Recieved
from Mike
and Aunt Joyce
on May 25, 1994

THE NEW OXFORD
ANNOTATED BIBLE

New Revised Standard Version

The New
Oxford Annotated

CONTRIBUTORS

BERNHARD W. ANDERSON   GEORGE W. ANDERSON
WILLIAM A. BEARDSLEE   JOHN BRECK   WALTER BRUEGGEMANN
MARY C. CALLAWAY   JOHN J. COLLINS   ROBERT C. DENTAN
WALTER J. HARRELSON   WILLIAM L. HOLLADAY
LESLIE J. HOPPE   PHILIP J. KING   BURKE O. LONG
BRUCE M. METZGER   CAREY A. MOORE   ROLAND E. MURPHY
PHEME PERKINS   JOHN REUMANN   JAMES A. SANDERS

ADDITIONAL CONTRIBUTORS TO PREVIOUS EDITIONS

†GEORGES A. BARROIS   DEMETRIOS J. CONSTANTELOS
†FLOYD V. FILSON   VICTOR R. GOLD   R. LANSING HICKS
†ARTHUR JEFFERY   SHERMAN E. JOHNSON   †JOHN KNOX
†HERBERT G. MAY   DONALD G. MILLER   †WARREN A. QUANBECK
†H. H. ROWLEY   †R. B. Y. SCOTT   WILLIAM F. STINESPRING
SAMUEL TERRIEN   ELWYN E. TILDEN   †LUTHER A. WEIGLE

BIBLE

with the
Apocryphal / Deuterocanonical Books

Edited by

BRUCE M. METZGER   ROLAND E. MURPHY

NEW REVISED
STANDARD VERSION

New York
OXFORD UNIVERSITY PRESS

OXFORD UNIVERSITY PRESS

Oxford   New York   Toronto
Delhi   Bombay   Calcutta   Madras   Karachi
Petaling Jaya   Singapore   Hong Kong   Tokyo
Nairobi   Dar es Salaam   Cape Town
Melbourne   Auckland

and associated companies in
Berlin   Ibadan

Published by Oxford University Press, Inc.
200 Madison Avenue, New York, New York 10016

Oxford is a registered trademark of Oxford University Press

4 6 8 10 9 7 5

Printed in the United States of America

Contents

Contents

Note: In accordance with traditional practice, the pages of the separate Testaments and of the Apocryphal/Deuterocanonical Books are numbered independently; thus, the Introductions to Genesis, to Tobit, and to Matthew are respectively the first page of the Hebrew Scriptures, the first page of the Apocryphal/Deuterocanonical Books, and the first page of the New Testament. As an aid to the reader, the page numbers in each case are distinguished by the abbreviations OT, AP, and NT. The reference materials at the back of the book continue the numbering of the pages of the New Testament, but without the distinguishing letters.

The Editors' Preface

The recent publication of the New Revised Standard Version of the Bible makes it opportune not only to adjust the annotations to the wording of the new version, but also to introduce many other changes and improvements as well. Besides the preface to the New Revised Standard Version entitled "To the Reader," the reader will find two introductory articles, one on the number and sequence of the books of the Bible, and the other on how to make the fullest use of the present annotated edition. These were drawn up by Bruce Metzger.

All of the general articles, as well as the introductions to and annotations on the several books of the Bible and of the Apocryphal/Deuterocanonical Books, have been reviewed, either by the original contributor or, in most instances, by another scholar. In some cases relatively few changes were called for; in others, where many advances have occurred since the publication of the original Oxford Annotated Bible in 1962 (for example, in the realm of the archaeology of Bible lands), an entirely new article has been prepared.

In the following list of contributors the name of the original writer is followed by the name of the reviewer; when the reviewer was also the original writer, the name occurs only once. It is understood that the editors have also introduced a certain number of modifications throughout.

THE OLD TESTAMENT

Genesis through Deuteronomy, BERNHARD W. ANDERSON; Joshua through Ruth, ROBERT C. DENTAN/ LESLIE J. HOPPE; 1 Samuel through 2 Chronicles, WILLIAM F. STINESPRING/ BURKE O. LONG; Ezra and Nehemiah, †ARTHUR JEFFERY/ JOHN J. COLLINS; Esther, †ARTHUR JEFFERY/ CAREY A. MOORE; Job, SAMUEL TERRIEN/ ROLAND E. MURPHY; Psalms, ROBERT C. DENTAN; Proverbs through Song of Solomon, †R. B. Y. SCOTT/ ROLAND E. MURPHY; Isaiah and Jeremiah, VICTOR R. GOLD/WILLIAM L. HOLLADAY; Lamentations, †R. B. Y. SCOTT/ ROLAND E. MURPHY; Ezekiel, VICTOR R. GOLD/ WILLIAM L. HOLLADAY; Daniel, †ARTHUR JEFFERY/ JOHN J. COLLINS; and Hosea through Malachi, R. LANSING HICKS/ WALTER BRUEGGEMANN.

THE APOCRYPHAL/DEUTEROCANONICAL BOOKS

1 Esdras, WALTER J. HARRELSON; 2 Esdras, BRUCE M. METZGER/ WALTER J. HARRELSON; Tobit and Judith, ROBERT C. DENTAN/ CAREY A. MOORE; Esther (Greek),

†FLOYD V. FILSON/ CAREY A. MOORE; Wisdom of Solomon, †FLOYD V. FILSON/ ROLAND E. MURPHY; Ecclesiasticus (Sirach), BRUCE M. METZGER/ ROLAND E. MURPHY; Baruch and the Letter of Jeremiah, †HERBERT G. MAY/ JAMES A. SANDERS; The Additions to the Book of Daniel, and the Prayer of Manasseh, BRUCE M. METZGER/ JAMES A. SANDERS; 1 and 2 Maccabees, SHERMAN E. JOHNSON/ MARY C. CALLAWAY; 3 Maccabees, DEMETRIOS J. CONSTANTELOS/ JOHN BRECK; 4 Maccabees, SHERMAN E. JOHNSON/ JOHN BRECK; and Psalm 151, BRUCE M. METZGER/ JAMES A. SANDERS. The introductory article on all these books and texts is by BRUCE M. METZGER, and the general article on "The Additions to the Greek Book of Daniel" is by JAMES A. SANDERS.

THE NEW TESTAMENT

Matthew through Luke, ELWYN E. TILDEN/ BRUCE M. METZGER; John, DONALD G. MILLER/ BRUCE M. METZGER; Acts, SHERMAN E. JOHNSON/ BRUCE M. METZGER; Romans through 2 Corinthians, †JOHN KNOX/ JOHN REUMANN; Galatians, BRUCE M. METZGER/ JOHN REUMANN; Ephesians and Colossians, BRUCE M. METZGER/ WILLIAM A. BEARDSLEE; Philippians, †JOHN KNOX/ WILLIAM A. BEARDSLEE; 1 Thessalonians through Philemon, †WARREN A. QUANBECK/ WILLIAM A. BEARDSLEE; Hebrews. BRUCE M. METZGER/ PHEME PERKINS; James through 2 Peter, †WARREN A. QUANBECK/ PHEME PERKINS; 1 John through 3 John. DONALD G. MILLER/ PHEME PERKINS; Jude, † WARREN A. QUANBECK/ PHEME PERKINS; and Revelation, BRUCE M. METZGER.

GENERAL AND SPECIAL ARTICLES

The several articles that appear throughout the volume were written by the following contributors: "Introduction to the Old Testament," †HERBERT G. MAY/ ROLAND E. MURPHY; "The Pentateuch," BERNHARD W. ANDERSON; "The Historical Books," BURKE O. LONG; "The Poetical Books or 'Writings,' " ROLAND E. MURPHY; "The Prophetical Books," WALTER BRUEGGEMANN; "Introduction to the Apocryphal/Deuterocanonical Books," "Introduction to the New Testament," "The Narrative Books—Gospels and Acts," and "Letters/Epistles in the New Testament," BRUCE M. METZGER; "Apocalyptic Literature," BRUCE M. METZGER and JOHN J. COLLINS; "Modern Approaches to Biblical Study," ROLAND E. MURPHY; "Characteristics of Hebrew Poetry," GEORGE W. ANDERSON; "Literary Forms in the Gospels," BRUCE M. METZGER; "Survey of the Geography, History, and Archaeology of the Bible Lands," †GEORGES A. BARROIS/ PHILIP J. KING; and "English Versions of the Bible," †LUTHER A. WEIGLE/ BRUCE M. METZGER.

The editors express their gratitude to all who are mentioned above, as well as to the staff and artisans of the Oxford University Press, who, under the careful guidance of Donald Kraus, have assisted in producing this annotated edition of the Holy Scriptures. May its varied helps continue to prove useful in leading readers to a deeper understanding of the meaning of the Word of God.

BRUCE M. METZGER
ROLAND E. MURPHY

Acknowledgments

Many people contributed to this volume. Among them were: Herbert J. Addison, Elizabeth M. Edman, Kelly Gallagher, June Gunden, Deborah Jenks, Susan Kraus, Estella Pate, Leslie Phillips, Lynn Stanley, Mildred and Frederick Tripp. The publisher expresses thanks to these and all others who had a part in bringing this work to completion.

To the Reader

This preface is addressed to you by the Committee of translators, who wish to explain, as briefly as possible, the origin and character of our work. The publication of our revision is yet another step in the long, continual process of making the Bible available in the form of the English language that is most widely current in our day. To summarize in a single sentence: the New Revised Standard Version of the Bible is an authorized revision of the Revised Standard Version, published in 1952, which was a revision of the American Standard Version, published in 1901, which, in turn, embodied earlier revisions of the King James Version, published in 1611.

In the course of time, the King James Version came to be regarded as "the Authorized Version." With good reason it has been termed "the noblest monument of English prose," and it has entered, as no other book has, into the making of the personal character and the public institutions of the English-speaking peoples. We owe to it an incalculable debt.

Yet the King James Version has serious defects. By the middle of the nineteenth century, the development of biblical studies and the discovery of many biblical manuscripts more ancient than those on which the King James Version was based made it apparent that these defects were so many as to call for revision. The task was begun, by authority of the Church of England, in 1870. The (British) Revised Version of the Bible was published in 1881–1885; and the American Standard Version, its variant embodying the preferences of the American scholars associated with the work, was published, as was mentioned above, in 1901. In 1928 the copyright of the latter was acquired by the International Council of Religious Education and thus passed into the ownership of the Churches of the United States and Canada that were associated in this Council through their boards of education and publication.

The Council appointed a committee of scholars to have charge of the text of the American Standard Version and to undertake inquiry concerning the need for further revision. After studying the questions whether or not revision should be undertaken, and if so, what its nature and extent should be, in 1937 the Council authorized a revision. The scholars who served as members of the Committee worked in two sections, one dealing with the Old Testament and one with the New Testament. In 1946 the Revised Standard Version of the New Testament was published. The publication of the Revised Standard Version of the Bible, containing the Old and New Testaments, took place on September 30, 1952. A translation of the Apocryphal/Deuterocanonical Books of the Old Testament followed in 1957. In 1977 this collection was issued in an expanded edition, containing three additional texts received by Eastern Orthodox communions (3 and 4 Maccabees and Psalm 151). Thereafter the Revised Standard Version gained the distinction of being officially authorized for use by all major Christian churches: Protestant, Anglican, Roman Catholic, and Eastern Orthodox.

The Revised Standard Version Bible Committee is a continuing body, comprising about thirty members, both men and women. Ecumenical in representation, it includes scholars affiliated with various Protestant denominations, as well as several Roman Catholic members, an Eastern Orthodox member, and a Jewish member who serves in the Old Testament section. For a period of time the Committee included several members from Canada and from England.

Because no translation of the Bible is perfect or is acceptable to all groups of readers, and because discoveries of older manuscripts and further investigation of linguistic features of the text continue to become available, renderings of the Bible have proliferated. During the years following the publication of the Revised Standard Version, twenty-six other English translations and revisions of the Bible were produced by committees and by individual scholars—not to mention twenty-five other translations and revisions of the New Testament alone. One of the latter was the second edition of the RSV New Testament, issued in 1971, twenty-five years after its initial publication.

Following the publication of the RSV Old Testament in 1952, significant advances were made in the discovery and interpretation of documents in Semitic languages related to Hebrew. In addition to the information that had become available in the late 1940s from the Dead Sea texts of Isaiah and Habakkuk, subsequent acquisitions from the same area brought to light many other early copies of all the books of the Hebrew Scriptures (except Esther), though most of these copies are fragmentary. During the same period early Greek manuscript copies of books of the New Testament also became available.

In order to take these discoveries into account, along with recent studies of documents in Semitic languages related to Hebrew, in 1974 the Policies Committee of the Revised Standard Version, which is a standing committee of the National Council of the Churches of Christ in the U.S.A., authorized the preparation of a revision of the entire RSV Bible.

For the Old Testament the Committee has made use of the *Biblia Hebraica Stuttgartensia* (1977; ed. sec. emendata, 1983). This is an edition of the Hebrew and Aramaic text as current early in the Christian era and fixed by Jewish scholars (the "Masoretes") of the sixth to the ninth centuries. The vowel signs, which were added by the Masoretes, are accepted in the main, but where a more probable and convincing reading can be obtained by assuming different vowels, this has been done. No notes are given in such cases, because the vowel points are less ancient and reliable than the consonants. When an alternative reading given by the Masoretes is translated in a footnote, this is identified by the words "Another reading is."

Departures from the consonantal text of the best manuscripts have been made only where it seems clear that errors in copying had been made before the text was standardized. Most of the corrections adopted are based on the ancient versions (translations into Greek, Aramaic, Syriac, and Latin), which were made prior to the time of the work of the Masoretes and which therefore may reflect earlier forms of the Hebrew text. In such instances a footnote specifies the version or versions from which the correction has been derived and also gives a translation of the Masoretic Text. Where it was deemed appropriate to do so, information is supplied in footnotes from subsidiary Jewish traditions concerning other textual readings (the *Tiqqune Sopherim,* "emendations of the scribes"). These are identified in the footnotes as "Ancient Heb tradition."

Occasionally it is evident that the text has suffered in transmission and that none of the versions provides a satisfactory restoration. Here we can only follow the best judgment of competent scholars as to the most probable reconstruction of the original text. Such reconstructions are indicated in footnotes by the abbreviation Cn ("Correction"), and a translation of the Masoretic Text is added.

For the Apocryphal/Deuterocanonical Books of the Old Testament the Committee has made use of a number of texts. For most of these books the basic Greek text from which the present translation was made is the edition of the Septuagint prepared by Alfred Rahlfs and published by the Württemberg Bible Society (Stuttgart, 1935). For several of the books the more recently published individual volumes of the Göttingen Septuagint project were utilized. For the book of Tobit it was decided to follow the form of the Greek text found in codex Sinaiticus (supported as it is by evidence from Qumran); where this text is defective, it was supplemented and corrected by other Greek manuscripts. For the three Additions to Daniel (namely, Susanna, the Prayer of Azariah and the Song of the Three Jews, and Bel and the Dragon) the Committee continued to use the Greek version attributed to Theodotion (the so-called "Theodotion-Daniel"). In translating Ecclesiasticus (Sirach), while constant reference was made to the Hebrew fragments of a large portion of this book (those discovered at Qumran and Masada as well as those recovered from the Cairo Geniza), the Committee generally followed the Greek text (including verse numbers) published by Joseph Ziegler in the Göttingen Septuagint (1965). But in many places the Committee has translated the Hebrew text when this provides a reading that is clearly superior to the Greek; the Syriac and Latin versions were also consulted throughout and occasionally adopted. The basic text adopted in rendering 2 Esdras is the Latin version given in *Biblia Sacra,* edited by Robert Weber (Stuttgart, 1971). This was supplemented by consulting the Latin text as edited by R. L. Bensly (1895) and by Bruno Violet (1910), as well as by taking into account the several Oriental versions of 2 Esdras, namely, the Syriac, Ethiopic, Arabic (two forms, referred to as Arabic 1 and Arabic 2), Armenian, and Georgian versions. Finally, since the Additions to the Book of Esther are disjointed and quite unintelligible as they stand in most editions of the Apocrypha, we have provided them with their original context by translating the whole of the Greek version of Esther from Robert Hanhart's Göttingen edition (1983).

For the New Testament the Committee has based its work on the most recent edition of *The Greek New Testament,* prepared by an interconfessional and international committee and published by the United Bible Societies (1966; 3rd ed. corrected, 1983; information concerning changes to be introduced into the critical apparatus of the forthcoming 4th edition was available to the Committee). As in that edition, double brackets are used to enclose a few passages that are generally regarded to be later additions to the text, but which we have retained because of their evident antiquity and their importance in the textual tradition. Only in very rare instances have we replaced the text or the punctuation of the Bible Societies' edition by an alternative that seemed to us to be superior. Here and there in the footnotes the phrase, "Other ancient authorities read," identifies alternative readings preserved by Greek manuscripts and early versions. In both Testaments, alternative renderings of the text are indicated by the word "Or."

As for the style of English adopted for the present revision, among the mandates given to the Committee in 1980 by the Division of Education and Ministry of the National Council of Churches of Christ (which now holds the copyright of the RSV Bible) was the directive to continue in the tradition of the King James Bible, but to introduce such changes as are warranted on the basis of accuracy, clarity, euphony, and current English usage. Within the constraints set by the original texts and by the mandates of the Division, the Committee has followed the maxim, "As literal as possible, as free as necessary." As a consequence, the New Revised Standard Version (NRSV) remains essentially a literal translation. Paraphrastic renderings have been adopted only sparingly, and then chiefly to compensate for a deficiency in the English language—the lack of a common gender third person singular pronoun.

During the almost half a century since the publication of the RSV, many in the churches have become sensitive to the danger of linguistic sexism arising from the inherent bias of the English language towards the masculine gender, a bias that in the case of the Bible has often restricted or obscured the meaning of the original text. The mandates from the Division specified that, in references to men and women, masculine-oriented language should be eliminated as far as this can be done without altering passages that reflect the historical situation of ancient patriarchal culture. As can be appreciated, more than once the Committee found that the several mandates stood in tension and even in conflict. The various concerns had to be balanced case by case in order to provide a faithful and acceptable rendering without using contrived English. Only very occasionally has the pronoun "he" or "him" been retained in passages where the reference may have been to a woman as well as to a man; for example, in several legal texts in Leviticus and Deuteronomy. In such instances of formal, legal language, the options of either putting the passage in the plural or of introducing additional nouns to avoid masculine pronouns in English seemed to the Committee to obscure the historic structure and literary character of the original. In the vast majority of cases, however, inclusiveness has been attained by simple rephrasing or by introducing plural forms when this does not distort the meaning of the passage. Of course, in narrative and in parable no attempt was made to generalize the sex of individual persons.

Another aspect of style will be detected by readers who compare the more stately English rendering of the Old Testament with the less formal rendering adopted for the New Testament. For example, the traditional distinction between *shall* and *will* in English has been retained in the Old Testament as appropriate in rendering a document that embodies what may be termed the classic form of Hebrew, while in the New Testament the abandonment of such distinctions in the usage of the future tense in English reflects the more colloquial nature of the koine Greek used by most New Testament authors except when they are quoting the Old Testament.

Careful readers will notice that here and there in the Old Testament the word LORD (or in certain cases GOD) is printed in capital letters. This represents the traditional manner in English versions of rendering the Divine Name, the "Tetragrammaton" (see the notes on Exodus 3.14, 15), following the precedent of the ancient Greek and Latin translators and the long established practice in the reading of the Hebrew Scriptures in the synagogue. While it is almost if not quite certain that the Name was originally pronounced "Yahweh," this pronunciation was not indicated when the Masoretes added vowel sounds to the consonantal Hebrew text. To the four consonants YHWH of the Name, which had come to be regarded as too sacred to be pronounced, they attached vowel signs indicating that in its place should be read the Hebrew word *Adonai* meaning "Lord" (or *Elohim* meaning "God"). Ancient Greek translators employed the word *Kyrios* ("Lord") for the Name. The Vulgate likewise used the Latin word *Dominus* ("Lord"). The form "Jehovah" is of late medieval origin; it is a combination of the consonants of the Divine Name and the vowels attached to it by the Masoretes but belonging to an entirely different word. Although the American Standard Version (1901) had used "Jehovah" to render the Tetragrammaton (the sound of Y being represented by J and the sound of W by V, as in Latin), for two reasons the Committees that produced the RSV and the NRSV returned to the more familiar usage of the King James Version. (1) The word "Jehovah" does not accurately represent any form of the Name ever used in Hebrew. (2) The use of any proper name for the one and only God, as though there were other gods from whom the true God had to be distinguished, began to be discontinued in Judaism before the Christian era and is inappropriate for the universal faith of the Christian Church.

It will be seen that in the Psalms and in other prayers addressed to God the archaic second person singular pronouns *(thee, thou, thine)* and verb forms *(art, hast, hadst)* are no longer used. Although some readers may regret this change, it should be pointed out that in the original languages neither the Old Testament nor the New makes any linguistic distinction between addressing a human being and addressing the Deity. Furthermore, in the tradition of the King James Version one will not expect to find the use of capital letters for pronouns that refer to the Deity—such capitalization is an unnecessary innovation that has only recently been introduced into a few English translations of the Bible. Finally, we have left to the discretion of the licensed publishers such matters as section headings, cross-references, and clues to the pronunciation of proper names.

This new version seeks to preserve all that is best in the English Bible as it has been known and used through the years. It is intended for use in public reading and congregational worship, as well as in private study, instruction, and meditation. We have resisted the temptation to introduce terms and phrases that merely reflect current moods, and have tried to put the message of the Scriptures in simple, enduring words and expressions that are worthy to stand in the great tradition of the King James Bible and its predecessors.

In traditional Judaism and Christianity, the Bible has been more than a historical document to be preserved or a classic of literature to be cherished and admired; it is recognized as the unique record of God's dealings with people over the ages. The Old Testament sets forth the call of a special people to enter into covenant relation with the God of justice and steadfast love and to bring God's law to the nations. The New Testament records the life and work of Jesus Christ, the one in whom "the Word became flesh," as well as describes the rise and spread of the early Christian Church. The Bible carries its full message, not to those who regard it simply as a noble literary heritage of the past or who wish to use it to enhance political purposes and advance otherwise desirable goals, but to all persons and communities who read it so that they may discern and understand what God is saying to them. That message must not be disguised in phrases that are no longer clear, or hidden under words that have changed or lost their meaning; it must be presented in language that is direct and plain and meaningful to people today. It is the hope and prayer of the translators that this version of the Bible may continue to hold a large place in congregational life and to speak to all readers, young and old alike, helping them to understand and believe and respond to its message.

For the Committee,
BRUCE M. METZGER

How to Use This
Study Bible Profitably

The following comments are chiefly arranged under the headings: "Kinds of Helps in this Edition," "Finding What the Text Says," and "Finding What the Text Means." Under the first heading are listed the several kinds of helps bearing on the major divisions of Scripture and the smaller collections within those divisions; helps bearing on individual books; and helps bearing on particular passages. Under the other two headings are mentioned practical hints as to interpreting any piece of ancient literature. Following these general comments are two further sections, "How to Look up References" and "Other Explanations."

I. KINDS OF HELPS IN THIS EDITION

1. General Introductions and Articles

At some stage, and preferably sooner rather than later, the student should read the "Introduction to the Old Testament" (p. xxxi OT), and/or the "Introduction to the New Testament" (p. iii NT). These give general information about the contents, original language(s), manuscripts, and canon of each Testament.

More specialized information about the literary form of the several parts of the Bible is provided by the articles in the Old Testament on "The Pentateuch" (p. xxxv OT), "The Historical Books" (p. 267 OT), "The Poetical Books or 'Writings' " (p. 623 OT), and "The Prophetical Books" (p. 862 OT), and in the New Testament by "The Narrative Books—Gospels and Acts" (p. viii NT), "Letters/Epistles in the New Testament" (p. 204 NT), and "Apocalyptic Literature" (p. 362 NT). A variety of other helps follow the close of the New Testament. These are essays on "Modern Approaches to Biblical Study" (p. 388), "Characteristics of Hebrew Poetry"

(p. 392), "Literary Forms in the Gospels" (p. 397), "English Versions of the Bible" (p. 400), and "Survey of the Geography, History, and Archaeology of the Bible Lands." (p.407).

Tables of Measures and Weights in the Bible, Chronological Tables of Rulers (now located at places in the text where they are most useful), an Index to the Annotations, and a set of Maps of Bible Lands (with an index to the maps) round out the wide variety of supplementary helps.

Standing between the Old and New Testaments, the Apocryphal/Deuterocanonical Books are provided with a general Introduction. Here one finds information on the meaning of the words "apocrypha" and "deuterocanonical," the kinds of literature in these books, divergent attitudes among Christian churches towards them, the pervasive influence of these books, and a brief discussion of other apocryphal and pseudepigraphical literature.

2. Introductions to Individual Books

For the study of a given book of the Bible, one should begin with the Introduction to that book. This provides necessary information about its author, themes, date (when known), and major sections. A more detailed consecutive outline of the book is provided by the boldface annotations at the foot of each page. By glancing through these one can obtain an idea of the overall structure and content.

The boldface headings are also useful in another way. When looking up an unfamiliar verse, attention given to such headings will alert the reader to the general drift of the passage and the larger unit of which it is a part, and thus provide guidance in understanding the verse in relation to the wider context.

3. Annotations

When examining individual paragraphs or shorter sections, help in understanding the passage will be found in the separate annotations. Annotations on a passage as a whole precede those on individual verses within the passage. Here italics designate words that are quoted verbatim from the biblical text.

4. Translators' Footnotes

The notes that belong to the New Revised Standard Version (designated by an italic letter and printed in smaller type at the foot of the second column of the biblical text) provide basically two kinds of information: divergent readings and alternative renderings. Here the phrase "Other ancient authorities read" means that the reading (i.e. wording) of the passage is different in various manuscripts and early versions, while the word "Or" signifies that the Hebrew or Greek text permits an alternative rendering besides the one given in the text.

5. Parallels in Synoptic Gospels

In the annotations on the Synoptic Gospels (Matthew, Mark, and Luke) the cross-references to parallel passages stand in parentheses immediately after the boldface heading of a new paragraph. For example, at the account of Jesus' transfiguration in Mk 9.2–8 the references to Mt 17.1–8 and Lk 9.28–36 indicate where the Synoptic parallels are to be found; similar information is given at the passages in Matthew and Luke. A careful comparison of such parallel texts will reveal the special emphases of each Synoptic account.

The best commentary on the Bible is often some other passage in the Bible; therefore it cannot be too strongly urged that, for the fullest comprehension of any one passage, all of the cross-references should be consulted. Particular attention should be given to the cross-references that end with "n." (as, for example, "see 5.24 n."); these involve biblical text along with annotation.

6. Maps

The hasty reader often overlooks the help that maps can provide. The set of specially prepared maps, with an index, found at the close of the volume should be consulted not only in connection with the general article on the geography of the Bible lands (p. 407), but also in order to obtain more specific orientation concerning countries, cities, towns, and places mentioned in the biblical text.

II. FINDING WHAT THE TEXT SAYS

1. Variant Readings and Alternative Renderings

First one should notice whether the passage involves a variant reading and/or an alternative rendering (for the distinction, see above at I.4). For example, in Rev 13.18 the footnote of the NRSV informs the reader that some Greek manuscripts give 616 rather than 666 as the number of the beast, while in the Lord's Prayer "Give us our bread for tomorrow" is an alternative rendering of the Greek text of Mt 6.11. Following 1 Sam 10.27 a footnote indicates that one of the Dead Sea Scrolls from Qumran provides the text of an additional paragraph, as given in the English translation.

2. Typographical Clues

Attention should be paid to the typography of the NRSV. Not only is the text arranged in paragraphs, but also one finds here and there a white line (a blank space); this indicates a still greater break in connection between two paragraphs than one normally expects.

Another typographical device, the strophic arrangement of the text, will alert the reader to the presence of semi-poetic or rhythmic material standing within a passage of prose (see the opening chapters of Genesis, or parts of the prophetical books). In the New Testament some passages (such as Phil 2.6–11; 1 Tim 2.5–6; 3.16; 2 Tim 2.11–13) may have been adapted from early Christian hymnody.

Likewise, not only are New Testament quotations and echoes from the Old Testament set off within quotation marks (see, for example, Heb 7.1–3), but also here and there in 1 Corinthians the presence of quotation marks helps to identify what seem to be questions and statements previously made by members of the church at Corinth, to which Paul now responds (as in 1 Cor 7.1; 8.1,4,8; 10.23).

3. Source and Context

One must pay attention not only to what is said, but also by whom it is said, to whom it is said, at what time, under what circumstances, what precedes, and what follows. There is the thoughtless habit of quoting all parts of the Bible as of equal value, whether they are the words of the Lord or the words of Bildad the Shuhite and Zophar the Naamathite in the book of Job, who are afterwards represented as condemned and contradicted by God. In other words, not everything contained in the Bible is affirmed by the Bible. For example, the behavior of some persons in Scripture is positively disgraceful; accounts of such behavior are recorded as a warning, not as an example (e.g. David with Uriah's wife, Bathsheba). Again, some statements are included in the biblical text not as truth to be believed, but as error to be rejected (e.g. "There is no God" in Pss 14.1 and 53.1).

4. What Is Not Said

Occasionally it is appropriate to consider not only what the text says, but also what it does not say. For example, 1 Jn 1.5 does not say that God is *a* light, as though one light among several other lights. Nor does it say that God is *the* light, as standing in relation to created beings. But it says simply, "God is light." Again, one should not read Jesus' denunciations against the scribes and Pharisees in Matthew ch 23 as though directed against *all* scribes and Pharisees, but only against those who are further identified in one or another particular.

III. FINDING WHAT THE TEXT MEANS

1. Intended Meaning

One should begin with the common-sense view that a text means what the author meant. For example, when one reads in 2 Tim 3.16, "All scripture is inspired by God," for the author "all scripture" denoted the Jewish scriptures—what we call the Old Testament. It was only subsequently that the expression "scripture" came to be applied also to writings of the New Testament (2 Pet 3.15–16).

2. Cultural Context

No word of the Bible was spoken or written in a cultural vacuum; every part of it was culturally conditioned. Thomas Aquinas wrote concerning the Bible, *Omnia ex Deo, omnia ex hominibus* ("All is from God, all is from human authors"). In other words, the mystery of Scripture is that the word of God is transmitted in words of human authors who were culturally conditioned. Therefore in order to ascertain the meaning of a passage one must take into account the background and thought-patterns of ancient Near Eastern culture. For example, ancient Hebrew psychology sometimes located the seat of the innermost emotions in the kidneys, translated "reins" in King James English (as in Ps 16.7, "My reins instruct me in the night watches"); here contemporary English translations substitute "heart" as more comprehensible to modern readers.

3. Literary Genres

One must also take into account the presence and intent of specific literary genres and forms. The recognition of different formats of different books, and of passages within a book, will assist one to understand them accordingly; prose is prose, poetry is poetry, proverb is proverb, story is story, parable is parable. (See also "Literary Forms in the Gospels," p. 397.)

4. Visions and Symbols

If a passage is identified as a visionary experience, one must seek the meaning behind what is said. In Ezekiel's vision of the valley of dry bones (Ezek 37) and in many parts of the book of Revelation (for example, the Four Horsemen of the Apocalypse, 6.1–8), the descriptions are not descriptions of real occurrences, but of symbols of real occurrences.

5. Natural Meaning

What might be called "the principle of simplicity" in interpreting the Bible is aptly set forth by Calvin in a comment on Gal 4.22 about allegorizing. "Let us know," he writes, "that the true meaning of Scripture is the natural and obvious meaning. . . . Let us not only neglect as doubtful, but boldly set aside as deadly corruptions, those pretended expositions that lead us away from the natural meaning." It is significant that Calvin uses the adjective "natural" rather than "literal," for the natural meaning of a text is sometimes figurative rather than literal. As Luther once cautioned, one should not conclude from the psalmist's statement, "Under his [God's] wings you will find refuge," that God has feathers!

IV. HOW TO LOOK UP REFERENCES

Jn 3.16	means the Gospel according to John, chapter 3, verse 16.
Isa 9.1,6	means Isaiah, chapter 9, verse 1 and verse 6.
Rev 22.1–5	means Revelation, chapter 22, verses 1 to 5 inclusive.
Acts 8.4; 11.19	means Acts, chapter 8, verse 4; and chapter 11, verse 19 (of the same book).
See Heb 3.2n.	means there is an annotation on Hebrews, chapter 3, verse 2. Both the Scripture passage and the annotation should be consulted.
v. 4a; v. 7b	means the first part of verse 4; the second part of verse 7.

When a book of the Bible is referred to within an annotation on that book, the name of the book is not repeated. For example, a reference to Ps 23.1 among the annotations elsewhere in the book of Psalms is written 23.1, not Ps. 23.1.

The best commentary on the Bible is often some other text in the Bible, it therefore cannot be too strongly urged that, for the fullest comprehension of any one passage, all of the cross-references should be looked up.

V. OTHER EXPLANATIONS

1. The Divine Name

In the translation of that part of the Old Testament that is the Hebrew Scriptures, the words LORD and GOD (printed in capital and small capital letters) represent the Divine name, YHWH (conventionally vocalized "Yahweh"). For further information, see p. xiii above.

2. Typographical Distinctions

In the annotations, transliterated Hebrew, Aramaic, or Greek words are enclosed within quotation marks; italics are used to designate the words that are quoted verbatim from the translated Scripture text. In the footnotes belonging to the New Revised Standard Version (designated by italic superscript letters, and standing at the foot of the second column of the Scripture text on any given page) alternative readings and renderings (for the distinction see I.4 above) are printed in italics.

3. Calendar Designations

For dates given in the present edition, the editors have decided to retain the designations B.C. ("before Christ") and A.D. (*anno domini*, "in the year of the Lord"), in preference to the designations B.C.E. ("before the common era") and C.E. ("common era"). These latter designations have not yet attained wide recognition among readers and would require frequent explanation in order to avoid confusion. The use of B.C. and A.D. is in the interest of ease of understanding, and is not meant to convey a theological position or preference.

The Number and Sequence of the Books of the Bible

THE OLD TESTAMENT

According to Jewish usage the twenty-four books of the Hebrew Scriptures fall into three divisions: the Law, the Prophets, and the Writings. The Prophets are divided into the Former Prophets and the Latter Prophets (the terms "former" and "latter" refer to their position in the list, and have no reference to date of composition). The books of Samuel, Kings, Chronicles, Ezra-Nehemiah were not divided by the Jews until the close of the Middle Ages. The twelve Minor Prophets are treated as one book. The two subdivisions of the Prophets therefore contain four books each. In most editions of the Hebrew Bible the sequence of the books is as follows:

The Law (five books):
Genesis, Exodus, Leviticus, Numbers, Deuteronomy.

The Prophets (eight books):
Former Prophets: Joshua, Judges, Samuel (1 and 2), Kings (1 and 2).
Latter Prophets: Isaiah, Jeremiah, Ezekiel, the Twelve (=Hosea, Joel, Amos, Obadiah, Jonah, Micah, Nahum, Habakkuk, Zephaniah, Haggai, Zechariah, Malachi).

The Writings (eleven books):
Psalms, Proverbs, Job, Song of Solomon, Ruth, Lamentations, Ecclesiastes, Esther, Daniel, Ezra-Nehemiah, Chronicles (1 and 2).

In Protestant editions of the Bible the Old Testament follows the Hebrew text as regards content, but the books in the second and third divisions are rearranged in sequence and several are divided, making a total of thirty-nine.

In Roman Catholic editions the Old Testament contains the rearranged thirty-nine books of the Hebrew Scriptures plus seven others that are current in the official Latin Vulgate Bible and that Protestants include among the Apocrypha. The order of these forty-six books in Vulgate manuscripts varies greatly; in fact, the manuscripts that have been examined disclose more than two hundred different ways of arranging the books. In current editions of Roman Catholic Bibles (including the Douay Version, the Jerusalem Bible, and the New American Bible), Tobit and Judith stand after Nehemiah; 1 and 2 Maccabees, after Esther (except in the Douay Version, in which these books conclude the Old Testament); Wisdom and Ecclesiasticus, after the Song of Solomon; and Baruch, with the Letter of

Jeremiah as ch. 6, after Lamentations. Furthermore, the books of Esther and Daniel are expanded by several additional chapters and parts of chapters, which Protestants regard as apocryphal. They comprise six Additions to the book of Esther and the following three supplements to the book of Daniel: the Prayer of Azariah and the Song of the Three Jews, Susanna, and Bel and the Dragon.

In the original Douay Version of 1609–1610 an appendix after the close of the Old Testament contains three other books, 3 and 4 Esdras (called 1 and 2 Esdras by Protestants) and the Prayer of Manasseh. These are regarded as apocryphal by Roman Catholics as well as by Protestants. It is curious that in the Geneva Bible of 1560, widely used by the Puritans, the Prayer of Manasseh is included in the Old Testament between 2 Chronicles and Ezra, though in the table of contents it is designated as apocryphal.

The Greek Orthodox Church, which uses the Greek Septuagint Version as its official text, has generally been accustomed to follow the longer canon of the Old Testament, including in this case also the 151st Psalm and 3 Maccabees. The Seventh Ecumenical Council held at Nicaea in 787 and the Council convened by Basil in Constantinople in 869 quote certain Apocrypha as authoritative. On the other hand, writers who raised the issue concerning the limits of the canon, such as John of Damascus and Nicephorus, express views that coincide with those of Athanasius, who adhered to the Hebrew canon. In the Schism of 1054 the Apocrypha were not an issue, though they became such during the Protestant Reformation. At that time a short-lived attempt was made by Cyril Lucar, Patriarch of Constantinople, to promote the adoption of the Hebrew canon in the Greek Church. Subsequently, however, the Synod of Jerusalem (1672) condemned Cyril and expressly designated the books of Tobit, Judith, Ecclesiasticus (Sirach), and Wisdom as canonical.

By way of summary, the Roman Catholic Church and the Greek Orthodox Church agree in regarding as authoritative certain books that they call deuterocanonical and that Protestants call apocryphal. In addition the following books are considered apocryphal by Protestants and Roman Catholics, but are in the Greek canon when indicated.

1 Esdras (= Esdras A in the Greek canon; 3 Esdras in Appendix to Latin Vulgate).

2 Esdras (= 4 Esdras in Appendix to Latin Vulgate).

Prayer of Manasseh (in the Greek canon; in Appendix to Latin Vulgate).

Psalm 151 and 3 Maccabees (in the Greek canon; 4 Maccabees in Appendix Greek canon).

In order to set forth clearly the several differences in usage among the Churches, the New Revised Standard Version presents the text of the Apocryphal/ Deuterocanonical Books in four successive groupings, as follows: *(a)* Books and Additions to Esther and Daniel that are in Roman Catholic, Greek, and Slavonic Bibles; *(b)* Books in the Greek and Slavonic Bibles, not in the Roman Catholic canon; *(c)* In the Slavonic Bible and the Latin Vulgate Appendix; *(d)* In an Appendix to the Greek Bible.

One can readily understand, therefore, why the reader does not find, for example, 3 and 4 Maccabees directly following 1 and 2 Maccabees.

THE NEW TESTAMENT

The number and the sequence of the twenty-seven books of the New Testament are the same in Protestant, Roman Catholic, and Eastern Orthodox Bibles.

A comparison of the extant manuscripts of the New Testament discloses that the early Church was accustomed to arrange them in four groups: (1) the Gospels, (2) the Acts and the General, or Catholic, Letters (that is, the seven letters which bear the names of James, Peter, John, and Jude), (3) the Pauline Letters, (4) the Apocalypse (as in the fifth-century codex Alexandrinus and many other manuscripts). Sometimes the Pauline Letters precede the Acts and General Letters, thus placing first the books which had earliest obtained canonical authority (as in the fourth-century codex Sinaiticus and the sixth-century codex Fuldensis).

Within each of the four groups there was a great variety of order. In the early Western Church the Gospel sequence most commonly followed was that of Matthew, John, Luke, Mark—namely, the Gospels attributed to apostles preceding those attributed to disciples of the apostles. The Letter to the Hebrews had no fixed place; sometimes it stood at the end of the Pauline Letters, sometimes between Paul's Letters to churches and those to individuals (that is, between 2 Thessalonians and 1 Timothy), and occasionally after Romans (as in the third-century Chester Beatty Papyrus) or after Galatians (as in an ancestor of the fourth-century codex Vaticanus, as is disclosed by the section numbers in Vaticanus). In the West the Letters of Peter were frequently placed first among the General Letters.

The current order of the Pauline and the General Letters seems to have been drawn up roughly in accord with length, the longest one in each group (Paul to churches, Paul to individuals, and General Letters) standing first and the shortest one, last.

Alphabetical Listing of the
Books of the Bible
(including the Apocryphal/Deuterocanonical Books of the Old Testament)

OT = Old Testament AP = Apocryphal/Deuterocanonical Books NT = New Testament

The Acts	160NT	James	331NT	Numbers	163OT
The Additions		Jeremiah	960OT	Obadiah	1183OT
to Esther (Gk)	41AP	Jeremiah,		1 Peter	337NT
Amos	1170OT	Letter of	169AP	2 Peter	344NT
Azariah,		Job	625OT	Philemon	314NT
Prayer of	174AP	Joel	1163OT	Philippians	279NT
Baruch	161AP	John	124NT	Prayer of	
Bel and the		1 John	349NT	Azariah	174AP
Dragon	183AP	2 John	355NT	Prayer of	
1 Chronicles	503OT	3 John	357NT	Manasseh	281AP
2 Chronicles	538OT	Jonah	1186OT	Proverbs	802OT
Colossians	285NT	Joshua	270OT	Psalm 151	284AP
1 Corinthians	229NT	Jude	359NT	Psalms	674OT
2 Corinthians	249NT	Judges	300OT	Revelation	364NT
Daniel	1126OT	Judith	20AP	Romans	208NT
Deuteronomy	217OT	1 Kings	423OT	Ruth	332OT
Ecclesiastes	841OT	2 Kings	463OT	1 Samuel	340OT
Ecclesiasticus	86AP	Lamentations	1074OT	2 Samuel	384OT
Ephesians	272NT	Letter of		Sirach, Wisdom of	
1 Esdras	259AP	Jeremiah	169AP	Jesus Son of	86AP
2 Esdras	300AP	Leviticus	125OT	Song of Solomon	853OT
Esther,	612OT	Luke	76NT	Song of the	
Esther, (Gk)		1 Maccabees	186AP	Three Jews	174AP
Additions to	41AP	2 Maccabees	228AP	Susanna	179AP
Exodus	69OT	3 Maccabees	285AP	1 Thessalonians	291NT
Ezekiel	1057OT	4 Maccabees	341AP	2 Thessalonians	296NT
Ezra	581OT	Malachi	1233OT	1 Timothy	300NT
Galatians	263NT	Manasseh,		2 Timothy	306NT
Genesis	1OT	Prayer of	281AP	Titus	311NT
Habakkuk	1205OT	Mark	47NT	Tobit	1AP
Haggai	1217OT	Matthew	1NT	Wisdom of	
Hebrews	316NT	Micah	1190OT	Solomon	57AP
Hosea	1148OT	Nahum	1200OT	Zechariah	1220OT
Isaiah	866OT	Nehemiah	594OT	Zephaniah	1211OT

Names and Order of the Books of the Old and New Testaments with the Apocryphal/Deuterocanonical Books

THE HEBREW SCRIPTURES

Genesis 1 OT	Job 625 OT
Exodus 69 OT	Psalms 674 OT
Leviticus 125 OT	Proverbs 802 OT
Numbers 163 OT	Ecclesiastes 841 OT
Deuteronomy 217 OT	Song of Solomon 853 OT
Joshua 270 OT	Isaiah 866 OT
Judges 300 OT	Jeremiah 960 OT
Ruth 332 OT	Lamentations 1047 OT
1 Samuel 340 OT	Ezekiel 1057 OT
(1 Kingdoms in Greek)	Daniel 1126 OT
2 Samuel 384 OT	Hosea 1148 OT
(2 Kingdoms in Greek)	Joel 1163 OT
1 Kings 423 OT	Amos 1170 OT
(3 Kingdoms in Greek)	Obadiah 1183 OT
2 Kings 463 OT	Jonah 1186 OT
(4 Kingdoms in Greek)	Micah 1190 OT
1 Chronicles 503 OT	Nahum 1200 OT
(1 Paralipomenon in Greek)	Habakkuk 1205 OT
2 Chronicles 538 OT	Zephaniah 1211 OT
(2 Paralipomenon in Greek)	Haggai 1217 OT
Ezra 581 OT	Zechariah 1220 OT
Nehemiah 594 OT } (= 2 Esdras in Greek)	Malachi 1233 OT
Esther 612 OT	

THE APOCRYPHAL/DEUTEROCANONICAL BOOKS

The Apocryphal/Deuterocanonical Books are listed here in four groupings, as follows:

(a) *Books and Additions to Esther and Daniel that are in the Roman Catholic, Greek, and Slavonic Bibles*

Tobit 1 AP

Judith 20 AP

The Additions to the Book of Esther

    (with a translation of the entire Greek text of Esther)   41 AP

Wisdom of Solomon   57 AP

Ecclesiasticus, or the Wisdom of Jesus Son of Sirach   86 AP

Baruch   161 AP

The Letter of Jeremiah (=Baruch ch. 6)   169 AP

The Additions to the Greek Book of Daniel   173 AP

    The Prayer of Azariah and the Song of the Three Jews   174 AP

    Susanna   179 AP

    Bel and the Dragon   183 AP

1 Maccabees   186 AP

2 Maccabees   228 AP

(b) *Books in the Greek and Slavonic Bibles; not in the Roman Catholic Canon*

1 Esdras (=2 Esdras in Slavonic=3 Esdras in Appendix to Vulgate)   259 AP

Prayer of Manasseh (in Appendix to Vulgate)   281 AP

Psalm 151, following Psalm 150 in the Greek Bible   283 AP

3 Maccabees   285 AP

(c) *In the Slavonic Bible and in the Latin Vulgate Appendix*

2 Esdras (=3 Esdras in Slavonic=4 Esdras in Vulgate Appendix)

(Note: In the Latin Vulgate, Ezra–Nehemiah=1 and 2 Esdras.)   300 AP

(d) *In an Appendix to the Greek Bible*

4 Maccabees   341 AP

THE NEW TESTAMENT

Matthew	1 NT	2 Thessalonians	296 NT
Mark	47 NT	1 Timothy	300 NT
Luke	76 NT	2 Timothy	306 NT
John	124 NT	Titus	311 NT
Acts of the Apostles	160 NT	Philemon	314 NT
Romans	208 NT	Hebrews	316 NT
1 Corinthians	229 NT	James	331 NT
2 Corinthians	249 NT	1 Peter	337 NT
Galatians	263 NT	2 Peter	344 NT
Ephesians	272 NT	1 John	349 NT
Philippians	279 NT	2 John	355 NT
Colossians	285 NT	3 John	357 NT
1 Thessalonians	291 NT	Jude	359 NT
		Revelation	364 NT

Abbreviations

The following abbreviations are used for the books of the Bible:

OLD TESTAMENT

Gen	Genesis	2 Chr	2 Chronicles	Dan	Daniel
Ex	Exodus	Ezra	Ezra	Hos	Hosea
Lev	Leviticus	Neh	Nehemiah	Joel	Joel
Num	Numbers	Esth	Esther	Am	Amos
Deut	Deuteronomy	Job	Job	Ob	Obadiah
Josh	Joshua	Ps	Psalms	Jon	Jonah
Judg	Judges	Prov	Proverbs	Mic	Micah
Ruth	Ruth	Eccl	Ecclesiastes	Nah	Nahum
1 Sam	1 Samuel	Song	Song of Solomon	Hab	Habakkuk
2 Sam	2 Samuel	Isa	Isaiah	Zeph	Zephaniah
1 Kings	1 Kings	Jer	Jeremiah	Hag	Haggai
2 Kings	2 Kings	Lam	Lamentations	Zech	Zechariah
1 Chr	1 Chronicles	Ezek	Ezekiel	Mal	Malachi

APOCRYPHAL / DEUTEROCANONICAL BOOKS

Tob	Tobit	Song of Thr	Prayer of Azariah and the Song of the Three Jews
Jdt	Judith		
Add Esth	Additions to Esther (Gk)	Sus	Susanna
Wis	Wisdom	Bel	Bel and the Dragon
Sir	Sirach (Ecclesiasticus)	1 Macc	1 Maccabees
Bar	Baruch	2 Macc	2 Maccabees
1 Esd	1 Esdras	3 Macc	3 Maccabees
2 Esd	2 Esdras	4 Macc	4 Maccabees
Let Jer	Letter of Jeremiah	Pr Man	Prayer of Manasseh

NEW TESTAMENT

Mt	Matthew	Acts	Acts of the Apostles	Gal	Galatians
Mk	Mark	Rom	Romans	Eph	Ephesians
Lk	Luke	1 Cor	1 Corinthians	Phil	Philippians
Jn	John	2 Cor	2 Corinthians	Col	Colossians

1 Thess	1 Thessalonians	Philem	Philemon	1 Jn	1 John
2 Thess	2 Thessalonians	Heb	Hebrews	2 Jn	2 John
1 Tim	1 Timothy	Jas	James	3 Jn	3 John
2 Tim	2 Timothy	1 Pet	1 Peter	Jude	Jude
Titus	Titus	2 Pet	2 Peter	Rev	Revelation

In the NRSV notes of the books of the Old Testament the following
abbreviations are used:

Ant.	Josephus, *Antiquities of the Jews*
Aram	Aramaic
Ch, chs	Chapter, chapters
Cn	Correction; made where the text has suffered in transmission and the versions provide no satisfactory restoration but where the Standard Bible Committee agrees with the judgment of competent scholars as to the most probable reconstruction of the original text.
Gk	Septuagint, Greek version of the Old Testament
Heb	Hebrew of the consonantal Masoretic Text of the Old Testament
Josephus	Flavius Josephus (Jewish historian, about A.D. 37 to about 95)
Macc	The book(s) of the Maccabees
Ms(s)	Manuscript(s)
MT	The Hebrew of the pointed Masoretic Text of the Old Testament
OL	Old Latin
Q Ms(s)	Manuscript(s) found at Qumran by the Dead Sea
Sam	Samaritan Hebrew text of the Old Testament
Syr	Syriac Version of the Old Testament
Syr H	Syriac Version of Origen's Hexapla
Tg	Targum
Vg	Vulgate, Latin Version of the Old Testament

THE HEBREW SCRIPTURES
COMMONLY CALLED
THE OLD TESTAMENT

New Revised Standard Version

Introduction to the Old Testament

Since the time of the early theologians of the church, Christians have adopted, for the Hebrew Scriptures and some additional Jewish writings, the phrase "Old Testament," in opposition to the "New Testament" (see Jer 31.31; Heb 8.8), which was itself not completed until the second century. The New Testament designation of the sacred books of the Jews was *graphē,* the Writing, either in the singular or plural form. At the time of Christ (see the Prologue to Sirach, already in the second century B.C., and also Luke 24.44), the Hebrew Scriptures seem to have been recognized as three-stepped: the Law, the Prophets, and the Writings (or Hagiographa). This threefold division is often referred to today by the acronym *TaNaK* (*Torah, Nebi'im* and *Kethubim,* the Hebrew words for the three groups). The traditional Christian division into historical, prophetical, and poetic derives from the Greek translation (the Septuagint) and tradition of the sacred books.

That the Bible was not written originally in English is a fact not always appreciated, and there are even those who are unaware of it. What we use is the translated Bible (see "English Versions of the Bible," pp. 400–406). The New Testament was written in Greek and the Old Testament in Hebrew, with the exception of parts of Daniel (2.4b–7.28) and Ezra (4.8–6.18; 7.12–26) and one verse in Jeremiah (10.11), which are in Aramaic. The translation of Hebrew and Aramaic presents distinctive and often difficult problems; they belong to the Semitic family of languages, to which Arabic, Assyrian and Babylonian, and Canaanite also belong, in contrast to Greek and English, which are Indo-European.

The Hebrew text of the Old Testament, like the Greek of the New (see "Introduction to the New Testament," pp. iii–vi NT), has suffered from copyist's errors and scribal emendations, some of which can be corrected in the light of the ancient versions (Greek, Latin, Syriac, and other early translations and collections). The notes in the New Revised Standard Version give examples of such corrections. The translators must also decide among variant readings preserved in the Hebrew manuscripts themselves, a task complicated by the recent discovery of much earlier manuscripts of the Hebrew Scriptures than had previously been known. Before about A.D. 100, there did not exist a single standard text of the various books of Scripture regarded as possessing sole authority (a *textus receptus*). Rather, as the ancient versions and the Dead Sea (Qumran) Scrolls afford evidence, there were variant recensions, or texts, of the same Old Testament book. It is true that there was already in existence a form of the Hebrew text that was to be edited later by the Jewish scholars known as Masoretes (from about A.D. 600 to the 10th century) and that is the standard text used today, but there were also variant forms of the text. In contrast to this evidence of early variant textual recensions, the biblical manuscripts found in the Wilderness of Judea at Wadi Murabba'at and belonging to the early part of the second century A.D. are strictly Masoretic (proto-Masoretic) in character, showing that by this time the standard text had been adopted. Although there are variants among the preserved manuscripts of the Masoretic Text, due largely to simple scribal errors, the Masoretic Text has been transmitted with remarkable accuracy. (See also "To the Reader," pp. ix–xiv.)

The books accepted as authoritative Scripture are spoken of as belonging to the "canon" (see "Introduction to the New Testament," pp. iii–vi NT) of Scripture. If by "canonical" one means that a book must be regarded as having a special authority, that it is holy and inspired, that it is one of a strictly limited number of books, and that there is a single, standard text, then one cannot speak of a canon of Old Testament Scripture before A.D. 100. Long before this, however, the Jews had their Scriptures.

The process by which the Jews became "the people of the Book" was gradual, and the development is shrouded in the mists of history and tradition. One might designate as their earliest Scripture "the book of the law" that was found in the temple at Jerusalem in 621 B.C. and used as the basis for King Josiah's reform (2 Kings chs 22–23). It is generally agreed that this "book of the law" is now incorporated within the present book of Deuteronomy (chs 5–26 [or perhaps 12–26] and ch 28). Another landmark is "the book of the law of Moses" brought by Ezra from Babylonia in 458 (398 ?) B.C. (Ezra 7.6–10, 14; Neh chs 8–10). This has been variously identified by scholars as the prototype or earlier form of that legislation that became the P Code in the Pentateuch, or the P Code itself, or the completed Pentateuch (see the Introduction to the Pentateuch, pp. xxxv–xxxvi OT). The date of the final compilation of the Pentateuch or Law, which was the first corpus or larger body of literature that came to be regarded as authoritative Scripture, is uncertain, but some have conservatively dated it to the time of the Exile in the sixth century. Although the story line goes from Genesis through Joshua (when the promise of the land is accomplished), those responsible for the final form of the Pentateuch highlighted the role of Moses as founding father and legislator. Genesis to Deuteronomy came to be regarded as Mosaic in origin, and the Law was limited to the Pentateuch (see Ex 19–40; Leviticus; Num 1–9). Certainly before the middle of the third century B.C., when according to tradition the Pentateuch was first translated into Greek, it had achieved a primary status as the Scripture of the Jews. It was to retain this primary position even after the body of Scripture had been enlarged to include the Prophets and the Writings. When in the second century B.C. the Samaritans finally separated from Judaism, they retained the Pentateuch as their sole Scripture, preserving it in a script derived from the old Hebrew script, which was revived in the Maccabean period.

Before the adoption of the Pentateuch as the Law of Moses, there had been compiled and edited in the spirit and diction of the Deuteronomic "school" the group of books consisting of Deuteronomy, Joshua, Judges, Samuel, and Kings (often called the Deuteronomistic History; see "The Historical Books," pp. 267–269 OT). This compilation, which was partly built up from earlier writings, may have occurred in two stages, i.e. shortly before the death of Josiah in 609 B.C. and during the Exile in the middle of the sixth century. Isaiah, Jeremiah, Ezekiel, and the book of the Twelve (or minor prophets) were edited and compiled during the post-exilic period. The process of compiling and editing these works is illustrated by the addition of chs 40–66 to the book of Isaiah and chs 9–14 to the prophecies of Zechariah, by the superscriptions giving information about the person and time of a given prophet, and by certain additions and changes here and there that are mentioned in the annotations in this volume. The editing was both an adaptation of the books of the prophets to the needs of the post-exilic period and a recognition of the relevance of the prophetic messages for contemporary as well as historic Israel.

Just when the prophetical books came to be regarded as a definitely limited body of Scripture is not clear. Ben Sira, the Jewish author of the apocryphal book of Ecclesiasticus, writing around

180 B.C., seems to have regarded his work as a continuation of the Prophets and the books of Wisdom. The grandson of Ben Sira refers in his Prologue to Ecclesiasticus to the threefold division of Scripture as the Law, the Prophets, and the other or rest of the books, though he does not necessarily imply a closed canon. The Prophets came to be divided into the Former Prophets (Joshua, Judges, Samuel, and Kings) and the Latter Prophets (Isaiah, Jeremiah, Ezekiel, and the book of the Twelve). Perhaps the association of the two divisions was based on allusions in the Former Prophets to prophetic figures and because of a tradition that they were composed largely by prophets. Such a tradition is suggested in 1 Chr 29.29; 2 Chr 9.29; 12.15; 13.22; 20.34; 26.22; 32.32.

In contrast to the Law and the Prophets, the books of the Writings contain less homogeneous materials. They were not edited in groups or combined as in the case of the other books, but circulated separately. This should be qualified by the recognition that 1 and 2 Chronicles, Ezra, and Nehemiah are a connected history, which some even regard as the work of a single author ("the Chronicler"; see "The Historical Books," pp. 267–269 OT). It was not until the time of the Christian era that the authority of the disputed books among the Writings (in particular, the Song of Songs and Ecclesiastes) was definitively settled among the Jews themselves. Later in the Christian era the five shortest of the Writings came to form a group called the "Scrolls" (*megilloth*): Song of Songs, Ruth, Lamentations, Ecclesiastes, and Esther. They were associated with certain Jewish festivals, when they were read publicly, such as the Song of Songs at Passover.

Among writings not included in the Hebrew canon are the so-called Apocrypha (see "Introduction to the Apocryphal/Deuterocanonical Books," pp. iii–xii AP). These were transmitted mainly in the Greek Septuagint tradition, and most of them were composed in the "intertestamental" era, the last two centuries B.C. They were not accepted in the Jewish tradition, in part because of the current conviction that the Old Testament canon was closed at the time of Ezra when prophetic revelation was supposed to have ceased (the book of Daniel is of course an exception to this). Later Christian groups accepted several of the apocrypha into their official Scriptures (see "Divergent Attitudes in the Christian Church" toward the Apocrypha, pp. v–vi AP).

The Pentateuch

The so-called "Five Books of Moses" are often designated as the Pentateuch, a Greek term referring to a work divided into "five scrolls." In Jewish circles these books are known by the Hebrew word Torah, a term that is frequently translated "law." For instance, the New Testament
makes reference to "the law" or "the prophets," two major parts of the Hebrew Bible (Mt 5.17;
Lk 24.44; Jn 1.45). This rendering is proper in the sense that the Torah includes many legal matters; for instance, the book of Leviticus is exclusively "law." The translation is too restrictive,
however, to do full justice to the meaning of the Hebrew word. In its larger sense the Hebrew
word *torah* refers to the "teaching" or even the "revelation" that God gives the people.

The Pentateuch or Torah is presented in the form of a narrative extending from the creation
of the universe to the eve of Israel's entrance into "the promised land." The comprehensive account has six major movements: the story of primeval times (Gen 1–11), of Israel's ancestors
(Gen 12–50), of the Exodus from Egypt (Ex 1–18), of the sojourn at Sinai (Ex 19 through
Leviticus to Num 10.11), of the wandering in the wilderness (the remainder of Numbers), and
of the entrance into the land of Canaan. The latter story, however, is only anticipated in the
book of Deuteronomy. With the death of Moses at the end of Deuteronomy the narrative breaks
off, only to be continued in the book of Joshua and subsequent books of the Hebrew Bible.

At various points along the way, the Torah story relates to what we would call "history."
This is especially the case in the book of Exodus, where the drama is enacted in the context of
the imperial politics of ancient Egypt, probably at the beginning of the Nineteenth Dynasty
(thirteenth century B.C.). The purpose of the narrators, however, is not to present a straightforward historical account but to confess faith in God. They do so by telling the story of how
the Holy God, Creator of heaven and earth, chooses to become involved in the life of a small
Hebrew people who were known in the ancient world as Israel. Storytelling or poetry is indispensable for confessing the faith of this "people of the LORD" (Judg 5.11, 13).

The heart of the story of the Pentateuch is found in the book of Exodus, which deals with
the exodus from Egypt and the sojourn at Mount Sinai. All Jewish tradition reaches back to
these "root experiences," which constitute the people's basic understanding of their own identity
and of the identity and character of God. At first these crucial experiences were related in story,
song, and proverb, but in the course of time, as the tradition was handed on in various circles
and reinterpreted for new situations, the Torah expanded and took written form. During the
monarchy it circulated in Old Epic literature; indeed, some scholars detect southern ("J" or Judean) and northern ("E" or Ephraimitic) literary versions. (These letters, derived originally from
the terms for God used in the different sources—"J" for *Jahveh* or *Yahweh,* "E" for *'elohim*—are
conventionally used as titles for these two strands of the tradition, as "P" and "D" are used for
the other two sources.) Eventually the old Israelite epic was edited by Priestly writers ("P"),

perhaps during the Babylonian Exile. Finally the book of Deuteronomy ("D"), which belongs with the Deuteronomistic History extending from Joshua through Second Kings, was inserted just after the conclusion of the Priestly version of the Torah, because it purports to be Moses' farewell address to the people.

Thus the Pentateuch in its manifold richness took shape over a long period of time. It preserves not only the tones that reverberated in the Mosaic period, but the overtones of meaning perceived by subsequent generations during Israel's historical pilgrimage. In tribute to his seminal influence, the whole Torah was ascribed to Moses (2 Chr 30.16; Lk 24.44), the leader to whom, Israel believed, God had spoken as to no other person (Deut 34.10–12).

Even though the Pentateuchal tradition evolved over a long period of time, it displays a striking overall unity in the midst of its great diversity. The leading theme is struck at a turning point, where the primeval history (Gen 1–11) moves into the story of Israel's ancestors (Gen 12–50). To the people represented by Abraham and Sarah, God makes a promise that has a threefold dimension: a numerous progeny, a land, and a relationship with God that will benefit other human families (Gen 12.1–3). At times the promise is put to the test, at times it is not even mentioned, at times events seem to negate it. In the book of Exodus, the promise is threatened by Pharaoh's mad design to destroy his own slave market. In the book of Leviticus, attention shifts to matters pertaining to cultic life. In the book of Numbers, the people are on the move again, though murmuring in doubt and disbelief. Finally, in the book of Deuteronomy, Moses preaches to the people about God's faithfulness to the covenant despite their infidelity, and about the people's obligations in response to God's acts of "steadfast love." When the book of Deuteronomy concludes, at the death of Moses, the promise is still on the way toward fulfillment. Thus the Torah is open-ended; it concludes with the people moving toward the horizon of God's future.

The priestly editors have enriched the whole Torah story by placing it in the scheme of the history of God's covenants. The first covenant is a covenant of creation, made with Noah and, through him, with all humanity, the non-human creatures, and the earth itself (Gen 9.1–17). The second, made with Abraham and Sarah, guarantees to the people of Israel the promises of land, posterity, and relationship with God (Gen 17.1–21). The third covenant, the one mediated by Moses at Mount Sinai (Ex 19–24), is regarded as a ratification and extension of the covenant with the ancestors (Ex 2.24). In the priestly view, which now dominates the whole Pentateuch, these are "everlasting covenants" that cannot be abrogated. Regardless of human weakness or failure, the covenant relationship with human beings universally and with Israel in particular endures, for it is grounded firmly in God's covenant loyalty (Hebrew *hesed*) or, in the appropriate NRSV rendering, God's "steadfast love."

Genesis

Genesis, meaning "origin," covers the time from creation to the Israelite sojourn in Egypt. The book falls naturally into two main sections. The primeval history (chs 1–11), which is universal in scope, tells how the blessing of God enabled humanity to multiply, diversify, and disperse on the face of the earth. The ancestral history (chs 12–50), on the other hand, deals with the limited family history of Israel's ancestors: Abraham and Sarah (chs 12–25), Isaac and Rebekah and their twin sons Esau and Jacob (chs 26–46), and Jacob's family, the chief member of which was Joseph (chs 37–50).

The primeval history reflects a "prehistorical" or mythical view of the movement from creation to the return of chaos in a catastrophic flood and the new beginning afterwards, while the ancestral history can be read, at least to some degree, in the context of the history of the Near East in the latter part of the second millennium (1500–1200 B.C.). The primary purpose of the book, however, is not to present straightforward history but to tell the dramatic story of God's dealings with the world and, in particular, to interpret Israel's special role in God's purpose.

Thus the migration of Abram and Abraham's family in response to God's promise (12.1–3) is the turning point in the unfolding story. God's creation had been marred by human violence that, under the judgment of God, threatened the earth with a return to pre-creation chaos. Out of this fallible human material, however, God gradually separated one family line, promising that the descendants of Abraham and Sarah would increase in number, receive a land, and have a relationship with God that would benefit other peoples (12.1–3). The promise in its threefold aspect is threatened by various experiences, such as Sarah's barrenness or Jacob's having to flee from the land of Canaan. Despite trials and tribulations, however, the people move toward the horizon of God's future, and when the book ends, in the time of Joseph's wise and benevolent administration of Egypt, the promise is pressing toward realization.

1 In the beginning when God created[a] the heavens and the earth, [2] the earth was a formless void and darkness covered the face of the deep, while a wind from God[b] swept over the face of the waters. [3] Then God said, "Let there be light"; and there was light. [4] And God saw that the light was good; and God separated the light from the darkness. [5] God called the light Day, and the darkness he called Night. And there was evening and there was morning, the first day.

6 And God said, "Let there be a dome in the midst of the waters, and let it separate the waters from the waters." [7] So God made the dome and separated the waters that were under the dome from the waters that were above the dome. And it was so. [8] God called the dome Sky. And there was evening and there was morning, the second day.

9 And God said, "Let the waters under the sky be gathered together into one place, and let the dry land appear." And it was so. [10] God called the dry land Earth, and the waters that were gathered together he called Seas. And God saw that it was good. [11] Then God said, "Let the earth put forth vegetation: plants yielding seed, and fruit trees of every kind on earth that bear fruit with the seed in it." And it was so. [12] The earth brought forth vegetation: plants yielding seed of every kind, and trees of every kind bearing fruit with the seed in it. And God saw that it was good. [13] And there was evening and there was morning, the third day.

14 And God said, "Let there be lights in the dome of the sky to separate the day from the night; and let them be for signs and for seasons and for days and years, [15] and let them be lights in the dome of the sky to give light upon the earth." And it was so. [16] God made the two great lights—the greater light to rule the day and the lesser light to rule the night—and the stars. [17] God set them in the dome of the sky to give light upon the earth, [18] to rule over the day and over the night, and to separate the light from the darkness. And God saw that it was good. [19] And there was evening and there was morning, the fourth day.

20 And God said, "Let the waters bring forth swarms of living creatures, and let birds fly above the earth across

a Or *when God began to create* or *In the beginning God created*   *b* Or *while the spirit of God* or *while a mighty wind*

1.1–2.3: The story of creation. Out of original chaos God created an orderly world, assigning a preeminent place to human beings. **1:** The traditional translation as an independent sentence, following the Greek Bible (Septuagint) of the 3rd cent. B.C., is defensible, in which case 1.1 is a thematic sentence, corresponding to the climactic summary of 2.1. Many, however, favor *When God began to create* (note *a*) . . . , taking the verse to be introductory to v. 2 (as above) or possibly to the first act of creation in v. 3 (compare 2.4b–7). The ancients believed the world originated from and was founded upon a watery abyss (*the deep;* compare Ps 24.2; 104.6), portrayed as a sea-monster in various myths (Isa 51.9). **3–5:** Creation by the word of God (Ps 33.6) expresses God's absolute sovereignty and anticipates the doctrine of creation out of nothing (2 Macc 7.28). *Light* burst forth first (2 Cor 4.6), even before the creation of the sun (vv. 14–18), and was *separated* from *night,* a remnant of uncreated darkness (v. 2). Since the Jewish day began with sundown, the order is *evening* and *morning.*

1.6–8: *A dome,* the sky was understood to be a solid expanse capable of separating the upper from the lower waters (Ex 20.4; Ps 148.4). See 7.11 n. **1.9–10:** The *Seas,* remnants of the watery chaos, were assigned boundaries at the edge of the earth (Ps 139.9; Prov 8.29), where they continue to menace God's creation (Ps 104.7–9). **11–12:** *Vegetation* was created only indirectly by God, whose creative command empowered mother earth to become fertile. **14–19:** The sun, moon, and stars are not divine powers that control human destiny, as was believed in antiquity, but are only *lights.* Implicitly, worship of the heavenly host is forbidden (Deut 4.19; Zeph 1.5). **1.20–23:** The creation of birds and fishes marks the first appearance of life on the earth. *Sea monsters,* Pss 74.13; 104.25–26. **24–25:** God's command for the earth to *bring forth* (a maternal verb, compare v. 20) suggests that

the dome of the sky." [21] So God created the great sea monsters and every living creature that moves, of every kind, with which the waters swarm, and every winged bird of every kind. And God saw that it was good. [22] God blessed them, saying, "Be fruitful and multiply and fill the waters in the seas, and let birds multiply on the earth." [23] And there was evening and there was morning, the fifth day.

[24] And God said, "Let the earth bring forth living creatures of every kind: cattle and creeping things and wild animals of the earth of every kind." And it was so. [25] God made the wild animals of the earth of every kind, and the cattle of every kind, and everything that creeps upon the ground of every kind. And God saw that it was good.

[26] Then God said, "Let us make humankind[c] in our image, according to our likeness; and let them have dominion over the fish of the sea, and over the birds of the air, and over the cattle, and over all the wild animals of the earth,[d] and over every creeping thing that creeps upon the earth."

[27] So God created humankind[c] in his
image,
in the image of God he created
them;[e]
male and female he created
them.

[28] God blessed them, and God said to them, "Be fruitful and multiply, and fill the earth and subdue it; and have dominion over the fish of the sea and over the birds of the air and over every living thing that moves upon the earth." [29] God said, "See, I have given you every plant yielding seed that is upon the face of all the earth, and every tree with seed in its fruit; you shall have them for food. [30] And to every beast of the earth, and to every bird of the air, and to everything that creeps on the earth, everything that has the breath of life, I have given every green plant for food." And it was so. [31] God saw everything that he had made, and indeed, it was very good. And there was evening and there was morning, the sixth day.

2 Thus the heavens and the earth were finished, and all their multitude. [2] And on the seventh day God finished the work that he had done, and he rested on the seventh day from all the work that he had done. [3] So God blessed the seventh day and hallowed it, because on it God rested from all the work that he had done in creation.

[4] These are the generations of the heavens and the earth when they were created.

---

c Heb *adam*   d Syr: Heb *and over all the earth*
e Heb *him*

---

the animals are immediately bound to *the ground* and only indirectly related to God, in contrast to human beings. **26–27:** The solemn divine declaration emphasizes humanity's supreme place at the climax of God's creative work.
1.26: The plural *us, our* (3.22; 11.7; Isa 6.8) probably refers to the divine beings who compose God's heavenly court (1 Kings 22.19; Job 1.6). *Image, likeness,* refer not to physical appearance but to relationship and activity. *Humankind* is commissioned to manifest God's rule on earth, on the analogy of a child who represents a parent (see 5.3). **27–28:** *Them,* literally "him" or "it" (see note *e*), referring to humankind; however, humanity is differentiated sexually, and to "them," *male and female,* God gives power to reproduce their kind and to exercise *dominion* over the

earth. Together men and women, made in the image of God, share the task of being God's stewards on earth (Ps 8.6–8). **29–30:** Human dominion is limited, as shown by the vegetarian requirement; in fact, there is to be no killing, a command that was relaxed in Noah's time (9.2–3). Human dominion, corresponding to God's rule, is to be benevolent and peaceful (compare Isa 11.6–8). **31:** *Very good* (vv. 4, 10, 12, etc.), corresponding perfectly to God's creative intention.
2.1–3: The verb *rested* (Hebrew "shabat") is the basis of the noun sabbath (Ex 31.12–17).
2.4a: Not the conclusion of the priestly creation story, but a separate caption introducing the following material, as elsewhere in Gen (e.g. 5.1; 6.9; 10.1).
2.4b–25: The creation of man and

In the day that the LORD God made the earth and the heavens, 5 when no plant of the field was yet in the earth and no herb of the field had yet sprung up—for the LORD God had not caused it to rain upon the earth, and there was no one to till the ground; 6 but a stream would rise from the earth, and water the whole face of the ground— 7 then the LORD God formed man from the dust of the ground, *f* and breathed into his nostrils the breath of life; and the man became a living being. 8 And the LORD God planted a garden in Eden, in the east; and there he put the man whom he had formed. 9 Out of the ground the LORD God made to grow every tree that is pleasant to the sight and good for food, the tree of life also in the midst of the garden, and the tree of the knowledge of good and evil.

10 A river flows out of Eden to water the garden, and from there it divides and becomes four branches. 11 The name of the first is Pishon; it is the one that flows around the whole land of Havilah, where there is gold; 12 and the gold of that land is good; bdellium and onyx stone are there. 13 The name of the second river is Gihon; it is the one that flows around the whole land of Cush. 14 The name of the third river is Tigris, which flows east of Assyria. And the fourth river is the Euphrates.

15 The LORD God took the man and put him in the garden of Eden to till it and keep it. 16 And the LORD God commanded the man, "You may freely eat of every tree of the garden; 17 but of the tree of the knowledge of good and evil you shall not eat, for in the day that you eat of it you shall die."

18 Then the LORD God said, "It is not good that the man should be alone; I will make him a helper as his partner." 19 So out of the ground the LORD God formed every animal of the field and every bird of the air, and brought them to the man to see what he would call them; and whatever the man called every living creature, that was its name. 20 The man gave names to all cattle, and to the birds of the air, and to every animal of the field; but for the man*g* there was not found a helper as his partner. 21 So the LORD God caused a deep sleep to fall upon the man, and he slept; then he took one of his ribs and closed up its place

f Or *formed a man* (Heb *adam*) *of dust from the ground* (Heb *adamah*)   *g* Or *for Adam*

---

woman. This is a different tradition from 1.1–2.3 as evidenced by the flowing style and the different order of events of creation. **6:** *Stream* probably refers to the moisture that welled up from the subterranean ocean, the source of fertility (49.25). **7:** The word-play on "'adam" (human being; here translated *man*) and "'adamah" (ground, soil) introduces a motif characteristic of this early tradition: the relation of humankind to the soil from which it was *formed,* as a potter molds clay (Jer 18.6). Human nature is not a duality of body and soul; rather God's *breath* animates the dust and it becomes *a living being* or psychophysical self (Ps 104.29; Job 34.14–15). **8–9:** *Eden,* meaning "delight," is a "garden of God" (Isa 51.3; Ezek 31.8–9; Joel 2.3). **9:** Ancients believed that the *tree of life* confers eternal life (3.22; see Prov 3.18; Rev 22.2, 14, 19), as *the tree of the knowledge of good and evil* confers wisdom (see 2 Sam 14.17; Isa 7.15). **2.10–14:** The rivers, springing from the subterranean ocean (v. 6), flowed out to the four corners of the known historical world, particularly the valley of the *Tigris* and *Euphrates* (Mesopotamia). **15–17:** *The man,* here, and often in the story, "'adam" is ambiguous. God put *him,* apparently a masculine being, in the garden to cultivate and take care of it; yet the prohibition against eating the forbidden fruit applies to the man and the woman inclusively (3.3). Gender distinction is not emphasized until vv. 21–25 where a different word is used to distinguish man ("'ish") from woman ("'ishah"). **2.18–21:** To be fully human one needs to be in relation to others who correspond to oneself. *Helper,* not in a relationship of subordination but of mutuality and interdependence (see 1.27–28 n.). **21–23:** Creation from the man's rib shows an affinity between man and woman such as is not possible between humans and animals. The affinity is expressed poetically in the jubilant cry of v. 23, with its word play on "man" ("'ish") and "woman" ("'ishah"). **24–25:** Sex is not regarded as evil but as a God-given impulse that draws a man and a woman together so that *they become one*

with flesh. 22 And the rib that the LORD God had taken from the man he made into a woman and brought her to the man. 23 Then the man said,

"This at last is bone of my bones
and flesh of my flesh;
this one shall be called Woman, *h*
for out of Man *i* this one was
taken."

24 Therefore a man leaves his father and his mother and clings to his wife, and they become one flesh. 25 And the man and his wife were both naked, and were not ashamed.

3 Now the serpent was more crafty than any other wild animal that the LORD God had made. He said to the woman, "Did God say, 'You shall not eat from any tree in the garden'?" 2 The woman said to the serpent, "We may eat of the fruit of the trees in the garden; 3 but God said, 'You shall not eat of the fruit of the tree that is in the middle of the garden, nor shall you touch it, or you shall die.' " 4 But the serpent said to the woman, "You will not die; 5 for God knows that when you eat of it your eyes will be opened, and you will be like God, *j* knowing good and evil." 6 So when the woman saw that the tree was good for food, and that it was a delight to the eyes, and that the tree was to be desired to make one wise, she took of its fruit and ate; and she also gave some to her husband, who was with her, and he ate. 7 Then the eyes of both were opened, and they knew that they were naked; and they sewed fig leaves together and made loincloths for themselves.

8 They heard the sound of the LORD God walking in the garden at the time of the evening breeze, and the man and his wife hid themselves from the presence of the LORD God among the trees of the garden. 9 But the LORD God called to the man, and said to him, "Where are you?" 10 He said, "I heard the sound of you in the garden, and I was afraid, because I was naked; and I hid myself." 11 He said, "Who told you that you were naked? Have you eaten from the tree of which I commanded you not to eat?" 12 The man said, "The woman whom you gave to be with me, she gave me fruit from the tree, and I ate." 13 Then the LORD God said to the woman, "What is this that you have done?" The woman said, "The serpent tricked me, and I ate." 14 The LORD God said to the serpent,

"Because you have done this,
cursed are you among all
animals
and among all wild creatures;
upon your belly you shall go,
and dust you shall eat
all the days of your life.
15 I will put enmity between you and
the woman,
and between your offspring and
hers;
he will strike your head,
and you will strike his heel."
16 To the woman he said,
"I will greatly increase your pangs
in childbearing;

---

h Heb *ishshah*   *i* Heb *ish*   *j* Or *gods*

---

flesh. The two were unashamedly naked, a symbol of their guiltless relation to God and to one another.

3.1–24: The temptation story. 3.1–7: The temptation begins with an insinuation of doubt (vv. 1–3), increases as suspicion is cast upon God's motive (vv. 4–5), and becomes irresistible when the couple sense the possibilities of freedom (v. 6). **1:** *The serpent,* one of the wild creatures, distinguished by uncanny wisdom (Mt 10.16); there is a hint of an evil power, hostile to God, out in the world. **5:** *Like God,* perhaps "like gods" (Septuagint), the divine beings of the heavenly court

(v. 22; 1.26 n.). *Knowing good and evil,* the entirety of knowledge; see 2.9 n. **7:** Bodily shame (2.25) symbolizes loss of an innocent, trusting relationship with God.

3.8–13: Guilt and anxiety prompt an attempt to hide from God (Ps 139.7–12), who is here picturesquely portrayed as strolling in the garden to enjoy the cool evening breeze. **14–15:** The curse contains an old explanation of why the serpent crawls rather than walks and why people are instinctively hostile to it. **16:** This divine judgment contains an old explanation of woman's pain in childbirth, her sexual *desire* for her husband (i.e. her motherly

in pain you shall bring forth
children,
yet your desire shall be for your
husband,
and he shall rule over you."
[17] And to the man[k] he said,
"Because you have listened to the
voice of your wife,
and have eaten of the tree
about which I commanded you,
'You shall not eat of it,'
cursed is the ground because of
you;
in toil you shall eat of it all the
days of your life;
[18] thorns and thistles it shall bring
forth for you;
and you shall eat the plants of
the field.
[19] By the sweat of your face
you shall eat bread
until you return to the ground,
for out of it you were taken;
you are dust,
and to dust you shall return."

20 The man named his wife Eve,[l] because she was the mother of all living.
[21] And the LORD God made garments of
skins for the man[m] and for his wife, and
clothed them.

22 Then the LORD God said, "See, the
man has become like one of us, knowing
good and evil; and now, he might reach
out his hand and take also from the tree
of life, and eat, and live forever"—
[23] therefore the LORD God sent him forth
from the garden of Eden, to till the
ground from which he was taken. [24] He
drove out the man; and at the east of the
garden of Eden he placed the cherubim,
and a sword flaming and turning to
guard the way to the tree of life.

4 Now the man knew his wife Eve,
and she conceived and bore Cain,
saying, "I have produced[n] a man with
the help of the LORD." [2] Next she bore his
brother Abel. Now Abel was a keeper of
sheep, and Cain a tiller of the ground. [3] In
the course of time Cain brought to the
LORD an offering of the fruit of the
ground, [4] and Abel for his part brought
of the firstlings of his flock, their fat portions. And the LORD had regard for Abel
and his offering, [5] but for Cain and his
offering he had no regard. So Cain was
very angry, and his countenance fell.
[6] The LORD said to Cain, "Why are you
angry, and why has your countenance
fallen? [7] If you do well, will you not be

k Or *to Adam*   l In Heb *Eve* resembles the
word for *living*   m Or *for Adam*
n The verb in Heb resembles the word for *Cain*

---

impulse, compare 30.1), and her subordinate
position to man in ancient society, in contrast
with the ideal equality of creation (see 1.27–
28 n.; 2.18–21 n.).

3.17–19: An explanation of why people
("'adam") have to struggle to eke out an existence from the ground ("'adamah"). Work is
not intrinsically evil (2.15) but it becomes *toil*
when relationship with the Creator is broken.
In v. 17 and elsewhere (2.20; 3.21) later editors took the Hebrew word "'adam" to be a
personal name (see note *k*), as in the genealogy of 5.1ff. The mortal nature of humanity
was implicit in the circumstances of its origin
from dust (2.7); because of human disobedience, God now makes death an inevitable fate
that haunts human beings throughout life until they *return to the ground* ("'adamah"). **21:**
Garments of skins, a sign of God's protective
care even in the time of judgment (4.15).

3.22: *Like one of us,* see 3.5 n. *The tree of life*
(2.9) does not figure in the temptation story,
which explicitly speaks of only one tree in the
center of the garden (3.3–6, 11–12, 17). **24:**
The cherubim, guardians of sacred areas
(1 Kings 8.6–7), were represented as winged
creatures like the Sphinx of Egypt, half-
human and half-lion (Ezek 41.18–19). A divine *sword* (compare Jer 47.6) was placed near
the cherubim to warn banished human beings
of the impossibility of overstepping their
creaturely bounds (compare Ezek 28.13–16).

4.1–26: Cain, Abel, and Seth. 1: The
verb translated *produced* (or "brought into being," "created") is a play on the name of Cain.
There is a suggestion that Eve's conception of
her first child was something marvellous:
God took part in her act of creation. **2–5:** The
story reflects the tension between farmers and
semi-nomads, two different ways of life that
are symbolized in the two types of offerings.
No reason is given for the acceptance of
Abel's offering (compare Ex 33.19). **7:** Perhaps the meaning is that Cain himself will be
accepted, even though his offering is not, if his
deed springs from the right motive. Sin is

accepted? And if you do not do well, sin is lurking at the door; its desire is for you, but you must master it."

8 Cain said to his brother Abel, "Let us go out to the field."*o* And when they were in the field, Cain rose up against his brother Abel, and killed him. 9 Then the LORD said to Cain, "Where is your brother Abel?" He said, "I do not know; am I my brother's keeper?" 10 And the LORD said, "What have you done? Listen; your brother's blood is crying out to me from the ground! 11 And now you are cursed from the ground, which has opened its mouth to receive your brother's blood from your hand. 12 When you till the ground, it will no longer yield to you its strength; you will be a fugitive and a wanderer on the earth." 13 Cain said to the LORD, "My punishment is greater than I can bear! 14 Today you have driven me away from the soil, and I shall be hidden from your face; I shall be a fugitive and a wanderer on the earth, and anyone who meets me may kill me." 15 Then the LORD said to him, "Not so!*p* Whoever kills Cain will suffer a sevenfold vengeance." And the LORD put a mark on Cain, so that no one who came upon him would kill him. 16 Then Cain went away from the presence of the LORD, and settled in the land of Nod,*q* east of Eden.

17 Cain knew his wife, and she conceived and bore Enoch; and he built a city, and named it Enoch after his son Enoch. 18 To Enoch was born Irad; and

Irad was the father of Mehujael, and Mehujael the father of Methushael, and Methushael the father of Lamech. 19 Lamech took two wives; the name of the one was Adah, and the name of the other Zillah. 20 Adah bore Jabal; he was the ancestor of those who live in tents and have livestock. 21 His brother's name was Jubal; he was the ancestor of all those who play the lyre and pipe. 22 Zillah bore Tubal-cain, who made all kinds of bronze and iron tools. The sister of Tubal-cain was Naamah.

23 Lamech said to his wives:

"Adah and Zillah, hear my voice;
    you wives of Lamech, listen to
        what I say:
I have killed a man for wounding
        me,
    a young man for striking me.
24 If Cain is avenged sevenfold,
    truly Lamech
        seventy-sevenfold."

25 Adam knew his wife again, and she bore a son and named him Seth, for she said, "God has appointed*r* for me another child instead of Abel, because Cain killed him." 26 To Seth also a son was born, and he named him Enosh. At that time people began to invoke the name of the LORD.

5 This is the list of the descendants of Adam. When God created human-

o Sam Gk Syr Compare Vg: MT lacks *Let us go out to the field*   *p* Gk Syr Vg: Heb *Therefore* *q* That is *Wandering*   *r* The verb in Heb resembles the word for *Seth*

---

pictured as a predatory animal, *lurking at the door.*
4.10–11: Blood is sacred to God, for it is the seat of life (Deut 12.23) and cries *from the ground* for vindication. **13–14:** Cain concludes that exile from the farmland is also exile from the LORD's *face,* i.e. protective presence, exposing him to blood revenge. **15:** The "mark of Cain" was a protective mark, perhaps a tattoo, signifying divine mercy.
4.17: Here Cain is not the ancestor of nomadic tribesmen (vv. 11–16) but the founder of sedentary culture. **19–22:** Cultural advance is evidenced by the three occupations of Lamech's sons: shepherds, musicians, and smiths. **23–24:** An ancient song, probably

once sung in praise of Lamech, is here quoted to illustrate the increase of violence from murder to measureless blood revenge. **25–26:** From Cain's genealogy the narrator returns to the sequel of Cain's banishment (vv. 11–16) and introduces the new line of Seth. **26b:** This tradition traces the worship of the LORD (Yahweh) back to the time of Adam's grandson (5.3), in contrast to other traditions which claim that the sacred name was introduced in Moses' time (Ex 3.13–15; 6.2–3).
5.1–32: The generations from Adam to Noah. This priestly tradition bridges the time from creation to the flood. **1:** *The list of descendants,* lit. "the book of generations,"

kind,[s] he made them[t] in the likeness of God. [2]Male and female he created them, and he blessed them and named them "Humankind"[s] when they were created.

[3] When Adam had lived one hundred thirty years, he became the father of a son in his likeness, according to his image, and named him Seth. [4]The days of Adam after he became the father of Seth were eight hundred years; and he had other sons and daughters. [5]Thus all the days that Adam lived were nine hundred thirty years; and he died.

[6] When Seth had lived one hundred five years, he became the father of Enosh. [7]Seth lived after the birth of Enosh eight hundred seven years, and had other sons and daughters. [8]Thus all the days of Seth were nine hundred twelve years; and he died.

[9] When Enosh had lived ninety years, he became the father of Kenan. [10]Enosh lived after the birth of Kenan eight hundred fifteen years, and had other sons and daughters. [11]Thus all the days of Enosh were nine hundred five years; and he died.

[12] When Kenan had lived seventy years, he became the father of Mahalalel. [13]Kenan lived after the birth of Mahalalel eight hundred and forty years, and had other sons and daughters. [14]Thus all the days of Kenan were nine hundred and ten years; and he died.

[15] When Mahalalel had lived sixty-five years, he became the father of Jared. [16]Mahalalel lived after the birth of Jared eight hundred thirty years, and had other sons and daughters. [17]Thus all the days of Mahalalel were eight hundred ninety-five years; and he died.

[18] When Jared had lived one hundred sixty-two years he became the father of Enoch. [19]Jared lived after the birth of Enoch eight hundred years, and had other sons and daughters. [20]Thus all the days of Jared were nine hundred sixty-two years; and he died.

[21] When Enoch had lived sixty-five years, he became the father of Methuselah. [22]Enoch walked with God after the birth of Methuselah three hundred years, and had other sons and daughters. [23]Thus all the days of Enoch were three hundred sixty-five years. [24]Enoch walked with God; then he was no more, because God took him.

[25] When Methuselah had lived one hundred eighty-seven years, he became the father of Lamech. [26]Methuselah lived after the birth of Lamech seven hundred eighty-two years, and had other sons and daughters. [27]Thus all the days of Methuselah were nine hundred sixty-nine years; and he died.

[28] When Lamech had lived one hundred eighty-two years, he became the father of a son; [29]he named him Noah, saying, "Out of the ground that the LORD has cursed this one shall bring us relief from our work and from the toil of our hands." [30]Lamech lived after the birth of Noah five hundred ninety-five years, and had other sons and daughters. [31]Thus all the days of Lamech were seven

s Heb *adam*   t Heb *him*

---

was evidently a separate source from which the writer drew genealogical data (6.9; 10.2; 11.10, 27; etc.). **1b–2:** Ambiguously the Hebrew word "'adam" refers to *humankind* inclusively, as in 1.26–28, and to a male individual, Adam, as in the succeeding genealogy (see 2.17 n.). **3:** The divine *likeness* (v. 1; see 1.26 n.) was continued in Adam's son Seth, born *in his likeness,* and thus was transmitted to succeeding generations without effacement (9.6). **4–32:** Ancient Babylonian tradition also reckons ten heroes before the flood but ascribes fantastically higher ages. In Hebrew tradition the ages decrease from 900–1000 (Adam to Noah) to 200–600 (Noah to Abraham), to 100–200 (Israel's ancestors), to the normal lifetime of 70 years (Ps 90.10). This list is somehow related to the genealogy of Cain (4.17–21) as shown by the resemblance of some of the names.

5.24: Babylonian tradition also reports that Enmeduranki, the seventh hero prior to the flood, was taken by God, i.e. translated (2 Kings 2.11). **29:** This verse, the only connection with the old epic traditions of Eden (3.17–19) and Cain and Abel, anticipates the new age inaugurated with Noah (9.20).

6.1–4: The birth of the Nephilim is re-

hundred seventy-seven years; and he died.

32 After Noah was five hundred years old, Noah became the father of Shem, Ham, and Japheth.

6 When people began to multiply on the face of the ground, and daughters were born to them, 2the sons of God saw that they were fair; and they took wives for themselves of all that they chose. 3Then the LORD said, "My spirit shall not abide*u* in mortals forever, for they are flesh; their days shall be one hundred twenty years." 4The Nephilim were on the earth in those days—and also afterward—when the sons of God went in to the daughters of humans, who bore children to them. These were the heroes that were of old, warriors of renown.

5 The LORD saw that the wickedness of humankind was great in the earth, and that every inclination of the thoughts of their hearts was only evil continually. 6And the LORD was sorry that he had made humankind on the earth, and it grieved him to his heart. 7So the LORD said, "I will blot out from the earth the human beings I have created—people together with animals and creeping things and birds of the air, for I am sorry that

I have made them." 8But Noah found favor in the sight of the LORD.

9 These are the descendants of Noah. Noah was a righteous man, blameless in his generation; Noah walked with God. 10And Noah had three sons, Shem, Ham, and Japheth.

11 Now the earth was corrupt in God's sight, and the earth was filled with violence. 12And God saw that the earth was corrupt; for all flesh had corrupted its ways upon the earth. 13And God said to Noah, "I have determined to make an end of all flesh, for the earth is filled with violence because of them; now I am going to destroy them along with the earth. 14Make yourself an ark of cypress*u* wood; make rooms in the ark, and cover it inside and out with pitch. 15This is how you are to make it: the length of the ark three hundred cubits, its width fifty cubits, and its height thirty cubits. 16Make a roof*v* for the ark, and finish it to a cubit above; and put the door of the ark in its side; make it with lower, second, and third decks. 17For my part, I am going to bring a flood of waters on the earth, to destroy from under heaven all

u Meaning of Heb uncertain   *v* Or *window*

---

lated to show the increase of violence (see 6.11), in this instance through the breaching of the boundaries separating heaven and earth. This old fragment of mythology connects immediately with chs 2–4. **1:** The narrator resumes the motif of the connection of *people* or human beings ("'adam") to the soil ("'adamah"), introduced in 2.4b–9. **2:** *The sons of God,* divine beings who belonged to the heavenly court (1.26 n.). **3:** Despite the lustful intrusion of divine beings into the human sphere, human beings did not become semi-divine (compare 3.22–24) but remained mortal creatures in whom the LORD's *spirit* dwells temporarily (see 2.7 n.). **4:** Originally the story accounted for the *Nephilim* (Num 13.33; Deut 2.10–11), people of gigantic stature whose superhuman power was thought to result from divine-human marriage.

6.5–8.22: The great flood. God's judgment took the form of a destructive flood and God's mercy was shown in saving a remnant, the seed of a new historical beginning. **5–8:** An introduction, belonging to the old literary

tradition found in 2.4b–3.24; 4.1–26; 6.1–4. **5:** The *heart* includes the will and reason, as evidenced by human capacity to decide between good and evil. **7:** The biblical account is superficially similar to the Babylonian Gilgamesh Epic, which also relates the story of a great flood. The biblical perspective, however, is basically different, for the flood was not the expression of polytheistic caprice but of God's judgment upon *the wickedness of humankind.* **9:** Noah was *a righteous man,* one who stood in right relationship to God (15.6).

6.11–22: A parallel version. Apparently an earlier and a later (priestly) tradition have been combined editorially, resulting in some duplications and discrepancies. **11–12:** The keynote of the dominant priestly version of the story: the earth, once described as "good" (1.31), is seen to be *corrupt* owing to human *violence* or willful, lawless deeds, beginning with rebellion in the garden. **14–16:** In the Babylonian epic too, the hero is commanded to build a houseboat, sealing it with pitch. **15:** The dimensions: about 450 × 75 × 45 feet.

flesh in which is the breath of life; everything that is on the earth shall die. 18 But I will establish my covenant with you; and you shall come into the ark, you, your sons, your wife, and your sons' wives with you. 19 And of every living thing, of all flesh, you shall bring two of every kind into the ark, to keep them alive with you; they shall be male and female. 20 Of the birds according to their kinds, and of the animals according to their kinds, of every creeping thing of the ground according to its kind, two of every kind shall come in to you, to keep them alive. 21 Also take with you every kind of food that is eaten, and store it up; and it shall serve as food for you and for them." 22 Noah did this; he did all that God commanded him.

7 Then the Lord said to Noah, "Go into the ark, you and all your household, for I have seen that you alone are righteous before me in this generation. 2 Take with you seven pairs of all clean animals, the male and its mate; and a pair of the animals that are not clean, the male and its mate; 3 and seven pairs of the birds of the air also, male and female, to keep their kind alive on the face of all the earth. 4 For in seven days I will send rain on the earth for forty days and forty nights; and every living thing that I have made I will blot out from the face of the ground." 5 And Noah did all that the Lord had commanded him.

6 Noah was six hundred years old when the flood of waters came on the earth. 7 And Noah with his sons and his wife and his sons' wives went into the ark to escape the waters of the flood. 8 Of clean animals, and of animals that are not clean, and of birds, and of everything that creeps on the ground, 9 two and two, male and female, went into the ark with Noah, as God had commanded Noah. 10 And after seven days the waters of the flood came on the earth.

11 In the six hundredth year of Noah's life, in the second month, on the seventeenth day of the month, on that day all the fountains of the great deep burst forth, and the windows of the heavens were opened. 12 The rain fell on the earth forty days and forty nights. 13 On the very same day Noah with his sons, Shem and Ham and Japheth, and Noah's wife and the three wives of his sons entered the ark, 14 they and every wild animal of every kind, and all domestic animals of every kind, and every creeping thing that creeps on the earth, and every bird of every kind—every bird, every winged creature. 15 They went into the ark with Noah, two and two of all flesh in which there was the breath of life. 16 And those that entered, male and female of all flesh, went in as God had commanded him; and the Lord shut him in.

17 The flood continued forty days on the earth; and the waters increased, and bore up the ark, and it rose high above the earth. 18 The waters swelled and increased greatly on the earth; and the ark floated on the face of the waters. 19 The waters swelled so mightily on the earth that all the high mountains under the whole heaven were covered; 20 the waters swelled above the mountains, covering them fifteen cubits deep. 21 And all flesh

---

7.1–10: Essentially a continuation of the early tradition (6.5–8). **2–3:** On clean and unclean animals, see Lev ch 11. (The priestly version mentions two animals of every sort [v. 9; 6.19], presuming that the clean-unclean distinction was introduced at Sinai.) **4:** The flood was caused by heavy rainfall, lasting *forty days and forty nights* (v. 12; compare the difference in the priestly version, v. 24). **11–24:** Largely from priestly tradition. **11:** Here the flood was not caused by a rain storm but is a cosmic catastrophe resulting from opening the *windows of the heavens* (fixed in the firmament) and the upsurging of *the fountains of the great deep* (or the subterranean watery chaos; see 1.6–8 n.). Thus the earth was threatened with a return to pre-creation chaos (1.2). **15:** The animals went in *two and two* (6.19; see 7.2 n.).

7.18–20: The waters covered *all the high mountains,* threatening a confluence of the upper and lower waters (1.6). Archaeological evidence suggests that traditions of a prehistoric flood covering the whole earth are heightened versions of local inundations, e.g. in the Tigris-Euphrates basin.

died that moved on the earth, birds, domestic animals, wild animals, all swarming creatures that swarm on the earth, and all human beings; 22everything on dry land in whose nostrils was the breath of life died. 23He blotted out every living thing that was on the face of the ground, human beings and animals and creeping things and birds of the air; they were blotted out from the earth. Only Noah was left, and those that were with him in the ark. 24And the waters swelled on the earth for one hundred fifty days.

8 But God remembered Noah and all the wild animals and all the domestic animals that were with him in the ark. And God made a wind blow over the earth, and the waters subsided; 2the fountains of the deep and the windows of the heavens were closed, the rain from the heavens was restrained, 3and the waters gradually receded from the earth. At the end of one hundred fifty days the waters had abated; 4and in the seventh month, on the seventeenth day of the month, the ark came to rest on the mountains of Ararat. 5The waters continued to abate until the tenth month; in the tenth month, on the first day of the month, the tops of the mountains appeared.

6 At the end of forty days Noah opened the window of the ark that he had made 7and sent out the raven; and it went to and fro until the waters were dried up from the earth. 8Then he sent out the dove from him, to see if the waters had subsided from the face of the ground; 9but the dove found no place to set its foot, and it returned to him to the ark, for the waters were still on the face of the whole earth. So he put out his hand and took it and brought it into the ark with him. 10He waited another seven days, and again he sent out the dove from the ark; 11and the dove came back to him in the evening, and there in its beak was a freshly plucked olive leaf; so Noah knew that the waters had subsided from the earth. 12Then he waited another seven days, and sent out the dove; and it did not return to him any more.

13 In the six hundred first year, in the first month, the first day of the month, the waters were dried up from the earth; and Noah removed the covering of the ark, and looked, and saw that the face of the ground was drying. 14In the second month, on the twenty-seventh day of the month, the earth was dry. 15Then God said to Noah, 16"Go out of the ark, you and your wife, and your sons and your sons' wives with you. 17Bring out with you every living thing that is with you of all flesh—birds and animals and every creeping thing that creeps on the earth—so that they may abound on the earth, and be fruitful and multiply on the earth." 18So Noah went out with his sons and his wife and his sons' wives. 19And every animal, every creeping thing, and every bird, everything that moves on the earth, went out of the ark by families.

20 Then Noah built an altar to the LORD, and took of every clean animal and of every clean bird, and offered

---

8.1–5: In the main a continuation of the dominant priestly tradition. 1: The dramatic turning point of the story: *God remembered Noah* and the remnant of humans and animals with him. The waters of chaos, which have peaked, are driven back by a wind (compare the *wind from God* of 1.2) sent by God. 4: In the Babylonian epic the boat also rested on a mountain. *Ararat* (2 Kings 19.37; Jer 51.27) is the name of a region in Armenia.
8.6–12: Essentially from the early tradition. In the Babylonian epic the hero sent out two birds, a dove and a swallow, each of which came back; the third, a raven, did not return.

8.13–19: A continuation of the priestly account with its theme of the greening of the earth (1.11–12). A new creation begins in which, as at the first, non-human creatures have their proper place (1.20–24).
8.20–22: The early tradition relates that Noah sacrificed *burnt offerings* (Lev ch 1) of clean animals (see 7.2–3 n.). In the Babylonian epic the hero offered sacrifices and "the gods smelled [compare v. 21] the goodly savor." For the curse, compare 3.17. Despite the *evil inclination of the human heart* (6.5), God's covenant faithfulness will be expressed in the regularities of nature, *seedtime and harvest*, etc.

burnt offerings on the altar. 21 And when the LORD smelled the pleasing odor, the LORD said in his heart, "I will never again curse the ground because of humankind, for the inclination of the human heart is evil from youth; nor will I ever again destroy every living creature as I have done.

22 As long as the earth endures,
   seedtime and harvest, cold and
      heat,
   summer and winter, day and
      night,
   shall not cease."

9 God blessed Noah and his sons, and said to them, "Be fruitful and multiply, and fill the earth. 2 The fear and dread of you shall rest on every animal of the earth, and on every bird of the air, on everything that creeps on the ground, and on all the fish of the sea; into your hand they are delivered. 3 Every moving thing that lives shall be food for you; and just as I gave you the green plants, I give you everything. 4 Only, you shall not eat flesh with its life, that is, its blood. 5 For your own lifeblood I will surely require a reckoning: from every animal I will require it and from human beings, each one for the blood of another, I will require a reckoning for human life.

6 Whoever sheds the blood of a
      human,
   by a human shall that person's
      blood be shed;
   for in his own image
   God made humankind.

7 And you, be fruitful and multiply, abound on the earth and multiply in it."

8 Then God said to Noah and to his sons with him, 9 "As for me, I am establishing my covenant with you and your descendants after you, 10 and with every living creature that is with you, the birds, the domestic animals, and every animal of the earth with you, as many as came out of the ark. *w* 11 I establish my covenant with you, that never again shall all flesh be cut off by the waters of a flood, and never again shall there be a flood to destroy the earth." 12 God said, "This is the sign of the covenant that I make between me and you and every living creature that is with you, for all future generations: 13 I have set my bow in the clouds, and it shall be a sign of the covenant between me and the earth. 14 When I bring clouds over the earth and the bow is seen in the clouds, 15 I will remember my covenant that is between me and you and every living creature of all flesh; and the waters shall never again become a flood to destroy all flesh. 16 When the bow is in the clouds, I will see it and remember the everlasting covenant between God and every living creature of all flesh that is on the earth." 17 God said to Noah, "This is the sign of the covenant that I have established between me and all flesh that is on the earth."

w Gk: Heb adds *every animal of the earth*

9.1–17: God's covenant with Noah includes all human and non-human creatures under divine promise and law. **1**: The new age opens with a renewal of the blessing given to human beings, as well as non-human creatures (8.17), at creation (1.22, 28). **3–6**: The command to exercise dominion (1.28–30) is qualified by permission to eat animal flesh, but not with *its life,* i.e. *its blood* (see 4.10–11 n.). The principle is reverence for life, God's gift, symbolized by blood (Lev 17.11). **6**: The violence that had corrupted the earth (6.11) is restrained by a very old law against murder, the validity of which is grounded in the creation: human beings are made in God's *image* (1.26–27). The laws given to

Noah are binding not only on Israel but on all humanity (Acts 15.20; 21.25).
   9.8–11: The preservation of the natural order from the powers of chaos is guaranteed by a *covenant* (see 17.2 n.). Unlike later covenants (ch 17; Ex ch 24), the covenant with Noah is universal in scope, for Noah's three sons (6.10; 9.18–19) are regarded as the ancestors of all the nations (see ch 10). This is also an ecological covenant, for it is made with *every living creature,* including birds and animals (vv. 10, 12, 15), and with *the earth* itself (v. 13). **13**: Ancients imagined the rainbow as the weapon (bow) of the Divine Warrior from which the lightnings of arrows were shot (Ps 7.12–13; Hab 3.9–11). The place-

18 The sons of Noah who went out of the ark were Shem, Ham, and Japheth. Ham was the father of Canaan. ¹⁹These three were the sons of Noah; and from these the whole earth was peopled.

20 Noah, a man of the soil, was the first to plant a vineyard. ²¹He drank some of the wine and became drunk, and he lay uncovered in his tent. ²²And Ham, the father of Canaan, saw the nakedness of his father, and told his two brothers outside. ²³Then Shem and Japheth took a garment, laid it on both their shoulders, and walked backward and covered the nakedness of their father; their faces were turned away, and they did not see their father's nakedness. ²⁴When Noah awoke from his wine and knew what his youngest son had done to him, ²⁵he said,

"Cursed be Canaan;
   lowest of slaves shall he be to
      his brothers."
²⁶He also said,
"Blessed by the LORD my God be
      Shem;
   and let Canaan be his slave.
²⁷ May God make space for ˣ
      Japheth,
   and let him live in the tents of
      Shem;
   and let Canaan be his slave."

28 After the flood Noah lived three hundred fifty years. ²⁹All the days of Noah were nine hundred fifty years; and he died.

10 These are the descendants of Noah's sons, Shem, Ham, and Japheth; children were born to them after the flood.

2 The descendants of Japheth: Gomer, Magog, Madai, Javan, Tubal, Meshech, and Tiras. ³The descendants of Gomer: Ashkenaz, Riphath, and Togarmah. ⁴The descendants of Javan: Elishah, Tarshish, Kittim, and Rodanim. ʸ ⁵From these the coastland peoples spread. These are the descendants of Japheth ᶻ in their lands, with their own language, by their families, in their nations.

6 The descendants of Ham: Cush, Egypt, Put, and Canaan. ⁷The descendants of Cush: Seba, Havilah, Sabtah, Raamah, and Sabteca. The descendants of Raamah: Sheba and Dedan. ⁸Cush became the father of Nimrod; he was the first on earth to become a mighty warrior. ⁹He was a mighty hunter before the LORD; therefore it is said, "Like Nimrod a mighty hunter before the LORD." ¹⁰The beginning of his kingdom was Babel, Erech, and Accad, all of them in the land of Shinar. ¹¹From that land he went into

x Heb *yapht*, a play on *Japheth*   y Heb Mss Sam Gk See 1 Chr 1.7: MT *Dodanim*
z Compare verses 20, 31. Heb lacks *These are the descendants of Japheth*

---

ment of this weapon in the heavens is a *sign*, or visible token, that God's wrath has abated. **9.18–27: Noah's curse upon Canaan. 20:** In the new age, Noah was the first to engage in agriculture. His success fulfilled the prophecy made at his birth (5.29). **22:** Since the curse was later put on Canaan rather than on Ham (v. 25), it is likely that Canaan was the actor originally. **24:** Here Noah's *youngest son* is clearly Canaan, not Ham as in v. 22. **25:** The curse implies that Canaan's subjugation to Israel was the result of Canaanite sexual practices (Lev 18.24–30). **26:** *Shem*, 10.21. **27:** *Japheth*, 10.2–5. The verse may refer to the Philistines, one of the sea-peoples who dwelt *in the tents of Shem*, i.e. conquered the coast of Canaan. **9.28–29:** The remainder of Noah's life and his death. **10.1–32: The table of the nations** provides a background of world history for the call of Abraham (ch 12). **1:** This list, which connects with 5.32, was probably drawn from "the book of generations" (5.1). The original unity of humanity is represented by the view that all the nations originated from Noah's three sons (9.19). Although the various "families" were separated by language and land (vv. 5, 20, 31), the present list is arranged primarily on the basis of political rather than ethnic considerations. **2–5:** *The descendants of Japheth* (9.27) had their political center in Asia Minor, the former territory of the Hittites (Heth, v. 15). The spread of the *coastland peoples,* including the Philistines (see 9.27 n.), reflects population movements in the Aegean area about 1200 B.C. **10.6–20:** *The descendants of Ham* lived in the Egyptian orbit. *Canaan* is included because it was nominally under Egyptian control from 1500–1200 B.C. **8–12:** An old fragment of tradition relates how Nimrod, a

Assyria, and built Nineveh, Rehoboth-ir, Calah, and 12 Resen between Nineveh and Calah; that is the great city. 13 Egypt became the father of Ludim, Anamim, Lehabim, Naphtuhim, 14 Pathrusim, Casluhim, and Caphtorim, from which the Philistines come. *a*

15 Canaan became the father of Sidon his firstborn, and Heth, 16 and the Jebusites, the Amorites, the Girgashites, 17 the Hivites, the Arkites, the Sinites, 18 the Arvadites, the Zemarites, and the Hamathites. Afterward the families of the Canaanites spread abroad. 19 And the territory of the Canaanites extended from Sidon, in the direction of Gerar, as far as Gaza, and in the direction of Sodom, Gomorrah, Admah, and Zeboiim, as far as Lasha. 20 These are the descendants of Ham, by their families, their languages, their lands, and their nations.

21 To Shem also, the father of all the children of Eber, the elder brother of Japheth, children were born. 22 The descendants of Shem: Elam, Asshur, Arpachshad, Lud, and Aram. 23 The descendants of Aram: Uz, Hul, Gether, and Mash. 24 Arpachshad became the father of Shelah; and Shelah became the father of Eber. 25 To Eber were born two sons: the name of the one was Peleg, *b* for in his days the earth was divided, and his brother's name was Joktan. 26 Joktan became the father of Almodad, Sheleph,

Hazarmaveth, Jerah, 27 Hadoram, Uzal, Diklah, 28 Obal, Abimael, Sheba, 29 Ophir, Havilah, and Jobab; all these were the descendants of Joktan. 30 The territory in which they lived extended from Mesha in the direction of Sephar, the hill country of the east. 31 These are the descendants of Shem, by their families, their languages, their lands, and their nations.

32 These are the families of Noah's sons, according to their genealogies, in their nations; and from these the nations spread abroad on the earth after the flood.

11 Now the whole earth had one language and the same words. 2 And as they migrated from the east, *c* they came upon a plain in the land of Shinar and settled there. 3 And they said to one another, "Come, let us make bricks, and burn them thoroughly." And they had brick for stone, and bitumen for mortar. 4 Then they said, "Come, let us build ourselves a city, and a tower with its top in the heavens, and let us make a name for ourselves; otherwise we shall be scattered abroad upon the face of the whole earth." 5 The LORD came down to

a Cn: Heb *Casluhim, from which the Philistines come, and Caphtorim* *b* That is *Division* *c* Or *migrated eastward*

---

successful warrior, built a kingdom in *Shinar* (Babylonia) and Assyria.

10.15–20: *Heth* (the Hittites), who once established a powerful empire in Asia Minor, disappeared as a world power in the twelfth century B.C. Here they are mentioned along with other Canaanite peoples; e.g. the *Jebusites* (located around Jerusalem), the *Amorites* (natives of the Palestinian hill country), the *Hivites* (perhaps Horites or Hurrians; see 34.2).

10.21–31: *Shem* is the father of Semitic peoples, *the children of Eber*, that is, all "Hebrews," including those Hebrews who later became Israel. During 1500–1200 B.C. waves of Hebrews came into Syria-Palestine and eventually established states, such as *Aram* or Syria (v. 23), Moab, Edom, and Israel.

11.1–9: The tower of Babel. God frustrated another attempt to overreach human

limitations (compare 3.22–24) by scattering the peoples and confusing their tongues. **1:** *One language,* compare 10.5, 20, 31. Clearly this tradition is independent of, and parallel to, the table of nations. **2:** *Shinar,* 10.10. The *plain* is the Tigris-Euphrates basin. **4:** In the eyes of nomads Mesopotamian city culture was characterized by the ziggurat, a pyramidal temple tower whose summit was believed to be the gateway to heaven, the realm of the gods. **6:** 3.22. **7:** *Let us,* see 1.26 n. **8:** Motivated by a Promethean desire for unity, fame, and security (v. 4), the enterprise ended in misunderstanding and thus arose the various language groups. **9:** *Babel,* meaning "Gate of God," is here interpreted by the Hebrew verb "confuse" (see note *d*). The story, now told to portray divine judgment upon the continuing sin of humanity, once explained the origin of languages and the cultural glory

see the city and the tower, which mortals had built. 6 And the LORD said, "Look, they are one people, and they have all one language; and this is only the beginning of what they will do; nothing that they propose to do will now be impossible for them. 7 Come, let us go down, and confuse their language there, so that they will not understand one another's speech." 8 So the LORD scattered them abroad from there over the face of all the earth, and they left off building the city. 9 Therefore it was called Babel, because there the LORD confused*d* the language of all the earth; and from there the LORD scattered them abroad over the face of all the earth.

10 These are the descendants of Shem. When Shem was one hundred years old, he became the father of Arpachshad two years after the flood; 11 and Shem lived after the birth of Arpachshad five hundred years, and had other sons and daughters.

12 When Arpachshad had lived thirty-five years, he became the father of Shelah; 13 and Arpachshad lived after the birth of Shelah four hundred three years, and had other sons and daughters.

14 When Shelah had lived thirty years, he became the father of Eber; 15 and Shelah lived after the birth of Eber four hundred three years, and had other sons and daughters.

16 When Eber had lived thirty-four years, he became the father of Peleg; 17 and Eber lived after the birth of Peleg four hundred thirty years, and had other sons and daughters.

18 When Peleg had lived thirty years, he became the father of Reu; 19 and Peleg lived after the birth of Reu two hundred nine years, and had other sons and daughters.

20 When Reu had lived thirty-two years, he became the father of Serug; 21 and Reu lived after the birth of Serug two hundred seven years, and had other sons and daughters.

22 When Serug had lived thirty years, he became the father of Nahor; 23 and Serug lived after the birth of Nahor two hundred years, and had other sons and daughters.

24 When Nahor had lived twenty-nine years, he became the father of Terah; 25 and Nahor lived after the birth of Terah one hundred nineteen years, and had other sons and daughters.

26 When Terah had lived seventy years, he became the father of Abram, Nahor, and Haran.

27 Now these are the descendants of Terah. Terah was the father of Abram, Nahor, and Haran; and Haran was the father of Lot. 28 Haran died before his father Terah in the land of his birth, in Ur of the Chaldeans. 29 Abram and Nahor took wives; the name of Abram's wife was Sarai, and the name of Nahor's wife was Milcah. She was the daughter of Haran the father of Milcah and Iscah. 30 Now Sarai was barren; she had no child.

31 Terah took his son Abram and his grandson Lot son of Haran, and his daughter-in-law Sarai, his son Abram's wife, and they went out together from Ur of the Chaldeans to go into the land of Canaan; but when they came to Ha-

d Heb *balal*, meaning *to confuse*

of Babylon, the center of Hammurabi's empire.

11.10–32: Genealogies to Abraham. The line of Shem (10.21–31) leads to Terah's three sons (v. 26) and narrows down to Abraham, showing how God chose Abraham and his seed from all the families of the earth. **10:** Here begin further quotations from the book of genealogies (5.1).
11.27: *Abram*, see 17.5 n. **30:** *Sarai*, see 17.15 n. **31:** *Haran*, in northwest Mesopota-

mia, was Abraham's ancestral home according to 24.10 (compare 29.4). Extra-biblical sources show that several of the names in the genealogy, e.g. Peleg (v. 16), Serug (v. 20), Terah and Nahor (v. 24), were place-names in this region, variously called Paddan-aram or Aram-naharaim. The migration from Mesopotamia into Canaan may be part of population movements in the second millennium B.C.

ran, they settled there. 32 The days of Terah were two hundred five years; and Terah died in Haran.

12 Now the LORD said to Abram, "Go from your country and your kindred and your father's house to the land that I will show you. 2 I will make of you a great nation, and I will bless you, and make your name great, so that you will be a blessing. 3 I will bless those who bless you, and the one who curses you I will curse; and in you all the families of the earth shall be blessed."*e*

4 So Abram went, as the LORD had told him; and Lot went with him. Abram was seventy-five years old when he departed from Haran. 5 Abram took his wife Sarai and his brother's son Lot, and all the possessions that they had gathered, and the persons whom they had acquired in Haran; and they set forth to go to the land of Canaan. When they had come to the land of Canaan, 6 Abram passed through the land to the place at Shechem, to the oak*f* of Moreh. At that time the Canaanites were in the land. 7 Then the LORD appeared to Abram, and said, "To your offspring*g* I will give this land." So he built there an altar to the LORD, who had appeared to him. 8 From there he moved on to the hill country on the east of Bethel, and pitched his tent, with Bethel on the west and Ai on the east; and there he built an altar to the LORD and invoked the name of the LORD. 9 And Abram journeyed on by stages toward the Negeb.

10 Now there was a famine in the land. So Abram went down to Egypt to reside there as an alien, for the famine was severe in the land. 11 When he was about to enter Egypt, he said to his wife Sarai, "I know well that you are a woman beautiful in appearance; 12 and when the Egyptians see you, they will say, 'This is his wife'; then they will kill me, but they will let you live. 13 Say you are my sister, so that it may go well with me because of you, and that my life may be spared on your account." 14 When Abram entered Egypt the Egyptians saw that the woman was very beautiful. 15 When the officials of Pharaoh saw her, they praised her to Pharaoh. And the woman was taken into Pharaoh's house. 16 And for her sake he dealt well with Abram; and he had sheep, oxen, male donkeys, male and female slaves, female donkeys, and camels.

17 But the LORD afflicted Pharaoh and his house with great plagues because of Sarai, Abram's wife. 18 So Pharaoh called Abram, and said, "What is this you have done to me? Why did you not tell me that she was your wife? 19 Why did you say, 'She is my sister,' so that I took her for my wife? Now then, here is your wife, take her, and be gone." 20 And Pharaoh gave his men orders concerning him; and they set him on the way, with his wife and all that he had.

e Or by you all the families of the earth shall bless themselves   f Or terebinth   g Heb seed

---

12.1–9: God's call of Abraham is sketched against the background of a broken, divided humanity. **1–3:** Israel, represented by Abraham and Sarah, is chosen to play a decisive role in God's historical purpose (Isa 19.24; 51.2; 49.6; Rom 4.13). The promise includes receiving a land, becoming a numerous people, and having a relationship with God that will benefit other human families. **1:** *Your country,* see 11.31 n. **2:** The language—*great nation* instead of "people"—reflects the national consciousness of the early monarchy, when the story was retold. **3:** By breaking ties of land and kindred (v. 1) and responding to God's summons into a new world, Abraham typifies the person of faith

(Heb 11.8). **6:** *Shechem,* located at the commercial crossroads of Canaan in the pass between Mount Ebal and Mount Gerizim, was a flourishing Canaanite city in the second millennium B.C. (Judg ch 9). Nearby was *the oak of Moreh* ("oracle giver"), a sacred tree (35.4; Deut 11.30; Josh 24.26; Judg 9.37).

12.10–13.1: Sarah in jeopardy. God's promise of a posterity was temporarily eclipsed in Egypt where Sarah, the ancestress of Israel, was almost taken into Pharaoh's harem. **13:** Sarah was Abraham's *sister,* i.e. half-sister, according to 20.12. The narrative does not moralize about the white lie but rather portrays the LORD's rescue of Sarah from the jeopardy into which Abraham's self-

13 So Abram went up from Egypt, he and his wife, and all that he had, and Lot with him, into the Negeb. 2 Now Abram was very rich in livestock, in silver, and in gold. 3 He journeyed on by stages from the Negeb as far as Bethel, to the place where his tent had been at the beginning, between Bethel and Ai, 4 to the place where he had made an altar at the first; and there Abram called on the name of the LORD. 5 Now Lot, who went with Abram, also had flocks and herds and tents, 6 so that the land could not support both of them living together; for their possessions were so great that they could not live together, 7 and there was strife between the herders of Abram's livestock and the herders of Lot's livestock. At that time the Canaanites and the Perizzites lived in the land.

8 Then Abram said to Lot, "Let there be no strife between you and me, and between your herders and my herders; for we are kindred. 9 Is not the whole land before you? Separate yourself from me. If you take the left hand, then I will go to the right; or if you take the right hand, then I will go to the left." 10 Lot looked about him, and saw that the plain of the Jordan was well watered everywhere like the garden of the LORD, like the land of Egypt, in the direction of Zoar; this was before the LORD had destroyed Sodom and Gomorrah. 11 So Lot chose for himself all the plain of the Jordan, and Lot journeyed eastward; thus they separated from each other. 12 Abram settled in the land of Canaan, while Lot settled among the cities of the Plain and moved his tent as far as Sodom. 13 Now the people of Sodom were wicked, great sinners against the LORD.

14 The LORD said to Abram, after Lot had separated from him, "Raise your eyes now, and look from the place where you are, northward and southward and eastward and westward; 15 for all the land that you see I will give to you and to your offspring[h] forever. 16 I will make your offspring like the dust of the earth; so that if one can count the dust of the earth, your offspring also can be counted. 17 Rise up, walk through the length and the breadth of the land, for I will give it to you." 18 So Abram moved his tent, and came and settled by the oaks[i] of Mamre, which are at Hebron; and there he built an altar to the LORD.

14 In the days of King Amraphel of Shinar, King Arioch of Ellasar, King Chedorlaomer of Elam, and King Tidal of Goiim, 2 these kings made war with King Bera of Sodom, King Birsha of Gomorrah, King Shinab of Admah, King Shemeber of Zeboiim, and the king of Bela (that is, Zoar). 3 All these joined forces in the Valley of Siddim (that is, the Dead Sea).[j] 4 Twelve years they had served Chedorlaomer, but in the thirteenth year they rebelled. 5 In the fourteenth year Chedorlaomer and the kings

h Heb *seed*   *i* Or *terebinths*   *j* Heb *Salt Sea*

---

interest had placed her (compare ch 20; 26.1–11).

13.2–18: Abraham and Lot. The LORD's promise of the land hung in the balance of a decision, but providentially Lot, the ancestor of Moab, chose the region around Sodom. **5–7:** Israel's ancestors are represented as living a semi-nomadic life in the midst of the Canaanites; thus adequate pasture land was vital to the *herders*. **13.10:** The fertility of the *plain of the Jordan* is compared to *the garden of the* LORD (2.8–10) and the Nile valley. *Zoar,* 19.20–22. *Sodom and Gomorrah,* 19.1–29. **11:** Lot thus became the ancestor of Moab and Ammon (19.30–38). **14–17:** The promise of 12.7 is renewed. **18:** *Mamre* was an ancient sacred place, slightly north of Hebron, with which Abraham was associated.

14.1–24: An alliance of four eastern kings was defeated by Abraham's small forces. This independent tradition, which pictures Abraham as a military hero in the field of world politics, deviates from the usual portrayal of Israel's ancestors as peaceful semi-nomads. **1:** None of the kings can be identified with certainty. *Shinar,* 10.10. *Ellasar,* i.e. Larsa. **2–4:** The rebellious city-kings held sway in the region of the Dead Sea before its lower basin was covered with water. **5–7:** The object of the invasion may have been to secure the trade routes to Egypt and southern Arabia.

who were with him came and subdued
the Rephaim in Ashteroth-karnaim, the
Zuzim in Ham, the Emim in Shaveh-
kiriathaim, 6and the Horites in the hill
country of Seir as far as El-paran on the
edge of the wilderness; 7then they turned
back and came to En-mishpat (that is,
Kadesh), and subdued all the country of
the Amalekites, and also the Amorites
who lived in Hazazon-tamar. 8Then the
king of Sodom, the king of Gomorrah,
the king of Admah, the king of Zeboiim,
and the king of Bela (that is, Zoar) went
out, and they joined battle in the Valley
of Siddim 9with King Chedorlaomer of
Elam, King Tidal of Goiim, King Amraphel of Shinar, and King Arioch of Ellasar, four kings against five. 10Now the
Valley of Siddim was full of bitumen
pits; and as the kings of Sodom and Gomorrah fled, some fell into them, and the
rest fled to the hill country. 11So the enemy took all the goods of Sodom and
Gomorrah, and all their provisions, and
went their way; 12they also took Lot, the
son of Abram's brother, who lived in
Sodom, and his goods, and departed.

13 Then one who had escaped came
and told Abram the Hebrew, who was
living by the oaks*k* of Mamre the Amorite, brother of Eshcol and of Aner; these
were allies of Abram. 14When Abram
heard that his nephew had been taken
captive, he led forth his trained men,
born in his house, three hundred eighteen of them, and went in pursuit as far
as Dan. 15He divided his forces against
them by night, he and his servants, and
routed them and pursued them to Ho-
bah, north of Damascus. 16Then he
brought back all the goods, and also
brought back his nephew Lot with his
goods, and the women and the people.

17 After his return from the defeat of
Chedorlaomer and the kings who were
with him, the king of Sodom went out
to meet him at the Valley of Shaveh (that
is, the King's Valley). 18And King Melchizedek of Salem brought out bread and
wine; he was priest of God Most High.*l*
19He blessed him and said,

"Blessed be Abram by God Most
    High,*l*
  maker of heaven and earth;
20 and blessed be God Most High,*l*
  who has delivered your enemies
    into your hand!"

And Abram gave him one tenth of everything. 21Then the king of Sodom said
to Abram, "Give me the persons, but
take the goods for yourself." 22But
Abram said to the king of Sodom, "I
have sworn to the LORD, God Most
High,*l* maker of heaven and earth, 23that
I would not take a thread or a sandal-
thong or anything that is yours, so that
you might not say, 'I have made Abram
rich.' 24I will take nothing but what the
young men have eaten, and the share of
the men who went with me—Aner, Eshcol, and Mamre. Let them take their
share."

15 After these things the word of the
LORD came to Abram in a vision,
"Do not be afraid, Abram, I am your
shield; your reward shall be very great."

k Or *terebinths*   *l* Heb *El Elyon*

14.13: *Mamre* here is the name of a person
(compare 13.18). **14**: *Dan* was known as Laish
in the early period (Judg 18.29). **17**: *The King's
Valley,* near Jerusalem (2 Sam 18.18). **18**: *Salem* is a name for Jerusalem (Ps 76.2), where
in pre-Israelite times *God Most High* (El
Elyon) was worshiped as the high god of the
pantheon. **19–20**: Melchizedek, the priest of
the Canaanite cult, blesses Abraham in the
name of his god, *maker of heaven and earth.* **22**:
Here *God Most High* is identified with the
LORD, the God of Israel (Num 24.16; Ps 46.4).
The mysterious Melchizedek was later interpreted messianically (Ps 110.4; Heb 7.1–17).
15.1–21: **The covenant with Abraham
and Sarah.** Even though Abraham as yet had
no heir, God reaffirmed the promise. **1**: *Your
shield,* i.e. divine protector (Ps 28.7; 33.20).
Abraham's reward (compare 14.21–24) will
be his numerous posterity (13.14–17). This
verse may reflect an ancient custom that allowed a slave to be adopted as heir in case of
childlessness. **6**: *He believed,* i.e. put his trust
in the divine promise even when there was no
concrete evidence to support it. *Righteousness,*
a right relationship with God (6.9; 7.1).

2 But Abram said, "O Lord GOD, what will you give me, for I continue childless, and the heir of my house is Eliezer of Damascus?"*m* 3 And Abram said, "You have given me no offspring, and so a slave born in my house is to be my heir." 4 But the word of the LORD came to him, "This man shall not be your heir; no one but your very own issue shall be your heir." 5 He brought him outside and said, "Look toward heaven and count the stars, if you are able to count them." Then he said to him, "So shall your descendants be." 6 And he believed the LORD; and the LORD *n* reckoned it to him as righteousness.

7 Then he said to him, "I am the LORD who brought you from Ur of the Chaldeans, to give you this land to possess." 8 But he said, "O Lord GOD, how am I to know that I shall possess it?" 9 He said to him, "Bring me a heifer three years old, a female goat three years old, a ram three years old, a turtledove, and a young pigeon." 10 He brought him all these and cut them in two, laying each half over against the other; but he did not cut the birds in two. 11 And when birds of prey came down on the carcasses, Abram drove them away.

12 As the sun was going down, a deep sleep fell upon Abram, and a deep and terrifying darkness descended upon him. 13 Then the LORD *n* said to Abram, "Know this for certain, that your offspring shall be aliens in a land that is not theirs, and shall be slaves there, and they shall be oppressed for four hundred years; 14 but I will bring judgment on the nation that they serve, and afterward they shall come out with great possessions. 15 As for yourself, you shall go to your ancestors in peace; you shall be buried in a good old age. 16 And they shall come back here in the fourth generation; for the iniquity of the Amorites is not yet complete."

17 When the sun had gone down and it was dark, a smoking fire pot and a flaming torch passed between these pieces. 18 On that day the LORD made a covenant with Abram, saying, "To your descendants I give this land, from the river of Egypt to the great river, the river Euphrates, 19 the land of the Kenites, the Kenizzites, the Kadmonites, 20 the Hittites, the Perizzites, the Rephaim, 21 the Amorites, the Canaanites, the Girgashites, and the Jebusites."

16 Now Sarai, Abram's wife, bore him no children. She had an Egyptian slave-girl whose name was Hagar, 2 and Sarai said to Abram, "You see that the LORD has prevented me from bearing children; go in to my slave-girl; it may be that I shall obtain children by her." And Abram listened to the voice of Sarai. 3 So, after Abram had lived ten years in the land of Canaan, Sarai, Abram's wife, took Hagar the Egyptian, her slave-girl, and gave her to her husband Abram as a wife. 4 He went in to

m Meaning of Heb uncertain   *n* Heb *ho*

15.7–12: The covenant ceremony described in these verses and vv. 17–18 comes from early tradition, as evidenced by the archaic ritual of making a covenant by cutting animals in two (Jer 34 17) and passing between the parts. The meaning is no longer clear; according to one view the parties invoke dismemberment if they fail to keep the covenant oath. **12:** Abraham fell into *a deep sleep* (compare 2.21), in which he received the revelation (Job 4.13; 33.15). **13–16:** An inserted tradition explains the delay in the fulfillment of the promise, referring to the Egyptian oppression for 400 years (Ex 12.40) and the victorious Exodus. *The iniquity of the Amorites* (10.16), i.e. Canaanites, was the sexual corruption that led to their downfall (see 9.25 n.).

15.17–18: Continuation of vv. 7–12. The presence of God as a covenant party is symbolized by fire (see Ex 3.2 n.) passing between the pieces. *A covenant,* see 17.2 n. The ideal boundaries of the promised land, from the *river of Egypt* to the *Euphrates,* were those of David's empire (Deut 11.24; 2 Sam 8.3; compare 1 Kings 4.21; 8.65).

16.1–16: The birth of Ishmael. 2: In antiquity both barrenness and fertility were traced to God (20.17 18; 30.2; 33.5; 1 Sam 1.6). According to ancient custom, a wife could give her maid to her husband and claim the child as her own (30.3, 9). **4–6:** Though

Hagar, and she conceived; and when she saw that she had conceived, she looked with contempt on her mistress. 5 Then Sarai said to Abram, "May the wrong done to me be on you! I gave my slave-girl to your embrace, and when she saw that she had conceived, she looked on me with contempt. May the LORD judge between you and me!" 6 But Abram said to Sarai, "Your slave-girl is in your power; do to her as you please." Then Sarai dealt harshly with her, and she ran away from her.

7 The angel of the LORD found her by a spring of water in the wilderness, the spring on the way to Shur. 8 And he said, "Hagar, slave-girl of Sarai, where have you come from and where are you going?" She said, "I am running away from my mistress Sarai." 9 The angel of the LORD said to her, "Return to your mistress, and submit to her." 10 The angel of the LORD also said to her, "I will so greatly multiply your offspring that they cannot be counted for multitude." 11 And the angel of the LORD said to her,

"Now you have conceived and
   shall bear a son;
you shall call him Ishmael, *o*
for the LORD has given heed to
   your affliction.

12 He shall be a wild ass of a man,
   with his hand against everyone,
   and everyone's hand against
      him;
   and he shall live at odds with all
      his kin."

13 So she named the LORD who spoke to her, "You are El-roi";*p* for she said, "Have I really seen God and remained alive after seeing him?"*q* 14 Therefore the well was called Beer-lahai-roi;*r* it lies between Kadesh and Bered.

15 Hagar bore Abram a son; and Abram named his son, whom Hagar bore, Ishmael. 16 Abram was eighty-six years old when Hagar bore him*s* Ishmael.

17 When Abram was ninety-nine years old, the LORD appeared to Abram, and said to him, "I am God Almighty;*t* walk before me, and be blameless. 2 And I will make my covenant between me and you, and will make you exceedingly numerous." 3 Then Abram fell on his face; and God said to him, 4 "As for me, this is my covenant with

o That is *God hears*   *p* Perhaps *God of seeing*
or *God who sees*   *q* Meaning of Heb uncertain
r That is *the Well of the Living One who sees me*
s Heb *Abram*   *t* Traditional rendering of Heb
El Shaddai

socially inferior, Hagar felt superior to Sarai and threatened to take her mistress's place (Prov 30.23) as the ancestress of Israel.
16.7: Here *the angel of the* LORD is not a heavenly being subordinate to God but the LORD (Yahweh) in earthly manifestation, as is clear from v. 13 (compare 21.17, 19; Ex 14.19). **12:** *A wild ass of a man* describes the bedouin freedom of the Ishmaelites in the southern wilderness (25.16–18). **13:** *God of seeing* (see note *p*) was the name of the deity of the sacred place, now identified with Israel's God. On Hagar's question, compare Ex 33.20; Judg 6.22–23; 13.22.
17.1–27: The everlasting covenant. This account from the priestly tradition is a parallel version of the Abrahamic covenant given in the early tradition (15.7–21). **1:** *God Almighty* (El Shaddai), meaning "God, the One of the Mountains," was a divine name current in the pre-Mosaic period (Ex 6.2–3), perhaps brought from Mesopotamia into Palestine. **2:** *Covenant* is a term of relationship

between a superior and an inferior party, the former "making" or "establishing" (v. 7) the bond. God's covenant guarantees an *exceedingly numerous* posterity, one of the divine promises (12.2; see below vv. 7–8). **5:** A new name signifies a new relationship or status (see 32.28). *Abraham,* a dialectical variant of *Abram,* means "the [divine] ancestor is exalted"; here the name is explained by its similarity to the Hebrew for *ancestor of a multitude,* referring to the nations whose ancestry was traced to Abraham (v. 16; 28.3; 35.11; 48.4), e.g. Edomites and Ishmaelites. **7:** Like the covenant with Noah (9.8–17), this is *an everlasting covenant* (vv. 13, 19), one that lasts in perpetuity because it is grounded in the sovereign will of God, not in human behavior. **7–8:** *To be God to you,* this covenant not only assures a new relationship with God, to be realized fully in the Exodus and the Sinai revelation (Ex 6.2–8), but unconditionally guarantees the promise of the land of Canaan as *a perpetual holding.*

you: You shall be the ancestor of a multitude of nations. [5] No longer shall your name be Abram, *u* but your name shall be Abraham; *v* for I have made you the ancestor of a multitude of nations. [6] I will make you exceedingly fruitful; and I will make nations of you, and kings shall come from you. [7] I will establish my covenant between me and you, and your offspring after you throughout their generations, for an everlasting covenant, to be God to you and to your offspring *w* after you. [8] And I will give to you, and to your offspring after you, the land where you are now an alien, all the land of Canaan, for a perpetual holding; and I will be their God."

[9] God said to Abraham, "As for you, you shall keep my covenant, you and your offspring after you throughout their generations. [10] This is my covenant, which you shall keep, between me and you and your offspring after you: Every male among you shall be circumcised. [11] You shall circumcise the flesh of your foreskins, and it shall be a sign of the covenant between me and you. [12] Throughout your generations every male among you shall be circumcised when he is eight days old, including the slave born in your house and the one bought with your money from any foreigner who is not of your offspring. [13] Both the slave born in your house and the one bought with your money must be circumcised. So shall my covenant be in your flesh an everlasting covenant. [14] Any uncircumcised male who is not circumcised in the flesh of his foreskin shall be cut off from his people; he has broken my covenant."

[15] God said to Abraham, "As for Sarai your wife, you shall not call her Sarai, but Sarah shall be her name. [16] I will bless

her, and moreover I will give you a son by her. I will bless her, and she shall give rise to nations; kings of peoples shall come from her." [17] Then Abraham fell on his face and laughed, and said to himself, "Can a child be born to a man who is a hundred years old? Can Sarah, who is ninety years old, bear a child?" [18] And Abraham said to God, "O that Ishmael might live in your sight!" [19] God said, "No, but your wife Sarah shall bear you a son, and you shall name him Isaac. *x* I will establish my covenant with him as an everlasting covenant for his offspring after him. [20] As for Ishmael, I have heard you; I will bless him and make him fruitful and exceedingly numerous; he shall be the father of twelve princes, and I will make him a great nation. [21] But my covenant I will establish with Isaac, whom Sarah shall bear to you at this season next year." [22] And when he had finished talking with him, God went up from Abraham.

[23] Then Abraham took his son Ishmael and all the slaves born in his house or bought with his money, every male among the men of Abraham's house, and he circumcised the flesh of their foreskins that very day, as God had said to him. [24] Abraham was ninety-nine years old when he was circumcised in the flesh of his foreskin. [25] And his son Ishmael was thirteen years old when he was circumcised in the flesh of his foreskin. [26] That very day Abraham and his son Ishmael were circumcised; [27] and all the men of his house, slaves born in the house and those bought with money from a foreigner, were circumcised with him.

u That is exalted ancestor *v Here taken to mean ancestor of a multitude* *w Heb seed* *x That is he laughs*

17.9–14: To *keep* the covenant involves the practice of circumcision, an ancient rite that was practiced by some of Israel's neighbors and whose origin is explained by various traditions (Ex 4.24–26; Josh 5.2–9). Binding only on males, circumcision is an external *sign* (v. 11) of membership in the covenant community. Unlike the universal Noachic covenant (see 9.1–19 n.), this *everlasting covenant* pertains only to the descendants of Abraham and Sarah.

17.15: *Sarah,* meaning "princess," is a variant of *Sarai.* See v. 5 n. **17:** 18.11–15. **18–20:** 16.10–12.

18 The LORD appeared to Abraham[y] by the oaks[z] of Mamre, as he sat at the entrance of his tent in the heat of the day. 2He looked up and saw three men standing near him. When he saw them, he ran from the tent entrance to meet them, and bowed down to the ground. 3He said, "My lord, if I find favor with you, do not pass by your servant. 4Let a little water be brought, and wash your feet, and rest yourselves under the tree. 5Let me bring a little bread, that you may refresh yourselves, and after that you may pass on—since you have come to your servant." So they said, "Do as you have said." 6And Abraham hastened into the tent to Sarah, and said, "Make ready quickly three measures[a] of choice flour, knead it, and make cakes." 7Abraham ran to the herd, and took a calf, tender and good, and gave it to the servant, who hastened to prepare it. 8Then he took curds and milk and the calf that he had prepared, and set it before them; and he stood by them under the tree while they ate.

9 They said to him, "Where is your wife Sarah?" And he said, "There, in the tent." 10Then one said, "I will surely return to you in due season, and your wife Sarah shall have a son." And Sarah was listening at the tent entrance behind him. 11Now Abraham and Sarah were old, advanced in age; it had ceased to be with Sarah after the manner of women. 12So Sarah laughed to herself, saying, "After I have grown old, and my husband is old, shall I have pleasure?" 13The LORD said to Abraham, "Why did Sarah laugh, and say, 'Shall I indeed bear a child, now that I am old?' 14Is anything too wonderful for the LORD? At the set time I will return to you, in due season, and Sarah shall have a son." 15But Sarah denied, saying, "I did not laugh"; for she was afraid. He said, "Oh yes, you did laugh."

16 Then the men set out from there, and they looked toward Sodom; and Abraham went with them to set them on their way. 17The LORD said, "Shall I hide from Abraham what I am about to do, 18seeing that Abraham shall become a great and mighty nation, and all the nations of the earth shall be blessed in him?[b] 19No, for I have chosen[c] him, that he may charge his children and his household after him to keep the way of the LORD by doing righteousness and justice; so that the LORD may bring about for Abraham what he has promised him." 20Then the LORD said, "How great is the outcry against Sodom and Gomorrah and how very grave their sin! 21I must go down and see whether they have done altogether according to the outcry that has come to me; and if not, I will know."

22 So the men turned from there, and went toward Sodom, while Abraham remained standing before the LORD.[d] 23Then Abraham came near and said,

y Heb *him*   z Or *terebinths*   a Heb *seahs*
b Or *and all the nations of the earth shall bless themselves by him*   c Heb *known*   d Another ancient tradition reads *while the LORD remained standing before Abraham*

18.1–15: The LORD's visit to Abraham and Sarah. 1: The oaks of *Mamre,* see 13.18 n. **2–8:** A fine description of oriental courtesy and hospitality. When the visitors appeared at the noontime siesta, Abraham did not recognize them as divine beings (Heb 13.2). The relation of the three visitors to the LORD or Yahweh (v. 1) is difficult. All three *angels* (19.1) may represent the LORD (see 16.7 n.); thus the plurality becomes a single person in vv. 10, 13. On the other hand, v. 22 and 19.1 suggest that the LORD is one of the three, the other two being attendants. **18.9–15:** The narrator stresses the incredibility of God's promise. **12:** Sarah's laughter arises from the absurd disproportion between the divine promise and the human possibilities. Other traditions also play upon the name of Isaac, meaning "he laughs" (compare 17.17–19; 21.6). **18.16–33: Abraham's intercession for Sodom and Gomorrah. 17–19:** Because Abraham is chosen for a special role (12.13), he is taken into the LORD's counsel, for Sodom will become an example to future generations (Isa 1.9). **18.23–33:** Abraham diplomatically questions the justice of God in allowing innocent people to be punished with the guilty. The first example in the Bible of expostulation

"Will you indeed sweep away the righteous with the wicked? 24 Suppose there are fifty righteous within the city; will you then sweep away the place and not forgive it for the fifty righteous who are in it? 25 Far be it from you to do such a thing, to slay the righteous with the wicked, so that the righteous fare as the wicked! Far be that from you! Shall not the Judge of all the earth do what is just?" 26 And the LORD said, "If I find at Sodom fifty righteous in the city, I will forgive the whole place for their sake." 27 Abraham answered, "Let me take it upon myself to speak to the Lord, I who am but dust and ashes. 28 Suppose five of the fifty righteous are lacking? Will you destroy the whole city for lack of five?" And he said, "I will not destroy it if I find forty-five there." 29 Again he spoke to him, "Suppose forty are found there." He answered, "For the sake of forty I will not do it." 30 Then he said, "Oh do not let the Lord be angry if I speak. Suppose thirty are found there." He answered, "I will not do it, if I find thirty there." 31 He said, "Let me take it upon myself to speak to the Lord. Suppose twenty are found there." He answered, "For the sake of twenty I will not destroy it." 32 Then he said, "Oh do not let the Lord be angry if I speak just once more. Suppose ten are found there." He answered, "For the sake of ten I will not destroy it." 33 And the LORD went his way, when he had finished speaking to Abraham; and Abraham returned to his place.

19 The two angels came to Sodom in the evening, and Lot was sitting in the gateway of Sodom. When Lot saw them, he rose to meet them, and bowed down with his face to the ground. 2 He said, "Please, my lords, turn aside to your servant's house and spend the night, and wash your feet; then you can rise early and go on your way." They said, "No; we will spend the night in the square." 3 But he urged them strongly; so they turned aside to him and entered his house; and he made them a feast, and baked unleavened bread, and they ate. 4 But before they lay down, the men of the city, the men of Sodom, both young and old, all the people to the last man, surrounded the house; 5 and they called to Lot, "Where are the men who came to you tonight? Bring them out to us, so that we may know them." 6 Lot went out of the door to the men, shut the door after him, 7 and said, "I beg you, my brothers, do not act so wickedly. 8 Look, I have two daughters who have not known a man; let me bring them out to you, and do to them as you please; only do nothing to these men, for they have come under the shelter of my roof." 9 But they replied, "Stand back!" And they said, "This fellow came here as an alien, and he would play the judge! Now we will deal worse with you than with them." Then they pressed hard against the man Lot, and came near the door to break it down. 10 But the men inside reached out their hands and brought Lot into the house with them, and shut the door. 11 And they struck with blindness the men who were at the door of the house, both small and great, so that they were unable to find the door.

12 Then the men said to Lot, "Have you anyone else here? Sons-in-law, sons, daughters, or anyone you have in the city—bring them out of the place. 13 For we are about to destroy this place, because the outcry against its people has become great before the LORD, and the

---

with God; see also Jeremiah's "confessions," the prophecy of Habakkuk, the book of Job. **19.1–38: The destruction of Sodom and Gomorrah** impressed itself deeply upon later generations as an example of God's total judgment upon appalling wickedness (Deut 29.23; Isa 1.9; Jer 49.18; Am 4.11). **1:** *Two angels,* see 18.2–8 n. **4–11:** Compare the crime of Gibeah (Judg 19.22–30). The episode is told to illustrate the sexual excesses of Canaanites. **5:** *Know* refers to sexual relations (v. 8), here homosexual ("sodomy"). **8:** Once guests had eaten in his house, Lot felt he had to obey the law of oriental hospitality which guaranteed protection. Thus his proposal to hand over his daughters showed his determination to put first his obligation as a host.

LORD has sent us to destroy it." [14]So Lot went out and said to his sons-in-law, who were to marry his daughters, "Up, get out of this place; for the LORD is about to destroy the city." But he seemed to his sons-in-law to be jesting.

15 When morning dawned, the angels urged Lot, saying, "Get up, take your wife and your two daughters who are here, or else you will be consumed in the punishment of the city." [16]But he lingered; so the men seized him and his wife and his two daughters by the hand, the LORD being merciful to him, and they brought him out and left him outside the city. [17]When they had brought them outside, they[e] said, "Flee for your life; do not look back or stop anywhere in the Plain; flee to the hills, or else you will be consumed." [18]And Lot said to them, "Oh, no, my lords; [19]your servant has found favor with you, and you have shown me great kindness in saving my life; but I cannot flee to the hills, for fear the disaster will overtake me and I die. [20]Look, that city is near enough to flee to, and it is a little one. Let me escape there—is it not a little one?—and my life will be saved!" [21]He said to him, "Very well, I grant you this favor too, and will not overthrow the city of which you have spoken. [22]Hurry, escape there, for I can do nothing until you arrive there." Therefore the city was called Zoar.[f] [23]The sun had risen on the earth when Lot came to Zoar.

24 Then the LORD rained on Sodom and Gomorrah sulfur and fire from the LORD out of heaven; [25]and he overthrew those cities, and all the Plain, and all the inhabitants of the cities, and what grew on the ground. [26]But Lot's wife, behind him, looked back, and she became a pillar of salt.

27 Abraham went early in the morning to the place where he had stood before the LORD; [28]and he looked down toward Sodom and Gomorrah and toward all the land of the Plain and saw the smoke of the land going up like the smoke of a furnace.

29 So it was that, when God destroyed the cities of the Plain, God remembered Abraham, and sent Lot out of the midst of the overthrow, when he overthrew the cities in which Lot had settled.

30 Now Lot went up out of Zoar and settled in the hills with his two daughters, for he was afraid to stay in Zoar; so he lived in a cave with his two daughters. [31]And the firstborn said to the younger, "Our father is old, and there is not a man on earth to come in to us after the manner of all the world. [32]Come, let us make our father drink wine, and we will lie with him, so that we may preserve offspring through our father." [33]So they made their father drink wine that night; and the firstborn went in, and lay with her father; he did not know when she lay down or when she rose. [34]On the next day, the firstborn said to the younger, "Look, I lay last night with my father; let us make him drink wine tonight also; then you go in and lie with him, so that we may preserve offspring through our father." [35]So they made their father drink wine that night also; and the younger rose, and lay with him; and he did not know when she lay down or when she rose. [36]Thus both the daughters of Lot became pregnant by their father. [37]The firstborn bore a son, and named him Moab; he is the ancestor of the Moabites to this day. [38]The younger also bore a son and named him Ben-

e Gk Syr Vg: Heb *he*   *f* That is *Little*

19.20–22: *Zoar,* meaning "small," was a town at the southern end of the Dead Sea which survived the calamity. **24:** *Sulfur and fire,* a memory of a catastrophe in remote times when seismic activity and the explosion of subterranean gases changed the face of the area, which was formerly fertile (13.10). **26:**

An old tradition to account for bizarre salt formations in the area such as may be seen today on Jebel Usdum.
19.30–38: A story which explains the origin of the Moabites and Ammonites, neighbors of Israel.

ammi; he is the ancestor of the Ammonites to this day.

20 From there Abraham journeyed toward the region of the Negeb, and settled between Kadesh and Shur. While residing in Gerar as an alien, 2 Abraham said of his wife Sarah, "She is my sister." And King Abimelech of Gerar sent and took Sarah. 3 But God came to Abimelech in a dream by night, and said to him, "You are about to die because of the woman whom you have taken; for she is a married woman." 4 Now Abimelech had not approached her; so he said, "Lord, will you destroy an innocent people? 5 Did he not himself say to me, 'She is my sister'? And she herself said, 'He is my brother.' I did this in the integrity of my heart and the innocence of my hands." 6 Then God said to him in the dream, "Yes, I know that you did this in the integrity of your heart; furthermore it was I who kept you from sinning against me. Therefore I did not let you touch her. 7 Now then, return the man's wife; for he is a prophet, and he will pray for you and you shall live. But if you do not restore her, know that you shall surely die, you and all that are yours."

8 So Abimelech rose early in the morning, and called all his servants and told them all these things; and the men were very much afraid. 9 Then Abimelech called Abraham, and said to him, "What have you done to us? How have I sinned against you, that you have brought such great guilt on me and my kingdom? You have done things to me that ought not to be done." 10 And Abimelech said to Abraham, "What were you thinking of, that you did this thing?"

11 Abraham said, "I did it because I thought, There is no fear of God at all in this place, and they will kill me because of my wife. 12 Besides, she is indeed my sister, the daughter of my father but not the daughter of my mother; and she became my wife. 13 And when God caused me to wander from my father's house, I said to her, 'This is the kindness you must do me: at every place to which we come, say of me, He is my brother.' " 14 Then Abimelech took sheep and oxen, and male and female slaves, and gave them to Abraham, and restored his wife Sarah to him. 15 Abimelech said, "My land is before you; settle where it pleases you." 16 To Sarah he said, "Look, I have given your brother a thousand pieces of silver; it is your exoneration before all who are with you; you are completely vindicated." 17 Then Abraham prayed to God; and God healed Abimelech, and also healed his wife and female slaves so that they bore children. 18 For the LORD had closed fast all the wombs of the house of Abimelech because of Sarah, Abraham's wife.

21 The LORD dealt with Sarah as he had said, and the LORD did for Sarah as he had promised. 2 Sarah conceived and bore Abraham a son in his old age, at the time of which God had spoken to him. 3 Abraham gave the name Isaac to his son whom Sarah bore him. 4 And Abraham circumcised his son Isaac when he was eight days old, as God had commanded him. 5 Abraham was a hundred years old when his son Isaac was born to him. 6 Now Sarah said, "God has brought laughter for me; everyone who hears will laugh with me." 7 And she

---

20.1–18: Abraham and Sarah in Gerar. This story parallels that of 12.10–20 (compare 26.6–11). Here, however, the narrator is more concerned with the ethical problems involved. **1:** *From there,* i.e. Mamre (18.1). **2:** Compare 12.11–13. **3–7:** Ethically sensitive, the narrator insists that Abimelech was innocent, for he did not go near Sarah. **7:** To Abraham is attributed the intercessory role of *a prophet* (18.22–33; compare Num 12.13; 21.7; 1 Sam 12.19–23).

20.11–12: Abraham's excuses. Marriage with a half-sister was permitted in ancient times (2 Sam 13.13) but later was forbidden (Lev 18.9, 11; 20.17). **16:** *Exoneration,* i.e. a gift to induce everyone to overlook the injury done to Sarah.

21.1–21: Isaac and Ishmael. Although Isaac was designated to continue Abraham's line, Ishmael too was promised a great future. **4:** On circumcision, see 17.9–14 n. **6:** See 18.9–15 n.

said, "Who would ever have said to Abraham that Sarah would nurse children? Yet I have borne him a son in his old age."

8 The child grew, and was weaned; and Abraham made a great feast on the day that Isaac was weaned. 9 But Sarah saw the son of Hagar the Egyptian, whom she had borne to Abraham, playing with her son Isaac. *g* 10 So she said to Abraham, "Cast out this slave woman with her son; for the son of this slave woman shall not inherit along with my son Isaac." 11 The matter was very distressing to Abraham on account of his son. 12 But God said to Abraham, "Do not be distressed because of the boy and because of your slave woman; whatever Sarah says to you, do as she tells you, for it is through Isaac that offspring shall be named for you. 13 As for the son of the slave woman, I will make a nation of him also, because he is your offspring." 14 So Abraham rose early in the morning, and took bread and a skin of water, and gave it to Hagar, putting it on her shoulder, along with the child, and sent her away. And she departed, and wandered about in the wilderness of Beer-sheba.

15 When the water in the skin was gone, she cast the child under one of the bushes. 16 Then she went and sat down opposite him a good way off, about the distance of a bowshot; for she said, "Do not let me look on the death of the child." And as she sat opposite him, she lifted up her voice and wept. 17 And God heard the voice of the boy; and the angel of God called to Hagar from heaven, and said to her, "What troubles you, Hagar? Do not be afraid; for God has heard the voice of the boy where he is. 18 Come, lift up the boy and hold him fast with your hand, for I will make a great nation of him." 19 Then God opened her eyes and she saw a well of water. She went, and filled the skin with water, and gave the boy a drink.

20 God was with the boy, and he grew up; he lived in the wilderness, and became an expert with the bow. 21 He lived in the wilderness of Paran; and his mother got a wife for him from the land of Egypt.

22 At that time Abimelech, with Phicol the commander of his army, said to Abraham, "God is with you in all that you do; 23 now therefore swear to me here by God that you will not deal falsely with me or with my offspring or with my posterity, but as I have dealt loyally with you, you will deal with me and with the land where you have resided as an alien." 24 And Abraham said, "I swear it."

25 When Abraham complained to Abimelech about a well of water that Abimelech's servants had seized, 26 Abimelech said, "I do not know who has done this; you did not tell me, and I have not heard of it until today." 27 So Abraham took sheep and oxen and gave

g Gk Vg: Heb lacks *with her son Isaac*

21.9–10: The jealous mother could not stand seeing the two boys on the same level, even at play. **11–14:** Compare Abraham's different attitude in the parallel story (ch 16). **14:** *Beer-sheba* is the locale of the Isaac stories, just as Abraham is associated primarily with Mamre or Hebron (see 13.18 n.).
21.17: *The angel of God,* see 16.7 n. *God has heard,* a play on the name Ishmael, meaning "God hears" (16.11). **20:** Although Ishmael was not the heir of the promise, *God was with the boy,* destining him to be the ancestor of bedouin tribes of the southern wilderness (16.12). Muslims trace their ancestry to Abraham through Ishmael.
21.22–34: Abraham's dispute with **Abimelech.** This story contains two traditional explanations of the name Beer-sheba. According to one, Abimelech guaranteed Abraham's loyalty by an oath. Hence Beer-sheba means "Well of the oath" (v. 31). According to the other (vv. 25–26, 28–30), a dispute over a well resulted in a covenant, seven ewe-lambs being taken in witness. Thus the alternate meaning, "Well of seven." **33:** *The Everlasting God* ("El Olam") is an ancient divine name, once associated with the pre-Israelite sanctuary of Beer-sheba, which Israel adopted as a title for the LORD (Isa 40.28). **34:** *The land of the Philistines* is an anachronism, for the Philistines came into Canaan after 1200 B.C. (see 10.2–5 n.).

them to Abimelech, and the two men made a covenant. 28 Abraham set apart seven ewe lambs of the flock. 29 And Abimelech said to Abraham, "What is the meaning of these seven ewe lambs that you have set apart?" 30 He said, "These seven ewe lambs you shall accept from my hand, in order that you may be a witness for me that I dug this well." 31 Therefore that place was called Beer-sheba;*h* because there both of them swore an oath. 32 When they had made a covenant at Beer-sheba, Abimelech, with Phicol the commander of his army, left and returned to the land of the Philistines. 33 Abraham*i* planted a tamarisk tree in Beer-sheba, and called there on the name of the LORD, the Everlasting God.*j* 34 And Abraham resided as an alien many days in the land of the Philistines.

22 After these things God tested Abraham. He said to him, "Abraham!" And he said, "Here I am." 2 He said, "Take your son, your only son Isaac, whom you love, and go to the land of Moriah, and offer him there as a burnt offering on one of the mountains that I shall show you." 3 So Abraham rose early in the morning, saddled his donkey, and took two of his young men with him, and his son Isaac; he cut the wood for the burnt offering, and set out and went to the place in the distance that God had shown him. 4 On the third day Abraham looked up and saw the place far away. 5 Then Abraham said to his young men, "Stay here with the donkey; the boy and I will go over there; we will worship, and then we will come back to you." 6 Abraham took the wood of the burnt offering and laid it on his son Isaac, and he himself

carried the fire and the knife. So the two of them walked on together. 7 Isaac said to his father Abraham, "Father!" And he said, "Here I am, my son." He said, "The fire and the wood are here, but where is the lamb for a burnt offering?" 8 Abraham said, "God himself will provide the lamb for a burnt offering, my son." So the two of them walked on together.

9 When they came to the place that God had shown him, Abraham built an altar there and laid the wood in order. He bound his son Isaac, and laid him on the altar, on top of the wood. 10 Then Abraham reached out his hand and took the knife to kill*k* his son. 11 But the angel of the LORD called to him from heaven, and said, "Abraham, Abraham!" And he said, "Here I am." 12 He said, "Do not lay your hand on the boy or do anything to him; for now I know that you fear God, since you have not withheld your son, your only son, from me." 13 And Abraham looked up and saw a ram, caught in a thicket by its horns. Abraham went and took the ram and offered it up as a burnt offering instead of his son. 14 So Abraham called that place "The LORD will provide";*l* as it is said to this day, "On the mount of the LORD it shall be provided."*m*

15 The angel of the LORD called to Abraham a second time from heaven, 16 and said, "By myself I have sworn, says the LORD: Because you have done this, and have not withheld your son,

h That is *Well of seven* or *Well of the oath*
i Heb *He*   *j* Or *the* LORD, *El Olam*   *k* Or *to slaughter*   *l* Or *will see*; Heb traditionally transliterated *Jehovah Jireh*   *m* Or *he shall be seen*

22.1–19: The testing of Abraham. In its oldest form this story was told to show that the Deity surrendered a claim upon the life of the firstborn and provided an animal for a substitute (Ex 13.2, 11–16; 22.29; 34.19–20). In its present context the story portrays another threat to the divine promise: God asks Abraham to sacrifice his only heir, the child of the promise, in whom the people's future would be realized. **1:** *Tested,* i.e. put under

trial to see whether he would obey in faith (12.4; compare Heb 11.17–19). **2:** The mountain in *the land of Moriah* is unknown. In 2 Chr 3.1 it is identified with Jerusalem. Samaritan tradition locates the scene on Mount Gerizim (Shechem; compare 12.6), three days' journey (22.4) from Beer-sheba (v. 19; 21.33). **22.14:** *Provide,* see v. 8. **15–18:** Since Abraham survives the test, God renews the promise to him and his descendants (see 12.1–3).

your only son, ¹⁷I will indeed bless you, and I will make your offspring as numerous as the stars of heaven and as the sand that is on the seashore. And your offspring shall possess the gate of their enemies, ¹⁸and by your offspring shall all the nations of the earth gain blessing for themselves, because you have obeyed my voice." ¹⁹So Abraham returned to his young men, and they arose and went together to Beer-sheba; and Abraham lived at Beer-sheba.

20 Now after these things it was told Abraham, "Milcah also has borne children, to your brother Nahor: ²¹Uz the firstborn, Buz his brother, Kemuel the father of Aram, ²²Chesed, Hazo, Pildash, Jidlaph, and Bethuel." ²³Bethuel became the father of Rebekah. These eight Milcah bore to Nahor, Abraham's brother. ²⁴Moreover, his concubine, whose name was Reumah, bore Tebah, Gaham, Tahash, and Maacah.

23 Sarah lived one hundred twenty-seven years; this was the length of Sarah's life. ²And Sarah died at Kiriath-arba (that is, Hebron) in the land of Canaan; and Abraham went in to mourn for Sarah and to weep for her. ³Abraham rose up from beside his dead, and said to the Hittites, ⁴"I am a stranger and an alien residing among you; give me property among you for a burying place, so that I may bury my dead out of my sight." ⁵The Hittites answered Abraham, ⁶"Hear us, my lord; you are a mighty prince among us. Bury your dead in the choicest of our burial places; none of us will withhold from you any burial ground for burying your dead." ⁷Abraham rose and bowed to the Hittites, the people of the land. ⁸He said to them, "If you are willing that I should bury my dead out of my sight, hear me,

and entreat for me Ephron son of Zohar, ⁹so that he may give me the cave of Machpelah, which he owns; it is at the end of his field. For the full price let him give it to me in your presence as a possession for a burying place." ¹⁰Now Ephron was sitting among the Hittites; and Ephron the Hittite answered Abraham in the hearing of the Hittites, of all who went in at the gate of his city, ¹¹"No, my lord, hear me; I give you the field, and I give you the cave that is in it; in the presence of my people I give it to you; bury your dead." ¹²Then Abraham bowed down before the people of the land. ¹³He said to Ephron in the hearing of the people of the land, "If you only will listen to me! I will give the price of the field; accept it from me, so that I may bury my dead there." ¹⁴Ephron answered Abraham, ¹⁵"My lord, listen to me; a piece of land worth four hundred shekels of silver—what is that between you and me? Bury your dead." ¹⁶Abraham agreed with Ephron; and Abraham weighed out for Ephron the silver that he had named in the hearing of the Hittites, four hundred shekels of silver, according to the weights current among the merchants.

17 So the field of Ephron in Machpelah, which was to the east of Mamre, the field with the cave that was in it and all the trees that were in the field, throughout its whole area, passed ¹⁸to Abraham as a possession in the presence of the Hittites, in the presence of all who went in at the gate of his city. ¹⁹After this, Abraham buried Sarah his wife in the cave of the field of Machpelah facing Mamre (that is, Hebron) in the land of Canaan. ²⁰The field and the cave that is in it passed from the Hittites into Abraham's possession as a burying place.

22.20–24: The descendants of Abraham's brother Nahor. See ch 24.

23.1–20: Abraham's purchase of a family burial place. 2: *Kiriath-arba,* the older name of Hebron (Josh 14.15; 15.13; Judg 1.10). **3:** *The Hittites* belonged at that time to the pre-Israelite population, *the people of the land* (v. 7; see 10.15 n.). **4–16:** Legal transac-

tions were handled by elders at the city gate (v. 10). **9:** Abraham insists on payment of *the full price* in order to obtain legal title to the land.

23.19: The cave of Machpelah was the tomb of Abraham and Sarah (25.9–10), Isaac (35.27–29) and Rebekah (49.31), Jacob (50.13) and Leah (49.31).

24 Now Abraham was old, well advanced in years; and the LORD had blessed Abraham in all things. 2 Abraham said to his servant, the oldest of his house, who had charge of all that he had, "Put your hand under my thigh 3 and I will make you swear by the LORD, the God of heaven and earth, that you will not get a wife for my son from the daughters of the Canaanites, among whom I live, 4 but will go to my country and to my kindred and get a wife for my son Isaac." 5 The servant said to him, "Perhaps the woman may not be willing to follow me to this land; must I then take your son back to the land from which you came?" 6 Abraham said to him, "See to it that you do not take my son back there. 7 The LORD, the God of heaven, who took me from my father's house and from the land of my birth, and who spoke to me and swore to me, 'To your offspring I will give this land,' he will send his angel before you, and you shall take a wife for my son from there. 8 But if the woman is not willing to follow you, then you will be free from this oath of mine; only you must not take my son back there." 9 So the servant put his hand under the thigh of Abraham his master and swore to him concerning this matter.

10 Then the servant took ten of his master's camels and departed, taking all kinds of choice gifts from his master; and he set out and went to Aram-naharaim, to the city of Nahor. 11 He made the camels kneel down outside the city by the well of water; it was toward evening, the time when women go out to draw water.

12 And he said, "O LORD, God of my master Abraham, please grant me success today and show steadfast love to my master Abraham. 13 I am standing here by the spring of water, and the daughters of the townspeople are coming out to draw water. 14 Let the girl to whom I shall say, 'Please offer your jar that I may drink,' and who shall say, 'Drink, and I will water your camels'—let her be the one whom you have appointed for your servant Isaac. By this I shall know that you have shown steadfast love to my master."

15 Before he had finished speaking, there was Rebekah, who was born to Bethuel son of Milcah, the wife of Nahor, Abraham's brother, coming out with her water jar on her shoulder. 16 The girl was very fair to look upon, a virgin, whom no man had known. She went down to the spring, filled her jar, and came up. 17 Then the servant ran to meet her and said, "Please let me sip a little water from your jar." 18 "Drink, my lord," she said, and quickly lowered her jar upon her hand and gave him a drink. 19 When she had finished giving him a drink, she said, "I will draw for your camels also, until they have finished drinking." 20 So she quickly emptied her jar into the trough and ran again to the well to draw, and she drew for all his camels. 21 The man gazed at her in silence to learn whether or not the LORD had made his journey successful.

22 When the camels had finished drinking, the man took a gold nose-ring weighing a half shekel, and two bracelets for her arms weighing ten gold shekels,

---

24.1–67: Finding a wife for Isaac among kinfolk in Haran. **2:** The servant was perhaps Abraham's major-domo, Eliezer (15.2). Putting the hand under the thigh, an old form of oath taking (47.29), reflected the view that the fountain of reproductivity was sacred to the Deity (see 16.2 n.). **3:** Aloofness from the Canaanites was based upon fear of the corrupting influence of Canaanite culture (Ex 34.15–16; Deut 7.3–4). **7:** *His angel,* see 16.7 n.

24.10: *The city of Nahor,* near Haran (see 11.31 n.). **12:** *Steadfast love:* NRSV's translation of a Hebrew term ("ḥesed") that signifies the loyalty arising from a relationship (e.g. friendship; see 1 Sam 20.8), which motivates the stronger party to show favor or give help to the weaker. Applied to God's covenant with human beings, it means benevolent action, loyalty manifest in deeds, gracious favor. **14:** The story assumes that the young woman had been *appointed* by the LORD to be Isaac's wife; therefore events unfold according to divine providence (see vv. 21, 26–27, 50, 56). **15:** 22.20–23.

23 and said, "Tell me whose daughter you are. Is there room in your father's house for us to spend the night?" 24 She said to him, "I am the daughter of Bethuel son of Milcah, whom she bore to Nahor." 25 She added, "We have plenty of straw and fodder and a place to spend the night." 26 The man bowed his head and worshiped the LORD 27 and said, "Blessed be the LORD, the God of my master Abraham, who has not forsaken his steadfast love and his faithfulness toward my master. As for me, the LORD has led me on the way to the house of my master's kin."

28 Then the girl ran and told her mother's household about these things. 29 Rebekah had a brother whose name was Laban; and Laban ran out to the man, to the spring. 30 As soon as he had seen the nose-ring, and the bracelets on his sister's arms, and when he heard the words of his sister Rebekah, "Thus the man spoke to me," he went to the man; and there he was, standing by the camels at the spring. 31 He said, "Come in, O blessed of the LORD. Why do you stand outside when I have prepared the house and a place for the camels?" 32 So the man came into the house; and Laban unloaded the camels, and gave him straw and fodder for the camels, and water to wash his feet and the feet of the men who were with him. 33 Then food was set before him to eat; but he said, "I will not eat until I have told my errand." He said, "Speak on."

34 So he said, "I am Abraham's servant. 35 The LORD has greatly blessed my master, and he has become wealthy; he has given him flocks and herds, silver and gold, male and female slaves, camels and donkeys. 36 And Sarah my master's wife bore a son to my master when she was old; and he has given him all that he has. 37 My master made me swear, saying, 'You shall not take a wife for my son from the daughters of the Canaanites, in whose land I live; 38 but you shall go to my father's house, to my kindred, and get a wife for my son.' 39 I said to my master, 'Perhaps the woman will not follow me.' 40 But he said to me, 'The LORD,

before whom I walk, will send his angel with you and make your way successful. You shall get a wife for my son from my kindred, from my father's house. 41 Then you will be free from my oath, when you come to my kindred; even if they will not give her to you, you will be free from my oath.'

42 "I came today to the spring, and said, 'O LORD, the God of my master Abraham, if now you will only make successful the way I am going! 43 I am standing here by the spring of water; let the young woman who comes out to draw, to whom I shall say, "Please give me a little water from your jar to drink," 44 and who will say to me, "Drink, and I will draw for your camels also"—let her be the woman whom the LORD has appointed for my master's son.'

45 "Before I had finished speaking in my heart, there was Rebekah coming out with her water jar on her shoulder; and she went down to the spring, and drew. I said to her, 'Please let me drink.' 46 She quickly let down her jar from her shoulder, and said, 'Drink, and I will also water your camels.' So I drank, and she also watered the camels. 47 Then I asked her, 'Whose daughter are you?' She said, 'The daughter of Bethuel, Nahor's son, whom Milcah bore to him.' So I put the ring on her nose, and the bracelets on her arms. 48 Then I bowed my head and worshiped the LORD, and blessed the LORD, the God of my master Abraham, who had led me by the right way to obtain the daughter of my master's kinsman for his son. 49 Now then, if you will deal loyally and truly with my master, tell me; and if not, tell me, so that I may turn either to the right hand or to the left."

50 Then Laban and Bethuel answered, "The thing comes from the LORD; we cannot speak to you anything bad or good. 51 Look, Rebekah is before you, take her and go, and let her be the wife of your master's son, as the LORD has spoken."

52 When Abraham's servant heard their words, he bowed himself to the ground before the LORD. 53 And the servant brought out jewelry of silver and of

gold, and garments, and gave them to Rebekah; he also gave to her brother and to her mother costly ornaments. [54]Then he and the men who were with him ate and drank, and they spent the night there. When they rose in the morning, he said, "Send me back to my master." [55]Her brother and her mother said, "Let the girl remain with us a while, at least ten days; after that she may go." [56]But he said to them, "Do not delay me, since the LORD has made my journey successful; let me go that I may go to my master." [57]They said, "We will call the girl, and ask her." [58]And they called Rebekah, and said to her, "Will you go with this man?" She said, "I will." [59]So they sent away their sister Rebekah and her nurse along with Abraham's servant and his men. [60]And they blessed Rebekah and said to her,

"May you, our sister, become
thousands of myriads;
may your offspring gain
possession
of the gates of their foes."

[61]Then Rebekah and her maids rose up, mounted the camels, and followed the man; thus the servant took Rebekah, and went his way.

[62]Now Isaac had come from[n] Beer-lahai-roi, and was settled in the Negeb. [63]Isaac went out in the evening to walk[o] in the field; and looking up, he saw camels coming. [64]And Rebekah looked up, and when she saw Isaac, she slipped quickly from the camel, [65]and said to the servant, "Who is the man over there, walking in the field to meet us?" The servant said, "It is my master." So she took her veil and covered herself. [66]And the servant told Isaac all the things that he had done. [67]Then Isaac brought her into his mother Sarah's tent. He took Rebekah, and she became his wife; and he loved her. So Isaac was comforted after his mother's death.

25 Abraham took another wife, whose name was Keturah. [2]She bore him Zimran, Jokshan, Medan, Midian, Ishbak, and Shuah. [3]Jokshan was the father of Sheba and Dedan. The sons of Dedan were Asshurim, Letushim, and Leummim. [4]The sons of Midian were Ephah, Epher, Hanoch, Abida, and Eldaah. All these were the children of Keturah. [5]Abraham gave all he had to Isaac. [6]But to the sons of his concubines Abraham gave gifts, while he was still living, and he sent them away from his son Isaac, eastward to the east country.

[7]This is the length of Abraham's life, one hundred seventy-five years. [8]Abraham breathed his last and died in a good old age, an old man and full of years, and was gathered to his people. [9]His sons Isaac and Ishmael buried him in the cave of Machpelah, in the field of Ephron son of Zohar the Hittite, east of Mamre, [10]the field that Abraham purchased from the Hittites. There Abraham was buried, with his wife Sarah. [11]After the death of Abraham God blessed his son Isaac. And Isaac settled at Beer-lahai-roi.

[12]These are the descendants of Ishmael, Abraham's son, whom Hagar the Egyptian, Sarah's slave-girl, bore to Abraham. [13]These are the names of the sons of Ishmael, named in the order of their birth: Nebaioth, the firstborn of Ishmael; and Kedar, Adbeel, Mibsam, [14]Mishma, Dumah, Massa, [15]Hadad, Tema, Jetur, Naphish, and Kedemah. [16]These are the sons of Ishmael and these are their names, by their villages and by their encampments, twelve princes according to their tribes. [17](This is the length of the life of Ishmael, one hundred thirty-seven years; he breathed his last and died, and was gathered to his people.) [18]They settled from Havilah to

n Syr Tg: Heb *from coming to*   o Meaning of Heb word is uncertain

24.62: *Beer-lahai-roi*, 16.14.
25.1–18: **The death of Abraham. 1–6:** The ancestry of Arabic tribes, including Midian (Ex 2.15b–22; 18.1), is traced to Abraham through his other wife, Keturah. **9:** See ch 23.

25.12–18: An excerpt from the book of generations (5.1). **16:** Like later Israel, the Ishmaelites were organized into twelve tribes, each with a tribal prince. **18:** 16.7, 14.

Shur, which is opposite Egypt in the direction of Assyria; he settled down*p* alongside of*q* all his people.

19 These are the descendants of Isaac, Abraham's son: Abraham was the father of Isaac, 20 and Isaac was forty years old when he married Rebekah, daughter of Bethuel the Aramean of Paddan-aram, sister of Laban the Aramean. 21 Isaac prayed to the LORD for his wife, because she was barren; and the LORD granted his prayer, and his wife Rebekah conceived. 22 The children struggled together within her; and she said, "If it is to be this way, why do I live?"*r* So she went to inquire of the LORD. 23 And the LORD said to her,

"Two nations are in your womb,
  and two peoples born of you
    shall be divided;
the one shall be stronger than the
    other,
  the elder shall serve the
    younger."

24 When her time to give birth was at hand, there were twins in her womb. 25 The first came out red, all his body like a hairy mantle; so they named him Esau. 26 Afterward his brother came out, with his hand gripping Esau's heel; so he was named Jacob.*s* Isaac was sixty years old when she bore them.

27 When the boys grew up, Esau was a skillful hunter, a man of the field, while Jacob was a quiet man, living in tents. 28 Isaac loved Esau, because he was fond of game; but Rebekah loved Jacob.

29 Once when Jacob was cooking a stew, Esau came in from the field, and he was famished. 30 Esau said to Jacob, "Let me eat some of that red stuff, for I am famished!" (Therefore he was called Edom.*t*) 31 Jacob said, "First sell me your birthright." 32 Esau said, "I am about to die; of what use is a birthright to me?" 33 Jacob said, "Swear to me first."*u* So he swore to him, and sold his birthright to Jacob. 34 Then Jacob gave Esau bread and lentil stew, and he ate and drank, and rose and went his way. Thus Esau despised his birthright.

26 Now there was a famine in the land, besides the former famine that had occurred in the days of Abraham. And Isaac went to Gerar, to King Abimelech of the Philistines. 2 The LORD appeared to Isaac*v* and said, "Do not go down to Egypt; settle in the land that I shall show you. 3 Reside in this land as an alien, and I will be with you, and will bless you; for to you and to your descendants I will give all these lands, and I will fulfill the oath that I swore to your father Abraham. 4 I will make your offspring as numerous as the stars of heaven, and will give to your offspring all these lands; and all the nations of the earth shall gain blessing for themselves through your offspring, 5 because Abraham obeyed my voice and kept my charge, my commandments, my statutes, and my laws."

6 So Isaac settled in Gerar. 7 When the men of the place asked him about his

p Heb *he fell*   *q* Or *down in opposition to*
r Syr: Meaning of Heb uncertain   *s* That is
He takes by the heel or *He supplants*   *t* That is
Red   *u* Heb *today*   *v* Heb *him*

25.19–34: The rivalry of Jacob (Israel) and Esau (Edom). 20: *Paddan-aram,* see 11.31 n. **21:** See 16.2 n. **22–23:** Rebekah went to a sanctuary *to inquire of the* LORD and received the oracular answer in v. 23. **25:** The Hebrew word *red* ("'admoni") is a play on the word Edom ("'edom"; v. 30); *hairy* ("se'ar") is a play on Seir, the region of the Edomites (32.3). **26:** *Jacob* is interpreted by a play on the Hebrew word for "heel," i.e. "he takes by the heel" or "he supplants" (Hos 12.3; Jer 9.4). **25.27–28:** The two boys typify the hunter and the shepherd, two rival ways of life (4.2). **31–34:** The *birthright* refers to the rights of the eldest son: leadership of the family and a double share of the inheritance (Deut 21.15–17). The caricature of Esau as a dull person, easily outwitted on an empty stomach, is intended to explain why Israel gained ascendancy over Edom (2 Sam 8.12–14; 2 Chr 25.11–24) even though the latter became a nation first (36.31–39). **26.1–34: Stories about Isaac. 1:** *The former famine,* 12.10. *The Philistines,* see 21.34 n.

wife, he said, "She is my sister"; for he was afraid to say, "My wife," thinking, "or else the men of the place might kill me for the sake of Rebekah, because she is attractive in appearance." 8 When Isaac had been there a long time, King Abimelech of the Philistines looked out of a window and saw him fondling his wife Rebekah. 9 So Abimelech called for Isaac, and said, "So she is your wife! Why then did you say, 'She is my sister'?" Isaac said to him, "Because I thought I might die because of her." 10 Abimelech said, "What is this you have done to us? One of the people might easily have lain with your wife, and you would have brought guilt upon us." 11 So Abimelech warned all the people, saying, "Whoever touches this man or his wife shall be put to death."

12 Isaac sowed seed in that land, and in the same year reaped a hundredfold. The LORD blessed him, 13 and the man became rich; he prospered more and more until he became very wealthy. 14 He had possessions of flocks and herds, and a great household, so that the Philistines envied him. 15 (Now the Philistines had stopped up and filled with earth all the wells that his father's servants had dug in the days of his father Abraham.) 16 And Abimelech said to Isaac, "Go away from us; you have become too powerful for us."

17 So Isaac departed from there and camped in the valley of Gerar and settled there. 18 Isaac dug again the wells of water that had been dug in the days of his father Abraham; for the Philistines had stopped them up after the death of Abraham; and he gave them the names that his father had given them. 19 But when Isaac's servants dug in the valley and found there a well of spring water, 20 the herders of Gerar quarreled with Isaac's herders, saying, "The water is ours." So he called the well Esek, *w* because they contended with him. 21 Then they dug another well, and they quarreled over that one also; so he called it Sitnah. *x* 22 He moved from there and dug another well, and they did not quarrel over it; so he called it Rehoboth, *y* saying, "Now the LORD has made room for us, and we shall be fruitful in the land."

23 From there he went up to Beersheba. 24 And that very night the LORD appeared to him and said, "I am the God of your father Abraham; do not be afraid, for I am with you and will bless you and make your offspring numerous for my servant Abraham's sake." 25 So he built an altar there, called on the name of the LORD, and pitched his tent there. And there Isaac's servants dug a well.

26 Then Abimelech went to him from Gerar, with Ahuzzath his adviser and Phicol the commander of his army. 27 Isaac said to them, "Why have you come to me, seeing that you hate me and have sent me away from you?" 28 They said, "We see plainly that the LORD has been with you; so we say, let there be an oath between you and us, and let us make a covenant with you 29 so that you will do us no harm, just as we have not touched you and have done to you nothing but good and have sent you away in peace. You are now the blessed of the LORD." 30 So he made them a feast, and they ate and drank. 31 In the morning they rose early and exchanged oaths; and Isaac set them on their way, and they departed from him in peace. 32 That same day Isaac's servants came and told him about the well that they had dug, and said to him, "We have found water!" 33 He

w That is *Contention*   x That is *Enmity*
y That is *Broad places* or *Room*

3–5: The promise, first given to Abraham (12.2–3, 7), is reaffirmed to Isaac. 7–11: 12.10–20 and ch 20. 26.12–33: Isaac is portrayed as a seminomad who settled down long enough to raise crops (v. 12) but also moved about to find pasturage and water for his flocks.

26.24: *The God of your father Abraham,* the God of the ancestors is not known in general or abstract terms but by concrete, historical relations to particular persons (Ex 3.6). 28–30: Compare 21.22–24. 32–33: *Shibah,* another explanation of the name of Isaac's shrine, Beer-sheba (see 21.22–34 n.).

called it Shibah;[z] therefore the name of the city is Beer-sheba[a] to this day.

34 When Esau was forty years old, he married Judith daughter of Beeri the Hittite, and Basemath daughter of Elon the Hittite; [35]and they made life bitter for Isaac and Rebekah.

27 When Isaac was old and his eyes were dim so that he could not see, he called his elder son Esau and said to him, "My son"; and he answered, "Here I am." [2]He said, "See, I am old; I do not know the day of my death. [3]Now then, take your weapons, your quiver and your bow, and go out to the field, and hunt game for me. [4]Then prepare for me savory food, such as I like, and bring it to me to eat, so that I may bless you before I die."

5 Now Rebekah was listening when Isaac spoke to his son Esau. So when Esau went to the field to hunt for game and bring it, [6]Rebekah said to her son Jacob, "I heard your father say to your brother Esau, [7]'Bring me game, and prepare for me savory food to eat, that I may bless you before the LORD before I die.' [8]Now therefore, my son, obey my word as I command you. [9]Go to the flock, and get me two choice kids, so that I may prepare from them savory food for your father, such as he likes; [10]and you shall take it to your father to eat, so that he may bless you before he dies." [11]But Jacob said to his mother Rebekah, "Look, my brother Esau is a hairy man, and I am a man of smooth skin. [12]Perhaps my father will feel me, and I shall seem to be mocking him, and bring a curse on myself and not a blessing." [13]His mother said to him, "Let your curse be on me, my son; only obey my word, and go, get them for me." [14]So he went and got them and brought them to his mother; and his mother prepared savory food, such as his father loved. [15]Then Rebekah took the best garments of her elder son Esau, which were with her in the house, and put them on her younger son Jacob; [16]and she put the skins of the kids on his hands and on the smooth part of his neck. [17]Then she handed the savory food, and the bread that she had prepared, to her son Jacob.

18 So he went in to his father, and said, "My father"; and he said, "Here I am; who are you, my son?" [19]Jacob said to his father, "I am Esau your firstborn. I have done as you told me; now sit up and eat of my game, so that you may bless me." [20]But Isaac said to his son, "How is it that you have found it so quickly, my son?" He answered, "Because the LORD your God granted me success." [21]Then Isaac said to Jacob, "Come near, that I may feel you, my son, to know whether you are really my son Esau or not." [22]So Jacob went up to his father Isaac, who felt him and said, "The voice is Jacob's voice, but the hands are the hands of Esau." [23]He did not recognize him, because his hands were hairy like his brother Esau's hands; so he blessed him. [24]He said, "Are you really my son Esau?" He answered, "I am." [25]Then he said, "Bring it to me, that I may eat of my son's game and bless you." So he brought it to him, and he ate; and he brought him wine, and he drank. [26]Then his father Isaac said to him, "Come near and kiss me, my son." [27]So he came near and kissed him; and he smelled the smell of his garments, and blessed him, and said,

"Ah, the smell of my son
    is like the smell of a field that
        the LORD has blessed.

z A word resembling the word for *oath*
a That is *Well of the oath* or *Well of seven*

27.1–45: Jacob cheats Esau out of the blessing. Here begins a cycle of stories about Jacob, the founder of the sanctuary of Bethel. **3**: For Esau as a hunter, see 25.27–34, an episode to which Jacob makes no reference. **4**: Death-bed blessings were important in the life and literature of ancient peoples (48.8–20;

49.1–28; Deut ch 33; Josh ch 23). It was believed that the blessing, like the curse (v. 12), released a power that effectively determined the character and destiny of the recipient (Num ch 22–24).

27.5–29: The elements of deceit and even outright lying (v. 20) mark this as a popular

28 May God give you of the dew of
    heaven,
  and of the fatness of the earth,
  and plenty of grain and wine.
29 Let peoples serve you,
  and nations bow down to you.
  Be lord over your brothers,
  and may your mother's sons
    bow down to you.
  Cursed be everyone who curses
    you,
  and blessed be everyone who
    blesses you!"

30 As soon as Isaac had finished blessing Jacob, when Jacob had scarcely gone
out from the presence of his father Isaac,
his brother Esau came in from his hunting. 31 He also prepared savory food, and
brought it to his father. And he said to his
father, "Let my father sit up and eat of his
son's game, so that you may bless me."
32 His father Isaac said to him, "Who are
you?" He answered, "I am your firstborn
son, Esau." 33 Then Isaac trembled violently, and said, "Who was it then that
hunted game and brought it to me, and
I ate it all[b] before you came, and I have
blessed him?—yes, and blessed he shall
be!" 34 When Esau heard his father's
words, he cried out with an exceedingly
great and bitter cry, and said to his father, "Bless me, me also, father!" 35 But
he said, "Your brother came deceitfully,
and he has taken away your blessing."
36 Esau said, "Is he not rightly named Jacob?[c] For he has supplanted me these
two times. He took away my birthright;
and look, now he has taken away my
blessing." Then he said, "Have you not
reserved a blessing for me?" 37 Isaac answered Esau, "I have already made him
your lord, and I have given him all his
brothers as servants, and with grain and

wine I have sustained him. What then
can I do for you, my son?" 38 Esau said to
his father, "Have you only one blessing,
father? Bless me, me also, father!" And
Esau lifted up his voice and wept.

39 Then his father Isaac answered
him:
  "See, away from[d] the fatness of
    the earth shall your home
    be,
  and away from[e] the dew of
    heaven on high.
40 By your sword you shall live,
  and you shall serve your
    brother;
  but when you break loose,[f]
    you shall break his yoke from
    your neck."

41 Now Esau hated Jacob because of
the blessing with which his father had
blessed him, and Esau said to himself,
"The days of mourning for my father are
approaching; then I will kill my brother
Jacob." 42 But the words of her elder son
Esau were told to Rebekah; so she sent
and called her younger son Jacob and said
to him, "Your brother Esau is consoling
himself by planning to kill you. 43 Now
therefore, my son, obey my voice; flee at
once to my brother Laban in Haran,
44 and stay with him a while, until your
brother's fury turns away— 45 until your
brother's anger against you turns away,
and he forgets what you have done to
him; then I will send, and bring you back
from there. Why should I lose both of
you in one day?"

46 Then Rebekah said to Isaac, "I am
weary of my life because of the Hittite
women. If Jacob marries one of the Hit-

b Cn: Heb *of all*   c That is *He supplants* or *He
takes by the heel*   d Or *See, of*   e Or *and of*
f Meaning of Heb uncertain

story; the sequel shows that later Jacob reaped
the consequences of his action.
  27.34: To appreciate the pathos of the
scene it must be remembered that the spoken
blessing, like an arrow shot toward its goal,
was believed to release a power which could
not be retracted (see v. 33). **36:** *He has supplanted me,* see 25.26 n. **39:** This blessing in-
verts the meaning of the same words in v. 28,

for the land of Edom was not fertile. **40:** In
David's time Edom was subjugated (2 Sam
8.12–14) but it revolted under Solomon
(1 Kings 11.14–22, 25; compare 2 Kings
8.20–22).
  **27.46–28.22: Jacob's departure for
Aram and his dream at Bethel.**
  27.46: *Hittite women,* i.e. Canaanites (see
23.3 n. and 24.3 n.).

tite women such as these, one of the women of the land, what good will my life be to me?"

28 Then Isaac called Jacob and blessed him, and charged him, "You shall not marry one of the Canaanite women. 2 Go at once to Paddan-aram to the house of Bethuel, your mother's father; and take as wife from there one of the daughters of Laban, your mother's brother. 3 May God Almighty*g* bless you and make you fruitful and numerous, that you may become a company of peoples. 4 May he give to you the blessing of Abraham, to you and to your offspring with you, so that you may take possession of the land where you now live as an alien—land that God gave to Abraham." 5 Thus Isaac sent Jacob away; and he went to Paddan-aram, to Laban son of Bethuel the Aramean, the brother of Rebekah, Jacob's and Esau's mother.

6 Now Esau saw that Isaac had blessed Jacob and sent him away to Paddan-aram to take a wife from there, and that as he blessed him he charged him, "You shall not marry one of the Canaanite women," 7 and that Jacob had obeyed his father and his mother and gone to Paddan-aram. 8 So when Esau saw that the Canaanite women did not please his father Isaac, 9 Esau went to Ishmael and took Mahalath daughter of Abraham's son Ishmael, and sister of Nebaioth, to be his wife in addition to the wives he had.

10 Jacob left Beer-sheba and went toward Haran. 11 He came to a certain place and stayed there for the night, because the sun had set. Taking one of the stones of the place, he put it under his head and lay down in that place. 12 And he dreamed that there was a ladder*h* set up on the earth, the top of it reaching to heaven; and the angels of God were ascending and descending on it. 13 And the LORD stood beside him*i* and said, "I am the LORD, the God of Abraham your father and the God of Isaac; the land on which you lie I will give to you and to your offspring; 14 and your offspring shall be like the dust of the earth, and you shall spread abroad to the west and to the east and to the north and to the south; and all the families of the earth shall be blessed*j* in you and in your offspring. 15 Know that I am with you and will keep you wherever you go, and will bring you back to this land; for I will not leave you until I have done what I have promised you." 16 Then Jacob woke from his sleep and said, "Surely the LORD is in this place—and I did not know it!" 17 And he was afraid, and said, "How awesome is this place! This is none other than the house of God, and this is the gate of heaven."

18 So Jacob rose early in the morning, and he took the stone that he had put under his head and set it up for a pillar and poured oil on the top of it. 19 He called that place Bethel;*k* but the name of the city was Luz at the first. 20 Then Jacob

*g Traditional rendering of Heb El Shaddai
h Or stairway or ramp   i Or stood above it
j Or shall bless themselves
k That is House of God*

28.2: *Paddan-aram,* see 11.31 n. **3:** *God Almighty,* see 17.1 n. **11:** Bethel was at this time unsettled. According to ancient belief, oracles could be received by sleeping in a holy place (1 Sam ch 3). **28.12:** Angels in a company are mentioned here and in 32.1–2 (compare 16.7; 21.17), suggesting the view of a retinue surrounding the heavenly King. **13–15:** The Deity, identified as *the God of Abraham* and *the God of Isaac* (see 26.24 n.), renews the promise (12.2–3, 7). **17:** An explanation of the name of Beth-el (*house of God*). *The gate of heaven* suggests the ancient view that a sanctuary was a place where God came down to meet the people, like Babel or "gate of God" (11.1–9). **28.18:** The *pillar* was a sacred stone, often found at ancient sanctuaries (Josh 24.26). Anointing the stone made it holy, that is, set it apart for the Deity (v. 22). **21:** The personal relationship with God indicated in the expression *my God,* is characteristic of the religion of Israel's ancestors. The family leader chooses God in response to a personal revelation. **22:** The story explains the origin of the northern sanctuary at Bethel which flour-

made a vow, saying, "If God will be with me, and will keep me in this way that I go, and will give me bread to eat and clothing to wear, ²¹ so that I come again to my father's house in peace, then the LORD shall be my God, ²² and this stone, which I have set up for a pillar, shall be God's house; and of all that you give me I will surely give one tenth to you."

29 Then Jacob went on his journey, and came to the land of the people of the east. ² As he looked, he saw a well in the field and three flocks of sheep lying there beside it; for out of that well the flocks were watered. The stone on the well's mouth was large, ³ and when all the flocks were gathered there, the shepherds would roll the stone from the mouth of the well, and water the sheep, and put the stone back in its place on the mouth of the well.

4 Jacob said to them, "My brothers, where do you come from?" They said, "We are from Haran." ⁵ He said to them, "Do you know Laban son of Nahor?" They said, "We do." ⁶ He said to them, "Is it well with him?" "Yes," they replied, "and here is his daughter Rachel, coming with the sheep." ⁷ He said, "Look, it is still broad daylight; it is not time for the animals to be gathered together. Water the sheep, and go, pasture them." ⁸ But they said, "We cannot until all the flocks are gathered together, and the stone is rolled from the mouth of the well; then we water the sheep."

9 While he was still speaking with them, Rachel came with her father's sheep; for she kept them. ¹⁰ Now when Jacob saw Rachel, the daughter of his mother's brother Laban, and the sheep of his mother's brother Laban, Jacob went up and rolled the stone from the well's mouth, and watered the flock of his

mother's brother Laban. ¹¹ Then Jacob kissed Rachel, and wept aloud. ¹² And Jacob told Rachel that he was her father's kinsman, and that he was Rebekah's son; and she ran and told her father.

13 When Laban heard the news about his sister's son Jacob, he ran to meet him; he embraced him and kissed him, and brought him to his house. Jacob^{*l*} told Laban all these things, ¹⁴ and Laban said to him, "Surely you are my bone and my flesh!" And he stayed with him a month.

15 Then Laban said to Jacob, "Because you are my kinsman, should you therefore serve me for nothing? Tell me, what shall your wages be?" ¹⁶ Now Laban had two daughters; the name of the elder was Leah, and the name of the younger was Rachel. ¹⁷ Leah's eyes were lovely, ^{*m*} and Rachel was graceful and beautiful. ¹⁸ Jacob loved Rachel; so he said, "I will serve you seven years for your younger daughter Rachel." ¹⁹ Laban said, "It is better that I give her to you than that I should give her to any other man; stay with me." ²⁰ So Jacob served seven years for Rachel, and they seemed to him but a few days because of the love he had for her.

21 Then Jacob said to Laban, "Give me my wife that I may go in to her, for my time is completed." ²² So Laban gathered together all the people of the place, and made a feast. ²³ But in the evening he took his daughter Leah and brought her to Jacob; and he went in to her. ²⁴ (Laban gave his maid Zilpah to his daughter Leah to be her maid.) ²⁵ When morning came, it was Leah! And Jacob said to Laban, "What is this you have done to me? Did I not serve with you for Rachel? Why then have you deceived me?" ²⁶ La-

l Heb *He*   *m* Meaning of Heb uncertain

---

ished from the time of Jeroboam I (1 Kings 12.26–29) to its destruction by Josiah (2 Kings 23.15). **29.1–31.55: Jacob's success in Haran. 29.1:** *The people of the east,* a general expression (11.2), is applied here to the Arameans. **8:** Local custom prevented use of the well until all rightful parties could be there to get

their fair share; thus the stone covering was bigger than one person could lift (vv. 2–3). **29.18:** Jacob asks for Rachel as a reward for service (compare Josh 15.16–17; 1 Sam 17.25; 18.17) instead of paying the usual marriage price (Ex 22.16–17; Deut 22.29). **23–25:** The exchange could be made because the bride was brought veiled to the bridegroom

ban said, "This is not done in our country—giving the younger before the firstborn. 27 Complete the week of this one, and we will give you the other also in return for serving me another seven years." 28 Jacob did so, and completed her week; then Laban gave him his daughter Rachel as a wife. 29 (Laban gave his maid Bilhah to his daughter Rachel to be her maid.) 30 So Jacob went in to Rachel also, and he loved Rachel more than Leah. He served Laban*n* for another seven years.

31 When the LORD saw that Leah was unloved, he opened her womb; but Rachel was barren. 32 Leah conceived and bore a son, and she named him Reuben;*o* for she said, "Because the LORD has looked on my affliction; surely now my husband will love me." 33 She conceived again and bore a son, and said, "Because the LORD has heard*p* that I am hated, he has given me this son also"; and she named him Simeon. 34 Again she conceived and bore a son, and said, "Now this time my husband will be joined*q* to me, because I have borne him three sons"; therefore he was named Levi. 35 She conceived again and bore a son, and said, "This time I will praise*r* the LORD"; therefore she named him Judah; then she ceased bearing.

30 When Rachel saw that she bore Jacob no children, she envied her sister; and she said to Jacob, "Give me children, or I shall die!" 2 Jacob became very angry with Rachel and said, "Am I in the place of God, who has withheld from you the fruit of the womb?" 3 Then she said, "Here is my maid Bilhah; go in to her, that she may bear upon my knees and that I too may have children through her." 4 So she gave him her maid Bilhah as a wife; and Jacob went in to her. 5 And Bilhah conceived and bore Jacob a son.

6 Then Rachel said, "God has judged me, and has also heard my voice and given me a son"; therefore she named him Dan.*s* 7 Rachel's maid Bilhah conceived again and bore Jacob a second son. 8 Then Rachel said, "With mighty wrestlings I have wrestled*t* with my sister, and have prevailed"; so she named him Naphtali.

9 When Leah saw that she had ceased bearing children, she took her maid Zilpah and gave her to Jacob as a wife. 10 Then Leah's maid Zilpah bore Jacob a son. 11 And Leah said, "Good fortune!" so she named him Gad.*u* 12 Leah's maid Zilpah bore Jacob a second son. 13 And Leah said, "Happy am I! For the women will call me happy"; so she named him Asher.*v*

14 In the days of wheat harvest Reuben went and found mandrakes in the field, and brought them to his mother Leah. Then Rachel said to Leah, "Please give me some of your son's mandrakes." 15 But she said to her, "Is it a small matter that you have taken away my husband? Would you take away my son's mandrakes also?" Rachel said, "Then he may lie with you tonight for your son's mandrakes." 16 When Jacob came from the field in the evening, Leah went out to meet him, and said, "You must come in to me; for I have hired you with my son's mandrakes." So he lay with her that night. 17 And God heeded Leah, and she conceived and bore Jacob a fifth son. 18 Leah said, "God has given me my hire*w* because I gave my maid to my husband"; so she named him Issachar. 19 And Leah conceived again, and she bore Jacob a sixth son. 20 Then Leah said, "God has endowed me with a good dowry; now

n Heb *him*   *o* That is *See, a son*
p Heb *shama*   *q* Heb *lawah*   *r* Heb *hodah*
s That is *He judged*   *t* Heb *niphtal*   *u* That is *Fortune*   *v* That is *Happy*   *w* Heb *sakar*

(24.65). **27:** *The week* refers to the week of marriage festivity (Judg 14.12).

29.31–30.24: Jacob's eleven sons (for Benjamin's birth, see 35.16–18). The fanciful name-explanations, based on Hebrew wordplays (see notes *o* to *y*), reflect the rivalry of the two wives for Jacob's affection. The story assumes the traditional twelve-tribe pattern of the Israelite confederacy.

30.2–3: See 16.2 n. **6:** *God has judged me,* i.e. has gotten justice for me.

30.14: In antiquity *mandrakes,* roots of a

my husband will honor[x] me, because I have borne him six sons"; so she named him Zebulun. 21 Afterwards she bore a daughter, and named her Dinah.

22 Then God remembered Rachel, and God heeded her and opened her womb. 23 She conceived and bore a son, and said, "God has taken away my reproach"; 24 and she named him Joseph,[y] saying, "May the LORD add to me another son!"

25 When Rachel had borne Joseph, Jacob said to Laban, "Send me away, that I may go to my own home and country. 26 Give me my wives and my children for whom I have served you, and let me go; for you know very well the service I have given you." 27 But Laban said to him, "If you will allow me to say so, I have learned by divination that the LORD has blessed me because of you; 28 name your wages, and I will give it." 29 Jacob said to him, "You yourself know how I have served you, and how your cattle have fared with me. 30 For you had little before I came, and it has increased abundantly; and the LORD has blessed you wherever I turned. But now when shall I provide for my own household also?" 31 He said, "What shall I give you?" Jacob said, "You shall not give me anything; if you will do this for me, I will again feed your flock and keep it: 32 let me pass through all your flock today, removing from it every speckled and spotted sheep and every black lamb, and the spotted and speckled among the goats; and such shall be my wages. 33 So my honesty will answer for me later, when you come to look into my wages with you. Every one that is not speckled and spotted among the goats and black among the lambs, if found with me, shall be counted stolen." 34 Laban said, "Good! Let it be as you have said." 35 But that day Laban re-

moved the male goats that were striped and spotted, and all the female goats that were speckled and spotted, every one that had white on it, and every lamb that was black, and put them in charge of his sons; 36 and he set a distance of three days' journey between himself and Jacob, while Jacob was pasturing the rest of Laban's flock.

37 Then Jacob took fresh rods of poplar and almond and plane, and peeled white streaks in them, exposing the white of the rods. 38 He set the rods that he had peeled in front of the flocks in the troughs, that is, the watering places, where the flocks came to drink. And since they bred when they came to drink, 39 the flocks bred in front of the rods, and so the flocks produced young that were striped, speckled, and spotted. 40 Jacob separated the lambs, and set the faces of the flocks toward the striped and the completely black animals in the flock of Laban; and he put his own droves apart, and did not put them with Laban's flock. 41 Whenever the stronger of the flock were breeding, Jacob laid the rods in the troughs before the eyes of the flock, that they might breed among the rods, 42 but for the feebler of the flock he did not lay them there; so the feebler were Laban's, and the stronger Jacob's. 43 Thus the man grew exceedingly rich, and had large flocks, and male and female slaves, and camels and donkeys.

31 Now Jacob heard that the sons of Laban were saying, "Jacob has taken all that was our father's; he has gained all this wealth from what belonged to our father." 2 And Jacob saw that Laban did not regard him as favorably as he did before. 3 Then the LORD said to Jacob, "Return to the land of your

x Heb *zabal*   y That is *He adds*

---

potato-like plant, were thought to have aphrodisiac properties that stimulated conception.
30.32–36: Since striped or speckled coloration was unusual, Laban seemingly had nothing to lose. **37–40:** Ancient cattle-breeders believed that the female, at the time of conception, was influenced by visual impressions which affect the color of the offspring. Jacob produced striped animals by putting striped sticks before the females' eyes while they were breeding.

ancestors and to your kindred, and I will be with you." ⁴So Jacob sent and called Rachel and Leah into the field where his flock was, ⁵and said to them, "I see that your father does not regard me as favorably as he did before. But the God of my father has been with me. ⁶You know that I have served your father with all my strength; ⁷yet your father has cheated me and changed my wages ten times, but God did not permit him to harm me. ⁸If he said, 'The speckled shall be your wages,' then all the flock bore speckled; and if he said, 'The striped shall be your wages,' then all the flock bore striped. ⁹Thus God has taken away the livestock of your father, and given them to me.

10 During the mating of the flock I once had a dream in which I looked up and saw that the male goats that leaped upon the flock were striped, speckled, and mottled. ¹¹Then the angel of God said to me in the dream, 'Jacob,' and I said, 'Here I am!' ¹²And he said, 'Look up and see that all the goats that leap on the flock are striped, speckled, and mottled; for I have seen all that Laban is doing to you. ¹³I am the God of Bethel,ᶻ where you anointed a pillar and made a vow to me. Now leave this land at once and return to the land of your birth.' " ¹⁴Then Rachel and Leah answered him, "Is there any portion or inheritance left to us in our father's house? ¹⁵Are we not regarded by him as foreigners? For he has sold us, and he has been using up the money given for us. ¹⁶All the property that God has taken away from our father belongs to us and to our children; now then, do whatever God has said to you."

17 So Jacob arose, and set his children and his wives on camels; ¹⁸and he drove away all his livestock, all the property that he had gained, the livestock in his possession that he had acquired in Paddan-aram, to go to his father Isaac in the land of Canaan.

19 Now Laban had gone to shear his sheep, and Rachel stole her father's household gods. ²⁰And Jacob deceived Laban the Aramean, in that he did not tell him that he intended to flee. ²¹So he fled with all that he had; starting out he crossed the Euphrates,ᵃ and set his face toward the hill country of Gilead.

22 On the third day Laban was told that Jacob had fled. ²³So he took his kinsfolk with him and pursued him for seven days until he caught up with him in the hill country of Gilead. ²⁴But God came to Laban the Aramean in a dream by night, and said to him, "Take heed that you say not a word to Jacob, either good or bad."

25 Laban overtook Jacob. Now Jacob had pitched his tent in the hill country, and Laban with his kinsfolk camped in the hill country of Gilead. ²⁶Laban said to Jacob, "What have you done? You have deceived me, and carried away my daughters like captives of the sword. ²⁷Why did you flee secretly and deceive me and not tell me? I would have sent you away with mirth and songs, with tambourine and lyre. ²⁸And why did you not permit me to kiss my sons and my daughters farewell? What you have done is foolish. ²⁹It is in my power to do you harm; but the God of your father spoke to me last night, saying, 'Take heed that you speak to Jacob neither good nor bad.' ³⁰Even though you had to go because you longed greatly for your father's house, why did you steal my gods?" ³¹Jacob answered Laban, "Because I was afraid, for I thought that you would take your daughters from me by force. ³²But anyone with whom you find your gods shall not live. In the presence of our kinsfolk, point out what I have that is yours, and take it." Now Jacob did

ᶻ Cn: Meaning of Heb uncertain   ᵃ Heb *the
river*

31.4–16: Jacob discusses the situation with his wives because in ancient society they legally belonged to their *father's house* (v. 14), and were part of the property (Ruth 4.5, 10).

31.19: Possession of the *household gods* (1 Sam 19.13–17), according to ancient custom, insured leadership of the family and legitimated any claim on the property.

not know that Rachel had stolen the gods. *b*

33 So Laban went into Jacob's tent, and into Leah's tent, and into the tent of the two maids, but he did not find them. And he went out of Leah's tent, and entered Rachel's. 34 Now Rachel had taken the household gods and put them in the camel's saddle, and sat on them. Laban felt all about in the tent, but did not find them. 35 And she said to her father, "Let not my lord be angry that I cannot rise before you, for the way of women is upon me." So he searched, but did not find the household gods.

36 Then Jacob became angry, and upbraided Laban. Jacob said to Laban, "What is my offense? What is my sin, that you have hotly pursued me? 37 Although you have felt about through all my goods, what have you found of all your household goods? Set it here before my kinsfolk and your kinsfolk, so that they may decide between us two. 38 These twenty years I have been with you; your ewes and your female goats have not miscarried, and I have not eaten the rams of your flocks. 39 That which was torn by wild beasts I did not bring to you; I bore the loss of it myself; of my hand you required it, whether stolen by day or stolen by night. 40 It was like this with me: by day the heat consumed me, and the cold by night, and my sleep fled from my eyes. 41 These twenty years I have been in your house; I served you fourteen years for your two daughters, and six years for your flock, and you have changed my wages ten times. 42 If the God of my father, the God of Abra-

ham and the Fear *c* of Isaac, had not been on my side, surely now you would have sent me away empty-handed. God saw my affliction and the labor of my hands, and rebuked you last night."

43 Then Laban answered and said to Jacob, "The daughters are my daughters, the children are my children, the flocks are my flocks, and all that you see is mine. But what can I do today about these daughters of mine, or about their children whom they have borne? 44 Come now, let us make a covenant, you and I; and let it be a witness between you and me." 45 So Jacob took a stone, and set it up as a pillar. 46 And Jacob said to his kinsfolk, "Gather stones," and they took stones, and made a heap; and they ate there by the heap. 47 Laban called it Jegar-sahadutha: *d* but Jacob called it Galeed. *e* 48 Laban said, "This heap is a witness between you and me today." Therefore he called it Galeed, 49 and the pillar *f* Mizpah, *g* for he said, "The Lord watch between you and me, when we are absent one from the other. 50 If you illtreat my daughters, or if you take wives in addition to my daughters, though no one else is with us, remember that God is witness between you and me."

51 Then Laban said to Jacob, "See this heap and see the pillar, which I have set between you and me. 52 This heap is a witness, and the pillar is a witness, that I will not pass beyond this heap to you, and you will not pass beyond this heap

b Heb *them*   *c* Meaning of Heb uncertain
d In Aramaic *The heap of witness*   *e* In Hebrew
The heap of witness   *f* Compare Sam: MT lacks
the pillar   *g* That is *Watchpost*

31.35: The narrator ridicules the idols upon which Rachel sat in her time of "uncleanness" (Lev 15.19–23). **42**: *The Fear of Isaac* (perhaps, the "Kinsman" of Isaac), an old epithet for the God of the ancestors that was appropriated as a title for Israel's God.
31.43: Laban's argument presupposes the legality of a type of marriage in which the wife stays in her father's household and the husband must leave his family (Judg 14.2–3; 15.1). **44**: The story reflects a boundary covenant between Arameans and Israelites, both

of whom laid claim to the area in Transjordan (v. 52). **46**: *They ate there,* a reference to the covenant meal (v. 54) at which, it was believed, the Deity was present. **47**: The stone-heap is given two names (see notes *d* and *e*), one in Laban's language (Aramaic) and one in Jacob's (Hebrew). **49**: The "Mizpah benediction" is a prayer that the Deity would oversee the treaty (since neither Jacob nor Laban could trust each other) and guarantee that both parties live up to the contract.

and this pillar to me, for harm. [53] May the God of Abraham and the God of Nahor"—the God of their father—"judge between us." So Jacob swore by the Fear[h] of his father Isaac, [54] and Jacob offered a sacrifice on the height and called his kinsfolk to eat bread; and they ate bread and tarried all night in the hill country.

[55][i] Early in the morning Laban rose up, and kissed his grandchildren and his daughters and blessed them; then he departed and returned home.

32 Jacob went on his way and the angels of God met him; [2] and when Jacob saw them he said, "This is God's camp!" So he called that place Mahanaim.[j]

[3] Jacob sent messengers before him to his brother Esau in the land of Seir, the country of Edom, [4] instructing them, "Thus you shall say to my lord Esau: Thus says your servant Jacob, 'I have lived with Laban as an alien, and stayed until now; [5] and I have oxen, donkeys, flocks, male and female slaves; and I have sent to tell my lord, in order that I may find favor in your sight.' "

[6] The messengers returned to Jacob, saying, "We came to your brother Esau, and he is coming to meet you, and four hundred men are with him." [7] Then Jacob was greatly afraid and distressed; and he divided the people that were with him, and the flocks and herds and camels, into two companies, [8] thinking, "If Esau comes to the one company and destroys it, then the company that is left will escape."

[9] And Jacob said, "O God of my father Abraham and God of my father Isaac, O Lord who said to me, 'Return to your country and to your kindred, and I will do you good,' [10] I am not worthy of the least of all the steadfast love and all the faithfulness that you have shown to

your servant, for with only my staff I crossed this Jordan; and now I have become two companies. [11] Deliver me, please, from the hand of my brother, from the hand of Esau, for I am afraid of him; he may come and kill us all, the mothers with the children. [12] Yet you have said, 'I will surely do you good, and make your offspring as the sand of the sea, which cannot be counted because of their number.' "

[13] So he spent that night there, and from what he had with him he took a present for his brother Esau, [14] two hundred female goats and twenty male goats, two hundred ewes and twenty rams, [15] thirty milch camels and their colts, forty cows and ten bulls, twenty female donkeys and ten male donkeys. [16] These he delivered into the hand of his servants, every drove by itself, and said to his servants, "Pass on ahead of me, and put a space between drove and drove." [17] He instructed the foremost, "When Esau my brother meets you, and asks you, 'To whom do you belong? Where are you going? And whose are these ahead of you?' [18] then you shall say, 'They belong to your servant Jacob; they are a present sent to my lord Esau; and moreover he is behind us.' " [19] He likewise instructed the second and the third and all who followed the droves, "You shall say the same thing to Esau when you meet him, [20] and you shall say, 'Moreover your servant Jacob is behind us.' " For he thought, "I may appease him with the present that goes ahead of me, and afterwards I shall see his face; perhaps he will accept me." [21] So the present passed on ahead of him; and he himself spent that night in the camp.

[22] The same night he got up and took

h Meaning of Heb uncertain   *i* Ch 32.1 in Heb   *j* Here taken to mean *Two camps*

32.1–33.20: Jacob's reconciliation with Esau. Before re-entering the promised land, Jacob underwent a struggle which chastened his self-confidence and prepared him for a new relation with Esau.
32.1–2: *Angels,* see 28.12 n. *Mahanaim,*

here explained by a wordplay (*God's camp*), was important in later Israelite history (2 Sam 2.8–9; 17.24–29; 1 Kings 4.14). **3:** *Seir,* the region of Edom in which apparently Esau had already settled (see 25.25 n.; 36.8). **3–21:** Although uncertain how his brother felt toward

his two wives, his two maids, and his eleven children, and crossed the ford of the Jabbok. 23 He took them and sent them across the stream, and likewise everything that he had. 24 Jacob was left alone; and a man wrestled with him until daybreak. 25 When the man saw that he did not prevail against Jacob, he struck him on the hip socket; and Jacob's hip was put out of joint as he wrestled with him. 26 Then he said, "Let me go, for the day is breaking." But Jacob said, "I will not let you go, unless you bless me." 27 So he said to him, "What is your name?" And he said, "Jacob." 28 Then the man*k* said, "You shall no longer be called Jacob, but Israel,*l* for you have striven with God and with humans,*m* and have prevailed." 29 Then Jacob asked him, "Please tell me your name." But he said, "Why is it that you ask my name?" And there he blessed him. 30 So Jacob called the place Peniel,*n* saying, "For I have seen God face to face, and yet my life is preserved." 31 The sun rose upon him as he passed Penuel, limping because of his hip. 32 Therefore to this day the Israelites do not eat the thigh muscle that is on the hip socket, because he struck Jacob on the hip socket at the thigh muscle.

33 Now Jacob looked up and saw Esau coming, and four hundred men with him. So he divided the children among Leah and Rachel and the two maids. 2 He put the maids with their children in front, then Leah with her children, and Rachel and Joseph last of all.

3 He himself went on ahead of them, bowing himself to the ground seven times, until he came near his brother. 4 But Esau ran to meet him, and embraced him, and fell on his neck and kissed him, and they wept. 5 When Esau looked up and saw the women and children, he said, "Who are these with you?" Jacob said, "The children whom God has graciously given your servant." 6 Then the maids drew near, they and their children, and bowed down; 7 Leah likewise and her children drew near and bowed down; and finally Joseph and Rachel drew near, and they bowed down. 8 Esau said, "What do you mean by all this company that I met?" Jacob answered, "To find favor with my lord." 9 But Esau said, "I have enough, my brother; keep what you have for yourself." 10 Jacob said, "No, please; if I find favor with you, then accept my present from my hand; for truly to see your face is like seeing the face of God—since you have received me with such favor. 11 Please accept my gift that is brought to you, because God has dealt graciously with me, and because I have everything I want." So he urged him, and he took it.

12 Then Esau said, "Let us journey on our way, and I will go alongside you." 13 But Jacob said to him, "My lord knows that the children are frail and that the flocks and herds, which are nursing, are

k Heb *he*   *l* That is *The one who strives with God* or *God strives*   *m* Or *with divine and human beings*   *n* That is *The face of God*

---

him after twenty years, the resourceful Jacob believed that he could get the situation into hand, either by saving half of his camp (vv. 6–8) or by winning Esau over with impressive gifts (vv. 13–21). **32.25:** Owing to his Herculean strength (29.10; compare 28.18), Jacob was winning the contest until his opponent sprained Jacob's hip. **26:** The divine being had to vanish before sunrise—a mark of the antiquity of the tradition. **27:** In antiquity it was believed that selfhood was expressed in the name given a person (compare v. 29). **28:** Jacob's new name signified a new self: no longer was he the Supplanter (25.26; 27.36) but *Israel*

(35.10), which probably means "God rules." This name, which later designated the tribal confederacy (see 33.20 n.), is interpreted to mean "The one who strives with God" (Hos 12.3–4). *And with humans* refers to Jacob's strife with Esau and Laban. **29:** The divine being refuses lest Jacob, by possessing the name, gain power over him (compare Ex 3.13–14; Judg 13.17). **30:** Jacob had feared to see Esau's face (v. 20; 33.10), but instead saw God *face to face* and was allowed to live (16.13; Ex 33.20). **32:** An explanation of the Israelite taboo against eating the corresponding muscle of an animal. **33.10:** *Like seeing the face of God,* who at

a care to me; and if they are overdriven for one day, all the flocks will die. 14 Let my lord pass on ahead of his servant, and I will lead on slowly, according to the pace of the cattle that are before me and according to the pace of the children, until I come to my lord in Seir."

15 So Esau said, "Let me leave with you some of the people who are with me." But he said, "Why should my lord be so kind to me?" 16 So Esau returned that day on his way to Seir. 17 But Jacob journeyed to Succoth, *o* and built himself a house, and made booths for his cattle; therefore the place is called Succoth.

18 Jacob came safely to the city of Shechem, which is in the land of Canaan, on his way from Paddan-aram; and he camped before the city. 19 And from the sons of Hamor, Shechem's father, he bought for one hundred pieces of money*p* the plot of land on which he had pitched his tent. 20 There he erected an altar and called it El-Elohe-Israel. *q*

34 Now Dinah the daughter of Leah, whom she had borne to Jacob, went out to visit the women of the region. 2 When Shechem son of Hamor the Hivite, prince of the region, saw her, he seized her and lay with her by force. 3 And his soul was drawn to Dinah daughter of Jacob; he loved the girl, and spoke tenderly to her. 4 So Shechem spoke to his father Hamor, saying, "Get me this girl to be my wife."

5 Now Jacob heard that Shechem*r* had defiled his daughter Dinah; but his sons were with his cattle in the field, so Jacob held his peace until they came.

6 And Hamor the father of Shechem went out to Jacob to speak with him, 7 just as the sons of Jacob came in from the field. When they heard of it, the men were indignant and very angry, because he had committed an outrage in Israel by lying with Jacob's daughter, for such a thing ought not to be done.

8 But Hamor spoke with them, saying, "The heart of my son Shechem longs for your daughter; please give her to him in marriage. 9 Make marriages with us; give your daughters to us, and take our daughters for yourselves. 10 You shall live with us; and the land shall be open to you; live and trade in it, and get property in it." 11 Shechem also said to her father and to her brothers, "Let me find favor with you, and whatever you say to me I will give. 12 Put the marriage present and gift as high as you like, and I will give whatever you ask me; only give me the girl to be my wife."

13 The sons of Jacob answered Shechem and his father Hamor deceitfully, because he had defiled their sister Dinah. 14 They said to them, "We cannot do this thing, to give our sister to one who is uncircumcised, for that would be a disgrace to us. 15 Only on this condition will we consent to you: that you will become as we are and every male among you be circumcised. 16 Then we will give our daughters to you, and we will take your daughters for ourselves, and we will live

o That is *Booths*   *p* Heb *one hundred qesitah*
q That is *God, the God of Israel*   *r* Heb *he*

Penuel (or Peniel) also proved to be gracious (32.30,31). *Shechem,* see 12.6 n. **19:** *The sons of Hamor* were the ruling clan of the city (Judg 9.28), of Hivite (perhaps Horite or Hurrian) extraction (see 36.20 n.). *The plot of land* later became the traditional burial place of Joseph (Josh 24.32). **20:** The worship of *El-Elohe-Israel* (see note *q*) apparently preceded the later establishment of an Israelite tribal confederacy, presumably at Shechem (Josh ch 24), when El (the Semitic word for "God") was succeeded by the LORD (Yahweh), the God of Israel.
34.1–31: Shechem's violation of Dinah and its consequences. **1:** Dinah is mentioned

elsewhere only in 30.21 and 46.15. *The women of the region,* i.e. Canaanite women. **2:** Here *Shechem* is the name of a person (33.19). The story portrays, in the guise of individuals, relations between the Canaanite city and early Hebrew tribes. *The son of Hamor,* see 33.19 n. **7:** *Committed an outrage in Israel* is an old expression for a crime affecting the whole tribal community of Israel (Deut 22.21; Josh 7.15; Judg 19.23–24; 20.6, 10). **12:** *Marriage present,* Ex 22.16–17; Deut 22.29.
34.14–15: Circumcision, see 17.9–14 n. **19:** The fact that Shechem *was most honored of all his family* and *the prince of the region* (v. 2)

among you and become one people.
[17] But if you will not listen to us and be
circumcised, then we will take our
daughter and be gone."

18 Their words pleased Hamor and
Hamor's son Shechem. [19] And the young
man did not delay to do the thing, because he was delighted with Jacob's
daughter. Now he was the most honored
of all his family. [20] So Hamor and his son
Shechem came to the gate of their city
and spoke to the men of their city, saying, [21] "These people are friendly with
us; let them live in the land and trade in
it, for the land is large enough for them;
let us take their daughters in marriage,
and let us give them our daughters.
[22] Only on this condition will they agree
to live among us, to become one people:
that every male among us be circumcised
as they are circumcised. [23] Will not their
livestock, their property, and all their animals be ours? Only let us agree with
them, and they will live among us."
[24] And all who went out of the city gate
heeded Hamor and his son Shechem; and
every male was circumcised, all who
went out of the gate of his city.

25 On the third day, when they were
still in pain, two of the sons of Jacob,
Simeon and Levi, Dinah's brothers, took
their swords and came against the city
unawares, and killed all the males.
[26] They killed Hamor and his son Shechem with the sword, and took Dinah
out of Shechem's house, and went away.
[27] And the other sons of Jacob came upon
the slain, and plundered the city, because
their sister had been defiled. [28] They took
their flocks and their herds, their donkeys, and whatever was in the city and in
the field. [29] All their wealth, all their little
ones and their wives, all that was in the
houses, they captured and made their
prey. [30] Then Jacob said to Simeon and
Levi, "You have brought trouble on me
by making me odious to the inhabitants
of the land, the Canaanites and the Perizzites; my numbers are few, and if they
gather themselves against me and attack
me, I shall be destroyed, both I and my
household." [31] But they said, "Should
our sister be treated like a whore?"

35 God said to Jacob, "Arise, go up
to Bethel, and settle there. Make
an altar there to the God who appeared
to you when you fled from your brother
Esau." [2] So Jacob said to his household
and to all who were with him, "Put away
the foreign gods that are among you, and
purify yourselves, and change your
clothes; [3] then come, let us go up to Bethel, that I may make an altar there to the
God who answered me in the day of my
distress and has been with me wherever
I have gone." [4] So they gave to Jacob all
the foreign gods that they had, and the
rings that were in their ears; and Jacob
hid them under the oak that was near
Shechem.

5 As they journeyed, a terror from
God fell upon the cities all around them,
so that no one pursued them. [6] Jacob
came to Luz (that is, Bethel), which is in
the land of Canaan, he and all the people
who were with him, [7] and there he built

---

shows the ascendancy of this Canaanite city
in the pre-Israelite period. **22–23:** *One people,*
i.e. a kindred-group in which the Shechemites would have the leadership.
    34.25–26: Simeon and Levi took the ini-
tiative because they were full brothers of Dinah. **30:** The violent action, which threatened
the good relations of Jacob's family with the
Canaanites, reflects events that forced Simeon and Levi out of the area and led to their
decline in power (49.5–7).
    **35.1–29: Jacob's journey from She-
chem to Mamre. 1:** *Bethel,* 28.19–22. Verses
1–4 may reflect the custom of making a pil-

grimage to the Bethel sanctuary (compare
Judg 20.26–27). **2:** The worshipers had to
undergo ceremonial purification, which involved changing garments and ablutions (Ex
19.10; Josh 7.13), and to renounce *foreign gods*
(Josh 24.14–18, 23), including the household
gods (31.19). **4:** The earrings and other magical amulets belonged to foreign idolatry (Ex
32.2–3; Judg 8.24). *The oak,* i.e. of Moreh
(12.6). **5:** *A terror from God,* an expression
derived from ancient holy war (Ex 23.27;
Josh 10.10), was a mysterious panic that paralyzed the enemy. **6–7:** 28.18–22.

an altar and called the place El-bethel, *s* because it was there that God had revealed himself to him when he fled from his brother. 8 And Deborah, Rebekah's nurse, died, and she was buried under an oak below Bethel. So it was called Allonbacuth. *t*

9 God appeared to Jacob again when he came from Paddan-aram, and he blessed him. 10 God said to him, "Your name is Jacob; no longer shall you be called Jacob, but Israel shall be your name." So he was called Israel. 11 God said to him, "I am God Almighty: *u* be fruitful and multiply; a nation and a company of nations shall come from you, and kings shall spring from you. 12 The land that I gave to Abraham and Isaac I will give to you, and I will give the land to your offspring after you." 13 Then God went up from him at the place where he had spoken with him. 14 Jacob set up a pillar in the place where he had spoken with him, a pillar of stone; and he poured out a drink offering on it, and poured oil on it. 15 So Jacob called the place where God had spoken with him Bethel.

16 Then they journeyed from Bethel; and when they were still some distance from Ephrath, Rachel was in childbirth, and she had hard labor. 17 When she was in her hard labor, the midwife said to her, "Do not be afraid; for now you will have another son." 18 As her soul was departing (for she died), she named him Ben-oni; *v* but his father called him Benjamin. *w* 19 So Rachel died, and she was buried on the way to Ephrath (that is,

Bethlehem), 20 and Jacob set up a pillar at her grave; it is the pillar of Rachel's tomb, which is there to this day. 21 Israel journeyed on, and pitched his tent beyond the tower of Eder.

22 While Israel lived in that land, Reuben went and lay with Bilhah his father's concubine; and Israel heard of it.

Now the sons of Jacob were twelve. 23 The sons of Leah: Reuben (Jacob's firstborn), Simeon, Levi, Judah, Issachar, and Zebulun. 24 The sons of Rachel: Joseph and Benjamin. 25 The sons of Bilhah, Rachel's maid: Dan and Naphtali. 26 The sons of Zilpah, Leah's maid: Gad and Asher. These were the sons of Jacob who were born to him in Paddan-aram.

27 Jacob came to his father Isaac at Mamre, or Kiriath-arba (that is, Hebron), where Abraham and Isaac had resided as aliens. 28 Now the days of Isaac were one hundred eighty years. 29 And Isaac breathed his last; he died and was gathered to his people, old and full of days; and his sons Esau and Jacob buried him.

36 These are the descendants of Esau (that is, Edom). 2 Esau took his wives from the Canaanites: Adah daughter of Elon the Hittite, Oholibamah daughter of Anah son *x* of Zibeon the Hivite, 3 and Basemath, Ishmael's daughter, sister of Nebaioth. 4 Adah bore Eliphaz to Esau; Basemath bore Reuel;

s That is *God of Bethel* *t* That is *Oak of weeping* *u* Traditional rendering of Heb *El Shaddai* *v* That is *Son of my sorrow* *w* That is *Son of the right hand* or *Son of the South* *x* Sam Gk Syr: Heb *daughter*

35.9–13: Another account of Jacob's receiving a new name (compare 32.24–30) and God's promise to him (28.13–15). **11:** *God Almighty,* see 17.1 n. **14–15:** A tradition parallel to 28.18–22.

35.16: *Ephrath,* see v. 19 n. **18:** Dying in childbirth, Rachel gave an ominous name (see note *v*) to the child (see 1 Sam 4.21), though Jacob changed the name to a propitious one. *Benjamin* (see note *w*) refers either to the right hand as a symbol of power and fortune or to the tribe's position south of Ephraim. **19:** Here and in 48.7 (Ruth 4.11; Mic 5.2) Ephrath is identified with Bethle-

hem. Another tradition located Rachel's grave in Benjaminite territory north of Jerusalem (1 Sam 10.2; Jer 31.15). **21–22:** This fragmentary account apparently related an incident that resulted in Reuben's loss of prestige as the firstborn son (49.3–4). **21:** From this point onward the name *Israel* is often used to refer to Jacob (32.28).

36.1–43: Edomite lists. Some of this material (vv. 1–8, 9–14, 15–19, 40–43) was apparently drawn from the book of generations (5.1). **1:** *Esau (Edom),* 25.21–28. **2–3:** Esau's wives, see 26.34; 28.8–9. **8:** *Seir,* or Edom (see 25.25 n.; 27.39–40).

5 and Oholibamah bore Jeush, Jalam, and Korah. These are the sons of Esau who were born to him in the land of Canaan.

6 Then Esau took his wives, his sons, his daughters, and all the members of his household, his cattle, all his livestock, and all the property he had acquired in the land of Canaan; and he moved to a land some distance from his brother Jacob. 7 For their possessions were too great for them to live together; the land where they were staying could not support them because of their livestock. 8 So Esau settled in the hill country of Seir; Esau is Edom.

9 These are the descendants of Esau, ancestor of the Edomites, in the hill country of Seir. 10 These are the names of Esau's sons: Eliphaz son of Adah the wife of Esau; Reuel, the son of Esau's wife Basemath. 11 The sons of Eliphaz were Teman, Omar, Zepho, Gatam, and Kenaz. 12 (Timna was a concubine of Eliphaz, Esau's son; she bore Amalek to Eliphaz.) These were the sons of Adah, Esau's wife. 13 These were the sons of Reuel: Nahath, Zerah, Shammah, and Mizzah. These were the sons of Esau's wife, Basemath. 14 These were the sons of Esau's wife Oholibamah, daughter of Anah son*y* of Zibeon: she bore to Esau Jeush, Jalam, and Korah.

15 These are the clans*z* of the sons of Esau. The sons of Eliphaz the firstborn of Esau: the clans*z* Teman, Omar, Zepho, Kenaz, 16 Korah, Gatam, and Amalek; these are the clans*z* of Eliphaz in the land of Edom; they are the sons of Adah. 17 These are the sons of Esau's son Reuel: the clans*z* Nahath, Zerah, Shammah, and Mizzah; these are the clans*z* of Reuel in the land of Edom; they are the sons of Esau's wife Basemath. 18 These are the sons of Esau's wife Oholibamah: the clans*z* Jeush, Jalam, and Korah; these are the clans*z* born of Esau's wife Oholibamah, the daughter of Anah. 19 These are

the sons of Esau (that is, Edom), and these are their clans.*z*

20 These are the sons of Seir the Horite, the inhabitants of the land: Lotan, Shobal, Zibeon, Anah, 21 Dishon, Ezer, and Dishan; these are the clans*z* of the Horites, the sons of Seir in the land of Edom. 22 The sons of Lotan were Hori and Heman; and Lotan's sister was Timna. 23 These are the sons of Shobal: Alvan, Manahath, Ebal, Shepho, and Onam. 24 These are the sons of Zibeon: Aiah and Anah; he is the Anah who found the springs*a* in the wilderness, as he pastured the donkeys of his father Zibeon. 25 These are the children of Anah: Dishon and Oholibamah daughter of Anah. 26 These are the sons of Dishon: Hemdan, Eshban, Ithran, and Cheran. 27 These are the sons of Ezer: Bilhan, Zaavan, and Akan. 28 These are the sons of Dishan: Uz and Aran. 29 These are the clans*z* of the Horites: the clans*z* Lotan, Shobal, Zibeon, Anah, 30 Dishon, Ezer, and Dishan; these are the clans*z* of the Horites, clan by clan*b* in the land of Seir.

31 These are the kings who reigned in the land of Edom, before any king reigned over the Israelites. 32 Bela son of Beor reigned in Edom, the name of his city being Dinhabah. 33 Bela died, and Jobab son of Zerah of Bozrah succeeded him as king. 34 Jobab died, and Husham of the land of the Temanites succeeded him as king. 35 Husham died, and Hadad son of Bedad, who defeated Midian in the country of Moab, succeeded him as king, the name of his city being Avith. 36 Hadad died, and Samlah of Masrekah succeeded him as king. 37 Samlah died, and Shaul of Rehoboth on the Euphrates succeeded him as king. 38 Shaul died, and Baal-hanan son of Achbor succeeded

y Gk Syr: Heb *daughter*   *z* Or *chiefs*
a Meaning of Heb uncertain   *b* Or *chief by chief*

36.20: *Horite* refers to the Hurrians, a non-Semitic people who migrated into Mesopotamia about 2000 B.C. and later formed an important element of the Canaanite population. **36.31–39:** Edom became a monarchy long (perhaps 150 years) before Israel did (Num 20.14), a circumstance reflected in the tradition that Esau was the older brother although Jacob gained the ascendancy.

him as king. [39] Baal-hanan son of Achbor died, and Hadar succeeded him as king, the name of his city being Pau; his wife's name was Mehetabel, the daughter of Matred, daughter of Me-zahab.

40 These are the names of the clans[c] of Esau, according to their families and their localities by their names: the clans[c] Timna, Alvah, Jetheth, [41] Oholibamah, Elah, Pinon, [42] Kenaz, Teman, Mibzar, [43] Magdiel, and Iram; these are the clans[c] of Edom (that is, Esau, the father of Edom), according to their settlements in the land that they held.

37 Jacob settled in the land where his father had lived as an alien, the land of Canaan. [2] This is the story of the family of Jacob.

Joseph, being seventeen years old, was shepherding the flock with his brothers; he was a helper to the sons of Bilhah and Zilpah, his father's wives; and Joseph brought a bad report of them to their father. [3] Now Israel loved Joseph more than any other of his children, because he was the son of his old age; and he had made him a long robe with sleeves.[d] [4] But when his brothers saw that their father loved him more than all his brothers, they hated him, and could not speak peaceably to him.

5 Once Joseph had a dream, and when he told it to his brothers, they hated him even more. [6] He said to them, "Listen to this dream that I dreamed. [7] There we were, binding sheaves in the field. Suddenly my sheaf rose and stood upright; then your sheaves gathered around it, and bowed down to my sheaf." [8] His brothers said to him, "Are you indeed to reign over us? Are you

indeed to have dominion over us?" So they hated him even more because of his dreams and his words.

9 He had another dream, and told it to his brothers, saying, "Look, I have had another dream: the sun, the moon, and eleven stars were bowing down to me." [10] But when he told it to his father and to his brothers, his father rebuked him, and said to him, "What kind of dream is this that you have had? Shall we indeed come, I and your mother and your brothers, and bow to the ground before you?" [11] So his brothers were jealous of him, but his father kept the matter in mind.

12 Now his brothers went to pasture their father's flock near Shechem. [13] And Israel said to Joseph, "Are not your brothers pasturing the flock at Shechem? Come, I will send you to them." He answered, "Here I am." [14] So he said to him, "Go now, see if it is well with your brothers and with the flock; and bring word back to me." So he sent him from the valley of Hebron.

He came to Shechem, [15] and a man found him wandering in the fields; the man asked him, "What are you seeking?" [16] "I am seeking my brothers," he said; "tell me, please, where they are pasturing the flock." [17] The man said, "They have gone away, for I heard them say, 'Let us go to Dothan.' " So Joseph went after his brothers, and found them at Dothan. [18] They saw him from a distance, and before he came near to them, they conspired to kill him. [19] They said to one

c Or *chiefs*   d Traditional rendering (compare Gk): *a coat of many colors*; Meaning of Heb uncertain

---

37.1–36: Joseph is sold into slavery. The narratives about Joseph, found in chs 37; 39–47; and 50, constitute a single literary form or short story. **2:** The Joseph story is regarded as part of the history of Jacob, who died at the end of the saga (49.33). *The sons of Bilhah and Zilpah,* 30.1–13. **3:** *A long robe with sleeves* was a luxurious robe (2 Sam 13.18–19), different from the ordinary sleeveless tunic that reached to the knees. It would be impossible to undertake any manual labor

while wearing it. The garment may also symbolize a royal claim.

37.5–11: Joseph's two dreams were prophetic of his future elevation in Egypt (42.6; 50.18). **9:** *Eleven stars,* apparently the eleven constellations that ancients pictured in animal form.

37.12: The previous scene is laid in Hebron (35.27), several days' journey from Jacob's pasture land *near Shechem* (33.18–20). **20:** *The pits* were open cisterns for storing rain water

another, "Here comes this dreamer. ²⁰Come now, let us kill him and throw him into one of the pits; then we shall say that a wild animal has devoured him, and we shall see what will become of his dreams." ²¹But when Reuben heard it, he delivered him out of their hands, saying, "Let us not take his life." ²²Reuben said to them, "Shed no blood; throw him into this pit here in the wilderness, but lay no hand on him"—that he might rescue him out of their hand and restore him to his father. ²³So when Joseph came to his brothers, they stripped him of his robe, the long robe with sleeves*e* that he wore; ²⁴and they took him and threw him into a pit. The pit was empty; there was no water in it.

25 Then they sat down to eat; and looking up they saw a caravan of Ishmaelites coming from Gilead, with their camels carrying gum, balm, and resin, on their way to carry it down to Egypt. ²⁶Then Judah said to his brothers, "What profit is it if we kill our brother and conceal his blood? ²⁷Come, let us sell him to the Ishmaelites, and not lay our hands on him, for he is our brother, our own flesh." And his brothers agreed. ²⁸When some Midianite traders passed by, they drew Joseph up, lifting him out of the pit, and sold him to the Ishmaelites for twenty pieces of silver. And they took Joseph to Egypt.

29 When Reuben returned to the pit and saw that Joseph was not in the pit, he tore his clothes. ³⁰He returned to his brothers, and said, "The boy is gone; and I, where can I turn?" ³¹Then they took Joseph's robe, slaughtered a goat, and dipped the robe in the blood. ³²They had the long robe with sleeves*e* taken to their father, and they said, "This we have found; see now whether it is your son's robe or not." ³³He recognized it, and said, "It is my son's robe! A wild animal has devoured him; Joseph is without doubt torn to pieces." ³⁴Then Jacob tore his garments, and put sackcloth on his loins, and mourned for his son many days. ³⁵All his sons and all his daughters sought to comfort him; but he refused to be comforted, and said, "No, I shall go down to Sheol to my son, mourning." Thus his father bewailed him. ³⁶Meanwhile the Midianites had sold him in Egypt to Potiphar, one of Pharaoh's officials, the captain of the guard.

38 It happened at that time that Judah went down from his brothers and settled near a certain Adullamite whose name was Hirah. ²There Judah saw the daughter of a certain Canaanite whose name was Shua; he married her and went in to her. ³She conceived and bore a son; and he named him Er. ⁴Again she conceived and bore a son whom she named Onan. ⁵Yet again she bore a son, and she named him Shelah. She*f* was in Chezib when she bore him. ⁶Judah took a wife for Er his firstborn; her name was Tamar. ⁷But Er, Judah's firstborn, was

e See note on 37.3   *f* Gk: Heb *He*

(Jer 38.6). **21**: *Reuben* may be a scribal mistake for Judah, Joseph's advocate in v. 26. **22**: The advice of Reuben and Judah reflects the ancient belief that blood cannot be "concealed" (v. 26) but cries out for requital (see 4.10–11 n.).

37.25–28: Apparently, two traditions have been combined: according to one, Joseph was sold to *Ishmaelites;* according to the other, he was kidnapped by *Midianite traders*. Dothan (a few miles north of Shechem), where Joseph found his brothers (v. 17), lay on the trade route from Syria to Egypt.

37.35: *Sheol*, the underworld to which, it was believed, a person's shade went at death (2 Sam 12.23; Ps 115.17). Since this after-life

was a shadowy existence, Jacob's going to his son there was not a comforting expectation. **36**: *Potiphar* is a form of Potiphera, the name of the Egyptian priest of 41.45 and 46.20. It has been thought they may be the same person, one tradition designating him as captain of the guard, and another as a priest.

38.1–30: **Judah and Tamar.** This chapter, an interlude in the Joseph story, deals with a woman's part in the LORD's promise that Abraham will have a great posterity. **1**: *Adullamite,* a resident of Adullam near Bethlehem. **2**: The marriage with Shuah's daughter reflects territorial expansion of the tribe of Judah and the consequent intermarriage with Canaanites. **7**: The early and childless death of

wicked in the sight of the LORD, and the LORD put him to death. 8 Then Judah said to Onan, "Go in to your brother's wife and perform the duty of a brother-in-law to her; raise up offspring for your brother." 9 But since Onan knew that the offspring would not be his, he spilled his semen on the ground whenever he went in to his brother's wife, so that he would not give offspring to his brother. 10 What he did was displeasing in the sight of the LORD, and he put him to death also. 11 Then Judah said to his daughter-in-law Tamar, "Remain a widow in your father's house until my son Shelah grows up"—for he feared that he too would die, like his brothers. So Tamar went to live in her father's house.

12 In course of time the wife of Judah, Shua's daughter, died; when Judah's time of mourning was over,*g* he went up to Timnah to his sheepshearers, he and his friend Hirah the Adullamite. 13 When Tamar was told, "Your father-in-law is going up to Timnah to shear his sheep," 14 she put off her widow's garments, put on a veil, wrapped herself up, and sat down at the entrance to Enaim, which is on the road to Timnah. She saw that Shelah was grown up, yet she had not been given to him in marriage. 15 When Judah saw her, he thought her to be a prostitute, for she had covered her face. 16 He went over to her at the road side, and said, "Come, let me come in to you," for he did not know that she was his daughter-in-law. She said, "What will you give me, that you may come in

to me?" 17 He answered, "I will send you a kid from the flock." And she said, "Only if you give me a pledge, until you send it." 18 He said, "What pledge shall I give you?" She replied, "Your signet and your cord, and the staff that is in your hand." So he gave them to her, and went in to her, and she conceived by him. 19 Then she got up and went away, and taking off her veil she put on the garments of her widowhood.

20 When Judah sent the kid by his friend the Adullamite, to recover the pledge from the woman, he could not find her. 21 He asked the townspeople, "Where is the temple prostitute who was at Enaim by the wayside?" But they said, "No prostitute has been here." 22 So he returned to Judah, and said, "I have not found her; moreover the townspeople said, 'No prostitute has been here.'" 23 Judah replied, "Let her keep the things as her own, otherwise we will be laughed at; you see, I sent this kid, and you could not find her."

24 About three months later Judah was told, "Your daughter-in-law Tamar has played the whore; moreover she is pregnant as a result of whoredom." And Judah said, "Bring her out, and let her be burned." 25 As she was being brought out, she sent word to her father-in-law, "It was the owner of these who made me pregnant." And she said, "Take note, please, whose these are, the signet and the cord and the staff." 26 Then Judah ac-

g Heb *when Judah was comforted*

Er is attributed to a divine act, almost demonic in character. **8:** According to the ancient widespread custom of levirate marriage (Deut 25.5–10; compare Ruth 4.1–12), *the duty of a brother-in-law* was to raise up a male descendant for his deceased brother and thus perpetuate his name and inheritance. **11:** Judah apparently feared that the death of his two sons resulted from Tamar's sinister power. A widow was supposed to return to her *father's house* (Ruth 1.8–9; Lev 22.13).

38.14: Suspecting that Judah's promise (v. 11) was insincere, Tamar took steps to make him perform the levirate duty. **15:** Tamar was taken to be a cult prostitute, a de-

votee of the mother-goddess Ishtar. Prostitution was connected with the worship of the nature gods of fertility (Deut 23.18; 1 Kings 14.24; 2 Kings 23.7; Hos 4.13; Am 2.7). **18:** The signet was a ring or cylinder, often suspended around the neck by a cord and used to stamp one's "signature."

38.24: In Israel stoning was the usual punishment for adultery (Deut 22.23–24; compare Jn 8.5), although burning was prescribed for exceptional cases (Lev 21.9). **26:** Tamar is singled out for approval, for judged by the levirate obligation, she was *more in the right* (see 6.9 n.) than Judah. Within the limitations of patriarchal society, she acted boldly to

knowledged them and said, "She is more in the right than I, since I did not give her to my son Shelah." And he did not lie with her again.

27 When the time of her delivery came, there were twins in her womb. 28 While she was in labor, one put out a hand; and the midwife took and bound on his hand a crimson thread, saying, "This one came out first." 29 But just then he drew back his hand, and out came his brother; and she said, "What a breach you have made for yourself!" Therefore he was named Perez. *h* 30 Afterward his brother came out with the crimson thread on his hand; and he was named Zerah. *i*

39 Now Joseph was taken down to Egypt, and Potiphar, an officer of Pharaoh, the captain of the guard, an Egyptian, bought him from the Ishmaelites who had brought him down there. 2 The LORD was with Joseph, and he became a successful man; he was in the house of his Egyptian master. 3 His master saw that the LORD was with him, and that the LORD caused all that he did to prosper in his hands. 4 So Joseph found favor in his sight and attended him; he made him overseer of his house and put him in charge of all that he had. 5 From the time that he made him overseer in his house and over all that he had, the LORD blessed the Egyptian's house for Joseph's sake; the blessing of the LORD was on all that he had, in house and field. 6 So he left all that he had in Joseph's charge; and, with him there, he had no concern for anything but the food that he ate.

Now Joseph was handsome and good-looking. 7 And after a time his master's wife cast her eyes on Joseph and said, "Lie with me." 8 But he refused and said to his master's wife, "Look, with me here, my master has no concern about anything in the house, and he has put everything that he has in my hand. 9 He is not greater in this house than I am, nor has he kept back anything from me except yourself, because you are his wife. How then could I do this great wickedness, and sin against God?" 10 And although she spoke to Joseph day after day, he would not consent to lie beside her or to be with her. 11 One day, however, when he went into the house to do his work, and while no one else was in the house, 12 she caught hold of his garment, saying, "Lie with me!" But he left his garment in her hand, and fled and ran outside. 13 When she saw that he had left his garment in her hand and had fled outside, 14 she called out to the members of her household and said to them, "See, my husband*j* has brought among us a Hebrew to insult us! He came in to me to lie with me, and I cried out with a loud voice; 15 and when he heard me raise my voice and cry out, he left his garment beside me, and fled outside." 16 Then she kept his garment by her until his master came home, 17 and she told him the same story, saying, "The Hebrew servant, whom you have brought among us, came in to me to insult me; 18 but as soon as I raised my voice and cried out, he left his garment beside me, and fled outside."

19 When his master heard the words that his wife spoke to him, saying, "This is the way your servant treated me," he became enraged. 20 And Joseph's master took him and put him into the prison, the

h That is *A breach*   *i* That is *Brightness*; perhaps alluding to the crimson thread
j Heb *he*

---

achieve justice. **27–30**: The birth of the twins (25.21–26) portrays the rivalry of Perez and Zerah, two clans of Judah (Num 26.19–22) who were partially Canaanite. Perez, the firstborn, was an ancestor of David (Ruth 4.18–22).

39.1–23: Joseph's success, temptation, and imprisonment. 1: This story continues from ch 37, and follows the tradition about the Ishmaelites (37.25). **5**: 30.27–30. **6**: For ritual reasons Potiphar took charge of his own food (43.32).

39.7–20: The Egyptian "Tale of Two Brothers" also tells how a man rejected the advances of his brother's wife, who then laid false accusations against him and almost brought about his death at the hands of his brother.

place where the king's prisoners were confined; he remained there in prison. 21 But the LORD was with Joseph and showed him steadfast love; he gave him favor in the sight of the chief jailer. 22 The chief jailer committed to Joseph's care all the prisoners who were in the prison, and whatever was done there, he was the one who did it. 23 The chief jailer paid no heed to anything that was in Joseph's care, because the LORD was with him; and whatever he did, the LORD made it prosper.

40 Some time after this, the cup-bearer of the king of Egypt and his baker offended their lord the king of Egypt. 2 Pharaoh was angry with his two officers, the chief cupbearer and the chief baker, 3 and he put them in custody in the house of the captain of the guard, in the prison where Joseph was confined. 4 The captain of the guard charged Joseph with them, and he waited on them; and they continued for some time in custody. 5 One night they both dreamed—the cupbearer and the baker of the king of Egypt, who were confined in the prison—each his own dream, and each dream with its own meaning. 6 When Joseph came to them in the morning, he saw that they were troubled. 7 So he asked Pharaoh's officers, who were with him in custody in his master's house, "Why are your faces downcast today?" 8 They said to him, "We have had dreams, and there is no one to interpret them." And Joseph said to them, "Do not interpretations belong to God? Please tell them to me."

9 So the chief cupbearer told his dream to Joseph, and said to him, "In my dream there was a vine before me, 10 and on the vine there were three branches. As soon as it budded, its blossoms came out and the clusters ripened into grapes. 11 Pharaoh's cup was in my hand; and I took the grapes and pressed them into Pharaoh's cup, and placed the cup in Pharaoh's hand." 12 Then Joseph said to him, "This is its interpretation: the three branches are three days; 13 within three days Pharaoh will lift up your head and restore you to your office; and you shall place Pharaoh's cup in his hand, just as you used to do when you were his cup-bearer. 14 But remember me when it is well with you; please do me the kindness to make mention of me to Pharaoh, and so get me out of this place. 15 For in fact I was stolen out of the land of the He-brews; and here also I have done nothing that they should have put me into the dungeon."

16 When the chief baker saw that the interpretation was favorable, he said to Joseph, "I also had a dream: there were three cake baskets on my head, 17 and in the uppermost basket there were all sorts of baked food for Pharaoh, but the birds were eating it out of the basket on my head." 18 And Joseph answered, "This is its interpretation: the three baskets are three days; 19 within three days Pharaoh will lift up your head—from you!—and hang you on a pole; and the birds will eat the flesh from you."

20 On the third day, which was Phar-aoh's birthday, he made a feast for all his servants, and lifted up the head of the chief cupbearer and the head of the chief baker among his servants. 21 He restored the chief cupbearer to his cupbearing, and he placed the cup in Pharaoh's hand; 22 but the chief baker he hanged, just as

---

39.21: Strangely Joseph, a slave, was not executed for alleged adultery, for *the LORD was with . . . him,* not only in success (v. 2) but also in adversity.

40.1–23: Joseph, the interpreter of dreams. 5: Since, according to ancient belief, dreams were a channel of divine communica-tion (1 Sam 28.6), the wise interpreter of dreams could discern the course of the future

(37.5–10; Dan 2.26–28). **6–8**: Professional interpreters of dreams were unnecessary (41.16), for *interpretations belong to God* who knows and controls the events of the future.

40.13: *Lift up your head,* i.e. graciously free you from prison (2 Kings 25.27). The same phrase, with extreme irony, is applied to the baker's fate in v. 19. **15**: *Stolen,* 37.28.

41.1–57: Joseph's elevation as a result of

Joseph had interpreted to them. 23 Yet the chief cupbearer did not remember Joseph, but forgot him.

41 After two whole years, Pharaoh dreamed that he was standing by the Nile, 2 and there came up out of the Nile seven sleek and fat cows, and they grazed in the reed grass. 3 Then seven other cows, ugly and thin, came up out of the Nile after them, and stood by the other cows on the bank of the Nile. 4 The ugly and thin cows ate up the seven sleek and fat cows. And Pharaoh awoke. 5 Then he fell asleep and dreamed a second time; seven ears of grain, plump and good, were growing on one stalk. 6 Then seven ears, thin and blighted by the east wind, sprouted after them. 7 The thin ears swallowed up the seven plump and full ears. Pharaoh awoke, and it was a dream. 8 In the morning his spirit was troubled; so he sent and called for all the magicians of Egypt and all its wise men. Pharaoh told them his dreams, but there was no one who could interpret them to Pharaoh.

9 Then the chief cupbearer said to Pharaoh, "I remember my faults today. 10 Once Pharaoh was angry with his servants, and put me and the chief baker in custody in the house of the captain of the guard. 11 We dreamed on the same night, he and I, each having a dream with its own meaning. 12 A young Hebrew was there with us, a servant of the captain of the guard. When we told him, he interpreted our dreams to us, giving an interpretation to each according to his dream. 13 As he interpreted to us, so it turned out; I was restored to my office, and the baker was hanged."

14 Then Pharaoh sent for Joseph, and he was hurriedly brought out of the dungeon. When he had shaved himself and changed his clothes, he came in before Pharaoh. 15 And Pharaoh said to Joseph, "I have had a dream, and there is no one who can interpret it. I have heard it said of you that when you hear a dream you can interpret it." 16 Joseph answered Pharaoh, "It is not I; God will give Pharaoh a favorable answer." 17 Then Pharaoh said to Joseph, "In my dream I was standing on the banks of the Nile; 18 and seven cows, fat and sleek, came up out of the Nile and fed in the reed grass. 19 Then seven other cows came up after them, poor, very ugly, and thin. Never had I seen such ugly ones in all the land of Egypt. 20 The thin and ugly cows ate up the first seven fat cows, 21 but when they had eaten them no one would have known that they had done so, for they were still as ugly as before. Then I awoke. 22 I fell asleep a second time*k* and I saw in my dream seven ears of grain, full and good, growing on one stalk, 23 and seven ears, withered, thin, and blighted by the east wind, sprouting after them; 24 and the thin ears swallowed up the seven good ears. But when I told it to the magicians, there was no one who could explain it to me."

25 Then Joseph said to Pharaoh, "Pharaoh's dreams are one and the same; God has revealed to Pharaoh what he is about to do. 26 The seven good cows are seven years, and the seven good ears are seven years; the dreams are one. 27 The seven lean and ugly cows that came up after them are seven years, as are the seven empty ears blighted by the east wind. They are seven years of famine. 28 It is as I told Pharaoh; God has shown to Pharaoh what he is about to do. 29 There will come seven years of great plenty throughout all the land of Egypt. 30 After them there will arise seven years of fam-

k Gk Syr Vg: Heb lacks *I fell asleep a second time*

---

his successful interpretation of Pharaoh's dreams. **1–2**: Egypt's fertility, symbolized by the sacred cows, was dependent upon the Nile. **6**: *The east wind,* the sirocco, a burning wind from the desert which withers vegetation (Hos 13.15). **8**: The narrator intends to demonstrate the superiority of Israel's God over heathen magic and wisdom (Ex 8.18–19; 9.11; Dan 2.2–19; 5.8, 15–28).

41.16: Joseph denies having any occult art and ascribes his skill solely to God (see 40.6–8 n.).

ine, and all the plenty will be forgotten in the land of Egypt; the famine will consume the land. 31 The plenty will no longer be known in the land because of the famine that will follow, for it will be very grievous. 32 And the doubling of Pharaoh's dream means that the thing is fixed by God, and God will shortly bring it about. 33 Now therefore let Pharaoh select a man who is discerning and wise, and set him over the land of Egypt. 34 Let Pharaoh proceed to appoint overseers over the land, and take one-fifth of the produce of the land of Egypt during the seven plenteous years. 35 Let them gather all the food of these good years that are coming, and lay up grain under the authority of Pharaoh for food in the cities, and let them keep it. 36 That food shall be a reserve for the land against the seven years of famine that are to befall the land of Egypt, so that the land may not perish through the famine."

37 The proposal pleased Pharaoh and all his servants. 38 Pharaoh said to his servants, "Can we find anyone else like this—one in whom is the spirit of God?" 39 So Pharaoh said to Joseph, "Since God has shown you all this, there is no one so discerning and wise as you. 40 You shall be over my house, and all my people shall order themselves as you command; only with regard to the throne will I be greater than you." 41 And Pharaoh said to Joseph, "See, I have set you over all the land of Egypt." 42 Removing his signet ring from his hand, Pharaoh put it on Joseph's hand; he arrayed him in garments of fine linen, and put a gold chain around his neck. 43 He had him ride in the chariot of his second-in-command; and they cried out in front of him, "Bow the knee!" *l* Thus he set him over all the land of Egypt. 44 Moreover Pharaoh said to Joseph, "I am Pharaoh, and without your consent no one shall lift up hand or foot in all the land of Egypt." 45 Pharaoh gave Joseph the name Zaphenath-paneah; and he gave him Asenath daughter of Potiphera, priest of On, as his wife. Thus Joseph gained authority over the land of Egypt.

46 Joseph was thirty years old when he entered the service of Pharaoh king of Egypt. And Joseph went out from the presence of Pharaoh, and went through all the land of Egypt. 47 During the seven plenteous years the earth produced abundantly. 48 He gathered up all the food of the seven years when there was plenty *m* in the land of Egypt, and stored up food in the cities; he stored up in every city the food from the fields around it. 49 So Joseph stored up grain in such abundance—like the sand of the sea—that he stopped measuring it; it was beyond measure.

50 Before the years of famine came, Joseph had two sons, whom Asenath daughter of Potiphera, priest of On, bore to him. 51 Joseph named the firstborn Manasseh, *n* "For," he said, "God has made me forget all my hardship and all my father's house." 52 The second he

l Abrek, apparently an Egyptian word similar in sound to the Hebrew word meaning *to kneel*
m Sam Gk: MT *the seven years that were*
n That is *Making to forget*

41.32: Two dreams with the same meaning (v. 25) show that the event is *fixed* or predestined by God. Note that this sense of God's overruling sovereignty does not evoke a fatalistic resignation but a practical plan of action (vv. 33–36).
41.38: *The spirit of God,* the source of extraordinary powers (Ex 31.3; Num 27.18; Dan 5.11, 14). **39–41:** Joseph was made prime minister, second only to Pharaoh in authority. Possibly the story reflects the period of Hyksos ascendancy in Egypt (about 1720–1550 B.C.), when the land was under pro-Semitic rule. At that time conditions would have been favorable for a Hebrew to rise to such a position of leadership. **42:** *His signet ring* (see 38.18 n.) empowered Joseph to act as Pharaoh's representative (compare Esth 3.10; 8.2). **45:** The installation rites, typically Egyptian, culminated with the bestowal of an Egyptian name. Joseph's adoption into the Egyptian court is further indicated by his marriage into the leading priesthood of *On* or Heliopolis. *Potiphera,* see 37.36 n.
41.46: Joseph's slavery and imprisonment lasted thirteen years (37.2–3).

named Ephraim,[o] "For God has made me fruitful in the land of my misfortunes."

53 The seven years of plenty that prevailed in the land of Egypt came to an end; [54]and the seven years of famine began to come, just as Joseph had said. There was famine in every country, but throughout the land of Egypt there was bread. [55]When all the land of Egypt was famished, the people cried to Pharaoh for bread. Pharaoh said to all the Egyptians, "Go to Joseph; what he says to you, do." [56]And since the famine had spread over all the land, Joseph opened all the storehouses,[p] and sold to the Egyptians, for the famine was severe in the land of Egypt. [57]Moreover, all the world came to Joseph in Egypt to buy grain, because the famine became severe throughout the world.

42 When Jacob learned that there was grain in Egypt, he said to his sons, "Why do you keep looking at one another? [2]I have heard," he said, "that there is grain in Egypt; go down and buy grain for us there, that we may live and not die." [3]So ten of Joseph's brothers went down to buy grain in Egypt. [4]But Jacob did not send Joseph's brother Benjamin with his brothers, for he feared that harm might come to him. [5]Thus the sons of Israel were among the other people who came to buy grain, for the famine had reached the land of Canaan.

6 Now Joseph was governor over the land; it was he who sold to all the people of the land. And Joseph's brothers came and bowed themselves before him with their faces to the ground. [7]When Joseph saw his brothers, he recognized them, but he treated them like strangers and spoke harshly to them. "Where do you come from?" he said. They said, "From the land of Canaan, to buy food." [8]Although Joseph had recognized his brothers, they did not recognize him. [9]Joseph also remembered the dreams that he had dreamed about them. He said to them, "You are spies; you have come to see the nakedness of the land!" [10]They said to him, "No, my lord; your servants have come to buy food. [11]We are all sons of one man; we are honest men; your servants have never been spies." [12]But he said to them, "No, you have come to see the nakedness of the land!" [13]They said, "We, your servants, are twelve brothers, the sons of a certain man in the land of Canaan; the youngest, however, is now with our father, and one is no more." [14]But Joseph said to them, "It is just as I have said to you; you are spies! [15]Here is how you shall be tested: as Pharaoh lives, you shall not leave this place unless your youngest brother comes here! [16]Let one of you go and bring your brother, while the rest of you remain in prison, in order that your words may be tested, whether there is truth in you; or else, as Pharaoh lives, surely you are spies." [17]And he put them all together in prison for three days.

18 On the third day Joseph said to them, "Do this and you will live, for I fear God: [19]if you are honest men, let one of your brothers stay here where you are imprisoned. The rest of you shall go and carry grain for the famine of your households, [20]and bring your youngest brother to me. Thus your words will be verified, and you shall not die." And they agreed to do so. [21]They said to one another, "Alas, we are paying the penalty for what we did to our brother; we saw

o From a Hebrew word meaning *to be fruitful*
p Gk Vg Compare Syr: Heb *opened all that was in* (or, *among*) *them*

42.1–38: Joseph's brothers journey to Egypt during the famine. 5: See 12.10. **6**: *Bowed themselves before him,* thereby unwittingly fulfilling the prediction of Joseph's dream (v. 9; see 37.5–11 n.).

42.9–14: The charge of espionage was natural, for Egypt's frontier, facing Canaan, was vulnerable to attack (Ex 1.10). **15–17**: The "testing" involved not only the verification of the brothers' words but also a discipline of suffering which would purge the evil of their hearts (compare v. 21). *As Pharaoh lives,* an oath in the name of Pharaoh, who was revered as divine in Egypt.

42.21–22: Once again the brothers have to announce to their father that misfortune has

his anguish when he pleaded with us, but we would not listen. That is why this anguish has come upon us." 22 Then Reuben answered them, "Did I not tell you not to wrong the boy? But you would not listen. So now there comes a reckoning for his blood." 23 They did not know that Joseph understood them, since he spoke with them through an interpreter. 24 He turned away from them and wept; then he returned and spoke to them. And he picked out Simeon and had him bound before their eyes. 25 Joseph then gave orders to fill their bags with grain, to return every man's money to his sack, and to give them provisions for their journey. This was done for them.

26 They loaded their donkeys with their grain, and departed. 27 When one of them opened his sack to give his donkey fodder at the lodging place, he saw his money at the top of the sack. 28 He said to his brothers, "My money has been put back; here it is in my sack!" At this they lost heart and turned trembling to one another, saying, "What is this that God has done to us?"

29 When they came to their father Jacob in the land of Canaan, they told him all that had happened to them, saying, 30 "The man, the lord of the land, spoke harshly to us, and charged us with spying on the land. 31 But we said to him, 'We are honest men, we are not spies. 32 We are twelve brothers, sons of our father; one is no more, and the youngest is now with our father in the land of Canaan.' 33 Then the man, the lord of the land, said to us, 'By this I shall know that you are honest men: leave one of your brothers with me, take grain for the famine of your households, and go your way. 34 Bring your youngest brother to me, and I shall know that you are not spies but honest men. Then I will release your brother to you, and you may trade in the land.'"

35 As they were emptying their sacks, there in each one's sack was his bag of money. When they and their father saw their bundles of money, they were dismayed. 36 And their father Jacob said to them, "I am the one you have bereaved of children: Joseph is no more, and Simeon is no more, and now you would take Benjamin. All this has happened to me!" 37 Then Reuben said to his father, "You may kill my two sons if I do not bring him back to you. Put him in my hands, and I will bring him back to you." 38 But he said, "My son shall not go down with you, for his brother is dead, and he alone is left. If harm should come to him on the journey that you are to make, you would bring down my gray hairs with sorrow to Sheol."

43 Now the famine was severe in the land. 2 And when they had eaten up the grain that they had brought from Egypt, their father said to them, "Go again, buy us a little more food." 3 But Judah said to him, "The man solemnly warned us, saying, 'You shall not see my face unless your brother is with you.' 4 If you will send our brother with us, we will go down and buy you food; 5 but if you will not send him, we will not go down, for the man said to us, 'You shall not see my face, unless your brother is with you.'" 6 Israel said, "Why did you treat me so badly as to tell the man that you had another brother?" 7 They replied, "The man questioned us carefully about ourselves and our kindred, saying, 'Is your father still alive? Have you another brother?' What we told him was in answer to these questions. Could we in any way know that he would say, 'Bring your brother down'?" 8 Then Judah said to his father Israel, "Send the boy with me, and let us be on our way, so that we may live and not die—you and we and also our little ones. 9 I myself will be surety for him; you can hold me accountable

---

befallen one of his sons (compare ch 37). Though they do not yet recognize Joseph, the similarity of the two situations evokes a feeling of guilt for their former behavior. **28:**

They sense that divine retribution is behind the mysterious events.

42.38: *Sheol,* see 37.35 n.; Eccl 9.10; Ps 6.5.

for him. If I do not bring him back to you and set him before you, then let me bear the blame forever. 10 If we had not delayed, we would now have returned twice."

11 Then their father Israel said to them, "If it must be so, then do this: take some of the choice fruits of the land in your bags, and carry them down as a present to the man—a little balm and a little honey, gum, resin, pistachio nuts, and almonds. 12 Take double the money with you. Carry back with you the money that was returned in the top of your sacks; perhaps it was an oversight. 13 Take your brother also, and be on your way again to the man; 14 may God Almighty*q* grant you mercy before the man, so that he may send back your other brother and Benjamin. As for me, if I am bereaved of my children, I am bereaved." 15 So the men took the present, and they took double the money with them, as well as Benjamin. Then they went on their way down to Egypt, and stood before Joseph.

16 When Joseph saw Benjamin with them, he said to the steward of his house, "Bring the men into the house, and slaughter an animal and make ready, for the men are to dine with me at noon." 17 The man did as Joseph said, and brought the men to Joseph's house. 18 Now the men were afraid because they were brought to Joseph's house, and they said, "It is because of the money, replaced in our sacks the first time, that we have been brought in, so that he may have an opportunity to fall upon us, to make slaves of us and take our donkeys." 19 So they went up to the steward of Joseph's house and spoke with him at the entrance to the house. 20 They said, "Oh, my lord, we came down the first time to buy food; 21 and when we came to the lodging place we opened our sacks, and there was each one's money in the top of his sack, our money in full weight. So we have brought it back with us. 22 Moreover we have brought down with us additional money to buy food. We do not know who put our money in our sacks." 23 He replied, "Rest assured, do not be afraid; your God and the God of your father must have put treasure in your sacks for you; I received your money." Then he brought Simeon out to them. 24 When the steward*r* had brought the men into Joseph's house, and given them water, and they had washed their feet, and when he had given their donkeys fodder, 25 they made the present ready for Joseph's coming at noon, for they had heard that they would dine there. 26 When Joseph came home, they brought him the present that they had carried into the house, and bowed to the ground before him. 27 He inquired about their welfare, and said, "Is your father well, the old man of whom you spoke? Is he still alive?" 28 They said, "Your servant our father is well; he is still alive." And they bowed their heads and did obeisance. 29 Then he looked up and saw his brother Benjamin, his mother's son, and said, "Is this your youngest brother, of whom you spoke to me? God be gracious to you, my son!" 30 With that, Joseph hurried out, because he was overcome with affection for his brother, and he was about to weep. So he went into a private room and wept there. 31 Then he washed his face and came out; and controlling himself he said, "Serve the meal." 32 They served him by himself, and them by themselves, and the Egyptians who ate with him by themselves,

q Traditional rendering of Heb *El Shaddai*
r Heb *the man*

43.1–34: The second journey to Egypt. 1–2: Simeon, left as a hostage in Egypt (vv. 14, 23), is apparently forgotten, for the brothers return only when more grain is needed. **3–7:** 42.29–34. **43.14:** *God Almighty,* see 17.1 n. **23:** The steward's words again stress the fundamental motif of the story: the working of divine providence. **43.29–30:** Joseph *was overcome with affection* for Benjamin, his only full brother through Rachel. **32:** Laws of ritual purity required that Egyptians eat apart from foreigners.

because the Egyptians could not eat with the Hebrews, for that is an abomination to the Egyptians. 33 When they were seated before him, the firstborn according to his birthright and the youngest according to his youth, the men looked at one another in amazement. 34 Portions were taken to them from Joseph's table, but Benjamin's portion was five times as much as any of theirs. So they drank and were merry with him.

44 Then he commanded the steward of his house, "Fill the men's sacks with food, as much as they can carry, and put each man's money in the top of his sack. 2 Put my cup, the silver cup, in the top of the sack of the youngest, with his money for the grain." And he did as Joseph told him. 3 As soon as the morning was light, the men were sent away with their donkeys. 4 When they had gone only a short distance from the city, Joseph said to his steward, "Go, follow after the men; and when you overtake them, say to them, 'Why have you returned evil for good? Why have you stolen my silver cup?' 5 Is it not from this that my lord drinks? Does he not indeed use it for divination? You have done wrong in doing this.' "

6 When he overtook them, he repeated these words to them. 7 They said to him, "Why does my lord speak such words as these? Far be it from your servants that they should do such a thing! 8 Look, the money that we found at the top of our sacks, we brought back to you from the land of Canaan; why then would we steal silver or gold from your lord's house? 9 Should it be found with

any one of your servants, let him die; moreover the rest of us will become my lord's slaves." 10 He said, "Even so; in accordance with your words, let it be: he with whom it is found shall become my slave, but the rest of you shall go free." 11 Then each one quickly lowered his sack to the ground, and each opened his sack. 12 He searched, beginning with the eldest and ending with the youngest; and the cup was found in Benjamin's sack. 13 At this they tore their clothes. Then each one loaded his donkey, and they returned to the city.

14 Judah and his brothers came to Joseph's house while he was still there; and they fell to the ground before him. 15 Joseph said to them, "What deed is this that you have done? Do you not know that one such as I can practice divination?" 16 And Judah said, "What can we say to my lord? What can we speak? How can we clear ourselves? God has found out the guilt of your servants; here we are then, my lord's slaves, both we and also the one in whose possession the cup has been found." 17 But he said, "Far be it from me that I should do so! Only the one in whose possession the cup was found shall be my slave; but as for you, go up in peace to your father."

18 Then Judah stepped up to him and said, "O my lord, let your servant please speak a word in my lord's ears, and do not be angry with your servant; for you are like Pharaoh himself. 19 My lord asked his servants, saying, 'Have you a

s Gk Compare Vg: Heb lacks *Why have you stolen my silver cup?*

44.1–34: Joseph puts his brothers to a final test. 1–2: The reference to the money harks back to the same motif in 42.25–28; here the real object of interest is Joseph's cup (v. 5). **5:** The cup was a sacred vessel used for divination, i.e. for magical prediction by observing the effects created when objects were thrown into the water contained therein. **44.15:** Having been initiated into Egyptian wisdom, Joseph can claim to practice divination, by which means the theft was discovered. **16:** Since the brothers acknowledge their

collective guilt, Judah's words may refer not only to the theft but to their treatment of Joseph in his youth. **17:** Joseph tests his brothers to see whether, as in his case once, they will let Benjamin go into slavery and return to their father to justify the loss of another of his sons.

44.18–34: Judah's speech, one of the finest prose pieces from Israel's early tradition, summarizes and epitomizes the whole sequence of events. **20:** The tragedy of Joseph's supposed death heightens the pathos; for of

father or a brother?' 20 And we said to my lord, 'We have a father, an old man, and a young brother, the child of his old age. His brother is dead; he alone is left of his mother's children, and his father loves him.' 21 Then you said to your servants, 'Bring him down to me, so that I may set my eyes on him.' 22 We said to my lord, 'The boy cannot leave his father, for if he should leave his father, his father would die.' 23 Then you said to your servants, 'Unless your youngest brother comes down with you, you shall see my face no more.' 24 When we went back to your servant my father we told him the words of my lord. 25 And when our father said, 'Go again, buy us a little food,' 26 we said, 'We cannot go down. Only if our youngest brother goes with us, will we go down; for we cannot see the man's face unless our youngest brother is with us.' 27 Then your servant my father said to us, 'You know that my wife bore me two sons; 28 one left me, and I said, Surely he has been torn to pieces; and I have never seen him since. 29 If you take this one also from me, and harm comes to him, you will bring down my gray hairs in sorrow to Sheol.' 30 Now therefore, when I come to your servant my father and the boy is not with us, then, as his life is bound up in the boy's life, 31 when he sees that the boy is not with us, he will die; and your servants will bring down the gray hairs of your servant our father with sorrow to Sheol. 32 For your servant became surety for the boy to my father, saying, 'If I do not bring him back to you, then I will bear the blame in the sight of my father all my life.' 33 Now therefore, please let your servant remain as a slave to my lord in place of the boy; and let the boy go back with his brothers. 34 For how can I go back to my father if the boy is not with me? I fear to see the suffering that would come upon my father."

45 Then Joseph could no longer control himself before all those who stood by him, and he cried out, "Send everyone away from me." So no one stayed with him when Joseph made himself known to his brothers. 2 And he wept so loudly that the Egyptians heard it, and the household of Pharaoh heard it. 3 Joseph said to his brothers, "I am Joseph. Is my father still alive?" But his brothers could not answer him, so dismayed were they at his presence.

4 Then Joseph said to his brothers, "Come closer to me." And they came closer. He said, "I am your brother, Joseph, whom you sold into Egypt. 5 And now do not be distressed, or angry with yourselves, because you sold me here; for God sent me before you to preserve life. 6 For the famine has been in the land these two years; and there are five more years in which there will be neither plowing nor harvest. 7 God sent me before you to preserve for you a remnant on earth, and to keep alive for you many survivors. 8 So it was not you who sent me here, but God; he has made me a father to Pharaoh, and lord of all his house and ruler over all the land of Egypt. 9 Hurry and go up to my father and say to him, 'Thus says your son Joseph, God has made me lord of all Egypt; come down to me, do not delay. 10 You shall settle in the land of Goshen, and you

---

Jacob's two sons by Rachel, only Benjamin is left and Jacob's *life is bound up in the boy's life* (v. 30).

45.1–28: Joseph makes himself known to his brothers. 7–8: This passage sets forth the central theme of the Joseph story: in a hidden manner, events were directed in accordance with divine, not human, purpose. God graciously contrives to bring good out of evil; for the brothers, in selling Joseph into slavery, had unwittingly carried out God's will. **7**: Through Joseph God acted *to preserve*

life (v. 5; 50.20), not only the life of famine-stricken Egyptians but also that of *a remnant,* that is, the family which is the bearer of the promise given to Abraham (12.2–3; 50.24). **8**: *A father to Pharaoh,* a title of the chief minister (Isa 22.21; compare 1 Macc 11.32). **10**: *The land of Goshen,* the present Wadi Tumilat, a narrow strip of grazing land in the Delta. Since the settlers would be *near* Joseph, the assumption is that Pharaoh's capital was in the Delta region, which was the case during the Hyksos period (see 41.39–41 n.).

shall be near me, you and your children and your children's children, as well as your flocks, your herds, and all that you have. 11 I will provide for you there— since there are five more years of famine to come—so that you and your household, and all that you have, will not come to poverty.' 12 And now your eyes and the eyes of my brother Benjamin see that it is my own mouth that speaks to you. 13 You must tell my father how greatly I am honored in Egypt, and all that you have seen. Hurry and bring my father down here." 14 Then he fell upon his brother Benjamin's neck and wept, while Benjamin wept upon his neck. 15 And he kissed all his brothers and wept upon them; and after that his brothers talked with him.

16 When the report was heard in Pharaoh's house, "Joseph's brothers have come," Pharaoh and his servants were pleased. 17 Pharaoh said to Joseph, "Say to your brothers, 'Do this: load your animals and go back to the land of Canaan. 18 Take your father and your households and come to me, so that I may give you the best of the land of Egypt, and you may enjoy the fat of the land.' 19 You are further charged to say, 'Do this: take wagons from the land of Egypt for your little ones and for your wives, and bring your father, and come. 20 Give no thought to your possessions, for the best of all the land of Egypt is yours.' "

21 The sons of Israel did so. Joseph gave them wagons according to the instruction of Pharaoh, and he gave them provisions for the journey. 22 To each one of them he gave a set of garments; but to Benjamin he gave three hundred pieces of silver and five sets of garments. 23 To his father he sent the following: ten donkeys loaded with the good things of Egypt, and ten female donkeys loaded with grain, bread, and provision for his father on the journey. 24 Then he sent his brothers on their way, and as they were leaving he said to them, "Do not quarrel[t] along the way."

25 So they went up out of Egypt and came to their father Jacob in the land of Canaan. 26 And they told him, "Joseph is still alive! He is even ruler over all the land of Egypt." He was stunned; he could not believe them. 27 But when they told him all the words of Joseph that he had said to them, and when he saw the wagons that Joseph had sent to carry him, the spirit of their father Jacob revived. 28 Israel said, "Enough! My son Joseph is still alive. I must go and see him before I die."

46 When Israel set out on his journey with all that he had and came to Beer-sheba, he offered sacrifices to the God of his father Isaac. 2 God spoke to Israel in visions of the night, and said, "Jacob, Jacob." And he said, "Here I am." 3 Then he said, "I am God,[u] the God of your father; do not be afraid to go down to Egypt, for I will make of you a great nation there. 4 I myself will go down with you to Egypt, and I will also bring you up again; and Joseph's own hand shall close your eyes."

5 Then Jacob set out from Beersheba; and the sons of Israel carried their father Jacob, their little ones, and their

t Or *be agitated*   *u* Heb *the God*

45.16–20: According to Egyptian sources, it was not unusual for Pharaoh to permit Asiatics to settle in his country in time of famine.
46.1–27: Jacob's migration to Egypt. 1: From Hebron (37.14) Jacob went first to *Beersheba,* the shrine associated with Isaac (26.23–25; 28.10). *The God of your father,* see 26.24 n. **2–4:** The descent into Egypt, which was to have decisive significance for Israel's history, was prompted not merely by Jacob's desire to see his long-lost son (45.28) but by divine revelation *in visions of the night.* **3:** As on a previous occasion (28.13–15), before Jacob left the land of the promise to go to a foreign land, God renewed the promise to make him *a great nation* (see 12.2 n.; 18.18) in Egypt (Ex 1.7). **4:** Only Jacob's corpse was returned from Egypt (50.4–14). However, according to Hebraic corporate thinking, the words *bring you up again* were fulfilled, for the father lived on in the person of his sons.

wives, in the wagons that Pharaoh had sent to carry him. 6 They also took their livestock and the goods that they had acquired in the land of Canaan, and they came into Egypt, Jacob and all his offspring with him, 7 his sons, and his sons' sons with him, his daughters, and his sons' daughters; all his offspring he brought with him into Egypt.

8 Now these are the names of the Israelites, Jacob and his offspring, who came to Egypt. Reuben, Jacob's firstborn, 9 and the children of Reuben: Hanoch, Pallu, Hezron, and Carmi. 10 The children of Simeon: Jemuel, Jamin, Ohad, Jachin, Zohar, and Shaul, *v* the son of a Canaanite woman. 11 The children of Levi: Gershon, Kohath, and Merari. 12 The children of Judah: Er, Onan, Shelah, Perez, and Zerah (but Er and Onan died in the land of Canaan); and the children of Perez were Hezron and Hamul. 13 The children of Issachar: Tola, Puvah, Jashub, *w* and Shimron. 14 The children of Zebulun: Sered, Elon, and Jahleel 15 (these are the sons of Leah, whom she bore to Jacob in Paddan-aram, together with his daughter Dinah; in all his sons and his daughters numbered thirty-three). 16 The children of Gad: Ziphion, Haggi, Shuni, Ezbon, Eri, Arodi, and Areli. 17 The children of Asher: Imnah, Ishvah, Ishvi, Beriah, and their sister Serah. The children of Beriah: Heber and Malchiel 18 (these are the children of Zilpah, whom Laban gave to his daughter Leah; and these she bore to Jacob—sixteen persons). 19 The children of Jacob's wife Rachel: Joseph and Benjamin. 20 To Joseph in the land of Egypt were born Manasseh and Ephraim, whom Asenath daughter of Potiphera, priest of

On, bore to him. 21 The children of Benjamin: Bela, Becher, Ashbel, Gera, Naaman, Ehi, Rosh, Muppim, Huppim, and Ard 22 (these are the children of Rachel, who were born to Jacob—fourteen persons in all). 23 The children of Dan: Hashum. *x* 24 The children of Naphtali: Jahzeel, Guni, Jezer, and Shillem 25 (these are the children of Bilhah, whom Laban gave to his daughter Rachel, and these she bore to Jacob—seven persons in all). 26 All the persons belonging to Jacob who came into Egypt, who were his own offspring, not including the wives of his sons, were sixty-six persons in all. 27 The children of Joseph, who were born to him in Egypt, were two; all the persons of the house of Jacob who came into Egypt were seventy.

28 Israel *y* sent Judah ahead to Joseph to lead the way before him into Goshen. When they came to the land of Goshen, 29 Joseph made ready his chariot and went up to meet his father Israel in Goshen. He presented himself to him, fell on his neck, and wept on his neck a good while. 30 Israel said to Joseph, "I can die now, having seen for myself that you are still alive." 31 Joseph said to his brothers and to his father's household, "I will go up and tell Pharaoh, and will say to him, 'My brothers and my father's household, who were in the land of Canaan, have come to me. 32 The men are shepherds, for they have been keepers of livestock; and they have brought their flocks, and their herds, and all that they have.' 33 When Pharaoh calls you, and says, 'What is your occupation?' 34 you shall

v Or *Saul*   *w* Compare Sam Gk Num 26.24 1 Chr 7.1: MT *Iob*   *x* Gk: Heb *Hushim* *y* Heb *He*

46.8–27: This section, from a separate priestly tradition, contains a list of Jacob's descendants, based on the traditional number seventy (v. 27; Ex 1.5; Deut 10.22). Most of the names of the ancestral clan leaders are found in the priestly list in Num ch 26. 27: *Seventy,* the author includes Joseph and his two sons born in Egypt, as well as Jacob himself.

46.28–47.12: Jacob and his sons settle in Egypt. **46.28:** Judah is sent ahead because he is the chief spokesperson (37.26; 43.3–10; 44.18–34). *Goshen,* see 45.10 n. **31–34:** Desiring to have his relatives near him in the Delta, Joseph advised his brothers to testify that they were shepherds; since this occupation was abominable to Egyptians in the interior

say, 'Your servants have been keepers of livestock from our youth even until now, both we and our ancestors'—in order that you may settle in the land of Goshen, because all shepherds are abhorrent to the Egyptians."

47 So Joseph went and told Pharaoh, "My father and my brothers, with their flocks and herds and all that they possess, have come from the land of Canaan; they are now in the land of Goshen." 2 From among his brothers he took five men and presented them to Pharaoh. 3 Pharaoh said to his brothers, "What is your occupation?" And they said to Pharaoh, "Your servants are shepherds, as our ancestors were." 4 They said to Pharaoh, "We have come to reside as aliens in the land; for there is no pasture for your servants' flocks because the famine is severe in the land of Canaan. Now, we ask you, let your servants settle in the land of Goshen." 5 Then Pharaoh said to Joseph, "Your father and your brothers have come to you. 6 The land of Egypt is before you; settle your father and your brothers in the best part of the land; let them live in the land of Goshen; and if you know that there are capable men among them, put them in charge of my livestock."

7 Then Joseph brought in his father Jacob, and presented him before Pharaoh, and Jacob blessed Pharaoh. 8 Pharaoh said to Jacob, "How many are the years of your life?" 9 Jacob said to Pharaoh, "The years of my earthly sojourn are one hundred thirty; few and hard have been the years of my life. They do not compare with the years of the life of my ancestors during their long sojourn."

10 Then Jacob blessed Pharaoh, and went out from the presence of Pharaoh. 11 Joseph settled his father and his brothers, and granted them a holding in the land of Egypt, in the best part of the land, in the land of Rameses, as Pharaoh had instructed. 12 And Joseph provided his father, his brothers, and all his father's household with food, according to the number of their dependents.

13 Now there was no food in all the land, for the famine was very severe. The land of Egypt and the land of Canaan languished because of the famine. 14 Joseph collected all the money to be found in the land of Egypt and in the land of Canaan, in exchange for the grain that they bought; and Joseph brought the money into Pharaoh's house. 15 When the money from the land of Egypt and from the land of Canaan was spent, all the Egyptians came to Joseph, and said, "Give us food! Why should we die before your eyes? For our money is gone." 16 And Joseph answered, "Give me your livestock, and I will give you food in exchange for your livestock, if your money is gone." 17 So they brought their livestock to Joseph; and Joseph gave them food in exchange for the horses, the flocks, the herds, and the donkeys. That year he supplied them with food in exchange for all their livestock. 18 When that year was ended, they came to him the following year, and said to him, "We can not hide from my lord that our money is all spent; and the herds of cattle are my lord's. There is nothing left in the sight of my lord but our bodies and our lands. 19 Shall we die before your eyes, both we and our land? Buy us and our

---

(v. 34), Pharaoh would see the wisdom of setting them apart in the land of Goshen.

47.1–6: Egyptian sources testify that pharaohs possessed large herds and gave much attention to cattle-breeding. **47.7–12:** According to this priestly tradition, *Jacob blessed Pharaoh,* presumably with the blessing of welfare and long life. **9:** Jacob's statement that his years had been *few and hard* reflects the view that there was an increasing shortening and troubling of human life (see

5.4–32 n.). **11:** *The land of Rameses* (= Goshen), was named after Rameses II (see Ex 1.8, 11 n.). **47.13–26: Joseph's agrarian program,** involving a change in the Egyptian system of land-tenure. **14:** First, the people spent all their money for grain (compare 41.56). **15–17:** Next, in their desperation they exchanged all their cattle for food. **18–19:** Finally, they offered themselves and their lands to Pharaoh.

land in exchange for food. We with our land will become slaves to Pharaoh; just give us seed, so that we may live and not die, and that the land may not become desolate." 20 So Joseph bought all the land of Egypt for Pharaoh. All the Egyptians sold their fields, because the famine was severe upon them; and the land became Pharaoh's. 21 As for the people, he made slaves of them*z* from one end of Egypt to the other. 22 Only the land of the priests he did not buy; for the priests had a fixed allowance from Pharaoh, and lived on the allowance that Pharaoh gave them; therefore they did not sell their land. 23 Then Joseph said to the people, "Now that I have this day bought you and your land for Pharaoh, here is seed for you; sow the land. 24 And at the harvests you shall give one-fifth to Pharaoh, and four-fifths shall be your own, as seed for the field and as food for yourselves and your households, and as food for your little ones." 25 They said, "You have saved our lives; may it please my lord, we will be slaves to Pharaoh." 26 So Joseph made it a statute concerning the land of Egypt, and it stands to this day, that Pharaoh should have the fifth. The land of the priests alone did not become Pharaoh's.

27 Thus Israel settled in the land of Egypt, in the region of Goshen; and they gained possessions in it, and were fruitful and multiplied exceedingly. 28 Jacob lived in the land of Egypt seventeen years; so the days of Jacob, the years of his life, were one hundred forty-seven years.

29 When the time of Israel's death drew near, he called his son Joseph and said to him, "If I have found favor with you, put your hand under my thigh and promise to deal loyally and truly with me. Do not bury me in Egypt. 30 When I lie down with my ancestors, carry me out of Egypt and bury me in their burial place." He answered, "I will do as you have said." 31 And he said, "Swear to me"; and he swore to him. Then Israel bowed himself on the head of his bed.

48 After this Joseph was told, "Your father is ill." So he took with him his two sons, Manasseh and Ephraim. 2 When Jacob was told, "Your son Joseph has come to you," he*a* summoned his strength and sat up in bed. 3 And Jacob said to Joseph, "God Almighty*b* appeared to me at Luz in the land of Canaan, and he blessed me, 4 and said to me, 'I am going to make you fruitful and increase your numbers; I will make of you a company of peoples, and will give this land to your offspring after you for a perpetual holding.' 5 Therefore your two sons, who were born to you in the land of Egypt before I came to you in Egypt, are now mine; Ephraim and Manasseh shall be mine, just as Reuben and Simeon are. 6 As for the offspring born to you after them, they shall be yours. They shall be recorded under the names of their brothers with regard to their inheritance. 7 For when I came from Paddan, Rachel, alas, died in the land of Canaan

z Sam Gk Compare Vg: MT *He removed them to the cities*   *a* Heb *Israel*   *b* Traditional rendering of Heb *El Shaddai*

47.20–26: The result was that former land-owners became tenants of Pharaoh, farming the land for him and paying him one-fifth of the produce as tax (v. 24). Temple lands were excepted (v. 22). **25:** The narrator does not intend to sanction absolutism but only to praise Joseph for his wisdom in delivering the people. **47.27–48.22: Jacob's adoption and blessing of Ephraim and Manasseh. 47.29:** *Put your hand under my thigh,* see 24.2 n. **30–31:** Joseph binds himself by oath to bury Jacob in his ancestors' *burial place,* i.e.

Machpelah (ch 23; 49.29–30; 50.12–13). *Israel* [i.e. Jacob, see 35.21 n.] *bowed himself on the head of his bed* (compare Heb 11.21), a gesture of reverence or gratitude (1 Kings 1.47). **48.3–4:** Jacob's adoption and blessing of Joseph's two sons are based on the divine promise given at Luz, or Bethel (35.9–13). **5–6:** By adopting his grandsons, Jacob gives them status equal to his eldest sons, Reuben and Simeon. The narrative accounts for the circumstance that "the house of Joseph" (Josh 17.17; 18.5; Judg 1.23, 35) came to be divided into two tribes, Manasseh and Ephraim, each

on the way, while there was still some distance to go to Ephrath; and I buried her there on the way to Ephrath" (that is, Bethlehem).

8 When Israel saw Joseph's sons, he said, "Who are these?" 9 Joseph said to his father, "They are my sons, whom God has given me here." And he said, "Bring them to me, please, that I may bless them." 10 Now the eyes of Israel were dim with age, and he could not see well. So Joseph brought them near him; and he kissed them and embraced them. 11 Israel said to Joseph, "I did not expect to see your face; and here God has let me see your children also." 12 Then Joseph removed them from his father's knees, *c* and he bowed himself with his face to the earth. 13 Joseph took them both, Ephraim in his right hand toward Israel's left, and Manasseh in his left hand toward Israel's right, and brought them near him. 14 But Israel stretched out his right hand and laid it on the head of Ephraim, who was the younger, and his left hand on the head of Manasseh, crossing his hands, for Manasseh was the firstborn. 15 He blessed Joseph, and said,

"The God before whom my
    ancestors Abraham and
    Isaac walked,
the God who has been my
    shepherd all my life to this
    day,

16 the angel who has redeemed me
    from all harm, bless the
    boys;
and in them let my name be
    perpetuated, and the name
    of my ancestors Abraham
    and Isaac;
and let them grow into a
    multitude on the earth."

17 When Joseph saw that his father laid his right hand on the head of Ephraim, it displeased him; so he took his father's hand, to remove it from Ephraim's head to Manasseh's head. 18 Joseph said to his father, "Not so, my father! Since this one is the firstborn, put your right hand on his head." 19 But his father refused, and said, "I know, my son, I know; he also shall become a people, and he also shall be great. Nevertheless his younger brother shall be greater than he, and his offspring shall become a multitude of nations." 20 So he blessed them that day, saying,

"By you *d* Israel will invoke
    blessings, saying,
'God make you *e* like Ephraim and
    like Manasseh.'"

So he put Ephraim ahead of Manasseh. 21 Then Israel said to Joseph, "I am about to die, but God will be with you and will

c Heb *from his knees*   *d* *you* here is singular in Heb   *e* *you* here is singular in Heb

claiming full rank with the other tribes (49.22–26). **7**: 35.16–20. *Paddan*(-aram), see 11.31 n.

48.12: An adoption ceremony may be suggested by the boys' having been between or on Jacob's knees. **13–14**: Joseph brings Manasseh, his firstborn, for the blessing of the right hand, but Jacob crosses his hands and puts his right hand on Ephraim, thereby giving him precedence. The narrator appeals to the ancient belief in the efficacy of the deathbed blessing (see 27.4 n.) to account for two facts: (1) Manasseh and Ephraim, located in the central hill country, were powerful tribes in early Israelite history; (2) the latter, during the period of the Judges and the early monarchy, gained pre-eminence over the "firstborn" tribe that once ranked first in leadership. **15–16**: Jacob invokes God by a threefold

description: the God before whom his ancestors *walked* (17.1; 24.40), who *has been* his *shepherd* (Ps 23.1) throughout his life, who *redeemed* him from evil (Isa 48.20). *The angel,* see 16.7 n. According to ancient belief, *the name,* that is, the psychic life of the father, was to be perpetuated in the two boys.

48.20: Another version of the blessing (compare vv. 15–16). *Israel* here refers not to Jacob but to the people (34.7). *Like Ephraim and like Manasseh,* i.e. as fruitful with offspring as these tribes. **21**: Joseph died in Egypt and his bones were brought back to Canaan (see 46.4 n.). **22**: In Hebrew *one portion* (or "shoulder") is a play on the name Shechem, the important city which lay in the territory of Joseph (see 12.6 n.). *With my sword and with my bow* reflects a different tradition from that in 33.19–20, which reports Jacob's peaceful

bring you again to the land of your ancestors. 22I now give to you one portion*f* more than to your brothers, the portion*f* that I took from the hand of the Amorites with my sword and with my bow."

49 Then Jacob called his sons, and said: "Gather around, that I may tell you what will happen to you in days to come.

2 Assemble and hear, O sons of
  Jacob;
  listen to Israel your father.

3 Reuben, you are my firstborn,
  my might and the first fruits of
    my vigor,
  excelling in rank and excelling
    in power.

4 Unstable as water, you shall no
    longer excel
  because you went up onto your
    father's bed;
  then you defiled it—you*g* went
    up onto my couch!

5 Simeon and Levi are brothers;
  weapons of violence are their
    swords.

6 May I never come into their
    council;
  may I not be joined to their
    company—

for in their anger they killed men,
  and at their whim they
    hamstrung oxen.

7 Cursed be their anger, for it is
    fierce,
  and their wrath, for it is cruel!
  I will divide them in Jacob,
  and scatter them in Israel.

8 Judah, your brothers shall praise
    you;
  your hand shall be on the neck
    of your enemies;
  your father's sons shall bow
    down before you.

9 Judah is a lion's whelp;
  from the prey, my son, you
    have gone up.
  He crouches down, he stretches
    out like a lion,
  like a lioness—who dares rouse
    him up?

10 The scepter shall not depart from
    Judah,
  nor the ruler's staff from
    between his feet,
  until tribute comes to him;*h*

---

f Or *mountain slope* (Heb *shekem*, a play on the name of the town and district of Shechem)
g Gk Syr Tg: Heb *he*
h Or *until Shiloh comes* or *until he comes to Shiloh* or (with Syr) *until he comes to whom it belongs*

---

coming to Shechem, and ch 34, which describes his protest against his sons' attack upon the city (34.25–30). *Amorites,* see 10.15–20 n.
49.1–28: Jacob's blessing on his twelve sons. This poem, apparently dating from the time of David (see vv. 8–12), portrays the character of the tribes in the person of their ancestor (Deut ch 33). Although regarded as Jacob's death-bed blessing (v. 28), this is not an altogether adequate description, for the poet sometimes speaks words of censure or even curse (e.g. v. 7). On death-bed blessings, see 27.4 n. **1–2:** The language in the future tense, *what shall happen to you in days to come,* is relevant to the Judah oracle (vv. 8–12); the other oracles, however, describe chiefly past events or present circumstances. **3–4:** Reuben, the firstborn, whose territory lay east of the Dead Sea, was once a leading

tribe but in early times was overcome by the Moabites (Judg 5.15–16; Deut 33.6). *4: Your father's bed,* a reference to the ancestor's act of immorality (35.22), which typifies the tribe's moral weakness and instability. **5–7:** Simeon and Levi are considered together, for they led in the attack against Shechem with *weapons of violence* (34.25–30). Levi, once a full tribe, came to be a priestly class (Ex 32.26–29; Deut 10.8–9). Simeon was eventually absorbed into the tribe of Judah.
49.8–12: This oracle reflects a situation, like that in David's time, when Judah had pre-eminence over the tribes. **10:** The first part of the verse portrays Judah as a sovereign; the second part, however, is very obscure (see note *h*). The verse has been interpreted to mean that after the kingdom of Judah has lasted an indefinite time there will arise a messianic ruler who will command *the*

and the obedience of the peoples
is his.

11 Binding his foal to the vine
and his donkey's colt to the
choice vine,
he washes his garments in wine
and his robe in the blood of
grapes;

12 his eyes are darker than wine,
and his teeth whiter than milk.

13 Zebulun shall settle at the shore of
the sea;
he shall be a haven for ships,
and his border shall be at Sidon.

14 Issachar is a strong donkey,
lying down between the
sheepfolds;

15 he saw that a resting place was
good,
and that the land was pleasant;
so he bowed his shoulder to the
burden,
and became a slave at forced
labor.

16 Dan shall judge his people
as one of the tribes of Israel.

17 Dan shall be a snake by the
roadside,
a viper along the path,
that bites the horse's heels
so that its rider falls backward.

18 I wait for your salvation, O LORD.

19 Gad shall be raided by raiders,
but he shall raid at their heels.

20 Asher's[i] food shall be rich,
and he shall provide royal
delicacies.

21 Naphtali is a doe let loose
that bears lovely fawns.[j]

22 Joseph is a fruitful bough,
a fruitful bough by a spring;
his branches run over the wall. [k]

23 The archers fiercely attacked him;
they shot at him and pressed
him hard.

24 Yet his bow remained taut,
and his arms[l] were made agile
by the hands of the Mighty One
of Jacob,
by the name of the Shepherd,
the Rock of Israel,

25 by the God of your father, who
will help you,
by the Almighty[m] who will
bless you
with blessings of heaven above,
blessings of the deep that lies
beneath,
blessings of the breasts and of
the womb.

i Gk Vg Syr: Heb *From Asher*   *j* Or *that gives*
beautiful words   *k* Meaning of Heb uncertain
l Heb *the arms of his hands*
m Traditional rendering of Heb *Shaddai*

---

obedience of the peoples (Num 24.17; Isa 11.1–
9). **11–12:** A picture of the marvelous fertility
that will ensue. **13:** Zebulun will have a favorable position, no longer shut up in the interior
(Josh 19.10–16) but with access to the Mediterranean Sea. Expansion into Asher's territory is assumed.
49.14–15: Issachar is compared to a domesticated beast of burden, contented with
a comfortable land and willing to surrender
political independence in subservience to the
Canaanites. **16–18:** Dan will rise to full tribal
prestige by judging (i.e. getting justice for)
his people. The Hebrew verb "judge" involves a play on the word Dan. The comparison with *a snake by the roadside* portrays the
insidious warfare of a small tribe in its rise to

power. **19:** Gad, settled east of the Jordan just
above Reuben, is cited for bravery in repel-
ling Ammonite and desert marauders (Judg
ch 11). **20:** Asher's land, situated on the coastal strip between Mount Carmel and Phoenicia, was so rich that it yielded *royal delicacies*
(Deut 33.24). **21:** The comparison of Naphtali
to *a doe let loose* suggests the idea of freedom,
agility, and vitality (compare Deut 33.23).
49.22–26: A picture of the prosperity and
strength of the populous tribe of Joseph, apparently harking back to a time before "the
house of Joseph" was divided into the tribes
of Manasseh and Ephraim, as in Deut 33.13–
17 (see Gen ch 48). **24:** *The Mighty One of
Jacob*, a title of "the God of the ancestors"
(v. 25a; Isa 1.24; 49.26). **25:** *The Almighty*, see

26 The blessings of your father
    are stronger than the blessings
        of the eternal mountains,
    the bounties*n* of the everlasting
        hills;
    may they be on the head of
        Joseph,
        on the brow of him who was
           set apart from his brothers.

27 Benjamin is a ravenous wolf,
    in the morning devouring the
        prey,
    and at evening dividing the
        spoil."

28 All these are the twelve tribes of Israel, and this is what their father said to them when he blessed them, blessing each one of them with a suitable blessing. 29 Then he charged them, saying to them, "I am about to be gathered to my people. Bury me with my ancestors—in the cave in the field of Ephron the Hittite, 30 in the cave in the field at Machpelah, near Mamre, in the land of Canaan, in the field that Abraham bought from Ephron the Hittite as a burial site. 31 There Abraham and his wife Sarah were buried; there Isaac and his wife Rebekah were buried; and there I buried Leah— 32 the field and the cave that is in it were purchased from the Hittites." 33 When Jacob ended his charge to his sons, he drew up his feet into the bed, breathed his last, and was gathered to his people.

50 Then Joseph threw himself on his father's face and wept over him and kissed him. 2 Joseph commanded the physicians in his service to embalm his father. So the physicians embalmed Isra-

el; 3 they spent forty days in doing this, for that is the time required for embalming. And the Egyptians wept for him seventy days.

4 When the days of weeping for him were past, Joseph addressed the household of Pharaoh, "If now I have found favor with you, please speak to Pharaoh as follows: 5 My father made me swear an oath; he said, 'I am about to die. In the tomb that I hewed out for myself in the land of Canaan, there you shall bury me.' Now therefore let me go up, so that I may bury my father; then I will return." 6 Pharaoh answered, "Go up, and bury your father, as he made you swear to do."

7 So Joseph went up to bury his father. With him went up all the servants of Pharaoh, the elders of his household, and all the elders of the land of Egypt, 8 as well as all the household of Joseph, his brothers, and his father's household. Only their children, their flocks, and their herds were left in the land of Goshen. 9 Both chariots and charioteers went up with him. It was a very great company. 10 When they came to the threshing floor of Atad, which is beyond the Jordan, they held there a very great and sorrowful lamentation; and he observed a time of mourning for his father seven days. 11 When the Canaanite inhabitants of the land saw the mourning on the threshing floor of Atad, they said, "This is a grievous mourning on the part of the Egyptians." Therefore the place was named Abel-mizraim;*o* it is beyond the Jordan. 12 Thus his sons did for him

n Cn Compare Gk: Heb of my progenitors to the boundaries   o That is mourning (or meadow) of Egypt

---

17.1 n. *Blessings of heaven,* i.e. rain, dew, sun. *The deep that lies beneath* (Deut 33.13), an allusion to the subterranean ocean (see 1.2, 6), believed to be the source of fertility (see 2.6 n.). **26:** The ancestral blessing surpasses even the majesty and fertility of the hills of Ephraim. *Set apart* by prestige and position.
    49.29–50.26: The death of Jacob and the final days of Joseph.
    49.30–32: *Machpelah,* see ch 23.

50.2–3: Embalming, an ancient Egyptian custom, was necessary if Jacob's body was to be carried back to Canaan. Egyptians are said to have mourned for a monarch seventy-two days; thus, out of respect for Joseph, Jacob was given a royal funeral. **5:** According to this fragment of tradition, Jacob had hewed out a tomb for himself east of the Jordan (v. 10) and was buried there rather than at Machpelah (vv. 12–13). This explains why the funeral

as he had instructed them. 13 They carried him to the land of Canaan and buried him in the cave of the field at Machpelah, the field near Mamre, which Abraham bought as a burial site from Ephron the Hittite. 14 After he had buried his father, Joseph returned to Egypt with his brothers and all who had gone up with him to bury his father.

15 Realizing that their father was dead, Joseph's brothers said, "What if Joseph still bears a grudge against us and pays us back in full for all the wrong that we did to him?" 16 So they approached[p] Joseph, saying, "Your father gave this instruction before he died, 17 'Say to Joseph: I beg you, forgive the crime of your brothers and the wrong they did in harming you.' Now therefore please forgive the crime of the servants of the God of your father." Joseph wept when they spoke to him. 18 Then his brothers also wept,[q] fell down before him, and said, "We are here as your slaves." 19 But Joseph said to them, "Do not be afraid! Am I in the place of God? 20 Even though you intended to do harm to me, God intend-

ed it for good, in order to preserve a numerous people, as he is doing today. 21 So have no fear; I myself will provide for you and your little ones." In this way he reassured them, speaking kindly to them.

22 So Joseph remained in Egypt, he and his father's household; and Joseph lived one hundred ten years. 23 Joseph saw Ephraim's children of the third generation; the children of Machir son of Manasseh were also born on Joseph's knees.

24 Then Joseph said to his brothers, "I am about to die; but God will surely come to you, and bring you up out of this land to the land that he swore to Abraham, to Isaac, and to Jacob." 25 So Joseph made the Israelites swear, saying, "When God comes to you, you shall carry up my bones from here." 26 And Joseph died, being one hundred ten years old; he was embalmed and placed in a coffin in Egypt.

p Gk Syr: Heb *they commanded*   *q* Cn: Heb *also came*

---

cortege detoured to Transjordan (vv. 10–11), though a main road from Egypt led along the coast to Beer-sheba.

50.18: A recapitulation of the motif introduced at the beginning of the Joseph story (see 37.5–11 n.). **19–20:** The heart and climax of the Joseph story: Joseph asserts that only God can forgive and heal human guilt, and he testifies to God's overruling providence that has already turned evil purposes to a good end

(45.4–7). **23:** The children of *Machir,* Joseph's grandson, were *born on Joseph's knees,* i.e. adopted as his descendants. Machir was the ancestor of a warlike clan of Manasseh that laid claim to Gilead (Num 32.39–40; Deut 3.15; Judg 5.14). **24:** An anticipation of the Exodus, based on the promise to Israel's ancestors. **25–26:** See Ex 13.19. According to tradition, Joseph was buried in Shechem (Josh 24.32; compare Gen 33.19; Acts 7.16).

Exodus

In this history of Israel's religious traditions, the two crucial "root experiences" were the exodus from Egypt and the revelation at Mount Sinai. The book of Exodus bears witness to the meaning of these seminal experiences: God's action to liberate a band of slaves from bondage and to make them a community, bound in covenant with their liberating God.

Although Egyptian records make no reference to the border incident of the flight of slaves into the Sinaitic wilderness, there can be little doubt that the story rests upon actual historical occurrences. Various lines of evidence point to the period of the 19th Dynasty (about 1350–1200 B.C.) as the most probable historical setting (see Ex 1.8 n.). The story unfolds against the background of Egyptian imperialism that motivated ambitious pharaohs to use Hebrew slaves as pawns in their scheme of world politics.

The book of Exodus discloses an editorial interweaving of traditions (Old Epic and Priestly) which preserve both the original Mosaic tradition and the interpretations of subsequent generations (see introduction to "The Pentateuch," pp. xxxv–xxxvi OT). The dramatic presentation falls into two major sections: (1) Israel's deliverance from Egyptian bondage and the pilgrimage to Sinai (chs 1–18) and (2) Israel's sojourn at Sinai, where the covenant was made and laws governing life and worship were promulgated (chs 19–40).

At the center of these memorable experiences stood Moses, in whose name the final edition of the whole account was issued. He was called to be the prophetic interpreter of God's liberating action and the priestly mediator of the covenant between God and people. Across the whole evolving tradition falls his massive shadow. Indeed, it was Moses who laid down the spiritual foundations upon which later generations built. Great religious reforms, such as those that occurred in the time of Elijah (1 Kings 19) or of King Josiah (2 Kings 22–23), were regarded as a return to the Mosaic source of Israel's faith.

1 These are the names of the sons of Israel who came to Egypt with Jacob, each with his household: 2Reuben, Simeon, Levi, and Judah, 3Issachar, Zebulun, and Benjamin, 4Dan and Naphtali, Gad and Asher. 5The total number of people born to Jacob was seventy. Joseph was already in Egypt. 6Then Joseph died, and all his brothers, and that whole generation. 7But the Israelites were fruitful and prolific; they multiplied and grew exceedingly strong, so that the land was filled with them.

8 Now a new king arose over Egypt, who did not know Joseph. 9He said to his people, "Look, the Israelite people are more numerous and more powerful than we. 10Come, let us deal shrewdly with them, or they will increase and, in the event of war, join our enemies and fight against us and escape from the land." 11Therefore they set taskmasters over them to oppress them with forced labor. They built supply cities, Pithom and Rameses, for Pharaoh. 12But the more they were oppressed, the more they multiplied and spread, so that the Egyptians came to dread the Israelites. 13The Egyptians became ruthless in imposing tasks on the Israelites, 14and made their lives bitter with hard service in mortar and brick and in every kind of field labor.

They were ruthless in all the tasks that they imposed on them.

15 The king of Egypt said to the Hebrew midwives, one of whom was named Shiphrah and the other Puah, 16"When you act as midwives to the Hebrew women, and see them on the birthstool, if it is a boy, kill him; but if it is a girl, she shall live." 17But the midwives feared God; they did not do as the king of Egypt commanded them, but they let the boys live. 18So the king of Egypt summoned the midwives and said to them, "Why have you done this, and allowed the boys to live?" 19The midwives said to Pharaoh, "Because the Hebrew women are not like the Egyptian women; for they are vigorous and give birth before the midwife comes to them." 20So God dealt well with the midwives; and the people multiplied and became very strong. 21And because the midwives feared God, he gave them families. 22Then Pharaoh commanded all his people, "Every boy that is born to the Hebrews*a* you shall throw into the Nile, but you shall let every girl live."

2 Now a man from the house of Levi went and married a Levite woman. 2The woman conceived and bore a son;

a Sam Gk Tg: Heb lacks *to the Hebrews*

1.1–22: Israel's bondage in Egypt. In spite of oppression, Abraham and Sarah's descendants multiplied and prospered, in fulfillment of the divine promise (Gen 12.2; 15.5). **1–7:** Gen 35.23–26; 50.26. **5:** *Total number of people . . . was seventy,* Gen 46.8–27; Deut 10.22. The book of Exodus reflects the memory of decisive events with which Israel as a people identified itself in faith. The tribal confederacy was formed later and embraced tribes that had not been in Egypt (Josh ch 24). **6:** Over four centuries elapsed since Joseph's death (12.40; compare Gen 15.13). **7:** The promise concerning Abraham's numerous posterity was being fulfilled (Gen 17.1–8; see Ex 12.37 n.). *The land,* see Gen 45.10 n.; 47.11 n.

1.8: Probably the allusion is to the new regime at the beginning of the 19th Dynasty under Seti I (1308–1290 B.C.) and Rameses II (1290–1224 B.C.). Hoping to regain Egypt's lost Asiatic empire, the pharaohs moved their capital from Thebes, where it had been during the 18th Dynasty, to the Delta. **9–10:** The presence of the Hebrews on Egypt's frontier was regarded as a security risk. **11:** *Supply cities,* an allusion to the fortification of the area. The new capital, *Rameses* (Zoan; Ps 78.12), was the former Hyksos capital (Avaris or Tanis) of Joseph's time (see Gen 45.10 n.; Num 13.22). As in the case of the pyramids, the work was carried out with the corvée, or forced labor gangs (compare 1 Kings 5.13).

1.15: *Hebrew,* an older and broader term than Israelite (see Gen 10.21–31 n.), was often used when foreigners spoke to or about Abraham's people (Gen 39.14, 17; 40.15). **19:** See vv. 7, 12.

2.1–22: The infancy and early career of Moses. 1: It was probably the Joseph tribes that took part in the Exodus, although elements of the tribe of Levi were also in Egypt. **2–10:** Aspects of this story are paralleled in

and when she saw that he was a fine baby, she hid him three months. ³When she could hide him no longer she got a papyrus basket for him, and plastered it with bitumen and pitch; she put the child in it and placed it among the reeds on the bank of the river. ⁴His sister stood at a distance, to see what would happen to him.

5 The daughter of Pharaoh came down to bathe at the river, while her attendants walked beside the river. She saw the basket among the reeds and sent her maid to bring it. ⁶When she opened it, she saw the child. He was crying, and she took pity on him, "This must be one of the Hebrews' children," she said. ⁷Then his sister said to Pharaoh's daughter, "Shall I go and get you a nurse from the Hebrew women to nurse the child for you?" ⁸Pharaoh's daughter said to her, "Yes." So the girl went and called the child's mother. ⁹Pharaoh's daughter said to her, "Take this child and nurse it for me, and I will give you your wages." So the woman took the child and nursed it. ¹⁰When the child grew up, she brought him to Pharaoh's daughter, and she took him as her son. She named him Moses,ᵇ "because," she said, "I drew him out*c* of the water."

11 One day, after Moses had grown up, he went out to his people and saw their forced labor. He saw an Egyptian beating a Hebrew, one of his kinsfolk. ¹²He looked this way and that, and seeing no one he killed the Egyptian and hid

him in the sand. ¹³When he went out the next day, he saw two Hebrews fighting; and he said to the one who was in the wrong, "Why do you strike your fellow Hebrew?" ¹⁴He answered, "Who made you a ruler and judge over us? Do you mean to kill me as you killed the Egyptian?" Then Moses was afraid and thought, "Surely the thing is known." ¹⁵When Pharaoh heard of it, he sought to kill Moses.

But Moses fled from Pharaoh. He settled in the land of Midian, and sat down by a well. ¹⁶The priest of Midian had seven daughters. They came to draw water, and filled the troughs to water their father's flock. ¹⁷But some shepherds came and drove them away. Moses got up and came to their defense and watered their flock. ¹⁸When they returned to their father Reuel, he said, "How is it that you have come back so soon today?" ¹⁹They said, "An Egyptian helped us against the shepherds; he even drew water for us and watered the flock." ²⁰He said to his daughters, "Where is he? Why did you leave the man? Invite him to break bread." ²¹Moses agreed to stay with the man, and he gave Moses his daughter Zipporah in marriage. ²²She bore a son, and he named him Gershom; for he said, "I have been an alienᵈ residing in a foreign land."

23 After a long time the king of Egypt died. The Israelites groaned under their

b Heb *Mosheh*   *c* Heb *mashah*   *d* Heb *ger*

---

the legends of other national heroes, e.g. Sargon of Agade (about 2600 B.C.) who in infancy was saved from danger by being put in a basket of rushes sealed with pitch and floated on the river. **4:** Moses' sister was Miriam (15.20; Num 26.59). **10:** The name *Moses,* from an Egyptian word meaning "to beget a child" and perhaps once joined with the name of an Egyptian deity (compare the name Thut-mose), is here explained by a Hebrew verb meaning "to draw out." The narrator sees divine providence at work, causing the evil design of Pharaoh to serve God's purpose. **2.11–14:** In spite of his Egyptian upbring-

ing Moses identified himself with his people (Heb 11.24–25). **15:** The Midianites (or Kenites, Judg 1.16) were distant blood relatives of Israel (Gen 25.2). **18:** "The priest of Midian" (v. 16) is usually called either Jethro (3.1; 4.18; 18.1) or Hobab (Num 10.29; Judg 4.11). *Reuel* was apparently his father (Num 10.29).

2.23–4.17: **The call of Moses.** In Midian the God of Israel's ancestors appeared to Moses and summoned him to take the lead in delivering Israel.

2.23: *The king* was probably Seti I (see 1.8 n.). The Israelites hoped that their condition would improve under the new regime, but Rameses II continued the oppressive

slavery, and cried out. Out of the slavery their cry for help rose up to God. 24 God heard their groaning, and God remembered his covenant with Abraham, Isaac, and Jacob. 25 God looked upon the Israelites, and God took notice of them.

3 Moses was keeping the flock of his father-in-law Jethro, the priest of Midian; he led his flock beyond the wilderness, and came to Horeb, the mountain of God. 2 There the angel of the LORD appeared to him in a flame of fire out of a bush; he looked, and the bush was blazing, yet it was not consumed. 3 Then Moses said, "I must turn aside and look at this great sight, and see why the bush is not burned up." 4 When the LORD saw that he had turned aside to see, God called to him out of the bush, "Moses, Moses!" And he said, "Here I am." 5 Then he said, "Come no closer! Remove the sandals from your feet, for the place on which you are standing is holy ground." 6 He said further, "I am the God of your father, the God of Abraham, the God of Isaac, and the God of Jacob." And Moses hid his face, for he was afraid to look at God.

7 Then the LORD said, "I have observed the misery of my people who are in Egypt; I have heard their cry on account of their taskmasters. Indeed, I know their sufferings, 8 and I have come down to deliver them from the Egyptians, and to bring them up out of that land to a good and broad land, a land flowing with milk and honey, to the country of the Canaanites, the Hittites, the Amorites, the Perizzites, the Hivites, and the Jebusites. 9 The cry of the Israelites has now come to me; I have also seen how the Egyptians oppress them. 10 So come, I will send you to Pharaoh to bring my people, the Israelites, out of Egypt." 11 But Moses said to God, "Who am I that I should go to Pharaoh, and bring the Israelites out of Egypt?" 12 He said, "I will be with you; and this shall be the sign for you that it is I who sent you: when you have brought the people out of Egypt, you shall worship God on this mountain."

13 But Moses said to God, "If I come to the Israelites and say to them, 'The God of your ancestors has sent me to you,' and they ask me, 'What is his name?' what shall I say to them?" 14 God

---

building program. **24:** Concerning the covenant with Israel's ancestors, see Gen 12.1–3; 17.1–14; 26.2–5.

3.1–6: The theophany of the bush. **1:** *The mountain of God,* called both Horeb and Sinai, was probably a Midianite sacred place (see v. 5 n.). Its location is unknown, but tradition places it in the eastern part of the Sinaitic Peninsula. **2:** *The angel of the LORD,* see Gen 16.7 n. *Fire* was conceived to be the form of the divine appearance (Gen 15.17; Ex 19.18; Ps 104.3–4; Ezek 1.27). **5:** Moses unexpectedly found himself in a holy place (see v. 1 n.; Gen 28.16–17). The removal of sandals before entering a holy place was an ancient custom (Josh 5.15). **6:** *The God of your father,* see Gen 26.24 n. The vision of God veiled in fire aroused dread (33.20), for divine holiness was experienced as a mysterious power that threatened human existence (19.10–13).

3.7–12: The divine commission. **8:** Canaan was *a land flowing with milk and honey,* foods that made it a paradise in the eyes of semi-nomads. On the pre-Israelite peoples, see Gen 10.15–20; Num 13.29. **11–12:** The

first of Moses' four objections (v. 13; 4.1, 10). God's word will be confirmed by a *sign* (compare Isa 7.10–17), i.e. the return of Israel to Sinai for worship. A sign may be an extraordinary wonder (4.1–9) or an ordinary phenomenon. What makes it significant, and therefore miraculous, is that God's presence and power are disclosed to the eyes of faith.

3.13–15: Moses' second question assumes a polytheistic environment; thus he must know the identity of the God who is dealing with him. On the *name,* see Gen 32.27 n. **14:** I AM WHO I AM is an etymology of the cultic name for the God of Israel, YHWH, probably pronounced Yahweh. (The NRSV, following ancient synagogue practice, substitutes "the LORD"; see To the Reader, p. xiii). YHWH is treated as a verbal form derived from "to be" and formulated in the first person because God is the speaker. Actually YHWH is a third person form and may mean "He causes to be." The name does not indicate God's eternal being but God's action and presence in historical affairs. **15:** The name is here introduced for the first time (6.2–3;

said to Moses, "I AM WHO I AM."*e* He said further, "Thus you shall say to the Israelites, 'I AM has sent me to you.'" ¹⁵God also said to Moses, "Thus you shall say to the Israelites, 'The LORD,*f* the God of your ancestors, the God of Abraham, the God of Isaac, and the God of Jacob, has sent me to you':

This is my name forever,
and this my title for all
generations.

¹⁶Go and assemble the elders of Israel, and say to them, 'The LORD, the God of your ancestors, the God of Abraham, of Isaac, and of Jacob, has appeared to me, saying: I have given heed to you and to what has been done to you in Egypt. ¹⁷I declare that I will bring you up out of the misery of Egypt, to the land of the Canaanites, the Hittites, the Amorites, the Perizzites, the Hivites, and the Jebusites, a land flowing with milk and honey.' ¹⁸They will listen to your voice; and you and the elders of Israel shall go to the king of Egypt and say to him, 'The LORD, the God of the Hebrews, has met with us; let us now go a three days' journey into the wilderness, so that we may sacrifice to the LORD our God.' ¹⁹I know, however, that the king of Egypt will not let you go unless compelled by a mighty hand.*g* ²⁰So I will stretch out my hand and strike Egypt with all my wonders that I will perform in it; after that he will let you go. ²¹I will bring this people into such favor with the Egyptians that, when you go, you will not go empty-handed; ²²each woman shall ask her neighbor and any woman living in the neighbor's house for jewelry of silver and of gold, and clothing, and you shall put them on your sons and on your daughters; and so you shall plunder the Egyptians."

4 Then Moses answered, "But suppose they do not believe me or listen to me, but say, 'The LORD did not appear to you.'" ²The LORD said to him, "What is that in your hand?" He said, "A staff." ³And he said, "Throw it on the ground." So he threw the staff on the ground, and it became a snake; and Moses drew back from it. ⁴Then the LORD said to Moses, "Reach out your hand, and seize it by the tail"—so he reached out his hand and grasped it, and it became a staff in his hand— ⁵"so that they may believe that the LORD, the God of their ancestors, the God of Abraham, the God of Isaac, and the God of Jacob, has appeared to you."

6 Again, the LORD said to him, "Put your hand inside your cloak." He put his hand into his cloak; and when he took it out, his hand was leprous,*h* as white as snow. ⁷Then God said, "Put your hand back into your cloak"—so he put his hand back into his cloak, and when he took it out, it was restored like the rest of his body— ⁸"If they will not believe you or heed the first sign, they may believe the second sign. ⁹If they will not believe even these two signs or heed you, you shall take some water from the Nile and pour it on the dry ground; and the water that you shall take from the Nile will become blood on the dry ground."

10 But Moses said to the LORD, "O my Lord, I have never been eloquent, neither in the past nor even now that you have spoken to your servant; but I am slow of speech and slow of tongue." ¹¹Then the LORD said to him, "Who gives speech to mortals? Who makes them mute or deaf, seeing or

e Or *I AM WHAT I AM* or *I WILL BE WHAT I WILL BE*
f The word "LORD" when spelled with capital letters stands for the divine name, *YHWH,* which is here connected with the verb *hayah,* "to be"   *g* Gk Vg: Heb *no, not by a mighty hand*   *h* A term for several skin diseases; precise meaning uncertain

---

compare Gen 4.26b n.). **21–22:** See 11.2–3; 12.35–36.

4.1–9: This narrative reflects superstitious magic which flourished in Egypt and claims that Moses was given power to excel in these "secret arts" (7.11; 8.18–19; 9.11). **3:** 7.8–12.

Serpent magic was practiced in Egypt from ancient times. The sign was the reverse of a trick whereby a snake is made rigid by hypnotism, so that it can be picked up by the tail. **4.10–17:** Aaron is designated as Moses' aide. **11:** Jer 1.6. In Hebraic thought human

blind? Is it not I, the LORD? 12 Now go, and I will be with your mouth and teach you what you are to speak." 13 But he said, "O my Lord, please send someone else." 14 Then the anger of the LORD was kindled against Moses and he said, "What of your brother Aaron, the Levite? I know that he can speak fluently; even now he is coming out to meet you, and when he sees you his heart will be glad. 15 You shall speak to him and put the words in his mouth; and I will be with your mouth and with his mouth, and will teach you what you shall do. 16 He indeed shall speak for you to the people; he shall serve as a mouth for you, and you shall serve as God for him. 17 Take in your hand this staff, with which you shall perform the signs."

18 Moses went back to his father-in-law Jethro and said to him, "Please let me go back to my kindred in Egypt and see whether they are still living." And Jethro said to Moses, "Go in peace." 19 The LORD said to Moses in Midian, "Go back to Egypt; for all those who were seeking your life are dead." 20 So Moses took his wife and his sons, put them on a donkey and went back to the land of Egypt; and Moses carried the staff of God in his hand.

21 And the LORD said to Moses, "When you go back to Egypt, see that you perform before Pharaoh all the wonders that I have put in your power; but

I will harden his heart, so that he will not let the people go. 22 Then you shall say to Pharaoh, 'Thus says the LORD: Israel is my firstborn son. 23 I said to you, "Let my son go that he may worship me." But you refused to let him go; now I will kill your firstborn son.' "

24 On the way, at a place where they spent the night, the LORD met him and tried to kill him. 25 But Zipporah took a flint and cut off her son's foreskin, and touched Moses'*i* feet with it, and said, "Truly you are a bridegroom of blood to me!" 26 So he let him alone. It was then she said, "A bridegroom of blood by circumcision."

27 The LORD said to Aaron, "Go into the wilderness to meet Moses." So he went; and he met him at the mountain of God and kissed him. 28 Moses told Aaron all the words of the LORD with which he had sent him, and all the signs with which he had charged him. 29 Then Moses and Aaron went and assembled all the elders of the Israelites. 30 Aaron spoke all the words that the LORD had spoken to Moses, and performed the signs in the sight of the people. 31 The people believed; and when they heard that the LORD had given heed to the Israelites and that he had seen their misery, they bowed down and worshiped.

5 Afterward Moses and Aaron went to Pharaoh and said, "Thus says the

i Heb *his*

---

conditions were not ascribed to secondary causes but to God whose will is sovereign in all things (Deut 32.39). **16:** The relation between God and God's prophetic spokesperson is analogous to the relation between Moses and Aaron (7.1; compare 16.9). **4.18–31: Moses returns to Egypt to arouse the faith of his people. 20:** Only one of Moses' sons has been mentioned so far (2.22); see 18.3–4. **21:** Even Pharaoh's stubbornness, which paradoxically was the expression of his own free will (8.15, 32; 9.34), was foreknown (3.19) and foreordained by God, thus indicating divine sovereignty in historical affairs (compare Isa 6.10). **22:** *Israel* (the people) is the LORD's *firstborn son* among the nations, a pre-eminent rank based upon divine adoption or election (Jer 31.9; Hos 11.1). **23:** 11.5; 12.29–34. **24–26:** An archaic

tradition that traces the origin of circumcision (compare Gen 17.9–14) to the Midianite wife of Moses. **4.24:** This verse reflects ancient belief in demonic attack (see Gen 38.7 n.), warded off by the timely performance of the rite. Originally circumcision was a puberty or marriage rite; *bridegroom of blood* (v. 26) is perhaps an old expression for a young man who was circumcised before marriage. **25:** Here it is assumed that the circumcision of the infant son was efficacious for Moses, who was evidently uncircumcised. *Feet,* a euphemism for the genitals (Isa 7.20). **27:** *The mountain of God* (3.1), which Moses had already left (4.20). **5.1–6.1: The first audience with Pharaoh fails. 1:** The petition is for a leave of absence to make a pilgrimage to the sacred mountain (v. 3; 3.18). **2:** The contemptuous

LORD, the God of Israel, 'Let my people go, so that they may celebrate a festival to me in the wilderness.' " 2But Pharaoh said, "Who is the LORD, that I should heed him and let Israel go? I do not know the LORD, and I will not let Israel go." 3Then they said, "The God of the Hebrews has revealed himself to us; let us go a three days' journey into the wilderness to sacrifice to the LORD our God, or he will fall upon us with pestilence or sword." 4But the king of Egypt said to them, "Moses and Aaron, why are you taking the people away from their work? Get to your labors!" 5Pharaoh continued, "Now they are more numerous than the people of the land *j* and yet you want them to stop working!" 6That same day Pharaoh commanded the taskmasters of the people, as well as their supervisors, 7"You shall no longer give the people straw to make bricks, as before; let them go and gather straw for themselves. 8But you shall require of them the same quantity of bricks as they have made previously; do not diminish it, for they are lazy; that is why they cry, 'Let us go and offer sacrifice to our God.' 9Let heavier work be laid on them; then they will labor at it and pay no attention to deceptive words."

10 So the taskmasters and the supervisors of the people went out and said to the people, "Thus says Pharaoh, 'I will not give you straw. 11Go and get straw yourselves, wherever you can find it; but your work will not be lessened in the least.' " 12So the people scattered throughout the land of Egypt, to gather stubble for straw. 13The taskmasters were urgent, saying, "Complete your work, the same daily assignment as when you were given straw." 14And the supervisors of the Israelites, whom Pharaoh's taskmasters had set over them, were beaten, and were asked, "Why did you not finish the required quantity of bricks yesterday and today, as you did before?"

15 Then the Israelite supervisors came to Pharaoh and cried, "Why do you treat your servants like this? 16No straw is given to your servants, yet they say to us, 'Make bricks!' Look how your servants are beaten! You are unjust to your own people."*k* 17He said, "You are lazy, lazy; that is why you say, 'Let us go and sacrifice to the LORD.' 18Go now, and work; for no straw shall be given you, but you shall still deliver the same number of bricks." 19The Israelite supervisors saw that they were in trouble when they were told, "You shall not lessen your daily number of bricks." 20As they left Pharaoh, they came upon Moses and Aaron who were waiting to meet them. 21They said to them, "The LORD look upon you and judge! You have brought us into bad odor with Pharaoh and his officials, and have put a sword in their hand to kill us."

22 Then Moses turned again to the LORD and said, "O LORD, why have you mistreated this people? Why did you ever send me? 23Since I first came to Pharaoh to speak in your name, he has mistreated this people, and you have done nothing at all to deliver your people."

6 Then the LORD said to Moses, "Now you shall see what I will do to Pharaoh: Indeed, by a mighty hand he will let them go; by a mighty hand he will drive them out of his land."

2 God also spoke to Moses and said to him: "I am the LORD. 3I appeared to

---

j Sam: Heb *The people of the land are now many*
k Gk Compare Syr Vg: Heb *beaten, and the sin of your people*

Pharaoh, whose absolute power was enforced by his deification in Egyptian religion, knew many gods; but the LORD (YHWH) was unheard of and a request made in the name of this deity carried no authority. **9:** Compare 1 Kings 12.1–11. Pharaoh treats the request for a three-day journey as *deceptive words,* i.e. a ruse to leave the country permanently.

5.16: The supervisors diplomatically suggest that the fault lies with Pharaoh's subordinates, i.e. *your own people.* **21:** *The LORD . . . judge,* i.e. give Moses and Aaron due justice for worsening their plight

6.2–7.7: The call of Moses and the appointment of Aaron are recapitulated in a priestly version (compare 3.1–4.17).

Abraham, Isaac, and Jacob as God Almighty,[l] but by my name 'The Lord'[m] I did not make myself known to them. 4 I also established my covenant with them, to give them the land of Canaan, the land in which they resided as aliens. 5 I have also heard the groaning of the Israelites whom the Egyptians are holding as slaves, and I have remembered my covenant. 6 Say therefore to the Israelites, 'I am the Lord, and I will free you from the burdens of the Egyptians and deliver you from slavery to them. I will redeem you with an outstretched arm and with mighty acts of judgment. 7 I will take you as my people, and I will be your God. You shall know that I am the Lord your God, who has freed you from the burdens of the Egyptians. 8 I will bring you into the land that I swore to give to Abraham, Isaac, and Jacob; I will give it to you for a possession. I am the Lord.'" 9 Moses told this to the Israelites; but they would not listen to Moses, because of their broken spirit and their cruel slavery.

10 Then the Lord spoke to Moses, 11 "Go and tell Pharaoh king of Egypt to let the Israelites go out of his land." 12 But Moses spoke to the Lord, "The Israelites have not listened to me; how then shall Pharaoh listen to me, poor speaker that I am?"[n] 13 Thus the Lord spoke to Moses and Aaron, and gave them orders regarding the Israelites and Pharaoh king of Egypt, charging them to free the Israelites from the land of Egypt.

14 The following are the heads of their ancestral houses: the sons of Reuben, the firstborn of Israel: Hanoch, Pallu, Hezron, and Carmi; these are the families of Reuben. 15 The sons of Simeon: Jemuel, Jamin, Ohad, Jachin, Zohar, and Shaul,[o] the son of a Canaanite woman; these are the families of Simeon.

16 The following are the names of the sons of Levi according to their genealogies: Gershon,[p] Kohath, and Merari, and the length of Levi's life was one hundred thirty-seven years. 17 The sons of Gershon:[p] Libni and Shimei, by their families. 18 The sons of Kohath: Amram, Izhar, Hebron, and Uzziel, and the length of Kohath's life was one hundred thirty-three years. 19 The sons of Merari: Mahli and Mushi. These are the families of the Levites according to their genealogies. 20 Amram married Jochebed his father's sister and she bore him Aaron and Moses, and the length of Amram's life was one hundred thirty-seven years. 21 The sons of Izhar: Korah, Nepheg, and Zichri. 22 The sons of Uzziel: Mishael, Elzaphan, and Sithri. 23 Aaron married Elisheba, daughter of Amminadab and sister of Nahshon, and she bore him Nadab, Abihu, Eleazar, and Ithamar. 24 The sons of Korah: Assir, Elkanah, and Abiasaph; these are the families of the Korahites. 25 Aaron's son Eleazar married one of the daughters of Putiel, and she bore him Phinehas. These are the heads of the ancestral houses of the Levites by their families.

26 It was this same Aaron and Moses to whom the Lord said, "Bring the Israelites out of the land of Egypt, company by company." 27 It was they who spoke to Pharaoh king of Egypt to bring the Israelites out of Egypt, the same Moses and Aaron.

28 On the day when the Lord spoke to Moses in the land of Egypt, 29 he said to him, "I am the Lord; tell Pharaoh king of Egypt all that I am speaking to you."

l Traditional rendering of Heb *El Shaddai* *m* Heb *YHWH*; see note at 3.15 *n* Heb *me? I am uncircumcised of lips* *o* Or *Saul* *p* Also spelled *Gershom*; see 2.22

6.2–3: See 3.14 n. The Lord (YHWH), the same God who led Israel's ancestors, was formerly known by another name, El Shaddai (see Gen 17.1 n.). 4: The *covenant* is the Abrahamic covenant (Gen ch 17), which guaranteed possession of the land of Canaan (v. 8). 7: *My people . . . your God,* a succinct statement of the meaning of the covenant relation (Hos 2.23; Jer 31.33). 6.12: See 4.10–16. The answer to Moses' question, interrupted by vv. 14–27, is found in v. 28 to 7.2.

30 But Moses said in the LORD's presence, "Since I am a poor speaker, *q* why would Pharaoh listen to me?"

7 The LORD said to Moses, "See, I have made you like God to Pharaoh, and your brother Aaron shall be your prophet. 2 You shall speak all that I command you, and your brother Aaron shall tell Pharaoh to let the Israelites go out of his land. 3 But I will harden Pharaoh's heart, and I will multiply my signs and wonders in the land of Egypt. 4 When Pharaoh does not listen to you, I will lay my hand upon Egypt and bring my people the Israelites, company by company, out of the land of Egypt by great acts of judgment. 5 The Egyptians shall know that I am the LORD, when I stretch out my hand against Egypt and bring the Israelites out from among them." 6 Moses and Aaron did so; they did just as the LORD commanded them. 7 Moses was eighty years old and Aaron eighty-three when they spoke to Pharaoh.

8 The LORD said to Moses and Aaron, 9 "When Pharaoh says to you, 'Perform a wonder,' then you shall say to Aaron, 'Take your staff and throw it down before Pharaoh, and it will become a snake.'" 10 So Moses and Aaron went to Pharaoh and did as the LORD had commanded; Aaron threw down his staff before Pharaoh and his officials, and it became a snake. 11 Then Pharaoh summoned the wise men and the sorcerers; and they also, the magicians of Egypt, did the same by their secret arts.

12 Each one threw down his staff, and they became snakes; but Aaron's staff swallowed up theirs. 13 Still Pharaoh's heart was hardened, and he would not listen to them, as the LORD had said.

14 Then the LORD said to Moses, "Pharaoh's heart is hardened; he refuses to let the people go. 15 Go to Pharaoh in the morning, as he is going out to the water; stand by at the river bank to meet him, and take in your hand the staff that was turned into a snake. 16 Say to him, 'The LORD, the God of the Hebrews, sent me to you to say, "Let my people go, so that they may worship me in the wilderness." But until now you have not listened.' 17 Thus says the LORD, "By this you shall know that I am the LORD." See, with the staff that is in my hand I will strike the water that is in the Nile, and it shall be turned to blood. 18 The fish in the river shall die, the river itself shall stink, and the Egyptians shall be unable to drink water from the Nile.'" 19 The LORD said to Moses, "Say to Aaron, 'Take your staff and stretch out your hand over the waters of Egypt—over its rivers, its canals, and its ponds, and all its pools of water—so that they may become blood; and there shall be blood throughout the whole land of Egypt, even in vessels of wood and in vessels of stone.'"

20 Moses and Aaron did just as the LORD commanded. In the sight of Phar-

q Heb *am uncircumcised of lips*; see 6.12

6.14–27: The purpose of the genealogy of Aaron and Moses is to trace priestly lineage from Levi, Jacob's son, to Aaron and through Aaron's third son, Eleazar, to Phinehas (Num ch 3). **14–16:** Gen 46.8–11. *Ancestral houses,* see Num 1.2 4 n. **28–30:** Continuation from v. 12.

7.1: See 4.16 n. **2:** See 5.1 where both Moses and Aaron are to speak to the king; here the role of Aaron, the priest, is stressed. **3–4:** See 4.21 n. The *mighty acts of judgment* (6.6) are the signs and wonders.

7.8–11.10: **The ten plagues** (Pss 78.44–51; 105.28–36).

7.8–13: The preface to the contest with Pharaoh is drawn from priestly tradition

(compare 4.1–5). Here Aaron is the chief actor and his rod—not Moses'—is turned into a snake.

7.14–24: **First plague:** the pollution of the Nile. **15:** Moses was to meet Pharaoh *at the river bank* with a challenge to Egyptian existence; for the Nile, believed to have its source in the subterranean ocean (see Gen 2.6 n.), was the source of life and fertility. **17–18:** The plague of blood apparently reflects a natural phenomenon of Egypt: namely, the reddish color of the Nile at its height in the summer owing to red particles of earth or perhaps minute organisms. **19:** The tendency to enhance the tradition is seen in the facts that here the rod is Aaron's (see 4.14; 7.2) and that all

aoh and of his officials he lifted up the staff and struck the water in the river, and all the water in the river was turned into blood, 21 and the fish in the river died. The river stank so that the Egyptians could not drink its water, and there was blood throughout the whole land of Egypt. 22 But the magicians of Egypt did the same by their secret arts; so Pharaoh's heart remained hardened, and he would not listen to them; as the LORD had said. 23 Pharaoh turned and went into his house, and he did not take even this to heart. 24 And all the Egyptians had to dig along the Nile for water to drink, for they could not drink the water of the river.

25 Seven days passed after the LORD had struck the Nile.

8 *r* Then the LORD said to Moses, "Go to Pharaoh and say to him, 'Thus says the LORD: Let my people go, so that they may worship me. 2 If you refuse to let them go, I will plague your whole country with frogs. 3 The river shall swarm with frogs; they shall come up into your palace, into your bedchamber and your bed, and into the houses of your officials and of your people, *s* and into your ovens and your kneading bowls. 4 The frogs shall come up on you and on your people and on all your officials.' " 5 *t* And the LORD said to Moses, "Say to Aaron, 'Stretch out your hand with your staff over the rivers, the canals, and the pools, and make frogs come up on the land of Egypt.' " 6 So Aaron stretched out his hand over the waters of Egypt; and the frogs came up and covered the land of Egypt. 7 But the magicians did the same by their secret arts,

and brought frogs up on the land of Egypt.

8 Then Pharaoh called Moses and Aaron, and said, "Pray to the LORD to take away the frogs from me and my people, and I will let the people go to sacrifice to the LORD." 9 Moses said to Pharaoh, "Kindly tell me when I am to pray for you and for your officials and for your people, that the frogs may be removed from you and your houses and be left only in the Nile." 10 And he said, "Tomorrow." Moses said, "As you say! So that you may know that there is no one like the LORD our God, 11 the frogs shall leave you and your houses and your officials and your people; they shall be left only in the Nile." 12 Then Moses and Aaron went out from Pharaoh; and Moses cried out to the LORD concerning the frogs that he had brought upon Pharaoh. *u* 13 And the LORD did as Moses requested: the frogs died in the houses, the courtyards, and the fields. 14 And they gathered them together in heaps, and the land stank. 15 But when Pharaoh saw that there was a respite, he hardened his heart, and would not listen to them, just as the LORD had said.

16 Then the LORD said to Moses, "Say to Aaron, 'Stretch out your staff and strike the dust of the earth, so that it may become gnats throughout the whole land of Egypt.' " 17 And they did so; Aaron stretched out his hand with his staff and struck the dust of the earth, and gnats came on humans and animals alike;

r Ch 7.26 in Heb   *s* Gk: Heb *upon your people*
t Ch 8.1 in Heb   *u* Or *frogs, as he had agreed*
with Pharaoh

the water of Egypt is said to have been polluted. **22**: See v. 11.

7.25–8.15: Second plague: frogs.

8.3: The mud of the Nile, after the seasonal overflowing, was a natural place for frogs to generate. Egypt has been spared more frequent occurrence of this pestilence by the frog-eating bird, the ibis. **5–7**: Another unit of tradition that extols Aaron the priest (7.19).

8.8: *Pray to the LORD,* for the first time

Pharaoh momentarily recognizes Israel's God. **9–11**: To enhance the wonder, Moses promises to pray that the scourge cease at a designated time.

8.16–32: Third and fourth plagues: gnats and flies. **16**: From ancient times stinging gnats or mosquitoes have plagued Egypt, especially in the autumn. The Nile, receding from its overflow, leaves stagnant pools of water in which the insects breed. On Aaron's rod, see 7.19; 8.5–7. **19**: This time the magi-

all the dust of the earth turned into gnats throughout the whole land of Egypt. [18]The magicians tried to produce gnats by their secret arts, but they could not. There were gnats on both humans and animals. [19]And the magicians said to Pharaoh, "This is the finger of God!" But Pharaoh's heart was hardened, and he would not listen to them, just as the LORD had said.

20 Then the LORD said to Moses, "Rise early in the morning and present yourself before Pharaoh, as he goes out to the water, and say to him, 'Thus says the LORD: Let my people go, so that they may worship me. [21]For if you will not let my people go, I will send swarms of flies on you, your officials, and your people, and into your houses; and the houses of the Egyptians shall be filled with swarms of flies; so also the land where they live. [22]But on that day I will set apart the land of Goshen, where my people live, so that no swarms of flies shall be there, that you may know that I the LORD am in this land. [23]Thus I will make a distinction[v] between my people and your people. This sign shall appear tomorrow.' " [24]The LORD did so, and great swarms of flies came into the house of Pharaoh and into his officials' houses; in all of Egypt the land was ruined because of the flies.

25 Then Pharaoh summoned Moses and Aaron, and said, "Go, sacrifice to your God within the land." [26]But Moses said, "It would not be right to do so; for the sacrifices that we offer to the LORD our God are offensive to the Egyptians. If we offer in the sight of the Egyptians sacrifices that are offensive to them, will they not stone us? [27]We must go a three days' journey into the wilderness and sacrifice to the LORD our God as he commands us." [28]So Pharaoh said, "I will let you go to sacrifice to the LORD your God in the wilderness, provided you do not go very far away. Pray for me." [29]Then Moses said, "As soon as I leave you, I will pray to the LORD that the swarms of flies may depart tomorrow from Pharaoh, from his officials, and from his people; only do not let Pharaoh again deal falsely by not letting the people go to sacrifice to the LORD."

30 So Moses went out from Pharaoh and prayed to the LORD. [31]And the LORD did as Moses asked: he removed the swarms of flies from Pharaoh, from his officials, and from his people; not one remained. [32]But Pharaoh hardened his heart this time also, and would not let the people go.

9 Then the LORD said to Moses, "Go to Pharaoh, and say to him, 'Thus says the LORD, the God of the Hebrews: Let my people go, so that they may worship me. [2]For if you refuse to let them go and still hold them, [3]the hand of the LORD will strike with a deadly pestilence your livestock in the field: the horses, the donkeys, the camels, the herds, and the flocks. [4]But the LORD will make a distinction between the livestock of Israel and the livestock of Egypt, so that nothing shall die of all that belongs to the Israelites.' " [5]The LORD set a time, saying, "Tomorrow the LORD will do this thing in the land." [6]And on the next day the LORD did so; all the livestock of the Egyptians died, but of the livestock of the Israelites not one died. [7]Pharaoh inquired and found that not one of the live-

v Gk Vg: Heb *will set redemption*

---

cians are unable to match the feat and confess that *this is the finger of God* (31.18; Ps 8.3; Lk 11.20).

8.20–32: The fourth plague is probably a variant of vv. 16–19. **22–23:** The sign was not just the coming of myriads of flies but the isolation of Goshen so that the scourge did not affect the Hebrews. **26:** Egypt had strong taboos against the religious practices of foreigners (Gen 43.32). **28:** Pharaoh's further concession—permission to go just beyond the border—reflects his suspicion of Moses' intention (5.9).

9.1–12: Fifth and sixth plagues: cattle plague and boils. **3:** This plague, perhaps anthrax, seems to have resulted from conditions created by former plagues: disease spread by mosquitoes or flies. **4:** See 8.22–23 n. **5:** This time Pharaoh was given twenty-four hours' notice.

stock of the Israelites was dead. But the heart of Pharaoh was hardened, and he would not let the people go.

8 Then the LORD said to Moses and Aaron, "Take handfuls of soot from the kiln, and let Moses throw it in the air in the sight of Pharaoh. 9 It shall become fine dust all over the land of Egypt, and shall cause festering boils on humans and animals throughout the whole land of Egypt." 10 So they took soot from the kiln, and stood before Pharaoh, and Moses threw it in the air, and it caused festering boils on humans and animals. 11 The magicians could not stand before Moses because of the boils, for the boils afflicted the magicians as well as all the Egyptians. 12 But the LORD hardened the heart of Pharaoh, and he would not listen to them, just as the LORD had spoken to Moses.

13 Then the LORD said to Moses, "Rise up early in the morning and present yourself before Pharaoh, and say to him, 'Thus says the LORD, the God of the Hebrews: Let my people go, so that they may worship me. 14 For this time I will send all my plagues upon you yourself, and upon your officials, and upon your people, so that you may know that there is no one like me in all the earth. 15 For by now I could have stretched out my hand and struck you and your people with pestilence, and you would have been cut off from the earth. 16 But this is why I have let you live: to show you my power, and to make my name resound through all the earth. 17 You are still exalting yourself against my people, and will not let them go. 18 Tomorrow at this time I will cause the heaviest hail to fall that has ever fallen in Egypt from the day it was founded until now. 19 Send, therefore, and have your livestock and everything that you have in the open field brought to a secure place; every human or animal that is in the open field and is not brought under shelter will die when the hail comes down upon them.' " 20 Those officials of Pharaoh who feared the word of the LORD hurried their slaves and livestock off to a secure place. 21 Those who did not regard the word of the LORD left their slaves and livestock in the open field.

22 The LORD said to Moses, "Stretch out your hand toward heaven so that hail may fall on the whole land of Egypt, on humans and animals and all the plants of the field in the land of Egypt." 23 Then Moses stretched out his staff toward heaven, and the LORD sent thunder and hail, and fire came down on the earth. And the LORD rained hail on the land of Egypt; 24 there was hail with fire flashing continually in the midst of it, such heavy hail as had never fallen in all the land of Egypt since it became a nation. 25 The hail struck down everything that was in the open field throughout all the land of Egypt, both human and animal; the hail also struck down all the plants of the field, and shattered every tree in the field. 26 Only in the land of Goshen, where the Israelites were, there was no hail.

27 Then Pharaoh summoned Moses and Aaron, and said to them, "This time I have sinned; the LORD is in the right, and I and my people are in the wrong. 28 Pray to the LORD! Enough of God's

---

9.8–12: The sixth plague, boils or a similar skin outbreak (Deut 28.27), is parallel to the previous one, so far as the cattle are concerned.

9.13–35: **Seventh plague:** hail and thunderstorm. 14: *All my plagues upon you yourself* is not clear but may refer to unleashing the full fury of heaven: thunder, hail, rain, and lightning (vv. 23, 33). 15–16: It is explained that the ineffectiveness of the plagues up to this point is not due to the LORD's weakness but to patient determination to demonstrate divine sovereignty (compare Rom 9.17). 26: See vv. 6–7; 8.22–23 n.

9.27: For the first time Pharaoh confesses that he is beaten, hoping to appease the foreign deity without making further concessions. 29: The plagues bear witness to the fact that *the earth is the LORD's* (19.5; Ps 24.1), for the powers of nature serve God's purpose. 31–32: This parenthetical remark explains why there were still plants for the locusts to eat during the next plague and, incidentally, dates the seventh plague around the middle of

thunder and hail! I will let you go; you need stay no longer." 29 Moses said to him, "As soon as I have gone out of the city, I will stretch out my hands to the LORD; the thunder will cease, and there will be no more hail, so that you may know that the earth is the LORD's. 30 But as for you and your officials, I know that you do not yet fear the LORD God." 31 (Now the flax and the barley were ruined, for the barley was in the ear and the flax was in bud. 32 But the wheat and the spelt were not ruined, for they are late in coming up.) 33 So Moses left Pharaoh, went out of the city, and stretched out his hands to the LORD; then the thunder and the hail ceased, and the rain no longer poured down on the earth. 34 But when Pharaoh saw that the rain and the hail and the thunder had ceased, he sinned once more and hardened his heart, he and his officials. 35 So the heart of Pharaoh was hardened, and he would not let the Israelites go, just as the LORD had spoken through Moses.

10 Then the LORD said to Moses, "Go to Pharaoh; for I have hardened his heart and the heart of his officials, in order that I may show these signs of mine among them, 2 and that you may tell your children and grandchildren how I have made fools of the Egyptians and what signs I have done among them—so that you may know that I am the LORD."

3 So Moses and Aaron went to Pharaoh, and said to him, "Thus says the LORD, the God of the Hebrews, 'How long will you refuse to humble yourself before me? Let my people go, so that they may worship me. 4 For if you refuse to let my people go, tomorrow I will bring locusts into your country. 5 They shall cover the surface of the land, so that no one will be able to see the land. They

shall devour the last remnant left you after the hail, and they shall devour every tree of yours that grows in the field. 6 They shall fill your houses, and the houses of all your officials and of all the Egyptians—something that neither your parents nor your grandparents have seen, from the day they came on earth to this day.'" Then he turned and went out from Pharaoh.

7 Pharaoh's officials said to him, "How long shall this fellow be a snare to us? Let the people go, so that they may worship the LORD their God; do you not yet understand that Egypt is ruined?" 8 So Moses and Aaron were brought back to Pharaoh, and he said to them, "Go, worship the LORD your God! But which ones are to go?" 9 Moses said, "We will go with our young and our old; we will go with our sons and daughters and with our flocks and herds, because we have the LORD's festival to celebrate." 10 He said to them, "The LORD indeed will be with you, if ever I let your little ones go with you! Plainly, you have some evil purpose in mind. 11 No, never! Your men may go and worship the LORD, for that is what you are asking." And they were driven out from Pharaoh's presence.

12 Then the LORD said to Moses, "Stretch out your hand over the land of Egypt, so that the locusts may come upon it and eat every plant in the land, all that the hail has left." 13 So Moses stretched out his staff over the land of Egypt, and the LORD brought an east wind upon the land all that day and all that night; when morning came, the east wind had brought the locusts. 14 The locusts came upon all the land of Egypt and settled on the whole country of Egypt, such a dense swarm of locusts as had

---

January, when the crops begin to mature. **10.1–20: Eighth plague:** locusts. **4–6:** Clouds of locusts were a familiar pestilence to farmers in the ancient Near East (see Joel). **10.7–11:** At the insistence of his impressed courtiers, Pharaoh attempts to negotiate before the twenty-four hour deadline (9.5); he

offers to let only the men go, since adult males took part in the religious rites (23.17; 34.23; Deut 16.16). **13:** The miracle rests upon a natural phenomenon (14.21; Num 11.31); an *east wind* brought the locusts and, when the wind shifted, a *west wind* drove them into the Red Sea (v. 19).

never been before, nor ever shall be again. [15]They covered the surface of the whole land, so that the land was black; and they ate all the plants in the land and all the fruit of the trees that the hail had left; nothing green was left, no tree, no plant in the field, in all the land of Egypt. [16]Pharaoh hurriedly summoned Moses and Aaron and said, "I have sinned against the LORD your God, and against you. [17]Do forgive my sin just this once, and pray to the LORD your God that at the least he remove this deadly thing from me." [18]So he went out from Pharaoh and prayed to the LORD. [19]The LORD changed the wind into a very strong west wind, which lifted the locusts and drove them into the Red Sea;[w] not a single locust was left in all the country of Egypt. [20]But the LORD hardened Pharaoh's heart, and he would not let the Israelites go.

21 Then the LORD said to Moses, "Stretch out your hand toward heaven so that there may be darkness over the land of Egypt, a darkness that can be felt." [22]So Moses stretched out his hand toward heaven, and there was dense darkness in all the land of Egypt for three days. [23]People could not see one another, and for three days they could not move from where they were; but all the Israelites had light where they lived. [24]Then Pharaoh summoned Moses, and said, "Go, worship the LORD. Only your flocks and your herds shall remain behind. Even your children may go with you." [25]But Moses said, "You must also let us have sacrifices and burnt offerings to sacrifice to the LORD our God. [26]Our livestock also must go with us; not a hoof shall be left behind, for we must choose some of them for the worship of the LORD our God, and we will not know what to use to worship the LORD until we

arrive there." [27]But the LORD hardened Pharaoh's heart, and he was unwilling to let them go. [28]Then Pharaoh said to him, "Get away from me! Take care that you do not see my face again, for on the day you see my face you shall die." [29]Moses said, "Just as you say! I will never see your face again."

11 The LORD said to Moses, "I will bring one more plague upon Pharaoh and upon Egypt; afterwards he will let you go from here; indeed, when he lets you go, he will drive you away. [2]Tell the people that every man is to ask his neighbor and every woman is to ask her neighbor for objects of silver and gold." [3]The LORD gave the people favor in the sight of the Egyptians. Moreover, Moses himself was a man of great importance in the land of Egypt, in the sight of Pharaoh's officials and in the sight of the people.

4 Moses said, "Thus says the LORD: About midnight I will go out through Egypt. [5]Every firstborn in the land of Egypt shall die, from the firstborn of Pharaoh who sits on his throne to the firstborn of the female slave who is behind the handmill, and all the firstborn of the livestock. [6]Then there will be a loud cry throughout the whole land of Egypt, such as has never been or will ever be again. [7]But not a dog shall growl at any of the Israelites—not at people, not at animals—so that you may know that the LORD makes a distinction between Egypt and Israel. [8]Then all these officials of yours shall come down to me, and bow low to me, saying, 'Leave us, you and all the people who follow you.' After that I will leave." And in hot anger he left Pharaoh.

9 The LORD said to Moses, "Pharaoh

w Or Sea of Reeds

10.21–29: Ninth plague: dense darkness. **21:** *A darkness that can be felt* aptly describes conditions created by the hot wind, the "khamsin," which blows in from the desert during the spring (March–May), bringing with it so much dust and sand that the air is darkened and breathing becomes difficult.

23b: Compare 8.22–23; 9.6–7, 26.

11.1–10: The announcement of the final plague, the death of the firstborn (concluded in 12.29–32). **2–3:** The "spoliation of the Egyptians" (3.21–22; 12.35–36) is explained as evidence of the favor which Moses and the Israelites had gained in Egypt.

will not listen to you, in order that my wonders may be multiplied in the land of Egypt." 10 Moses and Aaron performed all these wonders before Pharaoh; but the LORD hardened Pharaoh's heart, and he did not let the people of Israel go out of his land.

12 The LORD said to Moses and Aaron in the land of Egypt: 2 This month shall mark for you the beginning of months; it shall be the first month of the year for you. 3 Tell the whole congregation of Israel that on the tenth of this month they are to take a lamb for each family, a lamb for each household. 4 If a household is too small for a whole lamb, it shall join its closest neighbor in obtaining one; the lamb shall be divided in proportion to the number of people who eat of it. 5 Your lamb shall be without blemish, a year-old male; you may take it from the sheep or from the goats. 6 You shall keep it until the fourteenth day of this month; then the whole assembled congregation of Israel shall slaughter it at twilight. 7 They shall take some of the blood and put it on the two doorposts and the lintel of the houses in which they eat it. 8 They shall eat the lamb that same night; they shall eat it roasted over the fire with unleavened bread and bitter herbs. 9 Do not eat any of it raw or boiled in water, but roasted over the fire, with its head, legs, and inner organs. 10 You shall let none of it remain until the morning; anything that remains until the morning you shall burn. 11 This is how you shall eat it: your loins girded, your sandals on your feet, and your staff in your hand; and you shall eat it hurriedly. It is the passover of the LORD. 12 For I will pass through the land of Egypt that night, and I will strike down every firstborn in the land of Egypt, both human beings and animals; on all the gods of Egypt I will execute judgments: I am the LORD. 13 The blood shall be a sign for you on the houses where you live: when I see the blood, I will pass over you, and no plague shall destroy you when I strike the land of Egypt.

14 This day shall be a day of remembrance for you. You shall celebrate it as a festival to the LORD; throughout your generations you shall observe it as a perpetual ordinance. 15 Seven days you shall eat unleavened bread; on the first day you shall remove leaven from your houses, for whoever eats leavened bread from the first day until the seventh day shall be cut off from Israel. 16 On the first day you shall hold a solemn assembly, and on the seventh day a solemn assembly; no work shall be done on those days;

12.1–28: The festivals of passover and unleavened bread. 1–13 (and vv. 43–49): This is priestly tradition concerning the passover, an ancient nomadic spring festival which Israel reinterpreted as a memorial of the LORD's deliverance of the people from Egypt (Deut 16.1–8; Num 9.1–14; Ezek 45.21–25). **2:** *This month* refers to Nisan (March-April) which in the post-exilic ecclesiastical calendar was *the beginning of months* (see Lev 23.5, 23–25 n.). According to the older agricultural calendar, the new year began in the autumn (Ex 23.16; 34.22). **3–4:** Priestly tradition assumes that Israel in Egypt was already an organized *congregation* under the leadership of tribal princes (16.22). *Household,* see Num 1.2–4 n. The passover was a nocturnal festival, celebrated during full moon (v. 8; see Isa 30.29). **7:** Blood, regarded as the Deity's portion of the sacrifice (Lev 1.5), was smeared on the doorposts and the lintel, the holy places of the house (21.6; Deut 6.9), as a protection against the destroyer (vv. 22–23; see 4.24 n.). **11:** The feast must be eaten in readiness for the march, in commemoration of Israel's hasty exodus. **11–13:** Here *passover* is interpreted from a verb meaning "to pass over," referring to the LORD's passing over Israelite houses during the plague of the firstborn (vv. 24–27). **12.14–20:** The festival of unleavened bread, originally an agricultural festival held at the time of barley harvest, was also converted into an historical commemoration and came to be closely connected with the passover (Deut 16.1–8; Ezek 45.21–25). **14:** The passover was celebrated on the 14th of Nisan (v. 6); *this day* refers to the 15th (Lev 23.6; Num 28.17). The seven day festival is regarded as a continuation of the passover. **15:** The absence of leaven (yeast) is interpreted as due to hasty preparations for flight (vv. 34, 39; Deut 16.3). Originally leaven, owing to its fermenting or corrupting power (23.18; Mt 16.6; 1 Cor 5.7), was regarded as a ritually

only what everyone must eat, that alone may be prepared by you. 17 You shall observe the festival of unleavened bread, for on this very day I brought your companies out of the land of Egypt: you shall observe this day throughout your generations as a perpetual ordinance. 18 In the first month, from the evening of the fourteenth day until the evening of the twenty-first day, you shall eat unleavened bread. 19 For seven days no leaven shall be found in your houses; for whoever eats what is leavened shall be cut off from the congregation of Israel, whether an alien or a native of the land. 20 You shall eat nothing leavened; in all your settlements you shall eat unleavened bread.

21 Then Moses called all the elders of Israel and said to them, "Go, select lambs for your families, and slaughter the passover lamb. 22 Take a bunch of hyssop, dip it in the blood that is in the basin, and touch the lintel and the two doorposts with the blood in the basin. None of you shall go outside the door of your house until morning. 23 For the LORD will pass through to strike down the Egyptians; when he sees the blood on the lintel and on the two doorposts, the LORD will pass over that door and will not allow the destroyer to enter your houses to strike you down. 24 You shall observe this rite as a perpetual ordinance for you and your children. 25 When you come to the land that the LORD will give you, as he has promised, you shall keep this observance. 26 And when your children ask you, 'What do you mean by this observance?' 27 you shall say, 'It is the passover sacrifice to the LORD, for he passed over the houses of the Israelites in Egypt, when he struck down the Egyptians but

spared our houses.'" And the people bowed down and worshiped.

28 The Israelites went and did just as the LORD had commanded Moses and Aaron.

29 At midnight the LORD struck down all the firstborn in the land of Egypt, from the firstborn of Pharaoh who sat on his throne to the firstborn of the prisoner who was in the dungeon, and all the firstborn of the livestock. 30 Pharaoh arose in the night, he and all his officials and all the Egyptians; and there was a loud cry in Egypt, for there was not a house without someone dead. 31 Then he summoned Moses and Aaron in the night, and said, "Rise up, go away from my people, both you and the Israelites! Go, worship the LORD, as you said. 32 Take your flocks and your herds, as you said, and be gone. And bring a blessing on me too!"

33 The Egyptians urged the people to hasten their departure from the land, for they said, "We shall all be dead." 34 So the people took their dough before it was leavened, with their kneading bowls wrapped up in their cloaks on their shoulders. 35 The Israelites had done as Moses told them; they had asked the Egyptians for jewelry of silver and gold, and for clothing, 36 and the LORD had given the people favor in the sight of the Egyptians, so that they let them have what they asked. And so they plundered the Egyptians.

37 The Israelites journeyed from Rameses to Succoth, about six hundred thousand men on foot, besides children. 38 A mixed crowd also went up with them, and livestock in great numbers, both flocks and herds. 39 They baked unleavened cakes of the dough that they

unclean substance (compare Lev 2.11) which could contaminate the whole harvest. **18**: So closely is the festival combined with the passover that it is said to begin on the evening of the 14th, i.e. the night of the passover (see v. 14).

12.21–28: An older tradition concerning the passover. **22**: See v. 7 n. *Hyssop,* the fo-

liage of an aromatic plant. Because of its presumed magical powers, it was used for ritual purposes (Lev 14.4; Num 19.6, 18; Ps 51.7). **23**: *The destroyer,* or the angel of death (2 Sam 24.16; Isa 37.36), was regarded as a manifestation of the LORD's power.

12.29–51: **Israel's departure from Egypt. 29–32**: The conclusion of the tenth

had brought out of Egypt; it was not leavened, because they were driven out of Egypt and could not wait, nor had they prepared any provisions for themselves.

40 The time that the Israelites had lived in Egypt was four hundred thirty years. 41 At the end of four hundred thirty years, on that very day, all the companies of the LORD went out from the land of Egypt. 42 That was for the LORD a night of vigil, to bring them out of the land of Egypt. That same night is a vigil to be kept for the LORD by all the Israelites throughout their generations.

43 The LORD said to Moses and Aaron: This is the ordinance for the passover: no foreigner shall eat of it, 44 but any slave who has been purchased may eat of it after he has been circumcised; 45 no bound or hired servant may eat of it. 46 It shall be eaten in one house; you shall not take any of the animal outside the house, and you shall not break any of its bones. 47 The whole congregation of Israel shall celebrate it. 48 If an alien who resides with you wants to celebrate the passover to the LORD, all his males shall be circumcised; then he may draw near to celebrate it; he shall be regarded as a native of the land. But no uncircumcised person shall eat of it; 49 there shall be one

law for the native and for the alien who resides among you.

50 All the Israelites did just as the LORD had commanded Moses and Aaron. 51 That very day the LORD brought the Israelites out of the land of Egypt, company by company.

13 The LORD said to Moses: 2 Consecrate to me all the firstborn; whatever is the first to open the womb among the Israelites, of human beings and animals, is mine.

3 Moses said to the people, "Remember this day on which you came out of Egypt, out of the house of slavery, because the LORD brought you out from there by strength of hand; no leavened bread shall be eaten. 4 Today, in the month of Abib, you are going out. 5 When the LORD brings you into the land of the Canaanites, the Hittites, the Amorites, the Hivites, and the Jebusites, which he swore to your ancestors to give you, a land flowing with milk and honey, you shall keep this observance in this month. 6 Seven days you shall eat unleavened bread, and on the seventh day there shall be a festival to the LORD. 7 Unleavened bread shall be eaten for seven days; no leavened bread shall be seen in your possession, and no leaven shall be seen among you in all your territory. 8 You

---

plague (11.1–10). **33–34:** See v. 15 n. **35–36:** See 3.21–22 and 11.2–3.

12.37: Rameses (1.11) and Succoth (13.20) were the starting places on Israel's itinerary (Num 33.5). *Six hundred thousand men on foot* (Num 11.21), in addition to women and children, is an exaggeration, for neither the land of Goshen nor the southern Palestinian wilderness could have supported so large a population (at least two and a half million). The number apparently reflects the census list in Num 1.17–46. The *mixed crowd* (Num 11.4) included other "Hebrews" (see 1.15 n.) or rootless people. **40:** If the four hundred and thirty years (see Gen 15.13; Acts 7.6 n.; Gal 3.17 n.) covers the total time of the Egyptian sojourn, then the descent into Egypt would have coincided with the Hyksos invasion (about 1720 B.C.; see Gen 45.10 n.) and the Exodus occurred during the reign of Rameses II, about 1290 B.C. (see 1.8 n.). **42:** The *night of vigil* refers to the passover.

12.43–49: A supplement to the priestly tradition about the passover (12.1–13). A *foreigner* (v. 43), a *bound or hired servant* (v. 45) are excluded on the ground that they are related to other gods; however, the purchased slave who becomes a part of the family (v. 44) and the alien who resides permanently within Israel may eat the passover, if the *one law* of circumcision is kept (Gen 17.9–14).

13.1–16: The consecration of the firstborn. 2: According to ancient belief, the devotion of the firstborn of humans and of beasts to God, the giver of fertility, was necessary for continuing increase and well-being (22.29b–30; Lev 27.26–27; Num 3.13; 8.17–18; 18.15). **3–10:** Old tradition about the festival of unleavened bread (compare the parallel priestly version, 12.14–20). **4:** *Abib,* the older name for the month of the Exodus (23.15; see 12.2 n.). **5:** See 3.8. **8:** In later times people could tell *what the LORD did for me when I came out of Egypt,* for in worship the redemptive

shall tell your child on that day, 'It is because of what the LORD did for me when I came out of Egypt.' ⁹It shall serve for you as a sign on your hand and as a reminder on your forehead, so that the teaching of the LORD may be on your lips; for with a strong hand the LORD brought you out of Egypt. ¹⁰You shall keep this ordinance at its proper time from year to year.

11 "When the LORD has brought you into the land of the Canaanites, as he swore to you and your ancestors, and has given it to you, ¹²you shall set apart to the LORD all that first opens the womb. All the firstborn of your livestock that are males shall be the LORD's. ¹³But every firstborn donkey you shall redeem with a sheep; if you do not redeem it, you must break its neck. Every firstborn male among your children you shall redeem. ¹⁴When in the future your child asks you, 'What does this mean?' you shall answer, 'By strength of hand the LORD brought us out of Egypt, from the house of slavery. ¹⁵When Pharaoh stubbornly refused to let us go, the LORD killed all the firstborn in the land of Egypt, from human firstborn to the firstborn of animals. Therefore I sacrifice to the LORD every male that first opens the womb, but every firstborn of my sons I redeem.' ¹⁶It shall serve as a sign on your hand and as an emblem* on your fore-

head that by strength of hand the LORD brought us out of Egypt."

17 When Pharaoh let the people go, God did not lead them by way of the land of the Philistines, although that was nearer; for God thought, "If the people face war, they may change their minds and return to Egypt." ¹⁸So God led the people by the roundabout way of the wilderness toward the Red Sea.ʸ The Israelites went up out of the land of Egypt prepared for battle. ¹⁹And Moses took with him the bones of Joseph who had required a solemn oath of the Israelites, saying, "God will surely take notice of you, and then you must carry my bones with you from here." ²⁰They set out from Succoth, and camped at Etham, on the edge of the wilderness. ²¹The LORD went in front of them in a pillar of cloud by day, to lead them along the way, and in a pillar of fire by night, to give them light, so that they might travel by day and by night. ²²Neither the pillar of cloud by day nor the pillar of fire by night left its place in front of the people.

14 Then the LORD said to Moses: ²Tell the Israelites to turn back and camp in front of Pi-hahiroth, between Migdol and the sea, in front of Baal-zephon; you shall camp opposite it,

---

x Or *as a frontlet*; Meaning of Heb uncertain
y Or *Sea of Reeds*

---

event was made present (12.26–27; see Deut 5.2–3 n.). **9**: See Deut 6.8.

13.11–16: An old tradition about the consecration of the firstborn. **13**: Unclean animals, of which the donkey is typical (Lev ch 11; Deut ch 14), may be redeemed by substituting a lamb. In early times the custom arose of substituting an animal for the human firstborn (34.19–20; compare Gen 22.13), although pagan human sacrifice persisted (1 Kings 16.34; 2 Kings 16.3; Ezek 20.26; Mic 6.7). **14–15**: The practice, rooted in ancient fertility beliefs, is here reinterpreted in the light of the Exodus.

13.17–14.31: **Israel's deliverance.**

13.17–18: *Philistines,* see Gen 21.34 n. The route mentioned was the main military road into Canaan. To avoid attack, the people were providentially led *by the roundabout way*

of the wilderness. On the *Red Sea,* see 14.2 n. **19**: See Gen 50.25–26 n. **21–22**: The *pillar of cloud* and the *pillar of fire* may reflect the ancient custom of carrying a burning brazier at the head of a marching army or caravan to indicate the line of march by day and night. Whatever the nature of the phenomenon originally, cloud and fire have become traditional ways of expressing God's presence and guidance (see 3.2 n.; 19.9; 33.9; 40.34–38; 1 Kings 8.10–11).

14.2: The places mentioned, like Etham (13.20), were probably Egyptian frontier fortresses. Apparently the Israelites were unable to break through and had to *turn back,* with the result that they were trapped (v. 3) between the water barrier and the Egyptian forces. *The sea,* known in Hebrew as the "Sea of Reeds," was not the Red Sea itself but a

by the sea. ³Pharaoh will say of the Israelites, 'They are wandering aimlessly in the land; the wilderness has closed in on them.' ⁴I will harden Pharaoh's heart, and he will pursue them, so that I will gain glory for myself over Pharaoh and all his army; and the Egyptians shall know that I am the LORD. And they did so.

5 When the king of Egypt was told that the people had fled, the minds of Pharaoh and his officials were changed toward the people, and they said, "What have we done, letting Israel leave our service?" ⁶So he had his chariot made ready, and took his army with him; ⁷he took six hundred picked chariots and all the other chariots of Egypt with officers over all of them. ⁸The LORD hardened the heart of Pharaoh king of Egypt and he pursued the Israelites, who were going out boldly. ⁹The Egyptians pursued them, all Pharaoh's horses and chariots, his chariot drivers and his army; they overtook them camped by the sea, by Pi-hahiroth, in front of Baal-zephon.

10 As Pharaoh drew near, the Israelites looked back, and there were the Egyptians advancing on them. In great fear the Israelites cried out to the LORD. ¹¹They said to Moses, "Was it because there were no graves in Egypt that you have taken us away to die in the wilderness? What have you done to us, bringing us out of Egypt? ¹²Is this not the very thing we told you in Egypt, 'Let us alone and let us serve the Egyptians'? For it would have been better for us to serve the Egyptians than to die in the wilderness." ¹³But Moses said to the people, "Do not be afraid, stand firm, and see the deliverance that the LORD will accomplish for you today; for the Egyptians whom you see today you shall never see again. ¹⁴The LORD will fight for you, and you have only to keep still."

15 Then the LORD said to Moses, "Why do you cry out to me? Tell the Israelites to go forward. ¹⁶But you lift up your staff, and stretch out your hand over the sea and divide it, that the Israelites may go into the sea on dry ground. ¹⁷Then I will harden the hearts of the Egyptians so that they will go in after them; and so I will gain glory for myself over Pharaoh and all his army, his chariots, and his chariot drivers. ¹⁸And the Egyptians shall know that I am the LORD, when I have gained glory for myself over Pharaoh, his chariots, and his chariot drivers."

19 The angel of God who was going before the Israelite army moved and went behind them; and the pillar of cloud moved from in front of them and took its place behind them. ²⁰It came between the army of Egypt and the army of Israel. And so the cloud was there with the darkness, and it lit up the night; one did not come near the other all night.

21 Then Moses stretched out his hand over the sea. The LORD drove the sea back by a strong east wind all night, and turned the sea into dry land; and the waters were divided. ²²The Israelites went into the sea on dry ground, the waters forming a wall for them on their right and on their left. ²³The Egyptians pursued, and went into the sea after them, all of Pharaoh's horses, chariots, and chariot drivers. ²⁴At the morning watch the LORD in the pillar of fire and cloud looked down upon the Egyptian army, and threw the Egyptian army into panic.

---

shallow body of water farther north, perhaps in the area of Lake Timsah.
14.11–12: See 15.24 n. **13–14:** Viewed in faith, the victory was a mighty act of the LORD who was fighting for the people in a contest with the powerful Pharaoh (v. 25). *Deliverance,* see Gen 49.18.
14.19–20: One tradition expresses the divine presence as *the angel of God* (see Gen 16.7 n.), another as the shining pillar of cloud (v. 24; see 13.21–22 n.). **21–29:** The divine victory was rooted in a natural phenomenon: during a storm the shallow waters were driven back by *a strong east wind* (v. 21), making it possible for the Israelites to cross on foot. Egyptian chariots, however, were mired in the mud and engulfed by the returning waters. Tradition heightened the miracle by at-

25 He clogged[z] their chariot wheels so that they turned with difficulty. The Egyptians said, "Let us flee from the Israelites, for the LORD is fighting for them against Egypt."

26 Then the LORD said to Moses, "Stretch out your hand over the sea, so that the water may come back upon the Egyptians, upon their chariots and chariot drivers." 27 So Moses stretched out his hand over the sea, and at dawn the sea returned to its normal depth. As the Egyptians fled before it, the LORD tossed the Egyptians into the sea. 28 The waters returned and covered the chariots and the chariot drivers, the entire army of Pharaoh that had followed them into the sea; not one of them remained. 29 But the Israelites walked on dry ground through the sea, the waters forming a wall for them on their right and on their left.

30 Thus the LORD saved Israel that day from the Egyptians; and Israel saw the Egyptians dead on the seashore. 31 Israel saw the great work that the LORD did against the Egyptians. So the people feared the LORD and believed in the LORD and in his servant Moses.

15 Then Moses and the Israelites sang this song to the LORD:
"I will sing to the LORD, for he
has triumphed gloriously;
horse and rider he has thrown
into the sea.
2 The LORD is my strength and my
might,[a]
and he has become my
salvation;
this is my God, and I will praise
him,
my father's God, and I will
exalt him.
3 The LORD is a warrior;
the LORD is his name.

4 "Pharaoh's chariots and his army
he cast into the sea;
his picked officers were sunk in
the Red Sea.[b]
5 The floods covered them;
they went down into the depths
like a stone.
6 Your right hand, O LORD,
glorious in power—
your right hand, O LORD,
shattered the enemy.
7 In the greatness of your majesty
you overthrew your
adversaries;
you sent out your fury, it
consumed them like
stubble.
8 At the blast of your nostrils the
waters piled up,
the floods stood up in a heap;
the deeps congealed in the heart
of the sea.
9 The enemy said, 'I will pursue, I
will overtake,
I will divide the spoil, my desire
shall have its fill of them.
I will draw my sword, my hand
shall destroy them.'
10 You blew with your wind, the sea
covered them;
they sank like lead in the
mighty waters.

11 "Who is like you, O LORD, among
the gods?
Who is like you, majestic in
holiness,
awesome in splendor, doing
wonders?

---

z Sam Gk Syr: MT *removed*   a Or *song*
b Or *Sea of Reeds*

---

tributing it to Moses' wonder-working rod (vv. 16, 21a, 26–27) and by saying that the waters stood up like walls (vv. 22b, 29b).
15.1–21: Two songs of praise which celebrate the LORD's deliverance of the people. **1:** The song of Moses (vv. 1–18) is introduced by quoting the ancient song of Miriam (v. 21). **2:** See 14.12–14 n. *My father's God*

refers to "the God of the ancestors" (3.6). **3:** In this and the following verses, the poet uses the ancient Near Eastern metaphor of the Divine Warrior (Ps 24.8) to portray the LORD's saving action in behalf of Israel (14.14, 25). **4–10:** Recital of the Divine Warrior's victory at the Sea (Ps 78.12–13). **8–10:** The language is influenced by the ancient myth of a

12 You stretched out your right
  hand,
    the earth swallowed them.

13 "In your steadfast love you led the
    people whom you
    redeemed;
  you guided them by your
    strength to your holy
    abode.
14 The peoples heard, they trembled;
    pangs seized the inhabitants of
    Philistia.
15 Then the chiefs of Edom were
    dismayed;
  trembling seized the leaders of
    Moab;
  all the inhabitants of Canaan
    melted away.
16 Terror and dread fell upon them;
  by the might of your arm, they
    became still as a stone
  until your people, O Lord, passed
    by,
  until the people whom you
    acquired passed by.
17 You brought them in and planted
    them on the mountain of
    your own possession,
  the place, O Lord, that you
    made your abode,
  the sanctuary, O Lord, that
    your hands have
    established

18 The Lord will reign forever and
    ever."

19 When the horses of Pharaoh with
his chariots and his chariot drivers went
into the sea, the Lord brought back the
waters of the sea upon them; but the Israelites walked through the sea on dry
ground.

20 Then the prophet Miriam, Aaron's
sister, took a tambourine in her hand;
and all the women went out after her
with tambourines and with dancing.
21 And Miriam sang to them:

"Sing to the Lord, for he has
    triumphed gloriously;
  horse and rider he has thrown into
    the sea."

22 Then Moses ordered Israel to set
out from the Red Sea,*c* and they went
into the wilderness of Shur. They went
three days in the wilderness and found no
water. 23 When they came to Marah, they
could not drink the water of Marah because it was bitter. That is why it was
called Marah.*d* 24 And the people complained against Moses, saying, "What
shall we drink?" 25 He cried out to the
Lord; and the Lord showed him a piece
of wood;*e* he threw it into the water, and
the water became sweet.

There the Lord*f* made for them a stat-

---

c Or *Sea of Reeds*   *d* That is *Bitterness*
e Or *a tree*   *f* Heb *he*

---

divine battle against Sea, the chaotic power
hostile to God's rule (see Ps 77.16–19; 114.3–
6; Hab 3.8 note *l*).
  15.11: Another motif of ancient hymns: the
incomparability of the Lord *among the gods*
who compose the heavenly council (Pss 86.8·
89.7–8; Gen 1.26 n.). **13–17:** The guidance
into the land of Canaan. **13:** *Your holy abode,*
a mountain in Canaan (Ps 78.54). **14:** *Philistia*
was settled by the Philistines (Gen 21.32 n.)
about 1175 B.C.; hence the poem was written
afterwards, though probably before the monarchy. **15:** See Num 20.18–21; 21.13. **16:** *Acquired,* perhaps "created" (see Gen 4.1). **17:**
Canaan is described as the mythical cosmic
mountain, Zaphon, the location of God's
abode and *sanctuary* (see Ps 48.1–3 n.).
  15.19–21: In song and dance, Miriam and
her companions lead the people in praising

God for deliverance. Miriam (Num 26.59;
Mic 6.4), doubtless a dominant figure in early
tradition, is called a *prophet* because of her
ecstatic rousing of devotion to the Lord
(compare Judg 4.4). **21:** The Song of Miriam,
one of the oldest poetic couplets in the Old
Testament, may have been composed by an
eyewitness of the event.
  15.22–16.36: Crises in the wilderness. In
times of need, when faith was put to the test,
Israel perceived signs of the Lord's care and
protection.
  15.22: *The wilderness of Shur,* identified
with the wilderness of Etham in Num 33.8,
was on the border of Egypt. **24:** Israel's continual complaining in the wilderness is a dominant theme of the tradition (16.2–3; 17.3;
32.1–4, 25; Num 11.4–6; 12.1–2; 14.2–3;
16.13–14; 20.2–13; 21.4–5). **25:** It was be-

ute and an ordinance and there he put them to the test. 26 He said, "If you will listen carefully to the voice of the LORD your God, and do what is right in his sight, and give heed to his commandments and keep all his statutes, I will not bring upon you any of the diseases that I brought upon the Egyptians; for I am the LORD who heals you."

27 Then they came to Elim, where there were twelve springs of water and seventy palm trees; and they camped there by the water.

16 The whole congregation of the Israelites set out from Elim; and Israel came to the wilderness of Sin, which is between Elim and Sinai, on the fifteenth day of the second month after they had departed from the land of Egypt. 2 The whole congregation of the Israelites complained against Moses and Aaron in the wilderness. 3 The Israelites said to them, "If only we had died by the hand of the LORD in the land of Egypt, when we sat by the fleshpots and ate our fill of bread; for you have brought us out into this wilderness to kill this whole assembly with hunger."

4 Then the LORD said to Moses, "I am going to rain bread from heaven for you, and each day the people shall go out and gather enough for that day. In that way I will test them, whether they will follow my instruction or not. 5 On the sixth day, when they prepare what they bring in, it will be twice as much as they gather on other days." 6 So Moses and Aaron said to all the Israelites, "In the evening you shall know that it was the LORD who brought you out of the land of Egypt, 7 and in the morning you shall see the glory of the LORD, because he has heard your complaining against the LORD. For what are we, that you complain against us?" 8 And Moses said, "When the LORD gives you meat to eat in the evening and your fill of bread in the morning, because the LORD has heard the complaining that you utter against him—what are we? Your complaining is not against us but against the LORD."

9 Then Moses said to Aaron, "Say to the whole congregation of the Israelites, 'Draw near to the LORD, for he has heard your complaining.'" 10 And as Aaron spoke to the whole congregation of the Israelites, they looked toward the wilderness, and the glory of the LORD appeared in the cloud. 11 The LORD spoke to Moses and said, 12 "I have heard the complaining of the Israelites; say to them, 'At twilight you shall eat meat, and in the morning you shall have your fill of bread; then you shall know that I am the LORD your God.'"

13 In the evening quails came up and covered the camp; and in the morning there was a layer of dew around the camp. 14 When the layer of dew lifted, there on the surface of the wilderness was a fine flaky substance, as fine as frost on the ground. 15 When the Israelites saw it, they said to one another, "What is it?" *g* For they did not know what it was. Mo-

g Or *"It is manna"* (Heb *man hu,* see verse 31)

---

lieved that the leaves or bark of certain trees had magical properties for sweetening or "healing" water (2 Kings 2.21). **26:** *Diseases,* i.e. the Egyptian plagues. *Who heals you,* Num 21.4–9; Deut 7.15; Ps 103.3.

16.1–36: The provision of food in the wilderness. **1:** *The wilderness of Sin* (17.1; Num 33.11–12), probably on the Sinaitic Peninsula. **3:** The murmuring wanderers preferred the seasoned food of *the fleshpots* of Egypt to the precarious freedom of the wilderness. **4:** *Test* their faith by providing only a portion sufficient for one day (see Deut 8.3, 16; Mt 6.11). **5:** See vv. 22–30. **6–7:** *In the evening* when the quails come; *in the morning* when the manna is found (vv. 8, 12). In the priestly view, *the glory of the LORD* was an envelope of light (associated with the pillar of cloud and fire; see 13.21–22 n.) which veiled God's being. Though human beings could not see the Deity they could behold the glory that signified God's presence (40.34; Num 14.10b, 22; 16.19; Ezek 11.23).

16.9–10: *Draw near to the LORD,* see vv. 33–34 n. An early tradition concerning the provision of bread (v. 15). **13:** On the quails, see Num 11.1–35. **14:** The description here (see also v. 31 and Num 11.7–9) corresponds fairly closely to the "honey-dew" excretion of two scale-insects that feed on the twigs of the

ses said to them, "It is the bread that the LORD has given you to eat. 16 This is what the LORD has commanded: 'Gather as much of it as each of you needs, an omer to a person according to the number of persons, all providing for those in their own tents.'" 17 The Israelites did so, some gathering more, some less. 18 But when they measured it with an omer, those who gathered much had nothing over, and those who gathered little had no shortage; they gathered as much as each of them needed. 19 And Moses said to them, "Let no one leave any of it over until morning." 20 But they did not listen to Moses; some left part of it until morning, and it bred worms and became foul. And Moses was angry with them. 21 Morning by morning they gathered it, as much as each needed; but when the sun grew hot, it melted.

22 On the sixth day they gathered twice as much food, two omers apiece. When all the leaders of the congregation came and told Moses, 23 he said to them, "This is what the LORD has commanded: 'Tomorrow is a day of solemn rest, a holy sabbath to the LORD; bake what you want to bake and boil what you want to boil, and all that is left over put aside to be kept until morning.'" 24 So they put it aside until morning, as Moses commanded them; and it did not become foul, and there were no worms in it. 25 Moses said, "Eat it today, for today is a sabbath to the LORD; today you will not find it in the field. 26 Six days you shall gather it; but on the seventh day, which is a sabbath, there will be none."

27 On the seventh day some of the people went out to gather, and they found none. 28 The LORD said to Moses, "How long will you refuse to keep my commandments and instructions? 29 See!

The LORD has given you the sabbath, therefore on the sixth day he gives you food for two days; each of you stay where you are; do not leave your place on the seventh day." 30 So the people rested on the seventh day.

31 The house of Israel called it manna; it was like coriander seed, white, and the taste of it was like wafers made with honey. 32 Moses said, "This is what the LORD has commanded: 'Let an omer of it be kept throughout your generations, in order that they may see the food with which I fed you in the wilderness, when I brought you out of the land of Egypt.'" 33 And Moses said to Aaron, "Take a jar, and put an omer of manna in it, and place it before the LORD, to be kept throughout your generations." 34 As the LORD commanded Moses, so Aaron placed it before the covenant, *h* for safekeeping. 35 The Israelites ate manna forty years, until they came to a habitable land; they ate manna, until they came to the border of the land of Canaan. 36 An omer is a tenth of an ephah.

17 From the wilderness of Sin the whole congregation of the Israelites journeyed by stages, as the LORD commanded. They camped at Rephidim, but there was no water for the people to drink. 2 The people quarreled with Moses, and said, "Give us water to drink." Moses said to them, "Why do you quarrel with me? Why do you test the LORD?" 3 But the people thirsted there for water; and the people complained against Moses and said, "Why did you bring us out of Egypt, to kill us and our children and livestock with thirst?" 4 So Moses cried out to the LORD, "What shall I do with this people? They are almost

h Or *treaty* or *testimony*; Heb *eduth*

tamarisk tree. **15:** The name *manna* (v. 31), is explained by an expression meaning "What is it?" For persons of faith the answer was that the natural phenomenon was *bread that the LORD has given.* **22–36:** The provision of manna is the occasion for the insertion of priestly teaching on the sabbath, *a day of solemn rest* (31.15; 35.2). **33–34:** *Before the LORD,* i.e. before the ark, sometimes desig-

nated in priestly tradition by its chief contents, *the covenant* as represented by the tablets of law (27.21; Lev 16.13; Num 17.4). **17.1–16: Other trying experiences in the wilderness. 1–7:** Israel's thirst was quenched with water from the rock (compare Num 20.2–13). **1:** *By stages,* see Num 33.1–49. **2–3:** See 15.24 n. *Test the LORD,* i.e. demand proof that God was in their midst

ready to stone me." 5 The LORD said to Moses, "Go on ahead of the people, and take some of the elders of Israel with you; take in your hand the staff with which you struck the Nile, and go. 6 I will be standing there in front of you on the rock at Horeb. Strike the rock, and water will come out of it, so that the people may drink." Moses did so, in the sight of the elders of Israel. 7 He called the place Massah*i* and Meribah,*j* because the Israelites quarreled and tested the LORD, saying, "Is the LORD among us or not?"

8 Then Amalek came and fought with Israel at Rephidim. 9 Moses said to Joshua, "Choose some men for us and go out, fight with Amalek. Tomorrow I will stand on the top of the hill with the staff of God in my hand." 10 So Joshua did as Moses told him, and fought with Amalek, while Moses, Aaron, and Hur went up to the top of the hill. 11 Whenever Moses held up his hand, Israel prevailed; and whenever he lowered his hand, Amalek prevailed. 12 But Moses' hands grew weary; so they took a stone and put it under him, and he sat on it. Aaron and Hur held up his hands, one on one side, and the other on the other side; so his hands were steady until the sun set. 13 And Joshua defeated Amalek and his people with the sword.

14 Then the LORD said to Moses, "Write this as a reminder in a book and recite it in the hearing of Joshua: I will utterly blot out the remembrance of Amalek from under heaven." 15 And Moses built an altar and called it, The LORD is my banner. 16 He said, "A hand upon the banner of the LORD *k* The LORD will have war with Amalek from generation to generation."

18 Jethro, the priest of Midian, Moses' father-in-law, heard of all that God had done for Moses and for his people Israel, how the LORD had brought Israel out of Egypt. 2 After Moses had sent away his wife Zipporah, his father-in-law Jethro took her back, 3 along with her two sons. The name of the one was Gershom (for he said, "I have been an alien*l* in a foreign land"), 4 and the name of the other, Eliezer*m* (for he said, "The God of my father was my help, and delivered me from the sword of Pharaoh"). 5 Jethro, Moses' father-in-law, came into the wilderness where Moses was encamped at the mountain of God, bringing Moses' sons and wife to him. 6 He sent word to Moses, "I, your father-in-

---

i That is *Test*   *j* That is *Quarrel*
k Cn: Meaning of Heb uncertain   *l* Heb *ger*
m Heb *Eli,* my God; *ezer,* help

(v. 7b). **6:** Water lies below the limestone surface in the region of Sinai. **7:** The place is named both *Massah* from the Hebrew verb "test" and *Meribah* from the verb "find fault"—names that became memorials of Israel's faithlessness (Deut 6.16; 9.22; 33.8; Ps 95.8). Meribah was one of the springs at Kadesh (Num 20.13; 27.14; Deut 32.51). Marah (15.23) and Massah were evidently springs at the same oasis. Some traditions in 15.23–18.27 come from this oasis south of Beersheba (see Num 13.26 n.).

17.8–15: The battle with the Amalekites. **8:** The Amalekites, a fierce desert tribe, claimed control of the wilderness in the region of Kadesh (Gen 14.7; Num 13.29; 14.25). **9–13:** *Choose some men for us* implies holy war (v. 16) with a select group (compare Judg ch 7). The young warrior, Joshua, mentioned for the first time, headed the Israelite army. Moses, however, led the battle from a hilltop and ensured victory by the power of his rod and outstretched arms and perhaps by the power of the curse (Num 22.4–6). **10:** *Hur,* elsewhere mentioned only in 24.14. **14:** *Utterly blot out,* i.e. the foe will be subjected to the sacrificial ban, a practice of holy war. **16:** The bitter feud with Amalek persisted (Num 24.20; Deut 25.17–19; 1 Sam 15.7–8; 27.8; ch 30) until the foe was exterminated during the reign of Hezekiah (1 Chr 4.41–43).

18.1–27: Jethro's visit. The priest of Midian celebrated a sacred meal and counseled Moses about the administration of law. **1:** *Jethro,* see 2.18 n. **2–4:** Zipporah and her sons (2.21–22) apparently had been sent back from Egypt to Midian. **5:** The narrative is out of order, for Israel reached *the mountain of God* later (19.2). **9–12:** This passage may imply that the priest of Midian was already a worshiper of the LORD (see 3.1 n.). As the priest of the cult, Jethro came to rejoice in the LORD's great deeds and to officiate at a cultic

law Jethro, am coming to you, with your wife and her two sons." ⁷Moses went out to meet his father-in-law; he bowed down and kissed him; each asked after the other's welfare, and they went into the tent. ⁸Then Moses told his father-in-law all that the LORD had done to Pharaoh and to the Egyptians for Israel's sake, all the hardship that had beset them on the way, and how the LORD had delivered them. ⁹Jethro rejoiced for all the good that the LORD had done to Israel, in delivering them from the Egyptians.

10 Jethro said, "Blessed be the LORD, who has delivered you from the Egyptians and from Pharaoh. ¹¹Now I know that the LORD is greater than all gods, because he delivered the people from the Egyptians,ⁿ when they dealt arrogantly with them." ¹²And Jethro, Moses' father-in-law, brought a burnt offering and sacrifices to God; and Aaron came with all the elders of Israel to eat bread with Moses' father-in-law in the presence of God.

13 The next day Moses sat as judge for the people, while the people stood around him from morning until evening. ¹⁴When Moses' father-in-law saw all that he was doing for the people, he said, "What is this that you are doing for the people? Why do you sit alone, while all the people stand around you from morning until evening?" ¹⁵Moses said to his father-in-law, "Because the people come to me to inquire of God. ¹⁶When they have a dispute, they come to me and I decide between one person and another, and I make known to them the statutes and instructions of God." ¹⁷Moses' father-in-law said to him, "What you are doing is not good. ¹⁸You will surely wear yourself out, both you and these people with you. For the task is too heavy for you; you cannot do it alone. ¹⁹Now listen to me. I will give you counsel, and God be with you! You should represent the people before God, and you should bring their cases before God; ²⁰teach them the statutes and instructions and make known to them the way they are to go and the things they are to do. ²¹You should also look for able men among all the people, men who fear God, are trustworthy, and hate dishonest gain; set such men over them as officers over thousands, hundreds, fifties and tens. ²²Let them sit as judges for the people at all times; let them bring every important case to you, but decide every minor case themselves. So it will be easier for you, and they will bear the burden with you. ²³If you do this, and God so commands you, then you will be able to endure, and all these people will go to their home in peace."

24 So Moses listened to his father-in-law and did all that he had said. ²⁵Moses chose able men from all Israel and appointed them as heads over the people, as officers over thousands, hundreds, fifties, and tens. ²⁶And they judged the people at all times; hard cases they brought to Moses, but any minor case they decided themselves. ²⁷Then Moses let his father-in-law depart, and he went off to his own country.

19 On the third new moon after the Israelites had gone out of the land of Egypt, on that very day, they came into the wilderness of Sinai. ²They had

---

ⁿ The clause *because . . . Egyptians* has been transposed from verse 10

---

celebration. **12:** *Eat bread,* an allusion to a sacred meal *in the presence of God* (24.9–11). Moses was not invited, perhaps because he had already been initiated into the cult (3.1–6).

18.13–27: Jethro's plan for the reorganization of legal administration (compare Deut 1.9–18). **13:** Like a bedouin chief, Moses acted as judge in the people's disputes (2 Sam 15.1–6). **15–16:** *Inquire of God,* i.e. seek a verdict by oracle (Judg 4.4–5). **21–22:** Moses was to deal with cases without legal precedent which required a special oracle (compare Deut 17.8–13); ordinary cases were to be handled by lay leaders (Num 11.16–17, 24–25) or appointed judges (compare Deut 16.18–20). *Officers over thousands,* see Num 1.17–46 n.

19.1–25: The theophany at Sinai (20.18–21). At the sacred mountain the LORD offered to make a covenant with Israel. **2:** *Sinai,* see

journeyed from Rephidim, entered the wilderness of Sinai, and camped in the wilderness; Israel camped there in front of the mountain. ³Then Moses went up to God; the LORD called to him from the mountain, saying, "Thus you shall say to the house of Jacob, and tell the Israelites: ⁴You have seen what I did to the Egyptians, and how I bore you on eagles' wings and brought you to myself. ⁵Now therefore, if you obey my voice and keep my covenant, you shall be my treasured possession out of all the peoples. Indeed, the whole earth is mine, ⁶but you shall be for me a priestly kingdom and a holy nation. These are the words that you shall speak to the Israelites."

7 So Moses came, summoned the elders of the people, and set before them all these words that the LORD had commanded him. ⁸The people all answered as one: "Everything that the LORD has spoken we will do." Moses reported the words of the people to the LORD. ⁹Then the LORD said to Moses, "I am going to come to you in a dense cloud, in order that the people may hear when I speak with you and so trust you ever after."

When Moses had told the words of the people to the LORD, ¹⁰the LORD said to Moses: "Go to the people and consecrate them today and tomorrow. Have them wash their clothes ¹¹and prepare for the third day, because on the third day the LORD will come down upon Mount Sinai in the sight of all the people. ¹²You shall set limits for the people all around, saying, 'Be careful not to go up the mountain or to touch the edge of it. Any who touch the mountain shall be put to death. ¹³No hand shall touch them, but they shall be stoned or shot with arrows;ᵒ whether animal or human being, they shall not live.' When the trumpet sounds a long blast, they may go up on the mountain." ¹⁴So Moses went down from the mountain to the people. He consecrated the people, and they washed their clothes. ¹⁵And he said to the people, "Prepare for the third day; do not go near a woman."

16 On the morning of the third day there was thunder and lightning, as well as a thick cloud on the mountain, and a blast of a trumpet so loud that all the people who were in the camp trembled. ¹⁷Moses brought the people out of the camp to meet God. They took their stand at the foot of the mountain. ¹⁸Now Mount Sinai was wrapped in smoke, because the LORD had descended upon it in

o Heb lacks *with arrows*

3.1 n. **3:** The account assumes that the LORD's dwelling-place is in heaven, from which God "descended" to the mountain top for meeting with the people and their representatives (24.9–11). Compare the similar view reflected in the Babylonian temple-tower (Gen 11.1–9). **4–6:** The "eagles' wings" passage is formulated in liturgical style and employs the metaphor of an eagle carrying its young on its powerful wings (Deut 32.11–12). **5:** On Israel's side, the covenant rests upon a condition, *if you obey my voice*—an allusion to the covenant laws to be given. *My treasured possession:* a metaphor for God's special claim upon Israel. The God to whom all the earth belongs (Ex 9.29b; Ps 24.1) chose Israel for a special role as "the people of God" (Deut 7.6; 14.2; 26.18). **6:** That which is holy is set apart as belonging to the holy God; thus Israel is to be *a priestly kingdom,* consecrated for service to God (see Isa 61.6; 1 Pet 2.5, 9).

19.7–8: Compare 24.7. **9:** This tradition stresses Moses' role as the covenant mediator who represents God to the people and the people before God (20.19; 24.1–2, 9–11; Deut 5.2–27). **10–15:** In this tradition all the people are to prepare to take part in the covenant ceremony (24.3–8). **12:** The setting of bounds so that the people do not come near the mountain (v. 21) reflects the ancient view of holiness as a mysterious, threatening power with which the mountain is charged (see 3.6 n.; 2 Sam 6.6–9). No hand may touch the offender who has become affected with the contagion of holiness (Lev 6.27–28). **14–15:** Washing or changing of garments (Gen 35.2) and sexual abstinence (1 Sam 21.4–6) were forms of ceremonial purification.

19.16–19: The theophany is portrayed primarily in the imagery of a violent thunderstorm (Judg 5.4–5; Pss 18.7–15; 29.3–9; etc.). This traditional language—"earthquake, wind, and fire" (1 Kings 19.11–13)—depicts the wonder and majesty of God's revelation. **16:** The trumpet (v. 13) was sounded on cultic occasions (2 Sam 6.15).

fire; the smoke went up like the smoke of a kiln, while the whole mountain shook violently. 19 As the blast of the trumpet grew louder and louder, Moses would speak and God would answer him in thunder. 20 When the LORD descended upon Mount Sinai, to the top of the mountain, the LORD summoned Moses to the top of the mountain, and Moses went up. 21 Then the LORD said to Moses, "Go down and warn the people not to break through to the LORD to look; otherwise many of them will perish. 22 Even the priests who approach the LORD must consecrate themselves or the LORD will break out against them." 23 Moses said to the LORD, "The people are not permitted to come up to Mount Sinai; for you yourself warned us, saying, 'Set limits around the mountain and keep it holy.' " 24 The LORD said to him, "Go down, and come up bringing Aaron with you; but do not let either the priests or the people break through to come up to the LORD; otherwise he will break out against them." 25 So Moses went down to the people and told them.

20 Then God spoke all these words: 2 I am the LORD your God, who brought you out of the land of Egypt, out of the house of slavery; 3 you shall have no other gods before[p] me.

4 You shall not make for yourself an idol, whether in the form of anything that is in heaven above, or that is on the earth beneath, or that is in the water under the earth. 5 You shall not bow down to them or worship them; for I the LORD your God am a jealous God, punishing children for the iniquity of parents, to the third and the fourth generation of those who reject me, 6 but showing steadfast love to the thousandth generation[q] of those who love me and keep my commandments.

7 You shall not make wrongful use of the name of the LORD your God, for the LORD will not acquit anyone who misuses his name.

8 Remember the sabbath day, and keep it holy. 9 Six days you shall labor and do all your work. 10 But the seventh day is a sabbath to the LORD your God; you shall not do any work—you, your son or your daughter, your male or female slave, your livestock, or the alien resident in your towns. 11 For in six days the LORD made heaven and earth, the sea, and all that is in them, but rested the seventh day; therefore the LORD blessed the sabbath day and consecrated it.

12 Honor your father and your mother, so that your days may be long in the land that the LORD your God is giving you.

13 You shall not murder.[r]

14 You shall not commit adultery.

15 You shall not steal.

p Or *besides* *q* Or *to thousands* *r* Or *kill*

20.1–17: **The Ten Commandments,** the epitome of duties toward God and neighbor. **1**: *These words,* i.e. "the ten words" or the Decalogue (34.28; Deut 4.13; 10.4). Originally each commandment was a short utterance (see vv. 13, 14, 15), lacking the explanatory comments found, e.g., in vv. 5, 6, 9–11. **2**: Jewish tradition considers this to be the first commandment. Actually it is a preface that summarizes the meaning of the Exodus, thus setting law within the context of God's redemptive action. **3**: The first commandment asserts that for Israel there shall be no other gods, because the LORD is *a jealous God* (v. 5; 34.14) who will tolerate no rivals for the people's devotion.

20.4–6: Imageless worship of the LORD made Israel's faith unique in the ancient world where natural powers were personified and statues of them (animal or human) were worshiped. Some interpreters consider vv. 3–6 as one commandment and divide v. 17 into two commandments. **7**: The third commandment prohibits the misuse of the LORD's name in magic, divination, or false swearing (Lev 19.12). It reflects the ancient view that knowledge of the name could be used to exert magical control (see Gen 32.27, 29 n.).

20.8–11: Keeping the sabbath *holy* means to observe it as a day separated from others, a segment of time belonging especially to God. **10**: 16.22–30. **11**: Compare Deut 5.15. **12**: 21.15, 17; Deut 27.16. **13**: This commandment forbids murder (see Gen 9.5, 6 n.), not

16 You shall not bear false witness against your neighbor.

17 You shall not covet your neighbor's house; you shall not covet your neighbor's wife, or male or female slave, or ox, or donkey, or anything that belongs to your neighbor.

18 When all the people witnessed the thunder and lightning, the sound of the trumpet, and the mountain smoking, they were afraid[s] and trembled and stood at a distance, 19 and said to Moses, "You speak to us, and we will listen; but do not let God speak to us, or we will die." 20 Moses said to the people, "Do not be afraid; for God has come only to test you and to put the fear of him upon you so that you do not sin." 21 Then the people stood at a distance, while Moses drew near to the thick darkness where God was.

22 The LORD said to Moses: Thus you shall say to the Israelites: "You have seen for yourselves that I spoke with you from heaven. 23 You shall not make gods of silver alongside me, nor shall you make for yourselves gods of gold. 24 You need make for me only an altar of earth and sacrifice on it your burnt offerings and your offerings of well-being, your sheep and your oxen; in every place where I cause my name to be remem-

bered I will come to you and bless you. 25 But if you make for me an altar of stone, do not build it of hewn stones; for if you use a chisel upon it you profane it. 26 You shall not go up by steps to my altar, so that your nakedness may not be exposed on it."

21 These are the ordinances that you shall set before them:

2 When you buy a male Hebrew slave, he shall serve six years, but in the seventh he shall go out a free person, without debt. 3 If he comes in single, he shall go out single; if he comes in married, then his wife shall go out with him. 4 If his master gives him a wife and she bears him sons or daughters, the wife and her children shall be her master's and he shall go out alone. 5 But if the slave declares, "I love my master, my wife, and my children; I will not go out a free person," 6 then his master shall bring him before God.[t] He shall be brought to the door or the doorpost; and his master shall pierce his ear with an awl; and he shall serve him for life.

7 When a man sells his daughter as a slave, she shall not go out as the male slaves do. 8 If she does not please her mas-

s Sam Gk Syr Vg: MT *they saw*
t Or *to the judges*

---

the forms of killing authorized for Israel, e.g. war or capital punishment. **16**: This law demands telling the truth in a lawsuit involving the neighbor (23.1; Deut 19.15–21; 1 Kings 21.8–14).

20.17: Some regard the first sentence as a separate commandment; however, *neighbor's house* probably includes what is enumerated in the second part of the verse: wife, male or female slave, etc.

20.18–21: The conclusion to the theophany scene (ch 19). The people request that Moses be the covenant mediator (see 19.9 n.) so that they need not hear God's law directly (compare Deut 5.4–5).

20.22–23.33: The Covenant Code. These laws are largely neutral in regard to Israelite faith and presuppose a settled agricultural society. They reflect a situation after Israel's settlement in Canaan, when prevailing laws were borrowed and adapted to the covenant tradition.

20.22–26: Cultic regulations. **23**: See

20.4–6 n. **24–26**: The Israelite altar, in contrast to pagan models, is to be the simplest kind and is to be built wherever the LORD causes his *name to be remembered,* i.e. chooses to be manifest. Contrast the reform demanded in Deut 12.5–14.

21.1–11: The rights of a slave (compare Deut 15.12–18). **1**: *Ordinances* refers to laws formulated (usually in the third person) to deal with various cases, in contrast to the apodictic or unconditional law of the Israelite theocracy (e.g. the Decalogue). These case laws, though displaying Israel's peculiar humanitarian concern, reflect the agricultural way of life in Canaan (e.g. 22.5–6) and are similar in style and content to other legal codes of the ancient Near East. **2**: *Hebrew,* see Ex 1.15 n. An Israelite could go into servitude because of debts (Ex 22.1; Lev 25.39; 2 Kings 4.1). **6**: *Before God,* i.e. the legal act had to be performed at the sacred doorpost of the house (see 12.7 n.). **7–11**: The rights of a female slave or concubine (compare Deut 15.12, 17).

ter, who designated her for himself, then he shall let her be redeemed; he shall have no right to sell her to a foreign people, since he has dealt unfairly with her. 9 If he designates her for his son, he shall deal with her as with a daughter. 10 If he takes another wife to himself, he shall not diminish the food, clothing, or marital rights of the first wife. *u* 11 And if he does not do these three things for her, she shall go out without debt, without payment of money.

12 Whoever strikes a person mortally shall be put to death. 13 If it was not premeditated, but came about by an act of God, then I will appoint for you a place to which the killer may flee. 14 But if someone willfully attacks and kills another by treachery, you shall take the killer from my altar for execution.

15 Whoever strikes father or mother shall be put to death.

16 Whoever kidnaps a person, whether that person has been sold or is still held in possession, shall be put to death.

17 Whoever curses father or mother shall be put to death.

18 When individuals quarrel and one strikes the other with a stone or fist so that the injured party, though not dead, is confined to bed, 19 but recovers and walks around outside with the help of a staff, then the assailant shall be free of liability, except to pay for the loss of time, and to arrange for full recovery.

20 When a slaveowner strikes a male or female slave with a rod and the slave dies immediately, the owner shall be punished. 21 But if the slave survives a day or two, there is no punishment; for the slave is the owner's property.

22 When people who are fighting injure a pregnant woman so that there is a miscarriage, and yet no further harm follows, the one responsible shall be fined what the woman's husband demands, paying as much as the judges determine. 23 If any harm follows, then you shall give life for life, 24 eye for eye, tooth for tooth, hand for hand, foot for foot, 25 burn for burn, wound for wound, stripe for stripe.

26 When a slaveowner strikes the eye of a male or female slave, destroying it, the owner shall let the slave go, a free person, to compensate for the eye. 27 If the owner knocks out a tooth of a male or female slave, the slave shall be let go, a free person, to compensate for the tooth.

28 When an ox gores a man or a woman to death, the ox shall be stoned, and its flesh shall not be eaten; but the owner of the ox shall not be liable. 29 If the ox has been accustomed to gore in the past, and its owner has been warned but has not restrained it, and it kills a man or a woman, the ox shall be stoned, and its owner also shall be put to death. 30 If a ransom is imposed on the owner, then the owner shall pay whatever is imposed for the redemption of the victim's life. 31 If it gores a boy or a girl, the owner shall be dealt with according to this same rule. 32 If the ox gores a male or female slave, the owner shall pay to the slaveowner thirty shekels of silver, and the ox shall be stoned.

33 If someone leaves a pit open, or digs a pit and does not cover it, and an ox or a donkey falls into it, 34 the owner of

u Heb *of her*

8: *Redeemed,* i.e. by a relative or another buyer who pays the purchase price.

21.12–32: Laws protecting human beings. 12–14: A distinction is drawn between intentional and unintentional murder. As protection from the swift justice of the blood-avenger, the killer is guaranteed asylum (Num 35.12; Deut 4.41–43; 19.1–13; Josh ch 20), so that the case may be adjudicated soberly by legal authorities. The asylum in ancient times was at the altar (1 Kings 2.28–34).

21.17: The curse, according to ancient belief, released an inexorable power (Num 22.6), thus making it as serious to curse parents as to strike them. **22–25:** This "lex talionis," or "law of retaliation," (see Lev 24.20) was not a warrant for unrestrained vengeance but a limitation upon measureless revenge.

the pit shall make restitution, giving money to its owner, but keeping the dead animal.

35 If someone's ox hurts the ox of another, so that it dies, then they shall sell the live ox and divide the price of it; and the dead animal they shall also divide. 36 But if it was known that the ox was accustomed to gore in the past, and its owner has not restrained it, the owner shall restore ox for ox, but keep the dead animal.

22 *v* When someone steals an ox or a sheep, and slaughters it or sells it, the thief shall pay five oxen for an ox, and four sheep for a sheep. *w* The thief shall make restitution, but if unable to do so, shall be sold for the theft. 4 When the animal, whether ox or donkey or sheep, is found alive in the thief's possession, the thief shall pay double.

2 *x* If a thief is found breaking in, and is beaten to death, no bloodguilt is incurred; 3 but if it happens after sunrise, bloodguilt is incurred.

5 When someone causes a field or vineyard to be grazed over, or lets livestock loose to graze in someone else's field, restitution shall be made from the best in the owner's field or vineyard.

6 When fire breaks out and catches in thorns so that the stacked grain or the standing grain or the field is consumed, the one who started the fire shall make full restitution.

7 When someone delivers to a neighbor money or goods for safekeeping, and they are stolen from the neighbor's house, then the thief, if caught, shall pay double. 8 If the thief is not caught, the owner of the house shall be brought before God, *y* to determine whether or not the owner had laid hands on the neighbor's goods.

9 In any case of disputed ownership involving ox, donkey, sheep, clothing, or any other loss, of which one party says, "This is mine," the case of both parties shall come before God; *y* the one whom God condemns *z* shall pay double to the other.

10 When someone delivers to another a donkey, ox, sheep, or any other animal for safekeeping, and it dies or is injured or is carried off, without anyone seeing it, 11 an oath before the LORD shall decide between the two of them that the one has not laid hands on the property of the other; the owner shall accept the oath, and no restitution shall be made. 12 But if it was stolen, restitution shall be made to its owner. 13 If it was mangled by beasts, let it be brought as evidence; restitution shall not be made for the mangled remains.

14 When someone borrows an animal from another and it is injured or dies, the owner not being present, full restitution shall be made. 15 If the owner was present, there shall be no restitution; if it was hired, only the hiring fee is due.

16 When a man seduces a virgin who is not engaged to be married, and lies with her, he shall give the bride-price for her and make her his wife. 17 But if her father refuses to give her to him, he shall pay an amount equal to the bride-price for virgins.

v Ch 21.37 in Heb   *w* Verses 2, 3, and 4 rearranged thus: 3b, 4, 2, 3a   *x* Ch 22.1 in Heb   *y* Or *before the judges*   *z* Or *the judges condemn*

21.33–22.17: Laws dealing with property.
21.33–36: These laws establish responsibility in cases of carelessness.
22.1–4: Case laws regulating stealing. **2–3:** These verses may mean that if caught in the act (at night), the invader may be slain with impunity, but if slain in broad daylight, there is blood guilt. **5–6:** Cases of neglect.

22.7–15: Cases involving trusteeship. **9:** *Before God* (v. 8), i.e. to the sanctuary (possibly to the doorpost; 21.6) for an oracular decision or the sacred oath (v. 11; 1 Kings 8.31–32).
22.16–17: This law is included here because it deals with a financial matter, the *bride-price* (Deut 22.29). Laws concerning sexual relations are found in Deut 22.13–30.

18 You shall not permit a female sorcerer to live.

19 Whoever lies with an animal shall be put to death.

20 Whoever sacrifices to any god, other than the LORD alone, shall be devoted to destruction.

21 You shall not wrong or oppress a resident alien, for you were aliens in the land of Egypt. 22 You shall not abuse any widow or orphan. 23 If you do abuse them, when they cry out to me, I will surely heed their cry; 24 my wrath will burn, and I will kill you with the sword, and your wives shall become widows and your children orphans.

25 If you lend money to my people, to the poor among you, you shall not deal with them as a creditor; you shall not exact interest from them. 26 If you take your neighbor's cloak in pawn, you shall restore it before the sun goes down; 27 for it may be your neighbor's only clothing to use as cover; in what else shall that person sleep? And if your neighbor cries out to me, I will listen, for I am compassionate.

28 You shall not revile God, or curse a leader of your people.

29 You shall not delay to make offerings from the fullness of your harvest and from the outflow of your presses. *a*

The firstborn of your sons you shall give to me. 30 You shall do the same with your oxen and with your sheep: seven days it shall remain with its mother; on the eighth day you shall give it to me.

31 You shall be people consecrated to me; therefore you shall not eat any meat that is mangled by beasts in the field; you shall throw it to the dogs.

23 You shall not spread a false report. You shall not join hands with the wicked to act as a malicious witness. 2 You shall not follow a majority in wrongdoing; when you bear witness in a lawsuit, you shall not side with the majority so as to pervert justice; 3 nor shall you be partial to the poor in a lawsuit.

4 When you come upon your enemy's ox or donkey going astray, you shall bring it back.

5 When you see the donkey of one who hates you lying under its burden and you would hold back from setting it free, you must help to set it free. *a*

6 You shall not pervert the justice due to your poor in their lawsuits. 7 Keep far from a false charge, and do not kill the innocent and those in the right, for I will not acquit the guilty. 8 You shall take no bribe, for a bribe blinds the officials, and subverts the cause of those who are in the right.

9 You shall not oppress a resident alien; you know the heart of an alien, for you were aliens in the land of Egypt.

10 For six years you shall sow your land and gather in its yield; 11 but the seventh year you shall let it rest and lie fallow, so that the poor of your people may eat; and what they leave the wild animals may eat. You shall do the same with your vineyard, and with your olive orchard.

a Meaning of Heb uncertain

22.18–23.9: Miscellaneous social and cultic laws.
22.18–20: The laws of vv. 18–20 (compare 21.12, 15–17) are in the unconditional style of the Decalogue. **20:** Compare 20.3; Deut 13.12–18. **21–27:** Israel's God is the protector of the legally defenseless: the resident alien (sojourner), orphan, widow, and poor. **25:** Being a farming people, Israel frowned upon the mercantile way of life (Hos 12.7–8) and specifically upon the exaction of interest from another Israelite (Lev 25.35–38). **26:** A loan with a garment as security

could only be for the day, lest one of the poor suffer (Deut 24.12–13; Am 2.8).
22.28: Lev 24.15–16; 2 Sam 16.9; 1 Kings 2.8–9; 21.10. **29–30:** See 13.2 n. **31:** Flesh torn by beasts was regarded as unclean because it was not properly drained of blood (Lev 7.24; 17.15).
23.1–9: Laws expounding Israel's sense of justice. **4–5:** Justice extends even to helping *your enemy* (Deut 22.1–4).
23.10–19: A cultic calendar (34.18–26; Lev 23.1–44; Deut 16.1–17). **10–11:** See Lev 25.2–7. **12:** Here the observance of the sab-

12 Six days you shall do your work, but on the seventh day you shall rest, so that your ox and your donkey may have relief, and your homeborn slave and the resident alien may be refreshed. 13 Be attentive to all that I have said to you. Do not invoke the names of other gods; do not let them be heard on your lips.

14 Three times in the year you shall hold a festival for me. 15 You shall observe the festival of unleavened bread; as I commanded you, you shall eat unleavened bread for seven days at the appointed time in the month of Abib, for in it you came out of Egypt.

No one shall appear before me empty-handed.

16 You shall observe the festival of harvest, of the first fruits of your labor, of what you sow in the field. You shall observe the festival of ingathering at the end of the year, when you gather in from the field the fruit of your labor. 17 Three times in the year all your males shall appear before the Lord GOD.

18 You shall not offer the blood of my sacrifice with anything leavened, or let the fat of my festival remain until the morning.

19 The choicest of the first fruits of your ground you shall bring into the house of the LORD your God.

You shall not boil a kid in its mother's milk.

20 I am going to send an angel in front of you, to guard you on the way and to bring you to the place that I have prepared. 21 Be attentive to him and listen to his voice; do not rebel against him, for he will not pardon your transgression; for my name is in him.

22 But if you listen attentively to his voice and do all that I say, then I will be an enemy to your enemies and a foe to your foes.

23 When my angel goes in front of you, and brings you to the Amorites, the Hittites, the Perizzites, the Canaanites, the Hivites, and the Jebusites, and I blot them out, 24 you shall not bow down to their gods, or worship them, or follow their practices, but you shall utterly demolish them and break their pillars in pieces. 25 You shall worship the LORD your God, and I*b* will bless your bread and your water; and I will take sickness away from among you. 26 No one shall miscarry or be barren in your land; I will fulfill the number of your days. 27 I will send my terror in front of you, and will throw into confusion all the people against whom you shall come, and I will make all your enemies turn their backs to you. 28 And I will send the pestilence*c* in front of you, which shall drive out the Hivites, the Canaanites, and the Hittites from before you. 29 I will not drive them out from before you in one year, or the land would become desolate and the wild animals would multiply against you. 30 Little by little I will drive them out from before you, until you have increased and possess the land. 31 I will set your borders from the Red Sea*d* to the sea of the Philistines, and from the wilderness to the Euphrates; for I will hand over to you the inhabitants of the land, and you shall drive them out before you. 32 You shall make no covenant with them

b Gk Vg: Heb *he*   *c* Or *hornets*: Meaning of Heb uncertain   *d* Or *Sea of Reeds*

---

bath is based upon humanitarian concern (compare 20.11). **14–17**: This law reflects the practice of making a pilgrimage to the central sanctuary of the tribal confederacy (1 Sam 1.3, 21). **15**: *Empty-handed,* i.e. without a gift of the first fruits of the barley harvest. **16**: The *festival of harvest,* i.e. the festival of weeks (or pentecost, see Lev 23.15–21 n.) which was celebrated at the time of the wheat harvest (June). The third festival, *the festival of ingathering,* or festival of booths, was celebrated *at the* *end of the year* (autumn), according to the old agricultural calendar (see 12.2 n.), when fruit, grapes, and olives were harvested. **17**: According to ancient practice, men were the chief participants in the cult (34.23; see 10.7–11 n.). **18–19**: 34.25–26. The prohibition against boiling a kid in its mother's milk (Deut 14.21) seems to be a protest against a Canaanite method of preparing a sacrifice. **23.20–33**: **The conclusion to the Covenant Code** (beginning 20.22). **20–21**: The

and their gods. ³³They shall not live in your land, or they will make you sin against me; for if you worship their gods, it will surely be a snare to you.

24 Then he said to Moses, "Come up to the LORD, you and Aaron, Nadab, and Abihu, and seventy of the elders of Israel, and worship at a distance. ²Moses alone shall come near the LORD; but the others shall not come near, and the people shall not come up with him."

3 Moses came and told the people all the words of the LORD and all the ordinances; and all the people answered with one voice, and said, "All the words that the LORD has spoken we will do." ⁴And Moses wrote down all the words of the LORD. He rose early in the morning, and built an altar at the foot of the mountain, and set up twelve pillars, corresponding to the twelve tribes of Israel. ⁵He sent young men of the people of Israel, who offered burnt offerings and sacrificed oxen as offerings of well-being to the LORD. ⁶Moses took half of the blood and put it in basins, and half of the blood he dashed against the altar. ⁷Then he took the book of the covenant, and read it in the hearing of the people; and they said,

"All that the LORD has spoken we will do, and we will be obedient." ⁸Moses took the blood and dashed it on the people, and said, "See the blood of the covenant that the LORD has made with you in accordance with all these words."

9 Then Moses and Aaron, Nadab, and Abihu, and seventy of the elders of Israel went up, ¹⁰and they saw the God of Israel. Under his feet there was something like a pavement of sapphire stone, like the very heaven for clearness. ¹¹God^e did not lay his hand on the chief men of the people of Israel; also they beheld God, and they ate and drank.

12 The LORD said to Moses, "Come up to me on the mountain, and wait there; and I will give you the tablets of stone, with the law and the commandment, which I have written for their instruction." ¹³So Moses set out with his assistant Joshua, and Moses went up into the mountain of God. ¹⁴To the elders he had said, "Wait here for us, until we come to you again; for Aaron and Hur are with you; whoever has a dispute may go to them."

e Heb *He*

---

angel is the LORD in person (14.19; see Gen 16.7 n.). On *the name*, see Gen 32.27 n. **27–28**: Here the language of "holy war" is used. *Terror*, Gen 35.5 n. *Pestilence* apparently refers to the panic aroused in holy war (Deut 7.20; Josh 24.12 n.).

24.1–18: **The ceremony of covenant ratification. 1–2**: This tradition is continued in vv. 9–11. *Moses alone*, an indication of Moses' special role as covenant mediator (19.9; 20.19). **3–8**: The first version of the covenant ceremony stresses the people's participation (19.10–15). **3**: *The words*, i.e. the Decalogue; *the ordinances*, i.e. the laws of the Covenant Code (see 21.1 n.). **4**: The participation of all the people is symbolized by *twelve pillars*, one for each tribe. **5**: On the types of sacrifice, see Lev chs 1 and 3. **6–8**: The ritual dramatizes the uniting of the two parties: the LORD, whose presence is represented by the altar, and the people. Compare the ancient covenant ceremony found in Gen ch 15. **7**: *The book of the covenant* (Josh 24.25–26) apparently contained the covenant laws, and here is tacit-

ly identified with *the words* and *the ordinances* (v. 3). **8**: *The blood of the covenant* (compare Mt 26.28; 1 Cor 11.25) reflects the ancient view that blood was efficacious in establishing community between God and human beings (see Lev 1.5 n.).

24.9–11: The second version of the covenant ceremony (continuing vv. 1–2). **9**: The people did not take part but were represented by the seventy *elders* or *chief men*. Moses, the covenant mediator, was accompanied by the priestly family, Aaron, Nadab, and Abihu (6.14–25; Lev 10.1–3). **10**: The leaders did not see God directly; they saw only the lower part of the heavenly throne-room—the sapphire pavement (the firmament) above which the LORD was enthroned (compare Isa 6.1; Ezek 1.1, 26–28). **11**: Unharmed by divine holiness (see 3.6 n.), the leaders partook of the covenant meal (18.12).

24.12–14: A separate tradition about the gift of *the tablets of stone* on which the Decalogue was written (32.15; 34.28; Deut 9.9, 11, 15). **14**: 18.16. This verse sets the stage for the

15 Then Moses went up on the mountain, and the cloud covered the mountain. [16] The glory of the LORD settled on Mount Sinai, and the cloud covered it for six days; on the seventh day he called to Moses out of the cloud. [17] Now the appearance of the glory of the LORD was like a devouring fire on the top of the mountain in the sight of the people of Israel. [18] Moses entered the cloud, and went up on the mountain. Moses was on the mountain for forty days and forty nights.

25 The LORD said to Moses: [2] Tell the Israelites to take for me an offering; from all whose hearts prompt them to give you shall receive the offering for me. [3] This is the offering that you shall receive from them: gold, silver, and bronze, [4] blue, purple, and crimson yarns and fine linen, goats' hair, [5] tanned rams' skins, fine leather,*f* acacia wood, [6] oil for the lamps, spices for the anointing oil and for the fragrant incense, [7] onyx stones and gems to be set in the ephod and for the breastpiece. [8] And have them make me a sanctuary, so that I may dwell among them. [9] In accordance with all that I show you concerning the pattern of the tabernacle and of all its furniture, so you shall make it.

10 They shall make an ark of acacia wood; it shall be two and a half cubits long, a cubit and a half wide, and a cubit and a half high. [11] You shall overlay it with pure gold, inside and outside you shall overlay it, and you shall make a molding of gold upon it all around. [12] You shall cast four rings of gold for it and put them on its four feet, two rings on the one side of it, and two rings on the other side. [13] You shall make poles of acacia wood, and overlay them with gold. [14] And you shall put the poles into the rings on the sides of the ark, by which to carry the ark. [15] The poles shall remain in the rings of the ark; they shall not be taken from it. [16] You shall put into the ark the covenant*g* that I shall give you.

17 Then you shall make a mercy seat*h* of pure gold; two cubits and a half shall be its length, and a cubit and a half its width. [18] You shall make two cherubim of gold; you shall make them of hammered work, at the two ends of the mercy seat. *i* [19] Make one cherub at the one end, and one cherub at the other; of one piece with the mercy seat*i* you shall make the cherubim at its two ends. [20] The cherubim shall spread out their wings above, overshadowing the mercy seat*i* with their wings. They shall face one to another; the faces of the cherubim shall be turned toward the mercy seat. *i* [21] You shall put the mercy seat*i* on the top of the ark; and in the ark you shall put the

---

f Meaning of Heb uncertain   g Or *treaty*, or *testimony*; Heb *eduth*   h Or *a cover*   i Or *the cover*

---

episode of ch 32. *Hur*, see 17.10 n. **15–18:** This theophany introduces the priestly material of chs 25–31, which may have replaced the early tradition about Moses making the ark and putting the tablets of law in it (Deut 10.1–5). *The glory*, see 16.6–7 n.

25.1–40: The ark, the table, and the lampstand. 1–9: The request for a free-will offering for making the tabernacle and its equipment (35.4–29). **7:** *The ephod* and *the breastpiece*, 28.6–12, 13–30. **8:** *A sanctuary*, i.e. a tabernacle, will be the sign that God, the heavenly LORD, is present among the people.

25.10–22: Specifications for the ark. In the following sections, the priestly account moves from the center to the periphery; i.e. it begins with the most important item, the ark, and moves outward to the court. Comparable to ancient Arabic palladia, the ark was a portable wooden chest that served to guide Israel in wandering (Num 10.33), to lead in war (Num 10.35–36), and to be a medium for oracles (1 Sam ch 3). It was regarded as a throne-seat above which the LORD was invisibly enthroned (1 Sam 4.4; 2 Sam 6.2; 2 Kings 19.15). In contrast to the tent of meeting (33.7–11), it signified the divine nearness. Constructed by Moses himself, according to tradition (Deut 10.3), it was stationed at Shiloh during the days of the tribal confederacy (1 Sam chs 3–6) and was eventually brought by David to Jerusalem (2 Sam ch 6). **10:** The dimensions are about 45 × 27 × 27 inches. **16:** *The covenant*, i.e. the tablets of law (24.12). Hence the chest is called "the ark of the covenant" (v. 22) in view of its chief contents. **17:** The *mercy seat* (Lev 16.2, 13–15) was the cover of the ark. As the footstool of the LORD's

covenant*j* that I shall give you. 22 There I will meet with you, and from above the mercy seat,*k* from between the two cherubim that are on the ark of the covenant,*j* I will deliver to you all my commands for the Israelites.

23 You shall make a table of acacia wood, two cubits long, one cubit wide, and a cubit and a half high. 24 You shall overlay it with pure gold, and make a molding of gold around it. 25 You shall make around it a rim a handbreadth wide, and a molding of gold around the rim. 26 You shall make for it four rings of gold, and fasten the rings to the four corners at its four legs. 27 The rings that hold the poles used for carrying the table shall be close to the rim. 28 You shall make the poles of acacia wood, and overlay them with gold, and the table shall be carried with these. 29 You shall make its plates and dishes for incense, and its flagons and bowls with which to pour drink offerings; you shall make them of pure gold. 30 And you shall set the bread of the Presence on the table before me always.

31 You shall make a lampstand of pure gold. The base and the shaft of the lampstand shall be made of hammered work; its cups, its calyxes, and its petals shall be of one piece with it; 32 and there shall be six branches going out of its sides, three branches of the lampstand out of one side of it and three branches of the lampstand out of the other side of it; 33 three cups shaped like almond blossoms, each with calyx and petals, on one branch, and three cups shaped like almond blossoms, each with calyx and petals, on the other branch—so for the

six branches going out of the lampstand. 34 On the lampstand itself there shall be four cups shaped like almond blossoms, each with its calyxes and petals. 35 There shall be a calyx of one piece with it under the first pair of branches, a calyx of one piece with it under the next pair of branches, and a calyx of one piece with it under the last pair of branches—so for the six branches that go out of the lampstand. 36 Their calyxes and their branches shall be of one piece with it, the whole of it one hammered piece of pure gold. 37 You shall make the seven lamps for it; and the lamps shall be set up so as to give light on the space in front of it. 38 Its snuffers and trays shall be of pure gold. 39 It, and all these utensils, shall be made from a talent of pure gold. 40 And see that you make them according to the pattern for them, which is being shown you on the mountain.

26 Moreover you shall make the tabernacle with ten curtains of fine twisted linen, and blue, purple, and crimson yarns; you shall make them with cherubim skillfully worked into them. 2 The length of each curtain shall be twenty-eight cubits, and the width of each curtain four cubits; all the curtains shall be of the same size. 3 Five curtains shall be joined to one another; and the other five curtains shall be joined to one another. 4 You shall make loops of blue on the edge of the outermost curtain in the first set; and likewise you shall make loops on the edge of the outermost cur-

j Or *treaty*, or *testimony*; Heb *eduth*   *k* Or *the cover*

---

throne (1 Chr 28.2; Ps 132.7), it was regarded as the place where God meets the priestly representative of the people (v. 22). **18–20:** In antiquity cherubim (see Gen 3.24 n.) were adornments of a throne.

25.23–30: The table for the holy bread and the sacred vessels (1 Kings 7.48). **30:** The *bread of the Presence* or "holy bread" (1 Sam 21.4, 6) was bread placed before God as a sacrificial offering (Num 4.7; Lev 24.5–9; 1 Chr 9.32; Mt 12.4).

25.31–40: The seven-branched golden candlestick was to illumine the interior of the

holy place (30.7–8; 1 Kings 7.49). **40:** Ancients believed that earthly temples and their cultic equipment were made according to the *pattern* or prototype of heavenly models (v. 9; 26.30; 27.8).

26.1–37: The pattern of the tabernacle. This account blends the ancient tradition of the tent of meeting (33.7–11; Num 11.16–17, 24–30) and the later view of the structure and adornments of Solomon's temple (1 Kings ch 6; Ezek chs 40–43). Hence it is called "the tabernacle of the tent of meeting" (39.32; 40.6, 29). **1–6:** The interior was to consist

tain in the second set. 5 You shall make fifty loops on the one curtain, and you shall make fifty loops on the edge of the curtain that is in the second set; the loops shall be opposite one another. 6 You shall make fifty clasps of gold, and join the curtains to one another with the clasps, so that the tabernacle may be one whole.

7 You shall also make curtains of goats' hair for a tent over the tabernacle; you shall make eleven curtains. 8 The length of each curtain shall be thirty cubits, and the width of each curtain four cubits; the eleven curtains shall be of the same size. 9 You shall join five curtains by themselves, and six curtains by themselves, and the sixth curtain you shall double over at the front of the tent. 10 You shall make fifty loops on the edge of the curtain that is outermost in one set, and fifty loops on the edge of the curtain that is outermost in the second set.

11 You shall make fifty clasps of bronze, and put the clasps into the loops, and join the tent together, so that it may be one whole. 12 The part that remains of the curtains of the tent, the half curtain that remains, shall hang over the back of the tabernacle. 13 The cubit on the one side, and the cubit on the other side, of what remains in the length of the curtains of the tent, shall hang over the sides of the tabernacle, on this side and that side, to cover it. 14 You shall make for the tent a covering of tanned rams' skins and an outer covering of fine leather.[*l*]

15 You shall make upright frames of acacia wood for the tabernacle. 16 Ten cubits shall be the length of a frame, and a cubit and a half the width of each frame. 17 There shall be two pegs in each frame to fit the frames together; you shall make these for all the frames of the tabernacle. 18 You shall make the frames for the tab-

ernacle: twenty frames for the south side; 19 and you shall make forty bases of silver under the twenty frames, two bases under the first frame for its two pegs, and two bases under the next frame for its two pegs; 20 and for the second side of the tabernacle, on the north side twenty frames, 21 and their forty bases of silver, two bases under the first frame, and two bases under the next frame; 22 and for the rear of the tabernacle westward you shall make six frames. 23 You shall make two frames for corners of the tabernacle in the rear; 24 they shall be separate beneath, but joined at the top, at the first ring; it shall be the same with both of them; they shall form the two corners. 25 And so there shall be eight frames, with their bases of silver, sixteen bases; two bases under the first frame, and two bases under the next frame.

26 You shall make bars of acacia wood, five for the frames of the one side of the tabernacle, 27 and five bars for the frames of the other side of the tabernacle, and five bars for the frames of the side of the tabernacle at the rear westward. 28 The middle bar, halfway up the frames, shall pass through from end to end. 29 You shall overlay the frames with gold, and shall make their rings of gold to hold the bars; and you shall overlay the bars with gold. 30 Then you shall erect the tabernacle according to the plan for it that you were shown on the mountain.

31 You shall make a curtain of blue, purple, and crimson yarns, and of fine twisted linen; it shall be made with cherubim skillfully worked into it. 32 You shall hang it on four pillars of acacia overlaid with gold, which have hooks of

l Meaning of Heb uncertain

---

of ten richly decorated curtains (compare 1 Kings 6.29).

26.7–14: The oldest phase of the priestly tradition recalls the tent made of goat's hair (v. 7) and reddened ram's skins (v. 14), like the ancient, Arabic red-leather shrines which were also supported on desert acacia wood. The ancient tradition concerning the tent is combined with the design of Solomon's temple, with the result that the tent is conceived as a covering for the tabernacle, i.e. *a tent over the tabernacle* (v. 7; 39.33; 40.19).

26.15–30: The wooden framework is about 45 × 15 × 15 feet (1 Kings 6.2).

26.31–34: The *most holy place* (Holy of Holies) was to be separated from the *holy place*

gold and rest on four bases of silver.
33 You shall hang the curtain under the
clasps, and bring the ark of the covenant *m* in there, within the curtain; and
the curtain shall separate for you the holy
place from the most holy. 34 You shall
put the mercy seat *n* on the ark of the
covenant *m* in the most holy place. 35 You
shall set the table outside the curtain, and
the lampstand on the south side of the
tabernacle opposite the table; and you
shall put the table on the north side.

36 You shall make a screen for the
entrance of the tent, of blue, purple, and
crimson yarns, and of fine twisted linen,
embroidered with needlework. 37 You
shall make for the screen five pillars of
acacia, and overlay them with gold; their
hooks shall be of gold, and you shall cast
five bases of bronze for them.

27 You shall make the altar of acacia
wood, five cubits long and five
cubits wide; the altar shall be square, and
it shall be three cubits high. 2 You shall
make horns for it on its four corners; its
horns shall be of one piece with it, and
you shall overlay it with bronze. 3 You
shall make pots for it to receive its ashes,
and shovels and basins and forks and firepans; you shall make all its utensils of
bronze. 4 You shall also make for it a
grating, a network of bronze; and on the
net you shall make four bronze rings at
its four corners. 5 You shall set it under
the ledge of the altar so that the net shall
extend halfway down the altar. 6 You
shall make poles for the altar, poles of
acacia wood, and overlay them with
bronze; 7 the poles shall be put through
the rings, so that the poles shall be on the
two sides of the altar when it is carried.
8 You shall make it hollow, with boards.
They shall be made just as you were
shown on the mountain.

9 You shall make the court of the tabernacle. On the south side the court shall
have hangings of fine twisted linen one
hundred cubits long for that side; 10 its
twenty pillars and their twenty bases
shall be of bronze, but the hooks of the
pillars and their bands shall be of silver.
11 Likewise for its length on the north
side there shall be hangings one hundred
cubits long, their pillars twenty and their
bases twenty, of bronze, but the hooks of
the pillars and their bands shall be of silver. 12 For the width of the court on the
west side there shall be fifty cubits of
hangings, with ten pillars and ten bases.
13 The width of the court on the front to
the east shall be fifty cubits. 14 There shall
be fifteen cubits of hangings on the one
side, with three pillars and three bases.
15 There shall be fifteen cubits of hangings on the other side, with three pillars
and three bases. 16 For the gate of the
court there shall be a screen twenty cubits long, of blue, purple, and crimson
yarns, and of fine twisted linen, embroidered with needlework; it shall have four
pillars and with them four bases. 17 All
the pillars around the court shall be banded with silver; their hooks shall be of
silver, and their bases of bronze. 18 The
length of the court shall be one hundred
cubits, the width fifty, and the height
five cubits, with hangings of fine twisted
linen and bases of bronze. 19 All the utensils of the tabernacle for every use, and all
its pegs and all the pegs of the court, shall
be of bronze.

20 You shall further command the Israelites to bring you pure oil of beaten
olives for the light, so that a lamp may be
set up to burn regularly. 21 In the tent of

---

m Or *treaty*, or *testimony*; Heb *eduth*   *n* Or *the
cover*

(sanctuary) by a rich curtain, as in Solomon's
temple.
27.1–21: The altar and the court. From
the tabernacle, the writer turns to the surrounding sacred area. **1–8:** The central object
in the court was the altar of burnt offering,
where the main sacrificial services took place.
In comparison with the crude altar of 20.25,
this altar, overlaid with bronze (1 Kings

8.64), indicates a considerable amount of Canaanite influence.
27.9–19: *The court of the tabernacle* was a
sacred enclosure (150 × 75 feet), a common
feature of ancient temple plans. **20–21:** The
lamp (25.31–37; Lev 24.1–3) was to burn
continually as a sign of God's presence. *The
covenant,* see 16.33–34 n.

meeting, outside the curtain that is before the covenant,*o* Aaron and his sons shall tend it from evening to morning before the LORD. It shall be a perpetual ordinance to be observed throughout their generations by the Israelites.

28 Then bring near to you your brother Aaron, and his sons with him, from among the Israelites, to serve me as priests—Aaron and Aaron's sons, Nadab and Abihu, Eleazar and Ithamar. 2 You shall make sacred vestments for the glorious adornment of your brother Aaron. 3 And you shall speak to all who have ability, whom I have endowed with skill, that they make Aaron's vestments to consecrate him for my priesthood. 4 These are the vestments that they shall make: a breastpiece, an ephod, a robe, a checkered tunic, a turban, and a sash. When they make these sacred vestments for your brother Aaron and his sons to serve me as priests, 5 they shall use gold, blue, purple, and crimson yarns, and fine linen.

6 They shall make the ephod of gold, of blue, purple, and crimson yarns, and of fine twisted linen, skillfully worked. 7 It shall have two shoulder-pieces attached to its two edges, so that it may be joined together. 8 The decorated band on it shall be of the same workmanship and materials, of gold, of blue, purple, and crimson yarns, and of fine twisted linen. 9 You shall take two onyx stones, and engrave on them the names of the sons of Israel, 10 six of their names on the one stone, and the names of the remaining six on the other stone, in the order of their birth. 11 As a gem-cutter engraves sig-

nets, so you shall engrave the two stones with the names of the sons of Israel; you shall mount them in settings of gold filigree. 12 You shall set the two stones on the shoulder-pieces of the ephod, as stones of remembrance for the sons of Israel; and Aaron shall bear their names before the LORD on his two shoulders for remembrance. 13 You shall make settings of gold filigree, 14 and two chains of pure gold, twisted like cords; and you shall attach the corded chains to the settings.

15 You shall make a breastpiece of judgment, in skilled work; you shall make it in the style of the ephod; of gold, of blue and purple and crimson yarns, and of fine twisted linen you shall make it. 16 It shall be square and doubled, a span in length and a span in width. 17 You shall set in it four rows of stones. A row of carnelian,*p* chrysolite, and emerald shall be the first row; 18 and the second row a turquoise, a sapphire*q* and a moonstone; 19 and the third row a jacinth, an agate, and an amethyst; 20 and the fourth row a beryl, an onyx, and a jasper; they shall be set in gold filigree. 21 There shall be twelve stones with names corresponding to the names of the sons of Israel; they shall be like signets, each engraved with its name, for the twelve tribes. 22 You shall make for the breastpiece chains of pure gold, twisted like cords; 23 and you shall make for the breastpiece two rings of gold, and put the two rings on the two edges of the breastpiece. 24 You shall put

o Or *treaty,* or *testimony;* Heb *eduth*
p The identity of several of these stones is uncertain   *q* Or *lapis lazuli*

28.1–43: Priestly vestments. 1–5: Early in Israelite history the tribe of Levi became a priestly class (Gen 49.5–7; Num 1.47–54). Apparently Levites first exercised their priestly office at Kadesh (Deut 33.8–10). Those of Aaron's line (Ex 6.14–25; compare Num 8.5–26; Deut 18.1 n.) are here designated to officiate at the altar of the central shrine (i.e. the tabernacle and later the Jerusalem temple). **28.6–12:** The *ephod* harks back to early cultic practice at the central sanctuary of Shiloh (1 Sam 2.18–19, 28; 14.3). Sometimes

scorned as an idolatrous object, perhaps a garment for an idol (Judg 8.27; 17.5; 18.4), it is usually thought of as a linen apron worn by a priest (2 Sam 6.14–15) and used in connection with the sacred lot (see v. 30). The engraved stones on each shoulder-piece (vv. 9–12) symbolize the priest's intercessory function on behalf of the twelve tribes (v. 29). **28.15–30:** Hanging from the shoulder-pieces was *the breastpiece of judgment,* a pouch which contained the sacred lots, *Urim and Thummim* (v. 30). The priestly lot was used to

the two cords of gold in the two rings at the edges of the breastpiece; 25 the two ends of the two cords you shall attach to the two settings, and so attach it in front to the shoulder-pieces of the ephod. 26 You shall make two rings of gold, and put them at the two ends of the breastpiece, on its inside edge next to the ephod. 27 You shall make two rings of gold, and attach them in front to the lower part of the two shoulder-pieces of the ephod, at its joining above the decorated band of the ephod. 28 The breastpiece shall be bound by its rings to the rings of the ephod with a blue cord, so that it may lie on the decorated band of the ephod, and so that the breastpiece shall not come loose from the ephod. 29 So Aaron shall bear the names of the sons of Israel in the breastpiece of judgment on his heart when he goes into the holy place, for a continual remembrance before the LORD. 30 In the breastpiece of judgment you shall put the Urim and the Thummim, and they shall be on Aaron's heart when he goes in before the LORD; thus Aaron shall bear the judgment of the Israelites on his heart before the LORD continually.

31 You shall make the robe of the ephod all of blue. 32 It shall have an opening for the head in the middle of it, with a woven binding around the opening, like the opening in a coat of mail,ʳ so that it may not be torn. 33 On its lower hem you shall make pomegranates of blue, purple, and crimson yarns, all around the lower hem, with bells of gold between them all around—34 a golden bell and a pomegranate alternating all around the lower hem of the robe. 35 Aaron shall wear it when he ministers, and its sound shall be heard when he goes into the holy place before the LORD, and when he comes out, so that he may not die.

36 You shall make a rosette of pure gold, and engrave on it, like the engraving of a signet, "Holy to the LORD." 37 You shall fasten it on the turban with a blue cord; it shall be on the front of the turban. 38 It shall be on Aaron's forehead, and Aaron shall take on himself any guilt incurred in the holy offering that the Israelites consecrate as their sacred donations; it shall always be on his forehead, in order that they may find favor before the LORD.

39 You shall make the checkered tunic of fine linen, and you shall make a turban of fine linen, and you shall make a sash embroidered with needlework.

40 For Aaron's sons you shall make tunics and sashes and headdresses; you shall make them for their glorious adornment. 41 You shall put them on your brother Aaron, and on his sons with him, and shall anoint them and ordain them and consecrate them, so that they may serve me as priests. 42 You shall make for them linen undergarments to cover their naked flesh; they shall reach from the hips to the thighs; 43 Aaron and his sons shall wear them when they go into the tent of meeting, or when they come near the altar to minister in the holy place; or they will bring guilt on themselves and die. This shall be a perpetual ordinance for him and for his descendants after him.

29 Now this is what you shall do to them to consecrate them, so that

---

r Meaning of Heb uncertain

---

obtain an oracular decision (Lev 8.8; Num 27.21; Deut 33.8; 1 Sam 14.41–42; 23.6–13). **28.31–34:** A short garment, the blue *robe of the ephod,* was worn under the ephod. The bells were once thought to protect the priest from demonic attack, *so that he may not die* when he enters the holy place (v. 35). **36–38:** The head piece, *a rosette of pure gold* fastened to a turban, symbolizes the regal splendor of the priest (Ezek 21.26; Zech 3.5). **39:** Underneath the blue robe the priest wore a coat or long tunic with sleeves, gathered at the waist with a sash.

29.1–46: The service for the ordination of the priests. See Lev ch 8 where these directions are carried out. **1–9:** The anointing of the high priest (Lev 16.32; Ps 133.2) follows the ancient rite of anointing the king's head with oil, thus making him "the LORD's anointed" (1 Sam 24.6; Ps 2.2). Priestly tradition traces the line of the high priesthood from Aaron, to whom the priestly office be-

they may serve me as priests. Take one young bull and two rams without blemish, 2and unleavened bread, unleavened cakes mixed with oil, and unleavened wafers spread with oil. You shall make them of choice wheat flour. 3You shall put them in one basket and bring them in the basket, and bring the bull and the two rams. 4You shall bring Aaron and his sons to the entrance of the tent of meeting, and wash them with water. 5Then you shall take the vestments, and put on Aaron the tunic and the robe of the ephod, and the ephod, and the breastpiece, and gird him with the decorated band of the ephod; 6and you shall set the turban on his head, and put the holy diadem on the turban. 7You shall take the anointing oil, and pour it on his head and anoint him. 8Then you shall bring his sons, and put tunics on them, 9and you shall gird them with sashes*s* and tie headdresses on them; and the priesthood shall be theirs by a perpetual ordinance. You shall then ordain Aaron and his sons.

10 You shall bring the bull in front of the tent of meeting. Aaron and his sons shall lay their hands on the head of the bull, 11and you shall slaughter the bull before the Lord, at the entrance of the tent of meeting, 12and shall take some of the blood of the bull and put it on the horns of the altar with your finger, and all the rest of the blood you shall pour out at the base of the altar. 13You shall take all the fat that covers the entrails, and the appendage of the liver, and the two kidneys with the fat that is on them, and turn them into smoke on the altar. 14But

the flesh of the bull, and its skin, and its dung, you shall burn with fire outside the camp; it is a sin offering.

15 Then you shall take one of the rams, and Aaron and his sons shall lay their hands on the head of the ram, 16and you shall slaughter the ram, and shall take its blood and dash it against all sides of the altar. 17Then you shall cut the ram into its parts, and wash its entrails and its legs, and put them with its parts and its head, 18and turn the whole ram into smoke on the altar; it is a burnt offering to the Lord; it is a pleasing odor, an offering by fire to the Lord.

19 You shall take the other ram; and Aaron and his sons shall lay their hands on the head of the ram, 20and you shall slaughter the ram, and take some of its blood and put it on the lobe of Aaron's right ear and on the lobes of the right ears of his sons, and on the thumbs of their right hands, and on the big toes of their right feet, and dash the rest of the blood against all sides of the altar. 21Then you shall take some of the blood that is on the altar, and some of the anointing oil, and sprinkle it on Aaron and his vestments and on his sons and his sons' vestments with him; then he and his vestments shall be holy, as well as his sons and his sons' vestments.

22 You shall also take the fat of the ram, the fat tail, the fat that covers the entrails, the appendage of the liver, the two kidneys with the fat that is on them, and the right thigh (for it is a ram of

---

s Gk: Heb *sashes, Aaron and his sons*

---

longs *by a perpetual ordinance* (v. 9b). **10–14**: The sin offering for the priests (Lev 4.1–12). **29.10**: The laying on of hands signifies identification with the sacrificial victim. **12**: The horns of the altar (27.2) were its most sacred parts (1 Kings 1.50; Am 3.14). On the efficacy of blood for the expiation of sin, see Lev 1.5 n. **15–18**: One of the rams (v. 1), as distinguished from the ram of ordination (v. 22), was for *burnt offering* (Lev ch 1). *Pleasing odor,* see Lev 1.9 n. **19–34**: The installation sacrifice was essentially an offering of well-being (Lev ch 3). **20**: Touching the blood

upon the ears, hands, and feet consecrates the whole person for the office. **24**: Putting these things in the hands of the priests signified that they were authorized to receive their portions of the offerings (1 Sam 2.12–17). To "fill the hands" (literal Hebrew) was an old expression for investment with priestly prerogatives (Judg 17.5; 1 Kings 13.33). The *elevation offering* (vv. 26–28; Lev 7.29–36) refers to the act of moving the sacrifice toward and away from the altar, to symbolize presenting the gift to God and receiving it back as a portion. **31–34**: Further instructions for the ordination

ordination), 23 and one loaf of bread, one cake of bread made with oil, and one wafer, out of the basket of unleavened bread that is before the LORD; 24 and you shall place all these on the palms of Aaron and on the palms of his sons, and raise them as an elevation offering before the LORD. 25 Then you shall take them from their hands, and turn them into smoke on the altar on top of the burnt offering of pleasing odor before the LORD; it is an offering by fire to the LORD.

26 You shall take the breast of the ram of Aaron's ordination and raise it as an elevation offering before the LORD; and it shall be your portion. 27 You shall consecrate the breast that was raised as an elevation offering and the thigh that was raised as an elevation offering from the ram of ordination, from that which belonged to Aaron and his sons. 28 These things shall be a perpetual ordinance for Aaron and his sons from the Israelites, for this is an offering; and it shall be an offering by the Israelites from their sacrifice of offerings of well-being, their offering to the LORD.

29 The sacred vestments of Aaron shall be passed on to his sons after him; they shall be anointed in them and ordained in them. 30 The son who is priest in his place shall wear them seven days, when he comes into the tent of meeting to minister in the holy place.

31 You shall take the ram of ordination, and boil its flesh in a holy place; 32 and Aaron and his sons shall eat the flesh of the ram and the bread that is in the basket, at the entrance of the tent of meeting. 33 They themselves shall eat the food by which atonement is made, to ordain and consecrate them, but no one else shall eat of them, because they are holy. 34 If any of the flesh for the ordina-

tion, or of the bread, remains until the morning, then you shall burn the remainder with fire; it shall not be eaten, because it is holy.

35 Thus you shall do to Aaron and to his sons, just as I have commanded you; through seven days you shall ordain them. 36 Also every day you shall offer a bull as a sin offering for atonement. Also you shall offer a sin offering for the altar, when you make atonement for it, and shall anoint it, to consecrate it. 37 Seven days you shall make atonement for the altar, and consecrate it, and the altar shall be most holy; whatever touches the altar shall become holy.

38 Now this is what you shall offer on the altar: two lambs a year old regularly each day. 39 One lamb you shall offer in the morning, and the other lamb you shall offer in the evening; 40 and with the first lamb one-tenth of a measure of choice flour mixed with one-fourth of a hin of beaten oil, and one-fourth of a hin of wine for a drink offering. 41 And the other lamb you shall offer in the evening, and shall offer with it a grain offering and its drink offering, as in the morning, for a pleasing odor, an offering by fire to the LORD. 42 It shall be a regular burnt offering throughout your generations at the entrance of the tent of meeting before the LORD, where I will meet with you, to speak to you there. 43 I will meet with the Israelites there, and it shall be sanctified by my glory; 44 I will consecrate the tent of meeting and the altar; Aaron also and his sons I will consecrate, to serve me as priests. 45 I will dwell among the Israelites, and I will be their God. 46 And they shall know that I am the LORD their God, who brought them out of the land of Egypt that I might dwell among them; I am the LORD their God.

---

offering of well-being. *No one else,* i.e. no layperson.

29.35–37: The seven day ordination ceremony. Owing to the efficacy of blood (v. 12), the sin offering is for *atonement,* i.e. it produces a "covering" for sin and sanctifies the priest. The offering is also said to make atonement for the altar, i.e. to cleanse and consecrate it with blood (Ezek 43.18–27). In a

deeper sense atonement actualizes divine forgiveness and reconciliation (32.30–32).
29.38–42: The daily burnt offering, compare Num 28.3–8; Ezek 46.13–15. *One tenth of a measure,* a tenth of an ephah, see Num 15.2–10. **43–46:** God's meeting with the people. **43:** *It,* the door of the tent of meeting (v. 42). *Glory,* see 16.7 n. **45:** See 6.7 n.; 25.8; 40.34.

30 You shall make an altar on which to offer incense; you shall make it of acacia wood. ²It shall be one cubit long, and one cubit wide; it shall be square, and shall be two cubits high; its horns shall be of one piece with it. ³You shall overlay it with pure gold, its top, and its sides all around and its horns; and you shall make for it a molding of gold all around. ⁴And you shall make two golden rings for it; under its molding on two opposite sides of it you shall make them, and they shall hold the poles with which to carry it. ⁵You shall make the poles of acacia wood, and overlay them with gold. ⁶You shall place it in front of the curtain that is above the ark of the covenant,ᵗ in front of the mercy seatᵘ that is over the covenant,ᵗ where I will meet with you. ⁷Aaron shall offer fragrant incense on it; every morning when he dresses the lamps he shall offer it, ⁸and when Aaron sets up the lamps in the evening, he shall offer it, a regular incense offering before the LORD throughout your generations. ⁹You shall not offer unholy incense on it, or a burnt offering, or a grain offering; and you shall not pour a drink offering on it. ¹⁰Once a year Aaron shall perform the rite of atonement on its horns. Throughout your generations he shall perform the atonement for it once a year with the blood of the atoning sin offering. It is most holy to the LORD.

11 The LORD spoke to Moses: ¹²When you take a census of the Israelites to register them, at registration all of them shall give a ransom for their lives to the LORD, so that no plague may come upon them for being registered. ¹³This is what each one who is registered shall give: half a shekel according to the shekel of the sanctuary (the shekel is twenty gerahs), half a shekel as an offering to the LORD. ¹⁴Each one who is registered, from twenty years old and upward, shall give the LORD's offering. ¹⁵The rich shall not give more, and the poor shall not give less, than the half shekel, when you bring this offering to the LORD to make atonement for your lives. ¹⁶You shall take the atonement money from the Israelites and shall designate it for the service of the tent of meeting; before the LORD it will be a reminder to the Israelites of the ransom given for your lives.

17 The LORD spoke to Moses: ¹⁸You shall make a bronze basin with a bronze stand for washing. You shall put it between the tent of meeting and the altar, and you shall put water in it; ¹⁹with the waterᵛ Aaron and his sons shall wash their hands and their feet. ²⁰When they go into the tent of meeting, or when they come near the altar to minister, to make an offering by fire to the LORD, they shall wash with water, so that they may not die. ²¹They shall wash their hands and their feet, so that they may not die: it shall be a perpetual ordinance for them, for him and for his descendants throughout their generations.

22 The LORD spoke to Moses: ²³Take the finest spices: of liquid myrrh five hundred shekels, and of sweet-smelling cinnamon half as much, that is, two hundred fifty, and two hundred fifty of aromatic cane, ²⁴and five hundred of cassia—measured by the sanctuary shekel—and a hin of olive oil; ²⁵and you shall make of these a sacred anointing oil blended as by the perfumer; it shall be a

t Or treaty, or *testimony;* Heb *eduth*   *u* Or *the cover*   *v* Heb *it*

30.1–31.18: Other priestly matters.
30.1–10: The offering of incense was an ancient cultic practice (Isa 1.13), probably taken over from the Canaanites. **10:** On atonement for the altar see 29.36–37. *Once a year,* i.e. on the day of atonement (Lev ch 16).
30.11–16: The tax for the support of the sanctuary. **11–12:** The census, reported in Num ch 1, was originally for military pur-

poses. Fearing that God's wrath would be manifested against a census (2 Sam ch 24), the people paid a fee (compare 2 Chr 24.6, 9; Mt 17.24–27) as a *ransom* or *atonement* (v. 16). **13:** *The shekel of the sanctuary* (Lev 5.15; 27.25; Num 3.47; 18.16; Ezek 45.12) based on the older Phoenician or Hebrew measurement, was heavier than the Babylonian shekel used in post-exilic times. Here it is explained that

holy anointing oil. 26 With it you shall anoint the tent of meeting and the ark of the covenant, *w* 27 and the table and all its utensils, and the lampstand and its utensils, and the altar of incense, 28 and the altar of burnt offering with all its utensils, and the basin with its stand; 29 you shall consecrate them, so that they may be most holy; whatever touches them will become holy. 30 You shall anoint Aaron and his sons, and consecrate them, in order that they may serve me as priests. 31 You shall say to the Israelites, "This shall be my holy anointing oil throughout your generations. 32 It shall not be used in any ordinary anointing of the body, and you shall make no other like it in composition; it is holy, and it shall be holy to you. 33 Whoever compounds any like it or whoever puts any of it on an unqualified person shall be cut off from the people."

34 The LORD said to Moses: Take sweet spices, stacte, and onycha, and galbanum, sweet spices with pure frankincense (an equal part of each), 35 and make an incense blended as by the perfumer, seasoned with salt, pure and holy; 36 and you shall beat some of it into powder, and put part of it before the covenant *w* in the tent of meeting where I shall meet with you; it shall be for you most holy. 37 When you make incense according to this composition, you shall not make it for yourselves; it shall be regarded by you as holy to the LORD. 38 Whoever makes any like it to use as perfume shall be cut off from the people.

31 The LORD spoke to Moses: 2 See, I have called by name Bezalel son of Uri son of Hur, of the tribe of Judah: 3 and I have filled him with divine spirit, *x* with ability, intelligence, and knowledge in every kind of craft, 4 to devise artistic designs, to work in gold, silver, and bronze, 5 in cutting stones for setting, and in carving wood, in every kind of craft. 6 Moreover, I have appointed with him Oholiab son of Ahisamach, of the tribe of Dan; and I have given skill to all the skillful, so that they may make all that I have commanded you: 7 the tent of meeting, and the ark of the covenant, *w* and the mercy seat *y* that is on it, and all the furnishings of the tent, 8 the table and its utensils, and the pure lampstand with all its utensils, and the altar of incense, 9 and the altar of burnt offering with all its utensils, and the basin with its stand, 10 and the finely worked vestments, the holy vestments for the priest Aaron and the vestments of his sons, for their service as priests, 11 and the anointing oil and the fragrant incense for the holy place. They shall do just as I have commanded you.

12 The LORD said to Moses: 13 You yourself are to speak to the Israelites: "You shall keep my sabbaths, for this is a sign between me and you throughout your generations, given in order that you may know that I, the LORD, sanctify you. 14 You shall keep the sabbath, because it is holy for you; everyone who profanes it shall be put to death; whoever does any work on it shall be cut off from among the people. 15 Six days shall work be done, but the seventh day is a sabbath of solemn rest, holy to the LORD; whoever does any work on the sabbath day shall be put to death. 16 Therefore the Israelites shall keep the sabbath, observing the sabbath throughout their generations, as a perpetual covenant. 17 It is a sign forever between me and the people of Israel that

w Or treaty, or *testimony*; Heb *eduth*
x Or *with the spirit of God*   *y* Or *the cover*

---

the sanctuary tax is to be paid in the older weight. **17–21**: The *bronze basin* (1 Kings 7.38) was for ritual ablutions.
30.34–38: The formula for the incense, as for the holy oil (vv. 22–25), was a priestly secret. *Stacte*, an oil of myrrh. *Onycha*, a spice from a mollusk found in the Red Sea. *Galbanum*, an aromatic resin from Asiatic plants.

Frankincense, a fragrant gum resin from certain trees.
31.1–11: The appointment of craftsmen. *Bezalel*, 1 Chr 2.18–20. *Oholiab* is otherwise unknown.
31.12–17: The sabbath, anticipated by priestly tradition in 16.22–30, is here formally instituted at Sinai. **18**: A transitional verse

in six days the LORD made heaven and earth, and on the seventh day he rested, and was refreshed."

18 When God[z] finished speaking with Moses on Mount Sinai, he gave him the two tablets of the covenant,[a] tablets of stone, written with the finger of God.

32 When the people saw that Moses delayed to come down from the mountain, the people gathered around Aaron, and said to him, "Come, make gods for us, who shall go before us; as for this Moses, the man who brought us up out of the land of Egypt, we do not know what has become of him." ²Aaron said to them, "Take off the gold rings that are on the ears of your wives, your sons, and your daughters, and bring them to me." ³So all the people took off the gold rings from their ears, and brought them to Aaron. ⁴He took the gold from them, formed it in a mold,[b] and cast an image of a calf; and they said, "These are your gods, O Israel, who brought you up out of the land of Egypt!" ⁵When Aaron saw this, he built an altar before it; and Aaron made proclamation and said, "Tomorrow shall be a festival to the LORD." ⁶They rose early the next day, and offered burnt offerings and brought sacrifices of well-being; and the people sat down to eat and drink, and rose up to revel.

7 The LORD said to Moses, "Go down at once! Your people, whom you brought up out of the land of Egypt, have acted perversely; ⁸they have been quick to turn aside from the way that I commanded them; they have cast for themselves an image of a calf, and have worshiped it and sacrificed to it, and said, 'These are your gods, O Israel, who brought you up out of the land of Egypt!' " ⁹The LORD said to Moses, "I have seen this people, how stiff-necked they are. ¹⁰Now let me alone, so that my wrath may burn hot against them and I may consume them; and of you I will make a great nation."

11 But Moses implored the LORD his God, and said, "O LORD, why does your wrath burn hot against your people, whom you brought out of the land of Egypt with great power and with a mighty hand? ¹²Why should the Egyptians say, 'It was with evil intent that he brought them out to kill them in the mountains, and to consume them from the face of the earth'? Turn from your fierce wrath; change your mind and do not bring disaster on your people. ¹³Remember Abraham, Isaac, and Israel, your servants, how you swore to them by your own self, saying to them, 'I will multiply your descendants like the stars of heaven, and all this land that I have promised I will give to your descendants, and they shall inherit it forever.' " ¹⁴And the LORD changed his mind about the disaster that he planned to bring on his people.

15 Then Moses turned and went

z Heb *he*   *a* Or *treaty*, or *testimony*; Heb *eduth*
b Or *fashioned it with a graving tool*; Meaning of Heb uncertain

which resumes the narrative from 24.18. The inserted block of priestly tradition (chs 25–31) was supposedly delivered to Moses on Mount Sinai.

32.1–35: The breaking of the covenant. During Moses' absence the rebellious people chose Aaron as their leader and worshiped a golden bull. **1:** Moses stayed on the mountain top forty days and forty nights, a round number for an indefinitely long time (1 Kings 19.8; Mt 4.2). *Gods who shall go before us,* i.e. visible symbols of the divine presence as in pagan idolatry (see 20.4–6 n.). **2–3:** On the golden earrings, see Gen 35.4 n. The *calf,* or young bull, was a symbol of fertility in the nature-religions of the ancient Near East (compare 1 Kings 12.28; Hos 8.5). **6:** The eating and drinking accompanied *a festival to the LORD,* a dedication of the new cultic symbol (2 Sam 6.17–19).

32.7–14: Moses' first intercession (Num 14.13–19). **14:** *The LORD changed his mind:* God is not bound inflexibly to an announced plan but is free to change a course of action in a manner consistent with the divine purpose (Gen 6.5–6; Am 7.3, 6).

32.15: *Tablets of the covenant,* see 25.16 n. **19:** The breaking of the tablets symbolized that the covenant relationship had been broken. **20:** Moses subjected the people to a trial

down from the mountain, carrying the two tablets of the covenant*c* in his hands, tablets that were written on both sides, written on the front and on the back. [16] The tablets were the work of God, and the writing was the writing of God, engraved upon the tablets. [17] When Joshua heard the noise of the people as they shouted, he said to Moses, "There is a noise of war in the camp." [18] But he said,

> "It is not the sound made by
>   victors,
> or the sound made by losers;
> it is the sound of revelers that I
>   hear."

[19] As soon as he came near the camp and saw the calf and the dancing, Moses' anger burned hot, and he threw the tablets from his hands and broke them at the foot of the mountain. [20] He took the calf that they had made, burned it with fire, ground it to powder, scattered it on the water, and made the Israelites drink it.

21 Moses said to Aaron, "What did this people do to you that you have brought so great a sin upon them?" [22] And Aaron said, "Do not let the anger of my lord burn hot; you know the people, that they are bent on evil. [23] They said to me, 'Make us gods, who shall go before us; as for this Moses, the man who brought us up out of the land of Egypt, we do not know what has become of him.' [24] So I said to them, 'Whoever has gold, take it off'; so they gave it to me, and I threw it into the fire, and out came this calf!"

25 When Moses saw that the people were running wild (for Aaron had let them run wild, to the derision of their enemies), [26] then Moses stood in the gate of the camp, and said, "Who is on the LORD's side? Come to me!" And all the sons of Levi gathered around him. [27] He said to them, "Thus says the LORD, the God of Israel, 'Put your sword on your side, each of you! Go back and forth from gate to gate throughout the camp, and each of you kill your brother, your friend, and your neighbor.' " [28] The sons of Levi did as Moses commanded, and about three thousand of the people fell on that day. [29] Moses said, "Today you have ordained yourselves*d* for the service of the LORD, each one at the cost of a son or a brother, and so have brought a blessing on yourselves this day."

30 On the next day Moses said to the people, "You have sinned a great sin. But now I will go up to the LORD; perhaps I can make atonement for your sin." [31] So Moses returned to the LORD and said, "Alas, this people has sinned a great sin; they have made for themselves gods of gold. [32] But now, if you will only forgive their sin—but if not, blot me out of the book that you have written." [33] But the LORD said to Moses, "Whoever has sinned against me I will blot out of my book. [34] But now go, lead the people to the place about which I have spoken to you; see, my angel shall go in front of you. Nevertheless, when the day comes for punishment, I will punish them for their sin."

35 Then the LORD sent a plague on the

---

*c Or treaty, or testimony; Heb eduth   d Gk Vg
Compare Tg: Heb Today ordain yourselves*

---

by ordeal (Num 5.16–28). Those who suffered ill effects from drinking the water and pulverized metal were regarded as guilty and fell in a plague (v. 35). **21–24:** The rebuke of Aaron (see Num ch 12) stands in contrast to his priestly prestige and intercessory role as described in chs 25–31. **24:** Aaron feebly disclaims responsibility by saying that he did not make the calf: it emerged from the fire by itself. **32.25–29:** A separate tradition about how the Levites (see 28.1–5 n.) were consecrated to the priesthood (compare Num 25.10–13). Instead of being consecrated by a ritual ceremony (ch 29), the Levites *ordained* themselves by their zeal, that is, their passionate loyalty to the LORD (1 Kings 19.10; 2 Kings 10.16) despite social or family bonds. **30–35:** Moses' second intercession. **30:** *Make atonement,* i.e. obtain forgiveness (v. 32). **32:** *The book* is the register of the members of the theocratic community (Ps 69.28; Isa 4.3; Dan 12.1; Mal 3.16). **34:** On the *angel,* see Gen 16.7 n.; Ex 23.20.

people, because they made the calf—the one that Aaron made.

33 The LORD said to Moses, "Go, leave this place, you and the people whom you have brought up out of the land of Egypt, and go to the land of which I swore to Abraham, Isaac, and Jacob, saying, 'To your descendants I will give it.' 2 I will send an angel before you, and I will drive out the Canaanites, the Amorites, the Hittites, the Perizzites, the Hivites, and the Jebusites. 3 Go up to a land flowing with milk and honey; but I will not go up among you, or I would consume you on the way, for you are a stiff-necked people."

4 When the people heard these harsh words, they mourned, and no one put on ornaments. 5 For the LORD had said to Moses, "Say to the Israelites, 'You are a stiff-necked people; if for a single moment I should go up among you, I would consume you. So now take off your ornaments, and I will decide what to do to you.' " 6 Therefore the Israelites stripped themselves of their ornaments, from Mount Horeb onward.

7 Now Moses used to take the tent and pitch it outside the camp, far off from the camp; he called it the tent of meeting. And everyone who sought the LORD would go out to the tent of meeting, which was outside the camp. 8 Whenever Moses went out to the tent, all the people would rise and stand, each of them, at the entrance of their tents and watch Moses until he had gone into the tent. 9 When Moses entered the tent, the pillar of cloud would descend and stand at the entrance of the tent, and the LORD would speak with Moses. 10 When all the people saw the pillar of cloud standing at the entrance of the tent, all the people would rise and bow down, all of them, at the entrance of their tent. 11 Thus the LORD used to speak to Moses face to face, as one speaks to a friend. Then he would return to the camp; but his young assistant, Joshua son of Nun, would not leave the tent.

12 Moses said to the LORD, "See, you have said to me, 'Bring up this people'; but you have not let me know whom you will send with me. Yet you have

---

33.1–23: The LORD's guidance. Moses seeks assurance that God will accompany the people, despite their folly and sin. **1–6:** The accompanying angel is the LORD's representative or alter ego (32.34), showing that the people will not be God-forsaken. The LORD will not, however, accompany the sinful people directly, lest divine holiness consume them. **4–6:** The people removed their ornaments as a sign of contrition. **7–11:** An old tradition about the tent of meeting. **33.7:** *The tent* was portable, like ancient Arabic tent-shrines (see 26.7–14 n.). Unlike the priestly tabernacle, which was centrally located (25.8; Num 2.2), the tent was pitched *far off from the camp*. Originally perhaps the tent was a place of tribal assembly and an oracle-place, both ideas being implied in the term *meeting*. It was, however, chiefly a tent of revelation to Moses (Num 11.16–17, 24–40; 12.1–8; Deut 31.14–15; compare Ex 29.42–46). *Sought the LORD,* i.e. for oracular decisions (18.15–16). **8–9:** While the ark symbolized the nearness and presence of the LORD (see 25.10–22 n.), the tent signified divine distance and transcendence; hence the LORD used to *descend* from time to time to meet with Moses. In priestly tradition the

two views are combined by saying that the ark was placed within the "tent of covenant" (30.36; Num 9.15; 17.7–8). **11:** Moses' mediatorial role (19.9; 20.19) is indicated by the fact that the LORD used to speak to him *face to face, as one speaks to a friend* (Num 12.7–8; Deut 34.10–12). Here Joshua, rather than Aaron the priest, is the custodian of the tent. **33.12–16:** Moses' intercession. **14:** *My presence* (literally, "face"), perhaps a reference to the ark. Enthroned on the ark, the LORD goes before the people and gives them *rest* (Num 10.33). **16:** Israel is a unique people because it undertakes a special historical pilgrimage with the LORD leading them into the future (Num 23.9). **17–23:** These verses anticipate the theophany of 34.5–9. **18:** Having asked for a display of God's *ways* (v. 13) or manner of action in the world, Moses now asks for more: for a manifestation of God's *glory,* i.e. the visible radiance and majesty of the Godhead (see 16.7 n.). **19:** The proclamation of the divine name, the LORD (Yahweh; see 3.14 n.), was tantamount to a disclosure of the character or identity of God (see Gen 32.27 n.). Divine freedom is emphasized. God is free to act according to God's will, unbound by external hindrance or necessity

said, 'I know you by name, and you have also found favor in my sight.' 13 Now if I have found favor in your sight, show me your ways, so that I may know you and find favor in your sight. Consider too that this nation is your people." 14 He said, "My presence will go with you, and I will give you rest." 15 And he said to him, "If your presence will not go, do not carry us up from here. 16 For how shall it be known that I have found favor in your sight, I and your people, unless you go with us? In this way, we shall be distinct, I and your people, from every people on the face of the earth."

17 The LORD said to Moses, "I will do the very thing that you have asked; for you have found favor in my sight, and I know you by name." 18 Moses said, "Show me your glory, I pray." 19 And he said, "I will make all my goodness pass before you, and will proclaim before you the name, 'The LORD'; *e* and I will be gracious to whom I will be gracious, and will show mercy on whom I will show mercy. 20 But," he said, "you cannot see my face; for no one shall see me and live." 21 And the LORD continued, "See, there is a place by me where you shall stand on the rock; 22 and while my glory passes by I will put you in a cleft of the rock, and I will cover you with my hand until I have passed by; 23 then I will take away my hand, and you shall see my back; but my face shall not be seen."

34 The LORD said to Moses, "Cut two tablets of stone like the former ones, and I will write on the tablets the words that were on the former tab-

lets, which you broke. 2 Be ready in the morning, and come up in the morning to Mount Sinai and present yourself there to me, on the top of the mountain. 3 No one shall come up with you, and do not let anyone be seen throughout all the mountain; and do not let flocks or herds graze in front of that mountain." 4 So Moses cut two tablets of stone like the former ones; and he rose early in the morning and went up on Mount Sinai, as the LORD had commanded him, and took in his hand the two tablets of stone. 5 The LORD descended in the cloud and stood with him there, and proclaimed the name, "The LORD." *e* 6 The LORD passed before him, and proclaimed,

"The LORD, the LORD,
a God merciful and gracious,
slow to anger,
and abounding in steadfast love
    and faithfulness,
7 keeping steadfast love for the
    thousandth generation, *f*
forgiving iniquity and
    transgression and sin,
yet by no means clearing the
    guilty,
but visiting the iniquity of the
    parents
upon the children
and the children's children,
to the third and the fourth
    generation."

8 And Moses quickly bowed his head toward the earth, and worshiped. 9 He said,

e Heb *YHWH*; see note at 3.15   *f* Or *for thousands*

(compare Rom 9.15); God's action is not capricious, however, but is the expression of divine *goodness* (34.6–7; see 32.14 n.). **22**: On the cave or *cleft of the rock,* see 1 Kings 19.9–18. **23**: Although employing bold anthropomorphisms (the LORD's *hand* and *back),* the narrator stresses that God remains hidden (v. 20), even when most palpably present.
34.1–35: The renewal of the covenant, symbolized by the rewriting of the commandments. **1–4**: The second *tablets* were to contain *the words that were on the former tablets* (24.12–14; compare Deut 10.1–5). Their reissue, however, provides an opportunity for

the editor to introduce a cultic set of laws (vv. 12–16). **5–9**: The theophany is anticipated in 33.17–23. **6–7**: While the LORD *passed before* him (see 33.22 n.; compare 1 Kings 19.11–12), Moses heard the self-disclosure of the LORD's identity: Yahweh is above all the God of *steadfast love* (see Gen 24.12 n.), though covenant loyalty does not exclude divine judgment upon sin. This summary, echoed in various places in the Old Testament (Num 14.18; Neh 9.17, 31; Ps 103.8; Jer 32.18; Jon 4.2), is probably an old cultic confession.

"If now I have found favor in your sight, O Lord, I pray, let the Lord go with us. Although this is a stiff-necked people, pardon our iniquity and our sin, and take us for your inheritance."

10 He said: I hereby make a covenant. Before all your people I will perform marvels, such as have not been performed in all the earth or in any nation; and all the people among whom you live shall see the work of the LORD; for it is an awesome thing that I will do with you.

11 Observe what I command you today. See, I will drive out before you the Amorites, the Canaanites, the Hittites, the Perizzites, the Hivites, and the Jebusites. 12 Take care not to make a covenant with the inhabitants of the land to which you are going, or it will become a snare among you. 13 You shall tear down their altars, break their pillars, and cut down their sacred poles *g* 14 (for you shall worship no other god, because the LORD, whose name is Jealous, is a jealous God). 15 You shall not make a covenant with the inhabitants of the land, for when they prostitute themselves to their gods and sacrifice to their gods, someone among them will invite you, and you will eat of the sacrifice. 16 And you will take wives from among their daughters for your sons, and their daughters who prostitute themselves to their gods will make your sons also prostitute themselves to their gods.

17 You shall not make cast idols.

18 You shall keep the festival of unleavened bread. Seven days you shall eat unleavened bread, as I commanded you, at the time appointed in the month of Abib; for in the month of Abib you came out from Egypt.

19 All that first opens the womb is mine, all your male *h* livestock, the firstborn of cow and sheep. 20 The firstborn of a donkey you shall redeem with a lamb, or if you will not redeem it you shall break its neck. All the firstborn of your sons you shall redeem.

No one shall appear before me empty-handed.

21 Six days you shall work, but on the seventh day you shall rest; even in plowing time and in harvest time you shall rest. 22 You shall observe the festival of weeks, the first fruits of wheat harvest, and the festival of ingathering at the turn of the year. 23 Three times in the year all your males shall appear before the LORD God, the God of Israel. 24 For I will cast out nations before you, and enlarge your borders; no one shall covet your land when you go up to appear before the LORD your God three times in the year.

25 You shall not offer the blood of my sacrifice with leaven, and the sacrifice of the festival of the passover shall not be left until the morning.

g Heb *Asherim*   *h* Gk Theodotion Vg Tg: Meaning of Heb uncertain

---

34.10–28: This may be another tradition about the making of a covenant (v. 10), parallel to that of chs 19–24. In the present context, however, it is understood as a renewal of the covenant after it was broken by the people (ch 32). **11–16:** Intolerance of pagan forms of worship was motivated by fear of the seductive power of idolatry (see 23.24). **13:** The *pillars* were upright stones which stood near Baal shrines; the *sacred poles* (or "Asherim") symbolized Asherah, the mother goddess of Canaanite religion (Judg 2.13). **14:** Religious exclusivism is derived from the fundamental conviction of Mosaic faith; the total claim of Israel's God upon the people's loyalty (20.3). The LORD's *name* (or character) is *Jealous,* i.e. the LORD will tolerate no rivals for Israel's devotion (20.5; Deut 4.24). **34.18–24:** A cultic calendar. The laws concerning the three annual festivals (vv. 18, 22–23) are paralleled in 23.14–17. On the redemption of the firstborn (vv. 19–20), see 13.13 n. **24:** The property will be protected while the pilgrims are on their way to the central sanctuary. **34.25–26:** 23.18–19. **27–28:** Moses' special role as covenant mediator (see 19.9 n.) is shown by the fact that the covenant is made with him and, through him, with Israel. *These words,* a reference to the preceding cultic laws. Some have attempted to arrange these into a decalogue (compare v. 28b).

26 The best of the first fruits of your ground you shall bring to the house of the LORD your God.

You shall not boil a kid in its mother's milk.

27 The LORD said to Moses: Write these words; in accordance with these words I have made a covenant with you and with Israel. 28 He was there with the LORD forty days and forty nights; he neither ate bread nor drank water. And he wrote on the tablets the words of the covenant, the ten commandments. *i*

29 Moses came down from Mount Sinai. As he came down from the mountain with the two tablets of the covenant *j* in his hand, Moses did not know that the skin of his face shone because he had been talking with God. 30 When Aaron and all the Israelites saw Moses, the skin of his face was shining, and they were afraid to come near him. 31 But Moses called to them; and Aaron and all the leaders of the congregation returned to him, and Moses spoke with them. 32 Afterward all the Israelites came near, and he gave them in commandment all that the LORD had spoken with him on Mount Sinai. 33 When Moses had finished speaking with them, he put a veil on his face; 34 but whenever Moses went in before the LORD to speak with him, he would take the veil off, until he came out; and when he came out, and told the Israelites what he had been commanded, 35 the Israelites would see the face of Moses, that the skin of his face was shining; and Moses would put the veil on his face again, until he went in to speak with him.

35 Moses assembled all the congregation of the Israelites and said to them: These are the things that the LORD has commanded you to do:

2 Six days shall work be done, but on the seventh day you shall have a holy sabbath of solemn rest to the LORD; whoever does any work on it shall be put to death. 3 You shall kindle no fire in all your dwellings on the sabbath day.

4 Moses said to all the congregation of the Israelites: This is the thing that the LORD has commanded: 5 Take from among you an offering to the LORD; let whoever is of a generous heart bring the LORD's offering: gold, silver, and bronze; 6 blue, purple, and crimson yarns, and fine linen; goats' hair, 7 tanned rams' skins, and fine leather; *k* acacia wood, 8 oil for the light, spices for the anointing oil and for the fragrant incense, 9 and onyx stones and gems to be set in the ephod and the breastpiece.

10 All who are skillful among you shall come and make all that the LORD has commanded: the tabernacle, 11 its tent and its covering, its clasps and its frames, its bars, its pillars, and its bases; 12 the ark with its poles, the mercy seat, *l* and the curtain for the screen; 13 the table with its poles and all its utensils, and the bread of the Presence; 14 the lampstand also for the light, with its utensils and its lamps, and the oil for the light; 15 and the altar of incense, with its poles, and the anointing oil and the fragrant incense, and the screen for the entrance, the entrance of the tabernacle; 16 the altar of burnt offering, with its grating of bronze, its poles, and all its utensils, the basin with its stand; 17 the hangings of the court, its pillars and its bases, and the screen for the gate of the court; 18 the pegs of the tabernacle and the pegs of the court, and their

---

i Heb *words*   *j* Or *treaty*, or *testimony*; Heb
eduth   *k* Meaning of Heb uncertain
l Or *the cover*

Probably, however, the editor has blended two covenant traditions: one based on the decalogue and the other on this set of ritual laws. **29–35**: According to priestly tradition, the radiant glory of the LORD so transfigured Moses' face (compare Mt 17.1–7) that he had to wear a veil (2 Cor 3.7–18).

Chapters 35–40: The establishment of the cult. This priestly section shows how the instructions given to Moses in chs 25–31 were carried out.
35.1–3: See 31.12–17. The sabbath law is placed first so as to restrict work on the tabernacle. **4–29**: An expansion of 25.1–9.

cords; [19] the finely worked vestments for ministering in the holy place, the holy vestments for the priest Aaron, and the vestments of his sons, for their service as priests.

20 Then all the congregation of the Israelites withdrew from the presence of Moses. [21] And they came, everyone whose heart was stirred, and everyone whose spirit was willing, and brought the LORD's offering to be used for the tent of meeting, and for all its service, and for the sacred vestments. [22] So they came, both men and women; all who were of a willing heart brought brooches and earrings and signet rings and pendants, all sorts of gold objects, everyone bringing an offering of gold to the LORD. [23] And everyone who possessed blue or purple or crimson yarn or fine linen or goats' hair or tanned rams' skins or fine leather, [m] brought them. [24] Everyone who could make an offering of silver or bronze brought it as the LORD's offering; and everyone who possessed acacia wood of any use in the work, brought it. [25] All the skillful women spun with their hands, and brought what they had spun in blue and purple and crimson yarns and fine linen; [26] all the women whose hearts moved them to use their skill spun the goats' hair. [27] And the leaders brought onyx stones and gems to be set in the ephod and the breastpiece, [28] and spices and oil for the light, and for the anointing oil, and for the fragrant incense. [29] All the Israelite men and women whose hearts made them willing to bring anything for the work that the LORD had commanded by Moses to be done, brought it as a freewill offering to the LORD.

30 Then Moses said to the Israelites: See, the LORD has called by name Bezalel son of Uri son of Hur, of the tribe of Judah; [31] he has filled him with divine spirit, [n] with skill, intelligence, and knowledge in every kind of craft, [32] to devise artistic designs, to work in gold, silver, and bronze, [33] in cutting stones for setting, and in carving wood, in every

kind of craft. [34] And he has inspired him to teach, both him and Oholiab son of Ahisamach, of the tribe of Dan. [35] He has filled them with skill to do every kind of work done by an artisan or by a designer or by an embroiderer in blue, purple, and crimson yarns, and in fine linen, or by a weaver—by any sort of artisan or skilled designer.

36 Bezalel and Oholiab and every skillful one to whom the LORD has given skill and understanding to know how to do any work in the construction of the sanctuary shall work in accordance with all that the LORD has commanded.

2 Moses then called Bezalel and Oholiab and every skillful one to whom the LORD had given skill, everyone whose heart was stirred to come to do the work; [3] and they received from Moses all the freewill offerings that the Israelites had brought for doing the work on the sanctuary. They still kept bringing him freewill offerings every morning, [4] so that all the artisans who were doing every sort of task on the sanctuary came, each from the task being performed, [5] and said to Moses, "The people are bringing much more than enough for doing the work that the LORD has commanded us to do." [6] So Moses gave command, and word was proclaimed throughout the camp: "No man or woman is to make anything else as an offering for the sanctuary." So the people were restrained from bringing; [7] for what they had already brought was more than enough to do all the work.

8 All those with skill among the workers made the tabernacle with ten curtains; they were made of fine twisted linen, and blue, purple, and crimson yarns, with cherubim skillfully worked into them. [9] The length of each curtain was twenty-eight cubits, and the width of each curtain four cubits; all the curtains were of the same size.

m Meaning of Heb uncertain   *n* Or *the spirit
of God*

35.30–36.7: See 31.1–11.   36.8–38: See 26.1–37.

10 He joined five curtains to one another, and the other five curtains he joined to one another. 11 He made loops of blue on the edge of the outermost curtain of the first set; likewise he made them on the edge of the outermost curtain of the second set; 12 he made fifty loops on the one curtain, and he made fifty loops on the edge of the curtain that was in the second set; the loops were opposite one another. 13 And he made fifty clasps of gold, and joined the curtains one to the other with clasps; so the tabernacle was one whole.

14 He also made curtains of goats' hair for a tent over the tabernacle; he made eleven curtains. 15 The length of each curtain was thirty cubits, and the width of each curtain four cubits; the eleven curtains were of the same size. 16 He joined five curtains by themselves, and six curtains by themselves. 17 He made fifty loops on the edge of the outermost curtain of the one set, and fifty loops on the edge of the other connecting curtain. 18 He made fifty clasps of bronze to join the tent together so that it might be one whole. 19 And he made for the tent a covering of tanned rams' skins and an outer covering of fine leather.*o*

20 Then he made the upright frames for the tabernacle of acacia wood. 21 Ten cubits was the length of a frame, and a cubit and a half the width of each frame. 22 Each frame had two pegs for fitting together; he did this for all the frames of the tabernacle. 23 The frames for the tabernacle he made in this way: twenty frames for the south side; 24 and he made forty bases of silver under the twenty frames, two bases under the first frame for its two pegs, and two bases under the next frame for its two pegs. 25 For the second side of the tabernacle, on the north side, he made twenty frames 26 and their forty bases of silver, two bases under the first frame and two bases under the next frame. 27 For the rear of the tabernacle westward he made six frames. 28 He made two frames for corners of the tabernacle in the rear. 29 They were separate beneath, but joined at the top, at the first ring; he made two of them in this way, for the two corners. 30 There were eight frames with their bases of silver: sixteen bases, under every frame two bases.

31 He made bars of acacia wood, five for the frames of the one side of the tabernacle, 32 and five bars for the frames of the other side of the tabernacle, and five bars for the frames of the tabernacle at the rear westward. 33 He made the middle bar to pass through from end to end halfway up the frames. 34 And he overlaid the frames with gold, and made rings of gold for them to hold the bars, and overlaid the bars with gold.

35 He made the curtain of blue, purple, and crimson yarns, and fine twisted linen, with cherubim skillfully worked into it. 36 For it he made four pillars of acacia, and overlaid them with gold; their hooks were of gold, and he cast for them four bases of silver. 37 He also made a screen for the entrance to the tent, of blue, purple, and crimson yarns, and fine twisted linen, embroidered with needlework; 38 and its five pillars with their hooks. He overlaid their capitals and their bases with gold, but their five bases were of bronze.

37 Bezalel made the ark of acacia wood; it was two and a half cubits long, a cubit and a half wide, and a cubit and a half high. 2 He overlaid it with pure gold inside and outside, and made a molding of gold around it. 3 He cast for it four rings of gold for its four feet, two rings on its one side and two rings on its other side. 4 He made poles of acacia wood, and overlaid them with gold, 5 and put the poles into the rings on the sides of the ark, to carry the ark. 6 He made a mercy seat*p* of pure gold; two cubits and a half was its length, and a cubit and a half its width. 7 He made two cherubim of hammered gold; at the two

o Meaning of Heb uncertain   *p* Or *a cover*

37.1–9: See 25.10–22.

ends of the mercy seat*q* he made them, [8]one cherub at the one end, and one cherub at the other end; of one piece with the mercy seat*q* he made the cherubim at its two ends. [9]The cherubim spread out their wings above, overshadowing the mercy seat*q* with their wings. They faced one another; the faces of the cherubim were turned toward the mercy seat. *q*

10 He also made the table of acacia wood, two cubits long, one cubit wide, and a cubit and a half high. [11]He overlaid it with pure gold, and made a molding of gold around it. [12]He made around it a rim a handbreadth wide, and made a molding of gold around the rim. [13]He cast for it four rings of gold, and fastened the rings to the four corners at its four legs. [14]The rings that held the poles used for carrying the table were close to the rim. [15]He made the poles of acacia wood to carry the table, and overlaid them with gold. [16]And he made the vessels of pure gold that were to be on the table, its plates and dishes for incense, and its bowls and flagons with which to pour drink offerings.

17 He also made the lampstand of pure gold. The base and the shaft of the lampstand were made of hammered work; its cups, its calyxes, and its petals were of one piece with it. [18]There were six branches going out of its sides, three branches of the lampstand out of one side of it and three branches of the lampstand out of the other side of it; [19]three cups shaped like almond blossoms, each with calyx and petals, on one branch, and three cups shaped like almond blossoms, each with calyx and petals, on the other branch—so for the six branches going out of the lampstand. [20]On the lampstand itself there were four cups shaped like almond blossoms, each with its calyxes and petals. [21]There was a calyx of one piece with it under the first pair of branches, a calyx of one piece with it under the next pair of branches, and a

calyx of one piece with it under the last pair of branches. [22]Their calyxes and their branches were of one piece with it, the whole of it one hammered piece of pure gold. [23]He made its seven lamps and its snuffers and its trays of pure gold. [24]He made it and all its utensils of a talent of pure gold.

25 He made the altar of incense of acacia wood, one cubit long, and one cubit wide; it was square, and was two cubits high; its horns were of one piece with it. [26]He overlaid it with pure gold, its top, and its sides all around, and its horns; and he made for it a molding of gold all around, [27]and made two golden rings for it under its molding, on two opposite sides of it, to hold the poles with which to carry it. [28]And he made the poles of acacia wood, and overlaid them with gold.

29 He made the holy anointing oil also, and the pure fragrant incense, blended as by the perfumer.

38 He made the altar of burnt offering also of acacia wood; it was five cubits long, and five cubits wide; it was square, and three cubits high. [2]He made horns for it on its four corners; its horns were of one piece with it, and he overlaid it with bronze. [3]He made all the utensils of the altar, the pots, the shovels, the basins, the forks, and the firepans: all its utensils he made of bronze. [4]He made for the altar a grating, a network of bronze, under its ledge, extending halfway down. [5]He cast four rings on the four corners of the bronze grating to hold the poles; [6]he made the poles of acacia wood, and overlaid them with bronze. [7]And he put the poles through the rings on the sides of the altar, to carry it with them; he made it hollow, with boards.

8 He made the basin of bronze with its stand of bronze, from the mirrors of

q Or *the cover*

37.10–16: See 25.23–30. **37.17–24**: See 25.31–40. **37.25–28**: See 30.1–10. **29**: See 30.22–38. **38.1–7**: See 27.1–8. **8**: See 30.17–21.

the women who served at the entrance to the tent of meeting.

9 He made the court; for the south side the hangings of the court were of fine twisted linen, one hundred cubits long; [10] its twenty pillars and their twenty bases were of bronze, but the hooks of the pillars and their bands were of silver. [11] For the north side there were hangings one hundred cubits long; its twenty pillars and their twenty bases were of bronze, but the hooks of the pillars and their bands were of silver. [12] For the west side there were hangings fifty cubits long, with ten pillars and ten bases; the hooks of the pillars and their bands were of silver. [13] And for the front to the east, fifty cubits. [14] The hangings for one side of the gate were fifteen cubits, with three pillars and three bases. [15] And so for the other side; on each side of the gate of the court were hangings of fifteen cubits, with three pillars and three bases. [16] All the hangings around the court were of fine twisted linen. [17] The bases for the pillars were of bronze, but the hooks of the pillars and their bands were of silver; the overlaying of their capitals was also of silver, and all the pillars of the court were banded with silver. [18] The screen for the entrance to the court was embroidered with needlework in blue, purple, and crimson yarns and fine twisted linen. It was twenty cubits long and, along the width of it, five cubits high, corresponding to the hangings of the court. [19] There were four pillars; their four bases were of bronze, their hooks of silver, and the overlaying of their capitals and their bands of silver. [20] All the pegs for the tabernacle and for the court all around were of bronze.

21 These are the records of the tabernacle, the tabernacle of the covenant,[r] which were drawn up at the command-ment of Moses, the work of the Levites being under the direction of Ithamar son of the priest Aaron. [22] Bezalel son of Uri son of Hur, of the tribe of Judah, made all that the LORD commanded Moses; [23] and with him was Oholiab son of Ahisamach, of the tribe of Dan, engraver, designer, and embroiderer in blue, purple, and crimson yarns, and in fine linen.

24 All the gold that was used for the work, in all the construction of the sanctuary, the gold from the offering, was twenty-nine talents and seven hundred thirty shekels, measured by the sanctuary shekel. [25] The silver from those of the congregation who were counted was one hundred talents and one thousand seven hundred seventy-five shekels, measured by the sanctuary shekel; [26] a beka a head (that is, half a shekel, measured by the sanctuary shekel), for everyone who was counted in the census, from twenty years old and upward, for six hundred three thousand, five hundred fifty men. [27] The hundred talents of silver were for casting the bases of the sanctuary, and the bases of the curtain; one hundred bases for the hundred talents, a talent for a base. [28] Of the thousand seven hundred seventy-five shekels he made hooks for the pillars, and overlaid their capitals and made bands for them. [29] The bronze that was contributed was seventy talents, and two thousand four hundred shekels; [30] with it he made the bases for the entrance of the tent of meeting, the bronze altar and the bronze grating for it and all the utensils of the altar, [31] the bases all around the court, and the bases of the gate of the court, all the pegs of the tabernacle, and all the pegs around the court.

39 Of the blue, purple, and crimson yarns they made finely worked vestments, for ministering in the holy

r Or *treaty*, or *testimony*; Heb *eduth*

38.9–20: See 27.9–19.
38.21–31: This is a supplement which presupposes the appointment of Ithamar as head of the Levites (Num ch 3; 4.33) and the Israelite census (Num ch 1). **21:** *The tabernacle of the covenant,* a phrase indicating that the sanctu-ary contained the ark and ten commandments. **24–26:** The silver tax is computed at the rate set forth in 30.11–16 and with the figures of the later census in mind (see Num 1.17–46 n.), the result being that the figures are highly exaggerated.

place; they made the sacred vestments for Aaron; as the Lord had commanded Moses.

2 He made the ephod of gold, of blue, purple, and crimson yarns, and of fine twisted linen. 3 Gold leaf was hammered out and cut into threads to work into the blue, purple, and crimson yarns and into the fine twisted linen, in skilled design. 4 They made for the ephod shoulder-pieces, joined to it at its two edges. 5 The decorated band on it was of the same materials and workmanship, of gold, of blue, purple, and crimson yarns, and of fine twisted linen; as the Lord had commanded Moses.

6 The onyx stones were prepared, enclosed in settings of gold filigree and engraved like the engravings of a signet, according to the names of the sons of Israel. 7 He set them on the shoulder-pieces of the ephod, to be stones of remembrance for the sons of Israel; as the Lord had commanded Moses.

8 He made the breastpiece, in skilled work, like the work of the ephod, of gold, of blue, purple, and crimson yarns, and of fine twisted linen. 9 It was square; the breastpiece was made double, a span in length and a span in width when doubled. 10 They set in it four rows of stones. A row of carnelian, *s* chrysolite, and emerald was the first row; 11 and the second row, a turquoise, a sapphire, *t* and a moonstone; 12 and the third row, a jacinth, an agate, and an amethyst; 13 and the fourth row, a beryl, an onyx, and a jasper; they were enclosed in settings of gold filigree. 14 There were twelve stones with names corresponding to the names of the sons of Israel; they were like signets, each engraved with its name, for the twelve tribes. 15 They made on the breastpiece chains of pure gold, twisted like cords; 16 and they made two settings of gold filigree and two gold rings, and put the two rings on the two edges of the breastpiece; 17 and they put the two cords of gold in the two rings at the edges of the breastpiece. 18 Two ends of the two

cords they had attached to the two settings of filigree; in this way they attached it in front to the shoulder-pieces of the ephod. 19 Then they made two rings of gold, and put them at the two ends of the breastpiece, on its inside edge next to the ephod. 20 They made two rings of gold, and attached them in front to the lower part of the two shoulder-pieces of the ephod, at its joining above the decorated band of the ephod. 21 They bound the breastpiece by its rings to the rings of the ephod with a blue cord, so that it should lie on the decorated band of the ephod, and that the breastpiece should not come loose from the ephod; as the Lord had commanded Moses.

22 He also made the robe of the ephod woven all of blue yarn; 23 and the opening of the robe in the middle of it was like the opening in a coat of mail, *u* with a binding around the opening, so that it might not be torn. 24 On the lower hem of the robe they made pomegranates of blue, purple, and crimson yarns, and of fine twisted linen. 25 They also made bells of pure gold, and put the bells between the pomegranates on the lower hem of the robe all around, between the pomegranates; 26 a bell and a pomegranate, a bell and a pomegranate all around on the lower hem of the robe for ministering; as the Lord had commanded Moses.

27 They also made the tunics, woven of fine linen, for Aaron and his sons, 28 and the turban of fine linen, and the headdresses of fine linen, and the linen undergarments of fine twisted linen, 29 and the sash of fine twisted linen, and of blue, purple, and crimson yarns, embroidered with needlework; as the Lord had commanded Moses.

30 They made the rosette of the holy diadem of pure gold, and wrote on it an inscription, like the engraving of a signet, "Holy to the Lord." 31 They tied to it a blue cord, to fasten it on the turban

s The identification of several of these stones is uncertain   *t* Or *lapis lazuli*   *u* Meaning of Heb uncertain

39.1–31: See 28.1–43.

above; as the LORD had commanded Moses.

32 In this way all the work of the tabernacle of the tent of meeting was finished; the Israelites had done everything just as the LORD had commanded Moses. 33 Then they brought the tabernacle to Moses, the tent and all its utensils, its hooks, its frames, its bars, its pillars, and its bases; 34 the covering of tanned rams' skins and the covering of fine leather,*v* and the curtain for the screen; 35 the ark of the covenant*w* with its poles and the mercy seat;*x* 36 the table with all its utensils, and the bread of the Presence; 37 the pure lampstand with its lamps set on it and all its utensils, and the oil for the light; 38 the golden altar, the anointing oil and the fragrant incense, and the screen for the entrance of the tent; 39 the bronze altar, and its grating of bronze, its poles, and all its utensils; the basin with its stand; 40 the hangings of the court, its pillars, and its bases, and the screen for the gate of the court, its cords, and its pegs; and all the utensils for the service of the tabernacle, for the tent of meeting; 41 the finely worked vestments for ministering in the holy place, the sacred vestments for the priest Aaron, and the vestments of his sons to serve as priests. 42 The Israelites had done all of the work just as the LORD had commanded Moses. 43 When Moses saw that they had done all the work just as the LORD had commanded, he blessed them.

40 The LORD spoke to Moses: 2 On the first day of the first month you shall set up the tabernacle of the tent of meeting. 3 You shall put in it the ark of the covenant,*w* and you shall screen the ark with the curtain. 4 You shall bring in the table, and arrange its setting; and you shall bring in the lampstand, and set up its lamps. 5 You shall put the golden altar for incense before the ark of the covenant,*w* and set up the screen for the entrance of the tabernacle. 6 You shall set the altar of burnt offering before the entrance of the tabernacle of the tent of meeting, 7 and place the basin between the tent of meeting and the altar, and put water in it. 8 You shall set up the court all around, and hang up the screen for the gate of the court. 9 Then you shall take the anointing oil, and anoint the tabernacle and all that is in it, and consecrate it and all its furniture, so that it shall become holy. 10 You shall also anoint the altar of burnt offering and all its utensils, and consecrate the altar, so that the altar shall be most holy. 11 You shall also anoint the basin with its stand, and consecrate it. 12 Then you shall bring Aaron and his sons to the entrance of the tent of meeting, and shall wash them with water, 13 and put on Aaron the sacred vestments, and you shall anoint him and consecrate him, so that he may serve me as priest. 14 You shall bring his sons also and put tunics on them, 15 and anoint them, as you anointed their father, that they may serve me as priests: and their anointing shall admit them to a perpetual priesthood throughout all generations to come.

16 Moses did everything just as the LORD had commanded him. 17 In the first month in the second year, on the first day of the month, the tabernacle was set up. 18 Moses set up the tabernacle; he laid its bases, and set up its frames, and put in its poles, and raised up its pillars; 19 and he spread the tent over the tabernacle, and put the covering of the tent over it; as the LORD had commanded Moses. 20 He took the covenant*w* and put it into the ark, and put the poles on the ark, and set the mercy seat*x* above the ark; 21 and he brought

v Meaning of Heb uncertain   *w* Or *treaty*, or *testimony*; Heb *eduth*   *x* Or *the cover*

39.32–43: Completion of the work of *the tabernacle of the tent of meeting.* On the blending of the tabernacle and tent traditions (vv. 33, 40), compare 26.7–14.

40.1–33: The erection and furnishing of the tabernacle according to previous instructions. **9–15:** See 30.26–30. **17:** According to priestly chronology, the workers erected the tabernacle nine months after the arrival at Sinai (19.1).

the ark into the tabernacle, and set up the curtain for screening, and screened the ark of the covenant;*y* as the LORD had commanded Moses. 22 He put the table in the tent of meeting, on the north side of the tabernacle, outside the curtain, 23 and set the bread in order on it before the LORD; as the LORD had commanded Moses. 24 He put the lampstand in the tent of meeting, opposite the table on the south side of the tabernacle, 25 and set up the lamps before the LORD; as the LORD had commanded Moses. 26 He put the golden altar in the tent of meeting before the curtain, 27 and offered fragrant incense on it; as the LORD had commanded Moses. 28 He also put in place the screen for the entrance of the tabernacle. 29 He set the altar of burnt offering at the entrance of the tabernacle of the tent of meeting, and offered on it the burnt offering and the grain offering as the LORD had commanded Moses. 30 He set the basin between the tent of meeting and the altar, and put water in it for washing, 31 with which Moses and Aaron and his sons washed their hands and their feet. 32 When they went into the tent of meeting, and when they approached the altar, they washed; as the LORD had commanded Moses. 33 He set up the court around the tabernacle and the altar, and put up the screen at the gate of the court. So Moses finished the work.

34 Then the cloud covered the tent of meeting, and the glory of the LORD filled the tabernacle. 35 Moses was not able to enter the tent of meeting because the cloud settled upon it, and the glory of the LORD filled the tabernacle. 36 Whenever the cloud was taken up from the tabernacle, the Israelites would set out on each stage of their journey; 37 but if the cloud was not taken up, then they did not set out until the day that it was taken up. 38 For the cloud of the LORD was on the tabernacle by day, and fire was in the cloud*z* by night, before the eyes of all the house of Israel at each stage of their journey.

y Or *treaty*, or *testimony*; Heb *eduth*   *z* Heb *it*

40.34–38: These concluding verses hark back to 25.8 (see 29.43–46). *The cloud* and *the glory* are signs of God's tabernacling presence (see 16.6–7 n.). **35**: 1 Kings 8.10–11; Isa 6.3, 4. **36–38**: The ancient symbols of the pillar of cloud by day and the pillar of fire by night (13.21–22) are connected with the tabernacle on the view that it was a portable sanctuary that accompanied Israel on its journey (see Num 9.15–23).

Leviticus

Leviticus is pre-eminently a book of worship. The English title, derived from the Greek and Latin versions of the Hebrew Bible, refers to the Levitical priests who were set apart to minister at the sanctuary. In early rabbinic tradition the book is called "the Priests' Manual." It falls into six parts: (1) laws dealing with sacrifices (chs 1–7); (2) the consecration of the priests to their office (chs 8–10); (3) laws setting forth the distinction between clean and unclean (chs 11–15); (4) the ceremony for the annual day of atonement (ch 16); (5) laws to govern Israel's life as a holy people (chs 17–26); (6) an appendix on religious vows (ch 27).

Although this section of the Pentateuch has become a separate book, actually it is a continuation of the priestly tradition at the end of the book of Exodus (chs 25–31; 35–40). Moreover, the same tradition extends without interruption through the first ten chapters of Numbers. It is generally agreed that this priestly material in its present form comes from a relatively late period (see introduction to "The Pentateuch," pp. xxxv–xxxvi OT). The compiler, however, has relied upon independent source materials, such as the so-called Holiness Code (chs 17–26), and upon numerous traditions that reach back to ancient times.

Through the various rituals and laws there breathes the conviction that the holy God "tabernacles" in the midst of the people during their historical pilgrimage (Ex 40.34–38). The nearness of God not only accentuates the people's sense of sin but prompts them to seek God in sacrificial services of worship. For, according to the priestly witness, God has provided the means of grace whereby the people, forgiven and restored, may live in the presence of the holy God, avoiding those things that contaminate their health and well-being, and doing those things that make them a holy people, separated for a divine service in the world.

1 The LORD summoned Moses and spoke to him from the tent of meeting, saying: ²Speak to the people of Israel and say to them: When any of you bring an offering of livestock to the LORD, you shall bring your offering from the herd or from the flock.

3 If the offering is a burnt offering from the herd, you shall offer a male without blemish; you shall bring it to the entrance of the tent of meeting, for acceptance in your behalf before the LORD. ⁴You shall lay your hand on the head of the burnt offering, and it shall be acceptable in your behalf as atonement for you. ⁵The bull shall be slaughtered before the LORD; and Aaron's sons the priests shall offer the blood, dashing the blood against all sides of the altar that is at the entrance of the tent of meeting. ⁶The burnt offering shall be flayed and cut up into its parts. ⁷The sons of the priest Aaron shall put fire on the altar and arrange wood on the fire. ⁸Aaron's sons the priests shall arrange the parts, with the head and the suet, on the wood that is on the fire on the altar; ⁹but its entrails and its legs shall be washed with water. Then the priest shall turn the whole into smoke on the altar as a burnt offering, an offering by fire of pleasing odor to the LORD.

10 If your gift for a burnt offering is from the flock, from the sheep or goats, your offering shall be a male without blemish. ¹¹It shall be slaughtered on the north side of the altar before the LORD, and Aaron's sons the priests shall dash its blood against all sides of the altar. ¹²It shall be cut up into its parts, with its head and its suet, and the priest shall arrange them on the wood that is on the fire on the altar; ¹³but the entrails and the legs shall be washed with water. Then the priest shall offer the whole and turn it into smoke on the altar; it is a burnt offering, an offering by fire of pleasing odor to the LORD.

14 If your offering to the LORD is a burnt offering of birds, you shall choose your offering from turtledoves or pigeons. ¹⁵The priest shall bring it to the altar and wring off its head, and turn it into smoke on the altar; and its blood shall be drained out against the side of the altar. ¹⁶He shall remove its crop with its contents *a* and throw it at the east side of the altar, in the place for ashes. ¹⁷He shall tear it open by its wings without severing it. Then the priest shall turn it into smoke on the altar, on the wood that is on the fire; it is a burnt offering, an offering by fire of pleasing odor to the LORD.

2 When anyone presents a grain offering to the LORD, the offering shall be of choice flour; the worshiper shall pour

a Meaning of Heb uncertain

1.1–17: Burnt offerings. 1: In the priestly view the laws of Leviticus were delivered to Moses from *the tent of meeting* or tabernacle (Ex 25.22; 26.1–37) during the wilderness sojourn (7.37–38; Ex 40.16–38). **2:** The sacrifices dealt with in this chapter are regarded as offerings or gifts to God out of the worshiper's substance. **3–9:** A sacrifice from the herd. **3:** The burnt offering, the chief daily offering (6.9), was one in which the whole animal was burnt on the altar as an act of praise and adoration. *To the entrance,* i.e. at the great altar in the court (Ex ch 38; 40.6). **4:** The worshiper symbolically identifies with the sacrifice presented (see Ex 29.10 n.). According to the ancient notion of substitutionary sacrifice, the animal sacrifice makes *atonement* for, or "puts a cover" over, the sin of the worshiper (see Ex 29.35–37 n.). **5:** Blood, the seat of the mystery of life (17.11; Deut 12.23; Gen 9.4), was held to be peculiarly sacred to God. Therefore, on the principle of the sacrifice of life for life, the shedding of blood was efficacious in forgiving sin and reconciling persons to God. The act of throwing the blood against the altar symbolizes God's participation in the atonement ceremony (see Ex 24.6–8 n.). On *Aaron's sons the priests,* see Ex 28.1–5 n. **9:** *A pleasing odor,* a traditional expression for an offering acceptable to God (see Gen 8.20–22 n.; compare Eph 5.2).

1.10–13: A sacrifice from the flock. **14–17:** The offering to be made by poor people who cannot afford a sacrifice from the herd or flock.

2.1–16: Grain offerings. 1: As Abel's offering was from the flock, so Cain's was a typical grain offering from "the fruit of the ground" (Gen 4.3–5). Both types were expressions of gratitude and praise. Grain offer-

oil on it, and put frankincense on it, ²and bring it to Aaron's sons the priests. After taking from it a handful of the choice flour and oil, with all its frankincense, the priest shall turn this token portion into smoke on the altar, an offering by fire of pleasing odor to the LORD. ³And what is left of the grain offering shall be for Aaron and his sons, a most holy part of the offerings by fire to the LORD.

4 When you present a grain offering baked in the oven, it shall be of choice flour: unleavened cakes mixed with oil, or unleavened wafers spread with oil. ⁵If your offering is grain prepared on a griddle, it shall be of choice flour mixed with oil, unleavened; ⁶break it in pieces, and pour oil on it; it is a grain offering. ⁷If your offering is grain prepared in a pan, it shall be made of choice flour in oil. ⁸You shall bring to the LORD the grain offering that is prepared in any of these ways; and when it is presented to the priest, he shall take it to the altar. ⁹The priest shall remove from the grain offering its token portion and turn this into smoke on the altar, an offering by fire of pleasing odor to the LORD. ¹⁰And what is left of the grain offering shall be for Aaron and his sons; it is a most holy part of the offerings by fire to the LORD.

11 No grain offering that you bring to the LORD shall be made with leaven, for you must not turn any leaven or honey into smoke as an offering by fire to the LORD. ¹²You may bring them to the LORD as an offering of choice products, but they shall not be offered on the altar for a pleasing odor. ¹³You shall not omit from your grain offerings the salt of the covenant with your God; with all your offerings you shall offer salt.

14 If you bring a grain offering of first fruits to the LORD, you shall bring as the grain offering of your first fruits coarse new grain from fresh ears, parched with fire. ¹⁵You shall add oil to it and lay frankincense on it; it is a grain offering. ¹⁶And the priest shall turn a token portion of it into smoke—some of the coarse grain and oil with all its frankincense; it is an offering by fire to the LORD.

3 If the offering is a sacrifice of well-being, if you offer an animal of the herd, whether male or female, you shall offer one without blemish before the LORD. ²You shall lay your hand on the head of the offering and slaughter it at the entrance of the tent of meeting; and Aaron's sons the priests shall dash the blood against all sides of the altar. ³You shall offer from the sacrifice of well-being, as an offering by fire to the LORD, the fat that covers the entrails and all the fat that is around the entrails; ⁴the two kidneys with the fat that is on them at the loins, and the appendage of the liver, which he shall remove with the kidneys. ⁵Then Aaron's sons shall turn these into smoke on the altar, with the burnt offering that

---

ings often accompanied an animal sacrifice (7.11–14; 8.26; 9.4; Num 15.1–10). *Oil,* i.e. olive oil. *Frankincense,* Ex 30.34. **2–3**: Part of the offering is a *token portion,* i.e. it memorializes the worshiper before God. The unburned remainder is *most holy* because, being consecrated to God, only the priests could eat it. In Israel it was not believed that sacrifices satisfied the physical needs of the Deity (compare Ps 50.9–13). Rather, sacrifice supported the priests (see Ex 29.24 n.; 1 Sam 2.13–17), who, in eating their portion, identified themselves with worshipers in their approach to God. **2.11**: Honey was forbidden because, like leaven (see Ex 12.15 n.), it was associated with foods that ferment. **12**: *Choice products,* or "first fruits" (vv. 14–16; see Ex 23.19; 34.26). **13**: *Salt of the covenant* (Num 18.19; 2 Chr 13.5) reflects the oriental practice of making a covenant by eating a meal seasoned with salt. Here salt symbolizes the covenant relation upon which the whole sacrificial system rests.

3.1–17: **Offerings of well-being.** While the burnt offering was a sacrifice of praise (ch 1), the offering of well-being, also an ancient type of sacrifice (Ex 24.5; Deut 12.7, 18; 1 Sam 9.11–14, 22–24), was a covenant meal in which the worshiper was sacramentally related to the LORD and to the community of Israelites. **1–5**: Sacrifice from the herd (1.3–9). In this case only certain parts (suet, kidneys) are burned; the rest is consumed by priests and people in a communion meal.

is on the wood on the fire, as an offering by fire of pleasing odor to the LORD.

6 If your offering for a sacrifice of well-being to the LORD is from the flock, male or female, you shall offer one without blemish. 7 If you present a sheep as your offering, you shall bring it before the LORD 8 and lay your hand on the head of the offering. It shall be slaughtered before the tent of meeting, and Aaron's sons shall dash its blood against all sides of the altar. 9 You shall present its fat from the sacrifice of well-being, as an offering by fire to the LORD: the whole broad tail, which shall be removed close to the backbone, the fat that covers the entrails, and all the fat that is around the entrails; 10 the two kidneys with the fat that is on them at the loins, and the appendage of the liver, which you shall remove with the kidneys. 11 Then the priest shall turn these into smoke on the altar as a food offering by fire to the LORD.

12 If your offering is a goat, you shall bring it before the LORD 13 and lay your hand on its head; it shall be slaughtered before the tent of meeting; and the sons of Aaron shall dash its blood against all sides of the altar. 14 You shall present as your offering from it, as an offering by fire to the LORD, the fat that covers the entrails, and all the fat that is around the entrails; 15 the two kidneys with the fat that is on them at the loins, and the appendage of the liver, which you shall remove with the kidneys. 16 Then the priest shall turn these into smoke on the altar as a food offering by fire for a pleasing odor.

All fat is the LORD's. 17 It shall be a perpetual statute throughout your generations, in all your settlements: you must not eat any fat or any blood.

4 The LORD spoke to Moses, saying, 2 Speak to the people of Israel, saying: When anyone sins unintentionally in any of the LORD's commandments about things not to be done, and does any one of them:

3 If it is the anointed priest who sins, thus bringing guilt on the people, he shall offer for the sin that he has committed a bull of the herd without blemish as a sin offering to the LORD. 4 He shall bring the bull to the entrance of the tent of meeting before the LORD and lay his hand on the head of the bull; the bull shall be slaughtered before the LORD. 5 The anointed priest shall take some of the blood of the bull and bring it into the tent of meeting. 6 The priest shall dip his finger in the blood and sprinkle some of the blood seven times before the LORD in front of the curtain of the sanctuary. 7 The priest shall put some of the blood on the horns of the altar of fragrant incense that is in the tent of meeting before the LORD; and the rest of the blood of the bull he shall pour out at the base of the altar of burnt offering, which is at the entrance of the tent of meeting. 8 He shall remove all the fat from the bull of sin offering: the fat that covers the entrails and all the fat that is around the entrails; 9 the two kidneys with the fat that is on them at the loins; and the appendage of the liver, which he shall remove with the kidneys, 10 just as these are removed from the ox of the sacrifice of well-being. The priest shall turn them into

---

3.6–17: Sacrifice from the flock (compare 1.10–13). 16: Fat, like blood, was held to be God's portion of the sacrifice (7.22–27; Ex 23.18).

4.1–5.13: **The sin offering.** This is a sacrifice of repentance for sin that has broken one's relation to God and has endangered the welfare of the community.

4.2: The sacrifice is efficacious only for one who sins *unintentionally*, i.e. who inadvertently offends God's holiness. The sacrifice must be accompanied by confession (5.5).

Priestly tradition provides no expiation for sin committed deliberately (see Num 15.30 n.; Heb 5.2 n.). 3–12: Offering for the high priest (compare Ex 29.10–14). 4: Compare 1.3–4. 6: *The curtain of the sanctuary,* see Ex 26.31–33 n. 7: *The altar of fragrant incense* was inside the tabernacle (Ex 30.1–10). 8–12: Only the sacred, vital parts are to be sacrificed; the remainder, being contaminated by the sin of the offerer who presented it (v. 4), must be burned outside the camp.

smoke upon the altar of burnt offering. [11] But the skin of the bull and all its flesh, as well as its head, its legs, its entrails, and its dung— [12] all the rest of the bull— he shall carry out to a clean place outside the camp, to the ash heap, and shall burn it on a wood fire; at the ash heap it shall be burned.

13 If the whole congregation of Israel errs unintentionally and the matter escapes the notice of the assembly, and they do any one of the things that by the LORD's commandments ought not to be done and incur guilt; [14] when the sin that they have committed becomes known, the assembly shall offer a bull of the herd for a sin offering and bring it before the tent of meeting. [15] The elders of the congregation shall lay their hands on the head of the bull before the LORD, and the bull shall be slaughtered before the LORD. [16] The anointed priest shall bring some of the blood of the bull into the tent of meeting, [17] and the priest shall dip his finger in the blood and sprinkle it seven times before the LORD, in front of the curtain. [18] He shall put some of the blood on the horns of the altar that is before the LORD in the tent of meeting; and the rest of the blood he shall pour out at the base of the altar of burnt offering that is at the entrance of the tent of meeting. [19] He shall remove all its fat and turn it into smoke on the altar. [20] He shall do with the bull just as is done with the bull of sin offering; he shall do the same with this. The priest shall make atonement for them, and they shall be forgiven. [21] He shall carry the bull outside the camp, and burn it as he burned the first bull; it is the sin offering for the assembly.

22 When a ruler sins, doing unintentionally any one of all the things that by commandments of the LORD his God ought not to be done and incurs guilt, [23] once the sin that he has committed is made known to him, he shall bring as his offering a male goat without blemish.

[24] He shall lay his hand on the head of the goat; it shall be slaughtered at the spot where the burnt offering is slaughtered before the LORD; it is a sin offering. [25] The priest shall take some of the blood of the sin offering with his finger and put it on the horns of the altar of burnt offering, and pour out the rest of its blood at the base of the altar of burnt offering. [26] All its fat he shall turn into smoke on the altar, like the fat of the sacrifice of well-being. Thus the priest shall make atonement on his behalf for his sin, and he shall be forgiven.

27 If anyone of the ordinary people among you sins unintentionally in doing any one of the things that by the LORD's commandments ought not to be done and incurs guilt, [28] when the sin that you have committed is made known to you, you shall bring a female goat without blemish as your offering, for the sin that you have committed. [29] You shall lay your hand on the head of the sin offering; and the sin offering shall be slaughtered at the place of the burnt offering. [30] The priest shall take some of its blood with his finger and put it on the horns of the altar of burnt offering, and he shall pour out the rest of its blood at the base of the altar. [31] He shall remove all its fat, as the fat is removed from the offering of well-being, and the priest shall turn it into smoke on the altar for a pleasing odor to the LORD. Thus the priest shall make atonement on your behalf, and you shall be forgiven.

32 If the offering you bring as a sin offering is a sheep, you shall bring a female without blemish. [33] You shall lay your hand on the head of the sin offering; and it shall be slaughtered as a sin offering at the spot where the burnt offering is slaughtered. [34] The priest shall take some of the blood of the sin offering with his finger and put it on the horns of the altar of burnt offering, and pour out the rest of its blood at the base of the altar.

4.13–21: Offering for the whole congregation (compare Num 15.22–26). **15:** *The elders* are the representatives of the people. **20:** *Make* atonement (vv. 26, 31, 35), see 1.4 n.

4.22–26: Offering for a ruler. **27–35:** Offering for an ordinary person.

35 You shall remove all its fat, as the fat of the sheep is removed from the sacrifice of well-being, and the priest shall turn it into smoke on the altar, with the offerings by fire to the LORD. Thus the priest shall make atonement on your behalf for the sin that you have committed, and you shall be forgiven.

5 When any of you sin in that you have heard a public adjuration to testify and—though able to testify as one who has seen or learned of the matter—does not speak up, you are subject to punishment. 2 Or when any of you touch any unclean thing—whether the carcass of an unclean beast or the carcass of unclean livestock or the carcass of an unclean swarming thing—and are unaware of it, you have become unclean, and are guilty. 3 Or when you touch human uncleanness—any uncleanness by which one can become unclean—and are unaware of it, when you come to know it, you shall be guilty. 4 Or when any of you utter aloud a rash oath for a bad or a good purpose, whatever people utter in an oath, and are unaware of it, when you come to know it, you shall in any of these be guilty. 5 When you realize your guilt in any of these, you shall confess the sin that you have committed. 6 And you shall bring to the LORD, as your penalty for the sin that you have committed, a female from the flock, a sheep or a goat, as a sin offering; and the priest shall make atonement on your behalf for your sin.

7 But if you cannot afford a sheep, you shall bring to the LORD, as your penalty for the sin that you have committed, two turtledoves or two pigeons, one for a sin offering and the other for a burnt offering. 8 You shall bring them to the priest, who shall offer first the one for the sin offering, wringing its head at the nape without severing it. 9 He shall sprinkle some of the blood of the sin offering on the side of the altar, while the rest of the blood shall be drained out at the base of the altar; it is a sin offering. 10 And the second he shall offer for a burnt offering according to the regulation. Thus the priest shall make atonement on your behalf for the sin that you have committed, and you shall be forgiven.

11 But if you cannot afford two turtledoves or two pigeons, you shall bring as your offering for the sin that you have committed one-tenth of an ephah of choice flour for a sin offering; you shall not put oil on it or lay frankincense on it, for it is a sin offering. 12 You shall bring it to the priest, and the priest shall scoop up a handful of it as its memorial portion, and turn this into smoke on the altar, with the offerings by fire to the LORD; it is a sin offering. 13 Thus the priest shall make atonement on your behalf for whichever of these sins you have committed, and you shall be forgiven. Like the grain offering, the rest shall be for the priest.

14 The LORD spoke to Moses, saying: 15 When any of you commit a trespass and sin unintentionally in any of the holy things of the LORD, you shall bring, as your guilt offering to the LORD, a ram without blemish from the flock, convertible into silver by the sanctuary shekel; it is a guilt offering. 16 And you shall make restitution for the holy thing in which you were remiss, and shall add one-fifth to it and give it to the priest.

5.1–4: Cases which require a sin offering. 1: The refusal to testify as a witness could lead to a miscarriage of justice. 2–3: For laws about uncleanness, see chs 11–15; Num 19.11–13. 5: Confession of sin must precede the rite, for sacrifice is not a magical means of atonement.
5.7–13: A supplement to 4.27–35, which covers the case of a poor person who cannot afford a more costly animal (compare 1.14–15).

5.14–6.7: **The guilt offering.** This type of offering is prescribed for offenses against God and the community that require that restitution accompany the sacrifice.
5.15–16: These verses deal with the withholding of the LORD's *holy things,* i.e. the offerings and tithes that are due God. On *the sanctuary shekel,* see Ex 30.13 n. 17–19: The case of unwitting disobedience is similar to cases requiring a sin offering (4.27–35; compare 7.7).

The priest shall make atonement on your behalf with the ram of the guilt offering, and you shall be forgiven.

17 If any of you sin without knowing it, doing any of the things that by the LORD's commandments ought not to be done, you have incurred guilt, and are subject to punishment. 18 You shall bring to the priest a ram without blemish from the flock, or the equivalent, as a guilt offering; and the priest shall make atonement on your behalf for the error that you committed unintentionally, and you shall be forgiven. 19 It is a guilt offering; you have incurred guilt before the LORD.

6 b The LORD spoke to Moses, saying: 2 When any of you sin and commit a trespass against the LORD by deceiving a neighbor in a matter of a deposit or a pledge, or by robbery, or if you have defrauded a neighbor, 3 or have found something lost and lied about it—if you swear falsely regarding any of the various things that one may do and sin thereby— 4 when you have sinned and realize your guilt, and would restore what you took by robbery or by fraud or the deposit that was committed to you, or the lost thing that you found, 5 or anything else about which you have sworn falsely, you shall repay the principal amount and shall add one-fifth to it. You shall pay it to its owner when you realize your guilt. 6 And you shall bring to the priest, as your guilt offering to the LORD, a ram without blemish from the flock, or its equivalent, for a guilt offering. 7 The priest shall make atonement on your behalf before the LORD, and you shall be forgiven for any of the things that one may do and incur guilt thereby.

8 c The LORD spoke to Moses, saying: 9 Command Aaron and his sons, saying:

This is the ritual of the burnt offering. The burnt offering itself shall remain on the hearth upon the altar all night until the morning, while the fire on the altar shall be kept burning. 10 The priest shall put on his linen vestments after putting on his linen undergarments next to his body; and he shall take up the ashes to which the fire has reduced the burnt offering on the altar, and place them beside the altar. 11 Then he shall take off his vestments and put on other garments, and carry the ashes out to a clean place outside the camp. 12 The fire on the altar shall be kept burning; it shall not go out. Every morning the priest shall add wood to it, lay out the burnt offering on it, and turn into smoke the fat pieces of the offerings of well-being. 13 A perpetual fire shall be kept burning on the altar; it shall not go out.

14 This is the ritual of the grain offering: The sons of Aaron shall offer it before the LORD, in front of the altar. 15 They shall take from it a handful of the choice flour and oil of the grain offering, with all the frankincense that is on the offering, and they shall turn its memorial portion into smoke on the altar as a pleasing odor to the LORD. 16 Aaron and his sons shall eat what is left of it; it shall be eaten as unleavened cakes in a holy place; in the court of the tent of meeting they shall eat it. 17 It shall not be baked with leaven. I have given it as their portion of my offerings by fire; it is most holy, like the sin offering and the guilt offering. 18 Every male among the descendants of Aaron shall eat of it, as their perpetual due throughout your generations, from

b Ch 5.20 in Heb   c Ch 6.1 in Heb

6.1–7: Cases involving damage against another person (compare Ex 22.7–15). This law is supplemented by the legislation in Num 5.5–10.

6.8–7.38: **Instructions to priests concerning sacrifices.**

6.9–13: A supplement to 1.3–17. 9: The ever-burning fire on the altar (vv. 12–13) symbolizes Israel's perpetual service of the LORD. The burnt offering was sacrificed in the morning and evening according to Ex 29.38–42; Num 28.3–8 (but compare the pre-exilic practice reflected in 2 Kings 16.15). 10: For the priestly dress, see Ex 28.40–43. 14–18: Supplement to ch 2. 17: *It is most holy,* see 5.15–16 n. 18: Since the priests eat their portion in the holy place in a state of ritual purity (v. 16), holiness can be transferred to anyone who touches them (see Ex 19.12 n.).

the LORD's offerings by fire; anything that touches them shall become holy.

19 The LORD spoke to Moses, saying: 20This is the offering that Aaron and his sons shall offer to the LORD on the day when he is anointed: one-tenth of an ephah of choice flour as a regular offering, half of it in the morning and half in the evening. 21It shall be made with oil on a griddle; you shall bring it well soaked, as a grain offering of baked*d* pieces, and you shall present it as a pleasing odor to the LORD. 22And so the priest, anointed from among Aaron's descendants as a successor, shall prepare it; it is the LORD's—a perpetual due—to be turned entirely into smoke. 23Every grain offering of a priest shall be wholly burned; it shall not be eaten.

24 The LORD spoke to Moses, saying: 25Speak to Aaron and his sons, saying: This is the ritual of the sin offering. The sin offering shall be slaughtered before the LORD at the spot where the burnt offering is slaughtered; it is most holy. 26The priest who offers it as a sin offering shall eat of it; it shall be eaten in a holy place, in the court of the tent of meeting. 27Whatever touches its flesh shall become holy; and when any of its blood is spattered on a garment, you shall wash the bespattered part in a holy place. 28An earthen vessel in which it was boiled shall be broken; but if it is boiled in a bronze vessel, that shall be scoured and rinsed in water. 29Every male among the priests shall eat of it; it is most holy. 30But no sin offering shall be eaten from which any blood is brought into the tent of meeting for atonement in the holy place; it shall be burned with fire.

7 This is the ritual of the guilt offering. It is most holy; 2at the spot where the burnt offering is slaughtered, they shall slaughter the guilt offering, and its blood shall be dashed against all sides of the altar. 3All its fat shall be offered: the broad tail, the fat that covers the entrails, 4the two kidneys with the fat that is on them at the loins, and the appendage of the liver, which shall be removed with the kidneys. 5The priest shall turn them into smoke on the altar as an offering by fire to the LORD; it is a guilt offering. 6Every male among the priests shall eat of it; it shall be eaten in a holy place; it is most holy.

7 The guilt offering is like the sin offering, there is the same ritual for them; the priest who makes atonement with it shall have it. 8So, too, the priest who offers anyone's burnt offering shall keep the skin of the burnt offering that he has offered. 9And every grain offering baked in the oven, and all that is prepared in a pan or on a griddle, shall belong to the priest who offers it. 10But every other grain offering, mixed with oil or dry, shall belong to all the sons of Aaron equally.

11 This is the ritual of the sacrifice of the offering of well-being that one may offer to the LORD. 12If you offer it for thanksgiving, you shall offer with the thank offering unleavened cakes mixed with oil, unleavened wafers spread with oil, and cakes of choice flour well soaked in oil. 13With your thanksgiving sacrifice of well-being you shall bring your offering with cakes of leavened bread. 14From this you shall offer one cake from each offering, as a gift to the LORD; it shall belong to the priest who dashes the blood of the offering of well-being. 15And the flesh of your thanksgiving sacrifice of well-being shall be eaten on the day it is offered; you shall not leave any of it until morning. 16But if the sacrifice you offer is a votive offering or a freewill offering, it shall be eaten on the day that

d Meaning of Heb uncertain

6.19–23: Supplement to the law concerning the ordination of priests (Ex ch 29). 24–30: Supplement to 4.1–5.13. 27–28: These verses reflect the ancient view of holiness as something transferable by contact (v. 18).

Holiness can be scoured off a bronze vessel; but an earthen vessel, because it is absorbent, must be destroyed.
7.1–10: Supplement to 5.14–6.7.
7.11–36: Supplement to 3.1–17. 12–14:

you offer your sacrifice, and what is left of it shall be eaten the next day; 17but what is left of the flesh of the sacrifice shall be burned up on the third day. 18If any of the flesh of your sacrifice of well-being is eaten on the third day, it shall not be acceptable, nor shall it be credited to the one who offers it; it shall be an abomination, and the one who eats of it shall incur guilt.

19 Flesh that touches any unclean thing shall not be eaten; it shall be burned up. As for other flesh, all who are clean may eat such flesh. 20But those who eat flesh from the LORD's sacrifice of well-being while in a state of uncleanness shall be cut off from their kin. 21When any one of you touches any unclean thing— human uncleanness or an unclean animal or any unclean creature—and then eats flesh from the LORD's sacrifice of well-being, you shall be cut off from your kin.

22 The LORD spoke to Moses, saying: 23Speak to the people of Israel, saying: You shall eat no fat of ox or sheep or goat. 24The fat of an animal that died or was torn by wild animals may be put to any use, but you must not eat it. 25If any one of you eats the fat from an animal of which an offering by fire may be made to the LORD, you who eat it shall be cut off from your kin. 26You must not eat any blood whatever, either of bird or of animal, in any of your settlements. 27Any one of you who eats any blood shall be cut off from your kin.

28 The LORD spoke to Moses, saying: 29Speak to the people of Israel, saying: Any one of you who would offer to the Lord your sacrifice of well-being must yourself bring to the LORD your offering from your sacrifice of well-being.

30Your own hands shall bring the LORD's offering by fire; you shall bring the fat with the breast, so that the breast may be raised as an elevation offering before the LORD. 31The priest shall turn the fat into smoke on the altar, but the breast shall belong to Aaron and his sons. 32And the right thigh from your sacrifices of well-being you shall give to the priest as an offering; 33the one among the sons of Aaron who offers the blood and fat of the offering of well-being shall have the right thigh for a portion. 34For I have taken the breast of the elevation offering, and the thigh that is offered, from the people of Israel, from their sacrifices of well-being, and have given them to Aaron the priest and to his sons, as a perpetual due from the people of Israel. 35This is the portion allotted to Aaron and to his sons from the offerings made by fire to the LORD, once they have been brought forward to serve the LORD as priests; 36these the LORD commanded to be given them, when he anointed them, as a perpetual due from the people of Israel throughout their generations.

37 This is the ritual of the burnt offering, the grain offering, the sin offering, the guilt offering, the offering of ordination, and the sacrifice of well-being, 38which the LORD commanded Moses on Mount Sinai, when he commanded the people of Israel to bring their offerings to the LORD, in the wilderness of Sinai.

8 The LORD spoke to Moses, saying: 2Take Aaron and his sons with him, the vestments, the anointing oil, the bull of sin offering, the two rams, and the basket of unleavened bread; 3and assemble the whole congregation at the entrance of the tent of meeting. 4And Mo-

---

The offering of well-being may be accompanied by a grain offering (see 2.1 n.) for a thanksgiving. **16**: *Votive offering,* see ch 27.
7.19: *Any unclean thing,* see ch 11. **20**: Conditions of personal uncleanness are described in chs 12–15. *Cut off from their kin,* a reference to the death penalty. **23**: 3.16. **24**: See Ex 22.31 n. **26–27**: On this prohibition, see 1.5 n. **28–36**: The priests' portion of the offering of well-being. **30**: On the *elevation*

offering, see Ex 29.24 n. **34**: *A perpetual due,* see 2.2–3 n.
8.1–10.20: **Investiture and induction of priests.** These chapters are intended to follow Ex 25–40, which deal with the erection of the tabernacle and making of vestments for priestly service in the sanctuary.
8.1–36: **The ordination of the priests** (based on Ex ch 29). **2**: On the relation of the Aaronic order to the Levites as a whole, see

ses did as the LORD commanded him. When the congregation was assembled at the entrance of the tent of meeting, [5] Moses said to the congregation, "This is what the LORD has commanded to be done."

6 Then Moses brought Aaron and his sons forward, and washed them with water. [7] He put the tunic on him, fastened the sash around him, clothed him with the robe, and put the ephod on him. He then put the decorated band of the ephod around him, tying the ephod to him with it. [8] He placed the breastpiece on him, and in the breastpiece he put the Urim and the Thummim. [9] And he set the turban on his head, and on the turban, in front, he set the golden ornament, the holy crown, as the LORD commanded Moses.

10 Then Moses took the anointing oil and anointed the tabernacle and all that was in it, and consecrated them. [11] He sprinkled some of it on the altar seven times, and anointed the altar and all its utensils, and the basin and its base, to consecrate them. [12] He poured some of the anointing oil on Aaron's head and anointed him, to consecrate him. [13] And Moses brought forward Aaron's sons, and clothed them with tunics, and fastened sashes around them, and tied headdresses on them, as the LORD commanded Moses.

14 He led forward the bull of sin offering; and Aaron and his sons laid their hands upon the head of the bull of sin offering, [15] and it was slaughtered. Moses took the blood and with his finger put some on each of the horns of the altar, purifying the altar; then he poured out the blood at the base of the altar. Thus he consecrated it, to make atonement for it.

[16] Moses took all the fat that was around the entrails, and the appendage of the liver, and the two kidneys with their fat, and turned them into smoke on the altar. [17] But the bull itself, its skin and flesh and its dung, he burned with fire outside the camp, as the LORD commanded Moses.

18 Then he brought forward the ram of burnt offering. Aaron and his sons laid their hands on the head of the ram, [19] and it was slaughtered. Moses dashed the blood against all sides of the altar. [20] The ram was cut into its parts, and Moses turned into smoke the head and the parts and the suet. [21] And after the entrails and the legs were washed with water, Moses turned into smoke the whole ram on the altar; it was a burnt offering for a pleasing odor, an offering by fire to the LORD, as the LORD commanded Moses.

22 Then he brought forward the second ram, the ram of ordination. Aaron and his sons laid their hands on the head of the ram, [23] and it was slaughtered. Moses took some of its blood and put it on the lobe of Aaron's right ear and on the thumb of his right hand and on the big toe of his right foot. [24] After Aaron's sons were brought forward, Moses put some of the blood on the lobes of their right ears and on the thumbs of their right hands and on the big toes of their right feet; and Moses dashed the rest of the blood against all sides of the altar. [25] He took the fat—the broad tail, all the fat that was around the entrails, the appendage of the liver, and the two kidneys with their fat—and the right thigh. [26] From the basket of unleavened bread that was before the LORD, he took one cake of unleavened bread, one cake of bread with oil, and one wafer, and placed

---

Num 3.5–10 n. **5–9**: Ex 29.4–6. The priestly regalia is described in Ex ch 28.
8.10–13: The ceremony of anointment (Ex 29.7–9). On the anointing of tabernacle and furnishings, see Ex 30.26–30; 40.9–15. **14–17**: Compare Ex 29.10–14; the ceremony began with a special sin offering for the priests (4.1–12). **15**: On atonement for the altar, see Ex

29.35–37 n. **18–21**: The second ordination sacrifice was a burnt offering (1.10–13; Ex 29.15–17). **22–35**: Compare Ex 29.19–37. The climax of the service was the sacrifice of *the ram of ordination,* i.e. an offering of wellbeing or communion meal (vv. 31–35). **26**: The offering of well-being was accompanied by a grain offering (7.12–14). **27–28**: In this

them on the fat and on the right thigh. 27 He placed all these on the palms of Aaron and on the palms of his sons, and raised them as an elevation offering before the LORD. 28 Then Moses took them from their hands and turned them into smoke on the altar with the burnt offering. This was an ordination offering for a pleasing odor, an offering by fire to the LORD. 29 Moses took the breast and raised it as an elevation offering before the LORD; it was Moses' portion of the ram of ordination, as the LORD commanded Moses.

30 Then Moses took some of the anointing oil and some of the blood that was on the altar and sprinkled them on Aaron and his vestments, and also on his sons and their vestments. Thus he consecrated Aaron and his vestments, and also his sons and their vestments.

31 And Moses said to Aaron and his sons, "Boil the flesh at the entrance of the tent of meeting, and eat it there with the bread that is in the basket of ordination offerings, as I was commanded, 'Aaron and his sons shall eat it'; 32 and what remains of the flesh and the bread you shall burn with fire. 33 You shall not go outside the entrance of the tent of meeting for seven days, until the day when your period of ordination is completed. For it will take seven days to ordain you; 34 as has been done today, the LORD has commanded to be done to make atonement for you. 35 You shall remain at the entrance of the tent of meeting day and night for seven days, keeping the LORD's charge so that you do not die; for so I am commanded." 36 Aaron and his sons did all the things that the LORD commanded through Moses.

9 On the eighth day Moses summoned Aaron and his sons and the elders of Israel. 2 He said to Aaron, "Take a bull calf for a sin offering and a ram for a burnt offering, without blemish, and offer them before the LORD. 3 And say to the people of Israel, 'Take a male goat for a sin offering; a calf and a lamb, yearlings without blemish, for a burnt offering; 4 and an ox and a ram for an offering of well-being to sacrifice before the LORD; and a grain offering mixed with oil. For today the LORD will appear to you.'" 5 They brought what Moses commanded to the front of the tent of meeting; and the whole congregation drew near and stood before the LORD. 6 And Moses said, "This is the thing that the LORD commanded you to do, so that the glory of the LORD may appear to you." 7 Then Moses said to Aaron, "Draw near to the altar and sacrifice your sin offering and your burnt offering, and make atonement for yourself and for the people; and sacrifice the offering of the people, and make atonement for them; as the LORD has commanded."

8 Aaron drew near to the altar, and slaughtered the calf of the sin offering, which was for himself. 9 The sons of Aaron presented the blood to him, and he dipped his finger in the blood and put it on the horns of the altar; and the rest of the blood he poured out at the base of the altar. 10 But the fat, the kidneys, and the appendage of the liver from the sin offering he turned into smoke on the altar, as the LORD commanded Moses; 11 and the flesh and the skin he burned with fire outside the camp.

12 Then he slaughtered the burnt offering. Aaron's sons brought him the blood, and he dashed it against all sides of the altar. 13 And they brought him the burnt offering piece by piece, and the head, which he turned into smoke on the altar. 14 He washed the entrails and the legs and, with the burnt offering, turned them into smoke on the altar.

15 Next he presented the people's of-

---

case the *elevation offering* (see Ex 29.24 n.) did not go to the priests as, in the usual offering of well-being (7.28–36), but to Moses the officiating priest.
9.1–24: The commencement of Aaron's high priesthood. 1: *The eighth day,* i.e.

at the end of the seven day ordination ceremony (8.33). **2–7:** Preparations for an assembly of the whole congregation for worship.
9.8–14: Aaron offered a sin offering (4.1–12) and a burnt offering (1.3–13) to make atonement for himself.

fering. He took the goat of the sin offering that was for the people, and slaughtered it, and presented it as a sin offering like the first one. [16]He presented the burnt offering, and sacrificed it according to regulation. [17]He presented the grain offering, and, taking a handful of it, he turned it into smoke on the altar, in addition to the burnt offering of the morning.

18 He slaughtered the ox and the ram as a sacrifice of well-being for the people. Aaron's sons brought him the blood, which he dashed against all sides of the altar, [19]and the fat of the ox and of the ram—the broad tail, the fat that covers the entrails, the two kidneys and the fat on them, [e] and the appendage of the liver. [20]They first laid the fat on the breasts, and the fat was turned into smoke on the altar; [21]and the breasts and the right thigh Aaron raised as an elevation offering before the LORD, as Moses had commanded.

22 Aaron lifted his hands toward the people and blessed them; and he came down after sacrificing the sin offering, the burnt offering, and the offering of well-being. [23]Moses and Aaron entered the tent of meeting, and then came out and blessed the people; and the glory of the LORD appeared to all the people. [24]Fire came out from the LORD and consumed the burnt offering and the fat on the altar; and when all the people saw it, they shouted and fell on their faces.

10 Now Aaron's sons, Nadab and Abihu, each took his censer, put fire in it, and laid incense on it; and they offered unholy fire before the LORD, such as he had not commanded them. [2]And fire came out from the presence of the LORD and consumed them, and they died before the LORD. [3]Then Moses said to Aaron, "This is what the LORD meant when he said,

'Through those who are near me
I will show myself holy,
and before all the people
I will be glorified.' "

And Aaron was silent.

4 Moses summoned Mishael and Elzaphan, sons of Uzziel the uncle of Aaron, and said to them, "Come forward, and carry your kinsmen away from the front of the sanctuary to a place outside the camp." [5]They came forward and carried them by their tunics out of the camp, as Moses had ordered. [6]And Moses said to Aaron and to his sons Eleazar and Ithamar, "Do not dishevel your hair, and do not tear your vestments, or you will die and wrath will strike all the congregation; but your kindred, the whole house of Israel, may mourn the burning that the LORD has sent. [7]You shall not go outside the entrance of the tent of meeting, or you will die; for the anointing oil of the LORD is on you." And they did as Moses had ordered.

8 And the LORD spoke to Aaron: [9]Drink no wine or strong drink, neither you nor your sons, when you enter the tent of meeting, that you may not die; it is a statute forever throughout your generations. [10]You are to distinguish be-

e Gk: Heb *the broad tail, and that which covers, and the kidneys*

9.15–21: The people's sin offering, burnt and grain offering, and offering of well-being were presented according to the ordinances of chs 1–7. **22:** On the Aaronic blessing, see Num 6.24–26 n. *He came down,* i.e. from the steps leading up to the great altar (Ezek 43.17). **24:** Judg 6.21.
10.1–20: The fate of Nadab and Abihu. Two of Aaron's sons were destroyed for their sin in making an unauthorized offering before the LORD. **1:** *Unholy fire* refers to incense which was offered in presumptuous defiance of the rules of the theocratic community (Ex 30.34–38; Num ch 16). **2:** See 9.24 n. **3:** *Those who are near me,* i.e. the priests who have access to the holy place. The story explains why the priestly line was traced through Aaron's third son, Eleazar (Ex 6.23–25). **4:** *Mishael* and *Elzaphan,* cousins of Moses (Ex 6.22). **5:** *Their tunics,* i.e. priestly garments (Ex 28.4, 39–40). **6:** Aaron and his sons must abstain from signs of mourning, i.e. rending the garments and letting the hair hang loose. The reason is that they are still in a state of ritual purity (v. 7; see 21.10–12). **9:** Ezek 44.21. **10–11:** See chs 11–15.

tween the holy and the common, and between the unclean and the clean; [11] and you are to teach the people of Israel all the statutes that the LORD has spoken to them through Moses.

12 Moses spoke to Aaron and to his remaining sons, Eleazar and Ithamar: Take the grain offering that is left from the LORD's offerings by fire, and eat it unleavened beside the altar, for it is most holy; [13] you shall eat it in a holy place, because it is your due and your sons' due, from the offerings by fire to the LORD; for so I am commanded. [14] But the breast that is elevated and the thigh that is raised, you and your sons and daughters as well may eat in any clean place; for they have been assigned to you and your children from the sacrifices of the offerings of well-being of the people of Israel. [15] The thigh that is raised and the breast that is elevated they shall bring, together with the offerings by fire of the fat, to raise for an elevation offering before the LORD; they are to be your due and that of your children forever, as the LORD has commanded.

16 Then Moses made inquiry about the goat of the sin offering, and—it had already been burned! He was angry with Eleazar and Ithamar, Aaron's remaining sons, and said, [17] "Why did you not eat the sin offering in the sacred area? For it is most holy, and God[f] has given it to you that you may remove the guilt of the congregation, to make atonement on their behalf before the LORD. [18] Its blood was not brought into the inner part of the sanctuary. You should certainly have eaten it in the sanctuary, as I commanded." [19] And Aaron spoke to Moses, "See,

today they offered their sin offering and their burnt offering before the LORD; and yet such things as these have befallen me! If I had eaten the sin offering today, would it have been agreeable to the LORD?" [20] And when Moses heard that, he agreed.

11 The LORD spoke to Moses and Aaron, saying to them: [2] Speak to the people of Israel, saying:

From among all the land animals, these are the creatures that you may eat. [3] Any animal that has divided hoofs and is cleft-footed and chews the cud—such you may eat. [4] But among those that chew the cud or have divided hoofs, you shall not eat the following: the camel, for even though it chews the cud, it does not have divided hoofs; it is unclean for you. [5] The rock badger, for even though it chews the cud, it does not have divided hoofs; it is unclean for you. [6] The hare, for even though it chews the cud, it does not have divided hoofs; it is unclean for you. [7] The pig, for even though it has divided hoofs and is cleft-footed, it does not chew the cud; it is unclean for you. [8] Of their flesh you shall not eat, and their carcasses you shall not touch; they are unclean for you.

9 These you may eat, of all that are in the waters. Everything in the waters that has fins and scales, whether in the seas or in the streams—such you may eat. [10] But anything in the seas or the streams that does not have fins and scales, of the swarming creatures in the waters and among all the other living creatures that

f Heb *he*

10.12–15: 7.28–36. The grain offering (v. 12) accompanied the offering of well-being (7.11–14). 16–20: When Aaron and his sons offered their first sin offering after Nadab and Abihu were struck down during the performance of their priestly duties, they failed to eat their rightful portion (see 2.2–3 n.; 6.26), fearing that the LORD was not disposed to accept their atonement for the people.
11.1–47: **Clean and unclean animals** (Deut 14.3–20). The laws of chs 11–15 come after the ordination tradition (chs 8–10) be- cause one of the tasks of the priests was to make a distinction between clean and unclean (10.10). This distinction is not based merely on sanitary or hygienic considerations, in the modern sense that "cleanliness is next to godliness." That which is unclean is ritually impure and therefore the opposite of holy. **5–6:** Strictly, these animals do not chew the cud but only appear to. **7:** *The pig* came to be regarded as the animal that was particularly unclean (1 Macc 1.47).

are in the waters—they are detestable to you [11] and detestable they shall remain. Of their flesh you shall not eat, and their carcasses you shall regard as detestable. [12] Everything in the waters that does not have fins and scales is detestable to you.

13 These you shall regard as detestable among the birds. They shall not be eaten; they are an abomination: the eagle, the vulture, the osprey, [14] the buzzard, the kite of any kind; [15] every raven of any kind; [16] the ostrich, the nighthawk, the sea gull, the hawk of any kind; [17] the little owl, the cormorant, the great owl, [18] the water hen, the desert owl, [g] the carrion vulture, [19] the stork, the heron of any kind, the hoopoe, and the bat. [h]

20 All winged insects that walk upon all fours are detestable to you. [21] But among the winged insects that walk on all fours you may eat those that have jointed legs above their feet, with which to leap on the ground. [22] Of them you may eat: the locust according to its kind, the bald locust according to its kind, the cricket according to its kind, and the grasshopper according to its kind. [23] But all other winged insects that have four feet are detestable to you.

24 By these you shall become unclean; whoever touches the carcass of any of them shall be unclean until the evening, [25] and whoever carries any part of the carcass of any of them shall wash his clothes and be unclean until the evening. [26] Every animal that has divided hoofs but is not cleft-footed or does not chew the cud is unclean for you; everyone who touches one of them shall be unclean. [27] All that walk on their paws, among the animals that walk on all fours, are unclean for you; whoever touches the carcass of any of them shall be unclean until the evening, [28] and the one who carries the carcass shall wash his clothes and be unclean until the evening; they are unclean for you.

29 These are unclean for you among the creatures that swarm upon the earth: the weasel, the mouse, the great lizard according to its kind, [30] the gecko, the land crocodile, the lizard, the sand lizard, and the chameleon. [31] These are unclean for you among all that swarm; whoever touches one of them when they are dead shall be unclean until the evening. [32] And anything upon which any of them falls when they are dead shall be unclean, whether an article of wood or cloth or skin or sacking, any article that is used for any purpose; it shall be dipped into water, and it shall be unclean until the evening, and then it shall be clean. [33] And if any of them falls into any earthen vessel, all that is in it shall be unclean, and you shall break the vessel. [34] Any food that could be eaten shall be unclean if water from any such vessel comes upon it; and any liquid that could be drunk shall be unclean if it was in any such vessel. [35] Everything on which any part of the carcass falls shall be unclean; whether an oven or stove, it shall be broken in pieces; they are unclean, and shall remain unclean for you. [36] But a spring or a cistern holding water shall be clean, while whatever touches the carcass in it shall be unclean. [37] If any part of their carcass falls upon any seed set aside for sowing, it is clean; [38] but if water is put on the seed and any part of their carcass falls on it, it is unclean for you.

39 If an animal of which you may eat dies, anyone who touches its carcass shall be unclean until the evening. [40] Those who eat of its carcass shall wash their clothes and be unclean until the evening; and those who carry the carcass shall wash their clothes and be unclean until the evening.

41 All creatures that swarm upon the earth are detestable; they shall not be eaten. [42] Whatever moves on its belly, and whatever moves on all fours, or whatev-

g Or *pelican*   h Identification of several of the birds in verses 13–19 is uncertain

---

11.24–38: An unclean animal, when dead, transmits uncleanness at the touch but it may be safely handled while alive. Further, the contagion of uncleanness, like that of its opposite, holiness (6.27–28), affects objects, vessels, etc.

er has many feet, all the creatures that swarm upon the earth, you shall not eat; for they are detestable. 43 You shall not make yourselves detestable with any creature that swarms; you shall not defile yourselves with them, and so become unclean. 44 For I am the LORD your God; sanctify yourselves therefore, and be holy, for I am holy. You shall not defile yourselves with any swarming creature that moves on the earth. 45 For I am the LORD who brought you up from the land of Egypt, to be your God; you shall be holy, for I am holy.

46 This is the law pertaining to land animal and bird and every living creature that moves through the waters and every creature that swarms upon the earth, 47 to make a distinction between the unclean and the clean, and between the living creature that may be eaten and the living creature that may not be eaten.

12 The LORD spoke to Moses, saying: 2 Speak to the people of Israel, saying:

If a woman conceives and bears a male child, she shall be ceremonially unclean seven days; as at the time of her menstruation, she shall be unclean. 3 On the eighth day the flesh of his foreskin shall be circumcised. 4 Her time of blood purification shall be thirty-three days; she shall not touch any holy thing, or come into the sanctuary, until the days of her purification are completed. 5 If she bears a female child, she shall be unclean two weeks, as in her menstruation; her time of blood purification shall be sixty-six days.

6 When the days of her purification are completed, whether for a son or for a daughter, she shall bring to the priest at the entrance of the tent of meeting a lamb in its first year for a burnt offering, and a pigeon or a turtledove for a sin offering. 7 He shall offer it before the LORD, and make atonement on her behalf; then she shall be clean from her flow of blood. This is the law for her who bears a child, male or female. 8 If she cannot afford a sheep, she shall take two turtledoves or two pigeons, one for a burnt offering and the other for a sin offering; and the priest shall make atonement on her behalf, and she shall be clean.

13 The LORD spoke to Moses and Aaron, saying:

2 When a person has on the skin of his body a swelling or an eruption or a spot, and it turns into a leprous*i* disease on the skin of his body, he shall be brought to Aaron the priest or to one of his sons the priests. 3 The priest shall examine the disease on the skin of his body, and if the hair in the diseased area has turned white and the disease appears to be deeper than the skin of his body, it is a leprous*i* disease; after the priest has examined him he shall pronounce him ceremonially unclean. 4 But if the spot is white in the skin of his body, and appears no deeper than the skin, and the hair in it has not turned white, the priest shall confine the diseased person for seven days. 5 The priest

i A term for several skin diseases; precise meaning uncertain

---

11.44–45: The basis for these laws is not irrational taboo but the covenant relationship, which sets Israel apart for the service of God (Ex 19.3–6). As a holy and consecrated people (Ex 22.31), Israelites must avoid all impurity in order that the holy God may tabernacle in their midst (15.31; 18.1–5; 20.22–26; 26.11–12).
12.1–8: **Purification of a woman after childbirth. 2**: This law does not imply that sex is evil—a view completely foreign to Israel's thought; rather, the woman becomes unclean because of her bodily discharge (v. 7; see 15.16–18 n.). **6**: The *sin offering* is prescribed for inadvertent sin (4.27–35). **8**: See 5.7–13 n.; Lk 2.24.
13.1–59: **The diagnosis of leprosy.** In this chapter and the next, the word translated "leprosy" is a generic term that includes various skin diseases (including what is called leprosy today), as well as blemishes affecting garments and buildings. **1–8**: A suspicious skin eruption. Tubercular leprosy begins with reddish patches and progresses into nodules and deformities; anesthetic leprosy paralyzes the nerves so that the limbs are numb and eventually lifeless. Whatever type is in mind, the law emphasizes the need for

shall examine him on the seventh day, and if he sees that the disease is checked and the disease has not spread in the skin, then the priest shall confine him seven days more. 6 The priest shall examine him again on the seventh day, and if the disease has abated and the disease has not spread in the skin, the priest shall pronounce him clean; it is only an eruption; and he shall wash his clothes, and be clean. 7 But if the eruption spreads in the skin after he has shown himself to the priest for his cleansing, he shall appear again before the priest. 8 The priest shall make an examination, and if the eruption has spread in the skin, the priest shall pronounce him unclean; it is a leprous[j] disease.

9 When a person contracts a leprous[j] disease, he shall be brought to the priest. 10 The priest shall make an examination, and if there is a white swelling in the skin that has turned the hair white, and there is quick raw flesh in the swelling, 11 it is a chronic leprous[j] disease in the skin of his body. The priest shall pronounce him unclean; he shall not confine him, for he is unclean. 12 But if the disease breaks out in the skin, so that it covers all the skin of the diseased person from head to foot, so far as the priest can see, 13 then the priest shall make an examination, and if the disease has covered all his body, he shall pronounce him clean of the disease; since it has all turned white, he is clean. 14 But if raw flesh ever appears on him, he shall be unclean; 15 the priest shall examine the raw flesh and pronounce him unclean. Raw flesh is unclean, for it is a leprous[j] disease. 16 But if the raw flesh again turns white, he shall come to the priest; 17 the priest shall examine him, and if the disease has turned white, the priest shall pronounce the diseased person clean. He is clean.

18 When there is on the skin of one's body a boil that has healed, 19 and in the place of the boil there appears a white swelling or a reddish-white spot, it shall be shown to the priest. 20 The priest shall make an examination, and if it appears deeper than the skin and its hair has turned white, the priest shall pronounce him unclean; this is a leprous[j] disease, broken out in the boil. 21 But if the priest examines it and the hair on it is not white, nor is it deeper than the skin but has abated, the priest shall confine him seven days. 22 If it spreads in the skin, the priest shall pronounce him unclean; it is diseased. 23 But if the spot remains in one place and does not spread, it is the scar of the boil; the priest shall pronounce him clean.

24 Or, when the body has a burn on the skin and the raw flesh of the burn becomes a spot, reddish-white or white, 25 the priest shall examine it. If the hair in the spot has turned white and it appears deeper than the skin, it is a leprous[j] disease; it has broken out in the burn, and the priest shall pronounce him unclean. This is a leprous[j] disease. 26 But if the priest examines it and the hair in the spot is not white, and it is no deeper than the skin but has abated, the priest shall confine him seven days. 27 The priest shall examine him the seventh day; if it is spreading in the skin, the priest shall pronounce him unclean. This is a leprous[j] disease. 28 But if the spot remains in one place and does not spread in the skin but has abated, it is a swelling from the burn, and the priest shall pronounce him clean; for it is the scar of the burn.

29 When a man or woman has a disease on the head or in the beard, 30 the priest shall examine the disease. If it appears deeper than the skin and the hair in it is yellow and thin, the priest shall pro-

j A term for several skin diseases; precise meaning uncertain

---

observation so that the disease may be detected in its early stages.

13.9–17: In the case of *chronic leprous disease,* quarantine for further examination (v. 11) is pointless.

13.18–23: A boil in which leprosy may break out. **24–28:** A burn that may be infected with leprosy.

13.29–37: An itching disease, perhaps ringworm.

nounce him unclean; it is an itch, a leprous[k] disease of the head or the beard. 31 If the priest examines the itching disease, and it appears no deeper than the skin and there is no black hair in it, the priest shall confine the person with the itching disease for seven days. 32 On the seventh day the priest shall examine the itch; if the itch has not spread, and there is no yellow hair in it, and the itch appears to be no deeper than the skin, 33 he shall shave, but the itch he shall not shave. The priest shall confine the person with the itch for seven days more. 34 On the seventh day the priest shall examine the itch; if the itch has not spread in the skin and it appears to be no deeper than the skin, the priest shall pronounce him clean. He shall wash his clothes and be clean. 35 But if the itch spreads in the skin after he was pronounced clean, 36 the priest shall examine him. If the itch has spread in the skin, the priest need not seek for the yellow hair; he is unclean. 37 But if in his eyes the itch is checked, and black hair has grown in it, the itch is healed, he is clean; and the priest shall pronounce him clean.

38 When a man or a woman has spots on the skin of the body, white spots, 39 the priest shall make an examination, and if the spots on the skin of the body are of a dull white, it is a rash that has broken out on the skin; he is clean.

40 If anyone loses the hair from his head, he is bald but he is clean. 41 If he loses the hair from his forehead and temples, he has baldness of the forehead but he is clean. 42 But if there is on the bald head or the bald forehead a reddish-white diseased spot, it is a leprous[k] disease breaking out on his bald head or his bald forehead. 43 The priest shall examine him; if the diseased swelling is reddish-white on his bald head or on his bald forehead, which resembles a leprous[k]

disease in the skin of the body, 44 he is leprous,[k] he is unclean. The priest shall pronounce him unclean; the disease is on his head.

45 The person who has the leprous[k] disease shall wear torn clothes and let the hair of his head be disheveled; and he shall cover his upper lip and cry out, "Unclean, unclean." 46 He shall remain unclean as long as he has the disease; he is unclean. He shall live alone; his dwelling shall be outside the camp.

47 Concerning clothing: when a leprous[k] disease appears in it, in woolen or linen cloth, 48 in warp or woof of linen or wool, or in a skin or in anything made of skin, 49 if the disease shows greenish or reddish in the garment, whether in warp or woof or in skin or in anything made of skin, it is a leprous[k] disease and shall be shown to the priest. 50 The priest shall examine the disease, and put the diseased article aside for seven days. 51 He shall examine the disease on the seventh day. If the disease has spread in the cloth, in warp or woof, or in the skin, whatever be the use of the skin, this is a spreading leprous[k] disease; it is unclean. 52 He shall burn the clothing, whether diseased in warp or woof, woolen or linen, or anything of skin, for it is a spreading leprous[k] disease; it shall be burned in fire.

53 If the priest makes an examination, and the disease has not spread in the clothing, in warp or woof or in anything of skin, 54 the priest shall command them to wash the article in which the disease appears, and he shall put it aside seven days more. 55 The priest shall examine the diseased article after it has been washed. If the diseased spot has not changed color, though the disease has not spread, it is unclean; you shall burn

k A term for several skin diseases; precise meaning uncertain

---

13:38–39: A vesicular skin disease. **40–44**: Falling hair, unless accompanied by other symptoms, is not necessarily a sign of leprosy. **45–46**: If truly a leper, the individual must appear like a mourner (10.6) and must go into isolation (Job 2.7–8). The judgment of being *unclean* is based upon the belief that when one is ritually impure (see ch 11 n.) defilement could be transmitted to others. Even after being cured, a leper was not "clean" until ritually purified (ch 14). **47–59**: *Leprous disease* in cloth refers to mold or mildew.

it in fire, whether the leprous[1] spot is on the inside or on the outside.

56 If the priest makes an examination, and the disease has abated after it is washed, he shall tear the spot out of the cloth, in warp or woof, or out of skin. [57]If it appears again in the garment, in warp or woof, or in anything of skin, it is spreading; you shall burn with fire that in which the disease appears. [58]But the cloth, warp or woof, or anything of skin from which the disease disappears when you have washed it, shall then be washed a second time, and it shall be clean.

59 This is the ritual for a leprous[1] disease in a cloth of wool or linen, either in warp or woof, or in anything of skin, to decide whether it is clean or unclean.

14 The LORD spoke to Moses, saying: [2]This shall be the ritual for the leprous[1] person at the time of his cleansing:

He shall be brought to the priest; [3]the priest shall go out of the camp, and the priest shall make an examination. If the disease is healed in the leprous[1] person, [4]the priest shall command that two living clean birds and cedarwood and crimson yarn and hyssop be brought for the one who is to be cleansed. [5]The priest shall command that one of the birds be slaughtered over fresh water in an earthen vessel. [6]He shall take the living bird with the cedarwood and the crimson yarn and the hyssop, and dip them and the living bird in the blood of the bird that was slaughtered over the fresh water. [7]He shall sprinkle it seven times upon the one who is to be cleansed of the leprous[1] disease; then he shall pronounce him clean, and he shall let the living bird go into the open field. [8]The one who is to be cleansed shall wash his clothes, and shave off all his hair, and bathe himself in water, and he shall be clean. After that he shall come into the camp, but shall live outside his tent seven days. [9]On the seventh day he shall shave all his hair: of head, beard, eyebrows; he shall shave all his hair. Then he shall wash his clothes, and bathe his body in water, and he shall be clean.

10 On the eighth day he shall take two male lambs without blemish, and one ewe lamb in its first year without blemish, and a grain offering of three-tenths of an ephah of choice flour mixed with oil, and one log[m] of oil. [11]The priest who cleanses shall set the person to be cleansed, along with these things, before the LORD, at the entrance of the tent of meeting. [12]The priest shall take one of the lambs, and offer it as a guilt offering, along with the log[m] of oil, and raise them as an elevation offering before the LORD. [13]He shall slaughter the lamb in the place where the sin offering and the burnt offering are slaughtered in the holy place; for the guilt offering, like the sin offering, belongs to the priest: it is most holy. [14]The priest shall take some of the blood of the guilt offering and put it on the lobe of the right ear of the one to be cleansed, and on the thumb of the right hand, and on the big toe of the right foot. [15]The priest shall take some of the log[m] of oil and pour it into the palm of his own left hand, [16]and dip his right finger in the oil that is in his left hand and sprinkle some oil with his finger seven times before the LORD. [17]Some of the oil that remains in his hand the priest shall put on the lobe of the right ear of the one to be cleansed, and on the thumb of the right hand, and on the big toe of the right foot, on top of the blood of the guilt offering. [18]The rest of the oil that is in the priest's hand he shall put on the head of the one to be cleansed. Then the priest shall make atonement on his behalf before the LORD: [19]the priest shall offer the sin offering, to

---

[1] A term for several skin diseases; precise meaning uncertain   [m] A liquid measure

---

14.1–57: The cleansing of leprosy. 2: Mk 1.44. **4–8:** This ceremony has archaic elements that elude explanation. **4:** *Hyssop,* see Ex 12.22 n. **7:** Perhaps the freeing of the living bird symbolizes the carrying away of the leper's uncleanness. **8:** On the ritual washing of garments, see 16.23–24; Ex 19.10.

14.10–20: The offering of appropriate sac-

make atonement for the one to be cleansed from his uncleanness. Afterward he shall slaughter the burnt offering; [20] and the priest shall offer the burnt offering and the grain offering on the altar. Thus the priest shall make atonement on his behalf and he shall be clean. 21 But if he is poor and cannot afford so much, he shall take one male lamb for a guilt offering to be elevated, to make atonement on his behalf, and one-tenth of an ephah of choice flour mixed with oil for a grain offering and a log[n] of oil; [22] also two turtledoves or two pigeons, such as he can afford, one for a sin offering and the other for a burnt offering. [23] On the eighth day he shall bring them for his cleansing to the priest, to the entrance of the tent of meeting, before the LORD; [24] and the priest shall take the lamb of the guilt offering and the log[n] of oil, and the priest shall raise them as an elevation offering before the LORD. [25] The priest shall slaughter the lamb of the guilt offering and shall take some of the blood of the guilt offering, and put it on the lobe of the right ear of the one to be cleansed, and on the thumb of the right hand, and on the big toe of the right foot. [26] The priest shall pour some of the oil into the palm of his own left hand, [27] and shall sprinkle with his right finger some of the oil that is in his left hand seven times before the LORD. [28] The priest shall put some of the oil that is in his hand on the lobe of the right ear of the one to be cleansed, and on the thumb of the right hand, and the big toe of the right foot, where the blood of the guilt offering was placed. [29] The rest of the oil that is in the priest's hand he shall put on the head of the one to be cleansed, to make atonement on his behalf before the LORD. [30] And he shall offer, of the turtledoves or pigeons such as he can afford, [31] one[o] for a sin offering and the other for a burnt offering, along with a grain offering; and the priest shall make atonement before

the LORD on behalf of the one being cleansed. [32] This is the ritual for the one who has a leprous[p] disease, who cannot afford the offerings for his cleansing.

33 The LORD spoke to Moses and Aaron, saying: 34 When you come into the land of Canaan, which I give you for a possession, and I put a leprous[p] disease in a house in the land of your possession, [35] the owner of the house shall come and tell the priest, saying, "There seems to me to be some sort of disease in my house." [36] The priest shall command that they empty the house before the priest goes to examine the disease, or all that is in the house will become unclean; and afterward the priest shall go in to inspect the house. [37] He shall examine the disease; if the disease is in the walls of the house with greenish or reddish spots, and if it appears to be deeper than the surface, [38] the priest shall go outside to the door of the house and shut up the house seven days. [39] The priest shall come again on the seventh day and make an inspection; if the disease has spread in the walls of the house, [40] the priest shall command that the stones in which the disease appears be taken out and thrown into an unclean place outside the city. [41] He shall have the inside of the house scraped thoroughly, and the plaster that is scraped off shall be dumped in an unclean place outside the city. [42] They shall take other stones and put them in the place of those stones, and take other plaster and plaster the house.

43 If the disease breaks out again in the house, after he has taken out the stones and scraped the house and plastered it, [44] the priest shall go and make inspection; if the disease has spread in the house, it is a spreading leprous[p] disease

n A liquid measure   *o* Gk Syr: Heb *afford,*
[31]*such as he can afford, one*   *p* A term for
several skin diseases; precise meaning uncertain

---

rifices. **12**: In this case the guilt offering, like the sin offering (v. 19; 5.1–6), is prescribed for an inadvertent offence. *Elevation offering,* see Ex 29.24 n. **21–32**: See 5.7–13 n. **33–57**: Leprosy in houses. The "disease" appears to be mold or rot.

in the house; it is unclean. 45 He shall have the house torn down, its stones and timber and all the plaster of the house, and taken outside the city to an unclean place. 46 All who enter the house while it is shut up shall be unclean until the evening; 47 and all who sleep in the house shall wash their clothes; and all who eat in the house shall wash their clothes.

48 If the priest comes and makes an inspection, and the disease has not spread in the house after the house was plastered, the priest shall pronounce the house clean; the disease is healed. 49 For the cleansing of the house he shall take two birds, with cedarwood and crimson yarn and hyssop, 50 and shall slaughter one of the birds over fresh water in an earthen vessel, 51 and shall take the cedarwood and the hyssop and the crimson yarn, along with the living bird, and dip them in the blood of the slaughtered bird and the fresh water, and sprinkle the house seven times. 52 Thus he shall cleanse the house with the blood of the bird, and with the fresh water, and with the living bird, and with the cedarwood and hyssop and crimson yarn; 53 and he shall let the living bird go out of the city into the open field; so he shall make atonement for the house, and it shall be clean.

54 This is the ritual for any leprous*q* disease: for an itch, 55 for leprous*q* diseases in clothing and houses, 56 and for a swelling or an eruption or a spot, 57 to determine when it is unclean and when it is clean. This is the ritual for leprous*q* diseases.

15 The LORD spoke to Moses and Aaron, saying: 2 Speak to the people of Israel and say to them:

When any man has a discharge from his member,*r* his discharge makes him ceremonially unclean. 3 The uncleanness of his discharge is this: whether his member*r* flows with his discharge, or his

member*r* is stopped from discharging, it is uncleanness for him. 4 Every bed on which the one with the discharge lies shall be unclean; and everything on which he sits shall be unclean. 5 Anyone who touches his bed shall wash his clothes, and bathe in water, and be unclean until the evening. 6 All who sit on anything on which the one with the discharge has sat shall wash their clothes, and bathe in water, and be unclean until the evening. 7 All who touch the body of the one with the discharge shall wash their clothes, and bathe in water, and be unclean until the evening. 8 If the one with the discharge spits on persons who are clean, then they shall wash their clothes, and bathe in water, and be unclean until the evening. 9 Any saddle on which the one with the discharge rides shall be unclean. 10 All who touch anything that was under him shall be unclean until the evening, and all who carry such a thing shall wash their clothes, and bathe in water, and be unclean until the evening. 11 All those whom the one with the discharge touches without his having rinsed his hands in water shall wash their clothes, and bathe in water, and be unclean until the evening. 12 Any earthen vessel that the one with the discharge touches shall be broken; and every vessel of wood shall be rinsed in water.

13 When the one with a discharge is cleansed of his discharge, he shall count seven days for his cleansing; he shall wash his clothes and bathe his body in fresh water, and he shall be clean. 14 On the eighth day he shall take two turtledoves or two pigeons and come before the LORD to the entrance of the tent of meeting and give them to the priest. 15 The priest shall offer them, one for a sin offering and the other for a burnt

q A term for several skin diseases; precise meaning uncertain   *r* Heb *flesh*

15.1–32: Bodily discharges. Various bodily emissions, it was believed, produced ritual uncleanness that defiled the holy tabernacle in the midst of the people (v. 31). **3–12:**

Unclean persons contaminate anything or anyone they touch (11.24–38). **15.13–15:** The prescribed sacrifices to make atonement for uncleanness. On the sin

offering; and the priest shall make atonement on his behalf before the LORD for his discharge.

16 If a man has an emission of semen, he shall bathe his whole body in water, and be unclean until the evening. 17 Everything made of cloth or of skin on which the semen falls shall be washed with water, and be unclean until the evening. 18 If a man lies with a woman and has an emission of semen, both of them shall bathe in water, and be unclean until the evening.

19 When a woman has a discharge of blood that is her regular discharge from her body, she shall be in her impurity for seven days, and whoever touches her shall be unclean until the evening. 20 Everything upon which she lies during her impurity shall be unclean; everything also upon which she sits shall be unclean. 21 Whoever touches her bed shall wash his clothes, and bathe in water, and be unclean until the evening. 22 Whoever touches anything upon which she sits shall wash his clothes, and bathe in water, and be unclean until the evening; 23 whether it is the bed or anything upon which she sits, when he touches it he shall be unclean until the evening. 24 If any man lies with her, and her impurity falls on him, he shall be unclean seven days; and every bed on which he lies shall be unclean.

25 If a woman has a discharge of blood for many days, not at the time of her impurity, or if she has a discharge beyond the time of her impurity, all the days of the discharge she shall continue in uncleanness; as in the days of her impurity, she shall be unclean. 26 Every bed on which she lies during all the days of her discharge shall be treated as the bed of her impurity; and everything on which she sits shall be unclean, as in the uncleanness of her impurity. 27 Whoever touches these things shall be unclean, and shall wash his clothes, and bathe in water, and be unclean until the evening. 28 If she is cleansed of her discharge, she shall count seven days, and after that she shall be clean. 29 On the eighth day she shall take two turtledoves or two pigeons and bring them to the priest to the entrance of the tent of meeting. 30 The priest shall offer one for a sin offering and the other for a burnt offering; and the priest shall make atonement on her behalf before the LORD for her unclean discharge.

31 Thus you shall keep the people of Israel separate from their uncleanness, so that they do not die in their uncleanness by defiling my tabernacle that is in their midst.

32 This is the ritual for those who have a discharge: for him who has an emission of semen, becoming unclean thereby, 33 for her who is in the infirmity of her period, for anyone, male or female, who has a discharge, and for the man who lies with a woman who is unclean.

16 The LORD spoke to Moses after the death of the two sons of Aaron, when they drew near before the LORD and died. 2 The LORD said to Moses:

Tell your brother Aaron not to come

---

offering, see 4.2 n. **16–18**: It is not suggested that such secretions are evil or that sex is taboo. On holy occasions, such as worship (Ex 19.15) or the conduct of holy war (1 Sam 21.4–6), sexual abstinence was required. **15.19–30**: See ch 12. **31**: The people of Israel must be holy, for the LORD, whose tabernacle is in their midst, is holy (compare 19.2; see 11.44–45 n.).

16.1–34: **The ritual for the day of atonement.** Priestly tradition has preserved an ancient ritual, which has been elaborated during years of cultic usage, as shown by the composite character of this chapter. **1–5**: Preparatory instructions. **1**: After the block of laws on uncleanness (chs 11–15), the account resumes from ch 10. **2**: The high priest is to go into *the sanctuary inside the curtain* (Ex 26.31–35) only once a year to make atonement for priests and people (vv. 30, 34). *The mercy seat*, see Ex 25.17 n. *The cloud*, see Ex 40.34–38 n. **4**: Compare Ex ch 28. **6–10**: A short version of the ritual for the day of atonement. Two goats are chosen to bear symbolically the sins of the people. The one chosen *for the LORD* is to be sacrificed as a sin offering

just at any time into the sanctuary inside the curtain before the mercy seat[s] that is upon the ark, or he will die; for I appear in the cloud upon the mercy seat.[s] 3 Thus shall Aaron come into the holy place: with a young bull for a sin offering and a ram for a burnt offering. 4 He shall put on the holy linen tunic, and shall have the linen undergarments next to his body, fasten the linen sash, and wear the linen turban; these are the holy vestments. He shall bathe his body in water, and then put them on. 5 He shall take from the congregation of the people of Israel two male goats for a sin offering, and one ram for a burnt offering.

6 Aaron shall offer the bull as a sin offering for himself, and shall make atonement for himself and for his house. 7 He shall take the two goats and set them before the LORD at the entrance of the tent of meeting; 8 and Aaron shall cast lots on the two goats, one lot for the LORD and the other lot for Azazel.[t] 9 Aaron shall present the goat on which the lot fell for the LORD, and offer it as a sin offering; 10 but the goat on which the lot fell for Azazel[t] shall be presented alive before the LORD to make atonement over it, that it may be sent away into the wilderness to Azazel.[t]

11 Aaron shall present the bull as a sin offering for himself, and shall make atonement for himself and for his house; he shall slaughter the bull as a sin offering for himself. 12 He shall take a censer full of coals of fire from the altar before the LORD, and two handfuls of crushed sweet incense, and he shall bring it inside the curtain 13 and put the incense on the fire before the LORD, that the cloud of the incense may cover the mercy seat[s] that is upon the covenant,[u] or he will die. 14 He shall take some of the blood of the

bull, and sprinkle it with his finger on the front of the mercy seat,[s] and before the mercy seat[s] he shall sprinkle the blood with his finger seven times.

15 He shall slaughter the goat of the sin offering that is for the people and bring its blood inside the curtain, and do with its blood as he did with the blood of the bull, sprinkling it upon the mercy seat[s] and before the mercy seat.[s] 16 Thus he shall make atonement for the sanctuary, because of the uncleannesses of the people of Israel, and because of their transgressions, all their sins; and so he shall do for the tent of meeting, which remains with them in the midst of their uncleannesses. 17 No one shall be in the tent of meeting from the time he enters to make atonement in the sanctuary until he comes out and has made atonement for himself and for his house and for all the assembly of Israel. 18 Then he shall go out to the altar that is before the LORD and make atonement on its behalf, and shall take some of the blood of the bull and of the blood of the goat, and put it on each of the horns of the altar. 19 He shall sprinkle some of the blood on it with his finger seven times, and cleanse it and hallow it from the uncleannesses of the people of Israel.

20 When he has finished atoning for the holy place and the tent of meeting and the altar, he shall present the live goat. 21 Then Aaron shall lay both his hands on the head of the live goat, and confess over it all the iniquities of the people of Israel, and all their transgressions, all their sins, putting them on the head of the goat, and sending it away into the wilderness by means of someone

---

s Or *the cover*   t Traditionally rendered *a scapegoat*   u Or *treaty*, or *testament*; Heb *eduth*

(v. 15); the other is to be driven into the wilderness *to Azazel,* an evil spirit or desert demon (10.4; compare Lev 17.7; Isa 34.14). **16.11–28**: A more elaborate version of the ritual. The high priest enters the Holy of Holies once to make atonement for the priests (vv. 11–14) and once for the people (vv. 15–22). **13**: The cloud of incense is to cover the

mercy seat in order to protect the priest from beholding God (Ex 33.20). **16–19**: The high priest is to make atonement for the tabernacle and altar (Ex 29.35–37) because of the people's sins and uncleannesses. Priestly theology is deeply concerned about the presence of the holy God in the midst of a sinful people. **20–22**: This ritual symbolizes the transfer of

designated for the task. *v* 22 The goat shall bear on itself all their iniquities to a barren region; and the goat shall be set free in the wilderness.

23 Then Aaron shall enter the tent of meeting, and shall take off the linen vestments that he put on when he went into the holy place, and shall leave them there. 24 He shall bathe his body in water in a holy place, and put on his vestments; then he shall come out and offer his burnt offering and the burnt offering of the people, making atonement for himself and for the people. 25 The fat of the sin offering he shall turn into smoke on the altar. 26 The one who sets the goat free for Azazel *w* shall wash his clothes and bathe his body in water, and afterward may come into the camp. 27 The bull of the sin offering and the goat of the sin offering, whose blood was brought in to make atonement in the holy place, shall be taken outside the camp; their skin and their flesh and their dung shall be consumed in fire. 28 The one who burns them shall wash his clothes and bathe his body in water, and afterward may come into the camp.

29 This shall be a statute to you forever: In the seventh month, on the tenth day of the month, you shall deny yourselves, *x* and shall do no work, neither the citizen nor the alien who resides among you. 30 For on this day atonement shall be made for you, to cleanse you; from all your sins you shall be clean before the LORD. 31 It is a sabbath of complete rest to you, and you shall deny yourselves; *x* it is a statute forever. 32 The priest who is anointed and consecrated as priest in his father's place shall make atonement, wearing the linen vestments, the holy vestments. 33 He shall make atonement for the sanctuary, and he shall make atonement for the tent of meeting and for the altar, and he shall make atonement for the priests and for all the people of the assembly. 34 This shall be an everlasting statute for you, to make atonement for the people of Israel once in the year for all their sins. And Moses did as the LORD had commanded him.

17 The LORD spoke to Moses: 2 Speak to Aaron and his sons and to all the people of Israel and say to them: This is what the LORD has commanded. 3 If anyone of the house of Israel slaughters an ox or a lamb or a goat in the camp, or slaughters it outside the camp, 4 and does not bring it to the entrance of the tent of meeting, to present it as an offering to the LORD before the tabernacle of the LORD, he shall be held guilty of bloodshed; he has shed blood, and he shall be cut off from the people. 5 This is in order that the people of Israel may bring their sacrifices that they offer in the open field, that they may bring them to the LORD, to the priest at the entrance of the tent of meeting, and offer them as sacrifices of well-being to the LORD. 6 The priest shall dash the blood against the altar of the LORD at the entrance of the tent of meeting, and turn the fat into smoke as a pleasing odor to the LORD, 7 so that they may no longer offer their sacrifices for goat-demons, to whom

v Meaning of Heb uncertain   *w* Traditionally rendered *a scapegoat*   *x* Or *shall fast*

---

the people's sins to the animal which then carries the sins off into the wilderness (see Heb 9.1–10.18 n.).

16.23–28: Contact with holiness, like contact with its opposite, uncleanness, requires ceremonial change of garments and ritual ablutions. The ceremony concludes with a burnt offering for priest and people. **29–34:** A statutory requirement that the day of atonement shall be observed annually as *a sabbath of complete rest.*

Chapters 17–26: The Holiness Code. These chapters constitute a corpus of laws often called the Holiness Code because of the dominant theme: Israel must be holy as God is holy.

17.1–16: Restrictions upon the slaughter of animals. 3–7: The former practice of sacrifice at every legitimate place where the LORD was worshiped (Ex 20.24–25) is modified by this old tradition that requires that animals for food be sacrificed at the central sanctuary (Deut 12.15–28) as an offering of well-being (ch 3). **4:** The reason given in v. 7 is that sacrifices in the open fields had been offered to *goat-demons* or "satyrs" (Isa 34.14;

they prostitute themselves. This shall be a statute forever to them throughout their generations.

8 And say to them further: Anyone of the house of Israel or of the aliens who reside among them who offers a burnt offering or sacrifice, 9 and does not bring it to the entrance of the tent of meeting, to sacrifice it to the LORD, shall be cut off from the people.

10 If anyone of the house of Israel or of the aliens who reside among them eats any blood, I will set my face against that person who eats blood, and will cut that person off from the people. 11 For the life of the flesh is in the blood; and I have given it to you for making atonement for your lives on the altar; for, as life, it is the blood that makes atonement. 12 Therefore I have said to the people of Israel: No person among you shall eat blood, nor shall any alien who resides among you eat blood. 13 And anyone of the people of Israel, or of the aliens who reside among them, who hunts down an animal or bird that may be eaten shall pour out its blood and cover it with earth.

14 For the life of every creature—its blood is its life; therefore I have said to the people of Israel: You shall not eat the blood of any creature, for the life of every creature is its blood; whoever eats it shall be cut off. 15 All persons, citizens or aliens, who eat what dies of itself or what has been torn by wild animals, shall wash their clothes, and bathe themselves in water, and be unclean until the evening; then they shall be clean. 16 But if they do not wash themselves or bathe their body, they shall bear their guilt.

18 The LORD spoke to Moses, saying:

2 Speak to the people of Israel and say to them: I am the LORD your God. 3 You shall not do as they do in the land of Egypt, where you lived, and you shall not do as they do in the land of Canaan, to which I am bringing you. You shall not follow their statutes. 4 My ordinances you shall observe and my statutes you shall keep, following them: I am the LORD your God. 5 You shall keep my statutes and my ordinances; by doing so one shall live: I am the LORD.

6 None of you shall approach anyone near of kin to uncover nakedness: I am the LORD. 7 You shall not uncover the nakedness of your father, which is the nakedness of your mother; she is your mother, you shall not uncover her nakedness. 8 You shall not uncover the nakedness of your father's wife; it is the nakedness of your father. 9 You shall not uncover the nakedness of your sister, your father's daughter or your mother's daughter, whether born at home or born abroad. 10 You shall not uncover the nakedness of your son's daughter or of your daughter's daughter, for their nakedness is your own nakedness. 11 You shall not uncover the nakedness of your father's wife's daughter, begotten by your father, since she is your sister. 12 You shall not uncover the nakedness of your father's sister; she is your father's flesh. 13 You shall not uncover the nakedness of your mother's sister, for she is your mother's flesh. 14 You shall not uncover the nakedness of your father's brother, that is, you shall not approach his wife; she is your aunt. 15 You shall not uncover the nakedness of your daughter-in-law: she is your son's wife; you shall not uncover her nakedness. 16 You shall not uncover the nakedness of your brother's wife; it is your brother's nakedness. 17 You shall not uncover the nakedness of a woman and her daughter, and

---

Deut 32.17; 2 Chr 11.15). **9:** *Cut off* (v. 4), see 7.20 n.

17.10–14: Prohibition against eating blood. **11:** The basis of the priestly sacrificial system (see 1.5 n.). **15:** See Ex 22.31 n.

18.1–30: Forbidden sexual relations. 1–5: As a holy people, set apart for special rela-

tion to the LORD, Israel must not imitate the practices of other peoples (vv. 24–29; 11.44–45 n.).

18.6–18: An old list of twelve sexual prohibitions (compare the twelve curses in Deut ch 27). **16:** The levirate marriage was an exception to this rule (see Gen 38.8 n.). **21:**

you shall not take^y her son's daughter or her daughter's daughter to uncover her nakedness; they are your^z flesh; it is depravity. ¹⁸And you shall not take^y a woman as a rival to her sister, uncovering her nakedness while her sister is still alive.

19 You shall not approach a woman to uncover her nakedness while she is in her menstrual uncleanness. ²⁰You shall not have sexual relations with your kinsman's wife, and defile yourself with her. ²¹You shall not give any of your offspring to sacrifice them^a to Molech, and so profane the name of your God: I am the LORD. ²²You shall not lie with a male as with a woman; it is an abomination. ²³You shall not have sexual relations with any animal and defile yourself with it, nor shall any woman give herself to an animal to have sexual relations with it: it is perversion.

24 Do not defile yourselves in any of these ways, for by all these practices the nations I am casting out before you have defiled themselves. ²⁵Thus the land became defiled; and I punished it for its iniquity, and the land vomited out its inhabitants. ²⁶But you shall keep my statutes and my ordinances and commit none of these abominations, either the citizen or the alien who resides among you ²⁷(for the inhabitants of the land, who were before you, committed all of these abominations, and the land became defiled); ²⁸otherwise the land will vomit you out for defiling it, as it vomited out the nation that was before you. ²⁹For whoever commits any of these abominations shall be cut off from their people. ³⁰So keep my charge not to commit any of these abominations that were done be-fore you, and not to defile yourselves by them: I am the LORD your God.

19 The LORD spoke to Moses, saying:

2 Speak to all the congregation of the people of Israel and say to them: You shall be holy, for I the LORD your God am holy. ³You shall each revere your mother and father, and you shall keep my sabbaths: I am the LORD your God. ⁴Do not turn to idols or make cast images for yourselves: I am the LORD your God.

5 When you offer a sacrifice of well-being to the LORD, offer it in such a way that it is acceptable on your behalf. ⁶It shall be eaten on the same day you offer it, or on the next day; and anything left over until the third day shall be consumed in fire. ⁷If it is eaten at all on the third day, it is an abomination; it will not be acceptable. ⁸All who eat it shall be subject to punishment, because they have profaned what is holy to the LORD; and any such person shall be cut off from the people.

9 When you reap the harvest of your land, you shall not reap to the very edges of your field, or gather the gleanings of your harvest. ¹⁰You shall not strip your vineyard bare, or gather the fallen grapes of your vineyard; you shall leave them for the poor and the alien: I am the LORD your God.

11 You shall not steal; you shall not deal falsely; and you shall not lie to one another. ¹²And you shall not swear falsely by my name, profaning the name of your God: I am the LORD.

y Or *marry* z Gk: Heb lacks *your*
a Heb *to pass them over*

On the pagan rite of child sacrifice to *Molech,* the Ammonite deity (1 Kings 11.7), see Deut 18.10 n.

18.26–28: Although the laws of Leviticus are placed in the ancient setting of Mount Sinai, this passage clearly presupposes a time after the occupation of Canaan (vv. 25, 27).

19.1–37: The life of holiness. This chapter represents a fine blending of cultic requirements and ethical obligations, as expressed classically in the Ten Commandments (see vv. 9–10, 13–16). **2:** The keynote of the chapter and of the whole Holiness Code (chs 17–26). The people of Israel have been separated for a special covenant relationship with the God who liberated them from Egypt (v. 36). Israel's holiness, therefore, is derived from relationship to the holy God, it is not an intrinsic quality of their own life. **5–8:** See ch 3.

13 You shall not defraud your neighbor; you shall not steal; and you shall not keep for yourself the wages of a laborer until morning. [14] You shall not revile the deaf or put a stumbling block before the blind; you shall fear your God: I am the Lord.

15 You shall not render an unjust judgment; you shall not be partial to the poor or defer to the great: with justice you shall judge your neighbor. [16] You shall not go around as a slanderer[b] among your people, and you shall not profit by the blood[c] of your neighbor: I am the Lord.

17 You shall not hate in your heart anyone of your kin; you shall reprove your neighbor, or you will incur guilt yourself. [18] You shall not take vengeance or bear a grudge against any of your people, but you shall love your neighbor as yourself: I am the Lord.

19 You shall keep my statutes. You shall not let your animals breed with a different kind; you shall not sow your field with two kinds of seed; nor shall you put on a garment made of two different materials.

20 If a man has sexual relations with a woman who is a slave, designated for another man but not ransomed or given her freedom, an inquiry shall be held. They shall not be put to death, since she has not been freed; [21] but he shall bring a guilt offering for himself to the Lord, at the entrance of the tent of meeting, a ram as guilt offering. [22] And the priest shall make atonement for him with the ram of guilt offering before the Lord for his sin that he committed; and the sin he committed shall be forgiven him.

23 When you come into the land and plant all kinds of trees for food, then you shall regard their fruit as forbidden;[d]

three years it shall be forbidden[e] to you, it must not be eaten. [24] In the fourth year all their fruit shall be set apart for rejoicing in the Lord. [25] But in the fifth year you may eat of their fruit, that their yield may be increased for you: I am the Lord your God.

26 You shall not eat anything with its blood. You shall not practice augury or witchcraft. [27] You shall not round off the hair on your temples or mar the edges of your beard. [28] You shall not make any gashes in your flesh for the dead or tattoo any marks upon you: I am the Lord.

29 Do not profane your daughter by making her a prostitute, that the land not become prostituted and full of depravity. [30] You shall keep my sabbaths and reverence my sanctuary: I am the Lord.

31 Do not turn to mediums or wizards; do not seek them out, to be defiled by them: I am the Lord your God.

32 You shall rise before the aged, and defer to the old; and you shall fear your God: I am the Lord.

33 When an alien resides with you in your land, you shall not oppress the alien. [34] The alien who resides with you shall be to you as the citizen among you; you shall love the alien as yourself, for you were aliens in the land of Egypt: I am the Lord your God.

35 You shall not cheat in measuring length, weight, or quantity. [36] You shall have honest balances, honest weights, an honest ephah, and an honest hin: I am the Lord your God, who brought you out of the land of Egypt. [37] You shall keep all my statutes and all my ordinances, and observe them: I am the Lord.

b Meaning of Heb uncertain   *c* Heb *stand against the blood*   *d* Heb *as their uncircumcision*
e Heb *uncircumcision*

19.17–18: Previous ethical injunctions come to a climax in this law, the source of the "second" commandment quoted in the New Testament (Mk 12.31). *Neighbor* here means a fellow-Israelite (see Deut 15.3 n.); however, the law to *love your neighbor as yourself* is extended to include the resident alien in vv. 33–34. **19:** See Deut 22.9–11 n. **20–22:** The reason for this legal clemency is that the slave-woman is regarded as another man's property, i.e. his concubine (Ex 21.7–11). **23–25:** This law may reflect the ancient custom of propitiating the fertility powers of the soil, though here the practice is redefined in terms of Israel's faith.

19.26–31: Most of the laws in this section

20 The LORD spoke to Moses, saying: ² Say further to the people of Israel:

Any of the people of Israel, or of the aliens who reside in Israel, who give any of their offspring to Molech shall be put to death; the people of the land shall stone them to death. ³ I myself will set my face against them, and will cut them off from the people, because they have given of their offspring to Molech, defiling my sanctuary and profaning my holy name. ⁴ And if the people of the land should ever close their eyes to them, when they give of their offspring to Molech, and do not put them to death, ⁵ I myself will set my face against them and against their family, and will cut them off from among their people, them and all who follow them in prostituting themselves to Molech.

6 If any turn to mediums and wizards, prostituting themselves to them, I will set my face against them, and will cut them off from the people. ⁷ Consecrate yourselves therefore, and be holy; for I am the LORD your God. ⁸ Keep my statutes, and observe them; I am the LORD; I sanctify you. ⁹ All who curse father or mother shall be put to death; having cursed father or mother, their blood is upon them.

10 If a man commits adultery with the wife of*f* his neighbor, both the adulterer and the adulteress shall be put to death. ¹¹ The man who lies with his father's wife has uncovered his father's nakedness; both of them shall be put to death; their blood is upon them. ¹² If a man lies with his daughter-in-law, both of them shall be put to death; they have committed perversion, their blood is upon them. ¹³ If a man lies with a male as with a woman, both of them have committed an abomination; they shall be put to death; their blood is upon them. ¹⁴ If a man takes a wife and her mother also, it is depravity; they shall be burned to death, both he and they, that there may be no depravity among you. ¹⁵ If a man has sexual relations with an animal, he shall be put to death; and you shall kill the animal. ¹⁶ If a woman approaches any animal and has sexual relations with it, you shall kill the woman and the animal; they shall be put to death, their blood is upon them.

17 If a man takes his sister, a daughter of his father or a daughter of his mother, and sees her nakedness, and she sees his nakedness, it is a disgrace, and they shall be cut off in the sight of their people; he has uncovered his sister's nakedness, he shall be subject to punishment. ¹⁸ If a man lies with a woman having her sickness and uncovers her nakedness, he has laid bare her flow and she has laid bare her flow of blood; both of them shall be cut off from their people. ¹⁹ You shall not uncover the nakedness of your mother's sister or of your father's sister, for that is to lay bare one's own flesh; they shall be subject to punishment. ²⁰ If a man lies with his uncle's wife, he has uncovered his uncle's nakedness; they shall be subject to punishment; they shall die childless. ²¹ If a man takes his brother's wife, it is impurity; he has uncovered his brother's nakedness; they shall be childless.

22 You shall keep all my statutes and all my ordinances, and observe them, so that the land to which I bring you to settle in may not vomit you out. ²³ You shall not follow the practices of the nation that I am driving out before you.

f Heb repeats *if a man commits adultery with the wife of*

---

are protests against the practices of other peoples: magic and witchcraft (v. 26), heathen mourning customs (vv. 27–28; see 21.5 n.), sacred prostitution (v. 29), and necromancy (v. 31; 1 Sam ch 28; 2 Kings 21.6; 23.24).

20.1–27: Penalties for violating the rules of the theocratic community. The severity of punishment is based on the conviction that Israel is to be a holy people, separated from others by its manner of life and worship (vv. 7–8, 22–26). **2–5:** On Molech worship, see 18.21 n. **9:** See Ex 21.17 n.

20.10–21: The death penalty is prescribed for the sexual offenses dealt with in ch 18.

20.22–24: This appeal is similar to the conclusion of the Covenant Code (Ex 23.23–

Because they did all these things, I abhorred them. 24 But I have said to you: You shall inherit their land, and I will give it to you to possess, a land flowing with milk and honey. I am the LORD your God; I have separated you from the peoples. 25 You shall therefore make a distinction between the clean animal and the unclean, and between the unclean bird and the clean; you shall not bring abomination on yourselves by animal or by bird or by anything with which the ground teems, which I have set apart for you to hold unclean. 26 You shall be holy to me; for I the LORD am holy, and I have separated you from the other peoples to be mine.

27 A man or a woman who is a medium or a wizard shall be put to death; they shall be stoned to death, their blood is upon them.

21 The LORD said to Moses: Speak to the priests, the sons of Aaron, and say to them:

No one shall defile himself for a dead person among his relatives, 2 except for his nearest kin: his mother, his father, his son, his daughter, his brother; 3 likewise, for a virgin sister, close to him because she has had no husband, he may defile himself for her. 4 But he shall not defile himself as a husband among his people and so profane himself. 5 They shall not make bald spots upon their heads, or shave off the edges of their beards, or make any gashes in their flesh. 6 They shall be holy to their God, and not profane the name of their God; for they offer the LORD's offerings by fire, the food of

their God; therefore they shall be holy. 7 They shall not marry a prostitute or a woman who has been defiled; neither shall they marry a woman divorced from her husband. For they are holy to their God, 8 and you shall treat them as holy, since they offer the food of your God; they shall be holy to you, for I the LORD, I who sanctify you, am holy. 9 When the daughter of a priest profanes herself through prostitution, she profanes her father; she shall be burned to death.

10 The priest who is exalted above his fellows, on whose head the anointing oil has been poured and who has been consecrated to wear the vestments, shall not dishevel his hair, nor tear his vestments. 11 He shall not go where there is a dead body; he shall not defile himself even for his father or mother. 12 He shall not go outside the sanctuary and thus profane the sanctuary of his God; for the consecration of the anointing oil of his God is upon him: I am the LORD. 13 He shall marry only a woman who is a virgin. 14 A widow, or a divorced woman, or a woman who has been defiled, a prostitute, these he shall not marry. He shall marry a virgin of his own kin, 15 that he may not profane his offspring among his kin; for I am the LORD; I sanctify him.

16 The LORD spoke to Moses, saying: 17 Speak to Aaron and say: No one of your offspring throughout their generations who has a blemish may approach to offer the food of his God. 18 For no one who has a blemish shall draw near, one who is blind or lame, or one who has a mutilated face or a limb too long, 19 or

---

33) and to the exhortations of Deuteronomy. **25**: See ch 11. **26**: See 19.2 n. Much of the material in this chapter duplicates laws found elsewhere in the Holiness Code (e.g. chs 18 and 19), indicating that the editor has compiled various independent traditions. **27**: See 19.31. *Put to death,* compare 1 Sam 28.9.

21.1–24: Instructions to the priests. Behavior of the priests is governed by the fact that they have a special status: they are *holy to their God* (v. 6), i.e. separated for a special divine service. **1–15**: Rules concerning mourning and marriage. **1–3**: Mourning rites

were believed to be ritually defiling, owing to association with a dead body (10.6; Ezek 24.15–18). **5**: 19.27–28. Shaving the head and bodily mutilation were common mourning customs in antiquity (Deut 14.1; Am 8.10; Isa 22.12; Jer 16.6; Mic 1.16). **6**: *The food of their God* faintly echoes the ancient notion of sacrifice as the offering of food for the deity (see 2.2–3 n.); here the language is merely traditional. **10–12**: See 10.6 n.

21.16–23: Just as the sacrificial offering must be unblemished (22.17–25), so the priest who offers it must be without bodily

one who has a broken foot or a broken hand, 20 or a hunchback, or a dwarf, or a man with a blemish in his eyes or an itching disease or scabs or crushed testicles. 21 No descendant of Aaron the priest who has a blemish shall come near to offer the LORD's offerings by fire; since he has a blemish, he shall not come near to offer the food of his God. 22 He may eat the food of his God, of the most holy as well as of the holy. 23 But he shall not come near the curtain or approach the altar, because he has a blemish, that he may not profane my sanctuaries; for I am the LORD; I sanctify them. 24 Thus Moses spoke to Aaron and to his sons and to all the people of Israel.

22 The LORD spoke to Moses, saying: 2 Direct Aaron and his sons to deal carefully with the sacred donations of the people of Israel, which they dedicate to me, so that they may not profane my holy name; I am the LORD. 3 Say to them: If anyone among all your offspring throughout your generations comes near the sacred donations, which the people of Israel dedicate to the LORD, while he is in a state of uncleanness, that person shall be cut off from my presence: I am the LORD. 4 No one of Aaron's offspring who has a leprous*g* disease or suffers a discharge may eat of the sacred donations until he is clean. Whoever touches anything made unclean by a corpse or a man who has had an emission of semen, 5 and whoever touches any swarming thing by which he may be made unclean or any human being by whom he may be made unclean— whatever his uncleanness may be— 6 the person who touches any such shall be unclean until evening and shall not eat of the sacred donations unless he has washed his body in water. 7 When the

sun sets he shall be clean; and afterward he may eat of the sacred donations, for they are his food. 8 That which died or was torn by wild animals he shall not eat, becoming unclean by it: I am the LORD. 9 They shall keep my charge, so that they may not incur guilt and die in the sanctuary*h* for having profaned it: I am the LORD; I sanctify them.

10 No lay person shall eat of the sacred donations. No bound or hired servant of the priest shall eat of the sacred donations; 11 but if a priest acquires anyone by purchase, the person may eat of them; and those that are born in his house may eat of his food. 12 If a priest's daughter marries a layman, she shall not eat of the offering of the sacred donations; 13 but if a priest's daughter is widowed or divorced, without offspring, and returns to her father's house, as in her youth, she may eat of her father's food. No lay person shall eat of it. 14 If a man eats of the sacred donation unintentionally, he shall add one-fifth of its value to it, and give the sacred donation to the priest. 15 No one shall profane the sacred donations of the people of Israel, which they offer to the LORD, 16 causing them to bear guilt requiring a guilt offering, by eating their sacred donations: for I am the LORD; I sanctify them.

17 The LORD spoke to Moses, saying: 18 Speak to Aaron and his sons and all the people of Israel and say to them: When anyone of the house of Israel or of the aliens residing in Israel presents an offering, whether in payment of a vow or as a freewill offering that is offered to the LORD as a burnt offering, 19 to be accept-

g A term for several skin diseases; precise meaning uncertain   *h* Vg: Heb *incur guilt for it and die in it*

---

defect. Unnatural deformity or disfigurement is frowned upon, for human beings are God's creation, made in the divine image (Gen 5.1–3).
22.1–9: Priests may not partake of *the sacred donations,* i.e. the consecrated portion of the sacrifice which is their due (see 2.2–3 n.),

while in a state of ritual uncleanness as defined in the laws of chs 11–15. **10–16:** Further definition of who may eat the consecrated portions. *No lay person,* i.e. nobody outside the priest's immediate household (which included his slaves; see Ex 12.43–49 n.). **17–25:** Directions about acceptable sacrifices.

able in your behalf it shall be a male without blemish, of the cattle or the sheep or the goats. ²⁰ You shall not offer anything that has a blemish, for it will not be acceptable in your behalf.

21 When anyone offers a sacrifice of well-being to the LORD, in fulfillment of a vow or as a freewill offering, from the herd or from the flock, to be acceptable it must be perfect; there shall be no blemish in it. ²² Anything blind, or injured, or maimed, or having a discharge or an itch or scabs—these you shall not offer to the LORD or put any of them on the altar as offerings by fire to the LORD. ²³ An ox or a lamb that has a limb too long or too short you may present for a freewill offering; but it will not be accepted for a vow. ²⁴ Any animal that has its testicles bruised or crushed or torn or cut, you shall not offer to the LORD; such you shall not do within your land, ²⁵ nor shall you accept any such animals from a foreigner to offer as food to your God; since they are mutilated, with a blemish in them, they shall not be accepted in your behalf.

26 The LORD spoke to Moses, saying: ²⁷ When an ox or a sheep or a goat is born, it shall remain seven days with its mother, and from the eighth day on it shall be acceptable as the LORD's offering by fire. ²⁸ But you shall not slaughter, from the herd or the flock, an animal with its young on the same day. ²⁹ When you sacrifice a thanksgiving offering to the LORD, you shall sacrifice it so that it may be acceptable in your behalf. ³⁰ It shall be eaten on the same day; you shall not leave any of it until morning: I am the LORD.

31 Thus you shall keep my commandments and observe them: I am the LORD. ³² You shall not profane my holy name, that I may be sanctified among the people of Israel: I am the LORD; I sanctify you, ³³ I who brought you out of the land of Egypt to be your God: I am the LORD.

23 The LORD spoke to Moses, saying: ² Speak to the people of Israel and say to them: These are the appointed festivals of the LORD that you shall proclaim as holy convocations, my appointed festivals.

3 Six days shall work be done; but the seventh day is a sabbath of complete rest, a holy convocation; you shall do no work: it is a sabbath to the LORD throughout your settlements.

4 These are the appointed festivals of the LORD, the holy convocations, which you shall celebrate at the time appointed for them. ⁵ In the first month, on the fourteenth day of the month, at twilight,ⁱ there shall be a passover offering to the LORD, ⁶ and on the fifteenth day of the same month is the festival of unleavened bread to the LORD; seven days you shall eat unleavened bread. ⁷ On the first day you shall have a holy convocation; you shall not work at your occupations. ⁸ For seven days you shall present the LORD's offerings by fire; on the seventh day there shall be a holy convocation: you shall not work at your occupations.

9 The LORD spoke to Moses: ¹⁰ Speak to the people of Israel and say to them: When you enter the land that I am giving you and you reap its harvest, you shall bring the sheaf of the first fruits of your harvest to the priest. ¹¹ He shall raise the sheaf before the LORD, that you may find acceptance; on the day after the sabbath the priest shall raise it. ¹² On the day when you raise the sheaf, you shall offer a lamb a year old, without blemish, as a burnt offering to the LORD. ¹³ And the grain offering with it shall be two-tenths of an ephah of choice flour mixed with

i Heb *between the two evenings*

23.1–44: **The sacred calendar. 2:** There shall be a proclamation by trumpets (Num 10.10) to announce the appointed festivals. **3:** Ex 16.23. **4:** The calendar of sacred festivals is paralleled in Ex 23.14–17; 34.18–24; Deut 16.1–17. **5–8:** Passover and the festival of un-leavened bread are treated here (as in other calendars) as two phases of one celebration (see Ex 12.14–20 n.). *The first month,* see Ex 12.2 n.

23.9–14: The offering of the first fruits (Deut 26.5–10) took place the day after the concluding sabbath of the festival of unleavened

oil, an offering by fire of pleasing odor to the LORD; and the drink offering with it shall be of wine, one-fourth of a hin. [14] You shall eat no bread or parched grain or fresh ears until that very day, until you have brought the offering of your God: it is a statute forever throughout your generations in all your settlements.

15 And from the day after the sabbath, from the day on which you bring the sheaf of the elevation offering, you shall count off seven weeks; they shall be complete. [16] You shall count until the day after the seventh sabbath, fifty days; then you shall present an offering of new grain to the LORD. [17] You shall bring from your settlements two loaves of bread as an elevation offering, each made of two-tenths of an ephah; they shall be of choice flour, baked with leaven, as first fruits to the LORD. [18] You shall present with the bread seven lambs a year old without blemish, one young bull, and two rams; they shall be a burnt offering to the LORD, along with their grain offering and their drink offerings, an offering by fire of pleasing odor to the LORD. [19] You shall also offer one male goat for a sin offering, and two male lambs a year old as a sacrifice of well-being. [20] The priest shall raise them with the bread of the first fruits as an elevation offering before the LORD, together with the two lambs; they shall be holy to the LORD for the priest. [21] On that same day you shall make proclamation; you shall hold a holy convocation; you shall not work at your occupations. This is a statute forever in all your settlements throughout your generations.

22 When you reap the harvest of your land, you shall not reap to the very edges of your field, or gather the gleanings of your harvest; you shall leave them for the poor and for the alien: I am the LORD your God.

23 The LORD spoke to Moses, saying: [24] Speak to the people of Israel, saying: In the seventh month, on the first day of the month, you shall observe a day of complete rest, a holy convocation commemorated with trumpet blasts. [25] You shall not work at your occupations; and you shall present the LORD's offering by fire.

26 The LORD spoke to Moses, saying: [27] Now, the tenth day of this seventh month is the day of atonement; it shall be a holy convocation for you: you shall deny yourselves[j] and present the LORD's offering by fire; [28] and you shall do no work during that entire day; for it is a day of atonement, to make atonement on your behalf before the LORD your God. [29] For anyone who does not practice self-denial[k] during that entire day shall be cut off from the people. [30] And anyone who does any work during that entire day, such a one I will destroy from the midst of the people. [31] You shall do no work: it is a statute forever throughout your generations in all your settlements. [32] It shall be to you a sabbath of complete rest, and you shall deny yourselves;[j] on the ninth day of the month at evening, from evening to evening you shall keep your sabbath.

33 The LORD spoke to Moses, saying: [34] Speak to the people of Israel, saying: On the fifteenth day of this seventh month, and lasting seven days, there shall be the festival of booths[l] to the LORD. [35] The first day shall be a holy convocation; you shall not work at your occupations. [36] Seven days you shall present the LORD's offerings by fire; on the

j Or *shall fast*   *k* Or *does not fast*
l Or *tabernacles*: Heb *succoth*

---

bread (v. 11; compare vv. 7–8), i.e. at the beginning of the barley harvest in April. **11:** *Raise the sheaf,* see Ex 29.24 n.

23.15–21: Since the festival of weeks, celebrated at the time of the wheat harvest, was held fifty days after the festival of unleavened bread, it came to be called Pentecost (based on a Greek word meaning "fifty"). **23–25:** The festival of trumpets or New Year. This falls at the beginning of *the seventh month* (September–October), according to the ecclesiastical calendar (see Ex 12.2 n.).

23.26–32: The day of atonement (ch 16) occurs during the same month. **33–36** (sup-

eighth day you shall observe a holy convocation and present the LORD's offerings by fire; it is a solemn assembly; you shall not work at your occupations.

37 These are the appointed festivals of the LORD, which you shall celebrate as times of holy convocation, for presenting to the LORD offerings by fire—burnt offerings and grain offerings, sacrifices and drink offerings, each on its proper day— [38] apart from the sabbaths of the LORD, and apart from your gifts, and apart from all your votive offerings, and apart from all your freewill offerings, which you give to the LORD.

39 Now, the fifteenth day of the seventh month, when you have gathered in the produce of the land, you shall keep the festival of the LORD, lasting seven days; a complete rest on the first day, and a complete rest on the eighth day. [40] On the first day you shall take the fruit of majestic[m] trees, branches of palm trees, boughs of leafy trees, and willows of the brook; and you shall rejoice before the LORD your God for seven days. [41] You shall keep it as a festival to the LORD seven days in the year; you shall keep it in the seventh month as a statute forever throughout your generations. [42] You shall live in booths for seven days; all that are citizens in Israel shall live in booths, [43] so that your generations may know that I made the people of Israel live in booths when I brought them out of the land of Egypt: I am the LORD your God.

44 Thus Moses declared to the people of Israel the appointed festivals of the LORD.

24 The LORD spoke to Moses, saying: [2] Command the people of Israel to bring you pure oil of beaten olives for the lamp, that a light may be kept burning regularly. [3] Aaron shall set it up in the tent of meeting, outside the curtain of the covenant,[n] to burn from evening to morning before the LORD regularly; it shall be a statute forever throughout your generations. [4] He shall set up the lamps on the lampstand of pure gold[o] before the LORD regularly.

5 You shall take choice flour, and bake twelve loaves of it; two-tenths of an ephah shall be in each loaf. [6] You shall place them in two rows, six in a row, on the table of pure gold.[p] [7] You shall put pure frankincense with each row, to be a token offering for the bread, as an offering by fire to the LORD. [8] Every sabbath day Aaron shall set them in order before the LORD regularly as a commitment of the people of Israel, as a covenant forever. [9] They shall be for Aaron and his descendants, who shall eat them in a holy place, for they are most holy portions for him from the offerings by fire to the LORD, a perpetual due.

10 A man whose mother was an Israelite and whose father was an Egyptian came out among the people of Israel; and the Israelite woman's son and a certain Israelite began fighting in the camp. [11] The Israelite woman's son blasphemed the Name in a curse. And they brought him to Moses—now his mother's name was Shelomith, daughter of Dibri, of the tribe of Dan— [12] and they put him in custody, until the decision of the LORD should be made clear to them.

13 The LORD said to Moses, saying: [14] Take the blasphemer outside the camp; and let all who were within hearing lay their hands on his head, and let the whole congregation stone him. [15] And speak to the people of Israel, saying: Anyone who curses God shall bear the sin. [16] One who

---

m Meaning of Heb uncertain   *n* Or *treaty*, or *testament*; Heb *eduth*   *o* Heb *pure lampstand*   *p* Heb *pure table*

---

plemented in vv. 39–43): The festival of booths or thanksgiving was held at the time of the autumn ingathering.

23.37–38: See further Num chs 28 and 29 for the sacrifices to be offered at the holy convocations.

24.1–23: Various priestly laws. 1–4: Oil for the sanctuary lamp (Ex 27.20–21). **5–9:** The bread of the Presence, see Ex 25.23–30.

24.10–14: This incident serves as a setting for the laws that follow (see v. 23). *The Name,* a substitute for the sacred name of Israel's

blasphemes the name of the LORD shall be put to death; the whole congregation shall stone the blasphemer. Aliens as well as citizens, when they blaspheme the Name, shall be put to death. 17 Anyone who kills a human being shall be put to death. 18 Anyone who kills an animal shall make restitution for it, life for life. 19 Anyone who maims another shall suffer the same injury in return: 20 fracture for fracture, eye for eye, tooth for tooth; the injury inflicted is the injury to be suffered. 21 One who kills an animal shall make restitution for it; but one who kills a human being shall be put to death. 22 You shall have one law for the alien and for the citizen: for I am the LORD your God. 23 Moses spoke thus to the people of Israel; and they took the blasphemer outside the camp, and stoned him to death. The people of Israel did as the LORD had commanded Moses.

25 The LORD spoke to Moses on Mount Sinai, saying: 2 Speak to the people of Israel and say to them: When you enter the land that I am giving you, the land shall observe a sabbath for the LORD. 3 Six years you shall sow your field, and six years you shall prune your vineyard, and gather in their yield; 4 but in the seventh year there shall be a sabbath of complete rest for the land, a sabbath for the LORD: you shall not sow your field or prune your vineyard. 5 You shall not reap the aftergrowth of your

harvest or gather the grapes of your unpruned vine: it shall be a year of complete rest for the land. 6 You may eat what the land yields during its sabbath—you, your male and female slaves, your hired and your bound laborers who live with you; 7 for your livestock also, and for the wild animals in your land all its yield shall be for food.

8 You shall count off seven weeks *q* of years, seven times seven years, so that the period of seven weeks of years gives forty-nine years. 9 Then you shall have the trumpet sounded loud; on the tenth day of the seventh month—on the day of atonement—you shall have the trumpet sounded throughout all your land. 10 And you shall hallow the fiftieth year and you shall proclaim liberty throughout the land to all its inhabitants. It shall be a jubilee for you: you shall return, every one of you, to your property and every one of you to your family. 11 That fiftieth year shall be a jubilee for you: you shall not sow, or reap the aftergrowth, or harvest the unpruned vines. 12 For it is a jubilee; it shall be holy to you: you shall eat only what the field itself produces.

13 In this year of jubilee you shall return, every one of you, to your property. 14 When you make a sale to your neighbor or buy from your neighbor, you shall not cheat one another. 15 When you

q Or *sabbaths*

God (see Ex 3.14 n.). On cursing God, see Ex 20.7; 22.28. **12:** Compare Num 15.32–36. **13:** It is held that blasphemy pollutes the community. Stoning, a communal mode of execution, is the means of purifying the evil from the midst of the people (Deut 17.2–7; 21.18–21). **15–22:** An independent set of laws. **20:** For the "lex talionis" or "law of retaliation" (the law of "an eye for an eye, and a tooth for a tooth"), see Ex 21.22–25 n.

25.1–55: The sabbatical year and the year of jubilee. These laws attempt to prevent economic exploitation by stressing that the ownership of the land is vested in God, rather than human beings. **2–7:** On the seventh year *the land shall observe a sabbath,* i.e. lie fallow (Ex 23.10–11). Although this custom may be rooted ultimately in appeasement of the powers of the soil (see 19.23–25 n.), here

it signifies that the LORD owns the land and gives it to the people of Israel for stewardship (v. 2). **6–7:** These verses appear to modify the law by saying that, although there shall be no reaping or storing (v. 5), the crop that grows of itself can be taken (compare v. 12). Evidently it was difficult to enforce the law (see 26.34–35). It was enforced, however, in the Maccabean period (1 Macc 6.49, 53).

25.8–24: After seven sabbatical years, i.e. on the fiftieth year, comes the year of jubilee, named after the Hebrew word "yobhel" (ram's horn) which was blown in proclamation. **10:** Leases are to expire and all persons are to return to their ancestral estate. **13–17:** Although Israelites shall not sell their property in perpetuity (v. 23), they may lease it for farming. **14:** *Neighbor,* see 19.17–18 n. **15–16:** The rent is to be reckoned at the approximat-

buy from your neighbor, you shall pay only for the number of years since the jubilee; the seller shall charge you only for the remaining crop years. 16 If the years are more, you shall increase the price, and if the years are fewer, you shall diminish the price; for it is a certain number of harvests that are being sold to you. 17 You shall not cheat one another, but you shall fear your God; for I am the LORD your God.

18 You shall observe my statutes and faithfully keep my ordinances, so that you may live on the land securely. 19 The land will yield its fruit, and you will eat your fill and live on it securely. 20 Should you ask, What shall we eat in the seventh year, if we may not sow or gather in our crop? 21 I will order my blessing for you in the sixth year, so that it will yield a crop for three years. 22 When you sow in the eighth year, you will be eating from the old crop; until the ninth year, when its produce comes in, you shall eat the old. 23 The land shall not be sold in perpetuity, for the land is mine; with me you are but aliens and tenants. 24 Throughout the land that you hold, you shall provide for the redemption of the land.

25 If anyone of your kin falls into difficulty and sells a piece of property, then the next of kin shall come and redeem what the relative has sold. 26 If the person has no one to redeem it, but then prospers and finds sufficient means to do so, 27 the years since its sale shall be computed and the difference shall be refunded to the person to whom it was sold, and the property shall be returned. 28 But if there is not sufficient means to recover it, what

was sold shall remain with the purchaser until the year of jubilee; in the jubilee it shall be released, and the property shall be returned.

29 If anyone sells a dwelling house in a walled city, it may be redeemed until a year has elapsed since its sale; the right of redemption shall be one year. 30 If it is not redeemed before a full year has elapsed, a house that is in a walled city shall pass in perpetuity to the purchaser, throughout the generations; it shall not be released in the jubilee. 31 But houses in villages that have no walls around them shall be classed as open country; they may be redeemed, and they shall be released in the jubilee. 32 As for the cities of the Levites, the Levites shall forever have the right of redemption of the houses in the cities belonging to them. 33 Such property as may be redeemed from the Levites—houses sold in a city belonging to them—shall be released in the jubilee; because the houses in the cities of the Levites are their possession among the people of Israel. 34 But the open land around their cities may not be sold; for that is their possession for all time.

35 If any of your kin fall into difficulty and become dependent on you,*r* you shall support them; they shall live with you as though resident aliens. 36 Do not take interest in advance or otherwise make a profit from them, but fear your God; let them live with you. 37 You shall not lend them your money at interest taken in advance, or provide them food at a profit. 38 I am the LORD your God,

r Meaning of Heb uncertain

---

ed value of the crops in the remaining years of the jubilee period. **20–22**: A law dealing with the sabbatical years that fall during the jubilee period. **23**: A statement of the theological premise of the program: Israelites are *aliens and tenants* on land that does not belong to them by right but that the LORD has given them as an inheritance. Thus the land is not private property, to be bought and sold speculatively. Although there is no evidence that the jubilee program was ever carried out, the law opposes foreign conceptions of prop-

erty that resulted in the swallowing up of ancestral holdings (1 Kings 21.3; Isa 5.8). **25.25–55**: Laws dealing with property and its redemption. **25–28**: Even before the year of jubilee the land may be redeemed. **25**: The *next of kin* (lit., "redeemer") is the relative who upholds the rights of a family member, e.g. by blood revenge (Num 35.12) or the redemption of property (Ruth ch 4). **29–34**: An exception to cover the case of urban houses as distinguished from houses in villages. **32–34**: Num 35.1–8.

who brought you out of the land of Egypt, to give you the land of Canaan, to be your God.

39 If any who are dependent on you become so impoverished that they sell themselves to you, you shall not make them serve as slaves. 40 They shall remain with you as hired or bound laborers. They shall serve with you until the year of the jubilee. 41 Then they and their children with them shall be free from your authority; they shall go back to their own family and return to their ancestral property. 42 For they are my servants, whom I brought out of the land of Egypt; they shall not be sold as slaves are sold. 43 You shall not rule over them with harshness, but shall fear your God. 44 As for the male and female slaves whom you may have, it is from the nations around you that you may acquire male and female slaves. 45 You may also acquire them from among the aliens residing with you, and from their families that are with you, who have been born in your land; and they may be your property. 46 You may keep them as a possession for your children after you, for them to inherit as property. These you may treat as slaves, but as for your fellow Israelites, no one shall rule over the other with harshness.

47 If resident aliens among you prosper, and if any of your kin fall into difficulty with one of them and sell themselves to an alien, or to a branch of the alien's family, 48 after they have sold themselves they shall have the right of redemption; one of their brothers may redeem them, 49 or their uncle or their uncle's son may redeem them, or anyone of their family who is of their own flesh may redeem them; or if they prosper they may redeem themselves. 50 They shall compute with the purchaser the total from the year when they sold themselves to the alien until the jubilee year; the price of the sale shall be applied to the number of years: the time they were with the owner shall be rated as the time of a hired laborer. 51 If many years remain, they shall pay for their redemption in proportion to the purchase price; 52 and if few years remain until the jubilee year, they shall compute thus: according to the years involved they shall make payment for their redemption. 53 As a laborer hired by the year they shall be under the alien's authority, who shall not, however, rule with harshness over them in your sight. 54 And if they have not been redeemed in any of these ways, they and their children with them shall go free in the jubilee year. 55 For to me the people of Israel are servants; they are my servants whom I brought out from the land of Egypt: I am the LORD your God.

26 You shall make for yourselves no idols and erect no carved images or pillars, and you shall not place figured stones in your land, to worship at them; for I am the LORD your God. 2 You shall keep my sabbaths and reverence my sanctuary: I am the LORD.

3 If you follow my statutes and keep my commandments and observe them faithfully, 4 I will give you your rains in their season, and the land shall yield its produce, and the trees of the field shall yield their fruit. 5 Your threshing shall overtake the vintage, and the vintage shall overtake the sowing; you shall eat your bread to the full, and live securely in your land. 6 And I will grant peace in the land, and you shall lie down, and no one shall make you afraid; I will remove dangerous animals from the land, and no

---

25.35–38: See Ex 22.25 n. **39–55:** Israelites may sell themselves as hired servants but shall never become slaves. **40:** Compare Ex 21.1–6; Deut 15.12–18. **44–46:** It is permissible, however, to make slaves of non-Israelites, since, according to the ancient way of thinking, they are outside the boundaries of the covenant community.
25.48–49: See v. 25 n. **55:** This verse gives the theological basis for the prohibition against enslaving Israelites (vv. 42–43).

26.1–46: The two ways. The Holiness Code (chs 17–26), conceived as the LORD's address to the people through Moses (17.1–2), concludes with divine warnings and promises. See also the concluding addresses in Ex 23.20–33 and Deut ch 28. **1:** A prohibition against the cultic objects of Canaanite

sword shall go through your land. 7 You shall give chase to your enemies, and they shall fall before you by the sword. 8 Five of you shall give chase to a hundred, and a hundred of you shall give chase to ten thousand; your enemies shall fall before you by the sword. 9 I will look with favor upon you and make you fruitful and multiply you; and I will maintain my covenant with you. 10 You shall eat old grain long stored, and you shall have to clear out the old to make way for the new. 11 I will place my dwelling in your midst, and I shall not abhor you. 12 And I will walk among you, and will be your God, and you shall be my people. 13 I am the LORD your God who brought you out of the land of Egypt, to be their slaves no more; I have broken the bars of your yoke and made you walk erect.

14 But if you will not obey me, and do not observe all these commandments, 15 if you spurn my statutes, and abhor my ordinances, so that you will not observe all my commandments, and you break my covenant, 16 I in turn will do this to you: I will bring terror on you; consumption and fever that waste the eyes and cause life to pine away. You shall sow your seed in vain, for your enemies shall eat it. 17 I will set my face against you, and you shall be struck down by your enemies; your foes shall rule over you, and you shall flee though no one pursues you. 18 And if in spite of this you will not obey me, I will continue to punish you sevenfold for your sins. 19 I will break your proud glory, and I will make your sky like iron and your earth like copper. 20 Your strength shall be spent to no purpose: your land shall not yield its produce, and the trees of the land shall not yield their fruit.

21 If you continue hostile to me, and will not obey me, I will continue to

plague you sevenfold for your sins. 22 I will let loose wild animals against you, and they shall bereave you of your children and destroy your livestock; they shall make you few in number, and your roads shall be deserted.

23 If in spite of these punishments you have not turned back to me, but continue hostile to me, 24 then I too will continue hostile to you: I myself will strike you sevenfold for your sins. 25 I will bring the sword against you, executing vengeance for the covenant; and if you withdraw within your cities, I will send pestilence among you, and you shall be delivered into enemy hands. 26 When I break your staff of bread, ten women shall bake your bread in a single oven, and they shall dole out your bread by weight; and though you eat, you shall not be satisfied.

27 But if, despite this, you disobey me, and continue hostile to me, 28 I will continue hostile to you in fury; I in turn will punish you myself sevenfold for your sins. 29 You shall eat the flesh of your sons, and you shall eat the flesh of your daughters. 30 I will destroy your high places and cut down your incense altars; I will heap your carcasses on the carcasses of your idols. I will abhor you. 31 I will lay your cities waste, will make your sanctuaries desolate, and I will not smell your pleasing odors. 32 I will devastate the land, so that your enemies who come to settle in it shall be appalled at it. 33 And you I will scatter among the nations, and I will unsheathe the sword against you; your land shall be a desolation, and your cities a waste.

34 Then the land shall enjoy[s] its sabbath years as long as it lies desolate, while you are in the land of your enemies; then

[s] Or *make up for*

---

religion (see Ex 34.13 n.). **11–13:** The fundamental theme of the Holiness Code: the holy God has come to dwell in the midst of a sinful people. **12:** Ex 6.7.
26.26: In the time of scarcity ten women will have to bake in one oven, rather than each in her own house. **29:** Deut 28.53–57.

32–33: These and following verses indicate familiarity with the policy of deporting conquered peoples, a policy used effectively by the Assyrians (2 Kings ch 17), and anticipate the conquest of Judah and the exile of the people into Babylonia in 597–587 B.C. (vv. 34, 38, 39, 41, 43–44).

the land shall rest, and enjoy[t] its sabbath years. 35 As long as it lies desolate, it shall have the rest it did not have on your sabbaths when you were living on it. 36 And as for those of you who survive, I will send faintness into their hearts in the lands of their enemies; the sound of a driven leaf shall put them to flight, and they shall flee as one flees from the sword, and they shall fall though no one pursues. 37 They shall stumble over one another, as if to escape a sword, though no one pursues; and you shall have no power to stand against your enemies. 38 You shall perish among the nations, and the land of your enemies shall devour you. 39 And those of you who survive shall languish in the land of your enemies because of their iniquities; also they shall languish because of the iniquities of their ancestors.

40 But if they confess their iniquity and the iniquity of their ancestors, in that they committed treachery against me and, moreover, that they continued hostile to me— 41 so that I, in turn, continued hostile to them and brought them into the land of their enemies; if then their uncircumcised heart is humbled and they make amends for their iniquity, 42 then will I remember my covenant with Jacob; I will remember also my covenant with Isaac and also my covenant with Abraham, and I will remember the land. 43 For the land shall be deserted by them, and enjoy[t] its sabbath years by lying desolate without them, while they shall make amends for their iniquity, because they dared to spurn my ordinances, and they abhorred my statutes. 44 Yet for all that, when they are in the land of their enemies, I will not spurn them, or abhor them so as to destroy them utterly and break my covenant with them; for I am the LORD their God; 45 but I will remember in their favor the covenant with their ancestors whom I brought out of the land of Egypt in the sight of the nations, to be their God: I am the LORD.

46 These are the statutes and ordinances and laws that the LORD established between himself and the people of Israel on Mount Sinai through Moses.

27 The LORD spoke to Moses, saying: 2 Speak to the people of Israel and say to them: When a person makes an explicit vow to the LORD concerning the equivalent for a human being, 3 the equivalent for a male shall be: from twenty to sixty years of age the equivalent shall be fifty shekels of silver by the sanctuary shekel. 4 If the person is a female, the equivalent is thirty shekels. 5 If the age is from five to twenty years of age, the equivalent is twenty shekels for a male and ten shekels for a female. 6 If the age is from one month to five years, the equivalent for a male is five shekels of silver, and for a female the equivalent is three shekels of silver. 7 And if the person is sixty years old or over, then the equivalent for a male is fifteen shekels, and for a female ten shekels. 8 If any cannot afford the equivalent, they shall be brought before the priest and the priest shall assess them; the priest shall assess them according to what each one making a vow can afford.

9 If it concerns an animal that may be brought as an offering to the LORD, any such that may be given to the LORD shall be holy. 10 Another shall not be exchanged or substituted for it, either good for bad or bad for good; and if one animal is substituted for another, both that one and its substitute shall be holy. 11 If it concerns any unclean animal that may

---

t Or *make up for*

26.34–35: See 25.6–7 n. **41:** *Uncircumcised heart,* a heart that is sealed or unresponsive to the LORD's will (Jer 4.4).

27.1–34: An appendix dealing with religious vows (compare Pss 56.12; 116.14). See further Num ch 30. **1–8:** Persons dedicated to the LORD's service (compare 1 Sam 1.11) may be freed from the vow by the payment of a monetary substitute. **3:** *The sanctuary shekel,* see Ex 30.13 n. **9–13:** If an animal appropriate for sacrifice is vowed, it cannot be redeemed.

not be brought as an offering to the LORD, the animal shall be presented before the priest. [12] The priest shall assess it: whether good or bad, according to the assessment of the priest, so it shall be. [13] But if it is to be redeemed, one-fifth must be added to the assessment.

14 If a person consecrates a house to the LORD, the priest shall assess it: whether good or bad, as the priest assesses it, so it shall stand. [15] And if the one who consecrates the house wishes to redeem it, one-fifth shall be added to its assessed value, and it shall revert to the original owner.

16 If a person consecrates to the LORD any inherited landholding, its assessment shall be in accordance with its seed requirements: fifty shekels of silver to a homer of barley seed. [17] If the person consecrates the field as of the year of jubilee, that assessment shall stand; [18] but if the field is consecrated after the jubilee, the priest shall compute the price for it according to the years that remain until the year of jubilee, and the assessment shall be reduced. [19] And if the one who consecrates the field wishes to redeem it, then one-fifth shall be added to its assessed value, and it shall revert to the original owner; [20] but if the field is not redeemed, or if it has been sold to someone else, it shall no longer be redeemable. [21] But when the field is released in the jubilee, it shall be holy to the LORD as a devoted field; it becomes the priest's holding. [22] If someone consecrates to the LORD a field that has been purchased, which is not a part of the inherited landholding, [23] the priest shall compute for it the proportionate assessment up to the year of jubilee, and the assessment shall be paid as of that day, a sacred donation to the LORD. [24] In the year of jubilee the field shall return to the one from whom it was bought, whose holding the land is. [25] All assessments shall be by the sanctuary shekel: twenty gerahs shall make a shekel.

26 A firstling of animals, however, which as a firstling belongs to the LORD, cannot be consecrated by anyone; whether ox or sheep, it is the LORD's. [27] If it is an unclean animal, it shall be ransomed at its assessment, with one-fifth added; if it is not redeemed, it shall be sold at its assessment.

28 Nothing that a person owns that has been devoted to destruction for the LORD, be it human or animal, or inherited landholding, may be sold or redeemed; every devoted thing is most holy to the LORD. [29] No human beings who have been devoted to destruction can be ransomed; they shall be put to death.

30 All tithes from the land, whether the seed from the ground or the fruit from the tree, are the LORD's; they are holy to the LORD. [31] If persons wish to redeem any of their tithes, they must add one-fifth to them. [32] All tithes of herd and flock, every tenth one that passes under the shepherd's staff, shall be holy to the LORD. [33] Let no one inquire whether it is good or bad, or make substitution for it; if one makes substitution for it, then both it and the substitute shall be holy and cannot be redeemed.

34 These are the commandments that the LORD gave to Moses for the people of Israel on Mount Sinai.

27.14–25: Houses and lands vowed to the LORD are redeemable. 17–18: On this manner of valuation, see 25.15–16. 21: *Devoted,* i.e. set apart as belonging to the LORD. Booty taken in holy war was devoted and could not be appropriated for common use (Josh 6.19). 27.26: Firstlings of clean animals cannot be dedicated since they already belong to the LORD (see Ex 13.2 n.). 28: *Devoted thing,* see v. 21 n. 29: Compare 1 Sam ch 15. 30–33: On tithes (Mal 3.8–10). 32–33: The words *pass under the shepherd's staff* refer to the manner of counting animals (Jer 33.13).

Numbers

The English title "Numbers," derived from the Greek (*Arithmoi*) and Latin (*Numeri*) translations, is based primarily on the numbering or census of the people related in chs 1–4 and again in ch 26. If one follows this lead, the book falls into two main divisions: (1) chs 1–25 relate how the entire first Exodus generation, which murmured against God in the wilderness and refused to take the land, died off, except for Moses and the two faithful spies, Joshua and Caleb; (2) chs 26–36, beginning with another census, deal with the new generation that God led toward the promised land under Moses' leadership.

The narratives portray a people who found that, in the strange providence of God, the journey from promise to fulfillment led round about by way of the wilderness. In the Hebrew Bible the book is appropriately called "In the Wilderness," referring to the long period, traditionally forty years (33.38; Am 5.25), spent in the wilderness. Much of the time the people apparently stayed at an oasis known as Kadesh-barnea, where they arrived after the march from Sinai (see 10.11 and 20.1). The story is told in such a way as to reflect later struggles during the time of the monarchy, the Babylonian exile, and even later.

These narratives do not idealize the wilderness period. Again and again the people complained, sensing the contrast between the relative security of slavery in Egypt and the precarious insecurity of freedom in the wilderness. There were power struggles among the leaders, raising the question as to who speaks for God. There were crises that threatened faith in God's presence in their midst and God's guidance into the future. Yet these narratives are infused with the conviction that, despite the people's blindness and rebelliousness, God was faithful to promises made to Israel's ancestors. In the wilderness the people were being disciplined so that they might know their utter dependence upon the God who liberated them from bondage and thus be strengthened for the challenges of life in a new land.

1 The LORD spoke to Moses in the wilderness of Sinai, in the tent of meeting, on the first day of the second month, in the second year after they had come out of the land of Egypt, saying: 2 Take a census of the whole congregation of Israelites, in their clans, by ancestral houses, according to the number of names, every male individually; 3 from twenty years old and upward, everyone in Israel able to go to war. You and Aaron shall enroll them, company by company. 4 A man from each tribe shall be with you, each man the head of his ancestral house. 5 These are the names of the men who shall assist you:

> From Reuben, Elizur son of
> Shedeur.
6 From Simeon, Shelumiel son of
> Zurishaddai.
7 From Judah, Nahshon son of
> Amminadab.
8 From Issachar, Nethanel son of
> Zuar.
9 From Zebulun, Eliab son of
> Helon.
10 From the sons of Joseph:
> from Ephraim, Elishama son of
> Ammihud;
> from Manasseh, Gamaliel son of
> Pedahzur.
11 From Benjamin, Abidan son of
> Gideoni.
12 From Dan, Ahiezer son of
> Ammishaddai.
13 From Asher, Pagiel son of
> Ochran.
14 From Gad, Eliasaph son of Deuel.
15 From Naphtali, Ahira son of
> Enan.

16 These were the ones chosen from the congregation, the leaders of their ancestral tribes, the heads of the divisions of Israel.

17 Moses and Aaron took these men who had been designated by name, 18 and on the first day of the second month they assembled the whole congregation together. They registered themselves in their clans, by their ancestral houses, according to the number of names from twenty years old and upward, individually, 19 as the LORD commanded Moses. So he enrolled them in the wilderness of Sinai.

20 The descendants of Reuben, Israel's firstborn, their lineage, in their clans, by their ancestral houses, according to the number of names, individually, every male from twenty years old and upward, everyone able to go to war: 21 those enrolled of the tribe of Reuben were forty-six thousand five hundred.

22 The descendants of Simeon, their lineage, in their clans, by their ancestral houses, those of them that were numbered, according to the number of names, individually, every male from twenty years old and upward, everyone able to go to war: 23 those enrolled of the tribe of Simeon were fifty-nine thousand three hundred.

24 The descendants of Gad, their lineage, in their clans, by their ancestral houses, according to the number of the names, from twenty years old and upward, everyone able to go to war: 25 those enrolled of the tribe of Gad were forty-five thousand six hundred fifty.

26 The descendants of Judah, their

1.1–54: Census of the tribes. Moses is commanded to number all Israelites, making an exception of the tribe of Levi. **1:** One month had elapsed since the construction of the tabernacle (Ex 40.2, 17). **2–4:** See Ex 30.11–12 n. Military service was one of the chief duties involved in membership in the tribal assembly (ch 32). Strictly, a tribe included a number of *clans,* and a clan included several *ancestral houses* or family groups. **5–15:** The twelvefold structure of the tribal league, reflected in this old name-list, possibly goes back to Joshua's time (Josh ch 24); probably, however, this notion reflects social organization known in the time of David or may even be an ideal scheme. No longer a landed tribe, Levi is not mentioned here (see Ex 28.1–5 n.), but is later assigned a priestly role (vv. 47–54). The twelvefold pattern is maintained by counting the "house of Joseph" as two tribes: Manasseh and Ephraim. On the mention of Ephraim before Manasseh, see Gen 48.13–14 n. and contrast 26.28; 34.23–24. **1.17–46:** The census total of 603,550 (compare the slightly lower figure in 26.51) is extremely high (see Ex 12.37 n.). It has been

lineage, in their clans, by their ancestral houses, according to the number of names, from twenty years old and upward, everyone able to go to war: 27those enrolled of the tribe of Judah were seventy-four thousand six hundred.

28 The descendants of Issachar, their lineage, in their clans, by their ancestral houses, according to the number of names, from twenty years old and upward, everyone able to go to war: 29those enrolled of the tribe of Issachar were fifty-four thousand four hundred.

30 The descendants of Zebulun, their lineage, in their clans, by their ancestral houses, according to the number of names, from twenty years old and upward, everyone able to go to war: 31those enrolled of the tribe of Zebulun were fifty-seven thousand four hundred.

32 The descendants of Joseph, namely, the descendants of Ephraim, their lineage, in their clans, by their ancestral houses, according to the number of names, from twenty years old and upward, everyone able to go to war: 33those enrolled of the tribe of Ephraim were forty thousand five hundred.

34 The descendants of Manasseh, their lineage, in their clans, by their ancestral houses, according to the number of names, from twenty years old and upward, everyone able to go to war: 35those enrolled of the tribe of Manasseh were thirty-two thousand two hundred.

36 The descendants of Benjamin, their lineage, in their clans, by their ancestral houses, according to the number of names, from twenty years old and upward, everyone able to go to war: 37those enrolled of the tribe of Benjamin were thirty-five thousand four hundred.

38 The descendants of Dan, their lineage, in their clans, by their ancestral houses, according to the number of names, from twenty years old and upward, everyone able to go to war: 39those enrolled of the tribe of Dan were sixty-two thousand seven hundred.

40 The descendants of Asher, their lineage, in their clans, by their ancestral houses, according to the number of names, from twenty years old and upward, everyone able to go to war: 41those enrolled of the tribe of Asher were forty-one thousand five hundred.

42 The descendants of Naphtali, their lineage, in their clans, by their ancestral houses, according to the number of names, from twenty years old and upward, everyone able to go to war: 43those enrolled of the tribe of Naphtali were fifty-three thousand four hundred.

44 These are those who were enrolled, whom Moses and Aaron enrolled with the help of the leaders of Israel, twelve men, each representing his ancestral house. 45So the whole number of the Israelites, by their ancestral houses, from twenty years old and upward, everyone able to go to war in Israel— 46their whole number was six hundred three thousand five hundred fifty. 47The Levites, however, were not numbered by their ancestral tribe along with them.

18 The LORD had said to Moses: 49Only the tribe of Levi you shall not enroll, and you shall not take a census of them with the other Israelites. 50Rather you shall appoint the Levites over the

---

suggested that the Hebrew word translated "thousand" (vv. 21, 23, etc.) is an old term for a subsection of a tribe (31.14), based on the procedures for military muster employed by other ancient peoples, and that the original number follows "thousand" in each case, e.g. Reuben had forty-six tribal subsections with a total of five hundred men (v. 21). This reduces the total to 5,550. It is alleged that the present impossibly high figures were read back from the time of the monarchy when the military unit actually included one thousand soldiers. In any case, priestly tradition about the numerous population (see 11.21 n.) is based on the belief that the LORD marvelously increased Abraham's descendants (see Ex 1.7 n.) and miraculously supported this great people in the wilderness.

1.47–54: See 3.5–10 n. **50:** *Covenant,* see Ex 16.33–34 n.

tabernacle of the covenant, *a* and over all its equipment, and over all that belongs to it; they are to carry the tabernacle and all its equipment, and they shall tend it, and shall camp around the tabernacle. 51 When the tabernacle is to set out, the Levites shall take it down; and when the tabernacle is to be pitched, the Levites shall set it up. And any outsider who comes near shall be put to death. 52 The other Israelites shall camp in their respective regimental camps, by companies; 53 but the Levites shall camp around the tabernacle of the covenant, *a* that there may be no wrath on the congregation of the Israelites; and the Levites shall perform the guard duty of the tabernacle of the covenant. *a* 54 The Israelites did so; they did just as the LORD commanded Moses.

2 The LORD spoke to Moses and Aaron, saying: 2 The Israelites shall camp each in their respective regiments, under ensigns by their ancestral houses; they shall camp facing the tent of meeting on every side. 3 Those to camp on the east side toward the sunrise shall be of the regimental encampment of Judah by companies. The leader of the people of Judah shall be Nahshon son of Amminadab, 4 with a company as enrolled of seventy-four thousand six hundred. 5 Those to camp next to him shall be the tribe of Issachar. The leader of the Issacharites shall be Nethanel son of Zuar, 6 with a company as enrolled of fifty-four thousand four hundred. 7 Then the tribe of Zebulun: The leader of the Zebulunites shall be Eliab son of Helon, 8 with a company as enrolled of fifty-seven thousand four hundred. 9 The total enrollment of the camp of Judah, by compa-

nies, is one hundred eighty-six thousand four hundred. They shall set out first on the march.

10 On the south side shall be the regimental encampment of Reuben by companies. The leader of the Reubenites shall be Elizur son of Shedeur, 11 with a company as enrolled of forty-six thousand five hundred. 12 And those to camp next to him shall be the tribe of Simeon. The leader of the Simeonites shall be Shelumiel son of Zurishaddai, 13 with a company as enrolled of fifty-nine thousand three hundred. 14 Then the tribe of Gad: The leader of the Gadites shall be Eliasaph son of Reuel, 15 with a company as enrolled of forty-five thousand six hundred fifty. 16 The total enrollment of the camp of Reuben, by companies, is one hundred fifty-one thousand four hundred fifty. They shall set out second.

17 The tent of meeting, with the camp of the Levites, shall set out in the center of the camps; they shall set out just as they camp, each in position, by their regiments.

18 On the west side shall be the regimental encampment of Ephraim by companies. The leader of the people of Ephraim shall be Elishama son of Ammihud, 19 with a company as enrolled of forty thousand five hundred. 20 Next to him shall be the tribe of Manasseh. The leader of the people of Manasseh shall be Gamaliel son of Pedahzur, 21 with a company as enrolled of thirty-two thousand two hundred. 22 Then the tribe of Benjamin: The leader of the Benjaminites shall be Abidan son of Gideoni, 23 with a company as enrolled of thirty-five thousand

a Or *treaty*, or *testimony*; Heb *eduth*

2.1–34: The arrangement of the tribes while encamped or on the march (compare 10.13–28). **2:** Perhaps with the festival of tabernacles in mind (Lev 23.33–36), the priestly writer conceives the congregation as arranged symmetrically around the tent of meeting. In the oldest tradition, the tent was not in the center but on the outside of the camp (Ex 33.7–11). **3–31:** The Levites are to encamp immediately around the tabernacle

court (Ex 27.9–19), to protect and transport the sacred shrine (v. 17; 1.47–54). Farther out the twelve tribes are to encamp, three on each side. The arrangement expresses the idea of the tabernacling presence of the LORD in the midst of his people (Ex 25.8). **3–4:** Judah is assigned the favored position, on the east of the camp. **9:** Each division marches in assigned order, with Judah in the lead. **17:** During the march, the Levites are to be flanked by two

four hundred. 24 The total enrollment of the camp of Ephraim, by companies, is one hundred eight thousand one hundred. They shall set out third on the march.

25 On the north side shall be the regimental encampment of Dan by companies. The leader of the Danites shall be Ahiezer son of Ammishaddai, 26 with a company as enrolled of sixty-two thousand seven hundred. 27 Those to camp next to him shall be the tribe of Asher. The leader of the Asherites shall be Pagiel son of Ochran, 28 with a company as enrolled of forty-one thousand five hundred. 29 Then the tribe of Naphtali: The leader of the Naphtalites shall be Ahira son of Enan, 30 with a company as enrolled of fifty-three thousand four hundred. 31 The total enrollment of the camp of Dan is one hundred fifty-seven thousand six hundred. They shall set out last, by companies. b

32 This was the enrollment of the Israelites by their ancestral houses; the total enrollment in the camps by their companies was six hundred three thousand five hundred fifty. 33 Just as the LORD had commanded Moses, the Levites were not enrolled among the other Israelites.

34 The Israelites did just as the LORD had commanded Moses: They camped by regiments, and they set out the same way, everyone by clans, according to ancestral houses.

3 This is the lineage of Aaron and Moses at the time when the LORD spoke with Moses on Mount Sinai. 2 These are the names of the sons of Aaron: Nadab the firstborn, and Abihu, Eleazar, and Ithamar; 3 these are the names of the sons of Aaron, the anointed priests, whom he ordained to minister as priests. 4 Nadab and Abihu died before the LORD when they offered illicit fire before the LORD in the wilderness of Sinai, and they had no children. Eleazar and Ithamar served as priests in the lifetime of their father Aaron.

5 Then the LORD spoke to Moses, saying: 6 Bring the tribe of Levi near, and set them before Aaron the priest, so that they may assist him. 7 They shall perform duties for him and for the whole congregation in front of the tent of meeting, doing service at the tabernacle; 8 they shall be in charge of all the furnishings of the tent of meeting, and attend to the duties for the Israelites as they do service at the tabernacle. 9 You shall give the Levites to Aaron and his descendants; they are unreservedly given to him from among the Israelites. 10 But you shall make a register of Aaron and his descendants; it is they who shall attend to the priesthood, and any outsider who comes near shall be put to death.

11 Then the LORD spoke to Moses, saying: 12 I hereby accept the Levites from among the Israelites as substitutes for all the firstborn that open the womb among the Israelites. The Levites shall be mine, 13 for all the firstborn are mine; when I killed all the firstborn in the land of Egypt, I consecrated for my own all the firstborn in Israel, both human and animal; they shall be mine. I am the LORD.

14 Then the LORD spoke to Moses in the wilderness of Sinai, saying: 15 Enroll the Levites by ancestral houses and by

b Compare verses 9, 16, 24: Heb by their regiments

---

divisions on both the front and the rear. 18–24: On Ephraim's position of leadership, see 1.5–15 n.

3.1–51: The Levites are set apart for special service. 1–3: Ex 6.23–25. 4: Lev ch 10. 5–10: Moses and Aaron both belong to the tribe of Levi. Within the larger circle of Levites, however, a distinction is made between Aaron and his descendants and other Levites (see Ex 28.1–5 n.). Aaron's descendants exercise the chief priestly functions in the sanctuary, while the other Levites are given to the Aaronic order as priestly assistants.

3.11–13: A reinterpretation of the ancient law found in Ex 13.2; 22.29b–30; 34.19–20. Instead of having to go into lifelong religious service, the firstborn are redeemed by a "sacrifice" (see Ex 13.13 n.), for the LORD claims the Levites in their stead (Num 8.15–19). 14–39: 1.47–54; ch 4. 15: The Levitical census is

clans. You shall enroll every male from a month old and upward. 16 So Moses enrolled them according to the word of the LORD, as he was commanded. 17 The following were the sons of Levi, by their names: Gershon, Kohath, and Merari. 18 These are the names of the sons of Gershon by their clans: Libni and Shimei. 19 The sons of Kohath by their clans: Amram, Izhar, Hebron, and Uzziel. 20 The sons of Merari by their clans: Mahli and Mushi. These are the clans of the Levites, by their ancestral houses.

21 To Gershon belonged the clan of the Libnites and the clan of the Shimeites; these were the clans of the Gershonites. 22 Their enrollment, counting all the males from a month old and upward, was seven thousand five hundred. 23 The clans of the Gershonites were to camp behind the tabernacle on the west, 24 with Eliasaph son of Lael as head of the ancestral house of the Gershonites. 25 The responsibility of the sons of Gershon in the tent of meeting was to be the tabernacle, the tent with its covering, the screen for the entrance of the tent of meeting, 26 the hangings of the court, the screen for the entrance of the court that is around the tabernacle and the altar, and its cords—all the service pertaining to these.

27 To Kohath belonged the clan of the Amramites, the clan of the Izharites, the clan of the Hebronites, and the clan of the Uzzielites; these are the clans of the Kohathites. 28 Counting all the males, from a month old and upward, there were eight thousand six hundred, attending to the duties of the sanctuary. 29 The clans of the Kohathites were to camp on the south side of the tabernacle, 30 with Elizaphan son of Uzziel as head of the ancestral house of the clans of the Kohathites. 31 Their responsibility was to be the ark, the table, the lampstand, the altars, the vessels of the sanctuary with which the priests minister, and the screen—all the service pertaining to these. 32 Eleazar son of Aaron the priest was to be chief over the leaders of the Levites, and to have oversight of those who had charge of the sanctuary.

33 To Merari belonged the clan of the Mahlites and the clan of the Mushites: these are the clans of Merari. 34 Their enrollment, counting all the males from a month old and upward, was six thousand two hundred. 35 The head of the ancestral house of the clans of Merari was Zuriel son of Abihail; they were to camp on the north side of the tabernacle. 36 The responsibility assigned to the sons of Merari was to be the frames of the tabernacle, the bars, the pillars, the bases, and all their accessories—all the service pertaining to these; 37 also the pillars of the court all around, with their bases and pegs and cords.

38 Those who were to camp in front of the tabernacle on the east—in front of

---

not for military purposes (compare 1.2–3) but is associated with the firstborn who are redeemable from the age of *a month* (vv. 40–51). **17–20:** For *the clans of the Levites,* see Ex 6.16–19.
3.25–26: 4.21–28. The Gershonites have custody over the tent covering (Ex 26.7–14), the screen for the entrance to the tent (Ex 26.36), the hangings of the court (Ex 27.9), and the screen for the gate of the court (Ex 27.16). **31–32:** 4.4–15. The Kohathites have charge of the most holy objects (4.4–20), i.e. the ark, *the table* of the bread of the Presence, *the lampstand* (Ex ch 25), the bronze and gold *altars* (Ex 27.1–8; 30.1–10), the sacred *vessels* (Ex 30.17–21; 31.7–11), and *the screen* (Num 4.5; Ex 26.31–35). *Eleazar,* Aaron's oldest surviving son.

3.36–37: 4.29–33. The Merarites are responsible for the tabernacle framework (Ex 26.15–30) and the supports for the court (Ex 27.9–19). **38:** Aaron and his sons encamp on the east, the favored location (see 2.3–4 n.), symbolizing their priestly pre-eminence. *Any outsider* (3.10), i.e. a layperson. **39:** The numbers given in ch 3 actually total 22,300. **40–51:** Another Israelite census is taken to determine the number of firstborn males, in contrast to men of military age (1.2–3). **41:** See 3.11–13 n. The substitution of Levites' cattle for the people's firstlings of cattle modifies the law of the firstborn even further. **43:** A tally disclosed an excess of 273 Israelite firstborn over the number of Levites (but see v. 39 n.). **46–48:** Since the Levites were a ransom for Israelite males on a one to one basis,

the tent of meeting toward the east—
were Moses and Aaron and Aaron's
sons, having charge of the rites within
the sanctuary, whatever had to be done
for the Israelites; and any outsider who
came near was to be put to death. 39 The
total enrollment of the Levites whom
Moses and Aaron enrolled at the commandment of the LORD, by their clans,
all the males from a month old and upward, was twenty-two thousand.

40 Then the LORD said to Moses: Enroll all the firstborn males of the Israelites, from a month old and upward, and
count their names. 41 But you shall accept
the Levites for me— I am the LORD—as
substitutes for all the firstborn among the
Israelites, and the livestock of the Levites
as substitutes for all the firstborn among
the livestock of the Israelites. 42 So Moses
enrolled all the firstborn among the Israelites, as the LORD commanded him.
43 The total enrollment, all the firstborn
males from a month old and upward,
counting the number of names, was
twenty-two thousand two hundred
seventy-three.

44 Then the LORD spoke to Moses,
saying: 45 Accept the Levites as substitutes for all the firstborn among the Israelites, and the livestock of the Levites as
substitutes for their livestock; and the Levites shall be mine. I am the LORD. 46 As
the price of redemption of the two hundred seventy-three of the firstborn of the
Israelites, over and above the number of
the Levites, 47 you shall accept five shekels apiece, reckoning by the shekel of the
sanctuary, a shekel of twenty gerahs.
48 Give to Aaron and his sons the money
by which the excess number of them is
redeemed. 49 So Moses took the redemption money from those who were over
and above those redeemed by the Le-

vites; 50 from the firstborn of the Israelites he took the money, one thousand
three hundred sixty-five shekels, reckoned by the shekel of the sanctuary;
51 and Moses gave the redemption money
to Aaron and his sons, according to the
word of the LORD, as the LORD had commanded Moses.

4 The LORD spoke to Moses and Aaron, saying: 2 Take a census of the
Kohathites separate from the other Levites, by their clans and their ancestral
houses, 3 from thirty years old up to fifty
years old, all who qualify to do work
relating to the tent of meeting. 4 The service of the Kohathites relating to the tent
of meeting concerns the most holy
things.

5 When the camp is to set out, Aaron
and his sons shall go in and take down the
screening curtain, and cover the ark of
the covenant*c* with it; 6 then they shall
put on it a covering of fine leather,*d* and
spread over that a cloth all of blue, and
shall put its poles in place. 7 Over the
table of the bread of the Presence they
shall spread a blue cloth, and put on it the
plates, the dishes for incense, the bowls,
and the flagons for the drink offering; the
regular bread also shall be on it; 8 then
they shall spread over them a crimson
cloth, and cover it with a covering of fine
leather,*d* and shall put its poles in place.
9 They shall take a blue cloth, and cover
the lampstand for the light, with its
lamps, its snuffers, its trays, and all the
vessels for oil with which it is supplied;
10 and they shall put it with all its utensils
in a covering of fine leather,*d* and put it
on the carrying frame. 11 Over the golden
altar they shall spread a blue cloth, and

---

c Or *treaty,* or *testimony;* Heb *eduth*
d Meaning of Heb uncertain

---

the additional males had to be redeemed by
monetary payment. *The shekel of the sanctuary,*
see Ex 30.13 n.
4.1–49: Another Levitical census. 2–3:
This census, in contrast with 3 15, is to determine the number of Levites of the age required for the priestly duties (compare 8.23–
25) that are defined in vv. 4–33. **4–20:** The

Kohathites have a special distinction among
the Levites, for they have charge of *the most
holy things* and are therefore under the immediate supervision of Eleazar (see 3.31–32 n.).
5: *The screening curtain,* Ex 26.31–35. **7:** *The
regular bread,* i.e. the bread of the Presence that
is continually on the table (Ex 25.30).

cover it with a covering of fine leather,*e* and shall put its poles in place; 12 and they shall take all the utensils of the service that are used in the sanctuary, and put them in a blue cloth, and cover them with a covering of fine leather,*e* and put them on the carrying frame. 13 They shall take away the ashes from the altar, and spread a purple cloth over it; 14 and they shall put on it all the utensils of the altar, which are used for the service there, the firepans, the forks, the shovels, and the basins, all the utensils of the altar; and they shall spread on it a covering of fine leather,*e* and shall put its poles in place. 15 When Aaron and his sons have finished covering the sanctuary and all the furnishings of the sanctuary, as the camp sets out, after that the Kohathites shall come to carry these, but they must not touch the holy things, or they will die. These are the things of the tent of meeting that the Kohathites are to carry.

16 Eleazar son of Aaron the priest shall have charge of the oil for the light, the fragrant incense, the regular grain offering, and the anointing oil, the oversight of all the tabernacle and all that is in it, in the sanctuary and in its utensils.

17 Then the LORD spoke to Moses and Aaron, saying: 18 You must not let the tribe of the clans of the Kohathites be destroyed from among the Levites. 19 This is how you must deal with them in order that they may live and not die when they come near to the most holy things: Aaron and his sons shall go in and assign each to a particular task or burden. 20 But the Kohathites*f* must not go in to look on the holy things even for a moment; otherwise they will die.

21 Then the LORD spoke to Moses, saying: 22 Take a census of the Gershonites also, by their ancestral houses and by their clans; 23 from thirty years old up to fifty years old you shall enroll them, all who qualify to do work in the tent of meeting. 24 This is the service of the clans of the Gershonites, in serving and bearing burdens: 25 They shall carry the curtains of the tabernacle, and the tent of meeting with its covering, and the outer covering of fine leather*e* that is on top of it, and the screen for the entrance of the tent of meeting, 26 and the hangings of the court, and the screen for the entrance of the gate of the court that is around the tabernacle and the altar, and their cords, and all the equipment for their service; and they shall do all that needs to be done with regard to them. 27 All the service of the Gershonites shall be at the command of Aaron and his sons, in all that they are to carry, and in all that they have to do; and you shall assign to their charge all that they are to carry. 28 This is the service of the clans of the Gershonites relating to the tent of meeting, and their responsibilities are to be under the oversight of Ithamar son of Aaron the priest.

29 As for the Merarites, you shall enroll them by their clans and their ancestral houses; 30 from thirty years old up to fifty years old you shall enroll them, everyone who qualifies to do the work of the tent of meeting. 31 This is what they are charged to carry, as the whole of their service in the tent of meeting: the frames of the tabernacle, with its bars, pillars, and bases, 32 and the pillars of the court all around with their bases, pegs, and cords, with all their equipment and all their related service; and you shall assign by name the objects that they are required to carry. 33 This is the service of the clans of the Merarites, the whole of their service relating to the tent of meeting, under the hand of Ithamar son of Aaron the priest.

34 So Moses and Aaron and the leaders of the congregation enrolled the Kohathites, by their clans and their ancestral

e Meaning of Heb uncertain   *f* Heb *they*

4.15–20: The precautions taken by Aaron and his sons are explained on the assumption that the Kohathites must not see or touch the holy objects, lest divine holiness destroy them (compare 1 Sam ch 5–6; 2 Sam 6.6–11). 4.21–28: For the responsibility of the Ger-

houses, 35 from thirty years old up to fifty years old, everyone who qualified for work relating to the tent of meeting; 36 and their enrollment by clans was two thousand seven hundred fifty. 37 This was the enrollment of the clans of the Kohathites, all who served at the tent of meeting, whom Moses and Aaron enrolled according to the commandment of the LORD by Moses.

38 The enrollment of the Gershonites, by their clans and their ancestral houses, 39 from thirty years old up to fifty years old, everyone who qualified for work relating to the tent of meeting— 40 their enrollment by their clans and their ancestral houses was two thousand six hundred thirty. 41 This was the enrollment of the clans of the Gershonites, all who served at the tent of meeting, whom Moses and Aaron enrolled according to the commandment of the LORD.

42 The enrollment of the clans of the Merarites, by their clans and their ancestral houses, 43 from thirty years old up to fifty years old, everyone who qualified for work relating to the tent of meeting— 44 their enrollment by their clans was three thousand two hundred. 45 This is the enrollment of the clans of the Merarites, whom Moses and Aaron enrolled according to the commandment of the LORD by Moses.

46 All those who were enrolled of the Levites, whom Moses and Aaron and the leaders of Israel enrolled, by their clans and their ancestral houses, 47 from thirty years old up to fifty years old, everyone who qualified to do the work of service and the work of bearing burdens relating to the tent of meeting, 48 their enrollment was eight thousand five hundred eighty. 49 According to the commandment of the LORD through Moses they were appoint- ed to their several tasks of serving or carrying; thus they were enrolled by him, as the LORD commanded Moses.

5 The LORD spoke to Moses, saying: 2 Command the Israelites to put out of the camp everyone who is leprous, *g* or has a discharge, and everyone who is unclean through contact with a corpse; 3 you shall put out both male and female, putting them outside the camp; they must not defile their camp, where I dwell among them. 4 The Israelites did so, putting them outside the camp; as the LORD had spoken to Moses, so the Israelites did.

5 The LORD spoke to Moses, saying: 6 Speak to the Israelites: When a man or a woman wrongs another, breaking faith with the LORD, that person incurs guilt 7 and shall confess the sin that has been committed. The person shall make full restitution for the wrong, adding one fifth to it, and giving it to the one who was wronged. 8 If the injured party has no next of kin to whom restitution may be made for the wrong, the restitution for wrong shall go to the LORD for the priest, in addition to the ram of atonement with which atonement is made for the guilty party. 9 Among all the sacred donations of the Israelites, every gift that they bring to the priest shall be his. 10 The sacred donations of all are their own; whatever anyone gives to the priest shall be his.

11 The LORD spoke to Moses, saying: 12 Speak to the Israelites and say to them: If any man's wife goes astray and is unfaithful to him, 13 if a man has had intercourse with her but it is hidden from her

g A term for several skin diseases; precise meaning uncertain

---

shonites, see 3.25–26. **29–33**: 3.36–37. **48**: Compare the number of male Levites in 3.39.

5.1–6.21: Various instructions to the people.

5.1–4: The sanctity of the tabernacle is the reason for excluding unclean people, such as lepers (Lev 13.46), those having a bodily discharge (Lev ch 15), or those defiled by contact with the dead (Lev 21.1–12). **5–10**: This case law, supplemental to Lev 6.1–7 (Ex 22.7–15), deals with a situation in which there is no *next of kin* (see Lev 25.25 n.). **5.11–31**: Another case law, dealing with a woman suspected of adultery. **13–14**: This old law deals with a case not covered elsewhere (e.g. Lev 20.10): no witness can testify *since*

husband, so that she is undetected though she has defiled herself, and there is no witness against her since she was not caught in the act; 14 if a spirit of jealousy comes on him, and he is jealous of his wife who has defiled herself; or if a spirit of jealousy comes on him, and he is jealous of his wife, though she has not defiled herself; 15 then the man shall bring his wife to the priest. And he shall bring the offering required for her, one-tenth of an ephah of barley flour. He shall pour no oil on it and put no frankincense on it, for it is a grain offering of jealousy, a grain offering of remembrance, bringing iniquity to remembrance.

16 Then the priest shall bring her near, and set her before the LORD; 17 the priest shall take holy water in an earthen vessel, and take some of the dust that is on the floor of the tabernacle and put it into the water. 18 The priest shall set the woman before the LORD, dishevel the woman's hair, and place in her hands the grain offering of remembrance, which is the grain offering of jealousy. In his own hand the priest shall have the water of bitterness that brings the curse. 19 Then the priest shall make her take an oath, saying, "If no man has lain with you, if you have not turned aside to uncleanness while under your husband's authority, be immune to this water of bitterness that brings the curse. 20 But if you have gone astray while under your husband's authority, if you have defiled yourself and some man other than your husband has had intercourse with you," 21 —let the priest make the woman take the oath of the curse and say to the woman—"the LORD make you an execration and an oath among your people, when the LORD makes your uterus drop, your womb discharge; 22 now may this water that brings the curse enter your bowels and make your womb discharge, your uterus drop!" And the woman shall say, "Amen. Amen."

23 Then the priest shall put these curses in writing, and wash them off into the water of bitterness. 24 He shall make the woman drink the water of bitterness that brings the curse, and the water that brings the curse shall enter her and cause bitter pain. 25 The priest shall take the grain offering of jealousy out of the woman's hand, and shall elevate the grain offering before the LORD and bring it to the altar; 26 and the priest shall take a handful of the grain offering, as its memorial portion, and turn it into smoke on the altar, and afterward shall make the woman drink the water. 27 When he has made her drink the water, then, if she has defiled herself and has been unfaithful to her husband, the water that brings the curse shall enter into her and cause bitter pain, and her womb shall discharge, her uterus drop, and the woman shall become an execration among her people. 28 But if the woman has not defiled herself and is clean, then she shall be immune and be able to conceive children.

29 This is the law in cases of jealousy, when a wife, while under her husband's authority, goes astray and defiles herself, 30 or when a spirit of jealousy comes on a man and he is jealous of his wife; then he shall set the woman before the LORD, and the priest shall apply this entire law to her. 31 The man shall be free from iniquity, but the woman shall bear her iniquity.

6 The LORD spoke to Moses, saying: 2 Speak to the Israelites and say to them: When either men or women make

---

she was not caught in the act. **15**: The grain offering (Lev ch 2) of jealousy is intended to bring the case directly to God.
5.16–28: A trial by ordeal, a common practice among ancient peoples. **16**: Before the LORD, i.e. to the sanctuary (Ex 22.9). **19–22**: The oath of execration, compare Ex 22.10–11.
5.23–28: The oath is followed by the drinking of the water that brings the curse (v. 18). It was believed that if a person were guilty this potion would have effects which would signify the LORD's verdict of judgment (Ex 32.20, 35). **26**: Memorial portion, see Lev 2.2–3 n.

6.1–21: The vow of a nazirite. **2**: Nazirites are holy persons who have taken the vow to separate themselves to the LORD (Judg 13.5;

a special vow, the vow of a nazirite, [h] to separate themselves to the LORD, [3] they shall separate themselves from wine and strong drink; they shall drink no wine vinegar or other vinegar, and shall not drink any grape juice or eat grapes, fresh or dried. [4] All their days as nazirites [i] they shall eat nothing that is produced by the grapevine, not even the seeds or the skins.

5 All the days of their nazirite vow no razor shall come upon the head; until the time is completed for which they separate themselves to the LORD, they shall be holy; they shall let the locks of the head grow long.

6 All the days that they separate themselves to the LORD they shall not go near a corpse. [7] Even if their father or mother, brother or sister, should die, they may not defile themselves; because their consecration to God is upon the head. [8] All their days as nazirites [i] they are holy to the LORD.

9 If someone dies very suddenly nearby, defiling the consecrated head, then they shall shave the head on the day of their cleansing; on the seventh day they shall shave it. [10] On the eighth day they shall bring two turtledoves or two young pigeons to the priest at the entrance of the tent of meeting, [11] and the priest shall offer one as a sin offering and the other as a burnt offering, and make atonement for them, because they incurred guilt by reason of the corpse. They shall sanctify the head that same day, [12] and separate themselves to the LORD for their days as nazirites, [i] and bring a male lamb a year old as a guilt offering. The former time shall be void, because the consecrated head was defiled.

13 This is the law for the nazirites [i] when the time of their consecration has been completed: they shall be brought to the entrance of the tent of meeting, [14] and they shall offer their gift to the LORD, one male lamb a year old without blemish as a burnt offering, one ewe lamb a year old without blemish as a sin offering, one ram without blemish as an offering of well-being, [15] and a basket of unleavened bread, cakes of choice flour mixed with oil and unleavened wafers spread with oil, with their grain offering and their drink offerings. [16] The priest shall present them before the LORD and offer their sin offering and burnt offering, [17] and shall offer the ram as a sacrifice of well-being to the LORD, with the basket of unleavened bread; the priest also shall make the accompanying grain offering and drink offering. [18] Then the nazirites [i] shall shave the consecrated head at the entrance of the tent of meeting, and shall take the hair from the consecrated head and put it on the fire under the sacrifice of well-being. [19] The priest shall take the shoulder of the ram, when it is boiled, and one unleavened cake out of the basket, and one unleavened wafer, and shall put them in the palms of the nazirites, [i] after they have shaved the consecrated head. [20] Then the priest shall elevate them as an elevation offering before the LORD; they are a holy portion for the priest, together with the breast that is elevated and the thigh that is offered. After that the nazirites [i] may drink wine.

21 This is the law for the nazirites [i] who take a vow. Their offering to the LORD must be in accordance with the nazirite [h] vow, apart from what else they

h That is *one separated* or *one consecrated*
i That is *those separated* or *those consecrated*

1 Sam 1.11; Am 2.11–12). **3–5:** Abstinence from the fruit of the vine (Jer ch 35; compare Lk 1.15), as well as from vinegar made from soured intoxicants, dramatizes Israel's separation from Canaanite culture (compare Gen 9.21). For this restriction, as well as the law that *no razor shall come upon the head,* see the Samson story (especially Judg 13.4–5). **6–7:** Compare Lev 21.1–12.

6.9–12: Accidental contact with a corpse defiles the *consecrated head,* i.e. the long locks which evidence one's holy separation (v. 7b) so that the vow must be initiated again. **6.13–20:** The law for terminating the vow. **14–17:** For the kinds of sacrifice, see Lev chs 1–7. **18:** Hair offerings were not uncommon in ancient religion. **20:** *Elevation offering,* see Ex 29.24 n.

can afford. In accordance with whatever vow they take, so they shall do, following the law for their consecration.

22 The LORD spoke to Moses, saying: 23 Speak to Aaron and his sons, saying, Thus you shall bless the Israelites: You shall say to them,

24 The LORD bless you and keep you;
25 the LORD make his face to shine
upon you, and be gracious
to you;
26 the LORD lift up his countenance
upon you, and give you
peace.

27 So they shall put my name on the Israelites, and I will bless them.

7 On the day when Moses had finished setting up the tabernacle, and had anointed and consecrated it with all its furnishings, and had anointed and consecrated the altar with all its utensils, 2 the leaders of Israel, heads of their ancestral houses, the leaders of the tribes, who were over those who were enrolled, made offerings. 3 They brought their offerings before the LORD, six covered wagons and twelve oxen, a wagon for every two of the leaders, and for each one an ox; they presented them before the tabernacle. 4 Then the LORD said to Moses: 5 Accept these from them, that they may be used in doing the service of the tent of meeting, and give them to the Levites, to each according to his service. 6 So Moses took the wagons and the oxen, and gave them to the Levites. 7 Two wagons and four oxen he gave to the Gershonites, according to their service; 8 and four wagons and eight oxen he gave to the Merarites, according to their service, under the direction of Ithamar son of Aaron the priest. 9 But to the Kohathites he gave none, because they were charged with the care of the holy things that had to be carried on the shoulders.

10 The leaders also presented offerings for the dedication of the altar at the time when it was anointed; the leaders presented their offering before the altar. 11 The LORD said to Moses: They shall present their offerings, one leader each day, for the dedication of the altar.

12 The one who presented his offering the first day was Nahshon son of Amminadab, of the tribe of Judah; 13 his offering was one silver plate weighing one hundred thirty shekels, one silver basin weighing seventy shekels, according to the shekel of the sanctuary, both of them full of choice flour mixed with oil for a grain offering; 14 one golden dish weighing ten shekels, full of incense; 15 one young bull, one ram, one male lamb a year old, for a burnt offering; 16 one male goat for a sin offering; 17 and for the sacrifice of well-being, two oxen, five rams, five male goats, and five male lambs a year old. This was the offering of Nahshon son of Amminadab.

18 On the second day Nethanel son of Zuar, the leader of Issachar, presented an offering; 19 he presented for his offering one silver plate weighing one hundred thirty shekels, one silver basin weighing seventy shekels, according to the shekel of the sanctuary, both of them full of choice flour mixed with oil for a grain offering; 20 one golden dish weighing ten

---

6.22–27: The Aaronic benediction. This blessing, whose wording is very ancient, was undoubtedly used in the Jerusalem temple, usually at the conclusion of a service of worship (see Lev 9.22). **25–26:** The shining *face* (or "presence"; see Ex 33.14 n.) signifies that the holy God graciously turns toward the people in concern and favor (Pss 4.6; 31.16; 80.3); the expression *lift up his countenance* also signifies divine favor (Pss 44.3; 89.15). *Peace,* i.e. well–being, welfare, both material and spiritual. **27:** On the meaning of the *name,* see Gen 32.27 n.; Ex 3.13–15.

7.1–89: Offerings from the tribal lead- ers for the service of the tabernacle and the dedication of the altar. **1:** The date suddenly shifts back to Ex 40.17 (compare Num 1.1), showing that a new block of tradition begins here. **2–9:** The leaders present wagons for transporting the tabernacle equipment, though the Kohathites, for the sake of precaution, had to carry the *holy things* on the shoulder (4.1–15).

7.12–83: The tribal names are taken from the old list found in 1.5–15 (compare 10.14–27). The dedication offering presented on successive days was the same.

shekels, full of incense; [21] one young bull, one ram, one male lamb a year old, as a burnt offering; [22] one male goat as a sin offering; [23] and for the sacrifice of well-being, two oxen, five rams, five male goats, and five male lambs a year old. This was the offering of Nethanel son of Zuar.

24 On the third day Eliab son of Helon, the leader of the Zebulunites: [25] his offering was one silver plate weighing one hundred thirty shekels, one silver basin weighing seventy shekels, according to the shekel of the sanctuary, both of them full of choice flour mixed with oil for a grain offering; [26] one golden dish weighing ten shekels, full of incense; [27] one young bull, one ram, one male lamb a year old, for a burnt offering; [28] one male goat for a sin offering; [29] and for the sacrifice of well-being, two oxen, five rams, five male goats, and five male lambs a year old. This was the offering of Eliab son of Helon.

30 On the fourth day Elizur son of Shedeur, the leader of the Reubenites: [31] his offering was one silver plate weighing one hundred thirty shekels, one silver basin weighing seventy shekels, according to the shekel of the sanctuary, both of them full of choice flour mixed with oil for a grain offering; [32] one golden dish weighing ten shekels, full of incense; [33] one young bull, one ram, one male lamb a year old, for a burnt offering; [34] one male goat for a sin offering; [35] and for the sacrifice of well-being, two oxen, five rams, five male goats, and five male lambs a year old. This was the offering of Elizur son of Shedeur.

36 On the fifth day Shelumiel son of Zurishaddai, the leader of the Simeonites: [37] his offering was one silver plate weighing one hundred thirty shekels, one silver basin weighing seventy shekels, according to the shekel of the sanctuary, both of them full of choice flour mixed with oil for a grain offering; [38] one golden dish weighing ten shekels, full of incense; [39] one young bull, one ram, one male lamb a year old, for a burnt offering; [40] one male goat for a sin offering; [41] and for the sacrifice of well-being, two

oxen, five rams, five male goats, and five male lambs a year old. This was the offering of Shelumiel son of Zurishaddai.

42 On the sixth day Eliasaph son of Deuel, the leader of the Gadites: [43] his offering was one silver plate weighing one hundred thirty shekels, one silver basin weighing seventy shekels, according to the shekel of the sanctuary, both of them full of choice flour mixed with oil for a grain offering; [44] one golden dish weighing ten shekels, full of incense; [45] one young bull, one ram, one male lamb a year old, for a burnt offering; [46] one male goat for a sin offering; [47] and for the sacrifice of well-being, two oxen, five rams, five male goats, and five male lambs a year old. This was the offering of Eliasaph son of Deuel.

48 On the seventh day Elishama son of Ammihud, the leader of the Ephraimites: [49] his offering was one silver plate weighing one hundred thirty shekels, one silver basin weighing seventy shekels, according to the shekel of the sanctuary, both of them full of choice flour mixed with oil for a grain offering; [50] one golden dish weighing ten shekels, full of incense; [51] one young bull, one ram, one male lamb a year old, for a burnt offering; [52] one male goat for a sin offering; [53] and for the sacrifice of well-being, two oxen, five rams, five male goats, and five male lambs a year old. This was the offering of Elishama son of Ammihud.

54 On the eighth day Gamaliel son of Pedahzur, the leader of the Manassites: [55] his offering was one silver plate weighing one hundred thirty shekels, one silver basin weighing seventy shekels, according to the shekel of the sanctuary, both of them full of choice flour mixed with oil for a grain offering; [56] one golden dish weighing ten shekels, full of incense; [57] one young bull, one ram, one male lamb a year old, for a burnt offering; [58] one male goat for a sin offering; [59] and for the sacrifice of well-being, two oxen, five rams, five male goats, and five male lambs a year old. This was the offering of Gamaliel son of Pedahzur.

60 On the ninth day Abidan son of Gideoni, the leader of the Benjaminites:

61 his offering was one silver plate weighing one hundred thirty shekels, one silver basin weighing seventy shekels, according to the shekel of the sanctuary, both of them full of choice flour mixed with oil for a grain offering; 62 one golden dish weighing ten shekels, full of incense; 63 one young bull, one ram, one male lamb a year old, for a burnt offering; 64 one male goat for a sin offering; 65 and for the sacrifice of well-being, two oxen, five rams, five male goats, and five male lambs a year old. This was the offering of Abidan son of Gideoni.

66 On the tenth day Ahiezer son of Ammishaddai, the leader of the Danites: 67 his offering was one silver plate weighing one hundred thirty shekels, one silver basin weighing seventy shekels, according to the shekel of the sanctuary, both of them full of choice flour mixed with oil for a grain offering; 68 one golden dish weighing ten shekels, full of incense; 69 one young bull, one ram, one male lamb a year old, for a burnt offering; 70 one male goat for a sin offering; 71 and for the sacrifice of well-being, two oxen, five rams, five male goats, and five male lambs a year old. This was the offering of Ahiezer son of Ammishaddai.

72 On the eleventh day Pagiel son of Ochran, the leader of the Asherites: 73 his offering was one silver plate weighing one hundred thirty shekels, one silver basin weighing seventy shekels, according to the shekel of the sanctuary, both of them full of choice flour mixed with oil for a grain offering; 74 one golden dish weighing ten shekels, full of incense; 75 one young bull, one ram, one male lamb a year old, for a burnt offering; 76 one male goat for a sin offering; 77 and for the sacrifice of well-being, two oxen, five rams, five male goats, and five male lambs a year old. This was the offering of Pagiel son of Ochran.

78 On the twelfth day Ahira son of Enan, the leader of the Naphtalites: 79 his offering was one silver plate weighing one hundred thirty shekels, one silver basin weighing seventy shekels, according to the shekel of the sanctuary, both of them full of choice flour mixed with oil for a grain offering; 80 one golden dish weighing ten shekels, full of incense; 81 one young bull, one ram, one male lamb a year old, for a burnt offering; 82 one male goat for a sin offering; 83 and for the sacrifice of well-being, two oxen, five rams, five male goats, and five male lambs a year old. This was the offering of Ahira son of Enan.

84 This was the dedication offering for the altar, at the time when it was anointed, from the leaders of Israel: twelve silver plates, twelve silver basins, twelve golden dishes, 85 each silver plate weighing one hundred thirty shekels and each basin seventy, all the silver of the vessels two thousand four hundred shekels according to the shekel of the sanctuary, 86 the twelve golden dishes, full of incense, weighing ten shekels apiece according to the shekel of the sanctuary, all the gold of the dishes being one hundred twenty shekels; 87 all the livestock for the burnt offering twelve bulls, twelve rams, twelve male lambs a year old, with their grain offering; and twelve male goats for a sin offering; 88 and all the livestock for the sacrifice of well-being twenty-four bulls, the rams sixty, the male goats sixty, the male lambs a year old sixty. This was the dedication offering for the altar, after it was anointed.

89 When Moses went into the tent of meeting to speak with the LORD,*j* he would hear the voice speaking to him from above the mercy seat*k* that was on the ark of the covenant*l* from between the two cherubim; thus it spoke to him.

8 The LORD spoke to Moses, saying: 2 Speak to Aaron and say to him:

j Heb *him*   *k* Or *the cover*   *l* Or *treaty*, or *testimony*; Heb *eduth*

7.85: *The shekel of the sanctuary*, see Ex 30.13 n. 89: Ex 25.22.

8.1–26: **The consecration of the Levites. 1–4**: Ex 25.31–40; 27.20–21; Lev 24.2–

When you set up the lamps, the seven lamps shall give light in front of the lampstand. ³Aaron did so; he set up its lamps to give light in front of the lampstand, as the LORD had commanded Moses. ⁴Now this was how the lampstand was made, out of hammered work of gold. From its base to its flowers, it was hammered work; according to the pattern that the LORD had shown Moses, so he made the lampstand.

5 The LORD spoke to Moses, saying: ⁶Take the Levites from among the Israelites and cleanse them. ⁷Thus you shall do to them, to cleanse them: sprinkle the water of purification on them, have them shave their whole body with a razor and wash their clothes, and so cleanse themselves. ⁸Then let them take a young bull and its grain offering of choice flour mixed with oil, and you shall take another young bull for a sin offering. ⁹You shall bring the Levites before the tent of meeting, and assemble the whole congregation of the Israelites. ¹⁰When you bring the Levites before the LORD, the Israelites shall lay their hands on the Levites, ¹¹and Aaron shall present the Levites before the LORD as an elevation offering from the Israelites, that they may do the service of the LORD. ¹²The Levites shall lay their hands on the heads of the bulls, and he shall offer the one for a sin offering and the other for a burnt offering to the LORD, to make atonement for the Levites. ¹³Then you shall have the Levites stand before Aaron and his sons, and you shall present them as an elevation offering to the LORD.

14 Thus you shall separate the Levites from among the other Israelites, and the Levites shall be mine. ¹⁵Thereafter the Levites may go in to do service at the tent of meeting, once you have cleansed them and presented them as an elevation offering. ¹⁶For they are unreservedly given to me from among the Israelites; I have taken them for myself, in place of all that open the womb, the firstborn of all the Israelites. ¹⁷For all the firstborn among the Israelites are mine, both human and animal. On the day that I struck down all the firstborn in the land of Egypt I consecrated them for myself, ¹⁸but I have taken the Levites in place of all the firstborn among the Israelites. ¹⁹Moreover, I have given the Levites as a gift to Aaron and his sons from among the Israelites, to do the service for the Israelites at the tent of meeting, and to make atonement for the Israelites, in order that there may be no plague among the Israelites for coming too close to the sanctuary.

20 Moses and Aaron and the whole congregation of the Israelites did with the Levites accordingly; the Israelites did with the Levites just as the LORD had commanded Moses concerning them. ²¹The Levites purified themselves from sin and washed their clothes; then Aaron presented them as an elevation offering before the LORD, and Aaron made atonement for them to cleanse them. ²²Thereafter the Levites went in to do their service in the tent of meeting in attendance on Aaron and his sons. As the LORD had commanded Moses concerning the Levites, so they did with them.

23 The LORD spoke to Moses, saying: ²⁴This applies to the Levites: from twenty-five years old and upward they shall begin to do duty in the service of the

---

4. The lighting of the lamps is the prerogative of the Aaronic priests. **4**: *The pattern, see* Ex 25.9.

8.5–22: The Levites are to be consecrated for their office by a special purification ceremony (compare the ordination of the Aaronic priests, Lev ch 8). **7**: *The water of purification* (compare Ezek 36.25) refers to ceremonial cleansing. Shaving the hair and washing garments were rites of purification. **10–11**: By the laying on of hands (Lev 1.4) the people identify themselves with the Levites who are "sac-

rificed" instead of their firstborn (vv. 16–17; 3,13). *Elevation offering* is used symbolically to indicate that the Levites belong to Aaron and his sons just as the elevation sacrifice belongs to the priests (see Ex 29.24 n.). **12**: *Make atonement,* see Ex 29.35–37 n.

8.16–19: 3.5–13. The Levites, by their position in the camp (1.52–53; 3.38), shielded the people from the dreadful effects of holiness, which could cause a *plague* or other calamity (1 Sam chs 5–6). **24**: 4.3.

tent of meeting; ²⁵ and from the age of fifty years they shall retire from the duty of the service and serve no more. ²⁶ They may assist their brothers in the tent of meeting in carrying out their duties, but they shall perform no service. Thus you shall do with the Levites in assigning their duties.

9 The LORD spoke to Moses in the wilderness of Sinai, in the first month of the second year after they had come out of the land of Egypt, saying: ² Let the Israelites keep the passover at its appointed time. ³ On the fourteenth day of this month, at twilight, ^m you shall keep it at its appointed time; according to all its statutes and all its regulations you shall keep it. ⁴ So Moses told the Israelites that they should keep the passover. ⁵ They kept the passover in the first month, on the fourteenth day of the month, at twilight, ^m in the wilderness of Sinai. Just as the LORD had commanded Moses, so the Israelites did. ⁶ Now there were certain people who were unclean through touching a corpse, so that they could not keep the passover on that day. They came before Moses and Aaron on that day, ⁷ and said to him, "Although we are unclean through touching a corpse, why must we be kept from presenting the LORD's offering at its appointed time among the Israelites?" ⁸ Moses spoke to them, "Wait, so that I may hear what the LORD will command concerning you."

9 The LORD spoke to Moses, saying: ¹⁰ Speak to the Israelites, saying: Anyone of you or your descendants who is unclean through touching a corpse, or is away on a journey, shall still keep the passover to the LORD. ¹¹ In the second month on the fourteenth day, at twilight, ^m they shall keep it; they shall eat it with unleavened bread and bitter herbs. ¹² They shall leave none of it until morning, nor break a bone of it; according to all the statute for the passover they shall keep it. ¹³ But anyone who is clean and is not on a journey, and yet refrains from keeping the passover, shall be cut off from the people for not presenting the LORD's offering at its appointed time; such a one shall bear the consequences for the sin. ¹⁴ Any alien residing among you who wishes to keep the passover to the LORD shall do so according to the statute of the passover and according to its regulation; you shall have one statute for both the resident alien and the native.

15 On the day the tabernacle was set up, the cloud covered the tabernacle, the tent of the covenant; ⁿ and from evening until morning it was over the tabernacle, having the appearance of fire. ¹⁶ It was always so: the cloud covered it by day ^o and the appearance of fire by night. ¹⁷ Whenever the cloud lifted from over the tent, then the Israelites would set out; and in the place where the cloud settled down, there the Israelites would camp. ¹⁸ At the command of the LORD the Israelites would set out, and at the command of the LORD they would camp. As long as the cloud rested over the tabernacle, they would remain in camp. ¹⁹ Even when the cloud continued over the tabernacle many days, the Israelites would keep the charge of the LORD, and would not set out. ²⁰ Sometimes the cloud

---

m Heb *between the two evenings*   *n* Or *treaty*, or *testimony*; Heb *eduth*   *o* Gk Syr Vg: Heb lacks *by day*

9.1–10.10: Other events and instructions. 1–14: Supplement to the laws concerning the passover. **1**: Here, as in 7.1, the month is that of Ex 40.17 rather than the date assumed for chs 1–6. **2–4**: Ex 12.1–13, 21–27. **6–7**: On uncleanness through touching a corpse, see Lev 21.1–12. **8**: This verse shows that Israel's law was not fixed and unchangeable, but developed as the people faced new situations (15.23).

9.10: The case of one who is *away on a journey* apparently presupposes settlement in Canaan and goes beyond the case at hand (v. 7). **11**: In these cases, the passover may be kept one month late. **13**: *Cut off*, see Lev 7.20 n. **9.15–23**: On the fiery cloud, see Ex 14.24; 40.34–38 n. These verses (compare 14.14) presuppose Israel's march, rather than the sojourn at Sinai and thus anticipate 10.11–13. **15**: Ex 40.34–38.

would remain a few days over the tabernacle, and according to the command of the LORD they would remain in camp; then according to the command of the LORD they would set out. ²¹Sometimes the cloud would remain from evening until morning; and when the cloud lifted in the morning, they would set out, or if it continued for a day and a night, when the cloud lifted they would set out. ²²Whether it was two days, or a month, or a longer time, that the cloud continued over the tabernacle, resting upon it, the Israelites would remain in camp and would not set out; but when it lifted they would set out. ²³At the command of the LORD they would camp, and at the command of the LORD they would set out. They kept the charge of the LORD, at the command of the LORD by Moses.

10 The LORD spoke to Moses, saying: ²Make two silver trumpets; you shall make them of hammered work; and you shall use them for summoning the congregation, and for breaking camp. ³When both are blown, the whole congregation shall assemble before you at the entrance of the tent of meeting. ⁴But if only one is blown, then the leaders, the heads of the tribes of Israel, shall assemble before you. ⁵When you blow an alarm, the camps on the east side shall set out; ⁶when you blow a second alarm, the camps on the south side shall set out. An alarm is to be blown whenever they are to set out. ⁷But when the assembly is to be gathered, you shall blow, but you shall not sound an alarm. ⁸The sons of Aaron, the priests, shall blow the trumpets; this shall be a perpetual institution for you throughout your generations. ⁹When you go to war in your land against the adversary who oppresses you, you shall sound an alarm with the trumpets, so that you may be remem-

bered before the LORD your God and be saved from your enemies. ¹⁰Also on your days of rejoicing, at your appointed festivals, and at the beginnings of your months, you shall blow the trumpets over your burnt offerings and over your sacrifices of well-being; they shall serve as a reminder on your behalf before the LORD your God: I am the LORD your God.

11 In the second year, in the second month, on the twentieth day of the month, the cloud lifted from over the tabernacle of the covenant.ᵖ ¹²Then the Israelites set out by stages from the wilderness of Sinai, and the cloud settled down in the wilderness of Paran. ¹³They set out for the first time at the command of the LORD by Moses. ¹⁴The standard of the camp of Judah set out first, company by company, and over the whole company was Nahshon son of Amminadab. ¹⁵Over the company of the tribe of Issachar was Nethanel son of Zuar; ¹⁶and over the company of the tribe of Zebulun was Eliab son of Helon.

17 Then the tabernacle was taken down, and the Gershonites and the Merarites, who carried the tabernacle, set out. ¹⁸Next the standard of the camp of Reuben set out, company by company; and over the whole company was Elizur son of Shedeur. ¹⁹Over the company of the tribe of Simeon was Shelumiel son of Zurishaddai, ²⁰and over the company of the tribe of Gad was Eliasaph son of Deuel.

21 Then the Kohathites, who carried the holy things, set out; and the tabernacle was set up before their arrival. ²²Next the standard of the Ephraimite camp set out, company by company, and over the whole company was Elishama son of

p Or *treaty,* or *testimony;* Heb *eduth*

10.1–10: Sounding the trumpets. **5–7:** An *alarm* was a military call (v. 9; Am 3.6), as distinguished from the trumpet that proclaims assembly (Lev ch 23, especially v. 24). **10:** *The beginnings of your months,* see 28.11–15 n. **10.11–36: Israel on the march.** The peo-ple departed from Sinai with signs of the LORD's accompanying presence. **11:** The date was eleven months after the arrival at Sinai (Ex 19.1) or nineteen days after the census (Num 1.1). **12:** *By stages,* see ch 33. **13–28:** On the order of march and Levitical tasks, see chs 2–3.

Ammihud. 23 Over the company of the tribe of Manasseh was Gamaliel son of Pedahzur, 24 and over the company of the tribe of Benjamin was Abidan son of Gideoni.

25 Then the standard of the camp of Dan, acting as the rear guard of all the camps, set out, company by company, and over the whole company was Ahiezer son of Ammishaddai. 26 Over the company of the tribe of Asher was Pagiel son of Ochran, 27 and over the company of the tribe of Naphtali was Ahira son of Enan. 28 This was the order of march of the Israelites, company by company, when they set out.

29 Moses said to Hobab son of Reuel the Midianite, Moses' father-in-law, "We are setting out for the place of which the LORD said, 'I will give it to you'; come with us, and we will treat you well; for the LORD has promised good to Israel." 30 But he said to him, "I will not go, but I will go back to my own land and to my kindred." 31 He said, "Do not leave us, for you know where we should camp in the wilderness, and you will serve as eyes for us. 32 Moreover, if you go with us, whatever good the LORD does for us, the same we will do for you."

33 So they set out from the mount of the LORD three days' journey with the ark of the covenant of the LORD going before them three days' journey, to seek out a resting place for them, 34 the cloud of the LORD being over them by day when they set out from the camp.

35 Whenever the ark set out, Moses would say,

"Arise, O LORD, let your enemies
be scattered,
and your foes flee before you."
36 And whenever it came to rest, he would say,

"Return, O LORD of the ten
thousand thousands of
Israel." *q*

11 Now when the people complained in the hearing of the LORD about their misfortunes, the LORD heard it and his anger was kindled. Then the fire of the LORD burned against them, and consumed some outlying parts of the camp. 2 But the people cried out to Moses; and Moses prayed to the LORD, and the fire abated. 3 So that place was called Taberah, *r* because the fire of the LORD burned against them.

4 The rabble among them had a strong craving; and the Israelites also wept again, and said, "If only we had meat to eat! 5 We remember the fish we used to eat in Egypt for nothing, the cucumbers, the melons, the leeks, the onions, and the garlic; 6 but now our strength is dried up, and there is nothing at all but this manna to look at."

7 Now the manna was like coriander seed, and its color was like the color of gum resin. 8 The people went around and gathered it, ground it in mills or beat it in mortars, then boiled it in pots and made cakes of it; and the taste of it was like the taste of cakes baked with oil. 9 When the dew fell on the camp in the night, the manna would fall with it.

q Meaning of Heb uncertain   *r* That is *Burning*

10.29–36: Here old epic traditions reappear after a long, unbroken sequence of priestly material (Ex chs 35–40; Lev; Num 1.1–10.28). 29: *Hobab,* otherwise known as Jethro (Ex 2.18 n.), was leader of the Kenites, a Midianite clan who were allies during Israel's early history (Ex ch 18; Judg 1.16; 4.11; 5.24). 10.33–36: Ancient tradition about the ark (see Ex 25.10–22 n.). 33: This verse should not be interpreted to mean that the ark went ahead *three days' journey;* rather, it went *before* them (compare v. 21) during this time of travel from Sinai. 35–36: The "song to the ark," a very ancient piece of poetry, reflects the view that the ark was a throne upon which the LORD, the Divine Warrior, was seated invisibly while waging holy war (Josh chs 3–4; 1 Sam 4.3–22; compare Ps 24.7–10). 11.1–35: **Israel's complaining in the wilderness. 1:** See Ex 15.24 n. The *fire of the LORD* may refer to some natural phenomenon like lightning that was interpreted as God's consuming holiness (compare Lev 10.2). 4–15: Not satisfied with the manna, the people craved seasoned meat dishes such as they once

10 Moses heard the people weeping throughout their families, all at the entrances of their tents. Then the LORD became very angry, and Moses was displeased. ¹¹So Moses said to the LORD, "Why have you treated your servant so badly? Why have I not found favor in your sight, that you lay the burden of all this people on me? ¹²Did I conceive all this people? Did I give birth to them, that you should say to me, 'Carry them in your bosom, as a nurse carries a sucking child,' to the land that you promised on oath to their ancestors? ¹³Where am I to get meat to give to all this people? For they come weeping to me and say, 'Give us meat to eat!' ¹⁴I am not able to carry all this people alone, for they are too heavy for me. ¹⁵If this is the way you are going to treat me, put me to death at once—if I have found favor in your sight—and do not let me see my misery."

16 So the LORD said to Moses, "Gather for me seventy of the elders of Israel, whom you know to be the elders of the people and officers over them; bring them to the tent of meeting, and have them take their place there with you. ¹⁷I will come down and talk with you there; and I will take some of the spirit that is on you and put it on them; and they shall bear the burden of the people along with you so that you will not bear it all by yourself. ¹⁸And say to the people: Consecrate yourselves for tomorrow, and

you shall eat meat; for you have wailed in the hearing of the LORD, saying, 'If only we had meat to eat! Surely it was better for us in Egypt.' Therefore the LORD will give you meat, and you shall eat. ¹⁹You shall eat not only one day, or two days, or five days, or ten days, or twenty days, ²⁰but for a whole month—until it comes out of your nostrils and becomes loathsome to you—because you have rejected the LORD who is among you, and have wailed before him, saying, 'Why did we ever leave Egypt?' " ²¹But Moses said, "The people I am with number six hundred thousand on foot; and you say, 'I will give them meat, that they may eat for a whole month'! ²²Are there enough flocks and herds to slaughter for them? Are there enough fish in the sea to catch for them?" ²³The LORD said to Moses, "Is the LORD's power limited?ˢ Now you shall see whether my word will come true for you or not."

24 So Moses went out and told the people the words of the LORD; and he gathered seventy elders of the people, and placed them all around the tent. ²⁵Then the LORD came down in the cloud and spoke to him, and took some of the spirit that was on him and put it on the seventy elders; and when the spirit rested upon them, they prophesied. But they did not do so again.

26 Two men remained in the camp,

ˢ Heb LORD's hand too short?

enjoyed in Egypt (Ex 16.3). **4**: *Rabble,* see Ex 12.38 n. *Wept again,* see Ex 16.2–3. **7–9**: *The manna,* see Ex 16.14–21, 31.
11.11–15: In this chapter two motifs have been woven together: one deals with the miraculous provision of wilderness food, the other with Moses' heavy responsibilities as leader (vv. 11–12, 14–17, 24–30). **11–12**: On Moses' impatience, see Ex 17.2–4. **14**: On the *too heavy* burden, see Ex 18.17–18.
11.16–17 (also **24–25**): Moses' burden was lightened by investing seventy elders with power to assist him (Ex 18.21–23; see Deut 1.9–18). This account presupposes the old tradition about the *tent of meeting* (Ex 33.7–11). **17**: Moses is regarded as a charismatic leader, endowed with the divine spirit (com-

pare 24.2; Judg 11.29). The transfer of a portion of the spirit to the elders (2 Kings 2.9–10) indicates that the latter are subordinate to Moses, who stands in a close relation to God (Ex 33.11). **18–23**: Continuation of the food tradition. **21**: Ex 12.37. The story stresses the LORD's miraculous power to support an extremely large population in the wilderness. **25**: *They prophesied,* because the divine spirit was put upon them. Such ecstatic prophecy, familiar in Israel's early prophetic movement (1 Sam 10.6, 10–13; 1 Kings 22.6, 10–12), probably shows Canaanite influence.
11.26–30: Two other elders received the gift of prophecy even though they stayed in the camp. **26**: Here it is assumed that the tent of meeting was outside the camp (Ex 33.7; see

one named Eldad, and the other named Medad, and the spirit rested on them; they were among those registered, but they had not gone out to the tent, and so they prophesied in the camp. 27 And a young man ran and told Moses, "Eldad and Medad are prophesying in the camp." 28 And Joshua son of Nun, the assistant of Moses, one of his chosen men,[t] said, "My lord Moses, stop them!" 29 But Moses said to him, "Are you jealous for my sake? Would that all the LORD's people were prophets, and that the LORD would put his spirit on them!" 30 And Moses and the elders of Israel returned to the camp.

31 Then a wind went out from the LORD, and it brought quails from the sea and let them fall beside the camp, about a day's journey on this side and a day's journey on the other side, all around the camp, about two cubits deep on the ground. 32 So the people worked all that day and night and all the next day, gathering the quails; the least anyone gathered was ten homers; and they spread them out for themselves all around the camp. 33 But while the meat was still between their teeth, before it was consumed, the anger of the LORD was kindled against the people, and the LORD struck the people with a very great plague. 34 So that place was called Kibroth-hattaavah,[u] because there they buried the people who had the craving. 35 From Kibroth-hattaavah the people journeyed to Hazeroth.

12 While they were at Hazeroth, Miriam and Aaron spoke against Moses because of the Cushite woman whom he had married (for he had indeed married a Cushite woman); 2 and they said, "Has the LORD spoken only through Moses? Has he not spoken through us also?" And the LORD heard it. 3 Now the man Moses was very humble,[v] more so than anyone else on the face of the earth. 4 Suddenly the LORD said to Moses, Aaron, and Miriam, "Come out, you three, to the tent of meeting." So the three of them came out. 5 Then the LORD came down in a pillar of cloud, and stood at the entrance of the tent, and called Aaron and Miriam; and they both came forward. 6 And he said, "Hear my words:

When there are prophets among
    you,
  I the LORD make myself known
    to them in visions;
  I speak to them in dreams.
7 Not so with my servant Moses;
  he is entrusted with all my
    house.
8 With him I speak face to face—clearly,
    not in riddles;
  and he beholds the form of the
    LORD.

Why then were you not afraid to speak against my servant Moses?" 9 And the an-

t Or *of Moses from his youth*   u That is *Graves of craving*   v Or *devout*

Num 2.2 n.). *Registered,* i.e. enrolled as representatives of the community.
11.31–35: Continuation of vv. 18–23. **31–35:** Like the manna, the quails (Ex 16.13) are a natural phenomenon of the desert. They migrate over the region in great numbers and, when exhausted, are easily caught. **31:** *A wind,* sent by God (compare Ex 14.21), brought the quails from *the sea,* the Gulf of Aqaba. **33:** A sickness which resulted from eating quails is interpreted as divine judgment upon the people's faithlessness. A *plague* was an act of God; see Lev 26.21; Deut 28.27; 1 Sam 4.8.
12.1–16: A dispute over Moses' authority as leader. 1: Miriam, mentioned here and

in 20.1 and 26.59 was undoubtedly a major leader in ancient Israel (see Ex 15.19–21 n.; Mic 6.4). Together Miriam and Aaron criticized Moses for marrying *the Cushite woman,* a reference to Zipporah, a Midianite (Ex 2.21). The term Cushite apparently includes Midianites and other Arabic peoples (Hab 3.7). The story reflects a power struggle in the community, perhaps during the monarchy. **2:** The question is directed against Moses' position as covenant mediator and leader of the people (Ex 19.9; 33.11). **3:** *Humble,* i.e. before God (Ex 3.11); see note *v.* This verse is an age-old stumbling block to the belief that Moses wrote the whole Pentateuch. **4:** See 11.26 n. **6–8:** Normally the LORD com-

ger of the LORD was kindled against them, and he departed.

10 When the cloud went away from over the tent, Miriam had become leprous,ʷ as white as snow. And Aaron turned towards Miriam and saw that she was leprous. ¹¹ Then Aaron said to Moses, "Oh, my lord, do not punish usˣ for a sin that we have so foolishly committed. ¹² Do not let her be like one stillborn, whose flesh is half consumed when it comes out of its mother's womb." ¹³ And Moses cried to the LORD, "O God, please heal her." ¹⁴ But the LORD said to Moses, "If her father had but spit in her face, would she not bear her shame for seven days? Let her be shut out of the camp for seven days, and after that she may be brought in again." ¹⁵ So Miriam was shut out of the camp for seven days; and the people did not set out on the march until Miriam had been brought in again. ¹⁶ After that the people set out from Hazeroth, and camped in the wilderness of Paran.

13 The LORD said to Moses, ² "Send men to spy out the land of Canaan, which I am giving to the Israelites; from each of their ancestral tribes you shall send a man, every one a leader among them." ³ So Moses sent them from the wilderness of Paran, according to the command of the LORD, all of them leading men among the Israelites. ⁴ These were their names: From the tribe of Reuben, Shammua son of Zaccur; ⁵ from the tribe of Simeon, Shaphat son of Hori;

⁶ from the tribe of Judah, Caleb son of Jephunneh; ⁷ from the tribe of Issachar, Igal son of Joseph; ⁸ from the tribe of Ephraim, Hoshea son of Nun; ⁹ from the tribe of Benjamin, Palti son of Raphu; ¹⁰ from the tribe of Zebulun, Gaddiel son of Sodi; ¹¹ from the tribe of Joseph (that is, from the tribe of Manasseh), Gaddi son of Susi; ¹² from the tribe of Dan, Ammiel son of Gemalli; ¹³ from the tribe of Asher, Sethur son of Michael; ¹⁴ from the tribe of Naphtali, Nahbi son of Vophsi; ¹⁵ from the tribe of Gad, Geuel son of Machi. ¹⁶ These were the names of the men whom Moses sent to spy out the land. And Moses changed the name of Hoshea son of Nun to Joshua.

17 Moses sent them to spy out the land of Canaan, and said to them, "Go up there into the Negeb, and go up into the hill country, ¹⁸ and see what the land is like, and whether the people who live in it are strong or weak, whether they are few or many, ¹⁹ and whether the land they live in is good or bad, and whether the towns that they live in are unwalled or fortified, ²⁰ and whether the land is rich or poor, and whether there are trees in it or not. Be bold, and bring some of the fruit of the land." Now it was the season of the first ripe grapes.

21 So they went up and spied out the land from the wilderness of Zin to Re-

ʷ A term for several skin diseases; precise meaning uncertain   ˣ Heb *do not lay sin upon us*

---

municates with prophets indirectly through dreams or visions (see Deut 13.1 n.), but to Moses God speaks directly (see Ex 33.11 n.); he even *beholds the form of the LORD* (compare Ex 33.17–23).
12.10: *Leprous,* probably not leprosy but a skin disease of some sort (see Lev 13.1–8; Deut 24.8–9). **11–13:** Strangely, no explanation is given for the punishment falling only upon Miriam, even though Aaron and Moses show concern for her. **14:** The punishment is severe: the defilement of being spat upon by her father, the sign of a curse (Deut 25.9). Since Miriam was known to be a prophet (Ex 15.20), the story may reflect fear of a challenge to Moses' prophetic leadership (see

11.16–29) or tension between charismatic freedom and priestly order. **16:** The early traditions in 10.29–12.15 assume that Israel had not yet reached Paran (10.12).
13.1–33: The reconnaissance of the land. Twelve scouts brought back a majority report that formidable obstacles stood in the way of taking Canaan. **1–16:** Priestly tradition gives the names of the spies. The leaders are not the same as those of 1.5–15. **8:** *Hoshea,* an alternate form of Joshua (v. 16).
13.21: This priestly comment supposes that the spies surveyed the whole of Canaan from the area of Kadesh (see v. 26 n.) to Rehob (2 Sam 10.6) near Laish or Dan. *Lebo-hamath,* see 34.1–12 n. **22:** The mixed population of

hob, near Lebo-hamath. 22 They went up into the Negeb, and came to Hebron; and Ahiman, Sheshai, and Talmai, the Anakites, were there. (Hebron was built seven years before Zoan in Egypt.) 23 And they came to the Wadi Eshcol, and cut down from there a branch with a single cluster of grapes, and they carried it on a pole between two of them. They also brought some pomegranates and figs. 24 That place was called the Wadi Eshcol, *y* because of the cluster that the Israelites cut down from there.

25 At the end of forty days they returned from spying out the land. 26 And they came to Moses and Aaron and to all the congregation of the Israelites in the wilderness of Paran, at Kadesh; they brought back word to them and to all the congregation, and showed them the fruit of the land. 27 And they told him, "We came to the land to which you sent us; it flows with milk and honey, and this is its fruit. 28 Yet the people who live in the land are strong, and the towns are fortified and very large; and besides, we saw the descendants of Anak there. 29 The Amalekites live in the land of the Negeb; the Hittites, the Jebusites, and the Amorites live in the hill country; and the Canaanites live by the sea, and along the Jordan."

30 But Caleb quieted the people before Moses, and said, "Let us go up at once and occupy it, for we are well able to overcome it." 31 Then the men who had gone up with him said, "We are not able to go up against this people, for they are stronger than we." 32 So they brought to the Israelites an unfavorable report of the land that they had spied out, saying, "The land that we have gone through as spies is a land that devours its inhabitants; and all the people that we saw in it are of great size. 33 There we saw the Nephilim (the Anakites come from the Nephilim); and to ourselves we seemed like grasshoppers, and so we seemed to them."

14 Then all the congregation raised a loud cry, and the people wept that night. 2 And all the Israelites complained against Moses and Aaron; the whole congregation said to them, "Would that we had died in the land of Egypt! Or would that we had died in this wilderness! 3 Why is the LORD bringing us into this land to fall by the sword? Our wives and our little ones will become booty; would it not be better for us to go back to Egypt?" 4 So they said to one another, "Let us choose a captain, and go back to Egypt."

5 Then Moses and Aaron fell on their faces before all the assembly of the congregation of the Israelites. 6 And Joshua son of Nun and Caleb son of Jephunneh, who were among those who had spied out the land, tore their clothes 7 and said to all the congregation of the Israelites, "The land that we went through as spies is an exceedingly good land. 8 If the LORD is pleased with us, he will bring us into this land and give it to us, a land that flows with milk and honey. 9 Only, do not rebel against the LORD; and do not fear the people of the land, for they are no more than bread for us; their protec-

---

y That is *Cluster*

Canaan included the tribes descended from Anak, regarded as an unusually tall people (vv. 32–33). An archaeological comment dates the history of Hebron in relation to the founding of Zoan or Tanis, rebuilt as the Hyksos capital about 1700 B.C. (see Ex 1.11 n.). **23:** *The Wadi Eshcol* was in the vicinity of Hebron, a famous grape-producing area. **26:** *Kadesh* (-barnea), an oasis on the border between the wilderness of Paran and the wilderness of Zin. The traditions found in 10.11–21.3 deal with this oasis (see Ex 17.7 n.), where Israel spent most of the forty year sojourn in the wilderness (see Introduction). **27:** *Milk and honey,* see Ex 3.8 n. **29:** *Amalekites,* see Ex 17.8 n. For the pre-Israelite inhabitants, see Ex 3.8. **33:** *Nephilim,* see Gen 6.4 n.

14.1–45: **Decision to attack.** The Israelites attempted a foolhardy attack on Canaan from the south and were repulsed. **1–4:** See Ex 16.3 n.

14.5–6: According to priestly tradition, Joshua sided with Caleb in appraising the situation (compare 13.30). **10:** On the LORD's *glory,* see Ex 16.6–7 n. **11:** The *signs* (see Ex

tion is removed from them, and the LORD is with us; do not fear them." [10] But the whole congregation threatened to stone them.

Then the glory of the LORD appeared at the tent of meeting to all the Israelites. [11] And the LORD said to Moses, "How long will this people despise me? And how long will they refuse to believe in me, in spite of all the signs that I have done among them? [12] I will strike them with pestilence and disinherit them, and I will make of you a nation greater and mightier than they."

13 But Moses said to the LORD, "Then the Egyptians will hear of it, for in your might you brought up this people from among them, [14] and they will tell the inhabitants of this land. They have heard that you, O LORD, are in the midst of this people; for you, O LORD, are seen face to face, and your cloud stands over them and you go in front of them, in a pillar of cloud by day and in a pillar of fire by night. [15] Now if you kill this people all at one time, then the nations who have heard about you will say, [16] 'It is because the LORD was not able to bring this people into the land he swore to give them that he has slaughtered them in the wilderness.' [17] And now, therefore, let the power of the LORD be great in the way that you promised when you spoke, saying,

18 'The LORD is slow to anger,
    and abounding in steadfast love,
    forgiving iniquity and
        transgression,
    but by no means clearing the
        guilty,
    visiting the iniquity of the parents

upon the children
    to the third and the fourth
        generation.'
[19] Forgive the iniquity of this people according to the greatness of your steadfast love, just as you have pardoned this people, from Egypt even until now."

20 Then the LORD said, "I do forgive, just as you have asked; [21] nevertheless—as I live, and as all the earth shall be filled with the glory of the LORD— [22] none of the people who have seen my glory and the signs that I did in Egypt and in the wilderness, and yet have tested me these ten times and have not obeyed my voice, [23] shall see the land that I swore to give to their ancestors; none of those who despised me shall see it. [24] But my servant Caleb, because he has a different spirit and has followed me wholeheartedly, I will bring into the land into which he went, and his descendants shall possess it. [25] Now, since the Amalekites and the Canaanites live in the valleys, turn tomorrow and set out for the wilderness by the way to the Red Sea."z

26 And the LORD spoke to Moses and to Aaron, saying: [27] How long shall this wicked congregation complain against me? I have heard the complaints of the Israelites, which they complain against me. [28] Say to them, "As I live," says the LORD, "I will do to you the very things I heard you say: [29] your dead bodies shall fall in this very wilderness; and of all your number, included in the census, from twenty years old and upward, who have complained against me, [30] not one of you shall come into the land in which

z Or *Sea of Reeds*

3.11–12 n.) were ambiguous evidences, not proofs of God's presence. **12:** Ex 32.9–10. **14.13–19:** In his intercession, Moses again appeals to the LORD's honor and gracious ways (Ex 32.11–13). **15–16:** Other nations, owing to limitations of understanding, will conclude that the LORD does not have the power of deity (Deut 32.26–27), i.e. the power to deliver the people (1 Kings 18.20–40). **18:** A quotation from an old liturgical summary (see Ex 34.5–8). **20:** On the LORD's change of mind, see Ex 32.14 n. **22–23:** Since

divine mercy does not exclude divine judgment (v.18b), the verdict is that none of the present rebellious generation will enter Canaan. **24:** Caleb is expected in view of 13.30. Joshua is not mentioned because the account of his siding with Caleb (vv. 5–6, 26–38) comes from another tradition. **14.26–38:** This priestly tradition interprets the forty years in the wilderness as a time of divine judgment. **29:** *Twenty years,* the age for beginning military service (1.3).

I swore to settle you, except Caleb son of Jephunneh and Joshua son of Nun. 31 But your little ones, who you said would become booty, I will bring in, and they shall know the land that you have despised. 32 But as for you, your dead bodies shall fall in this wilderness. 33 And your children shall be shepherds in the wilderness for forty years, and shall suffer for your faithlessness, until the last of your dead bodies lies in the wilderness. 34 According to the number of the days in which you spied out the land, forty days, for every day a year, you shall bear your iniquity, forty years, and you shall know my displeasure." 35 I the LORD have spoken; surely I will do thus to all this wicked congregation gathered together against me: in this wilderness they shall come to a full end, and there they shall die.

36 And the men whom Moses sent to spy out the land, who returned and made all the congregation complain against him by bringing a bad report about the land— 37 the men who brought an unfavorable report about the land died by a plague before the LORD. 38 But Joshua son of Nun and Caleb son of Jephunneh alone remained alive, of those men who went to spy out the land.

39 When Moses told these words to all the Israelites, the people mourned greatly. 40 They rose early in the morning and went up to the heights of the hill country, saying, "Here we are. We will go up to the place that the LORD has promised, for we have sinned." 41 But Moses said, "Why do you continue to transgress the command of the LORD? That will not succeed. 42 Do not go up, for the LORD is not with you; do not let yourselves be struck down before your enemies. 43 For the Amalekites and the Canaanites will confront you there, and you shall fall by the sword; because you have turned back from following the LORD, the LORD will not be with you." 44 But they presumed to go up to the heights of the hill country, even though the ark of the covenant of the LORD, and Moses, had not left the camp. 45 Then the Amalekites and the Canaanites who lived in that hill country came down and defeated them, pursuing them as far as Hormah.

15 The LORD spoke to Moses, saying: 2 Speak to the Israelites and say to them: When you come into the land you are to inhabit, which I am giving you, 3 and you make an offering by fire to the LORD from the herd or from the flock—whether a burnt offering or a sacrifice, to fulfill a vow or as a freewill offering or at your appointed festivals—to make a pleasing odor for the LORD, 4 then whoever presents such an offering to the LORD shall present also a grain offering, one-tenth of an ephah of choice flour, mixed with one-fourth of a hin of oil. 5 Moreover, you shall offer one-fourth of a hin of wine as a drink offering with the burnt offering or the sacrifice, for each lamb. 6 For a ram, you shall offer a grain offering, two-tenths of an ephah of choice flour mixed with one-third of a hin of oil; 7 and as a drink offering you shall offer one-third of a hin of wine, a pleasing odor to the LORD. 8 When you offer a bull as a burnt offering or a sacrifice, to fulfill a vow or as an offering of well-being to the LORD, 9 then you shall present with the bull a grain offering, three-tenths of an ephah of choice flour, mixed with half a hin of oil, 10 and you shall present as a drink offering half a hin

---

14.39–45: These verses preserve the memory of an abortive attempt to penetrate Canaan from the south. **42:** The failure is interpreted by the words *the LORD is not with you,* i.e. the ark did not go with them into battle (v. 44). **45:** The bitterness of this and other battles is the basis of the ancient vow against the Amalekites (Ex 17.8–16).

15.1–41: Cultic regulations. The narrative is interrupted to introduce priestly matters (chs 15–19) and thereby to continue the main line of priestly tradition (see 10.29–36 n.). **1–16:** This law supplements the legislation of Leviticus by prescribing the grain (Lev ch 2) and drink offerings that are to accompany a burnt offering (Lev ch 1) or offering of well-being (Lev ch 3). **3:** *A pleasing odor,* see Lev 1.9 n. **16:** See Ex 12.43–49 n.

of wine, as an offering by fire, a pleasing odor to the LORD.

11 Thus it shall be done for each ox or ram, or for each of the male lambs or the kids. 12 According to the number that you offer, so you shall do with each and every one. 13 Every native Israelite shall do these things in this way, in presenting an offering by fire, a pleasing odor to the LORD. 14 An alien who lives with you, or who takes up permanent residence among you, and wishes to offer an offering by fire, a pleasing odor to the LORD, shall do as you do. 15 As for the assembly, there shall be for both you and the resident alien a single statute, a perpetual statute throughout your generations; you and the alien shall be alike before the LORD. 16 You and the alien who resides with you shall have the same law and the same ordinance.

17 The LORD spoke to Moses, saying: 18 Speak to the Israelites and say to them: After you come into the land to which I am bringing you, 19 whenever you eat of the bread of the land, you shall present a donation to the LORD. 20 From your first batch of dough you shall present a loaf as a donation; you shall present it just as you present a donation from the threshing floor. 21 Throughout your generations you shall give to the LORD a donation from the first of your batch of dough.

22 But if you unintentionally fail to observe all these commandments that the LORD has spoken to Moses— 23 everything that the LORD has commanded you by Moses, from the day the LORD gave commandment and thereafter, throughout your generations— 24 then if it was done unintentionally without the knowl-edge of the congregation, the whole congregation shall offer one young bull for a burnt offering, a pleasing odor to the LORD, together with its grain offering and its drink offering, according to the ordinance, and one male goat for a sin offering. 25 The priest shall make atonement for all the congregation of the Israelites, and they shall be forgiven; it was unintentional, and they have brought their offering, an offering by fire to the LORD, and their sin offering before the LORD, for their error. 26 All the congregation of the Israelites shall be forgiven, as well as the aliens residing among them, because the whole people was involved in the error

27 An individual who sins unintentionally shall present a female goat a year old for a sin offering. 28 And the priest shall make atonement before the LORD for the one who commits an error, when it is unintentional, to make atonement for the person, who then shall be forgiven. 29 For both the native among the Israelites and the alien residing among them—you shall have the same law for anyone who acts in error. 30 But whoever acts high-handedly, whether a native or an alien, affronts the LORD, and shall be cut off from among the people. 31 Because of having despised the word of the LORD and broken his commandment, such a person shall be utterly cut off and bear the guilt.

32 When the Israelites were in the wilderness, they found a man gathering sticks on the sabbath day. 33 Those who found him gathering sticks brought him to Moses, Aaron, and to the whole congregation. 34 They put him in custody, because it was not clear what should be

---

15.17–21: An offering of the first fruits (Lev 23.9–14). 22–31: A supplement to Lev ch 4. 23: Mosaic law includes the original commandments and supplementary laws, thus indicating the development of law as new situations were faced (see 9.8 n.). 25: *Make atonement,* see Lev 1.4 n. 30: Sin committed *high-handedly,* in contrast to sin committed inadvertently or in ignorance (vv. 25, 26, 27), is that which is done deliberately or defiantly (Ps 19.12–13). For such sin there is no atonement in priestly legislation. *Cut off,* see Lev 7.20 n.

15.32–36: The death penalty had already been prescribed for violating the sabbath (Ex 31.14–15; 35.2). The man was put in custody until, by divine oracle, it was determined whether his action actually came under the

done to him. 35 Then the LORD said to Moses, "The man shall be put to death; all the congregation shall stone him outside the camp." 36 The whole congregation brought him outside the camp and stoned him to death, just as the LORD had commanded Moses.

37 The LORD said to Moses: 38 Speak to the Israelites, and tell them to make fringes on the corners of their garments throughout their generations and to put a blue cord on the fringe at each corner. 39 You have the fringe so that, when you see it, you will remember all the commandments of the LORD and do them, and not follow the lust of your own heart and your own eyes. 40 So you shall remember and do all my commandments, and you shall be holy to your God. 41 I am the LORD your God, who brought you out of the land of Egypt, to be your God: I am the LORD your God.

16 Now Korah son of Izhar son of Kohath son of Levi, along with Dathan and Abiram sons of Eliab, and On son of Peleth—descendants of Reuben—took 2 two hundred fifty Israelite men, leaders of the congregation, chosen from the assembly, well-known men,*a* and they confronted Moses. 3 They assembled against Moses and against Aaron, and said to them, "You have gone too far! All the congregation are holy, everyone of them, and the LORD is among them. So why then do you exalt yourselves above the assembly of the LORD?" 4 When Moses heard it, he fell on his face. 5 Then he said to Korah and all his company, "In the morning the LORD will make known who is his, and who is holy, and who will be allowed to approach him; the one whom he will

choose he will allow to approach him. 6 Do this: take censers, Korah and all your*b* company, 7 and tomorrow put fire in them, and lay incense on them before the LORD; and the man whom the LORD chooses shall be the holy one. You Levites have gone too far!" 8 Then Moses said to Korah, "Hear now, you Levites! 9 Is it too little for you that the God of Israel has separated you from the congregation of Israel, to allow you to approach him in order to perform the duties of the LORD's tabernacle, and to stand before the congregation and serve them? 10 He has allowed you to approach him, and all your brother Levites with you; yet you seek the priesthood as well! 11 Therefore you and all your company have gathered together against the LORD. What is Aaron that you rail against him?"

12 Moses sent for Dathan and Abiram sons of Eliab; but they said, "We will not come! 13 Is it too little that you have brought us up out of a land flowing with milk and honey to kill us in the wilderness, that you must also lord it over us? 14 It is clear you have not brought us into a land flowing with milk and honey, or given us an inheritance of fields and vineyards. Would you put out the eyes of these men? We will not come!"

15 Moses was very angry and said to the LORD, "Pay no attention to their offering. I have not taken one donkey from them, and I have not harmed any one of them." 16 And Moses said to Korah, "As for you and all your company, be present tomorrow before the LORD, you and

a Cn: Heb *and they confronted Moses, and two hundred fifty men . . . well-known men* *b* Heb *his*

---

law (see Lev 24.12). **37–41:** The ancient custom of wearing *fringes* is reinterpreted as a reminder of God's law.

16.1–50: Revolts against Moses. These stories recall struggles that threatened the community during the sojourn at Kadesh, overlaid with the memory of later conflicts in the history of the priesthood. **3–11:** The rebellion led by Korah, of Levitical descent (v. 1a; Ex 6.21). Supported by other Levites (vv. 7,

8, 10), he challenged the subordination of some Levites to the Aaronite order (3.5–10; 8.5–22) and sought full priestly status (vv. 9–10, 40). **5:** A divine sign will determine who will be *allowed to approach* the LORD, i.e. the altar. **6–7:** Compare Lev 10.1–3.

16.12–15: According to an older literary tradition, a civil revolt was led by members of the tribe of Reuben who were dissatisfied with Moses' leadership (vv. 1b–2). **13:** Here

they and Aaron; [17] and let each one of you take his censer, and put incense on it, and each one of you present his censer before the LORD, two hundred fifty censers; you also, and Aaron, each his censer." [18] So each man took his censer, and they put fire in the censers and laid incense on them, and they stood at the entrance of the tent of meeting with Moses and Aaron. [19] Then Korah assembled the whole congregation against them at the entrance of the tent of meeting. And the glory of the LORD appeared to the whole congregation.

20 Then the LORD spoke to Moses and to Aaron, saying: [21] Separate yourselves from this congregation, so that I may consume them in a moment. [22] They fell on their faces, and said, "O God, the God of the spirits of all flesh, shall one person sin and you become angry with the whole congregation?"

23 And the LORD spoke to Moses, saying: [24] Say to the congregation: Get away from the dwellings of Korah, Dathan, and Abiram. [25] So Moses got up and went to Dathan and Abiram; the elders of Israel followed him. [26] He said to the congregation, "Turn away from the tents of these wicked men, and touch nothing of theirs, or you will be swept away for all their sins." [27] So they got away from the dwellings of Korah, Dathan, and Abiram; and Dathan and Abiram came out and stood at the entrance of their tents, together with their wives, their children, and their little ones. [28] And Moses said, "This is how you shall know that the LORD has sent me to do all these works; it has not been of my own accord: [29] If these people die a natural death, or if a natural fate comes on them, then the LORD has not sent me. [30] But if the LORD creates something new, and the ground opens its mouth and swallows them up, with all that belongs to them, and they go down alive into Sheol, then you shall know that these men have despised the LORD."

31 As soon as he finished speaking all these words, the ground under them was split apart. [32] The earth opened its mouth and swallowed them up, along with their households—everyone who belonged to Korah and all their goods. [33] So they with all that belonged to them went down alive into Sheol; the earth closed over them, and they perished from the midst of the assembly. [34] All Israel around them fled at their outcry, for they said, "The earth will swallow us too!" [35] And fire came out from the LORD and consumed the two hundred fifty men offering the incense.

36 *c* Then the LORD spoke to Moses, saying: [37] Tell Eleazar son of Aaron the priest to take the censers out of the blaze; then scatter the fire far and wide. [38] For the censers of these sinners have become holy at the cost of their lives. Make them into hammered plates as a covering for the altar, for they presented them before the LORD and they became holy. Thus they shall be a sign to the Israelites. [39] So Eleazar the priest took the bronze censers that had been presented by those who were burned; and they were hammered out as a covering for the altar— [40] a reminder to the Israelites that no outsider, who is not of the descendants of Aaron, shall approach to offer incense before the

c Ch 17.1 in Heb

Egypt, in contrast to the wilderness, is *a land flowing with milk and honey* (see Ex 3.8 n.). **14:** *Put out the eyes,* i.e. blind with deceit. **16.16–24:** Continuation of the Korah story (vv. 3–11). **19:** Apparently the Levites' revolt had popular support. **24:** An editorial revision which combines the two revolts. **16.25–34:** A continuation of the story of the civil revolt (vv. 12–15). **27a:** See v. 24 n. **30:** *Sheol,* see Gen 37.35 n. **31–33:** These verses express the ancient conception of corporate guilt which involves a person's whole family (Josh 7.22–26). **32:** *Everyone who belonged to Korah,* see v. 24 n. **16.35–40:** Continuation of the Korah story (vv. 16–24). **35:** The two hundred and fifty men were Levites, sons of Korah (vv. 2b, 17). **37–38:** On the bronze covering for the altar, see Ex 27.2. **40:** This verse states the purpose of the Korah story: to uphold the preroga-

LORD, so as not to become like Korah and his company—just as the LORD had said to him through Moses.

41 On the next day, however, the whole congregation of the Israelites rebelled against Moses and against Aaron, saying, "You have killed the people of the LORD." 42 And when the congregation had assembled against them, Moses and Aaron turned toward the tent of meeting; the cloud had covered it and the glory of the LORD appeared. 43 Then Moses and Aaron came to the front of the tent of meeting, 44 and the LORD spoke to Moses, saying, 45 "Get away from this congregation, so that I may consume them in a moment." And they fell on their faces. 46 Moses said to Aaron, "Take your censer, put fire on it from the altar and lay incense on it, and carry it quickly to the congregation and make atonement for them. For wrath has gone out from the LORD; the plague has begun." 47 So Aaron took it as Moses had ordered, and ran into the middle of the assembly, where the plague had already begun among the people. He put on the incense, and made atonement for the people. 48 He stood between the dead and the living; and the plague was stopped. 49 Those who died by the plague were fourteen thousand seven hundred, besides those who died in the affair of Korah. 50 When the plague was stopped, Aaron returned to Moses at the entrance of the tent of meeting.

17 *d* The LORD spoke to Moses, saying: 2 Speak to the Israelites, and get twelve staffs from them, one for each ancestral house, from all the leaders of their ancestral houses. Write each man's name on his staff, 3 and write Aaron's name on the staff of Levi. For there shall be one staff for the head of each ancestral house. 4 Place them in the tent of meeting before the covenant, *e* where I meet with you. 5 And the staff of the man whom I choose shall sprout; thus I will put a stop to the complaints of the Israelites that they continually make against you. 6 Moses spoke to the Israelites; and all their leaders gave him staffs, one for each leader, according to their ancestral houses, twelve staffs; and the staff of Aaron was among theirs. 7 So Moses placed the staffs before the LORD in the tent of the covenant. *e*

8 When Moses went into the tent of the covenant *e* on the next day, the staff of Aaron for the house of Levi had sprouted. It put forth buds, produced blossoms, and bore ripe almonds. 9 Then Moses brought out all the staffs from before the LORD to all the Israelites; and they looked, and each man took his staff. 10 And the LORD said to Moses, "Put back the staff of Aaron before the covenant, *e* to be kept as a warning to rebels, so that you may make an end of their complaints against me, or else they will die." 11 Moses did so; just as the LORD commanded him, so he did.

12 The Israelites said to Moses, "We are perishing; we are lost, all of us are lost! 13 Everyone who approaches the tabernacle of the LORD will die. Are we all to perish?"

18 The LORD said to Aaron: You and your sons and your ancestral house with you shall bear responsibility

d Ch 17.16 in Heb *e* Or *treaty*, or *testimony*; Heb *eduth*

---

tives of the Aaronite priesthood. **42**: *Cloud . . . glory,* see Ex 40.34–38 n.

17.1–13: Aaron's budding rod demonstrates the special status of the tribe of Levi, of which he is the leader. **2**: *Ancestral house,* here means "tribe" (see 1.2–4 n.). The rod or staff was a symbol of tribal authority. **3**: On Aaron's Levitical descent, see Ex 6.14–25. **4**: *The covenant,* see Ex 16.33–34 n. **5**: The people's complaints, related in 16.41–50.

17.8: Folk traditions of other peoples contain stories of blossoming rods, clubs, or spears. The sign signifies the LORD's choice of Aaron (v. 5) as leader of the whole priestly tribe of Levi. **10**: See Ex 25.16 n.

18.1–32: The responsibility of Aaronic priests and other Levites. This chapter comes naturally after the preceding one and gives the answer to the people's question in 17.13. **1**: *Bear responsibility,* i.e. bear the consequences for any cultic impropriety (Ex 28.38). All Levites are responsible for the

for offenses connected with the sanctuary, while you and your sons alone shall bear responsibility for offenses connected with the priesthood. ²So bring with you also your brothers of the tribe of Levi, your ancestral tribe, in order that they may be joined to you, and serve you while you and your sons with you are in front of the tent of the covenant.*f* ³They shall perform duties for you and for the whole tent. But they must not approach either the utensils of the sanctuary or the altar, otherwise both they and you will die. ⁴They are attached to you in order to perform the duties of the tent of meeting, for all the service of the tent; no outsider shall approach you. ⁵You yourselves shall perform the duties of the sanctuary and the duties of the altar, so that wrath may never again come upon the Israelites. ⁶It is I who now take your brother Levites from among the Israelites; they are now yours as a gift, dedicated to the LORD, to perform the service of the tent of meeting. ⁷But you and your sons with you shall diligently perform your priestly duties in all that concerns the altar and the area behind the curtain. I give your priesthood as a gift;*g* any outsider who approaches shall be put to death.

8 The LORD spoke to Aaron: I have given you charge of the offerings made to me, all the holy gifts of the Israelites; I have given them to you and your sons as a priestly portion due you in perpetuity. ⁹This shall be yours from the most holy things, reserved from the fire: every offering of theirs that they render to me as a most holy thing, whether grain offering, sin offering, or guilt offering, shall belong to you and your sons. ¹⁰As a most holy thing you shall eat it; every male may eat it; it shall be holy to you.

¹¹This also is yours: I have given to you, together with your sons and daughters, as a perpetual due, whatever is set aside from the gifts of all the elevation offerings of the Israelites; everyone who is clean in your house may eat them. ¹²All the best of the oil and all the best of the wine and of the grain, the choice produce that they give to the LORD, I have given to you. ¹³The first fruits of all that is in their land, which they bring to the LORD, shall be yours; everyone who is clean in your house may eat of it. ¹⁴Every devoted thing in Israel shall be yours. ¹⁵The first issue of the womb of all creatures, human and animal, which is offered to the LORD, shall be yours; but the firstborn of human beings you shall redeem, and the firstborn of unclean animals you shall redeem. ¹⁶Their redemption price, reckoned from one month of age, you shall fix at five shekels of silver, according to the shekel of the sanctuary (that is, twenty gerahs). ¹⁷But the firstborn of a cow, or the firstborn of a sheep, or the firstborn of a goat, you shall not redeem; they are holy. You shall dash their blood on the altar, and shall turn their fat into smoke as an offering by fire for a pleasing odor to the LORD; ¹⁸but their flesh shall be yours, just as the breast that is elevated and as the right thigh are yours. ¹⁹All the holy offerings that the Israelites present to the LORD I have given to you, together with your sons and daughters, as a perpetual due; it is a covenant of salt forever before the LORD for you and your descendants as well. ²⁰Then the LORD said to Aaron: You shall have no allotment in their land, nor shall you have any share

f Or *treaty*, or *testimony*; Heb *eduth*   *g* Heb *as a service of gift*

sanctuary; only Aaron's sons are responsible for the priesthood (v. 7), because they minister at the altar and guard the purity of the sanctuary. **2–6:** Other Levites are assistants to the Aaronic order (3.5–10; 8.5–22). **7:** *Area behind the curtain,* or in the Holy of Holies. **18.8–20:** Since the tribe of Levi has no land inheritance, the Aaronic priests are to be supported from the offerings (v. 20; see Lev 2.2–3 n.). **9–10:** A definition of *the most holy things*

which belong only to the Aaronic priests. For the types of sacrifice, see Lev chs 1–7. **11–19:** The priest and his family may eat the *holy things,* i.e. the *elevation offerings* (e.g. of offerings of well-being, Lev 7.28–36), offerings of the *first fruits* (Num 15.17–21), or every *devoted thing,* such as the firstborn of clean animals (Lev 27.26–27) **14–18:** Ex 13.11–13; 34.19–20. **19:** *A covenant of salt,* see Lev 2.13 n.

among them; I am your share and your possession among the Israelites.

21 To the Levites I have given every tithe in Israel for a possession in return for the service that they perform, the service in the tent of meeting. 22 From now on the Israelites shall no longer approach the tent of meeting, or else they will incur guilt and die. 23 But the Levites shall perform the service of the tent of meeting, and they shall bear responsibility for their own offenses; it shall be a perpetual statute throughout your generations. But among the Israelites they shall have no allotment, 24 because I have given to the Levites as their portion the tithe of the Israelites, which they set apart as an offering to the LORD. Therefore I have said of them that they shall have no allotment among the Israelites.

25 Then the LORD spoke to Moses, saying: 26 You shall speak to the Levites, saying: When you receive from the Israelites the tithe that I have given you from them for your portion, you shall set apart an offering from it to the LORD, a tithe of the tithe. 27 It shall be reckoned to you as your gift, the same as the grain of the threshing floor and the fullness of the wine press. 28 Thus you also shall set apart an offering to the LORD from all the tithes that you receive from the Israelites; and from them you shall give the LORD's offering to the priest Aaron. 29 Out of all the gifts to you, you shall set apart every offering due to the LORD; the best of all of them is the part to be consecrated. 30 Say also to them: When you have set apart the best of it, then the rest shall be reckoned to the Levites as produce of the threshing floor, and as produce of the wine press. 31 You may eat it in any place, you and your households; for it is your payment for your service in the tent of

meeting. 32 You shall incur no guilt by reason of it, when you have offered the best of it. But you shall not profane the holy gifts of the Israelites, on pain of death.

19 The LORD spoke to Moses and Aaron, saying: 2 This is a statute of the law that the LORD has commanded: Tell the Israelites to bring you a red heifer without defect, in which there is no blemish and on which no yoke has been laid. 3 You shall give it to the priest Eleazar, and it shall be taken outside the camp and slaughtered in his presence. 4 The priest Eleazar shall take some of its blood with his finger and sprinkle it seven times towards the front of the tent of meeting. 5 Then the heifer shall be burned in his sight; its skin, its flesh, and its blood, with its dung, shall be burned. 6 The priest shall take cedarwood, hyssop, and crimson material, and throw them into the fire in which the heifer is burning. 7 Then the priest shall wash his clothes and bathe his body in water, and afterwards he may come into the camp; but the priest shall remain unclean until evening. 8 The one who burns the heifer*h* shall wash his clothes in water and bathe his body in water; he shall remain unclean until evening. 9 Then someone who is clean shall gather up the ashes of the heifer, and deposit them outside the camp in a clean place; and they shall be kept for the congregation of the Israelites for the water for cleansing. It is a purification offering. 10 The one who gathers the ashes of the heifer shall wash his clothes and be unclean until evening.

This shall be a perpetual statute for the Israelites and for the alien residing among them. 11 Those who touch the

h Heb *it*

18.21–24: The rest of the Levites receive all tithes as their due (Lev 27.30–33). 25–32: The assistant Levites are to pay a *tithe of the tithe* to the Aaronic priests.
19.1–22: **Rites for purifying a person who is defiled by a corpse. 1–10:** The purpose of this ancient rite is to prepare ceremonially the ashes used in the *water for cleansing*

(vv. 9, 12–13). 4: The sacred number, the sacred blood, and the sacred tent of meeting combine to make the rite efficacious. 6: Compare Lev 14.4.
19.11–22: The ceremonial cleansing. 11: On defilement by the dead, see Lev 21.1–12. 12: *The water,* referred to in v. 9. 13: The defilement was not only dangerous to the individu-

dead body of any human being shall be unclean seven days. 12 They shall purify themselves with the water on the third day and on the seventh day, and so be clean; but if they do not purify themselves on the third day and on the seventh day, they will not become clean. 13 All who touch a corpse, the body of a human being who has died, and do not purify themselves, defile the tabernacle of the LORD; such persons shall be cut off from Israel. Since water for cleansing was not dashed on them, they remain unclean; their uncleanness is still on them.

14 This is the law when someone dies in a tent: everyone who comes into the tent, and everyone who is in the tent, shall be unclean seven days. 15 And every open vessel with no cover fastened on it is unclean. 16 Whoever in the open field touches one who has been killed by a sword, or who has died naturally, *i* or a human bone, or a grave, shall be unclean seven days. 17 For the unclean they shall take some ashes of the burnt purification offering, and running water shall be added in a vessel; 18 then a clean person shall take hyssop, dip it in the water, and sprinkle it on the tent, on all the furnishings, on the persons who were there, and on whoever touched the bone, the slain, the corpse, or the grave. 19 The clean person shall sprinkle the unclean ones on the third day and on the seventh day, thus purifying them on the seventh day. Then they shall wash their clothes and bathe themselves in water, and at evening they shall be clean. 20 Any who are unclean but do not purify themselves, those persons shall be cut off from the assembly, for they have defiled the sanctuary of the LORD. Since the water for cleansing has not been dashed on them, they are unclean.

21 It shall be a perpetual statute for them. The one who sprinkles the water for cleansing shall wash his clothes, and whoever touches the water for cleansing shall be unclean until evening. 22 Whatever the unclean person touches shall be unclean, and anyone who touches it shall be unclean until evening.

20 The Israelites, the whole congregation, came into the wilderness of Zin in the first month, and the people stayed in Kadesh. Miriam died there, and was buried there.

2 Now there was no water for the congregation; so they gathered together against Moses and against Aaron. 3 The people quarreled with Moses and said, "Would that we had died when our kindred died before the LORD! 4 Why have you brought the assembly of the LORD into this wilderness for us and our livestock to die here? 5 Why have you brought us up out of Egypt, to bring us to this wretched place? It is no place for grain, or figs, or vines, or pomegranates; and there is no water to drink." 6 Then Moses and Aaron went away from the assembly to the entrance of the tent of meeting; they fell on their faces, and the glory of the LORD appeared to them. 7 The LORD spoke to Moses, saying: 8 Take the staff, and assemble the congregation, you and your brother Aaron, and command the rock before their eyes to yield its water. Thus you shall bring water out of the rock for them; thus you shall provide drink for the congregation and their livestock.

9 So Moses took the staff from before the LORD, as he had commanded him. 10 Moses and Aaron gathered the assem-

---

i Heb lacks *naturally*

---

al but to the community, owing to the holiness of the tabernacle. **14–16**: Uncleanness, like the stench of death, could contaminate a tent or even an open vessel. **17**: *Some ashes,* see v. 9. *Hyssop,* see Ex 12.22 n.

20.1–29: Departure from Kadesh. Having failed to enter Canaan from the south (chs 13–14), Israel planned to detour through Transjordan in order to attack Canaan from the east. **1**: A summary that condenses Israel's long sojourn at Kadesh (see 13.26 n.). The year of *the first month* has dropped out. The whole generation, condemned to wander in the wilderness for about forty years (14.20–35), passed away. On the role of Miriam see comments on 12.1–16. **2–13**: The Meribah incident is paralleled in Ex 17.1–7. **3**: *Our kindred,* i.e. Dathan and Abiram (ch 16).

bly together before the rock, and he said to them, "Listen, you rebels, shall we bring water for you out of this rock?" ¹¹Then Moses lifted up his hand and struck the rock twice with his staff; water came out abundantly, and the congregation and their livestock drank. ¹²But the LORD said to Moses and Aaron, "Because you did not trust in me, to show my holiness before the eyes of the Israelites, therefore you shall not bring this assembly into the land that I have given them." ¹³These are the waters of Meribah,^j where the people of Israel quarreled with the LORD, and by which he showed his holiness.

14 Moses sent messengers from Kadesh to the king of Edom, "Thus says your brother Israel: You know all the adversity that has befallen us: ¹⁵how our ancestors went down to Egypt, and we lived in Egypt a long time; and the Egyptians oppressed us and our ancestors; ¹⁶and when we cried to the LORD, he heard our voice, and sent an angel and brought us out of Egypt; and here we are in Kadesh, a town on the edge of your territory. ¹⁷Now let us pass through your land. We will not pass through field or vineyard, or drink water from any well; we will go along the King's Highway, not turning aside to the right hand or to the left until we have passed through your territory."

18 But Edom said to him, "You shall not pass through, or we will come out with the sword against you." ¹⁹The Israelites said to him, "We will stay on the highway; and if we drink of your water,

we and our livestock, then we will pay for it. It is only a small matter; just let us pass through on foot." ²⁰But he said, "You shall not pass through." And Edom came out against them with a large force, heavily armed. ²¹Thus Edom refused to give Israel passage through their territory; so Israel turned away from them.

22 They set out from Kadesh, and the Israelites, the whole congregation, came to Mount Hor. ²³Then the LORD said to Moses and Aaron at Mount Hor, on the border of the land of Edom, ²⁴"Let Aaron be gathered to his people. For he shall not enter the land that I have given to the Israelites, because you rebelled against my command at the waters of Meribah. ²⁵Take Aaron and his son Eleazar, and bring them up Mount Hor; ²⁶strip Aaron of his vestments, and put them on his son Eleazar. But Aaron shall be gathered to his people,^k and shall die there." ²⁷Moses did as the LORD had commanded; they went up Mount Hor in the sight of the whole congregation. ²⁸Moses stripped Aaron of his vestments, and put them on his son Eleazar; and Aaron died there on the top of the mountain. Moses and Eleazar came down from the mountain. ²⁹When all the congregation saw that Aaron had died, all the house of Israel mourned for Aaron thirty days.

21 When the Canaanite, the king of Arad, who lived in the Negeb, heard that Israel was coming by the way of Atharim, he fought against Israel and

j That is *Quarrel*   k Heb lacks *to his people*

20.12: Although the preceding narrative is silent about Moses' disbelief, it is implied that he failed to interpret the giving of water as a sign from the LORD (Deut 32.50–52). **13**: See Ex 17.7 n. **14–21**: The request for passage through Edom. **14**: *Your brother*, Gen 25.24–26. By the thirteenth century B.C. Edom, one of the "Hebrew" groups (see Gen 10.21–31 n.), had developed a non-hereditary monarchy (Gen 36.31–39). **16**: *Angel*, see Ex 23.20–21 n. **17**: *The King's Highway* was the main Transjordanian route from Ezion-geber on the Gulf of Aqaba to Syria.

20.22–29: The death of Aaron. **24**: See 20.12 n. *Gathered to his people*, Gen 25.8. **26**: *Vestments*, i.e. of the high priest (Ex ch 28; Lev 8.7–9).

21.1–35: **Events along the way. 1–3**: Another account of the battle of Hormah (14.39–45; compare Judg 1.16–17). **1**: *The way of Atharim*, probably the name of a pass. **2–3**: A vow to wage holy war (see Ex 17.14 n.). This is apparently another tradition of an attempt to penetrate Canaan from the south; note that the people marched north from Kadesh. **4–9**: The serpent scourge. **4**: Leaving Mount Hor

took some of them captive. [2] Then Israel made a vow to the LORD and said, "If you will indeed give this people into our hands, then we will utterly destroy their towns." [3] The LORD listened to the voice of Israel, and handed over the Canaanites; and they utterly destroyed them and their towns; so the place was called Hormah. [l]

4 From Mount Hor they set out by the way to the Red Sea, [m] to go around the land of Edom; but the people became impatient on the way. [5] The people spoke against God and against Moses, "Why have you brought us up out of Egypt to die in the wilderness? For there is no food and no water, and we detest this miserable food." [6] Then the LORD sent poisonous [n] serpents among the people, and they bit the people, so that many Israelites died. [7] The people came to Moses and said, "We have sinned by speaking against the LORD and against you; pray to the LORD to take away the serpents from us." So Moses prayed for the people. [8] And the LORD said to Moses, "Make a poisonous [o] serpent, and set it on a pole; and everyone who is bitten shall look at it and live." [9] So Moses made a serpent of bronze, and put it upon a pole; and whenever a serpent bit someone, that person would look at the serpent of bronze and live.

10 The Israelites set out, and camped in Oboth. [11] They set out from Oboth, and camped at Iye-abarim, in the wilderness bordering Moab toward the sunrise. [12] From there they set out, and camped in the Wadi Zered. [13] From there they set out, and camped on the other side of the Arnon, in [p] the wilderness that extends from the boundary of the Amorites; for the Arnon is the boundary of Moab, between Moab and the Amorites. [14] Wherefore it is said in the Book of the Wars of the LORD,

"Waheb in Suphah and the wadis.
The Arnon [15] and the slopes of the
     wadis
that extend to the seat of Ar,
and lie along the border of
     Moab." [q]

16 From there they continued to Beer; [r] that is the well of which the LORD said to Moses, "Gather the people together, and I will give them water." [17] Then Israel sang this song:

"Spring up, O well!—Sing to
     it!—
[18] the well that the leaders sank,
that the nobles of the people dug,
with the scepter, with the staff."

From the wilderness to Mattanah, [19] from Mattanah to Nahaliel, from Nahaliel to Bamoth, [20] and from Bamoth to the valley lying in the region of Moab by the top of Pisgah that overlooks the wasteland. [s]

21 Then Israel sent messengers to King Sihon of the Amorites, saying, [22] "Let me pass through your land; we will not turn aside into field or vineyard; we will not drink the water of any well; we will go by the King's Highway until we have passed through your territory." [23] But Sihon would not allow Israel to pass through his territory. Sihon gath-

l Heb *Destruction*   m Or *Sea of Reeds*
n Or *fiery*; Heb *seraphim*   o Or *fiery*; Heb
seraph   p Gk: Heb *which is in*   q Meaning
of Heb uncertain   r That is *Well*
s Or *Jeshimon*

(20.27), the people turned south toward the Red Sea (Gulf of Aqaba), i.e. toward Eziongeber (33.35; Deut 2.1–8). **5:** See Ex 16.3 n. **6:** An attack by *poisonous serpents* was interpreted as divine judgment upon the people's rebellion. **8–9:** These verses echo serpent magic, as practiced e.g. in ancient Egypt. The *serpent of bronze* (Nehushtan) was an object of popular worship during the Israelite monarchy (2 Kings 18.4).

21.10–20: The itinerary of the march (see ch 33; Deut ch 2). **14–15:** A quotation from an ancient poetic collection, *the Book of the Wars of the LORD* (compare Josh 10.13; 2 Sam 1.18). Ar, a chief city of Moab (v. 28). **16–18:** The "Song of the Well," another ancient poetic fragment, is quoted here because water was miraculously provided along the way.

21.21–32: The defeat of the petty kingdom of the Amorites just north of Moab. **22:** The

ered all his people together, and went out against Israel to the wilderness; he came to Jahaz, and fought against Israel. 24 Israel put him to the sword, and took possession of his land from the Arnon to the Jabbok, as far as to the Ammonites; for the boundary of the Ammonites was strong. 25 Israel took all these towns, and Israel settled in all the towns of the Amorites, in Heshbon, and in all its villages. 26 For Heshbon was the city of King Sihon of the Amorites, who had fought against the former king of Moab and captured all his land as far as the Arnon. 27 Therefore the ballad singers say,

"Come to Heshbon, let it be built;
    let the city of Sihon be
        established.
28 For fire came out from Heshbon,
        flame from the city of Sihon.
    It devoured Ar of Moab,
        and swallowed up*t* the heights
            of the Arnon.
29 Woe to you, O Moab!
        You are undone, O people of
            Chemosh!
    He has made his sons fugitives,
        and his daughters captives,
    to an Amorite king, Sihon.
30 So their posterity perished
        from Heshbon*u* to Dibon,
        and we laid waste until fire
            spread to Medeba."*v*

31 Thus Israel settled in the land of the Amorites. 32 Moses sent to spy out Jazer; and they captured its villages, and dispossessed the Amorites who were there.

33 Then they turned and went up the road to Bashan; and King Og of Bashan came out against them, he and all his people, to battle at Edrei. 34 But the LORD said to Moses, "Do not be afraid of him; for I have given him into your hand, with all his people, and all his land. You shall do to him as you did to King Sihon of the Amorites, who ruled in Heshbon." 35 So they killed him, his sons, and all his people, until there was no survivor left; and they took possession of his land.

22 The Israelites set out, and camped in the plains of Moab across the Jordan from Jericho. 2 Now Balak son of Zippor saw all that Israel had done to the Amorites. 3 Moab was in great dread of the people, because they were so numerous; Moab was overcome with fear of the people of Israel. 4 And Moab said to the elders of Midian, "This horde will now lick up all that is around us, as an ox licks up the grass of the field." Now Balak son of Zippor was king of Moab at that time. 5 He sent messengers to Balaam son of Beor at Pethor, which is on the Euphrates, in the land of Amaw,*w* to summon him, saying, "A people has come out of Egypt; they have spread over the face of the earth, and they have settled next to me. 6 Come now, curse this people for me, since they are stronger than I; perhaps I shall be able to defeat them and drive them from the land; for I know that whomever you bless is blessed, and whomever you curse is cursed."

t Gk: Heb *and the lords of*   *u* Gk: Heb *we have shot at them; Heshbon has perished*   *v* Compare Sam Gk: Meaning of MT uncertain   *w* Or *land of his kinsfolk*

---

request (compare 20.14–21) was made at the boundary river, Arnon (v. 13). **27**: An ancient taunt song, quoted to show that Sihon had captured former Moabite territory, including the city of Heshbon. **29**: *Chemosh,* god of the Moabites.

21.33–35: *Bashan,* a petty kingdom north of the Jabbok river. *Og,* Deut 3.1–7.

22.1–40: **Balak and Balaam.** Fearful of the triumphant Israelites, the Moabite king invited a Mesopotamian diviner to put a curse on the invaders. **4**: On Moab's league with Midian, see ch 31.

22.5–20: Balak's two missions to obtain the services of Balaam who was in *Pethor,* located south of Carchemish *which is on the Euphrates.* In antiquity Babylonia was famed for the art of divination. **6**: In antiquity it was believed that the curse—the opposite of the blessing (see Gen 27.4 n.)—releases a negative power that shapes future events. **8**: Throughout chs 22–24 the conviction is expressed that a foreign priest-diviner, though not a member of the covenant community, was obedient to the LORD's will and that nothing could prevent the fulfillment of the

7 So the elders of Moab and the elders of Midian departed with the fees for divination in their hand; and they came to Balaam, and gave him Balak's message. [8] He said to them, "Stay here tonight, and I will bring back word to you, just as the LORD speaks to me"; so the officials of Moab stayed with Balaam. [9] God came to Balaam and said, "Who are these men with you?" [10] Balaam said to God, "King Balak son of Zippor of Moab, has sent me this message: [11] 'A people has come out of Egypt and has spread over the face of the earth; now come, curse them for me; perhaps I shall be able to fight against them and drive them out.' " [12] God said to Balaam, "You shall not go with them; you shall not curse the people, for they are blessed." [13] So Balaam rose in the morning, and said to the officials of Balak, "Go to your own land, for the LORD has refused to let me go with you." [14] So the officials of Moab rose and went to Balak, and said, "Balaam refuses to come with us."

15 Once again Balak sent officials, more numerous and more distinguished than these. [16] They came to Balaam and said to him, "Thus says Balak son of Zippor: 'Do not let anything hinder you from coming to me; [17] for I will surely do you great honor, and whatever you say to me I will do; come, curse this people for me.' " [18] But Balaam replied to the servants of Balak, "Although Balak were to give me his house full of silver and gold, I could not go beyond the command of the LORD my God, to do less or more. [19] You remain here, as the others did, so that I may learn what more the LORD may say to me." [20] That night God came to Balaam and said to him, "If the men have come to summon you, get up and go with them; but do only what I tell you to do." [21] So Balaam got up in the morning, saddled his donkey, and went with the officials of Moab.

22 God's anger was kindled because he was going, and the angel of the LORD took his stand in the road as his adversary. Now he was riding on the donkey, and his two servants were with him. [23] The donkey saw the angel of the LORD standing in the road, with a drawn sword in his hand; so the donkey turned off the road, and went into the field; and Balaam struck the donkey, to turn it back onto the road. [24] Then the angel of the LORD stood in a narrow path between the vineyards, with a wall on either side. [25] When the donkey saw the angel of the LORD, it scraped against the wall, and scraped Balaam's foot against the wall; so he struck it again. [26] Then the angel of the LORD went ahead, and stood in a narrow place, where there was no way to turn either to the right or to the left. [27] When the donkey saw the angel of the LORD, it lay down under Balaam; and Balaam's anger was kindled, and he struck the donkey with his staff. [28] Then the LORD opened the mouth of the donkey, and it said to Balaam, "What have I done to you, that you have struck me these three times?" [29] Balaam said to the donkey, "Because you have made a fool of me! I wish I had a sword in my hand! I would kill you right now!" [30] But the donkey said to Balaam, "Am I not your donkey, which you have ridden all your life to this day? Have I been in the habit of treating you this way?" And he said, "No."

31 Then the LORD opened the eyes of Balaam, and he saw the angel of the LORD standing in the road, with his drawn sword in his hand; and he bowed down, falling on his face. [32] The angel of the LORD said to him, "Why have you struck your donkey these three times? I have come out as an adversary, because

---

divine purpose for Israel. **13**: It is implied that the LORD spoke to him in a dream (vv. 8, 19–20).

22.21–35: The folk-story of Balaam's donkey, with its miraculously talking animal and unsophisticated, humorous irony, is ap-

parently another tradition. **22**: Note the contrast between this verse and the previous permission for Balaam to go (vv. 15–20). *The angel of the LORD,* see Gen 16.7 n. **23**: Josh 5.13.

your way is perverse[x] before me. 33 The donkey saw me, and turned away from me these three times. If it had not turned away from me, surely just now I would have killed you and let it live." 34 Then Balaam said to the angel of the LORD, "I have sinned, for I did not know that you were standing in the road to oppose me. Now therefore, if it is displeasing to you, I will return home." 35 The angel of the LORD said to Balaam, "Go with the men; but speak only what I tell you to speak." So Balaam went on with the officials of Balak.

36 When Balak heard that Balaam had come, he went out to meet him at Ir-moab, on the boundary formed by the Arnon, at the farthest point of the boundary. 37 Balak said to Balaam, "Did I not send to summon you? Why did you not come to me? Am I not able to honor you?" 38 Balaam said to Balak, "I have come to you now, but do I have power to say just anything? The word God puts in my mouth, that is what I must say." 39 Then Balaam went with Balak, and they came to Kiriath-huzoth. 40 Balak sacrificed oxen and sheep, and sent them to Balaam and to the officials who were with him.

41 On the next day Balak took Balaam and brought him up to Bamoth-baal; and from there he could see part of the people of Israel.[y] 1 Then Balaam said to Balak, "Build me seven altars here, and prepare seven bulls and seven rams for me." 2 Balak did as Balaam had said; and Balak and Balaam offered a bull and a ram on each altar. 3 Then Balaam said to Balak, "Stay here beside your burnt offerings while I go aside. Perhaps the LORD will come to meet me. Whatever he shows me I will tell you." And he went to a bare height.

4 Then God met Balaam; and Balaam said to him, "I have arranged the seven altars, and have offered a bull and a ram on each altar." 5 The LORD put a word in Balaam's mouth, and said, "Return to Balak, and this is what you must say." 6 So he returned to Balak,[z] who was standing beside his burnt offerings with all the officials of Moab. 7 Then Balaam[a] uttered his oracle, saying:

"Balak has brought me from Aram,
    the king of Moab from the eastern mountains:
'Come, curse Jacob for me;
    Come, denounce Israel!'
8 How can I curse whom God has not cursed?
    How can I denounce those whom the LORD has not denounced?
9 For from the top of the crags I see him,
    from the hills I behold him;
Here is a people living alone,
    and not reckoning itself among the nations!
10 Who can count the dust of Jacob,
    or number the dust-cloud[b] of Israel?
Let me die the death of the upright,
    and let my end be like his!"

11 Then Balak said to Balaam, "What have you done to me? I brought you to curse my enemies, but now you have

x Meaning of Heb uncertain   y Heb lacks *of Israel*   z Heb *him*   a Heb *he*   b Or *fourth part*

22.34–35: The incident serves to remind Balaam that he must speak only what the LORD bids him (compare 1 Kings 22.14), regardless of gifts or honors (v. 38). **22.41–24.25: The four oracles of Balaam,** probably dating in their present form from the early Israelite monarchy (24.17–19). **23.1–6:** Babylonian diviners resorted to this kind of sacrificial ceremony to obtain an omen. **7–10:** The first oracle. **7:** *Aram,* i.e. "Aram of the Two Rivers" or Mesopotamia (Gen 24.10). **9:** Israel is not a nation like other nations (compare 1 Sam 8.4–22) but a people set apart for a special destiny. **10:** Balaam prays that, by not cursing God's people, he may come to the end of his days as a righteous man.

23.11–12: See 22.20. The narrative is written with a touch of humor. **13:** Not wishing to defy God, Balak supposes that the oracle

done nothing but bless them." 12He answered, "Must I not take care to say what the Lord puts into my mouth?"

13 So Balak said to him, "Come with me to another place from which you may see them; you shall see only part of them, and shall not see them all; then curse them for me from there." 14So he took him to the field of Zophim, to the top of Pisgah. He built seven altars, and offered a bull and a ram on each altar. 15Balaam said to Balak, "Stand here beside your burnt offerings, while I meet the Lord over there. 16The Lord met Balaam, put a word into his mouth, and said, "Return to Balak, and this is what you shall say." 17When he came to him, he was standing beside his burnt offerings with the officials of Moab. Balak said to him, "What has the Lord said?" 18Then Balaam uttered his oracle, saying:

"Rise, Balak, and hear;
  listen to me, O son of Zippor:
19 God is not a human being, that he
    should lie,
  or a mortal, that he should
    change his mind.
Has he promised, and will he not
    do it?
Has he spoken, and will he not
    fulfill it?
20 See, I received a command to
    bless;
  he has blessed, and I cannot
    revoke it.
21 He has not beheld misfortune in
    Jacob;
  nor has he seen trouble in Israel.
The Lord their God is with them,
  acclaimed as a king among
    them.
22 God, who brings them out of
    Egypt,

is like the horns of a wild ox for
    them.
23 Surely there is no enchantment
    against Jacob,
  no divination against Israel;
  now it shall be said of Jacob and
    Israel,
  'See what God has done!'
24 Look, a people rising up like a
    lioness,
  and rousing itself like a lion!
It does not lie down until it has
    eaten the prey
  and drunk the blood of the
    slain."

25 Then Balak said to Balaam, "Do not curse them at all, and do not bless them at all." 26But Balaam answered Balak, "Did I not tell you, 'Whatever the Lord says, that is what I must do'?"

27 So Balak said to Balaam, "Come now, I will take you to another place; perhaps it will please God that you may curse them for me from there." 28So Balak took Balaam to the top of Peor, which overlooks the wasteland.[c] 29Balaam said to Balak, "Build me seven altars here, and prepare seven bulls and seven rams for me." 30So Balak did as Balaam had said, and offered a bull and a ram on each altar.

24 Now Balaam saw that it pleased the Lord to bless Israel, so he did not go, as at other times, to look for omens, but set his face toward the wilderness. 2Balaam looked up and saw Israel camping tribe by tribe. Then the spirit of God came upon him, 3and he uttered his oracle, saying:

"The oracle of Balaam son of
    Beor,

c Or *overlooks Jeshimon*

---

may be different if Balaam views the people from another perspective. **14:** *Pisgah,* see Deut 34.1 n.

23.18–24: The second oracle. **19:** God's change of mind displays consistency of purpose, unlike human beings who manifest deceit and caprice (see Ex 32.14 n.). **21:** The Lord's kingship is an ancient poetic motif (Ex

15.18). **22:** *The horns of a wild ox,* a symbol of brute strength. **23:** Since Israel's God cannot be coerced by magical techniques, divination is ineffective in altering Israel's destiny. **28:** *Peor,* see 25.3 n.

24.1: Balaam begins to abandon his technique of divination. **3–9:** The third oracle. **3–4:** These verses portray the ecstasy during

the oracle of the man whose eye
is clear, [d]
4 the oracle of one who hears the
words of God,
who sees the vision of the
Almighty, [e]
who falls down, but with eyes
uncovered:
5 how fair are your tents, O Jacob,
your encampments, O Israel!
6 Like palm groves that stretch far
away,
like gardens beside a river,
like aloes that the LORD has
planted,
like cedar trees beside the
waters.
7 Water shall flow from his buckets,
and his seed shall have abundant
water,
his king shall be higher than
Agag,
and his kingdom shall be
exalted.
8 God who brings him out of
Egypt,
is like the horns of a wild ox for
him;
he shall devour the nations that are
his foes
and break their bones.
He shall strike with his
arrows. [f]
9 He crouched, he lay down like a
lion,
and like a lioness; who will
rouse him up?
Blessed is everyone who blesses
you,
and cursed is everyone who
curses you."
10 Then Balak's anger was kindled
against Balaam, and he struck his hands

together. Balak said to Balaam, "I summoned you to curse my enemies, but
instead you have blessed them these three
times. 11 Now be off with you! Go home!
I said, 'I will reward you richly,' but the
LORD has denied you any reward."
12 And Balaam said to Balak, "Did I not
tell your messengers whom you sent to
me, 13 'If Balak should give me his house
full of silver and gold, I would not be
able to go beyond the word of the LORD,
to do either good or bad of my own will;
what the LORD says, that is what I will
say'? 14 So now, I am going to my people;
let me advise you what this people will
do to your people in days to come."
15 So he uttered his oracle, saying:
"The oracle of Balaam son of
Beor,
the oracle of the man whose eye
is clear, [d]
16 the oracle of one who hears the
words of God,
and knows the knowledge of
the Most High, [g]
who sees the vision of the
Almighty, [e]
who falls down, but with his
eyes uncovered:
17 I see him, but not now;
I behold him, but not near—
a star shall come out of Jacob,
and a scepter shall rise out of
Israel;
it shall crush the borderlands [h] of
Moab,
and the territory [i] of all the
Shethites.

d Or *closed* or *open*   e Traditional rendering of
Heb *Shaddai*   f Meaning of Heb uncertain
g Or *of Elyon*   h Or *forehead*   i Some Mss
read *skull*

---

which the oracle was received. *The Almighty,*
see Gen 17.1 n. **7**: If *Agag* refers to the Amalekite king of Saul's time (1 Sam 15.8), the
poem must have been composed in the early
monarchy. **9b**: Gen 12.3.
24.10: Striking the hands together was a
gesture of anger and reproach (Job 27.23). **14**:
Balaam not only defies Balak's command to

get out but gives his last oracle without the
preparatory rites of divination (23.1–6;
24.1 n.). **15–19**: The fourth oracle. *The Most
High,* see Gen 14.18 n. **17–19**: What he sees is
not now but in the future: Israel will be victorious over Moab and Edom, a prophecy that
was realized in the time of David (2 Sam 8.2,
13–14). *A star, a scepter,* these royal symbols

18 Edom will become a possession,
   Seir a possession of its
      enemies, ʲ
   while Israel does valiantly.
19 One out of Jacob shall rule,
   and destroy the survivors of Ir."

20 Then he looked on Amalek, and uttered his oracle, saying:
   "First among the nations was
      Amalek,
   but its end is to perish forever."

21 Then he looked on the Kenite, and uttered his oracle, saying:
   "Enduring is your dwelling place,
   and your nest is set in the rock;
22 yet Kain is destined for burning.
   How long shall Asshur take you
      away captive?"

23 Again he uttered his oracle, saying:
   "Alas, who shall live when God
      does this?
24 But ships shall come from
      Kittim
   and shall afflict Asshur and Eber;
   and he also shall perish forever."

25 Then Balaam got up and went back to his place, and Balak also went his way.

25 While Israel was staying at Shittim, the people began to have sexual relations with the women of Moab. ²These invited the people to the sacrifices of their gods, and the people ate and bowed down to their gods. ³Thus Israel yoked itself to the Baal of Peor, and the Lord's anger was kindled against Israel. ⁴The Lord said to Moses, "Take all the chiefs of the people, and impale them in the sun before the Lord, in order that the fierce anger of the Lord may turn away from Israel." ⁵And Moses said to the judges of Israel, "Each of you shall kill any of your people who have yoked themselves to the Baal of Peor."

6 Just then one of the Israelites came and brought a Midianite woman into his family, in the sight of Moses and in the sight of the whole congregation of the Israelites, while they were weeping at the entrance of the tent of meeting. ⁷When Phinehas son of Eleazar, son of Aaron the priest, saw it, he got up and left the congregation. Taking a spear in his hand, ⁸he went after the Israelite man into the tent, and pierced the two of them, the Israelite and the woman, through the belly. So the plague was stopped among the people of Israel. ⁹Nevertheless those that died by the plague were twenty-four thousand.

10 The Lord spoke to Moses, saying: ¹¹"Phinehas son of Eleazar, son of Aaron the priest, has turned back my wrath from the Israelites by manifesting such zeal among them on my behalf that in my jealousy I did not consume the Israelites. ¹²Therefore say, 'I hereby grant him my covenant of peace. ¹³It shall be for him and for his descendants after him a

ʲ Heb *Seir, its enemies, a possession*

---

fit a leader like David whose empire embraced the promised land (see Gen 49.10 n.). **18:** *Seir,* a synonym for Edom.
24.20–24: Supplementary oracles. **20:** Ex 17.14–16. **21–24:** The meaning of these verses is obscure, owing to the uncertainty of the names.
25.1–18: Incidents of apostasy in Moab. The protest against intermarriage with foreign peoples was based on the fear of the corruption of Israel's faith (Deut 7.1–5). **1:** *Shittim* (Abel-shittim), opposite Jericho. **2:** The Moabite women lured Israelites into idolatrous rites. *Ate,* a reference to a sacred meal in connection with the sacrifices (compare Ex 32.6). **3:** *Baal,* the Canaanite god of storm and fertility who was worshiped at the

cult center of Peor or Beth-peor (23.28; Deut 3.29). **4:** The execution of the *chiefs* was an expiation for the people, according to the ancient conception of corporate guilt (compare 2 Sam 21.1–6). **5:** *Judges,* Ex 18.25–26. **6–18:** Phinehas' zeal, compare Ex 32.25–29. **6:** The story assumes an attempt to seduce Israelites into intermarriage with Midianites, with the result that divine judgment came in the form of a plague (v. 9). In 31.16 Balaam is blamed for this incident.
25.11: *Zealous,* or jealous, see Ex 34.14 n.; 1 Kings 19.10. **12:** A *covenant of peace* (Ezek 34.25; 37.26) is one of welfare and well-being, based on right relation with God and harmonious relations in the community. **13:** The Aaronic line, traced through Phinehas

covenant of perpetual priesthood, because he was zealous for his God, and made atonement for the Israelites.' "

14 The name of the slain Israelite man, who was killed with the Midianite woman, was Zimri son of Salu, head of an ancestral house belonging to the Simeonites. 15 The name of the Midianite woman who was killed was Cozbi daughter of Zur, who was the head of a clan, an ancestral house in Midian.

16 The LORD said to Moses, 17 "Harass the Midianites, and defeat them; 18 for they have harassed you by the trickery with which they deceived you in the affair of Peor, and in the affair of Cozbi, the daughter of a leader of Midian, their sister; she was killed on the day of the plague that resulted from Peor."

26 After the plague the LORD said to Moses and to Eleazar son of Aaron the priest, 2 "Take a census of the whole congregation of the Israelites, from twenty years old and upward, by their ancestral houses, everyone in Israel able to go to war." 3 Moses and Eleazar the priest spoke with them in the plains of Moab by the Jordan opposite Jericho, saying, 4 "Take a census of the people,*k* from twenty years old and upward," as the LORD commanded Moses.

The Israelites, who came out of the land of Egypt, were:

5 Reuben, the firstborn of Israel. The descendants of Reuben: of Hanoch, the clan of the Hanochites; of Pallu, the clan of the Palluites; 6 of Hezron, the clan of the Hezronites; of Carmi, the clan of the Carmites. 7 These are the clans of the Reubenites; the number of those enrolled was forty-three thousand seven hundred thirty. 8 And the descendants of Pallu: Eliab. 9 The descendants of Eliab: Nemuel, Dathan, and Abiram. These are the same Dathan and Abiram, chosen from the congregation, who rebelled against Moses and Aaron in the company of Korah, when they rebelled against the LORD, 10 and the earth opened its mouth and swallowed them up along with Korah, when that company died, when the fire devoured two hundred fifty men; and they became a warning. 11 Notwithstanding, the sons of Korah did not die.

12 The descendants of Simeon by their clans: of Nemuel, the clan of the Nemuelites; of Jamin, the clan of the Jaminites; of Jachin, the clan of the Jachinites; 13 of Zerah, the clan of the Zerahites; of Shaul, the clan of the Shaulites.*l* 14 These are the clans of the Simeonites, twenty-two thousand two hundred.

15 The children of Gad by their clans: of Zephon, the clan of the Zephonites; of Haggi, the clan of the Haggites; of Shuni, the clan of the Shunites; 16 of Ozni, the clan of the Oznites; of Eri, the clan of the Erites; 17 of Arod, the clan of the Arodites; of Areli, the clan of the Arelites. 18 These are the clans of the Gadites: the number of those enrolled was forty thousand five hundred.

19 The sons of Judah: Er and Onan; Er and Onan died in the land of Canaan. 20 The descendants of Judah by their clans were: of Shelah, the clan of the Shelanites; of Perez, the clan of the Perezites; of Zerah, the clan of the Zerahites. 21 The descendants of Perez were: of Hezron, the clan of the Hezronites; of Hamul, the clan of the Hamulites. 22 These are the clans of Judah: the number of those enrolled was seventy-six thousand five hundred.

23 The descendants of Issachar by their clans: of Tola, the clan of the Tolaites; of Puvah, the clan of the Punites; 24 of Jashub, the clan of the Jashubites; of Shimron, the clan of the Shimronites. 25 These are the clans of Issachar: sixty-four thousand three hundred enrolled.

26 The descendants of Zebulun by their clans: of Sered, the clan of the Seredites; of Elon, the clan of the Elonites; of Jahleel, the clan of the Jahleelites.

k Heb lacks *take a census of the people*: Compare verse 2   *l* Or *Saul . . . Saulites*

(Ex 6.25), is guaranteed *a perpetual priesthood,* i.e. one that continues in perpetuity. **16–18:** See ch 31.

26.1–65: A second census is taken to ascertain the strength of the tribes and to allot the land. **5–50:** See 1.5–15 n. **9–11:** See ch 16.

27 These are the clans of the Zebulunites; the number of those enrolled was sixty thousand five hundred.

28 The sons of Joseph by their clans: Manasseh and Ephraim. 29 The descendants of Manasseh: of Machir, the clan of the Machirites; and Machir was the father of Gilead; of Gilead, the clan of the Gileadites. 30 These are the descendants of Gilead: of Iezer, the clan of the Iezerites; of Helek, the clan of the Helekites; 31 and of Asriel, the clan of the Asrielites; and of Shechem, the clan of the Shechemites; 32 and of Shemida, the clan of the Shemidaites; and of Hepher, the clan of the Hepherites. 33 Now Zelophehad son of Hepher had no sons, but daughters: and the names of the daughters of Zelophehad were Mahlah, Noah, Hoglah, Milcah, and Tirzah. 34 These are the clans of Manasseh; the number of those enrolled was fifty-two thousand seven hundred.

35 These are the descendants of Ephraim according to their clans: of Shuthelah, the clan of the Shuthelahites; of Becher, the clan of the Becherites; of Tahan, the clan of the Tahanites. 36 And these are the descendants of Shuthelah: of Eran, the clan of the Eranites. 37 These are the clans of the Ephraimites: the number of those enrolled was thirty-two thousand five hundred. These are the descendants of Joseph by their clans.

38 The descendants of Benjamin by their clans: of Bela, the clan of the Belaites; of Ashbel, the clan of the Ashbelites; of Ahiram, the clan of the Ahiramites; 39 of Shephupham, the clan of the Shuphamites; of Hupham, the clan of the Huphamites. 40 And the sons of Bela were Ard and Naaman: of Ard, the clan of the Ardites; of Naaman, the clan of the Naamites. 41 These are the descendants of Benjamin by their clans; the number of those enrolled was forty-five thousand six hundred.

42 These are the descendants of Dan by their clans: of Shuham, the clan of the Shuhamites. These are the clans of Dan by their clans. 43 All the clans of the Shuhamites: sixty-four thousand four hundred enrolled.

44 The descendants of Asher by their families: of Imnah, the clan of the Imnites; of Ishvi, the clan of the Ishvites; of Beriah, the clan of the Beriites. 45 Of the descendants of Beriah: of Heber, the clan of the Heberites; of Malchiel, the clan of the Malchielites. 46 And the name of the daughter of Asher was Serah. 47 These are the clans of the Asherites: the number of those enrolled was fifty-three thousand four hundred.

48 The descendants of Naphtali by their clans: of Jahzeel, the clan of the Jahzeelites; of Guni, the clan of the Gunites; 49 of Jezer, the clan of the Jezerites; of Shillem, the clan of the Shillemites. 50 These are the Naphtalites[m] by their clans: the number of those enrolled was forty-five thousand four hundred.

51 This was the number of the Israelites enrolled: six hundred and one thousand seven hundred thirty.

52 The LORD spoke to Moses, saying: 53 To these the land shall be apportioned for inheritance according to the number of names. 54 To a large tribe you shall give a large inheritance, and to a small tribe you shall give a small inheritance; every tribe shall be given its inheritance according to its enrollment. 55 But the land shall be apportioned by lot; according to the names of their ancestral tribes they shall inherit. 56 Their inheritance shall be apportioned according to lot between the larger and the smaller.

m Heb *clans of Naphtali*

26.12–14: Simeon suffered the greatest loss during the wilderness sojourn (compare 1.23; see Gen 49.5–7 n.).
26.28: Manasseh's priority over Ephraim (compare 1.10) points to a very old historical situation (Gen 48.13–22) and suggests that this list is older than that of ch 1.
26.51: See 1.17–46 n. The priestly writer emphasizes God's marvelous power to sustain the people despite the almost complete change of population (vv. 63–65). 52–56: The census provided a basis for allotting land

57 This is the enrollment of the Levites by their clans: of Gershon, the clan of the Gershonites; of Kohath, the clan of the Kohathites; of Merari, the clan of the Merarites. 58These are the clans of Levi: the clan of the Libnites, the clan of the Hebronites, the clan of the Mahlites, the clan of the Mushites, the clan of the Korahites. Now Kohath was the father of Amram. 59The name of Amram's wife was Jochebed daughter of Levi, who was born to Levi in Egypt; and she bore to Amram: Aaron, Moses, and their sister Miriam. 60To Aaron were born Nadab, Abihu, Eleazar, and Ithamar. 61But Nadab and Abihu died when they offered illicit fire before the LORD. 62The number of those enrolled was twenty-three thousand, every male one month old and up; for they were not enrolled among the Israelites because there was no allotment given to them among the Israelites.

63 These were those enrolled by Moses and Eleazar the priest, who enrolled the Israelites in the plains of Moab by the Jordan opposite Jericho. 64Among these there was not one of those enrolled by Moses and Aaron the priest, who had enrolled the Israelites in the wilderness of Sinai. 65For the LORD had said of them, "They shall die in the wilderness." Not one of them was left, except Caleb son of Jephunneh and Joshua son of Nun.

27 Then the daughters of Zelophehad came forward. Zelophehad was son of Hepher son of Gilead son of Machir son of Manasseh son of Joseph, a member of the Manassite clans. The names of his daughters were: Mahlah, Noah, Hoglah, Milcah, and Tirzah. 2They stood before Moses, Eleazar the priest, the leaders, and all the congregation, at the entrance of the tent of meeting, and they said, 3"Our father died in the wilderness; he was not among the company of those who gathered themselves together against the LORD in the company of Korah, but died for his own sin; and he had no sons. 4Why should the name of our father be taken away from his clan because he had no son? Give to us a possession among our father's brothers."

5 Moses brought their case before the LORD. 6And the LORD spoke to Moses, saying: 7The daughters of Zelophehad are right in what they are saying; you shall indeed let them possess an inheritance among their father's brothers and pass the inheritance of their father on to them. 8You shall also say to the Israelites, "If a man dies, and has no son, then you shall pass his inheritance on to his daughter. 9If he has no daughter, then you shall give his inheritance to his brothers. 10If he has no brothers, then you shall give his inheritance to his father's brothers. 11And if his father has no brothers, then you shall give his inheritance to the nearest kinsman of his clan, and he shall possess it. It shall be for the Israelites a statute and ordinance, as the LORD commanded Moses."

12 The LORD said to Moses, "Go up this mountain of the Abarim range, and see the land that I have given to the Israelites. 13When you have seen it, you also shall be gathered to your people, as your brother Aaron was, 14because you rebelled against my word in the wilderness of Zin when the congregation quarreled with me.*ⁿ* You did not show my holi-

---

n Heb lacks *with me*

---

after the conquest. **57–62:** The Levites, the landless tribe, are numbered separately. **61:** See Lev ch 10.

27.1–11: Inheritance of property by women. 1: 26.33. **2–4:** In Israel a family inheritance was to remain in the family (36.7; 1 Kings 21.3), thereby perpetuating the ancestral name. The request of the daughters of Zelophehad was unusual in that, according to ancient law, normally women did not inherit

property. **5–11:** Another example of how law developed as unprecedented situations were faced (see 9.8 n.). Compare also ch 36.

27.12–23: Joshua commissioned to succeed Moses (Deut 31.14–15, 23). **12:** *Abarim*, a hilly region in which Mount Nebo, the scene of Moses' death, was located (Deut ch 34). **13:** 20.22–29. **14:** 20.12–13. *Meribathkadesh*, see Ex 17.7 n. **18–20:** Like Moses, Joshua was held to be a charismatic leader,

ness before their eyes at the waters."
(These are the waters of Meribath-ka-desh in the wilderness of Zin.) 15 Moses
spoke to the LORD, saying, 16 "Let the
LORD, the God of the spirits of all flesh,
appoint someone over the congregation
17 who shall go out before them and come
in before them, who shall lead them out
and bring them in, so that the congrega-tion of the LORD may not be like sheep
without a shepherd." 18 So the LORD said
to Moses, "Take Joshua son of Nun, a
man in whom is the spirit, and lay your
hand upon him; 19 have him stand before
Eleazar the priest and all the congrega-tion, and commission him in their sight.
20 You shall give him some of your au-thority, so that all the congregation of
the Israelites may obey. 21 But he shall
stand before Eleazar the priest, who shall
inquire for him by the decision of the
Urim before the LORD; at his word they
shall go out, and at his word they shall
come in, both he and all the Israelites
with him, the whole congregation." 22 So
Moses did as the LORD commanded him.
He took Joshua and had him stand before
Eleazar the priest and the whole congre-gation; 23 he laid his hands on him and
commissioned him—as the LORD had di-rected through Moses.

28 The LORD spoke to Moses, say-ing: 2 Command the Israelites,
and say to them: My offering, the food
for my offerings by fire, my pleasing
odor, you shall take care to offer to me
at its appointed time. 3 And you shall say
to them, This is the offering by fire that
you shall offer to the LORD: two male
lambs a year old without blemish, daily,
as a regular offering. 4 One lamb you
shall offer in the morning, and the other
lamb you shall offer at twilight° 5 also
one-tenth of an ephah of choice flour for
a grain offering, mixed with one-fourth

of a hin of beaten oil. 6 It is a regular burnt
offering, ordained at Mount Sinai for a
pleasing odor, an offering by fire to the
LORD. 7 Its drink offering shall be one-fourth of a hin for each lamb; in the sanc-tuary you shall pour out a drink offering
of strong drink to the LORD. 8 The other
lamb you shall offer at twilight° with a
grain offering and a drink offering like
the one in the morning; you shall offer it
as an offering by fire, a pleasing odor to
the LORD.

9 On the sabbath day: two male
lambs a year old without blemish, and
two-tenths of an ephah of choice flour
for a grain offering, mixed with oil, and
its drink offering— 10 this is the burnt
offering for every sabbath, in addition to
the regular burnt offering and its drink
offering.

11 At the beginnings of your months
you shall offer a burnt offering to the
LORD: two young bulls, one ram, seven
male lambs a year old without blemish;
12 also three-tenths of an ephah of choice
flour for a grain offering, mixed with oil,
for each bull; and two-tenths of choice
flour for a grain offering, mixed with oil,
for the one ram; 13 and one-tenth of
choice flour mixed with oil as a grain
offering for every lamb—a burnt offer-ing of pleasing odor, an offering by fire
to the LORD. 14 Their drink offerings shall
be half a hin of wine for a bull, one-third
of a hin for a ram, and one-fourth of a hin
for a lamb. This is the burnt offering of
every month throughout the months of
the year. 15 And there shall be one male
goat for a sin offering to the LORD; it shall
be offered in addition to the regular
burnt offering and its drink offering.

16 On the fourteenth day of the first

---

o Heb *between the two evenings*

---

a man in whom is the spirit (11.17; 24.2). **21:**
Urim, see Ex 28.15–30 n. At Joshua's word,
given in obedience to the sacred oracle, the
people shall *go out* to battle and *come in* the
camp (v. 17; Deut 3.28).
28.1–29.40: Offerings for various occa-sions.
28.2: *The food . . . my pleasing odor,* see Lev

1.9 n. and 21.6 n. **3–8:** The daily sacrifice (Ex
29.38–42). **9–10:** The sabbath offering is
mentioned only here in the Pentateuch.
28.11–15: Since the year was based on a
lunar calendar, a festival was held at each new
moon (10.10; 1 Sam 20.5; Isa 1.14; Am 8.5).
28.16–25: Unleavened bread (Ex 12.1–27;
Lev 23.5–8).

month there shall be a passover offering to the LORD. [17] And on the fifteenth day of this month is a festival; seven days shall unleavened bread be eaten. [18] On the first day there shall be a holy convocation. You shall not work at your occupations. [19] You shall offer an offering by fire, a burnt offering to the LORD: two young bulls, one ram, and seven male lambs a year old; see that they are without blemish. [20] Their grain offering shall be of choice flour mixed with oil: three-tenths of an ephah shall you offer for a bull, and two-tenths for a ram; [21] one-tenth shall you offer for each of the seven lambs; [22] also one male goat for a sin offering, to make atonement for you. [23] You shall offer these in addition to the burnt offering of the morning, which belongs to the regular burnt offering. [24] In the same way you shall offer daily, for seven days, the food of an offering by fire, a pleasing odor to the LORD; it shall be offered in addition to the regular burnt offering and its drink offering. [25] And on the seventh day you shall have a holy convocation; you shall not work at your occupations.

26 On the day of the first fruits, when you offer a grain offering of new grain to the LORD at your festival of weeks, you shall have a holy convocation; you shall not work at your occupations. [27] You shall offer a burnt offering, a pleasing odor to the LORD: two young bulls, one ram, seven male lambs a year old. [28] Their grain offering shall be of choice flour mixed with oil, three-tenths of an ephah for each bull, two-tenths for one ram, [29] one-tenth for each of the seven lambs; [30] with one male goat, to make atonement for you. [31] In addition to the regular burnt offering with its grain offering, you shall offer them and their drink offering. They shall be without blemish.

29 On the first day of the seventh month you shall have a holy con-

vocation; you shall not work at your occupations. It is a day for you to blow the trumpets, [2] and you shall offer a burnt offering, a pleasing odor to the LORD: one young bull, one ram, seven male lambs a year old without blemish. [3] Their grain offering shall be of choice flour mixed with oil, three-tenths of one ephah for the bull, two-tenths for the ram, [4] and one-tenth for each of the seven lambs; [5] with one male goat for a sin offering, to make atonement for you. [6] These are in addition to the burnt offering of the new moon and its grain offering, and the regular burnt offering and its grain offering, and their drink offerings, according to the ordinance for them, a pleasing odor, an offering by fire to the LORD.

7 On the tenth day of this seventh month you shall have a holy convocation, and deny yourselves;[p] you shall do no work. [8] You shall offer a burnt offering to the LORD, a pleasing odor: one young bull, one ram, seven male lambs a year old. They shall be without blemish. [9] Their grain offering shall be of choice flour mixed with oil, three-tenths of an ephah for the bull, two-tenths for the one ram, [10] one-tenth for each of the seven lambs; [11] with one male goat for a sin offering, in addition to the sin offering of atonement, and the regular burnt offering and its grain offering, and their drink offerings.

12 On the fifteenth day of the seventh month you shall have a holy convocation; you shall not work at your occupations. You shall celebrate a festival to the LORD seven days. [13] You shall offer a burnt offering, an offering by fire, a pleasing odor to the LORD: thirteen young bulls, two rams, fourteen male lambs a year old. They shall be without blemish. [14] Their grain offering shall be of choice flour mixed with oil, three-tenths of an ephah for each of the thirteen

p Or *and fast*

28.26–31: Pentecost (Ex 23.16; 34.22; Lev 23.15–21).
29.1–6: New Year (Lev 23.23–25). 7–11: Day of atonement (Lev 16.29–34; 23.26–32).

29.12–38: The offering for the festival of booths (Lev 23.33–36) exceeds that of any other convocation.
30.1–16: **On vows made by women. 2:**

bulls, two-tenths for each of the two rams, [15] and one-tenth for each of the fourteen lambs; [16] also one male goat for a sin offering, in addition to the regular burnt offering, its grain offering and its drink offering.

17 On the second day: twelve young bulls, two rams, fourteen male lambs a year old without blemish, [18] with the grain offering and the drink offerings for the bulls, for the rams, and for the lambs, as prescribed in accordance with their number; [19] also one male goat for a sin offering, in addition to the regular burnt offering and its grain offering, and their drink offerings.

20 On the third day: eleven bulls, two rams, fourteen male lambs a year old without blemish, [21] with the grain offering and the drink offerings for the bulls, for the rams, and for the lambs, as prescribed in accordance with their number; [22] also one male goat for a sin offering, in addition to the regular burnt offering and its grain offering and its drink offering.

23 On the fourth day: ten bulls, two rams, fourteen male lambs a year old without blemish, [24] with the grain offering and the drink offerings for the bulls, for the rams, and for the lambs, as prescribed in accordance with their number; [25] also one male goat for a sin offering, in addition to the regular burnt offering, its grain offering and its drink offering.

26 On the fifth day: nine bulls, two rams, fourteen male lambs a year old without blemish, [27] with the grain offering and the drink offerings for the bulls, for the rams, and for the lambs, as prescribed in accordance with their number; [28] also one male goat for a sin offering, in addition to the regular burnt offering and its grain offering and its drink offering.

29 On the sixth day: eight bulls, two rams, fourteen male lambs a year old without blemish, [30] with the grain offering and the drink offerings for the bulls, for the rams, and for the lambs, as pre-scribed in accordance with their number; [31] also one male goat for a sin offering, in addition to the regular burnt offering, its grain offering, and its drink offerings.

32 On the seventh day: seven bulls, two rams, fourteen male lambs a year old without blemish, [33] with the grain offering and the drink offerings for the bulls, for the rams, and for the lambs, as prescribed in accordance with their number; [34] also one male goat for a sin offering, besides the regular burnt offering, its grain offering, and its drink offering.

35 On the eighth day you shall have a solemn assembly; you shall not work at your occupations. [36] You shall offer a burnt offering, an offering by fire, a pleasing odor to the LORD: one bull, one ram, seven male lambs a year old without blemish, [37] and the grain offering and the drink offerings for the bull, for the ram, and for the lambs, as prescribed in accordance with their number; [38] also one male goat for a sin offering, in addition to the regular burnt offering and its grain offering and its drink offering.

39 These you shall offer to the LORD at your appointed festivals, in addition to your votive offerings and your freewill offerings, as your burnt offerings, your grain offerings, your drink offerings, and your offerings of well-being.

40 [q] So Moses told the Israelites everything just as the LORD had commanded Moses.

30 Then Moses said to the heads of the tribes of the Israelites: This is what the LORD has commanded. 2 When a man makes a vow to the LORD, or swears an oath to bind himself by a pledge, he shall not break his word; he shall do according to all that proceeds out of his mouth.

3 When a woman makes a vow to the LORD, or binds herself by a pledge, while

q Ch 30.1 in Heb

Vows made by men (Gen 28.20–22; Judg 11.30–31) are absolutely binding (compare Lev ch 27; Deut 23.21–23).

30.3–15: These cases reflect a society in which a woman was subordinate to the man of the family. He could nullify her vow if he

within her father's house, in her youth, [4] and her father hears of her vow or her pledge by which she has bound herself, and says nothing to her; then all her vows shall stand, and any pledge by which she has bound herself shall stand. [5] But if her father expresses disapproval to her at the time that he hears of it, no vow of hers, and no pledge by which she has bound herself, shall stand; and the LORD will forgive her, because her father had expressed to her his disapproval.

[6] If she marries, while obligated by her vows or any thoughtless utterance of her lips by which she has bound herself, [7] and her husband hears of it and says nothing to her at the time that he hears, then her vows shall stand, and her pledges by which she has bound herself shall stand. [8] But if, at the time that her husband hears of it, he expresses disapproval to her, then he shall nullify the vow by which she was obligated, or the thoughtless utterance of her lips, by which she bound herself; and the LORD will forgive her. [9] (But every vow of a widow or of a divorced woman, by which she has bound herself, shall be binding upon her.) [10] And if she made a vow in her husband's house, or bound herself by a pledge with an oath, [11] and her husband heard it and said nothing to her, and did not express disapproval to her, then all her vows shall stand, and any pledge by which she bound herself shall stand. [12] But if her husband nullifies them at the time that he hears them, then whatever proceeds out of her lips concerning her vows, or concerning her pledge of herself, shall not stand. Her husband has nullified them, and the LORD will forgive her. [13] Any vow or any binding oath to deny herself,[r] her husband may allow to stand, or her husband may nullify. [14] But if her husband says nothing to her from day to day,[s] then he validates all her vows, or all her pledges, by which she is obligated; he has validated them, because he said nothing to her at the time that he heard of them. [15] But if he nullifies them some time after he has heard of them, then he shall bear her guilt.

16 These are the statutes that the LORD commanded Moses concerning a husband and his wife, and a father and his daughter while she is still young and in her father's house.

31 The LORD spoke to Moses, saying, [2] "Avenge the Israelites on the Midianites; afterward you shall be gathered to your people." [3] So Moses said to the people, "Arm some of your number for the war, so that they may go against Midian, to execute the LORD's vengeance on Midian. [4] You shall send a thousand from each of the tribes of Israel to the war." [5] So out of the thousands of Israel, a thousand from each tribe were conscripted, twelve thousand armed for battle. [6] Moses sent them to the war, a thousand from each tribe, along with Phinehas son of Eleazar the priest,[t] with the vessels of the sanctuary and the trumpets for sounding the alarm in his hand. [7] They did battle against Midian, as the LORD had commanded Moses, and killed every male. [8] They killed the kings of Midian: Evi, Rekem, Zur, Hur, and Reba, the five kings of Midian, in addition to others who were slain by them;

r Or *to fast*   s Or *from that day to the next*
t Gk: Heb adds *to the war*

---

felt that it was rash or thoughtless. **9:** The exception is a vow made by a woman when widowed or divorced. **13:** *Deny herself,* i.e. by a vow of abstinence or discipline.
31.1–54: Holy war against Midian. The reason for this war was the corrupting influence of Midianites at Peor (25.16–18). **3:** *The LORD's vengeance,* i.e. holy war in which the enemy was placed under the sacrificial ban (Ex 17.8–16). **6:** Phinehas was sent instead of Eleazar, for the high priest, more than other priests, had to avoid contact with the dead (Lev 21.10–15). Strangely, no mention is made of the ark going into battle. **8:** In contrast to early tradition that portrayed Balaam as an advocate for Israel (chs 22–24), this priestly tradition holds him responsible for the apostasy at Peor (v. 16). **7:** Despite this annihilation, the Midianites were a formidable foe in a later period (Judg ch 6).

and they also killed Balaam son of Beor with the sword. 9 The Israelites took the women of Midian and their little ones captive; and they took all their cattle, their flocks, and all their goods as booty. 10 All their towns where they had settled, and all their encampments, they burned, 11 but they took all the spoil and all the booty, both people and animals. 12 Then they brought the captives and the booty and the spoil to Moses, to Eleazar the priest, and to the congregation of the Israelites, at the camp on the plains of Moab by the Jordan at Jericho.

13 Moses, Eleazar the priest, and all the leaders of the congregation went to meet them outside the camp. 14 Moses became angry with the officers of the army, the commanders of thousands and the commanders of hundreds, who had come from service in the war. 15 Moses said to them, "Have you allowed all the women to live? 16 These women here, on Balaam's advice, made the Israelites act treacherously against the LORD in the affair of Peor, so that the plague came among the congregation of the LORD. 17 Now therefore, kill every male among the little ones, and kill every woman who has known a man by sleeping with him. 18 But all the young girls who have not known a man by sleeping with him, keep alive for yourselves. 19 Camp outside the camp seven days; whoever of you has killed any person or touched a corpse, purify yourselves and your captives on the third and on the seventh day. 20 You shall purify every garment, every article of skin, everything made of goats' hair, and every article of wood."

21 Eleazar the priest said to the troops who had gone to battle: "This is the statute of the law that the LORD has commanded Moses: 22 gold, silver, bronze, iron, tin, and lead— 23 everything that can withstand fire, shall be passed through fire, and it shall be clean. Nevertheless it shall also be purified with the water for purification; and whatever cannot withstand fire, shall be passed through the water. 24 You must wash your clothes on the seventh day, and you shall be clean; afterward you may come into the camp."

25 The LORD spoke to Moses, saying, 26 "You and Eleazar the priest and the heads of the ancestral houses of the congregation make an inventory of the booty captured, both human and animal. 27 Divide the booty into two parts, between the warriors who went out to battle and all the congregation. 28 From the share of the warriors who went out to battle, set aside as tribute for the LORD, one item out of every five hundred, whether persons, oxen, donkeys, sheep, or goats. 29 Take it from their half and give it to Eleazar the priest as an offering to the LORD. 30 But from the Israelites' half you shall take one out of every fifty, whether persons, oxen, donkeys, sheep, or goats—all the animals—and give them to the Levites who have charge of the tabernacle of the LORD."

31 Then Moses and Eleazar the priest did as the LORD had commanded Moses:

32 The booty remaining from the spoil that the troops had taken totaled six hundred seventy-five thousand sheep, 33 seventy-two thousand oxen, 34 sixty-one thousand donkeys, 35 and thirty-two thousand persons in all, women who had not known a man by sleeping with him.

36 The half-share, the portion of those who had gone out to war, was in number three hundred thirty-seven

---

31.9–12: According to the ideology of holy war, the enemy was offered as a sacrifice to the LORD (1 Sam ch 15).

31.19–24: A major concern of the story is the removal of uncleanness through contact with the dead. See ch 19 for the procedure for ceremonial purification. **20:** These articles could absorb uncleanness (Lev 11.24–38).

22–23: Non-absorbent articles had to be subjected to fire as well as to *the water for purification* (19.9).

31.25–30: The law for the distribution of booty. **27:** 1 Sam 30.24–25. **28–30:** The warriors were to contribute one-fifth of one percent from their half of the booty to the high priest; this offering was regarded as *the tribute*

thousand five hundred sheep and goats, 37 and the LORD's tribute of sheep and goats was six hundred seventy-five. 38 The oxen were thirty-six thousand, of which the LORD's tribute was seventy-two. 39 The donkeys were thirty thousand five hundred, of which the LORD's tribute was sixty-one. 40 The persons were sixteen thousand, of which the LORD's tribute was thirty-two persons. 41 Moses gave the tribute, the offering for the LORD, to Eleazar the priest, as the LORD had commanded Moses.

42 As for the Israelites' half, which Moses separated from that of the troops, 43 the congregation's half was three hundred thirty-seven thousand five hundred sheep and goats, 44 thirty-six thousand oxen, 45 thirty thousand five hundred donkeys, 46 and sixteen thousand persons. 47 From the Israelites' half Moses took one of every fifty, both of persons and of animals, and gave them to the Levites who had charge of the tabernacle of the LORD; as the LORD had commanded Moses.

48 Then the officers who were over the thousands of the army, the commanders of thousands and the commanders of hundreds, approached Moses, 49 and said to Moses, "Your servants have counted the warriors who are under our command, and not one of us is missing. 50 And we have brought the LORD's offering, what each of us found, articles of gold, armlets and bracelets, signet rings, earrings, and pendants, to make atonement for ourselves before the LORD." 51 Moses and Eleazar the priest received the gold from them, all in the form of crafted articles. 52 And all the gold of the offering that they offered to the LORD, from the commanders of thousands and the commanders of hundreds, was sixteen thousand seven hundred fifty shekels. 53 (The troops had all taken plunder for themselves.) 54 So Moses and Eleazar the priest received the gold from the commanders of thousands and of hundreds, and brought it into the tent of meeting as a memorial for the Israelites before the LORD.

32 Now the Reubenites and the Gadites owned a very great number of cattle. When they saw that the land of Jazer and the land of Gilead was a good place for cattle, 2 the Gadites and the Reubenites came and spoke to Moses, to Eleazar the priest, and to the leaders of the congregation, saying, 3 "Ataroth, Dibon, Jazer, Nimrah, Heshbon, Elealeh, Sebam, Nebo, and Beon— 4 the land that the LORD subdued before the congregation of Israel—is a land for cattle; and your servants have cattle." 5 They continued, "If we have found favor in your sight, let this land be given to your servants for a possession; do not make us cross the Jordan."

6 But Moses said to the Gadites and to the Reubenites, "Shall your brothers go to war while you sit here? 7 Why will you discourage the hearts of the Israelites from going over into the land that the LORD has given them? 8 Your fathers did this, when I sent them from Kadesh-barnea to see the land. 9 When they went up to the Wadi Eshcol and saw the land, they discouraged the hearts of the Israelites from going into the land that the LORD had given them. 10 The LORD's anger was kindled on that day and he swore, saying, 11 'Surely none of the people who came up out of Egypt, from twenty years old and upward, shall see the land that I swore to give to Abraham, to Isaac, and to Jacob, because they have not unreservedly followed me— 12 none except Caleb son of Jephunneh the Kenizzite and Joshua son of Nun, for they

---

for the LORD (vv. 32–41). The people were to give two percent from their half to the Levites (vv. 42–47).
31.50: *Make atonement* (see Lev 1.4 n.), i.e. on account of ritual defilement by the dead.
32.1–42: The allotment of land in **Transjordan** to Reuben, Gad, and Manasseh, on the condition that they help the other tribes in the battle for Canaan (Deut 3.12–22; Josh 13.8–32). **1:** *Jazer*, on the boundary of Ammon (21.24). **6:** Compare Judg 5.16–17. **32.8–13:** A summary of chs 13–14.

have unreservedly followed the LORD.'
13 And the LORD's anger was kindled
against Israel, and he made them wander
in the wilderness for forty years, until all
the generation that had done evil in the
sight of the LORD had disappeared.
14 And now you, a brood of sinners, have
risen in place of your fathers, to increase
the LORD's fierce anger against Israel!
15 If you turn away from following him,
he will again abandon them in the wilderness; and you will destroy all this
people."

16 Then they came up to him and
said, "We will build sheepfolds here for
our flocks, and towns for our little ones,
17 but we will take up arms as a van-
guard^u before the Israelites, until we
have brought them to their place. Meanwhile our little ones will stay in the fortified towns because of the inhabitants of
the land. 18 We will not return to our
homes until all the Israelites have obtained their inheritance. 19 We will not
inherit with them on the other side of the
Jordan and beyond, because our inheritance has come to us on this side of the
Jordan to the east."

20 So Moses said to them, "If you do
this—if you take up arms to go before
the LORD for the war, 21 and all those of
you who bear arms cross the Jordan before the LORD, until he has driven out his
enemies from before him 22 and the land
is subdued before the LORD—then after
that you may return and be free of obligation to the LORD and to Israel, and this
land shall be your possession before the
LORD. 23 But if you do not do this, you
have sinned against the LORD; and be sure
your sin will find you out. 24 Build towns
for your little ones, and folds for your
flocks; but do what you have promised."

25 Then the Gadites and the Reubenites said to Moses, "Your servants will
do as my lord commands. 26 Our little
ones, our wives, our flocks, and all our
livestock shall remain there in the towns
of Gilead; 27 but your servants will cross
over, everyone armed for war, to do battle for the LORD, just as my lord orders."

28 So Moses gave command concerning them to Eleazar the priest, to Joshua
son of Nun, and to the heads of the ancestral houses of the Israelite tribes.
29 And Moses said to them, "If the Gadites and the Reubenites, everyone armed
for battle before the LORD, will cross
over the Jordan with you and the land
shall be subdued before you, then you
shall give them the land of Gilead for a
possession; 30 but if they will not cross
over with you armed, they shall have
possessions among you in the land of
Canaan." 31 The Gadites and the Reubenites answered, "As the LORD has spoken
to your servants, so we will do. 32 We
will cross over armed before the LORD
into the land of Canaan, but the possession of our inheritance shall remain with
us on this side of^v the Jordan."

33 Moses gave to them—to the Gadites and to the Reubenites and to the half-
tribe of Manasseh son of Joseph—the
kingdom of King Sihon of the Amorites
and the kingdom of King Og of Bashan,
the land and its towns, with the territories of the surrounding towns. 34 And the
Gadites rebuilt Dibon, Ataroth, Aroer,
35 Atroth-shophan, Jazer, Jogbehah,
36 Beth-nimrah, and Beth-haran, fortified cities, and folds for sheep. 37 And the
Reubenites rebuilt Heshbon, Elealeh,
Kiriathaim, 38 Nebo, and Baal-meon
(some names being changed), and Sibmah; and they gave names to the towns
that they rebuilt. 39 The descendants of
Machir son of Manasseh went to Gilead,

u Cn: Heb hurrying   v Heb beyond

32.20–23: This is conceived as a holy war
which imposed a sacred obligation upon all
the tribal confederacy (Deut 33 21).
32.33: The half-tribe of Manasseh, i.e. the
part that located east of the Jordan as distinguished from the part that settled in the region around Shechem (Josh 17.1–3). 34–36:

Gad's territory included part of the kingdom
of Sihon (21.24), i.e. from Heshbon northward into Gilead (Josh 13.24–28). 37–38:
Reuben's territory extended from Heshbon
south to the Arnon, the frontier of Moab
(Josh 13.15–23). 39–42: Machir (26.29; see
Gen 50.23 n.) received the territory of King

captured it, and dispossessed the Amorites who were there; ⁴⁰so Moses gave Gilead to Machir son of Manasseh, and he settled there. ⁴¹Jair son of Manasseh went and captured their villages, and renamed them Havvoth-jair. *w* ⁴²And Nobah went and captured Kenath and its villages, and renamed it Nobah after himself.

33 These are the stages by which the Israelites went out of the land of Egypt in military formation under the leadership of Moses and Aaron. ²Moses wrote down their starting points, stage by stage, by command of the LORD; and these are their stages according to their starting places. ³They set out from Rameses in the first month, on the fifteenth day of the first month; on the day after the passover the Israelites went out boldly in the sight of all the Egyptians, ⁴while the Egyptians were burying all their firstborn, whom the LORD had struck down among them. The LORD executed judgments even against their gods.

5 So the Israelites set out from Rameses, and camped at Succoth. ⁶They set out from Succoth, and camped at Etham, which is on the edge of the wilderness. ⁷They set out from Etham, and turned back to Pi-hahiroth, which faces Baal-zephon; and they camped before Migdol. ⁸They set out from Pi-hahiroth, passed through the sea into the wilderness, went a three days' journey in the wilderness of Etham, and camped at Marah. ⁹They set out from Marah and came to Elim; at Elim there were twelve springs of water and seventy palm trees, and they camped there. ¹⁰They set out from Elim and camped by the Red Sea. *x* ¹¹They set out from the Red Sea*x* and camped in the wilderness of Sin. ¹²They set out from the wilderness of Sin and camped at Dophkah. ¹³They set out from Dophkah and camped at Alush. ¹⁴They set out

from Alush and camped at Rephidim, where there was no water for the people to drink. ¹⁵They set out from Rephidim and camped in the wilderness of Sinai. ¹⁶They set out from the wilderness of Sinai and camped at Kibroth-hattaavah. ¹⁷They set out from Kibroth-hattaavah and camped at Hazeroth. ¹⁸They set out from Hazeroth and camped at Rithmah. ¹⁹They set out from Rithmah and camped at Rimmon-perez. ²⁰They set out from Rimmon-perez and camped at Libnah. ²¹They set out from Libnah and camped at Rissah. ²²They set out from Rissah and camped at Kehelathah. ²³They set out from Kehelathah and camped at Mount Shepher. ²⁴They set out from Mount Shepher and camped at Haradah. ²⁵They set out from Haradah and camped at Makheloth. ²⁶They set out from Makheloth and camped at Tahath. ²⁷They set out from Tahath and camped at Terah. ²⁸They set out from Terah and camped at Mithkah. ²⁹They set out from Mithkah and camped at Hashmonah. ³⁰They set out from Hashmonah and camped at Moseroth. ³¹They set out from Moseroth and camped at Bene-jaakan. ³²They set out from Bene-jaakan and camped at Hor-haggidgad. ³³They set out from Hor-haggidgad and camped at Jotbathah. ³⁴They set out from Jotbathah and camped at Abronah. ³⁵They set out from Abronah and camped at Ezion-geber. ³⁶They set out from Ezion-geber and camped in the wilderness of Zin (that is, Kadesh). ³⁷They set out from Kadesh and camped at Mount Hor, on the edge of the land of Edom.

38 Aaron the priest went up Mount Hor at the command of the LORD and died there in the fortieth year after the Israelites had come out of the land of

w That is *the villages of Jair*   *x* Or *Sea of Reeds*

Og (21.33), which included the rest of Gilead and all of Bashan (Josh 13.29–31). **41:** *Havvoth-jair,* 1 Kings 4.13.

33.1–49: Israel's itinerary from Egypt to the border of Canaan.

33.5–15: From the land of Goshen to Sinai

(Ex 12.37–19.2). **8:** *Etham,* otherwise designated as Shur (Ex 15.22).

33.16–36: From Sinai to Kadesh-barnea (Num 10.11–20.1).

33.37–49: From Kadesh to the plains of Moab (20.22–22.1). The details of this phase

Egypt, on the first day of the fifth month. 39 Aaron was one hundred twenty-three years old when he died on Mount Hor.

40 The Canaanite, the king of Arad, who lived in the Negeb in the land of Canaan, heard of the coming of the Israelites.

41 They set out from Mount Hor and camped at Zalmonah. 42 They set out from Zalmonah and camped at Punon. 43 They set out from Punon and camped at Oboth. 44 They set out from Oboth and camped at Iye-abarim, in the territory of Moab. 45 They set out from Iyim and camped at Dibon-gad. 46 They set out from Dibon-gad and camped at Almon-diblathaim. 47 They set out from Almon-diblathaim and camped in the mountains of Abarim, before Nebo. 48 They set out from the mountains of Abarim and camped in the plains of Moab by the Jordan at Jericho; 49 they camped by the Jordan from Beth-jeshimoth as far as Abel-shittim in the plains of Moab.

50 In the plains of Moab by the Jordan at Jericho, the LORD spoke to Moses, saying: 51 Speak to the Israelites, and say to them: When you cross over the Jordan into the land of Canaan, 52 you shall drive out all the inhabitants of the land from before you, destroy all their figured stones, destroy all their cast images, and demolish all their high places. 53 You shall take possession of the land and settle in it, for I have given you the land to possess. 54 You shall apportion the land by lot according to your clans; to a large one you shall give a large inheritance, and to a small one you shall give a small inheritance; the inheritance shall belong to the person on whom the lot falls; ac-

cording to your ancestral tribes you shall inherit. 55 But if you do not drive out the inhabitants of the land from before you, then those whom you let remain shall be as barbs in your eyes and thorns in your sides; they shall trouble you in the land where you are settling. 56 And I will do to you as I thought to do to them.

34 The LORD spoke to Moses, saying: 2 Command the Israelites, and say to them: When you enter the land of Canaan (this is the land that shall fall to you for an inheritance, the land of Canaan, defined by its boundaries), 3 your south sector shall extend from the wilderness of Zin along the side of Edom. Your southern boundary shall begin from the end of the Dead Sea y on the east; 4 your boundary shall turn south of the ascent of Akrabbim, and cross to Zin, and its outer limit shall be south of Kadesh-barnea; then it shall go on to Hazar-addar, and cross to Azmon; 5 the boundary shall turn from Azmon to the Wadi of Egypt, and its termination shall be at the Sea.

6 For the western boundary, you shall have the Great Sea and its z coast; this shall be your western boundary.

7 This shall be your northern boundary: from the Great Sea you shall mark out your line to Mount Hor; 8 from Mount Hor you shall mark it out to Lebo-hamath, and the outer limit of the boundary shall be at Zedad; 9 then the boundary shall extend to Ziphron, and its end shall be at Hazar-enan; this shall be your northern boundary.

10 You shall mark out your eastern boundary from Hazar-enan to Shepham; 11 and the boundary shall continue down

y Heb *Salt Sea*   *z* Syr: Heb lacks *its*

---

of the itinerary, including the journey to Ezion-geber (v. 35), are not clear.

33.50–56: Warnings against the Canaanite cult. See Ex 23.23–33; Lev ch 26.

34.1–29: Ideal boundaries of the Promised Land (compare Josh chs 13–19; Ezek 47.13–20). **1–12:** The extent of the land, south to north, is essentially from *the Wadi of Egypt* (v. 5) to *Lebo-hamath* (v. 8), near Riblah. Isra-

el's territory did not extend so far north until the time of David (2 Sam 8.3–14; 1 Kings 8.65). Probably the tradition has visualized the past in the light of David, the "star" who arose out of Jacob (see 24.17–19 n.) to conquer the last of the opposing forces, thereby establishing in Israel's possession *the land of Canaan, defined by its boundaries* (v. 2).

from Shepham to Riblah on the east side of Ain; and the boundary shall go down, and reach the eastern slope of the sea of Chinnereth; 12 and the boundary shall go down to the Jordan, and its end shall be at the Dead Sea. *a* This shall be your land with its boundaries all around.

13 Moses commanded the Israelites, saying: This is the land that you shall inherit by lot, which the LORD has commanded to give to the nine tribes and to the half-tribe; 14 for the tribe of the Reubenites by their ancestral houses and the tribe of the Gadites by their ancestral houses have taken their inheritance, and also the half-tribe of Manasseh; 15 the two tribes and the half-tribe have taken their inheritance beyond the Jordan at Jericho eastward, toward the sunrise.

16 The LORD spoke to Moses, saying: 17 These are the names of the men who shall apportion the land to you for inheritance: the priest Eleazar and Joshua son of Nun. 18 You shall take one leader of every tribe to apportion the land for inheritance. 19 These are the names of the men: Of the tribe of Judah, Caleb son of Jephunneh. 20 Of the tribe of the Simeonites, Shemuel son of Ammihud. 21 Of the tribe of Benjamin, Elidad son of Chislon. 22 Of the tribe of the Danites a leader, Bukki son of Jogli. 23 Of the Josephites: of the tribe of the Manassites a leader, Hanniel son of Ephod, 24 and of the tribe of the Ephraimites a leader, Kemuel son of Shiphtan. 25 Of the tribe of the Zebulunites a leader, Eli-zaphan son of Parnach. 26 Of the tribe of the Issacharites a leader, Paltiel son of Azzan. 27 And of the tribe of the Asherites a leader, Ahihud son of Shelomi. 28 Of the tribe of the Naphtalites a leader, Pedahel son of Ammihud. 29 These were the ones whom the LORD commanded to apportion the inheritance for the Israelites in the land of Canaan.

35 In the plains of Moab by the Jordan at Jericho, the LORD spoke to Moses, saying: 2 Command the Israelites to give, from the inheritance that they possess, towns for the Levites to live in; you shall also give to the Levites pasture lands surrounding the towns. 3 The towns shall be theirs to live in, and their pasture lands shall be for their cattle, for their livestock, and for all their animals. 4 The pasture lands of the towns, which you shall give to the Levites, shall reach from the wall of the town outward a thousand cubits all around. 5 You shall measure, outside the town, for the east side two thousand cubits, for the south side two thousand cubits, for the west side two thousand cubits, and for the north side two thousand cubits, with the town in the middle; this shall belong to them as pasture land for their towns.

6 The towns that you give to the Levites shall include the six cities of refuge, where you shall permit a slayer to flee, and in addition to them you shall give forty-two towns. 7 The towns that you give to the Levites shall total forty-eight, with their pasture lands. 8 And as for the towns that you shall give from the possession of the Israelites, from the larger tribes you shall take many, and from the smaller tribes you shall take few; each, in proportion to the inheritance that it obtains, shall give of its towns to the Levites.

9 The LORD spoke to Moses, saying: 10 Speak to the Israelites, and say to them: When you cross the Jordan into the land

a Heb *Salt Sea*

34.14–15: Jericho is said to be east, presupposing a standpoint within Canaan rather than in Transjordan.

35.1–34: Plans for Levitical cities and cities of refuge. 1–8: Special cities were allotted because the Levites were not entitled to a tribal inheritance (Lev 25.32–34; Josh ch 21; 1 Chr 6.54–81).

35.9–15: The six cities of refuge (Deut 4.41–43; 19.1–13; Josh ch 20) represent an attempt to restrain the tribal law of blood revenge so that a killer might receive a trial (v. 12). Early legislation stipulated that a person might seek asylum from the *avenger* or next of kin (see Lev 25.25 n.), whose duty was to uphold family rights by killing the murderer or a relative (see Ex 21.12–14 n.).

35.16–34: This case law distinguishes be-

of Canaan, [11]then you shall select cities to be cities of refuge for you, so that a slayer who kills a person without intent may flee there. [12]The cities shall be for you a refuge from the avenger, so that the slayer may not die until there is a trial before the congregation.

13 The cities that you designate shall be six cities of refuge for you: [14]you shall designate three cities beyond the Jordan, and three cities in the land of Canaan, to be cities of refuge. [15]These six cities shall serve as refuge for the Israelites, for the resident or transient alien among them, so that anyone who kills a person without intent may flee there.

16 But anyone who strikes another with an iron object, and death ensues, is a murderer; the murderer shall be put to death. [17]Or anyone who strikes another with a stone in hand that could cause death, and death ensues, is a murderer; the murderer shall be put to death. [18]Or anyone who strikes another with a weapon of wood in hand that could cause death, and death ensues, is a murderer; the murderer shall be put to death. [19]The avenger of blood is the one who shall put the murderer to death; when they meet, the avenger of blood shall execute the sentence. [20]Likewise, if someone pushes another from hatred, or hurls something at another, lying in wait, and death ensues, [21]or in enmity strikes another with the hand, and death ensues, then the one who struck the blow shall be put to death; that person is a murderer; the avenger of blood shall put the murderer to death, when they meet.

22 But if someone pushes another suddenly without enmity, or hurls any object without lying in wait, [23]or, while handling any stone that could cause death, unintentionally[b] drops it on another and death ensues, though they were not enemies, and no harm was in-tended, [24]then the congregation shall judge between the slayer and the avenger of blood, in accordance with these ordinances; [25]and the congregation shall rescue the slayer from the avenger of blood. Then the congregation shall send the slayer back to the original city of refuge. The slayer shall live in it until the death of the high priest who was anointed with the holy oil. [26]But if the slayer shall at any time go outside the bounds of the original city of refuge, [27]and is found by the avenger of blood outside the bounds of the city of refuge, and is killed by the avenger, no bloodguilt shall be incurred. [28]For the slayer must remain in the city of refuge until the death of the high priest; but after the death of the high priest the slayer may return home.

29 These things shall be a statute and ordinance for you throughout your generations wherever you live.

30 If anyone kills another, the murderer shall be put to death on the evidence of witnesses; but no one shall be put to death on the testimony of a single witness. [31]Moreover you shall accept no ransom for the life of a murderer who is subject to the death penalty; a murderer must be put to death. [32]Nor shall you accept ransom for one who has fled to a city of refuge, enabling the fugitive to return to live in the land before the death of the high priest. [33]You shall not pollute the land in which you live; for blood pollutes the land, and no expiation can be made for the land, for the blood that is shed in it, except by the blood of the one who shed it. [34]You shall not defile the land in which you live, in which I also dwell; for I the LORD dwell among the Israelites.

36

The heads of the ancestral houses of the clans of the descendants of

b Heb *without seeing*

---

tween premeditated and unpremeditated murder (manslaughter). As a concession to ancient blood revenge, the tribunal lets the avenger exercise the duty of the next of kin if the killing was premeditated (vv. 19, 21).

35.33–34: Murder is a heinous offense because *blood pollutes the land* (Gen 4.10–11) in the midst of which the LORD dwells or "tabernacles." Therefore only the blood of the murderer can expiate the crime (Deut 19.10, 13).

Gilead son of Machir son of Manasseh, of the Josephite clans, came forward and spoke in the presence of Moses and the leaders, the heads of the ancestral houses of the Israelites; ²they said, "The LORD commanded my lord to give the land for inheritance by lot to the Israelites; and my lord was commanded by the LORD to give the inheritance of our brother Zelophehad to his daughters. ³But if they are married into another Israelite tribe, then their inheritance will be taken from the inheritance of our ancestors and added to the inheritance of the tribe into which they marry; so it will be taken away from the allotted portion of our inheritance. ⁴And when the jubilee of the Israelites comes, then their inheritance will be added to the inheritance of the tribe into which they have married; and their inheritance will be taken from the inheritance of our ancestral tribe."

5 Then Moses commanded the Israelites according to the word of the LORD, saying, "The descendants of the tribe of Joseph are right in what they are saying. ⁶This is what the LORD commands concerning the daughters of Zelophehad, 'Let them marry whom they think best; only it must be into a clan of their father's tribe that they are married, ⁷so that no inheritance of the Israelites shall be transferred from one tribe to another; for all Israelites shall retain the inheritance of their ancestral tribes. ⁸Every daughter who possesses an inheritance in any tribe of the Israelites shall marry one from the clan of her father's tribe, so that all Israelites may continue to possess their ancestral inheritance. ⁹No inheritance shall be transferred from one tribe to another; for each of the tribes of the Israelites shall retain its own inheritance.'"

10 The daughters of Zelophehad did as the LORD had commanded Moses. ¹¹Mahlah, Tirzah, Hoglah, Milcah, and Noah, the daughters of Zelophehad, married sons of their father's brothers. ¹²They were married into the clans of the descendants of Manasseh son of Joseph, and their inheritance remained in the tribe of their father's clan.

13 These are the commandments and the ordinances that the LORD commanded through Moses to the Israelites in the plains of Moab by the Jordan at Jericho.

36.1–12: Tribal property must be maintained intact. This law supplements 27.1–11 by stipulating that a woman who is allowed to inherit property must marry within her tribe. **1:** 26.28–34. **4:** On the jubilee year, see Lev 25.8–55. **6:** The law prevents tribal intermarriage only when the woman is an heiress.

36.13: A concluding statement covering all the laws given in Moab (22.1–36.12).

Deuteronomy

The basic theme of Deuteronomy, meaning the "second law," is the reaffirmation of the covenant between God and the people of Israel. Here the legal tradition of the book of Exodus (for example, the Decalogue or the Covenant Code) is not just repeated; it is reinterpreted in contemporary terms, so that the promises and demands of the covenant were brought near to every worshiping Israelite.

At the end of the book of Numbers Israel is encamped in the plains of Moab, prepared for an attack upon Canaan from the east. Deuteronomy purports to be Moses' farewell address to the people in which he rehearses the mighty acts of the LORD, solemnly warns of the temptations of the new ways of Canaan, and pleads for loyalty to and love of God as the condition for life in the promised land.

Actually Deuteronomy contains not one address by Moses, but three. The first is found in 1.6 to 4.40; the second in chs 5–28; and the third in chs 29 and 30. The remaining chapters (31–34) pick up the story where it was left at the end of Numbers (see introduction to "The Pentateuch").

A distinctive teaching of Deuteronomy is that the worship of the LORD is to be centralized in one place, so that the syncretism characteristic of local shrines may be eliminated (ch 12). When Deuteronomy was published, the Jerusalem temple was regarded as the central sanctuary. Indeed, Deuteronomy was probably the "book of the law" (2 Kings 22.11) that prompted Josiah's sweeping religious reform in 621 B.C. (2 Kings chs 22–23) and led to the revision of the history found in Joshua, Judges, Samuel, and Kings, often called the Deuteronomistic History. Although Deuteronomy rests upon ancient tradition, fundamentally it is a rediscovery and reinterpretation of Mosaic teaching in the light of later historical experience.

1 These are the words that Moses spoke to all Israel beyond the Jordan—in the wilderness, on the plain opposite Suph, between Paran and Tophel, Laban, Hazeroth, and Di-zahab. ²(By the way of Mount Seir it takes eleven days to reach Kadesh-barnea from Horeb.) ³In the fortieth year, on the first day of the eleventh month, Moses spoke to the Israelites just as the LORD had commanded him to speak to them. ⁴This was after he had defeated King Sihon of the Amorites, who reigned in Heshbon, and King Og of Bashan, who reigned in Ashtaroth and*ᵃ* in Edrei. ⁵Beyond the Jordan in the land of Moab, Moses undertook to expound this law as follows:

6 The LORD our God spoke to us at Horeb, saying, "You have stayed long enough at this mountain. ⁷Resume your journey, and go into the hill country of the Amorites as well as into the neighboring regions—the Arabah, the hill country, the Shephelah, the Negeb, and the seacoast—the land of the Canaanites and the Lebanon, as far as the great river, the river Euphrates. ⁸See, I have set the land before you; go in and take possession of the land that I*ᵇ* swore to your ancestors, to Abraham, to Isaac, and to Jacob, to give to them and to their descendants after them."

9 At that time I said to you, "I am unable by myself to bear you. ¹⁰The LORD your God has multiplied you, so that today you are as numerous as the stars of heaven. ¹¹May the LORD, the God of your ancestors, increase you a thousand times more and bless you, as he has promised you! ¹²But how can I bear the heavy burden of your disputes all by myself? ¹³Choose for each of your tribes individuals who are wise, discerning, and reputable to be your leaders." ¹⁴You answered me, "The plan you have proposed is a good one." ¹⁵So I took the leaders of your tribes, wise and reputable individuals, and installed them as leaders over you, commanders of thousands, commanders of hundreds, commanders of fifties, commanders of tens, and officials, throughout your tribes. ¹⁶I charged your judges at that time: "Give the members of your community a fair hearing, and judge rightly between one person and another, whether citizen or resident alien. ¹⁷You must not be partial in judging: hear out the small and the great alike; you shall not be intimidated by anyone, for the judgment is God's. Any case that is too hard for you, bring to me, and I will hear it." ¹⁸So I charged you at that time with all the things that you should do.

19 Then, just as the LORD our God had ordered us, we set out from Horeb and went through all that great and terrible wilderness that you saw, on the way to the hill country of the Amorites, until we reached Kadesh-barnea. ²⁰I said to you, "You have reached the hill country of the Amorites, which the LORD our God is giving us. ²¹See, the LORD your God has given the land to you; go up, take possession, as the LORD, the God of

a Gk Syr Vg Compare Josh 12.4: Heb lacks *and*
b Sam Gk: MT *the* LORD

1.1–3.29: Historical review. Moses rehearses events since the departure from Sinai (Horeb) to show how the LORD marvelously guided the people of Israel in the wilderness. **1.1–5:** Introduction to the first address (1.6–4.40), which was given *beyond the Jordan,* i.e. in the plains of Moab (Num 33.48; 36.13). **1–2:** The places mentioned refer to the wilderness journey. **3:** *Fortieth year,* counting from the Exodus (Ex 19.1; 40.17). **4:** Num 21.21–35. **7:** *Amorites, Canaanites,* see Num 13.29 n. The Israelite empire extended ideally to the Euphrates (Gen 15.18), the northern limit of David's conquests (2 Sam 8.3).

1.9–18: In this composite account (compare vv. 9–12 with Num 11.14–17 and vv. 13–17 with Ex 18.13–27) there is no reference to Jethro's initiative to lighten Moses' burden. **11:** *The God of your ancestors,* see Gen 26.24 n. **1.19–46:** From Horeb to Kadesh: a summary of the spies' reconnaissance of the land (Num ch 13), the people's complaining about the LORD's inability to fulfill the promise to Israel's ancestors (Num 14.1–38), and the abortive attempt to penetrate Canaan from the south (Num 14.39–45; compare 21.1–3).

your ancestors, has promised you; do not fear or be dismayed."

22 All of you came to me and said, "Let us send men ahead of us to explore the land for us and bring back a report to us regarding the route by which we should go up and the cities we will come to." 23 The plan seemed good to me, and I selected twelve of you, one from each tribe. 24 They set out and went up into the hill country, and when they reached the Valley of Eshcol they spied it out 25 and gathered some of the land's produce, which they brought down to us. They brought back a report to us, and said, "It is a good land that the LORD our God is giving us."

26 But you were unwilling to go up. You rebelled against the command of the LORD your God; 27 you grumbled in your tents and said, "It is because the LORD hates us that he has brought us out of the land of Egypt, to hand us over to the Amorites to destroy us. 28 Where are we headed? Our kindred have made our hearts melt by reporting, 'The people are stronger and taller than we; the cities are large and fortified up to heaven! We actually saw there the offspring of the Anakim!' " 29 I said to you, "Have no dread or fear of them. 30 The LORD your God, who goes before you, is the one who will fight for you, just as he did for you in Egypt before your very eyes, 31 and in the wilderness, where you saw how the LORD your God carried you, just as one carries a child, all the way that you traveled until you reached this place. 32 But in spite of this, you have no trust in the LORD your God, 33 who goes before you on the way to seek out a place for you to camp, in fire by night, and in the cloud by day, to show you the route you should take."

34 When the LORD heard your words, he was wrathful and swore: 35 "Not one of these—not one of this evil

generation—shall see the good land that I swore to give to your ancestors, 36 except Caleb son of Jephunneh. He shall see it, and to him and to his descendants I will give the land on which he set foot, because of his complete fidelity to the LORD." 37 Even with me the LORD was angry on your account, saying, "You also shall not enter there. 38 Joshua son of Nun, your assistant, shall enter there; encourage him, for he is the one who will secure Israel's possession of it. 39 And as for your little ones, who you thought would become booty, your children, who today do not yet know right from wrong, they shall enter there; to them I will give it, and they shall take possession of it. 40 But as for you, journey back into the wilderness, in the direction of the Red Sea."*c*

41 You answered me, "We have sinned against the LORD! We are ready to go up and fight, just as the LORD our God commanded us." So all of you strapped on your battle gear, and thought it easy to go up into the hill country. 42 The LORD said to me, "Say to them, 'Do not go up and do not fight, for I am not in the midst of you; otherwise you will be defeated by your enemies.' " 43 Although I told you, you would not listen. You rebelled against the command of the LORD and presumptuously went up into the hill country. 44 The Amorites who lived in that hill country then came out against you and chased you as bees do. They beat you down in Seir as far as Hormah. 45 When you returned and wept before the LORD, the LORD would neither heed your voice nor pay you any attention.

46 After you had stayed at Kadesh as many days as you did, 1 we journeyed back into the wilderness, in the direction of the Red Sea,*c* as the

c Or *Sea of Reeds*

1.28: *Anakim,* see Num 13.22, 33 n. **30**: Ex 14.14. **33**: *Fire . . . cloud,* see Ex 13.21–22 n. **37**: Here Moses is not punished for his own sin (Num 20.10–13; 27.12–23), but vicariously bears the divine wrath on Israel's account.

2.1–25: The circuit via Transjordan. **1–8a**: Num 20.14–21. From Kadesh Israel turned

LORD had told me and skirted Mount Seir for many days. [2] Then the LORD said to me: [3] "You have been skirting this hill country long enough. Head north, [4] and charge the people as follows: You are about to pass through the territory of your kindred, the descendants of Esau, who live in Seir. They will be afraid of you, so, be very careful [5] not to engage in battle with them, for I will not give you even so much as a foot's length of their land, since I have given Mount Seir to Esau as a possession. [6] You shall purchase food from them for money, so that you may eat; and you shall also buy water from them for money, so that you may drink. [7] Surely the LORD your God has blessed you in all your undertakings; he knows your going through this great wilderness. These forty years the LORD your God has been with you; you have lacked nothing." [8] So we passed by our kin, the descendants of Esau who live in Seir, leaving behind the route of the Arabah, and leaving behind Elath and Eziongeber.

When we had headed out along the route of the wilderness of Moab, [9] the LORD said to me: "Do not harass Moab or engage them in battle, for I will not give you any of its land as a possession, since I have given Ar as a possession to the descendants of Lot." [10] (The Emim— a large and numerous people, as tall as the Anakim—had formerly inhabited it. [11] Like the Anakim, they are usually reckoned as Rephaim, though the Moabites call them Emim. [12] Moreover, the Horim had formerly inhabited Seir, but the descendants of Esau dispossessed them, destroying them and settling in their place, as Israel has done in the land that the LORD gave them as a possession.) [13] "Now then, proceed to cross over the Wadi Zered."

So we crossed over the Wadi Zered.

[14] And the length of time we had traveled from Kadesh-barnea until we crossed the Wadi Zered was thirty-eight years, until the entire generation of warriors had perished from the camp, as the LORD had sworn concerning them. [15] Indeed, the LORD's own hand was against them, to root them out from the camp, until all had perished.

[16] Just as soon as all the warriors had died off from among the people, [17] the LORD spoke to me, saying, [18] "Today you are going to cross the boundary of Moab at Ar. [19] When you approach the frontier of the Ammonites, do not harass them or engage them in battle, for I will not give the land of the Ammonites to you as a possession, because I have given it to the descendants of Lot." [20] (It also is usually reckoned as a land of Rephaim. Rephaim formerly inhabited it, though the Ammonites call them Zamzummim, [21] a strong and numerous people, as tall as the Anakim. But the LORD destroyed them from before the Ammonites so that they could dispossess them and settle in their place. [22] He did the same for the descendants of Esau, who live in Seir, by destroying the Horim before them so that they could dispossess them and settle in their place even to this day. [23] As for the Avvim, who had lived in settlements in the vicinity of Gaza, the Caphtorim, who came from Caphtor, destroyed them and settled in their place.) [24] "Proceed on your journey and cross the Wadi Arnon. See, I have handed over to you King Sihon the Amorite of Heshbon, and his land. Begin to take possession by engaging him in battle. [25] This day I will begin to put the dread and fear of you upon the peoples everywhere under heaven; when they hear report of you, they will tremble and be in anguish because of you."

[26] So I sent messengers from the wil-

---

south through the Arabah to the Gulf of Aqaba in order to go around Edom (Num 21.4; compare 33.37–49). **4:** *Descendants of Esau,* Gen 36.1.

2.8b–25: Num 21.4–20. Turning along

the brook Zered (boundary of Edom), Israel detoured via the Moabite wilderness toward the Amorite kingdom of Sihon. **9:** Moab and Ammon (v. 19) were traditionally related through Lot (Gen 19.36–38). **10–11:** *Emim,*

derness of Kedemoth to King Sihon of Heshbon with the following terms of peace: 27 "If you let me pass through your land, I will travel only along the road; I will turn aside neither to the right nor to the left. 28 You shall sell me food for money, so that I may eat, and supply me water for money, so that I may drink. Only allow me to pass through on foot— 29 just as the descendants of Esau who live in Seir have done for me and likewise the Moabites who live in Ar— until I cross the Jordan into the land that the LORD our God is giving us." 30 But King Sihon of Heshbon was not willing to let us pass through, for the LORD your God had hardened his spirit and made his heart defiant in order to hand him over to you, as he has now done.

31 The LORD said to me, "See, I have begun to give Sihon and his land over to you. Begin now to take possession of his land." 32 So when Sihon came out against us, he and all his people for battle at Jahaz, 33 the LORD our God gave him over to us; and we struck him down, along with his offspring and all his people. 34 At that time we captured all his towns, and in each town we utterly destroyed men, women, and children. We left not a single survivor. 35 Only the livestock we kept as spoil for ourselves, as well as the plunder of the towns that we had captured. 36 From Aroer on the edge of the Wadi Arnon (including the town that is in the wadi itself) as far as Gilead, there was no citadel too high for us. The LORD our God gave everything to us. 37 You did not encroach, however, on the land of the Ammonites, avoiding the whole upper region of the Wadi Jabbok as well as the towns of the hill country, just as*d* the LORD our God had charged.

3 When we headed up the road to Bashan, King Og of Bashan came out against us, he and all his people, for battle at Edrei. 2 The LORD said to me, "Do not fear him, for I have handed him over to you, along with his people and his land. Do to him as you did to King Sihon of the Amorites, who reigned in Heshbon." 3 So the LORD our God also handed over to us King Og of Bashan and all his people. We struck him down until not a single survivor was left. 4 At that time we captured all his towns; there was no citadel that we did not take from them— sixty towns, the whole region of Argob, the kingdom of Og in Bashan. 5 All these were fortress towns with high walls, double gates, and bars, besides a great many villages. 6 And we utterly destroyed them, as we had done to King Sihon of Heshbon, in each city utterly destroying men, women, and children. 7 But all the livestock and the plunder of the towns we kept as spoil for ourselves.

8 So at that time we took from the two kings of the Amorites the land beyond the Jordan, from the Wadi Arnon to Mount Hermon 9 (the Sidonians call Hermon Sirion, while the Amorites call it Senir), 10 all the towns of the tableland, the whole of Gilead, and all of Bashan, as far as Salecah and Edrei, towns of Og's kingdom in Bashan. 11 (Now only King Og of Bashan was left of the remnant of the Rephaim. In fact his bed, an iron bed, can still be seen in Rabbah of the Ammonites. By the common cubit it is nine cubits long and four cubits wide.) 12 As for the land that we took possession of at that time, I gave to the Reubenites and Gadites the territory north of Aroer, *e*

d Gk Tg: Heb *and all* *e* Heb *territory from Aroer*

Rephaim (compare v. 20; 3.11–13), names reflecting the legendary view that the aborigines were giants. *Horim,* see Gen 36.20 n. **23:** *Caphtor,* i.e. Crete. This verse refers to the conquest of the coastal plain by "sea-peoples" shortly after 1200 B.C. (see Gen 10.2–5 n.).
2.26–37: The victory over Sihon (Num 21.21–32), whose capital was at Heshbon. **30:** See Ex 4.21 n. **37:** The river Jabbok makes a

wide bend south and thus forms the western border of Ammon (3.16).
3.1–11: The victory over Bashan (Num 21.33–35). **1:** *Edrei,* on the extreme south border of *Bashan.* **11:** The oversized bed of Og, one of the legendary Rephaim (2.10–11), was a "museum piece" in Rabbah, a city on the Ammonite border.
3.12–22: The allotment of tribal territories

that is on the edge of the Wadi Arnon, as well as half the hill country of Gilead with its towns, 13 and I gave to the half-tribe of Manasseh the rest of Gilead and all of Bashan, Og's kingdom. (The whole region of Argob: all that portion of Bashan used to be called a land of Rephaim; 14 Jair the Manassite acquired the whole region of Argob as far as the border of the Geshurites and the Maacathites, and he named them—that is, Bashan—after himself, Havvoth-jair,*f* as it is to this day.) 15 To Machir I gave Gilead. 16 And to the Reubenites and the Gadites I gave the territory from Gilead as far as the Wadi Arnon, with the middle of the wadi as a boundary, and up to the Jabbok, the wadi being boundary of the Ammonites; 17 the Arabah also, with the Jordan and its banks, from Chinnereth down to the sea of the Arabah, the Dead Sea,*g* with the lower slopes of Pisgah on the east.

18 At that time, I charged you as follows: "Although the LORD your God has given you this land to occupy, all your troops shall cross over armed as the vanguard of your Israelite kin. 19 Only your wives, your children, and your livestock—I know that you have much livestock—shall stay behind in the towns that I have given to you. 20 When the LORD gives rest to your kindred, as to you, and they too have occupied the land that the LORD your God is giving them beyond the Jordan, then each of you may return to the property that I have given to you." 21 And I charged Joshua as well at that time, saying: "Your own eyes have seen everything that the LORD your

God has done to these two kings; so the LORD will do to all the kingdoms into which you are about to cross. 22 Do not fear them, for it is the LORD your God who fights for you."

23 At that time, too, I entreated the LORD, saying: 24 "O Lord GOD, you have only begun to show your servant your greatness and your might; what god in heaven or on earth can perform deeds and mighty acts like yours! 25 Let me cross over to see the good land beyond the Jordan, that good hill country and the Lebanon." 26 But the LORD was angry with me on your account and would not heed me. The LORD said to me, "Enough from you! Never speak to me of this matter again! 27 Go up to the top of Pisgah and look around you to the west, to the north, to the south, and to the east. Look well, for you shall not cross over this Jordan. 28 But charge Joshua, and encourage and strengthen him, because it is he who shall cross over at the head of this people and who shall secure their possession of the land that you will see." 29 So we remained in the valley opposite Beth-peor.

4 So now, Israel, give heed to the statutes and ordinances that I am teaching you to observe, so that you may live to enter and occupy the land that the LORD, the God of your ancestors, is giving you. 2 You must neither add anything to what I command you nor take away anything from it, but keep the commandments of the LORD your God with which I am charging you. 3 You have

f That is Settlement of Jair   g Heb Salt Sea

---

in Transjordan (Num ch 32; Josh ch 13). **14:** Num 32.41. **17:** The territory included the eastern part of the Jordan Valley or Arabah. **22:** The whole story of the Exodus, wilderness journey, and invasion of Canaan is governed by the convictions of holy war: the LORD fights for the people of Israel; faith is response to God's dynamic presence in the events (20.1–20).

3.23–29: Num 27.12–23. **24:** The incomparability of the LORD, who performs *mighty acts* of salvation, is an ancient theme of

hymnic praise (Ex 15.11; Ps 89.5–8). **26:** On Moses' vicarious suffering, see 1.37 n. **27:** Mount *Pisgah,* see 34.1 n.

4.1–40: Conclusion to the first address. The preceding recital of what the LORD had done for the people is the basis of Moses' appeal for faithful obedience. **1–8:** The incident of Peor (Num 25.1–9) teaches that obedience to God's law is the condition for life in Canaan and a testimony to the wisdom which the LORD graciously imparts.

seen for yourselves what the LORD did with regard to the Baal of Peor—how the LORD your God destroyed from among you everyone who followed the Baal of Peor, 4while those of you who held fast to the LORD your God are all alive today.

5 See, just as the LORD my God has charged me, I now teach you statutes and ordinances for you to observe in the land that you are about to enter and occupy. 6You must observe them diligently, for this will show your wisdom and discernment to the peoples, who, when they hear all these statutes, will say, "Surely this great nation is a wise and discerning people!" 7For what other great nation has a god so near to it as the LORD our God is whenever we call to him? 8And what other great nation has statutes and ordinances as just as this entire law that I am setting before you today?

9 But take care and watch yourselves closely, so as neither to forget the things that your eyes have seen nor to let them slip from your mind all the days of your life; make them known to your children and your children's children— 10how you once stood before the LORD your God at Horeb, when the LORD said to me, "Assemble the people for me, and I will let them hear my words, so that they may learn to fear me as long as they live on the earth, and may teach their children so"; 11you approached and stood at the foot of the mountain while the mountain was blazing up to the very heavens, shrouded in dark clouds. 12Then the LORD spoke to you out of the fire. You heard the sound of words but saw no form; there was only a voice. 13He declared to you his covenant, which he charged you to observe, that is, the ten commandments;*h* and he wrote them on two stone tablets. 14And the

LORD charged me at that time to teach you statutes and ordinances for you to observe in the land that you are about to cross into and occupy.

15 Since you saw no form when the LORD spoke to you at Horeb out of the fire, take care and watch yourselves closely, 16so that you do not act corruptly by making an idol for yourselves, in the form of any figure—the likeness of male or female, 17the likeness of any animal that is on the earth, the likeness of any winged bird that flies in the air, 18the likeness of anything that creeps on the ground, the likeness of any fish that is in the water under the earth. 19And when you look up to the heavens and see the sun, the moon, and the stars, all the host of heaven, do not be led astray and bow down to them and serve them, things that the LORD your God has allotted to all the peoples everywhere under heaven. 20But the LORD has taken you and brought you out of the iron-smelter, out of Egypt, to become a people of his very own possession, as you are now.

21 The LORD was angry with me because of you, and he vowed that I should not cross the Jordan and that I should not enter the good land that the LORD your God is giving for your possession. 22For I am going to die in this land without crossing over the Jordan, but you are going to cross over to take possession of that good land. 23So be careful not to forget the covenant that the LORD your God made with you, and not to make for yourselves an idol in the form of anything that the LORD your God has forbidden you. 24For the LORD your God is a devouring fire, a jealous God.

25 When you have had children and children's children, and become compla-

h Heb *the ten words*

4.9–14: The revelation at Sinai-Horeb (Ex chs 19–20) should be a constant reminder to *fear,* i.e. reverence, the LORD. **12:** At Horeb Israel heard the LORD's voice but *saw no form*—a warning against idolatry and image worship (vv. 15–18).
4.15–40: The great exhortation. **16–18:** In

pagan religions, gods were represented in both human and animal form. The imageless worship of the invisible God was a fundamental tenet of Mosaic faith (Ex 20.4). Here it is grounded in the covenant (v. 23) made by the Lord of history and creation (v. 32). **24:** *A jealous God,* see Ex 34.14 n.

cent in the land, if you act corruptly by making an idol in the form of anything, thus doing what is evil in the sight of the LORD your God, and provoking him to anger, 26 I call heaven and earth to witness against you today that you will soon utterly perish from the land that you are crossing the Jordan to occupy; you will not live long on it, but will be utterly destroyed. 27 The LORD will scatter you among the peoples; only a few of you will be left among the nations where the LORD will lead you. 28 There you will serve other gods made by human hands, objects of wood and stone that neither see, nor hear, nor eat, nor smell. 29 From there you will seek the LORD your God, and you will find him if you search after him with all your heart and soul. 30 In your distress, when all these things have happened to you in time to come, you will return to the LORD your God and heed him. 31 Because the LORD your God is a merciful God, he will neither abandon you nor destroy you; he will not forget the covenant with your ancestors that he swore to them.

32 For ask now about former ages, long before your own, ever since the day that God created human beings on the earth; ask from one end of heaven to the other: has anything so great as this ever happened or has its like ever been heard of? 33 Has any people ever heard the voice of a god speaking out of a fire, as you have heard, and lived? 34 Or has any god ever attempted to go and take a nation for himself from the midst of another nation, by trials, by signs and wonders, by war, by a mighty hand and an outstretched arm, and by terrifying displays of power, as the LORD your God did for you in Egypt before your very eyes? 35 To you it was shown so that you would acknowledge that the LORD is God; there is no other besides him. 36 From heaven he made you hear his voice to discipline you. On earth he showed you his great fire, while you heard his words coming out of the fire. 37 And because he loved your ancestors, he chose their descendants after them. He brought you out of Egypt with his own presence, by his great power, 38 driving out before you nations greater and mightier than yourselves, to bring you in, giving you their land for a possession, as it is still today. 39 So acknowledge today and take to heart that the LORD is God in heaven above and on the earth beneath; there is no other. 40 Keep his statutes and his commandments, which I am commanding you today for your own well-being and that of your descendants after you, so that you may long remain in the land that the LORD your God is giving you for all time.

41 Then Moses set apart on the east side of the Jordan three cities 42 to which a homicide could flee, someone who unintentionally kills another person, the two not having been at enmity before; the homicide could flee to one of these cities and live: 43 Bezer in the wilderness on the tableland belonging to the Reubenites, Ramoth in Gilead belonging to the Gadites, and Golan in Bashan belonging to the Manassites.

44 This is the law that Moses set before the Israelites. 45 These are the decrees and the statutes and ordinances that Mo-

4.27–29: These verses allude to the exile of conquered populations, a policy used effectively by Assyrians and Babylonians. 31: Divine wrath is a temporary reaction to specific situations; God is fundamentally and unchangeably *a merciful God* (Ex 34.6–7).
4.32–36: The events that have happened in Israel's historical experience (i.e. Exodus and Sinai) show that the LORD only is God; for this people *there is no other besides him* (see Ex 20.3 n.). 33: Ex 20.18–26; 33.20. 36: *Disci*-pline *you,* Ex 20.20. 37: A new element is added to the Pentateuchal tradition: the LORD's choice of Israel was based upon divine love (Hos 11.1). Israel's obedience, therefore, should be motivated by a responding love (6.4–5). *Presence,* Ex 33.14.
4.41–43: **An appendix** dealing with cities of refuge (see ch 19).
4.44–26.19 **(and ch 28): Moses' second address. 44–49: Introduction.**
5.1–33: **The giving of the law at Sinai.**

ses spoke to the Israelites when they had come out of Egypt, [46]beyond the Jordan in the valley opposite Beth-peor, in the land of King Sihon of the Amorites, who reigned at Heshbon, whom Moses and the Israelites defeated when they came out of Egypt. [47]They occupied his land and the land of King Og of Bashan, the two kings of the Amorites on the eastern side of the Jordan: [48]from Aroer, which is on the edge of the Wadi Arnon, as far as Mount Sirion[i] (that is, Hermon), [49]together with all the Arabah on the east side of the Jordan as far as the Sea of the Arabah, under the slopes of Pisgah.

5 Moses convened all Israel, and said to them:

Hear, O Israel, the statutes and ordinances that I am addressing to you today; you shall learn them and observe them diligently. [2]The LORD our God made a covenant with us at Horeb. [3]Not with our ancestors did the LORD make this covenant, but with us, who are all of us here alive today. [4]The LORD spoke with you face to face at the mountain, out of the fire. [5](At that time I was standing between the LORD and you to declare to you the words[j] of the LORD; for you were afraid because of the fire and did not go up the mountain.) And he said:

[6] I am the LORD your God, who brought you out of the land of Egypt, out of the house of slavery; [7]you shall have no other gods before[k] me.

[8] You shall not make for yourself an idol, whether in the form of anything that is in heaven above, or that is on the earth beneath, or that is in the water under the earth. [9]You shall not bow down to them or worship them; for I the LORD your God am a jealous God, punishing children for the iniquity of parents, to the third and fourth generation of those who

reject me, [10]but showing steadfast love to the thousandth generation[l] of those who love me and keep my commandments.

[11] You shall not make wrongful use of the name of the LORD your God, for the LORD will not acquit anyone who misuses his name.

[12] Observe the sabbath day and keep it holy, as the LORD your God commanded you. [13]Six days you shall labor and do all your work. [14]But the seventh day is a sabbath to the LORD your God; you shall not do any work—you, or your son or your daughter, or your male or female slave, or your ox or your donkey, or any of your livestock, or the resident alien in your towns, so that your male and female slave may rest as well as you. [15]Remember that you were a slave in the land of Egypt, and the LORD your God brought you out from there with a mighty hand and an outstretched arm; therefore the LORD your God commanded you to keep the sabbath day.

[16] Honor your father and your mother, as the LORD your God commanded you, so that your days may be long and that it may go well with you in the land that the LORD your God is giving you.

[17] You shall not murder.[m]

[18] Neither shall you commit adultery.

[19] Neither shall you steal.

[20] Neither shall you bear false witness against your neighbor.

[21] Neither shall you covet your neighbor's wife.

Neither shall you desire your neighbor's house, or field, or male or female slave, or ox, or donkey, or anything that belongs to your neighbor.

i Syr: Heb *Sion*　　*j* Q Mss Sam Gk Syr Vg Tg: MT *word*　　*k* Or *besides*　　*l* Or *to thousands*　　*m* Or *kill*

1: *Hear,* a frequently repeated verb, indicating that Deuteronomy is a sermon that interprets Israel's covenant responsibilities. **2–3**: The making of the covenant (Ex ch 24) was not just a past ceremony involving another generation but it is a contemporary covenant *with*

us, who are all of us here alive today (see Ex 13.8 n.). The language may reflect a liturgy in which the covenant was periodically recalled and renewed (26.16–19; 31.10–11).

5.6–21: This version of the Decalogue differs only slightly from that in Ex 20.2–17.

22 These words the LORD spoke with a loud voice to your whole assembly at the mountain, out of the fire, the cloud, and the thick darkness, and he added no more. He wrote them on two stone tablets, and gave them to me. 23 When you heard the voice out of the darkness, while the mountain was burning with fire, you approached me, all the heads of your tribes and your elders; 24 and you said, "Look, the LORD our God has shown us his glory and greatness, and we have heard his voice out of the fire. Today we have seen that God may speak to someone and the person may still live. 25 So now why should we die? For this great fire will consume us; if we hear the voice of the LORD our God any longer, we shall die. 26 For who is there of all flesh that has heard the voice of the living God speaking out of fire, as we have, and remained alive? 27 Go near, you yourself, and hear all that the LORD our God will say. Then tell us everything that the LORD our God tells you, and we will listen and do it."

28 The LORD heard your words when you spoke to me, and the LORD said to me: "I have heard the words of this people, which they have spoken to you; they are right in all that they have spoken. 29 If only they had such a mind as this, to fear me and to keep all my commandments always, so that it might go well with them and with their children forever! 30 Go say to them, 'Return to your tents.' 31 But you, stand here by me, and I will tell you all the commandments, the statutes and the ordinances, that you shall teach them, so that they may do them in the land that I am giving them to possess." 32 You must therefore be careful to do as the LORD your God has commanded you; you shall not turn to the right or to the left. 33 You must follow exactly the path that the LORD your God has commanded you, so that you may live, and that it may go well with you, and that you may live long in the land that you are to possess.

6 Now this is the commandment—the statutes and the ordinances—that the LORD your God charged me to teach you to observe in the land that you are about to cross into and occupy, 2 so that you and your children and your children's children may fear the LORD your God all the days of your life, and keep all his decrees and his commandments that I am commanding you, so that your days may be long. 3 Hear therefore, O Israel, and observe them diligently, so that it may go well with you, and so that you may multiply greatly in a land flowing with milk and honey, as the LORD, the God of your ancestors, has promised you.

4 Hear, O Israel: The LORD is our

5.22–23: While the Decalogue was given directly to the people (vv. 4–5; compare 4.10–13), the rest of the laws were mediated to the people through Moses (v. 31; 4.14). 23–27: 4.33; Ex 20.18–21. 31: Moses is no mere legislator but is a teacher or expositor of God's will (1.5). Hence the *statutes and the ordinances* (chs 12–26; compare Ex 20.23–23.19) are expressed in a sermonic appeal to do God's will in the concrete situations of life.

6.1–25: **The meaning of the first commandment. 1:** In this chapter Moses deals with *the commandment;* in a later section he explains *the statutes* and *the ordinances* (see 5.31 n.). **3:** A characteristic Deuteronomic note: reverent obedience will result in divine blessings of long life, fruitfulness, and welfare (5.33; 6.18–19). Thus the promise made to

Israel's ancestors will be fulfilled (Gen 12.1–7; Ex 3.16–17). *Milk and honey,* see Ex 3.8 n. **4–9:** In Jewish tradition these verses are known as the Shema, from the first word in the Hebrew ("shema'"), which means "*Hear.*" **4–5:** The great commandment (Mk 12.29–30) is essentially a restatement of the first commandment of the Decalogue in positive form. There are not many gods, for *the LORD alone* is sovereign and unique; thus Israel is to have only one loyalty. *Heart* (mind, will), *soul* (self, vital being), *might* express the idea of loving God (see 4.37 n.) with the full measure of one's devotion. **8–9:** Putting this law on the hand, forehead, and doorpost (compare Ex 13.9) signifies that it is to be *in your heart* (v. 6), i.e. constantly thought about and acted upon (Ps 1.2).

6.10–19: See ch 8. **13:** The love of God is

God, the LORD alone. [n] 5 You shall love the LORD your God with all your heart, and with all your soul, and with all your might. 6 Keep these words that I am commanding you today in your heart. 7 Recite them to your children and talk about them when you are at home and when you are away, when you lie down and when you rise. 8 Bind them as a sign on your hand, fix them as an emblem [o] on your forehead, 9 and write them on the doorposts of your house and on your gates.

10 When the LORD your God has brought you into the land that he swore to your ancestors, to Abraham, to Isaac, and to Jacob, to give you—a land with fine, large cities that you did not build, 11 houses filled with all sorts of goods that you did not fill, hewn cisterns that you did not hew, vineyards and olive groves that you did not plant—and when you have eaten your fill, 12 take care that you do not forget the LORD, who brought you out of the land of Egypt, out of the house of slavery. 13 The LORD your God you shall fear; him you shall serve, and by his name alone you shall swear. 14 Do not follow other gods, any of the gods of the peoples who are all around you, 15 because the LORD your God, who is present with you, is a jealous God. The anger of the LORD your God would be kindled against you and he would destroy you from the face of the earth.

16 Do not put the LORD your God to the test, as you tested him at Massah. 17 You must diligently keep the commandments of the LORD your God, and his decrees, and his statutes that he has commanded you. 18 Do what is right and good in the sight of the LORD, so that it may go well with you, and so that you may go in and occupy the good land that the LORD swore to your ancestors to give you, 19 thrusting out all your enemies from before you, as the LORD has promised.

20 When your children ask you in time to come, "What is the meaning of the decrees and the statutes and the ordinances that the LORD our God has commanded you?" 21 then you shall say to your children, "We were Pharaoh's slaves in Egypt, but the LORD brought us out of Egypt with a mighty hand. 22 The LORD displayed before our eyes great and awesome signs and wonders against Egypt, against Pharaoh and all his household. 23 He brought us out from there in order to bring us in, to give us the land that he promised on oath to our ancestors. 24 Then the LORD commanded us to observe all these statutes, to fear the LORD our God, for our lasting good, so as to keep us alive, as is now the case. 25 If we diligently observe this entire commandment before the LORD our God, as he has commanded us, we will be in the right."

7 When the LORD your God brings you into the land that you are about to enter and occupy, and he clears away many nations before you—the Hittites, the Girgashites, the Amorites, the Canaanites, the Perizzites, the Hivites, and the Jebusites, seven nations mightier and more numerous than you— 2 and when the LORD your God gives them over to you and you defeat them, then you must utterly destroy them. Make no covenant with them and show them no mercy.

n Or *The* LORD *our God is one* LORD, or *The* LORD *our God, the* LORD *is one,* or *The* LORD *is our God, the* LORD *is one*   o Or *as a frontlet*

---

blended with fear, i.e. reverence before God's holy majesty (4.9–15). **15:** Divine jealousy is associated with divine wrath (4.24; Josh 24.19–20); both are expressions of the LORD's holiness which will not tolerate idolatry. **16:** See Ex 17.2–7; compare Mt 4.7.

6.20–25: Compare v. 7; Ex 13.14–16. Children are to be taught the marvelous story of God's redeeming acts which is the background and basis of the law. **21–23:** See 26.5–10 n.

7.1–26: Life in Canaan. Having dealt with the heart of the law (the Decalogue and the great commandment; chs 5–6), Moses now interprets what it means to be God's people in the new land. **1–5:** This holy war is

3 Do not intermarry with them, giving your daughters to their sons or taking their daughters for your sons, 4 for that would turn away your children from following me, to serve other gods. Then the anger of the LORD would be kindled against you, and he would destroy you quickly. 5 But this is how you must deal with them: break down their altars, smash their pillars, hew down their sacred poles,[p] and burn their idols with fire. 6 For you are a people holy to the LORD your God; the LORD your God has chosen you out of all the peoples on earth to be his people, his treasured possession.

7 It was not because you were more numerous than any other people that the LORD set his heart on you and chose you—for you were the fewest of all peoples. 8 It was because the LORD loved you and kept the oath that he swore to your ancestors, that the LORD has brought you out with a mighty hand, and redeemed you from the house of slavery, from the hand of Pharaoh king of Egypt. 9 Know therefore that the LORD your God is God, the faithful God who maintains covenant loyalty with those who love him and keep his commandments, to a thousand generations, 10 and who repays in their own person those who reject him. He does not delay but repays in their own person those who reject him. 11 Therefore, observe diligently the commandment—the statutes, and the ordinances—that I am commanding you today.

12 If you heed these ordinances, by diligently observing them, the LORD your God will maintain with you the covenant loyalty that he swore to your ancestors; 13 he will love you, bless you, and multiply you; he will bless the fruit of your womb and the fruit of your ground, your grain and your wine and your oil, the increase of your cattle and the issue of your flock, in the land that he swore to your ancestors to give you. 14 You shall be the most blessed of peoples, with neither sterility nor barrenness among you or your livestock. 15 The LORD will turn away from you every illness; all the dread diseases of Egypt that you experienced, he will not inflict on you, but he will lay them on all who hate you. 16 You shall devour all the peoples that the LORD your God is giving over to you, showing them no pity; you shall not serve their gods, for that would be a snare to you.

17 If you say to yourself, "These nations are more numerous than I; how can I dispossess them?" 18 do not be afraid of them. Just remember what the LORD your God did to Pharaoh and to all Egypt, 19 the great trials that your eyes saw, the signs and wonders, the mighty hand and the outstretched arm by which the LORD your God brought you out. The LORD your God will do the same to all the peoples of whom you are afraid. 20 Moreover, the LORD your God will send the pestilence[q] against them, until even the survivors and the fugitives are destroyed. 21 Have no dread of them, for the LORD your God, who is present with you, is a great and awesome God. 22 The

p Heb *Asherim* *q* Or *hornets*: Meaning of Heb uncertain

---

based on the fear of the corrupting influence of Canaanite culture (v. 16; Ex 23.23–33; 34.11–16). **1**: On the pre-Israelite peoples, see Gen 10.15–20. **2**: *Utterly destroy,* see Josh 6.17 n. **5**: *Pillars, sacred poles* (see Ex 34.13 n.).

7.6–16: An uncompromising attitude toward the nations is required because Israel is *a people holy to the LORD* (Ex 19.5–6), i.e. a people separated for a special service to God (see Lev 19.2 n.); therefore Israel is not to be a nation like other nations (Num 23.9; 1 Sam 8.4–22). **7–8**: The election of Israel is not based on the people's greatness or goodness (9.4–6) but upon the LORD's gracious love and faithfulness to the promise made to their ancestors. **9–10**: Ex 34.6–7. *Covenant loyalty* ("steadfast love"), see Gen 24.12 n. **12–14**: The blessings of fertility do not come from the nature gods of Canaan, but are bestowed by Israel's God who delivered the people from Egyptian bondage (Hos ch 2). **15**: *Dread diseases,* a reference to the plagues of Egypt.

7.17–26: Israel should not fear mighty nations, for, according to the conviction of holy

LORD your God will clear away these nations before you little by little; you will not be able to make a quick end of them, otherwise the wild animals would become too numerous for you. 23 But the LORD your God will give them over to you, and throw them into great panic, until they are destroyed. 24 He will hand their kings over to you and you shall blot out their name from under heaven; no one will be able to stand against you, until you have destroyed them. 25 The images of their gods you shall burn with fire. Do not covet the silver or the gold that is on them and take it for yourself, because you could be ensnared by it; for it is abhorrent to the LORD your God. 26 Do not bring an abhorrent thing into your house, or you will be set apart for destruction like it. You must utterly detest and abhor it, for it is set apart for destruction.

8 This entire commandment that I command you today you must diligently observe, so that you may live and increase, and go in and occupy the land that the LORD promised on oath to your ancestors. 2 Remember the long way that the LORD your God has led you these forty years in the wilderness, in order to humble you, testing you to know what was in your heart, whether or not you would keep his commandments. 3 He humbled you by letting you hunger, then by feeding you with manna, with which neither you nor your ancestors were acquainted, in order to make you understand that one does not live by bread alone, but by every word that comes from the mouth of the LORD. *r* 4 The clothes on your back did not wear out and your feet did not swell these forty years. 5 Know then in your heart that

as a parent disciplines a child so the LORD your God disciplines you. 6 Therefore keep the commandments of the LORD your God, by walking in his ways and by fearing him. 7 For the LORD your God is bringing you into a good land, a land with flowing streams, with springs and underground waters welling up in valleys and hills, 8 a land of wheat and barley, of vines and fig trees and pomegranates, a land of olive trees and honey, 9 a land where you may eat bread without scarcity, where you will lack nothing, a land whose stones are iron and from whose hills you may mine copper. 10 You shall eat your fill and bless the LORD your God for the good land that he has given you.

11 Take care that you do not forget the LORD your God, by failing to keep his commandments, his ordinances, and his statutes, which I am commanding you today. 12 When you have eaten your fill and have built fine houses and live in them, 13 and when your herds and flocks have multiplied, and your silver and gold is multiplied, and all that you have is multiplied, 14 then do not exalt yourself, forgetting the LORD your God, who brought you out of the land of Egypt, out of the house of slavery, 15 who led you through the great and terrible wilderness, an arid wasteland with poisonous *s* snakes and scorpions. He made water flow for you from flint rock, 16 and fed you in the wilderness with manna that your ancestors did not know, to humble you and to test you, and in the end to do you good. 17 Do not say to yourself, "My power and the might of

r Or *by anything that the LORD decrees*
s Or *fiery*; Heb *seraph*

war, *the LORD your God is present with you* (20.1–4). These verses echo ideas found in Ex 23.20–33. **20:** *Pestilence*, see Ex 23.27–28 n.; Josh 24.12. **22:** Ex 23.29–30.

8.1–20: The temptation to pride and self-sufficiency. Moses warns the people that success in Canaan will tempt them to forget the wilderness lesson of complete dependence upon God's mercies. **1–10:** An appeal to Israel's memory: in the wilderness

God cared for the people daily (Ex 12.37–17.16; Num chs 11–14). **3:** *Manna*, see Ex ch 16; Num 11.7, 8. **5:** Suffering is here interpreted as discipline, analogous to a parent's correction of a child (Hos ch 11; Heb 12.3–11). God's purpose is to humble the people's pride and to test the quality of their faith.

8.11–20: The peril of prosperity. **15:** *Poisonous snakes*, Num 21.6–9. *Water . . . from flint rock*, Num 20.2–13.

my own hand have gotten me this wealth." [18] But remember the LORD your God, for it is he who gives you power to get wealth, so that he may confirm his covenant that he swore to your ancestors, as he is doing today. [19] If you do forget the LORD your God and follow other gods to serve and worship them, I solemnly warn you today that you shall surely perish. [20] Like the nations that the LORD is destroying before you, so shall you perish, because you would not obey the voice of the LORD your God.

9 Hear, O Israel! You are about to cross the Jordan today, to go in and dispossess nations larger and mightier than you, great cities, fortified to the heavens, [2] a strong and tall people, the offspring of the Anakim, whom you know. You have heard it said of them, "Who can stand up to the Anakim?" [3] Know then today that the LORD your God is the one who crosses over before you as a devouring fire; he will defeat them and subdue them before you, so that you may dispossess and destroy them quickly, as the LORD has promised you.

4 When the LORD your God thrusts them out before you, do not say to yourself, "It is because of my righteousness that the LORD has brought me in to occupy this land"; it is rather because of the wickedness of these nations that the LORD is dispossessing them before you. [5] It is not because of your righteousness or the uprightness of your heart that you are going in to occupy their land; but because of the wickedness of these nations the LORD your God is dispossessing them before you, in order to fulfill the promise that the LORD made on oath to your ancestors, to Abraham, to Isaac, and to Jacob.

6 Know, then, that the LORD your God is not giving you this good land to occupy because of your righteousness; for you are a stubborn people. [7] Remember and do not forget how you provoked the LORD your God to wrath in the wilderness; you have been rebellious against the LORD from the day you came out of the land of Egypt until you came to this place.

8 Even at Horeb you provoked the LORD to wrath, and the LORD was so angry with you that he was ready to destroy you. [9] When I went up the mountain to receive the stone tablets, the tablets of the covenant that the LORD made with you, I remained on the mountain forty days and forty nights; I neither ate bread nor drank water. [10] And the LORD gave me the two stone tablets written with the finger of God; on them were all the words that the LORD had spoken to you at the mountain out of the fire on the day of the assembly. [11] At the end of forty days and forty nights the LORD gave me the two stone tablets, the tablets of the covenant. [12] Then the LORD said to me, "Get up, go down quickly from here, for your people whom you have brought from Egypt have acted corruptly. They have been quick to turn from the way that I commanded them; they have cast an image for themselves." [13] Furthermore the LORD said to me, "I have seen that this people is indeed a stubborn people. [14] Let me alone that I may destroy them and blot out their name from under heaven; and I will make of you a nation mightier and more numerous than they."

15 So I turned and went down from the mountain, while the mountain was ablaze; the two tablets of the covenant were in my two hands. [16] Then I saw that you had indeed sinned against the LORD your God, by casting for yourselves an

---

9.1–10.11: The temptation to self-righteousness. God does not give the land to the people of Israel as a reward for righteousness, for in the wilderness they proved to be a rebellious people.
9.2: See Num 13.22 n. **4–5:** Victory will be given in the holy war because (negatively)

Canaan has been corrupted by pagan religions and because (positively) the LORD is faithful to the promise made to Israel's ancestors.
9.6–24: The historical record shows that Israel has been rebellious ever since the Exodus (Ezek 20.5–8). **8–10:** Ex 24.12–18; 31.18. **11–21:** Ex ch 32. **22:** Num 11.1–3; Ex 17.1–7;

image of a calf; you had been quick to turn from the way that the LORD had commanded you. 17 So I took hold of the two tablets and flung them from my two hands, smashing them before your eyes. 18 Then I lay prostrate before the LORD as before, forty days and forty nights; I neither ate bread nor drank water, because of all the sin you had committed, provoking the LORD by doing what was evil in his sight. 19 For I was afraid that the anger that the LORD bore against you was so fierce that he would destroy you. But the LORD listened to me that time also. 20 The LORD was so angry with Aaron that he was ready to destroy him, but I interceded also on behalf of Aaron at that same time. 21 Then I took the sinful thing you had made, the calf, and burned it with fire and crushed it, grinding it thoroughly, until it was reduced to dust; and I threw the dust of it into the stream that runs down the mountain.

22 At Taberah also, and at Massah, and at Kibroth-hattaavah, you provoked the LORD to wrath. 23 And when the LORD sent you from Kadesh-barnea, saying, "Go up and occupy the land that I have given you," you rebelled against the command of the LORD your God, neither trusting him nor obeying him. 24 You have been rebellious against the LORD as long as he has*t* known you.

25 Throughout the forty days and forty nights that I lay prostrate before the LORD when the LORD intended to destroy you, 26 I prayed to the LORD and said, "Lord GOD, do not destroy the people who are your very own possession, whom you redeemed in your greatness, whom you brought out of Egypt with a mighty hand. 27 Remember your servants, Abraham, Isaac, and Jacob; pay no

attention to the stubbornness of this people, their wickedness and their sin, 28 otherwise the land from which you have brought us might say, 'Because the LORD was not able to bring them into the land that he promised them, and because he hated them, he has brought them out to let them die in the wilderness.' 29 For they are the people of your very own possession, whom you brought out by your great power and by your outstretched arm."

10 At that time the LORD said to me, "Carve out two tablets of stone like the former ones, and come up to me on the mountain, and make an ark of wood. 2 I will write on the tablets the words that were on the former tablets, which you smashed, and you shall put them in the ark." 3 So I made an ark of acacia wood, cut two tablets of stone like the former ones, and went up the mountain with the two tablets in my hand. 4 Then he wrote on the tablets the same words as before, the ten commandments*u* that the LORD had spoken to you on the mountain out of the fire on the day of the assembly; and the LORD gave them to me. 5 So I turned and came down from the mountain, and put the tablets in the ark that I had made; and there they are, as the LORD commanded me.

6 (The Israelites journeyed from Beeroth-bene-jaakan*v* to Moserah. There Aaron died, and there he was buried; his son Eleazar succeeded him as priest. 7 From there they journeyed to Gudgodah, and from Gudgodah to Jotbathah, a land with flowing streams. 8 At that time the LORD set apart the tribe of

t Sam Gk: MT *I have*　　*u* Heb *the ten words*
v Or *the wells of the Bene-jaakan*

and Num 11.31–34. **23:** Num chs 13–14. **9.25–29:** A paraphrase of Ex 32.11–14. In Deuteronomy Moses is portrayed as the ideal prophet (34.10–12) who intercedes for the people and who suffers on their behalf (1.37; compare Isa ch 53). **10.1–11:** The second ascent of the mountain (Ex 34.1–4, 27–28). **1–3:** These verses rest on an ancient tradition that Moses made the ark and put the stone tablets in it (1 Kings 8.9; see Ex 24.15–18 n.). **6–9:** An editorial insertion which in vv. 6–7 apparently quotes from a wilderness itinerary (compare Num 33.30–38). **6:** Num 20.22–29. **8:** Ex 32.25–29. The Levites' role is to bear the ark (Num 4.4–15), to *minister,* i.e. conduct the sacrificial services (Num ch 18), and to *bless* the people (Num 6.22–27).

Levi to carry the ark of the covenant of the LORD, to stand before the LORD to minister to him, and to bless in his name, to this day. 9 Therefore Levi has no allotment or inheritance with his kindred; the LORD is his inheritance, as the LORD your God promised him.)

10 I stayed on the mountain forty days and forty nights, as I had done the first time. And once again the LORD listened to me. The LORD was unwilling to destroy you. 11 The LORD said to me, "Get up, go on your journey at the head of the people, that they may go in and occupy the land that I swore to their ancestors to give them."

12 So now, O Israel, what does the LORD your God require of you? Only to fear the LORD your God, to walk in all his ways, to love him, to serve the LORD your God with all your heart and with all your soul, 13 and to keep the commandments of the LORD your God*w* and his decrees that I am commanding you today, for your own well-being. 14 Although heaven and the heaven of heavens belong to the LORD your God, the earth with all that is in it, 15 yet the LORD set his heart in love on your ancestors alone and chose you, their descendants after them, out of all the peoples, as it is today. 16 Circumcise, then, the foreskin of your heart, and do not be stubborn any longer. 17 For the LORD your God is God of gods and Lord of lords, the great God, mighty and awesome, who is not partial and takes no bribe, 18 who executes justice for the orphan and the widow, and who loves the strangers, providing them food and clothing. 19 You shall also love the stranger, for you were strangers in the land of Egypt. 20 You shall fear the LORD your God; him alone you shall worship; to him you shall hold fast, and by his name you shall swear. 21 He is your praise; he is your God, who has done for you these great and awesome things that your own eyes have seen. 22 Your ancestors went down to Egypt seventy persons; and now the LORD your God has made you as numerous as the stars in heaven.

11 You shall love the LORD your God, therefore, and keep his charge, his decrees, his ordinances, and his commandments always. 2 Remember today that it was not your children (who have not known or seen the discipline of the LORD your God), but it is you who must acknowledge his greatness, his mighty hand and his outstretched arm, 3 his signs and his deeds that he did in Egypt to Pharaoh, the king of Egypt, and to all his land; 4 what he did to the Egyptian army, to their horses and chariots, how he made the water of the Red Sea*x* flow over them as they pursued you, so that the LORD has destroyed them to this day; 5 what he did to you in the wilderness, until you came to this place; 6 and what he did to Dathan and Abiram, sons of Eliab son of Reuben, how in the midst of all Israel the earth opened its mouth and swallowed them up, along with their households, their tents, and every living being in their company; 7 for it is your own eyes that

w Q Ms Gk Syr: MT lacks *your God*
x Or *Sea of Reeds*

10.12–11.32. What the LORD requires. This section is the climax and conclusion of the historical review found in ch 8 and 9.1–10.11.

10.12: God's gracious dealings form the background and presupposition of God's requirements (compare Mic 6.1–8). **16:** *Circumcise the . . . heart* means to open the mind, to direct the will toward God (see Lev 26.41 n.). **17–18:** Divine justice is disclosed in God's impartiality and defense of the legally helpless. *The stranger,* who resided within the cove-nant community without tribal status, was in danger of exploitation. **19:** Ex 22.21; 23.9; Lev 19.34. This verse implies the second great commandment of Lev 19.17–18. **22:** *Seventy persons,* Ex 1.5.

11.1–26: Loyalty to the covenant is the condition for life in Canaan. **2:** The word *today* appears frequently in Deuteronomy to emphasize the contemporaneity of the covenant demands and promises (see 5.2–3 n.). *Discipline,* see 8.5 n. **6:** The address follows the early tradition of Num ch 16 concerning

have seen every great deed that the LORD did.

8 Keep, then, this entire commandment that I am commanding you today, so that you may have strength to go in and occupy the land that you are crossing over to occupy, 9and so that you may live long in the land that the LORD swore to your ancestors to give them and to their descendants, a land flowing with milk and honey. 10For the land that you are about to enter to occupy is not like the land of Egypt, from which you have come, where you sow your seed and irrigate by foot like a vegetable garden. 11But the land that you are crossing over to occupy is a land of hills and valleys, watered by rain from the sky, 12a land that the LORD your God looks after. The eyes of the LORD your God are always on it, from the beginning of the year to the end of the year.

13 If you will only heed his every commandment*y* that I am commanding you today—loving the LORD your God, and serving him with all your heart and with all your soul— 14then he*z* will give the rain for your land in its season, the early rain and the later rain, and you will gather in your grain, your wine, and your oil; 15and he*z* will give grass in your fields for your livestock, and you will eat your fill. 16Take care, or you will be seduced into turning away, serving other gods and worshiping them, 17for then the anger of the LORD will be kindled against you and he will shut up the heavens, so that there will be no rain and the land will yield no fruit; then you will perish quickly off the good land that the LORD is giving you.

18 You shall put these words of mine in your heart and soul, and you shall bind them as a sign on your hand, and fix them as an emblem*a* on your forehead. 19Teach them to your children, talking about them when you are at home and when you are away, when you lie down and when you rise. 20Write them on the doorposts of your house and on your gates, 21so that your days and the days of your children may be multiplied in the land that the LORD swore to your ancestors to give them, as long as the heavens are above the earth.

22 If you will diligently observe this entire commandment that I am commanding you, loving the LORD your God, walking in all his ways, and holding fast to him, 23then the LORD will drive out all these nations before you, and you will dispossess nations larger and mightier than yourselves. 24Every place on which you set foot shall be yours; your territory shall extend from the wilderness to the Lebanon and from the River, the river Euphrates, to the Western Sea. 25No one will be able to stand against you; the LORD your God will put the fear and dread of you on all the land on which you set foot, as he promised you.

26 See, I am setting before you today a blessing and a curse: 27the blessing, if you obey the commandments of the LORD your God that I am commanding you today; 28and the curse, if you do not obey the commandments of the LORD your God, but turn from the way that I

y Compare Gk: Heb *my commandments*
z Sam Gk Vg: MT *I*   *a* Or *as a frontlet*

---

the revolt of Dathan and Abiram; note the silence about Korah's rebellion (Num 16.3–11). **11.10–12:** The Nile valley must be irrigated through human effort; Palestine, however, is dependent upon seasonal rainfall. This difference is mentioned to show Israel's dependence upon the LORD, who gives and withholds rain (Am 4.7–9). **14:** The *early rain* comes at the end of the summer drought (October–November); the *later rain* comes in the spring (March–April). **16–17:** See 7.12–14 n.

11.18–21: 6.6–9. **24:** The territory is described in terms of the ideal limits of David's empire (see 1.7 n.). *The Western Sea,* the Mediterranean. **26–32:** The two ways (see ch 28; 30.15–20). **26:** *Blessing . . . curse,* an echo of an ancient ceremony of covenant renewal, perhaps inaugurated at Shechem (v. 29, see ch 27). The people stand in a time of solemn decision, facing the alternatives of

am commanding you today, to follow other gods that you have not known. 29 When the LORD your God has brought you into the land that you are entering to occupy, you shall set the blessing on Mount Gerizim and the curse on Mount Ebal. 30 As you know, they are beyond the Jordan, some distance to the west, in the land of the Canaanites who live in the Arabah, opposite Gilgal, beside the oak*b* of Moreh.

31 When you cross the Jordan to go in to occupy the land that the LORD your God is giving you, and when you occupy it and live in it, 32 you must diligently observe all the statutes and ordinances that I am setting before you today.

12 These are the statutes and ordinances that you must diligently observe in the land that the LORD, the God of your ancestors, has given you to occupy all the days that you live on the earth.

2 You must demolish completely all the places where the nations whom you are about to dispossess served their gods, on the mountain heights, on the hills, and under every leafy tree. 3 Break down their altars, smash their pillars, burn their sacred poles*c* with fire, and hew down the idols of their gods, and thus blot out their name from their places. 4 You shall not worship the LORD your God in such ways. 5 But you shall seek the place that

the LORD your God will choose out of all your tribes as his habitation to put his name there. You shall go there, 6 bringing there your burnt offerings and your sacrifices, your tithes and your donations, your votive gifts, your freewill offerings, and the firstlings of your herds and flocks. 7 And you shall eat there in the presence of the LORD your God, you and your households together, rejoicing in all the undertakings in which the LORD your God has blessed you.

8 You shall not act as we are acting here today, all of us according to our own desires, 9 for you have not yet come into the rest and the possession that the LORD your God is giving you. 10 When you cross over the Jordan and live in the land that the LORD your God is allotting to you, and when he gives you rest from your enemies all around so that you live in safety, 11 then you shall bring everything that I command you to the place that the LORD your God will choose as a dwelling for his name: your burnt offerings and your sacrifices, your tithes and your donations, and all your choice votive gifts that you vow to the LORD. 12 And you shall rejoice before the LORD your God, you together with your sons and your daughters, your male and fe-

b Gk Syr: Compare Gen 12.6; Heb *oaks* or *terebinths*   *c* Heb *Asherim*

divine blessing or divine judgment. **29–30:** See ch 27. *The oak of Moreh* was near Shechem (see Gen 12.6 n.).

12.1–31: The centralization of worship. Israel is to worship at the central sanctuary chosen by the LORD and to make sacrifices only at this place. **1:** Here begins the exposition of *the statutes and the ordinances* (see 5.31 n.) found in chs 12–26. **3:** See Ex 34.13 n. **4:** The following law concerning the centralization of worship is intended to guard Israel from the paganism that flourished in local shrines (vv. 29–31). **5–7:** During the period of the tribal confederacy pilgrimages were made to Shiloh (Josh 18.1; 1 Sam 1.3–28), and under the leadership of David Jerusalem became the central sanctuary of Israel (2 Sam ch 6). During this whole period pilgrimages to the sanctuary did not preclude sacrificing at any altar in the land (Gen 12.7;

1 Sam 10.8; 1 Kings 3.2, 4). The present stringent law, providing for only one place for sacrificial worship, was the basis of the great reform carried out by King Josiah (2 Kings chs 22–23). **5:** God's dwelling place is in heaven (1 Kings 8.27), but the LORD's *name*—signifying God's personal identity (see Gen 32.27 n.)—dwells in the sanctuary (v. 11) and in this sense God is present there. This "name" theology, a characteristic of the Deuteronomic perspective, deals with the paradox of divine distance and nearness, or transcendence and immanence. **6:** For the types of sacrifice, see Lev chs 1–7. *Votive gifts,* Lev ch 27; Num ch 30. *Eat there in the presence of the LORD* refers to certain offerings that must be eaten at the sanctuary, such as the tithe (14.22–28) or the offering of well-being (Lev ch 3). **12:** *The Levites,* see 18.1 n.

male slaves, and the Levites who reside in your towns (since they have no allotment or inheritance with you).

13 Take care that you do not offer your burnt offerings at any place you happen to see. 14 But only at the place that the Lord will choose in one of your tribes—there you shall offer your burnt offerings and there you shall do everything I command you.

15 Yet whenever you desire you may slaughter and eat meat within any of your towns, according to the blessing that the Lord your God has given you; the unclean and the clean may eat of it, as they would of gazelle or deer. 16 The blood, however, you must not eat; you shall pour it out on the ground like water. 17 Nor may you eat within your towns the tithe of your grain, your wine, and your oil, the firstlings of your herds and your flocks, any of your votive gifts that you vow, your freewill offerings, or your donations; 18 these you shall eat in the presence of the Lord your God at the place that the Lord your God will choose, you together with your son and your daughter, your male and female slaves, and the Levites resident in your towns, rejoicing in the presence of the Lord your God in all your undertakings. 19 Take care that you do not neglect the Levite as long as you live in your land.

20 When the Lord your God enlarges your territory, as he has promised you, and you say, "I am going to eat some meat," because you wish to eat meat, you may eat meat whenever you have the desire. 21 If the place where the Lord your God will choose to put his name is too far from you, and you slaughter as I have commanded you any of your herd or flock that the Lord has given you, then you may eat within your towns whenever you desire. 22 Indeed, just as gazelle or deer is eaten, so you may eat it; the unclean and the clean alike may eat it. 23 Only be sure that you do not eat the blood; for the blood is the life, and you shall not eat the life with the meat. 24 Do not eat it; you shall pour it out on the ground like water. 25 Do not eat it, so that all may go well with you and your children after you, because you do what is right in the sight of the Lord. 26 But the sacred donations that are due from you, and your votive gifts, you shall bring to the place that the Lord will choose. 27 You shall present your burnt offerings, both the meat and the blood, on the altar of the Lord your God; the blood of your other sacrifices shall be poured out beside *d* the altar of the Lord your God, but the meat you may eat.

28 Be careful to obey all these words that I command you today, *e* so that it may go well with you and with your children after you forever, because you will be doing what is good and right in the sight of the Lord your God.

29 When the Lord your God has cut off before you the nations whom you are about to enter to dispossess them, when you have dispossessed them and live in their land, 30 take care that you are not snared into imitating them, after they have been destroyed before you: do not inquire concerning their gods, saying, "How did these nations worship their gods? I also want to do the same." 31 You must not do the same for the Lord your God, because every abhorrent thing that the Lord hates they have done for their gods. They would even burn their sons and their daughters in the fire to their

d Or *on*   *e* Gk Sam Syr: MT lacks *today*

12.15–28: Now that sacrifice is made only at the central sanctuary, a distinction is drawn between sacrifice and slaughter of animals for food, thus modifying earlier legislation (Lev 17.1–9). **15–16:** When eating slaughtered meat in a town, it is not necessary to observe the laws of ceremonial cleanness (Lev 7.19–21), for slaughtered meat may be considered in the same category as game (e.g. gazelle or deer). However, the ancient prohibition against eating blood must be maintained (vv. 23–24; see Gen 9.3–4; Lev 17.10–11). **17–18:** *Sacred donations* (v. 26), which belong by right, or by dedication, to the Lord, must be eaten at the central sanctuary.

12.31: *Every abhorrent thing*, e.g. sacred

gods. [32][f] You must diligently observe everything that I command you; do not add to it or take anything from it.

13 [g] If prophets or those who divine by dreams appear among you and promise you omens or portents, [2] and the omens or the portents declared by them take place, and they say, "Let us follow other gods" (whom you have not known) "and let us serve them," [3] you must not heed the words of those prophets or those who divine by dreams; for the LORD your God is testing you, to know whether you indeed love the LORD your God with all your heart and soul. [4] The LORD your God you shall follow, him alone you shall fear, his commandments you shall keep, his voice you shall obey, him you shall serve, and to him you shall hold fast. [5] But those prophets or those who divine by dreams shall be put to death for having spoken treason against the LORD your God—who brought you out of the land of Egypt and redeemed you from the house of slavery—to turn you from the way in which the LORD your God commanded you to walk. So you shall purge the evil from your midst.

6 If anyone secretly entices you—even if it is your brother, your father's son or[h] your mother's son, or your own son or daughter, or the wife you embrace, or your most intimate friend—saying, "Let us go worship other gods," whom neither you nor your ancestors have known, [7] any of the gods of the peoples that are around you, whether near you or far away from you, from one end of the earth to the other, [8] you must not yield to or heed any such persons. Show them no pity or compassion and do not shield them. [9] But you shall surely kill them; your own hand shall be first against them to execute them, and afterwards the hand of all the people. [10] Stone them to death for trying to turn you away from the LORD your God, who brought you out of the land of Egypt, out of the house of slavery. [11] Then all Israel shall hear and be afraid, and never again do any such wickedness.

12 If you hear it said about one of the towns that the LORD your God is giving you to live in, [13] that scoundrels from among you have gone out and led the inhabitants of the town astray, saying, "Let us go and worship other gods," whom you have not known, [14] then you shall inquire and make a thorough investigation. If the charge is established that such an abhorrent thing has been done among you, [15] you shall put the inhabitants of that town to the sword, utterly destroying it and everything in it—even putting its livestock to the sword. [16] All of its spoil you shall gather into its public square; then burn the town and all its spoil with fire, as a whole burnt offering to the LORD your God. It shall remain a perpetual ruin, never to be rebuilt. [17] Do not let anything devoted to destruction

f Ch 13.1 in Heb   g Ch 13.2 in Heb
h Sam Gk Compare Tg: MT lacks *your father's son or*

---

prostitution (23.17–18) and ordeal by fire (see 18.10 n.).

12.32–13.18: Warning against idolatry. Subversives who entice their fellow Israelites into the worship of other gods must be eliminated from the covenant community.

13.1–5: The false prophet (Jer 23.9–32; Ezek ch 13). **1:** Dreams were regarded as a medium of divine communication to a prophet (Num 12.6; 22.20; 1 Sam 3.15; 28.6; Jer 23.25). This law makes it clear that *omens or portents* (see Ex 3.11–12 n.) are not in themselves proof that God has spoken, for God may give false prophets power to perform wonders in order to test the people's faith (compare 8.2). A miracle is not significant unless it prompts faith in the God whom Israel has *known* (vv. 6, 13) in its historical experience, that is, the God of Exodus and Sinai. **5:** The punishment is severe because idolatry contaminates the health or holiness of the community.

13.6–11: The same punishment applies to close friends or family members who entice one into idolatry. **10:** Stoning is prescribed because this is a communal mode of purging the evil that threatens the community (17.2–7).

13.12–18: When a city has turned to idolatry, it must be put under the sacrificial ban

stick to your hand, so that the LORD may turn from his fierce anger and show you compassion, and in his compassion multiply you, as he swore to your ancestors, [18] if you obey the voice of the LORD your God by keeping all his commandments that I am commanding you today, doing what is right in the sight of the LORD your God.

14 You are children of the LORD your God. You must not lacerate yourselves or shave your forelocks for the dead. [2] For you are a people holy to the LORD your God; it is you the LORD has chosen out of all the peoples on earth to be his people, his treasured possession.

3 You shall not eat any abhorrent thing. [4] These are the animals you may eat: the ox, the sheep, the goat, [5] the deer, the gazelle, the roebuck, the wild goat, the ibex, the antelope, and the mountain-sheep. [6] Any animal that divides the hoof and has the hoof cleft in two, and chews the cud, among the animals, you may eat. [7] Yet of those that chew the cud or have the hoof cleft you shall not eat these: the camel, the hare, and the rock badger, because they chew the cud but do not divide the hoof; they are unclean for you. [8] And the pig, because it divides the hoof but does not chew the cud, is unclean for you. You shall not eat their meat, and you shall not touch their carcasses.

9 Of all that live in water you may eat these: whatever has fins and scales you may eat. [10] And whatever does not have fins and scales you shall not eat; it is unclean for you.

11 You may eat any clean birds. [12] But these are the ones that you shall not eat: the eagle, the vulture, the osprey, [13] the buzzard, the kite, of any kind; [14] every

raven of any kind; [15] the ostrich, the nighthawk, the sea gull, the hawk, of any kind; [16] the little owl and the great owl, the water hen [17] and the desert owl, *i* the carrion vulture and the cormorant, [18] the stork, the heron, of any kind; the hoopoe and the bat. *j* [19] And all winged insects are unclean for you; they shall not be eaten. [20] You may eat any clean winged creature.

21 You shall not eat anything that dies of itself; you may give it to aliens residing in your towns for them to eat, or you may sell it to a foreigner. For you are a people holy to the LORD your God.

You shall not boil a kid in its mother's milk.

22 Set apart a tithe of all the yield of your seed that is brought in yearly from the field. [23] In the presence of the LORD your God, in the place that he will choose as a dwelling for his name, you shall eat the tithe of your grain, your wine, and your oil, as well as the firstlings of your herd and flock, so that you may learn to fear the LORD your God always. [24] But if, when the LORD your God has blessed you, the distance is so great that you are unable to transport it, because the place where the LORD your God will choose to set his name is too far away from you, [25] then you may turn it into money. With the money secure in hand, go to the place that the LORD your God will choose; [26] spend the money for whatever you wish—oxen, sheep, wine, strong drink, or whatever you desire. And you shall eat there in the presence of the LORD your God, you and your household rejoicing

i Or *pelican*   *j* Identification of several of the birds in verses 12-18 is uncertain

---

and consumed as *a whole burnt offering.* **17:** *Anything devoted* (see Lev 27.21 n., 28–29), such as people, cattle, spoil, are holy, for they belong to the LORD's sacrifice (20.10–18). **14.1–15.23: The lifestyle of a holy people** (see Lev 19.2 n.). In this section old laws are reinterpreted to enforce the contemporaneity of God's covenant demands. **14.1:** Prohibition against pagan customs of mourning (see Lev 19.28; 21.5). **3–21:** Clean

animals that are permitted for food (Lev 11.2–23). **21:** On the first law cited, see Ex 22.31 n. and the regulation for slaughter in Deut 12.15–28. On the second law, see Ex 23.18–19 n.

14.22–29: Israel is steward of the land which belongs to the LORD; therefore, tithes of produce must be offered annually at the harvest festival (16.9–12). **24–26:** This is a special Deuteronomic provision, arising out

together. 27 As for the Levites resident in your towns, do not neglect them, because they have no allotment or inheritance with you.

28 Every third year you shall bring out the full tithe of your produce for that year, and store it within your towns; 29 the Levites, because they have no allotment or inheritance with you, as well as the resident aliens, the orphans, and the widows in your towns, may come and eat their fill so that the LORD your God may bless you in all the work that you undertake.

15 Every seventh year you shall grant a remission of debts. 2 And this is the manner of the remission: every creditor shall remit the claim that is held against a neighbor, not exacting it of a neighbor who is a member of the community, because the LORD's remission has been proclaimed. 3 Of a foreigner you may exact it, but you must remit your claim on whatever any member of your community owes you. 4 There will, however, be no one in need among you, because the LORD is sure to bless you in the land that the LORD your God is giving you as a possession to occupy, 5 if only you will obey the LORD your God by diligently observing this entire commandment that I command you today. 6 When the LORD your God has blessed you, as he promised you, you will lend to many nations, but you will not borrow; you will rule over many nations, but they will not rule over you.

7 If there is among you anyone in need, a member of your community in any of your towns within the land that the LORD your God is giving you, do not be hard-hearted or tight-fisted toward your needy neighbor. 8 You should rather open your hand, willingly lending enough to meet the need, whatever it may be. 9 Be careful that you do not entertain a mean thought, thinking, "The seventh year, the year of remission, is near," and therefore view your needy neighbor with hostility and give nothing; your neighbor might cry to the LORD against you, and you would incur guilt. 10 Give liberally and be ungrudging when you do so, for on this account the LORD your God will bless you in all your work and in all that you undertake. 11 Since there will never cease to be some in need on the earth, I therefore command you, "Open your hand to the poor and needy neighbor in your land."

12 If a member of your community, whether a Hebrew man or a Hebrew woman, is sold*k* to you and works for you six years, in the seventh year you shall set that person free. 13 And when you send a male slave*l* out from you a free person, you shall not send him out empty-handed. 14 Provide liberally out of your flock, your threshing floor, and your wine press, thus giving to him some of the bounty with which the LORD your God has blessed you. 15 Remember that you were a slave in the land of Egypt, and the LORD your God redeemed you; for this reason I lay this command upon you today. 16 But if he says to you, "I will not go out from you," because he loves you and your

k Or sells himself or herself   l Heb him

---

of centralization of worship. **27–29**: It was impractical for all Levites to serve at the central sanctuary; therefore special consideration was to be given to the town Levites (see 18.1 n.; 26.12–15). This law modifies the previous provision that an annual tithe be brought to the sanctuary for support of the Levites (Num 18.21–32).

15.1–11: The year of release. In Ex 23.10–11 the sabbatical year is for the sake of the poor; in Lev 25.1–7 it is a fallow year for the sake of the land. Here the sabbatical year, like the jubilee year (Lev 25.8–55), is a time for remission of debts. **3**: The law does not apply to a *foreigner* who visits for business (v. 6). A *neighbor* (v. 2) is a fellow member of the covenant community (see Lev 19.17–18 n.). **11**: If God's will were fully obeyed, there would be no poverty (vv. 3–5); but until that time comes *there will never cease to be poor people* (compare Jn 12.8) and the law must remain in force.

15.12–18: The law on Hebrew servitude (Ex 21.2–11). **13–15**: An expansion, showing Deuteronomy's ethical fervor (compare Lev 25.42–43). **17**: Unlike the older law, male and

household, since he is well off with you, [17]then you shall take an awl and thrust it through his earlobe into the door, and he shall be your slave[m] forever.

You shall do the same with regard to your female slave.[n]

18 Do not consider it a hardship when you send them out from you free persons, because for six years they have given you services worth the wages of hired laborers; and the LORD your God will bless you in all that you do.

19 Every firstling male born of your herd and flock you shall consecrate to the LORD your God; you shall not do work with your firstling ox nor shear the firstling of your flock. [20]You shall eat it, you together with your household, in the presence of the LORD your God year by year at the place that the LORD will choose. [21]But if it has any defect—any serious defect, such as lameness or blindness—you shall not sacrifice it to the LORD your God; [22]within your towns you may eat it, the unclean and the clean alike, as you would a gazelle or deer. [23]Its blood, however, you must not eat; you shall pour it out on the ground like water.

16 Observe the month[o] of Abib by keeping the passover for the LORD your God, for in the month of Abib the LORD your God brought you out of Egypt by night. [2]You shall offer the passover sacrifice for the LORD your God, from the flock and the herd, at the place that the LORD will choose as a dwelling for his name. [3]You must not eat with it anything leavened. For seven days you shall eat unleavened bread with it—the bread of affliction—because you came out of the land of Egypt in great haste, so that all the days of your life you may remember the day of your departure from the land of Egypt. [4]No leaven shall be seen with you in all your territory for seven days; and none of the meat of what you slaughter on the evening of the first day shall remain until morning. [5]You are not permitted to offer the passover sacrifice within any of your towns that the LORD your God is giving you. [6]But at the place that the LORD your God will choose as a dwelling for his name, only there shall you offer the passover sacrifice, in the evening at sunset, the time of day when you departed from Egypt. [7]You shall cook it and eat it at the place that the LORD your God will choose; the next morning you may go back to your tents. [8]For six days you shall continue to eat unleavened bread, and on the seventh day there shall be a solemn assembly for the LORD your God, when you shall do no work.

9 You shall count seven weeks; begin to count the seven weeks from the time the sickle is first put to the standing grain. [10]Then you shall keep the festival of weeks for the LORD your God, contributing a freewill offering in proportion to the blessing that you have received from the LORD your God. [11]Rejoice before the LORD your God— you and your sons and your daughters, your male and female slaves, the Levites resident in your towns, as well as the strangers, the orphans, and the widows who are among you—at the place that the LORD your God will choose as a dwelling for his name. [12]Remember that

m Or *bondman*   n Or *bondwoman*   o Or *new moon*

---

female slaves are put on the same level. **18:** Obeying the law should be easy (compare Jer 34.8–16; Neh 5.5).

15.19–23: The old law on the sacrifice of firstlings (see Ex 13.2 n.) is adapted to the requirement of the central sanctuary (12.15–28).

16.1–17; A festal calendar (Ex 23.14–17; 34.18–24; Lev ch 23; Num chs 28–29). **1–8:** The festival of the passover and unleavened bread (Ex 12.1–27; 13.3–10; 23.15, 18; 34.18, 25; Lev 23.5–8; Num 28.16–25). On the relation between these festivals, see Ex 12.14–20 n. Here the passover must be eaten at the central sanctuary, rather than in the towns according to previous custom. **1:** *Abib,* see Ex 13.4 n.

16.9 12: The festival of weeks (Ex 23.16; 34.22; Lev 23.15–16; Num 28.26), a harvest festival held in June (in the New Testament

you were a slave in Egypt, and diligently observe these statutes.

13 You shall keep the festival of booths[p] for seven days, when you have gathered in the produce from your threshing floor and your wine press. [14]Rejoice during your festival, you and your sons and your daughters, your male and female slaves, as well as the Levites, the strangers, the orphans, and the widows resident in your towns. [15]Seven days you shall keep the festival for the LORD your God at the place that the LORD will choose; for the LORD your God will bless you in all your produce and in all your undertakings, and you shall surely celebrate.

16 Three times a year all your males shall appear before the LORD your God at the place that he will choose: at the festival of unleavened bread, at the festival of weeks, and at the festival of booths.[p] They shall not appear before the LORD empty-handed; [17]all shall give as they are able, according to the blessing of the LORD your God that he has given you.

18 You shall appoint judges and officials throughout your tribes, in all your towns that the LORD your God is giving you, and they shall render just decisions for the people. [19]You must not distort justice; you must not show partiality; and you must not accept bribes, for a bribe blinds the eyes of the wise and subverts the cause of those who are in the right. [20]Justice, and only justice, you shall pursue, so that you may live and occupy the land that the LORD your God is giving you.

21 You shall not plant any tree as a sacred pole[q] beside the altar that you make for the LORD your God; [22]nor shall you set up a stone pillar—things that the LORD your God hates.

17 You must not sacrifice to the LORD your God an ox or a sheep that has a defect, anything seriously wrong; for that is abhorrent to the LORD your God.

2 If there is found among you, in one of your towns that the LORD your God is giving you, a man or woman who does what is evil in the sight of the LORD your God, and transgresses his covenant [3]by going to serve other gods and worshiping them—whether the sun or the moon or any of the host of heaven, which I have forbidden— [4]and if it is reported to you or you hear of it, and you make a thorough inquiry, and the charge is proved true that such an abhorrent thing has occurred in Israel, [5]then you shall bring out to your gates that man or that woman who has committed this crime and you shall stone the man or woman to death. [6]On the evidence of two or three witnesses the death sentence shall be executed; a person must not be put to death on the evidence of only one witness. [7]The hands of the witnesses shall be the first raised against the person to execute the death penalty, and afterward the hands of all the people. So you shall purge the evil from your midst.

8 If a judicial decision is too difficult

p Or *tabernacles*; Heb *succoth*   q Heb *Asherah*

Pentecost, Acts 2.1; 20.16; 1 Cor 16.8). **13–15:** The festival of booths, in the King James Version "tabernacles" (Ex 23.16; 34.22; Lev 23.33–43), the autumn thanksgiving festival. **16–17:** Summary of the pilgrimage festivals (Ex 23.17; 34.23). **16.18–17.20: Laws dealing with justice and religion. 16.18–20:** The administration of justice. This law reflects a developed society in which responsibility for legal administration was delegated to appointed officials in every town. **18:** *Judges and officials,* 1.13–17. **19:** Ex

23.6–8. **21–22:** Prohibition against Canaanite cultic installations (7.5; see Ex 34.13 n.). **17.1:** An unblemished sacrifice is frequently required in cultic laws (15.21; Lev 22.17–25). **2–7:** The penalty for idolatry is the same as for enticement to idolatry (ch 13). **3:** See 4.19. **6:** 19.15; Num 35.30. **17.8–13:** A supreme tribunal is to adjudicate cases that are too difficult for local judges (16.18). Older legislation had dealt with premeditated and unpremeditated killing (Ex 21.12–14), bodily *assault* (Ex 21.18–27), and other legal rights, and had provided that cases

for you to make between one kind of bloodshed and another, one kind of legal right and another, or one kind of assault and another—any such matters of dispute in your towns—then you shall immediately go up to the place that the LORD your God will choose, 9 where you shall consult with the levitical priests and the judge who is in office in those days; they shall announce to you the decision in the case. 10 Carry out exactly the decision that they announce to you from the place that the LORD will choose, diligently observing everything they instruct you. 11 You must carry out fully the law that they interpret for you or the ruling that they announce to you; do not turn aside from the decision that they announce to you, either to the right or to the left. 12 As for anyone who presumes to disobey the priest appointed to minister there to the LORD your God, or the judge, that person shall die. So you shall purge the evil from Israel. 13 All the people will hear and be afraid, and will not act presumptuously again.

14 When you have come into the land that the LORD your God is giving you, and have taken possession of it and settled in it, and you say, "I will set a king over me, like all the nations that are around me," 15 you may indeed set over you a king whom the LORD your God will choose. One of your own community you may set as king over you; you are not permitted to put a foreigner over you, who is not of your own community. 16 Even so, he must not acquire many horses for himself, or return the people to Egypt in order to acquire more horses, since the LORD has said to you, "You must never return that way again." 17 And he must not acquire many wives for himself, or else his heart will turn away; also silver and gold he must not acquire in great quantity for himself. 18 When he has taken the throne of his kingdom, he shall have a copy of this law written for him in the presence of the levitical priests. 19 It shall remain with him and he shall read in it all the days of his life, so that he may learn to fear the LORD his God, diligently observing all the words of this law and these statutes, 20 neither exalting himself above other members of the community nor turning aside from the commandment, either to the right or to the left, so that he and his descendants may reign long over his kingdom in Israel.

18 The levitical priests, the whole tribe of Levi, shall have no allotment or inheritance within Israel. They may eat the sacrifices that are the LORD's

---

of doubt should be brought "before God" (Ex 22.9; compare 18.15–16; 33.7). Here the old legal practice is adapted to the law of the central sanctuary. **9**: *The levitical priests* are those who function in the sanctuary in distinction from town Levites (see 18.1 n.). *The judge,* perhaps the lay chief justice (19.17). One of the priests (see v. 12) was the ecclesiastical chief justice. The tribunal set up by Jehoshaphat was composed of lay and clerical judges (2 Chr 19.5–11). **14–20**: Restrictions upon the Israelite king. **17.14–15**: In the view of the Israelite tribal confederacy, perhaps formed at Shechem (Josh ch 24), kingship was alien to the theocracy (Judg 8.22–23). Israel's monarchy represented an attempt to be *like all the nations,* whose kings claimed absolute power (1 Sam 8.4–22). **16–18**: These verses allude to "the ways of the king" (compare 1 Sam 8.10–18) exemplified by Solomon, e.g. his trade in horses (1 Kings 10.26–29). **18**: *This law,* i.e. the law of Deuteronomy, which was kept in the custody of levitical priests at the sanctuary (2 Kings 22.8). **19–20**: The king, elected from among his own people (v. 15), is subject to God's law, like any other citizen. Compare the admonitions to the "prince" (the future kings of the restored line of David) in Ezek 45.7–9.

18.1–22: **The proper worship of God. 1–8**: Rights of the levitical priests. **1**: This law applies to *the whole tribe of Levi* (see Ex 28.1–5 n.), i.e. Levites who officiated at the central sanctuary ("the priests, the sons of Levi," 21.5) and those who functioned as teaching priests in the towns (12.18–19; 14.27, 29). Thus Deuteronomy adds a further qualification to priestly legislation which had distinguished only between Aaronic Levites who officiated at the altar and other Levites who were assistants in the sanctuary (see Num

portion[r] 2but they shall have no inheritance among the other members of the community; the LORD is their inheritance, as he promised them.

3 This shall be the priests' due from the people, from those offering a sacrifice, whether an ox or a sheep: they shall give to the priest the shoulder, the two jowls, and the stomach. 4The first fruits of your grain, your wine, and your oil, as well as the first of the fleece of your sheep, you shall give him. 5For the LORD your God has chosen Levi[s] out of all your tribes, to stand and minister in the name of the LORD, him and his sons for all time.

6 If a Levite leaves any of your towns, from wherever he has been residing in Israel, and comes to the place that the LORD will choose (and he may come whenever he wishes), 7then he may minister in the name of the LORD his God, like all his fellow-Levites who stand to minister there before the LORD. 8They shall have equal portions to eat, even though they have income from the sale of family possessions.[r]

9 When you come into the land that the LORD your God is giving you, you must not learn to imitate the abhorrent practices of those nations. 10No one shall be found among you who makes a son or daughter pass through fire, or who practices divination, or is a soothsayer, or an augur, or a sorcerer, 11or one who casts spells, or who consults ghosts or spirits, or who seeks oracles from the dead. 12For whoever does these things is ab-

horrent to the LORD; it is because of such abhorrent practices that the LORD your God is driving them out before you. 13You must remain completely loyal to the LORD your God. 14Although these nations that you are about to dispossess do give heed to soothsayers and diviners, as for you, the LORD your God does not permit you to do so.

15 The LORD your God will raise up for you a prophet[t] like me from among your own people; you shall heed such a prophet.[u] 16This is what you requested of the LORD your God at Horeb on the day of the assembly when you said: "If I hear the voice of the LORD my God any more, or ever again see this great fire, I will die." 17Then the LORD replied to me: "They are right in what they have said. 18I will raise up for them a prophet[t] like you from among their own people; I will put my words in the mouth of the prophet,[v] who shall speak to them everything that I command. 19Anyone who does not heed the words that the prophet[w] shall speak in my name, I myself will hold accountable. 20But any prophet who speaks in the name of other gods, or who presumes to speak in my name a word that I have not commanded the prophet to speak—that prophet shall die." 21You may say to yourself, "How can we recognize a word that the LORD has not spoken?" 22If a prophet speaks in the name

r Meaning of Heb uncertain   s Heb him
t Or prophets   u Or such prophets
v Or mouths of the prophets   w Heb he

---

ch 18). **2:** As a landless tribe, Levites are entitled to support from the sacrifices, i.e. to receive portions of the *offerings by fire* (see Lev 2.2–3 n.) and of the first fruits. **3–4:** Lev 6.14–18; 7.28–36; Num 18.8–19. **6–8:** Town Levites, whose former role was changed by the centralization of worship, may take part in the services at the central sanctuary. This provision, however, proved impracticable in Josiah's time (2 Kings 23.8–9). **18.9–14:** Prohibition against pagan superstition and magic. **10:** The meaning of the first practice is uncertain but probably refers to an ordeal of passing through the fire as a test of devotion to Molech, the god of Ammon

(12.31). This pagan rite is frequently mentioned in the Old Testament (Lev 18.21; 2 Kings 16.3; 21.6; Jer 7.31; 19.5; 32.35, etc.). **18.15–22:** Israelites are not to resort to pagan divination (see the foregoing law), for the LORD will raise up a prophet to reveal the divine will. **15:** *Like me,* Moses is regarded as the fountainhead of prophecy and the prototype of the true prophet (34.10–11). **16–17:** 5.23–31. **20–22:** The test of the true prophet (13.1–5) is that prophecy made by such a person will be brought to fulfillment according to God's purpose (1 Kings 22.26–28; Jer ch 28). **19.1–21: The administration of justice.**

of the LORD but the thing does not take place or prove true, it is a word that the LORD has not spoken. The prophet has spoken it presumptuously; do not be frightened by it.

19 When the LORD your God has cut off the nations whose land the LORD your God is giving you, and you have dispossessed them and settled in their towns and in their houses, 2 you shall set apart three cities in the land that the LORD your God is giving you to possess. 3 You shall calculate the distances *x* and divide into three regions the land that the LORD your God gives you as a possession, so that any homicide can flee to one of them.

4 Now this is the case of a homicide who might flee there and live, that is, someone who has killed another person unintentionally when the two had not been at enmity before: 5 Suppose someone goes into the forest with another to cut wood, and when one of them swings the ax to cut down a tree, the head slips from the handle and strikes the other person who then dies; the killer may flee to one of these cities and live. 6 But if the distance is too great, the avenger of blood in hot anger might pursue and overtake and put the killer to death, although a death sentence was not deserved, since the two had not been at enmity before. 7 Therefore I command you: You shall set apart three cities.

8 If the LORD your God enlarges your territory, as he swore to your ancestors—and he will give you all the land that he promised your ancestors to give you, 9 provided you diligently observe this entire commandment that I command you today, by loving the LORD your God and walking always in his ways—then you shall add three more cities to these three, 10 so that the blood of an innocent person may not be shed in the land that the LORD your God is giving you as an inheritance, thereby bringing bloodguilt upon you.

11 But if someone at enmity with another lies in wait and attacks and takes the life of that person, and flees into one of these cities, 12 then the elders of the killer's city shall send to have the culprit taken from there and handed over to the avenger of blood to be put to death. 13 Show no pity; you shall purge the guilt of innocent blood from Israel, so that it may go well with you.

14 You must not move your neighbor's boundary marker, set up by former generations, on the property that will be allotted to you in the land that the LORD your God is giving you to possess.

15 A single witness shall not suffice to convict a person of any crime or wrongdoing in connection with any offense that may be committed. Only on the evidence of two or three witnesses shall a charge be sustained. 16 If a malicious witness comes forward to accuse someone of wrongdoing, 17 then both parties to the dispute shall appear before the LORD, before the priests and the judges who are in office in those days, 18 and the judges shall make a thorough inquiry. If the witness is a false witness, having testified falsely against another, 19 then you shall do to the false witness just as the false witness had meant to do to the other. So you shall purge the evil from your midst. 20 The rest shall hear and be afraid, and a crime such as this shall never again be committed among you. 21 Show no pity: life for life, eye for eye, tooth for tooth, hand for hand, foot for foot.

x Or prepare roads to them

1–13: Cities of refuge are to be established to limit the ancient tribal law of blood revenge (Num ch 35). 1–2: Three cities are to be set apart in Canaan, besides the three in Transjordan (4.41–43). 4–6 (11–13): Num 35.16–28. 8–10: Num ch 35 and Josh ch 20 provide for only six cities.
19.14: This verse preserves an ancient law

prohibiting the removal of a property *boundary marker* (Isa 5.8; Hos 5.10; Prov 22.28). 15–21: The evidence of witnesses. 15: 17.6; Num 35.30. 16: Malicious witnessing is prohibited in the Decalogue (Ex 20.16; compare 23.1; Lev 19.16) 17: The case must come *before the LORD,* i.e. to the supreme tribunal (17.8–13). 21: *Life for life,* the "lex talionis," or "law of

20 When you go out to war against your enemies, and see horses and chariots, an army larger than your own, you shall not be afraid of them; for the LORD your God is with you, who brought you up from the land of Egypt. 2 Before you engage in battle, the priest shall come forward and speak to the troops, 3 and shall say to them: "Hear, O Israel! Today you are drawing near to do battle against your enemies. Do not lose heart, or be afraid, or panic, or be in dread of them; 4 for it is the LORD your God who goes with you, to fight for you against your enemies, to give you victory." 5 Then the officials shall address the troops, saying, "Has anyone built a new house but not dedicated it? He should go back to his house, or he might die in the battle and another dedicate it. 6 Has anyone planted a vineyard but not yet enjoyed its fruit? He should go back to his house, or he might die in the battle and another be first to enjoy its fruit. 7 Has anyone become engaged to a woman but not yet married her? He should go back to his house, or he might die in the battle and another marry her." 8 The officials shall continue to address the troops, saying, "Is anyone afraid or disheartened? He should go back to his house, or he might cause the heart of his comrades to melt like his own." 9 When the officials have finished addressing the troops, then the commanders shall take charge of them.

10 When you draw near to a town to fight against it, offer it terms of peace. 11 If it accepts your terms of peace and surrenders to you, then all the people in it shall serve you at forced labor. 12 If it does not submit to you peacefully, but makes war against you, then you shall besiege it; 13 and when the LORD your God gives it into your hand, you shall put all its males to the sword. 14 You may, however, take as your booty the women, the children, livestock, and everything else in the town, all its spoil. You may enjoy the spoil of your enemies, which the LORD your God has given you. 15 Thus you shall treat all the towns that are very far from you, which are not towns of the nations here. 16 But as for the towns of these peoples that the LORD your God is giving you as an inheritance, you must not let anything that breathes remain alive. 17 You shall annihilate them—the Hittites and the Amorites, the Canaanites and the Perizzites, the Hivites and the Jebusites—just as the LORD your God has commanded, 18 so that they may not teach you to do all the abhorrent things that they do for their gods, and you thus sin against the LORD your God.

19 If you besiege a town for a long time, making war against it in order to take it, you must not destroy its trees by wielding an ax against them. Although you may take food from them, you must not cut them down. Are trees in the field human beings that they should come under siege from you? 20 You may destroy

retaliation," is quoted to emphasize that the judges shall determine a penalty appropriate to the crime (see Ex 21.22–25 n.).
20.1–20: Rules for waging holy war. The conception of holy war, rooted in ancient wilderness experiences, provides the basis for the Deuteronomic understanding of the conquest (2.33–35; 3.3–7, 18–22; 7.1–5; 11.22–25). **1–4:** The premise is that holy war is not a human enterprise, like the wars fought by kings with trained soldiers and impressive *horses and chariots,* but is an action in which the LORD is personally engaged and in which the people respond with zealous devotion (Judg ch 5). **4:** Ex 14.14, 25. **6:** The fruit of a vineyard could not be *enjoyed* (or, put to common use) until the fifth year (Lev 19.23–25). **7:** 24.5. **8:** Holy war is waged with selected warriors. The size of the army is not important, and above all the *afraid or disheartened* must be sifted out (Judg 7.2–3).
20.10–15: In war against a non-Palestinian city, booty may be taken (see Num ch 31). **15–18:** A Palestinian city, however, must be utterly destroyed, i.e. put under the sacrificial ban, lest the inhabitants corrupt Israel's faith (9.1–6). Since the condemned city is a holy sacrifice to the LORD (compare 13.12–18), no booty can be taken (Josh ch 7). **19–20:** This stipulation limits wanton destruction of natural resources which, unlike the city and its booty, are gifts from the LORD.

only the trees that you know do not produce food; you may cut them down for use in building siegeworks against the town that makes war with you, until it falls.

21 If, in the land that the LORD your God is giving you to possess, a body is found lying in open country, and it is not known who struck the person down, ²then your elders and your judges shall come out to measure the distances to the towns that are near the body. ³The elders of the town nearest the body shall take a heifer that has never been worked, one that has not pulled in the yoke; ⁴the elders of that town shall bring the heifer down to a wadi with running water, which is neither plowed nor sown, and shall break the heifer's neck there in the wadi. ⁵Then the priests, the sons of Levi, shall come forward, for the LORD your God has chosen them to minister to him and to pronounce blessings in the name of the LORD, and by their decision all cases of dispute and assault shall be settled. ⁶All the elders of that town nearest the body shall wash their hands over the heifer whose neck was broken in the wadi, ⁷and they shall declare: "Our hands did not shed this blood, nor were we witnesses to it. ⁸Absolve, O LORD, your people Israel, whom you redeemed; do not let the guilt of innocent blood remain in the midst of your people Israel." Then they will be absolved of bloodguilt. ⁹So you shall purge the guilt of innocent blood from your midst, be-cause you must do what is right in the sight of the LORD.

10 When you go out to war against your enemies, and the LORD your God hands them over to you and you take them captive, ¹¹suppose you see among the captives a beautiful woman whom you desire and want to marry, ¹²and so you bring her home to your house: she shall shave her head, pare her nails, ¹³discard her captive's garb, and shall remain in your house a full month, mourning for her father and mother; after that you may go in to her and be her husband, and she shall be your wife. ¹⁴But if you are not satisfied with her, you shall let her go free and not sell her for money. You must not treat her as a slave, since you have dishonored her.

15 If a man has two wives, one of them loved and the other disliked, and if both the loved and the disliked have borne him sons, the firstborn being the son of the one who is disliked, ¹⁶then on the day when he wills his possessions to his sons, he is not permitted to treat the son of the loved as the firstborn in preference to the son of the disliked, who is the firstborn. ¹⁷He must acknowledge as firstborn the son of the one who is disliked, giving him a double portion*y* of all that he has; since he is the first issue of his virility, the right of the firstborn is his.

18 If someone has a stubborn and rebellious son who will not obey his father

y Heb *two-thirds*

21.1–23.14: Miscellaneous laws.
21.1–9: A law for the expiation of murder when the slayer is unknown. This ancient rite rests on the belief that bloodshed pollutes the land (vv. 8–9; 19.10, 13) and that it is the responsibility of the community to seek atonement for the crime. **2:** *The elders* (see v. 20 n.) represent the adjacent cities; the *judges* are apparently the Levites from the supreme tribunal of the sanctuary (v. 5; compare 17.8–13). **3–4:** The ritual is in some respects similar to that of Num 19.2–10 (compare Lev 14.4–7). In this case, the victim is a substitutionary sacrifice for the guilty party (see Lev 1.4 n.). **5:** *The priests, the sons of Levi,* i.e. the levitical priests from the central sanctuary rather than the town Levites (see 18.1 n.). **6–8:** The elders, representing the people, absolve themselves by the symbolic washing of hands and by taking a solemn oath. **21.10–14:** A supplement to the law on holy war against a non-Palestinian city (20.10–15) that deals with treatment of female captives (compare Num 31.18). **15–20:** Two old case laws dealing with family affairs. **21.15–17:** In antiquity it was believed that the right of the firstborn was inalienable (Gen 25.29–34). This law puts that right above family rivalry or preference. **18–21:** Like other ancient family laws (Ex 21.15, 17; Lev 20.9; compare Deut 27.16), this law is severe. It is designed to protect the family, the

and mother, who does not heed them when they discipline him, [19] then his father and his mother shall take hold of him and bring him out to the elders of his town at the gate of that place. [20] They shall say to the elders of his town, "This son of ours is stubborn and rebellious. He will not obey us. He is a glutton and a drunkard." [21] Then all the men of the town shall stone him to death. So you shall purge the evil from your midst; and all Israel will hear, and be afraid.

22 When someone is convicted of a crime punishable by death and is executed, and you hang him on a tree, [23] his corpse must not remain all night upon the tree; you shall bury him that same day, for anyone hung on a tree is under God's curse. You must not defile the land that the LORD your God is giving you for possession.

22 You shall not watch your neighbor's ox or sheep straying away and ignore them; you shall take them back to their owner. [2] If the owner does not reside near you or you do not know who the owner is, you shall bring it to your own house, and it shall remain with you until the owner claims it; then you shall return it. [3] You shall do the same with a neighbor's donkey; you shall do the same with a neighbor's garment; and you shall do the same with anything else that your neighbor loses and you find. You may not withhold your help.

4 You shall not see your neighbor's donkey or ox fallen on the road and ignore it; you shall help to lift it up.

5 A woman shall not wear a man's apparel, nor shall a man put on a woman's garment; for whoever does such things is abhorrent to the LORD your God.

6 If you come on a bird's nest, in any tree or on the ground, with fledglings or eggs, with the mother sitting on the fledglings or on the eggs, you shall not take the mother with the young. [7] Let the mother go, taking only the young for yourself, in order that it may go well with you and you may live long.

8 When you build a new house, you shall make a parapet for your roof; otherwise you might have bloodguilt on your house, if anyone should fall from it.

9 You shall not sow your vineyard with a second kind of seed, or the whole yield will have to be forfeited, both the crop that you have sown and the yield of the vineyard itself.

10 You shall not plow with an ox and a donkey yoked together.

11 You shall not wear clothes made of wool and linen woven together.

12 You shall make tassels on the four corners of the cloak with which you cover yourself.

13 Suppose a man marries a woman, but after going in to her, he dislikes her [14] and makes up charges against her, slandering her by saying, "I married this woman; but when I lay with her, I did not find evidence of her virginity." [15] The father of the young woman and her mother shall then submit the evidence of the young woman's virginity to the elders of the city at the gate. [16] The father of the young woman shall say to

---

basic social unit of Israelite society. **19–20:** *The elders* sat at the city gate (Ruth 4.1–12) where they acted as a judicial council in legal matters. This case is an application of the fifth commandment (Ex 20.12). **22–23:** Hanging (impaling) a criminal on a tree after execution was regarded as the greatest disgrace (Josh 8.29; 10.26–27; 2 Sam 4.12). Only criminals who were regarded as accursed were subjected to this ignominy.

22.1–4: Ex 23.4–5. **5:** Apparently the purpose of this law was to prevent Israelites from taking part in Canaanite rites where worshipers simulated change of sex. **6:** People should exercise a reverent concern for God's creatures (compare 20.19–20), taking only what is needed for livelihood. **8:** This safety law was designed for ancient flat-roofed houses. **9–11:** The mixing of kinds was believed to be a violation of the differences that God has ordained (v. 5; Lev 19.19). **12:** Num 15.37–41.

22.13–30: Laws governing sexual relations. These laws (also 21.15–21), written in the casuistic (case-law) style of some laws in the Covenant Code (Ex 21.1–22.17; see Ex 21.1 n.), are apparently quoted from an older source, since they show no trace of Deutero-

the elders: "I gave my daughter in marriage to this man but he dislikes her; [17]now he has made up charges against her, saying, 'I did not find evidence of your daughter's virginity.' But here is the evidence of my daughter's virginity." Then they shall spread out the cloth before the elders of the town. [18]The elders of that town shall take the man and punish him; [19]they shall fine him one hundred shekels of silver (which they shall give to the young woman's father) because he has slandered a virgin of Israel. She shall remain his wife; he shall not be permitted to divorce her as long as he lives.

20 If, however, this charge is true, that evidence of the young woman's virginity was not found, [21]then they shall bring the young woman out to the entrance of her father's house and the men of her town shall stone her to death, because she committed a disgraceful act in Israel by prostituting herself in her father's house. So you shall purge the evil from your midst.

22 If a man is caught lying with the wife of another man, both of them shall die, the man who lay with the woman as well as the woman. So you shall purge the evil from Israel.

23 If there is a young woman, a virgin already engaged to be married, and a man meets her in the town and lies with her, [24]you shall bring both of them to the gate of that town and stone them to death, the young woman because she did not cry for help in the town and the man because he violated his neighbor's wife. So you shall purge the evil from your midst.

25 But if the man meets the engaged woman in the open country, and the man seizes her and lies with her, then only the man who lay with her shall die. [26]You shall do nothing to the young woman; the young woman has not committed an offense punishable by death, because this case is like that of someone who attacks and murders a neighbor. [27]Since he found her in the open country, the engaged woman may have cried for help, but there was no one to rescue her.

28 If a man meets a virgin who is not engaged, and seizes her and lies with her, and they are caught in the act, [29]the man who lay with her shall give fifty shekels of silver to the young woman's father, and she shall become his wife. Because he violated her he shall not be permitted to divorce her as long as he lives.

30[z] A man shall not marry his father's wife, thereby violating his father's rights.[a]

23 No one whose testicles are crushed or whose penis is cut off shall be admitted to the assembly of the LORD.

2 Those born of an illicit union shall not be admitted to the assembly of the LORD. Even to the tenth generation, none of their descendants shall be admitted to the assembly of the LORD.

3 No Ammonite or Moabite shall be admitted to the assembly of the LORD. Even to the tenth generation, none of their descendants shall be admitted to the assembly of the LORD, [4]because they did not meet you with food and water on your journey out of Egypt, and because

z Ch 23.1 in Heb   a Heb *uncovering his father's skirt*

___

nomic style. **13–21**: The proof of virginity. **15**: On *the elders*, see 21.19–20 n. **21**: On the expression *committed a disgraceful act in Israel,* see Gen 34.7 n.

22.22–27: Adultery, a violation of the seventh commandment, was punishable by death (Lev 18.20; 20.10). Intercourse with a betrothed virgin is considered as adultery because the woman is already, in effect, another man's wife. **28–29**: The money payment to be given to the father is the marriage present

(Ex 22.16–17). **30**: *Father's wife,* i.e. stepmother (27.20; Lev 18.8; 20.11).

23.1–8: Exclusions from the assembly of the LORD (33.5; Judg 20.2). **1–2**: Only those who are unblemished can "present themselves before God" (Josh 24.1). This excludes eunuchs (compare Lev 21.17–23), who in ancient pagan religions were temple priests, and perhaps also excludes children born of incestuous union (Lev 18.6–18). **3–6**: 2.9–25. The account in Num chs 21–22 does not mention

they hired against you Balaam son of Beor, from Pethor of Mesopotamia, to curse you. 5 (Yet the LORD your God refused to heed Balaam; the LORD your God turned the curse into a blessing for you, because the LORD your God loved you.) 6 You shall never promote their welfare or their prosperity as long as you live.

7 You shall not abhor any of the Edomites, for they are your kin. You shall not abhor any of the Egyptians, because you were an alien residing in their land. 8 The children of the third generation that are born to them may be admitted to the assembly of the LORD.

9 When you are encamped against your enemies you shall guard against any impropriety.

10 If one of you becomes unclean because of a nocturnal emission, then he shall go outside the camp; he must not come within the camp. 11 When evening comes, he shall wash himself with water, and when the sun has set, he may come back into the camp.

12 You shall have a designated area outside the camp to which you shall go. 13 With your utensils you shall have a trowel; when you relieve yourself outside, you shall dig a hole with it and then cover up your excrement. 14 Because the LORD your God travels along with your camp, to save you and to hand over your enemies to you, therefore your camp must be holy, so that he may not see anything indecent among you and turn away from you.

15 Slaves who have escaped to you from their owners shall not be given back to them. 16 They shall reside with you, in your midst, in any place they choose in any one of your towns, wherever they please; you shall not oppress them.

17 None of the daughters of Israel shall be a temple prostitute; none of the sons of Israel shall be a temple prostitute. 18 You shall not bring the fee of a prostitute or the wages of a male prostitute[b] into the house of the LORD your God in payment for any vow, for both of these are abhorrent to the LORD your God.

19 You shall not charge interest on loans to another Israelite, interest on money, interest on provisions, interest on anything that is lent. 20 On loans to a foreigner you may charge interest, but on loans to another Israelite you may not charge interest, so that the LORD your God may bless you in all your undertakings in the land that you are about to enter and possess.

21 If you make a vow to the LORD your God, do not postpone fulfilling it; for the LORD your God will surely require it of you, and you would incur guilt. 22 But if you refrain from vowing, you will not incur guilt. 23 Whatever your lips utter you must diligently perform, just as you have freely vowed to the LORD your God with your own mouth.

24 If you go into your neighbor's vineyard, you may eat your fill of grapes, as many as you wish, but you shall not put any in a container.

25 If you go into your neighbor's standing grain, you may pluck the ears with your hand, but you shall not put a sickle to your neighbor's standing grain.

b Heb *a dog*

Ammonite hostility against Israel. **7:** Gen 25.24–26; 36.1. **23.9–14:** A supplement to the rules for holy war (see ch 20). Ceremonial cleanness is demanded because *the LORD your God travels along with your camp.* **10–11:** Lev 15.16–17. **12–14:** The camp must be clean not just in the sense of being sanitary but in the sense of being *holy* (see Lev 11.1–47 n.). **23.15–25.19: Laws dealing with humanitarian and religious obligations.**

23.15–16: In contrast with this humane law, the Code of Hammurabi decreed death as the penalty for sheltering a fugitive slave. **17–18:** A strict prohibition against sacred prostitution (see Gen 38.15 n.). **19–20:** See Ex 22.25 n.; Lev 25.35–38; Deut 15.1–11. Loans within Israel, as distinguished from loans to outsiders, were usually occasioned by financial desperation and therefore became means of oppressing a fellow-Israelite. **21–23:** See Lev ch 27. **24–25:** Neighbors' goodwill

24 Suppose a man enters into marriage with a woman, but she does not please him because he finds something objectionable about her, and so he writes her a certificate of divorce, puts it in her hand, and sends her out of his house; she then leaves his house ²and goes off to become another man's wife. ³Then suppose the second man dislikes her, writes her a bill of divorce, puts it in her hand, and sends her out of his house (or the second man who married her dies); ⁴her first husband, who sent her away, is not permitted to take her again to be his wife after she has been defiled; for that would be abhorrent to the LORD, and you shall not bring guilt on the land that the LORD your God is giving you as a possession.

5 When a man is newly married, he shall not go out with the army or be charged with any related duty. He shall be free at home one year, to be happy with the wife whom he has married.

6 No one shall take a mill or an upper millstone in pledge, for that would be taking a life in pledge.

7 If someone is caught kidnaping another Israelite, enslaving or selling the Israelite, then that kidnaper shall die. So you shall purge the evil from your midst.

8 Guard against an outbreak of a leprous ᶜ skin disease by being very careful; you shall carefully observe whatever the levitical priests instruct you, just as I have commanded them. ⁹Remember what the LORD your God did to Miriam on your journey out of Egypt.

10 When you make your neighbor a loan of any kind, you shall not go into the house to take the pledge. ¹¹You shall wait outside, while the person to whom you are making the loan brings the pledge out to you. ¹²If the person is poor, you shall not sleep in the garment given you as ᵈ the pledge. ¹³You shall give the pledge back by sunset, so that your neighbor may sleep in the cloak and bless you; and it will be to your credit before the LORD your God.

14 You shall not withhold the wages of poor and needy laborers, whether other Israelites or aliens who reside in your land in one of your towns. ¹⁵You shall pay them their wages daily before sunset, because they are poor and their livelihood depends on them; otherwise they might cry to the LORD against you, and you would incur guilt.

16 Parents shall not be put to death for their children, nor shall children be put to death for their parents; only for their own crimes may persons be put to death.

17 You shall not deprive a resident alien or an orphan of justice; you shall not take a widow's garment in pledge. ¹⁸Remember that you were a slave in Egypt and the LORD your God redeemed you from there; therefore I command you to do this.

19 When you reap your harvest in your field and forget a sheaf in the field, you shall not go back to get it; it shall be

---

c A term for several skin diseases; precise meaning uncertain   *d* Heb lacks *the garment given you as*

---

should not be presumed upon to the point of stealing from their harvest.
24.1–4: An old case law dealing with remarriage after divorce. Initiative for divorce rested with the husband, who, however, could not act without following a legal procedure, including the formulation of a *certificate of divorce* (Jer 3.8), stating the grounds. **5**: Another of the rules for holy war (20.1–20; 21.10–14; 23.9–14). **6**: A limitation upon the creditor. Since daily life depends upon bread, the mill cannot be taken as security for a loan (vv. 10–13). **7**: Ex 21.16. **8–9**: See Lev chs 13–14. *Remember . . . Miriam,* see comments on Num 12.1–15.

24.10–13: See Ex 22.26 n. (compare Am 2.8; Prov 22.27). **14–15**: Lev 19.13. Israel's humanitarianism is based on the conviction that the LORD is the vindicator of the weak and helpless. **16**: This law modifies the ancient belief that guilt affected a whole social group, especially the family (Num 16.31–33; Josh 7.24–25; 2 Sam 21.1–9). Although Deuteronomy does not reject the solidarity of the community in guilt (21.1–9; compare 13.12–18), it puts in the foreground the principle of individual responsibility (Jer 31.29; Ezek ch 18). See 2 Kings 14.6 for the application of this law. **17–18**: 15.15. **19–22**: Lev 19.9–10; 23.22.

left for the alien, the orphan, and the widow, so that the LORD your God may bless you in all your undertakings. 20 When you beat your olive trees, do not strip what is left; it shall be for the alien, the orphan, and the widow.

21 When you gather the grapes of your vineyard, do not glean what is left; it shall be for the alien, the orphan, and the widow. 22 Remember that you were a slave in the land of Egypt; therefore I am commanding you to do this.

25 Suppose two persons have a dispute and enter into litigation, and the judges decide between them, declaring one to be in the right and the other to be in the wrong. 2 If the one in the wrong deserves to be flogged, the judge shall make that person lie down and be beaten in his presence with the number of lashes proportionate to the offense. 3 Forty lashes may be given but not more; if more lashes than these are given, your neighbor will be degraded in your sight.

4 You shall not muzzle an ox while it is treading out the grain.

5 When brothers reside together, and one of them dies and has no son, the wife of the deceased shall not be married outside the family to a stranger. Her husband's brother shall go in to her, taking her in marriage, and performing the duty of a husband's brother to her, 6 and the firstborn whom she bears shall succeed to the name of the deceased brother, so that his name may not be blotted out of Israel. 7 But if the man has no desire to marry his brother's widow, then his brother's widow shall go up to the elders at the gate and say, "My husband's brother refuses to perpetuate his brother's name in Israel; he will not perform the duty of a husband's brother to me." 8 Then the elders of his town shall summon him and speak to him. If he persists, saying, "I have no desire to marry her," 9 then his brother's wife shall go up to him in the presence of the elders, pull his sandal off his foot, spit in his face, and declare, "This is what is done to the man who does not build up his brother's house." 10 Throughout Israel his family shall be known as "the house of him whose sandal was pulled off."

11 If men get into a fight with one another, and the wife of one intervenes to rescue her husband from the grip of his opponent by reaching out and seizing his genitals, 12 you shall cut off her hand; show no pity.

13 You shall not have in your bag two kinds of weights, large and small. 14 You shall not have in your house two kinds of measures, large and small. 15 You shall have only a full and honest weight; you shall have only a full and honest measure, so that your days may be long in the land that the LORD your God is giving you. 16 For all who do such things, all who act dishonestly, are abhorrent to the LORD your God.

17 Remember what Amalek did to you on your journey out of Egypt, 18 how he attacked you on the way, when you were faint and weary, and struck down all who lagged behind you; he did not fear God. 19 Therefore when the LORD your God has given you rest from all your enemies on every hand, in the land that the LORD your God is giving you as an inheritance to possess, you shall blot out the remembrance of Amalek from under heaven; do not forget.

26 When you have come into the land that the LORD your God is giving you as an inheritance to possess, and you possess it, and settle in it, 2 you shall take some of the first of all the fruit

---

25.1–3: An old case law which imposes judicial restrictions on corporal punishment. *The judges,* 16.18. **4:** A law in the humane spirit of 22.6–7. **5–10:** An old case law on levirate marriage (see Gen 38.8 n.). **6:** Since a man's name, according to ancient thought, was the bearer of his person, a father lived on in his son (Gen 48.15–16). **7:** As in the case of Tamar (Gen ch 38), the wife had the obligation to see that *the duty of a husband's brother* (or levir) was performed. **9–10:** For the symbolic action of removing the sandal, in a more complicated levirate marriage situation, see Ruth 4.7 n.

25.11–12: Another law (compare the preceding one) intended to make it possible for

of the ground, which you harvest from the land that the Lord your God is giving you, and you shall put it in a basket and go to the place that the Lord your God will choose as a dwelling for his name. 3 You shall go to the priest who is in office at that time, and say to him, "Today I declare to the Lord your God that I have come into the land that the Lord swore to our ancestors to give us." 4 When the priest takes the basket from your hand and sets it down before the altar of the Lord your God, 5 you shall make this response before the Lord your God: "A wandering Aramean was my ancestor; he went down into Egypt and lived there as an alien, few in number, and there he became a great nation, mighty and populous. 6 When the Egyptians treated us harshly and afflicted us, by imposing hard labor on us, 7 we cried to the Lord, the God of our ancestors; the Lord heard our voice and saw our affliction, our toil, and our oppression. 8 The Lord brought us out of Egypt with a mighty hand and an outstretched arm, with a terrifying display of power, and with signs and wonders; 9 and he brought us into this place and gave us this land, a land flowing with milk and honey. 10 So now I bring the first of the fruit of the ground that you, O Lord, have given me." You shall set it down before the Lord your God and bow down before the Lord your God. 11 Then you, together with the Levites and the aliens who reside among you, shall celebrate with all the bounty that the Lord your God has given to you and to your house.

12 When you have finished paying all the tithe of your produce in the third year (which is the year of the tithe), giving it to the Levites, the aliens, the orphans, and the widows, so that they may eat their fill within your towns, 13 then you shall say before the Lord your God: "I have removed the sacred portion from the house, and I have given it to the Levites, the resident aliens, the orphans, and the widows, in accordance with your entire commandment that you commanded me; I have neither transgressed nor forgotten any of your commandments: 14 I have not eaten of it while in mourning; I have not removed any of it while I was unclean; and I have not offered any of it to the dead. I have obeyed the Lord my God, doing just as you commanded me. 15 Look down from your holy habitation, from heaven, and bless your people Israel and the ground that you have given us, as you swore to our ancestors—a land flowing with milk and honey."

16 This very day the Lord your God is commanding you to observe these statutes and ordinances; so observe them diligently with all your heart and with all your soul. 17 Today you have obtained the Lord's agreement: to be your God; and for you to walk in his ways, to keep his statutes, his commandments, and his ordinances, and to obey him. 18 Today the Lord has obtained your agreement:

---

a man to produce offspring, in this case by protecting his genitals, the spring of fertility. **13–16:** Compare Lev 19.35–36. **17–19:** Holy war against Amalek (see Ex 17.8–15).

26.1–19: Concluding liturgies and exhortation. This chapter anticipates the climax and conclusion of Moses' main address (ch 28). **1–11:** A liturgy for the presentation of the first fruits at the central sanctuary. **2–3:** The occasion is the harvest pilgrimage festival, the feast of weeks (16.9–12), when the worshiper thanks God for the gift of the land and for a bounteous harvest. **5–10:** A basic confession of faith in the form of a story. **5:** A *wandering Aramean,* a reference to Jacob's semi-nomadic life. **6:** *Us:* the plural pronouns show that the worshiper identifies with the community of faith in making this recitation.

26.12–15: A liturgy for *the year of the tithe.* In the worshiper's home town *the sacred portion* has been set aside according to the ordinance of 14.28–29. **13:** *Before the Lord,* at the central sanctuary. **14:** The worshiper has not defiled the holy offering by touching it in a state of ritual uncleanness while mourning (Num 19.11–16) or by bringing it to a tomb as a funerary offering.

26.16–19: This exhortation concludes the exposition of the *statutes* and *ordinances* contained in chs 12–26. The language here may echo a ceremony of covenant renewal at the central sanctuary when the law was publicly

to be his treasured people, as he promised you, and to keep his commandments; [19] for him to set you high above all nations that he has made, in praise and in fame and in honor; and for you to be a people holy to the LORD your God, as he promised.

27 Then Moses and the elders of Israel charged all the people as follows: Keep the entire commandment that I am commanding you today. [2] On the day that you cross over the Jordan into the land that the LORD your God is giving you, you shall set up large stones and cover them with plaster. [3] You shall write on them all the words of this law when you have crossed over, to enter the land that the LORD your God is giving you, a land flowing with milk and honey, as the LORD, the God of your ancestors, promised you. [4] So when you have crossed over the Jordan, you shall set up these stones, about which I am commanding you today, on Mount Ebal, and you shall cover them with plaster. [5] And you shall build an altar there to the LORD your God, an altar of stones on which you have not used an iron tool. [6] You must build the altar of the LORD your God of unhewn[e] stones. Then offer up burnt offerings on it to the LORD your God, [7] make sacrifices of well-being, and eat them there, rejoicing before the LORD

your God. [8] You shall write on the stones all the words of this law very clearly.

9 Then Moses and the levitical priests spoke to all Israel, saying: Keep silence and hear, O Israel! This very day you have become the people of the LORD your God. [10] Therefore obey the LORD your God, observing his commandments and his statutes that I am commanding you today.

11 The same day Moses charged the people as follows: [12] When you have crossed over the Jordan, these shall stand on Mount Gerizim for the blessing of the people: Simeon, Levi, Judah, Issachar, Joseph, and Benjamin. [13] And these shall stand on Mount Ebal for the curse: Reuben, Gad, Asher, Zebulun, Dan, and Naphtali. [14] Then the Levites shall declare in a loud voice to all the Israelites:

15 "Cursed be anyone who makes an idol or casts an image, anything abhorrent to the LORD, the work of an artisan, and sets it up in secret." All the people shall respond, saying, "Amen!"

16 "Cursed be anyone who dishonors father or mother." All the people shall say, "Amen!"

17 "Cursed be anyone who moves a neighbor's boundary marker." All the people shall say, "Amen!"

e Heb *whole*

---

read and the covenant relation between the LORD and Israel was renewed (see 31.10–11 n.). *This very day,* i.e. the day of the ceremony, the LORD restates the demands of the covenant (v. 16), *today* the people reaffirm that the LORD alone is their God (v. 17), and *today* the promise is renewed that Israel is the LORD's *treasured people* (vv. 18–19; see Ex 19.5).

27.1–26: The Shechem ceremony. A cultic ceremony is to be inaugurated which will solemnly dramatize Israel's covenant responsibilities (11.26–32). **1:** Here Moses is referred to in the third person. This chapter stands independently and interrupts Moses' address (chs 5–26 and ch 28). **4:** *Mount Ebal* and Mount Gerizim (vv. 12, 13) overlook the city of Shechem (see Gen 12.6 n.). Contrary to the law of ch 12, an altar of unhewn stones (Ex 20.25) is to be set up at a place other than

the central sanctuary, an evidence of the antiquity and independence of this tradition. **7:** *Eat . . . rejoicing,* a reference to the communion meal in connection with the *sacrifices of well-being* (see Lev ch 3).

27.11–13: These verses deal with an ancient ceremony at Shechem, instituted at the command of Moses according to Josh 8.30–35. The antiquity of the tribal list is seen in the facts that Levi is listed as a tribe (Gen 49.5–7) and that the division of the house of Joseph into Manasseh and Ephraim has not yet occurred (Gen 49.22–26).

27.15–26: The ritual of the blessing (v. 12; 28.3–6) is not preserved in this fragmentary record; only the ceremony of the curse is given (v. 13). The presiding Levites read a curse and in response the congregation says *Amen,* i.e. in full knowledge of the divine disapproval expressed in the twelve curses (correspond-

18 "Cursed be anyone who misleads a blind person on the road." All the people shall say, "Amen!"

19 "Cursed be anyone who deprives the alien, the orphan, and the widow of justice." All the people shall say, "Amen!"

20 "Cursed be anyone who lies with his father's wife, because he has violated his father's rights."*f* All the people shall say, "Amen!"

21 "Cursed be anyone who lies with any animal." All the people shall say, "Amen!"

22 "Cursed be anyone who lies with his sister, whether the daughter of his father or the daughter of his mother." All the people shall say, "Amen!"

23 "Cursed be anyone who lies with his mother-in-law." All the people shall say, "Amen!"

24 "Cursed be anyone who strikes down a neighbor in secret." All the people shall say, "Amen!"

25 "Cursed be anyone who takes a bribe to shed innocent blood." All the people shall say, "Amen!"

26 "Cursed be anyone who does not uphold the words of this law by observing them." All the people shall say, "Amen!"

28 If you will only obey the LORD your God, by diligently observing all his commandments that I am commanding you today, the LORD your God will set you high above all the nations of the earth; 2all these blessings shall come upon you and overtake you, if you obey the LORD your God:

3 Blessed shall you be in the city, and blessed shall you be in the field.

4 Blessed shall be the fruit of your womb, the fruit of your ground, and the fruit of your livestock, both the increase of your cattle and the issue of your flock.

5 Blessed shall be your basket and your kneading bowl.

6 Blessed shall you be when you come in, and blessed shall you be when you go out.

7 The LORD will cause your enemies who rise against you to be defeated before you; they shall come out against you one way, and flee before you seven ways. 8The LORD will command the blessing upon you in your barns, and in all that you undertake; he will bless you in the land that the LORD your God is giving you. 9The LORD will establish you as his holy people, as he has sworn to you, if you keep the commandments of the LORD your God and walk in his ways. 10All the peoples of the earth shall see that you are called by the name of the LORD, and they shall be afraid of you. 11The LORD will make you abound in prosperity, in the fruit of your womb, in the fruit of your livestock, and in the fruit of your ground in the land that the LORD swore to your ancestors to give you. 12The LORD will open for you his rich storehouse, the heavens, to give the rain of your land in its season and to bless all your undertakings. You will lend to many nations, but you will not borrow. 13The LORD will make you the head, and not the tail; you shall be only at the top, and not at the bottom—if you obey the commandments of the LORD your God, which I am commanding you today, by diligently observing them, 14and if you do not turn aside from any of the words that I am commanding you today, either to the right or to the left, following other gods to serve them.

15 But if you will not obey the LORD

f Heb *uncovered his father's skirt*

---

ing to the twelve tribes), the people solemnly take upon themselves the covenant responsibilities (Jer 11.3). The twelve curses are very old laws which, with the exception of the last two, are paralleled in other codes of the Pentateuch.

28.1–68. The conclusion of Moses' second address. Moses calls the people to decision by announcing the blessings which accompany obedience and the curses that fall upon disobedience. **1**: Continuation from 26.16–19. **3–6**: These six blessings, formulated in short, emphatic style, were perhaps part of the old covenant ceremony preserved in fragmentary form in 27.11–26.

28.7–14: A commentary on the meaning

your God by diligently observing all his commandments and decrees, which I am commanding you today, then all these curses shall come upon you and overtake you:

16 Cursed shall you be in the city, and cursed shall you be in the field.

17 Cursed shall be your basket and your kneading bowl.

18 Cursed shall be the fruit of your womb, the fruit of your ground, the increase of your cattle and the issue of your flock.

19 Cursed shall you be when you come in, and cursed shall you be when you go out.

20 The LORD will send upon you disaster, panic, and frustration in everything you attempt to do, until you are destroyed and perish quickly, on account of the evil of your deeds, because you have forsaken me. 21 The LORD will make the pestilence cling to you until it has consumed you off the land that you arc entering to possess. 22 The LORD will afflict you with consumption, fever, inflammation, with fiery heat and drought, and with blight and mildew; they shall pursue you until you perish. 23 The sky over your head shall be bronze, and the earth under you iron. 24 The LORD will change the rain of your land into powder, and only dust shall come down upon you from the sky until you are destroyed.

25 The LORD will cause you to be defeated before your enemies; you shall go out against them one way and flee before them seven ways. You shall become an object of horror to all the kingdoms of the earth. 26 Your corpses shall be food for every bird of the air and animal of the earth, and there shall be no one to frighten them away. 27 The LORD will afflict you with the boils of Egypt, with ulcers, scurvy, and itch, of which you cannot be healed. 28 The LORD will afflict you with madness, blindness, and confusion of mind; 29 you shall grope about at noon as blind people grope in darkness, but you shall be unable to find your way; and you shall be continually abused and robbed, without anyone to help. 30 You shall become engaged to a woman, but another man shall lie with her. You shall build a house, but not live in it. You shall plant a vineyard, but not enjoy its fruit. 31 Your ox shall be butchered before your eyes, but you shall not eat of it. Your donkey shall be stolen in front of you, and shall not be restored to you. Your sheep shall be given to your enemies, without anyone to help you. 32 Your sons and daughters shall be given to another people, while you look on; you will strain your eyes looking for them all day but be powerless to do anything. 33 A people whom you do not know shall eat up the fruit of your ground and of all your labors; you shall be continually abused and crushed, 34 and driven mad by the sight that your eyes shall see. 35 The LORD will strike you on the knees and on the legs with grievous boils of which you cannot be healed, from the sole of your foot to the crown of your head. 36 The LORD will bring you, and the king whom you set over you, to a nation that neither you nor your ancestors have known, where you shall serve other gods, of wood and stone. 37 You shall become an object of horror, a proverb, and a byword among all the peoples where the LORD will lead you.

38 You shall carry much seed into the field but shall gather little in, for the locust shall consume it. 39 You shall plant vineyards and dress them, but you shall neither drink the wine nor gather the

---

of the blessings. **16–19**: These six curses, parallel to the six blessings, likewise echo the old Shechem ceremony.

28.20–46: A commentary on the curses. Perhaps it was the reading of this chapter which had a profound effect upon King Josiah (2 Kings 22.11–13). **27**: A reference to the Egyptian plagues. **30**: 20.5–7.

28.36–37: Conquest by a strange *nation* was experienced in 721 B.C. when the Assyrians destroyed the Northern Kingdom and deported its population into *captivity* (v. 41).

grapes, for the worm shall eat them.
40 You shall have olive trees throughout
all your territory, but you shall not
anoint yourself with the oil, for your olives shall drop off. 41 You shall have sons
and daughters, but they shall not remain
yours, for they shall go into captivity.
42 All your trees and the fruit of your
ground the cicada shall take over.
43 Aliens residing among you shall ascend
above you higher and higher, while you
shall descend lower and lower. 44 They
shall lend to you but you shall not lend
to them; they shall be the head and you
shall be the tail.

45 All these curses shall come upon
you, pursuing and overtaking you until
you are destroyed, because you did not
obey the LORD your God, by observing
the commandments and the decrees that
he commanded you. 46 They shall be
among you and your descendants as a
sign and a portent forever.

47 Because you did not serve the
LORD your God joyfully and with gladness of heart for the abundance of everything, 48 therefore you shall serve your
enemies whom the LORD will send
against you, in hunger and thirst, in nakedness and lack of everything. He will
put an iron yoke on your neck until he
has destroyed you. 49 The LORD will
bring a nation from far away, from the
end of the earth, to swoop down on you
like an eagle, a nation whose language
you do not understand, 50 a grim-faced
nation showing no respect to the old or
favor to the young. 51 It shall consume
the fruit of your livestock and the fruit of
your ground until you are destroyed,
leaving you neither grain, wine, and oil,
nor the increase of your cattle and the
issue of your flock, until it has made you
perish. 52 It shall besiege you in all your
towns until your high and fortified walls,
in which you trusted, come down
throughout your land; it shall besiege

you in all your towns throughout the
land that the LORD your God has given
you. 53 In the desperate straits to which
the enemy siege reduces you, you will
eat the fruit of your womb, the flesh of
your own sons and daughters whom the
LORD your God has given you. 54 Even
the most refined and gentle of men
among you will begrudge food to his
own brother, to the wife whom he embraces, and to the last of his remaining
children, 55 giving to none of them any of
the flesh of his children whom he is eating, because nothing else remains to him,
in the desperate straits to which the enemy siege will reduce you in all your
towns. 56 She who is the most refined and
gentle among you, so gentle and refined
that she does not venture to set the sole
of her foot on the ground, will begrudge
food to the husband whom she embraces, to her own son, and to her own
daughter, 57 begrudging even the after-
birth that comes out from between her
thighs, and the children that she bears,
because she is eating them in secret for
lack of anything else, in the desperate
straits to which the enemy siege will reduce you in your towns.

58 If you do not diligently observe all
the words of this law that are written in
this book, fearing this glorious and awesome name, the LORD your God, 59 then
the LORD will overwhelm both you and
your offspring with severe and lasting
afflictions and grievous and lasting mala-
dies. 60 He will bring back upon you all
the diseases of Egypt, of which you were
in dread, and they shall cling to you.
61 Every other malady and affliction,
even though not recorded in the book of
this law, the LORD will inflict on you
until you are destroyed. 62 Although once
you were as numerous as the stars in
heaven, you shall be left few in number,
because you did not obey the LORD your
God. 63 And just as the LORD took delight

28.47–68: A further expansion of the
meaning of the curses. These verses apparently reflect the tragedy that befell Judah when
the Babylonians laid siege to fortified cities

(v. 52), conquered Jerusalem, and between
597 and 587 took many people into captivity.
53–57: Lev 26.29.

in making you prosperous and numerous, so the LORD will take delight in bringing you to ruin and destruction; you shall be plucked off the land that you are entering to possess. 64 The LORD will scatter you among all peoples, from one end of the earth to the other; and there you shall serve other gods, of wood and stone, which neither you nor your ancestors have known. 65 Among those nations you shall find no ease, no resting place for the sole of your foot. There the LORD will give you a trembling heart, failing eyes, and a languishing spirit. 66 Your life shall hang in doubt before you; night and day you shall be in dread, with no assurance of your life. 67 In the morning you shall say, "If only it were evening!" and at evening you shall say, "If only it were morning!"—because of the dread that your heart shall feel and the sights that your eyes shall see. 68 The LORD will bring you back in ships to Egypt, by a route that I promised you would never see again; and there you shall offer yourselves for sale to your enemies as male and female slaves, but there will be no buyer.

29 g These are the words of the covenant that the LORD commanded Moses to make with the Israelites in the land of Moab, in addition to the covenant that he had made with them at Horeb.

2 h Moses summoned all Israel and said to them: You have seen all that the LORD did before your eyes in the land of Egypt, to Pharaoh and to all his servants and to all his land, 3 the great trials that your eyes saw, the signs, and those great wonders. 4 But to this day the LORD has not given you a mind to understand, or eyes to see, or ears to hear. 5 I have led you

forty years in the wilderness. The clothes on your back have not worn out, and the sandals on your feet have not worn out; 6 you have not eaten bread, and you have not drunk wine or strong drink—so that you may know that I am the LORD your God. 7 When you came to this place, King Sihon of Heshbon and King Og of Bashan came out against us for battle, but we defeated them. 8 We took their land and gave it as an inheritance to the Reubenites, the Gadites, and the half-tribe of Manasseh. 9 Therefore diligently observe the words of this covenant, in order that you may succeed i in everything that you do.

10 You stand assembled today, all of you, before the LORD your God—the leaders of your tribes, j your elders, and your officials, all the men of Israel, 11 your children, your women, and the aliens who are in your camp, both those who cut your wood and those who draw your water— 12 to enter into the covenant of the LORD your God, sworn by an oath, which the LORD your God is making with you today; 13 in order that he may establish you today as his people, and that he may be your God, as he promised you and as he swore to your ancestors, to Abraham, to Isaac, and to Jacob. 14 I am making this covenant, sworn by an oath, not only with you who stand here with us today before the LORD our God, 15 but also with those who are not here with us today. 16 You know how we lived in the land of Egypt, and how we came through the midst of the nations through which you passed.

g Ch 28.69 in Heb   h Ch 29.1 in Heb
i Or *deal wisely*   j Gk Syr: Heb *your leaders,
your tribes*

28.68: The disaster is interpreted as a return to Egyptian bondage, more pathetic than the former slavery. **29.1–30.20: Moses' third address.** In this supplement Moses again exhorts Israel to renew the covenant and warns of the disastrous consequences of disobedience. **29.1:** *The covenant* in Moab is not the original covenant but a renewal of it. **2–9:** As in the other addresses, the covenant is based on a recital of the LORD's mighty acts (compare Ex 19.3–6; Josh 24.2–13). **29.10–14:** See 26.16–19 n. **14:** On the contemporaneity of the covenant, see 5.2–3 n. **18–19:** The community is responsible for rooting out the infectious poison of idolatry lest the *moist* (fertile and wholesome) people be swept away with the *dry* (ch 13).

17 You have seen their detestable things, the filthy idols of wood and stone, of silver and gold, that were among them. 18 It may be that there is among you a man or woman, or a family or tribe, whose heart is already turning away from the LORD our God to serve the gods of those nations. It may be that there is among you a root sprouting poisonous and bitter growth. 19 All who hear the words of this oath and bless themselves, thinking in their hearts, "We are safe even though we go our own stubborn ways" (thus bringing disaster on moist and dry alike)*k*— 20 the LORD will be unwilling to pardon them, for the LORD's anger and passion will smoke against them. All the curses written in this book will descend on them, and the LORD will blot out their names from under heaven. 21 The LORD will single them out from all the tribes of Israel for calamity, in accordance with all the curses of the covenant written in this book of the law. 22 The next generation, your children who rise up after you, as well as the foreigner who comes from a distant country, will see the devastation of that land and the afflictions with which the LORD has afflicted it— 23 all its soil burned out by sulfur and salt, nothing planted, nothing sprouting, unable to support any vegetation, like the destruction of Sodom and Gomorrah, Admah and Zeboiim, which the LORD destroyed in his fierce anger— 24 they and indeed all the nations will wonder, "Why has the LORD done thus to this land? What caused this great display of anger?" 25 They will conclude, "It is because they abandoned the covenant of the LORD, the God of their ancestors, which he made with them when he brought them out of the land of Egypt. 26 They turned and served other gods, worshiping them, gods whom they had not known and whom he had not allotted to them; 27 so the anger of the LORD was kindled against that land, bringing on it every curse written in this book. 28 The LORD uprooted them from their land in anger, fury, and great wrath, and cast them into another land, as is now the case." 29 The secret things belong to the LORD our God, but the revealed things belong to us and to our children forever, to observe all the words of this law.

30 When all these things have happened to you, the blessings and the curses that I have set before you, if you call them to mind among all the nations where the LORD your God has driven you, 2 and return to the LORD your God, and you and your children obey him with all your heart and with all your soul, just as I am commanding you today, 3 then the LORD your God will restore your fortunes and have compassion on you, gathering you again from all the peoples among whom the LORD your God has scattered you. 4 Even if you are exiled to the ends of the world,*l* from there the LORD your God will gather you, and from there he will bring you back. 5 The LORD your God will bring you into the land that your ancestors possessed, and you will possess it; he will make you more prosperous and numerous than your ancestors.

6 Moreover, the LORD your God will circumcise your heart and the heart of your descendants, so that you will love the LORD your God with all your heart and with all your soul, in order that you may live. 7 The LORD your God will put all these curses on your enemies and on the adversaries who took advantage of you. 8 Then you shall again obey the LORD, observing all his commandments

k Meaning of Heb uncertain   *l* Heb *of heaven*

29.20: See Ex 34.14 n. 23: See Gen ch 19. 29: *The secret things* refer to the divine wisdom beyond the human ken; *the revealed things* are the teachings set forth in Deuteronomy.

30.1–10: This passage presupposes that Israel is already in exile and anticipates the time when the LORD will restore the people. 2: *Return to the LORD,* that is, repent—an act that involves turning away from unworthy loyalties and turning the heart (see 6.4–5 n.) to God (v. 10). 6: God will work the inner change that makes it possible to fulfill the law

that I am commanding you today, [9] and the LORD your God will make you abundantly prosperous in all your undertakings, in the fruit of your body, in the fruit of your livestock, and in the fruit of your soil. For the LORD will again take delight in prospering you, just as he delighted in prospering your ancestors, [10] when you obey the LORD your God by observing his commandments and decrees that are written in this book of the law, because you turn to the LORD your God with all your heart and with all your soul.

11 Surely, this commandment that I am commanding you today is not too hard for you, nor is it too far away. [12] It is not in heaven, that you should say, "Who will go up to heaven for us, and get it for us so that we may hear it and observe it?" [13] Neither is it beyond the sea, that you should say, "Who will cross to the other side of the sea for us, and get it for us so that we may hear it and observe it?" [14] No, the word is very near to you; it is in your mouth and in your heart for you to observe.

15 See, I have set before you today life and prosperity, death and adversity. [16] If you obey the commandments of the LORD your God[m] that I am commanding you today, by loving the LORD your God, walking in his ways, and observing his commandments, decrees, and ordinances, then you shall live and become numerous, and the LORD your God will bless you in the land that you are entering to possess. [17] But if your heart turns away and you do not hear, but are led astray to bow down to other gods and serve them, [18] I declare to you today that you shall perish; you shall not live long in the land that you are crossing the Jordan to enter and possess. [19] I call heaven and earth to witness against you today that I have set before you life and death, blessings and curses. Choose life so that you and your descendants may live, [20] loving the LORD your God, obeying him, and holding fast to him; for that means life to you and length of days, so that you may live in the land that the LORD swore to give to your ancestors, to Abraham, to Isaac, and to Jacob.

31 When Moses had finished speaking all[n] these words to all Israel, [2] he said to them: "I am now one hundred twenty years old. I am no longer able to get about, and the LORD has told me, 'You shall not cross over this Jordan.' [3] The LORD your God himself will cross over before you. He will destroy these nations before you, and you shall dispossess them. Joshua also will cross over before you, as the LORD promised. [4] The LORD will do to them as he did to Sihon and Og, the kings of the Amorites, and to their land, when he destroyed them. [5] The LORD will give them over to you and you shall deal with them in full accord with the command that I have given to you. [6] Be strong and bold; have no fear or dread of them, because it is the LORD your God who goes with you; he will not fail you or forsake you."

7 Then Moses summoned Joshua and said to him in the sight of all Israel: "Be strong and bold, for you are the one who will go with this people into the land that the LORD has sworn to their ancestors to give them; and you will put them in possession of it. [8] It is the LORD who goes before you. He will be with you; he will

m Gk: Heb lacks *If you obey the commandments of the* LORD *your God*   n Q Ms Gk: MT *Moses went and spoke*

---

of love. On circumcision of the heart, see 10.16 n. **9**: 28.3–6.

30.11–14: The covenant demand is not beyond human reach or understanding but has been graciously revealed (29.29) and, in the service of covenant renewal (29.1), *the word is very near to you*. **15–20**: The two ways, between which Israel must decide. Such a challenge formed a climactic moment in ceremonies of covenant renewal (26.16–27.26; Josh 24.14–15). **20**: *Life* is not the mere extension of days. It is loving, obeying, and cleaving to the LORD instead of pursuing the ways of idolatry.

31.1–29: **Concluding events of Moses' life**. **1–8**: Moses' final charge to Joshua and the people. **2–3**: Compare 1.37–38; 3.18–28. **7–8**: Josh 1.1–9.

not fail you or forsake you. Do not fear or be dismayed."

9 Then Moses wrote down this law, and gave it to the priests, the sons of Levi, who carried the ark of the covenant of the LORD, and to all the elders of Israel. 10 Moses commanded them: "Every seventh year, in the scheduled year of remission, during the festival of booths, [o] 11 when all Israel comes to appear before the LORD your God at the place that he will choose, you shall read this law before all Israel in their hearing. 12 Assemble the people—men, women, and children, as well as the aliens residing in your towns—so that they may hear and learn to fear the LORD your God and to observe diligently all the words of this law, 13 and so that their children, who have not known it, may hear and learn to fear the LORD your God, as long as you live in the land that you are crossing over the Jordan to possess."

14 The LORD said to Moses, "Your time to die is near; call Joshua and present yourselves in the tent of meeting, so that I may commission him." So Moses and Joshua went and presented themselves in the tent of meeting, 15 and the LORD appeared at the tent in a pillar of cloud; the pillar of cloud stood at the entrance to the tent.

16 The LORD said to Moses, "Soon you will lie down with your ancestors. Then this people will begin to prostitute themselves to the foreign gods in their midst, the gods of the land into which they are going; they will forsake me, breaking my covenant that I have made with them. 17 My anger will be kindled against them in that day. I will forsake them and hide my face from them; they will become easy prey, and many terrible troubles will come upon them. In that day they will say, 'Have not these troubles come upon us because our God is not in our midst?' 18 On that day I will surely hide my face on account of all the evil they have done by turning to other gods. 19 Now therefore write this song, and teach it to the Israelites; put it in their mouths, in order that this song may be a witness for me against the Israelites. 20 For when I have brought them into the land flowing with milk and honey, which I promised on oath to their ancestors, and they have eaten their fill and grown fat, they will turn to other gods and serve them, despising me and breaking my covenant. 21 And when many terrible troubles come upon them, this song will confront them as a witness, because it will not be lost from the mouths of their descendants. For I know what they are inclined to do even now, before I have brought them into the land that I promised them on oath." 22 That very day Moses wrote this song and taught it to the Israelites.

23 Then the LORD commissioned Joshua son of Nun and said, "Be strong and bold, for you shall bring the Israelites into the land that I promised them; I will be with you."

24 When Moses had finished writing down in a book the words of this law to the very end, 25 Moses commanded the Levites who carried the ark of the covenant of the LORD, saying, 26 "Take this book of the law and put it beside the ark of the covenant of the LORD your God; let it remain there as a witness against

o Or tabernacles; Heb succoth

31.9–13: A covenant ceremony, to be held every seven years, is instituted. 9: The belief is expressed that the whole Deuteronomic law was written by Moses. 10–11: Every seventh year or sabbatical year (15.1–11) the Deuteronomic law is to be read at the central sanctuary during the festival of booths (16.13–15). On this occasion a ceremony of covenant renewal is presumably to take place. 31.14–23: The commissioning of Joshua (Num 27.12–23). 14–15: These verses connect with the old epic tradition in Ex 33.7–11. 16–22: Moses is commanded to write a song which will confront the people as a witness when they turn to other gods. The song is found in ch 32. 23: Continuation of vv. 14–15.

31.24–27: These verses resume the thought of vv. 9–13. 26: This book of the law, i.e. Deuteronomy.

you. ²⁷For I know well how rebellious and stubborn you are. If you already have been so rebellious toward the LORD while I am still alive among you, how much more after my death! ²⁸Assemble to me all the elders of your tribes and your officials, so that I may recite these words in their hearing and call heaven and earth to witness against them. ²⁹For I know that after my death you will surely act corruptly, turning aside from the way that I have commanded you. In time to come trouble will befall you, because you will do what is evil in the sight of the LORD, provoking him to anger through the work of your hands."

30 Then Moses recited the words of this song, to the very end, in the hearing of the whole assembly of Israel:

32 Give ear, O heavens, and I will speak;
  let the earth hear the words of
    my mouth.
² May my teaching drop like the
    rain,
  my speech condense like the
    dew;
  like gentle rain on grass,
  like showers on new growth.
³ For I will proclaim the name of
    the LORD;
  ascribe greatness to our God!

⁴ The Rock, his work is perfect,
  and all his ways are just.
A faithful God, without deceit,
  just and upright is he;
⁵ yet his degenerate children have
    dealt falsely with him, *p*
  a perverse and crooked
    generation.
⁶ Do you thus repay the LORD,
  O foolish and senseless people?

Is not he your father, who created
    you,
  who made you and established
    you?
⁷ Remember the days of old,
  consider the years long past;
ask your father, and he will
    inform you;
  your elders, and they will tell
    you.
⁸ When the Most High *q*
    apportioned the nations,
  when he divided humankind,
he fixed the boundaries of the
    peoples
  according to the number of the
    gods; *r*
⁹ the LORD's own portion was his
    people,
  Jacob his allotted share.

¹⁰ He sustained *s* him in a desert
    land,
  in a howling wilderness waste;
he shielded him, cared for him,
  guarded him as the apple of his
    eye.
¹¹ As an eagle stirs up its nest,
  and hovers over its young;
as it spreads its wings, takes them
    up,
  and bears them aloft on its
    pinions,
¹² the LORD alone guided him;
  no foreign god was with him.
¹³ He set him atop the heights of the
    land,
  and fed him with *t* produce of
    the field;

p Meaning of Heb uncertain   q Traditional
rendering of Heb *Elyon*   r Q Ms Compare Gk
Tg: MT *the Israelites*   s Sam Gk Compare Tg:
MT *found*   t Sam Gk Syr Tg: MT *he ate*

31.28–29: *These words,* probably a reference to the following Song of Moses, not to the book of the law.
31.30–32.47: **The Song of Moses.** This psalm, which contrasts God's faithfulness with Israel's faithlessness, contains elements of a "covenant lawsuit" or legal controversy (Mic 6.1–8; Jer 2.1–13). Probably it dates from the early monarchy.

32.1–3: Introductory appeal (compare Isa 1.2; Mic 1.2). **4–6**: The integrity of God's ways is contrasted to the perversity of Israel. *Rock,* an ancient epithet for God (vv. 15, 18), suggesting stability and dependability.
32.7–14: A recital of the LORD's saving deeds in *the days of old.* **8**: *Gods,* the divine beings who belong to the heavenly court (see Gen 1.26 n.). To these heavenly beings the

he nursed him with honey from
the crags,
with oil from flinty rock;
14 curds from the herd, and milk
from the flock,
with fat of lambs and rams;
Bashan bulls and goats,
together with the choicest
wheat—
you drank fine wine from the
blood of grapes.
15 Jacob ate his fill;ᵘ
Jeshurun grew fat, and kicked.
You grew fat, bloated, and
gorged!
He abandoned God who made
him,
and scoffed at the Rock of his
salvation.
16 They made him jealous with
strange gods,
with abhorrent things they
provoked him.
17 They sacrificed to demons, not
God,
to deities they had never
known,
to new ones recently arrived,
whom your ancestors had not
feared.
18 You were unmindful of the Rock
that bore you;ᵛ
you forgot the God who gave
you birth.

19 The LORD saw it, and was
jealousʷ
he spurnedˣ his sons and
daughters.
20 He said: I will hide my face from
them,
I will see what their end will be;
for they are a perverse generation,

children in whom there is no
faithfulness.
21 They made me jealous with what
is no god,
provoked me with their idols.
So I will make them jealous with
what is no people,
provoke them with a foolish
nation.
22 For a fire is kindled by my anger,
and burns to the depths of
Sheol;
it devours the earth and its
increase,
and sets on fire the foundations
of the mountains.
23 I will heap disasters upon them,
spend my arrows against them:
24 wasting hunger,
burning consumption,
bitter pestilence.
The teeth of beasts I will send
against them,
with venom of things crawling
in the dust.
25 In the street the sword shall
bereave,
and in the chambers terror,
for young man and woman alike,
nursing child and old gray head.
26 I thought to scatter themʸ
and blot out the memory of
them from humankind;
27 but I feared provocation by the
enemy,
for their adversaries might
misunderstand
and say, "Our hand is triumphant;

u Q Mss Sam Gk: MT lacks *Jacob ate his fill*
v Or *that begot you*   *w* Q Mss Gk: MT
lacks *was jealous*   *x* Cn: Heb *he spurned
because of provocation*   *y* Gk: Meaning of
Heb uncertain

LORD delegated authority to govern other nations, but Israel was claimed as *the LORD's own
portion.* **11–12**: Ex 19.4.
   32.15–18: Israel's rebellious forgetfulness
of the LORD. **15**: Like a well-fed animal, Israel
rebelled against its master. *Jeshurun* ("the
Upright One"), a term applied ironically to
Israel (33.5, 26). **16**: *Jealous,* see Ex 34.14 n.
17: *Demons,* a reference to the gods of Ca-

naan (Ps 106.37–38), who are actually *not God.*
   32.19–27: The LORD's righteous indignation. **21**: *No people,* a wordplay on *no god.* **22**:
The poet envisions the foundations of the
earth crumbling under the fire of divine
wrath. According to ancient belief, the *foundations of the mountains* which supported the
firmament were sunk in the subterranean
ocean (Ps 46.2–3). *Sheol,* see Gen 37.35 n.

it was not the LORD who did all
this."

28 They are a nation void of sense;
there is no understanding in
them.
29 If they were wise, they would
understand this;
they would discern what the
end would be.
30 How could one have routed a
thousand,
and two put a myriad to flight,
unless their Rock had sold them,
the LORD had given them up?
31 Indeed their rock is not like our
Rock;
our enemies are fools. *z*
32 Their vine comes from the
vinestock of Sodom,
from the vineyards of
Gomorrah;
their grapes are grapes of poison,
their clusters are bitter;
33 their wine is the poison of
serpents,
the cruel venom of asps.

34 Is not this laid up in store with
me,
sealed up in my treasuries?
35 Vengeance is mine, and
recompense,
for the time when their foot
shall slip;
because the day of their calamity is
at hand,
their doom comes swiftly.

36 Indeed the LORD will vindicate his
people,

have compassion on his servants,
when he sees that their power is
gone,
neither bond nor free remaining.
37 Then he will say: Where are their
gods,
the rock in which they took
refuge,
38 who ate the fat of their sacrifices,
and drank the wine of their
libations?
Let them rise up and help you,
let them be your protection!

39 See now that I, even I, am he;
there is no god besides me.
I kill and I make alive;
I wound and I heal;
and no one can deliver from my
hand.
40 For I lift up my hand to heaven,
and swear: As I live forever,
41 when I whet my flashing sword,
and my hand takes hold on
judgment;
I will take vengeance on my
adversaries,
and will repay those who hate
me.
42 I will make my arrows drunk with
blood,
and my sword shall devour
flesh—
with the blood of the slain and the
captives,
from the long-haired enemy.
43 Praise, O heavens, *a* his people,
worship him, all you gods! *b*

---

z Gk: Meaning of Heb uncertain   *a* Q Ms Gk:
MT *nations*   *b* Q Ms Gk: MT lacks this line

32.28–33: The corruption of the nations.
In vv. 28–29 *they* apparently refers to Israel
and in vv. 31–33 *their* refers to the enemies.
Israel cannot perceive that disaster has befall-
en them because *their Rock* abandoned them.
The rock (god) of the enemies (vv. 37–38),
however, cannot be compared to Israel's
God. **32**: *Their vine* (i.e. of the enemies) is
likened to the proverbial fruit of Sodom and
Gomorrah: outwardly beautiful but inwardly
worthless.

32.34–43: Because of the corruption of
the nations the LORD will finally overthrow
the wicked nations and vindicate the people of
Israel. **34**: *This,* the nations' corruption
(vv. 32–33). **35**: *Vengeance* means both judgment upon the oppressor and vindication of
the oppressed (v. 36). *Their foot . . . their
doom,* the reference is to Israel's enemies. **37–
38**: See v. 17 n. **39**: *Kill . . . make alive,* i.e.
God is sovereign in all things according to the
poet's faith (Ex 4.11 n.; Isa 45.5–7). **43**: Con-

For he will avenge the blood of
his children, [c]
and take vengeance on his
adversaries;
he will repay those who hate
him, [d]
and cleanse the land for his
people. [e]

44 Moses came and recited all the
words of this song in the hearing of the
people, he and Joshua [f] son of Nun.
[45] When Moses had finished reciting all
these words to all Israel, [46] he said to
them: "Take to heart all the words that
I am giving in witness against you today;
give them as a command to your children, so that they may diligently observe
all the words of this law. [47] This is no
trifling matter for you, but rather your
very life; through it you may live long in
the land that you are crossing over the
Jordan to possess."

48 On that very day the LORD addressed Moses as follows: [49] "Ascend this
mountain of the Abarim, Mount Nebo,
which is in the land of Moab, across from
Jericho, and view the land of Canaan,
which I am giving to the Israelites for a
possession; [50] you shall die there on the
mountain that you ascend and shall be
gathered to your kin, as your brother
Aaron died on Mount Hor and was gathered to his kin; [51] because both of you
broke faith with me among the Israelites
at the waters of Meribath-kadesh in the
wilderness of Zin, by failing to maintain
my holiness among the Israelites. [52] Although you may view the land from a

distance, you shall not enter it—the land
that I am giving to the Israelites."

33 This is the blessing with which
Moses, the man of God, blessed
the Israelites before his death. [2] He said:
The LORD came from Sinai,
and dawned from Seir upon
us; [g]
he shone forth from Mount
Paran.
With him were myriads of holy
ones; [h]
at his right, a host of his
own. [i]
3 Indeed, O favorite among [j]
peoples,
all his holy ones were in your
charge;
they marched at your heels,
accepted direction from you.
4 Moses charged us with the law,
as a possession for the assembly
of Jacob.
5 There arose a king in Jeshurun,
when the leaders of the people
assembled—
the united tribes of Israel.

6 May Reuben live, and not die
out,
even though his numbers are
few.

[c] Q Ms Gk: MT *his servants*   [d] Q Ms Gk:
MT lacks this line   [e] Q Ms Sam Gk Vg: MT
his land his people   [f] Sam Gk Syr Vg: MT
Hoshea   [g] Gk Syr Vg Compare Tg: Hcb
upon them   [h] Cn Compare Gk Sam Syr Vg:
MT *He came from Ribeboth-kodesh,*
[i] Cn Compare Gk: meaning of Heb uncertain
[j] Or *O lover of the*

---

cluding hymn of praise. **44–47**: The conclusion of 31.16–29.
32.48–52: **Moses' vision.** Moses is commanded to ascend Mount Nebo (Num
20.10–13; 27.12–14).
33.1–29: **The blessing of Moses.** Before
going to his death, Moses gives his final blessing to the Israelite tribes (see introduction to
the blessing of Jacob, Gen ch 49). **1**: On death-
bed blessings, see Gen 27.4 n. This poem assumes that the tribes are already settled in
Palestine rather than looking forward to the
settlement. In its present form it probably
comes from the early period of the monar-

chy, though it may reflect slightly earlier tribal circumstances. Simeon, for example, is not
mentioned, perhaps because the tribe had already disappeared (Gen 49.5–7). **2–5**: Introduction: the theophany *from Sinai* (see Ps
68.17; Hab 3.3), here associated with the land
of Edom (Seir), as in Judg 5.4. **5**: *Jeshurun,* see
32.15 n. The context suggests the enthronement of the LORD as king in an assembly of
the united tribes of Israel (Josh 24.1; Judg 20.2).
6: Reuben, who once had the leadership of the
firstborn (see Gen 49.3–4 n.), is apparently
threatened with extinction, owing to military
pressures in Transjordan.

7 And this he said of Judah:
   O LORD, give heed to Judah,
      and bring him to his people;
   strengthen his hands for him, *k*
      and be a help against his
      adversaries.

8 And of Levi he said:
   Give to Levi *l* your Thummim,
      and your Urim to your loyal
      one,
   whom you tested at Massah,
      with whom you contended at
      the waters of Meribah;
9 who said of his father and mother,
      "I regard them not";
   he ignored his kin,
      and did not acknowledge his
      children.
   For they observed your word,
      and kept your covenant.
10 They teach Jacob your ordinances,
      and Israel your law;
   they place incense before you,
      and whole burnt offerings on
      your altar.
11 Bless, O LORD, his substance,
      and accept the work of his
      hands;
   crush the loins of his adversaries,
      of those that hate him, so that
      they do not rise again.

12 Of Benjamin he said:
   The beloved of the LORD rests in
      safety—
   the High God *m* surrounds him all
      day long—
   the beloved *n* rests between his
      shoulders.

13 And of Joseph he said:
   Blessed by the LORD be his land,
      with the choice gifts of heaven
      above,
   and of the deep that lies
      beneath;
14 with the choice fruits of the sun,
      and the rich yield of the
      months;
15 with the finest produce of the
      ancient mountains,
   and the abundance of the
      everlasting hills;
16 with the choice gifts of the earth
      and its fullness,
   and the favor of the one who
      dwells on Sinai. *o*
   Let these come on the head of
      Joseph,
   on the brow of the prince
      among his brothers.
17 A firstborn *p* bull—majesty is his!
   His horns are the horns of a
      wild ox;
   with them he gores the peoples,
      driving them to *q* the ends of the
      earth;
   such are the myriads of Ephraim,
      such the thousands of Manasseh.

18 And of Zebulun he said:
   Rejoice, Zebulun, in your going
      out;
   and Issachar, in your tents.

k Cn: Heb *with his hands he contended*   *l* Q Ms
Gk: MT lacks *Give to Levi*   *m* Heb *above him*
n Heb *he*   *o* Cn: Heb *in the bush*   *p* Q Ms
Gk Syr Vg: MT *His firstborn*   *q* Cn: Heb *the
peoples, together*

33.7: It is desired that Judah, in sore trouble because of an adversary (perhaps the Philistines), should be helped by other tribes. **8–11**: Levi, once a warlike tribe (Gen 49.5–7), is to receive the prerogatives of priesthood, namely teaching and officiating at the altar. **8**: *Urim* and *Thummim,* see Ex 28.15–30 n. Apparently the levitical priesthood originated at Kadesh, i.e. *Massah* and *Meribah* (Ex 17.1–7; Num 20.2–13). **9**: On Levi's zealous covenant loyalty see Ex 32.25–29. **33.13–17**: The wish is that *Joseph* may be blessed with the bounties of nature and with invincible military strength (compare Gen 49.25–26). **16**: *Dwells on Sinai,* a reference to the LORD's revelation there (Ex 3.1–6). *Prince,* at the time of the poem Joseph enjoyed even greater prestige than Judah (v. 7). **17**: *Ephraim and Manasseh,* the two tribes that composed "the house of Joseph" (see Gen 48.13–14 n.). **18–19**: *Zebulun* and *Issachar* will enjoy great affluence owing to the resources of the Mediterranean and Sea of Galilee (Gen 49.13).

19 They call peoples to the mountain;
    there they offer the right
      sacrifices;
    for they suck the affluence of the
      seas
    and the hidden treasures of the
      sand.

20 And of Gad he said:
    Blessed be the enlargement of
      Gad!
    Gad lives like a lion;
    he tears at arm and scalp.
21 He chose the best for himself,
    for there a commander's
      allotment was reserved;
    he came at the head of the people,
    he executed the justice of the
      LORD,
    and his ordinances for Israel.

22 And of Dan he said:
    Dan is a lion's whelp
    that leaps forth from Bashan.

23 And of Naphtali he said:
    O Naphtali, sated with favor,
    full of the blessing of the LORD,
    possess the west and the south.

24 And of Asher he said:
    Most blessed of sons be Asher;
    may he be the favorite of his
      brothers,
    and may he dip his foot in oil.
25 Your bars are iron and bronze;
    and as your days, so is your
      strength.

26 There is none like God,
    O Jeshurun,

who rides through the heavens
    to your help,
    majestic through the skies.
27 He subdues the ancient gods,*r*
    shatters*s* the forces of old;*t*
    he drove out the enemy before
      you,
    and said, "Destroy!"
28 So Israel lives in safety,
    untroubled is Jacob's abode*u*
    in a land of grain and wine,
    where the heavens drop down
      dew.
29 Happy are you, O Israel! Who is
      like you,
    a people saved by the LORD,
    the shield of your help,
    and the sword of your triumph!
    Your enemies shall come fawning
      to you,
    and you shall tread on their
      backs.

34 Then Moses went up from the plains of Moab to Mount Nebo, to the top of Pisgah, which is opposite Jericho, and the LORD showed him the whole land: Gilead as far as Dan, 2all Naphtali, the land of Ephraim and Manasseh, all the land of Judah as far as the Western Sea, 3the Negeb, and the Plain—that is, the valley of Jericho, the city of palm trees—as far as Zoar. 4The LORD said to him, "This is the land of which I swore to Abraham, to Isaac, and to Jacob, saying, 'I will give it to your descendants'; I have let you see it with

r Or *The eternal God is a dwelling place*
s Cn: Heb *from underneath*   *t* Or *the everlasting arms*   *u* Or *fountain*

33.20–21: *Gad* occupied the best tableland in Transjordan but aided the other tribes in the occupation of Canaan (See Num ch 32). **22:** *Dan*, vigorous as *a lion's whelp*, has already migrated from its former position north of Judah to the base of Mount Hermon (Judg ch 18). **23:** *Naphtali*, located in the region of the Sea of Galilee and the Ginnesar plain. **24–25:** *Asher*, located below Phoenicia, is to be strong and prosperous. **26–29:** A concluding ascription of praise to the LORD who has given Israel victory and security in

the bounteous land. **26:** *Jeshurun*, see 32.15 n.
    34.1–12: The death of Moses. This chapter resumes the story from the end of Numbers, after Moses' "Deuteronomic" addresses to Israel. **1:** Two traditions about the place of Moses' death are included here: *Mount Nebo* is in Transjordan east of Jericho; *Mount Pisgah* is a peak in the same range, slightly west. **2–3:** From this lofty height Moses looks northward to the Sea of Galilee (area of Dan and Naphtali), to the Western Sea (Mediterranean), south to the Negeb (southern wilder-

your eyes, but you shall not cross over there." ⁵Then Moses, the servant of the LORD, died there in the land of Moab, at the LORD's command. ⁶He was buried in a valley in the land of Moab, opposite Beth-peor, but no one knows his burial place to this day. ⁷Moses was one hundred twenty years old when he died; his sight was unimpaired and his vigor had not abated. ⁸The Israelites wept for Moses in the plains of Moab thirty days; then the period of mourning for Moses was ended.

9 Joshua son of Nun was full of the spirit of wisdom, because Moses had laid his hands on him; and the Israelites obeyed him, doing as the LORD had commanded Moses.

10 Never since has there arisen a prophet in Israel like Moses, whom the LORD knew face to face. ¹¹He was unequaled for all the signs and wonders that the LORD sent him to perform in the land of Egypt, against Pharaoh and all his servants and his entire land, ¹²and for all the mighty deeds and all the terrifying displays of power that Moses performed in the sight of all Israel.

---

ness of Judah) and the Jordan valley as far south as *Zoar* (once located at the end of the Dead Sea; Gen 14.2). **6:** The Hebrew text may mean that the LORD secretly buried Moses, showing the marvelous disappearance of God's prophet (compare 2 Kings 2.11–12). **9:** Compare Num 27.18–23. **10–11:** For the judgment that Moses was the greatest of Israel's prophets, see 18.15–22; Num 12.6–8; compare 11.24–30; Hos 12.13.

The Historical Books

Historical writings claim to be bound by the facts, and intend to record people, places, and events as they really were and as they, implicitly or explicitly, explain and give significance to the writer's own times.

The books in the Hebrew Scriptures that are usually labeled historical in this sense are Joshua, Judges, 1–2 Samuel, 1–2 Kings, 1–2 Chronicles, Ezra, and Nehemiah. Such grouping and descriptive nomenclature, however, are modern and, to a certain extent, misleading. In the original divisions of the Hebrew Bible, the books of Joshua through Kings belong to the "Former Prophets," and the rest appear as part of "The Writings." And there is little reason on literary grounds to exclude the books of Genesis and Exodus from the rubric of historical literature. Moreover, biblical authors and editors, whose work in some cases was in process of formation over several centuries, were less concerned with documentable fact than modern historians would require themselves to be. Often a writer developed high drama, including miracles and divine interventions; the narrator sometimes wove textures of metaphor and imaginative description, often adding aggrandizing elements that lent legendary and archetypal proportions to characters and themes.

In short, the historical books in the Hebrew Scriptures resist modern categories of either history or fiction; rather, they are theological and didactic historiography. The writings present different and sometimes overlapping versions of ancient Israel's past, each one shaped by the conviction that God was committed to intimate dealings with the descendants of Abraham, and each one asserting that this reality of faith and religion was of continuing relevance to the author and ancient reader. This perspective, which was so important in creating these books, is still held by religious Jews, Christians, and Muslims, who look to these historical books for divine instruction, whether informed by perspectives developed in the formative rabbinic writings, the New Testament and early Christian documents, or in the Koran.

Most scholars assume that Israel's earliest historians drew upon oral traditions as well as written sources when constructing a connected narrative of Israel's past. It is now difficult to distinguish one from the other. We may suppose that besides legends, folktales, and poetic materials, a writer would have used a variety of historical stories, such as Judg 9.1–21 and 1 Sam 11.1–11, and many lists, such as ancestral genealogy (Gen 10.1–32), tribal chiefs (Gen 36.40–43), rulers or "judges" (Judg 10.1–5), kings (1 Chr 1.43–51), itineraries (Num 33.5–7), royal officials (1 Kings 4.2–6; 2 Sam 23.24–39), populations (Neh 11.3–36), and towns (Josh chs 15–19). Writers associated with the monarchy, roughly 1000–600 B.C., would have used various administrative documents, including letters (Ezra 4.11–16) and reports of settlements and military campaigns (Judg 1.16–21; 1 Kings 14.25–26), palace and temple records (1 Kings 4.7–17; 7.1–12) and the like. If ancient Israel followed the ways of other ancient Near Eastern monarchies, its scribes would also have produced inscriptions that commemorated and praised the religio-political accomplishments of various kings. We may suppose that prior to the production of such books as 1–2 Kings, royal scribes had shaped these various materials into connected records, perhaps chronicle-like narratives. Some such documents may be referred to by the repeated citations of sources throughout the books of Kings and Chronicles, such as "Book of the Annals of the Kings of Judah/Israel" (1 Kings 14.29; 16.5), "Book of the Acts of Solomon" (1 Kings 11.41), "the record of the seer Samuel" (1 Chr 29.29), and "The Commentary on the Book of the Kings" (2 Chr 24.27). Finally, we may mention the historical memoir, which seems to lie behind a book such as Nehemiah.

Many scholars believe that the earliest example of theological and didactic historiography in the Bible is what can be termed the "Court History of David" (2 Samuel 9–20 and 1 Kings 1–2), which is now a major element in the canonical Books of Samuel and Kings. Here a variety of sources were assembled into a literary unity and given a consistent theological point of view. Commonly admired for its literary qualities, this version of history deals with the domestic and political troubles of David, and tells finally how Solomon, contrary to the usual rules of primogeniture, comes to occupy the Davidic throne. The outcome, which was constantly in danger because of the twists and turns of love, envy, greed, and struggles for power, finally assures the continued vitality of God's promise to be with his covenant people through the longevity of the Davidic dynasty.

Later, this "court history" was joined to narratives and records of kings who ruled after David, and, supplemented with already formed materials relating to pre-monarchical times (now visible in the books of Joshua and Judges and Samuel), resulted in a new, expanded version of Israel's past. This massive composition is known as the "Deuteronomistic History," and, according to most scholars, includes all the books from Joshua through 2 Kings, with the Book of Deuteronomy serving as an introduction and flagship for the work's governing theological perspectives.

This composition, which may have existed in at least two versions dating from the seventh and sixth centuries B.C., portrays Israel's life from entrance into the land to exile in Babylon. The writer shapes this political history into a story of constant struggle to remain obedient to God's teaching (*torah*), to be God's covenant people, and so inherit divine blessings. In such a theological context, entrance into the homeland is fraught with temptation and transgression, but tempered by God's steady willingness to forgive and start afresh. Beginning with Saul (1 Sam 11) and ending when the last of the rulers in David's line goes into exile (2 Kings 25.1–7), the monarchy is also seen as problematic. It is a threat to God's rule, and finally becomes a moral, political, and religious failure. Destruction of Jerusalem and the temple, and dissolution of God's people, are divine punishments necessary to set things right. Not even the righteousness of an occasionally admired king, such as Hezekiah (2 Kings 18–20) or Josiah (2 Kings 23.26), could forestall or turn it aside. The writer allows just a hint of hopefulness to intrude into this bleak picture by noting at the end that Jehoiachin, the last of the Davidic line, was treated kindly by his Babylonian captor (2 Kings 25.27–30).

A second major historical work was written at the time of Israel's restoration from exile, about the fifth or fourth century B.C. Called the "Chronicler's History," this writing encompassed 1–2 Chronicles and may also have included the books of Ezra and Nehemiah, though this opinion has been recently disputed. Working in much the same way as earlier historians, the Chronicler selected and shaped materials from a variety of sources, many of which are cited as written works (e.g. 1 Chr 29.29; 2 Chr 13.22; 26.22). Presupposing the earlier Deuteronomistic

History, this author for the most part repeated, rewrote, and supplemented large portions of 1 Sam 31–2 Kings 25, while also drawing upon other portions of the Bible.

The result is a version of Israel's history that parallels Genesis–2 Kings, running from Adam through the Davidic monarchy and down to the Persian period, when Cyrus decreed that exiled Israelites might return to their homeland. The Chronicler's vision is both idealized and practical. A chain of genealogy leads to the house of David, and from this point David emerges as the archetypal king and religious figure. He rules in obedience to God and establishes authentic divine service, the temple, its priesthood, singers, prayers, and offerings. Sacred space is carefully demarcated, hence showing God's power as effective and available in the kingdom. Subsequent kings fall from this standard in a procession of evil and good rulers, until exile, as divine punishment, brings closure to Israel's monarchy. In its place, it is hinted that for the future there will be a newly constituted, priest-directed people, living in God's regulated commonwealth, and focussed on the holy space and time of temple and worship. The books of Ezra and Nehemiah then chronicle the rebuilding of Jerusalem and the successful creation of such a theocratic order.

It is commonly said that the writing of history began with the Greeks. The examples of historical works in the Bible oblige us to recognize the extent to which modern habits of historical thinking and writing also have their beginnings in the Hebraic world of the Bible's historians.

Joshua

The book of Joshua is part of the larger story of Israel's life in its land. Optimism pervades the opening scenes of that story as told in this book: the successful settlement of Canaan by the Israelite tribes under Joshua. It takes on a more ominous character as it continues in the book of Judges, and reaches its climax with the founding of the Israelite monarchy in the books of Samuel. It comes to its tragic end with the books of Kings that narrate the fall of the Israelite kingdoms. Because the theological perspectives of the book of Deuteronomy undergird this narrative, the books of Joshua to 2 Kings are known as the Deuteronomistic History of Israel (see "The Historical Books," pp. 267–269 OT).

The book of Joshua simplifies what was the long and complex process by which the Israelite tribes came to settle in Canaan. The story does not reflect the actual course of events. Some details are missing; others are rearranged. Archaeological excavations, supplemented by sociological analysis, have helped reconstruct the history of the settlement period. All this has made it clear that the book of Joshua, using an idealized historical narrative, intends to describe Israel, past and future, its relationship with God, and the kind of society it wished to be.

The Canaan that the Israelite tribes entered following the Exodus was in a state of social and political turmoil. Their entrance into the region west of the Jordan river was a catalyst that brought a new social reality into being. The socio-economic system of the Canaanite city-states concentrated power in the upper classes to the detriment of the peasants. That system began to disintegrate under the pressure brought to bear by the new immigrants. The Israelite tribes that experienced a great liberation in Egypt entered Canaan as harbingers of a new society. The book of Joshua is the story of how an obedient Israel under God's chosen leader can bring into existence a society based on justice and freedom.

The book of Joshua falls into three parts. (a) Chapters 1–12 describe the settlement of the Israelite tribes in Canaan as the result of a successful military campaign led by Joshua against the Canaanites. The bulk of this narrative deals with central Canaan, followed by stories about campaigns in the south in 10.28–43, while 11.1–15 covers victories in the north. (b) Chapters 13–21 report the distribution of the land among the victorious tribes. These geographical lists probably come from the period of the Israelite monarchy. Here they serve to describe the extent of the Israelite occupation of Canaan. (c) The book concludes with three stories (chs 22–24) that focus on the loyalty that Israelite tribes owe to their God who has given them the land they now occupy.

The violence described in chs 1–12 has led some readers to ignore or spiritualize this book. There have been attempts to explain away the harshness of these stories by showing that they do not reflect the actual historical circumstances of Israelite tribes' conflicts with the Canaanites. Still it is true that ancient Israel did acquire the land, in part, through violent means; the violence that took place during the settlement period was evil. What the book of Joshua affirms is that God's purpose for Israel was served even by this evil. The aim of the book was not to edify but to move its readers to obedience. For ancient Israel this obedience was an act of faith in the God who brings good out of evil.

1 After the death of Moses the servant of the LORD, the LORD spoke to Joshua son of Nun, Moses' assistant, saying, 2"My servant Moses is dead. Now proceed to cross the Jordan, you and all this people, into the land that I am giving to them, to the Israelites. 3 Every place that the sole of your foot will tread upon I have given to you, as I promised to Moses. 4 From the wilderness and the Lebanon as far as the great river, the river Euphrates, all the land of the Hittites, to the Great Sea in the west shall be your territory. 5 No one shall be able to stand against you all the days of your life. As I was with Moses, so I will be with you; I will not fail you or forsake you. 6 Be strong and courageous; for you shall put this people in possession of the land that I swore to their ancestors to give them. 7 Only be strong and very courageous, being careful to act in accordance with all the law that my servant Moses commanded you; do not turn from it to the right hand or to the left, so that you may be successful wherever you go. 8 This book of the law shall not depart out of your mouth; you shall meditate on it day and night, so that you may be careful to act in accordance with all that is written in it. For then you shall make your way prosperous, and then you shall be successful. 9 I hereby command you: Be strong and courageous; do not be frightened or dismayed, for the LORD your God is with you wherever you go."

10 Then Joshua commanded the officers of the people, 11 "Pass through the camp, and command the people: 'Prepare your provisions; for in three days you are to cross over the Jordan, to go in to take possession of the land that the LORD your God gives you to possess.'"

12 To the Reubenites, the Gadites, and the half-tribe of Manasseh Joshua said, 13 "Remember the word that Moses the servant of the LORD commanded you, saying, 'The LORD your God is providing you a place of rest, and will give you this land.' 14 Your wives, your little ones, and your livestock shall remain in the land that Moses gave you beyond the Jordan. But all the warriors among you shall cross over armed before your kindred and shall help them, 15 until the LORD gives rest to your kindred as well as to you, and they too take possession of the land that the LORD your God is giving them. Then you shall return to your own land and take possession of it, the land that Moses the servant of the LORD gave you beyond the Jordan to the east."

16 They answered Joshua: "All that you have commanded us we will do, and wherever you send us we will go. 17 Just as we obeyed Moses in all things, so we will obey you. Only may the LORD your God be with you, as he was with Moses! 18 Whoever rebels against your orders and disobeys your words, whatever you command, shall be put to death. Only be strong and courageous."

2 Then Joshua son of Nun sent two men secretly from Shittim as spies,

---

1.1–9: The LORD's address to Joshua. The death of Moses provided the signal for entrance of the Israelites into Canaan, since God did not allow Moses to enter the land (Deut 32.48–52). **2:** The *Jordan* river was the natural eastern boundary of Canaan. **4:** The idealized boundaries of the Israelite settlement were *the wilderness* of Sinai to the south and east, the *Lebanon* mountain range to the northwest, the *river Euphrates* to the east, and the *Great Sea,* the Mediterranean Sea, to the west. **8:** Obedience to *this book of the law* (Deuteronomy) is the only requirement for the success of the project that the Israelites are to undertake.

1.10–18: Joshua's address to the people.

Joshua instructs the Israelites to prepare for the entrance into Canaan. **14:** *Beyond the Jordan,* the region east of the Jordan where these tribes settled (Num 32). **15:** After the Transjordanian groups helped the other Israelites acquire the land west of the Jordan, they were to return to the territory *that Moses the servant of the LORD gave them.*

2.1–24: Joshua's spies at Jericho. The men whom Joshua sent to reconnoiter the situation find a friendly reception in the house of Rahab, to whom they promise safety when the city is taken. In Heb 11.31 Rahab is counted as one of the heroes of faith. **1:** *Shittim* was the site of the Israelite camp east of the Jordan and northeast of the Dead Sea; its precise loca-

saying, "Go, view the land, especially Jericho." So they went, and entered the house of a prostitute whose name was Rahab, and spent the night there. 2 The king of Jericho was told, "Some Israelites have come here tonight to search out the land." 3 Then the king of Jericho sent orders to Rahab, "Bring out the men who have come to you, who entered your house, for they have come only to search out the whole land." 4 But the woman took the two men and hid them. Then she said, "True, the men came to me, but I did not know where they came from. 5 And when it was time to close the gate at dark, the men went out. Where the men went I do not know. Pursue them quickly, for you can overtake them." 6 She had, however, brought them up to the roof and hidden them with the stalks of flax that she had laid out on the roof. 7 So the men pursued them on the way to the Jordan as far as the fords. As soon as the pursuers had gone out, the gate was shut.

8 Before they went to sleep, she came up to them on the roof 9 and said to the men: "I know that the LORD has given you the land, and that dread of you has fallen on us, and that all the inhabitants of the land melt in fear before you. 10 For we have heard how the LORD dried up the water of the Red Sea*a* before you when you came out of Egypt, and what you did to the two kings of the Amorites that were beyond the Jordan, to Sihon and Og, whom you utterly destroyed. 11 As soon as we heard it, our hearts melted, and there was no courage left in any of us because of you. The LORD your God is indeed God in heaven above and on earth below. 12 Now then, since I have dealt kindly with you, swear to me by the LORD that you in turn will deal kindly with my family. Give me a sign of good faith 13 that you will spare my father and mother, my brothers and sisters, and all

who belong to them, and deliver our lives from death." 14 The men said to her, "Our life for yours! If you do not tell this business of ours, then we will deal kindly and faithfully with you when the LORD gives us the land."

15 Then she let them down by a rope through the window, for her house was on the outer side of the city wall and she resided within the wall itself. 16 She said to them, "Go toward the hill country, so that the pursuers may not come upon you. Hide yourselves there three days, until the pursuers have returned; then afterward you may go your way." 17 The men said to her, "We will be released from this oath that you have made us swear to you 18 if we invade the land and you do not tie this crimson cord in the window through which you let us down, and you do not gather into your house your father and mother, your brothers, and all your family. 19 If any of you go out of the doors of your house into the street, they shall be responsible for their own death, and we shall be innocent; but if a hand is laid upon any who are with you in the house, we shall bear the responsibility for their death. 20 But if you tell this business of ours, then we shall be released from this oath that you made us swear to you." 21 She said, "According to your words, so be it." She sent them away and they departed. Then she tied the crimson cord in the window.

22 They departed and went into the hill country and stayed there three days, until the pursuers returned. The pursuers had searched all along the way and found nothing. 23 Then the two men came down again from the hill country. They crossed over, came to Joshua son of Nun, and told him all that had happened to them. 24 They said to Joshua, "Truly

a Or Sea of Reeds

---

tion is uncertain. *Jericho* is twenty-three miles east of Jerusalem; its location in the lower Jordan Valley near several fords in the river made it a gateway to Canaan from the east. The earliest remains on the site date from the

10th–8th millennia B.C. Archaeology has shown that in the 13th century B.C., when it was likely that the Israelites entered Canaan under Joshua, Jericho was an unfortified village.

the LORD has given all the land into our hands; moreover all the inhabitants of the land melt in fear before us."

3 Early in the morning Joshua rose and set out from Shittim with all the Israelites, and they came to the Jordan. They camped there before crossing over. 2At the end of three days the officers went through the camp 3and commanded the people, "When you see the ark of the covenant of the LORD your God being carried by the levitical priests, then you shall set out from your place. Follow it, 4so that you may know the way you should go, for you have not passed this way before. Yet there shall be a space between you and it, a distance of about two thousand cubits; do not come any nearer to it." 5Then Joshua said to the people, "Sanctify yourselves; for tomorrow the LORD will do wonders among you." 6To the priests Joshua said, "Take up the ark of the covenant, and pass on in front of the people." So they took up the ark of the covenant and went in front of the people.

7 The LORD said to Joshua, "This day I will begin to exalt you in the sight of all Israel, so that they may know that I will be with you as I was with Moses. 8You are the one who shall command the priests who bear the ark of the covenant, 'When you come to the edge of the waters of the Jordan, you shall stand still in the Jordan.'" 9Joshua then said to the Israelites, "Draw near and hear the words of the LORD your God." 10Joshua said, "By this you shall know that among you is the living God who without fail will drive out from before you

the Canaanites, Hittites, Hivites, Perizzites, Girgashites, Amorites, and Jebusites: 11the ark of the covenant of the Lord of all the earth is going to pass before you into the Jordan. 12So now select twelve men from the tribes of Israel, one from each tribe. 13When the soles of the feet of the priests who bear the ark of the LORD, the Lord of all the earth, rest in the waters of the Jordan, the waters of the Jordan flowing from above shall be cut off; they shall stand in a single heap."

14 When the people set out from their tents to cross over the Jordan, the priests bearing the ark of the covenant were in front of the people. 15Now the Jordan overflows all its banks throughout the time of harvest. So when those who bore the ark had come to the Jordan, and the feet of the priests bearing the ark were dipped in the edge of the water, 16the waters flowing from above stood still, rising up in a single heap far off at Adam, the city that is beside Zarethan, while those flowing toward the sea of the Arabah, the Dead Sea, *b* were wholly cut off. Then the people crossed over opposite Jericho. 17While all Israel were crossing over on dry ground, the priests who bore the ark of the covenant of the LORD stood on dry ground in the middle of the Jordan, until the entire nation finished crossing over the Jordan.

4 When the entire nation had finished crossing over the Jordan, the LORD said to Joshua: 2"Select twelve men from the people, one from each tribe, 3and

b Heb *Salt Sea*

3.1–17: The crossing of the Jordan. The crossing of the river took on the appearance of a liturgical ceremony. **3:** *The ark of the covenant* is the sign of God's presence among the tribes; it can be carried only by priests. **4:** The people were to keep their distance. *Two thousand cubits,* about three thousand feet. **5:** *Sanctify yourselves* reflects the cultic background of this story. It requires the Israelites to render themselves ritually pure.

3.10: A stereotypical listing of the pre-Israelite population of Canaan. It is difficult to identify the individual groups specifically.

16: *Adam,* eighteen miles north of Jericho, is probably Tell ed-Damiyeh. *Zarethan* is farther north. *The Dead Sea* is about one third suspended salts. This makes it impossible for plant or animal life to survive in it.

4.1–5.1: A monument commemorating the crossing. The chapter preserves two traditions about memorial stones set up to commemorate the crossing of the Jordan by the Israelites. One describes the memorial stones set up at Gilgal (4.1–3, 6–7, 8b, 20); the other describes stones set up in the bed of the river (4.4–5, 8a, 9, 15–19).

command them, 'Take twelve stones from here out of the middle of the Jordan, from the place where the priests' feet stood, carry them over with you, and lay them down in the place where you camp tonight.' " 4Then Joshua summoned the twelve men from the Israelites, whom he had appointed, one from each tribe. 5Joshua said to them, "Pass on before the ark of the LORD your God into the middle of the Jordan, and each of you take up a stone on his shoulder, one for each of the tribes of the Israelites, 6so that this may be a sign among you. When your children ask in time to come, 'What do those stones mean to you?' 7then you shall tell them that the waters of the Jordan were cut off in front of the ark of the covenant of the LORD. When it crossed over the Jordan, the waters of the Jordan were cut off. So these stones shall be to the Israelites a memorial forever."

8 The Israelites did as Joshua commanded. They took up twelve stones out of the middle of the Jordan, according to the number of the tribes of the Israelites, as the LORD told Joshua, carried them over with them to the place where they camped, and laid them down there. 9(Joshua set up twelve stones in the middle of the Jordan, in the place where the feet of the priests bearing the ark of the covenant had stood; and they are there to this day.)

10 The priests who bore the ark remained standing in the middle of the Jordan, until everything was finished that the LORD commanded Joshua to tell the people, according to all that Moses had commanded Joshua. The people crossed over in haste. 11As soon as all the people had finished crossing over, the ark of the LORD, and the priests, crossed over in front of the people. 12The Reubenites, the Gadites, and the half-tribe of Manasseh crossed over armed before the Israelites, as Moses had ordered them. 13About forty thousand armed for war crossed over before the LORD to the plains of Jericho for battle.

14 On that day the LORD exalted Joshua in the sight of all Israel; and they stood in awe of him, as they had stood in awe of Moses, all the days of his life.

15 The LORD said to Joshua, 16"Command the priests who bear the ark of the covenant,*c* to come up out of the Jordan." 17Joshua therefore commanded the priests, "Come up out of the Jordan." 18When the priests bearing the ark of the covenant of the LORD came up from the middle of the Jordan, and the soles of the priests' feet touched dry ground, the waters of the Jordan returned to their place and overflowed all its banks, as before.

19 The people came up out of the Jordan on the tenth day of the first month, and they camped in Gilgal on the east border of Jericho. 20Those twelve stones, which they had taken out of the Jordan, Joshua set up in Gilgal, 21saying to the Israelites, "When your children ask their parents in time to come, 'What do these stones mean?' 22then you shall let your children know, 'Israel crossed over the Jordan here on dry ground.' 23For the LORD your God dried up the waters of the Jordan for you until you crossed over, as the LORD your God did to the Red Sea,*d* which he dried up for us until we crossed over, 24so that all the peoples of the earth may know that the hand of the LORD is mighty, and so that you may fear the LORD your God forever."

5 When all the kings of the Amorites beyond the Jordan to the west, and all the kings of the Canaanites by the sea, heard that the LORD had dried up the waters of the Jordan for the Israelites until they had crossed over, their hearts melted, and there was no longer any spirit in them, because of the Israelites.

c Or *treaty*, or *testimony*; Heb *eduth*   *d* Or *Sea of Reeds*

4.19: *The first month* was Abib (March–April), later called Nisan. *Gilgal,* probably Khirbet Mefjar, is about one mile from Jericho. It became an important Israelite shrine.

Here the people made Saul their king (1 Sam 11.15); later David and his rebellious subjects were reconciled here (2 Sam 19.15, 40).

2 At that time the LORD said to Joshua, "Make flint knives and circumcise the Israelites a second time." ³So Joshua made flint knives, and circumcised the Israelites at Gibeath-haaraloth. *ᵉ* ⁴This is the reason why Joshua circumcised them: all the males of the people who came out of Egypt, all the warriors, had died during the journey through the wilderness after they had come out of Egypt. ⁵Although all the people who came out had been circumcised, yet all the people born on the journey through the wilderness after they had come out of Egypt had not been circumcised. ⁶For the Israelites traveled forty years in the wilderness, until all the nation, the warriors who came out of Egypt, perished, not having listened to the voice of the LORD. To them the LORD swore that he would not let them see the land that he had sworn to their ancestors to give us, a land flowing with milk and honey. ⁷So it was their children, whom he raised up in their place, that Joshua circumcised; for they were uncircumcised, because they had not been circumcised on the way.

8 When the circumcising of all the nation was done, they remained in their places in the camp until they were healed. ⁹The LORD said to Joshua, "Today I have rolled away from you the disgrace of Egypt." And so that place is called Gilgalᶠ to this day.

10 While the Israelites were camped in Gilgal they kept the passover in the evening on the fourteenth day of the month in the plains of Jericho. ¹¹On the day after the passover, on that very day, they ate the produce of the land, unleavened cakes and parched grain. ¹²The manna ceased on the day they ate the produce of the land, and the Israelites no longer had manna; they ate the crops of the land of Canaan that year.

13 Once when Joshua was by Jericho, he looked up and saw a man standing before him with a drawn sword in his hand. Joshua went to him and said to him, "Are you one of us, or one of our adversaries?" ¹⁴He replied, "Neither; but as commander of the army of the LORD I have now come." And Joshua fell on his face to the earth and worshiped, and he said to him, "What do you command your servant, my lord?" ¹⁵The commander of the army of the LORD said to Joshua, "Remove the sandals from your feet, for the place where you stand is holy." And Joshua did so.

6 Now Jericho was shut up inside and out because of the Israelites; no one came out and no one went in. ²The LORD said to Joshua, "See, I have handed Jericho over to you, along with its king and soldiers. ³You shall march around the city, all the warriors circling the city once. Thus you shall do for six days, ⁴with seven priests bearing seven trumpets of rams' horns before the ark. On

e That is *the Hill of the Foreskins*   *f* Related to Heb *galal* to roll

---

5.2–12: The first passover in the land.
2–9: Ex 12.48 asserts that circumcision was required of all males who were to celebrate the passover. **2:** At one time circumcision was a common practice in the ancient Near East. It became a sign of Israel's unique relationship with God (Gen 17.11–13).
5.8–9: The precise nature of the *disgrace* is unclear; it may refer to the abject social status of the Hebrew slaves in Egypt. **11–12:** The first passover celebrates Israel's new life in the land. **12:** *The manna* was no longer necessary, since Israel could now live off the fruit of its new land.
5.13–15: A theophany. This passage belongs to a tradition that remembered the victory over Jericho as the result of a military victory. *The commander of the army of the LORD* did not give complete support to Joshua because God remained free to become Israel's adversary (see Hos 13.8; Lam 2.5). **15:** This is a direct quotation from Ex 3.5. The allusion to the call of Moses shows that Joshua was the divinely appointed successor of Moses.
6.1–14: The siege of Jericho. Archaeology does not allow this passage to be read as a factual account of events connected with the entrance of the Israelite tribes into Canaan. Excavations at Tell es-Sultan (ancient Jericho) have shown that in the 13th century B.C. an unfortified village occupied the site. The text reads like a description of the

the seventh day you shall march around the city seven times, the priests blowing the trumpets. 5 When they make a long blast with the ram's horn, as soon as you hear the sound of the trumpet, then all the people shall shout with a great shout; and the wall of the city will fall down flat, and all the people shall charge straight ahead." 6 So Joshua son of Nun summoned the priests and said to them, "Take up the ark of the covenant, and have seven priests carry seven trumpets of rams' horns in front of the ark of the LORD." 7 To the people he said, "Go forward and march around the city; have the armed men pass on before the ark of the LORD."

8 As Joshua had commanded the people, the seven priests carrying the seven trumpets of rams' horns before the LORD went forward, blowing the trumpets, with the ark of the covenant of the LORD following them. 9 And the armed men went before the priests who blew the trumpets; the rear guard came after the ark, while the trumpets blew continually. 10 To the people Joshua gave this command: "You shall not shout or let your voice be heard, nor shall you utter a word, until the day I tell you to shout. Then you shall shout." 11 So the ark of the LORD went around the city, circling it once; and they came into the camp, and spent the night in the camp.

12 Then Joshua rose early in the morning, and the priests took up the ark of the LORD. 13 The seven priests carrying the seven trumpets of rams' horns before the ark of the LORD passed on, blowing the trumpets continually. The armed men went before them, and the rear guard came after the ark of the LORD, while the trumpets blew continually. 14 On the second day they marched around the city once and then returned to the camp. They did this for six days.

15 On the seventh day they rose early, at dawn, and marched around the city in the same manner seven times. It was only on that day that they marched around the city seven times. 16 And at the seventh time, when the priests had blown the trumpets, Joshua said to the people, "Shout! For the LORD has given you the city. 17 The city and all that is in it shall be devoted to the LORD for destruction. Only Rahab the prostitute and all who are with her in her house shall live because she hid the messengers we sent. 18 As for you, keep away from the things devoted to destruction, so as not to covet*g* and take any of the devoted things and make the camp of Israel an object for destruction, bringing trouble upon it. 19 But all silver and gold, and vessels of bronze and iron, are sacred to the LORD; they shall go into the treasury of the LORD." 20 So the people shouted, and the trumpets were blown. As soon as the people heard the sound of the trumpets, they raised a great shout, and the wall fell down flat; so the people charged straight ahead into the city and captured it. 21 Then they devoted to destruction by the edge of the sword all in the city, both men and women, young and old, oxen, sheep, and donkeys.

22 Joshua said to the two men who had spied out the land, "Go into the prostitute's house, and bring the woman out of it and all who belong to her, as you swore to her." 23 So the young men who had been spies went in and brought Rahab out, along with her father, her mother, her brothers, and all who belonged to her—they brought all her kindred out—

g Gk: Heb *devote to destruction* Compare 7.21

later liturgical celebration of what must have been a conflict over the spring that watered the plains of Jericho. **6.4:** The sacred number *seven* occurs repeatedly in this chapter.

6.15–27: The fall of Jericho. 17: *Devoted . . . for destruction* is a fixed formula meaning that *the city and all that is in it* are to be separated from ordinary use by a ritual act. This text may contain memories of how warfare was conducted before the monarchy, but it is not certain that all the details found here or in Deut 20 were ever in use. The complete devastation of the city represents the radical break between the old Canaanite society and the new Israelite one.

6.25: *Ever since* may show that people who

and set them outside the camp of Israel.
24 They burned down the city, and everything in it; only the silver and gold, and the vessels of bronze and iron, they put into the treasury of the house of the LORD. 25 But Rahab the prostitute, with her family and all who belonged to her, Joshua spared. Her family[h] has lived in Israel ever since. For she hid the messengers whom Joshua sent to spy out Jericho.

26 Joshua then pronounced this oath, saying,

"Cursed before the LORD be
    anyone who tries
  to build this city—this Jericho!
At the cost of his firstborn he shall
    lay its foundation,
  and at the cost of his youngest
    he shall set up its gates!"

27 So the LORD was with Joshua; and his fame was in all the land.

7 But the Israelites broke faith in regard to the devoted things: Achan son of Carmi son of Zabdi son of Zerah, of the tribe of Judah, took some of the devoted things; and the anger of the LORD burned against the Israelites.

2 Joshua sent men from Jericho to Ai, which is near Beth-aven, east of Bethel, and said to them, "Go up and spy out the land." And the men went up and spied out Ai. 3 Then they returned to Joshua and said to him, "Not all the people need go up; about two or three thousand men should go up and attack Ai. Since they are so few, do not make the whole people toil up there." 4 So about three thousand of the people went up there; and they fled

before the men of Ai. 5 The men of Ai killed about thirty-six of them, chasing them from outside the gate as far as Shebarim and killing them on the slope. The hearts of the people melted and turned to water.

6 Then Joshua tore his clothes, and fell to the ground on his face before the ark of the LORD until the evening, he and the elders of Israel; and they put dust on their heads. 7 Joshua said, "Ah, Lord GOD! Why have you brought this people across the Jordan at all, to hand us over to the Amorites so as to destroy us? Would that we had been content to settle beyond the Jordan! 8 O Lord, what can I say, now that Israel has turned their backs to their enemies! 9 The Canaanites and all the inhabitants of the land will hear of it, and surround us, and cut off our name from the earth. Then what will you do for your great name?"

10 The LORD said to Joshua, "Stand up! Why have you fallen upon your face? 11 Israel has sinned; they have transgressed my covenant that I imposed on them. They have taken some of the devoted things; they have stolen, they have acted deceitfully, and they have put them among their own belongings. 12 Therefore the Israelites are unable to stand before their enemies; they turn their backs to their enemies, because they have become a thing devoted for destruction themselves. I will be with you no more, unless you destroy the devoted things from among you. 13 Proceed to sanctify the people, and say, 'Sanctify yourselves

h Heb She

claimed to be descendants of Rahab lived in Jericho at the time that this story was composed; it thus served to explain the survival of some Canaanites and their continued presence in Israel. **26:** The fulfillment of this curse is recorded in 1 Kings 16.34.
7.1–5: Defeat at Ai. 1: This verse serves to unite what were two separate events: the taking of Ai (7.2–5; 8.1–29) and the sin of Achan (7.6–26). **2:** The name Ai (modern et-Tell) means "ruin." The site, two miles east of Bethel, was uninhabited during the

settlement period. This is an exaggerated account of the Israelite capture of the unwalled village that occupied the site in the early Iron Age (around 1125 B.C.). **3–5:** The cause of Israel's defeat is a too-optimistic report on Ai's defenses.
7.6–26: Achan's sin. 11–13: The LORD reveals the true cause of Israel's defeat: the theft of booty from Jericho. The assumption here is that evil done by even a single individual has effects beyond that individual's person.

for tomorrow; for thus says the LORD, the God of Israel, "There are devoted things among you, O Israel; you will be unable to stand before your enemies until you take away the devoted things from among you." [14] In the morning therefore you shall come forward tribe by tribe. The tribe that the LORD takes shall come near by clans, the clan that the LORD takes shall come near by households, and the household that the LORD takes shall come near one by one. [15] And the one who is taken as having the devoted things shall be burned with fire, together with all that he has, for having transgressed the covenant of the LORD, and for having done an outrageous thing in Israel.' "

16 So Joshua rose early in the morning, and brought Israel near tribe by tribe, and the tribe of Judah was taken. [17] He brought near the clans of Judah, and the clan of the Zerahites was taken; and he brought near the clan of the Zerahites, family by family, [i] and Zabdi was taken. [18] And he brought near his household one by one, and Achan son of Carmi son of Zabdi son of Zerah, of the tribe of Judah, was taken. [19] Then Joshua said to Achan, "My son, give glory to the LORD God of Israel and make confession to him. Tell me now what you have done; do not hide it from me." [20] And Achan answered Joshua, "It is true; I am the one who sinned against the LORD God of Israel. This is what I did: [21] when I saw among the spoil a beautiful mantle from Shinar, and two hundred shekels of silver, and a bar of gold weighing fifty shekels, then I coveted them and took them. They now lie hidden in the ground inside my tent, with the silver underneath."

22 So Joshua sent messengers, and they ran to the tent; and there it was, hidden in his tent with the silver underneath. [23] They took them out of the tent and brought them to Joshua and all the Israelites; and they spread them out before the LORD. [24] Then Joshua and all Israel with him took Achan son of Zerah, with the silver, the mantle, and the bar of gold, with his sons and daughters, with his oxen, donkeys, and sheep, and his tent and all that he had; and they brought them up to the Valley of Achor. [25] Joshua said, "Why did you bring trouble on us? The LORD is bringing trouble on you today." And all Israel stoned him to death; they burned them with fire, cast stones on them, [26] and raised over him a great heap of stones that remains to this day. Then the LORD turned from his burning anger. Therefore that place to this day is called the Valley of Achor. [j]

8 Then the LORD said to Joshua, "Do not fear or be dismayed; take all the fighting men with you, and go up now to Ai. See, I have handed over to you the king of Ai with his people, his city, and his land. [2] You shall do to Ai and its king as you did to Jericho and its king; only its spoil and its livestock you may take as booty for yourselves. Set an ambush against the city, behind it."

3 So Joshua and all the fighting men set out to go up against Ai. Joshua chose thirty thousand warriors and sent them out by night [4] with the command, "You shall lie in ambush against the city, behind it; do not go very far from the city, but all of you stay alert. [5] I and all the people who are with me will approach the city. When they come out against us, as before, we shall flee from them. [6] They

---

i Mss Syr: MT *man by man*   *j* That is *Trouble*

7.14: *The LORD takes*, by the casting of lots. **16–25**: Once Achan took the booty, he set in motion a process that inevitably led to his destruction and threatened the very existence of Israel.

7.26: The precise location of the *Valley of Achor* is not certain; in 15.7 it describes the border between Benjamin and Judah. Achan belonged to the tribe of Judah. Originally this story may have been one symptom of the rivalry between Benjamin (Saul's tribe) and Judah (David's tribe).

8.1–29: **Victory at Ai.** This is a detailed and probable report of a military expedition. Except for a few references to God's role in the taking of the city (vv. 1, 7b, 18, 27), the account is without miraculous elements.

will come out after us until we have drawn them away from the city; for they will say, 'They are fleeing from us, as before.' While we flee from them, 7 you shall rise up from the ambush and seize the city; for the LORD your God will give it into your hand. 8 And when you have taken the city, you shall set the city on fire, doing as the LORD has ordered; see, I have commanded you." 9 So Joshua sent them out; and they went to the place of ambush, and lay between Bethel and Ai, to the west of Ai; but Joshua spent that night in the camp. *k*

10 In the morning Joshua rose early and mustered the people, and went up, with the elders of Israel, before the people to Ai. 11 All the fighting men who were with him went up, and drew near before the city, and camped on the north side of Ai, with a ravine between them and Ai. 12 Taking about five thousand men, he set them in ambush between Bethel and Ai, to the west of the city. 13 So they stationed the forces, the main encampment that was north of the city and its rear guard west of the city. But Joshua spent that night in the valley. 14 When the king of Ai saw this, he and all his people, the inhabitants of the city, hurried out early in the morning to the meeting place facing the Arabah to meet Israel in battle; but he did not know that there was an ambush against him behind the city. 15 And Joshua and all Israel made a pretense of being beaten before them, and fled in the direction of the wilderness. 16 So all the people who were in the city were called together to pursue them, and as they pursued Joshua they were drawn away from the city. 17 There was not a man left in Ai or Bethel who did not go out after Israel; they left the city open, and pursued Israel.

18 Then the LORD said to Joshua, "Stretch out the sword that is in your hand toward Ai; for I will give it into

your hand." And Joshua stretched out the sword that was in his hand toward the city. 19 As soon as he stretched out his hand, the troops in ambush rose quickly out of their place and rushed forward. They entered the city, took it, and at once set the city on fire. 20 So when the men of Ai looked back, the smoke of the city was rising to the sky. They had no power to flee this way or that, for the people who fled to the wilderness turned back against the pursuers. 21 When Joshua and all Israel saw that the ambush had taken the city and that the smoke of the city was rising, then they turned back and struck down the men of Ai. 22 And the others came out from the city against them; so they were surrounded by Israelites, some on one side, and some on the other; and Israel struck them down until no one was left who survived or escaped. 23 But the king of Ai was taken alive and brought to Joshua.

24 When Israel had finished slaughtering all the inhabitants of Ai in the open wilderness where they pursued them, and when all of them to the very last had fallen by the edge of the sword, all Israel returned to Ai, and attacked it with the edge of the sword. 25 The total of those who fell that day, both men and women, was twelve thousand—all the people of Ai. 26 For Joshua did not draw back his hand, with which he stretched out the sword, until he had utterly destroyed all the inhabitants of Ai. 27 Only the livestock and the spoil of that city Israel took as their booty, according to the word of the LORD that he had issued to Joshua. 28 So Joshua burned Ai, and made it forever a heap of ruins, as it is to this day. 29 And he hanged the king of Ai on a tree until evening; and at sunset Joshua commanded, and they took his body down from the tree, threw it down at the en-

k Heb *among the people*

8.9: *Bethel,* the modern Beitin, is eleven miles north of Jerusalem. Later it became one of the principal shrines of the Northern Kingdom (1 Kings 12.28–30).

8.26: *Utterly destroyed,* see 6.17 n. 29: *A great heap of stones* was a reminder that Israel acquired its land in part by force of arms (compare 4.9; 7.26; Judg 6.24; 1 Sam 6.18).

trance of the gate of the city, and raised over it a great heap of stones, which stands there to this day.

30 Then Joshua built on Mount Ebal an altar to the LORD, the God of Israel, 31just as Moses the servant of the LORD had commanded the Israelites, as it is written in the book of the law of Moses, "an altar of unhewn*l* stones, on which no iron tool has been used"; and they offered on it burnt offerings to the LORD, and sacrificed offerings of well-being. 32 And there, in the presence of the Israelites, Joshua*m* wrote on the stones a copy of the law of Moses, which he had written. 33 All Israel, alien as well as citizen, with their elders and officers and their judges, stood on opposite sides of the ark in front of the levitical priests who carried the ark of the covenant of the LORD, half of them in front of Mount Gerizim and half of them in front of Mount Ebal, as Moses the servant of the LORD had commanded at the first, that they should bless the people of Israel. 34 And afterward he read all the words of the law, blessings and curses, according to all that is written in the book of the law. 35 There was not a word of all that Moses commanded that Joshua did not read before all the assembly of Israel, and the women, and the little ones, and the aliens who resided among them.

9 Now when all the kings who were beyond the Jordan in the hill country and in the lowland all along the coast of the Great Sea toward Lebanon—the Hittites, the Amorites, the Canaanites, the Perizzites, the Hivites, and the Jebusites—heard of this, 2they gathered together with one accord to fight Joshua and Israel.

3 But when the inhabitants of Gibeon heard what Joshua had done to Jericho and to Ai, 4they on their part acted with cunning: they went and prepared provisions,*n* and took worn-out sacks for their donkeys, and wineskins, worn-out and torn and mended, 5with worn-out, patched sandals on their feet, and worn-out clothes; and all their provisions were dry and moldy. 6They went to Joshua in the camp at Gilgal, and said to him and to the Israelites, "We have come from a far country; so now make a treaty with us." 7But the Israelites said to the Hivites, "Perhaps you live among us; then how can we make a treaty with you?" 8They said to Joshua, "We are your servants." And Joshua said to them, "Who are you? And where do you come from?" 9They said to him, "Your servants have come from a very far country, because of the name of the LORD your God; for we have heard a report of him, of all that he did in Egypt, 10and of all that he did to the two kings of the Amorites who were beyond the Jordan, King Sihon of Heshbon, and King Og of Bashan who lived in Ashtaroth. 11So our elders and all the inhabitants of our country said to us, 'Take provisions in your hand for the journey; go to meet them, and say to them, "We are your servants; come now, make a treaty with us." ' 12Here is our bread; it was still warm when we took it from our houses as our food for the jour-

l Heb *whole*   *m* Heb *he*   *n* Cn: Meaning of Heb uncertain

---

8.30–35: The altar on Mount Ebal. The events narrated in ch 9 are the natural sequel to the story about the fall of Ai (note that 8.29 flows into 9.3). Traveling to Ebal required the tribes to make a twenty-mile trip from Ai to Ebal and then to retrace their steps to encamp at Gilgal (9.6). This narrative serves to portray Joshua as carrying out the command given to Moses in Deut 27.4–5 (compare 11.29–30). Joshua, who obeyed, is the foil to Achan, who did not. **30:** *Mount Ebal* is one of the two mountains (Gerizim being the other)

that flank the pass of Shechem in central Canaan.
9.1–27: The stratagem of the Gibeonites. The Israelite settlement in Canaan sometimes advanced through peaceful means, as this story shows. **3:** *Gibeon* is the modern el-Jib, about seven miles southwest of Ai. While Gibeon was an important city in the monarchic period, there is no archaeological evidence that the site was occupied when the Israelite tribes entered the region.
9.6: The Gibeonites pretend to be *from a far*

ney, on the day we set out to come to you, but now, see, it is dry and moldy; [13] these wineskins were new when we filled them, and see, they are burst; and these garments and sandals of ours are worn out from the very long journey." [14] So the leaders[o] partook of their provisions, and did not ask direction from the LORD. [15] And Joshua made peace with them, guaranteeing their lives by a treaty; and the leaders of the congregation swore an oath to them.

16 But when three days had passed after they had made a treaty with them, they heard that they were their neighbors and were living among them. [17] So the Israelites set out and reached their cities on the third day. Now their cities were Gibeon, Chephirah, Beeroth, and Kiriath-jearim. [18] But the Israelites did not attack them, because the leaders of the congregation had sworn to them by the LORD, the God of Israel. Then all the congregation murmured against the leaders. [19] But all the leaders said to all the congregation, "We have sworn to them by the LORD, the God of Israel, and now we must not touch them. [20] This is what we will do to them: We will let them live, so that wrath may not come upon us, because of the oath that we swore to them." [21] The leaders said to them, "Let them live." So they became hewers of wood and drawers of water for all the congregation, as the leaders had decided concerning them.

22 Joshua summoned them, and said to them, "Why did you deceive us, saying, 'We are very far from you,' while in fact you are living among us? [23] Now therefore you are cursed, and some of you shall always be slaves, hewers of wood and drawers of water for the house of my God." [24] They answered Joshua, "Because it was told to your servants for a certainty that the LORD your God had commanded his servant Moses to give you all the land, and to destroy all the inhabitants of the land before you; so we were in great fear for our lives because of you, and did this thing. [25] And now we are in your hand: do as it seems good and right in your sight to do to us." [26] This is what he did for them: he saved them from the Israelites; and they did not kill them. [27] But on that day Joshua made them hewers of wood and drawers of water for the congregation and for the altar of the LORD, to continue to this day, in the place that he should choose.

10 When King Adoni-zedek of Jerusalem heard how Joshua had taken Ai, and had utterly destroyed it, doing to Ai and its king as he had done to Jericho and its king, and how the inhabitants of Gibeon had made peace with Israel and were among them, [2] he[p] became greatly frightened, because Gibeon was a large city, like one of the royal cities, and was larger than Ai, and all its men were warriors. [3] So King Adoni-zedek of Jerusalem sent a message to King Hoham of Hebron, to King Piram of Jarmuth, to King Japhia of Lachish, and to King Debir of Eglon, saying, [4] "Come up and help me, and let us attack Gibeon; for it has made peace with Joshua and with the Israelites." [5] Then the five kings of the Amorites—the king of Jerusalem, the king of Hebron, the king of Jarmuth, the

---

o Gk: Heb *men*   *p* Heb *they*

---

country so as to take advantage of the more lenient treatment afforded to such people (Deut 20.15).

9.16: The Israelites discover the ruse of the Gibeonites, who readily admit their deception (vv. 24–25) because they know that Israel has to honor the pact between them (v. 20).

9.27: An explanation for the presence of non-Israelites in the service of Israelite sanctuaries. The story also explains the survival of some Canaanites despite the command to ex-

terminate them. *The place that he should choose* is the way that Deuteronomy alludes to Jerusalem in its legislation (e.g. Deut 12.5, 11, 14, 18).

10.1–43: Campaigns south of Gibeon. 1–15: The treaty between Gibeon and Israel incited the kings of five Amorite city-states to attack Gibeon. The Gibeonites call upon their Israelite allies, and Joshua defeats the coalition. **3:** The five city-states were to the south of Gibeon. **5:** Gen 10.16 lists *Amorites* as de-

king of Lachish, and the king of Eglon—gathered their forces, and went up with all their armies and camped against Gibeon, and made war against it.

6 And the Gibeonites sent to Joshua at the camp in Gilgal, saying, "Do not abandon your servants; come up to us quickly, and save us, and help us; for all the kings of the Amorites who live in the hill country are gathered against us." 7 So Joshua went up from Gilgal, he and all the fighting force with him, all the mighty warriors. 8 The LORD said to Joshua, "Do not fear them, for I have handed them over to you; not one of them shall stand before you." 9 So Joshua came upon them suddenly, having marched up all night from Gilgal. 10 And the LORD threw them into a panic before Israel, who inflicted a great slaughter on them at Gibeon, chased them by the way of the ascent of Beth-horon, and struck them down as far as Azekah and Makkedah. 11 As they fled before Israel, while they were going down the slope of Beth-horon, the LORD threw down huge stones from heaven on them as far as Azekah, and they died; there were more who died because of the hailstones than the Israelites killed with the sword.

12 On the day when the LORD gave the Amorites over to the Israelites, Joshua spoke to the LORD; and he said in the sight of Israel,

"Sun, stand still at Gibeon,
        and Moon, in the valley of
                Aijalon."

13 And the sun stood still, and the
        moon stopped,
until the nation took vengeance
        on their enemies.

Is this not written in the Book of Jashar? The sun stopped in midheaven, and did not hurry to set for about a whole day. 14 There has been no day like it before or since, when the LORD heeded a human voice; for the LORD fought for Israel.

15 Then Joshua returned, and all Israel with him, to the camp at Gilgal.

16 Meanwhile, these five kings fled and hid themselves in the cave at Makkedah. 17 And it was told Joshua, "The five kings have been found, hidden in the cave at Makkedah." 18 Joshua said, "Roll large stones against the mouth of the cave, and set men by it to guard them; 19 but do not stay there yourselves; pursue your enemies, and attack them from the rear. Do not let them enter their towns, for the LORD your God has given them into your hand." 20 When Joshua and the Israelites had finished inflicting a very great slaughter on them, until they were wiped out, and when the survivors had entered into the fortified towns, 21 all the people returned safe to Joshua in the camp at Makkedah; no one dared to speak *q* against any of the Israelites.

22 Then Joshua said, "Open the mouth of the cave, and bring those five kings out to me from the cave." 23 They did so, and brought the five kings out to him from the cave, the king of Jerusalem, the king of Hebron, the king of Jarmuth, the king of Lachish, and the king of Eglon. 24 When they brought the kings out to Joshua, Joshua summoned all the Israelites, and said to the chiefs of the warriors who had gone with him, "Come near, put your feet on the necks of these kings." Then they came near and put their feet on their necks. 25 And Joshua said to them, "Do not be afraid or dismayed; be strong and courageous; for thus the LORD will do to all the enemies against whom you fight." 26 Afterward Joshua struck them down and put them to death, and he hung them on five trees. And they hung on the trees until evening. 27 At sunset Joshua commanded, and they took them down from the trees and threw them into the cave where they

q Heb moved his tongue

---

scendants of Canaan. **11–14:** The miracle of the hailstones (v. 11) and of the sun (vv. 12–14) reflects the Israelite belief that *the LORD fought for Israel*. **13:** *The Book of Jashar,* no longer extant, appears to have been a collection of poetry that extolled Israel's military victories and heroes (compare 2 Sam 1.18).

10.16–27: The defeat and humiliation of

had hidden themselves; they set large stones against the mouth of the cave, which remain to this very day.

28 Joshua took Makkedah on that day, and struck it and its king with the edge of the sword; he utterly destroyed every person in it; he left no one remaining. And he did to the king of Makkedah as he had done to the king of Jericho.

29 Then Joshua passed on from Makkedah, and all Israel with him, to Libnah, and fought against Libnah. 30 The LORD gave it also and its king into the hand of Israel; and he struck it with the edge of the sword, and every person in it; he left no one remaining in it; and he did to its king as he had done to the king of Jericho.

31 Next Joshua passed on from Libnah, and all Israel with him, to Lachish, and laid siege to it, and assaulted it. 32 The LORD gave Lachish into the hand of Israel, and he took it on the second day, and struck it with the edge of the sword, and every person in it, as he had done to Libnah.

33 Then King Horam of Gezer came up to help Lachish; and Joshua struck him and his people, leaving him no survivors.

34 From Lachish Joshua passed on with all Israel to Eglon; and they laid siege to it, and assaulted it; 35 and they took it that day, and struck it with the edge of the sword; and every person in it he utterly destroyed that day, as he had done to Lachish.

36 Then Joshua went up with all Israel from Eglon to Hebron; they assaulted it, 37 and took it, and struck it with the edge of the sword, and its king and its towns, and every person in it; he left no one remaining, just as he had done to Eglon, and utterly destroyed it with every person in it.

38 Then Joshua, with all Israel, turned back to Debir and assaulted it, 39 and he took it with its king and all its towns; they struck them with the edge of the sword, and utterly destroyed every person in it; he left no one remaining; just as he had done to Hebron, and, as he had done to Libnah and its king, so he did to Debir and its king.

40 So Joshua defeated the whole land, the hill country and the Negeb and the lowland and the slopes, and all their kings; he left no one remaining, but utterly destroyed all that breathed, as the LORD God of Israel commanded. 41 And Joshua defeated them from Kadesh-barnea to Gaza, and all the country of Goshen, as far as Gibeon. 42 Joshua took all these kings and their land at one time, because the LORD God of Israel fought for Israel. 43 Then Joshua returned, and all Israel with him, to the camp at Gilgal.

11 When King Jabin of Hazor heard of this, he sent to King Jobab of Madon, to the king of Shimron, to the king of Achshaph, 2 and to the kings who were in the northern hill country, and in the Arabah south of Chinneroth, and in the lowland, and in Naphoth-dor on the west, 3 to the Canaanites in the east and the west, the Amorites, the Hittites, the Perizzites, and the Jebusites in the hill country, and the Hivites under Hermon

---

the five kings was complete because God assured Israel of victory (vv. 19,25). **27**: The final shape of this episode serves to explain a large heap of stones near Makkedah, whose precise location is unknown.

10.28–39: Archaeology does not support this summary of Joshua's southern campaign. Judg 1.8–15 ascribes the defeat of Jerusalem, Hebron, and Debir to Judah and Caleb. Most of these territories became part of Israel during the time of David and Solomon.

10.40–43: This summary credits Joshua for the acquisition of *the whole land*. This is a retroversion (or projection back) of the achievements of the Israelite monarchy into the premonarchic period.

11.1–23: **Israel's victories in the north. 1–15**: The source of this story was probably Judg 4–5. Archaeological evidence shows that Hazor, located nine miles north of the Sea of Galilee, was destroyed in the mid–13th century. Still there may be no connection between that destruction and Joshua's campaign. **1**: The name *Jabin* occurs here and in Judg 4 and Ps 83.9. The locations of *Madon, Shimron*, and *Achshaph* are unknown. **2**: *The Arabah* is the Jordan Valley; *Chinneroth,* the Sea of Galilee.

in the land of Mizpah. [4] They came out, with all their troops, a great army, in number like the sand on the seashore, with very many horses and chariots. [5] All these kings joined their forces, and came and camped together at the waters of Merom, to fight with Israel.

[6] And the LORD said to Joshua, "Do not be afraid of them, for tomorrow at this time I will hand over all of them, slain, to Israel; you shall hamstring their horses, and burn their chariots with fire." [7] So Joshua came suddenly upon them with all his fighting force, by the waters of Merom, and fell upon them. [8] And the LORD handed them over to Israel, who attacked them and chased them as far as Great Sidon and Misrephoth-maim, and eastward as far as the valley of Mizpeh. They struck them down, until they had left no one remaining. [9] And Joshua did to them as the LORD commanded him; he hamstrung their horses, and burned their chariots with fire.

[10] Joshua turned back at that time, and took Hazor, and struck its king down with the sword. Before that time Hazor was the head of all those kingdoms. [11] And they put to the sword all who were in it, utterly destroying them; there was no one left who breathed, and he burned Hazor with fire. [12] And all the towns of those kings, and all their kings, Joshua took, and struck them with the edge of the sword, utterly destroying them, as Moses the servant of the LORD had commanded. [13] But Israel burned none of the towns that stood on mounds except Hazor, which Joshua did burn. [14] All the spoil of these towns, and the livestock, the Israelites took for their booty; but all the people they struck down with the edge of the sword, until they had destroyed them, and they did not leave any who breathed. [15] As the LORD had commanded his servant Moses, so Moses commanded Joshua, and so

Joshua did; he left nothing undone of all that the LORD had commanded Moses.

[16] So Joshua took all that land: the hill country and all the Negeb and all the land of Goshen and the lowland and the Arabah and the hill country of Israel and its lowland, [17] from Mount Halak, which rises toward Seir, as far as Baal-gad in the valley of Lebanon below Mount Hermon. He took all their kings, struck them down, and put them to death. [18] Joshua made war a long time with all those kings. [19] There was not a town that made peace with the Israelites, except the Hivites, the inhabitants of Gibeon; all were taken in battle. [20] For it was the LORD's doing to harden their hearts so that they would come against Israel in battle, in order that they might be utterly destroyed, and might receive no mercy, but be exterminated, just as the LORD had commanded Moses.

[21] At that time Joshua came and wiped out the Anakim from the hill country, from Hebron, from Debir, from Anab, and from all the hill country of Judah, and from all the hill country of Israel; Joshua utterly destroyed them with their towns. [22] None of the Anakim was left in the land of the Israelites; some remained only in Gaza, in Gath, and in Ashdod. [23] So Joshua took the whole land, according to all that the LORD had spoken to Moses; and Joshua gave it for an inheritance to Israel according to their tribal allotments. And the land had rest from war.

12 Now these are the kings of the land, whom the Israelites defeated, whose land they occupied beyond the Jordan toward the east, from the Wadi Arnon to Mount Hermon, with all the Arabah eastward: [2] King Sihon of the Amorites who lived at Heshbon, and ruled from Aroer, which is on the edge of the Wadi Arnon, and from the middle of the valley as far as the river Jabbok, the

---

11.5: *Merom* is a city known from extrabiblical sources; its identification is not certain. **11.21:** *The Anakim,* a term for the pre-Israelite inhabitants of Canaan renowned for their size and strength (Deut 9.2). **23:** The

final summary to be followed by the distribution of the land. **12.1–24: A list of Israel's victories. 1–6:** Moses' exploits in Transjordan. This summary draws from Deut 2–3.

boundary of the Ammonites, that is, half of Gilead, 3 and the Arabah to the Sea of Chinneroth eastward, and in the direction of Beth-jeshimoth, to the sea of the Arabah, the Dead Sea,*r* southward to the foot of the slopes of Pisgah; 4 and King Og*s* of Bashan, one of the last of the Rephaim, who lived at Ashtaroth and at Edrei 5 and ruled over Mount Hermon and Salecah and all Bashan to the boundary of the Geshurites and the Maacathites, and over half of Gilead to the boundary of King Sihon of Heshbon. 6 Moses, the servant of the LORD, and the Israelites defeated them; and Moses the servant of the LORD gave their land for a possession to the Reubenites and the Gadites and the half-tribe of Manasseh.

7 The following are the kings of the land whom Joshua and the Israelites defeated on the west side of the Jordan, from Baal-gad in the valley of Lebanon to Mount Halak, that rises toward Seir (and Joshua gave their land to the tribes of Israel as a possession according to their allotments, 8 in the hill country, in the lowland, in the Arabah, in the slopes, in the wilderness, and in the Negeb, the land of the Hittites, Amorites, Canaanites, Perizzites, Hivites, and Jebusites):

9	the king of Jericho	one
	the king of Ai, which is next to Bethel	one
10	the king of Jerusalem	one
	the king of Hebron	one
11	the king of Jarmuth	one
	the king of Lachish	one
12	the king of Eglon	one
	the king of Gezer	one
13	the king of Debir	one
	the king of Geder	one
14	the king of Hormah	one
	the king of Arad	one
15	the king of Libnah	one
	the king of Adullam	one
16	the king of Makkedah	one
	the king of Bethel	one
17	the king of Tappuah	one
	the king of Hepher	one
18	the king of Aphek	one
	the king of Lasharon	one
19	the king of Madon	one
	the king of Hazor	one
20	the king of Shimron-meron	one
	the king of Achshaph	one
21	the king of Taanach	one
	the king of Megiddo	one
22	the king of Kedesh	one
	the king of Jokneam in Carmel	one
23	the king of Dor in Naphath-dor	one
	the king of Goiim in Galilee,*t*	one
24	the king of Tirzah	one

thirty-one kings in all.

13 Now Joshua was old and advanced in years; and the LORD said to him, "You are old and advanced in years, and very much of the land still remains to be possessed. 2 This is the land that still remains: all the regions of the Philistines, and all those of the Geshurites 3 (from the Shihor, which is east of Egypt, northward to the boundary of Ekron, it is reckoned as Canaanite; there are five rulers of the Philistines, those of Gaza, Ashdod, Ashkelon, Gath, and Ekron), and those of the Avvim, 4 in the south, all the land of the Canaanites, and Mearah that belongs to the Sidonians, to Aphek, to the boundary of the Amorites, 5 and the land of the Gebalites, and all Lebanon, toward the east, from Baal-gad

r Heb *Salt Sea* *s* Gk: Heb *the boundary of King Og* *t* Gk: Heb *Gilgal*

12.7–24: Joshua's victories west of the Jordan. Previous narratives mention few of these names; the list contains names of only five kings of cities located between Gibeon and Galilee. The narratives in Joshua do not describe a single battle in this region. This may show that people of this region became independent of the Canaanite regimes before the arrival of Joshua and the Israelites.

13.1–33: **The distribution of the land begins.** Chapters 13–21 focus exclusively on this subject. 1: The Israelite hegemony in Canaan was not the product of a total military conquest but of a complicated and gradual process. It was completed under David. 2–4: Israel controlled Philistia only briefly under David and Solomon.
13.5–7: Israel never ruled Phoenicia. 8–

below Mount Hermon to Lebo-hamath, 6 all the inhabitants of the hill country from Lebanon to Misrephoth-maim, even all the Sidonians. I will myself drive them out from before the Israelites; only allot the land to Israel for an inheritance, as I have commanded you. 7 Now therefore divide this land for an inheritance to the nine tribes and the half-tribe of Manasseh."

8 With the other half-tribe of Manasseh*u* the Reubenites and the Gadites received their inheritance, which Moses gave them, beyond the Jordan eastward, as Moses the servant of the Lord gave them: 9 from Aroer, which is on the edge of the Wadi Arnon, and the town that is in the middle of the valley, and all the tableland from*v* Medeba as far as Dibon; 10 and all the cities of King Sihon of the Amorites, who reigned in Heshbon, as far as the boundary of the Ammonites; 11 and Gilead, and the region of the Geshurites and Maacathites, and all Mount Hermon, and all Bashan to Salecah; 12 all the kingdom of Og in Bashan, who reigned in Ashtaroth and in Edrei (he alone was left of the survivors of the Rephaim); these Moses had defeated and driven out. 13 Yet the Israelites did not drive out the Geshurites or the Maacathites; but Geshur and Maacath live within Israel to this day.

14 To the tribe of Levi alone Moses gave no inheritance; the offerings by fire to the Lord God of Israel are their inheritance, as he said to them.

15 Moses gave an inheritance to the tribe of the Reubenites according to their clans. 16 Their territory was from Aroer, which is on the edge of the Wadi Arnon, and the town that is in the middle of the valley, and all the tableland by Medeba; 17 with Heshbon, and all its towns that are in the tableland; Dibon, and Bamothbaal, and Beth-baal-meon, 18 and Jahaz, and Kedemoth, and Mephaath, 19 and Kiriathaim, and Sibmah, and Zerethshahar on the hill of the valley, 20 and Beth-peor, and the slopes of Pisgah, and Beth-jeshimoth, 21 that is, all the towns of the tableland, and all the kingdom of King Sihon of the Amorites, who reigned in Heshbon, whom Moses defeated with the leaders of Midian, Evi and Rekem and Zur and Hur and Reba, as princes of Sihon, who lived in the land. 22 Along with the rest of those they put to death, the Israelites also put to the sword Balaam son of Beor, who practiced divination. 23 And the border of the Reubenites was the Jordan and its banks. This was the inheritance of the Reubenites, according to their families with their towns and villages.

24 Moses gave an inheritance also to the tribe of the Gadites, according to their families. 25 Their territory was Jazer, and all the towns of Gilead, and half the land of the Ammonites, to Aroer, which is east of Rabbah, 26 and from Heshbon to Ramath-mizpeh and Betonim, and from Mahanaim to the territory of Debir,*w* 27 and in the valley Beth-haram, Beth-nimrah, Succoth, and Zaphon, the rest of the kingdom of King Sihon of Heshbon, the Jordan and its banks, as far as the lower end of the Sea of Chinnereth, eastward beyond the Jordan. 28 This is the inheritance of the Gadites according to their clans, with their towns and villages.

29 Moses gave an inheritance to the half-tribe of Manasseh; it was allotted to the half-tribe of the Manassites according

u Cn: Heb *With it*   *v* Compare Gk: Heb lacks *from*   *w* Gk Syr Vg: Heb *Lidebir*

32: Israel seldom controlled this territory, though Deuteronomy considered it part of Israel (see Deut 3.12–17). In Ezek 48 the Transjordan is not part of the ideal Israel (see also Josh 22.10–34). **12:** Israel remembered *the Rephaim* as a race of giants (Deut 3.11) like the Anakim of 11.21.

13.13: The statement that *the Israelites did*

not drive out the Geshurites . . . is the first of a series of passages in this book and in Judges, derived from an ancient source, that indicate that the conquest was less thoroughgoing than the later editors of the books as a whole would have their readers understand (15.63; 16.10; 17.12–13; Judg 1.19, 21, 27–35). **33:** On *Levi,* see ch 21.

to their families. 30Their territory extended from Mahanaim, through all Bashan, the whole kingdom of King Og of Bashan, and all the settlements of Jair, which are in Bashan, sixty towns, 31and half of Gilead, and Ashtaroth, and Edrei, the towns of the kingdom of Og in Bashan; these were allotted to the people of Machir son of Manasseh according to their clans—for half the Machirites.

32 These are the inheritances that Moses distributed in the plains of Moab, beyond the Jordan east of Jericho. 33But to the tribe of Levi Moses gave no inheritance; the LORD God of Israel is their inheritance, as he said to them.

14 These are the inheritances that the Israelites received in the land of Canaan, which the priest Eleazar, and Joshua son of Nun, and the heads of the families of the tribes of the Israelites distributed to them. 2Their inheritance was by lot, as the LORD had commanded Moses for the nine and one-half tribes. 3For Moses had given an inheritance to the two and one-half tribes beyond the Jordan; but to the Levites he gave no inheritance among them. 4For the people of Joseph were two tribes, Manasseh and Ephraim; and no portion was given to the Levites in the land, but only towns to live in, with their pasture lands for their flocks and herds. 5The Israelites did as the LORD commanded Moses; they allotted the land.

6 Then the people of Judah came to Joshua at Gilgal; and Caleb son of Jephunneh the Kenizzite said to him, "You know what the LORD said to Moses the man of God in Kadesh-barnea concerning you and me. 7I was forty years old when Moses the servant of the LORD sent me from Kadesh-barnea to spy out the land; and I brought him an honest report. 8But my companions who went up with me made the heart of the people melt; yet I wholeheartedly followed the LORD my God. 9And Moses swore on that day, saying, 'Surely the land on which your foot has trodden shall be an inheritance for you and your children forever, because you have wholeheartedly followed the LORD my God.' 10And now, as you see, the LORD has kept me alive, as he said, these forty-five years since the time that the LORD spoke this word to Moses, while Israel was journeying through the wilderness; and here I am today, eighty-five years old. 11I am still as strong today as I was on the day that Moses sent me; my strength now is as my strength was then, for war, and for going and coming. 12So now give me this hill country of which the LORD spoke on that day; for you heard on that day how the Anakim were there, with great fortified cities; it may be that the LORD will be with me, and I shall drive them out, as the LORD said."

13 Then Joshua blessed him, and gave Hebron to Caleb son of Jephunneh for an inheritance. 14So Hebron became the inheritance of Caleb son of Jephunneh the Kenizzite to this day, because he wholeheartedly followed the LORD, the God of Israel. 15Now the name of Hebron formerly was Kiriath-arba;*x* this Arba was*y* the greatest man among the Anakim. And the land had rest from war.

15 The lot for the tribe of the people of Judah according to their families reached southward to the boundary

x That is *the city of Arba*   *y* Heb lacks *this Arba was*

14.1–5: Distribution of the land west of the Jordan. 4: *The people of Joseph were two tribes*, this reckoning, with the subtraction of Levi, makes a total of twelve still. Practically, however, Simeon became absorbed in Judah (see 19.9 n.), and each half-tribe of Ephraim and Manasseh came to count as a complete tribe.
14.6–15: Hebron, the inheritance of Caleb. The tribe of Judah eventually absorbed the Calebites. This tradition remembers that the claims of the Calebites on this region went back to Moses (Num 13.30; 14.24). **14**: *Hebron* is twenty miles south of Jerusalem; it was here that David was anointed king of Judah (2 Sam 2.4).
15.1–63: The territory assigned to Judah. 1–12: The boundaries of Judah described.

of Edom, to the wilderness of Zin at the farthest south. ²And their south boundary ran from the end of the Dead Sea,ᶻ from the bay that faces southward; ³it goes out southward of the ascent of Akrabbim, passes along to Zin, and goes up south of Kadesh-barnea, along by Hezron, up to Addar, makes a turn to Karka, ⁴passes along to Azmon, goes out by the Wadi of Egypt, and comes to its end at the sea. This shall be your south boundary. ⁵And the east boundary is the Dead Sea,ᶻ to the mouth of the Jordan. And the boundary on the north side runs from the bay of the sea at the mouth of the Jordan; ⁶and the boundary goes up to Beth-hoglah, and passes along north of Beth-arabah; and the boundary goes up to the Stone of Bohan, Reuben's son; ⁷and the boundary goes up to Debir from the Valley of Achor, and so northward, turning toward Gilgal, which is opposite the ascent of Adummim, which is on the south side of the valley; and the boundary passes along to the waters of En-shemesh, and ends at En-rogel; ⁸then the boundary goes up by the valley of the son of Hinnom at the southern slope of the Jebusites (that is, Jerusalem); and the boundary goes up to the top of the mountain that lies over against the valley of Hinnom, on the west, at the northern end of the valley of Rephaim; ⁹then the boundary extends from the top of the mountain to the spring of the Waters of Nephtoah, and from there to the towns of Mount Ephron; then the boundary bends around to Baalah (that is, Kiriath-jearim); ¹⁰and the boundary circles west of Baalah to Mount Seir, passes along to the northern slope of Mount Jearim (that is, Chesalon), and goes down to Beth-shemesh, and passes along by Timnah; ¹¹the boundary goes out to the slope of the hill north of Ekron, then the boundary bends around to Shikkeron, and passes along to Mount Baalah, and goes out to Jabneel; then the boundary comes to an end at the sea. ¹²And the west boundary was the Mediterranean with its coast. This is the boundary surrounding the people of Judah according to their families.

13 According to the commandment of the LORD to Joshua, he gave to Caleb son of Jephunneh a portion among the people of Judah, Kiriath-arba,ᵃ that is, Hebron (Arba was the father of Anak). ¹⁴And Caleb drove out from there the three sons of Anak: Sheshai, Ahiman, and Talmai, the descendants of Anak. ¹⁵From there he went up against the inhabitants of Debir; now the name of Debir formerly was Kiriath-sepher. ¹⁶And Caleb said, "Whoever attacks Kiriath-sepher and takes it, to him I will give my daughter Achsah as wife." ¹⁷Othniel son of Kenaz, the brother of Caleb, took it; and he gave him his daughter Achsah as wife. ¹⁸When she came to him, she urged him to ask her father for a field. As she dismounted from her donkey, Caleb said to her, "What do you wish?" ¹⁹She said to him, "Give me a present; since you have set me in the land of the Negeb, give me springs of water as well." So Caleb gave her the upper springs and the lower springs.

20 This is the inheritance of the tribe of the people of Judah according to their families. ²¹The towns belonging to the tribe of the people of Judah in the extreme South, toward the boundary of Edom, were Kabzeel, Eder, Jagur, ²²Kinah, Dimonah, Adadah, ²³Kedesh, Hazor, Ithnan, ²⁴Ziph, Telem, Bealoth, ²⁵Hazor-hadattah, Kerioth-hezron (that is, Hazor), ²⁶Amam, Shema, Moladah, ²⁷Hazar-gaddah, Heshmon, Beth-pelet, ²⁸Hazar-shual, Beer-sheba, Biziothiah, ²⁹Baalah, Iim, Ezem, ³⁰Eltolad, Chesil,

ᶻ Heb *Salt Sea*   ᵃ That is *the city of Arba*

15.13–19: Information about Caleb (compare 14.6–15). **16–19:** These verses are almost identical with Judg 1.11–15.
15.20–63: A list of Judah's towns by districts. The cities were divided into districts probably corresponding to the administrative districts of the kingdom of Judah. See 1 Kings 4.7–19 for a similar list.

Hormah, 31Ziklag, Madmannah, San-
sannah, 32Lebaoth, Shilhim, Ain, and
Rimmon: in all, twenty-nine towns,
with their villages.

33 And in the Lowland, Eshtaol, Zo-
rah, Ashnah, 34Zanoah, En-gannim,
Tappuah, Enam, 35Jarmuth, Adullam,
Socoh, Azekah, 36Shaaraim, Adithaim,
Gederah, Gederothaim: fourteen towns
with their villages.

37 Zenan, Hadashah, Migdal-gad,
38Dilan, Mizpeh, Jokthe-el, 39Lachish,
Bozkath, Eglon, 40Cabbon, Lahmam,
Chitlish, 41Gederoth, Beth-dagon, Naamah, and Makkedah: sixteen towns with
their villages.

42 Libnah, Ether, Ashan, 43Iphtah,
Ashnah, Nezib, 44Keilah, Achzib, and
Mareshah: nine towns with their vil-
lages.

45 Ekron, with its dependencies and
its villages; 46from Ekron to the sea, all
that were near Ashdod, with their villages.

47 Ashdod, its towns and its villages;
Gaza, its towns and its villages; to the
Wadi of Egypt, and the Great Sea with its
coast.

48 And in the hill country, Shamir,
Jattir, Socoh, 49Dannah, Kiriath-sannah
(that is, Debir), 50Anab, Eshtemoh,
Anim, 51Goshen, Holon, and Giloh:
eleven towns with their villages.

52 Arab, Dumah, Eshan, 53Janim,
Beth-tappuah, Aphekah, 54Humtah,
Kiriath-arba (that is, Hebron), and Zior:
nine towns with their villages.

55 Maon, Carmel, Ziph, Juttah,
56Jezreel, Jokdeam, Zanoah, 57Kain,
Gibeah, and Timnah: ten towns with
their villages.

58 Halhul, Beth-zur, Gedor, 59Maa-
rath, Beth-anoth, and Eltekon: six towns
with their villages.

60 Kiriath-baal (that is, Kiriath-

jearim), and Rabbah: two towns with
their villages.

61 In the wilderness, Beth-arabah,
Middin, Secacah, 62Nibshan, the City of
Salt, and En-gedi: six towns with their
villages.

63 But the people of Judah could not
drive out the Jebusites, the inhabitants of
Jerusalem; so the Jebusites live with the
people of Judah in Jerusalem to this day.

16 The allotment of the Josephites
went from the Jordan by Jericho,
east of the waters of Jericho, into the
wilderness, going up from Jericho into
the hill country to Bethel; 2then going
from Bethel to Luz, it passes along to
Ataroth, the territory of the Archites
3then it goes down westward to the terri
tory of the Japhletites, as far as the terr
tory of Lower Beth-horon, then to G
zer, and it ends at the sea.

4 The Josephites—Manasseh a
Ephraim—received their inheritance

5 The territory of the Ephraimite y
their families was as follows: the bou a-
ry of their inheritance on the east as
Ataroth-addar as far as Upper th-
horon, 6and the boundary goes om
there to the sea; on the north is Mi me-
thath; then on the east the bou dary
makes a turn toward Taanath- iloh,
and passes along beyond it on the ast to
Janoah, 7then it goes down from anoah
to Ataroth and to Naarah, and uches
Jericho, ending at the Jordan 8From
Tappuah the boundary goes we ward to
the Wadi Kanah, and ends a he sea.
Such is the inheritance of the t e of the
Ephraimites by their families together
with the towns that were set a rt for the
Ephraimites within the inher nce of the
Manassites, all those town with their
villages. 10They did not, ho ever, drive
out the Canaanites who live n Gezer: so
the Canaanites have lived v hin Ephra-

16.1–17.18: **The territory of the Joseph
tribes.** Since the Levites did not have a specific territory assigned to them (13.14), the tribe
of Joseph was able to split into the tribes of
Ephraim and Manasseh, named after Joseph's
two sons (Gen 41.50–52). These two tribes
occupied the central hill country north of Je-

rusalem with Ephraim in the south and Manasseh in the north. **1–4:** The southern
boundary ran from Jericho west to the Mediterranean. **5–10:** Delimitation of the boundary of Ephraim. **10:** This verse is very similar
to Judg 1.29.

im to this day but have been made to do forced labor.

17 Then allotment was made to the tribe of Manasseh, for he was the firstborn of Joseph. To Machir the firstborn of Manasseh, the father of Gilead, were allotted Gilead and Bashan, because he was a warrior. 2 And allotments were made to the rest of the tribe of Manasseh, by their families, Abiezer, Helek, Asriel, Shechem, Hepher, and Shemida; these were the male descendants of Manasseh son of Joseph, by their families.

3 Now Zelophehad son of Hepher son of Gilead son of Machir son of Manasseh had no sons, but only daughters; and these are the names of his daughters: Mahlah, Noah, Hoglah, Milcah, and Tirzah. 4 They came before the priest Eleazar and Joshua son of Nun and the leaders, and said, "The LORD commanded Moses to give us an inheritance along with our male kin." So according to the commandment of the LORD he gave them an inheritance among the kinsmen of their father. 5 Thus there fell to Manasseh ten portions, besides the land of Gilead and Bashan, which is on the other side of the Jordan, 6 because the daughters of Manasseh received an inheritance along with his sons. The land of Gilead was allotted to the rest of the Manassites.

7 The territory of Manasseh reached from Asher to Michmethath, which is east of Shechem; then the boundary goes along southward to the inhabitants of En-tappuah. 8 The land of Tappuah belonged to Manasseh, but the town of Tappuah on the boundary of Manasseh belonged to the Ephraimites. 9 Then the boundary went down to the Wadi Kanah. The towns here, to the south of the wadi, among the towns of Manasseh, belong to Ephraim. Then the boundary of Manasseh goes along the north side of the wadi and ends at the sea. 10 The land to the south is Ephraim's and that to the north is Manasseh's, with the sea forming its boundary; on the north Asher is reached, and on the east Issachar. 11 Within Issachar and Asher, Manasseh had Beth-shean and its villages, Ibleam and its villages, the inhabitants of Dor and its villages, the inhabitants of En-dor and its villages, the inhabitants of Taanach and its villages, and the inhabitants of Megiddo and its villages (the third is Naphath). *b* 12 Yet the Manassites could not take possession of those towns; but the Canaanites continued to live in that land. 13 But when the Israelites grew strong, they put the Canaanites to forced labor, but did not utterly drive them out.

14 The tribe of Joseph spoke to Joshua, saying, "Why have you given me but one lot and one portion as an inheritance, since we are a numerous people, whom all along the LORD has blessed?" 15 And Joshua said to them, "If you are a numerous people, go up to the forest, and clear ground there for yourselves in the land of the Perizzites and the Rephaim, since the hill country of Ephraim is too narrow for you." 16 The tribe of Joseph said, "The hill country is not enough for us; yet all the Canaanites who live in the plain have chariots of iron, both those in Bethshean and its villages and those in the Valley of Jezreel." 17 Then Joshua said to the house of Joseph, to Ephraim and Manasseh, "You are indeed a numerous people, and have great power; you shall not have one lot only, 18 but the hill country shall be yours, for though it is a forest, you shall clear it and possess it to its farthest borders; for you shall drive out

b Meaning of Heb uncertain

17.1–6: Arrangements for the clans of Manasseh. **2:** *The rest of the tribe,* those who had not already settled east of the Jordan (13.29–31). **7–13:** Delimitation of the boundary of Manasseh. **17.11–13:** These verses are closely parallel to Judg 1.27–28. **14–18:** The Joseph tribes demand and receive a double portion. **16:** The *Jezreel Valley* that ran from the Carmel range east to the Jordan was not under Israelite control. *Beth-Shean* is located in the southeast corner of the Jezreel Valley. According to this verse and Judg 1.27 Israel did not take this city. It did not fall into Israelite hands until the time of David and Solomon.

the Canaanites, though they have chariots of iron, and though they are strong."

18 Then the whole congregation of the Israelites assembled at Shiloh, and set up the tent of meeting there. The land lay subdued before them.

2 There remained among the Israelites seven tribes whose inheritance had not yet been apportioned. ³So Joshua said to the Israelites, "How long will you be slack about going in and taking possession of the land that the LORD, the God of your ancestors, has given you? ⁴Provide three men from each tribe, and I will send them out that they may begin to go throughout the land, writing a description of it with a view to their inheritances. Then come back to me. ⁵They shall divide it into seven portions, Judah continuing in its territory on the south, and the house of Joseph in their territory on the north. ⁶You shall describe the land in seven divisions and bring the description here to me; and I will cast lots for you here before the LORD our God. ⁷The Levites have no portion among you, for the priesthood of the LORD is their heritage; and Gad and Reuben and the half-tribe of Manasseh have received their inheritance beyond the Jordan eastward, which Moses the servant of the LORD gave them."

8 So the men started on their way; and Joshua charged those who went to write the description of the land, saying, "Go throughout the land and write a description of it, and come back to me; and I will cast lots for you here before the LORD in Shiloh." ⁹So the men went and traversed the land and set down in a book a description of it by towns in seven divisions; then they came back to Joshua in the camp at Shiloh, ¹⁰and Joshua cast lots for them in Shiloh before the LORD; and there Joshua apportioned the land to the Israelites, to each a portion.

11 The lot of the tribe of Benjamin according to its families came up, and the territory allotted to it fell between the tribe of Judah and the tribe of Joseph. ¹²On the north side their boundary began at the Jordan; then the boundary goes up to the slope of Jericho on the north, then up through the hill country westward; and it ends at the wilderness of Beth-aven. ¹³From there the boundary passes along southward in the direction of Luz, to the slope of Luz (that is, Bethel), then the boundary goes down to Ataroth-addar, on the mountain that lies south of Lower Beth-horon. ¹⁴Then the boundary goes in another direction, turning on the western side southward from the mountain that lies to the south, opposite Beth-horon, and it ends at Kiriath-baal (that is, Kiriath-jearim), a town belonging to the tribe of Judah. This forms the western side. ¹⁵The southern side begins at the outskirts of Kiriath-jearim; and the boundary goes from there to Ephron,ᶜ to the spring of the Waters of Nephtoah; ¹⁶then the boundary goes down to the border of the mountain that overlooks the valley of the son of Hinnom, which is at the north end of the valley of Rephaim; and it then goes down the valley of Hinnom, south of the slope of the Jebusites, and downward to En-rogel; ¹⁷then it bends in a northerly direction going on to En-shemesh, and from there goes to Geliloth, which is opposite the ascent of Adummim; then it goes down to the Stone of Bohan, Reuben's son; ¹⁸and passing on to the north of the slope of Beth-arabahᵈ it goes down to the Arabah; ¹⁹then the boundary passes on to the north of the slope of Beth-hoglah; and the boundary ends at the northern bay of the Dead Sea,ᵉ at the south end of the Jordan: this is the southern border. ²⁰The Jordan forms its

c Cn See 15.9. Heb *westward*   d Gk: Heb *to the slope over against the Arabah*   e Heb *Salt Sea*

18.1–19.51: The territory of other tribes. 1–10: General introduction to the allotment. **1:** *Shiloh,* an important Israelite sanctuary in the period before the monarchy (Judg 18.31; 1 Sam 4.3–4). It is twenty miles north of Jerusalem.

18.11–18: The territory of Benjamin is immediately north of Judah.

boundary on the eastern side. This is the inheritance of the tribe of Benjamin, according to its families, boundary by boundary all around.

21 Now the towns of the tribe of Benjamin according to their families were Jericho, Beth-hoglah, Emek-keziz, ²²Beth-arabah, Zemaraim, Bethel, ²³Avvim, Parah, Ophrah, ²⁴Chephar-ammoni, Ophni, and Geba—twelve towns with their villages: ²⁵Gibeon, Ramah, Beeroth, ²⁶Mizpeh, Chephirah, Mozah, ²⁷Rekem, Irpeel, Taralah, ²⁸Zela, Haeleph, Jebus*f* (that is, Jerusalem), Gibeah*g* and Kiriath-jearim*h*—fourteen towns with their villages. This is the inheritance of the tribe of Benjamin according to its families.

19 The second lot came out for Simeon, for the tribe of Simeon, according to its families; its inheritance lay within the inheritance of the tribe of Judah. ²It had for its inheritance Beer-sheba, Sheba, Moladah, ³Hazar-shual, Balah, Ezem, ⁴Eltolad, Bethul, Hormah, ⁵Ziklag, Beth-marcaboth, Hazar-susah, ⁶Beth-lebaoth, and Sharuhen—thirteen towns with their villages; ⁷Ain, Rimmon, Ether, and Ashan—four towns with their villages; ⁸together with all the villages all around these towns as far as Baalath-beer, Ramah of the Negeb. This was the inheritance of the tribe of Simeon according to its families. ⁹The inheritance of the tribe of Simeon formed part of the territory of Judah; because the portion of the tribe of Judah was too large for them, the tribe of Simeon obtained an inheritance within their inheritance.

10 The third lot came up for the tribe of Zebulun, according to its families. The boundary of its inheritance reached as far as Sarid; ¹¹then its boundary goes up westward, and on to Maralah, and touches Dabbesheth, then the wadi that is east of Jokneam; ¹²from Sarid it goes in the other direction eastward toward

the sunrise to the boundary of Chisloth-tabor; from there it goes to Daberath, then up to Japhia; ¹³from there it passes along on the east toward the sunrise to Gath-hepher, to Eth-kazin, and going on to Rimmon it bends toward Neah; ¹⁴then on the north the boundary makes a turn to Hannathon, and it ends at the valley of Iphtah-el; ¹⁵and Kattath, Nahalal, Shimron, Idalah, and Bethlehem—twelve towns with their villages. ¹⁶This is the inheritance of the tribe of Zebulun, according to its families—these towns with their villages.

17 The fourth lot came out for Issachar, for the tribe of Issachar, according to its families. ¹⁸Its territory included Jezreel, Chesulloth, Shunem, ¹⁹Hapharaim, Shion, Anaharath, ²⁰Rabbith, Kishion, Ebez, ²¹Remeth, En-gannim, En-haddah, Beth-pazzez; ²²the boundary also touches Tabor, Shahazumah, and Beth-shemesh, and its boundary ends at the Jordan—sixteen towns with their villages. ²³This is the inheritance of the tribe of Issachar, according to its families—the towns with their villages.

24 The fifth lot came out for the tribe of Asher according to its families. ²⁵Its boundary included Helkath, Hali, Beten, Achshaph, ²⁶Allammelech, Amad, and Mishal; on the west it touches Carmel and Shihor-libnath, ²⁷then it turns eastward, goes to Beth-dagon, and touches Zebulun and the valley of Iphtah-el northward to Beth-emek and Neiel; then it continues in the north to Cabul, ²⁸Ebron, Rehob, Hammon, Kanah, as far as Great Sidon; ²⁹then the boundary turns to Ramah, reaching to the fortified city of Tyre; then the boundary turns to Hosah, and it ends at the sea; Mahalab,*i* Achzib, ³⁰Ummah, Aphek,

---

f Gk Syr Vg: Heb *the Jebusite*   *g* Heb *Gibeath*
h Gk: Heb *Kiriath*   *i* Cn Compare Gk: Heb *Mehebel*

19.1–9: The territory of Simeon. **9:** A rationalization for the incorporation of Simeon into Judah.
   19.10–48: The territory of the tribes in Galilee. **10–16:** Zebulun. **17–23:** Issachar.
   19.24–31: Asher. The list includes some Phoenician cities like Tyre (v. 29) that were never under Israelite control.

and Rehob—twenty-two towns with their villages. 31 This is the inheritance of the tribe of Asher according to its families—these towns with their villages.

32 The sixth lot came out for the tribe of Naphtali, for the tribe of Naphtali, according to its families. 33 And its boundary ran from Heleph, from the oak in Zaanannim, and Adami-nekeb, and Jabneel, as far as Lakkum; and it ended at the Jordan; 34 then the boundary turns westward to Aznoth-tabor, and goes from there to Hukkok, touching Zebulun at the south, and Asher on the west, and Judah on the east at the Jordan. 35 The fortified towns are Ziddim, Zer, Hammath, Rakkath, Chinnereth, 36 Adamah, Ramah, Hazor, 37 Kedesh, Edrei, Enhazor, 38 Iron, Migdal-el, Horem, Bethanath, and Beth-shemesh—nineteen towns with their villages. 39 This is the inheritance of the tribe of Naphtali according to its families—the towns with their villages.

40 The seventh lot came out for the tribe of Dan, according to its families. 41 The territory of its inheritance included Zorah, Eshtaol, Ir-shemesh, 42 Shaalabbin, Aijalon, Ithlah, 43 Elon, Timnah, Ekron, 44 Eltekeh, Gibbethon, Baalath, 45 Jehud, Bene-berak, Gath-rimmon, 46 Me-jarkon, and Rakkon at the border opposite Joppa. 47 When the territory of the Danites was lost to them, the Danites went up and fought against Leshem, and after capturing it and putting it to the sword, they took possession of it and settled in it, calling Leshem, Dan, after their ancestor Dan. 48 This is the inheritance of the tribe of Dan, according to

their families—these towns with their villages.

49 When they had finished distributing the several territories of the land as inheritances, the Israelites gave an inheritance among them to Joshua son of Nun. 50 By command of the LORD they gave him the town that he asked for, Timnath-serah in the hill country of Ephraim; he rebuilt the town, and settled in it.

51 These are the inheritances that the priest Eleazar and Joshua son of Nun and the heads of the families of the tribes of the Israelites distributed by lot at Shiloh before the LORD, at the entrance of the tent of meeting. So they finished dividing the land.

20 Then the LORD spoke to Joshua, saying, 2 "Say to the Israelites, 'Appoint the cities of refuge, of which I spoke to you through Moses, 3 so that anyone who kills a person without intent or by mistake may flee there; they shall be for you a refuge from the avenger of blood. 4 The slayer shall flee to one of these cities and shall stand at the entrance of the gate of the city, and explain the case to the elders of that city; then the fugitive shall be taken into the city, and given a place, and shall remain with them. 5 And if the avenger of blood is in pursuit, they shall not give up the slayer, because the neighbor was killed by mistake, there having been no enmity between them before. 6 The slayer shall remain in that city until there is a trial before the congregation, until the death of the one who is high priest at the time: then the slayer may return home, to the town in which the deed was done.' "

19.32–39: Naphtali.
19.40–48: Dan. Ancient tradition located Dan in the south. **47**: Sometime before the time of the monarchy the Danites migrated to the north (see Judg 18). Judg 18.27 refers to *Leshem* as Laish. **49–51**: Conclusion. **49–50**: Joshua's personal allotment. **50**: *Timnathserah* is identified with Khirbet Tibnah, seventeen miles southwest of Shechem.

20.1–9: Joshua named cities of refuge that provided the right of asylum for one

accused of murder until the case was adjudicated. The passage shows how Joshua fulfilled what God had commanded Moses (Num 35.9–34) and Moses commanded Israel (Deut 19.1–13). **3**: *The avenger of blood* was the deceased's nearest relative. **4**: *The gate of the city* was where *the elders* of the city met to adjudicate disputes. Such gates were enclosed structures of more than one story with several rooms.

7 So they set apart Kedesh in Galilee in the hill country of Naphtali, and Shechem in the hill country of Ephraim, and Kiriath-arba (that is, Hebron) in the hill country of Judah. 8 And beyond the Jordan east of Jericho, they appointed Bezer in the wilderness on the tableland, from the tribe of Reuben, and Ramoth in Gilead, from the tribe of Gad, and Golan in Bashan, from the tribe of Manasseh. 9 These were the cities designated for all the Israelites, and for the aliens residing among them, that anyone who killed a person without intent could flee there, so as not to die by the hand of the avenger of blood, until there was a trial before the congregation.

21 Then the heads of the families of the Levites came to the priest Eleazar and to Joshua son of Nun and to the heads of the families of the tribes of the Israelites; 2 they said to them at Shiloh in the land of Canaan, "The LORD commanded through Moses that we be given towns to live in, along with their pasture lands for our livestock." 3 So by command of the LORD the Israelites gave to the Levites the following towns and pasture lands out of their inheritance.

4 The lot came out for the families of the Kohathites. So those Levites who were descendants of Aaron the priest received by lot thirteen towns from the tribes of Judah, Simeon, and Benjamin.

5 The rest of the Kohathites received by lot ten towns from the families of the tribe of Ephraim, from the tribe of Dan, and the half-tribe of Manasseh.

6 The Gershonites received by lot thirteen towns from the families of the tribe of Issachar, from the tribe of Asher, from the tribe of Naphtali, and from the half-tribe of Manasseh in Bashan.

7 The Merarites according to their families received twelve towns from the tribe of Reuben, the tribe of Gad, and the tribe of Zebulun.

8 These towns and their pasture lands the Israelites gave by lot to the Levites, as the LORD had commanded through Moses.

9 Out of the tribe of Judah and the tribe of Simeon they gave the following towns mentioned by name, 10 which went to the descendants of Aaron, one of the families of the Kohathites who belonged to the Levites, since the lot fell to them first. 11 They gave them Kiriath-arba (Arba being the father of Anak), that is Hebron, in the hill country of Judah, along with the pasture lands around it. 12 But the fields of the town and its villages had been given to Caleb son of Jephunneh as his holding.

13 To the descendants of Aaron the priest they gave Hebron, the city of refuge for the slayer, with its pasture lands, Libnah with its pasture lands, 14 Jattir with its pasture lands, Eshtemoa with its pasture lands, 15 Holon with its pasture lands, Debir with its pasture lands, 16 Ain with its pasture lands, Juttah with its pasture lands, and Beth-shemesh with its pasture lands—nine towns out of these two tribes. 17 Out of the tribe of Benjamin: Gibeon with its pasture lands, Geba with its pasture lands, 18 Anathoth with its pasture lands, and Almon with its pasture lands—four towns. 19 The towns of the descendants of Aaron—the priests—were thirteen in all, with their pasture lands.

20 As to the rest of the Kohathites belonging to the Kohathite families of the Levites, the towns allotted to them were out of the tribe of Ephraim. 21 To them were given Shechem, the city of refuge for the slayer, with its pasture lands in the hill country of Ephraim, Gezer with its pasture lands, 22 Kibzaim with its pasture lands, and Beth-horon with its pasture lands—four towns. 23 Out of the tribe of Dan: Elteke with its pasture lands, Gibbethon with its pasture

---

21.1–42: **The levitical cities.** Unlike previous statements in this book (13.14,33) that assert that the Levites received no land of their own, here (and in Num 35.1–8) the Levites receive forty-eight cities with the surrounding land for grazing. Most of the cities listed were not settled until 8th century.

lands, 24 Aijalon with its pasture lands, Gath-rimmon with its pasture lands— four towns. 25 Out of the half-tribe of Manasseh: Taanach with its pasture lands, and Gath-rimmon with its pasture lands—two towns. 26 The towns of the families of the rest of the Kohathites were ten in all, with their pasture lands.

27 To the Gershonites, one of the families of the Levites, were given out of the half-tribe of Manasseh, Golan in Bashan with its pasture lands, the city of refuge for the slayer, and Beeshterah with its pasture lands—two towns. 28 Out of the tribe of Issachar: Kishion with its pasture lands, Daberath with its pasture lands, 29 Jarmuth with its pasture lands, En-gannim with its pasture lands—four towns; 30 Out of the tribe of Asher: Mishal with its pasture lands, Abdon with its pasture lands, 31 Helkath with its pasture lands, and Rehob with its pasture lands—four towns. 32 Out of the tribe of Naphtali: Kedesh in Galilee with its pasture lands, the city of refuge for the slayer, Hammoth-dor with its pasture lands, and Kartan with its pasture lands—three towns. 33 The towns of the several families of the Gershonites were in all thirteen, with their pasture lands.

34 To the rest of the Levites—the Merarite families—were given out of the tribe of Zebulun: Jokneam with its pasture lands, Kartah with its pasture lands, 35 Dimnah with its pasture lands, Nahalal with its pasture lands—four towns. 36 Out of the tribe of Reuben: Bezer with its pasture lands, Jahzah with its pasture lands, 37 Kedemoth with its pasture lands, and Mephaath with its pasture lands—four towns. 38 Out of the tribe of Gad: Ramoth in Gilead with its pasture lands, the city of refuge for the slayer, Mahanaim with its pasture lands, 39 Heshbon with its pasture lands, Jazer with its pasture lands—four towns in all. 40 As for the towns of the several Merarite families, that is, the remainder of the families of the Levites, those allotted to them were twelve in all.

41 The towns of the Levites within the holdings of the Israelites were in all forty-eight towns with their pasture lands. 42 Each of these towns had its pasture lands around it; so it was with all these towns.

43 Thus the LORD gave to Israel all the land that he swore to their ancestors that he would give them; and having taken possession of it, they settled there. 44 And the LORD gave them rest on every side just as he had sworn to their ancestors; not one of all their enemies had withstood them, for the LORD had given all their enemies into their hands. 45 Not one of all the good promises that the LORD had made to the house of Israel had failed; all came to pass.

22 Then Joshua summoned the Reubenites, the Gadites, and the half-tribe of Manasseh, 2 and said to them, "You have observed all that Moses the servant of the LORD commanded you, and have obeyed me in all that I have commanded you; 3 you have not forsaken your kindred these many days, down to this day, but have been careful to keep the charge of the LORD your God. 4 And now the LORD your God has given rest to your kindred, as he promised them; therefore turn and go to your tents in the land where your possession lies, which Moses the servant of the LORD gave you on the other side of the Jordan. 5 Take good care to observe the commandment and instruction that Moses the servant of the LORD commanded you, to love the LORD your God, to walk in all his ways, to keep his commandments, and to hold fast to him, and to serve him with all your heart and with all your soul." 6 So Joshua blessed them and sent them away, and they went to their tents.

7 Now to the one half of the tribe of Manasseh Moses had given a possession in Bashan; but to the other half Joshua

---

21.43–45: **Another summary.** This conclusion to the whole of the preceding material is couched in the language of Deuteronomy.

22.1–34: **The Transjordanian tribes.** The ties binding the tribes together were not firm. Joshua's style of leadership helped to

had given a possession beside their fellow Israelites in the land west of the Jordan. And when Joshua sent them away to their tents and blessed them, 8 he said to them, "Go back to your tents with much wealth, and with very much livestock, with silver, gold, bronze, and iron, and with a great quantity of clothing; divide the spoil of your enemies with your kindred." 9 So the Reubenites and the Gadites and the half-tribe of Manasseh returned home, parting from the Israelites at Shiloh, which is in the land of Canaan, to go to the land of Gilead, their own land of which they had taken possession by command of the LORD through Moses.

10 When they came to the region*j* near the Jordan that lies in the land of Canaan, the Reubenites and the Gadites and the half-tribe of Manasseh built there an altar by the Jordan, an altar of great size. 11 The Israelites heard that the Reubenites and the Gadites and the half-tribe of Manasseh had built an altar at the frontier of the land of Canaan, in the region*k* near the Jordan, on the side that belongs to the Israelites. 12 And when the people of Israel heard of it, the whole assembly of the Israelites gathered at Shiloh, to make war against them.

13 Then the Israelites sent the priest Phinehas son of Eleazar to the Reubenites and the Gadites and the half-tribe of Manasseh, in the land of Gilead, 14 and with him ten chiefs, one from each of the tribal families of Israel, every one of them the head of a family among the clans of Israel. 15 They came to the Reubenites, the Gadites, and the half-tribe of Manasseh, in the land of Gilead, and they said to them, 16 "Thus says the whole congregation of the LORD, 'What is this treachery that you have committed against the God of Israel in turning away today from following the LORD, by building yourselves an altar today in rebellion against the LORD? 17 Have we not had enough of the sin at Peor from which even yet we have not cleansed ourselves, and for which a plague came upon the congregation of the LORD, 18 that you must turn away today from following the LORD! If you rebel against the LORD today, he will be angry with the whole congregation of Israel tomorrow. 19 But now, if your land is unclean, cross over into the LORD's land where the LORD's tabernacle now stands, and take for yourselves a possession among us; only do not rebel against the LORD, or rebel against us*l* by building yourselves an altar other than the altar of the LORD our God. 20 Did not Achan son of Zerah break faith in the matter of the devoted things, and wrath fell upon all the congregation of Israel? And he did not perish alone for his iniquity!' "

21 Then the Reubenites, the Gadites, and the half-tribe of Manasseh said in answer to the heads of the families of Israel, 22 "The LORD, God of gods! The LORD, God of gods! He knows; and let Israel itself know! If it was in rebellion or in breach of faith toward the LORD, do not spare us today 23 for building an altar to turn away from following the LORD; or if we did so to offer burnt offerings or grain offerings or offerings of well-being on it, may the LORD himself take vengeance. 24 No! We did it from fear that in time to come your children might say to our children, 'What have you to do with the LORD, the God of Israel? 25 For the LORD has made the Jordan a boundary between us and you, you Reubenites and Gadites; you have no portion in the

j Or *to Geliloth*   *k* Or *at Geliloth*
l Or *make rebels of us*

---

solve intertribal disputes, but such conflicts were a continuing problem. The issue here is the locus of legitimate worship, a central concern of Deuteronomy.

22.9–34: The central role of *Phineas* in dealing with this conflict may show that priestly circles edited this story. **12:** Holding strictly to the Deuteronomic law that forbade the offering of sacrifice anywhere except in the one central sanctuary (Deut 12.13,14), the other tribes apparently interpret the building of the altar as an act of disloyalty to Israel

Lord.' So your children might make our children cease to worship the Lord. 26 Therefore we said, 'Let us now build an altar, not for burnt offering, nor for sacrifice, 27 but to be a witness between us and you, and between the generations after us, that we do perform the service of the Lord in his presence with our burnt offerings and sacrifices and offerings of well-being; so that your children may never say to our children in time to come, "You have no portion in the Lord."' 28 And we thought, If this should be said to us or to our descendants in time to come, we could say, 'Look at this copy of the altar of the Lord, which our ancestors made, not for burnt offerings, nor for sacrifice, but to be a witness between us and you.' 29 Far be it from us that we should rebel against the Lord, and turn away this day from following the Lord by building an altar for burnt offering, grain offering, or sacrifice, other than the altar of the Lord our God that stands before his tabernacle!"

30 When the priest Phinehas and the chiefs of the congregation, the heads of the families of Israel who were with him, heard the words that the Reubenites and the Gadites and the Manassites spoke, they were satisfied. 31 The priest Phinehas son of Eleazar said to the Reubenites and the Gadites and the Manassites, "Today we know that the Lord is among us, because you have not committed this treachery against the Lord; now you have saved the Israelites from the hand of the Lord."

32 Then the priest Phinehas son of Eleazar and the chiefs returned from the Reubenites and the Gadites in the land of Gilead to the land of Canaan, to the Israelites, and brought back word to them. 33 The report pleased the Israelites; and the Israelites blessed God and spoke no more of making war against them, to destroy the land where the Reubenites and the Gadites were settled. 34 The Reubenites and the Gadites called the altar Witness;*m* "For," said they, "it is a witness between us that the Lord is God."

23 A long time afterward, when the Lord had given rest to Israel from all their enemies all around, and Joshua was old and well advanced in years, 2 Joshua summoned all Israel, their elders and heads, their judges and officers, and said to them, "I am now old and well advanced in years; 3 and you have seen all that the Lord your God has done to all these nations for your sake, for it is the Lord your God who has fought for you. 4 I have allotted to you as an inheritance for your tribes those nations that remain, along with all the nations that I have already cut off, from the Jordan to the Great Sea in the west. 5 The Lord your God will push them back before you, and drive them out of your sight; and you shall possess their land, as the Lord your God promised you. 6 Therefore be very steadfast to observe and do all that is written in the book of the law of Moses, turning aside from it neither to the right nor to the left, 7 so that you may not be mixed with these nations left here among you, or make mention of the names of their gods, or swear by them, or serve them, or bow yourselves down to them, 8 but hold fast to the Lord your God, as you have done to this day. 9 For the Lord has driven out before you great and strong nations; and as for you, no one has been able to withstand you to this day. 10 One of you puts to flight a thousand, since it is the Lord your God who fights for you, as he promised you. 11 Be very careful, therefore, to love the

m Cn Compare Syr: Heb lacks *Witness*

---

and to its God, and therefore prepare *to make war against them.* 17: *The sin at Peor,* Num 25.3–5.

22.20: *Achan,* 7.1. 26–27: It was not a real altar, but merely a memorial, *a witness.*

23.1–16: **Joshua's farewell admonitions.** Like other important leaders of Israel (Jacob, Gen 49; Moses, Deut 29–31; Samuel, 1 Sam 12; David, 1 Kings 2.1–9), Joshua delivers an address at the end of his life. Its tone is Deuteronomistic, stressing the fulfillment of God's promises and the need for Israel's obedience. This may have been the original ending of the book.

LORD your God. [12]For if you turn back, and join the survivors of these nations left here among you, and intermarry with them, so that you marry their women and they yours, [13]know assuredly that the LORD your God will not continue to drive out these nations before you; but they shall be a snare and a trap for you, a scourge on your sides, and thorns in your eyes, until you perish from this good land that the LORD your God has given you.

[14] "And now I am about to go the way of all the earth, and you know in your hearts and souls, all of you, that not one thing has failed of all the good things that the LORD your God promised concerning you; all have come to pass for you, not one of them has failed. [15]But just as all the good things that the LORD your God promised concerning you have been fulfilled for you, so the LORD will bring upon you all the bad things, until he has destroyed you from this good land that the LORD your God has given you. [16]If you transgress the covenant of the LORD your God, which he enjoined on you, and go and serve other gods and bow down to them, then the anger of the LORD will be kindled against you, and you shall perish quickly from the good land that he has given to you."

24 Then Joshua gathered all the tribes of Israel to Shechem, and summoned the elders, the heads, the judges, and the officers of Israel; and they presented themselves before God. [2]And Joshua said to all the people, "Thus says the LORD, the God of Israel: Long ago your ancestors—Terah and his sons Abraham and Nahor—lived beyond the Euphrates and served other gods. [3]Then I took your father Abraham from be-

yond the River and led him through all the land of Canaan and made his offspring many. I gave him Isaac; [4]and to Isaac I gave Jacob and Esau. I gave Esau the hill country of Seir to possess, but Jacob and his children went down to Egypt. [5]Then I sent Moses and Aaron, and I plagued Egypt with what I did in its midst; and afterwards I brought you out. [6]When I brought your ancestors out of Egypt, you came to the sea; and the Egyptians pursued your ancestors with chariots and horsemen to the Red Sea.[n] [7]When they cried out to the LORD, he put darkness between you and the Egyptians, and made the sea come upon them and cover them; and your eyes saw what I did to Egypt. Afterwards you lived in the wilderness a long time. [8]Then I brought you to the land of the Amorites, who lived on the other side of the Jordan; they fought with you, and I handed them over to you, and you took possession of their land, and I destroyed them before you. [9]Then King Balak son of Zippor of Moab, set out to fight against Israel. He sent and invited Balaam son of Beor to curse you, [10]but I would not listen to Balaam; therefore he blessed you; so I rescued you out of his hand. [11]When you went over the Jordan and came to Jericho, the citizens of Jericho fought against you, and also the Amorites, the Perizzites, the Canaanites, the Hittites, the Girgashites, the Hivites, and the Jebusites; and I handed them over to you. [12]I sent the hornet[o] ahead of you, which drove out before you the two kings of the Amorites; it was not by your sword or by your bow. [13]I gave you a land on

n Or *Sea of Reeds*   o Meaning of Heb uncertain

24.1–28: The covenant at Shechem. Joshua fulfills the commands of Moses in Deut 11, 27, 31. All Israel unite under Joshua's leadership in the service of the LORD. **1:** *Shechem* (Tell Balata) became an important Israelite cultic and political center. It was one of the few major cities in Canaan not described as destroyed by the Israelites. Apparently it was incorporated into Israel peacefully. Early

Israel then was a confederation of originally disparate groups unrelated by blood or common experience. **2–13:** This summary of God's actions on Israel's behalf does not mention Sinai or the divine guidance in the wilderness.

24.9–11: This text reflects a different memory of the Balaam incident (Num 22) and the fall of Jericho (Josh 6).

which you had not labored, and towns that you had not built, and you live in them; you eat the fruit of vineyards and oliveyards that you did not plant.

14 "Now therefore revere the LORD, and serve him in sincerity and in faithfulness; put away the gods that your ancestors served beyond the River and in Egypt, and serve the LORD. 15 Now if you are unwilling to serve the LORD, choose this day whom you will serve, whether the gods your ancestors served in the region beyond the River or the gods of the Amorites in whose land you are living; but as for me and my household, we will serve the LORD."

16 Then the people answered, "Far be it from us that we should forsake the LORD to serve other gods; 17 for it is the LORD our God who brought us and our ancestors up from the land of Egypt, out of the house of slavery, and who did those great signs in our sight. He protected us along all the way that we went, and among all the peoples through whom we passed; 18 and the LORD drove out before us all the peoples, the Amorites who lived in the land. Therefore we also will serve the LORD, for he is our God."

19 But Joshua said to the people, "You cannot serve the LORD, for he is a holy God. He is a jealous God; he will not forgive your transgressions or your sins. 20 If you forsake the LORD and serve foreign gods, then he will turn and do you harm, and consume you, after having done you good." 21 And the people said to Joshua, "No, we will serve the LORD!" 22 Then Joshua said to the people, "You are witnesses against yourselves that you have chosen the LORD, to serve him." And they said, "We are wit-

nesses." 23 He said, "Then put away the foreign gods that are among you, and incline your hearts to the LORD, the God of Israel." 24 The people said to Joshua, "The LORD our God we will serve, and him we will obey." 25 So Joshua made a covenant with the people that day, and made statutes and ordinances for them at Shechem. 26 Joshua wrote these words in the book of the law of God; and he took a large stone, and set it up there under the oak in the sanctuary of the LORD. 27 Joshua said to all the people, "See, this stone shall be a witness against us; for it has heard all the words of the LORD that he spoke to us; therefore it shall be a witness against you, if you deal falsely with your God." 28 So Joshua sent the people away to their inheritances.

29 After these things Joshua son of Nun, the servant of the LORD, died, being one hundred ten years old. 30 They buried him in his own inheritance at Timnath-serah, which is in the hill country of Ephraim, north of Mount Gaash.

31 Israel served the LORD all the days of Joshua, and all the days of the elders who outlived Joshua and had known all the work that the LORD did for Israel.

32 The bones of Joseph, which the Israelites had brought up from Egypt, were buried at Shechem, in the portion of ground that Jacob had bought from the children of Hamor, the father of Shechem, for one hundred pieces of money;p it became an inheritance of the descendants of Joseph.

33 Eleazar son of Aaron died; and they buried him at Gibeah, the town of his son Phinehas, which had been given him in the hill country of Ephraim.

p Heb *one hundred qesitah*

24.29–33: **Final notes. 29–31:** Joshua's death and burial. **32:** Joseph was reburied (see Gen 33.19; 50.25; Ex 13.19). **33:** The burial of Eleazar, a priest associated with Joshua (14.1).

Judges

This book tells the story of a period of transition for the Israelite tribes. The age of the great leaders of the past was gone. Both Moses and Joshua were dead. The age of the greatness under the rule of David was yet to come. Israel had leaders in this transitional period, but they were not like Moses, Joshua, or David. The judges were courageous, but they had their fears. Even their attitudes toward the God of Israel were not entirely commendable.

This is also the story of a new community emerging from disparate groups that were trying to create an entirely new pattern of life for their people. This new community was to be one in which all citizens had an equal range of opportunities. The Israelites rejected the absolutism of the Canaanite city-states with their oppressive political and social systems. This new people living in the highlands of central Canaan would serve only the LORD. The book of Judges shows that the creation of this new society was an immense struggle. In the midst of revolutionary social upheaval, the Israelites found support in their belief that they were ruled by the LORD who took the side of the lowly against their oppressors.

The traditional name of this book is The Book of Judges. Outside the introduction (2.16–19), however, the title "judge" appears only once and then it refers to the LORD (11.27). The human protagonists of these stories appear as servants of the LORD. To fulfill their task they receive "the spirit of the LORD" that leads them to exert great power in effecting the divine will as they lead in battle against the enemies of Israel. After defeating those enemies, the judges continued to exercise civil duties among their several clans and tribes.

This book is really a composite work, dealing with several clan and tribal heroes. The inclusion of their stories in the collection transforms these local champions into national figures. The hand of the Deuteronomistic editor is evident primarily in the texts that join one story with another. At one point this collection became part of a larger work that told the story of Israel in its land: the Deuteronomistic History (see pp. 267–269 OT). The author of that history found the stories of the judges illustrative of an important principle of the larger work: Israel's future is a product of the loyalty that Israel owes to its LORD.

1 After the death of Joshua, the Israelites inquired of the LORD, "Who shall go up first for us against the Canaanites, to fight against them?" ²The LORD said, "Judah shall go up. I hereby give the land into his hand." ³Judah said to his brother Simeon, "Come up with me into the territory allotted to me, that we may fight against the Canaanites; then I too will go with you into the territory allotted to you." So Simeon went with him. ⁴Then Judah went up and the LORD gave the Canaanites and the Perizzites into their hand; and they defeated ten thousand of them at Bezek. ⁵They came upon Adoni-bezek at Bezek, and fought against him, and defeated the Canaanites and the Perizzites. ⁶Adoni-bezek fled; but they pursued him, and caught him, and cut off his thumbs and big toes. ⁷Adoni-bezek said, "Seventy kings with their thumbs and big toes cut off used to pick up scraps under my table; as I have done, so God has paid me back." They brought him to Jerusalem, and he died there.

8 Then the people of Judah fought against Jerusalem and took it. They put it to the sword and set the city on fire. ⁹Afterward the people of Judah went down to fight against the Canaanites who lived in the hill country, in the Negeb, and in the lowland. ¹⁰Judah went against the Canaanites who lived in Hebron (the name of Hebron was formerly Kiriath-arba); and they defeated Sheshai and Ahiman and Talmai.

11 From there they went against the inhabitants of Debir (the name of Debir was formerly Kiriath-sepher). ¹²Then Caleb said, "Whoever attacks Kiriath-sepher and takes it, I will give him my daughter Achsah as wife." ¹³And Othniel son of Kenaz, Caleb's younger brother, took it; and he gave him his daughter Achsah as wife. ¹⁴When she came to him, she urged him to ask her father for a field. As she dismounted from her donkey, Caleb said to her, "What do you wish?" ¹⁵She said to him, "Give me a present; since you have set me in the land of the Negeb, give me also Gulloth-mayim."ᵃ So Caleb gave her Upper Gulloth and Lower Gulloth.

16 The descendants of Hobabᵇ the Kenite, Moses' father-in-law, went up with the people of Judah from the city of palms into the wilderness of Judah, which lies in the Negeb near Arad. Then they went and settled with the Amalekites.ᶜ ¹⁷Judah went with his brother Simeon, and they defeated the Canaanites who inhabited Zephath, and devoted it to destruction. So the city was called Hormah. ¹⁸Judah took Gaza with its territory, Ashkelon with its territory, and Ekron with its territory. ¹⁹The LORD was

a That is Basins of Water   _b Gk: Heb lacks
Hobab_   _c See 1 Sam 15.6: Heb people_

1.1–2.5: The period of the judges. Israel's acquisition of the land was incomplete because of its failures to obey the LORD's commands. The story of the judges details the effects of Israel's failures and how the LORD dealt with those failures.

1.1–21: The military conquest in the south (compare Josh 15) was not completely successful because of the Canaanite superiority in armaments. **3:** The tribe of _Judah_ in time absorbed _Simeon_, a tribe that played no significant role in the later history of Israel (see Josh 19.9 n.). **4:** The location of _Bezek_ is uncertain, though Khirbet Ibzik seventeen miles north of Shechem is a possibility. **1.8:** _Jerusalem_ is not an Israelite city in its only other appearance in Judges (19.10; see Josh 15.63). It was David who brought Jerusalem into the Israelite orbit (2 Sam 5.6–9). **10:** _Hebron,_ twenty miles south of Jerusalem. The continuous occupation of the area has made it impossible to locate the remains of the Bronze and Iron Age city. **11–15:** Josh 15.13–19. **11:** The identification of _Debir_ with Tell Beit Mirsim is not certain; Khirbet er-Rabub is another possibility. **1.16:** The Kenites were travelling smiths. _The city of palms_ is probably Jericho (as in 3.13). **17:** The name _Hormah_ derives from the Hebrew "ḥerem," the total destruction of a city and its inhabitants (see Josh 6.17 n.). **18:** _Gaza . . . Ashkelon . . . Ekron_ were three of the five principal cities of the Philistine confederation (see 14.19 n.). It was David who captured these Philistine cities. The Septuagint asserts that Judah did not capture these cities.

with Judah, and he took possession of the hill country, but could not drive out the inhabitants of the plain, because they had chariots of iron. 20 Hebron was given to Caleb, as Moses had said; and he drove out from it the three sons of Anak. 21 But the Benjaminites did not drive out the Jebusites who lived in Jerusalem; so the Jebusites have lived in Jerusalem among the Benjaminites to this day.

22 The house of Joseph also went up against Bethel; and the LORD was with them. 23 The house of Joseph sent out spies to Bethel (the name of the city was formerly Luz). 24 When the spies saw a man coming out of the city, they said to him, "Show us the way into the city, and we will deal kindly with you." 25 So he showed them the way into the city; and they put the city to the sword, but they let the man and all his family go. 26 So the man went to the land of the Hittites and built a city, and named it Luz; that is its name to this day.

27 Manasseh did not drive out the inhabitants of Beth-shean and its villages, or Taanach and its villages, or the inhabitants of Dor and its villages, or the inhabitants of Ibleam and its villages, or the inhabitants of Megiddo and its villages; but the Canaanites continued to live in that land. 28 When Israel grew strong, they put the Canaanites to forced labor, but did not in fact drive them out.

29 And Ephraim did not drive out the Canaanites who lived in Gezer; but the Canaanites lived among them in Gezer.

30 Zebulun did not drive out the inhabitants of Kitron, or the inhabitants of Nahalol; but the Canaanites lived among them, and became subject to forced labor.

31 Asher did not drive out the inhabitants of Acco, or the inhabitants of Sidon, or of Ahlab, or of Achzib, or of Helbah, or of Aphik, or of Rehob; 32 but the Asherites lived among the Canaanites, the inhabitants of the land; for they did not drive them out.

33 Naphtali did not drive out the inhabitants of Beth-shemesh, or the inhabitants of Beth-anath, but lived among the Canaanites, the inhabitants of the land; nevertheless the inhabitants of Beth-shemesh and of Beth-anath became subject to forced labor for them.

34 The Amorites pressed the Danites back into the hill country; they did not allow them to come down to the plain. 35 The Amorites continued to live in Har-heres, in Aijalon, and in Shaalbim, but the hand of the house of Joseph rested heavily on them, and they became subject to forced labor. 36 The border of the Amorites ran from the ascent of Akrabbim, from Sela and upward.

2 Now the angel of the LORD went up from Gilgal to Bochim, and said, "I brought you up from Egypt, and brought you into the land that I had promised to your ancestors. I said, 'I will never break my covenant with you. 2 For your part, do not make a covenant with the inhabitants of this land; tear down their altars.' But you have not obeyed my command. See what you have done! 3 So now I say, I will not drive them out before you; but they shall become adversaries*d* to you, and their gods shall be a snare to you." 4 When the angel of the LORD spoke these words to all the Israelites, the people lifted up their voices and wept. 5 So they named that place Bochim,*e* and there they sacrificed to the LORD.

6 When Joshua dismissed the people,

d OL Vg Compare Gk: Heb *sides*   *e* That is *Weepers*

1.22–29: The achievements of the Joseph tribes located in the central highlands north of Jerusalem (compare Josh chs 16–17). 30–36: The achievements of the tribes of Galilee (Josh chs 18–19). 2.1–5: Israel did not do as the LORD ordered and so it will suffer. 1: *The angel of the LORD* is a way of speaking about an appearance of God (compare Gen 16.7). *Bochim* occurs nowhere else in the Hebrew Bible; the Septuagint reads Bethel. 2.6–3.6: **Israel under the judges.** When good leaders die, the apostasy that they suppressed in Israel thrives again. 2.6–10: With the death of Joshua, Israel is bereft of the leadership that insured its fidelity

the Israelites all went to their own inheritances to take possession of the land. 7 The people worshiped the LORD all the days of Joshua, and all the days of the elders who outlived Joshua, who had seen all the great work that the LORD had done for Israel. 8 Joshua son of Nun, the servant of the LORD, died at the age of one hundred ten years. 9 So they buried him within the bounds of his inheritance in Timnath-heres, in the hill country of Ephraim, north of Mount Gaash. 10 Moreover, that whole generation was gathered to their ancestors, and another generation grew up after them, who did not know the LORD or the work that he had done for Israel.

11 Then the Israelites did what was evil in the sight of the LORD and worshiped the Baals; 12 and they abandoned the LORD, the God of their ancestors, who had brought them out of the land of Egypt; they followed other gods, from among the gods of the peoples who were all around them, and bowed down to them; and they provoked the LORD to anger. 13 They abandoned the LORD, and worshiped Baal and the Astartes. 14 So the anger of the LORD was kindled against Israel, and he gave them over to plunderers who plundered them, and he sold them into the power of their enemies all around, so that they could no longer withstand their enemies. 15 Whenever they marched out, the hand of the LORD was against them to bring misfortune, as the LORD had warned them and sworn to them; and they were in great distress.

16 Then the LORD raised up judges, who delivered them out of the power of those who plundered them. 17 Yet they did not listen even to their judges; for they lusted after other gods and bowed down to them. They soon turned aside

from the way in which their ancestors had walked, who had obeyed the commandments of the LORD; they did not follow their example. 18 Whenever the LORD raised up judges for them, the LORD was with the judge, and he delivered them from the hand of their enemies all the days of the judge; for the LORD would be moved to pity by their groaning because of those who persecuted and oppressed them. 19 But whenever the judge died, they would relapse and behave worse than their ancestors, following other gods, worshiping them and bowing down to them. They would not drop any of their practices or their stubborn ways. 20 So the anger of the LORD was kindled against Israel; and he said, "Because this people have transgressed my covenant that I commanded their ancestors, and have not obeyed my voice, 21 I will no longer drive out before them any of the nations that Joshua left when he died." 22 In order to test Israel, whether or not they would take care to walk in the way of the LORD as their ancestors did, 23 the LORD had left those nations, not driving them out at once, and had not handed them over to Joshua.

3 Now these are the nations that the LORD left to test all those in Israel who had no experience of any war in Canaan 2 (it was only that successive generations of Israelites might know war, to teach those who had no experience of it before): 3 the five lords of the Philistines, and all the Canaanites, and the Sidonians, and the Hivites who lived on Mount Lebanon, from Mount Baal-hermon as far as Lebo-hamath. 4 They were for the testing of Israel, to know whether Israel would obey the commandments of the LORD, which he commanded their ancestors by Moses. 5 So the Israelites lived among the Canaanites, the Hittites, the

---

to the LORD (see also Josh 24.29–31). **13:** *Baal* was the principal Canaanite deity. Fertility was his concern, so the service of this god was a constant temptation in an agrarian society like Israel. *Astarte* was the consort of Baal in the form of the fertility cult adopted by some Israelites.

2.16–19: The *judges* were leaders of the tribal militia whose victories gave them a principal role in the government of their tribes during their lifetime.
3.1–6: Marriage outside the Israelite community leads to weakening of Israel's bonds with the LORD.

Amorites, the Perizzites, the Hivites, and the Jebusites; 6 and they took their daughters as wives for themselves, and their own daughters they gave to their sons; and they worshiped their gods.

7 The Israelites did what was evil in the sight of the LORD, forgetting the LORD their God, and worshiping the Baals and the Asherahs. 8 Therefore the anger of the LORD was kindled against Israel, and he sold them into the hand of King Cushan-rishathaim of Aram-naharaim; and the Israelites served Cushan-rishathaim eight years. 9 But when the Israelites cried out to the LORD, the LORD raised up a deliverer for the Israelites, who delivered them, Othniel son of Kenaz, Caleb's younger brother. 10 The spirit of the LORD came upon him, and he judged Israel; he went out to war, and the LORD gave King Cushan-rishathaim of Aram into his hand; and his hand prevailed over Cushan-rishathaim. 11 So the land had rest forty years. Then Othniel son of Kenaz died.

12 The Israelites again did what was evil in the sight of the LORD; and the LORD strengthened King Eglon of Moab against Israel, because they had done what was evil in the sight of the LORD. 13 In alliance with the Ammonites and the Amalekites, he went and defeated Israel; and they took possession of the city of palms. 14 So the Israelites served King Eglon of Moab eighteen years.

15 But when the Israelites cried out to the LORD, the LORD raised up for them a deliverer, Ehud son of Gera, the Benjaminite, a left-handed man. The Israelites sent tribute by him to King Eglon of Moab. 16 Ehud made for himself a sword with two edges, a cubit in length; and he fastened it on his right thigh under his clothes. 17 Then he presented the tribute to King Eglon of Moab. Now Eglon was a very fat man. 18 When Ehud had finished presenting the tribute, he sent the people who carried the tribute on their way. 19 But he himself turned back at the sculptured stones near Gilgal, and said, "I have a secret message for you, O king." So the king said,*f* "Silence!" and all his attendants went out from his presence. 20 Ehud came to him, while he was sitting alone in his cool roof chamber, and said, "I have a message from God for you." So he rose from his seat. 21 Then Ehud reached with his left hand, took the sword from his right thigh, and thrust it into Eglon's*g* belly; 22 the hilt also went in after the blade, and the fat closed over the blade, for he did not draw the sword out of his belly; and the dirt came out. *h* 23 Then Ehud went out into the vestibule, *i* and closed the doors of the roof chamber on him, and locked them.

24 After he had gone, the servants came. When they saw that the doors of the roof chamber were locked, they thought, "He must be relieving himself*j* in the cool chamber." 25 So they waited until they were embarrassed. When he still did not open the doors of the roof chamber, they took the key and opened them. There was their lord lying dead on the floor.

26 Ehud escaped while they delayed, and passed beyond the sculptured stones, and escaped to Seirah. 27 When he arrived, he sounded the trumpet in the hill country of Ephraim; and the Israelites went down with him from the hill country, having him at their head. 28 He said

f Heb *he said*   *g* Heb *his*   *h* With Tg Vg: Meaning of Heb uncertain   *i* Meaning of Heb uncertain   *j* Heb *covering his feet*

3.7–11: Othniel. The hero of 1.12–15 appears here as a deliverer who defeated *Cushan-rishathaim* (Cushan "Double Trouble") of Aram "of the Two Rivers" (eastern Syria). **7:** *Asherah* was another female deity associated with Baal. **10:** *The spirit of the* LORD empowered Othniel. **11:** A type of editorial formula that occurs repeatedly in the book (see v. 30; 5.31; 8.28).

3.12–30: Ehud. The story of a Benjaminite hero who assassinated Eglon, the king of Moab, and delivered his tribe from Eglon's service. **12:** Moab, the country immediately east of the Dead Sea. **13:** *The city of palms,* Jericho. **19:** *Stones near Gilgal,* compare Josh 4.20.

to them, "Follow after me; for the LORD has given your enemies the Moabites into your hand." So they went down after him, and seized the fords of the Jordan against the Moabites, and allowed no one to cross over. 29 At that time they killed about ten thousand of the Moabites, all strong, able-bodied men; no one escaped. 30 So Moab was subdued that day under the hand of Israel. And the land had rest eighty years.

31 After him came Shamgar son of Anath, who killed six hundred of the Philistines with an oxgoad. He too delivered Israel.

4 The Israelites again did what was evil in the sight of the LORD, after Ehud died. 2 So the LORD sold them into the hand of King Jabin of Canaan, who reigned in Hazor; the commander of his army was Sisera, who lived in Harosheth-ha-goiim. 3 Then the Israelites cried out to the LORD for help; for he had nine hundred chariots of iron, and had oppressed the Israelites cruelly twenty years.

4 At that time Deborah, a prophetess, wife of Lappidoth, was judging Israel. 5 She used to sit under the palm of Deborah between Ramah and Bethel in the hill country of Ephraim; and the Israelites came up to her for judgment. 6 She sent and summoned Barak son of Abinoam from Kedesh in Naphtali, and said to him, "The LORD, the God of Israel, commands you, 'Go, take position at Mount Tabor, bringing ten thousand from the tribe of Naphtali and the tribe of Zebulun. 7 I will draw out Sisera, the general of Jabin's army, to meet you by the Wadi Kishon with his chariots and his troops; and I will give him into your hand.'" 8 Barak said to her, "If you will go with me, I will go; but if you will not go with me, I will not go." 9 And she said, "I will surely go with you; nevertheless, the road on which you are going will not lead to your glory, for the LORD will sell Sisera into the hand of a woman." Then Deborah got up and went with Barak to Kedesh. 10 Barak summoned Zebulun and Naphtali to Kedesh; and ten thousand warriors went up behind him; and Deborah went up with him.

11 Now Heber the Kenite had separated from the other Kenites, [k] that is, the descendants of Hobab the father-in-law of Moses, and had encamped as far away as Elon-bezaanannim, which is near Kedesh.

12 When Sisera was told that Barak son of Abinoam had gone up to Mount Tabor, 13 Sisera called out all his chariots, nine hundred chariots of iron, and all the troops who were with him, from Harosheth-ha-goiim to the Wadi Kishon. 14 Then Deborah said to Barak, "Up! For this is the day on which the LORD has given Sisera into your hand. The LORD is indeed going out before you." So Barak went down from Mount Tabor with ten thousand warriors following him. 15 And the LORD threw Sisera and all his chariots and all his army

---

[k] Heb *from the Kain*

3.31: Shamgar. This brief reference to Shamgar occurs here because the story of Deborah that follows mentions him (5.6). *The Philistines,* see 13.1 n.

4.1–5.31: Deborah. The prose story in ch 4 complements the older poetic account in ch 5. **4.2:** *Jabin* (see Josh 11.1; compare Ps 83.9) plays no direct role in this story. His title as *King . . . of Canaan* is inaccurate; Josh 11 calls him "king of Hazor." *Hazor* was an important city in Galilee because of its proximity to trade routes; it is nine miles north of the Sea of Galilee. *Sisera* is not a Semitic name. He may have been a Philistine in the service of Jabin. **3:** The number of chariots is unusually high and reflects the hopelessness of the situation. Such equipment gave the Canaanites an overwhelming advantage in battles waged on level ground, since the Israelites did not have chariot forces. **6:** *Mount Tabor* (1850 ft.) is a lone mountain standing in the northeast corner of the Jezreel Valley. **7:** The *Wadi Kishon* is a stream that flows westward through the Jezreel Valley that runs from the Carmel range to the Jordan. **10:** *Zebulun and Naphtali* are two tribes from Galilee. **11:** The *Kenites,* see 1.16 n.

into a panic[l] before Barak; Sisera got down from his chariot and fled away on foot, [16]while Barak pursued the chariots and the army to Harosheth-ha-goiim. All the army of Sisera fell by the sword; no one was left.

17 Now Sisera had fled away on foot to the tent of Jael wife of Heber the Kenite; for there was peace between King Jabin of Hazor and the clan of Heber the Kenite. [18]Jael came out to meet Sisera, and said to him, "Turn aside, my lord, turn aside to me; have no fear." So he turned aside to her into the tent, and she covered him with a rug. [19]Then he said to her, "Please give me a little water to drink; for I am thirsty." So she opened a skin of milk and gave him a drink and covered him. [20]He said to her, "Stand at the entrance of the tent, and if anybody comes and asks you, 'Is anyone here?' say, 'No.'" [21]But Jael wife of Heber took a tent peg, and took a hammer in her hand, and went softly to him and drove the peg into his temple, until it went down into the ground—he was lying fast asleep from weariness—and he died. [22]Then, as Barak came in pursuit of Sisera, Jael went out to meet him, and said to him, "Come, and I will show you the man whom you are seeking." So he went into her tent; and there was Sisera lying dead, with the tent peg in his temple.

23 So on that day God subdued King Jabin of Canaan before the Israelites. [24]Then the hand of the Israelites bore harder and harder on King Jabin of Canaan, until they destroyed King Jabin of Canaan.

5 Then Deborah and Barak son of Abinoam sang on that day, saying:
2 "When locks are long in Israel,
    when the people offer
       themselves willingly—
    bless[m] the Lord!

3 "Hear, O kings; give ear,
       O princes;
    to the Lord I will sing,
    I will make melody to the
       Lord, the God of Israel.

4 "Lord, when you went out from
       Seir,
    when you marched from the
       region of Edom,
  the earth trembled,
    and the heavens poured,
    the clouds indeed poured water.
5 The mountains quaked before the
       Lord, the One of Sinai,
    before the Lord, the God of
       Israel.

6 "In the days of Shamgar son of
       Anath,
    in the days of Jael, caravans
       ceased
    and travelers kept to the
       byways.
7 The peasantry prospered in Israel,
    they grew fat on plunder,
  because you arose, Deborah,
    arose as a mother in Israel.
8 When new gods were chosen,
    then war was in the gates.
  Was shield or spear to be seen
    among forty thousand in Israel?
9 My heart goes out to the
       commanders of Israel
  who offered themselves
       willingly among the people.
    Bless the Lord.

10 "Tell of it, you who ride on white
       donkeys,
    you who sit on rich carpets[n]
    and you who walk by the way.

l Heb adds *to the sword*; compare verse 16
m Or *You who offer yourselves willingly among the
people, bless*   *n* Meaning of Heb uncertain

5.2–31: The Song of Deborah may be the oldest part of the Hebrew Bible; it is also one of the most obscure. The poem celebrates the Lord's victory over Sisera, won because of a sudden downpour that made it impossible for chariots to maneuver. **4:** *From Seir . . . Edom.* The God who comes from the barren regions of the southeast showed an ability to give rain.

11 To the sound of musicians° at the
watering places,
there they repeat the triumphs
of the LORD,
the triumphs of his peasantry in
Israel.

"Then down to the gates marched
the people of the LORD.

12 "Awake, awake, Deborah!
Awake, awake, utter a song!
Arise, Barak, lead away your
captives,
O son of Abinoam.

13 Then down marched the remnant
of the noble;
the people of the LORD marched
down for him° against the
mighty.

14 From Ephraim they set out° into
the valley,°
following you, Benjamin, with
your kin;
from Machir marched down the
commanders,
and from Zebulun those who
bear the marshal's staff;

15 the chiefs of Issachar came with
Deborah,
and Issachar faithful to Barak;
into the valley they rushed out
at his heels
Among the clans of Reuben
there were great searchings of
heart.

16 Why did you tarry among the
sheepfolds,
to hear the piping for the flocks?
Among the clans of Reuben
there were great searchings of
heart.

17 Gilead stayed beyond the Jordan;
and Dan, why did he abide with
the ships?

Asher sat still at the coast of the
sea,
settling down by his landings.

18 Zebulun is a people that scorned
death;
Naphtali too, on the heights of
the field.

19 "The kings came, they fought;
then fought the kings of
Canaan,
at Taanach, by the waters of
Megiddo;
they got no spoils of silver.

20 The stars fought from heaven,
from their courses they fought
against Sisera.

21 The torrent Kishon swept them
away,
the onrushing torrent, the
torrent Kishon.
March on, my soul, with might!

22 "Then loud beat the horses' hoofs
with the galloping, galloping of
his steeds.

23 "Curse Meroz, says the angel of
the LORD,
curse bitterly its inhabitants,
because they did not come to the
help of the LORD,
to the help of the LORD against
the mighty.

24 "Most blessed of women be Jael,
the wife of Heber the Kenite,
of tent-dwelling women most
blessed.

25 He asked water and she gave him
milk,
she brought him curds in a
lordly bowl.

o Meaning of Heb uncertain   *p* Gk: Heb *me*
q Cn: Heb *From Ephraim their root*   *r* Gk: Heb
in Amalek

5.15–17: *Reuben . . . Gilead* (Gad) *. . . Dan*
and *Asher* do not join the battle. The tribes of
Judah, Simeon, and Levi receive no mention
in the poem. **19:** *Tanaach . . . Megiddo* were
two large cities that guarded the northern approaches to passes through Mount Carmel.
23: *Meroz* was a nearby Israelite village that
did not join in the battle. **24–27:** A non-
Israelite woman delivered the mortal blow to
Sisera.

26 She put her hand to the tent peg
and her right hand to the
workmen's mallet;
she struck Sisera a blow,
she crushed his head,
she shattered and pierced his
temple.
27 He sank, he fell,
he lay still at her feet;
at her feet he sank, he fell;
where he sank, there he fell
dead.

28 "Out of the window she peered,
the mother of Sisera gazed*s*
through the lattice:
'Why is his chariot so long in
coming?
Why tarry the hoofbeats of his
chariots?'
29 Her wisest ladies make answer,
indeed, she answers the question
herself:
30 'Are they not finding and dividing
the spoil?—
A girl or two for every man;
spoil of dyed stuffs for Sisera,
spoil of dyed stuffs
embroidered,
two pieces of dyed work
embroidered for my neck as
spoil?'

31 "So perish all your enemies,
O LORD!
But may your friends be like the
sun as it rises in its might."

And the land had rest forty years.

6 The Israelites did what was evil in
the sight of the LORD, and the LORD
gave them into the hand of Midian seven
years. 2The hand of Midian prevailed
over Israel; and because of Midian the
Israelites provided for themselves hiding
places in the mountains, caves and
strongholds. 3For whenever the Israelites put in seed, the Midianites and the
Amalekites and the people of the east
would come up against them. 4They
would encamp against them and destroy
the produce of the land, as far as the
neighborhood of Gaza, and leave no sustenance in Israel, and no sheep or ox or
donkey. 5For they and their livestock
would come up, and they would even
bring their tents, as thick as locusts; neither they nor their camels could be
counted; so they wasted the land as they
came in. 6Thus Israel was greatly impoverished because of Midian; and the Israelites cried out to the LORD for help.

7 When the Israelites cried to the
LORD on account of the Midianites, 8the
LORD sent a prophet to the Israelites; and
he said to them, "Thus says the LORD,
the God of Israel: I led you up from
Egypt, and brought you out of the house
of slavery; 9and I delivered you from the
hand of the Egyptians, and from the
hand of all who oppressed you, and
drove them out before you, and gave
you their land; 10and I said to you, 'I am

s Gk Compare Tg: Heb *exclaimed*

5.28–31: The final scene is a poignant description of the conflicting emotions felt by
the women who awaited the return of the
Canaanite army.
6.1–8.35: Gideon. A hero of the tribe of
Manasseh ends the oppression of his tribe by
the Midianites, who were confiscating the
food necessary for the survival of the people.
The assumption behind this story is that the
LORD guides the destinies of nations and
tribes. This insures the fulfillment of the divine will despite the intentions of human participants in these events.
6.1–40: Gideon's call. **2–6:** The plight of
the Israelite peasants. Behind this text may be
the harvest-season taxes imposed on peasants
by groups with the military power to make
such impositions. **2:** *Midian* was among the
descendants of Abraham (Gen 25.2–4; 1 Chr
1.32–33). **3:** *The Amalekites* were among the
tribes descended from Esau (Gen 36.12). *The
people of the east* had a role in prophetic oracles
(see Isa 11.14; Jer 49.28; Ezek 25.1–10). **4:**
Gaza, a Philistine city forty-five miles south
of Jaffa. **5:** The first appearance of the domesticated camel in the Bible. **11–24:** A message
from the LORD. The story about the establishment of a Yahwistic shrine at Ophrah (vv. 11a,
18–24) frames the commissioning of Gideon
to be a military leader (vv. 11b–17).
6.11: *The angel of the LORD,* see 2.1 n. *Abiezrite,* Josh 17.2; 1 Chr 7.18.

the LORD your God; you shall not pay reverence to the gods of the Amorites, in whose land you live.' But you have not given heed to my voice."

11 Now the angel of the LORD came and sat under the oak at Ophrah, which belonged to Joash the Abiezrite, as his son Gideon was beating out wheat in the wine press, to hide it from the Midianites. 12 The angel of the LORD appeared to him and said to him, "The LORD is with you, you mighty warrior." 13 Gideon answered him, "But sir, if the LORD is with us, why then has all this happened to us? And where are all his wonderful deeds that our ancestors recounted to us, saying, 'Did not the LORD bring us up from Egypt?' But now the LORD has cast us off, and given us into the hand of Midian." 14 Then the LORD turned to him and said, "Go in this might of yours and deliver Israel from the hand of Midian; I hereby commission you." 15 He responded, "But sir, how can I deliver Israel? My clan is the weakest in Manasseh, and I am the least in my family." 16 The LORD said to him, "But I will be with you, and you shall strike down the Midianites, every one of them." 17 Then he said to him, "If now I have found favor with you, then show me a sign that it is you who speak with me. 18 Do not depart from here until I come to you, and bring out my present, and set it before you." And he said, "I will stay until you return."

19 So Gideon went into his house and prepared a kid, and unleavened cakes from an ephah of flour; the meat he put in a basket, and the broth he put in a pot, and brought them to him under the oak and presented them. 20 The angel of God said to him, "Take the meat and the unleavened cakes, and put them on this rock, and pour out the broth." And he did so. 21 Then the angel of the LORD reached out the tip of the staff that was in his hand, and touched the meat and the unleavened cakes; and fire sprang up

from the rock and consumed the meat and the unleavened cakes; and the angel of the LORD vanished from his sight. 22 Then Gideon perceived that it was the angel of the LORD; and Gideon said, "Help me, Lord GOD! For I have seen the angel of the LORD face to face." 23 But the LORD said to him, "Peace be to you; do not fear, you shall not die." 24 Then Gideon built an altar there to the LORD, and called it, The LORD is peace. To this day it still stands at Ophrah, which belongs to the Abiezrites.

25 That night the LORD said to him, "Take your father's bull, the second bull seven years old, and pull down the altar of Baal that belongs to your father, and cut down the sacred pole[t] that is beside it; 26 and build an altar to the LORD your God on the top of the stronghold here, in proper order; then take the second bull, and offer it as a burnt offering with the wood of the sacred pole[t] that you shall cut down." 27 So Gideon took ten of his servants, and did as the LORD had told him; but because he was too afraid of his family and the townspeople to do it by day, he did it by night.

28 When the townspeople rose early in the morning, the altar of Baal was broken down, and the sacred pole[t] beside it was cut down, and the second bull was offered on the altar that had been built. 29 So they said to one another, "Who has done this?" After searching and inquiring, they were told, "Gideon son of Joash did it." 30 Then the townspeople said to Joash, "Bring out your son, so that he may die, for he has pulled down the altar of Baal and cut down the sacred pole[t] beside it." 31 But Joash said to all who were arrayed against him, "Will you contend for Baal? Or will you defend his cause? Whoever contends for him shall be put to death by morning. If he is a god, let him contend for himself,

---

t Heb *Asherah*

6.25–32: Gideon's Yahwistic credentials. Gideon became identified with Jerubbaal, the head of a clan that ruled Shechem (9.1–2).

Gideon's destruction of an altar to Baal made it possible to give an orthodox etymology to the name *Jerubbaal.* **25:** *Baal,* see 2.13 n.

because his altar has been pulled down."
[32] Therefore on that day Gideon[u] was called Jerubbaal, that is to say, "Let Baal contend against him," because he pulled down his altar.

33 Then all the Midianites and the Amalekites and the people of the east came together, and crossing the Jordan they encamped in the Valley of Jezreel. [34] But the spirit of the LORD took possession of Gideon; and he sounded the trumpet, and the Abiezrites were called out to follow him. [35] He sent messengers throughout all Manasseh, and they too were called out to follow him. He also sent messengers to Asher, Zebulun, and Naphtali, and they went up to meet them.

36 Then Gideon said to God, "In order to see whether you will deliver Israel by my hand, as you have said, [37] I am going to lay a fleece of wool on the threshing floor; if there is dew on the fleece alone, and it is dry on all the ground, then I shall know that you will deliver Israel by my hand, as you have said." [38] And it was so. When he rose early next morning and squeezed the fleece, he wrung enough dew from the fleece to fill a bowl with water. [39] Then Gideon said to God, "Do not let your anger burn against me, let me speak one more time; let me, please, make trial with the fleece just once more; let it be dry only on the fleece, and on all the ground let there be dew." [40] And God did so that night. It was dry on the fleece only, and on all the ground there was dew.

7 Then Jerubbaal (that is, Gideon) and all the troops that were with him rose early and encamped beside the spring of Harod; and the camp of Midian was north of them, below[v] the hill of Moreh, in the valley.

2 The LORD said to Gideon, "The troops with you are too many for me to give the Midianites into their hand. Israel would only take the credit away from me, saying, 'My own hand has delivered me.' [3] Now therefore proclaim this in the hearing of the troops, 'Whoever is fearful and trembling, let him return home.'" Thus Gideon sifted them out;[w] twenty-two thousand returned, and ten thousand remained.

4 Then the LORD said to Gideon, "The troops are still too many; take them down to the water and I will sift them out for you there. When I say, 'This one shall go with you,' he shall go with you; and when I say, 'This one shall not go with you,' he shall not go." [5] So he brought the troops down to the water; and the LORD said to Gideon, "All those who lap the water with their tongues, as a dog laps, you shall put to one side; all those who kneel down to drink, putting their hands to their mouths,[x] you shall put to the other side." [6] The number of those that lapped was three hundred; but all the rest of the troops knelt down to drink water. [7] Then the LORD said to Gideon, "With the three hundred that lapped I will deliver you, and give the Midianites into your hand. Let all the others go to their homes." [8] So he took the jars of the troops from their hands,[y] and their trumpets; and he sent all the rest of Israel back to their own tents, but retained the three hundred. The camp of Midian was below him in the valley.

9 That same night the LORD said to him, "Get up, attack the camp; for I have given it into your hand. [10] But if you fear to attack, go down to the camp with your servant Purah; [11] and you shall hear

u Heb *he*   *v* Heb *from*   *w* Cn: Heb *home, and depart from Mount Gilead'"*   *x* Heb places the words *putting their hands to their mouths* after the word *lapped* in verse 6   *y* Cn: Heb *So the people took provisions in their hands*

6.33–40: Gideon prepared to attack the Midianites. **33:** *Valley of Jezreel,* see 4.7 n. **34:** *The spirit of the LORD* empowered Gideon so the victory belongs to God. **36–40:** Similar tests are common in folklore.

7.1–22: Gideon's victory. **1:** *The spring of* *Harod* is at the foot of Mount Gilboa in the southeast of the Jezreel Valley. *The Hill of Moreh* is identified with Jebel ed-Dehi in the Jezreel Valley. **2–8:** The reduction of Gideon's forces made it impossible to ascribe the victory over Midian to human prowess.

what they say, and afterward your hands shall be strengthened to attack the camp." Then he went down with his servant Purah to the outposts of the armed men that were in the camp. 12 The Midianites and the Amalekites and all the people of the east lay along the valley as thick as locusts; and their camels were without number, countless as the sand on the seashore. 13 When Gideon arrived, there was a man telling a dream to his comrade; and he said, "I had a dream, and in it a cake of barley bread tumbled into the camp of Midian, and came to the tent, and struck it so that it fell; it turned upside down, and the tent collapsed." 14 And his comrade answered, "This is no other than the sword of Gideon son of Joash, a man of Israel; into his hand God has given Midian and all the army."

15 When Gideon heard the telling of the dream and its interpretation, he worshiped; and he returned to the camp of Israel, and said, "Get up; for the LORD has given the army of Midian into your hand." 16 After he divided the three hundred men into three companies, and put trumpets into the hands of all of them, and empty jars, with torches inside the jars, 17 he said to them, "Look at me, and do the same; when I come to the outskirts of the camp, do as I do. 18 When I blow the trumpet, I and all who are with me, then you also blow the trumpets around the whole camp, and shout, 'For the LORD and for Gideon!' "

19 So Gideon and the hundred who were with him came to the outskirts of the camp at the beginning of the middle watch, when they had just set the watch; and they blew the trumpets and smashed the jars that were in their hands. 20 So the three companies blew the trumpets and broke the jars, holding in their left hands the torches, and in their right hands the trumpets to blow; and they cried, "A sword for the LORD and for Gideon!" 21 Every man stood in his place all around the camp, and all the men in camp ran; they cried out and fled. 22 When they blew the three hundred trumpets, the LORD set every man's sword against his fellow and against all the army; and the army fled as far as Beth-shittah toward Zererah, z as far as the border of Abel-meholah, by Tabbath. 23 And the men of Israel were called out from Naphtali and from Asher and from all Manasseh, and they pursued after the Midianites.

24 Then Gideon sent messengers throughout all the hill country of Ephraim, saying, "Come down against the Midianites and seize the waters against them, as far as Beth-barah, and also the Jordan." So all the men of Ephraim were called out, and they seized the waters as far as Beth-barah, and also the Jordan. 25 They captured the two captains of Midian, Oreb and Zeeb; they killed Oreb at the rock of Oreb, and Zeeb they killed at the wine press of Zeeb, as they pursued the Midianites. They brought the heads of Oreb and Zeeb to Gideon beyond the Jordan.

8 Then the Ephraimites said to him, "What have you done to us, not to call us when you went to fight against the Midianites?" And they upbraided him violently. 2 So he said to them, "What have I done now in comparison with you? Is not the gleaning of the grapes of Ephraim better than the vintage of Abiezer? 3 God has given into your hands the captains of Midian, Oreb and Zeeb; what have I been able to do in comparison with you?" When he said this, their anger against him subsided.

4 Then Gideon came to the Jordan and crossed over, he and the three hundred who were with him, exhausted and

z Another reading is *Zeredah*

7.9–14: Dreams were believed to be a means of divine-human communication. 15–22: Gideon's forces surprised the Midianites. 7.23–8.3: The elimination of Midianite leaders consolidated Gideon's victory.

8.4–35: The battle east of the Jordan. 4–12: The elders of Succoth refuse to aid Gideon as he pursued the Midianite army. 5: *Succoth,* a Transjordanian city east of Shechem, identified with Tell Deir Alla.

famished. *a* 5 So he said to the people of Succoth, "Please give some loaves of bread to my followers, for they are exhausted, and I am pursuing Zebah and Zalmunna, the kings of Midian." 6 But the officials of Succoth said, "Do you already have in your possession the hands of Zebah and Zalmunna, that we should give bread to your army?" 7 Gideon replied, "Well then, when the LORD has given Zebah and Zalmunna into my hand, I will trample your flesh on the thorns of the wilderness and on briers." 8 From there he went up to Penuel, and made the same request of them; and the people of Penuel answered him as the people of Succoth had answered. 9 So he said to the people of Penuel, "When I come back victorious, I will break down this tower."

10 Now Zebah and Zalmunna were in Karkor with their army, about fifteen thousand men, all who were left of all the army of the people of the east; for one hundred twenty thousand men bearing arms had fallen. 11 So Gideon went up by the caravan route east of Nobah and Jogbehah, and attacked the army; for the army was off its guard. 12 Zebah and Zalmunna fled; and he pursued them and took the two kings of Midian, Zebah and Zalmunna, and threw all the army into a panic.

13 When Gideon son of Joash returned from the battle by the ascent of Heres, 14 he caught a young man, one of the people of Succoth, and questioned him; and he listed for him the officials and elders of Succoth, seventy-seven people. 15 Then he came to the people of Succoth, and said, "Here are Zebah and Zalmunna, about whom you taunted me, saying, 'Do you already have in your possession the hands of Zebah and Zalmunna, that we should give bread to your troops who are exhausted?' " 16 So he took the elders of the city and he took

thorns of the wilderness and briers and with them he trampled *b* the people of Succoth. 17 He also broke down the tower of Penuel, and killed the men of the city.

18 Then he said to Zebah and Zalmunna, "What about the men whom you killed at Tabor?" They answered, "As you are, so were they, every one of them; they resembled the sons of a king." 19 And he replied, "They were my brothers, the sons of my mother; as the LORD lives, if you had saved them alive, I would not kill you." 20 So he said to Jether his firstborn, "Go kill them!" But the boy did not draw his sword, for he was afraid, because he was still a boy. 21 Then Zebah and Zalmunna said, "You come and kill us; for as the man is, so is his strength." So Gideon proceeded to kill Zebah and Zalmunna; and he took the crescents that were on the necks of their camels.

22 Then the Israelites said to Gideon, "Rule over us, you and your son and your grandson also; for you have delivered us out of the hand of Midian." 23 Gideon said to them, "I will not rule over you, and my son will not rule over you; the LORD will rule over you." 24 Then Gideon said to them, "Let me make a request of you; each of you give me an earring he has taken as booty." (For the enemy *c* had golden earrings, because they were Ishmaelites.) 25 "We will willingly give them," they answered. So they spread a garment, and each threw into it an earring he had taken as booty. 26 The weight of the golden earrings that he requested was one thousand seven hundred shekels of gold (apart from the crescents and the pendants and the purple garments worn by the kings of Midian, and the collars that

a Gk: Heb *pursuing*   *b* With verse 7, Compare Gk: Heb *he taught*   *c* Heb *they*

8.13–21: Gideon took reprisals against the people who refused to support his army.
8.22–35: An epilogue. 22–23: Gideon's refusal reflected the ancient belief that only

the LORD was to rule over Israel. 24–28: The nature of the *ephod* is not certain; in Ex 28.6–14 and 39.2–7, it is a garment worn by the high priest. This passage suggests that the

were on the necks of their camels). ²⁷ Gideon made an ephod of it and put it in his town, in Ophrah; and all Israel prostituted themselves to it there, and it became a snare to Gideon and to his family. ²⁸ So Midian was subdued before the Israelites, and they lifted up their heads no more. So the land had rest forty years in the days of Gideon.

29 Jerubbaal son of Joash went to live in his own house. ³⁰ Now Gideon had seventy sons, his own offspring, for he had many wives. ³¹ His concubine who was in Shechem also bore him a son, and he named him Abimelech. ³² Then Gideon son of Joash died at a good old age, and was buried in the tomb of his father Joash at Ophrah of the Abiezrites.

33 As soon as Gideon died, the Israelites relapsed and prostituted themselves with the Baals, making Baal-berith their god. ³⁴ The Israelites did not remember the LORD their God, who had rescued them from the hand of all their enemies on every side; ³⁵ and they did not exhibit loyalty to the house of Jerubbaal (that is, Gideon) in return for all the good that he had done to Israel.

9 Now Abimelech son of Jerubbaal went to Shechem to his mother's kinsfolk and said to them and to the whole clan of his mother's family, ² "Say in the hearing of all the lords of Shechem, 'Which is better for you, that all seventy of the sons of Jerubbaal rule over you, or that one rule over you?' Remember also that I am your bone and your flesh." ³ So his mother's kinsfolk spoke all these words on his behalf in the hearing of all the lords of Shechem; and their hearts inclined to follow Abimelech, for they

said, "He is our brother." ⁴ They gave him seventy pieces of silver out of the temple of Baal-berith with which Abimelech hired worthless and reckless fellows, who followed him. ⁵ He went to his father's house at Ophrah, and killed his brothers the sons of Jerubbaal, seventy men, on one stone; but Jotham, the youngest son of Jerubbaal, survived, for he hid himself. ⁶ Then all the lords of Shechem and all Beth-millo came together, and they went and made Abimelech king, by the oak of the pillar^d at Shechem.

7 When it was told to Jotham, he went and stood on the top of Mount Gerizim, and cried aloud and said to them, "Listen to me, you lords of Shechem, so that God may listen to you.

⁸ The trees once went out
>   to anoint a king over
>   themselves.
> So they said to the olive tree,
>   'Reign over us.'
⁹ The olive tree answered them,
>   'Shall I stop producing my rich
>   oil
>   by which gods and mortals
>   are honored,
>   and go to sway over the
>   trees?'
¹⁰ Then the trees said to the fig tree,
>   'You come and reign over us.'
¹¹ But the fig tree answered them,
>   'Shall I stop producing my
>   sweetness
>   and my delicious fruit,
>   and go to sway over the
>   trees?'

d Cn: Meaning of Heb uncertain

---

ephod of Gideon was an image. **34**: Freed from the Midianite threat, Israel returns to its familiar faults.

9.1–57: Abimelech. A disinherited son of Gideon seizes the royal dignity that Gideon rejected. Abimelech's power was fragile and he fell in a revolt against his rule. This chapter contains no reference to the God of Israel. **1–6**: Abimelech eliminated all potential rivals and took the throne of Shechem. **4**: The god *Baal-berith* ("the Lord of the covenant") is

otherwise unknown. Since Shechem was the site of an Israelite covenant assembly (Josh 24.1–27), it may be that a hybrid form of Israelite religion was in vogue in the city. **6**: *The oak of the pillar,* Josh 24.26.

9.7–15: Jotham rebukes his elder brother by satirizing the monarchy. **7**: *Mount Gerizim* is just south of Shechem. **8**: *The olive tree* is the noblest tree, while *the bramble* (v. 14) is a useless bush.

12 Then the trees said to the vine,
  'You come and reign over us.'
13 But the vine said to them,
  'Shall I stop producing my wine
    that cheers gods and mortals,
    and go to sway over the
      trees?'
14 So all the trees said to the
    bramble,
  'You come and reign over us.'
15 And the bramble said to the trees,
  'If in good faith you are
    anointing me king over
    you,
  then come and take refuge in
    my shade;
  but if not, let fire come out of
    the bramble
  and devour the cedars of
    Lebanon.'

16 "Now therefore, if you acted in good faith and honor when you made Abimelech king, and if you have dealt well with Jerubbaal and his house, and have done to him as his actions deserved— 17 for my father fought for you, and risked his life, and rescued you from the hand of Midian; 18 but you have risen up against my father's house this day, and have killed his sons, seventy men on one stone, and have made Abimelech, the son of his slave woman, king over the lords of Shechem, because he is your kinsman— 19 if, I say, you have acted in good faith and honor with Jerubbaal and with his house this day, then rejoice in Abimelech, and let him also rejoice in you; 20 but if not, let fire come out from Abimelech, and devour the lords of Shechem, and Beth-millo; and let fire come out from the lords of Shechem, and from Beth-millo, and devour Abimelech." 21 Then Jotham ran away and fled, going to Beer, where he remained for fear of his brother Abimelech.

22 Abimelech ruled over Israel three years. 23 But God sent an evil spirit between Abimelech and the lords of Shechem; and the lords of Shechem dealt treacherously with Abimelech. 24 This happened so that the violence done to the seventy sons of Jerubbaal might be avenged[e] and their blood be laid on their brother Abimelech, who killed them, and on the lords of Shechem, who strengthened his hands to kill his brothers. 25 So, out of hostility to him, the lords of Shechem set ambushes on the mountain tops. They robbed all who passed by them along that way; and it was reported to Abimelech.

26 When Gaal son of Ebed moved into Shechem with his kinsfolk, the lords of Shechem put confidence in him. 27 They went out into the field and gathered the grapes from their vineyards, trod them, and celebrated. Then they went into the temple of their god, ate and drank, and ridiculed Abimelech. 28 Gaal son of Ebed said, "Who is Abimelech, and who are we of Shechem, that we should serve him? Did not the son of Jerubbaal and Zebul his officer serve the men of Hamor father of Shechem? Why then should we serve him? 29 If only this people were under my command! Then I would remove Abimelech; I would say[f] to him, 'Increase your army, and come out.' "

30 When Zebul the ruler of the city heard the words of Gaal son of Ebed, his anger was kindled. 31 He sent messengers to Abimelech at Arumah,[g] saying, "Look, Gaal son of Ebed and his kinsfolk have come to Shechem, and they are stirring up[h] the city against you. 32 Now therefore, go by night, you and the troops that are with you, and lie in wait in the fields. 33 Then early in the morning, as soon as the sun rises, get up and rush on the city; and when he and the troops that are with him come out

e Heb *might come*   f Gk: Heb *and he said*
g Cn See 9.41. Heb *Tormah*   h Cn: Heb *are besieging*

9.22–29: These conflicts show the tenuous hold that Abimelech had on Shechem.
9.30–49: Abimelech levels Shechem. Archaeology has revealed a massive destruction at Shechem in the 12th century.

against you, you may deal with them as best you can."

34 So Abimelech and all the troops with him got up by night and lay in wait against Shechem in four companies. [35] When Gaal son of Ebed went out and stood in the entrance of the gate of the city, Abimelech and the troops with him rose from the ambush. [36] And when Gaal saw them, he said to Zebul, "Look, people are coming down from the mountain tops!" And Zebul said to him, "The shadows on the mountains look like people to you." [37] Gaal spoke again and said, "Look, people are coming down from Tabbur-erez, and one company is coming from the direction of Elon-meonenim."[i] [38] Then Zebul said to him, "Where is your boast[j] now, you who said, 'Who is Abimelech, that we should serve him?' Are not these the troops you made light of? Go out now and fight with them." [39] So Gaal went out at the head of the lords of Shechem, and fought with Abimelech. [40] Abimelech chased him, and he fled before him. Many fell wounded, up to the entrance of the gate. [41] So Abimelech resided at Arumah; and Zebul drove out Gaal and his kinsfolk, so that they could not live on at Shechem.

42 On the following day the people went out into the fields. When Abimelech was told, [43] he took his troops and divided them into three companies, and lay in wait in the fields. When he looked and saw the people coming out of the city, he rose against them and killed them. [44] Abimelech and the company that was[k] with him rushed forward and stood at the entrance of the gate of the city, while the two companies rushed on all who were in the fields and killed them. [45] Abimelech fought against the city all that day; he took the city, and killed the people that were in it; and he razed the city and sowed it with salt.

46 When all the lords of the Tower of Shechem heard of it, they entered the stronghold of the temple of El-berith. [47] Abimelech was told that all the lords of the Tower of Shechem were gathered together. [48] So Abimelech went up to Mount Zalmon, he and all the troops that were with him. Abimelech took an ax in his hand, cut down a bundle of brushwood, and took it up and laid it on his shoulder. Then he said to the troops with him, "What you have seen me do, do quickly, as I have done." [49] So every one of the troops cut down a bundle and following Abimelech put it against the stronghold, and they set the stronghold on fire over them, so that all the people of the Tower of Shechem also died, about a thousand men and women.

50 Then Abimelech went to Thebez, and encamped against Thebez, and took it. [51] But there was a strong tower within the city, and all the men and women and all the lords of the city fled to it and shut themselves in; and they went to the roof of the tower. [52] Abimelech came to the tower, and fought against it, and came near to the entrance of the tower to burn it with fire. [53] But a certain woman threw an upper millstone on Abimelech's head, and crushed his skull. [54] Immediately he called to the young man who carried his armor and said to him, "Draw your sword and kill me, so people will not say about me, 'A woman killed him.'" So the young man thrust him through, and he died. [55] When the Israelites saw that Abimelech was dead, they all went home. [56] Thus God repaid Abimelech for the crime he committed against his father in killing his seventy brothers; [57] and God also made all the wickedness of the people of Shechem fall back on their heads, and on them came the curse of Jotham son of Jerubbaal.

10 After Abimelech, Tola son of Puah son of Dodo, a man of Issa-

---

i That is *Diviners' Oak*   *j* Heb *mouth*
k Vg and some Gk Mss: Heb *companies that were*

9.50–57: The end of Abimelech was a judgment on his attempt to reestablish the discredited Canaanite socio-political system.

10.1–5: **Tola and Jair.** Several judges remembered in this book (see 3.31; 12.8–15) have no specific deeds associated with their

char, who lived at Shamir in the hill country of Ephraim, rose to deliver Israel. 2 He judged Israel twenty-three years. Then he died, and was buried at Shamir.

3 After him came Jair the Gileadite, who judged Israel twenty-two years. 4 He had thirty sons who rode on thirty donkeys; and they had thirty towns, which are in the land of Gilead, and are called Havvoth-jair to this day. 5 Jair died, and was buried in Kamon.

6 The Israelites again did what was evil in the sight of the LORD, worshiping the Baals and the Astartes, the gods of Aram, the gods of Sidon, the gods of Moab, the gods of the Ammonites, and the gods of the Philistines. Thus they abandoned the LORD, and did not worship him. 7 So the anger of the LORD was kindled against Israel, and he sold them into the hand of the Philistines and into the hand of the Ammonites, 8 and they crushed and oppressed the Israelites that year. For eighteen years they oppressed all the Israelites that were beyond the Jordan in the land of the Amorites, which is in Gilead. 9 The Ammonites also crossed the Jordan to fight against Judah and against Benjamin and against the house of Ephraim; so that Israel was greatly distressed.

10 So the Israelites cried to the LORD, saying, "We have sinned against you, because we have abandoned our God and have worshiped the Baals." 11 And the LORD said to the Israelites, "Did I not deliver you[l] from the Egyptians and from the Amorites, from the Ammonites and from the Philistines? 12 The Sidonians also, and the Amalekites, and the Maonites, oppressed you; and you cried to me, and I delivered you out of their hand. 13 Yet you have abandoned me and worshiped other gods; therefore I will deliver you no more. 14 Go and cry to the gods whom you have chosen; let them deliver you in the time of your distress." 15 And the Israelites said to the LORD, "We have sinned; do to us whatever seems good to you; but deliver us this day!" 16 So they put away the foreign gods from among them and worshiped the LORD; and he could no longer bear to see Israel suffer.

17 Then the Ammonites were called to arms, and they encamped in Gilead; and the Israelites came together, and they encamped at Mizpah. 18 The commanders of the people of Gilead said to one another, "Who will begin the fight against the Ammonites? He shall be head over all the inhabitants of Gilead."

11 Now Jephthah the Gileadite, the son of a prostitute, was a mighty warrior. Gilead was the father of Jephthah. 2 Gilead's wife also bore him sons; and when his wife's sons grew up, they drove Jephthah away, saying to him, "You shall not inherit anything in our father's house; for you are the son of another woman." 3 Then Jephthah fled from his brothers and lived in the land of Tob. Outlaws collected around Jephthah and went raiding with him.

4 After a time the Ammonites made war against Israel. 5 And when the Am-

l Heb lacks *Did I not deliver you*

---

administration. Perhaps they were added to bring the total of judges named in the book to twelve.

10.6–12.7: Jephthah. The Transjordanian setting of this story is unique in Judges. The story unfolds as a series of crises. Though Jephthah saved Israel, he faced a personal tragedy because of his victory.

10.6–16: The Deuteronomistic introduction. Israel's problems resulted from its infidelity toward the LORD. **10:** This act of repentance in v. 16 is the only such act in the book.

13: God's refusal to deliver Israel showed the degree of Israel's failures. This was a harbinger of the tragedy to come, since God was no longer swayed by a claim of repentance.

10.17–11.11: The choice of Jephthah. The Israelites considered their problem to be a military one, so they chose an accomplished military leader. Because of Jephthah's previous experience with the Gileadites, he accepted their commission only on his terms. A ritual at Mizpah sealed the commission.

10.17: *The Ammonites* were an Aramean

monites made war against Israel, the elders of Gilead went to bring Jephthah from the land of Tob. 6 They said to Jephthah, "Come and be our commander, so that we may fight with the Ammonites." 7 But Jephthah said to the elders of Gilead, "Are you not the very ones who rejected me and drove me out of my father's house? So why do you come to me now when you are in trouble?" 8 The elders of Gilead said to Jephthah, "Nevertheless, we have now turned back to you, so that you may go with us and fight with the Ammonites, and become head over us, over all the inhabitants of Gilead." 9 Jephthah said to the elders of Gilead, "If you bring me home again to fight with the Ammonites, and the LORD gives them over to me, I will be your head." 10 And the elders of Gilead said to Jephthah, "The LORD will be witness between us; we will surely do as you say." 11 So Jephthah went with the elders of Gilead, and the people made him head and commander over them; and Jephthah spoke all his words before the LORD at Mizpah.

12 Then Jephthah sent messengers to the king of the Ammonites and said, "What is there between you and me, that you have come to me to fight against my land?" 13 The king of the Ammonites answered the messengers of Jephthah, "Because Israel, on coming from Egypt, took away my land from the Arnon to the Jabbok and to the Jordan; now therefore restore it peaceably." 14 Once again Jephthah sent messengers to the king of the Ammonites 15 and said to him: "Thus says Jephthah: Israel did not take away the land of Moab or the land of the Ammonites, 16 but when they came up from Egypt, Israel went through the wilderness to the Red Sea*m* and came to Kadesh. 17 Israel then sent messengers to the king of Edom, saying, 'Let us pass

through your land'; but the king of Edom would not listen. They also sent to the king of Moab, but he would not consent. So Israel remained at Kadesh. 18 Then they journeyed through the wilderness, went around the land of Edom and the land of Moab, arrived on the east side of the land of Moab, and camped on the other side of the Arnon. They did not enter the territory of Moab, for the Arnon was the boundary of Moab. 19 Israel then sent messengers to King Sihon of the Amorites, king of Heshbon; and Israel said to him, 'Let us pass through your land to our country.' 20 But Sihon did not trust Israel to pass through his territory; so Sihon gathered all his people together, and encamped at Jahaz, and fought with Israel. 21 Then the LORD, the God of Israel, gave Sihon and all his people into the hand of Israel, and they defeated them; so Israel occupied all the land of the Amorites, who inhabited that country. 22 They occupied all the territory of the Amorites from the Arnon to the Jabbok and from the wilderness to the Jordan. 23 So now the LORD, the God of Israel, has conquered the Amorites for the benefit of his people Israel. Do you intend to take their place? 24 Should you not possess what your god Chemosh gives you to possess? And should we not be the ones to possess everything that the LORD our God has conquered for our benefit? 25 Now are you any better than King Balak son of Zippor of Moab? Did he ever enter into conflict with Israel, or did he ever go to war with them? 26 While Israel lived in Heshbon and its villages, and in Aroer and its villages, and in all the towns that are along the Arnon, three hundred years, why did you not recover them within that time? 27 It is not I who have sinned against you, but you are the

m Or *Sea of Reeds*

group that settled along the Jabbok river in the 12th century. *Gilead* is the territory east of the Jordan river west of Ammonite territory. **11.12–28:** This is the only example in Judges of diplomacy as a way to resolve disputes. Religious beliefs and historical precedents were the bases of the claims of Israel and the counterclaims of Ammon. **17:** Num 20.14–21. **19–23:** Num 21.21–32. **24:** *Chemosh* was the national god of the Moabites. The god of the Ammonites was Milcom or Molech (1 Kings 11.5, 7). **25:** *Balak*, see Num

one who does me wrong by making war on me. Let the LORD, who is judge, decide today for the Israelites or for the Ammonites." 28 But the king of the Ammonites did not heed the message that Jephthah sent him.

29 Then the spirit of the LORD came upon Jephthah, and he passed through Gilead and Manasseh. He passed on to Mizpah of Gilead, and from Mizpah of Gilead he passed on to the Ammonites. 30 And Jephthah made a vow to the LORD, and said, "If you will give the Ammonites into my hand, 31 then whoever comes out of the doors of my house to meet me, when I return victorious from the Ammonites, shall be the LORD's, to be offered up by me as a burnt offering." 32 So Jephthah crossed over to the Ammonites to fight against them; and the LORD gave them into his hand. 33 He inflicted a massive defeat on them from Aroer to the neighborhood of Minnith, twenty towns, and as far as Abel-keramim. So the Ammonites were subdued before the people of Israel.

34 Then Jephthah came to his home at Mizpah; and there was his daughter coming out to meet him with timbrels and with dancing. She was his only child; he had no son or daughter except her. 35 When he saw her, he tore his clothes, and said, "Alas, my daughter! You have brought me very low; you have become the cause of great trouble to me. For I have opened my mouth to the LORD, and

I cannot take back my vow." 36 She said to him, "My father, if you have opened your mouth to the LORD, do to me according to what has gone out of your mouth, now that the LORD has given you vengeance against your enemies, the Ammonites." 37 And she said to her father, "Let this thing be done for me: Grant me two months, so that I may go and wander*ⁿ* on the mountains, and bewail my virginity, my companions and I." 38 "Go," he said and sent her away for two months. So she departed, she and her companions, and bewailed her virginity on the mountains. 39 At the end of two months, she returned to her father, who did with her according to the vow he had made. She had never slept with a man. So there arose an Israelite custom that 40 for four days every year the daughters of Israel would go out to lament the daughter of Jephthah the Gileadite.

12 The men of Ephraim were called to arms, and they crossed to Zaphon and said to Jephthah, "Why did you cross over to fight against the Ammonites, and did not call us to go with you? We will burn your house down over you!" 2 Jephthah said to them, "My people and I were engaged in conflict with the Ammonites who oppressed us*ᵒ* severely. But when I called you, you did

---

n Cn: Heb *go down*   *o* Gk OL, Syr H: Heb lacks *who oppressed us*

22–24. 28: The Ammonites reject the Israelite contention that the fact of conquest determines the legitimacy of a territorial claim. 11.29–40: The centerpiece of this story is Jephthah's vow; the outcome of the battle with Ammon is almost an aside. 29: The coming of *the spirit of the LORD* upon Jephthah should have been the guarantee of victory. 30–31: Jephthah promises a sacrifice in exchange for victory. The Hebrew text does not assume that the intended victim would be a human being. Given the arrangement of homes with courtyards that housed domesticated animals, it is likely that Jephthah assumed that one of these animals would be encountered first upon his return home. 35: While human sacrifice was not usually acceptable in ancient Israelite religion, neither was it unknown (2 Kings 16.3; Ezek 20.25–26, 31). 37: Because a Hebrew woman could suffer no greater disgrace than to die childless, Jephthah's daughter asks for time to *bewail* her *virginity*. 39: Similar rites elsewhere were disapproved; see Ezek 8.14; Zech 12.11. As the Israelite cult got an identity more detached from its Canaanite roots, practices such as human sacrifice and ritual mourning were censured (Jer 7.31; Mic 6.6–8; see Ps 106.37) and fell into disuse. 12.1–7: Perhaps the cause of the Ephraimites' complaints was that, not having participated in the war against Ammon, they would have no right to any spoils. 6: The word *shibboleth* means "ear of corn." Local variations in

not deliver me from their hand. [3]When
I saw that you would not deliver me, I
took my life in my hand, and crossed
over against the Ammonites, and the
LORD gave them into my hand. Why
then have you come up to me this day,
to fight against me?" [4]Then Jephthah
gathered all the men of Gilead and fought
with Ephraim; and the men of Gilead
defeated Ephraim, because they said,
"You are fugitives from Ephraim, you
Gileadites—in the heart of Ephraim and
Manasseh."[p] [5]Then the Gileadites took
the fords of the Jordan against the Ephraimites. Whenever one of the fugitives of
Ephraim said, "Let me go over," the
men of Gilead would say to him, "Are
you an Ephraimite?" When he said,
"No," [6]they said to him, "Then say
Shibboleth," and he said, "Sibboleth,"
for he could not pronounce it right. Then
they seized him and killed him at the
fords of the Jordan. Forty-two thousand
of the Ephraimites fell at that time.
7 Jephthah judged Israel six years.
Then Jephthah the Gileadite died, and
was buried in his town in Gilead.[q]
8 After him Ibzan of Bethlehem
judged Israel. [9]He had thirty sons. He
gave his thirty daughters in marriage
outside his clan and brought in thirty
young women from outside for his sons.
He judged Israel seven years. [10]Then Ibzan died, and was buried at Bethlehem.
11 After him Elon the Zebulunite
judged Israel; and he judged Israel ten
years. [12]Then Elon the Zebulunite died,
and was buried at Aijalon in the land of
Zebulun.
13 After him Abdon son of Hillel the
Pirathonite judged Israel. [14]He had forty
sons and thirty grandsons, who rode on
seventy donkeys; he judged Israel eight
years. [15]Then Abdon son of Hillel the
Pirathonite died, and was buried at Pirathon in the land of Ephraim, in the hill
country of the Amalekites.

13

The Israelites again did what was
evil in the sight of the LORD, and
the LORD gave them into the hand of the
Philistines forty years.
2 There was a certain man of Zorah,
of the tribe of the Danites, whose name
was Manoah. His wife was barren, having borne no children. [3]And the angel of
the LORD appeared to the woman and
said to her, "Although you are barren,
having borne no children, you shall conceive and bear a son. [4]Now be careful not
to drink wine or strong drink, or to eat
anything unclean, [5]for you shall conceive and bear a son. No razor is to come
on his head, for the boy shall be a nazirite[r] to God from birth. It is he who shall
begin to deliver Israel from the hand of
the Philistines." [6]Then the woman came
and told her husband, "A man of God
came to me, and his appearance was like

p Meaning of Heb uncertain: Gk omits *because . . . Manasseh* *q* Gk: Heb *in the towns of Gilead*
r That is *one separated* or *one consecrated*

---

pronunciation of common words aid in identifying the regional origin of the speaker (see Mt 26.73).

12.8–15: Ibzan, Elon, and Abdon. See 10.1–5 n. **8**: Since these judges were from the north, the *Bethlehem* mentioned here is probably the one in the territory of Zebulun (Josh 19.15).

13.1–16.31: Samson. The final deliverer described in the book does not fit the pattern set by his predecessors. He broke the vows he made to the LORD, married non-Israelite women, and never associated with other Israelites in his conflicts with the Philistines.

13.1–25: Samson's birth. **1**: This Deuteronomistic introduction helps to authenticate Samson as a judge. *The Philistines* were of Aegean-Asian origin. Prophetic tradition identified Caphtor as their precise place of origin (see Am 9.7; Jer 47.4). They entered Canaan shortly after the Israelites and were their chief rivals for hegemony in the region. **2**: *Zorah,* a town on the border between Judah and Dan (Josh 15.33; 19.41) identified with modern Sa'rah fifteen miles west of Jerusalem. Here the *Danites* are a southern tribe. Chapters 17–18 describe the migration of the tribe to the north. The Samson stories illustrate the reasons for the migration. **3**: *The angel of the LORD,* see 2.1 n. Any Israelite, male or female, who lived according to the norms of Num 6.1–21 was a *nazirite.* These norms reflect the values of a conservative Yahwistic lifestyle. **4**: The requirement that

that of an angel[s] of God, most awe-inspiring; I did not ask him where he came from, and he did not tell me his name; [7]but he said to me, 'You shall conceive and bear a son. So then drink no wine or strong drink, and eat nothing unclean, for the boy shall be a nazirite[t] to God from birth to the day of his death.' "

8 Then Manoah entreated the Lord, and said, "O, Lord, I pray, let the man of God whom you sent come to us again and teach us what we are to do concerning the boy who will be born." [9]God listened to Manoah, and the angel of God came again to the woman as she sat in the field; but her husband Manoah was not with her. [10]So the woman ran quickly and told her husband, "The man who came to me the other day has appeared to me." [11]Manoah got up and followed his wife, and came to the man and said to him, "Are you the man who spoke to this woman?" And he said, "I am." [12]Then Manoah said, "Now when your words come true, what is to be the boy's rule of life; what is he to do?" [13]The angel of the Lord said to Manoah, "Let the woman give heed to all that I said to her. [14]She may not eat of anything that comes from the vine. She is not to drink wine or strong drink, or eat any unclean thing. She is to observe everything that I commanded her."

15 Manoah said to the angel of the Lord, "Allow us to detain you, and prepare a kid for you." [16]The angel of the Lord said to Manoah, "If you detain me, I will not eat your food; but if you want to prepare a burnt offering, then offer it to the Lord." (For Manoah did not know that he was the angel of the Lord.) [17]Then Manoah said to the angel of the

Lord, "What is your name, so that we may honor you when your words come true?" [18]But the angel of the Lord said to him, "Why do you ask my name? It is too wonderful."

19 So Manoah took the kid with the grain offering, and offered it on the rock to the Lord, to him who works[u] wonders.[v] [20]When the flame went up toward heaven from the altar, the angel of the Lord ascended in the flame of the altar while Manoah and his wife looked on; and they fell on their faces to the ground. [21]The angel of the Lord did not appear again to Manoah and his wife. Then Manoah realized that it was the angel of the Lord. [22]And Manoah said to his wife, "We shall surely die, for we have seen God." [23]But his wife said to him, "If the Lord had meant to kill us, he would not have accepted a burnt offering and a grain offering at our hands, or shown us all these things, or now announced to us such things as these."

24 The woman bore a son, and named him Samson. The boy grew, and the Lord blessed him. [25]The spirit of the Lord began to stir him in Mahaneh-dan, between Zorah and Eshtaol.

14 Once Samson went down to Timnah, and at Timnah he saw a Philistine woman. [2]Then he came up, and told his father and mother, "I saw a Philistine woman at Timnah; now get her for me as my wife." [3]But his father and mother said to him, "Is there not a woman among your kin, or among all

---

s Or *the angel* *t* That is *one separated* or *one consecrated* *u* Gk Vg: Heb *and working* *v* Heb *wonders, while Manoah and his wife looked on*

---

the *nazirite* abstain from wine becomes prenatal care (also v. 7).

13.14: The cultivation of grapes represents the settled life and the consequent compromise with Canaanite culture.

13.24: The name *Samson* is related to the Hebrew word for sun. **25:** *The spirit of the Lord,* see 3.10 n. *Mahaneh-dan, between Zorah and Eshtaol,* the references to definite towns and places in these stories show that they are

part of a genuine local tradition. These sites were in the foothills southwest of Jerusalem.

14.1–20: Samson's wedding. The stories report Samson's naïveté toward women. It was a fatal flaw that led to his destruction. **1:** *Timnah* may be Tell el-Batashi, four miles north of Beth-Shemesh. **3:** Samson's marriage to a Philistine will lead to divided loyalties. **4:** An editorial comment reflecting the belief that God can use human folly to accom-

our[w] people, that you must go to take a wife from the uncircumcised Philistines?" But Samson said to his father, "Get her for me, because she pleases me." [4]His father and mother did not know that this was from the LORD; for he was seeking a pretext to act against the Philistines. At that time the Philistines had dominion over Israel.

5 Then Samson went down with his father and mother to Timnah. When he came to the vineyards of Timnah, suddenly a young lion roared at him. [6]The spirit of the LORD rushed on him, and he tore the lion apart barehanded as one might tear apart a kid. But he did not tell his father or his mother what he had done. [7]Then he went down and talked with the woman, and she pleased Samson. [8]After a while he returned to marry her, and he turned aside to see the carcass of the lion, and there was a swarm of bees in the body of the lion, and honey. [9]He scraped it out into his hands, and went on, eating as he went. When he came to his father and mother, he gave some to them, and they ate it. But he did not tell them that he had taken the honey from the carcass of the lion.

10 His father went down to the woman, and Samson made a feast there as the young men were accustomed to do. [11]When the people saw him, they brought thirty companions to be with him. [12]Samson said to them, "Let me now put a riddle to you. If you can explain it to me within the seven days of the feast, and find it out, then I will give you thirty linen garments and thirty festal garments. [13]But if you cannot explain it to me, then you shall give me thirty linen garments and thirty festal garments." So they said to him, "Ask your riddle; let us hear it." [14]He said to them,

"Out of the eater came something to eat.
Out of the strong came something sweet."

But for three days they could not explain the riddle.

15 On the fourth[x] day they said to Samson's wife, "Coax your husband to explain the riddle to us, or we will burn you and your father's house with fire. Have you invited us here to impoverish us?" [16]So Samson's wife wept before him, saying, "You hate me; you do not really love me. You have asked a riddle of my people, but you have not explained it to me." He said to her, "Look, I have not told my father or my mother. Why should I tell you?" [17]She wept before him the seven days that their feast lasted; and because she nagged him, on the seventh day he told her. Then she explained the riddle to her people. [18]The men of the town said to him on the seventh day before the sun went down,

"What is sweeter than honey?
What is stronger than a lion?"

And he said to them,

"If you had not plowed with my heifer,
you would not have found out my riddle."

[19]Then the spirit of the LORD rushed on him, and he went down to Ashkelon. He killed thirty men of the town, took their spoil, and gave the festal garments to those who had explained the riddle. In hot anger he went back to his father's house. [20]And Samson's wife was given to his companion, who had been his best man.

15 After a while, at the time of the wheat harvest, Samson went to

---

w Cn: Heb *my*   *x* Gk Syr: Heb *seventh*

---

plish the divine purpose. **6:** *The spirit of the LORD* equipped Samson to overcome any physical threat.

14.10–18: The hostility between the Danites and Philistines flared up because of verbal jousting between Samson and his guests. Samson proposed a riddle that could not be solved, or so he thought. Samson's bride betrayed him. **19:** *Ashkelon* on the southern Mediterranean coast was one of the five main Philistine cities. Samson robbed the people of this city to pay his debts to the people of Timnah.

15.1–20: Samson's revenge. Samson's anger after his wife was given to another man led him to destroy the Philistines' grain crop.

visit his wife, bringing along a kid. He said, "I want to go into my wife's room." But her father would not allow him to go in. ²Her father said, "I was sure that you had rejected her; so I gave her to your companion. Is not her younger sister prettier than she? Why not take her instead?" ³Samson said to them, "This time, when I do mischief to the Philistines, I will be without blame." ⁴So Samson went and caught three hundred foxes, and took some torches; and he turned the foxes*ʸ* tail to tail, and put a torch between each pair of tails. ⁵When he had set fire to the torches, he let the foxes go into the standing grain of the Philistines, and burned up the shocks and the standing grain, as well as the vineyards and*ᶻ* olive groves. ⁶Then the Philistines asked, "Who has done this?" And they said, "Samson, the son-in-law of the Timnite, because he has taken Samson's wife and given her to his companion." So the Philistines came up, and burned her and her father. ⁷Samson said to them, "If this is what you do, I swear I will not stop until I have taken revenge on you." ⁸He struck them down hip and thigh with great slaughter; and he went down and stayed in the cleft of the rock of Etam.

9 Then the Philistines came up and encamped in Judah, and made a raid on Lehi. ¹⁰The men of Judah said, "Why have you come up against us?" They said, "We have come up to bind Samson, to do to him as he did to us." ¹¹Then three thousand men of Judah went down to the cleft of the rock of Etam, and they said to Samson, "Do you not know that the Philistines are rulers over us? What then have you done to us?" He replied, "As they did to me, so I have done to them." ¹²They said to him, "We have come down to bind you, so that we may give you into the hands of the Philistines." Samson answered them, "Swear to me that you yourselves will not attack me." ¹³They said to him, "No, we will only bind you and give you into their hands; we will not kill you." So they bound him with two new ropes, and brought him up from the rock.

14 When he came to Lehi, the Philistines came shouting to meet him; and the spirit of the LORD rushed on him, and the ropes that were on his arms became like flax that has caught fire, and his bonds melted off his hands. ¹⁵Then he found a fresh jawbone of a donkey, reached down and took it, and with it he killed a thousand men. ¹⁶And Samson said,

"With the jawbone of a donkey,
  heaps upon heaps,
  with the jawbone of a donkey
  I have slain a thousand men."

¹⁷When he had finished speaking, he threw away the jawbone; and that place was called Ramath-lehi.*ᵃ*

18 By then he was very thirsty, and he called on the LORD, saying, "You have granted this great victory by the hand of your servant. Am I now to die of thirst, and fall into the hands of the uncircumcised?" ¹⁹So God split open the hollow place that is at Lehi, and water came from it. When he drank, his spirit returned, and he revived. Therefore it was named En-hakkore,*ᵇ* which is at Lehi to this day. ²⁰And he judged Israel in the days of the Philistines twenty years.

16 Once Samson went to Gaza, where he saw a prostitute and

ʸ Heb *them*   *ᶻ* Gk Tg Vg: Heb lacks *and*
ᵃ That is *The Hill of the Jawbone*   *ᵇ* That is
The Spring of the One who Called

This began a cycle of revenge. **1:** *A kid* was perhaps the usual gift for sexual intimacy (Gen 38.17). **8:** *Etam* cannot be located with certainty. It must have been in the south. *Hip and thigh,* a formulaic expression describing complete devastation.

15.9–13: The people of Judah refuse to provide Samson with sanctuary and even attempt to apprehend Samson for the Philis-

tines. **14:** *The spirit of the LORD,* see 14.6 n. **15:** The *jawbone of a donkey* was sometimes worked into a sickle; it could then serve as a weapon. **16:** The couplet involves a pun, for the same Hebrew word can mean both *donkey* and *heap.* **20:** An editorial comment to authenticate Samson as a judge.

16.1–31: Samson's death. Samson's weakness for women led to his betrayal and

went in to her. 2 The Gazites were told,[c] "Samson has come here." So they circled around and lay in wait for him all night at the city gate. They kept quiet all night, thinking, "Let us wait until the light of the morning; then we will kill him." 3 But Samson lay only until midnight. Then at midnight he rose up, took hold of the doors of the city gate and the two posts, pulled them up, bar and all, put them on his shoulders, and carried them to the top of the hill that is in front of Hebron.

4 After this he fell in love with a woman in the valley of Sorek, whose name was Delilah. 5 The lords of the Philistines came to her and said to her, "Coax him, and find out what makes his strength so great, and how we may overpower him, so that we may bind him in order to subdue him; and we will each give you eleven hundred pieces of silver." 6 So Delilah said to Samson, "Please tell me what makes your strength so great, and how you could be bound, so that one could subdue you." 7 Samson said to her, "If they bind me with seven fresh bowstrings that are not dried out, then I shall become weak, and be like anyone else." 8 Then the lords of the Philistines brought her seven fresh bowstrings that had not dried out, and she bound him with them. 9 While men were lying in wait in an inner chamber, she said to him, "The Philistines are upon you, Samson!" But he snapped the bowstrings, as a strand of fiber snaps when it touches the fire. So the secret of his strength was not known.

10 Then Delilah said to Samson, "You have mocked me and told me lies; please tell me how you could be bound." 11 He said to her, "If they bind me with new ropes that have not been used, then I shall become weak, and be like anyone else." 12 So Delilah took new ropes and bound him with them, and said to him, "The Philistines are upon you, Samson!" (The men lying in wait were in an inner chamber.) But he snapped the ropes off his arms like a thread.

13 Then Delilah said to Samson, "Until now you have mocked me and told me lies; tell me how you could be bound." He said to her, "If you weave the seven locks of my head with the web and make it tight with the pin, then I shall become weak, and be like anyone else." 14 So while he slept, Delilah took the seven locks of his head and wove them into the web,[d] and made them tight with the pin. Then she said to him, "The Philistines are upon you, Samson!" But he awoke from his sleep, and pulled away the pin, the loom, and the web.

15 Then she said to him, "How can you say, 'I love you,' when your heart is not with me? You have mocked me three times now and have not told me what makes your strength so great." 16 Finally, after she had nagged him with her words day after day, and pestered him, he was tired to death. 17 So he told her his whole secret, and said to her, "A razor has never come upon my head; for I have been a nazirite[e] to God from my mother's womb. If my head were shaved, then my strength would leave me; I would become weak, and be like anyone else."

18 When Delilah realized that he had told her his whole secret, she sent and called the lords of the Philistines, saying, "This time come up, for he has told his whole secret to me." Then the lords of the Philistines came up to her, and brought the money in their hands. 19 She

c Gk: Heb lacks *were told*   d Compare Gk: in verses 13-14, Heb lacks *and make it tight . . . into the web*   e That is *one separated* or *one consecrated*

---

death. **1**: *Gaza,* see 6.4 n. **3**: An exaggeration to make the Philistines look powerless. The distance from Gaza to Hebron is forty miles and involves an ascent of more than three thousand feet.

16.4–22: The Delilah incident. **4**: The *valley of Sorek* is the modern Wadi-es-Sarar that begins thirteen miles southwest of Jerusalem. The name *Delilah* may be related to an Arabic root that means "flirt." The text does not say that she was a Philistine. Living in the valley of Sorek would allow her to be either a Philistine or an Israelite. **13**: *The web* and *the pin* were parts of a loom.

let him fall asleep on her lap; and she called a man, and had him shave off the seven locks of his head. He began to weaken,[f] and his strength left him. [20] Then she said, "The Philistines are upon you, Samson!" When he awoke from his sleep, he thought, "I will go out as at other times, and shake myself free." But he did not know that the LORD had left him. [21] So the Philistines seized him and gouged out his eyes. They brought him down to Gaza and bound him with bronze shackles; and he ground at the mill in the prison. [22] But the hair of his head began to grow again after it had been shaved.

23 Now the lords of the Philistines gathered to offer a great sacrifice to their god Dagon, and to rejoice; for they said, "Our god has given Samson our enemy into our hand." [24] When the people saw him, they praised their god; for they said, "Our god has given our enemy into our hand, the ravager of our country, who has killed many of us." [25] And when their hearts were merry, they said, "Call Samson, and let him entertain us." So they called Samson out of the prison, and he performed for them. They made him stand between the pillars; [26] and Samson said to the attendant who held him by the hand, "Let me feel the pillars on which the house rests, so that I may lean against them." [27] Now the house was full of men and women; all the lords of the Philistines were there, and on the roof there were about three thousand men and women, who looked on while Samson performed.

28 Then Samson called to the LORD and said, "Lord GOD, remember me and strengthen me only this once, O God, so that with this one act of revenge I may pay back the Philistines for my two eyes."[g] [29] And Samson grasped the two middle pillars on which the house rested, and he leaned his weight against them, his right hand on the one and his left hand on the other. [30] Then Samson said, "Let me die with the Philistines." He strained with all his might; and the house fell on the lords and all the people who were in it. So those he killed at his death were more than those he had killed during his life. [31] Then his brothers and all his family came down and took him and brought him up and buried him between Zorah and Eshtaol in the tomb of his father Manoah. He had judged Israel twenty years.

17 There was a man in the hill country of Ephraim whose name was Micah. [2] He said to his mother, "The eleven hundred pieces of silver that were taken from you, about which you uttered a curse, and even spoke it in my hearing,—that silver is in my possession; I took it; but now I will return it to you."[h] And his mother said, "May my son be blessed by the LORD!" [3] Then he returned the eleven hundred pieces of silver to his mother; and his mother said, "I consecrate the silver to the LORD from my hand for my son, to make an idol of cast metal." [4] So when he returned the money to his mother, his mother took two hundred pieces of silver, and gave it to the silversmith, who made it into an idol of cast metal; and it was in the house of Micah. [5] This man Micah had a shrine,

f Gk: Heb *She began to torment him*   g Or *so that I may be avenged upon the Philistines for one of my two eyes*   h The words *but now I will return it to you* are transposed from the end of verse 3 in Heb

---

16.23–31: Samson's final act of vengeance. 23: *Dagon* was a Canaanite god adopted by the Philistines. 28–30: Samson's final words and deeds portray an honorable death.

17.1–21.25: **The failures of the Israelite tribes.** The final chapters of the book describe the self-destructive actions by the Israelites that almost succeeded in undoing the achievements of the judges. Chapters 17–18 describe cultic disintegration and chs 19–21 describe a civil war.

17.1–6: **The shrine of Micah. 1:** *The hill country of Ephraim* was the central highlands north of Bethel and south of Shechem. The story begins with an ironic twist: *Micah* is an abbreviation of a name that means "who is like the LORD"; an Israelite with such a name built a shrine with an image! **5:** An *ephod* was an elaborate priestly vestment. *Teraphim* were divinatory equipment of some type. **6:** An editorial comment that describes the end of the judges' period as disordered (21.25).

and he made an ephod and teraphim, and installed one of his sons, who became his priest. [6] In those days there was no king in Israel; all the people did what was right in their own eyes.

7 Now there was a young man of Bethlehem in Judah, of the clan of Judah. He was a Levite residing there. [8] This man left the town of Bethlehem in Judah, to live wherever he could find a place. He came to the house of Micah in the hill country of Ephraim to carry on his work.[i] [9] Micah said to him, "From where do you come?" He replied, "I am a Levite of Bethlehem in Judah, and I am going to live wherever I can find a place." [10] Then Micah said to him, "Stay with me, and be to me a father and a priest, and I will give you ten pieces of silver a year, a set of clothes, and your living."[j] [11] The Levite agreed to stay with the man; and the young man became to him like one of his sons. [12] So Micah installed the Levite, and the young man became his priest, and was in the house of Micah. [13] Then Micah said, "Now I know that the LORD will prosper me, because the Levite has become my priest."

18 In those days there was no king in Israel. And in those days the tribe of the Danites was seeking for itself a territory to live in; for until then no territory among the tribes of Israel had been allotted to them. [2] So the Danites sent five valiant men from the whole number of their clan, from Zorah and from Eshtaol, to spy out the land and to explore it; and they said to them, "Go, explore the land." When they came to the hill country of Ephraim, to the house of Micah, they stayed there. [3] While they were at Micah's house, they recognized the voice of the young Levite; so they went over and asked him, "Who brought you here? What are you doing in this place? What is your business here?" [4] He said to them, "Micah did such and such for me, and he hired me, and I have become his priest." [5] Then they said to him, "Inquire of God that we may know whether the mission we are undertaking will succeed." [6] The priest replied, "Go in peace. The mission you are on is under the eye of the LORD."

7 The five men went on, and when they came to Laish, they observed the people who were there living securely, after the manner of the Sidonians, quiet and unsuspecting, lacking[k] nothing on earth, and possessing wealth.[l] Furthermore, they were far from the Sidonians and had no dealings with Aram.[m] [8] When they came to their kinsfolk at Zorah and Eshtaol, they said to them, "What do you report?" [9] They said, "Come, let us go up against them; for we have seen the land, and it is very good. Will you do nothing? Do not be slow to go, but enter in and possess the land. [10] When you go, you will come to an unsuspecting people. The land is broad—God has indeed given it into your hands—a place where there is no lack of anything on earth."

11 Six hundred men of the Danite

---

i Or *Ephraim, continuing his journey*
j Heb *living, and the Levite went*
k Cn Compare 18.10: Meaning of Heb uncertain
l Meaning of Heb uncertain   m Symmachus: Heb *with anyone*

17.7–13: Micah's priest. While any male Israelite could function as a priest, the Levites were the preferred specialists for ritual activities. **7:** This *Bethlehem* (see 12.8 n.) was in the territory of Judah, five miles south of Jerusalem.

18.1–31: The migration of the tribe of Dan. The Danites could not control Philistine incursions into their territory in the south. They abandoned this region and migrated to the far north of Galilee. **1–6:** An oracle assures the Danites of divine support for their move. **5:** Here *to inquire of God* refers to divination.

18.7: *Laish* was a city at the foot of Mount Hermon. *Living securely* means "undefended." Archaeology has shown that the pre-Israelite city was unwalled. *Sidon* was a Phoenician city along the Mediterranean. It was too far from Laish to offer it effective protection. *Aram* was just to the north of Laish. Without defenses or effective alliances Laish was easy prey.

18.11–26: On their way to Laish, the Dan-

clan, armed with weapons of war, set out from Zorah and Eshtaol, [12] and went up and encamped at Kiriath-jearim in Judah. On this account that place is called Mahaneh-dan[n] to this day; it is west of Kiriath-jearim. [13] From there they passed on to the hill country of Ephraim, and came to the house of Micah.

14 Then the five men who had gone to spy out the land (that is, Laish) said to their comrades, "Do you know that in these buildings there are an ephod, teraphim, and an idol of cast metal? Now therefore consider what you will do." [15] So they turned in that direction and came to the house of the young Levite, at the home of Micah, and greeted him. [16] While the six hundred men of the Danites, armed with their weapons of war, stood by the entrance of the gate, [17] the five men who had gone to spy out the land proceeded to enter and take the idol of cast metal, the ephod, and the teraphim. [o] The priest was standing by the entrance of the gate with the six hundred men armed with weapons of war. [18] When the men went into Micah's house and took the idol of cast metal, the ephod, and the teraphim, the priest said to them, "What are you doing?" [19] They said to him, "Keep quiet! Put your hand over your mouth, and come with us, and be to us a father and a priest. Is it better for you to be priest to the house of one person, or to be priest to a tribe and clan in Israel?" [20] Then the priest accepted the offer. He took the ephod, the teraphim, and the idol, and went along with the people.

21 So they resumed their journey, putting the little ones, the livestock, and the goods in front of them. [22] When they were some distance from the home of Micah, the men who were in the houses near Micah's house were called out, and

they overtook the Danites. [23] They shouted to the Danites, who turned around and said to Micah, "What is the matter that you come with such a company?" [24] He replied, "You take my gods that I made, and the priest, and go away, and what have I left? How then can you ask me, 'What is the matter?'" [25] And the Danites said to him, "You had better not let your voice be heard among us or else hot-tempered fellows will attack you, and you will lose your life and the lives of your household." [26] Then the Danites went their way. When Micah saw that they were too strong for him, he turned and went back to his home.

27 The Danites, having taken what Micah had made, and the priest who belonged to him, came to Laish, to a people quiet and unsuspecting, put them to the sword, and burned down the city. [28] There was no deliverer, because it was far from Sidon and they had no dealings with Aram.[p] It was in the valley that belongs to Beth-rehob. They rebuilt the city, and lived in it. [29] They named the city Dan, after their ancestor Dan, who was born to Israel; but the name of the city was formerly Laish. [30] Then the Danites set up the idol for themselves. Jonathan son of Gershom, son of Moses,[q] and his sons were priests to the tribe of the Danites until the time the land went into captivity. [31] So they maintained as their own Micah's idol that he had made, as long as the house of God was at Shiloh.

19 In those days, when there was no king in Israel, a certain Levite, residing in the remote parts of the hill country of Ephraim, took to himself a

---

n That is *Camp of Dan*   o Compare 17.4, 5;
18.14: Heb *teraphim and the cast metal*
p Cn Compare verse 7: Heb *with anyone*
q Another reading is *son of Manasseh*

---

ites stole Micah's cultic objects and induced his priest to join them. **18.27–31:** The Danites took Laish, renamed it, and installed Micah's images there. **30:** A priesthood descended from Moses is associated with an illegitimate cult. **31:** A

Deuteronomistic indictment against the shrine at Dan that became an important center of worship for the northern kingdom (1 Kings 12.29). **19.1–9: The Levite and his concubine.** This incident set up a conflict that pitted the

concubine from Bethlehem in Judah.
[2] But his concubine became angry with[r]
him, and she went away from him to her
father's house at Bethlehem in Judah, and
was there some four months. [3] Then her
husband set out after her, to speak tenderly to her and bring her back. He had
with him his servant and a couple of donkeys. When he reached[s] her father's
house, the girl's father saw him and came
with joy to meet him. [4] His father-in-
law, the girl's father, made him stay, and
he remained with him three days; so they
ate and drank, and he[t] stayed there. [5] On
the fourth day they got up early in the
morning, and he prepared to go; but the
girl's father said to his son-in-law, "Fortify yourself with a bit of food, and after
that you may go." [6] So the two men sat
and ate and drank together; and the girl's
father said to the man, "Why not spend
the night and enjoy yourself?" [7] When the
man got up to go, his father-in-law kept
urging him until he spent the night there
again. [8] On the fifth day he got up early
in the morning to leave; and the girl's
father said, "Fortify yourself." So they
lingered[u] until the day declined, and the
two of them ate and drank. [v] [9] When the
man with his concubine and his servant
got up to leave, his father-in-law, the
girl's father, said to him, "Look, the day
has worn on until it is almost evening.
Spend the night. See, the day has drawn
to a close. Spend the night here and enjoy
yourself. Tomorrow you can get up early in the morning for your journey, and
go home."
10 But the man would not spend the
night; he got up and departed, and arrived opposite Jebus (that is, Jerusalem).
He had with him a couple of saddled
donkeys, and his concubine was with
him. [11] When they were near Jebus, the

day was far spent, and the servant said to
his master, "Come now, let us turn aside
to this city of the Jebusites, and spend the
night in it." [12] But his master said to him,
"We will not turn aside into a city of
foreigners, who do not belong to the
people of Israel; but we will continue on
to Gibeah." [13] Then he said to his servant,
"Come, let us try to reach one of these
places, and spend the night at Gibeah or
at Ramah." [14] So they passed on and went
their way; and the sun went down on
them near Gibeah, which belongs to
Benjamin. [15] They turned aside there, to
go in and spend the night at Gibeah. He
went in and sat down in the open square
of the city, but no one took them in to
spend the night.
16 Then at evening there was an old
man coming from his work in the field.
The man was from the hill country of
Ephraim, and he was residing in Gibeah.
(The people of the place were Benjaminites.) [17] When the old man looked up and
saw the wayfarer in the open square of
the city, he said, "Where are you going
and where do you come from?" [18] He
answered him, "We are passing from
Bethlehem in Judah to the remote parts
of the hill country of Ephraim, from
which I come. I went to Bethlehem in
Judah; and I am going to my home.[w]
Nobody has offered to take me in. [19] We
your servants have straw and fodder for
our donkeys, with bread and wine for
me and the woman and the young man
along with us. We need nothing more."
[20] The old man said, "Peace be to you. I
will care for all your wants; only do not

r Gk OL: Heb *prostituted herself against*
s Gk: Heb *she brought him*   t Compare verse 7
and Gk: Heb *they*   u Cn: Heb *Linger*
v Gk: Heb lacks *and drank*   w Gk Compare
19.29. Heb *to the house of the* Lord

tribal assembly against Benjamin. The book
of Judges then ends on a chaotic note. **2:** The
woman of Bethlehem was not the full legal
wife of the Levite, so the text calls her his
concubine. Because v. 3 calls the Levite her
husband, there must have been some
regularity about their relationship. It is the
woman who initiates the separation.

**19.10–30: The incident at Gibeah. 10–
21:** The Levites prefer to spend the night in an
Israelite rather than a Canaanite town. **12:**
Gibeah is identified with Tell el Ful, four
miles north of Jerusalem. It became Saul's
royal residence (1 Sam 15.34). **13:** *Ramah* is
identified with er-Ram, five miles north of
Jerusalem.

spend the night in the square." 21 So he brought him into his house, and fed the donkeys; they washed their feet, and ate and drank.

22 While they were enjoying themselves, the men of the city, a perverse lot, surrounded the house, and started pounding on the door. They said to the old man, the master of the house, "Bring out the man who came into your house, so that we may have intercourse with him." 23 And the man, the master of the house, went out to them and said to them, "No, my brothers, do not act so wickedly. Since this man is my guest, do not do this vile thing. 24 Here are my virgin daughter and his concubine; let me bring them out now. Ravish them and do whatever you want to them; but against this man do not do such a vile thing." 25 But the men would not listen to him. So the man seized his concubine, and put her out to them. They wantonly raped her, and abused her all through the night until the morning. And as the dawn began to break, they let her go. 26 As morning appeared, the woman came and fell down at the door of the man's house where her master was, until it was light.

27 In the morning her master got up, opened the doors of the house, and when he went out to go on his way, there was his concubine lying at the door of the house, with her hands on the threshold. 28 "Get up," he said to her, "we are going." But there was no answer. Then he put her on the donkey; and the man set out for his home. 29 When he had entered his house, he took a knife, and grasping his concubine he cut her into twelve pieces, limb by limb, and sent her throughout all the territory of Israel. 30 Then he commanded the men whom he sent, saying, "Thus shall you say to all the Israelites, 'Has such a thing ever happened*x* since the day that the Israelites came up from the land of Egypt until this day? Consider it, take counsel, and speak out.' "

20 Then all the Israelites came out, from Dan to Beer-sheba, including the land of Gilead, and the congregation assembled in one body before the LORD at Mizpah. 2 The chiefs of all the people, of all the tribes of Israel, presented themselves in the assembly of the people of God, four hundred thousand foot-soldiers bearing arms. 3 (Now the Benjaminites heard that the people of Israel had gone up to Mizpah.) And the Israelites said, "Tell us, how did this criminal act come about?" 4 The Levite, the husband of the woman who was murdered, answered, "I came to Gibeah that belongs to Benjamin, I and my concubine, to spend the night. 5 The lords of Gibeah rose up against me, and surrounded the house at night. They intended to kill me, and they raped my concubine until she died. 6 Then I took my concubine and cut her into pieces, and sent her throughout the whole extent of Israel's territory; for they have committed a vile outrage in Israel. 7 So now, you Israelites, all of you, give your advice and counsel here."

8 All the people got up as one, saying, "We will not any of us go to our tents, nor will any of us return to our houses. 9 But now this is what we will do to Gib-

---

x Compare Gk: Heb 30*And all who saw it said, "Such a thing has not happened or been seen*

19.22–26: To forestall his own rape by the men of Gibeah, the Levite offers them his concubine. The text implies that the woman died as a result of her night of horror. **27–30:** The Levite seeks redress of the crime by bringing the matter to the tribal authorities. **29:** In 1 Sam 11.7 Saul calls up the tribes of Israel in a strikingly similar fashion.

20.1–48: The war against Benjamin. 1–11: The tribal assembly condemns the crime against the Levite's concubine and takes steps to punish the guilty. **1:** *Dan* was the northernmost Israelite city, while *Beer-sheba* was the southernmost one. *Gilead* refers to the Trans-jordanian territories. *Mizpah* was the name of a few Israelite towns. The one intended here was probably the Mizpah of Benjamin (Josh 18.26). It is identified with Tell en-Nasbeh, eight miles north of Jerusalem.

eah: we will go up[y] against it by lot.
[10] We will take ten men of a hundred throughout all the tribes of Israel, and a hundred of a thousand, and a thousand of ten thousand, to bring provisions for the troops, who are going to repay[z] Gibeah of Benjamin for all the disgrace that they have done in Israel." [11] So all the men of Israel gathered against the city, united as one.

12 The tribes of Israel sent men through all the tribe of Benjamin, saying, "What crime is this that has been committed among you? [13] Now then, hand over those scoundrels in Gibeah, so that we may put them to death, and purge the evil from Israel." But the Benjaminites would not listen to their kinsfolk, the Israelites. [14] The Benjaminites came together out of the towns to Gibeah, to go out to battle against the Israelites. [15] On that day the Benjaminites mustered twenty-six thousand armed men from their towns, besides the inhabitants of Gibeah. [16] Of all this force, there were seven hundred picked men who were left-handed; every one could sling a stone at a hair, and not miss. [17] And the Israelites, apart from Benjamin, mustered four hundred thousand armed men, all of them warriors.

18 The Israelites proceeded to go up to Bethel, where they inquired of God, "Which of us shall go up first to battle against the Benjaminites?" And the LORD answered, "Judah shall go up first."

19 Then the Israelites got up in the morning, and encamped against Gibeah. [20] The Israelites went out to battle against Benjamin; and the Israelites drew up the battle line against them at Gibeah. [21] The Benjaminites came out of Gibeah, and struck down on that day twenty-two thousand of the Israelites. [23][a] The Israelites went up and wept before the LORD until the evening; and they inquired of

the LORD, "Shall we again draw near to battle against our kinsfolk the Benjaminites?" And the LORD said, "Go up against them." [22] The Israelites took courage, and again formed the battle line in the same place where they had formed it on the first day.

24 So the Israelites advanced against the Benjaminites the second day. [25] Benjamin moved out against them from Gibeah the second day, and struck down eighteen thousand of the Israelites, all of them armed men. [26] Then all the Israelites, the whole army, went back to Bethel and wept, sitting there before the LORD; they fasted that day until evening. Then they offered burnt offerings and sacrifices of well-being before the LORD. [27] And the Israelites inquired of the LORD (for the ark of the covenant of God was there in those days, [28] and Phinehas son of Eleazar, son of Aaron, ministered before it in those days), saying, "Shall we go out once more to battle against our kinsfolk the Benjaminites, or shall we desist?" The LORD answered, "Go up, for tomorrow I will give them into your hand."

29 So Israel stationed men in ambush around Gibeah. [30] Then the Israelites went up against the Benjaminites on the third day, and set themselves in array against Gibeah, as before. [31] When the Benjaminites went out against the army, they were drawn away from the city. As before they began to inflict casualties on the troops, along the main roads, one of which goes up to Bethel and the other to Gibeah, as well as in the open country, killing about thirty men of Israel. [32] The Benjaminites thought, "They are being routed before us, as previously." But the

y Gk: Heb lacks *we will go up*   z Compare Gk: Meaning of Heb uncertain   a Verses 22 and 23 are transposed

---

20.12–17: The Benjaminites refused to subject themselves to the authority of the tribal assembly.

20.18–28: The militia of the assembly experienced reversals in the field. *Bethel,* once known as Luz, had associations with both

Abraham (Gen 12.8; 13.3) and Jacob (Gen 28.10–22). It is near the modern Beitin, fourteen miles north of Jerusalem.

20.29–48: The Benjaminites were so thoroughly defeated that only a few survivors remained.

Israelites said, "Let us retreat and draw them away from the city toward the roads." 33 The main body of the Israelites drew back its battle line to Baal-tamar, while those Israelites who were in ambush rushed out of their place west*b* of Geba. 34 There came against Gibeah ten thousand picked men out of all Israel, and the battle was fierce. But the Benjaminites did not realize that disaster was close upon them.

35 The LORD defeated Benjamin before Israel; and the Israelites destroyed twenty-five thousand one hundred men of Benjamin that day, all of them armed.

36 Then the Benjaminites saw that they were defeated. *c*

The Israelites gave ground to Benjamin, because they trusted to the troops in ambush that they had stationed against Gibeah. 37 The troops in ambush rushed quickly upon Gibeah. Then they put the whole city to the sword. 38 Now the agreement between the main body of Israel and the men in ambush was that when they sent up a cloud of smoke out of the city 39 the main body of Israel should turn in battle. But Benjamin had begun to inflict casualties on the Israelites, killing about thirty of them; so they thought, "Surely they are defeated before us, as in the first battle." 40 But when the cloud, a column of smoke, began to rise out of the city, the Benjaminites looked behind them—and there was the whole city going up in smoke toward the sky! 41 Then the main body of Israel turned, and the Benjaminites were dismayed, for they saw that disaster was close upon them. 42 Therefore they turned away from the Israelites in the direction of the wilderness; but the battle overtook them, and those who came out of the city*d* were slaughtering them in between.*e* 43 Cutting down*f* the Benjaminites, they pursued them from Nohah*g* and trod them down as far as a place

east of Gibeah. 44 Eighteen thousand Benjaminites fell, all of them courageous fighters. 45 When they turned and fled toward the wilderness to the rock of Rimmon, five thousand of them were cut down on the main roads, and they were pursued as far as Gidom, and two thousand of them were slain. 46 So all who fell that day of Benjamin were twenty-five thousand arms-bearing men, all of them courageous fighters. 47 But six hundred turned and fled toward the wilderness to the rock of Rimmon, and remained at the rock of Rimmon for four months. 48 Meanwhile, the Israelites turned back against the Benjaminites, and put them to the sword—the city, the people, the animals, and all that remained. Also the remaining towns they set on fire.

21 Now the Israelites had sworn at Mizpah, "No one of us shall give his daughter in marriage to Benjamin." 2 And the people came to Bethel, and sat there until evening before God, and they lifted up their voices and wept bitterly. 3 They said, "O LORD, the God of Israel, why has it come to pass that today there should be one tribe lacking in Israel?" 4 On the next day, the people got up early, and built an altar there, and offered burnt offerings and sacrifices of well-being. 5 Then the Israelites said, "Which of all the tribes of Israel did not come up in the assembly to the LORD?" For a solemn oath had been taken concerning whoever did not come up to the LORD to Mizpah, saying, "That one shall be put to death." 6 But the Israelites had compassion for Benjamin their kin, and said, "One tribe is cut off from Israel this day. 7 What shall we do for wives for those

b Gk Vg: Heb *in the plain*   *c* This sentence is continued by verse 45.   *d* Compare Vg and some Gk Mss: Heb *cities*   *e* Compare Syr: Meaning of Heb uncertain   *f* Gk: Heb *Surrounding*   *g* Gk: Heb *pursued them at their resting place*

21.1–25: Repopulating Benjamin. The inept handling of the Gibeah incident led to the near extinction of the tribe of Benjamin. The other tribes resorted to murder, kidnap, and rape to repopulate the tribe. The final chapter of the book paints a pathetic picture of Israelite society. **1–14:** Revenge on Jabesh-gilead. **1:** The tribes swear not to intermarry with Benjamin, though that meant extinction for Benjamin. **5:** The oath here solves the

who are left, since we have sworn by the LORD that we will not give them any of our daughters as wives?"

8 Then they said, "Is there anyone from the tribes of Israel who did not come up to the LORD to Mizpah?" It turned out that no one from Jabesh-gilead had come to the camp, to the assembly. 9 For when the roll was called among the people, not one of the inhabitants of Jabesh-gilead was there. 10 So the congregation sent twelve thousand soldiers there and commanded them, "Go, put the inhabitants of Jabesh-gilead to the sword, including the women and the little ones. 11 This is what you shall do; every male and every woman that has lain with a male you shall devote to destruction." 12 And they found among the inhabitants of Jabesh-gilead four hundred young virgins who had never slept with a man and brought them to the camp at Shiloh, which is in the land of Canaan.

13 Then the whole congregation sent word to the Benjaminites who were at the rock of Rimmon, and proclaimed peace to them. 14 Benjamin returned at that time; and they gave them the women whom they had saved alive of the women of Jabesh-gilead; but they did not suffice for them.

15 The people had compassion on Benjamin because the LORD had made a breach in the tribes of Israel. 16 So the elders of the congregation said, "What shall we do for wives for those who are left, since there are no women left in Benjamin?" 17 And they said, "There must be heirs for the survivors of Benjamin, in order that a tribe may not be blotted out from Israel. 18 Yet we cannot give any of our daughters to them as wives." For the Israelites had sworn, "Cursed be anyone who gives a wife to Benjamin." 19 So they said, "Look, the yearly festival of the LORD is taking place at Shiloh, which is north of Bethel, on the east of the highway that goes up from Bethel to Shechem, and south of Lebonah." 20 And they instructed the Benjaminites, saying, "Go and lie in wait in the vineyards, 21 and watch; when the young women of Shiloh come out to dance in the dances, then come out of the vineyards and each of you carry off a wife for himself from the young women of Shiloh, and go to the land of Benjamin. 22 Then if their fathers or their brothers come to complain to us, we will say to them, 'Be generous and allow us to have them; because we did not capture in battle a wife for each man. But neither did you incur guilt by giving your daughters to them.' " 23 The Benjaminites did so; they took wives for each of them from the dancers whom they abducted. Then they went and returned to their territory, and rebuilt the towns, and lived in them. 24 So the Israelites departed from there at that time by tribes and families, and they went out from there to their own territories.

25 In those days there was no king in Israel; all the people did what was right in their own eyes.

---

problem resulting from the oath of v. 1. **9:** *Jabesh-gilead,* a town in the northern Transjordan region. It has been identified with Tell Abu Haraz and Tell el-Meqbereh. **12:** Four hundred virgins are spared and given to Benjamin as wives (v. 14). *Shiloh* is identified with the modern Seilun nine miles north of Bethel.

21.15–24: A second strategy for repopulating Benjamin allows them to abduct unprotected dancers among the women of Shiloh. **25:** A final verdict on the events of chs 19–21. There is no individual (except the concubine) or group whose conduct is not abominably immoral.

Ruth

In contrast to the story of the Israelite tribes in Joshua and Judges, the book of Ruth focuses on a single family. The story underscores the loyalty and fidelity that bind the family together. The book is highly entertaining because it skillfully moves winsome characters through an engaging plot. The mysterious ways of God are an important part of this story, which illustrates how God is at work in the lives of Naomi, Ruth, and Boaz. These individuals serve as models of faithful commitment to the God of Israel.

The Septuagint, the Greek version of the Hebrew Scriptures, is responsible for placing Ruth between Judges and 1 Samuel. Locating the book here disrupts the sequence of the Deuteronomistic History that begins with Joshua and ends with 2 Kings. Ruth shows no evidence of any Deuteronomistic editing. The editors of the Septuagint placed the book here because they considered it to reflect events during the time of the judges (1.1). More likely Ruth is a short story whose setting is the period before the monarchy.

It is difficult to be certain about a date for this book, yet ascertaining the dating is central to determining the reason for its composition. A date before the Exile leads to the view that the book intended to establish David's ancestry, to affirm the practice of levirate marriage (see Deut 25.5–10), or to commend the virtues of the book's protagonists to ancient Israel. A date after the Exile supports the conviction that the author wished to show that a non-Israelite could become a faithful worshiper of the LORD. This would counter the books of Ezra and Nehemiah, both of which consider intermarriage wrong (see Ezra chs 9 and 10; Neh 10.30). Like the books of Jonah and Isaiah chs 40–55, Ruth affirms that the concern of the LORD extends beyond the people of Israel to people of every nation.

1 In the days when the judges ruled, there was a famine in the land, and a certain man of Bethlehem in Judah went to live in the country of Moab, he and his wife and two sons. 2 The name of the man was Elimelech and the name of his wife Naomi, and the names of his two sons were Mahlon and Chilion; they were Ephrathites from Bethlehem in Judah. They went into the country of Moab and remained there. 3 But Elimelech, the husband of Naomi, died, and she was left with her two sons. 4 These took Moabite wives; the name of the one was Orpah and the name of the other Ruth. When they had lived there about ten years, 5 both Mahlon and Chilion also died, so that the woman was left without her two sons and her husband.

6 Then she started to return with her daughters-in-law from the country of Moab, for she had heard in the country of Moab that the LORD had considered his people and given them food. 7 So she set out from the place where she had been living, she and her two daughters-in-law, and they went on their way to go back to the land of Judah. 8 But Naomi said to her two daughters-in-law, "Go back each of you to your mother's house. May the LORD deal kindly with you, as you have dealt with the dead and with me. 9 The LORD grant that you may find security, each of you in the house of your husband." Then she kissed them, and they wept aloud. 10 They said to her, "No, we will return with you to your people." 11 But Naomi said, "Turn back, my daughters, why will you go with me? Do I still have sons in my womb that they may become your husbands? 12 Turn back, my daughters, go your way, for I am too old to have a husband.

Even if I thought there was hope for me, even if I should have a husband tonight and bear sons, 13 would you then wait until they were grown? Would you then refrain from marrying? No, my daughters, it has been far more bitter for me than for you, because the hand of the LORD has turned against me." 14 Then they wept aloud again. Orpah kissed her mother-in-law, but Ruth clung to her.

15 So she said, "See, your sister-in-law has gone back to her people and to her gods; return after your sister-in-law." 16 But Ruth said,

"Do not press me to leave you
 or to turn back from following
  you!
Where you go, I will go;
 where you lodge, I will lodge;
your people shall be my people,
 and your God my God.
17 Where you die, I will die—
 there will I be buried.
May the LORD do thus and so
  to me,
 and more as well,
 if even death parts me from you!"
18 When Naomi saw that she was determined to go with her, she said no more to her.

19 So the two of them went on until they came to Bethlehem. When they came to Bethlehem, the whole town was stirred because of them; and the women said, "Is this Naomi?" 20 She said to them,

"Call me no longer Naomi,*a*
 call me Mara,*b*
for the Almighty*c* has dealt
 bitterly with me.

---

a That is *Pleasant*   *b* That is *Bitter*
c Traditional rendering of Heb *Shaddai*

1.1–22: Ruth and Naomi. The two principal characters of the story are Naomi, an Israelite woman, and her Moabite daughter-in-law, Ruth. **1–5:** Naomi had to leave her homeland because of a famine. While in Moab she lost her husband and both sons to death. **1:** *Bethlehem* of Judah is five miles south of Jerusalem. It was the home of David's family (1 Sam 16.1–5). *Moab* occupied the territory east of the Dead Sea and south of the river Arnon. The Israelites believed that the Moabites were related to them through Lot (Gen 19.37), the nephew of Abraham. Conflicting claims to the same territory led to enmity between the two peoples (Deut 23.4).

1.6–18: Ruth chooses to stay with Naomi, who was returning to Judah.

1.19–22: Naomi's losses have embittered her.

21 I went away full,
  but the LORD has brought me
    back empty;
  why call me Naomi
    when the LORD has dealt harshly
      with[d] me,
  and the Almighty[e] has brought
    calamity upon me?"

22 So Naomi returned together with Ruth the Moabite, her daughter-in-law, who came back with her from the country of Moab. They came to Bethlehem at the beginning of the barley harvest.

2 Now Naomi had a kinsman on her husband's side, a prominent rich man, of the family of Elimelech, whose name was Boaz. 2 And Ruth the Moabite said to Naomi, "Let me go to the field and glean among the ears of grain, behind someone in whose sight I may find favor." She said to her, "Go, my daughter." 3 So she went. She came and gleaned in the field behind the reapers. As it happened, she came to the part of the field belonging to Boaz, who was of the family of Elimelech. 4 Just then Boaz came from Bethlehem. He said to the reapers, "The LORD be with you." They answered, "The LORD bless you." 5 Then Boaz said to his servant who was in charge of the reapers, "To whom does this young woman belong?" 6 The servant who was in charge of the reapers answered, "She is the Moabite who came back with Naomi from the country of Moab. 7 She said, 'Please, let me glean and gather among the sheaves behind the reapers.' So she came, and she has been on her feet from early this morning until now, without resting even for a moment."[f]

8 Then Boaz said to Ruth, "Now listen, my daughter, do not go to glean in another field or leave this one, but keep close to my young women. 9 Keep your eyes on the field that is being reaped, and follow behind them. I have ordered the young men not to bother you. If you get thirsty, go to the vessels and drink from what the young men have drawn." 10 Then she fell prostrate, with her face to the ground, and said to him, "Why have I found favor in your sight, that you should take notice of me, when I am a foreigner?" 11 But Boaz answered her, "All that you have done for your mother-in-law since the death of your husband has been fully told me, and how you left your father and mother and your native land and came to a people that you did not know before. 12 May the LORD reward you for your deeds, and may you have a full reward from the LORD, the God of Israel, under whose wings you have come for refuge!" 13 Then she said, "May I continue to find favor in your sight, my lord, for you have comforted me and spoken kindly to your servant, even though I am not one of your servants."

14 At mealtime Boaz said to her, "Come here, and eat some of this bread, and dip your morsel in the sour wine." So she sat beside the reapers, and he heaped up for her some parched grain. She ate until she was satisfied, and she had some left over. 15 When she got up to glean, Boaz instructed his young men, "Let her glean even among the standing sheaves, and do not reproach her. 16 You must also pull out some handfuls for her from the bundles, and leave them for her to glean, and do not rebuke her."

17 So she gleaned in the field until evening. Then she beat out what she had gleaned, and it was about an ephah of barley. 18 She picked it up and came into

d Or *has testified against*   e Traditional rendering of Heb *Shaddi*   f Compare Gk Vg: Meaning of Heb uncertain

---

2.1–23: Ruth and Boaz. A new character is introduced into the story. **1:** *Boaz* belonged to the family of Naomi's deceased husband. **2:** Israelite law made provision for the poor to provide for themselves by gleaning the fields after the reapers had finished their work. To ensure that there would be something for the poor, the law required farmers to leave a part of their harvest for gleaning by the poor (Lev 19.9–10; 23.22; Deut 24.19–22). **2.4–23:** Ruth attracts the attention of Boaz. **2.15–16:** Boaz gives instructions to ensure that Ruth would glean enough grain. **19–20:**

the town, and her mother-in-law saw how much she had gleaned. Then she took out and gave her what was left over after she herself had been satisfied. 19 Her mother-in-law said to her, "Where did you glean today? And where have you worked? Blessed be the man who took notice of you." So she told her mother-in-law with whom she had worked, and said, "The name of the man with whom I worked today is Boaz." 20 Then Naomi said to her daughter-in-law, "Blessed be he by the LORD, whose kindness has not forsaken the living or the dead!" Naomi also said to her, "The man is a relative of ours, one of our nearest kin."*g* 21 Then Ruth the Moabite said, "He even said to me, 'Stay close by my servants, until they have finished all my harvest.'" 22 Naomi said to Ruth, her daughter-in-law, "It is better, my daughter, that you go out with his young women, otherwise you might be bothered in another field." 23 So she stayed close to the young women of Boaz, gleaning until the end of the barley and wheat harvests; and she lived with her mother-in-law.

3 Naomi her mother-in-law said to her, "My daughter, I need to seek some security for you, so that it may be well with you. 2 Now here is our kinsman Boaz, with whose young women you have been working. See, he is winnowing barley tonight at the threshing floor. 3 Now wash and anoint yourself, and put on your best clothes and go down to the threshing floor; but do not make yourself known to the man until he has finished eating and drinking. 4 When he lies down, observe the place where he lies; then, go and uncover his feet and lie down; and he will tell you what to do." 5 She said to her, "All that you tell me I will do."

6 So she went down to the threshing floor and did just as her mother-in-law had instructed her. 7 When Boaz had eaten and drunk, and he was in a contented mood, he went to lie down at the end of the heap of grain. Then she came stealthily and uncovered his feet, and lay down. 8 At midnight the man was startled, and turned over, and there, lying at his feet, was a woman! 9 He said, "Who are you?" And she answered, "I am Ruth, your servant; spread your cloak over your servant, for you are next-of-kin."*g* 10 He said, "May you be blessed by the LORD, my daughter; this last instance of your loyalty is better than the first; you have not gone after young men, whether poor or rich. 11 And now, my daughter, do not be afraid, I will do for you all that you ask, for all the assembly of my people know that you are a worthy woman. 12 But now, though it is true that I am a near kinsman, there is another kinsman more closely related than I. 13 Remain this night, and in the morning, if he will act as next-of-kin*g* for you, good; let him do it. If he is not willing to act as next-of-kin*g* for you, then, as the LORD lives, I will act as next-of-kin for you. Lie down until the morning."

g Or one with the right to redeem

Naomi recognizes the kindness of Boaz toward Ruth as an act of divine providence.

3.1–18: Naomi's instructions. In the previous scene Ruth comes to the field of Boaz looking for food; she returns looking for a spouse. **1–5:** Naomi wants to help Ruth in finding a husband (see 1.9). She considers Boaz the obvious choice.

3.6–13: By following Naomi's instructions Ruth wins the favor of Boaz. **9:** To *spread* one's *cloak over* a woman means to take her for one's wife. There is a slight overtone of sexual intimacy in the passage, since *feet* (vv. 4, 8) in Hebrew can be a euphemism for "genitals." That the relationship between Ruth and Boaz is honorable according to the standards of their time, however, is shown by subsequent events. The *next-of-kin* had an obligation to protect the honor and property of the family (Lev 25.25; see Josh 20.3 n.). In addition, a surviving brother was to marry the wife of his deceased brother (Deut 25.5–6). The application of the latter law to Boaz is not obvious, since he was not Ruth's brother-in-law. **10–11:** Boaz asks God to bless Ruth because she was looking for a husband in accordance with Israelite custom. **12:** Boaz informs Ruth that there is another male relative who has a prior claim that must be respected.

14 So she lay at his feet until morning, but got up before one person could recognize another; for he said, "It must not be known that the woman came to the threshing floor." 15 Then he said, "Bring the cloak you are wearing and hold it out." So she held it, and he measured out six measures of barley, and put it on her back; then he went into the city. 16 She came to her mother-in-law, who said, "How did things go with you,*ʰ* my daughter?" Then she told her all that the man had done for her, 17 saying, "He gave me these six measures of barley, for he said, 'Do not go back to your mother-in-law empty-handed.'" 18 She replied, "Wait, my daughter, until you learn how the matter turns out, for the man will not rest, but will settle the matter today."

4 No sooner had Boaz gone up to the gate and sat down there than the next-of-kin,*ⁱ* of whom Boaz had spoken, came passing by. So Boaz said, "Come over, friend; sit down here." And he went over and sat down. 2 Then Boaz took ten men of the elders of the city, and said, "Sit down here"; so they sat down. 3 He then said to the next-of-kin,*ⁱ* "Naomi, who has come back from the country of Moab, is selling the parcel of land that belonged to our kinsman Elimelech. 4 So I thought I would tell you of it, and say: Buy it in the presence of those sitting here, and in the presence of the elders of my people. If you will redeem it, redeem it; but if you will not, tell me, so that I may know; for there is no one prior to you to redeem it, and I come after you." So he said, "I will redeem it." 5 Then Boaz said, "The day you acquire the field from the hand of Naomi, you are also acquiring Ruth*ʲ* the Moabite, the widow of the dead man, to maintain the dead man's name on his inheritance." 6 At this, the next-of-kin*ⁱ* said, "I cannot redeem it for myself without damaging my own inheritance. Take my right of redemption yourself, for I cannot redeem it."

7 Now this was the custom in former times in Israel concerning redeeming and exchanging: to confirm a transaction, the one took off a sandal and gave it to the other; this was the manner of attesting in Israel. 8 So when the next-of-kin*ⁱ* said to Boaz, "Acquire it for yourself," he took off his sandal. 9 Then Boaz said to the elders and all the people, "Today you are witnesses that I have acquired from the hand of Naomi all that belonged to Elimelech and all that belonged to Chilion and Mahlon. 10 I have also acquired Ruth the Moabite, the wife of Mahlon, to be my wife, to maintain the dead man's name on his inheritance, in order that the name of the dead may not be cut off from his kindred and from the gate of his native place; today you are witnesses." 11 Then all the people who were at the gate, along with the elders, said, "We are witnesses. May the LORD make the woman who is coming into your house like Rachel and Leah, who together built up the house of Israel. May you produce children in Ephrathah and bestow a name in Bethlehem; 12 and, through the children that the LORD will give you by this young woman, may your house be like the house of Perez, whom Tamar bore to Judah."

h Or *"Who are you,*   *i* Or *one with the right to redeem*   *j* OL Vg: Heb *from the hand of Naomi and from Ruth*

3.14–18: Naomi is certain that Ruth's status will be settled soon.

4.1–12: Boaz and Ruth will marry. 1–6: The unnamed rival to Boaz renounces the claims he had on Elimelech's property and household. Ruth is still young enough to have children, and any son she might bear would be considered a child and heir of her deceased husband. This would complicate matters concerning Elimelech's property for the next-of-kin's own heirs. **2:** *The elders of the city* were leading and influential men who interpreted traditional Israelite law and settled disputes.

4.7–12: Boaz asserts his claim to both Ruth and the property of Elimelech. **7:** The procedures described here do not follow the prescriptions of Deut 25.7–10. The precise way of handling such situations probably changed over time and differed regionally as well. **10:** *The gate,* see Josh 20.4 n. **12:** *Tamar's* marital situation was similar to that of Ruth (see Gen 38).

4.13–22: Ruth's child. 13–17: People rec-

13 So Boaz took Ruth and she became his wife. When they came together, the LORD made her conceive, and she bore a son. 14 Then the women said to Naomi, "Blessed be the LORD, who has not left you this day without next-of-kin;[k] and may his name be renowned in Israel! 15 He shall be to you a restorer of life and a nourisher of your old age; for your daughter-in-law who loves you, who is more to you than seven sons, has borne him." 16 Then Naomi took the child and laid him in her bosom, and became his nurse. 17 The women of the neighborhood gave him a name, saying, "A son has been born to Naomi." They named him Obed; he became the father of Jesse, the father of David.

18 Now these are the descendants of Perez: Perez became the father of Hezron, 19 Hezron of Ram, Ram of Amminadab, 20 Amminadab of Nahshon, Nahshon of Salmon, 21 Salmon of Boaz, Boaz of Obed, 22 Obed of Jesse, and Jesse of David.

k Or *one with the right to redeem*

---

ognize that the LORD was responsible for bringing Naomi's situation to a happy conclusion. 17: Ruth's son became the grandfather of David; the New Testament includes him among the ancestors of Jesus (Mt 1.4–6; Lk 3.32).

Chronological Tables
of Rulers

THE UNITED MONARCHY

The length of Saul's reign is not known; David and Solomon are each said to have ruled for forty years, which is often used as a general and somewhat indefinite number. Their reigns must have fallen between about 1020 and 922 (931) B.C., perhaps as follows:

Saul	1020–1000 B.C.
David	1000–961 (or 1000–965) B.C.
Solomon	961–922 (or 965–931) B.C.

THE DIVIDED MONARCHY, JUDAH AND ISRAEL

Problems of chronology of the kings of Israel and Judah permit no easy solution. The following tables are based upon two widely accepted systems of chronology, one developed by W. F. Albright, and the other by E. R. Thiele. The dates of Thiele's system (3rd ed.), which are enclosed within parentheses, are presented in simplified form apart from coregencies and simultaneous claims to the throne. The two columns show the chronological relationships of the kings of the two kingdoms.

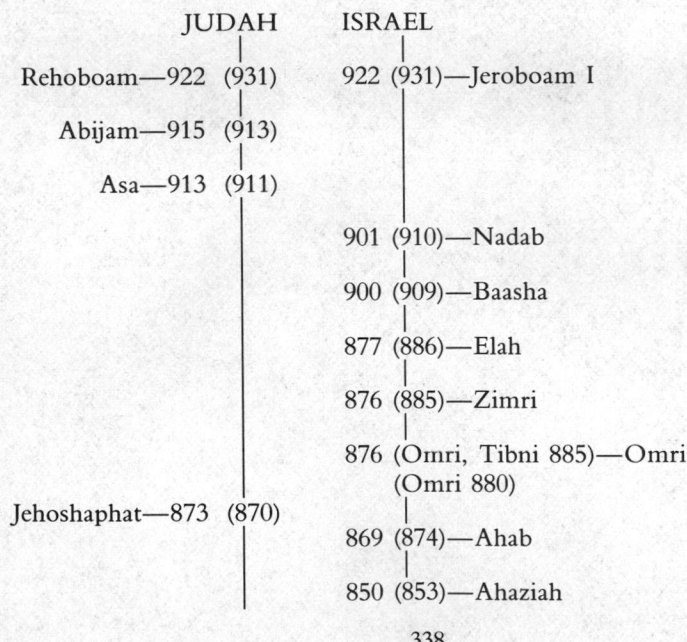

JUDAH	ISRAEL
Rehoboam—922 (931)	922 (931)—Jeroboam I
Abijam—915 (913)	
Asa—913 (911)	
	901 (910)—Nadab
	900 (909)—Baasha
	877 (886)—Elah
	876 (885)—Zimri
	876 (Omri, Tibni 885)—Omri (Omri 880)
Jehoshaphat—873 (870)	
	869 (874)—Ahab
	850 (853)—Ahaziah

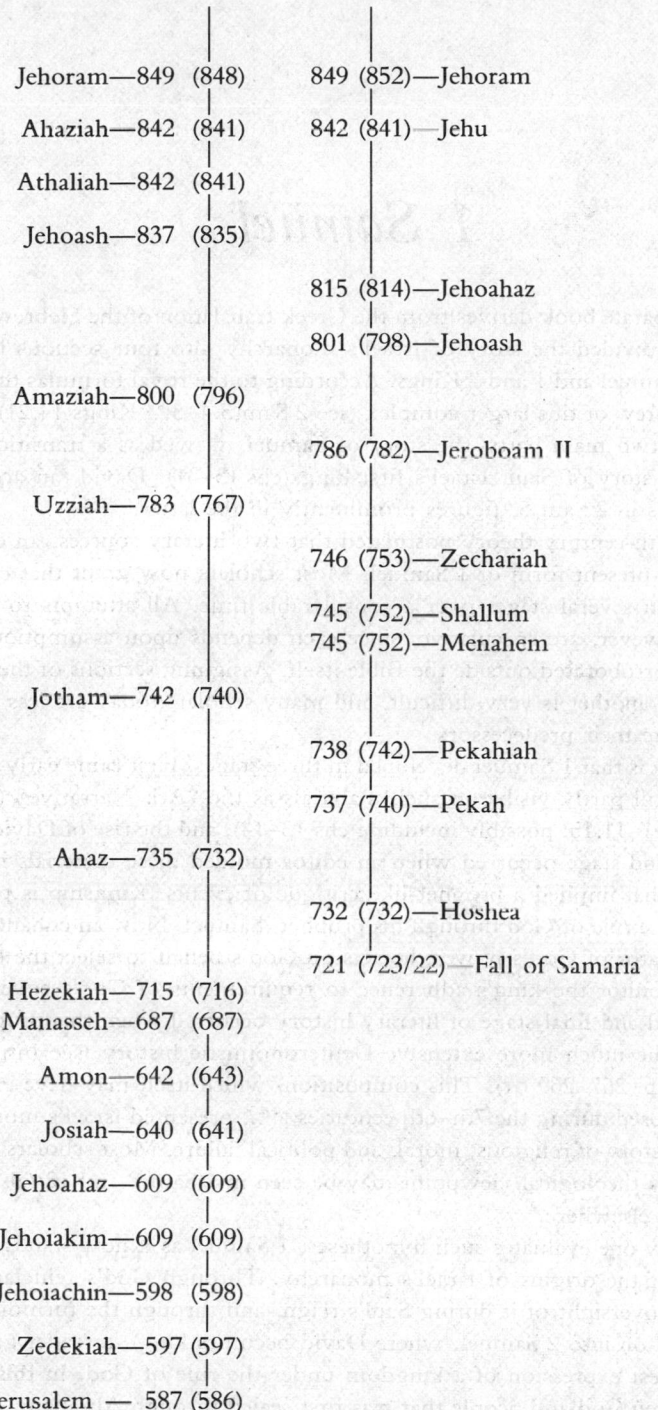

Jehoram—849 (848)   849 (852)—Jehoram

Ahaziah—842 (841)   842 (841)—Jehu

Athaliah—842 (841)

Jehoash—837 (835)

                                 815 (814)—Jehoahaz

                                 801 (798)—Jehoash

Amaziah—800 (796)

                                 786 (782)—Jeroboam II

Uzziah—783 (767)

                                 746 (753)—Zechariah

                                 745 (752)—Shallum
                                 745 (752)—Menahem

Jotham—742 (740)

                                 738 (742)—Pekahiah

                                 737 (740)—Pekah

Ahaz—735 (732)

                                 732 (732)—Hoshea

                                 721 (723/22)—Fall of Samaria

Hezekiah—715 (716)
Manasseh—687 (687)

Amon—642 (643)

Josiah—640 (641)

Jehoahaz—609 (609)

Jehoiakim—609 (609)

Jehoiachin—598 (598)

Zedekiah—597 (597)

Fall of Jerusalem — 587 (586)

1 Samuel

First Samuel as a separate book derives from the Greek translation of the Hebrew Scriptures (the Septuagint), which divided the story of Israel's monarchy into four sections now commonly known as 1 and 2 Samuel and 1 and 2 Kings. According to the royal formulas that mark literary divisions in the Hebrew of this larger complex (see 2 Sam 5.4–5; 1 Kings 14.21), the content of 1 Samuel falls into two main parts: the story of Samuel, viewed as a transition to monarchy (chs 1–12), and the story of Saul, Israel's first king (chs 13–31). David, an account of whose reign actually begins in 2 Sam 5, figures prominently in the latter.

A classic nineteenth-century theory postulated that two literary sources, an early and a later one, lay behind the present form of 1 Samuel. Most scholars now grant that composition and editing took place in several stages over a considerable time. All attempts to reconstruct this literary history, however, are speculative, since each depends upon assumptions and evidence that can rarely be corroborated outside the Bible itself. Assigning sections of the present biblical text to one stage or another is very difficult, and many scholars today are less confident about the results than were their predecessors.

One popular view is that 1 Samuel developed in three stages. First came early traditions about Samuel and Saul, still partly visible in such materials as the "Ark Narrative" (4.1–7.1), Saul's rise to the throne (9.1–11.15, possibly including chs 13–14), and the rise of David to prominence (chs 16–31). A second stage occurred when an editor molded these materials into a connected version of history that implied a prophet-like critique of events. Kingship is problematic, and must be set under the rule of God through his prophet, Samuel. Now an enhanced figure, Samuel is the sole mediator of God's power; he acts on God's behalf to select the king (8.1–10.27; 16.1–13) and to monitor the king's adherence to requirements of God's covenant with Israel (15.1–31). The third and final stage of literary history occurred when this prophetic story was incorporated into the much more extensive Deuteronomistic history (see Introduction to the Historical Books, pp. 267–269 ot). This composition, which itself may have existed in at least two versions composed during the 7th–6th centuries B.C., presented Israel's monarchy from beginning to end as a story of religious, moral, and political failure. Most scholars suggest that this historian's decidedly theological viewpoint may be seen in 1 Sam 8 and 12 (also 2 Sam 7), and in editorial touches elsewhere.

Regardless of how one evaluates such hypotheses, 1 Samuel as it now stands presents a fairly coherent account of the origins of Israel's monarchy. Through God's reluctant agreement to kingship, then full oversight of it during Saul's reign, and through the turmoil of David's rise to prominence (and on into 2 Samuel, where David becomes king), a reader grasps that David represents the highest expression of a kingdom under the rule of God. In this view, the covenantal bond between God and people that was first sealed through Abraham, Isaac, and Jacob, then refined through Moses, continues to define a nation whose roots should draw nourishment above all from religious ideals.

1 There was a certain man of Ramathaim, a Zuphite*ᵃ* from the hill country of Ephraim, whose name was Elkanah son of Jeroham son of Elihu son of Tohu son of Zuph, an Ephraimite. ²He had two wives; the name of the one was Hannah, and the name of the other Peninnah. Peninnah had children, but Hannah had no children.

3 Now this man used to go up year by year from his town to worship and to sacrifice to the LORD of hosts at Shiloh, where the two sons of Eli, Hophni and Phinehas, were priests of the LORD. ⁴On the day when Elkanah sacrificed, he would give portions to his wife Peninnah and to all her sons and daughters; ⁵but to Hannah he gave a double portion, *ᵇ* because he loved her, though the LORD had closed her womb. ⁶Her rival used to provoke her severely, to irritate her, because the LORD had closed her womb. ⁷So it went on year by year; as often as she went up to the house of the LORD, she used to provoke her. Therefore Hannah wept and would not eat. ⁸Her husband Elkanah said to her, "Hannah, why do you weep? Why do you not eat? Why is your heart sad? Am I not more to you than ten sons?"

9 After they had eaten and drunk at Shiloh, Hannah rose and presented herself before the LORD. *ᶜ* Now Eli the priest was sitting on the seat beside the doorpost of the temple of the LORD. ¹⁰She was deeply distressed and prayed to the LORD, and wept bitterly. ¹¹She made this vow: "O LORD of hosts, if only you will look on the misery of your servant, and remember me, and not forget your servant, but will give to your servant a male child, then I will set him before you as a nazirite*ᵈ* until the day of his death. He shall drink neither wine nor intoxicants, *ᵉ* and no razor shall touch his head."

12 As she continued praying before the LORD, Eli observed her mouth. ¹³Hannah was praying silently; only her lips moved, but her voice was not heard; therefore Eli thought she was drunk. ¹⁴So Eli said to her, "How long will you make a drunken spectacle of yourself? Put away your wine." ¹⁵But Hannah answered, "No, my lord, I am a woman deeply troubled; I have drunk neither wine nor strong drink, but I have been pouring out my soul before the LORD. ¹⁶Do not regard your servant as a worthless woman, for I have been speaking out of my great anxiety and vexation all this time." ¹⁷Then Eli answered, "Go in peace; the God of Israel grant the petition you have made to him." ¹⁸And she said, "Let your servant find favor in your sight." Then the woman went to her quarters, *ᶠ* ate and drank with her husband, *ᵍ* and her countenance was sad no longer. *ʰ*

a Compare Gk and 1 Chr 6.35–36: Heb *Ramathaim-zophim*   *b* Syr: Meaning of Heb uncertain   *c* Gk: Heb lacks *and presented herself before the LORD*   *d* That is *one separated* or *one consecrated*   *e* Cn Compare Gk Q Ms 1.22: MT *then I will give him to the LORD all the days of his life*   *f* Gk: Heb *went her way*   *g* Gk: Heb lacks *and drank with her husband*   *h* Gk: Meaning of Heb uncertain

1.1–28: The birth and consecration of Samuel. The story of a formerly barren woman who bears unusual offspring late in life as a special favor from God appears several times in the Bible. Besides Hannah, note Sarah (Gen 17.16–19), Rebekah (Gen 25.21–26), Rachel (Gen 29.31; 30.22–24), the mother of Samson (Judg 13.2–5), and Elizabeth (Lk 1.5–17). The unusual birth was thought to be symbolic of the importance of the person in later life. **1:** The Greek version begins thus: "There was a certain man of Ramah, a Zuphite of the hill country . . ."; comparison with v. 19 shows that Ramah was probably the familiar name of the place. **2:** *Two wives,* bigamy was not common in Old Testament times, but was allowed (Deut 21.15–17). **3:** The original meaning of the expression *LORD of hosts* is "LORD of armies" or "God of battles," referring to God's leadership of heavenly armies and those of his people in war. There was an important shrine at *Shiloh* in the days before the monarchy (Josh 18.1; Judg 21.19; Jer 7.12; Ps 78.60). We should read (with the Greek version), "Eli and his two sons"; it is evident from later verses that Eli had not retired.

1.9–18: Hannah went to the temple and

19 They rose early in the morning and worshiped before the Lord; then they went back to their house at Ramah. Elkanah knew his wife Hannah, and the Lord remembered her. 20 In due time Hannah conceived and bore a son. She named him Samuel, for she said, "I have asked him of the Lord."

21 The man Elkanah and all his household went up to offer to the Lord the yearly sacrifice, and to pay his vow. 22 But Hannah did not go up, for she said to her husband, "As soon as the child is weaned, I will bring him, that he may appear in the presence of the Lord, and remain there forever; I will offer him as a nazirite[i] for all time."[j] 23 Her husband Elkanah said to her, "Do what seems best to you, wait until you have weaned him; only—may the Lord establish his word."[k] So the woman remained and nursed her son, until she weaned him. 24 When she had weaned him, she took him up with her, along with a three-year-old bull,[l] an ephah of flour, and a skin of wine. She brought him to the house of the Lord at Shiloh; and the child was young. 25 Then they slaughtered the bull, and they brought the child to Eli. 26 And she said, "Oh, my lord! As you live, my lord, I am the woman who was standing here in your presence, praying to the Lord. 27 For this child I prayed; and the Lord has granted me the petition that I made to him. 28 Therefore I have lent him to the Lord; as long as he lives, he is given to the Lord."

She left him there for[m] the Lord.

2 Hannah prayed and said,
"My heart exults in the Lord;
   my strength is exalted in my God.[n]
My mouth derides my enemies,
   because I rejoice in my[o] victory.

2 "There is no Holy One like the Lord,
   no one besides you;
   there is no Rock like our God.
3 Talk no more so very proudly,
   let not arrogance come from your mouth;
for the Lord is a God of knowledge,
   and by him actions are weighed.
4 The bows of the mighty are broken,
   but the feeble gird on strength.

i That is one separated or *one consecrated*
j Cn Compare Q Ms: MT lacks *I will offer him as a nazirite for all time*   k MT: Q Ms Gk Compare Syr *that which goes out of your mouth*   l Q Ms Gk Syr: MT *three bulls*   m Gk (Compare Q Ms) and Gk at 2.11: MT *And he* (that is, Elkanah) *worshiped there before*   n Gk: Heb *the Lord*
o Q Ms: MT *your*

prayed for a son, promising to consecrate him to God's service if her prayer was granted.
1.19–28: The prayer was answered and the promise fulfilled. The Hebrew word for *asked* in v. 20 seems to be a play on the meaning of the name "Saul" rather than *Samuel,* which probably means "name of God," suggesting the boy's close affinity with God. There may be some confusion between Saul and Samuel here; but the word *asked* in Hebrew also means "borrowed," so perhaps this word should be connected with the word *lent* in v. 28; Hannah had begged or borrowed (same word in Hebrew) her son from God, so she *lent* him back to God, by whose grace he had been granted (see 2.20).
2.1–11: The song of Hannah. It was the custom of biblical editors to insert poems into prose books to increase artistic and religious appeal. The poems may be older or later than the contexts into which they are inserted. In this case the poem seems to be considerably later. It is really a psalm of national thanksgiving, and has a certain appropriateness at this point; it became the literary model for Mary's song of thanksgiving (the Magnificat) in the New Testament (Lk 1.46–55). **1:** *My strength,* literally "my horn," the figure of an animal tossing its head. The *I* of this psalm, as of many others, is the nation as well as the individual worshiper. The group and the individual are often identified in biblical poetry in a way strange to modern thinking. **2:** *Rock,* 2 Sam 22.2–3; Pss 18.2; 28.1; 62.2, 6; etc. **3:** The enemies of Israel are addressed. **4:** The reversal of fortune for the downtrodden and oppressed (also vv. 5, 8); compare Ps 113.7–9.

5 Those who were full have hired
    themselves out for bread,
  but those who were hungry are
    fat with spoil.
The barren has borne seven,
  but she who has many children
    is forlorn.
6 The LORD kills and brings to life;
  he brings down to Sheol and
    raises up.
7 The LORD makes poor and makes
    rich;
  he brings low, he also exalts.
8 He raises up the poor from the
    dust;
  he lifts the needy from the ash
    heap,
to make them sit with princes
  and inherit a seat of honor. *p*
For the pillars of the earth are the
    LORD's,
  and on them he has set the
    world.

9 "He will guard the feet of his
    faithful ones,
  but the wicked shall be cut off
    in darkness;
for not by might does one
    prevail.
10 The LORD! His adversaries shall be
    shattered;
  the Most High *q* will thunder in
    heaven.
The LORD will judge the ends of
    the earth;
  he will give strength to his
    king,

and exalt the power of his
    anointed."

11 Then Elkanah went home to Ramah, while the boy remained to minister to the LORD, in the presence of the priest Eli.

12 Now the sons of Eli were scoundrels; they had no regard for the LORD 13 or for the duties of the priests to the people. When anyone offered sacrifice, the priest's servant would come, while the meat was boiling, with a three-pronged fork in his hand, 14 and he would thrust it into the pan, or kettle, or caldron, or pot; all that the fork brought up the priest would take for himself. *r* This is what they did at Shiloh to all the Israelites who came there. 15 Moreover, before the fat was burned, the priest's servant would come and say to the one who was sacrificing, "Give meat for the priest to roast; for he will not accept boiled meat from you, but only raw." 16 And if the man said to him, "Let them burn the fat first, and then take whatever you wish," he would say, "No, you must give it now; if not, I will take it by force." 17 Thus the sin of the young men was very great in the sight of the LORD; for they treated the offerings of the LORD with contempt.

18 Samuel was ministering before the LORD, a boy wearing a linen ephod. 19 His mother used to make for him a

p Gk (Compare Q Ms) adds *He grants the vow of the one who vows, and blesses the years of the just* *q* Cn Heb *against him he* *r* Gk Syr Vg: Heb *with it*

2.5: *The barren has borne seven,* this line may have suggested the insertion of the poem at this place. *Seven* is merely symbolic of a sizable family; Hannah seems not to have had more than six children (see v. 21). **6–7:** God controls all of life. *Brings to life* may refer to birth rather than to resurrection from the dead; likewise the next line may refer to deep trouble or desperate injuries and recovery from them. *Sheol,* the place of the dead under the earth, like Hades among the Greeks (Isa 14.9–21); but the term is sometimes used of conditions near death (Pss 86.13; 88.3–7). **8:** *Pillars of the earth;* the earth was conceived of as a platform resting on great pillars. **9:** The

idea that the good would prosper and the wicked suffer ill fortune in this world was widespread in the post-exilic period (see Prov 3.9–10; 5.22–23); the books of Job and Ecclesiastes are a protest against this view of life. **10:** For God as the final judge of all, see Ps 98.9. *His king* and *his anointed* in this context could refer to the historical monarchy, but may also look forward to an ideal king of the future (see 10.1 n.). **2.12–26:** **The sons of Eli.** The evil conduct of the sons of Eli contrasts with the growing spirituality of Samuel (vv. 21, 26 and ch 3). **2.18:** The *linen ephod* was a light ceremonial

little robe and take it to him each year, when she went up with her husband to offer the yearly sacrifice. 20 Then Eli would bless Elkanah and his wife, and say, "May the LORD repay[s] you with children by this woman for the gift that she made to[t] the LORD"; and then they would return to their home.

21 And[u] the LORD took note of Hannah; she conceived and bore three sons and two daughters. And the boy Samuel grew up in the presence of the LORD.

22 Now Eli was very old. He heard all that his sons were doing to all Israel, and how they lay with the women who served at the entrance to the tent of meeting. 23 He said to them, "Why do you do such things? For I hear of your evil dealings from all these people. 24 No, my sons; it is not a good report that I hear the people of the LORD spreading abroad. 25 If one person sins against another, someone can intercede for the sinner with the LORD;[v] but if someone sins against the LORD, who can make intercession?" But they would not listen to the voice of their father; for it was the will of the LORD to kill them.

26 Now the boy Samuel continued to grow both in stature and in favor with the LORD and with the people.

27 A man of God came to Eli and said to him, "Thus the LORD has said, 'I revealed[w] myself to the family of your ancestor in Egypt when they were slaves[x] to the house of Pharaoh. 28 I chose him out of all the tribes of Israel to be my priest, to go up to my altar, to offer incense, to wear an ephod before me; and I gave to the family of your ancestor all my offerings by fire from the people of Israel. 29 Why then look with greedy eye[y] at my sacrifices and my offerings that I commanded, and honor your sons more than me by fattening yourselves on the choicest parts of every offering of my people Israel?' 30 Therefore the LORD the God of Israel declares: 'I promised that your family and the family of your ancestor should go in and out before me forever'; but now the LORD declares: 'Far be it from me; for those who honor me I will honor, and those who despise me shall be treated with contempt. 31 See, a time is coming when I will cut off your strength and the strength of your ancestor's family, so that no one in your family will live to old age. 32 Then in distress you will look with greedy eye[z] on all the prosperity that shall be bestowed upon Israel; and no one in your family shall ever live to old age. 33 The only one of you whom I shall not cut off from my altar shall be spared to weep out his[a] eyes and grieve his[b] heart; all the members of your household shall die by the sword.[c] 34 The fate of your two sons, Hophni and Phinehas, shall be the sign to you—both of them shall die on the same day. 35 I will

s Q Ms Gk: MT *give*   t Q Ms Gk: MT *for the petition that she asked of*   u Q Ms Gk: MT *When*   v Gk Compare Q Ms: MT *another, God will mediate for him*   w Gk Tg Syr: Heb *Did I reveal*   x Q Ms Gk: MT lacks *slaves*   y Q Ms Gk: MT *then kick*   z Q Ms Gk: MT *will kick*   a Q Ms Gk: MT *your*   b Q Ms Gk: Heb *your*   c Q Ms See Gk: MT *die like mortals*

garment covering only the front of the body (2 Sam 6.14); it is sometimes referred to as an apron. Later the priests wore a more elaborate ephod (Ex 28.5–14). **20:** See 1.19–28 n. **2.22:** The *tent of meeting* seems to be an error here, for the sanctuary at Shiloh was a building, not a tent. Perhaps there is some confusion with Ex 38.8. **25:** *It was the will of the LORD to kill them* is a theological way of saying that they were incorrigible, and therefore God had to punish them. **2.27–36: The condemnation of the house of Eli.** This is possibly an insertion designed to justify the exclusion of Abiathar and his descendants from the priesthood in favor of Zadok and his descendants (1 Kings 2.27, 35; see 3.11–14 n.). Eli was an ancestor of Abiathar, according to the writer of this passage, and the sins of the fathers (Eli and his sons) were visited upon the children (Deut 5.9). **27:** *Your ancestor* refers to Aaron. **28:** *An ephod,* see v. 18 n. **33:** *The only one of you* is Abiathar, the only one to escape at Nob (1 Sam 22.18–23; 1 Kings 2.26–27). **35:** The *faithful priest* is Zadok (2 Sam 8.17; 15.24; 1 Kings 1.8; 2.35). *My anointed* refers to the king, real or ideal (see v. 10 n.).

raise up for myself a faithful priest, who shall do according to what is in my heart and in my mind. I will build him a sure house, and he shall go in and out before my anointed one forever. 36 Everyone who is left in your family shall come to implore him for a piece of silver or a loaf of bread, and shall say, Please put me in one of the priest's places, that I may eat a morsel of bread.' "

3 Now the boy Samuel was ministering to the LORD under Eli. The word of the LORD was rare in those days; visions were not widespread.

2 At that time Eli, whose eyesight had begun to grow dim so that he could not see, was lying down in his room; 3 the lamp of God had not yet gone out, and Samuel was lying down in the temple of the LORD, where the ark of God was. 4 Then the LORD called, "Samuel! Samuel!"*d* and he said, "Here I am!" 5 and ran to Eli, and said, "Here I am, for you called me." But he said, "I did not call; lie down again." So he went and lay down. 6 The LORD called again, "Samuel!" Samuel got up and went to Eli, and said, "Here I am, for you called me." But he said, "I did not call, my son; lie down again." 7 Now Samuel did not yet know the LORD, and the word of the LORD had not yet been revealed to him. 8 The LORD called Samuel again, a third time. And he got up and went to Eli, and said, "Here I am, for you called me." Then Eli perceived that the LORD was calling the boy. 9 Therefore Eli said to Samuel, "Go, lie down; and if he calls you, you shall say, 'Speak, LORD, for your servant is listening.' " So Samuel went and lay down in his place.

10 Now the LORD came and stood there, calling as before, "Samuel! Samuel!" And Samuel said, "Speak, for your servant is listening." 11 Then the LORD said to Samuel, "See, I am about to do something in Israel that will make both ears of anyone who hears of it tingle. 12 On that day I will fulfill against Eli all that I have spoken concerning his house, from beginning to end. 13 For I have told him that I am about to punish his house forever, for the iniquity that he knew, because his sons were blaspheming God,*e* and he did not restrain them. 14 Therefore I swear to the house of Eli that the iniquity of Eli's house shall not be expiated by sacrifice or offering forever."

15 Samuel lay there until morning; then he opened the doors of the house of the LORD. Samuel was afraid to tell the vision to Eli. 16 But Eli called Samuel and said, "Samuel, my son." He said, "Here I am." 17 Eli said, "What was it that he told you? Do not hide it from me. May God do so to you and more also, if you hide anything from me of all that he told you." 18 So Samuel told him everything and hid nothing from him. Then he said, "It is the LORD; let him do what seems good to him."

19 As Samuel grew up, the LORD was with him and let none of his words fall to the ground. 20 And all Israel from Dan to Beer-sheba knew that Samuel was a trustworthy prophet of the LORD. 21 The LORD continued to appear at Shiloh, for the LORD revealed himself to Samuel at

d Q Ms Gk See 3.10: MT *the LORD called Samuel*
e Another reading is *for themselves*

3.1–4.1a: God's first revelation to Samuel. According to Jewish tradition Samuel was twelve years old at this time, the age of Jesus when he discoursed in the temple at Jerusalem (Lk 2.40–52). These beautiful stories of the birth and childhood of Samuel were in the mind of Luke as he began the writing of his Gospel (see 2.1–10 n.). **1:** *Word* and *vision* here are essentially the same thing—a revelation from God. **3:** *The lamp of God* burned all night (Ex 27.21); hence the

time was just before dawn. The *lamp* was near *the ark of God,* the portable shrine or chest symbolizing the presence and power of the Deity. Its early form was simple, but in later conception it became highly ornate (Ex 25.10–22; 37.1–9). The simple early form plays an important part in the story that follows in the next section. **3.11–14:** These verses are connected with 2.27–36 above. God chastens as well as offering salvation. **4.1a:** *Word* here probably means

4 Shiloh by the word of the LORD. ¹And the word of Samuel came to all Israel.

In those days the Philistines mustered for war against Israel,ᶠ and Israel went out to battle against them;ᵍ they encamped at Ebenezer, and the Philistines encamped at Aphek. ²The Philistines drew up in line against Israel, and when the battle was joined,ʰ Israel was defeated by the Philistines, who killed about four thousand men on the field of battle. ³When the troops came to the camp, the elders of Israel said, "Why has the LORD put us to rout today before the Philistines? Let us bring the ark of the covenant of the LORD here from Shiloh, so that he may come among us and save us from the power of our enemies." ⁴So the people sent to Shiloh, and brought from there the ark of the covenant of the LORD of hosts, who is enthroned on the cherubim. The two sons of Eli, Hophni and Phinehas, were there with the ark of the covenant of God.

5 When the ark of the covenant of the LORD came into the camp, all Israel gave a mighty shout, so that the earth resounded. ⁶When the Philistines heard the noise of the shouting, they said, "What does this great shouting in the camp of the Hebrews mean?" When they learned that the ark of the LORD had come to the camp, ⁷the Philistines were afraid; for they said, "Gods haveⁱ come into the camp." They also said, "Woe to us! For nothing like this has happened before. ⁸Woe to us! Who can deliver us from the power of these mighty gods? These are the gods who struck the Egyptians with every sort of plague in the wilderness. ⁹Take courage, and be men, O Philistines, in order not to become slaves to the Hebrews as they have been to you; be men and fight."

10 So the Philistines fought; Israel was defeated, and they fled, everyone to his home. There was a very great slaugh-

f Gk: Heb lacks *In those days the Philistines mustered for war against Israel* g Gk: Heb *against the Philistines* h Meaning of Heb uncertain i Or *A god has*

"reputation"; Samuel became known and trusted throughout all the land as a leader who spoke for God.

4.1b–7.2: The beginning of the war with the Philistines. The first sentence of this section of the Greek version tells us that *the Philistines* took the lead in the war by mustering their forces against the Israelites. For the origin of the Philistines, see Judg 13.1 n. At the time this story begins (about 1050 B.C.) the Philistines apparently had decided that they were strong enough to attack the Israelites and perhaps to take over the whole land. Thus begins for the Israelites a life-and-death struggle. Samuel is not mentioned in this section; one might almost say that *the ark,* rather than any person, is here the center of the author's attention.

4.1b–22: The defeat of Israel and capture of the ark. The strategy of the Philistines was to drive up the coast, then down the plain of Esdraelon to the Jordan river, thus cutting communication between the parts of Israel north and south of the plain. By the end of 1 Samuel this strategy has succeeded, though in the meantime there were attempts to penetrate Israelite territory farther south. **4.3–4:** *Ark of the covenant,* see 3.3 n. *The* LORD *of hosts,* see 1.3 n. *Who is enthroned on the*

cherubim, 2 Sam 6.2; 2 Kings 19.15; Pss 80.1; 99.1; etc. *Cherubim,* see Gen 3.24 n. In Phoenicia the king was sometimes represented as sitting on a throne supported by cherubim; the translators here suggest that the LORD God of Israel is similarly *enthroned.* Another interpretation is that God "dwells in the cherubim." There is no necessary implication that cherubim formed a part of this early, simple ark.

4.5–11: The loss of the ark. 6: The word *Hebrews* in the earlier parts of the Hebrew Bible is nearly always used by non-Israelites as a term of contempt (compare Gen 39.14; 43.32); the Israelites seldom use it of themselves. Only later could Jonah and Paul say proudly, "I am a Hebrew" (Jon 1.9; Phil 3.5). **7, 8:** The plural expression *gods* would seem to imply that the Philistines had no conception of how the Israelites worshiped only one God (but compare 5.7, 8, 10, 11). Since the plagues occurred in Egypt, the phrase "in the wilderness" may be a mistake, or possibly the author is making sport of the ignorance of the Philistines. **10:** Most ancient documents, biblical and non-biblical, are less exact with regard to numbers than are modern writings.

ter, for there fell of Israel thirty thousand foot soldiers. ¹¹ The ark of God was captured; and the two sons of Eli, Hophni and Phinehas, died.

12 A man of Benjamin ran from the battle line, and came to Shiloh the same day, with his clothes torn and with earth upon his head. ¹³ When he arrived, Eli was sitting upon his seat by the road watching, for his heart trembled for the ark of God. When the man came into the city and told the news, all the city cried out. ¹⁴ When Eli heard the sound of the outcry, he said, "What is this uproar?" Then the man came quickly and told Eli. ¹⁵ Now Eli was ninety-eight years old and his eyes were set, so that he could not see. ¹⁶ The man said to Eli, "I have just come from the battle; I fled from the battle today." He said, "How did it go, my son?" ¹⁷ The messenger replied, "Israel has fled before the Philistines, and there has also been a great slaughter among the troops; your two sons also, Hophni and Phinehas, are dead, and the ark of God has been captured." ¹⁸ When he mentioned the ark of God, Eli*ʲ* fell over backward from his seat by the side of the gate; and his neck was broken and he died, for he was an old man, and heavy. He had judged Israel forty years.

19 Now his daughter-in-law, the wife of Phinehas, was pregnant, about to give birth. When she heard the news that the ark of God was captured, and that her father-in-law and her husband were dead, she bowed and gave birth; for her labor pains overwhelmed her. ²⁰ As she was about to die, the women attending her said to her, "Do not be afraid, for you have borne a son." But she did not answer or give heed. ²¹ She named the child Ichabod, meaning, "The glory has departed from Israel," because the ark of God had been captured and because of her father-in-law and her husband. ²² She said, "The glory has departed from Israel, for the ark of God has been captured."

5 When the Philistines captured the ark of God, they brought it from Ebenezer to Ashdod; ² then the Philistines took the ark of God and brought it into the house of Dagon and placed it beside Dagon. ³ When the people of Ashdod rose early the next day, there was Dagon, fallen on his face to the ground before the ark of the LORD. So they took Dagon and put him back in his place. ⁴ But when they rose early on the next morning, Dagon had fallen on his face to the ground before the ark of the LORD, and the head of Dagon and both his hands were lying cut off upon the threshold; only the trunk of*ᵏ* Dagon was left to him. ⁵ This is why the priests of Dagon and all who enter the house of Dagon do

j Heb *he* *k* Heb lacks *the trunk of*

4.12–18: **The death of Eli. 13:** *By the road,* the reading of the Greek version, "by the gate" (presumably the door of the temple), is preferable (1.9). **18:** *He had judged Israel forty years,* added by the Deuteronomic editor on the assumption that practically every prominent leader in those days was a judge of all Israel (Judg 16.31).

4.19–22: **The birth of Ichabod.** The name *Ichabod* means "no glory" or "alas for the glory!" Many names in the Bible describe the circumstances surrounding the bearer's birth or give an omen; compare the naming of Cain (Gen 4.1), Isaac ("laugh," Gen 21.3, 6), the children of Jacob (Gen 30.6–24), Moses ("draw out," Ex 2.10), and Jesus ("save," Mt 1.21).

5.1–12: **The ark troubles the Philistines.** The early Israelites must have taken much delight in this story of the power of the symbol of their God. **1:** *Ashdod* was one of the five principal Philistine towns (pentapolis), along with Ashkelon, Ekron, Gath, and Gaza (Judg 14.19 n.). **2:** *House* here means temple. *Dagon,* the principal deity of the Philistines (Judg 16.23), appears to have been borrowed, at least in name, from some of the surrounding Semitic peoples. The root meaning of the word was formerly thought to be "fish" (the Philistines being near the sea), but is now considered to be "grain." Hence Dagon is reckoned among the "fertility" deities that have to do with agricultural productivity. **5:** Leaping over *the threshold* was a common practice in primitive religions (Zeph 1.9), the doorsill being regarded with superstitious awe (compare the modern custom of carrying a bride over the threshold). The origins of the custom are very ancient; hence the explanation given here can hardly be correct.

not step on the threshold of Dagon in Ashdod to this day.

6 The hand of the LORD was heavy upon the people of Ashdod, and he terrified and struck them with tumors, both in Ashdod and in its territory. 7 And when the inhabitants of Ashdod saw how things were, they said, "The ark of the God of Israel must not remain with us; for his hand is heavy on us and on our god Dagon." 8 So they sent and gathered together all the lords of the Philistines, and said, "What shall we do with the ark of the God of Israel?" The inhabitants of Gath replied, "Let the ark of God be moved on to us."*l* So they moved the ark of the God of Israel to Gath.*m* 9 But after they had brought it to Gath,*n* the hand of the LORD was against the city, causing a very great panic; he struck the inhabitants of the city, both young and old, so that tumors broke out on them. 10 So they sent the ark of the God of Israel*o* to Ekron. But when the ark of God came to Ekron, the people of Ekron cried out, "Why*p* have they brought around to us*q* the ark of the God of Israel to kill us*q* and our*r* people?" 11 They sent therefore and gathered together all the lords of the Philistines, and said, "Send away the ark of the God of Israel, and let it return to its own place, that it may not kill us and our people." For there was a deathly panic*s* throughout the whole city. The hand of God was very heavy there; 12 those who did not die were stricken with tumors, and the cry of the city went up to heaven.

6 The ark of the LORD was in the country of the Philistines seven

months. 2 Then the Philistines called for the priests and the diviners and said, "What shall we do with the ark of the LORD? Tell us what we should send with it to its place." 3 They said, "If you send away the ark of the God of Israel, do not send it empty, but by all means return him a guilt offering. Then you will be healed and will be ransomed;*t* will not his hand then turn from you?" 4 And they said, "What is the guilt offering that we shall return to him?" They answered, "Five gold tumors and five gold mice, according to the number of the lords of the Philistines; for the same plague was upon all of you and upon your lords. 5 So you must make images of your tumors and images of your mice that ravage the land, and give glory to the God of Israel; perhaps he will lighten his hand on you and your gods and your land. 6 Why should you harden your hearts as the Egyptians and Pharaoh hardened their hearts? After he had made fools of them, did they not let the people go, and they departed? 7 Now then, get ready a new cart and two milch cows that have never borne a yoke, and yoke the cows to the cart, but take their calves home, away from them. 8 Take the ark of the LORD and place it on the cart, and put in a box at its side the figures of gold, which you

l Gk Compare Q Ms: MT *They answered, "Let the ark of the God of Israel be brought around to Gath."*   *m* Gk: Heb lacks *to Gath*   *n* Q Ms: MT lacks *to Gath*   *o* Q Ms Gk: MT lacks *of Israel*   *p* Q Ms Gk: MT lacks *Why*   *q* Heb *me*   *r* Heb *my*   *s* Q Ms reads *a panic from the LORD*   *t* Q Ms Gk: MT *and it will be known to you*

5.6, 9, 12: These *tumors* are generally considered to have been the swellings of the bubonic plague. **7, 8, 10, 11**: The Philistines appear to understand that Israel has but one God.

6.1–7.2: The voluntary return of the ark. The Philistines, attributing the plague to the presence of the ark, make arrangements to return it to its former owners. **3**: The Philistines wish to make a guilt offering, to appease the God of Israel, who (so they thought) was angry. **4, 5**: *Five gold tumors and five gold mice* were prepared. We learn from the Greek ver-

sion (5.6 and 6.1) that the plague was accompanied by swarms of mice. Bubonic plague is usually spread by a flea carried by rats. The Philistines may have been correct in connecting the disease with these "mice" (probably rats). By sending away images of the troublesome objects, they hoped, by a process of sympathetic magic, to be rid of the troubles themselves; and they hoped that *the God of Israel* would be pleased by their generosity in making the images of gold. **6**: To *harden your hearts* means to "make your minds stubborn" (Ex 8.19, 32).

are returning to him as a guilt offering. Then send it off, and let it go its way. 9And watch; if it goes up on the way to its own land, to Beth-shemesh, then it is he who has done us this great harm; but if not, then we shall know that it is not his hand that struck us; it happened to us by chance."

10 The men did so; they took two milch cows and yoked them to the cart, and shut up their calves at home. 11They put the ark of the LORD on the cart, and the box with the gold mice and the images of their tumors. 12The cows went straight in the direction of Beth-shemesh along one highway, lowing as they went; they turned neither to the right nor to the left, and the lords of the Philistines went after them as far as the border of Beth-shemesh.

13 Now the people of Beth-shemesh were reaping their wheat harvest in the valley. When they looked up and saw the ark, they went with rejoicing to meet it. *u* 14The cart came into the field of Joshua of Beth-shemesh, and stopped there. A large stone was there; so they split up the wood of the cart and offered the cows as a burnt offering to the LORD. 15The Levites took down the ark of the LORD and the box that was beside it, in which were the gold objects, and set them upon the large stone. Then the people of Beth-shemesh offered burnt offerings and presented sacrifices on that day to the LORD. 16When the five lords of the Philistines saw it, they returned that day to Ekron.

17 These are the gold tumors, which the Philistines returned as a guilt offering to the LORD: one for Ashdod, one for Gaza, one for Ashkelon, one for Gath,

one for Ekron; 18also the gold mice, according to the number of all the cities of the Philistines belonging to the five lords, both fortified cities and unwalled villages. The great stone, beside which they set down the ark of the LORD, is a witness to this day in the field of Joshua of Beth-shemesh.

19 The descendants of Jeconiah did not rejoice with the people of Beth-shemesh when they greeted*v* the ark of the LORD; and he killed seventy men of them. *w* The people mourned because the LORD had made a great slaughter among the people. 20Then the people of Beth-shemesh said, "Who is able to stand before the LORD, this holy God? To whom shall he go so that we may be rid of him?" 21So they sent messengers to the inhabitants of Kiriath-jearim, saying, "The Philistines have returned the ark of the LORD. Come down and take it up to you." 1And the people of Kiriath-jearim came and took up the ark of the LORD, and brought it to the house of Abinadab on the hill. They consecrated his son, Eleazar, to have charge of the ark of the LORD.

2 From the day that the ark was lodged at Kiriath-jearim, a long time passed, some twenty years, and all the house of Israel lamented*x* after the LORD.

3 Then Samuel said to all the house of Israel, "If you are returning to the LORD with all your heart, then put away the foreign gods and the Astartes from among you. Direct your heart to the

u Gk: Heb *rejoiced to see it*   *v* Gk: Heb *And he killed some of the people of Beth-shemesh, because they looked into*   *w* Heb *killed seventy men, fifty thousand men*   *x* Meaning of Heb uncertain

6.9: *Beth-shemesh* was probably the nearest important Israelite town to Ekron, where the ark was prior to its return. **15**: This verse was inserted by a priestly editor to make the procedure conform to later requirements. The Levites did not begin to function as priestly assistants until later times. **17**: See 5.1 n.

6.19: The ungrammatical addition of the phrase "fifty thousand men" (see note *w*) shows how easily exaggerations could occur (see 4.10 n.). **21**: The ark was taken to *Kiriath-*

jearim probably because Shiloh had been destroyed in the meantime (Jer 7.12–14; 26.6–9).

7.2: The expression *lamented after the LORD* seems to mean, in the light of what follows, "implored the LORD for help."

7.3–17: **Samuel as judge of all Israel.** *Samuel* immediately comes to the fore, as the focus returns to his role in the transitional days leading to Israel's first king. **3–4**: Probably added by the Deuteronomic editor, who

LORD, and serve him only, and he will deliver you out of the hand of the Philistines." 4 So Israel put away the Baals and the Astartes, and they served the LORD only.

5 Then Samuel said, "Gather all Israel at Mizpah, and I will pray to the LORD for you." 6 So they gathered at Mizpah, and drew water and poured it out before the LORD. They fasted that day, and said, "We have sinned against the LORD." And Samuel judged the people of Israel at Mizpah.

7 When the Philistines heard that the people of Israel had gathered at Mizpah, the lords of the Philistines went up against Israel. And when the people of Israel heard of it they were afraid of the Philistines. 8 The people of Israel said to Samuel, "Do not cease to cry out to the LORD our God for us, and pray that he may save us from the hand of the Philistines." 9 So Samuel took a sucking lamb and offered it as a whole burnt offering to the LORD; Samuel cried out to the LORD for Israel, and the LORD answered him. 10 As Samuel was offering up the burnt offering, the Philistines drew near to attack Israel; but the LORD thundered with a mighty voice that day against the Philistines and threw them into confusion; and they were routed before Israel. 11 And the men of Israel went out of Mizpah and pursued the Philistines, and struck them down as far as beyond Bethcar.

12 Then Samuel took a stone and set it up between Mizpah and Jeshanah,*y* and named it Ebenezer;*z* for he said, "Thus far the LORD has helped us." 13 So the Philistines were subdued and did not again enter the territory of Israel; the hand of the LORD was against the Philistines all the days of Samuel. 14 The towns that the Philistines had taken from Israel were restored to Israel, from Ekron to Gath; and Israel recovered their territory from the hand of the Philistines. There was peace also between Israel and the Amorites.

15 Samuel judged Israel all the days of his life. 16 He went on a circuit year by year to Bethel, Gilgal, and Mizpah; and he judged Israel in all these places. 17 Then he would come back to Ramah, for his home was there; he administered justice there to Israel, and built there an altar to the LORD.

8 When Samuel became old, he made his sons judges over Israel. 2 The name of his firstborn son was Joel, and the name of his second, Abijah; they were judges in Beer-sheba. 3 Yet his sons did not follow in his ways, but turned aside after gain; they took bribes and perverted justice.

4 Then all the elders of Israel gathered together and came to Samuel at Ramah, 5 and said to him, "You are old and your sons do not follow in your ways; appoint

y Gk Syr: Heb *Shen*   *z* That is *Stone of Help*

---

felt that the worship of false gods was always the chief sin of the people (Judg 2.11–15; 3.7; 10.6; 13.1; etc.). *The Baals and the Astartes* (plural of Astarte, goddess of fertility) were the principal deities of the Canaanites, often said to have been worshiped by the Israelites when they departed from what the Deuteronomic writer considered to be the true faith. **7.5**: Continuation of v. 2. **6**: Libations of *water* were unusual (wine was used generally), but to people with a desert background water would be sufficiently precious. **9**: *A whole burnt offering,* one that is entirely consumed (Ex 29.18; Lev 8.21; Deut 13.16; 33.10). Most burnt offerings were of this type; the use of the word *whole* emphasizes the importance of the occasion. **10**: It is char-

acteristic of the Bible (and similar literary material) that battles are often won or lost by miraculous divine intervention. **13–14**: Most of the other potential enemies of Israel besides the Philistines could be classified as *Amorites.* The final sentence of v. 14 makes Samuel's success as a temporal ruler complete. But he was not really a "judge" or military hero like those in the Book of Judges; he was more like a circuit judge in the ordinary sense of the word (compare vv. 16–17). **8.1–22**: **The people request a king.** This section views the rise of kingship as especially problematic and associates it with earlier failings of the people. Samuel dislikes the request, but after seeking God's will, he finds that he must yield, even though God feels

for us, then, a king to govern us, like other nations." 6 But the thing displeased Samuel when they said, "Give us a king to govern us." Samuel prayed to the Lord, 7 and the Lord said to Samuel, "Listen to the voice of the people in all that they say to you; for they have not rejected you, but they have rejected me from being king over them. 8 Just as they have done to me, *a* from the day I brought them up out of Egypt to this day, forsaking me and serving other gods, so also they are doing to you. 9 Now then, listen to their voice; only— you shall solemnly warn them, and show them the ways of the king who shall reign over them."

10 So Samuel reported all the words of the Lord to the people who were asking him for a king. 11 He said, "These will be the ways of the king who will reign over you: he will take your sons and appoint them to his chariots and to be his horsemen, and to run before his chariots; 12 and he will appoint for himself commanders of thousands and commanders of fifties, and some to plow his ground and to reap his harvest, and to make his implements of war and the equipment of his chariots. 13 He will take your daughters to be perfumers and cooks and bakers. 14 He will take the best of your fields and vineyards and olive orchards and give them to his courtiers. 15 He will take one-tenth of your grain and of your vineyards and give it to his officers and his courtiers. 16 He will take your male and female slaves, and the best of your cattle *b* and donkeys, and put them to his work.

17 He will take one-tenth of your flocks, and you shall be his slaves. 18 And in that day you will cry out because of your king, whom you have chosen for yourselves; but the Lord will not answer you in that day."

19 But the people refused to listen to the voice of Samuel; they said, "No! but we are determined to have a king over us, 20 so that we also may be like other nations, and that our king may govern us and go out before us and fight our battles." 21 When Samuel had heard all the words of the people, he repeated them in the ears of the Lord. 22 The Lord said to Samuel, "Listen to their voice and set a king over them." Samuel then said to the people of Israel, "Each of you return home."

9 There was a man of Benjamin whose name was Kish son of Abiel son of Zeror son of Becorath son of Aphiah, a Benjaminite, a man of wealth. 2 He had a son whose name was Saul, a handsome young man. There was not a man among the people of Israel more handsome than he; he stood head and shoulders above everyone else.

3 Now the donkeys of Kish, Saul's father, had strayed. So Kish said to his son Saul, "Take one of the boys with you; go and look for the donkeys." 4 He passed through the hill country of Ephraim and passed through the land of Shalishah, but they did not find them. And they passed through the land of Shaalim, but they were not there. Then he passed

a Gk: Heb lacks *to me*   *b* Gk: Heb *young men*

---

offended. At God's behest Samuel delivers a severe lecture on the evils of kingship before yielding (vv. 9–18; compare Deut 17.14–20). The evils described here and in the Deuteronomy passage seem to be mainly those of the reign of Solomon, and it is probable that the resentment against the monarchy arose at this time and never ceased, becoming a part of the thought of many of the prophets and of the Deuteronomic writers.

9.1–10.16: The secret choice of Saul. Samuel appears here not as the judge or ruler of all Israel (7.5), but as a more modest figure, a local *man of God* (9.6), a highly respected *seer*

(9.11), that is, a clairvoyant (9.20; 10.2), and possibly a circuit judge. Some of this material may belong to the earliest layers of the book. Most noteworthy in this section is the idea that the kingship is a splendid thing, a blessing from God voluntarily bestowed, and not a concession to the improper desires of the people (9.16; 10.1). Samuel himself seems immensely pleased with the idea of the kingship and his part in creating it (9.19–24), in contrast to his attitude in ch 8. It will soon appear also that this material is friendly to Saul, in contrast to ch 13. **9.2:** *Young man* here means "man in the prime of life," for Saul had

through the land of Benjamin, but they did not find them.

5 When they came to the land of Zuph, Saul said to the boy who was with him, "Let us turn back, or my father will stop worrying about the donkeys and worry about us." ⁶But he said to him, "There is a man of God in this town; he is a man held in honor. Whatever he says always comes true. Let us go there now; perhaps he will tell us about the journey on which we have set out." ⁷Then Saul replied to the boy, "But if we go, what can we bring the man? For the bread in our sacks is gone, and there is no present to bring to the man of God. What have we?" ⁸The boy answered Saul again, "Here, I have with me a quarter shekel of silver; I will give it to the man of God, to tell us our way." ⁹(Formerly in Israel, anyone who went to inquire of God would say, "Come, let us go to the seer"; for the one who is now called a prophet was formerly called a seer.) ¹⁰Saul said to the boy, "Good; come, let us go." So they went to the town where the man of God was.

11 As they went up the hill to the town, they met some girls coming out to draw water, and said to them, "Is the seer here?" ¹²They answered, "Yes, there he is just ahead of you. Hurry; he has come just now to the town, because the people have a sacrifice today at the shrine. ¹³As soon as you enter the town, you will find him, before he goes up to the shrine to eat. For the people will not eat until he comes, since he must bless the sacrifice; afterward those eat who are invited. Now go up, for you will meet him immediately." ¹⁴So they went up to the town. As they were entering the town,

they saw Samuel coming out toward them on his way up to the shrine.

15 Now the day before Saul came, the LORD had revealed to Samuel: ¹⁶"Tomorrow about this time I will send to you a man from the land of Benjamin, and you shall anoint him to be ruler over my people Israel. He shall save my people from the hand of the Philistines; for I have seen the suffering of[c] my people, because their outcry has come to me." ¹⁷When Samuel saw Saul, the LORD told him, "Here is the man of whom I spoke to you. He it is who shall rule over my people." ¹⁸Then Saul approached Samuel inside the gate, and said, "Tell me, please, where is the house of the seer?" ¹⁹Samuel answered Saul, "I am the seer; go up before me to the shrine, for today you shall eat with me, and in the morning I will let you go and will tell you all that is on your mind. ²⁰As for your donkeys that were lost three days ago, give no further thought to them, for they have been found. And on whom is all Israel's desire fixed, if not on you and on all your ancestral house?" ²¹Saul answered, "I am only a Benjaminite, from the least of the tribes of Israel, and my family is the humblest of all the families of the tribe of Benjamin. Why then have you spoken to me in this way?"

22 Then Samuel took Saul and his servant-boy and brought them into the hall, and gave them a place at the head of those who had been invited, of whom there were about thirty. ²³And Samuel said to the cook, "Bring the portion I gave you, the one I asked you to put aside." ²⁴The cook took up the thigh and

c Gk: Heb lacks *the suffering of*

---

grown children. **6:** *This town* may be Ramah (7.17). **7:** Saul appears not to have heard of Samuel; contrast 4.1a and 7.15. **8:** There was no coinage in those days; *a quarter shekel of silver* was merely a bit of silver of a certain weight (about one-eighth oz.). **9:** A note by an editor, inserted after the word *seer* had gone out of use.

9.11–14: The *town* was apparently a small walled town, lying on the side of a hill, with only one gate; the spring was below, *the shrine* above. Samuel presided over the sacrificial meal as a sort of priest. Later, religious practices continued or developed at the high place shrines, and they were ordered to be destroyed (Deut 12.2–3; 2 Kings 23.8–9). **19:** Perhaps Saul had been brooding over the oppression by the Philistines. **25:** Roofs in that part of the world were flat, with protecting parapets, as they still are. Hence they were

what went with it[d] and set them before Saul. Samuel said, "See, what was kept is set before you. Eat; for it is set[e] before you at the appointed time, so that you might eat with the guests."[f]

So Saul ate with Samuel that day. 25 When they came down from the shrine into the town, a bed was spread for Saul[g] on the roof, and he lay down to sleep.[h] 26 Then at the break of dawn[i] Samuel called to Saul upon the roof, "Get up, so that I may send you on your way." Saul got up, and both he and Samuel went out into the street.

27 As they were going down to the outskirts of the town, Samuel said to Saul, "Tell the boy to go on before us, and when he has passed on, stop here yourself for a while, that I may make known to you the word of God."

10 1 Samuel took a vial of oil and poured it on his head, and kissed him; he said, "The Lord has anointed you ruler over his people Israel. You shall reign over the people of the Lord and you will save them from the hand of their enemies all around. Now this shall be the sign to you that the Lord has anointed you ruler[j] over his heritage: 2 When you depart from me today you will meet two men by Rachel's tomb in the territory of Benjamin at Zelzah; they will say to you, 'The donkeys that you went to seek are found, and now your father has stopped worrying about them and is worrying about you, saying: What shall I do about my son?' 3 Then you shall go on from there further and come to the oak of Tabor; three men going up to God

at Bethel will meet you there, one carrying three kids, another carrying three loaves of bread, and another carrying a skin of wine. 4 They will greet you and give you two loaves of bread, which you shall accept from them. 5 After that you shall come to Gibeath-elohim,[k] at the place where the Philistine garrison is; there, as you come to the town, you will meet a band of prophets coming down from the shrine with harp, tambourine, flute, and lyre playing in front of them; they will be in a prophetic frenzy. 6 Then the spirit of the Lord will possess you, and you will be in a prophetic frenzy along with them and be turned into a different person. 7 Now when these signs meet you, do whatever you see fit to do, for God is with you. 8 And you shall go down to Gilgal ahead of me; then I will come down to you to present burnt offerings and offer sacrifices of well-being. Seven days you shall wait, until I come to you and show you what you shall do."

9 As he turned away to leave Samuel, God gave him another heart; and all these signs were fulfilled that day. 10 When they were going from there[l] to Gibeah,[m] a band of prophets met him; and the spirit of God possessed him, and he fell into a prophetic frenzy along with them.

d Meaning of Heb uncertain   e Q Ms Gk: MT *it was kept*   f Cn: Heb *it was kept for you, saying, I have invited the people*   g Gk: Heb *and he spoke with Saul*   h Gk: Heb lacks *and he lay down to sleep*   i Gk: Heb *and they arose early and at break of dawn*   j Gk: Heb lacks *over his people Israel. You shall . . . anointed you ruler*   k Or *the Hill of God*   l Gk: Heb *they came there*   m Or *the hill*

---

and are well adapted for sleeping during the summer months, when no rain falls.

10.1: The unguent used in anointing kings was olive oil. Though priests (Ex 29.7) and prophets (1 Kings 19.16) were sometimes anointed, the ceremony was more relevant to the kingship, so that the king came to be called "the Lord's anointed" (16.6; 24.6), or simply "the anointed one." This title was applied to an ideal future king in the form "Messiah" in Hebrew, "Christos" in Greek (see Mt 1.16). **2:** This site of *Rachel's tomb* is in Benjamin, north of Jerusalem (Jer 31.15). Another tradition locates it south of Jerusalem near

Bethlehem, at the site shown to tourists today (Gen 35.16; 48.7; Mt 2.16–18). **3:** *Going up to God* means "going up to offer sacrifice." **5:** The mention of the *Philistine garrison* (or prefect) is a reminder of the task that lay before Saul. **6:** *To be in a prophetic frenzy . . . and be turned into a different person* means here to dance ecstatically and be out of one's head, in the fashion of the so-called ecstatic prophecy of those days (19.23–24). This was to be a sort of initiatory religious experience for the new king.

10.9: *God gave him another heart,* that is, made a new person of him. **11:** The change in

11 When all who knew him before saw how he prophesied with the prophets, the people said to one another, "What has come over the son of Kish? Is Saul also among the prophets?" 12 A man of the place answered, "And who is their father?" Therefore it became a proverb, "Is Saul also among the prophets?" 13 When his prophetic frenzy had ended, he went home. *n*

14 Saul's uncle said to him and to the boy, "Where did you go?" And he replied, "To seek the donkeys; and when we saw they were not to be found, we went to Samuel." 15 Saul's uncle said, "Tell me what Samuel said to you." 16 Saul said to his uncle, "He told us that the donkeys had been found." But about the matter of the kingship, of which Samuel had spoken, he did not tell him anything.

17 Samuel summoned the people to the LORD at Mizpah 18 and said to them, *o* "Thus says the LORD, the God of Israel, 'I brought up Israel out of Egypt, and I rescued you from the hand of the Egyptians and from the hand of all the kingdoms that were oppressing you.' 19 But today you have rejected your God, who saves you from all your calamities and your distresses; and you have said, 'No! but set a king over us.' Now therefore present yourselves before the LORD by your tribes and by your clans."

20 Then Samuel brought all the tribes of Israel near, and the tribe of Benjamin was taken by lot. 21 He brought the tribe of Benjamin near by its families, and the family of the Matrites was taken by lot. Finally he brought the family of the Matrites near man by man, *p* and Saul the son of Kish was taken by lot. But when they sought him, he could not be found. 22 So they inquired again of the LORD, "Did the man come here?" *q* and the LORD said, "See, he has hidden himself among the baggage." 23 Then they ran and brought him from there. When he took his stand among the people, he was head and shoulders taller than any of them. 24 Samuel said to all the people, "Do you see the one whom the LORD has chosen? There is no one like him among all the people." And all the people shouted, "Long live the king!"

25 Samuel told the people the rights and duties of the kingship; and he wrote them in a book and laid it up before the LORD. Then Samuel sent all the people back to their homes. 26 Saul also went to his home at Gibeah, and with him went warriors whose hearts God had touched. 27 But some worthless fellows said, "How can this man save us?" They despised him and brought him no present. But he held his peace.

Now Nahash, king of the Ammonites, had been grievously oppressing the Gadites and the Reubenites. He would gouge out the right eye of each of them and would not grant Israel a deliverer. No one was left of the Israelites across the Jordan whose right eye Nahash, king of the Ammonites, had not gouged out. But there were seven thousand men who had escaped from the Ammonites and had entered Jabesh-gilead. *r*

11 About a month later, *s* Nahash the Ammonite went up and besieged Jabesh-gilead; and all the men of Jabesh said to Nahash, "Make a treaty

n Cn: Heb *he came to the shrine* o Heb *to the people of Israel* p Gk: Heb lacks *Finally . . . man by man* q Gk: Heb *Is there yet a man to come here?* r Q Ms Compare Josephus, *Antiquities* VI.v.1 (68-71): MT lacks *Now Nahash . . . entered Jabesh-gilead.* s Q Ms Gk: MT lacks *About a month later*

Saul began to be noticed. **12**: The phrase, *And who is their father?* may have something to do with an honorific title applied to a great prophet. See 2 Kings 13.14; 2.12. **10.17–27: Saul chosen king by lot.** Samuel proceeds reluctantly (see v. 19), in contrast to 9.19–24. The story told here follows naturally upon ch 8, and has the same point of view. **21**: The clan of the *Matrites* is not mentioned elsewhere. **21–22**: These verses seem to reflect on Saul unfavorably.

10.25: This verse may be an editorial preparation for ch 12. The mention of *a book* is designed to show Samuel as a man of letters as well as priest, prophet, and judge. We know nothing more of this book. **26–27**:

with us, and we will serve you." ²But Nahash the Ammonite said to them, "On this condition I will make a treaty with you, namely that I gouge out everyone's right eye, and thus put disgrace upon all Israel." ³The elders of Jabesh said to him, "Give us seven days' respite that we may send messengers through all the territory of Israel. Then, if there is no one to save us, we will give ourselves up to you." ⁴When the messengers came to Gibeah of Saul, they reported the matter in the hearing of the people; and all the people wept aloud.

5 Now Saul was coming from the field behind the oxen; and Saul said, "What is the matter with the people, that they are weeping?" So they told him the message from the inhabitants of Jabesh. ⁶And the spirit of God came upon Saul in power when he heard these words, and his anger was greatly kindled. ⁷He took a yoke of oxen, and cut them in pieces and sent them throughout all the territory of Israel by messengers, saying, "Whoever does not come out after Saul and Samuel, so shall it be done to his oxen!" Then the dread of the LORD fell upon the people, and they came out as one. ⁸When he mustered them at Bezek, those from Israel were three hundred thousand, and those from Judah seventyᵗ thousand. ⁹They said to the messengers who had come, "Thus shall you say to the inhabitants of Jabesh-gilead: 'To-

morrow, by the time the sun is hot, you shall have deliverance.'" When the messengers came and told the inhabitants of Jabesh, they rejoiced. ¹⁰So the inhabitants of Jabesh said, "Tomorrow we will give ourselves up to you, and you may do to us whatever seems good to you." ¹¹The next day Saul put the people in three companies. At the morning watch they came into the camp and cut down the Ammonites until the heat of the day; and those who survived were scattered, so that no two of them were left together.

12 The people said to Samuel, "Who is it that said, 'Shall Saul reign over us?' Give them to us so that we may put them to death." ¹³But Saul said, "No one shall be put to death this day, for today the LORD has brought deliverance to Israel."

14 Samuel said to the people, "Come, let us go to Gilgal and there renew the kingship." ¹⁵So all the people went to Gilgal, and there they made Saul king before the LORD in Gilgal. There they sacrificed offerings of well-being before the LORD, and there Saul and all the Israelites rejoiced greatly.

12 Samuel said to all Israel, "I have listened to you in all that you have said to me, and have set a king over you. ²See, it is the king who leads you now; I am old and gray, but my sons are

---

ᵗ Q Ms Gk: MT *thirty*

These verses, which also may be editorial, reflect both favorable and unfavorable opinion about Saul.

11.1–15: Saul proves himself able to lead in battle and is publicly made king. This chapter logically follows 10.16, portraying the opportunity for which Saul had been advised to be ready (10.1). The Ammonites to the east (Gen 19.38; Num 21.24; Deut 2.37; Judg 10.9) were not so dangerous to national security as the Philistines, but a victory over them would show Saul capable of meeting the stronger enemy on the west, and also free Israel from having to fight on two fronts. **1:** *Jabesh-gilead,* an Israelite town east of the Jordan. **11.7:** Sending around the *pieces* of a sacrifice was a call to war (Judg 19.29). **8:** *Bezek* was

between Shechem and Beth-shan. On the numbers see 4.10 n. **11:** Their day began at sundown, hence *the next day* would mean "that evening" in our reckoning. Apparently the Israelites marched all night to attack by surprise in the morning.

12.1–25: Samuel's farewell address. Late material with Deuteronomic editing. This chapter logically follows 10.17–27 (compare ch 8), though there are certain adjustments to ch 11 (compare 12.12). Most ancient historical documents, biblical and nonbiblical alike, contain speeches written at a time later than the events of the narrative, but regarded as appropriate to the occasion. This chapter is an excellent example of such literary material. **2:** The misconduct of the sons is not mentioned here (compare 8.3). **2, 5:** The

with you. I have led you from my youth until this day. [3] Here I am; testify against me before the LORD and before his anointed. Whose ox have I taken? Or whose donkey have I taken? Or whom have I defrauded? Whom have I oppressed? Or from whose hand have I taken a bribe to blind my eyes with it? Testify against me[u] and I will restore it to you." [4] They said, "You have not defrauded us or oppressed us or taken anything from the hand of anyone." [5] He said to them, "The LORD is witness against you, and his anointed is witness this day, that you have not found anything in my hand." And they said, "He is witness."

6 Samuel said to the people, "The LORD is witness, who[v] appointed Moses and Aaron and brought your ancestors up out of the land of Egypt. [7] Now therefore take your stand, so that I may enter into judgment with you before the LORD, and I will declare to you[w] all the saving deeds of the LORD that he performed for you and for your ancestors. [8] When Jacob went into Egypt and the Egyptians oppressed them,[x] then your ancestors cried to the LORD and the LORD sent Moses and Aaron, who brought forth your ancestors out of Egypt, and settled them in this place. [9] But they forgot the LORD their God; and he sold them into the hand of Sisera, commander of the army of King Jabin of[y] Hazor, and into the hand of the Philistines, and into the hand of the king of Moab; and they fought against them. [10] Then they cried to the LORD, and said, 'We have sinned, because we have forsaken the LORD, and have served the Baals and the Astartes; but now rescue us out of the hand of our enemies, and we will serve you.' [11] And the LORD sent Jerubbaal and Barak,[z] and Jephthah, and Samson,[a] and rescued you out of the hand of your enemies on every side; and you lived in safety. [12] But when you saw that King Nahash of the Ammonites came against you, you said to me, 'No, but a king shall reign over us,' though the LORD your God was your king. [13] See, here is the king whom you have chosen, for whom you have asked; see, the LORD has set a king over you. [14] If you will fear the LORD and serve him and heed his voice and not rebel against the commandment of the LORD, and if both you and the king who reigns over you will follow the LORD your God, it will be well; [15] but if you will not heed the voice of the LORD, but rebel against the commandment of the LORD, then the hand of the LORD will be against you and your king.[b] [16] Now therefore take your stand and see this great thing that the LORD will do before your eyes. [17] Is it not the wheat harvest today? I will call upon the LORD, that he may send thunder and rain; and you shall know and see that the wickedness that you have done in the sight of the LORD is great in demanding a king for yourselves." [18] So Samuel called upon the LORD, and the LORD sent thunder and rain that day; and all the people greatly feared the LORD and Samuel.

19 All the people said to Samuel, "Pray to the LORD your God for your servants, so that we may not die; for we have added to all our sins the evil of de-

u Gk: Heb lacks *Testify against me*   v Gk: Heb lacks *is witness, who*   w Gk: Heb lacks *and I will declare to you*   x Gk: Heb lacks *and the Egyptians oppressed them*   y Gk: Heb lacks *Jabin king of*   z Gk Syr: Heb *Bedan*   a Gk: Heb *Samuel*   b Gk: Heb *and your ancestors*

---

expression *his anointed* refers to the king (see 10.1 n.).

12.9, 10, 11: Note the similarity to the Deuteronomic framework of the Book of Judges; e.g. Judg 4.2–3. *The Baals and the Astartes;* see 7.3–4 n. **14, 15, 24, 25:** This philosophy of divine retribution is more fully developed in Deut ch 28 and Judg 2.11–23. **16–19:** The author wishes to show that Sam-

uel was still in greater favor with God than was the idea of a king. *Rain* during the *wheat harvest* would be like "snow in summer" (Prov 26.1). This harvest took place in early summer, when no rain could be expected until autumn. Thus it is a miracle, the *thunder* making it more impressive. The miracle story is the writer's way of making a theological point.

manding a king for ourselves." 20 And Samuel said to the people, "Do not be afraid; you have done all this evil, yet do not turn aside from following the LORD, but serve the LORD with all your heart; 21 and do not turn aside after useless things that cannot profit or save, for they are uscless. 22 For the LORD will not cast away his people, for his great name's sake, because it has pleased the LORD to make you a people for himself. 23 Moreover as for me, far be it from me that I should sin against the LORD by ceasing to pray for you; and I will instruct you in the good and the right way. 24 Only fear the LORD, and serve him faithfully with all your heart; for consider what great things he has done for you. 25 But if you still do wickedly, you shall be swept away, both you and your king."

13 Saul was . . .*c* years old when he began to reign; and he reigned . . . and two*d* years over Israel.

2 Saul chose three thousand out of Israel; two thousand were with Saul in Michmash and the hill country of Bethel, and a thousand were with Jonathan in Gibeah of Benjamin; the rest of the people he sent home to their tents. 3 Jonathan defeated the garrison of the Philistines that was at Geba; and the Philistines heard of it. And Saul blew the trumpet throughout all the land, saying, "Let the Hebrews hear!" 4 When all Israel heard that Saul had defeated the garrison of the Philistines, and also that Israel had become odious to the Philistines, the people were called out to join Saul at Gilgal.

5 The Philistines mustered to fight with Israel, thirty thousand chariots, and six thousand horsemen, and troops like the sand on the seashore in multitude; they came up and encamped at Michmash, to the east of Beth-aven. 6 When the Israelites saw that they were in distress (for the troops were hard pressed), the people hid themselves in caves and in holes and in rocks and in tombs and in cisterns. 7 Some Hebrews crossed the Jordan to the land of Gad and Gilead. Saul was still at Gilgal, and all the people followed him trembling.

8 He waited seven days, the time appointed by Samuel; but Samuel did not come to Gilgal, and the people began to slip away from Saul.*e* 9 So Saul said, "Bring the burnt offering here to me, and the offerings of well-being." And he offered the burnt offering. 10 As soon as he had finished offering the burnt offering, Samuel arrived; and Saul went out to meet him and salute him. 11 Samuel said, "What have you done?" Saul replied, "When I saw that the people were slipping away from me, and that you did not come within the days appointed, and that the Philistines were mustering at Michmash, 12 I said, 'Now the Philistines will come down upon me at Gilgal, and I have not entreated the favor of the LORD'; so I forced myself, and offered the burnt offering." 13 Samuel said to Saul, "You have done foolishly; you have not kept the commandment of the LORD your God, which he commanded you. The LORD would have established your kingdom over Israel forever, 14 but now your kingdom will not continue; the

c The number is lacking in the Heb text (the verse is lacking in the Septuagint).   *d* *Two* is not the entire number; something has dropped out.   *e* Heb *him*

13.1–7a: Saul begins the war with the Philistines. Saul's son Jonathan makes the first move. **1:** Saul's age at this time is not known. Some suppose that he reigned twelve years; others put the figure at twenty-two (see note *d*). **5:** *Beth-aven,* probably an alternative name for Bethel. **7:** *Crossed the Jordan,* see Judg 12.5. *Gad and Gilead,* east of the Jordan.

13.7b–15a: Saul's ritual sin and rejection by God. This passage is more hostile to Saul than ch 15, another explanation of Saul's rejection. The thought is that the king had no right to exercise priestly functions, or at least Saul had no such right. Contrast the very different view of 14.31–35, where Saul definitely assists in priestly functions. Later David (2 Sam 6.12–19; 24.25) and Solomon (1 Kings 3.15) also exercised such functions. The *man after his* [God's] *own heart* (v. 14) is David. **8:** *The time appointed by Samuel,* see 10.8. No reason for Samuel's delay is given.

LORD has sought out a man after his own heart; and the LORD has appointed him to be ruler over his people, because you have not kept what the LORD commanded you." 15 And Samuel left and went on his way from Gilgal. *f* The rest of the people followed Saul to join the army; they went up from Gilgal toward Gibeah of Benjamin. *g*

Saul counted the people who were present with him, about six hundred men. 16 Saul, his son Jonathan, and the people who were present with them stayed in Geba of Benjamin; but the Philistines encamped at Michmash. 17 And raiders came out of the camp of the Philistines in three companies; one company turned toward Ophrah, to the land of Shual, 18 another company turned toward Beth-horon, and another company turned toward the mountain *h* that looks down upon the valley of Zeboim toward the wilderness.

19 Now there was no smith to be found throughout all the land of Israel; for the Philistines said, "The Hebrews must not make swords or spears for themselves"; 20 so all the Israelites went down to the Philistines to sharpen their plowshare, mattocks, axes, or sickles; *i* 21 The charge was two-thirds of a shekel *j* for the plowshares and for the mattocks, and one-third of a shekel for sharpening the axes and for setting the goads. *k* 22 So on the day of the battle neither sword nor spear was to be found in the possession of any of the people with Saul and Jonathan; but Saul and his son Jonathan had them.

23 Now a garrison of the Philistines had gone out to the pass of Michmash. 14 1 One day Jonathan son of Saul said to the young man who carried his armor, "Come, let us go over to the Philistine garrison on the other side." But he did not tell his father. 2 Saul was staying in the outskirts of Gibeah under the pomegranate tree that is at Migron; the troops that were with him were about six hundred men, 3 along with Ahijah son of Ahitub, Ichabod's brother, son of Phinehas son of Eli, the priest of the LORD in Shiloh, carrying an ephod. Now the people did not know that Jonathan had gone. 4 In the pass, *l* by which Jonathan tried to go over to the Philistine garrison, there was a rocky crag on one side and a rocky crag on the other; the name of the one was Bozez, and the name of the other Seneh. 5 One crag rose on the north in front of Michmash, and the other on the south in front of Geba.

6 Jonathan said to the young man who carried his armor, "Come, let us go over to the garrison of these uncircumcised; it may be that the LORD will act for us; for nothing can hinder the LORD from saving by many or by few." 7 His armor-bearer said to him, "Do all that your mind inclines to. *m* I am with you; as your mind is, so is mine." *n* 8 Then Jonathan

f Gk: Heb *went up from Gilgal to Gibeah of Benjamin*   *g* Gk: Heb lacks *The rest . . . of Benjamin*   *h* Cn Compare Gk: Heb *toward the border*   *i* Gk: Heb *plowshare*   *j* Heb *was a pim*   *k* Cn: Meaning of Heb uncertain   *l* Heb *Between the passes*   *m* Gk: Heb *Do all that is in your mind. Turn*   *n* Gk: Heb lacks *so is mine*

13.15b–14.52: Continuation of the Philistine War. The Israelites had scattered and only *six hundred* fighting men (a realistic estimate) remained with Saul. **19–22:** These verses are by a writer who conceived the situation to be worse than it really was. The age of iron was just beginning in Palestine; the Philistines were superior to the Israelites in material culture, but the latter overcame the deficiency and eventually won the war. A *shekel* was about eleven and one-half grams of silver. With v. 22 compare Judg 5.8.

14.1–5: Jonathan again takes the lead in forcing the fighting. **3:** *Carrying an ephod,* this was not the *linen ephod* of 2.18, but a box containing Urim and Thummim, the sacred objects for determining the divine will by lot (v. 41; see Ex 28.30 n.). **6–15:** Jonathan's bravery leads to victory. *Uncircumcised* (v. 6) and *Hebrews* (v. 11) are used as uncomplimentary terms (see 4.6 n.). The Israelites, along with the other Semitic peoples to the east, such as the Amorites, Ammonites, Moabites, and Edomites, practiced circumcision

said, "Now we will cross over to those men and will show ourselves to them. 9 If they say to us, 'Wait until we come to you,' then we will stand still in our place, and we will not go up to them. 10 But if they say, 'Come up to us,' then we will go up; for the LORD has given them into our hand. That will be the sign for us." 11 So both of them showed themselves to the garrison of the Philistines; and the Philistines said, "Look, Hebrews are coming out of the holes where they have hidden themselves." 12 The men of the garrison hailed Jonathan and his armor-bearer, saying, "Come up to us, and we will show you something." Jonathan said to his armor-bearer, "Come up after me; for the LORD has given them into the hand of Israel." 13 Then Jonathan climbed up on his hands and feet, with his armor-bearer following after him. The Philistines*o* fell before Jonathan, and his armor-bearer, coming after him, killed them. 14 In that first slaughter Jonathan and his armor-bearer killed about twenty men within an area about half a furrow long in an acre*p* of land. 15 There was a panic in the camp, in the field, and among all the people; the garrison and even the raiders trembled; the earth quaked; and it became a very great panic.

16 Saul's lookouts in Gibeah of Benjamin were watching as the multitude was surging back and forth.*q* 17 Then Saul said to the troops that were with him, "Call the roll and see who has gone from us." When they had called the roll, Jonathan and his armor-bearer were not there. 18 Saul said to Ahijah, "Bring the ark*r* of God here." For at that time the ark*r* of God went with the Israelites. 19 While Saul was talking to the priest, the tumult in the camp of the Philistines increased more and more; and Saul said to the priest, "Withdraw your hand." 20 Then Saul and all the people who were with him rallied and went into the battle; and every sword was against the other, so that there was very great confusion. 21 Now the Hebrews who previously had been with the Philistines and had gone up with them into the camp turned and joined the Israelites who were with Saul and Jonathan. 22 Likewise, when all the Israelites who had gone into hiding in the hill country of Ephraim heard that the Philistines were fleeing, they too followed closely after them in the battle. 23 So the LORD gave Israel the victory that day.

The battle passed beyond Beth-aven, and the troops with Saul numbered altogether about ten thousand men. The battle spread out over the hill country of Ephraim.

24 Now Saul committed a very rash act on that day.*s* He had laid an oath on the troops, saying, "Cursed be anyone who eats food before it is evening and I have been avenged on my enemies." So none of the troops tasted food. 25 All the troops*t* came upon a honeycomb; and there was honey on the ground. 26 When the troops came upon the honeycomb, the honey was dripping out; but they did not put their hands to their mouths, for they feared the oath. 27 But Jonathan had not heard his father charge the troops with the oath; so he extended the staff that was in his hand, and dipped the tip of it in the honeycomb, and put his hand to his mouth; and his eyes brightened. 28 Then one of the soldiers said, "Your father strictly charged the troops with an

---

o Heb *They*   *p* Heb *yoke*   *q* Gk: Heb *they
went and there*   *r* Gk *the ephod*   *s* Gk: Heb *The
Israelites were distressed that day*   *t* Heb *land*

(see Jer 9.25 n.). The non-Semitic Philistines, coming from the west (see 4.1b n.), had no such custom. **11:** *Hebrews,* see 4.6 n.
14.16–23: Saul capitalizes on the advantage gained by Jonathan. **18:** Instead of *the ark of God,* the Greek version reads "the ephod" (compare ephod vv. 3, 41). **19:** *Withdraw your hand,* the priest had begun the casting of the Urim and Thummim. **21:** *Hebrews* and *Israelites* are not necessarily identical peoples here. Perhaps we can say that not all Hebrews were Israelites, since the former was the broader term. **24–30:** Fasting was supposed to be pleasing to God. Saul hoped thus to further his cause, but he was remiss in not seeing that Jonathan was informed.

oath, saying, 'Cursed be anyone who eats food this day.' And so the troops are faint." 29 Then Jonathan said, "My father has troubled the land; see how my eyes have brightened because I tasted a little of this honey. 30 How much better if today the troops had eaten freely of the spoil taken from their enemies; for now the slaughter among the Philistines has not been great."

31 After they had struck down the Philistines that day from Michmash to Aijalon, the troops were very faint; 32 so the troops flew upon the spoil, and took sheep and oxen and calves, and slaughtered them on the ground; and the troops ate them with the blood. 33 Then it was reported to Saul, "Look, the troops are sinning against the LORD by eating with the blood." And he said, "You have dealt treacherously; roll a large stone before me here."*u* 34 Saul said, "Disperse yourselves among the troops, and say to them, 'Let all bring their oxen or their sheep, and slaughter them here, and eat; and do not sin against the LORD by eating with the blood.' " So all of the troops brought their oxen with them that night, and slaughtered them there. 35 And Saul built an altar to the LORD; it was the first altar that he built to the LORD.

36 Then Saul said, "Let us go down after the Philistines by night and despoil them until the morning light; let us not leave one of them." They said, "Do whatever seems good to you." But the priest said, "Let us draw near to God here." 37 So Saul inquired of God, "Shall I go down after the Philistines? Will you give them into the hand of Israel?" But he did not answer him that day. 38 Saul said, "Come here, all you leaders of the people; and let us find out how this sin has arisen today. 39 For as the LORD lives who

saves Israel, even if it is in my son Jonathan, he shall surely die!" But there was no one among all the people who answered him. 40 He said to all Israel, "You shall be on one side, and I and my son Jonathan will be on the other side." The people said to Saul, "Do what seems good to you." 41 Then Saul said, "O LORD God of Israel, why have you not answered your servant today? If this guilt is in me or in my son Jonathan, O LORD God of Israel, give Urim; but if this guilt is in your people Israel,*v* give Thummim." And Jonathan and Saul were indicated by the lot, but the people were cleared. 42 Then Saul said, "Cast the lot between me and my son Jonathan." And Jonathan was taken.

43 Then Saul said to Jonathan, "Tell me what you have done." Jonathan told him, "I tasted a little honey with the tip of the staff that was in my hand; here I am, I will die." 44 Saul said, "God do so to me and more also; you shall surely die, Jonathan!" 45 Then the people said to Saul, "Shall Jonathan die, who has accomplished this great victory in Israel? Far from it! As the LORD lives, not one hair of his head shall fall to the ground; for he has worked with God today." So the people ransomed Jonathan, and he did not die. 46 Then Saul withdrew from pursuing the Philistines; and the Philistines went to their own place.

47 When Saul had taken the kingship over Israel, he fought against all his enemies on every side—against Moab, against the Ammonites, against Edom, against the kings of Zobah, and against the Philistines; wherever he turned he routed them. 48 He did valiantly, and

u Gk: Heb *me this day*   *v* Vg Compare Gk: Heb *41 Saul said to the LORD, the God of Israel*

14.31–35: Saul rather than the priests seems to have been in charge of the religious rites (contrast the attitude expressed in 13.7b–15a). The law against partaking of the blood appears in Lev 19.26; Deut 12.16. **36–42:** Saul wanted to press his advantage, but the priest sensed that something was wrong and suggested use of the sacred lots, *Urim* and *Thum-*

mim. **43–46:** Presumably the ransom took the form of the substitution of an animal (Gen 22.13; Ex 13.13; 34.20). Verse 46 shows this point to be the end of a phase of the war. **14.47–48, 49–51, 52:** A series of three editorial notes, summarizing the public and private situation of Saul at this time and bringing this account into line with the earlier "judges"

struck down the Amalekites, and rescued Israel out of the hands of those who plundered them.

49 Now the sons of Saul were Jonathan, Ishvi, and Malchishua; and the names of his two daughters were these: the name of the firstborn was Merab, and the name of the younger, Michal. 50 The name of Saul's wife was Ahinoam daughter of Ahimaaz. And the name of the commander of his army was Abner son of Ner, Saul's uncle; 51 Kish was the father of Saul, and Ner the father of Abner was the son of Abiel.

52 There was hard fighting against the Philistines all the days of Saul; and when Saul saw any strong or valiant warrior, he took him into his service.

15 Samuel said to Saul, "The LORD sent me to anoint you king over his people Israel; now therefore listen to the words of the LORD. 2 Thus says the LORD of hosts, 'I will punish the Amalekites for what they did in opposing the Israelites when they came up out of Egypt. 3 Now go and attack Amalek, and utterly destroy all that they have; do not spare them, but kill both man and woman, child and infant, ox and sheep, camel and donkey.' "

4 So Saul summoned the people, and numbered them in Telaim, two hundred thousand foot soldiers, and ten thousand soldiers of Judah. 5 Saul came to the city of the Amalekites and lay in wait in the valley. 6 Saul said to the Kenites, "Go!

Leave! Withdraw from among the Amalekites, or I will destroy you with them; for you showed kindness to all the people of Israel when they came up out of Egypt." So the Kenites withdrew from the Amalekites. 7 Saul defeated the Amalekites, from Havilah as far as Shur, which is east of Egypt. 8 He took King Agag of the Amalekites alive, but utterly destroyed all the people with the edge of the sword. 9 Saul and the people spared Agag, and the best of the sheep and of the cattle and of the fatlings, and the lambs, and all that was valuable, and would not utterly destroy them; all that was despised and worthless they utterly destroyed.

10 The word of the LORD came to Samuel: 11 "I regret that I made Saul king, for he has turned back from following me, and has not carried out my commands." Samuel was angry; and he cried out to the LORD all night. 12 Samuel rose early in the morning to meet Saul, and Samuel was told, "Saul went to Carmel, where he set up a monument for himself, and on returning he passed on down to Gilgal." 13 When Samuel came to Saul, Saul said to him, "May you be blessed by the LORD; I have carried out the command of the LORD." 14 But Samuel said, "What then is this bleating of sheep in my ears, and the lowing of cattle that I hear?" 15 Saul said, "They have brought them from the Amalekites; for the people spared the best of the sheep and the cat-

---

who rescued Israel (see Judg chs 8, 9; 10.1–5). Observe how favorably Saul is presented, especially in the first note (vv. 47–48). **15.1–35: Another story of Saul's rejection.** Compare this section with 13.7b–15a and contrast with the preceding section. Samuel, not Saul, is the leading figure once more. Though Saul is king, and wins a military victory, he is depicted as a moral and religious reprobate, unworthy of the position he holds. **2:** *The Amalekites* is the name of a people traditionally descended from Esau (Gen 36.12). They were a wandering tribe from southern Canaan or northern Sinai which had been constantly troublesome to the Israelites (Deut 25.17–19; Ex 17.7–13; Judg 6.33). **3:** This verse seems to pick up the curse

recorded in Deut 25.19. *Utterly destroy* means "put under the ban" or "destroy as a type of religious sacrifice" (Deut 20.16–18). Both the Israelites and their neighbors attempted at times this type of holy war, but usually without complete success. The Amalekites are still numerous and troublesome to David in ch 30.

15.4: Exaggerated numbers are characteristic of many ancient documents. **5–7:** *The Kenites* were a clan partly with the Midianites and partly with the Amalekites. Moses' father-in-law seems to have belonged to this group (Judg 1.16; 4.11). The tradition of *kindness* referred to here appears in Num 10.29–32.

tle, to sacrifice to the LORD your God; but the rest we have utterly destroyed." 16 Then Samuel said to Saul, "Stop! I will tell you what the LORD said to me last night." He replied, "Speak."

17 Samuel said, "Though you are little in your own eyes, are you not the head of the tribes of Israel? The LORD anointed you king over Israel. 18 And the LORD sent you on a mission, and said, 'Go, utterly destroy the sinners, the Amalekites, and fight against them until they are consumed.' 19 Why then did you not obey the voice of the LORD? Why did you swoop down on the spoil, and do what was evil in the sight of the LORD?" 20 Saul said to Samuel, "I have obeyed the voice of the LORD, I have gone on the mission on which the LORD sent me, I have brought Agag the king of Amalek, and I have utterly destroyed the Amalekites. 21 But from the spoil the people took sheep and cattle, the best of the things devoted to destruction, to sacrifice to the LORD your God in Gilgal." 22 And Samuel said,

"Has the LORD as great delight in
        burnt offerings and
        sacrifices,
    as in obeying the voice of the
        LORD?
Surely, to obey is better than
        sacrifice,
    and to heed than the fat of
        rams.
23 For rebellion is no less a sin than
        divination,
    and stubbornness is like iniquity
        and idolatry.
Because you have rejected the
        word of the LORD,
    he has also rejected you from
        being king."

24 Saul said to Samuel, "I have sinned; for I have transgressed the commandment of the LORD and your words, because I feared the people and obeyed their voice. 25 Now therefore, I pray,

pardon my sin, and return with me, so that I may worship the LORD." 26 Samuel said to Saul, "I will not return with you; for you have rejected the word of the LORD, and the LORD has rejected you from being king over Israel." 27 As Samuel turned to go away, Saul caught hold of the hem of his robe, and it tore. 28 And Samuel said to him, "The LORD has torn the kingdom of Israel from you this very day, and has given it to a neighbor of yours, who is better than you. 29 Moreover the Glory of Israel will not recant *w* or change his mind; for he is not a mortal, that he should change his mind." 30 Then Saul *x* said, "I have sinned; yet honor me now before the elders of my people and before Israel, and return with me, so that I may worship the LORD your God." 31 So Samuel turned back after Saul; and Saul worshiped the LORD.

32 Then Samuel said, "Bring Agag king of the Amalekites here to me." And Agag came to him haltingly. *y* Agag said, "Surely this is the bitterness of death." *z*
33 But Samuel said,

"As your sword has made women
        childless,
    so your mother shall be childless
        among women."

And Samuel hewed Agag in pieces before the LORD in Gilgal.

34 Then Samuel went to Ramah; and Saul went up to his house in Gibeah of Saul. 35 Samuel did not see Saul again until the day of his death, but Samuel grieved over Saul. And the LORD was sorry that he had made Saul king over Israel.

16 The LORD said to Samuel, "How long will you grieve over Saul? I have rejected him from being king over Israel. Fill your horn with oil and set out; I will send you to Jesse the Bethlehemite,

w Q Ms Gk: MT *deceive*   x Heb *he*
y Cn Compare Gk: Meaning of Heb uncertain
z Q Ms Gk: MT *Surely the bitterness of death is past*

15.22: *To obey is better than sacrifice;* compare Hos 6.6. **23:** On the sin of *divination* see Deut 18.9–14. **28:** The *neighbor* is David. **35:**

There is a slight discrepancy between this verse and 19.24.
16.1–13: **The anointing of David.** This

for I have provided for myself a king among his sons." 2 Samuel said, "How can I go? If Saul hears of it, he will kill me." And the LORD said, "Take a heifer with you, and say, 'I have come to sacrifice to the LORD.' 3 Invite Jesse to the sacrifice, and I will show you what you shall do; and you shall anoint for me the one whom I name to you." 4 Samuel did what the LORD commanded, and came to Bethlehem. The elders of the city came to meet him trembling, and said, "Do you come peaceably?" 5 He said, "Peaceably; I have come to sacrifice to the LORD; sanctify yourselves and come with me to the sacrifice." And he sanctified Jesse and his sons and invited them to the sacrifice.

6 When they came, he looked on Eliab and thought, "Surely the LORD's anointed is now before the LORD." *a* 7 But the LORD said to Samuel, "Do not look on his appearance or on the height of his stature, because I have rejected him; for the LORD does not see as mortals see; they look on the outward appearance, but the LORD looks on the heart." 8 Then Jesse called Abinadab, and made him pass before Samuel. He said, "Neither has the LORD chosen this one." 9 Then Jesse made Shammah pass by. And he said, "Neither has the LORD chosen this one." 10 Jesse made seven of his sons pass before Samuel, and Samuel said to Jesse, "The LORD has not chosen any of these." 11 Samuel said to Jesse, "Are all your sons here?" And he said, "There remains yet the youngest, but he is keeping the sheep." And Samuel said to Jesse, "Send and bring him; for we will not sit down until he comes here." 12 He sent and brought him in. Now he was ruddy, and had beautiful eyes, and was handsome. The LORD said, "Rise and anoint him; for this

is the one." 13 Then Samuel took the horn of oil, and anointed him in the presence of his brothers; and the spirit of the LORD came mightily upon David from that day forward. Samuel then set out and went to Ramah.

14 Now the spirit of the LORD departed from Saul, and an evil spirit from the LORD tormented him. 15 And Saul's servants said to him, "See now, an evil spirit from God is tormenting you. 16 Let our lord now command the servants who attend you to look for someone who is skillful in playing the lyre; and when the evil spirit from God is upon you, he will play it, and you will feel better." 17 So Saul said to his servants, "Provide for me someone who can play well, and bring him to me." 18 One of the young men answered, "I have seen a son of Jesse the Bethlehemite who is skillful in playing, a man of valor, a warrior, prudent in speech, and a man of good presence; and the LORD is with him." 19 So Saul sent messengers to Jesse, and said, "Send me your son David who is with the sheep." 20 Jesse took a donkey loaded with bread, a skin of wine, and a kid, and sent them by his son David to Saul. 21 And David came to Saul, and entered his service. Saul loved him greatly, and he became his armor-bearer. 22 Saul sent to Jesse, saying, "Let David remain in my service, for he has found favor in my sight." 23 And whenever the evil spirit from God came upon Saul, David took the lyre and played it with his hand, and Saul would be relieved and feel better, and the evil spirit would depart from him.

17 Now the Philistines gathered their armies for battle; they were gathered at Socoh, which belongs to Ju-

a Heb *him*

---

story may be a counterpart to the anointing of Saul (see 10.1 n.). **5:** Consecration perhaps involved the ceremony of ritual washing **12:** The word *ruddy* (compare 17.42) is sometimes taken to mean that David was red-haired, but the reference is to his complexion (compare Song 5.10, 11).
16.14–23: David wins a position at the

court of Saul. This is a story of [how] became acquainted with David. [Ancient] the beginning of the story of S[aul ... are Lk] from mental illness, attri[buted ... relations] times to *an evil spirit* (1A problem) 11.24–26). **21–22:** No[...] between Saul and D[avid ...]
17.1–58: David

dah, and encamped between Socoh and Azekah, in Ephes-dammim. [2] Saul and the Israelites gathered and encamped in the valley of Elah, and formed ranks against the Philistines. [3] The Philistines stood on the mountain on the one side, and Israel stood on the mountain on the other side, with a valley between them. [4] And there came out from the camp of the Philistines a champion named Goliath, of Gath, whose height was six[b] cubits and a span. [5] He had a helmet of bronze on his head, and he was armed with a coat of mail; the weight of the coat was five thousand shekels of bronze. [6] He had greaves of bronze on his legs and a javelin of bronze slung between his shoulders. [7] The shaft of his spear was like a weaver's beam, and his spear's head weighed six hundred shekels of iron; and his shield-bearer went before him. [8] He stood and shouted to the ranks of Israel, "Why have you come out to draw up for battle? Am I not a Philistine, and are you not servants of Saul? Choose a man for yourselves, and let him come down to me. [9] If he is able to fight with me and kill me, then we will be your servants; but if I prevail against him and kill him, then you shall be our servants and serve us." [10] And the Philistine said, "Today I defy the ranks of Israel! Give me a man, that we may fight together." [11] When Saul and all Israel heard these words of the Philistine, they were dismayed and greatly afraid.

12 Now David was the son of an Ephrathite of Bethlehem in Judah, named Jesse, who had eight sons. In the days of Saul the man was already old and advanced in years.[c] [13] The three eldest sons of Jesse had followed Saul to the battle; the names of his three sons who went to the battle were Eliab the firstborn, and next to him Abinadab, and the third Shammah. [14] David was the youngest; the three eldest followed Saul, [15] but David went back and forth from Saul to feed his father's sheep at Bethlehem. [16] For forty days the Philistine came forward and took his stand, morning and evening.

17 Jesse said to his son David, "Take for your brothers an ephah of this parched grain and these ten loaves, and carry them quickly to the camp to your brothers; [18] also take these ten cheeses to the commander of their thousand. See how your brothers fare, and bring some token from them."

19 Now Saul, and they, and all the men of Israel, were in the valley of Elah, fighting with the Philistines. [20] David rose early in the morning, left the sheep with a keeper, took the provisions, and went as Jesse had commanded him. He came to the encampment as the army was going forth to the battle line, shouting the war cry. [21] Israel and the Philistines drew up for battle, army against army. [22] David left the things in charge of the keeper of the baggage, ran to the ranks, and went and greeted his brothers. [23] As he talked with them, the champion, the Philistine of Gath, Goliath by name, came up out of the ranks of the Philistines, and spoke the same words as before. And David heard him.

24 All the Israelites, when they saw the man, fled from him and were very much afraid. [25] The Israelites said, "Have

b MT: Q Ms Gk *four*   *c* Gk Syr: Heb *among men*

---

rises because, according to vv. 55–58, Saul ⸱⸱s not know David (16.21–23). This has ⸱⸱ taken as evidence that a variety of mate-⸱⸱ ⸱⸱ot always consistent in every detail, ⸱⸱ ⸱⸱ woven together (see Introduction). D⸱⸱ng to 2 Sam 21.19, *Goliath of Gath* this⸱⸱ a later time by Elhanan, one of Davi⸱⸱ors. It may be supposed that ⸱⸱ne erroneously attached to ⸱⸱hose name was unknown

(1 Chr 20.5). *Six cubits and a span,* about ten feet. **5:** *Five thousand shekels,* about one hundred and fifty lbs. **7:** *Six hundred shekels,* about nineteen lbs.

17.12: The abrupt break in the narrative at this point probably marks insertion of material from a different source.

17.32–40: Saul knows David well, in conformity with 16.14–23.

you seen this man who has come up? Surely he has come up to defy Israel. The king will greatly enrich the man who kills him, and will give him his daughter and make his family free in Israel." 26 David said to the men who stood by him, "What shall be done for the man who kills this Philistine, and takes away the reproach from Israel? For who is this uncircumcised Philistine that he should defy the armies of the living God?" 27 The people answered him in the same way, "So shall it be done for the man who kills him."

28 His eldest brother Eliab heard him talking to the men; and Eliab's anger was kindled against David. He said, "Why have you come down? With whom have you left those few sheep in the wilderness? I know your presumption and the evil of your heart; for you have come down just to see the battle." 29 David said, "What have I done now? It was only a question." 30 He turned away from him toward another and spoke in the same way; and the people answered him again as before.

31 When the words that David spoke were heard, they repeated them before Saul; and he sent for him. 32 David said to Saul, "Let no one's heart fail because of him; your servant will go and fight with this Philistine." 33 Saul said to David, "You are not able to go against this Philistine to fight with him; for you are just a boy, and he has been a warrior from his youth." 34 But David said to Saul, "Your servant used to keep sheep for his father; and whenever a lion or a bear came, and took a lamb from the flock, 35 I went after it and struck it down, rescuing the lamb from its mouth; and if it turned against me, I would catch it by the jaw, strike it down, and kill it. 36 Your servant has killed both lions and bears; and this uncircumcised Philistine shall be like one of them, since he has defied the armies of the living God." 37 David said, "The LORD, who saved me from the paw of the lion and from the paw of the bear, will save me from the hand of this Philistine." So Saul said to David, "Go, and may the LORD be with you!"

38 Saul clothed David with his armor; he put a bronze helmet on his head and clothed him with a coat of mail. 39 David strapped Saul's sword over the armor, and he tried in vain to walk, for he was not used to them. Then David said to Saul, "I cannot walk with these; for I am not used to them." So David removed them. 40 Then he took his staff in his hand, and chose five smooth stones from the wadi, and put them in his shepherd's bag, in the pouch; his sling was in his hand, and he drew near to the Philistine.

41 The Philistine came on and drew near to David, with his shield-bearer in front of him. 42 When the Philistine looked and saw David, he disdained him, for he was only a youth, ruddy and handsome in appearance. 43 The Philistine said to David, "Am I a dog, that you come to me with sticks?" And the Philistine cursed David by his gods. 44 The Philistine said to David, "Come to me, and I will give your flesh to the birds of the air and to the wild animals of the field." 45 But David said to the Philistine, "You come to me with sword and spear and javelin; but I come to you in the name of the LORD of hosts, the God of the armies of Israel, whom you have defied. 46 This very day the LORD will deliver you into my hand, and I will strike you down and cut off your head; and I will give the dead bodies of the Philistine army this very day to the birds of the air and to the wild animals of the earth, so that all the earth may know that there is a God in Israel, 47 and that all this assembly may know that the LORD does not save by sword and spear; for the battle is the LORD's and he will give you into our hand."

48 When the Philistine drew nearer to meet David, David ran quickly toward the battle line to meet the Philistine. 49 David put his hand in his bag, took out a stone, slung it, and struck the Philistine on his forehead; the stone sank into his forehead, and he fell face down on the ground.

50 So David prevailed over the Philistine with a sling and a stone, striking down the Philistine and killing him;

there was no sword in David's hand. [51] Then David ran and stood over the Philistine; he grasped his sword, drew it out of its sheath, and killed him; then he cut off his head with it.

When the Philistines saw that their champion was dead, they fled. [52] The troops of Israel and Judah rose up with a shout and pursued the Philistines as far as Gath[d] and the gates of Ekron, so that the wounded Philistines fell on the way from Shaaraim as far as Gath and Ekron. [53] The Israelites came back from chasing the Philistines, and they plundered their camp. [54] David took the head of the Philistine and brought it to Jerusalem; but he put his armor in his tent.

[55] When Saul saw David go out against the Philistine, he said to Abner, the commander of the army, "Abner, whose son is this young man?" Abner said, "As your soul lives, O king, I do not know." [56] The king said, "Inquire whose son the stripling is." [57] On David's return from killing the Philistine, Abner took him and brought him before Saul, with the head of the Philistine in his hand. [58] Saul said to him, "Whose son are you, young man?" And David answered, "I am the son of your servant Jesse the Bethlehemite."

18 When David[e] had finished speaking to Saul, the soul of Jonathan was bound to the soul of David, and Jonathan loved him as his own soul. [2] Saul took him that day and would not let him return to his father's house. [3] Then Jonathan made a covenant with David, because he loved him as his own soul. [4] Jonathan stripped himself of the robe that he was wearing, and gave it to David, and his armor, and even his sword and his bow and his belt. [5] David went out and was successful wherever Saul sent him; as a result, Saul set him

over the army. And all the people, even the servants of Saul, approved.

6 As they were coming home, when David returned from killing the Philistine, the women came out of all the towns of Israel, singing and dancing, to meet King Saul, with tambourines, with songs of joy, and with musical instruments.[f] [7] And the women sang to one another as they made merry,

"Saul has killed his thousands,
    and David his ten thousands."

[8] Saul was very angry, for this saying displeased him. He said, "They have ascribed to David ten thousands, and to me they have ascribed thousands; what more can he have but the kingdom?" [9] So Saul eyed David from that day on.

10 The next day an evil spirit from God rushed upon Saul, and he raved within his house, while David was playing the lyre, as he did day by day. Saul had his spear in his hand; [11] and Saul threw the spear, for he thought, "I will pin David to the wall." But David eluded him twice.

12 Saul was afraid of David, because the LORD was with him but had departed from Saul. [13] So Saul removed him from his presence, and made him a commander of a thousand; and David marched out and came in, leading the army. [14] David had success in all his undertakings; for the LORD was with him. [15] When Saul saw that he had great success, he stood in awe of him. [16] But all Israel and Judah loved David; for it was he who marched out and came in leading them.

17 Then Saul said to David, "Here is my elder daughter Merab; I will give her to you as a wife; only be valiant for me and fight the LORD's battles." For Saul thought, "I will not raise a hand against

---

d Gk Syr: Heb *Gai*   e Heb *he*
f Or *triangles,* or *three-stringed instruments*

17.55–58: Note how completely unknown David is to Saul, in contrast to vv. 32–40, but in entire consistency with 16.1–13. **18.1–30: Saul becomes jealous of David. 1–3:** A deep friendship arose between David and Jonathan.
18.6–9: Saul's first anger at David. **17–19:** This incident of *Merab* is lacking in some Greek texts.

him; let the Philistines deal with him."
18 David said to Saul, "Who am I and
who are my kinsfolk, my father's family
in Israel, that I should be son-in-law to
the king?" 19 But at the time when Saul's
daughter Merab should have been given
to David, she was given to Adriel the
Meholathite as a wife.

20 Now Saul's daughter Michal loved
David. Saul was told, and the thing
pleased him. 21 Saul thought, "Let me
give her to him that she may be a snare
for him and that the hand of the Philistines may be against him." Therefore
Saul said to David a second time, g "You
shall now be my son-in-law." 22 Saul
commanded his servants, "Speak to David in private and say, 'See, the king is
delighted with you, and all his servants
love you; now then, become the king's
son-in-law.'" 23 So Saul's servants reported these words to David in private.
And David said, "Does it seem to you a
little thing to become the king's son-in-
law, seeing that I am a poor man and of
no repute?" 24 The servants of Saul told
him, "This is what David said." 25 Then
Saul said, "Thus shall you say to David,
'The king desires no marriage present except a hundred foreskins of the Philistines, that he may be avenged on the
king's enemies.'" Now Saul planned to
make David fall by the hand of the Philistines. 26 When his servants told David
these words, David was well pleased to
be the king's son-in-law. Before the time
had expired, 27 David rose and went,
along with his men, and killed one hundred h of the Philistines; and David
brought their foreskins, which were giv-
en in full number to the king, that he
might become the king's son-in-law.
Saul gave him his daughter Michal as a
wife. 28 But when Saul realized that the
LORD was with David, and that Saul's
daughter Michal loved him, 29 Saul was

still more afraid of David. So Saul was
David's enemy from that time forward.

30 Then the commanders of the Philistines came out to battle; and as often as
they came out, David had more success
than all the servants of Saul, so that his
fame became very great.

19 Saul spoke with his son Jonathan
and with all his servants about
killing David. But Saul's son Jonathan
took great delight in David. 2 Jonathan
told David, "My father Saul is trying to
kill you; therefore be on guard tomorrow morning; stay in a secret place and
hide yourself. 3 I will go out and stand
beside my father in the field where you
are, and I will speak to my father about
you; if I learn anything I will tell you."
4 Jonathan spoke well of David to his father Saul, saying to him, "The king
should not sin against his servant David,
because he has not sinned against you,
and because his deeds have been of good
service to you; 5 for he took his life in his
hand when he attacked the Philistine, and
the LORD brought about a great victory
for all Israel. You saw it, and rejoiced;
why then will you sin against an innocent person by killing David without
cause?" 6 Saul heeded the voice of Jonathan; Saul swore, "As the LORD lives, he
shall not be put to death." 7 So Jonathan
called David and related all these things
to him. Jonathan then brought David to
Saul, and he was in his presence as before.

8 Again there was war, and David
went out to fight the Philistines. He
launched a heavy attack on them, so that
they fled before him. 9 Then an evil spirit
from the LORD came upon Saul, as he sat
in his house with his spear in his hand,
while David was playing music. 10 Saul

g Heb *by two*   h Gk Compare 2 Sam 3.14;
Heb *two hundred*

18.25: The *marriage present* is regarded by
some as a gift to the bride's family, by others
as a "price" paid for the bride. Probably the
custom combined both elements. **27:** Only
one hundred *foreskins* were required (v. 25);

this story is to show David's superior prowess.

19.1–10: Saul seeks David's life. 6–7:
Jonathan succeeds in restraining his father
temporarily. **8–10:** Compare 18.10–11.

sought to pin David to the wall with the spear; but he eluded Saul, so that he struck the spear into the wall. David fled and escaped that night.

11 Saul sent messengers to David's house to keep watch over him, planning to kill him in the morning. David's wife Michal told him, "If you do not save your life tonight, tomorrow you will be killed." 12 So Michal let David down through the window; he fled away and escaped. 13 Michal took an idol[i] and laid it on the bed; she put a net[j] of goats' hair on its head, and covered it with the clothes. 14 When Saul sent messengers to take David, she said, "He is sick." 15 Then Saul sent the messengers to see David for themselves. He said, "Bring him up to me in the bed, that I may kill him." 16 When the messengers came in, the idol[k] was in the bed, with the covering[j] of goats' hair on its head. 17 Saul said to Michal, "Why have you deceived me like this, and let my enemy go, so that he has escaped?" Michal answered Saul, "He said to me, 'Let me go; why should I kill you?'"

18 Now David fled and escaped; he came to Samuel at Ramah, and told him all that Saul had done to him. He and Samuel went and settled at Naioth. 19 Saul was told, "David is at Naioth in Ramah." 20 Then Saul sent messengers to take David. When they saw the company of the prophets in a frenzy, with Samuel standing in charge of[j] them, the spirit of God came upon the messengers of Saul, and they also fell into a prophetic frenzy. 21 When Saul was told, he sent other messengers, and they also fell into a frenzy. Saul sent messengers again the third time, and they also fell into a frenzy. 22 Then he himself went to Ramah. He came to the great well that is in Secu;[l] he asked, "Where are Samuel and David?" And someone said, "They are at Naioth in Ramah." 23 He went there, toward Naioth in Ramah; and the spirit of God came upon him. As he was going, he fell into a prophetic frenzy, until he came to Naioth in Ramah. 24 He too stripped off his clothes, and he too fell into a frenzy before Samuel. He lay naked all that day and all that night. Therefore it is said, "Is Saul also among the prophets?"

20 David fled from Naioth in Ramah. He came before Jonathan and said, "What have I done? What is my guilt? And what is my sin against your father that he is trying to take my life?" 2 He said to him, "Far from it! You shall not die. My father does nothing either great or small without disclosing it to me; and why should my father hide this from me? Never!" 3 But David also swore, "Your father knows well that you like me; and he thinks, 'Do not let Jonathan know this, or he will be grieved.' But truly, as the LORD lives and as you yourself live, there is but a step between me and death." 4 Then Jonathan said to David, "Whatever you say, I will do for you." 5 David said to Jonathan, "Tomorrow is the new moon, and I should not fail to sit with the king at the meal; but

i Heb *took the teraphim*   *j* Meaning of Heb uncertain   *k* Heb *the teraphim*   *l* Gk reads *to the well of the threshing floor on the bare height*

19.11–17: David forced to flee. This incident follows naturally upon the marriage in 18.20–29. David seems to be unaware of the danger (compare the preceding verses) until Michal tells him. **13:** *Idol,* see Gen 31.19 n.; Judg 17.5. The nature of the *idol* and *net* here is obscure.

19.18–24: Saul's ecstatic behavior. Probably an independent story. See another version of this matter (10.10–12). The nature of this *frenzy* (v. 20) as ecstatic dancing or whirling must be kept in mind (see 10.6 n.). The names *Secu* and *Naioth in Ramah* (v. 22)

seem to indicate locations within the town of Ramah.

20.1–42: An independent tradition of the break between Saul and David, incompatible with much of ch 19. David is represented as still a member of the king's household and Jonathan seems unaware of Saul's hatred of David. The break between Saul and David was so significant that many different stories about it were told. **1:** The first part of this verse is an editorial attempt to join it to ch 19. Actually, according to what follows, David had not yet left the court of Saul. **5:**

let me go, so that I may hide in the field until the third evening. ⁶If your father misses me at all, then say, 'David earnestly asked leave of me to run to Bethlehem his city; for there is a yearly sacrifice there for all the family.' ⁷If he says, 'Good!' it will be well with your servant; but if he is angry, then know that evil has been determined by him. ⁸Therefore deal kindly with your servant, for you have brought your servant into a sacred covenant*m* with you. But if there is guilt in me, kill me yourself; why should you bring me to your father?" ⁹Jonathan said, "Far be it from you! If I knew that it was decided by my father that evil should come upon you, would I not tell you?" ¹⁰Then David said to Jonathan, "Who will tell me if your father answers you harshly?" ¹¹Jonathan replied to David, "Come, let us go out into the field." So they both went out into the field.

12 Jonathan said to David, "By the LORD, the God of Israel! When I have sounded out my father, about this time tomorrow, or on the third day, if he is well disposed toward David, shall I not then send and disclose it to you? ¹³But if my father intends to do you harm, the LORD do so to Jonathan, and more also, if I do not disclose it to you, and send you away, so that you may go in safety. May the LORD be with you, as he has been with my father. ¹⁴If I am still alive, show me the faithful love of the LORD, but if I die,*n* ¹⁵never cut off your faithful love from my house, even if the LORD were to cut off every one of the enemies of David from the face of the earth." ¹⁶Thus Jonathan made a covenant with the house of David, saying, "May the LORD seek out the enemies of David." ¹⁷Jonathan made David swear again by his love for him; for he loved him as he loved his own life.

18 Jonathan said to him, "Tomorrow is the new moon; you will be missed, because your place will be empty. ¹⁹On the day after tomorrow, you shall go a long way down; go to the place where you hid yourself earlier, and remain beside the stone there.*n* ²⁰I will shoot three arrows to the side of it, as though I shot at a mark. ²¹Then I will send the boy, saying, 'Go, find the arrows.' If I say to the boy, 'Look, the arrows are on this side of you, collect them,' then you are to come, for, as the LORD lives, it is safe for you and there is no danger. ²²But if I say to the young man, 'Look, the arrows are beyond you,' then go; for the LORD has sent you away. ²³As for the matter about which you and I have spoken, the LORD is witness*o* between you and me forever."

24 So David hid himself in the field. When the new moon came, the king sat at the feast to eat. ²⁵The king sat upon his seat, as at other times, upon the seat by the wall. Jonathan stood, while Abner sat by Saul's side; but David's place was empty.

26 Saul did not say anything that day; for he thought, "Something has befallen him; he is not clean, surely he is not clean." ²⁷But on the second day, the day after the new moon, David's place was empty. And Saul said to his son Jonathan, "Why has the son of Jesse not come to the feast, either yesterday or today?" ²⁸Jonathan answered Saul, "David earnestly asked leave of me to go to Bethlehem; ²⁹he said, 'Let me go; for our family is holding a sacrifice in the city, and my brother has commanded me to be there. So now, if I have found favor in your sight, let me get away, and see my brothers.' For this reason he has not come to the king's table."

30 Then Saul's anger was kindled

---

m Heb *a covenant of the LORD*   *n* Meaning of Heb uncertain   *o* Gk: Heb lacks *witness*

The new moon was a festival day, when all members of the household were supposed to eat together. **8**: *Sacred covenant* is a reference to the deep friendship of David and Jonathan. See 18.1–4.

20.26: Some mishap could have caused temporary ritual uncleanness.
20.30: A popular form of cursing was then, and is now, to berate a person's mother; hence the words *your mother's nakedness,* as

against Jonathan. He said to him, "You son of a perverse, rebellious woman! Do I not know that you have chosen the son of Jesse to your own shame, and to the shame of your mother's nakedness? 31 For as long as the son of Jesse lives upon the earth, neither you nor your kingdom shall be established. Now send and bring him to me, for he shall surely die." 32 Then Jonathan answered his father Saul, "Why should he be put to death? What has he done?" 33 But Saul threw his spear at him to strike him; so Jonathan knew that it was the decision of his father to put David to death. 34 Jonathan rose from the table in fierce anger and ate no food on the second day of the month, for he was grieved for David, and because his father had disgraced him.

35 In the morning Jonathan went out into the field to the appointment with David, and with him was a little boy. 36 He said to the boy, "Run and find the arrows that I shoot." As the boy ran, he shot an arrow beyond him. 37 When the boy came to the place where Jonathan's arrow had fallen, Jonathan called after the boy and said, "Is the arrow not beyond you?" 38 Jonathan called after the boy, "Hurry, be quick, do not linger." So Jonathan's boy gathered up the arrows and came to his master. 39 But the boy knew nothing; only Jonathan and David knew the arrangement. 40 Jonathan gave his weapons to the boy and said to him, "Go and carry them to the city." 41 As soon as the boy had gone, David rose from beside the stone heap[p] and prostrated himself with his face to the ground. He bowed three times, and they kissed each other, and wept with each other; David wept the more.[q] 42 Then Jonathan said to David, "Go in peace, since both of us have sworn in the name of the LORD, saying, 'The LORD shall be between me and you, and between my descendants and your descendants, forever.'" He got up and left; and Jonathan went into the city.[r]

21 [s] David came to Nob to the priest Ahimelech. Ahimelech came trembling to meet David, and said to him, "Why are you alone, and no one with you?" 2 David said to the priest Ahimelech, "The king has charged me with a matter, and said to me, 'No one must know anything of the matter about which I send you, and with which I have charged you.' I have made an appointment[t] with the young men for such and such a place. 3 Now then, what have you at hand? Give me five loaves of bread, or whatever is here." 4 The priest answered David, "I have no ordinary bread at hand, only holy bread—provided that the young men have kept themselves from women." 5 David answered the priest, "Indeed women have been kept from us as always when I go on an expedition; the vessels of the young men are holy even when it is a common journey; how much more today will their vessels be holy?" 6 So the priest gave him the holy bread; for there was no bread there except the bread of the Presence, which is removed from before the LORD, to be replaced by hot bread on the day it is taken away.

7 Now a certain man of the servants of Saul was there that day, detained before the LORD; his name was Doeg the Edomite, the chief of Saul's shepherds.

8 David said to Ahimelech, "Is there

---

p Gk: Heb *from beside the south*
q Vg: Meaning of Heb uncertain   r This sentence is 21.1 in Heb   s Ch 21.2 in Heb
t Q Ms Vg Compare Gk: Meaning of MT uncertain

---

though she were a prostitute or something of that sort. **33**: Compare the similar treatment of David, 18.11 and 19.10.

21.1–9: David escapes to Nob. 6: On *the holy bread* or *the bread of the Presence* in later tradition, see Lev 24.5–9; on the use of this passage in the New Testament, see Mk 2.23–28. **7**: *Doeg* appears later as David's betrayer (22.9). He was *detained* to perform some sort of religious ceremony. **8**: David was not only alone, but unarmed, a fact testifying to the hastiness of his flight (contrast the story of more leisurely departure in ch 20). **9**: *The ephod* is here a box, not a linen apron (see 14.3 n.).

no spear or sword here with you? I did not bring my sword or my weapons with me, because the king's business required haste." 9 The priest said, "The sword of Goliath the Philistine, whom you killed in the valley of Elah, is here wrapped in a cloth behind the ephod; if you will take that, take it, for there is none here except that one." David said, "There is none like it; give it to me."

10 David rose and fled that day from Saul; he went to King Achish of Gath. 11 The servants of Achish said to him, "Is this not David the king of the land? Did they not sing to one another of him in dances,

'Saul has killed his thousands,
    and David his ten thousands'?"

12 David took these words to heart and was very much afraid of King Achish of Gath. 13 So he changed his behavior before them; he pretended to be mad when in their presence.*ᵘ* He scratched marks on the doors of the gate, and let his spittle run down his beard. 14 Achish said to his servants, "Look, you see the man is mad; why then have you brought him to me? 15 Do I lack madmen, that you have brought this fellow to play the madman in my presence? Shall this fellow come into my house?"

22 David left there and escaped to the cave of Adullam; when his brothers and all his father's house heard of it, they went down there to him. 2 Everyone who was in distress, and everyone who was in debt, and everyone who was discontented gathered to him; and he became captain over them. Those who were with him numbered about four hundred.

3 David went from there to Mizpeh of Moab. He said to the king of Moab, "Please let my father and mother come*ᵛ*

to you, until I know what God will do for me." 4 He left them with the king of Moab, and they stayed with him all the time that David was in the stronghold. 5 Then the prophet Gad said to David, "Do not remain in the stronghold; leave, and go into the land of Judah." So David left, and went into the forest of Hereth.

6 Saul heard that David and those who were with him had been located. Saul was sitting at Gibeah, under the tamarisk tree on the height, with his spear in his hand, and all his servants were standing around him. 7 Saul said to his servants who stood around him, "Hear now, you Benjaminites; will the son of Jesse give every one of you fields and vineyards, will he make you all commanders of thousands and commanders of hundreds? 8 Is that why all of you have conspired against me? No one discloses to me when my son makes a league with the son of Jesse, none of you is sorry for me or discloses to me that my son has stirred up my servant against me, to lie in wait, as he is doing today." 9 Doeg the Edomite, who was in charge of Saul's servants, answered, "I saw the son of Jesse coming to Nob, to Ahimelech son of Ahitub; 10 he inquired of the Lord for him, gave him provisions, and gave him the sword of Goliath the Philistine."

11 The king sent for the priest Ahimelech son of Ahitub and for all his father's house, the priests who were at Nob; and all of them came to the king. 12 Saul said, "Listen now, son of Ahitub." He answered, "Here I am, my lord." 13 Saul said to him, "Why have you conspired against me, you and the son of Jesse, by giving him bread and a sword, and by inquiring of God for him, so that he has

---

u Heb *in their hands*   *v* Syr Vg: Heb *come out*

21.10–15: David flees to Gath. Another version of this episode is found in ch 27. **11:** The purpose of this version of the story may have been to show that David and Achish had nothing to do with one another, in contrast to ch 27, where they get along very well.
22.1–23: David at Adullam; massacre of the priests of Nob. 1: *Adullam* was a place

southwest of Bethlehem, in David's home territory, so to speak. **2:** David collects a retinue of malcontents from his native Judah. **3–5:** The locations of *Mizpeh of Moab* and *Hereth* are unknown. On *the prophet Gad* see 2 Sam 24.12. **9:** *Doeg the Edomite,* a foreigner who had attached himself to Saul and held a good position under him (21.7).

risen against me, to lie in wait, as he is doing today?"

14 Then Ahimelech answered the king, "Who among all your servants is so faithful as David? He is the king's son-in-law, and is quick*w* to do your bidding, and is honored in your house. 15 Is today the first time that I have inquired of God for him? By no means! Do not let the king impute anything to his servant or to any member of my father's house; for your servant has known nothing of all this, much or little." 16 The king said, "You shall surely die, Ahimelech, you and all your father's house." 17 The king said to the guard who stood around him, "Turn and kill the priests of the LORD, because their hand also is with David; they knew that he fled, and did not disclose it to me." But the servants of the king would not raise their hand to attack the priests of the LORD. 18 Then the king said to Doeg, "You, Doeg, turn and attack the priests." Doeg the Edomite turned and attacked the priests; on that day he killed eighty-five who wore the linen ephod. 19 Nob, the city of the priests, he put to the sword; men and women, children and infants, oxen, donkeys, and sheep, he put to the sword.

20 But one of the sons of Ahimelech son of Ahitub, named Abiathar, escaped and fled after David. 21 Abiathar told David that Saul had killed the priests of the LORD. 22 David said to Abiathar, "I knew on that day, when Doeg the Edomite was there, that he would surely tell Saul. I am responsible*x* for the lives of all your father's house. 23 Stay with me, and do not be afraid; for the one who seeks my life seeks your life; you will be safe with me."

23 Now they told David, "The Philistines are fighting against Keilah, and are robbing the threshing floors." 2 David inquired of the LORD, "Shall I go and attack these Philistines?" The LORD said to David, "Go and attack the Philistines and save Keilah." 3 But David's men said to him, "Look, we are afraid here in Judah; how much more then if we go to Keilah against the armies of the Philistines?" 4 Then David inquired of the LORD again. The LORD answered him, "Yes, go down to Keilah; for I will give the Philistines into your hand." 5 So David and his men went to Keilah, fought with the Philistines, brought away their livestock, and dealt them a heavy defeat. Thus David rescued the inhabitants of Keilah.

6 When Abiathar son of Ahimelech fled to David at Keilah, he came down with an ephod in his hand. 7 Now it was told Saul that David had come to Keilah. And Saul said, "God has given*y* him into my hand; for he has shut himself in by entering a town that has gates and bars." 8 Saul summoned all the people to war, to go down to Keilah, to besiege David and his men. 9 When David learned that Saul was plotting evil against him, he said to the priest Abiathar, "Bring the ephod here." 10 David said, "O LORD, the God of Israel, your servant has heard that Saul seeks to come to Keilah, to destroy the city on my account. 11 And now, will*z* Saul come down as your servant has heard? O LORD, the God of Israel, I beseech you, tell your servant." The LORD said, "He will come down." 12 Then Da-

w Heb *and turns aside* *x* Gk Vg: Meaning of Heb uncertain *y* Gk Tg: Heb *made a stranger of* *z* Q Ms Compare Gk: MT *Will the men of Keilah surrender me into his hand? Will*

22.17: The soldiers of Saul refused to commit this atrocious deed against their fellow Israelites, and Saul did not have the authority to compel them. 18: *Doeg,* the foreigner, with loyalty to Saul but none to the Israelites, willingly did the deed. *Linen ephod,* see 2.18 n. 19: See 15.3 n. Compare Josh ch 7. 20–23: *Abiathar* was to play a very important part in the subsequent history of David.

23.1–14: **David's relief of Keilah. 1:** *Keilah* was a few miles south of Adullam (22.1). 3: From this verse it is clear that *Keilah* belonged neither to *Judah* nor to *the Philistines* at that time.

23.6: The *ephod* here is the box containing the sacred lots, as in 21.9 (see 14.3 n.); compare vv. 9–10.

23.15–24.22: **David spares Saul's life.**

vid said, "Will the men of Keilah surrender me and my men into the hand of Saul?" The LORD said, "They will surrender you." [13]Then David and his men, who were about six hundred, set out and left Keilah; they wandered wherever they could go. When Saul was told that David had escaped from Keilah, he gave up the expedition. [14]David remained in the strongholds in the wilderness, in the hill country of the Wilderness of Ziph. Saul sought him every day, but the LORD[a] did not give him into his hand.

[15] David was in the Wilderness of Ziph at Horesh when he learned that[b] Saul had come out to seek his life. [16]Saul's son Jonathan set out and came to David at Horesh; there he strengthened his hand through the LORD.[c] [17]He said to him, "Do not be afraid; for the hand of my father Saul shall not find you; you shall be king over Israel, and I shall be second to you; my father Saul also knows that this is so." [18]Then the two of them made a covenant before the LORD; David remained at Horesh, and Jonathan went home.

[19] Then some Ziphites went up to Saul at Gibeah and said, "David is hiding among us in the strongholds of Horesh, on the hill of Hachilah, which is south of Jeshimon. [20]Now, O king, whenever you wish to come down, do so; and our part will be to surrender him into the king's hand." [21]Saul said, "May you be blessed by the LORD for showing me compassion! [22]Go and make sure once more; find out exactly where he is, and who has seen him there; for I am told that he is very cunning. [23]Look around and learn all the hiding places where he lurks, and come back to me with sure information. Then I will go with you; and if he is in the land, I will search him out

among all the thousands of Judah." [24]So they set out and went to Ziph ahead of Saul.

David and his men were in the wilderness of Maon, in the Arabah to the south of Jeshimon. [25]Saul and his men went to search for him. When David was told, he went down to the rock and stayed in the wilderness of Maon. When Saul heard that, he pursued David into the wilderness of Maon. [26]Saul went on one side of the mountain, and David and his men on the other side of the mountain. David was hurrying to get away from Saul, while Saul and his men were closing in on David and his men to capture them. [27]Then a messenger came to Saul, saying, "Hurry and come; for the Philistines have made a raid on the land." [28]So Saul stopped pursuing David, and went against the Philistines; therefore that place was called the Rock of Escape.[d] [29e]David then went up from there, and lived in the strongholds of En-gedi.

24 When Saul returned from following the Philistines, he was told, "David is in the wilderness of En-gedi." [2]Then Saul took three thousand chosen men out of all Israel, and went to look for David and his men in the direction of the Rocks of the Wild Goats. [3]He came to the sheepfolds beside the road, where there was a cave; and Saul went in to relieve himself.[f] Now David and his men were sitting in the innermost parts of the cave. [4]The men of David said to him, "Here is the day of which the LORD said to you, 'I will give your enemy into your hand, and you shall do to him as it seems good to you.' " Then David went

---

a Q Ms Gk: MT *God*   b Or *saw that*
c Compare Q Ms Gk: MT *God*   d Or *Rock of Division*; Meaning of Heb uncertain
e Ch 24.1 in Heb   f Heb *to cover his feet*

Another version of this or a similar incident is found in ch 26. **14:** *Ziph* was a rocky area south of Hebron. *Every day*, in this context means "all the time" or "constantly." **19:** The names *Hachilah* and *Jeshimon*, which possibly mean "darkness" and "barren waste" respectively, suggest good places for hiding.
23.24: *Maon* was a little *south* of Ziph, Ha-

chilah, and *Jeshimon* (compare Josh 15.55). *Arabah* here simply means "desert," not the depression north and south of the Dead Sea. **27–29:** In ch 26 there is no interruption by the *Philistines,* nor does the scene shift to *En-gedi,* on the west shore of the Dead Sea.
24.2: *Rocks of the Wild Goats* is most descriptive of the character of the terrain.

and stealthily cut off a corner of Saul's cloak. 5 Afterward David was stricken to the heart because he had cut off a corner of Saul's cloak. 6 He said to his men, "The LORD forbid that I should do this thing to my lord, the LORD's anointed, to raise my hand against him; for he is the LORD's anointed." 7 So David scolded his men severely and did not permit them to attack Saul. Then Saul got up and left the cave, and went on his way.

8 Afterwards David also rose up and went out of the cave and called after Saul, "My lord the king!" When Saul looked behind him, David bowed with his face to the ground, and did obeisance. 9 David said to Saul, "Why do you listen to the words of those who say, 'David seeks to do you harm'? 10 This very day your eyes have seen how the LORD gave you into my hand in the cave; and some urged me to kill you, but I spared*g* you. I said, 'I will not raise my hand against my lord; for he is the LORD's anointed.' 11 See, my father, see the corner of your cloak in my hand; for by the fact that I cut off the corner of your cloak, and did not kill you, you may know for certain that there is no wrong or treason in my hands. I have not sinned against you, though you are hunting me to take my life. 12 May the LORD judge between me and you! May the LORD avenge me on you; but my hand shall not be against you. 13 As the ancient proverb says, 'Out of the wicked comes forth wickedness'; but my hand shall not be against you. 14 Against whom has the king of Israel come out? Whom do you pursue? A dead dog? A

single flea? 15 May the LORD therefore be judge, and give sentence between me and you. May he see to it, and plead my cause, and vindicate me against you."

16 When David had finished speaking these words to Saul, Saul said, "Is this your voice, my son David?" Saul lifted up his voice and wept. 17 He said to David, "You are more righteous than I; for you have repaid me good, whereas I have repaid you evil. 18 Today you have explained how you have dealt well with me, in that you did not kill me when the LORD put me into your hands. 19 For who has ever found an enemy, and sent the enemy safely away? So may the LORD reward you with good for what you have done to me this day. 20 Now I know that you shall surely be king, and that the kingdom of Israel shall be established in your hand. 21 Swear to me therefore by the LORD that you will not cut off my descendants after me, and that you will not wipe out my name from my father's house." 22 So David swore this to Saul. Then Saul went home; but David and his men went up to the stronghold.

25 Now Samuel died; and all Israel assembled and mourned for him. They buried him at his home in Ramah.

Then David got up and went down to the wilderness of Paran.

2 There was a man in Maon, whose property was in Carmel. The man was very rich; he had three thousand sheep and a thousand goats. He was shearing his sheep in Carmel. 3 Now the name of

g Gk Syr Tg Vg: Heb *it* (my eye) *spared*

24.14: Dogs were held in low esteem in those days; *a dead dog* would be less than nothing (2 Sam 9.8; 16.9). **20–21**: The incident comes to a close with Saul practically in tears, handing the kingdom to David (compare what Jonathan says in 23.17). In 26.22 nothing is said about David's becoming king. **25.1a**: **The death of Samuel.** The notation of an editor, perhaps, repeated in 28.3a. The brevity of the obituary is surprising. **25.1b–44**: **David, Nabal, and Abigail.** The story of how David obtained his second wife. **1b**: *The wilderness of Paran* is so far to the

south that it could hardly have had any connection with the movements of David at this time; the Greek version reads "the wilderness of Maon," and some modern translations follow this reading. **2**: This *Carmel* lay between Ziph and Maon (see 23.14 n. and 24 n.); it must not be confused with the famous Mount Carmel to the north near the seacoast. **3**: The *Calebite* clan had not yet been absorbed by the tribes of Judah (Josh 15.13–19; Judg 1.12–15). **11**: *Water,* the Greek version reads "wine," no doubt correctly (compare vv. 18 and 36).

the man was Nabal, and the name of his wife Abigail. The woman was clever and beautiful, but the man was surly and mean; he was a Calebite. 4David heard in the wilderness that Nabal was shearing his sheep. 5So David sent ten young men; and David said to the young men, "Go up to Carmel, and go to Nabal, and greet him in my name. 6Thus you shall salute him: 'Peace be to you, and peace be to your house, and peace be to all that you have. 7I hear that you have shearers; now your shepherds have been with us, and we did them no harm, and they missed nothing, all the time they were in Carmel. 8Ask your young men, and they will tell you. Therefore let my young men find favor in your sight; for we have come on a feast day. Please give whatever you have at hand to your servants and to your son David.' "

9 When David's young men came, they said all this to Nabal in the name of David; and then they waited. 10But Nabal answered David's servants, "Who is David? Who is the son of Jesse? There are many servants today who are breaking away from their masters. 11Shall I take my bread and my water and the meat that I have butchered for my shearers, and give it to men who come from I do not know where?" 12So David's young men turned away, and came back and told him all this. 13David said to his men, "Every man strap on his sword!" And every one of them strapped on his sword; David also strapped on his sword; and about four hundred men went up after David, while two hundred remained with the baggage.

14 But one of the young men told Abigail, Nabal's wife, "David sent messengers out of the wilderness to salute our master; and he shouted insults at them. 15Yet the men were very good to us, and we suffered no harm, and we never missed anything when we were in the fields, as long as we were with them;

16they were a wall to us both by night and by day, all the while we were with them keeping the sheep. 17Now therefore know this and consider what you should do; for evil has been decided against our master and against all his house; he is so ill-natured that no one can speak to him."

18 Then Abigail hurried and took two hundred loaves, two skins of wine, five sheep ready dressed, five measures of parched grain, one hundred clusters of raisins, and two hundred cakes of figs. She loaded them on donkeys 19and said to her young men, "Go on ahead of me; I am coming after you." But she did not tell her husband Nabal. 20As she rode on the donkey and came down under cover of the mountain, David and his men came down toward her; and she met them. 21Now David had said, "Surely it was in vain that I protected all that this fellow has in the wilderness, so that nothing was missed of all that belonged to him; but he has returned me evil for good. 22God do so to David*h* and more also, if by morning I leave so much as one male of all who belong to him."

23 When Abigail saw David, she hurried and alighted from the donkey, fell before David on her face, bowing to the ground. 24She fell at his feet and said, "Upon me alone, my lord, be the guilt; please let your servant speak in your ears, and hear the words of your servant. 25My lord, do not take seriously this ill-natured fellow, Nabal; for as his name is, so is he; Nabal*i* is his name, and folly is with him; but I, your servant, did not see the young men of my lord, whom you sent.

26 Now then, my lord, as the LORD lives, and as you yourself live, since the LORD has restrained you from bloodguilt and from taking vengeance with your

h Gk Compare Syr: Heb *the enemies of David*
i That is *Fool*

25.16: David and his men were outlaws, but of the "Robin Hood" type, even acting as a police force at times (compare v. 21).

25.26: The oath *as the LORD lives, and as you yourself live* may be translated more simply thus: "by the life of the LORD and by your

own hand, now let your enemies and those who seek to do evil to my lord be like Nabal. 27 And now let this present that your servant has brought to my lord be given to the young men who follow my lord. 28 Please forgive the trespass of your servant; for the LORD will certainly make my lord a sure house, because my lord is fighting the battles of the LORD; and evil shall not be found in you so long as you live. 29 If anyone should rise up to pursue you and to seek your life, the life of my lord shall be bound in the bundle of the living under the care of the LORD your God; but the lives of your enemies he shall sling out as from the hollow of a sling. 30 When the LORD has done to my lord according to all the good that he has spoken concerning you, and has appointed you prince over Israel, 31 my lord shall have no cause of grief, or pangs of conscience, for having shed blood without cause or for having saved himself. And when the LORD has dealt well with my lord, then remember your servant."

32 David said to Abigail, "Blessed be the LORD, the God of Israel, who sent you to meet me today! 33 Blessed be your good sense, and blessed be you, who have kept me today from bloodguilt and from avenging myself by my own hand! 34 For as surely as the LORD the God of Israel lives, who has restrained me from hurting you, unless you had hurried and come to meet me, truly by morning there would not have been left to Nabal so much as one male." 35 Then David received from her hand what she had brought him; he said to her, "Go up to your house in peace; see, I have heeded your voice, and I have granted your petition."

36 Abigail came to Nabal; he was holding a feast in his house, like the feast of a king. Nabal's heart was merry within him, for he was very drunk; so she told him nothing at all until the morning light. 37 In the morning, when the wine had gone out of Nabal, his wife told him these things, and his heart died within him; he became like a stone. 38 About ten days later the LORD struck Nabal, and he died.

39 When David heard that Nabal was dead, he said, "Blessed be the LORD who has judged the case of Nabal's insult to me, and has kept back his servant from evil; the LORD has returned the evildoing of Nabal upon his own head." Then David sent and wooed Abigail, to make her his wife. 40 When David's servants came to Abigail at Carmel, they said to her, "David has sent us to you to take you to him as his wife." 41 She rose and bowed down, with her face to the ground, and said, "Your servant is a slave to wash the feet of the servants of my lord." 42 Abigail got up hurriedly and rode away on a donkey; her five maids attended her. She went after the messengers of David and became his wife.

43 David also married Ahinoam of Jezreel; both of them became his wives. 44 Saul had given his daughter Michal, David's wife, to Palti son of Laish, who was from Gallim.

26 Then the Ziphites came to Saul at Gibeah, saying, "David is in hid-

---

own life." **29:** *The bundle of the living* is the precious package of those for whom the LORD cares. **30:** The writer anticipates the future events in crediting Abigail with knowing that David was going to be king of Israel.

25.37: Perhaps a stroke or a heart attack. **38:** In biblical theology life and death are in the hands of God. **41:** Note the formal obsequiousness of Abigail's words. **43–44:** David lost one wife and gained two more. By marrying Abigail, David consolidated his position with the powerful Calebite clan. *Ahino-* am was also a woman of the southern area; compare Josh 15.56 for the location of this *Jezreel* (not to be confused with the plain of Esdraelon). These marriages were a great help to David politically.

26.1–25: David spares Saul's life. See 23.15–24.22 for another version of this incident. **5–6:** *Abner, Joab,* and *Abishai* were destined to play important parts in the story of David's reign in 2 Samuel. *Ahimelech the Hittite* is not to be confused with Ahimelech the priest in ch 21. Hittites had settled in the

ing on the hill of Hachilah, which is opposite Jeshimon."[j] 2 So Saul rose and went down to the Wilderness of Ziph, with three thousand chosen men of Israel, to seek David in the Wilderness of Ziph. 3 Saul encamped on the hill of Hachilah, which is opposite Jeshimon[j] beside the road. But David remained in the wilderness. When he learned that Saul came after him into the wilderness, 4 David sent out spies, and learned that Saul had indeed arrived. 5 Then David set out and came to the place where Saul had encamped; and David saw the place where Saul lay, with Abner son of Ner, the commander of his army. Saul was lying within the encampment, while the army was encamped around him.

6 Then David said to Ahimelech the Hittite, and to Joab's brother Abishai son of Zeruiah, "Who will go down with me into the camp to Saul?" Abishai said, "I will go down with you." 7 So David and Abishai went to the army by night; there Saul lay sleeping within the encampment, with his spear stuck in the ground at his head; and Abner and the army lay around him. 8 Abishai said to David, "God has given your enemy into your hand today; now therefore let me pin him to the ground with one stroke of the spear; I will not strike him twice." 9 But David said to Abishai, "Do not destroy him; for who can raise his hand against the LORD's anointed, and be guiltless?" 10 David said, "As the LORD lives, the LORD will strike him down; or his day will come to die; or he will go down into battle and perish. 11 The LORD forbid that I should raise my hand against the LORD's anointed; but now take the spear that is at his head, and the water jar, and let us go." 12 So David took the spear that was at Saul's head and the water jar, and they went away. No one saw it, or knew it, nor did anyone awake; for they were all asleep, because a deep sleep from the LORD had fallen upon them.

13 Then David went over to the other side, and stood on top of a hill far away, with a great distance between them. 14 David called to the army and to Abner son of Ner, saying, "Abner! Will you not answer?" Then Abner replied, "Who are you that calls to the king?" 15 David said to Abner, "Are you not a man? Who is like you in Israel? Why then have you not kept watch over your lord the king? For one of the people came in to destroy your lord the king. 16 This thing that you have done is not good. As the LORD lives, you deserve to die, because you have not kept watch over your lord, the LORD's anointed. See now, where is the king's spear, or the water jar that was at his head?"

17 Saul recognized David's voice, and said, "Is this your voice, my son David?" David said, "It is my voice, my lord, O king." 18 And he added, "Why does my lord pursue his servant? For what have I done? What guilt is on my hands? 19 Now therefore let my lord the king hear the words of his servant. If it is the LORD who has stirred you up against me, may he accept an offering; but if it is mortals, may they be cursed before the LORD, for they have driven me out today from my share in the heritage of the LORD, saying, 'Go, serve other gods.' 20 Now therefore, do not let my blood fall to the ground, away from the presence of the LORD; for the king of Israel has come out to seek a single flea, like one who hunts a partridge in the mountains."

21 Then Saul said, "I have done wrong; come back, my son David, for I will never harm you again, because my life was precious in your sight today; I have been a fool, and have made a great

---

[j] Or *opposite the wasteland*

---

country in earlier times (Gen 23.7; Josh 1.4), coming from the north. The few who remained in the time of David joined with the Israelites, adopting Hebrew names (as Uriah the Hittite, 2 Sam 11.3). **9:** *The LORD's anointed,* see 10.1 n.

26.13–16: Even today the Arabs in the wilder parts of the country shout across great distances in this manner. **17:** See 24.16. **21–24:** The meeting ends on a note of reconciliation, but, to judge from the opening words of the next chapter, David trusted Saul not at all.

mistake." ²²David replied, "Here is the spear, O king! Let one of the young men come over and get it. ²³The LORD rewards everyone for his righteousness and his faithfulness; for the LORD gave you into my hand today, but I would not raise my hand against the LORD's anointed. ²⁴As your life was precious today in my sight, so may my life be precious in the sight of the LORD, and may he rescue me from all tribulation." ²⁵Then Saul said to David, "Blessed be you, my son David! You will do many things and will succeed in them." So David went his way, and Saul returned to his place.

27 David said in his heart, "I shall now perish one day by the hand of Saul; there is nothing better for me than to escape to the land of the Philistines; then Saul will despair of seeking me any longer within the borders of Israel, and I shall escape out of his hand." ²So David set out and went over, he and the six hundred men who were with him, to King Achish son of Maoch of Gath. ³David stayed with Achish at Gath, he and his troops, every man with his household, and David with his two wives, Ahinoam of Jezreel, and Abigail of Carmel, Nabal's widow. ⁴When Saul was told that David had fled to Gath, he no longer sought for him.

5 Then David said to Achish, "If I have found favor in your sight, let a place be given me in one of the country towns, so that I may live there; for why should your servant live in the royal city with you?" ⁶So that day Achish gave him Ziklag; therefore Ziklag has belonged to the kings of Judah to this day. ⁷The length of time that David lived in the country of the Philistines was one year and four months.

8 Now David and his men went up and made raids on the Geshurites, the Girzites, and the Amalekites; for these were the landed settlements from Telam^k on the way to Shur and on to the land of Egypt. ⁹David struck the land, leaving neither man nor woman alive, but took away the sheep, the oxen, the donkeys, the camels, and the clothing, and came back to Achish. ¹⁰When Achish asked, "Against whom^l have you made a raid today?" David would say, "Against the Negeb of Judah," or "Against the Negeb of the Jerahmeelites," or, "Against the Negeb of the Kenites." ¹¹David left neither man nor woman alive to be brought back to Gath, thinking, "They might tell about us, and say, 'David has done so and so.' " Such was his practice all the time he lived in the country of the Philistines. ¹²Achish trusted David, thinking, "He has made himself utterly abhorrent to his people Israel; therefore he shall always be my servant."

28 In those days the Philistines gathered their forces for war, to fight against Israel. Achish said to David,

k Compare Gk 15.4: Heb *from of old*   *l* Q Ms Gk Vg: MT lacks *whom*

27.1–12: David becomes a vassal of the Philistines. If David really became loyal to the Philistines, then he was a traitor to his own people; if he was not loyal to the Philistines, then he was a deceiver—so the dilemma is posed. In 27.10–12 he seems to be deceiving the Philistines, yet in 29.8 he seems to affirm his loyalty to Achish. Actually, the situation developed in such a way that he did not have to make a public choice (ch 29). **2:** David went to *Gath,* the very place from which Goliath was reputed to have come (17.4). **3:** Hope of better conditions for their wives and children was one of the reasons why David and his men took this step. **5–7:** *Ziklag* was located somewhere near the border between the territory of the Philistines and Judah.

27.8–12: David actually raided more distant and hostile peoples, such as *the Geshurites, the Girzites, and the Amalekites* (ch 15), while pretending that he had attacked his own or friendly peoples, such as *Judah, the Jerahmeelites,* or *the Kenites* (30.26–31), who lived in the *Negeb* (southern Palestine).

28.1–25: Saul consults the spirit of Samuel through the medium of Endor. This passage plots its action as an "aside," against the ominous background of gathering hostile forces (vv. 1, 4; 29.1). The scene is laid at Gilboa (v. 4) and nearby Endor (v. 7) on the night before the battle. **6:** The ordinary

"You know, of course, that you and your men are to go out with me in the army." [2]David said to Achish, "Very well, then you shall know what your servant can do." Achish said to David, "Very well, I will make you my bodyguard for life."

3 Now Samuel had died, and all Israel had mourned for him and buried him in Ramah, his own city. Saul had expelled the mediums and the wizards from the land. [4]The Philistines assembled, and came and encamped at Shunem. Saul gathered all Israel, and they encamped at Gilboa. [5]When Saul saw the army of the Philistines, he was afraid, and his heart trembled greatly. [6]When Saul inquired of the LORD, the LORD did not answer him, not by dreams, or by Urim, or by prophets. [7]Then Saul said to his servants, "Seek out for me a woman who is a medium, so that I may go to her and inquire of her." His servants said to him, "There is a medium at Endor."

8 So Saul disguised himself and put on other clothes and went there, he and two men with him. They came to the woman by night. And he said, "Consult a spirit for me, and bring up for me the one whom I name to you." [9]The woman said to him, "Surely you know what Saul has done, how he has cut off the mediums and the wizards from the land. Why then are you laying a snare for my life to bring about my death?" [10]But Saul swore to her by the LORD, "As the LORD lives, no punishment shall come upon you for this thing." [11]Then the woman said, "Whom shall I bring up for you?" He answered, "Bring up Samuel for me." [12]When the woman saw Samuel, she cried out with a loud voice; and the woman said to Saul, "Why have you deceived me? You are Saul!" [13]The king said to her, "Have no fear; what do you see?" The woman said to Saul, "I see a divine being*m* coming up out of the ground." [14]He said to her, "What is his appearance?" She said, "An old man is coming up; he is wrapped in a robe." So Saul knew that it was Samuel, and he bowed with his face to the ground, and did obeisance.

15 Then Samuel said to Saul, "Why have you disturbed me by bringing me up?" Saul answered, "I am in great distress, for the Philistines are warring against me, and God has turned away from me and answers me no more, either by prophets or by dreams; so I have summoned you to tell me what I should do." [16]Samuel said, "Why then do you ask me, since the LORD has turned from you and become your enemy? [17]The LORD has done to you just as he spoke by me; for the LORD has torn the kingdom out of your hand, and given it to your neighbor, David. [18]Because you did not obey the voice of the LORD, and did not carry out his fierce wrath against Amalek, therefore the LORD has done this thing to you today. [19]Moreover the LORD will give Israel along with you into the hands of the Philistines; and tomorrow you and your sons shall be with me; the LORD will also give the army of Israel into the hands of the Philistines."

20 Immediately Saul fell full length on the ground, filled with fear because of the words of Samuel; and there was no

m Or *a god*; or *gods*

---

means of finding God's will had failed Saul. On *Urim* ("and Thummim" understood) see 14.3 n. and 14.42.

28.9: Necromancy (consultation of the dead) continued to be practiced from time to time (Isa 8.19; 2 Kings 21.6) although forbidden by written law (Deut 18.10–11; Lev 19.31; 20.6, 27). Apparently it was already considered evil and a threat to true religion in Saul's time, and he had made an effort to stamp it out; but under pressure he himself relapsed into former practice. **12**: *Samuel*, probably an error for "Saul." **13**: The words *divine being* here mean a being from another world; that world was Sheol, the abode of the dead, conceived of as a great hollow place under the flat earth, hence the expression *coming up* (compare the word *up* in v. 11).

28.17–18: The writer or perhaps a later editor has added the prediction of David's kingship and a reference to the Amalek incident (compare ch 15).

28.20–25: Note the human kindness of the medium toward Saul.

strength in him, for he had eaten nothing all day and all night. 21 The woman came to Saul, and when she saw that he was terrified, she said to him, "Your servant has listened to you; I have taken my life in my hand, and have listened to what you have said to me. 22 Now therefore, you also listen to your servant; let me set a morsel of bread before you. Eat, that you may have strength when you go on your way." 23 He refused, and said, "I will not eat." But his servants, together with the woman, urged him; and he listened to their words. So he got up from the ground and sat on the bed. 24 Now the woman had a fatted calf in the house. She quickly slaughtered it, and she took flour, kneaded it, and baked unleavened cakes. 25 She put them before Saul and his servants, and they ate. Then they rose and went away that night.

29 Now the Philistines gathered all their forces at Aphek, while the Israelites were encamped by the fountain that is in Jezreel. 2 As the lords of the Philistines were passing on by hundreds and by thousands, and David and his men were passing on in the rear with Achish, 3 the commanders of the Philistines said, "What are these Hebrews doing here?" Achish said to the commanders of the Philistines, "Is this not David, the servant of King Saul of Israel, who has been with me now for days and years? Since he deserted to me I have found no fault in him to this day." 4 But the commanders of the Philistines were angry with him; and the commanders of the Philistines said to him, "Send the man back, so that he may return to the place that you have assigned to him; he shall not go down with us to battle, or

else he may become an adversary to us in the battle. For how could this fellow reconcile himself to his lord? Would it not be with the heads of the men here? 5 Is this not David, of whom they sing to one another in dances,

'Saul has killed his thousands,
and David his ten thousands'?"

6 Then Achish called David and said to him, "As the LORD lives, you have been honest, and to me it seems right that you should march out and in with me in the campaign; for I have found nothing wrong in you from the day of your coming to me until today. Nevertheless the lords do not approve of you. 7 So go back now; and go peaceably; do nothing to displease the lords of the Philistines." 8 David said to Achish, "But what have I done? What have you found in your servant from the day I entered your service until now, that I should not go and fight against the enemies of my lord the king?" 9 Achish replied to David, "I know that you are as blameless in my sight as an angel of God; nevertheless, the commanders of the Philistines have said, 'He shall not go up with us to the battle.' 10 Now then rise early in the morning, you and the servants of your lord who came with you, and go to the place that I appointed for you. As for the evil report, do not take it to heart, for you have done well before me. *n* Start early in the morning, and leave as soon as you have light." 11 So David set out with his men early in the morning, to return to the land of the Philistines. But the Philistines went up to Jezreel.

n Gk: Heb lacks *and go to the place . . . done well before me*

29.1–11: The services of David rejected by the Philistine army. On David's attitude toward the Philistines, see 27.1–12 n. **1:** *The Philistines gathered . . . at Aphek,* where they had captured the ark at the beginning of the war (4.1b). *The Israelites* were at *Jezreel,* a considerable distance away in the plain of Esdraelon (Esdraelon is the Greek form of Jezreel). **3:** *Hebrews,* see 4.6 n. **29.6:** According to the Hebrew text,

Achish shows his respect for David by swearing by David's God (compare *an angel of God* in v. 9). **11:** *The Philistines* quickly marched on *Jezreel,* about seventy-five miles away, where the Israelites were encamped. The latter then moved up Mount Gilboa, where they fortified themselves in preparation for the battle; the Philistines took up positions at Shunem (28.4).

30 Now when David and his men came to Ziklag on the third day, the Amalekites had made a raid on the Negeb and on Ziklag. They had attacked Ziklag, burned it down, 2and taken captive the women and all*o* who were in it, both small and great; they killed none of them, but carried them off, and went their way. 3When David and his men came to the city, they found it burned down, and their wives and sons and daughters taken captive. 4Then David and the people who were with him raised their voices and wept, until they had no more strength to weep. 5David's two wives also had been taken captive, Ahinoam of Jezreel, and Abigail the widow of Nabal of Carmel. 6David was in great danger; for the people spoke of stoning him, because all the people were bitter in spirit for their sons and daughters. But David strengthened himself in the LORD his God.

7 David said to the priest Abiathar son of Ahimelech, "Bring me the ephod." So Abiathar brought the ephod to David. 8David inquired of the LORD, "Shall I pursue this band? Shall I overtake them?" He answered him, "Pursue; for you shall surely overtake and shall surely rescue." 9So David set out, he and the six hundred men who were with him. They came to the Wadi Besor, where those stayed who were left behind. 10But David went on with the pursuit, he and four hundred men; two hundred stayed behind, too exhausted to cross the Wadi Besor.

11 In the open country they found an Egyptian, and brought him to David. They gave him bread and he ate, they gave him water to drink; 12they also gave him a piece of fig cake and two clusters of raisins. When he had eaten, his spirit revived; for he had not eaten bread or drunk water for three days and three nights. 13Then David said to him, "To whom do you belong? Where are you from?" He said, "I am a young man of Egypt, servant to an Amalekite. My master left me behind because I fell sick three days ago. 14We had made a raid on the Negeb of the Cherethites and on that which belongs to Judah and on the Negeb of Caleb; and we burned Ziklag down." 15David said to him, "Will you take me down to this raiding party?" He said, "Swear to me by God that you will not kill me, or hand me over to my master, and I will take you down to them."

16 When he had taken him down, they were spread out all over the ground, eating and drinking and dancing, because of the great amount of spoil they had taken from the land of the Philistines and from the land of Judah. 17David attacked them from twilight until the evening of the next day. Not one of them escaped, except four hundred young men, who mounted camels and fled. 18David recovered all that the Amalekites had taken; and David rescued his two wives. 19Nothing was missing, whether small or great, sons or daughters, spoil or anything that had been taken; David brought back everything. 20David also captured all the flocks and herds, which were driven ahead of the other cattle; people said, "This is David's spoil."

21 Then David came to the two hundred men who had been too exhausted to follow David, and who had been left at the Wadi Besor. They went out to meet David and to meet the people who were with him. When David drew near to the people he saluted them. 22Then all the corrupt and worthless fellows among the men who had gone with David said,

o Gk: Heb lacks *and all*

30.1–31: **An interlude: the burning of Ziklag and David's pursuit of the Amalekites. 1**: *Ziklag* was about eighty miles south of Aphek, a long two-day march through rough terrain. **6**: The men almost revolted, and David saved himself only by an appeal in the name of God.

30.7: *Abiathar,* see 22.20–23 n. *The ephod,* see 14.3 n. **9–10**: *The Wadi Besor,* south of Ziklag. See 27.8–12 n. **14**: Some interpret the word *Cherethites* as meaning "Cretans" and so equate them with the Philistines (see 4.1b–7.2 n.; compare Ezek 25.16; Zeph 2.5). **30.21–24**: David was not only fair-

"Because they did not go with us, we will not give them any of the spoil that we have recovered, except that each man may take his wife and children, and leave." 23 But David said, "You shall not do so, my brothers, with what the Lord has given us; he has preserved us and handed over to us the raiding party that attacked us. 24 Who would listen to you in this matter? For the share of the one who goes down into the battle shall be the same as the share of the one who stays by the baggage; they shall share alike." 25 From that day forward he made it a statute and an ordinance for Israel; it continues to the present day.

26 When David came to Ziklag, he sent part of the spoil to his friends, the elders of Judah, saying, "Here is a present for you from the spoil of the enemies of the Lord"; 27 it was for those in Bethel, in Ramoth of the Negeb, in Jattir, 28 in Aroer, in Siphmoth, in Eshtemoa, 29 in Racal, in the towns of the Jerahmeelites, in the towns of the Kenites, 30 in Hormah, in Bor-ashan, in Athach, 31 in Hebron, all the places where David and his men had roamed.

31 Now the Philistines fought against Israel; and the men of Israel fled before the Philistines, and many fell*p* on Mount Gilboa. 2 The Philistines overtook Saul and his sons; and the Philistines killed Jonathan and Abinadab and Malchishua, the sons of Saul. 3 The battle pressed hard upon Saul; the archers found him, and he was badly wounded by them. 4 Then Saul said to his armorbearer, "Draw your sword and thrust me through with it, so that these uncircumcised may not come and thrust me through, and make sport of me." But his armor-bearer was unwilling; for he was terrified. So Saul took his own sword and fell upon it. 5 When his armor-bearer saw that Saul was dead, he also fell upon his sword and died with him. 6 So Saul and his three sons and his armor-bearer and all his men died together on the same day. 7 When the men of Israel who were on the other side of the valley and those beyond the Jordan saw that the men of Israel had fled and that Saul and his sons were dead, they forsook their towns and fled; and the Philistines came and occupied them.

8 The next day, when the Philistines came to strip the dead, they found Saul and his three sons fallen on Mount Gilboa. 9 They cut off his head, stripped off his armor, and sent messengers throughout the land of the Philistines to carry the

p Heb *and they fell slain*

---

minded, but politically wise at this point; he retained everybody's support. **25:** This verse is editorial; David was not yet king of Judah, to say nothing of Israel (compare Num 31.27). **26–31:** All these places, including *Bethel* (not the familiar one farther north), were in Judah. It is little wonder that David was soon made king of Judah (2 Sam 2.4). He could now work to that end, having been freed of his Philistine entanglement.

31.1–13: The battle of Gilboa; death of Saul and his older sons. Following the interlude (30.1–31), this chapter returns to the main plot of the Philistine threat. No details of the battle are given. Obviously the Israelites fortified themselves near the summit of the hill and the Philistines stormed their position successfully. **1–3:** The Israelites were totally defeated, three of Saul's sons were killed, and Saul was *badly wounded.* **4:** *These uncircumcised,* see 14.6 n. **5:** Biblical suicides are rare; compare 2 Sam 17.23; 1 Kings 16.18; Mt 27.5.

31.7: The phrase *and those beyond the Jordan,* if it refers to the Philistines, seems to be in error (it is lacking in 1 Chr 10.7). There is no evidence that the Philistines occupied territory east of the Jordan. **10:** *Astarte,* see 7.3–4 n. Was this *temple* in *Beth-shan,* which the Philistines apparently had occupied, or back in Philistia? The story is told somewhat differently in 1 Chr 10.10. **11–13:** The men of *Jabeshgilead* now had opportunity to show their gratitude for what Saul had done for them (ch 11). With the burial in Jabesh, the story of Saul comes to an end, except for the later return of his bones to his native territory (2 Sam 21.12–14). According to some parts of the tradition, Saul came to a noble but tragic end, defending his people and land to the best of his ability. To other writers, probably later, however, he was utterly unwor-

good news to the houses of their idols and to the people. 10 They put his armor in the temple of Astarte;*q* and they fastened his body to the wall of Beth-shan. 11 But when the inhabitants of Jabesh-gilead heard what the Philistines had done to Saul, 12 all the valiant men set out, traveled all night long, and took the body of Saul and the bodies of his sons from the wall of Beth-shan. They came to Jabesh and burned them there. 13 Then they took their bones and buried them under the tamarisk tree in Jabesh, and fasted seven days.

q Heb plural

---

thy, entirely deserving of an unhappy fate. Note the similar conclusion of 1 Chr 10.13–14. Later writers, in order to glorify David, felt it necessary to depreciate Saul. David himself may have had no such attitude (compare the immediate sequel, 2 Sam ch 1).

2 Samuel

The second book of Samuel relates the rule of David, first as he gradually assumed control in Judah when Saul's claimants fell away (chs 1–4), and then (chs 5–24) as king over both Judah and the northern tribes (Israel). Most scholars believe that chs 9–20, together with 1 Kings 1–2, constituted originally an independent document. Commonly admired for its literary qualities as the "Court History" or the "Throne Succession Narrative," this story deals with the domestic and political troubles of David, and tells finally how Solomon assumed the throne, like Abraham's sons, at the expense of the rules of primogeniture (the requirement that the eldest surviving son should become king). The account in chs 21–24 somewhat interrupts this plot, and may have been inserted by another hand. The two books of Samuel were originally one book, and problems of composition are discussed in the Introduction to 1 Samuel.

1 After the death of Saul, when David had returned from defeating the Amalekites, David remained two days in Ziklag. 2 On the third day, a man came from Saul's camp, with his clothes torn and dirt on his head. When he came to David, he fell to the ground and did obeisance. 3 David said to him, "Where have you come from?" He said to him, "I have escaped from the camp of Israel." 4 David said to him, "How did things go? Tell me!" He answered, "The army fled from the battle, but also many of the army fell and died; and Saul and his son Jonathan also died." 5 Then David asked the young man who was reporting to him, "How do you know that Saul and his son Jonathan died?" 6 The young man reporting to him said, "I happened to be on Mount Gilboa; and there was Saul leaning on his spear, while the chariots and the horsemen drew close to him. 7 When he looked behind him, he saw me, and called to me. I answered, 'Here sir.' 8 And he said to me, 'Who are you?' I answered him, 'I am an Amalekite.' 9 He said to me, 'Come, stand over me and kill me; for convulsions have seized me, and yet my life still lingers.' 10 So I stood over him, and killed him, for I knew that he could not live after he had fallen. I took the crown that was on his head and the armlet that was on his arm, and I have brought them here to my lord."

11 Then David took hold of his clothes and tore them; and all the men who were with him did the same. 12 They mourned and wept, and fasted until evening for Saul and for his son Jonathan, and for the army of the Lord and for the house of Israel, because they had fallen by the sword. 13 David said to the young man who had reported to him, "Where do you come from?" He answered, "I am the son of a resident alien, an Amalekite." 14 David said to him, "Were you not afraid to lift your hand to destroy the Lord's anointed?" 15 Then David called one of the young men and said, "Come here and strike him down." So he struck him down and he died. 16 David said to him, "Your blood be on your head; for your own mouth has testified against you, saying, 'I have killed the Lord's anointed.'"

17 David intoned this lamentation over Saul and his son Jonathan. 18 (He ordered that The Song of the Bow*a* be taught to the people of Judah; it is written in the Book of Jashar.) He said:

19 Your glory, O Israel, lies slain
    upon your high places!
  How the mighty have fallen!

20 Tell it not in Gath,
    proclaim it not in the streets of
      Ashkelon;
  or the daughters of the Philistines
      will rejoice,
    the daughters of the
      uncircumcised will exult.

21 You mountains of Gilboa,
    let there be no dew or rain upon
      you,
    nor bounteous fields!*b*
  For there the shield of the mighty
      was defiled,

a Heb *that The Bow*   b Meaning of Heb uncertain

1.1–16: David learns of the death of Saul and Jonathan. 13: *Resident alien,* Hebrew "gēr," a technical term for a foreigner residing in Israel over a long period, deserving protection but not entitled to full civil rights (Ex 20.10; Deut 14.29). **1.14–16:** Note David's continued respect for the memory of Saul and the office he held (1 Sam 26.9, 11, 16, 24). David of course did not know that the Amalekite was lying. **1.17–27: David's elegy over Saul and Jonathan.** David is said to have played the lyre (1 Sam 16.23) and is traditionally connected with the composition of psalms (see superscriptions to Pss 3–6 and many others). This commemorative poem takes a positive view of Saul's heroism, in contrast to the harshly negative evaluation of him in some parts of 1 Samuel. **1.18:** *The Book of Jashar* was apparently a collection of poems from various sources (see Josh 10.13 n.). **20:** *Gath* and *Ashkelon* were Philistine cities. *The uncircumcised,* see 1 Sam 14.6 n.

the shield of Saul, anointed with
oil no more.

22 From the blood of the slain,
from the fat of the mighty,
the bow of Jonathan did not turn
back,
nor the sword of Saul return
empty.

23 Saul and Jonathan, beloved and
lovely!
In life and in death they were
not divided;
they were swifter than eagles,
they were stronger than lions.

24 O daughters of Israel, weep over
Saul,
who clothed you with crimson,
in luxury,
who put ornaments of gold on
your apparel.

25 How the mighty have fallen
in the midst of the battle!

Jonathan lies slain upon your high
places.

26 I am distressed for you, my
brother Jonathan;

greatly beloved were you to me;
your love to me was wonderful,
passing the love of women.

27 How the mighty have fallen,
and the weapons of war
perished!

2 After this David inquired of the
LORD, "Shall I go up into any of the
cities of Judah?" The LORD said to him,
"Go up." David said, "To which shall I
go up?" He said, "To Hebron." 2 So David went up there, along with his two
wives, Ahinoam of Jezreel, and Abigail
the widow of Nabal of Carmel. 3 David
brought up the men who were with him,
every one with his household; and they
settled in the towns of Hebron. 4 Then
the people of Judah came, and there they
anointed David king over the house of
Judah.

When they told David, "It was the
people of Jabesh-gilead who buried
Saul," 5 David sent messengers to the
people of Jabesh-gilead, and said to
them, "May you be blessed by the LORD,
because you showed this loyalty to Saul
your lord, and buried him! 6 Now may
the LORD show steadfast love and faithfulness to you! And I too will reward you
because you have done this thing.

2.1–11: **David becomes king of Judah. 1:**
Probably David *inquired of the LORD* by means
of the ephod and Urim and Thummim (see
1 Sam 14.3, 41; 30.7–8). It should occasion no
surprise that the tribe of *Judah* was able and
willing to set up a separate kingdom. In the
list of the tribes of Israel in Judg 5.13–18
Judah is not mentioned; it was apparently
leading a separate existence. For other indications of separateness, see 3.10; 5.5; 19.8–15,
40–43; 20.1–2. *Judah* might never have joined
the other tribes if the Philistine danger had
not arisen. **4b–7:** David showed again his
respect for the memory of Saul. When he
dropped the political hint in v. 7, he probably
did not know that Israel had a new king
(vv. 8–9), or the new king had not yet been
appointed.
 2.8: On the relation of *Abner* to Saul, see
1 Sam 14.50–51. The name *Ishbaal,* "man of
the lord," may have been changed in the Hebrew Mss (see note *c*) to *Ish-bosheth,* "man of
shame." The element "baal" might have sug-

gested the Canaanite god Baal who is particularly detested in the Deuteronomistic History
(see Introduction to the Historical Books) as
a rival of Israel's God (see Judg 2.11; 1 Kings
18.17–40). *Mahanaim* was the principal town
of Gilead; the Philistines were in control of
the territory of the Jordan. **9:** At the moment
Ishbaal was king over *Gilead* only, but was
theoretically king over *all* Israel. The other
places mentioned probably sent representatives. The word *Ashurites* really means "Assyrians," a manifest impossibility; probably
the correct form is "Asherites." **10–11:** As so
frequently in ancient works, biblical and non-
biblical, the figures seem doubtful. *Ishbaal*
could not have been *forty years old.* Though he
is mentioned in 1 Sam 14.49 (called "Ishvi"
there), he did not fight at Gilboa; therefore he
was probably a minor at this time. This circumstance would explain why he was so
completely under the control of Abner. If *he
reigned two years* only, then *David* could hardly have reigned more than *seven years* in He-

7 Therefore let your hands be strong, and be valiant; for Saul your lord is dead, and the house of Judah has anointed me king over them."

8 But Abner son of Ner, commander of Saul's army, had taken Ishbaal[c] son of Saul, and brought him over to Mahanaim. 9 He made him king over Gilead, the Ashurites, Jezreel, Ephraim, Benjamin, and over all Israel. 10 Ishbaal,[c] Saul's son, was forty years old when he began to reign over Israel, and he reigned two years. But the house of Judah followed David. 11 The time that David was king in Hebron over the house of Judah was seven years and six months.

12 Abner son of Ner, and the servants of Ishbaal[c] son of Saul, went out from Mahanaim to Gibeon. 13 Joab son of Zeruiah, and the servants of David, went out and met them at the pool of Gibeon. One group sat on one side of the pool, while the other sat on the other side of the pool. 14 Abner said to Joab, "Let the young men come forward and have a contest before us." Joab said, "Let them come forward." 15 So they came forward and were counted as they passed by, twelve for Benjamin and Ishbaal[c] son of Saul, and twelve of the servants of David. 16 Each grasped his opponent by the head, and thrust his sword in his opponent's side; so they fell down together. Therefore that place was called Helkath-hazzurim,[d] which is at Gibeon. 17 The battle was very fierce that day; and Abner and the men of Israel were beaten by the servants of David.

18 The three sons of Zeruiah were there, Joab, Abishai, and Asahel. Now Asahel was as swift of foot as a wild gazelle. 19 Asahel pursued Abner, turning neither to the right nor to the left as he followed him. 20 Then Abner looked back and said, "Is it you, Asahel?" He answered, "Yes, it is." 21 Abner said to him, "Turn to your right or to your left, and seize one of the young men, and take his spoil." But Asahel would not turn away from following him. 22 Abner said again to Asahel, "Turn away from following me; why should I strike you to the ground? How then could I show my face to your brother Joab?" 23 But he refused to turn away. So Abner struck him in the stomach with the butt of his spear, so that the spear came out at his back. He fell there, and died where he lay. And all those who came to the place where Asahel had fallen and died, stood still.

24 But Joab and Abishai pursued Abner. As the sun was going down they came to the hill of Ammah, which lies before Giah on the way to the wilderness of Gibeon. 25 The Benjaminites rallied around Abner and formed a single band; they took their stand on the top of a hill. 26 Then Abner called to Joab, "Is the sword to keep devouring forever? Do you not know that the end will be bitter? How long will it be before you order your people to turn from the pursuit of their kinsmen?" 27 Joab said, "As God lives, if you had not spoken, the people would have continued to pursue their kinsmen, not stopping until morning." 28 Joab sounded the trumpet and all the

c Gk Compare 1 Chr 8.33; 9.39: Heb *Ish-bosheth*, "man of shame"   d That is *Field of Sword-edges*

---

bron. We gain the impression from 5.6 that *David* left *Hebron* soon after the death of *Ishbaal*. Probably the latter reigned more than *two years* and the former less than *seven years* in Hebron.

2.12–32: War with Israel breaks out. *Gibeon* (Josh ch 9) was an important town about five miles northwest of Jerusalem. Apparently both sides were planning to occupy the place when their fighting men met there. Something like a tournament, the exact nature of which is not clear, was agreed upon, but this soon degenerated into a real military engagement. *Joab* was David's nephew; both Saul (v. 8) and David had appointed relatives to be their commanders-in-chief.

2.18: *Zeruiah* was a sister or half sister of David (1 Chr 2.13–16). Joab and Abishai (1 Sam 26.6–9) play important parts in the history that follows. **22–23:** The killing of *Asahel* by *Abner* brought about a blood feud between *Joab* and *Abner*.

2.24–28: Both commanders realized that they had gone too far and a truce was called.

people stopped; they no longer pursued Israel or engaged in battle any further. 29 Abner and his men traveled all that night through the Arabah; they crossed the Jordan, and, marching the whole forenoon, *e* they came to Mahanaim. 30 Joab returned from the pursuit of Abner; and when he had gathered all the people together, there were missing of David's servants nineteen men besides Asahel. 31 But the servants of David had killed of Benjamin three hundred sixty of Abner's men. 32 They took up Asahel and buried him in the tomb of his father, which was at Bethlehem. Joab and his men marched all night, and the day broke upon them at Hebron.

3 There was a long war between the house of Saul and the house of David; David grew stronger and stronger, while the house of Saul became weaker and weaker.

2 Sons were born to David at Hebron: his firstborn was Amnon, of Ahinoam of Jezreel; 3 his second, Chileab, of Abigail the widow of Nabal of Carmel; the third, Absalom son of Maacah, daughter of King Talmai of Geshur; 4 the fourth, Adonijah son of Haggith; the fifth, Shephatiah son of Abital; 5 and the sixth, Ithream, of David's wife Eglah. These were born to David in Hebron.

6 While there was war between the house of Saul and the house of David, Abner was making himself strong in the house of Saul. 7 Now Saul had a concubine whose name was Rizpah daughter of Aiah. And Ishbaal*f* said to Abner, "Why have you gone in to my father's concubine?" 8 The words of Ishbaal*g* made Abner very angry; he said, "Am I a dog's head for Judah? Today I keep showing loyalty to the house of your father Saul, to his brothers, and to his friends, and have not given you into the hand of David; and yet you charge me now with a crime concerning this woman. 9 So may God do to Abner and so may he add to it! For just what the Lord has sworn to David, that will I accomplish for him, 10 to transfer the kingdom from the house of Saul, and set up the throne of David over Israel and over Judah, from Dan to Beer-sheba." 11 And Ishbaal*f* could not answer Abner another word, because he feared him.

12 Abner sent messengers to David at Hebron, *h* saying, "To whom does the land belong? Make your covenant with me, and I will give you my support to bring all Israel over to you." 13 He said, "Good; I will make a covenant with you. But one thing I require of you: you shall never appear in my presence unless you bring Saul's daughter Michal when you come to see me." 14 Then David sent messengers to Saul's son Ishbaal, *i* saying, "Give me my wife Michal, to whom I became engaged at the price of one hundred foreskins of the Philistines." 15 Ish-

e Meaning of Heb uncertain   *f* Heb *And he*
g Gk Compare 1 Chr 8.33; 9.39: Heb *Ish-bosheth,*
"man of shame"   *h* Gk: Heb *where he was*
i Heb *Ish-bosheth*

29: *The Arabah* ("the desert") here means the Jordan valley north of the Dead Sea. The term is also used of the dry depression south of the Dead Sea. **30–32:** The initial victory of David's forces was symptomatic of what was to come (see 3.1).

3.1–39: Continuation of the war with Israel. 2–5: These valuable verses were inserted by an editor interested in family statistics.

3.6–11: Abner quarrels with Ishbaal. According to the usage of that day, a king's *concubines* were royal property that must be kept within the royal household. To aspire to have such a *concubine* (v. 7) was to aspire to the kingship, and that was treason (16.21–22;

1 Kings 2.22). The *dog,* being regarded as an unclean animal, served as a term of reproach; "a dead dog" or *a dog's head* (v. 8) were common expressions (9.8; 16.9; 1 Sam 17.43; 24.14). Evidently Abner realized that references to *Judah* and to *David* (vv. 8, 9, 10) would frighten *Ishbaal* into submission.

3.12–21: Abner negotiates with David. David's demand for the return of Michal (1 Sam 18.20–27) was not due to any lack of wives, but had the political purpose of strengthening David's claim to Saul's throne as a son-in-law of the late king. **19:** It was particularly necessary to obtain the consent of Benjamin, Saul's own tribe, which was most likely to object.

baal[j] sent and took her from her husband Paltiel the son of Laish. [16]But her husband went with her, weeping as he walked behind her all the way to Bahurim. Then Abner said to him, "Go back home!" So he went back.

17 Abner sent word to the elders of Israel, saying, "For some time past you have been seeking David as king over you. [18]Now then bring it about; for the LORD has promised David: Through my servant David I will save my people Israel from the hand of the Philistines, and from all their enemies." [19]Abner also spoke directly to the Benjaminites; then Abner went to tell David at Hebron all that Israel and the whole house of Benjamin were ready to do.

20 When Abner came with twenty men to David at Hebron, David made a feast for Abner and the men who were with him. [21]Abner said to David, "Let me go and rally all Israel to my lord the king, in order that they may make a covenant with you, and that you may reign over all that your heart desires." So David dismissed Abner, and he went away in peace.

22 Just then the servants of David arrived with Joab from a raid, bringing much spoil with them. But Abner was not with David at Hebron, for David[k] had dismissed him, and he had gone away in peace. [23]When Joab and all the army that was with him came, it was told Joab, "Abner son of Ner came to the king, and he has dismissed him, and he has gone away in peace." [24]Then Joab went to the king and said, "What have you done? Abner came to you; why did you dismiss him, so that he got away? [25]You know that Abner son of Ner came to deceive you, and to learn your com-

ings and goings and to learn all that you are doing."

26 When Joab came out from David's presence, he sent messengers after Abner, and they brought him back from the cistern of Sirah; but David did not know about it. [27]When Abner returned to Hebron, Joab took him aside in the gateway to speak with him privately, and there he stabbed him in the stomach. So he died for shedding[l] the blood of Asahel, Joab's[m] brother. [28]Afterward, when David heard of it, he said, "I and my kingdom are forever guiltless before the LORD for the blood of Abner son of Ner. [29]May the guilt[n] fall on the head of Joab, and on all his father's house; and may the house of Joab never be without one who has a discharge, or who is leprous,[o] or who holds a spindle, or who falls by the sword, or who lacks food!" [30]So Joab and his brother Abishai murdered Abner because he had killed their brother Asahel in the battle at Gibeon.

31 Then David said to Joab and to all the people who were with him, "Tear your clothes, and put on sackcloth, and mourn over Abner." And King David followed the bier. [32]They buried Abner at Hebron. The king lifted up his voice and wept at the grave of Abner, and all the people wept. [33]The king lamented for Abner, saying,

"Should Abner die as a fool dies?
[34] Your hands were not bound,
    your feet were not fettered;
as one falls before the wicked
    you have fallen."
And all the people wept over him again.

j Heb *Ish-bosheth*   *k* Heb *he*   *l* Heb lacks
shedding   *m* Heb *his*   *n* Heb *May it*   *o* A
term for several skin diseases; precise meaning
uncertain

3.22–39: The murder of Abner by Joab. *Joab* had two motives for this murder: (*a*) to avenge the death of *Asahel* (2.22–23); (*b*) to eliminate a rival for the post of commander-in-chief of David's army (19.13; 20.9–10).
3.28–39: *Joab* was too powerful (v. 39) for *David* to order him executed, as had been done in the case of the Amalekite (1.15). Yet

this murder greatly endangered David's position among the northern tribes. Therefore *David* did the best he could to make amends by invoking a series of curses upon *Joab* and his descendants (the *spindle* is a derogatory reference to women's work), by proclaiming public mourning throughout Judah, by himself taking the position of chief mourner at Abner's funeral and fasting all that day, then

35 Then all the people came to persuade David to eat something while it was still day; but David swore, saying, "So may God do to me, and more, if I taste bread or anything else before the sun goes down!" 36 All the people took notice of it, and it pleased them; just as everything the king did pleased all the people. 37 So all the people and all Israel understood that day that the king had no part in the killing of Abner son of Ner. 38 And the king said to his servants, "Do you not know that a prince and a great man has fallen this day in Israel? 39 Today I am powerless, even though anointed king; these men, the sons of Zeruiah, are too violent for me. The LORD pay back the one who does wickedly in accordance with his wickedness!"

4 When Saul's son Ishbaal*p* heard that Abner had died at Hebron, his courage failed, and all Israel was dismayed. 2 Saul's son had two captains of raiding bands; the name of the one was Baanah, and the name of the other Rechab. They were sons of Rimmon a Benjaminite from Beeroth—for Beeroth is considered to belong to Benjamin. 3 (Now the people of Beeroth had fled to Gittaim and are there as resident aliens to this day).

4 Saul's son Jonathan had a son who was crippled in his feet. He was five years old when the news about Saul and Jonathan came from Jezreel. His nurse picked him up and fled; and, in her haste to flee, it happened that he fell and became lame. His name was Mephibosheth. *q*

5 Now the sons of Rimmon the Bee-rothite, Rechab and Baanah, set out, and about the heat of the day they came to the house of Ishbaal, *r* while he was taking his noonday rest. 6 They came inside the house as though to take wheat, and they struck him in the stomach; then Rechab and his brother Baanah escaped. *s* 7 Now they had come into the house while he was lying on his couch in his bedchamber; they attacked him, killed him, and beheaded him. Then they took his head and traveled by way of the Arabah all night long. 8 They brought the head of Ishbaal *r* to David at Hebron and said to the king, "Here is the head of Ishbaal, *r* son of Saul, your enemy, who sought your life; the LORD has avenged my lord the king this day on Saul and on his off-spring."

9 David answered Rechab and his brother Baanah, the sons of Rimmon the Beerothite, "As the LORD lives, who has redeemed my life out of every adversity, 10 when the one who told me, 'See, Saul is dead,' thought he was bringing good news, I seized him and killed him at Ziklag—this was the reward I gave him for his news. 11 How much more then, when wicked men have killed a righteous man on his bed in his own house! And now shall I not require his blood at your hand, and destroy you from the earth?" 12 So David commanded the young men, and they killed them; they cut off their hands and feet, and hung their bodies beside the pool at Hebron. But the head

p Heb lacks *Ishbaal*   q In 1 Chr 8.34 and 9.40, *Merib-baal*   r Heb *Ish-bosheth*   s Meaning of Heb of verse 6 uncertain

---

continuing to praise Abner thereafter. The people of Israel were convinced of David's innocence (v. 37) of the murder, though naturally any idea of making David king of Israel was put aside for the time.

4.1–12: The murder of Ishbaal and David's punishment of the murderers. 2–3: *Beeroth* was originally a city under Gibeonite control (Josh 9.17); but the Gibeonite *people of Beeroth had fled to Gittaim,* probably during Saul's persecution (21.1–2, 5), thus making it possible for *Rimmon* and other Israelites to

settle in *Beeroth*. **4:** The original form of *Mephibosheth* ("he who spreads [?] shame") was Merib-baal ("he who strives for [?] the Lord," 1 Chr 8.34; 9.40). On reasons for the change, see 2.8 n. **8:** By saying *the LORD has avenged my lord,* the guilty pair try to justify what they have done.

4.9–12: David continues to show respect for the family of Saul and to punish anyone seeking to harm members of that family (1.14–16; 3.28–39).

of Ishbaal[t] they took and buried in the tomb of Abner at Hebron.

5 Then all the tribes of Israel came to David at Hebron, and said, "Look, we are your bone and flesh. [2]For some time, while Saul was king over us, it was you who led out Israel and brought it in. The LORD said to you: It is you who shall be shepherd of my people Israel, you who shall be ruler over Israel." [3]So all the elders of Israel came to the king at Hebron; and King David made a covenant with them at Hebron before the LORD, and they anointed David king over Israel. [4]David was thirty years old when he began to reign, and he reigned forty years. [5]At Hebron he reigned over Judah seven years and six months; and at Jerusalem he reigned over all Israel and Judah thirty-three years.

6 The king and his men marched to Jerusalem against the Jebusites, the inhabitants of the land, who said to David, "You will not come in here, even the blind and the lame will turn you back"—thinking, "David cannot come in here." [7]Nevertheless David took the stronghold of Zion, which is now the city of David. [8]David had said on that day, "Whoever would strike down the Jebusites, let him get up the water shaft to attack the lame and the blind, those

whom David hates."[u] Therefore it is said, "The blind and the lame shall not come into the house." [9]David occupied the stronghold, and named it the city of David. David built the city all around from the Millo inward. [10]And David became greater and greater, for the LORD, the God of hosts, was with him.

11 King Hiram of Tyre sent messengers to David, along with cedar trees, and carpenters and masons who built David a house. [12]David then perceived that the LORD had established him king over Israel, and that he had exalted his kingdom for the sake of his people Israel.

13 In Jerusalem, after he came from Hebron, David took more concubines and wives; and more sons and daughters were born to David. [14]These are the names of those who were born to him in Jerusalem: Shammua, Shobab, Nathan, Solomon, [15]Ibhar, Elishua, Nepheg, Japhia, [16]Elishama, Eliada, and Eliphelet.

17 When the Philistines heard that David had been anointed king over Israel, all the Philistines went up in search of David; but David heard about it and went down to the stronghold. [18]Now the Philistines had come and spread out

---

t Heb *Ish-bosheth*   *u* Another reading is *those who hate David*

5.1–16: David becomes king of all Israel and Judah; he captures Jerusalem and makes it his capital. 4–5: A chronological note by an editor which may be approximately correct, though David could hardly have been *thirty years old* at this time. *Forty years* is a favorite biblical expression meaning a reasonably long time. *Israel and Judah* are kept clearly separate in the writer's mind. **5.6–9:** The details of how David captured *Jerusalem* or *Zion,* the last town remaining to the Canaanites (locally called *Jebusites*), are not clear (compare 1 Chr 11.4–9 for a somewhat different account). The capture was important because it removed the last vestige of Canaanite power in the land, and also provided a place where David could establish a neutral capital, belonging to neither Israel nor Judah, but lying between the two. Later tradition assumed an earlier possession of the city; contrast Judg 1.8 and 1 Sam 17.54 with Judg 19.10–12. *The Millo,* see 1 Kings 9.15 n.

11: Apparently the same *Hiram* who much later helped Solomon in the building of the temple (1 Kings 5.1–12). **5.13–16:** Children of David. Editorial information similar to that in 3.2–5, though somewhat out of place, since *Solomon* at least was born later. Large harems were customary at that time (see Judg 8.30–31) and increased the prestige of the ruler. David, with at least twenty concubines and wives, was apparently never criticized on this score. It was quite otherwise with Solomon (1 Kings 11.1–4). **5.17–25: War with the Philistines.** The Philistines now realized that David was steering a course that would make him an enemy, and one to be feared (1 Sam chs 27 and 29); and David no longer pretended to be their ally. His success as king would depend on his ability to deal with the Philistines. **17:** It is not clear which *stronghold* is meant; some think Jerusalem, others Adullam (1 Sam 22.1). **18:** *The valley of Rephaim* was probably located

in the valley of Rephaim. ¹⁹David inquired of the LORD, "Shall I go up against the Philistines? Will you give them into my hand?" The LORD said to David, "Go up; for I will certainly give the Philistines into your hand." ²⁰So David came to Baal-perazim, and David defeated them there. He said, "The LORD has burst forth against*ᵛ* my enemies before me, like a bursting flood." Therefore that place is called Baal-perazim.*ʷ* ²¹The Philistines abandoned their idols there, and David and his men carried them away.

22 Once again the Philistines came up, and were spread out in the valley of Rephaim. ²³When David inquired of the LORD, he said, "You shall not go up; go around to their rear, and come upon them opposite the balsam trees. ²⁴When you hear the sound of marching in the tops of the balsam trees, then be on the alert; for then the LORD has gone out before you to strike down the army of the Philistines." ²⁵David did just as the LORD had commanded him; and he struck down the Philistines from Geba all the way to Gezer.

6 David again gathered all the chosen men of Israel, thirty thousand. ²David and all the people with him set out and went from Baale-judah, to bring up from there the ark of God, which is called by the name of the LORD of hosts who is enthroned on the cherubim. ³They carried the ark of God on a new cart, and brought it out of the house of Abinadab, which was on the hill. Uzzah and Ahio,*ˣ* the sons of Abinadab, were driving the new cart ⁴with the ark of God;*ʸ* and Ahio*ˣ* went in front of the ark. ⁵David and all the house of Israel were dancing before the LORD with all their might, with songs*ᶻ* and lyres and harps and tambourines and castanets and cymbals.

6 When they came to the threshing floor of Nacon, Uzzah reached out his hand to the ark of God and took hold of it, for the oxen shook it. ⁷The anger of the LORD was kindled against Uzzah; and God struck him there because he reached out his hand to the ark;*ᵃ* and he died there beside the ark of God. ⁸David was angry because the LORD had burst forth

---

v Heb *paraz*   *w* That is *Lord of Bursting Forth*
x Or *and his brother*   *y* Compare Gk: Heb *and brought it out of the house of Abinadab, which was on the hill with the ark of God*   *z* Q Ms Gk 1 Chr 13.8: Heb *fir-trees*   *a* 1 Chr 13.10 Compare Q Ms: Meaning of Heb uncertain

---

southwest of Jerusalem. **20:** In good oriental fashion, the writer makes a folkloristic pun on the place where David's enemies were defeated (compare Gen 38.29).

5.21: Formerly the Philistines had captured the sacred ark of the Israelites (1 Sam 4.11; 5.1); now the tables were turned, and the Philistines lost their *idols*. **22–25:** David defeated the Philistines again, pushing them back *from Geba,* near Jerusalem, *to Gezer,* near the border of their own territory. **24:** The rustling of the leaves *in the tops of the balsam trees* became *the sound of marching,* an omen taken as the signal to advance.

6.1–15: The bringing of the ark to Jerusalem. David wished to add to the prestige of Jerusalem by making it a religious, as well as a political and military, center. Hence it was particularly appropriate for him to bring the ark there, since it was the sacred object of the northern tribes and now the symbol of the national God. **2:** *Baale-judah* is either an error or another name for Kiriath-jearim, the place where the ark was left (1 Sam 6.21–7.2). *The LORD of hosts who is enthroned on the cherubim,* see 1 Sam 4.4 n. **5:** Some interpreters consider this verse a later embellishment, since the full orchestra seems more appropriate to the time after the temple had been built (compare 2 Chr 5.12–13; Ps 150.3–5).

6.6: *Uzzah* was apparently trying to steady the ark as *the oxen shook it.* **7:** At this point Uzzah died. Most ancient peoples attributed disaster to the anger of a deity. According to a later interpretation introduced here, Uzzah should not have touched the ark (see Num 4.15 n.). Even in these earlier times, the ark was regarded as very dangerous (1 Sam 5.11–12; 6.19). **10–11:** A *Gittite* means a person from Gath, but *Obed-edom* was not necessarily from the Philistine town. Gath means "wine press," and there were several towns by that name in Israelite territory. On the other hand, this man could have been a follower of *David* from the latter's Philistine days (1 Sam 27.2–4).

with an outburst upon Uzzah; so that place is called Perez-uzzah, *b* to this day. ⁹David was afraid of the LORD that day; he said, "How can the ark of the LORD come into my care?" ¹⁰So David was unwilling to take the ark of the LORD into his care in the city of David; instead David took it to the house of Obed-edom the Gittite. ¹¹The ark of the LORD remained in the house of Obed-edom the Gittite three months; and the LORD blessed Obed-edom and all his household.

12 It was told King David, "The LORD has blessed the household of Obed-edom and all that belongs to him, because of the ark of God." So David went and brought up the ark of God from the house of Obed-edom to the city of David with rejoicing; ¹³and when those who bore the ark of the LORD had gone six paces, he sacrificed an ox and a fatling. ¹⁴David danced before the LORD with all his might; David was girded with a linen ephod. ¹⁵So David and all the house of Israel brought up the ark of the LORD with shouting, and with the sound of the trumpet.

16 As the ark of the LORD came into the city of David, Michal daughter of Saul looked out of the window, and saw King David leaping and dancing before the LORD; and she despised him in her heart.

17 They brought in the ark of the LORD, and set it in its place, inside the tent that David had pitched for it; and David offered burnt offerings and offerings of well-being before the LORD. ¹⁸When David had finished offering the burnt offerings and the offerings of well-being, he blessed the people in the name of the LORD of hosts, ¹⁹and distributed food among all the people, the whole multitude of Israel, both men and women, to each a cake of bread, a portion of meat, *c* and a cake of raisins. Then all the people went back to their homes.

20 David returned to bless his household. But Michal the daughter of Saul came out to meet David, and said, "How the king of Israel honored himself today, uncovering himself today before the eyes of his servants' maids, as any vulgar fellow might shamelessly uncover himself!" ²¹David said to Michal, "It was before the LORD, who chose me in place of your father and all his household, to appoint me as prince over Israel, the people of the LORD, that I have danced before the LORD. ²²I will make myself yet more contemptible than this, and I will be abased in my own eyes; but by the maids of whom you have spoken, by them I shall be held in honor." ²³And Michal the daughter of Saul had no child to the day of her death.

7 Now when the king was settled in his house, and the LORD had given him rest from all his enemies around him, ²the king said to the prophet Nathan, "See now, I am living in a house of cedar, but the ark of God stays in a tent." ³Nathan said to the king, "Go, do all that you have in mind; for the LORD is with you."

4 But that same night the word of the

b That is *Bursting Out Against Uzzah*   *c* Vg: Meaning of Heb uncertain

---

6.12–15: The prosperity and peace of *the household of Obed-edom* were taken as signs that God's anger in connection with the ark had ceased (v. 12). Note that it was perfectly permissible for *David* to offer sacrifice (vv. 13, 17; compare 1 Sam 13.10–13); only later was this function restricted to priests. On religious dancing, see Ex 15.20 n.; 32.19; 1 Kings 18.26. On the *linen ephod,* see 1 Sam 2.18 n.
6.16–23: Michal's alienation. *Michal* may have been resentful at being torn away from her husband Paltiel (3.15–16), or at the discovery that she was only one among many wives of David, or at the decline of her family's fortunes (vv. 16, 20). **23:** It is not clear whether *Michal . . . had no child* because David put her away or because she was barren. Childlessness was considered a great misfortune; compare Gen 30.1; 1 Sam 1.6–11.
7.1–29: David wishes to build a temple, but God wills that he establish an everlasting dynasty. Like 1 Sam 2.27–36, this chapter is a late theological commentary

LORD came to Nathan: 5 Go and tell my servant David: Thus says the LORD: Are you the one to build me a house to live in? 6 I have not lived in a house since the day I brought up the people of Israel from Egypt to this day, but I have been moving about in a tent and a tabernacle. 7 Wherever I have moved about among all the people of Israel, did I ever speak a word with any of the tribal leaders*d* of Israel, whom I commanded to shepherd my people Israel, saying, "Why have you not built me a house of cedar?" 8 Now therefore thus you shall say to my servant David: Thus says the LORD of hosts: I took you from the pasture, from following the sheep to be prince over my people Israel; 9 and I have been with you wherever you went, and have cut off all your enemies from before you; and I will make for you a great name, like the name of the great ones of the earth. 10 And I will appoint a place for my people Israel and will plant them, so that they may live in their own place, and be disturbed no more; and evildoers shall afflict them no more, as formerly, 11 from the time that I appointed judges over my people Israel; and I will give you rest from all your enemies. Moreover the LORD declares to you that the LORD will make you a house. 12 When your days are fulfilled and you lie down with your ancestors, I will raise up your offspring after you, who shall come forth from your body, and I will establish his kingdom. 13 He shall build a house for my name, and I will establish the throne of his kingdom forever. 14 I will be a father to him, and he shall be a son to me. When he commits iniquity, I will punish him with a rod such as mortals use, with blows inflicted by human beings. 15 But I will not take*e* my steadfast love from him, as I

took it from Saul, whom I put away from before you. 16 Your house and your kingdom shall be made sure forever before me;*f* your throne shall be established forever. 17 In accordance with all these words and with all this vision, Nathan spoke to David.

18 Then King David went in and sat before the LORD, and said, "Who am I, O Lord GOD, and what is my house, that you have brought me thus far? 19 And yet this was a small thing in your eyes, O Lord GOD; you have spoken also of your servant's house for a great while to come. May this be instruction for the people,*g* O Lord GOD! 20 And what more can David say to you? For you know your servant, O Lord GOD! 21 Because of your promise, and according to your own heart, you have wrought all this greatness, so that your servant may know it. 22 Therefore you are great, O LORD God; for there is no one like you, and there is no God besides you, according to all that we have heard with our ears. 23 Who is like your people, like Israel? Is there another*h* nation on earth whose God went to redeem it as a people, and to make a name for himself, doing great and awesome things for them,*i* by driving out*j* before his people nations and their gods?*k* 24 And you established your people Israel for yourself to be your people forever; and you, O LORD, became their God. 25 And now, O LORD God, as for the word that you

d Or *any of the tribes*   *e* Gk Syr Vg
1 Chr 17.13: Heb *shall not depart*   *f* Gk Heb
Mss: MT *before you*; Compare 2 Sam 7.26, 29
g Meaning of Heb uncertain   *h* Gk: Heb *one*
i Heb *you*   *j* Gk 1 Chr 17.21: Heb *for your land*
k Cn: Heb *before your people, whom you redeemed for yourself from Egypt, nations and its gods*

inserted into an early source. Based in part on Ps 89 (compare Ps 132.11–12), the writer explains why David did not build the great Temple in Jerusalem, and at the same time shows the importance of David's dynasty to the religious meaning of ancient Israel. In v. 6 the writer ignores the temple at Shiloh (1 Sam 1.7; 3.3).

The key to the understanding of this chapter is the play on the various meanings of the word *house;* in vv. 1–2 it means "palace"; in vv. 5, 6, 7, 13 it means "temple"; in vv. 11, 16, 19, 25, 26, 27, 29 it means "dynasty"; in v. 18 it means "family status."

Historically the dynasty of David was not everlasting. It fell in 587 (586) B.C., probably

have spoken concerning your servant and concerning his house, confirm it forever; do as you have promised. 26 Thus your name will be magnified forever in the saying, 'The LORD of hosts is God over Israel'; and the house of your servant David will be established before you. 27 For you, O LORD of hosts, the God of Israel, have made this revelation to your servant, saying, 'I will build you a house'; therefore your servant has found courage to pray this prayer to you. 28 And now, O Lord GOD, you are God, and your words are true, and you have promised this good thing to your servant; 29 now therefore may it please you to bless the house of your servant, so that it may continue forever before you; for you, O Lord GOD, have spoken, and with your blessing shall the house of your servant be blessed forever."

8 Some time afterward, David attacked the Philistines and subdued them; David took Metheg-ammah out of the hand of the Philistines.

2 He also defeated the Moabites and, making them lie down on the ground, measured them off with a cord; he measured two lengths of cord for those who were to be put to death, and one length[l] for those who were to be spared. And the Moabites became servants to David and brought tribute.

3 David also struck down King Hadadezer son of Rehob of Zobah, as he went to restore his monument[m] at the river Euphrates. 4 David took from him

one thousand seven hundred horsemen, and twenty thousand foot soldiers. David hamstrung all the chariot horses, but left enough for a hundred chariots. 5 When the Arameans of Damascus came to help King Hadadezer of Zobah, David killed twenty-two thousand men of the Arameans. 6 Then David put garrisons among the Arameans of Damascus; and the Arameans became servants to David and brought tribute. The LORD gave victory to David wherever he went. 7 David took the gold shields that were carried by the servants of Hadadezer, and brought them to Jerusalem. 8 From Betah and from Berothai, towns of Hadadezer, King David took a great amount of bronze.

9 When King Toi of Hamath heard that David had defeated the whole army of Hadadezer, 10 Toi sent his son Joram to King David, to greet him and to congratulate him because he had fought against Hadadezer and defeated him. Now Hadadezer had often been at war with Toi. Joram brought with him articles of silver, gold, and bronze; 11 these also King David dedicated to the LORD, together with the silver and gold that he dedicated from all the nations he subdued, 12 from Edom, Moab, the Ammonites, the Philistines, Amalek, and from the spoil of King Hadadezer son of Rehob of Zobah.

13 David won a name for himself.

l Heb one full length   m Compare 1 Sam 15.12 and 2 Sam 18.18

some time before our author wrote. The language, however, is typical of courtly scribes who glorify a royal line.

8.1–18: Summary of David's wars and the administration of his kingdom. This chapter was contributed by a Deuteronomistic editor (see Introduction to 1 Samuel). It is not a mere digest of what follows, for it contains information not found elsewhere. **1:** This seems intended as a statement of the final defeat of the Philistines, compare 21.15–22 and 23.8–39 for some details. *Metheg-ammah,* 1 Chr 18.1 reads "Gath and its villages." **2:** The slaughter of two-thirds of the Moabites does not accord well with the friendly relations implied by 1 Sam 22.3–4. Perhaps

something happened to revive the old enmity (1 Sam 14.47). **3–8:** David defeats the Arameans (Syrians). This passage may be a partial summary of 10.6–19. *Zobah* (vv. 3, 5) lay north of Damascus (vv. 5, 6) and was attempting to extend its power to the *Euphrates* (v. 3). South of *Zobah,* Syria was controlled by *Damascus.* David defeated the coalition (v. 5), and thus controlled all of Syria. He kept enough *horses* for *a hundred chariots* only (v. 4). Solomon greatly expanded the use of horses and chariots (1 Kings 10.26–29). **8.9:** *Hamath* lay north of Zobah. David did not control this territory, but its king was friendly. **13–14:** David defeated Edom, and thus his little empire was enlarged to the

When he returned, he killed eighteen thousand Edomites[n] in the Valley of Salt. [14]He put garrisons in Edom; throughout all Edom he put garrisons, and all the Edomites became David's servants. And the LORD gave victory to David wherever he went.

15 So David reigned over all Israel; and David administered justice and equity to all his people. [16]Joab son of Zeruiah was over the army; Jehoshaphat son of Ahilud was recorder; [17]Zadok son of Ahitub and Ahimelech son of Abiathar were priests; Seraiah was secretary; [18]Benaiah son of Jehoiada was over[o] the Cherethites and the Pelethites; and David's sons were priests.

9 David asked, "Is there still anyone left of the house of Saul to whom I may show kindness for Jonathan's sake?" [2]Now there was a servant of the house of Saul whose name was Ziba, and he was summoned to David. The king said to him, "Are you Ziba?" And he said, "At your service!" [3]The king said, "Is there anyone remaining of the house of Saul to whom I may show the kindness of God?" Ziba said to the king, "There remains a son of Jonathan; he is crippled in

his feet." [4]The king said to him, "Where is he?" Ziba said to the king, "He is in the house of Machir son of Ammiel, at Lodebar." [5]Then King David sent and brought him from the house of Machir son of Ammiel, at Lo-debar. [6]Mephibosheth[p] son of Jonathan son of Saul came to David, and fell on his face and did obeisance. David said, "Mephibosheth!"[p] He answered, "I am your servant." [7]David said to him, "Do not be afraid, for I will show you kindness for the sake of your father Jonathan; I will restore to you all the land of your grandfather Saul, and you yourself shall eat at my table always." [8]He did obeisance and said, "What is your servant, that you should look upon a dead dog such as I?"

9 Then the king summoned Saul's servant Ziba, and said to him, "All that belonged to Saul and to all his house I have given to your master's grandson. [10]You and your sons and your servants shall till the land for him, and shall bring in the produce, so that your master's

n Gk: Heb *returned from striking down eighteen thousand Arameans* o Syr Tg Vg 20.23; 1 Chr 18.17: Heb lacks *was over*
p Or *Merib-baal*: See 4.4 note

south. In these campaigns the king did not always take to the field in person (11.1). *The Valley of Salt* is either the depression south of the Dead Sea, or a valley running southeast from Beer-sheba. **16–18**: Another version of this list is found in 20.23–26. *Zadok* appears here (v. 17) historically for the first time (1 Sam 2.35). His earlier history is not known, though later his ancestry is traced back to Aaron (1 Chr 6.3–8). *Ahimelech the son of Abiathar* should perhaps be reversed (1 Sam 22.20), though it is also possible that *Abiathar* had a son *Ahimelech* who acted in place of his aged father (1 Chr 18.16; 24.3, 6, 31). **18**: *The Cherethites and Pelethites* were foreign mercenaries; possibly the names are equivalent to "Cretans and Philistines" (15.18; 20.7, 23; 1 Kings 1.38, 44; see 1 Sam 30.14 n.). It is entirely possible that some of *David's sons were priests*, for the restriction of the priesthood to a certain family had not yet arisen. The parallels, however, in 20.26 and 1 Chr 18.17 have different readings.

9.1–20.26: The domestic and political troubles of David's reign. This material is, in view of its superb literary style, the prose

masterpiece of the Hebrew Scriptures. For the fullest appreciation the section should be read as a unit with its conclusion in 1 Kings chs 1–2.

9.1–13: David shows dutiful kindness to Mephibosheth. As preparation for this chapter, read 4.4 and 21.1–14; see comments on these passages. **1**: The word translated *kindness* contains the idea of obligation, "kindness required by loyalty to a covenant." *David* and *Jonathan* had sworn eternal loyalty to one another (1 Sam 18.1–4; 20.14–17). *David* was probably also fearful of reaction to the execution of seven of Saul's descendants at the request of the Gibeonites (see 21.11–14 n.). The king also needed to know whether any member of the former royal family remained who might serve as a focal point of rebellion. **3**: *The kindness of God* is a stronger expression of the *kindness* of v. 1, meaning "obligation under God." **4**: *Machir at Lodebar*, see 17.27. **7**: Probably one reason that Mephibosheth might be *afraid* was what had happened to two of his half brothers and five of his nephews (21.8–9).

9.8: *Dead dog*, see 1 Sam 24.14 n. **13**: David

grandson may have food to eat; but your master's grandson Mephibosheth[q] shall always eat at my table." Now Ziba had fifteen sons and twenty servants. [11] Then Ziba said to the king, "According to all that my lord the king commands his servant, so your servant will do." Mephibosheth[q] ate at David's[r] table, like one of the king's sons. [12] Mephibosheth[q] had a young son whose name was Mica. And all who lived in Ziba's house became Mephibosheth's[s] servants. [13] Mephibosheth[q] lived in Jerusalem, for he always ate at the king's table. Now he was lame in both his feet.

10 Some time afterward, the king of the Ammonites died, and his son Hanun succeeded him. [2] David said, "I will deal loyally with Hanun son of Nahash, just as his father dealt loyally with me." So David sent envoys to console him concerning his father. When David's envoys came into the land of the Ammonites, [3] the princes of the Ammonites said to their lord Hanun, "Do you really think that David is honoring your father just because he has sent messengers with condolences to you? Has not David sent his envoys to you to search the city, to spy it out, and to overthrow it?" [4] So Hanun seized David's envoys, shaved off half the beard of each, cut off their garments in the middle at their hips, and sent them away. [5] When David was told, he sent to meet them, for the men were greatly ashamed. The king said, "Remain at Jericho until your beards have grown, and then return."

6 When the Ammonites saw that they had become odious to David, the Ammonites sent and hired the Arameans of Beth-rehob and the Arameans of Zobah,

twenty thousand foot soldiers, as well as the king of Maacah, one thousand men, and the men of Tob, twelve thousand men. [7] When David heard of it, he sent Joab and all the army with the warriors. [8] The Ammonites came out and drew up in battle array at the entrance of the gate; but the Arameans of Zobah and of Rehob, and the men of Tob and Maacah, were by themselves in the open country.

9 When Joab saw that the battle was set against him both in front and in the rear, he chose some of the picked men of Israel, and arrayed them against the Arameans; [10] the rest of his men he put in the charge of his brother Abishai, and he arrayed them against the Ammonites. [11] He said, "If the Arameans are too strong for me, then you shall help me; but if the Ammonites are too strong for you, then I will come and help you. [12] Be strong, and let us be courageous for the sake of our people, and for the cities of our God; and may the LORD do what seems good to him." [13] So Joab and the people who were with him moved forward into battle against the Arameans; and they fled before him. [14] When the Ammonites saw that the Arameans fled, they likewise fled before Abishai, and entered the city. Then Joab returned from fighting against the Ammonites, and came to Jerusalem.

15 But when the Arameans saw that they had been defeated by Israel, they gathered themselves together. [16] Hadadezer sent and brought out the Arameans who were beyond the Euphrates; and they came to Helam, with Shobach the commander of the army of Hadadezer at

q Or *Merib-baal*: See 4.4 note   r Gk: Heb *my*
s Or *Merib-baal's*: See 4.4 note

---

may have been pained that Jonathan's son was such a pathetic cripple, and yet relieved that this scion of Saul could hardly be expected to head an insurrection. *Ziba* claimed later that Mephibosheth was disloyal (16.1–4), but the claim seems to have been false (19.24–30).

10.1–19: David defeats the Ammonites and Arameans (Syrians). 2: There is no record of the covenant or agreement between *David* and *Nahash* (but compare 17.27). The relations between Saul and Nahash had been

hostile (1 Sam ch 11). **3:** Perhaps Hanun's advisers were thinking of David's treatment of the Moabites (see 8.2 n.). **4:** Mutilation of the beard, the symbol of a man's honor, and forcible indecent exposure were the worst insults imaginable in those days.

10.6–19: The summary of these engagements in 8.3–8 omits mention of the part the Ammonites played in calling the *Arameans* (Syrians) into the conflict. **16:** Here *beyond the Euphrates* means "west of the Euphrates," as

their head. [17] When it was told David, he gathered all Israel together, and crossed the Jordan, and came to Helam. The Arameans arrayed themselves against David and fought with him. [18] The Arameans fled before Israel; and David killed of the Arameans seven hundred chariot teams, and forty thousand horsemen,[t] and wounded Shobach the commander of their army, so that he died there. [19] When all the kings who were servants of Hadadezer saw that they had been defeated by Israel, they made peace with Israel, and became subject to them. So the Arameans were afraid to help the Ammonites any more.

11 In the spring of the year, the time when kings go out to battle, David sent Joab with his officers and all Israel with him; they ravaged the Ammonites, and besieged Rabbah. But David remained at Jerusalem.

2 It happened, late one afternoon, when David rose from his couch and was walking about on the roof of the king's house, that he saw from the roof a woman bathing; the woman was very beautiful. [3] David sent someone to inquire about the woman. It was reported, "This is Bathsheba daughter of Eliam, the wife of Uriah the Hittite." [4] So David sent messengers to get her, and she came to him, and he lay with her. (Now she was purifying herself after her period.) Then she returned to her house. [5] The woman conceived; and she sent and told David, "I am pregnant."

6 So David sent word to Joab, "Send me Uriah the Hittite." And Joab sent Uriah to David. [7] When Uriah came to him, David asked how Joab and the people fared, and how the war was going. [8] Then David said to Uriah, "Go down to your house, and wash your feet." Uriah went out of the king's house, and there followed him a present from the king. [9] But Uriah slept at the entrance of the king's house with all the servants of his lord, and did not go down to his house. [10] When they told David, "Uriah did not go down to his house," David said to Uriah, "You have just come from a journey. Why did you not go down to your house?" [11] Uriah said to David, "The ark and Israel and Judah remain in booths;[u] and my lord Joab and the servants of my lord are camping in the open field; shall I then go to my house, to eat and to drink, and to lie with my wife? As you live, and as your soul lives, I will not do such a thing." [12] Then David said to Uriah, "Remain here today also, and tomorrow I will send you back." So Uriah remained in Jerusalem that day. On the next day, [13] David invited him to eat and

t 1 Chr 19.18 and some Gk Mss read *foot soldiers*
u Or *at Succoth*

---

in 1 Kings 4.24. David now takes to the field in person, in spite of the earlier warning in 21.17 (compare 12.28). **18**: On the numbers, see 1 Sam 4.10 n.

11.1–27: Second campaign against Ammon; David wrongs Uriah. At this point the author begins an intimate portrait of the domestic and political troubles of the royal family and the court, with all their sordidness and tragic consequences. The military history serves only as a framework in which to place the personal episodes in the private and public life of the king. The writer does not omit the sin of David with Bathsheba (as in 1 Chr 20.1–3) or gloss it over (as in 1 Sam 13.14). The consequences are so much a part of the story which follows that the cause must be clearly exhibited. Compare the superscription to Ps 51, added by an editor, who recognized the need for repentance and divine mercy in such a case. **1**: *His officers,* outstanding men of the royal bodyguard (23.8–39); *all Israel,* the army in general. **3**: *Uriah the Hittite,* see 1 Sam 26.6 n. **4**: See Lev 15.19–24 for the later codification of the law on the purification from a woman's *period.* **8**: The invitation *wash your feet* was standard custom after a journey (Gen 18.4; 19.2; 24.32). It may also be a euphemism for preparing for sexual intercourse, in view of David's secret plan (v. 11).

11.9–13: Continence was required of soldiers consecrated for war by religious sanction (1 Sam 21.4–5). Uriah refused to violate this taboo even when David made him drunk. **27**: Note the moral condemnation by the author in the name of God.

drink in his presence and made him drunk; and in the evening he went out to lie on his couch with the servants of his lord, but he did not go down to his house.

14 In the morning David wrote a letter to Joab, and sent it by the hand of Uriah. 15 In the letter he wrote, "Set Uriah in the forefront of the hardest fighting, and then draw back from him, so that he may be struck down and die." 16 As Joab was besieging the city, he assigned Uriah to the place where he knew there were valiant warriors. 17 The men of the city came out and fought with Joab; and some of the servants of David among the people fell. Uriah the Hittite was killed as well. 18 Then Joab sent and told David all the news about the fighting; 19 and he instructed the messenger, "When you have finished telling the king all the news about the fighting, 20 then, if the king's anger rises, and if he says to you, 'Why did you go so near the city to fight? Did you not know that they would shoot from the wall? 21 Who killed Abimelech son of Jerubbaal?*v* Did not a woman throw an upper millstone on him from the wall, so that he died at Thebez? Why did you go so near the wall?' then you shall say, 'Your servant Uriah the Hittite is dead too.'"

22 So the messenger went, and came and told David all that Joab had sent him to tell. 23 The messenger said to David, "The men gained an advantage over us, and came out against us in the field; but we drove them back to the entrance of the gate. 24 Then the archers shot at your servants from the wall; some of the king's servants are dead; and your servant Uriah the Hittite is dead also." 25 David said to the messenger, "Thus you shall say to Joab, 'Do not let this matter trouble you, for the sword devours now one and now another; press your attack on the city, and overthrow it.' And encourage him."

26 When the wife of Uriah heard that her husband was dead, she made lamentation for him. 27 When the mourning was over, David sent and brought her to his house, and she became his wife, and bore him a son.

But the thing that David had done displeased the LORD, 1 and the LORD sent Nathan to David. He came to him, and said to him, "There were two men in a certain city, the one rich and the other poor. 2 The rich man had very many flocks and herds; 3 but the poor man had nothing but one little ewe lamb, which he had bought. He brought it up, and it grew up with him and with his children; it used to eat of his meager fare, and drink from his cup, and lie in his bosom, and it was like a daughter to him. 4 Now there came a traveler to the rich man, and he was loath to take one of his own flock or herd to prepare for the wayfarer who had come to him, but he took the poor man's lamb, and prepared that for the guest who had come to him." 5 Then David's anger was greatly kindled against the man. He said to Nathan, "As the LORD lives, the man who has done this deserves to die; 6 he shall restore the lamb fourfold, because he did this thing, and because he had no pity."

7 Nathan said to David, "You are the man! Thus says the LORD, the God of Israel: I anointed you king over Israel, and I rescued you from the hand of Saul; 8 I gave you your master's house, and your master's wives into your bosom, and gave you the house of Israel and of Judah; and if that had been too little, I would have added as much more. 9 Why have you despised the word of the LORD, to do what is evil in his sight? You have

v Gk Syr Judg 7.1: Heb *Jerubbesheth*

12.1–25: Rebuke of David by the prophet Nathan; David's repentance and the birth of Solomon. 1–6: Nathan's famous parable of the ewe lamb. The prophet showed courage, but not originality; what David had done was wrong, even according to the standards of that day (see 11.27). Later Nathan became an active supporter of Bathsheba (1 Kings 1.5–14). **8:** There is no other hint that David took over Saul's wives, though such was the custom (16.21–22; 1 Kings 2.17–25).

struck down Uriah the Hittite with the sword, and have taken his wife to be your wife, and have killed him with the sword of the Ammonites. 10 Now therefore the sword shall never depart from your house, for you have despised me, and have taken the wife of Uriah the Hittite to be your wife. 11 Thus says the LORD: I will raise up trouble against you from within your own house; and I will take your wives before your eyes, and give them to your neighbor, and he shall lie with your wives in the sight of this very sun. 12 For you did it secretly; but I will do this thing before all Israel, and before the sun." 13 David said to Nathan, "I have sinned against the LORD." Nathan said to David, "Now the LORD has put away your sin; you shall not die. 14 Nevertheless, because by this deed you have utterly scorned the LORD, *w* the child that is born to you shall die." 15 Then Nathan went to his house.

The LORD struck the child that Uriah's wife bore to David, and it became very ill. 16 David therefore pleaded with God for the child; David fasted, and went in and lay all night on the ground. 17 The elders of his house stood beside him, urging him to rise from the ground; but he would not, nor did he eat food with them. 18 On the seventh day the child died. And the servants of David were afraid to tell him that the child was dead; for they said, "While the child was still alive, we spoke to him, and he did not listen to us; how then can we tell him the child is dead? He may do himself some harm." 19 But when David saw that his servants were whispering together, he perceived that the child was dead; and David said to his servants, "Is the child dead?" They said, "He is dead."

20 Then David rose from the ground, washed, anointed himself, and changed his clothes. He went into the house of the LORD, and worshiped; he then went to his own house; and when he asked, they set food before him and he ate. 21 Then his servants said to him, "What is this thing that you have done? You fasted and wept for the child while it was alive; but when the child died, you rose and ate food." 22 He said, "While the child was still alive, I fasted and wept; for I said, 'Who knows? The LORD may be gracious to me, and the child may live.' 23 But now he is dead; why should I fast? Can I bring him back again? I shall go to him, but he will not return to me."

24 Then David consoled his wife Bathsheba, and went to her, and lay with her; and she bore a son, and he named him Solomon. The LORD loved him, 25 and sent a message by the prophet Nathan; so he named him Jedidiah, *x* because of the LORD.

26 Now Joab fought against Rabbah of the Ammonites, and took the royal city. 27 Joab sent messengers to David, and said, "I have fought against Rabbah; moreover, I have taken the water city. 28 Now, then, gather the rest of the people together, and encamp against the city, and take it; or I myself will take the

w Ancient scribal tradition: Compare 1 Sam 25.22 note: Heb *scorned the enemies of the* LORD   *x* That is *Beloved of the* LORD

12.11: The prediction of *trouble . . . from within your own house* is very exact; it points to subsequent episodes in which David's sons carry out similar deeds (see chs 13 and 15). The *neighbor* is presumably Absalom (16.21–22). **13–14:** According to the idea of the "lex talionis" (law of exact retaliation, Ex 21.23–25; Lev 24.19–21; Deut 19.21) David should have died. Instead, divine judgment fell upon the child, according to the ideas of that day, as a special favor to David. The sin is *against the* LORD*;* it is God who sets the moral standard, not human beings.
12.23: This verse reflects the idea of Sheol, the cavity under the earth where all the dead go (see Gen 37.35 n.; 1 Sam 2.6–7 n.), and from which there is no return (Job 7.9–10).
12.26–31: Conclusion of the campaign against Ammon. David takes the Ammonite capital. **26:** *Rabbah of the Ammonites* was the full name of the city, called more briefly Rabbath-Ammon or simply Rabbah. **27:** *The water city* means the place protecting the water supply. **28–29:** Joab was always careful to keep himself subservient to *David.* The word *people* here refers to the army. **30:** *Milcom* (see note *y*), the god of the Ammonites, may be the better reading (compare 1 Kings 11.5); if

city, and it will be called by my name."
²⁹ So David gathered all the people together and went to Rabbah, and fought against it and took it. ³⁰ He took the crown of Milcom*y* from his head; the weight of it was a talent of gold, and in it was a precious stone; and it was placed on David's head. He also brought forth the spoil of the city, a very great amount. ³¹ He brought out the people who were in it, and set them to work with saws and iron picks and iron axes, or sent them to the brickworks. Thus he did to all the cities of the Ammonites. Then David and all the people returned to Jerusalem.

13 Some time passed. David's son Absalom had a beautiful sister whose name was Tamar; and David's son Amnon fell in love with her. ² Amnon was so tormented that he made himself ill because of his sister Tamar, for she was a virgin and it seemed impossible to Amnon to do anything to her. ³ But Amnon had a friend whose name was Jonadab, the son of David's brother Shimeah; and Jonadab was a very crafty man. ⁴ He said to him, "O son of the king, why are you so haggard morning after morning? Will you not tell me?" Amnon said to him, "I love Tamar, my brother Absalom's sister." ⁵ Jonadab said to him, "Lie down on your bed, and pretend to be ill; and when your father comes to see you, say to him, 'Let my sister Tamar come and give me something to eat, and prepare the food in my sight, so that I may see it and eat it from her hand.'" ⁶ So Amnon lay down, and pretended to be ill; and when the king came to see him,

Amnon said to the king, "Please let my sister Tamar come and make a couple of cakes in my sight, so that I may eat from her hand."

7 Then David sent home to Tamar, saying, "Go to your brother Amnon's house, and prepare food for him." ⁸ So Tamar went to her brother Amnon's house, where he was lying down. She took dough, kneaded it, made cakes in his sight, and baked the cakes. ⁹ Then she took the pan and set them*z* out before him, but he refused to eat. Amnon said, "Send out everyone from me." So everyone went out from him. ¹⁰ Then Amnon said to Tamar, "Bring the food into the chamber, so that I may eat from your hand." So Tamar took the cakes she had made, and brought them into the chamber to Amnon her brother. ¹¹ But when she brought them near him to eat, he took hold of her, and said to her, "Come, lie with me, my sister." ¹² She answered him, "No, my brother, do not force me; for such a thing is not done in Israel; do not do anything so vile! ¹³ As for me, where could I carry my shame? And as for you, you would be as one of the scoundrels in Israel. Now therefore, I beg you, speak to the king; for he will not withhold me from you." ¹⁴ But he would not listen to her; and being stronger than she, he forced her and lay with her.

15 Then Amnon was seized with a very great loathing for her; indeed, his

y Gk See 1 Kings 11.5, 33: Heb *their kings*
z Heb *and poured*

---

so, the reference is to an image of the god. *The weight* (*a talent,* about sixty-five lbs.) may be exaggerated.

13.1–20.26: The story of Absalom. The ensuing events show what was meant by *trouble . . . from within your own house* (12.11). The actions of Absalom brought forth a series of political crises, with decisive effect on the future of the nation. The deaths of Amnon and Absalom opened the way for Solomon (12.24) to become king, though he was originally far down in the line of succession (3.2–5; 1 Kings ch 1).

13.1–39: Tamar raped by Amnon; Absalom has Amnon murdered and flees from the court. 1: Tamar was a full sister of Absalom, a half sister of Amnon. The latter, being the eldest son of David (3.2), would normally have succeeded David as king. Chileab, the second son (3.3), seems to have died young, leaving Absalom as the next in line of succession to the throne; this fact must be borne in mind as one considers the deeds of Absalom.

13.13: At that time a man could marry his half sister, though later this practice was for-

loathing was even greater than the lust he had felt for her. Amnon said to her, "Get out!" 16 But she said to him, "No, my brother;*a* for this wrong in sending me away is greater than the other that you did to me." But he would not listen to her. 17 He called the young man who served him and said, "Put this woman out of my presence, and bolt the door after her." 18 (Now she was wearing a long robe with sleeves; for this is how the virgin daughters of the king were clothed in earlier times.*b*) So his servant put her out, and bolted the door after her. 19 But Tamar put ashes on her head, and tore the long robe that she was wearing; she put her hand on her head, and went away, crying aloud as she went.

20 Her brother Absalom said to her, "Has Amnon your brother been with you? Be quiet for now, my sister; he is your brother; do not take this to heart." So Tamar remained, a desolate woman, in her brother Absalom's house. 21 When King David heard of all these things, he became very angry, but he would not punish his son Amnon, because he loved him, for he was his firstborn.*c* 22 But Absalom spoke to Amnon neither good nor bad; for Absalom hated Amnon, because he had raped his sister Tamar.

23 After two full years Absalom had sheepshearers at Baal-hazor, which is near Ephraim, and Absalom invited all the king's sons. 24 Absalom came to the king, and said, "Your servant has sheepshearers; will the king and his servants please go with your servant?" 25 But the king said to Absalom, "No, my son, let us not all go, or else we will be burdensome to you." He pressed him, but he would not go but gave him his blessing. 26 Then Absalom said, "If not, please let my brother Amnon go with us." The king said to him, "Why should he go with you?" 27 But Absalom pressed him until he let Amnon and all the king's sons

go with him. Absalom made a feast like a king's feast.*d* 28 Then Absalom commanded his servants, "Watch when Amnon's heart is merry with wine, and when I say to you, 'Strike Amnon,' then kill him. Do not be afraid; have I not myself commanded you? Be courageous and valiant." 29 So the servants of Absalom did to Amnon as Absalom had commanded. Then all the king's sons rose, and each mounted his mule and fled.

30 While they were on the way, the report came to David that Absalom had killed all the king's sons, and not one of them was left. 31 The king rose, tore his garments, and lay on the ground; and all his servants who were standing by tore their garments. 32 But Jonadab, the son of David's brother Shimeah, said, "Let not my lord suppose that they have killed all the young men the king's sons; Amnon alone is dead. This has been determined by Absalom from the day Amnon*e* raped his sister Tamar. 33 Now therefore, do not let my lord the king take it to heart, as if all the king's sons were dead; for Amnon alone is dead."

34 But Absalom fled. When the young man who kept watch looked up, he saw many people coming from the Horonaim road*f* by the side of the mountain. 35 Jonadab said to the king, "See, the king's sons have come; as your servant said, so it has come about." 36 As soon as he had finished speaking, the king's sons arrived, and raised their voices and wept; and the king and all his servants also wept very bitterly.

37 But Absalom fled, and went to Talmai son of Ammihud, king of Ge-

a Cn Compare Gk Vg: Meaning of Heb uncertain   *b* Cn: Heb *were clothed in robes*   *c* Q Ms Gk: MT lacks *but he would not punish . . . firstborn*   *d* Gk Compare Q Ms: MT lacks *Absalom made a feast like a king's feast*   *e* Heb *he*   *f* Cn Compare Gk: Heb *the road behind him*

bidden (Lev 18.9). **19**: On these signs of grief see 15.32; 2 Kings 5.7; Esth 4.1; Jer 2.37. **13.23–29**: Festivities at the time of sheepshearing were usual (1 Sam 25.4–13). **37**: *Ab-salom fled* to the land of his maternal grandfather (3.3). As an Aramean kingdom (15.8), *Geshur* was under the military control of David (8.3–8; 10.6–19).

shur. David mourned for his son day after day. 38 Absalom, having fled to Geshur, stayed there three years. 39 And the heart of[g] the king went out, yearning for Absalom; for he was now consoled over the death of Amnon.

14 Now Joab son of Zeruiah perceived that the king's mind was on Absalom. 2 Joab sent to Tekoa and brought from there a wise woman. He said to her, "Pretend to be a mourner; put on mourning garments, do not anoint yourself with oil, but behave like a woman who has been mourning many days for the dead. 3 Go to the king and speak to him as follows." And Joab put the words into her mouth.

4 When the woman of Tekoa came to the king, she fell on her face to the ground and did obeisance, and said, "Help, O king!" 5 The king asked her, "What is your trouble?" She answered, "Alas, I am a widow; my husband is dead. 6 Your servant had two sons, and they fought with one another in the field; there was no one to part them, and one struck the other and killed him. 7 Now the whole family has risen against your servant. They say, 'Give up the man who struck his brother, so that we may kill him for the life of his brother whom he murdered, even if we destroy the heir as well.' Thus they would quench my one remaining ember, and leave to my husband neither name nor remnant on the face of the earth."

8 Then the king said to the woman, "Go to your house, and I will give orders concerning you." 9 The woman of Tekoa said to the king, "On me be the guilt, my lord the king, and on my father's house; let the king and his throne be guiltless." 10 The king said, "If anyone says anything to you, bring him to me, and he shall never touch you again." 11 Then she said, "Please, may the king keep the

Lord your God in mind, so that the avenger of blood may kill no more, and my son not be destroyed." He said, "As the Lord lives, not one hair of your son shall fall to the ground."

12 Then the woman said, "Please let your servant speak a word to my lord the king." He said, "Speak." 13 The woman said, "Why then have you planned such a thing against the people of God? For in giving this decision the king convicts himself, inasmuch as the king does not bring his banished one home again. 14 We must all die; we are like water spilled on the ground, which cannot be gathered up. But God will not take away a life; he will devise plans so as not to keep an outcast banished forever from his presence.[h] 15 Now I have come to say this to my lord the king because the people have made me afraid; your servant thought, 'I will speak to the king; it may be that the king will perform the request of his servant. 16 For the king will hear, and deliver his servant from the hand of the man who would cut both me and my son off from the heritage of God.' 17 Your servant thought, 'The word of my lord the king will set me at rest'; for my lord the king is like the angel of God, discerning good and evil. The Lord your God be with you!"

18 Then the king answered the woman, "Do not withhold from me anything I ask you." The woman said, "Let my lord the king speak." 19 The king said, "Is the hand of Joab with you in all this?" The woman answered and said, "As surely as you live, my lord the king, one cannot turn right or left from anything that my lord the king has said. For it was your servant Joab who commanded me; it was he who put all these words into the

g Q Ms Gk: MT *And David*   h Meaning of Heb uncertain

14.1–33: Joab brings about the return of Absalom. This incident is another manifestation of the power and influence of *Joab,* who had a way of taking things into his own hands, though he carefully remained subservient to David (v. 22 and 12.28). **2:** *Tekoa,* a few miles south of Bethlehem, and thus in the home territory of David and Joab.
14.7: The use of the word *heir* by *the woman* is a good indication that Absalom was expected to succeed David as king.

mouth of your servant. 20 In order to change the course of affairs your servant Joab did this. But my lord has wisdom like the wisdom of the angel of God to know all things that are on the earth."

21 Then the king said to Joab, "Very well, I grant this; go, bring back the young man Absalom." 22 Joab prostrated himself with his face to the ground and did obeisance, and blessed the king; and Joab said, "Today your servant knows that I have found favor in your sight, my lord the king, in that the king has granted the request of his servant." 23 So Joab set off, went to Geshur, and brought Absalom to Jerusalem. 24 The king said, "Let him go to his own house; he is not to come into my presence." So Absalom went to his own house, and did not come into the king's presence.

25 Now in all Israel there was no one to be praised so much for his beauty as Absalom; from the sole of his foot to the crown of his head there was no blemish in him. 26 When he cut the hair of his head (for at the end of every year he used to cut it; when it was heavy on him, he cut it), he weighed the hair of his head, two hundred shekels by the king's weight. 27 There were born to Absalom three sons, and one daughter whose name was Tamar; she was a beautiful woman.

28 So Absalom lived two full years in Jerusalem, without coming into the king's presence. 29 Then Absalom sent for Joab to send him to the king; but Joab would not come to him. He sent a second time, but Joab would not come. 30 Then he said to his servants, "Look, Joab's field is next to mine, and he has barley there; go and set it on fire." So Absalom's servants set the field on fire. 31 Then Joab rose and went to Absalom at his house, and said to him, "Why have your servants set my field on fire?" 32 Absalom answered Joab, "Look, I sent word to you: Come here, that I may send you to the king with the question, 'Why have I come from Geshur? It would be better for me to be there still.' Now let me go into the king's presence; if there is guilt in me, let him kill me!" 33 Then Joab went to the king and told him; and he summoned Absalom. So he came to the king and prostrated himself with his face to the ground before the king; and the king kissed Absalom.

15 After this Absalom got himself a chariot and horses, and fifty men to run ahead of him. 2 Absalom used to rise early and stand beside the road into the gate; and when anyone brought a suit before the king for judgment, Absalom would call out and say, "From what city are you?" When the person said, "Your servant is of such and such a tribe in Israel," 3 Absalom would say, "See, your claims are good and right; but there is no one deputed by the king to hear you." 4 Absalom said moreover, "If only I were judge in the land! Then all who had a suit or cause might come to me, and I would

---

14.26: There were at least two standards of *weight* at that time, the common and the royal, or *the king's weight.* The latter was somewhat heavier; *two hundred shekels* would have been roughly five pounds. **27:** Absalom seems to have named his daughter after his ravished sister, Tamar. The information about sons should be contrasted with 18.18. **28:** It was very likely during this period that Absalom began to plot revolt. Some have blamed David for delaying the reconciliation too long.

15.1–37: Absalom revolts and David flees from Jerusalem. Since *Absalom* was the natural heir to the throne (see 14.7 n.), it may be asked why he revolted during David's lifetime. Primogeniture (the requirement that the eldest surviving son should become king) was the normal practice in Israel (2 Sam 2.8–10; Deut 21.15–17), but there were exceptions (1 Chr 5.1–2; 26.10). David himself had established a new dynasty, and had failed to lay down a law of succession (1 Kings 1.27). Absalom, knowing that he had once been out of favor with his father, may have feared that something similar would happen again. Or perhaps, like Amnon, he simply could not wait for what he wanted so badly. **1–6:** *Chariot and horses, and fifty men,* compare 1 Kings 1.5. Absalom attacks another weakness of David's administration, the failure to establish a judiciary system, with consequent delay in the hearing of cases. David was better as a military leader than as a peacetime organizer.

give them justice." 5 Whenever people came near to do obeisance to him, he would put out his hand and take hold of them, and kiss them. 6 Thus Absalom did to every Israelite who came to the king for judgment; so Absalom stole the hearts of the people of Israel.

7 At the end of four*i* years Absalom said to the king, "Please let me go to Hebron and pay the vow that I have made to the LORD. 8 For your servant made a vow while I lived at Geshur in Aram: If the LORD will indeed bring me back to Jerusalem, then I will worship the LORD in Hebron."*j* 9 The king said to him, "Go in peace." So he got up, and went to Hebron. 10 But Absalom sent secret messengers throughout all the tribes of Israel, saying, "As soon as you hear the sound of the trumpet, then shout: Absalom has become king at Hebron!" 11 Two hundred men from Jerusalem went with Absalom; they were invited guests, and they went in their innocence, knowing nothing of the matter. 12 While Absalom was offering the sacrifices, he sent for*k* Ahithophel the Gilonite, David's counselor, from his city Giloh. The conspiracy grew in strength, and the people with Absalom kept increasing.

13 A messenger came to David, saying, "The hearts of the Israelites have gone after Absalom." 14 Then David said to all his officials who were with him at Jerusalem, "Get up! Let us flee, or there will be no escape for us from Absalom. Hurry, or he will soon overtake us, and bring disaster down upon us, and attack

the city with the edge of the sword." 15 The king's officials said to the king, "Your servants are ready to do whatever our lord the king decides." 16 So the king left, followed by all his household, except ten concubines whom he left behind to look after the house. 17 The king left, followed by all the people; and they stopped at the last house. 18 All his officials passed by him; and all the Cherethites, and all the Pelethites, and all the six hundred Gittites who had followed him from Gath, passed on before the king.

19 Then the king said to Ittai the Gittite, "Why are you also coming with us? Go back, and stay with the king; for you are a foreigner, and also an exile from your home. 20 You came only yesterday, and shall I today make you wander about with us, while I go wherever I can? Go back, and take your kinsfolk with you; and may the LORD show*l* steadfast love and faithfulness to you." 21 But Ittai answered the king, "As the LORD lives, and as my lord the king lives, wherever my lord the king may be, whether for death or for life, there also your servant will be." 22 David said to Ittai, "Go then, march on." So Ittai the Gittite marched on, with all his men and all the little ones who were with him. 23 The whole country wept aloud as all the people passed by; the king crossed the Wadi Kidron, and all the people moved on toward the wilderness.

i Gk Syr: Heb *forty*   *j* Gk Mss: Heb lacks *in Hebron*   *k* Or *he sent*   *l* Gk Compare 2.6: Heb lacks *may the LORD show*

15.7–12: The conspiracy was carefully nurtured for a period of *four years*. Absalom could have worshiped in Jerusalem, but he had probably found disaffection in Hebron over the loss of status as the capital city. *Ahithophel the Gilonite* was perhaps the grandfather of Bathsheba (compare 11.3 with 23.34). If so, he may have resented David's conduct with his granddaughter in spite of the higher status which it brought to the family. **15.13–18:** David flees Jerusalem, perhaps feeling that he would have an advantage by fighting in the open country. He probably also suspected the loyalty of some of those around him; by leaving, he could find out who would follow him regardless of circumstances, and at the same time he could leave behind his agents to work secretly for him in the city. See the superscription to Ps 3. *Ten concubines,* see 5.13–16 n. *Cherethites and Pelethites,* see 8.18 n. *Gittites* (rather than "Gathites") is the proper form designating people *from Gath*. It is rather surprising to find so many Philistines in David's army. **23:** The weeping of *the people* indicates that *the king* retained much popular support. *The Wadi Kidron* marked the eastern boundary of the city.

24 Abiathar came up, and Zadok also, with all the Levites, carrying the ark of the covenant of God. They set down the ark of God, until the people had all passed out of the city. 25 Then the king said to Zadok, "Carry the ark of God back into the city. If I find favor in the eyes of the LORD, he will bring me back and let me see both it and the place where it stays. 26 But if he says, 'I take no pleasure in you,' here I am, let him do to me what seems good to him." 27 The king also said to the priest Zadok, "Look, *m* go back to the city in peace, you and Abiathar, *n* with your two sons, Ahimaaz your son, and Jonathan son of Abiathar. 28 See, I will wait at the fords of the wilderness until word comes from you to inform me." 29 So Zadok and Abiathar carried the ark of God back to Jerusalem, and they remained there.

30 But David went up the ascent of the Mount of Olives, weeping as he went, with his head covered and walking barefoot; and all the people who were with him covered their heads and went up, weeping as they went. 31 David was told that Ahithophel was among the conspirators with Absalom. And David said, "O LORD, I pray you, turn the counsel of Ahithophel into foolishness."

32 When David came to the summit, where God was worshiped, Hushai the Archite came to meet him with his coat torn and earth on his head. 33 David said to him, "If you go on with me, you will be a burden to me. 34 But if you return to the city and say to Absalom, 'I will be your servant, O king; as I have been your father's servant in time past, so now I will be your servant,' then you will defeat for me the counsel of Ahithophel. 35 The priests Zadok and Abiathar will be with you there. So whatever you hear from the king's house, tell it to the priests Zadok and Abiathar. 36 Their two sons are with them there, Zadok's son Ahimaaz and Abiathar's son Jonathan; and by them you shall report to me everything you hear." 37 So Hushai, David's friend, came into the city, just as Absalom was entering Jerusalem.

16 When David had passed a little beyond the summit, Ziba the servant of Mephibosheth *o* met him, with a couple of donkeys saddled, carrying two hundred loaves of bread, one hundred bunches of raisins, one hundred of summer fruits, and one skin of wine. 2 The king said to Ziba, "Why have you brought these?" Ziba answered, "The donkeys are for the king's household to ride, the bread and summer fruit for the young men to eat, and the wine is for those to drink who faint in the wilderness." 3 The king said, "And where is your master's son?" Ziba said to the king, "He remains in Jerusalem; for he said, 'Today the house of Israel will give me back my grandfather's kingdom.'" 4 Then the king said to Ziba, "All that belonged to Mephibosheth *o* is now yours." Ziba said, "I do obeisance; let me find favor in your sight, my lord the king."

m Gk: Heb *Are you a seer* or *Do you see?*
n Cn: Heb lacks *and Abiathar*
o Or *Merib-baal:* See 4.4 note

15.24–37: David's leaving *the ark* in Jerusalem was a fair indication that he expected to return. It was also good diplomacy to have the loyal priests in the city (vv. 35–36). On *Zadok,* see 8.17 n. *Abiathar,* a priest closely associated with David (see 1 Sam 23.6–11; 2 Sam 20.25). *Hushai the Archite, David's friend* (i.e. "royal counselor," an official title, 1 Kings 4.5), came from an originally Canaanite group settled near Bethel (Josh 16.2). Under David's inclusive policy, various non-Israelite groups (Hittites, Philistines, Canaan-ites, and the like) became naturalized. These people seem to have been particularly loyal to David, perhaps partly because of his policies and personality, partly because he himself was not an Israelite (19.11–15).
16.1–23: David in flight; Absalom takes over Jerusalem. 1–4: *Mephibosheth,* see ch 9 and notes there. *Ziba* was probably lying for his own advantage; see Mephibosheth's defense in 19.24–30. David, perhaps disturbed by the reference to the dynasty of Saul, here makes too hasty a decision.

5 When King David came to Bahurim, a man of the family of the house of Saul came out whose name was Shimei son of Gera; he came out cursing. 6 He threw stones at David and at all the servants of King David; now all the people and all the warriors were on his right and on his left. 7 Shimei shouted while he cursed, "Out! Out! Murderer! Scoundrel! 8 The LORD has avenged on all of you the blood of the house of Saul, in whose place you have reigned; and the LORD has given the kingdom into the hand of your son Absalom. See, disaster has overtaken you; for you are a man of blood."

9 Then Abishai son of Zeruiah said to the king, "Why should this dead dog curse my lord the king? Let me go over and take off his head." 10 But the king said, "What have I to do with you, you sons of Zeruiah? If he is cursing because the LORD has said to him, 'Curse David,' who then shall say, 'Why have you done so?' " 11 David said to Abishai and to all his servants, "My own son seeks my life; how much more now may this Benjaminite! Let him alone, and let him curse; for the LORD has bidden him. 12 It may be that the LORD will look on my distress, *p* and the LORD will repay me with good for this cursing of me today." 13 So David and his men went on the road, while Shimei went along on the hillside opposite him and cursed as he went, throwing stones and flinging dust at him. 14 The king and all the people who were with him arrived weary at the Jordan; *q* and there he refreshed himself.

15 Now Absalom and all the Israel-ites *r* came to Jerusalem; Ahithophel was with him. 16 When Hushai the Architc, David's friend, came to Absalom, Hushai said to Absalom, "Long live the king! Long live the king!" 17 Absalom said to Hushai, "Is this your loyalty to your friend? Why did you not go with your friend?" 18 Hushai said to Absalom, "No; but the one whom the LORD and this people and all the Israelites have chosen, his I will be, and with him I will remain. 19 Moreover, whom should I serve? Should it not be his son? Just as I have served your father, so I will serve you."

20 Then Absalom said to Ahithophel, "Give us your counsel; what shall we do?" 21 Ahithophel said to Absalom, "Go in to your father's concubines, the ones he has left to look after the house; and all Israel will hear that you have made yourself odious to your father, and the hands of all who are with you will be strengthened." 22 So they pitched a tent for Absalom upon the roof; and Absalom went in to his father's concubines in the sight of all Israel. 23 Now in those days the counsel that Ahithophel gave was as if one consulted the oracle *s* of God; so all the counsel of Ahithophel was esteemed, both by David and by Absalom.

17 Moreover Ahithophel said to Absalom, "Let me choose twelve thousand men, and I will set out and pursue David tonight. 2 I will come upon him while he is weary and discouraged, and throw him into a panic; and all the

p Gk Vg: Hcb *iniquity*   q Gk: Heb lacks *at the Jordan*   r Gk: Heb *all the people, the men of Israel*   s Heb *word*

16.5–14: The disaffection in Judah over loss of the capital has already been noted (see 15.7–12 n.); here we see the disaffection in Israel over loss of the ruling dynasty (v. 8). This time, however, David keeps a cool head and refuses to be pushed into a hasty judgment. For the result, see 19.16–23; compare 1 Kings 2.8–9. *This dead dog,* see 3.8 n. On the undue haste of a *son of Zeruiah* with the sword, see 3.39.
16.20–23: The *concubines* were royal property; hence taking them over publicly was a sensational way of showing the people that Absalom had assumed the office and prerogatives of kingship (3.7; 15.16; 1 Kings 2.17–25).
17.1–23: Hushai frustrates the plan of Ahithophel. 1–14: The plan of Ahithophel would probably have led to success for Absalom. The plan of Hushai gave the advantage to David by enabling him to collect his forces for a formal battle in which experience and skill would be decisive. Recognition of these facts led to the suicide of Ahithophel (v. 23). The last sentence of v. 14 reminds the reader that all things are *ordained* by God.

people who are with him will flee. I will strike down only the king, 3and I will bring all the people back to you as a bride comes home to her husband. You seek the life of only one man,*t* and all the people will be at peace." 4The advice pleased Absalom and all the elders of Israel.

5 Then Absalom said, "Call Hushai the Archite also, and let us hear too what he has to say." 6When Hushai came to Absalom, Absalom said to him, "This is what Ahithophel has said; shall we do as he advises? If not, you tell us." 7Then Hushai said to Absalom, "This time the counsel that Ahithophel has given is not good." 8Hushai continued, "You know that your father and his men are warriors, and that they are enraged, like a bear robbed of her cubs in the field. Besides, your father is expert in war; he will not spend the night with the troops. 9Even now he has hidden himself in one of the pits, or in some other place. And when some of our troops*u* fall at the first attack, whoever hears it will say, 'There has been a slaughter among the troops who follow Absalom.' 10Then even the valiant warrior, whose heart is like the heart of a lion, will utterly melt with fear; for all Israel knows that your father is a warrior, and that those who are with him are valiant warriors. 11But my counsel is that all Israel be gathered to you, from Dan to Beer-sheba, like the sand by the sea for multitude, and that you go to battle in person. 12So we shall come upon him in whatever place he may be found, and we shall light on him as the dew falls on the ground; and he will not survive, nor will any of those with him. 13If he withdraws into a city, then all Israel will bring ropes to that city, and we shall drag it into the valley, until not even a pebble is to be found there." 14Absalom and all the men of Israel said, "The counsel of Hushai the Archite is better

than the counsel of Ahithophel." For the LORD had ordained to defeat the good counsel of Ahithophel, so that the LORD might bring ruin on Absalom.

15 Then Hushai said to the priests Zadok and Abiathar, "Thus and so did Ahithophel counsel Absalom and the elders of Israel; and thus and so I have counseled. 16Therefore send quickly and tell David, 'Do not lodge tonight at the fords of the wilderness, but by all means cross over; otherwise the king and all the people who are with him will be swallowed up.' " 17Jonathan and Ahimaaz were waiting at En-rogel; a servant-girl used to go and tell them, and they would go and tell King David; for they could not risk being seen entering the city. 18But a boy saw them, and told Absalom; so both of them went away quickly, and came to the house of a man at Bahurim, who had a well in his courtyard; and they went down into it. 19The man's wife took a covering, stretched it over the well's mouth, and spread out grain on it; and nothing was known of it. 20When Absalom's servants came to the woman at the house, they said, "Where are Ahimaaz and Jonathan?" The woman said to them, "They have crossed over the brook*v* of water." And when they had searched and could not find them, they returned to Jerusalem.

21 After they had gone, the men came up out of the well, and went and told King David. They said to David, "Go and cross the water quickly; for thus and so has Ahithophel counseled against you." 22So David and all the people who were with him set out and crossed the Jordan; by daybreak not one was left who had not crossed the Jordan.

23 When Ahithophel saw that his counsel was not followed, he saddled his

t Gk: Heb *like the return of the whole (is) the man whom you seek   u* Gk Mss: Heb *some of them
v* Meaning of Heb uncertain

17.15–22: David is informed and crosses the Jordan. Hushai did not know which way the decision had gone; hence his recommendation to cross the river immediately. *En-*

rogel is a spring near the southeast corner of Jerusalem, now called Job's Well. *Bahurim* is east of the Mt. of Olives. **23:** On biblical suicides, see 1 Sam 31.5 n.

donkey and went off home to his own city. He set his house in order, and hanged himself; he died and was buried in the tomb of his father.

24 Then David came to Mahanaim, while Absalom crossed the Jordan with all the men of Israel. 25 Now Absalom had set Amasa over the army in the place of Joab. Amasa was the son of a man named Ithra the Ishmaelite,ʷ who had married Abigal daughter of Nahash, sister of Zeruiah, Joab's mother. 26 The Israelites and Absalom encamped in the land of Gilead.

27 When David came to Mahanaim, Shobi son of Nahash from Rabbah of the Ammonites, and Machir son of Ammiel from Lo-debar, and Barzillai the Gileadite from Rogelim, 28 brought beds, basins, and earthen vessels, wheat, barley, meal, parched grain, beans and lentils,ˣ 29 honey and curds, sheep, and cheese from the herd, for David and the people with him to eat; for they said, "The troops are hungry and weary and thirsty in the wilderness."

18 Then David mustered the men who were with him, and set over them commanders of thousands and commanders of hundreds. 2 And David divided the army into three groups:ʸ one third under the command of Joab, one third under the command of Abishai son of Zeruiah, Joab's brother, and one third under the command of Ittai the Gittite. The king said to the men, "I myself will also go out with you." 3 But the men said, "You shall not go out. For if we flee, they will not care about us. If half of us die, they will not care about us. But you are worth ten thousand of us;ᶻ therefore it is better that you send us help from the city." 4 The king said to them, "Whatever seems best to you I will do." So the king stood at the side of the gate, while all the army marched out by hundreds and by thousands. 5 The king ordered Joab and Abishai and Ittai, saying, "Deal gently for my sake with the young man Absalom." And all the people heard when the king gave orders to all the commanders concerning Absalom.

6 So the army went out into the field against Israel; and the battle was fought in the forest of Ephraim. 7 The men of Israel were defeated there by the servants of David, and the slaughter there was great on that day, twenty thousand men. 8 The battle spread over the face of all the country; and the forest claimed more victims that day than the sword.

9 Absalom happened to meet the servants of David. Absalom was riding on his mule, and the mule went under the thick branches of a great oak. His head caught fast in the oak, and he was left hangingᵃ between heaven and earth, while the mule that was under him went on. 10 A man saw it, and told Joab, "I saw Absalom hanging in an oak." 11 Joab said

w 1 Chr 2.17: Heb *Israelite*   x Heb *and lentils and parched grain*   y Gk: Heb *sent forth the army*   z Gk Vg Symmachus: Heb *for now there are ten thousand such as w*   a Gk Syr Tg: Heb *was put*

17.24–29: **David in Transjordan.** *Amasa* was a cousin of Joab; both of them were nephews of David and cousins of Absalom (1 Chr 2.13–17). The *Nahash* in v. 25 (probably an error) is not the same person as the *Nahash* of v. 27. On the latter as a friend of David, see 10.2. *Shobi*, unlike his brother Hanun, was friendly to David. It has even been proposed that David had deposed Hanun after the events of ch 10, putting *Shobi* in his place. *Machir*, see 9.4. The Aramean name of *Barzillai* shows him to have been a non-Israelite; see the comment on Hushai, 15.24–37 n.

18.1–8: **The battle in the forest of Ephraim. 3–4**: For an earlier agreement that David should remain out of the front line of battle, see 21.17; for an exception, see 12.28–29. **5**: In spite of Absalom's complete apostasy, David still loved him and wished that his life might be saved **6–8**: As might have been expected, Absalom's hastily assembled forces were no match for David's standing army under skilled and experienced commanders. Great numbers of the raw recruits were driven into *the forest*, a trackless jungle, from which there was no escape. The number *twenty thousand* is an overgenerous estimate; there was no counting before or after the battle (see 12.30 n. and 1 Sam 4.10 n.).

18.9–18: **The death of Absalom.** There is a popular notion, based on 14.26, that Absalom was caught by his hair. The present

to the man who told him, "What, you saw him! Why then did you not strike him there to the ground? I would have been glad to give you ten pieces of silver and a belt." 12 But the man said to Joab, "Even if I felt in my hand the weight of a thousand pieces of silver, I would not raise my hand against the king's son; for in our hearing the king commanded you and Abishai and Ittai, saying: For my sake protect the young man Absalom! 13 On the other hand, if I had dealt treacherously against his life *b* (and there is nothing hidden from the king), then you yourself would have stood aloof." 14 Joab said, "I will not waste time like this with you." He took three spears in his hand, and thrust them into the heart of Absalom, while he was still alive in the oak. 15 And ten young men, Joab's armorbearers, surrounded Absalom and struck him, and killed him.

16 Then Joab sounded the trumpet, and the troops came back from pursuing Israel, for Joab restrained the troops. 17 They took Absalom, threw him into a great pit in the forest, and raised over him a very great heap of stones. Meanwhile all the Israelites fled to their homes. 18 Now Absalom in his lifetime had taken and set up for himself a pillar that is in the King's Valley, for he said, "I have no son to keep my name in remembrance"; he called the pillar by his own name. It is called Absalom's Monument to this day.

19 Then Ahimaaz son of Zadok said, "Let me run, and carry tidings to the king that the LORD has delivered him from the power of his enemies." 20 Joab said to him, "You are not to carry tidings today; you may carry tidings another day, but today you shall not do so, because the king's son is dead." 21 Then Joab said to a Cushite, "Go, tell the king what you have seen." The Cushite bowed before Joab, and ran. 22 Then Ahimaaz son of Zadok said again to Joab, "Come what may, let me also run after the Cushite." And Joab said, "Why will you run, my son, seeing that you have no reward *c* for the tidings?" 23 "Come what may," he said, "I will run." So he said to him, "Run." Then Ahimaaz ran by the way of the Plain, and outran the Cushite.

24 Now David was sitting between the two gates. The sentinel went up to the roof of the gate by the wall, and when he looked up, he saw a man running alone. 25 The sentinel shouted and told the king. The king said, "If he is alone, there are tidings in his mouth." He kept coming, and drew near. 26 Then the sentinel saw another man running; and the sentinel called to the gatekeeper and said, "See, another man running alone!" The king said, "He also is bringing tidings." 27 The sentinel said, "I think the running of the first one is like the running of Ahimaaz son of Zadok." The king said, "He is a good man, and comes with good tidings."

28 Then Ahimaaz cried out to the king, "All is well!" He prostrated himself before the king with his face to the ground, and said, "Blessed be the LORD your God, who has delivered up the men

b Another reading is *at the risk of my life* *c* Meaning of Heb uncertain

text implies that his whole *head* was caught (v. 9). Joab never withheld a fatal blow when he felt that it would be to his own or David's advantage (3.27; 20.10). **16–18:** These verses show that Absalom's tomb was very different from what he had intended; he had prepared for himself a showy monument near Jerusalem. The statement about his having *no son* appears to be inconsistent with 14.27. The monument commonly called "Absalom's Tomb," still standing today in the Kidron valley, is of Hellenistic or Roman date, and has no connection with the original *Absalom's Monument.*

18.19–33: The grief of David. The story in these verses provides the reason why some commentators believe *Ahimaaz son of Zadok* to be the author. But there is no persuasive reason to assume that an author (such as Ahimaaz here) would have put himself into the story. Ahimaaz was too excited to remain behind, yet when he faced *the king* he lacked the courage to tell the whole story, as Joab had foreseen.

who raised their hand against my lord the king." 29 The king said, "Is it well with the young man Absalom?" Ahimaaz answered, "When Joab sent your servant,*d* I saw a great tumult, but I do not know what it was." 30 The king said, "Turn aside, and stand here." So he turned aside, and stood still.

31 Then the Cushite came; and the Cushite said, "Good tidings for my lord the king! For the LORD has vindicated you this day, delivering you from the power of all who rose up against you." 32 The king said to the Cushite, "Is it well with the young man Absalom?" The Cushite answered, "May the enemies of my lord the king, and all who rise up to do you harm, be like that young man."

33*e* The king was deeply moved, and went up to the chamber over the gate, and wept; and as he went, he said, "O my son Absalom, my son, my son Absalom! Would I had died instead of you, O Absalom, my son, my son!"

19 It was told Joab, "The king is weeping and mourning for Absalom." 2 So the victory that day was turned into mourning for all the troops; for the troops heard that day, "The king is grieving for his son." 3 The troops stole into the city that day as soldiers steal in who are ashamed when they flee in battle. 4 The king covered his face, and the king cried with a loud voice, "O my son Absalom, O Absalom, my son, my son!" 5 Then Joab came into the house to the king, and said, "Today you have covered with shame the faces of all your officers who have saved your life today, and the lives of your sons and your daughters, and the lives of your wives and your concubines, 6 for love of those who hate you and for hatred of those who love you. You have made it clear today that commanders and officers are nothing to you; for I perceive that if Absalom were alive and all of us were dead today, then you would be pleased. 7 So go out at once and speak kindly to your servants; for I swear by the LORD, if you do not go, not a man will stay with you this night; and this will be worse for you than any disaster that has come upon you from your youth until now." 8 Then the king got up and took his seat in the gate. The troops were all told, "See, the king is sitting in the gate"; and all the troops came before the king.

Meanwhile, all the Israelites had fled to their homes. 9 All the people were disputing throughout all the tribes of Israel, saying, "The king delivered us from the hand of our enemies, and saved us from the hand of the Philistines; and now he has fled out of the land because of Absalom. 10 But Absalom, whom we anointed over us, is dead in battle. Now therefore why do you say nothing about bringing the king back?"

11 King David sent this message to the priests Zadok and Abiathar, "Say to the elders of Judah, 'Why should you be the last to bring the king back to his house? The talk of all Israel has come to the king.*f* 12 You are my kin, you are my bone and my flesh; why then should you be the last to bring back the king?' 13 And say to Amasa, 'Are you not my bone and my flesh? So may God do to me, and more, if you are not the commander of my army from now on, in place of Joab.' " 14 Amasa*g* swayed the hearts of all the people of Judah as one, and they

d Heb *the king's servant, your servant*
e Ch 19.1 in Heb   *f* Gk: Heb *to the king, to his house*   *g* Heb *He*

18.31–33: It remained for the Ethiopian slave (*the Cushite;* compare Jer 13.23; 38.7; 39.16–18) to reveal the true state of affairs. In his personal grief, David momentarily forgot his military victory and his public duties. **19.1–43: David attempts to restore a divided and disorganized nation. 1–8a:** It was only Joab who could speak realistically to the king about his royal obligations. **8b–10:** The Israelites made the first overtures to renew allegiance to David. **19.11–15:** *David* appeals to the Judahites on the basis of tribal loyalty and by giving Joab's position to *Amasa.* On Amasa's kinship to David and Joab, see 17.24–29 n. David no doubt resented Joab's killing of Absalom (18.14) and also remembered the case of Abner (3.27–29).

sent word to the king, "Return, both you and all your servants." 15 So the king came back to the Jordan; and Judah came to Gilgal to meet the king and to bring him over the Jordan.

16 Shimei son of Gera, the Benjaminite, from Bahurim, hurried to come down with the people of Judah to meet King David; 17 with him were a thousand people from Benjamin. And Ziba, the servant of the house of Saul, with his fifteen sons and his twenty servants, rushed down to the Jordan ahead of the king, 18 while the crossing was taking place, *h* to bring over the king's household, and to do his pleasure.

Shimei son of Gera fell down before the king, as he was about to cross the Jordan, 19 and said to the king, "May my lord not hold me guilty or remember how your servant did wrong on the day my lord the king left Jerusalem; may the king not bear it in mind. 20 For your servant knows that I have sinned; therefore, see, I have come this day, the first of all the house of Joseph to come down to meet my lord the king." 21 Abishai son of Zeruiah answered, "Shall not Shimei be put to death for this, because he cursed the LORD's anointed?" 22 But David said, "What have I to do with you, you sons of Zeruiah, that you should today become an adversary to me? Shall anyone be put to death in Israel this day? For do I not know that I am this day king over Israel?" 23 The king said to Shimei, "You shall not die." And the king gave him his oath.

24 Mephibosheth *i* grandson of Saul came down to meet the king; he had not taken care of his feet, or trimmed his beard, or washed his clothes, from the day the king left until the day he came back in safety. 25 When he came from Jerusalem to meet the king, the king said to him, "Why did you not go with me, Mephibosheth?" *i* 26 He answered, "My lord, O king, my servant deceived me; for your servant said to him, 'Saddle a donkey for me, *j* so that I may ride on it and go with the king.' For your servant is lame. 27 He has slandered your servant to my lord the king. But my lord the king is like the angel of God; do therefore what seems good to you. 28 For all my father's house were doomed to death before my lord the king; but you set your servant among those who eat at your table. What further right have I, then, to appeal to the king?" 29 The king said to him, "Why speak any more of your affairs? I have decided: you and Ziba shall divide the land." 30 Mephibosheth *i* said to the king, "Let him take it all, since my lord the king has arrived home safely."

31 Now Barzillai the Gileadite had come down from Rogelim; he went on with the king to the Jordan, to escort him over the Jordan. 32 Barzillai was a very aged man, eighty years old. He had provided the king with food while he stayed at Mahanaim, for he was a very wealthy man. 33 The king said to Barzillai, "Come over with me, and I will provide for you in Jerusalem at my side." 34 But Barzillai said to the king, "How many years have I still to live, that I should go up with the king to Jerusalem? 35 Today I am eighty years old; can I discern what is pleasant and what is not? Can your servant taste what he eats or what he drinks? Can I still listen to the voice of singing men and singing women? Why then should your servant be an added burden to my lord the king? 36 Your ser-

h Cn: Heb *the ford crossed*   *i* Or *Merib-baal*: See 4.4 note   *j* Gk Syr Vg:Heb *said, I will saddle a donkey for myself*

19.16–23: *Shimei* and *Ziba* rushed to pay homage to *David* (compare 16.1–14). *David* forgave *Shimei*. *House of Joseph* (v. 20), another name for the northern tribes or Israel.

19.24–30: See ch 9 and 16.1–5. David had made a hasty decision in Ziba's case against Mephibosheth. Now, confronted by both parties, *the king* seemed unable to make a firm decision, though Mephibosheth's evidences of loyalty (v. 24) and self-abnegation (vv. 28, 30) appear genuine.

19.31–40α: It is usually assumed that *Chimham* was the son of *Barzillai*, though the text does not specifically so state (compare 1 Kings 2.7). **40b**: The word *all* is something of an exaggeration, signifying that by this

vant will go a little way over the Jordan with the king. Why should the king recompense me with such a reward? 37 Please let your servant return, so that I may die in my own town, near the graves of my father and my mother. But here is your servant Chimham; let him go over with my lord the king; and do for him whatever seems good to you." 38 The king answered, "Chimham shall go over with me, and I will do for him whatever seems good to you; and all that you desire of me I will do for you." 39 Then all the people crossed over the Jordan, and the king crossed over; the king kissed Barzillai and blessed him, and he returned to his own home. 40 The king went on to Gilgal, and Chimham went on with him; all the people of Judah, and also half the people of Israel, brought the king on his way.

41 Then all the people of Israel came to the king, and said to him, "Why have our kindred the people of Judah stolen you away, and brought the king and his household over the Jordan, and all David's men with him?" 42 All the people of Judah answered the people of Israel, "Because the king is near of kin to us. Why then are you angry over this matter? Have we eaten at all at the king's expense? Or has he given us any gift?" 43 But the people of Israel answered the people of Judah, "We have ten shares in the king, and in David also we have more than you. Why then did you despise us? Were we not the first to speak of bringing back our king?" But the words of the people of Judah were fiercer than the words of the people of Israel.

20 Now a scoundrel named Sheba son of Bichri, a Benjaminite, happened to be there. He sounded the trumpet and cried out,

"We have no portion in David,
no share in the son of Jesse!
Everyone to your tents, O Israel!"

2 So all the people of Israel withdrew from David and followed Sheba son of Bichri; but the people of Judah followed their king steadfastly from the Jordan to Jerusalem.

3 David came to his house at Jerusalem; and the king took the ten concubines whom he had left to look after the house, and put them in a house under guard, and provided for them, but did not go in to them. So they were shut up until the day of their death, living as if in widowhood.

4 Then the king said to Amasa, "Call the men of Judah together to me within three days, and be here yourself." 5 So Amasa went to summon Judah; but he delayed beyond the set time that had been appointed him. 6 David said to Abishai, "Now Sheba son of Bichri will do us more harm than Absalom; take your lord's servants and pursue him, or he will find fortified cities for himself, and escape from us." 7 Joab's men went out after him, along with the Cherethites, the Pelethites, and all the warriors; they went out from Jerusalem to pursue Sheba son of Bichri. 8 When they were at the large stone that is in Gibeon, Amasa came to meet them. Now Joab was wearing a soldier's garment and over it was a belt with a sword in its sheath fastened at his waist; as he went forward it fell out. 9 Joab said to Amasa, "Is it well with you, my brother?" And Joab took Amasa by the beard with his right hand to kiss him. 10 But Amasa did not notice

---

time Judah was solidly in favor of David, whereas Israel was hardly more than half persuaded (compare 20.2). **41–43**: The antagonism between *Israel* and *Judah* breaks out again. In the light of this passage, it can be seen that the final division of the kingdom after the death of Solomon (1 Kings 12.16–20) had its roots in an old cleavage.

20.1–22: New revolt in Israel; Joab murders Amasa, regains his position, and quells the revolt. **1–2**: It is significant that *Sheba* came from Benjamin, the most disaffected tribe (compare 16.5, 8). The war cry is repeated in 1 Kings 12.16. **3**: See 6.23 n. **4–13**: The delay of *Amasa* showed David that the new commander was not equal to Joab; but the king was too proud to approach Joab directly. Hence he called upon Abishai, Joab's brother, knowing that Joab would soon take the initiative (note the order of names in

the sword in Joab's hand; Joab struck him in the belly so that his entrails poured out on the ground, and he died. He did not strike a second blow.

Then Joab and his brother Abishai pursued Sheba son of Bichri. 11 And one of Joab's men took his stand by Amasa, and said, "Whoever favors Joab, and whoever is for David, let him follow Joab." 12 Amasa lay wallowing in his blood on the highway, and the man saw that all the people were stopping. Since he saw that all who came by him were stopping, he carried Amasa from the highway into a field, and threw a garment over him. 13 Once he was removed from the highway, all the people went on after Joab to pursue Sheba son of Bichri.

14 Sheba*k* passed through all the tribes of Israel to Abel of Beth-maacah;*l* and all the Bichrites*m* assembled, and followed him inside. 15 Joab's forces*n* came and besieged him in Abel of Beth-maacah; they threw up a siege ramp against the city, and it stood against the rampart. Joab's forces were battering the wall to break it down. 16 Then a wise woman called from the city, "Listen! Listen! Tell Joab, 'Come here, I want to speak to you.' " 17 He came near her; and the woman said, "Are you Joab?" He answered, "I am." Then she said to him, "Listen to the words of your servant." He answered, "I am listening." 18 Then she said, "They used to say in the old days, 'Let them inquire at Abel'; and so they would settle a matter. 19 I am one of those who are peaceable and faithful in Israel; you seek to destroy a city that is a mother in Israel; why will you swallow up the heritage of the LORD?" 20 Joab answered, "Far be it from me, far be it, that I should swallow up or destroy! 21 That is not the case! But a man of the hill country of Ephraim, called Sheba son of Bichri, has lifted up his hand against King David; give him up alone, and I will withdraw from the city." The woman said to Joab, "His head shall be thrown over the wall to you." 22 Then the woman went to all the people with her wise plan. And they cut off the head of Sheba son of Bichri, and threw it out to Joab. So he blew the trumpet, and they dispersed from the city, and all went to their homes, while Joab returned to Jerusalem to the king.

23 Now Joab was in command of all the army of Israel;*o* Benaiah son of Jehoiada was in command of the Cherethites and the Pelethites; 24 Adoram was in charge of the forced labor; Jehoshaphat son of Ahilud was the recorder; 25 Sheva was secretary; Zadok and Abiathar were priests; 26 and Ira the Jairite was also David's priest.

21 Now there was a famine in the days of David for three years, year after year; and David inquired of the LORD. The LORD said, "There is bloodguilt on Saul and on his house, because he put the Gibeonites to death." 2 So the

k Heb *He*   *l* Compare 20.15: Heb *and Beth-maacah*   *m* Compare Gk Vg: Heb *Berites*   *n* Heb *They Israel*   *o* Cn: Heb *Joab to all the army,*

---

v. 10b and the name of Joab alone in v. 13). On the *Cherethites, the Pelethites* (v. 7), see 8.18 n.

20.14–22: Joab was now in complete command of the army and of the situation. Joab showed his energy and efficiency in pursuing the rebel to this remote spot. **14:** The *Bichrites* were the members of Sheba's own clan (v. 1), perhaps the only followers remaining to him. **18:** The meaning of this verse is not clear, but it would seem that the town had a reputation for wisdom and the settling of disputes. **19:** *A city that is a mother in Israel* means one that had dependent villages called "daughters" (compare *villages*, literally "daughters," in Num 21.25, 32; Josh 15.45; Judg 11.26). **22:** The story continues in 1 Kings chs 1–2.

20.23–26: A list of David's officials, added by an editor; it seems to be another version of the similar list in 8.16–18 (see note there). Perhaps it was inserted here to emphasize that *Joab* was again in command. *Adoram* (or Adoniram) kept his position under Solomon (1 Kings 4.6). The institution of *forced labor* was one of the causes of the final division of the kingdom (1 Kings 12.18–19). *Ira the Jairite* was from Gilead, a territory which had welcomed *David* when he was in flight before Absalom (compare Num 32.40–41).

21.1–24.25: This section is something of

king called the Gibeonites and spoke to them. (Now the Gibeonites were not of the people of Israel, but of the remnant of the Amorites; although the people of Israel had sworn to spare them, Saul had tried to wipe them out in his zeal for the people of Israel and Judah.) 3David said to the Gibeonites, "What shall I do for you? How shall I make expiation, that you may bless the heritage of the LORD?" 4The Gibeonites said to him, "It is not a matter of silver or gold between us and Saul or his house; neither is it for us to put anyone to death in Israel." He said, "What do you say that I should do for you?" 5They said to the king, "The man who consumed us and planned to destroy us, so that we should have no place in all the territory of Israel— 6let seven of his sons be handed over to us, and we will impale them before the LORD at Gibeon on the mountain of the LORD."*p* The king said, "I will hand them over."

7 But the king spared Mephibosheth,*q* the son of Saul's son Jonathan, because of the oath of the LORD that was between them, between David and Jonathan son of Saul. 8The king took the two sons of Rizpah daughter of Aiah, whom she bore to Saul, Armoni and Mephibo-

sheth;*q* and the five sons of Merab*r* daughter of Saul, whom she bore to Adriel son of Barzillai the Meholathite; 9he gave them into the hands of the Gibeonites, and they impaled them on the mountain before the LORD. The seven of them perished together. They were put to death in the first days of harvest, at the beginning of barley harvest.

10 Then Rizpah the daughter of Aiah took sackcloth, and spread it on a rock for herself, from the beginning of harvest until rain fell on them from the heavens; she did not allow the birds of the air to come on the bodies*s* by day, or the wild animals by night. 11When David was told what Rizpah daughter of Aiah, the concubine of Saul, had done, 12David went and took the bones of Saul and the bones of his son Jonathan from the people of Jabesh-gilead, who had stolen them from the public square of Bethshan, where the Philistines had hung them up, on the day the Philistines killed Saul on Gilboa. 13He brought up from there the bones of Saul and the bones of

p Cn Compare Gk and 21.9: Heb *at Gibeah of Saul, the chosen of the LORD*   *q* Or *Merib-baal:* See 4.4 note   *r* Two Heb Mss Syr Compare Gk: MT *Michal*   *s* Heb *them*

---

an appendix to 2 Samuel, and some items relate to earlier material in the larger story. The principles by which these traditions have been ordered are unclear.
21.1–14: The famine and the execution of the descendants of Saul. This passage seems to relate to the early part of David's reign. **1:** There is no record of the occasion upon which *Saul . . . put the Gibeonites to death.* Apparently Saul was highly intolerant of non-Israelite elements in his kingdom, in contrast to the inclusive policy of David (see 15.24–37 n.). *David* heard a complaint from *the Gibeonites,* which he took as the voice of God, connected with the *famine.* The notion of *bloodguilt* was based on the "lex talionis" (see 12.13–14 n.). **2:** For the story of the original contact with *the Gibeonites,* see Josh 9.3–27. The pre-Israelite inhabitants of Palestine are sometimes called Canaanites, sometimes *Amorites.* **6:** The number *seven* was thought to have special significance.
21.7: This verse makes an editorial link with 9.1–13. **8:** Only *two sons* of Saul were

found; the other *five* were grandsons. On *Rizpah,* see 3.7 The *Mephibosheth* of this verse is not the same person as the son of Jonathan in v. 7 and ch 9.
21.10: The *beginning of harvest* was in late April or May; rain does not fall in Palestine from that time until late autumn; hence the bodies were exposed all summer. The law of Deut 21.22–23 was not yet in effect; the long exposure (an added insult to the deceased; compare 1 Sam 31.10–13) was doubtless a part of the expiation which David was cooperating with the Gibeonites to bring about. **11–14:** The amazing vigil of Rizpah had doubtless excited much sympathy. David began to feel that the expiation had been overdone, and that it was time to do something favorable to the memory of Saul to avoid repercussions from partisans of the late king. Also, the beginning of rain had marked the end of the famine. For these reasons, the great public mass burial, including the bones of Saul and Jonathan (1 Sam 31.11–13), seemed in order.

his son Jonathan; and they gathered the bones of those who had been impaled. [14] They buried the bones of Saul and of his son Jonathan in the land of Benjamin in Zela, in the tomb of his father Kish; they did all that the king commanded. After that, God heeded supplications for the land.

15 The Philistines went to war again with Israel, and David went down together with his servants. They fought against the Philistines, and David grew weary. [16] Ishbi-benob, one of the descendants of the giants, whose spear weighed three hundred shekels of bronze, and who was fitted out with new weapons, [t] said he would kill David. [17] But Abishai son of Zeruiah came to his aid, and attacked the Philistine and killed him. Then David's men swore to him, "You shall not go out with us to battle any longer, so that you do not quench the lamp of Israel."

18 After this a battle took place with the Philistines, at Gob; then Sibbecai the Hushathite killed Saph, who was one of the descendants of the giants. [19] Then there was another battle with the Philistines at Gob; and Elhanan son of Jaare-oregim, the Bethlehemite, killed Goliath the Gittite, the shaft of whose spear was like a weaver's beam. [20] There was again war at Gath, where there was a man of great size, who had six fingers on each hand, and six toes on each foot, twenty-four in number; he too was descended from the giants. [21] When he taunted Israel, Jonathan son of David's brother Shimei, killed him. [22] These four were descended from the giants in Gath; they fell by the hands of David and his servants.

22 David spoke to the LORD the words of this song on the day when the LORD delivered him from the hand of all his enemies, and from the hand of Saul. [2] He said:

The LORD is my rock, my
    fortress, and my deliverer,
3   my God, my rock, in whom I
    take refuge,
my shield and the horn of my
    salvation,
my stronghold and my refuge,
my savior; you save me from
    violence.
4 I call upon the LORD, who is
    worthy to be praised,
    and I am saved from my
    enemies.

5 For the waves of death
    encompassed me,
    the torrents of perdition assailed
    me;
6 the cords of Sheol entangled me,
    the snares of death confronted
    me.

7 In my distress I called upon the
    LORD;
    to my God I called.
From his temple he heard my
    voice,
    and my cry came to his ears.

8 Then the earth reeled and rocked;
    the foundations of the heavens
    trembled

t Heb *was belted anew*

21.15–22: Exploits in the war against the Philistines. This brief fragment (probably with 23.8–39) may be part of the now partially lost story of David's victory over the Philistines. The tradition that there were giants among the Philistines here comes into full view. Another reflection of this idea is in 1 Sam ch 17. Some have supposed that these giants were the descendants of the Anakim, who had been defeated earlier by the Israelites (see Num 13.22 n.; Deut 1.28).

21.16: Compare the figure of *three hundred shekels* with that in 1 Sam 17.7. **17:** For other references to *Abishai,* see 2.18; 16.9–10; 19.21–22; 20.6–7. On the arrangement that David should stay behind, see 11.1; 12.28; 18.3–4. **19:** On the problem raised by the statement that *Elhanan . . . killed Goliath,* see 1 Sam 17.4 n. and 1 Chr 20.5 n. **22:** The preceding verses give no instance of *giants who fell by the hands of David;* perhaps these words were influenced by 1 Sam ch 17.

and quaked, because he was
angry.

9 Smoke went up from his nostrils,
and devouring fire from his
mouth;
glowing coals flamed forth from
him.

10 He bowed the heavens, and came
down;
thick darkness was under his
feet.

11 He rode on a cherub, and flew;
he was seen upon the wings of
the wind.

12 He made darkness around him a
canopy,
thick clouds, a gathering of
water.

13 Out of the brightness before him
coals of fire flamed forth.

14 The LORD thundered from heaven;
the Most High uttered his
voice.

15 He sent out arrows, and scattered
them
—lightning, and routed them.

16 Then the channels of the sea were
seen,
the foundations of the world
were laid bare
at the rebuke of the LORD,
at the blast of the breath of his
nostrils.

17 He reached from on high, he
took me,
he drew me out of mighty
waters.

18 He delivered me from my strong
enemy,
from those who hated me;
for they were too mighty for
me.

19 They came upon me in the day of
my calamity,
but the LORD was my stay.

20 He brought me out into a broad
place;

he delivered me, because he
delighted in me.

21 The LORD rewarded me according
to my righteousness;
according to the cleanness of my
hands he recompensed me.

22 For I have kept the ways of the
LORD,
and have not wickedly departed
from my God.

23 For all his ordinances were before
me,
and from his statutes I did not
turn aside.

24 I was blameless before him,
and I kept myself from guilt.

25 Therefore the LORD has
recompensed me according
to my righteousness,
according to my cleanness in his
sight.

26 With the loyal you show yourself
loyal;
with the blameless you show
yourself blameless;

27 with the pure you show yourself
pure,
and with the crooked you show
yourself perverse.

28 You deliver a humble people,
but your eyes are upon the
haughty to bring them
down.

29 Indeed, you are my lamp,
O LORD,
the LORD lightens my darkness.

30 By you I can crush a troop,
and by my God I can leap over
a wall.

31 This God—his way is perfect;
the promise of the LORD proves
true;
he is a shield for all who take
refuge in him.

32 For who is God, but the LORD?

22.1–51: A hymn of praise. It was the custom of certain writers and editors to insert poems into prose books for artistic and religious effect (see 1 Sam 2.1–10 n.). The present example is also preserved as Ps 18, to which reference should be made.

And who is a rock, except our
God?

33 The God who has girded me with
strength[u]
has opened wide my path. [v]

34 He made my[w] feet like the feet of
deer,
and set me secure on the
heights.

35 He trains my hands for war,
so that my arms can bend a
bow of bronze.

36 You have given me the shield of
your salvation,
and your help[x] has made me
great.

37 You have made me stride freely,
and my feet do not slip;

38 I pursued my enemies and
destroyed them,
and did not turn back until they
were consumed.

39 I consumed them; I struck them
down, so that they did not
rise;
they fell under my feet.

40 For you girded me with strength
for the battle;
you made my assailants sink
under me.

41 You made my enemies turn their
backs to me,
those who hated me, and I
destroyed them.

42 They looked, but there was no
one to save them;
they cried to the LORD, but he
did not answer them.

43 I beat them fine like the dust of
the earth,
I crushed them and stamped
them down like the mire of
the streets.

44 You delivered me from strife with
the peoples;[y]
you kept me as the head of the
nations;

people whom I had not known
served me.

45 Foreigners came cringing to me;
as soon as they heard of me,
they obeyed me.

46 Foreigners lost heart,
and came trembling out of their
strongholds.

47 The LORD lives! Blessed be my
rock,
and exalted be my God, the
rock of my salvation,

48 the God who gave me vengeance
and brought down peoples
under me,

49 who brought me out from my
enemies;
you exalted me above my
adversaries,
you delivered me from the
violent.

50 For this I will extol you, O LORD,
among the nations,
and sing praises to your name.

51 He is a tower of salvation for his
king,
and shows steadfast love to his
anointed,
to David and his descendants
forever.

23 Now these are the last words of
David:
The oracle of David, son of Jesse,
the oracle of the man whom
God exalted, [z]
the anointed of the God of Jacob,
the favorite of the Strong One
of Israel:

u Q Ms Gk Syr Vg Compare Ps 18.32: MT *God
is my strong refuge*   v Meaning of Heb
uncertain   w Another reading is *his*
x Q Ms: MT *your answering*   y Gk: Heb *from
strife with my people*   z Q Ms: MT *who was
raised on high*

23.1–7: Another hymn of praise, represented as *the last words of David* (see Gen
48.21–49.2 and Deut 33.1). This psalm is a
late composition in the style of Ps 1 and Prov

4.10–19, with an introduction suggestive of
the oracles of Balaam (Num 24.3–4, 15–16).
The theme of the perpetuity of the *house* (dynasty) of David (v. 5) has already appeared in

2 The spirit of the LORD speaks
through me,
his word is upon my tongue.
3 The God of Israel has spoken,
the Rock of Israel has said to
me:
One who rules over people justly,
ruling in the fear of God,
4 is like the light of morning,
like the sun rising on a cloudless
morning,
gleaming from the rain on the
grassy land.

5 Is not my house like this with
God?
For he has made with me an
everlasting covenant,
ordered in all things and secure.
Will he not cause to prosper
all my help and my desire?
6 But the godless are* all like thorns
that are thrown away;
for they cannot be picked up
with the hand;
7 to touch them one uses an iron
bar
or the shaft of a spear.
And they are entirely consumed
in fire on the spot. *b*

8 These are the names of the warriors
whom David had: Josheb-basshebeth a
Tahchemonite; he was chief of the
Three;*c* he wielded his spear*d* against
eight hundred whom he killed at one
time.
9 Next to him among the three warriors was Eleazar son of Dodo son of
Ahohi. He was with David when they
defied the Philistines who were gathered

there for battle. The Israelites withdrew,
10 but he stood his ground. He struck
down the Philistines until his arm grew
weary, though his hand clung to the
sword. The LORD brought about a great
victory that day. Then the people came
back to him—but only to strip the dead.
11 Next to him was Shammah son of
Agee, the Hararite. The Philistines gathered together at Lehi, where there was a
plot of ground full of lentils; and the
army fled from the Philistines. 12 But he
took his stand in the middle of the plot,
defended it, and killed the Philistines;
and the LORD brought about a great victory.
13 Towards the beginning of harvest
three of the thirty*e* chiefs went down to
join David at the cave of Adullam, while
a band of Philistines was encamped in the
valley of Rephaim. 14 David was then in
the stronghold; and the garrison of the
Philistines was then at Bethlehem. 15 David said longingly, "O that someone
would give me water to drink from the
well of Bethlehem that is by the gate!"
16 Then the three warriors broke through
the camp of the Philistines, drew water
from the well of Bethlehem that was by
the gate, and brought it to David. But he
would not drink of it; he poured it out to
the LORD, 17 for he said, "The LORD forbid that I should do this. Can I drink the
blood of the men who went at the risk of
their lives?" Therefore he would not
drink it. The three warriors did these
things.

a Heb *But worthlessness*   *b* Heb *in sitting*
c Gk Vg Compare 1 Chr 11.11: Meaning of Heb
uncertain   *d* 1 Chr 11.11: Meaning of Heb
uncertain   *e* Heb adds *head*

---

ch 7 (see the concluding note there). *The
unwinted of the God of Jacob,* see 1 Sam 2.10 n.
and 10.1 n. The text of the last two verses is
very corrupt, so that the exact meaning is not
certain. It is also not certain whether the
poem is complete as it stands or is only a
fragment.
**23.8–39: Other exploits in the war
against the Philistines, with a roster of
warriors.** This passage relates to 21.15–22. It
is also preserved in 1 Chr 11.11–47 at a more

logical place in the narrative, but with some
variation in the names and details. Here there
are two orders of heroes, the order of *the
Three* (v. 8) and the order of *the Thirty* (v. 18).
8–11: *Josheb-basshebeth a Tahchemonite* is an error of a copyist; 1 Chr 11.11 has *Jashobeam,
son of Hachmoni.* It has been proposed that
the man's original name was Ishbaal (see
2.8 n.).
23.13–17: This story is slightly misplaced,
since it concerns *three of the thirty,* not *the*

18 Now Abishai son of Zeruiah, the brother of Joab, was chief of the Thirty.*f* With his spear he fought against three hundred men and killed them, and won a name beside the Three. 19 He was the most renowned of the Thirty,*g* and became their commander; but he did not attain to the Three.

20 Benaiah son of Jehoiada was a valiant warrior*h* from Kabzeel, a doer of great deeds; he struck down two sons of Ariel*i* of Moab. He also went down and killed a lion in a pit on a day when snow had fallen. 21 And he killed an Egyptian, a handsome man. The Egyptian had a spear in his hand; but Benaiah went against him with a staff, snatched the spear out of the Egyptian's hand, and killed him with his own spear. 22 Such were the things Benaiah son of Jehoiada did, and won a name beside the three warriors. 23 He was renowned among the Thirty, but he did not attain to the Three. And David put him in charge of his bodyguard.

24 Among the Thirty were Asahel brother of Joab; Elhanan son of Dodo of Bethlehem; 25 Shammah of Harod; Elika of Harod; 26 Helez the Paltite; Ira son of Ikkesh of Tekoa; 27 Abiezer of Anathoth; Mebunnai the Hushathite; 28 Zalmon the Ahohite; Maharai of Netophah; 29 Heleb son of Baanah of Netophah; Ittai son of Ribai of Gibeah of the Benjaminites; 30 Benaiah of Pirathon; Hiddai of the torrents of Gaash; 31 Abi-albon the Arbathite; Azmaveth of Bahurim; 32 Eliahba of Shaalbon; the sons of Jashen: Jonathan 33 son of*j* Shammah the Hararite; Ahiam son of Sharar the Hararite; 34 Eliphelet son of Ahasbai of Maacah; Eliam son of Ahithophel the Gilonite; 35 Hezro*k* of Carmel; Paarai the Arbite; 36 Igal son of Nathan of Zobah; Bani the Gadite; 37 Zelek the Ammonite; Naharai of Beeroth, the armor-bearer of Joab son of Zeruiah; 38 Ira the Ithrite; Gareb the Ithrite; 39 Uriah the Hittite—thirty-seven in all.

24 Again the anger of the LORD was kindled against Israel, and he incited David against them, saying, "Go, count the people of Israel and Judah." 2 So the king said to Joab and the com-

f Two Heb Mss Syr: MT *Three*
g Syr Compare 1 Chr 11.25: Heb *Was he the most renowned of the Three?* *h* Another reading is *the son of Ish-hai* *i* Gk: Heb lacks *sons of* *j* Gk: Heb lacks *son of* *k* Another reading is *Hezrai*

Three of the preceding verses. *The Thirty* are introduced in v. 18. **18–23**: On *Abishai,* see 2.18; 10.10; 16.9–10; 18.2; 19.21–22; 20.6. On *Benaiah,* see 8.18; 20.23. On the *bodyguard,* see 8.18 n. *Joab* is not listed here; he had a special status as commander-in-chief (20.23). **24–39**: There is some confusion in the list, and more than *thirty* names occur. **23.24**: *Asahel brother of Joab* was killed while David was king of Judah only (2.23); either this name was included on an honorary basis, or the organization of *the Thirty* arose in the early days of David's kingship. On *Elhanan,* see 21.19 n. **23.33**: *Shammah the Hararite* seems to be repeated from v. 11. **39**: The inclusion of *Uriah the Hittite* is noteworthy (compare 11.15–17). The number *thirty-seven* seems to be intended to include all the names mentioned in vv. 8–39, though they total only thirty-six. It has been suggested that Joab, as commander-in-chief, was included in the reckoning. **24.1–25**: **The census, the plague, and the building of the altar.** This passage, along with 21.1–14, reflects the belief that natural calamity is caused by the wrath of God against human sin, and that this wrath must be appeased before the calamity can be stopped (compare Lk 13.1–4; Rom 1.18–23). **1**: 1 Chr 21.1 substitutes "Satan" for *the* LORD in this verse, a significant theological change. Some have wondered why it should have been considered sinful to *count the people of Israel and Judah.* The reason may have been that taking a census was deemed an infringement upon the prerogatives of their God, the sole arbiter of the destinies of the nation and its people. **3**: Joab expressed the popular fear in the matter, but he had to yield and oversee the counting in person (v. 4). **5–7**: Those who took the census began at *Aroer* (Deut 2.36; Josh 13.9) on the east side of the Dead Sea, went north to the limits of the country, then returned on the west side to *Beer-sheba.* If *Kadesh, Sidon,* and *Tyre* were included, the census must have been taken after David's Syrian campaigns (8.3–12; 10.15–19); but these cities were not truly in Israel or Judah,

manders of the army,[l] who were with him, "Go through all the tribes of Israel, from Dan to Beer-sheba, and take a census of the people, so that I may know how many there are." 3 But Joab said to the king, "May the LORD your God increase the number of the people a hundredfold, while the eyes of my lord the king can still see it! But why does my lord the king want to do this?" 4 But the king's word prevailed against Joab and the commanders of the army. So Joab and the commanders of the army went out from the presence of the king to take a census of the people of Israel. 5 They crossed the Jordan, and began from[m] Aroer and from the city that is in the middle of the valley, toward Gad and on to Jazer. 6 Then they came to Gilead, and to Kadesh in the land of the Hittites;[n] and they came to Dan, and from Dan[o] they went around to Sidon, 7 and came to the fortress of Tyre and to all the cities of the Hivites and Canaanites; and they went out to the Negeb of Judah at Beer-sheba. 8 So when they had gone through all the land, they came back to Jerusalem at the end of nine months and twenty days. 9 Joab reported to the king the number of those who had been recorded: in Israel there were eight hundred thousand soldiers able to draw the sword, and those of Judah were five hundred thousand.

10 But afterward, David was stricken to the heart because he had numbered the people. David said to the LORD, "I have sinned greatly in what I have done. But now, O LORD, I pray you, take away the guilt of your servant; for I have done very foolishly." 11 When David rose in the morning, the word of the LORD came to the prophet Gad, David's seer, saying, 12 "Go and say to David: Thus says the LORD: Three things I offer[p] you; choose

one of them, and I will do it to you." 13 So Gad came to David and told him; he asked him, "Shall three[q] years of famine come to you on your land? Or will you flee three months before your foes while they pursue you? Or shall there be three days' pestilence in your land? Now consider, and decide what answer I shall return to the one who sent me." 14 Then David said to Gad, "I am in great distress; let us fall into the hand of the LORD, for his mercy is great; but let me not fall into human hands."

15 So the LORD sent a pestilence on Israel from that morning until the appointed time; and seventy thousand of the people died, from Dan to Beer-sheba. 16 But when the angel stretched out his hand toward Jerusalem to destroy it, the LORD relented concerning the evil, and said to the angel who was bringing destruction among the people, "It is enough; now stay your hand." The angel of the LORD was then by the threshing floor of Araunah the Jebusite. 17 When David saw the angel who was destroying the people, he said to the LORD, "I alone have sinned, and I alone have done wickedly; but these sheep, what have they done? Let your hand, I pray, be against me and against my father's house."

18 That day Gad came to David and said to him, "Go up and erect an altar to the LORD on the threshing floor of Araunah the Jebusite." 19 Following Gad's instructions, David went up, as the LORD had commanded. 20 When Araunah looked down, he saw the king and his

---

l 1 Chn 21.2 Gk: Heb to Joab the commander of the army   m Gk Mss: Heb encamped in Aroer south of   n Gk: Heb to the land of Tahtim-hodshi o Cn Compare Gk: Heb they came to Dan-jaan and   p Or hold over   q 1 Chr 21.12 Gk: Heb seven

---

even when under the control of David. **9:** The numbers here given are round numbers and incredibly high (1,300,000 fighting men alone); see 1 Sam 4.10 n.

24.10–14: We might say today that "David had an uneasy conscience" (v. 10). Compare the appearance of *Gad* in 1 Sam 22.5. **16:** The

angel here is not specifically the angel of death, but simply a "messenger" (the original meaning of the word *angel*) of the LORD doing the LORD's will. **17:** The people are like innocent sheep led to slaughter; David feels that the evil should fall upon him.

24.18–25: *The threshing floor of Araunah the*

servants coming toward him; and Araunah went out and prostrated himself before the king with his face to the ground. 21 Araunah said, "Why has my lord the king come to his servant?" David said, "To buy the threshing floor from you in order to build an altar to the LORD, so that the plague may be averted from the people." 22 Then Araunah said to David, "Let my lord the king take and offer up what seems good to him; here are the oxen for the burnt offering, and the threshing sledges and the yokes of the oxen for the wood. 23 All this, O king, Araunah gives to the king." And Araunah said to the king, "May the LORD your God respond favorably to you."

24 But the king said to Araunah, "No, but I will buy them from you for a price; I will not offer burnt offerings to the LORD my God that cost me nothing." So David bought the threshing floor and the oxen for fifty shekels of silver. 25 David built there an altar to the LORD, and offered burnt offerings and offerings of well-being. So the LORD answered his supplication for the land, and the plague was averted from Israel.

Jebusite later became the site of Solomon's temple (1 Chr 22.1; 2 Chr 3.1). The acquisition of this holy place, the building of an altar, and a sacrifice there were regarded as sufficient expiation to avert the plague from Israel (v. 25). *Fifty shekels of silver*, about 24 oz. (see 1 Sam 9.8 n.).

1 Kings

The two books of Kings, like those of Samuel, were originally one. They continue the story of monarchy begun in 1–2 Samuel, and give a consecutive account of the Israelite kingdoms from the death of David and accession of Solomon, to the Exile (see "Survey of . . . Bible Lands," §§ 13 ff.). First Kings begins with the enthronement of Solomon and death of David (chs 1–2), recounts the reign of Solomon (chs 3–11) and the sometimes rivalrous, sometimes allied, kingdoms of the divided monarchy through the reigns of Ahab of Israel (the northern kingdom) and Jehoshaphat of Judah in the south (chs 12–22).

The books of Kings resulted from a long process of collection, writing, editing, and revising of diverse materials that themselves were based upon written documents (see, e.g., "Book of the Acts of Solomon," 1 Kings 11.41) or oral traditions (e.g. the stories of Elijah and Elisha, 1 Kings 17–19; 2 Kings 2–8). Most scholars now view the books of Kings as part of the Deuteronomistic History (see Introduction to the Historical Books). A widely shared opinion holds that 1 and 2 Kings were first cast into a pre-exilic version of this Deuteronomistic history (some suggest Josiah's time, others Hezekiah's), and later revised after 587 B.C. in light of the Judean exile, when the high hopes that had been vested in King Josiah (or perhaps Hezekiah) as religious reformer had come to nothing. Such elements as the formulaic introductory and concluding summaries that demarcate the regnal period of most kings (e.g. 1 Kings 14.21–22, 29–31), theological speeches and prayers (e.g. 1 Kings 2.1–4; 8.22–40), and sermon-like commentaries on events (e.g. 1 Kings 9.3–9; 11.9–13; 2 Kings 17.7–20) are usually attributed to these Deuteronomistic hands. However, disagreement on the details of such analysis persists.

Although the subject of 1–2 Kings is political history, its theme is the moral and religious failure that eventually led to the loss of national identity and autonomy (see 2 Kings 25). Each king is evaluated by how well he upheld the primacy of God and God's temple in Jerusalem or—more usually—how he failed in this responsibility, and in this departed from the ways of David. The people also are frequently condemned. Political and social turmoil, even national defeat, were thereby taken to result from such deficiencies. In particular, 1 Kings recounts the story of Solomon's glory and his downfall in apostasy (chs 3–11). The subsequent breakup of Solomon's rule leads to a long story of divided kingdoms whose troubles are rooted not only in his transgressions (11.9–13), but in those of Jeroboam as well, who, as the first king of the north, fell away from God's (and David's) ways (chs 11–13). The Deuteronomistic writer's voice is clearly heard through certain prophets (especially Elijah, chs 17–19), who zealously guard the absolute priority of being devoted exclusively to the God of Israel.

1 King David was old and advanced in years; and although they covered him with clothes, he could not get warm. 2 So his servants said to him, "Let a young virgin be sought for my lord the king, and let her wait on the king, and be his attendant; let her lie in your bosom, so that my lord the king may be warm." 3 So they searched for a beautiful girl throughout all the territory of Israel, and found Abishag the Shunammite, and brought her to the king. 4 The girl was very beautiful. She became the king's attendant and served him, but the king did not know her sexually.

5 Now Adonijah son of Haggith exalted himself, saying, "I will be king"; he prepared for himself chariots and horsemen, and fifty men to run before him. 6 His father had never at any time displeased him by asking, "Why have you done thus and so?" He was also a very handsome man, and he was born next after Absalom. 7 He conferred with Joab son of Zeruiah and with the priest Abiathar, and they supported Adonijah. 8 But the priest Zadok, and Benaiah son of Jehoiada, and the prophet Nathan, and Shimei, and Rei, and David's own warriors did not side with Adonijah.

9 Adonijah sacrificed sheep, oxen, and fatted cattle by the stone Zoheleth, which is beside En-rogel, and he invited all his brothers, the king's sons, and all the royal officials of Judah, 10 but he did not invite the prophet Nathan or Benaiah or the warriors or his brother Solomon. 11 Then Nathan said to Bathsheba,

Solomon's mother, "Have you not heard that Adonijah son of Haggith has become king and our lord David does not know it? 12 Now therefore come, let me give you advice, so that you may save your own life and the life of your son Solomon. 13 Go in at once to King David, and say to him, 'Did you not, my lord the king, swear to your servant, saying: Your son Solomon shall succeed me as king, and he shall sit on my throne? Why then is Adonijah king?' 14 Then while you are still there speaking with the king, I will come in after you and confirm your words."

15 So Bathsheba went to the king in his room. The king was very old; Abishag the Shunammite was attending the king. 16 Bathsheba bowed and did obeisance to the king, and the king said, "What do you wish?" 17 She said to him, "My lord, you swore to your servant by the LORD your God, saying: Your son Solomon shall succeed me as king, and he shall sit on my throne. 18 But now suddenly Adonijah has become king, though you, my lord the king, do not know it. 19 He has sacrificed oxen, fatted cattle, and sheep in abundance, and has invited all the children of the king, the priest Abiathar, and Joab the commander of the army; but your servant Solomon he has not invited. 20 But you, my lord the king—the eyes of all Israel are on you to tell them who shall sit on the throne of my lord the king after him. 21 Otherwise it will come to pass, when my lord the king sleeps with his ancestors, that

---

1.1–53: The struggle for the succession and Solomon's attainment of the kingship. This chapter continues the "Throne Succession Narrative," left off at 2 Sam 20 (see Introduction to 2 Samuel). A number of years had intervened and *King David was old and advanced in years* (v. 1). **3:** *Abishag the Shunammite,* i.e. from Shunem in the plain of Esdraelon near Mount Gilboa (1 Sam 28.4). **1.5–10:** *Adonijah,* as David's eldest living son (Chileab seems to have died young, 2 Sam 3.3–4), naturally assumed that he would succeed his father. This seems to have been taken for granted also by the older followers of David, such as *Joab* and *Abiathar*

(1 Sam 22.20–23; 2 Sam 2.13). Apparently, however, there was no strict law of primogeniture at that time (compare v. 20), and hence the younger followers of David, such as *Zadok* (2 Sam 8.17), *Benaiah* (2 Sam 8.18), and *Nathan* (2 Sam 12.1) were plotting to install Solomon in spite of his disadvantage in age. The group favoring Solomon had the advantage of including David's bodyguards. **9:** *En-rogel,* now known as Job's Well, a source of water southeast of the city, was considered a sacred place, appropriate for affairs of this kind (2 Sam 17.17). **1.11–27:** It is surprising to find Nathan, who once so eloquently denounced David for

my son Solomon and I will be counted offenders."

22 While she was still speaking with the king, the prophet Nathan came in. 23 The king was told, "Here is the prophet Nathan." When he came in before the king, he did obeisance to the king, with his face to the ground. 24 Nathan said, "My lord the king, have you said, 'Adonijah shall succeed me as king, and he shall sit on my throne'? 25 For today he has gone down and has sacrificed oxen, fatted cattle, and sheep in abundance, and has invited all the king's children, Joab the commander[a] of the army, and the priest Abiathar, who are now eating and drinking before him, and saying, 'Long live King Adonijah!' 26 But he did not invite me, your servant, and the priest Zadok, and Benaiah son of Jehoiada, and your servant Solomon. 27 Has this thing been brought about by my lord the king and you have not let your servants know who should sit on the throne of my lord the king after him?"

28 King David answered, "Summon Bathsheba to me." So she came into the king's presence, and stood before the king. 29 The king swore, saying, "As the LORD lives, who has saved my life from every adversity, 30 as I swore to you by the LORD, the God of Israel, 'Your son Solomon shall succeed me as king, and he shall sit on my throne in my place,' so will I do this day." 31 Then Bathsheba bowed with her face to the ground, and did obeisance to the king, and said, "May my lord King David live forever!"

32 King David said, "Summon to me the priest Zadok, the prophet Nathan, and Benaiah son of Jehoiada." When they came before the king, 33 the king said to them, "Take with you the servants of your lord, and have my son Solomon ride on my own mule, and bring him down to Gihon. 34 There let the priest Zadok and the prophet Nathan anoint him king over Israel; then blow the trumpet, and say, 'Long live King Solomon!' 35 You shall go up following him. Let him enter and sit on my throne; he shall be king in my place; for I have appointed him to be ruler over Israel and over Judah." 36 Benaiah son of Jehoiada answered the king, "Amen! May the LORD, the God of my lord the king, so ordain. 37 As the LORD has been with my lord the king, so may he be with Solomon, and make his throne greater than the throne of my lord King David."

38 So the priest Zadok, the prophet Nathan, and Benaiah son of Jehoiada, and the Cherethites and the Pelethites, went down and had Solomon ride on King David's mule, and led him to Gihon. 39 There the priest Zadok took the horn of oil from the tent and anointed Solomon. Then they blew the trumpet, and all the people said, "Long live King Solomon!" 40 And all the people went up following him, playing on pipes and rejoicing with great joy, so that the earth quaked at their noise.

41 Adonijah and all the guests who were with him heard it as they finished feasting. When Joab heard the sound of the trumpet, he said, "Why is the city in an uproar?" 42 While he was still speaking, Jonathan son of the priest Abiathar arrived. Adonijah said, "Come in, for you are a worthy man and surely you bring good news." 43 Jonathan answered Adonijah, "No, for our lord King David has made Solomon king; 44 the king has

a Gk: Heb *the commanders*

his affair with Bathsheba (2 Sam 12.1–15), now on Bathsheba's side. **28–31**: Bathsheba had apparently gone out while Nathan was speaking. **1.32–37**: When Bathsheba came back, Nathan left and hence had to be recalled. **33**: *Gihon*, now called the Virgin's Fountain or Mary's Spring, is a short distance north of En-rogel (v. 9), out of sight of the latter because of a curve in the valley (Kidron) in which they lie, but well within earshot (compare v. 41). Gihon for long was the principal source of water for the city of Jerusalem (2 Chr 32.30), and hence was a place of special significance. **38–40**: *Cherethites* and *Pelethites*, see 1 Sam 30.14 n.; 2 Sam 8.18; 15.18. On the ceremony of anointing, see 1 Sam 10.1 n. **1.41–48**: *Jonathan son of the priest Abiathar,*

sent with him the priest Zadok, the prophet Nathan, and Benaiah son of Jehoiada, and the Cherethites and the Pelethites; and they had him ride on the king's mule; 45 the priest Zadok and the prophet Nathan have anointed him king at Gihon; and they have gone up from there rejoicing, so that the city is in an uproar. This is the noise that you heard. 46 Solomon now sits on the royal throne. 47 Moreover the king's servants came to congratulate our lord King David, saying, 'May God make the name of Solomon more famous than yours, and make his throne greater than your throne.' The king bowed in worship on the bed 48 and went on to pray thus, 'Blessed be the LORD, the God of Israel, who today has granted one of my offspring*b* to sit on my throne and permitted me to witness it.' "

49 Then all the guests of Adonijah got up trembling and went their own ways. 50 Adonijah, fearing Solomon, got up and went to grasp the horns of the altar. 51 Solomon was informed, "Adonijah is afraid of King Solomon; see, he has laid hold of the horns of the altar, saying, 'Let King Solomon swear to me first that he will not kill his servant with the sword.' " 52 So Solomon responded, "If he proves to be a worthy man, not one of his hairs shall fall to the ground; but if wickedness is found in him, he shall die." 53 Then King Solomon sent to have him brought down from the altar. He came to do obeisance to King Solomon; and Solomon said to him, "Go home."

2 When David's time to die drew near, he charged his son Solomon, saying: 2 "I am about to go the way of all the earth. Be strong, be courageous, 3 and keep the charge of the LORD your God, walking in his ways and keeping his statutes, his commandments, his ordinances, and his testimonies, as it is written in the law of Moses, so that you may prosper in all that you do and wherever you turn. 4 Then the LORD will establish his word that he spoke concerning me: 'If your heirs take heed to their way, to walk before me in faithfulness with all their heart and with all their soul, there shall not fail you a successor on the throne of Israel.'

5 "Moreover you know also what Joab son of Zeruiah did to me, how he dealt with the two commanders of the armies of Israel, Abner son of Ner, and Amasa son of Jether, whom he murdered, retaliating in time of peace for blood that had been shed in war, and putting the blood of war on the belt around his waist, and on the sandals on his feet. 6 Act therefore according to your wisdom, but do not let his gray head go down to Sheol in peace. 7 Deal loyally, however, with the sons of Barzillai the Gileadite, and let them be among those who eat at your table; for with such loyalty they met me when I fled from your brother Absalom. 8 There is also with you Shimei son of Gera, the Benjaminite from Bahurim, who cursed me with a terrible curse on the day when I went to Mahanaim; but when he came down to meet me at the Jordan, I swore to him by the LORD, 'I will not put you to death with the sword.' 9 Therefore do not hold him guiltless, for you are a wise man; you will know what you ought to do to him, and you must bring his gray head down with blood to Sheol."

b Gk: Heb *one*

---

see 2 Sam 15.27, 36; 17.17–21. **49–53**: *The horns of the altar* were projections resembling horns at the four corners of an altar (Ex 29.12; 30.10; Lev 4.7; Ps 118.27). An altar was sacred, so that a person touching it was not supposed to be slain; but the taboo was not always honored (Ex 21.14; 1 Kings 2.28–34). **2.1–46**: **The death of David and the elimination of rivals to the reign of Solo**mon. **1–4**: The first example of the viewpoint of the Deuteronomist (see Introduction). Solomon is warned that he must follow *the law of Moses* (i.e. the Deuteronomic law) or national ruin will result. See the Introduction to Deuteronomy and compare Deut 4.40, 44–45; 5.1; 11.1–17; 17.14–20. **2.5–9**: When *Joab* (2 Sam 3.27; 20.10) and *Shimei* (2 Sam 16.5–14; 19.18–23) committed

10 Then David slept with his ancestors, and was buried in the city of David. [11] The time that David reigned over Israel was forty years; he reigned seven years in Hebron, and thirty-three years in Jerusalem. [12] So Solomon sat on the throne of his father David; and his kingdom was firmly established.

13 Then Adonijah son of Haggith came to Bathsheba, Solomon's mother. She asked, "Do you come peaceably?" He said, "Peaceably." [14] Then he said, "May I have a word with you?" She said, "Go on." [15] He said, "You know that the kingdom was mine, and that all Israel expected me to reign; however, the kingdom has turned about and become my brother's, for it was his from the LORD. [16] And now I have one request to make of you; do not refuse me." She said to him, "Go on." [17] He said, "Please ask King Solomon—he will not refuse you—to give me Abishag the Shunammite as my wife." [18] Bathsheba said, "Very well; I will speak to the king on your behalf."

19 So Bathsheba went to King Solomon, to speak to him on behalf of Adonijah. The king rose to meet her, and bowed down to her; then he sat on his throne, and had a throne brought for the king's mother, and she sat on his right. [20] Then she said, "I have one small request to make of you; do not refuse me." And the king said to her, "Make your request, my mother; for I will not refuse you." [21] She said, "Let Abishag the Shunammite be given to your brother Adonijah as his wife." [22] King Solomon answered his mother, "And why do you ask Abishag the Shunammite for Adonijah? Ask for him the kingdom as well! For he is my elder brother; ask not only for him but also for the priest Abiathar

and for Joab son of Zeruiah!" [23] Then King Solomon swore by the LORD, "So may God do to me, and more also, for Adonijah has devised this scheme at the risk of his life! [24] Now therefore as the LORD lives, who has established me and placed me on the throne of my father David, and who has made me a house as he promised, today Adonijah shall be put to death." [25] So King Solomon sent Benaiah son of Jehoiada; he struck him down, and he died.

26 The king said to the priest Abiathar, "Go to Anathoth, to your estate; for you deserve death. But I will not at this time put you to death, because you carried the ark of the Lord GOD before my father David, and because you shared in all the hardships my father endured." [27] So Solomon banished Abiathar from being priest to the LORD, thus fulfilling the word of the LORD that he had spoken concerning the house of Eli in Shiloh.

28 When the news came to Joab—for Joab had supported Adonijah though he had not supported Absalom—Joab fled to the tent of the LORD and grasped the horns of the altar. [29] When it was told King Solomon, "Joab has fled to the tent of the LORD and now is beside the altar," Solomon sent Benaiah son of Jehoiada, saying, "Go, strike him down." [30] So Benaiah came to the tent of the LORD and said to him, "The king commands, 'Come out.'" But he said, "No, I will die here." Then Benaiah brought the king word again, saying, "Thus said Joab, and thus he answered me." [31] The king replied to him, "Do as he has said, strike him down and bury him; and thus take away from me and from my father's house the guilt for the blood that Joab shed without cause. [32] The LORD will

---

the offenses referred to here, David was in such precarious political situations that he did not dare have these men put to death. On the friendship of Barzillai, see 2 Sam 17.27–29; 19.31–39.

2.13–25: An excuse is found to eliminate Adonijah. The members of the king's harem were considered royal property, to be passed on to the next king (compare v. 22). *Adonijah*

made himself vulnerable by his request (2 Sam 3.6–11; 16.21–22).

2.26–27: Since *Abiathar* was a priest, and hence a sacrosanct personality, Solomon possibly did not want to execute him, although the narrator suggests that Abiathar's loyalty to David is the reason. **27:** See 1 Sam 2.27–36. **28–35:** *Solomon* had orders from *David* to do away with *Joab,* and he proceeds quickly to

bring back his bloody deeds on his own head, because, without the knowledge of my father David, he attacked and killed with the sword two men more righteous and better than himself, Abner son of Ner, commander of the army of Israel, and Amasa son of Jether, commander of the army of Judah. 33 So shall their blood come back on the head of Joab and on the head of his descendants forever; but to David, and to his descendants, and to his house, and to his throne, there shall be peace from the LORD forevermore." 34 Then Benaiah son of Jehoiada went up and struck him down and killed him; and he was buried at his own house near the wilderness. 35 The king put Benaiah son of Jehoiada over the army in his place, and the king put the priest Zadok in the place of Abiathar.

36 Then the king sent and summoned Shimei, and said to him, "Build yourself a house in Jerusalem, and live there, and do not go out from there to any place whatever. 37 For on the day you go out, and cross the Wadi Kidron, know for certain that you shall die; your blood shall be on your own head." 38 And Shimei said to the king, "The sentence is fair; as my lord the king has said, so will your servant do." So Shimei lived in Jerusalem many days.

39 But it happened at the end of three years that two of Shimei's slaves ran away to King Achish son of Maacah of Gath. When it was told Shimei, "Your slaves are in Gath," 40 Shimei arose and saddled a donkey, and went to Achish in Gath, to search for his slaves; Shimei went and brought his slaves from Gath. 41 When Solomon was told that Shimei had gone from Jerusalem to Gath and returned, 42 the king sent and summoned Shimei, and said to him, "Did I not make you swear by the LORD, and solemnly adjure you, saying, 'Know for certain that on the day you go out and go to any place whatever, you shall die'? And you said to me, 'The sentence is fair; I accept.' 43 Why then have you not kept your oath to the LORD and the commandment with which I charged you?" 44 The king also said to Shimei, "You know in your own heart all the evil that you did to my father David; so the LORD will bring back your evil on your own head. 45 But King Solomon shall be blessed, and the throne of David shall be established before the LORD forever." 46 Then the king commanded Benaiah son of Jehoiada; and he went out and struck him down, and he died.

So the kingdom was established in the hand of Solomon.

3 Solomon made a marriage alliance with Pharaoh king of Egypt; he took Pharaoh's daughter and brought her into the city of David, until he had finished building his own house and the house of the LORD and the wall around Jerusalem. 2 The people were sacrificing at the high places, however, because no house had yet been built for the name of the LORD.

3 Solomon loved the LORD, walking in the statutes of his father David; only, he sacrificed and offered incense at the high places. 4 The king went to Gibeon to sacrifice there, for that was the principal

---

carry out the orders (see 1.49–53 n.). The ever ready *Benaiah* becomes the new commander.

2.36–46: *Shimei* had not joined the party of Adonijah (1.8); hence *Solomon* by agreement placed him in protective custody instead of executing him. *Shimei,* however, violated the agreement, and Benaiah had the welcome task of performing another execution.

3.1–28: Solomon in a dream prays for wisdom and receives it; he exercises it in judgment. The reign of Solomon now begins. One might say that the story begins and ends with wisdom (ch 3 and 11.41), but includes folly (ch 11). **1–2:** The identity of the pharaoh whose daughter Solomon married is uncertain. Verse 2 is Deuteronomistic and apologetic, since to worship anywhere except at the temple was contrary to the editor's principles (Deut 12.1–14). Worship at *the high places* (hilltop shrines) was not outlawed until the time of Josiah (2 Kings 23.8). Solomon himself worshiped at *the principal high place* at *Gibeon* (v. 4).

3.3–15: On the story of how *Gibeon* became a part of the community of Israel, see

high place; Solomon used to offer a thousand burnt offerings on that altar. 5 At Gibeon the LORD appeared to Solomon in a dream by night; and God said, "Ask what I should give you." 6 And Solomon said, "You have shown great and steadfast love to your servant my father David, because he walked before you in faithfulness, in righteousness, and in uprightness of heart toward you; and you have kept for him this great and steadfast love, and have given him a son to sit on his throne today. 7 And now, O LORD my God, you have made your servant king in place of my father David, although I am only a little child; I do not know how to go out or come in. 8 And your servant is in the midst of the people whom you have chosen, a great people, so numerous they cannot be numbered or counted. 9 Give your servant therefore an understanding mind to govern your people, able to discern between good and evil; for who can govern this your great people?"

10 It pleased the Lord that Solomon had asked this. 11 God said to him, "Because you have asked this, and have not asked for yourself long life or riches, or for the life of your enemies, but have asked for yourself understanding to discern what is right, 12 I now do according to your word. Indeed I give you a wise and discerning mind; no one like you has been before you and no one like you shall arise after you. 13 I give you also what you have not asked, both riches and honor all your life; no other king shall compare with you. 14 If you will walk in my ways, keeping my statutes and my commandments, as your father David walked, then I will lengthen your life."

15 Then Solomon awoke; it had been a dream. He came to Jerusalem where he stood before the ark of the covenant of the LORD. He offered up burnt offerings and offerings of well-being, and provided a feast for all his servants.

16 Later, two women who were prostitutes came to the king and stood before him. 17 The one woman said, "Please, my lord, this woman and I live in the same house; and I gave birth while she was in the house. 18 Then on the third day after I gave birth, this woman also gave birth. We were together; there was no one else with us in the house, only the two of us were in the house. 19 Then this woman's son died in the night, because she lay on him. 20 She got up in the middle of the night and took my son from beside me while your servant slept. She laid him at her breast, and laid her dead son at my breast. 21 When I rose in the morning to nurse my son, I saw that he was dead; but when I looked at him closely in the morning, clearly it was not the son I had borne." 22 But the other woman said, "No, the living son is mine, and the dead son is yours." The first said, "No, the dead son is yours, and the living son is mine." So they argued before the king.

23 Then the king said, "The one says, 'This is my son that is alive, and your son is dead'; while the other says, 'Not so! Your son is dead, and my son is the living one.'" 24 So the king said, "Bring me a sword," and they brought a sword before the king. 25 The king said, "Divide the living boy in two; then give half to the one, and half to the other." 26 But the woman whose son was alive said to the king—because compassion for her son burned within her—"Please, my lord, give her the living boy; certainly do not kill him!" The other said, "It shall be neither mine nor yours; divide it." 27 Then the king responded: "Give the first woman the living boy; do not kill him. She is his mother." 28 All Israel heard of the judgment that the king had rendered; and they stood in awe of the king, because they perceived that the wisdom of God was in him, to execute justice.

Josh ch 9; compare 2 Sam 21.2. **7:** The expression *only a little child* denotes humility, Solomon was probably about twenty years old. **14:** See v. 6. **15:** See 2 Sam 6.17–18. **16–28:** This is the most famous of the stories of Solomon's wisdom.

4 King Solomon was king over all Israel, ²and these were his high officials: Azariah son of Zadok was the priest; ³Elihoreph and Ahijah sons of Shisha were secretaries; Jehoshaphat son of Ahilud was recorder; ⁴Benaiah son of Jehoiada was in command of the army; Zadok and Abiathar were priests; ⁵Azariah son of Nathan was over the officials; Zabud son of Nathan was priest and king's friend; ⁶Ahishar was in charge of the palace; and Adoniram son of Abda was in charge of the forced labor.

7 Solomon had twelve officials over all Israel, who provided food for the king and his household; each one had to make provision for one month in the year. ⁸These were their names: Ben-hur, in the hill country of Ephraim; ⁹Ben-deker, in Makaz, Shaalbim, Beth-shemesh, and Elon-beth-hanan; ¹⁰Ben-hesed, in Arubboth (to him belonged Socoh and all the land of Hepher); ¹¹Ben-abinadab, in all Naphath-dor (he had Taphath, Solomon's daughter, as his wife); ¹²Baana son of Ahilud, in Taanach, Megiddo, and all Beth-shean, which is beside Zarethan below Jezreel, and from Beth-shean to Abel-meholah, as far as the other side of Jokmeam; ¹³Ben-geber, in Ramothgilead (he had the villages of Jair son of Manasseh, which are in Gilead, and he had the region of Argob, which is in Bashan, sixty great cities with walls and bronze bars); ¹⁴Ahinadab son of Iddo, in Mahanaim; ¹⁵Ahimaaz, in Naphtali (he had taken Basemath, Solomon's daughter, as his wife); ¹⁶Baana son of Hushai, in Asher and Bealoth; ¹⁷Jehoshaphat son of Paruah, in Issachar; ¹⁸Shimei son of Ela, in Benjamin; ¹⁹Geber son of Uri, in the land of Gilead, the country of King Sihon of the Amorites and of King Og of Bashan. And there was one official in the land of Judah.

20 Judah and Israel were as numerous as the sand by the sea; they ate and drank and were happy. ²¹ᶜSolomon was sovereign over all the kingdoms from the Euphrates to the land of the Philistines, even to the border of Egypt; they brought tribute and served Solomon all the days of his life.

22 Solomon's provision for one day was thirty cors of choice flour, and sixty cors of meal, ²³ten fat oxen, and twenty pasture-fed cattle, one hundred sheep, besides deer, gazelles, roebucks, and fatted fowl. ²⁴For he had dominion over all the region west of the Euphrates from Tiphsah to Gaza, over all the kings west of the Euphrates; and he had peace on all sides. ²⁵During Solomon's lifetime Judah and Israel lived in safety, from Dan even to Beer-sheba, all of them under their vines and fig trees. ²⁶Solomon also had forty thousand stalls of horses for his chariots, and twelve thousand horsemen. ²⁷Those officials supplied provisions for King Solomon and for all who came to King Solomon's table, each one in his month; they let nothing be lacking. ²⁸They also brought to the required place barley and straw for the horses and swift steeds, each according to his charge.

29 God gave Solomon very great wisdom, discernment, and breadth of understanding as vast as the sand on the seashore, ³⁰so that Solomon's wisdom surpassed the wisdom of all the people of

c Ch 5.1 in Heb

4.1–28: Solomon's organization of his kingdom. 1–6: This list of high officials from the court records is now somewhat disordered. *Azariah son of Zadok* was probably not *the priest* (v. 2); that honor belonged to (another?) Zadok (2.35) alone, *Abiathar (v. 4)* having been deposed (2.27).

4.7–19: Solomon's new administrative districts did not conform to the old tribal boundaries. Possibly this was one of the reasons for the revolt at the end of Solomon's reign, especially since Judah (v. 19) seems to have been exempt from the taxation mentioned in v. 7.

4.20–21: In the end, things turned out not to be so *happy*. Also, during Solomon's reign, the empire stretching *from the Euphrates . . . to the border of Egypt* began to fall away. Compare ch 11.

4.22–28: *Solomon's . . . stalls of horses for his chariots* (v. 26; compare 10.26–29) may have been similar to those of the time of Ahab found in the excavations at Megiddo.

4.29–34: **Further remarks on Solo-**

the east, and all the wisdom of Egypt. [31] He was wiser than anyone else, wiser than Ethan the Ezrahite, and Heman, Calcol, and Darda, children of Mahol; his fame spread throughout all the surrounding nations. [32] He composed three thousand proverbs, and his songs numbered a thousand and five. [33] He would speak of trees, from the cedar that is in the Lebanon to the hyssop that grows in the wall; he would speak of animals, and birds, and reptiles, and fish. [34] People came from all the nations to hear the wisdom of Solomon; they came from all the kings of the earth who had heard of his wisdom.

5 [d] Now King Hiram of Tyre sent his servants to Solomon, when he heard that they had anointed him king in place of his father; for Hiram had always been a friend to David. [2] Solomon sent word to Hiram, saying, [3] "You know that my father David could not build a house for the name of the LORD his God because of the warfare with which his enemies surrounded him, until the LORD put them under the soles of his feet. [e] [4] But now the LORD my God has given me rest on every side; there is neither adversary nor misfortune. [5] So I intend to build a house for the name of the LORD my God, as the LORD said to my father David, 'Your son, whom I will set on your throne in your place, shall build the house for my name.' [6] Therefore command that cedars from the Lebanon be cut for me. My servants will join your servants, and I will give you whatever wages you set for your servants; for you know that there is no one among us who knows how to cut timber like the Sidonians."

7 When Hiram heard the words of Solomon, he rejoiced greatly, and said, "Blessed be the LORD today, who has given to David a wise son to be over this great people." [8] Hiram sent word to Solomon, "I have heard the message that you have sent to me; I will fulfill all your needs in the matter of cedar and cypress timber. [9] My servants shall bring it down to the sea from the Lebanon; I will make it into rafts to go by sea to the place you indicate. I will have them broken up there for you to take away. And you shall meet my needs by providing food for my household." [10] So Hiram supplied Solomon's every need for timber of cedar and cypress. [11] Solomon in turn gave Hiram twenty thousand cors of wheat as food for his household, and twenty cors of fine oil. Solomon gave this to Hiram year by year. [12] So the LORD gave Solomon wisdom, as he promised him. There was peace between Hiram and Solomon; and the two of them made a treaty.

13 King Solomon conscripted forced labor out of all Israel; the levy numbered thirty thousand men. [14] He sent them to the Lebanon, ten thousand a month in shifts; they would be a month in the Leb-

d Ch 5.15 in Heb   e Gk Tg Vg: Heb *my feet* or *his feet*

---

mon's wisdom. In the biblical tradition, wisdom became Solomonic, just as laws were Mosaic, and psalms Davidic (compare Prov 1.1; Eccl 1.1, 12). 31: On *Ethan* and *Heman,* compare the titles of Pss 88 and 89 32: On Solomon's reputation as a maker of *proverbs,* see Prov 1.1. With regard to the attribution of *songs* to Solomon, compare Pss 72; 127; and Song 1.1. The number *a thousand and five* suggests "The Thousand and One Nights" of Arabic literature or the number of Solomon's wives and concubines (11.3).
  5.1–18: **Preparations for the building of the temple. 1–6:** On the friendship of *King Hiram of Tyre* with David, see 2 Sam 5.11–12. *The Sidonians* (v. 6), a term used here for all the people later called Phoenicians. From their two chief cities, Tyre and Sidon, they carried on a vast maritime enterprise, two of their chief articles of commerce being cedar wood and purple dye. About a century after the time of Solomon they established their most famous colony, Carthage in North Africa. 11: *Twenty thousand cors of wheat,* estimated to equal 125,000 bushels; *twenty thousand cors* of liquid would equal over a million gallons, according to some computations. *Fine oil,* oil extracted from olives by beating, considered the best method for quality production.
  5.13–18: One of the causes of the later disruption of the kingdom was *forced labor*

anon and two months at home; Adoniram was in charge of the forced labor. 15 Solomon also had seventy thousand laborers and eighty thousand stonecutters in the hill country, 16 besides Solomon's three thousand three hundred supervisors who were over the work, having charge of the people who did the work. 17 At the king's command, they quarried out great, costly stones in order to lay the foundation of the house with dressed stones. 18 So Solomon's builders and Hiram's builders and the Gebalites did the stonecutting and prepared the timber and the stone to build the house.

6 In the four hundred eightieth year after the Israelites came out of the land of Egypt, in the fourth year of Solomon's reign over Israel, in the month of Ziv, which is the second month, he began to build the house of the LORD. 2 The house that King Solomon built for the LORD was sixty cubits long, twenty cubits wide, and thirty cubits high. 3 The vestibule in front of the nave of the house was twenty cubits wide, across the width of the house. Its depth was ten cubits in front of the house. 4 For the house he made windows with recessed frames.*f* 5 He also built a structure against the wall of the house, running around the walls of the house, both the nave and the inner sanctuary; and he made side chambers all around. 6 The lowest story*g* was five cubits wide, the middle one was six cubits wide, and the third was seven cubits wide; for around the outside of the house

he made offsets on the wall in order that the supporting beams should not be inserted into the walls of the house. 7 The house was built with stone finished at the quarry, so that neither hammer nor ax nor any tool of iron was heard in the temple while it was being built. 8 The entrance for the middle story was on the south side of the house: one went up by winding stairs to the middle story, and from the middle story to the third. 9 So he built the house, and finished it; he roofed the house with beams and planks of cedar. 10 He built the structure against the whole house, each story*h* five cubits high, and it was joined to the house with timbers of cedar.

11 Now the word of the LORD came to Solomon, 12 "Concerning this house that you are building, if you will walk in my statutes, obey my ordinances, and keep all my commandments by walking in them, then I will establish my promise with you, which I made to your father David. 13 I will dwell among the children of Israel, and will not forsake my people Israel."

14 So Solomon built the house, and finished it. 15 He lined the walls of the house on the inside with boards of cedar; from the floor of the house to the rafters of the ceiling, he covered them on the inside with wood; and he covered the floor of the house with boards of cy-

f Gk: Meaning of Heb uncertain   *g* Gk: Heb structure   *h* Heb lacks *each story*

(12.4). *Adoniram,* also called Adoram and Hadoram (2 Sam 20.24; 2 Chr 10.18); compare 4.6; 12.18. *Gebalites,* people from Gebal, called Byblos by the Greeks, famous for its trade in Egyptian papyrus, the ancient form of paper. From the Greek name our word "Bible" is indirectly derived.

6.1–38: The building of the temple. 1: *The fourth year of Solomon's reign* would be approximately 960 B.C. *The four hundred eightieth year* before this (1440 B.C.) is now considered too early as a date for the Exodus and is possibly an artificially rounded number. *The month of Ziv* came in the spring (April–May). **2–6:** The *cubit,* approximately eighteen inches. Thus the temple itself had a rectangular floor plan of about 90 × 30 ft., and

was about 45 ft. high. It consisted of two rooms: *the nave* or main room, which was about 60 ft. long (v. 17); and *the inner sanctuary* or holy of holies, a perfect cube about 30 ft. on a side (v. 20). In addition there was *the vestibule in front of the nave,* which, like the rest of the temple, was 30 ft. wide and about 15 ft. deep (v. 3). *Side chambers* surrounded the nave and the inner sanctuary, but not the vestibule. **6.7:** There is no reason to doubt that the *stone* was *finished at the quarry,* though the writer has exaggerated the silence of the building operations. **8–10:** A further description of the side chambers.

6.14–22: No doubt the decorations of the temple were artistically intricate; the amount

press. 16 He built twenty cubits of the rear of the house with boards of cedar from the floor to the rafters, and he built this within as an inner sanctuary, as the most holy place. 17 The house, that is, the nave in front of the inner sanctuary, was forty cubits long. 18 The cedar within the house had carvings of gourds and open flowers; all was cedar, no stone was seen. 19 The inner sanctuary he prepared in the innermost part of the house, to set there the ark of the covenant of the LORD. 20 The interior of the inner sanctuary was twenty cubits long, twenty cubits wide, and twenty cubits high; he overlaid it with pure gold. He also overlaid the altar with cedar. *i* 21 Solomon overlaid the inside of the house with pure gold, then he drew chains of gold across, in front of the inner sanctuary, and overlaid it with gold. 22 Next he overlaid the whole house with gold, in order that the whole house might be perfect; even the whole altar that belonged to the inner sanctuary he overlaid with gold.

23 In the inner sanctuary he made two cherubim of olivewood, each ten cubits high. 24 Five cubits was the length of one wing of the cherub, and five cubits the length of the other wing of the cherub; it was ten cubits from the tip of one wing to the tip of the other. 25 The other cherub also measured ten cubits; both cherubim had the same measure and the same form. 26 The height of one cherub was ten cubits, and so was that of the other cherub. 27 He put the cherubim in the innermost part of the house; the wings of the cherubim were spread out so that a wing of one was touching the one wall, and a wing of the other cherub was touching the other wall; their other wings toward the center of the house were touching wing to wing. 28 He also overlaid the cherubim with gold.

29 He carved the walls of the house all around about with carved engravings of cherubim, palm trees, and open flowers, in the inner and outer rooms. 30 The floor of the house he overlaid with gold, in the inner and outer rooms.

31 For the entrance to the inner sanctuary he made doors of olivewood; the lintel and the doorposts were five-sided. *i* 32 He covered the two doors of olivewood with carvings of cherubim, palm trees, and open flowers; he overlaid them with gold, and spread gold on the cherubim and on the palm trees.

33 So also he made for the entrance to the nave doorposts of olivewood, four-sided each, 34 and two doors of cypress wood; the two leaves of the one door were folding, and the two leaves of the other door were folding. 35 He carved cherubim, palm trees, and open flowers, overlaying them with gold evenly applied upon the carved work. 36 He built the inner court with three courses of dressed stone to one course of cedar beams.

37 In the fourth year the foundation of the house of the LORD was laid, in the month of Ziv. 38 In the eleventh year, in the month of Bul, which is the eighth month, the house was finished in all its parts, and according to all its specifications. He was seven years in building it.

7 Solomon was building his own house thirteen years, and he finished his entire house.

2 He built the House of the Forest of the Lebanon one hundred cubits long, fifty cubits wide, and thirty cubits high, built on four rows of cedar pillars, with cedar beams on the pillars. 3 It was roofed with cedar on the forty-five rafters, fifteen in each row, which were on the pillars. 4 There were window frames in the three rows, facing each other in the three

i Meaning of Heb uncertain

---

of gold, however, has probably been somewhat exaggerated. On *the ark of the covenant,* see 1 Sam 3.3 n.

6.23–28: On *the cherubim,* see 1 Sam 4.4 n. **37–38:** *The month of Bul* was in the autumn (October–November).

7.1–51: Details of Solomon's building operations. 1: See 3.1. The palace and administrative complex took *thirteen years* to build, the temple only seven (6.38); compare 9.10.

rows. [5] All the doorways and doorposts had four-sided frames, opposite, facing each other in the three rows.

6 He made the Hall of Pillars fifty cubits long and thirty cubits wide. There was a porch in front with pillars, and a canopy in front of them.

7 He made the Hall of the Throne where he was to pronounce judgment, the Hall of Justice, covered with cedar from floor to floor.

8 His own house where he would reside, in the other court back of the hall, was of the same construction. Solomon also made a house like this hall for Pharaoh's daughter, whom he had taken in marriage.

9 All these were made of costly stones, cut according to measure, sawed with saws, back and front, from the foundation to the coping, and from outside to the great court. [10] The foundation was of costly stones, huge stones, stones of eight and ten cubits. [11] There were costly stones above, cut to measure, and cedarwood. [12] The great court had three courses of dressed stone to one layer of cedar beams all around; so had the inner court of the house of the LORD, and the vestibule of the house.

13 Now King Solomon invited and received Hiram from Tyre. [14] He was the son of a widow of the tribe of Naphtali, whose father, a man of Tyre, had been an artisan in bronze; he was full of skill, intelligence, and knowledge in working bronze. He came to King Solomon, and did all his work.

15 He cast two pillars of bronze. Eighteen cubits was the height of the one, and a cord of twelve cubits would encircle it; the second pillar was the same.[j] [16] He also made two capitals of molten bronze, to set on the tops of the pillars; the height of the one capital was five cubits, and the height of the other capital was five cubits. [17] There were nets of checker work with wreaths of chain work for the capitals on the tops of the pillars; seven[k] for the one capital, and seven[k] for the other capital. [18] He made the columns with two rows around each latticework to cover the capitals that were above the pomegranates; he did the same with the other capital. [19] Now the capitals that were on the tops of the pillars in the vestibule were of lily-work, four cubits high. [20] The capitals were on the two pillars and also above the rounded projection that was beside the latticework; there were two hundred pomegranates in rows all around; and so with the other capital. [21] He set up the pillars at the vestibule of the temple; he set up the pillar on the south and called it Jachin; and he set up the pillar on the north and called it Boaz. [22] On the tops of the pillars was lily-work. Thus the work of the pillars was finished.

23 Then he made the molten sea; it was round, ten cubits from brim to brim, and five cubits high. A line of thirty cubits would encircle it completely. [24] Under its brim were panels all around it, each of ten cubits, surrounding the sea; there were two rows of panels, cast when it was cast. [25] It stood on twelve oxen, three facing north, three facing west, three facing south, and three facing

j Cn: Heb *and a cord of twelve cubits encircled the second pillar*; Compare Jer 52.21   *k* Heb: Gk *a net*

7.6–12: Details of the palace and administrative complex lying immediately south of the temple. **13–14:** The artisan *Hiram from Tyre* is not to be confused with the king of the same name; compare 2 Chr 2.13, where the name is given as Huram-abi. **7.15–22:** In front of the vestibule were set up the two great freestanding *pillars of bronze* called *Jachin* and *Boaz*. The names mean something like "God establishes" and "He comes with power." The pillars themselves doubtless had a symbolic significance (e.g. "trees of life") which now eludes us. On their role at ceremonial occasions, see 2 Kings 11.14; 23.3. **7.23–26:** *The molten sea* was a huge bowl or tank supported on *twelve oxen*. Its capacity was *two thousand baths* (the bath was a liquid measure equal to about six gallons). According to 2 Chr 4.6 the purpose *was for the priests to wash in,* though it is difficult to imagine just how the ablutions were accomplished, since the rim of the tank was some ten feet above the pavement. As with the pillars, there may

east; the sea was set on them. The hindquarters of each were toward the inside. 26 Its thickness was a handbreadth; its brim was made like the brim of a cup, like the flower of a lily; it held two thousand baths. 1

27 He also made the ten stands of bronze; each stand was four cubits long, four cubits wide, and three cubits high. 28 This was the construction of the stands: they had borders; the borders were within the frames; 29 on the borders that were set in the frames were lions, oxen, and cherubim. On the frames, both above and below the lions and oxen, there were wreaths of beveled work. 30 Each stand had four bronze wheels and axles of bronze; at the four corners were supports for a basin. The supports were cast with wreaths at the side of each. 31 Its opening was within the crown whose height was one cubit; its opening was round, as a pedestal is made; it was a cubit and a half wide. At its opening there were carvings; its borders were four-sided, not round. 32 The four wheels were underneath the borders; the axles of the wheels were in the stands; and the height of a wheel was a cubit and a half. 33 The wheels were made like a chariot wheel; their axles, their rims, their spokes, and their hubs were all cast. 34 There were four supports at the four corners of each stand; the supports were of one piece with the stands. 35 On the top of the stand there was a round band half a cubit high; on the top of the stand, its stays and its borders were of one piece with it. 36 On the surfaces of its stays and on its borders he carved cherubim, lions, and palm trees, where each had space, with wreaths all around. 37 In this way he made the ten stands; all of them were cast alike, with the same size and the same form.

38 He made ten basins of bronze; each

basin held forty baths, 1 each basin measured four cubits; there was a basin for each of the ten stands. 39 He set five of the stands on the south side of the house, and five on the north side of the house; he set the sea on the southeast corner of the house.

40 Hiram also made the pots, the shovels, and the basins. So Hiram finished all the work that he did for King Solomon on the house of the LORD: 41 the two pillars, the two bowls of the capitals that were on the tops of the pillars, the two latticeworks to cover the two bowls of the capitals that were on the tops of the pillars; 42 the four hundred pomegranates for the two latticeworks, two rows of pomegranates for each latticework, to cover the two bowls of the capitals that were on the pillars; 43 the ten stands, the ten basins on the stands; 44 the one sea, and the twelve oxen underneath the sea.

45 The pots, the shovels, and the basins, all these vessels that Hiram made for King Solomon for the house of the LORD were of burnished bronze. 46 In the plain of the Jordan the king cast them, in the clay ground between Succoth and Zarethan. 47 Solomon left all the vessels unweighed, because there were so many of them; the weight of the bronze was not determined.

48 So Solomon made all the vessels that were in the house of the LORD: the golden altar, the golden table for the bread of the Presence, 49 the lampstands of pure gold, five on the south side and five on the north, in front of the inner sanctuary; the flowers, the lamps, and the tongs, of gold; 50 the cups, snuffers, basins, dishes for incense, and firepans, of pure gold; the sockets for the doors of the innermost part of the house, the most

---

1 A Heb measure of volume

---

have been also a symbolic significance (such as "life-giving water"; compare the artificial lakes near Egyptian temples). The *twelve oxen* were really couchant bulls, symbols of fertility and power in many religions of the ancient Near East. Their arrangement according to

the four points of the compass suggested universal dominion.

7.27–39: *The ten stands,* highly ornamented bronze wagons on each of which was mounted one of the *ten basins.*

7.48–50: In *the nave* were *the golden altar*

holy place, and for the doors of the nave of the temple, of gold.

51 Thus all the work that King Solomon did on the house of the LORD was finished. Solomon brought in the things that his father David had dedicated, the silver, the gold, and the vessels, and stored them in the treasuries of the house of the LORD.

8 Then Solomon assembled the elders of Israel and all the heads of the tribes, the leaders of the ancestral houses of the Israelites, before King Solomon in Jerusalem, to bring up the ark of the covenant of the LORD out of the city of David, which is Zion. ²All the people of Israel assembled to King Solomon at the festival in the month Ethanim, which is the seventh month. ³And all the elders of Israel came, and the priests carried the ark. ⁴So they brought up the ark of the LORD, the tent of meeting, and all the holy vessels that were in the tent; the priests and the Levites brought them up. ⁵King Solomon and all the congregation of Israel, who had assembled before him, were with him before the ark, sacrificing so many sheep and oxen that they could not be counted or numbered. ⁶Then the priests brought the ark of the covenant of the LORD to its place, in the inner sanctuary of the house, in the most holy place, underneath the wings of the cherubim. ⁷For the cherubim spread out their wings over the place of the ark, so that the cherubim made a covering above the ark and

its poles. ⁸The poles were so long that the ends of the poles were seen from the holy place in front of the inner sanctuary; but they could not be seen from outside; they are there to this day. ⁹There was nothing in the ark except the two tablets of stone that Moses had placed there at Horeb, where the LORD made a covenant with the Israelites, when they came out of the land of Egypt. ¹⁰And when the priests came out of the holy place, a cloud filled the house of the LORD, ¹¹so that the priests could not stand to minister because of the cloud; for the glory of the LORD filled the house of the LORD.

12 Then Solomon said,

"The LORD has said that he would
 dwell in thick darkness.
13 I have built you an exalted house,
 a place for you to dwell in
 forever."

14 Then the king turned around and blessed all the assembly of Israel, while all the assembly of Israel stood. ¹⁵He said, "Blessed be the LORD, the God of Israel, who with his hand has fulfilled what he promised with his mouth to my father David, saying, ¹⁶'Since the day that I brought my people Israel out of Egypt, I have not chosen a city from any of the tribes of Israel in which to build a house, that my name might be there; but I chose David to be over my people Israel.' ¹⁷My father David had it in mind to build a house for the name of the LORD, the God of Israel. ¹⁸But the LORD said to

---

for incense, the *table for the bread of the Presence* (or showbread, see Lev 24.5–9), and the ten *lampstands,* plus smaller objects. *The lampstands* were simple, each being a rod resting on a base, with a small bowl at the top for oil and wick; not to be confused with the lampstand described in Ex 25.31–40.

8.1–66: The dedication of the temple.
1–11: Bringing *the ark of the covenant* into the temple (2 Sam 6.12–15). *The city of David,* or *Zion,* was south of the temple area, in the southeastern part of the city. *The month Ethanim* or *Tishri* occurred in the autumn (September–October). The dedication of the temple was, therefore, postponed for eleven months (6.38), in order to make it a part of the autumnal new-year festival, the religious high point of the year.

8.7: *The cherubim,* see 6.23–28. The ark was carried by *the poles* (v. 8); the writer here possibly wrote or compiled this material during the time before the destruction of the temple, as is seen from the words *they are there to this day.*
8.12–21: Solomon's address to the people. It is impossible to tell how much of this comes from Solomon and how much from the Deuteronomistic writer. The *thick darkness* (v. 12) refers to the fact that the inner sanctuary had no windows, being in total darkness except when the door was opened on rare occasions. According to the Greek version, the poetic prologue (vv. 12–13) was taken from the Book of Jashar (see Josh 10.13 n.).

my father David, 'You did well to consider building a house for my name; [19]nevertheless you shall not build the house, but your son who shall be born to you shall build the house for my name.' [20]Now the LORD has upheld the promise that he made; for I have risen in the place of my father David; I sit on the throne of Israel, as the LORD promised, and have built the house for the name of the LORD, the God of Israel. [21]There I have provided a place for the ark, in which is the covenant of the LORD that he made with our ancestors when he brought them out of the land of Egypt."

22 Then Solomon stood before the altar of the LORD in the presence of all the assembly of Israel, and spread out his hands to heaven. [23]He said, "O LORD, God of Israel, there is no God like you in heaven above or on earth beneath, keeping covenant and steadfast love for your servants who walk before you with all their heart, [24]the covenant that you kept for your servant my father David as you declared to him; you promised with your mouth and have this day fulfilled with your hand. [25]Therefore, O LORD, God of Israel, keep for your servant my father David that which you promised him, saying, 'There shall never fail you a successor before me to sit on the throne of Israel, if only your children look to their way, to walk before me as you have walked before me.' [26]Therefore, O God of Israel, let your word be confirmed, which you promised to your servant my father David.

27 "But will God indeed dwell on the earth? Even heaven and the highest heaven cannot contain you, much less this house that I have built! [28]Regard your servant's prayer and his plea, O LORD my God, heeding the cry and the prayer that your servant prays to you today; [29]that your eyes may be open night and day toward this house, the place of which you said, 'My name shall be there,' that you may heed the prayer that your servant prays toward this place. [30]Hear the plea of your servant and of your people Israel when they pray toward this place; O hear in heaven your dwelling place; heed and forgive.

31 "If someone sins against a neighbor and is given an oath to swear, and comes and swears before your altar in this house, [32]then hear in heaven, and act, and judge your servants, condemning the guilty by bringing their conduct on their own head, and vindicating the righteous by rewarding them according to their righteousness.

33 "When your people Israel, having sinned against you, are defeated before an enemy but turn again to you, confess your name, pray and plead with you in this house, [34]then hear in heaven, forgive the sin of your people Israel, and bring them again to the land that you gave to their ancestors.

35 "When heaven is shut up and there is no rain because they have sinned against you, and then they pray toward this place, confess your name, and turn from their sin, because you punish[m] them, [36]then hear in heaven, and forgive the sin of your servants, your people Israel, when you teach them the good way in which they should walk; and grant rain on your land, which you have given to your people as an inheritance.

37 "If there is famine in the land, if there is plague, blight, mildew, locust, or caterpillar; if their enemy besieges them in any[n] of their cities; whatever plague, whatever sickness there is; [38]whatever prayer, whatever plea there is from any individual or from all your people Israel, all knowing the afflictions of their own hearts so that they stretch out their hands toward this house; [39]then hear in heaven your dwelling place, for-

m Or *when you answer*   *n* Gk Syr: Heb *in the land*

8.22–40: Solomon's prayer of dedication. This prayer, entirely in the spirit of the book of Deuteronomy, was composed mostly by the Deuteronomist. The phrase *and bring them again to the land that you gave to their ancestors* (v. 34) was possibly added during the Exile.

give, act, and render to all whose hearts you know—according to all their ways, for only you know what is in every human heart— 40 so that they may fear you all the days that they live in the land that you gave to our ancestors.

41 "Likewise when a foreigner, who is not of your people Israel, comes from a distant land because of your name 42 —for they shall hear of your great name, your mighty hand, and your outstretched arm—when a foreigner comes and prays toward this house, 43 then hear in heaven your dwelling place, and do according to all that the foreigner calls to you, so that all the peoples of the earth may know your name and fear you, as do your people Israel, and so that they may know that your name has been invoked on this house that I have built.

44 "If your people go out to battle against their enemy, by whatever way you shall send them, and they pray to the LORD toward the city that you have chosen and the house that I have built for your name, 45 then hear in heaven their prayer and their plea, and maintain their cause.

46 "If they sin against you—for there is no one who does not sin—and you are angry with them and give them to an enemy, so that they are carried away captive to the land of the enemy, far off or near; 47 yet if they come to their senses in the land to which they have been taken captive, and repent, and plead with you in the land of their captors, saying, 'We have sinned, and have done wrong; we have acted wickedly'; 48 if they repent with all their heart and soul in the land of their enemies, who took them captive, and pray to you toward their land, which you gave to their ancestors, the city that you have chosen, and the house that I have built for your name; 49 then hear in heaven your dwelling place their prayer and their plea, maintain their cause 50 and

forgive your people who have sinned against you, and all their transgressions that they have committed against you; and grant them compassion in the sight of their captors, so that they may have compassion on them 51 (for they are your people and heritage, which you brought out of Egypt, from the midst of the iron-smelter). 52 Let your eyes be open to the plea of your servant, and to the plea of your people Israel, listening to them whenever they call to you. 53 For you have separated them from among all the peoples of the earth, to be your heritage, just as you promised through Moses, your servant, when you brought our ancestors out of Egypt, O Lord GOD."

54 Now when Solomon finished offering all this prayer and this plea to the LORD, he arose from facing the altar of the LORD, where he had knelt with hands outstretched toward heaven; 55 he stood and blessed all the assembly of Israel with a loud voice:

56 "Blessed be the LORD, who has given rest to his people Israel according to all that he promised; not one word has failed of all his good promise, which he spoke through his servant Moses. 57 The LORD our God be with us, as he was with our ancestors; may he not leave us or abandon us, 58 but incline our hearts to him, to walk in all his ways, and to keep his commandments, his statutes, and his ordinances, which he commanded our ancestors. 59 Let these words of mine, with which I pleaded before the LORD, be near to the LORD our God day and night, and may he maintain the cause of his servant and the cause of his people Israel, as each day requires; 60 so that all the peoples of the earth may know that the LORD is God; there is no other. 61 Therefore devote yourselves completely to the LORD our God, walking in his statutes and keeping his commandments, as at this day."

8.41–53: An expansion of the prayer by a writer during the Exile. The many references to being *carried away captive to the land of*

the enemy, especially in vv. 46–53, should be noticed.

8.62–66: This would have been a natural

62 Then the king, and all Israel with him, offered sacrifice before the LORD. 63 Solomon offered as sacrifices of well-being to the LORD twenty-two thousand oxen and one hundred twenty thousand sheep. So the king and all the people of Israel dedicated the house of the LORD. 64 The same day the king consecrated the middle of the court that was in front of the house of the LORD; for there he offered the burnt offerings and the grain offerings and the fat pieces of the sacrifices of well-being, because the bronze altar that was before the LORD was too small to receive the burnt offerings and the grain offerings and the fat pieces of the sacrifices of well-being.

65 So Solomon held the festival at that time, and all Israel with him—a great assembly, people from Lebo-hamath to the Wadi of Egypt—before the LORD our God, seven days. *o* 66 On the eighth day he sent the people away; and they blessed the king, and went to their tents, joyful and in good spirits because of all the goodness that the LORD had shown to his servant David and to his people Israel.

9 When Solomon had finished building the house of the LORD and the king's house and all that Solomon desired to build, 2 the LORD appeared to Solomon a second time, as he had appeared to him at Gibeon. 3 The LORD said to him, "I have heard your prayer and your plea, which you made before me; I have consecrated this house that you have built, and put my name there forever; my eyes and my heart will be there for all time. 4 As for you, if you will walk before me, as David your father walked, with integrity of heart and uprightness, doing according to all that I have commanded you, and keeping my statutes and my ordinances, 5 then I will establish your royal throne over Israel forever, as I promised your father David, saying, 'There shall not fail you a successor on the throne of Israel.'

6 "If you turn aside from following me, you or your children, and do not keep my commandments and my statutes that I have set before you, but go and serve other gods and worship them, 7 then I will cut Israel off from the land that I have given them; and the house that I have consecrated for my name I will cast out of my sight; and Israel will become a proverb and a taunt among all peoples. 8 This house will become a heap of ruins; *p* everyone passing by it will be astonished, and will hiss; and they will say, 'Why has the LORD done such a thing to this land and to this house?' 9 Then they will say, 'Because they have forsaken the LORD their God, who brought their ancestors out of the land of Egypt, and embraced other gods, worshiping them and serving them; therefore the LORD has brought this disaster upon them.'"

10 At the end of twenty years, in which Solomon had built the two houses, the house of the LORD and the king's house, 11 King Hiram of Tyre having supplied Solomon with cedar and cypress timber and gold, as much as he desired, King Solomon gave to Hiram twenty cities in the land of Galilee. 12 But when Hiram came from Tyre to see the cities that Solomon had given him, they

o Compare Gk: Heb *seven days and seven days, fourteen days* *p* Syr Old Latin: Heb *will become high*

---

conclusion of the chapter by a writer who worked during the time before the destruction of the temple. In v. 64 *the bronze altar* in front of the temple is specifically mentioned for the first time, though it is alluded to in vv. 22 and 54.

9.1–9: Solomon's vision. A thoroughly Deuteronomic composition, perhaps by a writer during the Exile, who lived after the temple had *become a heap of ruins* (v. 8), and could point out the sinfulness that had led to national ruin (see Introduction).

9.10–14: An interlude; a story told to explain the origin of a name. The meaning of *Cabul* is not known; "like nothing" is as good a conjecture as any. On the historical side, Hiram's supplying of timber is irrelevant; Solomon had paid for that (5.11). It would seem that Solomon was in financial difficulties and was forced to cede *twenty cities* to

did not please him. 13 Therefore he said, "What kind of cities are these that you have given me, my brother?" So they are called the land of Cabul*q* to this day. 14 But Hiram had sent to the king one hundred twenty talents of gold.

15 This is the account of the forced labor that King Solomon conscripted to build the house of the LORD and his own house, the Millo and the wall of Jerusalem, Hazor, Megiddo, Gezer 16 (Pharaoh king of Egypt had gone up and captured Gezer and burned it down, had killed the Canaanites who lived in the city, and had given it as dowry to his daughter, Solomon's wife; 17 so Solomon rebuilt Gezer), Lower Beth-horon, 18 Baalath, Tamar in the wilderness, within the land, 19 as well as all of Solomon's storage cities, the cities for his chariots, the cities for his cavalry, and whatever Solomon desired to build, in Jerusalem, in Lebanon, and in all the land of his dominion. 20 All the people who were left of the Amorites, the Hittites, the Perizzites, the Hivites, and the Jebusites, who were not of the people of Israel— 21 their descendants who were still left in the land, whom the Israelites were unable to destroy completely—these Solomon conscripted for slave labor, and so they are to this day. 22 But of the Israelites Solomon made no slaves; they were the soldiers, they were his officials, his commanders, his captains, and the commanders of his chariotry and cavalry.

23 These were the chief officers who were over Solomon's work: five hundred fifty, who had charge of the people who carried on the work.

24 But Pharaoh's daughter went up from the city of David to her own house that Solomon had built for her; then he built the Millo.

25 Three times a year Solomon used to offer up burnt offerings and sacrifices of well-being on the altar that he built for the LORD, offering incense*r* before the LORD. So he completed the house.

26 King Solomon built a fleet of ships at Ezion-geber, which is near Eloth on the shore of the Red Sea,*s* in the land of Edom. 27 Hiram sent his servants with the fleet, sailors who were familiar with the sea, together with the servants of Solomon. 28 They went to Ophir, and imported from there four hundred twenty talents of gold, which they delivered to King Solomon.

10 When the queen of Sheba heard of the fame of Solomon, (fame due to*t* the name of the LORD), she came to test him with hard questions. 2 She came to Jerusalem with a very great retinue, with camels bearing spices, and very much gold, and precious stones; and when she came to Solomon, she told him all that was on her mind. 3 Solomon answered all her questions; there was nothing hidden from the king that he

q Perhaps meaning *a land good for nothing*
r Gk: Heb *offering incense with it that was*
s Or *Sea of Reeds*   *t* Meaning of Heb uncertain

Hiram for a cash consideration (a gold talent would have weighed about seventy-five pounds).
9.15–22: The forced levy (see 5.13). *The Millo* (also v. 24; 2 Sam 5.9) is usually interpreted as an earthwork south of the temple area; the word probably means "a filling." Verse 22 seems to be contradicted by 5.13 and 12.4.
9.23–28: Miscellaneous details. 24: Compare v. 16 and 3.1. **25:** In those early days the king still exercised priestly functions. **26–28:** *Red Sea* here refers to the Gulf of Aqaba. The location of *Ophir* is unknown; it may have been in southern Arabia (compare 10.11 and Gen 10.29). Excavations

at Ezion-geber have disclosed the refineries in which copper from the Arabah mines was smelted.
10.1–29: The visit of the queen of Sheba; Solomon's wealth and business enterprises. 1–5: *Sheba* is usually considered to have been located in southwest Arabia, modern Yemen. Some, however, think that this queen came from a colony of Sheba in northern Arabia, where a number of queens are known to have ruled. Practical wisdom, proverbial sayings, and contests of wits were characteristic of the biblical world; compare 4.29–34 and the Old Testament books of "wisdom," such as Job and Proverbs. The word for *spirit* (v. 5) can be translated

could not explain to her. [4] When the queen of Sheba had observed all the wisdom of Solomon, the house that he had built, [5] the food of his table, the seating of his officials, and the attendance of his servants, their clothing, his valets, and his burnt offerings that he offered at the house of the LORD, there was no more spirit in her.

6 So she said to the king, "The report was true that I heard in my own land of your accomplishments and of your wisdom, [7] but I did not believe the reports until I came and my own eyes had seen it. Not even half had been told me; your wisdom and prosperity far surpass the report that I had heard. [8] Happy are your wives! [u] Happy are these your servants, who continually attend you and hear your wisdom! [9] Blessed be the LORD your God, who has delighted in you and set you on the throne of Israel! Because the LORD loved Israel forever, he has made you king to execute justice and righteousness." [10] Then she gave the king one hundred twenty talents of gold, a great quantity of spices, and precious stones; never again did spices come in such quantity as that which the queen of Sheba gave to King Solomon.

11 Moreover, the fleet of Hiram, which carried gold from Ophir, brought from Ophir a great quantity of almug wood and precious stones. [12] From the almug wood the king made supports for the house of the LORD, and for the king's house, lyres also and harps for the singers; no such almug wood has come or been seen to this day.

13 Meanwhile King Solomon gave to the queen of Sheba every desire that she expressed, as well as what he gave her out of Solomon's royal bounty. Then she returned to her own land, with her servants.

14 The weight of gold that came to Solomon in one year was six hundred sixty-six talents of gold, [15] besides that which came from the traders and from the business of the merchants, and from all the kings of Arabia and the governors of the land. [16] King Solomon made two hundred large shields of beaten gold; six hundred shekels of gold went into each large shield. [17] He made three hundred shields of beaten gold; three minas of gold went into each shield; and the king put them in the House of the Forest of Lebanon. [18] The king also made a great ivory throne, and overlaid it with the finest gold. [19] The throne had six steps. The top of the throne was rounded in the back, and on each side of the seat were arm rests and two lions standing beside the arm rests, [20] while twelve lions were standing, one on each end of a step on the six steps. Nothing like it was ever made in any kingdom. [21] All King Solomon's drinking vessels were of gold, and all the vessels of the House of the Forest of Lebanon were of pure gold; none were of silver—it was not considered as anything in the days of Solomon. [22] For the king had a fleet of ships of Tarshish at sea with the fleet of Hiram. Once every three years the fleet of ships of Tarshish used to come bringing gold, silver, ivory, apes, and peacocks. [v]

23 Thus King Solomon excelled all the kings of the earth in riches and in wisdom. [24] The whole earth sought the presence of Solomon to hear his wisdom, which God had put into his mind. [25] Every one of them brought a present, objects of silver and gold, garments, weaponry, spices, horses, and mules, so much year by year.

u Gk Syr: Heb *men*   v Or *baboons*

"breath"—the queen was breathless with amazement. **10:** On the weight of *talents of gold,* see 9.14 n.

10.11–12: An interlude, related to 9.26–28. *Almug wood* is now often called sandalwood. **13:** The Ethiopians (Abyssinians) have a tradition according to which *every desire that she expressed* included an heir, who later became king of Ethiopia, thus establishing the true dynasty of David for that country. **16–17:** If there were fifty shekels in a mina (Ezek 45.12), the smaller shields weighed one hundred and fifty shekels.

10.22: *Ships of Tarshish,* i.e. ships capable of making long voyages; compare vv. 11–12; 9.26–28; Isa 23.1. **26–29:** These verses de-

26 Solomon gathered together chariots and horses; he had fourteen hundred chariots and twelve thousand horses, which he stationed in the chariot cities and with the king in Jerusalem. 27 The king made silver as common in Jerusalem as stones, and he made cedars as numerous as the sycamores of the Shephelah. 28 Solomon's import of horses was from Egypt and Kue, and the king's traders received them from Kue at a price. 29 A chariot could be imported from Egypt for six hundred shekels of silver, and a horse for one hundred fifty; so through the king's traders they were exported to all the kings of the Hittites and the kings of Aram.

11 King Solomon loved many foreign women along with the daughter of Pharaoh: Moabite, Ammonite, Edomite, Sidonian, and Hittite women, 2 from the nations concerning which the LORD had said to the Israelites, "You shall not enter into marriage with them, neither shall they with you; for they will surely incline your heart to follow their gods"; Solomon clung to these in love. 3 Among his wives were seven hundred princesses and three hundred concubines; and his wives turned away his heart. 4 For when Solomon was old, his wives turned away his heart after other gods; and his heart was not true to the LORD his God, as was the heart of his father David. 5 For Solomon followed Astarte the goddess of the Sidonians, and Milcom the abomination of the Ammonites. 6 So Solomon did what was evil in the sight of the LORD, and did not completely follow the LORD, as his father David had done. 7 Then Solomon built a high place for Chemosh the abomination of Moab, and for Molech the abomination of the Ammonites, on the mountain east of Jerusalem. 8 He did the same for all his foreign wives, who offered incense and sacrificed to their gods.

9 Then the LORD was angry with Solomon, because his heart had turned away from the LORD, the God of Israel, who had appeared to him twice, 10 and had commanded him concerning this matter, that he should not follow other gods; but he did not observe what the LORD commanded. 11 Therefore the LORD said to Solomon, "Since this has been your mind and you have not kept my covenant and my statutes that I have commanded you, I will surely tear the kingdom from you and give it to your servant. 12 Yet for the sake of your father David I will not do it in your lifetime; I will tear it out of the hand of your son. 13 I will not, however, tear away the entire kingdom; I will give one tribe to your son, for the sake of my servant David and for the sake of Jerusalem, which I have chosen."

14 Then the LORD raised up an adversary against Solomon, Hadad the Edomite; he was of the royal house in Edom. 15 For when David was in Edom, and Joab the commander of the army went up to bury the dead, he killed every male in Edom 16 (for Joab and all Israel remained there six months, until he had eliminated every male in Edom); 17 but Hadad fled to Egypt with some Edomites who were servants of his father. He was a young boy at that time. 18 They set out from Midian and came to Paran; they took people with them from Paran and came to Egypt, to Pharaoh king of Egypt, who gave him a house, assigned

---

scribe Solomon's very large commerce in *horses* and *chariots*. *Kue* was Cilicia, in southeast Asia Minor. *The kings of the Hittites* ruled small principalities in what is now northern Syria.

11.1–43: The dark side of Solomon's reign. The Deuteronomistic writer knew the tragedies that followed Solomon (vv. 12–13) and sought to probe their causes. In this connection, Deut 17.14–20, from the same school of writers, should be carefully considered. **1–13:** Judgment on Solomon. Compare Deut 7.1–4. On *Astarte,* see 1 Sam 7.3–4 n. On the *high place,* see 3.1–2 n.

11.14–40: Adversaries of Solomon.

11.14–22: Hadad the Edomite. **15:** *David* had conquered *Edom* (*every male* is of course an exaggeration). Near the beginning of Solomon's reign (v. 21), *Hadad the Edomite* revolted rather successfully (v. 25). **23–25:** Re-

him an allowance of food, and gave him land. ¹⁹Hadad found great favor in the sight of Pharaoh, so that he gave him his sister-in-law for a wife, the sister of Queen Tahpenes. ²⁰The sister of Tahpenes gave birth by him to his son Genubath, whom Tahpenes weaned in Pharaoh's house; Genubath was in Pharaoh's house among the children of Pharaoh. ²¹When Hadad heard in Egypt that David slept with his ancestors and that Joab the commander of the army was dead, Hadad said to Pharaoh, "Let me depart, that I may go to my own country." ²²But Pharaoh said to him, "What do you lack with me that you now seek to go to your own country?" And he said, "No, do let me go."

23 God raised up another adversary against Solomon, *ʷ* Rezon son of Eliada, who had fled from his master, King Hadadezer of Zobah. ²⁴He gathered followers around him and became leader of a marauding band, after the slaughter by David; they went to Damascus, settled there, and made him king in Damascus. ²⁵He was an adversary of Israel all the days of Solomon, making trouble as Hadad did; he despised Israel and reigned over Aram.

26 Jeroboam son of Nebat, an Ephraimite of Zeredah, a servant of Solomon, whose mother's name was Zeruah, a widow, rebelled against the king. ²⁷The following was the reason he rebelled against the king. Solomon built the Millo, and closed up the gap in the wall*ˣ* of the city of his father David. ²⁸The man Jeroboam was very able, and when Solomon saw that the young man was industrious he gave him charge over all the forced labor of the house of Joseph. ²⁹About that time, when Jeroboam was leaving Jerusalem, the prophet Ahijah

the Shilonite found him on the road. Ahijah had clothed himself with a new garment. The two of them were alone in the open country ³⁰when Ahijah laid hold of the new garment he was wearing and tore it into twelve pieces. ³¹He then said to Jeroboam: Take for yourself ten pieces; for thus says the LORD, the God of Israel, "See, I am about to tear the kingdom from the hand of Solomon, and will give you ten tribes. ³²One tribe will remain his, for the sake of my servant David and for the sake of Jerusalem, the city that I have chosen out of all the tribes of Israel. ³³This is because he has*ʸ* forsaken me, worshiped Astarte the goddess of the Sidonians, Chemosh the god of Moab, and Milcom the god of the Ammonites, and has*ʸ* not walked in my ways, doing what is right in my sight and keeping my statutes and my ordinances, as his father David did. ³⁴Nevertheless I will not take the whole kingdom away from him but will make him ruler all the days of his life, for the sake of my servant David whom I chose and who did keep my commandments and my statutes; ³⁵but I will take the kingdom away from his son and give it to you— that is, the ten tribes. ³⁶Yet to his son I will give one tribe, so that my servant David may always have a lamp before me in Jerusalem, the city where I have chosen to put my name. ³⁷I will take you, and you shall reign over all that your soul desires; you shall be king over Israel. ³⁸If you will listen to all that I command you, walk in my ways, and do what is right in my sight by keeping my statutes and my commandments, as David my servant did, I will be with you, and will build you an enduring house, as

w Heb *him*   *x* Heb lacks *in the wall*
y Gk Syr Vg: Heb *they have*

---

zon of *Aram* (Syria). On David's conquest of *King Hadadezer of Zobah* and of the Syrians (Arameans) of *Damascus,* see 2 Sam 8.3–6; 10.15–19. *Rezon* set up a new kingdom in *Damascus* and defied Solomon. Thus the empire of *David* began to melt away under Solomon.

11.26–40: Jeroboam of Israel. The greatest danger, namely internal revolt, came from *Jeroboam son of Nebat. The prophet Ahijah the Shilonite* (of Shiloh) expected that *Jeroboam* would be more loyal to *the God of Israel* than *Solomon* had been. The *lamp* (v. 36) was a symbol of the permanence of the Davidic dy-

I built for David, and I will give Israel to you. ³⁹For this reason I will punish the descendants of David, but not forever." ⁴⁰Solomon sought therefore to kill Jeroboam; but Jeroboam promptly fled to Egypt, to King Shishak of Egypt, and remained in Egypt until the death of Solomon.

41 Now the rest of the acts of Solomon, all that he did as well as his wisdom, are they not written in the Book of the Acts of Solomon? ⁴²The time that Solomon reigned in Jerusalem over all Israel was forty years. ⁴³Solomon slept with his ancestors and was buried in the city of his father David; and his son Rehoboam succeeded him.

12 Rehoboam went to Shechem, for all Israel had come to Shechem to make him king. ²When Jeroboam son of Nebat heard of it (for he was still in Egypt, where he had fled from King Solomon), then Jeroboam returned from ᶻ Egypt. ³And they sent and called him; and Jeroboam and all the assembly of Israel came and said to Rehoboam, ⁴"Your father made our yoke heavy. Now therefore lighten the hard service of your father and his heavy yoke that he placed on us, and we will serve you." ⁵He said to them, "Go away for three days, then come again to me." So the people went away.

6 Then King Rehoboam took counsel with the older men who had attended his father Solomon while he was still alive, saying, "How do you advise me to answer this people?" ⁷They answered him, "If you will be a servant to this people today and serve them, and speak good words to them when you answer them, then they will be your servants forever." ⁸But he disregarded the advice that the older men gave him, and consulted with the young men who had grown up with him and now attended him. ⁹He said to them, "What do you advise that we answer this people who have said to me, 'Lighten the yoke that your father put on us'?" ¹⁰The young men who had grown up with him said to him, "Thus you should say to this people who spoke to you, 'Your father made our yoke heavy, but you must lighten it for us'; thus you should say to them, 'My little finger is thicker than my father's loins. ¹¹Now, whereas my father laid on you a heavy yoke, I will add to your yoke. My father disciplined you with whips, but I will discipline you with scorpions.' "

12 So Jeroboam and all the people came to Rehoboam the third day, as the king had said, "Come to me again the third day." ¹³The king answered the people harshly. He disregarded the advice that the older men had given him ¹⁴and spoke to them according to the advice of the young men, "My father made your yoke heavy, but I will add to your yoke; my father disciplined you with whips, but I will discipline you with scorpions." ¹⁵So the king did not listen to the people, because it was a turn of affairs brought about by the LORD that he might fulfill his word, which the LORD had spoken by Ahijah the Shilonite to Jeroboam son of Nebat.

16 When all Israel saw that the king would not listen to them, the people answered the king,

"What share do we have in David?
    We have no inheritance in the
        son of Jesse.
To your tents, O Israel!

z Gk Vg Compare 2 Chr 10.2: Heb *lived in*

---

nasty (2 Sam 21.17). *King Shishak of Egypt* was probably the less friendly successor of the pharaoh whose daughter was married to *Solomon* (3.1). **41–43:** Conclusion. On *the Book of the Acts of Solomon,* see the Introduction. The figure *forty years* is an approximation; compare the same figure for David in 2.11.
12.1–33: The kingdom divided; Jeroboam's idolatry. 1–5: *Shechem* was at that time the chief town of the northern tribes (Josh 24.1, 32), where *Rehoboam* needed to be confirmed as *king* in order to hold the allegiance of these tribes. **11:** The word *scorpions* designates here stinging whips, much more cruel than ordinary *whips.*
12.12–15: *Rehoboam* takes the tragically wrong advice of *the young men.* The attitude of the writer in v. 15 is explained in 11.33.

Look now to your own house, O David."

So Israel went away to their tents. [17] But Rehoboam reigned over the Israelites who were living in the towns of Judah. [18] When King Rehoboam sent Adoram, who was taskmaster over the forced labor, all Israel stoned him to death. King Rehoboam then hurriedly mounted his chariot to flee to Jerusalem. [19] So Israel has been in rebellion against the house of David to this day.

20 When all Israel heard that Jeroboam had returned, they sent and called him to the assembly and made him king over all Israel. There was no one who followed the house of David, except the tribe of Judah alone.

21 When Rehoboam came to Jerusalem, he assembled all the house of Judah and the tribe of Benjamin, one hundred eighty thousand chosen troops to fight against the house of Israel, to restore the kingdom to Rehoboam son of Solomon. [22] But the word of God came to Shemaiah the man of God: [23] Say to King Rehoboam of Judah, son of Solomon, and to all the house of Judah and Benjamin, and to the rest of the people, [24] "Thus says the LORD, You shall not go up or fight against your kindred the people of Israel. Let everyone go home, for this thing is from me." So they heeded the word of the LORD and went home again, according to the word of the LORD.

25 Then Jeroboam built Shechem in the hill country of Ephraim, and resided there; he went out from there and built Penuel. [26] Then Jeroboam said to himself, "Now the kingdom may well revert to the house of David. [27] If this people continues to go up to offer sacrifices in the house of the LORD at Jerusalem, the heart of this people will turn again to their master, King Rehoboam of Judah; they will kill me and return to King Rehoboam of Judah." [28] So the king took counsel, and made two calves of gold. He said to the people, [a] "You have gone up to Jerusalem long enough. Here are your gods, O Israel, who brought you up out of the land of Egypt." [29] He set one in Bethel, and the other he put in Dan. [30] And this thing became a sin, for the people went to worship before the one at Bethel and before the other as far as Dan. [b] [31] He also made houses [c] on high places, and appointed priests from among all the people, who were not Levites. [32] Jeroboam appointed a festival on the fifteenth day of the eighth month like the festival that was in Judah, and he offered sacrifices on the altar; so he did in Bethel, sacrificing to the calves that he had made. And he placed in Bethel the priests of the high places that he had made. [33] He went up to the altar that he had made in Bethel on the fifteenth day in the eighth month, in the month that he alone had devised; he appointed a festival for the people of Israel, and he went up to the altar to offer incense.

13 While Jeroboam was standing by the altar to offer incense, a man of

---

a Gk: Heb *to them*   *b* Compare Gk: Heb *went to the one as far as Dan*   *c* Gk Vg Compare 13.32: Heb *a house*

12.16: On the readiness of the northern tribes to withdraw from Judah, compare 2 Sam 20.1, where even the wording is similar. **17–20.** Certain Israelites remained in *Judah;* and when *Rehoboam* attempted to enforce his oppressive policy in the north, *Jeroboam* was promptly made *king over all Israel. Judah* alone remained under *Rehoboam.* **21–24:** According to these verses, the prophetic party of *Judah,* represented by *Shemaiah,* favored the division, hoping that the north would be more loyal to the LORD (see 11.26–40 n.). *Benjamin* in vv. 21, 23 is editorial, to support the idea that there were exactly ten tribes in the north and two in the south. Actually, Benjamin may have been split in the division. **12.25–33:** *Jeroboam* straightway led his people into worse apostasy than that of the south. *Jeroboam* intended the *two calves of gold* (really bulls) to represent the LORD (*gods* in v. 28 should probably be "God"); according to the Deuteronomistic writer, this was very wrong, since the bull was also the symbol of Baal, and all images were forbidden (see 7.23–26 n.; Deut 5.8–9). Jeroboam also had a political purpose in attempting to keep his people from going to Jerusalem.

God came out of Judah by the word of the LORD to Bethel 2 and proclaimed against the altar by the word of the LORD, and said, "O altar, altar, thus says the LORD: 'A son shall be born to the house of David, Josiah by name; and he shall sacrifice on you the priests of the high places who offer incense on you, and human bones shall be burned on you.'" 3 He gave a sign the same day, saying, "This is the sign that the LORD has spoken: 'The altar shall be torn down, and the ashes that are on it shall be poured out.'" 4 When the king heard what the man of God cried out against the altar at Bethel, Jeroboam stretched out his hand from the altar, saying, "Seize him!" But the hand that he stretched out against him withered so that he could not draw it back to himself. 5 The altar also was torn down, and the ashes poured out from the altar, according to the sign that the man of God had given by the word of the LORD. 6 The king said to the man of God, "Entreat now the favor of the LORD your God, and pray for me, so that my hand may be restored to me." So the man of God entreated the LORD; and the king's hand was restored to him, and became as it was before. 7 Then the king said to the man of God, "Come home with me and dine, and I will give you a gift." 8 But the man of God said to the king, "If you give me half your kingdom, I will not go in with you; nor will I eat food or drink water in this place. 9 For thus I was commanded by the word of the LORD: You shall not eat food, or drink water, or return by the way that you came." 10 So he went another way, and did not return by the way that he had come to Bethel.

11 Now there lived an old prophet in Bethel. One of his sons came and told him all that the man of God had done that day in Bethel; the words also that he had spoken to the king, they told to their father. 12 Their father said to them, "Which way did he go?" And his sons showed him the way that the man of God who came from Judah had gone. 13 Then he said to his sons, "Saddle a donkey for me." So they saddled a donkey for him, and he mounted it. 14 He went after the man of God, and found him sitting under an oak tree. He said to him, "Are you the man of God who came from Judah?" He answered, "I am." 15 Then he said to him, "Come home with me and eat some food." 16 But he said, "I cannot return with you, or go in with you; nor will I eat food or drink water with you in this place; 17 for it was said to me by the word of the LORD: You shall not eat food or drink water there, or return by the way that you came." 18 Then the other*d* said to him, "I also am a prophet as you are, and an angel spoke to me by the word of the LORD: Bring him back with you into your house so that he may eat food and drink water." But he was deceiving him. 19 Then the man of God*d* went back with him, and ate food and drank water in his house.

20 As they were sitting at the table, the word of the LORD came to the prophet who had brought him back; 21 and he proclaimed to the man of God who came from Judah, "Thus says the LORD: Because you have disobeyed the word of the LORD, and have not kept the commandment that the LORD your God commanded you, 22 but have come back and have eaten food and drunk water in the place of which he said to you, 'Eat no food, and drink no water,' your body shall not come to your ancestral tomb."

d Heb *he*

13.1–34: **The prophets turn against Jeroboam.** This chapter continues the condemnation of Jeroboam begun in ch 12. To *eat food, or drink water* (vv. 9, 22) in Israel would have implied approval of what was happening there. There are two definite indications of date in the story: the territory of Israel was not called *Samaria* (v. 32) until after that kingdom fell in 721 B.C. (2 Kings 17.24); Josiah (v. 2) died in 609 B.C. The writer of about 600 B.C. is trying to evaluate the conditions existing about three hundred years earlier.

23 After the man of God[e] had eaten food and had drunk, they saddled for him a donkey belonging to the prophet who had brought him back. 24 Then as he went away, a lion met him on the road and killed him. His body was thrown in the road, and the donkey stood beside it; the lion also stood beside the body. 25 People passed by and saw the body thrown in the road, with the lion standing by the body. And they came and told it in the town where the old prophet lived.

26 When the prophet who had brought him back from the way heard of it, he said, "It is the man of God who disobeyed the word of the LORD; therefore the LORD has given him to the lion, which has torn him and killed him according to the word that the LORD spoke to him." 27 Then he said to his sons, "Saddle a donkey for me." So they saddled one, 28 and he went and found the body thrown in the road, with the donkey and the lion standing beside the body. The lion had not eaten the body or attacked the donkey. 29 The prophet took up the body of the man of God, laid it on the donkey, and brought it back to the city,[f] to mourn and to bury him. 30 He laid the body in his own grave; and they mourned over him, saying, "Alas, my brother!" 31 After he had buried him, he said to his sons, "When I die, bury me in the grave in which the man of God is buried; lay my bones beside his bones. 32 For the saying that he proclaimed by the word of the LORD against the altar in Bethel, and against all the houses of the high places that are in the cities of Samaria, shall surely come to pass."

33 Even after this event Jeroboam did not turn from his evil way, but made priests for the high places again from among all the people; any who wanted to be priests he consecrated for the high places. 34 This matter became sin to the house of Jeroboam, so as to cut it off and to destroy it from the face of the earth.

14 At that time Abijah son of Jeroboam fell sick. 2 Jeroboam said to his wife, "Go, disguise yourself, so that it will not be known that you are the wife of Jeroboam, and go to Shiloh; for the prophet Ahijah is there, who said of me that I should be king over this people. 3 Take with you ten loaves, some cakes, and a jar of honey, and go to him; he will tell you what shall happen to the child."

4 Jeroboam's wife did so; she set out and went to Shiloh, and came to the house of Ahijah. Now Ahijah could not see, for his eyes were dim because of his age. 5 But the LORD said to Ahijah, "The wife of Jeroboam is coming to inquire of you concerning her son; for he is sick. Thus and thus you shall say to her."

When she came, she pretended to be another woman. 6 But when Ahijah heard the sound of her feet, as she came in at the door, he said, "Come in, wife of Jeroboam; why do you pretend to be another? For I am charged with heavy tidings for you. 7 Go, tell Jeroboam, 'Thus says the LORD, the God of Israel: Because I exalted you from among the people, made you leader over my people Israel, 8 and tore the kingdom away from the house of David to give it to you; yet you have not been like my servant David, who kept my commandments and followed me with all his heart, doing only that which was right in my sight, 9 but you have done evil above all those who were before you and have gone and made

e Heb *he*   f Gk: Heb *he came to the town of the old prophet*

14.1–31: Conclusion of the reigns of Jeroboam and Rehoboam. From this point to the downfall of the northern kingdom (2 Kings 17.6), the writer skillfully weaves back and forth from one kingdom to the other in order to tell the story of both kingdoms as nearly continuously as possible. **1–16:** *Ahijah,* who had encouraged *Jeroboam* to revolt in the first place (11.29–31), now also turns against the king in bitter disappointment. The writer knew how short-lived was the *house* (dynasty) *of Jeroboam* (15.25–30) and how the whole northern kingdom fell in 721 B.C. (v. 15), and explained these somber events in terms of religious apostasy (vv. 9, 16).

for yourself other gods, and cast images, provoking me to anger, and have thrust me behind your back; 10 therefore, I will bring evil upon the house of Jeroboam. I will cut off from Jeroboam every male, both bond and free in Israel, and will consume the house of Jeroboam, just as one burns up dung until it is all gone. 11 Anyone belonging to Jeroboam who dies in the city, the dogs shall eat; and anyone who dies in the open country, the birds of the air shall eat; for the LORD has spoken.' 12 Therefore set out, go to your house. When your feet enter the city, the child shall die. 13 All Israel shall mourn for him and bury him; for he alone of Jeroboam's family shall come to the grave, because in him there is found something pleasing to the LORD, the God of Israel, in the house of Jeroboam. 14 Moreover the LORD will raise up for himself a king over Israel, who shall cut off the house of Jeroboam today, even right now! g

15 "The LORD will strike Israel, as a reed is shaken in the water; he will root up Israel out of this good land that he gave to their ancestors, and scatter them beyond the Euphrates, because they have made their sacred poles, h provoking the LORD to anger. 16 He will give Israel up because of the sins of Jeroboam, which he sinned and which he caused Israel to commit."

17 Then Jeroboam's wife got up and went away, and she came to Tirzah. As she came to the threshold of the house, the child died. 18 All Israel buried him and mourned for him, according to the word of the LORD, which he spoke by his servant the prophet Ahijah.

19 Now the rest of the acts of Jeroboam, how he warred and how he reigned, are written in the Book of the Annals of the Kings of Israel. 20 The time that Jeroboam reigned was twenty-two years;

then he slept with his ancestors, and his son Nadab succeeded him.

21 Now Rehoboam son of Solomon reigned in Judah. Rehoboam was forty-one years old when he began to reign, and he reigned seventeen years in Jerusalem, the city that the LORD had chosen out of all the tribes of Israel, to put his name there. His mother's name was Naamah the Ammonite. 22 Judah did what was evil in the sight of the LORD; they provoked him to jealousy with their sins that they committed, more than all that their ancestors had done. 23 For they also built for themselves high places, pillars, and sacred poles h on every high hill and under every green tree; 24 there were also male temple prostitutes in the land. They committed all the abominations of the nations that the LORD drove out before the people of Israel.

25 In the fifth year of King Rehoboam, King Shishak of Egypt came up against Jerusalem; 26 he took away the treasures of the house of the LORD and the treasures of the king's house; he took everything. He also took away all the shields of gold that Solomon had made; 27 so King Rehoboam made shields of bronze instead, and committed them to the hands of the officers of the guard, who kept the door of the king's house. 28 As often as the king went into the house of the LORD, the guard carried them and brought them back to the guardroom.

29 Now the rest of the acts of Rehoboam, and all that he did, are they not written in the Book of the Annals of the Kings of Judah? 30 There was war between Rehoboam and Jeroboam continually. 31 Rehoboam slept with his ancestors and was buried with his ancestors in the city of David. His mother's name

g Meaning of Heb uncertain   h Heb *Asherim*

14.15: *Sacred poles* were symbols of the Canaanite fertility goddess Asherah, who is mentioned about forty times in the Hebrew Scriptures as a temptation to the Israelites. **17**: *Tirzah* was *Jeroboam's* place of residence after Shechem (12.25), thus becoming the capital of Israel (15.33) until Samaria was built (16.24).

14.21–31: The reign of Rehoboam. Judah at this time did no better than Israel, according to the writer (vv. 22–24). According to Egyptian sources, *Shishak* (v. 25) also invad-

was Naamah the Ammonite. His son Abijam succeeded him.

15 Now in the eighteenth year of King Jeroboam son of Nebat, Abijam began to reign over Judah. 2He reigned for three years in Jerusalem. His mother's name was Maacah daughter of Abishalom. 3He committed all the sins that his father did before him; his heart was not true to the LORD his God, like the heart of his father David. 4Nevertheless for David's sake the LORD his God gave him a lamp in Jerusalem, setting up his son after him, and establishing Jerusalem; 5because David did what was right in the sight of the LORD, and did not turn aside from anything that he commanded him all the days of his life, except in the matter of Uriah the Hittite. 6The war begun between Rehoboam and Jeroboam continued all the days of his life. 7The rest of the acts of Abijam, and all that he did, are they not written in the Book of the Annals of the Kings of Judah? There was war between Abijam and Jeroboam. 8Abijam slept with his ancestors, and they buried him in the city of David. Then his son Asa succeeded him.

9 In the twentieth year of King Jeroboam of Israel, Asa began to reign over Judah; 10he reigned forty-one years in Jerusalem. His mother's name was Maacah daughter of Abishalom. 11Asa did what was right in the sight of the LORD, as his father David had done. 12He put away the male temple prostitutes out of the land, and removed all the idols that his ancestors had made. 13He also removed his mother Maacah from being queen mother, because she had made an abominable image for Asherah; Asa cut down her image and burned it at the Wadi Kidron. 14But the high places were not taken away. Nevertheless the heart of Asa was true to the LORD all his days.

15He brought into the house of the LORD the votive gifts of his father and his own votive gifts—silver, gold, and utensils.
16 There was war between Asa and King Baasha of Israel all their days. 17King Baasha of Israel went up against Judah, and built Ramah, to prevent anyone from going out or coming in to King Asa of Judah. 18Then Asa took all the silver and the gold that were left in the treasures of the house of the LORD and the treasures of the king's house, and gave them into the hands of his servants. King Asa sent them to King Ben-hadad son of Tabrimmon son of Hezion of Aram, who resided in Damascus, saying, 19"Let there be an alliance between me and you, like that between my father and your father: I am sending you a present of silver and gold; go, break your alliance with King Baasha of Israel, so that he may withdraw from me." 20Ben-hadad listened to King Asa, and sent the commanders of his armies against the cities of Israel. He conquered Ijon, Dan, Abel-beth-maacah, and all Chinneroth, with all the land of Naphtali. 21When Baasha heard of it, he stopped building Ramah and lived in Tirzah. 22Then King Asa made a proclamation to all Judah, none was exempt: they carried away the stones of Ramah and its timber, with which Baasha had been building; with them King Asa built Geba of Benjamin and Mizpah. 23Now the rest of all the acts of Asa, all his power, all that he did, and the cities that he built, are they not written in the Book of the Annals of the Kings of Judah? But in his old age he was diseased in his feet. 24Then Asa slept with his ancestors, and was buried with his ancestors in the city of his father David; his son Jehoshaphat succeeded him.

25 Nadab son of Jeroboam began to reign over Israel in the second year of

---

ed Israel; a fragment of his stela has been found in the excavations at Megiddo.

15.1–8: Abijam of Judah was little better than his father. Verse 6 is out of place; compare 14.30.

15.9–24: Asa of Judah. This king is cred-

ited with reforms and given a fairly clean bill of religious health. **13:** *Asherah*, see 14.15 n. **18–22:** Asa carried on continual war with Israel, and bribed *Ben-hadad . . . of Aram* (Syria) to change sides, thus tipping the balance in favor of Judah.

King Asa of Judah; he reigned over Israel two years. 26 He did what was evil in the sight of the LORD, walking in the way of his ancestor and in the sin that he caused Israel to commit.

27 Baasha son of Ahijah, of the house of Issachar, conspired against him; and Baasha struck him down at Gibbethon, which belonged to the Philistines; for Nadab and all Israel were laying siege to Gibbethon. 28 So Baasha killed Nadab[i] in the third year of King Asa of Judah, and succeeded him. 29 As soon as he was king, he killed all the house of Jeroboam; he left to the house of Jeroboam not one that breathed, until he had destroyed it, according to the word of the LORD that he spoke by his servant Ahijah the Shilonite— 30 because of the sins of Jeroboam that he committed and that he caused Israel to commit, and because of the anger to which he provoked the LORD, the God of Israel.

31 Now the rest of the acts of Nadab, and all that he did, are they not written in the Book of the Annals of the Kings of Israel? 32 There was war between Asa and King Baasha of Israel all their days.

33 In the third year of King Asa of Judah, Baasha son of Ahijah began to reign over all Israel at Tirzah; he reigned twenty-four years. 34 He did what was evil in the sight of the LORD, walking in the way of Jeroboam and in the sin that he caused Israel to commit.

16 The word of the LORD came to Jehu son of Hanani against Baasha, saying, 2 "Since I exalted you out of the dust and made you leader over my people Israel, and you have walked in the way of Jeroboam, and have caused my people Israel to sin, provoking me to anger with their sins, 3 therefore, I will consume Baasha and his house, and I will make your house like the house of Jeroboam son of Nebat. 4 Anyone belonging

to Baasha who dies in the city the dogs shall eat; and anyone of his who dies in the field the birds of the air shall eat."

5 Now the rest of the acts of Baasha, what he did, and his power, are they not written in the Book of the Annals of the Kings of Israel? 6 Baasha slept with his ancestors, and was buried at Tirzah; and his son Elah succeeded him. 7 Moreover the word of the LORD came by the prophet Jehu son of Hanani against Baasha and his house, both because of all the evil that he did in the sight of the LORD, provoking him to anger with the work of his hands, in being like the house of Jeroboam, and also because he destroyed it.

8 In the twenty-sixth year of King Asa of Judah, Elah son of Baasha began to reign over Israel in Tirzah; he reigned two years. 9 But his servant Zimri, commander of half his chariots, conspired against him. When he was at Tirzah, drinking himself drunk in the house of Arza, who was in charge of the palace at Tirzah, 10 Zimri came in and struck him down and killed him, in the twenty-seventh year of King Asa of Judah, and succeeded him.

11 When he began to reign, as soon as he had seated himself on his throne, he killed all the house of Baasha; he did not leave him a single male of his kindred or his friends. 12 Thus Zimri destroyed all the house of Baasha, according to the word of the LORD, which he spoke against Baasha by the prophet Jehu— 13 because of all the sins of Baasha and the sins of his son Elah that they committed, and that they caused Israel to commit, provoking the LORD God of Israel to anger with their idols. 14 Now the rest of the acts of Elah, and all that he did, are they not written in the Book of the Annals of the Kings of Israel?

i Heb *him*

15.25–32: **Nadab of Israel** and his assassination. Here begins a series of assassinations of kings of Israel, interpreted by the writer as a just judgment against the northern kingdom (vv. 29–30); see 14.1–16 n.

15.33–16.7: **Baasha of Israel.** A new

prophet, *Jehu son of Hanani,* takes up the theme of condemnation already made familiar by Ahijah the Shilonite.

16.8–14: **Elah of Israel** and his assassination by *Zimri.* Note the now familiar evaluation in v. 13.

15 In the twenty-seventh year of King Asa of Judah, Zimri reigned seven days in Tirzah. Now the troops were encamped against Gibbethon, which belonged to the Philistines, 16 and the troops who were encamped heard it said, "Zimri has conspired, and he has killed the king"; therefore all Israel made Omri, the commander of the army, king over Israel that day in the camp. 17 So Omri went up from Gibbethon, and all Israel with him, and they besieged Tirzah. 18 When Zimri saw that the city was taken, he went into the citadel of the king's house; he burned down the king's house over himself with fire, and died— 19 because of the sins that he committed, doing evil in the sight of the LORD, walking in the way of Jeroboam, and for the sin that he committed, causing Israel to sin. 20 Now the rest of the acts of Zimri, and the conspiracy that he made, are they not written in the Book of the Annals of the Kings of Israel?

21 Then the people of Israel were divided into two parts; half of the people followed Tibni son of Ginath, to make him king, and half followed Omri. 22 But the people who followed Omri overcame the people who followed Tibni son of Ginath; so Tibni died, and Omri became king. 23 In the thirty-first year of King Asa of Judah, Omri began to reign over Israel; he reigned for twelve years, six of them in Tirzah.

24 He bought the hill of Samaria from Shemer for two talents of silver; he fortified the hill, and called the city that he built, Samaria, after the name of Shemer, the owner of the hill.

25 Omri did what was evil in the sight of the LORD; he did more evil than all who were before him. 26 For he walked in all the way of Jeroboam son of Nebat, and in the sins that he caused Israel to commit, provoking the LORD, the God of Israel, to anger by their idols. 27 Now the rest of the acts of Omri that he did, and the power that he showed, are they not written in the Book of the Annals of the Kings of Israel? 28 Omri slept with his ancestors, and was buried in Samaria; his son Ahab succeeded him.

29 In the thirty-eighth year of King Asa of Judah, Ahab son of Omri began to reign over Israel; Ahab son of Omri reigned over Israel in Samaria twenty-two years. 30 Ahab son of Omri did evil in the sight of the LORD more than all who were before him.

31 And as if it had been a light thing for him to walk in the sins of Jeroboam son of Nebat, he took as his wife Jezebel daughter of King Ethbaal of the Sidonians, and went and served Baal, and worshiped him. 32 He erected an altar for Baal in the house of Baal, which he built in Samaria. 33 Ahab also made a sacred pole.[j] Ahab did more to provoke the anger of the LORD, the God of Israel, than had all the kings of Israel who were before him. 34 In his days Hiel of Bethel built Jericho; he laid its foundation at the cost of Abiram his firstborn, and set up its gates at the cost of his youngest son

j Heb *Asherah*

16.15–20: **Zimri of Israel** is attacked by *Omri* and commits suicide.
16.21–28: **Omri of Israel.** *Omri* was an able and important king, as we know from extra-biblical sources. The Moabite Stone tells us that he subjugated Moab, and the Assyrian records refer to Israel as "the land of the house of Omri" long after Omri's descendants had ceased to rule. The Deuteronomistic writer, interested chiefly in religious matters, lists among the accomplishments of Omri only the founding of Samaria as the new capital. Otherwise, the king is roundly condemned, no doubt chiefly because he married his son Ahab to Jezebel, the daughter of the king of the neighboring Phoenicians (*Sidonians*). This was good political policy, but it turned out to be religiously disastrous.
16.29–34: **Introduction to the reign of Ahab in Israel.** This reign is treated in great detail (17.1–22.40) because of the religious crisis that ensued. *The Sidonians,* see 5.1–6 n. *Sacred pole,* see 14.15 n. **34:** The bodies of the children were buried under the foundations to bring good luck to the building project. Such practices were regarded as abominable by the

Segub, according to the word of the LORD, which he spoke by Joshua son of Nun.

17 Now Elijah the Tishbite, of Tishbe[k] in Gilead, said to Ahab, "As the LORD the God of Israel lives, before whom I stand, there shall be neither dew nor rain these years, except by my word." 2 The word of the LORD came to him, saying, 3 "Go from here and turn eastward, and hide yourself by the Wadi Cherith, which is east of the Jordan. 4 You shall drink from the wadi, and I have commanded the ravens to feed you there." 5 So he went and did according to the word of the LORD; he went and lived by the Wadi Cherith, which is east of the Jordan. 6 The ravens brought him bread and meat in the morning, and bread and meat in the evening; and he drank from the wadi. 7 But after a while the wadi dried up, because there was no rain in the land.

8 Then the word of the LORD came to him, saying, 9 "Go now to Zarephath, which belongs to Sidon, and live there; for I have commanded a widow there to feed you." 10 So he set out and went to Zarephath. When he came to the gate of the town, a widow was there gathering sticks; he called to her and said, "Bring me a little water in a vessel, so that I may drink." 11 As she was going to bring it, he called to her and said, "Bring me a morsel of bread in your hand." 12 But she said, "As the LORD your God lives, I have nothing baked, only a handful of meal in a jar, and a little oil in a jug; I am now gathering a couple of sticks, so that I may go home and prepare it for myself and my son, that we may eat it, and die." 13 Elijah said to her, "Do not be afraid; go and do as you have said; but first make me a little cake of it and bring it to me, and afterwards make something for yourself and your son. 14 For thus says the LORD the God of Israel: The jar of meal will not be emptied and the jug of oil will not fail until the day that the LORD sends rain on the earth." 15 She went and did as Elijah said, so that she as well as he and her household ate for many days. 16 The jar of meal was not emptied, neither did the jug of oil fail, according to the word of the LORD that he spoke by Elijah.

17 After this the son of the woman, the mistress of the house, became ill; his illness was so severe that there was no breath left in him. 18 She then said to Elijah, "What have you against me, O man of God? You have come to me to bring my sin to remembrance, and to cause the death of my son!" 19 But he said to her, "Give me your son." He took him from her bosom, carried him up into the upper chamber where he was lodging, and laid him on his own bed. 20 He cried out to

k Gk: Heb *of the settlers*

Deuteronomists; hence the curse in Josh 6.26. The last clause of v. 34 brings the account into conformity with the passage in Joshua. **17.1–24: The beginning of the story of Elijah.** 1 Kings ch 17 to 2 Kings ch 10 is mostly a detailed account of affairs in the northern kingdom, featuring the prophets Elijah and Elisha, the reign of Ahab, and the destruction of the dynasty of Omri. These stories are based on northern source material, possibly brought to Judah by refugees from Israel. It must be remembered that all final redaction was done in Judah, and therefore we are fortunate in having this northern material. The element of the miraculous in the stories is an integral part of the writer's intent to dramatize the power of God through prophets' words and actions. **1–7:** The story opens abruptly, leading some to believe that the beginning, with a proper introduction of Elijah, has been lost. **1:** The Canaanite (or Phoenician) god Baal (16.31–32) was held by his worshipers to be the one who controlled the *rain.* Elijah intended to show that his God, *the LORD the God of Israel,* was the one who really controlled the *rain.* **3:** The words *east of the Jordan* (also v. 5) probably refer to territory outside of Ahab's jurisdiction. Ahab had been eagerly seeking to apprehend the prophet in order to find some reconciliation with him to end the drought (18.10). **17.8–16:** *Zarephath,* on the Phoenician coast, was definitely in territory beyond the control of Ahab. **17–24:** Some have argued that the child was not really dead, and hence that no miracle was involved. This is beside

the LORD, "O LORD my God, have you brought calamity even upon the widow with whom I am staying, by killing her son?" 21 Then he stretched himself upon the child three times, and cried out to the LORD, "O LORD my God, let this child's life come into him again." 22 The LORD listened to the voice of Elijah; the life of the child came into him again, and he revived. 23 Elijah took the child, brought him down from the upper chamber into the house, and gave him to his mother; then Elijah said, "See, your son is alive." 24 So the woman said to Elijah, "Now I know that you are a man of God, and that the word of the LORD in your mouth is truth."

18 After many days the word of the LORD came to Elijah, in the third year of the drought,¹ saying, "Go, present yourself to Ahab; I will send rain on the earth." 2 So Elijah went to present himself to Ahab. The famine was severe in Samaria. 3 Ahab summoned Obadiah, who was in charge of the palace. (Now Obadiah revered the LORD greatly; 4 when Jezebel was killing off the prophets of the LORD, Obadiah took a hundred prophets, hid them fifty to a cave, and provided them with bread and water.) 5 Then Ahab said to Obadiah, "Go through the land to all the springs of water and to all the wadis; perhaps we may find grass to keep the horses and mules alive, and not lose some of the animals." 6 So they divided the land between them to pass through it; Ahab went in one direction by himself, and Obadiah went in another direction by himself.

7 As Obadiah was on the way, Elijah met him; Obadiah recognized him, fell on his face, and said, "Is it you, my lord Elijah?" 8 He answered him, "It is I. Go, tell your lord that Elijah is here." 9 And he said, "How have I sinned, that you would hand your servant over to Ahab, to kill me? 10 As the LORD your God lives, there is no nation or kingdom to which my lord has not sent to seek you; and when they would say, 'He is not here,' he would require an oath of the kingdom or nation, that they had not found you. 11 But now you say, 'Go, tell your lord that Elijah is here.' 12 As soon as I have gone from you, the spirit of the LORD will carry you I know not where; so, when I come and tell Ahab and he cannot find you, he will kill me, although I your servant have revered the LORD from my youth. 13 Has it not been told my lord what I did when Jezebel killed the prophets of the LORD, how I hid a hundred of the LORD's prophets fifty to a cave, and provided them with bread and water? 14 Yet now you say, 'Go, tell your lord that Elijah is here'; he will surely kill me." 15 Elijah said, "As the LORD of hosts lives, before whom I stand, I will surely show myself to him today." 16 So Obadiah went to meet Ahab, and told him; and Ahab went to meet Elijah.

17 When Ahab saw Elijah, Ahab said to him, "Is it you, you troubler of Israel?" 18 He answered, "I have not troubled Israel; but you have, and your father's house, because you have forsaken the commandments of the LORD and followed the Baals. 19 Now therefore have all Israel assemble for me at Mount Carmel, with the four hundred fifty proph-

---

1 Heb lacks *of the drought*

---

the point. The writer meant to portray a powerful God and a worthy prophet (compare 2 Kings 4.32–37; Acts 20.9–12).

18.1–46: The contest on Mount Carmel. 1–6: The drought is about to end. The question is, Which God withholds and sends the rain: Yahweh (the LORD) or the great Baal, called Baal of the Heavens by his followers? The very name of *Elijah,* meaning "Yah [short form of Yahweh, or the LORD] is God," proclaims his faith. *Obadiah* means "Servant of Yahweh." *Jezebel* has been per-

secuting *the prophets of the LORD* (Yahweh). **18.7–16:** *Ahab* has been hunting *Elijah* everywhere, but now the prophet volunteers to meet the king (see 17.1–7 n.). **17–19:** *Elijah* is quick to point out that the fault lies with *Ahab* in not recognizing the LORD as the supreme God, and in allowing his wife *Jezebel* to propagate her religion in Israel. *Asherah,* one of the consorts of *Baal* (see 14.15 n.). *The Baals* may have been local versions of the great sky-god, *Baal;* thus there was one *Baal,* but also many *Baals.*

ets of Baal and the four hundred prophets of Asherah, who eat at Jezebel's table."

20 So Ahab sent to all the Israelites, and assembled the prophets at Mount Carmel. 21 Elijah then came near to all the people, and said, "How long will you go limping with two different opinions? If the LORD is God, follow him; but if Baal, then follow him." The people did not answer him a word. 22 Then Elijah said to the people, "I, even I only, am left a prophet of the LORD; but Baal's prophets number four hundred fifty. 23 Let two bulls be given to us; let them choose one bull for themselves, cut it in pieces, and lay it on the wood, but put no fire to it; I will prepare the other bull and lay it on the wood, but put no fire to it. 24 Then you call on the name of your god and I will call on the name of the LORD; the god who answers by fire is indeed God." All the people answered, "Well spoken!" 25 Then Elijah said to the prophets of Baal, "Choose for yourselves one bull and prepare it first, for you are many; then call on the name of your god, but put no fire to it." 26 So they took the bull that was given them, prepared it, and called on the name of Baal from morning until noon, crying, "O Baal, answer us!" But there was no voice, and no answer. They limped about the altar that they had made. 27 At noon Elijah mocked them, saying, "Cry aloud! Surely he is a god; either he is meditating, or he has wandered away, or he is on a journey, or perhaps he is asleep and must be awakened." 28 Then they cried aloud and, as was their custom, they cut themselves with swords and lances until the blood gushed out over them. 29 As midday passed, they raved on until the time of the offering of the oblation, but there was no voice, no answer, and no response.

30 Then Elijah said to all the people, "Come closer to me"; and all the people came closer to him. First he repaired the altar of the LORD that had been thrown down; 31 Elijah took twelve stones, according to the number of the tribes of the sons of Jacob, to whom the word of the LORD came, saying, "Israel shall be your name"; 32 with the stones he built an altar in the name of the LORD. Then he made a trench around the altar, large enough to contain two measures of seed. 33 Next he put the wood in order, cut the bull in pieces, and laid it on the wood. He said, "Fill four jars with water and pour it on the burnt offering and on the wood." 34 Then he said, "Do it a second time"; and they did it a second time. Again he said, "Do it a third time"; and they did it a third time, 35 so that the water ran all around the altar, and filled the trench also with water.

36 At the time of the offering of the oblation, the prophet Elijah came near and said, "O LORD, God of Abraham, Isaac, and Israel, let it be known this day that you are God in Israel, that I am your servant, and that I have done all these things at your bidding. 37 Answer me, O LORD, answer me, so that this people may know that you, O LORD, are God, and that you have turned their hearts back." 38 Then the fire of the LORD fell

---

18.20–29: The contest begins; the Baal worshipers test their god first. **26:** *They limped about the altar,* i.e. they performed a kind of limping dance, bending first one knee and then the other. This form of ritual is well known from a number of sources (compare Ps 26.6–7). **27:** One of the sharpest satires on non-Hebraic religion ever penned. *He has wandered away* is probably a euphemism for attending to natural needs. **28:** Ritualistic gashing of one's self was fairly common (compare Deut 14.1; Lev 19.28; Hos 7.14; Jer 16.6; 41.5; 47.5). **29:** *The time of the offering of the oblation* was 3 p.m. (compare 2 Kings 16.15; Acts 3.1).

18.30–35: Verses 31–32a are based on Gen 35.10. **38:** Some wish to rationalize *the fire of the LORD* by calling it lightning preceding the rain; but it must be borne in mind that the ancient writer intended to describe a miracle. **40:** The slaughter of *the prophets of Baal* is sometimes interpreted as a vast human sacrifice to the LORD. The people of the time and the (slightly later) writer saw the struggle between Baal and the LORD as one of life and death.

and consumed the burnt offering, the wood, the stones, and the dust, and even licked up the water that was in the trench. 39 When all the people saw it, they fell on their faces and said, "The LORD indeed is God; the LORD indeed is God." 40 Elijah said to them, "Seize the prophets of Baal; do not let one of them escape." Then they seized them; and Elijah brought them down to the Wadi Kishon, and killed them there.

41 Elijah said to Ahab, "Go up, eat and drink; for there is a sound of rushing rain." 42 So Ahab went up to eat and to drink. Elijah went up to the top of Carmel; there he bowed himself down upon the earth and put his face between his knees. 43 He said to his servant, "Go up now, look toward the sea." He went up and looked, and said, "There is nothing." Then he said, "Go again seven times." 44 At the seventh time he said, "Look, a little cloud no bigger than a person's hand is rising out of the sea." Then he said, "Go say to Ahab, 'Harness your chariot and go down before the rain stops you.'" 45 In a little while the heavens grew black with clouds and wind; there was a heavy rain. Ahab rode off and went to Jezreel. 46 But the hand of the LORD was on Elijah; he girded up his loins and ran in front of Ahab to the entrance of Jezreel.

19 Ahab told Jezebel all that Elijah had done, and how he had killed all the prophets with the sword. 2 Then Jezebel sent a messenger to Elijah, saying, "So may the gods do to me, and more also, if I do not make your life like the life of one of them by this time tomorrow." 3 Then he was afraid; he got up and fled for his life, and came to Beersheba, which belongs to Judah; he left his servant there.

4 But he himself went a day's journey into the wilderness, and came and sat down under a solitary broom tree. He asked that he might die: "It is enough; now, O LORD, take away my life, for I am no better than my ancestors." 5 Then he lay down under the broom tree and fell asleep. Suddenly an angel touched him and said to him, "Get up and eat." 6 He looked, and there at his head was a cake baked on hot stones, and a jar of water. He ate and drank, and lay down again. 7 The angel of the LORD came a second time, touched him, and said, "Get up and eat, otherwise the journey will be too much for you." 8 He got up, and ate and drank; then he went in the strength of that food forty days and forty nights to Horeb the mount of God. 9 At that place he came to a cave, and spent the night there.

Then the word of the LORD came to him, saying, "What are you doing here, Elijah?" 10 He answered, "I have been very zealous for the LORD, the God of hosts; for the Israelites have forsaken your covenant, thrown down your altars, and killed your prophets with the sword. I alone am left, and they are seeking my life, to take it away."

11 He said, "Go out and stand on the mountain before the LORD, for the LORD

18.41–46: *Eat and drink,* i.e. break the fast which was in effect during the religious ceremonies. The end of the drought was supposed to be final proof that the LORD, not Baal, ruled the heavens, and controlled the destinies of the Israelites. The town of *Jezreel,* near Mount Gilboa (1 Sam 29.1; 31.1), was used by *Ahab* as a second place of residence (21.1), the primary capital being Samaria (16.24; 20.43; 21.1). *Elijah* ran seventeen miles before the *chariot* of *Ahab,* in a high state of excitement, to herald what he considered a complete victory over the forces of Baal. 19.1–18: **The revelation to Elijah on Mount Horeb. 1–3**: The power of *Jezebel*

very quickly turned the tables, and Elijah *was afraid.* Suddenly, it seems, he is at *Beer-sheba,* one hundred and thirty miles south of Jezreel, well within *Judah.* 4–8: By miraculous divine help, the prophet arrived at *Horeb,* the place where the LORD revealed the law to Moses, according to northern (Israelite) tradition— called Sinai in the southern (Judahite) literature. The round number, *forty days,* simply indicates a great distance (compare Ex 34.28). The traditional site of Mount Horeb, probably not intended here, is some two hundred miles to the south. 19.9–18: Compare Ex 33.17–23. 12: The *sound of sheer silence* was the demanding voice

is about to pass by." Now there was a great wind, so strong that it was splitting mountains and breaking rocks in pieces before the Lord, but the Lord was not in the wind; and after the wind an earthquake, but the Lord was not in the earthquake; 12 and after the earthquake a fire, but the Lord was not in the fire; and after the fire a sound of sheer silence. 13 When Elijah heard it, he wrapped his face in his mantle and went out and stood at the entrance of the cave. Then there came a voice to him that said, "What are you doing here, Elijah?" 14 He answered, "I have been very zealous for the Lord, the God of hosts; for the Israelites have forsaken your covenant, thrown down your altars, and killed your prophets with the sword. I alone am left, and they are seeking my life, to take it away." 15 Then the Lord said to him, "Go, return on your way to the wilderness of Damascus; when you arrive, you shall anoint Hazael as king over Aram. 16 Also you shall anoint Jehu son of Nimshi as king over Israel; and you shall anoint Elisha son of Shaphat of Abel-meholah as prophet in your place. 17 Whoever escapes from the sword of Hazael, Jehu shall kill; and whoever escapes from the sword of Jehu, Elisha shall kill. 18 Yet I will leave seven thousand in Israel, all the knees that have not bowed to Baal, and every mouth that has not kissed him."

19 So he set out from there, and found Elisha son of Shaphat, who was plowing. There were twelve yoke of oxen ahead of him, and he was with the twelfth. Elijah passed by him and threw his mantle over him. 20 He left the oxen, ran after Elijah, and said, "Let me kiss my father and my mother, and then I will follow you." Then Elijah*m* said to him, "Go back again; for what have I done to you?" 21 He returned from following him, took the yoke of oxen, and slaughtered them; using the equipment from the oxen, he boiled their flesh, and gave it to the people, and they ate. Then he set out and followed Elijah, and became his servant.

20 King Ben-hadad of Aram gathered all his army together; thirty-two kings were with him, along with horses and chariots. He marched against Samaria, laid siege to it, and attacked it. 2 Then he sent messengers into the city to King Ahab of Israel, and said to him: "Thus says Ben-hadad: 3 Your silver and gold are mine; your fairest wives and children also are mine." 4 The king of Israel answered, "As you say, my lord, O king, I am yours, and all that I have." 5 The messengers came again and said: "Thus says Ben-hadad: I sent to you, saying, 'Deliver to me your silver and gold, your wives and children'; 6 nevertheless I will send my servants to you tomorrow about this time, and they shall search your house and the houses of your servants, and lay hands on whatever pleases them,*n* and take it away."

7 Then the king of Israel called all the elders of the land, and said, "Look now!

m Heb *he*   *n* Gk Syr Vg: Heb *you*

of God. Elijah was willing to oblige, for he was *very zealous*. **15**: *The wilderness of Damascus,* the present-day Syrian Desert. **18**: The words *seven thousand* express the idea of the righteous remnant, which appears again in Am 5.15; Isa 10.20; 11.11. *Kissed* the Baal image, see Hos 13.2.

19.19–21: **The call of Elisha.** *Elijah* returns from Horeb and proceeds to carry out first the third command of the "sound of sheer silence." He did not actually "anoint" *Elisha* (v. 16), but *threw his mantle over him.* The end of v. 20 means, "Go, and return to me, for I have done something very important to you." For similar extemporaneous

sacrifices (v. 21), compare 1 Sam 6.14; 2 Sam 24.22–25. It was *Elisha,* not *Elijah,* who carried out the first and second injunctions of the "sound" (2 Kings 8.13; 9.1–3).

20.1–43: **Ahab's wars with the Arameans; prophetic opposition to his policy.** Verses 1–34 come from a source favorable to *Ahab,* in sharp contrast to vv. 35–43 and to chs 17–19 and ch 21; Elijah and Elisha disappear for a moment. *Ahab* appears as an able and popular ruler; his *people* stand by him in adversity (v. 8), and the prophets of the Lord support him (vv. 13–14, 22); to a defeated enemy he shows noble generosity (vv. 32, 34). **1–6**: *Kings* here no doubt means, as often

See how this man is seeking trouble; for he sent to me for my wives, my children, my silver, and my gold; and I did not refuse him." 8 Then all the elders and all the people said to him, "Do not listen or consent." 9 So he said to the messengers of Ben-hadad, "Tell my lord the king: All that you first demanded of your servant I will do; but this thing I cannot do." The messengers left and brought him word again. 10 Ben-hadad sent to him and said, "The gods do so to me, and more also, if the dust of Samaria will provide a handful for each of the people who follow me." 11 The king of Israel answered, "Tell him: One who puts on armor should not brag like one who takes it off." 12 When Ben-hadad heard this message—now he had been drinking with the kings in the booths—he said to his men, "Take your positions!" And they took their positions against the city.

13 Then a certain prophet came up to King Ahab of Israel and said, "Thus says the LORD, Have you seen all this great multitude? Look, I will give it into your hand today; and you shall know that I am the LORD." 14 Ahab said, "By whom?" He said, "Thus says the LORD, By the young men who serve the district governors." Then he said, "Who shall begin the battle?" He answered, "You." 15 Then he mustered the young men who serve the district governors, two hundred thirty-two; after them he mustered all the people of Israel, seven thousand.

16 They went out at noon, while Ben-hadad was drinking himself drunk in the booths, he and the thirty-two kings allied with him. 17 The young men who serve the district governors went out first. Ben-hadad had sent out scouts,° and they reported to him, "Men have come out from Samaria." 18 He said, "If they have come out for peace, take them alive; if they have come out for war, take them alive."

19 But these had already come out of the city: the young men who serve the district governors, and the army that followed them. 20 Each killed his man; the Arameans fled and Israel pursued them, but King Ben-hadad of Aram escaped on a horse with the cavalry. 21 The king of Israel went out, attacked the horses and chariots, and defeated the Arameans with a great slaughter.

22 Then the prophet approached the king of Israel and said to him, "Come, strengthen yourself, and consider well what you have to do; for in the spring the king of Aram will come up against you."

23 The servants of the king of Aram said to him, "Their gods are gods of the hills, and so they were stronger than we; but let us fight against them in the plain, and surely we shall be stronger than they. 24 Also do this: remove the kings, each from his post, and put commanders in place of them; 25 and muster an army like the army that you have lost, horse for horse, and chariot for chariot; then we will fight against them in the plain, and surely we shall be stronger than they." He heeded their voice, and did so.

26 In the spring Ben-hadad mustered the Arameans and went up to Aphek to fight against Israel. 27 After the Israelites had been mustered and provisioned, they went out to engage them; the people of Israel encamped opposite them like two little flocks of goats, while the Arameans filled the country. 28 A man of God approached and said to the king of Israel, "Thus says the LORD: Because the Arameans have said, 'The LORD is a god of the hills but he is not a god of the valleys,' therefore I will give all this great

° Heb lacks *scouts*

---

in the Bible, rulers of small independent towns. The second demand (v. 6) was more drastic even than the first (v. 3), amounting to the privilege of unlimited looting.
20.11: Ahab quotes a proverb to the effect that as long as there is a battle to be fought, one should not boast; the battle may go either way.
20.24: *The kings* were figureheads, the *commanders* experienced military men. **29**: The

multitude into your hand, and you shall know that I am the LORD." 29 They encamped opposite one another seven days. Then on the seventh day the battle began; the Israelites killed one hundred thousand Aramean foot soldiers in one day. 30 The rest fled into the city of Aphek; and the wall fell on twenty-seven thousand men that were left.

Ben-hadad also fled, and entered the city to hide. 31 His servants said to him, "Look, we have heard that the kings of the house of Israel are merciful kings; let us put sackcloth around our waists and ropes on our heads, and go out to the king of Israel; perhaps he will spare your life." 32 So they tied sackcloth around their waists, put ropes on their heads, went to the king of Israel, and said, "Your servant Ben-hadad says, 'Please let me live.'" And he said, "Is he still alive? He is my brother." 33 Now the men were watching for an omen; they quickly took it up from him and said, "Yes, Ben-hadad is your brother." Then he said, "Go and bring him." So Ben-hadad came out to him; and he had him come up into the chariot. 34 Ben-hadad*p* said to him, "I will restore the towns that my father took from your father; and you may establish bazaars for yourself in Damascus, as my father did in Samaria." The king of Israel responded,*q* "I will let you go on those terms." So he made a treaty with him and let him go.

35 At the command of the LORD a certain member of a company of prophets*r* said to another, "Strike me!" But the man refused to strike him. 36 Then he said to him, "Because you have not obeyed the voice of the LORD, as soon as you have left me, a lion will kill you." And when he had left him, a lion met him and

killed him. 37 Then he found another man and said, "Strike me!" So the man hit him, striking and wounding him. 38 Then the prophet departed, and waited for the king along the road, disguising himself with a bandage over his eyes. 39 As the king passed by, he cried to the king and said, "Your servant went out into the thick of the battle; then a soldier turned and brought a man to me, and said, 'Guard this man; if he is missing, your life shall be given for his life, or else you shall pay a talent of silver.' 40 While your servant was busy here and there, he was gone." The king of Israel said to him, "So shall your judgment be; you yourself have decided it." 41 Then he quickly took the bandage away from his eyes. The king of Israel recognized him as one of the prophets. 42 Then he said to him, "Thus says the LORD, 'Because you have let the man go whom I had devoted to destruction, therefore your life shall be for his life, and your people for his people.'" 43 The king of Israel set out toward home, resentful and sullen, and came to Samaria.

21 Later the following events took place: Naboth the Jezreelite had a vineyard in Jezreel, beside the palace of King Ahab of Samaria. 2 And Ahab said to Naboth, "Give me your vineyard, so that I may have it for a vegetable garden, because it is near my house; I will give you a better vineyard for it; or, if it seems good to you, I will give you its value in money." 3 But Naboth said to Ahab, "The LORD forbid that I should give you my ancestral inheritance." 4 Ahab went home resentful and sullen because of

---

p Heb *He responded*   q Heb lacks *The king of Israel responded*   r Heb *of the sons of the prophets*

---

figure *one hundred thousand Aramean foot soldiers* is not to be taken numerically; the meaning is "a large number."

20.35–43: The change of tone in this section should be noted. Ahab is violently criticized for what seems a noble act in v. 34. **39:** *A talent of silver,* roughly seventy-five lbs., would have been an enormous fine in those days. **42:** On the ban or "herem," a practice

whereby something or someone is *devoted to destruction* as a religious act, see Josh 6.17 n.; 1 Sam 15.3 n.

21.1–29: Naboth's vineyard. In this chapter Elijah and the source concerned with him return. **1–4:** *Naboth* could not give up his *ancestral inheritance* because by well-established legal and religious custom, ancestral property must remain in the family in

what Naboth the Jezreelite had said to him; for he had said, "I will not give you my ancestral inheritance." He lay down on his bed, turned away his face, and would not eat.

5 His wife Jezebel came to him and said, "Why are you so depressed that you will not eat?" 6 He said to her, "Because I spoke to Naboth the Jezreelite and said to him, 'Give me your vineyard for money; or else, if you prefer, I will give you another vineyard for it'; but he answered, 'I will not give you my vineyard.'" 7 His wife Jezebel said to him, "Do you now govern Israel? Get up, eat some food, and be cheerful; I will give you the vineyard of Naboth the Jezreelite."

8 So she wrote letters in Ahab's name and sealed them with his seal; she sent the letters to the elders and the nobles who lived with Naboth in his city. 9 She wrote in the letters, "Proclaim a fast, and seat Naboth at the head of the assembly; 10 seat two scoundrels opposite him, and have them bring a charge against him, saying, 'You have cursed God and the king.' Then take him out, and stone him to death." 11 The men of his city, the elders and the nobles who lived in his city, did as Jezebel had sent word to them. Just as it was written in the letters that she had sent to them, 12 they proclaimed a fast and seated Naboth at the head of the assembly. 13 The two scoundrels came in and sat opposite him; and the scoundrels brought a charge against Naboth, in the presence of the people, saying, "Naboth cursed God and the king." So they took him outside the city, and stoned him to death. 14 Then they sent to Jezebel, saying, "Naboth has been stoned; he is dead."

15 As soon as Jezebel heard that Naboth had been stoned and was dead, Jezebel said to Ahab, "Go, take possession of the vineyard of Naboth the Jezreelite, which he refused to give you for money; for Naboth is not alive, but dead." 16 As soon as Ahab heard that Naboth was dead, Ahab set out to go down to the vineyard of Naboth the Jezreelite, to take possession of it.

17 Then the word of the LORD came to Elijah the Tishbite, saying: 18 Go down to meet King Ahab of Israel, who rules[s] in Samaria; he is now in the vineyard of Naboth, where he has gone to take possession. 19 You shall say to him, "Thus says the LORD: Have you killed, and also taken possession?" You shall say to him, "Thus says the LORD: In the place where dogs licked up the blood of Naboth, dogs will also lick up your blood."

20 Ahab said to Elijah, "Have you found me, O my enemy?" He answered, "I have found you. Because you have sold yourself to do what is evil in the sight of the LORD, 21 I will bring disaster on you; I will consume you, and will cut off from Ahab every male, bond or free, in Israel; 22 and I will make your house like the house of Jeroboam son of Nebat, and like the house of Baasha son of Ahijah, because you have provoked me to anger and have caused Israel to sin. 23 Also concerning Jezebel the LORD said, 'The dogs shall eat Jezebel within the bounds of Jezreel.' 24 Anyone belonging to Ahab who dies in the city the dogs shall eat; and anyone of his who dies in the open country the birds of the air shall eat."

25 (Indeed, there was no one like

_s_ Heb _who is_

perpetuity (Lev 25.10, 13–17, 23–24, 34). Ahab was *resentful and sullen* because he knew that *Naboth* was legally and religiously right. **5–7**: *Jezebel* had no real respect for the laws and religion of Israel.
21.8–14: For *a fast* as part of the solemn assembly when a serious problem was to be considered, see Judg 20.26; 1 Sam 7.6; 14.24. To *seat Naboth at the head of the assembly* prob-

ably means that he, as a prominent citizen, was to preside. *Two* witnesses and *death* by stoning, see Deut 17.5–6; 19.15. *Naboth cursed God and the king,* compare Ex 22.28; Lev 24.10–16.
21.22–24: Probably an expansion of Elijah's speech by the Deuteronomistic editor (compare 13.33–34; 14.10–11; 16.1–4; 2 Kings 9.35–36). **25–26**: The evaluation by

Ahab, who sold himself to do what was evil in the sight of the LORD, urged on by his wife Jezebel. 26 He acted most abominably in going after idols, as the Amorites had done, whom the LORD drove out before the Israelites.)

27 When Ahab heard those words, he tore his clothes and put sackcloth over his bare flesh; he fasted, lay in the sackcloth, and went about dejectedly. 28 Then the word of the LORD came to Elijah the Tishbite: 29 "Have you seen how Ahab has humbled himself before me? Because he has humbled himself before me, I will not bring the disaster in his days; but in his son's days I will bring the disaster on his house."

22 For three years Aram and Israel continued without war. 2 But in the third year King Jehoshaphat of Judah came down to the king of Israel. 3 The king of Israel said to his servants, "Do you know that Ramoth-gilead belongs to us, yet we are doing nothing to take it out of the hand of the king of Aram?" 4 He said to Jehoshaphat, "Will you go with me to battle at Ramoth-gilead?" Jehoshaphat replied to the king of Israel, "I am as you are; my people are your people, my horses are your horses."

5 But Jehoshaphat also said to the king of Israel, "Inquire first for the word of the LORD." 6 Then the king of Israel gathered the prophets together, about four hundred of them, and said to them, "Shall I go to battle against Ramoth-gilead, or shall I refrain?" They said, "Go up; for the LORD will give it into the hand of the king." 7 But Jehoshaphat said, "Is there no other prophet of the LORD here of whom we may inquire?" 8 The king of Israel said to Jehoshaphat, "There is still one other by whom we may inquire of the LORD, Micaiah son of Imlah; but I hate him, for he never prophesies anything favorable about me, but only disaster." Jehoshaphat said, "Let the king not say such a thing." 9 Then the king of Israel summoned an officer and said, "Bring quickly Micaiah son of Imlah." 10 Now the king of Israel and King Jehoshaphat of Judah were sitting on their thrones, arrayed in their robes, at the threshing floor at the entrance of the gate of Samaria; and all the prophets were prophesying before them. 11 Zedekiah son of Chenaanah made for himself horns of iron, and he said, "Thus says the LORD: With these you shall gore the Arameans until they are destroyed." 12 All the prophets were prophesying the same and saying, "Go up to Ramoth-gilead and triumph; the LORD will give it into the hand of the king."

13 The messenger who had gone to summon Micaiah said to him, "Look, the words of the prophets with one accord are favorable to the king; let your word be like the word of one of them, and speak favorably." 14 But Micaiah said, "As the LORD lives, whatever the LORD says to me, that I will speak."

15 When he had come to the king, the king said to him, "Micaiah, shall we go

---

the writer, breaking the connection between vv. 24 and 27. *Amorites,* see 2 Sam 21.2 n. **29:** This verse, in the light of what happened, appraises the situation more realistically than vv. 21 and 24; it was perhaps added by a writer during the Exile.

22.1–53: Ahab dies in battle; reign of Jehoshaphat in Judah; accession of Ahaziah in Israel. Verses 1–40 resume the story of the Aramean wars from ch 20. **1–4:** During the *three years* of peace between themselves, *Aram and Israel* had formed a military alliance and had successfully fought, together with other allies, against the invading Assyrians at Qarqar (853 B.C.), thus preserving both king-doms. Ahab now formed an alliance with *Jehoshaphat,* marrying his daughter Athaliah to the son of *the king of Judah* (2 Kings 8.18, 26); for he had quarreled with *the king of Aram* over the possession of the town of *Ramoth-gilead,* east of the Jordan (4.13; Deut 4.43).

22.5–12: The need to *inquire* of one's god before a battle was widely felt in ancient times (1 Sam 23.2). In this source *the prophets* of the LORD are mostly on the side of Ahab (compare 20.13, 28). Yet there is one *prophet of the LORD,* not Elijah, but *Micaiah,* who appears only here and is hostile. *Horns of iron,* compare Deut 33.17; Zech 1.18–21. **15:** *Micaiah* mimics the supporters of *the king.*

to Ramoth-gilead to battle, or shall we refrain?" He answered him, "Go up and triumph; the LORD will give it into the hand of the king." 16 But the king said to him, "How many times must I make you swear to tell me nothing but the truth in the name of the LORD?" 17 Then Micaiah[t] said, "I saw all Israel scattered on the mountains, like sheep that have no shepherd; and the LORD said, 'These have no master; let each one go home in peace.' " 18 The king of Israel said to Jehoshaphat, "Did I not tell you that he would not prophesy anything favorable about me, but only disaster?"

19 Then Micaiah[t] said, "Therefore hear the word of the LORD: I saw the LORD sitting on his throne, with all the host of heaven standing beside him to the right and to the left of him. 20 And the LORD said, 'Who will entice Ahab, so that he may go up and fall at Ramoth-gilead?' Then one said one thing, and another said another, 21 until a spirit came forward and stood before the LORD, saying, 'I will entice him.' 22 'How?' the LORD asked him. He replied, 'I will go out and be a lying spirit in the mouth of all his prophets.' Then the LORD[t] said, 'You are to entice him, and you shall succeed; go out and do it.' 23 So you see, the LORD has put a lying spirit in the mouth of all these your prophets; the LORD has decreed disaster for you."

24 Then Zedekiah son of Chenaanah came up to Micaiah, slapped him on the cheek, and said, "Which way did the spirit of the LORD pass from me to speak to you?" 25 Micaiah replied, "You will find out on that day when you go in to hide in an inner chamber." 26 The king of Israel then ordered, "Take Micaiah, and return him to Amon the governor of the city and to Joash the king's son, 27 and

say, 'Thus says the king: Put this fellow in prison, and feed him on reduced rations of bread and water until I come in peace.' " 28 Micaiah said, "If you return in peace, the LORD has not spoken by me." And he said, "Hear, you peoples, all of you!"

29 So the king of Israel and King Jehoshaphat of Judah went up to Ramoth-gilead. 30 The king of Israel said to Jehoshaphat, "I will disguise myself and go into battle, but you wear your robes." So the king of Israel disguised himself and went into battle. 31 Now the king of Aram had commanded the thirty-two captains of his chariots, "Fight with no one small or great, but only with the king of Israel." 32 When the captains of the chariots saw Jehoshaphat, they said, "It is surely the king of Israel." So they turned to fight against him; and Jehoshaphat cried out. 33 When the captains of the chariots saw that it was not the king of Israel, they turned back from pursuing him. 34 But a certain man drew his bow and unknowingly struck the king of Israel between the scale armor and the breastplate; so he said to the driver of his chariot, "Turn around, and carry me out of the battle, for I am wounded." 35 The battle grew hot that day, and the king was propped up in his chariot facing the Arameans, until at evening he died; the blood from the wound had flowed into the bottom of the chariot. 36 Then about sunset a shout went through the army, "Every man to his city, and every man to his country!"

37 So the king died, and was brought to Samaria; they buried the king in Samaria. 38 They washed the chariot by the pool of Samaria; the dogs licked up his

t Heb *he*

22.19–23: The celestial scene portrayed here is unique in the earlier literature of Israel. The *lying spirit,* here still under the control of the LORD; compare the "evil spirit" of 1 Sam 16.14, and Satan in Zech 3.1–2; Job chs 1–2. 26: Ahab put the prisoner in charge of *the king's son* (one of his own sons) for maximum security. 27–28: *In peace,* i.e. "victorious."

22.30: The *disguise* shows that Ahab was fearful that Micaiah might be right. 31: *Thirty-two captains,* compare 20.1, 24. 22.38: This verse is editorial, with reference to 21.19, though Naboth died in Jezreel. The addition of *the prostitutes* symbolizes the fertility cult, a prominent feature of the Baal religion that included ritual sexual inter-

blood, and the prostitutes washed themselves in it, *u* according to the word of the LORD that he had spoken. ³⁹ Now the rest of the acts of Ahab, and all that he did, and the ivory house that he built, and all the cities that he built, are they not written in the Book of the Annals of the Kings of Israel? ⁴⁰ So Ahab slept with his ancestors; and his son Ahaziah succeeded him.

41 Jehoshaphat son of Asa began to reign over Judah in the fourth year of King Ahab of Israel. ⁴² Jehoshaphat was thirty-five years old when he began to reign, and he reigned twenty-five years in Jerusalem. His mother's name was Azubah daughter of Shilhi. ⁴³ He walked in all the way of his father Asa; he did not turn aside from it, doing what was right in the sight of the LORD; yet the high places were not taken away, and the people still sacrificed and offered incense on the high places. ⁴⁴ Jehoshaphat also made peace with the king of Israel.

45 Now the rest of the acts of Jehoshaphat, and his power that he showed, and how he waged war, are they not written in the Book of the Annals of the Kings of Judah? ⁴⁶ The remnant of the male temple prostitutes who were still in the land in the days of his father Asa, he exterminated.

47 There was no king in Edom; a deputy was king. ⁴⁸ Jehoshaphat made ships of the Tarshish type to go to Ophir for gold; but they did not go, for the ships were wrecked at Ezion-geber. ⁴⁹ Then Ahaziah son of Ahab said to Jehoshaphat, "Let my servants go with your servants in the ships," but Jehoshaphat was not willing. ⁵⁰ Jehoshaphat slept with his ancestors and was buried with his ancestors in the city of his father David; his son Jehoram succeeded him.

51 Ahaziah son of Ahab began to reign over Israel in Samaria in the seventeenth year of King Jehoshaphat of Judah; he reigned two years over Israel. ⁵² He did what was evil in the sight of the LORD, and walked in the way of his father and mother, and in the way of Jeroboam son of Nebat, who caused Israel to sin. ⁵³ He served Baal and worshiped him; he provoked the LORD, the God of Israel, to anger, just as his father had done.

u Heb lacks *in it*

---

course. **39**: *The ivory house,* a palace in Samaria, decorated with carved ivory inlay and containing furniture so decorated. Some of these decorative inlays have been found in the archaeological excavation at Samaria (compare 10.22; Am 3.15; 6.4; Ps 45.8). **41–46**: *Jehoshaphat of Judah* is credited with certain reforms and given qualified approval by the Deuteronomistic editor, as was his father *Asa* (15.9–14).

22.47–50: Jehoshaphat controlled Edom (2 Chr ch 20), and wished to imitate Solomon in maritime operations (compare 9.26–28; 10.22). *Ships of the Tarshish type,* see 10.22 n. **48–49**: For a different version, see 2 Chr 20.35–37. **51–53**: *Ahaziah* is condemned by the Deuteronomistic editor, as are all the kings of *Israel.* The account of his reign is continued immediately in 2 Kings ch 1.

2 Kings

Since 1 and 2 Kings are really one book, the problems and sources common to both are discussed in the Introduction to 1 Kings. Second Kings continues the story of the Hebrew monarchies. Chapters 1–17 describe the period from the reigns of Ahaziah of Israel and Jehoshaphat of Judah until the fall of Samaria and the end of the kingdom of Israel in 721 B.C. Chapters 18–25 relate the story of the kingdom of Judah from the fall of the kingdom of Israel to the fall of Judah with the capture and destruction of Jerusalem by Nebuchadnezzar in 586 B.C., ending with a brief account of the governorship of Gedaliah and the elevation of King Jehoiachin in exile. The book thus covers the period from the middle of the ninth century to near the middle of the sixth century (see "Survey of . . . Bible Lands," §§ 15–17 and 27). The fall of both Israel and Judah is interpreted in terms of the judgment of the LORD.

1 After the death of Ahab, Moab rebelled against Israel.

2 Ahaziah had fallen through the lattice in his upper chamber in Samaria, and lay injured; so he sent messengers, telling them, "Go, inquire of Baal-zebub, the god of Ekron, whether I shall recover from this injury." ³But the angel of the LORD said to Elijah the Tishbite, "Get up, go to meet the messengers of the king of Samaria, and say to them, 'Is it because there is no God in Israel that you are going to inquire of Baal-zebub, the god of Ekron?' ⁴Now therefore thus says the LORD, 'You shall not leave the bed to which you have gone, but you shall surely die.' " So Elijah went.

5 The messengers returned to the king, who said to them, "Why have you returned?" ⁶They answered him, "There came a man to meet us, who said to us, 'Go back to the king who sent you, and say to him: Thus says the LORD: Is it because there is no God in Israel that you are sending to inquire of Baal-zebub, the god of Ekron? Therefore you shall not leave the bed to which you have gone, but shall surely die.' " ⁷He said to them, "What sort of man was he who came to meet you and told you these things?" ⁸They answered him, "A hairy man, with a leather belt around his waist." He said, "It is Elijah the Tishbite."

9 Then the king sent to him a captain of fifty with his fifty men. He went up to Elijah, who was sitting on the top of a hill, and said to him, "O man of God, the king says, 'Come down.' " ¹⁰But Elijah answered the captain of fifty, "If I am a man of God, let fire come down from heaven and consume you and your fif-

ty." Then fire came down from heaven, and consumed him and his fifty.

11 Again the king sent to him another captain of fifty with his fifty. He went up*a* and said to him, "O man of God, this is the king's order: Come down quickly!" ¹²But Elijah answered them, "If I am a man of God, let fire come down from heaven and consume you and your fifty." Then the fire of God came down from heaven and consumed him and his fifty.

13 Again the king sent the captain of a third fifty with his fifty. So the third captain of fifty went up, and came and fell on his knees before Elijah, and entreated him, "O man of God, please let my life, and the life of these fifty servants of yours, be precious in your sight. ¹⁴Look, fire came down from heaven and consumed the two former captains of fifty men with their fifties; but now let my life be precious in your sight." ¹⁵Then the angel of the LORD said to Elijah, "Go down with him; do not be afraid of him." So he set out and went down with him to the king, ¹⁶and said to him, "Thus says the LORD: Because you have sent messengers to inquire of Baal-zebub, the god of Ekron,—is it because there is no God in Israel to inquire of his word?—therefore you shall not leave the bed to which you have gone, but you shall surely die."

17 So he died according to the word of the LORD that Elijah had spoken. His brother,*b* Jehoram succeeded him as king in the second year of King Jehoram

---

a Gk Compare verses 9, 13: Heb *He answered*
b Gk Syr: Heb lacks *His brother*

1.1–18: The reappearance of Elijah, the death of Ahaziah, and the accession of Jehoram of Israel. 1: This information is probably out of place here (compare 3.5). **2–5:** *The lattice,* some sort of protective fence or grating, in this case failed to perform its function. *Baal-zebub* means "lord of flies"; it is perhaps a mocking distortion of "ba'al zbl," a Canaanite name meaning either "lord of the divine abode," or "Baal the Prince," one of the names and manifestations of the great Baal, the rival of the God of Israel (see

1 Kings 18.1–6 n., 17–19 n.). Later, the name of this deity became one synonym for the leader of evil forces in the Gospels (Mt 10.25; 12.24; Mk 3.22; Lk 11.15–19). If Ahaziah was a son of Jezebel, it is not surprising that he preferred Baal to the LORD. **1.9–16:** As before (1 Kings 18.38), the power of God acting through Elijah is symbolized by fire. **17:** *Jehoram,* son of Ahab and king of Israel, must not be confused with his brother-in-law, *Jehoram son of King Jehoshaphat of Judah* (8.16, 25).

son of Jehoshaphat of Judah, because Ahaziah had no son. 18 Now the rest of the acts of Ahaziah that he did, are they not written in the Book of the Annals of the Kings of Israel?

2 Now when the LORD was about to take Elijah up to heaven by a whirlwind, Elijah and Elisha were on their way from Gilgal. 2 Elijah said to Elisha, "Stay here; for the LORD has sent me as far as Bethel." But Elisha said, "As the LORD lives, and as you yourself live, I will not leave you." So they went down to Bethel. 3 The company of prophets‹ who were in Bethel came out to Elisha, and said to him, "Do you know that today the LORD will take your master away from you?" And he said, "Yes, I know; keep silent."

4 Elijah said to him, "Elisha, stay here; for the LORD has sent me to Jericho." But he said, "As the LORD lives, and as you yourself live, I will not leave you." So they came to Jericho. 5 The company of prophets‹ who were at Jericho drew near to Elisha, and said to him, "Do you know that today the LORD will take your master away from you?" And he answered, "Yes, I know; be silent."

6 Then Elijah said to him, "Stay here; for the LORD has sent me to the Jordan." But he said, "As the LORD lives, and as you yourself live, I will not leave you." So the two of them went on. 7 Fifty men of the company of prophets‹ also went, and stood at some distance from them, as they both were standing by the Jordan. 8 Then Elijah took his mantle and rolled it up, and struck the water; the water was parted to the one side and to the other, until the two of them crossed on dry ground.

9 When they had crossed, Elijah said to Elisha, "Tell me what I may do for you, before I am taken from you." Elisha said, "Please let me inherit a double share of your spirit." 10 He responded, "You have asked a hard thing; yet, if you see me as I am being taken from you, it will be granted you; if not, it will not." 11 As they continued walking and talking, a chariot of fire and horses of fire separated the two of them, and Elijah ascended in a whirlwind into heaven. 12 Elisha kept watching and crying out, "Father, father! The chariots of Israel and its horsemen!" But when he could no longer see him, he grasped his own clothes and tore them in two pieces.

13 He picked up the mantle of Elijah that had fallen from him, and went back and stood on the bank of the Jordan. 14 He took the mantle of Elijah that had fallen from him, and struck the water, saying, "Where is the LORD, the God of Elijah?" When he had struck the water, the water was parted to the one side and to the other, and Elisha went over.

15 When the company of prophets‹ who were at Jericho saw him at a distance, they declared, "The spirit of Elijah rests on Elisha." They came to meet him and bowed to the ground before him. 16 They said to him, "See now, we have fifty strong men among your servants; please let them go and seek your master; it may be that the spirit of the LORD has caught him up and thrown him down on some mountain or into some valley." He responded, "No, do not send them." 17 But when they urged him until he was ashamed, he said, "Send them." So they

c Heb *sons of the prophets*

2.1–25: Elijah, taken up to heaven, is succeeded by Elisha. As always, the power and greatness of Elijah are expressed by the ancient writer in terms of legend and miracle. According to biblical tradition, only two persons, Enoch (Gen 5.24) and Elijah, were worthy to be taken up by God without having to die. **1–2:** This *Gilgal* was north of *Bethel.* **3:** The expression *the company of prophets* means "members of the prophetic order"; see 1 Sam 10.5. **11:** The *chariot of fire and horses of fire* continue the symbolism of fire (see 1.9–16 n.). **12:** Elisha means perhaps that Elijah was more important and more powerful than *chariots* and *horsemen;* in 13.14 Elisha receives the same compliment. *Father* as the title of a man of religion is a very old usage (Judg 17.10). **13–14:** See 1 Kings 19.19. **15:** Elisha is acknowledged leader by *the company of prophets.*

sent fifty men who searched for three days but did not find him. 18 When they came back to him (he had remained at Jericho), he said to them, "Did I not say to you, Do not go?"

19 Now the people of the city said to Elisha, "The location of this city is good, as my lord sees; but the water is bad, and the land is unfruitful." 20 He said, "Bring me a new bowl, and put salt in it." So they brought it to him. 21 Then he went to the spring of water and threw the salt into it, and said, "Thus says the LORD, I have made this water wholesome; from now on neither death nor miscarriage shall come from it." 22 So the water has been wholesome to this day, according to the word that Elisha spoke.

23 He went up from there to Bethel; and while he was going up on the way, some small boys came out of the city and jeered at him, saying, "Go away, baldhead! Go away, baldhead!" 24 When he turned around and saw them, he cursed them in the name of the LORD. Then two she-bears came out of the woods and mauled forty-two of the boys. 25 From there he went on to Mount Carmel, and then returned to Samaria.

3 In the eighteenth year of King Jehoshaphat of Judah, Jehoram son of Ahab became king over Israel in Samaria; he reigned twelve years. 2 He did what was evil in the sight of the LORD, though not like his father and mother, for he removed the pillar of Baal that his father had made. 3 Nevertheless he clung to the sin of Jeroboam son of Nebat, which he caused Israel to commit; he did not depart from it.

4 Now King Mesha of Moab was a sheep breeder, who used to deliver to the king of Israel one hundred thousand lambs, and the wool of one hundred thousand rams. 5 But when Ahab died, the king of Moab rebelled against the king of Israel. 6 So King Jehoram marched out of Samaria at that time and mustered all Israel. 7 As he went he sent word to King Jehoshaphat of Judah, "The king of Moab has rebelled against me; will you go with me to battle against Moab?" He answered, "I will; I am with you, my people are your people, my horses are your horses." 8 Then he asked, "By which way shall we march?" Jehoram answered, "By the way of the wilderness of Edom."

9 So the king of Israel, the king of Judah, and the king of Edom set out; and when they had made a roundabout march of seven days, there was no water for the army or for the animals that were with them. 10 Then the king of Israel said, "Alas! The LORD has summoned us, three kings, only to be handed over to Moab." 11 But Jehoshaphat said, "Is there no prophet of the LORD here, through whom we may inquire of the LORD?" Then one of the servants of the king of Israel answered, "Elisha son of Shaphat, who used to pour water on the hands of Elijah, is here." 12 Jehoshaphat said, "The word of the LORD is with him." So the

---

2.19–22: The God-given power of Elisha is attested by a miracle. Today, the finest spring in Jericho is sometimes called Elisha's Fountain. **23–25:** Not all ancient writers, to say nothing of modern, would have told a story like this to inculcate respect for a prophet. On *forty-two* as a number of ill-omen, compare 10.14; Rev 11.2; 13.5.
3.1–27: **The war with Moab. 1–3:** *In the eighteenth year of King Jehoshaphat* does not agree with 1.17 ("second year of King Jehoram . . . of Judah"). The data come from two irreconcilable chronological systems. **4–8:** The famous "Moabite Stone," discovered in 1868, gives an account of the war from the standpoint of *King Mesha of Moab*. In naming *King Jehoram,* the author disregards the brief reign of Ahaziah (vv. 5–6; compare 1 Kings 22.40). The naming of *King Jehoshaphat of Judah* is based on the synchronism in v. 1, not that in 1.17. It was necessary for *the king of Israel* to get the cooperation of the *king of Judah* in order to march through the territory west of the Dead Sea on the way to *the wilderness of Edom.*
3.9–12: The plan was to strike *Moab* from the south through *Edom.* At that time the *king of Edom* was a vassal of *the king of Judah,* not a king in his own right (1 Kings 22.47). *Used to pour water on the hands of Elijah* (when he washed), i.e. *Elisha* had waited on his mentor like a servant.

king of Israel and Jehoshaphat and the king of Edom went down to him.

13 Elisha said to the king of Israel, "What have I to do with you? Go to your father's prophets or to your mother's." But the king of Israel said to him, "No; it is the LORD who has summoned us, three kings, only to be handed over to Moab." 14 Elisha said, "As the LORD of hosts lives, whom I serve, were it not that I have regard for King Jehoshaphat of Judah, I would give you neither a look nor a glance. 15 But get me a musician." And then, while the musician was playing, the power of the LORD came on him. 16 And he said, "Thus says the LORD, 'I will make this wadi full of pools.' 17 For thus says the LORD, 'You shall see neither wind nor rain, but the wadi shall be filled with water, so that you shall drink, you, your cattle, and your animals.' 18 This is only a trifle in the sight of the LORD, for he will also hand Moab over to you. 19 You shall conquer every fortified city and every choice city; every good tree you shall fell, all springs of water you shall stop up, and every good piece of land you shall ruin with stones." 20 The next day, about the time of the morning offering, suddenly water began to flow from the direction of Edom, until the country was filled with water.

21 When all the Moabites heard that the kings had come up to fight against them, all who were able to put on armor, from the youngest to the oldest, were called out and were drawn up at the frontier. 22 When they rose early in the morning, and the sun shone upon the water, the Moabites saw the water opposite them as red as blood. 23 They said, "This is blood; the kings must have fought together, and killed one another. Now then, Moab, to the spoil!" 24 But when they came to the camp of Israel, the Israelites rose up and attacked the Moabites, who fled before them; as they entered Moab they continued the attack. *d* 25 The cities they overturned, and on every good piece of land everyone threw a stone, until it was covered; every spring of water they stopped up, and every good tree they felled. Only at Kirhareseth did the stone walls remain, until the slingers surrounded and attacked it. 26 When the king of Moab saw that the battle was going against him, he took with him seven hundred swordsmen to break through, opposite the king of Edom; but they could not. 27 Then he took his firstborn son who was to succeed him, and offered him as a burnt offering on the wall. And great wrath came upon Israel, so they withdrew from him and returned to their own land.

4 Now the wife of a member of the company of prophets *e* cried to Elisha, "Your servant my husband is dead; and you know that your servant feared

d Compare Gk Syr: Meaning of Heb uncertain
e Heb *the sons of the prophets*

3.13–20: The phrase *to your father's prophets or to your mother's* (prophets of Baal) indicates that Jehoram was a son of Jezebel. **15:** The *musician* was used to induce a trance, out of which the prophet could give his oracle (1 Sam 10.5–6; 19.20–24). **16:** The *wadi* may have been "the brook Zered" (Deut 2.13). **22:** *Red as blood,* colored by the red sandstone of Edom (see Gen 25.30 note *t*).
3.27: Human sacrifice, common in many ancient religions, was not unknown among the people of Israel and Judah (Ex 22.29–30; Judg 11.30–31, 39; 1 Kings 16.34), though they learned that it was wrong (Gen 22.12; Ex 34.20; Deut 18.10). *The king of Moab* in his terrible extremity made the supreme sacrifice of *his firstborn son,* and the forces of Israel were

so impressed and so filled with fear of the *great wrath* (perhaps of Chemosh, the god of Moab) that they gave up the victory that lay within their grasp and hastily *returned to their own land.* Two later kings of Judah are condemned for participating in the horrible practice (16.3; 21.6).
4.1–8.6: An interlude on the miracles of Elisha. See 2.19–22 n. In ancient times, miracle stories were considered to be one of the best ways of portraying the importance of a religious leader. Two differences from Elijah should be noted: *Elisha* belonged to and worked with *the company of prophets* (members of the prophetic order living in communities); he was often in the company of the king and the army (compare ch 3).

the LORD, but a creditor has come to take my two children as slaves." 2 Elisha said to her, "What shall I do for you? Tell me, what do you have in the house?" She answered, "Your servant has nothing in the house, except a jar of oil." 3 He said, "Go outside, borrow vessels from all your neighbors, empty vessels and not just a few. 4 Then go in, and shut the door behind you and your children, and start pouring into all these vessels; when each is full, set it aside." 5 So she left him and shut the door behind her and her children; they kept bringing vessels to her, and she kept pouring. 6 When the vessels were full, she said to her son, "Bring me another vessel." But he said to her, "There are no more." Then the oil stopped flowing. 7 She came and told the man of God, and he said, "Go sell the oil and pay your debts, and you and your children can live on the rest."

8 One day Elisha was passing through Shunem, where a wealthy woman lived, who urged him to have a meal. So whenever he passed that way, he would stop there for a meal. 9 She said to her husband, "Look, I am sure that this man who regularly passes our way is a holy man of God. 10 Let us make a small roof chamber with walls, and put there for him a bed, a table, a chair, and a lamp, so that he can stay there whenever he comes to us."

11 One day when he came there, he went up to the chamber and lay down there. 12 He said to his servant Gehazi, "Call the Shunammite woman." When he had called her, she stood before him. 13 He said to him, "Say to her, Since you have taken all this trouble for us, what may be done for you? Would you have a word spoken on your behalf to the king or to the commander of the army?" She answered, "I live among my own people." 14 He said, "What then may be done for her?" Gehazi answered, "Well, she has no son, and her husband is old." 15 He said, "Call her." When he had called her, she stood at the door. 16 He said, "At this season, in due time, you shall embrace a son." She replied, "No, my lord, O man of God; do not deceive your servant."

17 The woman conceived and bore a son at that season, in due time, as Elisha had declared to her.

18 When the child was older, he went out one day to his father among the reapers. 19 He complained to his father, "Oh, my head, my head!" The father said to his servant, "Carry him to his mother." 20 He carried him and brought him to his mother; the child sat on her lap until noon, and he died. 21 She went up and laid him on the bed of the man of God, closed the door on him, and left. 22 Then she called to her husband, and said, "Send me one of the servants and one of the donkeys, so that I may quickly go to the man of God and come back again." 23 He said, "Why go to him today? It is neither new moon nor sabbath." She said, "It will be all right." 24 Then she saddled the donkey and said to her servant, "Urge the animal on; do not hold back for me unless I tell you." 25 So she set out, and came to the man of God at Mount Carmel.

When the man of God saw her coming, he said to Gehazi his servant, "Look, there is the Shunammite woman; 26 run at once to meet her, and say to her, Are you all right? Is your husband all right? Is the child all right?" She answered, "It is all right." 27 When she came to the man of God at the mountain, she caught hold

---

4.1–7: **The jar of oil.** This story is a parallel to Elijah's miracle in 1 Kings 17.14–16. Taking *children* as *slaves* for debt was legal in Israel (Ex 21.7). Later, in Judah, the practice was modified somewhat (Deut 15.12–18; Lev 25.39–46), at least in theory (compare Jer 34.8–16).

4.8–37: **The son restored to life.** The first part of this story (vv. 8–17) exhibits the favorite theme of the birth of a child late in life to a hitherto barren woman (see 1 Sam ch 1 n.). The second part of the story (vv. 18–37) parallels the account of Elijah's resuscitation of a child in 1 Kings 17.17–24.

4.23: *It is neither new moon nor sabbath,* it was considered more propitious to visit a prophet on holy days.

of his feet. Gehazi approached to push her away. But the man of God said, "Let her alone, for she is in bitter distress; the LORD has hidden it from me and has not told me." 28 Then she said, "Did I ask my lord for a son? Did I not say, Do not mislead me?" 29 He said to Gehazi, "Gird up your loins, and take my staff in your hand, and go. If you meet anyone, give no greeting, and if anyone greets you, do not answer; and lay my staff on the face of the child." 30 Then the mother of the child said, "As the LORD lives, and as you yourself live, I will not leave without you." So he rose up and followed her. 31 Gehazi went on ahead and laid the staff on the face of the child, but there was no sound or sign of life. He came back to meet him and told him, "The child has not awakened."

32 When Elisha came into the house, he saw the child lying dead on his bed. 33 So he went in and closed the door on the two of them, and prayed to the LORD. 34 Then he got up on the bed[f] and lay upon the child, putting his mouth upon his mouth, his eyes upon his eyes, and his hands upon his hands; and while he lay bent over him, the flesh of the child became warm. 35 He got down, walked once to and fro in the room, then got up again and bent over him; the child sneezed seven times, and the child opened his eyes. 36 Elisha[g] summoned Gehazi and said, "Call the Shunammite woman." So he called her. When she came to him, he said, "Take your son." 37 She came and fell at his feet, bowing to the ground; then she took her son and left.

38 When Elisha returned to Gilgal, there was a famine in the land. As the company of prophets was[h] sitting before him, he said to his servant, "Put the large

pot on, and make some stew for the company of prophets."[i] 39 One of them went out into the field to gather herbs; he found a wild vine and gathered from it a lapful of wild gourds, and came and cut them up into the pot of stew, not knowing what they were. 40 They served some for the men to eat. But while they were eating the stew, they cried out, "O man of God, there is death in the pot!" They could not eat it. 41 He said, "Then bring some flour." He threw it into the pot, and said, "Serve the people and let them eat." And there was nothing harmful in the pot.

42 A man came from Baal-shalishah, bringing food from the first fruits to the man of God: twenty loaves of barley and fresh ears of grain in his sack. Elisha said, "Give it to the people and let them eat." 43 But his servant said, "How can I set this before a hundred people?" So he repeated, "Give it to the people and let them eat, for thus says the LORD, 'They shall eat and have some left.'" 44 He set it before them, they ate, and had some left, according to the word of the LORD.

5 Naaman, commander of the army of the king of Aram, was a great man and in high favor with his master, because by him the LORD had given victory to Aram. The man, though a mighty warrior, suffered from leprosy.[j] 2 Now the Arameans on one of their raids had taken a young girl captive from the land of Israel, and she served Naaman's wife. 3 She said to her mistress, "If only my lord were with the prophet who is in Samaria! He would cure him of his lepro-

---

[f] Heb lacks *on the bed*   [g] Heb *he*
[h] Heb *sons of the prophets were*   [i] Heb *sons of the prophets*
[j] A term for several skin diseases; precise meaning uncertain

---

4 38–41: The spoiled pot of stew. Compare 2.19–22. **42–44: The twenty loaves.** The most striking parallels to this miracle are to be found in the New Testament, Mt 14.13–21; 15.32–38.

5.1–27: The curing of the leprosy of Naaman. 1–7: Neither the name of *the king of Aram* nor that of *the king of Israel* is men-

tioned. The story assumes that the Arameans held the upper hand at this time. The *ten talents of silver* would have weighed 750 lbs.; the *six thousand shekels of gold* would have weighed about 150 lbs. There is no way to estimate what the equivalent worth today would be, but a considerable fortune is intended.

sy."ᵏ ⁴So Naamanˡ went in and told his lord just what the girl from the land of Israel had said. ⁵And the king of Aram said, "Go then, and I will send along a letter to the king of Israel."

He went, taking with him ten talents of silver, six thousand shekels of gold, and ten sets of garments. ⁶He brought the letter to the king of Israel, which read, "When this letter reaches you, know that I have sent to you my servant Naaman, that you may cure him of his leprosy."ᵏ ⁷When the king of Israel read the letter, he tore his clothes and said, "Am I God, to give death or life, that this man sends word to me to cure a man of his leprosy?ᵏ Just look and see how he is trying to pick a quarrel with me."

8 But when Elisha the man of God heard that the king of Israel had torn his clothes, he sent a message to the king, "Why have you torn your clothes? Let him come to me, that he may learn that there is a prophet in Israel." ⁹So Naaman came with his horses and chariots, and halted at the entrance of Elisha's house. ¹⁰Elisha sent a messenger to him, saying, "Go, wash in the Jordan seven times, and your flesh shall be restored and you shall be clean." ¹¹But Naaman became angry and went away, saying, "I thought that for me he would surely come out, and stand and call on the name of the LORD his God, and would wave his hand over the spot, and cure the leprosy!ᵏ ¹²Are not Abanaᵐ and Pharpar, the rivers of Damascus, better than all the waters of Israel? Could I not wash in them, and be clean?" He turned and went away in a rage. ¹³But his servants approached and said to him, "Father, if the prophet had commanded you to do something difficult, would you not have done it? How much more, when all he said to you was, 'Wash, and be clean'?" ¹⁴So he went

down and immersed himself seven times in the Jordan, according to the word of the man of God; his flesh was restored like the flesh of a young boy, and he was clean.

15 Then he returned to the man of God, he and all his company; he came and stood before him and said, "Now I know that there is no God in all the earth except in Israel; please accept a present from your servant." ¹⁶But he said, "As the LORD lives, whom I serve, I will accept nothing!" He urged him to accept, but he refused. ¹⁷Then Naaman said, "If not, please let two mule-loads of earth be given to your servant; for your servant will no longer offer burnt offering or sacrifice to any god except the LORD. ¹⁸But may the LORD pardon your servant on one count: when my master goes into the house of Rimmon to worship there, leaning on my arm, and I bow down in the house of Rimmon, when I do bow down in the house of Rimmon, may the LORD pardon your servant on this one count." ¹⁹He said to him, "Go in peace."

But when Naaman had gone from him a short distance, ²⁰Gehazi, the servant of Elisha the man of God, thought, "My master has let that Aramean Naaman off too lightly by not accepting from him what he offered. As the LORD lives, I will run after him and get something out of him." ²¹So Gehazi went after Naaman. When Naaman saw someone running after him, he jumped down from the chariot to meet him and said, "Is everything all right?" ²²He replied, "Yes, but my master has sent me to say, 'Two members of a company of prophetsⁿ have just come to me from the hill country of Ephraim; please give them a talent of sil-

k A term for several skin diseases; precise meaning uncertain   *l* Heb *he*   *m* Another reading is *Amana*   *n* Heb *sons of the prophets*

5.14: For another description of the disease of leprosy, compare Lev ch 13. The New Testament has a number of examples of the healing of this or a similar disease (Mt 8.2–3; Mk 1.40–42; Lk 5.12–13; compare Mt 11.5; Lk 7.22).

5.15–19a: *Naaman* would have preferred to worship the LORD alone thenceforth if that

would have been possible. He asked to take home *two mule-loads of earth* from Israel, the idea being that a god could not be worshiped apart from his own land. If he should bow down to *Rimmon* (another name for Hadad, the chief god of Aram), it would only be because he was forced to do so.

ver and two changes of clothing.'" 23 Naaman said, "Please accept two talents." He urged him, and tied up two talents of silver in two bags, with two changes of clothing, and gave them to two of his servants, who carried them in front of Gehazi. *o* 24 When he came to the citadel, he took the bags*p* from them, and stored them inside; he dismissed the men, and they left.

25 He went in and stood before his master; and Elisha said to him, "Where have you been, Gehazi?" He answered, "Your servant has not gone anywhere at all." 26 But he said to him, "Did I not go with you in spirit when someone left his chariot to meet you? Is this a time to accept money and to accept clothing, olive orchards and vineyards, sheep and oxen, and male and female slaves? 27 Therefore the leprosy*q* of Naaman shall cling to you, and to your descendants forever." So he left his presence leprous,*q* as white as snow.

6 Now the company of prophets*r* said to Elisha, "As you see, the place where we live under your charge is too small for us. 2 Let us go to the Jordan, and let us collect logs there, one for each of us, and build a place there for us to live." He answered, "Do so." 3 Then one of them said, "Please come with your servants." And he answered, "I will." 4 So he went with them. When they came to the Jordan, they cut down trees. 5 But as one was felling a log, his ax head fell into the water; he cried out, "Alas, master! It was borrowed." 6 Then the man of God said, "Where did it fall?" When he showed him the place, he cut off a stick, and threw it in there, and made the iron float. 7 He said, "Pick it up." So he reached out his hand and took it.

8 Once when the king of Aram was at war with Israel, he took counsel with his officers. He said, "At such and such a place shall be my camp." 9 But the man of God sent word to the king of Israel, "Take care not to pass this place, because the Arameans are going down there." 10 The king of Israel sent word to the place of which the man of God spoke. More than once or twice he warned such a place*s* so that it was on the alert.

11 The mind of the king of Aram was greatly perturbed because of this; he called his officers and said to them, "Now tell me who among us sides with the king of Israel?" 12 Then one of his officers said, "No one, my lord king. It is Elisha, the prophet in Israel, who tells the king of Israel the words that you speak in your bedchamber." 13 He said, "Go and find where he is; I will send and seize him." He was told, "He is in Dothan." 14 So he sent horses and chariots there and a great army; they came by night, and surrounded the city.

15 When an attendant of the man of God rose early in the morning and went out, an army with horses and chariots was all around the city. His servant said, "Alas, master! What shall we do?" 16 He replied, "Do not be afraid, for there are more with us than there are with them." 17 Then Elisha prayed: "O LORD, please open his eyes that he may see." So the LORD opened the eyes of the servant, and he saw; the mountain was full of horses and chariots of fire all around Elisha. 18 When the Arameans*t* came down against him, Elisha prayed to the LORD, and said, "Strike this people, please, with blindness." So he struck them with

o Heb *him*   *p* Heb lacks *the bags*   *q* A term for several skin diseases; precise meaning uncertain   *r* Heb *sons of the prophets*   *s* Heb *warned it*   *t* Heb *they*

5.26: Elisha caught the culprit by extrasensory perception (*in spirit*). This trait comes out more strongly in the next chapter. 27: For another case of *leprosy* as punishment, compare 15.4–5.

6.1–7: **The iron ax head.** See 4.1–8.6 n.
6.8–23: **The Aramean army blinded and captured.** As in 5.1–7 neither *the king of Aram* nor *the king of Israel* is named. Verses 8–14 portray Elisha's extraordinary skill in the use of extrasensory perception. **10:** The meaning is that the *king of Israel sent* someone else to investigate. *Dothan* (v. 13) was about ten miles north of *Samaria* (v. 19).

6.17: The *horses and chariots of fire* remind one of the stories about Elijah (2.11; 1 Kings

blindness as Elisha had asked. ¹⁹Elisha said to them, "This is not the way, and this is not the city; follow me, and I will bring you to the man whom you seek." And he led them to Samaria.

20 As soon as they entered Samaria, Elisha said, "O Lord, open the eyes of these men so that they may see." The Lord opened their eyes, and they saw that they were inside Samaria. ²¹When the king of Israel saw them he said to Elisha, "Father, shall I kill them? Shall I kill them?" ²²He answered, "No! Did you capture with your sword and your bow those whom you want to kill? Set food and water before them so that they may eat and drink; and let them go to their master." ²³So he prepared for them a great feast; after they ate and drank, he sent them on their way, and they went to their master. And the Arameans no longer came raiding into the land of Israel.

24 Some time later King Ben-hadad of Aram mustered his entire army; he marched against Samaria and laid siege to it. ²⁵As the siege continued, famine in Samaria became so great that a donkey's head was sold for eighty shekels of silver, and one-fourth of a kab of dove's dung for five shekels of silver. ²⁶Now as the king of Israel was walking on the city wall, a woman cried out to him, "Help, my lord king!" ²⁷He said, "No! Let the Lord help you. How can I help you? From the threshing floor or from the wine press?" ²⁸But then the king asked her, "What is your complaint?" She answered, "This woman said to me, 'Give up your son; we will eat him today, and we will eat my son tomorrow.' ²⁹So we cooked my son and ate him. The next day I said to her, 'Give up your son and

we will eat him.' But she has hidden her son." ³⁰When the king heard the words of the woman he tore his clothes—now since he was walking on the city wall, the people could see that he had sackcloth on his body underneath— ³¹and he said, "So may God do to me, and more, if the head of Elisha son of Shaphat stays on his shoulders today." ³²So he dispatched a man from his presence.

Now Elisha was sitting in his house, and the elders were sitting with him. Before the messenger arrived, Elisha said to the elders, "Are you aware that this murderer has sent someone to take off my head? When the messenger comes, see that you shut the door and hold it closed against him. Is not the sound of his master's feet behind him?" ³³While he was still speaking with them, the king*u* came down to him and said, "This trouble is from the Lord! Why should I hope in the Lord any longer?" ¹But Elisha said, "Hear the word of the Lord: thus says the Lord, Tomorrow about this time a measure of choice meal shall be sold for a shekel, and two measures of barley for a shekel, at the gate of Samaria." ²Then the captain on whose hand the king leaned said to the man of God, "Even if the Lord were to make windows in the sky, could such a thing happen?" But he said, "You shall see it with your own eyes, but you shall not eat from it."

3 Now there were four leprous*v* men outside the city gate, who said to one another, "Why should we sit here until we die? ⁴If we say, 'Let us enter the city,'

u See 7.2: Heb *messenger*   *v* A term for several skin diseases; precise meaning uncertain

---

18.38). **23:** *The Arameans* were either so pleased by the hospitality of the king or so frightened by the power of the prophet that they momentarily gave up the war. **6.24–7.20: Ben-hadad's siege of Samaria repulsed by divine intervention. 24:** The mention of the name of the *king of Aram* does not establish the chronology, since there were probably two Ben-hadads during these times, and the name of the king of Israel

is not mentioned. **25:** These prices were enormously high. *Eighty shekels* would have been about a pound, *a kab* about a quart. **28–29:** See Deut 28.54–57; Lam 2.20; 4.10; Ezek 5.10. **31:** For some unknown reason, the king blamed *Elisha*, though in vv. 8–23 king and prophet were on the best of terms. **6.32–7.2:** Elisha predicted a quick reversal of fortune; a *captain* doubted and his doom was predicted by the prophet.

the famine is in the city, and we shall die there; but if we sit here, we shall also die. Therefore, let us desert to the Aramean camp; if they spare our lives, we shall live; and if they kill us, we shall but die." 5 So they arose at twilight to go to the Aramean camp; but when they came to the edge of the Aramean camp, there was no one there at all. 6 For the Lord had caused the Aramean army to hear the sound of chariots, and of horses, the sound of a great army, so that they said to one another, "The king of Israel has hired the kings of the Hittites and the kings of Egypt to fight against us." 7 So they fled away in the twilight and abandoned their tents, their horses, and their donkeys leaving the camp just as it was, and fled for their lives. 8 When these leprous *w* men had come to the edge of the camp, they went into a tent, ate and drank, carried off silver, gold, and clothing, and went and hid them. Then they came back, entered another tent, carried off things from it, and went and hid them.

9 Then they said to one another, "What we are doing is wrong. This is a day of good news; if we are silent and wait until the morning light, we will be found guilty; therefore let us go and tell the king's household." 10 So they came and called to the gatekeepers of the city, and told them, "We went to the Aramean camp, but there was no one to be seen or heard there, nothing but the horses tied, the donkeys tied, and the tents as they were." 11 Then the gatekeepers called out and proclaimed it to the king's household. 12 The king got up in the night, and said to his servants, "I will tell you what the Arameans have prepared against us. They know that we are starving; so they have left the camp to hide themselves in the open country, thinking, 'When they come out of the city, we

shall take them alive and get into the city.'" 13 One of his servants said, "Let some men take five of the remaining horses, since those left here will suffer the fate of the whole multitude of Israel that have perished already; *x* let us send and find out." 14 So they took two mounted men, and the king sent them after the Aramean army, saying, "Go and find out." 15 So they went after them as far as the Jordan; the whole way was littered with garments and equipment that the Arameans had thrown away in their haste. So the messengers returned, and told the king.

16 Then the people went out, and plundered the camp of the Arameans. So a measure of choice meal was sold for a shekel, and two measures of barley for a shekel, according to the word of the Lord. 17 Now the king had appointed the captain on whose hand he leaned to have charge of the gate; the people trampled him to death in the gate, just as the man of God had said when the king came down to him. 18 For when the man of God had said to the king, "Two measures of barley shall be sold for a shekel, and a measure of choice meal for a shekel, about this time tomorrow in the gate of Samaria," 19 the captain had answered the man of God, "Even if the Lord were to make windows in the sky, could such a thing happen?" And he had answered, "You shall see it with your own eyes, but you shall not eat from it." 20 It did indeed happen to him; the people trampled him to death in the gate.

8 Now Elisha had said to the woman whose son he had restored to life, "Get up and go with your household, and settle wherever you can; for the Lord has called for a famine, and it will

w A term for several skin diseases; precise meaning uncertain   *x* Compare Gk Syr Vg: Meaning of Heb uncertain

---

7.3–8: The lepers do not seem to be overly ill from their disease. It is the Lord alone who routs the enemy; no fighting is recorded. The word here translated *Egypt* probably refers to a small country to the north, in the same

general area as *the Hittites* (likewise in 1 Kings 10.28–29).

8.1–6: Elisha again helps the woman whose child he had saved. This section may be a continuation of 4.8–37. **1–2:** Isaac also

come on the land for seven years." ²So the woman got up and did according to the word of the man of God; she went with her household and settled in the land of the Philistines seven years. ³At the end of the seven years, when the woman returned from the land of the Philistines, she set out to appeal to the king for her house and her land. ⁴Now the king was talking with Gehazi the servant of the man of God, saying, "Tell me all the great things that Elisha has done." ⁵While he was telling the king how Elisha had restored a dead person to life, the woman whose son he had restored to life appealed to the king for her house and her land. Gehazi said, "My lord king, here is the woman, and here is her son whom Elisha restored to life." ⁶When the king questioned the woman, she told him. So the king appointed an official for her, saying, "Restore all that was hers, together with all the revenue of the fields from the day that she left the land until now."

7 Elisha went to Damascus while King Ben-hadad of Aram was ill. When it was told him, "The man of God has come here," ⁸the king said to Hazael, "Take a present with you and go to meet the man of God. Inquire of the LORD through him, whether I shall recover from this illness." ⁹So Hazael went to meet him, taking a present with him, all kinds of goods of Damascus, forty camel loads. When he entered and stood before him, he said, "Your son King Ben-hadad of Aram has sent me to you, saying, 'Shall I recover from this illness?' " ¹⁰Eli-

sha said to him, "Go, say to him, 'You shall certainly recover'; but the LORD has shown me that he shall certainly die." ¹¹He fixed his gaze and stared at him, until he was ashamed. Then the man of God wept. ¹²Hazael asked, "Why does my lord weep?" He answered, "Because I know the evil that you will do to the people of Israel; you will set their fortresses on fire, you will kill their young men with the sword, dash in pieces their little ones, and rip up their pregnant women." ¹³Hazael said, "What is your servant, who is a mere dog, that he should do this great thing?" Elisha answered, "The LORD has shown me that you are to be king over Aram." ¹⁴Then he left Elisha, and went to his master Ben-hadad,ʸ who said to him, "What did Elisha say to you?" And he answered, "He told me that you would certainly recover." ¹⁵But the next day he took the bed-cover and dipped it in water and spread it over the king's face, until he died. And Hazael succeeded him.

16 In the fifth year of King Joram son of Ahab of Israel,ᶻ Jehoram son of King Jehoshaphat of Judah began to reign. ¹⁷He was thirty-two years old when he became king, and he reigned eight years in Jerusalem. ¹⁸He walked in the way of the kings of Israel, as the house of Ahab had done, for the daughter of Ahab was his wife. He did what was evil in the sight of the LORD. ¹⁹Yet the LORD would not destroy Judah, for the sake of his

y Heb lacks *Ben-hadad*   z Gk Syr: Heb adds *Jehoshaphat being king of Judah,*

---

went to *the land of the Philistines* to avoid *a famine* (Gen 26.1); compare Gen 12.10, where the place of refuge is Egypt. Drought and famine were common (4.38; 1 Kings 17.1). **3:** Apparently the caretakers were trying to usurp the property. **4:** The leprosy of *Gehazi* (5.27) is ignored here. Perhaps it was a mild case (7.3, 8), or perhaps the tradition from which this story comes knew nothing of it. **8.7–15: Elisha foments revolution in Syria.** Elijah was able to carry out only one (the third) of the three divine commands given him at Horeb (1 Kings 19.15–16). *Elisha* now proceeds to carry out the first command.

The idea behind this is that Israel needed to be punished for its sins, and that Hazael was divinely ordained to do this work. The prophet was sorrowful over this doleful necessity (vv. 11–12), but nevertheless felt that God's will must be done (compare Hos 13.16).

8.16–24: The reign of Jehoram (Joram) of Judah. 16: This synchronism with *Joram* (Jehoram) *of Israel* agrees with that in 3.1, not with that in 1.17 (see 3.1–3 n.). **18:** See 1 Kings 22.1–4 n. The adverse moral evaluation of the Deuteronomist is probably owing to the marriage. **19:** On the *lamp* as a

servant David, since he had promised to give a lamp to him and to his descendants forever.

20 In his days Edom revolted against the rule of Judah, and set up a king of their own. 21 Then Joram crossed over to Zair with all his chariots. He set out by night and attacked the Edomites and their chariot commanders who had surrounded him;*a* but his army fled home. 22 So Edom has been in revolt against the rule of Judah to this day. Libnah also revolted at the same time. 23 Now the rest of the acts of Joram, and all that he did, are they not written in the Book of the Annals of the Kings of Judah? 24 So Joram slept with his ancestors, and was buried with them in the city of David; his son Ahaziah succeeded him.

25 In the twelfth year of King Joram son of Ahab of Israel, Ahaziah son of King Jehoram of Judah began to reign. 26 Ahaziah was twenty-two years old when he began to reign; he reigned one year in Jerusalem. His mother's name was Athaliah, a granddaughter of King Omri of Israel. 27 He also walked in the way of the house of Ahab, doing what was evil in the sight of the LORD, as the house of Ahab had done, for he was son-in-law to the house of Ahab.

28 He went with Joram son of Ahab to wage war against King Hazael of Aram at Ramoth-gilead, where the Arameans wounded Joram. 29 King Joram returned to be healed in Jezreel of the wounds that the Arameans had inflicted on him at Ramah, when he fought against King Hazael of Aram. King Ahaziah son of Jehoram of Judah went down to see Joram son of Ahab in Jezreel, because he was wounded.

9 Then the prophet Elisha called a member of the company of prophets*b* and said to him, "Gird up your loins; take this flask of oil in your hand, and go to Ramoth-gilead. 2 When you arrive, look there for Jehu son of Jehoshaphat, son of Nimshi; go in and get him to leave his companions, and take him into an inner chamber. 3 Then take the flask of oil, pour it on his head, and say, 'Thus says the LORD: I anoint you king over Israel.' Then open the door and flee; do not linger."

4 So the young man, the young prophet, went to Ramoth-gilead. 5 He arrived while the commanders of the army were in council, and he announced, "I have a message for you, commander." "For which one of us?" asked Jehu. "For you, commander." 6 So Jehu*c* got up and went inside; the young man poured the oil on his head, saying to him, "Thus says the LORD the God of Israel: I anoint you king over the people of the LORD, over Israel. 7 You shall strike down the house of your master Ahab, so that I may avenge on Jezebel the blood of my servants the prophets, and the blood of all the servants of the LORD. 8 For the whole house of Ahab shall perish; I will cut off from Ahab every male, bond or free, in Israel. 9 I will make the house of Ahab like the house of Jeroboam son of Nebat, and like the house of Baasha son of Ahijah. 10 The dogs shall eat Jezebel in the territory of Jezreel, and no one shall bury

a Meaning of Heb uncertain   *b* Heb *sons of the prophets*   *c* Heb *he*

---

symbol of the permanence of the Davidic dynasty, see 2 Sam 21.17; 1 Kings 11.36; 15.4. **20:** See 3.9–12 n.

8.25–29: The reign of Ahaziah of Judah. This *Ahaziah* must not be confused with his uncle, Ahaziah of Israel (1.2–18; 1 Kings 22.40, 51–53). His relationship to *the house of Ahab* was enough to condemn him in the eyes of the editor. The term *son-in-law* (v. 27) would apply more accurately to Ahaziah's father. Ahaziah himself was actually related

by blood *to the house of Ahab. Ramoth-gilead*, see 1 Kings 22.29–36. Like *Ahab, Joram* sought help from the *king of Judah.* The visit of *Ahaziah* to *Joram* set the stage for the catastrophic events of the next chapter. (*Ramah* is the same as Ramoth-gilead.)

9.1–37: Elisha foments the revolution of Jehu; assassination of Joram, Ahaziah, and Jezebel. Elisha now proceeds to carry out the second and only unfulfilled command given him at Horeb (see 8.7–15 n.).

her." Then he opened the door and fled.

11 When Jehu came back to his master's officers, they said to him, "Is everything all right? Why did that madman come to you?" He answered them, "You know the sort and how they babble." 12 They said, "Liar! Come on, tell us!" So he said, "This is just what he said to me: 'Thus says the LORD, I anoint you king over Israel.' " 13 Then hurriedly they all took their cloaks and spread them for him on the bare*d* steps; and they blew the trumpet, and proclaimed, "Jehu is king."

14 Thus Jehu son of Jehoshaphat son of Nimshi conspired against Joram. Joram with all Israel had been on guard at Ramoth-gilead against King Hazael of Aram; 15 but King Joram had returned to be healed in Jezreel of the wounds that the Arameans had inflicted on him, when he fought against King Hazael of Aram. So Jehu said, "If this is your wish, then let no one slip out of the city to go and tell the news in Jezreel." 16 Then Jehu mounted his chariot and went to Jezreel, where Joram was lying ill. King Ahaziah of Judah had come down to visit Joram.

17 In Jezreel, the sentinel standing on the tower spied the company of Jehu arriving, and said, "I see a company." Joram said, "Take a horseman; send him to meet them, and let him say, 'Is it peace?' " 18 So the horseman went to meet him; he said, "Thus says the king, 'Is it peace?' " Jehu responded, "What have you to do with peace? Fall in behind me." The sentinel reported, saying, "The messenger reached them, but he is not coming back." 19 Then he sent out a second horseman, who came to them and said, "Thus says the king, 'Is it peace?' " Jehu answered, "What have you to do with peace? Fall in behind me." 20 Again the sentinel reported, "He reached them, but he is not coming back. It looks like the driving of Jehu son of Nimshi; for he drives like a maniac."

21 Joram said, "Get ready." And they got his chariot ready. Then King Joram of Israel and King Ahaziah of Judah set out, each in his chariot, and went to meet Jehu; they met him at the property of Naboth the Jezreelite. 22 When Joram saw Jehu, he said, "Is it peace, Jehu?" He answered, "What peace can there be, so long as the many whoredoms and sorceries of your mother Jezebel continue?" 23 Then Joram reined about and fled, saying to Ahaziah, "Treason, Ahaziah!" 24 Jehu drew his bow with all his strength, and shot Joram between the shoulders, so that the arrow pierced his heart; and he sank in his chariot. 25 Jehu said to his aide Bidkar, "Lift him out, and throw him on the plot of ground belonging to Naboth the Jezreelite; for remember, when you and I rode side by side behind his father Ahab how the LORD uttered this oracle against him: 26 'For the blood of Naboth and for the blood of his children that I saw yesterday, says the LORD, I swear I will repay you on this very plot of ground.' Now therefore lift him out and throw him on the plot of ground, in accordance with the word of the LORD."

27 When King Ahaziah of Judah saw this, he fled in the direction of Beth-haggan. Jehu pursued him, saying, "Shoot him also!" And they shot him*e* in the chariot at the ascent to Gur, which is by Ibleam. Then he fled to Megiddo, and died there. 28 His officers carried him in a chariot to Jerusalem, and buried him in his tomb with his ancestors in the city of David.

29 In the eleventh year of Joram son of Ahab, Ahaziah began to reign over Judah.

30 When Jehu came to Jezreel, Jezebel heard of it; she painted her eyes, and

d Meaning of Heb uncertain   *e* Syr Vg Compare Gk: Heb lacks *and they shot him*

9.16: Compare 8.29. *Jezreel,* see 1 Kings 18.41–46 n. **26:** Compare 1 Kings 21.17–19, 28–29.

9.29: This verse may be a correction of 8.25. **31:** *Zimri* was infamous for his brutal assassinations (1 Kings 16.8–12). **37:** The editor may have added this verse to the oracle in 1 Kings 21.23.

adorned her head, and looked out of the window. 31 As Jehu entered the gate, she said, "Is it peace, Zimri, murderer of your master?" 32 He looked up to the window and said, "Who is on my side? Who?" Two or three eunuchs looked out at him. 33 He said, "Throw her down." So they threw her down; some of her blood spattered on the wall and on the horses, which trampled on her. 34 Then he went in and ate and drank; he said, "See to that cursed woman and bury her; for she is a king's daughter." 35 But when they went to bury her, they found no more of her than the skull and the feet and the palms of her hands. 36 When they came back and told him, he said, "This is the word of the LORD, which he spoke by his servant Elijah the Tishbite, 'In the territory of Jezreel the dogs shall eat the flesh of Jezebel; 37 the corpse of Jezebel shall be like dung on the field in the territory of Jezreel, so that no one can say, This is Jezebel.' "

10 Now Ahab had seventy sons in Samaria. So Jehu wrote letters and sent them to Samaria, to the rulers of Jezreel,*f* to the elders, and to the guardians of the sons of*g* Ahab, saying, 2 "Since your master's sons are with you and you have at your disposal chariots and horses, a fortified city, and weapons, 3 select the son of your master who is the best qualified, set him on his father's throne, and fight for your master's house." 4 But they were utterly terrified and said, "Look, two kings could not withstand him; how then can we stand?" 5 So the steward of the palace, and the governor of the city, along with the elders and the guardians, sent word to Jehu: "We are your servants; we will do anything you say. We will not make anyone king; do whatever you think right." 6 Then he wrote them a second letter, saying, "If you are on my side, and if you

are ready to obey me, take the heads of your master's sons and come to me at Jezreel tomorrow at this time." Now the king's sons, seventy persons, were with the leaders of the city, who were charged with their upbringing. 7 When the letter reached them, they took the king's sons and killed them, seventy persons; they put their heads in baskets and sent them to him at Jezreel. 8 When the messenger came and told him, "They have brought the heads of the king's sons," he said, "Lay them in two heaps at the entrance of the gate until the morning." 9 Then in the morning when he went out, he stood and said to all the people, "You are innocent. It was I who conspired against my master and killed him; but who struck down all these? 10 Know then that there shall fall to the earth nothing of the word of the LORD, which the LORD spoke concerning the house of Ahab; for the LORD has done what he said through his servant Elijah." 11 So Jehu killed all who were left of the house of Ahab in Jezreel, all his leaders, close friends, and priests, until he left him no survivor.

12 Then he set out and went to Samaria. On the way, when he was at Beth-eked of the Shepherds, 13 Jehu met relatives of King Ahaziah of Judah and said, "Who are you?" They answered, "We are kin of Ahaziah; we have come down to visit the royal princes and the sons of the queen mother." 14 He said, "Take them alive." They took them alive, and slaughtered them at the pit of Beth-eked, forty-two in all; he spared none of them.

15 When he left there, he met Jehonadab son of Rechab coming to meet him; he greeted him, and said to him, "Is your heart as true to mine as mine is to

f Or *of the city*; Vg Compare Gk   *g* Gk. Heb lacks *of the sons of*

10.1–36: Continuation of the purge; massacre of the princes of Israel, the princes of Judah, and the worshipers of Baal; decline of the power of Israel. 1: *Seventy sons* probably includes grandsons (compare, however, Judg 8.30; 9.5; 12.13).

10.10–11: Compare 1 Kings 21.21. **15:** *Jehonadab* (or Jonadab) *son of Rechab* was the leader of that group, called the Rechabites, which, some scholars believe, fiercely maintained the old desert way of life, believing that only thus could they properly worship

yours?"*h* Jehonadab answered, "It is."
Jehu said, *i* "If it is, give me your hand."
So he gave him his hand. Jehu took him
up with him into the chariot. 16 He said,
"Come with me, and see my zeal for the
LORD." So he*j* had him ride in his chariot. 17 When he came to Samaria, he killed
all who were left to Ahab in Samaria,
until he had wiped them out, according
to the word of the LORD that he spoke to
Elijah.

18 Then Jehu assembled all the people
and said to them, "Ahab offered Baal
small service; but Jehu will offer much
more. 19 Now therefore summon to me
all the prophets of Baal, all his worshipers, and all his priests; let none be missing, for I have a great sacrifice to offer to
Baal; whoever is missing shall not live."
But Jehu was acting with cunning in order to destroy the worshipers of Baal.
20 Jehu decreed, "Sanctify a solemn assembly for Baal." So they proclaimed it.
21 Jehu sent word throughout all Israel; all
the worshipers of Baal came, so that
there was no one left who did not come.
They entered the temple of Baal, until
the temple of Baal was filled from wall to
wall. 22 He said to the keeper of the wardrobe, "Bring out the vestments for all the
worshipers of Baal." So he brought out
the vestments for them. 23 Then Jehu entered the temple of Baal with Jehonadab
son of Rechab; he said to the worshipers
of Baal, "Search and see that there is no
worshiper of the LORD here among you,
but only worshipers of Baal." 24 Then
they proceeded to offer sacrifices and
burnt offerings.

Now Jehu had stationed eighty men
outside, saying, "Whoever allows any of
those to escape whom I deliver into your

hands shall forfeit his life." 25 As soon as
he had finished presenting the burnt offering, Jehu said to the guards and to the
officers, "Come in and kill them; let no
one escape." So they put them to the
sword. The guards and the officers threw
them out, and then went into the citadel
of the temple of Baal. 26 They brought
out the pillar*k* that was in the temple of
Baal, and burned it. 27 Then they demolished the pillar of Baal, and destroyed the
temple of Baal, and made it a latrine to
this day.

28 Thus Jehu wiped out Baal from
Israel. 29 But Jehu did not turn aside from
the sins of Jeroboam son of Nebat, which
he caused Israel to commit—the golden
calves that were in Bethel and in Dan.
30 The LORD said to Jehu, "Because you
have done well in carrying out what I
consider right, and in accordance with all
that was in my heart have dealt with the
house of Ahab, your sons of the fourth
generation shall sit on the throne of Israel." 31 But Jehu was not careful to follow
the law of the LORD the God of Israel
with all his heart; he did not turn from
the sins of Jeroboam, which he caused
Israel to commit.

32 In those days the LORD began to
trim off parts of Israel. Hazael defeated
them throughout the territory of Israel:
33 from the Jordan eastward, all the land
of Gilead, the Gadites, the Reubenites,
and the Manassites, from Aroer, which is
by the Wadi Arnon, that is, Gilead and
Bashan. 34 Now the rest of the acts of
Jehu, all that he did, and all his power,

h Gk: Heb *Is it right with your heart, as my heart is
with your heart?*   *i* Gk: Heb lacks *Jehu said*
j Gk Syr Tg: Heb *they*   *k* Gk Vg Syr Tg:
Heb *pillars*

the LORD (1 Chr 2.55; Jer ch 35). The nazirites were of similar type, though they practiced as individuals (Judg 13.4–5; Am 2.11–
12; Num 6.1–21).
  10.18–27: In slaughtering *the prophets of
Baal,* with *his priests* and *worshipers, Jehu* was
following in the footsteps of Elijah (compare
v. 25 with 1 Kings 18.40).
  10.28–31: The verses belong to the formulaic summaries (see Introduction to

1 Kings). A later prophet condemns this
butchery by Jehu (Hos 1.4–5). **32–36:** On the
successes of *Hazael* against *the territory of Israel,* see the prediction in 8.12. After the death
of so many leaders, the defensive position of
Israel was naturally weakened. It was probably during this time that Moab, south of the
Arnon, also attacked Israel again and won
final freedom from vassalage, as related on
the Moabite Stone (see 3.4–8 n.).

are they not written in the Book of the Annals of the Kings of Israel? 35 So Jehu slept with his ancestors, and they buried him in Samaria. His son Jehoahaz succeeded him. 36 The time that Jehu reigned over Israel in Samaria was twenty-eight years.

11 Now when Athaliah, Ahaziah's mother, saw that her son was dead, she set about to destroy all the royal family. 2 But Jehosheba, King Joram's daughter, Ahaziah's sister, took Joash son of Ahaziah, and stole him away from among the king's children who were about to be killed; she put*l* him and his nurse in a bedroom. Thus she*m* hid him from Athaliah, so that he was not killed; 3 he remained with her six years, hidden in the house of the LORD, while Athaliah reigned over the land.

4 But in the seventh year Jehoiada summoned the captains of the Carites and of the guards and had them come to him in the house of the LORD. He made a covenant with them and put them under oath in the house of the LORD; then he showed them the king's son. 5 He commanded them, "This is what you are to do: one-third of you, those who go off duty on the sabbath and guard the king's house 6 (another third being at the gate Sur and a third at the gate behind the guards), shall guard the palace; 7 and your two divisions that come on duty in force on the sabbath and guard the house of the LORD*n* 8 shall surround the king, each with weapons in hand; and whoever ap-

proaches the ranks is to be killed. Be with the king in his comings and goings."

9 The captains did according to all that the priest Jehoiada commanded; each brought his men who were to go off duty on the sabbath, with those who were to come on duty on the sabbath, and came to the priest Jehoiada. 10 The priest delivered to the captains the spears and shields that had been King David's, which were in the house of the LORD; 11 the guards stood, every man with his weapons in his hand, from the south side of the house to the north side of the house, around the altar and the house, to guard the king on every side. 12 Then he brought out the king's son, put the crown on him, and gave him the covenant;*o* they proclaimed him king, and anointed him; they clapped their hands and shouted, "Long live the king!"

13 When Athaliah heard the noise of the guard and of the people, she went into the house of the LORD to the people; 14 when she looked, there was the king standing by the pillar, according to custom, with the captains and the trumpeters beside the king, and all the people of the land rejoicing and blowing trumpets. Athaliah tore her clothes and cried, "Treason! Treason!" 15 Then the priest Jehoiada commanded the captains who were set over the army, "Bring her out

l With 2 Chr 22.11: Heb lacks *she put*
m Gk Syr Vg Compare 2 Chr 22.11: Heb *they*
n Heb *the* LORD *to the king*   *o* Or *treaty* or
testimony; Heb *eduth*

11.1–21: The usurpation of Athaliah in Judah, the revolt against her, and the coronation of Joash (Jehoash). Jehu's attempt to eliminate the family of Ahab and the worship of Baal from Israel resulted in the temporary seizure of the throne of Judah by one who was both a descendant of Ahab and a worshiper of Baal (v. 18). **2–3:** *Jehosheba* was not a daughter of Athaliah; hence she was only *Ahaziah's* half *sister.* According to 2 Chr 22.11 she was the wife of Jehoiada the priest (vv. 4, 9). *Joash* (Jehoash, 12.1) was the grandson of *Athaliah* as well as of *King Joram;* but while he was *hidden in the house of the* LORD, he was taught the ways of the LORD. **4:**

The revolt was led by Jehoiada, the priest of the LORD (v. 9). *The Carites* were mercenaries; the word may be only an error for Cherethites (see 1 Sam 30.14 n.; 2 Sam 8.16–18 n.). **11.12:** The word *covenant* or testimony (see note *o*) may refer to a book or document; or it may be an error for "armlet," a royal symbol in 2 Sam 1.10. **14:** *The pillar* (before the temple), see 1 Kings 7.15–22 n. Much debated is the social and political significance of *the people of the land* (see also 23.35; 25.19); perhaps they were free persons who had remained loyal to the old Yahwistic traditions.

between the ranks, and kill with the sword anyone who follows her." For the priest said, "Let her not be killed in the house of the LORD." 16 So they laid hands on her; she went through the horses' entrance to the king's house, and there she was put to death.

17 Jehoiada made a covenant between the LORD and the king and people, that they should be the LORD's people; also between the king and the people. 18 Then all the people of the land went to the house of Baal, and tore it down; his altars and his images they broke in pieces, and they killed Mattan, the priest of Baal, before the altars. The priest posted guards over the house of the LORD. 19 He took the captains, the Carites, the guards, and all the people of the land; then they brought the king down from the house of the LORD, marching through the gate of the guards to the king's house. He took his seat on the throne of the kings. 20 So all the people of the land rejoiced; and the city was quiet after Athaliah had been killed with the sword at the king's house.

21 *p* Jehoash *q* was seven years old when he began to reign.

12 In the seventh year of Jehu, Jehoash began to reign; he reigned forty years in Jerusalem. His mother's name was Zibiah of Beer-sheba. 2 Jehoash did what was right in the sight of the LORD all his days, because the priest Jehoiada instructed him. 3 Nevertheless the high places were not taken away; the people continued to sacrifice and make offerings on the high places.

4 Jehoash said to the priests, "All the money offered as sacred donations that is brought into the house of the LORD, the money for which each person is assessed—the money from the assessment of persons—and the money from the voluntary offerings brought into the house of the LORD, 5 let the priests receive from each of the donors; and let them repair the house wherever any need of repairs is discovered." 6 But by the twenty-third year of King Jehoash the priests had made no repairs on the house. 7 Therefore King Jehoash summoned the priest Jehoiada with the other priests and said to them, "Why are you not repairing the house? Now therefore do not accept any more money from your donors but hand it over for the repair of the house." 8 So the priests agreed that they would neither accept more money from the people nor repair the house.

9 Then the priest Jehoiada took a chest, made a hole in its lid, and set it beside the altar on the right side as one entered the house of the LORD; the priests who guarded the threshold put in it all the money that was brought into the house of the LORD. 10 Whenever they saw that there was a great deal of money in the chest, the king's secretary and the high priest went up, counted the money that was found in the house of the LORD, and tied it up in bags. 11 They would give the money that was weighed out into the hands of the workers who had the oversight of the house of the LORD; then they paid it out to the carpenters and the builders who worked on the house of the LORD, 12 to the masons and the stonecut-

p Ch 12.1 in Heb   *q* Another spelling is *Joash*; see verse 19

12.1–21: Reign of Jehoash of Judah; difficulties with repairs to the temple; attack by the Arameans; assassination of Jehoash. 1–3: During the childhood of the king, *the priest Jehoiada instructed him,* being no doubt the real power behind the throne. **4–8:** According to the writer, the controversy over the repair of the temple was the fault of the priests rather than of the king. The identity of the *donors* (vv. 5, 7) is not clear. **12.9–16:** A new method of collection resulted in enough funds for repairs, but not enough for replacement of the valuable furnishings probably lost during Athaliah's regime (2 Chr 24.7). *Jehoiada* might properly be called "chief priest," but the term *high priest* is post-exilic and here editorial. For a story about apostasy on the part of Joash, and his assassination of Jehoiada's son, see 2 Chr 24.17–22. **19–21:** According to 2 Chr 24.23–27, *Joash* (Jehoash) was defeated and assassinated because of his apostasy. In the end he turned out to be a true grandson of Athaliah (see 11.2–3 n.).

ters, as well as to buy timber and quarried stone for making repairs on the house of the LORD, as well as for any outlay for repairs of the house. 13 But for the house of the LORD no basins of silver, snuffers, bowls, trumpets, or any vessels of gold, or of silver, were made from the money that was brought into the house of the LORD, 14 for that was given to the workers who were repairing the house of the LORD with it. 15 They did not ask an accounting from those into whose hand they delivered the money to pay out to the workers, for they dealt honestly. 16 The money from the guilt offerings and the money from the sin offerings was not brought into the house of the LORD; it belonged to the priests.

17 At that time King Hazael of Aram went up, fought against Gath, and took it. But when Hazael set his face to go up against Jerusalem, 18 King Jehoash of Judah took all the votive gifts that Jehoshaphat, Jehoram, and Ahaziah, his ancestors, the kings of Judah, had dedicated, as well as his own votive gifts, all the gold that was found in the treasuries of the house of the LORD and of the king's house, and sent these to King Hazael of Aram. Then Hazael withdrew from Jerusalem.

19 Now the rest of the acts of Joash, and all that he did, are they not written in the Book of the Annals of the Kings of Judah? 20 His servants arose, devised a conspiracy, and killed Joash in the house of Millo, on the way that goes down to Silla. 21 It was Jozacar son of Shimeath and Jehozabad son of Shomer, his servants, who struck him down, so that he died. He was buried with his ancestors in the city of David; then his son Amaziah succeeded him.

13 In the twenty-third year of King Joash son of Ahaziah of Judah, Jehoahaz son of Jehu began to reign over Israel in Samaria; he reigned seventeen years. 2 He did what was evil in the sight of the LORD, and followed the sins of Jeroboam son of Nebat, which he caused Israel to sin; he did not depart from them. 3 The anger of the LORD was kindled against Israel, so that he gave them repeatedly into the hand of King Hazael of Aram, then into the hand of Benhadad son of Hazael. 4 But Jehoahaz entreated the LORD, and the LORD heeded him; for he saw the oppression of Israel, how the king of Aram oppressed them. 5 Therefore the LORD gave Israel a savior, so that they escaped from the hand of the Arameans; and the people of Israel lived in their homes as formerly. 6 Nevertheless they did not depart from the sins of the house of Jeroboam, which he caused Israel to sin, but walked[r] in them; the sacred pole[s] also remained in Samaria. 7 So Jehoahaz was left with an army of not more than fifty horsemen, ten chariots and ten thousand footmen; for the king of Aram had destroyed them and made them like the dust at threshing. 8 Now the rest of the acts of Jehoahaz and all that he did, including his might, are they not written in the Book of the Annals of the Kings of Israel? 9 So Jehoahaz slept with his ancestors, and they buried him in Samaria; then his son Joash succeeded him.

10 In the thirty-seventh year of King Joash of Judah, Jehoash son of Jehoahaz began to reign over Israel in Samaria; he reigned sixteen years. 11 He also did what was evil in the sight of the LORD; he did not depart from all the sins of Jeroboam son of Nebat, which he caused Israel to sin, but he walked in them. 12 Now the rest of the acts of Joash, and all that he did, as well as the might with which he fought against King Amaziah of Judah, are they not written in the Book of the Annals of the Kings of Israel? 13 So Joash

r Gk Syr Tg Vg: Heb *he walked*
s Heb *Asherah*

13.1–25: **The reigns of Jehoahaz and Jehoash of Israel; the death of Elisha. 3:** Compare 8.12. **4–5:** Compare the "framework" of Judges (see Introduction to Judges).

6: *The sacred pole,* here a symbol of the goddess (see 1 Kings 14.15 n.; 18.17–19 n.).

13.13: This Jeroboam is Jeroboam II, to be distinguished from Jeroboam I mentioned in

slept with his ancestors, and Jeroboam sat upon his throne; Joash was buried in Samaria with the kings of Israel.

14 Now when Elisha had fallen sick with the illness of which he was to die, King Joash of Israel went down to him, and wept before him, crying, "My father, my father! The chariots of Israel and its horsemen!" ¹⁵Elisha said to him, "Take a bow and arrows"; so he took a bow and arrows. ¹⁶Then he said to the king of Israel, "Draw the bow"; and he drew it. Elisha laid his hands on the king's hands. ¹⁷Then he said, "Open the window eastward"; and he opened it. Elisha said, "Shoot"; and he shot. Then he said, "The LORD's arrow of victory, the arrow of victory over Aram! For you shall fight the Arameans in Aphek until you have made an end of them." ¹⁸He continued, "Take the arrows"; and he took them. He said to the king of Israel, "Strike the ground with them"; he struck three times, and stopped. ¹⁹Then the man of God was angry with him, and said, "You should have struck five or six times; then you would have struck down Aram until you had made an end of it, but now you will strike down Aram only three times."

20 So Elisha died, and they buried him. Now bands of Moabites used to invade the land in the spring of the year. ²¹As a man was being buried, a marauding band was seen and the man was thrown into the grave of Elisha; as soon as the man touched the bones of Elisha, he came to life and stood on his feet. 22 Now King Hazael of Aram op-

pressed Israel all the days of Jehoahaz. ²³But the LORD was gracious to them and had compassion on them; he turned toward them, because of his covenant with Abraham, Isaac, and Jacob, and would not destroy them; nor has he banished them from his presence until now.

24 When King Hazael of Aram died, his son Ben-hadad succeeded him. ²⁵Then Jehoash son of Jehoahaz took again from Ben-hadad son of Hazael the towns that he had taken from his father Jehoahaz in war. Three times Joash defeated him and recovered the towns of Israel.

14 In the second year of King Joash son of Joahaz of Israel, King Amaziah son of Joash of Judah, began to reign. ²He was twenty-five years old when he began to reign, and he reigned twenty-nine years in Jerusalem. His mother's name was Jehoaddin of Jerusalem. ³He did what was right in the sight of the LORD, yet not like his ancestor David; in all things he did as his father Joash had done. ⁴But the high places were not removed; the people still sacrificed and made offerings on the high places. ⁵As soon as the royal power was firmly in his hand he killed his servants who had murdered his father the king. ⁶But he did not put to death the children of the murderers; according to what is written in the book of the law of Moses, where the LORD commanded, "The parents shall not be put to death for the children, or the children be put to death for the parents; but all shall be put to death for their own sins."

v. 11. **14**: Note the friendly relations between the prophet and the grandson of Jehu; on the words of Joash, see 2.12 n. **18**: The striking *three times* refers to the three victories in v. 25. **13.20–21**: On the miracle, see 4.1–8.6 n. **23**: A theological anticipation of vv. 24–25, added by an editor who, though friendly to Israel, yet knew that in the end the kingdom had fallen. **24–25**: This *Ben-hadad* (also in v. 3) is the third of that name, possibly the fourth, mentioned in these pages (compare 1 Kings 15.18–21; 20.1; 2 Kings 6.24; 8.7). **14.1–22**: **Warfare between Amaziah of Judah and Jehoash of Israel. 1**: The reigns of

Joash (Jehoash) *of Israel* and *Joash* (Jehoash) *of Judah* overlapped by about two years. Compare the earlier overlapping and more nearly coincident reigns of the two brothers-in-law, Joram (Jehoram) of Israel and Joram (Jehoram) of Judah (see 1.17; 8.16, 25—exhibiting two different chronological systems; see p. 339 OT. **6**: *The law,* Deut 24.16; compare Jer 31.29–30; Ezek 18.2–4, 20; earlier theory and practice may be seen from Ex 20.5; Deut 5.9–10; Josh 7.24–25; 1 Kings 21.21. **7**: *The Valley of Salt,* probably the depression south of the Dead Sea. On the relations between Judah and the *Edomites,* compare 8.20–22.

7 He killed ten thousand Edomites in the Valley of Salt and took Sela by storm; he called it Jokthe-el, which is its name to this day.

8 Then Amaziah sent messengers to King Jehoash son of Jehoahaz, son of Jehu, of Israel, saying, "Come, let us look one another in the face." 9 King Jehoash of Israel sent word to King Amaziah of Judah, "A thornbush on Lebanon sent to a cedar on Lebanon, saying, 'Give your daughter to my son for a wife'; but a wild animal of Lebanon passed by and trampled down the thornbush. 10 You have indeed defeated Edom, and your heart has lifted you up. Be content with your glory, and stay at home; for why should you provoke trouble so that you fall, you and Judah with you?"

11 But Amaziah would not listen. So King Jehoash of Israel went up; he and King Amaziah of Judah faced one another in battle at Beth-shemesh, which belongs to Judah. 12 Judah was defeated by Israel; everyone fled home. 13 King Jehoash of Israel captured King Amaziah of Judah son of Jehoash, son of Ahaziah, at Beth-shemesh; he came to Jerusalem, and broke down the wall of Jerusalem from the Ephraim Gate to the Corner Gate, a distance of four hundred cubits. 14 He seized all the gold and silver, and all the vessels that were found in the house of the LORD and in the treasuries of the king's house, as well as hostages; then he returned to Samaria.

15 Now the rest of the acts that Jehoash did, his might, and how he fought with King Amaziah of Judah, are they not written in the Book of the Annals of the Kings of Israel? 16 Jehoash slept with his ancestors, and was buried in Samaria with the kings of Israel; then his son Jeroboam succeeded him.

17 King Amaziah son of Joash of Judah lived fifteen years after the death of King Jehoash son of Jehoahaz of Israel. 18 Now the rest of the deeds of Amaziah, are they not written in the Book of the Annals of the Kings of Judah? 19 They made a conspiracy against him in Jerusalem, and he fled to Lachish. But they sent after him to Lachish, and killed him there. 20 They brought him on horses; he was buried in Jerusalem with his ancestors in the city of David. 21 All the people of Judah took Azariah, who was sixteen years old, and made him king to succeed his father Amaziah. 22 He rebuilt Elath and restored it to Judah, after King Amaziah[t] slept with his ancestors.

23 In the fifteenth year of King Amaziah son of Joash of Judah, King Jeroboam son of Joash of Israel began to reign in Samaria; he reigned forty-one years. 24 He did what was evil in the sight of the LORD; he did not depart from all the sins of Jeroboam son of Nebat, which he caused Israel to sin. 25 He restored the border of Israel from Lebo-hamath as far as the Sea of the Arabah, according to the word of the LORD, the God of Israel, which he spoke by his servant Jonah son of Amittai, the prophet, who was from

t Heb *the king*

14.8–10: See Judg 9.8–15. The analogy in the fable here should not be pressed in every detail. In general, the idea is that a miserable little *thornbush* tried to make itself equal to a *cedar*, and was badly *trampled* for its trouble. *Jehoash* was both *cedar* and *wild animal.* **13:** *Four hundred cubits,* about two hundred yards. *The Ephraim Gate* would naturally be in the north wall. *The Corner Gate* was probably near the northwest angle of the wall. **14:** Compare 12.18. Apparently the captured king (v. 13) was released and *hostages* were taken instead.

14.15–16: These verses duplicate 13.12–13. **19:** Perhaps Amaziah was assassinated by those resentful of the punishment he had meted out to his father's assassins (v. 5). **21:** *Azariah,* also called Uzziah (2 Chr 26.1). **22:** The restoration of Elath as a seaport of Judah was possible because Edom had again been subdued (v. 7; see 8.20–22; 1 Kings 9.26–28; 22.47–50).

14.23–29: The reign of Jeroboam II was long and prosperous. **25:** Jeroboam is here said to have ruled from the Dead Sea (*Sea of the Arabah*) as far north as Solomon's limit (1 Kings 8.65). His prophetic supporter was *Jonah son of Amittai,* whose name was long afterwards given to the hero of the Book of Jonah (see Jon 1.1). **26–27:** These verses

Gath-hepher. 26 For the LORD saw that the distress of Israel was very bitter; there was no one left, bond or free, and no one to help Israel. 27 But the LORD had not said that he would blot out the name of Israel from under heaven, so he saved them by the hand of Jeroboam son of Joash.

28 Now the rest of the acts of Jeroboam, and all that he did, and his might, how he fought, and how he recovered for Israel Damascus and Hamath, which had belonged to Judah, are they not written in the Book of the Annals of the Kings of Israel? 29 Jeroboam slept with his ancestors, the kings of Israel; his son Zechariah succeeded him.

15 In the twenty-seventh year of King Jeroboam of Israel King Azariah son of Amaziah of Judah began to reign. 2 He was sixteen years old when he began to reign, and he reigned fifty-two years in Jerusalem. His mother's name was Jecoliah of Jerusalem. 3 He did what was right in the sight of the LORD, just as his father Amaziah had done. 4 Nevertheless the high places were not taken away; the people still sacrificed and made offerings on the high places. 5 The LORD struck the king, so that he was leprous*u* to the day of his death, and lived in a separate house. Jotham the king's son was in charge of the palace, governing the people of the land. 6 Now the rest of the acts of Azariah, and all that he did,

are they not written in the Book of the Annals of the Kings of Judah? 7 Azariah slept with his ancestors; they buried him with his ancestors in the city of David; his son Jotham succeeded him.

8 In the thirty-eighth year of King Azariah of Judah, Zechariah son of Jeroboam reigned over Israel in Samaria six months. 9 He did what was evil in the sight of the LORD, as his ancestors had done. He did not depart from the sins of Jeroboam son of Nebat, which he caused Israel to sin. 10 Shallum son of Jabesh conspired against him, and struck him down in public and killed him, and reigned in place of him. 11 Now the rest of the deeds of Zechariah are written in the Book of the Annals of the Kings of Israel. 12 This was the promise of the LORD that he gave to Jehu, "Your sons shall sit on the throne of Israel to the fourth generation." And so it happened.

13 Shallum son of Jabesh began to reign in the thirty-ninth year of King Uzziah of Judah; he reigned one month in Samaria. 14 Then Menahem son of Gadi came up from Tirzah and came to Samaria; he struck down Shallum son of Jabesh in Samaria and killed him; he reigned in place of him. 15 Now the rest of the deeds of Shallum, including the conspiracy that he made, are written in

u A term for several skin diseases; precise meaning uncertain

---

show sympathy for Israel and are inconsistent with the usual attitudes of the Deuteronomist (see Introduction to 1 Kings and contrast v. 24). Contrast the attitudes of Amos (Am 7.11) and Hosea (Hos 1.4–5; 10.7, 15; 13.16), who prophesied during the reign of *Jeroboam* (see Introductions to Amos and Hosea). **28:** This verse has been badly damaged in transmission; some scholars propose to read something like this: "how he fought with Damascus and how he averted the wrath of the LORD from Israel . . ." **15.1–7: The reign of Azariah (Uzziah) in Judah.** The reign of *Azariah* (called Uzziah in vv. 13, 30, 32, 34, and elsewhere) was long and prosperous, like that of *Jeroboam* II in Israel (see 14.23–24 and additional data in 2 Chr 26.6–15). **5:** Because *the king was leprous,* his

son *Jotham* acted as regent (*was in charge of the palace*) during the latter years of the reign. **7:** On his death see Isa 6.1. A limestone inscription found at Jerusalem bears the inscription: "Hither were brought the bones of Uzziah, King of Judah: not to be opened" (date 1st cent. A.D.). **15.8–31: The reigns of Zechariah, Shallum, Menahem, Pekahiah, and Pekah in Israel. 8–12:** *Zechariah* was the last of the dynasty of *Jehu* (v. 12; compare 10.30). His assassination began a series of revolts and counter-revolts like those preceding the reign of Omri (1 Kings ch 16). **13–16:** *Shallum* was quickly murdered by *Menahem.* On the custom of disemboweling pregnant women, see 8.12; Hos 13.16; Am 1.13.

the Book of the Annals of the Kings of Israel. 16 At that time Menahem sacked Tiphsah, all who were in it and its territory from Tirzah on; because they did not open it to him, he sacked it. He ripped open all the pregnant women in it.

17 In the thirty-ninth year of King Azariah of Judah, Menahem son of Gadi began to reign over Israel; he reigned ten years in Samaria. 18 He did what was evil in the sight of the LORD; he did not depart all his days from any of the sins of Jeroboam son of Nebat, which he caused Israel to sin. 19 King Pul of Assyria came against the land; Menahem gave Pul a thousand talents of silver, so that he might help him confirm his hold on the royal power. 20 Menahem exacted the money from Israel, that is, from all the wealthy, fifty shekels of silver from each one, to give to the king of Assyria. So the king of Assyria turned back, and did not stay there in the land. 21 Now the rest of the deeds of Menahem, and all that he did, are they not written in the Book of the Annals of the Kings of Israel? 22 Menahem slept with his ancestors, and his son Pekahiah succeeded him.

23 In the fiftieth year of King Azariah of Judah, Pekahiah son of Menahem began to reign over Israel in Samaria; he reigned two years. 24 He did what was evil in the sight of the LORD; he did not turn away from the sins of Jeroboam son of Nebat, which he caused Israel to sin. 25 Pekah son of Remaliah, his captain,

conspired against him with fifty of the Gileadites, and attacked him in Samaria, in the citadel of the palace along with Argob and Arieh; he killed him, and reigned in place of him. 26 Now the rest of the deeds of Pekahiah, and all that he did, are written in the Book of the Annals of the Kings of Israel.

27 In the fifty-second year of King Azariah of Judah, Pekah son of Remaliah began to reign over Israel in Samaria; he reigned twenty years. 28 He did what was evil in the sight of the LORD; he did not depart from the sins of Jeroboam son of Nebat, which he caused Israel to sin.

29 In the days of King Pekah of Israel, King Tiglath-pileser of Assyria came and captured Ijon, Abel-beth-maacah, Janoah, Kedesh, Hazor, Gilead, and Galilee, all the land of Naphtali; and he carried the people captive to Assyria. 30 Then Hoshea son of Elah made a conspiracy against Pekah son of Remaliah, attacked him, and killed him; he reigned in place of him, in the twentieth year of Jotham son of Uzziah. 31 Now the rest of the acts of Pekah, and all that he did, are written in the Book of the Annals of the Kings of Israel.

32 In the second year of King Pekah son of Remaliah of Israel, King Jotham son of Uzziah of Judah began to reign. 33 He was twenty-five years old when he began to reign and reigned sixteen years in Jerusalem. His mother's name was Jerusha daughter of Zadok. 34 He did what

---

15.17–22: Menahem's reign. Israel, tottering internally, was an easy prey to outside attack by a reviving Assyria. **19–20:** *Pul* was another name of Tiglath-pileser III, *king of Assyria*. Menahem was forced to pay a huge sum to be allowed to hold his throne (roughly, a talent weighs 75 lbs.; it contains 3000 shekels. In order to amass 1000 talents, 60,000 of *the wealthy* were each taxed *fifty shekels,* about 20 oz., *of silver.*). The time was 738 B.C. Israel was henceforth a vassal state. **23–26:** Pekahiah's reign. **27–31:** Pekah's reign. **27:** The total of *twenty years* is much too long for the reign of Pekah (see 14.1 n.). **15.29:** A somewhat garbled list, containing first the names of certain towns, followed by

three names of larger areas: *Gilead, and Galilee, all the land of Naphtali.* Two or three lists may have been combined. In any case, all the names belong to the northern part of the country, which was ravaged by Tiglath-pileser in his campaigns of 733–732 B.C., when he punished Pekah for anti-Assyrian plotting (v. 37; 16.5, 7–8) and also put an end to the kingdom of Aram (Syria) by capturing Damascus (16.9). We notice the beginning of Israelite deportation in this verse (compare 17.6). **30:** Tiglath-pileser in his own "Annals" claims to have had a hand in the overthrow of Pekah by Hoshea. Thus Hoshea began with a pro-Assyrian policy. **15.32–38:** **The reign of Jotham in Ju-**

was right in the sight of the LORD, just as his father Uzziah had done. 35 Nevertheless the high places were not removed; the people still sacrificed and made offerings on the high places. He built the upper gate of the house of the LORD. 36 Now the rest of the acts of Jotham, and all that he did, are they not written in the Book of the Annals of the Kings of Judah? 37 In those days the LORD began to send King Rezin of Aram and Pekah son of Remaliah against Judah. 38 Jotham slept with his ancestors, and was buried with his ancestors in the city of David, his ancestor; his son Ahaz succeeded him.

16 In the seventeenth year of Pekah son of Remaliah, King Ahaz son of Jotham of Judah began to reign. 2 Ahaz was twenty years old when he began to reign; he reigned sixteen years in Jerusalem. He did not do what was right in the sight of the LORD his God, as his ancestor David had done, 3 but he walked in the way of the kings of Israel. He even made his son pass through fire, according to the abominable practices of the nations whom the LORD drove out before the people of Israel. 4 He sacrificed and made offerings on the high places, on the hills, and under every green tree.

5 Then King Rezin of Aram and King Pekah son of Remaliah of Israel came up to wage war on Jerusalem; they besieged Ahaz but could not conquer him. 6 At that time the king of Edom *v* recovered Elath for Edom, *w* and drove the Judeans

from Elath; and the Edomites came to Elath, where they live to this day. 7 Ahaz sent messengers to King Tiglath-pileser of Assyria, saying, "I am your servant and your son. Come up, and rescue me from the hand of the king of Aram and from the hand of the king of Israel, who are attacking me." 8 Ahaz also took the silver and gold found in the house of the LORD and in the treasures of the king's house, and sent a present to the king of Assyria. 9 The king of Assyria listened to him; the king of Assyria marched up against Damascus, and took it, carrying its people captive to Kir; then he killed Rezin.

10 When King Ahaz went to Damascus to meet King Tiglath-pileser of Assyria, he saw the altar that was at Damascus. King Ahaz sent to the priest Uriah a model of the altar, and its pattern, exact in all its details. 11 The priest Uriah built the altar; in accordance with all that King Ahaz had sent from Damascus, just so did the priest Uriah build it, before King Ahaz arrived from Damascus. 12 When the king came from Damascus, the king viewed the altar. Then the king drew near to the altar, went up on it, 13 and offered his burnt offering and his grain offering, poured his drink offering, and dashed the blood of his offerings of well-

v Cn: Heb *King Rezin of Aram*   *w* Cn: Heb *Aram*

---

dah. For futher details of his reign, see 2 Chr 27.1–9. **37:** The idea of this attack was to force Judah to join a coalition against Assyria (see v. 29 n. and 16.5–9).
   16.1–20: The reign of Ahaz in Judah. 1–4: *Ahaz* is condemned as unusually bad by the Deuteronomistic editor in comparison with other kings of Judah. Worst of all, he revived the barbarous custom of human sacrifice (see 3.27 n.). **5:** See 15.37 n.; compare also Isa 7.1–17; 8.1–8a. **6:** *Edom* took advantage of the situation to throw off the control of Judah (see 14.22 n.). **7–9:** *Ahaz* disregarded the advice of Isaiah (Isa 7.4, 16–17; 8.4–8a) and called on *Tiglath-pileser* for help, sending an enormous gift. *The king of Assyria,* glad to be paid for what he had intended to do anyhow, took Damascus and devastated Israel

(see 15.29 n.). *Kir* was the place in Mesopotamia from which some of the Arameans (Syrians) had come originally (Am 1.5; 9.7).
   16.10–16: *Ahaz went to Damascus* to pay homage to his overlord, the *King . . . of Assyria;* while there he saw an altar, probably of Assyrian type, which he liked. He had this altar duplicated and placed before the temple in Jerusalem. Ahaz was probably thus paying his respects to a foreign religion, though the matter is not absolutely certain. *The priest Uriah* is probably the same person as the Uriah of Isa 8.2, and hence could hardly have been wholly disloyal to the LORD. **17–18:** It was necessary for Ahaz to dismantle some of the costly bronze equipment of the temple in order to pay the heavy tribute to Assyria. *The stands,* see 1 Kings 7.27–37. *The sea* and the

being against the altar. 14The bronze altar that was before the LORD he removed from the front of the house, from the place between his altar and the house of the LORD, and put it on the north side of his altar. 15King Ahaz commanded the priest Uriah, saying, "Upon the great altar offer the morning burnt offering, and the evening grain offering, and the king's burnt offering, and his grain offering, with the burnt offering of all the people of the land, their grain offering, and their drink offering; then dash against it all the blood of the burnt offering, and all the blood of the sacrifice; but the bronze altar shall be for me to inquire by." 16The priest Uriah did everything that King Ahaz commanded.

17 Then King Ahaz cut off the frames of the stands, and removed the laver from them; he removed the sea from the bronze oxen that were under it, and put it on a pediment of stone. 18The covered portal for use on the sabbath that had been built inside the palace, and the outer entrance for the king he removed from[x] the house of the LORD. He did this because of the king of Assyria. 19Now the rest of the acts of Ahaz that he did, are they not written in the Book of the Annals of the Kings of Judah? 20Ahaz slept with his ancestors, and was buried with his ancestors in the city of David; his son Hezekiah succeeded him.

17 In the twelfth year of King Ahaz of Judah, Hoshea son of Elah began to reign in Samaria over Israel; he reigned nine years. 2He did what was evil in the sight of the LORD, yet not like the kings of Israel who were before him. 3King Shalmaneser of Assyria came up against him; Hoshea became his vassal, and paid him tribute. 4But the king of Assyria found treachery in Hoshea; for he had sent messengers to King So of Egypt, and offered no tribute to the king of Assyria, as he had done year by year; therefore the king of Assyria confined him and imprisoned him.

5 Then the king of Assyria invaded all the land and came to Samaria; for three years he besieged it. 6In the ninth year of Hoshea the king of Assyria captured Samaria; he carried the Israelites away to Assyria. He placed them in Halah, on the Habor, the river of Gozan, and in the cities of the Medes.

7 This occurred because the people of Israel had sinned against the LORD their God, who had brought them up out of the land of Egypt from under the hand of Pharaoh king of Egypt. They had worshiped other gods 8and walked in the customs of the nations whom the LORD drove out before the people of Israel, and in the customs that the kings of Israel had introduced.[y] 9The people of Israel secretly did things that were not right against the LORD their God. They built

x Cn: Heb lacks *from*   y Meaning of Heb uncertain

bronze oxen that were under it, 1 Kings 7.23–26. The words *the covered portal for use on the sabbath* represent a Hebrew phrase of uncertain meaning.
17.1–41: The end of Israel, with reasons for the catastrophe; the origin of the Samaritans. 1: *In the twelfth year of King Ahaz* is probably an erroneous synchronism (compare 15.30; see p. 339 OT). **2:** Perhaps the writer treated Hoshea more gently because of his tragic position as the last king of Israel. **3:** This was *Shalmaneser* V (727–722 B.C.). **4–5:** See 15.30 n. *Hoshea* now foolishly and disastrously plotted with Egypt (compare Hos 7.11); he left the capital, perhaps to plead for mercy, but was imprisoned; the siege of *Samaria* continued *for three years.* **6:** In the mean-

time Shalmaneser died and was succeeded by Sargon II (Isa 20.1), who captured the city and deported, according to his own records, 27,290 inhabitants to faraway places. Thus ended for all time the kingdom of Israel (721 B.C.).
17.7–18: These verses are the most important in the entire book (1 and 2 Kings) for the understanding of the theological and ethical viewpoint of the Deuteronomists (see Introduction to 1 Kings). The Israelites were considered sinners in the hands of an angry God (v. 18). Israel, not Assyria, was blamed. Playing politics with Egypt and Assyria was of no avail. Israel could have been saved by a proper attitude toward God and internal religious reform, but the warnings of the prophets

for themselves high places at all their towns, from watchtower to fortified city; ¹⁰they set up for themselves pillars and sacred poles *z* on every high hill and under every green tree; ¹¹there they made offerings on all the high places, as the nations did whom the LORD carried away before them. They did wicked things, provoking the LORD to anger; ¹²they served idols, of which the LORD had said to them, "You shall not do this." ¹³Yet the LORD warned Israel and Judah by every prophet and every seer, saying, "Turn from your evil ways and keep my commandments and my statutes, in accordance with all the law that I commanded your ancestors and that I sent to you by my servants the prophets." ¹⁴They would not listen but were stubborn, as their ancestors had been, who did not believe in the LORD their God. ¹⁵They despised his statutes, and his covenant that he made with their ancestors, and the warnings that he gave them. They went after false idols and became false; they followed the nations that were around them, concerning whom the LORD had commanded them that they should not do as they did. ¹⁶They rejected all the commandments of the LORD their God and made for themselves cast images of two calves; they made a sacred pole, *a* worshiped all the host of heaven, and served Baal. ¹⁷They made their sons and their daughters pass through fire; they used divination and augury; and they sold themselves to do evil in the sight of the LORD, provoking him to anger. ¹⁸Therefore the LORD was very angry with Israel and removed them out of his sight; none was left but the tribe of Judah alone.

19 Judah also did not keep the commandments of the LORD their God but

walked in the customs that Israel had introduced. ²⁰The LORD rejected all the descendants of Israel; he punished them and gave them into the hand of plunderers, until he had banished them from his presence.

21 When he had torn Israel from the house of David, they made Jeroboam son of Nebat king. Jeroboam drove Israel from following the LORD and made them commit great sin. ²²The people of Israel continued in all the sins that Jeroboam committed; they did not depart from them ²³until the LORD removed Israel out of his sight, as he had foretold through all his servants the prophets. So Israel was exiled from their own land to Assyria until this day.

24 The king of Assyria brought people from Babylon, Cuthah, Avva, Hamath, and Sepharvaim, and placed them in the cities of Samaria in place of the people of Israel; they took possession of Samaria, and settled in its cities. ²⁵When they first settled there, they did not worship the LORD; therefore the LORD sent lions among them, which killed some of them. ²⁶So the king of Assyria was told, "The nations that you have carried away and placed in the cities of Samaria do not know the law of the god of the land; therefore he has sent lions among them; they are killing them, because they do not know the law of the god of the land." ²⁷Then the king of Assyria commanded, "Send there one of the priests whom you carried away from there; let him *b* go and live there, and teach them the law of the god of the land." ²⁸So one of the priests whom they had carried away from Samaria came and lived in Bethel; he taught

z Heb *Asherim*   *a* Heb *Asherah*   *b* Syr Vg: Heb *them*

were disregarded (vv. 13–14). *Sacred pole* (vv. 10, 16), see 1 Kings 14.15 n.; 18.17–19 n. On the worship of the heavenly bodies (*host of heaven*, v. 16), see 21.5; 23.4–5. **17**: Human sacrifice, see 3.27 n.

17.19–20: By a writer, perhaps a second Deuteronomist, who knew the fate of the southern kingdom. **21–23**: A concluding

summary of vv. 7–18. **24–28**: Sargon's own record confirms v. 24 thus: "[The cities] I set up again and made more populous than before. People from lands which I had taken I settled there." Later Assyrian kings continued this policy. The land was no longer called Israel, but *Samaria* (v. 24). The people were Samaritans (v. 29), not Israelites.

them how they should worship the LORD.

29 But every nation still made gods of its own and put them in the shrines of the high places that the people of Samaria had made, every nation in the cities in which they lived; 30 the people of Babylon made Succoth-benoth, the people of Cuth made Nergal, the people of Hamath made Ashima; 31 the Avvites made Nibhaz and Tartak; the Sepharvites burned their children in the fire to Adrammelech and Anammelech, the gods of Sepharvaim. 32 They also worshiped the LORD and appointed from among themselves all sorts of people as priests of the high places, who sacrificed for them in the shrines of the high places. 33 So they worshiped the LORD but also served their own gods, after the manner of the nations from among whom they had been carried away. 34 To this day they continue to practice their former customs.

They do not worship the LORD and they do not follow the statutes or the ordinances or the law or the commandment that the LORD commanded the children of Jacob, whom he named Israel. 35 The LORD had made a covenant with them and commanded them, "You shall not worship other gods or bow yourselves to them or serve them or sacrifice to them, 36 but you shall worship the LORD, who brought you out of the land of Egypt with great power and with an outstretched arm; you shall bow yourselves to him, and to him you shall sacrifice. 37 The statutes and the ordinances and the law and the commandment that he wrote for you, you shall always be careful to observe. You shall not worship other gods; 38 you shall not forget the covenant that I have made with you. You shall not worship other gods, 39 but you shall worship the LORD your God; he will deliver you out of the hand of all your enemies." 40 They would not listen, however, but they continued to practice their former custom.

41 So these nations worshiped the LORD, but also served their carved images; to this day their children and their children's children continue to do as their ancestors did.

18 In the third year of King Hoshea son of Elah of Israel, Hezekiah son of King Ahaz of Judah began to reign. 2 He was twenty-five years old when he began to reign; he reigned

---

17.29–34a: Thus it came about that the worship of the LORD became contaminated by various foreign cults. *To this day* (v. 34a) presumably means the time of the Deuteronomist, a century after the fall of Israel; *their former customs* refers to their religions practiced before they were brought to Samaria.
17.34b–40: An addition by a late editor who wished to make it plain that the Samaritans were not to be credited with any proper worship of the LORD whatever—they were to be considered beyond the pale of religiously proper persons. This attitude toward the Samaritans continued into New Testament times (Ezra 4.1–3; Lk 10.33; 17.16–18; Jn 4.9; 8.48). **41:** This verse summarizes vv. 29–34a.
18.1–20.21: **The reign of Hezekiah.** *Hezekiah* was almost as highly favored by the Deuteronomistic writer (18.5) as Josiah (23.25). For this reason, perhaps, a comparatively large amount of space is allotted to Hezekiah's reign. The reign was also significant because of political events such as the attack of *Sennacherib* (18.13), and because of

the activities of *the prophet Isaiah* (19.2). These three chapters present many exegetical problems, but the main outlines of the story seem to be clear: Hezekiah revolted against Assyria (18.7) and Judah was severely punished (701 B.C.). Sennacherib himself reported that forty-six of the fortified cities and "countless small villages" were taken, while Hezekiah was shut up in Jerusalem "like a bird in a cage" (compare Isa 1.7–8). Jerusalem escaped capture only by the payment of a huge sum for indemnity (18.13–16). Judah was forced to remain a subservient vassal of Assyria. Some scholars believe that Sennacherib attacked again several years later and was repulsed; solid proof of this view (called "the two-campaign theory") is lacking. 18.13–20.19 is repeated in Isa chs 36–39 with certain omissions and additions. For additional comments see notes there.
18.1–12: **The accession of Hezekiah, his reforms. 1:** *In the third year of King Hoshea* is probably too early. The Deuteronomistic author wishes to place the fall of Samaria during

twenty-nine years in Jerusalem. His mother's name was Abi daughter of Zechariah. [3] He did what was right in the sight of the LORD just as his ancestor David had done. [4] He removed the high places, broke down the pillars, and cut down the sacred pole. [c] He broke in pieces the bronze serpent that Moses had made, for until those days the people of Israel had made offerings to it; it was called Nehushtan. [5] He trusted in the LORD the God of Israel; so that there was no one like him among all the kings of Judah after him, or among those who were before him. [6] For he held fast to the LORD; he did not depart from following him but kept the commandments that the LORD commanded Moses. [7] The LORD was with him; wherever he went, he prospered. He rebelled against the king of Assyria and would not serve him. [8] He attacked the Philistines as far as Gaza and its territory, from watchtower to fortified city.

9 In the fourth year of King Hezekiah, which was the seventh year of King Hoshea son of Elah of Israel, King Shalmaneser of Assyria came up against Samaria, besieged it, [10] and at the end of three years, took it. In the sixth year of Hezekiah, which was the ninth year of King Hoshea of Israel, Samaria was taken. [11] The king of Assyria carried the Israelites away to Assyria, settled them in Halah, on the Habor, the river of Gozan, and in the cities of the Medes, [12] because they did not obey the voice of the LORD their God but transgressed his covenant—all that Moses the servant of the LORD had commanded; they neither listened nor obeyed.

13 In the fourteenth year of King Hezekiah, King Sennacherib of Assyria came up against all the fortified cities of Judah and captured them. [14] King Hezekiah of Judah sent to the king of Assyria at Lachish, saying, "I have done wrong; withdraw from me; whatever you impose on me I will bear." The king of Assyria demanded of King Hezekiah of Judah three hundred talents of silver and thirty talents of gold. [15] Hezekiah gave him all the silver that was found in the house of the LORD and in the treasuries of the king's house. [16] At that time Hezekiah stripped the gold from the doors of the temple of the LORD, and from the doorposts that King Hezekiah of Judah had overlaid and gave it to the king of Assyria. [17] The king of Assyria sent the Tartan, the Rabsaris, and the Rabshakeh with a great army from Lachish to King Hezekiah at Jerusalem. They went up and came to Jerusalem. When they arrived, they came and stood by the conduit of the upper pool, which is on the highway to the Fuller's Field. [18] When they called for the king, there came out to them Eliakim son of Hilkiah, who was in charge of the palace, and Shebnah the secretary, and Joah son of Asaph, the recorder.

19 The Rabshakeh said to them, "Say to Hezekiah: Thus says the great king, the king of Assyria: On what do you base this confidence of yours? [20] Do you think that mere words are strategy and power

c Heb *Asherah*

---

the reign of Hezekiah (vv. 9–10), but modern chronologists tend to place this catastrophe in the reign of Ahaz, in conformity with v. 13 (fourteenth yr. of Hezekiah = about 701 B.C., first yr. = about 715 B.C.). **18.4:** *The sacred pole,* see 1 Kings 14.15 n. *The bronze serpent,* see Num 21.6–9; this object doubtless had negative associations, since the serpent was one of the symbols of the Baal religion. **9–12:** The fall of Samaria; largely a repetition of 17.5–6. **18.13–37: The attack of Sennacherib.**

See Isa ch 36. These verses present a highly dramatized account of the historical incident. Verses 14–16 are lacking from the account in Isa ch 36. **17–36:** If this narrative belongs to the first (and only historically certain) attack of Sennacherib, it should come before v. 14. Otherwise, it may be referred to the hypothetical second campaign some twelve or thirteen years later. **18.17–18:** Three Assyrian officials, *the Tartan, the Rabsaris, and the Rabshakeh* confer with three Judahite officials.

for war? On whom do you now rely, that you have rebelled against me? 21 See, you are relying now on Egypt, that broken reed of a staff, which will pierce the hand of anyone who leans on it. Such is Pharaoh king of Egypt to all who rely on him. 22 But if you say to me, 'We rely on the LORD our God,' is it not he whose high places and altars Hezekiah has removed, saying to Judah and to Jerusalem, 'You shall worship before this altar in Jerusalem'? 23 Come now, make a wager with my master the king of Assyria: I will give you two thousand horses, if you are able on your part to set riders on them. 24 How then can you repulse a single captain among the least of my master's servants, when you rely on Egypt for chariots and for horsemen? 25 Moreover, is it without the LORD that I have come up against this place to destroy it? The LORD said to me, Go up against this land, and destroy it."

26 Then Eliakim son of Hilkiah, and Shebnah, and Joah said to the Rabshakeh, "Please speak to your servants in the Aramaic language, for we understand it; do not speak to us in the language of Judah within the hearing of the people who are on the wall." 27 But the Rabshakeh said to them, "Has my master sent me to speak these words to your master and to you, and not to the people sitting on the wall, who are doomed with you to eat their own dung and to drink their own urine?"

28 Then the Rabshakeh stood and called out in a loud voice in the language of Judah, "Hear the word of the great king, the king of Assyria! 29 Thus says the king: 'Do not let Hezekiah deceive you, for he will not be able to deliver you out of my hand. 30 Do not let Hezekiah make you rely on the LORD by saying, The

LORD will surely deliver us, and this city will not be given into the hand of the king of Assyria.' 31 Do not listen to Hezekiah; for thus says the king of Assyria: 'Make your peace with me and come out to me; then every one of you will eat from your own vine and your own fig tree, and drink water from your own cistern, 32 until I come and take you away to a land like your own land, a land of grain and wine, a land of bread and vineyards, a land of olive oil and honey, that you may live and not die. Do not listen to Hezekiah when he misleads you by saying, The LORD will deliver us. 33 Has any of the gods of the nations ever delivered its land out of the hand of the king of Assyria? 34 Where are the gods of Hamath and Arpad? Where are the gods of Sepharvaim, Hena, and Ivvah? Have they delivered Samaria out of my hand? 35 Who among all the gods of the countries have delivered their countries out of my hand, that the LORD should deliver Jerusalem out of my hand?' "

36 But the people were silent and answered him not a word, for the king's command was, "Do not answer him." 37 Then Eliakim son of Hilkiah, who was in charge of the palace, and Shebna the secretary, and Joah son of Asaph, the recorder, came to Hezekiah with their clothes torn and told him the words of the Rabshakeh.

19 When King Hezekiah heard it, he tore his clothes, covered himself with sackcloth, and went into the house of the LORD. 2 And he sent Eliakim, who was in charge of the palace, and Shebna the secretary, and the senior priests, covered with sackcloth, to the prophet Isaiah son of Amoz. 3 They said to him, "Thus says Hezekiah, This day is a day of distress, of rebuke, and of disgrace; children

---

18.26: After the Exile *the Aramaic language* (the language of Syria) came into common use among the Jews of Palestine (see Neh 8.7–8 n.), and thus was the medium in which Jesus spoke; *the language of Judah* here means what is now called Hebrew, which was largely superseded by Aramaic.

19.1–37: **Hezekiah consults Isaiah.** See

Isa ch 37. Sennacherib departs without capturing Jerusalem and is assassinated by his sons. Legend and history seem to be interwoven here. In neither campaign (if there were two) did Judah really win a victory; Jerusalem merely escaped capture. It is questionable whether *the prophet Isaiah* predicted the assassination of Sennacherib (v. 7), uttered a taunt

have come to the birth, and there is no strength to bring them forth. ⁴It may be that the LORD your God heard all the words of the Rabshakeh, whom his master the king of Assyria has sent to mock the living God, and will rebuke the words that the LORD your God has heard; therefore lift up your prayer for the remnant that is left." ⁵When the servants of King Hezekiah came to Isaiah, ⁶Isaiah said to them, "Say to your master, 'Thus says the LORD: Do not be afraid because of the words that you have heard, with which the servants of the king of Assyria have reviled me. ⁷I myself will put a spirit in him, so that he shall hear a rumor and return to his own land; I will cause him to fall by the sword in his own land.'"

8 The Rabshakeh returned, and found the king of Assyria fighting against Libnah; for he had heard that the king had left Lachish. ⁹When the king*d* heard concerning King Tirhakah of Ethiopia,*e* "See, he has set out to fight against you," he sent messengers again to Hezekiah, saying, ¹⁰"Thus shall you speak to King Hezekiah of Judah: Do not let your God on whom you rely deceive you by promising that Jerusalem will not be given into the hand of the king of Assyria. ¹¹See, you have heard what the kings of Assyria have done to all lands, destroying them utterly. Shall you be delivered? ¹²Have the gods of the nations delivered them, the nations that my predecessors destroyed, Gozan, Haran, Rezeph, and the people of Eden who were in Telassar? ¹³Where is the king of Hamath, the king of Arpad, the king of the city of Sepharvaim, the king of Hena, or the king of Ivvah?"

14 Hezekiah received the letter from the hand of the messengers and read it; then Hezekiah went up to the house of the LORD and spread it before the LORD.

¹⁵And Hezekiah prayed before the LORD, and said: "O LORD the God of Israel, who are enthroned above the cherubim, you are God, you alone, of all the kingdoms of the earth; you have made heaven and earth. ¹⁶Incline your ear, O LORD, and hear; open your eyes, O LORD, and see; hear the words of Sennacherib, which he has sent to mock the living God. ¹⁷Truly, O LORD, the kings of Assyria have laid waste the nations and their lands, ¹⁸and have hurled their gods into the fire, though they were no gods but the work of human hands—wood and stone—and so they were destroyed. ¹⁹So now, O LORD our God, save us, I pray you, from his hand, so that all the kingdoms of the earth may know that you, O LORD, are God alone."

20 Then Isaiah son of Amoz sent to Hezekiah, saying, "Thus says the LORD, the God of Israel: I have heard your prayer to me about King Sennacherib of Assyria. ²¹This is the word that the LORD has spoken concerning him:

She despises you, she scorns
    you—
    virgin daughter Zion;
she tosses her head—behind your
    back,
    daughter Jerusalem.

²² Whom have you mocked and
    reviled?
    Against whom have you raised
    your voice
and haughtily lifted your eyes?
    Against the Holy One of Israel!
²³ By your messengers you have
    mocked the Lord,
    and you have said, 'With my
    many chariots
I have gone up the heights of the
    mountains,

d Heb *he*   *e* Or Nubia; Heb *Cush*

---

song or ode of derision against the Assyrian king (vv. 20–28), or proclaimed the inviolability of the city (v. 34). Tradition tended to turn an escape into a victory.

19.9: *King Tirhakah of Ethiopia* did not become king or pharaoh of Egypt until 690 or

688 B.C.; hence some have regarded mention of this name as proof of a second campaign (see 18.17–36 n.). Others regard the name as an error, or point out that *Tirhakah* was a general long before he became king.

19.20–28: For similar anti-Assyrian mate-

to the far recesses of Lebanon;
  I felled its tallest cedars,
    its choicest cypresses;
  I entered its farthest retreat,
    its densest forest.
24 I dug wells
    and drank foreign waters,
  I dried up with the sole of my
      foot
    all the streams of Egypt.'

25 Have you not heard
    that I determined it long ago?
  I planned from days of old
    what now I bring to pass,
  that you should make fortified
      cities
    crash into heaps of ruins,
26 while their inhabitants, shorn of
      strength,
    are dismayed and confounded;
  they have become like plants of
      the field
    and like tender grass,
  like grass on the housetops,
    blighted before it is grown.

27 "But I know your rising*f* and
      your sitting,
    your going out and coming in,
    and your raging against me.
28 Because you have raged against
      me
    and your arrogance has come to
      my ears,
  I will put my hook in your nose
    and my bit in your mouth;

I will turn you back on the way
  by which you came.

29 "And this shall be the sign for you:
This year you shall eat what grows of
itself, and in the second year what
springs from that; then in the third year
sow, reap, plant vineyards, and eat their
fruit. 30 The surviving remnant of the
house of Judah shall again take root
downward, and bear fruit upward; 31 for
from Jerusalem a remnant shall go out,
and from Mount Zion a band of survivors. The zeal of the LORD of hosts will
do this.

32 "Therefore thus says the LORD
concerning the king of Assyria: He shall
not come into this city, shoot an arrow
there, come before it with a shield, or
cast up a siege ramp against it. 33 By the
way that he came, by the same he shall
return; he shall not come into this city,
says the LORD. 34 For I will defend this
city to save it, for my own sake and for
the sake of my servant David."

35 That very night the angel of the
LORD set out and struck down one hundred eighty-five thousand in the camp of
the Assyrians; when morning dawned,
they were all dead bodies. 36 Then King
Sennacherib of Assyria left, went home,
and lived at Nineveh. 37 As he was worshiping in the house of his god Nisroch,
his sons Adrammelech and Sharezer
killed him with the sword, and they es-

f Gk Compare Isa 37.27 Q Ms: MT lacks *rising*

---

rial, also attributed to *Isaiah,* compare Isa
10.12–19; 14.24–27. **21:** *Daughter Zion* and
daughter Jerusalem, terms of endearment.
19.29–31: These verses, though not very
clear, seem to be more modestly realistic than
the rest of the chapter in expecting only *a
remnant* and *a band of survivors* to carry forward the work of the LORD. **32–34:** Contrast
these verses with the preceding. It has been
claimed, on the basis of this passage and the
parallel in Isa 37.33–35, that Isaiah taught the
doctrine of "the inviolability of Zion," i.e.
that Jerusalem could never be taken (compare
20.6). Many interpreters doubt this claim,
with very good reason.

19.35: This defeat has been discussed without much justification in relation to a tradition by Herodotus (*Hist.,* II.141) that *the Assyrians* suffered a defeat on the borders of
Egypt because their bowstrings and other
leather equipment were chewed by a sudden
onslaught of field mice. Mice may be carriers
of the plague (see 1 Sam 6.4–5 n.). **37:** The
murder did not take place until the year 681,
twenty years after the first (and only?) campaign against Jerusalem, and at least eight
years after the hypothetical second campaign.
The god *Nisroch* has not been identified with
certainty. *Ararat* is Armenia (compare Gen
8.4).

caped into the land of Ararat. His son Esar-haddon succeeded him.

20 In those days Hezekiah became sick and was at the point of death. The prophet Isaiah son of Amoz came to him, and said to him, "Thus says the LORD: Set your house in order, for you shall die; you shall not recover." ²Then Hezekiah turned his face to the wall and prayed to the LORD: ³"Remember now, O LORD, I implore you, how I have walked before you in faithfulness with a whole heart, and have done what is good in your sight." Hezekiah wept bitterly. ⁴Before Isaiah had gone out of the middle court, the word of the LORD came to him: ⁵"Turn back, and say to Hezekiah prince of my people, Thus says the LORD, the God of your ancestor David: I have heard your prayer, I have seen your tears; indeed, I will heal you; on the third day you shall go up to the house of the LORD. ⁶I will add fifteen years to your life. I will deliver you and this city out of the hand of the king of Assyria; I will defend this city for my own sake and for my servant David's sake." ⁷Then Isaiah said, "Bring a lump of figs. Let them take it and apply it to the boil, so that he may recover."

8 Hezekiah said to Isaiah, "What shall be the sign that the LORD will heal me, and that I shall go up to the house of the LORD on the third day?" ⁹Isaiah said, "This is the sign to you from the LORD, that the LORD will do the thing that he has promised: the shadow has now advanced ten intervals; shall it retreat ten intervals?" ¹⁰Hezekiah answered, "It is normal for the shadow to lengthen ten intervals; rather let the shadow retreat ten intervals." ¹¹The prophet Isaiah cried to the LORD; and he brought the shadow back the ten intervals, by which the sun*g* had declined on the dial of Ahaz.

12 At that time King Merodach-baladan son of Baladan of Babylon sent envoys with letters and a present to Hezekiah, for he had heard that Hezekiah had been sick. ¹³Hezekiah welcomed them;*h* he showed them all his treasure house, the silver, the gold, the spices, the precious oil, his armory, all that was found in his storehouses; there was nothing in his house or in all his realm that Hezekiah did not show them. ¹⁴Then the prophet Isaiah came to King Hezekiah, and said to him, "What did these men say? From where did they come to you?" Hezekiah answered, "They have come from a far country, from Babylon." ¹⁵He said, "What have they seen in your house?" Hezekiah answered, "They have seen all that is in my house; there is nothing in my storehouses that I did not show them."

16 Then Isaiah said to Hezekiah, "Hear the word of the LORD: ¹⁷Days are coming when all that is in your house, and that which your ancestors have stored up until this day, shall be carried

g Syr See Isa 38.8 and Tg: Heb *it* *h* Gk Vg Syr: Heb *When Hezekiah heard about them*

20.1–11: Hezekiah's illness and recovery. See Isa ch 38. **1:** *In those days* is only a vague indication of time. Actually, this chapter (except for vv. 20–21) is out of order, for both stories in it belong before the attack of Sennacherib in 701 B.C. (compare v. 6; Merodach-baladan ruled in Babylon, 722–710 and 703–702). **7:** The use of *a lump of figs* as a poultice was widespread in Bible times. **20.8–10:** Compare Josh 10.12–13 and 1 Kings ch 17 n. Though it is really the earth which moves, we still say, "The sun rises." **20.12–19: Merodach-baladan's embassy.** See Isa 39.1–8. The Babylonian name of *Merodach-baladan* was Marduk-apaliddina. This mission was a part of the plotting against Assyria to which *Isaiah* was opposed and which led to the disastrous attack of 701 B.C., discussed above (18.7, 21; Isa 7.17; 8.7–8; 20.1–6; 31.1–3). **16–19:** Babylon and Egypt were both plotting; Isaiah was a neutralist, opposed to both; he was equally opposed to an alliance with Assyria (see 16.7–9 n.); but after Judah became an Assyrian vassal, he advocated loyalty to the obligation. The point of these verses, that Isaiah rebuked Hezekiah, and warned against the danger of invasion, would seem to be correct; the anachronistic reference to the Babylonian captivity, which took place more than a century later, brings this story into the Deuteronomist's aim to explain the catastrophe of exile.

to Babylon; nothing shall be left, says the LORD. 18 Some of your own sons who are born to you shall be taken away; they shall be eunuchs in the palace of the king of Babylon." 19 Then Hezekiah said to Isaiah, "The word of the LORD that you have spoken is good." For he thought, "Why not, if there will be peace and security in my days?"

20 The rest of the deeds of Hezekiah, all his power, how he made the pool and the conduit and brought water into the city, are they not written in the Book of the Annals of the Kings of Judah? 21 Hezekiah slept with his ancestors; and his son Manasseh succeeded him.

21 Manasseh was twelve years old when he began to reign; he reigned fifty-five years in Jerusalem. His mother's name was Hephzibah. 2 He did what was evil in the sight of the LORD, following the abominable practices of the nations that the LORD drove out before the people of Israel. 3 For he rebuilt the high places that his father Hezekiah had destroyed; he erected altars for Baal, made a sacred pole, *i* as King Ahab of Israel had done, worshiped all the host of heaven, and served them. 4 He built altars in the house of the LORD, of which the LORD had said, "In Jerusalem I will put my name." 5 He built altars for all the host of heaven in the two courts of the house of the LORD. 6 He made his son pass through fire; he practiced soothsaying and augury, and dealt with mediums

and with wizards. He did much evil in the sight of the LORD, provoking him to anger. 7 The carved image of Asherah that he had made he set in the house of which the LORD said to David and to his son Solomon, "In this house, and in Jerusalem, which I have chosen out of all the tribes of Israel, I will put my name forever; 8 I will not cause the feet of Israel to wander any more out of the land that I gave to their ancestors, if only they will be careful to do according to all that I have commanded them, and according to all the law that my servant Moses commanded them." 9 But they did not listen; Manasseh misled them to do more evil than the nations had done that the LORD destroyed before the people of Israel.

10 The LORD said by his servants the prophets, 11 "Because King Manasseh of Judah has committed these abominations, has done things more wicked than all that the Amorites did, who were before him, and has caused Judah also to sin with his idols; 12 therefore thus says the LORD, the God of Israel, I am bringing upon Jerusalem and Judah such evil that the ears of everyone who hears of it will tingle. 13 I will stretch over Jerusalem the measuring line for Samaria, and the plummet for the house of Ahab; I will wipe Jerusalem as one wipes a dish, wiping it and turning it upside down. 14 I will

i Heb *Asherah*

20.20–21: Deuteronomistic conclusion to the reign of Hezekiah, with an additional note about *the pool and the conduit,* improvements in the water supply of Jerusalem in preparation for possible attack (compare Isa 22.8b–11). This conduit, with an identifying inscription, has been found and is now popularly called Hezekiah's Tunnel, or the Siloam Tunnel (compare 2 Chr 32.30). It runs from Gihon (see 1 Kings 1.33 n.), which was outside the city wall, to the Pool of Siloam, which was inside the wall. Extending 1700 feet through solid rock, this tunnel was a remarkable engineering achievement in its time.
21.1–26: The wicked reigns of Manasseh and Amon. According to the Deuteronomists, this was the worst period of aposta-

sy in Judah. Any good done by Hezekiah was quickly undone by Manasseh, who probably reigned longer than any other king of Israel or Judah, though the figure of *fifty-five years* appears to be exaggerated (see p. 339 OT). For a somewhat different interpretation, see 2 Chr ch 33. **3:** *Sacred pole,* see 1 Kings 14.15 n. Child sacrifice, see 3.27 n. **7–9:** Compare Deut 12.5, 29–31; 17.3; 18.9–14.
21.10–15: This section was written after the fall of *Jerusalem* in 587–6 B.C. and during the Babylonian captivity (see Introduction to 1 Kings). This writer blames the fall of Jerusalem and Judah on the apostasy of Manasseh. *Amorites,* see 2 Sam 21.2 n. *The measuring line . . . and the plummet,* see Isa 34.11; Lam 2.8;

cast off the remnant of my heritage, and give them into the hand of their enemies; they shall become a prey and a spoil to all their enemies, 15 because they have done what is evil in my sight and have provoked me to anger, since the day their ancestors came out of Egypt, even to this day."

16 Moreover Manasseh shed very much innocent blood, until he had filled Jerusalem from one end to another, besides the sin that he caused Judah to sin so that they did what was evil in the sight of the LORD.

17 Now the rest of the acts of Manasseh, all that he did, and the sin that he committed, are they not written in the Book of the Annals of the Kings of Judah? 18 Manasseh slept with his ancestors, and was buried in the garden of his house, in the garden of Uzza. His son Amon succeeded him.

19 Amon was twenty-two years old when he began to reign; he reigned two years in Jerusalem. His mother's name was Meshullemeth daughter of Haruz of Jotbah. 20 He did what was evil in the sight of the LORD, as his father Manasseh had done. 21 He walked in all the way in which his father walked, served the idols that his father served, and worshiped them; 22 he abandoned the LORD, the God of his ancestors, and did not walk in the way of the LORD. 23 The servants of Amon conspired against him, and killed the king in his house. 24 But the people of the land killed all those who had conspired against King Amon, and the people of the land made his son Josiah king in place of him. 25 Now the rest of the acts of Amon that he did, are they not written in the Book of the Annals of the Kings of Judah? 26 He was buried in his tomb in the garden of Uzza; then his son Josiah succeeded him.

22 Josiah was eight years old when he began to reign; he reigned thirty-one years in Jerusalem. His mother's name was Jedidah daughter of Adaiah of Bozkath. 2 He did what was right in the sight of the LORD, and walked in all the way of his father David; he did not turn aside to the right or to the left.

3 In the eighteenth year of King Josiah, the king sent Shaphan son of Azaliah, son of Meshullam, the secretary, to the house of the LORD, saying, 4 "Go up to the high priest Hilkiah, and have him count the entire sum of the money that has been brought into the house of the LORD, which the keepers of the threshold have collected from the people; 5 let it be given into the hand of the workers who have the oversight of the house of the LORD; let them give it to the workers who are at the house of the LORD, repairing the house, 6 that is, to the carpenters, to the builders, to the masons; and let them use it to buy timber and quarried stone to repair the house. 7 But no accounting shall be asked from them for the money that is delivered into their hand, for they deal honestly."

8 The high priest Hilkiah said to Shaphan the secretary, "I have found the book of the law in the house of the LORD." When Hilkiah gave the book to Shaphan, he read it. 9 Then Shaphan the secretary came to the king, and reported to the king, "Your servants have emptied

---

Am 7.7–9. **16:** *Manasseh* actually persecuted the followers of the LORD, practically driving them underground.
21.19–26: A popular reaction and revolt set in, ending in the assassination of Amon early in his reign, and the accession of Josiah.
22.1–20: The beginning of the reign of Josiah; the repairing of the temple and the finding of the book of the law. 1–2: Besides David, *Josiah* is the favorite king of the Deuteronomists because of his reforms (see ch 23). **3–7:** *The eighteenth year of King Josiah,* 621 B.C. On the collection of *money*

(unminted metal) to repair the temple, compare 12.4–16. **8–10:** This *book of the law,* really a scroll, was apparently found in a collection box (12.9) or in some rubbish about to be removed from the temple. It is also possible that *Hilkiah* had known about it and brought it forth intentionally at this time. Most scholars suppose that the scroll contained the earliest form of our present book of Deuteronomy (see Introduction to Deuteronomy), as subsequent references in this and the following chapter will show.

out the money that was found in the house, and have delivered it into the hand of the workers who have oversight of the house of the LORD." [10]Shaphan the secretary informed the king, "The priest Hilkiah has given me a book." Shaphan then read it aloud to the king.

[11] When the king heard the words of the book of the law, he tore his clothes. [12]Then the king commanded the priest Hilkiah, Ahikam son of Shaphan, Achbor son of Micaiah, Shaphan the secretary, and the king's servant Asaiah, saying, [13]"Go, inquire of the LORD for me, for the people, and for all Judah, concerning the words of this book that has been found; for great is the wrath of the LORD that is kindled against us, because our ancestors did not obey the words of this book, to do according to all that is written concerning us."

[14] So the priest Hilkiah, Ahikam, Achbor, Shaphan, and Asaiah went to the prophetess Huldah the wife of Shallum son of Tikvah, son of Harhas, keeper of the wardrobe; she resided in Jerusalem in the Second Quarter, where they consulted her. [15]She declared to them, "Thus says the LORD, the God of Israel: Tell the man who sent you to me, [16]Thus says the LORD, I will indeed bring disaster on this place and on its inhabitants — all the words of the book that the king of Judah has read. [17]Because they have abandoned me and have made offerings to other gods, so that they have provoked me to anger with all the work of their hands, therefore my wrath will be kindled against this place, and it will not be quenched. [18]But as to the king of Ju-

dah, who sent you to inquire of the LORD, thus shall you say to him, Thus says the LORD, the God of Israel: Regarding the words that you have heard, [19]because your heart was penitent, and you humbled yourself before the LORD, when you heard how I spoke against this place, and against its inhabitants, that they should become a desolation and a curse, and because you have torn your clothes and wept before me, I also have heard you, says the LORD. [20]Therefore, I will gather you to your ancestors, and you shall be gathered to your grave in peace; your eyes shall not see all the disaster that I will bring on this place." They took the message back to the king.

23 Then the king directed that all the elders of Judah and Jerusalem should be gathered to him. [2]The king went up to the house of the LORD, and with him went all the people of Judah, all the inhabitants of Jerusalem, the priests, the prophets, and all the people, both small and great; he read in their hearing all the words of the book of the covenant that had been found in the house of the LORD. [3]The king stood by the pillar and made a covenant before the LORD, to follow the LORD, keeping his commandments, his decrees, and his statutes, with all his heart and all his soul, to perform the words of this covenant that were written in this book. All the people joined in the covenant.

[4] The king commanded the high priest Hilkiah, the priests of the second order, and the guardians of the threshold, to bring out of the temple of the LORD all the vessels made for Baal, for

22.11–13: The consternation of *the king* and his reference to *the wrath of the LORD* suggest that the scroll contained such words as those of Deut 6.13–15 and 28.15–24. To *inquire of the LORD* in an emergency by consulting a prophet was common practice (see 1 Kings 22.5–12 n.). 16–17: These words seem to reflect the Deuteronomist's view of the fall of Jerusalem. 23.1–30: The reforms and death of Josiah. 1–3: The adoption of this program not only meant a religious reform, but also a dec-

laration of political independence from Assyria, which was now in its last days as a great power (Nineveh fell in 612 B.C.). 2: *The book of the covenant,* contrast 22.8, 11. The idea of *the covenant,* or agreement, between the LORD and his people is one of the most important theological elements in the Hebrew Scriptures (see Ex 24.7–8; Deut 29.1; Jer 11.3–4; 30.22 and many other similar passages). *The pillar,* see 11.14; 1 Kings 7.15–22 n. 23.4–14: Josiah tried valiantly to get rid of the pagan cults in accordance with Deut

Asherah, and for all the host of heaven; he burned them outside Jerusalem in the fields of the Kidron, and carried their ashes to Bethel. 5He deposed the idolatrous priests whom the kings of Judah had ordained to make offerings in the high places at the cities of Judah and around Jerusalem; those also who made offerings to Baal, to the sun, the moon, the constellations, and all the host of the heavens. 6He brought out the image of*j* Asherah from the house of the LORD, outside Jerusalem, to the Wadi Kidron, burned it at the Wadi Kidron, beat it to dust and threw the dust of it upon the graves of the common people. 7He broke down the houses of the male temple prostitutes that were in the house of the LORD, where the women did weaving for Asherah. 8He brought all the priests out of the towns of Judah, and defiled the high places where the priests had made offerings, from Geba to Beer-sheba; he broke down the high places of the gates that were at the entrance of the gate of Joshua the governor of the city, which were on the left at the gate of the city. 9The priests of the high places, however, did not come up to the altar of the LORD in Jerusalem, but ate unleavened bread among their kindred. 10He defiled Topheth, which is in the valley of Benhinnom, so that no one would make a son or a daughter pass through fire as an offering to Molech. 11He removed the horses that the kings of Judah had dedicated to the sun, at the entrance to the house of the LORD, by the chamber of the eunuch Nathan-melech, which was in the precincts;*k* then he burned the chariots of the sun with fire. 12The altars on the roof of the upper chamber of Ahaz, which the kings of Judah had made, and the altars that Manasseh had made in the two courts of the house of the LORD, he pulled down from there and broke in

pieces, and threw the rubble into the Wadi Kidron. 13The king defiled the high places that were east of Jerusalem, to the south of the Mount of Destruction, which King Solomon of Israel had built for Astarte the abomination of the Sidonians, for Chemosh the abomination of Moab, and for Milcom the abomination of the Ammonites. 14He broke the pillars in pieces, cut down the sacred poles,*l* and covered the sites with human bones.

15 Moreover, the altar at Bethel, the high place erected by Jeroboam son of Nebat, who caused Israel to sin—he pulled down that altar along with the high place. He burned the high place, crushing it to dust; he also burned the sacred pole.*m* 16As Josiah turned, he saw the tombs there on the mount; and he sent and took the bones out of the tombs, and burned them on the altar, and defiled it, according to the word of the LORD that the man of God proclaimed,*n* when Jeroboam stood by the altar at the festival; he turned and looked up at the tomb of the man of God who had predicted these things. 17Then he said, "What is that monument that I see?" The people of the city told him, "It is the tomb of the man of God who came from Judah and predicted these things that you have done against the altar at Bethel." 18He said, "Let him rest; let no one move his bones." So they let his bones alone, with the bones of the prophet who came out of Samaria. 19Moreover, Josiah removed all the shrines of the high places that were in the towns of Samaria, which kings of Israel had made, provoking the LORD to anger; he did to them just as he had done at Bethel. 20He slaughtered on the altars

j Heb lacks *image of*   *k* Meaning of Heb uncertain   *l* Heb *Asherim*   *m* Heb *Asherah*
n Gk: Heb *proclaimed, who had predicted these things*

15.19–18.22. He even attempted to abolish all worship of the LORD outside of the temple in Jerusalem (at *the high places,* v. 8, compare Deut ch 12). The consequently unemployed local priests (called Levites in Deuteronomy) were supposed to join the temple staff in Jeru-

salem (Deut 18.6–8), but this arrangement did not succeed (v. 9). **17–18:** See 1 Kings ch 13 (and notes there). *Samaria* is probably an error for Bethel (compare 1 Kings 13.11, 31).

all the priests of the high places who were there, and burned human bones on them. Then he returned to Jerusalem.

21 The king commanded all the people, "Keep the passover to the LORD your God as prescribed in this book of the covenant." 22 No such passover had been kept since the days of the judges who judged Israel, or during all the days of the kings of Israel or of the kings of Judah; 23 but in the eighteenth year of King Josiah this passover was kept to the LORD in Jerusalem.

24 Moreover Josiah put away the mediums, wizards, teraphim, ⁰ idols, and all the abominations that were seen in the land of Judah and in Jerusalem, so that he established the words of the law that were written in the book that the priest Hilkiah had found in the house of the LORD. 25 Before him there was no king like him, who turned to the LORD with all his heart, with all his soul, and with all his might, according to all the law of Moses; nor did any like him arise after him.

26 Still the LORD did not turn from the fierceness of his great wrath, by which his anger was kindled against Judah, because of all the provocations with which Manasseh had provoked him. 27 The LORD said, "I will remove Judah also out of my sight, as I have removed Israel; and I will reject this city that I have chosen, Jerusalem, and the house of which I said, My name shall be there."

28 Now the rest of the acts of Josiah, and all that he did, are they not written in the Book of the Annals of the Kings of Judah? 29 In his days Pharaoh Neco king of Egypt went up to the king of Assyria to the river Euphrates. King Josiah went to meet him; but when Pharaoh Neco met him at Megiddo, he killed him. 30 His servants carried him dead in a chariot from Megiddo, brought him to Jerusalem, and buried him in his own tomb. The people of the land took Jehoahaz son of Josiah, anointed him, and made him king in place of his father.

31 Jehoahaz was twenty-three years old when he began to reign; he reigned three months in Jerusalem. His mother's name was Hamutal daughter of Jeremiah of Libnah. 32 He did what was evil in the sight of the LORD, just as his ancestors had done. 33 Pharaoh Neco confined him at Riblah in the land of Hamath, so that he might not reign in Jerusalem, and imposed tribute on the land of one hundred talents of silver and a talent of gold. 34 Pharaoh Neco made Eliakim son of Josiah king in place of his father Josiah, and changed his name to Jehoiakim. But he took Jehoahaz away; he came to Egypt, and died there. 35 Jehoiakim gave the silver and the gold to Pharaoh, but he taxed the land in order to meet Pharaoh's demand for money. He exacted the silver and the gold from the people of the land, from all according to their assessment, to give it to Pharaoh Neco.

36 Jehoiakim was twenty-five years old when he began to reign; he reigned eleven years in Jerusalem. His mother's name was Zebidah daughter of Pedaiah

⁰ Or *household gods*

23.21–23: The great *passover*, the climax of the reform. **24–25:** Opinion differs as to whether the author ended the account just before or just after the death of Josiah in 609. **25:** Compare the evaluation of Hezekiah in 18.5. **26–27:** These verses have been ascribed to an author writing during the exile (see 21.10–15 n.). **28–30:** Some scholars believe an author writing before the destruction of the temple ended the story of the monarchy here. Compare the fuller account of Josiah's death in 2 Chr 35.20–24. Nineveh had fallen in 612 before the Medes and Chaldeans, but Assyria was still fighting; Egypt was the ally of Assyria, hence *Pharaoh Neco king of Egypt went up to the king of Assyria* to help him. Josiah, now in revolt against Assyria, felt it necessary to oppose Neco. Josiah lost his life, and Judah became a vassal of Egypt.

23.31–37: The reign of Jehoahaz and the beginning of the reign of Jehoiakim. The people's choice for king (v. 30) was *Jehoahaz* or Shallum (1 Chr 3.15; Jer 22.11), a younger son of Josiah; but *Pharaoh Neco* was now master. He deposed *Jehoahaz,* levied heavy *tribute,* and made *Eliakim,* an older son, king, changing his name to *Jehoiakim* as a symbol of his vassalage.

of Rumah. 37 He did what was evil in the sight of the Lord, just as all his ancestors had done.

24 In his days King Nebuchadnezzar of Babylon came up; Jehoiakim became his servant for three years; then he turned and rebelled against him. 2 The Lord sent against him bands of the Chaldeans, bands of the Arameans, bands of the Moabites, and bands of the Ammonites; he sent them against Judah to destroy it, according to the word of the Lord that he spoke by his servants the prophets. 3 Surely this came upon Judah at the command of the Lord, to remove them out of his sight, for the sins of Manasseh, for all that he had committed, 4 and also for the innocent blood that he had shed; for he filled Jerusalem with innocent blood, and the Lord was not willing to pardon. 5 Now the rest of the deeds of Jehoiakim, and all that he did, are they not written in the Book of the Annals of the Kings of Judah? 6 So Jehoiakim slept with his ancestors; then his son Jehoiachin succeeded him. 7 The king of Egypt did not come again out of his land, for the king of Babylon had taken over all that belonged to the king of Egypt from the Wadi of Egypt to the River Euphrates.

8 Jehoiachin was eighteen years old when he began to reign; he reigned three months in Jerusalem. His mother's name was Nehushta daughter of Elnathan of Jerusalem. 9 He did what was evil in the sight of the Lord, just as his father had done.

10 At that time the servants of King Nebuchadnezzar of Babylon came up to Jerusalem, and the city was besieged. 11 King Nebuchadnezzar of Babylon came to the city, while his servants were besieging it; 12 King Jehoiachin of Judah gave himself up to the king of Babylon, himself, his mother, his servants, his officers, and his palace officials. The king of Babylon took him prisoner in the eighth year of his reign.

13 He carried off all the treasures of the house of the Lord, and the treasures of the king's house; he cut in pieces all the vessels of gold in the temple of the Lord, which King Solomon of Israel had made, all this as the Lord had foretold. 14 He carried away all Jerusalem, all the officials, all the warriors, ten thousand captives, all the artisans and the smiths; no one remained, except the poorest people of the land. 15 He carried away Jehoiachin to Babylon; the king's mother, the king's wives, his officials, and the elite of the land, he took into captivity from Jerusalem to Babylon. 16 The king of Babylon brought captive to Babylon all the men of valor, seven thousand, the artisans and the smiths, one thousand, all of them strong and fit for war. 17 The king of Babylon made Mattaniah, Jehoiachin's uncle, king in his place, and changed his name to Zedekiah.

18 Zedekiah was twenty-one years

24.1–17: **The first fall of Jerusalem and the first deportation (reigns of Jehoiakim and Jehoiachin). 1–7:** The Chaldeans (the Neo-Babylonians) defeated the Assyrians and Egyptians at the battle of Carchemish in 605 B.C. (Jer 46.2). This event removed Egyptian control of Judah, which thereupon came under the dominance of the Chaldeans, but revolted after *three years* (v. 1). Verses 3–4 are the comment of the Deuteronomist. **8:** Jehoiakim died (598 B.C.) before he could be punished, and his young son Jehoiachin (also called Jeconiah, 1 Chr 3.16, and Coniah, Jer 22.24) took the tottering throne. **24.10–17:** Jehoiachin almost immediately surrendered to *Nebuchadnezzar* (more correct form, Nebuchadrezzar, Jer 21.2 and frequent-

ly in Jeremiah), on March 16, 597 B.C., in the seventh (Jer 52.28) year of Nebuchadrezzar, as related in the Babylonian records. According to v. 14, *ten thousand captives* were deported; according to v. 16, eight thousand; Jer 52.28 says 3,023, which perhaps does not count women and children, or perhaps it is a more accurate figure. Tablets found in Babylon make reference to Jehoiachin and five sons (compare 1 Chr 3.17–18). His uncle *Mattaniah,* another son of Josiah (1 Chr 3.15), was made *king in his place,* and given a new name, *Zedekiah,* as a token of vassalage (see 23.31–37 n.). **24.18–25.21: The reign of Zedekiah; second fall of Jerusalem and its destruction; the second deportation.** This section

old when he began to reign; he reigned eleven years in Jerusalem. His mother's name was Hamutal daughter of Jeremiah of Libnah. [19]He did what was evil in the sight of the LORD, just as Jehoiakim had done. [20]Indeed, Jerusalem and Judah so angered the LORD that he expelled them from his presence.

Zedekiah rebelled against the king of

25 Babylon. [1]And in the ninth year of his reign, in the tenth month, on the tenth day of the month, King Nebuchadnezzar of Babylon came with all his army against Jerusalem, and laid siege to it; they built siegeworks against it all around. [2]So the city was besieged until the eleventh year of King Zedekiah. [3]On the ninth day of the fourth month the famine became so severe in the city that there was no food for the people of the land. [4]Then a breach was made in the city wall;[p] the king with all the soldiers fled[q] by night by the way of the gate between the two walls, by the king's garden, though the Chaldeans were all around the city. They went in the direction of the Arabah. [5]But the army of the Chaldeans pursued the king, and overtook him in the plains of Jericho; all his army was scattered, deserting him. [6]Then they captured the king and brought him up to the king of Babylon at Riblah, who passed sentence on him. [7]They slaughtered the sons of Zedekiah before his eyes, then put out the eyes of Zedekiah; they bound him in fetters and took him to Babylon.

[8] In the fifth month, on the seventh day of the month—which was the nineteenth year of King Nebuchadnezzar,

king of Babylon—Nebuzaradan, the captain of the bodyguard, a servant of the king of Babylon, came to Jerusalem. [9]He burned the house of the LORD, the king's house, and all the houses of Jerusalem; every great house he burned down. [10]All the army of the Chaldeans who were with the captain of the guard broke down the walls around Jerusalem. [11]Nebuzaradan the captain of the guard carried into exile the rest of the people who were left in the city and the deserters who had defected to the king of Babylon—all the rest of the population. [12]But the captain of the guard left some of the poorest people of the land to be vinedressers and tillers of the soil.

[13] The bronze pillars that were in the house of the LORD, as well as the stands and the bronze sea that were in the house of the LORD, the Chaldeans broke in pieces, and carried the bronze to Babylon. [14]They took away the pots, the shovels, the snuffers, the dishes for incense, and all the bronze vessels used in the temple service, [15]as well as the firepans and the basins. What was made of gold the captain of the guard took away for the gold, and what was made of silver, for the silver. [16]As for the two pillars, the one sea, and the stands, which Solomon had made for the house of the LORD, the bronze of all these vessels was beyond weighing. [17]The height of the one pillar was eighteen cubits, and on it was a bronze capital; the height of the capital was three cubits; latticework and

p Heb lacks *wall*   *q* Gk Compare Jer 39.4; 52.7: Heb lacks *the king* and lacks *fled*

---

is repeated in Jer ch 52 with certain changes; 25.1–12 is found in Jer 39.1–10 with changes; compare the parallel in 2 Chr 36.11–21. **24.18:** More information about Zedekiah is found in Jer chs 21; 24; 27; 29; 32; 37; 38; Ezek 17.11–21.

24.20b–25.7: The sad story hastens to its woeful conclusion. In spite of his oath of allegiance (2 Chr 36.13; Ezek 17.13), *Zedekiah* began to plot with Egypt and other nations against the Chaldeans (Jer 27.3–7; Ezek 17.15), and *rebelled against the king of Babylon.*

Jerusalem was besieged, and after terrible privation (25.3; Deut 28.52–57; Lam 4.10), fell in *the eleventh year of King Zedekiah,* 587 or 586 B.C. The king was tortured and carried *to Babylon,* where he died (Jer 52.11).

25.8–12: The temple, the palace, *and all the houses of Jerusalem* were *burned,* and the walls were broken down. A second deportation was carried out (see 24.10–17 n.), the number taken being only 832, according to Jer 52.29. **13–17:** All the valuable equipment of the temple was carried to Babylon.

pomegranates, all of bronze, were on the capital all around. The second pillar had the same, with the latticework.

18 The captain of the guard took the chief priest Seraiah, the second priest Zephaniah, and the three guardians of the threshold; 19 from the city he took an officer who had been in command of the soldiers, and five men of the king's council who were found in the city; the secretary who was the commander of the army who mustered the people of the land; and sixty men of the people of the land who were found in the city. 20 Nebuzaradan the captain of the guard took them, and brought them to the king of Babylon at Riblah. 21 The king of Babylon struck them down and put them to death at Riblah in the land of Hamath. So Judah went into exile out of its land.

22 He appointed Gedaliah son of Ahikam son of Shaphan as governor over the people who remained in the land of Judah, whom King Nebuchadnezzar of Babylon had left. 23 Now when all the captains of the forces and their men heard that the king of Babylon had appointed Gedaliah as governor, they came with their men to Gedaliah at Mizpah, namely, Ishmael son of Nethaniah, Johanan son of Kareah, Seraiah son of Tanhumeth the Netophathite, and Jaazaniah son of the Maacathite. 24 Gedaliah swore to them and their men, saying, "Do not be afraid because of the Chaldean officials; live in the land, serve the king of Babylon, and it shall be well with you." 25 But in the seventh month, Ishmael son of Nethaniah son of Elishama, of the royal family, came with ten men; they struck down Gedaliah so that he died, along with the Judeans and Chaldeans who were with him at Mizpah. 26 Then all the people, high and low[r] and the captains of the forces set out and went to Egypt; for they were afraid of the Chaldeans.

27 In the thirty-seventh year of the exile of King Jehoiachin of Judah, in the twelfth month, on the twenty-seventh day of the month, King Evil-merodach of Babylon, in the year that he began to reign, released King Jehoiachin of Judah from prison; 28 he spoke kindly to him, and gave him a seat above the other seats of the kings who were with him in Babylon. 29 So Jehoiachin put aside his prison clothes. Every day of his life he dined regularly in the king's presence. 30 For his allowance, a regular allowance was given him by the king, a portion every day, as long as he lived.

r Or young and old

25.18–21: All remaining leaders were executed. **22–26:** This section does not appear in Jer ch 52; there is, however, an expanded account in Jer 40.7–41.18. In Jer 40.6 and chs 42–43, the prophet is brought into the story. It is a cause for amazement that Jeremiah is not mentioned even once in 2 Kings, whereas the book of Jeremiah contains a wealth of additional information about the last days of Judah. **25.22–26: The governorship of Gedaliah.** **22–24:** The abolition of the monarchy and the appointment of *Gedaliah* as a native governor under Chaldean control should have meant peace and stability for Judah. Gedaliah's father, *Ahikam,* had been a trusted adviser of Josiah (22.12) and a friend of Jeremiah (Jer 26.24). *Gedaliah* himself was well disposed toward the prophet (Jer 39.14; 40.6), and enjoyed the confidence of his fellow citizens (Jer 40.11–12). **25–26:** The dastardly assassination of this good man by *Ishmael,* a member of the deposed *royal family,* brought about utter chaos and ruin (Jer 40.13–41.18); see the evaluation of the situation by the Deuteronomic editor of the book of Jeremiah in Jer 44.1–14. **25.27–30: Jehoiachin in exile.** Also found in Jer 52.31–34. From exile, the Deuteronomist reports that Jehoiachin was still safe, and was enjoying favored treatment by *King Evil-merodach of Babylon.* The report has been partially confirmed by archaeological researchers. The writer may have used this information to end the book with a note of modest hope, as though to say (in spite of 24.9): the Davidic dynasty has not been snuffed out.

1 Chronicles

The two books of Chronicles, like 1 and 2 Samuel and 1 and 2 Kings, were originally one book; this combined book of Chronicles formed the original conclusion to the Hebrew Scriptures. In the ancient Greek version of the Bible, however, 1 and 2 Chronicles were grouped together with Samuel and Kings and called, "Miscellanies [Paralipomena, i.e. matters omitted] concerning the Kings of Judah." This arrangement persists in most modern non-Jewish translations today, despite the fact that it falsifies the original ordering of the Hebrew Scriptures.

Chronicles tells the story of Israel from the creation of the world down to the beginnings of the Persian Empire. Summarizing the period from Adam to the reign of King Saul in genealogies (1 Chr 1–9), the writer deals extensively with David's reign (chs 10–29). Second Chronicles begins with Solomon's rule and carries the account through destruction and exile, until a decree from King Cyrus of Persia, allows a return to the land (2 Chr 36.22–23). Many scholars believe that 1 and 2 Chronicles were originally part of a larger work, written in the fifth or fourth century B.C., which also included Ezra and Nehemiah. The matter is disputed, however, and some recent scholars have strongly urged that Ezra and Nehemiah be treated as independent works (see Introduction to the Historical Books; Introduction to Ezra).

The Chronicler drew mainly, but selectively, upon 1 Sam 31–2 Kings 25. Other materials were used in briefer fashion, including many unknown sources, some of which may have overlapped with writings mentioned in 1 and 2 Kings (see 1 Kings 14.29), and many others that apparently did not, such as "the records of the seer Samuel . . . the prophet Nathan . . . the seer Gad" (1 Chr 29.29), "the story of the prophet Iddo" (2 Chr 13.22), "The Commentary on the Book of the Kings" (2 Chr 24.27), or a writing by Isaiah on the reign of Uzziah (2 Chr 26.22). It is difficult to tell whether these citations point to real or fictional sources. What seems certain is that the Chronicler rewrote, quoted, edited, and supplemented material from the Pentateuch, Samuel, and Kings without attribution. The result is a parallel version of Israel's history from Genesis through Kings, and a theological account of monarchical times that differs markedly from the Deuteronomistic History (see Introduction to 1 Kings; Introduction to the Historical Books).

The Davidic dynasty, the socio-political shape of his kingdom, the holy temple in Jerusalem with its authorized attendants, Levites, and temple singers; the surety of rewards and punishments linked to obedience or disobedience; attentiveness to God's presence in temple service—these are the themes that the Chronicler used to support consciousness of God through memory of a glorious, but failed, past. For this author, the bond between God and the true Israel, and the sacrality of those times, were symbolized above all in David and his establishment of a divinely empowered kingdom with its center in God's sacred space (temple). Even though the monarchy after David eventually came to a bad end, the writer allows us to glimpse a future that opens onto restoration and the survival of God's people (2 Chr 36.23).

1 Adam, Seth, Enosh; [2]Kenan, Mahalalel, Jared; [3]Enoch, Methuselah, Lamech; [4]Noah, Shem, Ham, and Japheth.

5 The descendants of Japheth: Gomer, Magog, Madai, Javan, Tubal, Meshech, and Tiras. [6]The descendants of Gomer: Ashkenaz, Diphath,[a] and Togarmah. [7]The descendants of Javan: Elishah, Tarshish, Kittim, and Rodanim.[b]

8 The descendants of Ham: Cush, Egypt, Put, and Canaan. [9]The descendants of Cush: Seba, Havilah, Sabta, Raama, and Sabteca. The descendants of Raamah: Sheba and Dedan. [10]Cush became the father of Nimrod; he was the first to be a mighty one on the earth.

11 Egypt became the father of Ludim, Anamim, Lehabim, Naphtuhim, [12]Pathrusim, Casluhim, and Caphtorim, from whom the Philistines come.[c]

13 Canaan became the father of Sidon his firstborn, and Heth, [14]and the Jebusites, the Amorites, the Girgashites, [15]the Hivites, the Arkites, the Sinites, [16]the Arvadites, the Zemarites, and the Hamathites.

17 The descendants of Shem: Elam, Asshur, Arpachshad, Lud, Aram, Uz, Hul, Gether, and Meshech.[d] [18]Arpachshad became the father of Shelah; and Shelah became the father of Eber. [19]To Eber were born two sons: the name of the one was Peleg (for in his days the earth was divided), and the name of his

brother Joktan. [20]Joktan became the father of Almodad, Sheleph, Hazarmaveth, Jerah, [21]Hadoram, Uzal, Diklah, [22]Ebal, Abimael, Sheba, [23]Ophir, Havilah, and Jobab; all these were the descendants of Joktan.

24 Shem, Arpachshad, Shelah; [25]Eber, Peleg, Reu; [26]Serug, Nahor, Terah; [27]Abram, that is, Abraham.

28 The sons of Abraham: Isaac and Ishmael. [29]These are their genealogies: the firstborn of Ishmael, Nebaioth; and Kedar, Adbeel, Mibsam, [30]Mishma, Dumah, Massa, Hadad, Tema, [31]Jetur, Naphish, and Kedemah. These are the sons of Ishmael. [32]The sons of Keturah, Abraham's concubine: she bore Zimran, Jokshan, Medan, Midian, Ishbak, and Shuah. The sons of Jokshan: Sheba and Dedan. [33]The sons of Midian: Ephah, Epher, Hanoch, Abida, and Eldaah. All these were the descendants of Keturah.

34 Abraham became the father of Isaac. The sons of Isaac: Esau and Israel. [35]The sons of Esau: Eliphaz, Reuel, Jeush, Jalam, and Korah. [36]The sons of Eliphaz: Teman, Omar, Zephi, Gatam, Kenaz, Timna, and Amalek. [37]The sons of Reuel: Nahath, Zerah, Shammah, and Mizzah.

38 The sons of Seir: Lotan, Shobal,

a Gen 10.3 *Ripath*; See Gk Vg   b Gen 10.4 *Dodanim*; See Syr Vg   c Heb *Casluhim, from which the Philistines come, Caphtorim*; See Am 9.7, Jer 47.4   d *Mash* in Gen 10.23

1.1–54: The place of Abraham and his descendants among the nations. 1–4: Summary of Gen ch 5. **5–7:** The sons of Japheth, or the Japhethites, in the genealogies represent in general the ancestors of those people whom today we call Indo-Europeans; for example, *Javan* (lit., Ionia) refers to the Greeks (compare Gen 10.2–4).

1.8–16: Today the term "Hamitic" is restricted largely to the peoples and languages of ancient Egypt; the Canaanites (*Canaan*) and *the Amorites* are now classified as Semites, and the Hittites (*Heth*) are sometimes classified among the Indo-Europeans. *Cush* sometimes refers to Ethiopia (v. 8), sometimes to a location in Mesopotamia (v. 10). Compare Gen 10.6–20.

1.17–27: *The descendants of Shem,* or the Semites. The Hebrews (*Eber*) belonged to this

group, and from it emerged *Abram* or *Abraham.* Such modern terms as "Hamitic" and "Semitic" have linguistic, rather than racial, connotations. Ancient classifications (such as *descendants of Shem*) were based on considerations of geography and cultural (not racial) affinity. Verses 17–23 are condensed from Gen 10.21–31, vv. 24–27 from Gen 11.10–26.

1.28–33: From *Abraham* sprang both *Isaac,* the progenitor of the Edomites and Israelites (v. 34), and *Ishmael,* traditionally, since Josephus (*Antiq.*I.xii.2), the progenitor of the Arabs (compare Gen 25.1–4, 12–18). **34–37:** The Chronicler does not use the personal name Jacob at all, but only the community term, *Israel.* Esau (Edom) is given special attention as the brother of *Israel.*

1.38–42: *Seir* is another name for Edom

Zibeon, Anah, Dishon, Ezer, and Dishan. [39] The sons of Lotan: Hori and Homam; and Lotan's sister was Timna. [40] The sons of Shobal: Alian, Manahath, Ebal, Shephi, and Onam. The sons of Zibeon: Aiah and Anah. [41] The sons of Anah: Dishon. The sons of Dishon: Hamran, Eshban, Ithran, and Cheran. [42] The sons of Ezer: Bilhan, Zaavan, and Jaakan.[e] The sons of Dishan:[f] Uz and Aran.

43 These are the kings who reigned in the land of Edom before any king reigned over the Israelites: Bela son of Beor, whose city was called Dinhabah. [44] When Bela died, Jobab son of Zerah of Bozrah succeeded him. [45] When Jobab died, Husham of the land of the Temanites succeeded him. [46] When Husham died, Hadad son of Bedad, who defeated Midian in the country of Moab, succeeded him; and the name of his city was Avith. [47] When Hadad died, Samlah of Masrekah succeeded him. [48] When Samlah died, Shaul[g] of Rehoboth on the Euphrates succeeded him. [49] When Shaul[g] died, Baal-hanan son of Achbor succeeded him. [50] When Baal-hanan died, Hadad succeeded him; the name of his city was Pai, and his wife's name Mehetabel daughter of Matred, daughter of Mezahab. [51] And Hadad died.

The clans[h] of Edom were: clans[h] Timna, Aliah,[i] Jetheth, [52] Oholibamah, Elah, Pinon, [53] Kenaz, Teman, Mibzar, [54] Magdiel, and Iram; these are the clans[h] of Edom.

2 These are the sons of Israel: Reuben, Simeon, Levi, Judah, Issachar, Zebulun, [2] Dan, Joseph, Benjamin, Naphtali, Gad, and Asher. [3] The sons of Judah: Er, Onan, and Shelah; these three the Canaanite woman Bath-shua bore to him. Now Er, Judah's firstborn, was wicked in the sight of the LORD, and he put him to death. [4] His daughter-in-law

Tamar also bore him Perez and Zerah. Judah had five sons in all.

5 The sons of Perez: Hezron and Hamul. [6] The sons of Zerah: Zimri, Ethan, Heman, Calcol, and Dara,[j] five in all. [7] The sons of Carmi: Achar, the troubler of Israel, who transgressed in the matter of the devoted thing; [8] and Ethan's son was Azariah.

9 The sons of Hezron, who were born to him: Jerahmeel, Ram, and Chelubai. [10] Ram became the father of Amminadab, and Amminadab became the father of Nahshon, prince of the sons of Judah. [11] Nahshon became the father of Salma, Salma of Boaz, [12] Boaz of Obed, Obed of Jesse. [13] Jesse became the father of Eliab his firstborn, Abinadab the second, Shimea the third, [14] Nethanel the fourth, Raddai the fifth, [15] Ozem the sixth, David the seventh; [16] and their sisters were Zeruiah and Abigail. The sons of Zeruiah: Abishai, Joab, and Asahel, three. [17] Abigail bore Amasa, and the father of Amasa was Jether the Ishmaelite.

18 Caleb son of Hezron had children by his wife Azubah, and by Jerioth; these were her sons: Jesher, Shobab, and Ardon. [19] When Azubah died, Caleb married Ephrath, who bore him Hur. [20] Hur became the father of Uri, and Uri became the father of Bezalel.

21 Afterward Hezron went in to the daughter of Machir father of Gilead, whom he married when he was sixty years old; and she bore him Segub; [22] and Segub became the father of Jair, who had twenty-three towns in the land of Gilead. [23] But Geshur and Aram took from them Havvoth-jair, Kenath and its villages, sixty towns. All these were descendants of Machir, father of Gilead. [24] After the death of Hezron, in Caleb-ephrathah,

e Or *and Akan*; See Gen 36.27   *f* See 1.38: Heb *Dishon*   *g* Or *Saul*   *h* Or *chiefs* *i* Or *Alvah*; See Gen 36.40   *j* Or *Darda*; Compare Syr Tg some Gk Mss; See 1 Kings 4.31

(Gen 36.8, 20–30). **43–54:** *The kings . . . of Edom,* compare Gen 36.31–43.

2.1–55: Descendants of Judah. The chief

sources are Gen chs 35; 38; 46; Num 1.7; 32.41–42; Josh chs 7; 14; 15; Ruth 4.18–22; 1 Sam 27.10.

Abijah wife of Hezron bore him Ashhur, father of Tekoa.

25 The sons of Jerahmeel, the firstborn of Hezron: Ram his firstborn, Bunah, Oren, Ozem, and Ahijah. 26 Jerahmeel also had another wife, whose name was Atarah; she was the mother of Onam. 27 The sons of Ram, the firstborn of Jerahmeel: Maaz, Jamin, and Eker. 28 The sons of Onam: Shammai and Jada. The sons of Shammai: Nadab and Abishur. 29 The name of Abishur's wife was Abihail, and she bore him Ahban and Molid. 30 The sons of Nadab: Seled and Appaim; and Seled died childless. 31 The son*k* of Appaim: Ishi. The son*k* of Ishi: Sheshan. The son*k* of Sheshan: Ahlai. 32 The sons of Jada, Shammai's brother: Jether and Jonathan; and Jether died childless. 33 The sons of Jonathan: Peleth and Zaza. These were the descendants of Jerahmeel. 34 Now Sheshan had no sons, only daughters; but Sheshan had an Egyptian slave, whose name was Jarha. 35 So Sheshan gave his daughter in marriage to his slave Jarha; and she bore him Attai. 36 Attai became the father of Nathan, and Nathan of Zabad. 37 Zabad became the father of Ephlal, and Ephlal of Obed. 38 Obed became the father of Jehu, and Jehu of Azariah. 39 Azariah became the father of Helez, and Helez of Eleasah. 40 Eleasah became the father of Sismai, and Sismai of Shallum. 41 Shallum became the father of Jekamiah, and Jekamiah of Elishama.

42 The sons of Caleb brother of Jerahmeel: Mesha*l* his firstborn, who was father of Ziph. The sons of Mareshah father of Hebron. 43 The sons of Hebron: Korah, Tappuah, Rekem, and Shema. 44 Shema became father of Raham, father of Jorkeam; and Rekem became the father of Shammai. 45 The son of Shammai: Maon; and Maon was the father of Bethzur. 46 Ephah also, Caleb's concubine, bore Haran, Moza, and Gazez; and Ha-

ran became the father of Gazez. 47 The sons of Jahdai: Regem, Jotham, Geshan, Pelet, Ephah, and Shaaph. 48 Maacah, Caleb's concubine, bore Sheber and Tirhanah. 49 She also bore Shaaph father of Madmannah, Sheva father of Machbenah and father of Gibea; and the daughter of Caleb was Achsah. 50 These were the descendants of Caleb.

The sons*m* of Hur the firstborn of Ephrathah: Shobal father of Kiriath-jearim, 51 Salma father of Bethlehem, and Hareph father of Beth-gader. 52 Shobal father of Kiriath-jearim had other sons: Haroeh, half of the Menuhoth. 53 And the families of Kiriath-jearim: the Ithrites, the Puthites, the Shumathites, and the Mishraites; from these came the Zorathites and the Eshtaolites. 54 The sons of Salma: Bethlehem, the Netophathites, Atroth-beth-joab, and half of the Manahathites, the Zorites. 55 The families also of the scribes that lived at Jabez: the Tirathites, the Shimeathites, and the Sucathites. These are the Kenites who came from Hammath, father of the house of Rechab.

3 These are the sons of David who were born to him in Hebron: the firstborn Amnon, by Ahinoam the Jezreelite; the second Daniel, by Abigail the Carmelite; 2 the third Absalom, son of Maacah, daughter of King Talmai of Geshur; the fourth Adonijah, son of Haggith; 3 the fifth Shephatiah, by Abital; the sixth Ithream, by his wife Eglah; 4 six were born to him in Hebron, where he reigned for seven years and six months. And he reigned thirty-three years in Jerusalem. 5 These were born to him in Jerusalem: Shimea, Shobab, Nathan, and Solomon, four by Bath-shua, daughter of Ammiel; 6 then Ibhar, Elishama, Eliphelet, 7 Nogah, Nepheg, Japhia,

k Heb *sons*   *l* Gk reads *Mareshah*
m Gk Vg: Heb *son*

3.1–24: The descendants of David are traced down to the Chronicler's own time. These were very important people, even though the dynasty of David had not been restored (compare 2 Sam 3.2–5; 5.13–16; 13.1). Non-biblical sources were also utilized.

8 Elishama, Eliada, and Eliphelet, nine. 9 All these were David's sons, besides the sons of the concubines; and Tamar was their sister.

10 The descendants of Solomon: Rehoboam, Abijah his son, Asa his son, Jehoshaphat his son, 11 Joram his son, Ahaziah his son, Joash his son, 12 Amaziah his son, Azariah his son, Jotham his son, 13 Ahaz his son, Hezekiah his son, Manasseh his son, 14 Amon his son, Josiah his son. 15 The sons of Josiah: Johanan the firstborn, the second Jehoiakim, the third Zedekiah, the fourth Shallum. 16 The descendants of Jehoiakim: Jeconiah his son, Zedekiah his son; 17 and the sons of Jeconiah, the captive: Shealtiel his son, 18 Malchiram, Pedaiah, Shenazzar, Jekamiah, Hoshama, and Nedabiah; 19 The sons of Pedaiah: Zerubbabel and Shimei; and the sons of Zerubbabel: Meshullam and Hananiah, and Shelomith was their sister; 20 and Hashubah, Ohel, Berechiah, Hasadiah, and Jushab-hesed, five. 21 The sons of Hananiah: Pelatiah and Jeshaiah, his son[n] Rephaiah, his son[n] Arnan, his son[n] Obadiah, his son[n] Shecaniah. 22 The son[o] of Shecaniah: Shemaiah. And the sons of Shemaiah: Hattush, Igal, Bariah, Neariah, and Shaphat, six. 23 The sons of Neariah: Elioenai, Hizkiah, and Azrikam, three. 24 The sons of Elioenai: Hodaviah, Eliashib, Pelaiah, Akkub, Johanan, Delaiah, and Anani, seven.

4 The sons of Judah: Perez, Hezron, Carmi, Hur, and Shobal. 2 Reaiah son of Shobal became the father of Jahath, and Jahath became the father of Ahumai and Lahad. These were the families of the Zorathites. 3 These were the sons[p] of Etam: Jezreel, Ishma, and Idbash; and the name of their sister was Hazzelelponi, 4 and Penuel was the father of Gedor, and Ezer the father of Hushah. These were the sons of Hur, the firstborn of Ephrathah, the father of Bethlehem.

5 Ashhur father of Tekoa had two wives, Helah and Naarah; 6 Naarah bore him Ahuzzam, Hepher, Temeni, and Haahashtari.[q] These were the sons of Naarah. 7 The sons of Helah: Zereth, Izhar,[r] and Ethnan. 8 Koz became the father of Anub, Zobebah, and the families of Aharhel son of Harum. 9 Jabez was honored more than his brothers; and his mother named him Jabez, saying, "Because I bore him in pain." 10 Jabez called on the God of Israel, saying, "Oh that you would bless me and enlarge my border, and that your hand might be with me, and that you would keep me from hurt and harm!" And God granted what he asked. 11 Chelub the brother of Shuhah became the father of Mehir, who was the father of Eshton. 12 Eshton became the father of Beth-rapha, Paseah, and Tehinnah the father of Ir-nahash. These are the men of Recah. 13 The sons of Kenaz: Othniel and Seraiah; and the sons of Othniel: Hathath and Meonothai.[s] 14 Meonothai became the father of Ophrah; and Seraiah became the father of Joab father of Ge-harashim,[t] so-called because they were artisans. 15 The sons of Caleb son of Jephunneh: Iru, Elah, and Naam; and the son[o] of Elah: Kenaz. 16 The sons of Jehallelel: Ziph, Ziphah, Tiria, and Asarel. 17 The sons of Ezrah: Jether, Mered, Epher, and Jalon. These are the sons of Bithiah, daughter of Pharaoh, whom Mered married;[u] and she conceived and bore[v] Miriam, Shammai, and Ishbah father of Eshtemoa. 18 And his Judean wife bore Jered father of Gedor, Heber father of Soco, and Jekuthiel father of Zanoah. 19 The sons of the wife

n Gk Compare Syr Vg: Heb *sons of*
o Heb *sons*   p Gk Compare Vg: Heb *the father*
q Or *Ahashtari*   r Another reading is *Zohar*
s Gk Vg: Heb lacks *and Meonothai*   t That is *Valley of artisans*   u The clause: *These are . . . married* is transposed from verse 18
v Heb lacks *and bore*

4.1–43: The descendants of Judah and Simeon. The writer takes a genealogical excursion into the early days of the tribes of Judah and Simeon (compare Josh chs 15 and 19). These two tribes were closely related (Gen 29.33, 35), and Simeon was finally absorbed into Judah and lost its identity (note its absence in Deut ch 33).

of Hodiah, the sister of Naham, were the fathers of Keilah the Garmite and Eshtemoa the Maacathite. 20 The sons of Shimon: Amnon, Rinnah, Ben-hanan, and Tilon. The sons of Ishi: Zoheth and Benzoheth. 21 The sons of Shelah son of Judah: Er father of Lecah, Laadah father of Mareshah, and the families of the guild of linen workers at Beth-ashbea; 22 and Jokim, and the men of Cozeba, and Joash, and Saraph, who married into Moab but returned to Lehem *w* (now the records *x* are ancient). 23 These were the potters and inhabitants of Netaim and Gederah; they lived there with the king in his service.

24 The sons of Simeon: Nemuel, Jamin, Jarib, Zerah, Shaul; *y* 25 Shallum was his son, Mibsam his son, Mishma his son. 26 The sons of Mishma: Hammuel his son, Zaccur his son, Shimei his son. 27 Shimei had sixteen sons and six daughters; but his brothers did not have many children, nor did all their family multiply like the Judeans. 28 They lived in Beersheba, Moladah, Hazar-shual, 29 Bilhah, Ezem, Tolad, 30 Bethuel, Hormah, Ziklag, 31 Beth-marcaboth, Hazar-susim, Beth-biri, and Shaaraim. These were their towns until David became king. 32 And their villages were Etam, Ain, Rimmon, Tochen, and Ashan, five towns, 33 along with all their villages that were around these towns as far as Baal. These were their settlements. And they kept a genealogical record.

34 Meshobab, Jamlech, Joshah son of Amaziah, 35 Joel, Jehu son of Joshibiah son of Seraiah son of Asiel, 36 Elioenai, Jaakobah, Jeshohaiah, Asaiah, Adiel, Jesimiel, Benaiah, 37 Ziza son of Shiphi son of Allon son of Jedaiah son of Shimri son of Shemaiah— 38 these mentioned by name were leaders in their families, and their clans increased greatly. 39 They journeyed to the entrance of Gedor, to the east side of the valley, to seek pasture for their flocks, 40 where they found rich, good pasture, and the land was very broad, quiet, and peaceful; for the former inhabitants there belonged to Ham. 41 These, registered by name, came in the days of King Hezekiah of Judah, and attacked their tents and the Meunim who were found there, and exterminated them to this day, and settled in their place, because there was pasture there for their flocks. 42 And some of them, five hundred men of the Simeonites, went to Mount Seir, having as their leaders Pelatiah, Neariah, Rephaiah, and Uzziel, sons of Ishi; 43 they destroyed the remnant of the Amalekites that had escaped, and they have lived there to this day.

5 The sons of Reuben the firstborn of Israel. (He was the firstborn, but because he defiled his father's bed his birthright was given to the sons of Joseph son of Israel, so that he is not enrolled in the genealogy according to the birthright; 2 though Judah became prominent among his brothers and a ruler came from him, yet the birthright belonged to Joseph.) 3 The sons of Reuben, the firstborn of Israel: Hanoch, Pallu, Hezron, and Carmi. 4 The sons of Joel: Shemaiah his son, Gog his son, Shimei his son, 5 Micah his son, Reaiah his son, Baal his son, 6 Beerah his son, whom King Tilgath-pilneser of Assyria carried away into exile; he was a chieftain of the Reubenites. 7 And his kindred by their families, when the genealogy of their generations was reckoned: the chief, Jeiel, and Zechariah, 8 and Bela son of Azaz, son of Shema, son of Joel, who lived in Aroer, as far as Nebo and Baal-meon. 9 He also lived to the east as far as the beginning of the desert this side of the Euphrates, be-

w Vg Compare Gk: Heb *and Jashubi-lahem*
x Or *matters*   *y* Or *Saul*

5.1–26: **The descendants of Reuben, Gad, and the half-tribe of Manasseh,** the tribes located east of the Jordan. **1–10:** *Reuben* was traditionally connected with *Judah* (Gen 29.32, 35), but by remaining east of the Jordan lost power and influence (Gen 35.22; Deut 33.6). The Joseph tribes (*sons of Joseph*) became dominant (Gen 49.22–26; Deut 33.13–17), yet from *Judah* sprang *a ruler,* David, the greatest figure of the past, who became in other literature one prototype of the Messiah to come.

cause their cattle had multiplied in the land of Gilead. ¹⁰And in the days of Saul they made war on the Hagrites, who fell by their hand; and they lived in their tents throughout all the region east of Gilead.

11 The sons of Gad lived beside them in the land of Bashan as far as Salecah: ¹²Joel the chief, Shapham the second, Janai, and Shaphat in Bashan. ¹³And their kindred according to their clans: Michael, Meshullam, Sheba, Jorai, Jacan, Zia, and Eber, seven. ¹⁴These were the sons of Abihail son of Huri, son of Jaroah, son of Gilead, son of Michael, son of Jeshishai, son of Jahdo, son of Buz; ¹⁵Ahi son of Abdiel, son of Guni, was chief in their clan; ¹⁶and they lived in Gilead, in Bashan and in its towns, and in all the pasture lands of Sharon to their limits. ¹⁷All of these were enrolled by genealogies in the days of King Jotham of Judah, and in the days of King Jeroboam of Israel.

18 The Reubenites, the Gadites, and the half-tribe of Manasseh had valiant warriors, who carried shield and sword, and drew the bow, expert in war, forty-four thousand seven hundred sixty, ready for service. ¹⁹They made war on the Hagrites, Jetur, Naphish, and Nodab; ²⁰and when they received help against them, the Hagrites and all who were with them were given into their hands, for they cried to God in the battle, and he granted their entreaty because they trusted in him. ²¹They captured their livestock: fifty thousand of their camels, two hundred fifty thousand sheep, two thousand donkeys, and one hundred thousand captives. ²²Many fell

slain, because the war was of God. And they lived in their territory until the exile.

23 The members of the half-tribe of Manasseh lived in the land; they were very numerous from Bashan to Baalhermon, Senir, and Mount Hermon. ²⁴These were the heads of their clans: Epher,ᶻ Ishi, Eliel, Azriel, Jeremiah, Hodaviah, and Jahdiel, mighty warriors, famous men, heads of their clans. ²⁵But they transgressed against the God of their ancestors, and prostituted themselves to the gods of the peoples of the land, whom God had destroyed before them. ²⁶So the God of Israel stirred up the spirit of King Pul of Assyria, the spirit of King Tilgath-pilneser of Assyria, and he carried them away, namely, the Reubenites, the Gadites, and the half-tribe of Manasseh, and brought them to Halah, Habor, Hara, and the river Gozan, to this day.

6 ᵃ The sons of Levi: Gershom,ᵇ Kohath, and Merari. ²The sons of Kohath: Amram, Izhar, Hebron, and Uzziel. ³The children of Amram: Aaron, Moses, and Miriam. The sons of Aaron: Nadab, Abihu, Eleazar, and Ithamar. ⁴Eleazar became the father of Phinehas, Phinehas of Abishua, ⁵Abishua of Bukki, Bukki of Uzzi, ⁶Uzzi of Zerahiah, Zerahiah of Meraioth, ⁷Meraioth of Amariah, Amariah of Ahitub, ⁸Ahitub of Zadok, Zadok of Ahimaaz, ⁹Ahimaaz of Azariah, Azariah of Johanan, ¹⁰and Johanan of Azariah (it was he who served as priest in the house that Solomon built in

z Gk Vg: Heb *and Epher*   a Ch 5.27 in Heb
b Heb *Gershon*, variant of *Gershom*; See 6.16

5.11–17: *Gad* was just north of Reuben and seems to have been stronger and more important (Deut 33.20–21). **18–22**: In this story of the wars of the Transjordanian tribes with their neighbors, we see the tendency to express the greatness of God's people in terms of military might, with figures somewhat exaggerated.
  5.23–26: To the north of Gad was the *half-tribe of Manasseh*, i.e. the part of Manasseh that remained east of the Jordan (Num 32.33–42). All these tribes were carried into

exile by the Assyrians, according to the tradition utilized in these verses (compare 2 Kings 15.29).
  6.1–81: **The lineage of the Levites. 1–5**: The Levites were an ancient warlike tribe (Gen 49.5–7) that became a priestly caste (Deut 33.8–10; Judg 17.9–13; 18.19). The line running through Zadok (v. 8; compare 2 Sam 8.17) produced the chief priests, who are here traced down to the Exile (compare Gen 46.11; Ex 6.18–19; Num 26.59–60).

Jerusalem). [11] Azariah became the father of Amariah, Amariah of Ahitub, [12] Ahitub of Zadok, Zadok of Shallum, [13] Shallum of Hilkiah, Hilkiah of Azariah, [14] Azariah of Seraiah, Seraiah of Jehozadak; [15] and Jehozadak went into exile when the LORD sent Judah and Jerusalem into exile by the hand of Nebuchadnezzar.

[16c] The sons of Levi: Gershom, Kohath, and Merari. [17] These are the names of the sons of Gershom: Libni and Shimei. [18] The sons of Kohath: Amram, Izhar, Hebron, and Uzziel. [19] The sons of Merari: Mahli and Mushi. These are the clans of the Levites according to their ancestry. [20] Of Gershom: Libni his son, Jahath his son, Zimmah his son, [21] Joah his son, Iddo his son, Zerah his son, Jeatherai his son. [22] The sons of Kohath: Amminadab his son, Korah his son, Assir his son, [23] Elkanah his son, Ebiasaph his son, Assir his son, [24] Tahath his son, Uriel his son, Uzziah his son, and Shaul his son. [25] The sons of Elkanah: Amasai and Ahimoth, [26] Elkanah his son, Zophai his son, Nahath his son, [27] Eliab his son, Jeroham his son, Elkanah his son. [28] The sons of Samuel: Joel[d] his firstborn, the second Abijah.[e] [29] The sons of Merari: Mahli, Libni his son, Shimei his son, Uzzah his son, [30] Shimea his son, Haggiah his son, and Asaiah his son.

31 These are the men whom David put in charge of the service of song in the house of the LORD, after the ark came to rest there. [32] They ministered with song before the tabernacle of the tent of meeting, until Solomon had built the house of the LORD in Jerusalem; and they performed their service in due order. [33] These are the men who served; and their sons were: Of the Kohathites: He-

man, the singer, son of Joel, son of Samuel, [34] son of Elkanah, son of Jeroham, son of Eliel, son of Toah, [35] son of Zuph, son of Elkanah, son of Mahath, son of Amasai, [36] son of Elkanah, son of Joel, son of Azariah, son of Zephaniah, [37] son of Tahath, son of Assir, son of Ebiasaph, son of Korah, [38] son of Izhar, son of Kohath, son of Levi, son of Israel; [39] and his brother Asaph, who stood on his right, namely, Asaph son of Berechiah, son of Shimea, [40] son of Michael, son of Baaseiah, son of Malchijah, [41] son of Ethni, son of Zerah, son of Adaiah, [42] son of Ethan, son of Zimmah, son of Shimei, [43] son of Jahath, son of Gershom, son of Levi. [44] On the left were their kindred the sons of Merari: Ethan son of Kishi, son of Abdi, son of Malluch, [45] son of Hashabiah, son of Amaziah, son of Hilkiah, [46] son of Amzi, son of Bani, son of Shemer, [47] son of Mahli, son of Mushi, son of Merari, son of Levi; [48] and their kindred the Levites were appointed for all the service of the tabernacle of the house of God.

49 But Aaron and his sons made offerings on the altar of burnt offering and on the altar of incense, doing all the work of the most holy place, to make atonement for Israel, according to all that Moses the servant of God had commanded. [50] These are the sons of Aaron: Eleazar his son, Phinehas his son, Abishua his son, [51] Bukki his son, Uzzi his son, Zerahiah his son, [52] Meraioth his son, Amariah his son, Ahitub his son, [53] Zadok his son, Ahimaaz his son.

54 These are their dwelling places ac-

c Ch 6.1 in Heb   *d* Gk Syr Compare verse 33 and 1 Sam 8.2: Heb lacks *Joel*   *e* Heb reads *Vashni, and Abijah* for *the second Abijah,* taking *the second* as a proper name

---

6.16–48: Other Levites, not of the privileged priestly family, performed auxiliary duties in the temple, such as singing (*the service of song,* v. 31). *David* is credited with instituting the musical service of the temple as it existed in the Chronicler's own day, somewhat as Moses is credited with all the laws in Deuteronomy. Prominent family names associated with these singers were *Heman*

(v. 33), *Korah* (v. 37), *Asaph* (v. 39), and *Ethan* (v. 44); see the titles to Pss 73–83; 88–89.

6.49–53: To be a priest in the Chronicler's time it was necessary to be a descendant of Aaron; to be a high priest, one had to be a descendant of Zadok.

6.54–81: Territory allotted to Levitical clans before the Levites had become a priestly

cording to their settlements within their borders: to the sons of Aaron of the families of Kohathites—for the lot fell to them first— ⁵⁵ to them they gave Hebron in the land of Judah and its surrounding pasture lands, ⁵⁶ but the fields of the city and its villages they gave to Caleb son of Jephunneh. ⁵⁷ To the sons of Aaron they gave the cities of refuge: Hebron, Libnah with its pasture lands, Jattir, Eshtemoa with its pasture lands, ⁵⁸ Hilen*f* with its pasture lands, Debir with its pasture lands, ⁵⁹ Ashan with its pasture lands, and Beth-shemesh with its pasture lands. ⁶⁰ From the tribe of Benjamin, Geba with its pasture lands, Alemeth with its pasture lands, and Anathoth with its pasture lands. All their towns throughout their families were thirteen.

61 To the rest of the Kohathites were given by lot out of the family of the tribe, out of the half-tribe, the half of Manasseh, ten towns. ⁶² To the Gershomites according to their families were allotted thirteen towns out of the tribes of Issachar, Asher, Naphtali, and Manasseh in Bashan. ⁶³ To the Merarites according to their families were allotted twelve towns out of the tribes of Reuben, Gad, and Zebulun. ⁶⁴ So the people of Israel gave the Levites the towns with their pasture lands. ⁶⁵ They also gave them by lot out of the tribes of Judah, Simeon, and Benjamin these towns that are mentioned by name.

66 And some of the families of the sons of Kohath had towns of their territory out of the tribe of Ephraim. ⁶⁷ They were given the cities of refuge: Shechem with its pasture lands in the hill country of Ephraim, Gezer with its pasture lands, ⁶⁸ Jokmeam with its pasture lands, Beth-horon with its pasture lands, ⁶⁹ Aijalon with its pasture lands, Gath-rimmon with its pasture lands; ⁷⁰ and out of the half-tribe of Manasseh, Aner with its

pasture lands, and Bileam with its pasture lands, for the rest of the families of the Kohathites.

71 To the Gershomites: out of the half-tribe of Manasseh: Golan in Bashan with its pasture lands and Ashtaroth with its pasture lands; ⁷² and out of the tribe of Issachar: Kedesh with its pasture lands, Daberath*g* with its pasture lands, ⁷³ Ramoth with its pasture lands, and Anem with its pasture lands; ⁷⁴ out of the tribe of Asher: Mashal with its pasture lands, Abdon with its pasture lands, ⁷⁵ Hukok with its pasture lands, and Rehob with its pasture lands; ⁷⁶ and out of the tribe of Naphtali: Kedesh in Galilee with its pasture lands, Hammon with its pasture lands, and Kiriathaim with its pasture lands. ⁷⁷ To the rest of the Merarites out of the tribe of Zebulun: Rimmono with its pasture lands, Tabor with its pasture lands, ⁷⁸ and across the Jordan from Jericho, on the east side of the Jordan, out of the tribe of Reuben: Bezer in the steppe with its pasture lands, Jahzah with its pasture lands, ⁷⁹ Kedemoth with its pasture lands, and Mephaath with its pasture lands; ⁸⁰ and out of the tribe of Gad: Ramoth in Gilead with its pasture lands, Mahanaim with its pasture lands, ⁸¹ Heshbon with its pasture lands, and Jazer with its pasture lands.

7 The sons*h* of Issachar: Tola, Puah, Jashub, and Shimron, four. ² The sons of Tola: Uzzi, Rephaiah, Jeriel, Jahmai, Ibsam, and Shemuel, heads of their ancestral houses, namely of Tola, mighty warriors of their generations, their number in the days of David being twenty-two thousand six hundred. ³ The son*i* of Uzzi: Izrahiah. And the sons of Izrahiah: Michael, Obadiah, Joel, and Isshiah, five, all of them chiefs; ⁴ and along

---

f Other readings *Hilez, Holon*; See Josh 21.15
g Or *Dobrath*   *h* Syr Compare Vg: Heb *And to the sons*   *i* Heb *sons*

---

caste supported by other Israelites and thus needing no land (Num 26.62). The main source is Josh ch 21.

7.1–40: Descendants of Issachar, Benjamin, Naphtali, Ephraim, and Asher. A

swift genealogical glance at the northern tribes, which were perhaps of interest to the writer mainly because they served David (v. 2).

with them, by their generations, according to their ancestral houses, were units of the fighting force, thirty-six thousand, for they had many wives and sons. 5 Their kindred belonging to all the families of Issachar were in all eighty-seven thousand mighty warriors, enrolled by genealogy.

6 The sons of Benjamin: Bela, Becher, and Jediael, three. 7 The sons of Bela: Ezbon, Uzzi, Uzziel, Jerimoth, and Iri, five, heads of ancestral houses, mighty warriors; and their enrollment by genealogies was twenty-two thousand thirty-four. 8 The sons of Becher: Zemirah, Joash, Eliezer, Elioenai, Omri, Jeremoth, Abijah, Anathoth, and Alemeth. All these were the sons of Becher; 9 and their enrollment by genealogies, according to their generations, as heads of their ancestral houses, mighty warriors, was twenty thousand two hundred. 10 The sons of Jediael: Bilhan. And the sons of Bilhan: Jeush, Benjamin, Ehud, Chenaanah, Zethan, Tarshish, and Ahishahar. 11 All these were the sons of Jediael according to the heads of their ancestral houses, mighty warriors, seventeen thousand two hundred, ready for service in war. 12 And Shuppim and Huppim were the sons of Ir, Hushim the son *j* of Aher.

13 The descendants of Naphtali: Jahziel, Guni, Jezer, and Shallum, the descendants of Bilhah.

14 The sons of Manasseh: Asriel, whom his Aramean concubine bore; she bore Machir the father of Gilead. 15 And Machir took a wife for Huppim and for Shuppim. The name of his sister was Maacah. And the name of the second was Zelophehad; and Zelophehad had daughters. 16 Maacah the wife of Machir bore a son, and she named him Peresh; the name of his brother was Sheresh; and his sons were Ulam and Rekem. 17 The son *j* of Ulam: Bedan. These were the sons of Gilead son of Machir, son of Ma-

nasseh. 18 And his sister Hammolecheth bore Ishhod, Abiezer, and Mahlah. 19 The sons of Shemida were Ahian, Shechem, Likhi, and Aniam.

20 The sons of Ephraim: Shuthelah, and Bered his son, Tahath his son, Eleadah his son, Tahath his son, 21 Zabad his son, Shuthelah his son, and Ezer and Elead. Now the people of Gath, who were born in the land, killed them, because they came down to raid their cattle. 22 And their father Ephraim mourned many days, and his brothers came to comfort him. 23 Ephraim *k* went in to his wife, and she conceived and bore a son; and he named him Beriah, because disaster *l* had befallen his house. 24 His daughter was Sheerah, who built both Lower and Upper Beth-horon, and Uzzen-sheerah. 25 Rephah was his son, Resheph his son, Telah his son, Tahan his son, 26 Ladan his son, Ammihud his son, Elishama his son, 27 Nun *m* his son, Joshua his son. 28 Their possessions and settlements were Bethel and its towns, and eastward Naaran, and westward Gezer and its towns, Shechem and its towns, as far as Ayyah and its towns; 29 also along the borders of the Manassites, Beth-shean and its towns, Taanach and its towns, Megiddo and its towns, Dor and its towns. In these lived the sons of Joseph son of Israel.

30 The sons of Asher: Imnah, Ishvah, Ishvi, Beriah, and their sister Serah. 31 The sons of Beriah: Heber and Malchiel, who was the father of Birzaith. 32 Heber became the father of Japhlet, Shomer, Hotham, and their sister Shua. 33 The sons of Japhlet: Pasach, Bimhal, and Ashvath. These are the sons of Japhlet. 34 The sons of Shemer: Ahi, Rohgah, Hubbah, and Aram. 35 The sons of Helem *n* his brother: Zophah, Imna, Shelesh, and Amal. 36 The sons of Zophah:

j Heb *sons*   *k* Heb *He*   *l* Heb *beraah*
m Here spelled *Non*; see Ex 33.11
n Or *Hotham*; see 7.32

7.6: *Benjamin,* which appears again in the next chapter, may have been confused with Zebulun, which strangely fails to appear here in its proper place (compare Gen 46.17–27; Num 26.23–50; much non-canonical material is also used).

Suah, Harnepher, Shual, Beri, Imrah, 37 Bezer, Hod, Shamma, Shilshah, Ithran, and Beera. 38 The sons of Jether: Jephunneh, Pispa, and Ara. 39 The sons of Ulla: Arah, Hanniel, and Rizia. 40 All of these were men of Asher, heads of ancestral houses, select mighty warriors, chief of the princes. Their number enrolled by genealogies, for service in war, was twenty-six thousand men.

8 Benjamin became the father of Bela his firstborn, Ashbel the second, Aharah the third, 2 Nohah the fourth, and Rapha the fifth. 3 And Bela had sons: Addar, Gera, Abihud, *o* 4 Abishua, Naaman, Ahoah, 5 Gera, Shephuphan, and Huram. 6 These are the sons of Ehud (they were heads of ancestral houses of the inhabitants of Geba, and they were carried into exile to Manahath): 7 Naaman, *p* Ahijah, and Gera, that is, Heglam, *q* who became the father of Uzza and Ahihud. 8 And Shaharaim had sons in the country of Moab after he had sent away his wives Hushim and Baara. 9 He had sons by his wife Hodesh: Jobab, Zibia, Mesha, Malcam, 10 Jeuz, Sachia, and Mirmah. These were his sons, heads of ancestral houses. 11 He also had sons by Hushim: Abitub and Elpaal. 12 The sons of Elpaal: Eber, Misham, and Shemed, who built Ono and Lod with its towns, 13 and Beriah and Shema (they were heads of ancestral houses of the inhabitants of Aijalon, who put to flight the inhabitants of Gath); 14 and Ahio, Shashak, and Jeremoth. 15 Zebadiah, Arad, Eder, 16 Michael, Ishpah, and Joha were sons of Beriah. 17 Zebadiah, Meshullam, Hizki, Heber, 18 Ishmerai, Izliah, and Jobab were the sons of Elpaal. 19 Jakim, Zichri, Zabdi, 20 Elienai, Zillethai, Eliel, 21 Adaiah, Beraiah, and Shimrath were the sons of Shimei. 22 Ishpan, Eber, Eliel, 23 Abdon, Zichri, Hanan, 24 Hananiah,

Elam, Anthothijah, 25 Iphdeiah, and Penuel were the sons of Shashak. 26 Shamsherai, Shehariah, Athaliah, 27 Jaareshiah, Elijah, and Zichri were the sons of Jeroham. 28 These were the heads of ancestral houses, according to their generations, chiefs. These lived in Jerusalem.

29 Jeiel *r* the father of Gibeon lived in Gibeon, and the name of his wife was Maacah. 30 His firstborn son: Abdon, then Zur, Kish, Baal, *s* Nadab, 31 Gedor, Ahio, Zecher, 32 and Mikloth, who became the father of Shimeah. Now these also lived opposite their kindred in Jerusalem, with their kindred. 33 Ner became the father of Kish, Kish of Saul, *t* Saul *t* of Jonathan, Malchishua, Abinadab, and Esh-baal; 34 and the son of Jonathan was Merib-baal; and Merib-baal became the father of Micah. 35 The sons of Micah: Pithon, Melech, Tarea, and Ahaz. 36 Ahaz became the father of Jehoaddah; and Jehoaddah became the father of Alemeth, Azmaveth, and Zimri; Zimri became the father of Moza. 37 Moza became the father of Binea; Raphah was his son, Eleasah his son, Azel his son. 38 Azel had six sons, and these are their names: Azrikam, Bocheru, Ishmael, Sheariah, Obadiah, and Hanan; all these were the sons of Azel. 39 The sons of his brother Eshek: Ulam his firstborn, Jeush the second, and Eliphelet the third. 40 The sons of Ulam were mighty warriors, archers, having many children and grandchildren, one hundred fifty. All these were Benjaminites.

9 So all Israel was enrolled by genealogies; and these are written in the Book of the Kings of Israel. And Judah was taken into exile in Babylon because

o Or *father of Ehud*; see 8.6   *p* Heb *and Naaman*   *q* Or *he carried them into exile*
r Compare 9.35: Heb lacks *Jeiel*   *s* Gk Ms adds *Ner*; Compare 8.33 and 9.36   *t* Or *Shaul*

8.1–40: The descendants of Benjamin. *Benjamin* was given special treatment because *Jerusalem* (v. 28) belonged traditionally to this tribe (Josh 18.28), and perhaps because from it came *Saul* (v. 33), who, though disliked by the Chronicler (10.13–14), was nevertheless the first king of Israel and the predecessor of

David. Verse 28 is usually taken to mean that there were numerous Benjaminites in Jerusalem in the Chronicler's own time. **33–34:** The names *Esh-baal* and *Meribbaal* are correctly transmitted (see 2 Sam 2.8 n. and 4.4 n.), **9.1–44: The genealogies of families in Jerusalem after the Exile. 1:** *The Book of the*

of their unfaithfulness. 2 Now the first to live again in their possessions in their towns were Israelites, priests, Levites, and temple servants.

3 And some of the people of Judah, Benjamin, Ephraim, and Manasseh lived in Jerusalem: 4 Uthai son of Ammihud, son of Omri, son of Imri, son of Bani, from the sons of Perez son of Judah. 5 And of the Shilonites: Asaiah the first-born, and his sons. 6 Of the sons of Zerah: Jeuel and their kin, six hundred ninety. 7 Of the Benjaminites: Sallu son of Meshullam, son of Hodaviah, son of Hassenuah, 8 Ibneiah son of Jeroham, Elah son of Uzzi, son of Michri, and Meshullam son of Shephatiah, son of Reuel, son of Ibnijah; 9 and their kindred according to their generations, nine hundred fifty-six. All these were heads of families according to their ancestral houses.

10 Of the priests: Jedaiah, Jehoiarib, Jachin, 11 and Azariah son of Hilkiah, son of Meshullam, son of Zadok, son of Meraioth, son of Ahitub, the chief officer of the house of God; 12 and Adaiah son of Jeroham, son of Pashhur, son of Malchijah, and Maasai son of Adiel, son of Jahzerah, son of Meshullam, son of Meshillemith, son of Immer; 13 besides their kindred, heads of their ancestral houses, one thousand seven hundred sixty, qualified for the work of the service of the house of God.

14 Of the Levites: Shemaiah son of Hasshub, son of Azrikam, son of Hashabiah, of the sons of Merari; 15 and Bakbakkar, Heresh, Galal, and Mattaniah son of Mica, son of Zichri, son of Asaph; 16 and Obadiah son of Shemaiah, son of Galal, son of Jeduthun, and Berechiah son of Asa, son of Elkanah, who lived in the villages of the Netophathites.

17 The gatekeepers were: Shallum, Akkub, Talmon, Ahiman; and their kindred Shallum was the chief, 18 stationed previously in the king's gate on the east side. These were the gatekeepers of the camp of the Levites. 19 Shallum son of Kore, son of Ebiasaph, son of Korah, and his kindred of his ancestral house, the Korahites, were in charge of the work of the service, guardians of the thresholds of the tent, as their ancestors had been in charge of the camp of the LORD, guardians of the entrance. 20 And Phinehas son of Eleazar was chief over them in former times; the LORD was with him. 21 Zechariah son of Meshelemiah was gatekeeper at the entrance of the tent of meeting. 22 All these, who were chosen as gatekeepers at the thresholds, were two hundred twelve. They were enrolled by genealogies in their villages. David and the seer Samuel established them in their office of trust. 23 So they and their descendants were in charge of the gates of the house of the LORD, that is, the house of the tent, as guards. 24 The gatekeepers were on the four sides, east, west, north, and south; 25 and their kindred who were in their villages were obliged to come in every seven days, in turn, to be with them; 26 for the four chief gatekeepers, who were Levites, were in charge of the chambers and the treasures of the house of God. 27 And they would spend the night near the house of God; for on them lay the duty of watching, and they had charge of opening it every morning.

Kings of Israel is not to be confused with the canonical books of Kings. When the northern kingdom fell, the southern kingdom appropriated the name *Israel* (Mic 1.13–15; 2.7; 3.1, 9–10). For a further comment on *their unfaithfulness,* see 2 Chr 36.11–21. **2–16:** These verses have an inexact parallel in Neh 11.3–19, though it is not claimed there that *people of . . . Ephraim and Manasseh dwelt in Jerusalem.* In earlier times people of these tribes did not live in Jerusalem, but in the Chronicler's time representatives of "all Israel" (the post-exilic Jews) did live there.

9.17–34: In these paragraphs is developed the idea that *David* established (with the help of *Samuel*) all the arrangements of the temple and its services which were carried on into post-exilic times. If 1 Kings chs 6–7 are true, then David did not actually accomplish the building of the temple; he must perforce have utilized *the tent of meeting* (or "tabernacle"), the portable shrine that preceded the temple.

28 Some of them had charge of the utensils of service, for they were required to count them when they were brought in and taken out. 29 Others of them were appointed over the furniture, and over all the holy utensils, also over the choice flour, the wine, the oil, the incense, and the spices. 30 Others, of the sons of the priests, prepared the mixing of the spices, 31 and Mattithiah, one of the Levites, the firstborn of Shallum the Korahite, was in charge of making the flat cakes. 32 Also some of their kindred of the Kohathites had charge of the rows of bread, to prepare them for each sabbath.

33 Now these are the singers, the heads of ancestral houses of the Levites, living in the chambers of the temple free from other service, for they were on duty day and night. 34 These were heads of ancestral houses of the Levites, according to their generations; these leaders lived in Jerusalem.

35 In Gibeon lived the father of Gibeon, Jeiel, and the name of his wife was Maacah. 36 His firstborn son was Abdon, then Zur, Kish, Baal, Ner, Nadab, 37 Gedor, Ahio, Zechariah, and Mikloth; 38 and Mikloth became the father of Shimeam; and these also lived opposite their kindred in Jerusalem, with their kindred. 39 Ner became the father of Kish, Kish of Saul, Saul of Jonathan, Malchishua, Abinadab, and Esh-baal; 40 and the son of Jonathan was Meribbaal; and Merib-baal became the father of Micah. 41 The sons of Micah: Pithon, Melech, Tahrea, and Ahaz;*u* 42 and Ahaz became the father of Jarah, and Jarah of Alemeth, Azmaveth, and Zimri; and Zimri became the father of Moza. 43 Moza became the father of Binea; and Rephaiah was his son, Eleasah his son, Azel his son. 44 Azel had six sons, and these are their names: Azrikam, Boche-

ru, Ishmael, Sheariah, Obadiah, and Hanan; these were the sons of Azel.

10 Now the Philistines fought against Israel; and the men of Israel fled before the Philistines, and fell slain on Mount Gilboa. 2 The Philistines overtook Saul and his sons; and the Philistines killed Jonathan and Abinadab and Malchishua, sons of Saul. 3 The battle pressed hard on Saul; and the archers found him, and he was wounded by the archers. 4 Then Saul said to his armorbearer, "Draw your sword, and thrust me through with it, so that these uncircumcised may not come and make sport of me." But his armor-bearer was unwilling, for he was terrified. So Saul took his own sword and fell on it. 5 When his armor-bearer saw that Saul was dead, he also fell on his sword and died. 6 Thus Saul died; he and his three sons and all his house died together. 7 When all the men of Israel who were in the valley saw that the army*v* had fled and that Saul and his sons were dead, they abandoned their towns and fled; and the Philistines came and occupied them.

8 The next day when the Philistines came to strip the dead, they found Saul and his sons fallen on Mount Gilboa. 9 They stripped him and took his head and his armor, and sent messengers throughout the land of the Philistines to carry the good news to their idols and to the people. 10 They put his armor in the temple of their gods, and fastened his head in the temple of Dagon. 11 But when all Jabesh-gilead heard everything that the Philistines had done to Saul, 12 all the valiant warriors got up and took away the body of Saul and the bodies of his sons, and brought them to Jabesh. Then

u Compare 8.35: Heb lacks *and Ahaz*
v Heb *they*

9.35–44: The genealogy of the family of Saul is repeated here from 8.29–38 in anticipation of the next chapter.
10.1–14: **Saul, the unfaithful predecessor of David. 1–12:** These verses follow

closely their source, 1 Sam 31.1–13, except in v. 10. Compare also v. 12 in the two passages. In spite of his dislike of Saul, the Chronicler hesitated to relate the disgraceful exposure of the bodies.

they buried their bones under the oak in Jabesh, and fasted seven days.

13 So Saul died for his unfaithfulness; he was unfaithful to the LORD in that he did not keep the command of the LORD; moreover, he had consulted a medium, seeking guidance, 14 and did not seek guidance from the LORD. Therefore the LORD *w* put him to death and turned the kingdom over to David son of Jesse.

11 Then all Israel gathered together to David at Hebron and said, "See, we are your bone and flesh. 2 For some time now, even while Saul was king, it was you who commanded the army of Israel. The LORD your God said to you: It is you who shall be shepherd of my people Israel, you who shall be ruler over my people Israel." 3 So all the elders of Israel came to the king at Hebron, and David made a covenant with them at Hebron before the LORD. And they anointed David king over Israel, according to the word of the LORD by Samuel.

4 David and all Israel marched to Jerusalem, that is Jebus, where the Jebusites were, the inhabitants of the land. 5 The inhabitants of Jebus said to David, "You will not come in here." Nevertheless David took the stronghold of Zion, now the city of David. 6 David had said, "Whoever attacks the Jebusites first shall be chief and commander." And Joab son of Zeruiah went up first, so he became chief. 7 David resided in the stronghold; therefore it was called the city of David. 8 He built the city all around, from the Millo in complete circuit; and Joab repaired the rest of the city. 9 And David became greater and greater, for the LORD of hosts was with him.

10 Now these are the chiefs of David's warriors, who gave him strong support in his kingdom, together with all Israel, to make him king, according to the word of the LORD concerning Israel. 11 This is an account of David's mighty warriors: Jashobeam, son of Hachmoni, *x* was chief of the Three; *y* he wielded his spear against three hundred whom he killed at one time.

12 And next to him among the three warriors was Eleazar son of Dodo, the Ahohite. 13 He was with David at Pasdammim when the Philistines were gathered there for battle. There was a plot of ground full of barley. Now the people had fled from the Philistines, 14 but he and David took their stand in the middle of the plot, defended it, and killed the Philistines; and the LORD saved them by a great victory.

15 Three of the thirty chiefs went down to the rock to David at the cave of Adullam, while the army of Philistines was encamped in the valley of Rephaim. 16 David was then in the stronghold; and the garrison of the Philistines was then at Bethlehem. 17 David said longingly, "O that someone would give me water to drink from the well of Bethlehem that is by the gate!" 18 Then the Three broke through the camp of the Philistines, and drew water from the well of Bethlehem that was by the gate, and they brought it to David. But David would not drink of it; he poured it out to the LORD, 19 and said, "My God forbid that I should do this. Can I drink the blood of these men? For at the risk of their lives they brought

w Heb *he* *x* Or *a Hachmonite* *y* Compare 2 Sam 23.8: Heb *Thirty* or *captains*

10.13–14: The Chronicler brings *Saul* into his story mainly in contrast with *David* to the greater glory of the latter (see 1 Sam 31.10–13 n.). *Medium,* 1 Sam 28.3–25.
11.1–47: The accession of David, the capture of Jerusalem, and a list of David's mighty warriors. 1–3: Omitting all the problems and troubles of 2 Sam chs 1–4, the Chronicler sees *David* quickly and easily anointed king over all *Israel.* The source is

2 Sam 5.1–3, except the last phrase, which is condensed from 1 Sam 16.1–13. **4–9:** The source is 2 Sam 5.6–10. The additional information on *Joab* presented here is generally considered to be historical (see 2 Sam 5.6–9 n.).
11.10–47: The main source of this list is 2 Sam 23.8–39, though additional names of unknown origin are also present.

it." Therefore he would not drink it. The three warriors did these things.

20 Now Abishai, *z* the brother of Joab, was chief of the Thirty. *a* With his spear he fought against three hundred and killed them, and won a name beside the Three. 21 He was the most renowned *b* of the Thirty, *a* and became their commander; but he did not attain to the Three.

22 Benaiah son of Jehoiada was a valiant man *c* of Kabzeel, a doer of great deeds; he struck down two sons of *d* Ariel of Moab. He also went down and killed a lion in a pit on a day when snow had fallen. 23 And he killed an Egyptian, a man of great stature, five cubits tall. The Egyptian had in his hand a spear like a weaver's beam; but Benaiah went against him with a staff, snatched the spear out of the Egyptian's hand, and killed him with his own spear. 24 Such were the things Benaiah son of Jehoiada did, and he won a name beside the three warriors. 25 He was renowned among the Thirty, but he did not attain to the Three. And David put him in charge of his bodyguard.

26 The warriors of the armies were Asahel brother of Joab, Elhanan son of Dodo of Bethlehem, 27 Shammoth of Harod, *e* Helez the Pelonite, 28 Ira son of Ikkesh of Tekoa, Abiezer of Anathoth, 29 Sibbecai the Hushathite, Ilai the Ahohite, 30 Maharai of Netophah, Heled son of Baanah of Netophah, 31 Ithai son of Ribai of Gibeah of the Benjaminites, Benaiah of Pirathon, 32 Hurai of the wadis of Gaash, Abiel the Arbathite, 33 Azmaveth of Baharum, Eliahba of Shaalbon, 34 Hashem *f* the Gizonite, Jonathan son of Shagee the Hararite, 35 Ahiam son of Sachar the Hararite, Eliphal son of Ur, 36 Hepher the Mecherathite, Ahijah the Pelonite, 37 Hezro of Carmel, Naarai son of Ezbai, 38 Joel the brother of Nathan, Mibhar son of Hagri, 39 Zelek the Am-monite, Naharai of Beeroth, the armor-bearer of Joab son of Zeruiah, 40 Ira the Ithrite, Gareb the Ithrite, 41 Uriah the Hittite, Zabad son of Ahlai, 42 Adina son of Shiza the Reubenite, a leader of the Reubenites, and thirty with him, 43 Hanan son of Maacah, and Joshaphat the Mithnite, 44 Uzzia the Ashterathite, Shama and Jeiel sons of Hotham the Aroer-ite, 45 Jediael son of Shimri, and his brother Joha the Tizite, 46 Eliel the Maha-vite, and Jeribai and Joshaviah sons of Elnaam, and Ithmah the Moabite, 47 Eli-el, and Obed, and Jaasiel the Mezobaite.

12 The following are those who came to David at Ziklag, while he could not move about freely because of Saul son of Kish; they were among the mighty warriors who helped him in war. 2 They were archers, and could shoot arrows and sling stones with either the right hand or the left; they were Benja-minites, Saul's kindred. 3 The chief was Ahiezer, then Joash, both sons of Shema-ah of Gibeah; also Jeziel and Pelet sons of Azmaveth; Beracah, Jehu of Anathoth, 4 Ishmaiah of Gibeon, a warrior among the Thirty and a leader over the Thirty; Jeremiah, *g* Jahaziel, Johanan, Jozabad of Gederah, 5 Eluzai, *h* Jerimoth, Bealiah, Shemariah, Shephatiah the Haruphite; 6 Elkanah, Isshiah, Azarel, Joezer, and Ja-shobeam, the Korahites; 7 and Joelah and Zebadiah, sons of Jeroham of Gedor.

8 From the Gadites there went over to David at the stronghold in the wilderness mighty and experienced warriors, expert with shield and spear, whose faces were like the faces of lions, and who were

z Gk Vg Tg Compare 2 Sam 23.18: Heb *Abshai*
a Syr: Heb *Three*   *b* Compare 2 Sam 23.19: Heb *more renowned among the two*   *c* Syr: Heb *the son of a valiant man*   *d* See 2 Sam 23.20: Heb lacks *sons of*   *e* Compare 2 Sam 23.25: Heb *the Harorite*   *f* Compare Gk and 2 Sam 23.32: Heb *the sons of Hashem*
g Heb verse 5   *h* Heb verse 6

12.1–40: David's ability to attract men of valor. A description of David's army. David is linked to the past by historical allusions, yet he is also in other sources the prototype of the ideal future king or Messiah. **1–7:** Compare v. 1 with 1 Sam 27.5–12; v. 2 with Judg 3.15 and 20.15–16; v. 4 with 2 Sam 23.18–19. **12.8–15:** Compare 1 Sam 26.1–3 and Deut 33.20–21. These *Gadites,* like *David* himself,

swift as gazelles on the mountains: 9 Ezer the chief, Obadiah second, Eliab third, 10 Mishmannah fourth, Jeremiah fifth, 11 Attai sixth, Eliel seventh, 12 Johanan eighth, Elzabad ninth, 13 Jeremiah tenth, Machbannai eleventh. 14 These Gadites were officers of the army, the least equal to a hundred and the greatest to a thousand. 15 These are the men who crossed the Jordan in the first month, when it was overflowing all its banks, and put to flight all those in the valleys, to the east and to the west.

16 Some Benjaminites and Judahites came to the stronghold to David. 17 David went out to meet them and said to them, "If you have come to me in friendship, to help me, then my heart will be knit to you; but if you have come to betray me to my adversaries, though my hands have done no wrong, then may the God of our ancestors see and give judgment." 18 Then the spirit came upon Amasai, chief of the Thirty, and he said,

"We are yours, O David;
    and with you, O son of Jesse!
Peace, peace to you,
    and peace to the one who helps
        you!
    For your God is the one who
        helps you."

Then David received them, and made them officers of his troops.

19 Some of the Manassites deserted to David when he came with the Philistines for the battle against Saul. (Yet he did not help them, for the rulers of the Philistines took counsel and sent him away, saying, "He will desert to his master Saul at the cost of our heads.") 20 As he went to Ziklag these Manassites deserted to him: Adnah, Jozabad, Jediael, Michael, Jozabad, Elihu, and Zillethai, chiefs of the thousands in Manasseh. 21 They helped David against the band of raiders,[i] for they were all warriors and commanders in the army. 22 Indeed from day to day people kept coming to David to help him, until there was a great army, like an army of God.

23 These are the numbers of the divisions of the armed troops who came to David in Hebron to turn the kingdom of Saul over to him, according to the word of the LORD. 24 The people of Judah bearing shield and spear numbered six thousand eight hundred armed troops. 25 Of the Simeonites, mighty warriors, seven thousand one hundred. 26 Of the Levites four thousand six hundred. 27 Jehoiada, leader of the house of Aaron, and with him three thousand seven hundred. 28 Zadok, a young warrior, and twenty-two commanders from his own ancestral house. 29 Of the Benjaminites, the kindred of Saul, three thousand, of whom the majority had continued to keep their allegiance to the house of Saul. 30 Of the Ephraimites, twenty thousand eight hundred, mighty warriors, notables in their ancestral houses. 31 Of the half-tribe of Manasseh, eighteen thousand, who were expressly named to come and make David king. 32 Of Issachar, those who had understanding of the times, to know what Israel ought to do, two hundred chiefs, and all their kindred under their command. 33 Of Zebulun, fifty thousand seasoned troops, equipped for battle with all the weapons of war, to help David[j] with singleness of purpose. 34 Of Naphtali, a thousand commanders, with whom there were thirty-seven thousand armed with shield and spear. 35 Of the Danites, twenty-eight thousand six hundred equipped for battle. 36 Of Asher, forty thousand seasoned troops ready for battle. 37 Of the Reubenites and Gadites and the half-tribe of Manasseh from be-

i Or *as officers of his troops*   *j* Gk: Heb lacks *David*

---

are here highly idealized. They are as much soldiers of the future as of the past.
12.19–22: The Chronicler's revised version of 1 Sam chs 29–30. That *Manassites deserted to David* in this manner is not attested elsewhere.
12.23–37: The writer now returns to the enthronement scene of 11.1–3. The total

yond the Jordan, one hundred twenty thousand armed with all the weapons of war.

38 All these, warriors arrayed in battle order, came to Hebron with full intent to make David king over all Israel; likewise all the rest of Israel were of a single mind to make David king. 39 They were there with David for three days, eating and drinking, for their kindred had provided for them. 40 And also their neighbors, from as far away as Issachar and Zebulun and Naphtali, came bringing food on donkeys, camels, mules, and oxen—abundant provisions of meal, cakes of figs, clusters of raisins, wine, oil, oxen, and sheep, for there was joy in Israel.

13 David consulted with the commanders of the thousands and of the hundreds, with every leader. 2 David said to the whole assembly of Israel, "If it seems good to you, and if it is the will of the LORD our God, let us send abroad to our kindred who remain in all the land of Israel, including the priests and Levites in the cities that have pasture lands, that they may come together to us. 3 Then let us bring again the ark of our God to us; for we did not turn to it in the days of Saul." 4 The whole assembly agreed to do so, for the thing pleased all the people.

5 So David assembled all Israel from the Shihor of Egypt to Lebo-hamath, to bring the ark of God from Kiriath-jearim. 6 And David and all Israel went up to Baalah, that is, to Kiriath-jearim, which belongs to Judah, to bring up from there the ark of God, the LORD, who is enthroned on the cherubim, which is called by his *k* name. 7 They carried the ark of God on a new cart, from the house of Abinadab, and Uzzah and Ahio *l* were driving the cart. 8 David and all Israel were dancing before God with all their might, with song and lyres and harps and tambourines and cymbals and trumpets.

9 When they came to the threshing floor of Chidon, Uzzah put out his hand to hold the ark, for the oxen shook it. 10 The anger of the LORD was kindled against Uzzah; he struck him down because he put out his hand to the ark; and he died there before God. 11 David was angry because the LORD had burst out against Uzzah; so that place is called Perez-uzzah *m* to this day. 12 David was afraid of God that day; he said, "How can I bring the ark of God into my care?" 13 So David did not take the ark into his care into the city of David; he took it instead to the house of Obed-edom the Gittite. 14 The ark of God remained with the household of Obed-edom in his house three months, and the LORD blessed the household of Obed-edom and all that he had.

14 King Hiram of Tyre sent messengers to David, along with cedar logs, and masons and carpenters to build a house for him. 2 David then perceived that the LORD had established him as king over Israel, and that his kingdom was highly exalted for the sake of his people Israel.

3 David took more wives in Jerusalem, and David became the father of more sons and daughters. 4 These are the names of the children whom he had in Jerusalem: Shammua, Shobab, and Na-

k Heb lacks *his*   *l* Or *and his brother*
m That is *Bursting Out Against Uzzah*

number of fighting men listed here is fantastic, historically speaking.

13.1–14: David's concern to bring the ark to Jerusalem. The Chronicler partly reverses the order of 2 Samuel, placing his version of 2 Sam 6.2–11, which is the source of the present chapter, ahead of his treatment of 2 Sam 5.11–25 (in ch 14). 2 Sam 6.12–19 is then utilized in 15.25–16.3. The purpose of this rearrangement is probably to show the new David's primary concern with cultic rather than military matters, in spite of his skill in the latter (compare 14.17). **1–4:** In the times of the historical *David,* the distinction between *priests* and *Levites* did not exist. **5:** *The Shihor of Egypt,* the eastern branch of the Nile delta; *Lebo-hamath* was in Syria. **6:** *The cherubim,* see 1 Sam 4.4 n.

14.1–17: David's family and his defeat of the Philistines. The source is 2 Sam 5.11–

than; Solomon, [5]Ibhar, Elishua, and El-
pelet; [6]Nogah, Nepheg, and Japhia;
[7]Elishama, Beeliada, and Eliphelet.

8 When the Philistines heard that David had been anointed king over all Israel, all the Philistines went up in search of
David; and David heard of it and went
out against them. [9]Now the Philistines
had come and made a raid in the valley
of Rephaim. [10]David inquired of God,
"Shall I go up against the Philistines?
Will you give them into my hand?" The
Lord said to him, "Go up, and I will give
them into your hand." [11]So he went up
to Baal-perazim, and David defeated
them there. David said, "God has burst
out[n] against my enemies by my hand,
like a bursting flood." Therefore that
place is called Baal-perazim.[o] [12]They
abandoned their gods there, and at David's command they were burned.

13 Once again the Philistines made a
raid in the valley. [14]When David again
inquired of God, God said to him, "You
shall not go up after them; go around and
come on them opposite the balsam trees.
[15]When you hear the sound of marching
in the tops of the balsam trees, then go
out to battle; for God has gone out before
you to strike down the army of the Philistines." [16]David did as God had commanded him, and they struck down the
Philistine army from Gibeon to Gezer.
[17]The fame of David went out into all
lands, and the Lord brought the fear of
him on all nations.

15 David[p] built houses for himself
in the city of David, and he prepared a place for the ark of God and
pitched a tent for it. [2]Then David commanded that no one but the Levites were
to carry the ark of God, for the Lord had
chosen them to carry the ark of the Lord
and to minister to him forever. [3]David
assembled all Israel in Jerusalem to bring
up the ark of the Lord to its place, which
he had prepared for it. [4]Then David
gathered together the descendants of
Aaron and the Levites: [5]of the sons of
Kohath, Uriel the chief, with one hundred twenty of his kindred; [6]of the sons
of Merari, Asaiah the chief, with two
hundred twenty of his kindred; [7]of the
sons of Gershom, Joel the chief, with one
hundred thirty of his kindred; [8]of the
sons of Elizaphan, Shemaiah the chief,
with two hundred of his kindred; [9]of the
sons of Hebron, Eliel the chief, with
eighty of his kindred; [10]of the sons of
Uzziel, Amminadab the chief, with one
hundred twelve of his kindred.

11 David summoned the priests Zadok and Abiathar, and the Levites Uriel,
Asaiah, Joel, Shemaiah, Eliel, and Amminadab. [12]He said to them, "You are
the heads of families of the Levites; sanctify yourselves, you and your kindred, so
that you may bring up the ark of the
Lord, the God of Israel, to the place that
I have prepared for it. [13]Because you did
not carry it the first time,[q] the Lord our
God burst out against us, because we did
not give it proper care." [14]So the priests
and the Levites sanctified themselves to
bring up the ark of the Lord, the God of
Israel. [15]And the Levites carried the ark
of God on their shoulders with the poles,
as Moses had commanded according to
the word of the Lord.

16 David also commanded the chiefs
of the Levites to appoint their kindred as
the singers to play on musical instruments, on harps and lyres and cymbals,
to raise loud sounds of joy. [17]So the Levites appointed Heman son of Joel; and
of his kindred Asaph son of Berechiah;
and of the sons of Merari, their kindred,

---

n Heb *paraz*   *o* That is *Lord of Bursting Out*
p Heb *He*   *q* Meaning of Heb uncertain

---

25. The Chronicler adds v. 17 to portray *David* (see 12.23–37 n.) as a military figure
feared throughout the world.
**15.1–29: The ark is brought to Jerusa-
lem.** The Chronicler returns to the picture of
David as establishing and overseeing temple
worship (see ch 13 n.). **1–15:** Note the emphasis on the Levites, who did not exist as a
special class in the time of the historical *David;* they are not even mentioned in 2 Sam
ch 6.
15.16–24: The musical arrangements here
set forth were largely drawn from the practice
in the Chronicler's own day. These names

Ethan son of Kushaiah; [18] and with them their kindred of the second order, Zechariah, Jaaziel, Shemiramoth, Jehiel, Unni, Eliab, Benaiah, Maaseiah, Mattithiah, Eliphelehu, and Mikneiah, and the gatekeepers Obed-edom and Jeiel. [19] The singers Heman, Asaph, and Ethan were to sound bronze cymbals; [20] Zechariah, Aziel, Shemiramoth, Jehiel, Unni, Eliab, Maaseiah, and Benaiah were to play harps according to Alamoth; [21] but Mattithiah, Eliphelehu, Mikneiah, Obededom, Jeiel, and Azaziah were to lead with lyres according to the Sheminith. [22] Chenaniah, leader of the Levites in music, was to direct the music, for he understood it. [23] Berechiah and Elkanah were to be gatekeepers for the ark. [24] Shebaniah, Joshaphat, Nethanel, Amasai, Zechariah, Benaiah, and Eliezer, the priests, were to blow the trumpets before the ark of God. Obed-edom and Jehiah also were to be gatekeepers for the ark.

25 So David and the elders of Israel, and the commanders of the thousands, went to bring up the ark of the covenant of the LORD from the house of Obededom with rejoicing. [26] And because God helped the Levites who were carrying the ark of the covenant of the LORD, they sacrificed seven bulls and seven rams. [27] David was clothed with a robe of fine linen, as also were all the Levites who were carrying the ark, and the singers, and Chenaniah the leader of the music of the singers; and David wore a linen ephod. [28] So all Israel brought up the ark of the covenant of the LORD with shouting, to the sound of the horn, trumpets, and cymbals, and made loud music on harps and lyres.

29 As the ark of the covenant of the LORD came to the city of David, Michal daughter of Saul looked out of the window, and saw King David leaping and dancing; and she despised him in her heart.

16 They brought in the ark of God, and set it inside the tent that David had pitched for it; and they offered burnt offerings and offerings of well-being before God. [2] When David had finished offering the burnt offerings and the offerings of well-being, he blessed the people in the name of the LORD; [3] and he distributed to every person in Israel—man and woman alike—to each a loaf of bread, a portion of meat,[r] and a cake of raisins.

4 He appointed certain of the Levites as ministers before the ark of the LORD, to invoke, to thank, and to praise the LORD, the God of Israel. [5] Asaph was the chief, and second to him Zechariah, Jeiel, Shemiramoth, Jehiel, Mattithiah, Eliab, Benaiah, Obed-edom, and Jeiel, with harps and lyres; Asaph was to sound the cymbals, [6] and the priests Benaiah and Jahaziel were to blow trumpets regularly, before the ark of the covenant of God.

7 Then on that day David first appointed the singing of praises to the LORD by Asaph and his kindred.

8 O give thanks to the LORD, call on
　　his name,
　　make known his deeds among
　　　the peoples.
9 Sing to him, sing praises to him,
　　tell of all his wonderful works.
10 Glory in his holy name;
　　let the hearts of those who seek
　　　the LORD rejoice.

r Compare Gk Syr Vg: Meaning of Heb uncertain

had become traditional in the musical guild. *Sheminith* (v. 21), a musical term; the exact meaning is unknown.
15.25–28: Based upon 2 Sam 6.12–15, with the addition of Levites, singers, more musical instruments, and more clothing on the person of *David* (see 1 Sam 2.18 n.). **29:** A close approximation of 2 Sam 6.16.
16.1–43: A service of dedication, and further arrangements. A continuation of

ch 15. **1–3:** Compare 2 Sam 6.17–19 as source. All reference to 2 Sam 6.20–23 is omitted. This idealized David would not be guilty of indecent exposure (see 15.25–28 n.) or of unseemly disputes with members of his harem.
16.8–36: A composite psalm inserted for artistic effect, perhaps by an editor (see 1 Sam 2.1–10 n.; 2 Sam ch 22 n.; 23.1–7 n.). The sources of the present composition are

11 Seek the LORD and his strength,
 seek his presence continually.
12 Remember the wonderful works
 he has done,
 his miracles, and the judgments
 he uttered,
13 O offspring of his servant Israel, *s*
 children of Jacob, his chosen ones.

14 He is the LORD our God;
 his judgments are in all the earth.
15 Remember his covenant forever,
 the word that he commanded,
 for a thousand generations,
16 the covenant that he made with
 Abraham,
 his sworn promise to Isaac,
17 which he confirmed to Jacob as a
 statute,
 to Israel as an everlasting
 covenant,
18 saying, "To you I will give the
 land of Canaan
 as your portion for an
 inheritance."

19 When they were few in number,
 of little account, and strangers
 in the land, *t*
20 wandering from nation to nation,
 from one kingdom to another
 people,
21 he allowed no one to oppress
 them;
 he rebuked kings on their
 account,
22 saying, "Do not touch my
 anointed ones;
 do my prophets no harm."

23 Sing to the LORD, all the earth.
 Tell of his salvation from day to
 day.
24 Declare his glory among the
 nations,
 his marvelous works among all
 the peoples.
25 For great is the LORD, and greatly
 to be praised;
 he is to be revered above all gods.

26 For all the gods of the peoples are
 idols,
 but the LORD made the heavens.
27 Honor and majesty are before him;
 strength and joy are in his place.

28 Ascribe to the LORD, O families of
 the peoples,
 ascribe to the LORD glory and
 strength.
29 Ascribe to the LORD the glory due
 his name;
 bring an offering, and come
 before him.
 Worship the LORD in holy
 splendor;
30 tremble before him, all the
 earth.
 The world is firmly established;
 it shall never be moved.
31 Let the heavens be glad, and let
 the earth rejoice,
 and let them say among the
 nations, "The LORD is
 king!"
32 Let the sea roar, and all that
 fills it;
 let the field exult, and
 everything in it.
33 Then shall the trees of the forest
 sing for joy
 before the LORD, for he comes
 to judge the earth.
34 O give thanks to the LORD, for he
 is good;
 for his steadfast love endures
 forever.

35 Say also:
 "Save us, O God of our salvation,
 and gather and rescue us from
 among the nations,
 that we may give thanks to your
 holy name,
 and glory in your praise.
36 Blessed be the LORD, the God of
 Israel,
 from everlasting to everlasting."

s Another reading is *Abraham* (compare Ps
105.6)   *t* Heb *in it*

Ps 105.1–15 (vv. 8–22); Ps 96.1–13 (vv. 23–
33); Ps 106.1, 47–48 (vv. 34–36).

Then all the people said "Amen!" and praised the LORD.

37 David left Asaph and his kinsfolk there before the ark of the covenant of the LORD to minister regularly before the ark as each day required, 38 and also Obed-edom and his*u* sixty-eight kinsfolk; while Obed-edom son of Jeduthun and Hosah were to be gatekeepers. 39 And he left the priest Zadok and his kindred the priests before the tabernacle of the LORD in the high place that was at Gibeon, 40 to offer burnt offerings to the LORD on the altar of burnt offering regularly, morning and evening, according to all that is written in the law of the LORD that he commanded Israel, 41 With them were Heman and Jeduthun, and the rest of those chosen and expressly named to render thanks to the LORD, for his steadfast love endures forever. 42 Heman and Jeduthun had with them trumpets and cymbals for the music, and instruments for sacred song. The sons of Jeduthun were appointed to the gate.

43 Then all the people departed to their homes, and David went home to bless his household.

17 Now when David settled in his house, David said to the prophet Nathan, "I am living in a house of cedar, but the ark of the covenant of the LORD is under a tent." 2 Nathan said to David, "Do all that you have in mind, for God is with you."

3 But that same night the word of the LORD came to Nathan, saying: 4 Go and tell my servant David: Thus says the LORD: You shall not build me a house to live in. 5 For I have not lived in a house since the day I brought out Israel to this very day, but I have lived in a tent and a tabernacle. *v* 6 Wherever I have moved about among all Israel, did I ever speak a word with any of the judges of Israel, whom I commanded to shepherd my people, saying, Why have you not built me a house of cedar? 7 Now therefore thus you shall say to my servant David: Thus says the LORD of hosts: I took you from the pasture, from following the sheep, to be ruler over my people Israel; 8 and I have been with you wherever you went, and have cut off all your enemies before you; and I will make for you a name, like the name of the great ones of the earth. 9 I will appoint a place for my people Israel, and will plant them, so that they may live in their own place, and be disturbed no more; and evildoers shall wear them down no more, as they did formerly, 10 from the time that I appointed judges over my people Israel; and I will subdue all your enemies.

Moreover I declare to you that the LORD will build you a house. 11 When your days are fulfilled to go to be with your ancestors, I will raise up your offspring after you, one of your own sons, and I will establish his kingdom. 12 He shall build a house for me, and I will establish his throne forever. 13 I will be a father to him, and he shall be a son to me. I will not take my steadfast love from him, as I took it from him who was before you, 14 but I will confirm him in my house and in my kingdom forever, and his throne shall be established forever. 15 In accordance with all these words and all this vision, Nathan spoke to David.

16 Then King David went in and sat before the LORD, and said, "Who am I, O LORD God, and what is my house, that you have brought me thus far? 17 And even this was a small thing in your sight, O God; you have also spoken of your servant's house for a great while to come. You regard me as someone of high rank, *w* O LORD God! 18 And what more can David say to you for honoring your servant? You know your servant. 19 For your servant's sake, O LORD, and

u Gk Syr Vg: Heb *their*   *v* Gk 2 Sam 7.6: Heb *but I have been from tent to tent and from tabernacle*   *w* Meaning of Heb uncertain

16.37–42: A continuation of vv. 4–6; similar to 15.1–24.

16.43: Taken from 2 Sam 6.19b–20a; a continuation of v. 3.

17.1–27: Why David himself did not build the temple. This chapter follows closely its source, 2 Sam ch 7.

according to your own heart, you have done all these great deeds, making known all these great things. 20 There is no one like you, O LORD, and there is no God besides you, according to all that we have heard with our ears. 21 Who is like your people Israel, one nation on the earth whom God went to redeem to be his people, making for yourself a name for great and terrible things, in driving out nations before your people whom you redeemed from Egypt? 22 And you made your people Israel to be your people forever; and you, O LORD, became their God.

23 "And now, O LORD, as for the word that you have spoken concerning your servant and concerning his house, let it be established forever, and do as you have promised. 24 Thus your name will be established and magnified forever in the saying, 'The LORD of hosts, the God of Israel, is Israel's God'; and the house of your servant David will be established in your presence. 25 For you, my God, have revealed to your servant that you will build a house for him; therefore your servant has found it possible to pray before you. 26 And now, O LORD, you are God, and you have promised this good thing to your servant; 27 therefore may it please you to bless the house of your servant, that it may continue forever before you. For you, O LORD, have blessed and are blessed *x* forever."

18 Some time afterward, David attacked the Philistines and subdued them; he took Gath and its villages from the Philistines.

2 He defeated Moab, and the Moabites became subject to David and brought tribute.

3 David also struck down King Hadadezer of Zobah, toward Hamath, *y* as he went to set up a monument at the river Euphrates. 4 David took from him one thousand chariots, seven thousand cavalry, and twenty thousand foot soldiers. David hamstrung all the chariot horses, but left one hundred of them. 5 When the Arameans of Damascus came to help King Hadadezer of Zobah, David killed twenty-two thousand Arameans. 6 Then David put garrisons *z* in Aram of Damascus; and the Arameans became subject to David, and brought tribute. The LORD gave victory to David wherever he went. 7 David took the gold shields that were carried by the servants of Hadadezer, and brought them to Jerusalem. 8 From Tibhath and from Cun, cities of Hadadezer, David took a vast quantity of bronze; with it Solomon made the bronze sea and the pillars and the vessels of bronze.

9 When King Tou of Hamath heard that David had defeated the whole army of King Hadadezer of Zobah, 10 he sent his son Hadoram to King David, to greet him and to congratulate him, because he had fought against Hadadezer and defeated him. Now Hadadezer had often been at war with Tou. He sent all sorts of articles of gold, of silver, and of bronze; 11 these also King David dedicated to the LORD, together with the silver and gold that he had carried off from all the nations, from Edom, Moab, the Ammonites, the Philistines, and Amalek.

12 Abishai son of Zeruiah killed eighteen thousand Edomites in the Valley of Salt. 13 He put garrisons in Edom; and all the Edomites became subject to David. And the LORD gave victory to David wherever he went.

14 So David reigned over all Israel;

x Or *and it is blessed*   *y* Meaning of Heb uncertain   *z* Gk Vg 2 Sam 8.6 Compare Syr: Heb lacks *garrisons*

18.1–17: David's further military prowess. This chapter and the two following resume the theme of ch 14 (see note there). The idealized *David* was both a great military leader and a noble founder of religious institutions. **1–13:** *David* defeats *the Philistines, Moabites, Arameans* (Syrians), and *Edomites.* 2 Sam 8.1–14 is the source.

18.14–17: The source is 2 Sam 8.15–18. **17:** *David's sons* are changed from priests (see

and he administered justice and equity to all his people. 15 Joab son of Zeruiah was over the army; Jehoshaphat son of Ahilud was recorder; 16 Zadok son of Ahitub and Ahimelech son of Abiathar were priests; Shavsha was secretary; 17 Benaiah son of Jehoiada was over the Cherethites and the Pelethites; and David's sons were the chief officials in the service of the king.

19 Some time afterward, King Nahash of the Ammonites died, and his son succeeded him. 2 David said, "I will deal loyally with Hanun son of Nahash, for his father dealt loyally with me." So David sent messengers to console him concerning his father. When David's servants came to Hanun in the land of the Ammonites, to console him, 3 the officials of the Ammonites said to Hanun, "Do you think, because David has sent consolers to you, that he is honoring your father? Have not his servants come to you to search and to overthrow and to spy out the land?" 4 So Hanun seized David's servants, shaved them, cut off their garments in the middle at their hips, and sent them away; 5 and they departed. When David was told about the men, he sent messengers to them, for they felt greatly humiliated. The king said, "Remain at Jericho until your beards have grown, and then return."

6 When the Ammonites saw that they had made themselves odious to David, Hanun and the Ammonites sent a thousand talents of silver to hire chariots and cavalry from Mesopotamia, from Aram-maacah and from Zobah. 7 They hired thirty-two thousand chariots and the king of Maacah with his army, who came and camped before Medeba. And the Ammonites were mustered from their cities and came to battle. 8 When David heard of it, he sent Joab and all the army of the warriors. 9 The Ammonites came out and drew up in battle array at the entrance of the city, and the kings who had come were by themselves in the open country.

10 When Joab saw that the line of battle was set against him both in front and in the rear, he chose some of the picked men of Israel and arrayed them against the Arameans; 11 the rest of his troops he put in the charge of his brother Abishai, and they were arrayed against the Ammonites. 12 He said, "If the Arameans are too strong for me, then you shall help me; but if the Ammonites are too strong for you, then I will help you. 13 Be strong, and let us be courageous for our people and for the cities of our God; and may the LORD do what seems good to him." 14 So Joab and the troops who were with him advanced toward the Arameans for battle; and they fled before him. 15 When the Ammonites saw that the Arameans fled, they likewise fled before Abishai, Joab's brother, and entered the city. Then Joab came to Jerusalem.

16 But when the Arameans saw that they had been defeated by Israel, they sent messengers and brought out the Arameans who were beyond the Euphrates, with Shophach the commander of the army of Hadadezer at their head. 17 When David was informed, he gathered all Israel together, crossed the Jordan, came to them, and drew up his forces against them. When David set the battle in array against the Arameans, they fought with him. 18 The Arameans fled before Israel; and David killed seven thousand Aramean charioteers and forty thousand foot soldiers, and also killed Shophach the commander of their army. 19 When the servants of Hadadezer saw that they had been defeated by Israel, they made peace with David, and became subject to him. So the Arameans were not willing to help the Ammonites any more.

---

2 Sam 8.18 n.) to *chief officials,* for the Chronicler believed only descendants of Aaron could be priests.

19.1–19: David defeats the Ammonites and their Aramean (Syrian) allies. The

Chronicler omits 2 Sam ch 9, having already declared (10.6) that the whole house of Saul had disappeared. The theme of David's military prowess is continued, the source being 2 Sam ch 10 with a few changes.

20 In the spring of the year, the time when kings go out to battle, Joab led out the army, ravaged the country of the Ammonites, and came and besieged Rabbah. But David remained at Jerusalem. Joab attacked Rabbah, and overthrew it. ²David took the crown of Milcom*a* from his head; he found that it weighed a talent of gold, and in it was a precious stone; and it was placed on David's head. He also brought out the booty of the city, a very great amount. ³He brought out the people who were in it, and set them to work*b* with saws and iron picks and axes.*c* Thus David did to all the cities of the Ammonites. Then David and all the people returned to Jerusalem.

4 After this, war broke out with the Philistines at Gezer; then Sibbecai the Hushathite killed Sippai, who was one of the descendants of the giants; and the Philistines were subdued. ⁵Again there was war with the Philistines; and Elhanan son of Jair killed Lahmi the brother of Goliath the Gittite, the shaft of whose spear was like a weaver's beam. ⁶Again there was war at Gath, where there was a man of great size, who had six fingers on each hand, and six toes on each foot, twenty-four in number; he also was descended from the giants. ⁷When he taunted Israel, Jonathan son of Shimea,

David's brother, killed him. ⁸These were descended from the giants in Gath; they fell by the hand of David and his servants.

21 Satan stood up against Israel, and incited David to count the people of Israel. ²So David said to Joab and the commanders of the army, "Go, number Israel, from Beer-sheba to Dan, and bring me a report, so that I may know their number." ³But Joab said, "May the LORD increase the number of his people a hundredfold! Are they not, my lord the king, all of them my lord's servants? Why then should my lord require this? Why should he bring guilt on Israel?" ⁴But the king's word prevailed against Joab. So Joab departed and went throughout all Israel, and came back to Jerusalem. ⁵Joab gave the total count of the people to David. In all Israel there were one million one hundred thousand men who drew the sword, and in Judah four hundred seventy thousand who drew the sword. ⁶But he did not include Levi and Benjamin in the numbering, for the king's command was abhorrent to Joab.

7 But God was displeased with this

a Gk Vg See 1 Kings 11.5, 33: MT *of their king*
b Compare 2 Sam 12.31: Heb *and he sawed*
c Compare 2 Sam 12.31: Heb *saws*

20.1–8: The conquest of Ammon; further wars with the Philistines. 1: The first two sentences of this verse come from 2 Sam 11.1; to get the last sentence the Chronicler turned to 2 Sam 12.26, thus eliminating the story of David's adultery with Bathsheba, his murder of her husband, and the prophet's rebuke of the king for his conduct. **2–3**: The source is 2 Sam 12.30–31 (see 2 Sam 12.30 n.). **20.4–8**: The writer now jumps to 2 Sam 21.18–22 for source material, thus eliminating a whole series of discreditable incidents: the violent and immoral conduct of David's sons, Absalom's rebellion, David's flight and weak sentimentality, the disaffection of the northern tribes, Joab's control over David, the execution of more of Saul's descendants. **5**: The addition of the words *Lahmi the brother of* would seem to be for the purpose of resolving the conflict between 1 Sam ch 17 and

2 Sam 21.19 as to who killed Goliath (see 1 Sam 17.4 n.).
    21.1–22.1: The census, the plague, and the acquisition of a site for the sanctuary. This story, based upon 2 Sam ch 24, though not altogether to the credit of David, was used by the Chronicler as the introduction to the next section of his book because the outcome of the trouble in the story was the fixing of the temple site, and the Chronicler from here on is concerned with David as the real founder of the temple. **1**: *Satan* (or "an adversary") replaces "the anger of the LORD" of 2 Sam 24.1. At this time, perhaps some five hundred years since the writing of the earlier account, a figure called "the *satan*" (the "adversary" or "accuser") was associated with evil and misfortune, but not as an enemy of God (see Job 1–2; Zech 3.1–2 and compare 1 Kings 22.19–22). **21.5**: The numbers are somewhat different

thing, and he struck Israel. [8]David said to God, "I have sinned greatly in that I have done this thing. But now, I pray you, take away the guilt of your servant; for I have done very foolishly." [9]The LORD spoke to Gad, David's seer, saying, [10]"Go and say to David, 'Thus says the LORD: Three things I offer you; choose one of them, so that I may do it to you.'" [11]So Gad came to David and said to him, "Thus says the LORD, 'Take your choice: [12]either three years of famine; or three months of devastation by your foes, while the sword of your enemies overtakes you; or three days of the sword of the LORD, pestilence on the land, and the angel of the LORD destroying throughout all the territory of Israel.' Now decide what answer I shall return to the one who sent me." [13]Then David said to Gad, "I am in great distress; let me fall into the hand of the LORD, for his mercy is very great; but let me not fall into human hands."

14 So the LORD sent a pestilence on Israel; and seventy thousand persons fell in Israel. [15]And God sent an angel to Jerusalem to destroy it; but when he was about to destroy it, the LORD took note and relented concerning the calamity; he said to the destroying angel, "Enough! Stay your hand." The angel of the LORD was then standing by the threshing floor of Ornan the Jebusite. [16]David looked up and saw the angel of the LORD standing between earth and heaven, and in his hand a drawn sword stretched out over Jerusalem. Then David and the elders, clothed in sackcloth, fell on their faces. [17]And David said to God, "Was it not I who gave the command to count the people? It is I who have sinned and done very wickedly. But these sheep, what have they done? Let your hand, I pray, O LORD my God, be against me and against my father's house; but do not let your people be plagued!"

18 Then the angel of the LORD commanded Gad to tell David that he should go up and erect an altar to the LORD on the threshing floor of Ornan the Jebusite. [19]So David went up following Gad's instructions, which he had spoken in the name of the LORD. [20]Ornan turned and saw the angel; and while his four sons who were with him hid themselves, Ornan continued to thresh wheat. [21]As David came to Ornan, Ornan looked and saw David; he went out from the threshing floor, and did obeisance to David with his face to the ground. [22]David said to Ornan, "Give me the site of the threshing floor that I may build on it an altar to the LORD—give it to me at its full price—so that the plague may be averted from the people." [23]Then Ornan said to David, "Take it; and let my lord the king do what seems good to him; see, I present the oxen for burnt offerings, and the threshing sledges for the wood, and the wheat for a grain offering. I give it all." [24]But King David said to Ornan, "No; I will buy them for the full price. I will not take for the LORD what is yours, nor offer burnt offerings that cost me nothing." [25]So David paid Ornan six hundred shekels of gold by weight for the site. [26]David built there an altar to the LORD and presented burnt offerings and offerings of well-being. He called upon the LORD, and he answered him with fire from heaven on the altar of burnt offering. [27]Then the LORD commanded the angel, and he put his sword back into its sheath.

28 At that time, when David saw that the LORD had answered him at the threshing floor of Ornan the Jebusite, he made his sacrifices there. [29]For the tabernacle of the LORD, which Moses had

---

in 2 Sam 24.9. **6:** *Levi,* compare Num 1.47–49. **15:** *Ornan,* Araunah in 2 Sam 24.16–24. **20:** The *four sons* do not appear in 2 Sam 24.19–20.

21.25: The modest price of fifty shekels (about 20 oz.) of silver (see 2 Sam 24.18–25 n.) is here raised to the very large sum of

six hundred shekels of gold, which would weigh about 15 lbs. The author is telling his readers that no price was too great for this precious site. **26:** The *fire from heaven* is lacking in the original account; it is a symbol of God's strong approval (compare 1 Kings 18.36–39).

21.28–30: These verses, not in 2 Sam

made in the wilderness, and the altar of burnt offering were at that time in the high place at Gibeon; ³⁰but David could not go before it to inquire of God, for he was afraid of the sword of the angel of the LORD. ¹Then David said, "Here shall be the house of the LORD God and here the altar of burnt offering for Israel."

22 2 David gave orders to gather together the aliens who were residing in the land of Israel, and he set stonecutters to prepare dressed stones for building the house of God. ³David also provided great stores of iron for nails for the doors of the gates and for clamps, as well as bronze in quantities beyond weighing, ⁴and cedar logs without number—for the Sidonians and Tyrians brought great quantities of cedar to David. ⁵For David said, "My son Solomon is young and inexperienced, and the house that is to be built for the LORD must be exceedingly magnificent, famous and glorified throughout all lands; I will therefore make preparation for it." So David provided materials in great quantity before his death.

6 Then he called for his son Solomon and charged him to build a house for the LORD, the God of Israel. ⁷David said to Solomon, "My son, I had planned to build a house to the name of the LORD my God. ⁸But the word of the LORD came to me, saying, 'You have shed much blood and have waged great wars; you shall not build a house to my name, because you have shed so much blood in my sight on

the earth. ⁹See, a son shall be born to you; he shall be a man of peace. I will give him peace from all his enemies on every side; for his name shall be Solomon,ᵈ and I will give peaceᵉ and quiet to Israel in his days. ¹⁰He shall build a house for my name. He shall be a son to me, and I will be a father to him, and I will establish his royal throne in Israel forever.' ¹¹Now, my son, the LORD be with you, so that you may succeed in building the house of the LORD your God, as he has spoken concerning you. ¹²Only, may the LORD grant you discretion and understanding, so that when he gives you charge over Israel you may keep the law of the LORD your God. ¹³Then you will prosper if you are careful to observe the statutes and the ordinances that the LORD commanded Moses for Israel. Be strong and of good courage. Do not be afraid or dismayed. ¹⁴With great pains I have provided for the house of the LORD one hundred thousand talents of gold, one million talents of silver, and bronze and iron beyond weighing, for there is so much of it; timber and stone too I have provided. To these you must add more. ¹⁵You have an abundance of workers: stonecutters, masons, carpenters, and all kinds of artisans without number, skilled in working ¹⁶gold, silver, bronze, and iron. Now begin the work, and the LORD be with you."

17 David also commanded all the

d Heb *Shelomoh*   e Heb *shalom*

---

ch 24, are designed to show that there is now only one true place of sacrifice. **22.1:** A continuation of the preceding verses. The source, 2 Sam ch 24, identified the site as that of *the altar,* but said nothing about *the house of the LORD God.*

22.2–19: David makes preparations for the construction of the temple. According to the earlier record, David may have thought of building a temple (1 Kings 5.2–6), but it was Solomon who made and executed the plans, with the aid of Hiram of Tyre (1 Kings chs 5–7). The Chronicler regarded the temple as Judaism's greatest institution, and David as the nation's most important human figure.

This figure and this institution must, therefore, be brought together. **2:** The reference to forced labor of *aliens* only may be based on 1 Kings 9.22, which overlooks 1 Kings 5.13; 11.28; 12.4. **3–5:** See 1 Kings ch 5, where this preparatory activity is credited to Solomon. **22.6–12:** Compare 1 Kings 8.17–21. **8:** *Blood and . . . wars,* compare 1 Kings 5.3. **13:** Compare 1 Kings 2.2. **14:** The amount of *gold* and *silver* is fantastically large (about 3750 tons of gold and 37,500 tons of silver). It is a figure of speech expressing the inestimable preciousness of the temple as a religious institution. Compare the more modest figures in 1 Kings 9.14, 28; 10.10, 14.

leaders of Israel to help his son Solomon, saying, 18 "Is not the LORD your God with you? Has he not given you peace on every side? For he has delivered the inhabitants of the land into my hand; and the land is subdued before the LORD and his people. 19 Now set your mind and heart to seek the LORD your God. Go and build the sanctuary of the LORD God so that the ark of the covenant of the LORD and the holy vessels of God may be brought into a house built for the name of the LORD."

23 When David was old and full of days, he made his son Solomon king over Israel.

2 David assembled all the leaders of Israel and the priests and the Levites. 3 The Levites, thirty years old and upward, were counted, and the total was thirty-eight thousand. 4 "Twenty-four thousand of these," David said, "shall have charge of the work in the house of the LORD, six thousand shall be officers and judges, 5 four thousand gatekeepers, and four thousand shall offer praises to the LORD with the instruments that I have made for praise." 6 And David organized them in divisions corresponding to the sons of Levi: Gershon,*f* Kohath, and Merari.

7 The sons of Gershon*g* were Ladan and Shimei. 8 The sons of Ladan: Jehiel the chief, Zetham, and Joel, three. 9 The sons of Shimei: Shelomoth, Haziel, and Haran, three. These were the heads of families of Ladan. 10 And the sons of Shimei: Jahath, Zina, Jeush, and Beriah. These four were the sons of Shimei. 11 Jahath was the chief, and Zizah the second; but Jeush and Beriah did not have many

sons, so they were enrolled as a single family.

12 The sons of Kohath: Amram, Izhar, Hebron, and Uzziel, four. 13 The sons of Amram: Aaron and Moses. Aaron was set apart to consecrate the most holy things, so that he and his sons forever should make offerings before the LORD, and minister to him and pronounce blessings in his name forever; 14 but as for Moses the man of God, his sons were to be reckoned among the tribe of Levi. 15 The sons of Moses: Gershom and Eliezer. 16 The sons of Gershom: Shebuel the chief. 17 The sons of Eliezer: Rehabiah the chief; Eliezer had no other sons, but the sons of Rehabiah were very numerous. 18 The sons of Izhar: Shelomith the chief. 19 The sons of Hebron: Jeriah the chief, Amariah the second, Jahaziel the third, and Jekameam the fourth. 20 The sons of Uzziel: Micah the chief and Isshiah the second.

21 The sons of Merari: Mahli and Mushi. The sons of Mahli: Eleazar and Kish. 22 Eleazar died having no sons, but only daughters; their kindred, the sons of Kish, married them. 23 The sons of Mushi: Mahli, Eder, and Jeremoth, three.

24 These were the sons of Levi by their ancestral houses, the heads of families as they were enrolled according to the number of the names of the individuals from twenty years old and upward who were to do the work for the service of the house of the LORD. 25 For David said, "The LORD, the God of Israel, has given rest to his people; and he resides in Jerusalem forever. 26 And so the Levites

f Or *Gershom*; See 1 Chr 6.1, note, and 23.15
g Vg Compare Gk Syr: Heb *to the Gershonite*

23.1–32: David organizes the Levites and assigns them their duties. In the time of the historical David there was no special class of Levites (see notes on ch 6). But to the Chronicler, more than five hundred years later, they were very important; hence they must be brought into relationship with David. **1:** This single verse moves swiftly all the way from 1 Kings 1.1 to 1 Kings 2.1. **2–6:** On the use of men *thirty years old and upward,*

compare Num 4.3; this was apparently older practice (but see Num 8.24). When more elaborate ritual required more men, the age for beginning service was apparently lowered to *twenty years,* as in vv. 24, 27. On *the instruments,* see ch 26, and compare Am 6.5.
23.7–21: These verses were probably inspired by Num chs 3; 4; 8, at least in part.
23.24–32: The new duties of *the Levites* are here more carefully defined.

no longer need to carry the tabernacle or any of the things for its service"— 27 for according to the last words of David these were the number of the Levites from twenty years old and upward— 28 "but their duty shall be to assist the descendants of Aaron for the service of the house of the LORD, having the care of the courts and the chambers, the cleansing of all that is holy, and any work for the service of the house of God; 29 to assist also with the rows of bread, the choice flour for the grain offering, the wafers of unleavened bread, the baked offering, the offering mixed with oil, and all measures of quantity or size. 30 And they shall stand every morning, thanking and praising the LORD, and likewise at evening, 31 and whenever burnt offerings are offered to the LORD on sabbaths, new moons, and appointed festivals, according to the number required of them, regularly before the LORD. 32 Thus they shall keep charge of the tent of meeting and the sanctuary, and shall attend the descendants of Aaron, their kindred, for the service of the house of the LORD."

24 The divisions of the descendants of Aaron were these. The sons of Aaron: Nadab, Abihu, Eleazar, and Ithamar. 2 But Nadab and Abihu died before their father, and had no sons; so Eleazar and Ithamar became the priests. 3 Along with Zadok of the sons of Eleazar, and Ahimelech of the sons of Ithamar, David organized them according to the appointed duties in their service. 4 Since more chief men were found among the sons of Eleazar than among the sons of Ithamar, they organized them under sixteen heads of ancestral houses of the sons of Eleazar, and eight of the sons of Ithamar. 5 They organized them by lot, all alike, for there were officers of the sanctuary and officers of God among both the sons of Eleazar and the sons of Ithamar. 6 The scribe Shemaiah son of Nethanel, a Levite, recorded them in the presence of the king, and the officers, and Zadok the priest, and Ahimelech son of Abiathar, and the heads of ancestral houses of the priests and of the Levites; one ancestral house being chosen for Eleazar and one chosen for Ithamar.

7 The first lot fell to Jehoiarib, the second to Jedaiah, 8 the third to Harim, the fourth to Seorim, 9 the fifth to Malchijah, the sixth to Mijamin, 10 the seventh to Hakkoz, the eighth to Abijah, 11 the ninth to Jeshua, the tenth to Shecaniah, 12 the eleventh to Eliashib, the twelfth to Jakim, 13 the thirteenth to Huppah, the fourteenth to Jeshebeab, 14 the fifteenth to Bilgah, the sixteenth to Immer, 15 the seventeenth to Hezir, the eighteenth to Happizzez, 16 the nineteenth to Pethahiah, the twentieth to Jehezkel, 17 the twenty-first to Jachin, the twenty-second to Gamul, 18 the twenty-third to Delaiah, the twenty-fourth to Maaziah. 19 These had as their appointed duty in their service to enter the house of the LORD according to the procedure established for them by their ancestor Aaron, as the LORD God of Israel had commanded him.

20 And of the rest of the sons of Levi: of the sons of Amram, Shubael; of the sons of Shubael, Jehdeiah. 21 Of Rehabiah: of the sons of Rehabiah, Isshiah the chief. 22 Of the Izharites, Shelomoth; of the sons of Shelomoth, Jahath. 23 The sons of Hebron:*h* Jeriah the chief,*i* Amariah the second, Jahaziel the third, Jekameam the fourth. 24 The sons of Uzziel,

h See 23.19: Heb lacks *Hebron*   *i* See 23.19: Heb lacks *the chief*

24.1–31: David organizes the priests. David organized not only the Levites (ch 23), but also the priests. Thus the highest religious authority was lodged in David. **1–6:** See Num 3.2–4. **3:** *Zadok . . . and Ahimelech,* this should read "Zadok . . . and Abiathar"; compare vv. 6, 31; 18.16; see also 2 Sam 8.16–18 n. **4:** The priests, at least in theory, were organized into twenty-four groups or divisions.
24.7–19: A detailed exposition of the twenty-four divisions as they were or should have been in the Chronicler's own day. On the division of Abijah, see Lk 1.5.
24.20–31: A supplementary list of Levites, having some names in common with 23.7–23.

Micah; of the sons of Micah, Shamir. 25 The brother of Micah, Isshiah; of the sons of Isshiah, Zechariah. 26 The sons of Merari: Mahli and Mushi. The sons of Jaaziah: Beno.*j* 27 The sons of Merari: of Jaaziah, Beno,*j* Shoham, Zaccur, and Ibri. 28 Of Mahli: Eleazar, who had no sons. 29 Of Kish, the sons of Kish: Jerahmeel. 30 The sons of Mushi: Mahli, Eder, and Jerimoth. These were the sons of the Levites according to their ancestral houses. 31 These also cast lots corresponding to their kindred, the descendants of Aaron, in the presence of King David, Zadok, Ahimelech, and the heads of ancestral houses of the priests and of the Levites, the chief as well as the youngest brother.

25 David and the officers of the army also set apart for the service the sons of Asaph, and of Heman, and of Jeduthun, who should prophesy with lyres, harps, and cymbals. The list of those who did the work and of their duties was: 2 Of the sons of Asaph: Zaccur, Joseph, Nethaniah, and Asarelah, sons of Asaph, under the direction of Asaph, who prophesied under the direction of the king. 3 Of Jeduthun, the sons of Jeduthun: Gedaliah, Zeri, Jeshaiah, Shimei,*k* Hashabiah, and Mattithiah, six, under the direction of their father Jeduthun, who prophesied with the lyre in thanksgiving and praise to the LORD. 4 Of Heman, the sons of Heman: Bukkiah, Mattaniah, Uzziel, Shebuel, and Jerimoth, Hananiah, Hanani, Eliathah, Giddalti, and Romamti-ezer, Joshbekashah, Mallothi, Hothir, Mahazioth. 5 All these were the sons of Heman the king's seer, according to the promise of God to exalt

him; for God had given Heman fourteen sons and three daughters. 6 They were all under the direction of their father for the music in the house of the LORD with cymbals, harps, and lyres for the service of the house of God. Asaph, Jeduthun, and Heman were under the order of the king. 7 They and their kindred, who were trained in singing to the LORD, all of whom were skillful, numbered two hundred eighty-eight. 8 And they cast lots for their duties, small and great, teacher and pupil alike.

9 The first lot fell for Asaph to Joseph; the second to Gedaliah, to him and his brothers and his sons, twelve; 10 the third to Zaccur, his sons and his brothers, twelve; 11 the fourth to Izri, his sons and his brothers, twelve; 12 the fifth to Nethaniah, his sons and his brothers, twelve; 13 the sixth to Bukkiah, his sons and his brothers, twelve; 14 the seventh to Jesarelah,*l* his sons and his brothers, twelve; 15 the eighth to Jeshaiah, his sons and his brothers, twelve; 16 the ninth to Mattaniah, his sons and his brothers, twelve; 17 the tenth to Shimei, his sons and his brothers, twelve; 18 the eleventh to Azarel, his sons and his brothers, twelve; 19 the twelfth to Hashabiah, his sons and his brothers, twelve; 20 to the thirteenth, Shubael, his sons and his brothers, twelve; 21 to the fourteenth, Mattithiah, his sons and his brothers, twelve; 22 to the fifteenth, to Jeremoth, his sons and his brothers, twelve; 23 to the sixteenth, to Hananiah, his sons and his brothers,

j Or *his son*: Meaning of Heb uncertain
k One Ms: Gk: MT lacks *Shimei*
l Or *Asarelah*; see 25.2

25.1–31: David organizes the musicians. 1–8: Corresponding to the twenty-four divisions of priests (ch 24) are projected twenty-four divisions of musicians, arranged under three great names, *Asaph, Heman,* and *Jeduthun* (compare 6.31–48; 15.16–24; 16.4–7, 37–42; 23.5). Here Jeduthun takes the place of Ethan in some of the other lists. According to 23.5 the total number of musicians was thought to be four thousand but here (v. 7) only *two hundred and eighty-eight* (24 × 12) are considered. For a more primitive use of music

by a *seer* (v. 5) to aid him to *prophesy* (v. 1), see 2 Kings 3.15; compare 1 Sam 10.5. **4:** *Hananiah . . . Mahazioth,* these nine names seem artificially formed, for when put together and translated they suggest a series of phrases often used in prayer: "Be gracious, O LORD, be gracious to me; thou art my God, whom I magnify and exalt, my help when in trouble; I have fulfilled (or spoken), he has increased visions."

25.9–31: This is a highly artificial alternating arrangement of the names of vv. 2–4.

twelve; [24] to the seventeenth, to Joshbe-kashah, his sons and his brothers, twelve; [25] to the eighteenth, to Hanani, his sons and his brothers, twelve; [26] to the nine-teenth, to Mallothi, his sons and his brothers, twelve; [27] to the twentieth, to Eliathah, his sons and his brothers, twelve; [28] to the twenty-first, to Hothir, his sons and his brothers, twelve; [29] to the twenty-second, to Giddalti, his sons and his brothers, twelve; [30] to the twenty-third, to Mahazioth, his sons and his brothers, twelve; [31] to the twenty-fourth, to Romamti-ezer, his sons and his brothers, twelve.

26 As for the divisions of the gate-keepers: of the Korahites, Me-shelemiah son of Kore, of the sons of Asaph. [2] Meshelemiah had sons: Zechari-ah the firstborn, Jediael the second, Zeb-adiah the third, Jathniel the fourth, [3] Elam the fifth, Jehohanan the sixth, Eli-ehoenai the seventh. [4] Obed-edom had sons: Shemaiah the firstborn, Jehozabad the second, Joah the third, Sachar the fourth, Nethanel the fifth, [5] Ammiel the sixth, Issachar the seventh, Peullethai the eighth; for God blessed him. [6] Also to his son Shemaiah sons were born who exercised authority in their ancestral houses, for they were men of great ability. [7] The sons of Shemaiah: Othni, Rapha-el, Obed, and Elzabad, whose brothers were able men, Elihu and Semachiah. [8] All these, sons of Obed-edom with their sons and brothers, were able men qualified for the service; sixty-two of Obed-edom. [9] Meshelemiah had sons and brothers, able men, eighteen. [10] Ho-sah, of the sons of Merari, had sons: Shimri the chief (for though he was not the firstborn, his father made him chief), [11] Hilkiah the second, Tebaliah the third,

Zechariah the fourth: all the sons and brothers of Hosah totaled thirteen.

12 These divisions of the gatekeepers, corresponding to their leaders, had du-ties, just as their kindred did, ministering in the house of the LORD; [13] and they cast lots by ancestral houses, small and great alike, for their gates. [14] The lot for the east fell to Shelemiah. They cast lots also for his son Zechariah, a prudent counsel-or, and his lot came out for the north. [15] Obed-edom's came out for the south, and to his sons was allotted the store-house. [16] For Shuppim and Hosah it came out for the west, at the gate of Shalle-cheth on the ascending road. Guard cor-responded to guard. [17] On the east there were six Levites each day, [m] on the north four each day, on the south four each day, as well as two and two at the store-house; [18] and for the colonnade [n] on the west there were four at the road and two at the colonnade. [n] [19] These were the divi-sions of the gatekeepers among the Ko-rahites and the sons of Merari.

20 And of the Levites, Ahijah had charge of the treasuries of the house of God and the treasuries of the dedicated gifts. [21] The sons of Ladan, the sons of the Gershonites belonging to Ladan, the heads of families belonging to Ladan the Gershonite: Jehieli. [o]

22 The sons of Jehieli, Zetham and his brother Joel, were in charge of the trea-suries of the house of the LORD. [23] Of the Amramites, the Izharites, the Hebron-ites, and the Uzzielites: [24] Shebuel son of Gershom, son of Moses, was chief offi-cer in charge of the treasuries. [25] His brothers: from Eliezer were his son Re-

m Gk: Heb lacks *each day*   *n* Heb *parbar*: meaning uncertain   *o* The Hebrew text of verse 21 is confused

---

26.1–32: Organization of the gate-keepers, treasurers, and other functionar-ies. The Chronicler concludes the presenta-tion of the organization of the Levites in the service of the temple (compare 9.17–27; 15.23–24; 16.37–42; Ps 84.10). **1–11:** It is ob-vious from such names as *Korahites* (6.37), *Asaph* (6.39), and *Obed-edom* (15.21) that the gatekeepers were considered to be closely re-lated to the musicians. All told there were four thousand (23.5), though only ninety-three (62 + 18 + 13, vv. 8, 9, 11) are consid-ered here.

26.12–19: The Chronicler describes the temple at the time of writing, mentioning the four sides on the *east, north, south,* and *west;* also the gates (v. 13), even giving the name of one (v. 16). **26:** Compare Num 31.48–54.

habiah, his son Jeshaiah, his son Joram,
his son Zichri, and his son Shelomoth.
26 This Shelomoth and his brothers were
in charge of all the treasuries of the dedicated gifts that King David, and the
heads of families, and the officers of
the thousands and the hundreds, and the
commanders of the army, had dedicated.
27 From booty won in battles they dedicated gifts for the maintenance of the
house of the LORD. 28 Also all that Samuel
the seer, and Saul son of Kish, and Abner
son of Ner, and Joab son of Zeruiah had
dedicated—all dedicated gifts were in the
care of Shelomoth[p] and his brothers.
29 Of the Izharites, Chenaniah and
his sons were appointed to outside duties
for Israel, as officers and judges. 30 Of the
Hebronites, Hashabiah and his brothers,
one thousand seven hundred men of ability, had the oversight of Israel west of the
Jordan for all the work of the LORD and
for the service of the king. 31 Of the Hebronites, Jerijah was chief of the Hebronites. (In the fortieth year of David's reign
search was made, of whatever genealogy
or family, and men of great ability
among them were found at Jazer in Gilead.) 32 King David appointed him and his
brothers, two thousand seven hundred
men of ability, heads of families, to have
the oversight of the Reubenites, the Gadites, and the half-tribe of the Manassites
for everything pertaining to God and for
the affairs of the king.

27 This is the list of the people of
Israel, the heads of families, the
commanders of the thousands and the
hundreds, and their officers who served
the king in all matters concerning the
divisions that came and went, month after month throughout the year, each division numbering twenty-four thousand:
2 Jashobeam son of Zabdiel was in
charge of the first division in the first
month; in his division were twenty-four
thousand. 3 He was a descendant of Perez, and was chief of all the commanders
of the army for the first month. 4 Dodai
the Ahohite was in charge of the division
of the second month; Mikloth was the
chief officer of his division. In his division were twenty-four thousand. 5 The
third commander, for the third month,
was Benaiah son of the priest Jehoiada, as
chief; in his division were twenty-four
thousand. 6 This is the Benaiah who was
a mighty man of the Thirty and in command of the Thirty; his son Ammizabad
was in charge of his division.[q] 7 Asahel
brother of Joab was fourth, for the fourth
month, and his son Zebadiah after him;
in his division were twenty-four thousand. 8 The fifth commander, for the fifth
month, was Shamhuth, the Izrahite; in
his division were twenty-four thousand.
9 Sixth, for the sixth month, was Ira son
of Ikkesh the Tekoite; in his division
were twenty-four thousand. 10 Seventh,
for the seventh month, was Helez the
Pelonite, of the Ephraimites; in his division were twenty-four thousand.
11 Eighth, for the eighth month, was Sibbecai the Hushathite, of the Zerahites; in
his division were twenty-four thousand.
12 Ninth, for the ninth month, was Abiezer of Anathoth, a Benjaminite; in his
division were twenty-four thousand.
13 Tenth, for the tenth month, was Maharai of Netophah, of the Zerahites; in

p Gk Compare 26.28: Heb *Shelomith*
q Gk Vg: Heb *Ammizabad was his division*

26.29–32: Those appointed to outside duties were to look after the interests of the religious establishment in all the rest of the idealized nation.

27.1–34: **David organizes military and civil affairs.** The arbitrary numerical schemes already applied to the temple officials (23.4–5; 24.4; 25.9–31) are extended to the military and other groups. **1**: The total number of the king's bodyguard, 288,000 (12 × 24,000) is historically unrealistic, but entirely in line with the Chronicler's idealization of David's kingdom (see 22.14 n. and contrast the modest figure in 2 Sam 15.18). **2–4**: Compare 11.11–12; 2 Sam 23.8–9. **5**: Compare 11.22; 2 Sam 23.20–23.

27.7: *Asahel* was killed early in his career (2 Sam 2.18–23); hence *his son Zebadiah* had to take his place.

his division were twenty-four thousand. ¹⁴ Eleventh, for the eleventh month, was Benaiah of Pirathon, of the Ephraimites; in his division were twenty-four thousand. ¹⁵ Twelfth, for the twelfth month, was Heldai the Netophathite, of Othniel; in his division were twenty-four thousand.

16 Over the tribes of Israel, for the Reubenites, Eliezer son of Zichri was chief officer; for the Simeonites, Shephatiah son of Maacah; ¹⁷ for Levi, Hashabiah son of Kemuel; for Aaron, Zadok; ¹⁸ for Judah, Elihu, one of David's brothers; for Issachar, Omri son of Michael; ¹⁹ for Zebulun, Ishmaiah son of Obadiah; for Naphtali, Jerimoth son of Azriel; ²⁰ for the Ephraimites, Hoshea son of Azaziah; for the half-tribe of Manasseh, Joel son of Pedaiah; ²¹ for the half-tribe of Manasseh in Gilead, Iddo son of Zechariah; for Benjamin, Jaasiel son of Abner; ²² for Dan, Azarel son of Jeroham. These were the leaders of the tribes of Israel. ²³ David did not count those below twenty years of age, for the LORD had promised to make Israel as numerous as the stars of heaven. ²⁴ Joab son of Zeruiah began to count them, but did not finish; yet wrath came upon Israel for this, and the number was not entered into the account of the Annals of King David.

25 Over the king's treasuries was Azmaveth son of Adiel. Over the treasuries in the country, in the cities, in the villages and in the towers, was Jonathan son of Uzziah. ²⁶ Over those who did the work of the field, tilling the soil, was Ezri son of Chelub. ²⁷ Over the vineyards was Shimei the Ramathite. Over the produce of the vineyards for the wine cellars was Zabdi the Shiphmite. ²⁸ Over the olive and sycamore trees in the Shephelah was Baal-hanan the Gederite. Over the stores of oil was Joash. ²⁹ Over the herds that pastured in Sharon was Shitrai the Sharonite. Over the herds in the valleys was Shaphat son of Adlai. ³⁰ Over the camels was Obil the Ishmaelite. Over the donkeys was Jehdeiah the Meronothite. Over the flocks was Jaziz the Hagrite. ³¹ All these were stewards of King David's property.

32 Jonathan, David's uncle, was a counselor, being a man of understanding and a scribe; Jehiel son of Hachmoni attended the king's sons. ³³ Ahithophel was the king's counselor, and Hushai the Archite was the king's friend. ³⁴ After Ahithophel came Jehoiada son of Benaiah, and Abiathar. Joab was commander of the king's army.

28 David assembled at Jerusalem all the officials of Israel, the officials of the tribes, the officers of the divisions that served the king, the commanders of the thousands, the commanders of the hundreds, the stewards of all the property and cattle of the king and his sons, together with the palace officials, the mighty warriors, and all the warriors. ² Then King David rose to his feet and said: "Hear me, my brothers and my people. I had planned to build a house of rest for the ark of the covenant of the LORD, for the footstool of our God; and I made preparations for building. ³ But God said to me, 'You shall not build a

---

27.16–22: The traditional tribal divisions, obsolete in the Chronicler's day, are regarded as important in this ideal past.
27.23–24: Two references to the census of ch 21. 23: Compare 22.17.
27.32–34: No *Jonathan* who was David's uncle is known elsewhere; perhaps the name here is reminiscent of Saul's son (1 Sam 18.1–4). This *Jehiel* is otherwise unknown. *Ahithophel* (2 Sam 15.31; 16.23; 17.23) and *Hushai the Archite* (2 Sam 15.32–37; 16.16–19; 17.5–16) were well-known historical characters. *Jehoiada son of Benaiah,* the reverse of the names in v. 5; probably correct, as boys were often named after their grandfathers. *Abiathar* is doubtless the priest (compare 1 Sam 22.20–23). *Joab,* see 2 Sam 2.12–17 n.
28.1–21: **David transmits the final plans for the temple to Solomon. 1–8:** Compare 22.2–19 and 23.1, where the new David's resolve to name Solomon his successor and to turn the temple plans over to him is already set forth (see 22.2–19 n.). Here a great assembly is called to carry out the resolve. **2:** Compare Ps 132. **3:** See 22.6–12 n. **6–8:** Compare 17.11–14; 22.9–10; Deut 4.5.

house for my name, for you are a warrior and have shed blood.' 4 Yet the LORD God of Israel chose me from all my ancestral house to be king over Israel forever; for he chose Judah as leader, and in the house of Judah my father's house, and among my father's sons he took delight in making me king over all Israel. 5 And of all my sons, for the LORD has given me many, he has chosen my son Solomon to sit upon the throne of the kingdom of the LORD over Israel. 6 He said to me, 'It is your son Solomon who shall build my house and my courts, for I have chosen him to be a son to me, and I will be a father to him. 7 I will establish his kingdom forever if he continues resolute in keeping my commandments and my ordinances, as he is today.' 8 Now therefore in the sight of all Israel, the assembly of the LORD, and in the hearing of our God, observe and search out all the commandments of the LORD your God; that you may possess this good land, and leave it for an inheritance to your children after you forever.

9 "And you, my son Solomon, know the God of your father, and serve him with single mind and willing heart; for the LORD searches every mind, and understands every plan and thought. If you seek him, he will be found by you; but if you forsake him, he will abandon you forever. 10 Take heed now, for the LORD has chosen you to build a house as the sanctuary; be strong, and act."

11 Then David gave his son Solomon the plan of the vestibule of the temple, and of its houses, its treasuries, its upper rooms, and its inner chambers, and of the room for the mercy seat;*r* 12 and the plan of all that he had in mind: for the courts of the house of the LORD, all the surrounding chambers, the treasuries of the house of God, and the treasuries

for dedicated gifts; 13 for the divisions of the priests and of the Levites, and all the work of the service in the house of the LORD; for all the vessels for the service in the house of the LORD, 14 the weight of gold for all golden vessels for each service, the weight of silver vessels for each service, 15 the weight of the golden lampstands and their lamps, the weight of gold for each lampstand and its lamps, the weight of silver for a lampstand and its lamps, according to the use of each in the service, 16 the weight of gold for each table for the rows of bread, the silver for the silver tables, 17 and pure gold for the forks, the basins, and the cups; for the golden bowls and the weight of each; for the silver bowls and the weight of each; 18 for the altar of incense made of refined gold, and its weight; also his plan for the golden chariot of the cherubim that spread their wings and covered the ark of the covenant of the LORD.

19 "All this, in writing at the LORD's direction, he made clear to me—the plan of all the works."

20 David said further to his son Solomon, "Be strong and of good courage, and act. Do not be afraid or dismayed; for the LORD God, my God, is with you. He will not fail you or forsake you, until all the work for the service of the house of the LORD is finished. 21 Here are the divisions of the priests and the Levites for all the service of the house of God; and with you in all the work will be every volunteer who has skill for any kind of service; also the officers and all the people will be wholly at your command."

29 King David said to the whole assembly, "My son Solomon, whom alone God has chosen, is young

r Or *the cover*

28.9–10: Compare Deut 4.25–31.
28.11–19: The Chronicler makes the point that David is much greater than Solomon, and that all these things were done in accordance with the divine will. *The golden chariot* may have been suggested by Ps 18.10 and Ezek 1.4–21. 19: David is said to have re-

ceived plans for the temple in much the same way as Moses traditionally received plans for the tabernacle (Ex ch 25). It is not clear whether *the writing* is thought of as done by David or the LORD (compare Ex 24.4, 12; 31.18; 34.1, 27–28; Deut 5.22).
29.1–30: **The investiture of Solomon**

and inexperienced, and the work is great; for the temple[s] will not be for mortals but for the LORD God. 2So I have provided for the house of my God, so far as I was able, the gold for the things of gold, the silver for the things of silver, and the bronze for the things of bronze, the iron for the things of iron, and wood for the things of wood, besides great quantities of onyx and stones for setting, antimony, colored stones, all sorts of precious stones, and marble in abundance. 3Moreover, in addition to all that I have provided for the holy house, I have a treasure of my own of gold and silver, and because of my devotion to the house of my God I give it to the house of my God: 4three thousand talents of gold, of the gold of Ophir, and seven thousand talents of refined silver, for overlaying the walls of the house, 5and for all the work to be done by artisans, gold for the things of gold and silver for the things of silver. Who then will offer willingly, consecrating themselves today to the LORD?"

6 Then the leaders of ancestral houses made their freewill offerings, as did also the leaders of the tribes, the commanders of the thousands and of the hundreds, and the officers over the king's work. 7They gave for the service of the house of God five thousand talents and ten thousand darics of gold, ten thousand talents of silver, eighteen thousand talents of bronze, and one hundred thousand talents of iron. 8Whoever had precious stones gave them to the treasury of the house of the LORD, into the care of Jehiel the Gershonite. 9Then the people rejoiced because these had given willingly, for with single mind they had offered freely to the LORD; King David also rejoiced greatly.

10 Then David blessed the LORD in the presence of all the assembly; David said: "Blessed are you, O LORD, the God of our ancestor Israel, forever and ever. 11Yours, O LORD, are the greatness, the power, the glory, the victory, and the majesty; for all that is in the heavens and on the earth is yours; yours is the kingdom, O LORD, and you are exalted as head above all. 12Riches and honor come from you, and you rule over all. In your hand are power and might; and it is in your hand to make great and to give strength to all. 13And now, our God, we give thanks to you and praise your glorious name.

14 "But who am I, and what is my people, that we should be able to make this freewill offering? For all things come from you, and of your own have we given you. 15For we are aliens and transients before you, as were all our ancestors; our days on the earth are like a shadow, and there is no hope. 16O LORD our God, all this abundance that we have provided for building you a house for your holy name comes from your hand and is all your own. 17I know, my God, that you search the heart, and take pleasure in uprightness; in the uprightness of my heart I have freely offered all these things, and now I have seen your people, who are present here, offering freely and joyously to you. 18O LORD, the God of Abraham, Isaac, and Israel, our ancestors, keep forever such purposes and thoughts in the hearts of your people, and direct their hearts toward you. 19Grant to my son Solomon that with single mind he may keep your commandments, your decrees, and your statutes, performing all

[s] Heb *fortress*

and the death of David. 2–5: Compare 22.14–16. **6–9:** The call for a freewill offering again follows the tradition of Moses (see 28.19 n.; compare Ex 25.1–9; 35.4–29). *Darics* were originally coins of Persian origin; compare Ezra 8.27.
29.10–19: David's prayer of farewell, a beautiful composition of the Chronicler, which parallels, segment by segment, the psalmic anthology in 16.8–36. **22:** The phrase *the second time* is used in reference to 23.1. The scene here is a kind of ratification of what David had already done because God had determined it (v. 1). Note the part played by *Zadok* in 1 Kings ch 1.
29.23: The equivalent of this verse in the

of them, and that he may build the temple[t] for which I have made provision."

20 Then David said to the whole assembly, "Bless the LORD your God." And all the assembly blessed the LORD, the God of their ancestors, and bowed their heads and prostrated themselves before the LORD and the king. 21 On the next day they offered sacrifices and burnt offerings to the LORD, a thousand bulls, a thousand rams, and a thousand lambs, with their libations, and sacrifices in abundance for all Israel; 22 and they ate and drank before the LORD on that day with great joy.

They made David's son Solomon king a second time; they anointed him as the LORD's prince, and Zadok as priest. 23 Then Solomon sat on the throne of the LORD, succeeding his father David as king; he prospered, and all Israel obeyed him. 24 All the leaders and the mighty warriors, and also all the sons of King David, pledged their allegiance to King Solomon. 25 The LORD highly exalted Solomon in the sight of all Israel, and bestowed upon him such royal majesty as had not been on any king before him in Israel.

26 Thus David son of Jesse reigned over all Israel. 27 The period that he reigned over Israel was forty years; he reigned seven years in Hebron, and thirty-three years in Jerusalem. 28 He died in a good old age, full of days, riches, and honor; and his son Solomon succeeded him. 29 Now the acts of King David, from first to last, are written in the records of the seer Samuel, and in the records of the prophet Nathan, and in the records of the seer Gad, 30 with accounts of all his rule and his might and of the events that befell him and Israel and all the kingdoms of the earth.

t Heb *fortress*

earlier account (1 Kings 2.12) comes after the death of David. **27:** See 1 Kings 2.11. **28:** Compare the adulatory quality of this verse with 1 Kings 2.10.

29.29–30: On the nature of these sources, see the Introduction. The phrase *the kingdoms of the earth* refers to the surrounding peoples with which David came into contact, such as the Philistines, the Arameans, the Ammonites, the Moabites, and the Edomites.

2 Chronicles

The character of 2 Chronicles is discussed in the Introduction to 1 Chronicles, since 1 and 2 Chronicles were originally a single book. Chapters 1–9 of 2 Chronicles present the reign of Solomon, who is an idealized figure nearly as great as David. Chapters 10–36 cover the subsequent years of the divided kingdoms down to exile of the southern kingdom. As little notice as possible is given to the northern kingdom, which was irredeemably false to God, and hence not God's true Israel. Instead, the Chronicler pictures Judah as a kingdom with holy space (the temple) and holy servants (the priests, prophets, and temple singers) at its center, and kings who are supposed to rule under stringent obligations to God. National calamity comes precisely because the Judean kings after Solomon, despite the righteous actions of a few of them, finally and utterly fail in their religious duty. The book ends with Judah in exile, but not without hope. Cyrus of Persia, under God's prompting, decrees that a new temple is to be built in Jerusalem and that God's people may go up, exodus-like, to their homeland (2 Chr 36.22–23). These words of hope and anticipation are the conclusion of the Hebrew Scriptures.

1 Solomon son of David established himself in his kingdom; the LORD his God was with him and made him exceedingly great.

2 Solomon summoned all Israel, the commanders of the thousands and of the hundreds, the judges, and all the leaders of all Israel, the heads of families. 3 Then Solomon, and the whole assembly with him, went to the high place that was at Gibeon; for God's tent of meeting, which Moses the servant of the LORD had made in the wilderness, was there. 4 (But David had brought the ark of God up from Kiriath-jearim to the place that David had prepared for it; for he had pitched a tent for it in Jerusalem.) 5 Moreover the bronze altar that Bezalel son of Uri, son of Hur, had made, was there in front of the tabernacle of the LORD. And Solomon and the assembly inquired at it. 6 Solomon went up there to the bronze altar before the LORD, which was at the tent of meeting, and offered a thousand burnt offerings on it.

7 That night God appeared to Solomon, and said to him, "Ask what I should give you." 8 Solomon said to God, "You have shown great and steadfast love to my father David, and have made me succeed him as king. 9 O LORD God, let your promise to my father David now be fulfilled, for you have made me king over a people as numerous as the dust of the earth. 10 Give me now wisdom and knowledge to go out and come in before this people, for who can rule this great people of yours?" 11 God answered Solomon, "Because this was in your heart, and you have not asked for possessions, wealth, honor, or the life of those who hate you, and have not even asked for long life, but have asked for wisdom and knowledge for yourself that you may rule my people over whom I have made you king, 12 wisdom and knowledge are granted to you. I will also give you riches, possessions, and honor, such as none of the kings had who were before you, and none after you shall have the like." 13 So Solomon came from*a* the high place at Gibeon, from the tent of meeting, to Jerusalem. And he reigned over Israel.

14 Solomon gathered together chariots and horses; he had fourteen hundred chariots and twelve thousand horses, which he stationed in the chariot cities and with the king in Jerusalem. 15 The king made silver and gold as common in Jerusalem as stone, and he made cedar as plentiful as the sycamore of the Shephelah. 16 Solomon's horses were imported from Egypt and Kue; the king's traders received them from Kue at the prevailing price. 17 They imported from Egypt, and then exported, a chariot for six hundred shekels of silver, and a horse for one hundred fifty; so through them these were exported to all the kings of the Hittites and the kings of Aram.

2*b* Solomon decided to build a temple for the name of the LORD, and a royal palace for himself. 2*c* Solomon conscripted seventy thousand laborers and eighty thousand stonecutters in the hill country, with three thousand six hundred to oversee them.

3 Solomon sent word to King Huram

---

a Gk Vg: Heb *to*   *b* Ch 1.18 in Heb
c Ch 2.1 in Heb

1.1–17: Solomon receives wisdom. 2–6: The author passes over the machinations and murders of 1 Kings chs 1–2, also the marriage alliance with Egypt of 1 Kings 3.1–2, and begins the account of Solomon with a religious act, the visit to the high place at Gibeon. The ark of God is introduced into the story (contrast 1 Kings 3.3–15) to legitimate a visit to a high place (compare 1 Kings 3.3; 2 Kings 23.5; 1 Chr 21.29). *Bezalel* and *the bronze altar,* compare Ex 27.1–2; 31.1; 1 Chr 2.20.

1.7–13: This episode, very creditable to Solomon, is here abridged from 1 Kings 3.3–15; the statement that it was a dream has been removed. Also removed is the story of the judgment between the two prostitutes (1 Kings 3.16–28). **9:** *People as numerous as the dust of the earth,* compare 1 Kings 4.20. **14–17:** This section is taken from 1 Kings 10.26–29. **2.1–18: Solomon prepares to build the temple. 1–10:** This is a rewritten version of 1 Kings 5.1–6, 11; note *Huram* instead of Hi-

of Tyre: "Once you dealt with my father David and sent him cedar to build himself a house to live in. 4I am now about to build a house for the name of the LORD my God and dedicate it to him for offering fragrant incense before him, and for the regular offering of the rows of bread, and for burnt offerings morning and evening, on the sabbaths and the new moons and the appointed festivals of the LORD our God, as ordained forever for Israel. 5The house that I am about to build will be great, for our God is greater than other gods. 6But who is able to build him a house, since heaven, even highest heaven, cannot contain him? Who am I to build a house for him, except as a place to make offerings before him? 7So now send me an artisan skilled to work in gold, silver, bronze, and iron, and in purple, crimson, and blue fabrics, trained also in engraving, to join the skilled workers who are with me in Judah and Jerusalem, whom my father David provided. 8Send me also cedar, cypress, and algum timber from Lebanon, for I know that your servants are skilled in cutting Lebanon timber. My servants will work with your servants 9to prepare timber for me in abundance, for the house I am about to build will be great and wonderful. 10I will provide for your servants, those who cut the timber, twenty thousand cors of crushed wheat, twenty thousand cors of barley, twenty thousand baths*d* of wine, and twenty thousand baths of oil."

11 Then King Huram of Tyre answered in a letter that he sent to Solomon, "Because the LORD loves his people

he has made you king over them." 12Huram also said, "Blessed be the LORD God of Israel, who made heaven and earth, who has given King David a wise son, endowed with discretion and understanding, who will build a temple for the LORD, and a royal palace for himself.

13 "I have dispatched Huram-abi, a skilled artisan, endowed with understanding, 14the son of one of the Danite women, his father a Tyrian. He is trained to work in gold, silver, bronze, iron, stone, and wood, and in purple, blue, and crimson fabrics and fine linen, and to do all sorts of engraving and execute any design that may be assigned him, with your artisans, the artisans of my lord, your father David. 15Now, as for the wheat, barley, oil, and wine, of which my lord has spoken, let him send them to his servants. 16We will cut whatever timber you need from Lebanon, and bring it to you as rafts by sea to Joppa; you will take it up to Jerusalem."

17 Then Solomon took a census of all the aliens who were residing in the land of Israel, after the census that his father David had taken; and there were found to be one hundred fifty-three thousand six hundred. 18Seventy thousand of them he assigned as laborers, eighty thousand as stonecutters in the hill country, and three thousand six hundred as overseers to make the people work.

3 Solomon began to build the house of the LORD in Jerusalem on Mount Moriah, where the LORD had appeared to his father David, at the place that David

---

d A Hebrew measure of volume

---

ram as the name of *the king of Tyre.* **11–12:** Expanded from 1 Kings 5.7, where the *letter* does not appear.

2.13–16: *Huram-abi,* see 1 Kings 7.13–14 n. Here his mother is from *Dan,* in 1 Kings 7.14 from the tribe of Naphtali (compare Ex 31.1–5). *Joppa* is not mentioned in 1 Kings 5.9. **17–18:** In the idealized Israel of the Chronicler (see Introduction to 1 Chronicles) *aliens* must of course perform the forced labor, though the historical arrangement was quite otherwise (compare 1 Kings 5.13–18, especially v. 13; also 1 Kings 9.22; 12.4).

3.1–17: Details of the building. This is a condensed and rewritten version of 1 Kings ch 6 and 1 Kings 7.15–22 (see notes there). **1:** The name *Moriah* appears only here and in Gen 22.2, where another location is probably to be understood. *Ornan the Jebusite,* in 2 Sam 24.16 "Araunah" (compare 1 Chr 21.15). **3:** *Cubits,* see 1 Kings 6.2–6 n. The *cubits of the old standard* were larger, being about twenty inches in length. **4:** In 1 Kings 6.2 the height of the main part of the building is given as "thirty cubits"; here the measurement *hundred and twenty cubits,* given as the height of the

had designated, on the threshing floor of Ornan the Jebusite. 2 He began to build on the second day of the second month of the fourth year of his reign. 3 These are Solomon's measurements *e* for building the house of God: the length, in cubits of the old standard, was sixty cubits, and the width twenty cubits. 4 The vestibule in front of the nave of the house was twenty cubits long, across the width of the house; *f* and its height was one hundred twenty cubits. He overlaid it on the inside with pure gold. 5 The nave he lined with cypress, covered it with fine gold, and made palms and chains on it. 6 He adorned the house with settings of precious stones. The gold was gold from Parvaim. 7 So he lined the house with gold—its beams, its thresholds, its walls, and its doors; and he carved cherubim on the walls.

8 He made the most holy place; its length, corresponding to the width of the house, was twenty cubits, and its width was twenty cubits; he overlaid it with six hundred talents of fine gold. 9 The weight of the nails was fifty shekels of gold. He overlaid the upper chambers with gold.

10 In the most holy place he made two carved cherubim and overlaid *g* them with gold. 11 The wings of the cherubim together extended twenty cubits: one wing of the one, five cubits long, touched the wall of the house, and its other wing, five cubits long, touched the wing of the other cherub; 12 and of this cherub, one wing, five cubits long, touched the wall of the house, and the other wing, also five cubits long, was joined to the wing of the first cherub. 13 The wings of these cherubim extended twenty cubits; the cherubim *h* stood on their feet, facing the nave. 14 And Solomon *i* made the curtain of blue and purple and crimson fabrics and fine linen, and worked cherubim into it.

15 In front of the house he made two pillars thirty-five cubits high, with a capital of five cubits on the top of each. 16 He made encircling *j* chains and put them on the tops of the pillars; and he made one hundred pomegranates, and put them on the chains. 17 He set up the pillars in front of the temple, one on the right, the other on the left; the one on the right he called Jachin, and the one on the left, Boaz.

4 He made an altar of bronze, twenty cubits long, twenty cubits wide, and ten cubits high. 2 Then he made the molten sea; it was round, ten cubits from rim to rim, and five cubits high. A line of thirty cubits would encircle it completely. 3 Under it were panels all around, each of ten cubits, surrounding the sea; there were two rows of panels, cast when it was cast. 4 It stood on twelve oxen, three facing north, three facing west, three facing south, and three facing east; the sea was set on them. The hindquarters of each were toward the inside. 5 Its thickness was a handbreadth; its rim was made like the rim of a cup, like the flower

e Syr: Heb *foundations*   *f* Compare
1 Kings 6.3: Meaning of Heb uncertain
g Heb *they overlaid*   *h* Heb *they*   *i* Heb *he*
j Cn: Heb *in the inner sanctuary*

vestibule, is either a textual error or a typical exaggeration of the Chronicler to emphasize the impressiveness of the ideal temple. **6:** The location of *Parvaim* is unknown; it may be Arabia (see Ophir, 1 Kings 9.28 n.). **3.8–9:** Compare 1 Kings 6.19–22, where much *gold* is already in evidence; but *six hundred talents* (about 45,000 lbs.) stands here only and fairly staggers the imagination, so impressive is the ideal temple. **10–14:** Abridged from 1 Kings 6.23–28. *Cherubim,* see 1 Sam 4.4 n. The description of *the curtain* before *the most holy place* (the inner room at the rear of the temple, sometimes called the

"holy of holies") is drawn from a similar feature of the tabernacle (Ex 26.31). The historical temple of Solomon had doors at this point (1 Kings 6.31). At the time of the Chronicler there may have been a curtain, as there was in New Testament times (Mt 27.51; Mk 15.38; Lk 23.45). **15–17:** *Two pillars . . . called Jachin and . . . Boaz,* see 1 Kings 7.15–22 n.

4.1–22: The temple equipment. This chapter is taken mainly from 1 Kings 7.23–51, omitting vv. 27–37. **1:** *The altar of bronze* is not mentioned in 1 Kings ch 7, but it appears in 1 Kings 8.64 and 2 Kings 16.14. **4.5–6:** On the idea *that the sea was for the*

of a lily; it held three thousand baths. *k*
⁶He also made ten basins in which to
wash, and set five on the right side, and
five on the left. In these they were to
rinse what was used for the burnt offering. The sea was for the priests to
wash in.

7 He made ten golden lampstands as
prescribed, and set them in the temple,
five on the south side and five on the
north. ⁸He also made ten tables and
placed them in the temple, five on the
right side and five on the left. And he
made one hundred basins of gold. ⁹He
made the court of the priests, and the
great court, and doors for the court; he
overlaid their doors with bronze. ¹⁰He
set the sea at the southeast corner of the
house.

11 And Huram made the pots, the
shovels, and the basins. Thus Huram finished the work that he did for King Solomon on the house of God: ¹²the two pillars, the bowls, and the two capitals on
the top of the pillars; and the two latticeworks to cover the two bowls of the capitals that were on the top of the pillars;
¹³the four hundred pomegranates for the
two latticeworks, two rows of pomegranates for each latticework, to cover
the two bowls of the capitals that were
on the pillars. ¹⁴He made the stands, the
basins on the stands, ¹⁵the one sea, and
the twelve oxen underneath it. ¹⁶The
pots, the shovels, the forks, and all the
equipment for these Huram-abi made of
burnished bronze for King Solomon for
the house of the LORD. ¹⁷In the plain of
the Jordan the king cast them, in the clay
ground between Succoth and Zeredah.
¹⁸Solomon made all these things in great
quantities, so that the weight of the
bronze was not determined.

19 So Solomon made all the things
that were in the house of God: the golden
altar, the tables for the bread of the Pres-

ence, ²⁰the lampstands and their lamps of
pure gold to burn before the inner sanctuary, as prescribed; ²¹the flowers, the
lamps, and the tongs, of purest gold;
²²the snuffers, basins, ladles, and firepans, of pure gold. As for the entrance to
the temple: the inner doors to the most
holy place and the doors of the nave of
the temple were of gold.

5 Thus all the work that Solomon did
for the house of the LORD was finished. Solomon brought in the things
that his father David had dedicated, and
stored the silver, the gold, and all the
vessels in the treasuries of the house of
God.

2 Then Solomon assembled the elders
of Israel and all the heads of the tribes, the
leaders of the ancestral houses of the people of Israel, in Jerusalem, to bring up the
ark of the covenant of the LORD out of the
city of David, which is Zion. ³And all
the Israelites assembled before the king at
the festival that is in the seventh month.
⁴And all the elders of Israel came, and the
Levites carried the ark. ⁵So they brought
up the ark, the tent of meeting, and all
the holy vessels that were in the tent; the
priests and the Levites brought them up.
⁶King Solomon and all the congregation
of Israel, who had assembled before him,
were before the ark, sacrificing so many
sheep and oxen that they could not be
numbered or counted. ⁷Then the priests
brought the ark of the covenant of the
LORD to its place, in the inner sanctuary
of the house, in the most holy place, underneath the wings of the cherubim. ⁸For
the cherubim spread out their wings over
the place of the ark, so that the cherubim
made a covering above the ark and its
poles. ⁹The poles were so long that the
ends of the poles were seen from the holy
place in front of the inner sanctuary; but

k A Hebrew measure of volume

---

priests to wash in, see 1 Kings 7.23–26 n.
(which has the more correct figure of "two
thousand" rather than *three thousand baths*).
  **5.1–14: Bringing the ark of the cove-
nant into the temple.** This is drawn mainly
from 1 Kings 8.1–11, with the insertion of a

section on the priests and the Levitical singers
(vv. 11b–13a), in which the Chronicler was
particularly interested (see Introduction to
1 Chronicles); compare 1 Chr chs 25–26. **4:**
The Levites are substituted for "the priests" of
the source (1 Kings 8.3).

they could not be seen from outside; they are there to this day. [10]There was nothing in the ark except the two tablets that Moses put there at Horeb, where the LORD made a covenant*l* with the people of Israel after they came out of Egypt.

11 Now when the priests came out of the holy place (for all the priests who were present had sanctified themselves, without regard to their divisions, [12]and all the levitical singers, Asaph, Heman, and Jeduthun, their sons and kindred, arrayed in fine linen, with cymbals, harps, and lyres, stood east of the altar with one hundred twenty priests who were trumpeters). [13]It was the duty of the trumpeters and singers to make themselves heard in unison in praise and thanksgiving to the LORD, and when the song was raised, with trumpets and cymbals and other musical instruments, in praise to the LORD,

"For he is good,
     for his steadfast love endures
          forever,"

the house, the house of the LORD, was filled with a cloud, [14]so that the priests could not stand to minister because of the cloud; for the glory of the LORD filled the house of God.

6 Then Solomon said, "The LORD has said that he would reside in thick darkness. [2]I have built you an exalted house, a place for you to reside in forever."

3 Then the king turned around and blessed all the assembly of Israel, while all the assembly of Israel stood. [4]And he said, "Blessed be the LORD, the God of Israel, who with his hand has fulfilled what he promised with his mouth to my father David, saying, [5]'Since the day that I brought my people out of the land of Egypt, I have not chosen a city from any of the tribes of Israel in which to build a house, so that my name might be there, and I chose no one as ruler over my people Israel; [6]but I have chosen Jerusalem in order that my name may be there, and I have chosen David to be over my people Israel.' [7]My father David had it in mind to build a house for the name of the LORD, the God of Israel. [8]But the LORD said to my father David, 'You did well to consider building a house for my name; [9]nevertheless you shall not build the house, but your son who shall be born to you shall build the house for my name.' [10]Now the LORD has fulfilled his promise that he made; for I have succeeded my father David, and sit on the throne of Israel, as the LORD promised, and have built the house for the name of the LORD, the God of Israel. [11]There I have set the ark, in which is the covenant of the LORD that he made with the people of Israel."

12 Then Solomon*m* stood before the altar of the LORD in the presence of the whole assembly of Israel, and spread out his hands. [13]Solomon had made a bronze platform five cubits long, five cubits wide, and three cubits high, and had set it in the court; and he stood on it. Then he knelt on his knees in the presence of the whole assembly of Israel, and spread out his hands toward heaven. [14]He said, "O LORD, God of Israel, there is no God like you, in heaven or on earth, keeping covenant in steadfast love with your servants who walk before you with all their heart— [15]you who have kept for your servant, my father David, what you promised to him. Indeed, you promised with your mouth and this day have ful-

l Heb lacks *a covenant*   *m* Heb *he*

6.1–42: Solomon's address to the people and his prayer of dedication. This whole chapter, with the exception of vv. 13 and 41–42, is taken with little change from 1 Kings 8.12–52 (see the notes there). Verses 4–11 are the address to the people and vv. 12–42 the prayer of dedication. The *bronze platform* of v. 13, added by the Chronicler, provides a proper place for *Solomon* to kneel; the regular place of prayer before the altar was sacred to the priests, according to usage at the time of the Chronicler. Verses 41–42, adapted from Ps 132.8–10, provide a fitting conclusion, showing the idealized Solomon to be a very devout and holy man, yet reminding the reader that *David* is, after all, the more important figure; *your anointed one* is King Solomon himself.

filled with your hand. [16] Therefore, O LORD, God of Israel, keep for your servant, my father David, that which you promised him, saying, 'There shall never fail you a successor before me to sit on the throne of Israel, if only your children keep to their way, to walk in my law as you have walked before me.' [17] Therefore, O LORD, God of Israel, let your word be confirmed, which you promised to your servant David.

18 "But will God indeed reside with mortals on earth? Even heaven and the highest heaven cannot contain you, how much less this house that I have built! [19] Regard your servant's prayer and his plea, O LORD my God, heeding the cry and the prayer that your servant prays to you. [20] May your eyes be open day and night toward this house, the place where you promised to set your name, and may you heed the prayer that your servant prays toward this place. [21] And hear the plea of your servant and of your people Israel, when they pray toward this place; may you hear from heaven your dwelling place; hear and forgive.

22 "If someone sins against another and is required to take an oath and comes and swears before your altar in this house, [23] may you hear from heaven, and act, and judge your servants, repaying the guilty by bringing their conduct on their own head, and vindicating those who are in the right by rewarding them in accordance with their righteousness.

24 "When your people Israel, having sinned against you, are defeated before an enemy but turn again to you, confess your name, pray and plead with you in this house, [25] may you hear from heaven, and forgive the sin of your people Israel, and bring them again to the land that you gave to them and to their ancestors.

26 "When heaven is shut up and there is no rain because they have sinned against you, and then they pray toward this place, confess your name, and turn from their sin, because you punish them, [27] may you hear in heaven, forgive the sin of your servants, your people Israel, when you teach them the good way in which they should walk; and send down rain upon your land, which you have given to your people as an inheritance.

28 "If there is famine in the land, if there is plague, blight, mildew, locust, or caterpillar; if their enemies besiege them in any of the settlements of the lands; whatever suffering, whatever sickness there is; [29] whatever prayer, whatever plea from any individual or from all your people Israel, all knowing their own suffering and their own sorrows so that they stretch out their hands toward this house; [30] may you hear from heaven, your dwelling place, forgive, and render to all whose heart you know, according to all their ways, for only you know the human heart. [31] Thus may they fear you and walk in your ways all the days that they live in the land that you gave to our ancestors.

32 "Likewise when foreigners, who are not of your people Israel, come from a distant land because of your great name, and your mighty hand, and your outstretched arm, when they come and pray toward this house, [33] may you hear from heaven your dwelling place, and do whatever the foreigners ask of you, in order that all the peoples of the earth may know your name and fear you, as do your people Israel, and that they may know that your name has been invoked on this house that I have built.

34 "If your people go out to battle against their enemies, by whatever way you shall send them, and they pray to you toward this city that you have chosen and the house that I have built for your name, [35] then hear from heaven their prayer and their plea, and maintain their cause.

36 "If they sin against you—for there is no one who does not sin—and you are angry with them and give them to an enemy, so that they are carried away captive to a land far or near; [37] then if they come to their senses in the land to which they have been taken captive, and repent, and plead with you in the land of their captivity, saying, 'We have sinned, and have done wrong; we have acted wickedly'; [38] if they repent with all their heart and soul in the land of their captivity, to

which they were taken captive, and pray toward their land, which you gave to their ancestors, the city that you have chosen, and the house that I have built for your name, ³⁹then hear from heaven your dwelling place their prayer and their pleas, maintain their cause and forgive your people who have sinned against you. ⁴⁰Now, O my God, let your eyes be open and your ears attentive to prayer from this place.

⁴¹ "Now rise up, O Lord God, and
go to your resting place,
you and the ark of your might.
Let your priests, O Lord God, be
clothed with salvation,
and let your faithful rejoice in
your goodness.
⁴² O Lord God, do not reject your
anointed one.
Remember your steadfast love
for your servant David."

7 When Solomon had ended his prayer, fire came down from heaven and consumed the burnt offering and the sacrifices; and the glory of the Lord filled the temple. ²The priests could not enter the house of the Lord, because the glory of the Lord filled the Lord's house. ³When all the people of Israel saw the fire come down and the glory of the Lord on the temple, they bowed down on the pavement with their faces to the ground, and worshiped and gave thanks to the Lord, saying,

"For he is good,
for his steadfast love endures
forever."

4 Then the king and all the people offered sacrifice before the Lord. ⁵King Solomon offered as a sacrifice twenty-two thousand oxen and one hundred twenty thousand sheep. So the king and all the people dedicated the house of God. ⁶The priests stood at their posts; the Levites also, with the instruments for music to the Lord that King David had made for giving thanks to the Lord—for his steadfast love endures forever—whenever David offered praises by their ministry. Opposite them the priests sounded trumpets; and all Israel stood.

7 Solomon consecrated the middle of the court that was in front of the house of the Lord; for there he offered the burnt offerings and the fat of the offerings of well-being because the bronze altar Solomon had made could not hold the burnt offering and the grain offering and the fat parts.

8 At that time Solomon held the festival for seven days, and all Israel with him, a very great congregation, from Lebo-hamath to the Wadi of Egypt. ⁹On the eighth day they held a solemn assembly; for they had observed the dedication of the altar seven days and the festival seven days. ¹⁰On the twenty-third day of the seventh month he sent the people away to their homes, joyful and in good spirits because of the goodness that the Lord had shown to David and to Solomon and to his people Israel.

11 Thus Solomon finished the house of the Lord and the king's house; all that Solomon had planned to do in the house of the Lord and in his own house he successfully accomplished.

12 Then the Lord appeared to Solomon in the night and said to him: "I have

---

7.1–22: The consecration of the sanctuary and the divine admonition. 1: *Fire . . . from heaven* (not in 1 Kings 8.54), see 1 Chr 21.26 n. and compare Lev 9.24. **2:** *The glory of the Lord,* compare 5.14 and see 1 Kings 8.11. **3:** The Chronicler added the obeisance of the people. For the liturgical words, see 5.13; Ps 136.1. **4–5:** Compare 1 Kings 8.62–63, the source. **6:** Compare 5.11–13, also 1 Chr ch 25; the Levitical musicians did not exist in Solomon's day.
7.7–8: Compare 1 Kings 8.64–65, the

source. **9–10:** The source, 1 Kings 8.66, has been drastically changed. The people are not sent away *on the eighth day,* but are kept on for another occasion, *a solemn assembly,* until *the twenty-third day* (compare Lev 23.36; Num 29.35–38; Neh 8.18).
7.11–22: This section, the divine admonition, is taken from 1 Kings 9.1–9, with little change except that vv. 13–15 were inserted to give a slightly more hopeful tone (compare 6.26–27, 37–39).

heard your prayer, and have chosen this place for myself as a house of sacrifice. ¹³When I shut up the heavens so that there is no rain, or command the locust to devour the land, or send pestilence among my people, ¹⁴if my people who are called by my name humble themselves, pray, seek my face, and turn from their wicked ways, then I will hear from heaven, and will forgive their sin and heal their land. ¹⁵Now my eyes will be open and my ears attentive to the prayer that is made in this place. ¹⁶For now I have chosen and consecrated this house so that my name may be there forever; my eyes and my heart will be there for all time. ¹⁷As for you, if you walk before me, as your father David walked, doing according to all that I have commanded you and keeping my statutes and my ordinances, ¹⁸then I will establish your royal throne, as I made covenant with your father David saying, 'You shall never lack a successor to rule over Israel.'

19 "But if you*ⁿ* turn aside and forsake my statutes and my commandments that I have set before you, and go and serve other gods and worship them, ²⁰then I will pluck you*ᵒ* up from the land that I have given you;*ᵒ* and this house, which I have consecrated for my name, I will cast out of my sight, and will make it a proverb and a byword among all peoples. ²¹And regarding this house, now exalted, everyone passing by will be astonished, and say, 'Why has the LORD done such a thing to this land and to this house?' ²²Then they will say, 'Because they abandoned the LORD the God of their ancestors who brought them out of the land of Egypt, and they adopted other gods, and worshiped them and served

them; therefore he has brought all this calamity upon them.'"

8 At the end of twenty years, during which Solomon had built the house of the LORD and his own house, ²Solomon rebuilt the cities that Huram had given to him, and settled the people of Israel in them.

3 Solomon went to Hamath-zobah, and captured it. ⁴He built Tadmor in the wilderness and all the storage towns that he built in Hamath. ⁵He also built Upper Beth-horon and Lower Beth-horon, fortified cities, with walls, gates, and bars, ⁶and Baalath, as well as all Solomon's storage towns, and all the towns for his chariots, the towns for his cavalry, and whatever Solomon desired to build, in Jerusalem, in Lebanon, and in all the land of his dominion. ⁷All the people who were left of the Hittites, the Amorites, the Perizzites, the Hivites, and the Jebusites, who were not of Israel, ⁸from their descendants who were still left in the land, whom the people of Israel had not destroyed—these Solomon conscripted for forced labor, as is still the case today. ⁹But of the people of Israel Solomon made no slaves for his work; they were soldiers, and his officers, the commanders of his chariotry and cavalry. ¹⁰These were the chief officers of King Solomon, two hundred fifty of them, who exercised authority over the people.

11 Solomon brought Pharaoh's daughter from the city of David to the house that he had built for her, for he said, "My wife shall not live in the house of King David of Israel, for the places to

n The word *you* in this verse is plural
o Heb *them*

8.1–18: Various activities of Solomon. This chapter is based on 1 Kings 9.10–28. **2:** In 1 Kings 9.10–14 *Solomon* ceded *the cities* to Hiram (here called *Huram*) to raise needed money; but here the cities go to Solomon. **3:** This campaign is otherwise unknown; it is perhaps an adaptation of 1 Chr 18.3. **4:** *Tadmor* is Palmyra, the great city of the Syrian desert, which could hardly be intended here; 1 Kings 9.18 reads "Tamar," a small place in southern Judah (Ezek 47.19; 48.28).

8.7–10: These verses follow 1 Kings 9.20–23 closely, except that in v. 10 a smaller number is given; usually the numbers of the Chronicler are larger (see 1 Chr 27.1 n.). **11:** Compare 1 Kings 3.1; 7.8; 9.24. The reason given for moving the residence of the woman is an addition by the Chronicler (compare Ezek 44.9).

8.12–15: These verses are an expansion of

which the ark of the LORD has come are holy."

12 Then Solomon offered up burnt offerings to the LORD on the altar of the LORD that he had built in front of the vestibule, 13 as the duty of each day required, offering according to the commandment of Moses for the sabbaths, the new moons, and the three annual festivals—the festival of unleavened bread, the festival of weeks, and the festival of booths. 14 According to the ordinance of his father David, he appointed the divisions of the priests for their service, and the Levites for their offices of praise and ministry alongside the priests as the duty of each day required, and the gatekeepers in their divisions for the several gates; for so David the man of God had commanded. 15 They did not turn away from what the king had commanded the priests and Levites regarding anything at all, or regarding the treasuries.

16 Thus all the work of Solomon was accomplished from*p* the day the foundation of the house of the LORD was laid until the house of the LORD was finished completely.

17 Then Solomon went to Eziongeber and Eloth on the shore of the sea, in the land of Edom. 18 Huram sent him, in the care of his servants, ships and servants familiar with the sea. They went to Ophir, together with the servants of Solomon, and imported from there four hundred fifty talents of gold and brought it to King Solomon.

9 When the queen of Sheba heard of the fame of Solomon, she came to Jerusalem to test him with hard questions, having a very great retinue and camels bearing spices and very much gold and precious stones. When she came to Solomon, she discussed with him all that was on her mind. 2 Solomon answered all her questions; there was noth-

ing hidden from Solomon that he could not explain to her. 3 When the queen of Sheba had observed the wisdom of Solomon, the house that he had built, 4 the food of his table, the seating of his officials, and the attendance of his servants, and their clothing, his valets, and their clothing, and his burnt offerings*q* that he offered at the house of the LORD, there was no more spirit left in her.

5 So she said to the king, "The report was true that I heard in my own land of your accomplishments and of your wisdom, 6 but I did not believe the*r* reports until I came and my own eyes saw it. Not even half of the greatness of your wisdom had been told to me; you far surpass the report that I had heard. 7 Happy are your people! Happy are these your servants, who continually attend you and hear your wisdom! 8 Blessed be the LORD your God, who has delighted in you and set you on his throne as king for the LORD your God. Because your God loved Israel and would establish them forever, he has made you king over them, that you may execute justice and righteousness." 9 Then she gave the king one hundred twenty talents of gold, a very great quantity of spices, and precious stones: there were no spices such as those that the queen of Sheba gave to King Solomon.

10 Moreover the servants of Huram and the servants of Solomon who brought gold from Ophir brought algum wood and precious stones. 11 From the algum wood, the king made steps*s* for the house of the LORD and for the king's house, lyres also and harps for the singers; there never was seen the like of them before in the land of Judah.

12 Meanwhile King Solomon granted the queen of Sheba every desire that

p Gk Syr Vg: Heb *to*   *q* Gk Syr Vg
1 Kings 10.5: Heb *ascent*   *r* Heb *their*
s Gk Vg: Meaning of Heb uncertain

---

1 Kings 9.25 to bring these acts of Solomon in line with the ideas of the Chronicler. In v. 12 it is made clear that Solomon remained outside the sanctuary *in front of the vestibule;* only the priests could go inside (see 1 Kings 9.25 n.). **16:** Added by the Chronicler. **17–18:** Taken with a few changes from 1 Kings 9.26–28 (see notes there).

9.1–31: The visit of the queen of Sheba; Solomon's wealth and grandeur; his death. 1–12: Taken with little change from 1 Kings 10.1–13 (see notes there).

she expressed, well beyond what she had brought to the king. Then she returned to her own land, with her servants.

13 The weight of gold that came to Solomon in one year was six hundred sixty-six talents of gold, 14 besides that which the traders and merchants brought; and all the kings of Arabia and the governors of the land brought gold and silver to Solomon. 15 King Solomon made two hundred large shields of beaten gold; six hundred shekels of beaten gold went into each large shield. 16 He made three hundred shields of beaten gold; three hundred shekels of gold went into each shield; and the king put them in the House of the Forest of Lebanon. 17 The king also made a great ivory throne, and overlaid it with pure gold. 18 The throne had six steps and a footstool of gold, which were attached to the throne, and on each side of the seat were arm rests and two lions standing beside the arm rests, 19 while twelve lions were standing, one on each end of a step on the six steps. The like of it was never made in any kingdom. 20 All King Solomon's drinking vessels were of gold, and all the vessels of the House of the Forest of Lebanon were of pure gold; silver was not considered as anything in the days of Solomon. 21 For the king's ships went to Tarshish with the servants of Huram; once every three years the ships of Tarshish used to come bringing gold, silver, ivory, apes, and peacocks. *t*

22 Thus King Solomon excelled all the kings of the earth in riches and in wisdom. 23 All the kings of the earth sought the presence of Solomon to hear his wisdom, which God had put into his mind. 24 Every one of them brought a present, objects of silver and gold, garments, weaponry, spices, horses, and mules, so much year by year. 25 Solomon had four thousand stalls for horses and chariots, and twelve thousand horses, which he stationed in the chariot cities and with the king in Jerusalem. 26 He ruled over all the kings from the Euphrates to the land of the Philistines, and to the border of Egypt. 27 The king made silver as common in Jerusalem as stone, and cedar as plentiful as the sycamore of the Shephelah. 28 Horses were imported for Solomon from Egypt and from all lands.

29 Now the rest of the acts of Solomon, from first to last, are they not written in the history of the prophet Nathan, and in the prophecy of Ahijah the Shilonite, and in the visions of the seer Iddo concerning Jeroboam son of Nebat? 30 Solomon reigned in Jerusalem over all Israel forty years. 31 Solomon slept with his ancestors and was buried in the city of his father David; and his son Rehoboam succeeded him.

10 Rehoboam went to Shechem, for all Israel had come to Shechem to make him king. 2 When Jeroboam son of Nebat heard of it (for he was in Egypt, where he had fled from King Solomon), then Jeroboam returned from Egypt. 3 They sent and called him; and Jeroboam and all Israel came and said to Rehoboam, 4 "Your father made our yoke heavy. Now therefore lighten the hard service of your father and his heavy yoke that he placed on us, and we will serve you." 5 He said to them, "Come to me again in three days." So the people went away.

6 Then King Rehoboam took counsel with the older men who had attended his father Solomon while he was still alive, saying, "How do you advise me to answer this people?" 7 They answered him, "If you will be kind to this people and please them, and speak good words to them, then they will be your servants

t Or *baboons*

9.13–28: From 1 Kings 10.14–28a with a few changes. With vv. 25–28, compare 1.14–17; 1 Kings 4.21, 26.

9.29–31: Taken with changes from 1 Kings 11.41–43, omitting all unfavorable aspects of Solomon's reign presented in 1 Kings 11.1–40 (see notes there). On the books mentioned in v. 29, see Introduction to 1 Chronicles.

10.1–19: **The division of the kingdom.** This chapter reproduces almost exactly 1 Kings 12.1–19, in spite of the fact that the

forever." 8 But he rejected the advice that the older men gave him, and consulted the young men who had grown up with him and now attended him. 9 He said to them, "What do you advise that we answer this people who have said to me, 'Lighten the yoke that your father put on us'?" 10 The young men who had grown up with him said to him, "Thus should you speak to the people who said to you, 'Your father made our yoke heavy, but you must lighten it for us'; tell them, 'My little finger is thicker than my father's loins. 11 Now, whereas my father laid on you a heavy yoke, I will add to your yoke. My father disciplined you with whips, but I will discipline you with scorpions.'"

12 So Jeroboam and all the people came to Rehoboam the third day, as the king had said, "Come to me again the third day." 13 The king answered them harshly. King Rehoboam rejected the advice of the older men; 14 he spoke to them in accordance with the advice of the young men, "My father made your yoke heavy, but I will add to it; my father disciplined you with whips, but I will discipline you with scorpions." 15 So the king did not listen to the people, because it was a turn of affairs brought about by God so that the LORD might fulfill his word, which he had spoken by Ahijah the Shilonite to Jeroboam son of Nebat.

16 When all Israel saw that the king would not listen to them, the people answered the king,

"What share do we have in David?
We have no inheritance in the
    son of Jesse.
Each of you to your tents,
    O Israel!

Look now to your own house,
    O David."

So all Israel departed to their tents. 17 But Rehoboam reigned over the people of Israel who were living in the cities of Judah. 18 When King Rehoboam sent Hadoram, who was taskmaster over the forced labor, the people of Israel stoned him to death. King Rehoboam hurriedly mounted his chariot to flee to Jerusalem. 19 So Israel has been in rebellion against the house of David to this day.

11 When Rehoboam came to Jerusalem, he assembled one hundred eighty thousand chosen troops of the house of Judah and Benjamin to fight against Israel, to restore the kingdom to Rehoboam. 2 But the word of the LORD came to Shemaiah the man of God: 3 Say to King Rehoboam of Judah, son of Solomon, and to all Israel in Judah and Benjamin, 4 "Thus says the LORD: You shall not go up or fight against your kindred. Let everyone return home, for this thing is from me." So they heeded the word of the LORD and turned back from the expedition against Jeroboam.

5 Rehoboam resided in Jerusalem, and he built cities for defense in Judah. 6 He built up Bethlehem, Etam, Tekoa, 7 Beth-zur, Soco, Adullam, 8 Gath, Mareshah, Ziph, 9 Adoraim, Lachish, Azekah, 10 Zorah, Aijalon, and Hebron, fortified cities that are in Judah and in Benjamin. 11 He made the fortresses strong, and put commanders in them, and stores of food, oil, and wine. 12 He also put large shields and spears in all the cities, and made them very strong. So he held Judah and Benjamin.

13 The priests and the Levites who were in all Israel presented themselves to

---

account reflects little credit on *Rehoboam;* 1 Kings 12.20, which reveals the weakness of Judah, is omitted.
11.1–23: The beginning of the reign of Rehoboam. 1–4: These verses follow closely 1 Kings 12.21–24, in spite of a tendency in favor of the northern kingdom. The phrase *all Israel in Judah and Benjamin,* lacking in 1 Kings 12.23, is intended to make clear that the southern kingdom was the true Israel. In

the older sources, "all Israel" means the northern tribes only (compare 2 Sam 2.9; 1 Kings 12.18, 20). After the fall of the northern kingdom, this name was gradually assumed by Judah. **5–12:** These verses, which have no parallel in 1 Kings, are added for the purpose of showing the strength of Judah; they are probably based on a good source. **10, 12:** *Benjamin,* see 1 Kings 12.21–24 n.
11.13–17: These verses, also without paral-

him from all their territories. 14 The Levites had left their common lands and their holdings and had come to Judah and Jerusalem, because Jeroboam and his sons had prevented them from serving as priests of the LORD, 15 and had appointed his own priests for the high places, and for the goat-demons, and for the calves that he had made. 16 Those who had set their hearts to seek the LORD God of Israel came after them from all the tribes of Israel to Jerusalem to sacrifice to the LORD, the God of their ancestors. 17 They strengthened the kingdom of Judah, and for three years they made Rehoboam son of Solomon secure, for they walked for three years in the way of David and Solomon.

18 Rehoboam took as his wife Mahalath daughter of Jerimoth son of David, and of Abihail daughter of Eliab son of Jesse. 19 She bore him sons: Jeush, Shemariah, and Zaham. 20 After her he took Maacah daughter of Absalom, who bore him Abijah, Attai, Ziza, and Shelomith. 21 Rehoboam loved Maacah daughter of Absalom more than all his other wives and concubines (he took eighteen wives and sixty concubines, and became the father of twenty-eight sons and sixty daughters). 22 Rehoboam appointed Abijah son of Maacah as chief prince among his brothers, for he intended to make him king. 23 He dealt wisely, and distributed some of his sons through all the districts of Judah and Benjamin, in all the fortified cities; he gave them abundant provisions, and found many wives for them.

12 When the rule of Rehoboam was established and he grew strong, he abandoned the law of the LORD, he and all Israel with him. 2 In the fifth year of King Rehoboam, because they had been unfaithful to the LORD, King Shishak of Egypt came up against Jerusalem 3 with twelve hundred chariots and sixty thousand cavalry. A countless army came with him from Egypt—Libyans, Sukkiim, and Ethiopians. *u* 4 He took the fortified cities of Judah and came as far as Jerusalem. 5 Then the prophet Shemaiah came to Rehoboam and to the officers of Judah, who had gathered at Jerusalem because of Shishak, and said to them, "Thus says the LORD: You abandoned me, so I have abandoned you to the hand of Shishak." 6 Then the officers of Israel and the king humbled themselves and said, "The LORD is in the right." 7 When the LORD saw that they humbled themselves, the word of the LORD came to Shemaiah, saying: "They have humbled themselves; I will not destroy them, but I will grant them some deliverance, and my wrath shall not be poured out on Jerusalem by the hand of Shishak. 8 Nevertheless they shall be his servants, so that they may know the difference between serving me and serving the kingdoms of other lands."

9 So King Shishak of Egypt came up against Jerusalem; he took away the treasures of the house of the LORD and the treasures of the king's house; he took everything. He also took away the shields of gold that Solomon had made; 10 but King Rehoboam made in place of them shields of bronze, and committed them to the hands of the officers of the guard, who kept the door of the king's house. 11 Whenever the king went into the house

u Or *Nubians*; Heb *Cushites*

---

lel in 1 Kings, are designed to show that the northern kingdom (*all Israel,* v. 13; see vv. 1–4 n.) was faithless to the LORD; hence *the priests and Levites* flocked to the south. Other good people also came, and things went well *for three years.*

11.18–23: In 1 Kings ch 11, Solomon is reproached for having a multitude of *wives and concubines.* The Chronicler omits this section and, in an account from an otherwise unknown source, transfers some of the onus to *Rehoboam.*

12.1–16: Rehoboam's sin; his punishment and death. The source is 1 Kings 14.21–31, with expansions to show that Shishak's invasion was the direct result of the apostasy of *Rehoboam* and *all Israel* (meaning Judah; compare v. 6 and see 11.1–4 n.). Complete destruction was averted by repentance (vv. 7, 12). See 1 Kings 14.21–31 n.

of the LORD, the guard would come along bearing them, and would then bring them back to the guardroom. 12 Because he humbled himself the wrath of the LORD turned from him, so as not to destroy them completely; moreover, conditions were good in Judah.

13 So King Rehoboam established himself in Jerusalem and reigned. Rehoboam was forty-one years old when he began to reign; he reigned seventeen years in Jerusalem, the city that the LORD had chosen out of all the tribes of Israel to put his name there. His mother's name was Naamah the Ammonite. 14 He did evil, for he did not set his heart to seek the LORD.

15 Now the acts of Rehoboam, from first to last, are they not written in the records of the prophet Shemaiah and of the seer Iddo, recorded by genealogy? There were continual wars between Rehoboam and Jeroboam. 16 Rehoboam slept with his ancestors and was buried in the city of David; and his son Abijah succeeded him.

13 In the eighteenth year of King Jeroboam, Abijah began to reign over Judah. 2 He reigned for three years in Jerusalem. His mother's name was Micaiah daughter of Uriel of Gibeah.

Now there was war between Abijah and Jeroboam. 3 Abijah engaged in battle, having an army of valiant warriors, four hundred thousand picked men; and Jeroboam drew up his line of battle against him with eight hundred thousand picked mighty warriors. 4 Then Abijah stood on the slope of Mount Zemaraim that is in the hill country of Ephraim, and said, "Listen to me, Jeroboam and all Israel! 5 Do you not know that the LORD God of Israel gave the kingship over Israel forever to David and his sons by a covenant of salt? 6 Yet Jeroboam son of Nebat, a servant of Solomon son of David, rose up and rebelled against his lord; 7 and certain worthless scoundrels gathered around him and defied Rehoboam son of Solomon, when Rehoboam was young and irresolute and could not withstand them.

8 "And now you think that you can withstand the kingdom of the LORD in the hand of the sons of David, because you are a great multitude and have with you the golden calves that Jeroboam made as gods for you. 9 Have you not driven out the priests of the LORD, the descendants of Aaron, and the Levites, and made priests for yourselves like the peoples of other lands? Whoever comes to be consecrated with a young bull or seven rams becomes a priest of what are no gods. 10 But as for us, the LORD is our God, and we have not abandoned him. We have priests ministering to the LORD who are descendants of Aaron, and Levites for their service. 11 They offer to the LORD every morning and every evening burnt offerings and fragrant incense, set out the rows of bread on the table of pure gold, and care for the golden lampstand so that its lamps may burn every evening; for we keep the charge of the LORD our God, but you have abandoned him. 12 See, God is with us at our head, and his priests have their battle trumpets to sound the call to battle against you. O Israelites, do not fight against the LORD, the God of your ancestors; for you cannot succeed."

13 Jeroboam had sent an ambush around to come on them from behind; thus his troops*v* were in front of Judah, and the ambush was behind them. 14 When Judah turned, the battle was in front of them and behind them. They cried out to the LORD, and the priests blew the trumpets. 15 Then the people of

v Heb *they*

13.1–22: The reign of Abijah; his great victory over the north. Verses 1–2 parallel 1 Kings 15.1–2, with slight differences in names; v. 22 may be compared with 1 Kings 15.7a. Except for these parallels, the chapter is apparently a free composition of the Chronicler, developed from 1 Kings 15.7b. The theme is not really a war between north and south, or Israel and Judah (note the exaggerated numbers); rather it is the perpetual conflict between apostasy and true worship in all times.

Judah raised the battle shout. And when the people of Judah shouted, God defeated Jeroboam and all Israel before Abijah and Judah. 16 The Israelites fled before Judah, and God gave them into their hands. 17 Abijah and his army defeated them with great slaughter; five hundred thousand picked men of Israel fell slain. 18 Thus the Israelites were subdued at that time, and the people of Judah prevailed, because they relied on the LORD, the God of their ancestors. 19 Abijah pursued Jeroboam, and took cities from him: Bethel with its villages and Jeshanah with its villages and Ephron*w* with its villages. 20 Jeroboam did not recover his power in the days of Abijah; the LORD struck him down, and he died. 21 But Abijah grew strong. He took fourteen wives, and became the father of twenty-two sons and sixteen daughters. 22 The rest of the acts of Abijah, his behavior and his deeds, are written in the story of the prophet Iddo.

14 *x* So Abijah slept with his ancestors, and they buried him in the city of David. His son Asa succeeded him. In his days the land had rest for ten years. 2 *y* Asa did what was good and right in the sight of the LORD his God. 3 He took away the foreign altars and the high places, broke down the pillars, hewed down the sacred poles, *z* 4 and commanded Judah to seek the LORD, the God of their ancestors, and to keep the law and the commandment. 5 He also removed from all the cities of Judah the high places and the incense altars. And the kingdom had rest under him. 6 He built fortified cities in Judah while the land had rest. He had no war in those years, for the LORD gave him peace. 7 He said to Judah, "Let us build these cities,

and surround them with walls and towers, gates and bars; the land is still ours because we have sought the LORD our God; we have sought him, and he has given us peace on every side." So they built and prospered. 8 Asa had an army of three hundred thousand from Judah, armed with large shields and spears, and two hundred eighty thousand troops from Benjamin who carried shields and drew bows; all these were mighty warriors.

9 Zerah the Ethiopian *a* came out against them with an army of a million men and three hundred chariots, and came as far as Mareshah. 10 Asa went out to meet him, and they drew up their lines of battle in the valley of Zephathah at Mareshah. 11 Asa cried to the LORD his God, "O LORD, there is no difference for you between helping the mighty and the weak. Help us, O LORD our God, for we rely on you, and in your name we have come against this multitude. O LORD, you are our God; let no mortal prevail against you." 12 So the LORD defeated the Ethiopians *b* before Asa and before Judah, and the Ethiopians *b* fled. 13 Asa and the army with him pursued them as far as Gerar, and the Ethiopians *b* fell until no one remained alive; for they were broken before the LORD and his army. The people of Judah *c* carried away a great quantity of booty. 14 They defeated all the cities around Gerar, for the fear of the LORD was on them. They plundered all the cities; for there was much plunder in them. 15 They also attacked the tents of

w Another reading is *Ephrain*   *x* Ch 13.23 in Heb   *y* Ch 14.1 in Heb   *z* Heb *Asherim*   *a* Or *Nubian*; Heb *Cushite*   *b* Or *Nubians*; Heb *Cushites*   *c* Heb *They*

14.1–15: The beginning of the reign of Asa, and his great victory over the Ethiopians. In 1 Kings only a relatively brief passage (1 Kings 15.9–24) deals with Asa; here the story is expanded to three chapters (chs 14; 15; 16). Asa is presented by the Chronicler as a man like David, great on the field of battle and diligent in religious reform. In the end, his religious loyalty flagged, and he had to be suitably punished. **1–5:** Roughly

parallel to 1 Kings 15.8–14. **6–8:** Added by the Chronicler from another source (compare 1 Kings 15.23; Jer 41.9).

14.9–15: This is another legendary battle, like that of ch 13, to show the piety of Asa and the power of the LORD. There is no hint of it elsewhere in the Bible. *The Ethiopian* may mean "the Egyptian" (see 2 Kings 19.9 n.) or "the Arabian," but no king named *Zerah* is otherwise known.

those who had livestock,[d] and carried away sheep and goats in abundance, and camels. Then they returned to Jerusalem.

15 The spirit of God came upon Azariah son of Oded. [2]He went out to meet Asa and said to him, "Hear me, Asa, and all Judah and Benjamin: The LORD is with you, while you are with him. If you seek him, he will be found by you, but if you abandon him, he will abandon you. [3]For a long time Israel was without the true God, and without a teaching priest, and without law; [4]but when in their distress they turned to the LORD, the God of Israel, and sought him, he was found by them. [5]In those times it was not safe for anyone to go or come, for great disturbances afflicted all the inhabitants of the lands. [6]They were broken in pieces, nation against nation and city against city, for God troubled them with every sort of distress. [7]But you, take courage! Do not let your hands be weak, for your work shall be rewarded."

8 When Asa heard these words, the prophecy of Azariah son of Oded,[e] he took courage, and put away the abominable idols from all the land of Judah and Benjamin and from the towns that he had taken in the hill country of Ephraim. He repaired the altar of the LORD that was in front of the vestibule of the house of the LORD.[f] [9]He gathered all Judah and Benjamin, and those from Ephraim, Manasseh, and Simeon who were residing as aliens with them, for great numbers had deserted to him from Israel when they saw that the LORD his God was with him. [10]They were gathered at Jerusalem in the third month of the fifteenth year of the reign of Asa. [11]They sacrificed to the LORD on that day, from the booty that they had brought, seven hundred oxen and seven thousand sheep. [12]They entered into a covenant to seek the LORD, the God of their ancestors, with all their

heart and with all their soul. [13]Whoever would not seek the LORD, the God of Israel, should be put to death, whether young or old, man or woman. [14]They took an oath to the LORD with a loud voice, and with shouting, and with trumpets, and with horns. [15]All Judah rejoiced over the oath; for they had sworn with all their heart, and had sought him with their whole desire, and he was found by them, and the LORD gave them rest all around.

16 King Asa even removed his mother Maacah from being queen mother because she had made an abominable image for Asherah. Asa cut down her image, crushed it, and burned it at the Wadi Kidron. [17]But the high places were not taken out of Israel. Nevertheless the heart of Asa was true all his days. [18]He brought into the house of God the votive gifts of his father and his own votive gifts— silver, gold, and utensils. [19]And there was no more war until the thirty-fifth year of the reign of Asa.

16 In the thirty-sixth year of the reign of Asa, King Baasha of Israel went up against Judah, and built Ramah, to prevent anyone from going out or coming into the territory of[g] King Asa of Judah. [2]Then Asa took silver and gold from the treasures of the house of the LORD and the king's house, and sent them to King Ben-hadad of Aram, who resided in Damascus, saying, [3]"Let there be an alliance between me and you, like that between my father and your father; I am sending to you silver and gold; go, break your alliance with King Baasha of Israel, so that he may withdraw from me." [4]Ben-hadad listened to King Asa, and sent the commanders of his armies

---

d Meaning of Heb uncertain   e Compare Syr Vg: Heb *the prophecy, the prophet Obed*   f Heb *the vestibule of the LORD*   g Heb lacks *the territory of*

15.1–19: Asa's reform. 1–5: In this passage, which has no parallel, Asa is pictured as a great reformer, like Hezekiah and Josiah at a later date. The prophet *Azariah son of Oded* is not mentioned elsewhere.
15.16–19: For a parallel, see 1 Kings 15.13–

15, which suggested Asa as a reformer. **16:** *Asherah,* see 1 Kings 14.15 n.
16.1–14: The war with Baasha; Asa's apostasy, punishment, and death. 1–6: Except for the date, these verses follow closely 1 Kings 15.17–22.

against the cities of Israel. They conquered Ijon, Dan, Abel-maim, and all the store-cities of Naphtali. 5 When Baasha heard of it, he stopped building Ramah, and let his work cease. 6 Then King Asa brought all Judah, and they carried away the stones of Ramah and its timber, with which Baasha had been building, and with them he built up Geba and Mizpah.

7 At that time the seer Hanani came to King Asa of Judah, and said to him, "Because you relied on the king of Aram, and did not rely on the LORD your God, the army of the king of Aram has escaped you. 8 Were not the Ethiopians*h* and the Libyans a huge army with exceedingly many chariots and cavalry? Yet because you relied on the LORD, he gave them into your hand. 9 For the eyes of the LORD range throughout the entire earth, to strengthen those whose heart is true to him. You have done foolishly in this; for from now on you will have wars." 10 Then Asa was angry with the seer, and put him in the stocks, in prison, for he was in a rage with him because of this. And Asa inflicted cruelties on some of the people at the same time.

11 The acts of Asa, from first to last, are written in the Book of the Kings of Judah and Israel. 12 In the thirty-ninth year of his reign Asa was diseased in his feet, and his disease became severe; yet even in his disease he did not seek the LORD, but sought help from physicians. 13 Then Asa slept with his ancestors, dying in the forty-first year of his reign.

14 They buried him in the tomb that he had hewn out for himself in the city of David. They laid him on a bier that had been filled with various kinds of spices prepared by the perfumer's art; and they made a very great fire in his honor.

17 His son Jehoshaphat succeeded him, and strengthened himself against Israel. 2 He placed forces in all the fortified cities of Judah, and set garrisons in the land of Judah, and in the cities of Ephraim that his father Asa had taken. 3 The LORD was with Jehoshaphat, because he walked in the earlier ways of his father;*i* he did not seek the Baals, 4 but sought the God of his father and walked in his commandments, and not according to the ways of Israel. 5 Therefore the LORD established the kingdom in his hand. All Judah brought tribute to Jehoshaphat, and he had great riches and honor. 6 His heart was courageous in the ways of the LORD; and furthermore he removed the high places and the sacred poles*j* from Judah.

7 In the third year of his reign he sent his officials, Ben-hail, Obadiah, Zechariah, Nethanel, and Micaiah, to teach in the cities of Judah. 8 With them were the Levites, Shemaiah, Nethaniah, Zebadiah, Asahel, Shemiramoth, Jehonathan, Adonijah, Tobijah, and Tob-adonijah; and with these Levites, the priests Elishama and Jehoram. 9 They taught in Judah, having the book of the law of the LORD

h Or *Nubians*; Heb *Cushites*   *i* Another reading is *his father David*   *j* Heb *Asherim*

16.7–10: This rebuke by the prophet and the unrepentant attitude of Asa were introduced here by the Chronicler to explain the later illness of the king (v. 12). **11–14:** 1 Kings 15.23–24a has been expanded by emphasis on the culpability of Asa and the grandeur of his funeral, in spite of his sin.

17.1–19: The initially good and prosperous reign of Jehoshaphat. Chapters 17–20 are taken up with the reign of this king, exhibiting, as also in the case of Asa (see 14.1–14 n.), a considerable expansion of the source material (found in 1 Kings 15.24b; 22.1–35a, 41–49). Jehoshaphat is represented as a ruler who did mostly good things, for which he

was rewarded, but also some bad things, for which he was punished. **1–6:** With the exception of v. 1a (1 Kings 15.24b), this comes largely from sources now lost or is the author's creation. **6:** This contradicts 20.33 and 1 Kings 22.43. *The sacred poles,* see 1 Kings 14.15 n.

17.7–9: No parallel. Here the Chronicler has the king send out a delegation of five princes, nine Levites, and two priests for religious education *among the people.* The preponderance of Levites is significant for the Chronicler's ideology. *The book of the law,* compare 2 Kings 22.8–13; Deut 17.18–20.

with them; they went around through all the cities of Judah and taught among the people.

10 The fear of the LORD fell on all the kingdoms of the lands around Judah, and they did not make war against Jehoshaphat. 11 Some of the Philistines brought Jehoshaphat presents, and silver for tribute; and the Arabs also brought him seven thousand seven hundred rams and seven thousand seven hundred male goats. 12 Jehoshaphat grew steadily greater. He built fortresses and storage cities in Judah. 13 He carried out great works in the cities of Judah. He had soldiers, mighty warriors, in Jerusalem. 14 This was the muster of them by ancestral houses: Of Judah, the commanders of the thousands: Adnah the commander, with three hundred thousand mighty warriors, 15 and next to him Jehohanan the commander, with two hundred eighty thousand, 16 and next to him Amasiah son of Zichri, a volunteer for the service of the LORD, with two hundred thousand mighty warriors. 17 Of Benjamin: Eliada, a mighty warrior, with two hundred thousand armed with bow and shield, 18 and next to him Jehozabad with one hundred eighty thousand armed for war. 19 These were in the service of the king, besides those whom the king had placed in the fortified cities throughout all Judah.

18 Now Jehoshaphat had great riches and honor; and he made a marriage alliance with Ahab. 2 After some years he went down to Ahab in Samaria. Ahab slaughtered an abundance of sheep and oxen for him and for the people who were with him, and induced him to go up against Ramoth-gilead. 3 King Ahab of Israel said to King Jehosh-aphat of Judah, "Will you go with me to Ramoth-gilead?" He answered him, "I am with you, my people are your people. We will be with you in the war."

4 But Jehoshaphat also said to the king of Israel, "Inquire first for the word of the LORD." 5 Then the king of Israel gathered the prophets together, four hundred of them, and said to them, "Shall we go to battle against Ramoth-gilead, or shall I refrain?" They said, "Go up; for God will give it into the hand of the king." 6 But Jehoshaphat said, "Is there no other prophet of the LORD here of whom we may inquire?" 7 The king of Israel said to Jehoshaphat, "There is still one other by whom we may inquire of the LORD, Micaiah son of Imlah; but I hate him, for he never prophesies anything favorable about me, but only disaster." Jehoshaphat said, "Let the king not say such a thing." 8 Then the king of Israel summoned an officer and said, "Bring quickly Micaiah son of Imlah." 9 Now the king of Israel and King Jehoshaphat of Judah were sitting on their thrones, arrayed in their robes; and they were sitting at the threshing floor at the entrance of the gate of Samaria; and all the prophets were prophesying before them. 10 Zedekiah son of Chenaanah made for himself horns of iron, and he said, "Thus says the LORD: With these you shall gore the Arameans until they are destroyed." 11 All the prophets were prophesying the same and saying, "Go up to Ramoth-gilead and triumph; the LORD will give it into the hand of the king."

12 The messenger who had gone to summon Micaiah said to him, "Look, the words of the prophets with one accord are favorable to the king; let your word be like the word of one of them,

17.10–19: Without parallel. The king was not only a man of religion, but also a man of military might. Since he is represented as having an army of 1,160,000 men, it is understandable that *the fear of the LORD fell on all the kingdoms of the lands around Judah.* The numbers are exaggerated.
18.1–34: **Jehoshaphat's ill-fated alliance with Ahab.** With the exception of vv. 1–2, this chapter is taken almost verbatim from 1 Kings 22.1–35a. It constitutes by far the largest body of material from the northern kingdom used by the Chronicler. It was inserted here probably because it reflects upon Ahab and prepares for the rebuke of Jehoshaphat in the next chapter (see 1 Kings ch 22 n.). **1:** Compare 2 Kings 8.18, 25–27. **2:** An expansion of 1 Kings 22.2.

and speak favorably." 13 But Micaiah said, "As the LORD lives, whatever my God says, that I will speak."

14 When he had come to the king, the king said to him, "Micaiah, shall we go to Ramoth-gilead to battle, or shall I refrain?" He answered, "Go up and triumph; they will be given into your hand." 15 But the king said to him, "How many times must I make you swear to tell me nothing but the truth in the name of the LORD?" 16 Then Micaiah*k* said, "I saw all Israel scattered on the mountains, like sheep without a shepherd; and the LORD said, 'These have no master; let each one go home in peace.' " 17 The king of Israel said to Jehoshaphat, "Did I not tell you that he would not prophesy anything favorable about me, but only disaster?"

18 Then Micaiah*k* said, "Therefore hear the word of the LORD: I saw the LORD sitting on his throne, with all the host of heaven standing to the right and to the left of him. 19 And the LORD said, 'Who will entice King Ahab of Israel, so that he may go up and fall at Ramoth-gilead?' Then one said one thing, and another said another, 20 until a spirit came forward and stood before the LORD, saying, 'I will entice him.' The LORD asked him, 'How?' 21 He replied, 'I will go out and be a lying spirit in the mouth of all his prophets.' Then the LORD*k* said, 'You are to entice him, and you shall succeed; go out and do it.' 22 So you see, the LORD has put a lying spirit in the mouth of these your prophets; the LORD has decreed disaster for you."

23 Then Zedekiah son of Chenaanah came up to Micaiah, slapped him on the cheek, and said, "Which way did the spirit of the LORD pass from me to speak to you?" 24 Micaiah replied, "You will find out on that day when you go in to hide in an inner chamber." 25 The king of Israel then ordered, "Take Micaiah, and return him to Amon the governor of the city and to Joash the king's son; 26 and say, 'Thus says the king: Put this fellow in prison, and feed him on reduced rations of bread and water until I return in peace.' " 27 Micaiah said, "If you return in peace, the LORD has not spoken by me." And he said, "Hear, you peoples, all of you!"

28 So the king of Israel and King Jehoshaphat of Judah went up to Ramoth-gilead. 29 The king of Israel said to Jehoshaphat, "I will disguise myself and go into battle, but you wear your robes." So the king of Israel disguised himself, and they went into battle. 30 Now the king of Aram had commanded the captains of his chariots, "Fight with no one small or great, but only with the king of Israel." 31 When the captains of the chariots saw Jehoshaphat, they said, "It is the king of Israel." So they turned to fight against him; and Jehoshaphat cried out, and the LORD helped him. God drew them away from him, 32 for when the captains of the chariots saw that it was not the king of Israel, they turned back from pursuing him. 33 But a certain man drew his bow and unknowingly struck the king of Israel between the scale armor and the breastplate; so he said to the driver of his chariot, "Turn around, and carry me out of the battle, for I am wounded." 34 The battle grew hot that day, and the king of Israel propped himself up in his chariot facing the Arameans until evening; then at sunset he died.

19 King Jehoshaphat of Judah returned in safety to his house in Jerusalem. 2 Jehu son of Hanani the seer went out to meet him and said to King Jehoshaphat, "Should you help the wicked and love those who hate the LORD? Because of this, wrath has gone out against you from the LORD. 3 Neverthe-

k Heb *he*

19.1–11: Jehoshaphat, having been rebuked by a prophet, rules wisely. 1–3: These verses were composed by the Chronicler, who felt it necessary to protest against the entanglement with Ahab. According to 1 Kings 16.1, *Jehu son of Hanani* prophesied nearly fifty years earlier. This may be the same man, or a person of the same name;

less, some good is found in you, for you destroyed the sacred poles[1] out of the land, and have set your heart to seek God."

4 Jehoshaphat resided at Jerusalem; then he went out again among the people, from Beer-sheba to the hill country of Ephraim, and brought them back to the LORD, the God of their ancestors. [5]He appointed judges in the land in all the fortified cities of Judah, city by city, [6]and said to the judges, "Consider what you are doing, for you judge not on behalf of human beings but on the LORD's behalf; he is with you in giving judgment. [7]Now, let the fear of the LORD be upon you; take care what you do, for there is no perversion of justice with the LORD our God, or partiality, or taking of bribes."

8 Moreover in Jerusalem Jehoshaphat appointed certain Levites and priests and heads of families of Israel, to give judgment for the LORD and to decide disputed cases. They had their seat at Jerusalem. [9]He charged them: "This is how you shall act: in the fear of the LORD, in faithfulness, and with your whole heart; [10]whenever a case comes to you from your kindred who live in their cities, concerning bloodshed, law or commandment, statutes or ordinances, then you shall instruct them, so that they may not incur guilt before the LORD and wrath may not come on you and your kindred. Do so, and you will not incur guilt. [11]See, Amariah the chief priest is over you in all matters of the LORD; and Zebadiah son of Ishmael, the governor of the house of Judah, in all the king's mat-

ters; and the Levites will serve you as officers. Deal courageously, and may the LORD be with the good!"

20 After this the Moabites and Ammonites, and with them some of the Meunites,[m] came against Jehoshaphat for battle. [2]Messengers[n] came and told Jehoshaphat, "A great multitude is coming against you from Edom,[o] from beyond the sea; already they are at Hazazon-tamar" (that is, En-gedi). [3]Jehoshaphat was afraid; he set himself to seek the LORD, and proclaimed a fast throughout all Judah. [4]Judah assembled to seek help from the LORD; from all the towns of Judah they came to seek the LORD.

5 Jehoshaphat stood in the assembly of Judah and Jerusalem, in the house of the LORD, before the new court, [6]and said, "O LORD, God of our ancestors, are you not God in heaven? Do you not rule over all the kingdoms of the nations? In your hand are power and might, so that no one is able to withstand you. [7]Did you not, O our God, drive out the inhabitants of this land before your people Israel, and give it forever to the descendants of your friend Abraham? [8]They have lived in it, and in it have built you a sanctuary for your name, saying, [9]'If disaster comes upon us, the sword, judgment,[p] or pestilence, or famine, we will stand before this house, and before you, for your name is in this house, and cry to you in our distress, and you will hear and

l Heb *Asheroth*   m Compare 26.7: Heb *Ammonites*   n Heb *They*   o One Ms: MT *Aram*   p Or *the sword of judgment*

---

compare the "Hanani" of 16.7. *The sacred poles,* see 1 Kings 14.15 n. **4–7:** No parallel in 1 Kings. The idea of the appointment of *judges* by this king may have been suggested to the Chronicler by the name *Jehoshaphat,* which means "the LORD judges." Compare Deut 1.16–17; 16.18.
19.8: Here *Israel* means Judah, the true Israel in the eyes of the Chronicler (see 11.1–4 n.). **11:** Note the prominence of *the Levites,* who did not exist as a separate class in the time of the historical *Jehoshaphat* (see 1 Chr

6.1–5 n.; 13.1–4 n.). Compare the procedure of judges in Deut 17.8–13.
20.1–37: The great victory over enemies of the true faith; Jehoshaphat's final mistake and consequent punishment. The Chronicler, having established the character of Jehoshaphat as a righteous king, attributes to him a victory of faith even greater than those of Abijah (ch 13) and Asa (ch 14). There is only a faint parallel in 2 Kings 3.4–27.
20.5: *The new court* was a feature of the temple of the Chronicler's time. **6:** This

save.' 10 See now, the people of Ammon, Moab, and Mount Seir, whom you would not let Israel invade when they came from the land of Egypt, and whom they avoided and did not destroy— 11 they reward us by coming to drive us out of your possession that you have given us to inherit. 12 O our God, will you not execute judgment upon them? For we are powerless against this great multitude that is coming against us. We do not know what to do, but our eyes are on you."

13 Meanwhile all Judah stood before the LORD, with their little ones, their wives, and their children. 14 Then the spirit of the LORD came upon Jahaziel son of Zechariah, son of Benaiah, son of Jeiel, son of Mattaniah, a Levite of the sons of Asaph, in the middle of the assembly. 15 He said, "Listen, all Judah and inhabitants of Jerusalem, and King Jehoshaphat: Thus says the LORD to you: 'Do not fear or be dismayed at this great multitude; for the battle is not yours but God's. 16 Tomorrow go down against them; they will come up by the ascent of Ziz; you will find them at the end of the valley, before the wilderness of Jeruel. 17 This battle is not for you to fight; take your position, stand still, and see the victory of the LORD on your behalf, O Judah and Jerusalem.' Do not fear or be dismayed; tomorrow go out against them, and the LORD will be with you."

18 Then Jehoshaphat bowed down with his face to the ground, and all Judah and the inhabitants of Jerusalem fell down before the LORD, worshiping the LORD. 19 And the Levites, of the Kohathites and the Korahites, stood up to praise the LORD, the God of Israel, with a very loud voice.

20 They rose early in the morning and went out into the wilderness of Te-koa; and as they went out, Jehoshaphat stood and said, "Listen to me, O Judah and inhabitants of Jerusalem! Believe in the LORD your God and you will be established; believe his prophets." 21 When he had taken counsel with the people, he appointed those who were to sing to the LORD and praise him in holy splendor, as they went before the army, saying,

"Give thanks to the LORD,
　for his steadfast love endures
　　forever."

22 As they began to sing and praise, the LORD set an ambush against the Ammonites, Moab, and Mount Seir, who had come against Judah, so that they were routed. 23 For the Ammonites and Moab attacked the inhabitants of Mount Seir, destroying them utterly; and when they had made an end of the inhabitants of Seir, they all helped to destroy one another.

24 When Judah came to the watchtower of the wilderness, they looked toward the multitude; they were corpses lying on the ground; no one had escaped. 25 When Jehoshaphat and his people came to take the booty from them, they found livestock*q* in great numbers, goods, clothing, and precious things, which they took for themselves until they could carry no more. They spent three days taking the booty, because of its abundance. 26 On the fourth day they assembled in the Valley of Beracah, for there they blessed the LORD; therefore that place has been called the Valley of Beracah*r* to this day. 27 Then all the people of Judah and Jerusalem, with Jehoshaphat at their head, returned to Jerusalem with joy, for the LORD had enabled them to rejoice over their enemies. 28 They came to Jerusalem, with harps and lyres and

q Gk: Heb *among them*　　*r* That is *Blessing*

---

highly-developed monotheism was characteristic of the theology of the Chronicler's time (compare 6.14). **14:** *A Levite* takes the lead in prophecy. **17:** In battles of this kind, God's people do not need to fight, since God alone is commander and warrior among human and divine forces (compare 13.15; 14.12;

Ezek 38.17–23). Jehoshaphat's army of 1,160,000 (see 17.10–19 n.) was unnecessary. **20.18–23:** The singing of *the Levites* seems to have been the decisive human factor in this victory (compare Judg 7.22). **24–30:** Compare Ezek ch 39.

trumpets, to the house of the LORD.
29 The fear of God came on all the kingdoms of the countries when they heard that the LORD had fought against the enemies of Israel. 30 And the realm of Jehoshaphat was quiet, for his God gave him rest all around.

31 So Jehoshaphat reigned over Judah. He was thirty-five years old when he began to reign; he reigned twenty-five years in Jerusalem. His mother's name was Azubah daughter of Shilhi. 32 He walked in the way of his father Asa and did not turn aside from it, doing what was right in the sight of the LORD. 33 Yet the high places were not removed; the people had not yet set their hearts upon the God of their ancestors.

34 Now the rest of the acts of Jehoshaphat, from first to last, are written in the Annals of Jehu son of Hanani, which are recorded in the Book of the Kings of Israel.

35 After this King Jehoshaphat of Judah joined with King Ahaziah of Israel, who did wickedly. 36 He joined him in building ships to go to Tarshish; they built the ships in Ezion-geber. 37 Then Eliezer son of Dodavahu of Mareshah prophesied against Jehoshaphat, saying, "Because you have joined with Ahaziah, the LORD will destroy what you have made." And the ships were wrecked and were not able to go to Tarshish.

21 Jehoshaphat slept with his ancestors and was buried with his ancestors in the city of David; his son Jehoram succeeded him. 2 He had brothers, the sons of Jehoshaphat: Azariah, Jehiel, Zechariah, Azariah, Michael, and Shephatiah; all these were the sons of King Jehoshaphat of Judah.*s* 3 Their father gave them many gifts, of silver, gold, and valuable possessions, together with fortified cities in Judah; but he gave the kingdom to Jehoram, because he was the firstborn. 4 When Jehoram had ascended the throne of his father and was established, he put all his brothers to the sword, and also some of the officials of Israel. 5 Jehoram was thirty-two years old when he began to reign; he reigned eight years in Jerusalem. 6 He walked in the way of the kings of Israel, as the house of Ahab had done; for the daughter of Ahab was his wife. He did what was evil in the sight of the LORD. 7 Yet the LORD would not destroy the house of David because of the covenant that he had made with David, and since he had promised to give a lamp to him and to his descendants forever.

8 In his days Edom revolted against the rule of Judah and set up a king of their own. 9 Then Jehoram crossed over with his commanders and all his chariots. He set out by night and attacked the Edomites, who had surrounded him and his chariot commanders. 10 So Edom has been in revolt against the rule of Judah to this day. At that time Libnah also revolted against his rule, because he had forsaken the LORD, the God of his ancestors.

11 Moreover he made high places in the hill country of Judah, and led the inhabitants of Jerusalem into unfaithfulness, and made Judah go astray. 12 A letter came to him from the prophet Elijah,

s Gk Syr: Heb *Israel*

20.31–33: Compare the source, 1 Kings 22.41–43. 34: The parallel is 1 Kings 22.45, with a difference in the name of the book. *Israel* here means Judah (see 11.1–4 n.).

20.35–37: The parallel is 1 Kings 22.44, 48–49, drastically rewritten to show that the loss of the ships was due to improper association with *Ahaziah*.

21.1–20: **Jehoram's wicked reign and his punishment.** *Jehoram* is already given a bad reputation in 2 Kings 8.18; the Chronicler makes him worse and increases his punishments. 1: The source is 1 Kings 22.50. 2–4: These verses, which have no known source, are probably based on historically correct information (compare 11.22–23; Judg 9.1–6). 5–7: Taken with changes from 2 Kings 8.17–19.

21.8–10: Taken from 2 Kings 8.20–22, with the addition of the last clause giving the Chronicler's explanation why *Jehoram* was having this trouble. 11–15: The *letter . . . from the prophet Elijah* is a creation of the Chronicler, used to sharpen the rebuke of Jehoram.

saying: "Thus says the LORD, the God of your father David: Because you have not walked in the ways of your father Jehoshaphat or in the ways of King Asa of Judah, 13but have walked in the way of the kings of Israel, and have led Judah and the inhabitants of Jerusalem into unfaithfulness, as the house of Ahab led Israel into unfaithfulness, and because you also have killed your brothers, members of your father's house, who were better than yourself, 14see, the LORD will bring a great plague on your people, your children, your wives, and all your possessions, 15and you yourself will have a severe sickness with a disease of your bowels, until your bowels come out, day after day, because of the disease."

16 The LORD aroused against Jehoram the anger of the Philistines and of the Arabs who are near the Ethiopians.*t* 17They came up against Judah, invaded it, and carried away all the possessions they found that belonged to the king's house, along with his sons and his wives, so that no son was left to him except Jehoahaz, his youngest son.

18 After all this the LORD struck him in his bowels with an incurable disease. 19In course of time, at the end of two years, his bowels came out because of the disease, and he died in great agony. His people made no fire in his honor, like the fires made for his ancestors. 20He was thirty-two years old when he began to reign; he reigned eight years in Jerusalem. He departed with no one's regret. They buried him in the city of David, but not in the tombs of the kings.

22 The inhabitants of Jerusalem made his youngest son Ahaziah king as his successor; for the troops who came with the Arabs to the camp had killed all the older sons. So Ahaziah son

of Jehoram reigned as king of Judah. 2Ahaziah was forty-two years old when he began to reign; he reigned one year in Jerusalem. His mother's name was Athaliah, a granddaughter of Omri. 3He also walked in the ways of the house of Ahab, for his mother was his counselor in doing wickedly. 4He did what was evil in the sight of the LORD, as the house of Ahab had done; for after the death of his father they were his counselors, to his ruin. 5He even followed their advice, and went with Jehoram son of King Ahab of Israel to make war against King Hazael of Aram at Ramoth-gilead. The Arameans wounded Joram, 6and he returned to be healed in Jezreel of the wounds that he had received at Ramah, when he fought King Hazael of Aram. And Ahaziah son of King Jehoram of Judah went down to see Joram son of Ahab in Jezreel, because he was sick.

7 But it was ordained by God that the downfall of Ahaziah should come about through his going to visit Joram. For when he came there he went out with Jehoram to meet Jehu son of Nimshi, whom the LORD had anointed to destroy the house of Ahab. 8When Jehu was executing judgment on the house of Ahab, he met the officials of Judah and the sons of Ahaziah's brothers, who attended Ahaziah, and he killed them. 9He searched for Ahaziah, who was captured while hiding in Samaria and was brought to Jehu, and put to death. They buried him, for they said, "He is the grandson of Jehoshaphat, who sought the LORD with all his heart." And the house of Ahaziah had no one able to rule the kingdom.

10 Now when Athaliah, Ahaziah's

t Or Nubians; Heb Cushites

21.16–17: More military punishment; not found in 2 Kings. *Arabs who are near the Ethiopians,* see 14.9–15 n. **18–19:** Personal punishment, to fulfill the prophecy in v. 15. Contrast the fine funeral of Asa, 16.14. **20:** The same as v. 5, with typical additions (compare 2 Kings 8.24a).
22.1–12: The brief but wicked reign of

Ahaziah; his punishment; the usurpation of Athaliah. 1–6: This is a slightly rewritten version of 2 Kings 8.24b–29, with the guilt of Ahaziah heightened. **2:** *Forty-two* is an error for "twenty-two"; compare 2 Kings 8.26.
22.7: A theological interpretation of v. 6 and of 2 Kings 9.21. **8–9:** Drawn from 2 Kings 9.27–28, but somewhat rewritten to

mother, saw that her son was dead, she set about to destroy all the royal family of the house of Judah. 11 But Jehoshabeath, the king's daughter, took Joash son of Ahaziah, and stole him away from among the king's children who were about to be killed; she put him and his nurse in a bedroom. Thus Jehoshabeath, daughter of King Jehoram and wife of the priest Jehoiada—because she was a sister of Ahaziah—hid him from Athaliah, so that she did not kill him; 12 he remained with them six years, hidden in the house of God, while Athaliah reigned over the land.

23 But in the seventh year Jehoiada took courage, and entered into a compact with the commanders of the hundreds, Azariah son of Jeroham, Ishmael son of Jehohanan, Azariah son of Obed, Maaseiah son of Adaiah, and Elishaphat son of Zichri. 2 They went around through Judah and gathered the Levites from all the towns of Judah, and the heads of families of Israel, and they came to Jerusalem. 3 Then the whole assembly made a covenant with the king in the house of God. Jehoiada*u* said to them, "Here is the king's son! Let him reign, as the LORD promised concerning the sons of David. 4 This is what you are to do: one third of you, priests and Levites, who come on duty on the sabbath, shall be gatekeepers, 5 one third shall be at the king's house, and one third at the Gate of the Foundation; and all the people shall be in the courts of the house of the LORD. 6 Do not let anyone enter the house of the LORD except the priests and ministering Levites; they may enter, for they are holy, but all the other*v* people shall observe the instructions of the LORD. 7 The Levites shall surround the king, each with his weapons in his hand;

and whoever enters the house shall be killed. Stay with the king in his comings and goings."

8 The Levites and all Judah did according to all that the priest Jehoiada commanded; each brought his men, who were to come on duty on the sabbath, with those who were to go off duty on the sabbath; for the priest Jehoiada did not dismiss the divisions. 9 The priest Jehoiada delivered to the captains the spears and the large and small shields that had been King David's, which were in the house of God; 10 and he set all the people as a guard for the king, everyone with weapon in hand, from the south side of the house to the north side of the house, around the altar and the house. 11 Then he brought out the king's son, put the crown on him, and gave him the covenant;*w* they proclaimed him king, and Jehoiada and his sons anointed him; and they shouted, "Long live the king!"

12 When Athaliah heard the noise of the people running and praising the king, she went into the house of the LORD to the people; 13 and when she looked, there was the king standing by his pillar at the entrance, and the captains and the trumpeters beside the king, and all the people of the land rejoicing and blowing trumpets, and the singers with their musical instruments leading in the celebration. Athaliah tore her clothes, and cried, "Treason! Treason!" 14 Then the priest Jehoiada brought out the captains who were set over the army, saying to them, "Bring her out between the ranks; anyone who follows her is to be put to the sword." For the priest said, "Do not put her to death in the house of the LORD."

u Heb *He*   *v* Heb lacks *other*   *w* Or *treaty,* or *testimony*; Heb *eduth*

---

make the end of Ahaziah less honorable, and to prepare for what follows. **10–12**: 2 Kings 11.1–3 with slight changes.

23.1–21: The revolt against Athaliah; her punishment by death and the enthronement of Joash. Taken from 2 Kings 11.4–20, with certain characteristic changes and additions with a view to making the re-

volt the work of the Levites and singers rather than of the military; see vv. 2–3, 6, 18–19, added by the Chronicler; in v. 19, even *the gatekeepers* have a share (compare 1 Chr ch 26). In v. 13 the insertion of *the singers with their musical instruments leading in the celebration* changes drastically the tense military atmosphere of 2 Kings ch 11.

15 So they laid hands on her; she went into the entrance of the Horse Gate of the king's house, and there they put her to death.

16 Jehoiada made a covenant between himself and all the people and the king that they should be the LORD's people. 17 Then all the people went to the house of Baal, and tore it down; his altars and his images they broke in pieces, and they killed Mattan, the priest of Baal, in front of the altars. 18 Jehoiada assigned the care of the house of the LORD to the levitical priests whom David had organized to be in charge of the house of the LORD, to offer burnt offerings to the LORD, as it is written in the law of Moses, with rejoicing and with singing, according to the order of David. 19 He stationed the gatekeepers at the gates of the house of the LORD so that no one should enter who was in any way unclean. 20 And he took the captains, the nobles, the governors of the people, and all the people of the land, and they brought the king down from the house of the LORD, marching through the upper gate to the king's house. They set the king on the royal throne. 21 So all the people of the land rejoiced, and the city was quiet after Athaliah had been killed with the sword.

24 Joash was seven years old when he began to reign; he reigned forty years in Jerusalem; his mother's name was Zibiah of Beer-sheba. 2 Joash did what was right in the sight of the LORD all the days of the priest Jehoiada. 3 Jehoiada got two wives for him, and he became the father of sons and daughters.

4 Some time afterward Joash decided to restore the house of the LORD. 5 He assembled the priests and the Levites and said to them, "Go out to the cities of Judah and gather money from all Israel to repair the house of your God, year by year; and see that you act quickly." But the Levites did not act quickly. 6 So the king summoned Jehoiada the chief, and said to him, "Why have you not required the Levites to bring in from Judah and Jerusalem the tax levied by Moses, the servant of the LORD, on^x the congregation of Israel for the tent of the covenant?"^y 7 For the children of Athaliah, that wicked woman, had broken into the house of God, and had even used all the dedicated things of the house of the LORD for the Baals.

8 So the king gave command, and they made a chest, and set it outside the gate of the house of the LORD. 9 A proclamation was made throughout Judah and Jerusalem to bring in for the LORD the tax that Moses the servant of God laid on Israel in the wilderness. 10 All the leaders and all the people rejoiced and brought their tax and dropped it into the chest until it was full. 11 Whenever the chest was brought to the king's officers by the Levites, when they saw that there was a large amount of money in it, the king's secretary and the officer of the chief priest would come and empty the chest and take it and return it to its place. So they did day after day, and collected money in abundance. 12 The king and Jehoiada gave it to those who had charge of the work of the house of the LORD, and they hired masons and carpenters to restore the house of the LORD, and also workers in iron and bronze to repair the house of the LORD. 13 So those who were engaged in the work labored, and the repairing went forward at their hands, and they restored the house of God to its

x Compare Vg: Heb *and* y Or *treaty*, or *testimony*; Heb *eduth*

24.1–27: The good beginning of Joash; his later apostasy and consequent punishment. Based upon 2 Kings ch 12. Joash is depicted by the Chronicler as being considerably worse than he appears to be in 2 Kings, probably because he was a grandson of Athaliah (see 2 Kings 12.19–21 n.). Such treatment also provides an explanation of Joash's military misfortune and his assassination, both unexplained in 2 Kings. **1–3:** 2 Kings 12.3 is omitted and v. 3 here is added. **4–7:** Rewritten to give *the Levites* a greater share, even in the delaying tactics (v. 5). Verse 7 is an added observation.

24.8–14: See 2 Kings 12.9–16 n.

proper condition and strengthened it. ¹⁴When they had finished, they brought the rest of the money to the king and Jehoiada, and with it were made utensils for the house of the LORD, utensils for the service and for the burnt offerings, and ladles, and vessels of gold and silver. They offered burnt offerings in the house of the LORD regularly all the days of Jehoiada.

15 But Jehoiada grew old and full of days, and died; he was one hundred thirty years old at his death. ¹⁶And they buried him in the city of David among the kings, because he had done good in Israel, and for God and his house.

17 Now after the death of Jehoiada the officials of Judah came and did obeisance to the king; then the king listened to them. ¹⁸They abandoned the house of the LORD, the God of their ancestors, and served the sacred poles*ᶻ* and the idols. And wrath came upon Judah and Jerusalem for this guilt of theirs. ¹⁹Yet he sent prophets among them to bring them back to the LORD; they testified against them, but they would not listen.

20 Then the spirit of God took possession of*ᵃ* Zechariah son of the priest Jehoiada; he stood above the people and said to them, "Thus says God: Why do you transgress the commandments of the LORD, so that you cannot prosper? Because you have forsaken the LORD, he has also forsaken you." ²¹But they conspired against him, and by command of the king they stoned him to death in the court of the house of the LORD. ²²King Joash did not remember the kindness that Jehoiada, Zechariah's father, had shown him, but killed his son. As he was dying, he said, "May the LORD see and avenge!"

23 At the end of the year the army of Aram came up against Joash. They came to Judah and Jerusalem, and destroyed all the officials of the people from among them, and sent all the booty they took to the king of Damascus. ²⁴Although the army of Aram had come with few men, the LORD delivered into their hand a very great army, because they had abandoned the LORD, the God of their ancestors. Thus they executed judgment on Joash.

25 When they had withdrawn, leaving him severely wounded, his servants conspired against him because of the blood of the son*ᵇ* of the priest Jehoiada, and they killed him on his bed. So he died; and they buried him in the city of David, but they did not bury him in the tombs of the kings. ²⁶Those who conspired against him were Zabad son of Shimeath the Ammonite, and Jehozabad son of Shimrith the Moabite. ²⁷Accounts of his sons, and of the many oracles against him, and of the rebuilding*ᶜ* of the house of God are written in the Commentary on the Book of the Kings. And his son Amaziah succeeded him.

25 Amaziah was twenty-five years old when he began to reign, and he reigned twenty-nine years in Jerusalem. His mother's name was Jehoaddan of Jerusalem. ²He did what was right in the sight of the LORD, yet not with a true heart. ³As soon as the royal power was firmly in his hand he killed his servants who had murdered his father the king. ⁴But he did not put their children to death, according to what is written in the law, in the book of Moses, where the LORD commanded, "The parents shall not be put to death for the children, or the children be put to death for the parents; but all shall be put to death for their own sins."

5 Amaziah assembled the people of

z Heb *Asherim*   *a* Heb *clothed itself with*
b Gk Vg: Heb *sons*   *c* Heb *founding*

24.15–22: This story, not attested elsewhere, prepares for what follows. Compare Lk 11.51. 23–24: A drastically rewritten form of 2 Kings 12.17–18. 25–27: A reinterpretation of 2 Kings 12.19–21. *The Commentary on the Book of the Kings* can no more be identified than can other such works referred to by the Chronicler (see Introduction to 1 Chronicles).
25.1–28: **Amaziah acts wrongfully and suffers punishment.** Based upon 2 Kings 14.2–14, 17–20. **1–4:** Not essentially different from 2 Kings 14.2–6.
25.5–13: An expansion of a single verse,

Judah, and set them by ancestral houses under commanders of the thousands and of the hundreds for all Judah and Benjamin. He mustered those twenty years old and upward, and found that they were three hundred thousand picked troops fit for war, able to handle spear and shield. 6 He also hired one hundred thousand mighty warriors from Israel for one hundred talents of silver. 7 But a man of God came to him and said, "O king, do not let the army of Israel go with you, for the LORD is not with Israel—all these Ephraimites. 8 Rather, go by yourself and act; be strong in battle, or God will fling you down before the enemy; for God has power to help or to overthrow." 9 Amaziah said to the man of God, "But what shall we do about the hundred talents that I have given to the army of Israel?" The man of God answered, "The LORD is able to give you much more than this." 10 Then Amaziah discharged the army that had come to him from Ephraim, letting them go home again. But they became very angry with Judah, and returned home in fierce anger.

11 Amaziah took courage, and led out his people; he went to the Valley of Salt, and struck down ten thousand men of Seir. 12 The people of Judah captured another ten thousand alive, took them to the top of Sela, and threw them down from the top of Sela, so that all of them were dashed to pieces. 13 But the men of the army whom Amaziah sent back, not letting them go with him to battle, fell on the cities of Judah from Samaria to Bethhoron; they killed three thousand people in them, and took much booty.

14 Now after Amaziah came from the slaughter of the Edomites, he brought the gods of the people of Seir, set them up as his gods, and worshiped them, making offerings to them. 15 The LORD was angry with Amaziah and sent to him

a prophet, who said to him, "Why have you resorted to a people's gods who could not deliver their own people from your hand?" 16 But as he was speaking the king *d* said to him, "Have we made you a royal counselor? Stop! Why should you be put to death?" So the prophet stopped, but said, "I know that God has determined to destroy you, because you have done this and have not listened to my advice."

17 Then King Amaziah of Judah took counsel and sent to King Joash son of Jehoahaz son of Jehu of Israel, saying, "Come, let us look one another in the face." 18 King Joash of Israel sent word to King Amaziah of Judah, "A thornbush on Lebanon sent to a cedar on Lebanon, saying, 'Give your daughter to my son for a wife'; but a wild animal of Lebanon passed by and trampled down the thornbush. 19 You say, 'See, I have defeated Edom,' and your heart has lifted you up in boastfulness. Now stay at home; why should you provoke trouble so that you fall, you and Judah with you?"

20 But Amaziah would not listen—it was God's doing, in order to hand them over, because they had sought the gods of Edom. 21 So King Joash of Israel went up; he and King Amaziah of Judah faced one another in battle at Beth-shemesh, which belongs to Judah. 22 Judah was defeated by Israel; everyone fled home. 23 King Joash of Israel captured King Amaziah of Judah, son of Joash, son of Ahaziah, at Beth-shemesh; he brought him to Jerusalem, and broke down the wall of Jerusalem from the Ephraim Gate to the Corner Gate, a distance of four hundred cubits. 24 He seized all the gold and silver, and all the vessels that were found in the house of God, and Obededom with them; he seized also the trea-

d Heb *he*

---

2 Kings 14.7. *Amaziah* did wrong in hiring troops from the northern kingdom; even while winning a victory over Edom, he was punished for his indiscretion (v. 13). *Samaria* and *Beth-horon* were not among the *cities of Judah;* there is some confusion here.

25.14–16: These verses, not found in 2 Kings, are the Chronicler's way of explaining what follows. **17–19:** See note on the parallel, 2 Kings 14.8–10.

25.20–24: Similar to 2 Kings 14.11–14, except for the theological explanation in

suries of the king's house, also hostages; then he returned to Samaria.

25 King Amaziah son of Joash of Judah, lived fifteen years after the death of King Joash son of Jehoahaz of Israel. 26 Now the rest of the deeds of Amaziah, from first to last, are they not written in the Book of the Kings of Judah and Israel? 27 From the time that Amaziah turned away from the LORD they made a conspiracy against him in Jerusalem, and he fled to Lachish. But they sent after him to Lachish, and killed him there. 28 They brought him back on horses; he was buried with his ancestors in the city of David.

26 Then all the people of Judah took Uzziah, who was sixteen years old, and made him king to succeed his father Amaziah. 2 He rebuilt Eloth and restored it to Judah, after the king slept with his ancestors. 3 Uzziah was sixteen years old when he began to reign, and he reigned fifty-two years in Jerusalem. His mother's name was Jecoliah of Jerusalem. 4 He did what was right in the sight of the LORD, just as his father Amaziah had done. 5 He set himself to seek God in the days of Zechariah, who instructed him in the fear of God; and as long as he sought the LORD, God made him prosper.

6 He went out and made war against the Philistines, and broke down the wall of Gath and the wall of Jabneh and the wall of Ashdod; he built cities in the territory of Ashdod and elsewhere among the Philistines. 7 God helped him against the Philistines, against the Arabs who lived in Gur-baal, and against the Meunites. 8 The Ammonites paid tribute to Uzziah, and his fame spread even to the border of Egypt, for he became very strong. 9 Moreover Uzziah built towers in Jerusalem at the Corner Gate, at the Valley Gate, and at the Angle, and fortified them. 10 He built towers in the wilderness and hewed out many cisterns, for he had large herds, both in the Shephelah and in the plain, and he had farmers and vinedressers in the hills and in the fertile lands, for he loved the soil. 11 Moreover Uzziah had an army of soldiers, fit for war, in divisions according to the numbers in the muster made by the secretary Jeiel and the officer Maaseiah, under the direction of Hananiah, one of the king's commanders. 12 The whole number of the heads of ancestral houses of mighty warriors was two thousand six hundred. 13 Under their command was an army of three hundred seven thousand five hundred, who could make war with mighty power, to help the king against the enemy. 14 Uzziah provided for all the army the shields, spears, helmets, coats of mail, bows, and stones for slinging. 15 In Jerusalem he set up machines, invented by skilled workers, on the towers and the corners for shooting arrows and large stones. And his fame spread far, for he was marvelously helped until he became strong.

16 But when he had become strong he grew proud, to his destruction. For he was false to the LORD his God, and entered the temple of the LORD to make offering on the altar of incense. 17 But the priest Azariah went in after him, with eighty priests of the LORD who were men of valor; 18 they withstood King Uzziah, and said to him, "It is not for you, Uzziah, to make offering to the LORD, but for the priests the descendants of Aaron,

---

v. 20b. **25–28:** Much the same as 2 Kings 14.17–20, except for the explanatory phrase *From the time that Amaziah turned away from the* LORD (v. 27).
26.1–23: Uzziah's good beginning and consequent prosperity; his subsequent sin and consequent punishment. The brief parallel is found in 2 Kings 14.21–22; 15.2–3, 5–7. The reign of Uzziah (called also Azariah in 2 Kings) was a long one (v. 3), and the account is greatly expanded here. **1–4:** Compare 2 Kings 14.21–22; 15.2–3. **5:** Added by the Chronicler. This Zechariah is otherwise unknown; he served the same function as Jehoiada in 24.2.
26.6–15: Without parallel in 2 Kings, but there is here valuable historical material from an unknown source.
26.16–20: A composition of the Chronicler, perhaps from a lost source, to explain the

who are consecrated to make offering. Go out of the sanctuary; for you have done wrong, and it will bring you no honor from the LORD God." 19 Then Uzziah was angry. Now he had a censer in his hand to make offering, and when he became angry with the priests a leprous[e] disease broke out on his forehead, in the presence of the priests in the house of the LORD, by the altar of incense. 20 When the chief priest Azariah, and all the priests, looked at him, he was leprous[e] in his forehead. They hurried him out, and he himself hurried to get out, because the LORD had struck him. 21 King Uzziah was leprous[e] to the day of his death, and being leprous[e] lived in a separate house, for he was excluded from the house of the LORD. His son Jotham was in charge of the palace of the king, governing the people of the land.

22 Now the rest of the acts of Uzziah, from first to last, the prophet Isaiah son of Amoz wrote. 23 Uzziah slept with his ancestors; they buried him near his ancestors in the burial field that belonged to the kings, for they said, "He is leprous."[e] His son Jotham succeeded him.

27 Jotham was twenty-five years old when he began to reign; he reigned sixteen years in Jerusalem. His mother's name was Jerushah daughter of Zadok. 2 He did what was right in the sight of the LORD just as his father Uzziah had done—only he did not invade the temple of the LORD. But the people still followed corrupt practices. 3 He built the upper gate of the house of the LORD, and did extensive building on the wall of Ophel. 4 Moreover he built cities in the hill country of Judah, and forts and towers on the wooded hills. 5 He fought with the king of the Ammonites and prevailed against them. The Ammonites gave him that year one hundred talents of silver, ten thousand cors of wheat and ten thousand of barley. The Ammonites paid him the same amount in the second and the third years. 6 So Jotham became strong because he ordered his ways before the LORD his God. 7 Now the rest of the acts of Jotham, and all his wars and his ways, are written in the Book of the Kings of Israel and Judah. 8 He was twenty-five years old when he began to reign; he reigned sixteen years in Jerusalem. 9 Jotham slept with his ancestors, and they buried him in the city of David; and his son Ahaz succeeded him.

28 Ahaz was twenty years old when he began to reign; he reigned sixteen years in Jerusalem. He did not do what was right in the sight of the LORD, as his ancestor David had done, 2 but he walked in the ways of the kings of Israel. He even made cast images for the Baals; 3 and he made offerings in the valley of the son of Hinnom, and made his sons pass through fire, according to the abominable practices of the nations whom the LORD drove out before the people of Israel. 4 He sacrificed and made offerings on the high places, on the hills, and under every green tree.

5 Therefore the LORD his God gave him into the hand of the king of Aram, who defeated him and took captive a

e A term for several skin diseases; precise meaning uncertain

---

leprosy of Uzziah. **21–23:** Parallel to 2 Kings 15.5–7, with the change of the name of a book in v. 22.

27.1–9: Jotham's goodness and consequent prosperity. This short chapter parallels 2 Kings 15.33–38, with a few significant changes. Jotham is represented by the Chronicler as a totally good and hence completely prosperous king. Verse 2b was added to show that the king was ritually correct; v. 2c to show that faults came from the people, not the king. Verses 4–6 were added to show the good king's military might; 2 Kings 15.37 was omitted because the Chronicler wished to present the reign of the idealized Jotham as being without any troubles whatever.

28.1–27: The enormous iniquity of Ahaz and the appropriate punishments which he received. Based upon 2 Kings ch 16 with drastic rewriting to intensify the unfavorable portrait already present in the source. To the Chronicler, the wholly good reign of Jotham was followed by the wholly bad reign of Ahaz. **1–4:** Not greatly changed from the source, 2 Kings 16.2–4.

great number of his people and brought them to Damascus. He was also given into the hand of the king of Israel, who defeated him with great slaughter. 6 Pekah son of Remaliah killed one hundred twenty thousand in Judah in one day, all of them valiant warriors, because they had abandoned the LORD, the God of their ancestors. 7 And Zichri, a mighty warrior of Ephraim, killed the king's son Maaseiah, Azrikam the commander of the palace, and Elkanah the next in authority to the king.

8 The people of Israel took captive two hundred thousand of their kin, women, sons, and daughters; they also took much booty from them and brought the booty to Samaria. 9 But a prophet of the LORD was there, whose name was Oded; he went out to meet the army that came to Samaria, and said to them, "Because the LORD, the God of your ancestors, was angry with Judah, he gave them into your hand, but you have killed them in a rage that has reached up to heaven. 10 Now you intend to subjugate the people of Judah and Jerusalem, male and female, as your slaves. But what have you except sins against the LORD your God? 11 Now hear me, and send back the captives whom you have taken from your kindred, for the fierce wrath of the LORD is upon you." 12 Moreover, certain chiefs of the Ephraimites, Azariah son of Johanan, Berechiah son of Meshillemoth, Jehizkiah son of Shallum, and Amasa son of Hadlai, stood up against those who were coming from the war, 13 and said to them, "You shall not bring the captives in here, for you propose to bring on us guilt against the LORD in addition to our present sins and guilt. For our guilt is already great, and there is fierce wrath against Israel." 14 So the warriors left the captives and the booty before the officials and all the assembly.

15 Then those who were mentioned by name got up and took the captives, and with the booty they clothed all that were naked among them; they clothed them, gave them sandals, provided them with food and drink, and anointed them; and carrying all the feeble among them on donkeys, they brought them to their kindred at Jericho, the city of palm trees. Then they returned to Samaria.

16 At that time King Ahaz sent to the king*f* of Assyria for help. 17 For the Edomites had again invaded and defeated Judah, and carried away captives. 18 And the Philistines had made raids on the cities in the Shephelah and the Negeb of Judah, and had taken Beth-shemesh, Aijalon, Gederoth, Soco with its villages, Timnah with its villages, and Gimzo with its villages; and they settled there. 19 For the LORD brought Judah low because of King Ahaz of Israel, for he had behaved without restraint in Judah and had been faithless to the LORD. 20 So King Tilgath-pilneser of Assyria came against him, and oppressed him instead of strengthening him. 21 For Ahaz plundered the house of the LORD and the houses of the king and of the officials, and gave tribute to the king of Assyria; but it did not help him.

22 In the time of his distress he became yet more faithless to the LORD—this same King Ahaz. 23 For he sacrificed to the gods of Damascus, which had defeated him, and said, "Because the gods of the kings of Aram helped them, I will sacrifice to them so that they may help me." But they were the ruin of him, and of all Israel. 24 Ahaz gathered together the utensils of the house of God, and cut in pieces the utensils of the house of God. He shut up the doors of the house of the LORD and made himself altars in every

f Gk Syr Vg Compare 2 Kings 16.7: Heb *kings*

28.5–15: A drastic rewriting and expansion of 2 Kings 16.5–6. Both *Aram* and *Israel* defeat *Judah* with tremendous slaughter, signalizing God's retributive justice. At the same time, the wicked northern kingdom is not allowed to retain any advantage from its punitive victory (vv. 8–15).
28.16–25: Somewhat briefer than the source, 2 Kings 16.7–18. The aim of the writer is to show that Ahaz did not profit even

corner of Jerusalem. 25 In every city of Judah he made high places to make offerings to other gods, provoking to anger the LORD, the God of his ancestors. 26 Now the rest of his acts and all his ways, from first to last, are written in the Book of the Kings of Judah and Israel. 27 Ahaz slept with his ancestors, and they buried him in the city, in Jerusalem; but they did not bring him into the tombs of the kings of Israel. His son Hezekiah succeeded him.

29 Hezekiah began to reign when he was twenty-five years old; he reigned twenty-nine years in Jerusalem. His mother's name was Abijah daughter of Zechariah. 2 He did what was right in the sight of the LORD, just as his ancestor David had done.

3 In the first year of his reign, in the first month, he opened the doors of the house of the LORD and repaired them. 4 He brought in the priests and the Levites and assembled them in the square on the east. 5 He said to them, "Listen to me, Levites! Sanctify yourselves, and sanctify the house of the LORD, the God of your ancestors, and carry out the filth from the holy place. 6 For our ancestors have been unfaithful and have done what was evil in the sight of the LORD our God; they have forsaken him, and have turned away their faces from the dwelling of the LORD, and turned their backs. 7 They also shut the doors of the vestibule and put out the lamps, and have not offered incense or made burnt offerings in the holy place to the God of Israel. 8 Therefore the wrath of the LORD came upon Judah and

Jerusalem, and he has made them an object of horror, of astonishment, and of hissing, as you see with your own eyes. 9 Our fathers have fallen by the sword and our sons and our daughters and our wives are in captivity for this. 10 Now it is in my heart to make a covenant with the LORD, the God of Israel, so that his fierce anger may turn away from us. 11 My sons, do not now be negligent, for the LORD has chosen you to stand in his presence to minister to him, and to be his ministers and make offerings to him."

12 Then the Levites arose, Mahath son of Amasai, and Joel son of Azariah, of the sons of the Kohathites; and of the sons of Merari, Kish son of Abdi, and Azariah son of Jehallelel; and of the Gershonites, Joah son of Zimmah, and Eden son of Joah; 13 and of the sons of Elizaphan, Shimri and Jeuel; and of the sons of Asaph, Zechariah and Mattaniah; 14 and of the sons of Heman, Jehuel and Shimei; and of the sons of Jeduthun, Shemaiah and Uzziel. 15 They gathered their brothers, sanctified themselves, and went in as the king had commanded, by the words of the LORD, to cleanse the house of the LORD. 16 The priests went into the inner part of the house of the LORD to cleanse it, and they brought out all the unclean things that they found in the temple of the LORD into the court of the house of the LORD; and the Levites took them and carried them out to the Wadi Kidron. 17 They began to sanctify on the first day of the first month, and on the eighth day of the month they came to the vestibule of the LORD; then for eight

temporarily from his unholy alliance with the Assyrians. **26–27:** Rewritten from 2 Kings 16.19–20. Verse 27 contradicts 2 Kings 16.20 with regard to the burial of the king; the evil Ahaz could not be buried in a royal tomb (compare 21.20; 24.25; 26.23).

29.1–36: Hezekiah begins his good reign by cleansing the temple. Second Kings considers Hezekiah an important and on the whole a good king; three chapters (chs 18–20) are devoted to him. The Chronicler makes him a reformer like the later Josiah, and devotes four chapters to him (chs 29–

32). **1–2:** Taken with little change from 2 Kings 18.2–3. **3–11:** In 28.24 the Chronicler had said that Ahaz closed down the temple entirely (in contrast to 2 Kings 16.10–16). Thus the way was prepared for the present passage, with its theme of reform carried out by *priests and Levites*—a favorite subject with the Chronicler. There is only the slightest affinity with 2 Kings 18.4–6. Verses 6–9 suggest liturgical confession during the Babylonian captivity (compare Zech 1.1–6).

29.12–19: Continuation of the reform; no parallel; the Chronicler presents an ideal re-

days they sanctified the house of the LORD, and on the sixteenth day of the first month they finished. 18 Then they went inside to King Hezekiah and said, "We have cleansed all the house of the LORD, the altar of burnt offering and all its utensils, and the table for the rows of bread and all its utensils. 19 All the utensils that King Ahaz repudiated during his reign when he was faithless, we have made ready and sanctified; see, they are in front of the altar of the LORD."

20 Then King Hezekiah rose early, assembled the officials of the city, and went up to the house of the LORD. 21 They brought seven bulls, seven rams, seven lambs, and seven male goats for a sin offering for the kingdom and for the sanctuary and for Judah. He commanded the priests the descendants of Aaron to offer them on the altar of the LORD. 22 So they slaughtered the bulls, and the priests received the blood and dashed it against the altar; they slaughtered the rams and their blood was dashed against the altar; they also slaughtered the lambs and their blood was dashed against the altar. 23 Then the male goats for the sin offering were brought to the king and the assembly; they laid their hands on them, 24 and the priests slaughtered them and made a sin offering with their blood at the altar, to make atonement for all Israel. For the king commanded that the burnt offering and the sin offering should be made for all Israel.

25 He stationed the Levites in the house of the LORD with cymbals, harps, and lyres, according to the commandment of David and of Gad the king's seer and of the prophet Nathan, for the commandment was from the LORD through his prophets. 26 The Levites stood with the instruments of David, and the priests with the trumpets. 27 Then Hezekiah commanded that the burnt offering be offered on the altar. When the burnt offering began, the song to the LORD began also, and the trumpets, accompanied by the instruments of King David of Israel. 28 The whole assembly worshiped, the singers sang, and the trumpeters sounded; all this continued until the burnt offering was finished. 29 When the offering was finished, the king and all who were present with him bowed down and worshiped. 30 King Hezekiah and the officials commanded the Levites to sing praises to the LORD with the words of David and of the seer Asaph. They sang praises with gladness, and they bowed down and worshiped.

31 Then Hezekiah said, "You have now consecrated yourselves to the LORD; come near, bring sacrifices and thank offerings to the house of the LORD." The assembly brought sacrifices and thank offerings; and all who were of a willing heart brought burnt offerings. 32 The number of the burnt offerings that the assembly brought was seventy bulls, one hundred rams, and two hundred lambs; all these were for a burnt offering to the LORD. 33 The consecrated offerings were six hundred bulls and three thousand sheep. 34 But the priests were too few and could not skin all the burnt offerings, so, until other priests had sanctified themselves, their kindred, the Levites, helped them until the work was finished for the Levites were more conscientious[g] than the priests in sanctifying themselves. 35 Besides the great number of burnt offerings there was the fat of the offerings of well-being, and there were the drink offerings for the burnt offerings. Thus the service of the house of the LORD was restored. 36 And Hezekiah and all the people rejoiced because of what God had done for the people; for the thing had come about suddenly.

g Heb *upright in heart*

---

form with emphasis on *the Levites*. **20–24:** The ideal sacrifice of rededication.

29.25–30: Note the great importance of the Levitical musicians (see Introduction to 1 Chronicles and compare 1 Chr ch 25). **31–36:** To the Chronicler, *the Levites were more conscientious than the priests in sanctifying themselves.*

30 Hezekiah sent word to all Israel and Judah, and wrote letters also to Ephraim and Manasseh, that they should come to the house of the LORD at Jerusalem, to keep the passover to the LORD the God of Israel. ²For the king and his officials and all the assembly in Jerusalem had taken counsel to keep the passover in the second month ³(for they could not keep it at its proper time because the priests had not sanctified themselves in sufficient number, nor had the people assembled in Jerusalem). ⁴The plan seemed right to the king and all the assembly. ⁵So they decreed to make a proclamation throughout all Israel, from Beer-sheba to Dan, that the people should come and keep the passover to the LORD the God of Israel, at Jerusalem; for they had not kept it in great numbers as prescribed. ⁶So couriers went throughout all Israel and Judah with letters from the king and his officials, as the king had commanded, saying, "O people of Israel, return to the LORD, the God of Abraham, Isaac, and Israel, so that he may turn again to the remnant of you who have escaped from the hand of the kings of Assyria. ⁷Do not be like your ancestors and your kindred, who were faithless to the LORD God of their ancestors, so that he made them a desolation, as you see. ⁸Do not now be stiff-necked as your ancestors were, but yield yourselves to the LORD and come to his sanctuary, which he has sanctified forever, and serve the LORD your God, so that his fierce anger may turn away from you. ⁹For as you return to the LORD, your kindred and your children will find compassion with their captors, and return to this land. For the LORD your God is gracious and merciful, and will not turn away his face from you, if you return to him."

10 So the couriers went from city to city through the country of Ephraim and Manasseh, and as far as Zebulun; but they laughed them to scorn, and mocked them. ¹¹Only a few from Asher, Manasseh, and Zebulun humbled themselves and came to Jerusalem. ¹²The hand of God was also on Judah to give them one heart to do what the king and the officials commanded by the word of the LORD.

13 Many people came together in Jerusalem to keep the festival of unleavened bread in the second month, a very large assembly. ¹⁴They set to work and removed the altars that were in Jerusalem, and all the altars for offering incense they took away and threw into the Wadi Kidron. ¹⁵They slaughtered the passover lamb on the fourteenth day of the second month. The priests and the Levites were ashamed, and they sanctified themselves and brought burnt offerings into the house of the LORD. ¹⁶They took their accustomed posts according to the law of Moses the man of God; the priests dashed the blood that they received[h] from the hands of the Levites. ¹⁷For there were many in the assembly who had not sanctified themselves; therefore the Levites had to slaughter the passover lamb for everyone who was not clean, to make it holy to the LORD. ¹⁸For a multitude of the people, many of them from Ephraim, Manasseh, Issachar, and Zebulun, had not cleansed themselves, yet they ate the passover otherwise than as prescribed. But Hezekiah prayed for them, saying, "The good LORD pardon all ¹⁹who set their hearts to seek God, the LORD the God of their ancestors, even though not in accordance with the sanctuary's rules of cleanness." ²⁰The LORD heard Hezekiah, and healed the people. ²¹The people of Israel who were present at Jerusalem kept the festival of unleavened bread seven days with great gladness; and the Levites and the priests praised the LORD day by day, accompanied by loud instruments for the LORD.

h Heb lacks *that they received*

30.1–27: **Hezekiah's great passover festival.** There is no hint of this in 2 Kings. It seems to be a kind of parallel to or preparation for Josiah's great passover in ch 35. Included is the prophetic hope of the return of the northern tribes to their former loyalty to Je-

22 Hezekiah spoke encouragingly to all the Levites who showed good skill in the service of the LORD. So the people ate the food of the festival for seven days, sacrificing offerings of well-being and giving thanks to the LORD the God of their ancestors.

23 Then the whole assembly agreed together to keep the festival for another seven days; so they kept it for another seven days with gladness. 24 For King Hezekiah of Judah gave the assembly a thousand bulls and seven thousand sheep for offerings, and the officials gave the assembly a thousand bulls and ten thousand sheep. The priests sanctified themselves in great numbers. 25 The whole assembly of Judah, the priests and the Levites, and the whole assembly that came out of Israel, and the resident aliens who came out of the land of Israel, and the resident aliens who lived in Judah, rejoiced. 26 There was great joy in Jerusalem, for since the time of Solomon son of King David of Israel there had been nothing like this in Jerusalem. 27 Then the priests and the Levites stood up and blessed the people, and their voice was heard; their prayer came to his holy dwelling in heaven.

31 Now when all this was finished, all Israel who were present went out to the cities of Judah and broke down the pillars, hewed down the sacred poles, *i* and pulled down the high places and the altars throughout all Judah and Benjamin, and in Ephraim and Manasseh, until they had destroyed them all. Then all the people of Israel returned to their cities, all to their individual properties.

2 Hezekiah appointed the divisions of the priests and of the Levites, division by division, everyone according to his service, the priests and the Levites, for burnt offerings and offerings of well-being, to minister in the gates of the camp of the LORD and to give thanks and praise. 3 The contribution of the king from his own possessions was for the burnt offerings: the burnt offerings of morning and evening, and the burnt offerings for the sabbaths, the new moons, and the appointed festivals, as it is written in the law of the LORD. 4 He commanded the people who lived in Jerusalem to give the portion due to the priests and the Levites, so that they might devote themselves to the law of the LORD. 5 As soon as the word spread, the people of Israel gave in abundance the first fruits of grain, wine, oil, honey, and of all the produce of the field; and they brought in abundantly the tithe of everything. 6 The people of Israel and Judah who lived in the cities of Judah also brought in the tithe of cattle and sheep, and the tithe of the dedicated things that had been consecrated to the LORD their God, and laid them in heaps. 7 In the third month they began to pile up the heaps, and finished them in the seventh month. 8 When Hezekiah and the officials came and saw the heaps, they blessed the LORD and his people Israel. 9 Hezekiah questioned the priests and the Levites about the heaps. 10 The chief priest Azariah, who was of the house of Zadok, answered him, "Since they began to bring the contributions into the house of the LORD, we have had enough to eat and have plenty to spare; for the LORD has blessed his people, so that we have this great supply left over."

i Heb *Asherim*

---

rusalem (compare Ezek 37.15–23). The tolerant laxity of vv. 17–19 is in contrast to the usual attitude of this writer.

31.1–21: Hezekiah's reform and his re-establishment of the priests and Levites. 1: Similar to 2 Kings 18.4 (see note there), except that here the reforming activity is extended to the territory of the former northern kingdom. Some regard this latter as historically correct, pointing out that the northern kingdom had fallen, and claiming that the Assyrians did not exercise strict control. There is also the possibility that we have here the hope of reunion referred to in ch 30 (see notes there).

31.2–10: As David established *the divisions of the priests and of the Levites,* so Hezekiah re-established them, and saw that they were

11 Then Hezekiah commanded them to prepare store-chambers in the house of the LORD; and they prepared them. 12Faithfully they brought in the contributions, the tithes and the dedicated things. The chief officer in charge of them was Conaniah the Levite, with his brother Shimei as second; 13while Jehiel, Azaziah, Nahath, Asahel, Jerimoth, Jozabad, Eliel, Ismachiah, Mahath, and Benaiah were overseers assisting Conaniah and his brother Shimei, by the appointment of King Hezekiah and of Azariah the chief officer of the house of God. 14Kore son of Imnah the Levite, keeper of the east gate, was in charge of the freewill offerings to God, to apportion the contribution reserved for the LORD and the most holy offerings. 15Eden, Miniamin, Jeshua, Shemaiah, Amariah, and Shecaniah were faithfully assisting him in the cities of the priests, to distribute the portions to their kindred, old and young alike, by divisions, 16except those enrolled by genealogy, males from three years old and upwards, all who entered the house of the LORD as the duty of each day required, for their service according to their offices, by their divisions. 17The enrollment of the priests was according to their ancestral houses; that of the Levites from twenty years old and upwards was according to their offices, by their divisions. 18The priests were enrolled with all their little children, their wives, their sons, and their daughters, the whole multitude; for they were faithful in keeping themselves holy. 19And for the descendants of Aaron, the priests, who were in the fields of common land belonging to their towns, town by town, the people designated by name were to distribute portions to every male among the priests and to everyone among the Levites who was enrolled.

20 Hezekiah did this throughout all Judah; he did what was good and right and faithful before the LORD his God.

21And every work that he undertook in the service of the house of God, and in accordance with the law and the commandments, to seek his God, he did with all his heart; and he prospered.

32 After these things and these acts of faithfulness, King Sennacherib of Assyria came and invaded Judah and encamped against the fortified cities, thinking to win them for himself. 2When Hezekiah saw that Sennacherib had come and intended to fight against Jerusalem, 3he planned with his officers and his warriors to stop the flow of the springs that were outside the city; and they helped him. 4A great many people were gathered, and they stopped all the springs and the wadi that flowed through the land, saying, "Why should the Assyrian kings come and find water in abundance?" 5Hezekiah*j* set to work resolutely and built up the entire wall that was broken down, and raised towers on it, *k* and outside it he built another wall; he also strengthened the Millo in the city of David, and made weapons and shields in abundance. 6He appointed combat commanders over the people, and gathered them together to him in the square at the gate of the city and spoke encouragingly to them, saying, 7"Be strong and of good courage. Do not be afraid or dismayed before the king of Assyria and all the horde that is with him; for there is one greater with us than with him. 8With him is an arm of flesh; but with us is the LORD our God, to help us and to fight our battles." The people were encouraged by the words of King Hezekiah of Judah.

9 After this, while King Sennacherib of Assyria was at Lachish with all his forces, he sent his servants to Jerusalem to King Hezekiah of Judah and to all the people of Judah that were in Jerusalem, saying, 10"Thus says King Sennacherib of Assyria: On what are you relying, that

j Heb *He*   *k* Vg: Heb *and raised on the towers*

provided for. 11–21: Compare 1 Chr 23.7–23. 32.1–33: The invasion of Sennacherib and the end of Hezekiah's reign. Based upon 2 Kings 18.13–20.21, with much omission and some addition. 1: From 2 Kings 18.13, with the change of "and took them" to

you undergo the siege of Jerusalem? ¹¹Is not Hezekiah misleading you, handing you over to die by famine and by thirst, when he tells you, 'The LORD our God will save us from the hand of the king of Assyria'? ¹²Was it not this same Hezekiah who took away his high places and his altars and commanded Judah and Jerusalem, saying, 'Before one altar you shall worship, and upon it you shall make your offerings'? ¹³Do you not know what I and my ancestors have done to all the peoples of other lands? Were the gods of the nations of those lands at all able to save their lands out of my hand? ¹⁴Who among all the gods of those nations that my ancestors utterly destroyed was able to save his people from my hand, that your God should be able to save you from my hand? ¹⁵Now therefore do not let Hezekiah deceive you or mislead you in this fashion, and do not believe him, for no god of any nation or kingdom has been able to save his people from my hand or from the hand of my ancestors. How much less will your God save you out of my hand!"

16 His servants said still more against the Lord GOD and against his servant Hezekiah. ¹⁷He also wrote letters to throw contempt on the LORD the God of Israel and to speak against him, saying, "Just as the gods of the nations in other lands did not rescue their people from my hands, so the God of Hezekiah will not rescue his people from my hand." ¹⁸They shouted it with a loud voice in the language of Judah to the people of Jerusalem who were on the wall, to frighten and terrify them, in order that they might take the city. ¹⁹They spoke of the God of Jerusalem as if he were like the gods of the peoples of the earth, which are the work of human hands.

20 Then King Hezekiah and the

prophet Isaiah son of Amoz prayed because of this and cried to heaven. ²¹And the LORD sent an angel who cut off all the mighty warriors and commanders and officers in the camp of the king of Assyria. So he returned in disgrace to his own land. When he came into the house of his god, some of his own sons struck him down there with the sword. ²²So the LORD saved Hezekiah and the inhabitants of Jerusalem from the hand of King Sennacherib of Assyria and from the hand of all his enemies; he gave them rest¹ on every side. ²³Many brought gifts to the LORD in Jerusalem and precious things to King Hezekiah of Judah, so that he was exalted in the sight of all nations from that time onward.

24 In those days Hezekiah became sick and was at the point of death. He prayed to the LORD, and he answered him and gave him a sign. ²⁵But Hezekiah did not respond according to the benefit done to him, for his heart was proud. Therefore wrath came upon him and upon Judah and Jerusalem. ²⁶Then Hezekiah humbled himself for the pride of his heart, both he and the inhabitants of Jerusalem, so that the wrath of the LORD did not come upon them in the days of Hezekiah.

27 Hezekiah had very great riches and honor; and he made for himself treasuries for silver, for gold, for precious stones, for spices, for shields, and for all kinds of costly objects; ²⁸storehouses also for the yield of grain, wine, and oil; and stalls for all kinds of cattle, and sheepfolds. ᵐ ²⁹He likewise provided cities for himself, and flocks and herds in abundance; for God had given him very great possessions. ³⁰This same Hezekiah closed the upper

l Gk Vg: Heb *guided them*   *m* Gk Vg: Heb *flocks for folds*

---

thinking to win them for himself. **2–8:** These preparations for the siege were added by the Chronicler from a source now lost. **32.9–19:** A greatly condensed version of 2 Kings 18.17–35. **32.20:** This single sentence is a condensation of the long passage, 2 Kings 19.1–34. **21:**

The Chronicler's version of 2 Kings 19.35–37. **22–23:** The Chronicler's comments on the preceding part of the chapter. **24–26:** A condensation of 2 Kings 20.1–19 (Isa chs 38–39), showing that Hezekiah was very good, but not quite perfect. **32.27–29:** Compare 2 Kings 20.13. **30:**

outlet of the waters of Gihon and directed them down to the west side of the city of David. Hezekiah prospered in all his works. 31 So also in the matter of the envoys of the officials of Babylon, who had been sent to him to inquire about the sign that had been done in the land, God left him to himself, in order to test him and to know all that was in his heart.

32 Now the rest of the acts of Hezekiah, and his good deeds, are written in the vision of the prophet Isaiah son of Amoz in the Book of the Kings of Judah and Israel. 33 Hezekiah slept with his ancestors, and they buried him on the ascent to the tombs of the descendants of David; and all Judah and the inhabitants of Jerusalem did him honor at his death. His son Manasseh succeeded him.

33 Manasseh was twelve years old when he began to reign; he reigned fifty-five years in Jerusalem. 2 He did what was evil in the sight of the LORD, according to the abominable practices of the nations whom the LORD drove out before the people of Israel. 3 For he rebuilt the high places that his father Hezekiah had pulled down, and erected altars to the Baals, made sacred poles, *n* worshiped all the host of heaven, and served them. 4 He built altars in the house of the LORD, of which the LORD had said, "In Jerusalem shall my name be forever." 5 He built altars for all the host of heaven in the two courts of the house of the LORD. 6 He made his son pass through fire in the valley of the son of Hinnom, practiced soothsaying and augury and sorcery, and dealt with medi-

ums and with wizards. He did much evil in the sight of the LORD, provoking him to anger. 7 The carved image of the idol that he had made he set in the house of God, of which God said to David and to his son Solomon, "In this house, and in Jerusalem, which I have chosen out of all the tribes of Israel, I will put my name forever; 8 I will never again remove the feet of Israel from the land that I appointed for your ancestors, if only they will be careful to do all that I have commanded them, all the law, the statutes, and the ordinances given through Moses." 9 Manasseh misled Judah and the inhabitants of Jerusalem, so that they did more evil than the nations whom the LORD had destroyed before the people of Israel.

10 The LORD spoke to Manasseh and to his people, but they gave no heed. 11 Therefore the LORD brought against them the commanders of the army of the king of Assyria, who took Manasseh captive in manacles, bound him with fetters, and brought him to Babylon. 12 While he was in distress he entreated the favor of the LORD his God and humbled himself greatly before the God of his ancestors. 13 He prayed to him, and God received his entreaty, heard his plea, and restored him again to Jerusalem and to his kingdom. Then Manasseh knew that the LORD indeed was God.

14 Afterward he built an outer wall for the city of David west of Gihon, in the valley, reaching the entrance at the Fish Gate; he carried it around Ophel,

n Heb *Asheroth*

Compare vv. 3–4 and see 2 Kings 20.20–21 n. **31:** Further comment on 2 Kings 20.12–19. **32–33:** Slightly expanded from 2 Kings 20.20–21. The reference to *the vision of the prophet Isaiah* is not necessarily to the canonical book of this prophet, which does, however, contain material on Hezekiah (Isa chs 36–39; compare Isa 1.1).
   33.1–25: The bad beginning of Manasseh; his punishment and repentance; the evil reign of Amon. This chapter is unusual in that it presents a king of Judah, Manasseh, as less evil than he appears in 2 Kings. Perhaps the Chronicler could not believe that Manas-

seh was wholly bad, since he reigned longer than any other king of Israel or Judah. **1–9:** These verses follow closely their source, 2 Kings 21.1–9.
   33.10–13: There is no hint of Manasseh's captivity and release in 2 Kings, though his name appears in Assyrian inscriptions as a vassal of Esarhaddon and Ashurbanipal, sometimes under suspicion. Thus the Babylonian captivity of Manasseh is historically possible, though the Chronicler uses it here as divine punishment to bring the sinning king to repentance. This repentance is somewhat doubtful in view of 2 Kings 21.10–17.

and raised it to a very great height. He also put commanders of the army in all the fortified cities in Judah. 15 He took away the foreign gods and the idol from the house of the LORD, and all the altars that he had built on the mountain of the house of the LORD and in Jerusalem, and he threw them out of the city. 16 He also restored the altar of the LORD and offered on it sacrifices of well-being and of thanksgiving; and he commanded Judah to serve the LORD the God of Israel. 17 The people, however, still sacrificed at the high places, but only to the LORD their God.

18 Now the rest of the acts of Manasseh, his prayer to his God, and the words of the seers who spoke to him in the name of the LORD God of Israel, these are in the Annals of the Kings of Israel. 19 His prayer, and how God received his entreaty, all his sin and his faithlessness, the sites on which he built high places and set up the sacred poles*o* and the images, before he humbled himself, these are written in the records of the seers.*p* 20 So Manasseh slept with his ancestors, and they buried him in his house. His son Amon succeeded him.

21 Amon was twenty-two years old when he began to reign; he reigned two years in Jerusalem. 22 He did what was evil in the sight of the LORD, as his father Manasseh had done. Amon sacrificed to all the images that his father Manasseh had made, and served them. 23 He did not humble himself before the LORD, as his father Manasseh had humbled himself, but this Amon incurred more and more

guilt. 24 His servants conspired against him and killed him in his house. 25 But the people of the land killed all those who had conspired against King Amon; and the people of the land made his son Josiah king to succeed him.

34 Josiah was eight years old when he began to reign; he reigned thirty-one years in Jerusalem. 2 He did what was right in the sight of the LORD, and walked in the ways of his ancestor David; he did not turn aside to the right or to the left. 3 For in the eighth year of his reign, while he was still a boy, he began to seek the God of his ancestor David, and in the twelfth year he began to purge Judah and Jerusalem of the high places, the sacred poles,*o* and the carved and the cast images. 4 In his presence they pulled down the altars of the Baals; he demolished the incense altars that stood above them. He broke down the sacred poles*o* and the carved and the cast images; he made dust of them and scattered it over the graves of those who had sacrificed to them. 5 He also burned the bones of the priests on their altars, and purged Judah and Jerusalem. 6 In the towns of Manasseh, Ephraim, and Simeon, and as far as Naphtali, in their ruins*q* all around, 7 he broke down the altars, beat the sacred poles*r* and the images into powder, and demolished all the incense altars throughout all the land of Israel. Then he returned to Jerusalem.

8 In the eighteenth year of his reign,

o Heb *Asherim*   *p* One Ms Gk: MT *of Hozai*
q Meaning of Heb uncertain   *r* Heb *Asherim*

33.14–17: None of these verses, except the last, has any parallel. Verse 17 shows that the reform fell short of perfection; compare 2 Kings 21.16, which is much stronger in condemnation.

33.18–20: These verses are a new version of 2 Kings 21.17–18 in the light of the Chronicler's own interpretation of the reign of Manasseh. **21–25:** From 2 Kings 21.19–24, and in agreement with the source in the low evaluation of Amon; v. 23 has been rewritten to conform to the new evaluation of Manasseh. For the (apocryphal) Prayer of Manasseh, see p. 282 AP.

34.1–33: The good reign of Josiah; his reforms and the finding of the book of the law. The Chronicler pictures Josiah as the great reformer, the best king since David and Solomon, and the last good king. 2 Kings 22.1–23.30 is in basic agreement on the character and importance of Josiah. **3–7:** These verses transfer the great reform from a time after the finding of the book of the law to a time near the beginning of Josiah's reign (compare 2 Kings 23.4–20). *The sacred poles,* see 1 Kings 14.15 n.

34.8–13: The repair of the temple. This paragraph follows 2 Kings 22.3–7, but with

when he had purged the land and the house, he sent Shaphan son of Azaliah, Maaseiah the governor of the city, and Joah son of Joahaz, the recorder, to repair the house of the LORD his God. 9 They came to the high priest Hilkiah and delivered the money that had been brought into the house of God, which the Levites, the keepers of the threshold, had collected from Manasseh and Ephraim and from all the remnant of Israel and from all Judah and Benjamin and from the inhabitants of Jerusalem. 10 They delivered it to the workers who had the oversight of the house of the LORD, and the workers who were working in the house of the LORD gave it for repairing and restoring the house. 11 They gave it to the carpenters and the builders to buy quarried stone, and timber for binders, and beams for the buildings that the kings of Judah had let go to ruin. 12 The people did the work faithfully. Over them were appointed the Levites Jahath and Obadiah, of the sons of Merari, along with Zechariah and Meshullam, of the sons of the Kohathites, to have oversight. Other Levites, all skillful with instruments of music, 13 were over the burden bearers and directed all who did work in every kind of service; and some of the Levites were scribes, and officials, and gatekeepers.

14 While they were bringing out the money that had been brought into the house of the LORD, the priest Hilkiah found the book of the law of the LORD given through Moses. 15 Hilkiah said to the secretary Shaphan, "I have found the book of the law in the house of the LORD"; and Hilkiah gave the book to Shaphan. 16 Shaphan brought the book to the king, and further reported to the king, "All that was committed to your servants they are doing. 17 They have emptied out the money that was found in the house of the LORD and have delivered it into the hand of the overseers and the

workers." 18 The secretary Shaphan informed the king, "The priest Hilkiah has given me a book." Shaphan then read it aloud to the king.

19 When the king heard the words of the law he tore his clothes. 20 Then the king commanded Hilkiah, Ahikam son of Shaphan, Abdon son of Micah, the secretary Shaphan, and the king's servant Asaiah: 21 "Go, inquire of the LORD for me and for those who are left in Israel and in Judah, concerning the words of the book that has been found; for the wrath of the LORD that is poured out on us is great, because our ancestors did not keep the word of the LORD, to act in accordance with all that is written in this book."

22 So Hilkiah and those whom the king had sent went to the prophet Huldah, the wife of Shallum son of Tokhath son of Hasrah, keeper of the wardrobe (who lived in Jerusalem in the Second Quarter) and spoke to her to that effect. 23 She declared to them, "Thus says the LORD, the God of Israel: Tell the man who sent you to me, 24 Thus says the LORD: I will indeed bring disaster upon this place and upon its inhabitants, all the curses that are written in the book that was read before the king of Judah. 25 Because they have forsaken me and have made offerings to other gods, so that they have provoked me to anger with all the works of their hands, my wrath will be poured out on this place and will not be quenched. 26 But as to the king of Judah, who sent you to inquire of the LORD, thus shall you say to him: Thus says the LORD, the God of Israel: Regarding the words that you have heard, 27 because your heart was penitent and you humbled yourself before God when you heard his words against this place and its inhabitants, and you have humbled yourself before me, and have torn your clothes and wept before me, I also have

more attention to the Levites, the musicians, and the help from the territory of the former northern kingdom.

34.14–18: The finding of the book of the

law. This account is not essentially different from its source, 2 Kings 22.8–10 (see note there).

34.19–28: Very similar to its source,

heard you, says the LORD. ²⁸I will gather you to your ancestors and you shall be gathered to your grave in peace; your eyes shall not see all the disaster that I will bring on this place and its inhabitants." They took the message back to the king.

29 Then the king sent word and gathered together all the elders of Judah and Jerusalem. ³⁰The king went up to the house of the LORD, with all the people of Judah, the inhabitants of Jerusalem, the priests and the Levites, all the people both great and small; he read in their hearing all the words of the book of the covenant that had been found in the house of the LORD. ³¹The king stood in his place and made a covenant before the LORD, to follow the LORD, keeping his commandments, his decrees, and his statutes, with all his heart and all his soul, to perform the words of the covenant that were written in this book. ³²Then he made all who were present in Jerusalem and in Benjamin pledge themselves to it. And the inhabitants of Jerusalem acted according to the covenant of God, the God of their ancestors. ³³Josiah took away all the abominations from all the territory that belonged to the people of Israel, and made all who were in Israel worship the LORD their God. All his days they did not turn away from following the LORD the God of their ancestors.

35 Josiah kept a passover to the LORD in Jerusalem; they slaughtered the passover lamb on the fourteenth day of the first month. ²He appointed the priests to their offices and encouraged them in the service of the house of the LORD. ³He said to the Levites who taught all Israel and who were holy to the LORD, "Put the holy ark in the house that Solomon son of David, king of Israel, built; you need no longer carry it on your shoulders. Now serve the LORD your God and his people Israel. ⁴Make preparations by your ancestral houses by your divisions, following the written directions of King David of Israel and the written directions of his son Solomon. ⁵Take position in the holy place according to the groupings of the ancestral houses of your kindred the people, and let there be Levites for each division of an ancestral house. ^s ⁶Slaughter the passover lamb, sanctify yourselves, and on behalf of your kindred make preparations, acting according to the word of the LORD by Moses."

7 Then Josiah contributed to the people, as passover offerings for all that were present, lambs and kids from the flock to the number of thirty thousand, and three thousand bulls; these were from the king's possessions. ⁸His officials contributed willingly to the people, to the priests, and to the Levites. Hilkiah, Zechariah, and Jehiel, the chief officers of the house of God, gave to the priests for the passover offerings two thousand six hundred lambs and kids and three hundred bulls. ⁹Conaniah also, and his brothers Shemaiah and Nethanel, and Hashabiah and Jeiel and Jozabad, the chiefs of the Levites, gave to the Levites for the passover offerings five thousand lambs and kids and five hundred bulls.

10 When the service had been prepared for, the priests stood in their place, and the Levites in their divisions according to the king's command. ¹¹They slaughtered the passover lamb, and the priests dashed the blood that they re-

s Meaning of Heb uncertain

---

2 Kings 22.11–20 (see notes there). **29–33**: Expanded from 2 Kings 23.1–3. **30**: *The Levites* are substituted for "the prophets." **32**: *Benjamin* is added. **33**: A kind of summary of 2 Kings 23.4–20.

35.1–27: Josiah's great passover celebration; his tragic mistake and consequent death. 1–6: 2 Kings 23.21–23 tells us that Josiah kept the greatest passover of any up to his time, but gives no details. The Chronicler seizes the opportunity to describe in detail what he regarded as the most important festival, as it was kept in his time, or as it ought to be kept in the future. Verse 4 refers to 1 Chr chs 24–26.

35.7–9: Compare 1 Chr 29.6–9. **10–15**: A prominent place is given to *the Levites* and *the singers* (see Introduction to 1 Chronicles).

ceived*t* from them, while the Levites did the skinning. 12 They set aside the burnt offerings so that they might distribute them according to the groupings of the ancestral houses of the people, to offer to the LORD, as it is written in the book of Moses. And they did the same with the bulls. 13 They roasted the passover lamb with fire according to the ordinance; and they boiled the holy offerings in pots, in caldrons, and in pans, and carried them quickly to all the people. 14 Afterward they made preparations for themselves and for the priests, because the priests the descendants of Aaron were occupied in offering the burnt offerings and the fat parts until night; so the Levites made preparations for themselves and for the priests, the descendants of Aaron. 15 The singers, the descendants of Asaph, were in their place according to the command of David, and Asaph, and Heman, and the king's seer Jeduthun. The gatekeepers were at each gate; they did not need to interrupt their service, for their kindred the Levites made preparations for them.

16 So all the service of the LORD was prepared that day, to keep the passover and to offer burnt offerings on the altar of the LORD, according to the command of King Josiah. 17 The people of Israel who were present kept the passover at that time, and the festival of unleavened bread seven days. 18 No passover like it had been kept in Israel since the days of the prophet Samuel; none of the kings of Israel had kept such a passover as was kept by Josiah, by the priests and the Levites, by all Judah and Israel who were present, and by the inhabitants of Jerusalem. 19 In the eighteenth year of the reign of Josiah this passover was kept.

20 After all this, when Josiah had set the temple in order, King Neco of Egypt went up to fight at Carchemish on the Euphrates, and Josiah went out against him. 21 But Neco*u* sent envoys to him, saying, "What have I to do with you, king of Judah? I am not coming against you today, but against the house with which I am at war; and God has commanded me to hurry. Cease opposing God, who is with me, so that he will not destroy you." 22 But Josiah would not turn away from him, but disguised himself in order to fight with him. He did not listen to the words of Neco from the mouth of God, but joined battle in the plain of Megiddo. 23 The archers shot King Josiah; and the king said to his servants, "Take me away, for I am badly wounded." 24 So his servants took him out of the chariot and carried him in his second chariot*v* and brought him to Jerusalem. There he died, and was buried in the tombs of his ancestors. All Judah and Jerusalem mourned for Josiah. 25 Jeremiah also uttered a lament for Josiah, and all the singing men and singing women have spoken of Josiah in their laments to this day. They made these a custom in Israel; they are recorded in the Laments. 26 Now the rest of the acts of Josiah and his faithful deeds in accordance with what is written in the law of the LORD, 27 and his acts, first and last, are

t Heb lacks *that they received*   *u* Heb *he*
v Or *the chariot of his deputy*

35.16–19: Compare 2 Kings 23.22–23. 20–24: A greatly expanded version of 2 Kings 23.29–30; some of the added details are doubtless historically correct. The most characteristic touch of the Chronicler is in describing *the words of Neco* as coming *from the mouth of God* (v. 22). Only thus could a writer who believed every misfortune a direct punishment from God explain the untimely death of the otherwise righteous Josiah. 35.25: Strangely enough the prophet *Jeremiah* is not mentioned in 2 Kings. The Chronicler here makes good the omission (compare Jer 1.2; 3.6; 22.10, 15). *The Laments* mentioned here are not to be confused with the canonical book of Lamentations. 26–27: A characteristic expansion of 2 Kings 23.28. 36.1–23: **The last agonies of the doomed nation.** The goodness of David, Solomon, Asa, Hezekiah, and Josiah was not sufficient, and the apostate nation was doomed (compare 34.24–28). The source is 2 Kings 23.30b–25.21, with some condensation and much rewriting near the end. 1–4:

written in the Book of the Kings of Israel
and Judah.

36 The people of the land took Jehoahaz son of Josiah and made him
king to succeed his father in Jerusalem.
²Jehoahaz was twenty-three years old
when he began to reign; he reigned three
months in Jerusalem. ³Then the king of
Egypt deposed him in Jerusalem and laid
on the land a tribute of one hundred talents of silver and one talent of gold. ⁴The
king of Egypt made his brother Eliakim
king over Judah and Jerusalem, and
changed his name to Jehoiakim; but
Neco took his brother Jehoahaz and carried him to Egypt.

5 Jehoiakim was twenty-five years
old when he began to reign; he reigned
eleven years in Jerusalem. He did what
was evil in the sight of the Lord his God.
⁶Against him King Nebuchadnezzar of
Babylon came up, and bound him with
fetters to take him to Babylon. ⁷Nebu-
chadnezzar also carried some of the vessels of the house of the Lord to Babylon
and put them in his palace in Babylon.
⁸Now the rest of the acts of Jehoiakim,
and the abominations that he did, and
what was found against him, are written
in the Book of the Kings of Israel and
Judah; and his son Jehoiachin succeeded
him.

9 Jehoiachin was eight years old when
he began to reign; he reigned three
months and ten days in Jerusalem. He
did what was evil in the sight of the
Lord. ¹⁰In the spring of the year King
Nebuchadnezzar sent and brought him

to Babylon, along with the precious vessels of the house of the Lord, and made
his brother Zedekiah king over Judah
and Jerusalem.

11 Zedekiah was twenty-one years
old when he began to reign; he reigned
eleven years in Jerusalem. ¹²He did what
was evil in the sight of the Lord his God.
He did not humble himself before the
prophet Jeremiah who spoke from the
mouth of the Lord. ¹³He also rebelled
against King Nebuchadnezzar, who had
made him swear by God; he stiffened his
neck and hardened his heart against turning to the Lord, the God of Israel. ¹⁴All
the leading priests and the people also
were exceedingly unfaithful, following
all the abominations of the nations; and
they polluted the house of the Lord that
he had consecrated in Jerusalem.

15 The Lord, the God of their ancestors, sent persistently to them by his
messengers, because he had compassion
on his people and on his dwelling place;
¹⁶but they kept mocking the messengers
of God, despising his words, and scoffing at his prophets, until the wrath of the
Lord against his people became so great
that there was no remedy.

17 Therefore he brought up against
them the king of the Chaldeans, who
killed their youths with the sword in the
house of their sanctuary, and had no
compassion on young man or young
woman, the aged or the feeble; he gave
them all into his hand. ¹⁸All the vessels
of the house of God, large and small, and
the treasures of the house of the Lord,

The ill-fated reign of Jehoahaz, taken with
slight condensation from 2 Kings 23.30b–34.
36.5–8: Much abridged from 2 Kings
23.36–24.6. In v. 6 Jehoiakim seems to be
taken *to Babylon,* contrary to the source,
2 Kings 24.6 (compare Dan 1.1–2). Verse 7 is
added, anticipating v. 10 (compare Dan 1.2).
9–10: The story of *Jehoiachin* is drastically
abridged from 2 Kings 24.8–17. The correct
age in v. 9 is not *eight years,* but eighteen years
(compare 2 Kings 24.8). Tablets from Babylonia disclose that by 592 B.C. he had five
sons.
36.11–13: Based on 2 Kings 24.18–20, but
with certain changes, such as the dropping of

the names of the king's mother and grandfather, the introduction of *the prophet Jeremiah*
(see 35.25 n. and compare Jer chs 37–39), and
the mention of the king's oath (compare Ezek
17.16). **14:** There is no parallel to this in
2 Kings, where the blame is laid chiefly on the
king (compare Jer ch 7; Ezek ch 8). The
Chronicler does not condemn the Levites and
musicians. **15–16:** The Chronicler is fond of
the theme of the unheeded prophet (compare
12.5–8; 15.1–8; 19.1–3; 21.12–15; 25.7–9,
15–17; Jer 26.20–24; 29.16–20; 35.14–15).
17–21: A condensation of 2 Kings 25.1–21,
with additional references to *the establishment
of the kingdom of Persia* (compare v. 22), the

and the treasures of the king and of his officials, all these he brought to Babylon. ¹⁹They burned the house of God, broke down the wall of Jerusalem, burned all its palaces with fire, and destroyed all its precious vessels. ²⁰He took into exile in Babylon those who had escaped from the sword, and they became servants to him and to his sons until the establishment of the kingdom of Persia, ²¹to fulfill the word of the LORD by the mouth of Jeremiah, until the land had made up for its sabbaths. All the days that it lay desolate it kept sabbath, to fulfill seventy years.

22 In the first year of King Cyrus of Persia, in fulfillment of the word of the LORD spoken by Jeremiah, the LORD stirred up the spirit of King Cyrus of Persia so that he sent a herald throughout all his kingdom and also declared in a written edict: ²³"Thus says King Cyrus of Persia: The LORD, the God of heaven, has given me all the kingdoms of the earth, and he has charged me to build him a house at Jerusalem, which is in Judah. Whoever is among you of all his people, may the LORD his God be with him! Let him go up."

---

prophecies of *Jeremiah* (compare Jer 25.11–12; 29.10), and the *sabbaths* (sabbatical years; compare Lev 25.1–7; 26.27–39). **22–23:** This note of hope is reproduced here from what immediately follows, Ezra 1.1–3, where it more properly belongs. If Chronicles originally belonged with Ezra and Nehemiah (see Introductions to 1 and 2 Chronicles), then

one might say that 1 and 2 Chronicles portrayed the downfall, and Ezra/Nehemiah the restoration, of God's people. In the arrangement of the Hebrew Bible, 2 Chronicles is the last book; thus this version of Israel's history, like 2 Kings 25.27–30, does not end on a note of doom (compare 2 Kings 25.27–30 and see note there).

Ezra

Anciently several books circulated under the name of Ezra, not only our Ezra and Nehemiah as a single book, but others now preserved in the Apocrypha and Pseudepigrapha. Our Ezra-Nehemiah is usually ascribed to the Chronicler (see Introduction to 1 Chronicles), prepared as a supplement to Chronicles on the basis of Hebrew and Aramaic documents, memoirs of Nehemiah, a memorial of Ezra, genealogies, and archives. Chronicles ends with the destruction of Jerusalem and the carrying away of treasure and captives. This supplement, the first verses of which appear also at the end of Chronicles, was written to tell how some returned from captivity and labored at restoring religion at a restored temple in a refortified Jerusalem (see "Survey of . . . Bible Lands," §18). During the exile religious interest had concentrated particularly on the laws associated with the name of Moses, and had fostered that exclusiveness which became so characteristic of Judaism. Thus the returned exiles were concerned not only with reconstruction of altar, temple, and city, but with social and religious problems, freeing the community of foreign elements, and establishing religious practice in stricter conformity to their understanding of Mosaic law. To the author the returned exiles were a godly remnant with a religious mission.

There seem to have been four stages of the return: (1) a return under Cyrus (about 538 B.C.) led by Sheshbazzar, who commenced rebuilding the temple but, under local opposition, had to leave it unfinished; (2) a return under Darius I (521–485) led by Zerubbabel and Jeshua, who also encountered opposition but, with encouragement from the prophets Haggai and Zechariah, completed the temple; (3) a group, under Artaxerxes I (464–423), led by Ezra, who brought a codification of Mosaic law; (4) another group, under Artaxerxes II (404–358), led by Nehemiah, who came twice under Artaxerxes I to build the walls of Jerusalem, still against opposition from the local groups, and to attempt to establish purity of community and worship. An alternate solution of difficult textual problems places Ezra's return under Artaxerxes II, after Nehemiah.

The text of Ezra and Nehemiah has been dislocated in transmission. It is probable that Neh 8 originally stood between Ezra 8 and 9 and that Neh 9.1–5 originally stood between Ezra 10.15 and 10.16. An editor, faced with this textual confusion, endeavored to ease it by supplying connecting sentences and by insertions, such as the name of Nehemiah in Neh 8.9 and Ezra in Neh 12.26,36.

1 In the first year of King Cyrus of Persia, in order that the word of the LORD by the mouth of Jeremiah might be accomplished, the LORD stirred up the spirit of King Cyrus of Persia so that he sent a herald throughout all his kingdom, and also in a written edict declared:

2 "Thus says King Cyrus of Persia: The LORD, the God of heaven, has given me all the kingdoms of the earth, and he has charged me to build him a house at Jerusalem in Judah. 3 Any of those among you who are of his people—may their God be with them!—are now permitted to go up to Jerusalem in Judah, and rebuild the house of the LORD, the God of Israel—he is the God who is in Jerusalem; 4 and let all survivors, in whatever place they reside, be assisted by the people of their place with silver and gold, with goods and with animals, besides freewill offerings for the house of God in Jerusalem."

5 The heads of the families of Judah and Benjamin, and the priests and the Levites—everyone whose spirit God had stirred—got ready to go up and rebuild the house of the LORD in Jerusalem. 6 All their neighbors aided them with silver vessels, with gold, with goods, with animals, and with valuable gifts, besides all that was freely offered. 7 King Cyrus himself brought out the vessels of the house of the LORD that Nebuchadnezzar had carried away from Jerusalem and placed in the house of his gods. 8 King Cyrus of Persia had them released into the charge of Mithredath the treasurer, who counted them out to Sheshbazzar the prince of Judah. 9 And this was the inventory: gold basins, thirty; silver basins, one thousand; knives,*a* twenty-

nine; 10 gold bowls, thirty; other silver bowls, four hundred ten; other vessels, one thousand; 11 the total of the gold and silver vessels was five thousand four hundred. All these Sheshbazzar brought up, when the exiles were brought up from Babylonia to Jerusalem.

2 Now these were the people of the province who came from those captive exiles whom King Nebuchadnezzar of Babylon had carried captive to Babylonia; they returned to Jerusalem and Judah, all to their own towns. 2 They came with Zerubbabel, Jeshua, Nehemiah, Seraiah, Reelaiah, Mordecai, Bilshan, Mispar, Bigvai, Rehum, and Baanah.

The number of the Israelite people: 3 the descendants of Parosh, two thousand one hundred seventy-two. 4 Of Shephatiah, three hundred seventy-two. 5 Of Arah, seven hundred seventy-five. 6 Of Pahath-moab, namely the descendants of Jeshua and Joab, two thousand eight hundred twelve. 7 Of Elam, one thousand two hundred fifty-four. 8 Of Zattu, nine hundred forty-five. 9 Of Zaccai, seven hundred sixty. 10 Of Bani, six hundred forty-two. 11 Of Bebai, six hundred twenty-three. 12 Of Azgad, one thousand two hundred twenty-two. 13 Of Adonikam, six hundred sixty-six. 14 Of Bigvai, two thousand fifty-six. 15 Of Adin, four hundred fifty-four. 16 Of Ater, namely of Hezekiah, ninety-eight. 17 Of Bezai, three hundred twenty-three. 18 Of Jorah, one hundred twelve. 19 Of Hashum, two hundred twenty-three. 20 Of Gibbar, ninety-five. 21 Of Bethlehem, one hundred twenty-three. 22 The people of Netophah, fifty-six. 23 Of Ana-

a Vg: Meaning of Heb uncertain

1.1–4: Cyrus' decree. His *first year* at Babylon was 538 B.C., and his proclamation seemed the fulfilment of Jer 29.10 (2 Chr 36.21) under divine direction (Isa 44.28; 45.1–3). **2–3:** In his inscriptions Cyrus shows interest in restoring temples. To him the LORD is the local deity of Jerusalem (6.12), whose temple was destroyed in 586 B.C. by Nebuchadnezzar (5.12).

1.5–11: Gifts for refurnishing the temple, including the return of the *vessels* taken by *Nebuchadnezzar* (2 Kings 25.14–16). **1.8:** *Mithredath* (Mithridates) was temple treasurer. *Sheshbazzar* is the Babylonian name of a Jewish court official. **9–11:** These numbers vary in different texts and are uncertain. **2.1–70: A census of the first return** (= Neh 7.6–73a), covering several groups: lead-

thoth, one hundred twenty-eight. 24 The descendants of Azmaveth, forty-two. 25 Of Kiriatharim, Chephirah, and Beeroth, seven hundred forty-three. 26 Of Ramah and Geba, six hundred twenty-one. 27 The people of Michmas, one hundred twenty-two. 28 Of Bethel and Ai, two hundred twenty-three. 29 The descendants of Nebo, fifty-two. 30 Of Magbish, one hundred fifty-six. 31 Of the other Elam, one thousand two hundred fifty-four. 32 Of Harim, three hundred twenty. 33 Of Lod, Hadid, and Ono, seven hundred twenty-five. 34 Of Jericho, three hundred forty-five. 35 Of Senaah, three thousand six hundred thirty.

36 The priests: the descendants of Jedaiah, of the house of Jeshua, nine hundred seventy-three. 37 Of Immer, one thousand fifty-two. 38 Of Pashhur, one thousand two hundred forty-seven. 39 Of Harim, one thousand seventeen.

40 The Levites: the descendants of Jeshua and Kadmiel, of the descendants of Hodaviah, seventy-four. 41 The singers: the descendants of Asaph, one hundred twenty-eight. 42 The descendants of the gatekeepers: of Shallum, of Ater, of Talmon, of Akkub, of Hatita, and of Shobai, in all one hundred thirty-nine.

43 The temple servants: the descendants of Ziha, Hasupha, Tabbaoth, 44 Keros, Siaha, Padon, 45 Lebanah, Hagabah, Akkub, 46 Hagab, Shamlai, Hanan, 47 Giddel, Gahar, Reaiah, 48 Rezin, Nekoda, Gazzam, 49 Uzza, Paseah, Besai, 50 Asnah, Meunim, Nephisim, 51 Bakbuk, Hakupha, Harhur, 52 Bazluth, Mehida, Harsha, 53 Barkos, Sisera, Temah, 54 Neziah, and Hatipha.

55 The descendants of Solomon's servants: Sotai, Hassophereth, Peruda, 56 Jaalah, Darkon, Giddel, 57 Shephatiah, Hattil, Pochereth-hazzebaim, and Ami.

58 All the temple servants and the descendants of Solomon's servants were three hundred ninety-two.

59 The following were those who came up from Tel-melah, Tel-harsha, Cherub, Addan, and Immer, though they could not prove their families or their descent, whether they belonged to Israel: 60 the descendants of Delaiah, Tobiah, and Nekoda, six hundred fifty-two. 61 Also, of the descendants of the priests: the descendants of Habaiah, Hakkoz, and Barzillai (who had married one of the daughters of Barzillai the Gileadite, and was called by their name). 62 These looked for their entries in the genealogical records, but they were not found there, and so they were excluded from the priesthood as unclean; 63 the governor told them that they were not to partake of the most holy food, until there should be a priest to consult Urim and Thummim.

64 The whole assembly together was forty-two thousand three hundred sixty, 65 besides their male and female servants, of whom there were seven thousand three hundred thirty-seven; and they had two hundred male and female singers. 66 They had seven hundred thirty-six horses, two hundred forty-five mules, 67 four hundred thirty-five camels, and six thousand seven hundred twenty donkeys.

68 As soon as they came to the house of the Lord in Jerusalem, some of the heads of families made freewill offerings for the house of God, to erect it on its site. 69 According to their resources they gave to the building fund sixty-one thousand darics of gold, five thousand minas of silver, and one hundred priestly robes.

70 The priests, the Levites, and some of the people lived in Jerusalem and its

---

ers, temple officials, servants, those of dubious genealogy, and animals.
2.36–39: There are only four priestly groups as against twenty-four in 1 Chr ch 24.
2.43, 58: *The temple servants* are the Nethinim (lit., "those given" [to the temple]; see 1 Chr 9.2; Josh 9.27 n.), many of whose names are non-Jewish. **63:** The *governor* is Sheshbazzar. *Urim and Thummim* are the sacred lots (1 Sam 14.41; Deut 33.8).
2.64: This total exceeds the numbers of the various groups; it doubtless includes others unmentioned. **69:** A *daric* was a Persian gold coin, a *mina* a Mesopotamian weight for precious metals. The amount differs from that mentioned in Neh 7.70–72.

vicinity;[b] and the singers, the gatekeepers, and the temple servants lived in their towns, and all Israel in their towns.

3 When the seventh month came, and the Israelites were in the towns, the people gathered together in Jerusalem. [2] Then Jeshua son of Jozadak, with his fellow priests, and Zerubbabel son of Shealtiel with his kin set out to build the altar of the God of Israel, to offer burnt offerings on it, as prescribed in the law of Moses the man of God. [3] They set up the altar on its foundation, because they were in dread of the neighboring peoples, and they offered burnt offerings upon it to the LORD, morning and evening. [4] And they kept the festival of booths,[c] as prescribed, and offered the daily burnt offerings by number according to the ordinance, as required for each day, [5] and after that the regular burnt offerings, the offerings at the new moon and at all the sacred festivals of the LORD, and the offerings of everyone who made a freewill offering to the LORD. [6] From the first day of the seventh month they began to offer burnt offerings to the LORD. But the foundation of the temple of the LORD was not yet laid. [7] So they gave money to the masons and the carpenters, and food, drink, and oil to the Sidonians and the Tyrians to bring cedar trees from Lebanon to the sea, to Joppa, according to the grant that they had from King Cyrus of Persia.

8 In the second year after their arrival at the house of God at Jerusalem, in the second month, Zerubbabel son of Shealtiel and Jeshua son of Jozadak made a beginning, together with the rest of their people, the priests and the Levites and all who had come to Jerusalem from the captivity. They appointed the Levites, from twenty years old and upward, to have the oversight of the work on the house of the LORD. [9] And Jeshua with his sons and his kin, and Kadmiel and his sons, Binnui and Hodaviah[d] along with the sons of Henadad, the Levites, their sons and kin, together took charge of the workers in the house of God.

10 When the builders laid the foundation of the temple of the LORD, the priests in their vestments were stationed to praise the LORD with trumpets, and the Levites, the sons of Asaph, with cymbals, according to the directions of King David of Israel; [11] and they sang responsively, praising and giving thanks to the LORD,

"For he is good,
for his steadfast love endures
forever toward Israel."

And all the people responded with a great shout when they praised the LORD, because the foundation of the house of the LORD was laid. [12] But many of the priests and Levites and heads of families, old people who had seen the first house on its foundations, wept with a loud voice when they saw this house, though many shouted aloud for joy, [13] so that the people could not distinguish the sound of the joyful shout from the sound of the people's weeping, for the people shouted so loudly that the sound was heard far away.

---

b 1 Esdras 5.46: Heb lacks *lived in Jerusalem and its vicinity*   *c* Or *tabernacles*; Heb *succoth*
d Compare 2.40; Neh 7.43; 1 Esdras 5.58: Heb *sons of Judah*

3.1–13: Rebuilding the temple. This had commenced under Sheshbazzar (5.14–16) but stopped, and recommenced now under Zerubbabel and Jeshua, leaders of the second return, who *built the altar* as center for the temple cultus. The *seventh month* would be Tishri (September-October) of 520 B.C., in the second year of Darius I (Hag 2.1–4). **2:** *The law of Moses* is not our Pentateuch, but the body of laws associated with Moses' name, especially in Deuteronomy. **3:** Once the altar was *set,* the religious life with *morning and evening* sacrifice, **4:** annual festivals such as *booths* (tabernacles), etc., could be recommenced, and then the temple built. **3.7:** Compare Solomon's preparations in 2 Chr 2.1–11. Minted *money* was known in the Persian period, though payment in kind was more common. **10–11:** The laying of the foundation was celebrated with music and praise (compare Ps 118). For the directions of David see 2 Chr 29.25–30.

4 When the adversaries of Judah and Benjamin heard that the returned exiles were building a temple to the LORD, the God of Israel, 2they approached Zerubbabel and the heads of families and said to them, "Let us build with you, for we worship your God as you do, and we have been sacrificing to him ever since the days of King Esarhaddon of Assyria who brought us here." 3But Zerubbabel, Jeshua, and the rest of the heads of families in Israel said to them, "You shall have no part with us in building a house to our God; but we alone will build to the LORD, the God of Israel, as King Cyrus of Persia has commanded us."

4 Then the people of the land discouraged the people of Judah, and made them afraid to build, 5and they bribed officials to frustrate their plan throughout the reign of King Cyrus of Persia and until the reign of King Darius of Persia.

6 In the reign of Ahasuerus, in his accession year, they wrote an accusation against the inhabitants of Judah and Jerusalem.

7 And in the days of Artaxerxes, Bishlam and Mithredath and Tabeel and the rest of their associates wrote to King Artaxerxes of Persia; the letter was written in Aramaic and translated.ᵉ 8Rehum the royal deputy and Shimshai the scribe wrote a letter against Jerusalem to King Artaxerxes as follows 9(then Rehum the royal deputy, Shimshai the scribe, and the rest of their associates, the judges, the envoys, the officials, the Persians, the people of Erech, the Babylonians, the people of Susa, that is, the Elamites, 10and the rest of the nations whom the great and noble Osnappar deported and settled in the cities of Samaria and in the rest of the province Beyond the River wrote—and now 11this is a copy of the letter that they sent):

"To King Artaxerxes: Your servants, the people of the province Beyond the River, send greeting. And now 12may it be known to the king that the Jews who came up from you to us have gone to Jerusalem. They are rebuilding that rebellious and wicked city; they are finishing the walls and repairing the foundations. 13Now may it be known to the king that, if this city is rebuilt and the walls finished, they will not pay tribute, custom, or toll, and the royal revenue will be reduced. 14Now because we share the salt of the palace and it is not fitting for us to witness the king's dishonor, therefore we send and inform the king, 15so that a search may be made in the annals of your ancestors. You will discover in the annals that this is a rebellious city, hurtful to kings and provinces, and that sedition was stirred up in it from long ago. On that account this city was laid waste. 16We make known to the king that, if this city is rebuilt and its walls finished, you will then have no possession in the province Beyond the River."

17 The king sent an answer: "To Re-

ᵉ Heb adds *in Aramaic*, indicating that 4.8-6.18 is in Aramaic. Another interpretation is *The letter was written in the Aramaic script and set forth in the Aramaic language*

4.1–6: Opposition to the rebuilding of the temple. The rejoicing alerted neighboring mixed groups who claimed fellowship in religion (2 Kings 17.24–28). *Esarhaddon* settled groups in the cities of Samaria in 676 B.C. **3:** The phrase *we alone* evidences the exclusiveness which turned these neighbors into adversaries. **5:** The *officials* would plead their case at court. *Cyrus* died in 529 B.C. *Darius* reigned 521–485. **6:** *Ahasuerus*, Xerxes I, reigned 485–464.

4.7–23: Opposition to the rebuilding of the city. This section is a digression on later opposition encountered by the Jews. Ar-

taxerxes reigned 464–423 B.C. The Aramaic letter to him is from his local officials, authorized by the *commander*, and written by the *scribe*. **10:** *Osnappar* apparently means Assurbanipal (668–630 B.C.). *Beyond the River* denotes the Syro-Palestinian region west of the Euphrates.

4.13: The revenues expected by the Persian treasury were the provincial *tribute*, the *custom* levies for provincial expenses, and *toll* for upkeep of the roads. **15:** For such books of *annals* see Esth 2.23; 6.1; Mal 3.16. *Laid waste*, see 2 Kings 24.13–15.

hum the royal deputy and Shimshai the scribe and the rest of their associates who live in Samaria and in the rest of the province Beyond the River, greeting. And now [18] the letter that you sent to us has been read in translation before me. [19] So I made a decree, and someone searched and discovered that this city has risen against kings from long ago, and that rebellion and sedition have been made in it. [20] Jerusalem has had mighty kings who ruled over the whole province Beyond the River, to whom tribute, custom, and toll were paid. [21] Therefore issue an order that these people be made to cease, and that this city not be rebuilt, until I make a decree. [22] Moreover, take care not to be slack in this matter; why should damage grow to the hurt of the king?"

23 Then when the copy of King Artaxerxes' letter was read before Rehum and the scribe Shimshai and their associates, they hurried to the Jews in Jerusalem and by force and power made them cease. [24] At that time the work on the house of God in Jerusalem stopped and was discontinued until the second year of the reign of King Darius of Persia.

5 Now the prophets, Haggai[f] and Zechariah son of Iddo, prophesied to the Jews who were in Judah and Jerusalem, in the name of the God of Israel who was over them. [2] Then Zerubbabel son of Shealtiel and Jeshua son of Jozadak set out to rebuild the house of God in Jerusalem; and with them were the prophets of God, helping them.

3 At the same time Tattenai the governor of the province Beyond the River and Shethar-bozenai and their associates came to them and spoke to them thus, "Who gave you a decree to build this house and to finish this structure?" [4] They[g] also asked them this, "What are the names of the men who are building

this building?" [5] But the eye of their God was upon the elders of the Jews, and they did not stop them until a report reached Darius and then answer was returned by letter in reply to it.

6 The copy of the letter that Tattenai the governor of the province Beyond the River and Shethar-bozenai and his associates the envoys who were in the province Beyond the River sent to King Darius; [7] they sent him a report, in which was written as follows: "To Darius the king, all peace! [8] May it be known to the king that we went to the province of Judah, to the house of the great God. It is being built of hewn stone, and timber is laid in the walls; this work is being done diligently and prospers in their hands. [9] Then we spoke to those elders and asked them, 'Who gave you a decree to build this house and to finish this structure?' [10] We also asked them their names, for your information, so that we might write down the names of the men at their head. [11] This was their reply to us: 'We are the servants of the God of heaven and earth, and we are rebuilding the house that was built many years ago, which a great king of Israel built and finished. [12] But because our ancestors had angered the God of heaven, he gave them into the hand of King Nebuchadnezzar of Babylon, the Chaldean, who destroyed this house and carried away the people to Babylonia. [13] However, King Cyrus of Babylon, in the first year of his reign, made a decree that this house of God should be rebuilt. [14] Moreover, the gold and silver vessels of the house of God, which Nebuchadnezzar had taken out of the temple in Jerusalem and had brought into the temple of Babylon, these King Cyrus took out of the temple of Babylon, and they were delivered to a man

f Aram adds *the prophet*   *g* Gk Syr: Aram *We*

4.24–6.22: The rebuilding continued. Verse 24 follows naturally on v. 5, taking us back to the reign of Darius, in whose second year the prophets Haggai and Zechariah encouraged the building (Hag 1.1–4; 2.1–4; Zech 4.9; 6.15). **5.3:** *Tattenai* appears in cunei-

form tablets as governor of the province *Beyond the River.*

5.11: The *great king* was Solomon. **14:** *The temple of Babylon* would be that of Marduk. **17:** *Royal archives*, the *king's treasury* of 7.20; compare 6.1.

named Sheshbazzar, whom he had made governor. 15 He said to him, "Take these vessels; go and put them in the temple in Jerusalem, and let the house of God be rebuilt on its site." 16 Then this Sheshbazzar came and laid the foundations of the house of God in Jerusalem; and from that time until now it has been under construction, and it is not yet finished.' 17 And now, if it seems good to the king, have a search made in the royal archives there in Babylon, to see whether a decree was issued by King Cyrus for the rebuilding of this house of God in Jerusalem. Let the king send us his pleasure in this matter."

6 Then King Darius made a decree, and they searched the archives where the documents were stored in Babylon. 2 But it was in Ecbatana, the capital in the province of Media, that a scroll was found on which this was written: "A record. 3 In the first year of his reign, King Cyrus issued a decree: Concerning the house of God at Jerusalem, let the house be rebuilt, the place where sacrifices are offered and burnt offerings are brought;*h* its height shall be sixty cubits and its width sixty cubits, 4 with three courses of hewn stones and one course of timber; let the cost be paid from the royal treasury. 5 Moreover, let the gold and silver vessels of the house of God, which Nebuchadnezzar took out of the temple in Jerusalem and brought to Babylon, be restored and brought back to the temple in Jerusalem, each to its place; you shall put them in the house of God."

6 "Now you, Tattenai, governor of the province Beyond the River, Shethar-bozenai, and you, their associates, the envoys in the province Beyond the River, keep away; 7 let the work on this house of God alone; let the governor of the Jews and the elders of the Jews rebuild this house of God on its site. 8 Moreover I make a decree regarding what you shall do for these elders of the Jews for the rebuilding of this house of God: the cost is to be paid to these people, in full and without delay, from the royal revenue, the tribute of the province Beyond the River. 9 Whatever is needed—young bulls, rams, or sheep for burnt offerings to the God of heaven, wheat, salt, wine, or oil, as the priests in Jerusalem require—let that be given to them day by day without fail, 10 so that they may offer pleasing sacrifices to the God of heaven, and pray for the life of the king and his children. 11 Furthermore I decree that if anyone alters this edict, a beam shall be pulled out of the house of the perpetrator, who then shall be impaled on it. The house shall be made a dunghill. 12 May the God who has established his name there overthrow any king or people that shall put forth a hand to alter this, or to destroy this house of God in Jerusalem. I, Darius, make a decree; let it be done with all diligence."

13 Then, according to the word sent by King Darius, Tattenai, the governor of the province Beyond the River, Shethar-bozenai, and their associates did with all diligence what King Darius had ordered. 14 So the elders of the Jews built and prospered, through the prophesying of the prophet Haggai and Zechariah son of Iddo. They finished their building by command of the God of Israel and by decree of Cyrus, Darius, and King Artaxerxes of Persia; 15 and this house was finished on the third day of the month of Adar, in the sixth year of the reign of King Darius.

h Meaning of Aram uncertain

6.2: *Ecbatana* was the king's summer residence. **7:** *Governor* probably means Zerubbabel (see Hag 2.21).
6.9: Provision is made for the burnt offering (Lev ch 1), grain offering (Lev 2.1–13), and drink offering (Lev 23.13). **11:** The punishment for interference was impalement, on which see Esth 2.23; 9.14; Herodotus, *Hist.* III, 159. To make a house a dunghill was the limit of contempt (2 Kings 10.27; Dan 2.5). **15:** *Adar,* the twelfth month (March-April). *The sixth year . . . of King Darius,* 516 B.C.; therefore *Artaxerxes* in v. 14 is a scribal interpolation on the ground of 4.7.

16 The people of Israel, the priests and the Levites, and the rest of the returned exiles, celebrated the dedication of this house of God with joy. 17 They offered at the dedication of this house of God one hundred bulls, two hundred rams, four hundred lambs, and as a sin offering for all Israel, twelve male goats, according to the number of the tribes of Israel. 18 Then they set the priests in their divisions and the Levites in their courses for the service of God at Jerusalem, as it is written in the book of Moses.

19 On the fourteenth day of the first month the returned exiles kept the passover. 20 For both the priests and the Levites had purified themselves; all of them were clean. So they killed the passover lamb for all the returned exiles, for their fellow priests, and for themselves. 21 It was eaten by the people of Israel who had returned from exile, and also by all who had joined them and separated themselves from the pollutions of the nations of the land to worship the LORD, the God of Israel. 22 With joy they celebrated the festival of unleavened bread seven days; for the LORD had made them joyful, and had turned the heart of the king of Assyria to them, so that he aided them in the work on the house of God, the God of Israel.

7 After this, in the reign of King Artaxerxes of Persia, Ezra son of Seraiah, son of Azariah, son of Hilkiah, 2 son of Shallum, son of Zadok, son of Ahitub, 3 son of Amariah, son of Azariah, son of Meraioth, 4 son of Zerahiah, son of Uzzi, son of Bukki, 5 son of Abishua, son of Phinehas, son of Eleazar, son of the chief priest Aaron— 6 this Ezra went up from Babylonia. He was a scribe skilled in the law of Moses that the LORD the God of Israel had given; and the king

granted him all that he asked, for the hand of the LORD his God was upon him.

7 Some of the people of Israel, and some of the priests and Levites, the singers and gatekeepers, and the temple servants also went up to Jerusalem, in the seventh year of King Artaxerxes. 8 They came to Jerusalem in the fifth month, which was in the seventh year of the king. 9 On the first day of the first month the journey up from Babylon was begun, and on the first day of the fifth month he came to Jerusalem, for the gracious hand of his God was upon him. 10 For Ezra had set his heart to study the law of the LORD, and to do it, and to teach the statutes and ordinances in Israel.

11 This is a copy of the letter that King Artaxerxes gave to the priest Ezra, the scribe, a scholar of the text of the commandments of the LORD and his statutes for Israel: 12 "Artaxerxes, king of kings, to the priest Ezra, the scribe of the law of the God of heaven: Peace. *i* And now 13 I decree that any of the people of Israel or their priests or Levites in my kingdom who freely offers to go to Jerusalem may go with you. 14 For you are sent by the king and his seven counselors to make inquiries about Judah and Jerusalem according to the law of your God, which is in your hand, 15 and also to convey the silver and gold that the king and his counselors have freely offered to the God of Israel, whose dwelling is in Jerusalem, 16 with all the silver and gold that you shall find in the whole province of Babylonia, and with the freewill offerings of the people and the priests, given willingly for the house of their God in Jerusalem. 17 With this money, then, you shall with all diligence buy bulls, rams,

i Syr Vg 1 Esdras 8.9: Aram *Perfect*

6.16–18: There was a dedication of the new temple as Solomon had dedicated his (1 Kings 8.5, 63), and only then could *priests and Levites* take up their *courses*. No such division of courses is contained in the Pentateuch. **6.19–22:** Now also the festivals (e.g. *the passover* and then *unleavened bread; Deut* 16.1–8), could be celebrated by officiants rit-

ually *clean. King of Assyria,* perhaps because Persia now ruled the former Assyria.

7.1–10.16: The history of Ezra. 7: *Seventh year of King Artaxerxes* I, 458 B.C. (See Introduction).

7.12–26: The king's letter is in Aramaic. **14:** For the *seven counselors* see Esth 1.14; Herodotus, *Hist.* III, 84. **20:** The *king's*

and lambs, and their grain offerings and their drink offerings, and you shall offer them on the altar of the house of your God in Jerusalem. 18 Whatever seems good to you and your colleagues to do with the rest of the silver and gold, you may do, according to the will of your God. 19 The vessels that have been given you for the service of the house of your God, you shall deliver before the God of Jerusalem. 20 And whatever else is required for the house of your God, which you are responsible for providing, you may provide out of the king's treasury.

21 "I, King Artaxerxes, decree to all the treasurers in the province Beyond the River: Whatever the priest Ezra, the scribe of the law of the God of heaven, requires of you, let it be done with all diligence, 22 up to one hundred talents of silver, one hundred cors of wheat, one hundred baths*j* of wine, one hundred baths*j* of oil, and unlimited salt. 23 Whatever is commanded by the God of heaven, let it be done with zeal for the house of the God of heaven, or wrath will come upon the realm of the king and his heirs. 24 We also notify you that it shall not be lawful to impose tribute, custom, or toll on any of the priests, the Levites, the singers, the doorkeepers, the temple servants, or other servants of this house of God.

25 "And you, Ezra, according to the God-given wisdom you possess, appoint magistrates and judges who may judge all the people in the province Beyond the River who know the laws of your God; and you shall teach those who do not know them. 26 All who will not obey the law of your God and the law of the king, let judgment be strictly executed on them, whether for death or for banishment or for confiscation of their goods or for imprisonment."

27 Blessed be the LORD, the God of our ancestors, who put such a thing as this into the heart of the king to glorify the house of the LORD in Jerusalem, 28 and who extended to me steadfast love before the king and his counselors, and before all the king's mighty officers. I took courage, for the hand of the LORD my God was upon me, and I gathered leaders from Israel to go up with me.

8 These are their family heads, and this is the genealogy of those who went up with me from Babylonia, in the reign of King Artaxerxes: 2 Of the descendants of Phinehas, Gershom. Of Ithamar, Daniel. Of David, Hattush, 3 of the descendants of Shecaniah. Of Parosh, Zechariah, with whom were registered one hundred fifty males. 4 Of the descendants of Pahath-moab, Eliehoenai son of Zerahiah, and with him two hundred males. 5 Of the descendants of Zattu,*k* Shecaniah son of Jahaziel, and with him three hundred males. 6 Of the descendants of Adin, Ebed son of Jonathan, and with him fifty males. 7 Of the descendants of Elam, Jeshaiah son of Athaliah, and with him seventy males. 8 Of the descendants of Shephatiah, Zebadiah son of Michael, and with him eighty males. 9 Of the descendants of Joab, Obadiah son of Jehiel, and with him two hundred eighteen males. 10 Of the descendants of Bani,*l* Shelomith son of Josiphiah, and with him one hundred sixty males. 11 Of the descendants of Bebai, Zechariah son of Bebai, and with him twenty-eight males. 12 Of the descendants of Azgad, Johanan son of Hakkatan, and with him one hundred ten males. 13 Of the descendants of Adonikam, those who came later, their names being Eliphelet, Jeuel, and Shemaiah, and with them sixty males. 14 Of the descendants of Bigvai,

j A Heb measure of volume
k Gk 1 Esdras 8.32: Heb lacks *of Zattu*
l Gk 1 Esdras 8.36: Heb lacks *Bani*

---

treasury must mean the provincial treasure house, see v. 21. **22:** For the equivalent of the *talent, cor,* and *bath,* see p. 424. **24:** Temple functionaries are exempted from taxation. **25:** The *wisdom* of Ezra is that of the new law book of v. 14, which he was empowered to enforce through *magistrates and judges,* with four types of penalty.

8.1–29: A list of priestly and lay clans who returned with Ezra.

Uthai and Zaccur, and with them seventy males.

15 I gathered them by the river that runs to Ahava, and there we camped three days. As I reviewed the people and the priests, I found there none of the descendants of Levi. [16] Then I sent for Eliezer, Ariel, Shemaiah, Elnathan, Jarib, Elnathan, Nathan, Zechariah, and Meshullam, who were leaders, and for Joiarib and Elnathan, who were wise, [17] and sent them to Iddo, the leader at the place called Casiphia, telling them what to say to Iddo and his colleagues the temple servants at Casiphia, namely, to send us ministers for the house of our God. [18] Since the gracious hand of our God was upon us, they brought us a man of discretion, of the descendants of Mahli son of Levi son of Israel, namely Sherebiah, with his sons and kin, eighteen; [19] also Hashabiah and with him Jeshaiah of the descendants of Merari, with his kin and their sons, twenty; [20] besides two hundred twenty of the temple servants, whom David and his officials had set apart to attend the Levites. These were all mentioned by name.

21 Then I proclaimed a fast there, at the river Ahava, that we might deny ourselves[m] before our God, to seek from him a safe journey for ourselves, our children, and all our possessions. [22] For I was ashamed to ask the king for a band of soldiers and cavalry to protect us against the enemy on our way, since we had told the king that the hand of our God is gracious to all who seek him, but his power and his wrath are against all who forsake him. [23] So we fasted and petitioned our God for this, and he listened to our entreaty.

24 Then I set apart twelve of the leading priests: Sherebiah, Hashabiah, and ten of their kin with them. [25] And I weighed out to them the silver and the gold and the vessels, the offering for the house of our God that the king, his counselors, his lords, and all Israel there present had offered; [26] I weighed out into their hand six hundred fifty talents of silver, and one hundred silver vessels worth . . . talents,[n] and one hundred talents of gold, [27] twenty gold bowls worth a thousand darics, and two vessels of fine polished bronze as precious as gold. [28] And I said to them, "You are holy to the LORD, and the vessels are holy; and the silver and the gold are a freewill offering to the LORD, the God of your ancestors. [29] Guard them and keep them until you weigh them before the chief priests and the Levites and the heads of families in Israel at Jerusalem, within the chambers of the house of the LORD." [30] So the priests and the Levites took over the silver, the gold, and the vessels as they were weighed out, to bring them to Jerusalem, to the house of our God.

31 Then we left the river Ahava on the twelfth day of the first month, to go to Jerusalem; the hand of our God was upon us, and he delivered us from the hand of the enemy and from ambushes along the way. [32] We came to Jerusalem and remained there three days. [33] On the fourth day, within the house of our God, the silver, the gold, and the vessels were weighed into the hands of the priest Meremoth son of Uriah, and with him was Eleazar son of Phinehas, and with them were the Levites, Jozabad son of Jeshua and Noadiah son of Binnui. [34] The total was counted and weighed, and the weight of everything was recorded.

35 At that time those who had come from captivity, the returned exiles, offered burnt offerings to the God of Israel, twelve bulls for all Israel, ninety-six rams, seventy-seven lambs, and as a sin offering twelve male goats; all this was a

---

m Or *might fast*   *n* The number of talents is lacking

8.15: The location of *Ahava* is unknown, but *the river* was doubtless a tributary of the Euphrates. The dearth of Levites parallels that of 2.40. **17:** *Casiphia* is unidentified.
8.21, 23: For preparatory fasting see Esth 4.16; 2 Chron 20.3; Jer 36.9. **28:** Only *holy* persons should handle *holy* things.
8.29: The *chambers* were the temple storehouses (10.6; Neh 10.37; 13.4–7).
9.1–15: The problem of mixed mar-

burnt offering to the LORD. [36] They also delivered the king's commissions to the king's satraps and to the governors of the province Beyond the River; and they supported the people and the house of God.

9 After these things had been done, the officials approached me and said, "The people of Israel, the priests, and the Levites have not separated themselves from the peoples of the lands with their abominations, from the Canaanites, the Hittites, the Perizzites, the Jebusites, the Ammonites, the Moabites, the Egyptians, and the Amorites. [2] For they have taken some of their daughters as wives for themselves and for their sons. Thus the holy seed has mixed itself with the peoples of the lands, and in this faithlessness the officials and leaders have led the way." [3] When I heard this, I tore my garment and my mantle, and pulled hair from my head and beard, and sat appalled. [4] Then all who trembled at the words of the God of Israel, because of the faithlessness of the returned exiles, gathered around me while I sat appalled until the evening sacrifice.

5 At the evening sacrifice I got up from my fasting, with my garments and my mantle torn, and fell on my knees, spread out my hands to the LORD my God, [6] and said,

"O my God, I am too ashamed and embarrassed to lift my face to you, my God, for our iniquities have risen higher than our heads, and our guilt has mounted up to the heavens. [7] From the days of our ancestors to this day we have been deep in guilt, and for our iniquities we, our kings, and our priests have been handed over to the kings of the lands, to the sword, to captivity, to plundering, and to utter shame, as is now the case. [8] But now for a brief moment favor has been shown by the LORD our God, who has left us a remnant, and given us a stake in his holy place, in order that he[o] may brighten our eyes and grant us a little sustenance in our slavery. [9] For we are slaves; yet our God has not forsaken us in our slavery, but has extended to us his steadfast love before the kings of Persia, to give us new life to set up the house of our God, to repair its ruins, and to give us a wall in Judea and Jerusalem.

10 "And now, our God, what shall we say after this? For we have forsaken your commandments, [11] which you commanded by your servants the prophets, saying, 'The land that you are entering to possess is a land unclean with the pollutions of the peoples of the lands, with their abominations. They have filled it from end to end with their uncleanness. [12] Therefore do not give your daughters to their sons, neither take their daughters for your sons, and never seek their peace or prosperity, so that you may be strong and eat the good of the land and leave it for an inheritance to your children forever.' [13] After all that has come upon us for our evil deeds and for our great guilt, seeing that you, our God, have punished us less than our iniquities deserved and have given us such a remnant as this, [14] shall we break your commandments again and intermarry with the peoples who practice these abominations? Would you not be angry with us until you destroy us without remnant or survivor? [15] O LORD, God of

o Heb *our God*

---

riages. The older codes had not specially discouraged intermarriages, but experience during the exile, which had developed the principle of exclusiveness, made this contamination of the *holy seed* a major problem for those who returned under Zerubbabel (4.3; 6.21), under Nehemiah (Neh 10.28–30; 13.3, 23–30), and now under Ezra, who found priests, Levites, and chief officials involved therein. **1**: This list of peoples is only a formula (Deut 7.1; Neh 13.23). **2**: *Holy seed,* because as dedicated to God they were separated (Ex 19.5, 6; Isa 62.12).

9.5–15: Ezra's confession. 7: *To this day,* for they were still subject to Persia (v. 9). **8**: *Remnant,* see vv. 13–15; Zech 8.12; Isa 28.5. **9**: *Kings,* i.e. Cyrus, Darius, Xerxes, and Artaxerxes. **11–12**: No such passage occurs in extant prophetic writings, but see Lev 18.24–30 and Deut 7.3.

Israel, you are just, but we have escaped as a remnant, as is now the case. Here we are before you in our guilt, though no one can face you because of this."

10 While Ezra prayed and made confession, weeping and throwing himself down before the house of God, a very great assembly of men, women, and children gathered to him out of Israel; the people also wept bitterly. 2 Shecaniah son of Jehiel, of the descendants of Elam, addressed Ezra, saying, "We have broken faith with our God and have married foreign women from the peoples of the land, but even now there is hope for Israel in spite of this. 3 So now let us make a covenant with our God to send away all these wives and their children, according to the counsel of my lord and of those who tremble at the commandment of our God; and let it be done according to the law. 4 Take action, for it is your duty, and we are with you; be strong, and do it." 5 Then Ezra stood up and made the leading priests, the Levites, and all Israel swear that they would do as had been said. So they swore.

6 Then Ezra withdrew from before the house of God, and went to the chamber of Jehohanan son of Eliashib, where he spent the night. *p* He did not eat bread or drink water, for he was mourning over the faithlessness of the exiles. 7 They made a proclamation throughout Judah and Jerusalem to all the returned exiles that they should assemble at Jerusalem, 8 and that if any did not come within three days, by order of the officials and the elders all their property should be forfeited, and they themselves banned from the congregation of the exiles.

9 Then all the people of Judah and Benjamin assembled at Jerusalem within the three days; it was the ninth month, on the twentieth day of the month. All the people sat in the open square before the house of God, trembling because of this matter and because of the heavy rain. 10 Then Ezra the priest stood up and said to them, "You have trespassed and married foreign women, and so increased the guilt of Israel. 11 Now make confession to the LORD the God of your ancestors, and do his will; separate yourselves from the peoples of the land and from the foreign wives." 12 Then all the assembly answered with a loud voice, "It is so; we must do as you have said. 13 But the people are many, and it is a time of heavy rain; we cannot stand in the open. Nor is this a task for one day or for two, for many of us have transgressed in this matter. 14 Let our officials represent the whole assembly, and let all in our towns who have taken foreign wives come at appointed times, and with them the elders and judges of every town, until the fierce wrath of our God on this account is averted from us." 15 Only Jonathan son of Asahel and Jahzeiah son of Tikvah opposed this, and Meshullam and Shabbethai the Levites supported them.

16 Then the returned exiles did so. Ezra the priest selected men, *q* heads of families, according to their families, each of them designated by name. On the first day of the tenth month they sat down to examine the matter. 17 By the first day of the first month they had come to the end

p 1 Esdras 9.2: Heb *where he went*
q 1 Esdra 9.16: Syr: Heb *And there were selected Ezra,*

10.1–44: Repentance of the people and divorce of gentile wives. 2: No extant law required divorce in such cases, so *Shechaniah* proposes a way out, a *hope* by a *covenant* (v. 3), suggesting by the words he uses for *married* and *foreign women* that these had not been true marriages. **3:** The *law* would be that of Deut 7.3. **4:** *It is your duty,* for the covenant oath must be administered by Ezra.
10.6: *Jehohanan* was grandson of Eliashib the high priest, and became himself high priest under Darius II. **9:** *Heavy rain* is usual in Chislev, *the ninth month* (November-December); so *the assembly* asks for local enforcement of the covenant.
10.16–17: The work commenced in Tebet (*the tenth month,* December-January) and was completed in Nisan (*the first month,* March-April).
10.18–44: Lists, which were probably derived from official archives. **19:** The *guilt offering* signified that the wrong had been righted.

of all the men who had married foreign women.

18 There were found of the descendants of the priests who had married foreign women, of the descendants of Jeshua son of Jozadak and his brothers: Maaseiah, Eliezer, Jarib, and Gedaliah. [19]They pledged themselves to send away their wives, and their guilt offering was a ram of the flock for their guilt. [20]Of the descendants of Immer: Hanani and Zebadiah. [21]Of the descendants of Harim: Maaseiah, Elijah, Shemaiah, Jehiel, and Uzziah. [22]Of the descendants of Pashhur: Elioenai, Maaseiah, Ishmael, Nethanel, Jozabad, and Elasah.

23 Of the Levites: Jozabad, Shimei, Kelaiah (that is, Kelita), Pethahiah, Judah, and Eliezer. [24]Of the singers: Eliashib. Of the gatekeepers: Shallum, Telem, and Uri.

25 And of Israel: of the descendants of Parosh: Ramiah, Izziah, Malchijah, Mijamin, Eleazar, Hashabiah,[r] and Benaiah. [26]Of the descendants of Elam: Mattaniah, Zechariah, Jehiel, Abdi, Jeremoth, and Elijah. [27]Of the descendants of Zattu: Elioenai, Eliashib, Mattaniah, Jeremoth, Zabad, and Aziza. [28]Of the descendants of Bebai: Jehohanan, Hananiah, Zabbai, and Athlai. [29]Of the descendants of Bani: Meshullam, Malluch, Adaiah, Jashub, Sheal, and Jeremoth. [30]Of the descendants of Pahathmoab: Adna, Chelal, Benaiah, Maaseiah, Mattaniah, Bezalel, Binnui, and Manasseh. [31]Of the descendants of Harim: Eliezer, Isshijah, Malchijah, Shemaiah, Shimeon, [32]Benjamin, Malluch, and Shemariah. [33]Of the descendants of Hashum: Mattenai, Mattattah, Zabad, Eliphelet, Jeremai, Manasseh, and Shimei. [34]Of the descendants of Bani: Maadai, Amram, Uel, [35]Benaiah, Bedeiah, Cheluhi, [36]Vaniah, Meremoth, Eliashib, [37]Mattaniah, Mattenai, and Jaasu. [38]Of the descendants of Binnui:[s] Shimei, [39]Shelemiah, Nathan, Adaiah, [40]Machnadebai, Shashai, Sharai, [41]Azarel, Shelemiah, Shemariah, [42]Shallum, Amariah, and Joseph. [43]Of the descendants of Nebo: Jeiel, Mattithiah, Zabad, Zebina, Jaddai, Joel, and Benaiah. [44]All these had married foreign women, and they sent them away with their children.[t]

r 1 Esdras 9.26 Gk: Heb *Malchijah*   s Gk: Heb *Bani, Binnui*   t 1 Esdras 9.36; Meaning of Heb uncertain

Nehemiah

Since the books of Ezra and Nehemiah are really one book and since parts of Nehemiah are apparently misplaced and belong in Ezra, problems of origin, structure, and historical background are discussed in the Introduction to Ezra. It is sufficient to say here that the book which we call Nehemiah relates the return of Nehemiah for two periods of governorship over Judah during the reign of Artaxerxes I (464–423 B.C.). The first return was in 445/444. Nehemiah rebuilt the walls of Jerusalem and instituted social and religious reforms. An outstanding aspect of Nehemiah's religious life was his dependence on God and his frequent prayers (1.4–10; 2.18; 4.4, 20; 5.9, 19; 6.14; 13.14, 31).

1 The words of Nehemiah son of Hacaliah. In the month of Chislev, in the twentieth year, while I was in Susa the capital, ²one of my brothers, Hanani, came with certain men from Judah; and I asked them about the Jews that survived, those who had escaped the captivity, and about Jerusalem. ³They replied, "The survivors there in the province who escaped captivity are in great trouble and shame; the wall of Jerusalem is broken down, and its gates have been destroyed by fire."

4 When I heard these words I sat down and wept, and mourned for days, fasting and praying before the God of heaven. ⁵I said, "O LORD God of heaven, the great and awesome God who keeps covenant and steadfast love with those who love him and keep his commandments; ⁶let your ear be attentive and your eyes open to hear the prayer of your servant that I now pray before you day and night for your servants, the people of Israel, confessing the sins of the people of Israel, which we have sinned against you. Both I and my family have sinned. ⁷We have offended you deeply, failing to keep the commandments, the statutes, and the ordinances that you commanded your servant Moses. ⁸Remember the word that you commanded your servant Moses, 'If you are unfaithful, I will scatter you among the peoples; ⁹but if you return to me and keep my commandments and do them, though your outcasts are under the farthest skies, I will gather them from there and bring them to the place at which I have chosen to establish my name.' ¹⁰They are your servants and your people, whom you redeemed by your great power and your strong hand. ¹¹O Lord, let your ear be attentive to the prayer of your servant, and to the prayer of your servants who delight in revering your name. Give success to your servant today, and grant him mercy in the sight of this man!"

At the time, I was cupbearer to the king.

2 In the month of Nisan, in the twentieth year of King Artaxerxes, when wine was served him, I carried the wine and gave it to the king. Now, I had never been sad in his presence before. ²So the king said to me, "Why is your face sad, since you are not sick? This can only be sadness of the heart." Then I was very much afraid. ³I said to the king, "May the king live forever! Why should my face not be sad, when the city, the place of my ancestors' graves, lies waste, and its gates have been destroyed by fire?" ⁴Then the king said to me, "What do you request?" So I prayed to the God of heaven. ⁵Then I said to the king, "If it pleases the king, and if your servant has found favor with you, I ask that you send me to Judah, to the city of my ancestors' graves, so that I may rebuild it." ⁶The king said to me (the queen also was sitting beside him), "How long will you be gone, and when will you return?" So it pleased the king to send me, and I set him a date. ⁷Then I said to the king, "If it pleases the king, let letters be given me to the governors of the province Beyond the River, that they may grant me passage until I arrive in Judah; ⁸and a letter to Asaph, the keeper of the king's forest,

---

1.1–7.5: **Nehemiah's memoirs. 1.1–11: Report from Jerusalem and Nehemiah's prayer.** Nehemiah was a palace servant of Artaxerxes I at *Susa* in Elam, winter residence of the Persian kings (Esth 1.2, 5; Dan 8.2). **1.1:** *Chislev,* November-December. The *twentieth year* of Artaxerxes was 445–444 B.C. The date here is problematic, as it is later than that given in 2.1. Perhaps it should be the nineteenth year. **2:** The *men from Judah* seem to have been visitors to the *capital*. **1.4:** *Mourned,* compare 9.3–5; Ezra 10.6. **6:** He makes confession for the people just as Ezra does. **8–9:** There is no such passage in our Pentateuch, but see Deut 30.1–5. The *place* is Jerusalem.

1.11: As *this man* means Artaxerxes, this verse belongs more appropriately after 2.4. *Cupbearer,* i.e. butler, who sampled and poured wine for the king, an honorable and privileged office.

2.1–20: Nehemiah's mission. 5: *Graves* was a polite understatement, for his plan was to restore the walls (vv. 8, 12–15).

directing him to give me timber to make beams for the gates of the temple fortress, and for the wall of the city, and for the house that I shall occupy." And the king granted me what I asked, for the gracious hand of my God was upon me.

9 Then I came to the governors of the province Beyond the River, and gave them the king's letters. Now the king had sent officers of the army and cavalry with me. 10 When Sanballat the Horonite and Tobiah the Ammonite official heard this, it displeased them greatly that someone had come to seek the welfare of the people of Israel.

11 So I came to Jerusalem and was there for three days. 12 Then I got up during the night, I and a few men with me; I told no one what my God had put into my heart to do for Jerusalem. The only animal I took was the animal I rode. 13 I went out by night by the Valley Gate past the Dragon's Spring and to the Dung Gate, and I inspected the walls of Jerusalem that had been broken down and its gates that had been destroyed by fire. 14 Then I went on to the Fountain Gate and to the King's Pool; but there was no place for the animal I was riding to continue. 15 So I went up by way of the valley by night and inspected the wall. Then I turned back and entered by the Valley Gate, and so returned. 16 The officials did not know where I had gone or what I was doing; I had not yet told the Jews, the priests, the nobles, the officials, and the rest that were to do the work.

17 Then I said to them, "You see the trouble we are in, how Jerusalem lies in ruins with its gates burned. Come, let us rebuild the wall of Jerusalem, so that we may no longer suffer disgrace." 18 I told

them that the hand of my God had been gracious upon me, and also the words that the king had spoken to me. Then they said, "Let us start building!" So they committed themselves to the common good. 19 But when Sanballat the Horonite and Tobiah the Ammonite official, and Geshem the Arab heard of it, they mocked and ridiculed us, saying, "What is this that you are doing? Are you rebelling against the king?" 20 Then I replied to them, "The God of heaven is the one who will give us success, and we his servants are going to start building; but you have no share or claim or historic right in Jerusalem."

3 Then the high priest Eliashib set to work with his fellow priests and rebuilt the Sheep Gate. They consecrated it and set up its doors; they consecrated it as far as the Tower of the Hundred and as far as the Tower of Hananel. 2 And the men of Jericho built next to him. And next to them *a* Zaccur son of Imri built.

3 The sons of Hassenaah built the Fish Gate; they laid its beams and set up its doors, its bolts, and its bars. 4 Next to them Meremoth son of Uriah son of Hakkoz made repairs. Next to them Meshullam son of Berechiah son of Meshezabel made repairs. Next to them Zadok son of Baana made repairs. 5 Next to them the Tekoites made repairs; but their nobles would not put their shoulders to the work of their Lord. *b*

6 Joiada son of Paseah and Meshullam son of Besodeiah repaired the Old Gate; they laid its beams and set up its doors, its bolts, and its bars. 7 Next to them repairs were made by Melatiah the Gibe-

a Heb *him*   *b* Or *lords*

2.8: *Fortress*, see 7.2; 1 Macc 13.52; Acts 21.37. 10: As local leaders had hindered the work under Zerubbabel (Ezra 4.3–24), so they try again. *Sanballat* is mentioned in the Elephantine papyri as governor of Samaria. Tobiah was apparently an Ammonite official in Persian service.
2.12–16: A secret inspection of the ruined walls by night, making a complete circuit of the walls, beginning and ending at the *Valley*

Gate. 19: His enemies suspect the building of the wall was a prelude to rebellion. *Geshem* (6.6) was king of Kedar.
3.1–32: **Work on the wall.** 1: *Eliashib* (12.22; 13.4) was grandson of that Jeshua who worked with Zerubbabel (12.10), and grandfather of the later high priest Jehohanan.
3.5: *The Tekoites* (Am 1.1) were not supported by their upper class families.

onite and Jadon the Meronothite—the men of Gibeon and of Mizpah—who were under the jurisdiction of*c* the governor of the province Beyond the River. [8]Next to them Uzziel son of Harhaiah, one of the goldsmiths, made repairs. Next to him Hananiah, one of the perfumers, made repairs; and they restored Jerusalem as far as the Broad Wall. [9]Next to them Rephaiah son of Hur, ruler of half the district of*d* Jerusalem, made repairs. [10]Next to them Jedaiah son of Harumaph made repairs opposite his house; and next to him Hattush son of Hashabneiah made repairs. [11]Malchijah son of Harim and Hasshub son of Pahath-moab repaired another section and the Tower of the Ovens. [12]Next to him Shallum son of Hallohesh, ruler of half the district of*d* Jerusalem, made repairs, he and his daughters.

13 Hanun and the inhabitants of Zanoah repaired the Valley Gate; they rebuilt it and set up its doors, its bolts, and its bars, and repaired a thousand cubits of the wall, as far as the Dung Gate.

14 Malchijah son of Rechab, ruler of the district of*e* Beth-haccherem, repaired the Dung Gate; he rebuilt it and set up its doors, its bolts, and its bars.

15 And Shallum son of Col-hozeh, ruler of the district of*e* Mizpah, repaired the Fountain Gate; he rebuilt it and covered it and set up its doors, its bolts, and its bars; and he built the wall of the Pool of Shelah of the king's garden, as far as the stairs that go down from the City of David. [16]After him Nehemiah son of Azbuk, ruler of half the district of*d* Bethzur, repaired from a point opposite the graves of David, as far as the artificial pool and the house of the warriors. [17]After him the Levites made repairs: Rehum son of Bani; next to him Hashabiah, ruler of half the district of*d* Keilah, made repairs for his district. [18]After him their kin made repairs: Binnui,*f* son of Henadad, ruler of half the district of*d* Keilah; [19]next to him Ezer son of Jeshua, ruler*g*

of Mizpah, repaired another section opposite the ascent to the armory at the Angle. [20]After him Baruch son of Zabbai repaired another section from the Angle to the door of the house of the high priest Eliashib. [21]After him Meremoth son of Uriah son of Hakkoz repaired another section from the door of the house of Eliashib to the end of the house of Eliashib. [22]After him the priests, the men of the surrounding area, made repairs. [23]After them Benjamin and Hasshub made repairs opposite their house. After them Azariah son of Maaseiah son of Ananiah made repairs beside his own house. [24]After him Binnui son of Henadad repaired another section, from the house of Azariah to the Angle and to the corner. [25]Palal son of Uzai repaired opposite the Angle and the tower projecting from the upper house of the king at the court of the guard. After him Pedaiah son of Parosh [26]and the temple servants living*h* on Ophel made repairs up to a point opposite the Water Gate on the east and the projecting tower. [27]After him the Tekoites repaired another section opposite the great projecting tower as far as the wall of Ophel.

28 Above the Horse Gate the priests made repairs, each one opposite his own house. [29]After them Zadok son of Immer made repairs opposite his own house. After him Shemaiah son of Shecaniah, the keeper of the East Gate, made repairs. [30]After him Hananiah son of Shelemiah and Hanun sixth son of Zalaph repaired another section. After him Meshullam son of Berechiah made repairs opposite his living quarters. [31]After him Malchijah, one of the goldsmiths, made repairs as far as the house of the temple servants and of the merchants, opposite the Muster Gate,*i* and to the

---

c Meaning of Heb uncertain   *d* Or *supervisor of half the portion assigned to*   *e* Or *supervisor of the portion assigned to*   *f* Gk Syr Compare verse 24, 10.9: Heb *Bavvai*   *g* Or *supervisor*   *h* Cn: Heb *were living*   *i* Or *Hammiphkad Gate*

3.15: *City of David,* the old Jebusite town on the hill (2 Sam 5.6–9). **16:** *Graves of David,*

2 Chr 32.33. The *house of the warriors* was the barracks.

upper room of the corner. [32] And between the upper room of the corner and the Sheep Gate the goldsmiths and the merchants made repairs.

4 [j] Now when Sanballat heard that we were building the wall, he was angry and greatly enraged, and he mocked the Jews. [2] He said in the presence of his associates and of the army of Samaria, "What are these feeble Jews doing? Will they restore things? Will they sacrifice? Will they finish it in a day? Will they revive the stones out of the heaps of rubbish—and burned ones at that?" [3] Tobiah the Ammonite was beside him, and he said, "That stone wall they are building—any fox going up on it would break it down!" [4] Hear, O our God, for we are despised; turn their taunt back on their own heads, and give them over as plunder in a land of captivity. [5] Do not cover their guilt, and do not let their sin be blotted out from your sight; for they have hurled insults in the face of the builders.

6 So we rebuilt the wall, and all the wall was joined together to half its height; for the people had a mind to work.

7 [k] But when Sanballat and Tobiah and the Arabs and the Ammonites and the Ashdodites heard that the repairing of the walls of Jerusalem was going forward and the gaps were beginning to be closed, they were very angry, [8] and all plotted together to come and fight against Jerusalem and to cause confusion in it. [9] So we prayed to our God, and set a guard as a protection against them day and night.

10 But Judah said, "The strength of the burden bearers is failing, and there is too much rubbish so that we are unable to work on the wall." [11] And our enemies said, "They will not know or see anything before we come upon them and kill them and stop the work." [12] When the Jews who lived near them came, they said to us ten times, "From all the places where they live[l] they will come up against us."[m] [13] So in the lowest parts of the space behind the wall, in open places, I stationed the people according to their families,[n] with their swords, their spears, and their bows. [14] After I looked these things over, I stood up and said to the nobles and the officials and the rest of the people, "Do not be afraid of them. Remember the LORD, who is great and awesome, and fight for your kin, your sons, your daughters, your wives, and your homes."

15 When our enemies heard that their plot was known to us, and that God had frustrated it, we all returned to the wall, each to his work. [16] From that day on, half of my servants worked on construction, and half held the spears, shields, bows, and body-armor; and the leaders posted themselves behind the whole house of Judah, [17] who were building the wall. The burden bearers carried their loads in such a way that each labored on the work with one hand and with the other held a weapon. [18] And each of the builders had his sword strapped at his side while he built. The man who sounded the trumpet was beside me. [19] And I said to the nobles, the officials, and the rest of the people, "The work is great and widely spread out, and we are separated far from one another on the wall. [20] Rally to us wherever you hear the sound of the trumpet. Our God will fight for us."

21 So we labored at the work, and half of them held the spears from break of dawn until the stars came out. [22] I also said to the people at that time, "Let every man and his servant pass the night inside Jerusalem, so that they may be a guard

j Ch 3.33 in Heb   *k* Ch 4.1 in Heb
l Cn: Heb *you return*   *m* Compare Gk Syr:
Meaning of Heb uncertain   *n* Meaning of
Heb uncertain

4.1–7.5: Troubles for the builders. 4.1– 9: Trouble from Sanballat and his associates. **4–5**: Words of ill omen must be countered by words of ill omen. **10–14**: To trouble from without was added trouble from within. **16**: By *my servants* he would mean his personal, better armed, bodyguard. **21**: Night work was hardly possible in those days.

for us by night and may labor by day."
23 So neither I nor my brothers nor my servants nor the men of the guard who followed me ever took off our clothes; each kept his weapon in his right hand. *o*

5 Now there was a great outcry of the people and of their wives against their Jewish kin. 2 For there were those who said, "With our sons and our daughters, we are many; we must get grain, so that we may eat and stay alive." 3 There were also those who said, "We are having to pledge our fields, our vineyards, and our houses in order to get grain during the famine." 4 And there were those who said, "We are having to borrow money on our fields and vineyards to pay the king's tax. 5 Now our flesh is the same as that of our kindred; our children are the same as their children; and yet we are forcing our sons and daughters to be slaves, and some of our daughters have been ravished; we are powerless, and our fields and vineyards now belong to others."

6 I was very angry when I heard their outcry and these complaints. 7 After thinking it over, I brought charges against the nobles and the officials; I said to them, "You are all taking interest from your own people." And I called a great assembly to deal with them, 8 and said to them, "As far as we were able, we have bought back our Jewish kindred who had been sold to other nations; but now you are selling your own kin, who must then be bought back by us!" They were silent, and could not find a word to say. 9 So I said, "The thing that you are doing is not good. Should you not walk in the fear of our God, to prevent the taunts of the nations our enemies? 10 Moreover I and my brothers and my servants are lending them money and grain. Let us stop this taking of interest. 11 Restore to them, this very day, their

fields, their vineyards, their olive orchards, and their houses, and the interest on money, grain, wine, and oil that you have been exacting from them." 12 Then they said, "We will restore everything and demand nothing more from them. We will do as you say." And I called the priests, and made them take an oath to do as they had promised. 13 I also shook out the fold of my garment and said, "So may God shake out everyone from house and from property who does not perform this promise. Thus may they be shaken out and emptied." And all the assembly said, "Amen," and praised the LORD. And the people did as they had promised.

14 Moreover from the time that I was appointed to be their governor in the land of Judah, from the twentieth year to the thirty-second year of King Artaxerxes, twelve years, neither I nor my brothers ate the food allowance of the governor. 15 The former governors who were before me laid heavy burdens on the people, and took food and wine from them, besides forty shekels of silver. Even their servants lorded it over the people. But I did not do so, because of the fear of God. 16 Indeed, I devoted myself to the work on this wall, and acquired no land; and all my servants were gathered there for the work. 17 Moreover there were at my table one hundred fifty people, Jews and officials, besides those who came to us from the nations around us. 18 Now that which was prepared for one day was one ox and six choice sheep; also fowls were prepared for me, and every ten days skins of wine in abundance; yet with all this I did not demand the food allowance of the governor, because of the heavy burden of labor on the people. 19 Re-

o Cn: Heb *each his weapon the water*

5.1–13: Economic ills. Concentration on the walls had led to economic crisis. 3: *Pledge*, i.e. giving as security. 4: *The king's tax* was the tribute due to Persia.

5.7–10: It was legal to take *interest* from non-Jews but not from fellow Jews (Deut

23.20). 11: The *interest* (literally one hundredth) is usually taken as meaning one percent monthly. 13: *Fold* refers to that part of the outer garment which served as a pocket.

5.14–19: Nehemiah's apologia for his *twelve years* as *governor*.

member for my good, O my God, all that I have done for this people.

6 Now when it was reported to Sanballat and Tobiah and to Geshem the Arab and to the rest of our enemies that I had built the wall and that there was no gap left in it (though up to that time I had not set up the doors in the gates), 2 Sanballat and Geshem sent to me, saying, "Come and let us meet together in one of the villages in the plain of Ono." But they intended to do me harm. 3 So I sent messengers to them, saying, "I am doing a great work and I cannot come down. Why should the work stop while I leave it to come down to you?" 4 They sent to me four times in this way, and I answered them in the same manner. 5 In the same way Sanballat for the fifth time sent his servant to me with an open letter in his hand. 6 In it was written, "It is reported among the nations—and Geshem p also says it—that you and the Jews intend to rebel; that is why you are building the wall; and according to this report you wish to become their king. 7 You have also set up prophets to proclaim in Jerusalem concerning you, 'There is a king in Judah!' And now it will be reported to the king according to these words. So come, therefore, and let us confer together." 8 Then I sent to him, saying, "No such things as you say have been done; you are inventing them out of your own mind" 9—for they all wanted to frighten us, thinking, "Their hands will drop from the work, and it will not be done." But now, O God, strengthen my hands.

10 One day when I went into the house of Shemaiah son of Delaiah son of Mehetabel, who was confined to his house, he said, "Let us meet together in the house of God, within the temple, and let us close the doors of the temple, for they are coming to kill you; indeed, tonight they are coming to kill you." 11 But I said, "Should a man like me run away? Would a man like me go into the temple to save his life? I will not go in!" 12 Then I perceived and saw that God had not sent him at all, but he had pronounced the prophecy against me because Tobiah and Sanballat had hired him. 13 He was hired for this purpose, to intimidate me and make me sin by acting in this way, and so they could give me a bad name, in order to taunt me. 14 Remember Tobiah and Sanballat, O my God, according to these things that they did, and also the prophetess Noadiah and the rest of the prophets who wanted to make me afraid.

15 So the wall was finished on the twenty-fifth day of the month Elul, in fifty-two days. 16 And when all our enemies heard of it, all the nations around us were afraid q and fell greatly in their own esteem; for they perceived that this work had been accomplished with the help of our God. 17 Moreover in those days the nobles of Judah sent many letters to Tobiah, and Tobiah's letters came to them. 18 For many in Judah were bound by oath to him, because he was the son-in-law of Shecaniah son of Arah: and his son Jehohanan had married the daughter of Meshullam son of Berechiah. 19 Also they spoke of his good deeds in my presence, and reported my words to him. And Tobiah sent letters to intimidate me.

7 Now when the wall had been built and I had set up the doors, and the gatekeepers, the singers, and the Levites had been appointed, 2 I gave my brother Hanani charge over Jerusalem, along with Hananiah the commander of the citadel—for he was a faithful man and feared God more than many. 3 And I said

p Heb Gashmu   q Another reading is saw

6.1–19: Finishing the wall, in spite of plots. 2: They try to lure Nehemiah to Ono near Lydda to harm him. 5–7: Failing this, they try to frighten him by threatening to report him to the king, mentioning prophets because they were known as fomenters of rebellion (Jer 28.1–4). 6.10: Shemaiah is the enemy within the camp. 13: That prophets could be hired illustrates Zech 13.2–6. 15: Elul was the sixth month (August–September).
7.1–5: Setting up the guards. 2: Hanani, see 1.2. The citadel is the fortress (2.8). 3: The gates are to be open from sunrise to sunset. At

to them, "The gates of Jerusalem are not to be opened until the sun is hot; while the gatekeepers*r* are still standing guard, let them shut and bar the doors. Appoint guards from among the inhabitants of Jerusalem, some at their watch posts, and others before their own houses." 4 The city was wide and large, but the people within it were few and no houses had been built.

5 Then my God put it into my mind to assemble the nobles and the officials and the people to be enrolled by genealogy. And I found the book of the genealogy of those who were the first to come back, and I found the following written in it:

6 These are the people of the province who came up out of the captivity of those exiles whom King Nebuchadnezzar of Babylon had carried into exile; they returned to Jerusalem and Judah, each to his town. 7 They came with Zerubbabel, Jeshua, Nehemiah, Azariah, Raamiah, Nahamani, Mordecai, Bilshan, Mispereth, Bigvai, Nehum, Baanah.

The number of the Israelite people: 8 the descendants of Parosh, two thousand one hundred seventy-two. 9 Of Shephatiah, three hundred seventy-two. 10 Of Arah, six hundred fifty-two. 11 Of Pahath-moab, namely the descendants of Jeshua and Joab, two thousand eight hundred eighteen. 12 Of Elam, one thousand two hundred fifty-four. 13 Of Zattu, eight hundred forty-five. 14 Of Zaccai, seven hundred sixty. 15 Of Binnui, six hundred forty-eight. 16 Of Bebai, six hundred twenty-eight. 17 Of Azgad, two thousand three hundred twenty-two. 18 Of Adonikam, six hundred sixty-seven. 19 Of Bigvai, two thousand sixty-seven. 20 Of Adin, six hundred fifty-five. 21 Of Ater, namely of Hezekiah, ninety-eight. 22 Of Hashum, three hundred twenty-eight. 23 Of Bezai, three hundred twenty-four. 24 Of Hariph, one hundred twelve. 25 Of Gibeon, ninety-five. 26 The people of Bethlehem and Netophah, one hundred eighty-eight. 27 Of Anathoth, one hundred twenty-eight. 28 Of Beth-azmaveth, forty-two. 29 Of Kiriath-jearim, Chephirah, and Beeroth, seven hundred forty-three. 30 Of Ramah and Geba, six hundred twenty-one. 31 Of Michmas, one hundred twenty-two. 32 Of Bethel and Ai, one hundred twenty-three. 33 Of the other Nebo, fifty-two. 34 The descendants of the other Elam, one thousand two hundred fifty-four. 35 Of Harim, three hundred twenty. 36 Of Jericho, three hundred forty-five. 37 Of Lod, Hadid, and Ono, seven hundred twenty-one. 38 Of Senaah, three thousand nine hundred thirty.

39 The priests: the descendants of Jedaiah, namely the house of Jeshua, nine hundred seventy-three. 40 Of Immer, one thousand fifty-two. 41 Of Pashhur, one thousand two hundred forty-seven. 42 Of Harim, one thousand seventeen.

43 The Levites: the descendants of Jeshua, namely of Kadmiel of the descendants of Hodevah, seventy-four. 44 The singers: the descendants of Asaph, one hundred forty-eight. 45 The gatekeepers: the descendants of Shallum, of Ater, of Talmon, of Akkub, of Hatita, of Shobai, one hundred thirty-eight.

46 The temple servants: the descendants of Ziha, of Hasupha, of Tabbaoth, 47 of Keros, of Sia, of Padon, 48 of Lebana, of Hagaba, of Shalmai, 49 of Hanan, of Giddel, of Gahar, 50 of Reaiah, of Rezin, of Nekoda, 51 of Gazzam, of Uzza, of Paseah, 52 of Besai, of Meunim, of Nephushesim, 53 of Bakbuk, of Hakupha, of Harhur, 54 of Bazlith, of Mehida, of Harsha, 55 of Barkos, of Sisera, of Temah, 56 of Neziah, of Hatipha.

57 The descendants of Solomon's servants: of Sotai, of Sophereth, of Perida,

r Heb *while they*

v. 5a the memoirs of Nehemiah break off to be picked up again after chapter 10. Verse 5b is a transition verse to introduce the genealogies of vv. 6–73a which parallel Ezra 2.1–70.

7.6–73: **A census of the first return.** See Ezra 2.1–70 n. **73b:** A connecting verse based on Ezra 3.1.

58 of Jaala, of Darkon, of Giddel, 59 of Shephatiah, of Hattil, of Pochereth-hazzebaim, of Amon.

60 All the temple servants and the descendants of Solomon's servants were three hundred ninety-two.

61 The following were those who came up from Tel-melah, Tel-harsha, Cherub, Addon, and Immer, but they could not prove their ancestral houses or their descent, whether they belonged to Israel: 62 the descendants of Delaiah, of Tobiah, of Nekoda, six hundred forty-two. 63 Also, of the priests: the descendants of Hobaiah, of Hakkoz, of Barzillai (who had married one of the daughters of Barzillai the Gileadite and was called by their name). 64 These sought their registration among those enrolled in the genealogies, but it was not found there, so they were excluded from the priesthood as unclean; 65 the governor told them that they were not to partake of the most holy food, until a priest with Urim and Thummim should come.

66 The whole assembly together was forty-two thousand three hundred sixty, 67 besides their male and female slaves, of whom there were seven thousand three hundred thirty-seven; and they had two hundred forty-five singers, male and female. 68 They had seven hundred thirty-six horses, two hundred forty-five mules, *s* 69 four hundred thirty-five camels, and six thousand seven hundred twenty donkeys.

70 Now some of the heads of ancestral houses contributed to the work. The governor gave to the treasury one thousand darics of gold, fifty basins, and five hundred thirty priestly robes. 71 And some of the heads of ancestral houses gave into the building fund twenty thousand darics of gold and two thousand two hundred minas of silver. 72 And what the rest of the people gave was twenty thousand darics of gold, two thousand minas of silver, and sixty-seven priestly robes.

73 So the priests, the Levites, the gatekeepers, the singers, some of the people, the temple servants, and all Israel settled in their towns.

When the seventh month came—the people of Israel being settled in their 8 towns— 1 all the people gathered together into the square before the Water Gate. They told the scribe Ezra to bring the book of the law of Moses, which the LORD had given to Israel. 2 Accordingly, the priest Ezra brought the law before the assembly, both men and women and all who could hear with understanding. This was on the first day of the seventh month. 3 He read from it facing the square before the Water Gate from early morning until midday, in the presence of the men and the women and those who could understand; and the ears of all the people were attentive to the book of the law. 4 The scribe Ezra stood on a wooden platform that had been made for the purpose; and beside him stood Mattithiah, Shema, Anaiah, Uriah, Hilkiah, and Maaseiah on his right hand; and Pedaiah, Mishael, Malchijah, Hashum, Hash-baddanah, Zechariah, and Meshullam on his left hand. 5 And Ezra opened the book in the sight of all the people, for he was standing above all the people; and when he opened it, all the people stood up. 6 Then Ezra blessed the LORD, the great God, and all the people answered, "Amen, Amen," lifting up their hands. Then they bowed their heads and worshiped the LORD with their faces to the ground. 7 Also Jeshua, Bani, Sherebiah, Jamin, Akkub, Shabbethai, Hodiah, Maaseiah, Kelita, Azariah, Joza-

s Ezra 2.66 and the margins of some Hebrew Mss: MT lacks *They had . . . forty-five mules*

8.1–9.37: Continuation of the Ezra story. 8.1–12: Reading the book of the law. 1: *The Water Gate* is doubtless that of 3.26. For Ezra's *book of the law* see Ezra 7.6, 10, 14. The citations from it in vv. 14, 15 (and Ezra 9.11, 12) are not in our Pentateuch, but see Lev 23.40–42. **2:** *The seventh month* was Tishri (September-October), the *first day* of which was a day of convocation (Num 29.1). **4:** *Wooden platform,* lit. "tower." **7–8:** Since the *book* was in Hebrew, it was interpreted to the people in the more familiar Aramaic. The

bad, Hanan, Pelaiah, the Levites,[t] helped the people to understand the law, while the people remained in their places. [8] So they read from the book, from the law of God, with interpretation. They gave the sense, so that the people understood the reading.

9 And Nehemiah, who was the governor, and Ezra the priest and scribe, and the Levites who taught the people said to all the people, "This day is holy to the LORD your God; do not mourn or weep." For all the people wept when they heard the words of the law. [10] Then he said to them, "Go your way, eat the fat and drink sweet wine and send portions of them to those for whom nothing is prepared, for this day is holy to our LORD; and do not be grieved, for the joy of the LORD is your strength." [11] So the Levites stilled all the people, saying, "Be quiet, for this day is holy; do not be grieved." [12] And all the people went their way to eat and drink and to send portions and to make great rejoicing, because they had understood the words that were declared to them.

13 On the second day the heads of ancestral houses of all the people, with the priests and the Levites, came together to the scribe Ezra in order to study the words of the law. [14] And they found it written in the law, which the LORD had commanded by Moses, that the people of Israel should live in booths[u] during the festival of the seventh month, [15] and that they should publish and proclaim in all their towns and in Jerusalem as follows, "Go out to the hills and bring branches of olive, wild olive, myrtle, palm, and other leafy trees to make booths,[u] as it is written." [16] So the people went out and brought them, and made booths[u] for themselves, each on the roofs of their houses, and in their courts and in the courts of the house of God, and in the square at the Water Gate and in the square at the Gate of Ephraim. [17] And all the assembly of those who had returned from the captivity made booths[u] and lived in them; for from the days of Jeshua son of Nun to that day the people of Israel had not done so. And there was very great rejoicing. [18] And day by day, from the first day to the last day, he read from the book of the law of God. They kept the festival seven days; and on the eighth day there was a solemn assembly, according to the ordinance.

9 Now on the twenty-fourth day of this month the people of Israel were assembled with fasting and in sackcloth, and with earth on their heads.[v] [2] Then those of Israelite descent separated themselves from all foreigners, and stood and confessed their sins and the iniquities of their ancestors. [3] They stood up in their place and read from the book of the law of the LORD their God for a fourth part of the day, and for another fourth they made confession and worshiped the LORD their God. [4] Then Jeshua, Bani, Kadmiel, Shebaniah, Bunni, Sherebiah, Bani, and Chenani stood on the stairs of the Levites and cried out with a loud voice to the LORD their God. [5] Then the Levites, Jeshua, Kadmiel, Bani, Hashabneiah, Sherebiah, Hodiah, Shebaniah, and Pethahiah, said, "Stand up and bless the LORD your God from everlasting to everlasting. Blessed be your glorious

t 1 Esdras 9.48 Vg: Heb and the Levites
u Or tabernacles; Heb succoth   v Heb on them

---

name Nehemiah in v. 9 is a scribal insertion. **10–12**: A feast is appointed to celebrate the event.

8.13–18: Celebration of the festival of booths. Reading the law made them aware that this month of Tishri (September-October) was the time for the festival of booths, the Mosaic regulations for which are in Lev 23.33–43. **15**: The hills must mean the hilly country around Jerusalem. **16**: The booths were apparently both domestic and communal. **17**: Jeshua son of Nun, that is, from the time of the Conquest of the land.

9.1–37: The great confession. 1–5: The setting. **2**: The confession was something in which non-Jews might not participate. **3**: They read, i.e. they listened to the reading. **4**: The stairs possibly belonged to the pulpit of 8.4.

name, which is exalted above all blessing and praise."

6 And Ezra said:[w] "You are the LORD, you alone; you have made heaven, the heaven of heavens, with all their host, the earth and all that is on it, the seas and all that is in them. To all of them you give life, and the host of heaven worships you. [7] You are the LORD, the God who chose Abram and brought him out of Ur of the Chaldeans and gave him the name Abraham; [8] and you found his heart faithful before you, and made with him a covenant to give to his descendants the land of the Canaanite, the Hittite, the Amorite, the Perizzite, the Jebusite, and the Girgashite; and you have fulfilled your promise, for you are righteous.

9 "And you saw the distress of our ancestors in Egypt and heard their cry at the Red Sea.[x] [10] You performed signs and wonders against Pharaoh and all his servants and all the people of his land, for you knew that they acted insolently against our ancestors. You made a name for yourself, which remains to this day. [11] And you divided the sea before them, so that they passed through the sea on dry land, but you threw their pursuers into the depths, like a stone into mighty waters. [12] Moreover, you led them by day with a pillar of cloud, and by night with a pillar of fire, to give them light on the way in which they should go. [13] You came down also upon Mount Sinai, and spoke with them from heaven, and gave them right ordinances and true laws, good statutes and commandments, [14] and you made known your holy sabbath to them and gave them commandments and statutes and a law through your servant Moses. [15] For their hunger you gave them bread from heaven, and for their thirst you brought water for them out of the rock, and you told them to go in to possess the land that you swore to give them.

16 "But they and our ancestors acted presumptuously and stiffened their necks and did not obey your commandments; [17] they refused to obey, and were not mindful of the wonders that you performed among them; but they stiffened their necks and determined to return to their slavery in Egypt. But you are a God ready to forgive, gracious and merciful, slow to anger and abounding in steadfast love, and you did not forsake them. [18] Even when they had cast an image of a calf for themselves and said, 'This is your God who brought you up out of Egypt,' and had committed great blasphemies, [19] you in your great mercies did not forsake them in the wilderness; the pillar of cloud that led them in the way did not leave them by day, nor the pillar of fire by night that gave them light on the way by which they should go. [20] You gave your good spirit to instruct them, and did not withhold your manna from their mouths, and gave them water for their thirst. [21] Forty years you sustained them in the wilderness so that they lacked nothing; their clothes did not wear out and their feet did not swell. [22] And you gave them kingdoms and peoples, and allotted to them every corner,[y] so they took possession of the land of King Sihon of Heshbon and the land of King Og of Bashan. [23] You multiplied their descendants like the stars of heaven, and brought them into the land that you had told their ancestors to enter and possess. [24] So the descendants went in and possessed the land, and you subdued before them the inhabitants of the land, the Canaanites, and gave them into their hands, with their kings and the peoples of the land, to do with them as they pleased. [25] And they captured fortress cities and a rich land, and took possession of houses filled with all sorts of goods, hewn cisterns, vineyards, olive orchards,

---

w Gk: Heb lacks *And Ezra said*   *x* Or *Sea of Reeds*   *y* Meaning of Heb uncertain

---

9.6–37: Ezra's confession and complaint. **17:** Num 14.4 says only that they suggested appointing such a leader.

9.20a: *You gave your good spirit,* perhaps a reference to the appointment of elders (Num 11.17–29).

and fruit trees in abundance; so they ate, and were filled and became fat, and delighted themselves in your great goodness.

26 "Nevertheless they were disobedient and rebelled against you and cast your law behind their backs and killed your prophets, who had warned them in order to turn them back to you, and they committed great blasphemies. 27 Therefore you gave them into the hands of their enemies, who made them suffer. Then in the time of their suffering they cried out to you and you heard them from heaven, and according to your great mercies you gave them saviors who saved them from the hands of their enemies. 28 But after they had rest, they again did evil before you, and you abandoned them to the hands of their enemies, so that they had dominion over them; yet when they turned and cried to you, you heard from heaven, and many times you rescued them according to your mercies. 29 And you warned them in order to turn them back to your law. Yet they acted presumptuously and did not obey your commandments, but sinned against your ordinances, by the observance of which a person shall live. They turned a stubborn shoulder and stiffened their neck and would not obey. 30 Many years you were patient with them, and warned them by your spirit through your prophets; yet they would not listen. Therefore you handed them over to the peoples of the lands. 31 Nevertheless, in your great mercies you did not make an end of them or forsake them, for you are a gracious and merciful God.

32 "Now therefore, our God—the great and mighty and awesome God, keeping covenant and steadfast love—do not treat lightly all the hardship that has come upon us, upon our kings, our officials, our priests, our prophets, our ancestors, and all your people, since the time of the kings of Assyria until today. 33 You have been just in all that has come upon us, for you have dealt faithfully and we have acted wickedly; 34 our kings, our officials, our priests, and our ancestors have not kept your law or heeded the commandments and the warnings that you gave them. 35 Even in their own kingdom, and in the great goodness you bestowed on them, and in the large and rich land that you set before them, they did not serve you and did not turn from their wicked works. 36 Here we are, slaves to this day—slaves in the land that you gave to our ancestors to enjoy its fruit and its good gifts. 37 Its rich yield goes to the kings whom you have set over us because of our sins; they have power also over our bodies and over our livestock at their pleasure, and we are in great distress."

38 z Because of all this we make a firm agreement in writing, and on that sealed document are inscribed the names of our officials, our Levites, and our priests.

10 a Upon the sealed document are the names of Nehemiah the governor, son of Hacaliah, and Zedekiah; 2 Seraiah, Azariah, Jeremiah, 3 Pashhur, Amariah, Malchijah, 4 Hattush, Shebaniah, Malluch, 5 Harim, Meremoth, Obadiah, 6 Daniel, Ginnethon, Baruch, 7 Meshullam, Abijah, Mijamin, 8 Maaziah, Bilgai, Shemaiah; these are the priests. 9 And the Levites: Jeshua son of Azaniah, Binnui of the sons of Henadad, Kadmiel; 10 and their associates, Shebaniah, Hodiah, Kelita, Pelaiah, Hanan, 11 Mica, Rehob, Hashabiah, 12 Zaccur, Sherebiah, Shebaniah, 13 Hodiah, Bani, Beninu. 14 The leaders of the people: Parosh, Pahath-moab, Elam, Zattu, Bani, 15 Bunni, Azgad, Bebai, 16 Adonijah, Bigvai, Adin, 17 Ater, Hezekiah, Azzur, 18 Hodiah, Hashum, Bezai, 19 Hariph,

z Ch 10.1 in Heb   a Ch 10.2 in Heb

9.27: The *saviors* here are the judges (Judg 2.16). 30: *By your spirit*, see Zech 7.12; 2 Chr 24.20.
9.38–10.39: **The covenant to support** God's house; a document belonging to the Nehemiah story. 9.38: *Because of all this* refers to ch 13.

Anathoth, Nebai, 20 Magpiash, Meshullam, Hezir, 21 Meshezabel, Zadok, Jaddua, 22 Pelatiah, Hanan, Anaiah, 23 Hoshea, Hananiah, Hasshub, 24 Hallohesh, Pilha, Shobek, 25 Rehum, Hashabnah, Maaseiah, 26 Ahiah, Hanan, Anan, 27 Malluch, Harim, and Baanah.

28 The rest of the people, the priests, the Levites, the gatekeepers, the singers, the temple servants, and all who have separated themselves from the peoples of the lands to adhere to the law of God, their wives, their sons, their daughters, all who have knowledge and understanding, 29 join with their kin, their nobles, and enter into a curse and an oath to walk in God's law, which was given by Moses the servant of God, and to observe and do all the commandments of the LORD our Lord and his ordinances and his statutes. 30 We will not give our daughters to the peoples of the land or take their daughters for our sons; 31 and if the peoples of the land bring in merchandise or any grain on the sabbath day to sell, we will not buy it from them on the sabbath or on a holy day; and we will forego the crops of the seventh year and the exaction of every debt.

32 We also lay on ourselves the obligation to charge ourselves yearly one-third of a shekel for the service of the house of our God: 33 for the rows of bread, the regular grain offering, the regular burnt offering, the sabbaths, the new moons, the appointed festivals, the sacred donations, and the sin offerings to make atonement for Israel, and for all the work of the house of our God. 34 We have also cast lots among the priests, the Levites, and the people, for the wood offering, to bring it into the house of our God, by ancestral houses, at appointed times, year by year, to burn on the altar of the Lord our God, as it is written in the law. 35 We obligate ourselves to bring the first fruits of our soil and the first fruits of all fruit of every tree, year by year, to the house of the LORD; 36 also to bring to the house of our God, to the priests who minister in the house of our God, the firstborn of our sons and of our livestock, as it is written in the law, and the firstlings of our herds and of our flocks; 37 and to bring the first of our dough, and our contributions, the fruit of every tree, the wine and the oil, to the priests, to the chambers of the house of our God; and to bring to the Levites the tithes from our soil, for it is the Levites who collect the tithes in all our rural towns. 38 And the priest, the descendant of Aaron, shall be with the Levites when the Levites receive the tithes; and the Levites shall bring up a tithe of the tithes to the house of our God, to the chambers of the storehouse. 39 For the people of Israel and the sons of Levi shall bring the contribution of grain, wine, and oil to the storerooms where the vessels of the sanctuary are, and where the priests that minister, and the gatekeepers and the singers are. We will not neglect the house of our God.

11 Now the leaders of the people lived in Jerusalem; and the rest of the people cast lots to bring one out of ten to live in the holy city Jerusalem, while nine-tenths remained in the other towns. 2 And the people blessed all those who willingly offered to live in Jerusalem.

3 These are the leaders of the province who lived in Jerusalem; but in the towns of Judah all lived on their property in their towns: Israel, the priests, the Levites, the temple servants, and the de-

---

10.28: *Separated themselves,* see 9.2; 13.3; and Ezra 6.21. 31: *Sabbath day,* see 13.15–18. *A holy day* was treated as a sabbath. For the *seventh year* see Ex 23.10, and for the *debt,* Deut 15.1–3.
10.32: The temple tax was voluntary; later it became an obligatory half-shekel tax (Ex 30.13; Mt 17.24). 34: Casting *lots* was a recognized procedure (11.1; 1 Chr 25.8; 1 Sam

14.40–45; Acts 1.23–26). *The wood offering* was for the altar fires (13.31; Lev 6.12).
10.35–36: The law on *first fruits,* Ex 22.29; 23.19; Num 18.15–18. 37: *Tithes,* see Lev 27.30; Num 18.25–32. *The chambers* were the temple store-rooms (13.13; Ezra 8.29; 10.6).
11.1–13.31: **Continuation of Nehemiah's memoirs,** picking up from 7.5a. 11.1–36: **A census list.**

scendants of Solomon's servants. 4 And in Jerusalem lived some of the Judahites and of the Benjaminites. Of the Judahites: Athaiah son of Uzziah son of Zechariah son of Amariah son of Shephatiah son of Mahalalel, of the descendants of Perez; 5 and Maaseiah son of Baruch son of Col-hozeh son of Hazaiah son of Adaiah son of Joiarib son of Zechariah son of the Shilonite. 6 All the descendants of Perez who lived in Jerusalem were four hundred sixty-eight valiant warriors.

7 And these are the Benjaminites: Sallu son of Meshullam son of Joed son of Pedaiah son of Kolaiah son of Maaseiah son of Ithiel son of Jeshaiah. 8 And his brothers[b] Gabbai, Sallai: nine hundred twenty-eight. 9 Joel son of Zichri was their overseer; and Judah son of Hassenuah was second in charge of the city.

10 Of the priests: Jedaiah son of Joiarib, Jachin, 11 Seraiah son of Hilkiah son of Meshullam son of Zadok son of Meraioth son of Ahitub, officer of the house of God, 12 and their associates who did the work of the house, eight hundred twenty-two; and Adaiah son of Jeroham son of Pelaliah son of Amzi son of Zechariah son of Pashhur son of Malchijah, 13 and his associates, heads of ancestral houses, two hundred forty-two; and Amashsai son of Azarel son of Ahzai son of Meshillemoth son of Immer, 14 and their associates, valiant warriors, one hundred twenty-eight; their overseer was Zabdiel son of Haggedolim.

15 And of the Levites: Shemaiah son of Hasshub son of Azrikam son of Hashabiah son of Bunni; 16 and Shabbethai and Jozabad, of the leaders of the Levites, who were over the outside work of the house of God; 17 and Mattaniah son of Mica son of Zabdi son of Asaph, who was the leader to begin the thanksgiving in prayer, and Bakbukiah, the second among his associates; and Abda son of Shammua son of Galal son of Jeduthun.

18 All the Levites in the holy city were two hundred eighty-four.

19 The gatekeepers, Akkub, Talmon and their associates, who kept watch at the gates, were one hundred seventy-two. 20 And the rest of Israel, and of the priests and the Levites, were in all the towns of Judah, all of them in their inheritance. 21 But the temple servants lived on Ophel; and Ziha and Gishpa were over the temple servants.

22 The overseer of the Levites in Jerusalem was Uzzi son of Bani son of Hashabiah son of Mattaniah son of Mica, of the descendants of Asaph, the singers, in charge of the work of the house of God. 23 For there was a command from the king concerning them, and a settled provision for the singers, as was required every day. 24 And Pethahiah son of Meshezabel, of the descendants of Zerah son of Judah, was at the king's hand in all matters concerning the people.

25 And as for the villages, with their fields, some of the people of Judah lived in Kiriath-arba and its villages, and in Dibon and its villages, and in Jekabzeel and its villages, 26 and in Jeshua and in Moladah and Beth-pelet, 27 in Hazar-shual, in Beer-sheba and its villages, 28 in Ziklag, in Meconah and its villages, 29 in En-rimmon, in Zorah, in Jarmuth, 30 Zanoah, Adullam, and their villages, Lachish and its fields, and Azekah and its villages. So they camped from Beer-sheba to the valley of Hinnom. 31 The people of Benjamin also lived from Geba onward, at Michmash, Aija, Bethel and its villages, 32 Anathoth, Nob, Ananiah, 33 Hazor, Ramah, Gittaim, 34 Hadid, Zeboim, Neballat, 35 Lod, and Ono, the valley of artisans. 36 And certain divisions of the Levites in Judah were joined to Benjamin.

12 These are the priests and the Levites who came up with Zerub-

b Gk Mss: Heb *And after him*

11.9, 14, 22: There was an overseer over each group. **24:** *At the king's hand,* i.e. he was one of the Persian monarch's famous "eyes" reporting on Judean affairs.

11.25–36: Census of the village distribution. **12.1–26: An appendix** to the census lists, see Ezra 2.36–40. Verses 9 and 24 suggest

babel son of Shealtiel, and Jeshua: Seraiah, Jeremiah, Ezra, ²Amariah, Malluch, Hattush, ³Shecaniah, Rehum, Meremoth, ⁴Iddo, Ginnethoi, Abijah, ⁵Mijamin, Maadiah, Bilgah, ⁶Shemaiah, Joiarib, Jedaiah, ⁷Sallu, Amok, Hilkiah, Jedaiah. These were the leaders of the priests and of their associates in the days of Jeshua.

8 And the Levites: Jeshua, Binnui, Kadmiel, Sherebiah, Judah, and Mattaniah, who with his associates was in charge of the songs of thanksgiving. ⁹And Bakbukiah and Unno their associates stood opposite them in the service. ¹⁰Jeshua was the father of Joiakim, Joiakim the father of Eliashib, Eliashib the father of Joiada, ¹¹Joiada the father of Jonathan, and Jonathan the father of Jaddua.

12 In the days of Joiakim the priests, heads of ancestral houses, were: of Seraiah, Meraiah; of Jeremiah, Hananiah; ¹³of Ezra, Meshullam; of Amariah, Jehohanan; ¹⁴of Malluchi, Jonathan; of Shebaniah, Joseph; ¹⁵of Harim, Adna; of Meraioth, Helkai; ¹⁶of Iddo, Zechariah; of Ginnethon, Meshullam; ¹⁷of Abijah, Zichri; of Miniamin, of Moadiah, Piltai; ¹⁸of Bilgah, Shammua; of Shemaiah, Jehonathan; ¹⁹of Joiarib, Mattenai; of Jedaiah, Uzzi; ²⁰of Sallai, Kallai; of Amok, Eber; ²¹of Hilkiah, Hashabiah; of Jedaiah, Nethanel.

22 As for the Levites, in the days of Eliashib, Joiada, Johanan, and Jaddua, there were recorded the heads of ancestral houses; also the priests until the reign of Darius the Persian. ²³The Levites, heads of ancestral houses, were recorded in the Book of the Annals until the days of Johanan son of Eliashib. ²⁴And the leaders of the Levites: Hashabiah, Sherebiah, and Jeshua son of Kadmiel, with their associates over against them, to

praise and to give thanks, according to the commandment of David the man of God, section opposite to section. ²⁵Mattaniah, Bakbukiah, Obadiah, Meshullam, Talmon, and Akkub were gatekeepers standing guard at the storehouses of the gates. ²⁶These were in the days of Joiakim son of Jeshua son of Jozadak, and in the days of the governor Nehemiah and of the priest Ezra, the scribe.

27 Now at the dedication of the wall of Jerusalem they sought out the Levites in all their places, to bring them to Jerusalem to celebrate the dedication with rejoicing, with thanksgivings and with singing, with cymbals, harps, and lyres. ²⁸The companies of the singers gathered together from the circuit around Jerusalem and from the villages of the Netophathites; ²⁹also from Beth-gilgal and from the region of Geba and Azmaveth; for the singers had built for themselves villages around Jerusalem. ³⁰And the priests and the Levites purified themselves; and they purified the people and the gates and the wall.

31 Then I brought the leaders of Judah up onto the wall, and appointed two great companies that gave thanks and went in procession. One went to the right on the wall to the Dung Gate; ³²and after them went Hoshaiah and half the officials of Judah, ³³and Azariah, Ezra, Meshullam, ³⁴Judah, Benjamin, Shemaiah, and Jeremiah, ³⁵and some of the young priests with trumpets: Zechariah son of Jonathan son of Shemaiah son of Mattaniah son of Micaiah son of Zaccur son of Asaph; ³⁶and his kindred, Shemaiah, Azarel, Milalai, Gilalai, Maai, Nethanel, Judah, and Hanani, with the musical instruments of David the man of God; and the scribe Ezra went in front of them. ³⁷At the Fountain Gate, in front of

---

antiphonal singing. **23**: This does not necessarily mean our book of Chronicles. **26**: The phrase *the priest Ezra, the scribe* is an editorial insertion.

12.27–43: The dedication of the walls, with music, purification, procession, and sacrifice. **28**: People from Netophath, southwest

of Jerusalem (1 Chr 9.16; Ezra 2.22). **31**: *Two great companies,* for they marched in opposite directions and met at the temple.

12.36: They used the same kinds of instruments which tradition said belonged to David's time. The reference to Ezra here and in v. 33 is an editorial insertion.

them, they went straight up by the stairs of the city of David, at the ascent of the wall, above the house of David, to the Water Gate on the east.

38 The other company of those who gave thanks went to the left,[c] and I followed them with half of the people on the wall, above the Tower of the Ovens, to the Broad Wall, 39and above the Gate of Ephraim, and by the Old Gate, and by the Fish Gate and the Tower of Hananel and the Tower of the Hundred, to the Sheep Gate; and they came to a halt at the Gate of the Guard. 40So both companies of those who gave thanks stood in the house of God, and I and half of the officials with me; 41and the priests Eliakim, Maaseiah, Miniamin, Micaiah, Eliocnai, Zechariah, and Hananiah, with trumpets; 42and Maaseiah, Shemaiah, Eleazar, Uzzi, Jehohanan, Malchijah, Elam, and Ezer. And the singers sang with Jezrahiah as their leader. 43They offered great sacrifices that day and rejoiced, for God had made them rejoice with great joy; the women and children also rejoiced. The joy of Jerusalem was heard far away.

44 On that day men were appointed over the chambers for the stores, the contributions, the first fruits, and the tithes, to gather into them the portions required by the law for the priests and for the Levites from the fields belonging to the towns; for Judah rejoiced over the priests and the Levites who ministered. 45They performed the service of their God and the service of purification, as did the singers and the gatekeepers, according to the command of David and his son Solomon. 46For in the days of David and Asaph long ago there was a leader of the singers, and there were songs of praise and thanksgiving to God.

47In the days of Zerubbabel and in the days of Nehemiah all Israel gave the daily portions for the singers and the gatekeepers. They set apart that which was for the Levites; and the Levites set apart that which was for the descendants of Aaron.

13 On that day they read from the book of Moses in the hearing of the people; and in it was found written that no Ammonite or Moabite should ever enter the assembly of God, 2because they did not meet the Israelites with bread and water, but hired Balaam against them to curse them—yet our God turned the curse into a blessing. 3When the people heard the law, they separated from Israel all those of foreign descent.

4 Now before this, the priest Eliashib, who was appointed over the chambers of the house of our God, and who was related to Tobiah, 5prepared for Tobiah a large room where they had previously put the grain offering, the frankincense, the vessels, and the tithes of grain, wine, and oil, which were given by commandment to the Levites, singers, and gatekeepers, and the contributions for the priests. 6While this was taking place I was not in Jerusalem, for in the thirty-second year of King Artaxerxes of Babylon I went to the king. After some time I asked leave of the king 7and returned to Jerusalem. I then discovered the wrong that Eliashib had done on behalf of Tobiah, preparing a room for him in the courts of the house of God. 8And I was very angry, and I threw all the household furniture of Tobiah out of the room. 9Then I gave orders and they cleansed the chambers, and I brought back the

---

c Cn: Heb *opposite*

12.43: *Great sacrifices,* see Ezra 6.17.
12.44–47: **Arrangements for temple revenues. 45**: See 2 Chr 8.14, referring back to 1 Chr chs 23–26.
13.1–31: **Nehemiah's second administration.** Verses 1–3 are introductory to connect with 8.18. **1**: *Written,* compare Deut 23.3–5. **2**: *Balaam,* see Num chs 22–24. **4–9**: The ejection of *Tobiah* an Ammonite (2.10).

4–5: *Eliashib* is the high priest of 3.1, 20; 12.22; Ezra 10.6, and who was connected by marriage with Sanballat (v. 28); in defiance of the law against Ammonites (v. 1), he had allotted Tobiah a temple *room.* **6**: Artaxerxes' *thirty-second year,* 433 B.C. (5.14); as he died in 423 the *some time* falls between those two years. **9**: That he *gave orders* proves that his return was with royal authority.

vessels of the house of God, with the grain offering and the frankincense.

10 I also found out that the portions of the Levites had not been given to them; so that the Levites and the singers, who had conducted the service, had gone back to their fields. 11 So I remonstrated with the officials and said, "Why is the house of God forsaken?" And I gathered them together and set them in their stations. 12 Then all Judah brought the tithe of the grain, wine, and oil into the storehouses. 13 And I appointed as treasurers over the storehouses the priest Shelemiah, the scribe Zadok, and Pedaiah of the Levites, and as their assistant Hanan son of Zaccur son of Mattaniah, for they were considered faithful; and their duty was to distribute to their associates. 14 Remember me, O my God, concerning this, and do not wipe out my good deeds that I have done for the house of my God and for his service.

15 In those days I saw in Judah people treading wine presses on the sabbath, and bringing in heaps of grain and loading them on donkeys; and also wine, grapes, figs, and all kinds of burdens, which they brought into Jerusalem on the sabbath day; and I warned them at that time against selling food. 16 Tyrians also, who lived in the city, brought in fish and all kinds of merchandise and sold them on the sabbath to the people of Judah, and in Jerusalem. 17 Then I remonstrated with the nobles of Judah and said to them, "What is this evil thing that you are doing, profaning the sabbath day? 18 Did not your ancestors act in this way, and did not our God bring all this disaster on us and on this city? Yet you bring more wrath on Israel by profaning the sabbath."

19 When it began to be dark at the gates of Jerusalem before the sabbath, I commanded that the doors should be shut and gave orders that they should not be opened until after the sabbath. And I set some of my servants over the gates, to prevent any burden from being brought in on the sabbath day. 20 Then the merchants and sellers of all kinds of merchandise spent the night outside Jerusalem once or twice. 21 But I warned them and said to them, "Why do you spend the night in front of the wall? If you do so again, I will lay hands on you." From that time on they did not come on the sabbath. 22 And I commanded the Levites that they should purify themselves and come and guard the gates, to keep the sabbath day holy. Remember this also in my favor, O my God, and spare me according to the greatness of your steadfast love.

23 In those days also I saw Jews who had married women of Ashdod, Ammon, and Moab; 24 and half of their children spoke the language of Ashdod, and they could not speak the language of Judah, but spoke the language of various peoples. 25 And I contended with them and cursed them and beat some of them and pulled out their hair; and I made them take an oath in the name of God, saying, "You shall not give your daughters to their sons, or take their daughters for your sons or for yourselves. 26 Did not King Solomon of Israel sin on account of such women? Among the many nations there was no king like him, and he was beloved by his God, and God made him king over all Israel; nevertheless, foreign women made even him to sin. 27 Shall we then listen to you and do all this great evil and act treacherously against our God by marrying foreign women?"

28 And one of the sons of Jehoiada, son of the high priest Eliashib, was the

---

13.10–14: Restoration of tithes to the temple staff. **14:** *Do not wipe out,* see 4.5. **15–22:** Instances of sabbath breaking.
13.19–20: Shutting the gates prevented produce from entering the city, a restriction which merchants sought to nullify by trading *outside* the walls. **23–30:** The problem of mixed marriages (10.30; Ezra 9.1–2; 10.2). **25:** Such physical violence is common in Oriental history. **26:** *Solomon,* 1 Kings 11.1–8.

son-in-law of Sanballat the Horonite; I chased him away from me. ²⁹Remember them, O my God, because they have defiled the priesthood, the covenant of the priests and the Levites.

30 Thus I cleansed them from every-

thing foreign, and I established the duties of the priests and Levites, each in his work; ³¹and I provided for the wood offering, at appointed times, and for the first fruits. Remember me, O my God, for good.

Esther

In the Hebrew Bible, Esther comes last of the five scrolls (megillōth) that are read at the great festivals of the Jewish year. It is the scroll for Purim. In the Greek Bible, Esther has its place among the historical books, and has been lengthened by the addition of passages (totalling 107 verses) intended to make it a more religious book. The King James Version included these, "The Additions to the Book of Esther," among the Apocrypha.

Esther is the fascinating tale that provides the "historical" basis for a non-Mosaic (and, probably, originally a pagan) festival. Thanks to the wisdom of Mordecai and the courageous efforts of Queen Esther, wife of the Persian King Ahasuerus (or Xerxes I), Jews throughout the Achaemenid empire were saved from a subtly planned anti-Semitic pogrom, with the result that henceforth all Jews must celebrate that deliverance on the fourteenth and fifteenth of the month of Adar (in February-March).

Simple though its story and purpose are, no other book of the Hebrew Bible has received such mixed reviews from good, God-fearing people—both Jews and Christians—as Esther. Some have criticized the book for what it contains; others, for what it lacks.

The Persian king, for instance, is mentioned 190 times, but the God of Israel, not once—nor are such basic Jewish practices and institutions as the Law, covenant, prayer, dietary regulations, or Jerusalem. However, other religious books, in varying degrees, also share many of these same "deficiencies," notably, such wisdom books as Proverbs, Job, and Ecclesiastes.

The book has frequently been faulted for its moral tone. Not only are such basic Judaic values as kindness, mercy, and forgiveness lacking; but, as many Jews and Christians have lamented, the story evidences a vengeful, bloodthirsty, and chauvinistic spirit. Intrigue, deceit, and hatred abound regardless of whether the spotlight is on Haman, Esther, Mordecai, or on their enemies. The Hebrew account, however, in contrast to its Greek version, is not anti-Gentile.

Small wonder, then, that the book's canonicity was much contested by certain Jews as late as the fourth century A.D. and by a number of Greek church fathers centuries after that.

Nonetheless, Esther is still a religious work. For although the Deity is not seen or even heard on its stage, God is standing in the wings, following the drama and arranging the props for a successful resolution of the play. Certain religious concepts are clearly presupposed in the Hebrew text, notably the concept of Providence (4.14; 5.1–6) and the efficacy of fasting (4.16) and, by implication, of prayer (4.16). The author is saying that, as in the case of Mordecai and Esther, Providence can be relied upon to reverse the ill-fortunes that beset individuals or the nation—provided that such leaders and their followers actively do their part, acting wisely and courageously.

In spite of the story's historical trappings and its author's familiarity with Persian history, customs, vocabulary, and names, the book of Esther is essentially fiction, created out of three originally separate and independent tales: a harem tale involving a certain Vashti, a court intrigue/deliverance tale featuring a Mordecai, and a success/deliverance tale starring an Esther, or Hadassah. They were combined to provide the explanation for the origins of a non-Mosaic and, probably, pagan festival the real origins of which scholars have, thus far, been unable to trace.

In its present form, Esther dates to the early Hellenistic Period, that is, shortly prior to the Maccabean Period (167–135 B.C.), when antagonism toward Gentiles ran high among Jews. Esther may have been composed in the Eastern Dispersion, but Palestine is a distinct possibility.

1 This happened in the days of Ahasuerus, the same Ahasuerus who ruled over one hundred twenty-seven provinces from India to Ethiopia.*a* 2 In those days when King Ahasuerus sat on his royal throne in the citadel of Susa, 3 in the third year of his reign, he gave a banquet for all his officials and ministers. The army of Persia and Media and the nobles and governors of the provinces were present, 4 while he displayed the great wealth of his kingdom and the splendor and pomp of his majesty for many days, one hundred eighty days in all.

5 When these days were completed, the king gave for all the people present in the citadel of Susa, both great and small, a banquet lasting for seven days, in the court of the garden of the king's palace. 6 There were white cotton curtains and blue hangings tied with cords of fine linen and purple to silver rings*b* and marble pillars. There were couches of gold and silver on a mosaic pavement of porphyry, marble, mother-of-pearl, and colored stones. 7 Drinks were served in golden goblets, goblets of different kinds, and the royal wine was lavished according to the bounty of the king. 8 Drinking was by flagons, without restraint; for the king had given orders to all the officials of his palace to do as each one desired. 9 Furthermore, Queen Vashti gave a banquet for the women in the palace of King Ahasuerus.

10 On the seventh day, when the king was merry with wine, he commanded Mehuman, Biztha, Harbona, Bigtha and Abagtha, Zethar and Carkas, the seven eunuchs who attended him, 11 to bring Queen Vashti before the king, wearing the royal crown, in order to show the peoples and the officials her beauty; for she was fair to behold. 12 But Queen Vashti refused to come at the king's command conveyed by the eunuchs. At this the king was enraged, and his anger burned within him.

13 Then the king consulted the sages who knew the laws*c* (for this was the king's procedure toward all who were versed in law and custom, 14 and those next to him were Carshena, Shethar, Admatha, Tarshish, Meres, Marsena, and Memucan, the seven officials of Persia and Media, who had access to the king, and sat first in the kingdom): 15 "According to the law, what is to be done to Queen Vashti because she has not performed the command of King Ahasuerus conveyed by the eunuchs?" 16 Then Memucan said in the presence of the king and the officials, "Not only has Queen Vashti done wrong to the king, but also to all the officials and all the peoples who are in all the provinces of King Ahasuerus. 17 For this deed of the queen will be made known to all women, causing them to look with contempt on their husbands, since they will say, 'King Ahasuerus commanded Queen Vashti to be brought before him, and she did not come.' 18 This very day the noble ladies of Persia and Media who have heard of the queen's behavior will rebel against*d*

---

a Or *Nubia*; Heb *Cush*   *b* Or *rods*
c Cn: Heb *times*   *d* Cn: Heb *will tell*

1.1–9: Ahasuerus' feast. *Ahasuerus* (Ezra 4.6; Dan 9.1) is Xerxes I (485–464 B.C.), whose Persian empire, from *India*, i.e. the Indus valley, to *Ethiopia*, the modern Nubia, included some twenty satrapies (Herodotus, *Hist.* III, 89), subdivided into *provinces*. **2:** Persepolis was his capital; *the citadel* ("acropolis") *of Susa*, his winter residence in Elam, was located two hundred miles northeast of Babylon. **3:** Greek writers mention fabulous feasts given by Persian kings. **5:** Excavations at *Susa* and at Persepolis have uncovered such a *court*. **9:** *Vashti* is, in all likelihood, a Persian name, and figures in the original harem tale.

Xerxes' *queen* was Amestris, a Persian (Herodotus, *Hist.* VII, 61). **1.10–2.4: Vashti's fall.** Apart from deposing Vashti so as to clear the way for Esther, the entire story is told for comedic effect. The names here, as elsewhere in this book, are attempts to reproduce Persian names. **11:** *Crown* indicates that she was to appear in royal attire. **1.12:** *Vashti refused . . . the king's command* to come before him while, ironically, Esther refused to stay away (4.11), which resulted in quite different consequences. **13:** *Sages,* probably his seven privy counselors (Ezra 7.14;

the king's officials, and there will be no end of contempt and wrath! 19 If it pleases the king, let a royal order go out from him, and let it be written among the laws of the Persians and the Medes so that it may not be altered, that Vashti is never again to come before King Ahasuerus; and let the king give her royal position to another who is better than she. 20 So when the decree made by the king is proclaimed throughout all his kingdom, vast as it is, all women will give honor to their husbands, high and low alike."

21 This advice pleased the king and the officials, and the king did as Memucan proposed; 22 he sent letters to all the royal provinces, to every province in its own script and to every people in its own language, declaring that every man should be master in his own house. *e*

2 After these things, when the anger of King Ahasuerus had abated, he remembered Vashti and what she had done and what had been decreed against her. 2 Then the king's servants who attended him said, "Let beautiful young virgins be sought out for the king. 3 And let the king appoint commissioners in all the provinces of his kingdom to gather all the beautiful young virgins to the harem in the citadel of Susa under custody of Hegai, the king's eunuch, who is in charge of the women; let their cosmetic treatments be given them. 4 And let the girl who pleases the king be queen instead of Vashti." This pleased the king, and he did so.

5 Now there was a Jew in the citadel of Susa whose name was Mordecai son of Jair son of Shimei son of Kish, a Benjaminite. 6 Kish*f* had been carried away from Jerusalem among the captives carried away with King Jeconiah of Judah, whom King Nebuchadnezzar of Babylon had carried away. 7 Mordecai*g* had brought up Hadassah, that is Esther, his cousin, for she had neither father nor mother; the girl was fair and beautiful, and when her father and her mother died, Mordecai adopted her as his own daughter. 8 So when the king's order and his edict were proclaimed, and when many young women were gathered in the citadel of Susa in custody of Hegai, Esther also was taken into the king's palace and put in custody of Hegai, who had charge of the women. 9 The girl pleased him and won his favor, and he quickly provided her with her cosmetic treatments and her portion of food, and with seven chosen maids from the king's palace, and advanced her and her maids to the best place in the harem. 10 Esther did not reveal her people or kindred, for Mordecai had charged her not to tell. 11 Every day Mordecai would walk around in front of the court of the harem, to learn how Esther was and how she fared.

12 The turn came for each girl to go

e Heb adds *and speak according to the language of his people* *f* Heb *a Benjamite* 6 *who* *g* Heb *He*

Herodotus, *Hist.* III, 31, 84). **19**: This belief in the immutability of Medo-Persian law appears in 8.8; Dan 6.8. **22**: *To every people in its own language* is a rhetorical exaggeration here (as in 3.12 and 8.9). Aramaic was normally used for such official correspondence.

2.2: *Servants,* i.e. his pages, not his counselors. **3, 9**: *Harem,* lit. "house of women." **2.5–23**: **Mordecai and Esther.** They were cousins with names derived, perhaps, from the Babylonian deities Marduk and Ishtar (or, possibly, Persian "stara," star), who also were cousins. That, like Saul, he was a Benjaminite, is emphasized to prepare readers for the enmity of Haman, an Amalekite of the Agag family, Saul's enemy (see 3.1–15 n.).

The Babylonian name "Marduka" was borne by a number of individuals, including an accountant from Susa who lived during the early years of Xerxes or the last years of his father. **6**: *Jeconiah* is Jehoiachin (2 Kings 24.6); so Mordecai was about one hundred and twenty years old. **7**: *Hadassah* ("myrtle") was Esther's Hebrew name. He was her foster-father (Isa 49.23). **10**: Esther's concealment of her race is a necessary literary device (v. 20); but the Greek version is at pains to show that as queen she strictly followed the law, including dietary restrictions. **11**: Mordecai's closeness to the *harem* suggests that he was a minor official, perhaps a eunuch gatekeeper (vv. 19, 21; 6.10).

in to King Ahasuerus, after being twelve months under the regulations for the women, since this was the regular period of their cosmetic treatment, six months with oil of myrrh and six months with perfumes and cosmetics for women. 13 When the girl went in to the king she was given whatever she asked for to take with her from the harem to the king's palace. 14 In the evening she went in; then in the morning she came back to the second harem in custody of Shaashgaz, the king's eunuch, who was in charge of the concubines; she did not go in to the king again, unless the king delighted in her and she was summoned by name.

15 When the turn came for Esther daughter of Abihail the uncle of Mordecai, who had adopted her as his own daughter, to go in to the king, she asked for nothing except what Hegai the king's eunuch, who had charge of the women, advised. Now Esther was admired by all who saw her. 16 When Esther was taken to King Ahasuerus in his royal palace in the tenth month, which is the month of Tebeth, in the seventh year of his reign, 17 the king loved Esther more than all the other women; of all the virgins she won his favor and devotion, so that he set the royal crown on her head and made her queen instead of Vashti. 18 Then the king gave a great banquet to all his officials and ministers—"Esther's banquet." He also granted a holiday[h] to the provinces, and gave gifts with royal liberality.

19 When the virgins were being gathered together,[i] Mordecai was sitting at the king's gate. 20 Now Esther had not revealed her kindred or her people, as Mordecai had charged her; for Esther obeyed Mordecai just as when she was brought up by him. 21 In those days, while Mordecai was sitting at the king's gate, Bigthan and Teresh, two of the king's eunuchs, who guarded the threshold, became angry and conspired to assassinate[j] King Ahasuerus. 22 But the matter came to the knowledge of Mordecai, and he told it to Queen Esther, and Esther told the king in the name of Mordecai. 23 When the affair was investigated and found to be so, both the men were hanged on the gallows. It was recorded in the book of the annals in the presence of the king.

3 After these things King Ahasuerus promoted Haman son of Hammedatha the Agagite, and advanced him and set his seat above all the officials who were with him. 2 And all the king's servants who were at the king's gate bowed down and did obeisance to Haman; for the king had so commanded concerning him. But Mordecai did not bow down or do obeisance. 3 Then the king's servants who were at the king's gate said to Mordecai, "Why do you disobey the king's command?" 4 When they spoke to him day after day and he would not listen to them, they told Haman, in order to see whether Mordecai's words would avail; for he had told them that he was a Jew. 5 When Haman saw that Mordecai did not bow down or do obeisance to him, Haman was infuriated. 6 But he thought it beneath him to lay hands on Mordecai alone. So, having been told who Mordecai's people were, Haman plotted to de-

h Or an amnesty   i Heb adds a second time
j Heb to lay hands on

2.12: *Their cosmetic treatment,* apparently by massage. **15:** *Abihail,* in the Hebrew Bible, can be either a male (Num 3.35; 1 Chr 5.14) or female (1 Chr 2.29) name. **16:** *Tebeth* (December-January), as all the month-names in this book, is a Babylonian name. **18:** This *banquet* being a wedding feast, the king favors his subjects with largesse. **21:** *The threshold,* that of the bed-chamber. Xerxes actually fell victim to such a conspiracy of bed-chamber servants, as have other Oriental potentates. **23:** *On the gallows,* lit. "on a tree," i.e. impale-ment, the normal form of punishment in Persia for political offenders. In the Greek version (12.6), Haman's hostility toward Mordecai begins with his reporting of this plot. *Annals,* such records kept by court scribes in Persia are mentioned by Greek writers.

3.1–15: Haman and Mordecai. For the Benjaminite-Agagite enmity see 1 Sam 15.7–9. *Haman* was *advanced* to be grand vizier, to whom lower officials must make obeisance, an honor which Mordecai as a Benjaminite

stroy all the Jews, the people of Mordecai, throughout the whole kingdom of Ahasuerus.

7 In the first month, which is the month of Nisan, in the twelfth year of King Ahasuerus, they cast Pur—which means "the lot"—before Haman for the day and for the month, and the lot fell on the thirteenth day*k* of the twelfth month, which is the month of Adar. 8 Then Haman said to King Ahasuerus, "There is a certain people scattered and separated among the peoples in all the provinces of your kingdom; their laws are different from those of every other people, and they do not keep the king's laws, so that it is not appropriate for the king to tolerate them. 9 If it pleases the king, let a decree be issued for their destruction, and I will pay ten thousand talents of silver into the hands of those who have charge of the king's business, so that they may put it into the king's treasuries." 10 So the king took his signet ring from his hand and gave it to Haman son of Hammedatha the Agagite, the enemy of the Jews. 11 The king said to Haman, "The money is given to you, and the people as well, to do with them as it seems good to you."

12 Then the king's secretaries were summoned on the thirteenth day of the first month, and an edict, according to all that Haman commanded, was written to the king's satraps and to the governors over all the provinces and to the officials of all the peoples, to every province in its own script and every people in its own language; it was written in the name of King Ahasuerus and sealed with the king's ring. 13 Letters were sent by couriers to all the king's provinces, giving or-

ders to destroy, to kill, and to annihilate all Jews, young and old, women and children, in one day, the thirteenth day of the twelfth month, which is the month of Adar, and to plunder their goods. 14 A copy of the document was to be issued as a decree in every province by proclamation, calling on all the peoples to be ready for that day. 15 The couriers went quickly by order of the king, and the decree was issued in the citadel of Susa. The king and Haman sat down to drink; but the city of Susa was thrown into confusion.

4 When Mordecai learned all that had been done, Mordecai tore his clothes and put on sackcloth and ashes, and went through the city, wailing with a loud and bitter cry; 2 he went up to the entrance of the king's gate, for no one might enter the king's gate clothed with sackcloth. 3 In every province, wherever the king's command and his decree came, there was great mourning among the Jews, with fasting and weeping and lamenting, and most of them lay in sackcloth and ashes.

4 When Esther's maids and her eunuchs came and told her, the queen was deeply distressed; she sent garments to clothe Mordecai, so that he might take off his sackcloth; but he would not accept them. 5 Then Esther called for Hathach, one of the king's eunuchs, who had been appointed to attend her, and ordered him to go to Mordecai to learn what was happening and why. 6 Hathach went out to Mordecai in the open square of the city in front of the king's gate, 7 and Mordecai told him all that had happened to him,

k Cn Compare Gk and verse 13 below: Heb the twelfth month

would not pay an Agagite (see 2.5–23 n.; 9.16 n.). **6**: Haman's indignation naturally suggested a pogrom of *all the Jews.*
3.7: *Pur* is an Akkadian word for lot, and casting lots was not irreligious (1 Sam 14.42; Acts 1.26; Herodotus, *Hist.* III, 128). The lot was to decide an appropriate date for the pogrom. **8**: *Their laws are different,* i.e. the Mosaic law. **9**: To clinch his argument Haman offers a fabulous sum as a bribe. **10**: *His signet*

ring gave validity to documents (8.2, 8; Gen 41.42). **11**: The king refuses the bribe but authorizes the pogrom. **13**: The *couriers* carrying the royal edict were the famous post service organized by Cyrus. The apocryphal Additions to the Book of Esther purport to give the text of the documents. **14**: The *copy* was the local public distribution of the *edict* (v.12).
4.1–17: **The appeal to Esther. 1**: These are traditional Oriental manifestations of

and the exact sum of money that Haman had promised to pay into the king's treasuries for the destruction of the Jews. [8] Mordecai also gave him a copy of the written decree issued in Susa for their destruction, that he might show it to Esther, explain it to her, and charge her to go to the king to make supplication to him and entreat him for her people.

9 Hathach went and told Esther what Mordecai had said. [10] Then Esther spoke to Hathach and gave him a message for Mordecai, saying, [11] "All the king's servants and the people of the king's provinces know that if any man or woman goes to the king inside the inner court without being called, there is but one law—all alike are to be put to death. Only if the king holds out the golden scepter to someone, may that person live. I myself have not been called to come in to the king for thirty days." [12] When they told Mordecai what Esther had said, [13] Mordecai told them to reply to Esther, "Do not think that in the king's palace you will escape any more than all the other Jews. [14] For if you keep silence at such a time as this, relief and deliverance will rise for the Jews from another quarter, but you and your father's family will perish. Who knows? Perhaps you have come to royal dignity for just such a time as this." [15] Then Esther said in reply to Mordecai, [16] "Go, gather all the Jews to be found in Susa, and hold a fast on my behalf, and neither eat nor drink for three days, night or day. I and my maids will also fast as you do.

After that I will go to the king, though it is against the law; and if I perish, I perish." [17] Mordecai then went away and did everything as Esther had ordered him.

5 On the third day Esther put on her royal robes and stood in the inner court of the king's palace, opposite the king's hall. The king was sitting on his royal throne inside the palace opposite the entrance to the palace. [2] As soon as the king saw Queen Esther standing in the court, she won his favor and he held out to her the golden scepter that was in his hand. Then Esther approached and touched the top of the scepter. [3] The king said to her, "What is it, Queen Esther? What is your request? It shall be given you, even to the half of my kingdom." [4] Then Esther said, "If it pleases the king, let the king and Haman come today to a banquet that I have prepared for the king." [5] Then the king said, "Bring Haman quickly, so that we may do as Esther desires." So the king and Haman came to the banquet that Esther had prepared. [6] While they were drinking wine, the king said to Esther, "What is your petition? It shall be granted you. And what is your request? Even to the half of my kingdom, it shall be fulfilled." [7] Then Esther said, "This is my petition and request: [8] If I have won the king's favor, and if it pleases the king to grant my petition and fulfill my request, let the king and Haman come tomorrow to the banquet that I will prepare for them, and then I will do as the king has said."

---

grief and mourning, which render one ritually unclean, so Mordecai (v. 2) could not *enter the king's gate.*

4.11: Interdiction to enter without being summoned was a security precaution. **14:** *From another quarter* and *Perhaps you have come . . . for just such a time . . .* probably refer to Divine providence. **16:** *Hold a fast,* the only religious activity mentioned in the book (9.31); fasting in the Hebrew Bible is usually accompanied by prayer. The apocryphal Additions to the Book of Esther insert here the prayers of Esther and Mordecai, which greatly enhance the religious character of the Greek version.

5.1–8: Esther before the king. 1: By appearing unbidden she had violated custom; so she *stood* at the *inner court* to await the monarch's reaction. **2:** *She won his favor,* in the Greek version (Addition D), the king is initially hostile, but God miraculously changes his attitude, thereby making that the climax of the entire book. **3, 6:** *The half of my kingdom* (7.2) is a customary hyperbole (compare Mk 6.23). **4, 8:** No subtle reason need be sought for the first and second *banquet,* nor for Haman's being the only guest; these are but literary embellishments needful to the story.

9 Haman went out that day happy and in good spirits. But when Haman saw Mordecai in the king's gate, and observed that he neither rose nor trembled before him, he was infuriated with Mordecai; 10 nevertheless Haman restrained himself and went home. Then he sent and called for his friends and his wife Zeresh, 11 and Haman recounted to them the splendor of his riches, the number of his sons, all the promotions with which the king had honored him, and how he had advanced him above the officials and the ministers of the king. 12 Haman added, "Even Queen Esther let no one but myself come with the king to the banquet that she prepared. Tomorrow also I am invited by her, together with the king. 13 Yet all this does me no good so long as I see the Jew Mordecai sitting at the king's gate." 14 Then his wife Zeresh and all his friends said to him, "Let a gallows fifty cubits high be made, and in the morning tell the king to have Mordecai hanged on it; then go with the king to the banquet in good spirits." This advice pleased Haman, and he had the gallows made.

6 On that night the king could not sleep, and he gave orders to bring the book of records, the annals, and they were read to the king. 2 It was found written how Mordecai had told about Bigthana and Teresh, two of the king's eunuchs, who guarded the threshold, and who had conspired to assassinate[l] King Ahasuerus. 3 Then the king said, "What honor or distinction has been bestowed on Mordecai for this?" The king's servants who attended him said, "Nothing has been done for him." 4 The king said, "Who is in the court?" Now Haman had just entered the outer court of the king's palace to speak to the king about having Mordecai hanged on the gallows that he had prepared for him. 5 So the king's servants told him, "Haman is there, standing in the court." The king said, "Let him come in." 6 So Haman came in, and the king said to him, "What shall be done for the man whom the king wishes to honor?" Haman said to himself, "Whom would the king wish to honor more than me?" 7 So Haman said to the king, "For the man whom the king wishes to honor, 8 let royal robes be brought, which the king has worn, and a horse that the king has ridden, with a royal crown on its head. 9 Let the robes and the horse be handed over to one of the king's most noble officials; let him[m] robe the man whom the king wishes to honor, and let him[m] conduct the man on horseback through the open square of the city, proclaiming before him: 'Thus shall it be done for the man whom the king wishes to honor.' " 10 Then the king said to Haman, "Quickly, take the robes and the horse, as you have said, and do so to the Jew Mordecai who sits at the king's gate. Leave out nothing that you have mentioned." 11 So Haman took the robes and the horse and robed Mordecai and led him riding through the open square

l Heb *to lay hands on*   *m* Heb *them*

5.9–14: Haman's exaltation and chagrin. 10: Haman's boasting before *his friends* will make his subsequent fall even more ironic. They are the *advisers* of 6.13. **14:** *Gallows,* see 2.23 n. The *fifty cubits,* i.e. over seventy feet, is hyperbole for literary effect. Haman could prepare this, but only the monarch could condemn anyone to death.

6.1–14: Mordecai's triumph. 1: Whether the king's sleeplessness was simply a literary device common to many Near Eastern tales or an expression of divine Providence is debated by scholars. In the Greek version, God is the cause of it. **3:** Herodotus (*Hist.* VIII, 85, 90) records how Persian monarchs kept records of notable services rendered. **4:** The inquiry, *Who is in the court?* merely means, What official is currently in attendance? Haman's eagerness drove him to early attendance, and his superior rank gave him precedence. **8:** The bestowal of *robes* of honor was common in ancient times (Gen 41.42; 1 Sam 18.4), and is still practiced. The *royal crown* here was an ornament for the horse, such as is pictured on monuments. **10:** *Do so to . . . Mordecai,* this is one of the many dramatic reversals in the book, and it clearly foreshadows the ultimate triumph of Mordecai and his people over their enemies.

of the city, proclaiming, "Thus shall it be done for the man whom the king wishes to honor."

12 Then Mordecai returned to the king's gate, but Haman hurried to his house, mourning and with his head covered. 13 When Haman told his wife Zeresh and all his friends everything that had happened to him, his advisers and his wife Zeresh said to him, "If Mordecai, before whom your downfall has begun, is of the Jewish people, you will not prevail against him, but will surely fall before him."

14 While they were still talking with him, the king's eunuchs arrived and hurried Haman off to the banquet that Esther had prepared. 1 So the king and Haman went in to feast with Queen Esther. 2 On the second day, as they were drinking wine, the king again said to Esther, "What is your petition, Queen Esther? It shall be granted you. And what is your request? Even to the half of my kingdom, it shall be fulfilled." 3 Then Queen Esther answered, "If I have won your favor, O king, and if it pleases the king, let my life be given me—that is my petition—and the lives of my people— that is my request. 4 For we have been sold, I and my people, to be destroyed, to be killed, and to be annihilated. If we had been sold merely as slaves, men and women, I would have held my peace; but no enemy can compensate for this damage to the king." [n] 5 Then King Ahasuerus said to Queen Esther, "Who is he, and where is he, who has presumed to do this?" 6 Esther said, "A foe and enemy, this wicked Haman!" Then Haman was

terrified before the king and the queen. 7 The king rose from the feast in wrath and went into the palace garden, but Haman stayed to beg his life from Queen Esther, for he saw that the king had determined to destroy him. 8 When the king returned from the palace garden to the banquet hall, Haman had thrown himself on the couch where Esther was reclining; and the king said, "Will he even assault the queen in my presence, in my own house?" As the words left the mouth of the king, they covered Haman's face. 9 Then Harbona, one of the eunuchs in attendance on the king, said, "Look, the very gallows that Haman has prepared for Mordecai, whose word saved the king, stands at Haman's house, fifty cubits high." And the king said, "Hang him on that." 10 So they hanged Haman on the gallows that he had prepared for Mordecai. Then the anger of the king abated.

8 On that day King Ahasuerus gave to Queen Esther the house of Haman, the enemy of the Jews; and Mordecai came before the king, for Esther had told what he was to her. 2 Then the king took off his signet ring, which he had taken from Haman, and gave it to Mordecai. So Esther set Mordecai over the house of Haman.

3 Then Esther spoke again to the king; she fell at his feet, weeping and pleading with him to avert the evil design of Haman the Agagite and the plot that he had devised against the Jews. 4 The king held out the golden scepter to

[n] Meaning of Heb uncertain

6.12: *Mourning*, i.e. weeping with chagrin. To cover the head was a sign of grief (2 Sam 15.30; Jer 14.4). 14: It was a Near Eastern custom for guests to be brought to an entertainment (Lk 14.17).
7.1–10: **Haman's fall. 4**: *We have been sold* is a reference to Haman's bribe in 3.9. Had Haman planned to sell the Jews into slavery to enrich the royal treasury, Esther would not have intervened; but planned destruction changes the situation. 6: *Enemy* is used to recall the ancient enmity of Agag and the Amalekites (see 3.1–15 n.). 8: To throw one-

self as a suppliant at someone's feet was a common custom (8.3; 1 Sam 25.23, 24; 2 Kings 4.27; Mk 7.25), but even touching the queen's couch was a violation of harem law. Whether covering the face of the condemned was a Persian practice is unknown; it was certainly a Greco-Roman custom. 9: It was inevitable that some servant would think of this as "poetic justice."
8.1–17: **The rise of Mordecai and the revocation of the edict. 1**: Herodotus (*Hist.* III, 129) mentions the property of criminals being confiscated. The transference of the sig-

Esther, 5and Esther rose and stood before the king. She said, "If it pleases the king, and if I have won his favor, and if the thing seems right before the king, and I have his approval, let an order be written to revoke the letters devised by Haman son of Hammedatha the Agagite, which he wrote giving orders to destroy the Jews who are in all the provinces of the king. 6For how can I bear to see the calamity that is coming on my people? Or how can I bear to see the destruction of my kindred?" 7Then King Ahasuerus said to Queen Esther and to the Jew Mordecai, "See, I have given Esther the house of Haman, and they have hanged him on the gallows, because he plotted to lay hands on the Jews. 8You may write as you please with regard to the Jews, in the name of the king, and seal it with the king's ring; for an edict written in the name of the king and sealed with the king's ring cannot be revoked."

9 The king's secretaries were summoned at that time, in the third month, which is the month of Sivan, on the twenty-third day; and an edict was written, according to all that Mordecai commanded, to the Jews and to the satraps and the governors and the officials of the provinces from India to Ethiopia,*o* one hundred twenty-seven provinces, to every province in its own script and to every people in its own language, and also to the Jews in their script and their language. 10He wrote letters in the name of King Ahasuerus, sealed them with the king's ring, and sent them by mounted couriers riding on fast steeds bred from the royal herd.*p* 11By these letters the king allowed the Jews who were in every

city to assemble and defend their lives, to destroy, to kill, and to annihilate any armed force of any people or province that might attack them, with their children and women, and to plunder their goods 12on a single day throughout all the provinces of King Ahasuerus, on the thirteenth day of the twelfth month, which is the month of Adar. 13A copy of the writ was to be issued as a decree in every province and published to all peoples, and the Jews were to be ready on that day to take revenge on their enemies. 14So the couriers, mounted on their swift royal steeds, hurried out, urged by the king's command. The decree was issued in the citadel of Susa.

15 Then Mordecai went out from the presence of the king, wearing royal robes of blue and white, with a great golden crown and a mantle of fine linen and purple, while the city of Susa shouted and rejoiced. 16For the Jews there was light and gladness, joy and honor. 17In every province and in every city, wherever the king's command and his edict came, there was gladness and joy among the Jews, a festival and a holiday. Furthermore, many of the peoples of the country professed to be Jews, because the fear of the Jews had fallen upon them.

9 Now in the twelfth month, which is the month of Adar, on the thirteenth day, when the king's command and edict were about to be executed, on the very day when the enemies of the Jews hoped to gain power over them, but which had been changed to a day when the Jews

o Or *Nubia*; Heb *Cush*   *p* Meaning of Heb uncertain

---

net ring indicates promotion of Mordecai as grand vizier. **5**: Esther subtly suggests that the edict was not really his but *devised by Haman.* **8**: The king rejects this subterfuge, accepts the responsibility, but neutralizes the edict by new instructions.

8.9: *Sivan* is the hebraized name for the Babylonian month of May–June. **10**: Greek writers mention the *steeds* of the Persian postal system. **12**: The date is carefully given, for it is important for the Purim festival. After v. 12 the apocryphal Additions to the Book of

Esther insert the text of the king's letter. **15**: Mordecai appears in his state robes, *crown* here being the viziral turban. **17**: *A festival and a holiday* contrast with the fasting and lamentation of 4.3. The improbability of many Persians becoming proselytes leads some to translate the verb as "joined themselves to," i.e. took the part of the Jews (9.27).

9.1–32: **Destruction of the enemies and inauguration of the feast of Purim. 1**: The important date is again given (compare 8.12). **3–5**: *The royal officials,* anxious to please the

would gain power over their foes, 2the Jews gathered in their cities throughout all the provinces of King Ahasuerus to lay hands on those who had sought their ruin; and no one could withstand them, because the fear of them had fallen upon all peoples. 3All the officials of the provinces, the satraps and the governors, and the royal officials were supporting the Jews, because the fear of Mordecai had fallen upon them. 4For Mordecai was powerful in the king's house, and his fame spread throughout all the provinces as the man Mordecai grew more and more powerful. 5So the Jews struck down all their enemies with the sword, slaughtering, and destroying them, and did as they pleased to those who hated them. 6In the citadel of Susa the Jews killed and destroyed five hundred people. 7They killed Parshandatha, Dalphon, Aspatha, 8Poratha, Adalia, Aridatha, 9Parmashta, Arisai, Aridai, Vaizatha, 10the ten sons of Haman son of Hammedatha, the enemy of the Jews; but they did not touch the plunder.

11 That very day the number of those killed in the citadel of Susa was reported to the king. 12The king said to Queen Esther, "In the citadel of Susa the Jews have killed five hundred people and also the ten sons of Haman. What have they done in the rest of the king's provinces? Now what is your petition? It shall be granted you. And what further is your request? It shall be fulfilled." 13Esther said, "If it pleases the king, let the Jews who are in Susa be allowed tomorrow also to do according to this day's edict, and let the ten sons of Haman be hanged on the gallows." 14So the king commanded this to be done; a decree was issued in Susa, and the ten sons of Haman were hanged. 15The Jews who were in Susa gathered also on the fourteenth day of the month of Adar and they killed three hundred persons in Susa; but they did not touch the plunder.

16 Now the other Jews who were in the king's provinces also gathered to defend their lives, and gained relief from their enemies, and killed seventy-five thousand of those who hated them; but they laid no hands on the plunder. 17This was on the thirteenth day of the month of Adar, and on the fourteenth day they rested and made that a day of feasting and gladness.

18 But the Jews who were in Susa gathered on the thirteenth day and on the fourteenth, and rested on the fifteenth day, making that a day of feasting and gladness. 19Therefore the Jews of the villages, who live in the open towns, hold the fourteenth day of the month of Adar as a day for gladness and feasting, a holiday on which they send gifts of food to one another.

20 Mordecai recorded these things, and sent letters to all the Jews who were in all the provinces of King Ahasuerus, both near and far, 21enjoining them that they should keep the fourteenth day of the month Adar and also the fifteenth day of the same month, year by year, 22as the days on which the Jews gained relief from their enemies, and as the month that had been turned for them from sor-

new vizier, *were supporting the Jews.* **5:** *Slaughtering . . . and destroying them* represents one more dramatic incidence of the sudden changes so characteristic of the book. **7:** The names of Haman's sons, represented as Persian names, are written peculiarly in the Hebrew text, favoring cryptic interpretations. *Parshandatha* is an authentic Persian name; it has been found on a carved seal from that general period. **9.13:** It is unclear whether Haman's ten sons were hanged on the gallows out of revenge or as a deterrent or as an expression of peripeteia (sudden turn of fortune). **15:** The second day's slaughter is probably a device to explain why the city Jews had a two-day Purim celebration but the country Jews only once. **16:** The emphasis here, as in vv. 10, 15, that the Jews did not *plunder* seems a reminder of the Agag story, where they plundered to their own hurt (1 Sam 15.9). **19, 22:** *Gifts of food,* i.e. dainties (Neh 8.10).

9.20–32: The official institution of Purim (possibly an addition to the book from a later source). Mordecai as vizier takes the initiative; Esther with queenly authority confirms the arrangements, so Mordecai writes to the Jewish communities and Esther follows with

row into gladness and from mourning
into a holiday; that they should make
them days of feasting and gladness, days
for sending gifts of food to one another
and presents to the poor. 23 So the Jews
adopted as a custom what they had begun to do, as Mordecai had written to
them.

24 Haman son of Hammedatha the
Agagite, the enemy of all the Jews, had
plotted against the Jews to destroy them,
and had cast Pur—that is "the lot"—to
crush and destroy them; 25 but when Esther came before the king, he gave orders
in writing that the wicked plot that he
had devised against the Jews should
come upon his own head, and that he and
his sons should be hanged on the gallows. 26 Therefore these days are called
Purim, from the word Pur. Thus because
of all that was written in this letter, and
of what they had faced in this matter, and
of what had happened to them, 27 the
Jews established and accepted as a custom for themselves and their descendants
and all who joined them, that without
fail they would continue to observe these
two days every year, as it was written and at the time appointed. 28 These
days should be remembered and kept
throughout every generation, in every
family, province, and city; and these
days of Purim should never fall into dis-

use among the Jews, nor should the commemoration of these days cease among
their descendants.

29 Queen Esther daughter of Abihail,
along with the Jew Mordecai, gave full
written authority, confirming this second letter about Purim. 30 Letters were
sent wishing peace and security to all the
Jews, to the one hundred twenty-seven
provinces of the kingdom of Ahasuerus,
31 and giving orders that these days of
Purim should be observed at their appointed seasons, as the Jew Mordecai and
Queen Esther enjoined on the Jews, just
as they had laid down for themselves and
for their descendants regulations concerning their fasts and their lamentations. 32 The command of Queen Esther
fixed these practices of Purim, and it was
recorded in writing.

10 King Ahasuerus laid tribute on
the land and on the islands of the
sea. 2 All the acts of his power and might,
and the full account of the high honor of
Mordecai, to which the king advanced
him, are they not written in the annals of
the kings of Media and Persia? 3 For Mordecai the Jew was next in rank to King
Ahasuerus, and he was powerful among
the Jews and popular with his many kindred, for he sought the good of his people and interceded for the welfare of all
his descendants.

a letter. This emphasis on the *written* word is
doubtless intended to legalize a festival not
mentioned in the Torah.
10.1–3: Conclusion. A paragraph on the
continued greatness of Xerxes and fame of

Mordecai. Following 10.3 the apocryphal
Additions to the Book of Esther append Mor-
decai's interpretation of the dream (Addition
F) that he had had at the beginning of the
Greek version (Addition A).

The Poetical Books
or "Writings"

The poetical books of the Old Testament coincide only in part with the "Writings," the third and final part of the Hebrew canon (see "Introduction to the Old Testament," pp. xxxi–xxxiii OT). Neither term is exact: poetry is scattered throughout the Old Testament, and writings can refer to the entire Bible. In the Jewish canon, the Writings include such disparate books as Chronicles and Psalms, as well as the so-called wisdom literature. This third part is perhaps the most interesting because it contains such a wide variety of literature. The Psalms, many dating from before the exile, were finally collected and used in worship in the second temple after the exile. First and Second Chronicles constitute a history that replicates in a sense the books of Samuel and Kings, but interprets Israel's history for the post-exilic community. Ezra and Nehemiah fill out the details of the return from the exile and the establishment of Judah. Ruth and Esther are stories of valiant women from different periods (Ruth 1.1, "In the days when the judges ruled . . ."; Esth 1.1, "In the days of Ahasuerus . . ."). Both of them were later included in the *megilloth* or five scrolls that were used for Jewish liturgical festivals: Ruth (Weeks), Esther (Purim)—and to them were added Song of Solomon (Passover), Lamentations (destruction of Temple), and Ecclesiastes (Tabernacles). Because Daniel was written late in the post-exilic period (*ca.* 165, the time of the Maccabees), it too was included among the Writings.

In more modern times the term "wisdom literature" has become a popular designation for three books among the Writings: Job, Proverbs, and Ecclesiastes (and, outside the Hebrew canon, Sirach and the Wisdom of Solomon among the apocryphal/deuterocanonical books). The influence of this literature has also extended to other parts of the Bible, such as "wisdom" Psalms like Ps 37. As the Psalms were generally attributed to David, so these books were assigned to Solomon as author (Prov 1.1; Eccl 1.1; Song 1.1; see 1 Kings 4.29–34). But it is generally agreed that they are the products of Israelite sages. The precise settings in which they worked are hard to determine. Some were associated with the court (e.g. Prov 25.1, "the officials of King Hezekiah," a Jerusalem monarch of Isaiah's time). Qoheleth and Ben Sira were recognized as teachers (Eccl 12.9; Sir 51.23). The role of the family parent and tribal elder in the education of youth must also be recognized (Prov 10.1). Although one may presume that some sort of "school" (court? scribal?) existed, on the analogy of the educational institutions known in the Fertile Crescent, we have no firsthand evidence of this. Most of the wisdom literature was written in the post-exilic period (parts of Proverbs and Job being the only exception).

Wisdom (Hebrew *ḥokhmah*) has its own favorite but limited vocabulary (e.g. understanding, *binah;* discipline, *musar*), and its own emphasis (e.g. proper speech and conduct; the connection between wisdom/prosperity and folly/destruction). A characteristic feature is a certain linguistic style, such as the short saying or aphorism (two or at times three lines in parallelism), a command or admonition (usually with a motive clause). There are also lengthy wisdom poems, especially the 22-line (the number of letters in the Hebrew alphabet) poem, as in Prov 2.1–22; 31.10–31; Sir 1.11–30. Each book can have its special traits, such as the narrative framework to the book of Job (1.1–2.22; 42.7–17), or the reflections in Ecclesiastes (e.g. 2.18–26), or the "praise of the fathers" in Sir chs 44–50. This broad range of literary style is united by the dominant concern of wisdom, which deals with daily experience and how to cope with it.

During the twentieth century many examples of this kind of wisdom have been recovered from ancient Mesopotamia and especially from Egypt. Many of the "wise men" of Egypt are known by name, such as Ptah-hotep and Amenemope (the influence of the latter upon Prov 22.17ff. is generally admitted). It is significant that the wisdom of Solomon was said to surpass "all the wisdom of Egypt" (1 Kings 4.30); the Israelites were aware of the achievements of their neighbors in this area. Some have described the wisdom enterprise as a search for order, according to the model of the Egyptian *ma'at* (justice, or order) established by the divinity—and the pharaoh was divine. It is simpler to see wisdom as the teaching that the sages propagated as a result of their dialogue with reality, the lessons that they considered necessary for the attainment of a happy and prosperous life. The modern reader will be particularly impressed by the fact that the teaching of the sages was far from monolithic. The contrast of Proverbs with Job and Ecclesiastes is a striking illustration. Proverbs presents a serene program for prosperous living; Job wrestles with the suffering of the just; Qoheleth cannot understand the "work of God" (8.17; 11.5). Furthermore, Sirach is a traditionalist, but the Wisdom of Solomon (1.15) proclaims a message of blessed immortality.

There are no wisdom books in the New Testament, but the style of the Letter of James approximates Old Testament wisdom. More importantly, the sayings and parables of Jesus continue the wisdom tradition of his forebears. It is not for nothing that in the Gospels he is frequently called "teacher."

Job

The Book of Job does not explain the mystery of suffering or "justify the ways of God" with human beings, but it does probe the depths of faith in the midst of suffering. An ancient folktale about a saintly Job (Ezek 14.14,20) has been used as a setting (1.1–2.13; 42.7–17) for a discussion between Job and his three friends (3.1–31.40), a statement by Elihu (32.1–37.24, regarded by some as a later insertion), and the speeches of the LORD (38.1–42.6). The date of the work is uncertain (perhaps around the time of the Exile, in the sixth century B.C.).

The folktale introduces a profound question on the lips of Satan (a Hebrew word meaning "adversary" or "accuser," not the devil of New Testament thought): "Does Job fear God for nothing?" (1.9) That is, does Job serve God because it is profitable, because God has been good to him? In the prologue, Job is totally resigned to God (1.21; 2.10). Then the author stages the great dispute. Job curses the day he was born (ch 3), and the debate is on (chs 4–31, where Job and his friends alternate in speeches which gradually become acrimonious; the text seems disturbed in chs 25–27, for Zophar disappears). The friends give lectures to Job in defence of the orthodox opinion about divine justice. Job oscillates between despair (e.g. 7.11–21; 9.14–24) and faith (19.23–27); he replies to the friends, and only Job addresses God directly. He closes out the dialogue with bravado and self-confidence (31.35–37). The intervention of Elihu (chs 32–37) is in the style of the friends. Finally the LORD speaks from the whirlwind, asking Job "impossible" questions and drawing him ever deeper into mystery, until Job has the experience of an immediate communion with God (42.5, "now my eye sees you").

In the poetic language of the book, God is at work in the universe, even "to bring rain on a land where no one lives" (38.26), and God is fully aware of evil (personified by the monsters Behemoth and Leviathan, 40.15–41.34). At the same time the LORD cares for Job so much that he reveals himself personally and shares with him the vision of cosmic responsibilities. A God who confesses his burdens to a human being is a God who is profoundly involved in human destiny. He is also a God who respects human independence, who wishes to have the *free* gift of human service, and towards that end there is the testing of human beings. The restoration of Job in 42.7ff. is not the point of the book; it is merely the recognition of Job's integrity, and in line with divine generosity.

1 There was once a man in the land of Uz whose name was Job. That man was blameless and upright, one who feared God and turned away from evil. 2 There were born to him seven sons and three daughters. 3 He had seven thousand sheep, three thousand camels, five hundred yoke of oxen, five hundred donkeys, and very many servants; so that this man was the greatest of all the people of the east. 4 His sons used to go and hold feasts in one another's houses in turn; and they would send and invite their three sisters to eat and drink with them. 5 And when the feast days had run their course, Job would send and sanctify them, and he would rise early in the morning and offer burnt offerings according to the number of them all; for Job said, "It may be that my children have sinned, and cursed God in their hearts." This is what Job always did.

6 One day the heavenly beings[a] came to present themselves before the LORD, and Satan[b] also came among them. 7 The LORD said to Satan,[b] "Where have you come from?" Satan[b] answered the LORD, "From going to and fro on the earth, and from walking up and down on it." 8 The LORD said to Satan,[b] "Have you considered my servant Job? There is no one like him on the earth, a blameless and upright man who fears God and turns away from evil." 9 Then Satan[b] answered the LORD, "Does Job fear God for nothing? 10 Have you not put a fence around him and his house and all that he has, on every side? You have blessed the work of his hands, and his possessions have increased in the land. 11 But stretch out your hand now, and touch all that he has, and he will curse you to your face." 12 The LORD said to Satan,[b] "Very well, all that he has is in your power; only do not stretch out your hand against him!" So Satan[b] went out from the presence of the LORD.

13 One day when his sons and daughters were eating and drinking wine in the eldest brother's house, 14 a messenger came to Job and said, "The oxen were plowing and the donkeys were feeding beside them, 15 and the Sabeans fell on them and carried them off, and killed the servants with the edge of the sword; I alone have escaped to tell you." 16 While he was still speaking, another came and said, "The fire of God fell from heaven and burned up the sheep and the servants, and consumed them; I alone have escaped to tell you." 17 While he was still speaking, another came and said, "The Chaldeans formed three columns, made a raid on the camels and carried them off, and killed the servants with the edge of the sword; I alone have escaped to tell you." 18 While he was still speaking, another came and said, "Your sons and daughters were eating and drinking wine in their eldest brother's house, 19 and suddenly a great wind came across the desert, struck the four corners of the house, and it fell on the young people, and they are dead; I alone have escaped to tell you."

20 Then Job arose, tore his robe,

a Heb sons of God   b Or the Accuser; Heb ha-satan

1.1–2.13: **The prologue.** A blameless man is deprived of wealth, posterity and health, but keeps his faith in God. 1: *The land of Uz* is perhaps Edom.

1.6–7: The scene shifts to the heavenly court, where the LORD musters out the *heavenly beings,* among them *Satan.* His function is to patrol *the earth;* he is not yet the demonic personification of later Judaism (compare 1 Chr 21.1) and Christianity. 9: *Does Job fear God for nothing?* One of the most fundamental questions in the Bible: Do human beings serve God because of what they receive from God? 11–12: Satan's proposal puts God in a no-win situation. If God refuses, it looks as though he fears there may be a basis to Satan's claim; if God accepts, he comes out of it looking heartless.

1.13–19: Conversation is the style in which this legend describes activity: in this case, Job's loss of property and children. 20–21: Although a foreigner, Job is portrayed as a worshiper of the LORD, since he is the protagonist of the author. *Shall I return there:* there is correspondence between mother's womb and "mother earth." 22: The sinlessness of Job is explicitly noted, and this is fundamental to the book.

shaved his head, and fell on the ground and worshiped. [21]He said, "Naked I came from my mother's womb, and naked shall I return there; the LORD gave, and the LORD has taken away; blessed be the name of the LORD."

22 In all this Job did not sin or charge God with wrongdoing.

2 One day the heavenly beings[c] came to present themselves before the LORD, and Satan[d] also came among them to present himself before the LORD. [2]The LORD said to Satan,[d] "Where have you come from?" Satan[e] answered the LORD, "From going to and fro on the earth, and from walking up and down on it." [3]The LORD said to Satan,[d] "Have you considered my servant Job? There is no one like him on the earth, a blameless and upright man who fears God and turns away from evil. He still persists in his integrity, although you incited me against him, to destroy him for no reason." [4]Then Satan[d] answered the LORD, "Skin for skin! All that people have they will give to save their lives.[f] [5]But stretch out your hand now and touch his bone and his flesh, and he will curse you to your face." [6]The LORD said to Satan,[d] "Very well, he is in your power; only spare his life."

7 So Satan[d] went out from the presence of the LORD, and inflicted loathsome sores on Job from the sole of his foot to the crown of his head. [8]Job[g] took a potsherd with which to scrape himself, and sat among the ashes.

9 Then his wife said to him, "Do you still persist in your integrity? Curse[h] God, and die." [10]But he said to her, "You speak as any foolish woman would speak. Shall we receive the good at the hand of God, and not receive the bad?" In all this Job did not sin with his lips.

11 Now when Job's three friends heard of all these troubles that had come upon him, each of them set out from his home—Eliphaz the Temanite, Bildad the Shuhite, and Zophar the Naamathite. They met together to go and console and comfort him. [12]When they saw him from a distance, they did not recognize him, and they raised their voices and wept aloud; they tore their robes and threw dust in the air upon their heads. [13]They sat with him on the ground seven days and seven nights, and no one spoke a word to him, for they saw that his suffering was very great.

3 After this Job opened his mouth and cursed the day of his birth. [2]Job said:
3 "Let the day perish in which I was
      born,
   and the night that said,
   'A man-child is conceived.'
4 Let that day be darkness!
   May God above not seek it,
   or light shine on it.
5 Let gloom and deep darkness
      claim it.
   Let clouds settle upon it;
   let the blackness of the day
      terrify it.
6 That night—let thick darkness
      seize it!
   let it not rejoice among the days
      of the year;
   let it not come into the number
      of the months.
7 Yes, let that night be barren;
   let no joyful cry be heard[i] in it.
8 Let those curse it who curse the
      Sea,[j]
   those who are skilled to rouse
      up Leviathan.

c Heb *sons of God*   d Or *the Accuser*; Heb *ha-satan*   e Or *The Accuser*; Heb *ha-satan*
f Or *All that the man has he will give for his life*
g Heb *He*   h Heb *Bless*   i Heb *come*
j Cn: Heb *day*

2.4: *Skin for skin*: the image comes from bartering hides. **7**: *Sores*, not necessarily leprosy (Hansen's disease) but a skin ailment, otherwise unknown.
2.9: *Curse*, literally "bless," a euphemism. Job's wife still believed in his *integrity* (see 4.6 n.) but wishes to shorten his torture. **11**: Like Job, his friends are also non-Israelites, apparently from northwest Arabia.
3.1–26: **Job's curse of the day/night of his birth**. **8**: Job wants skilled cursers, who control Sea (see 7.12; Isa 51.10; Hab 3.8) and Leviathan (see 41.1; Ps 104.26), the monsters of the deep, that personify chaos.

9 Let the stars of its dawn be dark;
 let it hope for light, but have
  none;
 may it not see the eyelids of the
  morning—
10 because it did not shut the doors
  of my mother's womb,
 and hide trouble from my eyes.

11 "Why did I not die at birth,
 come forth from the womb and
  expire?
12 Why were there knees to receive
  me,
 or breasts for me to suck?
13 Now I would be lying down and
  quiet;
 I would be asleep; then I would
  be at rest
14 with kings and counselors of the
  earth
 who rebuild ruins for
  themselves,
15 or with princes who have gold,
 who fill their houses with silver.
16 Or why was I not buried like a
  stillborn child,
 like an infant that never sees the
  light?
17 There the wicked cease from
  troubling,
 and there the weary are at rest.
18 There the prisoners are at ease
  together;
 they do not hear the voice of
  the taskmaster.
19 The small and the great are there,
 and the slaves are free from
  their masters.

20 "Why is light given to one in
  misery,
 and life to the bitter in soul,
21 who long for death, but it does
  not come,

and dig for it more than for
  hidden treasures;
22 who rejoice exceedingly,
 and are glad when they find the
  grave?
23 Why is light given to one who
  cannot see the way,
 whom God has fenced in?
24 For my sighing comes like[k] my
  bread,
 and my groanings are poured
  out like water.
25 Truly the thing that I fear comes
  upon me,
 and what I dread befalls me.
26 I am not at ease, nor am I quiet;
 I have no rest; but trouble
  comes."

4 Then Eliphaz the Temanite answered:
2 "If one ventures a word with you,
  will you be offended?
 But who can keep from
  speaking?
3 See, you have instructed many;
 you have strengthened the weak
  hands.
4 Your words have supported those
  who were stumbling,
 and you have made firm the
  feeble knees.
5 But now it has come to you, and
  you are impatient;
 it touches you, and you are
  dismayed.
6 Is not your fear of God your
  confidence,
 and the integrity of your ways
  your hope?

7 "Think now, who that was
  innocent ever perished?

k Heb *before*

3.11: *Why?* A question continually repeated here and throughout the book.
3.23: *God,* Hebrew "Eloah," the only explicit mention in this poem. Job, like many Semites, accepted God as the cause of both good and evil (disaster, calamities, etc.); see 2.10; Isa 45.7; Am 3.6; Lam 3.38).

4.1–5.27: **First discourse of Eliphaz. 1–4:** A courteous and gentle beginning. The poet insists on the sincerity of Job's comforters, who present the traditional orthodox teaching on retribution (4.7–11).
4.6: Job's *integrity* (Hebrew word related to *blameless* in 1.1; see 2.9 n.) is not yet ques-

Or where were the upright cut
off?

8 As I have seen, those who plow
iniquity
and sow trouble reap the same.

9 By the breath of God they perish,
and by the blast of his anger
they are consumed.

10 The roar of the lion, the voice of
the fierce lion,
and the teeth of the young lions
are broken.

11 The strong lion perishes for lack
of prey,
and the whelps of the lioness are
scattered.

12 "Now a word came stealing to
me,
my ear received the whisper of
it.

13 Amid thoughts from visions of the
night,
when deep sleep falls on
mortals,

14 dread came upon me, and
trembling,
which made all my bones shake.

15 A spirit glided past my face;
the hair of my flesh bristled.

16 It stood still,
but I could not discern its
appearance.
A form was before my eyes;
there was silence, then I heard a
voice:

17 'Can mortals be righteous before[l]
God?
Can human beings be pure
before[l] their Maker?

18 Even in his servants he puts no
trust,
and his angels he charges with
error;

19 how much more those who live in
houses of clay,

whose foundation is in the dust,
who are crushed like a moth.

20 Between morning and evening
they are destroyed;
they perish forever without any
regarding it.

21 Their tent-cord is plucked up
within them,
and they die devoid of wisdom.'

5 "Call now; is there anyone who
will answer you?
To which of the holy ones will
you turn?

2 Surely vexation kills the fool,
and jealousy slays the simple.

3 I have seen fools taking root,
but suddenly I cursed their
dwelling.

4 Their children are far from safety,
they are crushed in the gate,
and there is no one to deliver
them.

5 The hungry eat their harvest,
and they take it even out of the
thorns;[m]
and the thirsty[n] pant after their
wealth.

6 For misery does not come from
the earth,
nor does trouble sprout from
the ground;

7 but human beings are born to
trouble
just as sparks[o] fly upward.

8 "As for me, I would seek God,
and to God I would commit my
cause.

9 He does great things and
unsearchable,

l Or *more than*   m Meaning of Heb uncertain
n Aquila Symmachus Syr Vg: Heb *snare*
o Or *birds;* Heb *sons of Resheph*

---

tioned. **12–16:** Eliphaz appeals to a *vision,* a
spirit, as his source of authority; normally the
sages appealed to experience and tradition.
4.17–21: This reasoning is intended to
make it easy for Job to admit that he is sinful
(and thus preserve the divine justice). *Tent-*

cord: an image for sudden disaster that comes
to the wicked.
5.1: *The holy ones,* or sons of God (see also
1.6; 15.15; Ex 15.11 n.; Ps 82.1 n.), can not
help Job since they are at fault (4.18).
5.8–16: Eliphaz draws a picture of the God

marvelous things without
number.

10 He gives rain on the earth
and sends waters on the fields;

11 he sets on high those who are
lowly,
and those who mourn are lifted
to safety.

12 He frustrates the devices of the
crafty,
so that their hands achieve no
success.

13 He takes the wise in their own
craftiness;
and the schemes of the wily are
brought to a quick end.

14 They meet with darkness in the
daytime,
and grope at noonday as in the
night.

15 But he saves the needy from the
sword of their mouth,
from the hand of the mighty.

16 So the poor have hope,
and injustice shuts its mouth.

17 "How happy is the one whom
God reproves;
therefore do not despise the
discipline of the Almighty. *p*

18 For he wounds, but he binds up;
he strikes, but his hands heal.

19 He will deliver you from six
troubles;
in seven no harm shall touch
you.

20 In famine he will redeem you
from death,
and in war from the power of
the sword.

21 You shall be hidden from the
scourge of the tongue,
and shall not fear destruction
when it comes.

22 At destruction and famine you
shall laugh,
and shall not fear the wild
animals of the earth.

23 For you shall be in league with the
stones of the field,
and the wild animals shall be at
peace with you.

24 You shall know that your tent is
safe,
you shall inspect your fold and
miss nothing.

25 You shall know that your
descendants will be many,
and your offspring like the grass
of the earth.

26 You shall come to your grave in
ripe old age,
as a shock of grain comes up to
the threshing floor in its
season.

27 See, we have searched this out; it
is true.
Hear, and know it for yourself."

6 Then Job answered:
2 "O that my vexation were
weighed,
and all my calamity laid in the
balances!

3 For then it would be heavier than
the sand of the sea;
therefore my words have been
rash.

4 For the arrows of the Almighty*p*
are in me;
my spirit drinks their poison;
the terrors of God are arrayed
against me.

5 Does the wild ass bray over its
grass,
or the ox low over its fodder?

p Traditional rendering of Heb *Shaddai*

to whom Job should appeal. **17:** Suffering is
discipline (Hebrew "musar"), which will lead
to Job's healing and eventual prosperity. Discipline was seen as a means to conversion
(Prov 3.11–12; Job 36.5–15).
6.1–7.21: Reply of Job. The orthodox explanation cannot be valid in Job's case, as the
reader knows from chs 1–2. His *calamity* exceeds all ordinary misfortunes, and his relationship to *the Almighty* seems destroyed.
6.8–10: Death would be preferable in
these circumstances; at least Job could have
the consolation of not having *denied the words
of the Holy One* (this text is obscure).

6 Can that which is tasteless be
 eaten without salt,
 or is there any flavor in the
 juice of mallows?*q*
7 My appetite refuses to touch them;
 they are like food that is
 loathsome to me. *q*

8 "O that I might have my request,
 and that God would grant my
 desire;
9 that it would please God to crush
 me,
 that he would let loose his hand
 and cut me off!
10 This would be my consolation;
 I would even exult*q* in
 unrelenting pain;
 for I have not denied the words
 of the Holy One.
11 What is my strength, that I should
 wait?
 And what is my end, that I
 should be patient?
12 Is my strength the strength of
 stones,
 or is my flesh bronze?
13 In truth I have no help in me,
 and any resource is driven from
 me.

14 "Those who withhold*r* kindness
 from a friend
 forsake the fear of the
 Almighty. *s*
15 My companions are treacherous
 like a torrent-bed,
 like freshets that pass away,
16 that run dark with ice,
 turbid with melting snow.
17 In time of heat they disappear;
 when it is hot, they vanish from
 their place.
18 The caravans turn aside from their
 course;
 they go up into the waste, and
 perish.

19 The caravans of Tema look,
 the travelers of Sheba hope.
20 They are disappointed because
 they were confident;
 they come there and are
 confounded.
21 Such you have now become to
 me;*t*
 you see my calamity, and are
 afraid.
22 Have I said, 'Make me a gift'?
 Or, 'From your wealth offer a
 bribe for me'?
23 Or, 'Save me from an opponent's
 hand'?
 Or, 'Ransom me from the hand
 of oppressors'?

24 "Teach me, and I will be silent;
 make me understand how I have
 gone wrong.
25 How forceful are honest words!
 But your reproof, what does it
 reprove?
26 Do you think that you can
 reprove words,
 as if the speech of the desperate
 were wind?
27 You would even cast lots over the
 orphan,
 and bargain over your friend.

28 "But now, be pleased to look at
 me;
 for I will not lie to your face.
29 Turn, I pray, let no wrong be
 done.
 Turn now, my vindication is at
 stake.
30 Is there any wrong on my tongue?
 Cannot my taste discern
 calamity?

q Meaning of Heb uncertain   *r* Syr Vg
Compare Tg: Meaning of Heb uncertain
s Traditional rendering of Heb *Shaddai*
t Cn Compare Gk Syr: Meaning of Heb
uncertain

6.14–23: He condemns the three friends,
likening them to *treacherous* gullies that contain no water.

6.24–30: He further challenges them to
show how he has *erred,* to convict him of *any
wrong.*

7 "Do not human beings have a
hard service on earth,
and are not their days like the
days of a laborer?
2 Like a slave who longs for the
shadow,
and like laborers who look for
their wages,
3 so I am allotted months of
emptiness,
and nights of misery are
apportioned to me.
4 When I lie down I say, 'When
shall I rise?'
But the night is long,
and I am full of tossing until
dawn.
5 My flesh is clothed with worms
and dirt;
my skin hardens, then breaks
out again.
6 My days are swifter than a
weaver's shuttle,
and come to their end without
hope. *u*

7 "Remember that my life is a
breath;
my eye will never again see
good.
8 The eye that beholds me will see
me no more;
while your eyes are upon me, I
shall be gone.
9 As the cloud fades and vanishes,
so those who go down to Sheol
do not come up;

10 they return no more to their
houses,
nor do their places know them
any more.

11 "Therefore I will not restrain my
mouth;
I will speak in the anguish of
my spirit;
I will complain in the bitterness
of my soul.
12 Am I the Sea, or the Dragon,
that you set a guard over me?
13 When I say, 'My bed will comfort
me,
my couch will ease my
complaint,'
14 then you scare me with dreams
and terrify me with visions,
15 so that I would choose strangling
and death rather than this body.
16 I loathe my life; I would not live
forever.
Let me alone, for my days are a
breath.
17 What are human beings, that you
make so much of them,
that you set your mind on
them,
18 visit them every morning,
test them every moment?
19 Will you not look away from me
for a while,
let me alone until I swallow my
spittle?

u Or as the thread runs out

7.1–6: A classic description of the hard life
that mortals (and Job in particular, v. 5) are
subject to. **7–11:** Job addresses God directly;
he will do this frequently, whereas the friends
never do so. They address their lectures to
Job. Job appeals to divine compassion with
the implied mockery that God will act only
when it is too late: *I shall be gone* (see also
v. 21). The statement about *Sheol*, repeated in
several ways elsewhere (10.21–22), is a clear
indication that Job does entertain the idea of
a blessed immortality (compare 14.7–22).
7.12: Were Job *the Sea, or the Dragon* (see
3.8 n.), he could understand the intensity of
God's supervision, which is described in the
following verses. **16:** But Job is a mere *breath*,
a common biblical theme (e.g., Ps 144.3–4).
7.17–18: A parody of Ps 8. While the
psalmist praised the creator who assigns to
insignificant mortals a place of pre-eminence
in nature, Job ironically prefers to receive
minimal attention. **20:** Job hypothetically admits his *sin* in order to provoke God who is
acting unlike a merciful God by making Job
his *target*. **21:** Compare v. 8. *You will seek me*,
i.e. you will grope in the darkness after me,
but I shall not be. Job threatens the Almighty
with his own non-being! He at once reaffirms
his former trust in a loving God and sarcasti-
cally implies the frustration of that love.

20 If I sin, what do I do to you, you
      watcher of humanity?
    Why have you made me your
      target?
    Why have I become a burden to
      you?
21 Why do you not pardon my
      transgression
    and take away my iniquity?
    For now I shall lie in the earth;
      you will seek me, but I shall not
      be."

8 Then Bildad the Shuhite answered:
  2 "How long will you say these
      things,
    and the words of your mouth
      be a great wind?
  3 Does God pervert justice?
    Or does the Almighty*v* pervert
      the right?
  4 If your children sinned against
      him,
    he delivered them into the
      power of their
      transgression.
  5 If you will seek God
    and make supplication to the
      Almighty,*v*
  6 if you are pure and upright,
    surely then he will rouse himself
      for you
    and restore to you your rightful
      place.
  7 Though your beginning was
      small,
    your latter days will be very
      great.

  8 "For inquire now of bygone
      generations,

and consider what their
      ancestors have found;
  9 for we are but of yesterday, and
      we know nothing,
    for our days on earth are but a
      shadow.
10 Will they not teach you and tell
      you
    and utter words out of their
      understanding?

11 "Can papyrus grow where there is
      no marsh?
    Can reeds flourish where there
      is no water?
12 While yet in flower and not cut
      down,
    they wither before any other
      plant.
13 Such are the paths of all who
      forget God;
    the hope of the godless shall
      perish.
14 Their confidence is gossamer,
    a spider's house their trust.
15 If one leans against its house, it
      will not stand;
    if one lays hold of it, it will not
      endure.
16 The wicked thrive*w* before the
      sun,
    and their shoots spread over the
      garden.
17 Their roots twine around the
      stoneheap;
    they live among the rocks.*x*
18 If they are destroyed from their
      place,

v Traditional rendering of Heb *Shaddai*
w Heb *He thrives*   *x* Gk Vg: Meaning of Heb
uncertain

8.1–22: First discourse of Bildad. 1–3:
The debate heats up, and Bildad implies that
Job accuses God of perverting *justice.* 4: He
also implies that Job suffers for the sin of his
children, whose *transgression* has *the power* to
destroy the collective personality of the clan
to which they belonged.

8.5–10: If only Job would *make supplication*
(i.e. admit his sinfulness), and be *upright*
(ironically, this word, "yashar," is used of Job

in the prologue (1.1, 8), he would be restored.
Such is the traditional teaching of the *ances-*
tors.

8.11–22: Bildad uses *plant* imagery to describe the fate of the *godless* (the text is very
uncertain) and assures Job of God's acceptance of a *blameless person* (again an echo of the
prologue, where Job is described as blameless, 1.1, 8!).

then it will deny them, saying,
'I have never seen you.'

19 See, these are their happy ways,^y
and out of the earth still others
will spring.

20 "See, God will not reject a
blameless person,
nor take the hand of evildoers.
21 He will yet fill your mouth with
laughter,
and your lips with shouts of
joy.
22 Those who hate you will be
clothed with shame,
and the tent of the wicked will
be no more."

9 Then Job answered:
2 "Indeed I know that this is so;
but how can a mortal be just
before God?
3 If one wished to contend with
him,
one could not answer him once
in a thousand.
4 He is wise in heart, and mighty in
strength
—who has resisted him, and
succeeded?—
5 he who removes mountains, and
they do not know it,
when he overturns them in his
anger;
6 who shakes the earth out of its
place,
and its pillars tremble;
7 who commands the sun, and it
does not rise;
who seals up the stars;
8 who alone stretched out the
heavens
and trampled the waves of the
Sea;^z

9 who made the Bear and Orion,
the Pleiades and the chambers of
the south;
10 who does great things beyond
understanding,
and marvelous things without
number.
11 Look, he passes by me, and I do
not see him;
he moves on, but I do not
perceive him.
12 He snatches away; who can stop
him?
Who will say to him, 'What are
you doing?'

13 "God will not turn back his anger;
the helpers of Rahab bowed
beneath him.
14 How then can I answer him,
choosing my words with him?
15 Though I am innocent, I cannot
answer him;
I must appeal for mercy to my
accuser. ^a
16 If I summoned him and he
answered me,
I do not believe that he would
listen to my voice.
17 For he crushes me with a tempest,
and multiplies my wounds
without cause;
18 he will not let me get my breath,
but fills me with bitterness.
19 If it is a contest of strength, he is
the strong one!
If it is a matter of justice, who
can summon him?^b
20 Though I am innocent, my own
mouth would condemn me;
though I am blameless, he
would prove me perverse.

---

y Meaning of Heb uncertain   *z* Or *trampled
the back of the sea dragon*   *a* Or *for my right*
b Compare Gk: Heb *me*

9.1–10.22: Reply of Job. There is not
much logical sequence in the argumentation;
here Job replies (compare 9.2 with 4.17) to
Eliphaz rather than to Bildad. **9.3–28:** In a
lawsuit, the creator has the advantage of
strength. There is no human or divine judge
of God (who is implicitly compared to a thief

or kidnapper, v. 12), a thought that anticipates the theme of the umpire (see vv. 33–
35).
   9.13: *Rahab* is the dragon of chaos defeated
in the beginning (26.23; Ps 89.10; Isa 51.9).
14–24: Job describes the futility of his appeal;
he crushes me with a tempest (this is in contrast

21 I am blameless; I do not know
        myself;
    I loathe my life.
22 It is all one; therefore I say,
        he destroys both the blameless
        and the wicked.
23 When disaster brings sudden
        death,
    he mocks at the calamity^c of
        the innocent.
24 The earth is given into the hand of
        the wicked;
    he covers the eyes of its
        judges—
    if it is not he, who then is it?

25 "My days are swifter than a
        runner;
    they flee away, they see no
        good.
26 They go by like skiffs of reed,
    like an eagle swooping on the
        prey.
27 If I say, 'I will forget my
        complaint;
    I will put off my sad
        countenance and be of good
        cheer,'
28 I become afraid of all my
        suffering,
    for I know you will not hold
        me innocent.
29 I shall be condemned;
    why then do I labor in vain?
30 If I wash myself with soap
    and cleanse my hands with lye,
31 yet you will plunge me into filth,
    and my own clothes will abhor
        me.
32 For he is not a mortal, as I am,
        that I might answer him,
    that we should come to trial
        together.
33 There is no umpire^d between us,
    who might lay his hand on us
        both.

34 If he would take his rod away
        from me,
    and not let dread of him terrify
        me,
35 then I would speak without fear of
        him,
    for I know I am not what I am
        thought to be.^e

10 "I loathe my life;
    I will give free utterance to
        my complaint;
    I will speak in the bitterness of
        my soul.
2 I will say to God, Do not
        condemn me;
    let me know why you contend
        against me.
3 Does it seem good to you to
        oppress,
    to despise the work of your
        hands
    and favor the schemes of the
        wicked?
4 Do you have eyes of flesh?
    Do you see as humans see?
5 Are your days like the days of
        mortals,
    or your years like human years,
6 that you seek out my iniquity
    and search for my sin,
7 although you know that I am not
        guilty,
    and there is no one to deliver
        out of your hand?
8 Your hands fashioned and made
        me;
    and now you turn and destroy
        me.^f
9 Remember that you fashioned me
        like clay;

c Meaning of Heb uncertain   d Another
reading is *Would that there were an umpire*
e Cn: Heb *for I am not so in myself*
f Cn Compare Gk Syr: Heb *made me together all
around, and you destroy me*

to the comparatively gentle theophany of
chs 38–41!). But he accuses God of rank injustice in vv. 22–24.
    9.25–35: Job returns to the theme of fleet-
ing existence (vv. 25–26) and entertains three
possibilities: to *forget* his *complaint*, to attempt

to *cleanse* himself, and the possibility of an
umpire between them. But none of these
works.
    10.1–12: Job prepares a tender address to
God: Can all that divine care in creation be
nullified by the present treatment of him?

and will you turn me to dust
  again?
10 Did you not pour me out like
  milk
  and curdle me like cheese?
11 You clothed me with skin and
  flesh,
  and knit me together with bones
  and sinews.
12 You have granted me life and
  steadfast love,
  and your care has preserved my
  spirit.
13 Yet these things you hid in your
  heart;
  I know that this was your
  purpose.
14 If I sin, you watch me,
  and do not acquit me of my
  iniquity.
15 If I am wicked, woe to me!
  If I am righteous, I cannot lift
  up my head,
  for I am filled with disgrace
  and look upon my affliction.
16 Bold as a lion you hunt me;
  you repeat your exploits against
  me.
17 You renew your witnesses against
  me,
  and increase your vexation
  toward me;
  you bring fresh troops against
  me.*g*

18 "Why did you bring me forth
  from the womb?
  Would that I had died before
  any eye had seen me,
19 and were as though I had not
  been,
  carried from the womb to the
  grave.
20 Are not the days of my life few?*h*
  Let me alone, that I may find a
  little comfort*i*

21 before I go, never to return,
  to the land of gloom and deep
  darkness,
22 the land of gloom*j* and chaos,
  where light is like darkness."

11 Then Zophar the Naamathite answered:
2 "Should a multitude of words go
  unanswered,
  and should one full of talk be
  vindicated?
3 Should your babble put others to
  silence,
  and when you mock, shall no
  one shame you?
4 For you say, 'My conduct*k* is
  pure,
  and I am clean in God's*l* sight.'
5 But oh, that God would speak,
  and open his lips to you,
6 and that he would tell you the
  secrets of wisdom!
  For wisdom is many-sided.*m*
  Know then that God exacts of you
  less than your guilt
  deserves.

7 "Can you find out the deep things
  of God?
  Can you find out the limit of
  the Almighty?*n*
8 It is higher than heaven*o*—what
  can you do?
  Deeper than Sheol—what can
  you know?
9 Its measure is longer than the
  earth,

g Cn Compare Gk: Heb *toward me; changes
and a troop are with me*   *h* Cn Compare
Gk Syr: Heb *Are not my days few? Let him
cease!*   *i* Heb *that I may brighten up a
little*   *j* Heb *gloom as darkness, deep
darkness*   *k* Gk: Heb *teaching*   *l* Heb *your*
m Meaning of Heb uncertain
n Traditional rendering of Heb *Shaddai*
o Heb *The heights of heaven*

10.13–22: He is actually being hounded by
God (v. 16), and he can only conclude with
the eternal question, Why? If God would only
let him *alone,* he might find *comfort* in that *land
of gloom and deep darkness,* Sheol (see also
14.13).

11.1–20: First discourse of Zophar. 1–6:
The third friend cannot restrain his theological passion. He turns aside Job's claim (v. 4)
and accuses him of *guilt* (v. 6), but the *secrets
of wisdom* will not be what Zophar thinks (see
42.7).

and broader than the sea.
10 If he passes through, and
imprisons,
and assembles for judgment,
who can hinder him?
11 For he knows those who are
worthless;
when he sees iniquity, will he
not consider it?
12 But a stupid person will get
understanding,
when a wild ass is born
human. *p*

13 "If you direct your heart rightly,
you will stretch out your hands
toward him.
14 If iniquity is in your hand, put it
far away,
and do not let wickedness reside
in your tents.
15 Surely then you will lift up your
face without blemish;
you will be secure, and will not
fear.
16 You will forget your misery;
you will remember it as waters
that have passed away.
17 And your life will be brighter than
the noonday;
its darkness will be like the
morning.
18 And you will have confidence,
because there is hope;
you will be protected *q* and take
your rest in safety.
19 You will lie down, and no one
will make you afraid;
many will entreat your favor.
20 But the eyes of the wicked will
fail;
all way of escape will be lost to
them,

and their hope is to breathe their
last."

12 Then Job answered:
2 "No doubt you are the
people,
and wisdom will die with you.
3 But I have understanding as well
as you;
I am not inferior to you.
Who does not know such things
as these?
4 I am a laughingstock to my
friends;
I, who called upon God and he
answered me,
a just and blameless man, I am a
laughingstock.
5 Those at ease have contempt for
misfortune, *p*
but it is ready for those whose
feet are unstable.
6 The tents of robbers are at peace,
and those who provoke God are
secure,
who bring their god in their
hands. *r*

7 "But ask the animals, and they
will teach you;
the birds of the air, and they
will tell you;
8 ask the plants of the earth, *s* and
they will teach you;
and the fish of the sea will
declare to you.
9 Who among all these does not
know

p Meaning of Heb uncertain   q Or *you will
look around*   r Or *whom God brought forth
by his hand*; Meaning of Heb uncertain
s Or *speak to the earth*

11.7–12: This poem on God's inscrutable
deep things, or mystery, ends with insulting
implications for Job (even if the meaning of
v. 12 is obscure).
11.13–20: This address assumes that Job's
misery is due to his *iniquity;* if he repents he
will be restored. It is precisely the integrity of
Job that prohibits him from espousing the
short-sighted theology of the friends.
12.1–14.22: Reply of Job. 12.3–6: Job answers with sarcasm and contrasts his situation
(*a laughingstock*) with *those who provoke God.*
6: *Who bring their god in their hand.* This seems
to mean that they make power their god.
12.7–25: Even the beasts know that God's
unrestrained and irresistible omnipotence is

that the hand of the LORD has
   done this?
10 In his hand is the life of every
      living thing
   and the breath of every human
      being.
11 Does not the ear test words
   as the palate tastes food?
12 Is wisdom with the aged,
   and understanding in length of
      days?

13 "With God[t] are wisdom and
      strength;
   he has counsel and
      understanding.
14 If he tears down, no one can
      rebuild;
   if he shuts someone in, no one
      can open up.
15 If he withholds the waters, they
      dry up;
   if he sends them out, they
      overwhelm the land.
16 With him are strength and
      wisdom;
   the deceived and the deceiver
      are his.
17 He leads counselors away stripped,
   and makes fools of judges.
18 He looses the sash of kings,
   and binds a waistcloth on their
      loins.
19 He leads priests away stripped,
   and overthrows the mighty.
20 He deprives of speech those who
      are trusted,
   and takes away the discernment
      of the elders.
21 He pours contempt on princes,
   and looses the belt of the
      strong.
22 He uncovers the deeps out of
      darkness,
   and brings deep darkness to light.
23 He makes nations great, then
      destroys them;

he enlarges nations, then leads
      them away.
24 He strips understanding from the
      leaders[u] of the earth,
   and makes them wander in a
      pathless waste.
25 They grope in the dark without
      light;
   he makes them stagger like a
      drunkard.

13 "Look, my eye has seen all
      this,
   my ear has heard and
      understood it.
2 What you know, I also know;
   I am not inferior to you.
3 But I would speak to the
      Almighty,[v]
   and I desire to argue my case
      with God.
4 As for you, you whitewash with
      lies;
   all of you are worthless
      physicians.
5 If you would only keep silent,
   that would be your wisdom!
6 Hear now my reasoning,
   and listen to the pleadings of
      my lips.
7 Will you speak falsely for God,
   and speak deceitfully for him?
8 Will you show partiality toward
      him,
   will you plead the case for God?
9 Will it be well with you when he
      searches you out?
   Or can you deceive him, as one
      person deceives another?
10 He will surely rebuke you
   if in secret you show partiality.
11 Will not his majesty terrify you,
   and the dread of him fall upon
      you?

---

t Heb *him*   *u* Heb adds *of the people*
v Traditional rendering of Heb *Shaddai*

---

behind everything; God has *strength and wisdom,* but he *overthrows* the leaders of the world.

13.1–12: Job attacks the friends as speaking *falsely for God* and offering *proverbs of ashes.*

12 Your maxims are proverbs of
      ashes,
   your defenses are defenses of
      clay.

13 "Let me have silence, and I will
      speak,
   and let come on me what may.
14 I will take my flesh in my teeth,
   and put my life in my hand. *w*
15 See, he will kill me; I have no
      hope; *x*
   but I will defend my ways to
      his face.
16 This will be my salvation,
   that the godless shall not come
      before him.
17 Listen carefully to my words,
   and let my declaration be in
      your ears.
18 I have indeed prepared my case;
   I know that I shall be
      vindicated.
19 Who is there that will contend
      with me?
   For then I would be silent and
      die.
20 Only grant two things to me,
   then I will not hide myself from
      your face:
21 withdraw your hand far from me,
   and do not let dread of you
      terrify me.
22 Then call, and I will answer;
   or let me speak, and you reply
      to me.
23 How many are my iniquities and
      my sins?
   Make me know my
      transgression and my sin.
24 Why do you hide your face,

and count me as your enemy?
25 Will you frighten a windblown
      leaf
   and pursue dry chaff?
26 For you write bitter things against
      me,
   and make me reap *y* the
      iniquities of my youth.
27 You put my feet in the stocks,
   and watch all my paths;
   you set a bound to the soles of
      my feet.
28 One wastes away like a rotten
      thing,
   like a garment that is
      moth-eaten.

14 "A mortal, born of woman,
      few of days and full of
      trouble,
2 comes up like a flower and
      withers,
   flees like a shadow and does not
      last.
3 Do you fix your eyes on such a
      one?
   Do you bring me into judgment
      with you?
4 Who can bring a clean thing out
      of an unclean?
   No one can.
5 Since their days are determined,
   and the number of their months
      is known to you,
   and you have appointed the
      bounds that they cannot
      pass,

w Gk: Heb *Why should I take . . . in my hand?*
x Or *Though he kill me, yet I will trust in him*
y Heb *inherit*

13.13–19: He has *prepared* his *case,* and despite the risk he is taking (vv. 14–15), he will
state it boldly, certain that he *shall be vindicated.* **15:** *I have no hope,* is the reading of the
consonantal text, contrary to the traditional
"yet will I trust in him" (compare v. 18, and
see note *x*). **16:** Job's passionate desire to *come
before* God, his would-be slayer, testifies to
his innocence, for *the godless* would not dare
do this. **19:** The question is a challenge to God
to engage in a lawsuit.

13.20–28: Despite his bold words, Job
speaks rather gently, asking for *two things:*
perhaps the withdrawal of the divine *hand* and
dread. **23:** Job is willing to learn from God
whether his *sins* (see v. 26) are the cause of his
misery. This seems to be rhetorical flourish
(compare 31.35–36); he has done no wrong
that would call for such treatment (see 1.1, 8,
22).

14.1–22: A classic poem on the human
condition: *few of days and full of trouble,* even

6 look away from them, and
desist,^z
that they may enjoy, like
laborers, their days.

7 "For there is hope for a tree,
if it is cut down, that it will
sprout again,
and that its shoots will not
cease.
8 Though its root grows old in the
earth,
and its stump dies in the
ground,
9 yet at the scent of water it will
bud
and put forth branches like a
young plant.
10 But mortals die, and are laid low;
humans expire, and where are
they?
11 As waters fail from a lake,
and a river wastes away and
dries up,
12 so mortals lie down and do not
rise again;
until the heavens are no more,
they will not awake
or be roused out of their sleep.
13 Oh that you would hide me in
Sheol,
that you would conceal me until
your wrath is past,
that you would appoint me a set
time, and remember me!
14 If mortals die, will they live again?
All the days of my service I
would wait
until my release should come.
15 You would call, and I would
answer you;
you would long for the work of
your hands.
16 For then you would not^a number
my steps,

you would not keep watch over
my sin;
17 my transgression would be sealed
up in a bag,
and you would cover over my
iniquity.

18 "But the mountain falls and
crumbles away,
and the rock is removed from
its place;
19 the waters wear away the stones;
the torrents wash away the soil
of the earth;
so you destroy the hope of
mortals.
20 You prevail forever against them,
and they pass away;
you change their countenance,
and send them away.
21 Their children come to honor, and
they do not know it;
they are brought low, and it
goes unnoticed.
22 They feel only the pain of their
own bodies,
and mourn only for
themselves."

15 Then Eliphaz the Temanite answered:
2 "Should the wise answer with
windy knowledge,
and fill themselves with the east
wind?
3 Should they argue in unprofitable
talk,
or in words with which they
can do no good?
4 But you are doing away with the
fear of God,

---

z Cn: Heb *that they may desist*   *a* Syr: Heb
lacks *not*

---

sinful (v. 4); why bother to *fix your eyes on
such a one?* (see 7.19). **7–22:** *There is hope for
a tree* to live again, but not for *mortals*.
14.13–14: Job toys with the idea of respite
in Sheol, hidden from divine *wrath* that would
eventually pass away. Finally God would *call*
him and *cover over his iniquity!* **18–22:** This is

merely a flight of fancy; reality is otherwise,
for God destroys *the hope of mortals.*
15.1–35: Second discourse of Eliphaz.
A second round of discussion now begins.
2–3: This time the speaker dispenses completely with soothing formulas of introduction (contrast 4.1–4). **4:** The insecurity of tra-

and hindering meditation before
God.
5 For your iniquity teaches your
mouth,
and you choose the tongue of
the crafty.
6 Your own mouth condemns you,
and not I;
your own lips testify against
you.

7 "Are you the firstborn of the
human race?
Were you brought forth before
the hills?
8 Have you listened in the council of
God?
And do you limit wisdom to
yourself?
9 What do you know that we do
not know?
What do you understand that is
not clear to us?
10 The gray-haired and the aged are
on our side,
those older than your father.
11 Are the consolations of God too
small for you,
or the word that deals gently
with you?
12 Why does your heart carry you
away,
and why do your eyes flash, *b*
13 so that you turn your spirit against
God,
and let such words go out of
your mouth?
14 What are mortals, that they can be
clean?
Or those born of woman, that
they can be righteous?
15 God puts no trust even in his holy
ones,

and the heavens are not clean in
his sight;
16 how much less one who is
abominable and corrupt,
one who drinks iniquity like
water!

17 "I will show you; listen to me;
what I have seen I will
declare—
18 what sages have told,
and their ancestors have not
hidden,
19 to whom alone the land was
given,
and no stranger passed among
them.
20 The wicked writhe in pain all their
days,
through all the years that are
laid up for the ruthless.
21 Terrifying sounds are in their ears;
in prosperity the destroyer will
come upon them.
22 They despair of returning from
darkness,
and they are destined for the
sword.
23 They wander abroad for bread,
saying, 'Where is it?'
They know that a day of
darkness is ready at hand;
24 distress and anguish terrify them;
they prevail against them, like a
king prepared for battle.
25 Because they stretched out their
hands against God,
and bid defiance to the
Almighty, *c*
26 running stubbornly against him

b Meaning of Heb uncertain
c Traditional rendering of Heb *Shaddai*

ditional wisdom faces the courage of
non-conformism.
15.7–14: The following questions are designed to ridicule Job. The *firstborn of the human race* is not a reference to Adam of Gen
1–3. More likely, it is a reference to a myth
about a First Man, who existed before the
earth, had a share in the divine *council* of the
heavenly court, and was born *before the hills*.
Such a person would have *wisdom* not available to others. The myth resembles partly the
story of Lady Wisdom in Prov 8.22–31. **11:**
The consolations of God, such words as Eliphaz
uttered in 4.15ff. **14–16:** Compare 4.17–19.
15.17–35: A typical statement concerning
the retribution visited upon *the wicked.*

with a thick-bossed shield;
27 because they have covered their
    faces with their fat,
    and gathered fat upon their
      loins,
28 they will live in desolate cities,
    in houses that no one should
      inhabit,
    houses destined to become heaps
      of ruins;
29 they will not be rich, and their
      wealth will not endure,
    nor will they strike root in the
      earth;*d*
30 they will not escape from
      darkness;
    the flame will dry up their
      shoots,
    and their blossom*e* will be
      swept away*f* by the wind.
31 Let them not trust in emptiness,
      deceiving themselves;
    for emptiness will be their
      recompense.
32 It will be paid in full before their
      time,
    and their branch will not be
      green.
33 They will shake off their unripe
      grape, like the vine,
    and cast off their blossoms, like
      the olive tree.
34 For the company of the godless is
      barren,
    and fire consumes the tents of
      bribery.
35 They conceive mischief and bring
      forth evil
    and their heart prepares deceit."

16 Then Job answered:
2 "I have heard many such
      things;
    miserable comforters are you
      all.
3 Have windy words no limit?
    Or what provokes you that you
      keep on talking?

4 I also could talk as you do,
    if you were in my place;
    I could join words together against
      you,
    and shake my head at you.
5 I could encourage you with my
      mouth,
    and the solace of my lips would
      assuage your pain.

6 "If I speak, my pain is not
      assuaged,
    and if I forbear, how much of it
      leaves me?
7 Surely now God has worn me out;
    he has*g* made desolate all my
      company.
8 And he has*g* shriveled me up,
    which is a witness against me;
    my leanness has risen up against
      me,
    and it testifies to my face.
9 He has torn me in his wrath, and
      hated me;
    he has gnashed his teeth at me;
    my adversary sharpens his eyes
      against me.
10 They have gaped at me with their
      mouths;
    they have struck me insolently
      on the cheek;
    they mass themselves together
      against me.
11 God gives me up to the ungodly,
    and casts me into the hands of
      the wicked.
12 I was at ease, and he broke me in
      two;
    he seized me by the neck and
      dashed me to pieces;
    he set me up as his target;
13   his archers surround me.
He slashes open my kidneys, and
      shows no mercy;

---

d Vg:Meaning of Heb uncertain   *e* Gk: Heb
mouth   *f* Cn: Heb *will depart*   *g* Heb *you have*

16.1–17.16: Reply of Job. 16.1–17: In the
face of God's continuing hostility (vv. 6–16),
Job reaffirms his innocence (v. 17).
    16.18: His "murder" must be avenged; per-
sonified *earth* is invited to assist in the avenging (compare Gen 4.10–11). **19–22:** The identity of the *witness* is disputed: either God or
someone else *in heaven.* In the first case Job is

he pours out my gall on the
   ground.

14 He bursts upon me again and
   again;
   he rushes at me like a warrior.

15 I have sewed sackcloth upon my
   skin,
   and have laid my strength in the
   dust.

16 My face is red with weeping,
   and deep darkness is on my
   eyelids,

17 though there is no violence in my
   hands,
   and my prayer is pure.

18 "O earth, do not cover my blood;
   let my outcry find no resting
   place.

19 Even now, in fact, my witness is
   in heaven,
   and he that vouches for me is
   on high.

20 My friends scorn me;
   my eye pours out tears to God,

21 that he would maintain the right
   of a mortal with God,
   as[h] one does for a neighbor.

22 For when a few years have come,
   I shall go the way from which I
   shall not return.

17 My spirit is broken, my days
   are extinct,
   the grave is ready for me.

2 Surely there are mockers around
   me,
   and my eye dwells on their
   provocation.

3 "Lay down a pledge for me with
   yourself;
   who is there that will give
   surety for me?

4 Since you have closed their minds
   to understanding,

therefore you will not let them
   triumph.

5 Those who denounce friends for
   reward—
   the eyes of their children will
   fail.

6 "He has made me a byword of the
   peoples,
   and I am one before whom
   people spit.

7 My eye has grown dim from
   grief,
   and all my members are like a
   shadow.

8 The upright are appalled at this,
   and the innocent stir themselves
   up against the godless.

9 Yet the righteous hold to their
   way,
   and they that have clean hands
   grow stronger and stronger.

10 But you, come back now, all of
   you,
   and I shall not find a sensible
   person among you.

11 My days are past, my plans are
   broken off,
   the desires of my heart.

12 They make night into day;
   'The light,' they say, 'is near to
   the darkness.'[i]

13 If I look for Sheol as my house,
   if I spread my couch in
   darkness,

14 if I say to the Pit, 'You are my
   father,'
   and to the worm, 'My mother,'
   or 'My sister,'

15 where then is my hope?
   Who will see my hope?

16 Will it go down to the bars of
   Sheol?

h Syr Vg Tg: Heb and   i Meaning of Heb
uncertain

---

appealing to God against God: *to maintain the
right of a mortal against God.* Otherwise the one
who *vouches* for him could be an angel (see
5.1; 33.23, and also 9.33).
   17.1–10: He describes his piteous condition; no one *will give surety for* him, especially
his three friends (v. 10).
   17.11–16: Job turns to the thought of death,
as frequently in ending his replies (7.21;
10.21–33).

Shall we descend together into
the dust?"

18 Then Bildad the Shuhite answered:
2 "How long will you hunt for
words?
Consider, and then we shall
speak.
3 Why are we counted as cattle?
Why are we stupid in your
sight?
4 You who tear yourself in your
anger—
shall the earth be forsaken
because of you,
or the rock be removed out of
its place?

5 "Surely the light of the wicked is
put out,
and the flame of their fire does
not shine.
6 The light is dark in their tent,
and the lamp above them is put
out.
7 Their strong steps are shortened,
and their own schemes throw
them down.
8 For they are thrust into a net by
their own feet,
and they walk into a pitfall.
9 A trap seizes them by the heel;
a snare lays hold of them.
10 A rope is hid for them in the
ground,
a trap for them in the path.
11 Terrors frighten them on every
side,
and chase them at their heels.
12 Their strength is consumed by
hunger,*j*
and calamity is ready for their
stumbling.
13 By disease their skin is
consumed,*k*

the firstborn of Death consumes
their limbs.
14 They are torn from the tent in
which they trusted,
and are brought to the king of
terrors.
15 In their tents nothing remains;
sulfur is scattered upon their
habitations.
16 Their roots dry up beneath,
and their branches wither above.
17 Their memory perishes from the
earth,
and they have no name in the
street.
18 They are thrust from light into
darkness,
and driven out of the world.
19 They have no offspring or
descendant among their
people,
and no survivor where they
used to live.
20 They of the west are appalled at
their fate,
and horror seizes those of the
east.
21 Surely such are the dwellings of
the ungodly,
such is the place of those who
do not know God."

19 Then Job answered:
2 "How long will you
torment me,
and break me in pieces with
words?
3 These ten times you have cast
reproach upon me;
are you not ashamed to wrong
me?
4 And even if it is true that I have
erred,

*j Or Disaster is hungry for them   k Cn: Heb It
consumes the limbs of his skin*

18.1–21: Second discourse of Bildad. It
deals with the fate of the wicked (see 15.17–
35), specifying that a disease, the *firstborn of
Death,* is in store for them, a foretaste of their
ultimate fate, when they will be brought before the *king of terrors,* or Death personified.
19.1–29: Reply of Job. 2–6: The friends
do not understand, and *reproach* him *ten times*
(i.e. repeatedly; see Gen 31.7). **4:** Job does not
admit his *error,* or sin (he has asserted his

my error remains with me.
5 If indeed you magnify yourselves
against me,
and make my humiliation an
argument against me,
6 know then that God has put me in
the wrong,
and closed his net around me.
7 Even when I cry out, 'Violence!' I
am not answered;
I call aloud, but there is no
justice.
8 He has walled up my way so that
I cannot pass,
and he has set darkness upon
my paths.
9 He has stripped my glory from
me,
and taken the crown from my
head.
10 He breaks me down on every side,
and I am gone,
he has uprooted my hope like a
tree.
11 He has kindled his wrath against
me,
and counts me as his adversary.
12 His troops come on together;
they have thrown up
siegeworks[1] against me,
and encamp around my tent.

13 "He has put my family far from
me,
and my acquaintances are
wholly estranged from me.
14 My relatives and my close friends
have failed me;
15 the guests in my house have
forgotten me;

my serving girls count me as a
stranger;
I have become an alien in their
eyes.
16 I call to my servant, but he gives
me no answer;
I must myself plead with him.
17 My breath is repulsive to my wife;
I am loathsome to my own
family.
18 Even young children despise me;
when I rise, they talk against
me.
19 All my intimate friends abhor me,
and those whom I loved have
turned against me.
20 My bones cling to my skin and to
my flesh,
and I have escaped by the skin
of my teeth.
21 Have pity on me, have pity on
me, O you my friends,
for the hand of God has touched
me!
22 Why do you, like God, pursue
me,
never satisfied with my flesh?

23 "O that my words were written
down!
O that they were inscribed in a
book!
24 O that with an iron pen and with
lead
they were engraved on a rock
forever!
25 For I know that my Redeemer[m]
lives,

1 Cn: Heb *their way*   m Or *Vindicator*

innocence often enough, 9.21; 10.7; 16.17);
his point is that his conduct is a matter between him and God, and not for them to
judge (in v. 22 he accuses them of pursuing
him, *like God*). 7–12: Once more he describes
God's attack (see 16.7–14).
19.13–22: Society has expelled Job from its
ranks, even the most intimate circle of household, family, and friends. This ostracism is
typical, and the author of the narrative does
not feel obliged to follow the details of the
prologue (1.16–19).

19.23–24: Job wishes that his *words* were
made permanent (does this refer to his whole
argument, or to what he is about to say?)
They are to be inscribed *in a book* (a scroll or
a tablet of clay), *engraved on a rock forever,* as
some inscriptions were engraved on a cliff at
the Nahr el-Kelb in Lebanon by military conquerors over many centuries.
19.25–27: There is no certain translation
of these difficult, probably textually corrupt,
verses. What is clear is that Job makes an act
of faith in a *Redeemer* (better, *Vindicator,* as in

and that at the last heⁿ will
stand upon the earth;^o

26 and after my skin has been thus
destroyed,
then in^p my flesh I shall see
God,^q

27 whom I shall see on my side,^r
and my eyes shall behold, and
not another.
My heart faints within me!

28 If you say, 'How we will
persecute him!'
and, 'The root of the matter is
found in him';

29 be afraid of the sword,
for wrath brings the punishment
of the sword,
so that you may know there is a
judgment."

20 Then Zophar the Naamathite answered:

2 "Pay attention! My thoughts urge
me to answer,
because of the agitation within
me.

3 I hear censure that insults me,
and a spirit beyond my
understanding answers me.

4 Do you not know this from of
old,
ever since mortals were placed
on earth,

5 that the exulting of the wicked is
short,
and the joy of the godless is but
for a moment?

6 Even though they mount up high
as the heavens,
and their head reaches to the
clouds,

7 they will perish forever like their
own dung;
those who have seen them will
say, 'Where are they?'

8 They will fly away like a dream,
and not be found;
they will be chased away like a
vision of the night.

9 The eye that saw them will see
them no more,
nor will their place behold them
any longer.

10 Their children will seek the favor
of the poor,
and their hands will give back
their wealth.

11 Their bodies, once full of youth,
will lie down in the dust with
them.

12 "Though wickedness is sweet in
their mouth,
though they hide it under their
tongues,

13 though they are loath to let it go,
and hold it in their mouths,

14 yet their food is turned in their
stomachs;
it is the venom of asps within
them.

n Or *that he the Last*   o Heb *dust*
p Or *without*   q Meaning of Heb of this
verse uncertain   r Or *for myself*

note *m*), and that he claims he shall see God
with his eyes (affirmed three times). The vindicator (Hebrew "gō'ēl") is a member of
one's kinfolk who will avenge one's honor,
or stand good for one's debts (compare Lev
25.25; Deut 25.5–10; Ruth 2.20).
There is some dispute whether the "gō'ēl"
is to be identified with God, or a mediator
(see 9.33–35), or a heavenly witness (see
16.19–21). In any case Job makes the magnificent affirmation that his justice will be recognized (therefore, ultimately by God). When
will this be? The meaning of v. 26 is too uncertain a one on which to base a firm conclusion, but the rendering of this verse in the
NRSV would allow the possibility of a resur-
rected Job. The doctrine of the resurrection,
however, appears late in Hebrew thought (see
Dan 12.1–3, probably in Maccabean times),
and nowhere in the rest of the book does Job
seriously consider the possibility (see ch 14).
Nevertheless, one must not lose sight of this
emphasis on being vindicated on high, and
upon seeing God (a theme that will be all
important in the theophany of chs 38–42).
20.1–29: Second discourse of Zophar.
3: The *censure* Zophar hears is in part 19.28–
29. **4–29:** Another description of the fate of
the wicked.

15 They swallow down riches and
    vomit them up again;
    God casts them out of their
    bellies.

16 They will suck the poison of asps;
    the tongue of a viper will kill
    them.

17 They will not look on the rivers,
    the streams flowing with honey
    and curds.

18 They will give back the fruit of
    their toil,
    and will not swallow it down;
    from the profit of their trading
    they will get no enjoyment.

19 For they have crushed and
    abandoned the poor,
    they have seized a house that
    they did not build.

20 "They knew no quiet in their
    bellies;
    in their greed they let nothing
    escape.

21 There was nothing left after they
    had eaten;
    therefore their prosperity will
    not endure.

22 In full sufficiency they will be in
    distress;
    all the force of misery will come
    upon them.

23 To fill their belly to the full
    God^s will send his fierce anger
    into them,
    and rain it upon them as their
    food.^{t}

24 They will flee from an iron
    weapon;
    a bronze arrow will strike them
    through.

25 It is drawn forth and comes out of
    their body,
    and the glittering point comes
    out of their gall;
    terrors come upon them.

26 Utter darkness is laid up for their
    treasures;

a fire fanned by no one will
    devour them;
    what is left in their tent will be
    consumed.

27 The heavens will reveal their
    iniquity,
    and the earth will rise up against
    them.

28 The possessions of their house will
    be carried away,
    dragged off in the day of
    God's^u wrath.

29 This is the portion of the wicked
    from God,
    the heritage decreed for them by
    God."

21 Then Job answered:
   2 "Listen carefully to my
      words,
    and let this be your consolation.

3 Bear with me, and I will speak;
    then after I have spoken, mock
    on.

4 As for me, is my complaint
    addressed to mortals?
    Why should I not be impatient?

5 Look at me, and be appalled,
    and lay your hand upon your
    mouth.

6 When I think of it I am dismayed,
    and shuddering seizes my flesh.

7 Why do the wicked live on,
    reach old age, and grow mighty
    in power?

8 Their children are established in
    their presence,
    and their offspring before their
    eyes.

9 Their houses are safe from fear,
    and no rod of God is upon
    them.

10 Their bull breeds without fail;
    their cow calves and never
    miscarries.

s Heb *he*　　*t* Cn: Meaning of Heb uncertain
u Heb *his*

21.1–34: Reply of Job. This discourse is
directed to the friends only, and he points out
that the wicked remain unpunished (vv. 7–

13), though they repudiate God (vv. 14–18;
compare v. 15 with 1.9); the traditional belief
in hereditary guilt (collective responsibility)

11 They send out their little ones like
a flock,
and their children dance around.
12 They sing to the tambourine and
the lyre,
and rejoice to the sound of the
pipe.
13 They spend their days in
prosperity,
and in peace they go down to
Sheol.
14 They say to God, 'Leave us alone!
We do not desire to know your
ways.
15 What is the Almighty,ᵛ that we
should serve him?
And what profit do we get if
we pray to him?'
16 Is not their prosperity indeed their
own achievement?ʷ
The plans of the wicked are
repugnant to me.

17 "How often is the lamp of the
wicked put out?
How often does calamity come
upon them?
How often does Godˣ distribute
pains in his anger?
18 How often are they like straw
before the wind,
and like chaff that the storm
carries away?
19 You say, 'God stores up their
iniquity for their children.'
Let it be paid back to them, so
that they may know it.
20 Let their own eyes see their
destruction,
and let them drink of the wrath
of the Almighty.ᵛ
21 For what do they care for their
household after them,
when the number of their
months is cut off?
22 Will any teach God knowledge,
seeing that he judges those that
are on high?

23 One dies in full prosperity,
being wholly at ease and secure,
24 his loins full of milk
and the marrow of his bones
moist.
25 Another dies in bitterness of soul,
never having tasted of good.
26 They lie down alike in the dust,
and the worms cover them.

27 "Oh, I know your thoughts,
and your schemes to wrong me.
28 For you say, 'Where is the house
of the prince?
Where is the tent in which the
wicked lived?'
29 Have you not asked those who
travel the roads,
and do you not accept their
testimony,
30 that the wicked are spared in the
day of calamity,
and are rescued in the day of
wrath?
31 Who declares their way to their
face,
and who repays them for what
they have done?
32 When they are carried to the
grave,
a watch is kept over their tomb.
33 The clods of the valley are sweet
to them;
everyone will follow after,
and those who went before are
innumerable.
34 How then will you comfort me
with empty nothings?
There is nothing left of your
answers but falsehood."

22 Then Eliphaz the Temanite answered:
2 "Can a mortal be of use to God?
Can even the wisest be of
service to him?

ᵛ Traditional rendering of Heb *Shaddai*
ʷ Heb *in their hand*   ˣ Heb *he*

is ineffective; finally, the honorable funeral of
the wicked (vv. 27–34) is a travesty of justice.
22.1–30: Third discourse of Eliphaz.

The third cycle of speeches begins, but there
may be some disorder in chs 24–27, since Zophar is not portrayed as speaking. **2–3:** No

3 Is it any pleasure to the
   Almighty[y] if you are
   righteous,
   or is it gain to him if you make
   your ways blameless?
4 Is it for your piety that he
   reproves you,
   and enters into judgment with
   you?
5 Is not your wickedness great?
   There is no end to your
   iniquities.
6 For you have exacted pledges
   from your family for no
   reason,
   and stripped the naked of their
   clothing.
7 You have given no water to the
   weary to drink,
   and you have withheld bread
   from the hungry.
8 The powerful possess the land,
   and the favored live in it.
9 You have sent widows away
   empty-handed,
   and the arms of the orphans you
   have crushed.[z]
10 Therefore snares are around you,
   and sudden terror overwhelms
   you,
11 or darkness so that you cannot see;
   a flood of water covers you.

12 "Is not God high in the heavens?
   See the highest stars, how lofty
   they are!
13 Therefore you say, 'What does
   God know?
   Can he judge through the deep
   darkness?
14 Thick clouds enwrap him, so that
   he does not see,
   and he walks on the dome of
   heaven.'
15 Will you keep to the old way

that the wicked have trod?
16 They were snatched away before
   their time;
   their foundation was washed
   away by a flood.
17 They said to God, 'Leave us
   alone,'
   and 'What can the Almighty[y]
   do to us?'[a]
18 Yet he filled their houses with
   good things—
   but the plans of the wicked are
   repugnant to me.
19 The righteous see it and are glad;
   the innocent laugh them to
   scorn,
20 saying, 'Surely our adversaries are
   cut off,
   and what they left, the fire has
   consumed.'

21 "Agree with God,[b] and be at
   peace;
   in this way good will come to
   you.
22 Receive instruction from his
   mouth,
   and lay up his words in your
   heart.
23 If you return to the Almighty,[y]
   you will be restored,
   if you remove unrighteousness
   from your tents,
24 if you treat gold like dust,
   and gold of Ophir like the
   stones of the torrent-bed,
25 and if the Almighty[y] is your gold
   and your precious silver,
26 then you will delight yourself in
   the Almighty,[y]
   and lift up your face to God.

[y] Traditional rendering of Heb *Shaddai*
[z] Gk Syr Tg Vg: Heb *were crushed*   [a] Gk Syr:
Heb *them*   [b] Heb *him*

---

one is of any *use* to God, no matter how wise
or *righteous;* hence it can be presumed that the
divine judgment on Job is exact and justified.
Eliphaz proceeds to show that it is so by de-
tailing Job's wrongdoings (vv. 6–11).
22.12–30: He draws a picture of a tran-

scendent and impartial God of justice and invites Job to *return to the Almighty.* God will
hear his prayer (of repentance) and he will be
restored. This, of course, Job cannot do because of his very integrity.

27 You will pray to him, and he will
   hear you,
      and you will pay your vows.
28 You will decide on a matter, and
   it will be established for
   you,
      and light will shine on your
      ways.
29 When others are humiliated, you
   say it is pride;
      for he saves the humble.
30 He will deliver even those who are
   guilty;
      they will escape because of the
      cleanness of your hands." *c*

23 Then Job answered:
   2 "Today also my complaint
   is bitter; *d*
      his *e* hand is heavy despite my
      groaning.
3 Oh, that I knew where I might
   find him,
      that I might come even to his
      dwelling!
4 I would lay my case before him,
   and fill my mouth with
   arguments.
5 I would learn what he would
   answer me,
      and understand what he would
      say to me.
6 Would he contend with me in the
   greatness of his power?
      No; but he would give heed to
      me.
7 There an upright person could
   reason with him,
      and I should be acquitted
      forever by my judge.

8 "If I go forward, he is not there;
      or backward, I cannot perceive
      him;

9 on the left he hides, and I cannot
   behold him;
      I turn *f* to the right, but I
      cannot see him.
10 But he knows the way that I take;
   when he has tested me, I shall
   come out like gold.
11 My foot has held fast to his steps;
   I have kept his way and have
      not turned aside.
12 I have not departed from the
   commandment of his lips;
      I have treasured in *g* my bosom
      the words of his mouth.
13 But he stands alone and who can
   dissuade him?
      What he desires, that he does.
14 For he will complete what he
   appoints for me;
      and many such things are in his
      mind.
15 Therefore I am terrified at his
   presence;
      when I consider, I am in dread
      of him.
16 God has made my heart faint;
   the Almighty *h* has terrified me;
17 If only I could vanish in darkness,
   and thick darkness would cover
   my face! *i*

24 "Why are times not kept by the
   Almighty, *h*
      and why do those who know
      him never see his days?
2 The wicked *j* remove landmarks;
      they seize flocks and pasture
      them.

c Meaning of Heb uncertain   *d* Syr Vg Tg:
Heb *rebellious*   *e* Gk Syr: Heb *my*
f Syr Vg: Heb *he turns*   *g* Gk Vg: Heb *from*
h Traditional rendering of Heb *Shaddai*
i Or *But I am not destroyed by the darkness; he has
concealed the thick darkness from me*
j Gk: Heb *they*

23.1–24.17: **Reply of Job. 23.1–17:** Another search for God's presence; now Job
wants to *lay* his *case* before God, confident
that *he would give heed* (in contrast to 9.32–34;
13.20–21). But he cannot find God, despite
his own fidelity (vv. 11–12), and once more
he experiences *dread*. **13:** *He stands alone,* literally, "he is in one," perhaps denoting divine
sovereignty (see Deut 6.4).
   24.1–17: Job accuses God of not having
times for adjudication, and catalogues the sins
of the wicked. Meanwhile, the *poor* suffer,
but *God pays no attention to their prayer.*

3 They drive away the donkey of
the orphan;
they take the widow's ox for a
pledge.
4 They thrust the needy off the
road;
the poor of the earth all hide
themselves.
5 Like wild asses in the desert
they go out to their toil,
scavenging in the wasteland
food for their young.
6 They reap in a field not their own
and they glean in the vineyard
of the wicked.
7 They lie all night naked, without
clothing,
and have no covering in the
cold.
8 They are wet with the rain of the
mountains,
and cling to the rock for want
of shelter.

9 "There are those who snatch the
orphan child from the
breast,
and take as a pledge the infant
of the poor.
10 They go about naked, without
clothing;
though hungry, they carry the
sheaves;
11 between their terraces^k they press
out oil;
they tread the wine presses, but
suffer thirst.
12 From the city the dying groan,
and the throat of the wounded
cries for help;
yet God pays no attention to
their prayer.

13 "There are those who rebel against
the light,

who are not acquainted with its
ways,
and do not stay in its paths.
14 The murderer rises at dusk
to kill the poor and needy,
and in the night is like a thief.
15 The eye of the adulterer also waits
for the twilight,
saying, 'No eye will see me';
and he disguises his face.
16 In the dark they dig through
houses;
by day they shut themselves up;
they do not know the light.
17 For deep darkness is morning to
all of them;
for they are friends with the
terrors of deep darkness.

18 "Swift are they on the face of the
waters;
their portion in the land is
cursed;
no treader turns toward their
vineyards.
19 Drought and heat snatch away the
snow waters;
so does Sheol those who have
sinned.
20 The womb forgets them;
the worm finds them sweet;
they are no longer remembered;
so wickedness is broken like a
tree.
21 "They harm^l the childless
woman,
and do no good to the widow.
22 Yet God^m prolongs the life of the
mighty by his power;
they rise up when they despair
of life.

k Meaning of Heb uncertain   l Gk Tg: Heb feed
on or associate with   m Heb he

24.18–25: Not only are there textual problems in these verses, but they are also difficult
to interpret and seem out of character if spoken by Job, since they appear to affirm the
doctrine advocated by the friends: divine
punishment for the wicked. The various solutions proposed, however (to attribute them to
Bildad, or to Zophar, or to add "you say" in
v. 18, as in the RSV), remain quite hypothetical. The wicked seem to be described in
vv. 18–21; God's justice may be delayed
(vv. 22–23), but it never fails (vv. 24–25)

23 He gives them security, and they
are supported;
his eyes are upon their ways.
24 They are exalted a little while, and
then are gone;
they wither and fade like the
mallow;[n]
they are cut off like the heads of
grain.
25 If it is not so, who will prove me
a liar,
and show that there is nothing
in what I say?"

25 Then Bildad the Shuhite answered:
2 "Dominion and fear are with
God;[o]
he makes peace in his high
heaven.
3 Is there any number to his armies?
Upon whom does his light not
arise?
4 How then can a mortal be
righteous before God?
How can one born of woman
be pure?
5 If even the moon is not bright
and the stars are not pure in his
sight,
6 how much less a mortal, who is a
maggot,

and a human being, who is a
worm!"

26 Then Job answered:
2 "How you have helped one
who has no power!
How you have assisted the arm
that has no strength!
3 How you have counseled one who
has no wisdom,
and given much good advice!
4 With whose help have you uttered
words,
and whose spirit has come forth
from you?
5 The shades below tremble,
the waters and their inhabitants.
6 Sheol is naked before God,
and Abaddon has no covering.
7 He stretches out Zaphon[p] over the
void,
and hangs the earth upon
nothing.
8 He binds up the waters in his
thick clouds,
and the cloud is not torn open
by them.
9 He covers the face of the full
moon,

---

n Gk: Heb *like all others*   *o* Heb *him*
p Or *the North*

25.1–6: Third discourse of Bildad. This
is a suspiciously short speech, and some join
it with 26.5–14 in order to fill out the usual
pattern of complete speeches. There is no textual warrant for this. The theme is the uncleanness and abasement of mortals, which
had already been developed by Eliphaz (4.17–
21; 15.14–16) and Zophar (11.5–12); Job also
subscribed to this (9.2–12; 12.9–25; 14.4),
without using it for self-accusation.
26.1–14: Reply of Job. 2–4: Job's reply to
Bildad is harsh and sarcastic. He leaves unnamed the *spirit* that *has come forth from* Bildad,
but the implication is that it is evil or unworthy. **5–14:** A hymn about divine omnipotence. While such a theme is not impossible
in the mouth of Job, it has been ignored by
him in the third round of the discussion, to be
replaced by the theme of divine indifference.
Neither is it directly related to the immediate
context (vv. 2–4 and 27.2). Hence some

attribute the poem to Bildad, but without
textual evidence. **5–6:** The *shades* are those
who inhabit the *waters* that are found in
Sheol/Abaddon. They are not beyond the divine reach (12.22; Ps 139.8; Prov 15.11; see
also Am 9.2), but they enjoy no loving contact with God (Ps 6.5).
26.7: *Zaphon* means "north," and is usually associated with the dwelling of the gods
according to Canaanite mythology. The
mountain of their assembly (see Isa 14.13) is
comparable to the Greek Mt. Olympus; here,
however, it is said to be stretched out (like a
tent?) *over the void* (Hebrew "tōhū," mentioned in Gen 1.2). This picture differs from
the usual description, according to which the
earth is built on pillars (9.6) that seem to be
fixed in the abyss; here the earth *hangs upon
nothing.* **8:** Despite the weight of the *waters,
the cloud is not torn open by them.* **9:** God has his
own *cloud,* which covers *the face of the full moon*

and spreads over it his cloud.
10 He has described a circle on the
     face of the waters,
   at the boundary between light
     and darkness.
11 The pillars of heaven tremble,
   and are astounded at his rebuke.
12 By his power he stilled the Sea;
   by his understanding he struck
     down Rahab.
13 By his wind the heavens were
     made fair;
   his hand pierced the fleeing
     serpent.
14 These are indeed but the outskirts
     of his ways;
   and how small a whisper do we
     hear of him!
   But the thunder of his power
     who can understand?"

27 Job again took up his discourse
   and said:
2 "As God lives, who has taken
     away my right,
   and the Almighty, *q* who has
     made my soul bitter,
3 as long as my breath is in me
   and the spirit of God is in my
     nostrils,
4 my lips will not speak falsehood,
   and my tongue will not utter
     deceit.

5 Far be it from me to say that you
     are right;
   until I die I will not put away
     my integrity from me.
6 I hold fast my righteousness, and
     will not let it go;
   my heart does not reproach me
     for any of my days.

7 "May my enemy be like the
     wicked,
   and may my opponent be like
     the unrighteous.
8 For what is the hope of the
     godless when God cuts
     them off,
   when God takes away their
     lives?
9 Will God hear their cry
   when trouble comes upon them?
10 Will they take delight in the
     Almighty?*q*
   Will they call upon God at all
     times?
11 I will teach you concerning the
     hand of God;
   that which is with the
     Almighty*q* I will not
     conceal.
12 All of you have seen it yourselves;

q Traditional rendering of Heb *Shaddai*

(or perhaps his throne?). This prevents him
from being seen, and it is no wonder that Job
looked for him in vain (ch 23). **10:** The *circle*
is the God-given limit to the *waters* of chaos
on the horizon, which is *the boundary between
light and darkness.*
26.11. The *pillars of heaven* are supports for the
firmament or sky, and perhaps designate the
mountains on the horizon; they *tremble,* because of the divine *rebuke* of the waters, described in v. 12. *Sea* is personified as in 3.8,
where it is paired with Leviathan instead of
with *Rahab* (see 9.13 n.). These are the waters
of chaos (38.8–11). **13:** The description of the
conflict with chaos continues in the mention
of piercing *the fleeing serpent,* who is coupled
with Leviathan in Isa 27.1. **14:** *Outskirts of his
ways* is perhaps the outline of the divine plan
for the cosmos, and mortals merely catch *a
whisper* of *the thunder of his power* (see also
37.2, 4).

27.1–23: Job's Reply. 1: *Discourse:* Hebrew "mashal" probably indicates a very solemn statement, such as the oath that Job is
about to take (2–6). **2:** *As God lives* indicates
that Job is invoking the very God he has accused, to curse him if his words are not true.
Parallel to this is an oath by *the Almighty*
("Shaddai"). Job is really forcing God's hand.
3: *As long as* Job lives, to the very end, he will
uphold his oath. **5:** *Far be it from me* is another
type of oath formula (compare 2 Sam 20.20).
My integrity, since Job can not deny his conscience (nor the verdict of the Lord in 1.8!).
6: *Righteousness* is not to be mistaken for self-
righteousness.
27.7–23: Some contend that the "lost"
third speech of Zophar is in these verses, but
in the Hebrew text they are the words of Job.
7–10: An imprecation against an *enemy* (not
explicitly identified). Job relies, somewhat illogically, on divine justice against *the godless.*

why then have you become
altogether vain?

13 "This is the portion of the wicked
with God,
and the heritage that oppressors
receive from the
Almighty:*r*
14 If their children are multiplied, it
is for the sword;
and their offspring have not
enough to eat.
15 Those who survive them the
pestilence buries,
and their widows make no
lamentation.
16 Though they heap up silver like
dust,
and pile up clothing like clay—
17 they may pile it up, but the just
will wear it,
and the innocent will divide the
silver.
18 They build their houses like nests,
like booths made by sentinels of
the vineyard.
19 They go to bed with wealth, but
will do so no more;
they open their eyes, and it is
gone.
20 Terrors overtake them like a
flood;
in the night a whirlwind carries
them off.
21 The east wind lifts them up and
they are gone;
it sweeps them out of their
place.
22 It*s* hurls at them without pity;
they flee from its*t* power in
headlong flight.

23 It*s* claps its*t* hands at them,
and hisses at them from its*t*
place.

28 "Surely there is a mine for
silver,
and a place for gold to be
refined.
2 Iron is taken out of the earth,
and copper is smelted from ore.
3 Miners put*u* an end to darkness,
and search out to the farthest
bound
the ore in gloom and deep
darkness.
4 They open shafts in a valley away
from human habitation;
they are forgotten by travelers,
they sway suspended, remote
from people.
5 As for the earth, out of it comes
bread;
but underneath it is turned up as
by fire.
6 Its stones are the place of
sapphires,*v*
and its dust contains gold.

7 "That path no bird of prey knows,
and the falcon's eye has not seen
it.
8 The proud wild animals have not
trodden it;
the lion has not passed over it.

9 "They put their hand to the flinty
rock,

r Traditional rendering of Heb *Shaddai*
s Or *He* (that is God)   *t* Or *his*   *u* Heb *He
puts*   *v* Or *lapis lazuli*

27.13–23: Typical of the descriptions of
the portion of the wicked with God. Since it does
not fit with Job's viewpoint (the wicked are
not punished; 21.7–34), it is difficult to explain; perhaps it belongs to one of the friends,
or is intended as an imprecation upon Job's
enemies.
**28.1–28: Hymn on the inaccessibility
of wisdom.** This is not prefaced by the usual
introduction, and it is difficult to consider it
as a continuation of the previous chapter. It is

not addressed to anyone, and it is out of character with the usual complaints of Job. Hence
it is often called a "bridge," or "interlude,"
but neither is helpful. The point is nevertheless clear: One can find precious stones that
are deep in the earth, but no one can find
wisdom (see also vv. 12, 20) because it is with
God, who alone knows the *way* to wisdom
(v. 23). In fact, his vision is perfect and hence
he *sees everything* (v. 24), and seems to have
put wisdom somewhere (v. 27).

and overturn mountains by the
roots.

10 They cut out channels in the
rocks,
and their eyes see every precious
thing.

11 The sources of the rivers they
probe;*w*
hidden things they bring to
light.

12 "But where shall wisdom be
found?
And where is the place of
understanding?

13 Mortals do not know the way to
it,*x*
and it is not found in the land
of the living.

14 The deep says, 'It is not in me,'
and the sea says, 'It is not with
me.'

15 It cannot be gotten for gold,
and silver cannot be weighed
out as its price.

16 It cannot be valued in the gold of
Ophir,
in precious onyx or sapphire.*y*

17 Gold and glass cannot equal it,
nor can it be exchanged for
jewels of fine gold.

18 No mention shall be made of coral
or of crystal;
the price of wisdom is above
pearls.

19 The chrysolite of Ethiopia*z* cannot
compare with it,
nor can it be valued in pure
gold.

20 "Where then does wisdom come
from?
And where is the place of
understanding?

21 It is hidden from the eyes of all
living,
and concealed from the birds of
the air.

22 Abaddon and Death say,
'We have heard a rumor of it
with our ears.'

23 "God understands the way to it,
and he knows its place.

24 For he looks to the ends of the
earth,
and sees everything under the
heavens.

25 When he gave to the wind its
weight,
and apportioned out the waters
by measure;

26 when he made a decree for the
rain,
and a way for the thunderbolt;

27 then he saw it and declared it;
he established it, and searched it
out.

28 And he said to humankind,
'Truly, the fear of the Lord, that
is wisdom;
and to depart from evil is
understanding.'"

29 Job again took up his discourse
and said:

2 "Oh, that I were as in the months
of old,
as in the days when God
watched over me;

3 when his lamp shone over my
head,
and by his light I walked
through darkness;

4 when I was in my prime,

---

w Gk Vg: Heb *bind* *x* Gk: Heb *its price*
y Or *lapis lazuli* *z* Or *Nubia*; Heb *Cush*

28.28: The final verse is in a certain tension
with the previous lines; for humankind, wisdom is fear of God (the sacred name YHWH
is not used) and avoidance of evil—a link
with 1.2, 8. Obviously this is not the transcendent wisdom described in vv. 1–27.
29.1–31.40: The final defense of Job.
Contrasting his happy past (ch 29) with the
present distress (ch 30), Job delivers his final
plea or oath of clearance (ch 31). It is in the
style of a soliloquy, for Job no longer addresses the friends. **1:** *Discourse*, as in 27.1,
there is the same solemn introduction ("mashal"). **2–6:** In former days Job enjoyed a
relationship with God, and the extravagant
imagery in v. 6 indicates his prosperity.

when the friendship of God was
upon my tent;

5 when the Almighty*a* was still
with me,
when my children were around
me;

6 when my steps were washed with
milk,
and the rock poured out for me
streams of oil!

7 When I went out to the gate of
the city,
when I took my seat in the
square,

8 the young men saw me and
withdrew,
and the aged rose up and stood;

9 the nobles refrained from talking,
and laid their hands on their
mouths;

10 the voices of princes were hushed,
and their tongues stuck to the
roof of their mouths.

11 When the ear heard, it
commended me,
and when the eye saw, it
approved;

12 because I delivered the poor who
cried,
and the orphan who had no
helper.

13 The blessing of the wretched came
upon me,
and I caused the widow's heart
to sing for joy.

14 I put on righteousness, and it
clothed me;
my justice was like a robe and a
turban.

15 I was eyes to the blind,
and feet to the lame.

16 I was a father to the needy,
and I championed the cause of
the stranger.

17 I broke the fangs of the
unrighteous,

and made them drop their prey
from their teeth.

18 Then I thought, 'I shall die in my
nest,
and I shall multiply my days
like the phoenix;*b*

19 my roots spread out to the waters,
with the dew all night on my
branches;

20 my glory was fresh with me,
and my bow ever new in my
hand.'

21 "They listened to me, and waited,
and kept silence for my counsel.

22 After I spoke they did not speak
again,
and my word dropped upon
them like dew.*c*

23 They waited for me as for the
rain;
they opened their mouths as for
the spring rain.

24 I smiled on them when they had
no confidence;
and the light of my countenance
they did not extinguish.*d*

25 I chose their way, and sat as chief,
and I lived like a king among
his troops,
like one who comforts
mourners.

30 "But now they make sport of
me,
those who are younger than I,
whose fathers I would have
disdained
to set with the dogs of my
flock.

2 What could I gain from the
strength of their hands?
All their vigor is gone.

a Traditional rendering of Heb *Shaddai*
b Or *like sand*   c Heb lacks *like dew*
d Meaning of Heb uncertain

29.7–10: He was honored in the community (see also vv. 21–25). **11–20:** His reputation was well deserved because of his care for the poor; he had no reason to think that his lot would change. **18:** *Like the phoenix,* an allusion to the myth that the phoenix rises to a new life from its ashes.

30.1–8: Job describes the treatment he now receives from those who are the outcasts of society.

3 Through want and hard hunger
   they gnaw the dry and desolate
      ground,
4 they pick mallow and the leaves of
      bushes,
   and to warm themselves the
      roots of broom.
5 They are driven out from society;
   people shout after them as after
      a thief.
6 In the gullies of wadis they must
      live,
   in holes in the ground, and in
      the rocks.
7 Among the bushes they bray;
   under the nettles they huddle
      together.
8 A senseless, disreputable brood,
   they have been whipped out of
      the land.

9 "And now they mock me in song;
   I am a byword to them.
10 They abhor me, they keep aloof
      from me;
   they do not hesitate to spit at
      the sight of me.
11 Because God has loosed my
      bowstring and humbled
      me,
   they have cast off restraint in
      my presence.
12 On my right hand the rabble rise
      up;
   they send me sprawling,
   and build roads for my ruin.
13 They break up my path,
   they promote my calamity;
   no one restrains^e them.
14 As through a wide breach they
      come;
   amid the crash they roll on.
15 Terrors are turned upon me;
   my honor is pursued as by the
      wind,
   and my prosperity has passed
      away like a cloud.

16 "And now my soul is poured out
      within me;
   days of affliction have taken
      hold of me.
17 The night racks my bones,
   and the pain that gnaws me
      takes no rest.
18 With violence he seizes my
      garment;^f
   he grasps me by^g the collar of
      my tunic.
19 He has cast me into the mire,
   and I have become like dust and
      ashes.
20 I cry to you and you do not
      answer me;
   I stand, and you merely look at
      me.
21 You have turned cruel to me;
   with the might of your hand
      you persecute me.
22 You lift me up on the wind, you
      make me ride on it,
   and you toss me about in the
      roar of the storm.
23 I know that you will bring me to
      death,
   and to the house appointed for
      all living.

24 "Surely one does not turn against
      the needy,^h
   when in disaster they cry for
      help.^i
25 Did I not weep for those whose
      day was hard?
   Was not my soul grieved for the
      poor?
26 But when I looked for good, evil
      came;
   and when I waited for light,
      darkness came.
27 My inward parts are in turmoil,
   and are never still;

e Cn: Heb *helps*   f Gk: Heb *my garment is*
disfigured   g Heb *like*   h Heb *ruin*
i Cn: Meaning of Heb uncertain

30.9–11: Opposition comes to him from
the *rabble*. But it is God who has broken his
bowstring (by undoing Job's bow, rendering
him powerless).

30.16–23: God, not humankind, is Job's
real enemy. 24–31: A piteous lament about
his present condition.

days of affliction come to meet
   me.
28 I go about in sunless gloom;
   I stand up in the assembly and
      cry for help.
29 I am a brother of jackals,
   and a companion of ostriches.
30 My skin turns black and falls from
      me,
   and my bones burn with heat.
31 My lyre is turned to mourning,
   and my pipe to the voice of
      those who weep.

31 "I have made a covenant with
   my eyes;
   how then could I look upon a
      virgin?
2 What would be my portion from
      God above,
   and my heritage from the
      Almighty*ʲ* on high?
3 Does not calamity befall the
      unrighteous,
   and disaster the workers of
      iniquity?
4 Does he not see my ways,
   and number all my steps?

5 "If I have walked with falsehood,
   and my foot has hurried to
      deceit—
6 let me be weighed in a just
      balance,
   and let God know my
      integrity!—
7 if my step has turned aside from
      the way,
   and my heart has followed my
      eyes,
   and if any spot has clung to my
      hands;
8 then let me sow, and another eat;
   and let what grows for me be
      rooted out.

9 "If my heart has been enticed by a
      woman,
   and I have lain in wait at my
      neighbor's door;
10 then let my wife grind for
      another,
   and let other men kneel over
      her.
11 For that would be a heinous
      crime;
   that would be a criminal
      offense;
12 for that would be a fire consuming
      down to Abaddon,
   and it would burn to the root
      all my harvest.

13 "If I have rejected the cause of my
      male or female slaves,
   when they brought a complaint
      against me;
14 what then shall I do when God
      rises up?
   When he makes inquiry, what
      shall I answer him?
15 Did not he who made me in the
      womb make them?
   And did not one fashion us in
      the womb?

16 "If I have withheld anything that
      the poor desired,
   or have caused the eyes of the
      widow to fail,
17 or have eaten my morsel alone,
   and the orphan has not eaten
      from it—
18 for from my youth I reared the
      orphan*ᵏ* like a father,
   and from my mother's womb I
      guided the widow*ˡ*—

j Traditional rendering of Heb *Shaddai*
k Heb *him*   *l* Heb *her*

31.1–40: A negative confession, consisting of a series of oaths calling down a curse if one has acted wrongly. **5:** *If* begins the customary form of the oath, to which the consequence is usually, but not always, added: If I have done so and so, then let so and so happen to me. The exact number of oaths Job takes is not easy to determine (between nine and sixteen). He mentions lust, dishonesty, adultery, treatment of slaves, mistreatment of the poor, avarice, idolatry (v. 27 may refer to throwing kisses with the *hand*), revenge, failure of hospitality, hypocrisy, and mistreatment of the land.

19 if I have seen anyone perish for
lack of clothing,
or a poor person without
covering,

20 whose loins have not blessed me,
and who was not warmed with
the fleece of my sheep;

21 if I have raised my hand against
the orphan,
because I saw I had supporters
at the gate;

22 then let my shoulder blade fall
from my shoulder,
and let my arm be broken from
its socket.

23 For I was in terror of calamity
from God,
and I could not have faced his
majesty.

24 "If I have made gold my trust,
or called fine gold my
confidence;

25 if I have rejoiced because my
wealth was great,
or because my hand had gotten
much;

26 if I have looked at the sun[m] when
it shone,
or the moon moving in
splendor,

27 and my heart has been secretly
enticed,
and my mouth has kissed my
hand;

28 this also would be an iniquity to
be punished by the judges,
for I should have been false to
God above.

29 "If I have rejoiced at the ruin of
those who hated me,
or exulted when evil overtook
them—

30 I have not let my mouth sin
by asking for their lives with a
curse—

31 if those of my tent ever said,

'O that we might be sated with
his flesh!'[n]—

32 the stranger has not lodged in the
street;
I have opened my doors to the
traveler—

33 if I have concealed my
transgressions as others
do,[o]
by hiding my iniquity in my
bosom,

34 because I stood in great fear of the
multitude,
and the contempt of families
terrified me,
so that I kept silence, and did
not go out of doors—

35 Oh, that I had one to hear me!
(Here is my signature! let the
Almighty[p] answer me!)
Oh, that I had the indictment
written by my adversary!

36 Surely I would carry it on my
shoulder;
I would bind it on me like a
crown;

37 I would give him an account of all
my steps;
like a prince I would approach
him.

38 "If my land has cried out against
me,
and its furrows have wept
together;

39 if I have eaten its yield without
payment,
and caused the death of its
owners;

40 let thorns grow instead of wheat,
and foul weeds instead of
barley."

The words of Job are ended.

m Heb *the light*   n Meaning of Heb
uncertain   o Or *as Adam did*   p Traditional
rendering of Heb *Shaddai*

31.35: With considerable bravado Job offers his *signature*, literally his "taw," the last
letter of the Hebrew alphabet, which was
shaped like a cross, meaning "mark," or signature. Any *indictment* of him would be merely a badge of honor, since he can give full
account of his life.

32 So these three men ceased to answer Job, because he was righteous in his own eyes. 2 Then Elihu son of Barachel the Buzite, of the family of Ram, became angry. He was angry at Job because he justified himself rather than God; 3 he was angry also at Job's three friends because they had found no answer, though they had declared Job to be in the wrong. *q* 4 Now Elihu had waited to speak to Job, because they were older than he. 5 But when Elihu saw that there was no answer in the mouths of these three men, he became angry.

6 Elihu son of Barachel the Buzite answered:
   "I am young in years,
      and you are aged;
   therefore I was timid and afraid
      to declare my opinion to you.
7 I said, 'Let days speak,
      and many years teach wisdom.'
8 But truly it is the spirit in a
         mortal,
      the breath of the Almighty, *r*
      that makes for
         understanding.
9 It is not the old *s* that are wise,
      nor the aged that understand
      what is right.

10 Therefore I say, 'Listen to me;
      let me also declare my opinion.'

11 "See, I waited for your words,
      I listened for your wise sayings,
      while you searched out what to
         say.
12 I gave you my attention,
      but there was in fact no one that
         confuted Job,
   no one among you that
      answered his words.
13 Yet do not say, 'We have found
         wisdom;
   God may vanquish him, not a
         human.'
14 He has not directed his words
         against me,
   and I will not answer him with
      your speeches.

15 "They are dismayed, they answer
      no more;
   they have not a word to say.
16 And am I to wait, because they do
      not speak,

---

q Another ancient tradition reads *answer, and had put God in the wrong*   *r* Traditional rendering of Heb *Shaddai*   *s* Gk Syr Vg: Heb *many*

32.1–37.24: Discourses of Elihu. Many commentators believe that these chapters, with their peculiar language (containing many Aramaic words) and different style (Elihu seems to quote Job's speeches and even the as-yet-undelivered discourse of the Lord), belong to another hand than that of the poet. In addition it seems to be interruptive: Job's final challenge (31.35–37) calls for the immediate manifestation of God (which occurs now in ch 38). The epilogue (42.7–17) ignores Elihu, which possibly indicates that he did not belong to the original cast of characters; there is, however, no textual evidence that the book ever existed without these chapters. It is possible to construe Elihu as the answer to Job's request for someone to hear him (31.35). Elihu does develop the revelatory function of suffering (33.19–33; 36.5–15); some would argue that his speeches serve to prepare dramatically, psychologically, and even theologically for the intervention of the Lord.
   32.1–33.7: Prose and Poetic introduc-tion. After the prose narrative, in which Elihu appears to be hotheaded and impatient (vv. 1–6), he launches into a bombastic and prolix introduction that continues to 33.7. **2:** *Elihu* means "My God is he." *Buzite* suggests that he was an Aramean living not far from Edom, for Buz in Gen 22.21 is a brother of Uz (see Job 1.1 n.) and an Aramean (compare Gen 11.26–32). *He justified himself rather than God:* Elihu quite correctly estimates the situation. Job has risen for the defense of self, not God. **3:** The reading in note *p* (*had put God in the wrong*) may be the original one, which scribes found offensive and corrected in the manuscripts. **5:** *No answer,* since the silence of the friends after Job's final words in ch 31 could be taken as an admission of defeat (see vv. 11–14).
   32.8: *The breath of the Almighty* need not refer to anything more than the breath of life (Gen 2.7), because in 33.4–7 Elihu does not claim to be any different from Job in his humanity. **18:** *I am full of words,* indeed!

because they stand there, and
answer no more?

17 I also will give my answer;
I also will declare my opinion.

18 For I am full of words;
the spirit within me constrains
me.

19 My heart is indeed like wine that
has no vent;
like new wineskins, it is ready
to burst.

20 I must speak, so that I may find
relief;
I must open my lips and
answer.

21 I will not show partiality to any
person
or use flattery toward anyone.

22 For I do not know how to
flatter—
or my Maker would soon put
an end to me!

33 "But now, hear my speech,
O Job,
and listen to all my words.

2 See, I open my mouth;
the tongue in my mouth speaks.

3 My words declare the uprightness
of my heart,
and what my lips know they
speak sincerely.

4 The spirit of God has made me,
and the breath of the Almighty^t^
gives me life.

5 Answer me, if you can;
set your words in order before
me; take your stand.

6 See, before God I am as you are;
I too was formed from a piece
of clay.

7 No fear of me need terrify you;
my pressure will not be heavy
on you.

8 "Surely, you have spoken in my
hearing,
and I have heard the sound of
your words.

9 You say, 'I am clean, without
transgression;
I am pure, and there is no
iniquity in me.

10 Look, he finds occasions against
me,
he counts me as his enemy;

11 he puts my feet in the stocks,
and watches all my paths.'

12 "But in this you are not right. I
will answer you:
God is greater than any mortal.

13 Why do you contend against him,
saying, 'He will answer none of
my^u^ words'?

14 For God speaks in one way,
and in two, though people do
not perceive it.

15 In a dream, in a vision of the
night,
when deep sleep falls on
mortals,
while they slumber on their
beds,

16 then he opens their ears,
and terrifies them with
warnings,

17 that he may turn them aside from
their deeds,
and keep them from pride,

18 to spare their souls from the Pit,
their lives from traversing the
River.

19 They are also chastened with pain
upon their beds,
and with continual strife in their
bones,

t Traditional rendering of Heb *Shaddai*
u Compare Gk: Heb *his*

33.1: An explicit reference to Job, and the following verses continue the bombast.
33.8–33: Elihu summarizes Job's claims and gives the rather simplistic answer that *God is greater than any mortal.* But he also argues that God does speak to humans *in a dream* and also by chastening them *with pain.* He also describes the intercessory role of *an angel, a mediator,* who will offer a *ransom,* so that the sufferer is healed. But this occurs apparently only on the condition that a person admits: *I sinned* (v. 27). This is obviously not appropriate to Job's situation, but Elihu rattles on (vv. 31–33).

20 so that their lives loathe bread,
    and their appetites dainty food.
21 Their flesh is so wasted away that
        it cannot be seen;
    and their bones, once invisible,
        now stick out.
22 Their souls draw near the Pit,
    and their lives to those who
        bring death.
23 Then, if there should be for one of
        them an angel,
    a mediator, one of a thousand,
    one who declares a person
        upright,
24 and he is gracious to that person,
        and says,
    'Deliver him from going down
        into the Pit;
    I have found a ransom;
25 let his flesh become fresh with
        youth;
    let him return to the days of his
        youthful vigor.'
26 Then he prays to God, and is
        accepted by him,
    he comes into his presence with
        joy,
    and God*ᵛ* repays him for his
        righteousness.
27   That person sings to others and
        says,
    'I sinned, and perverted what was
        right,
    and it was not paid back to me.
28 He has redeemed my soul from
        going down to the Pit,
    and my life shall see the light.'

29 "God indeed does all these things,
    twice, three times, with
        mortals,
30 to bring back their souls from the
        Pit,
    so that they may see the light of
        life.*ʷ*
31 Pay heed, Job, listen to me;

be silent, and I will speak.
32 If you have anything to say,
        answer me;
    speak, for I desire to justify
        you.
33 If not, listen to me;
    be silent, and I will teach you
        wisdom."

34 Then Elihu continued and said:
2 "Hear my words, you wise
        men,
    and give ear to me, you who
        know;
3 for the ear tests words
    as the palate tastes food.
4 Let us choose what is right;
    let us determine among
        ourselves what is good.
5 For Job has said, 'I am innocent,
    and God has taken away my
        right;
6 in spite of being right I am
        counted a liar;
    my wound is incurable, though
        I am without transgression.'
7 Who is there like Job,
    who drinks up scoffing like
        water,
8 who goes in company with
        evildoers
    and walks with the wicked?
9 For he has said, 'It profits one
        nothing
    to take delight in God.'

10 "Therefore, hear me, you who
        have sense,
    far be it from God that he
        should do wickedness,
    and from the Almighty*ˣ* that he
        should do wrong.

v Heb *he*
light of life   *w* Syr: Heb *to be lighted with the
light of life*   *x* Traditional rendering of Heb
Shaddai

34.1–37: Second discourse of Elihu.
Elihu addresses *wise men* and cites Job for
claiming to be *innocent* and charging that *God
has taken away* his *right* (v. 5). But it is incon-
ceivable that God *should do wrong.* Elihu then
gives a lengthy lecture on the divine exercise
of justice in governing the world. He concludes (vv. 31–37) that the *wise* will properly
condemn Job.

11 For according to their deeds he
will repay them,
and according to their ways he
will make it befall them.
12 Of a truth, God will not do
wickedly,
and the Almighty^y will not
pervert justice.
13 Who gave him charge over the
earth
and who laid on him^z the
whole world?
14 If he should take back his spirit^a
to himself,
and gather to himself his breath,
15 all flesh would perish together,
and all mortals return to dust.

16 "If you have understanding, hear
this;
listen to what I say.
17 Shall one who hates justice
govern?
Will you condemn one who is
righteous and mighty,
18 who says to a king, 'You
scoundrel!'
and to princes, 'You wicked
men!';
19 who shows no partiality to nobles,
nor regards the rich more than
the poor,
for they are all the work of his
hands?
20 In a moment they die;
at midnight the people are
shaken and pass away,
and the mighty are taken away
by no human hand.

21 "For his eyes are upon the ways of
mortals,
and he sees all their steps.
22 There is no gloom or deep
darkness
where evildoers may hide
themselves.
23 For he has not appointed a time^b
for anyone
to go before God in judgment.
24 He shatters the mighty without
investigation,
and sets others in their place.

25 Thus, knowing their works,
he overturns them in the night,
and they are crushed.
26 He strikes them for their
wickedness
while others look on,
27 because they turned aside from
following him,
and had no regard for any of his
ways,
28 so that they caused the cry of the
poor to come to him,
and he heard the cry of the
afflicted—
29 When he is quiet, who can
condemn?
When he hides his face, who can
behold him,
whether it be a nation or an
individual?—
30 so that the godless should not
reign,
or those who ensnare the
people.

31 "For has anyone said to God,
'I have endured punishment; I
will not offend any more;
32 teach me what I do not see;
if I have done iniquity, I will do
it no more'?
33 Will he then pay back to suit you,
because you reject it?
For you must choose, and not I;
therefore declare what you
know.^c
34 Those who have sense will say to
me,
and the wise who hear me will
say,
35 'Job speaks without knowledge,
his words are without insight.'
36 Would that Job were tried to the
limit,
because his answers are those of
the wicked.
37 For he adds rebellion to his sin;
he claps his hands among us,

y Traditional rendering of Heb. *Shaddai*
z Heb lacks *on him* a Heb *his heart his spirit*
b Cn: Heb *yet* c Meaning of Heb of
verses 29-33 uncertain

and multiplies his words against
God."

35 Elihu continued and said:
2 "Do you think this to be
just?
You say, 'I am in the right
before God.'
3 If you ask, 'What advantage have
I?
How am I better off than if I
had sinned?'
4 I will answer you
and your friends with you.
5 Look at the heavens and see;
observe the clouds, which are
higher than you.
6 If you have sinned, what do you
accomplish against him?
And if your transgressions are
multiplied, what do you do
to him?
7 If you are righteous, what do you
give to him;
or what does he receive from
your hand?
8 Your wickedness affects others
like you,
and your righteousness, other
human beings.

9 "Because of the multitude of
oppressions people cry out;
they call for help because of the
arm of the mighty.
10 But no one says, 'Where is God
my Maker,
who gives strength in the night,
11 who teaches us more than the
animals of the earth,
and makes us wiser than the
birds of the air?'
12 There they cry out, but he does
not answer,

because of the pride of
evildoers.
13 Surely God does not hear an
empty cry,
nor does the Almighty*d* regard
it.
14 How much less when you say that
you do not see him,
that the case is before him, and
you are waiting for him!
15 And now, because his anger does
not punish,
and he does not greatly heed
transgression, *e*
16 Job opens his mouth in empty
talk,
he multiplies words without
knowledge."

36 Elihu continued and said:
2 "Bear with me a little, and I
will show you,
for I have yet something to say
on God's behalf.
3 I will bring my knowledge from
far away,
and ascribe righteousness to my
Maker.
4 For truly my words are not false;
one who is perfect in knowledge
is with you.

5 "Surely God is mighty and does
not despise any;
he is mighty in strength of
understanding.
6 He does not keep the wicked
alive,
but gives the afflicted their
right.

d Traditional rendering of Heb *Shaddai*
e Theodotion Symmachus Compare Vg:
Meaning of Heb uncertain

35.1–16: Third discourse of Elihu. He
claims to *answer* (v. 4) Job's argument to be *in
the right* (v. 2). God is transcendent, and receives nothing from Job's *hand* (v. 7). When
there is no answer to the cry of the oppressed,
it is *because of the pride of evildoers* (v. 12). Job's
case is even less worthy, *empty talk* (v. 16). **10:**

Strength in the night is literally "songs in the
night."
36.1–37.24: Fourth discourse of Elihu.
He defends divine justice, which distinguishes between *the wicked* and *the afflicted*
(v. 6). Affliction is an *instruction* that will lead
to *prosperity* (v. 11), if heeded; see v. 15. But

7 He does not withdraw his eyes
        from the righteous,
    but with kings on the throne
        he sets them forever, and they
        are exalted.
8 And if they are bound in fetters
        and caught in the cords of
        affliction,
9 then he declares to them their
        work
    and their transgressions, that
        they are behaving
        arrogantly.
10 He opens their ears to instruction,
    and commands that they return
        from iniquity.
11 If they listen, and serve him,
    they complete their days in
        prosperity,
    and their years in pleasantness.
12 But if they do not listen, they
        shall perish by the sword,
    and die without knowledge.

13 "The godless in heart cherish
        anger;
    they do not cry for help when
        he binds them.
14 They die in their youth,
    and their life ends in shame.*f*
15 He delivers the afflicted by their
        affliction,
    and opens their ear by adversity.
16 He also allured you out of distress
    into a broad place where there
        was no constraint,
    and what was set on your table
        was full of fatness.

17 "But you are obsessed with the
        case of the wicked;
    judgment and justice seize you.
18 Beware that wrath does not entice
        you into scoffing,

and do not let the greatness of
        the ransom turn you aside.
19 Will your cry avail to keep you
        from distress,
    or will all the force of your
        strength?
20 Do not long for the night,
    when peoples are cut off in their
        place.
21 Beware! Do not turn to iniquity;
    because of that you have been
        tried by affliction.
22 See, God is exalted in his power;
    who is a teacher like him?
23 Who has prescribed for him his
        way,
    or who can say, 'You have done
        wrong'?

24 "Remember to extol his work,
    of which mortals have sung.
25 All people have looked on it;
    everyone watches it from far
        away.
26 Surely God is great, and we do
        not know him;
    the number of his years is
        unsearchable.
27 For he draws up the drops of
        water;
    he distills*g* his mist in rain,
28 which the skies pour down
    and drop upon mortals
        abundantly.
29 Can anyone understand the
        spreading of the clouds,
    the thunderings of his pavilion?
30 See, he scatters his lightning
        around him
    and covers the roots of the sea.
31 For by these he governs peoples;

f Heb *ends among the temple prostitutes*
g Cn: Heb *they distill*

Job's situation is different (vv. 17–23): his iniquity is the reason he has been *tried by affliction* (v. 21). He urges Job to praise God and
launches into a hymn to the divine power.
God is the sovereign ruler of nature; his purpose and his benevolence appear in the majestic unfolding of the seasons: autumn (36.27–
33, *rain* and *clouds*), winter (37.1–13, *snow* and

ice), and summer (37.14–22, *hot* and *south
wind*). Throughout the hymn, thunder and
lightning and clouds have a major role, and in
37.14–22 Elihu speaks more personally to Job
in the style of the speeches of the Lord in
chs 38–41. **37.24**: A typical wisdom ending;
compare Job 28.28; Prov 1.7.

he gives food in abundance.
32 He covers his hands with the
lightning,
and commands it to strike the
mark.
33 Its crashing[h] tells about him;
he is jealous[h] with anger against
iniquity.

37 "At this also my heart
trembles,
and leaps out of its place.
2 Listen, listen to the thunder of his
voice
and the rumbling that comes
from his mouth.
3 Under the whole heaven he lets it
loose,
and his lightning to the corners
of the earth.
4 After it his voice roars;
he thunders with his majestic
voice
and he does not restrain the
lightnings[i] when his voice
is heard.
5 God thunders wondrously with
his voice;
he does great things that we
cannot comprehend.
6 For to the snow he says, 'Fall on
the earth';
and the shower of rain, his
heavy shower of rain,
7 serves as a sign on everyone's
hand,
so that all whom he has made
may know it.[j]
8 Then the animals go into their
lairs
and remain in their dens.
9 From its chamber comes the
whirlwind,
and cold from the scattering
winds.
10 By the breath of God ice is given,
and the broad waters are frozen
fast.
11 He loads the thick cloud with
moisture;
the clouds scatter his lightning.
12 They turn round and round by his
guidance,

to accomplish all that he
commands them
on the face of the habitable
world.
13 Whether for correction, or for his
land,
or for love, he causes it to
happen.
14 "Hear this, O Job;
stop and consider the wondrous
works of God.
15 Do you know how God lays his
command upon them,
and causes the lightning of his
cloud to shine?
16 Do you know the balancings of
the clouds,
the wondrous works of the one
whose knowledge is
perfect,
17 you whose garments are hot
when the earth is still because of
the south wind?
18 Can you, like him, spread out the
skies,
hard as a molten mirror?
19 Teach us what we shall say to
him;
we cannot draw up our case
because of darkness.
20 Should he be told that I want to
speak?
Did anyone ever wish to be
swallowed up?
21 Now, no one can look on the
light
when it is bright in the skies,
when the wind has passed and
cleared them.
22 Out of the north comes golden
splendor;
around God is awesome
majesty.
23 The Almighty[k]—we cannot find
him;
he is great in power and justice,
and abundant righteousness he
will not violate.

h Meaning of Heb uncertain   i Heb them
j Meaning of Heb of verse 7 uncertain
k Traditional rendering of Heb Shaddai

24 Therefore mortals fear him;
    he does not regard any who are
      wise in their own conceit."

38 Then the LORD answered Job out
   of the whirlwind:
2 "Who is this that darkens counsel
    by words without
      knowledge?
3 Gird up your loins like a man,
    I will question you, and you
      shall declare to me.

4 "Where were you when I laid the
    foundation of the earth?
   Tell me, if you have
    understanding.
5 Who determined its
    measurements—surely you
    know!
   Or who stretched the line upon
    it?
6 On what were its bases sunk,
    or who laid its cornerstone
7 when the morning stars sang
    together
   and all the heavenly beings[l]
    shouted for joy?

8 "Or who shut in the sea with
    doors
   when it burst out from the
    womb?—
9 when I made the clouds its
    garment,
   and thick darkness its swaddling
    band,
10 and prescribed bounds for it,
    and set bars and doors,

11 and said, 'Thus far shall you
    come, and no farther,
   and here shall your proud waves
    be stopped'?

12 "Have you commanded the
    morning since your days
    began,
   and caused the dawn to know
    its place,
13 so that it might take hold of the
    skirts of the earth,
   and the wicked be shaken out of
    it?
14 It is changed like clay under the
    seal,
   and it is dyed[m] like a garment.
15 Light is withheld from the
    wicked,
   and their uplifted arm is broken.

16 "Have you entered into the
    springs of the sea,
   or walked in the recesses of the
    deep?
17 Have the gates of death been
    revealed to you,
   or have you seen the gates of
    deep darkness?
18 Have you comprehended the
    expanse of the earth?
   Declare, if you know all this.

19 "Where is the way to the dwelling
    of light,

l Heb *sons of God*   m Cn: Heb *and they stand
forth*

**38.1–42.6: The voice from the whirl-
wind.** There are two remarkable speeches of
the LORD, and two replies from Job.
**38.1–40.2: First speech of the LORD.
38.1:** *The whirlwind,* a frequent setting of
theophanies, i.e. divine appearances (Nah
1.3; Zech 9.14; Ps 18.7–15; 50.3; Ezek 1.4;
Hab ch 3). **2–3:** Throughout, Job has asked
why misfortune happened (e.g. 3.11, 16, 20;
13.24). The LORD now offers him the right to
challenge the divine rule, but first he has
questions to put to Job. These questions seem
irrelevant and unexpected, yet playful (only
38.21 displays sharp irony). Their apparent
purpose is to lead Job into the mystery of
God's creation, of which Job and his suffering
form only one part. The questions range
through the whole of creation: *earth* (vv. 4–
7); *sea* (vv. 8–11); *dawn* (vv. 12–15); Sheol
(*the deep, death,* vv. 16–18); *light* and *darkness*
(vv. 19–21); weather phenomena (vv. 22–
30); the constellations (vv. 31–33); *clouds* and
water (vv. 34–38); the providential care of the
animal world (from *lion* to *eagle,* 38.39–
39.40). **7:** *Heavenly beings,* see 1.6 n.; here
they are associated with the *morning stars* on
the festive occasion of the creative actions.
8: *Sea,* see 3.8 n.

and where is the place of
darkness,
20 that you may take it to its
territory
and that you may discern the
paths to its home?
21 Surely you know, for you were
born then,
and the number of your days is
great!

22 "Have you entered the storehouses
of the snow,
or have you seen the
storehouses of the hail,
23 which I have reserved for the time
of trouble,
for the day of battle and war?
24 What is the way to the place
where the light is
distributed,
or where the east wind is
scattered upon the earth?

25 "Who has cut a channel for the
torrents of rain,
and a way for the thunderbolt,
26 to bring rain on a land where no
one lives,
on the desert, which is empty of
human life,
27 to satisfy the waste and desolate
land,
and to make the ground put
forth grass?

28 "Has the rain a father,
or who has begotten the drops
of dew?
29 From whose womb did the ice
come forth,
and who has given birth to the
hoarfrost of heaven?
30 The waters become hard like
stone,

and the face of the deep is
frozen.

31 "Can you bind the chains of the
Pleiades,
or loose the cords of Orion?
32 Can you lead forth the Mazzaroth
in their season,
or can you guide the Bear with
its children?
33 Do you know the ordinances of
the heavens?
Can you establish their rule on
the earth?

34 "Can you lift up your voice to the
clouds,
so that a flood of waters may
cover you?
35 Can you send forth lightnings, so
that they may go
and say to you, 'Here we are'?
36 Who has put wisdom in the
inward parts,[n]
or given understanding to the
mind?[n]
37 Who has the wisdom to number
the clouds?
Or who can tilt the waterskins
of the heavens,
38 when the dust runs into a mass
and the clods cling together?

39 "Can you hunt the prey for the
lion,
or satisfy the appetite of the
young lions,
40 when they crouch in their dens,
or lie in wait in their covert?
41 Who provides for the raven its
prey,
when its young ones cry to
God,
and wander about for lack of
food?

n Meaning of Heb uncertain

38.21: A truly sarcastic jibe; Job would have had to have been *born* before the creation, like Wisdom in Prov 8.22ff. **26:** Astonishingly, *rain* falls *on a land where no one lives;* such puzzling prodigality!

38.32: The determination of these stars is uncertain.
39.13–18: The text is uncertain; perhaps the *ostrich,* lacking *wisdom* (v. 17), is a point made against Job.

39

"Do you know when the
 mountain goats give birth?
 Do you observe the calving of
 the deer?
2 Can you number the months that
 they fulfill,
 and do you know the time
 when they give birth,
3 when they crouch to give birth to
 their offspring,
 and are delivered of their
 young?
4 Their young ones become strong,
 they grow up in the open;
 they go forth, and do not return
 to them.

5 "Who has let the wild ass go free?
 Who has loosed the bonds of
 the swift ass,
6 to which I have given the steppe
 for its home,
 the salt land for its dwelling
 place?
7 It scorns the tumult of the city;
 it does not hear the shouts of
 the driver.
8 It ranges the mountains as its
 pasture,
 and it searches after every green
 thing.

9 "Is the wild ox willing to serve
 you?
 Will it spend the night at your
 crib?
10 Can you tie it in the furrow with
 ropes,
 or will it harrow the valleys
 after you?
11 Will you depend on it because its
 strength is great,
 and will you hand over your
 labor to it?
12 Do you have faith in it that it will
 return,
 and bring your grain to your
 threshing floor?*o*

13 "The ostrich's wings flap wildly,
 though its pinions lack
 plumage.*p*
14 For it leaves its eggs to the earth,

and lets them be warmed on the
 ground,
15 forgetting that a foot may crush
 them,
 and that a wild animal may
 trample them.
16 It deals cruelly with its young, as
 if they were not its own;
 though its labor should be in
 vain, yet it has no fear;
17 because God has made it forget
 wisdom,
 and given it no share in
 understanding.
18 When it spreads its plumes aloft,*p*
 it laughs at the horse and its
 rider.

19 "Do you give the horse its might?
 Do you clothe its neck with
 mane?
20 Do you make it leap like the
 locust?
 Its majestic snorting is terrible.
21 It paws*q* violently, exults
 mightily;
 it goes out to meet the
 weapons.
22 It laughs at fear, and is not
 dismayed;
 it does not turn back from the
 sword.
23 Upon it rattle the quiver,
 the flashing spear, and the
 javelin.
24 With fierceness and rage it
 swallows the ground;
 it cannot stand still at the sound
 of the trumpet.
25 When the trumpet sounds, it says
 'Aha!'
 From a distance it smells the
 battle,
 the thunder of the captains, and
 the shouting.

26 "Is it by your wisdom that the
 hawk soars,
 and spreads its wings toward
 the south?

o Heb *your grain and your threshing floor*
p Meaning of Heb uncertain   *q* Gk Syr Vg:
Heb *they dig*

27 Is it at your command that the
      eagle mounts up
   and makes its nest on high?
28 It lives on the rock and makes its
      home
   in the fastness of the rocky crag.
29 From there it spies the prey;
   its eyes see it from far away.
30 Its young ones suck up blood;
   and where the slain are, there
      it is."

40 And the LORD said to Job:
   2 "Shall a faultfinder contend
      with the Almighty?*r*
   Anyone who argues with God
      must respond."

3 Then Job answered the LORD:
4 "See, I am of small account; what
      shall I answer you?
   I lay my hand on my mouth.
5 I have spoken once, and I will not
      answer;
   twice, but will proceed no
      further."

6 Then the LORD answered Job out of
the whirlwind:
7 "Gird up your loins like a man;
   I will question you, and you
      declare to me.
8 Will you even put me in the
      wrong?
   Will you condemn me that you
      may be justified?
9 Have you an arm like God,
   and can you thunder with a
      voice like his?

10 "Deck yourself with majesty and
      dignity;

clothe yourself with glory and
      splendor.
11 Pour out the overflowings of your
      anger,
   and look on all who are proud,
      and abase them.
12 Look on all who are proud, and
      bring them low;
   tread down the wicked where
      they stand.
13 Hide them all in the dust together;
   bind their faces in the world
      below.*s*
14 Then I will also acknowledge to
      you
   that your own right hand can
      give you victory.

15 "Look at Behemoth,
   which I made just as I made
      you;
   it eats grass like an ox.
16 Its strength is in its loins,
   and its power in the muscles of
      its belly.
17 It makes its tail stiff like a cedar;
   the sinews of its thighs are knit
      together.
18 Its bones are tubes of bronze,
   its limbs like bars of iron.

19 "It is the first of the great acts of
      God—
   only its Maker can approach it
      with the sword.
20 For the mountains yield food for it
   where all the wild animals play.
21 Under the lotus plants it lies,

r Traditional rendering of Heb *Shaddai*
s Heb *the hidden place*

40.3–5: Job's reply. Job offers an elaborate statement—of silence. Is this evasive, or humility, or defiance? He does not say that he has sinned (see 42.7).
40.6–41.34: Second speech of the LORD. 8: *Will you condemn me that you may be justified?* The issue is finally joined, and the LORD seems to say in vv. 9–14 that if Job can take on the role of king (v. 10) in a divine style, and can manage the world, especially the

wicked, then the LORD will *acknowledge* his *victory.* This is an invitation to usurp God's role! But it is a safe one, since Job cannot and will not accept it.
40.15–24: *Behemoth* is not a mere hippopotamus but a primeval monster (v. 19), a mythical symbol of chaos and evil. **15:** *Which I made just as I made you,* Job and Behemoth have that in common, at least.

in the covert of the reeds and in
the marsh.

22 The lotus trees cover it for shade;
the willows of the wadi
surround it.

23 Even if the river is turbulent, it is
not frightened;
it is confident though Jordan
rushes against its mouth.

24 Can one take it with hooks*t*
or pierce its nose with a snare?

41 *u* "Can you draw out Leviathan*v*
with a fishhook,
or press down its tongue with a
cord?

2 Can you put a rope in its nose,
or pierce its jaw with a hook?

3 Will it make many supplications to
you?
Will it speak soft words to you?

4 Will it make a covenant with you
to be taken as your servant
forever?

5 Will you play with it as with a
bird,
or will you put it on leash for
your girls?

6 Will traders bargain over it?
Will they divide it up among
the merchants?

7 Can you fill its skin with
harpoons,
or its head with fishing spears?

8 Lay hands on it;
think of the battle; you will not
do it again!

9*w* Any hope of capturing it*x* will be
disappointed;
were not even the gods*y*
overwhelmed at the sight of
it?

10 No one is so fierce as to dare to
stir it up.
Who can stand before it?*z*

11 Who can confront it*z* and be
safe?*a*

—under the whole heaven,
who?*b*

12 "I will not keep silence concerning
its limbs,
or its mighty strength, or its
splendid frame.

13 Who can strip off its outer
garment?
Who can penetrate its double
coat of mail?*c*

14 Who can open the doors of its
face?
There is terror all around its
teeth.

15 Its back*d* is made of shields in
rows,
shut up closely as with a seal.

16 One is so near to another
that no air can come between
them.

17 They are joined one to another;
they clasp each other and cannot
be separated.

18 Its sneezes flash forth light,
and its eyes are like the eyelids
of the dawn.

19 From its mouth go flaming
torches;
sparks of fire leap out.

20 Out of its nostrils comes smoke,
as from a boiling pot and
burning rushes.

21 Its breath kindles coals,
and a flame comes out of its
mouth.

22 In its neck abides strength,
and terror dances before it.

23 The folds of its flesh cling
together;
it is firmly cast and immovable.

24 Its heart is as hard as stone,

t Cn: Heb *in his eyes*   *u* Ch 40.25 in Heb
v Or *the crocodile*   *w* Ch 41.1 in Heb
x Heb *of it*   *y* Cn Compare Symmachus Syr.
Heb *one is*   *z* Heb *me*   *a* Gk: Heb *that I
shall repay*   *b* Heb *to me*   *c* Gk: Heb *bridle*
d Cn Compare Gk Vg: Heb *pride*

41.1–34: *Leviathan* is not an ordinary crocodile, but the sea-monster that personifies
chaos (3.8; 26.13; Ps 74.14). It is only a play-
thing in the eyes of God, as the psalmist also
indicates (Ps 104.26).

as hard as the lower millstone.
25 When it raises itself up the gods
are afraid;
at the crashing they are beside
themselves.
26 Though the sword reaches it, it
does not avail,
nor does the spear, the dart, or
the javelin.
27 It counts iron as straw,
and bronze as rotten wood.
28 The arrow cannot make it flee;
slingstones, for it, are turned to
chaff.
29 Clubs are counted as chaff;
it laughs at the rattle of javelins.
30 Its underparts are like sharp
potsherds;
it spreads itself like a threshing
sledge on the mire.
31 It makes the deep boil like a pot;
it makes the sea like a pot of
ointment.
32 It leaves a shining wake behind it;
one would think the deep to be
white-haired.
33 On earth it has no equal,
a creature without fear.
34 It surveys everything that is lofty;
it is king over all that are
proud."

42 Then Job answered the LORD:
2 "I know that you can do all
things,
and that no purpose of yours
can be thwarted.

3 'Who is this that hides counsel
without knowledge?'
Therefore I have uttered what I
did not understand,
things too wonderful for me,
which I did not know.
4 'Hear, and I will speak;
I will question you, and you
declare to me.'
5 I had heard of you by the hearing
of the ear,
but now my eye sees you;
6 therefore I despise myself,
and repent in dust and ashes."

7 After the LORD had spoken these
words to Job, the LORD said to Eliphaz
the Temanite: "My wrath is kindled
against you and against your two friends;
for you have not spoken of me what is
right, as my servant Job has. 8 Now
therefore take seven bulls and seven
rams, and go to my servant Job, and offer up for yourselves a burnt offering;
and my servant Job shall pray for you,
for I will accept his prayer not to deal
with you according to your folly; for you
have not spoken of me what is right, as
my servant Job has done." 9 So Eliphaz
the Temanite and Bildad the Shuhite and
Zophar the Naamathite went and did
what the LORD had told them; and the
LORD accepted Job's prayer.

10 And the LORD restored the fortunes of Job when he had prayed for his
friends; and the LORD gave Job twice as

42.1–6: Job's reply. Job acknowledges
the divine *purpose* (v. 2) and reacts to two
quotations of the LORD's words: 38.2 (v. 3a)
and 40.7 (v. 4) by stating his ignorance (v. 3b)
and his experience of God (v. 5). A hearsay
understanding of the LORD has now yielded
to a vision (compare 19.26–27) or experience
that satisfies Job. He receives no explanation
of his predicament; it is enough that his creator has come to him personally. **6:** The
meaning is not clear. *I despise myself,* but no
object of the verb is indicated in the Hebrew.
Repent, a verb that is often used to indicate a
change of mind on the LORD's part (Ex 32.14;
Jer 18.8, 10). Here it does not mean repentance for sin (see vv. 7–8, where Job is said to

have spoken *what is right*). *In dust and ashes,* in
the sense that this figure expresses his weakness and humanity "since I am but dust and
ashes" (see also Gen 18.27; Job 30.19).
42.7–17: **The epilogue.** The style, language, and situation ties in with 1.1–2.13.
7–8: The verdict of the LORD is clearly in
favor of Job, against the friends, who are to
ask Job to be their intercessor; quite a turn-
around! **9–16:** Job's restoration follows upon
his intercession. It is a misreading to consider
the restoration the point of the story. The
author would not deny that God also rewards
people, but this is at the divine pleasure. **14:**
Jemimah: Dove; *Kezizah:* cassia, or cinnamon; *Keren-happuch:* horn of eye cosmetic.

much as he had before. ¹¹Then there came to him all his brothers and sisters and all who had known him before, and they ate bread with him in his house; they showed him sympathy and comforted him for all the evil that the LORD had brought upon him; and each of them gave him a piece of money*e* and a gold ring. ¹²The LORD blessed the latter days of Job more than his beginning; and he had fourteen thousand sheep, six thousand camels, a thousand yoke of oxen, and a thousand donkeys. ¹³He also had seven sons and three daughters. ¹⁴He named the first Jemimah, the second Keziah, and the third Keren-happuch. ¹⁵In all the land there were no women so beautiful as Job's daughters; and their father gave them an inheritance along with their brothers. ¹⁶After this Job lived one hundred and forty years, and saw his children, and his children's children, four generations. ¹⁷And Job died, old and full of days.

e Heb *a qesitah*

The Psalms

The living spirit of any religion shines through most clearly in the hymns by which its adherents bring before God their troubles and fears, their hopes, aspirations, and reasons for confidence. The Psalter is the hymnal of ancient Israel, compiled from older collections of lyrics for use in the temple of Zerubbabel (Ezra 5.2; Hag 1.14). Most of the psalms were probably composed to accompany acts of worship in the temple and may be classified as follows: *Hymns* (acts of praise suitable for any occasion and including the sub-types *Enthronement Hymns,* celebrating the Lord's kingship, and *Songs of Zion,* expressing devotion to the Holy City); *Laments* (in which an individual seeks deliverance from illness or false accusation, or the nation asks for help in time of distress); *Songs of Trust* (in which individuals express their confidence in God's readiness to help); *Thanksgivings* (in which individuals express their gratitude for deliverance); *Sacred History* (in which the nation recounts the story of God's dealings with it); *Royal Psalms* (designed to be used for occasions such as a coronation or royal wedding); *Wisdom Psalms* (which are meditations on life and the ways of God); and *Liturgies* (which are often of mixed type and were composed for some special cultic or historical occasion. One suggested cultic occasion is a regularly occurring ceremony for renewing the covenant). In imitation of the Pentateuch the Psalter is divided into five books (see 41.13 n.).

The titles or superscriptions of the psalms contain a variety of items. They frequently include musical directions, such as "To the choirmaster: with stringed instruments" (Pss 4; 6; etc.). The meaning of a number of technical terms is no longer definitely known; for example, Shiggaion (Ps 7), Miktam (Pss 16; 56–60), Maskil (Pss 42; 44; 52–55; etc.), Sheminith (Pss 6; 12), Gittith (Pss 8; 81; 84), Muth-Labben (Ps 9), Alamoth (Ps 46), Mahalath (Ps 53), Mahalath Leannoth (Ps 88), Shushan-Eduth (Ps 60). The first three may refer to types of psalms, and the rest, preceded by "According to," may be the names of tunes, as also The Hind of the Dawn (Ps 22), Lilies (Pss 45; 69), The Dove on the Far-off Terebinths (Ps 56), and Do Not Destroy (Pss 57–59; 75). References to the descendants of Korah (Pss 42; 44–49; 84; etc.), Jeduthun (Pss 39; 62; 77), Asaph (Pss 50; 73–83), Heman (Ps 88), and Ethan (Ps 89) point to the work of professional Levitical musicians (1 Chr 15.16–22; Neh 12.41–46). Song of Ascents (Pss 120–134) may mean "pilgrim song," that is, of ascent to Jerusalem. The ascription of nearly half of the psalms to David is testimony to the regard in which the great singer of Israel was held. The tradition that David wrote *all* the psalms is much later than the book itself. Interior evidence shows that the Psalter is the product of many minds during many centuries (see, for example, Pss 72.1; 82.1; 98.1; 102.1). For the significance of the frequently recurring word, "Selah," see 3.2 n.

The book of Psalms reflects many aspects of the religious experience of Israel. Its intrinsic spiritual depth and beauty have made it from earliest times a treasury of resources for public and private devotion in both Judaism and Christianity.

BOOK I

Psalm 1

1 Happy are those
     who do not follow the advice of
          the wicked,
  or take the path that sinners tread,
     or sit in the seat of scoffers;
2 but their delight is in the law of
          the LORD,
     and on his law they meditate
          day and night.
3 They are like trees
     planted by streams of water,
  which yield their fruit in its
          season,
     and their leaves do not wither.
  In all that they do, they prosper.

4 The wicked are not so,
     but are like chaff that the wind
          drives away.
5 Therefore the wicked will not
          stand in the judgment,
     nor sinners in the congregation
          of the righteous;
6 for the LORD watches over the
          way of the righteous,
     but the way of the wicked will
          perish.

Psalm 2

1 Why do the nations conspire,
     and the peoples plot in vain?
2 The kings of the earth set
          themselves,

and the rulers take counsel
          together,
  against the LORD and his
          anointed, saying,
3 "Let us burst their bonds asunder,
     and cast their cords from us."

4 He who sits in the heavens laughs;
     the LORD has them in derision.
5 Then he will speak to them in his
          wrath,
     and terrify them in his fury,
          saying,
6 "I have set my king on Zion, my
          holy hill."

7 I will tell of the decree of the
          LORD:
  He said to me, "You are my son;
     today I have begotten you.
8 Ask of me, and I will make the
          nations your heritage,
     and the ends of the earth your
          possession.
9 You shall break them with a rod
          of iron,
     and dash them in pieces like a
          potter's vessel."

10 Now therefore, O kings, be wise;
     be warned, O rulers of the
          earth.
11 Serve the LORD with fear,
     with trembling 12 kiss his feet, *a*

a Cn: Meaning of Heb of verses 11b and 12a is
uncertain

Ps 1: **The contrasting fate of the righteous and the wicked** (a wisdom psalm). Compare Jer 17.5–8. **1–3**: The prosperity of the righteous. **1**: *Happy* is the conventional translation of a Hebrew expression meaning literally "the happinesses of." **2**: Compare Ps 119. **4–6**: The disastrous end of the wicked. **4**: In threshing, the crushed sheaves were tossed into the air, where the wind blew away the lighter *chaff*.

Ps 2: **The LORD gives universal dominion to his king** (a royal psalm, composed for a coronation). **1–3**: Israel's subject peoples plot rebellion against the new king. **2**: The

word *anointed* in Hebrew is literally "messiah," one of the titles of an Israelite king; after the extinction of the Hebrew monarchy, this became a name for the ideal king of a future, hoped-for restoration, and the psalm was reinterpreted accordingly (compare Acts 4.25–29).
2.4–9: The newly-enthroned king quotes God's promise of universal rule. **7**: *You are my son . . . begotten you*, a formula of adoption whereby the king became God's son (compare 2 Sam 7.14; Ps 89.26–27; also Acts 13.33). **10–12**: Rebellious rulers are warned to submit.

or he will be angry, and you will
perish in the way;
for his wrath is quickly kindled.

Happy are all who take refuge
in him.

Psalm 3

*A Psalm of David, when he fled from
his son Absalom.*

1 O LORD, how many are my foes!
Many are rising against me;
2 many are saying to me,
"There is no help for you*b* in
God."   *Selah*

3 But you, O LORD, are a shield
around me,
my glory, and the one who lifts
up my head.
4 I cry aloud to the LORD,
and he answers me from his
holy hill.   *Selah*

5 I lie down and sleep;
I wake again, for the LORD
sustains me.
6 I am not afraid of ten thousands of
people
who have set themselves against
me all around.

7 Rise up, O LORD!
Deliver me, O my God!
For you strike all my enemies on
the cheek;
you break the teeth of the
wicked.

8 Deliverance belongs to the LORD;

may your blessing be on your
people!   *Selah*

Psalm 4

*To the leader: with stringed instruments.
A Psalm of David.*

1 Answer me when I call, O God of
my right!
You gave me room when I was
in distress.
Be gracious to me, and hear my
prayer.

2 How long, you people, shall my
honor suffer shame?
How long will you love vain
words, and seek after lies?
Selah
3 But know that the LORD has set
apart the faithful for
himself;
the LORD hears when I call to
him.

4 When you are disturbed,*c* do not
sin;
ponder it on your beds, and be
silent.   *Selah*
5 Offer right sacrifices,
and put your trust in the LORD.

6 There are many who say, "O that
we might see some good!
Let the light of your face shine
on us, O LORD!"
7 You have put gladness in my heart
more than when their grain and
wine abound.

b Syr: Heb *him*   *c* Or *are angry*

Ps 3: **Prayer for deliverance from personal enemies** (a lament). **1–2**: The psalmist's complaint. **2**: *Selah* is a liturgical direction; it may indicate that there should be an
instrumental interlude at this point in the
singing of the psalm. **3–4**: Expression of
trust, a regular element in psalms which are
laments. **5–6**: It is perhaps after having spent
the night in the temple that the psalmist receives assurance that the LORD's attitude is
still favorable. **7–8**: Concluding prayer for
the psalmist and the nation.

Ps 4: **Prayer for deliverance from personal enemies** (a lament). **1**: Cry for help.
2–4: A rebuke for those who falsely accuse
the psalmist of wrong-doing. The charge is
probably a formal accusation to which the
accused replies by a public declaration of innocence (compare 26.4–7). **2, 4**: *Selah*, see
3.2 n. **5–8**: The accused is assured of the
LORD's help. The ceremony in which this
psalm was used evidently included sacrifice
(v. 5) and possibly spending the night in the
temple (v. 8; compare 3.5).

8 I will both lie down and sleep in
     peace;
   for you alone, O LORD, make
     me lie down in safety.

Psalm 5

To the leader: for the flutes.
A Psalm of David.

1 Give ear to my words, O LORD;
   give heed to my sighing.
2 Listen to the sound of my cry,
   my King and my God,
   for to you I pray.
3 O LORD, in the morning you hear
     my voice;
   in the morning I plead my case
     to you, and watch.

4 For you are not a God who
     delights in wickedness;
   evil will not sojourn with you.
5 The boastful will not stand before
     your eyes;
   you hate all evildoers.
6 You destroy those who speak lies;
   the LORD abhors the
     bloodthirsty and deceitful.

7 But I, through the abundance of
     your steadfast love,
   will enter your house,
   I will bow down toward your
     holy temple
   in awe of you.
8 Lead me, O LORD, in your
     righteousness
   because of my enemies;
   make your way straight
     before me.

9 For there is no truth in their
     mouths;
   their hearts are destruction;

their throats are open graves;
   they flatter with their tongues.
10 Make them bear their guilt,
     O God;
   let them fall by their own
     counsels;
   because of their many
     transgressions cast them
     out,
   for they have rebelled against
     you.

11 But let all who take refuge in you
     rejoice;
   let them ever sing for joy.
   Spread your protection over them,
   so that those who love your
     name may exult in you.
12 For you bless the righteous,
     O LORD;
   you cover them with favor as
     with a shield.

Psalm 6

To the leader: with stringed instruments;
according to The Sheminith.
A Psalm of David.

1 O LORD, do not rebuke me in
     your anger,
   or discipline me in your wrath.
2 Be gracious to me, O LORD, for I
     am languishing;
   O LORD, heal me, for my bones
     are shaking with terror.
3 My soul also is struck with terror,
   while you, O LORD—how long?

4 Turn, O LORD, save my life;
   deliver me for the sake of your
     steadfast love.
5 For in death there is no
     remembrance of you;

Ps 5: Prayer for deliverance from personal enemies (a lament). **1–2:** Cry for help. **3–7:** Expression of trust (compare 3.3–4). **3:** *In the morning . . . I plead my case to you* probably refers to a ceremonial plea at the end of a night's vigil (compare 3.5; 4.8). **5.8–12:** Concluding prayer. **5.9–10:** See 4.2–4 n.

Ps 6: Prayer for healing from severe illness (a lament). In the church liturgy this is the first of seven Penitential Psalms (6; 32; 38; 51; 102; 130; 143). **1–5:** Cry for help. **5:** *Sheol*, the place of the dead, where people retain only the faintest semblance of life (compare 88.5–6; Job 3.13–19; see Gen 37.35 n.).

in Sheol who can give you
praise?

6 I am weary with my moaning;
every night I flood my bed with
tears;
I drench my couch with my
weeping.
7 My eyes waste away because of
grief;
they grow weak because of all
my foes.

8 Depart from me, all you workers
of evil,
for the LORD has heard the
sound of my weeping.
9 The LORD has heard my
supplication;
the LORD accepts my prayer.
10 All my enemies shall be ashamed
and struck with terror;
they shall turn back, and in a
moment be put to shame.

Psalm 7

*A Shiggaion of David, which he sang to the
LORD concerning Cush, a Benjaminite.*

1 O LORD my God, in you I take
refuge;
save me from all my pursuers,
and deliver me,
2 or like a lion they will tear me
apart;
they will drag me away, with
no one to rescue.

3 O LORD my God, if I have done
this,
if there is wrong in my hands,
4 if I have repaid my ally with harm

or plundered my foe without
cause,
5 then let the enemy pursue and
overtake me,
trample my life to the ground,
and lay my soul in the dust.
Selah

6 Rise up, O LORD, in your anger;
lift yourself up against the fury
of my enemies;
awake, O my God;*d* you have
appointed a judgment.
7 Let the assembly of the peoples be
gathered around you,
and over it take your seat*e* on
high.
8 The LORD judges the peoples;
judge me, O LORD, according to
my righteousness
and according to the integrity
that is in me.

9 O let the evil of the wicked come
to an end,
but establish the righteous,
you who test the minds and
hearts,
O righteous God.
10 God is my shield,
who saves the upright in heart.
11 God is a righteous judge,
and a God who has indignation
every day.

12 If one does not repent, God*f* will
whet his sword;
he has bent and strung his bow;
13 he has prepared his deadly
weapons,

d Or *awake for me*   *e* Cn: Heb *return*
f Heb *he*

6.6–7: The psalmist's weakened condition. 8–10: The petitioner receives assurance that the prayer has been heard. 8: The psalmist puts the blame for the illness on enemies, perhaps believing they have uttered a curse (compare Ps 102.8).
Ps 7: **Prayer for deliverance from personal enemies** (a lament). 1–2: Cry for help. 3–5: Protestation of innocence (see 4.2–4 n.).

3: *This,* the specific crime of which the psalmist was accused. 5: *Selah,* see 3.2 n.
7.6–9: Prayer for God's intervention in the psalmist's trial. 10–16: Expression of trust in God's righteous dealing (compare 3.3–4). 17: The psalmist promises to give formal thanks to God, probably by offering a thanksgiving sacrifice (compare 66.13–15).

making his arrows fiery shafts.

14 See how they conceive evil,
    and are pregnant with mischief,
    and bring forth lies.

15 They make a pit, digging it out,
    and fall into the hole that they
      have made.

16 Their mischief returns upon their
      own heads,
    and on their own heads their
      violence descends.

17 I will give to the LORD the thanks
    due to his righteousness,
    and sing praise to the name of
      the LORD, the Most High.

Psalm 8

*To the leader: according to The Gittith.
A Psalm of David.*

1 O LORD, our Sovereign,
    how majestic is your name in all
      the earth!

You have set your glory above the
    heavens.

2   Out of the mouths of babes and
      infants
you have founded a bulwark
      because of your foes,
    to silence the enemy and the
      avenger.

3 When I look at your heavens, the
      work of your fingers,
the moon and the stars that you
    have established;

4 what are human beings that you
    are mindful of them,

mortals[g] that you care for
    them?

5 Yet you have made them a little
    lower than God,[h]
    and crowned them with glory
      and honor.

6 You have given them dominion
      over the works of your
      hands;
you have put all things under
      their feet,

7 all sheep and oxen,
    and also the beasts of the field,

8 the birds of the air, and the fish of
      the sea,
    whatever passes along the paths
      of the seas.

9 O LORD, our Sovereign,
    how majestic is your name in all
      the earth!

Psalm 9

*To the leader: according to Muth-labben.
A Psalm of David.*

1 I will give thanks to the LORD
    with my whole heart;
    I will tell of all your wonderful
      deeds.

2 I will be glad and exult in you;
    I will sing praise to your name,
      O Most High.

3 When my enemies turned back,

g Heb *ben adam*, lit. *son of man*   *h* Or *than the
divine beings* or *angels*: Heb *elohim*

**Ps 8: Hymn celebrating God's glory
and the God-given dignity of human beings. 1–4:** The glory of God as manifest in the
night sky and in the songs of children. **4:** See
144.3–4; Job 7.17–18 n. *Mortals,* literally "son
of man." See note *g.* In the New Testament,
this is taken to be a title of the Messiah (see,
for example, Heb 2.5–8; Mt 10.23).
8.5–8: God has given human beings a
share in his own dignity by conferring on
them dominion over the rest of creation (Gen
1.26). **9:** The opening verse repeated as a refrain.

**Pss 9–10: Prayer for deliverance from
personal enemies** (a lament). This composition, which is printed as a single psalm in the
Septuagint, is constructed on the acrostic
principle, every second verse beginning with
a different successive letter of the Hebrew
alphabet, a circumstance which helps to account for the seeming lack of logical sequence. **1–2:** A vow in anticipation of deliverance (see 7.17 n.). **3–12:** Expression of
confidence, based on God's previous just
dealings (compare 3.3–4).

they stumbled and perished
before you.
4 For you have maintained my just
cause;
you have sat on the throne
giving righteous judgment.

5 You have rebuked the nations,
you have destroyed the
wicked;
you have blotted out their name
forever and ever.
6 The enemies have vanished in
everlasting ruins;
their cities you have rooted out;
the very memory of them has
perished.

7 But the LORD sits enthroned
forever,
he has established his throne for
judgment.
8 He judges the world with
righteousness;
he judges the peoples with
equity.

9 The LORD is a stronghold for the
oppressed,
a stronghold in times of trouble.
10 And those who know your name
put their trust in you,
for you, O LORD, have not
forsaken those who seek
you.

11 Sing praises to the LORD, who
dwells in Zion.
Declare his deeds among the
peoples.
12 For he who avenges blood is
mindful of them;
he does not forget the cry of the
afflicted.

13 Be gracious to me, O LORD.
See what I suffer from those
who hate me;
you are the one who lifts me up
from the gates of death,
14 so that I may recount all your
praises,
and, in the gates of daughter
Zion,
rejoice in your deliverance.

15 The nations have sunk in the pit
that they made;
in the net that they hid has their
own foot been caught.
16 The LORD has made himself
known, he has executed
judgment;
the wicked are snared in the
work of their own hands.
Higgaion. Selah

17 The wicked shall depart to Sheol,
all the nations that forget God.

18 For the needy shall not always be
forgotten,
nor the hope of the poor perish
forever.

19 Rise up, O LORD! Do not let
mortals prevail;
let the nations be judged before
you.
20 Put them in fear, O LORD;
let the nations know that they
are only human.   *Selah*

Psalm 10

1 Why, O LORD, do you stand far
off?
Why do you hide yourself in
times of trouble?

9.13–14: The psalmist's situation. 15–18: Renewed expression of confidence. 15: *The nations,* the psalmist's enemies. 16: The meaning of *Higgaion* is unknown. *Selah,* see 3.2 n. 17: *Sheol,* see 6.5 n.
9.19–10.15: Prayer that God may intervene against the wicked. 10.3–11: The greedy, haughty, irreligious, and tyrannical spirit of the psalmist's enemies. 4: They are not atheists, but they deny that God is concerned with moral retribution (compare v. 11 and 94.7). 16–18: Concluding expression of confidence.

2 In arrogance the wicked persecute
    the poor—
  let them be caught in the
    schemes they have devised.

3 For the wicked boast of the desires
    of their heart,
  those greedy for gain curse and
    renounce the LORD.
4 In the pride of their countenance
    the wicked say, "God will
    not seek it out";
  all their thoughts are, "There is
    no God."

5 Their ways prosper at all times;
  your judgments are on high, out
    of their sight;
  as for their foes, they scoff at
    them.
6 They think in their heart, "We
    shall not be moved;
  throughout all generations we
    shall not meet adversity."

7 Their mouths are filled with
    cursing and deceit and
    oppression;
  under their tongues are mischief
    and iniquity.
8 They sit in ambush in the villages;
  in hiding places they murder the
    innocent.

  Their eyes stealthily watch for the
    helpless;
9 they lurk in secret like a lion in
    its covert;
  they lurk that they may seize the
    poor;
  they seize the poor and drag
    them off in their net.

10 They stoop, they crouch,
  and the helpless fall by their
    might.
11 They think in their heart, "God
    has forgotten,

he has hidden his face, he will
    never see it."

12 Rise up, O LORD; O God, lift up
    your hand;
  do not forget the oppressed.
13 Why do the wicked renounce
    God,
  and say in their hearts, "You
    will not call us to account"?

14 But you do see! Indeed you note
    trouble and grief,
  that you may take it into your
    hands;
  the helpless commit themselves
    to you;
  you have been the helper of the
    orphan.

15 Break the arm of the wicked and
    evildoers;
  seek out their wickedness until
    you find none.
16 The LORD is king forever and
    ever;
  the nations shall perish from his
    land.

17 O LORD, you will hear the desire
    of the meek;
  you will strengthen their heart,
  you will incline your ear
18 to do justice for the orphan and
    the oppressed,
  so that those from earth may
    strike terror no more. *i*

Psalm 11

To the leader. Of David.

1 In the LORD I take refuge; how
    can you say to me,
  "Flee like a bird to the
    mountains;*j*

i Meaning of Heb uncertain   *j* Gk Syr Jerome
Tg: Heb *flee to your mountain, O bird*

Ps 11: Confidence in God's concern for justice. This type of psalm (the song of trust) probably developed as an expansion of the expression of trust that is a common feature of the laments (see 3.3–4 n.; compare 5.3–7; 7.10–16; 9.3–12). **1–3:** The psalmist rebukes

2 for look, the wicked bend the
    bow,
  they have fitted their arrow to
    the string,
  to shoot in the dark at the
    upright in heart.
3 If the foundations are destroyed,
  what can the righteous do?"

4 The LORD is in his holy temple;
  the LORD's throne is in heaven.
  His eyes behold, his gaze
    examines humankind.
5 The LORD tests the righteous and
    the wicked,
  and his soul hates the lover of
    violence.
6 On the wicked he will rain coals
    of fire and sulfur;
  a scorching wind shall be the
    portion of their cup.
7 For the LORD is righteous;
  he loves righteous deeds;
  the upright shall behold his face.

Psalm 12

*To the leader: according to The Sheminith.
A Psalm of David.*

1 Help, O LORD, for there is no
    longer anyone who is
    godly;
  the faithful have disappeared
    from humankind.
2 They utter lies to each other;
  with flattering lips and a double
    heart they speak.

3 May the LORD cut off all flattering
    lips,
  the tongue that makes great
    boasts,

4 those who say, "With our tongues
    we will prevail;
  our lips are our own—who is
    our master?"

5 "Because the poor are despoiled,
    because the needy groan,
  I will now rise up," says the
    LORD;
  "I will place them in the safety
    for which they long."
6 The promises of the LORD are
    promises that are pure,
  silver refined in a furnace on the
    ground,
  purified seven times.

7 You, O LORD, will protect us;
  you will guard us from this
    generation forever.
8 On every side the wicked prowl,
  as vileness is exalted among
    humankind.

Psalm 13

To the leader. A Psalm of David.

1 How long, O LORD? Will you
    forget me forever?
  How long will you hide your
    face from me?
2 How long must I bear pain[k] in
    my soul,
  and have sorrow in my heart all
    day long?
  How long shall my enemy be
    exalted over me?

3 Consider and answer me, O LORD
    my God!

k Syr: Heb *hold counsels*

---

those who say that only flight can provide
safety from one's enemies. **4–7:** One should
stand one's ground and depend on God for
vindication. **6:** *Cup,* Isa 51.17; see Lk 22.42.
  **Ps 12: Prayer for deliverance from per-
sonal enemies** (a lament). **1–4:** Cry for help.
The psalmist's trials are characteristic of the
faithlessness of the times. **5:** Perhaps spoken
by a priest or temple prophet as a formal act

of assurance to the suppliant. **6–8:** The choir
(which is perhaps implied in the use of *us*)
commends reliance on the promise just made
and prays that it may soon be carried out.
  **Ps 13: Prayer for deliverance from per-
sonal enemies** (a lament). **1–2:** The psalmist's complaint is given special emotional
force by the four-fold "How long?" **3–4:**
Prayer for help (see 4.2–4 n.). *Give light to my*

Give light to my eyes, or I will
  sleep the sleep of death,
4 and my enemy will say, "I have
  prevailed";
  my foes will rejoice because I
  am shaken.

5 But I trusted in your steadfast
  love;
  my heart shall rejoice in your
  salvation.
6 I will sing to the LORD,
  because he has dealt bountifully
  with me.

Psalm 14

To the leader. Of David.

1 Fools say in their hearts, "There is
  no God."
  They are corrupt, they do
  abominable deeds;
  there is no one who does good.

2 The LORD looks down from
  heaven on humankind
  to see if there are any who are
  wise,
  who seek after God.

3 They have all gone astray, they
  are all alike perverse;
  there is no one who does good,
  no, not one.

4 Have they no knowledge, all the
  evildoers
  who eat up my people as they
  eat bread,
  and do not call upon the LORD?

5 There they shall be in great terror,
  for God is with the company of
  the righteous.
6 You would confound the plans of
  the poor,
  but the LORD is their refuge.

7 O that deliverance for Israel would
  come from Zion!
  When the LORD restores the
  fortunes of his people,
  Jacob will rejoice; Israel will be
  glad.

Psalm 15

A Psalm of David.

1 O LORD, who may abide in your
  tent?
  Who may dwell on your holy
  hill?

2 Those who walk blamelessly, and
  do what is right,
  and speak the truth from their
  heart;
3 who do not slander with their
  tongue,
  and do no evil to their friends,
  nor take up a reproach against
  their neighbors;
4 in whose eyes the wicked are
  despised,
  but who honor those who fear
  the LORD;
  who stand by their oath even to
  their hurt;
5 who do not lend money at
  interest,

---

eyes, compare 38.10. **5**: Expression of trust (compare 3.3–4). **6**: The vow (see 7.17 n.). **Ps 14: Condemnation of a cynical and unrighteous age.** This psalm (which is almost identical with Ps 53) seems to be a variation on the typical lament; the psalmist generalizes personal troubles as characteristic of an evil generation (compare 12.1–4). **1–3**: The corruption of the age. **1**: *Fools,* not silly people, but those who are utterly corrupt in their moral character (compare 2 Sam 13.13; Prov 10.23). *There is no God,* see 10.4 n. **4–6**: A threat to the evil-doers, having the force of a curse. *Eat up my people,* compare Mic 3.2–3. **7**: A yearning for better times.

Ps 15: A liturgy for admission to the temple. 1: The question: Who shall be admitted to the worshiping congregation? **2–5**: The answer: Only those who have the requisite moral qualities. **5**: The prohibition of *interest* (Ex 22.25; Lev 25.35–37) has reference to charitable loans made for the relief of distress rather than to the purely business type of loan which became common in a later commercial age.

and do not take a bribe against
the innocent.

Those who do these things shall
never be moved.

Psalm 16

A Miktam of David.

1 Protect me, O God, for in you I
take refuge.
2 I say to the LORD, "You are my
Lord;
I have no good apart from
you."[l]

3 As for the holy ones in the land,
they are the noble,
in whom is all my delight.

4 Those who choose another god
multiply their sorrows;[m]
their drink offerings of blood I
will not pour out
or take their names upon my
lips.

5 The LORD is my chosen portion
and my cup;
you hold my lot.
6 The boundary lines have fallen for
me in pleasant places;
I have a goodly heritage.

7 I bless the LORD who gives me
counsel;
in the night also my heart
instructs me.
8 I keep the LORD always before me;
because he is at my right hand,
I shall not be moved.

9 Therefore my heart is glad, and
my soul rejoices;

my body also rests secure.
10 For you do not give me up to
Sheol,
or let your faithful one see the
Pit.

11 You show me the path of life.
In your presence there is fullness
of joy;
in your right hand are pleasures
forevermore.

Psalm 17

A Prayer of David.

1 Hear a just cause, O LORD; attend
to my cry;
give ear to my prayer from lips
free of deceit.
2 From you let my vindication
come;
let your eyes see the right.

3 If you try my heart, if you visit
me by night,
if you test me, you will find no
wickedness in me;
my mouth does not transgress.
4 As for what others do, by the
word of your lips
I have avoided the ways of the
violent.
5 My steps have held fast to your
paths;
my feet have not slipped.

6 I call upon you, for you will
answer me, O God;
incline your ear to me, hear my
words.

l Jerome Tg: Meaning of Heb uncertain
m Cn: Meaning of Heb uncertain

Ps 16: An act of personal faith in God's power to save (a song of trust; see Ps 11 n.). **1:** Prayer for deliverance from trouble. **2–8:** The psalmist makes an appeal to God based upon past devotion to God and to the community of the faithful.
16.9–11: The psalmist has complete confidence that God will not permit a devoted worshiper to perish. **10:** *Sheol*, see 6.5 n.; *the Pit* is a synonym. **11:** The psalmist will survive the ordeal and through God's help enjoy life's pleasures again.
Ps 17: Prayer for deliverance from personal enemies (a lament). **1:** Cry for vindication against false accusers (compare vv. 9–12). **3–5:** Protestation of innocence (see

7 Wondrously show your steadfast
love,
O savior of those who seek
refuge
from their adversaries at your
right hand.

8 Guard me as the apple of the eye;
hide me in the shadow of your
wings,
9 from the wicked who despoil me,
my deadly enemies who
surround me.
10 They close their hearts to pity;
with their mouths they speak
arrogantly.
11 They track me down;[n] now they
surround me;
they set their eyes to cast me to
the ground.
12 They are like a lion eager to tear,
like a young lion lurking in
ambush.

13 Rise up, O LORD, confront them,
overthrow them!
By your sword deliver my life
from the wicked,
14 from mortals—by your hand,
O LORD—
from mortals whose portion in
life is in this world.
May their bellies be filled with
what you have stored up
for them;
may their children have more
than enough;
may they leave something over
to their little ones.

15 As for me, I shall behold your
face in righteousness;

when I awake I shall be
satisfied, beholding your
likeness.

Psalm 18

*To the leader. A Psalm of David the
servant of the LORD, who addressed the
words of this song to the LORD on the day
when the LORD delivered him from the hand
of all his enemies, and from the hand of
Saul. He said:*

1 I love you, O LORD, my strength.
2 The LORD is my rock, my
fortress, and my deliverer,
my God, my rock in whom I
take refuge,
my shield, and the horn of my
salvation, my stronghold.
3 I call upon the LORD, who is
worthy to be praised,
so I shall be saved from my
enemies.

4 The cords of death encompassed
me;
the torrents of perdition
assailed me;
5 the cords of Sheol entangled me;
the snares of death confronted
me.

6 In my distress I called upon the
LORD;
to my God I cried for help.
From his temple he heard my
voice,
and my cry to him reached his
ears.

7 Then the earth reeled and rocked;

n One Ms Compare Syr: MT *Our steps*

4.2–4 n.). **3:** *By night,* see 3.5–6 n. **6–14:** The
psalmist's prayer renewed.
17.8: *Apple of the eye,* the pupil, the most
precious part (Zech 2.8). **10–12:** Description
of the accusers. **15:** Expression of confidence
(see 3.3–4 n.). *When I awake,* compare v. 3.
**Ps 18: A king gives thanks for a victory
in battle** (a royal thanksgiving). This psalm
is also found with minor variations in 2 Sam

ch 22. **1–3:** Confession of faith in God's readiness to help. **4–6:** Metaphorical description of
the psalmist's difficulties; for the actual situation see vv. 37–45. Narration of the troubles
from which the worshiper has been saved is
a regular feature of the thanksgiving psalms.
18.7–19: God's intervention in the battle.
The imagery associates the LORD's appearance with the phenomena of a thunderstorm

the foundations also of the
  mountains trembled
and quaked, because he was
  angry.
8 Smoke went up from his nostrils,
  and devouring fire from his
    mouth;
  glowing coals flamed forth from
    him.
9 He bowed the heavens, and came
    down;
  thick darkness was under his
    feet.
10 He rode on a cherub, and flew;
  he came swiftly upon the wings
    of the wind.
11 He made darkness his covering
    around him,
  his canopy thick clouds dark
    with water.
12 Out of the brightness before him
  there broke through his clouds
  hailstones and coals of fire.
13 The LORD also thundered in the
    heavens,
  and the Most High uttered his
    voice. *o*
14 And he sent out his arrows, and
    scattered them;
  he flashed forth lightnings, and
    routed them.
15 Then the channels of the sea were
    seen,
  and the foundations of the
    world were laid bare
  at your rebuke, O LORD,
  at the blast of the breath of your
    nostrils.

16 He reached down from on high,
    he took me;
  he drew me out of mighty
    waters.
17 He delivered me from my strong
    enemy,

and from those who hated me;
  for they were too mighty for
    me.
18 They confronted me in the day of
    my calamity;
  but the LORD was my support.
19 He brought me out into a broad
    place;
  he delivered me, because he
    delighted in me.

20 The LORD rewarded me according
    to my righteousness;
  according to the cleanness of my
    hands he recompensed me.
21 For I have kept the ways of the
    LORD,
  and have not wickedly departed
    from my God.
22 For all his ordinances were before
    me,
  and his statutes I did not put
    away from me.
23 I was blameless before him,
  and I kept myself from guilt.
24 Therefore the LORD has
    recompensed me according
    to my righteousness,
  according to the cleanness of my
    hands in his sight.

25 With the loyal you show yourself
    loyal;
  with the blameless you show
    yourself blameless;
26 with the pure you show yourself
    pure;
  and with the crooked you show
    yourself perverse.
27 For you deliver a humble people,
  but the haughty eyes you bring
    down.
28 It is you who light my lamp;

o Gk See 2 Sam 22.14: Heb adds *hailstones and
coals of fire*

(compare Judg 5.4–5; Ps 29.3–9). **8:** *Glowing
coals,* the lightning. **10:** *Cherub,* probably a
personification of the storm cloud (see Ezek
1.5 n.; Pss 68.33; 104.3). **13:** *His voice,* the
thunder, as in Ps 29. **15:** *The breath of your
nostrils,* the storm wind.

18.20–27: The king attributes the LORD's
favor toward him to his own previous loyalty
to God and his keeping the commandments.
28–30: An exuberant outburst of confidence
in what he can accomplish with God's help.

the LORD, my God, lights up
    my darkness.
29 By you I can crush a troop,
    and by my God I can leap over
      a wall.
30 This God—his way is perfect;
    the promise of the LORD proves
      true;
    he is a shield for all who take
      refuge in him.

31 For who is God except the LORD?
    And who is a rock besides our
      God?—
32 the God who girded me with
      strength,
    and made my way safe.
33 He made my feet like the feet of a
      deer,
    and set me secure on the
      heights.
34 He trains my hands for war,
    so that my arms can bend a
      bow of bronze.
35 You have given me the shield of
      your salvation,
    and your right hand has
      supported me;
    your help*p* has made me great.
36 You gave me a wide place for my
      steps under me,
    and my feet did not slip.
37 I pursued my enemies and
      overtook them;
    and did not turn back until they
      were consumed.
38 I struck them down, so that they
      were not able to rise;
    they fell under my feet.
39 For you girded me with strength
      for the battle;
    you made my assailants sink
      under me.
40 You made my enemies turn their
      backs to me,
    and those who hated me I
      destroyed.

41 They cried for help, but there was
      no one to save them;
    they cried to the LORD, but he
      did not answer them.
42 I beat them fine, like dust before
      the wind;
    I cast them out like the mire of
      the streets.

43 You delivered me from strife with
      the peoples;*q*
    you made me head of the
      nations;
    people whom I had not known
      served me.
44 As soon as they heard of me they
      obeyed me;
    foreigners came cringing to me.
45 Foreigners lost heart,
    and came trembling out of their
      strongholds.

46 The LORD lives! Blessed be my
      rock,
    and exalted be the God of my
      salvation,
47 the God who gave me vengeance
    and subdued peoples under me;
48 who delivered me from my
      enemies;
    indeed, you exalted me above
      my adversaries;
    you delivered me from the
      violent.

49 For this I will extol you, O LORD,
      among the nations,
    and sing praises to your name.
50 Great triumphs he gives to his
      king,
    and shows steadfast love to his
      anointed,
    to David and his descendants
      forever.

p Or *gentleness*   *q* Gk Tg: Heb *people*

18.31–42: A more prosaic account of the battle and the subsequent victory (compare vv. 17–19). **18.43–45:** The king became ruler of an empire at peace. **46–50:** Concluding hymn of praise and thanksgiving. **50:** This verse establishes the identity of the psalmist as a king of the Davidic dynasty. *His anointed,* see Ps 2.2 n. At least the nucleus of this psalm may go back to David.

Psalm 19

To the leader. A Psalm of David.

1 The heavens are telling the glory
        of God;
    and the firmament[r] proclaims
        his handiwork.
2 Day to day pours forth speech,
    and night to night declares
        knowledge.
3 There is no speech, nor are there
        words;
    their voice is not heard;
4 yet their voice[s] goes out through
        all the earth,
    and their words to the end of
        the world.

    In the heavens[t] he has set a tent
        for the sun,
5 which comes out like a
        bridegroom from his
        wedding canopy,
    and like a strong man runs its
        course with joy.
6 Its rising is from the end of the
        heavens,
    and its circuit to the end of
        them;
    and nothing is hid from its heat.

7 The law of the LORD is perfect,
        reviving the soul;
    the decrees of the LORD are sure,
        making wise the simple;
8 the precepts of the LORD are right,
        rejoicing the heart;
    the commandment of the LORD is
        clear,
        enlightening the eyes;

9 the fear of the LORD is pure,
        enduring forever;
    the ordinances of the LORD are
        true
    and righteous altogether.
10 More to be desired are they than
        gold,
    even much fine gold;
    sweeter also than honey,
        and drippings of the
        honeycomb.

11 Moreover by them is your servant
        warned;
    in keeping them there is great
        reward.
12 But who can detect their errors?
    Clear me from hidden faults.
13 Keep back your servant also from
        the insolent;[u]
    do not let them have dominion
        over me.
    Then I shall be blameless,
        and innocent of great
        transgression.

14 Let the words of my mouth and
        the meditation of my heart
    be acceptable to you,
    O LORD, my rock and my
        redeemer.

Psalm 20

To the leader. A Psalm of David.

1 The LORD answer you in the day
        of trouble!

r Or *dome*   s Gk Jerome Compare Syr: Heb
line   t Heb *In them*   u Or *from proud
thoughts*

Ps 19: Hymn to God as creator of nature and giver of the law. 1–6: The glory of God is manifested in the phenomena of the heavens and particularly in the might of the sun. **1–4a:** The sky and the successive days and nights are personified as members of a choir ceaselessly singing God's praises. **3:** *There is no speech . . .* i.e. they cannot be heard by human ears. **4b–6:** The skies provide a track along which the sun, like an athlete, runs its daily course.
19.7–14: Probably a later writer added these verses praising the revelation of God's will in the Mosaic law in order to counterbalance what seemed to him the almost pagan emphasis upon the revelation of God in nature (vv. 1–6). **7–9:** Six synonyms are used to describe the law (compare Ps 119). **9:** *Fear*, many scholars emend to "word" (see 119.11). **10:** Observance of the law is a joy, not a burden. **12–13:** The poet asks to be delivered from either accidental or deliberate violations of the commandments. **14:** A prayer that this hymn may be pleasing to God.

The name of the God of Jacob
  protect you!
2 May he send you help from the
    sanctuary,
  and give you support from
    Zion.
3 May he remember all your
    offerings,
  and regard with favor your
    burnt sacrifices.   *Selah*

4 May he grant you your heart's
    desire,
  and fulfill all your plans.
5 May we shout for joy over your
    victory,
  and in the name of our God set
    up our banners.
  May the Lord fulfill all your
    petitions.

6 Now I know that the Lord will
    help his anointed;
  he will answer him from his
    holy heaven
  with mighty victories by his
    right hand.
7 Some take pride in chariots, and
    some in horses,
  but our pride is in the name of
    the Lord our God.
8 They will collapse and fall,
  but we shall rise and stand
    upright.

9 Give victory to the king, O Lord;
  answer us when we call. *v*

Psalm 21

To the leader. A Psalm of David

1 In your strength the king rejoices,
    O Lord,

and in your help how greatly he
    exults!
2 You have given him his heart's
    desire,
  and have not withheld the
    request of his lips.   *Selah*
3 For you meet him with rich
    blessings;
  you set a crown of fine gold on
    his head.
4 He asked you for life; you gave it
    to him—
  length of days forever and ever.
5 His glory is great through your
    help;
  splendor and majesty you
    bestow on him.
6 You bestow on him blessings
    forever;
  you make him glad with the joy
    of your presence.
7 For the king trusts in the Lord,
  and through the steadfast love
    of the Most High he shall
    not be moved.

8 Your hand will find out all your
    enemies;
  your right hand will find out
    those who hate you.
9 You will make them like a fiery
    furnace
  when you appear.
  The Lord will swallow them up
    in his wrath,
  and fire will consume them.
10 You will destroy their offspring
    from the earth,
  and their children from among
    humankind.

v Gk: Heb *give victory, O Lord; let the King
answer us when we call*

**Ps 20: Prayer for the king's victory in
battle.** This psalm was probably composed
to accompany a sacrifice offered before the
battle was begun (v 3; compare 1 Sam 13.8–
15a). **1–5:** The prayer sung at the altar, or
while approaching it. **3:** *Selah,* see 3.2 n. **5:**
Between v. 5 and v. 6 some liturgical action
evidently occurred, probably the proclamation of an oracle of victory by a priest or

temple prophet (see 12.5 n.; compare also
21.8–12), which then inspired the expression
of confidence in vv. 6–8. **6:** *Anointed,* see
2.2 n. **9:** Concluding exclamatory prayer.
  **Ps 21: Thanksgiving after the king's
victory in battle.** This psalm is intentionally
paired with Ps 20. **1–7:** God is praised for
having answered the king's prayer. **2:** *Selah,*
see 3.2 n. **8–12:** An oracle promising a succes-

11 If they plan evil against you,
   if they devise mischief, they will
      not succeed.
12 For you will put them to flight;
   you will aim at their faces with
      your bows.

13 Be exalted, O LORD, in your
      strength!
   We will sing and praise your
      power.

Psalm 22

*To the leader: according to The Deer of the
Dawn. A Psalm of David.*

1 My God, my God, why have you
      forsaken me?
   Why are you so far from
      helping me, from the
      words of my groaning?
2 O my God, I cry by day, but you
      do not answer;
   and by night, but find no rest.

3 Yet you are holy,
   enthroned on the praises of
      Israel.
4 In you our ancestors trusted;
   they trusted, and you delivered
      them.
5 To you they cried, and were
      saved;
   in you they trusted, and were
      not put to shame.

6 But I am a worm, and not human;
   scorned by others, and despised
      by the people.
7 All who see me mock at me;

they make mouths at me, they
   shake their heads;
8 "Commit your cause to the LORD;
   let him deliver—
   let him rescue the one in whom
      he delights!"

9 Yet it was you who took me from
      the womb;
   you kept me safe on my
      mother's breast.
10 On you I was cast from my birth,
   and since my mother bore me
      you have been my God.
11 Do not be far from me,
   for trouble is near
   and there is no one to help.

12 Many bulls encircle me,
   strong bulls of Bashan surround
      me;
13 they open wide their mouths at
      me,
   like a ravening and roaring lion.

14 I am poured out like water,
   and all my bones are out of
      joint;
   my heart is like wax;
   it is melted within my breast;
15 my mouth[w] is dried up like a
      potsherd,
   and my tongue sticks to my
      jaws;
   you lay me in the dust of death.

16 For dogs are all around me;
   a company of evildoers
      encircles me.

w Cn: Heb *strength*

---

sion of victories, addressed to the king by a
priest or temple prophet (compare 20.5 n.).
13: Concluding exclamation of praise.
**Ps 22: Prayer for deliverance from
mortal illness** (a lament). **1–2:** Cry for help.
1: Quoted by Jesus on the cross (Mk 15.34).
3: God has helped his people in time past.
6–8: The psalmist's misery is aggravated by
the mockery of those who regard illness as
proof of God's displeasure.
22.9–11: God's help in former times in-

spires the psalmist to pray that it may continue in the present. **12–18:** Description of the
psalmist's condition. **12–13:** The detractors
behave like savage animals (see also v. 16,
although the meaning of the third line is obscure; and vv. 20–21). *Bulls of Bashan,* Am
4.1.
22.14–15, 17–18: A vivid account of the
poet's fever and resulting weakness. **16:** *Dogs,*
enemies (compare *lion, wild oxen,* v. 21). **18:**
The sick one is so nearly dead that neighbors

My hands and feet have
    shriveled;[x]
17 I can count all my bones.
    They stare and gloat over me;
18 they divide my clothes among
        themselves,
    and for my clothing they cast
        lots.

19 But you, O LORD, do not be far
        away!
    O my help, come quickly to
        my aid!
20 Deliver my soul from the sword,
    my life[y] from the power of the
        dog!
21 Save me from the mouth of the
        lion!

From the horns of the wild oxen
    you have rescued[z] me.
22 I will tell of your name to my
        brothers and sisters;[a]
    in the midst of the congregation
        I will praise you:
23 You who fear the LORD, praise
        him!
    All you offspring of Jacob,
        glorify him;
    stand in awe of him, all you
        offspring of Israel!
24 For he did not despise or abhor
        the affliction of the afflicted;
    he did not hide his face from me,[b]
    but heard when I[c] cried to him.

25 From you comes my praise in the
        great congregation;
    my vows I will pay before those
        who fear him.
26 The poor[d] shall eat and be
        satisfied;
    those who seek him shall praise
        the LORD.
    May your hearts live forever!

27 All the ends of the earth shall
        remember
    and turn to the LORD;
    and all the families of the nations
        shall worship before him.[e]
28 For dominion belongs to the
        LORD,
    and he rules over the nations.

29 To him,[f] indeed, shall all who
        sleep in[g] the earth bow
        down;
    before him shall bow all who go
        down to the dust,
    and I shall live for him.[h]
30 Posterity will serve him;
    future generations will be told
        about the Lord,
31 and[i] proclaim his deliverance to a
        people yet unborn,
    saying that he has done it.

Psalm 23

A Psalm of David.

1 The LORD is my shepherd, I shall
        not want.
2   He makes me lie down in green
        pastures;
    he leads me beside still waters;[j]
3   he restores my soul.[k]
    He leads me in right paths[l]
    for his name's sake.

4 Even though I walk through the
        darkest valley,[m]

x Meaning of Heb uncertain   y Heb *my only
one*   z Heb *answered*   a Or *kindred*
b Heb *him*   c Heb *he*   d Or *afflicted*
e Gk Syr Jerome: Heb *you*   f Cn: Heb *They
have eaten and*   g Cn: Heb *all the fat ones*
h Compare Gk Syr Vg: Heb *and he who cannot
keep himself alive*   i Compare Gk: Heb *it will
be told about the Lord to the generation,* [31]*they will
come and*   j Heb *waters of rest*   k Or *life*
l Or *paths of righteousness*   m Or *the valley of the
shadow of death*

---

and relatives have already begun to divide the
property. **19–21:** Prayer for healing and for
deliverance from slanderers.
    22.22–31: The psalmist vows, on recovery, to offer a formal thanksgiving *in the midst
of the congregation* at the temple (see 7.17 n.).

22: The vow (compare v. 25). **23–31:** The
hymn that will then be sung.
    **Ps 23: An expression of confidence in
God's protection** (a song of trust; see Ps
11 n.). **1–4.** The LORD is compared to a shepherd. **3:** *Soul* means vitality, the individual-

I fear no evil;
for you are with me;
your rod and your staff—
they comfort me.

5 You prepare a table before me
in the presence of my enemies;
you anoint my head with oil;
my cup overflows.
6 Surely[n] goodness and mercy[o]
shall follow me
all the days of my life,
and I shall dwell in the house of
the LORD
my whole life long.[p]

Psalm 24

Of David. A Psalm.

1 The earth is the LORD's and all
that is in it,
the world, and those who live
in it;
2 for he has founded it on the seas,
and established it on the rivers.

3 Who shall ascend the hill of the
LORD?
And who shall stand in his holy
place?
4 Those who have clean hands and
pure hearts,
who do not lift up their souls to
what is false,
and do not swear deceitfully.
5 They will receive blessing from
the LORD,

and vindication from the God of
their salvation.
6 Such is the company of those who
seek him,
who seek the face of the God of
Jacob.[q]   *Selah*

7 Lift up your heads, O gates!
and be lifted up, O ancient
doors!
that the King of glory may
come in.
8 Who is the King of glory?
The LORD, strong and mighty,
the LORD, mighty in battle.
9 Lift up your heads, O gates!
and be lifted up, O ancient
doors!
that the King of glory may
come in.
10 Who is this King of glory?
The LORD of hosts,
he is the King of glory.   *Selah*

Psalm 25

Of David.

1 To you, O LORD, I lift up my
soul.
2 O my God, in you I trust;
do not let me be put to shame;
do not let my enemies exult
over me.

---

n Or *Only*   *o* Or *kindness*   *p* Heb *for length of
days*   *q* Gk Syr: Heb *your face, O Jacob*

---

ized principle of life. *Right paths,* literally
"paths of rightness." **4:** *Darkest valley:* "shadow of death" is an ancient, but probably fanciful, rendering; compare 44.19; 107.10; Job
3.5; Isa 9.2; etc. where the same Hebrew expression occurs. **5–6:** The LORD is compared
to a gracious host. **6:** *Dwell in the house of the
LORD* means to worship in the temple (compare 27.4). *My whole life long,* Hebrew, "for
length of days."
**Ps 24: A liturgy on entering the sanc-
tuary,** probably used in connection with a
procession of the ark. **1–6:** Perhaps sung by a
choir within the temple gates. **1–2:** Acknowl-

edgment of the LORD as the creator. **3:** The
question: Who shall be admitted to the temple? (compare Ps 15). **4–6:** The answer: Only
those who have the requisite moral qualities.
6: *Selah,* see 3.2 n.
24.7–10: The choir outside the gates, presumably accompanied by the ark, now re-
quests to be admitted. **7:** *Heads,* lintels. **8–10:**
The King of glory . . . . , means the God of
Israel, whose presence was associated with
the ark (Num 10.35–36).
**Ps 25: Prayer for deliverance from
personal enemies** (a lament). In acrostic
form (compare Pss 9–10 n.), every successive

3 Do not let those who wait for you
      be put to shame;
  let them be ashamed who are
      wantonly treacherous.

4 Make me to know your ways,
      O Lord;
  teach me your paths.
5 Lead me in your truth, and teach
      me,
  for you are the God of my
      salvation;
  for you I wait all day long.

6 Be mindful of your mercy,
      O Lord, and of your
      steadfast love,
  for they have been from of old.
7 Do not remember the sins of my
      youth or my transgressions;
  according to your steadfast love
      remember me,
  for your goodness' sake,
      O Lord!

8 Good and upright is the Lord;
      therefore he instructs sinners in
      the way.
9 He leads the humble in what is
      right,
  and teaches the humble his way.
10 All the paths of the Lord are
      steadfast love and
      faithfulness,
  for those who keep his covenant
      and his decrees.

11 For your name's sake, O Lord,
      pardon my guilt, for it is great.
12 Who are they that fear the Lord?
  He will teach them the way that
      they should choose.

13 They will abide in prosperity,
      and their children shall possess
      the land.
14 The friendship of the Lord is for
      those who fear him,
  and he makes his covenant
      known to them.
15 My eyes are ever toward the
      Lord,
  for he will pluck my feet out of
      the net.

16 Turn to me and be gracious to
      me,
  for I am lonely and afflicted.
17 Relieve the troubles of my heart,
  and bring me^r out of my
      distress.
18 Consider my affliction and my
      trouble,
  and forgive all my sins.
19 Consider how many are my foes,
      and with what violent hatred
      they hate me.
20 O guard my life, and deliver me;
      do not let me be put to shame,
      for I take refuge in you.
21 May integrity and uprightness
      preserve me,
  for I wait for you.

22 Redeem Israel, O God,
  out of all its troubles.

Psalm 26

Of David.

1 Vindicate me, O Lord,
  for I have walked in my
      integrity,

r Or *The troubles of my heart are enlarged; bring me*

---

verse beginning with another letter of the Hebrew alphabet. This artificial pattern accounts for the absence of any clear, logical structure, although the psalm has most of the elements of the typical lament: cry for help (vv. 1–3), the psalmist's situation (vv. 18–19), protestation of present innocence in spite of sins in the past (v. 21, compare 7, 11, 18), expression of trust (vv. 8–15), prayer for vindication (vv. 16–20).
25.13: *Possess the land,* see 37.9, 11, 29; compare Deut 11.8–9. **22:** Probably a liturgical addition adapting an individual prayer to congregational use.
Ps 26: Prayer for deliverance from personal enemies (a lament). **1–3:** Cry for

and I have trusted in the LORD
without wavering.
2 Prove me, O LORD, and try me;
test my heart and mind.
3 For your steadfast love is before
my eyes,
and I walk in faithfulness to
you. *s*

4 I do not sit with the worthless,
nor do I consort with
hypocrites;
5 I hate the company of evildoers,
and will not sit with the
wicked.

6 I wash my hands in innocence,
and go around your altar,
O LORD,
7 singing aloud a song of
thanksgiving,
and telling all your wondrous
deeds.

8 O LORD, I love the house in
which you dwell,
and the place where your glory
abides.
9 Do not sweep me away with
sinners,
nor my life with the
bloodthirsty,
10 those in whose hands are evil
devices,
and whose right hands are full
of bribes.

11 But as for me, I walk in my
integrity;
redeem me, and be gracious
to me.
12 My foot stands on level ground;
in the great congregation I will
bless the LORD.

Psalm 27

Of David.

1 The LORD is my light and my
salvation;
whom shall I fear?
The LORD is the stronghold *t* of
my life;
of whom shall I be afraid?

2 When evildoers assail me
to devour my flesh—
my adversaries and foes—
they shall stumble and fall.

3 Though an army encamp against
me,
my heart shall not fear;
though war rise up against me,
yet I will be confident.

4 One thing I asked of the LORD,
that will I seek after:
to live in the house of the LORD
all the days of my life,
to behold the beauty of the LORD,
and to inquire in his temple.

5 For he will hide me in his shelter
in the day of trouble;
he will conceal me under the cover
of his tent;
he will set me high on a rock.

6 Now my head is lifted up
above my enemies all around
me,
and I will offer in his tent
sacrifices with shouts of joy;
I will sing and make melody to
the LORD.

7 Hear, O LORD, when I cry aloud,

s Or *in your faithfulness*   *t* Or *refuge*

---

vindication against an unjust charge (compare
1 Kings 8.31–32). **4–7:** Protestation of innocence (compare 4.2–4).
   26.6–7: The protest is dramatized in a
liturgical ceremony. *Wash my hands in innocence,* compare Deut 21.6–8; Ps 51.7. **8–11:**
Prayer for help. *Your glory abides,* Isa 4.5; Ezek
43.4–5. **12:** The vow (see 7.17 n.).

**Ps 27: An act of devotion and a prayer
for deliverance. 1–6:** Song of trust (see Ps
11 n.). **4:** *Live in the house of the LORD,* see
23.6 n. As with most of the psalms, the author is probably a temple functionary, a Levite. **6:** *Tent,* poetic name for the temple.
7–14: A lament.
   27.7–9: Cry for help. **10–12:** The psalm-

be gracious to me and answer
me!

8 "Come," my heart says, "seek his
face!"
Your face, Lord, do I seek.
9 Do not hide your face from me.

Do not turn your servant away in
anger,
you who have been my help.
Do not cast me off, do not
forsake me,
O God of my salvation!
10 If my father and mother forsake
me,
the Lord will take me up.

11 Teach me your way, O Lord,
and lead me on a level path
because of my enemies.
12 Do not give me up to the will of
my adversaries,
for false witnesses have risen
against me,
and they are breathing out
violence.

13 I believe that I shall see the
goodness of the Lord
in the land of the living.
14 Wait for the Lord;
be strong, and let your heart
take courage;
wait for the Lord!

Psalm 28

Of David.

1 To you, O Lord, I call;
my rock, do not refuse to
hear me,
for if you are silent to me,
I shall be like those who go
down to the Pit.
2 Hear the voice of my supplication,

as I cry to you for help,
as I lift up my hands
toward your most holy
sanctuary. *u*

3 Do not drag me away with the
wicked,
with those who are workers
of evil,
who speak peace with their
neighbors,
while mischief is in their hearts.
4 Repay them according to their
work,
and according to the evil of
their deeds;
repay them according to the work
of their hands;
render them their due reward.
5 Because they do not regard the
works of the Lord,
or the work of his hands,
he will break them down and
build them up no more.

6 Blessed be the Lord,
for he has heard the sound of
my pleadings.
7 The Lord is my strength and my
shield;
in him my heart trusts;
so I am helped, and my heart
exults,
and with my song I give thanks
to him.

8 The Lord is the strength of his
people;
he is the saving refuge of his
anointed.
9 O save your people, and bless
your heritage;

u Heb *your innermost sanctuary*

---

ist's situation. **13–14:** Expression of confidence. *In the land of the living,* i.e. "during my life."

Ps 28: Prayer for deliverance from personal enemies (a lament). **1–5:** Cry for vindication (compare 26.1–3). **1:** *The Pit,* i.e. Sheol (see 6.5 n.).

28.6–7: After v. 5, a priest or temple prophet probably delivered an oracle of assurance (compare 12.5 n.), to which these verses are the psalmist's grateful response. **8–9:** See 25.22 n. **8:** *Anointed,* see 2.2 n. **9:** *Shepherd,* Isa 40.11.

be their shepherd, and carry
them forever.

Psalm 29

A Psalm of David.

1 Ascribe to the LORD, O heavenly
beings,[v]
ascribe to the LORD glory and
strength.
2 Ascribe to the LORD the glory of
his name;
worship the LORD in holy
splendor.

3 The voice of the LORD is over the
waters;
the God of glory thunders,
the LORD, over mighty waters.
4 The voice of the LORD is
powerful;
the voice of the LORD is full of
majesty.

5 The voice of the LORD breaks the
cedars;
the LORD breaks the cedars of
Lebanon.
6 He makes Lebanon skip like a calf,
and Sirion like a young wild ox.

7 The voice of the LORD flashes
forth flames of fire.
8 The voice of the LORD shakes the
wilderness;
the LORD shakes the wilderness
of Kadesh.

9 The voice of the LORD causes the
oaks to whirl,[w]
and strips the forest bare;

and in his temple all say,
"Glory!"

10 The LORD sits enthroned over the
flood;
the LORD sits enthroned as king
forever.
11 May the LORD give strength to his
people!
May the LORD bless his people
with peace!

Psalm 30

*A Psalm. A Song at the dedication of the
temple. Of David.*

1 I will extol you, O LORD, for you
have drawn me up,
and did not let my foes rejoice
over me.
2 O LORD my God, I cried to you
for help,
and you have healed me.
3 O LORD, you brought up my soul
from Sheol,
restored me to life from among
those gone down to the
Pit.[x]

4 Sing praises to the LORD, O you
his faithful ones,
and give thanks to his holy
name.
5 For his anger is but for a moment;
his favor is for a lifetime.
Weeping may linger for the night,
but joy comes with the
morning.

v Heb *sons of gods*   *w* Or *causes the deer to calve*
x Or *that I should not go down to the Pit*

Ps 29: **Hymn to the God of the storm.**
1–2: Call to worship. **3–9**: The LORD's manifestation in the thunderstorm. **3**: *The voice of
the LORD,* the thunder. *The waters,* the Mediterranean, but with mythological overtones.
5: *Lebanon,* the principal mountains of Syria.
6: *Sirion,* Mount Hermon. **7**: *Flames of fire,* the
lightning. **8**: *Kadesh,* probably the Kadesh of
the wilderness wanderings (Num 20.1). **10–**
11: Above the tumult of the storm, the LORD
reigns in majestic peace.
Ps 30: **Thanksgiving for healing. 1–3**:
God is praised for the psalmist's recovery.
The psalm was probably sung in fulfillment
of a vow (see 7.17 n.). **3**: *Sheol . . . the Pit,* see
28.1 n. **4–5**: The congregation is invited to
join in giving thanks.

6 As for me, I said in my
prosperity,
"I shall never be moved."
7 By your favor, O LORD,
you had established me as a
strong mountain;
you hid your face;
I was dismayed.

8 To you, O LORD, I cried,
and to the LORD I made
supplication:
9 "What profit is there in my death,
if I go down to the Pit?
Will the dust praise you?
Will it tell of your faithfulness?
10 Hear, O LORD, and be gracious to
me!
O LORD, be my helper!"

11 You have turned my mourning
into dancing;
you have taken off my sackcloth
and clothed me with joy,
12 so that my soul*y* may praise you
and not be silent.
O LORD my God, I will give
thanks to you forever.

Psalm 31

To the leader. A Psalm of David.

1 In you, O LORD, I seek refuge;
do not let me ever be put to
shame;
in your righteousness deliver
me.
2 Incline your ear to me;
rescue me speedily.
Be a rock of refuge for me,
a strong fortress to save me.

3 You are indeed my rock and my
fortress;
for your name's sake lead me
and guide me,
4 take me out of the net that is
hidden for me,
for you are my refuge.
5 Into your hand I commit my
spirit;
you have redeemed me,
O LORD, faithful God.

6 You hate*z* those who pay regard
to worthless idols,
but I trust in the LORD.
7 I will exult and rejoice in your
steadfast love,
because you have seen my
affliction;
you have taken heed of my
adversities,
8 and have not delivered me into the
hand of the enemy;
you have set my feet in a broad
place.

9 Be gracious to me, O LORD, for I
am in distress;
my eye wastes away from grief,
my soul and body also.
10 For my life is spent with sorrow,
and my years with sighing;
my strength fails because of my
misery,*a*
and my bones waste away.

11 I am the scorn of all my
adversaries,

y Heb *that glory*   *z* One Heb Ms Gk Syr
Jerome: MT *I hate*   *a* Gk Syr: Heb *my iniquity*

30.6–12: The story of what had happened
(see 18.4–6 n.): Before becoming ill, the
psalmist had felt perfectly secure (vv. 6–7),
but, when illness came, turned to God in
prayer (vv. 8–10) and God had answered by
restoring health (vv. 11–12).
**Ps 31: Prayer for deliverance from per-
sonal enemies** (a lament). Verses 1–8 and
9–24 are parallel in form, both containing the

principal elements of a lament. **1–8:** Cry for
help (vv. 1–5), the psalmist's situation (v. 4),
expression of confidence (v. 5), protestation
of innocence (v. 6), grateful recognition of
God's help (vv. 7–8; perhaps this followed an
oracle of assurance; see 12.5 n.).
31.9–24: Cry for help (v. 9), the psalmist's
situation (vv. 10–13), expression of confidence (vv. 14, 19–20), prayer for vindication

a horror[b] to my neighbors,
an object of dread to my
acquaintances;
those who see me in the street
flee from me.

12 I have passed out of mind like one
who is dead;
I have become like a broken
vessel.

13 For I hear the whispering of
many—
terror all around!—
as they scheme together against
me,
as they plot to take my life.

14 But I trust in you, O Lord;
I say, "You are my God."

15 My times are in your hand;
deliver me from the hand of my
enemies and persecutors.

16 Let your face shine upon your
servant;
save me in your steadfast love.

17 Do not let me be put to shame,
O Lord,
for I call on you;
let the wicked be put to shame;
let them go dumbfounded to
Sheol.

18 Let the lying lips be stilled
that speak insolently against the
righteous
with pride and contempt.

19 O how abundant is your goodness
that you have laid up for those
who fear you,
and accomplished for those who
take refuge in you,
in the sight of everyone!

20 In the shelter of your presence you
hide them
from human plots;

you hold them safe under your
shelter
from contentious tongues.

21 Blessed be the Lord,
for he has wondrously shown
his steadfast love to me
when I was beset as a city under
siege.

22 I had said in my alarm,
"I am driven far[c] from your
sight."
But you heard my supplications
when I cried out to you for
help.

23 Love the Lord, all you his saints.
The Lord preserves the faithful,
but abundantly repays the one
who acts haughtily.

24 Be strong, and let your heart take
courage,
all you who wait for the Lord.

Psalm 32

Of David. A Maskil.

1 Happy are those whose
transgression is forgiven,
whose sin is covered.

2 Happy are those to whom the
Lord imputes no iniquity,
and in whose spirit there is no
deceit.

3 While I kept silence, my body
wasted away
through my groaning all day
long.

4 For day and night your hand was
heavy upon me;

b Cn: Heb *exceedingly*   c Another reading is
cut off

(vv. 15–18), grateful recognition of God's
help (vv. 21–24). **12**: *Broken vessel*, compare
Eccl 12.6. **13**: Jer 20.10. **23**: See 30.4 n.
   Ps 32: Thanksgiving for healing. 1–2:
God is praised for the psalmist's recovery. **1:**

Happy, see 1.1 n. Since disease was commonly regarded as punishment for sin, healing is
evidence that the *transgression is forgiven*. **3–5**:
The psalmist's experience (see 18.4–6 n.). **4**:
Selah, see 3.2 n.

my strength was dried up[d] as
 by the heat of summer.
     Selah

5 Then I acknowledged my sin to
 you,
 and I did not hide my iniquity;
I said, "I will confess my
 transgressions to the LORD,"
and you forgave the guilt of
 my sin.   *Selah*

6 Therefore let all who are faithful
 offer prayer to you;
at a time of distress,[e] the rush of
 mighty waters
shall not reach them.
7 You are a hiding place for me;
 you preserve me from trouble;
 you surround me with glad cries
  of deliverance. *Selah*

8 I will instruct you and teach you
 the way you should go;
 I will counsel you with my eye
  upon you.
9 Do not be like a horse or a mule,
 without understanding,
whose temper must be curbed
 with bit and bridle,
else it will not stay near you.

10 Many are the torments of the
 wicked,
 but steadfast love surrounds
  those who trust in the
  LORD
11 Be glad in the LORD and rejoice,
 O righteous,
 and shout for joy, all you
  upright in heart.

Psalm 33

1 Rejoice in the LORD, O you
 righteous.

Praise befits the upright.
2 Praise the LORD with the lyre;
 make melody to him with the
  harp of ten strings.
3 Sing to him a new song;
 play skillfully on the strings,
  with loud shouts.

4 For the word of the LORD is
 upright,
 and all his work is done in
  faithfulness.
5 He loves righteousness and justice;
 the earth is full of the steadfast
  love of the LORD.

6 By the word of the LORD the
 heavens were made,
 and all their host by the breath
  of his mouth.
7 He gathered the waters of the sea
 as in a bottle;
 he put the deeps in storehouses.

8 Let all the earth fear the LORD;
 let all the inhabitants of the
  world stand in awe of him.
9 For he spoke, and it came to be;
 he commanded, and it stood
  firm.

10 The LORD brings the counsel of
 the nations to nothing;
 he frustrates the plans of the
  peoples.
11 The counsel of the LORD stands
 forever,
 the thoughts of his heart to all
  generations.
12 Happy is the nation whose God is
 the LORD,
 the people whom he has chosen
  as his heritage.

d Meaning of Heb uncertain *e* Cn: Heb *at a
time of finding only*

32.5: Healing came only after acknowledgment of sin. **6–11:** The psalmist commends to the congregation similar faith in God (vv. 6–7, 10) and obedience to God's will (vv. 8–9).
Ps 33: Hymn to God as creator and lord of history. **1–3:** Call to worship. **4–5:** The character of Israel's God. **6–9:** The LORD as creator. The emphasis upon the divine word in vv. 6 and 9 reflects Gen 1.3–31.
 33.10–19: The LORD rules over the destinies of nations.

13 The Lord looks down from
heaven;
  he sees all humankind.
14 From where he sits enthroned he
watches
  all the inhabitants of the earth—
15 he who fashions the hearts of
them all,
  and observes all their deeds.
16 A king is not saved by his great
army;
  a warrior is not delivered by his
great strength.
17 The war horse is a vain hope for
victory,
  and by its great might it cannot
save.

18 Truly the eye of the Lord is on
those who fear him,
  on those who hope in his
steadfast love,
19 to deliver their soul from death,
  and to keep them alive in
famine.

20 Our soul waits for the Lord;
  he is our help and shield.
21 Our heart is glad in him,
  because we trust in his holy
name.
22 Let your steadfast love, O Lord,
be upon us,
  even as we hope in you.

Psalm 34

*Of David, when he feigned madness before
Abimelech, so that he drove him out, and he
went away.*

1 I will bless the Lord at all times;
  his praise shall continually be in
my mouth.

2 My soul makes its boast in the
Lord;
  let the humble hear and be glad.
3 O magnify the Lord with me,
  and let us exalt his name
together.

4 I sought the Lord, and he
answered me,
  and delivered me from all my
fears.
5 Look to him, and be radiant;
  so your*f* faces shall never be
ashamed.
6 This poor soul cried, and was
heard by the Lord,
  and was saved from every
trouble.
7 The angel of the Lord encamps
around those who fear him, and
delivers them.
8 O taste and see that the Lord is
good;
  happy are those who take refuge
in him.
9 O fear the Lord, you his holy
ones,
  for those who fear him have no
want.
10 The young lions suffer want and
hunger,
  but those who seek the Lord
lack no good thing.

11 Come, O children, listen to me;
  I will teach you the fear of the
Lord.
12 Which of you desires life,
  and covets many days to enjoy
good?
13 Keep your tongue from evil,

f Gk Syr Jerome: Heb *their*

33.20–22: Israel puts her complete trust
in him.
**Ps 34: Thanksgiving for deliverance
from trouble.** Like Pss 9–10 and 25, this
psalm is an alphabetical acrostic. **1–3:** A brief
hymn of praise. **4–6:** The narrative of the
psalmist's experience (see 18.4–6 n.; compare
30.6–12; 32.3–5).

34.7–22: He commends to the congregation the same kind of faith in God, and assures
them that the Lord will never be found wanting (compare 32.6–11). The style is that of the
teachers of wisdom (compare Pss 1 and 37;
also v. 11 with Prov 1.8; 2.1). **12–14:** 1 Pet
3.10–12. **20:** Jn 19.36.

Blank

---

and your lips from speaking
deceit.
14 Depart from evil, and do good;
seek peace, and pursue it.

15 The eyes of the LORD are on the
righteous,
and his ears are open to their
cry.
16 The face of the LORD is against
evildoers,
to cut off the remembrance of
them from the earth.
17 When the righteous cry for help,
the LORD hears,
and rescues them from all their
troubles.
18 The LORD is near to the
brokenhearted,
and saves the crushed in spirit.

19 Many are the afflictions of the
righteous,
but the LORD rescues them from
them all.
20 He keeps all their bones;
not one of them will be broken.
21 Evil brings death to the wicked,
and those who hate the
righteous will be
condemned.
22 The LORD redeems the life of his
servants;
none of those who take refuge
in him will be condemned.

Psalm 35

Of David.

1 Contend, O LORD, with those
who contend with me;
fight against those who fight
against me!
2 Take hold of shield and buckler,
and rise up to help me!
3 Draw the spear and javelin
against my pursuers;
say to my soul,
"I am your salvation."

4 Let them be put to shame and
dishonor
who seek after my life.
Let them be turned back and
confounded
who devise evil against me.
5 Let them be like chaff before the
wind,
with the angel of the LORD
driving them on.
6 Let their way be dark and
slippery,
with the angel of the LORD
pursuing them.
7 For without cause they hid their
net^g for me;
without cause they dug a pit^h
for my life.
8 Let ruin come on them unawares.
And let the net that they hid
ensnare them;
let them fall in it—to their ruin.

9 Then my soul shall rejoice in the
LORD,
exulting in his deliverance.
10 All my bones shall say,
"O LORD, who is like you?
You deliver the weak
from those too strong for them,
the weak and needy from those
who despoil them."

11 Malicious witnesses rise up;
they ask me about things I do
not know.
12 They repay me evil for good;
my soul is forlorn.

g Heb *a pit, their net*   h The word *pit* is
transposed from the preceding line

Ps 35: **Prayer for deliverance from
personal enemies** (a lament). As in Ps 31,
the elements of the lament occur more than
once, so that vv. 1–10, 11–18, and 19–28 can
be treated as separate units. **1–10:** The first
lament. **1–6:** Cry for vindication and vengeance. **7:** The psalmist's situation. **8:** Prayer
for vengeance renewed.
35.9–10: The vow (see 7.17 n.). **11–18:**
The second lament. **11–16:** The psalmist's sit-

13 But as for me, when they were
    sick,
  I wore sackcloth;
  I afflicted myself with fasting.
  I prayed with head bowed*ⁱ* on my
    bosom,
14   as though I grieved for a friend
    or a brother;
  I went about as one who laments
    for a mother,
  bowed down and in mourning.

15 But at my stumbling they
    gathered in glee,
  they gathered together against
    me;
  ruffians whom I did not know
    tore at me without ceasing;
16 they impiously mocked more and
    more,*ʲ*
  gnashing at me with their teeth.

17 How long, O LORD, will you look
    on?
  Rescue me from their ravages,
  my life from the lions!
18 Then I will thank you in the great
    congregation;
  in the mighty throng I will
    praise you.

19 Do not let my treacherous enemies
    rejoice over me,
  or those who hate me without
    cause wink the eye.
20 For they do not speak peace,
  but they conceive deceitful
    words
  against those who are quiet in
    the land.
21 They open wide their mouths
    against me;
  they say, "Aha, Aha,
    our eyes have seen it."

22 You have seen, O LORD; do not
    be silent!

  O Lord, do not be far from me!
23 Wake up! Bestir yourself for my
    defense,
  for my cause, my God and my
    Lord!
24 Vindicate me, O LORD, my God,
  according to your righteousness,
  and do not let them rejoice
    over me.
25 Do not let them say to
    themselves,
  "Aha, we have our heart's
    desire."
  Do not let them say, "We have
    swallowed you*ᵏ* up."

26 Let all those who rejoice at my
    calamity
  be put to shame and confusion;
  let those who exalt themselves
    against me
  be clothed with shame and
    dishonor.

27 Let those who desire my
    vindication
  shout for joy and be glad,
  and say evermore,
  "Great is the LORD,
  who delights in the welfare of
    his servant."
28 Then my tongue shall tell of your
    righteousness
  and of your praise all day long.

Psalm 36

*To the leader. Of David, the servant
of the LORD.*

1 Transgression speaks to the
    wicked
  deep in their hearts;
  there is no fear of God

---

i Or *My prayer turned back*   *j* Cn Compare
Gk: Heb *like the profanest of mockers of a cake*
k Heb *him*

---

uation: persecution without cause. As in
some other laments, it is not clear whether the
psalmist's principal trouble arises from enemies or from a sickness of which they have
taken advantage. **17**: Cry for help. **18**: The
vow (see 7.17 n.). **19–29**: The third lament.
35.19: Cry for help. **20–21**: The psalmist's
situation. **22–27**: Prayer for vindication. **28**:
The vow (compare v. 18).
    Ps 36: **A psalm of mixed type**: vv. 1–4

before their eyes.
2 For they flatter themselves in their
own eyes
that their iniquity cannot be
found out and hated.
3 The words of their mouths are
mischief and deceit;
they have ceased to act wisely
and do good.
4 They plot mischief while on their
beds;
they are set on a way that is
not good;
they do not reject evil.

5 Your steadfast love, O LORD,
extends to the heavens,
your faithfulness to the clouds.
6 Your righteousness is like the
mighty mountains,
your judgments are like the
great deep;
you save humans and animals
alike, O LORD.

7 How precious is your steadfast
love, O God!
All people may take refuge in
the shadow of your wings.
8 They feast on the abundance of
your house,
and you give them drink from
the river of your delights.
9 For with you is the fountain of
life;
in your light we see light.

10 O continue your steadfast love to
those who know you,
and your salvation to the
upright of heart!

11 Do not let the foot of the arrogant
tread on me,
or the hand of the wicked drive
me away.
12 There the evildoers lie prostrate;
they are thrust down, unable
to rise.

Psalm 37

Of David.

1 Do not fret because of the wicked;
do not be envious of
wrongdoers,
2 for they will soon fade like the
grass,
and wither like the green herb.

3 Trust in the LORD, and do good;
so you will live in the land, and
enjoy security.
4 Take delight in the LORD,
and he will give you the desires
of your heart.

5 Commit your way to the LORD;
trust in him, and he will act.
6 He will make your vindication
shine like the light,
and the justice of your cause like
the noonday.

7 Be still before the LORD, and wait
patiently for him;
do not fret over those who
prosper in their way,
over those who carry out evil
devices.

8 Refrain from anger, and forsake
wrath.

are in the type of a wisdom psalm, vv. 5–9,
a hymn; vv. 10–12, a prayer in the style of a
lament. Since the last section seems to determine the character of the whole, the psalm
should probably be classified as a liturgy of
lament. 1–4: The character of the wicked. 1:
The wicked are inspired by *transgression* just as
a prophet is inspired by the Spirit of God.
36.5–9: The character of God, who is a
source of blessing to the righteous and provides them with a refuge. *Fountain of life,*
compare Jer 2.13. 10–11: Prayer for deliverance. 12: Assurance of having been heard.
**Ps 37: The certainty of retribution for
the wicked** (a wisdom psalm). The argument is directed to those discouraged by the
injustices which apparently dominate the
world. The acrostic form of the psalm (see
Pss 9–10 n.) explains the lack of a clear outline
or logical progression of thought. 3: *The land,*
Palestine (vv. 9, 11, 22, 29; compare Deut
11.8–32).

Do not fret—it leads only to
evil.

9 For the wicked shall be cut off,
but those who wait for the
LORD shall inherit the land.

10 Yet a little while, and the wicked
will be no more;
though you look diligently for
their place, they will not
be there.

11 But the meek shall inherit the
land,
and delight themselves in
abundant prosperity.

12 The wicked plot against the
righteous,
and gnash their teeth at them;

13 but the LORD laughs at the
wicked,
for he sees that their day is
coming.

14 The wicked draw the sword and
bend their bows
to bring down the poor and
needy,
to kill those who walk
uprightly;

15 their sword shall enter their own
heart,
and their bows shall be broken.

16 Better is a little that the righteous
person has
than the abundance of many
wicked.

17 For the arms of the wicked shall
be broken,
but the LORD upholds the
righteous.

18 The LORD knows the days of the
blameless,
and their heritage will abide
forever;

19 they are not put to shame in evil
times,
in the days of famine they have
abundance.

20 But the wicked perish,
and the enemies of the LORD are
like the glory of the
pastures;
they vanish—like smoke they
vanish away.

21 The wicked borrow, and do not
pay back,
but the righteous are generous
and keep giving;

22 for those blessed by the LORD shall
inherit the land,
but those cursed by him shall be
cut off.

23 Our steps*l* are made firm by the
LORD,
when he delights in our*m* way;

24 though we stumble,*n* we*o* shall
not fall headlong,
for the LORD holds us*p* by the
hand.

25 I have been young, and now am
old,
yet I have not seen the righteous
forsaken
or their children begging bread.

26 They are ever giving liberally and
lending,
and their children become a
blessing.

27 Depart from evil, and do good;
so you shall abide forever.

28 For the LORD loves justice;
he will not forsake his faithful
ones.

l Heb *a man's steps*　　*m* Heb *his*　　*n* Heb *he
stumbles*　　*o* Heb *he*　　*p* Heb *him*

37.10: The heart of the argument: Do not
be impatient! Retribution will come soon,
even though it is not evident now (compare
vv. 35–36). 11: Compare Mt 5.5.

37.25: Compare Job 4.7. 34: Patient waiting for the LORD to act is the proper attitude,
not querulous anxiety (compare v. 9; 38.15;
62.1, 5; 130.5; Isa 40.31).

The righteous shall be kept safe
forever,
but the children of the wicked
shall be cut off.
29 The righteous shall inherit the
land,
and live in it forever.

30 The mouths of the righteous utter
wisdom,
and their tongues speak justice.
31 The law of their God is in their
hearts;
their steps do not slip.

32 The wicked watch for the
righteous,
and seek to kill them.
33 The LORD will not abandon them
to their power,
or let them be condemned when
they are brought to trial.

34 Wait for the LORD, and keep to his
way,
and he will exalt you to inherit
the land;
you will look on the destruction
of the wicked.

35 I have seen the wicked oppressing,
and towering like a cedar of
Lebanon. *q*
36 Again I*r* passed by, and they were
no more;
though I sought them, they
could not be found.

37 Mark the blameless, and behold
the upright,
for there is posterity for the
peaceable.
38 But transgressors shall be
altogether destroyed;

the posterity of the wicked shall
be cut off.
39 The salvation of the righteous is
from the LORD;
he is their refuge in the time of
trouble.
40 The LORD helps them and rescues
them;
he rescues them from the
wicked, and saves them,
because they take refuge in him.

Psalm 38

*A Psalm of David, for the memorial
offering.*

1 O LORD, do not rebuke me in
your anger,
or discipline me in your wrath.
2 For your arrows have sunk into
me,
and your hand has come down
on me.

3 There is no soundness in my flesh
because of your indignation;
there is no health in my bones
because of my sin.
4 For my iniquities have gone over
my head;
they weigh like a burden too
heavy for me.

5 My wounds grow foul and fester
because of my foolishness;
6 I am utterly bowed down and
prostrate;
all day long I go around
mourning.

q Gk: Meaning of Heb uncertain   *r* Gk Syr
Jerome: Heb *he*

Ps 38: Prayer for healing in sickness (a lament). The psalmist is mainly concerned with an attack of some grave disease, but also is troubled by enemies who are taking advantage of it. Their attitude was made more plausible by the common belief that illness was a punishment for sin and therefore an indica-tion that God was against the sick person (compare Ps 22). **Title:** *Memorial offering,* Lev 2.1–10; compare Lev 24.7. **1:** Cry for help. **2–20:** The psalmist's situation. **2:** *Your arrows,* Job 6.4; 16.12, 13. **3–10:** The nature of the disease. **3–4:** The psalmist acknowledges that the disease must be a deserved punish-

7 For my loins are filled with
burning,
and there is no soundness in my
flesh.
8 I am utterly spent and crushed;
I groan because of the tumult of
my heart.

9 O Lord, all my longing is known
to you;
my sighing is not hidden from
you.
10 My heart throbs, my strength
fails me;
as for the light of my eyes—it
also has gone from me.
11 My friends and companions stand
aloof from my affliction,
and my neighbors stand far off.

12 Those who seek my life lay their
snares;
those who seek to hurt me
speak of ruin,
and meditate treachery all day
long.

13 But I am like the deaf, I do not
hear;
like the mute, who cannot
speak.
14 Truly, I am like one who does not
hear,
and in whose mouth is no
retort.

15 But it is for you, O Lord, that I
wait;
it is you, O Lord my God,
who will answer.
16 For I pray, "Only do not let them
rejoice over me,

those who boast against me
when my foot slips."

17 For I am ready to fall,
and my pain is ever with me.
18 I confess my iniquity;
I am sorry for my sin.
19 Those who are my foes without
cause[s] are mighty,
and many are those who hate
me wrongfully.
20 Those who render me evil for
good
are my adversaries because I
follow after good.

21 Do not forsake me, O Lord;
O my God, do not be far from
me;
22 make haste to help me,
O Lord, my salvation.

Psalm 39

*To the leader: to Jeduthun.
A Psalm of David.*

1 I said, "I will guard my ways
that I may not sin with my
tongue;
I will keep a muzzle on my mouth
as long as the wicked are in my
presence."
2 I was silent and still;
I held my peace to no avail;
my distress grew worse,
3 my heart became hot within
me.
While I mused, the fire burned;
then I spoke with my tongue:

s Q Ms: MT *my living foes*

ment for having committed some sin, perhaps unwittingly (compare v. 18). **11:** Former
friends avoid the writer as one forsaken by
God. **12:** Enemies are circulating lies about the
psalmist (compare vv. 19–20).
38.13–16: Not really fretful, the writer is
waiting for God to act (compare 37.34). **18:**
By confessing to be a sinner, the psalmist
hopes to be forgiven and healed (compare
32.3–5). **21–22:** Cry for help.

Ps 39: Prayer for healing in sickness (a
lament). **1–3:** During a severe illness, by
which faith in God's good-will has been almost shaken, the psalmist has not made any
public complaint, not wishing to give encouragement to *the wicked* (i.e. the skeptical),
who deny God's concern for justice and
goodness (compare 10.4 and Ps 14). But now
the psalmist can be silent no longer: a complaint must be made, if only to God in pri-

4 "LORD, let me know my end,
  and what is the measure of my
    days;
  let me know how fleeting my
    life is.
5 You have made my days a few
    handbreadths,
  and my lifetime is as nothing in
    your sight.
  Surely everyone stands as a mere
    breath.   *Selah*
6 Surely everyone goes about like
    a shadow.
  Surely for nothing they are in
    turmoil;
  they heap up, and do not know
    who will gather.

7 "And now, O Lord, what do I
    wait for?
  My hope is in you.
8 Deliver me from all my
    transgressions.
  Do not make me the scorn of
    the fool.
9 I am silent; I do not open my
    mouth,
  for it is you who have done it.
10 Remove your stroke from me;
  I am worn down by the blows *t*
    of your hand.

11 "You chastise mortals
    in punishment for sin,
  consuming like a moth what is
    dear to them;
  surely everyone is a mere
    breath.   *Selah*

12 "Hear my prayer, O LORD,
    and give ear to my cry;

  do not hold your peace at my
    tears.
  For I am your passing guest,
    an alien, like all my forebears.
13 Turn your gaze away from me,
    that I may smile again,
  before I depart and am no
    more."

Psalm 40

To the leader. Of David. A Psalm.

1 I waited patiently for the LORD;
  he inclined to me and heard
    my cry.
2 He drew me up from the desolate
    pit, *u*
  out of the miry bog,
  and set my feet upon a rock,
    making my steps secure.
3 He put a new song in my mouth,
    a song of praise to our God.
  Many will see and fear,
    and put their trust in the LORD.

4 Happy are those who make
    the LORD their trust,
  who do not turn to the proud,
    to those who go astray after
      false gods.
5 You have multiplied, O LORD my
    God,
  your wondrous deeds and your
    thoughts toward us;
  none can compare with you.
  Were I to proclaim and tell of
    them,
  they would be more than can be
    counted.

t Heb *hostility*   *u* Cn: Heb *pit of tumult*

---

vate. **4–6:** The writer knows that all human
life is short. **5:** *Selah,* see 3.2 n. *Handbreadths,*
a measure of about three inches (four fingers);
compare Ezek 40.5; 43.13.
39.7–13: Prayer for healing. **8:** *Deliver . . .
from . . . transgressions,* i.e. heal the diseases
they have caused (compare 31.1). **12–13:** The
psalmist knows that, in any case, life will be
short, but hopes for at least one more respite,
brief though it may be.

**Ps 40: Thanksgiving for deliverance
from trouble, together with a prayer for
help.** A composite psalm: vv. 1–11 are a
thanksgiving; vv. 12–17, a lament. Perhaps
two originally independent units have been
combined into a liturgy. **1–3:** The psalmist's
experience (compare 18.4–6). The nature of
the trouble is not specified. **4–10:** Fulfillment
of the vow (see 7.17 n. and 22.22–31 n.).

6 Sacrifice and offering you do not
desire,
but you have given me an open
ear. *v*
Burnt offering and sin offering
you have not required.
7 Then I said, "Here I am;
in the scroll of the book it is
written of me. *w*
8 I delight to do your will, O my
God;
your law is within my heart."

9 I have told the glad news of
deliverance
in the great congregation;
see, I have not restrained my lips,
as you know, O LORD.
10 I have not hidden your saving help
within my heart,
I have spoken of your
faithfulness and your
salvation;
I have not concealed your steadfast
love and your faithfulness
from the great congregation.

11 Do not, O LORD, withhold
your mercy from me;
let your steadfast love and your
faithfulness
keep me safe forever.
12 For evils have encompassed me
without number;
my iniquities have overtaken me,
until I cannot see;
they are more than the hairs of
my head,
and my heart fails me.

13 Be pleased, O LORD, to deliver
me;

O LORD, make haste to help
me.
14 Let all those be put to shame and
confusion
who seek to snatch away my
life;
let those be turned back and
brought to dishonor
who desire my hurt.
15 Let those be appalled because of
their shame
who say to me, "Aha, Aha!"

16 But may all who seek you
rejoice and be glad in you;
may those who love your
salvation
say continually, "Great is the
LORD!"
17 As for me, I am poor and needy,
but the Lord takes thought for
me.
You are my help and my
deliverer;
do not delay, O my God.

Psalm 41

To the leader. A Psalm of David.

1 Happy are those who consider the
poor; *x*
the LORD delivers them in the
day of trouble.
2 The LORD protects them and keeps
them alive;
they are called happy in the
land.
You do not give them up to the
will of their enemies.

v Heb *ears you have dug for me*   *w* Meaning of
Heb uncertain   *x* Or *weak*

40.6–8: Rather than offer a formal sacrifice of thanksgiving, the writer promises to
do God's will. Quoted in Heb 10.5–7. **6:**
Compare 50.8–13; 51.16, 17; Am 5.21–24;
Hos 6.6. **7:** *Scroll of the book,* compare 56.8;
139.16. **9:** Compare 22.22. **11:** Prayer for continuing help. **12–17:** A lament. **12:** The psalmist's situation.

40.13–17: Prayer for deliverance. This
passage is almost identical with Ps 70.
Ps 41: Prayer for healing from sickness
(a lament). **1–3:** Act of faith, in the style of the
wisdom writers (compare Ps 1).
41.4–9: The psalmist's situation: Mortally
ill, with enemies who are glad of it; even a
dear friend has become an enemy. **10–12:**

3 The Lord sustains them on their
    sickbed;
  in their illness you heal all their
    infirmities. *y*

4 As for me, I said, "O Lord, be
    gracious to me;
  heal me, for I have sinned
    against you."
5 My enemies wonder in malice
  when I will die, and my name
    perish.
6 And when they come to see me,
    they utter empty words,
  while their hearts gather
    mischief;
  when they go out, they tell it
    abroad.
7 All who hate me whisper together
    about me;
  they imagine the worst for me.

8 They think that a deadly thing has
    fastened on me,
  that I will not rise again from
    where I lie.
9 Even my bosom friend in whom I
    trusted,
  who ate of my bread, has lifted
    the heel against me.
10 But you, O Lord, be gracious
    to me,
  and raise me up, that I may
    repay them.

11 By this I know that you are
    pleased with me;
  because my enemy has not
    triumphed over me.
12 But you have upheld me because
    of my integrity,

and set me in your presence
    forever.

13 Blessed be the Lord, the God of
    Israel,
  from everlasting to everlasting.
    Amen and Amen.

BOOK II

Psalm 42

To the leader. A Maskil of the Korahites.

1 As a deer longs for flowing
    streams,
  so my soul longs for you,
    O God.
2 My soul thirsts for God,
  for the living God.
  When shall I come and behold
    the face of God?
3 My tears have been my food
  day and night,
  while people say to me
    continually,
    "Where is your God?"

4 These things I remember,
  as I pour out my soul:
  how I went with the throng, *z*
    and led them in procession to
    the house of God,
  with glad shouts and songs of
    thanksgiving,
  a multitude keeping festival.
5 Why are you cast down, O my
    soul,
  and why are you disquieted
    within me?

y Heb *you change all his bed*   *z* Meaning of
Heb uncertain

Prayer for deliverance. **13:** A doxology (no
part of the psalm) marking the end of the first
of the five books into which the Psalter was
divided in imitation of the Pentateuch (compare 72.18–20; 89.52; 106.48).
    **Pss 42–43: Prayer for healing in prep-
aration for a pilgrimage** (a lament). These
two psalms are a single lyric consisting of
three stanzas with a refrain (42.5, 11; 43.5).
The author, who lives in the far north of

Palestine near Mount Hermon and the
sources of the Jordan (42.6–7), has been prevented by illness (42.10) from making the
accustomed pilgrimage to Jerusalem (42.4;
43.3–4). **42.1–4:** The psalmist's love for the
temple and for God's presence there. **3:** People regard the sickness as evidence that God
has forsaken the sufferer (compare 22.6–8).
    42.5–10: The writer's present situation:
sick and far from the temple. **6:** *Mount Mizar,*

Hope in God; for I shall again
    praise him,
  my help [6]and my God.

My soul is cast down within me;
    therefore I remember you
  from the land of Jordan and of
      Hermon,
    from Mount Mizar.
7 Deep calls to deep
    at the thunder of your cataracts;
  all your waves and your billows
    have gone over me.
8 By day the LORD commands his
      steadfast love,
    and at night his song is with
      me,
    a prayer to the God of my life.

9 I say to God, my rock,
    "Why have you forgotten me?
  Why must I walk about
      mournfully
    because the enemy oppresses
      me?"
10 As with a deadly wound in my
      body,
    my adversaries taunt me,
  while they say to me continually,
    "Where is your God?"

11 Why are you cast down, O my
      soul,
    and why are you disquieted
      within me?
  Hope in God; for I shall again
      praise him,
    my help and my God.

Psalm 43

1 Vindicate me, O God, and defend
      my cause
    against an ungodly people;
  from those who are deceitful and
      unjust
    deliver me!

2 For you are the God in whom I
      take refuge;
    why have you cast me off?
  Why must I walk about
      mournfully
    because of the oppression of the
      enemy?

3 O send out your light and your
      truth;
    let them lead me;
  let them bring me to your holy
      hill
    and to your dwelling.
4 Then I will go to the altar of God,
    to God my exceeding joy;
  and I will praise you with the
      harp,
    O God, my God.

5 Why are you cast down, O my
      soul,
    and why are you disquieted
      within me?
  Hope in God; for I shall again
      praise him,
    my help and my God.

Psalm 44

To the leader. Of the Korahites. A Maskil.

1 We have heard with our ears,
      O God,
    our ancestors have told us,
  what deeds you performed in their
      days,
    in the days of old:
2 you with your own hand drove
      out the nations,
    but them you planted;
  you afflicted the peoples,
    but them you set free;
3 for not by their own sword did
      they win the land,
    nor did their own arm give
      them victory;
  but your right hand, and your
      arm,

---

probably near Mount *Hermon,* but exact location is unknown.
43.1–4: Prayer that, by the healing of the
disease, the psalmist may be vindicated as a

righteous person and enabled to go to Jerusalem.
**Ps 44: Prayer for deliverance from na-
tional enemies** (a group lament). Israel has

and the light of your
countenance,
for you delighted in them.

4 You are my King and my God;
you command[a] victories for
Jacob.
5 Through you we push down our
foes;
through your name we tread
down our assailants.
6 For not in my bow do I trust,
nor can my sword save me.
7 But you have saved us from our
foes,
and have put to confusion those
who hate us.
8 In God we have boasted
continually,
and we will give thanks to your
name forever.   *Selah*

9 Yet you have rejected us and
abased us,
and have not gone out with our
armies.
10 You made us turn back from
the foe,
and our enemies have gotten
spoil.
11 You have made us like sheep for
slaughter,
and have scattered us among
the nations.
12 You have sold your people for a
trifle,
demanding no high price for
them.

13 You have made us the taunt of
our neighbors,
the derision and scorn of those
around us.

14 You have made us a byword
among the nations,
a laughingstock[b] among the
peoples.
15 All day long my disgrace is before
me,
and shame has covered my face
16 at the words of the taunters and
revilers,
at the sight of the enemy and
the avenger.

17 All this has come upon us,
yet we have not forgotten you,
or been false to your covenant.
18 Our heart has not turned back,
nor have our steps departed
from your way,
19 yet you have broken us in the
haunt of jackals,
and covered us with deep
darkness.

20 If we had forgotten the name of
our God,
or spread out our hands to a
strange god,
21 would not God discover this?
For he knows the secrets of the
heart.
22 Because of you we are being killed
all day long,
and accounted as sheep for the
slaughter.

23 Rouse yourself! Why do you
sleep, O Lord?
Awake, do not cast us off
forever!
24 Why do you hide your face?

a Gk Syr. Heb *You are my King, O God; command*
b Heb *a shaking of the head*

suffered a humiliating defeat at the hand of
some unnamed foreign nation. **1–8:** God is
reminded of the victories he had given Israel
in previous times, from the days of the conquest onward. **3:** Deut 8.17; Judg 7.2. **8:**
Selah, see 3.2 n.
  44.9–16: Now God has allowed Israel to
be defeated and shamed. **11:** *You . . . have scattered us among the nations* seems appropriate
only to the period after the Babylonian Exile,
a circumstance which helps to date the poem.
Some have put it as late as the Maccabean Age
(second century B.C.).
  44.17–22: Defeat has not been the result of
sin or apostasy, but of unshaken loyalty to the
God of Israel. **23–26:** Cry for help (compare
78.65; Isa 51.9). The language emphasizes the
strong emotions of the speaker.

Why do you forget our
    affliction and oppression?
25 For we sink down to the dust;
    our bodies cling to the ground.
26 Rise up, come to our help.
    Redeem us for the sake of your
    steadfast love.

Psalm 45

*To the leader: according to Lilies. Of the
Korahites. A Maskil. A love song.*

1 My heart overflows with a goodly
    theme;
    I address my verses to the king;
    my tongue is like the pen of a
    ready scribe.

2 You are the most handsome of
    men;
    grace is poured upon your lips;
    therefore God has blessed you
    forever.
3 Gird your sword on your thigh,
    O mighty one,
    in your glory and majesty.

4 In your majesty ride on
    victoriously
    for the cause of truth and to
    defend* the right;
    let your right hand teach you
    dread deeds.
5 Your arrows are sharp
    in the heart of the king's
    enemies;
    the peoples fall under you.

6 Your throne, O God,*d* endures
    forever and ever.
    Your royal scepter is a scepter
    of equity;

7   you love righteousness and hate
    wickedness.
Therefore God, your God, has
    anointed you
    with the oil of gladness beyond
    your companions;
8   your robes are all fragrant with
    myrrh and aloes and cassia.
From ivory palaces stringed
    instruments make you glad;
9   daughters of kings are among
    your ladies of honor;
    at your right hand stands the
    queen in gold of Ophir.

10 Hear, O daughter, consider and
    incline your ear;
    forget your people and your
    father's house,
11   and the king will desire your
    beauty.
Since he is your lord, bow to him;
12   the people*e* of Tyre will seek
    your favor with gifts,
    the richest of the people 13 with
    all kinds of wealth.

The princess is decked in her
    chamber with gold-woven
    robes;*f*
14   in many-colored robes she is led
    to the king;
    behind her the virgins, her
    companions, follow.
15 With joy and gladness they are
    led along
    as they enter the palace of the
    king.

c Cn: Heb *and the meekness of*   *d* Or *Your
throne is a throne of God, it*   *e* Heb *daughter*
f Or *people. 13All glorious is the princess within,
gold embroidery is her clothing*

Ps 45: An ode for a royal wedding. 1:
Introduction. The author is identified as a
professional writer (*a ready scribe*), presumably a court poet. **2–9:** The king is addressed
in flattering language.
    45.6: *Your throne, O God:* the king seems
to be addressed as "God." This is without
parallel in the Hebrew Scriptures, but common in surrounding nations. Compare Heb
1.8. **8:** The mention of *ivory palaces* (1 Kings

22.39; Am 3.15) shows that the bridegroom
was a king of the kingdom of Israel.
    45.10–13a: The queen, probably a Phoe-
nician (note the mention of *Tyre* in v. 12), is
instructed to forget her own ancestry and be
loyal to her husband.
    45.13b–15: The wedding procession. **16–
17:** The poet promises the king (the pronouns,
in Hebrew, are masculine; see note *g*) successful progeny and literary immortality.

16 In the place of ancestors you,
  O king,*g* shall have sons;
 you will make them princes in
  all the earth.
17 I will cause your name to be
  celebrated in all generations;
 therefore the peoples will praise
  you forever and ever.

Psalm 46

*To the leader. Of the Korahites. According
to Alamoth. A Song.*

1 God is our refuge and strength,
 a very present*h* help in trouble.
2 Therefore we will not fear, though
  the earth should change,
 though the mountains shake in
  the heart of the sea;
3 though its waters roar and foam,
 though the mountains tremble
  with its tumult. *Selah*

4 There is a river whose streams
  make glad the city of God,
 the holy habitation of the
  Most High.
5 God is in the midst of the city;*i* it
  shall not be moved;
 God will help it when the
  morning dawns.
6 The nations are in an uproar, the
  kingdoms totter;
 he utters his voice, the earth
  melts.
7 The LORD of hosts is with us;
 the God of Jacob is our refuge.*j*
     Selah

8 Come, behold the works of the
  LORD;
 see what desolations he has
  brought on the earth.
9 He makes wars cease to the end of
  the earth;
 he breaks the bow, and shatters
  the spear;
 he burns the shields with fire.
10 "Be still, and know that I am
  God!
 I am exalted among the nations,
 I am exalted in the earth."
11 The LORD of hosts is with us;
 the God of Jacob is our refuge.*j*
     Selah

Psalm 47

To the leader. Of the Korahites. A Psalm.

1 Clap your hands, all you peoples;
 shout to God with loud songs
  of joy.
2 For the LORD, the Most High, is
  awesome,
 a great king over all the earth.
3 He subdued peoples under us,
 and nations under our feet.
4 He chose our heritage for us,
 the pride of Jacob whom he
  loves. *Selah*

5 God has gone up with a shout,
 the LORD with the sound of a
  trumpet.

g Heb lacks *O king* *h* Or *well proved*
i Heb *of it* *j* Or *fortress*

**Ps 46: A song of Zion celebrating
God's ultimate victory over the nations.**
1–3: God will preserve his people even during
the cosmic tumults of the latter days (compare Joel 3.16). **1:** This verse inspired Luther's
hymn, "A Mighty Fortress." **3:** *Selah,* see
3.2 n. **4–7:** Jerusalem, God's dwelling place
on earth, will stand secure. This emphasis
upon the Holy City leads to the classification
of this and some other psalms as "songs of
Zion" (for the term, see 137.3).
46.4: *A river,* compare Isa 33.21; Ezek
47.1–12; Zech 14.8; Rev 22.1–2. **7:** A refrain

(compare v. 11); it has probably dropped out
by accident after v. 3. **8–11:** The establishment of God's kingdom will bring peace to
the earth (compare Isa 2.4).
Ps 47: A hymn celebrating God's enthronement as king of all nations. 1–4:
Summons to all the world to praise the God
of Israel as king. **4:** *Selah,* see 3.2 n. **5–9:**
These words were evidently composed to accompany a religious ceremony, probably
connected with the ark, which dramatized the
LORD's kingship (compare 24.7–10 and
68.17–18). **7:** The word translated "psalm" is

6 Sing praises to God, sing praises;
   sing praises to our King, sing
      praises.
7 For God is the king of all the
      earth;
   sing praises with a psalm. *k*

8 God is king over the nations;
   God sits on his holy throne.
9 The princes of the peoples gather
      as the people of the God of
      Abraham.
   For the shields of the earth belong
      to God;
   he is highly exalted.

Psalm 48

A Song. A Psalm of the Korahites.

1 Great is the LORD and greatly to
      be praised
   in the city of our God.
   His holy mountain, 2beautiful in
      elevation,
   is the joy of all the earth,
   Mount Zion, in the far north,
      the city of the great King.
3 Within its citadels God
   has shown himself a sure
      defense.

4 Then the kings assembled,
   they came on together.
5 As soon as they saw it, they were
      astounded;
   they were in panic, they took to
      flight;
6 trembling took hold of them
      there,
   pains as of a woman in labor,

7 as when an east wind shatters
   the ships of Tarshish.
8 As we have heard, so have we
      seen
   in the city of the LORD of hosts,
   in the city of our God,
   which God establishes forever.
                              Selah

9 We ponder your steadfast love,
      O God,
   in the midst of your temple.
10 Your name, O God, like your
      praise,
   reaches to the ends of the earth.
   Your right hand is filled with
      victory.
11 Let Mount Zion be glad,
   let the towns[l] of Judah rejoice
   because of your judgments.

12 Walk about Zion, go all around it,
   count its towers,
13 consider well its ramparts;
   go through its citadels,
   that you may tell the next
      generation
14 that this is God,
   our God forever and ever.
   He will be our guide forever.

Psalm 49

To the leader. Of the Korahites. A Psalm.

1 Hear this, all you peoples;
   give ear, all inhabitants of the
      world,
2 both low and high,
   rich and poor together.

---

k Heb *Maskil*   *l* Heb *daughters*

Maskil; see Introduction. **8**: *Throne,* Isa 6.1. **9**: *Shields,* rulers (compare 89.18 n.).
  Ps 48: A song celebrating the beauty and security of Zion. 1–3: The LORD is praised as the God of Jerusalem. **2**: *Mount Zion,* one of the hills upon which Jerusalem is built. *In the far north,* a curious phrase which apparently identifies the Israelite holy mountain with the Canaanite mountain of the gods (called Zaphon or "North" in the Ras Shamra tablets). **4–8**: When, in the last days, the heathen unite to attack God's city, they will be routed (compare Ezek chs 38–39; Zech chs 12; 14; Rev 20.9–10). **7**: *Ships of Tarshish,* see 1 Kings 10.22 n. **8**: *Selah,* see 3.2 n.
  48.9–11: All of this causes anticipatory rejoicing in the ceremonies of the temple. **12–14**: Call for a procession about the city walls (Ps 24.7–10).
  Ps 49: Meditation on the transience of life and wealth (a wisdom psalm). **1**: The poet summons an audience. **5–6**: The theme:

3 My mouth shall speak wisdom;
the meditation of my heart shall
be understanding.
4 I will incline my ear to a proverb;
I will solve my riddle to the
music of the harp.

5 Why should I fear in times of
trouble,
when the iniquity of my
persecutors surrounds me,
6 those who trust in their wealth
and boast of the abundance of
their riches?
7 Truly, no ransom avails for one's
life, *m*
there is no price one can give to
God for it.
8 For the ransom of life is costly,
and can never suffice
9 that one should live on forever
and never see the grave. *n*

10 When we look at the wise, they
die;
fool and dolt perish together
and leave their wealth to others.
11 Their graves*o* are their homes
forever,
their dwelling places to all
generations,
though they named lands their
own.
12 Mortals cannot abide in their
pomp;
they are like the animals that
perish.

13 Such is the fate of the foolhardy,
the end of those*p* who are
pleased with their lot. *Selah*
14 Like sheep they are appointed for
Sheol;
Death shall be their shepherd;

straight to the grave they
descend, *q*
and their form shall waste away;
Sheol shall be their home. *r*
15 But God will ransom my soul
from the power of Sheol,
for he will receive me. *Selah*

16 Do not be afraid when some
become rich,
when the wealth of their houses
increases.
17 For when they die they will carry
nothing away;
their wealth will not go down
after them.
18 Though in their lifetime they
count themselves happy
—for you are praised when you
do well for yourself—
19 they*s* will go to the company of
their ancestors,
who will never again see the
light.
20 Mortals cannot abide in their
pomp;
they are like the animals that
perish

Psalm 50

A Psalm of Asaph.

1 The mighty one, God the LORD,
speaks and summons the earth
from the rising of the sun to its
setting.
2 Out of Zion, the perfection of
beauty,
God shines forth.

m Another reading is *no one can ransom a brother*
n Heb *the pit*   o Gk Syr Compare Tg: Heb
their inward (thought)   p Tg: Heb *after them*
q Cn: Heb *the upright shall have dominion over
them in the morning*   r Meaning of Heb
uncertain   s Cn: Heb *you*

Why should one be afraid of the wealthy and
powerful? **7–12**: People do not have power
over the disposition of their own lives, nor
can they take their wealth with them.
49.9: *The Pit,* Sheol (see 6.5 n.). **12**: A refrain; compare v. 20. **13–14**: Death is the end
even for the arrogant. **13**: *Selah,* see 3.2 n. **15**:

This might express the psalmist's assurance of
personal immortality, but is perhaps better
understood merely as confidence in being delivered from present trouble (89.48; Hos
13.14). **16–20**: The audience is instructed
therefore not to be afraid of the wealthy.
Ps 50: A liturgy of divine judgment.

3 Our God comes and does not keep
    silence,
      before him is a devouring fire,
      and a mighty tempest all around
      him.
4 He calls to the heavens above
      and to the earth, that he may
      judge his people:
5 "Gather to me my faithful ones,
      who made a covenant with me
      by sacrifice!"
6 The heavens declare his
      righteousness,
      for God himself is judge.   *Selah*

7 "Hear, O my people, and I will
      speak,
      O Israel, I will testify against
      you.
      I am God, your God.
8 Not for your sacrifices do I rebuke
      you;
      your burnt offerings are
      continually before me.
9 I will not accept a bull from your
      house,
      or goats from your folds.
10 For every wild animal of the forest
      is mine,
      the cattle on a thousand hills.
11 I know all the birds of the air,*t*
      and all that moves in the field is
      mine.

12 "If I were hungry, I would not tell
      you,
      for the world and all that is in it
      is mine.
13 Do I eat the flesh of bulls,
      or drink the blood of goats?
14 Offer to God a sacrifice of
      thanksgiving,*u*
      and pay your vows to the
      Most High.

15 Call on me in the day of trouble;
      I will deliver you, and you shall
      glorify me."

16 But to the wicked God says:
      "What right have you to recite
      my statutes,
      or take my covenant on your
      lips?
17 For you hate discipline,
      and you cast my words behind
      you.
18 You make friends with a thief
      when you see one,
      and you keep company with
      adulterers.

19 "You give your mouth free rein
      for evil,
      and your tongue frames deceit.
20 You sit and speak against your
      kin;
      you slander your own mother's
      child.
21 These things you have done and I
      have been silent;
      you thought that I was one just
      like yourself.
      But now I rebuke you, and lay the
      charge before you.

22 "Mark this, then, you who forget
      God,
      or I will tear you apart, and
      there will be no one to
      deliver.
23 Those who bring thanksgiving as
      their sacrifice honor me;
      to those who go the right way*v*
      I will show the salvation of
      God."

t Gk Syr Tg: Heb *mountains*   *u* Or *make
thanksgiving your sacrifice to God   v* Hcb *who set
a way*

1–6: God coming to judge his people. 3:
Compare 18.8; Hab 3.4. 6: *Selah,* see 3.2 n.
50.7–23: The Lᴏʀᴅ's arraignment of the
nation. 8–13: They have brought sacrifices in
abundance, but this is not what God wants
(see 40.6 n.). 14–15: His demand is rather for
thanksgiving and prayer.

50.16–21: They have violated God's law
by tolerating evil (v. 18) and indulging in
slander (vv. 19–20). 16: The first line is probably an editorial addition, modifying the
poet's sweeping judgment on the nation as a
whole. 22–23: Concluding warning.

Psalm 51

To the leader. A Psalm of David, when the prophet Nathan came to him, after he had gone in to Bathsheba.

1 Have mercy on me, O God,
   according to your steadfast love;
 according to your abundant mercy
   blot out my transgressions.
2 Wash me thoroughly from my
     iniquity,
   and cleanse me from my sin.

3 For I know my transgressions,
   and my sin is ever before me.
4 Against you, you alone, have I
     sinned,
   and done what is evil in your
     sight,
 so that you are justified in your
     sentence
   and blameless when you pass
     judgment.
5 Indeed, I was born guilty,
   a sinner when my mother
     conceived me.

6 You desire truth in the inward
     being; *w*
   therefore teach me wisdom in
     my secret heart.
7 Purge me with hyssop, and I shall
     be clean;
   wash me, and I shall be whiter
     than snow.
8 Let me hear joy and gladness;
   let the bones that you have
     crushed rejoice.
9 Hide your face from my sins,
   and blot out all my iniquities.

10 Create in me a clean heart,
     O God,

and put a new and right *x* spirit
     within me.
11 Do not cast me away from your
     presence,
   and do not take your holy spirit
     from me.
12 Restore to me the joy of your
     salvation,
   and sustain in me a willing *y*
     spirit.

13 Then I will teach transgressors
     your ways,
   and sinners will return to you.
14 Deliver me from bloodshed,
     O God,
   O God of my salvation,
   and my tongue will sing aloud
     of your deliverance.

15 O Lord, open my lips,
   and my mouth will declare your
     praise.
16 For you have no delight in
     sacrifice;
   if I were to give a burnt
     offering, you would not be
     pleased.
17 The sacrifice acceptable to God *z* is
     a broken spirit;
   a broken and contrite heart,
     O God, you will not
     despise.

18 Do good to Zion in your good
     pleasure;
   rebuild the walls of Jerusalem,
19 then you will delight in right
     sacrifices,

---

w Meaning of Heb uncertain   *x* Or *steadfast*
y Or *generous*   *z* Or *My sacrifice, O God,*

Ps 51: Prayer for healing and moral renewal (a lament). **1–2:** Prayer for deliverance. Although v. 8 makes it clear that the psalmist's problem is one of illness, the main emphasis is upon restoration to moral, rather than merely physical, health. **3–5:** The psalmist confesses to having had a sinful nature even from the moment of conception.

51.6–12: Renewed prayer for deliverance. **7:** Purging with hyssop is a reference to some ceremony of sprinkling (Ex 12.22; Lev 14.51), but in a purely metaphorical sense. **51.13–17:** The vow (see 7.17 n.) is to instruct others and to praise and serve God rather than to offer a sacrifice (see 40.6 n.). **18–19:** A later addition, evidently from the exilic or

in burnt offerings and whole
burnt offerings;
then bulls will be offered on
your altar.

Psalm 52

*To the leader. A Maskil of David, when
Doeg the Edomite came to Saul and said to
him, "David has come to the house of
Ahimelech."*

1 Why do you boast, O mighty
one,
of mischief done against the
godly?[a]
All day long 2 you are plotting
destruction.
Your tongue is like a sharp razor,
you worker of treachery.
3 You love evil more than good,
and lying more than speaking
the truth.   *Selah*
4 You love all words that devour,
O deceitful tongue.

5 But God will break you down
forever;
he will snatch and tear you from
your tent;
he will uproot you from the
land of the living.   *Selah*
6 The righteous will see, and fear,
and will laugh at the evildoer,[b]
saying,
7 "See the one who would not take
refuge in God,
but trusted in abundant riches,
and sought refuge in wealth!"[c]

8 But I am like a green olive tree
in the house of God.
I trust in the steadfast love of God
forever and ever.

9 I will thank you forever,
because of what you have done.
In the presence of the faithful
I will proclaim[d] your name, for
it is good.

Psalm 53

*To the leader: according to Mahalath.
A Maskil of David.*

1 Fools say in their hearts, "There is
no God."
They are corrupt, they commit
abominable acts;
there is no one who does good.

2 God looks down from heaven on
humankind
to see if there are any who are
wise,
who seek after God.

3 They have all fallen away, they are
all alike perverse;
there is no one who does good,
no, not one.

4 Have they no knowledge, those
evildoers,
who eat up my people as they
eat bread,
and do not call upon God?

5 There they shall be in great terror,
in terror such as has not been.
For God will scatter the bones of
the ungodly;[e]

---

a Cn Compare Syr: Heb *the kindness of God*
b Heb *him*   c Syr Tg: Heb *in his destruction*
d Cn: Heb *wait for*   e Cn Compare Gk Syr:
Heb *him who encamps against you*

---

post-exilic period, designed to modify the
anti-sacrificial spirit of the preceding verses
and to adapt the psalm to liturgical use.
**Ps 52: God's imminent judgment
against a tyrant** (a lament). Prayer for deliverance in the form of a denunciation of the
psalmist's enemy. **1–4:** Character of the
psalmist's enemy. **3:** *Selah*, see 3.2 n.
52.5–7: Retribution is about to befall the

enemy. **8:** The psalmist is confident of being
delivered (compare 1.3). **9:** The vow (see
7.17 n.).
**Ps 53: Condemnation of a cynical and
unrighteous age.** This psalm is almost identical with Ps 14, except for the alteration of
the divine name Yahweh (the LORD) to Elohim (God).

they will be put to shame,*f* for
   God has rejected them.

6 O that deliverance for Israel would
   come from Zion!
When God restores the fortunes
   of his people,
Jacob will rejoice; Israel will be
   glad.

Psalm 54

*To the leader: with stringed instruments.
A Maskil of David, when the Ziphites went
and told Saul, "David is in hiding
among us."*

1 Save me, O God, by your name,
   and vindicate me by your
   might.
2 Hear my prayer, O God;
   give ear to the words of my
   mouth.

3 For the insolent have risen
   against me,
the ruthless seek my life;
they do not set God before
   them.   *Selah*

4 But surely, God is my helper;
the Lord is the upholder of*g* my
   life.
5 He will repay my enemies for
   their evil.
In your faithfulness, put an end
   to them.

6 With a freewill offering I will
   sacrifice to you;
I will give thanks to your name,
   O Lord, for it is good.
7 For he has delivered me from
   every trouble,
and my eye has looked in
   triumph on my enemies.

Psalm 55

*To the leader: with stringed instruments.
A Maskil of David.*

1 Give ear to my prayer, O God;
   do not hide yourself from my
   supplication.
2 Attend to me, and answer me;
   I am troubled in my complaint.
I am distraught 3by the noise of
   the enemy,
because of the clamor of the
   wicked.
For they bring*h* trouble upon me,
   and in anger they cherish
   enmity against me.

4 My heart is in anguish within me,
   the terrors of death have fallen
   upon me.
5 Fear and trembling come upon
   me,
and horror overwhelms me.
6 And I say, "O that I had wings
   like a dove!
I would fly away and be at rest;
7 truly, I would flee far away;
   I would lodge in the wilderness;
       Selah
8 I would hurry to find a shelter for
   myself
from the raging wind and
   tempest."

9 Confuse, O Lord, confound their
   speech;
for I see violence and strife in
   the city.
10 Day and night they go around it
   on its walls,

f Gk: Heb *you will put to shame*   *g* Gk Syr
Jerome: Heb *is of those who uphold* or *is
with those who uphold*   *h* Cn Compare Gk:
Heb *they cause to totter*

Ps 54: **Prayer for deliverance from
personal enemies** (a lament). **1:** Cry for help.
3: The psalmist's situation. *Selah,* see 3.2 n.
4–5: Expression of trust. **6–7:** The vow (see
7.17 n.). **6:** *Freewill offering,* Num 15.3.
  Ps 55: **Prayer for deliverance from**
personal enemies (a lament). **1–2:** Cry for
help. **3–14:** The psalmist's situation: the chief
enemy is a former friend (vv. 12–14, 20–21).
7: *Selah,* see 3.2 n. **15:** A curse upon the enemies. The violence of the language is probably an attempt to counteract the effect of a

and iniquity and trouble are
within it;
11   ruin is in its midst;
oppression and fraud
do not depart from its
marketplace.

12 It is not enemies who taunt me—
I could bear that;
it is not adversaries who deal
insolently with me—
I could hide from them.
13 But it is you, my equal,
my companion, my familiar
friend,
14 with whom I kept pleasant
company;
we walked in the house of God
with the throng.
15 Let death come upon them;
let them go down alive to
Sheol;
for evil is in their homes and in
their hearts.

16 But I call upon God,
and the LORD will save me.
17 Evening and morning and at noon
I utter my complaint and moan,
and he will hear my voice.
18 He will redeem me unharmed
from the battle that I wage,
for many are arrayed against
me.
19 God, who is enthroned from of
old,   *Selah*
will hear, and will humble
them—
because they do not change,
and do not fear God.

20 My companion laid hands on a
friend

and violated a covenant with
me[i]
21 with speech smoother than butter,
but with a heart set on war;
with words that were softer than
oil,
but in fact were drawn swords.

22 Cast your burden[j] on the LORD,
and he will sustain you;
he will never permit
the righteous to be moved.

23 But you, O God, will cast them
down
into the lowest pit;
the bloodthirsty and treacherous
shall not live out half their days.
But I will trust in you.

Psalm 56

*To the leader: according to The Dove on
Far-off Terebinths. Of David. A Miktam,
when the Philistines seized him in Gath.*

1 Be gracious to me, O God, for
people trample on me;
all day long foes oppress me;
2 my enemies trample on me all day
long,
for many fight against me.
O Most High, 3when I am afraid,
I put my trust in you.
4 In God, whose word I praise,
in God I trust; I am not afraid;
what can flesh do to me?

5 All day long they seek to injure
my cause;
all their thoughts are against me
for evil.

i Heb lacks *with me*   j Or *Cast what he has
given you*

---

curse believed to have been pronounced by
the enemies (compare Ps 58). *Alive to Sheol,*
compare Num 16.30.
    55.16–19: Expression of trust. **20–23:**
Complaint (vv. 20–21); confidence (vv. 22–
23).
    Ps 56: Prayer for deliverance from

personal enemies (a lament). **1–2, 5–6ab:**
The psalmist's situation. **3–4, 8–11:** Expression of trust.
    56.6c–7: Prayer for vindication. **7:** *The
peoples,* a metaphor for the psalmist's enemies. **8:** *Your record,* 40.7. **12–13:** The vow
(see 7.17 n.).

6 They stir up strife, they lurk,
   they watch my steps.
 As they hoped to have my life,
7   so repay[k] them for their crime;
   in wrath cast down the peoples,
     O God!

8 You have kept count of my
     tossings;
   put my tears in your bottle.
   Are they not in your record?
9 Then my enemies will retreat
   in the day when I call.
   This I know, that[l] God is for
     me.
10 In God, whose word I praise,
   in the LORD, whose word I
     praise,
11 in God I trust; I am not afraid.
   What can a mere mortal do to
     me?

12 My vows to you I must perform,
     O God;
   I will render thank offerings
     to you.
13 For you have delivered my soul
     from death,
   and my feet from falling,
 so that I may walk before God
   in the light of life.

Psalm 57

*To the leader: Do Not Destroy.
Of David. A Miktam, when he fled from
Saul, in the cave.*

1 Be merciful to me, O God, be
     merciful to me,
   for in you my soul takes refuge;
 in the shadow of your wings I
     will take refuge,
   until the destroying storms
     pass by.
2 I cry to God Most High,

to God who fulfills his purpose
     for me.
3 He will send from heaven and
     save me,
   he will put to shame those who
     trample on me.   *Selah*
 God will send forth his steadfast
   love and his faithfulness.

4 I lie down among lions
   that greedily devour[m] human
     prey;
 their teeth are spears and arrows,
   their tongues sharp swords.

5 Be exalted, O God, above the
     heavens.
   Let your glory be over all the
     earth.

6 They set a net for my steps;
   my soul was bowed down.
 They dug a pit in my path,
   but they have fallen into it
     themselves.   *Selah*
7 My heart is steadfast, O God,
   my heart is steadfast.
 I will sing and make melody.
8   Awake, my soul!
 Awake, O harp and lyre!
   I will awake the dawn.
9 I will give thanks to you, O Lord,
     among the peoples;
   I will sing praises to you among
     the nations.
10 For your steadfast love is as high
     as the heavens;
   your faithfulness extends to the
     clouds.

11 Be exalted, O God, above the
     heavens.
   Let your glory be over all the
     earth.

k Cn: Heb *rescue*   l Or *because*
m Cn: Heb *are aflame for*

**Ps 57: Prayer for deliverance from
personal enemies** (a lament). **1**: Cry for help.
2–3: Expression of trust. **3**: *Selah*, see 3.2 n.
4–6: The psalmist's situation, prayer, and
certainty of being heard.

57.5: A refrain (compare v. 11). **7–11**: The
customary vow (see 7.17 n.) in the form of a
thanksgiving sung in anticipation of deliverance (compare 22.22–31). These verses are
practically identical with 108.1–5.

Psalm 58

To the leader: Do Not Destroy. Of David.
A Miktam.

1 Do you indeed decree what is
    right, you gods?[n]
    Do you judge people fairly?
2 No, in your hearts you devise
    wrongs;
    your hands deal out violence on
    earth.

3 The wicked go astray from the
    womb;
    they err from their birth,
    speaking lies.
4 They have venom like the venom
    of a serpent,
    like the deaf adder that stops
    its ear,
5 so that it does not hear the voice
    of charmers
    or of the cunning enchanter.

6 O God, break the teeth in their
    mouths;
    tear out the fangs of the young
    lions, O LORD!
7 Let them vanish like water that
    runs away;
    like grass let them be trodden
    down[o] and wither.
8 Let them be like the snail that
    dissolves into slime;
    like the untimely birth that
    never sees the sun.
9 Sooner than your pots can feel the
    heat of thorns,
    whether green or ablaze, may he
    sweep them away!

10 The righteous will rejoice when
    they see vengeance done;

    they will bathe their feet in the
    blood of the wicked.
11 People will say, "Surely there is a
    reward for the righteous;
    surely there is a God who
    judges on earth."

Psalm 59

To the leader: Do Not Destroy. Of David.
A Miktam, when Saul ordered his house to
be watched in order to kill him.

1 Deliver me from my enemies,
    O my God;
    protect me from those who rise
    up against me.
2 Deliver me from those who work
    evil;
    from the bloodthirsty save me.

3 Even now they lie in wait for my
    life;
    the mighty stir up strife against
    me.
    For no transgression or sin of
    mine, O LORD,
4   for no fault of mine, they run
    and make ready.

    Rouse yourself, come to my help
    and see!
5   You, LORD God of hosts, are
    God of Israel.
    Awake to punish all the nations;
    spare none of those who
    treacherously plot evil.
                  Selah

6 Each evening they come back,
    howling like dogs

n Or *mighty lords*   *o* Cn: Meaning of Heb
uncertain

Ps 58: Curse upon enemies (a lament).
The violent language is probably to be explained as a curse that the psalmist utters to
counteract the effect of a curse the enemies are
believed to have pronounced previously
(compare Ps 109.17–19). **1–5:** The character
of the enemies. It is unclear whether they are
heathen *gods* or "mighty lords" (note *n*); the
translation of v. 1 is uncertain (compare
82.1). **6–9:** The curse. **10–11:** Deliverance is
certain.
    **Ps 59: Prayer for deliverance from
personal enemies** (a lament). **1–2:** Cry for
help. **3–7:** The psalmist's situation (and
prayer, vv. 4b–5). **5:** *Nations,* see 56.7 n.
Selah, see 3.2 n.

and prowling about the city.
7 There they are, bellowing with
    their mouths,
  with sharp words*p* on their
    lips—
  for "Who," they think,*q* "will
    hear us?"

8 But you laugh at them, O LORD;
  you hold all the nations in
    derision.
9 O my strength, I will watch for
    you;
  for you, O God, are my
    fortress.
10 My God in his steadfast love will
    meet me;
  my God will let me look in
    triumph on my enemies.

11 Do not kill them, or my people
    may forget;
  make them totter by your
    power, and bring them
    down,
  O Lord, our shield.
12 For the sin of their mouths, the
    words of their lips,
  let them be trapped in their
    pride.
  For the cursing and lies that they
    utter,
13 consume them in wrath;
  consume them until they are no
    more.
  Then it will be known to the ends
    of the earth
  that God rules over Jacob. *Selah*

14 Each evening they come back,
  howling like dogs
  and prowling about the city.
15 They roam about for food,
  and growl if they do not get
    their fill.

16 But I will sing of your might;
  I will sing aloud of your
    steadfast love in the
    morning.
  For you have been a fortress for
    me
  and a refuge in the day of my
    distress.
17 O my strength, I will sing praises
    to you,
  for you, O God, are my
    fortress,
  the God who shows me
    steadfast love.

Psalm 60

*To the leader: according to the
Lily of the Covenant. A Miktam of
David; for instruction; when he struggled with
Aram-naharaim and with Aram-zobah,
and when Joab on his return killed twelve
thousand Edomites in the Valley of Salt.*

1 O God, you have rejected us,
    broken our defenses;
  you have been angry; now
    restore us!
2 You have caused the land to
    quake; you have torn it
    open;
  repair the cracks in it, for it is
    tottering.
3 You have made your people suffer
    hard things;
  you have given us wine to drink
    that made us reel.

4 You have set up a banner for
    those who fear you,
  to rally to it out of bowshot.*r*
                              Selah
5 Give victory with your right
    hand, and answer us,*s*

p Heb *with swords*   q Heb lacks *they think*
r Gk Syr Jerome: Heb *because of the truth*
s Another reading is *me*

59.8–10: Expression of trust. 11–15: Appeal for vengeance. 12: *Cursing,* see Ps 58 n. 16–17: The vow (see 7.17 n.).
Ps 60: Prayer for deliverance from national enemies (a group lament). 1–3: The people's situation. They have suffered a humiliating defeat, probably by the Edomites (see v. 9). 4–5: Prayer for deliverance. 4: *Selah,* see 3.2 n.

so that those whom you love
may be rescued.

6 God has promised in his
sanctuary:[t]
"With exultation I will divide
up Shechem,
and portion out the Vale of
Succoth.
7 Gilead is mine, and Manasseh is
mine;
Ephraim is my helmet;
Judah is my scepter.
8 Moab is my washbasin;
on Edom I hurl my shoe;
over Philistia I shout in
triumph."

9 Who will bring me to the fortified
city?
Who will lead me to Edom?
10 Have you not rejected us, O God?
You do not go out, O God,
with our armies.
11 O grant us help against the foe,
for human help is worthless.
12 With God we shall do valiantly;
it is he who will tread down
our foes.

Psalm 61

*To the leader: with stringed instruments.
Of David.*

1 Hear my cry, O God;
listen to my prayer.
2 From the end of the earth I call
to you,
when my heart is faint.

Lead me to the rock
that is higher than I;
3 for you are my refuge,

a strong tower against the
enemy.

4 Let me abide in your tent forever,
find refuge under the shelter of
your wings. *Selah*
5 For you, O God, have heard my
vows;
you have given me the heritage
of those who fear your
name.

6 Prolong the life of the king;
may his years endure to all
generations!
7 May he be enthroned forever
before God;
appoint steadfast love and
faithfulness to watch
over him!

8 So I will always sing praises to
your name,
as I pay my vows day after day.

Psalm 62

*To the leader: according to Jeduthun.
A Psalm of David.*

1 For God alone my soul waits in
silence;
from him comes my salvation.
2 He alone is my rock and my
salvation,
my fortress; I shall never be
shaken.

3 How long will you assail a person,
will you batter your victim, all
of you,

---

t Or *by his holiness*

60.6–8: The answer prayed for in v. 5—a divine oracle, probably delivered by a priest or temple prophet. The places referred to are either Hebrew territories or parts of the one-time Hebrew empire under the United Monarchy. The God of Israel lays claim to them all. **9–12:** Prayer for victory. **9:** *Me,* probably the king. *The fortified city,* perhaps Sela, Edom's capital.

Ps 61: Prayer for protection (a lament). **1–2:** Cry for help. **2:** *From the end of the earth* seems to show that the psalmist did not live in Palestine. **3–5:** Expression of trust. **4:** *Your tent,* see 27.6 n. The psalmist is perhaps preparing to make a pilgrimage. *Selah,* see 3.2 n. **61.6–7:** Prayer for the king, the guarantor of safety for pilgrims and others. **8:** The vow (see 7.17 n.).

as you would a leaning wall, a
tottering fence?
4 Their only plan is to bring down a
person of prominence.
They take pleasure in falsehood;
they bless with their mouths,
but inwardly they curse. *Selah*

5 For God alone my soul waits in
silence,
for my hope is from him.
6 He alone is my rock and my
salvation,
my fortress; I shall not be
shaken.
7 On God rests my deliverance and
my honor;
my mighty rock, my refuge is
in God.

8 Trust in him at all times,
O people;
pour out your heart before him;
God is a refuge for us. *Selah*

9 Those of low estate are but a
breath,
those of high estate are a
delusion;
in the balances they go up;
they are together lighter than a
breath.
10 Put no confidence in extortion,
and set no vain hopes on
robbery;
if riches increase, do not set
your heart on them.

11 Once God has spoken;
twice have I heard this:
that power belongs to God,
12 and steadfast love belongs to
you, O Lord.

For you repay to all
according to their work.

Psalm 63

*A Psalm of David, when he was in the
Wilderness of Judah.*

1 O God, you are my God, I seek
you,
my soul thirsts for you;
my flesh faints for you,
as in a dry and weary land
where there is no water.
2 So I have looked upon you in the
sanctuary,
beholding your power and
glory.
3 Because your steadfast love is
better than life,
my lips will praise you.
4 So I will bless you as long as I
live;
I will lift up my hands and call
on your name.

5 My soul is satisfied as with a rich
feast, *u*
and my mouth praises you with
joyful lips
6 when I think of you on my bed,
and meditate on you in the
watches of the night;
7 for you have been my help,
and in the shadow of your
wings I sing for joy.
8 My soul clings to you;
your right hand upholds me.

9 But those who seek to destroy
my life
shall go down into the depths of
the earth;

u Heb *with fat and fatness*

Ps 62: Confidence in God's protection
(a song of trust; see Ps 11 n.). **1–2, 5–7**: God
is the psalmist's only help. **3–4**: The psalmist's situation: cursed by enemies. **8–12**: The
psalmist's compatriots are instructed to trust
in God also.
**Ps 63: Prayer for deliverance from
personal enemies** (a lament; though this
psalm may also be classified as a song of trust,
see vv. 9–10). The psalmist's almost mystical
delight in God's presence nearly obscures any
sense of need for personal security (compare
73.23–28). **1**: Compare 42.1–2. **6**: See 3.5–
6 n. The night was divided into three *watches*.
11: *The king*, see 61.6–7 n.

10 they shall be given over to the
     power of the sword,
   they shall be prey for jackals.
11 But the king shall rejoice in God;
   all who swear by him shall
     exult,
   for the mouths of liars will be
     stopped.

Psalm 64

To the leader. A Psalm of David.

1 Hear my voice, O God, in my
     complaint;
   preserve my life from the dread
     enemy.
2 Hide me from the secret plots of
     the wicked,
   from the scheming of evildoers,
3 who whet their tongues like
     swords,
   who aim bitter words like
     arrows,
4 shooting from ambush at the
     blameless;
   they shoot suddenly and
     without fear.
5 They hold fast to their evil
     purpose;
   they talk of laying snares
     secretly,
   thinking, "Who can see us?*v*
6   Who can search out our
     crimes?*w*
   We have thought out a cunningly
     conceived plot."
   For the human heart and mind
     are deep.

7 But God will shoot his arrow at
     them;
   they will be wounded suddenly.
8 Because of their tongue he will
     bring them to ruin;*x*
   all who see them will shake
     with horror.

9 Then everyone will fear;
   they will tell what God has
     brought about,
   and ponder what he has done.

10 Let the righteous rejoice in the
     Lord
   and take refuge in him.
   Let all the upright in heart glory.

Psalm 65

To the leader. A Psalm of David. A Song.

1 Praise is due to you,
     O God, in Zion;
   and to you shall vows be
     performed,
2   O you who answer prayer!
   To you all flesh shall come.
3 When deeds of iniquity
     overwhelm us,
   you forgive our transgressions.
4 Happy are those whom you
     choose and bring near
   to live in your courts.
   We shall be satisfied with the
     goodness of your house,
   your holy temple.

5 By awesome deeds you answer us
     with deliverance,
   O God of our salvation;
   you are the hope of all the ends of
     the earth
   and of the farthest seas.
6 By your*y* strength you established
     the mountains;
   you are girded with might.
7 You silence the roaring of the
     seas,
   the roaring of their waves,
   the tumult of the peoples.

v Syr: Heb *them*   *w* Cn: Heb *They search out
crimes*   *x* Cn: Heb *They will bring him to ruin,
their tongue being against them*   *y* Gk Jerome:
Heb *his*

**Ps 64: Prayer for deliverance from
personal enemies** (a lament). **1–2:** Cry for
help. **3–6:** The psalmist's situation. **3:** *Bitter
words,* perhaps magical curses (compare Ps
58 n.). **7–9:** Expression of confidence. **10:**
Concluding prayer.
     **Ps 65: Thanksgiving for a good har-
vest. 1–5:** It is good to gather at the temple

8 Those who live at earth's farthest
        bounds are awed by your
        signs;
    you make the gateways of the
        morning and the evening
        shout for joy.

9 You visit the earth and water it,
        you greatly enrich it;
    the river of God is full of water;
    you provide the people with
        grain,
        for so you have prepared it.
10 You water its furrows abundantly,
        settling its ridges,
    softening it with showers,
        and blessing its growth.
11 You crown the year with your
        bounty;
    your wagon tracks overflow
        with richness.
12 The pastures of the wilderness
        overflow,
    the hills gird themselves with
        joy,
13 the meadows clothe themselves
        with flocks,
    the valleys deck themselves with
        grain,
    they shout and sing together for
        joy.

Psalm 66

To the leader. A Song. A Psalm.

1 Make a joyful noise to God, all
        the earth;
2 sing the glory of his name;
        give to him glorious praise.
3 Say to God, "How awesome are
        your deeds!
    Because of your great power,
        your enemies cringe
        before you.
4 All the earth worships you;

    they sing praises to you,
        sing praises to your name."
                                Selah

5 Come and see what God has done:
        he is awesome in his deeds
        among mortals.
6 He turned the sea into dry land;
        they passed through the river
        on foot.
    There we rejoiced in him,
7   who rules by his might forever,
    whose eyes keep watch on the
        nations—
    let the rebellious not exalt
        themselves.   *Selah*

8 Bless our God, O peoples,
    let the sound of his praise be
        heard,
9 who has kept us among the living,
        and has not let our feet slip.
10 For you, O God, have tested us;
        you have tried us as silver is
        tried.
11 You brought us into the net;
        you laid burdens on our backs;
12 you let people ride over our heads;
        we went through fire and
        through water;
    yet you have brought us out to a
        spacious place. *z*

13 I will come into your house with
        burnt offerings;
    I will pay you my vows,
14 those that my lips uttered
    and my mouth promised when I
        was in trouble.
15 I will offer to you burnt offerings
        of fatlings,
    with the smoke of the sacrifice
        of rams;

z Cn Compare Gk Syr Jerome Tg: Heb *to a
saturation*

---

to sing God's praises. **6–8:** It was he who
created the world.
    65.9–13: And it is he who makes the earth
fertile. **9:** 104.13. *River of God,* see Ps 46.4 n.
    Ps 66: Liturgy of praise and thanksgiv-

ing. 1–12: Hymn in praise of God's might and
his care for his people. **4:** *Selah,* see 3.2 n.
    66.13–20: An individual presents a
thanksgiving sacrifice in fulfillment of a vow
(7.17). **13–15:** Announcement of the worship-

I will make an offering of bulls
and goats.   *Selah*

16 Come and hear, all you who fear
God,
and I will tell what he has done
for me.
17 I cried aloud to him,
and he was extolled with my
tongue.
18 If I had cherished iniquity in my
heart,
the Lord would not have
listened.
19 But truly God has listened;
he has given heed to the words
of my prayer.

20 Blessed be God,
because he has not rejected my
prayer
or removed his steadfast love
from me.

Psalm 67

*To the leader: with stringed instruments.
A Psalm. A Song.*

1 May God be gracious to us and
bless us
and make his face to shine upon
us,   *Selah*
2 that your way may be known
upon earth,
your saving power among all
nations.
3 Let the peoples praise you,
O God;
let all the peoples praise you.

4 Let the nations be glad and sing
for joy,
for you judge the peoples with
equity

and guide the nations upon
earth.   *Selah*
5 Let the peoples praise you,
O God;
let all the peoples praise you.

6 The earth has yielded its increase;
God, our God, has blessed us.
7 May God continue to bless us;
let all the ends of the earth
revere him.

Psalm 68

*To the leader. Of David. A Psalm.
A Song.*

1 Let God rise up, let his enemies be
scattered;
let those who hate him flee
before him.
2 As smoke is driven away, so drive
them away;
as wax melts before the fire,
let the wicked perish before
God.
3 But let the righteous be joyful;
let them exult before God;
let them be jubilant with joy.

4 Sing to God, sing praises to his
name;
lift up a song to him who rides
upon the clouds*a*—
his name is the LORD—
be exultant before him.

5 Father of orphans and protector of
widows
is God in his holy habitation.
6 God gives the desolate a home to
live in;

a Or *cast up a highway for him who rides through
the deserts*

---

er's purpose. **16–19:** The story of the writer's
experience (see 18.4–6 n.). **20:** Concluding
act of praise.
   **Ps 67: Thanksgiving for a good har-
vest** (see v. 6). **1–2:** Prayer that the blessing
may continue (compare Num 6.25). *Selah*,
see 3.2 n. **3–5:** May other nations know that

Israel's God is Lord of all! **6–7:** The occasion
of the psalm.
   **Ps 68: Liturgy for a festival celebra-
tion in the temple.** This is the most difficult
of the psalms to interpret, and there is no
general agreement either as to its meaning as
a whole or in many of its details. Some schol-

he leads out the prisoners to
  prosperity,
but the rebellious live in a
  parched land.

7 O God, when you went out
    before your people,
  when you marched through the
    wilderness,   *Selah*
8 the earth quaked, the heavens
    poured down rain
  at the presence of God, the God
    of Sinai,
  at the presence of God, the God
    of Israel.
9 Rain in abundance, O God, you
    showered abroad;
  you restored your heritage when
    it languished;
10 your flock found a dwelling in it;
  in your goodness, O God, you
    provided for the needy.

11 The Lord gives the command;
  great is the company of those[b]
    who bore the tidings:
12   "The kings of the armies, they
    flee, they flee!"
  The women at home divide the
    spoil,
13   though they stay among the
    sheepfolds—
  the wings of a dove covered with
    silver,
  its pinions with green gold.
14 When the Almighty[c] scattered
    kings there,
  snow fell on Zalmon.

15 O mighty mountain, mountain of
    Bashan;

O many-peaked mountain,
  mountain of Bashan!
16 Why do you look with envy,
    O many-peaked mountain,
  at the mount that God desired
    for his abode,
  where the LORD will reside
    forever?
17 With mighty chariotry, twice ten
    thousand,
  thousands upon thousands,
  the Lord came from Sinai into
    the holy place.[d]
18 You ascended the high mount,
  leading captives in your train
  and receiving gifts from people,
  even from those who rebel against
    the LORD God's abiding
    there.
19 Blessed be the Lord,
  who daily bears us up;
  God is our salvation.   *Selah*
20 Our God is a God of salvation,
  and to GOD, the Lord, belongs
    escape from death.

21 But God will shatter the heads of
    his enemies,
  the hairy crown of those who
    walk in their guilty ways.
22 The Lord said,
  "I will bring them back from
    Bashan,
  I will bring them back from the
    depths of the sea,
23 so that you may bathe[e] your feet
    in blood,

---

b Or *company of the women*   *c* Traditional
rendering of Heb *Shaddai*   *d* Cn: Heb *The
Lord among them Sinai in the holy* (place)
e Gk Syr Tg: Heb *shatter*

---

ars regard it as merely a collection of unrelated fragments. **1–3:** Prayer for God to manifest
himself in battle (compare v. 1 with Num
10.35). **4–6:** Praise to God as helper of the
helpless.
  68.7–10: God's care for his people in time
past (compare vv. 7–8 with Judg 5.4–5). **7:**
Selah, see 3.2 n. **11–14:** Announcement of a
great victory. **13:** *The wings of a dove . . . ,*
presumably describes some treasure found

among the spoil. **14:** An incident otherwise
unknown.
  68.15–16: Praise of the mount of God
(Zion). **15:** *Bashan*, a region east of the Sea of
Galilee. **17–18:** God ascends his throne in the
temple (compare 47.5). **19–20:** Praise of the
God who daily delivers the people.
  68.21: Expression of certainty that God
will give the victory.

so that the tongues of your dogs
    may have their share from
    the foe."

24 Your solemn processions are
    seen,*f* O God,
  the processions of my God, my
    King, into the sanctuary—
25 the singers in front, the musicians
    last,
  between them girls playing
    tambourines:
26 "Bless God in the great
    congregation,
  the LORD, O you who are of
    Israel's fountain!"
27 There is Benjamin, the least of
    them, in the lead,
  the princes of Judah in a body,
  the princes of Zebulun, the
    princes of Naphtali.

28 Summon your might, O God;
  show your strength, O God, as
    you have done for us
    before.
29 Because of your temple at
    Jerusalem
  kings bear gifts to you.
30 Rebuke the wild animals that live
    among the reeds,
  the herd of bulls with the calves
    of the peoples.
  Trample*g* under foot those who
    lust after tribute;
  scatter the peoples who delight
    in war.*h*
31 Let bronze be brought from
    Egypt;
  let Ethiopia*i* hasten to stretch
    out its hands to God.

32 Sing to God, O kingdoms of the
    earth;
  sing praises to the Lord,   *Selah*

33 O rider in the heavens, the ancient
    heavens;
  listen, he sends out his voice,
    his mighty voice.
34 Ascribe power to God,
  whose majesty is over Israel;
  and whose power is in the skies.
35 Awesome is God in his*j*
    sanctuary,
  the God of Israel;
  he gives power and strength to
    his people.

Blessed be God!

Psalm 69

To the leader: according to Lilies.
Of David.

1 Save me, O God,
  for the waters have come up to
    my neck.
2 I sink in deep mire,
  where there is no foothold;
  I have come into deep waters,
  and the flood sweeps over me.
3 I am weary with my crying;
  my throat is parched.
  My eyes grow dim
  with waiting for my God.

4 More in number than the hairs of
    my head
  are those who hate me without
    cause;
  many are those who would
    destroy me,
  my enemies who accuse me
    falsely.
  What I did not steal
    must I now restore?

f Or *have been seen*   *g* Cn: Heb *Trampling*
h Meaning of Heb of verse 30 is uncertain
i Or *Nubia*; Heb *Cush*   *j* Gk: Heb *from your*

68.24–27: A procession enters the temple
(see 24.7–10 n.). **27:** The reason for the mention of just these four tribes—the first two
from the south, the other two from Galilee—
is uncertain. Benjamin was the tribe of Saul;
Judah, of David.

68.28–31: Prayer for victory against
Egypt (*the wild animals . . . among the reeds,*
v. 30). **32–35:** Hymn to the God of heaven.
33: Compare 18.10–13 and 29.3–9.
**Ps 69: Prayer for deliverance from
personal enemies** (a lament). **1–4:** Cry for

5 O God, you know my folly;
   the wrongs I have done are not
      hidden from you.

6 Do not let those who hope in you
      be put to shame because of
         me,
   O Lord GOD of hosts;
   do not let those who seek you be
      dishonored because of me,
   O God of Israel.
7 It is for your sake that I have
      borne reproach,
   that shame has covered my face.
8 I have become a stranger to my
      kindred,
   an alien to my mother's
      children.

9 It is zeal for your house that has
      consumed me;
   the insults of those who insult
      you have fallen on me.
10 When I humbled my soul with
      fasting, *k*
   they insulted me for doing so.
11 When I made sackcloth my
      clothing,
   I became a byword to them.
12 I am the subject of gossip for
      those who sit in the gate,
   and the drunkards make songs
      about me.

13 But as for me, my prayer is to
      you, O LORD.
   At an acceptable time, O God,
   in the abundance of your
      steadfast love, answer me.
   With your faithful help 14 rescue
      me
   from sinking in the mire;
   let me be delivered from my
      enemies
   and from the deep waters.

15 Do not let the flood sweep over
      me,
   or the deep swallow me up,
   or the Pit close its mouth
      over me.

16 Answer me, O LORD, for your
      steadfast love is good;
   according to your abundant
      mercy, turn to me.
17 Do not hide your face from your
      servant,
   for I am in distress—make haste
      to answer me.
18 Draw near to me, redeem me,
   set me free because of my
      enemies.

19 You know the insults I receive,
   and my shame and dishonor;
   my foes are all known to you.
20 Insults have broken my heart,
   so that I am in despair.
   I looked for pity, but there was
      none;
   and for comforters, but I found
      none.
21 They gave me poison for food,
   and for my thirst they gave me
      vinegar to drink.

22 Let their table be a trap for them,
   a snare for their allies.
23 Let their eyes be darkened so that
      they cannot see,
   and make their loins tremble
      continually.
24 Pour out your indignation upon
      them,
   and let your burning anger
      overtake them.
25 May their camp be a desolation;

k Gk Syr: Heb *I wept, with fasting my soul*, or *I
made my soul mourn with fasting*

---

help. The language in vv. 2–3 is metaphorical. **5**: Confession of sin (compare 32.3–5;
51.3–5).
   69.6: The prayer continued. **7–21**: The situation: the psalmist may have been, like Haggai and Zechariah, a zealot for the rebuilding
of the temple after the Exile (compare v. 9

with vv. 35–36; see also Ezra 4.1–5, 23–24;
5.2–3), and thus have aroused opposition. **9**:
Quoted in Jn 2.17. **13–18**: Deep emotions
force the writer to interrupt the narrative
with a prayer. **21**: Quoted in all the Gospels
(Mt 27.34, 48; Mk 15.36; Lk 23.36; Jn 19.29).
   69.22–28: Curse upon all the enemies (see

let no one live in their tents.
26 For they persecute those whom
 you have struck down,
 and those whom you have
 wounded, they attack still
 more. *l*
27 Add guilt to their guilt;
 may they have no acquittal from
 you.
28 Let them be blotted out of the
 book of the living;
 let them not be enrolled among
 the righteous.
29 But I am lowly and in pain;
 let your salvation, O God,
 protect me.

30 I will praise the name of God with
 a song;
 I will magnify him with
 thanksgiving.
31 This will please the LORD more
 than an ox
 or a bull with horns and hoofs.
32 Let the oppressed see it and be
 glad;
 you who seek God, let your
 hearts revive.
33 For the LORD hears the needy,
 and does not despise his own
 that are in bonds.

34 Let heaven and earth praise him,
 the seas and everything that
 moves in them.
35 For God will save Zion
 and rebuild the cities of Judah;
 and his servants shall live *m* there
 and possess it;
36 the children of his servants shall
 inherit it,
 and those who love his name
 shall live in it.

Psalm 70

*To the leader. Of David, for the
memorial offering.*

1 Be pleased, O God, to deliver me.
 O LORD, make haste to help me!
2 Let those be put to shame and
 confusion
 who seek my life.
 Let those be turned back and
 brought to dishonor
 who desire to hurt me.
3 Let those who say, "Aha, Aha!"
 turn back because of their
 shame.

4 Let all who seek you
 rejoice and be glad in you.
 Let those who love your salvation
 say evermore, "God is great!"
5 But I am poor and needy;
 hasten to me, O God!
 You are my help and my
 deliverer;
 O LORD, do not delay!

Psalm 71

1 In you, O LORD, I take refuge;
 let me never be put to shame.
2 In your righteousness deliver me
 and rescue me;
 incline your ear to me and
 save me.
3 Be to me a rock of refuge,
 a strong fortress, *n* to save me,
 for you are my rock and my
 fortress.

---

l Gk Syr: Heb *recount the pain of*   *m* Syr: Heb
and they shall live   *n* Gk Compare 31.3: Heb *to
come continually you have commanded*

Ps 58 n.). **25:** Quoted in Acts 1.20. **29:** Exclamatory prayer.
 69.30–36: Thanksgiving for a favorable
answer. A priest or temple prophet may have
delivered an oracle of assurance between v. 29
and v. 30 (see 12.5 n. and 20.5 n.). **30–31:**
Like the authors of Pss 40 (vv. 6–8); 50
(vv. 8–13); and 51 (vv. 16–17), the psalmist
prefers to offer a hymn of thanksgiving. **35–**

36: If these verses are not a later addition, they
show the psalm to be post-exilic (compare
51.18–19).
 **Ps 70: Prayer for deliverance from
personal enemies** (a lament). This psalm is
practically identical with 40.13–17.
 **Ps 71: An aged worshiper's prayer for
deliverance from personal enemies** (a lament). **1–8:** Cry for help, mingled with ex-

4 Rescue me, O my God, from the
 hand of the wicked,
 from the grasp of the unjust and
 cruel.
5 For you, O Lord, are my hope,
 my trust, O Lord, from my
 youth.
6 Upon you I have leaned from my
 birth;
 it was you who took me from
 my mother's womb.
 My praise is continually of you.

7 I have been like a portent to
 many,
 but you are my strong refuge.
8 My mouth is filled with your
 praise,
 and with your glory all day
 long.
9 Do not cast me off in the time of
 old age;
 do not forsake me when my
 strength is spent.
10 For my enemies speak concerning
 me,
 and those who watch for my
 life consult together.
11 They say, "Pursue and seize that
 person
 whom God has forsaken,
 for there is no one to deliver."

12 O God, do not be far from me;
 O my God, make haste to
 help me!
13 Let my accusers be put to shame
 and consumed;
 let those who seek to hurt me
 be covered with scorn and
 disgrace.
14 But I will hope continually,
 and will praise you yet more
 and more.

15 My mouth will tell of your
 righteous acts,
 of your deeds of salvation all
 day long,
 though their number is past my
 knowledge.
16 I will come praising the mighty
 deeds of the Lord God,
 I will praise your righteousness,
 yours alone.

17 O God, from my youth you have
 taught me,
 and I still proclaim your
 wondrous deeds.
18 So even to old age and gray hairs,
 O God, do not forsake me,
 until I proclaim your might
 to all the generations to come. *o*
 Your power 19 and your
 righteousness, O God,
 reach the high heavens.

 You who have done great things,
 O God, who is like you?
20 You who have made me see many
 troubles and calamities
 will revive me again;
 from the depths of the earth
 you will bring me up again.
21 You will increase my honor,
 and comfort me once again.

22 I will also praise you with the
 harp
 for your faithfulness, O my
 God;
 I will sing praises to you with the
 lyre,
 O Holy One of Israel.
23 My lips will shout for joy
 when I sing praises to you;

o Gk Compare Syr: Heb *to a generation, to all that
come*

pressions of trust. **5–6:** Throughout the
whole of previous life, God has never failed.
71.9–11: The psalmist's situation: there are
violent enemies and age is a disadvantage
(compare v. 18).
71.12–13: Prayer for vindication. **14–24:**

The vow (see 7.17 n.). If the prayer is answered, the psalmist will use a talent for music (v. 22) to celebrate God's saving acts
(vv. 15–16) so that future generations will
know of them (v. 18). **20:** *From the depths of
the earth,* compare 9.13; 30.3.

my soul also, which you have
rescued.
24 All day long my tongue will talk
of your righteous help,
for those who tried to do me
harm
have been put to shame, and
disgraced.

Psalm 72

Of Solomon.

1 Give the king your justice,
O God,
and your righteousness to a
king's son.
2 May he judge your people with
righteousness,
and your poor with justice.
3 May the mountains yield
prosperity for the people,
and the hills, in righteousness.
4 May he defend the cause of the
poor of the people,
give deliverance to the needy,
and crush the oppressor.

5 May he live*p* while the sun
endures,
and as long as the moon,
throughout all generations.
6 May he be like rain that falls on
the mown grass,
like showers that water the
earth.
7 In his days may righteousness
flourish
and peace abound, until the
moon is no more.

8 May he have dominion from sea
to sea,

and from the River to the ends
of the earth.
9 May his foes*q* bow down before
him,
and his enemies lick the dust.
10 May the kings of Tarshish and of
the isles
render him tribute,
may the kings of Sheba and Seba
bring gifts.
11 May all kings fall down before
him,
all nations give him service.

12 For he delivers the needy when
they call,
the poor and those who have no
helper.
13 He has pity on the weak and the
needy,
and saves the lives of the needy.
14 From oppression and violence he
redeems their life;
and precious is their blood in his
sight.

15 Long may he live!
May gold of Sheba be given to
him.
May prayer be made for him
continually,
and blessings invoked for him
all day long.
16 May there be abundance of grain
in the land;
may it wave on the tops of the
mountains;
may its fruit be like Lebanon;

p Gk: Heb *may they fear you*   *q* Cn: Heb *those
who live in the wilderness*

**Ps 72: Prayer for God's blessing on the
king.** The occasion for the psalm may have
been a coronation or its annual commemoration. **1–4:** The king is to be the guarantor of
justice for the helpless (vv. 12–15). **5–6:**
These verses suggest the supernatural aura
which surrounded the person of the king in
the thought of the ancient Near East (even in
Israel he could be called "God's son," see 2.7,
possibly even "God," see 45.6). The health,

fertility, and success of the nation were bound
up inextricably in those of its monarch. **7:**
Righteousness and *peace,* as frequently elsewhere, mean "right conditions" and "prosperity."
72.8–11: The king's ideal universal empire. **8:** *The River,* the Euphrates (1 Kings
4.21). **10:** *Tarshish,* in the western Mediterranean, possibly in Spain. *Sheba and Seba,* in
south Arabia.

and may people blossom in the
cities
like the grass of the field.
17 May his name endure forever,
his fame continue as long as the
sun.
May all nations be blessed in
him; *r*
may they pronounce him
happy.

18 Blessed be the Lord, the God of
Israel,
who alone does wondrous
things.
19 Blessed be his glorious name
forever;
may his glory fill the whole
earth.
Amen and Amen.

20 The prayers of David son of Jesse
are ended.

BOOK III

Psalm 73

A Psalm of Asaph.

1 Truly God is good to the
upright, *s*
to those who are pure in heart.
2 But as for me, my feet had almost
stumbled;
my steps had nearly slipped.
3 For I was envious of the arrogant;
I saw the prosperity of the
wicked.

4 For they have no pain;
their bodies are sound and sleek.

5 They are not in trouble as others
are;
they are not plagued like other
people.
6 Therefore pride is their necklace;
violence covers them like a
garment.
7 Their eyes swell out with fatness;
their hearts overflow with
follies.
8 They scoff and speak with malice;
loftily they threaten oppression.
9 They set their mouths against
heaven,
and their tongues range over the
earth.

10 Therefore the people turn and
praise them, *t*
and find no fault in them. *u*
11 And they say, "How can God
know?
Is there knowledge in the
Most High?"
12 Such are the wicked;
always at ease, they increase in
riches.
13 All in vain I have kept my heart
clean
and washed my hands in
innocence.
14 For all day long I have been
plagued,
and am punished every
morning.

15 If I had said, "I will talk on in this
way,"

*r Or bless themselves by him   s Or good to Israel
t Cn: Heb his people return here   u Cn: Heb
abundant waters are drained by them*

72.12–14: The character of the king. 15–
17: The prayer concluded. 18–19: A doxology
(no part of the psalm) marking the end of
Book II of the Psalter (see 41.13 n.). 20: An
editorial colophon to one of the collections of
psalms now included in the Psalter.
  Ps 73: **Meditation on the justice of God**
(a wisdom psalm). The problem is the same
as that of the book of Job: How can one reconcile the belief that God is just with the
apparent inequities in his government of the

world? (compare Ps 37). 1: Both a confession
of faith and a thesis to be debated. 2–16: The
psalmist's experience and the grounds for
doubt. 2: The psalmist had nearly lost faith.
3–12: Cynical, wicked people seemed to
grow fat on their crimes (in sharp contradiction of such promises as Ps 1.4–6 or Deut
28.15–19; compare Job ch 21). 11: Their indifference to morality was grounded in a skeptical attitude toward God (10.4; 14.1). 13–16:
The psalmist's immediate, impulsive reac-

I would have been untrue to the
circle of your children.
16 But when I thought how to
understand this,
it seemed to me a wearisome
task,
17 until I went into the sanctuary
of God;
then I perceived their end.
18 Truly you set them in slippery
places;
you make them fall to ruin.
19 How they are destroyed in a
moment,
swept away utterly by terrors!
20 They are*v* like a dream when one
awakes;
on awaking you despise their
phantoms.

21 When my soul was embittered,
when I was pricked in heart,
22 I was stupid and ignorant;
I was like a brute beast toward
you.
23 Nevertheless I am continually
with you;
you hold my right hand.
24 You guide me with your counsel,
and afterward you will receive
me with honor.*w*
25 Whom have I in heaven but you?
And there is nothing on earth
that I desire other than you.
26 My flesh and my heart may fail,

but God is the strength*x* of my
heart and my portion
forever.

27 Indeed, those who are far from
you will perish;
you put an end to those who
are false to you.
28 But for me it is good to be near
God;
I have made the Lord GOD my
refuge,
to tell of all your works.

Psalm 74

A Maskil of Asaph.

1 O God, why do you cast us off
forever?
Why does your anger smoke
against the sheep of your
pasture?
2 Remember your congregation,
which you acquired long
ago,
which you redeemed to be the
tribe of your heritage.
Remember Mount Zion, where
you came to dwell.
3 Direct your steps to the perpetual
ruins;
the enemy has destroyed
everything in the sanctuary.

v Cn: Heb *Lord*   *w* Or *to glory*   *x* Heb *rock*

---

tion. **13**: If wickedness is not punished, why
be good? **14**: Efforts to be righteous brought
only suffering.
73.15: This was what the psalmist was
tempted to say, but had refrained from doing
so because of the effect it might have on others (39.1). **16**: Private cogitations brought no
answer to the problem. **17–28**: It was only on
going into the temple to seek God's help that
a satisfactory answer came. **17–22**: First of all
it became clear that the prosperity of the
wicked is only temporary (compare Ps 37).
73.21–22: The psalmist's previous attitude had been foolish and obtuse. **23–28**: In
the second place, and far more importantly,
the writer now realized that the righteous has
something that the wicked, however pros-

perous, could never enjoy—a sense of the
nearness of God (compare Ps 63; also 27.4).
Ps 74: **Prayer for deliverance from national enemies** (a group lament; compare Ps
44). **1–3**: Cry for help. **4–11**: The foe has devastated and burned the temple. If it were not
for the statement in v. 9 that there is no longer
any prophet, it would be natural to take the
situation as being that of the conquest by the
Babylonians in 587 B.C., but at that time, as
evidenced by the activity of Jeremiah and
Ezekiel, the prophetic movement was at its
height. Consequently the situation must be
some otherwise unknown event of the post-
exilic period (compare Isa 64.11). Some have
connected it with the Maccabean age, when
there were no prophets (see 1 Macc 4.46;

4 Your foes have roared within your
    holy place;
  they set up their emblems there.
5 At the upper entrance they hacked
    the wooden trellis with axes. *y*
6 And then, with hatchets and
    hammers,
  they smashed all its carved
    work.
7 They set your sanctuary on fire;
  they desecrated the dwelling
    place of your name,
  bringing it to the ground.
8 They said to themselves, "We will
    utterly subdue them";
  they burned all the meeting
    places of God in the land.

9 We do not see our emblems;
  there is no longer any prophet,
  and there is no one among us
    who knows how long.
10 How long, O God, is the foe to
    scoff?
  Is the enemy to revile your
    name forever?
11 Why do you hold back your hand;
  why do you keep your hand in *z*
    your bosom?

12 Yet God my King is from of old,
  working salvation in the earth.
13 You divided the sea by your
    might;
  you broke the heads of the
    dragons in the waters.
14 You crushed the heads of
    Leviathan;
  you gave him as food *a* for the
    creatures of the wilderness.
15 You cut openings for springs and
    torrents;
  you dried up ever-flowing
    streams.

16 Yours is the day, yours also the
    night;
  you established the luminaries *b*
    and the sun.
17 You have fixed all the bounds of
    the earth;
  you made summer and winter.

18 Remember this, O Lord, how the
    enemy scoffs,
  and an impious people reviles
    your name.
19 Do not deliver the soul of your
    dove to the wild animals;
  do not forget the life of your
    poor forever.

20 Have regard for your *c* covenant,
  for the dark places of the land
    are full of the haunts of
    violence.
21 Do not let the downtrodden be
    put to shame;
  let the poor and needy praise
    your name.
22 Rise up, O God, plead your cause;
  remember how the impious
    scoff at you all day long.
23 Do not forget the clamor of your
    foes,
  the uproar of your adversaries
    that goes up continually.

Psalm 75

*To the leader: Do Not Destroy. A Psalm of
Asaph. A Song.*

1 We give thanks to you, O God,
  we give thanks; your name is
    near.

y Cn Compare Gk Syr: Meaning of Heb
uncertain   *z* Cn: Heb *do you consume your right
hand from*   *a* Heb *food for the people*
b Or *moon;* Heb *light*   *c* Gk Syr: Heb *the*

14.41). **4:** *Emblems,* perhaps military standards.
  74.12–17: A hymn-like interlude, celebrating God as the creator. The allusion in
vv. 13–15 is to one of the popular accounts of
creation in which God was said to have subdued the monster of chaos (Leviathan or Rahab), a personification of the restless waters
of the sea, before beginning his constructive
activity (compare 89.10; Job 3.8 n.; 26.12–13;
Isa 27.1; 51.9). **18–23:** Prayer for deliverance.
19: *Your dove,* Israel.
  **Ps 75: National thanksgiving for God's
mighty acts. 1:** The verse shows that the

People tell of your wondrous
deeds.

2 At the set time that I appoint
I will judge with equity.
3 When the earth totters, with all its
inhabitants,
it is I who keep its pillars
steady.   *Selah*
4 I say to the boastful, "Do not
boast,"
and to the wicked, "Do not lift
up your horn;
5 do not lift up your horn on high,
or speak with insolent neck."

6 For not from the east or from
the west
and not from the wilderness
comes lifting up;
7 but it is God who executes
judgment,
putting down one and lifting up
another.
8 For in the hand of the LORD there
is a cup
with foaming wine, well mixed;
he will pour a draught from it,
and all the wicked of the earth
shall drain it down to the dregs.
9 But I will rejoice*d* forever;
I will sing praises to the God of
Jacob.

10 All the horns of the wicked I will
cut off,
but the horns of the righteous
shall be exalted.

Psalm 76

*To the leader: with stringed instruments.
A Psalm of Asaph. A Song.*

1 In Judah God is known,
his name is great in Israel.

2 His abode has been established in
Salem,
his dwelling place in Zion.
3 There he broke the flashing
arrows,
the shield, the sword, and the
weapons of war.   *Selah*

4 Glorious are you, more majestic
than the everlasting mountains. *e*
5 The stouthearted were stripped of
their spoil;
they sank into sleep;
none of the troops
was able to lift a hand.
6 At your rebuke, O God of Jacob,
both rider and horse lay
stunned.

7 But you indeed are awesome!
Who can stand before you
when once your anger is
roused?
8 From the heavens you uttered
judgment;
the earth feared and was still
9 when God rose up to establish
judgment,
to save all the oppressed of the
earth.   *Selah*

10 Human wrath serves only to
praise you,
when you bind the last bit of
your*f* wrath around you.
11 Make vows to the LORD your
God, and perform them;
let all who are around him bring
gifts
to the one who is awesome,
12 who cuts off the spirit of princes,

d Gk: Heb *declare*   *e* Gk: Heb *the mountains of
prey*   *f* Heb lacks *your*

---

basic theme is Israel's corporate (*we*) thankfulness, perhaps for a victory in battle. **2–5**: A divine oracle prophesying judgment for the wicked, probably spoken by a priest or temple prophet. **3**: *Selah*, see 3.2 n. **4–5**: *Horn*, symbol of strength and power.

75.6–8: A warning to all that the God of Israel controls the destinies of nations. **8**: *Cup*, see Isa 51.22–23; Jer 25.15; Lk 22.42 n. **9–10**: An individual, perhaps the king (see 60.9 n.) brings the thanksgiving liturgy to a conclusion in an act of praise and faith.

who inspires fear in the kings of
the earth.

Psalm 77

To the leader: according to Jeduthun.
Of Asaph. A Psalm.

1 I cry aloud to God,
aloud to God, that he may
hear me.

2 In the day of my trouble I seek
the Lord;
in the night my hand is
stretched out without
wearying;
my soul refuses to be
comforted.

3 I think of God, and I moan;
I meditate, and my spirit faints.
Selah

4 You keep my eyelids from
closing;
I am so troubled that I cannot
speak.

5 I consider the days of old,
and remember the years of long
ago.

6 I commune^g with my heart in the
night;
I meditate and search my
spirit:^{h}

7 "Will the Lord spurn forever,
and never again be favorable?

8 Has his steadfast love ceased
forever?
Are his promises at an end for
all time?

9 Has God forgotten to be gracious?

Has he in anger shut up his
compassion?"   *Selah*

10 And I say, "It is my grief
that the right hand of the Most
High has changed."

11 I will call to mind the deeds of the
LORD;
I will remember your wonders
of old.

12 I will meditate on all your work,
and muse on your mighty
deeds.

13 Your way, O God, is holy.
What god is so great as our
God?

14 You are the God who works
wonders;
you have displayed your might
among the peoples.

15 With your strong arm you
redeemed your people,
the descendants of Jacob and
Joseph.   *Selah*

16 When the waters saw you,
O God,
when the waters saw you, they
were afraid;
the very deep trembled.

17 The clouds poured out water;
the skies thundered;
your arrows flashed on every
side.

18 The crash of your thunder was in
the whirlwind;

g Gk Syr: Heb *My music*   *h* Syr Jerome: Heb
my spirit searches

**Ps 76: A song of Zion celebrating
God's ultimate victory over the nations**
(compare Ps 46). **1–3:** Announcement of
God's victory. **2:** *Salem,* a poetical name for
Jerusalem (compare Gen 14.18). **3:** Although
the verb is past, the reference is almost certainly to the great eschatological conflict of
the latter days (see 48.4–8 n.; 46.6, 9). *Selah,*
see 3.2 n.
76.4–9: Hymn to the victorious God of
Israel. **10:** God turns even evil intentions to his
own good purposes. **11–12:** The congregation
is exhorted to join in worshiping him.

**Ps 77: Prayer for deliverance from
personal trouble** (a lament). **1–6:** The
psalmist's miserable situation: the nature of
the difficulty is not specified. **3:** *Selah,* see
3.2 n. **7–10:** The mental agony is so intense as
to raise questions about God's justice and
love.
77.11–15: For encouragement, the psalmist recalls God's mighty works in the past.
16–20: Fragment of an ancient hymn praising
God for his work in creation (vv. 16–19) and
in the history of Israel (v. 20).

your lightnings lit up the world;
the earth trembled and shook.
19 Your way was through the sea,
your path, through the mighty
waters;
yet your footprints were unseen.
20 You led your people like a flock
by the hand of Moses and
Aaron.

Psalm 78

A Maskil of Asaph.

1 Give ear, O my people, to my
teaching;
incline your ears to the words of
my mouth.
2 I will open my mouth in a
parable;
I will utter dark sayings from
of old,
3 things that we have heard and
known,
that our ancestors have told us.
4 We will not hide them from their
children;
we will tell to the coming
generation
the glorious deeds of the LORD,
and his might,
and the wonders that he has
done.

5 He established a decree in Jacob,
and appointed a law in Israel,
which he commanded our
ancestors
to teach to their children;
6 that the next generation might
know them,
the children yet unborn,

and rise up and tell them to their
children,
7 so that they should set their
hope in God,
and not forget the works of God,
but keep his commandments;
8 and that they should not be like
their ancestors,
a stubborn and rebellious
generation,
a generation whose heart was not
steadfast,
whose spirit was not faithful
to God.

9 The Ephraimites, armed with*i* the
bow,
turned back on the day of
battle.
10 They did not keep God's
covenant,
but refused to walk according to
his law.
11 They forgot what he had done,
and the miracles that he had
shown them.
12 In the sight of their ancestors he
worked marvels
in the land of Egypt, in the
fields of Zoan.
13 He divided the sea and let them
pass through it,
and made the waters stand like
a heap.
14 In the daytime he led them with
a cloud,
and all night long with a fiery
light.

i Heb *armed with shooting*

Ps 78: **The story of God's great deeds
and his people's faithlessness.** One of a
group of psalms (105; 106; 135; 136), composed for use at the major festivals, which
simply recite the history of God's dealings
with Israel. This psalm emphasizes the disobedience and ingratitude of the people, especially noting the defection of the Ephraimites
(vv. 9–11) which led God to reject them in
favor of Judah (vv. 67–69). **1–4:** The poet addresses the congregation in the style of the
wisdom writers (compare 49.1–4). **2:** Quoted
in Mt 13.35. **5–8:** The giving of the law. **6:**
Deut 6.7.
78.9–11: It is uncertain what incident is
referred to. **12–53:** God's miraculous care
for his people during the Exodus and the wilderness wanderings, a review of events
in Exodus and Numbers, vv. 17–20 and
32–41 being interludes which describe the

15 He split rocks open in the
   wilderness,
   and gave them drink abundantly
   as from the deep.
16 He made streams come out of the
   rock,
   and caused waters to flow down
   like rivers.

17 Yet they sinned still more against
   him,
   rebelling against the Most High
   in the desert.
18 They tested God in their heart
   by demanding the food they
   craved.
19 They spoke against God, saying,
   "Can God spread a table in the
   wilderness?
20 Even though he struck the rock so
   that water gushed out
   and torrents overflowed,
   can he also give bread,
   or provide meat for his people?"

21 Therefore, when the LORD heard,
   he was full of rage;
   a fire was kindled against Jacob,
   his anger mounted against
   Israel,
22 because they had no faith in God,
   and did not trust his saving
   power.
23 Yet he commanded the skies
   above,
   and opened the doors of heaven;
24 he rained down on them manna
   to eat,
   and gave them the grain of
   heaven.
25 Mortals ate of the bread of angels;
   he sent them food in abundance.
26 He caused the east wind to blow
   in the heavens,
   and by his power he led out the
   south wind;
27 he rained flesh upon them like
   dust,

winged birds like the sand of
   the seas;
28 he let them fall within their camp,
   all around their dwellings.
29 And they ate and were well filled,
   for he gave them what they
   craved.
30 But before they had satisfied their
   craving,
   while the food was still in their
   mouths,
31 the anger of God rose against
   them
   and he killed the strongest of
   them,
   and laid low the flower of
   Israel.

32 In spite of all this they still sinned;
   they did not believe in his
   wonders.
33 So he made their days vanish like
   a breath,
   and their years in terror.
34 When he killed them, they sought
   for him;
   they repented and sought God
   earnestly.
35 They remembered that God was
   their rock,
   the Most High God their
   redeemer.
36 But they flattered him with their
   mouths;
   they lied to him with their
   tongues.
37 Their heart was not steadfast
   toward him;
   they were not true to his
   covenant.
38 Yet he, being compassionate,
   forgave their iniquity,
   and did not destroy them;
   often he restrained his anger,
   and did not stir up all his wrath.
39 He remembered that they were
   but flesh,

characteristic faithlessness of Israel. **12:** *Zoan,*
a royal Egyptian store city, is identical with
the Rameses of Ex 1.11. **15–16:** Ex 17.6; Num
20.10–13. **21–31:** Ex ch 16; Num ch 11. **24:**
Quoted in Jn 6.31. **25:** The manna is fanciful-
ly interpreted as *the bread of the angels.*

a wind that passes and does not
come again.
40 How often they rebelled against
him in the wilderness
and grieved him in the desert!
41 They tested God again and again,
and provoked the Holy One of
Israel.
42 They did not keep in mind his
power,
or the day when he redeemed
them from the foe;
43 when he displayed his signs in
Egypt,
and his miracles in the fields of
Zoan.
44 He turned their rivers to blood,
so that they could not drink of
their streams.
45 He sent among them swarms of
flies, which devoured them,
and frogs, which destroyed
them.
46 He gave their crops to the
caterpillar,
and the fruit of their labor to
the locust.
47 He destroyed their vines with hail,
and their sycamores with frost.
48 He gave over their cattle to the
hail,
and their flocks to thunderbolts.
49 He let loose on them his fierce
anger,
wrath, indignation, and distress,
a company of destroying angels.
50 He made a path for his anger;
he did not spare them from
death,
but gave their lives over to the
plague.
51 He struck all the firstborn in
Egypt,
the first issue of their strength in
the tents of Ham.
52 Then he led out his people like
sheep,

and guided them in the
wilderness like a flock.
53 He led them in safety, so that they
were not afraid;
but the sea overwhelmed their
enemies.
54 And he brought them to his holy
hill,
to the mountain that his right
hand had won.
55 He drove out nations before them;
he apportioned them for a
possession
and settled the tribes of Israel in
their tents.

56 Yet they tested the Most High
God,
and rebelled against him.
They did not observe his
decrees,
57 but turned away and were faithless
like their ancestors;
they twisted like a treacherous
bow.
58 For they provoked him to anger
with their high places;
they moved him to jealousy
with their idols.
59 When God heard, he was full of
wrath,
and he utterly rejected Israel.
60 He abandoned his dwelling at
Shiloh,
the tent where he dwelt among
mortals,
61 and delivered his power to
captivity,
his glory to the hand of the foe.
62 He gave his people to the sword,
and vented his wrath on his
heritage.
63 Fire devoured their young men,
and their girls had no marriage
song.
64 Their priests fell by the sword,

78.44–51: The plagues, Ex chs 7–12. **51:**
Ham, a poetic synonym for Egypt (Gen 10.6).
78.54–64: God's care for his people during the conquest and in the days of the Judges;
their continued faithlessness. **55:** *Nations,*
Deut 7.1. **60–64:** The destruction of the temple at *Shiloh* is mentioned elsewhere only in
Jer 7.12–14 and 26.6; the period was that of

and their widows made no
lamentation.
65 Then the Lord awoke as from
sleep,
like a warrior shouting because
of wine.
66 He put his adversaries to rout;
he put them to everlasting
disgrace.

67 He rejected the tent of Joseph,
he did not choose the tribe of
Ephraim;
68 but he chose the tribe of Judah,
Mount Zion, which he loves.
69 He built his sanctuary like the
high heavens,
like the earth, which he has
founded forever.
70 He chose his servant David,
and took him from the
sheepfolds;
71 from tending the nursing ewes he
brought him
to be the shepherd of his people
Jacob,
of Israel, his inheritance.
72 With upright heart he tended
them,
and guided them with skillful
hand.

Psalm 79

A Psalm of Asaph.

1 O God, the nations have come
into your inheritance;
they have defiled your holy
temple;
they have laid Jerusalem in
ruins.
2 They have given the bodies of
your servants

to the birds of the air for food,
the flesh of your faithful to the
wild animals of the earth.
3 They have poured out their blood
like water
all around Jerusalem,
and there was no one to bury
them.
4 We have become a taunt to our
neighbors,
mocked and derided by those
around us.

5 How long, O LORD? Will you be
angry forever?
Will your jealous wrath burn
like fire?
6 Pour out your anger on the
nations
that do not know you,
and on the kingdoms
that do not call on your name.
7 For they have devoured Jacob
and laid waste his habitation.

8 Do not remember against us the
iniquities of our ancestors;
let your compassion come
speedily to meet us,
for we are brought very low.
9 Help us, O God of our salvation,
for the glory of your name;
deliver us, and forgive our sins,
for your name's sake.
10 Why should the nations say,
"Where is their God?"
Let the avenging of the outpoured
blood of your servants
be known among the nations
before our eyes.

11 Let the groans of the prisoners
come before you;

---

the events related in 1 Sam 4.1–7.2. **65–66:**
Victories over the Philistines in the time of
Saul and David. **67–72:** The building of the
temple and the establishment of the Davidic
dynasty. The passage reflects the tension
between North and South which led to the
division of the nation into Judah and Israel

(Ephraim) and the eventual schism between
Jews and Samaritans.
Ps 79: Prayer for deliverance from national enemies (a group lament). The occasion is probably the same as that of Ps 74 (see
74.4–11 n.). **1–4:** The people's situation: they
are defeated and persecuted. **5–12:** Prayer for

according to your great power
   preserve those doomed to
   die.
12 Return sevenfold into the bosom
   of our neighbors
   the taunts with which they
   taunted you, O Lord!
13 Then we your people, the flock of
   your pasture,
   will give thanks to you forever;
   from generation to generation
   we will recount your
   praise.

Psalm 80

To the leader: on Lilies, a Covenant.
Of Asaph. A Psalm.

1 Give ear, O Shepherd of Israel,
   you who lead Joseph like a
   flock!
You who are enthroned upon the
   cherubim, shine forth
2   before Ephraim and Benjamin
   and Manasseh.
Stir up your might,
   and come to save us!

3 Restore us, O God;
   let your face shine, that we may
   be saved.

4 O Lord God of hosts,
   how long will you be angry
   with your people's prayers?
5 You have fed them with the bread
   of tears,
   and given them tears to drink in
   full measure.
6 You make us the scorn *j* of our
   neighbors;
   our enemies laugh among
   themselves.

7 Restore us, O God of hosts;
   let your face shine, that we may
   be saved.

8 You brought a vine out of Egypt;
   you drove out the nations and
   planted it.
9 You cleared the ground for it;
   it took deep root and filled the
   land.
10 The mountains were covered with
   its shade,
   the mighty cedars with its
   branches;
11 it sent out its branches to the sea,
   and its shoots to the River.
12 Why then have you broken down
   its walls,
   so that all who pass along the
   way pluck its fruit?
13 The boar from the forest ravages
   it,
   and all that move in the field
   feed on it.

14 Turn again, O God of hosts;
   look down from heaven, and
   see;
have regard for this vine,
15   the stock that your right hand
   planted. *k*
16 They have burned it with fire,
   they have cut it down; *l*
   may they perish at the rebuke of
   your countenance.
17 But let your hand be upon the one
   at your right hand,
   the one whom you made strong
   for yourself.

j Syr: Heb *strife*   *k* Heb adds *from verse 17*
and upon the one whom you made strong for yourself
l Cn: Heb *it is cut down*

---

deliverance and vengeance. **13**: The vow
(see 7.17 n.).
  Ps 80: **Prayer for deliverance from na-**
tional enemies (a group lament). **1–3**: Cry
for help. The tribes mentioned show that the
psalm is a product of the northern kingdom,
Israel (compare 78.67–68). **1**: *Cherubim,* see
1 Sam 4.4. **3**: A refrain (see vv. 7, 19).

80.8–13: Israel as a vine, once carefully
tended, but now forsaken (compare Isa 5.1–
7). **8**: *Nations,* see 78.55 n. **11**: *The River,* the
Euphrates (1 Kings 4.21).
  80.14–19: Prayer for deliverance. **17**: *The*
one at your right hand is a personification of
Israel.

18 Then we will never turn back
    from you;
    give us life, and we will call on
      your name.

19 Restore us, O LORD God of hosts;
    let your face shine, that we may
      be saved.

Psalm 81

*To the leader: according to The Gittith.
Of Asaph.*

1 Sing aloud to God our strength;
    shout for joy to the God of
      Jacob.
2 Raise a song, sound the
      tambourine,
    the sweet lyre with the harp.
3 Blow the trumpet at the new
      moon,
    at the full moon, on our
      festal day.
4 For it is a statute for Israel,
    an ordinance of the God of
      Jacob.
5 He made it a decree in Joseph,
    when he went out over*m* the
      land of Egypt.

    I hear a voice I had not known:
6 "I relieved your*n* shoulder of the
      burden;
    your*n* hands were freed from
      the basket.
7 In distress you called, and I
      rescued you;
    I answered you in the secret
      place of thunder;
    I tested you at the waters of
      Meribah.   *Selah*
8 Hear, O my people, while I
    admonish you;

    O Israel, if you would but listen
      to me!
9 There shall be no strange god
      among you;
    you shall not bow down to a
      foreign god.
10 I am the LORD your God,
    who brought you up out of the
      land of Egypt.
    Open your mouth wide and I
      will fill it.

11 "But my people did not listen to
      my voice;
    Israel would not submit to me.
12 So I gave them over to their
      stubborn hearts,
    to follow their own counsels.
13 O that my people would listen
      to me,
    that Israel would walk in my
      ways!
14 Then I would quickly subdue their
      enemies,
    and turn my hand against their
      foes.
15 Those who hate the LORD would
      cringe before him,
    and their doom would last
      forever.
16 I would feed you*o* with the finest
      of the wheat,
    and with honey from the rock I
      would satisfy you."

Psalm 82

A Psalm of Asaph.

1 God has taken his place in the
    divine council;

m Or *against*   *n* Heb *his*   *o* Cn Compare
verse 16b: Heb *he would feed him*

Ps 81: **Liturgy for a festival. 1–5a:**
Hymn-like summons to worship. **3:** *Festal
day,* probably festival of booths (Deut 16.13–
15).
    81.5b–16: A priest or temple prophet de-
livers a divine oracle in which God reminds
his people of what he has done for them in the
past (vv. 6–7); his demand for their loyalty
(vv. 8–10) and their disobedience to it

(vv. 11–12); their future prosperity depends
on their willingness now to change their ways
(vv. 13–16; compare 95.7b–11). **7:** *The secret
place of thunder,* Sinai. *Meribah,* Ex 17.7; Num
20.13. *Selah,* see 3.2 n.
    **Ps 82: Liturgy of the LORD's judgment
on pagan gods. 1:** Making use of a conception, common to the ancient Near East, that
the world is ruled by a council of gods (89.5–

in the midst of the gods he
holds judgment:
2 "How long will you judge
unjustly
and show partiality to the
wicked? *Selah*
3 Give justice to the weak and the
orphan;
maintain the right of the lowly
and the destitute.
4 Rescue the weak and the needy;
deliver them from the hand of
the wicked."

5 They have neither knowledge nor
understanding,
they walk around in darkness;
all the foundations of the earth
are shaken.

6 I say, "You are gods,
children of the Most High, all
of you;
7 nevertheless, you shall die like
mortals,
and fall like any prince."*p*

8 Rise up, O God, judge the earth;
for all the nations belong to
you!

Psalm 83

A Song. A Psalm of Asaph.

1 O God, do not keep silence;
do not hold your peace or be
still, O God!
2 Even now your enemies are in
tumult;
those who hate you have raised
their heads.
3 They lay crafty plans against your
people;

they consult together against
those you protect.
4 They say, "Come, let us wipe
them out as a nation;
let the name of Israel be
remembered no more."
5 They conspire with one accord;
against you they make a
covenant—
6 the tents of Edom and the
Ishmaelites,
Moab and the Hagrites,
7 Gebal and Ammon and Amalek,
Philistia with the inhabitants of
Tyre;
8 Assyria also has joined them;
they are the strong arm of the
children of Lot. *Selah*

9 Do to them as you did to Midian,
as to Sisera and Jabin at the
Wadi Kishon,
10 who were destroyed at En-dor,
who became dung for the
ground.
11 Make their nobles like Oreb and
Zeeb,
all their princes like Zebah and
Zalmunna,
12 who said, "Let us take the pastures
of God
for our own possession."

13 O my God, make them like
whirling dust, *q*
like chaff before the wind.
14 As fire consumes the forest,
as the flame sets the mountains
ablaze,
15 so pursue them with your tempest

p Or *fall as one man, O princes*   q Or *a
tumbleweed*

---

7), the poet (presumably a priest or temple
prophet) sees, in a vision, the God of Israel
standing up in the midst of the council and
pronouncing judgment upon all the other
members. **2–7:** Because they govern the earth
unjustly, they shall all perish like mere human
beings (Ezek 28.9). **2:** *Selah,* see 3.2 n. **6:**
Quoted in Jn 10.34. **8:** A prayer that the poet's
vision may be realized.

**Ps 83: Prayer for deliverance from na-
tional enemies** (a lament). **1:** Cry for help.
2–8: Israel's perilous situation, attacked by a
powerful group of foreign allies. **6–7:** These
are Israel's near neighbors. **8:** *Children of Lot,*
Moab and Edom (Gen 19.36–38; Deut 2.9).
Selah, see 3.2 n.
83.9–18: Prayer for victory. **9–11:** Gideon's decisive victory over Midian is recorded

and terrify them with your
hurricane.
16 Fill their faces with shame,
so that they may seek your
name, O LORD.
17 Let them be put to shame and
dismayed forever;
let them perish in disgrace.
18 Let them know that you alone,
whose name is the LORD,
are the Most High over all the
earth.

Psalm 84

To the leader: according to The Gittith.
Of the Korahites. A Psalm.

1 How lovely is your dwelling
place,
O LORD of hosts!
2 My soul longs, indeed it faints
for the courts of the LORD;
my heart and my flesh sing for
joy
to the living God.

3 Even the sparrow finds a home,
and the swallow a nest for
herself,
where she may lay her young,
at your altars, O LORD of hosts,
my King and my God.
4 Happy are those who live in your
house,
ever singing your praise. *Selah*

5 Happy are those whose strength is
in you,
in whose heart are the highways
to Zion.[r]
6 As they go through the valley of
Baca
they make it a place of springs;

the early rain also covers it with
pools.
7 They go from strength to
strength;
the God of gods will be seen
in Zion.

8 O LORD God of hosts, hear my
prayer;
give ear, O God of Jacob! *Selah*
9 Behold our shield, O God;
look on the face of your
anointed.

10 For a day in your courts is better
than a thousand elsewhere.
I would rather be a doorkeeper in
the house of my God
than live in the tents of
wickedness.
11 For the LORD God is a sun and
shield;
he bestows favor and honor.
No good thing does the LORD
withhold
from those who walk uprightly.
12 O LORD of hosts,
happy is everyone who trusts
in you.

Psalm 85

To the leader. Of the Korahites. A Psalm.

1 LORD, you were favorable to your
land;
you restored the fortunes of
Jacob.
2 You forgave the iniquity of your
people;
you pardoned all their sin. *Selah*
3 You withdrew all your wrath;

r Heb lacks *to Zion*

---

in Judg chs 6–8; the defeat of *Sisera and Jabin*
by Deborah and Barak in Judg chs 4–5. *Oreb
and Zeeb*, Judg 7.25. *Zebah and Zalmunna*,
Judg 8.21.
**Ps 84: Song praising Zion as the
longed-for goal of the pilgrim. 1–2:** Praise
for the temple. **3–4:** Envy of the birds and
servitors who live there. **4:** *Selah*, see 3.2 n.

84.5–7: The joys of the pilgrimage. **6:**
Baca, some unknown, desolate place through
which the pilgrims must go. **8–9:** Prayer for
the king (see 61.6–7 n.). **9:** *Anointed,* see
2.2 n. **10–12:** The superiority of life in the
temple to life anywhere else.
**Ps 85: Prayer for deliverance from na-
tional adversity** (a group lament). **1–3:**

you turned from your hot
anger.

4 Restore us again, O God of our
salvation,
and put away your indignation
toward us.
5 Will you be angry with us
forever?
Will you prolong your anger to
all generations?
6 Will you not revive us again,
so that your people may rejoice
in you?
7 Show us your steadfast love,
O LORD,
and grant us your salvation.

8 Let me hear what God the LORD
will speak,
for he will speak peace to his
people,
to his faithful, to those who
turn to him in their
hearts. *s*
9 Surely his salvation is at hand for
those who fear him,
that his glory may dwell in our
land.

10 Steadfast love and faithfulness
will meet;
righteousness and peace will kiss
each other.
11 Faithfulness will spring up from
the ground,
and righteousness will look
down from the sky.
12 The LORD will give what is good,
and our land will yield its
increase.
13 Righteousness will go before him,
and will make a path for his
steps.

Psalm 86

A Prayer of David.

1 Incline your ear, O LORD, and
answer me,
for I am poor and needy.
2 Preserve my life, for I am devoted
to you;
save your servant who trusts
in you.
You are my God; 3 be gracious to
me, O Lord,
for to you do I cry all day long.
4 Gladden the soul of your servant,
for to you, O Lord, I lift up my
soul.
5 For you, O Lord, are good and
forgiving,
abounding in steadfast love to
all who call on you.
6 Give ear, O LORD, to my prayer;
listen to my cry of supplication.
7 In the day of my trouble I call
on you,
for you will answer me.

8 There is none like you among the
gods, O Lord,
nor are there any works like
yours.
9 All the nations you have made
shall come
and bow down before you,
O Lord,
and shall glorify your name.
10 For you are great and do
wondrous things;
you alone are God.
11 Teach me your way, O LORD,
that I may walk in your truth;
give me an undivided heart to
revere your name.

s Gk: Heb *but let them not turn back to folly*

God's favor toward his people in time past.
4–7: Prayer that God's favor may be manifested again in their present difficulties (perhaps poor crops; compare v. 12).
   85.8–13: A priest or temple prophet delivers an oracle of assurance.

Ps 86: Prayer for deliverance from personal enemies (a lament). **1–7**: Cry for help.
   86.8–10: God is praised for his power. **11**: Prayer for guidance. **12–13**: Thanksgiving, spoken in confident anticipation of deliver-

12 I give thanks to you, O Lord my
        God, with my whole heart,
    and I will glorify your name
        forever.
13 For great is your steadfast love
        toward me;
    you have delivered my soul
        from the depths of Sheol.

14 O God, the insolent rise up
        against me;
    a band of ruffians seeks my life,
    and they do not set you before
        them.
15 But you, O Lord, are a God
        merciful and gracious,
    slow to anger and abounding in
        steadfast love and
        faithfulness.
16 Turn to me and be gracious to
        me;
    give your strength to your
        servant;
    save the child of your serving
        girl.
17 Show me a sign of your favor,
    so that those who hate me may
        see it and be put to shame,
    because you, Lord, have helped
        me and comforted me.

Psalm 87

Of the Korahites. A Psalm. A Song.

1 On the holy mount stands the city
        he founded;
2   the Lord loves the gates of Zion
        more than all the dwellings of
        Jacob.

3 Glorious things are spoken of you,
        O city of God.   *Selah*

4 Among those who know me I
        mention Rahab and
        Babylon;
    Philistia too, and Tyre, with
        Ethiopia *t*—
    "This one was born there," they
        say.

5 And of Zion it shall be said,
    "This one and that one were
        born in it";
    for the Most High himself will
        establish it.
6 The Lord records, as he registers
        the peoples,
    "This one was born there."
                                Selah

7 Singers and dancers alike say,
    "All my springs are in you."

Psalm 88

*A Song. A Psalm of the Korahites. To the
leader: according to Mahalath Leannoth.
A Maskil of Heman the Ezrahite.*

1 O Lord, God of my salvation,
    when, at night, I cry out in
        your presence,
2 let my prayer come before you;
    incline your ear to my cry.

3 For my soul is full of troubles,
    and my life draws near to Sheol.

t Or Nubia; Heb Cush

---

ance. **13**: *Sheol,* the abode of the dead. The
psalmist is sure God will not permit the
present situation to end in death.
   86.14: Prayer for preservation from ruthless enemies. **15–17**: Return to the mood of
supplication. **16**: *Child of your serving girl,* synonymous with *your servant.*
   Ps 87: **Song praising Zion as the mother of believers everywhere.** The text of this
psalm is damaged and disarranged. **1–3**: The
glory of Zion (Jerusalem). **3**: This verse provided the theme for Augustine's *City of God*
and was the inspiration of a popular hymn.

Selah, see 3.2 n. **4**: God's people, the citizens
of Zion, are found in every country. *Rahab,*
a poetical name for Egypt. **5**: The roll of
Zion's citizens is a long and proud one. **6**: As
God keeps the census records of the world, he
notes with special pleasure the citizens of
Zion. **7**: This verse seems to be only a fragment. *Springs,* sources of welfare.
   Ps 88: **Desperate prayer for healing in
sickness** (a lament). **1–2**: Cry for help.
   88.3–9: The psalmist's situation, probably one of lifelong illness (compare v. 15).
3–4: *Sheol . . . the Pit,* see 6.5 n. **5–6**: A classic

4 I am counted among those who
    go down to the Pit;
    I am like those who have no
    help,
5 like those forsaken among the
    dead,
    like the slain that lie in the
    grave,
    like those whom you remember
    no more,
    for they are cut off from your
    hand.
6 You have put me in the depths of
    the Pit,
    in the regions dark and deep.
7 Your wrath lies heavy upon me,
    and you overwhelm me with all
    your waves.   *Selah*

8 You have caused my companions
    to shun me;
    you have made me a thing of
    horror to them.
    I am shut in so that I cannot
    escape;
9   my eye grows dim through
    sorrow.
    Every day I call on you, O LORD;
    I spread out my hands to you.
10 Do you work wonders for the
    dead?
    Do the shades rise up to praise
    you?   *Selah*
11 Is your steadfast love declared in
    the grave,
    or your faithfulness in
    Abaddon?
12 Are your wonders known in the
    darkness,
    or your saving help in the land
    of forgetfulness?

13 But I, O LORD, cry out to you;
    in the morning my prayer
    comes before you.
14 O LORD, why do you cast me off?

Why do you hide your face
    from me?
15 Wretched and close to death from
    my youth up,
    I suffer your terrors; I am
    desperate. *u*
16 Your wrath has swept over me;
    your dread assaults destroy me.
17 They surround me like a flood all
    day long;
    from all sides they close in on
    me.
18 You have caused friend and
    neighbor to shun me;
    my companions are in darkness.

Psalm 89

A Maskil of Ethan the Ezrahite.

1 I will sing of your steadfast love,
    O LORD, *v* forever;
    with my mouth I will proclaim
    your faithfulness to all
    generations.
2 I declare that your steadfast love is
    established forever;
    your faithfulness is as firm as
    the heavens.

3 You said, "I have made a covenant
    with my chosen one,
    I have sworn to my servant
    David:
4 'I will establish your descendants
    forever,
    and build your throne for all
    generations.' "   *Selah*

5 Let the heavens praise your
    wonders, O LORD,
    your faithfulness in the assembly
    of the holy ones.

u Meaning of Heb uncertain   *v* Gk: Heb *the
steadfast love of the LORD*

---

description of Sheol (compare vv. 10–12; Job
10.21–22; Isa 38.18–19). **7:** *Selah*, see 3.2 n.
    88.10–18: Prayer for deliverance, notable
for the absence of even a spark of hopefulness.
11: *Abaddon*, see Job 26.6 n.
    Ps 89: A king prays for deliverance
from his enemies. **1–18:** Hymn extolling
God's power and faithfulness. **3–4:** His covenant with David recalled (compare vv. 19–
37; 2 Sam 7.16). **4:** *Selah*, see 3.2 n.
    89.5–7: See 82.1 n. **10:** *Rahab*, see 74.12–
17 n. **12:** *Tabor*, a mountain in the north of

6 For who in the skies can be
compared to the Lord?
Who among the heavenly beings
is like the Lord,
7 a God feared in the council of the
holy ones,
great and awesome*w* above all
that are around him?
8 O Lord God of hosts,
who is as mighty as you,
O Lord?
Your faithfulness surrounds
you.
9 You rule the raging of the sea;
when its waves rise, you still
them.
10 You crushed Rahab like a carcass;
you scattered your enemies with
your mighty arm.
11 The heavens are yours, the earth
also is yours;
the world and all that is in
it—you have founded
them.
12 The north and the south*x*—you
created them;
Tabor and Hermon joyously
praise your name.
13 You have a mighty arm;
strong is your hand, high your
right hand.
14 Righteousness and justice are the
foundation of your throne;
steadfast love and faithfulness go
before you.
15 Happy are the people who know
the festal shout,
who walk, O Lord, in the light
of your countenance;
16 they exult in your name all day
long,
and extol*y* your righteousness.
17 For you are the glory of their
strength;
by your favor our horn is
exalted.

18 For our shield belongs to the
Lord,
our king to the Holy One of
Israel.
19 Then you spoke in a vision to
your faithful one, and said:
"I have set the crown*z* on one
who is mighty,
I have exalted one chosen from
the people.
20 I have found my servant David;
with my holy oil I have
anointed him;
21 my hand shall always remain with
him;
my arm also shall strengthen
him.
22 The enemy shall not outwit him,
the wicked shall not humble
him.
23 I will crush his foes before him
and strike down those who hate
him.
24 My faithfulness and steadfast love
shall be with him;
and in my name his horn shall
be exalted.
25 I will set his hand on the sea
and his right hand on the rivers.
26 He shall cry to me, 'You are my
Father,
my God, and the Rock of my
salvation!'
27 I will make him the firstborn,
the highest of the kings of the
earth.
28 Forever I will keep my steadfast
love for him,
and my covenant with him will
stand firm.
29 I will establish his line forever,

w Gk Syr: Heb *greatly awesome*   *x* Or *Zaphon
and Yamin*   *y* Cn: Heb *are exalted in*
z Cn: Heb *help*

Palestine. *Hermon,* the highest mountain of
Syria. **17:** *Horn,* see 75.4–5 n. **18:** *Shield,* the
king (see 47.9 n.).
89.19–37: The terms of the unalterable
covenant that God had once established with
the Davidic dynasty. **19:** *Faithful one,* David
(compare vv. 3–4); alternatively, Nathan (see
2 Sam 7.4). **27:** *The firstborn,* compare 2.7.

and his throne as long as the
    heavens endure.
30 If his children forsake my law
    and do not walk according to
      my ordinances,
31 if they violate my statutes
    and do not keep my
      commandments,
32 then I will punish their
      transgression with the rod
    and their iniquity with scourges;
33 but I will not remove from him
      my steadfast love,
    or be false to my faithfulness.
34 I will not violate my covenant,
    or alter the word that went
      forth from my lips.
35 Once and for all I have sworn by
      my holiness;
    I will not lie to David.
36 His line shall continue forever,
    and his throne endure before me
      like the sun.
37 It shall be established forever like
      the moon,
    an enduring witness in the
      skies."   *Selah*

38 But now you have spurned and
      rejected him;
    you are full of wrath against
      your anointed.
39 You have renounced the covenant
      with your servant;
    you have defiled his crown in
      the dust.
40 You have broken through all
      his walls;
    you have laid his strongholds
      in ruins.
41 All who pass by plunder him;
    he has become the scorn of his
      neighbors.
42 You have exalted the right hand of
      his foes;

you have made all his enemies
    rejoice.
43 Moreover, you have turned back
      the edge of his sword,
    and you have not supported him
      in battle.
44 You have removed the scepter
      from his hand, *a*
    and hurled his throne to the
      ground.
45 You have cut short the days of his
      youth;
    you have covered him with
      shame.   *Selah*

46 How long, O LORD? Will you
    hide yourself forever?
    How long will your wrath burn
      like fire?
47 Remember how short my time
      is— *b*
    for what vanity you have
      created all mortals!
48 Who can live and never see death?
    Who can escape the power of
      Sheol?   *Selah*

49 Lord, where is your steadfast love
      of old,
    which by your faithfulness you
      swore to David?
50 Remember, O Lord, how your
      servant is taunted;
    how I bear in my bosom the
      insults of the peoples, *c*
51 with which your enemies taunt,
      O LORD,
    with which they taunted the
      footsteps of your anointed.

52 Blessed be the LORD forever.
    Amen and Amen.

a Cn: Heb *removed his cleanness*   *b* Meaning of
Heb uncertain   *c* Cn: Heb *bosom all of many
peoples*

89.38–45: The king has been defeated in
battle (v. 43), and it seems that God has forsaken the covenant. 38: *Anointed* (see v. 20;
2.2 n.).
    89.46–51: Prayer that God will remember
his promises and give victory to David's de-

scendant. 48: *Sheol,* see 6.5 n.; 49.15. 52: A
doxology (no part of the psalm) marking the
end of Book III of the Psalter (see 41.13 n.).
    Ps 90: **Prayer for deliverance from na-
tional adversity** (a group lament). 1–6:
Hymn-like introduction declaring God's

BOOK IV

Psalm 90

A Prayer of Moses, the man of God.

1 Lord, you have been our dwelling
place[d]
in all generations.
2 Before the mountains were
brought forth,
or ever you had formed the
earth and the world,
from everlasting to everlasting
you are God.

3 You turn us[e] back to dust,
and say, "Turn back, you
mortals."
4 For a thousand years in your sight
are like yesterday when it is
past,
or like a watch in the night.

5 You sweep them away; they are
like a dream,
like grass that is renewed in the
morning;
6 in the morning it flourishes and is
renewed;
in the evening it fades and
withers.

7 For we are consumed by your
anger;
by your wrath we are
overwhelmed.
8 You have set our iniquities
before you,
our secret sins in the light of
your countenance.

9 For all our days pass away under
your wrath;
our years come to an end[f] like
a sigh.
10 The days of our life are seventy
years,

or perhaps eighty, if we are
strong;
even then their span[g] is only toil
and trouble;
they are soon gone, and we fly
away.

11 Who considers the power of your
anger?
Your wrath is as great as the
fear that is due you.
12 So teach us to count our days
that we may gain a wise heart.

13 Turn, O Lord! How long?
Have compassion on your
servants!
14 Satisfy us in the morning with
your steadfast love,
so that we may rejoice and be
glad all our days.
15 Make us glad as many days as you
have afflicted us,
and as many years as we have
seen evil.
16 Let your work be manifest to your
servants,
and your glorious power to
their children.
17 Let the favor of the Lord our God
be upon us,
and prosper for us the work of
our hands—
O prosper the work of our
hands!

Psalm 91

1 You who live in the shelter of the
Most High,
who abide in the shadow of the
Almighty,[h]

---

d Another reading is *our refuge*
e Heb *humankind*   f Syr: Heb *we bring our
years to an end*   g Cn Compare Gk Syr Jerome
Tg: Heb *pride*   h Traditional rendering of
Heb *Shaddai*

---

eternity and the transience of human life
(compare Isa 40.6–8). **4:** *Watch,* see 63.6 n.
7–10: Brevity and sorrow of human existence.

90.11–12: Prayer that people may learn
wisdom from considering this. **13–17:** Prayer
that Israel may be delivered from its difficulties.

2 will say to the LORD, "My refuge
and my fortress;
my God, in whom I trust."
3 For he will deliver you from the
snare of the fowler
and from the deadly pestilence;
4 he will cover you with his
pinions,
and under his wings you will
find refuge;
his faithfulness is a shield and
buckler.
5 You will not fear the terror of the
night,
or the arrow that flies by day,
6 or the pestilence that stalks in
darkness,
or the destruction that wastes at
noonday.

7 A thousand may fall at your side,
ten thousand at your right hand,
but it will not come near you.
8 You will only look with your eyes
and see the punishment of the
wicked.

9 Because you have made the LORD
your refuge, *i*
the Most High your dwelling
place,
10 no evil shall befall you,
no scourge come near your tent.

11 For he will command his angels
concerning you
to guard you in all your ways.
12 On their hands they will bear
you up,
so that you will not dash your
foot against a stone.

13 You will tread on the lion and the
adder,
the young lion and the serpent
you will trample under
foot.

14 Those who love me, I will deliver;
I will protect those who know
my name.
15 When they call to me, I will
answer them;
I will be with them in trouble,
I will rescue them and honor
them.
16 With long life I will satisfy them,
and show them my salvation.

Psalm 92

A Psalm. A Song for the Sabbath Day.

1 It is good to give thanks to the
LORD,
to sing praises to your name,
O Most High;
2 to declare your steadfast love in
the morning,
and your faithfulness by night,
3 to the music of the lute and the
harp,
to the melody of the lyre.
4 For you, O LORD, have made me
glad by your work;
at the works of your hands I
sing for joy.

5 How great are your works,
O LORD!
Your thoughts are very deep!
6 The dullard cannot know,

i Cn: Heb *Because you,* LORD, *are my refuge; you
have made*

Ps 91: **Meditation on God as the protector of the faithful** (a wisdom psalm).
1–13: Those who trust in the LORD need have
no fear of any peril (121.2–8). Some of the
dangers mentioned are human foes; others are
demonic forces.
91.11–12: Quoted in Mt 4.6; Lk 4.10–11.
14–16: A divine oracle of assurance, probably
spoken by a priest or temple prophet.
Ps 92: **Thanksgiving after deliverance**
from personal enemies. **1–3**: Hymn-like introduction. **2**: *Morning . . . night,* compare the
morning and evening sacrifices (Ex 29.38–
42). **4**: The occasion (described more explicitly in vv. 10–11) is a desire to give thanks for
an answer to the psalmist's prayer for help.
92.5–9: The mystery and power of God.
10–11: The psalmist exults at being vindicated. *Horn,* see 75.4–5 n.

the stupid cannot understand
this:
7 though the wicked sprout like
grass
and all evildoers flourish,
they are doomed to destruction
forever,
8 but you, O LORD, are on high
forever.
9 For your enemies, O LORD,
for your enemies shall perish;
all evildoers shall be scattered.

10 But you have exalted my horn like
that of the wild ox;
you have poured over me[j]
fresh oil.
11 My eyes have seen the downfall of
my enemies;
my ears have heard the doom of
my evil assailants.

12 The righteous flourish like the
palm tree,
and grow like a cedar in
Lebanon.
13 They are planted in the house of
the LORD;
they flourish in the courts of
our God.
14 In old age they still produce fruit;
they are always green and full
of sap,
15 showing that the LORD is upright;
he is my rock, and there is no
unrighteousness in him.

Psalm 93

1 The LORD is king, he is robed in
majesty;

the LORD is robed, he is girded
with strength.
He has established the world; it
shall never be moved;
2 your throne is established from
of old;
you are from everlasting.

3 The floods have lifted up,
O LORD,
the floods have lifted up their
voice;
the floods lift up their roaring.
4 More majestic than the thunders
of mighty waters,
more majestic than the waves[k]
of the sea,
majestic on high is the LORD!

5 Your decrees are very sure;
holiness befits your house,
O LORD, forevermore.

Psalm 94

1 O LORD, you God of vengeance,
you God of vengeance, shine
forth!
2 Rise up, O judge of the earth;
give to the proud what they
deserve!
3 O LORD, how long shall the
wicked,
how long shall the wicked
exult?

4 They pour out their arrogant
words;

j Syr: Meaning of Heb uncertain   *k* Cn: Heb
majestic are the waves

92.12–15: The rewards of righteousness
(1.1–3). **13:** Compare 52.8.
Ps 93: Hymn extolling God as king.
This psalm begins a collection of hymns (Pss
93; 95–99) dealing with the kingly rule of the
God of Israel. They were apparently composed
for use in connection with some festival, perhaps the festival of booths, when the theme
of God's kingship (or "kingdom") was especially emphasized. Compare Ps 47, which is
closely related. **1–2:** The theme is stated. **3–4:**
God's rule is based upon his control over the

powers of chaos, symbolized by the waters of
the sea (74.12–17; 104.7–9; Job 38.8–11). **5:**
God is praised because his law offers dependable guidance and his temple is holy.
**Ps 94: A prayer for deliverance from
evil people** (a lament). Although this seems
to be originally the prayer of an individual
(note vv. 16–23), the enemies are described in
such general and indefinite terms (vv. 4–7,
20–21) that it was easily adapted to congregational use. **1–3:** Cry for help.
94.4–7: The condition of society. The

all the evildoers boast.
5 They crush your people, O LORD,
and afflict your heritage.
6 They kill the widow and the
stranger,
they murder the orphan,
7 and they say, "The LORD does
not see;
the God of Jacob does not
perceive."

8 Understand, O dullest of the
people;
fools, when will you be wise?
9 He who planted the ear, does he
not hear?
He who formed the eye, does he
not see?
10 He who disciplines the nations,
he who teaches knowledge to
humankind,
does he not chastise?
11 The LORD knows our thoughts,[1]
that they are but an empty
breath.

12 Happy are those whom you
discipline, O LORD,
and whom you teach out of
your law,
13 giving them respite from days of
trouble,
until a pit is dug for the wicked.
14 For the LORD will not forsake his
people;
he will not abandon his heritage;
15 for justice will return to the
righteous,
and all the upright in heart will
follow it.

16 Who rises up for me against the
wicked?

Who stands up for me against
evildoers?
17 If the LORD had not been my help,
my soul would soon have lived
in the land of silence.
18 When I thought, "My foot is
slipping,"
your steadfast love, O LORD,
held me up.
19 When the cares of my heart are
many,
your consolations cheer my
soul.
20 Can wicked rulers be allied
with you,
those who contrive mischief by
statute?
21 They band together against the life
of the righteous,
and condemn the innocent to
death.
22 But the LORD has become my
stronghold,
and my God the rock of my
refuge.
23 He will repay them for their
iniquity
and wipe them out for their
wickedness;
the LORD our God will wipe
them out.

Psalm 95

1 O come, let us sing to the LORD;
let us make a joyful noise to the
rock of our salvation!
2 Let us come into his presence with
thanksgiving;
let us make a joyful noise to
him with songs of praise!

---

1 Heb *the thoughts of humankind*

---

psalmist sees personal misfortunes as typical of the corruption of the age (compare 10.2–11; Ps 14). **94.8–15:** In the style of the wisdom writers, the psalmist appeals to the congregation to understand the ways of the LORD. **94.16:** Prayer for deliverance in the form of a rhetorical question. **17–23:** An expression of confidence, perhaps in response to an oracle of assurance delivered after v. 16 (see 12.5 n.).

17: *Land of silence,* Sheol (compare 115.17). **20–21:** The psalmist's enemies were persons in high official position in the state. **Ps 95: A liturgy of God's kingship.** This brief outline of a service opens with a hymn celebrating God's kingship (vv. 1–7a; see Ps 93 n.) and closes with an oracle, delivered by a priest or temple prophet, warning the congregation against disobeying God's laws (vv. 7b–11; compare 81.6–16). **1–2:**

3 For the Lord is a great God,
   and a great King above all gods.
4 In his hand are the depths of the
      earth;
   the heights of the mountains are
      his also.
5 The sea is his, for he made it,
   and the dry land, which his
      hands have formed.

6 O come, let us worship and bow
      down,
   let us kneel before the Lord,
      our Maker!
7 For he is our God,
   and we are the people of his
      pasture,
   and the sheep of his hand.

O that today you would listen to
      his voice!
8   Do not harden your hearts, as at
      Meribah,
   as on the day at Massah in the
      wilderness,
9 when your ancestors tested me,
   and put me to the proof, though
      they had seen my work.
10 For forty years I loathed that
      generation
   and said, "They are a people
      whose hearts go astray,
   and they do not regard my
      ways."
11 Therefore in my anger I swore,
   "They shall not enter my rest."

Psalm 96

1 O sing to the Lord a new song;
   sing to the Lord, all the earth.
2 Sing to the Lord, bless his name;

tell of his salvation from day
      to day.
3 Declare his glory among the
      nations,
   his marvelous works among all
      the peoples.
4 For great is the Lord, and greatly
      to be praised;
   he is to be revered above
      all gods.
5 For all the gods of the peoples are
      idols,
   but the Lord made the heavens.
6 Honor and majesty are before
      him;
   strength and beauty are in his
      sanctuary.

7 Ascribe to the Lord, O families of
      the peoples,
   ascribe to the Lord glory and
      strength.
8 Ascribe to the Lord the glory due
      his name;
   bring an offering, and come into
      his courts.
9 Worship the Lord in holy
      splendor;
   tremble before him, all the
      earth.

10 Say among the nations, "The
      Lord is king!
   The world is firmly established;
      it shall never be moved.
   He will judge the peoples with
      equity."
11 Let the heavens be glad, and let
      the earth rejoice;
   let the sea roar, and all that
      fills it;
12   let the field exult, and
      everything in it.

Summons to worship. **3–5:** God's rule based
upon the fact that he created the world. **6–7a:**
Renewed summons to worship.
   95.7b–11: Prophetic warning against disobedience. Quoted in Heb 3.7–11; 4.3–11.
The declaration that worship without obedience is displeasing to God is a cardinal principle of Old Testament religion (Pss 15; 24.3–6;
see 40.6 n.). **8:** *Meribah . . . Massah,* Ex 17.1–

7; Num 20.1–13; compare 106.32–33. **10–11:**
Num 14.33. *My rest,* the Promised Land.
   Ps 96: Hymn celebrating God's kingship. 1–3: Summons to worship. **4–6:** The
Lord is the powerful creator; all other gods
are only powerless images (115.3–8).
   96.7–13: Summons to all nations and to
the physical universe to join in his praise. **9:**
Holy splendor means ceremonial garments. **10:**

Then shall all the trees of the
forest sing for joy
13 before the LORD; for he is
coming,
for he is coming to judge the
earth.
He will judge the world with
righteousness,
and the peoples with his truth.

Psalm 97

1 The LORD is king! Let the earth
rejoice;
let the many coastlands be glad!
2 Clouds and thick darkness are all
around him;
righteousness and justice are the
foundation of his throne.
3 Fire goes before him,
and consumes his adversaries on
every side.
4 His lightnings light up the world;
the earth sees and trembles.
5 The mountains melt like wax
before the LORD,
before the Lord of all the earth.

6 The heavens proclaim his
righteousness;
and all the peoples behold his
glory.
7 All worshipers of images are put
to shame,
those who make their boast in
worthless idols;
all gods bow down before him.
8 Zion hears and is glad,
and the towns *m* of Judah rejoice,
because of your judgments,
O God.

9 For you, O LORD, are most high
over all the earth;
you are exalted far above
all gods.

10 The LORD loves those who hate *n*
evil;
he guards the lives of his
faithful;
he rescues them from the hand
of the wicked.
11 Light dawns *o* for the righteous,
and joy for the upright in heart.
12 Rejoice in the LORD, O you
righteous,
and give thanks to his holy
name!

Psalm 98

A Psalm.

1 O sing to the LORD a new song,
for he has done marvelous
things.
His right hand and his holy arm
have gotten him victory.
2 The LORD has made known his
victory;
he has revealed his vindication
in the sight of the nations.
3 He has remembered his steadfast
love and faithfulness
to the house of Israel.
All the ends of the earth have seen
the victory of our God.

4 Make a joyful noise to the LORD,
all the earth;

m Heb *daughters*   *n* Cn: Heb *You who love the*
LORD *hate*   *o* Gk Syr Jerome: Heb *is sown*

The kingship of the God of Israel. **13**: Compare Ps 98.4–9.
    Ps 97: **Hymn celebrating God's kingship**. **1**: Summons to worship the LORD as king. **2–6**: The manifestation of the LORD (18.7–15; 50.1–3). **7**: Idolaters will realize their folly (96.5). **8–9**: Israel's faith will be vindicated.
    97.10–11: Expression of confidence in the LORD's justice. **12**: Renewed call to worship.
    Ps 98: **Hymn proclaiming the future establishment of God's kingship on the**

earth. 1: Summons to worship. **2–3**: The LORD's triumph over all the powers that oppose him. Although the verbs are past tense, the reference is to a future event (compare 76.3; also Pss 46; 47; 48.4–8).
    98.4–9: Summons to all nations and to the physical universe to join in God's praise (96.7–13). The language in this and other of the "kingship" psalms is strikingly similar to that of many passages in Isa chs 40–55 (e.g. 44.23; 52.10; 55.12).

break forth into joyous song
and sing praises.
5 Sing praises to the LORD with the lyre,
with the lyre and the sound of melody.
6 With trumpets and the sound of the horn
make a joyful noise before the King, the LORD.

7 Let the sea roar, and all that fills it;
the world and those who live in it.
8 Let the floods clap their hands;
let the hills sing together for joy
9 at the presence of the LORD, for he is coming
to judge the earth.
He will judge the world with righteousness,
and the peoples with equity.

Psalm 99

1 The LORD is king; let the peoples tremble!
He sits enthroned upon the cherubim; let the earth quake!
2 The LORD is great in Zion;
he is exalted over all the peoples.
3 Let them praise your great and awesome name.
Holy is he!
4 Mighty King,[p] lover of justice,
you have established equity;
you have executed justice
and righteousness in Jacob.
5 Extol the LORD our God;
worship at his footstool.
Holy is he!

6 Moses and Aaron were among his priests,
Samuel also was among those who called on his name.
They cried to the LORD, and he answered them.
7 He spoke to them in the pillar of cloud;
they kept his decrees,
and the statutes that he gave them.

8 O LORD our God, you answered them;
you were a forgiving God to them,
but an avenger of their wrongdoings.
9 Extol the LORD our God,
and worship at his holy mountain;
for the LORD our God is holy.

Psalm 100

A Psalm of thanksgiving.

1 Make a joyful noise to the LORD, all the earth.
2 Worship the LORD with gladness;
come into his presence with singing
3 Know that the LORD is God.
It is he that made us, and we are his;[q]
we are his people, and the sheep of his pasture.

4 Enter his gates with thanksgiving,
and his courts with praise.
Give thanks to him, bless his name.

p Cn: Heb *And a king's strength* *q* Another reading is *and not we ourselves*

Ps 99: Hymn celebrating God's kingship. 1–3: The LORD God of Israel is the ruler of the earth. **3:** *Holy is he!* is the refrain (see v. 5 and v. 9). **4–5:** God's concern for justice. **6–9:** His fidelity toward his people.

Ps 100: Hymn calling on all nations to praise the LORD. While this psalm does not refer specifically to God as king, its mood is similar to that of the preceding "kingship" psalms (see Ps 93 n.) and it may be taken as a doxology for the collection.

5 For the LORD is good;
   his steadfast love endures
      forever,
   and his faithfulness to all
      generations.

Psalm 101

Of David. A Psalm.

1 I will sing of loyalty and of
      justice;
   to you, O LORD, I will sing.
2 I will study the way that is
      blameless.
   When shall I attain it?

   I will walk with integrity of heart
      within my house;
3 I will not set before my eyes
      anything that is base.

   I hate the work of those who
      fall away;
   it shall not cling to me.
4 Perverseness of heart shall be far
      from me;
   I will know nothing of evil.

5 One who secretly slanders a
      neighbor
   I will destroy.
   A haughty look and an arrogant
      heart
   I will not tolerate.

6 I will look with favor on the
      faithful in the land,
   so that they may live with me;
   whoever walks in the way that is
      blameless
   shall minister to me.

7 No one who practices deceit
   shall remain in my house;
   no one who utters lies
   shall continue in my presence.

8 Morning by morning I will
      destroy
   all the wicked in the land,
   cutting off all evildoers
   from the city of the LORD.

Psalm 102

*A prayer of one afflicted, when faint and
pleading before the LORD.*

1 Hear my prayer, O LORD;
   let my cry come to you.
2 Do not hide your face from me
   in the day of my distress.
   Incline your ear to me;
   answer me speedily in the day
      when I call.

3 For my days pass away like
      smoke,
   and my bones burn like a
      furnace.
4 My heart is stricken and withered
      like grass;
   I am too wasted to eat my
      bread.
5 Because of my loud groaning
   my bones cling to my skin.
6 I am like an owl of the wilderness,
   like a little owl of the waste
      places.
7 I lie awake;
   I am like a lonely bird on the
      housetop.
8 All day long my enemies taunt
      me;
   those who deride me use my
      name for a curse.
9 For I eat ashes like bread,
   and mingle tears with my drink,
10 because of your indignation and
      anger;
   for you have lifted me up and
      thrown me aside.
11 My days are like an evening
      shadow;
   I wither away like grass.

Ps 101: **A king pledges himself to rule justly.** This psalm may have been used as part of a coronation ceremony.
Ps 102: **Prayer for healing in sickness** (a lament). **1–2:** Cry for help. **3–11** (and **23**): The psalmist's situation: tormented by pain and accused by enemies of being forsaken by God.

12 But you, O Lord, are enthroned
        forever;
    your name endures to all
        generations.
13 You will rise up and have
        compassion on Zion,
    for it is time to favor it;
    the appointed time has come.
14 For your servants hold its stones
        dear,
    and have pity on its dust.
15 The nations will fear the name of
        the Lord,
    and all the kings of the earth
        your glory.
16 For the Lord will build up Zion;
    he will appear in his glory.
17 He will regard the prayer of the
        destitute,
    and will not despise their
        prayer.

18 Let this be recorded for a
        generation to come,
    so that a people yet unborn may
        praise the Lord:
19 that he looked down from his
        holy height,
    from heaven the Lord looked at
        the earth,
20 to hear the groans of the
        prisoners,
    to set free those who were
        doomed to die;
21 so that the name of the Lord may
        be declared in Zion,
    and his praise in Jerusalem,
22 when peoples gather together,
    and kingdoms, to worship the
        Lord.

23 He has broken my strength in
        midcourse;
    he has shortened my days.
24 "O my God," I say, "do not take
        me away
    at the mid-point of my life,
    you whose years endure
        throughout all generations."

25 Long ago you laid the foundation
        of the earth,
    and the heavens are the work of
        your hands.
26 They will perish, but you endure;
    they will all wear out like a
        garment.
    You change them like clothing,
        and they pass away;
27   but you are the same, and your
        years have no end.
28 The children of your servants shall
        live secure;
    their offspring shall be
        established in your
        presence.

Psalm 103

Of David.

1 Bless the Lord, O my soul,
    and all that is within me,
    bless his holy name.
2 Bless the Lord, O my soul,
    and do not forget all his
        benefits—
3 who forgives all your iniquity,
    who heals all your diseases,
4 who redeems your life from
        the Pit,
    who crowns you with steadfast
        love and mercy,

102.12–22: A hymn of praise to God; the words are perhaps quoted from a familiar song of Zion anticipating the future glory of Jerusalem. **24:** A prayer.
102.25–28: The psalmist concludes by again singing a fragment of a hymn, this time one in praise of God's eternity. **25–27:** Quoted in Heb 1.10–12. **27:** *You are the same,* literally, "You are he," compare Isa 43.10, 13, 25.
Ps 103: Thanksgiving for recovery

from sickness. This might be classified as a hymn, but vv. 1–5 suggest that the words, though general, are intended to express the emotion of a particular individual on a specific occasion. **1–5:** The whole being of the psalmist is summoned to praise God for what he has done; v. 3 and v. 4a show that the reference is to physical healing. **3:** *Forgives all your iniquity,* see 32.1 n. *The Pit,* see 16.10 n. *Your,* the psalmist's. An inner monologue,

5 who satisfies you with good as
   long as you live[r]
      so that your youth is renewed
      like the eagle's.

6 The LORD works vindication
      and justice for all who are
      oppressed.
7 He made known his ways to
   Moses,
      his acts to the people of Israel.
8 The LORD is merciful and
   gracious,
      slow to anger and abounding in
      steadfast love.
9 He will not always accuse,
      nor will he keep his anger
      forever.
10 He does not deal with us
      according to our sins,
      nor repay us according to our
      iniquities.
11 For as the heavens are high above
   the earth,
      so great is his steadfast love
      toward those who fear him;
12 as far as the east is from the west,
      so far he removes our
      transgressions from us.
13 As a father has compassion for his
   children,
      so the LORD has compassion for
      those who fear him.
14 For he knows how we were made;
      he remembers that we are dust.

15 As for mortals, their days are like
   grass;
      they flourish like a flower of the
      field;
16 for the wind passes over it, and it
   is gone,

and its place knows it no more.
17 But the steadfast love of the LORD
      is from everlasting to
      everlasting
      on those who fear him,
      and his righteousness to
      children's children,
18 to those who keep his covenant
      and remember to do his
      commandments.

19 The LORD has established his
      throne in the heavens,
      and his kingdom rules over all.
20 Bless the LORD, O you his angels,
      you mighty ones who do his
      bidding,
      obedient to his spoken word.
21 Bless the LORD, all his hosts,
      his ministers that do his will.
22 Bless the LORD, all his works,
      in all places of his dominion.
      Bless the LORD, O my soul.

Psalm 104

1 Bless the LORD, O my soul.
      O LORD my God, you are very
      great.
   You are clothed with honor and
      majesty,
2   wrapped in light as with a
      garment.
   You stretch out the heavens like
      a tent,
3   you set the beams of your[s]
         chambers on the waters,
   you make the clouds your[s]
      chariot,
      you ride on the wings of the
      wind,

r Meaning of Heb uncertain   s Heb *his*

---

the self ("soul") speaking to the self. **5:** *Like
the eagle's,* the vigor of the eagle was proverbial (Isa 40.31).
   103.6–18: The being of God, in his justice,
love, and eternity, compared with the frail
and transient nature of human beings. **15:**
Compare Isa 40.6–8.
   103.19–22: Conclusion, with a hymn-like

summons to all created things to join their
voices with that of the psalmist.
   Ps 104: Hymn to God the creator. 1: Address. The first line was probably added in
imitation of the preceding psalm (compare
also v. 35). **2–4:** The creation of the heavens.
The account is much more mythological than
that of Gen ch 1. **2:** *Light,* compare Gen 1.3;

⁴ you make the winds your¹
    messengers,
    fire and flame your¹ ministers.

⁵ You set the earth on its
    foundations,
    so that it shall never be shaken.
⁶ You cover it with the deep as
    with a garment;
    the waters stood above the
    mountains.
⁷ At your rebuke they flee;
    at the sound of your thunder
    they take to flight.
⁸ They rose up to the mountains,
    ran down to the valleys
    to the place that you appointed
    for them.
⁹ You set a boundary that they may
    not pass,
    so that they might not again
    cover the earth.

¹⁰ You make springs gush forth in
    the valleys;
    they flow between the hills,
¹¹ giving drink to every wild animal;
    the wild asses quench their
    thirst.
¹² By the streams" the birds of the
    air have their habitation;
    they sing among the branches.
¹³ From your lofty abode you water
    the mountains;
    the earth is satisfied with the
    fruit of your work.

¹⁴ You cause the grass to grow for
    the cattle,
    and plants for people to use,ᵛ
    to bring forth food from the earth,
¹⁵   and wine to gladden the human
    heart,
    oil to make the face shine,

and bread to strengthen the
    human heart.
¹⁶ The trees of the Lord are watered
    abundantly,
    the cedars of Lebanon that he
    planted.
¹⁷ In them the birds build their nests;
    the stork has its home in the fir
    trees.
¹⁸ The high mountains are for the
    wild goats;
    the rocks are a refuge for the
    coneys.
¹⁹ You have made the moon to mark
    the seasons;
    the sun knows its time for
    setting.
²⁰ You make darkness, and it is
    night,
    when all the animals of the
    forest come creeping out.
²¹ The young lions roar for their
    prey,
    seeking their food from God.
²² When the sun rises, they withdraw
    and lie down in their dens.
²³ People go out to their work
    and to their labor until the
    evening.

²⁴ O Lord, how manifold are your
    works!
    In wisdom you have made
    them all;
    the earth is full of your
    creatures.
²⁵ Yonder is the sea, great and wide,
    creeping things innumerable are
    there,
    living things both small and
    great.
²⁶ There go the ships,

t Heb *his*   *u* Heb *By them*   *v* Or *to cultivate*

---

1 Tim 6.16. **3**: See 18.10 n. **4**: Quoted in Heb
1.7.
    104.5–9: The creation of the earth. **7–9**:
God's victory over the waters, the symbol of
chaos (see 93.3–4 n.).
    104.10–18: God's continuing care for the
earth and its inhabitants, animal and human.

19: God's careful demarcation of the limits of
months and days. **24**: The poet's exclamation
of praise and admiration. **25–26**: The wonders of the sea. **26**: The monster *Leviathan* (see
74.12–17 n.) has become, for this author,
merely a harmless, sportive creature of God.

and Leviathan that you formed
to sport in it.

27 These all look to you
to give them their food in due
season;
28 when you give to them, they
gather it up;
when you open your hand, they
are filled with good things.
29 When you hide your face, they are
dismayed;
when you take away their
breath, they die
and return to their dust.
30 When you send forth your spirit, *w*
they are created;
and you renew the face of the
ground.

31 May the glory of the LORD endure
forever;
may the LORD rejoice in his
works—
32 who looks on the earth and it
trembles,
who touches the mountains and
they smoke.
33 I will sing to the LORD as long as
I live;
I will sing praise to my God
while I have being.
34 May my meditation be pleasing
to him,
for I rejoice in the LORD.
35 Let sinners be consumed from the
earth,
and let the wicked be no more.
Bless the LORD, O my soul.
Praise the LORD!

Psalm 105

1 O give thanks to the LORD, call on
his name,
make known his deeds among
the peoples.
2 Sing to him, sing praises to him;
tell of all his wonderful works.
3 Glory in his holy name;
let the hearts of those who seek
the LORD rejoice.
4 Seek the LORD and his strength;
seek his presence continually.
5 Remember the wonderful works
he has done,
his miracles, and the judgments
he uttered,
6 O offspring of his servant
Abraham, *x*
children of Jacob, his chosen
ones.

7 He is the LORD our God;
his judgments are in all the
earth.
8 He is mindful of his covenant
forever,
of the word that he
commanded, for a thousand
generations,
9 the covenant that he made with
Abraham,
his sworn promise to Isaac,
10 which he confirmed to Jacob as a
statute,
to Israel as an everlasting
covenant,

w Or *your breath*   *x* Another reading is *Israel*
(compare 1 Chr 16.13)

104.27–30: All living things depend on
God at every moment for their existence.
104.31: Concluding act of praise. **35:** The
psalmist prays for the restoration of the original, intended harmony of creation. Ancient
Hebrew culture tended to use personal and
concrete words (*sinners . . . wicked*) where we
should use the impersonal and abstract ("sin,"
"wickedness"). *Praise the LORD!* (Hebrew
"Hallelujah!") belongs to the beginning of the
following psalm (see 105 n.).
**Ps 105: The story of God's great deeds
on behalf of his people** (compare Ps 78).

This psalm, now paired with Ps 106, was
composed for use at one of the major festivals
and consists of a recital of the basic events
which created the nation of Israel; its mood is
one of grateful recollection. Both 105 and 106
at some point in their history were provided
with the ritual shout "Hallelujah!" (*Praise the
LORD!*) at the beginning and end (see
104.35 n.). **1–6:** Hymn-like introduction
summoning the congregation to praise and
thanksgiving.
105.7–11: The psalmist's theme: God's
faithfulness to his covenant with Israel's an-

11 saying, "To you I will give the
    land of Canaan
    as your portion for an
      inheritance."

12 When they were few in number,
    of little account, and strangers
      in it,
13 wandering from nation to nation,
    from one kingdom to another
      people,
14 he allowed no one to oppress
    them;
    he rebuked kings on their
      account,
15 saying, "Do not touch my
    anointed ones;
    do my prophets no harm."

16 When he summoned famine
    against the land,
    and broke every staff of bread,
17 he had sent a man ahead of them,
    Joseph, who was sold as a slave.
18 His feet were hurt with fetters,
    his neck was put in a collar of
      iron;
19 until what he had said came to
    pass,
    the word of the LORD kept
      testing him.
20 The king sent and released him;
    the ruler of the peoples set
      him free.
21 He made him lord of his house,
    and ruler of all his possessions,
22 to instruct[y] his officials at his
    pleasure,
    and to teach his elders wisdom.

23 Then Israel came to Egypt;
    Jacob lived as an alien in the
      land of Ham.
24 And the LORD made his people
    very fruitful,

    and made them stronger than
      their foes,
25 whose hearts he then turned to
    hate his people,
    to deal craftily with his servants.

26 He sent his servant Moses,
    and Aaron whom he had
      chosen.
27 They performed his signs among
    them,
    and miracles in the land of
      Ham.
28 He sent darkness, and made the
    land dark;
    they rebelled[z] against his
      words.
29 He turned their waters into blood,
    and caused their fish to die.
30 Their land swarmed with frogs,
    even in the chambers of their
      kings.
31 He spoke, and there came swarms
    of flies,
    and gnats throughout their
      country.
32 He gave them hail for rain,
    and lightning that flashed
      through their land.
33 He struck their vines and fig trees,
    and shattered the trees of their
      country.
34 He spoke, and the locusts came,
    and young locusts without
      number;
35 they devoured all the vegetation in
    their land,
    and ate up the fruit of their
      ground.
36 He struck down all the firstborn in
    their land,
    the first issue of all their
      strength.

y Gk Syr Jerome: Heb *to bind*   *z* Cn Compare
Gk Syr: Heb *they did not rebel*

---

cestors (Gen 15.18–21; 17.1–8; 26.1–5;
28.10–15). **12–41:** The narrative.
105.12–15: The story of Abraham, Isaac,
and Jacob. **14:** *Rebuked kings*, Gen 12.17; 20.3.
15: Only here are the most ancient ancestors
of Israel given the title *anointed ones*, presum-

ably because they are called *prophets* (compare
Gen 20.7; 1 Kings 19.16).
    105.16–22: The story of Joseph (Gen
chs 37; 39–50).
    105.23–38: The story of the Exodus (Ex
1.1–15.21). **27:** *Ham*, see 78.51 n.

37 Then he brought Israel*ᵃ* out with
　　silver and gold,
　　and there was no one among
　　　their tribes who stumbled.
38 Egypt was glad when they
　　departed,
　　for dread of them had fallen
　　　upon it.
39 He spread a cloud for a covering,
　　and fire to give light by night.
40 They asked, and he brought
　　quails,
　　and gave them food from
　　　heaven in abundance.
41 He opened the rock, and water
　　gushed out;
　　it flowed through the desert like
　　　a river.
42 For he remembered his holy
　　promise,
　　and Abraham, his servant.

43 So he brought his people out
　　with joy,
　　his chosen ones with singing.
44 He gave them the lands of the
　　nations,
　　and they took possession of the
　　　wealth of the peoples,
45 that they might keep his statutes
　　and observe his laws.
　　Praise the Lord!

Psalm 106

1 Praise the Lord!
　　O give thanks to the Lord, for
　　　he is good;
　　for his steadfast love endures
　　　forever.
2 Who can utter the mighty doings
　　of the Lord,

or declare all his praise?
3 Happy are those who observe
　　justice,
　　who do righteousness at all
　　　times.

4 Remember me, O Lord, when
　　you show favor to your
　　　people;
　　help me when you deliver them;
5 that I may see the prosperity of
　　your chosen ones,
　　that I may rejoice in the
　　　gladness of your nation,
　　that I may glory in your
　　　heritage.

6 Both we and our ancestors have
　　sinned;
　　we have committed iniquity,
　　have done wickedly.
7 Our ancestors, when they were in
　　Egypt,
　　did not consider your wonderful
　　　works;
　　they did not remember the
　　　abundance of your steadfast
　　　love,
　　but rebelled against the Most
　　　High*ᵇ* at the Red Sea.*ᶜ*
8 Yet he saved them for his name's
　　sake,
　　so that he might make known
　　　his mighty power.
9 He rebuked the Red Sea,*ᶜ* and it
　　became dry;
　　he led them through the deep as
　　　through a desert.
10 So he saved them from the hand
　　of the foe,

a Heb *them*　　*b* Cn Compare 78.17, 56: Heb
rebelled at the sea　　*c* Or *Sea of Reeds*

**Ps 106: The story of God's great deeds,
with confession of sin and prayer for
help.** In contrast to Ps 105, the mood of this
psalm is a somber one, stress being laid on the
perversity and obtuseness of the people. **1–2:**
Call to praise the Lord. **3:** The need for right-
doing on the part of his worshipers (105.45).
4–5: The psalmist prays for a share in Israel's
blessings when God restores prosperity to it
(compare v. 47).

106.6: The theme: Whatever God has
done, Israel has always been unfaithful (vv. 7,
13–14, 19, 21, 24–25, 28–29, 32, 34–39, 43);
nevertheless he has constantly forgiven its
disobedience and shown mercy (vv. 8, 15, 23,
30, 44–46). This thought encourages the
psalmist to offer the prayer in v. 47. **7–46:**
The narrative.
106.7–12: The story of the Exodus
(105.23–38). **13–33:** Incidents in the wilder-

and delivered them from the hand of the enemy.
11 The waters covered their adversaries;
not one of them was left.
12 Then they believed his words;
they sang his praise.

13 But they soon forgot his works;
they did not wait for his counsel.
14 But they had a wanton craving in the wilderness,
and put God to the test in the desert;
15 he gave them what they asked,
but sent a wasting disease among them.

16 They were jealous of Moses in the camp,
and of Aaron, the holy one of the LORD.
17 The earth opened and swallowed up Dathan,
and covered the faction of Abiram.
18 Fire also broke out in their company;
the flame burned up the wicked.

19 They made a calf at Horeb
and worshiped a cast image.
20 They exchanged the glory of God[d]
for the image of an ox that eats grass.
21 They forgot God, their Savior,
who had done great things in Egypt,
22 wondrous works in the land of Ham,
and awesome deeds by the Red Sea.[e]

23 Therefore he said he would destroy them—
had not Moses, his chosen one,
stood in the breach before him,
to turn away his wrath from destroying them.

24 Then they despised the pleasant land,
having no faith in his promise.
25 They grumbled in their tents,
and did not obey the voice of the LORD.
26 Therefore he raised his hand and swore to them
that he would make them fall in the wilderness,
27 and would disperse[f] their descendants among the nations,
scattering them over the lands.

28 Then they attached themselves to the Baal of Peor,
and ate sacrifices offered to the dead;
29 they provoked the LORD to anger with their deeds,
and a plague broke out among them.
30 Then Phinehas stood up and interceded,
and the plague was stopped.
31 And that has been reckoned to him as righteousness
from generation to generation forever.

32 They angered the LORD[g] at the waters of Meribah,

d Compare Gk Mss: Heb *exchanged their glory*
e Or *Sea of Reeds*   *f* Syr Compare Ezek 20.23:
Heb *cause to fall*   *g* Heb *him*

---

ness wanderings and at Sinai. **13–15:** The manna and quails (Num 11.4–6, 31–35). **16–18:** Dathan and Abiram's rebellion (Num ch 16). **106.19–23:** The golden *calf* (Ex ch 32). **19:** *Horeb*, alternative name for Mount Sinai. **24–27:** The refusal to enter Canaan (Num 14.1–35). **106.28–31:** Apostasy to *the Baal of Peor* (Num 25.1–13). **32–33:** *The waters of Meribah,* Num 20.2–13 (compare Ex 17.1–7).

and it went ill with Moses on
    their account;
33 for they made his spirit bitter,
    and he spoke words that were
      rash.

34 They did not destroy the peoples,
    as the LORD commanded them,
35 but they mingled with the nations
    and learned to do as they did.
36 They served their idols,
    which became a snare to them.
37 They sacrificed their sons
    and their daughters to the
      demons;
38 they poured out innocent blood,
    the blood of their sons and
      daughters,
    whom they sacrificed to the idols
      of Canaan;
    and the land was polluted with
      blood.
39 Thus they became unclean by their
    acts,
    and prostituted themselves in
      their doings.

40 Then the anger of the LORD was
    kindled against his people,
    and he abhorred his heritage;
41 he gave them into the hand of the
    nations,
    so that those who hated them
      ruled over them.
42 Their enemies oppressed them,
    and they were brought into
      subjection under their
      power.
43 Many times he delivered them,
    but they were rebellious in their
      purposes,
    and were brought low through
      their iniquity.

44 Nevertheless he regarded their
    distress
    when he heard their cry.
45 For their sake he remembered his
    covenant,
    and showed compassion
      according to the abundance
      of his steadfast love.
46 He caused them to be pitied
    by all who held them captive.

47 Save us, O LORD our God,
    and gather us from among the
      nations,
    that we may give thanks to your
      holy name
    and glory in your praise.

48 Blessed be the LORD, the God of
    Israel,
    from everlasting to everlasting.
And let all the people say,
    "Amen."
Praise the LORD!

BOOK V

Psalm 107

1 O give thanks to the LORD, for he
    is good;
    for his steadfast love endures
      forever.
2 Let the redeemed of the LORD
    say so,
    those he redeemed from trouble
3 and gathered in from the lands,
    from the east and from the
      west,
    from the north and from the
      south. *h*

h Cn: Heb *sea*

106.34–39: Israel's apostasies in the days
of the judges (Judg 2.11–19).
106.47: Prayer that Israel may be delivered
from its present distress. *Gather us from among
the nations* suggests that the psalm, in its
present form at least, dates from after the
Babylonian exile. **48:** A doxology (no original part of the psalm) marking the end of
Book IV of the Psalter (see 41.13 n.). *Praise*

the LORD! has been illogically attached to this
verse rather than the preceding verse (see n.
at beginning of Ps 105).
**Ps 107: A group thanksgiving for pil-
grims.** This psalm was perhaps sung by
groups of pilgrims who came to Jerusalem to
celebrate one of the festivals, offering thanks
for escape from various dangers. **1–3:** Call for
everyone to give thanks. **4–9:** Thanksgiving

4 Some wandered in desert wastes,
   finding no way to an inhabited
     town;
5 hungry and thirsty,
   their soul fainted within them.
6 Then they cried to the LORD in
   their trouble,
  and he delivered them from
   their distress;
7 he led them by a straight way,
   until they reached an inhabited
    town.
8 Let them thank the LORD for his
   steadfast love,
  for his wonderful works to
   humankind.
9 For he satisfies the thirsty,
  and the hungry he fills with
   good things.

10 Some sat in darkness and in
   gloom,
  prisoners in misery and in irons,
11 for they had rebelled against the
   words of God,
  and spurned the counsel of the
   Most High.
12 Their hearts were bowed down
   with hard labor;
  they fell down, with no one
   to help.
13 Then they cried to the LORD in
   their trouble,
  and he saved them from their
   distress;
14 he brought them out of darkness
   and gloom,
  and broke their bonds asunder.
15 Let them thank the LORD for his
   steadfast love,
  for his wonderful works to
   humankind.
16 For he shatters the doors of
   bronze,
  and cuts in two the bars of iron.

17 Some were sick*i* through their
   sinful ways,
  and because of their iniquities
   endured affliction;
18 they loathed any kind of food,
  and they drew near to the gates
   of death.
19 Then they cried to the LORD in
   their trouble,
  and he saved them from their
   distress;
20 he sent out his word and healed
   them,
  and delivered them from
   destruction.
21 Let them thank the LORD for his
   steadfast love,
  for his wonderful works to
   humankind.
22 And let them offer thanksgiving
   sacrifices,
  and tell of his deeds with songs
   of joy.

23 Some went down to the sea
   in ships,
  doing business on the mighty
   waters;
24 they saw the deeds of the LORD,
  his wondrous works in the
   deep.
25 For he commanded and raised the
   stormy wind,
  which lifted up the waves of
   the sea.
26 They mounted up to heaven, they
   went down to the depths;
  their courage melted away in
   their calamity;
27 they reeled and staggered like
   drunkards,
  and were at their wits' end.
28 Then they cried to the LORD in
   their trouble,

i Cn: Heb *fools*

---

of those who traveled across the desert. **8–9:**
A refrain (see vv. 15–16, 21–22, 31–32), appropriately varied for each occasion.
  107.10–16: Thanksgiving of those who
had been freed from prison.

107.17–22: Thanksgiving of those healed
from sickness. **22:** *Thanksgiving sacrifices,* see
7.17 n.
  107.23–32: Thanksgiving of those who
had traveled by sea.

and he brought them out from
their distress;

29 he made the storm be still,
and the waves of the sea were
hushed.

30 Then they were glad because they
had quiet,
and he brought them to their
desired haven.

31 Let them thank the LORD for his
steadfast love,
for his wonderful works to
humankind.

32 Let them extol him in the
congregation of the people,
and praise him in the assembly
of the elders.

33 He turns rivers into a desert,
springs of water into thirsty
ground,

34 a fruitful land into a salty waste,
because of the wickedness of its
inhabitants.

35 He turns a desert into pools of
water,
a parched land into springs of
water.

36 And there he lets the hungry live,
and they establish a town to
live in;

37 they sow fields, and plant
vineyards,
and get a fruitful yield.

38 By his blessing they multiply
greatly,
and he does not let their cattle
decrease.

39 When they are diminished and
brought low
through oppression, trouble,
and sorrow,

40 he pours contempt on princes
and makes them wander in
trackless wastes;

41 but he raises up the needy out of
distress,
and makes their families like
flocks.

42 The upright see it and are glad;
and all wickedness stops its
mouth.

43 Let those who are wise give heed
to these things,
and consider the steadfast love
of the LORD.

Psalm 108

A Song. A Psalm of David.

1 My heart is steadfast, O God, my
heart is steadfast;[j]
I will sing and make melody.
Awake, my soul![k]

2 Awake, O harp and lyre!
I will awake the dawn.

3 I will give thanks to you,
O LORD, among the
peoples,
and I will sing praises to you
among the nations.

4 For your steadfast love is higher
than the heavens,
and your faithfulness reaches to
the clouds.

5 Be exalted, O God, above the
heavens,
and let your glory be over all
the earth.

6 Give victory with your right
hand, and answer me,
so that those whom you love
may be rescued.

7 God has promised in his
sanctuary:[l]

j Heb Mss Gk Syr: MT lacks *my heart is steadfast*
k Compare 57.8: Heb *also my soul*
l Or *by his holiness*

107.33–43: Part of a hymn praising God
for his bounty; it was probably not originally
composed to go with the preceding verses,
but nevertheless provides a suitable unison
conclusion for the thanksgiving liturgy.
Ps 108: A liturgy of prayer for victory
over national enemies. This liturgy is entirely composed of portions of two other
psalms. Verses 1–5 are practically identical
with 57.7–11; vv. 6–13, with 60.5–12. For
details of interpretation see those psalms.
Ps 109: Prayer for deliverance from

"With exultation I will divide
up Shechem,
and portion out the Vale of
Succoth.
8 Gilead is mine; Manasseh is mine;
Ephraim is my helmet;
Judah is my scepter.
9 Moab is my washbasin;
on Edom I hurl my shoe;
over Philistia I shout in
triumph."

10 Who will bring me to the fortified
city?
Who will lead me to Edom?
11 Have you not rejected us,
O God?
You do not go out, O God,
with our armies.
12 O grant us help against the foe,
for human help is worthless.
13 With God we shall do valiantly;
it is he who will tread down
our foes.

Psalm 109

To the leader. Of David. A Psalm.

1 Do not be silent, O God of my
praise.
2 For wicked and deceitful mouths
are opened against me,
speaking against me with lying
tongues.
3 They beset me with words of
hate,
and attack me without cause.
4 In return for my love they
accuse me,
even while I make prayer for
them.*m*
5 So they reward me evil for good,
and hatred for my love.

6 They say,*n* "Appoint a wicked
man against him;
let an accuser stand on his
right.
7 When he is tried, let him be found
guilty;
let his prayer be counted as sin.
8 May his days be few;
may another seize his position.
9 May his children be orphans,
and his wife a widow.
10 May his children wander about
and beg;
may they be driven out of*o* the
ruins they inhabit.
11 May the creditor seize all that
he has;
may strangers plunder the fruits
of his toil.
12 May there be no one to do him a
kindness,
nor anyone to pity his orphaned
children.
13 May his posterity be cut off;
may his name be blotted out in
the second generation.
14 May the iniquity of his father*p* be
remembered before the
LORD,
and do not let the sin of his
mother be blotted out.
15 Let them be before the LORD
continually,
and may his*q* memory be cut
off from the earth.
16 For he did not remember to show
kindness,
but pursued the poor and needy
and the brokenhearted to their
death.

m Syr: Heb *I prayer*   *n* Heb lacks *They say*
o Gk: Heb *and seek*   *p* Cn: Heb *fathers*
q Gk: Heb *their*

personal enemies (a lament). **1:** Cry for help.
2–5: The psalmist's situation. He has been
cursed and falsely accused of crime (compare
vv. 22–25). **4–5:** In view of the temper of
vv. 6–19, it is important to note that the writer believed love for others to be the proper
attitude and was accustomed to pray for
them.

109.6–19: The curse. It is clear that the
purpose of the violent and repellent language
of this section is to provide a counter-curse
which will be effective against the black-
magic curses of the psalmist's enemies (note
vv. 17–19). Compare Ps 58. **8:** The second
line is quoted in Acts 1.20, though translated
differently.

17 He loved to curse; let curses come
   on him.
   He did not like blessing; may it
   be far from him.
18 He clothed himself with cursing as
   his coat,
   may it soak into his body like
   water,
   like oil into his bones.
19 May it be like a garment that he
   wraps around himself,
   like a belt that he wears every
   day."

20 May that be the reward of my
   accusers from the LORD,
   of those who speak evil against
   my life.
21 But you, O LORD my Lord,
   act on my behalf for your
   name's sake;
   because your steadfast love is
   good, deliver me.
22 For I am poor and needy,
   and my heart is pierced
   within me.
23 I am gone like a shadow at
   evening;
   I am shaken off like a locust.
24 My knees are weak through
   fasting;
   my body has become gaunt.
25 I am an object of scorn to my
   accusers;
   when they see me, they shake
   their heads.

26 Help me, O LORD my God!
   Save me according to your
   steadfast love.

27 Let them know that this is your
   hand;
   you, O LORD, have done it.
28 Let them curse, but you will bless.
   Let my assailants be put to
   shame;[r] may your servant
   be glad.
29 May my accusers be clothed with
   dishonor;
   may they be wrapped in their
   own shame as in a mantle.
30 With my mouth I will give great
   thanks to the LORD;
   I will praise him in the midst of
   the throng.
31 For he stands at the right hand of
   the needy,
   to save them from those who
   would condemn them to
   death.

Psalm 110

Of David. A Psalm.

1 The LORD says to my lord,
   "Sit at my right hand
   until I make your enemies your
   footstool."

2 The LORD sends out from Zion
   your mighty scepter.
   Rule in the midst of your foes.
3 Your people will offer themselves
   willingly
   on the day you lead your forces
   on the holy mountains.[s]
   From the womb of the morning,

---

r Gk: Heb *They have risen up and have been put to*
shame s Another reading is *in holy splendor*

109.20–29: Prayer for deliverance. The
poet does not share the popular, primitive
belief that a curse would be automatically efficacious apart from the personal intervention
of God. 30–31: The vow (see 7.17 n.).
**Ps 110: The LORD promises victory to
his king** (a royal psalm, probably composed
for a coronation; compare Ps 2). The Hebrew
text is unusually corrupt and the interpretation of many details extremely difficult. 1:
The king (*my lord*) is invited by Israel's God
(*the LORD*) to ascend the throne. This verse is
quoted frequently in the New Testament
(e.g. Mt 22.44; Acts 2.34; 1 Cor 15.25; Eph
1.20; Heb 1.3, 13), where the later Jewish
belief that it was composed by David in hon-
or of the Messiah is naturally assumed to be
correct.
110.2–7: The new king, who also bears the
dignity of priesthood, will thoroughly defeat
his enemies. 3: The meaning of the second
sentence can no longer be recovered with certainty. 4: Like all early Israelite kings, the new
king will have the privileges of a priest (com-

like dew, your youth[t] will
   come to you.
4 The Lord has sworn and will not
   change his mind,
"You are a priest forever
   according to the order of
   Melchizedek."[u]

5 The Lord is at your right hand;
   he will shatter kings on the day
   of his wrath.
6 He will execute judgment among
   the nations,
   filling them with corpses;
he will shatter heads
   over the wide earth.
7 He will drink from the stream by
   the path;
   therefore he will lift up his
   head.

Psalm 111

1 Praise the Lord!
I will give thanks to the Lord
   with my whole heart,
in the company of the upright,
   in the congregation.
2 Great are the works of the Lord,
   studied by all who delight
   in them.
3 Full of honor and majesty is
   his work,
   and his righteousness endures
   forever.
4 He has gained renown by his
   wonderful deeds;
   the Lord is gracious and
   merciful.
5 He provides food for those who
   fear him;

he is ever mindful of his
   covenant.
6 He has shown his people the
   power of his works,
   in giving them the heritage of
   the nations.
7 The works of his hands are
   faithful and just;
   all his precepts are trustworthy.
8 They are established forever and
   ever,
   to be performed with
   faithfulness and uprightness.
9 He sent redemption to his people;
   he has commanded his covenant
   forever.
   Holy and awesome is his name.
10 The fear of the Lord is the
   beginning of wisdom;
   all those who practice it[v] have a
   good understanding.
   His praise endures forever.

Psalm 112

1 Praise the Lord!
   Happy are those who fear
   the Lord,
   who greatly delight in his
   commandments.
2 Their descendants will be mighty
   in the land;
   the generation of the upright
   will be blessed.
3 Wealth and riches are in their
   houses,
   and their righteousness endures
   forever.

t Cn: Heb *the dew of your youth*   *u* Or *forever,*
a rightful king by my edict   *v* Gk Syr: Heb *them*

---

pare 2 Sam 8.18; 1 Kings 3.4). In this respect
he will be like Melchizedek, the ancient Canaanite priest-king of Jerusalem (Gen 14.18).
7: The meaning is not clear. The verse may be
only a fragment.
   **Ps 111: Hymn of praise to the Lord for
his great deeds,** especially for his fidelity to
the covenant. An acrostic psalm, every line
beginning with a successive letter of the alphabet. **1:** Like the two following psalms, it
begins with the ritual cry, *Praise the Lord!*

(Hallelujah!); see 104.35 n. **10:** Job 28.28; Prov
1.7.
   **Ps 112: The contrasting fate of the righ-
teous and the wicked** (a wisdom psalm). An
alphabetical acrostic like Ps 111, and similar
in theme to Ps 1. It is, however, more concerned with the rewards of righteousness
(vv. 1–9) than the punishment of ungodliness
(v. 10). **1:** *Praise the Lord,* see 111.1 n. **9:**
Quoted in part in 2 Cor 9.9. *Horn,* see 75.4–
5 n.

4 They rise in the darkness as a light
    for the upright;
  they are gracious, merciful, and
    righteous.
5 It is well with those who deal
    generously and lend,
  who conduct their affairs with
    justice.
6 For the righteous will never
    be moved;
  they will be remembered
    forever.
7 They are not afraid of evil tidings;
  their hearts are firm, secure in
    the LORD.
8 Their hearts are steady, they will
    not be afraid;
  in the end they will look in
    triumph on their foes.
9 They have distributed freely, they
    have given to the poor;
  their righteousness endures
    forever;
  their horn is exalted in honor.
10 The wicked see it and are angry;
  they gnash their teeth and melt
    away;
  the desire of the wicked comes
    to nothing.

Psalm 113

1 Praise the LORD!
  Praise, O servants of the LORD;
    praise the name of the LORD.

2 Blessed be the name of the LORD
    from this time on and
    forevermore.
3 From the rising of the sun to its
    setting

the name of the LORD is to be
    praised.
4 The LORD is high above all
    nations,
  and his glory above the heavens.

5 Who is like the LORD our God,
    who is seated on high,
6 who looks far down
    on the heavens and the earth?
7 He raises the poor from the dust,
    and lifts the needy from the ash
      heap,
8 to make them sit with princes,
    with the princes of his people.
9 He gives the barren woman a
      home,
  making her the joyous mother
    of children.
Praise the LORD!

Psalm 114

1 When Israel went out from Egypt,
    the house of Jacob from a
    people of strange language,
2 Judah became God's*w* sanctuary,
    Israel his dominion.

3 The sea looked and fled;
    Jordan turned back.
4 The mountains skipped like rams,
    the hills like lambs.

5 Why is it, O sea, that you flee?
    O Jordan, that you turn back?
6 O mountains, that you skip like
      rams?
    O hills, like lambs?

w Heb *his*

Ps 113: **Hymn celebrating the LORD as
helper of the humble.** Another "Hallelujah!" psalm (see 111.1 n.). In Jewish liturgical
tradition, Pss 113–118 constitute the so-called
"Egyptian Hallel," used in connection with
the great festivals. At the Passover, Pss 113–
114 are sung before the meal; 115–118 afterwards (compare Mt 26.30). **1**: Summons to
worship. **2–4**: Choral response.

113.5–9: Though the LORD is gloriously
exalted, he cares for the needy (1 Sam 2.4–8;
Lk 1.48–53).
Ps 114: **Hymn in praise of God's great
work in creating the nation.** Although its
form is unusual, vv. 7–8 show this psalm
should be classified as a hymn (see Ps 113 n.).
1–2: The events of the Exodus recalled. **3–6**:
The remarkable natural phenomena which

7 Tremble, O earth, at the presence
of the LORD,
at the presence of the God of
Jacob,
8 who turns the rock into a pool of
water,
the flint into a spring of water.

Psalm 115

1 Not to us, O LORD, not to us, but
to your name give glory,
for the sake of your steadfast
love and your faithfulness.
2 Why should the nations say,
"Where is their God?"

3 Our God is in the heavens;
he does whatever he pleases.
4 Their idols are silver and gold,
the work of human hands.
5 They have mouths, but do not
speak;
eyes, but do not see.
6 They have ears, but do not hear;
noses, but do not smell.
7 They have hands, but do not feel;
feet, but do not walk;
they make no sound in their
throats.
8 Those who make them are like
them;
so are all who trust in them.

9 O Israel, trust in the LORD!
He is their help and their shield.
10 O house of Aaron, trust in the
LORD!
He is their help and their shield.

11 You who fear the LORD, trust in
the LORD!
He is their help and their shield.

12 The LORD has been mindful of us;
he will bless us;
he will bless the house of Israel;
he will bless the house of
Aaron;
13 he will bless those who fear the
LORD,
both small and great.

14 May the LORD give you increase,
both you and your children.
15 May you be blessed by the LORD,
who made heaven and earth.

16 The heavens are the LORD's
heavens,
but the earth he has given to
human beings.
17 The dead do not praise the LORD,
nor do any that go down into
silence.
18 But we will bless the LORD
from this time on and
forevermore.
Praise the LORD!

Psalm 116

1 I love the LORD, because he has
heard
my voice and my supplications.
2 Because he inclined his ear to me,
therefore I will call on him as
long as I live.
3 The snares of death encompassed
me;

---

accompanied Israel's crossing of the Red Sea
and the Jordan. The past events of vv. 3–4
become contemporary in vv. 5–6.
114.7–8: The physical world summoned
to worship the LORD. *Who turns the rock into
a pool,* Ex 17.6; Num 20.11.
**Ps 115: Liturgy contrasting the LORD's
power with the impotence of heathen
gods.** Perhaps sung antiphonally as follows:
1–2: A choir ascribes glory to God alone. **3–8:**
A soloist proclaims that God is omnipotent;
idols have no life at all.

115.9–11: The choir therefore asks Israel to
trust in the LORD. **10:** *House of Aaron,* the
priests. **12–13:** The congregation confidently
responds. **14–15:** A priest pronounces a blessing. **16–18:** Concluding hymn of praise. **17:** In
early Israel it was believed that the dead in
Sheol (*silence*) were separated from God
(88.5–6). **18:** *Praise the LORD! Hallelujah!* (see
Ps 113 n.).
Ps 116: Thanksgiving for healing. 1–2:
Address to the congregation. **3–11:** The
psalmist's experience (see 18.4–6 n.). **3:** A de-

the pangs of Sheol laid hold
on me;
I suffered distress and anguish.
4 Then I called on the name of
the LORD:
"O LORD, I pray, save my life!"

5 Gracious is the LORD, and
righteous;
our God is merciful.
6 The LORD protects the simple;
when I was brought low, he
saved me.
7 Return, O my soul, to your rest,
for the LORD has dealt
bountifully with you.

8 For you have delivered my soul
from death,
my eyes from tears,
my feet from stumbling.
9 I walk before the LORD
in the land of the living.
10 I kept my faith, even when I said,
"I am greatly afflicted";
11 I said in my consternation,
"Everyone is a liar."

12 What shall I return to the LORD
for all his bounty to me?
13 I will lift up the cup of salvation
and call on the name of the
LORD,
14 I will pay my vows to the LORD
in the presence of all his people.
15 Precious in the sight of the LORD
is the death of his faithful ones.

16 O LORD, I am your servant;
I am your servant, the child of
your serving girl.
You have loosed my bonds.
17 I will offer to you a thanksgiving
sacrifice
and call on the name of the
LORD.
18 I will pay my vows to the LORD
in the presence of all his people,
19 in the courts of the house of
the LORD,
in your midst, O Jerusalem.
Praise the LORD!

Psalm 117

1 Praise the LORD, all you nations!
Extol him, all you peoples!
2 For great is his steadfast love
toward us,
and the faithfulness of the LORD
endures forever.
Praise the LORD!

Psalm 118

1 O give thanks to the LORD, for he
is good;
his steadfast love endures
forever!

2 Let Israel say,
"His steadfast love endures
forever."
3 Let the house of Aaron say,
"His steadfast love endures
forever."

---

scription of serious illness. *Sheol,* see 6.5 n.
5–8: Strong emotion leads the psalmist to
address God directly in describing the answer
to the prayer (v. 8).
116.8–9: The recovery described. **10–11:**
Even when most profoundly depressed, the
psalmist had trusted in God rather than human beings.
116.12–19: The fulfillment of the vow described (see 7.17 n.). **13:** A libation is offered
(Ex 29.40). **15:** *Precious . . . is the death* means
that such a death is rarely allowed to happen.
16: See 86.16 n. **19:** *Praise the LORD!* Hallelujah!
(see Ps 113 n.).

Ps 117: Doxology. 1: *Praise the LORD!* Hallelujah! (see Ps 113 n.). **2:** *Praise the LORD!*
properly belongs to the following psalm.
**Ps 118: Thanksgiving for deliverance in
battle.** The last of the "Egyptian Hallel"
psalms (see Ps 113 n.). While it is difficult to
be sure whether the language of vv. 10–14 is
literal or figurative, it is tenable that the
speaker is a king who has come to the temple
to offer thanks for a victory. **1–4:** Summons
to thanksgiving. **2–4:** *Let Israel say,* etc.
These are probably actual directions to various groups in the congregation. **3:** *House of
Aaron,* see 115.10 n.

4 Let those who fear the LORD say,
   "His steadfast love endures
      forever."

5 Out of my distress I called on the
      LORD;
   the LORD answered me and set
      me in a broad place.
6 With the LORD on my side I do
      not fear.
   What can mortals do to me?
7 The LORD is on my side to help
      me;
   I shall look in triumph on those
      who hate me.
8 It is better to take refuge in the
      LORD
   than to put confidence in
      mortals.
9 It is better to take refuge in the
      LORD
   than to put confidence in
      princes.

10 All nations surrounded me;
   in the name of the LORD I cut
      them off!
11 They surrounded me, surrounded
      me on every side;
   in the name of the LORD I cut
      them off!
12 They surrounded me like bees;
   they blazed* like a fire of
      thorns;
   in the name of the LORD I cut
      them off!
13 I was pushed hard,* so that I was
      falling,
   but the LORD helped me.
14 The LORD is my strength and my
      might;
   he has become my salvation.

15 There are glad songs of victory in
      the tents of the righteous:
   "The right hand of the LORD does
      valiantly;
16   the right hand of the LORD is
      exalted;
   the right hand of the LORD does
      valiantly."
17 I shall not die, but I shall live,
   and recount the deeds of the
      LORD.
18 The LORD has punished me
      severely,
   but he did not give me over to
      death.

19 Open to me the gates of
      righteousness,
   that I may enter through them
   and give thanks to the LORD.

20 This is the gate of the LORD;
   the righteous shall enter
      through it.

21 I thank you that you have
      answered me
   and have become my salvation.
22 The stone that the builders
      rejected
   has become the chief
      cornerstone.
23 This is the LORD's doing;
   it is marvelous in our eyes.
24 This is the day that the LORD has
      made;
   let us rejoice and be glad in it.*
25 Save us, we beseech you, O LORD!
   O LORD, we beseech you, give
      us success!

x Gk: Heb *were extinguished*   y Gk Syr
Jerome: Heb *You pushed me hard*   z Or *in him*

118.5–18: The psalmist's experience (see
18.4–6 n.). 6: Quoted in Heb 13.6. 10–13:
The desperateness of the king's situation. 14–
18: God gave him the victory.
118.19: He asks to be admitted at the temple
gates. The preceding ceremony evidently
took place outside. 20: A voice from within
replies that only the qualified may enter
(compare Ps 15 and 24.3–6). 21–22: He answers that God had borne witness to his character by delivering him (18.20–24) when others had given him up. 22–23: Frequently
quoted in the New Testament (e.g. Mt 21.42;
Acts 4.11; 1 Pet 2.7). 23–25: The choir joyously acknowledges what God has done. 25:
Save us, in Hebrew, is "Hoshianna" (Hosanna). This verse and the following are alluded
to in Mt 21.9 and parallel passages.

26 Blessed is the one who comes in
   the name of the LORD. *a*
   We bless you from the house of
   the LORD.
27 The LORD is God,
   and he has given us light.
   Bind the festal procession with
   branches,
   up to the horns of the altar. *b*

28 You are my God, and I will give
   thanks to you;
   you are my God, I will extol
   you.

29 O give thanks to the LORD, for he
   is good,
   for his steadfast love endures
   forever.

Psalm 119

1 Happy are those whose way is
   blameless,
   who walk in the law of the
   LORD.
2 Happy are those who keep his
   decrees,
   who seek him with their whole
   heart,
3 who also do no wrong,
   but walk in his ways.
4 You have commanded your
   precepts
   to be kept diligently.
5 O that my ways may be steadfast
   in keeping your statutes!
6 Then I shall not be put to shame,
   having my eyes fixed on all
   your commandments.
7 I will praise you with an upright
   heart,

when I learn your righteous
   ordinances.
8 I will observe your statutes;
   do not utterly forsake me.

9 How can young people keep their
   way pure?
   By guarding it according to
   your word.
10 With my whole heart I seek you;
   do not let me stray from your
   commandments.
11 I treasure your word in my heart,
   so that I may not sin against
   you.
12 Blessed are you, O LORD;
   teach me your statutes.
13 With my lips I declare
   all the ordinances of your
   mouth.
14 I delight in the way of your
   decrees
   as much as in all riches.
15 I will meditate on your precepts,
   and fix my eyes on your ways.
16 I will delight in your statutes;
   I will not forget your word.

17 Deal bountifully with your
   servant,
   so that I may live and observe
   your word.
18 Open my eyes, so that I may
   behold
   wondrous things out of your
   law.
19 I live as an alien in the land;
   do not hide your
   commandments from me.

a Or *Blessed in the name of the* LORD *is the one who
comes*   *b* Meaning of Heb uncertain

118.26–27: The suppliant is admitted with
a choral blessing. (The second sentence of
v. 27 is perhaps a liturgical direction.) **28:** He
makes his act of thanksgiving. **29:** The choir
begins a hymn of praise (compare Ps 136).
Ps 119: Meditation on the law of God.
The length of this psalm is the result of its
unusual, and highly artificial, structure. It is
an alphabetical acrostic (compare Pss 9–10;
25; 34; 37; 111; 112; 145) in which each stanza
consists of eight lines all beginning with the
same Hebrew letter; the twenty-two stanzas
use all the letters in turn. In addition almost
every line contains the word "law" or a synonym (e.g. "testimonies," "ways," "precepts"; compare Ps 19.7–14). The predominant mood of lament suggests that it may
have been composed as a prayer for deliverance from trouble, though the language may
be merely imitative and the whole a purely

20 My soul is consumed with longing
  for your ordinances at all times.
21 You rebuke the insolent, accursed
    ones,
  who wander from your
    commandments;
22 take away from me their scorn
    and contempt,
  for I have kept your decrees.
23 Even though princes sit plotting
    against me,
  your servant will meditate on
    your statutes.
24 Your decrees are my delight,
  they are my counselors.

25 My soul clings to the dust;
  revive me according to your
    word.
26 When I told of my ways, you
    answered me;
  teach me your statutes.
27 Make me understand the way of
    your precepts,
  and I will meditate on your
    wondrous works.
28 My soul melts away for sorrow;
  strengthen me according to your
    word.
29 Put false ways far from me;
  and graciously teach me
    your law.
30 I have chosen the way of
    faithfulness;
  I set your ordinances before me.
31 I cling to your decrees, O LORD;
  let me not be put to shame.
32 I run the way of your
    commandments,
  for you enlarge my
    understanding.

33 Teach me, O LORD, the way of
    your statutes,
  and I will observe it to the end.

34 Give me understanding, that I
    may keep your law
  and observe it with my whole
    heart.
35 Lead me in the path of your
    commandments,
  for I delight in it.
36 Turn my heart to your decrees,
  and not to selfish gain.
37 Turn my eyes from looking at
    vanities;
  give me life in your ways.
38 Confirm to your servant your
    promise,
  which is for those who fear
    you.
39 Turn away the disgrace that I
    dread,
  for your ordinances are good.
40 See, I have longed for your
    precepts;
  in your righteousness give
    me life.

41 Let your steadfast love come to
    me, O LORD,
  your salvation according to your
    promise.
42 Then I shall have an answer for
    those who taunt me,
  for I trust in your word.
43 Do not take the word of truth
    utterly out of my mouth,
  for my hope is in your
    ordinances.
44 I will keep your law continually,
  forever and ever.
45 I shall walk at liberty,
  for I have sought your
    precepts.
46 I will also speak of your decrees
    before kings,
  and shall not be put to shame;
47 I find my delight in your
    commandments,
  because I love them.

---

literary exercise in honor of the written law.
It is a very late composition.
119.1–8, 9–16: Prayer for help in observing the law. **1:** *Happy,* see 1.1 n.
119.17–24: Prayer for deliverance from en-
mies.

119.25–32: Declaration of fidelity to the
law.
119.33–40: Prayer to understand the law.
41–48: Prayer for an answer to taunters.

48 I revere your commandments,
which I love,
and I will meditate on your
statutes.

49 Remember your word to your
servant,
in which you have made
me hope.
50 This is my comfort in my distress,
that your promise gives me life.
51 The arrogant utterly deride me,
but I do not turn away from
your law.
52 When I think of your ordinances
from of old,
I take comfort, O LORD.
53 Hot indignation seizes me because
of the wicked,
those who forsake your law.
54 Your statutes have been my songs
wherever I make my home.
55 I remember your name in the
night, O LORD,
and keep your law.
56 This blessing has fallen to me,
for I have kept your precepts.

57 The LORD is my portion;
I promise to keep your words.
58 I implore your favor with all my
heart;
be gracious to me according to
your promise.
59 When I think of your ways,
I turn my feet to your decrees;
60 I hurry and do not delay
to keep your commandments.
61 Though the cords of the wicked
ensnare me,
I do not forget your law.
62 At midnight I rise to praise you,
because of your righteous
ordinances.
63 I am a companion of all who fear
you,
of those who keep your
precepts.
64 The earth, O LORD, is full of your
steadfast love;
teach me your statutes.

65 You have dealt well with your
servant,
O LORD, according to your
word.
66 Teach me good judgment and
knowledge,
for I believe in your
commandments.
67 Before I was humbled I went
astray,
but now I keep your word.
68 You are good and do good;
teach me your statutes.
69 The arrogant smear me with lies,
but with my whole heart I keep
your precepts.
70 Their hearts are fat and gross,
but I delight in your law.
71 It is good for me that I was
humbled,
so that I might learn your
statutes.
72 The law of your mouth is better
to me
than thousands of gold and
silver pieces.

73 Your hands have made and
fashioned me;
give me understanding that I
may learn your
commandments.
74 Those who fear you shall see me
and rejoice,
because I have hoped in your
word.
75 I know, O LORD, that your
judgments are right,
and that in faithfulness you have
humbled me.

119.49–56: Expression of confidence during persecution.
119.57–64: Declaration of devotion to God.

119.65–72: Recognition of the disciplinary value of suffering.
119.73–80: Acknowledgment of the justice of God's ways, and prayer for help.

76 Let your steadfast love become my
     comfort
  according to your promise to
     your servant.
77 Let your mercy come to me, that
     I may live;
  for your law is my delight.
78 Let the arrogant be put to shame,
  because they have subverted me
     with guile;
  as for me, I will meditate on
     your precepts.
79 Let those who fear you turn to
     me,
  so that they may know your
     decrees.
80 May my heart be blameless in
     your statutes,
  so that I may not be put to
     shame.

81 My soul languishes for your
     salvation;
  I hope in your word.
82 My eyes fail with watching for
     your promise;
  I ask, "When will you comfort
     me?"
83 For I have become like a wineskin
     in the smoke,
  yet I have not forgotten your
     statutes.
84 How long must your servant
     endure?
  When will you judge those who
     persecute me?
85 The arrogant have dug pitfalls
     for me;
  they flout your law.
86 All your commandments are
     enduring;
  I am persecuted without cause;
     help me!
87 They have almost made an end of
     me on earth;
  but I have not forsaken your
     precepts.
88 In your steadfast love spare my
     life,

so that I may keep the decrees
     of your mouth.
89 The LORD exists forever;
  your word is firmly fixed in
     heaven.
90 Your faithfulness endures to all
     generations;
  you have established the earth,
     and it stands fast.
91 By your appointment they stand
     today,
  for all things are your servants.
92 If your law had not been my
     delight,
  I would have perished in my
     misery.
93 I will never forget your precepts,
  for by them you have given
     me life.
94 I am yours; save me,
  for I have sought your precepts.
95 The wicked lie in wait to
     destroy me,
  but I consider your decrees.
96 I have seen a limit to all
     perfection,
  but your commandment is
     exceedingly broad.

97 Oh, how I love your law!
  It is my meditation all day long.
98 Your commandment makes me
     wiser than my enemies,
  for it is always with me.
99 I have more understanding than
     all my teachers,
  for your decrees are my
     meditation.
100 I understand more than the aged,
  for I keep your precepts.
101 I hold back my feet from every
     evil way,
  in order to keep your word.
102 I do not turn away from your
     ordinances,
  for you have taught me.
103 How sweet are your words to
     my taste,

119.81–88: Plea for deliverance from enemies. **89–96:** Faith in God's word.

119.97–104: The beauty and sweetness of
the law.

sweeter than honey to my
mouth!

104 Through your precepts I get
understanding;
therefore I hate every false way.

105 Your word is a lamp to my feet
and a light to my path.

106 I have sworn an oath and
confirmed it,
to observe your righteous
ordinances.

107 I am severely afflicted;
give me life, O LORD, according
to your word.

108 Accept my offerings of praise,
O LORD,
and teach me your ordinances.

109 I hold my life in my hand
continually,
but I do not forget your law.

110 The wicked have laid a snare
for me,
but I do not stray from your
precepts.

111 Your decrees are my heritage
forever;
they are the joy of my heart.

112 I incline my heart to perform
your statutes
forever, to the end.

113 I hate the double-minded,
but I love your law.

114 You are my hiding place and my
shield;
I hope in your word.

115 Go away from me, you evildoers,
that I may keep the
commandments of my
God.

116 Uphold me according to your
promise, that I may live,
and let me not be put to shame
in my hope.

117 Hold me up, that I may be safe

and have regard for your
statutes continually.

118 You spurn all who go astray
from your statutes;
for their cunning is in vain.

119 All the wicked of the earth you
count as dross;
therefore I love your decrees.

120 My flesh trembles for fear of
you,
and I am afraid of your
judgments.

121 I have done what is just and
right;
do not leave me to my
oppressors.

122 Guarantee your servant's
well-being;
do not let the godless oppress
me.

123 My eyes fail from watching for
your salvation,
and for the fulfillment of your
righteous promise.

124 Deal with your servant according
to your steadfast love,
and teach me your statutes.

125 I am your servant; give me
understanding,
so that I may know your
decrees.

126 It is time for the LORD to act,
for your law has been broken.

127 Truly I love your commandments
more than gold, more than
fine gold.

128 Truly I direct my steps by all
your precepts; *c*
I hate every false way.

129 Your decrees are wonderful;
therefore my soul keeps them.

130 The unfolding of your words
gives light;

c Gk Jerome: Meaning of Heb uncertain

119.105–112: Prayer for help.
119.113–120: Expression of confidence and
prayers for deliverance.

119.121–128: Declaration of juridical innocence and loyalty to the law.
119.129–136: Praise of the law and prayer

it imparts understanding to the
simple.

131 With open mouth I pant,
because I long for your
commandments.

132 Turn to me and be gracious to
me,
as is your custom toward those
who love your name.

133 Keep my steps steady according
to your promise,
and never let iniquity have
dominion over me.

134 Redeem me from human
oppression,
that I may keep your precepts.

135 Make your face shine upon your
servant,
and teach me your statutes.

136 My eyes shed streams of tears
because your law is not kept.

137 You are righteous, O Lord,
and your judgments are right.

138 You have appointed your decrees
in righteousness
and in all faithfulness.

139 My zeal consumes me
because my foes forget your
words.

140 Your promise is well tried,
and your servant loves it.

141 I am small and despised,
yet I do not forget your
precepts.

142 Your righteousness is an
everlasting righteousness,
and your law is the truth.

143 Trouble and anguish have come
upon me,
but your commandments are
my delight.

144 Your decrees are righteous
forever;
give me understanding that I
may live.

145 With my whole heart I cry;
answer me, O Lord.
I will keep your statutes.

146 I cry to you; save me,
that I may observe your decrees.

147 I rise before dawn and cry
for help;
I put my hope in your words.

148 My eyes are awake before each
watch of the night,
that I may meditate on your
promise.

149 In your steadfast love hear my
voice;
O Lord, in your justice
preserve my life.

150 Those who persecute me with
evil purpose draw near;
they are far from your law.

151 Yet you are near, O Lord,
and all your commandments
are true.

152 Long ago I learned from your
decrees
that you have established them
forever.

153 Look on my misery and rescue
me,
for I do not forget your law.

154 Plead my cause and redeem me;
give me life according to your
promise.

155 Salvation is far from the wicked,
for they do not seek your
statutes.

156 Great is your mercy, O Lord;
give me life according to your
justice.

157 Many are my persecutors and my
adversaries,
yet I do not swerve from your
decrees.

158 I look at the faithless with
disgust,
because they do not keep your
commands.

for deliverance from enemies. **137–144:** Acknowledgment of God's justice. **145–152:** Passionate cry to be saved from persecution.

119.153–160: Prayer for the preservation of the psalmist's life.

159 Consider how I love your
precepts;
preserve my life according to
your steadfast love.
160 The sum of your word is truth;
and every one of your righteous
ordinances endures forever.

161 Princes persecute me without
cause,
but my heart stands in awe of
your words.
162 I rejoice at your word
like one who finds great spoil.
163 I hate and abhor falsehood,
but I love your law.
164 Seven times a day I praise you
for your righteous ordinances.
165 Great peace have those who love
your law;
nothing can make them
stumble.
166 I hope for your salvation,
O Lord,
and I fulfill your
commandments.
167 My soul keeps your decrees;
I love them exceedingly.
168 I keep your precepts and decrees,
for all my ways are before you.

169 Let my cry come before you,
O Lord;
give me understanding
according to your word.
170 Let my supplication come before
you;
deliver me according to your
promise.
171 My lips will pour forth praise,
because you teach me your
statutes.
172 My tongue will sing of your
promise,

for all your commandments are
right.
173 Let your hand be ready to help
me,
for I have chosen your precepts.
174 I long for your salvation,
O Lord,
and your law is my delight.
175 Let me live that I may praise you,
and let your ordinances help
me.
176 I have gone astray like a lost
sheep; seek out your
servant,
for I do not forget your
commandments.

Psalm 120

A Song of Ascents.

1 In my distress I cry to the Lord,
that he may answer me:
2 "Deliver me, O Lord,
from lying lips,
from a deceitful tongue."

3 What shall be given to you?
And what more shall be done
to you,
you deceitful tongue?
4 A warrior's sharp arrows,
with glowing coals of the
broom tree!

5 Woe is me, that I am an alien in
Meshech,
that I must live among the tents
of Kedar.
6 Too long have I had my dwelling
among those who hate peace.
7 I am for peace;
but when I speak,
they are for war.

119.161–168: The psalmist's piety contrasted with the unjust actions of the persecutors.
119.169–176: If delivered, the poet vows to
sing the praise of God's law.
Ps 120: **An exile's prayer for deliverance from enemies** (a lament). 4: *The broom*

tree produces a specially hot, long-burning
fire. 5: *Meshech . . . Kedar,* remote regions in
Asia Minor and north Arabia. The psalmist
lives in such a place, far from the land of
Israel.

Psalm 121

A Song of Ascents.

1 I lift up my eyes to the hills—
   from where will my help come?
2 My help comes from the LORD,
   who made heaven and earth.

3 He will not let your foot be
       moved;
   he who keeps you will not
       slumber.
4 He who keeps Israel
   will neither slumber nor sleep.

5 The LORD is your keeper;
   the LORD is your shade at your
       right hand.
6 The sun shall not strike you
       by day,
   nor the moon by night.

7 The LORD will keep you from
       all evil;
   he will keep your life.
8 The LORD will keep
   your going out and your
       coming in
   from this time on and
       forevermore.

Psalm 122

A Song of Ascents. Of David.

1 I was glad when they said to me,
   "Let us go to the house of the
       LORD!"
2 Our feet are standing
   within your gates, O Jerusalem.

3 Jerusalem—built as a city
   that is bound firmly together.

4 To it the tribes go up,
       the tribes of the LORD,
   as was decreed for Israel,
       to give thanks to the name of
       the LORD.
5 For there the thrones for judgment
       were set up,
   the thrones of the house of
       David.

6 Pray for the peace of Jerusalem:
   "May they prosper who love
       you.
7 Peace be within your walls,
   and security within your
       towers."
8 For the sake of my relatives and
       friends
   I will say, "Peace be within
       you."
9 For the sake of the house of the
       LORD our God,
   I will seek your good.

Psalm 123

A Song of Ascents.

1 To you I lift up my eyes,
   O you who are enthroned in the
       heavens!
2 As the eyes of servants
       look to the hand of their master,
   as the eyes of a maid
       to the hand of her mistress,
   so our eyes look to the LORD
       our God,
   until he has mercy upon us.

3 Have mercy upon us, O LORD,
       have mercy upon us,
   for we have had more than
       enough of contempt.

Ps 121: A liturgy of blessing. 1: The
psalmist asks a rhetorical question. *The hills*
may be the "high places" where the baals, the
local fertility gods, were worshiped (2 Kings
23.5). **2–8:** A priest answers and pronounces
a blessing. **2:** Possibly *my* should be emended
to "your" (compare v. 3).
   **Ps 122: A song praising Zion as the
pilgrim's goal. 1–5:** Arrived in Jerusalem, a group of pilgrims admire its buildings
and the unity it symbolizes.
122.6–9: They pray for its continuing
prosperity.
**Ps 123: Prayer for deliverance from
enemies** (a group lament). **1–2:** An act of
humble submission to God's will. **3–4:**
Prayer for help. The speakers may represent
either Israel or some oppressed class or sect
within it.

4 Our soul has had more than its fill
  of the scorn of those who are
    at ease,
  of the contempt of the proud.

Psalm 124

A Song of Ascents. Of David.

1 If it had not been the LORD who
    was on our side
  —let Israel now say—
2 if it had not been the LORD who
    was on our side,
  when our enemies attacked us,
3 then they would have swallowed
    us up alive,
  when their anger was kindled
    against us;
4 then the flood would have swept
    us away,
  the torrent would have gone
    over us;
5 then over us would have gone
    the raging waters.

6 Blessed be the LORD,
  who has not given us
    as prey to their teeth.
7 We have escaped like a bird
    from the snare of the fowlers;
  the snare is broken,
    and we have escaped.

8 Our help is in the name of
    the LORD,
  who made heaven and earth.

Psalm 125

A Song of Ascents.

1 Those who trust in the LORD are
    like Mount Zion,
  which cannot be moved, but
    abides forever.
2 As the mountains surround
    Jerusalem,

so the LORD surrounds his
    people,
  from this time on and
    forevermore.
3 For the scepter of wickedness shall
    not rest
  on the land allotted to the
    righteous,
  so that the righteous might not
    stretch out
  their hands to do wrong.
4 Do good, O LORD, to those who
    are good,
  and to those who are upright in
    their hearts.
5 But those who turn aside to their
    own crooked ways
  the LORD will lead away with
    evildoers.
  Peace be upon Israel!

Psalm 126

A Song of Ascents.

1 When the LORD restored the
    fortunes of Zion, *d*
  we were like those who dream.
2 Then our mouth was filled with
    laughter,
  and our tongue with shouts
    of joy;
  then it was said among the
    nations,
  "The LORD has done great
    things for them."
3 The LORD has done great things
    for us,
  and we rejoiced.

4 Restore our fortunes, O LORD,
  like the watercourses in the
    Negeb.

---

d Or brought back those who returned to Zion

Ps 124: **Thanksgiving for a national deliverance. 4–5:** Compare 32.6; 69.1–2, 14–
15.
  Ps 125: **Prayer for deliverance from
national enemies** (a group lament). **1–3:** Expression of confidence. **4–5:** Prayer for help.

Ps 126: **Prayer for deliverance from
national misfortune. 1–3:** The joy inspired
by God's favor toward his people in former
times. **4–6:** Prayer that it may be granted
once again. **4:** *Negeb,* the arid region to the
south of Palestine.

5 May those who sow in tears
reap with shouts of joy.
6 Those who go out weeping,
bearing the seed for sowing,
shall come home with shouts of
joy,
carrying their sheaves.

Psalm 127

A Song of Ascents. Of Solomon.

1 Unless the LORD builds the house,
those who build it labor in vain.
Unless the LORD guards the city,
the guard keeps watch in vain.
2 It is in vain that you rise up early
and go late to rest,
eating the bread of anxious toil;
for he gives sleep to his
beloved. *e*

3 Sons are indeed a heritage from
the LORD,
the fruit of the womb a reward.
4 Like arrows in the hand of a
warrior
are the sons of one's youth.
5 Happy is the man who has
his quiver full of them.
He shall not be put to shame
when he speaks with his
enemies in the gate.

Psalm 128

A Song of Ascents.

1 Happy is everyone who fears the
LORD,
who walks in his ways.
2 You shall eat the fruit of the labor
of your hands;
you shall be happy, and it shall
go well with you.

3 Your wife will be like a fruitful
vine
within your house;
your children will be like olive
shoots
around your table.
4 Thus shall the man be blessed
who fears the LORD.

5 The LORD bless you from Zion.
May you see the prosperity of
Jerusalem
all the days of your life.
6 May you see your children's
children.
Peace be upon Israel!

Psalm 129

A Song of Ascents.

1 "Often have they attacked me
from my youth"
—let Israel now say—
2 "often have they attacked me from
my youth,
yet they have not prevailed
against me.
3 The plowers plowed on my back;
they made their furrows long."
4 The LORD is righteous;
he has cut the cords of the
wicked.
5 May all who hate Zion
be put to shame and turned
backward.
6 Let them be like the grass on the
housetops
that withers before it grows up,
7 with which reapers do not fill
their hands
or binders of sheaves their arms,
8 while those who pass by do not
say,

e Or *for he provides for his beloved during sleep*

Ps 127: **A safe home and a large family
are the LORD's gift** (a wisdom psalm). **1–2:**
Anxiety has no place in the life of the faithful
(compare Mt 6.25–34) **3–5:** The gift of
many stalwart sons makes a father feel secure.

Ps 128: **A large and prosperous family
is a reward for devotion to the LORD** (a

wisdom psalm). **1:** *Happy,* see 1.1 n.

Ps 129: **Prayer for deliverance from
national enemies** (a group lament). **1–4:**
God has preserved Israel in spite of all its
suffering in the past. **5–8:** May its present
enemies be destroyed!

"The blessing of the LORD be
    upon you!
We bless you in the name of the
    LORD!"

Psalm 130

A Song of Ascents.

1 Out of the depths I cry to you,
        O LORD.
2   Lord, hear my voice!
Let your ears be attentive
    to the voice of my supplications!

3 If you, O LORD, should mark
        iniquities,
    Lord, who could stand?
4 But there is forgiveness with you,
        so that you may be revered.

5 I wait for the LORD, my soul
        waits,
    and in his word I hope;
6 my soul waits for the Lord
        more than those who watch for
        the morning,
        more than those who watch for
        the morning.

7 O Israel, hope in the LORD!
    For with the LORD there is
        steadfast love,
    and with him is great power to
        redeem.
8 It is he who will redeem Israel
    from all its iniquities.

Psalm 131

A Song of Ascents. Of David.

1 O LORD, my heart is not lifted up,
    my eyes are not raised too high;

I do not occupy myself with
        things
    too great and too marvelous
        for me.
2 But I have calmed and quieted
        my soul,
    like a weaned child with its
        mother;
    my soul is like the weaned child
        that is with me.*f*

3 O Israel, hope in the LORD
    from this time on and
        forevermore.

Psalm 132

A Song of Ascents.

1 O LORD, remember in David's
        favor
    all the hardships he endured;
2 how he swore to the LORD
    and vowed to the Mighty One
        of Jacob,
3 "I will not enter my house
    or get into my bed;
4 I will not give sleep to my eyes
    or slumber to my eyelids,
5 until I find a place for the LORD,
    a dwelling place for the Mighty
        One of Jacob."

6 We heard of it in Ephrathah;
    we found it in the fields of Jaar.
7 "Let us go to his dwelling place;
    let us worship at his footstool."

8 Rise up, O LORD, and go to your
        resting place,

f Or *my soul within me is like a weaned child*

Ps 130: **Prayer for deliverance from
personal trouble** (a lament). **1–2:** Cry for
help. The psalmist's trouble is described only
in general terms. **3–6:** Because it is God's
nature to be merciful, the psalmist eagerly
awaits his help.
130.7–8: Israel should take the same attitude in its national difficulties.
Ps 131: **Act of humble submission to**

God's will and guidance (a song of trust;
see Ps 11 n.).
Ps 132: **Liturgy commemorating God's
choice of Zion and the Davidic dynasty. 1–5:** God is reminded of David's determination to provide him a sanctuary
(2 Sam 7.1–2).
132.6–10: These words evidently accompanied a dramatic ceremony that reenacted

you and the ark of your might.
9 Let your priests be clothed with
righteousness,
and let your faithful shout
for joy.
10 For your servant David's sake
do not turn away the face of
your anointed one.

11 The LORD swore to David a sure
oath
from which he will not turn
back:
"One of the sons of your body
I will set on your throne.
12 If your sons keep my covenant
and my decrees that I shall teach
them,
their sons also, forevermore,
shall sit on your throne."

13 For the LORD has chosen Zion;
he has desired it for his
habitation:
14 "This is my resting place
forever;
here I will reside, for I have
desired it.
15 I will abundantly bless its
provisions;
I will satisfy its poor with
bread.
16 Its priests I will clothe with
salvation,
and its faithful will shout for
joy.
17 There I will cause a horn to sprout
up for David;
I have prepared a lamp for my
anointed one.

18 His enemies I will clothe with
disgrace,
but on him, his crown will
gleam."

Psalm 133

A Song of Ascents.

1 How very good and pleasant
it is
when kindred live together in
unity!
2 It is like the precious oil on the
head,
running down upon the beard,
on the beard of Aaron,
running down over the collar of
his robes.
3 It is like the dew of Hermon,
which falls on the mountains of
Zion.
For there the LORD ordained his
blessing,
life forevermore.

Psalm 134

A Song of Ascents.

1 Come, bless the LORD, all you
servants of the LORD,
who stand by night in the house
of the LORD!
2 Lift up your hands to the holy
place,
and bless the LORD.

3 May the LORD, maker of heaven
and earth,
bless you from Zion.

---

the discovery of the ark by David and the procession by which he brought it to the sanctuary (2 Sam 6.2–15). **6**: *Ephrathah,* Bethlehem, David's city. *The fields of Jaar,* Kiriath-jearim, where the ark had been kept from Samuel's time until David became king in Jerusalem (1 Sam 7.1–2; 2 Chr 1.4).
132.11–18: A priest or temple prophet recites God's promise concerning Jerusalem and the dynasty.
Ps 133: **The joys of harmony in the**

family (a wisdom psalm). **1**: *Kindred,* literally "brothers." This may refer to the situation described in Deut 25.5. **2**: *Aaron,* the ancestor and type of the priests. Anointing with consecrated *oil* was part of the ordination ceremony (Ex 29.7). **3**: *Hermon,* the chief mountain of Syria.
Ps 134: **A liturgy of blessing. 1–2**: The priests are summoned to offer praise to the LORD. **3**: They bless the congregation.

Psalm 135

1 Praise the LORD!
Praise the name of the LORD;
give praise, O servants of
the LORD,
2 you that stand in the house of
the LORD,
in the courts of the house of
our God.
3 Praise the LORD, for the LORD
is good;
sing to his name, for he is
gracious.
4 For the LORD has chosen Jacob for
himself,
Israel as his own possession.

5 For I know that the LORD is great;
our Lord is above all gods.
6 Whatever the LORD pleases he
does,
in heaven and on earth,
in the seas and all deeps.
7 He it is who makes the clouds rise
at the end of the earth;
he makes lightnings for the rain
and brings out the wind from
his storehouses.

8 He it was who struck down the
firstborn of Egypt,
both human beings and animals;
9 he sent signs and wonders
into your midst, O Egypt,
against Pharaoh and all his
servants.
10 He struck down many nations
and killed mighty kings—
11 Sihon, king of the Amorites,
and Og, king of Bashan,
and all the kingdoms of
Canaan—

12 and gave their land as a heritage,
a heritage to his people Israel.

13 Your name, O LORD, endures
forever,
your renown, O LORD,
throughout all ages.
14 For the LORD will vindicate his
people,
and have compassion on his
servants.

15 The idols of the nations are silver
and gold,
the work of human hands.
16 They have mouths, but they do
not speak;
they have eyes, but they do
not see;
17 they have ears, but they do not
hear,
and there is no breath in their
mouths.
18 Those who make them
and all who trust them
shall become like them.

19 O house of Israel, bless the LORD!
O house of Aaron, bless the
LORD!
20 O house of Levi, bless the LORD!
You that fear the LORD, bless
the LORD!
21 Blessed be the LORD from Zion,
he who resides in Jerusalem.
Praise the LORD!

Psalm 136

1 O give thanks to the LORD, for he
is good,
for his steadfast love endures
forever.

Ps 135: **Hymn praising the LORD for his
mighty deeds. 1–4:** Call to worship. **5–7:**
The LORD's control of nature.
135.8–12: His work in the Exodus and the
conquest of Canaan. **11:** *Sihon . . . Og,* Num
21.21–35. **13–14:** A lyrical interlude.
135.15–18: The LORD's power contrasted
with the impotence of heathen gods (compare

115.3–8). **19–21:** Concluding summons to
worship. **19:** *House of Aaron . . . Levi,* the
priests and their assistants, the Levites.
Ps 136: **Thanksgiving for the LORD's
great deeds on behalf of his people.** The
second half of each verse is a congregational
response. **1–3:** Summons to give thanks.

2 O give thanks to the God of gods,
    for his steadfast love endures
        forever.
3 O give thanks to the Lord of
        lords,
    for his steadfast love endures
        forever;

4 who alone does great wonders,
    for his steadfast love endures
        forever;
5 who by understanding made the
        heavens,
    for his steadfast love endures
        forever;
6 who spread out the earth on the
        waters,
    for his steadfast love endures
        forever;
7 who made the great lights,
    for his steadfast love endures
        forever;
8 the sun to rule over the day,
    for his steadfast love endures
        forever;
9 the moon and stars to rule over
        the night,
    for his steadfast love endures
        forever;

10 who struck Egypt through their
        firstborn,
    for his steadfast love endures
        forever;
11 and brought Israel out from
        among them,
    for his steadfast love endures
        forever;
12 with a strong hand and an
        outstretched arm,
    for his steadfast love endures
        forever;
13 who divided the Red Sea*g* in two,
    for his steadfast love endures
        forever;
14 and made Israel pass through the
        midst of it,

for his steadfast love endures
        forever;
15 but overthrew Pharaoh and his
        army in the Red Sea,*g*
    for his steadfast love endures
        forever;
16 who led his people through the
        wilderness,
    for his steadfast love endures
        forever;
17 who struck down great kings,
    for his steadfast love endures
        forever;
18 and killed famous kings,
    for his steadfast love endures
        forever;
19 Sihon, king of the Amorites,
    for his steadfast love endures
        forever;
20 and Og, king of Bashan,
    for his steadfast love endures
        forever;
21 and gave their land as a heritage,
    for his steadfast love endures
        forever;
22 a heritage to his servant Israel,
    for his steadfast love endures
        forever.

23 It is he who remembered us in our
        low estate,
    for his steadfast love endures
        forever;
24 and rescued us from our foes,
    for his steadfast love endures
        forever;
25 who gives food to all flesh,
    for his steadfast love endures
        forever.

26 O give thanks to the God of
        heaven,
    for his steadfast love endures
        forever.

g Or *Sea of Reeds*

136.4–9: God's work in creation.
136.10–22: His work in the history of Israel. **10–15:** The Exodus. **16:** The wilderness wanderings. **17–22:** The conquest of Canaan.

Sihon . . . Og, Num 21.21–35. **23–25:** Recapitulation. **26:** Concluding summons to give thanks.

Psalm 137

1 By the rivers of Babylon—
    there we sat down and there
        we wept
    when we remembered Zion.
2 On the willows[h] there
    we hung up our harps.
3 For there our captors
    asked us for songs,
    and our tormentors asked for
        mirth, saying,
    "Sing us one of the songs of
        Zion!"

4 How could we sing the LORD's
        song
    in a foreign land?
5 If I forget you, O Jerusalem,
    let my right hand wither!
6 Let my tongue cling to the roof of
        my mouth,
    if I do not remember you,
    if I do not set Jerusalem
        above my highest joy.

7 Remember, O LORD, against the
        Edomites
    the day of Jerusalem's fall,
    how they said, "Tear it down!
        Tear it down!
    Down to its foundations!"
8 O daughter Babylon, you
        devastator![i]
    Happy shall they be who pay
        you back
    what you have done to us!
9 Happy shall they be who take
        your little ones
    and dash them against the rock!

Psalm 138

Of David.

1 I give you thanks, O LORD, with
        my whole heart;
    before the gods I sing your
        praise;
2 I bow down toward your holy
        temple
    and give thanks to your name
        for your steadfast love and
        your faithfulness;
    for you have exalted your name
        and your word
        above everything.[j]
3 On the day I called, you answered
        me,
    you increased my strength of
        soul.[k]

4 All the kings of the earth shall
        praise you, O LORD,
    for they have heard the words
        of your mouth.
5 They shall sing of the ways of the
        LORD,
    for great is the glory of the
        LORD.
6 For though the LORD is high, he
        regards the lowly;
    but the haughty he perceives
        from far away.

7 Though I walk in the midst of
        trouble,
    you preserve me against the
        wrath of my enemies;

---

h Or *poplars*   *i* Or *you who are devastated*
j Cn: Heb *you have exalted your word above all
your name*   *k* Syr Compare Gk Tg: Heb *you
made me arrogant in my soul with strength*

Ps 137: Prayer for vengeance on Israel's enemies (a lament). **1–6:** Lament over the holy city. **1–4:** It is difficult to sing the LORD's praise when an exile among foreigners. **1:** *Rivers of Babylon*, streams or canals of the Tigris and Euphrates rivers. **3:** *Songs of Zion*, see 46.4–7 n. **5–6:** The psalmist's solemn pledge.
137.7–9: Cry for revenge. **7:** *Edomites*, who helped the Babylonians sack Jerusalem in 587 (586) B.C. (Ob 10–14; 2 Kings 25.8–12). **8:** *Daughter Babylon*, literally, "daughter of Babylon," personification of the Babylonian people.
Ps 138: Thanksgiving for deliverance from trouble. 1–2: The psalmist in the temple court to offer thanks. **3:** The psalmist's experience (compare 18.37–45). **4–6:** A hymn of praise. **7–8:** An expression of faith.

you stretch out your hand,
and your right hand delivers
me.
8 The LORD will fulfill his purpose
for me;
your steadfast love, O LORD,
endures forever.
Do not forsake the work of
your hands.

Psalm 139

To the leader. Of David. A Psalm.

1 O LORD, you have searched me
and known me.
2 You know when I sit down and
when I rise up;
you discern my thoughts from
far away.
3 You search out my path and my
lying down,
and are acquainted with all
my ways.
4 Even before a word is on my
tongue,
O LORD, you know it
completely.
5 You hem me in, behind and
before,
and lay your hand upon me.
6 Such knowledge is too wonderful
for me;
it is so high that I cannot
attain it.

7 Where can I go from your spirit?
Or where can I flee from your
presence?
8 If I ascend to heaven, you are
there;
if I make my bed in Sheol, you
are there.

9 If I take the wings of the morning
and settle at the farthest limits
of the sea,
10 even there your hand shall
lead me,
and your right hand shall hold
me fast.
11 If I say, "Surely the darkness shall
cover me,
and the light around me become
night,"
12 even the darkness is not dark
to you;
the night is as bright as the day,
for darkness is as light to you.

13 For it was you who formed my
inward parts;
you knit me together in my
mother's womb.
14 I praise you, for I am fearfully and
wonderfully made.
Wonderful are your works;
that I know very well.
15 My frame was not hidden from
you,
when I was being made in secret,
intricately woven in the depths
of the earth.
16 Your eyes beheld my unformed
substance.
In your book were written
all the days that were formed
for me,
when none of them as yet
existed.
17 How weighty to me are your
thoughts, O God!
How vast is the sum of them!
18 I try to count them—they are
more than the sand;

Ps 139: **Prayer for deliverance from
personal enemies** (a lament). **1–18:** An appeal to the LORD, on the basis of his omniscience and universal power, to save the psalmist from the enemies of them both (see vv. 19–20). **1–6:** Everything the psalmist has ever done or thought is known to God.

139.7–12: There could have been no secret crimes, for God is with everyone, everywhere. **8:** *Sheol,* see 88.5–6 n. It is a new thought that God is in Sheol as well as in heaven (see 115.17 n.). **13–16:** God formed the embryo in the womb (poetically called *the depths of the earth*) and knew the psalmist's character from the moment of conception. **16:** *Your book,* compare 40.7 n. **17–18:** Exclamation of wonder.

I come to the end[l]—I am still
    with you.

19 O that you would kill the wicked,
    O God,
    and that the bloodthirsty would
      depart from me—
20 those who speak of you
    maliciously,
    and lift themselves up against
      you for evil![m]
21 Do I not hate those who hate you,
    O LORD?
    And do I not loathe those who
      rise up against you?
22 I hate them with perfect hatred;
    I count them my enemies.
23 Search me, O God, and know my
    heart;
    test me and know my thoughts.
24 See if there is any wicked[n] way
    in me,
    and lead me in the way
      everlasting.[o]

Psalm 140

To the leader. A Psalm of David.

1 Deliver me, O LORD, from
    evildoers;
    protect me from those who are
      violent,
2 who plan evil things in their
    minds
    and stir up wars continually.
3 They make their tongue sharp as a
    snake's,
    and under their lips is the
      venom of vipers.   *Selah*

4 Guard me, O LORD, from the
    hands of the wicked;
    protect me from the violent
    who have planned my downfall.

5 The arrogant have hidden a trap
    for me,
    and with cords they have spread
      a net,[p]
    along the road they have set
      snares for me.   *Selah*

6 I say to the LORD, "You are
    my God;
    give ear, O LORD, to the voice
      of my supplications."
7 O LORD, my Lord, my strong
    deliverer,
    you have covered my head in
      the day of battle.
8 Do not grant, O LORD, the desires
    of the wicked;
    do not further their evil plot.[q]
                  Selah

9 Those who surround me lift up
    their heads;[r]
    let the mischief of their lips
      overwhelm them!
10 Let burning coals fall on them!
    Let them be flung into pits, no
      more to rise!
11 Do not let the slanderer be
    established in the land;
    let evil speedily hunt down the
      violent!

12 I know that the LORD maintains
    the cause of the needy,
    and executes justice for the
      poor.
13 Surely the righteous shall give
    thanks to your name;

---

l Or *I awake*   *m* Cn: Meaning of Heb
uncertain   *n* Heb *hurtful*   *o* Or *the ancient
way*. Compare Jer 6.16   *p* Or *they have spread
cords as a net*   *q* Heb adds *they are exalted*
r Cn Compare Gk: Heb *those who surround me are
uplifted in head*; Heb divides verses 8 and 9
differently

139.19–24: Prayer for vindication and deliverance. **21:** The psalmist, a religious leader,
does not hesitate to identify personal enemies
as God's enemies also.
    **Ps 140: Prayer for deliverance from
personal enemies** (a lament). **1–2:** Cry for
help. **3–5:** The psalmist's situation. **3:** The

mention of *their tongue* and *lips* probably implies that they have pronounced a curse
against the psalmist (compare Pss 58 and
109). *Selah*, see 3.2 n.
    140.6–11: Prayer for deliverance. **7:** *Cov-
ered my head*, given protection. **12–13:** Expression of confidence.

the upright shall live in your
    presence.

Psalm 141

A Psalm of David.

1 I call upon you, O LORD; come
    quickly to me;
  give ear to my voice when I call
    to you.
2 Let my prayer be counted as
    incense before you,
  and the lifting up of my hands
    as an evening sacrifice.

3 Set a guard over my mouth,
    O LORD;
  keep watch over the door of
    my lips.
4 Do not turn my heart to any evil,
  to busy myself with wicked
    deeds
  in company with those who work
    iniquity;
  do not let me eat of their
    delicacies.

5 Let the righteous strike me;
  let the faithful correct me.
  Never let the oil of the wicked
    anoint my head,*s*
  for my prayer is continually*t*
    against their wicked deeds.
6 When they are given over to those
    who shall condemn them,
  then they shall learn that my
    words were pleasant.
7 Like a rock that one breaks apart
    and shatters on the land,
  so shall their bones be strewn at
    the mouth of Sheol.*u*

8 But my eyes are turned toward
    you, O GOD, my Lord;

in you I seek refuge; do not
    leave me defenseless.
9 Keep me from the trap that they
    have laid for me,
  and from the snares of
    evildoers.
10 Let the wicked fall into their
    own nets,
  while I alone escape.

Psalm 142

*A Maskil of David. When he was in the
cave. A Prayer.*

1 With my voice I cry to the LORD;
  with my voice I make
    supplication to the LORD.
2 I pour out my complaint before
    him;
  I tell my trouble before him.
3 When my spirit is faint,
  you know my way.

In the path where I walk
  they have hidden a trap for me.
4 Look on my right hand and see—
  there is no one who takes notice
    of me;
  no refuge remains to me;
  no one cares for me.

5 I cry to you, O LORD;
  I say, "You are my refuge,
  my portion in the land of the
    living."
6 Give heed to my cry,
  for I am brought very low.

Save me from my persecutors,
  for they are too strong for me.

s Gk: Meaning of Heb uncertain   *t* Cn: Heb
for continually and my prayer   *u* Meaning of
Heb of verses 5-7 is uncertain

Ps 141: **Prayer for deliverance from
personal enemies** (a lament). **1–2**: Cry for
help. *Evening sacrifice,* Ex 29.38–42; 1 Kings
18.36. **3–7**: A prayer that the writer may never compromise with the wicked, whose destruction is sure. **5**: *Oil,* compare 133.2. **7**:

Sheol, see 88.3–6 n. **8–10**: Prayer for deliverance.
    Ps 142: **Prayer for deliverance from
personal enemies** (a lament). **1–3a**: Cry for
help. **3b–4**: The psalmist's situation. **5–7**:
Prayer for deliverance.

7 Bring me out of prison,
   so that I may give thanks to
      your name.
   The righteous will surround me,
      for you will deal bountifully
         with me.

Psalm 143

A Psalm of David.

1 Hear my prayer, O LORD;
   give ear to my supplications in
      your faithfulness;
   answer me in your
      righteousness.
2 Do not enter into judgment with
      your servant,
   for no one living is righteous
      before you.

3 For the enemy has pursued me,
   crushing my life to the ground,
   making me sit in darkness like
      those long dead.
4 Therefore my spirit faints
      within me;
   my heart within me is appalled.

5 I remember the days of old,
   I think about all your deeds,
   I meditate on the works of your
      hands.
6 I stretch out my hands to you;
   my soul thirsts for you like a
      parched land.   *Selah*

7 Answer me quickly, O LORD;
   my spirit fails.
   Do not hide your face from me,
      or I shall be like those who go
         down to the Pit.
8 Let me hear of your steadfast love
      in the morning,
   for in you I put my trust.

Teach me the way I should go,
   for to you I lift up my soul.

9 Save me, O LORD, from my
      enemies;
   I have fled to you for refuge. *v*
10 Teach me to do your will,
      for you are my God.
   Let your good spirit lead me
      on a level path.

11 For your name's sake, O LORD,
      preserve my life.
   In your righteousness bring me
      out of trouble.
12 In your steadfast love cut off my
      enemies,
   and destroy all my adversaries,
      for I am your servant.

Psalm 144

Of David.

1 Blessed be the LORD, my rock,
   who trains my hands for war,
      and my fingers for battle;
2 my rock *w* and my fortress,
   my stronghold and my
      deliverer,
   my shield, in whom I take refuge,
   who subdues the peoples *x*
      under me.

3 O LORD, what are human beings
      that you regard them,
   or mortals that you think of
      them?
4 They are like a breath;
   their days are like a passing
      shadow.

v One Heb Ms Gk: MT *to you I have hidden*
w With 18.2 and 2 Sam 22.2: Heb *my steadfast
love* *x* Heb Mss Syr Aquila Jerome: MT *my
people*

**Ps 143: Prayer for deliverance from
personal enemies** (a lament). **1–2:** Cry for
vindication. **3–6:** The psalmist's situation. **6:**
Selah, see 3.2 n.
143.7–12: Prayer for deliverance. **7:** *Pit,*
see 16.10 n.
Ps 144: A king prays for deliverance

from his enemies (a royal lament). **1–2:**
Praise to God for his protecting power. **3–4:**
The weakness and inadequacy of human beings (compare 8.4; 90.5–6; 146.3–4).
144.5–8: Prayer for victory (compare
18.6–17). **9–10:** The vow (see 7.17 n.).

5 Bow your heavens, O Lord, and
            come down;
      touch the mountains so that
            they smoke.
6 Make the lightning flash and
            scatter them;
      send out your arrows and rout
            them.
7 Stretch out your hand from on
            high;
      set me free and rescue me from
            the mighty waters,
      from the hand of aliens,
8 whose mouths speak lies,
      and whose right hands are
            false.

9 I will sing a new song to you,
            O God;
      upon a ten-stringed harp I will
            play to you,
10 the one who gives victory to
            kings,
      who rescues his servant David.
11 Rescue me from the cruel sword,
      and deliver me from the hand of
            aliens,
      whose mouths speak lies,
      and whose right hands are false.

12 May our sons in their youth
      be like plants full grown,
      our daughters like corner pillars,
      cut for the building of a palace.
13 May our barns be filled,
      with produce of every kind;
      may our sheep increase by
            thousands,
      by tens of thousands in our
            fields,
14   and may our cattle be heavy
            with young.
      May there be no breach in the
            walls, *y* no exile,

and no cry of distress in our
      streets.
15 Happy are the people to whom
      such blessings fall;
      happy are the people whose
            God is the Lord.

Psalm 145

Praise. Of David.

1 I will extol you, my God and
            King,
      and bless your name forever
            and ever.
2 Every day I will bless you,
      and praise your name forever
            and ever.
3 Great is the Lord, and greatly to
            be praised;
      his greatness is unsearchable.

4 One generation shall laud your
            works to another,
      and shall declare your mighty
            acts.
5 On the glorious splendor of your
            majesty,
      and on your wondrous works, I
            will meditate.
6 The might of your awesome deeds
            shall be proclaimed,
      and I will declare your
            greatness.
7 They shall celebrate the fame of
            your abundant goodness,
      and shall sing aloud of your
            righteousness.

8 The Lord is gracious and
            merciful,
      slow to anger and abounding in
            steadfast love.

y Heb lacks *in the walls*

144.12–15: Prayer for a prosperous year,
perhaps originally a separate psalm.
**Ps 145: Hymn epitomizing the charac-
ter of the God of Israel.** An acrostic psalm,
each new verse beginning with the next in
order of the Hebrew alphabet (see Ps 119 n.);
the verse beginning with the Hebrew letter
"nun" appears as the third and fourth lines of
verse 13 (see note *b*). **1–3:** The psalmist's personal expression of praise. **4–7:** God's wonderful deeds.
145.8–9: The love of God (Ex 34.6; Num

9 The LORD is good to all,
and his compassion is over all
that he has made.

10 All your works shall give thanks
to you, O LORD,
and all your faithful shall
bless you.
11 They shall speak of the glory of
your kingdom,
and tell of your power,
12 to make known to all people
your*z* mighty deeds,
and the glorious splendor of
your*a* kingdom.
13 Your kingdom is an everlasting
kingdom,
and your dominion endures
throughout all generations.

The LORD is faithful in all his
words,
and gracious in all his deeds.*b*
14 The LORD upholds all who are
falling,
and raises up all who are bowed
down.
15 The eyes of all look to you,
and you give them their food in
due season.
16 You open your hand,
satisfying the desire of every
living thing.
17 The LORD is just in all his ways,
and kind in all his doings.
18 The LORD is near to all who call
on him,
to all who call on him in truth.
19 He fulfills the desire of all who
fear him;
he also hears their cry, and saves
them.
20 The LORD watches over all who
love him,

but all the wicked he will
destroy.
21 My mouth will speak the praise of
the LORD,
and all flesh will bless his holy
name forever and ever.

Psalm 146

1 Praise the LORD!
Praise the LORD, O my soul!
2 I will praise the LORD as long as
I live;
I will sing praises to my God all
my life long.

3 Do not put your trust in princes,
in mortals, in whom there is no
help.
4 When their breath departs, they
return to the earth;
on that very day their plans
perish.

5 Happy are those whose help is the
God of Jacob,
whose hope is in the LORD their
God,
6 who made heaven and earth,
the sea, and all that is in them;
who keeps faith forever;
7 who executes justice for the
oppressed;
who gives food to the hungry.

The LORD sets the prisoners free;
8 the LORD opens the eyes of
the blind.
The LORD lifts up those who are
bowed down;
the LORD loves the righteous.

z Gk Jerome Syr: Heb *his* *a* Heb *his*
b These two lines supplied by Q Ms Gk Syr

14.18; etc.). **10–13a:** The kingship of God (see
Ps 93 n.).
  145.13b–20: God's providential care for
his creatures. **21:** Return to the personal mood
of vv. 1–3.
  **Ps 146: Hymn praising God for his
help.** Pss 146–150 all begin and end with

"Hallelujah!" (see Ps 111.1 n.). **1–2:** Personal
expression of praise (compare 145.1–3). **3–4:**
The inadequacy of human beings (compare
144.3–4).
  146.5–9: The LORD gives help to all who
need it. **10:** Concluding expression of praise.

9 The LORD watches over the
strangers;
he upholds the orphan and the
widow,
but the way of the wicked he
brings to ruin.

10 The LORD will reign forever,
your God, O Zion, for all
generations.
Praise the LORD!

Psalm 147

1 Praise the LORD!
How good it is to sing praises to
our God;
for he is gracious, and a song of
praise is fitting.
2 The LORD builds up Jerusalem;
he gathers the outcasts of Israel.
3 He heals the brokenhearted,
and binds up their wounds.
4 He determines the number of
the stars;
he gives to all of them their
names.
5 Great is our Lord, and abundant
in power;
his understanding is beyond
measure.
6 The LORD lifts up the
downtrodden;
he casts the wicked to the
ground.

7 Sing to the LORD with
thanksgiving;
make melody to our God on
the lyre.
8 He covers the heavens with
clouds,
prepares rain for the earth,
makes grass grow on the hills.
9 He gives to the animals their food,
and to the young ravens when
they cry.

10 His delight is not in the strength
of the horse,
nor his pleasure in the speed of
a runner;[c]
11 but the LORD takes pleasure in
those who fear him,
in those who hope in his
steadfast love.

12 Praise the LORD, O Jerusalem!
Praise your God, O Zion!
13 For he strengthens the bars of
your gates;
he blesses your children within
you.
14 He grants peace[d] within your
borders;
he fills you with the finest of
wheat.
15 He sends out his command to the
earth;
his word runs swiftly.
16 He gives snow like wool;
he scatters frost like ashes.
17 He hurls down hail like crumbs—
who can stand before his cold?
18 He sends out his word, and melts
them;
he makes his wind blow, and
the waters flow.
19 He declares his word to Jacob,
his statutes and ordinances to
Israel.
20 He has not dealt thus with any
other nation;
they do not know his
ordinances.
Praise the LORD!

Psalm 148

1 Praise the LORD!
Praise the LORD from the heavens;
praise him in the heights!
2 Praise him, all his angels;
praise him, all his host!

c Heb *legs of a person*   d Or *prosperity*

**Ps 147: Hymn praising God for his uni-
versal power and providential care. 4**: Isa
40.26. **16–18**: Compare Job 37.9–11.

**Ps 148: Hymn calling upon all created
things to praise the LORD. 1–6**: All things in
heaven called to praise him.

3 Praise him, sun and moon;
    praise him, all you shining stars!
4 Praise him, you highest heavens,
    and you waters above the
        heavens!

5 Let them praise the name of
    the LORD,
    for he commanded and they
        were created.
6 He established them forever
    and ever;
    he fixed their bounds, which
        cannot be passed. *e*

7 Praise the LORD from the earth,
    you sea monsters and all deeps,
8 fire and hail, snow and frost,
    stormy wind fulfilling his
        command!

9 Mountains and all hills,
    fruit trees and all cedars!
10 Wild animals and all cattle,
    creeping things and flying birds!

11 Kings of the earth and all peoples,
    princes and all rulers of the
        earth!
12 Young men and women alike,
    old and young together!

13 Let them praise the name of
    the LORD,
    for his name alone is exalted;
    his glory is above earth and
        heaven.
14 He has raised up a horn for his
    people,
    praise for all his faithful,
    for the people of Israel who are
        close to him.
    Praise the LORD!

Psalm 149

1 Praise the LORD!
  Sing to the LORD a new song,
    his praise in the assembly of the
        faithful.
2 Let Israel be glad in its Maker;
    let the children of Zion rejoice
        in their King.
3 Let them praise his name with
    dancing,
    making melody to him with
        tambourine and lyre.
4 For the LORD takes pleasure in
    his people;
    he adorns the humble with
        victory.
5 Let the faithful exult in glory;
    let them sing for joy on their
        couches.
6 Let the high praises of God be in
    their throats
    and two-edged swords in their
        hands,
7 to execute vengeance on the
    nations
    and punishment on the peoples,
8 to bind their kings with fetters
    and their nobles with chains
        of iron,
9 to execute on them the judgment
    decreed.
    This is glory for all his faithful
        ones.
  Praise the LORD!

Psalm 150

1 Praise the LORD!
  Praise God in his sanctuary;
    praise him in his mighty
        firmament!*f*

---

e Or *he set a law that cannot pass away*
f Or *dome*

148.7–10: All animals, plants, and objects
on earth called to praise him. **11–14:** Everyone
summoned to join in the chorus. **14:** *Horn,* see
75.4–5 n.
Ps 149: Hymn to accompany a festival

dance. 5: *On their couches,* the meaning is
uncertain; perhaps reclining on couches was
part of the festival drama. **6–9:** The dance
was evidently of war-like character.
Ps 150: Doxology marking the end of

2 Praise him for his mighty deeds;
  praise him according to his
    surpassing greatness!

3 Praise him with trumpet sound;
  praise him with lute and harp!
4 Praise him with tambourine and
    dance;

praise him with strings and
    pipe!
5 Praise him with clanging cymbals;
  praise him with loud clashing
    cymbals!
6 Let everything that breathes praise
    the LORD!
  Praise the LORD!

---

the Psalter (compare 41.13). Verses 3–5 indicate the nature of the instrumental accompaniment to the psalms. **1:** *Praise the LORD!* see

Ps 146 n. **3:** *Lute,* a stringed instrument. **6:** The climax of the psalm and a fitting conclusion to the book of Psalms.

Proverbs

The book consists essentially of several collections, as the titles indicate: (1) "The proverbs of Solomon" (1.1–9.18), where 1.1 serves also as the title of the book. (2) "The proverbs of Solomon" (10.1–22.16), and also the Solomonic collection in 25.1–29.27, supposedly preserved by the efforts of the "men of Hezekiah" (25.1). (3) "The words of the wise" (22.17–23.11), a title preserved in the Septuagint but incorporated into the text of the Hebrew at 22.17. (4) "Sayings of the wise" in 24.23–24. (5) "The words of Agur" in 30.1ff., which are followed by a series of numerical sayings. (6) "The words of Lemuel" in 31.1–10. (7) Finally, the acrostic poem on the "good wife," 31.11–31.

While some of the sayings may very well be pre-exilic, the work was edited in the post-exilic period, and 1.1–6 serves as an introduction or key to the whole. The attribution to Solomon is due to his legendary wisdom (1 Kings 4.29–34). The "proverbs" consist of sayings, commands and admonitions, and long poems. The sayings are usually two lines in parallel thought (see "Characteristics of Hebrew Poetry," pp. 392–397). The commands and admonitions usually have a motive clause; the poems are characteristic of chs 1–9. The book is typical of the "wisdom" of the Old Testament (see "The Poetical Books or Writings," pp. 623–624 OT) and also of the ancient Near East, especially Egypt. In fact, most scholars agree that 22.17–23.11 is in some way dependent upon the Instruction of the Egyptian sage, Amen-em-ope (see 22.17 n.).

The purpose of the book (1.1–6) is to transmit the insights whereby a youth might learn to cope with life. These were gathered from the tradition of the elders (e.g. 4.1–4) and from experience and observation (e.g. 6.6–11). Certain key emphases appear in the process of moral formation: honesty, diligence, trustworthiness, docility (the ability to "hear" or obey), control of one's appetites, the cultivation of true and proper speech, the correct attitudes towards riches and poverty, etc. Some sayings are observations, but more often a moral ideal is inculcated (e.g., the contrast between the righteous and the wicked, especially in chs 10–15). The underlying presupposition is that "the fear of the LORD is the beginning of wisdom" (9.10; 1.7; 15.33).

The long wisdom poems in chs 1–9 differ in style from the short staccato sayings that predominate in the rest of the book. They are strongly hortatory, somewhat in the spirit of Deuteronomy. Chapters 1, 8, and 9 present a personification of Wisdom, a woman preaching to simple youth. She and her proclamation of "life" (8.35) stand in contrast to the "strange woman" or prostitute (7.10), and to Dame Folly (9.13). The sayings in chs 10ff. have not been collected in a haphazard manner. Catchwords, plays on words, alliteration and assonance, and content are at work to create a certain unity. Some sayings have been repeated in other collections (e.g. 21.9 = 25.24).

The teaching of the sages is marked by an optimistic view of retribution. Wisdom (generally equated with righteousness) brings success; folly (or wickedness) brings destruction. This is also the teaching of Deuteronomy and other biblical works. The sages were not, however, unaware of the limitations of wisdom, and of the ambiguities and mysteries of life (3.11–12). There was no wisdom that could prevail against the LORD (21.30), whose work is mysterious and definitive (16.1, 2, 9). But retribution remained a problem, as the books of Job and Qoheleth (Ecclesiastes) demonstrate.

1 The proverbs of Solomon son of David, king of Israel:

2 For learning about wisdom and instruction,
  for understanding words of insight,
3 for gaining instruction in wise dealing,
  righteousness, justice, and equity;
4 to teach shrewdness to the simple,
  knowledge and prudence to the young—
5 Let the wise also hear and gain in learning,
  and the discerning acquire skill,
6 to understand a proverb and a figure,
  the words of the wise and their riddles.

7 The fear of the LORD is the beginning of knowledge;
  fools despise wisdom and instruction.

8 Hear, my child, your father's instruction,
  and do not reject your mother's teaching;
9 for they are a fair garland for your head,
  and pendants for your neck.

10 My child, if sinners entice you,
  do not consent.
11 If they say, "Come with us, let us lie in wait for blood;
  let us wantonly ambush the innocent;
12 like Sheol let us swallow them alive
  and whole, like those who go down to the Pit.
13 We shall find all kinds of costly things;
  we shall fill our houses with booty.
14 Throw in your lot among us;
  we will all have one purse"—
15 my child, do not walk in their way,
  keep your foot from their paths;
16 for their feet run to evil,
  and they hurry to shed blood.
17 For in vain is the net baited
  while the bird is looking on;
18 yet they lie in wait—to kill themselves!
  and set an ambush—for their own lives!
19 Such is the end*a* of all who are greedy for gain;
  it takes away the life of its possessors.

20 Wisdom cries out in the street;

a Gk: Heb *ways*

1.1–9.18: These chapters contain mostly long poems, as opposed to short sayings, which serve as an introduction to the book. The style is hortatory (*hear, my child,* 1.8). The personification of Wisdom, in opposition to Dame Folly (ch 9) and the *loose woman* (2.16), is particularly noteworthy. She threatens and cajoles her followers. See the Introduction concerning the title in 1.1, and also in 10.1 and 25.1.
1.2–6: This is a complete sentence in Hebrew, which states the purpose of the sage: to awaken students and provide them with moral formation leading to their maturity. Wisdom is essentially practical, aiming at proper conduct, rather than theoretical. One will *understand a proverb* when one's life style is in harmony with it. **6:** *A figure* is a metaphor, parable, or allegory such as in vv. 20–33. A

riddle is a comparison or analogy that enforces a lesson when the hearer discerns its intention and is able to complete it (see 25.14; 26.7; 30.18–19).
1.7: A basic wisdom theme: An attitude of awe and reverence toward God is the necessary preliminary to knowledge—the first stage of enlightenment; compare 9.10; Job 28.28; Sir 1.14. *Fools* are not the stupid, but those whose conduct is reprehensible.
1.8–19: Home-training is a moral safeguard against the temptations of the wicked. **12:** *Sheol . . . the Pit,* common expressions for death and the place and state of the dead (see Gen 37.35 n.). **17:** This popular proverb affirms that the consequences of crime are too obvious to be missed.
1.20–33: Wisdom personified as a prophetess. She speaks publicly at the city

in the squares she raises her
voice.

21 At the busiest corner she cries out;
at the entrance of the city gates
she speaks:

22 "How long, O simple ones, will
you love being simple?
How long will scoffers delight in
their scoffing
and fools hate knowledge?

23 Give heed to my reproof;
I will pour out my thoughts to
you;
I will make my words known
to you.

24 Because I have called and you
refused,
have stretched out my hand and
no one heeded,

25 and because you have ignored all
my counsel
and would have none of my
reproof,

26 I also will laugh at your calamity;
I will mock when panic strikes
you,

27 when panic strikes you like a
storm,
and your calamity comes like a
whirlwind,
when distress and anguish come
upon you.

28 Then they will call upon me, but I
will not answer;
they will seek me diligently, but
will not find me.

29 Because they hated knowledge
and did not choose the fear of
the LORD,

30 would have none of my counsel,
and despised all my reproof,

31 therefore they shall eat the fruit of
their way
and be sated with their own
devices.

32 For waywardness kills the simple,

and the complacency of fools
destroys them;

33 but those who listen to me will be
secure
and will live at ease, without
dread of disaster."

2 My child, if you accept my words
and treasure up my
commandments within you,

2 making your ear attentive to
wisdom
and inclining your heart to
understanding;

3 if you indeed cry out for insight,
and raise your voice for
understanding;

4 if you seek it like silver,
and search for it as for hidden
treasures—

5 then you will understand the fear
of the LORD
and find the knowledge of God.

6 For the LORD gives wisdom;
from his mouth come
knowledge and
understanding;

7 he stores up sound wisdom for the
upright;
he is a shield to those who walk
blamelessly,

8 guarding the paths of justice
and preserving the way of his
faithful ones.

9 Then you will understand
righteousness and justice
and equity, every good path;

10 for wisdom will come into your
heart,
and knowledge will be pleasant
to your soul;

11 prudence will watch over you;
and understanding will
guard you.

12 It will save you from the way of
evil,

---

gates where she can find an audience. In language that echoes the prophets (for v. 24 compare Isa 65.2, 12; 66.4; Jer 7.13, 24–27) she pronounces dire threats against those who will not listen (vv. 26–32).

2.1–22: The fruits of the search for wisdom. A very compact poem in 22 lines (the number of letters in the Hebrew alphabet) expressing "if" (vv. 1, 4) and "then" (vv. 5, 9).

from those who speak
  perversely,
13 who forsake the paths of
  uprightness
 to walk in the ways of darkness,
14 who rejoice in doing evil
  and delight in the perverseness
  of evil;
15 those whose paths are crooked,
  and who are devious in their
  ways.

16 You will be saved from the loose[b]
  woman,
 from the adulteress with her
  smooth words,
17 who forsakes the partner of her
  youth
 and forgets her sacred covenant;
18 for her way[c] leads down to death,
  and her paths to the shades;
19 those who go to her never come
  back,
 nor do they regain the paths
  of life.

20 Therefore walk in the way of the
  good,
 and keep to the paths of the
  just.
21 For the upright will abide in the
  land,
 and the innocent will remain
  in it;
22 but the wicked will be cut off
  from the land,
 and the treacherous will be
  rooted out of it.

3 My child, do not forget my
  teaching,
 but let your heart keep my
  commandments;
2 for length of days and years of life
 and abundant welfare they will
  give you.

3 Do not let loyalty and faithfulness
  forsake you;
 bind them around your neck,
 write them on the tablet of your
  heart.
4 So you will find favor and good
  repute
 in the sight of God and of
  people.

5 Trust in the LORD with all your
  heart,
 and do not rely on your own
  insight.
6 In all your ways acknowledge
  him,
 and he will make straight your
  paths.
7 Do not be wise in your own eyes;
 fear the LORD, and turn away
  from evil.
8 It will be a healing for your flesh
 and a refreshment for your
  body.

9 Honor the LORD with your
  substance
 and with the first fruits of all
  your produce;
10 then your barns will be filled with
  plenty,
 and your vats will be bursting
  with wine.

11 My child, do not despise the
  LORD's discipline
 or be weary of his reproof,
12 for the LORD reproves the one he
  loves,
 as a father the son in whom he
  delights.

13 Happy are those who find
  wisdom,

b Heb *strange* c Cn: Heb *house*

2.16: *Loose woman*, literally, "strange woman"; although the identity is not certain, probably the wife of another (see ch 7). **17:** *Covenant of her God:* Marital faithfulness is a sacred obligation (Gen 2.24; Hos ch 2; Mal 2.14).

3.1–12: Six admonitions concerning conduct that leads to prosperity. Worthy of note is the idea that suffering is a *discipline*, a sign of divine love (vv. 11–12; see also Heb 12.5–6). **8:** *Body*, Hebrew "navel." **3.13–18: The incomparable value of wis-

and those who get
understanding,

14 for her income is better than
silver,
and her revenue better than
gold.

15 She is more precious than jewels,
and nothing you desire can
compare with her.

16 Long life is in her right hand;
in her left hand are riches and
honor.

17 Her ways are ways of
pleasantness,
and all her paths are peace.

18 She is a tree of life to those who
lay hold of her;
those who hold her fast are
called happy.

19 The LORD by wisdom founded the
earth;
by understanding he established
the heavens;

20 by his knowledge the deeps broke
open,
and the clouds drop down the
dew.

21 My child, do not let these escape
from your sight:
keep sound wisdom and
prudence,

22 and they will be life for your soul
and adornment for your neck.

23 Then you will walk on your way
securely
and your foot will not stumble.

24 If you sit down, *d* you will not be
afraid;
when you lie down, your sleep
will be sweet.

25 Do not be afraid of sudden panic,
or of the storm that strikes the
wicked;

26 for the LORD will be your
confidence

and will keep your foot from
being caught.

27 Do not withhold good from those
to whom it is due, *e*
when it is in your power to do
it.

28 Do not say to your neighbor,
"Go, and come again,
tomorrow I will give it"—when
you have it with you.

29 Do not plan harm against your
neighbor
who lives trustingly beside you.

30 Do not quarrel with anyone
without cause,
when no harm has been done
to you.

31 Do not envy the violent
and do not choose any of their
ways;

32 for the perverse are an
abomination to the LORD,
but the upright are in his
confidence.

33 The LORD's curse is on the house
of the wicked,
but he blesses the abode of the
righteous.

34 Toward the scorners he is
scornful,
but to the humble he shows
favor.

35 The wise will inherit honor,
but stubborn fools, disgrace.

4 Listen, children, to a father's
instruction,
and be attentive, that you may
gain *f* insight;

2 for I give you good precepts:
do not forsake my teaching.

---

d Gk: Heb *lie down*   *e* Heb *from its owners*
f Heb *know*

---

dom: she brings *life, riches and honors.* **19–20:**
Wisdom had a role in creation (8.22–31),
which is sustained by water from the deeps
(Ex 20.4) and dew from above.
3.21–35: An admonition followed by six

prohibitions, and a statement about the fate
of the righteous and wicked.
4.1–27: An urgent appeal to acquire wisdom, after the example of the teacher (vv. 1–
7), who describes life in terms of the two

3 When I was a son with my father,
      tender, and my mother's
         favorite,
4 he taught me, and said to me,
      "Let your heart hold fast my
         words;
      keep my commandments,
         and live.
5 Get wisdom; get insight: do not
      forget, nor turn away
      from the words of my mouth.
6 Do not forsake her, and she will
      keep you;
      love her, and she will guard
         you.
7 The beginning of wisdom is this:
      Get wisdom,
      and whatever else you get, get
         insight.
8 Prize her highly, and she will
      exalt you;
      she will honor you if you
         embrace her.
9 She will place on your head a fair
      garland;
      she will bestow on you a
         beautiful crown."

10 Hear, my child, and accept my
      words,
      that the years of your life may
         be many.
11 I have taught you the way of
      wisdom;
      I have led you in the paths of
         uprightness.
12 When you walk, your step will
      not be hampered;
      and if you run, you will not
         stumble.
13 Keep hold of instruction; do not
      let go;
      guard her, for she is your life.
14 Do not enter the path of the
      wicked,
      and do not walk in the way of
         evildoers.

15 Avoid it; do not go on it;
      turn away from it and pass on.
16 For they cannot sleep unless they
      have done wrong;
      they are robbed of sleep unless
      they have made someone
         stumble.
17 For they eat the bread of
      wickedness
      and drink the wine of violence.
18 But the path of the righteous is
      like the light of dawn,
      which shines brighter and
         brighter until full day.
19 The way of the wicked is like
      deep darkness;
      they do not know what they
         stumble over.
20 My child, be attentive to my
      words;
      incline your ear to my sayings.
21 Do not let them escape from
      your sight;
      keep them within your heart.
22 For they are life to those who
      find them,
      and healing to all their flesh.
23 Keep your heart with all vigilance,
      for from it flow the springs
         of life.
24 Put away from you crooked
      speech,
      and put devious talk far
         from you.
25 Let your eyes look directly
      forward,
      and your gaze be straight
         before you.
26 Keep straight the path of your
      feet,
      and all your ways will be sure.
27 Do not swerve to the right or to
      the left;
      turn your foot away from evil.

5 My child, be attentive to my
      wisdom;

ways: wisdom's *way* or *path* (vv. 11, 18), and the *path* or *way* of the wicked (vv. 14, 19).
  5.1–23: A warning (vv. 1–15) to avoid the *loose woman* who leads to *death* (2.16–19;

7.26–27; 9.18) is followed by encouragement to *rejoice* in one's own spouse (v. 15, *cistern*, v. 18, *fountain*).

incline your ear to my
understanding,
2 so that you may hold on to
prudence,
and your lips may guard
knowledge.
3 For the lips of a loose*g* woman
drip honey,
and her speech is smoother
than oil;
4 but in the end she is bitter as
wormwood,
sharp as a two-edged sword.
5 Her feet go down to death;
her steps follow the path to
Sheol.
6 She does not keep straight to the
path of life;
her ways wander, and she does
not know it.

7 And now, my child,*h* listen to
me,
and do not depart from the
words of my mouth.
8 Keep your way far from her,
and do not go near the door of
her house;
9 or you will give your honor to
others,
and your years to the merciless,
10 and strangers will take their fill of
your wealth,
and your labors will go to the
house of an alien;
11 and at the end of your life you
will groan,
when your flesh and body are
consumed,
12 and you say, "Oh, how I hated
discipline,
and my heart despised reproof!
13 I did not listen to the voice of my
teachers
or incline my ear to my
instructors.
14 Now I am at the point of utter
ruin
in the public assembly."

15 Drink water from your own
cistern,
flowing water from your own
well.
16 Should your springs be scattered
abroad,
streams of water in the streets?
17 Let them be for yourself alone,
and not for sharing with
strangers.
18 Let your fountain be blessed,
and rejoice in the wife of your
youth,
19 a lovely deer, a graceful doe.
May her breasts satisfy you at all
times;
may you be intoxicated always
by her love.
20 Why should you be intoxicated,
my son, by another woman
and embrace the bosom of an
adulteress?
21 For human ways are under the
eyes of the LORD,
and he examines all their paths.
22 The iniquities of the wicked
ensnare them,
and they are caught in the toils
of their sin.
23 They die for lack of discipline,
and because of their great folly
they are lost.

6 My child, if you have given your
pledge to your neighbor,
if you have bound yourself to
another,*i*
2 you are snared by the utterance of
your lips,*j*
caught by the words of your
mouth.
3 So do this, my child, and save
yourself,
for you have come into your
neighbor's power:
go, hurry,*k* and plead with your
neighbor.

g Heb *strange*   h Gk Vg: Heb *children*
i Or *a stranger*   j Cn Compare Gk Syr: Heb
the words of your mouth   k Or *humble yourself*

6.1–35: Warnings. 1–5: Do not offer your
property as collateral for your neighbor's
debts (apparently a source of trouble in Israelite society; see 11.15).

4 Give your eyes no sleep
  and your eyelids no slumber;
5 save yourself like a gazelle from
    the hunter,[l]
  like a bird from the hand of the
    fowler.

6 Go to the ant, you lazybones;
  consider its ways, and be wise.
7 Without having any chief
  or officer or ruler,
8 it prepares its food in summer,
  and gathers its sustenance in
    harvest.
9 How long will you lie there,
    O lazybones?
  When will you rise from your
    sleep?
10 A little sleep, a little slumber,
   a little folding of the hands
     to rest,
11 and poverty will come upon you
     like a robber,
   and want, like an armed
     warrior.

12 A scoundrel and a villain
   goes around with crooked
     speech,
13 winking the eyes, shuffling the
     feet,
   pointing the fingers,
14 with perverted mind devising evil,
   continually sowing discord;
15 on such a one calamity will
     descend suddenly;
   in a moment, damage beyond
     repair.

16 There are six things that the LORD
     hates,
   seven that are an abomination
     to him:
17 haughty eyes, a lying tongue,
   and hands that shed innocent
     blood,
18 a heart that devises wicked plans,
   feet that hurry to run to evil,
19 a lying witness who testifies
     falsely,
   and one who sows discord in a
     family.

20 My child, keep your father's
     commandment,
   and do not forsake your
     mother's teaching.
21 Bind them upon your heart
     always;
   tie them around your neck.
22 When you walk, they[m] will
     lead you;
   when you lie down, they[m] will
     watch over you;
   and when you awake, they[m]
     will talk with you.
23 For the commandment is a lamp
     and the teaching a light,
   and the reproofs of discipline are
     the way of life,
24 to preserve you from the wife of
     another,[n]
   from the smooth tongue of the
     adulteress.
25 Do not desire her beauty in your
     heart,
   and do not let her capture you
     with her eyelashes;
26 for a prostitute's fee is only a loaf
     of bread,[o]
   but the wife of another stalks a
     man's very life.
27 Can fire be carried in the bosom
     without burning one's clothes?
28 Or can one walk on hot coals
     without scorching the feet?
29 So is he who sleeps with his
     neighbor's wife;
   no one who touches her will go
     unpunished.

l Cn: Heb *from the hand*   *m* Heb *it*
n Gk: MT *the evil woman*   *o* Cn Compare Gk
Syr Vg Tg: Heb *for because of a harlot to a piece of
bread*

6.6–11: Avoid laziness (see 24.30–34), as even the ants can tell you. **12–15:** Beware of shady, disreputable characters. **16–19:** A typical numerical saying (30.14–31; Am 1.3–2.8) about seven ugly vices. **20–35:** Another warning against adultery, stressing the penalties one will have to pay (vv. 26, 31–34).

30 Thieves are not despised who
  steal only
   to satisfy their appetite when
    they are hungry.
31 Yet if they are caught, they will
  pay sevenfold;
   they will forfeit all the goods of
    their house.
32 But he who commits adultery has
  no sense;
   he who does it destroys himself.
33 He will get wounds and dishonor,
   and his disgrace will not be
    wiped away.
34 For jealousy arouses a husband's
  fury,
   and he shows no restraint when
    he takes revenge.
35 He will accept no compensation,
   and refuses a bribe no matter
    how great.

7 My child, keep my words
   and store up my
    commandments with you;
2 keep my commandments and live,
   keep my teachings as the apple
    of your eye;
3 bind them on your fingers,
   write them on the tablet of your
    heart.
4 Say to wisdom, "You are my
  sister,"
   and call insight your intimate
    friend,
5 that they may keep you from the
    loose*p* woman,
   from the adulteress with her
    smooth words.

6 For at the window of my house
   I looked out through my lattice,
7 and I saw among the simple ones,
   I observed among the youths,
   a young man without sense,
8 passing along the street near her
    corner,

taking the road to her house
9 in the twilight, in the evening,
   at the time of night and
    darkness.
10 Then a woman comes toward
  him,
   decked out like a prostitute,
    wily of heart. *q*
11 She is loud and wayward;
   her feet do not stay at home;
12 now in the street, now in the
    squares,
   and at every corner she lies
    in wait.
13 She seizes him and kisses him,
   and with impudent face she says
    to him:
14 "I had to offer sacrifices,
   and today I have paid my vows;
15 so now I have come out to
    meet you,
   to seek you eagerly, and I have
    found you!
16 I have decked my couch with
    coverings,
   colored spreads of Egyptian
    linen;
17 I have perfumed my bed with
    myrrh,
   aloes, and cinnamon.
18 Come, let us take our fill of love
    until morning;
   let us delight ourselves with
    love.
19 For my husband is not at home;
   he has gone on a long journey.
20 He took a bag of money with
  him;
   he will not come home until full
    moon."
21 With much seductive speech she
    persuades him;

p Heb *strange*   *q* Meaning of Heb uncertain

**7.1–27: Wisdom as safeguard against
adultery.** One is to prefer Lady Wisdom as
sister and *friend* (terms of endearment, in view

of marriage; compare Song 4.9–12). Hence
the teacher gives a lengthy description of how
easily one can be seduced (vv. 6–27).

with her smooth talk she
compels him.

22 Right away he follows her,
and goes like an ox to the
slaughter,
or bounds like a stag toward
the trap[r]

23 until an arrow pierces its
entrails.
He is like a bird rushing into a
snare,
not knowing that it will cost
him his life.

24 And now, my children, listen to
me,
and be attentive to the words of
my mouth.

25 Do not let your hearts turn aside
to her ways;
do not stray into her paths.

26 for many are those she has
laid low,
and numerous are her victims.

27 Her house is the way to Sheol,
going down to the chambers of
death.

8 Does not wisdom call,
and does not understanding raise
her voice?

2 On the heights, beside the way,
at the crossroads she takes her
stand;

3 beside the gates in front of the
town,
at the entrance of the portals she
cries out:

4 "To you, O people, I call,
and my cry is to all that live.

5 O simple ones, learn prudence;
acquire intelligence, you who
lack it.

6 Hear, for I will speak noble
things,
and from my lips will come
what is right;

7 for my mouth will utter truth;
wickedness is an abomination to
my lips.

8 All the words of my mouth are
righteous;
there is nothing twisted or
crooked in them.

9 They are all straight to one who
understands
and right to those who find
knowledge.

10 Take my instruction instead of
silver,
and knowledge rather than
choice gold;

11 for wisdom is better than jewels,
and all that you may desire
cannot compare with her.

12 I, wisdom, live with prudence,[s]
and I attain knowledge and
discretion.

13 The fear of the LORD is hatred
of evil.
Pride and arrogance and the way
of evil
and perverted speech I hate.

14 I have good advice and sound
wisdom;
I have insight, I have strength.

15 By me kings reign,
and rulers decree what is just;

16 by me rulers rule,
and nobles, all who govern
rightly.

17 I love those who love me,
and those who seek me
diligently find me.

18 Riches and honor are with me,
enduring wealth and prosperity.

19 My fruit is better than gold, even
fine gold,
and my yield than choice silver.

20 I walk in the way of
righteousness,

r Cn Compare Gk: Meaning of Heb uncertain
s Meaning of Heb uncertain

8.1–36: Lady Wisdom. She pronounces a public address (see 1.20–21), proclaiming her worth (vv. 6–11), her authority (vv. 12–16), and her rewards (vv. 17–21). In vv. 22–31 she describes her divine origin (vv. 22, 24, 26) before anything was created (this is asserted repeatedly in vv. 22–30). She was with God as a *master worker* (v. 30), and thus had some

along the paths of justice,
21 endowing with wealth those who
love me,
and filling their treasuries.
22 The LORD created me at the
beginning[t] of his work,[u]
the first of his acts of long ago.
23 Ages ago I was set up,
at the first, before the beginning
of the earth.
24 When there were no depths I was
brought forth,
when there were no springs
abounding with water.
25 Before the mountains had been
shaped,
before the hills, I was brought
forth—
26 when he had not yet made earth
and fields,[v]
or the world's first bits of soil.
27 When he established the heavens, I
was there,
when he drew a circle on the
face of the deep,
28 when he made firm the skies
above,
when he established the
fountains of the deep,
29 when he assigned to the sea its
limit,
so that the waters might not
transgress his command,
when he marked out the
foundations of the earth,
30 then I was beside him, like a
master worker;[w]
and I was daily his[x] delight,
rejoicing before him always,
31 rejoicing in his inhabited world
and delighting in the
human race.

32 And now, my children, listen to
me:

happy are those who keep
my ways.
33 Hear instruction and be wise,
and do not neglect it.
34 Happy is the one who listens to
me,
watching daily at my gates,
waiting beside my doors.
35 For whoever finds me finds life
and obtains favor from the
LORD;
36 but those who miss me injure
themselves;
all who hate me love death."

9 Wisdom has built her house,
she has hewn her seven pillars.
2 She has slaughtered her animals,
she has mixed her wine,
she has also set her table.
3 She has sent out her servant-girls,
she calls
from the highest places in
the town,
4 "You that are simple, turn in
here!"
To those without sense she says,
5 "Come, eat of my bread
and drink of the wine I have
mixed.
6 Lay aside immaturity,[y] and live,
and walk in the way of insight."

7 Whoever corrects a scoffer wins
abuse;
whoever rebukes the wicked
gets hurt.
8 A scoffer who is rebuked will only
hate you;
the wise, when rebuked, will
love you.

t Or *me as the beginning*   u Heb *way*
v Meaning of Heb uncertain   w Another
reading is *little child*   x Gk: Heb lacks *his*
y Or *simpleness*

unspecified role in the creative activity (Sir
1.9–10; Wis 7.22). Furthermore, she is all *delight, rejoicing* before God and *delighting in* his
human creatures.
8.32–36: The conclusion of the speech is
an offer of *life* in place of *death*.

9.1–18: Lady Wisdom invites the unwise,
or *simple*, to a banquet (vv. 1–6), while Dame
Folly invites them to her *stolen water*. Between the invitations are aphorisms about
scoffers and the wise. **10:** See 1.7 n.

9 Give instruction[z] to the wise, and
    they will become wiser still;
  teach the righteous and they will
    gain in learning.
10 The fear of the LORD is the
    beginning of wisdom,
  and the knowledge of the Holy
    One is insight.
11 For by me your days will be
    multiplied,
  and years will be added to your
    life.
12 If you are wise, you are wise for
    yourself;
  if you scoff, you alone will
    bear it.

13 The foolish woman is loud;
  she is ignorant and knows
    nothing.
14 She sits at the door of her house,
  on a seat at the high places of
    the town,
15 calling to those who pass by,
  who are going straight on their
    way,
16 "You who are simple, turn in
    here!"
  And to those without sense she
    says,
17 "Stolen water is sweet,
  and bread eaten in secret is
    pleasant."
18 But they do not know that the
    dead[a] are there,
  that her guests are in the depths
    of Sheol.

10 The proverbs of Solomon.

A wise child makes a glad father,
  but a foolish child is a mother's
    grief.
2 Treasures gained by wickedness
    do not profit,

  but righteousness delivers from
    death.
3 The LORD does not let the
    righteous go hungry,
  but he thwarts the craving of
    the wicked.
4 A slack hand causes poverty,
  but the hand of the diligent
    makes rich.
5 A child who gathers in summer is
    prudent,
  but a child who sleeps in harvest
    brings shame.
6 Blessings are on the head of the
    righteous,
  but the mouth of the wicked
    conceals violence.
7 The memory of the righteous is a
    blessing,
  but the name of the wicked
    will rot.
8 The wise of heart will heed
    commandments,
  but a babbling fool will come
    to ruin.
9 Whoever walks in integrity walks
    securely,
  but whoever follows perverse
    ways will be found out.
10 Whoever winks the eye causes
    trouble,
  but the one who rebukes boldly
    makes peace.[b]
11 The mouth of the righteous is a
    fountain of life,
  but the mouth of the wicked
    conceals violence.
12 Hatred stirs up strife,
  but love covers all offenses.
13 On the lips of one who has
    understanding wisdom
    is found,

z Heb lacks *instruction*   a Heb *shades*
b Gk· Heb *but a babbling fool will come to ruin*

**10.1–22.16: A collection of proverbial
sayings.** The discrete sayings in this collection are often bound together by catchwords and general context. There is a general
equation between wise and righteous on the
one hand, fool and wicked on the other. Anti-

thetical parallelism (see "Characteristics of
Hebrew Poetry," p. 392) is characteristic of
chs 10–15. See the Introduction, p. 802 OT.
    10.2: An early or unhappy *death* was con-
sidered as punishment for sin; hence long life
is the portion of the just or wise (3.2; 4.10;

but a rod is for the back of one
who lacks sense.

14 The wise lay up knowledge,
but the babbling of a fool brings
ruin near.

15 The wealth of the rich is their
fortress;
the poverty of the poor is their
ruin.

16 The wage of the righteous leads
to life,
the gain of the wicked to sin.

17 Whoever heeds instruction is on
the path to life,
but one who rejects a rebuke
goes astray.

18 Lying lips conceal hatred,
and whoever utters slander is
a fool.

19 When words are many,
transgression is not lacking,
but the prudent are restrained in
speech.

20 The tongue of the righteous is
choice silver;
the mind of the wicked is of
little worth.

21 The lips of the righteous feed
many,
but fools die for lack of sense.

22 The blessing of the LORD makes
rich,
and he adds no sorrow with it. *c*

23 Doing wrong is like sport to a
fool,
but wise conduct is pleasure to a
person of understanding.

24 What the wicked dread will come
upon them,
but the desire of the righteous
will be granted.

25 When the tempest passes, the
wicked are no more,

but the righteous are established
forever.

26 Like vinegar to the teeth, and
smoke to the eyes,
so are the lazy to their
employers.

27 The fear of the LORD prolongs life,
but the years of the wicked will
be short.

28 The hope of the righteous ends in
gladness,
but the expectation of the
wicked comes to nothing.

29 The way of the LORD is a
stronghold for the upright,
but destruction for evildoers.

30 The righteous will never be
removed,
but the wicked will not remain
in the land.

31 The mouth of the righteous brings
forth wisdom,
but the perverse tongue will be
cut off.

32 The lips of the righteous know
what is acceptable,
but the mouth of the wicked
what is perverse.

11 A false balance is an
abomination to the LORD,
but an accurate weight is his
delight.

2 When pride comes, then comes
disgrace;
but wisdom is with the humble.

3 The integrity of the upright guides
them,
but the crookedness of the
treacherous destroys them.

4 Riches do not profit in the day
of wrath,

c Or *and toil adds nothing to it*

9.11; 10.27). The perspective of the writer is
this-worldly; it must all happen here before
one dies and goes to Sheol; hence God must
show justice by rewarding the righteous and
the wicked (as in 10.3; see also Ps 37). This
doctrine of retribution appears frequently
here (compare 11.21) and in the rest of the
Bible, but the books of Job and Ecclesiastes
are a warning against over-simplification.
10.15: As it stands, this is a mere observation about the difference that *wealth* makes,
but it should be noted that in 18.11 it is considered to be only an imaginary protection;
see also 11.4. **22:** See note *c*; wealth comes
from the LORD's blessing, not from hard toil.
11.1: Compare Deut 25.13–16.

but righteousness delivers from
  death.
5 The righteousness of the blameless
    keeps their ways straight,
  but the wicked fall by their own
    wickedness.
6 The righteousness of the upright
    saves them,
  but the treacherous are taken
    captive by their schemes.
7 When the wicked die, their hope
    perishes,
  and the expectation of the
    godless comes to nothing.
8 The righteous are delivered from
    trouble,
  and the wicked get into it
    instead.
9 With their mouths the godless
    would destroy their
    neighbors,
  but by knowledge the righteous
    are delivered.
10 When it goes well with the
    righteous, the city rejoices;
  and when the wicked perish,
    there is jubilation.
11 By the blessing of the upright a
    city is exalted,
  but it is overthrown by the
    mouth of the wicked.
12 Whoever belittles another lacks
    sense,
  but an intelligent person remains
    silent.
13 A gossip goes about telling
    secrets,
  but one who is trustworthy in
    spirit keeps a confidence.
14 Where there is no guidance, a
    nation[d] falls,
  but in an abundance of
    counselors there is safety.
15 To guarantee loans for a stranger
    brings trouble,
  but there is safety in refusing
    to do so.
16 A gracious woman gets honor,

but she who hates virtue is
    covered with shame.[e]
  The timid become destitute,[f]
    but the aggressive gain riches.
17 Those who are kind reward
    themselves,
  but the cruel do themselves
    harm.
18 The wicked earn no real gain,
  but those who sow
    righteousness get a true
    reward.
19 Whoever is steadfast in
    righteousness will live,
  but whoever pursues evil will
    die.
20 Crooked minds are an
    abomination to the LORD,
  but those of blameless ways are
    his delight.
21 Be assured, the wicked will not go
    unpunished,
  but those who are righteous will
    escape.
22 Like a gold ring in a pig's snout
  is a beautiful woman without
    good sense.
23 The desire of the righteous ends
    only in good;
  the expectation of the wicked in
    wrath.
24 Some give freely, yet grow all the
    richer;
  others withhold what is due,
    and only suffer want.
25 A generous person will be
    enriched,
  and one who gives water will
    get water.
26 The people curse those who hold
    back grain,
  but a blessing is on the head of
    those who sell it.
27 Whoever diligently seeks good
    seeks favor,

d Or *an army*   e Compare Gk Syr: Heb lacks
but she . . . shame   f Gk: Heb lacks *The
timid . . . destitute*

11.15: Compare 6.1–5.
11.24: One of many paradoxical sayings.

25: *Water,* a precious commodity. **26:** *Hold
back grain,* for higher prices.

but evil comes to the one who
    searches for it.

28 Those who trust in their riches
    will wither,*g*
but the righteous will flourish
    like green leaves.

29 Those who trouble their
    households will inherit
    wind,
and the fool will be servant to
    the wise.

30 The fruit of the righteous is a tree
    of life,
but violence*h* takes lives away.

31 If the righteous are repaid on
    earth,
how much more the wicked and
    the sinner!

12 Whoever loves discipline loves
    knowledge,
but those who hate to be
    rebuked are stupid.

2 The good obtain favor from the
    LORD,
but those who devise evil he
    condemns.

3 No one finds security by
    wickedness,
but the root of the righteous
    will never be moved.

4 A good wife is the crown of her
    husband,
but she who brings shame is
    like rottenness in his bones.

5 The thoughts of the righteous
    are just;
the advice of the wicked is
    treacherous.

6 The words of the wicked are a
    deadly ambush,
but the speech of the upright
    delivers them.

7 The wicked are overthrown and
    are no more,
but the house of the righteous
    will stand.

8 One is commended for good
    sense,
but a perverse mind is despised.

9 Better to be despised and have a
    servant,
than to be self-important and
    lack food.

10 The righteous know the needs of
    their animals,
but the mercy of the wicked is
    cruel.

11 Those who till their land will have
    plenty of food,
but those who follow worthless
    pursuits have no sense.

12 The wicked covet the proceeds of
    wickedness,*i*
but the root of the righteous
    bears fruit.

13 The evil are ensnared by the
    transgression of their lips,
but the righteous escape from
    trouble.

14 From the fruit of the mouth one is
    filled with good things,
and manual labor has its reward.

15 Fools think their own way is
    right,
but the wise listen to advice.

16 Fools show their anger at once,
but the prudent ignore an insult.

17 Whoever speaks the truth gives
    honest evidence,
but a false witness speaks
    deceitfully.

18 Rash words are like sword thrusts,
but the tongue of the wise
    brings healing.

19 Truthful lips endure forever,
but a lying tongue lasts only a
    moment.

20 Deceit is in the mind of those who
    plan evil,
but those who counsel peace
    have joy.

g Cn: Heb *fall*   *h* Cn Compare Gk Syr: Heb *a
wise man*   *i* Or *covet the catch of the wicked*

12.1: The wise cultivate *discipline,* and listen
to criticism (see 13.1). **3:** *The root* holds a tree
firm, just as righteousness keeps the just person.

12.9: The "better" saying is a favorite locu-
tion of the sages; see also 15.16–17; 16.8;
17.1, etc. **10:** An oxymoron; what passes for
mercy with *the wicked* is *cruel.*

21 No harm happens to the
   righteous,
   but the wicked are filled with
      trouble.
22 Lying lips are an abomination to
   the LORD,
   but those who act faithfully are
      his delight.
23 One who is clever conceals
   knowledge,
   but the mind of a fool*j*
      broadcasts folly.
24 The hand of the diligent will rule,
   while the lazy will be put to
      forced labor.
25 Anxiety weighs down the human
   heart,
   but a good word cheers it up.
26 The righteous gives good advice
   to friends,*k*
   but the way of the wicked leads
      astray.
27 The lazy do not roast*l* their game,
   but the diligent obtain precious
      wealth.*l*
28 In the path of righteousness there
   is life,
   in walking its path there is no
      death.

13 A wise child loves discipline,*m*
   but a scoffer does not listen to
      rebuke.
2 From the fruit of their words
   good persons eat good
      things,
   but the desire of the treacherous
      is for wrongdoing.
3 Those who guard their mouths
   preserve their lives;
   those who open wide their lips
      come to ruin.
4 The appetite of the lazy craves,
   and gets nothing,
   while the appetite of the diligent
      is richly supplied.
5 The righteous hate falsehood,

but the wicked act shamefully
   and disgracefully.
6 Righteousness guards one whose
   way is upright,
   but sin overthrows the wicked.
7 Some pretend to be rich, yet have
   nothing;
   others pretend to be poor, yet
      have great wealth.
8 Wealth is a ransom for a person's
   life,
   but the poor get no threats.
9 The light of the righteous rejoices,
   but the lamp of the wicked
      goes out.
10 By insolence the heedless make
   strife,
   but wisdom is with those who
      take advice.
11 Wealth hastily gotten*n* will
   dwindle,
   but those who gather little by
      little will increase it.
12 Hope deferred makes the heart
   sick,
   but a desire fulfilled is a tree
      of life.
13 Those who despise the word bring
   destruction on themselves,
   but those who respect the
      commandment will be
      rewarded.
14 The teaching of the wise is a
   fountain of life,
   so that one may avoid the snares
      of death.
15 Good sense wins favor,
   but the way of the faithless is
      their ruin.*o*
16 The clever do all things
   intelligently,
   but the fool displays folly.

*j Heb the heart of fools   k Syr: Meaning of
Heb uncertain   l Meaning of Heb uncertain
m Cn: Heb A wise child the discipline of his father
n Gk Vg: Heb from vanity   o Cn Compare Gk
Syr Vg Tg: Heb is enduring*

13.2: Good words bring good results (see
12.14).
13.9: *Light* (a symbol of life, 1 Kings 15.4)
rejoices, i.e. burns brightly. **12:** *Tree of life,* as
in 3.18, a symbol of divine life and immortality in the ancient world, is a metaphor for life
itself. **13:** *Word* and *commandment* here mean
the teaching of the sages.

17 A bad messenger brings trouble,
   but a faithful envoy, healing.
18 Poverty and disgrace are for the
      one who ignores
      instruction,
   but one who heeds reproof is
      honored.
19 A desire realized is sweet to
      the soul,
   but to turn away from evil is an
      abomination to fools.
20 Whoever walks with the wise
      becomes wise,
   but the companion of fools
      suffers harm.
21 Misfortune pursues sinners,
   but prosperity rewards the
      righteous.
22 The good leave an inheritance to
      their children's children,
   but the sinner's wealth is laid up
      for the righteous.
23 The field of the poor may yield
      much food,
   but it is swept away through
      injustice.
24 Those who spare the rod hate
      their children,
   but those who love them are
      diligent to discipline them.
25 The righteous have enough to
      satisfy their appetite,
   but the belly of the wicked is
      empty.

14 The wise woman*p* builds her
      house,
   but the foolish tears it down
      with her own hands.
2 Those who walk uprightly fear the
      LORD,
   but one who is devious in
      conduct despises him.
3 The talk of fools is a rod for their
      backs,*q*
   but the lips of the wise preserve
      them.
4 Where there are no oxen, there is
      no grain;

abundant crops come by the
      strength of the ox.
5 A faithful witness does not lie,
   but a false witness breathes
      out lies.
6 A scoffer seeks wisdom in vain,
   but knowledge is easy for one
      who understands.
7 Leave the presence of a fool,
   for there you do not find words
      of knowledge.
8 It is the wisdom of the clever to
      understand where they go,
   but the folly of fools misleads.
9 Fools mock at the guilt offering,*r*
   but the upright enjoy God's
      favor.
10 The heart knows its own
      bitterness,
   and no stranger shares its joy.
11 The house of the wicked is
      destroyed,
   but the tent of the upright
      flourishes.
12 There is a way that seems right to
      a person,
   but its end is the way to death.*s*
13 Even in laughter the heart is sad,
   and the end of joy is grief.
14 The perverse get what their ways
      deserve,
   and the good, what their deeds
      deserve.*t*
15 The simple believe everything,
   but the clever consider
      their steps.
16 The wise are cautious and turn
      away from evil,
   but the fool throws off restraint
      and is careless.
17 One who is quick-tempered acts
      foolishly,
   and the schemer is hated.
18 The simple are adorned with*u*
      folly,

p Heb *Wisdom of women*   q Cn: Heb *a rod of
pride*   r Meaning of Heb uncertain
s Heb *ways of death*   t Cn: Heb *from upon him*
u Or *inherit*

14.1: See 24.3. **12**: Appearances can be
deceptive; see 16.25. **20**: A telling observation. **31**: See also 19.17.
15.8: Sincerity in worship is essential.

but the clever are crowned with
knowledge.

19 The evil bow down before the
good,
the wicked at the gates of the
righteous.

20 The poor are disliked even by
their neighbors,
but the rich have many friends.

21 Those who despise their neighbors
are sinners,
but happy are those who are
kind to the poor.

22 Do they not err that plan evil?
Those who plan good find
loyalty and faithfulness.

23 In all toil there is profit,
but mere talk leads only
to poverty.

24 The crown of the wise is their
wisdom, *v*
but folly is the garland *w* of
fools.

25 A truthful witness saves lives,
but one who utters lies is a
betrayer.

26 In the fear of the Lord one has
strong confidence,
and one's children will have a
refuge.

27 The fear of the Lord is a fountain
of life,
so that one may avoid the snares
of death.

28 The glory of a king is a multitude
of people;
without people a prince is
ruined.

29 Whoever is slow to anger has
great understanding,
but one who has a hasty temper
exalts folly.

30 A tranquil mind gives life to
the flesh,
but passion makes the bones
rot.

31 Those who oppress the poor insult
their Maker,
but those who are kind to the
needy honor him.

32 The wicked are overthrown by
their evildoing,

but the righteous find a refuge
in their integrity. *x*

33 Wisdom is at home in the mind of
one who has understanding,
but it is not *y* known in the
heart of fools.

34 Righteousness exalts a nation,
but sin is a reproach to any
people.

35 A servant who deals wisely has
the king's favor,
but his wrath falls on one who
acts shamefully.

15 A soft answer turns away
wrath,
but a harsh word stirs up anger.

2 The tongue of the wise dispenses
knowledge, *z*
but the mouths of fools pour
out folly.

3 The eyes of the Lord are in every
place,
keeping watch on the evil and
the good.

4 A gentle tongue is a tree of life,
but perverseness in it breaks the
spirit.

5 A fool despises a parent's
instruction,
but the one who heeds
admonition is prudent.

6 In the house of the righteous there
is much treasure,
but trouble befalls the income of
the wicked.

7 The lips of the wise spread
knowledge;
not so the minds of fools.

8 The sacrifice of the wicked is an
abomination to the Lord,
but the prayer of the upright is
his delight.

9 The way of the wicked is an
abomination to the Lord,
but he loves the one who
pursues righteousness.

10 There is severe discipline for one
who forsakes the way,

v Cn Compare Gk: Heb *riches*   *w* Cn: Heb *is
the folly*   *x* Gk Syr: Heb *in their death*
y Gk Syr: Heb lacks *not*   *z* Cn: Heb *makes
knowledge good*

but one who hates a rebuke
will die.

11 Sheol and Abaddon lie open
before the LORD,
how much more human hearts!

12 Scoffers do not like to be rebuked;
they will not go to the wise.

13 A glad heart makes a cheerful
countenance,
but by sorrow of heart the spirit
is broken.

14 The mind of one who has
understanding seeks
knowledge,
but the mouths of fools feed
on folly.

15 All the days of the poor are hard,
but a cheerful heart has a
continual feast.

16 Better is a little with the fear of
the LORD
than great treasure and trouble
with it.

17 Better is a dinner of vegetables
where love is
than a fatted ox and hatred
with it.

18 Those who are hot-tempered stir
up strife,
but those who are slow to anger
calm contention.

19 The way of the lazy is overgrown
with thorns,
but the path of the upright is a
level highway.

20 A wise child makes a glad father,
but the foolish despise their
mothers.

21 Folly is a joy to one who has no
sense,
but a person of understanding
walks straight ahead.

22 Without counsel, plans go wrong,
but with many advisers they
succeed.

23 To make an apt answer is a joy to
anyone,

and a word in season, how
good it is!

24 For the wise the path of life leads
upward,
in order to avoid Sheol below.

25 The LORD tears down the house of
the proud,
but maintains the widow's
boundaries.

26 Evil plans are an abomination to
the LORD,
but gracious words are pure.

27 Those who are greedy for unjust
gain make trouble for their
households,
but those who hate bribes
will live.

28 The mind of the righteous ponders
how to answer,
but the mouth of the wicked
pours out evil.

29 The LORD is far from the wicked,
but he hears the prayer of the
righteous.

30 The light of the eyes rejoices the
heart,
and good news refreshes
the body.

31 The ear that heeds wholesome
admonition
will lodge among the wise.

32 Those who ignore instruction
despise themselves,
but those who heed admonition
gain understanding.

33 The fear of the LORD is instruction
in wisdom,
and humility goes before honor.

16 The plans of the mind belong
to mortals,
but the answer of the tongue is
from the LORD.

2 All one's ways may be pure in
one's own eyes,
but the LORD weighs the spirit.

3 Commit your work to the LORD,

15.11: *Sheol and Abaddon* are the place and state (destruction) of the dead; see 1.12 n.
15.23: One of the goals of the sages. **30**:

Light of the eyes, i.e. of the one who brings *good news.*
16.1: Compare v. 9; expressed here is a

and your plans will be
    established.
4 The LORD has made everything for
        its purpose,
    even the wicked for the day of
        trouble.
5 All those who are arrogant are an
        abomination to the LORD;
    be assured, they will not go
        unpunished.
6 By loyalty and faithfulness
        iniquity is atoned for,
    and by the fear of the LORD one
        avoids evil.
7 When the ways of people please
        the LORD,
    he causes even their enemies to
        be at peace with them.
8 Better is a little with righteousness
    than large income with injustice.
9 The human mind plans the way,
    but the LORD directs the steps.
10 Inspired decisions are on the lips
        of a king;
    his mouth does not sin in
        judgment.
11 Honest balances and scales are
        the LORD's;
    all the weights in the bag are
        his work.
12 It is an abomination to kings to
        do evil,
    for the throne is established by
        righteousness.
13 Righteous lips are the delight of
        a king,
    and he loves those who speak
        what is right.
14 A king's wrath is a messenger
        of death,
    and whoever is wise will
        appease it.
15 In the light of a king's face there
        is life,
    and his favor is like the clouds
        that bring the spring rain.

16 How much better to get wisdom
        than gold!
    To get understanding is to be
        chosen rather than silver.
17 The highway of the upright
        avoids evil;
    those who guard their way
        preserve their lives.
18 Pride goes before destruction,
    and a haughty spirit before a
        fall.
19 It is better to be of a lowly spirit
        among the poor
    than to divide the spoil with
        the proud.
20 Those who are attentive to a
        matter will prosper,
    and happy are those who trust
        in the LORD.
21 The wise of heart is called
        perceptive,
    and pleasant speech increases
        persuasiveness.
22 Wisdom is a fountain of life to
        one who has it,
    but folly is the punishment of
        fools.
23 The mind of the wise makes their
        speech judicious,
    and adds persuasiveness to
        their lips.
24 Pleasant words are like a
        honeycomb,
    sweetness to the soul and health
        to the body.
25 Sometimes there is a way that
        seems to be right,
    but in the end it is the way
        to death.
26 The appetite of workers works for
        them;
    their hunger urges them on.
27 Scoundrels concoct evil,
    and their speech is like a
        scorching fire.
28 A perverse person spreads strife,

---

form of biblical determinism (the Bible honors human responsibility or free will, but does not try to harmonize it with the all-pervasive divine causality). From ch 14 on, several "LORD" sayings begin to appear. **2:** Ultimately we are unable to judge our own motives. **4:** Nothing escapes the LORD's purpose.
16.10: *Sin,* err. The ideal king was a just king. **11:** See 11.1. **25:** Also 14.12.

and a whisperer separates close
    friends.
29 The violent entice their neighbors,
    and lead them in a way that is
      not good.
30 One who winks the eyes plans[a]
    perverse things;
  one who compresses the lips
    brings evil to pass.
31 Gray hair is a crown of glory;
  it is gained in a righteous life.
32 One who is slow to anger is better
    than the mighty,
  and one whose temper is
    controlled than one who
      captures a city.
33 The lot is cast into the lap,
  but the decision is the LORD's
    alone.

17 Better is a dry morsel with
       quiet
  than a house full of feasting
    with strife.
2 A slave who deals wisely will rule
    over a child who acts
      shamefully,
  and will share the inheritance as
    one of the family.
3 The crucible is for silver, and the
    furnace is for gold,
  but the LORD tests the heart.
4 An evildoer listens to wicked lips;
  and a liar gives heed to a
    mischievous tongue.
5 Those who mock the poor insult
    their Maker;
  those who are glad at calamity
    will not go unpunished.
6 Grandchildren are the crown of
    the aged,
  and the glory of children is their
    parents.
7 Fine speech is not becoming to
    a fool;
  still less is false speech to a
    ruler.[b]

8 A bribe is like a magic stone in
    the eyes of those who give
      it;
  wherever they turn they
    prosper.
9 One who forgives an affront
    fosters friendship,
  but one who dwells on disputes
    will alienate a friend.
10 A rebuke strikes deeper into a
    discerning person
  than a hundred blows into a
    fool.
11 Evil people seek only rebellion,
  but a cruel messenger will be
    sent against them.
12 Better to meet a she-bear robbed
    of its cubs
  than to confront a fool
    immersed in folly.
13 Evil will not depart from the
    house
  of one who returns evil for
    good.
14 The beginning of strife is like
    letting out water;
  so stop before the quarrel
    breaks out.
15 One who justifies the wicked and
    one who condemns the
      righteous
  are both alike an abomination to
    the LORD.
16 Why should fools have a price
    in hand
  to buy wisdom, when they have
    no mind to learn?
17 A friend loves at all times,
  and kinsfolk are born to share
    adversity.
18 It is senseless to give a pledge,
  to become surety for a
    neighbor.

a Gk Syr Vg Tg: Heb *to plan*   *b* Or *a noble
person*

16.30: See 6.12–15. **31**: This reflects the
idea that the wise and virtuous person lives a
long life (*gray hair*). **33**: Nothing happens by
chance; even the fall of *the lot* is determined by
God.

17.5: See 14.31. **8**: *A bribe* works; elsewhere
bribery is condemned (17.23).
  17.17: *Is born for,* to provide help in *adversity* is what *kinsfolk* are for. **18**: See 11.5. **19**:
Builds a high threshold, an obscure metaphor,

19 One who loves transgression
    loves strife;
    one who builds a high threshold
    invites broken bones.
20 The crooked of mind do not
    prosper,
    and the perverse of tongue fall
    into calamity.
21 The one who begets a fool gets
    trouble;
    the parent of a fool has no joy.
22 A cheerful heart is a good
    medicine,
    but a downcast spirit dries up
    the bones.
23 The wicked accept a concealed
    bribe
    to pervert the ways of justice.
24 The discerning person looks to
    wisdom,
    but the eyes of a fool to the
    ends of the earth.
25 Foolish children are a grief to their
    father
    and bitterness to her who
    bore them.
26 To impose a fine on the innocent
    is not right,
    or to flog the noble for their
    integrity.
27 One who spares words is
    knowledgeable;
    one who is cool in spirit has
    understanding.
28 Even fools who keep silent are
    considered wise;
    when they close their lips, they
    are deemed intelligent.

18 The one who lives alone is
    self-indulgent,
    showing contempt for all who
    have sound judgment. *c*
2 A fool takes no pleasure in
    understanding,
    but only in expressing personal
    opinion.

3 When wickedness comes,
    contempt comes also;
    and with dishonor comes
    disgrace.
4 The words of the mouth are deep
    waters;
    the fountain of wisdom is a
    gushing stream.
5 It is not right to be partial to
    the guilty,
    or to subvert the innocent in
    judgment.
6 A fool's lips bring strife,
    and a fool's mouth invites a
    flogging.
7 The mouths of fools are their ruin,
    and their lips a snare to
    themselves.
8 The words of a whisperer are like
    delicious morsels;
    they go down into the inner
    parts of the body.
9 One who is slack in work
    is close kin to a vandal.
10 The name of the LORD is a strong
    tower;
    the righteous run into it and
    are safe.
11 The wealth of the rich is their
    strong city;
    in their imagination it is like a
    high wall.
12 Before destruction one's heart is
    haughty,
    but humility goes before honor.
13 If one gives answer before
    hearing,
    it is folly and shame.
14 The human spirit will endure
    sickness;
    but a broken spirit—who
    can bear?
15 An intelligent mind acquires
    knowledge,

c Meaning of Heb uncertain

---

perhaps indicating pride or ostentation.
17.24: *The eyes of a fool* are focused on unreal goals. **27–28:** Silence can be ambiguous.
18.4: *Deep waters,* an obscure metaphor
(profound, or murky?), and it is not clear if
the lines are antithetical or synonymous.
18.8: Slander listened to is not soon forgotten.

and the ear of the wise seeks
    knowledge.

16 A gift opens doors;
    it gives access to the great.

17 The one who first states a case
    seems right,
    until the other comes and
    cross-examines.

18 Casting the lot puts an end to
    disputes
    and decides between powerful
    contenders.

19 An ally offended is stronger than
    a city;*d*
    such quarreling is like the bars
    of a castle.

20 From the fruit of the mouth one's
    stomach is satisfied;
    the yield of the lips brings
    satisfaction.

21 Death and life are in the power of
    the tongue,
    and those who love it will eat
    its fruits.

22 He who finds a wife finds a
    good thing,
    and obtains favor from the
    Lord.

23 The poor use entreaties,
    but the rich answer roughly.

24 Some*e* friends play at friendship*f*
    but a true friend sticks closer
    than one's nearest kin.

19 Better the poor walking in
    integrity
    than one perverse of speech who
    is a fool.

2 Desire without knowledge is
    not good,
    and one who moves too
    hurriedly misses the way.

3 One's own folly leads to ruin,
    yet the heart rages against
    the Lord.

4 Wealth brings many friends,
    but the poor are left friendless.

5 A false witness will not go
    unpunished,
    and a liar will not escape.

6 Many seek the favor of the
    generous,
    and everyone is a friend to a
    giver of gifts.

7 If the poor are hated even by their
    kin,
    how much more are they
    shunned by their friends!
    When they call after them, they
    are not there.*g*

8 To get wisdom is to love oneself;
    to keep understanding is
    to prosper.

9 A false witness will not go
    unpunished,
    and the liar will perish.

10 It is not fitting for a fool to live
    in luxury,
    much less for a slave to rule
    over princes.

11 Those with good sense are slow
    to anger,
    and it is their glory to overlook
    an offense.

12 A king's anger is like the growling
    of a lion,
    but his favor is like dew on
    the grass.

13 A stupid child is ruin to a father,
    and a wife's quarreling is a
    continual dripping of rain.

14 House and wealth are inherited
    from parents,
    but a prudent wife is from
    the Lord.

15 Laziness brings on deep sleep;
    an idle person will suffer
    hunger.

16 Those who keep the
    commandment will live;
    those who are heedless of their
    ways will die.

17 Whoever is kind to the poor lends
    to the Lord,
    and will be repaid in full.

---

d Gk Syr Vg Tg: Meaning of Heb uncertain
e Syr Tg: Heb *A man of*   *f* Cn Compare Syr
Vg Tg: Meaning of Heb uncertain
g Meaning of Heb uncertain

18.17: Both sides of a question must be
heard. **18**: See 16.33. **19**: *Bars of a castle* prevent reconciliation; text is probably corrupt. **20**:
See 13.2.

18 Discipline your children while
there is hope;
do not set your heart on their
destruction.

19 A violent tempered person will
pay the penalty;
if you effect a rescue, you will
only have to do it again. *h*

20 Listen to advice and accept
instruction,
that you may gain wisdom for
the future.

21 The human mind may devise
many plans,
but it is the purpose of the
LORD that will be
established.

22 What is desirable in a person
is loyalty,
and it is better to be poor than
a liar.

23 The fear of the LORD is life indeed;
filled with it one rests secure
and suffers no harm.

24 The lazy person buries a hand in
the dish,
and will not even bring it back
to the mouth.

25 Strike a scoffer, and the simple
will learn prudence;
reprove the intelligent, and they
will gain knowledge.

26 Those who do violence to their
father and chase away their
mother
are children who cause shame
and bring reproach.

27 Cease straying, my child, from the
words of knowledge,
in order that you may hear
instruction.

28 A worthless witness mocks
at justice,
and the mouth of the wicked
devours iniquity.

29 Condemnation is ready for
scoffers,

and flogging for the backs of
fools.

20 Wine is a mocker, strong drink
a brawler,
and whoever is led astray by it
is not wise.

2 The dread anger of a king is like
the growling of a lion;
anyone who provokes him to
anger forfeits life itself.

3 It is honorable to refrain
from strife,
but every fool is quick to
quarrel.

4 The lazy person does not plow
in season;
harvest comes, and there is
nothing to be found.

5 The purposes in the human mind
are like deep water,
but the intelligent will draw
them out.

6 Many proclaim themselves loyal,
but who can find one worthy
of trust?

7 The righteous walk in integrity—
happy are the children who
follow them!

8 A king who sits on the throne of
judgment
winnows all evil with his eyes.

9 Who can say, "I have made my
heart clean;
I am pure from my sin"?

10 Diverse weights and diverse
measures
are both alike an abomination to
the LORD.

11 Even children make themselves
known by their acts,
by whether what they do is
pure and right.

12 The hearing ear and the seeing
eye—
the LORD has made them both.

13 Do not love sleep, or else you will
come to poverty;

h Meaning of Heb uncertain

19.17: See 14.31. **21:** See 16.1, 9. **24:** One is
too *lazy* even to eat.
20.8: *Winnows,* by his sharp judgment. **16:**
See 27.13; do not hesitate to exact a pledge;
see 6.1–5 n.

open your eyes, and you will
have plenty of bread.
14 "Bad, bad," says the buyer,
then goes away and boasts.
15 There is gold, and abundance of
costly stones;
but the lips informed by
knowledge are a
precious jewel.
16 Take the garment of one who has
given surety for a stranger;
seize the pledge given as surety
for foreigners.
17 Bread gained by deceit is sweet,
but afterward the mouth will be
full of gravel.
18 Plans are established by
taking advice;
wage war by following wise
guidance.
19 A gossip reveals secrets;
therefore do not associate with a
babbler.
20 If you curse father or mother,
your lamp will go out in utter
darkness.
21 An estate quickly acquired in the
beginning
will not be blessed in the end.
22 Do not say, "I will repay evil";
wait for the LORD, and he will
help you.
23 Differing weights are an
abomination to the LORD,
and false scales are not good.
24 All our steps are ordered by the
LORD;
how then can we understand
our own ways?
25 It is a snare for one to say rashly,
"It is holy,"
and begin to reflect only after
making a vow.
26 A wise king winnows the wicked,
and drives the wheel over them.
27 The human spirit is the lamp of
the LORD,

searching every inmost part.
28 Loyalty and faithfulness preserve
the king,
and his throne is upheld by
righteousness. *i*
29 The glory of youths is their
strength,
but the beauty of the aged is
their gray hair.
30 Blows that wound cleanse
away evil;
beatings make clean the
innermost parts.

21 The king's heart is a stream of
water in the hand of
the LORD;
he turns it wherever he will.
2 All deeds are right in the sight of
the doer,
but the LORD weighs the heart.
3 To do righteousness and justice
is more acceptable to the LORD
than sacrifice.
4 Haughty eyes and a proud heart—
the lamp of the wicked—are
sin.
5 The plans of the diligent lead
surely to abundance,
but everyone who is hasty
comes only to want.
6 The getting of treasures by a lying
tongue
is a fleeting vapor and a snare*j*
of death.
7 The violence of the wicked will
sweep them away,
because they refuse to do what
is just.
8 The way of the guilty is crooked,
but the conduct of the pure
is right.
9 It is better to live in a corner of
the housetop
than in a house shared with a
contentious wife.

i Gk: Heb *loyalty*   *j* Gk: Heb *seekers*

20.20: *Lamp will go out,* a metaphor for
misfortune or death (see also 13.9). **24**: See
19.21; Jer 10.23. **25**: See Eccl 5.1–6. **27**: *Lamp,*
a divine illumination within human beings.

21.1: Even the mind of a *king* is controlled
by God. **2**: See 16.2. **3**: See 1 Sam 15.22; Hos
6.6.

10 The souls of the wicked desire
    evil;
    their neighbors find no mercy in
      their eyes.

11 When a scoffer is punished, the
    simple become wiser;
    when the wise are instructed,
      they increase in knowledge.

12 The Righteous One observes the
    house of the wicked;
    he casts the wicked down to
      ruin.

13 If you close your ear to the cry of
    the poor,
    you will cry out and not be
      heard.

14 A gift in secret averts anger;
    and a concealed bribe in the
      bosom, strong wrath.

15 When justice is done, it is a joy to
    the righteous,
    but dismay to evildoers.

16 Whoever wanders from the way
    of understanding
    will rest in the assembly of
      the dead.

17 Whoever loves pleasure will suffer
    want;
    whoever loves wine and oil will
      not be rich.

18 The wicked is a ransom for the
    righteous,
    and the faithless for the upright.

19 It is better to live in a desert land
    than with a contentious and
      fretful wife.

20 Precious treasure remains*k* in the
    house of the wise,
    but the fool devours it.

21 Whoever pursues righteousness
    and kindness
    will find life*l* and honor.

22 One wise person went up against
    a city of warriors
    and brought down the
      stronghold in which they
      trusted.

23 To watch over mouth and tongue

is to keep out of trouble.

24 The proud, haughty person,
    named "Scoffer,"
    acts with arrogant pride.

25 The craving of the lazy person
    is fatal,
    for lazy hands refuse to labor.

26 All day long the wicked covet,*m*
    but the righteous give and do
      not hold back.

27 The sacrifice of the wicked is an
    abomination;
    how much more when brought
      with evil intent.

28 A false witness will perish,
    but a good listener will testify
      successfully.

29 The wicked put on a bold face,
    but the upright give thought
      to*n* their ways.

30 No wisdom, no understanding, no
    counsel,
    can avail against the Lord.

31 The horse is made ready for the
    day of battle,
    but the victory belongs to
      the Lord.

22 A good name is to be chosen
    rather than great riches,
    and favor is better than silver
      or gold.

2 The rich and the poor have this in
    common:
    the Lord is the maker of
      them all.

3 The clever see danger and hide;
    but the simple go on, and suffer
      for it.

4 The reward for humility and fear
    of the Lord
    is riches and honor and life.

5 Thorns and snares are in the way
    of the perverse;
    the cautious will keep far from
      them.

k Gk: Heb *and oil*   *l* Gk: Heb *life and
righteousness*   *m* Gk: Heb *all day long one covets
covetously*   *n* Another reading is *establish*

21.18: Obscure; compare 11.8.
21.30–31: Such sayings indicate that the
sages recognized the limitations of human
wisdom, in view of the Lord's own mysterious wisdom. **22.2**: Compare 29.13; the attitude to the poor in Proverbs is complex.

6 Train children in the right way,
   and when old, they will not
      stray.
7 The rich rules over the poor,
   and the borrower is the slave of
      the lender.
8 Whoever sows injustice will reap
      calamity,
   and the rod of anger will fail.
9 Those who are generous are
      blessed,
   for they share their bread with
      the poor.
10 Drive out a scoffer, and strife
      goes out;
   quarreling and abuse will cease.
11 Those who love a pure heart and
      are gracious in speech
   will have the king as a friend.
12 The eyes of the LORD keep watch
      over knowledge,
   but he overthrows the words of
      the faithless.
13 The lazy person says, "There is a
      lion outside!
   I shall be killed in the streets!"
14 The mouth of a loose[o] woman is
      a deep pit;
   he with whom the LORD is
      angry falls into it.
15 Folly is bound up in the heart of
      a boy,
   but the rod of discipline drives
      it far away.
16 Oppressing the poor in order to
      enrich oneself,
   and giving to the rich, will lead
      only to loss.

17   The words of the wise:

   Incline your ear and hear my
      words,[p]
   and apply your mind to my
      teaching;

18 for it will be pleasant if you keep
      them within you,
   if all of them are ready on
      your lips.
19 So that your trust may be in
      the LORD,
   I have made them known to
      you today—yes, to you.
20 Have I not written for you thirty
      sayings
   of admonition and knowledge,
21 to show you what is right and
      true,
   so that you may give a true
      answer to those who
      sent you?

22 Do not rob the poor because they
      are poor,
   or crush the afflicted at the gate;
23 for the LORD pleads their cause
   and despoils of life those who
      despoil them.
24 Make no friends with those given
      to anger,
   and do not associate with
      hotheads,
25 or you may learn their ways
   and entangle yourself in a snare.
26 Do not be one of those who give
      pledges,
   who become surety for debts.
27 If you have nothing with which
      to pay,
   why should your bed be taken
      from under you?
28 Do not remove the ancient
      landmark
   that your ancestors set up.
29 Do you see those who are skillful
      in their work?
   they will serve kings;

o Heb *strange*   *p* Cn Compare Gk: Heb *Incline
your ear, and hear the words of the wise*

**22.17–24.34: A second collection of pro-
verbial sayings.** See the Introduction; this
is another collection, entitled "The Words of
the Wise" (v. 17); it has many parallels to the
Instruction of the famous Egyptian sage,
Amen-em-ope, which has thirty chapters (see

22.20). There is a change in form from the
discrete sayings that characterized chs 10–21.
The topics are characteristic of (international)
wisdom literature: attitude to the poor, offering collateral, discipline of children, table
manners, abstemiousness, adultery, etc.

they will not serve common
people.

23 When you sit down to eat with
a ruler,
observe carefully what[q] is
before you,

2 and put a knife to your throat
if you have a big appetite.

3 Do not desire the ruler's[r]
delicacies,
for they are deceptive food.

4 Do not wear yourself out to
get rich;
be wise enough to desist.

5 When your eyes light upon it,
it is gone;
for suddenly it takes wings to
itself,
flying like an eagle toward
heaven.

6 Do not eat the bread of the stingy;
do not desire their delicacies;

7 for like a hair in the throat, so
are they.[s]
"Eat and drink!" they say to
you;
but they do not mean it.

8 You will vomit up the little you
have eaten,
and you will waste your
pleasant words.

9 Do not speak in the hearing of
a fool,
who will only despise the
wisdom of your words.

10 Do not remove an ancient
landmark
or encroach on the fields of
orphans,

11 for their redeemer is strong;
he will plead their cause
against you.

12 Apply your mind to instruction
and your ear to words of
knowledge.

13 Do not withhold discipline from
your children;
if you beat them with a rod,
they will not die.

14 If you beat them with the rod,
you will save their lives from
Sheol.

15 My child, if your heart is wise,
my heart too will be glad.

16 My soul will rejoice
when your lips speak what
is right.

17 Do not let your heart envy
sinners,
but always continue in the fear
of the LORD.

18 Surely there is a future,
and your hope will not be cut
off.

19 Hear, my child, and be wise,
and direct your mind in the
way.

20 Do not be among winebibbers,
or among gluttonous eaters
of meat;

21 for the drunkard and the glutton
will come to poverty,
and drowsiness will clothe them
with rags.

22 Listen to your father who
begot you,
and do not despise your mother
when she is old.

23 Buy truth, and do not sell it;
buy wisdom, instruction, and
understanding.

24 The father of the righteous will
greatly rejoice;
he who begets a wise son will
be glad in him.

25 Let your father and mother
be glad;
let her who bore you rejoice.

26 My child, give me your heart,
and let your eyes observe[t] my
ways.

27 For a prostitute is a deep pit;
an adulteress[u] is a narrow well.

28 She lies in wait like a robber
and increases the number of the
faithless.

29 Who has woe? Who has sorrow?

23.29-35: A vivid description of the ef-

---

q Or *who*   *r* Heb *his*   *s* Meaning of Heb
uncertain   *t* Another reading is *delight in*
u Heb *an alien woman*

Who has strife? Who has
     complaining?
Who has wounds without cause?
Who has redness of eyes?
30 Those who linger late over wine,
     those who keep trying
          mixed wines.
31 Do not look at wine when it is
          red,
     when it sparkles in the cup
     and goes down smoothly.
32 At the last it bites like a serpent,
     and stings like an adder.
33 Your eyes will see strange things,
     and your mind utter perverse
          things.
34 You will be like one who lies
          down in the midst of the
          sea,
     like one who lies on the top of
          a mast. *v*
35 "They struck me," you will say, *w*
     "but I was not hurt;
     they beat me, but I did not
          feel it.
When shall I awake?
I will seek another drink."

24 Do not envy the wicked,
          nor desire to be with them;
2 for their minds devise violence,
     and their lips talk of mischief.

3 By wisdom a house is built,
     and by understanding it is
          established;
4 by knowledge the rooms are filled
     with all precious and pleasant
          riches.
5 Wise warriors are mightier than
          strong ones, *x*
     and those who have knowledge
          than those who have
          strength;
6 for by wise guidance you can
          wage your war,
     and in abundance of counselors
          there is victory.

7 Wisdom is too high for fools;
     in the gate they do not open
          their mouths.

8 Whoever plans to do evil
     will be called a mischief-maker.
9 The devising of folly is sin,
     and the scoffer is an
          abomination to all.

10 If you faint in the day of
          adversity,
     your strength being small;
11 if you hold back from rescuing
          those taken away to death,
     those who go staggering to the
          slaughter;
12 if you say, "Look, we did not
          know this"—
     does not he who weighs the
          heart perceive it?
Does not he who keeps watch
          over your soul know it?
And will he not repay all
          according to their deeds?

13 My child, eat honey, for it is
          good,
     and the drippings of the
          honeycomb are sweet to
          your taste.
14 Know that wisdom is such to
          your soul;
     if you find it, you will find a
          future,
     and your hope will not be cut
          off.

15 Do not lie in wait like an outlaw
          against the home of the
          righteous;
     do no violence to the place
          where the righteous live;

v Meaning of Heb uncertain   *w* Gk Syr Vg
Tg: Heb lacks *you will say*   *x* Gk Compare
Syr Tg: Heb *A wise man is strength*

23.29–35: A vivid description of the effects of drunkenness.
24.10–12: A good person should intervene on behalf of victims of violence. **16:** *Seven times,* i.e. any number of times; the righteous will *rise again,* in contrast to *the wicked.* **17:** See Job 31.29.

16 for though they fall seven times,
   they will rise again;
but the wicked are overthrown
   by calamity.

17 Do not rejoice when your enemies
   fall,
and do not let your heart be
   glad when they stumble,
18 or else the LORD will see it and be
   displeased,
and turn away his anger
   from them.

19 Do not fret because of evildoers.
   Do not envy the wicked;
20 for the evil have no future;
   the lamp of the wicked will
      go out.

21 My child, fear the LORD and
   the king,
and do not disobey either of
   them;y
22 for disaster comes from them
   suddenly,
and who knows the ruin that
   both can bring?

23 These also are sayings of
   the wise:

Partiality in judging is not good.
24 Whoever says to the wicked,
   "You are innocent,"
will be cursed by peoples,
   abhorred by nations;
25 but those who rebuke the wicked
   will have delight,
and a good blessing will come
   upon them.
26 One who gives an honest answer
   gives a kiss on the lips.

27 Prepare your work outside,
   get everything ready for you in
      the field;
and after that build your house.

28 Do not be a witness against your
   neighbor without cause,
and do not deceive with your
   lips.
29 Do not say, "I will do to others as
   they have done to me;
I will pay them back for what
   they have done."

30 I passed by the field of one who
   was lazy,
by the vineyard of a stupid
   person;
31 and see, it was all overgrown with
   thorns;
the ground was covered with
   nettles,
and its stone wall was broken
   down.
32 Then I saw and considered it;
   I looked and received
      instruction.
33 A little sleep, a little slumber,
   a little folding of the hands
      to rest,
34 and poverty will come upon you
   like a robber,
and want, like an armed
   warrior.

25 These are other proverbs of Solomon that the officials of King Hezekiah of Judah copied.

2 It is the glory of God to conceal
   things,
but the glory of kings is to
   search things out.
3 Like the heavens for height, like
   the earth for depth,
so the mind of kings is
   unsearchable.
4 Take away the dross from the
   silver,
and the smith has material for
   a vessel;

y Gk: Heb *do not associate with those who change*

24.23: A title for another and short collection (see Introduction). **29**: See 24.17–18. **30–34**: See 6.6–11.
25.1–29.27: Another collection of "Proverbs of Solomon" (1.1; 10.1; see Introduction). **1**: *Officials of King Hezekiah* are scribes of this famous king (715–687 B.C.), and an indication that wisdom was cultivated at the

5 take away the wicked from the
presence of the king,
and his throne will be
established in righteousness.
6 Do not put yourself forward in
the king's presence
or stand in the place of the
great;
7 for it is better to be told, "Come
up here,"
than to be put lower in the
presence of a noble.

What your eyes have seen
8 do not hastily bring into court;
for*z* what will you do in the end,
when your neighbor puts you to
shame?
9 Argue your case with your
neighbor directly,
and do not disclose another's
secret;
10 or else someone who hears you
will bring shame upon you,
and your ill repute will have
no end.

11 A word fitly spoken
is like apples of gold in a setting
of silver.
12 Like a gold ring or an ornament
of gold
is a wise rebuke to a listening
ear.
13 Like the cold of snow in the time
of harvest
are faithful messengers to those
who send them;
they refresh the spirit of their
masters.
14 Like clouds and wind without rain
is one who boasts of a gift
never given.
15 With patience a ruler may be
persuaded,

and a soft tongue can break
bones.
16 If you have found honey, eat only
enough for you,
or else, having too much, you
will vomit it.
17 Let your foot be seldom in your
neighbor's house,
otherwise the neighbor will
become weary of you and
hate you.
18 Like a war club, a sword, or a
sharp arrow
is one who bears false witness
against a neighbor.
19 Like a bad tooth or a lame foot
is trust in a faithless person in
time of trouble.
20 Like vinegar on a wound*a*
is one who sings songs to a
heavy heart.
Like a moth in clothing or a
worm in wood,
sorrow gnaws at the human
heart. *b*
21 If your enemies are hungry, give
them bread to eat;
and if they are thirsty, give
them water to drink;
22 for you will heap coals of fire on
their heads,
and the LORD will reward you.
23 The north wind produces rain,
and a backbiting tongue, angry
looks.
24 It is better to live in a corner of
the housetop
than in a house shared with a
contentious wife.
25 Like cold water to a thirsty soul,

z Cn: Heb *or else*   *a* Gk: Heb *Like one who*
takes off a garment on a cold day, like vinegar on lye
b Gk Syr Tg: Heb lacks *Like a moth . . . human*
heart

---

royal court. **2–3**: Like God, the king questions but cannot be questioned. **6–7**: See Lk
14.7–11.
  25.7c–10: A caution against rash accusations. **11**: A goal of the sages: right and timely
expressions. **15b**: A "fitly spoken" paradox.
22: *Coals of fire*, are interpreted as (increased)
punishment, or as the remorse and hence conversion of the *enemies*. See also Rom 12.20.

so is good news from a far
country.

26 Like a muddied spring or a
polluted fountain
are the righteous who give way
before the wicked.

27 It is not good to eat much honey,
or to seek honor on top of
honor.

28 Like a city breached, without
walls,
is one who lacks self-control.

26 Like snow in summer or rain in
harvest,
so honor is not fitting for a
fool.

2 Like a sparrow in its flitting, like a
swallow in its flying,
an undeserved curse goes
nowhere.

3 A whip for the horse, a bridle for
the donkey,
and a rod for the back of fools.

4 Do not answer fools according to
their folly,
or you will be a fool yourself.

5 Answer fools according to their
folly,
or they will be wise in their
own eyes.

6 It is like cutting off one's foot and
drinking down violence,
to send a message by a fool.

7 The legs of a disabled person hang
limp;
so does a proverb in the mouth
of a fool.

8 It is like binding a stone in a sling
to give honor to a fool.

9 Like a thornbush brandished by
the hand of a drunkard
is a proverb in the mouth of
a fool.

10 Like an archer who wounds
everybody

is one who hires a passing fool
or drunkard. [c]

11 Like a dog that returns to its
vomit
is a fool who reverts to his
folly.

12 Do you see persons wise in their
own eyes?
There is more hope for fools
than for them.

13 The lazy person says, "There is a
lion in the road!
There is a lion in the streets!"

14 As a door turns on its hinges,
so does a lazy person in bed.

15 The lazy person buries a hand in
the dish,
and is too tired to bring it back
to the mouth.

16 The lazy person is wiser in
self-esteem
than seven who can answer
discreetly.

17 Like somebody who takes a
passing dog by the ears
is one who meddles in the
quarrel of another.

18 Like a maniac who shoots deadly
firebrands and arrows,

19 so is one who deceives a neighbor
and says, "I am only joking!"

20 For lack of wood the fire goes
out,
and where there is no
whisperer, quarreling
ceases.

21 As charcoal is to hot embers and
wood to fire,
so is a quarrelsome person for
kindling strife.

22 The words of a whisperer are like
delicious morsels;

_c_ Meaning of Heb uncertain

26.4–5: The ambiguity of sayings. *According to their folly* can mean "in fools' words" (v. 4) or "as fools deserve" (v. 5). **7**: *A fool* is by definition unable to profit from *a proverb.* **8**: A stone tied in a sling is useless.
26.9a: The metaphor is vague; it may indi-
cate uselessness (see vv. 7 8), or the use of a proverb in a hurtful way. **12**: The ambiguity of being *wise.*
26.22: See 18.1 n. **27**: A kind of "poetic justice"; see Ps 7.16; Eccl 10.8.

they go down into the inner
parts of the body.
23 Like the glaze[d] covering an
earthen vessel
are smooth[e] lips with an
evil heart.
24 An enemy dissembles in speaking
while harboring deceit within;
25 when an enemy speaks graciously,
do not believe it,
for there are seven abominations
concealed within;
26 though hatred is covered with
guile,
the enemy's wickedness will be
exposed in the assembly.
27 Whoever digs a pit will fall into it,
and a stone will come back on
the one who starts it
rolling.
28 A lying tongue hates its victims,
and a flattering mouth works
ruin.

27 Do not boast about tomorrow,
for you do not know what a
day may bring.
2 Let another praise you, and not
your own mouth—
a stranger, and not your own
lips.
3 A stone is heavy, and sand is
weighty,
but a fool's provocation is
heavier than both.
4 Wrath is cruel, anger is
overwhelming,
but who is able to stand before
jealousy?
5 Better is open rebuke
than hidden love.
6 Well meant are the wounds a
friend inflicts,
but profuse are the kisses of an
enemy.
7 The sated appetite spurns honey,
but to a ravenous appetite even
the bitter is sweet.
8 Like a bird that strays from
its nest

is one who strays from home.
9 Perfume and incense make the
heart glad,
but the soul is torn by trouble.[f]
10 Do not forsake your friend or the
friend of your parent;
do not go to the house of your
kindred in the day of your
calamity.
Better is a neighbor who is nearby
than kindred who are far away.
11 Be wise, my child, and make my
heart glad,
so that I may answer whoever
reproaches me.
12 The clever see danger and hide;
but the simple go on, and suffer
for it.
13 Take the garment of one who has
given surety for a stranger;
seize the pledge given as surety
for foreigners.[g]
14 Whoever blesses a neighbor with a
loud voice,
rising early in the morning,
will be counted as cursing.
15 A continual dripping on a
rainy day
and a contentious wife are alike;
16 to restrain her is to restrain
the wind
or to grasp oil in the right
hand.[h]
17 Iron sharpens iron,
and one person sharpens the
wits[i] of another.
18 Anyone who tends a fig tree will
eat its fruit,
and anyone who takes care of a
master will be honored.
19 Just as water reflects the face,
so one human heart reflects
another.

d Cn: Heb *silver of dross*   e Gk: Heb *burning*
f Gk: Heb *the sweetness of a friend is better than
one's own counsel*   g Vg and 20.16: Heb *for a
foreign woman*   h Meaning of Heb uncertain
i Heb *face*

27.6: *Profuse,* false. **7b:** A paradox, as in
25.15b. **10:** Line b is explained by lines cd.

27.13: See 20.16 n. **20:** See 15.11 n. Sheol is
personified as a dynamic power that devours

20 Sheol and Abaddon are never
satisfied,
and human eyes are never
satisfied.

21 The crucible is for silver, and the
furnace is for gold,
so a person is tested^j by being
praised.

22 Crush a fool in a mortar with a
pestle
along with crushed grain,
but the folly will not be
driven out.

23 Know well the condition of
your flocks,
and give attention to your
herds;

24 for riches do not last forever,
nor a crown for all generations.

25 When the grass is gone, and new
growth appears,
and the herbage of the
mountains is gathered,

26 the lambs will provide your
clothing,
and the goats the price of a
field;

27 there will be enough goats' milk
for your food,
for the food of your household
and nourishment for your
servant-girls.

28 The wicked flee when no one
pursues,
but the righteous are as bold as
a lion.

2 When a land rebels
it has many rulers;
but with an intelligent ruler
there is lasting order.^{k}

3 A ruler^l who oppresses the poor
is a beating rain that leaves
no food.

4 Those who forsake the law praise
the wicked,
but those who keep the law
struggle against them.

5 The evil do not understand justice,

but those who seek the Lord
understand it completely.

6 Better to be poor and walk in
integrity
than to be crooked in one's
ways even though rich.

7 Those who keep the law are wise
children,
but companions of gluttons
shame their parents.

8 One who augments wealth by
exorbitant interest
gathers it for another who is
kind to the poor.

9 When one will not listen to
the law,
even one's prayers are an
abomination.

10 Those who mislead the upright
into evil ways
will fall into pits of their own
making,
but the blameless will have a
goodly inheritance.

11 The rich is wise in self-esteem,
but an intelligent poor person
sees through the pose.

12 When the righteous triumph, there
is great glory,
but when the wicked prevail,
people go into hiding.

13 No one who conceals
transgressions will prosper,
but one who confesses and
forsakes them will obtain
mercy.

14 Happy is the one who is never
without fear,
but one who is hard-hearted
will fall into calamity.

15 Like a roaring lion or a charging
bear
is a wicked ruler over a poor
people.

16 A ruler who lacks understanding is
a cruel oppressor;

j Heb lacks *is tested*   *k* Meaning of Heb
uncertain   *l* Cn: Heb *A poor person*

individuals (see 30.16). **21:** See 17.3, where it
is the Lord who tests.

28.8: By the law of retribution, the oppressor should not be allowed to profit.

but one who hates unjust gain
will enjoy a long life.

17 If someone is burdened with the
blood of another,
let that killer be a fugitive
until death;
let no one offer assistance.

18 One who walks in integrity will
be safe,
but whoever follows crooked
ways will fall into the Pit. *m*

19 Anyone who tills the land will
have plenty of bread,
but one who follows worthless
pursuits will have plenty
of poverty.

20 The faithful will abound with
blessings,
but one who is in a hurry to be
rich will not go
unpunished.

21 To show partiality is not good—
yet for a piece of bread a person
may do wrong.

22 The miser is in a hurry to get rich
and does not know that loss is
sure to come.

23 Whoever rebukes a person will
afterward find more favor
than one who flatters with
the tongue.

24 Anyone who robs father or
mother
and says, "That is no crime,"
is partner to a thug.

25 The greedy person stirs up strife,
but whoever trusts in the Lord
will be enriched.

26 Those who trust in their own wits
are fools;
but those who walk in wisdom
come through safely.

27 Whoever gives to the poor will
lack nothing,
but one who turns a blind eye
will get many a curse.

28 When the wicked prevail, people
go into hiding;
but when they perish, the
righteous increase.

29 One who is often reproved, yet
remains stubborn,
will suddenly be broken beyond
healing.

2 When the righteous are in
authority, the people
rejoice;
but when the wicked rule, the
people groan.

3 A child who loves wisdom makes
a parent glad,
but to keep company with
prostitutes is to squander
one's substance.

4 By justice a king gives stability to
the land,
but one who makes heavy
exactions ruins it.

5 Whoever flatters a neighbor
is spreading a net for the
neighbor's feet.

6 In the transgression of the evil
there is a snare,
but the righteous sing and
rejoice.

7 The righteous know the rights of
the poor;
the wicked have no such
understanding.

8 Scoffers set a city aflame,
but the wise turn away wrath.

9 If the wise go to law with fools,
there is ranting and ridicule
without relief.

10 The bloodthirsty hate the
blameless,
and they seek the life of the
upright.

11 A fool gives full vent to anger,
but the wise quietly holds
it back.

12 If a ruler listens to falsehood,
all his officials will be wicked.

13 The poor and the oppressor have
this in common:
the Lord gives light to the eyes
of both.

m Syr: Heb *fall all at once*

29.13: See 22.2; Mt 5.45. **24**: Lev 5.1.

14 If a king judges the poor with
  equity,
    his throne will be established
    forever.
15 The rod and reproof give wisdom,
    but a mother is disgraced by a
    neglected child.
16 When the wicked are in authority,
  transgression increases,
    but the righteous will look upon
    their downfall.
17 Discipline your children, and they
  will give you rest;
    they will give delight to your
    heart.
18 Where there is no prophecy, the
  people cast off restraint,
    but happy are those who keep
    the law.
19 By mere words servants are not
  disciplined,
    for though they understand,
    they will not give heed.
20 Do you see someone who is hasty
  in speech?
    There is more hope for a fool
    than for anyone like that.
21 A slave pampered from childhood
  will come to a bad end. *n*
22 One given to anger stirs up strife,
  and the hothead causes much
    transgression.
23 A person's pride will bring
  humiliation,
    but one who is lowly in spirit
    will obtain honor.
24 To be a partner of a thief is to
  hate one's own life;
    one hears the victim's curse, but
    discloses nothing. *o*
25 The fear of others *p* lays a snare,
    but one who trusts in the LORD
    is secure.
26 Many seek the favor of a ruler,

but it is from the LORD that one
    gets justice.
27 The unjust are an abomination to
  the righteous,
    but the upright are an
    abomination to the wicked.

30 The words of Agur son of Jakeh.
  An oracle.

Thus says the man: I am weary,
  O God,
  I am weary, O God. How can I
  prevail? *q*
2 Surely I am too stupid to be
  human;
  I do not have human
  understanding.
3 I have not learned wisdom,
  nor have I knowledge of the
  holy ones. *r*
4 Who has ascended to heaven and
  come down?
  Who has gathered the wind in
  the hollow of the hand?
  Who has wrapped up the waters in
  a garment?
  Who has established all the ends
  of the earth?
  What is the person's name?
  And what is the name of the
  person's child?
  Surely you know!

5 Every word of God proves true;
  he is a shield to those who take
  refuge in him.
6 Do not add to his words,
  or else he will rebuke you, and
  you will be found a liar.

n Vg: Meaning of Heb uncertain   *o* Meaning
of Heb uncertain   *p* Or *human fear*   *q* Or *I
am spent*. Meaning of Heb uncertain
r Or *Holy One*

30.1–4: *The words of Agur* are a title (see
Introduction), but the extent of his sayings is
uncertain. **1:** *Oracle:* Others understand the
Hebrew to mean "from Massa," hence a non-
Israelite. The rest of v. 1 is all but untranslata-
ble. **2–3:** The writer acknowledges *human*

limitations. **4:** The questions suggest this is a
riddle, to which the answer would be "God."
30.7–9: The request is for truth, and
enough to live on. **11–14:** Four types of
sinners—the unfilial, self-satisfied, arrogant,
and avaricious.

7 Two things I ask of you;
    do not deny them to me before
        I die:
8 Remove far from me falsehood
        and lying;
    give me neither poverty nor
        riches;
    feed me with the food that
        I need,
9 or I shall be full, and deny you,
        and say, "Who is the LORD?"
    or I shall be poor, and steal,
        and profane the name of my
        God.

10 Do not slander a servant to a
        master,
    or the servant will curse you,
        and you will be held guilty.

11 There are those who curse their
        fathers
    and do not bless their mothers.
12 There are those who are pure in
        their own eyes
    yet are not cleansed of their
        filthiness.
13 There are those—how lofty are
        their eyes,
    how high their eyelids lift!
14 There are those whose teeth are
        swords,
    whose teeth are knives,
    to devour the poor from off the
        earth,
    the needy from among mortals.

15 The leech*s* has two daughters;
    "Give, give," they cry.
    Three things are never satisfied;
    four never say, "Enough":
16 Sheol, the barren womb,
    the earth ever thirsty for water,
    and the fire that never says,
        "Enough."*s*

17 The eye that mocks a father
    and scorns to obey a mother
will be pecked out by the ravens
        of the valley
    and eaten by the vultures.

18 Three things are too wonderful
        for me;
    four I do not understand:
19 the way of an eagle in the sky,
    the way of a snake on a rock,
    the way of a ship on the high seas,
    and the way of a man with a
        girl.

20 This is the way of an adulteress:
    she eats, and wipes her mouth,
    and says, "I have done no
        wrong."

21 Under three things the earth
        trembles;
    under four it cannot bear up:
22 a slave when he becomes king,
    and a fool when glutted
        with food;
23 an unloved woman when she gets
        a husband,
    and a maid when she succeeds
        her mistress.

24 Four things on earth are small,
    yet they are exceedingly wise:
25 the ants are a people without
        strength,
    yet they provide their food in
        the summer;
26 the badgers are a people without
        power,
    yet they make their homes in
        the rocks;
27 the locusts have no king,
    yet all of them march in rank;
28 the lizard*t* can be grasped in
        the hand,
    yet it is found in kings' palaces.

---

s Meaning of Heb uncertain   *t* Or *spider*

30.15–16: A numerical saying about four *things* that *are never satisfied*, to which the *two daughters* of the *leech* seem to be compared. **18–19:** Four wonders are compared to each other as regards the *way*, a mysterious and irrecoverable way. **30.20:** *Eats . . . . mouth*, a euphemism; this verse seems to be a moralizing interpretation

29 Three things are stately in their
    stride;
    four are stately in their gait:
30 the lion, which is mightiest among
    wild animals
    and does not turn back
    before any;
31 the strutting rooster,ᵘ the he-goat,
    and a king striding beforeᵛ
    his people.

32 If you have been foolish, exalting
    yourself,
    or if you have been devising
    evil,
    put your hand on your mouth.
33 For as pressing milk produces
    curds,
    and pressing the nose produces
    blood,
    so pressing anger produces
    strife.

31 The words of King Lemuel. An
oracle that his mother taught
him:

2 No, my son! No, son of my
    womb!
    No, son of my vows!
3 Do not give your strength to
    women,
    your ways to those who destroy
    kings.
4 It is not for kings, O Lemuel,
    it is not for kings to drink wine,
    or for rulers to desireʷ strong
    drink;
5 or else they will drink and forget
    what has been decreed,
    and will pervert the rights of all
    the afflicted.

6 Give strong drink to one who is
    perishing,
    and wine to those in bitter
    distress;
7 let them drink and forget their
    poverty,
    and remember their misery
    no more.
8 Speak out for those who cannot
    speak,
    for the rights of all the
    destitute.ˣ
9 Speak out, judge righteously,
    defend the rights of the poor
    and needy.

10 A capable wife who can find?
    She is far more precious than
    jewels.
11 The heart of her husband trusts
    in her,
    and he will have no lack of
    gain.
12 She does him good, and not harm,
    all the days of her life.
13 She seeks wool and flax,
    and works with willing hands.
14 She is like the ships of the
    merchant,
    she brings her food from
    far away.
15 She rises while it is still night
    and provides food for her
    household
    and tasks for her servant-girls.
16 She considers a field and buys it;
    with the fruit of her hands she
    plants a vineyard.

u Gk Syr Tg Compare Vg: Meaning of Heb
uncertain   *v* Meaning of Heb uncertain
w Cn: Heb *where*   *x* Heb *all children of passing
away*

---

of v. 19d. **21–23:** Four intolerable situations.
24–28: Four small creatures that can teach us
lessons. **29–31:** Four examples of stateliness.
  31.1–9: If *Lemuel* is not from Massa (here
translated as *oracle;* see 30.1), he is unidentified; the teaching of his *mother* warns him
against women and wine, and urges the ideals
of justice.
  31.10–31: An acrostic poem about the *capable wife.* **10:** The "finding" of a wife is a com-

mon theme (see 18.22), and so is the "finding" of wisdom (Job 28.12, 20). Is there a
relation between the two? Like wisdom, this
woman is *more precious than jewels* (so also
wisdom in 3.14–15; 8.10). The description
seems to stem from a male point of view.
There may be a play on "sophia" (Greek for
"wisdom") in the Hebrew text of verse 31,
which reads "sophiyyah" (*she looks well to*).

17 She girds herself with strength,
and makes her arms strong.
18 She perceives that her merchandise
is profitable.
Her lamp does not go out
at night.
19 She puts her hands to the distaff,
and her hands hold the
spindle.
20 She opens her hand to the poor,
and reaches out her hands to
the needy.
21 She is not afraid for her household
when it snows,
for all her household are clothed
in crimson.
22 She makes herself coverings;
her clothing is fine linen and
purple.
23 Her husband is known in the
city gates,
taking his seat among the elders
of the land.
24 She makes linen garments and sells
them;
she supplies the merchant with
sashes.

25 Strength and dignity are her
clothing,
and she laughs at the time
to come.
26 She opens her mouth with
wisdom,
and the teaching of kindness is
on her tongue.
27 She looks well to the ways of her
household,
and does not eat the bread of
idleness.
28 Her children rise up and call her
happy;
her husband too, and he
praises her:
29 "Many women have done
excellently,
but you surpass them all."
30 Charm is deceitful, and beauty
is vain,
but a woman who fears the
Lord is to be praised.
31 Give her a share in the fruit of her
hands,
and let her works praise her in
the city gates.

Ecclesiastes

The name of the book, and ultimately of the author, derives from the Hebrew *Qoheleth*. In Hebrew the root *qhl* has to do with an assembly or congregation (an "ecclesia," hence the Greek and Latin form, Ecclesiastes). The name, then, would seem to mean "leader of an assembly." The term "Preacher" goes back through Luther (*Prediger*) to Jerome (*concionator*), but the author is not a preacher, nor is this a series of sermons. The editor of the book, perhaps one of the author's students, is responsible for the superscription that identifies him as "son of David" (1.1), and hence the work was attributed to Solomon. But the late language (close to the style of the Mishnah) and the tenor of the work make this attribution virtually impossible; a date around 300 B.C. is likely. (The Mishnah is the part of the Talmud that contains laws and regulations. The collection was formed after the Hebrew Scriptures were largely completed.)

No definitive outline of the book has won a consensus. But there is a clear thematizing in 1.2 and its echo in 12.8, "vanity of vanities." This construction (like the similar "holy of holies") is Hebrew idiom for the superlative—thus, utter vanity. The Hebrew word used, *hebel,* means vapor, something unsubstantial, hence futile or vain, and it occurs 38 times in the Hebrew text of this book. This verdict is applied to several aspects of life: wisdom, toil, joy, etc., and all are found wanting. If Qoheleth cannot find immediate disadvantages, there is always the fact of death. Ultimately the wise man must die like the fool (2.16; 3.20; 9.10). The author of the epilogue in 12.9f. seems to have had a high regard for Qoheleth as a *teacher* (as Qoheleth is rendered in NRSV), though many interpreters regard 12.13–14 as a kind of "correction" to Qoheleth's radical thought. Because "joy" also plays a role in the book (2.26; 3.22; 9.7–9), some have interpreted Qoheleth very positively: enjoy the pleasures of life, the "gift" of God (5.19). These conclusions about joy, however, seem to be doubtful concessions. One never knows if joy will be meted out, or on what basis. The divine generosity cannot be comprehended: "Whether it is love or hate one does not know" (9.1), and "the work of God" is sheer mystery for humans (3.11; 8.17; 11.5). Qoheleth draws a grim picture of life (and death; see 12.1–7), but he held fast to the need of reverence for God (5.7; 7.18). Within the biblical canon his book fulfills a necessary role, warning against human hubris and preserving divine mystery.

1 ¹ The words of the Teacher,[a] the son of David, king in Jerusalem. ² Vanity of vanities, says the Teacher,[a] vanity of vanities! All is vanity.

³ What do people gain from all the toil
at which they toil under the sun?
⁴ A generation goes, and a generation comes,
but the earth remains forever.
⁵ The sun rises and the sun goes down,
and hurries to the place where it rises.
⁶ The wind blows to the south, and goes around to the north;
round and round goes the wind, and on its circuits the wind returns.
⁷ All streams run to the sea, but the sea is not full;
to the place where the streams flow,
there they continue to flow.
⁸ All things[b] are wearisome; more than one can express;
the eye is not satisfied with seeing, or the ear filled with hearing.
⁹ What has been is what will be, and what has been done is what will be done;
there is nothing new under the sun.
¹⁰ Is there a thing of which it is said, "See, this is new"?
It has already been, in the ages before us.
¹¹ The people of long ago are not remembered,

nor will there be any remembrance
of people yet to come
by those who come after them.

12 I, the Teacher,[a] when king over Israel in Jerusalem, ¹³ applied my mind to seek and to search out by wisdom all that is done under heaven; it is an unhappy business that God has given to human beings to be busy with. ¹⁴ I saw all the deeds that are done under the sun; and see, all is vanity and a chasing after wind.[c]

¹⁵ What is crooked cannot be made straight,
and what is lacking cannot be counted.

16 I said to myself, "I have acquired great wisdom, surpassing all who were over Jerusalem before me; and my mind has had great experience of wisdom and knowledge." ¹⁷ And I applied my mind to know wisdom and to know madness and folly. I perceived that this also is but a chasing after wind.[c]

¹⁸ For in much wisdom is much vexation,
and those who increase knowledge increase sorrow.

2 ¹ I said to myself, "Come now, I will make a test of pleasure; enjoy yourself." But again, this also was vanity. ²I said of laughter, "It is mad," and of pleasure, "What use is it?" ³I searched with my mind how to cheer my body with wine—my mind still guiding me with wisdom—and how to lay hold on folly,

---

a Heb *Qoheleth*, traditionally rendered *Preacher*
b Or *words*   c Or *a feeding on wind*. See Hos 12.1

1.1–11: Title and Thesis. For the superscription (verse 1) and themes (v. 2, compare 12.8), see the Introduction. The failure of toil (v. 3) is illustrated by a poem that exemplifies monotony and repetitiousness: *generation, sun, wind, streams*. Nothing is achieved, and similarly human desires are left unsatisfied (v. 8). Repetition marks human experience: *nothing new under the sun*, because there is no *remembrance* (v. 11).

1.12–18: The search for wisdom. These opening remarks anticipate Qoheleth's judgment about the futility of wisdom (compare 8.17). **18:** Both as a process and as achievement, *in much wisdom is much vexation*.

2.1–11: The experiment with pleasure. A preliminary judgment (vv. 1–2) is passed on an experiment that is described in great detail (vv. 3–11).

until I might see what was good for mortals to do under heaven during the few days of their life. 4I made great works; I built houses and planted vineyards for myself; 5I made myself gardens and parks, and planted in them all kinds of fruit trees. 6I made myself pools from which to water the forest of growing trees. 7I bought male and female slaves, and had slaves who were born in my house; I also had great possessions of herds and flocks, more than any who had been before me in Jerusalem. 8I also gathered for myself silver and gold and the treasure of kings and of the provinces; I got singers, both men and women, and delights of the flesh, and many concubines.*d*

9 So I became great and surpassed all who were before me in Jerusalem; also my wisdom remained with me. 10Whatever my eyes desired I did not keep from them; I kept my heart from no pleasure, for my heart found pleasure in all my toil, and this was my reward for all my toil. 11Then I considered all that my hands had done and the toil I had spent in doing it, and again, all was vanity and a chasing after wind,*e* and there was nothing to be gained under the sun.

12 So I turned to consider wisdom and madness and folly; for what can the one do who comes after the king? Only what has already been done. 13Then I saw that wisdom excels folly as light excels darkness.

14 The wise have eyes in their head,
  but fools walk in darkness.

Yet I perceived that the same fate befalls all of them. 15Then I said to myself, "What happens to the fool will happen to me also; why then have I been so very wise?" And I said to myself that this also

is vanity. 16For there is no enduring remembrance of the wise or of fools, seeing that in the days to come all will have been long forgotten. How can the wise die just like fools? 17So I hated life, because what is done under the sun was grievous to me; for all is vanity and a chasing after wind.*e*

18 I hated all my toil in which I had toiled under the sun, seeing that I must leave it to those who come after me 19—and who knows whether they will be wise or foolish? Yet they will be master of all for which I toiled and used my wisdom under the sun. This also is vanity. 20So I turned and gave my heart up to despair concerning all the toil of my labors under the sun, 21because sometimes one who has toiled with wisdom and knowledge and skill must leave all to be enjoyed by another who did not toil for it. This also is vanity and a great evil. 22What do mortals get from all the toil and strain with which they toil under the sun? 23For all their days are full of pain, and their work is a vexation; even at night their minds do not rest. This also is vanity.

24 There is nothing better for mortals than to eat and drink, and find enjoyment in their toil. This also, I saw, is from the hand of God; 25for apart from him*f* who can eat or who can have enjoyment? 26For to the one who pleases him God gives wisdom and knowledge and joy; but to the sinner he gives the work of gathering and heaping, only to give to one who pleases God. This also is vanity and a chasing after wind.*e*

---

d Meaning of Heb uncertain   *e* Or *a feeding
on wind.* See Hos 12.1   *f* Gk Syr: Heb *apart
from me*

2.10–11: *Pleasure* was achieved, but it is judged as vain.
2.12–23: A judgment on wisdom. 12: The translation is uncertain. **13–17:** Qoheleth rejects two sayings about the superiority of wisdom (vv. 13–14a), because there is one *fate* for all: death (vv. 14b–16).
2.24–26: The first of Qoheleth's re- signed conclusions about joy (3.12, 22; 5.17–18; 8.15; 9.7–9; 11.7–12). **26:** Joy is a gift of God, but God is quite arbitrary, dispensing it as he pleases. No moral evaluation is intended by *the one who pleases him,* or by *sinner* (i.e. one who misses the mark, a bungler; compare 9.18).

3 For everything there is a season, and a time for every matter under heaven:

2 a time to be born, and a time
to die;
a time to plant, and a time to
pluck up what is planted;
3 a time to kill, and a time to heal;
a time to break down, and a time
to build up;
4 a time to weep, and a time
to laugh;
a time to mourn, and a time
to dance;
5 a time to throw away stones, and
a time to gather stones
together;
a time to embrace, and a time to
refrain from embracing;
6 a time to seek, and a time to lose;
a time to keep, and a time to
throw away;
7 a time to tear, and a time to sew;
a time to keep silence, and a time
to speak;
8 a time to love, and a time to hate;
a time for war, and a time for
peace.

9 What gain have the workers from their toil? 10 I have seen the business that God has given to everyone to be busy with. 11 He has made everything suitable for its time; moreover he has put a sense of past and future into their minds, yet they cannot find out what God has done from the beginning to the end. 12 I know that there is nothing better for them than to be happy and enjoy themselves as long as they live; 13 moreover, it is God's gift that all should eat and drink and take pleasure in all their toil. 14 I know that whatever God does endures forever; nothing can be added to it, nor anything taken from it; God has done this, so that all should stand in awe before him. 15 That which is, already has been; that which is to be, already is; and God seeks out what has gone by. *g*

16 Moreover I saw under the sun that in the place of justice, wickedness was there, and in the place of righteousness, wickedness was there as well. 17 I said in my heart, God will judge the righteous and the wicked, for he has appointed a time for every matter, and for every work. 18 I said in my heart with regard to human beings that God is testing them to show that they are but animals. 19 For the fate of humans and the fate of animals is the same; as one dies, so dies the other. They all have the same breath, and humans have no advantage over the animals; for all is vanity. 20 All go to one place; all are from the dust, and all turn to dust again. 21 Who knows whether the human spirit goes upward and the spirit of animals goes downward to the earth? 22 So I saw that there is nothing better than that all should enjoy their work, for that is their lot; who can bring them to see what will be after them?

4 Again I saw all the oppressions that are practiced under the sun. Look, the tears of the oppressed—with no one to comfort them! On the side of their oppressors there was power—with no one to comfort them. 2 And I thought the dead, who have already died, more for-

g Heb *what is pursued*

3.1–15: Everything has its time, determined by God. 2: *To be born,* Hebrew "for giving birth."
3.11: God has so fashioned the human heart that it cannot understand what God is doing; hence Qoheleth offers another of his resigned conclusions (vv. 12–13; see also 2.26). **14–15:** Human beings can only stand in awe before God's mastery and mystery.
3.16–22: Human injustice is subject to divine judgment (compare 11.9). Qoheleth never denies that God is a judge; he denies that one can understand the divine judgment. It is meaningless, since humans and beasts resemble each other in that both have the same *fate,* death. **21:** *Who knows:* This is equivalent to a denial; it does not contradict 12.7. **22:** Since no one survives after death, the only good is to enjoy life (see 2.24).
4.1–6: Life's oppressions and inequalities. In view of pitiless oppression, the dead (and even more, those who have never been

tunate than the living, who are still alive; [3] but better than both is the one who has not yet been, and has not seen the evil deeds that are done under the sun.

4 Then I saw that all toil and all skill in work come from one person's envy of another. This also is vanity and a chasing after wind. [h]

5 Fools fold their hands
and consume their own flesh.

6 Better is a handful with quiet
than two handfuls with toil,
and a chasing after wind. [h]

7 Again, I saw vanity under the sun: [8] the case of solitary individuals, without sons or brothers; yet there is no end to all their toil, and their eyes are never satisfied with riches. "For whom am I toiling," they ask, "and depriving myself of pleasure?" This also is vanity and an unhappy business.

9 Two are better than one, because they have a good reward for their toil. [10] For if they fall, one will lift up the other; but woe to one who is alone and falls and does not have another to help. [11] Again, if two lie together, they keep warm; but how can one keep warm alone? [12] And though one might prevail against another, two will withstand one. A threefold cord is not quickly broken.

13 Better is a poor but wise youth than an old but foolish king, who will no longer take advice. [14] One can indeed come out of prison to reign, even though born poor in the kingdom. [15] I saw all the living who, moving about under the sun, follow that[i] youth who replaced the king;[j] [16] there was no end to all those people whom he led. Yet those who

come later will not rejoice in him. Surely this also is vanity and a chasing after wind. [h]

5 [k] Guard your steps when you go to the house of God; to draw near to listen is better than the sacrifice offered by fools; for they do not know how to keep from doing evil. [l] [2][m] Never be rash with your mouth, nor let your heart be quick to utter a word before God, for God is in heaven, and you upon earth; therefore let your words be few.

3 For dreams come with many cares, and a fool's voice with many words.

4 When you make a vow to God, do not delay fulfilling it; for he has no pleasure in fools. Fulfill what you vow. [5] It is better that you should not vow than that you should vow and not fulfill it. [6] Do not let your mouth lead you into sin, and do not say before the messenger that it was a mistake; why should God be angry at your words, and destroy the work of your hands?

7 With many dreams come vanities and a multitude of words;[n] but fear God.

8 If you see in a province the oppression of the poor and the violation of justice and right, do not be amazed at the matter; for the high official is watched by a higher, and there are yet higher ones over them. [9] But all things considered, this is an advantage for a land: a king for a plowed field. [n]

h Or *a feeding on wind.* See Hos 12.1
i Heb *the second*   j Heb *him*   k Ch 4.17 in Heb   l Cn: Heb *they do not know how to do evil*   m Ch 5.1 in Heb   n Meaning of Heb uncertain

---

born) are better off than the living, who are given to rivalry and *envy* (v. 4). **5–6**: These two sayings more or less cancel each other; neither laziness nor over-achievement is profitable. **5**: *Fold their hands,* i.e. do not work (Prov 6.10; 24.33).

4.7–12: The value of a friend. The futility of the solitary person (vv. 7–8) is illustrated by sayings that underscore the advantages of human companionship.

4.13–16: The impermanence of fame. This obscure story suggests that fortune and

fame are fickle; one ruler is replaced by another.

5.1–7: Advice on religious observance. Beware of wordy and insincere piety. **3**: 10.12–13. **4–6**: Vows are to be kept, or disaster can be expected.

5.8–9: Oppression is not surprising in view of the need to supervise officials. Hence kingship is an *advantage* to a country; the text of v. 9 is ambiguous.

5.10–6.9: The topic of possessions. 10–12: Since desires are insatiable, riches are dan-

10 The lover of money will not be satisfied with money; nor the lover of wealth, with gain. This also is vanity.

11 When goods increase, those who eat them increase; and what gain has their owner but to see them with his eyes?

12 Sweet is the sleep of laborers, whether they eat little or much; but the surfeit of the rich will not let them sleep.

13 There is a grievous ill that I have seen under the sun: riches were kept by their owners to their hurt, 14 and those riches were lost in a bad venture; though they are parents of children, they have nothing in their hands. 15 As they came from their mother's womb, so they shall go again, naked as they came; they shall take nothing for their toil, which they may carry away with their hands. 16 This also is a grievous ill: just as they came, so shall they go; and what gain do they have from toiling for the wind? 17 Besides, all their days they eat in darkness, in much vexation and sickness and resentment.

18 This is what I have seen to be good: it is fitting to eat and drink and find enjoyment in all the toil with which one toils under the sun the few days of the life God gives us; for this is our lot. 19 Likewise all to whom God gives wealth and possessions and whom he enables to enjoy them, and to accept their lot and find enjoyment in their toil—this is the gift of God. 20 For they will scarcely brood over the days of their lives, because God keeps them occupied with the joy of their hearts.

6 There is an evil that I have seen under the sun, and it lies heavy upon humankind: 2 those to whom God gives wealth, possessions, and honor, so that they lack nothing of all that they desire, yet God does not enable them to enjoy these things, but a stranger enjoys them. This is vanity; it is a grievous ill. 3 A man may beget a hundred children, and live many years; but however many are the days of his years, if he does not enjoy life's good things, or has no burial, I say that a stillborn child is better off than he. 4 For it comes into vanity and goes into darkness, and in darkness its name is covered; 5 moreover it has not seen the sun or known anything; yet it finds rest rather than he. 6 Even though he should live a thousand years twice over, yet enjoy no good—do not all go to one place?

7 All human toil is for the mouth, yet the appetite is not satisfied. 8 For what advantage have the wise over fools? And what do the poor have who know how to conduct themselves before the living? 9 Better is the sight of the eyes than the wandering of desire; this also is vanity and a chasing after wind. *o*

10 Whatever has come to be has already been named, and it is known what human beings are, and that they are not able to dispute with those who are stronger. 11 The more words, the more vanity, so how is one the better? 12 For who knows what is good for mortals while they live the few days of their vain life, which they pass like a shadow? For who can tell them what will be after them under the sun?

7 A good name is better than
 precious ointment,
   and the day of death, than the
     day of birth.
2 It is better to go to the house of
    mourning
   than to go to the house of
     feasting;

---

o Or *a feeding on wind.* See Hos 12.1

---

gerous. **13–17:** An example of the uncertainty of relying upon material possessions. **18–20:** A typical conclusion; compare 3.22.

6.1–6: Two examples of "evil": a *stranger* can come to enjoy the *possessions* of another (vv. 1–2); the *stillborn* is happier than one who fails to succeed in life (vv. 3–6). **7–9:** The problems begotten of unsatisfied desires.

6.10–12: The human condition. Things are determined by God, who has the upper hand. We *mortals* have recourse only to *words,* but we know not what is *good,* for we do not know what is to come.

7.1–22: Various comments. It is difficult to determine if this series of sayings is accepted or rejected (v. 6, *vanity*) by Qoheleth. **1:**

for this is the end of everyone,
 and the living will lay it
  to heart.
3 Sorrow is better than laughter,
 for by sadness of countenance
  the heart is made glad.
4 The heart of the wise is in the
  house of mourning;
 but the heart of fools is in the
  house of mirth.
5 It is better to hear the rebuke of
  the wise
 than to hear the song of fools.
6 For like the crackling of thorns
  under a pot,
 so is the laughter of fools;
 this also is vanity.
7 Surely oppression makes the wise
  foolish,
 and a bribe corrupts the heart.
8 Better is the end of a thing than
  its beginning;
 the patient in spirit are better
  than the proud in spirit.
9 Do not be quick to anger,
 for anger lodges in the bosom
  of fools.
10 Do not say, "Why were the
  former days better than
  these?"
 For it is not from wisdom that
  you ask this.
11 Wisdom is as good as an
  inheritance,
 an advantage to those who see
  the sun.
12 For the protection of wisdom is
  like the protection of
  money,
 and the advantage of knowledge
  is that wisdom gives life to
  the one who possesses it.
13 Consider the work of God;

who can make straight what he
  has made crooked?
14 In the day of prosperity be joyful, and in the day of adversity consider; God has made the one as well as the other, so that mortals may not find out anything that will come after them.

15 In my vain life I have seen everything; there are righteous people who perish in their righteousness, and there are wicked people who prolong their life in their evildoing. 16 Do not be too righteous, and do not act too wise; why should you destroy yourself? 17 Do not be too wicked, and do not be a fool; why should you die before your time? 18 It is good that you should take hold of the one, without letting go of the other; for the one who fears God shall succeed with both.

19 Wisdom gives strength to the wise more than ten rulers that are in a city.

20 Surely there is no one on earth so righteous as to do good without ever sinning.

21 Do not give heed to everything that people say, or you may hear your servant cursing you; 22 your heart knows that many times you have yourself cursed others.

23 All this I have tested by wisdom; I said, "I will be wise," but it was far from me. 24 That which is, is far off, and deep, very deep; who can find it out? 25 I turned my mind to know and to search out and to seek wisdom and the sum of things, and to know that wickedness is folly and that foolishness is madness. 26 I found more bitter than death the woman who is a trap, whose heart is snares and nets, whose hands are fetters; one who pleases God escapes her, but the sinner is taken by her. 27 See, this is what I found, says

---

Reputation (*a good name*) is never secure until *death* (*ointment* for burial).
7.5–7: Wisdom is better than folly, but *oppression* can make a fool of the wise person.
11–14: *Wisdom* has its advantages, but it cannot affect the *work of God*.
7.15–18: Two admonitions dealing with retribution. Because theories of retribution are failures, even contradictory, one should

hold on to both—but the God-fearer has the best chance.
7.23–8.9: **The meaning of existence is hidden.** After confessing his failure to attain wisdom, Qoheleth warns against the woman *who is a trap* (compare Prov 7). Although this is usually interpreted as misogynistic, the condemnation of both male and female appears in v. 29.

the Teacher,[p] adding one thing to another to find the sum, 28 which my mind has sought repeatedly, but I have not found. One man among a thousand I found, but a woman among all these I have not found. 29 See, this alone I found, that God made human beings straightforward, but they have devised many schemes.

8 Who is like the wise man?
And who knows the
interpretation of a thing?
Wisdom makes one's face shine,
and the hardness of one's
countenance is changed.

2 Keep[q] the king's command because of your sacred oath. 3 Do not be terrified; go from his presence, do not delay when the matter is unpleasant, for he does whatever he pleases. 4 For the word of the king is powerful, and who can say to him, "What are you doing?" 5 Whoever obeys a command will meet no harm, and the wise mind will know the time and way. 6 For every matter has its time and way, although the troubles of mortals lie heavy upon them. 7 Indeed, they do not know what is to be, for who can tell them how it will be? 8 No one has power over the wind[r] to restrain the wind,[r] or power over the day of death; there is no discharge from the battle, nor does wickedness deliver those who practice it. 9 All this I observed, applying my mind to all that is done under the sun, while one person exercises authority over another to the other's hurt.

10 Then I saw the wicked buried; they used to go in and out of the holy place, and were praised in the city where they had done such things.[s] This also is vanity. 11 Because sentence against an evil deed is not executed speedily, the human heart is fully set to do evil. 12 Though sinners do evil a hundred times and prolong their lives, yet I know that it will be well with those who fear God, because they stand in fear before him, 13 but it will not be well with the wicked, neither will they prolong their days like a shadow, because they do not stand in fear before God.

14 There is a vanity that takes place on earth, that there are righteous people who are treated according to the conduct of the wicked, and there are wicked people who are treated according to the conduct of the righteous. I said that this also is vanity. 15 So I commend enjoyment, for there is nothing better for people under the sun than to eat, and drink, and enjoy themselves, for this will go with them in their toil through the days of life that God gives them under the sun.

16 When I applied my mind to know wisdom, and to see the business that is done on earth, how one's eyes see sleep neither day nor night, 17 then I saw all the work of God, that no one can find out what is happening under the sun. However much they may toil in seeking, they will not find it out; even though those who are wise claim to know, they cannot find it out.

9 All this I laid to heart, examining it all, how the righteous and the wise and their deeds are in the hand of God;

p *Qoheleth,* traditionally rendered *Preacher*
q Heb *I keep*   r Or *breath*   s Meaning of Heb uncertain

8.1–4: The saying in v. 1 about *the wise man* is conditioned by the careful way in which he must deal with royalty.

8.5–9: The traditional wisdom given in v. 5 (*the wise mind will know the time and the way*) is severely modified by human ignorance, especially of *what is to be,* and by death; but neither does *wickedness* profit!

8.10–17: **Retribution.** Qoheleth describes the failure of retribution for the *wicked* (vv. 10–12a). He knows what the traditional teaching calls for (vv. 12b–13), but things do not work out that way (v. 14). Hence he repeats his familiar conclusion (v. 15), but denies that any one, even the *wise,* can know *the work of God* (vv. 16–17).

9.1–6: **A judgment about life.** One cannot distinguish *love or hate* on the part of God, since *the same fate comes to all.* While the hope of the *living* is suggested by the proverb in v. 4, there seems to be irony in the next verse: What do *the living know?* Only *that they will die!* This is hardly an advantage over the *dead* whose lot is described in vv. 5–6.

whether it is love or hate one does not know. Everything that confronts them [2]is vanity,[t] since the same fate comes to all, to the righteous and the wicked, to the good and the evil,[u] to the clean and the unclean, to those who sacrifice and those who do not sacrifice. As are the good, so are the sinners; those who swear are like those who shun an oath. [3]This is an evil in all that happens under the sun, that the same fate comes to everyone. Moreover, the hearts of all are full of evil; madness is in their hearts while they live, and after that they go to the dead. [4]But whoever is joined with all the living has hope, for a living dog is better than a dead lion. [5]The living know that they will die, but the dead know nothing; they have no more reward, and even the memory of them is lost. [6]Their love and their hate and their envy have already perished; never again will they have any share in all that happens under the sun.

[7] Go, eat your bread with enjoyment, and drink your wine with a merry heart; for God has long ago approved what you do. [8]Let your garments always be white; do not let oil be lacking on your head. [9]Enjoy life with the wife whom you love, all the days of your vain life that are given you under the sun, because that is your portion in life and in your toil at which you toil under the sun. [10]Whatever your hand finds to do, do with your might; for there is no work or thought or knowledge or wisdom in Sheol, to which you are going.

[11] Again I saw that under the sun the race is not to the swift, nor the battle to the strong, nor bread to the wise, nor riches to the intelligent, nor favor to the skillful; but time and chance happen to them all. [12]For no one can anticipate the time of disaster. Like fish taken in a cruel net, and like birds caught in a snare, so mortals are snared at a time of calamity, when it suddenly falls upon them.

13 I have also seen this example of wisdom under the sun, and it seemed great to me. [14]There was a little city with few people in it. A great king came against it and besieged it, building great siegeworks against it. [15]Now there was found in it a poor wise man, and he by his wisdom delivered the city. Yet no one remembered that poor man. [16]So I said, "Wisdom is better than might; yet the poor man's wisdom is despised, and his words are not heeded."

17 The quiet words of the wise are
       more to be heeded
    than the shouting of a ruler
       among fools.
18 Wisdom is better than weapons
       of war,
    but one bungler destroys much
       good.

10 Dead flies make the perfumer's ointment give off a foul odor;
    so a little folly outweighs wisdom and honor.
2 The heart of the wise inclines to
       the right,
    but the heart of a fool to the
       left.
3 Even when fools walk on the
       road, they lack sense,
    and show to everyone that they
       are fools.
4 If the anger of the ruler rises
       against you, do not leave
       your post,

t Syr Compare Gk: Heb *Everything that confronts them* [2]*is everything*   *u* Gk Syr Vg: Heb lacks *and the evil*

9.7–12: Qoheleth offers another of his conclusions: *Enjoy life* to the full, since there is nothing *in Sheol.* But *under the sun* (vv. 11–12) it is *time and chance* that ensnare mortals. **9.13–18: Wisdom and might.** The story of the one who *by his wisdom delivered the city* is nullified by the way he was treated (vv. 13–16). Similarly, the power of wisdom can be undone by *one bungler* (v. 18).

10.1–20: A series of observations. 1: This continues the idea of the vulnerability of wisdom expressed in 9.16. **2–3:** Although Qoheleth criticizes traditional wisdom, folly is not an option! (see vv. 12–15).

for calmness will undo great
offenses.

5 There is an evil that I have seen under the sun, as great an error as if it proceeded from the ruler: ⁶folly is set in
many high places, and the rich sit in a
low place. ⁷I have seen slaves on horseback, and princes walking on foot like
slaves.

8 Whoever digs a pit will fall into it;
   and whoever breaks through a
      wall will be bitten by
      a snake.
9 Whoever quarries stones will be
      hurt by them;
   and whoever splits logs will be
      endangered by them.
10 If the iron is blunt, and one does
      not whet the edge,
   then more strength must be
      exerted;
   but wisdom helps one to
      succeed.
11 If the snake bites before it is
      charmed,
   there is no advantage in a
      charmer.

12 Words spoken by the wise bring
      them favor,
   but the lips of fools consume
      them.
13 The words of their mouths begin
      in foolishness,
   and their talk ends in wicked
      madness;
14 yet fools talk on and on.
   No one knows what is to
      happen,
   and who can tell anyone what
      the future holds?
15 The toil of fools wears them out,
   for they do not even know the
      way to town.

16 Alas for you, O land, when your
      king is a servant, ᵛ
   and your princes feast in the
      morning!
17 Happy are you, O land, when
      your king is a nobleman,
   and your princes feast at the
      proper time—
   for strength, and not for
      drunkenness!
18 Through sloth the roof sinks in,
   and through indolence the house
      leaks.
19 Feasts are made for laughter;
      wine gladdens life,
   and money meets every need.
20 Do not curse the king, even in
      your thoughts,
   or curse the rich, even in your
      bedroom;
   for a bird of the air may carry
      your voice,
   or some winged creature tell
      the matter.

11 Send out your bread upon the
      waters,
   for after many days you will get
      it back.
2 Divide your means seven ways, or
      even eight,
   for you do not know what
      disaster may happen
      on earth.
3 When clouds are full,
   they empty rain on the earth;
   whether a tree falls to the south or
      to the north,
   in the place where the tree falls,
      there it will lie.
4 Whoever observes the wind will
      not sow;

v Or *a child*

10.4–7: One must deal calmly with a king (compare Prov 16.4), but this does not always bring success (vv. 5–7). 8–11: Accidents can happen (vv. 8–9), but the use of wisdom and talent may lessen their chances (vv. 10–11). 12–15: See 5.3; 10.2–3.
10.16–20: Reflections concerning royalty.

11.1–6: Sayings about the future and its uncertainty. 1: Take a chance (*send out your bread:* probably a commercial venture); you may receive a profit. 2: But don't put all your eggs in one basket. 3–4: Whether one counts on the future (*clouds . . . rain*), or real-izes that it is impenetrable (*there the tree will*

850
OT

and whoever regards the clouds will not reap.

5 Just as you do not know how the breath comes to the bones in the mother's womb, so you do not know the work of God, who makes everything.

6 In the morning sow your seed, and at evening do not let your hands be idle; for you do not know which will prosper, this or that, or whether both alike will be good.

7 Light is sweet, and it is pleasant for the eyes to see the sun.

8 Even those who live many years should rejoice in them all; yet let them remember that the days of darkness will be many. All that comes is vanity.

9 Rejoice, young man, while you are young, and let your heart cheer you in the days of your youth. Follow the inclination of your heart and the desire of your eyes, but know that for all these things God will bring you into judgment.

10 Banish anxiety from your mind, and put away pain from your body; for youth and the dawn of life are vanity.

12 Remember your creator in the days of your youth, before the days of trouble come, and the years draw near when you will say, "I have no pleasure in them"; 2before the sun and the light and the moon and the stars are darkened and the clouds return with*w* the rain; 3in the day when the guards of the house tremble, and the strong men are bent, and the women who grind cease

working because they are few, and those who look through the windows see dimly; 4when the doors on the street are shut, and the sound of the grinding is low, and one rises up at the sound of a bird, and all the daughters of song are brought low; 5when one is afraid of heights, and terrors are in the road; the almond tree blossoms, the grasshopper drags itself along*x* and desire fails; because all must go to their eternal home, and the mourners will go about the streets; 6before the silver cord is snapped,*y* and the golden bowl is broken, and the pitcher is broken at the fountain, and the wheel broken at the cistern, 7and the dust returns to the earth as it was, and the breath*z* returns to God who gave it. 8Vanity of vanities, says the Teacher;*a* all is vanity.

9 Besides being wise, the Teacher*a* also taught the people knowledge, weighing and studying and arranging many proverbs. 10The Teacher*a* sought to find pleasing words, and he wrote words of truth plainly.

11 The sayings of the wise are like goads, and like nails firmly fixed are the collected sayings that are given by one shepherd.*b* 12Of anything beyond these, my child, beware. Of making many

---

w Or *after*; Heb *'ahar*   x Or *is a burden*
y Syr Vg Compare Gk: Heb *is removed*
z Or *the spirit*   a *Qoheleth*, traditionally
rendered *Preacher*   b Meaning of Heb
uncertain

---

lie), one must act (v. 4). **5**: The unfathomable work of God (see also 3.11; 8.17). **6**: Despite one's ignorance, one must act.

11.7–12.8: Conclusion. 11.7–10: Qoheleth recommends joy and zest in life, but he does not remove the cloud: the coming *days of darkness . . . vanity . . . judgment . . . vanity*.

12.1–7: This famous description of the approach of old age and death (*the days of trouble*) is difficult to interpret. There is a suggestion of the approach of winter (of life) in v. 3 and the deterioration of a house (v. 4). But an allegorical approach has also been suggested: the *strong men* are the legs; the *guards* are the arms; the *women who grind* are the teeth; the *windows* are the eyes; etc. In any case, the

difficulties of old age seem to be reflected in vv. 4–5 (the *grasshopper* may be an image of one using crutches) and the picture of death is portrayed by the metaphors of v. 6. **7**: For the picture, see Gen 2.7; the body returns to the dust, and the *breath* of life is God's breath (not the "soul"), which sustains all living things (Ps 104.29–30).

12.9–14: Epilogue. 8: This is a deliberate repetition, or inclusion, of 1.2, the theme of the book. **9–12**: The author of these lines characterizes Qoheleth as a teacher with great literary style (*pleasing words*). He comments favorably on *the sayings of the wise*, as *given by one shepherd* (perhaps God), and advises his readers that they now have enough wisdom

books there is no end, and much study is a weariness of the flesh.

13 The end of the matter; all has been heard. Fear God, and keep his commandments; for that is the whole duty of everyone. 14 For God will bring every deed into judgment, including*c* every secret thing, whether good or evil.

c Or *into the judgment on*

---

books (including the present one), and no more need be added (v. 12). **13–14:** Some have conjectured that this final comment was added by an "orthodox" editor. But it is not hostile to Qoheleth; it goes beyond him, bringing in the *commandments,* and affirming *judgment* more positively than he ever does (compare 11.9; 3.17).

The Song of Solomon

This book, known also as the Song of Songs (see 1.1) or the Canticle (of Canticles), contains love poems, or songs, of Israel. The main speakers are a man and a woman, with the Daughters of Jerusalem serving as a foil for the woman's declarations (1.5, etc.). The work is dramatic, but not a drama with pre-arranged acts. Its unity is that of dialogue, though some have divided it into many unrelated poems. The most telling signs of unity are the frequent repetitions of words, such as 2.6–7 (compare 8.3–4; 3.5) and 2.8 (compare 2.17; 8.14), and also of scenes (e.g., 3.1–5; 5.2–8). Throughout the eight chapters a delicate mood of love and devotion is sustained. The woman (to whom most of the lines are to be attributed) and the man (portrayed both as king and shepherd) express their love for each other in poems that reflect desire, admiration, and boasting. Each delights in describing the physical charms of the other (4.1–7; 7.1–8). The only reference to marriage is in 3.11 (Solomon's wedding day). The term "bride" (4.8–5.1) is to be taken as part of the normal love language (as also "sister," 4.9, and 10) to designate the beloved. It is possible that these songs were originally oral compositions that came to be used in weddings. Here they have been assembled as a "book" of the Bible, perhaps by the Israelite sages because they underscore important values (mutuality and fidelity in love; compare Prov 5.15–20).

Both Jewish and Christian tradition found another level of meaning in the work: the love between God and God's people. Accordingly, the main characters would be the Lord and Israel (as for example in the Targum, the Aramaic paraphrase of the Hebrew Scriptures that was used in the synagogues for those who no longer understood Hebrew; the Song is the scroll that is read at the Passover festival), or Christ and the Church or the individual soul (as in the writings of Origen and Bernard of Clairvaux). The closest comparisons for the Song are the love poems of ancient Egypt. Allusions to divine marriage (as in Mesopotamian literature) are perhaps made in a few verses, but these belong to the prehistory of the text. As it stands, the collection of poems deals with human sexual love. It is impossible to date the Song with any certainty. The attribution to Solomon as author (1.1) is due to the mention of his name in 3.9, 11; 8.11–12, and perhaps to the claim made in 1 Kings 4.32.

OT

1

The Song of Songs, which is Solomon's.

2 Let him kiss me with the kisses of
his mouth!
For your love is better than wine,
3   your anointing oils are fragrant,
your name is perfume poured out;
therefore the maidens love you.
4 Draw me after you, let us make
haste.
The king has brought me into
his chambers.
We will exult and rejoice in you;
we will extol your love more
than wine;
rightly do they love you.

5 I am black and beautiful,
O daughters of Jerusalem,
like the tents of Kedar,
like the curtains of Solomon.
6 Do not gaze at me because I am
dark,
because the sun has gazed on
me.
My mother's sons were angry
with me;
they made me keeper of the
vineyards,
but my own vineyard I have
not kept!
7 Tell me, you whom my soul
loves,
where you pasture your flock,
where you make it lie down
at noon;
for why should I be like one who
is veiled
beside the flocks of your
companions?

8 If you do not know,
O fairest among women,
follow the tracks of the flock,
and pasture your kids
beside the shepherds' tents.

9 I compare you, my love,
to a mare among Pharaoh's
chariots.
10 Your cheeks are comely with
ornaments,
your neck with strings of
jewels.
11 We will make you ornaments
of gold,
studded with silver.

12 While the king was on his couch,
my nard gave forth its
fragrance.
13 My beloved is to me a bag of
myrrh
that lies between my breasts.
14 My beloved is to me a cluster of
henna blossoms
in the vineyards of En-gedi.

15 Ah, you are beautiful, my love;
ah, you are beautiful;
your eyes are doves.
16 Ah, you are beautiful, my
beloved,
truly lovely.
Our couch is green;
17 the beams of our house are
cedar,
our rafters*a* are pine.

2

I am a rose*b* of Sharon,
a lily of the valleys.

a Meaning of Heb uncertain   *b* Heb *crocus*

1.1: Title. The repetition of *song* is Hebrew
idiom for the superlative: the greatest song
(compare "king of kings"). On *Solomon,* see
the Introduction.
1.2–8: The woman's song. The woman
yearns for her lover, who is portrayed as a *king*
(vv. 2–4), explains her color to the *daughters*
(vv. 5–6), and seeks a rendezvous with her
lover, who is now described as a shepherd
(vv. 7–8).
1.9–2.7: The man's song. The man praises

her adornment (vv. 8–11), while she delights
in his presence (vv. 12–14). An exchange of
compliments follows (1.15–2.3). **14:** *En-gedi*
is an oasis on the western side of the Dead Sea.
2.1: The flower of Sharon is probably the
crocus that grows in the coastal plain of Sharon; the *lily* is the lotus flower.
2.4–7: The woman proclaims her love-
sickness and delivers an adjuration about love
to the daughters (see 3.5; 5.8; 8.4).

2 As a lily among brambles,
so is my love among maidens.

3 As an apple tree among the trees
of the wood,
so is my beloved among young
men.
With great delight I sat in his
shadow,
and his fruit was sweet to
my taste.

4 He brought me to the banqueting
house,
and his intention toward me
was love.

5 Sustain me with raisins,
refresh me with apples;
for I am faint with love.

6 O that his left hand were under
my head,
and that his right hand
embraced me!

7 I adjure you, O daughters of
Jerusalem,
by the gazelles or the wild does:
do not stir up or awaken love
until it is ready!

8 The voice of my beloved!
Look, he comes,
leaping upon the mountains,
bounding over the hills.

9 My beloved is like a gazelle
or a young stag.
Look, there he stands
behind our wall,
gazing in at the windows,
looking through the lattice.

10 My beloved speaks and says to
me:
"Arise, my love, my fair one,
and come away;

11 for now the winter is past,
the rain is over and gone.

12 The flowers appear on the earth;
the time of singing has come,
and the voice of the turtledove
is heard in our land.

13 The fig tree puts forth its figs,
and the vines are in blossom;
they give forth fragrance.
Arise, my love, my fair one,
and come away.

14 O my dove, in the clefts of the
rock,
in the covert of the cliff,
let me see your face,
let me hear your voice;
for your voice is sweet,
and your face is lovely.

15 Catch us the foxes,
the little foxes,
that ruin the vineyards—
for our vineyards are in
blossom."

16 My beloved is mine and I am his;
he pastures his flock among the
lilies.

17 Until the day breathes
and the shadows flee,
turn, my beloved, be like a gazelle
or a young stag on the cleft
mountains.[c]

3 Upon my bed at night
I sought him whom my soul
loves;
I sought him, but found him not;
I called him, but he gave no
answer.[d]

2 "I will rise now and go about
the city,
in the streets and in the squares;

c Or on the mountains of Bether: meaning of Heb
uncertain   d Gk: Heb lacks this line

2.8–17: The lover's spring visit. The
woman reminisces about a springtime
visit and invitation from her lover. To his
request to *hear* her *voice* (v. 14), she replies
with the enigmatic line about the *little
foxes* (v. 15), and invites him to the *mountains,* i.e. herself (v. 17).

3.1–5: The woman's dream. The wo-
man tells the *daughters* (v. 5; see also 2.7) of
a successful search (in dream or fantasy) for
her lover; compare 5.2–8.

I will seek him whom my soul
loves."
I sought him, but found him
not.
3 The sentinels found me,
as they went about in the city.
"Have you seen him whom my
soul loves?"
4 Scarcely had I passed them,
when I found him whom my
soul loves.
I held him, and would not let
him go
until I brought him into my
mother's house,
and into the chamber of her that
conceived me.
5 I adjure you, O daughters of
Jerusalem,
by the gazelles or the wild does:
do not stir up or awaken love
until it is ready!

6 What is that coming up from the
wilderness,
like a column of smoke,
perfumed with myrrh and
frankincense,
with all the fragrant powders of
the merchant?
7 Look, it is the litter of Solomon!
Around it are sixty mighty men
of the mighty men of Israel,
8 all equipped with swords
and expert in war,
each with his sword at his thigh
because of alarms by night.
9 King Solomon made himself a
palanquin
from the wood of Lebanon.
10 He made its posts of silver,
its back of gold, its seat of
purple;
its interior was inlaid with love. *e*

Daughters of Jerusalem,
11   come out.
Look, O daughters of Zion,
at King Solomon,
at the crown with which his
mother crowned him
on the day of his wedding,
on the day of the gladness of
his heart.

4 How beautiful you are, my love,
how very beautiful!
Your eyes are doves
behind your veil.
Your hair is like a flock of goats,
moving down the slopes of
Gilead.
2 Your teeth are like a flock of
shorn ewes
that have come up from the
washing,
all of which bear twins,
and not one among them is
bereaved.
3 Your lips are like a crimson
thread,
and your mouth is lovely.
Your cheeks are like halves of a
pomegranate
behind your veil.
4 Your neck is like the tower of
David,
built in courses;
on it hang a thousand bucklers,
all of them shields of warriors.
5 Your two breasts are like two
fawns,
twins of a gazelle,
that feed among the lilies.
6 Until the day breathes
and the shadows flee,

e Meaning of Heb uncertain

3.6–11: A wedding procession. Description of an ornate wedding procession. **11:** *King Solomon*, probably to be identified with the "king" of 1.4, 12.
4.1–5.1: The man's praise and invitation. There are several units: the man describes the physical charms of the woman (4.1–7; see also 6.5–7; 7.1–5); he invites her to make herself accessible (v. 8; compare 2.14), by leaving the *mountains;* he declares the effect her presence has upon him (4.9–11; compare 6.5); he utters the garden song (4.12–5.1).

I will hasten to the mountain of
    myrrh
  and the hill of frankincense.
7 You are altogether beautiful,
    my love;
  there is no flaw in you.
8 Come with me from Lebanon,
    my bride;
  come with me from Lebanon.
Depart*f* from the peak of Amana,
  from the peak of Senir and
    Hermon,
from the dens of lions,
  from the mountains of leopards.

9 You have ravished my heart, my
    sister, my bride,
  you have ravished my heart
    with a glance of your eyes,
  with one jewel of your necklace.
10 How sweet is your love, my
    sister, my bride!
  how much better is your love
    than wine,
  and the fragrance of your oils
    than any spice!
11 Your lips distill nectar, my bride;
  honey and milk are under your
    tongue;
  the scent of your garments is
    like the scent of Lebanon.
12 A garden locked is my sister,
    my bride,
  a garden locked, a fountain
    sealed.
13 Your channel*g* is an orchard of
    pomegranates
  with all choicest fruits,
  henna with nard,
14 nard and saffron, calamus and
    cinnamon,
  with all trees of frankincense,
myrrh and aloes,
  with all chief spices—

15 a garden fountain, a well of living
    water,
  and flowing streams from
    Lebanon.

16 Awake, O north wind,
  and come, O south wind!
Blow upon my garden
  that its fragrance may be wafted
    abroad.
Let my beloved come to his
    garden,
  and eat its choicest fruits.

5 I come to my garden, my sister,
    my bride;
  I gather my myrrh with my
    spice,
  I eat my honeycomb with my
    honey,
  I drink my wine with my milk.

Eat, friends, drink,
  and be drunk with love.

2 I slept, but my heart was awake.
Listen! my beloved is knocking.
"Open to me, my sister, my love,
  my dove, my perfect one;
for my head is wet with dew,
  my locks with the drops of the
    night."
3 I had put off my garment;
  how could I put it on again?
I had bathed my feet;
  how could I soil them?
4 My beloved thrust his hand into
    the opening,
  and my inmost being yearned
    for him.
5 I arose to open to my beloved,

f Or *Look*   *g* Meaning of Heb uncertain

4.9: *My sister, my bride,* characteristic of
this section; *sister* is a term of endearment in
Egyptian love-poetry; *bride* anticipates the
man's intention. **12**: The metaphors of *garden*
and *fountain* (see v. 15) designate the life-
giving qualities of the woman, and also her
fidelity (*locked, sealed*).

5.1: The man replies to the woman's invitation (4.16), and an unidentified voice (*Eat,
friends*) encourages them in their love.
5.2–6.3: **The woman's search.** Two
questions (5.9; 6.1) unite this section. **5.2–8**:
The woman relates another dream, similar to
3.1–5, but she fails to find her lover.

and my hands dripped with
   myrrh,
my fingers with liquid myrrh,
   upon the handles of the bolt.
6 I opened to my beloved,
   but my beloved had turned and
      was gone.
My soul failed me when he spoke.
I sought him, but did not find
      him;
   I called him, but he gave
      no answer.
7 Making their rounds in the city
   the sentinels found me;
they beat me, they wounded me,
   they took away my mantle,
   those sentinels of the walls.
8 I adjure you, O daughters of
      Jerusalem,
   if you find my beloved,
tell him this:
   I am faint with love.

9 What is your beloved more than
      another beloved,
   O fairest among women?
What is your beloved more than
      another beloved,
   that you thus adjure us?

10 My beloved is all radiant and
      ruddy,
   distinguished among ten
      thousand.
11 His head is the finest gold;
   his locks are wavy,
   black as a raven.
12 His eyes are like doves
   beside springs of water,
bathed in milk,
   fitly set. *h*
13 His cheeks are like beds of spices,
   yielding fragrance.
His lips are lilies,
   distilling liquid myrrh.

14 His arms are rounded gold,
   set with jewels.
His body is ivory work, *h*
   encrusted with sapphires. *i*
15 His legs are alabaster columns,
   set upon bases of gold.
His appearance is like Lebanon,
   choice as the cedars.
16 His speech is most sweet,
   and he is altogether desirable.
This is my beloved and this is
      my friend,
   O daughters of Jerusalem.

6 Where has your beloved gone,
   O fairest among women?
Which way has your beloved
      turned,
   that we may seek him with
      you?

2 My beloved has gone down to his
      garden,
   to the beds of spices,
to pasture his flock in the gardens,
   and to gather lilies.
3 I am my beloved's and my
      beloved is mine;
   he pastures his flock among the
      lilies.

4 You are beautiful as Tirzah,
      my love,
   comely as Jerusalem,
   terrible as an army with
      banners.
5 Turn away your eyes from me,
   for they overwhelm me!
Your hair is like a flock of goats,
   moving down the slopes of
      Gilead.

h Meaning of Heb uncertain   *i* Heb *lapis
lazuli*

5.9–16: In answer to the query of the
daughters she describes his physical charms.
6.1–3: When they too wish to seek him,
she tells them that he has never really been
lost, but has always been with his *garden,* herself.

6.4–10: **The man's song of praise.** He
praises her captivating beauty (compare 4.
1–3, 9), which is acknowledged even by the
royal harem; she is unique (vv. 8– 9). **4:** *Tirzah* was an early capital of the Northern
Kingdom; see 1 Kings 15.33.

6 Your teeth are like a flock of
    ewes,
  that have come up from the
    washing;
  all of them bear twins,
    and not one among them is
      bereaved.
7 Your cheeks are like halves of a
    pomegranate
  behind your veil.
8 There are sixty queens and eighty
    concubines,
  and maidens without number.
9 My dove, my perfect one, is the
    only one,
  the darling of her mother,
  flawless to her that bore her.
The maidens saw her and called
    her happy;
  the queens and concubines also,
    and they praised her.

10 "Who is this that looks forth like
    the dawn,
  fair as the moon, bright as
    the sun,
  terrible as an army with
    banners?"

11 I went down to the nut orchard,
  to look at the blossoms of the
    valley,
  to see whether the vines had
    budded,
  whether the pomegranates were
    in bloom.
12 Before I was aware, my fancy
    set me
  in a chariot beside my prince.*j*

13*k* Return, return, O Shulammite!
  Return, return, that we may
    look upon you.

Why should you look upon the
    Shulammite,
  as upon a dance before two
    armies?*l*

7 How graceful are your feet in
    sandals,
  O queenly maiden!
Your rounded thighs are like
    jewels,
  the work of a master hand.
2 Your navel is a rounded bowl
  that never lacks mixed wine.
Your belly is a heap of wheat,
  encircled with lilies.
3 Your two breasts are like two
    fawns,
  twins of a gazelle.
4 Your neck is like an ivory tower.
Your eyes are pools in Heshbon,
  by the gate of Bath-rabbim.
Your nose is like a tower of
    Lebanon,
  overlooking Damascus.
5 Your head crowns you like
    Carmel,
  and your flowing locks are like
    purple;
  a king is held captive in the
    tresses.*m*

6 How fair and pleasant you are,
  O loved one, delectable
    maiden!*n*
7 You are stately*o* as a palm tree,
  and your breasts are like its
    clusters.
8 I say I will climb the palm tree
  and lay hold of its branches.

j Cn: Meaning of Heb uncertain   *k* Ch 7.1 in
Heb   *l* Or *dance of Mahanaim*   *m* Meaning
of Heb uncertain   *n* Syr: Heb *in delights*
o Heb *This your stature is*

6.11–12: The woman visits the garden.
The woman seems to be referring to an
encounter in *the nut orchard* with her lover
(*my prince*), but the meaning of v. 12 is
uncertain.

**6.13–8.4: Praise of the woman and her
promise of love.** Unidentified persons
want to look at the *Shulammite* (in context

this mysterious term designates the woman),
but she seems to put them off (6.13b). Another description of her physical charms (see
also 4.1–7) is uttered (7.1–6), presumably
by the lover, who expresses his passionate
desires in vv. 7–9. She responds with a declaration of her love (7.10–8.4), ending with
the adjuration to the *daughters* (see 2.6–7).

Oh, may your breasts be like
    clusters of the vine,
    and the scent of your breath like
      apples,
9 and your kisses*p* like the best wine
    that goes down*q* smoothly,
    gliding over lips and teeth.*r*

10 I am my beloved's,
    and his desire is for me.
11 Come, my beloved,
    let us go forth into the fields,
    and lodge in the villages;
12 let us go out early to the
      vineyards,
    and see whether the vines have
      budded,
    whether the grape blossoms have
      opened
    and the pomegranates are
      in bloom.
    There I will give you my love.
13 The mandrakes give forth
      fragrance,
    and over our doors are all
      choice fruits,
    new as well as old,
    which I have laid up for you,
      O my beloved.

8 O that you were like a brother
      to me,
    who nursed at my mother's
      breast!
    If I met you outside, I would
      kiss you,
    and no one would despise me.
2 I would lead you and bring you
    into the house of my mother,
    and into the chamber of the one
      who bore me.*s*
    I would give you spiced wine
      to drink,
    the juice of my pomegranates.
3 O that his left hand were under
      my head,

    and that his right hand
      embraced me!
4 I adjure you, O daughters of
      Jerusalem,
    do not stir up or awaken love
    until it is ready!

5 Who is that coming up from the
      wilderness,
    leaning upon her beloved?

Under the apple tree I awakened
      you.
There your mother was in labor
      with you;
    there she who bore you was in
      labor.

6 Set me as a seal upon your heart,
    as a seal upon your arm;
for love is strong as death,
    passion fierce as the grave.
Its flashes are flashes of fire,
    a raging flame.
7 Many waters cannot quench love,
    neither can floods drown it.
If one offered for love
    all the wealth of his house,
    it would be utterly scorned.

8 We have a little sister,
    and she has no breasts.
What shall we do for our sister,
    on the day when she is
      spoken for?
9 If she is a wall,
    we will build upon her a
      battlement of silver;
but if she is a door,
    we will enclose her with boards
      of cedar.
10 I was a wall,

p Heb *palate*   *q* Heb *down for my lover*
r Gk Syr Vg: Heb *lips of sleepers*   *s* Gk Syr: Heb
my mother; she (or *you*) *will teach me*

**8.5–14: The lovers' vows and their final
exchange.** This consists of separate units: a
reference to a past encounter (v. 5); the stirring words about the power of *love* (vv. 6–7);
the reminiscence by the woman of the futile

concern of her brothers (vv. 8–10; compare
1.6); a boast concerning the *vineyard,* i.e. the
woman, more valuable than Solomon's harem (vv. 11–12); a final exchange between the
lovers (vv. 13–14, echoing 2.14, 17).

and my breasts were like
  towers;
then I was in his eyes
  as one who brings[t] peace.
11 Solomon had a vineyard at
  Baal-hamon;
he entrusted the vineyard to
  keepers;
each one was to bring for its
  fruit a thousand pieces of
  silver.
12 My vineyard, my very own, is
  for myself;
you, O Solomon, may have the
  thousand,

and the keepers of the fruit two
  hundred!

13 O you who dwell in the gardens,
  my companions are listening for
   your voice;
  let me hear it.

14 Make haste, my beloved,
  and be like a gazelle
or a young stag
  upon the mountains of spices!

t Or finds

The Prophetical Books

"The Prophets" constitutes the second part of the Hebrew Scriptures, following after the Torah ("law" or "teaching"). These two elements of the canon of sacred writings together constitute "the law and the prophets." In the arrangement of the Hebrew Scriptures, however, "the Prophets" includes the two sections of "Former Prophets" (Joshua, Judges, Samuel, and Kings) and "Latter Prophets" (Isaiah, Jeremiah, Ezekiel, and the Twelve). See the "Introduction to the Old Testament" (pp. xxxi–xxxiii OT) and the essay on "The Historical Books" (pp. 267–269 OT) for discussion of the Former Prophets and of the formation of the different collections of the Hebrew Scriptures. The book of Daniel, though grouped in English Bibles with the prophets, is in the Hebrew Scriptures part of the third division of "the Writings." In addition, it contains elements of a different type from the books of classic prophecy, namely apocalyptic; see the essays on "The Poetical Books or 'Writings'" (pp. 623–624 OT) and on "Apocalyptic Literature" (pp. 362–363 NT) for further discussion of Daniel. The following explication is concerned primarily with the collections of prophetic material that in the Hebrew Scriptures are called the Latter Prophets.

This prophetic canon stands in a tense relation to the Torah, sometimes complementing its teaching, and sometimes protesting against it. Whereas the Torah provides the foundation for biblical faith, the prophetic materials mark the move into the tensions of present-tense faith, and into an anticipation of a future outworking of God's powerful purpose. Thus the prophetic materials tilt the horizon of biblical faith towards a yet-unfulfilled future.

I. THE FORMATION OF THE PROPHETIC BOOKS

§1. There is no doubt that behind the prophetic literature were powerful, generative personalities called prophets. About some of the prophets (as, for instance, Jeremiah), we know a great deal. About many others we know nothing. We can conclude that they were, over a long period of time, passionate, extraordinary persons who refused a domesticated view of reality, and who dared to voice their peculiar discernment of their life among God's people under God's rule. They are a diverse lot and regularly mediate their faith and their urgings through their own lives and their own experiences. In various ways they are skillful users of language, having the capacity to shape perception of public reality in a rich variety of images and metaphors. Because they are imaginative and sometimes irreverent in their utterances, the speech of the prophets is daring and unpredictable, and often confronts and even affronts the present-day reader, much as it would have done to the original hearers. The prophets voice a restlessness about social reality that bespeaks the restless rule of God, and the opening of social reality for God's new possibilities.

These are no ordinary persons, though it is difficult to understand the roots of their extraordinariness. They reflect a peculiar psychodynamic that resists any easy explanation. Although they are highly individualized, they are persons who were capable of transcending themselves, in order to be available for a purpose larger than their own, and therefore to voice a message other than their own. As "messengers," the prophets are understood to be vehicles and channels for a word other than their own. They claimed to be and were perceived to be channels for the sovereign speech of God, albeit expressed through their own personal experiences.

The prophets are generative personalities. They did not, however, operate alone or in a social vacuum. They were deeply engaged in the socio-economic, political issues of their day. Moreover, they are regularly partisan figures, taking up pronounced and often unpopular positions in the conflicts of their communities. Thus they were partisan advocates for various social policies and social interests, and no doubt had important allies and supporters, as well as vigorous adversaries. Consequently, prophetic religion in this literary corpus is no easy or innocent enterprise, but demands a theological passion that carries with it determined and partisan sociopolitical advocacy. The prophets characteristically advocate just social relations and pure worship of God that together constitute an alternative to dominant modes of social relations and religion.

The prophets do not simply mirror the world around them; rather they exercise enormous imagination both in their discernment and in their articulation. They see what others do not see, and they dare to utter what others would not dare to utter. The prophets are not simply social analysts who issue moral and religious urgings; they are also artists who redescribe reality and who construe social experience in new and venturesome categories. For that reason much of the prophetic speech is couched in images and metaphors that require imaginative reception by the hearers.

§2. These generative personalities who evidence a remarkable extraordinariness, who voice partisan positions, and who redescribe reality in daring speech leave behind them a powerful and enduring influence. On the one hand, their powerful influence brings about the preservation of their words and the remembrance of their acts. Around these awesome figures there must have been loyal friends, allies, and disciples who remembered and cherished their words and who regarded them as having enduring authority and pertinence. It is this residue of remembered words that became the core of the several prophetic books.

On the other hand, the prophets not only left their words behind them; they also left behind a vision of reality that continued to inform and empower others after them. These others continued the rhetorical, interpretive tradition of the generative personalities, so that in the wake of the prophetic person, the prophetic tradition continued with new prophetic speech that was faithful to and congruent with the originary personality. Thus not all the speech in a prophetic book comes from the prophetic personality, but much of it comes from the ongoing interpretive tradition derived from that personality. The derivative prophetic tradition continues to be powerful and passionate, and continues to claim the authority of the original person. Consequently, all the words in a collection, even those from a derivative tradition, are ascribed to the prophet himself.

As the prophetic books now stand, they are not "authored" (in any modern sense) by the prophetic personality, but are shaped through a sustained, elongated editorial process by many hands who selected materials, arranged and juxtaposed materials to create a literary corpus. By such editorial work, oral speech became literary deposit. Through this process it is clear that we are no longer dealing with a prophetic person, but the person has now been superseded by the prophetic book, which now claims authority. The various prophetic books stand in various relations to the prophetic personalities behind them.

§3. The words of the prophetic personality have become a prophetic book through a traditioning process (see "Modern Approaches to Biblical Study," pp. 388–392). In the end, prophetic books have come to be regarded as "holy books," i.e. books of sacred scripture having enduring authority for a religious community. Clearly the initial prophetic person did not intend to create scripture, but scripture has resulted through the ongoing theological reflection of the religious community. On the one hand, this process of making a holy text (i.e. canonization) consists in the recognition that the literature has ongoing normative authority. The community finds the literature to be enduringly true. That is, words uttered in one context are found to be compelling and powerful in many subsequent contexts, and are therefore adjudged to be "oracles of God." On the other hand, this process of shaping the prophetic book is not done around literary motifs or historical events, but around theological themes and convictions. Thus the literature, begun with a concrete personality, in the end is thematized in a more or less intentional way around governing theological convictions of the community. The various pieces of rhetoric are pressed into the service of the theological community, which both adapts to its own norms and finds its norms changed by them. Characteristically the prophetic books are thematized around the notion of God's passionate purpose, God's severe judgment, and God's enduring promise. The prophet urges that the listening community come to terms with God's powerful identity that cannot be overridden or nullified.

II. THEMES OF PROPHETIC FAITH

We may identify three central aspects of the theological content of the prophetic books.

§1. The prophetic books are rooted in an older tradition of faith. The prophets and the ongoing prophetic tradition do not work *de novo,* but proceed from a remembered theological and textual tradition. Canonically, as the prophets come after the Torah, we may say that the prophets operate from the affirmations and convictions of the Torah. As the Bible is now put together, there is no doubt that the teaching of the prophetic books depends upon the Torah.

That general rootage, however, is enormously variegated, given the diversity of prophetic traditions. We may identify several dominant traditions that lie behind the prophets. First, the covenant tradition of Moses contributes decisively to the dominant patterns and structures of prophetic proclamation. This contribution includes not only the saving tradition of the exodus, which is variously taken up by the prophets, but also the ethical tradition of command, and the liturgical practice of covenant blessing and covenant curse.

Second, the prophets are variously informed by the priestly traditions of ancient Israel, with their concern for the reality of sin and guilt, right sacrifice, and the urgency of purification and cleanness. There is a modern propensity to contrast priestly and prophetic; against that propensity, however, it is clear that all of these traditions are present in and active for the prophets.

Third, the sapiential (wisdom) tradition reflected in the book of Proverbs is operative in the prophets. This tradition is focused on the disciplined discernment of life and especially of non-human life. The sapiential tradition is evident in the prophets through a variety of images and analogies held together by an affirmation of the coherence of creation under the rule of God.

§2. The prophets have a passionate and determined conviction about the future of their people. They are not, however, prognosticators, predictors, or fortune-tellers. Rather, they are convinced that the world is morally coherent and that God's purposes for the future of the world are powerful and will prevail. Moreover, the prophets, because they are rooted in the trusted traditions just identified, believe that God's intention for the future is indeed identifiable and can be known. The future is not a blind fate, nor a capricious intervention, nor an inscrutable mystery. The future can be reliably expected, given who God is and what God has willed and promised. When God's intention for the world is mocked or dismissed through arrogant practice and policy, trouble (judgment) is sure to come in the future. Beyond that sure trouble, however, God's resilient intention for well-being (i.e. fertility, peace, justice, righteousness, and joy) is also sure and will be established. The future is known and certain because the character of God is known and reliable.

§3. Situated between the several old, remembered traditions and the sure expectations for the future, the prophets are primarily advocates concerned for present-tense social relations and institutional practices and policies. They are convinced that God holds a righteous purpose for the present that grows out of God's old promises and commands that cannot be resisted. On the one hand, the present-tense life of the community must attend to the moral coherence of all of creation and must live (if it is to live at all) in terms of that moral coherence rooted in God's rule. On the other hand, the prophets are attentive to social pathologies and disorders that create injustice, pain, and suffering, and that lead to the pathos and absence of God. Thus the prophets, rooted in a past and open to God's future, believe that the present moment of life is a critical moment when life-and-death decisions must be made that bespeak loyalty to God and that impinge upon public power and purpose.

The prophetic books provide a resource and ground from which a passionate insistence is voiced in the present and a passionate advocacy is sounded. That insistence and advocacy are daring and radical, frequently subversive of all that is settled. While the books of the prophets as bodies of literature are finished and closed, their powerful voices continue to generate energy and courage and hope for obedience to God, obedience that is often costly and always life-giving.

Isaiah

The prophet Isaiah, the son of Amoz, proclaimed his message to Judah and Jerusalem from 742 until 701 B.C. (some believe until 687 B.C.), that critical period in which the Northern Kingdom was annexed to the Assyrian empire (2 Kings ch 17) while Judah lived uneasily in its shadow as a tributary (2 Chr 28.21). Nothing is known about the early life of the prophet, though it has been conjectured from certain aspects of his message and from Isa 6.1–8 that he may have been a priest.

Only chs 1–39 can be assigned to Isaiah's time, and even they contain later materials (see notes); it is generally accepted that chs 40–66 come from the time of Cyrus of Persia (539 B.C.) and later, as shown by the differences in historical background, literary style, and theological emphases. Chapters 1–39 begin with Isaiah's memoirs (1.1–12.6); they continue with oracles against foreign and domestic enemies (13.1–23.18), followed by the post-exilic "Isaiah Apocalypse" (24.1–27.13). Oracles generally concerned with Judah's intrigue with Egypt, its implications and consequences (28.1–32.20), are followed by a short collection of post-exilic eschatological oracles (33.1–35.10). An historical appendix (36.1–39.8) completes chs 1–39, in which there are other additions and some rearranging of oracles by post-exilic editors.

In the tradition of Amos, Hosea, and Micah, contemporaries whose work he seems to know, Isaiah attacks social injustice as that which is most indicative of Judah's tenuous relationship with God. He exhorts his hearers to place their confidence in their omnipotent God and to lead such public and private lives as manifest this. Thus justice and righteousness, teaching and word, and assurance of divine blessing upon the faithful and punishment upon the faithless are recurrent themes in his message from the Holy One of Israel to a proud and stubborn people.

Chapters 40–66, commonly called Second Isaiah (or Second and Third Isaiah), originated immediately before the fall of Babylon (October 29, 539 B.C.) to the armies of Cyrus, king of Persia, and during the generation following. The anonymous author of the first bipartite section (chs 40–55 [40–48; 49–55]) exults in joyful anticipation of exiled Judah's restoration to Palestine, for which Cyrus is God's precipitating agent (44.28). Second Isaiah emphasizes the significance of historical events in God's plan, a plan that extends from creation to redemption—and beyond. Blindness to God's way is a cardinal sin in Second Isaiah. The author's interest in cosmogony was unique up to his time; it is used to emphasize the concept of God as exclusive creator and lord of all, whose ultimate glorious manifestation will be accompanied by a new creation. A noteworthy feature of his prophecy is the songs of the Servant (see 42.1–4 n.).

This eschatological hope is shared with the author, or authors, of the second bipartite section (chs 56–66 [56–59 and 63–66; 60–62]). The contents of this section (sometimes called Third Isaiah) suggest a date between 530 and 510 B.C., perhaps contemporary with Haggai and Zechariah (520–518); chapters 60–62 may be later. Other concepts are also shared. Jacob and Israel have primarily religious, albeit national, significance. God's concern for the exiles in chs 40–55 is paralleled by the comforting assurance to Zion's afflicted in chs 56–66. But the direct "I—thou" relationship of Second Isaiah gives way to a more transcendent concept. In chs 56–66 one is confronted by the sobering realities of life in the restored community. The Servant-motif vanishes, and there is growing emphasis on cultic matters.

Together these theologically significant sections present a moving vision of the assured hope of God's people in a world whose times are in God's hands.

1 The vision of Isaiah son of Amoz,
which he saw concerning Judah and
Jerusalem in the days of Uzziah, Jotham,
Ahaz, and Hezekiah, kings of Judah.
2 Hear, O heavens, and listen,
O earth;
for the LORD has spoken:
I reared children and brought
them up,
but they have rebelled against
me.
3 The ox knows its owner,
and the donkey its master's crib;
but Israel does not know,
my people do not understand.

4 Ah, sinful nation,
people laden with iniquity,
offspring who do evil,
children who deal corruptly,
who have forsaken the LORD,
who have despised the Holy
One of Israel,
who are utterly estranged!

5 Why do you seek further beatings?
Why do you continue to rebel?
The whole head is sick,
and the whole heart faint.
6 From the sole of the foot even to
the head,
there is no soundness in it,
but bruises and sores

and bleeding wounds;
they have not been drained, or
bound up,
or softened with oil.

7 Your country lies desolate,
your cities are burned with fire;
in your very presence
aliens devour your land;
it is desolate, as overthrown by
foreigners.
8 And daughter Zion is left
like a booth in a vineyard,
like a shelter in a cucumber field,
like a besieged city.
9 If the LORD of hosts
had not left us a few survivors,
we would have been like Sodom,
and become like Gomorrah.

10 Hear the word of the LORD,
you rulers of Sodom!
Listen to the teaching of our God,
you people of Gomorrah!
11 What to me is the multitude of
your sacrifices?
says the LORD;
I have had enough of burnt
offerings of rams
and the fat of fed beasts;
I do not delight in the blood
of bulls,
or of lambs, or of goats.

1.1–5.24: Oracles against rebellious Judah. 1.1: Superscription. *Vision of Isaiah* (6.1–13; Jer ch 1; Ezek chs 1–3) identifies Isa chs 1–39 as God's message to Judah through the prophet. The name *Isaiah* means "The LORD [Yahweh] gives salvation." The latter part of the verse beginning with "in the days of" may be an editorial expansion.
1.2–31: First series of oracles, serving as a kind of prologue. **2–3:** Poetic exhortation reminiscent of God's address to the heavenly host in 40.1–2. *Children,* compare Jer 3.19–22. The biblical word *know* implies a profound, identifying comprehension of the right relationship with God; it is a recurring prophetic theme (Jer 1.5; Hos 2.20; 4.1, 6; 5.4).
1.4–9: An appeal to a people heedless of the significance of Judah's devastation by Tiglath-Pileser III (734–733 B.C.; 7.1–2) or

Sennacherib (701 B.C.; 36.1) and Jerusalem's isolation. **4:** Note the poetic parallelism: *nation, people; offspring, children.* The expression, *Holy One of Israel* (5.19, 24; 10.20; 12.6; 17.7; 29.19; 30.11, 12, 15; 37.23), emphasizes God's unapproachable separateness, which he has bridged by his gracious election of Israel as his people (Hos 8.1; Jer 3.20).
1.10–20: God's pronouncement concerning Judah's religious superficiality (Am 5.21–24; Jer 6.20). Judah may repent and return (Jer 7.5–7); the alternative is destruction (Jer 7.22–34). **10:** *Teaching,* the Hebrew word is "torah," which is frequently translated "law." On *Sodom* and *Gomorrah* see Gen 18.16–19.28; Jer 23.14; Ezek 16.46–58. **14:** *My soul,* a Hebrew idiom which in this context means "I" (compare Lev 26.11, 30). **16–17:** Compare Ex 22.21, 22; Am 5.6–7. **18:** *Argue,* as one argues a case before a judge (Job

12 When you come to appear
    before me,[a]
  who asked this from your hand?
  Trample my courts no more;
13 bringing offerings is futile;
  incense is an abomination to
    me.
  New moon and sabbath and
    calling of convocation—
  I cannot endure solemn
    assemblies with iniquity.
14 Your new moons and your
    appointed festivals
  my soul hates;
  they have become a burden to me,
  I am weary of bearing them.
15 When you stretch out your hands,
  I will hide my eyes from you;
  even though you make many
    prayers,
  I will not listen;
  your hands are full of blood.
16 Wash yourselves; make yourselves
    clean;
  remove the evil of your doings
    from before my eyes;
  cease to do evil,
17   learn to do good;
  seek justice,
    rescue the oppressed,
  defend the orphan,
    plead for the widow.

18 Come now, let us argue it out,
  says the LORD:
  though your sins are like scarlet,
    they shall be like snow;
  though they are red like crimson,
    they shall become like wool.
19 If you are willing and obedient,
  you shall eat the good of the
    land;
20 but if you refuse and rebel,
  you shall be devoured by the
    sword;

for the mouth of the LORD has
  spoken.

21 How the faithful city
  has become a whore!
  She that was full of justice,
  righteousness lodged in her—
  but now murderers!
22 Your silver has become dross,
  your wine is mixed with water.
23 Your princes are rebels
  and companions of thieves.
  Everyone loves a bribe
  and runs after gifts.
  They do not defend the orphan,
  and the widow's cause does not
    come before them.

24 Therefore says the Sovereign, the
  LORD of hosts, the Mighty
    One of Israel:
  Ah, I will pour out my wrath on
    my enemies,
  and avenge myself on my foes!
25 I will turn my hand against you;
  I will smelt away your dross as
    with lye
  and remove all your alloy.
26 And I will restore your judges as
    at the first,
  and your counselors as at the
    beginning.
  Afterward you shall be called the
    city of righteousness,
  the faithful city.

27 Zion shall be redeemed by justice,
  and those in her who repent, by
    righteousness.
28 But rebels and sinners shall be
    destroyed together,
  and those who forsake the LORD
    shall be consumed.

a Or *see my face*

23.7). *Scarlet* for wickedness (garments of
Babylon, Rev 17.4); *like snow,* white for purity (Rev 19.8).
  1.21–23: Lamentation over Jerusalem. **21:**
Whore, Jer 3.6–10; Ezek chs 16 and 23. *Justice*
and *righteousness* express Isaiah's ideal for the
people of God. **24:** *Mighty One of Israel* recalls

Israel's patriarchal traditions (49.26; Gen
49.24; Ps 132.2, 5). **25:** *As with lye,* or "thoroughly." **26:** Isaiah frequently uses symbolic
names (7.14; 8.1; 9.6; see also Jer 33.16; Ezek
48.35 n.). There will be a new creation; compare Am 9.11; Rev 3.12; 21.1–4.

29 For you shall be ashamed of
the oaks
in which you delighted;
and you shall blush for the
gardens
that you have chosen.
30 For you shall be like an oak
whose leaf withers,
and like a garden without water.
31 The strong shall become like
tinder,
and their work *b* like a spark;
they and their work shall burn
together,
with no one to quench them.

2 The word that Isaiah son of Amoz
saw concerning Judah and Jerusalem.

2 In days to come
the mountain of the LORD's
house
shall be established as the highest
of the mountains,
and shall be raised above
the hills;
all the nations shall stream to it.
3 Many peoples shall come and
say,
"Come, let us go up to the
mountain of the LORD,
to the house of the God of
Jacob;
that he may teach us his ways
and that we may walk in his
paths."
For out of Zion shall go forth
instruction,

and the word of the LORD from
Jerusalem.
4 He shall judge between the
nations,
and shall arbitrate for many
peoples;
they shall beat their swords into
plowshares,
and their spears into pruning
hooks;
nation shall not lift up sword
against nation,
neither shall they learn war
any more.

5 O house of Jacob,
come, let us walk
in the light of the LORD!
6 For you have forsaken the ways
of *c* your people,
O house of Jacob.
Indeed they are full of diviners *d*
from the east
and of soothsayers like the
Philistines,
and they clasp hands with
foreigners.
7 Their land is filled with silver
and gold,
and there is no end to their
treasures;
their land is filled with horses,
and there is no end to their
chariots.
8 Their land is filled with idols;

b Or *its makers*   *c* Heb lacks *the ways of*
d Cn: Heb lacks *of diviners*

1.29–31: Judah is faithless; the comparison
is based on one of Isaiah's rare references
to pagan religious practices; compare 57.5; Jer
2.27; Ezek 6.1–14.
   2.1: **Second superscription,** perhaps for
chs 2–4. *Word* connotes "message" (Jer 7.1;
11.1). **2–5**: **The new age,** involving the elevation of Zion, the acknowledgment of the
nations, and the age of peace. This oracle
(vv. 2–4) is also found in Mic 4.1–4. **3**: *Instruction,* i.e. "teaching" (1.10). **4**: The age of
peace will follow the judgment of the LORD
(compare 5.25; 30.27–28). **5**: Compare v. 3,
paraphrased in Mic 4.5.

2.6–22: **The day of the LORD.** This is
probably to be taken as three stanzas, vv. 6–
11, 12–17, 18–22. The first two have a similar
conclusion (compare vv. 11, 17), and it is
suggested that the third ended similarly, for
the present v. 22 is missing in the Septuagint
and is grammatically corrupt. **6–11**: Judgment
on idolatry. **6**: *Diviners* were forbidden in Israel (Ex 22.18; Lev 20.27; Deut 18.10–11;
compare 8.19; 1 Sam 28.8–25; Ezek 13.9).
The situation fits Uzziah's reign (2 Kings
15.1–7; 2 Chr ch 26). **7**: Judah's prosperity
(Deut 17.16–17; 1 Kings 10.14–29). **11**: *In that
day,* the day of the LORD, in which God judges

they bow down to the work of
their hands,
to what their own fingers
have made.

9 And so people are humbled,
and everyone is brought low—
do not forgive them!

10 Enter into the rock,
and hide in the dust
from the terror of the LORD,
and from the glory of his
majesty.

11 The haughty eyes of people shall
be brought low,
and the pride of everyone shall
be humbled;
and the LORD alone will be exalted
in that day.

12 For the LORD of hosts has a day
against all that is proud and
lofty,
against all that is lifted up
and high;*e*

13 against all the cedars of Lebanon,
lofty and lifted up;
and against all the oaks of
Bashan;

14 against all the high mountains,
and against all the lofty hills;

15 against every high tower,
and against every fortified wall;

16 against all the ships of Tarshish,
and against all the beautiful
craft.*f*

17 The haughtiness of people shall be
humbled,
and the pride of everyone shall
be brought low;
and the LORD alone will be
exalted on that day.

18 The idols shall utterly pass away.

19 Enter the caves of the rocks
and the holes of the ground,
from the terror of the LORD,

and from the glory of his
majesty,
when he rises to terrify the
earth.

20 On that day people will throw
away
to the moles and to the bats
their idols of silver and their idols
of gold,
which they made for themselves
to worship,

21 to enter the caverns of the rocks
and the clefts in the crags,
from the terror of the LORD,
and from the glory of his
majesty,
when he rises to terrify the
earth.

22 Turn away from mortals,
who have only breath in their
nostrils,
for of what account are they?

3 For now the Sovereign, the LORD
of hosts,
is taking away from Jerusalem
and from Judah
support and staff—
all support of bread,
and all support of water—

2 warrior and soldier,
judge and prophet,
diviner and elder,

3 captain of fifty
and dignitary,
counselor and skillful magician
and expert enchanter.

4 And I will make boys their
princes,
and babes shall rule over them.

5 The people will be oppressed,
everyone by another

e Cn Compare Gk: Heb *low*   *f* Compare Gk:
Meaning of Heb uncertain

his enemies and manifests his glory, is a
recurring prophetic theme (13.6; Am 5.18–
20; Jer 17.16–18; Ezek 30.3; Joel 1.15).
2.12–17: Pride and punishment. **13:** *Lebanon, Bashan,* Ezek 27.5–6; Jer 22.20. **16:** *Ships
of Tarshish,* the phrase may mean "refinery
fleet" (see 1 Kings 10.22 n.; Jer 10.9 n.).
2.18–22: Judgment on idolatry. **19:** The

innumerable *caves* in Palestine's limestone
hills are age-old places of refuge.
3.1–15: Anarchy in Jerusalem. 1–7:
Without leaders, society breaks down. **1:** *Support and staff,* everything which supports life,
including food and drink (economic resources), and perhaps also the functionaries in
vv. 2–3. **2–3:** Offices deemed necessary for

and everyone by a neighbor;
the youth will be insolent to
   the elder,
and the base to the honorable.

6 Someone will even seize a relative,
   a member of the clan, saying,
"You have a cloak;
   you shall be our leader,
and this heap of ruins
   shall be under your rule."
7 But the other will cry out on that
   day, saying,
"I will not be a healer;
   in my house there is neither
   bread nor cloak;
you shall not make me
   leader of the people."
8 For Jerusalem has stumbled
   and Judah has fallen,
because their speech and their
   deeds are against the LORD,
   defying his glorious presence.

9 The look on their faces bears
   witness against them;
they proclaim their sin like
   Sodom,
   they do not hide it.
Woe to them!
   For they have brought evil on
   themselves.
10 Tell the innocent how fortunate
   they are,
for they shall eat the fruit of
   their labors.
11 Woe to the guilty! How
   unfortunate they are,
for what their hands have done
   shall be done to them.
12 My people—children are their
   oppressors,
   and women rule over them.
O my people, your leaders
   mislead you,

and confuse the course of your
   paths.
13 The LORD rises to argue his case;
   he stands to judge the peoples.
14 The LORD enters into judgment
   with the elders and princes of
   his people:
It is you who have devoured the
   vineyard;
the spoil of the poor is in your
   houses.
15 What do you mean by crushing
   my people,
by grinding the face of the
   poor? says the Lord GOD of
   hosts.

16 The LORD said:
Because the daughters of Zion are
   haughty
and walk with outstretched
   necks,
glancing wantonly with
   their eyes,
mincing along as they go,
   tinkling with their feet;
17 the Lord will afflict with scabs
   the heads of the daughters of
   Zion,
and the LORD will lay bare their
   secret parts.

18 In that day the Lord will take away
the finery of the anklets, the headbands,
and the crescents; 19the pendants, the
bracelets, and the scarfs; 20the headdresses, the armlets, the sashes, the perfume boxes, and the amulets; 21the signet
rings and nose rings; 22the festal robes,
the mantles, the cloaks, and the hand-
bags; 23the garments of gauze, the linen
garments, the turbans, and the veils.
24 Instead of perfume there will be a
   stench;

---

the continuity and stability of the state. **4**: The
inexperienced and naive will rule. **5–6**: Civil
unrest will become open violence.
   3.8–12: A commentary on vv. 1–7. Judah's brazen sinfulness and rejection of God's
leadership has ruined the people.
   3.13–15: God will judge the corrupt

judges. *Elders,* primary administrators of justice (Ex 19.7; Josh 20.4; Deut 21.19–21).
Princes, royal appointees (1 Kings 4.2;
2 Kings 10.1; Jer 34.19). *Vineyard,* see 5.1–7.
   **3.16–4.1: The humiliation of Jerusa-
lem's women** (Am 4.1–3). **3.18–24**: Detailed expansion of v. 17. **25–4.1**: War's decima-

and instead of a sash, a rope;
and instead of well-set hair,
    baldness;
and instead of a rich robe, a
    binding of sackcloth;
instead of beauty, shame. *g*

25 Your men shall fall by the sword
    and your warriors in battle.
26 And her gates shall lament and
        mourn;
    ravaged, she shall sit upon the
        ground.

4 Seven women shall take hold of one
    man in that day, saying,
"We will eat our own bread and
    wear our own clothes;
just let us be called by your name;
    take away our disgrace."

2 On that day the branch of the LORD shall be beautiful and glorious, and the fruit of the land shall be the pride and glory of the survivors of Israel. 3 Whoever is left in Zion and remains in Jerusalem will be called holy, everyone who has been recorded for life in Jerusalem, 4 once the Lord has washed away the filth of the daughters of Zion and cleansed the bloodstains of Jerusalem from its midst by a spirit of judgment and by a spirit of burning. 5 Then the LORD will create over the whole site of Mount Zion and over its places of assembly a cloud by day and smoke and the shining of a flaming fire by night. Indeed over all the glory there will be a canopy. 6 It will serve as a pavilion, a shade by day from the heat, and a

refuge and a shelter from the storm and rain.

5 Let me sing for my beloved
    my love-song concerning his
        vineyard:
My beloved had a vineyard
    on a very fertile hill.
2 He dug it and cleared it of stones,
    and planted it with choice vines;
he built a watchtower in the midst
        of it,
    and hewed out a wine vat in it;
he expected it to yield grapes,
    but it yielded wild grapes.

3 And now, inhabitants of Jerusalem
    and people of Judah,
judge between me
    and my vineyard.
4 What more was there to do for
        my vineyard
    that I have not done in it?
When I expected it to yield grapes,
    why did it yield wild grapes?

5 And now I will tell you
    what I will do to my vineyard.
I will remove its hedge,
    and it shall be devoured;
I will break down its wall,
    and it shall be trampled down.
6 I will make it a waste;
    it shall not be pruned or hoed,

g Q Ms: MT lacks *shame*

tion of the male population forces the women to resort to desperate measures to preserve themselves and their self-respect. *Our disgrace* summarizes 3.16–4.1.
   4.2–6: Jerusalem's restoration. 2: *Branch,* the righteous remnant (3.10; compare the Messiah as Branch in 11.1; Jer 23.5); *fruit of the land,* a "return to paradise." **3:** *Recorded for life,* compare Ex 32.32; Mal 3.16; Dan 12.1; Rev 20.12, 15. **5:** *Smoke* and *flaming fire,* the signs of God's presence among his people at the Exodus (Ex 13.21–22; 40.34–38).
   5.1–7: Song of the vineyard (Hos 10.1; Jer 2.21; Ezek 19.10–14), an allegory. This unique didactic poem may have been composed for a celebration of the festival of

booths during Jotham's reign, the prophet imitating a vintage festival song. **1a:** Introduction to the poem. **2:** *Choice vines,* the Hebrew word ("soreq") means either red grapes, or grapes native to the valley of Sorek, west of Jerusalem.
   5.3–4 Judah's only possible answer would be judgment against the vineyard. Judah is asked to pass judgment on herself, much as Nathan through a parable had David pass judgment on himself (2 Sam 12.1–12). **7:** *Justice,* the faithful application of God's will to daily living. *Righteousness,* the living, dynamic relationship between the nation and God wherein the nation is spiritually and morally acceptable to God (1.27; 9.7; 16.5; 28.17).

and it shall be overgrown with
briers and thorns;
I will also command the clouds
that they rain no rain upon it.

7 For the vineyard of the LORD
of hosts
is the house of Israel,
and the people of Judah
are his pleasant planting;
he expected justice,
but saw bloodshed;
righteousness,
but heard a cry!
8 Ah, you who join house to house,
who add field to field,
until there is room for no one
but you,
and you are left to live alone
in the midst of the land!
9 The LORD of hosts has sworn in
my hearing:
Surely many houses shall be
desolate,
large and beautiful houses,
without inhabitant.
10 For ten acres of vineyard shall
yield but one bath,
and a homer of seed shall yield
a mere ephah. *h*

11 Ah, you who rise early in the
morning
in pursuit of strong drink,
who linger in the evening
to be inflamed by wine,
12 whose feasts consist of lyre and
harp,
tambourine and flute and wine,
but who do not regard the deeds
of the LORD,
or see the work of his hands!

13 Therefore my people go into exile
without knowledge;
their nobles are dying of hunger,
and their multitude is parched
with thirst.

14 Therefore Sheol has enlarged its
appetite
and opened its mouth beyond
measure;
the nobility of Jerusalem *i* and her
multitude go down,
her throng and all who exult
in her.
15 People are bowed down, everyone
is brought low,
and the eyes of the haughty are
humbled.
16 But the LORD of hosts is exalted
by justice,
and the Holy God shows
himself holy by
righteousness.
17 Then the lambs shall graze as in
their pasture,
fatlings and kids *j* shall feed
among the ruins.

18 Ah, you who drag iniquity along
with cords of falsehood,
who drag sin along as with cart
ropes,
19 who say, "Let him make haste,
let him speed his work
that we may see it;
let the plan of the Holy One of
Israel hasten to fulfillment,
that we may know it!"

h The Heb *bath, homer,* and *ephah* are measures
of quantity   *i* Heb *her nobility*
j Cn Compare Gk: Heb *aliens*

Righteousness and justice are naturally coupled (1.21) and grow out of the covenant relationship, the existence of which is assumed
(Ex chs 19–20). *A cry,* from the oppressed.
  5.8–24a: Six reproaches (vv. 8, 11, 18,
20, 21, 22; perhaps 10.1–4 is a seventh).
Compare Am 5.7, 18; 6.1; Jer 22.13. **8–10:**
Against covetousness (Mic 2.1–5, 8–9; Ex
20.17). *Bath,* about 6 U.S. gallons; *homer,* 6.5
bushels (see Ezek 45.11 n.); *ephah,* about two-
thirds of a bushel.
  5.11–12: Against carousing (Am 6.4–6).
13–17: *Knowledge,* compare *know,* 1.3. The
severity of Judah's punishment will require
the enlargement of *Sheol* (the underworld,
14.9–18). **16:** In all he does, God is just and
right.
  5.18–19: Against mocking God. **20:**

20 Ah, you who call evil good
    and good evil,
who put darkness for light
    and light for darkness,
who put bitter for sweet
    and sweet for bitter!
21 Ah, you who are wise in your
    own eyes,
    and shrewd in your own sight!
22 Ah, you who are heroes in
    drinking wine
    and valiant at mixing drink,
23 who acquit the guilty for a bribe,
    and deprive the innocent of their
    rights!
24 Therefore, as the tongue of fire
    devours the stubble,
    and as dry grass sinks down in
    the flame,
so their root will become rotten,
    and their blossom go up like
    dust;
for they have rejected the
    instruction of the LORD
    of hosts,
    and have despised the word of
    the Holy One of Israel.

25 Therefore the anger of the LORD
    was kindled against his
    people,
    and he stretched out his hand
    against them and struck
    them;
    the mountains quaked,
and their corpses were like refuse
    in the streets.
For all this his anger has not
    turned away,
    and his hand is stretched
    out still.

26 He will raise a signal for a nation
    far away,
    and whistle for a people at the
    ends of the earth;
Here they come, swiftly, speedily!
27 None of them is weary, none
    stumbles,
    none slumbers or sleeps,
not a loincloth is loose,
    not a sandal-thong broken;
28 their arrows are sharp,
    all their bows bent,
their horses' hoofs seem like flint,
    and their wheels like the
    whirlwind.
29 Their roaring is like a lion,
    like young lions they roar;
they growl and seize their prey,
    they carry it off, and no one
    can rescue.
30 They will roar over it on that day,
    like the roaring of the sea.
And if one look to the land—
    only darkness and distress;
and the light grows dark with
    clouds.

6 In the year that King Uzziah died, I saw the Lord sitting on a throne, high and lofty; and the hem of his robe filled the temple. 2 Seraphs were in attendance above him; each had six wings: with two they covered their faces, and with two they covered their feet, and with two they flew. 3 And one called to another and said:
    "Holy, holy, holy is the LORD of
    hosts;
    the whole earth is full of his
    glory."

Against moral depravity (32.5; Prov 17.15). **21:** Against conceit. **22–23:** Against bravado and bribery.
    5.24b–30: A judgment. These verses are probably displaced; for v. 25b compare 9.12b, 17b, 21b, 10.4b (see 9.8–10.4 n.). *Instruction,* here also in the sense of "teaching" (see 2.3 n.). The Assyrians (*a nation far away;* Jer 5.15, referring to Babylon) will be the executors of God's judgment.

6.1–13: The call of Isaiah. God's appearance is described in the setting of the Jerusalem temple (compare the description of the enthroned deity in 1 Kings 22.19–23; Ezek 1.4–2.1). **1:** *Year,* 742 B.C. **2:** *Seraphs,* possibly griffin-like creatures; compare the cherubim, also associated with the glory of the LORD (Ezek ch 1). **3:** Thrice-*holy* for emphasis (Jer 7.4). **5:** Before the holy God, a sinful person cannot stand (Ex 33.18–20).

4 The pivots[k] on the thresholds shook at the voices of those who called, and the house filled with smoke. 5 And I said: "Woe is me! I am lost, for I am a man of unclean lips, and I live among a people of unclean lips; yet my eyes have seen the King, the LORD of hosts!"

6 Then one of the seraphs flew to me, holding a live coal that had been taken from the altar with a pair of tongs. 7 The seraph[l] touched my mouth with it and said: "Now that this has touched your lips, your guilt has departed and your sin is blotted out." 8 Then I heard the voice of the Lord saying, "Whom shall I send, and who will go for us?" And I said, "Here am I; send me!" 9 And he said, "Go and say to this people:

'Keep listening, but do not
      comprehend;
keep looking, but do not
      understand.'

10 Make the mind of this people dull,
      and stop their ears,
      and shut their eyes,
so that they may not look with
      their eyes,
      and listen with their ears,
and comprehend with their minds,
      and turn and be healed."

11 Then I said, "How long,
      O Lord?" And he said:
"Until cities lie waste
      without inhabitant,
and houses without people,
      and the land is utterly desolate;

12 until the LORD sends everyone
      far away,
      and vast is the emptiness in the
      midst of the land.

13 Even if a tenth part remain in it,
      it will be burned again,
like a terebinth or an oak
      whose stump remains standing
      when it is felled."[k]
The holy seed is its stump.

7 In the days of Ahaz son of Jotham son of Uzziah, king of Judah, King Rezin of Aram and King Pekah son of Remaliah of Israel went up to attack Jerusalem, but could not mount an attack against it. 2 When the house of David heard that Aram had allied itself with Ephraim, the heart of Ahaz[m] and the heart of his people shook as the trees of the forest shake before the wind.

3 Then the LORD said to Isaiah, Go out to meet Ahaz, you and your son Shear-jashub,[n] at the end of the conduit of the upper pool on the highway to the Fuller's Field, 4 and say to him, Take heed, be quiet, do not fear, and do not let your heart be faint because of these two smoldering stumps of firebrands, because of the fierce anger of Rezin and Aram and the son of Remaliah. 5 Because Aram—with Ephraim and the son of Remaliah—has plotted evil against you, saying, 6 Let us go up against Judah and cut off Jerusalem[o] and conquer it for ourselves and make the son of Tabeel king in it; 7 therefore thus says the Lord GOD:

It shall not stand,
      and it shall not come to pass.

8 For the head of Aram is Damascus,

k Meaning of Heb uncertain   l Heb *He*
m Heb *his heart*   n That is *A remnant shall return*   o Heb *cut it off*

6.6–8: Cleansed by God's forgiving act, Isaiah may now speak for God. **9–12**: Compare Jer 1.10, 13–19. Verses 9b–10 are quoted in Mt 13.10–15; compare Mk 4.12; Lk 8.10; Jn 12.39–41; Acts 28.26–27. **13**: The last part of the verse is obscure and textually corrupt and perhaps should be restored to read, ". . . . like the terebinth [of the goddess] and the oak of Asherah, cast out with the pillar of the high places," that is, like the destroyed furnishings of a pagan high place.
7.1–8.15: **Isaiah and the Syro Ephrai-** mite War (734–733 B.C.). For the historical background see 2 Kings 16.1–20. **7.1–9**: **Sign of Shear-jashub. 1**: *Aram,* Syria. **2**: The continuation of the Davidic monarchy was threatened (see v. 6). **3**: *Shear-jashub,* "A remnant shall return"; assuming the worst eventuality, God's promise to David (2 Sam 7.8–16) will be preserved in the remnant (10.20–23). *Upper pool,* reservoir south of the Pool of Siloam. **6**: *Son of Tabeel,* perhaps a prince of Judah whose mother came from Tabeel, a region of northern Transjordan. **8–**

and the head of Damascus is
    Rezin.
(Within sixty-five years Ephraim will
be shattered, no longer a people.)
 9 The head of Ephraim is Samaria,
    and the head of Samaria is the
     son of Remaliah.
If you do not stand firm in faith,
    you shall not stand at all.

10 Again the LORD spoke to Ahaz,
saying, 11 Ask a sign of the LORD your
God; let it be deep as Sheol or high as
heaven. 12 But Ahaz said, I will not ask,
and I will not put the LORD to the test.
13 Then Isaiah*p* said: "Hear then,
O house of David! Is it too little for you
to weary mortals, that you weary my
God also? 14 Therefore the Lord himself
will give you a sign. Look, the young
woman*q* is with child and shall bear a
son, and shall name him Immanuel.*r*
15 He shall eat curds and honey by the
time he knows how to refuse the evil and
choose the good. 16 For before the child
knows how to refuse the evil and choose
the good, the land before whose two
kings you are in dread will be deserted.
17 The LORD will bring on you and on
your people and on your ancestral house
such days as have not come since the day
that Ephraim departed from Judah—the
king of Assyria."

18 On that day the LORD will whistle
for the fly that is at the sources of the
streams of Egypt, and for the bee that is
in the land of Assyria. 19 And they will all
come and settle in the steep ravines, and
in the clefts of the rocks, and on all the
thornbushes, and on all the pastures.
20 On that day the Lord will shave

with a razor hired beyond the River—
with the king of Assyria—the head and
the hair of the feet, and it will take off the
beard as well.

21 On that day one will keep alive a
young cow and two sheep, 22 and will eat
curds because of the abundance of milk
that they give; for everyone that is left in
the land shall eat curds and honey.

23 On that day every place where
there used to be a thousand vines, worth
a thousand shekels of silver, will become
briers and thorns. 24 With bow and arrows one will go there, for all the land
will be briers and thorns; 25 and as for all
the hills that used to be hoed with a hoe,
you will not go there for fear of briers
and thorns; but they will become a place
where cattle are let loose and where
sheep tread.

8 Then the LORD said to me, Take a
large tablet and write on it in common characters, "Belonging to Maher-
shalal-hash-baz,"*s* 2 and have it attested*t*
for me by reliable witnesses, the priest
Uriah and Zechariah son of Jeberechiah.
3 And I went to the prophetess, and she
conceived and bore a son. Then the LORD
said to me, Name him Maher-shalal-
hash-baz; 4 for before the child knows
how to call "My father" or "My mother," the wealth of Damascus and the
spoil of Samaria will be carried away by
the king of Assyria.

p Heb *he*   *q* Gk *the virgin*   *r* That is *God is
with us*   *s* That is *The spoil speeds, the prey
hastens*   *t* Q Ms Gk Syr: MT *and I caused to be
attested*

9a: These words are intended to reduce
Ahaz's panic (v. 2); behind Aram and Ephraim are simply—two men.
   7.10–17: Sign of Immanuel. 13: This expresses Isaiah's impatience. **14:** The sign is
Immanuel, "God with us"; a second (compare
vv. 3–9) assurance to the frightened, wavering Ahaz. *Young woman,* Hebrew " 'almah,"
feminine of " 'elem," young man (1 Sam
17.56; 20.22); the word appears in Gen 24.43;
Ex 2.8; Ps 68.25, and elsewhere, where it is
translated "young woman," "girl," "maid-
en." **15:** *Curds, honey,* simple foods for a child
being weaned; *good and evil,* age of moral discrimination. **18–25: Four threats** amplifying
v. 17. **20:** *Feet,* see Ex 4.25 n.
   **8.1–4: The sign of Maher-shalal-hash-
baz,** "The spoil speeds, the prey hastes"; Isaiah's third assurance to Ahaz. **1:** *Tablet,* of
wood. **2:** *Uriah,* 2 Kings 16.10–16. *Zechariah,*
perhaps Ahaz's father-in-law (2 Kings 18.2).
3: *Prophetess,* Isaiah's wife.

5 The LORD spoke to me again: 6 Because this people has refused the waters of Shiloah that flow gently, and melt in fear before[u] Rezin and the son of Remaliah; 7 therefore, the Lord is bringing up against it the mighty flood waters of the River, the king of Assyria and all his glory; it will rise above all its channels and overflow all its banks; 8 it will sweep on into Judah as a flood, and, pouring over, it will reach up to the neck; and its outspread wings will fill the breadth of your land, O Immanuel.

9 Band together, you peoples, and
be dismayed;
listen, all you far countries;
gird yourselves and be dismayed;
gird yourselves and be
dismayed!
10 Take counsel together, but it shall
be brought to naught;
speak a word, but it will not
stand,
for God is with us.[v]

11 For the LORD spoke thus to me while his hand was strong upon me, and warned me not to walk in the way of this people, saying: 12 Do not call conspiracy all that this people calls conspiracy, and do not fear what it fears, or be in dread. 13 But the LORD of hosts, him you shall regard as holy; let him be your fear, and let him be your dread. 14 He will become a sanctuary, a stone one strikes against; for both houses of Israel he will become a rock one stumbles over—a trap and a snare for the inhabitants of Jerusalem. 15 And many among them shall stumble; they shall fall and be broken; they shall be snared and taken.

16 Bind up the testimony, seal the teaching among my disciples. 17 I will wait for the LORD, who is hiding his face from the house of Jacob, and I will hope in him. 18 See, I and the children whom the LORD has given me are signs and portents in Israel from the LORD of hosts, who dwells on Mount Zion. 19 Now if people say to you, "Consult the ghosts and the familiar spirits that chirp and mutter; should not a people consult their gods, the dead on behalf of the living, 20 for teaching and for instruction?" Surely, those who speak like this will have no dawn! 21 They will pass through the land,[w] greatly distressed and hungry; when they are hungry, they will be enraged and will curse[x] their king and their gods. They will turn their faces upward, 22 or they will look to the earth, but will see only distress and darkness, the gloom of anguish; and they will be thrust into thick darkness.[y]

9[z] But there will be no gloom for those who were in anguish. In the former time he brought into contempt the land of Zebulun and the land of Naphtali, but in the latter time he will make glorious the way of the sea, the land beyond the Jordan, Galilee of the nations.
2[a] The people who walked in
darkness

---

u Cn: Meaning of Heb uncertain
v Heb *immanu el* w Heb *it* x Or *curse by*
y Meaning of Heb uncertain z Ch 8.23 in Heb a Ch 9.1 in Heb

8.5–10: Oracle of Shiloah and the Euphrates; Judah also is included in Assyria's sweep. *Shiloah,* a conduit flanking Ophel from the spring Gihon (see 1 Kings 1.33 n.) to the reservoir (7.3), is contrasted with the *River,* the great Euphrates. Ahaz's mighty ally, Assyria, will inundate tiny Judah, God's people. **9–10**: God is with his people (see 7.14 n.) to deliver them (Ps 46, esp. vv. 7, 11).
8.11–22: **The testimony and the teaching. 11–15**: God's ways are not the ways of human beings (55.8–9). **16**: *Bind, seal,* as one binds and seals a scroll (Jer 32.10). **18**: *Signs,* 7.3, 14; 8.1. **19–20**: Condemnation of superstition (2.6). For necromancy (consultation of the dead), see 1 Sam 28.7 n.
9.1: **Transitional verse** from doom to promise. *Zebulun, Naphtali,* and Issachar constituted later *Galilee;* these areas were annexed by Assyria in 733 B.C. (2 Kings 15.29). *Way of the sea,* one of the provinces set up by Assyria, around the sea-coast city of Dor south of Mount Carmel.
9.2–7: **The messianic king** (compare 11.1–9). Filled with borrowed phrases referring to the Davidic monarchy, this passage may have originally celebrated the accession of a Judean king, perhaps Hezekiah; in its

have seen a great light;
those who lived in a land of deep
    darkness—
on them light has shined.
3 You have multiplied the nation,
    you have increased its joy;
they rejoice before you
    as with joy at the harvest,
    as people exult when dividing
        plunder.
4 For the yoke of their burden,
    and the bar across their
        shoulders,
    the rod of their oppressor,
    you have broken as on the day
        of Midian.
5 For all the boots of the tramping
        warriors
    and all the garments rolled in
        blood
    shall be burned as fuel for
        the fire.
6 For a child has been born for us,
    a son given to us;
authority rests upon his shoulders;
    and he is named
Wonderful Counselor, Mighty
    God,
Everlasting Father, Prince of
    Peace.
7 His authority shall grow
        continually,
    and there shall be endless peace
for the throne of David and his
    kingdom.
    He will establish and uphold it
with justice and with
    righteousness
from this time onward and
    forevermore.

The zeal of the LORD of hosts will
    do this.

8 The Lord sent a word against
        Jacob,
    and it fell on Israel;
9 and all the people knew it—
    Ephraim and the inhabitants of
        Samaria—
    but in pride and arrogance of
        heart they said:
10 "The bricks have fallen,
    but we will build with dressed
        stones;
the sycamores have been cut
        down,
    but we will put cedars in their
        place."
11 So the LORD raised adversaries[b]
        against them,
    and stirred up their enemies,
12 the Arameans on the east and the
        Philistines on the west,
    and they devoured Israel with
        open mouth.
For all this his anger has not
        turned away;
    his hand is stretched out still.

13 The people did not turn to him
        who struck them,
    or seek the LORD of hosts.
14 So the LORD cut off from Israel
        head and tail,
    palm branch and reed in one
        day—
15 elders and dignitaries are the head,

b Cn: Heb *the adversaries of Rezin*

present context it describes the coming Messiah as the ideal king. **4:** *Midian,* Judg 7.15–25. **6:** *Authority,* symbol of power (see 22.21). *Mighty God,* divine in might. *Everlasting Father,* continuing fatherly love and care. *Prince of Peace,* the king who brings peace and prosperity. The king represents the best qualities of Israel's heroes (Ezek 37.25).

9.8–10.4: Ephraim's judgment an object lesson for Judah (five stanzas, including 5.24b–30; with the same refrain, 9.12, 17, 21;

10.4; 5.25; compare Jer 3.6–10; Ezek 16.44–58). **8–12:** Punishment for pride and unrepented wickedness. **8:** *Word,* more than a statement; it includes the potential and fact of accomplishment (55.10–11; Jer 23.18–20). **10:** *Bricks, sycamore,* for ordinary houses; *dressed stone, cedar* for palaces (Jer 22.7, 23).

9.13–17: Corrupt leaders misled their people (Jer 6.14). **18–21:** Moral decay consumes like a forest fire (Hos 7.6); civil war breaks out (2 Kings 15.23–31; 16.5). **20:** *The flesh of their*

and prophets who teach lies are
the tail;

16 for those who led this people led
them astray,
and those who were led by
them were left in
confusion.

17 That is why the Lord did not have
pity on[c] their young
people,
or compassion on their orphans
and widows;
for everyone was godless and an
evildoer,
and every mouth spoke folly.
For all this his anger has not
turned away,
his hand is stretched out still.

18 For wickedness burned like a fire,
consuming briers and thorns;
it kindled the thickets of the
forest,
and they swirled upward in a
column of smoke.

19 Through the wrath of the LORD of
hosts
the land was burned,
and the people became like fuel for
the fire;
no one spared another.

20 They gorged on the right, but still
were hungry,
and they devoured on the left,
but were not satisfied;
they devoured the flesh of their
own kindred;[d]

21 Manasseh devoured Ephraim, and
Ephraim Manasseh,
and together they were against
Judah.
For all this his anger has not
turned away;
his hand is stretched out still.

10 Ah, you who make iniquitous
decrees,

who write oppressive statutes,

2 to turn aside the needy from
justice
and to rob the poor of my
people of their right,
that widows may be your spoil,
and that you may make the
orphans your prey!

3 What will you do on the day of
punishment,
in the calamity that will come
from far away?
To whom will you flee for help,
and where will you leave your
wealth,

4 so as not to crouch among the
prisoners
or fall among the slain?
For all this his anger has not
turned away;
his hand is stretched out still.

5 Ah, Assyria, the rod of my
anger—
the club in their hands is
my fury!

6 Against a godless nation I
send him,
and against the people of my
wrath I command him,
to take spoil and seize plunder,
and to tread them down like the
mire of the streets.

7 But this is not what he intends,
nor does he have this in mind;
but it is in his heart to destroy,
and to cut off nations not a few.

8 For he says:
"Are not my commanders all
kings?

9 Is not Calno like Carchemish?
Is not Hamath like Arpad?
Is not Samaria like Damascus?

10 As my hand has reached to the
kingdoms of the idols

c Q Ms: MT *rejoice over*   d Or *arm*

---

own kindred; on cannibalism, see Jer 19.9.
Some treat this passage as a proverb. **10.1–4:**
Justice is miscarried (3.13–15; Jer 8.8).
  10.5–19: Woe, O Assyria! Unaware that

they were serving as God's instrument, the
powerful Assyrians were doomed by their
pride to destruction (Jer 25.8–14; 50.23). **9:** In
northern Syria, Tiglath-Pileser III captured

whose images were greater than
those of Jerusalem and
Samaria,
11 shall I not do to Jerusalem and
her idols
what I have done to Samaria
and her images?"

12 When the Lord has finished all his
work on Mount Zion and on Jerusalem,
he*e* will punish the arrogant boasting of
the king of Assyria and his haughty
pride. 13 For he says:
"By the strength of my hand I
have done it,
and by my wisdom, for I have
understanding;
I have removed the boundaries of
peoples,
and have plundered their
treasures;
like a bull I have brought down
those who sat on thrones.
14 My hand has found, like a nest,
the wealth of the peoples;
and as one gathers eggs that have
been forsaken,
so I have gathered all the earth;
and there was none that moved a
wing,
or opened its mouth, or
chirped."

15 Shall the ax vaunt itself over the
one who wields it,
or the saw magnify itself against
the one who handles it?

As if a rod should raise the one
who lifts it up,
or as if a staff should lift the one
who is not wood!
16 Therefore the Sovereign, the Lord
of hosts,
will send wasting sickness
among his stout warriors,
and under his glory a burning will
be kindled,
like the burning of fire.
17 The light of Israel will become
a fire,
and his Holy One a flame;
and it will burn and devour
his thorns and briers in one day.
18 The glory of his forest and his
fruitful land
the Lord will destroy, both soul
and body,
and it will be as when an invalid
wastes away.
19 The remnant of the trees of his
forest will be so few
that a child can write them
down.

20 On that day the remnant of Israel
and the survivors of the house of Jacob
will no more lean on the one who struck
them, but will lean on the Lord, the
Holy One of Israel, in truth. 21 A remnant will return, the remnant of Jacob, to
the mighty God. 22 For though your people Israel were like the sand of the sea,
only a remnant of them will return. De-

e Heb *I*

Calno (742 B.C.), *Carchemish, Hamath* (738),
Arpad (741), southern Syria, *Damascus* (732).
Menahem of Israel paid him tribute (2 Kings
15.19–20). **10–11:** To Assyria, the Lord was
another idol.
10.12: Prose summation of vv. 5–11, 13–
19. **13–14:** Assyria's boast. *Removed the boundaries,* to discourage rebellion, Assyria transplanted subject peoples.
10.15: Rhetorical question recalling v. 5
(45.9). **16–19:** Light of Israel, God's majestic
glory (2.10; 29.6; Ezek 1.26–28). God will
ravage Assyria like a forest fire.
**10.20–23: Only a remnant will return.
21:** *A remnant will return,* in Hebrew this is the

same as the name of Isaiah's son Shear-jashub;
in 7.3–4 it stands in an oracle of encouragement, but here in an oracle of doom. **22:** *Sand
of the sea* recalls God's oath to the patriarchs
(Gen 22.17; compare Rom 9.27). In Isaiah
(4.2–3; 6.13; 7.3; 28.5–6; 37.4; 37.31–32;
compare Mic 4.7; 5.2–9; Zeph 2.7) *remnant*
refers to those remaining after Judah's punishment, from whom a great people will
arise. During the Exile the remnant was the
deported people (Ezek 6.8–10; Jer 23.3; 31.7),
whom God would bring back and make
great. After the Exile Jewish faithlessness
evoked again the pre-exilic concept (Zech
8.11; Hag 1.12; Zech 14.2).

struction is decreed, overflowing with righteousness. 23For the Lord God of hosts will make a full end, as decreed, in all the earth.*f*

24 Therefore thus says the Lord God of hosts: O my people, who live in Zion, do not be afraid of the Assyrians when they beat you with a rod and lift up their staff against you as the Egyptians did. 25For in a very little while my indignation will come to an end, and my anger will be directed to their destruction. 26The Lord of hosts will wield a whip against them, as when he struck Midian at the rock of Oreb; his staff will be over the sea, and he will lift it as he did in Egypt. 27On that day his burden will be removed from your shoulder, and his yoke will be destroyed from your neck.

He has gone up from Rimmon,*g*
28   he has come to Aiath;
he has passed through Migron,
    at Michmash he stores his
       baggage;
29 they have crossed over the pass,
    at Geba they lodge for the night;
Ramah trembles,
    Gibeah of Saul has fled.
30 Cry aloud, O daughter Gallim!
Listen, O Laishah!
    Answer her, O Anathoth!
31 Madmenah is in flight,
    the inhabitants of Gebim flee
       for safety.
32 This very day he will halt at Nob,
he will shake his fist
    at the mount of daughter Zion,
    the hill of Jerusalem.

33 Look, the Sovereign, the Lord
       of hosts,

will lop the boughs with
    terrifying power;
the tallest trees will be cut down,
    and the lofty will be brought
       low.
34 He will hack down the thickets of
    the forest with an ax,
and Lebanon with its majestic
    trees*h* will fall.

11 A shoot shall come out from
    the stump of Jesse,
and a branch shall grow out of
    his roots.
2 The spirit of the Lord shall rest
    on him,
the spirit of wisdom and
    understanding,
the spirit of counsel and might,
    the spirit of knowledge and the
       fear of the Lord.
3 His delight shall be in the fear of
    the Lord.

He shall not judge by what his
    eyes see,
    or decide by what his ears hear;
4 but with righteousness he shall
       judge the poor,
    and decide with equity for the
       meek of the earth;
he shall strike the earth with the
    rod of his mouth,
    and with the breath of his lips
       he shall kill the wicked.
5 Righteousness shall be the belt
    around his waist,
    and faithfulness the belt around
       his loins.

f Or land   g Cn: Heb and his yoke from your
neck, and a yoke will be destroyed because of fatness
h Cn Compare Gk Vg: Heb with a majestic one

10.24–27c: Oracle of promise. *Oreb,* Judg 7.25; *rod,* Ex 14.16.
10.27d–32: The approach of the Assyrians. The invader (Tiglath-Pileser III or Sennacherib, 1.4–9 n.) approached from the north toward the outskirts of Jerusalem (Jer 6.1–3). This may be a "traditional" description of the northern invasion route; for a southern route, see Mic 1.10–15.
10:33–34: The Lord, the forester, will cut down Assyria.
11.1–9: The messianic king (compare 9.2–7). For the occasion of the original oracle, see 9.2–7 n. **1–3a:** The Messiah will manifest the characteristics of those who were great in Israel. **1:** *Jesse,* David's father (1 Sam 16.1–20). **2:** To these six "Gifts of the Spirit" the Septuagint adds "piety." **3b–5:** Wisdom and justice (5.7) were traditionally associated in the ideal king (1 Kings ch 3; Ps 72).

6 The wolf shall live with the lamb,
   the leopard shall lie down with
      the kid,
   the calf and the lion and the
      fatling together,
   and a little child shall lead them.
7 The cow and the bear shall graze,
   their young shall lie down
      together;
   and the lion shall eat straw like
      the ox.
8 The nursing child shall play over
   the hole of the asp,
   and the weaned child shall put
      its hand on the adder's den.
9 They will not hurt or destroy
   on all my holy mountain;
   for the earth will be full of the
      knowledge of the LORD
   as the waters cover the sea.

10 On that day the root of Jesse shall
stand as a signal to the peoples; the nations shall inquire of him, and his dwelling shall be glorious.

11 On that day the Lord will extend
his hand yet a second time to recover the
remnant that is left of his people, from
Assyria, from Egypt, from Pathros,
from Ethiopia, *i* from Elam, from Shinar, from Hamath, and from the coastlands of the sea.

12 He will raise a signal for the
   nations,
   and will assemble the outcasts of
      Israel,
   and gather the dispersed of Judah
      from the four corners of the
      earth.
13 The jealousy of Ephraim shall
   depart,

the hostility of Judah shall be
   cut off;
Ephraim shall not be jealous of
   Judah,
and Judah shall not be hostile
   towards Ephraim.
14 But they shall swoop down on the
   backs of the Philistines in
      the west,
   together they shall plunder the
      people of the east.
They shall put forth their hand
   against Edom and Moab,
   and the Ammonites shall
      obey them.
15 And the LORD will utterly destroy
   the tongue of the sea of Egypt;
   and will wave his hand over
      the River
   with his scorching wind;
   and will split it into seven
      channels,
   and make a way to cross on
      foot;
16 so there shall be a highway from
   Assyria
   for the remnant that is left of
      his people,
   as there was for Israel
      when they came up from the
      land of Egypt.

12 You will say in that day:
   I will give thanks to you,
      O LORD,
   for though you were angry
      with me,
   your anger turned away,
      and you comforted me.

i Or *Nubia*; Heb *Cush*

11.6–8: His reign will be "paradise regained"; the disorder of nature will be restored to its pristine harmony (Ezek 47.1–
12). 9: *My holy mountain*, 65.25; Ezek 20.40.
11.10–16: **The messianic age. 10:** *Root* is a
person, not the dynasty (v. 1). 11–16: Restored and reunited Israel takes vengeance
against her oppressors. The terminology and
mood of vv. 11–16 indicate a post-exilic date.
11: *Pathros*, Upper Egypt; *Shinar*, Babylonia;

coastlands, Aegean seacoast and islands. 15:
The tongue of the sea, the Red Sea (Ex ch 14);
River, Euphrates.
12.1–6: **Two songs** conclude Section I of
the book of Isaiah. (*a*) 1–3: Song of deliverance (compare Ps 116). 1a and 4a are liturgical
rubrics. 2b: Ex 15.2; Ps 118.14. (*b*) 4–6: Song
of thanksgiving. *Shout aloud and sing for joy*,
compare Zeph 3.14. *In your midst*, God in his
temple. *Holy One*, see 1.4.

2 Surely God is my salvation;
    I will trust, and will not
       be afraid,
  for the LORD GOD[j] is my strength
     and my might;
    he has become my salvation.

3 With joy you will draw water from
the wells of salvation. 4 And you will say
in that day:
  Give thanks to the LORD,
    call on his name;
  make known his deeds among the
     nations;
  proclaim that his name is
    exalted.

5 Sing praises to the LORD, for he
    has done gloriously;
  let this be known[k] in all the
    earth.
6 Shout aloud and sing for joy,
    O royal[l] Zion,
  for great in your midst is the
    Holy One of Israel.

13 The oracle concerning Babylon
that Isaiah son of Amoz saw.

2 On a bare hill raise a signal,
    cry aloud to them;
  wave the hand for them to enter
    the gates of the nobles.
3 I myself have commanded my
    consecrated ones,
  have summoned my warriors,
    my proudly exulting ones,
  to execute my anger.

4 Listen, a tumult on the mountains
    as of a great multitude!

Listen, an uproar of kingdoms,
    of nations gathering together!
  The LORD of hosts is mustering
    an army for battle.
5 They come from a distant land,
    from the end of the heavens,
  the LORD and the weapons of his
    indignation,
  to destroy the whole earth.

6 Wail, for the day of the LORD
    is near;
  it will come like destruction
    from the Almighty![m]
7 Therefore all hands will be feeble,
    and every human heart will
     melt,
8   and they will be dismayed.
Pangs and agony will seize them;
    they will be in anguish like a
     woman in labor.
They will look aghast at one
    another;
  their faces will be aflame.
9 See, the day of the LORD comes,
    cruel, with wrath and fierce
     anger,
to make the earth a desolation,
    and to destroy its sinners from
     it.
10 For the stars of the heavens and
    their constellations
  will not give their light;
  the sun will be dark at its rising,
    and the moon will not shed
     its light.
11 I will punish the world for its evil,

---

j Heb *for Yah, the* LORD   *k* Or *this is made
known*   *l* Or O *Inhabitant of*   *m* Traditional
rendering of Heb *Shaddai*

**13.1–23.18: Oracles against foreign na-
tions** (compare Jer chs 46–51; Ezek chs 25–
32); each is introduced by the word "oracle."
13.1–22: Oracle against Babylon. This
is clearly subsequent to Isaiah, when Babylon, whose power is here assumed, superseded Assyria; vv. 17–22 suggest a time
after Nebuchadrezzar's death in 562 B.C. **2:**
Gates of the nobles, perhaps the name of one of
Babylon's city gates (Jer 19.2). **3:** *My consecrated ones,* perhaps Persian soldiers serving

God's purposes (10.5–6; 45.1–4). **5:** *Distant
land,* perhaps Persia. *The whole earth,* the Babylonian empire.
13.6–16: The cataclysmic fall of Babylon,
attended by portents in the heavens and civil
disorder, is seen as the *day of the LORD* (compare 2.11). Before the Exile, the day of the
LORD marked the punishment of Israel (Am
8.9–10; Jer 30.5–7); after the Exile, it referred
to the punishment of Israel's oppressors, and
was again a day of hope for Israel (Am 5.18).

Proclamation
against Babylon

and the wicked for their
iniquity;
I will put an end to the pride of
the arrogant,
and lay low the insolence of
tyrants.
12 I will make mortals more rare
than fine gold,
and humans than the gold of
Ophir.
13 Therefore I will make the heavens
tremble,
and the earth will be shaken out
of its place,
at the wrath of the LORD of hosts
in the day of his fierce anger.
14 Like a hunted gazelle,
or like sheep with no one to
gather them,
all will turn to their own people,
and all will flee to their
own lands.
15 Whoever is found will be thrust
through,
and whoever is caught will fall
by the sword.
16 Their infants will be dashed
to pieces
before their eyes;
their houses will be plundered,
and their wives ravished.
17 See, I am stirring up the Medes
against them,
who have no regard for silver
and do not delight in gold.
18 Their bows will slaughter the
young men;
they will have no mercy on the
fruit of the womb;
their eyes will not pity children.
19 And Babylon, the glory of
kingdoms,
the splendor and pride of the
Chaldeans,

will be like Sodom and Gomorrah
when God overthrew them.
20 It will never be inhabited
or lived in for all generations;
Arabs will not pitch their
tents there,
shepherds will not make their
flocks lie down there.
21 But wild animals will lie
down there,
and its houses will be full of
howling creatures;
there ostriches will live,
and there goat-demons
will dance.
22 Hyenas will cry in its towers,
and jackals in the pleasant
palaces;
its time is close at hand,
and its days will not be
prolonged.

14 But the LORD will have compassion on Jacob and will again choose Israel, and will set them in their own land; and aliens will join them and attach themselves to the house of Jacob. 2 And the nations will take them and bring them to their place, and the house of Israel will possess the nations[n] as male and female slaves in the LORD's land; they will take captive those who were their captors, and rule over those who oppressed them.

3 When the LORD has given you rest from your pain and turmoil and the hard service with which you were made to serve, 4 you will take up this taunt against the king of Babylon:

How the oppressor has ceased!
How his insolence[o] has ceased!

n Heb *them*   *o* Q Ms Compare Gk Syr Vg:
Meaning of MT uncertain

13.17–19: *Medes*, people northwest of Persia who earlier were Babylon's allies against Assyria. **20–22**: Busy Babylon will become a wasteland inhabited only by animals (Jer 50.35–40). *Goat-demons*, see 34.14; Lev 17.7. **14.1–2**: **The return from Exile.** *Aliens*, proselytes to Judaism, referred to in post-exilic prophecy (Zech 8.20–22). **3–23**:

"**How are the mighty fallen!**" A mocking dirge against a tyrant; the tyrant may be a king (Nebuchadrezzar?), or a nation (Babylon?) whose fall would precede Judah's restoration. **3–4a**: Prose introduction. **14.4b–8**: After the tyrant's fall, peace settles over the oppressed peoples. **8**: Assyrian and Babylonian kings used great quantities of

884
OT

5 The LORD has broken the staff of
the wicked,
the scepter of rulers,
6 that struck down the peoples
in wrath
with unceasing blows,
that ruled the nations in anger
with unrelenting persecution.
7 The whole earth is at rest
and quiet;
they break forth into singing.
8 The cypresses exult over you,
the cedars of Lebanon, saying,
"Since you were laid low,
no one comes to cut us down."
9 Sheol beneath is stirred up
to meet you when you come;
it rouses the shades to greet you,
all who were leaders of the
earth;
it raises from their thrones
all who were kings of the
nations.
10 All of them will speak
and say to you:
"You too have become as weak
as we!
You have become like us!"
11 Your pomp is brought down
to Sheol,
and the sound of your harps;
maggots are the bed beneath you,
and worms are your covering.

12 How you are fallen from heaven,
O Day Star, son of Dawn!
How you are cut down to the
ground,
you who laid the nations low!
13 You said in your heart,
"I will ascend to heaven;
I will raise my throne
above the stars of God;

I will sit on the mount of
assembly
on the heights of Zaphon;*p*
14 I will ascend to the tops of the
clouds,
I will make myself like the
Most High."
15 But you are brought down to
Sheol,
to the depths of the Pit.
16 Those who see you will stare
at you,
and ponder over you:
"Is this the man who made the
earth tremble,
who shook kingdoms,
17 who made the world like a desert
and overthrew its cities,
who would not let his prisoners
go home?"
18 All the kings of the nations lie
in glory,
each in his own tomb;
19 but you are cast out, away from
your grave,
like loathsome carrion,*q*
clothed with the dead, those
pierced by the sword,
who go down to the stones of
the Pit,
like a corpse trampled
underfoot.
20 You will not be joined with them
in burial,
because you have destroyed
your land,
you have killed your people.

May the descendants of evildoers
nevermore be named!
21 Prepare slaughter for his sons

*p Or assembly in the far north   q Cn Compare
Gk: Heb like a loathed branch*

---

cedars for their palaces. **9–11**: The tyrant joins
earlier rulers in *Sheol* (see 5.14–17 n.; Ezek
32.17–22).
14.12–15: From aspirations for divinity
(compare Gen 11.4–8), he falls to the
anonymity of Sheol. Canaanite mythological
background is reflected in *Day Star* and *Dawn*
(Hebrew "Helal" and "Shahar," names of de-

ities), *mount of assembly* (of the gods, and located at Jebel 'Aqra, north of Ugarit), and
Most High, a title also applied to Baal Sha-
maim (Lord of Heaven).
14.15: *The Pit,* an alternative designation of
the underworld (Ezek 31.15–18). **16–19**: He
will have no greatness in death. **20–21**: Ignominy and hatred are his fate and heritage.

because of the guilt of their
father.[r]
Let them never rise to possess
the earth
or cover the face of the world
with cities.

22 I will rise up against them, says the
LORD of hosts, and will cut off from Babylon name and remnant, offspring and posterity, says the LORD. 23 And I will make it a possession of the hedgehog, and pools of water, and I will sweep it with the broom of destruction, says the LORD of hosts.

24 The LORD of hosts has sworn:
As I have designed,
so shall it be;
and as I have planned,
so shall it come to pass:
25 I will break the Assyrian in
my land,
and on my mountains trample
him under foot;
his yoke shall be removed
from them,
and his burden from their
shoulders.
26 This is the plan that is planned
concerning the whole earth;
and this is the hand that is
stretched out
over all the nations.
27 For the LORD of hosts has planned,
and who will annul it?
His hand is stretched out,
and who will turn it back?

28 In the year that King Ahaz died this oracle came:

29 Do not rejoice, all you Philistines,
that the rod that struck you
is broken,

for from the root of the snake will
come forth an adder,
and its fruit will be a flying
fiery serpent.
30 The firstborn of the poor will
graze,
and the needy lie down in
safety;
but I will make your root die
of famine,
and your remnant I[s] will kill.
31 Wail, O gate; cry, O city;
melt in fear, O Philistia, all
of you!
For smoke comes out of the
north,
and there is no straggler in
its ranks.

32 What will one answer the
messengers of the nation?
"The LORD has founded Zion,
and the needy among his people
will find refuge in her."

15 An oracle concerning Moab.

Because Ar is laid waste in a
night,
Moab is undone;
because Kir is laid waste in
a night,
Moab is undone.
2 Dibon[t] has gone up to the
temple,
to the high places to weep;
over Nebo and over Medeba
Moab wails.
On every head is baldness,
every beard is shorn;
3 in the streets they bind on
sackcloth;

r Syr Compare Gk: Heb *fathers*   s Q Ms Vg:
MT *he*   t Cn: Heb *the house and Dibon*

14.22–23: Prose conclusion.
14.24–27: Against Assyria (17.12–14; 30.27–33; 37.22–29); no power on earth can prevent God from accomplishing his purposes. **28–32: Against Philistia** (715 B.C.; Jer ch 47; Am 1.6–8). Though momentarily it appears that Philistia is secure, the foe *out of*

the north (Jer 1.13–15) will shortly destroy her, too.
15.1–16.14: Against Moab (Ezek 25.8–11; Am 2.1–3). Moab (Gen 19.30–37) lies prostrate under the invader's (probably Assyria) heel (vv. 1–9).

on the housetops and in the
squares
everyone wails and melts
in tears.
4 Heshbon and Elealeh cry out,
their voices are heard as far
as Jahaz;
therefore the loins of Moab
quiver;*u*
his soul trembles.
5 My heart cries out for Moab;
his fugitives flee to Zoar,
to Eglath-shelishiyah.
For at the ascent of Luhith
they go up weeping;
on the road to Horonaim
they raise a cry of destruction;
6 the waters of Nimrim
are a desolation;
the grass is withered, the new
growth fails,
the verdure is no more.
7 Therefore the abundance they
have gained
and what they have laid up
they carry away
over the Wadi of the Willows.
8 For a cry has gone
around the land of Moab;
the wailing reaches to Eglaim,
the wailing reaches to
Beer-elim.
9 For the waters of Dibon*v* are full
of blood;
yet I will bring upon Dibon*v*
even more—
a lion for those of Moab
who escape,
for the remnant of the land.

16 Send lambs
to the ruler of the land,
from Sela, by way of the desert,
to the mount of daughter Zion.
2 Like fluttering birds,
like scattered nestlings,
so are the daughters of Moab
at the fords of the Arnon.

3 "Give counsel,
grant justice;
make your shade like night
at the height of noon;
hide the outcasts,
do not betray the fugitive;
4 let the outcasts of Moab
settle among you;
be a refuge to them
from the destroyer."

When the oppressor is no more,
and destruction has ceased,
and marauders have vanished from
the land,
5 then a throne shall be established
in steadfast love
in the tent of David,
and on it shall sit in faithfulness
a ruler who seeks justice
and is swift to do what is right.

6 We have heard of the pride of
Moab
—how proud he is!—
of his arrogance, his pride, and his
insolence;
his boasts are false.
7 Therefore let Moab wail,
let everyone wail for Moab.
Mourn, utterly stricken,
for the raisin cakes of
Kir-hareseth.

8 For the fields of Heshbon
languish,
and the vines of Sibmah,
whose clusters once made drunk
the lords of the nations,
reached to Jazer
and strayed to the desert;
their shoots once spread abroad
and crossed over the sea.

u Cn Compare Gk Syr: Heb *the armed men of*
Moab cry aloud   *v* Q Ms Vg Compare Syr:
MT *Dimon*

16.1–5: Moabite refugees, appealing to the
mercies of the crown, seek sanctuary in Judah; since the thought seems out of place
here, the passage may be an insertion. *Lambs,*
a gift; compare earlier tribute Moab paid Judah (2 Kings 3.4).
16.6–12: A description of the disaster, especially the destruction of the vineyards (par-

9 Therefore I weep with the
weeping of Jazer
for the vines of Sibmah;
I drench you with my tears,
O Heshbon and Elealeh;
for the shout over your fruit
harvest
and your grain harvest
has ceased.
10 Joy and gladness are taken away
from the fruitful field;
and in the vineyards no songs
are sung,
no shouts are raised;
no treader treads out wine in
the presses;
the vintage-shout is hushed. *w*
11 Therefore my heart throbs like a
harp for Moab,
and my very soul for Kir-heres.
12 When Moab presents himself,
when he wearies himself upon the high
place, when he comes to his sanctuary to
pray, he will not prevail.
13 This was the word that the LORD
spoke concerning Moab in the past.
14 But now the LORD says, In three years,
like the years of a hired worker, the glory
of Moab will be brought into contempt,
in spite of all its great multitude; and
those who survive will be very few and
feeble.

17 An oracle concerning Damascus.

See, Damascus will cease to be
a city,
and will become a heap of ruins.
2 Her towns will be deserted
forever; *x*
they will be places for flocks,
which will lie down, and no one
will make them afraid.

3 The fortress will disappear from
Ephraim,
and the kingdom from
Damascus;
and the remnant of Aram will be
like the glory of the children
of Israel,
says the LORD of hosts.
4 On that day
the glory of Jacob will be
brought low,
and the fat of his flesh will
grow lean.
5 And it shall be as when reapers
gather standing grain
and their arms harvest the ears,
and as when one gleans the ears
of grain
in the Valley of Rephaim.
6 Gleanings will be left in it,
as when an olive tree is
beaten—
two or three berries
in the top of the highest bough,
four or five
on the branches of a fruit tree,
says the LORD God of
Israel.

7 On that day people will regard their
Maker, and their eyes will look to the
Holy One of Israel; 8 they will not have
regard for the altars, the work of their
hands, and they will not look to what
their own fingers have made, either the
sacred poles *y* or the altars of incense.
9 On that day their strong cities will
be like the deserted places of the Hivites

w Gk: Heb *I have hushed*   *x* Cn Compare Gk:
Heb *the cities of Aroer are deserted*
y Heb *Asherim*

---

allel, Jer 48.29–33). *Kir-hareseth* and *Kir-heres,*
the same as Kir, 15.1. **13–14**: In a short time,
despite intervening prosperity, Moab will
again be devastated.
**17.1–6: Against the Syro-Ephraimite
alliance** (about 734 B.C.–see 2 Kings 16.1–
20 n.; Isa 7.1–8.4). Two stanzas (1–3, 4–6)
reaffirm the ultimate defeat of *Damascus* and
Ephraim. The glory of . . . Israel refers to Sa-

maria; *Rephaim,* valley northwest of Jerusalem (2 Sam 5.18).
17.7–11: Against idolatry. 7–8: Eventually people will return to God (1.29–31). *Sacred poles,* cult images of Asherah (6.13 n.).
9–11: Those who forsake God for idols will be
displaced as Israel displaced the *Hivites* and
Amorites, original inhabitants of Palestine

and the Amorites, *z* which they deserted because of the children of Israel, and there will be desolation.

10 For you have forgotten the God of
  your salvation,
  and have not remembered the
    Rock of your refuge;
therefore, though you plant
    pleasant plants
  and set out slips of an alien god,
11 though you make them grow on
    the day that you plant
    them,
  and make them blossom in the
    morning that you sow;
yet the harvest will flee away
  in a day of grief and incurable
    pain.

12 Ah, the thunder of many peoples,
  they thunder like the thundering
    of the sea!
Ah, the roar of nations,
  they roar like the roaring of
    mighty waters!
13 The nations roar like the roaring
    of many waters,
  but he will rebuke them, and
    they will flee far away,
chased like chaff on the mountains
    before the wind
  and whirling dust before the
    storm.
14 At evening time, lo, terror!
  Before morning, they are no
    more.
This is the fate of those who
    despoil us,
  and the lot of those who
    plunder us.

18 Ah, land of whirring wings
  beyond the rivers of
  Ethiopia, *a*

2 sending ambassadors by the Nile
  in vessels of papyrus on the
    waters!
Go, you swift messengers,
  to a nation tall and smooth,
  to a people feared near and far,
  a nation mighty and
    conquering,
  whose land the rivers divide.

3 All you inhabitants of the world,
  you who live on the earth,
when a signal is raised on the
    mountains, look!
  When a trumpet is blown,
    listen!
4 For thus the LORD said to me:
I will quietly look from my
    dwelling
  like clear heat in sunshine,
  like a cloud of dew in the heat
    of harvest.
5 For before the harvest, when the
    blossom is over
  and the flower becomes a
    ripening grape,
he will cut off the shoots with
    pruning hooks,
  and the spreading branches he
    will hew away.
6 They shall all be left
  to the birds of prey of the
    mountains
  and to the animals of the earth.
And the birds of prey will
    summer on them,
  and all the animals of the earth
    will winter on them.

7 At that time gifts will be brought to

<hr/>

z Cn Compare Gk: Heb *places of the wood
and the highest bough*   *a* Or *Nubia*; Heb *Cush*

<hr/>

(Deut 7.1). *Plants, slips,* plants dedicated to Tammuz (compare Ezek 8.14–18). **12–14:** The appearance of a potentially destructive storm (perhaps Assyria's onslaught of 701 B.C.) is dispelled by God who defends his people.
18.1–20.6: Concerning Egypt and Ethiopia. 18.1–7: Concerning Ethiopia. The

occasion may be the invitation of Egypt and Ethiopia to participate in an anti-Assyrian plot, about 714 B.C. Using the figure of harvest, Isaiah cautions that since God determines the course of events human beings must wait. When ready, God will signal the beginning of revolt. *Land the rivers divide,* Egypt divided by the Nile.

the LORD of hosts from[b] a people tall and
smooth, from a people feared near and
far, a nation mighty and conquering,
whose land the rivers divide, to Mount
Zion, the place of the name of the LORD
of hosts.

19 An oracle concerning Egypt.

See, the LORD is riding on a
swift cloud
and comes to Egypt;
the idols of Egypt will tremble at
his presence,
and the heart of the Egyptians
will melt within them.
2 I will stir up Egyptians against
Egyptians,
and they will fight, one against
the other,
neighbor against neighbor,
city against city, kingdom
against kingdom;
3 the spirit of the Egyptians within
them will be emptied out,
and I will confound their plans;
they will consult the idols and the
spirits of the dead
and the ghosts and the familiar
spirits;
4 I will deliver the Egyptians
into the hand of a hard master;
a fierce king will rule over them,
says the Sovereign, the LORD
of hosts.

5 The waters of the Nile will be
dried up,
and the river will be parched
and dry;
6 its canals will become foul,
and the branches of Egypt's
Nile will diminish and
dry up,
reeds and rushes will rot away.

7 There will be bare places by
the Nile,
on the brink of the Nile;
and all that is sown by the Nile
will dry up,
be driven away, and be no
more.
8 Those who fish will mourn;
all who cast hooks in the Nile
will lament,
and those who spread nets on
the water will languish.
9 The workers in flax will be in
despair,
and the carders and those at the
loom will grow pale.
10 Its weavers will be dismayed,
and all who work for wages
will be grieved.

11 The princes of Zoan are utterly
foolish;
the wise counselors of Pharaoh
give stupid counsel.
How can you say to Pharaoh,
"I am one of the sages,
a descendant of ancient kings"?
12 Where now are your sages?
Let them tell you and make
known
what the LORD of hosts has
planned against Egypt.
13 The princes of Zoan have become
fools,
and the princes of Memphis
are deluded;
those who are the cornerstones of
its tribes
have led Egypt astray.
14 The LORD has poured into them[c]
a spirit of confusion;

b Q Ms Gk Vg: MT *of*   c Gk Compare Tg:
Heb *it*

19.1–15: **Against Egypt** (compare Ezek
chs 29–32). **1–4:** God's judgment is seen in
civil turbulence. The *hard master* of v. 4 may
be Pianchi, the founder of the Twenty-fifth
(Ethiopian) Dynasty, begun about 714 B.C.
19.5–10: *The Nile,* Egypt's lifeline, will
dry up, destroying her economy. People have

no control over natural catastrophes.
19.11–15: Isaiah taunts Egypt; if she, with
her vaunted wisdom, can devise schemes effecting national destinies, how could she have
overlooked God's plan for herself? For *Zoan*
and Memphis see Ezek 30.13–19 n. *Palm* and
reed represent rulers and the ruled (9.14).

and they have made Egypt stagger
    in all its doings
as a drunkard staggers around
    in vomit.
15 Neither head nor tail, palm branch
    or reed,
    will be able to do anything for
      Egypt.

16 On that day the Egyptians will be like women, and tremble with fear before the hand that the LORD of hosts raises against them. 17 And the land of Judah will become a terror to the Egyptians; everyone to whom it is mentioned will fear because of the plan that the LORD of hosts is planning against them.

18 On that day there will be five cities in the land of Egypt that speak the language of Canaan and swear allegiance to the LORD of hosts. One of these will be called the City of the Sun.

19 On that day there will be an altar to the LORD in the center of the land of Egypt, and a pillar to the LORD at its border. 20 It will be a sign and a witness to the LORD of hosts in the land of Egypt; when they cry to the LORD because of oppressors, he will send them a savior, and will defend and deliver them. 21 The LORD will make himself known to the Egyptians; and the Egyptians will know the LORD on that day, and will worship with sacrifice and burnt offering, and they will make vows to the LORD and perform them. 22 The LORD will strike Egypt, striking and healing; they will return to the LORD, and he will listen to their supplications and heal them.

23 On that day there will be a highway from Egypt to Assyria, and the Assyrian will come into Egypt, and the Egyptian into Assyria, and the Egyptians will worship with the Assyrians. 24 On that day Israel will be the third with Egypt and Assyria, a blessing in the midst of the earth, 25 whom the LORD of hosts has blessed, saying, "Blessed be Egypt my people, and Assyria the work of my hands, and Israel my heritage."

20 In the year that the commander-in-chief, who was sent by King Sargon of Assyria, came to Ashdod and fought against it and took it— 2 at that time the LORD had spoken to Isaiah son of Amoz, saying, "Go, and loose the sackcloth from your loins and take your sandals off your feet," and he had done so, walking naked and barefoot. 3 Then the LORD said, "Just as my servant Isaiah has walked naked and barefoot for three years as a sign and a portent against Egypt and Ethiopia,*d* 4 so shall the king of Assyria lead away the Egyptians as captives and the Ethiopians*e* as exiles, both the young and the old, naked and barefoot, with buttocks uncovered, to the shame of Egypt. 5 And they shall be dismayed and confounded because of Ethiopia*d* their hope and of Egypt their boast. 6 In that day the inhabitants of this coastland will say, 'See, this is what has happened to those in whom we hoped and to whom we fled for help and deliverance from the king of Assyria! And we, how shall we escape?' "

21 The oracle concerning the wilderness of the sea.

d Or *Nubia*; Heb *Cush*   *e* Or *Nubians*; Heb
Cushites

19.16–25: **Conversion of Egypt and Assyria** (five paragraphs, each beginning with *On that day*). 16–17: A later nationalistic paragraph emphasizing Judah's (and God's) eventual overthrow of Egypt (v. 12). 18: *Language of Canaan,* Jewish settlements in Egypt from the early sixth century on are known (Jer 44.1). *City of the sun,* perhaps Heliopolis (compare Jer 43.13); a city by this name was in Egypt (it is now a suburb of Cairo). 19–22: Egyptian proselytes will be treated as Israel, with punishment and mercy (*striking* and *healing*). 23, 24–25: Israel will become the mediator and blessing for the nations (Gen 12.3).
20.1–6: **Against Egypt.** Egypt failed to defend a co-conspirator, *Ashdod,* against Sargon's devastating attack, 711 B.C. Isaiah, naked like a prisoner, warns Egypt of its approaching captivity. 2: Apparently *sackcloth* was customarily worn by prophets (2 Kings 1.8; Zech 13.4–6).
21.1–10: **Against Babylon** (13.1). 1: *Wil-*

As whirlwinds in the Negeb
  sweep on,
  it comes from the desert,
  from a terrible land.
2 A stern vision is told to me;
  the betrayer betrays,
  and the destroyer destroys.
Go up, O Elam,
  lay siege, O Media;
  all the sighing she has caused
  I bring to an end.
3 Therefore my loins are filled
    with anguish;
  pangs have seized me,
  like the pangs of a woman
    in labor;
I am bowed down so that I
    cannot hear,
I am dismayed so that I
    cannot see.
4 My mind reels, horror has
    appalled me;
  the twilight I longed for
  has been turned for me into
    trembling.
5 They prepare the table,
  they spread the rugs,
  they eat, they drink.
Rise up, commanders,
  oil the shield!
6 For thus the Lord said to me:
  "Go, post a lookout,
  let him announce what he sees.
7 When he sees riders, horsemen
    in pairs,
  riders on donkeys, riders
    on camels,
  let him listen diligently,
  very diligently."

8 Then the watcher*f* called out:
  "Upon a watchtower I stand,
    O Lord,
  continually by day,
  and at my post I am stationed
    throughout the night.
9 Look, there they come, riders,
    horsemen in pairs!"
Then he responded,
  "Fallen, fallen is Babylon;
  and all the images of her gods
    lie shattered on the ground."
10 O my threshed and winnowed
    one,
  what I have heard from the
    LORD of hosts,
  the God of Israel, I announce
    to you.

11 The oracle concerning Dumah.

One is calling to me from Seir,
  "Sentinel, what of the night?
  Sentinel, what of the night?"
12 The sentinel says:
  "Morning comes, and also
    the night.
  If you will inquire, inquire;
    come back again."

13 The oracle concerning the
    desert plain.

In the scrub of the desert plain
    you will lodge,
  O caravans of Dedanites.
14 Bring water to the thirsty,
    meet the fugitive with bread,

f Q Ms: MT *a lion*

---

derness of the sea: this description is completely
obscure. The beginning of the poem should
perhaps be emended to read, "Words like
whirlwinds sweeping through the Negeb,
coming from the desert . . ." *Negeb,* southern
Palestine. **2–5**: Prepare for the attack! *Elam,*
Jer 49.34–39. *Media* aided Persia against Babylon (13.17–19). *Oil the shield,* to preserve the
leather's suppleness. **6–10**: The prophet (a
sentinel; Ezek 3.17) awaits the news of Babylon's fall (Jer 51.33; Rev 18.2). **9**: *On the
ground,* perhaps implying "to the underworld" (see Jer 17.12, 13 n.).

21.11–12: Concerning Edom, who
shared Ashdod's fate (20.1). The prophet suggests a time of deliverance (*morning*) followed
by renewed oppression (*night*). *Dumah,* a
town in Arabia; the word also means "silence" and here may be a symbolic name for
Edom (*Seir,* Gen 32.3).
21.13–17: Concerning Arabia. *Dedanites*
in northern Arabia (Ezek 25.13); *Tema* an oasis in northwestern Arabia (Jer 25.23); *Kedar*
in northern Arabia, apparently the aggressor
(Jer 49.28).

O inhabitants of the land
of Tema.
15 For they have fled from the
swords,
from the drawn sword,
from the bent bow,
and from the stress of battle.
16 For thus the Lord said to me:
Within a year, according to the years of
a hired worker, all the glory of Kedar
will come to an end; 17 and the remaining
bows of Kedar's warriors will be few; for
the LORD, the God of Israel, has spoken.

22 The oracle concerning the valley
of vision.

What do you mean that you have
gone up,
all of you, to the housetops,
2 you that are full of shoutings,
tumultuous city, exultant town?
Your slain are not slain by
the sword,
nor are they dead in battle.
3 Your rulers have all fled together;
they were captured without the
use of a bow. *g*
All of you who were found were
captured,
though they had fled far away. *h*
4 Therefore I said:
Look away from me,
let me weep bitter tears;
do not try to comfort me
for the destruction of my
beloved people.

5 For the Lord GOD of hosts has
a day

of tumult and trampling and
confusion
in the valley of vision,
a battering down of walls
and a cry for help to the
mountains.
6 Elam bore the quiver
with chariots and cavalry, *i*
and Kir uncovered the shield.
7 Your choicest valleys were full of
chariots,
and the cavalry took their stand
at the gates.
8 He has taken away the covering
of Judah.

On that day you looked to the weapons of the House of the Forest, 9 and you
saw that there were many breaches in the
city of David, and you collected the waters of the lower pool. 10 You counted the
houses of Jerusalem, and you broke
down the houses to fortify the wall.
11 You made a reservoir between the two
walls for the water of the old pool. But
you did not look to him who did it, or
have regard for him who planned it long
ago.

12 In that day the Lord GOD of hosts
called to weeping and
mourning,
to baldness and putting on
sackcloth;
13 but instead there was joy and
festivity,

*g Or without their bows   h Gk Syr Vg: Heb
fled from far away   i Meaning of Heb
uncertain*

**22.1–14: Warning to Jerusalem of ap-
proaching destruction.** The occasion may
have been Sargon's expedition of 711 B.C.
(20.1), which by-passed Judah, or the period
of civil strife in Assyria following Sargon's
death (705 B.C.). **1**: *Valley of vision*, the title is
taken from v. 5 and the reference perhaps to
the valley of Hinnom (compare Jer 7.30–34;
32.35). **4**: *My beloved people*, Jerusalem. **5–8a**:
The *day* of the Lord GOD is described as a day
of invasion (2.11; 13.6). *Elam* (21.2) and *Kir*
(region in southern Babylonia), supplying

mercenaries in Assyria's army. *Choicest valleys*, such as Valley of Rephaim and the
King's Valley (17.5; 2 Sam 18 18).
22.8b–11: Military preparedness does not
replace faithfulness to God. *House*, the royal
palace (Jer 22.13–14). *The city of David* was
the oldest (southeastern) section of Jerusalem
(2 Sam 5.7). *Collected the waters* may refer to
the excavation of Hezekiah's tunnel (2 Kings
20.20; 2 Chr 32.3–5). **12–14**: The words of the
revelers are turned against them (5.11–12).

killing oxen and slaughtering
sheep,
eating meat and drinking wine.
"Let us eat and drink,
for tomorrow we die."
14 The LORD of hosts has revealed
himself in my ears:
Surely this iniquity will not be
forgiven you until you die,
says the Lord GOD of hosts.

15 Thus says the Lord GOD of hosts:
Come, go to this steward, to Shebna,
who is master of the household, and say
to him: 16 What right do you have here?
Who are your relatives here, that you
have cut out a tomb here for yourself,
cutting a tomb on the height, and carving a habitation for yourself in the rock?
17 The LORD is about to hurl you away
violently, my fellow. He will seize firm
hold on you, 18 whirl you round and
round, and throw you like a ball into a
wide land; there you shall die, and there
your splendid chariots shall lie, O you
disgrace to your master's house! 19 I will
thrust you from your office, and you will
be pulled down from your post.

20 On that day I will call my servant
Eliakim son of Hilkiah, 21 and will clothe
him with your robe and bind your sash
on him. I will commit your authority to
his hand, and he shall be a father to the
inhabitants of Jerusalem and to the house
of Judah. 22 I will place on his shoulder
the key of the house of David; he shall
open, and no one shall shut; he shall shut,
and no one shall open. 23 I will fasten him
like a peg in a secure place, and he will
become a throne of honor to his ancestral
house. 24 And they will hang on him the
whole weight of his ancestral house, the
offspring and issue, every small vessel,
from the cups to all the flagons. 25 On

that day, says the LORD of hosts, the peg
that was fastened in a secure place will
give way; it will be cut down and fall,
and the load that was on it will perish, for
the LORD has spoken.

23 The oracle concerning Tyre.

Wail, O ships of Tarshish,
for your fortress is destroyed.*j*
When they came in from Cyprus
they learned of it.
2 Be still, O inhabitants of the coast,
O merchants of Sidon,
your messengers crossed over
the sea*k*
3 and were on the mighty waters;
your revenue was the grain of
Shihor,
the harvest of the Nile;
you were the merchant of the
nations.
4 Be ashamed, O Sidon, for the sea
has spoken,
the fortress of the sea, saying:
"I have neither labored nor given
birth,
I have neither reared young men
nor brought up young women."
5 When the report comes to Egypt,
they will be in anguish over the
report about Tyre.
6 Cross over to Tarshish—
wail, O inhabitants of the coast!
7 Is this your exultant city
whose origin is from days of
old,
whose feet carried her
to settle far away?

j Cn Compare verse 14: Heb *for it is destroyed,*
without houses   *k* Q Ms: MT *crossing over the*
sea, they replenished you

22.15–25: Against Shebna (Jer 20.1–6).
Hezekiah's major-domo apparently arrogat-
ed improper status to himself, and may have
been an instigator in the anti-Assyrian plot
preceding 711 B.C. (18.1–7 n.). Verses 24–25
suggest that Eliakim (36.3), a welcome replacement, failed to fulfill expectations.
23.1–18: The oracle concerning Sidon
(vv. 1–4, 12–14) is combined with a later one

against Tyre (vv. 5–11, 15–18). **1a:** Superscription for the chapter. **1b–c:** *Tarshish,* see
Jer 10.9 n. There were Phoenician colonies in
Cyprus. **3:** *Shihor,* the "waters of Horus," near
Zoan. **4:** Ruined Sidon is barren like the sea
without sailors (*young men*).
23.5–11: Fallen Tyre's commercial empire
reacts with alarm. With this section compare
Ezek ch 26; 27.1–9, 25–36; ch 28.

8 Who has planned this
        against Tyre, the bestower
            of crowns,
    whose merchants were princes,
        whose traders were the honored
            of the earth?
9 The LORD of hosts has planned
            it—
    to defile the pride of all glory,
    to shame all the honored of the
            earth.
10 Cross over to your own land,
        O ships of[^l] Tarshish;
    this is a harbor[^m] no more.
11 He has stretched out his hand over
            the sea,
    he has shaken the kingdoms;
    the LORD has given command
            concerning Canaan
        to destroy its fortresses.
12 He said:
    You will exult no longer,
        O oppressed virgin daughter
            Sidon;
    rise, cross over to Cyprus—
        even there you will have no
            rest.

13 Look at the land of the Chaldeans!
This is the people; it was not Assyria.
They destined Tyre for wild animals.
They erected their siege towers, they
tore down her palaces, they made her a
ruin.[^n]
14 Wail, O ships of Tarshish,
        for your fortress is destroyed.
15 From that day Tyre will be forgotten
for seventy years, the lifetime of one
king. At the end of seventy years, it will
happen to Tyre as in the song about the
prostitute:
16 Take a harp,
        go about the city,
        you forgotten prostitute!
    Make sweet melody,
        sing many songs,
        that you may be remembered.
17 At the end of seventy years, the LORD
will visit Tyre, and she will return to her
trade, and will prostitute herself with all
the kingdoms of the world on the face of
the earth. 18 Her merchandise and her
wages will be dedicated to the LORD; her
profits[^o] will not be stored or hoarded,
but her merchandise will supply abundant food and fine clothing for those
who live in the presence of the LORD.

24 Now the LORD is about to lay
        waste the earth and make
            it desolate,
    and he will twist its surface and
            scatter its inhabitants.
2 And it shall be, as with the
        people, so with the priest;
    as with the slave, so with his
            master;
    as with the maid, so with her
            mistress;
    as with the buyer, so with
            the seller;
    as with the lender, so with the
            borrower;

[^l]: Cn Compare Gk: Heb *like the Nile, daughter*
[^m]: Cn: Heb *restraint*
[^n]: Meaning of Heb uncertain
[^o]: Heb *it*

23.15–18: A late addition; compare the
restoration in the late oracles in Jer 48.47;
49.6, 39. *Seventy years,* Jer 25.11. **17:** *Prostitute
herself,* do anything for gain. **18:** Even these
sordid treasures will finally be dedicated to
God (18.7; 45.14).
    Chs 24–27: The "Isaiah Apocalypse."
These chapters, unrelated to their context, are
frequently called the "Isaiah Apocalypse" because of their use of eschatological themes
found in later apocalyptic writings (universal
judgment, eschatological banquet, heavenly
signs, and the like). One may regard the section as a transitional form between traditional
prophetic and apocalyptic materials, dating
between 540 and 425 B.C. The chapters contain a variety of types of materials, e.g. eschatological prophecy in four sections (24.1–
6, 16b–23; 25.6–10a; 26.20–27.1), four
apocalyptic poems of deliverance (24.7–16a;
25.1–5; 26.1–6; 27.2–11), oracles of doom
and triumph (26.20–27.1; 27.12–13; compare
25.10b–12), and a processional and an apocalyptic psalm (26.1–6; 27.2–11).
    24.1–6: Universal judgment (compare
vv. 16b–23). A picture of total destruction;
compare the flood of Noah. **1:** *Twist,* by *earthquake.* **2:** Expansion of Hos 4.9. **3:** *This word,*

as with the creditor, so with
the debtor.
3 The earth shall be utterly laid
waste and utterly despoiled;
for the LORD has spoken
this word.

4 The earth dries up and withers,
the world languishes and
withers;
the heavens languish together
with the earth.
5 The earth lies polluted
under its inhabitants;
for they have transgressed laws,
violated the statutes,
broken the everlasting covenant.
6 Therefore a curse devours
the earth,
and its inhabitants suffer for
their guilt;
therefore the inhabitants of the
earth dwindled,
and few people are left.
7 The wine dries up,
the vine languishes,
all the merry-hearted sigh.
8 The mirth of the timbrels is
stilled,
the noise of the jubilant
has ceased,
the mirth of the lyre is stilled.
9 No longer do they drink wine
with singing;
strong drink is bitter to those
who drink it.
10 The city of chaos is broken down,
every house is shut up so that
no one can enter.
11 There is an outcry in the streets
for lack of wine;
all joy has reached its eventide;

the gladness of the earth is
banished.
12 Desolation is left in the city,
the gates are battered into ruins.
13 For thus it shall be on the earth
and among the nations,
as when an olive tree is beaten,
as at the gleaning when the
grape harvest is ended.

14 They lift up their voices, they sing
for joy;
they shout from the west over
the majesty of the LORD.
15 Therefore in the east give glory to
the LORD;
in the coastlands of the sea
glorify the name of the
LORD, the God of Israel.
16 From the ends of the earth we
hear songs of praise,
of glory to the Righteous One.
But I say, I pine away,
I pine away. Woe is me!
For the treacherous deal
treacherously,
the treacherous deal very
treacherously.

17 Terror, and the pit, and the snare
are upon you, O inhabitant of
the earth!
18 Whoever flees at the sound of
the terror
shall fall into the pit;
and whoever climbs out of the pit
shall be caught in the snare.
For the windows of heaven are
opened,
and the foundations of the earth
tremble.
19 The earth is utterly broken,
the earth is torn asunder,

---

vv. 1–2. **5**: *Laws . . . statutes . . . covenant,* Isa 42.22–25. *Everlasting covenant,* perhaps a reference to the promise to Noah (Gen 9.1–17; compare Isa 54.9).
24.7–16a: Deliverance after destruction. 7–12: Happiness of the vintage festivals is stilled. **10:** *City of chaos,* a general picture of human society destined for God's judgment. **13–16a:** The coming triumph. *It,* the scat-

tered Jewish remnant (19.24; Ezek 38.12), which praises God, *the Righteous One,* for saving and vindicating Israel.
24.16b–23: Universal judgement. 16b–18b: This continues vv. 1–6. **17–18b:** This appears in Jer 48.43–44. **18c–23:** Reflection of Canaanite and Babylonian mythological background. *Windows of heaven,* source of downpours (Gen 7.11).

the earth is violently shaken.
20 The earth staggers like a drunkard,
    it sways like a hut;
  its transgression lies heavy upon
      it,
  and it falls, and will not
      rise again.

21 On that day the LORD will punish
    the host of heaven in heaven,
  and on earth the kings of
      the earth.
22 They will be gathered together
    like prisoners in a pit;
  they will be shut up in a prison,
  and after many days they will
      be punished.
23 Then the moon will be abashed,
    and the sun ashamed;
  for the LORD of hosts will reign
    on Mount Zion and in
      Jerusalem,
  and before his elders he will
      manifest his glory.

25 O LORD, you are my God;
    I will exalt you, I will praise
      your name;
  for you have done wonderful
      things,
  plans formed of old, faithful
      and sure.
2 For you have made the city
      a heap,
  the fortified city a ruin;
  the palace of aliens is a city
      no more,
  it will never be rebuilt.
3 Therefore strong peoples will
      glorify you;
  cities of ruthless nations will
      fear you.

4 For you have been a refuge to
      the poor,
  a refuge to the needy in their
      distress,
  a shelter from the rainstorm and
      a shade from the heat.
  When the blast of the ruthless was
      like a winter rainstorm,
5   the noise of aliens like heat in a
      dry place,
  you subdued the heat with the
      shade of clouds;
  the song of the ruthless
      was stilled.

6 On this mountain the LORD of
      hosts will make for all
      peoples
  a feast of rich food, a feast of
      well-aged wines,
  of rich food filled with marrow,
  of well-aged wines strained
      clear.
7 And he will destroy on this
      mountain
  the shroud that is cast over all
      peoples,
  the sheet that is spread over all
      nations;
  he will swallow up death
      forever.
8 Then the Lord GOD will wipe
      away the tears from all
      faces,
  and the disgrace of his people he
      will take away from all
      the earth,
  for the LORD has spoken.
9 It will be said on that day,
    Lo, this is our God; we have
      waited for him, so that he
      might save us.

24.21–22: *Host of heaven,* rebellious astral deities (Zeph 1.5; Jer 19.13) imprisoned in the *pit* (Isa 14.15; Rev 20.1–3). *Many days,* compare Rev 20.3, 7. **23:** *Moon . . . abashed, . . . sun ashamed* after losing their divine status (Jer 8.2; Deut 17.3). *Elders,* as with Moses (Ex 24.9–11, 12–16; compare Rev 4.4, 10–11). **25.1–5:** **Psalm of thanksgiving** (compare Ps 145). **2:** Identity of *the city* uncertain.

4: *Poor,* the helpless (perhaps Jews) contrasted with *strong peoples* (v. 3). **25.6–10a: Third eschatological section. 6:** *This mountain,* Zion. Cultic *feast,* (1 Sam 9.13), later a part of the Messianic expectation (Lk 14.15–24). **7–8:** *Destroy . . . shroud . . . sheet* (of mourning, or perhaps ignorance), *swallow up death,* phrases reminiscent of Canaanite mythology.

This is the LORD for whom we
    have waited;
  let us be glad and rejoice in his
    salvation.
10 For the hand of the LORD will rest
    on this mountain.

The Moabites shall be trodden
    down in their place
  as straw is trodden down in a
    dung-pit.
11 Though they spread out their
    hands in the midst of it,
  as swimmers spread out their
    hands to swim,
  their pride will be laid low
    despite the struggle*p* of
    their hands.
12 The high fortifications of his walls
    will be brought down,
  laid low, cast to the ground,
    even to the dust.

26 On that day this song will be
    sung in the land of Judah:
We have a strong city;
  he sets up victory
  like walls and bulwarks.
2 Open the gates,
  so that the righteous nation that
    keeps faith
  may enter in.
3 Those of steadfast mind you keep
    in peace—
  in peace because they trust
    in you.
4 Trust in the LORD forever,
  for in the LORD GOD *q*
  you have an everlasting rock.
5 For he has brought low
  the inhabitants of the height;
  the lofty city he lays low.

He lays it low to the ground,
  casts it to the dust.
6 The foot tramples it,
  the feet of the poor,
  the steps of the needy.

7 The way of the righteous is level;
  O Just One, you make smooth
    the path of the righteous.
8 In the path of your judgments,
  O LORD, we wait for you;
  your name and your renown
  are the soul's desire.
9 My soul yearns for you in
    the night,
  my spirit within me earnestly
    seeks you.
For when your judgments are in
    the earth,
  the inhabitants of the world
  learn righteousness.
10 If favor is shown to the wicked,
  they do not learn righteousness;
  in the land of uprightness they
    deal perversely
  and do not see the majesty of
    the LORD.
11 O LORD, your hand is lifted up,
  but they do not see it.
Let them see your zeal for your
    people, and be ashamed.
Let the fire for your adversaries
  consume them.
12 O LORD, you will ordain peace
    for us,
  for indeed, all that we have
    done, you have done for us.
13 O LORD our God,
  other lords besides you have
    ruled over us,

p Meaning of Heb uncertain   *q* Heb *in Yah,
the* LORD

25.10b–12: Oracle of doom. 10b: The
mention of *Moab* is unexpected, but anti-
Moabite feeling was high in the post-exilic
period.
  26.1–6: Song of victory (24.7–16a; 25.1–
5), a processional psalm, sung on entering
Jerusalem, the *strong city* (v. 1; compare Ps
24.7–10), celebrating God's *victory* (salvation)
over the enemies of Judah, the *righteous nation.*
5: *Low . . . height,* a familiar antithesis: the
proud are abased, the humble exalted (Ps
147.6; Lk 1.52).
  26.7–19: Apocalyptic psalm. 7: Proverb
of confidence in God's help (Ps 9.19; 18.25–
27; 34.21–22). **8:** Compare Rev 22.20. God's
chastisements are designed to benefit those
chastised. **10–14:** In good times or bad the
wicked fail to note God's guiding hand, but
God and the righteous will ultimately defeat
them. **14:** Compare v. 19.

but we acknowledge your
name alone.
14 The dead do not live;
shades do not rise—
because you have punished and
destroyed them,
and wiped out all memory
of them.
15 But you have increased the nation,
O Lord,
you have increased the nation;
you are glorified;
you have enlarged all the
borders of the land.

16 O Lord, in distress they sought
you,
they poured out a prayer[r]
when your chastening was
on them.
17 Like a woman with child,
who writhes and cries out in
her pangs
when she is near her time,
so were we because of you,
O Lord;
18 we were with child, we
writhed,
but we gave birth only to wind.
We have won no victories on
earth,
and no one is born to inhabit
the world.
19 Your dead shall live, their
corpses[s] shall rise.
O dwellers in the dust, awake
and sing for joy!
For your dew is a radiant dew,
and the earth will give birth to
those long dead.[t]

20 Come, my people, enter your
chambers,

and shut your doors behind
you;
hide yourselves for a little while
until the wrath is past.
21 For the Lord comes out from
his place
to punish the inhabitants of the
earth for their iniquity;
the earth will disclose the blood
shed on it,
and will no longer cover its
slain.

27 On that day the Lord with his
cruel and great and strong sword
will punish Leviathan the fleeing serpent,
Leviathan the twisting serpent, and he
will kill the dragon that is in the sea.

2 On that day:
A pleasant vineyard, sing about it!
3 I, the Lord, am its keeper;
every moment I water it.
I guard it night and day
so that no one can harm it;
4 I have no wrath.
If it gives me thorns and briers,
I will march to battle against it.
I will burn it up.
5 Or else let it cling to me for
protection,
let it make peace with me,
let it make peace with me.

6 In days to come[u] Jacob shall
take root,
Israel shall blossom and put
forth shoots,
and fill the whole world
with fruit.

r Meaning of Heb uncertain   s Cn Compare
Syr Tg: Heb *my corpse*   t Heb *to the shades*
u Heb *Those to come*

26.16–19: Without God, the people were in agony and helpless before their oppressors. Though as dead, they will be raised up by God, whose *radiant dew* will illumine the gloom of despair.
26.20–27.1: Fourth eschatological section, following logically the preceding petition, returns to the theme of judgment, and prepares for the return of the exiles referred to in 27.12–13. **20–21:** The people should await God's victory.
27.1: *Leviathan,* mythological sea monster, parallel with *dragon,* Hebrew "tannin" (compare Ps 74.13–14).
27.2–11: Fourth apocalyptic poem of deliverance. 2–5: God's vineyard; compare 5.1–7, where the same figure is used in another sense.

7 Has he struck them down as he
  struck down those who
  struck them?
  Or have they been killed as their
  killers were killed?
8 By expulsion, *v* by exile you
  struggled against them;
  with his fierce blast he removed
  them in the day of the
  east wind.
9 Therefore by this the guilt of
  Jacob will be expiated,
  and this will be the full fruit of
  the removal of his sin:
  when he makes all the stones of
  the altars
  like chalkstones crushed
  to pieces,
  no sacred poles *w* or incense
  altars will remain standing.
10 For the fortified city is solitary,
  a habitation deserted and
  forsaken, like the
  wilderness;
  the calves graze there,
  there they lie down, and strip its
  branches.
11 When its boughs are dry, they
  are broken;
  women come and make a fire
  of them.
  For this is a people without
  understanding;
  therefore he that made them
  will not have compassion
  on them,
  he that formed them will show
  them no favor.

12 On that day the LORD will thresh
from the channel of the Euphrates to the
Wadi of Egypt, and you will be gathered
one by one, O people of Israel. 13 And on
that day a great trumpet will be blown,
and those who were lost in the land of
Assyria and those who were driven out
to the land of Egypt will come and worship the LORD on the holy mountain at
Jerusalem.

28 Ah, the proud garland of the
  drunkards of Ephraim,
  and the fading flower of its
  glorious beauty,
  which is on the head of those
  bloated with rich food, of
  those overcome with wine!
2 See, the Lord has one who is
  mighty and strong;
  like a storm of hail, a destroying
  tempest,
  like a storm of mighty,
  overflowing waters;
  with his hand he will hurl them
  down to the earth.
3 Trampled under foot will be
  the proud garland of the
  drunkards of Ephraim.
4 And the fading flower of its
  glorious beauty,
  which is on the head of those
  bloated with rich food,
  will be like a first-ripe fig before
  the summer;
  whoever sees it, eats it up
  as soon as it comes to hand.

5 In that day the LORD of hosts will
  be a garland of glory,
  and a diadem of beauty, to the
  remnant of his people;

v Meaning of Heb uncertain   *w* Heb *Asherim*

27.7–11: Israel (*Jacob*) will be blessed after
all semblance of idolatry has been removed.
The enemy will be vanquished, his *city* destroyed (vv. 10–11; 24.10; 26.5).
27.12–13: **Concluding oracle of doom
and triumph. 12**: The figure of an eschatological harvest symbolizes the separation of
the wicked from the righteous (Joel 3.13; Mt
13.39; Rev 14.15–16). *Wadi of Egypt,* Wadi
el-'Arish, fifty miles southwest of Gaza. **13**:

The great trumpet, which summoned Israel for
solemn convocations (Num 10.2–10; Joel
2.15), will signal the final assembly of God's
elect (Mt 24.31; 1 Thess 4.16).
28.1–35.10: **Oracles concerning Judah
and Ephraim. 28.1–13**: **Against religious
leaders. 1–4**: An oracle concerning *Ephraim*
(Samaria), pronounced before Assyria's onslaught (2 Kings 17.5) and introducing the
longer (and later) oracle concerning Judah

6 and a spirit of justice to the one
who sits in judgment,
and strength to those who turn
back the battle at the gate.

7 These also reel with wine
and stagger with strong drink;
the priest and the prophet reel
with strong drink,
they are confused with wine,
they stagger with strong drink;
they err in vision,
they stumble in giving
judgment.
8 All tables are covered with
filthy vomit;
no place is clean.

9 "Whom will he teach knowledge,
and to whom will he explain
the message?
Those who are weaned from milk,
those taken from the breast?
10 For it is precept upon precept,
precept upon precept,
line upon line, line upon line,
here a little, there a little."ˣ

11 Truly, with stammering lip
and with alien tongue
he will speak to this people,
12 to whom he has said,
"This is rest;
give rest to the weary;
and this is repose";
yet they would not hear.
13 Therefore the word of the LORD
will be to them,
"Precept upon precept, precept
upon precept,

line upon line, line upon line,
here a little, there a little;"ˣ
in order that they may go, and fall
backward,
and be broken, and snared,
and taken.

14 Therefore hear the word of the
LORD, you scoffers
who rule this people in
Jerusalem.
15 Because you have said, "We have
made a covenant with
death,
and with Sheol we have an
agreement;
when the overwhelming scourge
passes through
it will not come to us;
for we have made lies our refuge,
and in falsehood we have taken
shelter";
16 therefore thus says the Lord GOD,
See, I am laying in Zion a
foundation stone,
a tested stone,
a precious cornerstone, a sure
foundation:
"One who trusts will not
panic."
17 And I will make justice the line,
and righteousness the plummet;
hail will sweep away the refuge
of lies,
and waters will overwhelm the
shelter.
18 Then your covenant with death
will be annulled,

ˣ Meaning of Heb of this verse uncertain

(for the occasion see 22.1–14 n.). **1**: *Garland,
walls of Samaria* (Am 3.9, 15). **4**: *First-ripe fig,*
just as quickly will Samaria be consumed.
5–6: A later interpolation; here *garland* is
God's blessing of the remnant.
28.7–13: This section continues vv. 1–4,
but is directed against Judah. The unrestrained hedonism of Samaria was paralleled
in Judah, whose intemperate religious leaders
were incapable of responsible guidance (Jer
13.12–14). *Priest* and *prophet,* presumably Isaiah's opponents (Jer 26.7–9), resented (compare vv. 9–10) Isaiah's condescending atti-

tude. *Tables,* for sacrificial feast (1 Sam
9.12–13).
28.11–13: If the people reject Isaiah, others
(the Assyrians) will address them (Jer 5.15).
28.14–22: **Against civil leaders. 14–15**:
The *scoffers* have forsaken God, their only defender, to seek foreign military aid against
Assyria (vv. 15–16; 30.1–7). **16–17a**: The
symbolic name of the foundation stone
(v. 16d; 1.26) proclaims salvation for those
who trust God. *Justice* and *righteousness* (1.21;
5.16) will characterize the new Jerusalem.
17b–22: The faithless will be destroyed by the

and your agreement with Sheol
will not stand;
when the overwhelming scourge
passes through
you will be beaten down by it.
19 As often as it passes through, it
will take you;
for morning by morning it will
pass through,
by day and by night;
and it will be sheer terror to
understand the message.
20 For the bed is too short to stretch
oneself on it,
and the covering too narrow to
wrap oneself in it.
21 For the LORD will rise up as on
Mount Perazim,
he will rage as in the valley of
Gibeon;
to do his deed—strange is his
deed!
and to work his work—alien is
his work!
22 Now therefore do not scoff,
or your bonds will be made
stronger;
for I have heard a decree of
destruction
from the Lord GOD of hosts
upon the whole land.

23 Listen, and hear my voice;
Pay attention, and hear my
speech.
24 Do those who plow for sowing
plow continually?
Do they continually open and
harrow their ground?
25 When they have leveled its
surface,
do they not scatter dill, sow
cummin,
and plant wheat in rows

and barley in its proper place,
and spelt as the border?
26 For they are well instructed;
their God teaches them.

27 Dill is not threshed with a
threshing sledge,
nor is a cart wheel rolled over
cummin;
but dill is beaten out with a stick,
and cummin with a rod.
28 Grain is crushed for bread,
but one does not thresh it
forever;
one drives the cart wheel and
horses over it,
but does not pulverize it.
29 This also comes from the LORD
of hosts;
he is wonderful in counsel,
and excellent in wisdom.

29 Ah, Ariel, Ariel,
the city where David
encamped!
Add year to year;
let the festivals run their round.
2 Yet I will distress Ariel,
and there shall be moaning and
lamentation,
and Jerusalem^y shall be to me
like an Ariel. ^{z}
3 And like David^a I will encamp
against you;
I will besiege you with towers
and raise siegeworks against
you.
4 Then deep from the earth you
shall speak,

_y_ Heb _she_   _z_ Probable meaning, _altar hearth_;
compare Ezek 43.15   _a_ Gk: Meaning of Heb
uncertain

storm of God's wrath. **20**: A proverb describing an irremediable situation. **21**: _Perazim_, 2 Sam 5.17–21; _Gibeon,_ Josh 10.10.
28.23–29: **Parable of the farmers,** patterned on contemporary wisdom literature. God, like a farmer, conducts his affairs according to plan (compare vv. 16–17a). **25**: _Spelt,_ less valuable than wheat. **27**: _Dill_ and

cummin, spices, too soft to thresh with a sledge.
29.1–8: **Judah's eventual restoration. 1**: _Ariel,_ probably meaning "altar hearth," an allusion to Jerusalem. **2**: _Like an Ariel,_ "like an altar-hearth": the entire city will be a burnt offering.

from low in the dust your
  words shall come;
your voice shall come from the
  ground like the voice of
  a ghost,
and your speech shall whisper
  out of the dust.

5 But the multitude of your foes[b]
  shall be like small dust,
and the multitude of tyrants like
  flying chaff.
And in an instant, suddenly,
6   you will be visited by the LORD
  of hosts
with thunder and earthquake and
  great noise,
  with whirlwind and tempest,
  and the flame of a
  devouring fire.
7 And the multitude of all the
  nations that fight against
  Ariel,
  all that fight against her and her
  stronghold, and who
  distress her,
  shall be like a dream, a vision
  of the night.
8 Just as when a hungry person
  dreams of eating
  and wakes up still hungry,
or a thirsty person dreams of
  drinking
  and wakes up faint, still thirsty,
so shall the multitude of all the
  nations be
  that fight against Mount Zion.

9 Stupefy yourselves and be in a
  stupor,
  blind yourselves and be blind!
Be drunk, but not from wine;
  stagger, but not from strong
  drink!

10 For the LORD has poured out
  upon you
  a spirit of deep sleep;
he has closed your eyes, you
  prophets,
and covered your heads, you
  seers.
11 The vision of all this has become
for you like the words of a sealed document. If it is given to those who can read,
with the command, "Read this," they
say, "We cannot, for it is sealed." 12 And
if it is given to those who cannot read,
saying, "Read this," they say, "We cannot read."

13 The Lord said:
Because these people draw near
  with their mouths
  and honor me with their lips,
  while their hearts are far
  from me,
and their worship of me is a
  human commandment
  learned by rote;
14 so I will again do
  amazing things with this people,
  shocking and amazing.
The wisdom of their wise shall
  perish,
  and the discernment of the
  discerning shall be hidden.

15 Ha! You who hide a plan too deep
  for the LORD,
  whose deeds are in the dark,
  and who say, "Who sees us?
  Who knows us?"
16 You turn things upside down!
  Shall the potter be regarded as
  the clay?

b Cn: Heb *strangers*

29.9–24: Spiritual insensibility; miscellaneous oracles. **9–12:** An independent section, related in thought to vv. 1–8 and 13–14.
9–10: As drunkards, Jerusalem's leaders are
incapable of moral discrimination. **11–12:** Explanatory later prose addition. *Sealed document,* a scroll (Jer 32.9–15).
  29.13–14: A new relationship with God,
based on a positive response to his acts (Ex
19.4–6), will replace Judah's superficial tradi-
tionalism (ironically called *wisdom* and *discernment;* compare 1 Cor 1.19). **15–16:** Judah's
leaders have usurped God's prerogatives by
plotting against Assyria (45.9; Jer 18.1–6;
Rom 9.20–21; Mt 10.24).

Shall the thing made say of its
maker,
"He did not make me";
or the thing formed say of the one
who formed it,
"He has no understanding"?

17 Shall not Lebanon in a very little
while
become a fruitful field,
and the fruitful field be regarded
as a forest?
18 On that day the deaf shall hear
the words of a scroll,
and out of their gloom and
darkness
the eyes of the blind shall see.
19 The meek shall obtain fresh joy in
the LORD,
and the neediest people shall
exult in the Holy One of
Israel.
20 For the tyrant shall be no more,
and the scoffer shall cease to be;
all those alert to do evil shall be
cut off—
21 those who cause a person to lose a
lawsuit,
who set a trap for the arbiter in
the gate,
and without grounds deny
justice to the one in the
right.

22 Therefore thus says the LORD,
who redeemed Abraham, concerning the
house of Jacob:
No longer shall Jacob be ashamed,
no longer shall his face
grow pale.

23 For when he sees his children,
the work of my hands, in
his midst,
they will sanctify my name;
they will sanctify the Holy One of
Jacob,
and will stand in awe of the
God of Israel.
24 And those who err in spirit will
come to understanding,
and those who grumble will
accept instruction.

30 Oh, rebellious children, says
the LORD,
who carry out a plan, but not
mine;
who make an alliance, but against
my will,
adding sin to sin;
2 who set out to go down to Egypt
without asking for my counsel,
to take refuge in the protection of
Pharaoh,
and to seek shelter in the
shadow of Egypt;
3 Therefore the protection of
Pharaoh shall become your
shame,
and the shelter in the shadow of
Egypt your humiliation.
4 For though his officials are at
Zoan
and his envoys reach Hanes,
5 everyone comes to shame
through a people that cannot
profit them,
that brings neither help nor profit,
but shame and disgrace.

29.17–24: Two later stanzas based on materials from chs 40–66 and reflecting that period. **17–21**: Israel's suffering will soon be
completed (35.1–2; 41.17). **19**: *Holy One of
Israel*, see 1.4 n.; 41.14, 16, 20, and elsewhere.
22–24: Abraham's God will restore repentant
Israel, who will accept God's instruction
(Ezek 36.22–32). Compare the references to
Abraham in 41.8; 51.2. This kind of reference
is later than Isaiah's time.
**30.1–7: Concerning the embassy sent
to Egypt** (about 703 B.C.) soliciting support

against Assyria, *a plan* which Isaiah considered rebellion against God (28.14–22; 29.15–
16). **1–5**: Since the pact is against God's
wishes, it will fail. *Zoan*, see Ezek 30.13–
19 n.; *Hanes* (Anusis), fifty miles south of
Memphis. **6a**: Obscure, perhaps emend to
read, "They carry, through the heat of the
Negeb, . . ." **7b**: Obscure, perhaps emend to
read, "Rahab, who shall be destroyed." *Rahab,* a mythological sea dragon vanquished in
cosmic combat (26.20–27.1; Job 26.12).

6 An oracle concerning the animals of
the Negeb.
Through a land of trouble and
distress,
of lioness and roaring^c lion,
of viper and flying serpent,
they carry their riches on the
backs of donkeys,
and their treasures on the humps
of camels,
to a people that cannot profit
them.
7 For Egypt's help is worthless and
empty,
therefore I have called her,
"Rahab who sits still."^d

8 Go now, write it before them on
a tablet,
and inscribe it in a book,
so that it may be for the time
to come
as a witness forever.
9 For they are a rebellious people,
faithless children,
children who will not hear
the instruction of the LORD;
10 who say to the seers, "Do not
see";
and to the prophets, "Do not
prophesy to us what is
right;
speak to us smooth things,
prophesy illusions,
11 leave the way, turn aside from
the path,
let us hear no more about the
Holy One of Israel."
12 Therefore thus says the Holy One
of Israel:
Because you reject this word,
and put your trust in oppression
and deceit,

and rely on them;
13 therefore this iniquity shall
become for you
like a break in a high wall,
bulging out, and about to
collapse,
whose crash comes suddenly, in
an instant;
14 its breaking is like that of a
potter's vessel
that is smashed so ruthlessly
that among its fragments not a
sherd is found
for taking fire from the hearth,
or dipping water out of the
cistern.

15 For thus said the Lord GOD, the
Holy One of Israel:
In returning and rest you shall
be saved;
in quietness and in trust shall be
your strength.
But you refused 16and said,
"No! We will flee upon horses"—
therefore you shall flee!
and, "We will ride upon swift
steeds"—
therefore your pursuers shall be
swift!
17 A thousand shall flee at the threat
of one,
at the threat of five you
shall flee,
until you are left
like a flagstaff on the top of
a mountain,
like a signal on a hill.

18 Therefore the LORD waits to be
gracious to you;

c Cn: Heb *from them*   *d* Meaning of Heb
uncertain

**30.8–17: Judah's connivance with
Egypt** against Assyria. **8–11:** Isaiah is to preserve his oracles in writing in preparation for
their ultimate fulfillment (8.16–18). *Tablet,*
see 8.1 n. *Book,* a scroll (see 29.11 n.). *Faithless children,* those who deny their status (1.2–
4). *Instruction,* Hebrew "torah," see 1.10 n.
Smooth things, pleasant, though false (Jer
28.8–9). **12–14:** Sentence of judgment; this

should probably follow vv. 15–17, since *this
word* (v. 12b) is recorded in v. 15. Judah's perverseness is a basic flaw which, under pressure, will burst and all will be lost.
30.15–17: Instead of demonstrating victorious faith, Judah seeks more tangible military devices, but these will be no source of
strength (7.3–9; 10.20–21).
30.18–26: Hope for the afflicted. A later

therefore he will rise up to show
mercy to you.
For the LORD is a God of justice;
blessed are all those who wait
for him.

19 Truly, O people in Zion, inhabitants of Jerusalem, you shall weep no more. He will surely be gracious to you at the sound of your cry; when he hears it, he will answer you. 20 Though the Lord may give you the bread of adversity and the water of affliction, yet your Teacher will not hide himself any more, but your eyes shall see your Teacher. 21 And when you turn to the right or when you turn to the left, your ears shall hear a word behind you, saying, "This is the way; walk in it." 22 Then you will defile your silver-covered idols and your gold-plated images. You will scatter them like filthy rags; you will say to them, "Away with you!"

23 He will give rain for the seed with which you sow the ground, and grain, the produce of the ground, which will be rich and plenteous. On that day your cattle will graze in broad pastures; 24 and the oxen and donkeys that till the ground will eat silage, which has been winnowed with shovel and fork. 25 On every lofty mountain and every high hill there will be brooks running with water—on a day of the great slaughter, when the towers fall. 26 Moreover the light of the moon will be like the light of the sun, and the light of the sun will be sevenfold, like the light of seven days, on the day when the LORD binds up the injuries of his people, and heals the wounds inflicted by his blow.

27 See, the name of the LORD comes
from far away,

burning with his anger, and in
thick rising smoke;*e*
his lips are full of indignation,
and his tongue is like a
devouring fire;
28 his breath is like an overflowing
stream
that reaches up to the neck—
to sift the nations with the sieve of
destruction,
and to place on the jaws of the
peoples a bridle that leads
them astray.

29 You shall have a song as in the night when a holy festival is kept; and gladness of heart, as when one sets out to the sound of the flute to go to the mountain of the LORD, to the Rock of Israel. 30 And the LORD will cause his majestic voice to be heard and the descending blow of his arm to be seen, in furious anger and a flame of devouring fire, with a cloudburst and tempest and hailstones. 31 The Assyrian will be terror-stricken at the voice of the LORD, when he strikes with his rod. 32 And every stroke of the staff of punishment that the LORD lays upon him will be to the sound of timbrels and lyres; battling with brandished arm he will fight with him. 33 For his burning place*f* has long been prepared; truly it is made ready for the king,*g* its pyre made deep and wide, with fire and wood in abundance; the breath of the LORD, like a stream of sulfur, kindles it.

31 Alas for those who go down to
Egypt for help
and who rely on horses,

e Meaning of Heb uncertain   f Or *Topheth*
g Or *Molech*

(perhaps Exilic) addition. Judah is not condemned for faithlessness, but encouraged to have patience under trial. Verses 19–26 are a commentary on v. 18 and include familiar, though later, imagery (Jer 31.10–14; Ezek 34.25–30). **20**: *Teacher,* with some Hebrew manuscripts, but probably to be read "teachers," i.e. prophets.
30.27–33: Oracle against Assyria (about

701 B.C.; see chs 36–37). The text is combined with a song of deliverance. **27–28, 30**: Like an approaching thunderstorm, God's fury will burst over Judah's enemies. **33**: *Burning place,* Topheth (see 2 Kings 23.10 n.). *For the king,* i.e. Molech (see Lev 18.21 n.).
31.1–3: Against Egypt (see 30.1–7). **1**: Judah's limited treasury precluded extensive military expenditures; her terrain was unfa-

who trust in chariots because they
    are many
and in horsemen because they
    are very strong,
but do not look to the Holy One
    of Israel
    or consult the LORD!
2 Yet he too is wise and brings
    disaster;
    he does not call back his words,
but will rise against the house of
    the evildoers,
    and against the helpers of those
      who work iniquity.
3 The Egyptians are human, and
    not God;
    their horses are flesh, and
      not spirit.
When the LORD stretches out
    his hand,
    the helper will stumble, and the
      one helped will fall,
    and they will all perish together.

4 For thus the LORD said to me,
As a lion or a young lion growls
    over its prey,
    and—when a band of shepherds
      is called out against it—
is not terrified by their shouting
    or daunted at their noise,
so the LORD of hosts will
    come down
    to fight upon Mount Zion and
      upon its hill.
5 Like birds hovering overhead, so
    the LORD of hosts
    will protect Jerusalem;
he will protect and deliver it,
    he will spare and rescue it.

6 Turn back to him whom you[h] have
deeply betrayed, O people of Israel. 7 For

on that day all of you shall throw away
your idols of silver and idols of gold,
which your hands have sinfully made for
you.
8 "Then the Assyrian shall fall by a
    sword, not of mortals;
    and a sword, not of humans,
      shall devour him;
he shall flee from the sword,
    and his young men shall be put
      to forced labor.
9 His rock shall pass away in terror,
    and his officers desert the
      standard in panic,"
says the LORD, whose fire is in
    Zion,
    and whose furnace is in
      Jerusalem.

32 See, a king will reign in
    righteousness,
    and princes will rule with
      justice.
2 Each will be like a hiding place
    from the wind,
    a covert from the tempest,
like streams of water in a
    dry place,
    like the shade of a great rock in
      a weary land.
3 Then the eyes of those who have
    sight will not be closed,
    and the ears of those who have
      hearing will listen.
4 The minds of the rash will have
    good judgment,
    and the tongues of stammerers
      will speak readily and
      distinctly.
5 A fool will no longer be called
    noble,

h Heb *they*

---

vorable for Egyptian chariots. **3**: Perhaps a
reference to the crossing of the Red Sea (Ex
14.26–31).
    31.4–9: Against Sennacherib (29.1–8;
37.21–25). **4–5**: God will defend Jerusalem.
6–7: A later addition (compare 2.20). **8–9**:
God's *sword* will destroy Assyria (Ezek
ch 21). *Fire, furnace,* a reference to the temple
altar.

32.1–8: The coming age of justice
(probably a later addition). Patterned on wisdom literature (28.23–29), this non-messianic
oracle describes Judah's coming rulers as
people of integrity, patience and high-
mindedness (Prov 8.15–21). Fools will no
longer receive attention (compare Prov 15.2,
7, 14, etc.).

nor a villain said to be
honorable.
6 For fools speak folly,
and their minds plot iniquity:
to practice ungodliness,
to utter error concerning
the LORD,
to leave the craving of the hungry
unsatisfied,
and to deprive the thirsty of
drink.
7 The villainies of villains are evil;
they devise wicked devices
to ruin the poor with lying words,
even when the plea of the needy
is right.
8 But those who are noble plan
noble things,
and by noble things they stand.

9 Rise up, you women who are at
ease, hear my voice;
you complacent daughters, listen
to my speech.
10 In little more than a year
you will shudder, you
complacent ones;
for the vintage will fail,
the fruit harvest will not come.
11 Tremble, you women who are
at ease,
shudder, you complacent ones;
strip, and make yourselves bare,
and put sackcloth on your loins.
12 Beat your breasts for the pleasant
fields,
for the fruitful vine,
13 for the soil of my people
growing up in thorns and briers;
yes, for all the joyous houses
in the jubilant city.
14 For the palace will be forsaken,

the populous city deserted;
the hill and the watchtower
will become dens forever,
the joy of wild asses,
a pasture for flocks;
15 until a spirit from on high is
poured out on us,
and the wilderness becomes a
fruitful field,
and the fruitful field is deemed
a forest.
16 Then justice will dwell in the
wilderness,
and righteousness abide in the
fruitful field.
17 The effect of righteousness will
be peace,
and the result of righteousness,
quietness and trust forever.
18 My people will abide in a peaceful
habitation,
in secure dwellings, and in quiet
resting places.
19 The forest will disappear
completely,*i*
and the city will be utterly
laid low.
20 Happy will you be who sow
beside every stream,
who let the ox and the donkey
range freely.

33 Ah, you destroyer,
who yourself have not been
destroyed;
you treacherous one,
with whom no one has dealt
treacherously!
When you have ceased to destroy,

i Cn: Heb *And it will hail when the forest comes
down*

**32.9–14: Against the complacency of
Judah's women** (3.16–4.1) who now celebrate vintage festival but in another year will
bemoan Judah's desolation. **11–12a:** Typical
signs of mourning (Jer 4.7–8; 16.1–9).
32.15–20: The age of the Spirit (11.2)
**will see the transformation of all cre-
ation.** The outpouring of the Spirit will produce a condition wherein human beings may
enjoy true happiness (compare Jer 31.31–34).

The section was probably added later (see
29.17–24).
33.1–24: A prophetic liturgy, an independent post-exilic addition, consisting of
entreaties and oracles (compare Pss 85; 46),
perhaps used in the temple service, led by a
prophet (Jer 14.20; 18.20). **1–6:** First section.
1: A prophetic reproach; *destroyer,* used of
Babylon in 21.2, here unidentified. **2:** A congregational prayer. *Our arm,* source of

you will be destroyed;
and when you have stopped
  dealing treacherously,
you will be dealt with
  treacherously.

2 O LORD, be gracious to us; we
  wait for you.
Be our arm every morning,
our salvation in the time of
  trouble.
3 At the sound of tumult, peoples
  fled;
before your majesty, nations
  scattered.
4 Spoil was gathered as the
  caterpillar gathers;
as locusts leap, they leaped[j]
  upon it.
5 The LORD is exalted, he dwells
  on high;
he filled Zion with justice and
  righteousness;
6 he will be the stability of your
  times,
abundance of salvation, wisdom,
  and knowledge;
the fear of the LORD is Zion's
  treasure.[k]

7 Listen! the valiant[j] cry in the
  streets;
the envoys of peace weep
  bitterly.
8 The highways are deserted,
travelers have quit the road.
The treaty is broken,
its oaths[l] are despised,
its obligation[m] is disregarded.
9 The land mourns and languishes;
Lebanon is confounded and
  withers away;
Sharon is like a desert;
and Bashan and Carmel shake
  off their leaves.

10 "Now I will arise," says the LORD,
  "now I will lift myself up;
now I will be exalted.
11 You conceive chaff, you bring
  forth stubble;
your breath is a fire that will
  consume you.
12 And the peoples will be as if
  burned to lime,
like thorns cut down, that are
  burned in the fire."

13 Hear, you who are far away, what
  I have done;
and you who are near,
  acknowledge my might.
14 The sinners in Zion are afraid;
trembling has seized the godless:
"Who among us can live with the
  devouring fire?
Who among us can live with
  everlasting flames?"
15 Those who walk righteously and
  speak uprightly,
who despise the gain of
  oppression,
who wave away a bribe instead of
  accepting it,
who stop their ears from
  hearing of bloodshed
and shut their eyes from looking
  on evil,
16 they will live on the heights;
their refuge will be the
  fortresses of rocks;
their food will be supplied, their
  water assured.

17 Your eyes will see the king in his
  beauty;
they will behold a land that
  stretches far away.

j Meaning of Heb uncertain   k Heb *his*
treasure; meaning of Heb uncertain   l Q Ms:
MT *cities*   m Or *everyone*

---

strength. **3–6**: Promise of God's victory and
restoration of Zion.
  33.7–16: Second section. **7–9**: A congregational lament. The conditions (insecurity,
distrust) are as critical as if normally forested
and fruitful regions became barren. **10–13**:

God commands the nations to acknowledge
his might. **14–16**: God's wrath will destroy
both *the sinners in Zion* and Zion's external
enemies. **15–16**: See Ps 15 and 24.3-5.
  33.17–24: Third section. **17–20**: The unpleasant past (28.11–13) will become a mem-

18 Your mind will muse on the
    terror:
    "Where is the one who counted?
    Where is the one who weighed
      the tribute?
    Where is the one who counted
      the towers?"
19 No longer will you see the
    insolent people,
    the people of an obscure speech
      that you cannot
      comprehend,
    stammering in a language that
      you cannot understand.
20 Look on Zion, the city of our
    appointed festivals!
    Your eyes will see Jerusalem,
    a quiet habitation, an
      immovable tent,
    whose stakes will never be
      pulled up,
    and none of whose ropes will
      be broken.
21 But there the LORD in majesty will
      be for us
    a place of broad rivers and
      streams,
    where no galley with oars can go,
      nor stately ship can pass.
22 For the LORD is our judge, the
      LORD is our ruler,
    the LORD is our king; he will
      save us.

23 Your rigging hangs loose;
    it cannot hold the mast firm in
      its place,
    or keep the sail spread out.

    Then prey and spoil in abundance
      will be divided;
    even the lame will fall to
      plundering.

24 And no inhabitant will say, "I am
      sick";
    the people who live there will
      be forgiven their iniquity.

34 Draw near, O nations, to hear;
    O peoples, give heed!
    Let the earth hear, and all that fills
      it;
    the world, and all that comes
      from it.
2 For the LORD is enraged against all
      the nations,
    and furious against all their
      hoards;
    he has doomed them, has given
      them over for slaughter.
3 Their slain shall be cast out,
    and the stench of their corpses
      shall rise;
    the mountains shall flow with
      their blood.
4 All the host of heaven shall rot
      away,
    and the skies roll up like a
      scroll.
    All their host shall wither
    like a leaf withering on a vine,
      or fruit withering on a fig tree.

5 When my sword has drunk its fill
      in the heavens,
    lo, it will descend upon Edom,
    upon the people I have doomed
      to judgment.
6 The LORD has a sword; it is sated
      with blood,
    it is gorged with fat,
    with the blood of lambs and
      goats,
    with the fat of the kidneys
      of rams.
    For the LORD has a sacrifice in
      Bozrah,

---

ory in the peaceful reign of the Messianic
king. **21–24**: The land will be amply watered
(Ezek 47.1–12), not for war fleets but to satisfy the people's needs.
    34.1–17: **The terrible end of God's ene-
mies,** another post-exilic addition. Compare
Ezek chs 38–39. **4**: For the destruction and fall
of *the host of heaven,* synonymous with God's

earthly enemies, see 24.21. **5**: The destruction
of Edom (see Jer 49.7–22) illustrates the fate
of all of God's enemies, as in Ob 15–16. **6**: For
the *sword* of the LORD, see Ezek ch 21. *Sacrifice,* compare Ezek 39.17–20. The animals
named are those usually sacrificed. *Bozrah,*
see Jer 49.7–22 n.
    34.8: *Day,* see 13.6; 27.2. **9–17**: Aftermath

a great slaughter in the land
of Edom.

7 Wild oxen shall fall with them,
and young steers with the
mighty bulls.
Their land shall be soaked with
blood,
and their soil made rich with
fat.

8 For the LORD has a day of
vengeance,
a year of vindication by Zion's
cause. *n*

9 And the streams of Edom*o* shall
be turned into pitch,
and her soil into sulfur;
her land shall become burning
pitch.

10 Night and day it shall not be
quenched;
its smoke shall go up forever.
From generation to generation it
shall lie waste;
no one shall pass through it
forever and ever.

11 But the hawk*p* and the hedgehog*p*
shall possess it;
the owl*p* and the raven shall live
in it.
He shall stretch the line of
confusion over it,
and the plummet of chaos over*q*
its nobles.

12 They shall name it No Kingdom
There,
and all its princes shall be
nothing.

13 Thorns shall grow over its
strongholds,
nettles and thistles in its
fortresses.
It shall be the haunt of jackals,
an abode for ostriches.

14 Wildcats shall meet with hyenas,
goat-demons shall call to
each other;
there too Lilith shall repose,
and find a place to rest.

15 There shall the owl nest
and lay and hatch and brood in
its shadow;
there too the buzzards shall gather,
each one with its mate.

16 Seek and read from the book of
the LORD:
Not one of these shall be
missing;
none shall be without its mate.
For the mouth of the LORD has
commanded,
and his spirit has gathered them.

17 He has cast the lot for them,
his hand has portioned it out to
them with the line;
they shall possess it forever,
from generation to generation
they shall live in it.

35 The wilderness and the dry
land shall be glad,
the desert shall rejoice and
blossom;
like the crocus ²it shall blossom
abundantly,
and rejoice with joy and
singing.
The glory of Lebanon shall be
given to it,
the majesty of Carmel and
Sharon.
They shall see the glory of the
LORD,
the majesty of our God.

n Or *of recompense by Zion's defender*
o Heb *her streams*   *p* Identification uncertain
q Heb lacks *over*

---

of God's assault. **9–10:** Edom suffers the punishment of Sodom and Gomorrah (13.19;
Gen 19.24). **11:** Her desolation will be as the
primordial chaos (Jer 4.23–28). **12:** *No Kingdom There,* a mocking name suggesting the
end of all who oppose God. **13–15:** *Lilith,* the
storm demon found in abandoned places, and

wild animals (13.19–22) haunt her ruins. **16:**
The book of the LORD, compare 4.3.
35.1–10: Zion restored. This probably
belonged originally to chs 40–66. **1–6a:** All
creation will see God's glory; the helpless exiles (*feeble knees* of exhausted prisoners; *blind;
speechless*) will receive new courage and hope.

3 Strengthen the weak hands,
and make firm the feeble knees.
4 Say to those who are of a fearful
heart,
"Be strong, do not fear!
Here is your God.
He will come with vengeance,
with terrible recompense.
He will come and save you."

5 Then the eyes of the blind shall
be opened,
and the ears of the deaf
unstopped;
6 then the lame shall leap like
a deer,
and the tongue of the speechless
sing for joy.
For waters shall break forth in the
wilderness,
and streams in the desert;
7 the burning sand shall become
a pool,
and the thirsty ground springs
of water;
the haunt of jackals shall become
a swamp,^{r}
the grass shall become reeds and
rushes.

8 A highway shall be there,
and it shall be called the
Holy Way;
the unclean shall not travel on it,^{s}
but it shall be for God's
people;^{t}
no traveler, not even fools, shall
go astray.
9 No lion shall be there,
nor shall any ravenous beast
come up on it;
they shall not be found there,
but the redeemed shall walk
there.

10 And the ransomed of the Lord
shall return,
and come to Zion with singing;
everlasting joy shall be upon their
heads;
they shall obtain joy and
gladness,
and sorrow and sighing shall
flee away.

36 In the fourteenth year of King
Hezekiah, King Sennacherib of
Assyria came up against all the fortified
cities of Judah and captured them. 2 The
king of Assyria sent the Rabshakeh from
Lachish to King Hezekiah at Jerusalem,
with a great army. He stood by the conduit of the upper pool on the highway to
the Fuller's Field. 3 And there came out to
him Eliakim son of Hilkiah, who was in
charge of the palace, and Shebna the secretary, and Joah son of Asaph, the recorder.

4 The Rabshakeh said to them, "Say
to Hezekiah: Thus says the great king,
the king of Assyria: On what do you base
this confidence of yours? 5 Do you think
that mere words are strategy and power
for war? On whom do you now rely,
that you have rebelled against me? 6 See,
you are relying on Egypt, that broken
reed of a staff, which will pierce the hand
of anyone who leans on it. Such is Pharaoh king of Egypt to all who rely on him.
7 But if you say to me, 'We rely on the
Lord our God,' is it not he whose high
places and altars Hezekiah has removed,
saying to Judah and to Jerusalem, 'You
shall worship before this altar'? 8 Come
now, make a wager with my master the
king of Assyria: I will give you two
thousand horses, if you are able on your

*r Cn: Heb in the haunt of jackals is her resting place
s Or pass it by   t Cn: Heb for them*

35.8–10: On *the Holy Way,* through a land
like paradise (11.6–9), they will *come to Zion*
to sing the praises of God, their deliverer.
36.1–39.8: **Historical appendix,** duplicated in 2 Kings 18.13–20.19, except for
38.9–20.
36.1–22: **The attack of Sennacherib**

(701 B.C.), see 2 Kings 18.13–27. **1**: *All the
fortified cities,* forty-six according to Sennacherib's account. **2**: *Rabshakeh,* Assyrian title meaning "chief steward." **3**: *Eliakim, Shebna,* compare 22.15–25. **6**: Compare 31.1–3;
Ezek 29.6. **7**: A reference to Hezekiah's reform, 2 Kings 18.4.

part to set riders on them. 9How then can you repulse a single captain among the least of my master's servants, when you rely on Egypt for chariots and for horsemen? 10Moreover, is it without the LORD that I have come up against this land to destroy it? The LORD said to me, Go up against this land, and destroy it."

11 Then Eliakim, Shebna, and Joah said to the Rabshakeh, "Please speak to your servants in Aramaic, for we understand it; do not speak to us in the language of Judah within the hearing of the people who are on the wall." 12But the Rabshakeh said, "Has my master sent me to speak these words to your master and to you, and not to the people sitting on the wall, who are doomed with you to eat their own dung and drink their own urine?"

13 Then the Rabshakeh stood and called out in a loud voice in the language of Judah, "Hear the words of the great king, the king of Assyria! 14Thus says the king: 'Do not let Hezekiah deceive you, for he will not be able to deliver you. 15Do not let Hezekiah make you rely on the LORD by saying, The LORD will surely deliver us; this city will not be given into the hand of the king of Assyria.' 16Do not listen to Hezekiah; for thus says the king of Assyria: 'Make your peace with me and come out to me; then everyone of you will eat from your own vine and your own fig tree and drink water from your own cistern, 17until I come and take you away to a land like your own land, a land of grain and wine, a land of bread and vineyards. 18Do not let Hezekiah mislead you by saying, The LORD will save us. Has any of the gods of the nations saved their land out of the hand of the king of Assyria? 19Where are the gods of Hamath and Arpad? Where are the gods of Sepharvaim? Have they delivered Samaria out of my hand? 20Who among all the gods of these countries have saved their countries out of my hand, that the LORD should save Jerusalem out of my hand?' "

21 But they were silent and answered him not a word, for the king's command was, "Do not answer him." 22Then Eliakim son of Hilkiah, who was in charge of the palace, and Shebna the secretary, and Joah son of Asaph, the recorder, came to Hezekiah with their clothes torn, and told him the words of the Rabshakeh.

37 When King Hezekiah heard it, he tore his clothes, covered himself with sackcloth, and went into the house of the LORD. 2And he sent Eliakim, who was in charge of the palace, and Shebna the secretary, and the senior priests, covered with sackcloth, to the prophet Isaiah son of Amoz. 3They said to him, "Thus says Hezekiah, This day is a day of distress, of rebuke, and of disgrace; children have come to the birth, and there is no strength to bring them forth. 4It may be that the LORD your God heard the words of the Rabshakeh, whom his master the king of Assyria has sent to mock the living God, and will rebuke the words that the LORD your God has heard; therefore lift up your prayer for the remnant that is left."

5 When the servants of King Hezekiah came to Isaiah, 6Isaiah said to them, "Say to your master, 'Thus says the LORD: Do not be afraid because of the words that you have heard, with which the servants of the king of Assyria have reviled me. 7I myself will put a spirit in him, so that he shall hear a rumor, and

---

36.11: *Aramaic* was the current diplomatic language. 12: The results of the projected siege. 13–20: A speech intended to demoralize Jerusalem's defenders. 19: *Hamath, Arpad,* see 10.9 n.; *Sepharvaim* (Sibraim, Ezek 47.16), a town between Hamath and Damascus, near Homs; *Samaria,* now an Assyrian province (2 Kings 17.5–6, 24). 37.1–35: **Hezekiah consults Isaiah.** See 2 Kings ch 19. 1: *Tore his clothes, covered himself with sackcloth,* signs of mourning and despair (15.3; Joel 2.12). For recourse to the temple in national crises, see Neh ch 9; Joel 1.13–14. 2: The delegation to Isaiah recalls the consultation of prophets by earlier kings (1 Kings 22.8–28; 2 Kings 1.9–17; 3.11–27). 5–7: Isaiah's reply reassures Hezekiah. *Rumor* of internal problems at home.

return to his own land; I will cause him to fall by the sword in his own land.' "

8 The Rabshakeh returned, and found the king of Assyria fighting against Libnah; for he had heard that the king had left Lachish. [9] Now the king[u] heard concerning King Tirhakah of Ethiopia,[v] "He has set out to fight against you." When he heard it, he sent messengers to Hezekiah, saying, [10] "Thus shall you speak to King Hezekiah of Judah: Do not let your God on whom you rely deceive you by promising that Jerusalem will not be given into the hand of the king of Assyria. [11] See, you have heard what the kings of Assyria have done to all lands, destroying them utterly. Shall you be delivered? [12] Have the gods of the nations delivered them, the nations that my predecessors destroyed, Gozan, Haran, Rezeph, and the people of Eden who were in Telassar? [13] Where is the king of Hamath, the king of Arpad, the king of the city of Sepharvaim, the king of Hena, or the king of Ivvah?''

14 Hezekiah received the letter from the hand of the messengers and read it; then Hezekiah went up to the house of the LORD and spread it before the LORD. [15] And Hezekiah prayed to the LORD, saying: [16] "O LORD of hosts, God of Israel, who are enthroned above the cherubim, you are God, you alone, of all the kingdoms of the earth; you have made heaven and earth. [17] Incline your ear, O LORD, and hear; open your eyes, O LORD, and see; hear all the words of Sennacherib, which he has sent to mock the living God. [18] Truly, O LORD, the kings of Assyria have laid waste all the nations and their lands, [19] and have

hurled their gods into the fire, though they were no gods, but the work of human hands—wood and stone—and so they were destroyed. [20] So now, O LORD our God, save us from his hand, so that all the kingdoms of the earth may know that you alone are the LORD."

21 Then Isaiah son of Amoz sent to Hezekiah, saying: "Thus says the LORD, the God of Israel: Because you have prayed to me concerning King Sennacherib of Assyria, [22] this is the word that the LORD has spoken concerning him:

> She despises you, she scorns
> you—
> virgin daughter Zion;
> she tosses her head—behind
> your back,
> daughter Jerusalem.

[23] Whom have you mocked and
> reviled?
> Against whom have you raised
> your voice
> and haughtily lifted your eyes?
> Against the Holy One of Israel!
[24] By your servants you have
> mocked the Lord,
> and you have said, 'With my
> many chariots
> I have gone up the heights of the
> mountains,
> to the far recesses of Lebanon;
> I felled its tallest cedars,
> its choicest cypresses;
> I came to its remotest height,
> its densest forest.
[25] I dug wells
> and drank waters,

---

u Heb *he* *v* Or *Nubia*; Heb *Cush*

37.8–20: A second account of Sennacherib's challenge to Hezekiah. **8:** *Libnah,* ten miles north of Lachish. **9:** *Tirhakah,* 2 Kings 19.9. **12:** Places in Mesopotamia: *Gozan,* on west tributary of Khabur River, east of Haran; *Haran,* on upper Balikh River (Gen 11.27–32); *Rezeph,* between Nineveh and northern Khabur; *Eden,* Bit Adini, on middle Euphrates (Ezek 27.23); *Telassar,* on middle Euphrates.

37.14: *Letter,* i.e. scroll. **16:** *Enthroned . . .,* temple imagery (1 Kings 8.6–7; Ezek 1.4–28). **17–20:** This recalls Sennacherib's claim, vv. 11–13; *no gods,* Jer 10.1–16. **21:** See vv. 33–35. **22–29:** Isaiah's challenge to Sennacherib (10.5–19). **37.22–29:** Isaiah taunts Sennacherib, reminding him that he has defied God, who has determined history's course and will frustrate Sennacherib's plans. **27–28:** It is preferable

I dried up with the sole of my
    foot
  all the streams of Egypt.'

26 Have you not heard
    that I determined it long ago?
  I planned from days of old
    what now I bring to pass,
  that you should make fortified
    cities
    crash into heaps of ruins,
27 while their inhabitants, shorn of
    strength,
  are dismayed and confounded;
  they have become like plants of
    the field
  and like tender grass,
  like grass on the housetops,
    blighted*w* before it is grown.

28 I know your rising up*x* and your
    sitting down,
  your going out and coming in,
  and your raging against me.
29 Because you have raged against
    me
  and your arrogance has come to
    my ears,
  I will put my hook in your nose
  and my bit in your mouth;
  I will turn you back on the way
  by which you came.

30 "And this shall be the sign for you: This year eat what grows of itself, and in the second year what springs from that; then in the third year sow, reap, plant vineyards, and eat their fruit. 31 The surviving remnant of the house of Judah shall again take root downward, and bear fruit upward; 32 for from Jerusalem a remnant shall go out, and from Mount Zion a band of survivors. The zeal of the LORD of hosts will do this.

33 "Therefore thus says the LORD concerning the king of Assyria: He shall not come into this city, shoot an arrow there, come before it with a shield, or cast up a siege ramp against it. 34 By the way that he came, by the same he shall return; he shall not come into this city, says the LORD. 35 For I will defend this city to save it, for my own sake and for the sake of my servant David."

36 Then the angel of the LORD set out and struck down one hundred eighty-five thousand in the camp of the Assyrians; when morning dawned, they were all dead bodies. 37 Then King Sennacherib of Assyria left, went home, and lived at Nineveh. 38 As he was worshiping in the house of his god Nisroch, his sons Adrammelech and Sharezer killed him with the sword, and they escaped into the land of Ararat. His son Esar-haddon succeeded him.

38 In those days Hezekiah became sick and was at the point of death. The prophet Isaiah son of Amoz came to him, and said to him, "Thus says the LORD: Set your house in order, for you shall die; you shall not recover." 2 Then Hezekiah turned his face to the wall, and prayed to the LORD: 3 "Remember now, O LORD, I implore you, how I have walked before you in faithfulness with a whole heart, and have done what is good in your sight." And Hezekiah wept bitterly.

4 Then the word of the LORD came to Isaiah: 5 "Go and say to Hezekiah, Thus

---

w With 2 Kings 19.26: Heb *field*   *x* Q Ms Gk: MT lacks *your rising up*

---

to read with the Dead Sea Scroll of Isaiah, ". . . housetops which is parched by the east wind; your rising and your sitting down I know . . ."
37.30–32: A second assurance to Hezekiah (see vv. 5–7), of which the sign will be a return to normal conditions by the third year. These verses belong between v. 35 and v. 36. *Remnant,* see 10.22 n. **33–35:** A third word of assurance to Hezekiah (compare v. 21). Verse 35 recalls Nathan's words to David (2 Sam 7.12–17).
37.36–38: Devastation of Sennacherib's army. 36: *Angel,* a figure and explanation for a plague (Ex 12.29; 2 Sam 24.15–17).
38.1–22: Hezekiah's illness and recovery. See 2 Kings ch 20. **3:** The form of Hezekiah's prayer for recovery is found in the

says the LORD, the God of your ancestor David: I have heard your prayer, I have seen your tears; I will add fifteen years to your life. ⁶I will deliver you and this city out of the hand of the king of Assyria, and defend this city.

7 "This is the sign to you from the LORD, that the LORD will do this thing that he has promised: ⁸See, I will make the shadow cast by the declining sun on the dial of Ahaz turn back ten steps." So the sun turned back on the dial the ten steps by which it had declined. *ʸ*

9 A writing of King Hezekiah of Judah, after he had been sick and had recovered from his sickness:

10 I said: In the noontide of my days
I must depart;
I am consigned to the gates of
Sheol
for the rest of my years.
11 I said, I shall not see the LORD
in the land of the living;
I shall look upon mortals no more
among the inhabitants of
the world.
12 My dwelling is plucked up and
removed from me
like a shepherd's tent;
like a weaver I have rolled up
my life;
he cuts me off from the loom;
from day to night you bring me
to an end; *ʸ*
13   I cry for help *ᶻ* until morning;
like a lion he breaks all my bones;
from day to night you bring me
to an end. *ʸ*

14 Like a swallow or a crane *ʸ* I
clamor,
I moan like a dove.
My eyes are weary with looking
upward.
O Lord, I am oppressed; be my
security!
15 But what can I say? For he has
spoken to me,
and he himself has done it.
All my sleep has fled *ᵃ*
because of the bitterness of
my soul.

16 O Lord, by these things people
live,
and in all these is the life of
my spirit. *ʸ*
Oh, restore me to health and
make me live!
17 Surely it was for my welfare
that I had great bitterness;
but you have held back *ᵇ* my life
from the pit of destruction,
for you have cast all my sins
behind your back.
18 For Sheol cannot thank you,
death cannot praise you;
those who go down to the Pit
cannot hope
for your faithfulness.
19 The living, the living, they
thank you,
as I do this day;
fathers make known to children
your faithfulness.

ʸ Meaning of Heb uncertain   *z* Cn: Meaning of Heb uncertain   *a* Cn Compare Syr: Heb *I will walk slowly all my years*   *b* Cn Compare Gk Vg: Heb *loved*

Psalms (compare Ps 6). **8**: *Dial,* literally "stairs" or "steps." Following the text of one ancient Hebrew manuscript (among the Dead Sea Scrolls) one may translate: " '. . . Behold, I shall turn back the shadow of the steps down which the sun has moved on the steps of the roof chamber of Ahaz your father. I will cause the sun to move backwards ten steps.' And the sun moved back ten steps on the steps down which the shadow had descended." **38.9–20**: This song, traditionally ascribed

to Hezekiah, is a liturgical thanksgiving for use when presenting in the temple a thank-offering for personal deliverance (Ps 32; 1 Sam 2.1–10; Jon 2.2–9). **9**: *Writing,* Hebrew "miktab," read "Miktam," as in Pss 56–60. **38.16ab**: Corrupted text; emend to read, "O LORD, with thee are the days of my life, thine alone is the life of my spirit." **17**: God's saving power is shown in recovery from illness and forgiveness of sins (Lk 5.17–26). **21–22**: These verses belong between vv. 6–7.

20 The LORD will save me,
    and we will sing to stringed
        instruments[c]
all the days of our lives,
    at the house of the LORD.

21 Now Isaiah had said, "Let them take a lump of figs, and apply it to the boil, so that he may recover." 22 Hezekiah also had said, "What is the sign that I shall go up to the house of the LORD?"

39 At that time King Merodachbaladan son of Baladan of Babylon sent envoys with letters and a present to Hezekiah, for he heard that he had been sick and had recovered. 2 Hezekiah welcomed them; he showed them his treasure house, the silver, the gold, the spices, the precious oil, his whole armory, all that was found in his storehouses. There was nothing in his house or in all his realm that Hezekiah did not show them. 3 Then the prophet Isaiah came to King Hezekiah and said to him, "What did these men say? From where did they come to you?" Hezekiah answered, "They have come to me from a far country, from Babylon." 4 He said, "What have they seen in your house?" Hezekiah answered, "They have seen all that is in my house; there is nothing in my storehouses that I did not show them."

5 Then Isaiah said to Hezekiah, "Hear the word of the LORD of hosts: 6 Days are coming when all that is in your house, and that which your ancestors have stored up until this day, shall be carried to Babylon; nothing shall be left, says the LORD. 7 Some of your own sons who are born to you shall be taken away; they shall be eunuchs in the palace of the king of Babylon." 8 Then Hezekiah said to Isaiah, "The word of the LORD that you have spoken is good." For he thought, "There will be peace and security in my days."

40 Comfort, O comfort my
        people,
    says your God.
2 Speak tenderly to Jerusalem,
    and cry to her
that she has served her term,
    that her penalty is paid,
that she has received from the
        LORD's hand
    double for all her sins.

3 A voice cries out:
"In the wilderness prepare the way
        of the LORD,
    make straight in the desert a
        highway for our God.
4 Every valley shall be lifted up,
    and every mountain and hill be
        made low;
the uneven ground shall become
        level,
    and the rough places a plain.
5 Then the glory of the LORD shall
        be revealed,
    and all people shall see it
        together,
    for the mouth of the LORD has
        spoken."

c Heb *my stringed instruments*

39.1–8: Merodach-baladan's embassy. See 2 Kings 20.12–19. **2:** The tribute paid Sennacherib (701 B.C.) depleted Hezekiah's treasury (2 Kings 18.14–16), much of which had been derived from Arabian trade (2 Chr 32.27–29). **3–4:** Isaiah suspected intrigue with the Babylonian revolutionary. **5–7:** The description fits the events of 597 B.C.; see 2 Kings 24.10–17. **8:** With this postponement of punishment, compare 1 Kings 21.27–29.

Chs 40–55: Book of the Consolation of Israel. 40.1–11: The prophet is called to announce God's coming. The background scene is the council of heaven, from which the voices come. **1–2:** Introduction. *Comfort,* for the Exile is nearly over (for similar repetitions, see 51.9; 52.1; 57.14). *My people . . . your God,* covenant words (Ex 19.4–6; Jer 11.5). *Served her term,* a reference to the Exile. *Double* may reflect Ex 22.7–8, or suggest that God exacted more from his people than was to be expected. **3–5:** Quoted in Lk 3.4–6; see also Mt 3.3; Mk 1.3; Jn 1.23. *The way of the LORD* (common figure in Second Isaiah; 42.16; 43.16, 19; 48.17; 49.11; 51.10), who comes to restore his people (35.1–10). *The glory of the LORD* shall return (Ezek 1.28; 10.18–19; 43.1–5).

6 A voice says, "Cry out!"
  And I said, "What shall I cry?"
  All people are grass,
     their constancy is like the flower
        of the field.
7 The grass withers, the flower
     fades,
  when the breath of the LORD
     blows upon it;
  surely the people are grass.
8 The grass withers, the flower
     fades;
  but the word of our God will
     stand forever.
9 Get you up to a high mountain,
     O Zion, herald of good
        tidings;*d*
  lift up your voice with strength,
     O Jerusalem, herald of good
        tidings,*e*
  lift it up, do not fear;
  say to the cities of Judah,
     "Here is your God!"
10 See, the Lord GOD comes with
      might,
   and his arm rules for him;
   his reward is with him,
   and his recompense before him.
11 He will feed his flock like a
      shepherd;
   he will gather the lambs in
      his arms,
   and carry them in his bosom,
   and gently lead the mother
      sheep.

12 Who has measured the waters in
      the hollow of his hand
   and marked off the heavens
      with a span,
   enclosed the dust of the earth in
      a measure,
   and weighed the mountains
      in scales
   and the hills in a balance?

13 Who has directed the spirit of
      the LORD,
   or as his counselor has
      instructed him?
14 Whom did he consult for his
      enlightenment,
   and who taught him the path of
      justice?
   Who taught him knowledge,
   and showed him the way of
      understanding?
15 Even the nations are like a drop
      from a bucket,
   and are accounted as dust on
      the scales;
   see, he takes up the isles like
      fine dust.
16 Lebanon would not provide fuel
      enough,
   nor are its animals enough for a
      burnt offering.
17 All the nations are as nothing
      before him;
   they are accounted by him as
      less than nothing and
      emptiness.

18 To whom then will you liken
      God,
   or what likeness compare
      with him?
19 An idol? —A workman casts it,
   and a goldsmith overlays it
      with gold,
   and casts for it silver chains.
20 As a gift one chooses mulberry
      wood*f*
   —wood that will not rot—
   then seeks out a skilled artisan
   to set up an image that will
      not topple.

d Or *O herald of good tidings to Zion*
e Or *O herald of good tidings to Jerusalem*
f Meaning of Heb uncertain

40.6–8: The *voice* of God's herald proclaims the immutability of God's word (9.8; 55.8–11) in contrast to all living things. Quoted in 1 Pet 1.24–25. **6a:** Compare Jer 1.4–8. **6b:** See 51.12 n. **7:** *Breath,* wind (Jer 4.11). **9–11:** *"Here is your God"* who comes in celestial grandeur, yet is compassionate (Ezek ch 34). See Acts 10.36; Rom 10.15; Rev 22.7. *Jerusalem,* used for Judah over thirty times in chs 40–55. *Arm,* symbol of power (33.2; 48.14; 51.5, 9; 52.10; 53.1).

40.12–31: Creator of the universe. 12: *Waters* (seas), *heavens, earth*—the world's three divisions. **13–14:** God is the source of all

21 Have you not known? Have you
    not heard?
  Has it not been told you from
    the beginning?
  Have you not understood from
    the foundations of the
    earth?
22 It is he who sits above the circle
    of the earth,
  and its inhabitants are like
    grasshoppers;
  who stretches out the heavens like
    a curtain,
  and spreads them like a tent to
    live in;
23 who brings princes to naught,
  and makes the rulers of the
    earth as nothing.

24 Scarcely are they planted, scarcely
    sown,
  scarcely has their stem taken
    root in the earth,
  when he blows upon them, and
    they wither,
  and the tempest carries them off
    like stubble.

25 To whom then will you compare
    me,
  or who is my equal? says the
    Holy One.
26 Lift up your eyes on high and see:
  Who created these?
  He who brings out their host and
    numbers them,
  calling them all by name;
  because he is great in strength,
    mighty in power,
  not one is missing.

27 Why do you say, O Jacob,
    and speak, O Israel,
  "My way is hidden from the
    LORD,
  and my right is disregarded by
    my God"?
28 Have you not known? Have you
    not heard?
  The LORD is the everlasting God,
    the Creator of the ends of
    the earth.
  He does not faint or grow weary;
    his understanding is
    unsearchable.
29 He gives power to the faint,
  and strengthens the powerless.
30 Even youths will faint and
    be weary,
  and the young will fall
    exhausted;
31 but those who wait for the LORD
    shall renew their strength,
  they shall mount up with wings
    like eagles,
  they shall run and not be weary,
  they shall walk and not faint.

41 Listen to me in silence,
    O coastlands;
  let the peoples renew their
    strength;
  let them approach, then let
    them speak;
  let us together draw near for
    judgment.

2 Who has roused a victor from
    the east,
  summoned him to his service?
  He delivers up nations to him,
    and tramples kings under foot;

---

knowledge and wisdom (Prov 8.22–31; Job
38.2–39.30). **15–17:** Before him all nations
must bow. **18–20:** Idols cannot be compared
with him (see 42.17; 45.16, 20; Jer 10.1–16).
40.21–24: God, the creator, is also Lord of
history (44.24–28; 51.9–10). *Circle,* the horizon; *tent,* the vault of heaven (Prov 8.27; Job
22.14).
40.25–27: God is incomparable (vv. 18–
20) and omniscient. *Created* (Gen 1.1), a word
used more often by Second Isaiah than by
other Old Testament writers (v. 28; 41.20;
42.5; 43.7, 15; 45.7, 8, 12, 18; 54.16). **26:** *Host*

of heaven (3.1). **28–31:** The omnipotent God
is concerned for the people he has created. **31:**
Wait expresses confidence that God will not
desert his people; a common concept from
this period on (49.23; Ps 25.3; 33.20).
41.1–42.4: **The trial of the nations. 1:**
The background scene is the law court (a
recurrent theme in chs 41–46; 48). The historical background is the victories of Cyrus of
Persia. *Coastlands,* Mediterranean lands and
islands (frequent in Second Isaiah, 40.15;
42.4; 49.1; 51.5). **2–4:** *Victor from the east,*
Cyrus. The God of history is calling Cyrus.

he makes them like dust with
   his sword,
   like driven stubble with his
     bow.
3 He pursues them and passes
   on safely,
   scarcely touching the path with
     his feet.
4 Who has performed and done this,
   calling the generations from the
     beginning?
   I, the LORD, am first,
   and will be with the last.
5 The coastlands have seen and
   are afraid,
   the ends of the earth tremble;
   they have drawn near and come.
6 Each one helps the other,
   saying to one another, "Take
     courage!"
7 The artisan encourages the
     goldsmith,
   and the one who smooths with
     the hammer encourages the
     one who strikes the anvil,
   saying of the soldering, "It is
     good";
   and they fasten it with nails so
     that it cannot be moved.
8 But you, Israel, my servant,
   Jacob, whom I have chosen,
   the offspring of Abraham,
     my friend;
9 you whom I took from the ends
     of the earth,
   and called from its farthest
     corners,
   saying to you, "You are my
     servant,
   I have chosen you and not cast
     you off";
10 do not fear, for I am with you,

do not be afraid, for I am your
     God;
I will strengthen you, I will help
     you,
I will uphold you with my
     victorious right hand.

11 Yes, all who are incensed
     against you
   shall be ashamed and disgraced;
   those who strive against you
   shall be as nothing and
     shall perish.
12 You shall seek those who contend
     with you,
   but you shall not find them;
   those who war against you
   shall be as nothing at all.
13 For I, the LORD your God,
   hold your right hand;
   it is I who say to you, "Do not
     fear,
   I will help you."

14 Do not fear, you worm Jacob,
   you insect*g* Israel!
   I will help you, says the LORD;
   your Redeemer is the Holy One
     of Israel.
15 Now, I will make of you a
     threshing sledge,
   sharp, new, and having teeth;
   you shall thresh the mountains and
     crush them,
   and you shall make the hills
     like chaff.
16 You shall winnow them and the
     wind shall carry them
     away,
   and the tempest shall scatter
     them.

g Syr: Heb *men of*

First . . . last, 43.10; 44.6; Rev 22.13. **5–7:** The
nations are as powerless as their gods (40.19–
22).
   41.8–10: *Israel, my servant* (Jer 30.10;
46.27–28), though punished for faithlessness,
will now be restored (44.1–5; 45.4; 48.10).
There is a different emphasis on "servant" in
the Servant Songs (see 42.1). **11–13:** The nations are judged.

41.14–16: With God's help Israel can thresh
the mountains into dust. *Redeemer,* the Hebrew term elsewhere may refer to the blood
avenger (e.g. Num 35.19), but here it describes the avenger of the oppressed and liberator of the LORD's people (43.14; 44.6; 47.4).
Mountains, hills, suggests Mesopotamian ziggurats (see Gen 11.4 n.).
   41.17–20: A rhapsody on God as Lord of

Then you shall rejoice in the
LORD;
in the Holy One of Israel you
shall glory.

17 When the poor and needy seek
water,
and there is none,
and their tongue is parched
with thirst,
I the LORD will answer them,
I the God of Israel will not
forsake them.
18 I will open rivers on the bare
heights,^h
and fountains in the midst of
the valleys;
I will make the wilderness a pool
of water,
and the dry land springs of
water.
19 I will put in the wilderness
the cedar,
the acacia, the myrtle, and
the olive;
I will set in the desert the cypress,
the plane and the pine together,
20 so that all may see and know,
all may consider and
understand,
that the hand of the LORD has
done this,
the Holy One of Israel has
created it.

21 Set forth your case, says the LORD;
bring your proofs, says the
King of Jacob.
22 Let them bring them, and tell us
what is to happen.
Tell us the former things, what
they are,
so that we may consider them,

and that we may know their
outcome;
or declare to us the things
to come.
23 Tell us what is to come hereafter,
that we may know that you
are gods;
do good, or do harm,
that we may be afraid and
terrified.
24 You, indeed, are nothing
and your work is nothing at all;
whoever chooses you is an
abomination.

25 I stirred up one from the north,
and he has come,
from the rising of the sun he
was summoned by name.ⁱ
He shall trample^j on rulers as on
mortar,
as the potter treads clay.
26 Who declared it from the
beginning, so that we
might know,
and beforehand, so that we
might say, "He is right"?
There was no one who declared it,
none who proclaimed,
none who heard your words.
27 I first have declared it to Zion,^k
and I give to Jerusalem a herald
of good tidings.
28 But when I look there is no one;
among these there is no
counselor
who, when I ask, gives an
answer.
29 No, they are all a delusion;
their works are nothing;
their images are empty wind.

h Or *trails*   i Cn Compare Q Ms Gk: MT *and
he shall call on my name*   j Cn: Heb *come*
k Cn: Heb *First to Zion—Behold, behold them*

nature (55.13; Ezek 47.12). *Holy One,* v. 16;
40.25; 1.4. **21–24:** The pagan nations are challenged to prove the validity of their claims
and their gods. *Former things* (46.9) have a
present significance. The nations have no defense. Verses 23–24 contain the first explicit
statement against the very existence of the
gods of other nations (43.8–13).

41.25–29: Second (compare vv. 11–13)
judgment against the nations. *From the north
. . . the rising of the sun,* from Persia to the
northeast (the first phrase earlier referred to
Assyria [14.31], and Babylonia [Jer 6.22]).
The reference here is again to Cyrus (see
vv. 2–4 n.). **29:** Compare v. 24; 40.18–20.

42 Here is my servant, whom I
uphold,
my chosen, in whom my soul
delights;
I have put my spirit upon him;
he will bring forth justice to the
nations.
2 He will not cry or lift up his
voice,
or make it heard in the street;
3 a bruised reed he will not break,
and a dimly burning wick he
will not quench;
he will faithfully bring forth
justice.
4 He will not grow faint or be
crushed
until he has established justice in
the earth;
and the coastlands wait for his
teaching. *l*

5 Thus says God, the LORD,
who created the heavens and
stretched them out,
who spread out the earth and
what comes from it,
who gives breath to the people
upon it
and spirit to those who walk
in it:
6 I am the LORD, I have called you
in righteousness,
I have taken you by the hand
and kept you;
I have given you as a covenant to
the people, *l*
a light to the nations,
7 to open the eyes that are blind,
to bring out the prisoners from
the dungeon,

from the prison those who sit in
darkness.
8 I am the LORD, that is my name;
my glory I give to no other,
nor my praise to idols.
9 See, the former things have come
to pass,
and new things I now declare;
before they spring forth,
I tell you of them.

10 Sing to the LORD a new song,
his praise from the end of the
earth!
Let the sea roar *m* and all that
fills it,
the coastlands and their
inhabitants.
11 Let the desert and its towns lift up
their voice,
the villages that Kedar inhabits;
let the inhabitants of Sela sing
for joy,
let them shout from the tops of
the mountains.
12 Let them give glory to the LORD,
and declare his praise in the
coastlands.
13 The LORD goes forth like a soldier,
like a warrior he stirs up his
fury;
he cries out, he shouts aloud,
he shows himself mighty against
his foes.

14 For a long time I have held my
peace,

*l Meaning of Heb uncertain   m Cn Compare
Ps 96.11; 98.7: Heb Those who go down to the sea*

42.1–4: The first Servant Song (49.1–6;
50.4–11; 52.13–53.12). Scholars hold differing views of the identity of the Servant in
these Songs. The position taken here is that
the Servant is the nation; others regard him as
an individual, and some as both. **The mission of the Servant.** The Servant is Israel
(*my chosen;* not Cyrus) who will, with all patience (vv. 3–4a), bring God's *teaching* (1.10)
and restore *justice* (5.16) to the nations. *Spirit,*
11.2. See Mt 12.18–21.

42.5–17: God's glorious victory. 5: God
is creator of all (40.21–22) and source of life
(Gen 2.7; Acts 17.24–25). **6–9:** God illumines
history. **6–7:** He has called Israel, his covenant people, to bring *light* to the nations groping in the darkness of ignorance (60.1–3; Lk
2.30–32; Acts 13.47; 26.23). **8:** LORD, Yahweh, the only God (41.23–24; Deut 4.23–24).
10–13: Song of victory (Pss 93; 96; 149). *Kedar*
(Jer 49.28–29) and *Sela* (see Jer 49.20) represent remote and isolated places; all the world

I have kept still and restrained
myself;
now I will cry out like a woman
in labor,
I will gasp and pant.
15 I will lay waste mountains and
hills,
and dry up all their herbage;
I will turn the rivers into islands,
and dry up the pools.
16 I will lead the blind
by a road they do not know,
by paths they have not known
I will guide them.
I will turn the darkness before
them into light,
the rough places into level
ground.
These are the things I will do,
and I will not forsake them.
17 They shall be turned back and
utterly put to shame—
those who trust in carved
images,
who say to cast images,
"You are our gods."

18 Listen, you that are deaf;
and you that are blind, look up
and see!
19 Who is blind but my servant,
or deaf like my messenger
whom I send?
Who is blind like my dedicated
one,
or blind like the servant of
the LORD?
20 He sees many things, but does[n]
not observe them;
his ears are open, but he does
not hear.
21 The LORD was pleased, for the
sake of his righteousness,

to magnify his teaching and
make it glorious.
22 But this is a people robbed and
plundered,
all of them are trapped in holes
and hidden in prisons;
they have become a prey with no
one to rescue,
a spoil with no one to say,
"Restore!"
23 Who among you will give heed
to this,
who will attend and listen for
the time to come?
24 Who gave up Jacob to the spoiler,
and Israel to the robbers?
Was it not the LORD, against
whom we have sinned,
in whose ways they would
not walk,
and whose law they would
not obey?
25 So he poured upon him the heat
of his anger
and the fury of war;
it set him on fire all around, but
he did not understand;
it burned him, but he did not
take it to heart.

43 But now thus says the LORD,
he who created you, O Jacob,
he who formed you, O Israel:
Do not fear, for I have redeemed
you;
I have called you by name, you
are mine.
2 When you pass through the
waters, I will be with you;
and through the rivers, they
shall not overwhelm you;

n Heb *You see many things but do*

---

praises God, the victorious warrior (Ex 15.1–
18; Judg 5.2–5).
42.14–17: God's intervention in history.
Long time, from creation (v. 5) to redemption
(vv. 14–16). **15:** Jer 4.23–28. **16:** *Blind,* (vv. 6–
7) recalling God's leading Israel safely
through the wilderness (41.17–20; Ex 13.21–
22).
42.18–43.7: Israel, the blind and deaf

servant. 18–25: Israel's judgment. **19–21:** To
be *blind* to God's will and way is, for Second
Isaiah, Israel's chief sin (compare 6.9–10).
23–25: God has punished Israel for its sins.
43.1–7: Israel's redemption (41.8–13). **1–
3b:** God is Israel's creator (44.2, 21, 24) and
redeemer (41.14; 48.17; 49.7). *By name,* Israel
is God's unique possession (45.3–4; Ex 19.5;
33.17). Through all dangers, God is with his

when you walk through fire you
    shall not be burned,
    and the flame shall not
      consume you.
3 For I am the LORD your God,
    the Holy One of Israel, your
      Savior.
I give Egypt as your ransom,
    Ethiopia*o* and Seba in exchange
      for you.
4 Because you are precious in
    my sight,
    and honored, and I love you,
I give people in return for you,
    nations in exchange for your
      life.
5 Do not fear, for I am with you;
    I will bring your offspring from
      the east,
    and from the west I will
      gather you;
6 I will say to the north, "Give
    them up,"
    and to the south, "Do not
      withhold;
bring my sons from far away
    and my daughters from the end
      of the earth—
7 everyone who is called by my
    name,
    whom I created for my glory,
    whom I formed and made."

8 Bring forth the people who are
    blind, yet have eyes,
    who are deaf, yet have ears!
9 Let all the nations gather together,
    and let the peoples assemble.
Who among them declared this,
    and foretold to us the former
      things?

Let them bring their witnesses to
    justify them,
    and let them hear and say, "It
      is true."
10 You are my witnesses, says the
    LORD,
    and my servant whom I have
      chosen,
so that you may know and
    believe me
    and understand that I am he.
Before me no god was formed,
    nor shall there be any after me.
11 I, I am the LORD,
    and besides me there is no
      savior.
12 I declared and saved and
    proclaimed,
    when there was no strange god
      among you;
    and you are my witnesses, says
      the LORD.
13 I am God, and also henceforth I
    am He;
    there is no one who can deliver
      from my hand;
    I work and who can hinder it?

14 Thus says the LORD,
    your Redeemer, the Holy One
      of Israel:
For your sake I will send to
    Babylon
    and break down all the bars,
    and the shouting of the
      Chaldeans will be turned to
      lamentation.*p*
15 I am the LORD, your Holy One,
    the Creator of Israel, your King.

o Or *Nubia*; Heb *Cush*   *p* Meaning of Heb
uncertain

---

people (Ps 66.12). **3c–7**: God's ransom of
Israel includes the nations of Africa (Egypt
and Ethiopia) and Arabia (Seba), all of which
Cyrus was expected to conquer.
    43.8–13: **The servant Israel is the
LORD's witness. 8**: Israel is capable of conversion. **9**: Who can defend the nations'
claims for their gods? **11–13**: Besides God
there is and will be no other god (41.23–24;
48.5).
    43.14–44.5: **The redemption and resto-
ration of Israel. 43.14–15**: The apposition of

Redeemer (v. 1; 54.5; 59.20; Jer 50.34) and
Holy One (41.14; 47.4; 48.17) is noteworthy.
Second Isaiah refers to God as Redeemer
more often than do other Old Testament
writers. God redeems Israel not for its merits
but because of his covenant relationship with
Israel. Other nations have kings, but God is
Israel's *King* (1 Sam 8.4–9). **16–17**: Allusion
to the passage through the Red Sea (Ex 14–
15). The "new (and greater) Exodus" is an
important theme in Second Isaiah (41.17–20;
42.16; compare Ex 14–15). **18–19**: *New thing,*

16 Thus says the LORD,
    who makes a way in the sea,
    a path in the mighty waters,
17 who brings out chariot and horse,
    army and warrior;
they lie down, they cannot rise,
    they are extinguished, quenched
    like a wick:
18 Do not remember the former
    things,
    or consider the things of old.
19 I am about to do a new thing;
    now it springs forth, do you not
    perceive it?
I will make a way in the
    wilderness
    and rivers in the desert.
20 The wild animals will honor me,
    the jackals and the ostriches;
for I give water in the wilderness,
    rivers in the desert,
to give drink to my chosen
    people,
21   the people whom I formed
      for myself
so that they might declare
    my praise.

22 Yet you did not call upon me,
    O Jacob;
but you have been weary of me,
    O Israel!
23 You have not brought me your
    sheep for burnt offerings,
    or honored me with your
    sacrifices.
I have not burdened you with
    offerings,
    or wearied you with
    frankincense.
24 You have not bought me sweet
    cane with money,
    or satisfied me with the fat of
    your sacrifices.

But you have burdened me with
    your sins;
    you have wearied me with your
    iniquities.

25 I, I am He
    who blots out your
    transgressions for my own
    sake,
    and I will not remember
    your sins.
26 Accuse me, let us go to trial;
    set forth your case, so that you
    may be proved right.
27 Your first ancestor sinned,
    and your interpreters
    transgressed against me.
28 Therefore I profaned the princes of
    the sanctuary,
    I delivered Jacob to utter
    destruction,
    and Israel to reviling.

44 But now hear, O Jacob
    my servant,
Israel whom I have chosen!
2 Thus says the LORD who made
    you,
who formed you in the womb
    and will help you:
Do not fear, O Jacob my servant,
    Jeshurun whom I have chosen.
3 For I will pour water on the
    thirsty land,
    and streams on the dry ground;
I will pour my spirit upon your
    descendants,
    and my blessing on your
    offspring.
4 They shall spring up like a green
    tamarisk,
    like willows by flowing
    streams.

---

the return of Israel to Palestine. *A way in the
wilderness.* This theme (40.3) is a part of the
Exodus story (vv. 16–17). **20–21:** *Water in the
wilderness,* see Ex 17.1–7.
    43.22–24: Israel has ignored and offended
God. *Sweet cane,* Ex 30.23; Jer 6.20. **25–28:**
The imagery of a lawcourt (41.1); God shows

himself gracious, yet just. *First ancestor,* Jacob
(Gen ch 27; Hos 12.2–4). *Interpreters,* perhaps
other patriarchal figures or prophets (1 Kings
13.11–32).
    44.1–2: *Jeshurun* (Deut 32.15; 33.5, 26), a
poetic name of endearment for Israel; it may
mean "dearest upright one." **3–4:** Israel, like

5 This one will say, "I am the
LORD's,"
another will be called by the
name of Jacob,
yet another will write on the
hand, "The LORD's,"
and adopt the name of Israel.

6 Thus says the LORD, the King
of Israel,
and his Redeemer, the LORD
of hosts:
I am the first and I am the last;
besides me there is no god.
7 Who is like me? Let them
proclaim it,
let them declare and set it forth
before me.
Who has announced from of old
the things to come?*q*
Let them tell us*r* what is yet
to be.
8 Do not fear, or be afraid;
have I not told you from of old
and declared it?
You are my witnesses!
Is there any god besides me?
There is no other rock; I know
not one.

9 All who make idols are nothing,
and the things they delight in do not
profit; their witnesses neither see nor
know. And so they will be put to shame.
10 Who would fashion a god or cast an
image that can do no good? 11 Look, all
its devotees shall be put to shame; the
artisans too are merely human. Let them
all assemble, let them stand up; they shall
be terrified, they shall all be put to
shame.
12 The ironsmith fashions it*s* and
works it over the coals, shaping it with
hammers, and forging it with his strong
arm; he becomes hungry and his strength

fails, he drinks no water and is faint.
13 The carpenter stretches a line, marks it
out with a stylus, fashions it with planes,
and marks it with a compass; he makes
it in human form, with human beauty, to
be set up in a shrine. 14 He cuts down
cedars or chooses a holm tree or an oak
and lets it grow strong among the trees
of the forest. He plants a cedar and the
rain nourishes it. 15 Then it can be used as
fuel. Part of it he takes and warms himself; he kindles a fire and bakes bread.
Then he makes a god and worships it,
makes it a carved image and bows down
before it. 16 Half of it he burns in the fire;
over this half he roasts meat, eats it and
is satisfied. He also warms himself and
says, "Ah, I am warm, I can feel the fire!"
17 The rest of it he makes into a god, his
idol, bows down to it and worships it; he
prays to it and says, "Save me, for you
are my god!"

18 They do not know, nor do they
comprehend; for their eyes are shut, so
that they cannot see, and their minds as
well, so that they cannot understand.
19 No one considers, nor is there knowledge or discernment to say, "Half of it I
burned in the fire; I also baked bread on
its coals, I roasted meat and have eaten.
Now shall I make the rest of it an abomination? Shall I fall down before a block
of wood?" 20 He feeds on ashes; a deluded
mind has led him astray, and he cannot
save himself or say, "Is not this thing in
my right hand a fraud?"

21 Remember these things, O Jacob,
and Israel, for you are my
servant;
I formed you, you are my servant;

*q Cn: Heb from my placing an eternal people and
things to come   r Tg: Heb them   s Cn: Heb
an ax*

---

the world of nature, will be transformed.
God's *spirit* brings new life. **5:** All people will
identify themselves with Israel and Israel's
God. *Write on the hand,* compare Deut 6.8.
44.6–8: God's uniqueness (compare
vv. 21–23). **6:** Note the apposition of *Redeemer* (43.14; 60.16) and LORD *of hosts* (first of

several references in Second Isaiah; 45.13;
47.4). *First . . . last,* 41.4; Rev 1.8, 17; 2.8;
22.13. **7:** Perhaps a reference to God's promises to the patriarchs (Gen 12.1–3). **8:** *Rock,*
17.10; Deut 32.4, 18; Ps 18.2.
44.9–20: Satire against idolatry (compare 40.18–20; 41.6–7). **9–11:** An idol is not

O Israel, you will not be
forgotten by me.
22 I have swept away your
transgressions like a cloud,
and your sins like mist;
return to me, for I have redeemed
you.

23 Sing, O heavens, for the LORD has
done it;
shout, O depths of the earth;
break forth into singing,
O mountains,
O forest, and every tree in it!
For the LORD has redeemed Jacob,
and will be glorified in Israel.

24 Thus says the LORD, your
Redeemer,
who formed you in the womb:
I am the LORD, who made all
things,
who alone stretched out the
heavens,
who by myself spread out
the earth;
25 who frustrates the omens of liars,
and makes fools of diviners;
who turns back the wise,
and makes their knowledge
foolish;
26 who confirms the word of his
servant,
and fulfills the prediction of his
messengers;
who says of Jerusalem, "It shall be
inhabited,"
and of the cities of Judah, "They
shall be rebuilt,
and I will raise up their ruins";

27 who says to the deep, "Be dry—
I will dry up your rivers";
28 who says of Cyrus, "He is my
shepherd,
and he shall carry out all my
purpose";
and who says of Jerusalem, "It
shall be rebuilt,"
and of the temple, "Your
foundation shall be laid."

45 Thus says the LORD to his
anointed, to Cyrus,
whose right hand I have grasped
to subdue nations before him
and strip kings of their robes,
to open doors before him—
and the gates shall not be closed:
2 I will go before you
and level the mountains,[t]
I will break in pieces the doors
of bronze
and cut through the bars of
iron,
3 I will give you the treasures of
darkness
and riches hidden in secret
places,
so that you may know that it is I,
the LORD,
the God of Israel, who call you
by your name.
4 For the sake of my servant Jacob,
and Israel my chosen,
I call you by your name,
I surname you, though you do
not know me.

t Q Ms Gk: MT *the swellings*

---

a god but an image. **13–14:** It is made by
human artisans. **14–20:** The same tree is used
for fuel and for carving a god; Israel should
note this ridiculous inconsistency.
**44.21–23: Israel is forgiven and re-
deemed.** (A reversion to the theme of vv. 6–
8). **22:** Like the morning sun, God, the Redeemer (41.14; 63.9), removes the sin which
beclouds Israel. **23:** Let all creation praise the
LORD! (Compare Jer 51.48; Rev 12.12; 18.20.)
**44.24–45.13: The commission of Cy-
rus. 44.24–28:** An historical-prophetic prologue
summarizing the preceding poems, and coming to a climax in God's designating *Cyrus* to
be his *shepherd* (term for king, Jer 23.4).
45.1–7: Charge to Cyrus, God's *anointed*
(i.e. messiah), is universal in scope, just as
Israel's mission is to all nations. **1:** The only
Old Testament passage in which "messiah"
refers to a non-Israelite. **2–3:** God has commissioned Cyrus and will assist him in accomplishing his mission (40.4). **4:** *My servant
. . . my chosen,* 41.8; 42.1; 44.1; 49.3, 5, 6.
5–7: Cyrus is unaware of his charge from

5 I am the LORD, and there is
no other;
besides me there is no god.
I arm you, though you do not
know me,
6 so that they may know, from the
rising of the sun
and from the west, that there is
no one besides me;
I am the LORD, and there is
no other.
7 I form light and create darkness,
I make weal and create woe;
I the LORD do all these things.

8 Shower, O heavens, from above,
and let the skies rain down
righteousness;
let the earth open, that salvation
may spring up, *u*
and let it cause righteousness to
sprout up also;
I the LORD have created it.

9 Woe to you who strive with your
Maker,
earthen vessels with the potter! *v*
Does the clay say to the one who
fashions it, "What are you
making"?
or "Your work has no handles"?
10 Woe to anyone who says to a
father, "What are you
begetting?"
or to a woman, "With what are
you in labor?"
11 Thus says the LORD,
the Holy One of Israel, and its
Maker:
Will you question me *w* about my
children,

or command me concerning the
work of my hands?
12 I made the earth,
and created humankind upon it;
it was my hands that stretched out
the heavens,
and I commanded all their host.
13 I have aroused Cyrus *x* in
righteousness,
and I will make all his paths
straight;
he shall build my city
and set my exiles free,
not for price or reward,
says the LORD of hosts.
14 Thus says the LORD:
The wealth of Egypt and the
merchandise of Ethiopia, *y*
and the Sabeans, tall of stature,
shall come over to you and
be yours,
they shall follow you;
they shall come over in chains
and bow down to you.
They will make supplication to
you, saying,
"God is with you alone, and
there is no other;
there is no god besides him."
15 Truly, you are a God who hides
himself,
O God of Israel, the Savior.
16 All of them are put to shame and
confounded,
the makers of idols go in
confusion together.
17 But Israel is saved by the LORD
with everlasting salvation;

u Q Ms: MT *that they may bring forth salvation*
v Cn: Heb *with the potsherds,* or *with the potters*
w Cn: Heb *Ask me of things to come*
x Heb *him*   *y* Or *Nubia;* Heb *Cush*

God, who alone determines history's course
and will reveal his uniqueness (44.6–8) to all.
8: A song exulting in the extension of salvation to all.
45.9–13: The sovereign power of God.
9–10: This is the only invective in Second
Isaiah; it is directed against those questioning
the propriety of Cyrus's messiahship (29.16).
9: Rom 9.20. **11**: *Children, . . . the work of my
hands,* Cyrus and his allies. As Lord of cre-

ation and history, God's action should be
unquestioned. **12**: 40.12–31; 44.24.
45.14–25: **The conversion of the nations** (2.2–4; 42.1–4; 55.3–5; Jer 16.19–21).
14–15: The nations' wealth will pour into Israel; they will acknowledge Israel's God. Earlier the nations listed here served as Israel's
ransom (43.3). **14**: 1 Cor 14.25. **16–17:** Idol-
makers are condemned (44.9–20), but Israel is
saved.

you shall not be put to shame or
    confounded
    to all eternity.

18 For thus says the LORD,
    who created the heavens
        (he is God!),
    who formed the earth and made it
        (he established it;
    he did not create it a chaos,
        he formed it to be inhabited!):
    I am the LORD, and there is
        no other.
19 I did not speak in secret,
    in a land of darkness;
    I did not say to the offspring
        of Jacob,
    "Seek me in chaos."
    I the LORD speak the truth,
        I declare what is right.

20 Assemble yourselves and come
        together,
    draw near, you survivors of the
        nations!
    They have no knowledge—
        those who carry about their
            wooden idols,
    and keep on praying to a god
        that cannot save.
21 Declare and present your case;
    let them take counsel together!
    Who told this long ago?
    Who declared it of old?
    Was it not I, the LORD?
        There is no other god besides
            me,
    a righteous God and a Savior;
        there is no one besides me.

22 Turn to me and be saved,
    all the ends of the earth!
    For I am God, and there is
        no other.

23 By myself I have sworn,
    from my mouth has gone forth
        in righteousness
    a word that shall not return:
    "To me every knee shall bow,
        every tongue shall swear."

24 Only in the LORD, it shall be said
        of me,
    are righteousness and strength;
    all who were incensed against him
        shall come to him and be
            ashamed.
25 In the LORD all the offspring of
        Israel
    shall triumph and glory.

46 Bel bows down, Nebo stoops,
    their idols are on beasts
        and cattle;
    these things you carry are loaded
        as burdens on weary animals.
2 They stoop, they bow down
        together;
    they cannot save the burden,
    but themselves go into
        captivity.

3 Listen to me, O house of Jacob,
    all the remnant of the house of
        Israel,
    who have been borne by me from
        your birth,
    carried from the womb;
4 even to your old age I am he,
    even when you turn gray I will
        carry you.
    I have made, and I will bear;
    I will carry and will save.

5 To whom will you liken me and
        make me equal,
    and compare me, as though we
        were alike?

45.18–25: The Creator has revealed himself to Israel. *Chaos,* the state of the world before creation (Gen 1.2). **20–21:** The nations' gods are powerless (44.8; Acts 15.18). **22–25:** Let all nations bow before God and sing his praises! **23:** Compare Rom 14.11; Phil 2.10–11.

46.1–13: **The LORD supports Israel. 1–4:** *Bel,* chief god of Babylon, Bel-Marduk; *Nebo,* the Babylonian god Nabu, son of Marduk (Jer 50.2). The gods must be borne and are incapable of protecting their devotees (37.12–13), but God bears (44.2) and protects his devotees. **5–7:** 40.18–20. **7:** Perhaps a ref-

6 Those who lavish gold from the
        purse,
    and weigh out silver in the
        scales—
    they hire a goldsmith, who makes
        it into a god;
    then they fall down and
        worship!
7 They lift it to their shoulders, they
        carry it,
    they set it in its place, and it
        stands there;
    it cannot move from its place.
    If one cries out to it, it does
        not answer
    or save anyone from trouble.

8 Remember this and consider, z
    recall it to mind, you
        transgressors,
9   remember the former things
        of old;
    for I am God, and there is no
        other;
    I am God, and there is no one
        like me,
10 declaring the end from the
        beginning
    and from ancient times things
        not yet done,
    saying, "My purpose shall stand,
    and I will fulfill my intention,"
11 calling a bird of prey from the
        east,
    the man for my purpose from a
        far country.
    I have spoken, and I will bring it
        to pass;
    I have planned, and I will do it.

12 Listen to me, you stubborn of
        heart,
    you who are far from
        deliverance:

13 I bring near my deliverance, it is
        not far off,
    and my salvation will not tarry;
    I will put salvation in Zion,
    for Israel my glory.

47 Come down and sit in the
        dust,
    virgin daughter Babylon!
    Sit on the ground without a
        throne,
    daughter Chaldea!
    For you shall no more be called
        tender and delicate.
2 Take the millstones and grind
        meal,
    remove your veil,
    strip off your robe, uncover
        your legs,
    pass through the rivers.
3 Your nakedness shall be
        uncovered,
    and your shame shall be seen.
    I will take vengeance,
    and I will spare no one.
4 Our Redeemer—the LORD of hosts
        is his name—
    is the Holy One of Israel.

5 Sit in silence, and go into
        darkness,
    daughter Chaldea!
    For you shall no more be called
        the mistress of kingdoms.
6 I was angry with my people,
    I profaned my heritage;
    I gave them into your hand,
    you showed them no mercy;
    on the aged you made your yoke
        exceedingly heavy.
7 You said, "I shall be mistress
        forever,"

z Meaning of Heb uncertain

---

erence to New Year's festival processions in
Babylon (45.20). **8–11:** 45.9–13.
    46.8: *This,* Israel's history (44.21). **9:**
41.22–29; 42.8–9. **11:** *Bird of prey,* a reference
to Cyrus (41.2; 44.28). **12–13:** Only God can
save; unbelievers take note!
    **47.1–15: Lamentation over Babylon. 1–
4:** God strips Babylon of her royalty. **1:** *Virgin*
daughter, unconquered (used of Jerusalem,
37.22; Israel, Jer 31.4; Egypt, Jer 46.11). **2:**
Grind meal, do the work of slaves. **3:** Hos
2.9–12; Jer 13.20–27; Nah 3.5. **5–7:** Babylon's pride condemned. She wrongly attributed her strength to herself; it came from
God (Jer 27.6–7; 25.12–14).

so that you did not lay these
    things to heart
    or remember their end.

8 Now therefore hear this, you
    lover of pleasures,
    who sit securely,
  who say in your heart,
    "I am, and there is no one
      besides me;
  I shall not sit as a widow
    or know the loss of children"—
9 both these things shall come
    upon you
    in a moment, in one day:
  the loss of children and
    widowhood
  shall come upon you in full
    measure,
  in spite of your many sorceries
    and the great power of your
    enchantments.

10 You felt secure in your
    wickedness;
    you said, "No one sees me."
  Your wisdom and your
    knowledge
  led you astray,
  and you said in your heart,
    "I am, and there is no one
      besides me."
11 But evil shall come upon you,
    which you cannot charm away;
  disaster shall fall upon you,
    which you will not be able to
      ward off;
  and ruin shall come on you
    suddenly,
    of which you know nothing.

12 Stand fast in your enchantments
    and your many sorceries,
    with which you have labored
      from your youth;

perhaps you may be able to
    succeed,
  perhaps you may inspire terror.
13 You are wearied with your many
    consultations;
    let those who study*a* the
      heavens
  stand up and save you,
    those who gaze at the stars,
  and at each new moon predict
    what*b* shall befall you.

14 See, they are like stubble,
    the fire consumes them;
  they cannot deliver themselves
    from the power of the flame.
  No coal for warming oneself is
    this,
    no fire to sit before!
15 Such to you are those with whom
    you have labored,
  who have trafficked with you
    from your youth;
  they all wander about in their
    own paths;
    there is no one to save you.

48 Hear this, O house of Jacob,
    who are called by the name of
      Israel,
  and who came forth from the
    loins*c* of Judah;
  who swear by the name of the
    Lord,
  and invoke the God of Israel,
    but not in truth or right.
2 For they call themselves after the
    holy city,
    and lean on the God of Israel;
    the Lord of hosts is his name.

---

a Meaning of Heb uncertain   *b* Gk Syr
Compare Vg: Heb *from what*   *c* Cn: Heb
waters

47.8–9: Babylon's sense of security is illusory. **8:** *I am . . .* , an arrogant presumption of divinity (42.8; 44.6). **9:** Jer 10.17–22; 15.5–9; Rev 18.8. **10–11:** *Wisdom* and *knowledge* (e.g. divination, astrology, magic; vv. 12–13) fail to reveal her approaching fall (*you know nothing*). **14–15:** Like stubble in a fire, so Babylon will be consumed; both vassals and allies will desert her (Jer 51.58).

    48.1–22: **Hear and see.** This chapter summarizes the first section of Second Isaiah, which it concludes. It emphasizes the control of God over history and his action within it. **1–2:** The close relationship between God and

3 The former things I declared
  long ago,
   they went out from my mouth
    and I made them known;
   then suddenly I did them and
    they came to pass.
4 Because I know that you are
  obstinate,
   and your neck is an iron sinew
   and your forehead brass,
5 I declared them to you from
  long ago,
   before they came to pass I
    announced them to you,
  so that you would not say, "My
   idol did them,
   my carved image and my cast
    image commanded them."

6 You have heard; now see all this;
   and will you not declare it?
  From this time forward I make
   you hear new things,
   hidden things that you have not
    known.
7 They are created now, not long
  ago;
   before today you have never
    heard of them,
   so that you could not say, "I
    already knew them."
8 You have never heard, you have
   never known,
   from of old your ear has not
    been opened.
  For I knew that you would deal
   very treacherously,
   and that from birth you were
    called a rebel.

9 For my name's sake I defer my
   anger,
   for the sake of my praise I
    restrain it for you,
   so that I may not cut you off.
10 See, I have refined you, but not
   like*d* silver;
   I have tested you in the furnace
    of adversity.
11 For my own sake, for my own
   sake, I do it,
   for why should my name*e* be
    profaned?
   My glory I will not give to
    another.

12 Listen to me, O Jacob,
   and Israel, whom I called:
  I am He; I am the first,
   and I am the last.
13 My hand laid the foundation of
   the earth,
   and my right hand spread out
    the heavens;
  when I summon them,
   they stand at attention.

14 Assemble, all of you, and hear!
   Who among them has declared
    these things?
  The LORD loves him;
   he shall perform his purpose on
    Babylon,
   and his arm shall be against the
    Chaldeans.
15 I, even I, have spoken and
   called him,

d Cn: Heb *with*   *e* Gk Old Latin: Heb *for why
should it*

---

his undeserving people (*not in truth or right*).
3–5: God's prophets announced future events
to preclude Israel's wrongly ascribing them to
other forces (*idol; cast image*). *Former things,*
41.22–29; 43.9–12; 45.21. Israel's obstinacy is
frequently deplored (Ex 32.9; Deut 9.13; Jer
5.5; 7.26).
   48.6–8: *New things,* Israel's deliverance by
Cyrus (43.18–19). God's goodness is renewed daily, but Israel's unfaithfulness is almost axiomatic (Ezek 2.6–8; Deut 32.5). **9–
10:** The glory of God is the purpose of Israel's

salvation (Ezek 20.22). *Furnace,* the Exile; earlier used to refer to the Egyptian bondage
(Deut 4.20; Jer 11.4). **12–13:** *I am He . . . the
first . . . the last,* emphasizing God's oneness,
uniqueness, and eternity (43.10; 44.6; Rev
1.17; 2.8; 22.13).
   48.14–15: *You,* Israel; *them,* idols; *him,* Cyrus. **16:** *It,* the creation. **17:** *Teaches* and *leads,*
description of God's historic relationship
with Israel (55.12; Ex 13.18; Deut 29.5; Ps
27.11). **18–19:** Ps 81.13–16. *Sand,* Gen 22.17.
20–22: The prophet sees Israel's deliverance

I have brought him, and he will
prosper in his way.
16 Draw near to me, hear this!
From the beginning I have not
spoken in secret,
from the time it came to be I
have been there.
And now the Lord GOD has sent
me and his spirit.

17 Thus says the LORD,
your Redeemer, the Holy One
of Israel:
I am the LORD your God,
who teaches you for your
own good,
who leads you in the way you
should go.
18 O that you had paid attention to
my commandments!
Then your prosperity would
have been like a river,
and your success like the waves
of the sea;
19 your offspring would have been
like the sand,
and your descendants like its
grains;
their name would never be cut off
or destroyed from before me.

20 Go out from Babylon, flee from
Chaldea,
declare this with a shout of joy,
proclaim it,
send it forth to the end of the
earth;
say, "The LORD has redeemed
his servant Jacob!"
21 They did not thirst when he led
them through the deserts;
he made water flow for them
from the rock;
he split open the rock and the
water gushed out.

22 "There is no peace," says the
LORD, "for the wicked."

49 Listen to me, O coastlands,
pay attention, you peoples
from far away!
The LORD called me before I
was born,
while I was in my mother's
womb he named me.
2 He made my mouth like a sharp
sword,
in the shadow of his hand he
hid me;
he made me a polished arrow,
in his quiver he hid me away.
3 And he said to me, "You are my
servant,
Israel, in whom I will be
glorified."
4 But I said, "I have labored in vain,
I have spent my strength for
nothing and vanity;
yet surely my cause is with
the LORD,
and my reward with my God."

5 And now the LORD says,
who formed me in the womb to
be his servant,
to bring Jacob back to him,
and that Israel might be
gathered to him,
for I am honored in the sight of
the LORD,
and my God has become my
strength—
6 he says,
"It is too light a thing that you
should be my servant
to raise up the tribes of Jacob
and to restore the survivors of
Israel;
I will give you as a light to the
nations,

---

as the Exodus (Ex 17.1–7; Jer 31.10). **22:**
57.21.
49.1–6: The second Servant Song (see
42.1–4 n.). **1–3:** The servant (Israel) speaks. **1:**
Listen, 41.1. *Called . . . named*, Jer 1.5; Gal
1.15. **2:** *Sharp sword*, Eph 6.17; Heb 4.12. **3–4:**

In serving God, Israel will be glorified.
Though his ministry appears futile, his reward is in God (1 Kings 19.4–18; Jer 15.15–
21). **5–6:** The life and mission of old and new
Israel are contrasted (compare Acts 13.47;
26.23). *Light*, 42.6.

that my salvation may reach to
the end of the earth."

7 Thus says the LORD,
the Redeemer of Israel and his
Holy One,
to one deeply despised, abhorred
by the nations,
the slave of rulers,
"Kings shall see and stand up,
princes, and they shall prostrate
themselves,
because of the LORD, who is
faithful,
the Holy One of Israel, who has
chosen you."

8 Thus says the LORD:
In a time of favor I have answered
you,
on a day of salvation I have
helped you;
I have kept you and given you
as a covenant to the people,*f*
to establish the land,
to apportion the desolate
heritages;
9 saying to the prisoners, "Come
out,"
to those who are in darkness,
"Show yourselves."
They shall feed along the ways,
on all the bare heights*g* shall be
their pasture;
10 they shall not hunger or thirst,
neither scorching wind nor sun
shall strike them down,
for he who has pity on them will
lead them,
and by springs of water will
guide them.
11 And I will turn all my mountains
into a road,
and my highways shall be
raised up.

12 Lo, these shall come from far
away,
and lo, these from the north and
from the west,
and these from the land of
Syene.*h*

13 Sing for joy, O heavens, and
exult, O earth;
break forth, O mountains, into
singing!
For the LORD has comforted his
people,
and will have compassion on his
suffering ones.

14 But Zion said, "The LORD has
forsaken me,
my Lord has forgotten me."
15 Can a woman forget her nursing
child,
or show no compassion for the
child of her womb?
Even these may forget,
yet I will not forget you.
16 See, I have inscribed you on the
palms of my hands;
your walls are continually
before me.
17 Your builders outdo your
destroyers,*i*
and those who laid you waste
go away from you.
18 Lift up your eyes all around
and see;
they all gather, they come to
you.
As I live, says the LORD,
you shall put all of them on like
an ornament,

f Meaning of Heb uncertain   *g* Or *the trails*
h Q Ms: MT *Sinim*   *i* Or *Your children come
swiftly; your destroyers*

49.7–26: Return and restoration. 7:
The Servant of the nations is served by them
(some take this verse with vv. 1–6). God's
historic faithfulness is Israel's assurance. **8–
9b:** Though in bondage as in Egypt (48.10),
they will be released and restored (compare
2 Cor 6.2). **9c–11:** The imagery of the Exodus
(48.20–22) is combined with that of Israel's
deliverance (40.3–4, 11). **12:** *Syene,* see Ezek
29.10 n.

49.13: Hymn praising God, the Comforter
(44.23). **14–16:** God protests his love for Israel
(Hos 2.14–23; Jer 31.20). *Inscribed you,* as a
tattoo. **18:** As if studded with jewels (Jer

and like a bride you shall bind
them on.

19 Surely your waste and your
desolate places
and your devastated land—
surely now you will be too
crowded for your
inhabitants,
and those who swallowed you
up will be far away.
20 The children born in the time of
your bereavement
will yet say in your hearing:
"The place is too crowded for me;
make room for me to settle."
21 Then you will say in your heart,
"Who has borne me these?
I was bereaved and barren,
exiled and put away—
so who has reared these?
I was left all alone—
where then have these come
from?"

22 Thus says the Lord GOD:
I will soon lift up my hand to the
nations,
and raise my signal to the
peoples;
and they shall bring your sons in
their bosom,
and your daughters shall be
carried on their shoulders.
23 Kings shall be your foster fathers,
and their queens your nursing
mothers.
With their faces to the ground
they shall bow down to
you,
and lick the dust of your feet.

Then you will know that I am the
LORD;
those who wait for me shall not
be put to shame.

24 Can the prey be taken from the
mighty,
or the captives of a tyrant[j] be
rescued?
25 But thus says the LORD:
Even the captives of the mighty
shall be taken,
and the prey of the tyrant be
rescued;
for I will contend with those who
contend with you,
and I will save your children.
26 I will make your oppressors eat
their own flesh,
and they shall be drunk with
their own blood as with
wine.
Then all flesh shall know
that I am the LORD your Savior,
and your Redeemer, the Mighty
One of Jacob.

50 Thus says the LORD:
Where is your mother's bill of
divorce
with which I put her away?
Or which of my creditors is it
to whom I have sold you?
No, because of your sins you
were sold,
and for your transgressions your
mother was put away.
2 Why was no one there when I
came?

---

j Q Ms Syr Vg: MT *of a righteous person*

2.32), ruined Jerusalem will be repopulated.
20–21: She (Zion) who had no hope for children will have more inhabitants than room for them (Jer 31.15–17). *Children*, those born during the Exile.
    49.22–23: A *signal* (fire, 30.17; Jer 6.1) announces, not an invasion (5.26; 13.2; 18.3), but the beginning of Judah's restoration (11.10, 12; 62.10). Those who oppress God's people will become their servants. **24–26**: God will free helpless captive Israel (9.20). All

the earth will acknowledge Israel's God (1.24; 60.16).
    50.1–3: **Covenant, faithfulness, and judgment. 1**: Exiled Israel is neither divorced (Deut 24.1–4) nor irretrievably sold into slavery (52.3; Ex 21.7) but punished for her unfaithfulness (Hos 2.4–9; Jer ch 3; Ezek ch 16). **2–3**: When God appeared in his mighty acts or spoke through his prophets, no one responded.

Why did no one answer when I
  called?
Is my hand shortened, that it
  cannot redeem?
Or have I no power to deliver?
By my rebuke I dry up the sea,
  I make the rivers a desert;
their fish stink for lack of water,
  and die of thirst. *k*
3 I clothe the heavens with
    blackness,
  and make sackcloth their
    covering.

4 The Lord GOD has given me
    the tongue of a teacher, *l*
  that I may know how to sustain
    the weary with a word.
  Morning by morning he
    wakens—
  wakens my ear
    to listen as those who are
    taught.
5 The Lord GOD has opened my ear,
    and I was not rebellious,
  I did not turn backward.
6 I gave my back to those who
    struck me,
  and my cheeks to those who
    pulled out the beard;
  I did not hide my face
    from insult and spitting.

7 The Lord GOD helps me;
    therefore I have not been
    disgraced;
  therefore I have set my face
    like flint,

and I know that I shall not be
  put to shame;
8 he who vindicates me is near.
  Who will contend with me?
  Let us stand up together.
  Who are my adversaries?
  Let them confront me.
9 It is the Lord GOD who helps me;
    who will declare me guilty?
  All of them will wear out like a
    garment;
  the moth will eat them up.

10 Who among you fears the LORD
    and obeys the voice of his
    servant,
  who walks in darkness
    and has no light,
  yet trusts in the name of the LORD
    and relies upon his God?
11 But all of you are kindlers of fire,
    lighters of firebrands. *m*
  Walk in the flame of your fire,
    and among the brands that you
    have kindled!
  This is what you shall have from
    my hand:
  you shall lie down in torment.

51 Listen to me, you that pursue
    righteousness,
  you that seek the LORD.
  Look to the rock from which you
    were hewn,

k Or *die on the thirsty ground*   *l* Cn: Heb *of
those who are taught*   *m* Syr: Heb *you gird
yourselves with firebrands*

50.4–11: The third Servant Song (see
42.1–4 n.). **4–6:** Taught by God (Jer 1.4–10),
the Servant conscientiously brings God's
comfort to his fellow Israelites (the *weary*)
who treat him despicably (52.13–53.12).
Here the prophet may be identifying himself
with the Servant (Jer 11.18; compare Lk 2.32;
Acts 13.47; 26.23). **7–9:** Using law court terminology, the Servant expresses unshakable
confidence that God will vindicate him (Jer
1.18–19; 17.17–18; Ezek 3.7–11; Rom 8.33).
10–11: God leads his Servant safely through
the darkness of his faithless people's rejection;
they who walk by their own lights will be
punished.

**51.1–16: Salvation for Abraham's chil-
dren. 1–8:** Past revelation and future salvation. **1–3:** As with his promises to Abraham
(Gen 12.1–3), God will fulfill his promised
salvation of Zion (49.20–21). **1:** *Rock, quarry,*
symbols of Israel's solidarity. **2:** The only reference to *Sarah* in the Old Testament outside
of Genesis. **3:** *Eden . . . garden of the LORD;* the
symbolism of creation appears often in biblical pictures of the new age. The end-time will
be a return to the ideal conditions in Eden (see
Ezek 36.35; 47.1–12). **4–6:** Salvation is imminent. **4:** *Teaching* (1.10) and *justice* (42.4) recall
First Isaiah (1.21; 5.7). **5:** *Arms,* 40.10. **6:** Mt
24.35.

and to the quarry from which
you were dug.
2 Look to Abraham your father
and to Sarah who bore you;
for he was but one when I called
him,
but I blessed him and made
him many.
3 For the LORD will comfort Zion;
he will comfort all her waste
places,
and will make her wilderness
like Eden,
her desert like the garden of
the LORD;
joy and gladness will be found
in her,
thanksgiving and the voice
of song.

4 Listen to me, my people,
and give heed to me, my nation;
for a teaching will go out from
me,
and my justice for a light to
the peoples.
5 I will bring near my deliverance
swiftly,
my salvation has gone out
and my arms will rule the
peoples;
the coastlands wait for me,
and for my arm they hope.
6 Lift up your eyes to the heavens,
and look at the earth beneath;
for the heavens will vanish like
smoke,
the earth will wear out like a
garment,
and those who live on it will die
like gnats;[n]
but my salvation will be forever,
and my deliverance will never
be ended.

7 Listen to me, you who know
righteousness,
you people who have my
teaching in your hearts;
do not fear the reproach of others,
and do not be dismayed when
they revile you.
8 For the moth will eat them up like
a garment,
and the worm will eat them
like wool;
but my deliverance will be
forever,
and my salvation to all
generations.

9 Awake, awake, put on strength,
O arm of the LORD!
Awake, as in days of old,
the generations of long ago!
Was it not you who cut Rahab in
pieces,
who pierced the dragon?
10 Was it not you who dried up
the sea,
the waters of the great deep;
who made the depths of the sea
a way
for the redeemed to cross over?
11 So the ransomed of the LORD shall
return,
and come to Zion with singing;
everlasting joy shall be upon their
heads;
they shall obtain joy and
gladness,
and sorrow and sighing shall
flee away.

12 I, I am he who comforts you;
why then are you afraid of a
mere mortal who must die,
a human being who fades like
grass?

n Or *in like manner*

51.9–16: Prayer of lament and assuring answer (for pattern of vv. 1–8, 9–16, compare 26.8–14, 16–21; 49.14–23, 24–26; Jer 15.10, 15–21). 9–11: Plea for God's intervention, using the Hebrew version of the primeval combat between God and chaos-monsters (*Rahab*, 30.7; *dragon* [Hebrew, "tannin"], 27.1; Ezek 29.3; *sea* [Hebrew, "yam"], Job 38.8; *deep* [Hebrew, "tehom"], Ezek 26.19; Job 38.16). 12–14: What mortal can compete with the

13 You have forgotten the LORD,
  your Maker,
   who stretched out the heavens
   and laid the foundations of
     the earth.
  You fear continually all day long
   because of the fury of the
     oppressor,
  who is bent on destruction.
   But where is the fury of the
     oppressor?
14 The oppressed shall speedily be
  released;
   they shall not die and go down
     to the Pit,
   nor shall they lack bread.
15 For I am the LORD your God,
   who stirs up the sea so that its
     waves roar—
   the LORD of hosts is his name.
16 I have put my words in your
  mouth,
   and hidden you in the shadow
     of my hand,
  stretching out*o* the heavens
   and laying the foundations of
     the earth,
   and saying to Zion, "You are
     my people."

17 Rouse yourself, rouse yourself!
   Stand up, O Jerusalem,
  you who have drunk at the hand
     of the LORD
   the cup of his wrath,
  who have drunk to the dregs
   the bowl of staggering.
18 There is no one to guide her
   among all the children she
     has borne;
  there is no one to take her by
     the hand
   among all the children she has
     brought up.

19 These two things have befallen
     you
   —who will grieve with you?—
  devastation and destruction,
     famine and sword—
   who will comfort you?*p*
20 Your children have fainted,
   they lie at the head of every
     street
  like an antelope in a net;
  they are full of the wrath of
     the LORD,
   the rebuke of your God.

21 Therefore hear this, you who are
     wounded,*q*
   who are drunk, but not
     with wine:
22 Thus says your Sovereign, the
     LORD,
   your God who pleads the cause
     of his people:
  See, I have taken from your hand
     the cup of staggering;
  you shall drink no more
   from the bowl of my wrath.
23 And I will put it into the hand of
     your tormentors,
   who have said to you,
  "Bow down, that we may walk
     on you";
  and you have made your back like
     the ground
   and like the street for them to
     walk on.

52 Awake, awake,
  put on your strength, O Zion!
 Put on your beautiful garments,
   O Jerusalem, the holy city;
 for the uncircumcised and the
     unclean

o Syr: Heb *planting*   *p* Q Ms Gk Syr Vg: MT
how may I comfort you?   *q* Or *humbled*

---

omnipotent and eternal God (Ps 90.5)? *Oppressed*, as prisoners. *Pit*, Sheol (14.9–11). **15–16**: *Your God . . . my people* reemphasizes the covenant relationship (43.1). **15**: Jer 31.35.
   51.17–52.12: God's kingship. 17–20: Jerusalem has drunk deeply of *the cup of* God's *wrath* (Jer 25.15–31) and lies exhausted,

depopulated, and destroyed. **21–23**: But now her oppressors will drink of it (41.1–42.4; 45.20–25).
   52.1–2: Jerusalem is bidden to arise (51.17). *The uncircumcised and the unclean* refers to foreigners (33.19) and ritually unclean Jews. **3–6**: A prose interpolation. *Name* is

shall enter you no more.
2 Shake yourself from the dust,
    rise up,
O captive[r] Jerusalem;
loose the bonds from your neck,
    O captive daughter Zion!

3 For thus says the LORD: You were
sold for nothing, and you shall be redeemed without money. 4 For thus says
the Lord GOD: Long ago, my people
went down into Egypt to reside there as
aliens; the Assyrian, too, has oppressed
them without cause. 5 Now therefore
what am I doing here, says the LORD,
seeing that my people are taken away
without cause? Their rulers howl, says
the LORD, and continually, all day long,
my name is despised. 6 Therefore my
people shall know my name; therefore in
that day they shall know that it is I who
speak; here am I.

7 How beautiful upon the
    mountains
  are the feet of the messenger
    who announces peace,
who brings good news,
  who announces salvation,
  who says to Zion, "Your God
    reigns."
8 Listen! Your sentinels lift up
    their voices,
  together they sing for joy;
for in plain sight they see
    the return of the LORD to Zion.
9 Break forth together into singing,
  you ruins of Jerusalem;
for the LORD has comforted his
    people,
  he has redeemed Jerusalem.
10 The LORD has bared his holy arm

before the eyes of all the
    nations;
and all the ends of the earth
    shall see
  the salvation of our God.

11 Depart, depart, go out from there!
  Touch no unclean thing;
go out from the midst of it, purify
    yourselves,
  you who carry the vessels of
    the LORD.
12 For you shall not go out in haste,
  and you shall not go in flight;
for the LORD will go before you,
  and the God of Israel will be
    your rear guard.

13 See, my servant shall prosper;
  he shall be exalted and lifted up,
  and shall be very high.
14 Just as there were many who were
    astonished at him[s]
  —so marred was his
    appearance, beyond human
    semblance,
  and his form beyond that of
    mortals—
15 so he shall startle[t] many nations;
  kings shall shut their mouths
    because of him;
for that which had not been told
    them they shall see,
  and that which they had not
    heard they shall
    contemplate.

53 Who has believed what we
    have heard?
  And to whom has the arm of

r Cn: Heb *rise up, sit*   s Syr Tg: Heb *you*
t Meaning of Heb uncertain

more than appellation; it involves the person
herself or himself (compare Ex 20.7). God
will not allow himself to be viewed with contempt by his people's oppressors: Egypt
(Ezek 29.9), Assyria (10.7–11), Babylon (Jer
50.29).
    52.7–8: Tensely, all creation awaits word
of God's decisive victory (Ps 125.2; 2 Sam
18.25–27; Nah 1.15; Rom 10.15). The *sentinels* see the victor returning (40.9). **11–12:**

Ritually clean, people and priests return home
in peace (Ex 13.21–22).
    52.13–53.12: The fourth Servant Song
(see 42.1–4 n.). **52.13–15:** God will exalt his
cruelly disfigured Servant (taken here to be
Israel) to the numbed astonishment of the
world's rulers (49.7, 23). **15:** Rom 15.21.
    53.1–3: A report beginning with rhetori-
cal questions (40.12; 50.8–10). The Servant's
background and appearance (52.14) are un-

the LORD been revealed?

2 For he grew up before him like a
young plant,
and like a root out of dry
ground;
he had no form or majesty that
we should look at him,
nothing in his appearance that
we should desire him.

3 He was despised and rejected
by others;
a man of suffering *u* and
acquainted with infirmity;
and as one from whom others
hide their faces *v*
he was despised, and we held
him of no account.

4 Surely he has borne our infirmities
and carried our diseases;
yet we accounted him stricken,
struck down by God, and
afflicted.

5 But he was wounded for our
transgressions,
crushed for our iniquities;
upon him was the punishment that
made us whole,
and by his bruises we are
healed.

6 All we like sheep have gone
astray;
we have all turned to our
own way,
and the LORD has laid on him
the iniquity of us all.

7 He was oppressed, and he was
afflicted,
yet he did not open his mouth;
like a lamb that is led to the
slaughter,

and like a sheep that before its
shearers is silent,
so he did not open his mouth.

8 By a perversion of justice he was
taken away.
Who could have imagined his
future?
For he was cut off from the land
of the living,
stricken for the transgression of
my people.

9 They made his grave with the
wicked
and his tomb *w* with the rich, *x*
although he had done no violence,
and there was no deceit in
his mouth.

10 Yet it was the will of the LORD to
crush him with pain. *y*
When you make his life an
offering for sin, *z*
he shall see his offspring, and
shall prolong his days;
through him the will of the LORD
shall prosper.

11 Out of his anguish he shall
see light; *a*
he shall find satisfaction through
his knowledge.
The righteous one, *b* my
servant, shall make many
righteous,
and he shall bear their iniquities.

12 Therefore I will allot him a
portion with the great,

u Or *a man of sorrows* v Or *as one who hides
his face from us* w Q Ms: MT *and in his death*
x Cn: Heb *with a rich person* y Or *by disease*;
meaning of Heb uncertain z Meaning of
Heb uncertain a Q Mss: MT lacks *light*
b Or *and he shall find satisfaction. Through his
knowledge, the righteous one*

distinguished; his person, rejected. **1:** John
12.38. **2:** *Young plant,* compare 11.1; *root,*
11.10, compare Jer 23.5; these are sometimes
considered Messianic allusions. **3:** Like a leper, he suffers painful loneliness and rejection
by the community (Job 19.13–19). **4–6:** By
the Servant's vicarious suffering, he restores
all people to God (Mt 8.17; 1 Pet 2.24–25).
Whole, general well-being ("peace," 48.18).
53.7–9: Unlike Jeremiah (11.18–12.6) or
Job, the Servant suffers silently. He is unjustly condemned (compare Lk 22.37), executed,
and ignominiously buried (compare Mt
27.57–60). See Acts 8.32–33. **9b:** Some
emend to read, "and his tomb with evildoers"
(others read "with demons [satyrs]"). **10–12:**
The Servant's suffering manifests God's
judgment (against sin) and mercy (upon sinful human beings). The Servant brings blessings to many. The poem is taken to describe

and he shall divide the spoil
with the strong;
because he poured out himself
to death,
and was numbered with the
transgressors;
yet he bore the sin of many,
and made intercession for the
transgressors.

54

Sing, O barren one who did
not bear;
burst into song and shout,
you who have not been in
labor!
For the children of the desolate
woman will be more
than the children of her that is
married, says the LORD.
2 Enlarge the site of your tent,
and let the curtains of your
habitations be stretched out;
do not hold back; lengthen
your cords
and strengthen your stakes.
3 For you will spread out to the
right and to the left,
and your descendants will
possess the nations
and will settle the desolate
towns.

4 Do not fear, for you will not be
ashamed;
do not be discouraged, for you
will not suffer disgrace;
for you will forget the shame of
your youth,
and the disgrace of your
widowhood you will
remember no more.
5 For your Maker is your husband,
the LORD of hosts is his name;

the Holy One of Israel is your
Redeemer,
the God of the whole earth he
is called.
6 For the LORD has called you
like a wife forsaken and grieved
in spirit,
like the wife of a man's youth
when she is cast off,
says your God.
7 For a brief moment I abandoned
you,
but with great compassion I will
gather you.
8 In overflowing wrath for a
moment
I hid my face from you,
but with everlasting love I will
have compassion on you,
says the LORD, your Redeemer.

9 This is like the days of Noah to
me:
Just as I swore that the waters
of Noah
would never again go over
the earth,
so I have sworn that I will not be
angry with you
and will not rebuke you.
10 For the mountains may depart
and the hills be removed,
but my steadfast love shall not
depart from you,
and my covenant of peace shall
not be removed,
says the LORD, who has
compassion on you.

11 O afflicted one, storm-tossed, and
not comforted,
I am about to set your stones in
antimony,

---

the purpose of God's people, the covenant community. **54.1–17: Song of assurance to Israel. 1–3**: 49.14–21. **1**: *Barren,* exilic Zion, deserted by God, her "husband" (Ezek ch 16); *married,* pre-exilic Zion (62.4; Gal 4.27). **4–8**: God, the faithful husband. **4**: *Shame . . . youth,* pre-exilic infidelity; *widowhood,* the Exile. **6**: God will not irrevocably reject his "wife" (Hos

2.19; 11.8–9). **7**: Reconciliation is imminent (Ps 27.10; 2 Cor 4.17–18). **8**: God's *everlasting love* for his people (43.4; Jer 31.3; a parent's love, 49.14–16; Jer 31.20; a husband's love, Jer 2.2; Ezek 16.8) is unmerited and unwaveringly faithful.
54.9–10: The everlasting covenant (Jer 32.40). **9**: Gen 8.21–22; 9.11–17. **10**: Jer 31.35–36; Mk 13.31; *covenant of peace,* 48.18;

and lay your foundations with
sapphires. *c*
12 I will make your pinnacles of
rubies,
your gates of jewels,
and all your wall of precious
stones.
13 All your children shall be taught
by the LORD,
and great shall be the prosperity
of your children.
14 In righteousness you shall be
established;
you shall be far from
oppression, for you shall
not fear;
and from terror, for it shall not
come near you.
15 If anyone stirs up strife,
it is not from me;
whoever stirs up strife with you
shall fall because of you.
16 See it is I who have created
the smith
who blows the fire of coals,
and produces a weapon fit for
its purpose;
I have also created the ravager
to destroy.
17 No weapon that is fashioned
against you shall prosper,
and you shall confute every
tongue that rises against
you in judgment.
This is the heritage of the servants
of the LORD
and their vindication from me,
says the LORD.

55 Ho, everyone who thirsts,
come to the waters;
and you that have no money,
come, buy and eat!

Come, buy wine and milk
without money and without
price.
2 Why do you spend your money
for that which is not bread,
and your labor for that which
does not satisfy?
Listen carefully to me, and eat
what is good,
and delight yourselves in
rich food.
3 Incline your ear, and come to me;
listen, so that you may live.
I will make with you an
everlasting covenant,
my steadfast, sure love for
David.
4 See, I made him a witness to
the peoples,
a leader and commander for
the peoples.
5 See, you shall call nations that you
do not know,
and nations that do not know
you shall run to you,
because of the LORD your God,
the Holy One of Israel,
for he has glorified you.

6 Seek the LORD while he may
be found,
call upon him while he is near;
7 let the wicked forsake their way,
and the unrighteous their
thoughts;
let them return to the LORD, that
he may have mercy on
them,
and to our God, for he will
abundantly pardon.

c Or *lapis lazuli*

Ezek 34.25. **11–14:** New Jerusalem (*afflicted one,* compare 51.21) is Paradise restored (Ezek 28.13–19; Jn 6.45; Rev 21.19). **15;** Commentary on v. 14. **17:** No one can assault God's *servants* with impunity (65.13–15).
55.1–13: A hymn of joy and triumph, celebrating the approaching consummation of Israel's restoration. This concludes the second section of Second Isaiah (see 48.1–22 n.).

1–2: Reminiscent of wisdom's invitation to a banquet (Prov 9.3–6; compare Jn 7.37); God's grace cannot be purchased. **3–5:** God's promise to David (2 Sam 7.4–17) continues in the *everlasting covenant* (54.9–10; Jer 33.19–26; compare Jer 31.31–34). Emphasis is on Israel's messianism, not Davidic messianism.
55.6–9: A call to repentance (Jer 29.12–14; Mt 3.2) and trust in God's inscrutable grace

8 For my thoughts are not your
   thoughts,
   nor are your ways my ways,
   says the LORD.
9 For as the heavens are higher than
   the earth,
   so are my ways higher than
   your ways
   and my thoughts than your
   thoughts.

10 For as the rain and the snow come
   down from heaven,
   and do not return there until
   they have watered the
   earth,
   making it bring forth and sprout,
   giving seed to the sower and
   bread to the eater,
11 so shall my word be that goes out
   from my mouth;
   it shall not return to me empty,
   but it shall accomplish that which
   I purpose,
   and succeed in the thing for
   which I sent it.

12 For you shall go out in joy,
   and be led back in peace;
   the mountains and the hills
   before you
   shall burst into song,
   and all the trees of the field shall
   clap their hands.
13 Instead of the thorn shall come up
   the cypress;
   instead of the brier shall come
   up the myrtle;
   and it shall be to the LORD for a
   memorial,
   for an everlasting sign that shall
   not be cut off.

56 Thus says the LORD:
   Maintain justice, and do what
   is right,
   for soon my salvation will come,
   and my deliverance be revealed.

2 Happy is the mortal who does
   this,
   the one who holds it fast,
   who keeps the sabbath, not
   profaning it,
   and refrains from doing any
   evil.

3 Do not let the foreigner joined to
   the LORD say,
   "The LORD will surely separate
   me from his people";
   and do not let the eunuch say,
   "I am just a dry tree."
4 For thus says the LORD:
   To the eunuchs who keep my
   sabbaths,
   who choose the things that
   please me
   and hold fast my covenant,
5 I will give, in my house and
   within my walls,
   a monument and a name
   better than sons and daughters;
   I will give them an everlasting
   name
   that shall not be cut off.

6 And the foreigners who join
   themselves to the LORD,
   to minister to him, to love the
   name of the LORD,
   and to be his servants,
   all who keep the sabbath, and do
   not profane it,
   and hold fast my covenant—

(Ps 103.11; Rom 11.33–36). **10–11**: As *rain* causes germination and ultimately provides sustenance, so does God's word (see 9.8 n.). **12–13**: The new Exodus (compare 43.16–21; 49.9–11) into an Eden-like land (see 51.3 n.; 41.18, 19; 44.3–4).

56.1–66.24: Miscellaneous post-restoration oracles (after 538 B.C. This section is sometimes called Third Isaiah; see Introduction, p. 866 OT).

56.1–8: A blessing on all who keep the sabbath (compare 58.13–14; Jer 17.19–27). **2:** *Happy is the mortal,* compare Ps 1.1; Jer 17.7; Mt 5.2–12. **3:** *Foreigner,* the reference is to the proselyte (see v. 6). **4–5:** God himself will honor faithful *eunuchs.* **6–7:** Faithful proselytes will present acceptable sacrifices; *house of prayer,* Mk 11.17; *for all peoples,* 60.1–14. **8:** God's community includes Israel and proselytes.

7 these I will bring to my holy
mountain,
and make them joyful in my
house of prayer;
their burnt offerings and their
sacrifices
will be accepted on my altar;
for my house shall be called a
house of prayer
for all peoples.
8 Thus says the Lord God,
who gathers the outcasts of
Israel,
I will gather others to them
besides those already gathered. *d*

9 All you wild animals,
all you wild animals in the
forest, come to devour!
10 Israel's *e* sentinels are blind,
they are all without knowledge;
they are all silent dogs
that cannot bark;
dreaming, lying down,
loving to slumber.
11 The dogs have a mighty appetite;
they never have enough.
The shepherds also have no
understanding;
they have all turned to their
own way,
to their own gain, one and all.
12 "Come," they say, "let us *f*
get wine;
let us fill ourselves with
strong drink.
And tomorrow will be like today,
great beyond measure."

57 The righteous perish,
and no one takes it to heart;
the devout are taken away,

while no one understands.
For the righteous are taken away
from calamity,
2 and they enter into peace;
those who walk uprightly
will rest on their couches.
3 But as for you, come here,
you children of a sorceress,
you offspring of an adulterer
and a whore. *g*
4 Whom are you mocking?
Against whom do you open
your mouth wide
and stick out your tongue?
Are you not children of
transgression,
the offspring of deceit—
5 you that burn with lust among
the oaks,
under every green tree;
you that slaughter your children in
the valleys,
under the clefts of the rocks?
6 Among the smooth stones of the
valley is your portion;
they, they, are your lot;
to them you have poured out a
drink offering,
you have brought a grain
offering.
Shall I be appeased for these
things?
7 Upon a high and lofty mountain
you have set your bed,
and there you went up to offer
sacrifice.
8 Behind the door and the doorpost
you have set up your symbol;

d Heb *besides his gathered ones*   *e* Heb *His*
f Q Ms Syr Vg Tg: MT *me*   *g* Heb *an
adulterer and she plays the whore*

56.9–12: **Against corrupt leaders. 9:**
Wild animals, nations (Jer 12.8–9; Ezek 39.17).
10: *Sentinels,* prophets (Ezek 3.17; 33.7). **11:**
Shepherds, rulers (Ezek 34.1; Zech 11.4–17).
57.1–13: Against idolatry. 1–2: The *righteous* die unnoticed by their hedonistic religious leaders. *Couches,* graves. **3–4:** *You,* i.e.
Jerusalem's apostates ("illegitimate children")
who abuse the righteous ("legitimate children"; (Jer 3.1–20; Ezek 16.1–63). **5–6:** The

prophet claims that the old fertility cults persist. *Oaks,* see 6.13 n.; *slaughter your children,*
Jer 19.5; *portion, lot,* the gods of the valleys
(v. 6), mountains (v. 7), house (v. 8), and
other shrines (vv. 9–10), rather than the true
God (Jer 10.16). **7–8:** Sexual immorality was
characteristic of these cults. **9–10:** *Molech,* see
Jer 7.31 n.; *to Sheol,* to the gods of the underworld (14.9–11). All such practices are futile
(Jer 2.25).

for, in deserting me, [h] you have
　　uncovered your bed,
　you have gone up to it,
　you have made it wide;
　and you have made a bargain for
　　yourself with them,
　you have loved their bed,
　you have gazed on their
　　nakedness. [i]

9 You journeyed to Molech [j] with
　　oil,
　and multiplied your perfumes;
　you sent your envoys far away,
　and sent down even to Sheol.

10 You grew weary from your many
　　wanderings,
　but you did not say, "It is
　　useless."
　You found your desire rekindled,
　and so you did not weaken.

11 Whom did you dread and fear
　　so that you lied,
　and did not remember me
　or give me a thought?
　Have I not kept silent and closed
　　my eyes, [k]
　and so you do not fear me?

12 I will concede your righteousness
　　and your works,
　but they will not help you.

13 When you cry out, let your
　　collection of idols deliver
　　you!
　The wind will carry them off,
　a breath will take them away.
　But whoever takes refuge in me
　　shall possess the land
　and inherit my holy mountain.

14 It shall be said,
　"Build up, build up, prepare
　　the way,
　remove every obstruction from
　　my people's way."

15 For thus says the high and
　　lofty one
　who inhabits eternity, whose
　　name is Holy:
　I dwell in the high and holy place,
　　and also with those who are
　　contrite and humble in
　　spirit,
　to revive the spirit of the humble,
　and to revive the heart of the
　　contrite.

16 For I will not continually accuse,
　　nor will I always be angry;
　for then the spirits would grow
　　faint before me,
　even the souls that I have made.

17 Because of their wicked
　　covetousness I was angry;
　I struck them, I hid and was
　　angry;
　but they kept turning back to
　　their own ways.

18 I have seen their ways, but I will
　　heal them;
　I will lead them and repay them
　　with comfort,
　creating for their mourners the
　　fruit of the lips. [h]

19 Peace, peace, to the far and the
　　near, says the LORD;
　and I will heal them.

20 But the wicked are like the
　　tossing sea
　that cannot keep still;
　its waters toss up mire and
　　mud.

21 There is no peace, says my God,
　　for the wicked.

58 Shout out, do not hold back!
　Lift up your voice like a
　　trumpet!

h Meaning of Heb uncertain　　i Or *their
phallus*; Heb *the hand*　　j Or *the king*
k Gk Vg: Heb *silent even for a long time*

57.11–13: God rebukes the idolators (42.8); their gods are impotent (42.17; Jer 2.20–28). **57.14–21: Poem of consolation. 14:** Compare 40.1–4. **15:** God is both distant and near. **16–21:** God's justifiable wrath is not unending. **17:** Israel continued to sin despite repeated punishments. **18:** Yet the LORD will heal him, by an act of grace. *Mourners,* a reference to those who repent. **19–21:** Israel will have peace, but the wicked will not (48.18; Eph 2.17).

58.1–14: The LORD does not desire fast-

Announce to my people their
rebellion,
to the house of Jacob their sins.
2 Yet day after day they seek me
and delight to know my ways,
as if they were a nation that
practiced righteousness
and did not forsake the
ordinance of their God;
they ask of me righteous
judgments,
they delight to draw near to
God.
3 "Why do we fast, but you do
not see?
Why humble ourselves, but you
do not notice?"
Look, you serve your own interest
on your fast day,
and oppress all your workers.
4 Look, you fast only to quarrel and
to fight
and to strike with a wicked fist.
Such fasting as you do today
will not make your voice heard
on high.
5 Is such the fast that I choose,
a day to humble oneself?
Is it to bow down the head like a
bulrush,
and to lie in sackcloth and
ashes?
Will you call this a fast,
a day acceptable to the LORD?

6 Is not this the fast that I choose:
to loose the bonds of injustice,
to undo the thongs of the yoke,
to let the oppressed go free,
and to break every yoke?
7 Is it not to share your bread with
the hungry,

and bring the homeless poor
into your house;
when you see the naked, to
cover them,
and not to hide yourself from
your own kin?
8 Then your light shall break forth
like the dawn,
and your healing shall spring
up quickly;
your vindicator¹ shall go
before you,
the glory of the LORD shall be
your rear guard.
9 Then you shall call, and the LORD
will answer;
you shall cry for help, and he
will say, Here I am.

If you remove the yoke from
among you,
the pointing of the finger, the
speaking of evil,
10 if you offer your food to the
hungry
and satisfy the needs of the
afflicted,
then your light shall rise in
the darkness
and your gloom be like the
noonday.
11 The LORD will guide you
continually,
and satisfy your needs in
parched places,
and make your bones strong;
and you shall be like a watered
garden,
like a spring of water,
whose waters never fail.

1 Or vindication

---

ing, but kindness and justice; compare Mic
6.6–8; Jas 1.27. **1–5**: Proper motivation is
necessary for the acceptance of fasting. **1**:
Trumpet, the trumpet was used to announce
a fast day (Joel 2.15; compare Ezek 33.3). **3**:
Fast day, Lev 23.26–32; Jer 36.9. **5**: *Sackcloth,*
worn by mourners (Ezek 7.18).
    58.6–9b: One's relationship to others re-
veals one's relationship to God (Lk 10.25–
37). When appropriate "fruits" are present

(social justice, mercy, sharing; Lk 3.8), God
will hear (1.10–20; Mt 25.34–40). *Light,*
42.6–7; *rear guard,* 52.12. **9c–12**: God will
give his people strength (66.14), abundance
(Jer 31.12) and auspicious circumstances for
reconstruction (44.26; 61.4). *Pointing of the
finger,* gesture of contempt (Prov 6.13). **13–
14**: Strict observance of *the sabbath* was increasingly emphasized in post-exilic Judaism
(56.2; Mt 12.1–8).

12 Your ancient ruins shall be rebuilt;
　you shall raise up the
　　foundations of many
　　generations;
　you shall be called the repairer of
　　the breach,
　the restorer of streets to live in.

13 If you refrain from trampling the
　　sabbath,
　from pursuing your own
　　interests on my holy day;
　if you call the sabbath a delight
　and the holy day of the LORD
　　honorable;
　if you honor it, not going your
　　own ways,
　serving your own interests, or
　　pursuing your own affairs;[m]
14 then you shall take delight in
　　the LORD,
　and I will make you ride upon
　　the heights of the earth;
　I will feed you with the heritage
　　of your ancestor Jacob,
　for the mouth of the LORD has
　　spoken.

59 See, the LORD's hand is not too
　　short to save,
　nor his ear too dull to hear.
2 Rather, your iniquities have been
　　barriers
　between you and your God,
　and your sins have hidden his face
　　from you
　so that he does not hear.
3 For your hands are defiled
　　with blood,
　and your fingers with iniquity;
　your lips have spoken lies,
　　your tongue mutters
　　wickedness.
4 No one brings suit justly,
　no one goes to law honestly;

they rely on empty pleas, they
　　speak lies,
　conceiving mischief and
　　begetting iniquity.
5 They hatch adders' eggs,
　and weave the spider's web;
　whoever eats their eggs dies,
　and the crushed egg hatches out
　　a viper.
6 Their webs cannot serve as
　　clothing;
　they cannot cover themselves
　　with what they make.
　Their works are works of iniquity,
　and deeds of violence are in
　　their hands.
7 Their feet run to evil,
　and they rush to shed innocent
　　blood;
　their thoughts are thoughts
　　of iniquity,
　desolation and destruction are in
　　their highways.
8 The way of peace they do not
　　know,
　and there is no justice in
　　their paths.
　Their roads they have made
　　crooked;
　no one who walks in them
　　knows peace.

9 Therefore justice is far from us,
　and righteousness does not
　　reach us;
　we wait for light, and lo! there is
　　darkness;
　and for brightness, but we walk
　　in gloom.
10 We grope like the blind along
　　a wall,
　groping like those who have
　　no eyes;

m Heb *or speaking words*

**59.1–21: Call to national repentance. 1–
4**: God is not impotent or deaf; Judah's transgressions have broken the covenant (50.1–2).
The people are totally given over to wickedness; injustice and dishonesty are rampant
(Rom 3.10–18). **5–8**: Graphic continuation of

vv. 3–4. The whole community is contaminated (Mt 23.13–36). Synonyms of "ways" in
vv. 7–8 (Rom 3.15–17) describe the blind al-
leys down which Judah is going.
　59.9–11: Consequently, Judah gropes in
the darkness of social and spiritual depravity

we stumble at noon as in the
twilight,
among the vigorous" as though
we were dead.
11 We all growl like bears;
like doves we moan mournfully.
We wait for justice, but there is
none;
for salvation, but it is far
from us.
12 For our transgressions before you
are many,
and our sins testify against us.
Our transgressions indeed are
with us,
and we know our iniquities:
13 transgressing, and denying the
LORD,
and turning away from
following our God,
talking oppression and revolt,
conceiving lying words and
uttering them from
the heart.
14 Justice is turned back,
and righteousness stands at a
distance;
for truth stumbles in the public
square,
and uprightness cannot enter.
15 Truth is lacking,
and whoever turns from evil is
despoiled.

The LORD saw it, and it displeased
him
that there was no justice.
16 He saw that there was no one,
and was appalled that there was
no one to intervene;
so his own arm brought him
victory,

and his righteousness upheld
him.
17 He put on righteousness like a
breastplate,
and a helmet of salvation on
his head;
he put on garments of vengeance
for clothing,
and wrapped himself in fury as
in a mantle.
18 According to their deeds, so will
he repay;
wrath to his adversaries, requital
to his enemies;
to the coastlands he will render
requital.
19 So those in the west shall fear the
name of the LORD,
and those in the east, his glory;
for he will come like a pent-up
stream
that the wind of the LORD
drives on.

20 And he will come to Zion as
Redeemer,
to those in Jacob who turn from
transgression, says the
LORD.
21 And as for me, this is my covenant
with them, says the LORD: my spirit that
is upon you, and my words that I have
put in your mouth, shall not depart out
of your mouth, or out of the mouths of
your children, or out of the mouths of
your children's children, says the LORD,
from now on and forever.

60 Arise, shine; for your light
has come,

n Meaning of Heb uncertain

(13.10; 50.10–11), fear, and foreboding
(38.14). **12–15b**: The community confesses
the magnitude of its offenses: rebellion (*transgressing*), faithlessness (*denying*), disobedience
(*turning away*); its integrity is wholly corrupted.
59.15c–17: With no human helper available, God himself intervened with every
means at his command (42.13; compare Eph

6.14–17). **18–20**: God brings judgment upon
all his enemies with irresistible fury and
might (30.27–28) to redeem repentant *Zion*
(Jerusalem; compare Rom 11.26). **21**: A prose
oracle assuring Judah of the abiding presence
of God's *spirit* and *words* (44.3–5).
**60.1–62.12: Poems on the glory of Jeru-
salem and of God's people,** reminiscent of
chapters 40–55. **60.1–22: Jerusalem's glori-**

and the glory of the LORD has
risen upon you.

2 For darkness shall cover the earth,
and thick darkness the peoples;
but the LORD will arise upon you,
and his glory will appear
over you.

3 Nations shall come to your light,
and kings to the brightness of
your dawn.

4 Lift up your eyes and look around;
they all gather together, they
come to you;
your sons shall come from
far away,
and your daughters shall be
carried on their nurses'
arms.

5 Then you shall see and be radiant;
your heart shall thrill and
rejoice, *o*
because the abundance of the sea
shall be brought to you,
the wealth of the nations shall
come to you.

6 A multitude of camels shall
cover you,
the young camels of Midian and
Ephah;
all those from Sheba shall come.
They shall bring gold and
frankincense,
and shall proclaim the praise of
the LORD.

7 All the flocks of Kedar shall be
gathered to you,
the rams of Nebaioth shall
minister to you;
they shall be acceptable on my
altar,
and I will glorify my glorious
house.

8 Who are these that fly like a
cloud,
and like doves to their
windows?

9 For the coastlands shall wait
for me,
the ships of Tarshish first,
to bring your children from far
away,
their silver and gold with them,
for the name of the LORD your
God,
and for the Holy One of Israel,
because he has glorified you.

10 Foreigners shall build up your
walls,
and their kings shall minister
to you;
for in my wrath I struck you
down,
but in my favor I have had
mercy on you.

11 Your gates shall always be open;
day and night they shall not
be shut,
so that nations shall bring you
their wealth,
with their kings led in
procession.

12 For the nation and kingdom
that will not serve you shall
perish;
those nations shall be utterly
laid waste.

13 The glory of Lebanon shall come
to you,
the cypress, the plane, and
the pine,
to beautify the place of my
sanctuary;
and I will glorify where my
feet rest.

o Heb *be enlarged*

---

ous restoration. 1–3: Fallen Zion bidden to
arise, shine (42.6–7), and reflect *the glory of the
LORD* (6.3; Ezek 1.4–28; 10.4), which will
attract all nations (66.18). **4–5**: Risen Zion
welcomes her children home (49.18ab, 22de);
her poverty is replaced by wealth (45.14;
61.6). **6–7**: Arabia's riches are brought by
camel caravan. *Midian* (see Ex 2.15 n.),
Ephah, Arab tribes east of Gulf of Aqaba;

Sheba (Gen 10.7); *Kedar* (21.16); *Nebaioth,* an
Arab tribe (Gen 25.13).
60.8–9: Some returning exiles (*children*)
and wealth from the west arrive by ship
(v. 8). *Tarshish* (see Jer 10.9 n.). **10–11**: Destroyed by foreigners (Jer 52.13–16), the new
Jerusalem will be built by foreigners (compare Rev 21.24–27). **13–14**: As before
(1 Kings 5.8–10) wood from *Lebanon* (41.19)

14 The descendants of those who
      oppressed you
   shall come bending low to you,
   and all who despised you
   shall bow down at your feet;
   they shall call you the City of
      the LORD,
      the Zion of the Holy One of
      Israel.
15 Whereas you have been forsaken
      and hated,
   with no one passing through,
   I will make you majestic forever,
      a joy from age to age.
16 You shall suck the milk of
      nations,
   you shall suck the breasts
      of kings;
   and you shall know that I, the
      LORD, am your Savior
   and your Redeemer, the Mighty
      One of Jacob.

17 Instead of bronze I will bring
      gold,
   instead of iron I will bring
      silver;
   instead of wood, bronze,
   instead of stones, iron.
   I will appoint Peace as your
      overseer
   and Righteousness as your
      taskmaster.
18 Violence shall no more be heard in
      your land,
   devastation or destruction
      within your borders;
   you shall call your walls Salvation,
   and your gates Praise.
19 The sun shall no longer be
      your light by day,
   nor for brightness shall the moon

   give light to you by night;*p*
   but the LORD will be your
      everlasting light,
   and your God will be your
      glory.
20 Your sun shall no more go down,
   or your moon withdraw itself;
   for the LORD will be your
      everlasting light,
   and your days of mourning shall
      be ended.
21 Your people shall all be righteous;
   they shall possess the land
      forever.
   They are the shoot that I planted,
      the work of my hands,
   so that I might be glorified.
22 The least of them shall become
      a clan,
   and the smallest one a mighty
      nation;
   I am the LORD;
   in its time I will accomplish
      it quickly.

61 The spirit of the Lord GOD is
      upon me,
   because the LORD has
      anointed me;
   he has sent me to bring good
      news to the oppressed,
   to bind up the brokenhearted,
   to proclaim liberty to the captives,
   and release to the prisoners;
2 to proclaim the year of the
      LORD's favor,
   and the day of vengeance of
      our God;
   to comfort all who mourn;
3 to provide for those who mourn
      in Zion—

p Q Ms Gk Old Latin Tg: MT lacks *by night*

---

and wealth from the nations will build the
temple and holy *City* (35.1–10; 49.26).
   60.15–16: Once forsaken (54.6–7), Zion
will be exalted (1.24; Ezek 16.1–63). **17–18:**
The new Jerusalem will surpass Solomon's
city in beauty and tranquillity.
   60.19–20: God's *glory* will perpetually illumine the joyful city (35.10; Rev 21.4). **21–**

22: In the divinely restored city, God will be
glorified (9.7; 54.1–2).
   61.1–11: The mission to Zion. This poem
recalls the Servant Songs of chs 42–53, especially 50.4–11. **1–3:** Lk 4.16–20; compare Mt
11.5; Lk 7.22. God sends his prophet to bring
encouragement to the exiled and oppressed;
he will make them mighty oaks (60.21; Jer

to give them a garland instead of ashes,
the oil of gladness instead of mourning,
the mantle of praise instead of a faint spirit.
They will be called oaks of righteousness,
the planting of the LORD, to display his glory.
4 They shall build up the ancient ruins,
they shall raise up the former devastations;
they shall repair the ruined cities,
the devastations of many generations.

5 Strangers shall stand and feed your flocks,
foreigners shall till your land and dress your vines;
6 but you shall be called priests of the LORD,
you shall be named ministers of our God;
you shall enjoy the wealth of the nations,
and in their riches you shall glory.
7 Because their[q] shame was double,
and dishonor was proclaimed as their lot,
therefore they shall possess a double portion;
everlasting joy shall be theirs.

8 For I the LORD love justice,
I hate robbery and wrongdoing;[r]
I will faithfully give them their recompense,
and I will make an everlasting covenant with them.
9 Their descendants shall be known among the nations,
and their offspring among the peoples;
all who see them shall acknowledge
that they are a people whom the LORD has blessed.
10 I will greatly rejoice in the LORD,
my whole being shall exult in my God;
for he has clothed me with the garments of salvation,
he has covered me with the robe of righteousness,
as a bridegroom decks himself with a garland,
and as a bride adorns herself with her jewels.
11 For as the earth brings forth its shoots,
and as a garden causes what is sown in it to spring up,
so the Lord GOD will cause righteousness and praise
to spring up before all the nations.

62 For Zion's sake I will not keep silent,
and for Jerusalem's sake I will not rest,
until her vindication shines out like the dawn,
and her salvation like a burning torch.

q Heb *your* *r* Or *robbery with a burnt offering*

17.8). *Spirit* is strongly emphasized in exilic and post-exilic writings (42.1; 59.21; Ezek 2.2). **2**: *Vengeance,* better, "rescue." **4–5**: 60.10–11. **6–7**: *Priests* (Ex 19.6). *Double portion,* for double punishment (40.2).

61.8–9: In Judah, all nations will see God's faithfulness and his blessing. *Justice,* 5.7, 16. *Everlasting covenant,* 54.9–10. **10–11**: The prophet identifies himself with Zion and rejoices in her salvation (compare vv. 1–3), which is as certain as the earth's producing vegetation.

62.1–12: **The glory of God's people. 1–3**: The prophet continues to proclaim Zion's approaching vindication. *New name* denotes a change in status (1.26; Jer 33.16; Ezek 48.35; compare Abram, Abraham, Gen 17.5 notes *u* and *v*). *Crown of beauty* (28.1–6). **4–5**: New

2 The nations shall see your
      vindication,
    and all the kings your glory;
    and you shall be called by a
      new name
      that the mouth of the LORD
        will give.
3 You shall be a crown of beauty in
      the hand of the LORD,
    and a royal diadem in the hand
      of your God.
4 You shall no more be termed
      Forsaken, *s*
    and your land shall no more be
      termed Desolate; *t*
    but you shall be called My Delight
      Is in Her, *u*
    and your land Married; *v*
    for the LORD delights in you,
    and your land shall be married.
5 For as a young man marries a
      young woman,
    so shall your builder *w* marry
      you,
    and as the bridegroom rejoices
      over the bride,
    so shall your God rejoice
      over you.
6 Upon your walls, O Jerusalem,
    I have posted sentinels;
    all day and all night
      they shall never be silent.
    You who remind the LORD,
      take no rest,
7 and give him no rest
    until he establishes Jerusalem
      and makes it renowned
        throughout the earth.
8 The LORD has sworn by his
      right hand
    and by his mighty arm:
    I will not again give your grain
      to be food for your enemies,

and foreigners shall not drink
      the wine
    for which you have labored;
9 but those who garner it shall eat it
    and praise the LORD,
    and those who gather it shall
      drink it
    in my holy courts.

10 Go through, go through the gates,
    prepare the way for the people;
    build up, build up the highway,
      clear it of stones,
    lift up an ensign over the
      peoples.
11 The LORD has proclaimed
    to the end of the earth:
    Say to daughter Zion,
      "See, your salvation comes;
    his reward is with him,
      and his recompense before
        him."
12 They shall be called, "The
      Holy People,
    The Redeemed of the LORD";
    and you shall be called,
      "Sought Out,
    A City Not Forsaken."

63 "Who is this that comes from
      Edom,
    from Bozrah in garments
      stained crimson?
    Who is this so splendidly robed,
      marching in his great might?"

    "It is I, announcing vindication,
      mighty to save."

2 "Why are your robes red,

s Heb *Azubah*   t Heb *Shemamah*
u Heb *Hephzibah*   v Heb *Beulah*
w Cn: Heb *your sons*

status, new names. **6–9**: *Sentinels,* probably
prophets (52.8; Ezek ch 33), to remind Jerusalem of her imminent salvation, the certainty
of which the LORD's oath (vv. 8–9) underscores. *Right hand* (Ex 15.6, 12). *Mighty arm*
(40.10; Deut 7.19). *Foreigners,* invaders, as the
Babylonians, and encroachers, as the Edomites (Ob 13).

62.10–12: A summary of the eschatological hopes described in chs 60–62. **10b**: 49.22.
11b: 40.10.

63.1–6: **Poem on divine vengeance. 1**:
The prophet (sentinel; 62.6 n.) challenges him
who approaches. *Edom, Bozrah,* symbolic of
God's opponents (34.5–7). **2–3**: Compare
Joel 3.13; Rev 14.19–20. **4–6**: *Day of ven-*

and your garments like theirs
who tread the wine press?"

3 "I have trodden the wine press
alone,
and from the peoples no one
was with me;
I trod them in my anger
and trampled them in my
wrath;
their juice spattered on my
garments,
and stained all my robes.

4 For the day of vengeance was in
my heart,
and the year for my redeeming
work had come.

5 I looked, but there was no helper;
I stared, but there was no one
to sustain me;
so my own arm brought me
victory,
and my wrath sustained me.

6 I trampled down peoples in my
anger,
I crushed them in my wrath,
and I poured out their lifeblood
on the earth."

7 I will recount the gracious deeds
of the LORD,
the praiseworthy acts of the
LORD,
because of all that the LORD has
done for us,
and the great favor to the house
of Israel
that he has shown them according
to his mercy,
according to the abundance of
his steadfast love.

8 For he said, "Surely they are
my people,
children who will not deal
falsely";

and he became their savior
9 in all their distress.
It was no messenger[x] or angel
but his presence that saved
them;[y]
in his love and in his pity he
redeemed them;
he lifted them up and carried
them all the days of old.

10 But they rebelled
and grieved his holy spirit;
therefore he became their enemy;
he himself fought against them.

11 Then they[z] remembered the days
of old,
of Moses his servant.[a]
Where is the one who brought
them up out of the sea
with the shepherds of his flock?
Where is the one who put
within them
his holy spirit,

12 who caused his glorious arm
to march at the right hand
of Moses,
who divided the waters before
them
to make for himself an
everlasting name,

13 who led them through the
depths?
Like a horse in the desert,
they did not stumble.

14 Like cattle that go down into
the valley,
the spirit of the LORD gave
them rest.
Thus you led your people,
to make for yourself a glorious
name.

x Gk: Heb *anguish*   y Or *savior. 9In all their
distress he was distressed; the angel of his presence
saved them;*   z Heb *he*   a Cn: Heb *his people*

---

geance and *year for my redeeming work* are synchronous.
63.7–64.12: Psalm of intercession. 63.7–14: Historical prologue recalling Israel's deliverance from Egypt. Israel is called (v. 8; Ex 4.22–23; 19.3–6), protected (v. 9a; Ex 12.1–32), exalted (v. 9b; Ezek ch 16), delivered (vv. 11–12; Ex 14.9–15.21; Jer 15.1), and safely led through Sinai into Canaan (vv. 13–14). Israel's rebellion necessitated God's opposing them (Jer 5.20–29). **7:** *Steadfast love* expresses God's continuing covenant loyalty despite Israel's faithlessness.
63.15–16: The prophet therefore hopefully

15 Look down from heaven and see,
   from your holy and glorious
      habitation.
   Where are your zeal and your
      might?
   The yearning of your heart and
      your compassion?
   They are withheld from me.
16 For you are our father,
   though Abraham does not
      know us
   and Israel does not acknowledge
      us;
   you, O LORD, are our father;
      our Redeemer from of old is
         your name.
17 Why, O LORD, do you make us
      stray from your ways
   and harden our heart, so that we
      do not fear you?
   Turn back for the sake of your
      servants,
   for the sake of the tribes that are
      your heritage.
18 Your holy people took possession
      for a little while;
   but now our adversaries have
      trampled down your
         sanctuary.
19 We have long been like those
      whom you do not rule,
   like those not called by your
      name.

64 O that you would tear open the
      heavens and come down,
   so that the mountains would
      quake at your presence—
2b as when fire kindles brushwood
   and the fire causes water to
      boil—
   to make your name known to
      your adversaries,
   so that the nations might
      tremble at your presence!

3 When you did awesome deeds that
      we did not expect,
   you came down, the mountains
      quaked at your presence.
4 From ages past no one has heard,
   no ear has perceived,
   no eye has seen any God besides
      you,
   who works for those who wait
      for him.
5 You meet those who gladly do
      right,
   those who remember you in
      your ways.
   But you were angry, and we
      sinned;
   because you hid yourself we
      transgressed.c
6 We have all become like one who
      is unclean,
   and all our righteous deeds are
      like a filthy cloth.
   We all fade like a leaf,
   and our iniquities, like the
      wind, take us away.
7 There is no one who calls on
      your name,
   or attempts to take hold of you;
   for you have hidden your face
      from us,
   and have deliveredd us into the
      hand of our iniquity.
8 Yet, O LORD, you are our Father;
   we are the clay, and you are
      our potter;
   we are all the work of your
      hand.
9 Do not be exceedingly angry,
      O LORD,
   and do not remember iniquity
      forever.

b Ch 64.1 in Heb   c Meaning of Heb
uncertain   d Gk Syr Old Latin Tg: Heb *melted*

petitions God, whose immortality is contrasted with the mortality of Israel's patriarchs (*Abraham, Israel* [Jacob]). **17–19**: Israel's
sinfulness begets sinfulness. The prophet
begs God to deliver his forsaken people from
their sins and to restore their ruined sanctuary.
   64.1–5b: A prayer that God should reveal

himself in power as in days of old (Ex 19.16–
18; Judg 5.4–5; Hab 3.3–15). Human beings
cannot approach God (Gen 11.1–9; Ex 33.17–
23), but God comes to them (Ex 19–20). **5c–
7**: The prophet, in behalf of his people, confesses their sin and hopelessness.
   64.8–12: In his final petition he pleads that
the LORD turn away from his anger and have

Now consider, we are all your
people.
10 Your holy cities have become a
wilderness,
Zion has become a wilderness,
Jerusalem a desolation.
11 Our holy and beautiful house,
where our ancestors praised
you,
has been burned by fire,
and all our pleasant places have
become ruins.
12 After all this, will you restrain
yourself, O LORD?
Will you keep silent, and punish
us so severely?

65 I was ready to be sought out
by those who did not ask,
to be found by those who did
not seek me.
I said, "Here I am, here I am,"
to a nation that did not call on
my name.
2 I held out my hands all day long
to a rebellious people,
who walk in a way that is not
good,
following their own devices;
3 a people who provoke me
to my face continually,
sacrificing in gardens
and offering incense on bricks;
4 who sit inside tombs,
and spend the night in secret
places;
who eat swine's flesh,
with broth of abominable things
in their vessels;
5 who say, "Keep to yourself,

do not come near me, for I am
too holy for you."
These are a smoke in my nostrils,
a fire that burns all day long.
6 See, it is written before me:
I will not keep silent, but I will
repay;
I will indeed repay into their laps
7 their*e* iniquities and their*e*
ancestors' iniquities
together,
says the LORD;
because they offered incense on
the mountains
and reviled me on the hills,
I will measure into their laps
full payment for their actions.
8 Thus says the LORD:
As the wine is found in the
cluster,
and they say, "Do not destroy
it,
for there is a blessing in it,"
so I will do for my servants' sake,
and not destroy them all.
9 I will bring forth descendants*f*
from Jacob,
and from Judah inheritors*g* of
my mountains;
my chosen shall inherit it,
and my servants shall settle
there.
10 Sharon shall become a pasture
for flocks,
and the Valley of Achor a place
for herds to lie down,

e Gk Syr: Heb *your*   *f* Or *a descendant*
g Or *an inheritor*

---

compassion on desolate Jerusalem and the destroyed temple. **10:** *Your holy cities,* the cities
are the LORD's, for the Promised Land is his.
12: *Will you keep silent . . . ?* Compare Ps 79.5;
85.5–7.
65.1–25: God's answer. 1–2: Not God,
but the people were silent. *Held out my hands,*
in a gesture of invitation (Rom 10.20–21). **3:**
Sacrificing in gardens, an allusion to nature-cult
practices. *Offering incense* was peculiarly associated with pagan worship, although also a
part of Israelite worship (Jer 1.16). **4:** *Sit inside
tombs,* for divination, to consult the dead

(29.4). *Spend the night in secret places,* in a
shrine to receive visions (compare 1 Kings
3.4–15). To *eat swine's flesh* was forbidden
(Deut 14.8). **5:** *I am too holy for you,* sanctified
by some idolatrous rite. **6–7:** God has noted
their faithlessness (4.3).
65.8–10: As good clusters of grapes are
separated from the bad, so God will separate
the righteous from the unrighteous (Mt
25.32–33). The LORD's *chosen,* his *servants*
(44.1), Israel, will inherit the land (57.13).
Sharon, the northern coastal plain. *Achor,* the
desolate region west of the Dead Sea (Josh

for my people who have
sought me.

11 But you who forsake the LORD,
who forget my holy mountain,
who set a table for Fortune
and fill cups of mixed wine for
Destiny;

12 I will destine you to the sword,
and all of you shall bow down
to the slaughter;
because, when I called, you did
not answer,
when I spoke, you did not
listen,
but you did what was evil in
my sight,
and chose what I did not
delight in.

13 Therefore thus says the Lord GOD:
My servants shall eat,
but you shall be hungry;
my servants shall drink,
but you shall be thirsty;
my servants shall rejoice,
but you shall be put to shame;

14 my servants shall sing for gladness
of heart,
but you shall cry out for pain
of heart,
and shall wail for anguish
of spirit.

15 You shall leave your name to my
chosen to use as a curse,
and the Lord GOD will put you
to death;
but to his servants he will give a
different name.

16 Then whoever invokes a blessing
in the land
shall bless by the God of
faithfulness,
and whoever takes an oath in
the land
shall swear by the God of
faithfulness;

because the former troubles are
forgotten
and are hidden from my sight.

17 For I am about to create new
heavens
and a new earth;
the former things shall not be
remembered
or come to mind.

18 But be glad and rejoice forever
in what I am creating;
for I am about to create Jerusalem
as a joy,
and its people as a delight.

19 I will rejoice in Jerusalem,
and delight in my people;
no more shall the sound of
weeping be heard in it,
or the cry of distress.

20 No more shall there be in it
an infant that lives but a
few days,
or an old person who does not
live out a lifetime;
for one who dies at a hundred
years will be considered a
youth,
and one who falls short of a
hundred will be considered
accursed.

21 They shall build houses and
inhabit them;
they shall plant vineyards and
eat their fruit.

22 They shall not build and another
inhabit;
they shall not plant and
another eat;
for like the days of a tree shall the
days of my people be,
and my chosen shall long enjoy
the work of their hands.

23 They shall not labor in vain,
or bear children for calamity;[h]

h Or *sudden terror*

7.24; see Hos 2.15 n.). **11–12**: In the Hebrew text *Fortune* and *Destiny* are Gad and Meni, Syrian gods of fate. **15–16**: Apostates' name, *a curse;* the righteous, a new name (62.2, 4). **65.17–19**: Heaven and earth will be trans- formed; God will rejoice with Jerusalem (62.5; contrast 64.10); her mourning is over (25.8; 2 Pet 3.13; Rev 21.4). **21–23**: Those in Jerusalem will live in happiness and security. *Tree,* Jer 17.8. *Calamity,* sudden misfortune

for they shall be offspring blessed
    by the LORD—
and their descendants as well.
24 Before they call I will answer,
    while they are yet speaking I
      will hear.
25 The wolf and the lamb shall feed
    together,
    the lion shall eat straw like
      the ox;
but the serpent—its food shall
    be dust!
They shall not hurt or destroy
    on all my holy mountain,
          says the LORD.

66 Thus says the LORD:
    Heaven is my throne
    and the earth is my footstool;
what is the house that you would
    build for me,
and what is my resting place?
2 All these things my hand has
    made,
    and so all these things
    are mine, *i*
          says the LORD.
But this is the one to whom I will
    look,
to the humble and contrite
    in spirit,
    who trembles at my word.

3 Whoever slaughters an ox is like
    one who kills a human
    being;
whoever sacrifices a lamb, like
    one who breaks a dog's
    neck;
whoever presents a grain offering,

like one who offers swine's
    blood;*j*
whoever makes a memorial
    offering of frankincense,
like one who blesses an
    idol.
These have chosen their own
    ways,
and in their abominations they
    take delight;
4 I also will choose to mock*k* them,
    and bring upon them what
    they fear;
because, when I called, no one
    answered,
when I spoke, they did not
    listen;
but they did what was evil in
    my sight,
and chose what did not
    please me.
5 Hear the word of the LORD,
    you who tremble at his word:
Your own people who hate you
    and reject you for my name's
    sake
have said, "Let the LORD be
    glorified,
so that we may see your joy";
but it is they who shall be put
    to shame.

6 Listen, an uproar from the city!
    A voice from the temple!
The voice of the LORD,
    dealing retribution to his
    enemies!

i Gk Syr: Heb *these things came to be*
j Meaning of Heb uncertain   *k* Or *to punish*

(Jer 15.8). **24–25**: In the new Jerusalem re-stored, all will be at peace (11.6–9). *My holy mountain,* see 11.9 n.; compare 27.13; 56.7; Joel 3.17.
    66.1–24: Concluding oracles. 1–6: On temple worship. **1–2**: A house made of creat-ed materials cannot contain the Creator and Lord of heaven and earth (compare 1 Kings 8.27), nor do mere acts suffice for devout humility (1.10–20). **3–4**: These verses are perhaps best taken as the continuation of the thought in vv. 1–2. The insertion of the word "like" (not in the Hebrew) interprets one pos-sible meaning of these verses. In accord with another interpretation read: "He who slaugh-ters an ox, he who kills a human being . . . ; these have chosen their own ways . . ." The first part of each line describes normally ac-ceptable practice, the second what is not. The "Canaanitish" practices of sacrificing humans (*kills a human being*), dogs (*breaks a dog's neck*), and swine (*offers swine's blood*) are abomina-tions. **5–6**: Vindication for true believers. God has returned to his temple (Ezek 43.1–5). **7–16**: On Jerusalem's restoration.

7 Before she was in labor
   she gave birth;
before her pain came upon her
   she delivered a son.
8 Who has heard of such a thing?
   Who has seen such things?
Shall a land be born in one day?
   Shall a nation be delivered in
      one moment?
Yet as soon as Zion was in labor
   she delivered her children.
9 Shall I open the womb and not
      deliver?
   says the LORD;
shall I, the one who delivers, shut
      the womb?
   says your God.

10 Rejoice with Jerusalem, and be
      glad for her,
   all you who love her;
rejoice with her in joy,
   all you who mourn over her—
11 that you may nurse and be
      satisfied
   from her consoling breast;
that you may drink deeply with
      delight
   from her glorious bosom.

12 For thus says the LORD:
I will extend prosperity to her like
      a river,
   and the wealth of the nations
      like an overflowing stream;
and you shall nurse and be carried
      on her arm,
   and dandled on her knees.
13 As a mother comforts her child,
   so I will comfort you;
you shall be comforted in
      Jerusalem.

14 You shall see, and your heart shall
      rejoice;
   your bodies[l] shall flourish like
      the grass;
and it shall be known that the
      hand of the LORD is with
      his servants,
   and his indignation is against
      his enemies.
15 For the LORD will come in fire,
   and his chariots like the
      whirlwind,
to pay back his anger in fury,
   and his rebuke in flames of fire.
16 For by fire will the LORD execute
      judgment,
   and by his sword, on all flesh;
   and those slain by the LORD
      shall be many.

17 Those who sanctify and purify
themselves to go into the gardens, following the one in the center, eating the
flesh of pigs, vermin, and rodents, shall
come to an end together, says the LORD.

18 For I know[m] their works and their
thoughts, and I am[n] coming to gather all
nations and tongues; and they shall come
and shall see my glory, 19 and I will set a
sign among them. From them I will send
survivors to the nations, to Tarshish,
Put,[o] and Lud—which draw the bow—
to Tubal and Javan, to the coastlands far
away that have not heard of my fame or
seen my glory; and they shall declare my
glory among the nations. 20 They shall
bring all your kindred from all the na-

---

l Heb *bones*   *m* Gk Syr: Heb lacks *know*
n Gk Syr Vg Tg: Heb *it is*   *o* Gk: Heb *Pul*

66.7–9: Jerusalem's rebirth is a divinely
wrought miracle. **10–11:** Restored "Mother"
Jerusalem can provide abundantly for all
(49.17–21).
   66.12–14: Continuation of the vision of
Jerusalem's prosperity and comfort. **15–16:**
With *fire* (29.6; 64.11) and *sword* (31.8; Ezek
ch 21), God will destroy his enemies. **17:**
Compare vv. 3–4. *In the center,* perhaps in a
pagan procession.

66.18–24: The return of Israel and the fate
of the wicked. **18–21:** God will reveal his glory to the nations. With Israel, they will assemble in Jerusalem, some Gentiles even becoming priests. **22–23:** A magnificent
climax: God's people, like the new creation,
will endure forever (Jer 31.34–36) and a
mighty chorus of praises will ceaselessly rise
to God's throne. **24:** Final reminder of the
eternal punishment of the wicked.

tions as an offering to the LORD, on horses, and in chariots, and in litters, and on mules, and on dromedaries, to my holy mountain Jerusalem, says the LORD, just as the Israelites bring a grain offering in a clean vessel to the house of the LORD. 21 And I will also take some of them as priests and as Levites, says the LORD.

22 For as the new heavens and the
new earth,
which I will make,
shall remain before me, says
the LORD;

so shall your descendants and
your name remain.
23 From new moon to new moon,
and from sabbath to sabbath,
all flesh shall come to worship
before me,
says the LORD.

24 And they shall go out and look at the dead bodies of the people who have rebelled against me; for their worm shall not die, their fire shall not be quenched, and they shall be an abhorrence to all flesh.

Jeremiah

Jeremiah was the son of Hilkiah, a priest at Anathoth (1.1); as such he may have been a descendant of the priest Abiathar, who was banished by Solomon to Anathoth (1 Kings 2.26, 27). His ministry began in 627 B.C. (1.2) or, by another interpretation, he was born in 627 B.C. (1.5). He died sometime after 587 in Egypt.

The book bearing his name consists essentially of a collection of oracles against Judah and Jerusalem, which he dictated to his aide Baruch (1.4–20.18, from the time of Josiah and Jehoiakim; 21.1–25.14, from the time of Zedekiah); Baruch's memoirs (chs 26–35; 36–45); and a group of oracles against the foreign nations (25.15–38; chs 46–51), together with an introduction (1.1–3) and an historical appendix (ch 52). Some other materials were added when the book was edited sometime after 560 B.C.

The material of the book is not in chronological order. Instead, there are many signs that the book was compiled over a period of time from smaller collections of oracles. One clue to smaller collections may be the first two scrolls of judgment oracles that Jeremiah is said to have dictated to Baruch (ch 36): the second scroll (36.32) could have formed the nucleus for the material now found in 2.1–20.18. Another small collection may have been the hopeful scroll (30.1–3), which in its original form probably contained much of what is now in chs 30–32. But the core of Baruch's memoirs (chs 37–44), dealing with the last few months of Jeremiah's life, does appear to be in chronological order.

The book of Jeremiah is noteworthy in that the present Hebrew text differs substantially from the Greek version (the Septuagint) in both content and order. Thus the Septuagint omits several passages (e.g. 33.14–26) and combines the oracles against the foreign nations into a single section following 25.14, though in a different order. In addition, there are many smaller differences from verse to verse. Remarkably, among the portions of the text of Jeremiah in Hebrew that are found among the Dead Sea Scrolls are not only those that reflect the standard Hebrew text but also those that reflect the text tradition represented by the Septuagint. It is likely, then, that these two text traditions represent the contrasting editorial work on the book of Jeremiah that took place in Egypt (the Septuagint tradition) and in Palestine or Babylon (the traditional Hebrew text).

Unique to the book of Jeremiah is the series of laments of the prophet that begin at 11.18 (see there). For no other prophet in the Hebrew Scriptures do we have a comparable reflection of the spiritual struggle with God. Unique too is the wealth of detail concerning the various trials endured by the prophet throughout his career, from the judicial hearing at the time of his temple sermon in 609 B.C. (ch 26) to the opposition he incurred at the end of his life in Egypt (ch 44).

The prophet Jeremiah was much concerned with rewards and punishment (compare 1.10), the recompense for good and evil, faithfulness and disobedience. He criticizes Judah for its worship of gods other than the LORD, with all the attendant evils in cult and daily life. God's covenant people must return to him. The judgment must come, but the ominous future (later, the unhappy present) would be replaced by a new and more enduring relationship with God.

1 The words of Jeremiah son of Hilkiah, of the priests who were in Anathoth in the land of Benjamin, ²to whom the word of the LORD came in the days of King Josiah son of Amon of Judah, in the thirteenth year of his reign. ³It came also in the days of King Jehoiakim son of Josiah of Judah, and until the end of the eleventh year of King Zedekiah son of Josiah of Judah, until the captivity of Jerusalem in the fifth month.

4 Now the word of the LORD came to me saying,

5 "Before I formed you in the
        womb I knew you,
    and before you were born I
        consecrated you;
    I appointed you a prophet to the
        nations."
⁶Then I said, "Ah, Lord GOD! Truly I do not know how to speak, for I am only a boy." ⁷But the LORD said to me,

    "Do not say, 'I am only a boy';
    for you shall go to all to whom I
        send you,
    and you shall speak whatever I
        command you,
8 Do not be afraid of them,
    for I am with you to deliver you,
                    says the LORD."
⁹Then the LORD put out his hand and touched my mouth; and the LORD said to me,

    "Now I have put my words in
        your mouth.
10 See, today I appoint you over
        nations and over kingdoms,

    to pluck up and to pull down,
    to destroy and to overthrow,
    to build and to plant."

11 The word of the LORD came to me, saying, "Jeremiah, what do you see?" And I said, "I see a branch of an almond tree."ᵃ ¹²Then the LORD said to me, "You have seen well, for I am watchingᵇ over my word to perform it." ¹³The word of the LORD came to me a second time, saying, "What do you see?" And I said, "I see a boiling pot, tilted away from the north."

14 Then the LORD said to me: Out of the north disaster shall break out on all the inhabitants of the land. ¹⁵For now I am calling all the tribes of the kingdoms of the north, says the LORD; and they shall come and all of them shall set their thrones at the entrance of the gates of Jerusalem, against all its surrounding walls and against all the cities of Judah. ¹⁶And I will utter my judgments against them, for all their wickedness in forsaking me; they have made offerings to other gods, and worshiped the works of their own hands. ¹⁷But you, gird up your loins; stand up and tell them everything that I command you. Do not break down before them, or I will break you before them. ¹⁸And I for my part have made you today a fortified city, an iron pillar, and a bronze wall, against the whole land—against the kings of Judah, its princes, its priests, and the people of

a Heb *shaqed*   *b* Heb *shoqed*

1.1–3: Superscription. 1: *Words,* i.e. "history." The name *Jeremiah* may mean "The LORD [Yahweh] exalts." *Priests . . . in Anathoth,* see Introduction. Anathoth is located at modern Ras Kharrubeh about two miles northeast of Jerusalem. **2:** *Thirteenth year of his reign,* 627 B.C. **3:** *Eleventh year of King Zedekiah,* 587 B.C.
1.4–19: Jeremiah's call and related visions. 4: *Word of the LORD,* characteristic expression in Jeremiah, emphasizing that his message is God's word. **5:** *Knew,* in the biblical sense, a profound and intimate knowledge. *To the nations,* Assyria, Babylonia, Egypt, Judah.
1.6–8: God's support will supplement Jeremiah's inexperience (*a boy*). **9:** Compare 15.19; Mt 10.19–20. **10:** God's word is a dynamic and vital force, not a static and symbolic figure (Isa 55.10–11).
1.11–12: Play on words (see notes *a* and *b*) to encourage the hesitant young prophet in the face of certain opposition. **13–14:** *Tilted away from the north:* this translation implies that the pot was spilling its hot contents toward the south, symbolizing the trouble which would come from the north. Another view is that the draft on the fire came from the north. Invaders into Palestine often came from *the north.* **17–19:** An expansion of the thought in vv. 4–8.

the land. ¹⁹They will fight against you; but they shall not prevail against you, for I am with you, says the LORD, to deliver you.

2 The word of the LORD came to me, saying: ²Go and proclaim in the hearing of Jerusalem, Thus says the LORD:

I remember the devotion of your
    youth,
    your love as a bride,
how you followed me in the
    wilderness,
    in a land not sown.
³ Israel was holy to the LORD,
    the first fruits of his harvest.
All who ate of it were held guilty;
    disaster came upon them,
                says the LORD.

4 Hear the word of the LORD, O house of Jacob, and all the families of the house of Israel. ⁵Thus says the LORD:

What wrong did your ancestors
    find in me
    that they went far from me,
and went after worthless things,
    and became worthless
        themselves?
⁶ They did not say, "Where is
    the LORD
    who brought us up from the
        land of Egypt,
who led us in the wilderness,
    in a land of deserts and pits,
in a land of drought and deep
        darkness,
in a land that no one passes
        through,
    where no one lives?"
⁷ I brought you into a plentiful land
    to eat its fruits and its good
        things.

But when you entered you defiled
    my land,
and made my heritage an
    abomination.
⁸ The priests did not say, "Where is
    the LORD?"
Those who handle the law did
    not know me;
the rulers*ᶜ* transgressed against
    me;
the prophets prophesied by
    Baal,
and went after things that do
    not profit.

⁹ Therefore once more I accuse you,
              says the LORD,
and I accuse your children's
    children.
¹⁰ Cross to the coasts of Cyprus
    and look,
send to Kedar and examine with
    care;
see if there has ever been such
    a thing.
¹¹ Has a nation changed its gods,
    even though they are no gods?
But my people have changed
    their glory
for something that does not
    profit.
¹² Be appalled, O heavens, at this,
    be shocked, be utterly desolate,
             says the LORD,
¹³ for my people have committed
    two evils:
they have forsaken me,
the fountain of living water,
    and dug out cisterns for
        themselves,
cracked cisterns
    that can hold no water.

c Heb *shepherds*

2.1–37: The apostasy of Israel. 1–3: God defended his *bride* (comparing the Sinai covenant with the marriage vow, see Hos 2.16) against all attempts to violate her (Amalekites, Canaanites, Philistines, etc.).
2.4–9: God remained unwaveringly faithful, despite Israel's rebellion. **8:** *Priests, rulers, prophets,* all Judah's leaders have failed to lead.

10–13: The LORD calls (upon the heavenly assembly; compare Isa 1.2; Mic 6.1) to witness the folly unprecedented (8.4) in both West (*Cyprus*) and East (*Kedar*) of a people who forsake the *fountain of living water* (Jn 4.10–15; 7.38) for the stagnant water at the bottom of leaky *cisterns.*

14 Is Israel a slave? Is he a homeborn
  servant?
 Why then has he become
  plunder?
15 The lions have roared against him,
  they have roared loudly.
 They have made his land a waste;
  his cities are in ruins, without
   inhabitant.
16 Moreover, the people of Memphis
  and Tahpanhes
 have broken the crown of
  your head.
17 Have you not brought this upon
  yourself
 by forsaking the LORD your
  God,
 while he led you in the way?
18 What then do you gain by going
  to Egypt,
 to drink the waters of the Nile?
 Or what do you gain by going to
  Assyria,
 to drink the waters of the
  Euphrates?
19 Your wickedness will punish you,
  and your apostasies will
   convict you.
 Know and see that it is evil
  and bitter
 for you to forsake the LORD
  your God,
 the fear of me is not in you,
   says the Lord GOD of hosts.

20 For long ago you broke your yoke
  and burst your bonds,
 and you said, "I will not serve!"
 On every high hill
  and under every green tree

you sprawled and played
 the whore.
21 Yet I planted you as a choice vine,
  from the purest stock.
 How then did you turn degenerate
  and become a wild vine?
22 Though you wash yourself with
  lye
 and use much soap,
 the stain of your guilt is still
  before me,
    says the Lord GOD.
23 How can you say, "I am not
  defiled,
 I have not gone after the Baals"?
 Look at your way in the valley;
  know what you have done—
 a restive young camel interlacing
  her tracks,
24 a wild ass at home in the
  wilderness,
 in her heat sniffing the wind!
 Who can restrain her lust?
 None who seek her need weary
  themselves;
 in her month they will find her.
25 Keep your feet from going unshod
  and your throat from thirst.
 But you said, "It is hopeless,
 for I have loved strangers,
  and after them I will go."

26 As a thief is shamed when caught,
  so the house of Israel shall be
   shamed—
 they, their kings, their officials,
  their priests, and their prophets,
27 who say to a tree, "You are
  my father,"

2.14–19: Israel has forsaken her covenant birthright of responsible freedom to become a slave of Assyria or Babylonia (*the lions*) and Egypt (*Memphis,* the capital of northern Egypt, fourteen miles south of Cairo) bringing disgrace (with v. 16b compare Isa 3.17; 7.20) upon themselves. *Tahpanhes,* Egyptian border fortress, also known as Baal-Zaphon, Greek Daphne, modern Tell Defneh. **16, 18**: Compare v. 36.
2.20–28: Unfaithful Israel is compared to a stubborn ox, to a wild vine (Isa 5.1–7; Hos

10.1) germinating from good seed but now inexplicably and irremediably worthless, to a nymphomaniacal prostitute (Hos 4.13) who restlessly seeks satisfaction from anyone and yet denies her guilt in spite of evidence (such as human sacrifice *in the valley*), and to a thief who is remorseful only when confronted by evidence of his wrong-doing (such as the *tree* or Asherah and the *stone* or sacred pillar, both cult furnishings). Let their gods now rescue Israel in her time of need!

and to a stone, "You gave
    me birth."
For they have turned their backs
    to me,
    and not their faces.
But in the time of their trouble
    they say,
    "Come and save us!"
28 But where are your gods
    that you made for yourself?
Let them come, if they can
    save you,
    in your time of trouble;
for you have as many gods
    as you have towns, O Judah.

29 Why do you complain against me?
    You have all rebelled against
    me,
                says the LORD.
30 In vain I have struck down
    your children;
    they accepted no correction.
Your own sword devoured your
    prophets
    like a ravening lion.
31 And you, O generation, behold
    the word of the LORD!*d*
Have I been a wilderness to Israel,
    or a land of thick darkness?
Why then do my people say, "We
    are free,
    we will come to you no more"?
32 Can a girl forget her ornaments,
    or a bride her attire?
Yet my people have forgotten me,
    days without number.

33 How well you direct your course
    to seek lovers!
So that even to wicked women
    you have taught your ways.
34 Also on your skirts is found
    the lifeblood of the innocent
    poor,

though you did not catch them
    breaking in.
Yet in spite of all these things*d*
35 you say, "I am innocent;
    surely his anger has turned
    from me."
Now I am bringing you to
    judgment
    for saying, "I have not sinned."
36 How lightly you gad about,
    changing your ways!
You shall be put to shame
    by Egypt
    as you were put to shame
    by Assyria.
37 From there also you will
    come away
with your hands on your head;
for the LORD has rejected those in
    whom you trust,
    and you will not prosper
    through them.

3 If*e* a man divorces his wife
    and she goes from him
and becomes another man's wife,
    will he return to her?
Would not such a land be greatly
    polluted?
You have played the whore with
    many lovers;
    and would you return to me?
             says the LORD.
2 Look up to the bare heights,*f*
    and see!
Where have you not been
    lain with?
By the waysides you have sat
    waiting for lovers,
    like a nomad in the wilderness.
You have polluted the land
    with your whoring and
    wickedness.

d Meaning of Heb uncertain   *e* Q Ms Gk
Syr: MT *Saying, If*   *f* Or *the trails*

2.29–31: Israel rejected God's correction and leadership, even killing his prophets (1 Kings 19.10; 2 Kings 21.16). **32**: See vv. 2–3. **33–37**: Undeniable evidence accuses her. In shame and sorrow (*hands on your head*), abandoned by her lovers (*Egypt* and *Assyria*), faithless Israel will stand alone and condemned before God.
3.1–4.4: Exhortations to repent. 3.1–5: Judah's sin surpasses anything envisioned by the law (Deut 24.1–4). Though there is a drought in the land (14.1–6) she does not

3 Therefore the showers have been
    withheld,
  and the spring rain has not
    come;
  yet you have the forehead of
    a whore,
  you refuse to be ashamed.
4 Have you not just now called to
    me,
  "My Father, you are the friend
    of my youth—
5 will he be angry forever,
  will he be indignant to the
    end?"
This is how you have spoken,
  but you have done all the evil
    that you could.

6 The LORD said to me in the days of
King Josiah: Have you seen what she did,
that faithless one, Israel, how she went
up on every high hill and under every
green tree, and played the whore there?
7 And I thought, "After she has done all
this she will return to me"; but she did
not return, and her false sister Judah saw
it. 8 She*g* saw that for all the adulteries of
that faithless one, Israel, I had sent her
away with a decree of divorce; yet her
false sister Judah did not fear, but she too
went and played the whore. 9 Because she
took her whoredom so lightly, she polluted the land, committing adultery with
stone and tree. 10 Yet for all this her false
sister Judah did not return to me with her
whole heart, but only in pretense, says
the LORD.

11 Then the LORD said to me: Faithless Israel has shown herself less guilty
than false Judah. 12 Go, and proclaim
these words toward the north, and say:
  Return, faithless Israel,
                says the LORD.

I will not look on you in anger,
  for I am merciful,
                says the LORD;
I will not be angry forever.
13 Only acknowledge your guilt,
  that you have rebelled against
    the LORD your God,
  and scattered your favors among
    strangers under every
    green tree,
  and have not obeyed my voice,
                says the LORD.
14 Return, O faithless children,
                says the LORD,
  for I am your master;
  I will take you, one from a city
    and two from a family,
  and I will bring you to Zion.

15 I will give you shepherds after my
own heart, who will feed you with
knowledge and understanding. 16 And
when you have multiplied and increased
in the land, in those days, says the LORD,
they shall no longer say, "The ark of the
covenant of the LORD." It shall not come
to mind, or be remembered, or missed;
nor shall another one be made. 17 At that
time Jerusalem shall be called the throne
of the LORD, and all nations shall gather
to it, to the presence of the LORD in Jerusalem, and they shall no longer stubbornly follow their own evil will. 18 In
those days the house of Judah shall join
the house of Israel, and together they
shall come from the land of the north to
the land that I gave your ancestors for a
heritage.

19 I thought
  how I would set you among
    my children,

g Q Ms Gk Mss Syr: MT *I*

---

abandon her indiscriminate harlotry (2.20);
she can expect no re-acceptance (contrast
vv. 6–13).
  3.6–11: Judah (the south) is more guilty
than Israel (the north). This prose section is
late, intruding on the sequence of poems. **12–
14**: Israel is invited to repent and return.
  3.15–18: Return from exile and reunion of

Israel and Judah under faithful rulers. This
oracle is probably later than the rest of the
section. **15**: *Shepherds,* rulers. **16–17**: Jerusalem will replace the *ark* as the symbol of the
throne of the LORD (14.21; 17.12).
  3.19–20: These verses continue vv. 12–
14. Judah's unfaithfulness prevented God's
achievement of his hope for his people.

and give you a pleasant land,
the most beautiful heritage of all
the nations.
And I thought you would call me,
My Father,
and would not turn from
following me.
20 Instead, as a faithless wife leaves
her husband,
so you have been faithless to
me, O house of Israel,
says the LORD.

21 A voice on the bare heights[h]
is heard,
the plaintive weeping of Israel's
children,
because they have perverted
their way,
they have forgotten the LORD
their God:
22 Return, O faithless children,
I will heal your faithlessness.

"Here we come to you;
for you are the LORD our God.
23 Truly the hills are[i] a delusion,
the orgies on the mountains.
Truly in the LORD our God
is the salvation of Israel.

24 "But from our youth the shameful
thing has devoured all for which our ancestors had labored, their flocks and their
herds, their sons and their daughters.
25 Let us lie down in our shame, and let
our dishonor cover us; for we have
sinned against the LORD our God, we and
our ancestors, from our youth even to
this day; and we have not obeyed the
voice of the LORD our God."

4 If you return, O Israel,
says the LORD,
if you return to me,
if you remove your abominations
from my presence,

and do not waver,
2 and if you swear, "As the LORD
lives!"
in truth, in justice, and in
uprightness,
then nations shall be blessed[j]
by him,
and by him they shall boast.

3 For thus says the LORD to the people
of Judah and to the inhabitants of Jerusalem:
Break up your fallow ground,
and do not sow among thorns.
4 Circumcise yourselves to the
LORD,
remove the foreskin of your
hearts,
O people of Judah and
inhabitants of Jerusalem,
or else my wrath will go forth
like fire,
and burn with no one to
quench it,
because of the evil of your
doings.

5 Declare in Judah, and proclaim in
Jerusalem, and say:
Blow the trumpet through the
land;
shout aloud[k] and say,
"Gather together, and let us go
into the fortified cities!"
6 Raise a standard toward Zion,
flee for safety, do not delay,
for I am bringing evil from the
north,
and a great destruction.
7 A lion has gone up from its
thicket,
a destroyer of nations has set
out;

h Or *the trails*   *i* Gk Syr Vg: Heb *Truly from
the hills is*   *j* Or *shall bless themselves*
k Or *shout, take your weapons*: Heb *shout, fill*
(*your hand*)

3.21–4.4: From the *heights,* sites of futile
idolatry, come cries of repentance (vv. 12–
14) and resolve to return to God (Hos 14.2–
3). The conditions of repentance are removal
of pagan shrines, recognition of God's exclusive claim by swearing in his name only
(4.2b), and cleansing of hearts. 3–4: Compare
Hos 10.12; Deut 10.16; Rom 2.25–29.
4.5–31: **The foe from the north,** a recurring theme of Jeremiah (1.13–14; 5.15–17;
6.1–5; etc.). 5–12: Sound the alarm! Muster
for the defense (6.1–8)! Like a beast of prey,

he has gone out from his place
to make your land a waste;
your cities will be ruins
without inhabitant.
8 Because of this put on sackcloth,
lament and wail:
"The fierce anger of the LORD
has not turned away from us."

9 On that day, says the LORD, courage shall fail the king and the officials; the priests shall be appalled and the prophets astounded. 10 Then I said, "Ah, Lord GOD, how utterly you have deceived this people and Jerusalem, saying, 'It shall be well with you,' even while the sword is at the throat!"

11 At that time it will be said to this people and to Jerusalem: A hot wind comes from me out of the bare heights *l* in the desert toward my poor people, not to winnow or cleanse— 12 a wind too strong for that. Now it is I who speak in judgment against them.
13 Look! He comes up like clouds,
his chariots like the whirlwind;
his horses are swifter than
eagles—
woe to us, for we are ruined!
14 O Jerusalem, wash your heart
clean of wickedness
so that you may be saved.
How long shall your evil schemes
lodge within you?
15 For a voice declares from Dan
and proclaims disaster from
Mount Ephraim.
16 Tell the nations, "Here they are!"
Proclaim against Jerusalem,
"Besiegers come from a distant
land;

they shout against the cities of
Judah.
17 They have closed in around her
like watchers of a field,
because she has rebelled
against me,
says the LORD.
18 Your ways and your doings
have brought this upon you.
This is your doom; how bitter it
is!
It has reached your very heart."

19 My anguish, my anguish! I writhe
in pain!
Oh, the walls of my heart!
My heart is beating wildly;
I cannot keep silent;
for I *m* hear the sound of the
trumpet,
the alarm of war.
20 Disaster overtakes disaster,
the whole land is laid waste.
Suddenly my tents are destroyed,
my curtains in a moment.
21 How long must I see the standard,
and hear the sound of the
trumpet?
22 "For my people are foolish,
they do not know me;
they are stupid children,
they have no understanding.
They are skilled in doing evil,
but do not know how to do
good."

23 I looked on the earth, and lo, it
was waste and void;
and to the heavens, and they
had no light.

l Or *the trails*   *m* Another reading is *for you,
O my soul,*

the foe approaches (5.6)! Courage will fail the leaders of the people who have ignored all warnings of impending doom (6.13–15; 14.13–16; 23.16–17). God's judgment will sweep over the land like the *hot wind* (the sirocco), desiccating everything before it (18.17). **13–18:** Swiftly, like the eagle and the stormwind, the chariotry and cavalry of the enemy approach. Communiqués trace this

advance from *Dan* (8.16), through *Mount Ephraim* (central Palestine), Benjamin (6.1) into Judah's heartland.
4.19–22: Though the people are foolish and stupid (5.2–3), the prophet laments the sudden disaster which has destroyed his beloved land like the striking of a tent (10.19–21).
4.23–28: In a vision, the prophet sees the

24 I looked on the mountains, and lo,
       they were quaking,
   and all the hills moved to
       and fro.
25 I looked, and lo, there was no one
       at all,
   and all the birds of the air
       had fled.
26 I looked, and lo, the fruitful land
       was a desert,
   and all its cities were laid in
       ruins
   before the LORD, before his
       fierce anger.
27   For thus says the LORD: The whole
land shall be a desolation; yet I will not
make a full end.
28 Because of this the earth shall
       mourn,
   and the heavens above grow
       black;
   for I have spoken, I have
       purposed;
   I have not relented nor will I
       turn back.

29 At the noise of horseman and
       archer
   every town takes to flight;
   they enter thickets; they climb
       among rocks;
   all the towns are forsaken,
   and no one lives in them.
30 And you, O desolate one,
   what do you mean that you dress
       in crimson,
   that you deck yourself with
       ornaments of gold,
   that you enlarge your eyes with
       paint?
   In vain you beautify yourself.

Your lovers despise you;
   they seek your life.
31 For I heard a cry as of a woman
       in labor,
   anguish as of one bringing forth
       her first child,
   the cry of daughter Zion gasping
       for breath,
   stretching out her hands,
   "Woe is me! I am fainting before
       killers!"

5 Run to and fro through the streets
       of Jerusalem,
   look around and take note!
Search its squares and see
   if you can find one person
who acts justly
   and seeks truth—
so that I may pardon Jerusalem. *n*
2 Although they say, "As the LORD
       lives,"
   yet they swear falsely.
3 O LORD, do your eyes not look
       for truth?
You have struck them,
   but they felt no anguish;
you have consumed them,
   but they refused to take
       correction.
They have made their faces harder
       than rock;
   they have refused to turn back.

4 Then I said, "These are only
       the poor,
   they have no sense;
for they do not know the way of
       the LORD,

n Heb *it*

---

terrifying results of God's irrevocable judgment (7.16; 15.1–4). As if struck by sudden
devastation, the earth has been returned to its
primeval state: waste and void (Gen 1.2).
   4.29–31: Like a rejected prostitute (3.2–3),
like a woman in the anguish of childbirth, like
a victim helpless before her murderer, Jerusalem, the daughter Zion, stretches out her
hands in futile appeal and suffers her death
throes—alone.

**5.1–6.30: The corruptions for which
judgment is coming. 5.1–9:** Jeremiah is
commanded to search carefully for someone
faithful (6.9–10). If one could be found, the
LORD would pardon (Gen 18.23–33). He seeks
among the "little people" (*poor*), but finds
none—perhaps they know no better. He
seeks among the privileged *rich,* but there is
none there either (compare Mt 19.23–25).
Therefore beasts of prey will be unleashed

the law of their God.
5 Let me go to the rich⁰
and speak to them;
surely they know the way of
the LORD,
the law of their God."
But they all alike had broken
the yoke,
they had burst the bonds.

6 Therefore a lion from the forest
shall kill them,
a wolf from the desert shall
destroy them.
A leopard is watching against
their cities;
everyone who goes out of them
shall be torn in pieces—
because their transgressions
are many,
their apostasies are great.

7 How can I pardon you?
Your children have forsaken
me,
and have sworn by those who
are no gods.
When I fed them to the full,
they committed adultery
and trooped to the houses of
prostitutes.
8 They were well-fed lusty stallions,
each neighing for his neighbor's
wife.
9 Shall I not punish them for these
things?
says the LORD;
and shall I not bring retribution
on a nation such as this?

10 Go up through her vine-rows
and destroy,
but do not make a full end;
strip away her branches,
for they are not the LORD's.
11 For the house of Israel and the
house of Judah

have been utterly faithless
to me,
says the LORD.
12 They have spoken falsely of
the LORD,
and have said, "He will do
nothing.
No evil will come upon us,
and we shall not see sword or
famine."
13 The prophets are nothing but
wind,
for the word is not in them.
Thus shall it be done to them!

14 Therefore thus says the LORD, the
God of hosts:
Because theyᵖ have spoken
this word,
I am now making my words in
your mouth a fire,
and this people wood, and the
fire shall devour them.
15 I am going to bring upon you
a nation from far away,
O house of Israel,
says the LORD.
It is an enduring nation,
it is an ancient nation,
a nation whose language you do
not know,
nor can you understand what
they say.
16 Their quiver is like an open tomb;
all of them are mighty warriors.
17 They shall eat up your harvest and
your food;
they shall eat up your sons and
your daughters;
they shall eat up your flocks and
your herds;
they shall eat up your vines and
your fig trees;
they shall destroy with the sword

o Or *the great*   *p* Heb *you*

---

against this wayward people (2.15; 4.7; Hab
1.8). **7–9**: In the face of rampant idolatry
(2.11) and immorality (v. 29; 9.9), what alternative does God have?

5.10–17: The LORD's choice vineyard has

grown wild and must be destroyed (2.20–21;
Isa 5.1–7). **12–17**: The judgment. The prophetic words which the people had ignored as
wind will be revealed as the consuming *word*
of God. God is not powerless (Pss 10.4; 14.1)

your fortified cities in which
you trust.

18 But even in those days, says the
LORD, I will not make a full end of you.
19 And when your people say, "Why has
the LORD our God done all these things
to us?" you shall say to them, "As you
have forsaken me and served foreign
gods in your land, so you shall serve
strangers in a land that is not yours."

20 Declare this in the house of Jacob,
  proclaim it in Judah:
21 Hear this, O foolish and senseless
    people,
  who have eyes, but do not see,
  who have ears, but do not hear.
22 Do you not fear me? says the
    LORD;
  Do you not tremble before me?
  I placed the sand as a boundary
    for the sea,
  a perpetual barrier that it cannot
    pass;
  though the waves toss, they
    cannot prevail,
  though they roar, they cannot
    pass over it.
23 But this people has a stubborn and
    rebellious heart;
  they have turned aside and
    gone away.
24 They do not say in their hearts,
  "Let us fear the LORD our God,
  who gives the rain in its season,
    the autumn rain and the
      spring rain,
  and keeps for us
    the weeks appointed for the
      harvest."

25 Your iniquities have turned
    these away,
  and your sins have deprived you
    of good.
26 For scoundrels are found among
    my people;
  they take over the goods of
    others.
  Like fowlers they set a trap;*q*
  they catch human beings.
27 Like a cage full of birds,
  their houses are full of
    treachery;
  therefore they have become great
    and rich,
28   they have grown fat and sleek.
  They know no limits in deeds of
    wickedness;
  they do not judge with justice
  the cause of the orphan, to make
    it prosper,
  and they do not defend the
    rights of the needy.
29 Shall I not punish them for these
    things?
              says the LORD,
  and shall I not bring retribution
    on a nation such as this?

30 An appalling and horrible thing
    has happened in the land:
31   the prophets prophesy falsely,
    and the priests rule as the
      prophets direct;*r*
  my people love to have it so,
    but what will you do when the
      end comes?

q Meaning of Heb uncertain   *r* Or *rule by*
their own authority

---

but will bring upon them the ravages of a
vicious invader. **18–19**: A recurring comment
(9.12–14; 16.10–13; 22.8–9), modifying Jeremiah's customary warnings of total destruction (vv. 16–17; 13.13–14).
  5.20–29: Judah's foolish stubbornness has
closed her eyes and ears to the manifest acts
of God (Prov 1.7; Isa 6.9–10; Mt 13.10–15).
22: Perhaps a reflection of the ancient myth of
a deity's conquest of the mighty monster, the
primordial sea (Job 38.8–11; Ps 104.6–9;
compare Isa 51.9–10). **24b**: The seven weeks

from Passover to Pentecost (Deut 16.9–10)
when rain could ruin the harvest (1 Sam
12.16–18). **26–29**: Judah must be punished
(v. 9) for her tolerance of those who grow
wealthy by defrauding the defenseless (Deut
24.17–18; Am 2.6–7) like fowlers catching
hapless birds in nets and putting them in baskets. **29**: Verse 9.
  5.30–31: The people heartily approve of
the perversity of priest and prophet (6.13–15;
23.9–22; Mic 3.5–8).

6 Flee for safety, O children of
      Benjamin,
   from the midst of Jerusalem!
  Blow the trumpet in Tekoa,
   and raise a signal on
      Beth-haccherem;
  for evil looms out of the north,
   and great destruction.
2 I have likened daughter Zion
   to the loveliest pasture. *s*
3 Shepherds with their flocks shall
      come against her.
  They shall pitch their tents
     around her;
  they shall pasture, all in
     their places.
4 "Prepare war against her;
   up, and let us attack at noon!"
  "Woe to us, for the day declines,
   the shadows of evening
     lengthen!"
5 "Up, and let us attack by night,
   and destroy her palaces!"
6 For thus says the LORD of hosts:
  Cut down her trees;
   cast up a siege ramp against
     Jerusalem.
  This is the city that must be
     punished; *t*
  there is nothing but oppression
     within her.
7 As a well keeps its water fresh,
   so she keeps fresh her
     wickedness;
  violence and destruction are heard
     within her;
  sickness and wounds are ever
     before me.
8 Take warning, O Jerusalem,
   or I shall turn from you in
     disgust,
  and make you a desolation,
   an uninhabited land.

9 Thus says the LORD of hosts:
  Glean *u* thoroughly as a vine
   the remnant of Israel;
  like a grape-gatherer, pass your
     hand again
   over its branches.

10 To whom shall I speak and give
     warning,
   that they may hear?
  See, their ears are closed, *v*
   they cannot listen.
  The word of the LORD is to them
     an object of scorn;
  they take no pleasure in it.
11 But I am full of the wrath of
     the LORD;
   I am weary of holding it in.

  Pour it out on the children in
     the street,
   and on the gatherings of young
     men as well;
  both husband and wife shall
     be taken,
   the old folk and the very aged.
12 Their houses shall be turned over
     to others,
   their fields and wives together;
  for I will stretch out my hand
     against the inhabitants of
     the land,
          says the LORD.

13 For from the least to the greatest
     of them,
   everyone is greedy for unjust
     gain;
  and from prophet to priest,
   everyone deals falsely.

*s Or I will destroy daughter Zion, the loveliest
pasture   t Or the city of license   u Cn: Heb
They shall glean   v Heb are uncircumcised*

6.1–8: The foe approaches from the north.
1: *Tekoa,* twelve miles south of Jerusalem;
Beth-haccherem, modern Ramet Rahel, two
miles south of Jerusalem. **3:** *Shepherds with
their flocks,* kings with their armies (compare
1.15; 12.10). **4:** *Prepare,* more literally, "sanctify" (see Joel 3.9 note *i*).
6.9–15: Jerusalem's obduracy. Jeremiah is

to search meticulously for a God-fearing person (5.1), but he finds none (20.7–18; Mic
7.1–2). God will pour out his wrath (Isa 5.25;
Ezek 6.14) upon the impenitent people, especially on the leaders who promised material
and spiritual well-being (*peace*) when none
was to be had (8.10–12; Ezek 13.10–11).

14 They have treated the wound of
    my people carelessly,
  saying, "Peace, peace,"
    when there is no peace.
15 They acted shamefully, they
      committed abomination;
  yet they were not ashamed,
    they did not know how to
      blush.
  Therefore they shall fall among
      those who fall;
  at the time that I punish them,
    they shall be overthrown,
              says the LORD.
16 Thus says the LORD:
  Stand at the crossroads, and look,
    and ask for the ancient paths,
  where the good way lies; and walk
    in it,
  and find rest for your souls.
  But they said, "We will not walk
    in it."
17 Also I raised up sentinels for you:
  "Give heed to the sound of the
    trumpet!"
  But they said, "We will not give
    heed."
18 Therefore hear, O nations,
  and know, O congregation,
    what will happen to them.
19 Hear, O earth; I am going to
    bring disaster on this
      people,
  the fruit of their schemes,
  because they have not given heed
    to my words;
  and as for my teaching, they
    have rejected it.
20 Of what use to me is frankincense
    that comes from Sheba,
  or sweet cane from a distant
    land?
  Your burnt offerings are not
    acceptable,

nor are your sacrifices pleasing
    to me.
21 Therefore thus says the LORD:
  See, I am laying before this people
    stumbling blocks against which
      they shall stumble;
  parents and children together,
    neighbor and friend shall perish.

22 Thus says the LORD:
  See, a people is coming from the
      land of the north,
  a great nation is stirring from
      the farthest parts of the
        earth.
23 They grasp the bow and the
      javelin,
  they are cruel and have no
      mercy,
  their sound is like the roaring
      sea;
  they ride on horses,
    equipped like a warrior for
      battle,
  against you, O daughter Zion!

24 "We have heard news of them,
    our hands fall helpless;
  anguish has taken hold of us,
    pain as of a woman in labor.
25 Do not go out into the field,
    or walk on the road;
  for the enemy has a sword,
    terror is on every side."

26 O my poor people, put on
      sackcloth,
    and roll in ashes;
  make mourning as for an
      only child,
    most bitter lamentation:
  for suddenly the destroyer
    will come upon us.

6.16–21: Jerusalem is without excuse. God gave his covenant and instruction (*paths, my teaching*) and sent his prophets (*sentinels, my words;* compare Hos 9.8) but both were ignored. The finest and rarest offerings (*frankincense, sweet cane*) are no substitute for faithfulness (7.21–23; Am 5.21–24).

Sheba, southwest Arabia.
6.22–26: The foe from the north (4.5–8) strikes terror into the hearts of the people of the invaded land (Am 8.10; Zech 12.10). **25:** *Terror is on every side,* a characteristic expression of Jeremiah to describe all-encompassing danger (20.3, 10; 46.5; 49.29; Lam 2.22).

27 I have made you a tester and a
refiner*w* among my people
so that you may know and test
their ways.
28 They are all stubbornly rebellious,
going about with slanders;
they are bronze and iron,
all of them act corruptly.
29 The bellows blow fiercely,
the lead is consumed by the fire;
in vain the refining goes on,
for the wicked are not removed.
30 They are called "rejected silver,"
for the LORD has rejected them.

7 The word that came to Jeremiah
from the LORD: 2 Stand in the gate of
the LORD's house, and proclaim there this
word, and say, Hear the word of the
LORD, all you people of Judah, you that
enter these gates to worship the LORD.
3 Thus says the LORD of hosts, the God of
Israel: Amend your ways and your doings, and let me dwell with you*x* in this
place. 4 Do not trust in these deceptive
words: "This is*y* the temple of the LORD,
the temple of the LORD, the temple of the
LORD."

5 For if you truly amend your ways
and your doings, if you truly act justly
one with another, 6 if you do not oppress
the alien, the orphan, and the widow, or
shed innocent blood in this place, and if
you do not go after other gods to your
own hurt, 7 then I will dwell with you in
this place, in the land that I gave of old
to your ancestors forever and ever.

8 Here you are, trusting in deceptive
words to no avail. 9 Will you steal, murder, commit adultery, swear falsely,
make offerings to Baal, and go after other gods that you have not known, 10 and
then come and stand before me in this
house, which is called by my name, and
say, "We are safe!"—only to go on doing
all these abominations? 11 Has this house,
which is called by my name, become a
den of robbers in your sight? You know,
I too am watching, says the LORD. 12 Go
now to my place that was in Shiloh,
where I made my name dwell at first, and
see what I did to it for the wickedness of
my people Israel. 13 And now, because
you have done all these things, says the
LORD, and when I spoke to you persistently, you did not listen, and when I
called you, you did not answer, 14 therefore I will do to the house that is called
by my name, in which you trust, and to
the place that I gave to you and to your
ancestors, just what I did to Shiloh.
15 And I will cast you out of my sight,
just as I cast out all your kinsfolk, all the
offspring of Ephraim.

16 As for you, do not pray for this
people, do not raise a cry or prayer on
their behalf, and do not intercede with
me, for I will not hear you. 17 Do you not
see what they are doing in the towns of
Judah and in the streets of Jerusalem?
18 The children gather wood, the fathers
kindle fire, and the women knead dough,
to make cakes for the queen of heaven;
and they pour out drink offerings to other gods, to provoke me to anger. 19 Is it

w Or *a fortress*   *x* Or *and I will let you dwell*
y Heb *They are*

6.27–30: Jeremiah, the assayer. One function of the prophetic office was to assay the
faithfulness (*silver*) of the people. Jeremiah
discovers that there is no precious metal to be
found.
7.1–15: **The temple sermon** (compare
26.4–6). **4**: The temple's presence was mistakenly interpreted as necessarily assuring
God's protection (Isa 31.4; for three-fold emphasis, compare 22.29; Isa 6.3). Jeremiah disagreed (Mic 3.12); a complete moral change
was required (vv. 5–6; compare Hos 4.2; Mic
6.8).
7.10–12: As *Shiloh* (7.12, eighteen miles
north of Jerusalem), the earlier central shrine,
was destroyed (around 1050 B.C. in the days
of Samuel; compare 1 Sam chs 4–6; Ps 78.56–
72), so also *this house*, desecrated by idolatry,
will be destroyed (vv. 10, 11; compare Mt
21.13). Immediately following this sermon,
Jeremiah was arrested (see 26.8).
7.16–8.3: **Abuses in worship. 16–20**: Because of Judah's general apostasy, God forbids Jeremiah to exercise one of the functions
of the prophetic office: intercession (7.16;
11.14; 15.1; compare Am 7.2, 5). *Queen of
heaven*, see 44.15–28 n.

I whom they provoke? says the LORD. Is it not themselves, to their own hurt? ²⁰ Therefore thus says the Lord GOD: My anger and my wrath shall be poured out on this place, on human beings and animals, on the trees of the field and the fruit of the ground; it will burn and not be quenched.

21 Thus says the LORD of hosts, the God of Israel: Add your burnt offerings to your sacrifices, and eat the flesh. ²² For in the day that I brought your ancestors out of the land of Egypt, I did not speak to them or command them concerning burnt offerings and sacrifices. ²³ But this command I gave them, "Obey my voice, and I will be your God, and you shall be my people; and walk only in the way that I command you, so that it may be well with you." ²⁴ Yet they did not obey or incline their ear, but, in the stubbornness of their evil will, they walked in their own counsels, and looked backward rather than forward. ²⁵ From the day that your ancestors came out of the land of Egypt until this day, I have persistently sent all my servants the prophets to them, day after day; ²⁶ yet they did not listen to me, or pay attention, but they stiffened their necks. They did worse than their ancestors did.

27 So you shall speak all these words to them, but they will not listen to you. You shall call to them, but they will not answer you. ²⁸ You shall say to them: This is the nation that did not obey the voice of the LORD their God, and did not accept discipline; truth has perished; it is cut off from their lips.

29 Cut off your hair and throw
      it away;
   raise a lamentation on the bare
      heights, ^z

for the LORD has rejected and
      forsaken
   the generation that provoked
      his wrath.

30 For the people of Judah have done evil in my sight, says the LORD; they have set their abominations in the house that is called by my name, defiling it. ³¹ And they go on building the high place^a of Topheth, which is in the valley of the son of Hinnom, to burn their sons and their daughters in the fire—which I did not command, nor did it come into my mind. ³² Therefore, the days are surely coming, says the LORD, when it will no more be called Topheth, or the valley of the son of Hinnom, but the valley of Slaughter: for they will bury in Topheth until there is no more room. ³³ The corpses of this people will be food for the birds of the air, and for the animals of the earth; and no one will frighten them away. ³⁴ And I will bring to an end the sound of mirth and gladness, the voice of the bride and bridegroom in the cities of Judah and in the streets of Jerusalem; for the land shall become a waste.

8 At that time, says the LORD, the bones of the kings of Judah, the bones of its officials, the bones of the priests, the bones of the prophets, and the bones of the inhabitants of Jerusalem shall be brought out of their tombs; ² and they shall be spread before the sun and the moon and all the host of heaven, which they have loved and served, which they have followed, and which they have inquired of and worshiped; and they shall not be gathered or buried; they shall be like dung on the surface of the ground. ³ Death shall be preferred to life by all the

---

z Or *the trails*   *a* Gk Tg: Heb *high places*

7.21–28: Faithfulness, not sacrifice, is required. Sacrifices are acceptable only when the right relationship exists between God and those who sacrifice (6.20; Ps 51.15–19). Until then, worshipers may as well eat the flesh of the *burnt offerings,* customarily incinerated (Lev ch 1), as well as the other offerings (Lev ch 3; 7.11–18).
   7.29–8.3: The fate of Judah. **29:** *Cut off your hair,* a sign of mourning (16.6; Mic 1.16).

31: The most gruesome of Israel's aberrations was the sacrifice of children (19.5; 32.35) on the burning platform (*Topheth,* 2 Kings 23.10). Strictly forbidden by God (Lev 18.21), it will eventually be recognized as murder. *Valley of the son of Hinnom,* southwest of the city joining the Kidron valley. **33:** Corpses as carrion were a frightful prospect. **34:** 16.9. **8.1–2:** Jeremiah here uses biting irony.

remnant that remains of this evil family in all the places where I have driven them, says the LORD of hosts.

4 You shall say to them, Thus says
the LORD:
When people fall, do they not get
up again?
If they go astray, do they not
turn back?
5 Why then has this people*b* turned
away
in perpetual backsliding?
They have held fast to deceit,
they have refused to return.
6 I have given heed and listened,
but they do not speak honestly;
no one repents of wickedness,
saying, "What have I done!"
All of them turn to their own
course,
like a horse plunging headlong
into battle.
7 Even the stork in the heavens
knows its times;
and the turtledove, swallow, and
crane*c*
observe the time of their
coming;
but my people do not know
the ordinance of the LORD.

8 How can you say, "We are wise,
and the law of the LORD is
with us,"
when, in fact, the false pen of
the scribes
has made it into a lie?
9 The wise shall be put to shame,
they shall be dismayed and
taken;
since they have rejected the word
of the LORD,
what wisdom is in them?

10 Therefore I will give their wives
to others
and their fields to conquerors,
because from the least to the
greatest
everyone is greedy for unjust
gain;
from prophet to priest
everyone deals falsely.
11 They have treated the wound of
my people carelessly,
saying, "Peace, peace,"
when there is no peace.
12 They acted shamefully, they
committed abomination;
yet they were not at all
ashamed,
they did not know how to
blush.
Therefore they shall fall among
those who fall;
at the time when I punish them,
they shall be overthrown,
says the LORD.
13 When I wanted to gather them,
says the LORD,
there are*d* no grapes on the
vine,
nor figs on the fig tree;
even the leaves are withered,
and what I gave them has
passed away from them.*c*

14 Why do we sit still?
Gather together, let us go into the
fortified cities
and perish there;
for the LORD our God has doomed
us to perish,
and has given us poisoned water
to drink,

b One Ms Gk: MT *this people, Jerusalem,*
c Meaning of Heb uncertain   *d* Or *I will make
an end of them, says the* LORD. *There are*

**8.4–10.25: Miscellaneous oracles. 8.4–
7:** Israel's incredible indifference (18.13–17).
Anyone who falls naturally gets up; the birds
follow their natural instincts; but Israel,
God's people, forgets God's law.
8.8–9: Word vs. law. Here Jeremiah contrasts his proclaimed *word of the* LORD with the
written tradition (*law*) misinterpreted by

those who administer it (*scribes;* compare 2.8).
10c–12: A doublet of 6.13–15. **13:** Judah, like
an unfruitful *vine* or *fig tree,* will be destroyed
(Lk 13.7).
8.14–17: Panic-stricken before the invader, the inhabitants of Judah seek temporary
defense in their fortresses, but there is no defense. *Poisoned water* (cup of God's wrath,

because we have sinned against
the LORD.
15 We look for peace, but find no
good,
for a time of healing, but there
is terror instead.

16 The snorting of their horses is
heard from Dan;
at the sound of the neighing of
their stallions
the whole land quakes.
They come and devour the land
and all that fills it,
the city and those who live in
it.
17 See, I am letting snakes loose
among you,
adders that cannot be charmed,
and they shall bite you,
says the LORD.

18 My joy is gone, grief is upon me,
my heart is sick.
19 Hark, the cry of my poor people
from far and wide in the land:
"Is the LORD not in Zion?
Is her King not in her?"
("Why have they provoked me to
anger with their images,
with their foreign idols?")
20 "The harvest is past, the summer
is ended,
and we are not saved."
21 For the hurt of my poor people I
am hurt,
I mourn, and dismay has taken
hold of me.

22 Is there no balm in Gilead?
Is there no physician there?
Why then has the health of my
poor people
not been restored?

9 *e* O that my head were a spring of
water,
and my eyes a fountain of tears,
so that I might weep day and
night
for the slain of my poor people!
2 *f* O that I had in the desert
a traveler's lodging place,
that I might leave my people
and go away from them!
For they are all adulterers,
a band of traitors.
3 They bend their tongues like
bows;
they have grown strong in the
land for falsehood, and not
for truth;
for they proceed from evil to evil,
and they do not know me, says
the LORD.

4 Beware of your neighbors,
and put no trust in any of
your kin;*g*
for all your kin*h* are supplanters,
and every neighbor goes around
like a slanderer.
5 They all deceive their neighbors,
and no one speaks the truth;
they have taught their tongues to
speak lies;
they commit iniquity and are
too weary to repent.*i*
6 Oppression upon oppression,
deceit*j* upon deceit!
They refuse to know me, says
the LORD.

7 Therefore thus says the LORD of
hosts:

e Ch 8.23 in Heb   *f* Ch 9.1 in Heb
g Heb *in a brother*   *h* Heb *for every brother*
i Cn Compare Gk: Heb *they weary themselves
with iniquity.* 6*Your dwelling*   *j* Cn: Heb *Your
dwelling in the midst of deceit*

9.15; Ps 75.8; Num ch 5). *Dan,* northernmost
point in Israel. *Snakes* (Eccl 10.11; Ps 58.4–5;
Num 21.4–9).
8.18–9.1: Lament over Judah. It distresses
Jeremiah to denounce his people. **20:** They
were perhaps suffering from drought (ch 14).
22: *My poor people,* Judah (4.11). *Balm in Gilead,* resin from the Styrax tree, produced espe-

cially in the north Transjordan region of Gilead, widely used for medicinal purposes
(46.11; Gen 37.25). **9.2–9:** The complete corruption of the people is the basis for this accusation. Here, instead of sympathy, God has
nothing but contempt for the lying, deceiving, untrustworthy nation (11.19–23; 12.6;
20.10). A remote stopping place in the desert

I will now refine and test them,
  for what else can I do with my
    sinful people?*k*
8 Their tongue is a deadly arrow;
  it speaks deceit through the
    mouth.
They all speak friendly words to
    their neighbors,
  but inwardly are planning to lay
    an ambush.
9 Shall I not punish them for these
    things? says the LORD;
  and shall I not bring retribution
    on a nation such as this?

10 Take up*l* weeping and wailing for
    the mountains,
  and a lamentation for the
    pastures of the wilderness,
because they are laid waste so that
    no one passes through,
  and the lowing of cattle is not
    heard;
both the birds of the air and the
    animals
  have fled and are gone.
11 I will make Jerusalem a heap
    of ruins,
  a lair of jackals;
and I will make the towns of
    Judah a desolation,
  without inhabitant.

12 Who is wise enough to understand
this? To whom has the mouth of the
LORD spoken, so that they may declare
it? Why is the land ruined and laid waste
like a wilderness, so that no one passes
through? 13 And the LORD says: Because
they have forsaken my law that I set before them, and have not obeyed my
voice, or walked in accordance with it,
14 but have stubbornly followed their
own hearts and have gone after the Baals,

as their ancestors taught them. 15 Therefore thus says the LORD of hosts, the God
of Israel: I am feeding this people with
wormwood, and giving them poisonous
water to drink. 16 I will scatter them
among nations that neither they nor their
ancestors have known; and I will send
the sword after them, until I have consumed them.

17 Thus says the LORD of hosts:
Consider, and call for the
    mourning women to come;
  send for the skilled women
    to come;
18 let them quickly raise a dirge
    over us,
  so that our eyes may run down
    with tears,
  and our eyelids flow with
    water.
19 For a sound of wailing is heard
    from Zion:
  "How we are ruined!
We are utterly shamed,
  because we have left the land,
  because they have cast down
    our dwellings."

20 Hear, O women, the word of
    the LORD,
  and let your ears receive the
    word of his mouth;
teach to your daughters a dirge,
  and each to her neighbor
    a lament.
21 "Death has come up into our
    windows,
  it has entered our palaces,
to cut off the children from
    the streets

---

k Or *my poor people*
l Gk Syr: Heb *I will take up*

---

would be preferable (1 Kings 19.3–4). **9**: 5.9,
29.
9.10–22: Lamentation over Zion. **12–16**: A
composite commentary on Jeremiah's oracle
in vv. 10–11, 17–22 (compare 5.18–19). *Poisonous water*, 8.14; 23.15; Ezek 23.31–34. Baal
worship was a human fabrication, not God's
revelation. **10–11, 17–22**: The stillness of death

shrouds the ruined land (7.34); only the howl
of jackals breaks the silence. As customary in
the ancient world, professional *mourning
women* are hired to bewail the fate of Zion. **21**:
The verse reflects the figure of speech of a
Canaanite mythological allusion to the god of
death.

and the young men from
the squares."

22 Speak! Thus says the LORD:
"Human corpses shall fall
like dung upon the open field,
like sheaves behind the reaper,
and no one shall gather them."

23 Thus says the LORD: Do not let the wise boast in their wisdom, do not let the mighty boast in their might, do not let the wealthy boast in their wealth; 24 but let those who boast boast in this, that they understand and know me, that I am the LORD; I act with steadfast love, justice, and righteousness in the earth, for in these things I delight, says the LORD.

25 The days are surely coming, says the LORD, when I will attend to all those who are circumcised only in the foreskin: 26 Egypt, Judah, Edom, the Ammonites, Moab, and all those with shaven temples who live in the desert. For all these nations are uncircumcised, and all the house of Israel is uncircumcised in heart.

10 Hear the word that the LORD speaks to you, O house of Israel.
2 Thus says the LORD:
Do not learn the way of the
nations,
or be dismayed at the signs of
the heavens;
for the nations are dismayed
at them.
3 For the customs of the peoples
are false:
a tree from the forest is cut down,
and worked with an ax by the
hands of an artisan;
4 people deck it with silver and
gold;

they fasten it with hammer
and nails
so that it cannot move.
5 Their idols*m* are like scarecrows in
a cucumber field,
and they cannot speak;
they have to be carried,
for they cannot walk.
Do not be afraid of them,
for they cannot do evil,
nor is it in them to do good.

6 There is none like you, O LORD;
you are great, and your name is
great in might.
7 Who would not fear you, O King
of the nations?
For that is your due;
among all the wise ones of the
nations
and in all their kingdoms
there is no one like you.
8 They are both stupid and foolish;
the instruction given by idols
is no better than wood!*n*
9 Beaten silver is brought from
Tarshish,
and gold from Uphaz.
They are the work of the artisan
and of the hands of the
goldsmith;
their clothing is blue and purple;
they are all the product of
skilled workers.
10 But the LORD is the true God;
he is the living God and the
everlasting King.
At his wrath the earth quakes,
and the nations cannot endure
his indignation.

m Heb *They*   *n* Meaning of Heb uncertain

9.23–24: True glory (1 Cor 1.31). The common goals of human beings are nothing (1 Kings 3.10–12) compared with the knowledge of God. **25–26:** Though physically circumcised, her neighbors will share with faithless Israel the lot of those not in right relationship with God (Deut 10.16; Rom 2.25–29). *Those with shaven temples,* Arab tribes (25.23).
10.1–16: God and the idols. This passage is an elaboration of an oracle by Jeremiah (compare Isa 44.9–20; Ps 115.3–8). **2:** *Signs of the heavens* (eclipses, comets, astrological observations) are meaningless. **4:** *People deck it with silver and gold,* compare the description of making an idol in Isa 40.18–20; 41.6–7. **5:** *They have to be carried,* this theme is effectively presented in Isa 46.1–7. *Cucumber field,* Isa 1.8. **9:** *Tarshish,* Sardinia, or Tartessus in southern Spain. *Uphaz,* unknown.

11 Thus shall you say to them: The gods who did not make the heavens and the earth shall perish from the earth and from under the heavens. *o*

12 It is he who made the earth by
    his power,
  who established the world by
    his wisdom,
  and by his understanding
    stretched out the heavens.
13 When he utters his voice, there is
    a tumult of waters in the
    heavens,
  and he makes the mist rise from
    the ends of the earth.
  He makes lightnings for the rain,
    and he brings out the wind
    from his storehouses.
14 Everyone is stupid and without
    knowledge;
  goldsmiths are all put to shame
    by their idols;
  for their images are false,
    and there is no breath in them.
15 They are worthless, a work of
    delusion;
  at the time of their punishment
    they shall perish.
16 Not like these is the LORD, *p* the
    portion of Jacob,
  for he is the one who formed
    all things,
  and Israel is the tribe of his
    inheritance;
  the LORD of hosts is his name.

17 Gather up your bundle from the
    ground,
  O you who live under siege!
18 For thus says the LORD:
  I am going to sling out the
    inhabitants of the land
    at this time,

and I will bring distress on them,
    so that they shall feel it.

19 Woe is me because of my hurt!
    My wound is severe.
  But I said, "Truly this is my
    punishment,
    and I must bear it."
20 My tent is destroyed,
    and all my cords are broken;
  my children have gone from me,
    and they are no more;
  there is no one to spread my tent
    again,
    and to set up my curtains.
21 For the shepherds are stupid,
    and do not inquire of the LORD;
  therefore they have not prospered,
    and all their flock is scattered.

22 Hear, a noise! Listen, it is
    coming—
  a great commotion from the
    land of the north
  to make the cities of Judah a
    desolation,
    a lair of jackals.

23 I know, O LORD, that the way of
    human beings is not in
    their control,
  that mortals as they walk cannot
    direct their steps.
24 Correct me, O LORD, but in
    just measure;
  not in your anger, or you will
    bring me to nothing.

25 Pour out your wrath on the
    nations that do not know
    you,

o This verse is in Aramaic   *p* Heb lacks *the
LORD*

10.11: An Aramaic gloss (unique in Jeremiah). **12–16**: He alone (*the portion of Jacob*; Lam 3.24) is worthy of worship (51.15–19). **10.17–22**: Prepare to leave! Related to but later than 9.10–22, this passage describes a siege (perhaps in 597 B.C.) which will not be lifted. In Exile, mother Zion will lose her children (Isa 49.14–23; 54.1–3) because of the stupidity of Judah's rulers (*shepherds*). God's role in political events is not to be overlooked. **22b**: 9.11. **10.23–25**: A prayer. The one who prays cites Prov 20.24 (v. 23, compare Rom 1.18–23) and Ps 79.6–7 (v. 25) to direct God's attention away from Judah and to neighboring nations instead.

and on the peoples that do not
  call on your name;
for they have devoured Jacob;
  they have devoured him and
    consumed him,
  and have laid waste his
    habitation.

11 The word that came to Jeremiah
from the LORD: ²Hear the words
of this covenant, and speak to the people
of Judah and the inhabitants of Jerusalem. ³You shall say to them, Thus says
the LORD, the God of Israel: Cursed be
anyone who does not heed the words of
this covenant, ⁴which I commanded
your ancestors when I brought them out
of the land of Egypt, from the iron-
smelter, saying, Listen to my voice, and
do all that I command you. So shall you
be my people, and I will be your God,
⁵that I may perform the oath that I swore
to your ancestors, to give them a land
flowing with milk and honey, as at this
day. Then I answered, "So be it, LORD."

6 And the LORD said to me: Proclaim
all these words in the cities of Judah, and
in the streets of Jerusalem: Hear the
words of this covenant and do them.
⁷For I solemnly warned your ancestors
when I brought them up out of the land
of Egypt, warning them persistently,
even to this day, saying, Obey my voice.
⁸Yet they did not obey or incline their
ear, but everyone walked in the stubbornness of an evil will. So I brought
upon them all the words of this covenant, which I commanded them to do,
but they did not.

9 And the LORD said to me: Conspiracy exists among the people of Judah and
the inhabitants of Jerusalem. ¹⁰They
have turned back to the iniquities of their

ancestors of old, who refused to heed my
words; they have gone after other gods
to serve them; the house of Israel and the
house of Judah have broken the covenant
that I made with their ancestors.
¹¹Therefore, thus says the LORD, assuredly I am going to bring disaster upon
them that they cannot escape; though
they cry out to me, I will not listen to
them. ¹²Then the cities of Judah and the
inhabitants of Jerusalem will go and cry
out to the gods to whom they make offerings, but they will never save them in
the time of their trouble. ¹³For your gods
have become as many as your towns,
O Judah; and as many as the streets of
Jerusalem are the altars you have set up
to shame, altars to make offerings to
Baal.

14 As for you, do not pray for this
people, or lift up a cry or prayer on their
behalf, for I will not listen when they call
to me in the time of their trouble. ¹⁵What
right has my beloved in my house, when
she has done vile deeds? Can vows*q* and
sacrificial flesh avert your doom? Can
you then exult? ¹⁶The LORD once called
you, "A green olive tree, fair with goodly fruit"; but with the roar of a great
tempest he will set fire to it, and its
branches will be consumed. ¹⁷The LORD
of hosts, who planted you, has pronounced evil against you, because of the
evil that the house of Israel and the house
of Judah have done, provoking me to
anger by making offerings to Baal.

18 It was the LORD who made it
    known to me, and I knew;
  then you showed me their
    evil deeds.

q Gk: Heb *Can many*

11.1–17: Jeremiah and the covenant. Jeremiah undoubtedly strongly supported Josiah's effort to eradicate practices of foreign
worship (2 Kings chs 22–23) and to return to
the provisions of the Mosaic covenant (v. 3,
this covenant; v. 10, *the covenant*). Prophetic
preaching was based upon the covenant relationship with God. Typically Deuteronomic
is the word *command* used of the covenant
(Deut 4.13; 6.17; etc.); note also other familiar Deuteronomic phrases: v. 3, Deut 27.26;
v. 4, Deut 4.20; v. 5, Deut 7.12–13; v. 8,
Deut 29.19. *Iron-smelter,* Deut 4.20; 1 Kings
8.51; Isa 48.10. Intercession for the apostate
people is useless (14.11–12), as are also their
rituals. **16:** The *green olive tree* is a symbol of
Israel (Hos 14.6).

19 But I was like a gentle lamb
    led to the slaughter.
And I did not know it was
        against me
    that they devised schemes,
        saying,
    "Let us destroy the tree with
        its fruit,
    let us cut him off from the land
        of the living,
    so that his name will no longer
        be remembered!"
20 But you, O LORD of hosts, who
        judge righteously,
    who try the heart and the mind,
    let me see your retribution upon
        them,
        for to you I have committed
        my cause.

21 Therefore thus says the LORD concerning the people of Anathoth, who
seek your life, and say, "You shall not
prophesy in the name of the LORD, or
you will die by our hand"— 22 therefore
thus says the LORD of hosts: I am going
to punish them; the young men shall die
by the sword; their sons and their daughters shall die by famine; 23 and not even a
remnant shall be left of them. For I will
bring disaster upon the people of Anathoth, the year of their punishment.

12 You will be in the right,
        O LORD,
    when I lay charges against you;
    but let me put my case to you.
Why does the way of the guilty
        prosper?
    Why do all who are treacherous
        thrive?

2 You plant them, and they take
        root;
    they grow and bring forth fruit;
    you are near in their mouths
        yet far from their hearts.
3 But you, O LORD, know me;
    You see me and test me—my
        heart is with you.
Pull them out like sheep for the
        slaughter,
    and set them apart for the day
        of slaughter.
4 How long will the land mourn,
    and the grass of every field
        wither?
For the wickedness of those who
        live in it
    the animals and the birds are
        swept away,
    and because people said, "He is
        blind to our ways."[r]

5 If you have raced with
        foot-runners and they have
        wearied you,
    how will you compete with
        horses?
And if in a safe land you fall
        down,
    how will you fare in the thickets
        of the Jordan?
6 For even your kinsfolk and your
        own family,
    even they have dealt
        treacherously with you;
    they are in full cry after you;
do not believe them,
    though they speak friendly
        words to you.

r Gk: Heb *to our future*

**11.18–12.6: Jeremiah's first personal la-
ment. A plot against the life of Jeremiah.**
The first of Jeremiah's six personal laments;
the others are 15.10–21; 17.14–18; 18.18–23;
20.7–13; 20.14–18. He learns that he is the
unwitting object (*gentle lamb;* compare Isa
53.7) of an assassination plot, and prays the
omniscient God for protection (17.10; Ps 26).
11.21–23: The origin of the grievance
against Jeremiah may be his identification of
certain fellow citizens with false prophets and
priests. In his characteristic phraseology

(5.12; 18.21; 19.15; 23.12) their end is foretold.
12.1–5: Against the background of the law
court, Jeremiah questions the then nearly universal idea that the wicked always suffer and
the righteous always prosper. The wicked,
ostensibly faithful, prosper (Job ch 21; Ps 73).
After Jeremiah's assertion of integrity (Ps
139.23–24), God, by two proverbs, informs
him that the present is but a preparation for
a more demanding future.

7 I have forsaken my house,
　I have abandoned my heritage;
　I have given the beloved of
　　my heart
　　into the hands of her enemies.
8 My heritage has become to me
　like a lion in the forest;
　she has lifted up her voice
　　against me—
　　therefore I hate her.
9 Is the hyena greedyˢ for my
　　heritage at my command?
　Are the birds of prey all
　　around her?
　Go, assemble all the wild animals;
　　bring them to devour her.
10 Many shepherds have destroyed
　　my vineyard,
　they have trampled down
　　my portion,
　they have made my pleasant
　　portion
　　a desolate wilderness.
11 They have made it a desolation;
　　desolate, it mourns to me.
　The whole land is made desolate,
　　but no one lays it to heart.
12 Upon all the bare heightsᵗ in
　　the desert
　　spoilers have come;
　for the sword of the LORD devours
　　from one end of the land to
　　the other;
　　no one shall be safe.
13 They have sown wheat and have
　　reaped thorns,
　they have tired themselves out
　　but profit nothing.
　They shall be ashamed of theirᵘ
　　harvests
　　because of the fierce anger of
　　the LORD.

14 Thus says the LORD concerning all my evil neighbors who touch the heritage that I have given my people Israel to inherit: I am about to pluck them up from their land, and I will pluck up the house of Judah from among them. ¹⁵ And after I have plucked them up, I will again have compassion on them, and I will bring them again to their heritage and to their land, everyone of them. ¹⁶ And then, if they will diligently learn the ways of my people, to swear by my name, "As the LORD lives," as they taught my people to swear by Baal, then they shall be built up in the midst of my people. ¹⁷ But if any nation will not listen, then I will completely uproot it and destroy it, says the LORD.

13 Thus said the LORD to me, "Go and buy yourself a linen loincloth, and put it on your loins, but do not dip it in water." ² So I bought a loincloth according to the word of the LORD, and put it on my loins. ³ And the word of the LORD came to me a second time, saying, ⁴ "Take the loincloth that you bought and are wearing, and go now to the Euphrates,ᵛ and hide it there in a cleft of the rock." ⁵ So I went, and hid it by the Euphrates,ʷ as the LORD commanded me. ⁶ And after many days the LORD said to me, "Go now to the Euphrates,ᵛ and take from there the loincloth that I commanded you to hide there." ⁷ Then I went to the Euphrates,ᵛ and dug, and I took the loincloth from the place where I had

---

s Cn: Heb *Is the hyena, the bird of prey*
t Or *the trails*　u Heb *your*　v Or *to Parah*;
Heb *perath*　w Or *by Parah*; Heb *perath*

12.7–13: God's lament. Using several figures (*beloved of my heart*, 11.15; *lion; birds of prey; vineyard*, 5.10; Isa 5.1–7), God laments the necessary ravaging of Judah (*my house*, Hos 9.15; *heritage*, 1 Sam 10.1) by the Babylonian-inspired raids of 601 B.C., led by several puppet rulers (*birds of prey, shepherds*), in reprisal for Jehoiakim's short-lived revolt (2 Kings 24.1–4).
12.14–17: Judah's neighbors. If Judah's

erstwhile enemies are converted (16.19–21; Ps 87), God will spare them when Judah is restored; otherwise they will be destroyed (25.12–38). **16:** *They,* a reference to the enemies.
13.1–11: The story of the loincloth. Jeremiah evidently associated the name of a village near his home town of Anathoth, Parah (Josh 18.23), where there are rocky cliffs (v. 4), with the Hebrew name of the Euphra-

hidden it. But now the loincloth was ruined; it was good for nothing.

8 Then the word of the LORD came to me: ⁹Thus says the LORD: Just so I will ruin the pride of Judah and the great pride of Jerusalem. ¹⁰This evil people, who refuse to hear my words, who stubbornly follow their own will and have gone after other gods to serve them and worship them, shall be like this loincloth, which is good for nothing. ¹¹For as the loincloth clings to one's loins, so I made the whole house of Israel and the whole house of Judah cling to me, says the LORD, in order that they might be for me a people, a name, a praise, and a glory. But they would not listen.

12 You shall speak to them this word: Thus says the LORD, the God of Israel: Every wine-jar should be filled with wine. And they will say to you, "Do you think we do not know that every wine-jar should be filled with wine?" ¹³Then you shall say to them: Thus says the LORD: I am about to fill all the inhabitants of this land—the kings who sit on David's throne, the priests, the prophets, and all the inhabitants of Jerusalem—with drunkenness. ¹⁴And I will dash them one against another, parents and children together, says the LORD. I will not pity or spare or have compassion when I destroy them.

15 Hear and give ear; do not be
    haughty,
    for the LORD has spoken.
16 Give glory to the LORD your God
    before he brings darkness,

and before your feet stumble
    on the mountains at twilight;
while you look for light,
    he turns it into gloom
    and makes it deep darkness.
17 But if you will not listen,
    my soul will weep in secret for
        your pride;
my eyes will weep bitterly and
    run down with tears,
    because the LORD's flock has
        been taken captive.

18 Say to the king and the queen
        mother;
    "Take a lowly seat,
for your beautiful crown
    has come down from your
        head."ˣ
19 The towns of the Negeb are shut
        up
    with no one to open them;
all Judah is taken into exile,
    wholly taken into exile.

20 Lift up your eyes and see
    those who come from the
        north.
Where is the flock that was
        given you,
    your beautiful flock?
21 What will you say when they set
        as head over you
    those whom you have trained
    to be your allies?
Will not pangs take hold of you,
    like those of a woman in labor?
22 And if you say in your heart,

---

x Gk Syr Vg: Meaning of Heb uncertain

---

tes (*Perath*). He buried a linen loincloth there to underscore the corrupting effect of Jehoiakim's pro-Babylonian foreign policy (2.18) and the accompanying religious syncretism in Judah (2 Kings 24.1–7).

13.12–14: The allegory of the wine-jar. Jeremiah uses a proverb (v. 12a) and the figure of drunkenness (v. 13; compare 25.15–16; Ezek 23.31) to describe the divine judgment.

13.15–17: The last opportunity for repentance is offered to wayward Judah; her captivity is imminent (v. 17).

13.18–19: Exile! Almost resignedly, Jeremiah notes the deposition of Jehoiachin (2 Kings 24.8, 15), the Babylonian military occupation of Judah (597 B.C.), and the first deportation (Jer 52.28).

13.20–27: Jerusalem's shame. Using a common biblical simile for judgment, the ravishing of a woman (Isa 47.2; here the untrustworthy shepherdess), Jeremiah describes the approaching rape (vv. 22, 26) of Jerusalem by Babylon. **20:** *North,* see 1.13–14 n.; 4.6. Her shamelessness and her shame are di-

"Why have these things come
          upon me?"
it is for the greatness of your
          iniquity
    that your skirts are lifted up,
    and you are violated.
23 Can Ethiopians*y* change their skin
          or leopards their spots?
    Then also you can do good
          who are accustomed to do evil.
24 I will scatter you*z* like chaff
          driven by the wind from
          the desert.
25 This is your lot,
          the portion I have measured out
          to you, says the LORD,
    because you have forgotten me
          and trusted in lies.
26 I myself will lift up your skirts
          over your face,
    and your shame will be seen.
27 I have seen your abominations,
          your adulteries and neighings,
          your shameless prostitutions
          on the hills of the countryside.
    Woe to you, O Jerusalem!
    How long will it be
    before you are made clean?

14 The word of the LORD that came
to Jeremiah concerning the
drought:
2 Judah mourns
    and her gates languish;
    they lie in gloom on the ground,
    and the cry of Jerusalem goes
          up.
3 Her nobles send their servants
          for water;
    they come to the cisterns,
    they find no water,
          they return with their
          vessels empty.
    They are ashamed and dismayed

and cover their heads,
4 because the ground is cracked.
    Because there has been no rain
          on the land
    the farmers are dismayed;
    they cover their heads.
5 Even the doe in the field forsakes
          her newborn fawn
    because there is no grass.
6 The wild asses stand on the bare
          heights,*a*
    they pant for air like jackals;
    their eyes fail
    because there is no herbage.

7 Although our iniquities testify
          against us,
    act, O LORD, for your name's
          sake;
    our apostasies indeed are many,
    and we have sinned against you.
8 O hope of Israel,
          its savior in time of trouble,
    why should you be like a stranger
          in the land,
    like a traveler turning aside for
          the night?
9 Why should you be like someone
          confused,
    like a mighty warrior who
          cannot give help?
    Yet you, O LORD, are in the midst
          of us,
    and we are called by your name;
          do not forsake us!

10 Thus says the LORD concerning
          this people:
    Truly they have loved to wander,
          they have not restrained
          their feet;

y Or *Nubians*; Heb *Cushites*   *z* Heb *them*
a Or *the trails*

rectly related. **23**: In her present state, she
cannot change herself (Hos 5.4; Jn 8.34; compare 3.22).
    **14.1–15.9: Lament over catastrophic
drought and the coming military defeat
of Jerusalem. 14.2–6:** By examples drawn
from city and country, forest and wilderness,
Jeremiah describes the drought.

14.7–9: Prostrate on the ground (v. 2); in
the words of a prayer for a Day of Repentance, the people confess their sins and, as
they are his covenant people, plead with God
to remember them.
    14.10–12: In their need, they remember
God; in their prosperity, they forget him. No
intercession or ritual acts will avert the divine

therefore the LORD does not
accept them,
now he will remember their
iniquity
and punish their sins.

11 The LORD said to me: Do not pray for the welfare of this people. 12 Although they fast, I do not hear their cry, and although they offer burnt offering and grain offering, I do not accept them; but by the sword, by famine, and by pestilence I consume them.

13 Then I said: "Ah, Lord GOD! Here are the prophets saying to them, 'You shall not see the sword, nor shall you have famine, but I will give you true peace in this place.'" 14 And the LORD said to me: The prophets are prophesying lies in my name; I did not send them, nor did I command them or speak to them. They are prophesying to you a lying vision, worthless divination, and the deceit of their own minds. 15 Therefore thus says the LORD concerning the prophets who prophesy in my name though I did not send them, and who say, "Sword and famine shall not come on this land": By sword and famine those prophets shall be consumed. 16 And the people to whom they prophesy shall be thrown out into the streets of Jerusalem, victims of famine and sword. There shall be no one to bury them—themselves, their wives, their sons, and their daughters. For I will pour out their wickedness upon them.

17 You shall say to them this word:
Let my eyes run down with tears
night and day,
and let them not cease,
for the virgin daughter—my
people—is struck down
with a crushing blow,

with a very grievous wound.
18 If I go out into the field,
look—those killed by the
sword!
And if I enter the city,
look—those sick with[b] famine!
For both prophet and priest ply
their trade throughout
the land,
and have no knowledge.

19 Have you completely rejected
Judah?
Does your heart loathe Zion?
Why have you struck us down
so that there is no healing for
us?
We look for peace, but find no
good;
for a time of healing, but there
is terror instead.
20 We acknowledge our wickedness,
O LORD,
the iniquity of our ancestors,
for we have sinned against you.
21 Do not spurn us, for your name's
sake;
do not dishonor your glorious
throne;
remember and do not break
your covenant with us.
22 Can any idols of the nations
bring rain?
Or can the heavens give
showers?
Is it not you, O LORD our God?
We set our hope on you,
for it is you who do all this.

15 Then the LORD said to me: Though Moses and Samuel stood before me, yet my heart would not turn toward this people. Send them out of my sight, and let them go! 2 And when they

b Heb *look—the sicknesses of*

---

punishment (7.16). **13–16:** *In this place,* in the temple. Jeremiah's effort to excuse the people's heedlessness because of misplaced reliance in the false prophets' beguiling message is rejected. The people have made their decision—the wrong one—and will be punished.

14.17–18: Here Jeremiah describes the calamity, using figures of war, drought, unburied dead, hunger, and confused spiritual leaders. **19–22:** More strongly than in vv. 10–12, the people plead their case, casting themselves wholly on God's mercy. **15.1–4:** Even if the people's greatest in-

say to you, "Where shall we go?" you shall say to them: Thus says the LORD:

Those destined for pestilence, to
pestilence,
and those destined for the
sword, to the sword;
those destined for famine, to
famine,
and those destined for captivity,
to captivity.

³ And I will appoint over them four kinds of destroyers, says the LORD: the sword to kill, the dogs to drag away, and the birds of the air and the wild animals of the earth to devour and destroy. ⁴ I will make them a horror to all the kingdoms of the earth because of what King Manasseh son of Hezekiah of Judah did in Jerusalem.

5 Who will have pity on you,
O Jerusalem,
or who will bemoan you?
Who will turn aside
to ask about your welfare?
6 You have rejected me, says
the LORD,
you are going backward;
so I have stretched out my hand
against you and destroyed
you—
I am weary of relenting.
7 I have winnowed them with a
winnowing fork
in the gates of the land;
I have bereaved them, I have
destroyed my people;
they did not turn from
their ways.

8 Their widows became more
numerous
than the sand of the seas;
I have brought against the mothers
of youths
a destroyer at noonday;
I have made anguish and terror
fall upon her suddenly.
9 She who bore seven has
languished;
she has swooned away;
her sun went down while it was
yet day;
she has been shamed and
disgraced.
And the rest of them I will give to
the sword
before their enemies,
says the LORD.

10 Woe is me, my mother, that you ever bore me, a man of strife and contention to the whole land! I have not lent, nor have I borrowed, yet all of them curse me. ¹¹ The LORD said: Surely I have intervened in your life *c* for good, surely I have imposed enemies on you in a time of trouble and in a time of distress. *d* ¹² Can iron and bronze break iron from the north?

13 Your wealth and your treasures I will give as plunder, without price, for all your sins, throughout all your territory. ¹⁴ I will make you serve your enemies in a land that you do not know, for in my

---

c Heb *intervened with you*   *d* Meaning of Heb uncertain

---

tercessors, *Moses* (Ex 32.11–14; Num 14.13–19; one of four references to Moses in prophetic literature [the others are Mic 6.4; Isa 63.11; Mal 4.4]) and *Samuel* (1 Sam 12.17–18), were to plead for them, their sentence of total destruction would not be withdrawn (17.16–17). *King Manasseh,* 2 Kings ch 21. **15.5–9: Jerusalem's end. 6:** *Weary of relenting,* compare Am 7.1–9. **7:** *Winnowed them,* a common threshing figure (see Isa 21.10; 27.12; Jer 51.2; Mt 3.12; Lk 3.17). **9:** *She who bore seven,* a sign of favor (Ruth 4.15;

1 Sam 2.5). Jerusalem will be bereaved (compare Hos 9.12).
15.10–21: Jeremiah's second personal lament, see 11.18–12.6 n. **10–18:** Jeremiah points to his ostracism and the continued rejection of the life-giving word (v. 16; Ezek 2.8–10; Jn 4.32–34). **15:** With the familiar phrase, *you know* (Pss 40.9; 139), Jeremiah prays God for retribution against his persecutors. **17:** The *hand* of God, often the symbol of inspiration (Isa 8.11; Ezek 3.14, 22), is here a symbol of burden.

anger a fire is kindled that shall burn forever.
15 O Lord, you know;
      remember me and visit me,
      and bring down retribution for
            me on my persecutors.
In your forbearance do not take
            me away;
      know that on your account I
            suffer insult.
16 Your words were found, and I
            ate them,
      and your words became to me
            a joy
      and the delight of my heart;
for I am called by your name,
      O Lord, God of hosts.
17 I did not sit in the company of
            merrymakers,
      nor did I rejoice;
under the weight of your hand I
            sat alone,
      for you had filled me with
            indignation.
18 Why is my pain unceasing,
      my wound incurable,
      refusing to be healed?
Truly, you are to me like a
            deceitful brook,
      like waters that fail.

19 Therefore thus says the Lord:
If you turn back, I will take
            you back,
      and you shall stand before me.
If you utter what is precious, and
            not what is worthless,
      you shall serve as my mouth.
It is they who will turn to you,
      not you who will turn to them.
20 And I will make you to this
            people
      a fortified wall of bronze;

they will fight against you,
      but they shall not prevail
            over you,
for I am with you
      to save you and deliver you,
                        says the Lord.
21 I will deliver you out of the hand
            of the wicked,
      and redeem you from the grasp
            of the ruthless.

16 The word of the Lord came to
me: 2 You shall not take a wife,
nor shall you have sons or daughters in
this place. 3 For thus says the Lord concerning the sons and daughters who are
born in this place, and concerning the
mothers who bear them and the fathers
who beget them in this land: 4 They shall
die of deadly diseases. They shall not be
lamented, nor shall they be buried; they
shall become like dung on the surface of
the ground. They shall perish by the
sword and by famine, and their dead
bodies shall become food for the birds of
the air and for the wild animals of the
earth.

5 For thus says the Lord: Do not enter the house of mourning, or go to lament, or bemoan them; for I have taken
away my peace from this people, says the
Lord, my steadfast love and mercy.
6 Both great and small shall die in this
land; they shall not be buried, and no one
shall lament for them; there shall be no
gashing, no shaving of the head for
them. 7 No one shall break bread*e* for the
mourner, to offer comfort for the dead;
nor shall anyone give them the cup of
consolation to drink for their fathers or
their mothers. 8 You shall not go into the
house of feasting to sit with them, to eat

e Two Mss Gk: MT *break for them*

15.19–21: God applies Jeremiah's own
message to himself; God promises no respite
from opposition, but promising him constant
support (1.18–19), urges Jeremiah to continue as his *mouth* (Ex 4.16).
16.1–9: Jeremiah's life, a symbol (Hos
1.2–9; Isa 8.3–4). **1–9:** An illustration of
15.17. **1–4:** In view of the impending doom,
Jeremiah relinquishes hope for home and

family as a sign of the judgment (Ezek 24.15–
27; 1 Cor 7.25–40).
16.5–7: Israel's faithlessness has nullified
God's obligation for covenant loyalty. All
symbols of mourning (*gashing,* 41.5; *shaving
of the head,* Am 8.10; Isa 22.12; compare Deut
14.1) are as futile as death is comfortless. **8–9:**
25.10.

and drink. 9For thus says the LORD of hosts, the God of Israel: I am going to banish from this place, in your days and before your eyes, the voice of mirth and the voice of gladness, the voice of the bridegroom and the voice of the bride.

10 And when you tell this people all these words, and they say to you, "Why has the LORD pronounced all this great evil against us? What is our iniquity? What is the sin that we have committed against the LORD our God?" 11then you shall say to them: It is because your ancestors have forsaken me, says the LORD, and have gone after other gods and have served and worshiped them, and have forsaken me and have not kept my law; 12and because you have behaved worse than your ancestors, for here you are, every one of you, following your stubborn evil will, refusing to listen to me. 13Therefore I will hurl you out of this land into a land that neither you nor your ancestors have known, and there you shall serve other gods day and night, for I will show you no favor.

14 Therefore, the days are surely coming, says the LORD, when it shall no longer be said, "As the LORD lives who brought the people of Israel up out of the land of Egypt," 15but "As the LORD lives who brought the people of Israel up out of the land of the north and out of all the lands where he had driven them." For I will bring them back to their own land that I gave to their ancestors.

16 I am now sending for many fishermen, says the LORD, and they shall catch them; and afterward I will send for many hunters, and they shall hunt them from every mountain and every hill, and out of the clefts of the rocks. 17For my eyes are on all their ways; they are not hidden from my presence, nor is their iniquity concealed from my sight. 18And*f* I will doubly repay their iniquity and their sin, because they have polluted my land with the carcasses of their detestable idols, and have filled my inheritance with their abominations.

19 O LORD, my strength and my
        stronghold,
    my refuge in the day of trouble,
to you shall the nations come
    from the ends of the earth
        and say:
Our ancestors have inherited
        nothing but lies,
    worthless things in which there
        is no profit.
20 Can mortals make for themselves
        gods?
    Such are no gods!

21 "Therefore I am surely going to teach them, this time I am going to teach them my power and my might, and they shall know that my name is the LORD."

17 The sin of Judah is written with an iron pen; with a diamond point it is engraved on the tablet of their hearts, and on the horns of their altars, 2while their children remember their altars and their sacred poles,*g* beside every green tree, and on the high hills, 3on the mountains in the open country. Your wealth and all your treasures I will give for spoil as the price of your sin*h* throughout all your territory. 4By your own act you shall lose the heritage that I gave you, and I will make you serve your

f Gk: Heb *And first*   *g* Heb *Asherim*   *h* Cn: Heb *spoil your high places for sin*

16.10–13: Judah's fate. As Judah has abandoned the true God, so she is condemned to worship false gods in exile (5.19). So Israel too will be alone (5.18–19).

16.14–15: Israel's homecoming. Their return, accomplished by God's actions, will be celebrated as the new Exodus (the passage duplicates 23.7–8; probably an editorial inclusion here).

16.16–18: No refuge for Judah. Judah cannot escape either Egypt or Babylonia (*fishermen* and *hunters;* Hab 1.14–15).

16.19–21: Conversion of the heathen (Isa 45.20–24; Mic 4.1–4). Recognizing the vanity of their idol-worship, all nations will join in worshiping Israel's omnipotent God.

17.1–4: Judah's sin. The practice of the Judeans indelibly and openly demonstrates

enemies in a land that you do not know,
for in my anger a fire is kindled[i] that
shall burn forever.

5 Thus says the LORD:
Cursed are those who trust in
mere mortals
and make mere flesh their
strength,
whose hearts turn away from
the LORD.
6 They shall be like a shrub in
the desert,
and shall not see when relief
comes.
They shall live in the parched
places of the wilderness,
in an uninhabited salt land.

7 Blessed are those who trust in
the LORD,
whose trust is the LORD.
8 They shall be like a tree planted
by water,
sending out its roots by
the stream.
It shall not fear when heat comes,
and its leaves shall stay green;
in the year of drought it is
not anxious,
and it does not cease to bear
fruit.

9 The heart is devious above all else;
it is perverse—
who can understand it?
10 I the LORD test the mind
and search the heart,
to give to all according to
their ways,
according to the fruit of
their doings.

11 Like the partridge hatching what
it did not lay,
so are all who amass wealth
unjustly;
in mid-life it will leave them,
and at their end they will prove
to be fools.

12 O glorious throne, exalted from
the beginning,
shrine of our sanctuary!
13 O hope of Israel! O LORD!
All who forsake you shall be
put to shame;
those who turn away from you[j]
shall be recorded in the
underworld,[k]
for they have forsaken the
fountain of living water,
the LORD.

14 Heal me, O LORD, and I shall
be healed;
save me, and I shall be saved;
for you are my praise.
15 See how they say to me,
"Where is the word of the
LORD?
Let it come!"
16 But I have not run away from
being a shepherd[l] in
your service,
nor have I desired the fatal day.
You know what came from my
lips;

i Two Mss Theodotion: *you kindled*
j Heb *me*   k Or *in the earth*
l Meaning of Heb uncertain

their attitudes. **1**: *Iron pen,* a diamond-tipped
stylus (Job 19.24). *Horns of their altar,* Ex
29.12.
17.5–8: **Trust in God.** The model is Ps 1.
The godless person is like a fruitless desert
plant; the godly person is like a fruitful, well-
watered tree (Ps 1.3; Prov 3.18).
17.9–10: **The heart is devious.** Only God
can really understand human beings (Rom
7.18–19); only God can therefore properly
judge human beings (1 Sam 16.7; Ps 62.12).

17.11: **A proverb,** perhaps referring to Jehoiakim (22.13; 2 Kings 23.35).
17.12–13: **God's ways are high.** *Throne,*
the ark in the temple (Isa 6.1). *Fountain,* 2.13.
17.14–18: **Jeremiah's third personal lament** (see 11.18–12.6 n.). Sick at heart because of his opponents' taunts, Jeremiah prays
for healing (Ps 6.2–3). He does not want to
see God's day of judgment (Am 5.18; Isa
2.11); but if this be the only means of vindication, let it descend in all its fury.

it was before your face.

17 Do not become a terror to me;
you are my refuge in the day
of disaster;
18 Let my persecutors be shamed,
but do not let me be shamed;
let them be dismayed,
but do not let me be dismayed;
bring on them the day of disaster;
destroy them with double
destruction!

19 Thus said the LORD to me: Go and stand in the People's Gate, by which the kings of Judah enter and by which they go out, and in all the gates of Jerusalem, 20 and say to them: Hear the word of the LORD, you kings of Judah, and all Judah, and all the inhabitants of Jerusalem, who enter by these gates. 21 Thus says the LORD: For the sake of your lives, take care that you do not bear a burden on the sabbath day or bring it in by the gates of Jerusalem. 22 And do not carry a burden out of your houses on the sabbath or do any work, but keep the sabbath day holy, as I commanded your ancestors. 23 Yet they did not listen or incline their ear; they stiffened their necks and would not hear or receive instruction.

24 But if you listen to me, says the LORD, and bring in no burden by the gates of this city on the sabbath day, but keep the sabbath day holy and do no work on it, 25 then there shall enter by the gates of this city kings *m* who sit on the throne of David, riding in chariots and on horses, they and their officials, the people of Judah and the inhabitants of Jerusalem; and this city shall be inhabited forever. 26 And people shall come from the towns of Judah and the places around Jerusalem, from the land of Benjamin, from the Shephelah, from the hill coun-

try, and from the Negeb, bringing burnt offerings and sacrifices, grain offerings and frankincense, and bringing thank offerings to the house of the LORD. 27 But if you do not listen to me, to keep the sabbath day holy, and to carry in no burden through the gates of Jerusalem on the sabbath day, then I will kindle a fire in its gates; it shall devour the palaces of Jerusalem and shall not be quenched.

18 The word that came to Jeremiah from the LORD: 2 "Come, go down to the potter's house, and there I will let you hear my words." 3 So I went down to the potter's house, and there he was working at his wheel. 4 The vessel he was making of clay was spoiled in the potter's hand, and he reworked it into another vessel, as seemed good to him.

5 Then the word of the LORD came to me: 6 Can I not do with you, O house of Israel, just as this potter has done? says the LORD. Just like the clay in the potter's hand, so are you in my hand, O house of Israel. 7 At one moment I may declare concerning a nation or a kingdom, that I will pluck up and break down and destroy it, 8 but if that nation, concerning which I have spoken, turns from its evil, I will change my mind about the disaster that I intended to bring on it. 9 And at another moment I may declare concerning a nation or a kingdom that I will build and plant it, 10 but if it does evil in my sight, not listening to my voice, then I will change my mind about the good that I had intended to do to it. 11 Now, therefore, say to the people of Judah and the inhabitants of Jerusalem: Thus says the LORD: Look, I am a potter shaping evil against you and devising a plan

m Cn: Heb *kings and officials*

17.19–27: Judah and the sabbath. Perhaps an insertion from the time of Nehemiah (compare Neh 13.15–22) dealing with the violation of the sabbath (Ex 23.12; compare Mt 12.1–8) as symptomatic of the people's general alienation from God (11.1–8; Am 8.5).

18.1–12: The allegory of the potter. Probably in Jerusalem's southern section, Jeremiah watched a *potter* shaping the inanimate clay. So God molds his people (Rom 9.20–24). God does not deal capriciously with them. The design for *evil* can be replaced by

against you. Turn now, all of you from your evil way, and amend your ways and your doings.

12 But they say, "It is no use! We will follow our own plans, and each of us will act according to the stubbornness of our evil will."

13 Therefore thus says the LORD:
Ask among the nations:
    Who has heard the like of this?
The virgin Israel has done
    a most horrible thing.
14 Does the snow of Lebanon leave
    the crags of Sirion?[n]
Do the mountain[o] waters
        run dry,[p]
    the cold flowing streams?
15 But my people have forgotten me,
    they burn offerings to a
        delusion;
they have stumbled[q] in their
        ways,
    in the ancient roads,
and have gone into bypaths,
    not the highway,
16 making their land a horror,
    a thing to be hissed at forever.
All who pass by it are horrified
    and shake their heads.
17 Like the wind from the east,
    I will scatter them before
        the enemy.
I will show them my back, not
        my face,
    in the day of their calamity.

18 Then they said, "Come, let us make plots against Jeremiah—for instruction shall not perish from the priest, nor counsel from the wise, nor the word from the prophet. Come, let us bring charges against him,[r] and let us not heed any of his words."

19 Give heed to me, O LORD,
    and listen to what my
        adversaries say!
20 Is evil a recompense for good?
    Yet they have dug a pit for
        my life.
Remember how I stood before
        you
    to speak good for them,
    to turn away your wrath
        from them.
21 Therefore give their children over
        to famine;
    hurl them out to the power of
        the sword,
let their wives become childless
        and widowed.
    May their men meet death by
        pestilence,
    their youths be slain by the
        sword in battle.
22 May a cry be heard from their
        houses,
    when you bring the marauder
        suddenly upon them!
For they have dug a pit to
        catch me,
    and laid snares for my feet.
23 Yet you, O LORD, know
    all their plotting to kill me.
Do not forgive their iniquity,
    do not blot out their sin from
        your sight.

n Cn: Heb *of the field*   o Cn: Heb *foreign*
p Cn: Heb *Are . . . plucked up?*   q Gk Syr Vg:
Heb *they made them stumble*   r Heb *strike him
with the tongue*

---

the design for *good*—if they repent. But they refuse.

18.13–17: Faithless Judah is condemned to endless walking. Judah has committed the nearly incredible folly of forsaking her God (2.10, 32). God will ignore them (2.27–28) and will scatter them as before the east wind (4.11; 13.24; 23.19). *Sirion,* Mount Hermon. *Mountain waters,* perhaps from the Anti-Lebanon Mountains or from Mount Hermon itself, such as the Pharpar and Abanah rivers or the sources of the Jordan. **16:** *A thing to be hissed at,* see Zeph 2.15 n.

18.18–23: Jeremiah's fourth personal lament (see 11.18–12.6 n.). *They* (identity unknown) plot against Jeremiah because of his attacks against the leaders (2.8; 8.8). Jeremiah defends his innocence and prays to God for the merciless, total destruction of his enemies and their families. Note the assignment of the instruction to the priests (v. 18).

Let them be tripped up before
 you;
 deal with them while you are
 angry.

19 Thus said the LORD: Go and buy
 a potter's earthenware jug. Take
with you[s] some of the elders of the people and some of the senior priests, ²and
go out to the valley of the son of Hinnom
at the entry of the Potsherd Gate, and
proclaim there the words that I tell you.
³You shall say: Hear the word of the
LORD, O kings of Judah and inhabitants
of Jerusalem. Thus says the LORD of
hosts, the God of Israel: I am going to
bring such disaster upon this place that
the ears of everyone who hears of it will
tingle. ⁴Because the people have forsak-
en me, and have profaned this place by
making offerings in it to other gods
whom neither they nor their ancestors
nor the kings of Judah have known; and
because they have filled this place with
the blood of the innocent, ⁵and gone on
building the high places of Baal to burn
their children in the fire as burnt offerings to Baal, which I did not command
or decree, nor did it enter my mind.
⁶Therefore the days are surely coming,
says the LORD, when this place shall no
more be called Topheth, or the valley of
the son of Hinnom, but the valley of
Slaughter. ⁷And in this place I will make
void the plans of Judah and Jerusalem,
and will make them fall by the sword
before their enemies, and by the hand of
those who seek their life. I will give their
dead bodies for food to the birds of the
air and to the wild animals of the earth.
⁸And I will make this city a horror, a
thing to be hissed at; everyone who
passes by it will be horrified and will hiss
because of all its disasters. ⁹And I will
make them eat the flesh of their sons and
the flesh of their daughters, and all shall
eat the flesh of their neighbors in the
siege, and in the distress with which their
enemies and those who seek their life afflict them.

10 Then you shall break the jug in the
sight of those who go with you, ¹¹and
shall say to them: Thus says the LORD of
hosts: So will I break this people and this
city, as one breaks a potter's vessel, so
that it can never be mended. In Topheth
they shall bury until there is no more
room to bury. ¹²Thus will I do to this
place, says the LORD, and to its inhabitants, making this city like Topheth.
¹³And the houses of Jerusalem and the
houses of the kings of Judah shall be defiled like the place of Topheth—all the
houses upon whose roofs offerings have
been made to the whole host of heaven,
and libations have been poured out to
other gods.

14 When Jeremiah came from Topheth, where the LORD had sent him to
prophesy, he stood in the court of the
LORD's house and said to all the people:
¹⁵Thus says the LORD of hosts, the God
of Israel: I am now bringing upon this
city and upon all its towns all the disaster
that I have pronounced against it, because they have stiffened their necks, refusing to hear my words.

20 Now the priest Pashhur son of
 Immer, who was chief officer in
the house of the LORD, heard Jeremiah
prophesying these things. ²Then Pash-

---

[s] Syr Tg Compare Gk: Heb lacks *take with you*

**19.1–20.6: The public persecution of
Jeremiah. 1–2:** Jeremiah buys an earthenware jug and goes to the Potsherd Gate (probably the later Dung Gate, Neh 2.13), taking
with him some sympathizers. **3–9:** Condemnation of the people for forsaking the LORD
and for worshiping idols, and worst of all, for
the offering of children to Baal (7.30–32).
With a play on the Hebrew words for *jug*
(vv. 2, 10) and *make void* (v. 7), Jeremiah announces the people's horrible fate, which will
make them resort to cannibalism. **8:** *A thing
to be hissed at,* see Zeph 2.15 n.
 19.10–15: The idol-worshiping city, like
the jug, will be irreparably smashed; the spec-
ter of death will hover over it as over *Topheth*
(see 7.31 n.). *Host of heaven,* 8.2; 2 Kings
21.3–5.
 20.1–6: To forestall a repetition of this
frightful pronouncement, Pashhur of the
temple police publicly punishes Jeremiah. On
his release (v. 3) he tells Pashhur that *Terror-*

hur struck the prophet Jeremiah, and put him in the stocks that were in the upper Benjamin Gate of the house of the LORD. ³The next morning when Pashhur released Jeremiah from the stocks, Jeremiah said to him, The LORD has named you not Pashhur but "Terror-all-around." ⁴For thus says the LORD: I am making you a terror to yourself and to all your friends; and they shall fall by the sword of their enemies while you look on. And I will give all Judah into the hand of the king of Babylon; he shall carry them captive to Babylon, and shall kill them with the sword. ⁵I will give all the wealth of this city, all its gains, all its prized belongings, and all the treasures of the kings of Judah into the hand of their enemies, who shall plunder them, and seize them, and carry them to Babylon. ⁶And you, Pashhur, and all who live in your house, shall go into captivity, and to Babylon you shall go; there you shall die, and there you shall be buried, you and all your friends, to whom you have prophesied falsely.

7 O LORD, you have enticed me,
    and I was enticed;
  you have overpowered me,
    and you have prevailed.
  I have become a laughingstock all
      day long;
    everyone mocks me.
8 For whenever I speak, I must
      cry out,
    I must shout, "Violence and
      destruction!"
  For the word of the LORD has
      become for me
    a reproach and derision all
      day long.
9 If I say, "I will not mention him,

or speak any more in his name,"
  then within me there is something
    like a burning fire
  shut up in my bones;
  I am weary with holding it in,
    and I cannot.
10 For I hear many whispering:
    "Terror is all around!
  Denounce him! Let us denounce
      him!"
    All my close friends
    are watching for me to stumble.
  "Perhaps he can be enticed,
    and we can prevail against him,
    and take our revenge on him."
11 But the LORD is with me like a
      dread warrior;
    therefore my persecutors will
      stumble,
    and they will not prevail.
  They will be greatly shamed,
    for they will not succeed.
  Their eternal dishonor
    will never be forgotten.
12 O LORD of hosts, you test the
      righteous,
    you see the heart and the mind;
  let me see your retribution
      upon them,
    for to you I have committed
      my cause.

13 Sing to the LORD;
    praise the LORD!
  For he has delivered the life of
      the needy
    from the hands of evildoers.

14 Cursed be the day
    on which I was born!
  The day when my mother bore
      me,
    let it not be blessed!

---

all-around (6.25; Ps 31.13) will be his name and lot, for he and his family will share the fate of the doomed city (25.8–11).
    20.7–13, 14–18: Jeremiah's fifth and sixth personal laments (see 11.18–12.6 n.). **7–9:** Almost blasphemously, Jeremiah accuses God of deceiving him and of exerting irresistible power over him. Regardless of its effect, Jeremiah cannot refrain from his proc-

lamation (Am 3.8; 1 Cor 9.16). **10–13:** Though his opponents plot his fall, Jeremiah expresses confidence in God's invincible protection of those who rely on him (*the needy*); he uses excerpts from liturgical hymns (compare Pss 6.9–10; 31.13; 109.30; 140.12–13).
    20.14–18: In this independent section, Jeremiah curses not God but his own existence (15.10; Job ch 3). In it one glimpses the inner

15 Cursed be the man
who brought the news to my
father, saying,
"A child is born to you, a son,"
making him very glad.
16 Let that man be like the cities
that the LORD overthrew
without pity;
let him hear a cry in the morning
and an alarm at noon,
17 because he did not kill me in
the womb;
so my mother would have been
my grave,
and her womb forever great.
18 Why did I come forth from
the womb
to see toil and sorrow,
and spend my days in shame?

21 This is the word that came to Jeremiah from the LORD, when King Zedekiah sent to him Pashhur son of Malchiah and the priest Zephaniah son of Maaseiah, saying, 2 "Please inquire of the LORD on our behalf, for King Nebuchadrezzar of Babylon is making war against us; perhaps the LORD will perform a wonderful deed for us, as he has often done, and will make him withdraw from us."

3 Then Jeremiah said to them: 4 Thus you shall say to Zedekiah: Thus says the LORD, the God of Israel: I am going to turn back the weapons of war that are in your hands and with which you are fighting against the king of Babylon and against the Chaldeans who are besieging you outside the walls; and I will bring them together into the center of this city. 5 I myself will fight against you with outstretched hand and mighty arm, in anger, in fury, and in great wrath. 6 And I will strike down the inhabitants of this city, both human beings and animals; they shall die of a great pestilence. 7 Afterward, says the LORD, I will give King Zedekiah of Judah, and his servants, and the people in this city—those who survive the pestilence, sword, and famine—into the hands of King Nebuchadrezzar of Babylon, into the hands of their enemies, into the hands of those who seek their lives. He shall strike them down with the edge of the sword; he shall not pity them, or spare them, or have compassion.

8 And to this people you shall say: Thus says the LORD: See, I am setting before you the way of life and the way of death. 9 Those who stay in this city shall die by the sword, by famine, and by pestilence; but those who go out and surrender to the Chaldeans who are besieging you shall live and shall have their lives as a prize of war. 10 For I have set my face against this city for evil and not for good, says the LORD: it shall be given into the hands of the king of Babylon, and he shall burn it with fire.

11 To the house of the king of Judah say: Hear the word of the LORD, 12 O house of David! Thus says the LORD:
Execute justice in the morning,
and deliver from the hand of the
oppressor
anyone who has been robbed,
or else my wrath will go forth
like fire,
and burn, with no one to
quench it,
because of your evil doings.

13 See, I am against you,
O inhabitant of the valley,

agony of God's prophet as he confronts the unheeding godlessness of his day.
21.1–24.10: Oracles from the time of Zedekiah. 21.1–10: Oracle against Zedekiah and Jerusalem. 1–7: For these verses compare 37.1–10. Although the accounts are similar, they evidently describe separate events. Pashhur, the son of Malchiah, appears also in ch 38. Zephaniah the priest was later executed at Riblah by Nebuchadrezzar (52.24–27). **5a:** Deut 4.34; 5.15. **5b:** Deut 29.28.
21.8–10: The only possibility for life is surrender; but Jerusalem will be destroyed (38.17).
21.11–23.8: Oracles concerning the royal house. 21.11–12: The king is to administer justice (1 Kings 3.9; Ps 72.1–4); if he does

O rock of the plain,
> says the LORD;
you who say, "Who can come
> down against us,
> or who can enter our places
> of refuge?"
14 I will punish you according to the
> fruit of your doings,
> says the LORD;
> I will kindle a fire in its forest,
> and it shall devour all that is
> around it.

22 Thus says the LORD: Go down to
the house of the king of Judah,
and speak there this word, 2and say:
Hear the word of the LORD, O King of
Judah sitting on the throne of David—
you, and your servants, and your people
who enter these gates. 3Thus says the
LORD: Act with justice and righteousness, and deliver from the hand of the
oppressor anyone who has been robbed.
And do no wrong or violence to the
alien, the orphan, and the widow, or
shed innocent blood in this place. 4For if
you will indeed obey this word, then
through the gates of this house shall enter
kings who sit on the throne of David,
riding in chariots and on horses, they,
and their servants, and their people. 5But
if you will not heed these words, I swear
by myself, says the LORD, that this house
shall become a desolation. 6For thus says
the LORD concerning the house of the
king of Judah:

You are like Gilead to me,
> like the summit of Lebanon;
but I swear that I will make you
> a desert,

an uninhabited city. [t]
7 I will prepare destroyers
> against you,
> all with their weapons;
> they shall cut down your choicest
> cedars
> and cast them into the fire.

8 And many nations will pass by this
city, and all of them will say one to another, "Why has the LORD dealt in this
way with that great city?" 9And they will
answer, "Because they abandoned the
covenant of the LORD their God, and
worshiped other gods and served them."

10 Do not weep for him who is
> dead,
> nor bemoan him;
> weep rather for him who
> goes away,
> for he shall return no more
> to see his native land.

11 For thus says the LORD concerning
Shallum son of King Josiah of Judah,
who succeeded his father Josiah, and
who went away from this place: He shall
return here no more, 12but in the place
where they have carried him captive he
shall die, and he shall never see this land
again.

13 Woe to him who builds his house
> by unrighteousness,
> and his upper rooms by
> injustice;
> who makes his neighbors work
> for nothing,
> and does not give them their
> wages;

---

t Cn: Heb *uninhabited cities*

---

evil, he will incur the wrath of God. **13–14**:
Jerusalem claims to be invulnerable to attack.
Forest suggests royal palace (1 Kings 7.2).
22.1–5: Expansion of 21.11–12 (compare
22.3 with 21.12). If the people repent, their
destruction will be averted and the Davidic
dynasty (*house*) preserved. **6–7**: Parallel to
21.13–14. Foresters (*destroyers*) with axes and
saws (*weapons*) will burn the "forest" (cedarwood palace), reducing its once beautiful site
to wasteland. **22.8–9**: A later comment (5.19; Deut

29.23–28; 1 Kings 9.8–9) referring to Jerusalem, not the palace.
22.10–30: **Oracles concerning Jehoahaz, Jehoiakim, and Jehoiachin, kings of
Judah. 10–12**: The dead Josiah was better off
than *Shallum* (personal name of Jehoahaz;
1 Chr 3.15), who in 609 B.C. was banished
(*went away*) to Egypt by Neco (2 Kings
23.33–34; 2 Chr 36.1–4; Ezek 19.4).
22.13–19: Jeremiah censures Jehoiakim for
irresponsibly expanding his palace (*house*) after Egyptian styles (v. 14). An administration

14 who says, "I will build myself a
     spacious house
  with large upper rooms,"
and who cuts out windows for it,
  paneling it with cedar,
  and painting it with vermilion.
15 Are you a king
  because you compete in cedar?
Did not your father eat and drink
  and do justice and
     righteousness?
Then it was well with him.
16 He judged the cause of the poor
     and needy;
  then it was well.
Is not this to know me?
  says the LORD.
17 But your eyes and heart
  are only on your dishonest gain,
for shedding innocent blood,
  and for practicing oppression
     and violence.

18 Therefore thus says the LORD concerning King Jehoiakim son of Josiah of Judah:
  They shall not lament for him,
     saying,
  "Alas, my brother!" or "Alas,
     sister!"
  They shall not lament for him,
     saying,
  "Alas, lord!" or "Alas, his
     majesty!"
19 With the burial of a donkey he
     shall be buried—
  dragged off and thrown out
  beyond the gates of
     Jerusalem.

20 Go up to Lebanon, and cry out,
  and lift up your voice in
     Bashan;
cry out from Abarim,
  for all your lovers are crushed.

21 I spoke to you in your prosperity,
  but you said, "I will not listen."
This has been your way from
     your youth,
  for you have not obeyed
     my voice.
22 The wind shall shepherd all your
     shepherds,
  and your lovers shall go into
     captivity;
then you will be ashamed and
     dismayed
  because of all your wickedness.
23 O inhabitant of Lebanon,
  nested among the cedars,
how you will groan *u* when pangs
     come upon you,
  pain as of a woman in labor!

24 As I live, says the LORD, even if King Coniah son of Jehoiakim of Judah were the signet ring on my right hand, even from there I would tear you off 25 and give you into the hands of those who seek your life, into the hands of those of whom you are afraid, even into the hands of King Nebuchadrezzar of Babylon and into the hands of the Chaldeans. 26 I will hurl you and the mother who bore you into another country, where you were not born, and there you shall die. 27 But they shall not return to the land to which they long to return.
28 Is this man Coniah a despised
     broken pot,
  a vessel no one wants?
Why are he and his offspring
     hurled out
  and cast away in a land that
     they do not know?
29 O land, land, land,
  hear the word of the LORD!

u Gk Vg Syr: Heb *will be pitied*

---

that will *do justice and righteousness,* not one housed *in cedar* (v. 15), is what makes the king (21.11–12; Mic 3.9–10). Jehoiakim should emulate his *father* Josiah whose death was mourned. His own death will be accompanied by indignities because of his misdeeds (36.30; 2 Kings 24.1–5).
22.20–23: From the highest peaks lamentations will rise over the desertion of *Jerusalem* by her allies (*lovers;* 3.1), the exile of her leaders (*shepherds;* 23.1), and her own terror (*inhabitant of Lebanon;* compare 21.13–14). **24–30**: The inexorable punishment of Jehoiachin (*King Coniah*) and his *mother* (13.18) is described by the symbol of authority (*signet ring;* Hag 2.23) and a *broken pot.* The threefold

30 Thus says the Lord:
  Record this man as childless,
    a man who shall not succeed in
      his days;
  for none of his offspring shall
      succeed
    in sitting on the throne of
      David,
    and ruling again in Judah.

23 Woe to the shepherds who destroy and scatter the sheep of my pasture! says the Lord. ²Therefore thus says the Lord, the God of Israel, concerning the shepherds who shepherd my people: It is you who have scattered my flock, and have driven them away, and you have not attended to them. So I will attend to you for your evil doings, says the Lord. ³Then I myself will gather the remnant of my flock out of all the lands where I have driven them, and I will bring them back to their fold, and they shall be fruitful and multiply. ⁴I will raise up shepherds over them who will shepherd them, and they shall not fear any longer, or be dismayed, nor shall any be missing, says the Lord.

5 The days are surely coming, says the Lord, when I will raise up for David a righteous Branch, and he shall reign as king and deal wisely, and shall execute justice and righteousness in the land. ⁶In his days Judah will be saved and Israel will live in safety. And this is the name by which he will be called: "The Lord is our righteousness."

7 Therefore, the days are surely coming, says the Lord, when it shall no longer be said, "As the Lord lives who brought the people of Israel up out of the land of Egypt," ⁸but "As the Lord lives who brought out and led the offspring of the house of Israel out of the land of the north and out of all the lands where he ᵛ had driven them." Then they shall live in their own land.

9 Concerning the prophets:
  My heart is crushed within me,
    all my bones shake;
  I have become like a drunkard,
    like one overcome by wine,
  because of the Lord
    and because of his holy words.
10 For the land is full of adulterers;
    because of the curse the land
      mourns,
    and the pastures of the
      wilderness are dried up.
  Their course has been evil,
    and their might is not right.
11 Both prophet and priest are
      ungodly;
    even in my house I have found
      their wickedness,
                  says the Lord.
12 Therefore their way shall be
      to them
    like slippery paths in the
      darkness,
    into which they shall be driven
      and fall;
  for I will bring disaster upon them
    in the year of their punishment,
                  says the Lord.
13 In the prophets of Samaria
    I saw a disgusting thing:
  they prophesied by Baal
    and led my people Israel astray.

ᵛ Gk: Heb *I*

---

address (v. 29a; Isa 6.3; Ezek 21.27) emphasizes the following oracle announcing that no descendant of Jehoiachin shall rule Judah.
  23.1–8: A messianic oracle. After reproaching Judah's rulers (*shepherds;* 22.22; Ezek ch 34) for scattering Judah (his *flock*), God promises to establish a righteous member (*branch;* Isa 11.1; Zech 3.8) of David's line over a restored Israel (30.9). He will rule responsibly before God as a king (Isa 9.2–7), not as a puppet (like Zedekiah). This messianic expectation differs from the later militant nationalism (16.14–15).
  23.9–40: Oracles concerning the prophets. 9–12: In the light of God's will, Jeremiah is distraught over the moral delinquency of *priest* and *prophet,* supposedly guardians of Israel's faith. **12:** By their own devices the prophets will be destroyed. **13–15:** In words they are worse than the prophets of Baal in Samaria; and in deeds, worse than the people of Sodom and Gomorrah.

14 But in the prophets of Jerusalem
I have seen a more shocking
thing:
they commit adultery and walk
in lies;
they strengthen the hands of
evildoers,
so that no one turns from
wickedness;
all of them have become like
Sodom to me,
and its inhabitants like
Gomorrah.

15 Therefore thus says the LORD of
hosts concerning the
prophets:
"I am going to make them eat
wormwood,
and give them poisoned water
to drink;
for from the prophets of Jerusalem
ungodliness has spread
throughout the land."

16 Thus says the LORD of hosts: Do not listen to the words of the prophets who prophesy to you; they are deluding you. They speak visions of their own minds, not from the mouth of the LORD. 17 They keep saying to those who despise the word of the LORD, "It shall be well with you"; and to all who stubbornly follow their own stubborn hearts, they say, "No calamity shall come upon you."

18 For who has stood in the council
of the LORD
so as to see and to hear his
word?
Who has given heed to his word
so as to proclaim it?
19 Look, the storm of the LORD!

Wrath has gone forth,
a whirling tempest;
it will burst upon the head of
the wicked.
20 The anger of the LORD will not
turn back
until he has executed and
accomplished
the intents of his mind.
In the latter days you will
understand it clearly.

21 I did not send the prophets,
yet they ran;
I did not speak to them,
yet they prophesied.
22 But if they had stood in my
council,
then they would have
proclaimed my words to
my people,
and they would have turned them
from their evil way,
and from the evil of their
doings.

23 Am I a God near by, says the LORD, and not a God far off? 24 Who can hide in secret places so that I cannot see them? says the LORD. Do I not fill heaven and earth? says the LORD. 25 I have heard what the prophets have said who prophesy lies in my name, saying, "I have dreamed, I have dreamed!" 26 How long? Will the hearts of the prophets ever turn back—those who prophesy lies, and who prophesy the deceit of their own heart? 27 They plan to make my people forget my name by their dreams that they tell one another, just as their ancestors forgot my name for Baal. 28 Let the prophet who has a dream tell the dream, but let the one who has my word speak

---

23.16–22: Turning from their deeds to their message, Jeremiah reproves the prophets for their assurance of well-being to those who flout God's word. Obviously they cannot be sent by God or speak for him (for the divine *council,* see 1 Kings 22.19–23; Isa 6.1–7; 40.1–2).
23.23–32: God's omnipresence precludes his being unaware of the prophets' misleading

his people by their claims of divine revelation through *dreams* (27.9; 29.8; Deut 13.3; compare 2.8b). **29:** God's living word is devastating as *fire* (5.14) and shattering as *hammer*. With biting sarcasm (*tell the dream*, v. 28; *steal my words*, v. 30) and threefold emphasis (*see*, vv. 30, 31, 32), Jeremiah proclaims God's condemnation of their pretensions.

my word faithfully. What has straw in common with wheat? says the LORD. 29 Is not my word like fire, says the LORD, and like a hammer that breaks a rock in pieces? 30 See, therefore, I am against the prophets, says the LORD, who steal my words from one another. 31 See, I am against the prophets, says the LORD, who use their own tongues and say, "Says the LORD." 32 See, I am against those who prophesy lying dreams, says the LORD, and who tell them, and who lead my people astray by their lies and their recklessness, when I did not send them or appoint them; so they do not profit this people at all, says the LORD.

33 When this people, or a prophet, or a priest asks you, "What is the burden of the LORD?" you shall say to them, "You are the burden,^w and I will cast you off, says the LORD." 34 And as for the prophet, priest, or the people who say, "The burden of the LORD," I will punish them and their households. 35 Thus shall you say to one another, among yourselves, "What has the LORD answered?" or "What has the LORD spoken?" 36 But "the burden of the LORD" you shall mention no more, for the burden is everyone's own word, and so you pervert the words of the living God, the LORD of hosts, our God. 37 Thus you shall ask the prophet, "What has the LORD answered you?" or "What has the LORD spoken?" 38 But if you say, "the burden of the LORD," thus says the LORD: Because you have said these words, "the burden of the LORD," when I sent to you, saying, You shall not say, "the burden of the LORD," 39 therefore, I will surely lift you up^x and cast you away from my presence, you and the city that I gave to you and your ancestors. 40 And I will bring upon you ev-

erlasting disgrace and perpetual shame, which shall not be forgotten.

24 The LORD showed me two baskets of figs placed before the temple of the LORD. This was after King Nebuchadrezzar of Babylon had taken into exile from Jerusalem King Jeconiah son of Jehoiakim of Judah, together with the officials of Judah, the artisans, and the smiths, and had brought them to Babylon. 2 One basket had very good figs, like first-ripe figs, but the other basket had very bad figs, so bad that they could not be eaten. 3 And the LORD said to me, "What do you see, Jeremiah?" I said, "Figs, the good figs very good, and the bad figs very bad, so bad that they cannot be eaten."

4 Then the word of the LORD came to me: 5 Thus says the LORD, the God of Israel: Like these good figs, so I will regard as good the exiles from Judah, whom I have sent away from this place to the land of the Chaldeans. 6 I will set my eyes upon them for good, and I will bring them back to this land. I will build them up, and not tear them down; I will plant them, and not pluck them up. 7 I will give them a heart to know that I am the LORD; and they shall be my people and I will be their God, for they shall return to me with their whole heart.

8 But thus says the LORD: Like the bad figs that are so bad they cannot be eaten, so will I treat King Zedekiah of Judah, his officials, the remnant of Jerusalem who remain in this land, and those who live in the land of Egypt. 9 I will make them a horror, an evil thing, to all the

w Gk Vg: Heb *What burden*   x Heb Mss Gk Vg: MT *forget you*

23.33: With bitter irony, Jeremiah uses a play on words. 34–40: This is a later commentary on v. 33, which it misunderstood, concentrating on the phrase "the burden of the LORD"; it is theologically non-Jeremianic.
24.1–10: **The vision of the baskets of figs.** Those remaining in Palestine (and Egypt) after 597 B.C. (*the bad figs*) smugly as-

sumed that they were favored over those deported (*the good figs*), who were considered the objects of God's wrath (29.15–19; Ezek 11.14–15). But the exiles will be returned (29.10–14) and will become a faithful nation, while the others will be destroyed for their wilful indifference to God's chastisement.

kingdoms of the earth—a disgrace, a by-word, a taunt, and a curse in all the places where I shall drive them. ¹⁰ And I will send sword, famine, and pestilence upon them, until they are utterly destroyed from the land that I gave to them and their ancestors.

25 The word that came to Jeremiah concerning all the people of Judah, in the fourth year of King Jehoiakim son of Josiah of Judah (that was the first year of King Nebuchadrezzar of Babylon), ² which the prophet Jeremiah spoke to all the people of Judah and all the inhabitants of Jerusalem: ³ For twenty-three years, from the thirteenth year of King Josiah son of Amon of Judah, to this day, the word of the LORD has come to me, and I have spoken persistently to you, but you have not listened. ⁴ And though the LORD persistently sent you all his servants the prophets, you have neither listened nor inclined your ears to hear ⁵ when they said, "Turn now, everyone of you, from your evil way and wicked doings, and you will remain upon the land that the LORD has given to you and your ancestors from of old and forever; ⁶ do not go after other gods to serve and worship them, and do not provoke me to anger with the work of your hands. Then I will do you no harm." ⁷ Yet you did not listen to me, says the LORD, and so you have provoked me to anger with the work of your hands to your own harm.

⁸ Therefore thus says the LORD of hosts: Because you have not obeyed my words, ⁹ I am going to send for all the tribes of the north, says the LORD, even for King Nebuchadrezzar of Babylon, my servant, and I will bring them against

this land and its inhabitants, and against all these nations around; I will utterly destroy them, and make them an object of horror and of hissing, and an everlasting disgrace.^{*y*} ¹⁰ And I will banish from them the sound of mirth and the sound of gladness, the voice of the bridegroom and the voice of the bride, the sound of the millstones and the light of the lamp. ¹¹ This whole land shall become a ruin and a waste, and these nations shall serve the king of Babylon seventy years. ¹² Then after seventy years are completed, I will punish the king of Babylon and that nation, the land of the Chaldeans, for their iniquity, says the LORD, making the land an everlasting waste. ¹³ I will bring upon that land all the words that I have uttered against it, everything written in this book, which Jeremiah prophesied against all the nations. ¹⁴ For many nations and great kings shall make slaves of them also; and I will repay them according to their deeds and the work of their hands.

¹⁵ For thus the LORD, the God of Israel, said to me: Take from my hand this cup of the wine of wrath, and make all the nations to whom I send you drink it. ¹⁶ They shall drink and stagger and go out of their minds because of the sword that I am sending among them.

¹⁷ So I took the cup from the LORD's hand, and made all the nations to whom the LORD sent me drink it: ¹⁸ Jerusalem and the towns of Judah, its kings and officials, to make them a desolation and a waste, an object of hissing and of cursing, as they are today; ¹⁹ Pharaoh king of Egypt, his servants, his officials, and all

y Gk Compare Syr: Heb *and everlasting desolations*

25.1–14: Babylon, God's instrument for punishment. This conclusion to Jeremiah's memoirs (36.1–4) was written after Nebuchadrezzar's victory over Neco of Egypt at Carchemish, in June, 605 B.C. Jeremiah's warnings to Judah are near fulfillment. The foe from the north (6.1) will devastate the apostate land. All manifestations of normal daily life will disappear. No one of this faithless generation will see the restoration "sev-

enty years" later (v. 12; compare Num 14.20–24). Though Judah's destruction is her punishment, wanton destruction does not go unpunished; ultimately Babylon herself will succumb to her enemies. **9**: *Hissing,* see Zeph 2.15 n.

25.15–38: The cup of wrath. 15–29: The vision originally introduced the now separated section of oracles against the nations (chs 46–51; compare 1.5). All nations, be-

his people; 20 all the mixed people; z all the kings of the land of Uz; all the kings of the land of the Philistines—Ashkelon, Gaza, Ekron, and the remnant of Ashdod; 21 Edom, Moab, and the Ammonites; 22 all the kings of Tyre, all the kings of Sidon, and the kings of the coastland across the sea; 23 Dedan, Tema, Buz, and all who have shaven temples; 24 all the kings of Arabia and all the kings of the mixed peoples z that live in the desert; 25 all the kings of Zimri, all the kings of Elam, and all the kings of Media; 26 all the kings of the north, far and near, one after another, and all the kingdoms of the world that are on the face of the earth. And after them the king of Sheshach a shall drink.

27 Then you shall say to them, Thus says the LORD of hosts, the God of Israel: Drink, get drunk and vomit, fall and rise no more, because of the sword that I am sending among you.

28 And if they refuse to accept the cup from your hand to drink, then you shall say to them: Thus says the LORD of hosts: You must drink! 29 See, I am beginning to bring disaster on the city that is called by my name, and how can you possibly avoid punishment? You shall not go unpunished, for I am summoning a sword against all the inhabitants of the earth, says the LORD of hosts.

30 You, therefore, shall prophesy against them all these words, and say to them:

The LORD will roar from on high,
    and from his holy habitation
        utter his voice;
he will roar mightily against
        his fold,
    and shout, like those who tread
        grapes,

against all the inhabitants of
        the earth.
31 The clamor will resound to the
        ends of the earth,
    for the LORD has an indictment
        against the nations;
he is entering into judgment with
        all flesh,
    and the guilty he will put to
        the sword,
                    says the LORD.

32 Thus says the LORD of hosts:
See, disaster is spreading
        from nation to nation,
and a great tempest is stirring
        from the farthest parts of
        the earth!

33 Those slain by the LORD on that day shall extend from one end of the earth to the other. They shall not be lamented, or gathered, or buried; they shall become dung on the surface of the ground.

34 Wail, you shepherds, and cry out;
    roll in ashes, you lords of
        the flock,
for the days of your slaughter
        have come—and your
        dispersions, z
    and you shall fall like a
        choice vessel.
35 Flight shall fail the shepherds,
    and there shall be no escape for
        the lords of the flock.
36 Hark! the cry of the shepherds,
    and the wail of the lords of
        the flock!
For the LORD is despoiling their
        pasture,

z Meaning of Heb uncertain   a *Sheshach* is a cryptogram for *Babel*, Babylon

cause of their transgressions (Am 1.3–3.2), must suffer God's wrath (vv. 27–29 continue v. 16). The *cup* as a symbol of God's judgment (note its equation with *sword*, v. 29) originated perhaps with Jeremiah (8.14; Isa 51.17; Ps 11.6). **26**: Babylon is written in the Hebrew text as *Sheshach;* this kind of cipher, in which the letters are substituted in reverse order of the Hebrew alphabet, is called "atbash" (see 51.1, 41 n.). **25.30–31**: The judgment is described in conventional figures: roaring (Am 1.2; Ps 46.6); vintage (Isa 16.9–10; 63.1–3); courtroom (12.1); sword (12.12). **32–33**: Distant foe (6.22) and heavy casualties (8.2; 16.4). **34–38**: The rulers (*shepherds, lords of the flock*)

37 and the peaceful folds are
devastated,
because of the fierce anger of
the LORD.
38 Like a lion he has left his covert;
for their land has become
a waste
because of the cruel sword,
and because of his fierce anger.

26 At the beginning of the reign of King Jehoiakim son of Josiah of Judah, this word came from the LORD: 2 Thus says the LORD: Stand in the court of the LORD's house, and speak to all the cities of Judah that come to worship in the house of the LORD; speak to them all the words that I command you; do not hold back a word. 3 It may be that they will listen, all of them, and will turn from their evil way, that I may change my mind about the disaster that I intend to bring on them because of their evil doings. 4 You shall say to them: Thus says the LORD: If you will not listen to me, to walk in my law that I have set before you, 5 and to heed the words of my servants the prophets whom I send to you urgently—though you have not heeded— 6 then I will make this house like Shiloh, and I will make this city a curse for all the nations of the earth.

7 The priests and the prophets and all the people heard Jeremiah speaking these words in the house of the LORD. 8 And when Jeremiah had finished speaking all that the LORD had commanded him to speak to all the people, then the priests and the prophets and all the people laid hold of him, saying, "You shall die! 9 Why have you prophesied in the name of the LORD, saying, 'This house shall be like Shiloh, and this city shall be desolate, without inhabitant'?" And all the people gathered around Jeremiah in the house of the LORD.

10 When the officials of Judah heard these things, they came up from the king's house to the house of the LORD and took their seat in the entry of the New Gate of the house of the LORD. 11 Then the priests and the prophets said to the officials and to all the people, "This man deserves the sentence of death because he has prophesied against this city, as you have heard with your own ears."

12 Then Jeremiah spoke to all the officials and all the people, saying, "It is the LORD who sent me to prophesy against this house and this city all the words you have heard. 13 Now therefore amend your ways and your doings, and obey the voice of the LORD your God, and the LORD will change his mind about the disaster that he has pronounced against you. 14 But as for me, here I am in your hands. Do with me as seems good and right to you. 15 Only know for certain that if you put me to death, you will be bringing innocent blood upon yourselves and upon this city and its inhabitants, for in truth the LORD sent me to you to speak all these words in your ears."

16 Then the officials and all the people said to the priests and the prophets, "This man does not deserve the sentence of death, for he has spoken to us in the name of the LORD our God." 17 And some of the elders of the land arose and said to all the assembled people, 18 "Micah of

---

are confused and desperate. **38**: The *lion* is either God, or Nebuchadrezzar (4.7).

26.1–35.19: Events and prophecies (largely from Baruch's memoirs); predictions of restoration.

26.1–24: The temple sermon. 1–6: The sermon (7.1–15) was delivered perhaps during the Festival of Booths (September–October), 609 B.C. Baruch's summation here contains elements of ch 7 and other sayings of Jeremiah (4.1–2; 18.7–11; 36.3).

26.7–19: Jeremiah's arrest and release. Frequently criticized by Jeremiah (2.8; 5.30–31; 6.13–14), priest and prophet alike would suffer from the temple's destruction. The royal officials assemble to hear the case. Jeremiah's dignified defense wins his release; the judges base their decision on a century-old precedent set by Hezekiah (715–687 B.C.) regarding the prophet Micah (Mic 3.12). *New Gate,* perhaps the Benjamin Gate, north of the temple (20.2; 2 Kings 15.35). Avenging of

Moresheth, who prophesied during the days of King Hezekiah of Judah, said to all the people of Judah: 'Thus says the LORD of hosts,

> Zion shall be plowed as a field;
> Jerusalem shall become a heap
>   of ruins,
>   and the mountain of the house a
>   wooded height.'

19 Did King Hezekiah of Judah and all Judah actually put him to death? Did he not fear the LORD and entreat the favor of the LORD, and did not the LORD change his mind about the disaster that he had pronounced against them? But we are about to bring great disaster on ourselves!"

20 There was another man prophesying in the name of the LORD, Uriah son of Shemaiah from Kiriath-jearim. He prophesied against this city and against this land in words exactly like those of Jeremiah. 21 And when King Jehoiakim, with all his warriors and all the officials, heard his words, the king sought to put him to death; but when Uriah heard of it, he was afraid and fled and escaped to Egypt. 22 Then King Jehoiakim sent[b] Elnathan son of Achbor and men with him to Egypt, 23 and they took Uriah from Egypt and brought him to King Jehoiakim, who struck him down with the sword and threw his dead body into the burial place of the common people.

24 But the hand of Ahikam son of Shaphan was with Jeremiah so that he was not given over into the hands of the people to be put to death.

27 In the beginning of the reign of King Zedekiah[c] son of Josiah of Judah, this word came to Jeremiah from the LORD. 2 Thus the LORD said to me: Make yourself a yoke of straps and bars, and put them on your neck. 3 Send word[d] to the king of Edom, the king of Moab, the king of the Ammonites, the king of Tyre, and the king of Sidon by the hand of the envoys who have come to Jerusalem to King Zedekiah of Judah. 4 Give them this charge for their masters: Thus says the LORD of hosts, the God of Israel: This is what you shall say to your masters: 5 It is I who by my great power and my outstretched arm have made the earth, with the people and animals that are on the earth, and I give it to whomever I please. 6 Now I have given all these lands into the hand of King Nebuchadnezzar of Babylon, my servant, and I have given him even the wild animals of the field to serve him. 7 All the nations shall serve him and his son and his grandson, until the time of his own land comes; then many nations and great kings shall make him their slave.

8 But if any nation or kingdom will not serve this king, Nebuchadnezzar of Babylon, and put its neck under the yoke of the king of Babylon, then I will punish that nation with the sword, with famine, and with pestilence, says the LORD, until I have completed its[e] destruction by his hand. 9 You, therefore, must not listen to your prophets, your diviners, your dreamers,[f] your soothsayers, or your sorcerers, who are saying to you, 'You shall not serve the king of Babylon.'

---

b Heb adds *men to Egypt*   c Another reading
is *Jehoiakim*   d Cn: Heb *send them*
e Heb *their*   f Gk Syr Vg: Heb *dreams*

---

innocent blood, Gen 4.10; 2 Sam 21.1–14; 1 Kings ch 21.
26.20–24: Martyrdom of Uriah. Baruch adds this story to illustrate Jeremiah's personal danger and fortunate official support (*Ahikam,* 2 Kings 22.12, 14). As Egypt's vassal (2 Kings 23.34–35), Jehoiakim had no difficulty arresting Uriah (*Elnathan,* 36.12, 25). His execution is unusual in the annals of Israelite prophetism (2 Chr 24.20–22; Mt 23.29–31).

27.1–28.17: The yoke of the king of Babylon. 27.1–11: The yoke of Babylon was imposed by God upon Judah and her neighbors (21.1–10; 32.3–5); thus their plans for rebellion are against God's will. The occasion for the conspiracy lay in a revolt within Nebuchadrezzar's army (December 595–January 594 B.C.) and the accession in Egypt of Psammetichus II (594). Perhaps responding to Jeremiah's warning, Zedekiah did not carry out the rebellion, and so Judah was spared during

10 For they are prophesying a lie to you, with the result that you will be removed far from your land; I will drive you out, and you will perish. 11 But any nation that will bring its neck under the yoke of the king of Babylon and serve him, I will leave on its own land, says the LORD, to till it and live there.

12 I spoke to King Zedekiah of Judah in the same way: Bring your necks under the yoke of the king of Babylon, and serve him and his people, and live. 13 Why should you and your people die by the sword, by famine, and by pestilence, as the LORD has spoken concerning any nation that will not serve the king of Babylon? 14 Do not listen to the words of the prophets who are telling you not to serve the king of Babylon, for they are prophesying a lie to you. 15 I have not sent them, says the LORD, but they are prophesying falsely in my name, with the result that I will drive you out and you will perish, you and the prophets who are prophesying to you.

16 Then I spoke to the priests and to all this people, saying, Thus says the LORD: Do not listen to the words of your prophets who are prophesying to you, saying, "The vessels of the LORD's house will soon be brought back from Babylon," for they are prophesying a lie to you. 17 Do not listen to them; serve the king of Babylon and live. Why should this city become a desolation? 18 If indeed they are prophets, and if the word of the LORD is with them, then let them intercede with the LORD of hosts, that the vessels left in the house of the LORD, in the house of the king of Judah, and in Jerusalem may not go to Babylon. 19 For thus says the LORD of hosts concerning the pillars, the sea, the stands, and the rest of the vessels that are left in this city, 20 which King Nebuchadnezzar of Babylon did not take away when he took into exile from Jerusalem to Babylon King Jeconiah son of Jehoiakim of Judah, and all the nobles of Judah and Jerusalem— 21 thus says the LORD of hosts, the God of Israel, concerning the vessels left in the house of the LORD, in the house of the king of Judah, and in Jerusalem: 22 They shall be carried to Babylon, and there they shall stay, until the day when I give attention to them, says the LORD. Then I will bring them up and restore them to this place.

28 In that same year, at the beginning of the reign of King Zedekiah of Judah, in the fifth month of the fourth year, the prophet Hananiah son of Azzur, from Gibeon, spoke to me in the house of the LORD, in the presence of the priests and all the people, saying, 2 "Thus says the LORD of hosts, the God of Israel: I have broken the yoke of the king of Babylon. 3 Within two years I will bring back to this place all the vessels of the LORD's house, which King Nebuchadnezzar of Babylon took away from this place and carried to Babylon. 4 I will also bring back to this place King Jeconiah son of Jehoiakim of Judah, and all the exiles from Judah who went to Babylon, says the LORD, for I will break the yoke of the king of Babylon."

5 Then the prophet Jeremiah spoke to the prophet Hananiah in the presence of the priests and all the people who were

---

Nebuchadrezzar's punitive campaign later that year. The form "Nebuchadnezzar" occurs in Jer 27.6–29.3; elsewhere in Jeremiah "Nebuchadrezzar" appears. The Babylonian form is Nabu-kudurru-usur.

27.12–15: Jeremiah repeats his warning: since God has not sent the prophets who are advising Zedekiah (14.14), they are unreliable guides.

27.16–22: He cautions the priests and the people against believing the baseless assurances of these prophets that the temple equip-

ment taken as booty in 597 B.C. would be shortly returned. Instead, they should pray that what they have (Jer 52.17; 2 Kings 25.13) will not also be taken to Babylon.

28.1–17: Jeremiah and Hananiah. In August 594 B.C., Hananiah faced Jeremiah in the temple and predicted total restoration within two years, breaking Jeremiah's yoke to emphasize his point. Jeremiah recalled that their predecessors had predicted chastisement (Deut 18.20–22; Mic 3.5–12). Because the situation was unchanged, Hananiah's prom-

standing in the house of the LORD; 6and the prophet Jeremiah said, "Amen! May the LORD do so; may the LORD fulfill the words that you have prophesied, and bring back to this place from Babylon the vessels of the house of the LORD, and all the exiles. 7But listen now to this word that I speak in your hearing and in the hearing of all the people. 8The prophets who preceded you and me from ancient times prophesied war, famine, and pestilence against many countries and great kingdoms. 9As for the prophet who prophesies peace, when the word of that prophet comes true, then it will be known that the LORD has truly sent the prophet."

10 Then the prophet Hananiah took the yoke from the neck of the prophet Jeremiah, and broke it. 11And Hananiah spoke in the presence of all the people, saying, "Thus says the LORD: This is how I will break the yoke of King Nebuchadnezzar of Babylon from the neck of all the nations within two years." At this, the prophet Jeremiah went his way.

12 Sometime after the prophet Hananiah had broken the yoke from the neck of the prophet Jeremiah, the word of the LORD came to Jeremiah: 13Go, tell Hananiah, Thus says the LORD: You have broken wooden bars only to forge iron bars in place of them! 14For thus says the LORD of hosts, the God of Israel: I have put an iron yoke on the neck of all these nations so that they may serve King Nebuchadnezzar of Babylon, and they shall indeed serve him; I have even given him the wild animals. 15And the prophet Jeremiah said to the prophet Hananiah, "Listen, Hananiah, the LORD has not sent you, and you made this people trust in a lie. 16Therefore thus says the LORD: I am going to send you off the face of the earth. Within this year you will be dead, because you have spoken rebellion against the LORD."

17 In that same year, in the seventh month, the prophet Hananiah died.

29 These are the words of the letter that the prophet Jeremiah sent from Jerusalem to the remaining elders among the exiles, and to the priests, the prophets, and all the people, whom Nebuchadnezzar had taken into exile from Jerusalem to Babylon. 2This was after King Jeconiah, and the queen mother, the court officials, the leaders of Judah and Jerusalem, the artisans, and the smiths had departed from Jerusalem. 3The letter was sent by the hand of Elasah son of Shaphan and Gemariah son of Hilkiah, whom King Zedekiah of Judah sent to Babylon to King Nebuchadnezzar of Babylon. It said: 4Thus says the LORD of hosts, the God of Israel, to all the exiles whom I have sent into exile from Jerusalem to Babylon: 5Build houses and live in them; plant gardens and eat what they produce. 6Take wives and have sons and daughters; take wives for your sons, and give your daughters in marriage, that they may bear sons and daughters; multiply there, and do not decrease. 7But seek the welfare of the city where I have sent you into exile, and pray to the LORD on its behalf, for in its welfare you will find your welfare. 8For thus says the LORD of hosts, the God of Israel: Do not let the prophets and the diviners who are among you deceive you, and do not listen to the dreams that they dream,*g* 9for it is a lie that they are prophesying to you in my name; I did not send them, says the LORD.

g Cn: Heb *your dreams that you cause to dream*

---

ise, though attractive, seemed very doubtful. Later Jeremiah used the symbolism of an iron yoke to emphasize that the Exile would continue (27.7). *Yoke,* placed on the neck of an ox, held on by leather straps.

29.1–32: Jeremiah's letters to Babylon. 1–23: Letter to the exiles. The exiles were being misled by the same baseless assurances of speedy return as those in Palestine (ch 27).

To counter this Jeremiah sent a letter by *Elasah* (perhaps the brother of Ahikam, 26.24) and *Gemariah* (36.10) to the *elders* of the people (Ezek 8.1; 14.1). His advice was revolutionary. They were to establish homes in Babylonia and even pray for the welfare of the state. God would be with them and would ultimately restore them (*seventy years,* 25.11; 27.7). Two colleagues of Hananiah, Ahab and

10 For thus says the LORD: Only when Babylon's seventy years are completed will I visit you, and I will fulfill to you my promise and bring you back to this place. 11 For surely I know the plans I have for you, says the LORD, plans for your welfare and not for harm, to give you a future with hope. 12 Then when you call upon me and come and pray to me, I will hear you. 13 When you search for me, you will find me; if you seek me with all your heart, 14 I will let you find me, says the LORD, and I will restore your fortunes and gather you from all the nations and all the places where I have driven you, says the LORD, and I will bring you back to the place from which I sent you into exile.

15 Because you have said, "The LORD has raised up prophets for us in Babylon," — 16 Thus says the LORD concerning the king who sits on the throne of David, and concerning all the people who live in this city, your kinsfolk who did not go out with you into exile: 17 Thus says the LORD of hosts, I am going to let loose on them sword, famine, and pestilence, and I will make them like rotten figs that are so bad they cannot be eaten. 18 I will pursue them with the sword, with famine, and with pestilence, and will make them a horror to all the kingdoms of the earth, to be an object of cursing, and horror, and hissing, and a derision among all the nations where I have driven them, 19 because they did not heed my words, says the LORD, when I persistently sent to you my servants the prophets, but they*h* would not listen, says the LORD. 20 But now, all you exiles whom I sent away from Jerusalem to Babylon, hear the word of the LORD: 21 Thus says the LORD of hosts, the God of Israel, concerning Ahab son of Kolaiah and Zedekiah son of Maaseiah, who are prophesying a lie to you in my name: I am going to deliver them into the hand of King Nebuchadrezzar of Babylon, and he shall kill them before your eyes. 22 And on account of them this curse shall be used by all the exiles from Judah in Babylon: "The LORD make you like Zedekiah and Ahab, whom the king of Babylon roasted in the fire," 23 because they have perpetrated outrage in Israel and have committed adultery with their neighbors' wives, and have spoken in my name lying words that I did not command them; I am the one who knows and bears witness, says the LORD.

24 To Shemaiah of Nehelam you shall say: 25 Thus says the LORD of hosts, the God of Israel: In your own name you sent a letter to all the people who are in Jerusalem, and to the priest Zephaniah son of Maaseiah, and to all the priests, saying, 26 The LORD himself has made you priest instead of the priest Jehoiada, so that there may be officers in the house of the LORD to control any madman who plays the prophet, to put him in the stocks and the collar. 27 So now why have you not rebuked Jeremiah of Anathoth who plays the prophet for you? 28 For he has actually sent to us in Babylon, saying, "It will be a long time; build houses and live in them, and plant gardens and eat what they produce."

29 The priest Zephaniah read this letter in the hearing of the prophet Jeremiah. 30 Then the word of the LORD came to Jeremiah: 31 Send to all the exiles, saying, Thus says the LORD concerning Shemaiah of Nehelam: Because Shemaiah has

h Syr: Heb *you*

Zedekiah (v. 21), were condemned by Jeremiah (compare Ezek ch 13). He predicted their execution; from the Babylonian point of view it would be for political subversion, not for the reason mentioned in v. 23. Verses 21–23 should follow v. 15; vv. 16–20 are an editorial comment on Jeremiah's letter.

29.24–32: Jeremiah and Shemaiah. Another "colleague" of Hananiah, Shemaiah, wrote from Babylon a scathing letter to the new temple overseer (compare 20.1), Zephaniah, charging him with dereliction of duty in not arresting Jeremiah for his letter (above). Instead Zephaniah (21.1; 37.3) read the letter to his friend Jeremiah. The latter sent another letter to the exiles, condemning Shemaiah and predicting that neither he nor his descendants would see the day of restoration (20.6).

prophesied to you, though I did not send him, and has led you to trust in a lie, [32]therefore thus says the LORD: I am going to punish Shemaiah of Nehelam and his descendants; he shall not have anyone living among this people to see[i] the good that I am going to do to my people, says the LORD, for he has spoken rebellion against the LORD.

30 The word that came to Jeremiah from the LORD: [2]Thus says the LORD, the God of Israel: Write in a book all the words that I have spoken to you. [3]For the days are surely coming, says the LORD, when I will restore the fortunes of my people, Israel and Judah, says the LORD, and I will bring them back to the land that I gave to their ancestors and they shall take possession of it.

[4] These are the words that the LORD spoke concerning Israel and Judah:

[5] Thus says the LORD:
We have heard a cry of panic,
of terror, and no peace.
[6] Ask now, and see,
can a man bear a child?
Why then do I see every man
with his hands on his loins like
a woman in labor?
Why has every face turned pale?
[7] Alas! that day is so great
there is none like it;
it is a time of distress for Jacob;
yet he shall be rescued from it.

[8] On that day, says the LORD of hosts, I will break the yoke from off his[j] neck, and I will burst his[j] bonds, and strangers shall no more make a servant of him. [9]But they shall serve the LORD their God and David their king, whom I will raise up for them.

[10] But as for you, have no fear, my
servant Jacob, says the
LORD,
and do not be dismayed,
O Israel;
for I am going to save you from
far away,
and your offspring from the
land of their captivity.
Jacob shall return and have quiet
and ease,
and no one shall make him
afraid.
[11] For I am with you, says the LORD,
to save you;
I will make an end of all the
nations
among which I scattered you,
but of you I will not make
an end.
I will chastise you in just measure,
and I will by no means leave
you unpunished.

[12] For thus says the LORD:
Your hurt is incurable,
your wound is grievous.
[13] There is no one to uphold your
cause,
no medicine for your wound,
no healing for you.
[14] All your lovers have forgotten
you;
they care nothing for you;
for I have dealt you the blow of
an enemy,
the punishment of a merciless
foe,
because your guilt is great,

i Gk: Heb *and he shall not see*   *j* Cn: Heb *your*

30.1–31.40: The Book of Consolation. **30.1–4:** Introduction. Initially collected by Baruch, some parts of chs 30–31 consist of early oracles addressed to the north (*Ephraim,* 31.6, 9, 18, 20; compare 3.12). These are expanded by Jeremiah and addressed to the south just before the fall of Jerusalem in 587 B.C. **30.5–9:** Oracles concerning Israel. After the travail (6.24) of God's judgment (*that day* [of the LORD], Am 5.18–20), God will break the yoke and a Davidic king will again rule Israel (23.5–6; Hos 3.5). **10–11:** *Have no fear,* common phrase in God's address to his people, Gen 15.1; Isa 35.4; Lk 2.10. *Not make an end,* a remnant will be left (see 4.27; 5.10, 18; compare 46.27–28). **30.12–17:** Though ostensibly incurably wounded (8.22; 14.17) and deserted (4.30; 13.21) because of her sinfulness, Israel will be healed (Hos 14.4) and her oppressors despoiled (9.25–26; 25.13–14).

because your sins are so
numerous.

15 Why do you cry out over your
hurt?
Your pain is incurable.
Because your guilt is great,
because your sins are so
numerous,
I have done these things to you.

16 Therefore all who devour you
shall be devoured,
and all your foes, everyone of
them, shall go into
captivity;
those who plunder you shall be
plundered,
and all who prey on you I will
make a prey.

17 For I will restore health to you,
and your wounds I will heal,
says the LORD,
because they have called you
an outcast:
"It is Zion; no one cares for
her!"

18 Thus says the LORD:
I am going to restore the fortunes
of the tents of Jacob,
and have compassion on his
dwellings;
the city shall be rebuilt upon its
mound,
and the citadel set on its
rightful site.

19 Out of them shall come
thanksgiving,
and the sound of merrymakers.
I will make them many, and they
shall not be few;
I will make them honored, and
they shall not be disdained.

20 Their children shall be as of old,
their congregation shall be
established before me;

and I will punish all who
oppress them.

21 Their prince shall be one of
their own,
their ruler shall come from their
midst;
I will bring him near, and he shall
approach me,
for who would otherwise dare
to approach me?
says the LORD.

22 And you shall be my people,
and I will be your God.

23 Look, the storm of the LORD!
Wrath has gone forth,
a whirling *k* tempest;
it will burst upon the head of
the wicked.

24 The fierce anger of the LORD will
not turn back
until he has executed and
accomplished
the intents of his mind.
In the latter days you will
understand this.

31 At that time, says the LORD, I will
be the God of all the families of
Israel, and they shall be my people.

2 Thus says the LORD:
The people who survived the
sword
found grace in the wilderness;
when Israel sought for rest,

3 the LORD appeared to him *l*
from far away. *m*
I have loved you with an
everlasting love;
therefore I have continued my
faithfulness to you.

4 Again I will build you, and you
shall be built,

k One Ms: Meaning of MT uncertain
l Gk: Heb me   m Or to him long ago

30.18–22: *The city* (Samaria or Jerusalem)
will be *rebuilt upon its mound;* it was customary
in antiquity to rebuild on the leveled rubble
of the preceding city, producing the present
flat-topped hills, called tells. **21:** *I will bring
him near* reassures the one who approaches

God (Ex 19.21; 33.20; Num 8.19). **22:** 7.23;
11.4; 24.7. **30.23–31.1:** The storm of the
LORD. **30.23–24:** = 23.19–20.
31.1: This emphasizes the inclusiveness of
the term *Israel;* compare 30.22. **2–6:** With a
formula taken from the Exodus narrative

O virgin Israel!
Again you shall take[n] your
tambourines,
and go forth in the dance of the
merrymakers.
5 Again you shall plant vineyards
on the mountains of Samaria;
the planters shall plant,
and shall enjoy the fruit.
6 For there shall be a day when
sentinels will call
in the hill country of Ephraim:
"Come, let us go up to Zion,
to the LORD our God."

7 For thus says the LORD:
Sing aloud with gladness for
Jacob,
and raise shouts for the chief of
the nations;
proclaim, give praise, and say,
"Save, O LORD, your people,
the remnant of Israel."
8 See, I am going to bring them
from the land of the north,
and gather them from the
farthest parts of the earth,
among them the blind and
the lame,
those with child and those in
labor, together;
a great company, they shall
return here.
9 With weeping they shall come,
and with consolations[o] I will
lead them back,
I will let them walk by brooks of
water,
in a straight path in which they
shall not stumble;
for I have become a father to
Israel,
and Ephraim is my firstborn.

10 Hear the word of the LORD,
O nations,
and declare it in the coastlands
far away;
say, "He who scattered Israel will
gather him,
and will keep him as a shepherd
a flock."
11 For the LORD has ransomed Jacob,
and has redeemed him from
hands too strong for him.
12 They shall come and sing aloud on
the height of Zion,
and they shall be radiant over
the goodness of the LORD,
over the grain, the wine, and
the oil,
and over the young of the flock
and the herd;
their life shall become like a
watered garden,
and they shall never languish
again.
13 Then shall the young women
rejoice in the dance,
and the young men and the old
shall be merry.
I will turn their mourning into
joy,
I will comfort them, and give
them gladness for sorrow.
14 I will give the priests their fill of
fatness,
and my people shall be satisfied
with my bounty,
says the LORD.

15 Thus says the LORD:
A voice is heard in Ramah,
lamentation and bitter weeping.

n Or *adorn yourself with*   *o* Gk Compare Vg
Tg: Heb *supplications*

(*found grace* [favor], Ex 33.12–17; compare Jer 23.7–8) and the emphatic historical-theological implications of *everlasting love* and *faithfulness*, God promises the joyful restoration of all Israel, including the renewal of pilgrimages to the Jerusalem temple (*Zion*; 41.5).

31.7–14: Homecoming. God will assemble the dispersed in their homeland (Isa 35.5–

10; Ps 23.2–3). *Israel . . . Ephraim is my first-born* (Ex 4.22), as Ephraim is restored, so is all Israel (including Judah; 2.3; 3.19). **10–11**: Praises to God will be raised far (*coastlands*, Ps 72.10–11; Isa 41.1, 5) and near for deliverance (*shepherd*, Isa 40.11; *redeemed*, Isa 48.20). **14**: Bountiful produce of the land will mark the new age of well-being (Isa 58.11).

31.15–20: *Rachel*, mother of Joseph and

Rachel is weeping for her children;
she refuses to be comforted for
her children,
because they are no more.
16 Thus says the LORD:
Keep your voice from weeping,
and your eyes from tears;
for there is a reward for your
work,
                    says the LORD:
they shall come back from the
land of the enemy;
17 there is hope for your future,
                    says the LORD:
your children shall come back to
their own country.

18 Indeed I heard Ephraim pleading:
"You disciplined me, and I took
the discipline;
I was like a calf untrained.
Bring me back, let me come back,
for you are the LORD my God.
19 For after I had turned away I
repented;
and after I was discovered, I
struck my thigh;
I was ashamed, and I was
dismayed
because I bore the disgrace of
my youth."
20 Is Ephraim my dear son?
Is he the child I delight in?
As often as I speak against him,
I still remember him.
Therefore I am deeply moved
for him;
I will surely have mercy on him,
                    says the LORD.

21 Set up road markers for yourself,
make yourself guideposts;

consider well the highway,
the road by which you went.
Return, O virgin Israel,
return to these your cities.
22 How long will you waver,
O faithless daughter?
For the LORD has created a new
thing on the earth:
a woman encompasses*p* a man.

23 Thus says the LORD of hosts, the
God of Israel: Once more they shall use
these words in the land of Judah and in
its towns when I restore their fortunes:
"The LORD bless you, O abode of
righteousness,
O holy hill!"
24 And Judah and all its towns shall live
there together, and the farmers and those
who wander*q* with their flocks.
25 I will satisfy the weary,
and all who are faint I will
replenish.
26 Thereupon I awoke and looked,
and my sleep was pleasant to me.
27 The days are surely coming, says
the LORD, when I will sow the house of
Israel and the house of Judah with the
seed of humans and the seed of animals.
28 And just as I have watched over them
to pluck up and break down, to overthrow, destroy, and bring evil, so I will
watch over them to build and to plant,
says the LORD. 29 In those days they shall
no longer say:
"The parents have eaten sour
grapes,
and the children's teeth are set
on edge."

p Meaning of Heb uncertain   *q* Cn Compare
Syr Vg Tg: Heb *and they shall wander*

Benjamin (Gen 30.22; 35.16–20; 1 Sam 10.2),
laments their (the northern tribes') exile (*Ramah*, 1 Sam 8.4). Matthew 2.18 depicts un-
mitigated grief, but here the verse introduces
the promise of restoration. **18**: *Ephraim* (Joseph's son, Gen 41.50–52) repents (3.22–25;
Hos 6.1–3). **21–22**: Israel is bidden to note the
road markers on her way to exile so that she can
find her way home again. *A new thing*, compare Isa 43.19. *A woman encompasses a man*, she

does not have to be protected by a man; compare Isa 11.6–9 for the reversal of the usual in
the new age. **23–40**: Restoration and the new
covenant. The following oracles are after
587 B.C.

31.23–30: As God depopulated and depleted Judah (1.10), he will also restore it
(Ezek 36.8–11) and reunite it with Israel
(vv. 2–14; Isa 11.11–16). **29–30**: This may
reflect the contemporary problem (Deut

30 But all shall die for their own sins; the teeth of everyone who eats sour grapes shall be set on edge.

31 The days are surely coming, says the LORD, when I will make a new covenant with the house of Israel and the house of Judah. 32 It will not be like the covenant that I made with their ancestors when I took them by the hand to bring them out of the land of Egypt—a covenant that they broke, though I was their husband,*r* says the LORD. 33 But this is the covenant that I will make with the house of Israel after those days, says the LORD: I will put my law within them, and I will write it on their hearts; and I will be their God, and they shall be my people. 34 No longer shall they teach one another, or say to each other, "Know the LORD," for they shall all know me, from the least of them to the greatest, says the LORD; for I will forgive their iniquity, and remember their sin no more.

35 Thus says the LORD,
who gives the sun for light by day
and the fixed order of the moon
and the stars for light
by night,
who stirs up the sea so that its
waves roar—
the LORD of hosts is his name:
36 If this fixed order were ever
to cease
from my presence, says the
LORD,
then also the offspring of Israel
would cease

to be a nation before me
forever.

37 Thus says the LORD:
If the heavens above can be
measured,
and the foundations of the earth
below can be explored,
then I will reject all the offspring
of Israel
because of all they have done,
says the LORD.

38 The days are surely coming, says the LORD, when the city shall be rebuilt for the LORD from the tower of Hananel to the Corner Gate. 39 And the measuring line shall go out farther, straight to the hill Gareb, and shall then turn to Goah. 40 The whole valley of the dead bodies and the ashes, and all the fields as far as the Wadi Kidron, to the corner of the Horse Gate toward the east, shall be sacred to the LORD. It shall never again be uprooted or overthrown.

32 The word that came to Jeremiah from the LORD in the tenth year of King Zedekiah of Judah, which was the eighteenth year of Nebuchadrezzar. 2 At that time the army of the king of Babylon was besieging Jerusalem, and the prophet Jeremiah was confined in the court of the guard that was in the palace of the king of Judah, 3 where King Zedekiah of Judah had confined him. Zedeki-

r Or master

24.16), discussed in some detail by Ezekiel (ch 18).
31.31–34: Using the oldest expression for covenant making ("to cut a covenant") and opposing what was an increasingly limited concept of the Sinai covenant, the prophet affirms that God will *make a new covenant* (32.38–40; Heb 8.8–12; 10.16–17) inscribed in the hearts of his people (17.1; Ezek 11.19; Hos 2.20).
31.35–37: In this passage, coming from a time after Jeremiah (compare 33.14–26), the seemingly eternal cycle of nature (5.22) is

used to emphasize God's assurance of Israel's continued existence (Isa 44.24; 54.9–10). **38–40:** This passage, again coming from after the time of Jeremiah (compare Zech 14.10–11), describes Jerusalem's four corners: northeast (*Hananel,* Neh 3.1), northwest (*Corner Gate,* 2 Kings 14.13), southeast and southwest (*Gareb, Goah,* both unidentified); and southern (Hinnom, 7.31–32) and eastern (Wadi Kidron, 2 Kings 23.4, 6) boundaries. *Horse Gate* (southeast corner), Neh 3.28.
32.1–44: Jeremiah purchases land in Anathoth. 1–5: For background, see ch 37

ah had said, "Why do you prophesy and say: Thus says the LORD: I am going to give this city into the hand of the king of Babylon, and he shall take it; [4] King Zedekiah of Judah shall not escape out of the hands of the Chaldeans, but shall surely be given into the hands of the king of Babylon, and shall speak with him face to face and see him eye to eye; [5] and he shall take Zedekiah to Babylon, and there he shall remain until I attend to him, says the LORD; though you fight against the Chaldeans, you shall not succeed?"

6 Jeremiah said, The word of the LORD came to me: [7] Hanamel son of your uncle Shallum is going to come to you and say, "Buy my field that is at Anathoth, for the right of redemption by purchase is yours." [8] Then my cousin Hanamel came to me in the court of the guard, in accordance with the word of the LORD, and said to me, "Buy my field that is at Anathoth in the land of Benjamin, for the right of possession and redemption is yours; buy it for yourself." Then I knew that this was the word of the LORD.

9 And I bought the field at Anathoth from my cousin Hanamel, and weighed out the money to him, seventeen shekels of silver. [10] I signed the deed, sealed it, got witnesses, and weighed the money on scales. [11] Then I took the sealed deed of purchase, containing the terms and conditions, and the open copy; [12] and I gave the deed of purchase to Baruch son of Neriah son of Mahseiah, in the presence of my cousin Hanamel, in the presence of the witnesses who signed the deed of purchase, and in the presence of all the Judeans who were sitting in the court of the guard. [13] In their presence I charged Baruch, saying, [14] Thus says the LORD of hosts, the God of Israel: Take these deeds, both this sealed deed of purchase and this open deed, and put them in an earthenware jar, in order that they may last for a long time. [15] For thus says the LORD of hosts, the God of Israel: Houses and fields and vineyards shall again be bought in this land.

16 After I had given the deed of purchase to Baruch son of Neriah, I prayed to the LORD, saying: [17] Ah Lord GOD! It is you who made the heavens and the earth by your great power and by your outstretched arm! Nothing is too hard for you. [18] You show steadfast love to the thousandth generation,[s] but repay the guilt of parents into the laps of their children after them, O great and mighty God whose name is the LORD of hosts, [19] great in counsel and mighty in deed; whose eyes are open to all the ways of mortals, rewarding all according to their ways and according to the fruit of their doings. [20] You showed signs and wonders in the land of Egypt, and to this day in Israel and among all humankind, and have made yourself a name that continues to this very day. [21] You brought your people Israel out of the land of Egypt

s Or *to thousands*

(date: early 587 B.C.). Chronologically the account should follow ch 37. Its being placed here emphasizes the validity of the preceding oracles concerning Judah's restoration. **6–15:** To prevent the loss of family property, Jeremiah's cousin, *Hanamel*, offered to sell his land to Jeremiah (Lev 25.25–28). This is the most detailed account of a business transaction in the Bible (compare Gen 23.1–16). *Seventeen shekels* (about seven ounces) refers to weight, not coins. The official copy of the deed, written on papyrus, was rolled up and sealed; *the open copy* was for easy reference.

Similar storage of deeds in earthen jars is known from Elephantine in Egypt. *Baruch,* Jeremiah's secretary, see ch 36. Jeremiah's purchase illustrates his confidence in the future of Judah.

32.16–44: This section deals with the theme of the future of Judah; some of it represents an editorial expansion. First is a formal prayer (vv. 16–25). After praising God's omnipotence (10.16; 27.5), his omniscience (17.10), and his wondrous deeds in Israel's behalf (11.5), Jeremiah questions the wisdom of his purchase in view of the circumstances

with signs and wonders, with a strong hand and outstretched arm, and with great terror; 22 and you gave them this land, which you swore to their ancestors to give them, a land flowing with milk and honey; 23 and they entered and took possession of it. But they did not obey your voice or follow your law; of all you commanded them to do, they did nothing. Therefore you have made all these disasters come upon them. 24 See, the siege ramps have been cast up against the city to take it, and the city, faced with sword, famine, and pestilence, has been given into the hands of the Chaldeans who are fighting against it. What you spoke has happened, as you yourself can see. 25 Yet you, O Lord God, have said to me, "Buy the field for money and get witnesses"—though the city has been given into the hands of the Chaldeans.

26 The word of the Lord came to Jeremiah: 27 See, I am the Lord, the God of all flesh; is anything too hard for me? 28 Therefore, thus says the Lord: I am going to give this city into the hands of the Chaldeans and into the hand of King Nebuchadrezzar of Babylon, and he shall take it. 29 The Chaldeans who are fighting against this city shall come, set it on fire, and burn it, with the houses on whose roofs offerings have been made to Baal and libations have been poured out to other gods, to provoke me to anger. 30 For the people of Israel and the people of Judah have done nothing but evil in my sight from their youth; the people of Israel have done nothing but provoke me to anger by the work of their hands, says the Lord. 31 This city has aroused my anger and wrath, from the day it was built until this day, so that I will remove it from my sight 32 because of all the evil of the people of Israel and the people of

Judah that they did to provoke me to anger—they, their kings and their officials, their priests and their prophets, the citizens of Judah and the inhabitants of Jerusalem. 33 They have turned their backs to me, not their faces; though I have taught them persistently, they would not listen and accept correction. 34 They set up their abominations in the house that bears my name, and defiled it. 35 They built the high places of Baal in the valley of the son of Hinnom, to offer up their sons and daughters to Molech, though I did not command them, nor did it enter my mind that they should do this abomination, causing Judah to sin.

36 Now therefore thus says the Lord, the God of Israel, concerning this city of which you say, "It is being given into the hand of the king of Babylon by the sword, by famine, and by pestilence": 37 See, I am going to gather them from all the lands to which I drove them in my anger and my wrath and in great indignation; I will bring them back to this place, and I will settle them in safety. 38 They shall be my people, and I will be their God. 39 I will give them one heart and one way, that they may fear me for all time, for their own good and the good of their children after them. 40 I will make an everlasting covenant with them, never to draw back from doing good to them, and I will put the fear of me in their hearts, so that they may not turn from me. 41 I will rejoice in doing good to them, and I will plant them in this land in faithfulness, with all my heart and all my soul.

42 For thus says the Lord: Just as I have brought all this great disaster upon this people, so I will bring upon them all the good fortune that I now promise them. 43 Fields shall be bought in this

---

(vv. 24–25). God's answer (vv. 26–44) summarizes Jeremiah's (and the Deuteronomic) interpretation of the contemporary critical events. After a résumé of Judah's idolatry (incense, wine and oil offered on the rooftops, 19.13; comparative sinfulness of Israel and Judah, 3.6–11; human sacrifices, 7.30–32) and her stubborn indifference to God's warn- ings (17.21–23), her imminent destruction (in conventional terms: sword, pestilence, and famine, 14.11–12; 21.7) is affirmed. The section concludes with assurances of restoration, first with reference to the new covenant (31.31–34) and then the exchange and holding of property (vv. 42–44).

land of which you are saying, It is a desolation, without human beings or animals; it has been given into the hands of the Chaldeans. ⁴⁴ Fields shall be bought for money, and deeds shall be signed and sealed and witnessed, in the land of Benjamin, in the places around Jerusalem, and in the cities of Judah, of the hill country, of the Shephelah, and of the Negeb; for I will restore their fortunes, says the LORD.

33 The word of the LORD came to Jeremiah a second time, while he was still confined in the court of the guard: ² Thus says the LORD who made the earth,ᵗ the LORD who formed it to establish it—the LORD is his name: ³ Call to me and I will answer you, and will tell you great and hidden things that you have not known. ⁴ For thus says the LORD, the God of Israel, concerning the houses of this city and the houses of the kings of Judah that were torn down to make a defense against the siege ramps and before the sword:ᵘ ⁵ The Chaldeans are coming in to fightᵛ and to fill them with the dead bodies of those whom I shall strike down in my anger and my wrath, for I have hidden my face from this city because of all their wickedness. ⁶ I am going to bring it recovery and healing; I will heal them and reveal to them abundanceᵘ of prosperity and security. ⁷ I will restore the fortunes of Judah and the fortunes of Israel, and rebuild them as they were at first. ⁸ I will cleanse them from all the guilt of their sin against me, and I will forgive all the guilt of their sin and rebellion against me. ⁹ And this cityᵂ shall be to me a name of joy, a praise and a glory before all the nations of the earth who shall hear of all the good that I do for them; they shall fear and tremble be-

cause of all the good and all the prosperity I provide for it.

10 Thus says the LORD: In this place of which you say, "It is a waste without human beings or animals," in the towns of Judah and the streets of Jerusalem that are desolate, without inhabitants, human or animal, there shall once more be heard ¹¹ the voice of mirth and the voice of gladness, the voice of the bridegroom and the voice of the bride, the voices of those who sing, as they bring thank offerings to the house of the LORD:

"Give thanks to the LORD of hosts,
   for the LORD is good,
   for his steadfast love endures
      forever!"

For I will restore the fortunes of the land as at first, says the LORD.

12 Thus says the LORD of hosts: In this place that is waste, without human beings or animals, and in all its towns there shall again be pasture for shepherds resting their flocks. ¹³ In the towns of the hill country, of the Shephelah, and of the Negeb, in the land of Benjamin, the places around Jerusalem, and in the towns of Judah, flocks shall again pass under the hands of the one who counts them, says the LORD.

14 The days are surely coming, says the LORD, when I will fulfill the promise I made to the house of Israel and the house of Judah. ¹⁵ In those days and at that time I will cause a righteous Branch to spring up for David; and he shall execute justice and righteousness in the land. ¹⁶ In those days Judah will be saved and Jerusalem will live in safety. And this is

---

t Gk: Heb *it*   *u* Meaning of Heb uncertain
v Cn: Heb *They are coming in to fight against the Chaldeans*   *w* Heb *And it*

33.1–13: God will rebuild the walls of Jerusalem. 1: Editorial connection with preceding chapter. **4–5:** In siege defense measures, there was demolition of houses built in the city wall to allow for the military needs of defenders (Isa 22.10). **6–8a:** After punishment (and repentance) will come rebuilding and restoration. **8b–9:** Generalizing expansion.

33.10–13: Two proclamations of salvation; vv. 10–11 reverse 7.34.
33.14–26: A variation of 23.5–6. This passage is lacking in the Septuagint, and is to be dated probably to 450–400 B.C. **14–18:** There will be not one but a succession of Davidic rulers (2 Sam 7.16; 1 Kings 9.5). The reference to *levitical priests* is unique in Jeremiah (Deut 18.1–5). **16:** The new *name* of Jerusalem

the name by which it will be called: "The LORD is our righteousness."

17 For thus says the LORD: David shall never lack a man to sit on the throne of the house of Israel, 18 and the levitical priests shall never lack a man in my presence to offer burnt offerings, to make grain offerings, and to make sacrifices for all time.

19 The word of the LORD came to Jeremiah: 20 Thus says the LORD: If any of you could break my covenant with the day and my covenant with the night, so that day and night would not come at their appointed time, 21 only then could my covenant with my servant David be broken, so that he would not have a son to reign on his throne, and my covenant with my ministers the Levites. 22 Just as the host of heaven cannot be numbered and the sands of the sea cannot be measured, so I will increase the offspring of my servant David, and the Levites who minister to me.

23 The word of the LORD came to Jeremiah: 24 Have you not observed how these people say, "The two families that the LORD chose have been rejected by him," and how they hold my people in such contempt that they no longer regard them as a nation? 25 Thus says the LORD: Only if I had not established my covenant with day and night and the ordinances of heaven and earth, 26 would I reject the offspring of Jacob and of my servant David and not choose any of his descendants as rulers over the offspring of Abraham, Isaac, and Jacob. For I will restore their fortunes, and will have mercy upon them.

34 The word that came to Jeremiah from the LORD, when King Neb-

uchadrezzar of Babylon and all his army and all the kingdoms of the earth and all the peoples under his dominion were fighting against Jerusalem and all its cities: 2 "Thus says the LORD, the God of Israel: Go and speak to King Zedekiah of Judah and say to him: Thus says the LORD: I am going to give this city into the hand of the king of Babylon, and he shall burn it with fire. 3 And you yourself shall not escape from his hand, but shall surely be captured and handed over to him; you shall see the king of Babylon eye to eye and speak with him face to face; and you shall go to Babylon. 4 Yet hear the word of the LORD, O King Zedekiah of Judah! Thus says the LORD concerning you: You shall not die by the sword; 5 you shall die in peace. And as spices were burned* for your ancestors, the earlier kings who preceded you, so they shall burn spices*y* for you and lament for you, saying, "Alas, lord!" For I have spoken the word, says the LORD.

6 Then the prophet Jeremiah spoke all these words to Zedekiah king of Judah, in Jerusalem, 7 when the army of the king of Babylon was fighting against Jerusalem and against all the cities of Judah that were left, Lachish and Azekah; for these were the only fortified cities of Judah that remained.

8 The word that came to Jeremiah from the LORD, after King Zedekiah had made a covenant with all the people in Jerusalem to make a proclamation of liberty to them, 9 that all should set free their Hebrew slaves, male and female, so that no one should hold another Judean in slavery. 10 And they obeyed, all the officials and all the people who had en-

x Heb *as there was burning*   *y* Heb *shall burn*

(see Isa 1.26 n.). **22**: A reinterpretation of Gen 22.17–18.

33.23–26: As night follows day (31.35–37; Gen 1.5; 8.22), God's promise to the patriarchs (compare Rom 4.13) and David will be kept.

34.1–7: **Warning to Zedekiah.** The siege of Jerusalem (January 588 B.C.) was but weeks away; only *Lachish* (twenty-three

miles southwest of Jerusalem) and *Azekah* (eleven miles north of Lachish) held out. Warning Zedekiah of Jerusalem's imminent destruction and his own captivity, Jeremiah assured Zedekiah of a peaceful death and appropriate funeral rites (but compare 52.8–11).

34.8–22: **Manumission of slaves and perfidy of the Jerusalemites.** Probably to simplify the problem of the domestic food

tered into the covenant that all would set free their slaves, male or female, so that they would not be enslaved again; they obeyed and set them free. [11] But afterward they turned around and took back the male and female slaves they had set free, and brought them again into subjection as slaves. [12] The word of the LORD came to Jeremiah from the LORD: [13] Thus says the LORD, the God of Israel: I myself made a covenant with your ancestors when I brought them out of the land of Egypt, out of the house of slavery, saying, [14] "Every seventh year each of you must set free any Hebrews who have been sold to you and have served you six years; you must set them free from your service." But your ancestors did not listen to me or incline their ears to me. [15] You yourselves recently repented and did what was right in my sight by proclaiming liberty to one another, and you made a covenant before me in the house that is called by my name; [16] but then you turned around and profaned my name when each of you took back your male and female slaves, whom you had set free according to their desire, and you brought them again into subjection to be your slaves. [17] Therefore, thus says the LORD: You have not obeyed me by granting a release to your neighbors and friends; I am going to grant a release to you, says the LORD—a release to the sword, to pestilence, and to famine. I will make you a horror to all the king-

doms of the earth. [18] And those who transgressed my covenant and did not keep the terms of the covenant that they made before me, I will make like[z] the calf when they cut it in two and passed between its parts: [19] the officials of Judah, the officials of Jerusalem, the eunuchs, the priests, and all the people of the land who passed between the parts of the calf [20] shall be handed over to their enemies and to those who seek their lives. Their corpses shall become food for the birds of the air and the wild animals of the earth. [21] And as for King Zedekiah of Judah and his officials, I will hand them over to their enemies and to those who seek their lives, to the army of the king of Babylon, which has withdrawn from you. [22] I am going to command, says the LORD, and will bring them back to this city; and they will fight against it, and take it, and burn it with fire. The towns of Judah I will make a desolation without inhabitant.

35 The word that came to Jeremiah from the LORD in the days of King Jehoiakim son of Josiah of Judah: [2] Go to the house of the Rechabites, and speak with them, and bring them to the house of the LORD, into one of the chambers; then offer them wine to drink. [3] So I took Jaazaniah son of Jeremiah son of Habazziniah, and his brothers, and all his

---

z Cn: Heb lacks *like*

---

supply, to make more people available for Jerusalem's defense, and to propitiate the LORD, Zedekiah proclaimed the manumission of all slaves. But when the Egyptian army approached (37.6–15), causing the Babylonians temporarily to lift the siege, the manumission was rescinded.

34.13–22: Making reference to Ex 21.2 and Deut 15.12, Jeremiah condemned this perfidy as one more example of notorious faithlessness. He assured the people of punishment like that implied in their covenant ritual (v. 18; Gen 15.9–17); the transgressor should suffer the same fate as the slaughtered animal, a practice known from non-biblical sources also. The use of the ancient formula "cut a covenant" is noteworthy here (in vv. 8,

13, 15, etc., "made" is literally "cut"; see 31.31 n.) rather than the Deuteronomic formula "command a covenant."

35.1–19: The symbol of the Rechabites. The Rechabites were a religious order, similar to the nazirites (Num 6.1–21), founded by Jonadab, son of Rechab, during the reign of Jehu (842–815 B.C.). Religiously devoted, they assisted Jehu in the purge accompanying the revolt against the dynasty of Omri (2 Kings 10.15–28). They held that the more sophisticated sedentary life in Canaan jeopardized the purity of the worship of the LORD. Consequently, returning to desert ways, they lived in tents, were shepherds, and abstained from wine. Jeremiah's association with them does not imply acceptance of their position,

sons, and the whole house of the Rechabites. 4I brought them to the house of the LORD into the chamber of the sons of Hanan son of Igdaliah, the man of God, which was near the chamber of the officials, above the chamber of Maaseiah son of Shallum, keeper of the threshold. 5Then I set before the Rechabites pitchers full of wine, and cups; and I said to them, "Have some wine." 6But they answered, "We will drink no wine, for our ancestor Jonadab son of Rechab commanded us, 'You shall never drink wine, neither you nor your children; 7nor shall you ever build a house, or sow seed; nor shall you plant a vineyard, or even own one; but you shall live in tents all your days, that you may live many days in the land where you reside.' 8We have obeyed the charge of our ancestor Jonadab son of Rechab in all that he commanded us, to drink no wine all our days, ourselves, our wives, our sons, or our daughters, 9and not to build houses to live in. We have no vineyard or field or seed; 10but we have lived in tents, and have obeyed and done all that our ancestor Jonadab commanded us. 11But when King Nebuchadrezzar of Babylon came up against the land, we said, 'Come, and let us go to Jerusalem for fear of the army of the Chaldeans and the army of the Arameans.' That is why we are living in Jerusalem."

12 Then the word of the LORD came to Jeremiah: 13Thus says the LORD of hosts, the God of Israel: Go and say to the people of Judah and the inhabitants of Jerusalem, Can you not learn a lesson and obey my words? says the LORD. 14The command has been carried out that Jonadab son of Rechab gave to his descendants to drink no wine; and they drink none to this day, for they have obeyed their ancestor's command. But I myself have spoken to you persistently,

and you have not obeyed me. 15I have sent to you all my servants the prophets, sending them persistently, saying, 'Turn now everyone of you from your evil way, and amend your doings, and do not go after other gods to serve them, and then you shall live in the land that I gave to you and your ancestors.' But you did not incline your ear or obey me. 16The descendants of Jonadab son of Rechab have carried out the command that their ancestor gave them, but this people has not obeyed me. 17Therefore, thus says the LORD, the God of hosts, the God of Israel: I am going to bring on Judah and on all the inhabitants of Jerusalem every disaster that I have pronounced against them; because I have spoken to them and they have not listened, I have called to them and they have not answered.

18 But to the house of the Rechabites Jeremiah said: Thus says the LORD of hosts, the God of Israel: Because you have obeyed the command of your ancestor Jonadab, and kept all his precepts, and done all that he commanded you, 19therefore thus says the LORD of hosts, the God of Israel: Jonadab son of Rechab shall not lack a descendant to stand before me for all time.

36 In the fourth year of King Jehoiakim son of Josiah of Judah, this word came to Jeremiah from the LORD: 2Take a scroll and write on it all the words that I have spoken to you against Israel and Judah and all the nations, from the day I spoke to you, from the days of Josiah until today. 3It may be that when the house of Judah hears of all the disasters that I intend to do to them, all of them may turn from their evil ways, so that I may forgive their iniquity and their sin.

4 Then Jeremiah called Baruch son of Neriah, and Baruch wrote on a scroll at

---

but only the approbation of their faithfulness to their principles, in contrast to Judah's faithlessness. The occasion is probably the crisis of 601 B.C. (see 12.7–13 n.). *Maaseiah,* perhaps the father of Zephaniah (compare 21.1).

36.1–32: Jehoiakim burns a scroll of

Jeremiah, and Jeremiah dictates another.
1–4: In the light of portentous contemporary events (25.1–14), Jeremiah commissioned *Baruch son of Neriah* and brother of Seraiah (32.12; 51.59), to transcribe his oracles (*words of the LORD*) on a papyrus scroll. *Fourth year,*

Jeremiah's dictation all the words of the LORD that he had spoken to him. 5 And Jeremiah ordered Baruch, saying, "I am prevented from entering the house of the LORD; 6 so you go yourself, and on a fast day in the hearing of the people in the LORD's house you shall read the words of the LORD from the scroll that you have written at my dictation. You shall read them also in the hearing of all the people of Judah who come up from their towns. 7 It may be that their plea will come before the LORD, and that all of them will turn from their evil ways, for great is the anger and wrath that the LORD has pronounced against this people." 8 And Baruch son of Neriah did all that the prophet Jeremiah ordered him about reading from the scroll the words of the LORD in the LORD's house.

9 In the fifth year of King Jehoiakim son of Josiah of Judah, in the ninth month, all the people in Jerusalem and all the people who came from the towns of Judah to Jerusalem proclaimed a fast before the LORD. 10 Then, in the hearing of all the people, Baruch read the words of Jeremiah from the scroll, in the house of the LORD, in the chamber of Gemariah son of Shaphan the secretary, which was in the upper court, at the entry of the New Gate of the LORD's house.

11 When Micaiah son of Gemariah son of Shaphan heard all the words of the LORD from the scroll, 12 he went down to the king's house, into the secretary's chamber; and all the officials were sitting there: Elishama the secretary, Delaiah son of Shemaiah, Elnathan son of Achbor, Gemariah son of Shaphan, Zedekiah son of Hananiah, and all the officials. 13 And Micaiah told them all the words

that he had heard, when Baruch read the scroll in the hearing of the people. 14 Then all the officials sent Jehudi son of Nethaniah son of Shelemiah son of Cushi to say to Baruch, "Bring the scroll that you read in the hearing of the people, and come." So Baruch son of Neriah took the scroll in his hand and came to them. 15 And they said to him, "Sit down and read it to us." So Baruch read it to them. 16 When they heard all the words, they turned to one another in alarm, and said to Baruch, "We certainly must report all these words to the king." 17 Then they questioned Baruch, "Tell us now, how did you write all these words? Was it at his dictation?" 18 Baruch answered them, "He dictated all these words to me, and I wrote them with ink on the scroll." 19 Then the officials said to Baruch, "Go and hide, you and Jeremiah, and let no one know where you are."

20 Leaving the scroll in the chamber of Elishama the secretary, they went to the court of the king; and they reported all the words to the king. 21 Then the king sent Jehudi to get the scroll, and he took it from the chamber of Elishama the secretary; and Jehudi read it to the king and all the officials who stood beside the king. 22 Now the king was sitting in his winter apartment (it was the ninth month), and there was a fire burning in the brazier before him. 23 As Jehudi read three or four columns, the king[a] would cut them off with a penknife and throw them into the fire in the brazier, until the entire scroll was consumed in the fire that was in the brazier. 24 Yet neither the king, nor any of his servants who heard

a Heb *he*

605 B.C. **5–10**: For an unknown reason, Jeremiah was barred from the temple; so, on the occasion of a fast proclaimed by Jehoiakim because of Nebuchadrezzar's advance against Ashkelon (in November, 604 B.C.), he instructed Baruch to read the scroll in his stead. *Shaphan,* a friend of Jeremiah (26.24).

36.11–19: Baruch was asked to re-read the

scroll before an assembly of royal officials. Deeply impressed, they resolved to inform Jehoiakim. Sensing the possibility of repercussions, they instructed Baruch and Jeremiah to go into hiding.

36.20–26: Jehoiakim ordered the scroll brought from the court scribe's office. Despite protests from some officials, he burned

all these words, was alarmed, nor did they tear their garments. 25 Even when Elnathan and Delaiah and Gemariah urged the king not to burn the scroll, he would not listen to them. 26 And the king commanded Jerahmeel the king's son and Seraiah son of Azriel and Shelemiah son of Abdeel to arrest the secretary Baruch and the prophet Jeremiah. But the LORD hid them.

27 Now, after the king had burned the scroll with the words that Baruch wrote at Jeremiah's dictation, the word of the LORD came to Jeremiah: 28 Take another scroll and write on it all the former words that were in the first scroll, which King Jehoiakim of Judah has burned. 29 And concerning King Jehoiakim of Judah you shall say: Thus says the LORD, You have dared to burn this scroll, saying, Why have you written in it that the king of Babylon will certainly come and destroy this land, and will cut off from it human beings and animals? 30 Therefore thus says the LORD concerning King Jehoiakim of Judah: He shall have no one to sit upon the throne of David, and his dead body shall be cast out to the heat by day and the frost by night. 31 And I will punish him and his offspring and his servants for their iniquity; I will bring on them, and on the inhabitants of Jerusalem, and on the people of Judah, all the disasters with which I have threatened them—but they would not listen.

32 Then Jeremiah took another scroll and gave it to the secretary Baruch son of Neriah, who wrote on it at Jeremiah's dictation all the words of the scroll that King Jehoiakim of Judah had burned in

the fire; and many similar words were added to them.

37 Zedekiah son of Josiah, whom King Nebuchadrezzar of Babylon made king in the land of Judah, succeeded Coniah son of Jehoiakim. 2 But neither he nor his servants nor the people of the land listened to the words of the LORD that he spoke through the prophet Jeremiah.

3 King Zedekiah sent Jehucal son of Shelemiah and the priest Zephaniah son of Maaseiah to the prophet Jeremiah saying, "Please pray for us to the LORD our God." 4 Now Jeremiah was still going in and out among the people, for he had not yet been put in prison. 5 Meanwhile, the army of Pharaoh had come out of Egypt; and when the Chaldeans who were besieging Jerusalem heard news of them, they withdrew from Jerusalem.

6 Then the word of the LORD came to the prophet Jeremiah: 7 Thus says the LORD, God of Israel: This is what the two of you shall say to the king of Judah, who sent you to me to inquire of me, Pharaoh's army, which set out to help you, is going to return to its own land, to Egypt. 8 And the Chaldeans shall return and fight against this city; they shall take it and burn it with fire. 9 Thus says the LORD: Do not deceive yourselves, saying, "The Chaldeans will surely go away from us," for they will not go away. 10 Even if you defeated the whole army of Chaldeans who are fighting against you, and there remained of them only wounded men in their tents, they would rise up and burn this city with fire.

11 Now when the Chaldean army

---

it as it was read, three or four columns at a time. Irritated by its contents, Jehoiakim ordered the arrest of Jeremiah and Baruch. *Penknife,* a knife used to sharpen the point of the scribes' reed pens.

36.27–32: Using the destruction of the scroll as a symbol, Jeremiah announced the ignominious death of Jehoiakim (22.18–19; 2 Kings 24.6–15) and dictated an expanded copy of the scroll (probably contained largely in chs 1–25).

37.1–44.30: The sufferings of Jeremiah

before, during, and after the fall of Jerusalem.

37.1–38.28: Jeremiah, Zedekiah, and the siege. 37.1–2: Editorial transition from ch 36 to 37. **3–10:** Shortly after his accession (spring, 588 B.C.), an army of Pharaoh Hophra (Apries) came up from Egypt to relieve besieged Jerusalem (34.21). The inhabitants concluded that a deliverance as in the days of Hezekiah (2 Kings 19.32–37) had occurred, but Jeremiah warned them that such optimism was wholly unwarranted.

had withdrawn from Jerusalem at the approach of Pharaoh's army, 12 Jeremiah set out from Jerusalem to go to the land of Benjamin to receive his share of property*b* among the people there. 13 When he reached the Benjamin Gate, a sentinel there named Irijah son of Shelemiah son of Hananiah arrested the prophet Jeremiah saying, "You are deserting to the Chaldeans." 14 And Jeremiah said, "That is a lie; I am not deserting to the Chaldeans." But Irijah would not listen to him, and arrested Jeremiah and brought him to the officials. 15 The officials were enraged at Jeremiah, and they beat him and imprisoned him in the house of the secretary Jonathan, for it had been made a prison. 16 Thus Jeremiah was put in the cistern house, in the cells, and remained there many days.

17 Then King Zedekiah sent for him, and received him. The king questioned him secretly in his house, and said, "Is there any word from the LORD?" Jeremiah said, "There is!" Then he said, "You shall be handed over to the king of Babylon." 18 Jeremiah also said to King Zedekiah, "What wrong have I done to you or your servants or this people, that you have put me in prison? 19 Where are your prophets who prophesied to you, saying, 'The king of Babylon will not come against you and against this land'? 20 Now please hear me, my lord king: be good enough to listen to my plea, and do not send me back to the house of the secretary Jonathan to die there." 21 So King Zedekiah gave orders, and they committed Jeremiah to the court of the guard; and a loaf of bread was given him

daily from the bakers' street, until all the bread of the city was gone. So Jeremiah remained in the court of the guard.

38 Now Shephatiah son of Mattan, Gedaliah son of Pashhur, Jucal son of Shelemiah, and Pashhur son of Malchiah heard the words that Jeremiah was saying to all the people, 2 Thus says the LORD, Those who stay in this city shall die by the sword, by famine, and by pestilence; but those who go out to the Chaldeans shall live; they shall have their lives as a prize of war, and live. 3 Thus says the LORD, This city shall surely be handed over to the army of the king of Babylon and be taken. 4 Then the officials said to the king, "This man ought to be put to death, because he is discouraging the soldiers who are left in this city, and all the people, by speaking such words to them. For this man is not seeking the welfare of this people, but their harm." 5 King Zedekiah said, "Here he is; he is in your hands; for the king is powerless against you." 6 So they took Jeremiah and threw him into the cistern of Malchiah, the king's son, which was in the court of the guard, letting Jeremiah down by ropes. Now there was no water in the cistern, but only mud, and Jeremiah sank in the mud.

7 Ebed-melech the Ethiopian,*c* a eunuch in the king's house, heard that they had put Jeremiah into the cistern. The king happened to be sitting at the Benjamin Gate, 8 So Ebed-melech left the king's house and spoke to the king,

b Meaning of Heb uncertain   *c* Or *Nubian*;
Heb *Cushite*

37.11–15: Leaving the city by the Benjamin Gate (20.2), Jeremiah was arrested on suspicion of desertion (38.18–19), perhaps based on oracles such as 21.1–10. **12:** *To receive his share;* see 32.6–15, which may suggest the occasion. **16–21:** In a secret interview, Zedekiah sought reassurance for his ill-advised revolt, but in vain (21.2). He modified Jeremiah's imprisonment to house arrest, and guaranteed him a minimum food ration as long as there were supplies. **38.1–13:** Jeremiah's continued insistence on surrender to Nebuchadrezzar was danger-

ous for Zedekiah's pro-Egyptian counselors (*Gedaliah son of Pashhur,* 20.1; *Jucal* = Jehucal, 37.3; *Pashhur,* 21.1). They persuaded the vacillating Zedekiah that Jeremiah was subverting the war effort (there is similar phraseology in a letter written eighteen months earlier, found in the excavations at Lachish). **6:** The *cistern* was nearly dry, indicating a time shortly before Nebuchadrezzar's final assault August 587 B.C. (52.5–7). Ironically, Jeremiah was delivered from murder at the hands of his countrymen by a foreign court official (who was not necessarily a *eunuch* physically, v. 7).

9 "My lord king, these men have acted wickedly in all they did to the prophet Jeremiah by throwing him into the cistern to die there of hunger, for there is no bread left in the city." 10 Then the king commanded Ebed-melech the Ethiopian,*d* "Take three men with you from here, and pull the prophet Jeremiah up from the cistern before he dies." 11 So Ebed-melech took the men with him and went to the house of the king, to a wardrobe of*e* the storehouse, and took from there old rags and worn-out clothes, which he let down to Jeremiah in the cistern by ropes. 12 Then Ebed-melech the Ethiopian*d* said to Jeremiah, "Just put the rags and clothes between your armpits and the ropes." Jeremiah did so. 13 Then they drew Jeremiah up by the ropes and pulled him out of the cistern. And Jeremiah remained in the court of the guard.

14 King Zedekiah sent for the prophet Jeremiah and received him at the third entrance of the temple of the LORD. The king said to Jeremiah, "I have something to ask you; do not hide anything from me." 15 Jeremiah said to Zedekiah, "If I tell you, you will put me to death, will you not? And if I give you advice, you will not listen to me." 16 So King Zedekiah swore an oath in secret to Jeremiah, "As the LORD lives, who gave us our lives, I will not put you to death or hand you over to these men who seek your life."

17 Then Jeremiah said to Zedekiah, "Thus says the LORD, the God of hosts, the God of Israel, If you will only surrender to the officials of the king of Babylon, then your life shall be spared, and this city shall not be burned with fire, and you and your house shall live. 18 But if you do not surrender to the officials of the king of Babylon, then this city shall be handed over to the Chaldeans, and they shall burn it with fire, and you yourself shall not escape from their hand." 19 King Zedekiah said to Jeremiah, "I am afraid of the Judeans who have deserted to the Chaldeans, for I might be handed over to them and they would abuse me." 20 Jeremiah said, "That will not happen. Just obey the voice of the LORD in what I say to you, and it shall go well with you, and your life shall be spared. 21 But if you are determined not to surrender, this is what the LORD has shown me— 22 a vision of all the women remaining in the house of the king of Judah being led out to the officials of the king of Babylon and saying,

'Your trusted friends have
        seduced you
    and have overcome you;
Now that your feet are stuck in
        the mud,
    they desert you.'

23 All your wives and your children shall be led out to the Chaldeans, and you yourself shall not escape from their hand, but shall be seized by the king of Babylon; and this city shall be burned with fire."

24 Then Zedekiah said to Jeremiah, "Do not let anyone else know of this conversation, or you will die. 25 If the officials should hear that I have spoken with you, and they should come and say to you, 'Just tell us what you said to the king; do not conceal it from us, or we will put you to death. What did the king say to you?' 26 then you shall say to them, 'I was presenting my plea to the king not to send me back to the house of Jonathan to die there.'" 27 All the officials did come to Jeremiah and questioned him; and he answered them in the very words

---

d Or Nubian; Heb Cushite   e Cn: Heb to under

38.14–28: Uncertain and fearful, Zedekiah again (37.17–20) summoned Jeremiah for a private interview (location of *third entrance* is unknown). Justifiably suspicious, Jeremiah exacted an oath from Zedekiah for his safety. He repeated his counsel of "surrender and live" (20.1–16; 21.4–10; 27.1–11) and emphasized it by relating a vision depicting the capture of the royal household and the burning of Jerusalem (*in the mud* recalls Jeremiah's own recent experience, 38.6). Jeremiah assured Zedekiah of safety from Judean deserters (39.9), if he surrendered. Zedekiah characteristically hesitated, exacted a promise of secrecy from Jeremiah, and returned him to house arrest (37.21).

the king had commanded. So they stopped questioning him, for the conversation had not been overheard. 28 And Jeremiah remained in the court of the guard until the day that Jerusalem was taken.

39 In the ninth year of King Zedekiah of Judah, in the tenth month, King Nebuchadrezzar of Babylon and all his army came against Jerusalem and besieged it; 2 in the eleventh year of Zedekiah, in the fourth month, on the ninth day of the month, a breach was made in the city. 3 When Jerusalem was taken,*f* all the officials of the king of Babylon came and sat in the middle gate: Nergal-sharezer, Samgar-nebo, Sarsechim the Rabsaris, Nergal-sharezer the Rabmag, with all the rest of the officials of the king of Babylon. 4 When King Zedekiah of Judah and all the soldiers saw them, they fled, going out of the city at night by way of the king's garden through the gate between the two walls; and they went toward the Arabah. 5 But the army of the Chaldeans pursued them, and overtook Zedekiah in the plains of Jericho; and when they had taken him, they brought him up to King Nebuchadrezzar of Babylon, at Riblah, in the land of Hamath; and he passed sentence on him. 6 The king of Babylon slaughtered the sons of Zedekiah at Riblah before his eyes; also the king of Babylon slaughtered all the nobles of Judah. 7 He put out the eyes of Zedekiah, and bound him in fetters to take him to Babylon. 8 The Chaldeans burned the king's house and the houses of the people, and broke down the walls of Jerusalem. 9 Then Nebuzaradan the captain of the guard exiled to Babylon the rest of the people who were left in the city, those who had deserted to him, and the people who re-

mained. 10 Nebuzaradan the captain of the guard left in the land of Judah some of the poor people who owned nothing, and gave them vineyards and fields at the same time.

11 King Nebuchadrezzar of Babylon gave command concerning Jeremiah through Nebuzaradan, the captain of the guard, saying, 12 "Take him, look after him well and do him no harm, but deal with him as he may ask you." 13 So Nebuzaradan the captain of the guard, Nebushazban the Rabsaris, Nergal-sharezer the Rabmag, and all the chief officers of the king of Babylon sent 14 and took Jeremiah from the court of the guard. They entrusted him to Gedaliah son of Ahikam son of Shaphan to be brought home. So he stayed with his own people.

15 The word of the LORD came to Jeremiah while he was confined in the court of the guard: 16 Go and say to Ebed-melech the Ethiopian:*g* Thus says the LORD of hosts, the God of Israel: I am going to fulfill my words against this city for evil and not for good, and they shall be accomplished in your presence on that day. 17 But I will save you on that day, says the LORD, and you shall not be handed over to those whom you dread. 18 For I will surely save you, and you shall not fall by the sword; but you shall have your life as a prize of war, because you have trusted in me, says the LORD.

40 The word that came to Jeremiah from the LORD after Nebuzaradan the captain of the guard had let him go from Ramah, when he took him bound in fetters along with all the captives of Jerusalem and Judah who were

---

f This clause has been transposed from 38.28
g Or *Nubian*; Heb *Cushite*

39.1–40.6: Jeremiah and the fall of Jerusalem. 39.1–14: Verses 1–10 summarize 52.4–16 (2 Kings 25.1–12) adding the names of Babylonian officials (v. 3). Alternative reading of v. 3: ". . . Nergal-sharezer the Simmagir, Nebushazban the chief court official, Nergal-sharezer the Rabmag . . ." Simmagir and Rabmag are titles of Babylonian

officials. According to 52.6–14, the sack of Jerusalem (v. 8) occurred a month after its capture.

39.15–18: This oracle, assuring Ebed-melech of his personal safety because of his trust in God, is evidently a continuation of 38.27.

40.1–6: Ramah (31.15) was a transit point

being exiled to Babylon. ²The captain of the guard took Jeremiah and said to him, "The LORD your God threatened this place with this disaster; ³and now the LORD has brought it about, and has done as he said, because all of you sinned against the LORD and did not obey his voice. Therefore this thing has come upon you. ⁴Now look, I have just released you today from the fetters on your hands. If you wish to come with me to Babylon, come, and I will take good care of you; but if you do not wish to come with me to Babylon, you need not come. See, the whole land is before you; go wherever you think it good and right to go. ⁵If you remain, *ʰ* then return to Gedaliah son of Ahikam son of Shaphan, whom the king of Babylon appointed governor of the towns of Judah, and stay with him among the people; or go wherever you think it right to go." So the captain of the guard gave him an allowance of food and a present, and let him go. ⁶Then Jeremiah went to Gedaliah son of Ahikam at Mizpah, and stayed with him among the people who were left in the land.

7 When all the leaders of the forces in the open country and their troops heard that the king of Babylon had appointed Gedaliah son of Ahikam governor in the land, and had committed to him men, women, and children, those of the poorest of the land who had not been taken into exile to Babylon, ⁸they went to Gedaliah at Mizpah—Ishmael son of Nethaniah, Johanan son of Kareah, Seraiah son of Tanhumeth, the sons of Ephai the Netophathite, Jezaniah son of the

Maacathite, they and their troops. ⁹Gedaliah son of Ahikam son of Shaphan swore to them and their troops, saying, "Do not be afraid to serve the Chaldeans. Stay in the land and serve the king of Babylon, and it shall go well with you. ¹⁰As for me, I am staying at Mizpah to represent you before the Chaldeans who come to us; but as for you, gather wine and summer fruits and oil, and store them in your vessels, and live in the towns that you have taken over." ¹¹Likewise, when all the Judeans who were in Moab and among the Ammonites and in Edom and in other lands heard that the king of Babylon had left a remnant in Judah and had appointed Gedaliah son of Ahikam son of Shaphan as governor over them, ¹²then all the Judeans returned from all the places to which they had been scattered and came to the land of Judah, to Gedaliah at Mizpah; and they gathered wine and summer fruits in great abundance.

13 Now Johanan son of Kareah and all the leaders of the forces in the open country came to Gedaliah at Mizpah ¹⁴and said to him, "Are you at all aware that Baalis king of the Ammonites has sent Ishmael son of Nethaniah to take your life?" But Gedaliah son of Ahikam would not believe them. ¹⁵Then Johanan son of Kareah spoke secretly to Gedaliah at Mizpah, "Please let me go and kill Ishmael son of Nethaniah, and no one else will know. Why should he take your life, so that all the Judeans who are gathered around you would be scattered, and

---

h Syr: Meaning of Heb uncertain

---

for deportees. For reasons unknown, Jeremiah was allowed to choose exile or residence in Judah. Choosing the latter, he was placed in the custody of Gedaliah, the newly-appointed governor of Judah with whose family Jeremiah had long been friendly (26.24; 36.10).
40.7–41.18: **The third revolt. 40.7–12:** Gedaliah, a member of a prominent Judean family (2 Kings 22.12–14), assured his countrymen that he would represent them before the Babylonians (*Chaldeans*) and urged them to return to their fields and cities. Benjamin seems to have been largely spared (32.1–8;

Neh ch 7) and *Mizpah* (probably present-day Nebi Samwil, five miles northwest of Jerusalem) became the provincial capital.
40.13–41.3: After a time (perhaps only two months) and with encouragement from Baalis of Ammon (for political reasons), Ishmael, a member of the royal family (as Gedaliah was not) and a super-patriot, plotted Gedaliah's assassination. When told of it, Gedaliah discredited the report; nevertheless he, his entourage, and the Babylonian garrison in Mizpah, were slain.

the remnant of Judah would perish?" [16] But Gedaliah son of Ahikam said to Johanan son of Kareah, "Do not do such a thing, for you are telling a lie about Ishmael."

41 In the seventh month, Ishmael son of Nethaniah son of Elishama, of the royal family, one of the chief officers of the king, came with ten men to Gedaliah son of Ahikam, at Mizpah. As they ate bread together there at Mizpah, [2] Ishmael son of Nethaniah and the ten men with him got up and struck down Gedaliah son of Ahikam son of Shaphan with the sword and killed him, because the king of Babylon had appointed him governor in the land. [3] Ishmael also killed all the Judeans who were with Gedaliah at Mizpah, and the Chaldean soldiers who happened to be there.

[4] On the day after the murder of Gedaliah, before anyone knew of it, [5] eighty men arrived from Shechem and Shiloh and Samaria, with their beards shaved and their clothes torn, and their bodies gashed, bringing grain offerings and incense to present at the temple of the LORD. [6] And Ishmael son of Nethaniah came out from Mizpah to meet them, weeping as he came. As he met them, he said to them, "Come to Gedaliah son of Ahikam." [7] When they reached the middle of the city, Ishmael son of Nethaniah and the men with him slaughtered them, and threw them[i] into a cistern. [8] But there were ten men among them who said to Ishmael, "Do not kill us, for we have stores of wheat, barley, oil, and honey hidden in the fields." So he refrained, and did not kill them along with their companions.

[9] Now the cistern into which Ishmael had thrown all the bodies of the men whom he had struck down was the large cistern[j] that King Asa had made for defense against King Baasha of Israel; Ishmael son of Nethaniah filled that cistern with those whom he had killed. [10] Then Ishmael took captive all the rest of the people who were in Mizpah, the king's daughters and all the people who were left at Mizpah, whom Nebuzaradan, the captain of the guard, had committed to Gedaliah son of Ahikam. Ishmael son of Nethaniah took them captive and set out to cross over to the Ammonites.

[11] But when Johanan son of Kareah and all the leaders of the forces with him heard of all the crimes that Ishmael son of Nethaniah had done, [12] they took all their men and went to fight against Ishmael son of Nethaniah. They came upon him at the great pool that is in Gibeon. [13] And when all the people who were with Ishmael saw Johanan son of Kareah and all the leaders of the forces with him, they were glad. [14] So all the people whom Ishmael had carried away captive from Mizpah turned around and came back, and went to Johanan son of Kareah. [15] But Ishmael son of Nethaniah escaped from Johanan with eight men, and went to the Ammonites. [16] Then Johanan son of Kareah and all the leaders of the forces with him took all the rest of the people whom Ishmael son of Nethaniah had carried away captive[k] from Mizpah after he had slain Gedaliah son of Ahikam— soldiers, women, children, and eunuchs, whom Johanan brought back from Gibeon.[l] [17] And they set out, and stopped at Geruth Chimham near Bethlehem, intending to go to Egypt [18] because of the

i Syr: Heb lacks *and threw them*; compare verse 9 *j* Gk: Heb *whom he had killed by the hand of Gedaliah* *k* Cn: Heb *whom he recovered from Ishmael son of Nethaniah* *l* Meaning of Heb uncertain

41.4–10: The next day Ishmael intercepted a group from the north on a pilgrimage to Jerusalem. Luring them into Mizpah, he slaughtered all but ten who bought their lives with *stores* of food. After casting the bodies into an old cistern (1 Kings 15.22), the assassins took the remaining people in Mizpah and headed for Ammon. **11–18:** An avenging group under Johanan overtook Ishmael at Gibeon (28.1; 2 Sam 2.13). In the melee, Ishmael and eight conspirators escaped to Ammon (40.14), but their captives were freed. Fearing Babylonian reprisals Johanan's group turned toward Egypt, encamping en route at *Geruth Chimham* (perhaps "Chimham's Inn") near Bethlehem.

Chaldeans; for they were afraid of them, because Ishmael son of Nethaniah had killed Gedaliah son of Ahikam, whom the king of Babylon had made governor over the land.

42 Then all the commanders of the forces, and Johanan son of Kareah and Azariah*m* son of Hoshaiah, and all the people from the least to the greatest, approached 2the prophet Jeremiah and said, "Be good enough to listen to our plea, and pray to the LORD your God for us—for all this remnant. For there are only a few of us left out of many, as your eyes can see. 3Let the LORD your God show us where we should go and what we should do." 4The prophet Jeremiah said to them, "Very well: I am going to pray to the LORD your God as you request, and whatever the LORD answers you I will tell you; I will keep nothing back from you." 5They in their turn said to Jeremiah, "May the LORD be a true and faithful witness against us if we do not act according to everything that the LORD your God sends us through you. 6Whether it is good or bad, we will obey the voice of the LORD our God to whom we are sending you, in order that it may go well with us when we obey the voice of the LORD our God."

7 At the end of ten days the word of the LORD came to Jeremiah. 8Then he summoned Johanan son of Kareah and all the commanders of the forces who were with him, and all the people from the least to the greatest, 9and said to them, "Thus says the LORD, the God of Israel, to whom you sent me to present your plea before him: 10If you will only remain in this land, then I will build you up and not pull you down; I will plant you, and not pluck you up; for I am sorry for the disaster that I have brought upon you. 11Do not be afraid of the king of Babylon, as you have been; do not be afraid of him, says the LORD, for I am with you, to save you and to rescue you from his hand. 12I will grant you mercy, and he will have mercy on you and restore you to your native soil. 13But if you continue to say, 'We will not stay in this land,' thus disobeying the voice of the LORD your God 14and saying, 'No, we will go to the land of Egypt, where we shall not see war, or hear the sound of the trumpet, or be hungry for bread, and there we will stay,' 15then hear the word of the LORD, O remnant of Judah. Thus says the LORD of hosts, the God of Israel: If you are determined to enter Egypt and go to settle there, 16then the sword that you fear shall overtake you there, in the land of Egypt; and the famine that you dread shall follow close after you into Egypt; and there you shall die. 17All the people who have determined to go to Egypt to settle there shall die by the sword, by famine, and by pestilence; they shall have no remnant or survivor from the disaster that I am bringing upon them.

18 "For thus says the LORD of hosts, the God of Israel: Just as my anger and my wrath were poured out on the inhabitants of Jerusalem, so my wrath will be poured out on you when you go to Egypt. You shall become an object of execration and horror, of cursing and ridicule. You shall see this place no more. 19The LORD has said to you, O remnant of Judah, Do not go to Egypt. Be well aware that I have warned you today 20that you have made a fatal mistake. For you yourselves sent me to the LORD your God, saying, 'Pray for us to the LORD our God, and whatever the LORD our God says, tell us and we will

m Gk: Heb *Jezaniah*

42.1–43.7: Flight to Egypt. 42.1–6: Jeremiah was evidently one of the captives freed by Johanan (41.16). He was asked to intercede (15.11) for the group, who were uncertain what they should do and where they should go. To remain could mean reprisals from Babylon (52.30); to flee would mean safety, though with a *de facto* admission of guilt. **7–22:** After ten days, Jeremiah brought the reply: to remain was God's will (29.1–14; 32.6–15) and they would receive God's blessing; to flee would only bring suffering upon the fugitives.

do it.' 21 So I have told you today, but you have not obeyed the voice of the LORD your God in anything that he sent me to tell you. 22 Be well aware, then, that you shall die by the sword, by famine, and by pestilence in the place where you desire to go and settle."

43 When Jeremiah finished speaking to all the people all these words of the LORD their God, with which the LORD their God had sent him to them, 2 Azariah son of Hoshaiah and Johanan son of Kareah and all the other insolent men said to Jeremiah, "You are telling a lie. The LORD our God did not send you to say, 'Do not go to Egypt to settle there'; 3 but Baruch son of Neriah is inciting you against us, to hand us over to the Chaldeans, in order that they may kill us or take us into exile in Babylon." 4 So Johanan son of Kareah and all the commanders of the forces and all the people did not obey the voice of the LORD, to stay in the land of Judah. 5 But Johanan son of Kareah and all the commanders of the forces took all the remnant of Judah who had returned to settle in the land of Judah from all the nations to which they had been driven— 6 the men, the women, the children, the princesses, and everyone whom Nebuzaradan the captain of the guard had left with Gedaliah son of Ahikam son of Shaphan; also the prophet Jeremiah and Baruch son of Neriah. 7 And they came into the land of Egypt, for they did not obey the voice of the LORD. And they arrived at Tahpanhes.

8 Then the word of the LORD came to Jeremiah in Tahpanhes: 9 Take some large stones in your hands, and bury them in the clay pavement *n* that is at the entrance to Pharaoh's palace in Tahpanhes. Let the Judeans see you do it, 10 and say to them, Thus says the LORD of hosts, the God of Israel: I am going to send and take my servant King Nebuchadrezzar of Babylon, and he *o* will set his throne above these stones that I have buried, and he will spread his royal canopy over them. 11 He shall come and ravage the land of Egypt, giving

those who are destined for
    pestilence, to pestilence,
and those who are destined for
    captivity, to captivity,
and those who are destined for
    the sword, to the sword.

12 He *p* shall kindle a fire in the temples of the gods of Egypt; and he shall burn them and carry them away captive; and he shall pick clean the land of Egypt, as a shepherd picks his cloak clean of vermin; and he shall depart from there safely. 13 He shall break the obelisks of Heliopolis, which is in the land of Egypt; and the temples of the gods of Egypt he shall burn with fire.

44 The word that came to Jeremiah for all the Judeans living in the land of Egypt, at Migdol, at Tahpanhes, at Memphis, and in the land of Pathros,

n Meaning of Heb uncertain   *o* Gk Syr: Heb *I*
p Gk Syr Vg: Heb *I*

43.1–7: Apparently the tension in the camp during the ten days that followed allowed those who favored the flight to prevail over the good intentions of those who wished to abide by Jeremiah's oracle. Jeremiah was accused of being unduly influenced by Baruch, and his oracle was not regarded as God's word. It has been suggested that 42.19–22 should come between verses 3 and 4 as Jeremiah's response to Azariah and his colleagues, since 42.19–22 may indicate that the decision had already been made to go to Egypt; Jeremiah reminded them of their earlier resolve and the danger of disobeying the LORD's will. They decided to go to Egypt,

taking Jeremiah and Baruch with them, perhaps to assure themselves of the continuing availability of oracles from God. **7:** *Tahpanhes,* see 2.14–19 n. **43.8–44.30: Jeremiah in Egypt. 43.8–13:** This oracle is intended to show that Egypt is no safe refuge from Nebuchadrezzar (*my servant,* 25.9; 27.6), who led a successful raid against Amasis (Ahmosis II) in 568/567 B.C., some years later than this oracle (46.13–26). *Picks his cloak clean,* literally "delouse," illustrates Jeremiah's low opinion of Egypt. *Heliopolis* (also called On; Gen 41.45), six miles northeast of Cairo; ancient center of the worship of sun-god Re (see Isa 19.18 n.). *Obelisks,*

2 Thus says the LORD of hosts, the God of Israel: You yourselves have seen all the disaster that I have brought on Jerusalem and on all the towns of Judah. Look at them; today they are a desolation, without an inhabitant in them, 3 because of the wickedness that they committed, provoking me to anger, in that they went to make offerings and serve other gods that they had not known, neither they, nor you, nor your ancestors. 4 Yet I persistently sent to you all my servants the prophets, saying, "I beg you not to do this abominable thing that I hate!" 5 But they did not listen or incline their ear, to turn from their wickedness and make no offerings to other gods. 6 So my wrath and my anger were poured out and kindled in the towns of Judah and in the streets of Jerusalem; and they became a waste and a desolation, as they still are today. 7 And now thus says the LORD God of hosts, the God of Israel: Why are you doing such great harm to yourselves, to cut off man and woman, child and infant, from the midst of Judah, leaving yourselves without a remnant? 8 Why do you provoke me to anger with the works of your hands, making offerings to other gods in the land of Egypt where you have come to settle? Will you be cut off and become an object of cursing and ridicule among all the nations of the earth? 9 Have you forgotten the crimes of your ancestors, of the kings of Judah, of their^q wives, your own crimes and those of your wives, which they committed in the land of Judah and in the streets of Jerusalem? 10 They have shown no contrition or fear to this day, nor have they walked in my law and my statutes that I set before you and before your ancestors.

11 Therefore thus says the LORD of hosts, the God of Israel: I am determined to bring disaster on you, to bring all Judah to an end. 12 I will take the remnant of Judah who are determined to come to the land of Egypt to settle, and they shall perish, everyone; in the land of Egypt they shall fall; by the sword and by famine they shall perish; from the least to the greatest, they shall die by the sword and by famine; and they shall become an object of execration and horror, of cursing and ridicule. 13 I will punish those who live in the land of Egypt, as I have punished Jerusalem, with the sword, with famine, and with pestilence, 14 so that none of the remnant of Judah who have come to settle in the land of Egypt shall escape or survive or return to the land of Judah. Although they long to go back to live there, they shall not go back, except some fugitives.

15 Then all the men who were aware that their wives had been making offerings to other gods, and all the women who stood by, a great assembly, all the people who lived in Pathros in the land of Egypt, answered Jeremiah: 16 "As for the word that you have spoken to us in the name of the LORD, we are not going to listen to you. 17 Instead, we will do everything that we have vowed, make offerings to the queen of heaven and pour out libations to her, just as we and our ancestors, our kings and our officials, used to do in the towns of Judah and in the streets of Jerusalem. We used to have plenty of food, and prospered, and saw no misfortune. 18 But from the time we stopped making offerings to the queen of heaven and pouring out liba-

_q_ Heb _his_

---

monuments, slightly tapered square granite shafts, capped by a pyramidion.

44.1–14: This oracle, addressed to the Jewish diaspora in Egypt, is Jeremiah's warning against repeating the mistakes of their fathers in Judah and suffering the consequences (compare 42.14–18). **1:** _Migdol,_ just north of present day Tell el-Heir, east of Tahpanhes (43.7). _Memphis,_ 2.16. _Pathros,_ "Land of the

South," upper (i.e. southern) Egypt, where perhaps there was already by this time a Jewish colony at Elephantine, known from discoveries of Aramaic papyri dating from the fifth century B.C.

44.15–28: The refugees return to the worship of the _queen of heaven_ (7.16–20), which was the Babylonian-Assyrian goddess Ishtar, goddess of the star Venus (compare the Ca-

tions to her, we have lacked everything and have perished by the sword and by famine." 19 And the women said,' "Indeed we will go on making offerings to the queen of heaven and pouring out libations to her; do you think that we made cakes for her, marked with her image, and poured out libations to her without our husbands' being involved?"

20 Then Jeremiah said to all the people, men and women, all the people who were giving him this answer: 21 "As for the offerings that you made in the towns of Judah and in the streets of Jerusalem, you and your ancestors, your kings and your officials, and the people of the land, did not the LORD remember them? Did it not come into his mind? 22 The LORD could no longer bear the sight of your evil doings, the abominations that you committed; therefore your land became a desolation and a waste and a curse, without inhabitant, as it is to this day. 23 It is because you burned offerings, and because you sinned against the LORD and did not obey the voice of the LORD or walk in his law and in his statutes and in his decrees, that this disaster has befallen you, as is still evident today."

24 Jeremiah said to all the people and all the women, "Hear the word of the LORD, all you Judeans who are in the land of Egypt, 25 Thus says the LORD of hosts, the God of Israel: You and your wives have accomplished in deeds what you declared in words, saying, 'We are determined to perform the vows that we have made, to make offerings to the queen of heaven and to pour out libations to her.' By all means, keep your vows and make your libations! 26 Therefore hear the word of the LORD, all you Judeans who live in the land of Egypt: Lo, I swear by my great name, says the LORD, that my name shall no longer be pronounced on the lips of any of the people of Judah in all the land of Egypt, saying, 'As the Lord GOD lives.' 27 I am going to watch over them for harm and not for good; all the people of Judah who are in the land of Egypt shall perish by the sword and by famine, until not one is left. 28 And those who escape the sword shall return from the land of Egypt to the land of Judah, few in number; and all the remnant of Judah, who have come to the land of Egypt to settle, shall know whose words will stand, mine or theirs! 29 This shall be the sign to you, says the LORD, that I am going to punish you in this place, in order that you may know that my words against you will surely be carried out: 30 Thus says the LORD, I am going to give Pharaoh Hophra, king of Egypt, into the hands of his enemies, those who seek his life, just as I gave King Zedekiah of Judah into the hand of King Nebuchadrezzar of Babylon, his enemy who sought his life."

45 The word that the prophet Jeremiah spoke to Baruch son of Neriah, when he wrote these words in a scroll at the dictation of Jeremiah, in the fourth year of King Jehoiakim son of Josiah of Judah: 2 Thus says the LORD, the God of Israel, to you, O Baruch: 3 You said, "Woe is me! The LORD has added sorrow to my pain; I am weary with my groaning, and I find no rest." 4 Thus you

r Compare Syr: Heb lacks *And the women said*

---

naanite Astarte, Greek Aphrodite, Roman Venus). First introduced, presumably, by Manasseh (2 Kings 21.1–18), suppressed by Josiah (2 Kings 23.4–14), and restored by Jehoiakim (2 Kings 23.36–24.7), the cult was especially popular among women, who had an inferior role in the cult of the LORD. Offerings included wine and star-shaped or crescent-shaped cakes or figurines bearing the *image* of the goddess. The cult persisted into the Christian centuries, and features of it were incorporated by the early Syrian church in the veneration of the Virgin.

44.29–30: *Pharaoh Hophra* (Apries, 588–569 B.C.; 37.5) was assassinated by Ahmosis II (Amasis, 569–526 B.C.), a former court official, co-regent for three years, and founder of the Twenty-seventh (Libyan) Dynasty. For similar *signs,* see Isa 7.11–17; Ex 3.12.

45.1–5: God's word to Baruch. *Baruch* ends his memoirs by recalling God's assurance to him of physical deliverance, given in

shall say to him, "Thus says the LORD: I am going to break down what I have built, and pluck up what I have planted—that is, the whole land. 5 And you, do you seek great things for yourself? Do not seek them; for I am going to bring disaster upon all flesh, says the LORD; but I will give you your life as a prize of war in every place to which you may go."

46 The word of the LORD that came to the prophet Jeremiah concerning the nations.

2 Concerning Egypt, about the army of Pharaoh Neco, king of Egypt, which was by the river Euphrates at Carchemish and which King Nebuchadrezzar of Babylon defeated in the fourth year of King Jehoiakim son of Josiah of Judah:

3 Prepare buckler and shield,
    and advance for battle!
4 Harness the horses;
    mount the steeds!
Take your stations with your
        helmets,
    whet your lances,
    put on your coats of mail!
5 Why do I see them terrified?
    They have fallen back;
their warriors are beaten down,
    and have fled in haste.
They do not look back—
    terror is all around!
                    says the LORD.
6 The swift cannot flee away,
    nor can the warrior escape;
in the north by the river Euphrates
    they have stumbled and fallen.

7 Who is this, rising like the Nile,
    like rivers whose waters surge?

8 Egypt rises like the Nile,
    like rivers whose waters surge.
It said, Let me rise, let me cover
        the earth,
    let me destroy cities and their
        inhabitants.
9 Advance, O horses,
    and dash madly, O chariots!
Let the warriors go forth:
    Ethiopia*s* and Put who carry
        the shield,
    the Ludim, who draw*t* the
        bow.
10 That day is the day of the Lord
        GOD of hosts,
    a day of retribution,
    to gain vindication from his
        foes.
The sword shall devour and
        be sated,
    and drink its fill of their blood.
For the Lord GOD of hosts holds
        a sacrifice
    in the land of the north by the
        river Euphrates.
11 Go up to Gilead, and take balm,
    O virgin daughter Egypt!
In vain you have used many
        medicines;
    there is no healing for you.
12 The nations have heard of
        your shame,
    and the earth is full of your cry;
for warrior has stumbled against
        warrior;
    both have fallen together.

13 The word that the LORD spoke to the prophet Jeremiah about the coming

---

s Or *Nubia*; Heb *Cush*   *t* Cn: Heb *who grasp, who draw*

---

the fourth year of Jehoiakim (605 B.C., compare 36.1). For *your life as a prize of war* see 39.18.

46.1–51.64: Oracles against foreign nations (compare Isa chs 13–23; Ezek chs 25–32). **46.1:** Introduction (1.2; 14.1); continues 25.15–38.

46.2–28: Against Egypt. 2–12: In May or June 605 B.C. Crown Prince Nebuchadrezzar of Babylon defeated Neco II of Egypt at Carchemish, on the northern Euphrates sixty miles west of Haran (Gen 11.31), the last capital of Assyria, and pursued him to the borders of Egypt. Like the Nile, Egypt was preparing to inundate the lands to the north. Instead she was grievously wounded, driven back, and humiliated before the nations. **9:** *Put,* Cyrene. *The Ludim,* the Lydians (in Asia Minor, Ezek 30.5).

46.13–26: This oracle may date from

of King Nebuchadrezzar of Babylon to attack the land of Egypt:

14 Declare in Egypt, and proclaim
in Migdol;
proclaim in Memphis and
Tahpanhes;
Say, "Take your stations and
be ready,
for the sword shall devour those
around you."
15 Why has Apis fled?ᵘ
Why did your bull not stand?
—because the LORD thrust
him down.
16 Your multitude stumbledᵛ and
fell,
and one said to another,ʷ
"Come, let us go back to our
own people
and to the land of our birth,
because of the destroying
sword."
17 Give Pharaoh, king of Egypt,
the name
"Braggart who missed his
chance."

18 As I live, says the King,
whose name is the LORD of
hosts,
one is coming
like Tabor among the
mountains,
and like Carmel by the sea.
19 Pack your bags for exile,
sheltered daughter Egypt!
For Memphis shall become a
waste,
a ruin, without inhabitant.

20 A beautiful heifer is Egypt—
a gadfly from the north lights
upon her.
21 Even her mercenaries in her midst
are like fatted calves;

they too have turned and fled
together,
they did not stand;
for the day of their calamity has
come upon them,
the time of their punishment.
22 She makes a sound like a snake
gliding away;
for her enemies march in force,
and come against her with axes,
like those who fell trees.
23 They shall cut down her forest,
says the LORD,
though it is impenetrable,
because they are more numerous
than locusts;
they are without number.
24 Daughter Egypt shall be put
to shame;
she shall be handed over to a
people from the north.

25 The LORD of hosts, the God of Israel, said: See, I am bringing punishment upon Amon of Thebes, and Pharaoh, and Egypt and her gods and her kings, upon Pharaoh and those who trust in him. 26 I will hand them over to those who seek their life, to King Nebuchadrezzar of Babylon and his officers. Afterward Egypt shall be inhabited as in the days of old, says the LORD.

27 But as for you, have no fear, my
servant Jacob,
and do not be dismayed,
O Israel;
for I am going to save you from
far away,
and your offspring from the
land of their captivity.

u Gk: Heb *Why was it swept away*
v Gk: Meaning of Heb uncertain   w Gk: Heb
and fell one to another and they said

588 B.C., when Egypt attempted to relieve the siege of Jerusalem (37.7). The poetic section refers to lower Egypt (*Memphis*); the prose to upper Egypt (*Thebes*). 15: *Apis,* bull-god of (southern) Egypt. *Tabor,* great mountain rising above the plain of Jezreel (Esdraelon; Josh 19.22).

46.18: *Carmel,* mountain at the end of the plain of Jezreel, projecting into the Mediterranean (Josh 19.26).
46.27–28: A doublet of 30.10–11, which here contrasts the destruction of Egypt (except 26b) and the reconstruction of all Israel.

Jacob shall return and have quiet
and ease,
and no one shall make him
afraid.
28 As for you, have no fear, my
servant Jacob,
says the LORD,
for I am with you.
I will make an end of all the
nations
among which I have banished
you,
but I will not make an end
of you!
I will chastise you in just measure,
and I will by no means leave
you unpunished.

47 The word of the LORD that came
to the prophet Jeremiah concerning the Philistines, before Pharaoh attacked Gaza:
2 Thus says the LORD:
See, waters are rising out of
the north
and shall become an
overflowing torrent;
they shall overflow the land and
all that fills it,
the city and those who live in
it.
People shall cry out,
and all the inhabitants of the
land shall wail.
3 At the noise of the stamping of
the hoofs of his stallions,
at the clatter of his chariots, at
the rumbling of their
wheels,
parents do not turn back for
children,

so feeble are their hands,
4 because of the day that is coming
to destroy all the Philistines,
to cut off from Tyre and Sidon
every helper that remains.
For the LORD is destroying the
Philistines,
the remnant of the coastland
of Caphtor.
5 Baldness has come upon Gaza,
Ashkelon is silenced.
O remnant of their power!ˣ
How long will you gash
yourselves?
6 Ah, sword of the LORD!
How long until you are quiet?
Put yourself into your scabbard,
rest and be still!
7 How can itʸ be quiet,
when the LORD has given it
an order?
Against Ashkelon and against the
seashore—
there he has appointed it.

48 Concerning Moab.

Thus says the LORD of hosts, the God of
Israel:
Alas for Nebo, it is laid waste!
Kiriathaim is put to shame, it
is taken;
the fortress is put to shame and
broken down;
2 the renown of Moab is no
more.
In Heshbon they planned evil
against her:

---

x Gk: Heb *their valley*   y Gk Vg: Heb *you*

47.1–7: **Against the Philistines.** Dis-
similar from other oracles against the Philistines (Isa 14.29–31; Ezek 25.15–17), this oracle may be associated with Nebuchadrezzar's
sack of Ashkelon in 604 B.C. (vv. 5, 7; 36.9).
The Phoenician cities, Tyre and Sidon, were
perhaps allied also with the Philistines (27.3).
The Philistines were related to the Indo-
European inhabitants of Crete (*Caphtor,* see
Judg 13.1 n.; 2 Sam 8.18 n.). **5:** For similar
signs of lamentation, compare 16.6; 41.5.

48.1–47: **Against Moab.** This long chapter appears to be a sequence of three oracles,
vv. 1–13, vv. 14–38, and vv. 39–45. Each of
them has been expanded by later editors,
especially vv. 14–38, that has been expanded by material from Isa 15–16. A possible
date for vv. 1–13 would be the threat from
Nebuchadrezzar in 605 B.C.; for vv. 14–
38, 601 B.C. (see 12.7–13 n.); for vv. 39–45,
594 B.C. (compare 27.1–3).
48.1–10: The advance of the enemy (per-

"Come, let us cut her off from
    being a nation!"
You also, O Madmen, shall be
    brought to silence;*z*
the sword shall pursue you.

3 Hark! a cry from Horonaim,
    "Desolation and great
    destruction!"
4 "Moab is destroyed!"
    her little ones cry out.
5 For at the ascent of Luhith
    they go*a* up weeping bitterly;
for at the descent of Horonaim
    they have heard the distressing
    cry of anguish.
6 Flee! Save yourselves!
    Be like a wild ass*b* in the desert!

7 Surely, because you trusted in
    your strongholds*c* and your
    treasures,
you also shall be taken;
Chemosh shall go out into exile,
    with his priests and his
    attendants.
8 The destroyer shall come upon
    every town,
    and no town shall escape;
the valley shall perish,
    and the plain shall be destroyed,
    as the LORD has spoken.

9 Set aside salt for Moab,
    for she will surely fall;
her towns shall become a
    desolation,
    with no inhabitant in them.
10 Accursed is the one who is slack in
doing the work of the LORD; and accursed is the one who keeps back the
sword from bloodshed.

11 Moab has been at ease from his
    youth,
    settled like wine*d* on its dregs;

he has not been emptied from
    vessel to vessel,
    nor has he gone into exile;
therefore his flavor has remained
    and his aroma is unspoiled.
12 Therefore, the time is surely coming, says the LORD, when I shall send to
him decanters to decant him, and empty
his vessels, and break his*e* jars in pieces.
13 Then Moab shall be ashamed of Chemosh, as the house of Israel was ashamed
of Bethel, their confidence.

14 How can you say, "We are heroes
    and mighty warriors"?
15 The destroyer of Moab and his
        towns has come up,
    and the choicest of his young
        men have gone down to
        slaughter,
    says the King, whose name is
        the LORD of hosts.
16 The calamity of Moab is near
        at hand
    and his doom approaches
        swiftly.
17 Mourn over him, all you his
        neighbors,
    and all who know his name;
say, "How the mighty scepter is
        broken,
    the glorious staff!"

18 Come down from glory,
    and sit on the parched ground,
    enthroned daughter Dibon!
For the destroyer of Moab has
    come up against you;
he has destroyed your
    strongholds.
19 Stand by the road and watch,
    you inhabitant of Aroer!

---

z The place-name *Madmen* sounds like the
Hebrew verb *to be silent*   *a* Cn: Heb *he goes*
b Gk Aquila: Heb *like Aroer*   *c* Gk: Heb *works*
d Heb lacks *like wine*   *e* Gk Aquila: Heb *their*

---

haps Nebuchadrezzar), against which Moab
is defenseless. **7**: For *Chemosh,* chief god of
Moab, see Judg 11.24 n. **13**: Though Moab
was not on the main trade-route or invasion-
route, she will not escape; her heroes will
quail before the enemy. Perhaps here *Bethel* is
the name of a contemporary god of Aramaic
origin, not the city.

Ask the man fleeing and the
woman escaping;
say, "What has happened?"
20 Moab is put to shame, for it is
broken down;
wail and cry!
Tell it by the Arnon,
that Moab is laid waste.

21 Judgment has come upon the tableland, upon Holon, and Jahzah, and Mephaath, 22 and Dibon, and Nebo, and Beth-diblathaim, 23 and Kiriathaim, and Beth-gamul, and Beth-meon, 24 and Kerioth, and Bozrah, and all the towns of the land of Moab, far and near. 25 The horn of Moab is cut off, and his arm is broken, says the LORD.
26 Make him drunk, because he magnified himself against the LORD; let Moab wallow in his vomit; he too shall become a laughingstock. 27 Israel was a laughingstock for you, though he was not caught among thieves; but whenever you spoke of him you shook your head!

28 Leave the towns, and live on
the rock,
O inhabitants of Moab!
Be like the dove that nests
on the sides of the mouth of
a gorge.
29 We have heard of the pride of
Moab—
he is very proud—
of his loftiness, his pride, and his
arrogance,
and the haughtiness of his heart.
30 I myself know his insolence, says
the LORD;
his boasts are false,
his deeds are false.
31 Therefore I wail for Moab;
I cry out for all Moab;
for the people of Kir-heres I
mourn.
32 More than for Jazer I weep for
you,

O vine of Sibmah!
Your branches crossed over the
sea,
reached as far as Jazer;*f*
upon your summer fruits and
your vintage
the destroyer has fallen.
33 Gladness and joy have been
taken away
from the fruitful land of Moab;
I have stopped the wine from the
wine presses;
no one treads them with shouts
of joy;
the shouting is not the shout
of joy.

34 Heshbon and Elealeh cry out;*g* as far as Jahaz they utter their voice, from Zoar to Horonaim and Eglath-shelishiyah. For even the waters of Nimrim have become desolate. 35 And I will bring to an end in Moab, says the LORD, those who offer sacrifice at a high place and make offerings to their gods. 36 Therefore my heart moans for Moab like a flute, and my heart moans like a flute for the people of Kir-heres; for the riches they gained have perished.
37 For every head is shaved and every beard cut off; on all the hands there are gashes, and on the loins sackcloth. 38 On all the housetops of Moab and in the squares there is nothing but lamentation; for I have broken Moab like a vessel that no one wants, says the LORD. 39 How it is broken! How they wail! How Moab has turned his back in shame! So Moab has become a derision and a horror to all his neighbors.
40 For thus says the LORD:
Look, he shall swoop down like
an eagle,
and spread his wings against
Moab;

f Two Mss and Isa 16.8: MT *the sea of Jazer*
g Cn: Heb *From the cry of Heshbon to Elealeh*

48.18–28: As Moab had spitefully derided Judah, may she now choke on her own spite! The desert is no safe refuge; as the wild dove (Ezek 7.16), she must take to the rocks.

48.29–38: Moab, once so proud, is humiliated before her conqueror. As a broken pot is discarded, so smitten Moab is accounted worthless.

41 the towns[h] shall be taken
and the strongholds seized.
The hearts of the warriors of
Moab, on that day,
shall be like the heart of a
woman in labor.
42 Moab shall be destroyed as
a people,
because he magnified himself
against the LORD.
43 Terror, pit, and trap
are before you, O inhabitants
of Moab!
says the LORD.
44 Everyone who flees from the
terror
shall fall into the pit,
and everyone who climbs out of
the pit
shall be caught in the trap.
For I will bring these things[i]
upon Moab
in the year of their punishment,
says the LORD.

45 In the shadow of Heshbon
fugitives stop exhausted;
for a fire has gone out from
Heshbon,
a flame from the house of
Sihon;
it has destroyed the forehead
of Moab,
the scalp of the people of
tumult.[j]
46 Woe to you, O Moab!
The people of Chemosh have
perished,
for your sons have been taken
captive,
and your daughters into
captivity.

47 Yet I will restore the fortunes
of Moab
in the latter days, says the
LORD.
Thus far is the judgment on
Moab.

49 Concerning the Ammonites.

Thus says the LORD:
Has Israel no sons?
Has he no heir?
Why then has Milcom
dispossessed Gad,
and his people settled in its
towns?
2 Therefore, the time is surely
coming,
says the LORD,
when I will sound the battle alarm
against Rabbah of the
Ammonites;
it shall become a desolate mound,
and its villages shall be burned
with fire;
then Israel shall dispossess those
who dispossessed him,
says the LORD.

3 Wail, O Heshbon, for Ai is laid
waste!
Cry out, O daughters[k] of
Rabbah!
Put on sackcloth,
lament, and slash yourselves
with whips![l]
For Milcom shall go into exile,
with his priests and his
attendants.

h Or *Kerioth*   i Gk Syr: Heb *bring upon it*
j Or *of Shaon*   k Or *villages*   l Cn: Meaning
of Heb uncertain

48.39–45: God's judgment against Moab.
47: After thoroughgoing devastation (23.12;
Isa 24.17–18) Moab will be restored.
49.1–6: **Against Ammon.** The occasion
for this comparatively mild oracle may be the
Ammonite sponsorship of Ishmael by King
Baalis (40.13–14), perhaps in the autumn of
587 B.C. Ammon, the northern "brother" of
Moab (Gen 19.30–38), had previously occupied Transjordanian territory claimed by Israel (Judg 10.6–12.6; 2 Kings 15.29) and was
Israelite territory under David (2 Sam 12.26–
31; Am 1.13–14 perhaps reflects their war for
independence). Ammon too must suffer for
its idolatry and violence. *Milcom,* Ammonite
national god (1 Kings 11.5, 33); *Rabbah,* capital of Ammon.

4 Why do you boast in your
strength?
Your strength is ebbing,
O faithless daughter.
You trusted in your treasures,
saying,
"Who will attack me?"
5 I am going to bring terror
upon you,
says the Lord GOD of hosts,
from all your neighbors,
and you will be scattered, each
headlong,
with no one to gather the
fugitives.
6 But afterward I will restore the fortunes of the Ammonites, says the LORD.

7 Concerning Edom.

Thus says the LORD of hosts:
Is there no longer wisdom in
Teman?
Has counsel perished from
the prudent?
Has their wisdom vanished?
8 Flee, turn back, get down low,
inhabitants of Dedan!
For I will bring the calamity of
Esau upon him,
the time when I punish him.
9 If grape-gatherers came to you,
would they not leave gleanings?
If thieves came by night,
even they would pillage only
what they wanted.
10 But as for me, I have stripped
Esau bare,
I have uncovered his hiding
places,
and he is not able to conceal
himself.
His offspring are destroyed,
his kinsfolk

and his neighbors; and he is
no more.
11 Leave your orphans, I will keep
them alive;
and let your widows trust in
me.
12 For thus says the LORD: If those
who do not deserve to drink the cup still
have to drink it, shall you be the one to
go unpunished? You shall not go unpunished; you must drink it. 13 For by myself
I have sworn, says the LORD, that Bozrah
shall become an object of horror and ridicule, a waste, and an object of cursing;
and all her towns shall be perpetual
wastes.
14 I have heard tidings from the
LORD,
and a messenger has been sent
among the nations:
"Gather yourselves together and
come against her,
and rise up for battle!"
15 For I will make you least among
the nations,
despised by humankind.
16 The terror you inspire
and the pride of your heart have
deceived you,
you who live in the clefts of
the rock, *m*
who hold the height of the hill.
Although you make your nest as
high as the eagle's,
from there I will bring you
down,
says the LORD.
17 Edom shall become an object of
horror; everyone who passes by it will be
horrified and will hiss because of all its
disasters. 18 As when Sodom and Gomorrah and their neighbors were over-

m Or of Sela

49.7–22: **Against Edom.** After 587 B.C.
the relationship between Israel and "brother"
Edom (Deut 23.7–8) deteriorated to consistent Jewish vengefulness because of Edom's
occupation of southern Judah (Lam 4.21–22;
Ezek 25.12–14), an occupation caused by
pressure from Arab tribes. Verses 7–8, 10–
11, 18–22 appear to be original to Jeremiah.

Verse 9 is a variation of Ob 5, and vv. 14–16
are a variation of Ob 1–4. The occasion for
Jeremiah's oracle may have been 594 B.C. (see
27.3). *Teman,* modern Tawilan, about three
miles east of Sela (Petra). *Bozrah,* a great fortress city in northern Edom.
49.17: *Hiss,* see Zeph 2.15 n.

thrown, says the LORD, no one shall live there, nor shall anyone settle in it. [19] Like a lion coming up from the thickets of the Jordan against a perennial pasture, I will suddenly chase Edom[n] away from it; and I will appoint over it whomever I choose.[o] For who is like me? Who can summon me? Who is the shepherd who can stand before me? [20] Therefore hear the plan that the LORD has made against Edom and the purposes that he has formed against the inhabitants of Teman: Surely the little ones of the flock shall be dragged away; surely their fold shall be appalled at their fate. [21] At the sound of their fall the earth shall tremble; the sound of their cry shall be heard at the Red Sea.[p] [22] Look, he shall mount up and swoop down like an eagle, and spread his wings against Bozrah, and the heart of the warriors of Edom in that day shall be like the heart of a woman in labor.

23 Concerning Damascus.

> Hamath and Arpad are
>   confounded,
>   for they have heard bad news;
> they melt in fear, they are
>   troubled like the sea[q]
>   that cannot be quiet.
> [24] Damascus has become feeble, she
>   turned to flee,
>   and panic seized her;
> anguish and sorrows have taken
>   hold of her,
>   as of a woman in labor.
> [25] How the famous city is forsaken,[r]
>   the joyful town![s]
> [26] Therefore her young men shall fall
>   in her squares,

> and all her soldiers shall be
>   destroyed in that day,
>   says the LORD of hosts.
> [27] And I will kindle a fire at the wall
>   of Damascus,
>   and it shall devour the
>   strongholds of Ben-hadad.

28 Concerning Kedar and the kingdoms of Hazor that King Nebuchadrezzar of Babylon defeated.

> Thus says the LORD:
> Rise up, advance against Kedar!
>   Destroy the people of the east!
> [29] Take their tents and their flocks,
>   their curtains and all their
>   goods;
>   carry off their camels for
>   yourselves,
> and a cry shall go up: "Terror is
>   all around!"
> [30] Flee, wander far away, hide in
>   deep places,
>   O inhabitants of Hazor!
>   says the LORD.
> For King Nebuchadrezzar of
>   Babylon
>   has made a plan against you
>   and formed a purpose against
>   you.
>
> [31] Rise up, advance against a nation
>   at ease,
>   that lives secure,
>   says the LORD,

n Heb *him*   o Or *and I will single out the choicest of his rams*: Meaning of Heb uncertain   p Or *Sea of Reeds*   q Cn: Heb *there is trouble in the sea*   r Vg: Heb *is not forsaken*   s Syr Vg Tg: Heb *the town of my joy*

49.19–21: Editorial revision of 50.44–46.

49.23–27: Against Damascus. The occasion for this composite oracle (compare v. 27; Am 1.4) may have been the raids by the Syrians after 601 B.C. (2 Kings 24.2). Damascus lost its independence with the capture of *Arpad* by Tiglath-Pileser III in 740 B.C., *Hamath* in 738, and Damascus in 732 (Isa 10.9; 37.13). *Ben-hadad*, 1 Kings 15.18, 20.

49.28–33: Against Kedar and Hazor. In mid-winter 599/598 B.C. Nebuchadrezzar led a successful expedition against the Arab tribes in the desert east of Syria-Palestine, which may have been the occasion for this oracle (9.26; 25.23–24). *Kedar* is a collective term for nomadic or semi-nomadic Arabs (2.10); *Hazor* is not the city in Galilee (Josh 11.1) but a collective term for sedentary Arabs, who lived at oases. Neither group had fortresses, and neither would escape attack from the wide-ranging Nebuchadrezzar.

that has no gates or bars,
that lives alone.
32 Their camels shall become booty,
their herds of cattle a spoil.
I will scatter to every wind
those who have shaven temples,
and I will bring calamity
against them from every side,
says the LORD.
33 Hazor shall become a lair of
jackals,
an everlasting waste;
no one shall live there,
nor shall anyone settle in it.

34 The word of the LORD that came to the prophet Jeremiah concerning Elam, at the beginning of the reign of King Zedekiah of Judah.
35 Thus says the LORD of hosts: I am going to break the bow of Elam, the mainstay of their might; 36 and I will bring upon Elam the four winds from the four quarters of heaven; and I will scatter them to all these winds, and there shall be no nation to which the exiles from Elam shall not come. 37 I will terrify Elam before their enemies, and before those who seek their life; I will bring disaster upon them, my fierce anger, says the LORD. I will send the sword after them, until I have consumed them; 38 and I will set my throne in Elam, and destroy their king and officials, says the LORD.
39 But in the latter days I will restore the fortunes of Elam, says the LORD.

50 The word that the LORD spoke concerning Babylon, concerning the land of the Chaldeans, by the prophet Jeremiah:
2 Declare among the nations
and proclaim,
set up a banner and proclaim,
do not conceal it, say:
Babylon is taken,
Bel is put to shame,
Merodach is dismayed.
Her images are put to shame,
her idols are dismayed.
3 For out of the north a nation has come up against her; it shall make her land a desolation, and no one shall live in it; both human beings and animals shall flee away.

4 In those days and in that time, says the LORD, the people of Israel shall come, they and the people of Judah together; they shall come weeping as they seek the LORD their God. 5 They shall ask the way to Zion, with faces turned toward it, and they shall come and join[t] themselves to the LORD by an everlasting covenant that will never be forgotten.

6 My people have been lost sheep; their shepherds have led them astray, turning them away on the mountains; from mountain to hill they have gone, they have forgotten their fold. 7 All who found them have devoured them, and

t Gk: Heb toward it. Come! They shall join

49.34–39: Against Elam. There is some evidence that Nebuchadrezzar attacked Elam, east of Babylonia, in the winter of 596 B.C.; if so, that attack might have been an occasion for this oracle. *Zedekiah* became king in March 597, when Jehoiachin was deposed. *Bow of Elam* reflects the prowess of Elamite archers (Isa 22.6). **39:** This verse (like 46.26; 48.47; and 49.6) is probably an editorial addition.
50.1–51.64: Against Babylon. This collection of oracles contains two main themes: the fall of Babylon (sometimes represented as accomplished, sometimes represented as still in the future) and the return of the exiles (compare 24.6; 29.10). The attitude toward Babylon is somewhat harsher (50.14, 24) than one finds elsewhere in the book (27.6; 43.10),

but is not unprecedented (e.g. 25.12–14). The oracles concerning Judah (Israel) are here marked with the letters (*a*) to (*f*). **50.1:** *By the prophet Jeremiah,* literally "by the hand of Jeremiah the prophet," an idiom found also in the superscriptions in Hag 1.1; Mal 1.1; contrast Jer 46.1; 49.34. **2–3:** The downfall of Babylon. The nation *out of the north,* patterned on Jeremiah's expression (4.6), may or may not refer to Cyrus and Persia, before whom Babylon fell in October 539 B.C. *Bel* (Baal), 51.44; Isa 46.1, originally chief god of Nippur, later identified in Babylon with the great cosmic god Marduk (*Merodach*).
50.4–5, 6–7: (*a*) An apostrophe on Israel's homecoming (31.7–9) and guilt (2.20; 23.1–2).

their enemies have said, "We are not guilty, because they have sinned against the LORD, the true pasture, the LORD, the hope of their ancestors."

8 Flee from Babylon, and go out of the land of the Chaldeans, and be like male goats leading the flock. ⁹For I am going to stir up and bring against Babylon a company of great nations from the land of the north; and they shall array themselves against her; from there she shall be taken. Their arrows are like the arrows of a skilled warrior who does not return empty-handed. ¹⁰Chaldea shall be plundered; all who plunder her shall be sated, says the LORD.

11 Though you rejoice, though
	you exult,
	O plunderers of my heritage,
	though you frisk about like a
		heifer on the grass,
	and neigh like stallions,
12 your mother shall be utterly shamed,
	and she who bore you shall
		be disgraced.
	Lo, she shall be the last of the
		nations,
	a wilderness, dry land, and
		a desert.
13 Because of the wrath of the LORD
	she shall not be inhabited,
	but shall be an utter desolation;
	everyone who passes by Babylon
		shall be appalled
	and hiss because of all her
		wounds.
14 Take up your positions around
		Babylon,
	all you that bend the bow;
	shoot at her, spare no arrows,
	for she has sinned against
		the LORD.
15 Raise a shout against her from
	all sides,

"She has surrendered;
	her bulwarks have fallen,
	her walls are thrown down."
For this is the vengeance of
	the LORD:
	take vengeance on her,
	do to her as she has done.
16 Cut off from Babylon the sower,
	and the wielder of the sickle in
		time of harvest;
	because of the destroying sword
	all of them shall return to their
		own people,
	and all of them shall flee to their
		own land.

17 Israel is a hunted sheep driven away by lions. First the king of Assyria devoured it, and now at the end King Nebuchadrezzar of Babylon has gnawed its bones. ¹⁸Therefore, thus says the LORD of hosts, the God of Israel: I am going to punish the king of Babylon and his land, as I punished the king of Assyria. ¹⁹I will restore Israel to its pasture, and it shall feed on Carmel and in Bashan, and on the hills of Ephraim and in Gilead its hunger shall be satisfied. ²⁰In those days and at that time, says the LORD, the iniquity of Israel shall be sought, and there shall be none; and the sins of Judah, and none shall be found; for I will pardon the remnant that I have spared.

21 Go up to the land of Merathaim;ᵘ
	go up against her,
	and attack the inhabitants of
		Pekodᵛ
	and utterly destroy the last
		of them,ʷ
				says the LORD;

u Or *of Double Rebellion*	v Or *of Punishment*
w Tg: Heb *destroy after them*

50.8–16: The residents are urged to flee before Babylon's approaching doom (13.14) and desolation (18.16). Not only is the city destroyed, but more damaging, in a way, is the destruction of food production. **13:** *Hiss,* see Zeph 2.15 n.

50.17–20: (*b*) Though successively subject to Assyria and Babylonia, Israel will be restored (31.4–5; 33.8) and Babylon, as Assyria earlier, will be destroyed (25.12).
50.21–32: God's judgment against Babylon. *Merathaim,* "Double Rebellion" is a play

do all that I have commanded
you.

22 The noise of battle is in the land,
and great destruction!
23 How the hammer of the whole
earth
is cut down and broken!
How Babylon has become
a horror among the nations!
24 You set a snare for yourself and
you were caught,
O Babylon,
but you did not know it;
you were discovered and seized,
because you challenged the LORD.
25 The LORD has opened his armory,
and brought out the weapons of
his wrath,
for the Lord GOD of hosts has a
task to do
in the land of the Chaldeans.
26 Come against her from every
quarter;
open her granaries;
pile her up like heaps of grain, and
destroy her utterly;
let nothing be left of her.
27 Kill all her bulls,
let them go down to the
slaughter.
Alas for them, their day has come,
the time of their punishment!

28 Listen! Fugitives and refugees
from the land of Babylon are coming to
declare in Zion the vengeance of the
LORD our God, vengeance for his temple.

29 Summon archers against Babylon,
all who bend the bow. Encamp all
around her; let no one escape. Repay her
according to her deeds; just as she has
done, do to her—for she has arrogantly
defied the LORD, the Holy One of Israel.
30 Therefore her young men shall fall in
her squares, and all her soldiers shall be
destroyed on that day, says the LORD.

31 I am against you, O arrogant one,
says the Lord GOD of hosts;
for your day has come,
the time when I will punish
you.
32 The arrogant one shall stumble
and fall,
with no one to raise him up,
and I will kindle a fire in his cities,
and it will devour everything
around him.

33 Thus says the LORD of hosts: The
people of Israel are oppressed, and so too
are the people of Judah; all their captors
have held them fast and refuse to let them
go. 34 Their Redeemer is strong; the
LORD of hosts is his name. He will surely
plead their cause, that he may give rest to
the earth, but unrest to the inhabitants of
Babylon.

35 A sword against the Chaldeans,
says the LORD,
and against the inhabitants of
Babylon,
and against her officials and
her sages!
36 A sword against the diviners,
so that they may become fools!
A sword against her warriors,
so that they may be destroyed!
37 A sword against her*x* horses and
against her*x* chariots,
and against all the foreign
troops in her midst,
so that they may become
women!
A sword against all her treasures,
that they may be plundered!
38 A drought*y* against her waters,
that they may be dried up!

x Cn: Heb *his*   *y* Another reading is *A sword*

---

on the name of southern Babylonia, *mât mar-*
râti, "Land of the Lagoons." *Pekod,* "Punishment," is a play on the name *Puqûdu,* an east
Babylonian tribe (Ezek 23.23). The writer derides Babylon as a smashed hammer and a
captured bird (compare 5.26–27). He sees the

destruction of the temple as an affront to God
which must and will be avenged (21.14; Am
2.2). **50.33–34:** (*c*) While Israel is helpless,
God, her *Redeemer* (Isa 47.4), will deliver her
and discomfit her oppressors. **35–37:** The or-

For it is a land of images,
and they go mad over idols.

39 Therefore wild animals shall live with hyenas in Babylon, *z* and ostriches shall inhabit her; she shall never again be peopled, or inhabited for all generations. 40 As when God overthrew Sodom and Gomorrah and their neighbors, says the LORD, so no one shall live there, nor shall anyone settle in her.

41 Look, a people is coming from
the north;
a mighty nation and many kings
are stirring from the farthest
parts of the earth.
42 They wield bow and spear,
they are cruel and have
no mercy.
The sound of them is like the
roaring sea;
they ride upon horses,
set in array as a warrior for battle,
against you, O daughter
Babylon!

43 The king of Babylon heard news
of them,
and his hands fell helpless;
anguish seized him,
pain like that of a woman
in labor.

44 Like a lion coming up from the thickets of the Jordan against a perennial pasture, I will suddenly chase them away from her; and I will appoint over her whomever I choose. *a* For who is like me? Who can summon me? Who is the shepherd who can stand before me? 45 Therefore hear the plan that the LORD has made against Babylon, and the purposes that he has formed against the land of the Chaldeans: Surely the little ones of the flock shall be dragged away; surely their *b* fold shall be appalled at their fate. 46 At the sound of the capture of Babylon the earth shall tremble, and her cry shall be heard among the nations.

51 Thus says the LORD:
I am going to stir up a
destructive wind *c*
against Babylon
and against the inhabitants of
Leb-qamai; *d*
2 and I will send winnowers to
Babylon,
and they shall winnow her.
They shall empty her land
when they come against her
from every side
on the day of trouble.
3 Let not the archer bend his bow,
and let him not array himself in
his coat of mail.
Do not spare her young men;
utterly destroy her entire army.
4 They shall fall down slain in the
land of the Chaldeans,
and wounded in her streets.
5 Israel and Judah have not been
forsaken
by their God, the LORD of
hosts,
though their land is full of guilt
before the Holy One of Israel.

z Heb lacks *in Babylon*   *a* Or *and I will single out the choicest of her rams*: Meaning of Heb uncertain   *b* Syr Gk Tg Compare 49.20: Heb lacks *their*   *c* Or *stir up the spirit of a destroyer*   *d* Leb-qamai is a cryptogram for *Kasdim,* Chaldea

---

acle of the sword. **38–40:** Babylon will lie as a desert, unproductive, and inhabited only by wild animals (Isa 34.13–14).
50.41–46: Babylon, once the foe from the north (6.22–24), now stands in dread anticipation of a foe from the north. As Edom could not escape her punishment (49.19–21), neither can Babylon. There is no king (*shepherd*) who can successfully defy God.
51.1–64: God's judgment against Babylon. 1–4: As grain is winnowed, so will Babylon be cut down and winnowed (see 15.7 n.). *Leb-qamai* is an atbash cipher (see 25.26 n.) for "Chaldea." **5–10:** (*d*) A note reminding Israel that while they have hope, Babylon's case is hopeless (50.33–34). Babylon was the golden cup from which the nations would drink of God's wrath (see 25.15–29 n.), but now that cup is smashed; she stands desolate (46.11; Ezek 27.27) as God delivers his people.

6 Flee from the midst of Babylon,
    save your lives, each of you!
  Do not perish because of her guilt,
    for this is the time of the LORD's
      vengeance;
  he is repaying her what is due.
7 Babylon was a golden cup in the
      LORD's hand,
    making all the earth drunken;
  the nations drank of her wine,
    and so the nations went mad.
8 Suddenly Babylon has fallen and
      is shattered;
    wail for her!
  Bring balm for her wound;
    perhaps she may be healed.
9 We tried to heal Babylon,
    but she could not be healed.
  Forsake her, and let each of us go
    to our own country;
  for her judgment has reached up
      to heaven
    and has been lifted up even to
      the skies.
10 The LORD has brought forth our
      vindication;
    come, let us declare in Zion
      the work of the LORD our God.

11 Sharpen the arrows!
    Fill the quivers!
  The LORD has stirred up the spirit of the
  kings of the Medes, because his purpose
  concerning Babylon is to destroy it, for
  that is the vengeance of the LORD, vengeance for his temple.
12 Raise a standard against the walls
      of Babylon;
    make the watch strong;
  post sentinels;
    prepare the ambushes;
  for the LORD has both planned
      and done
    what he spoke concerning the
      inhabitants of Babylon.
13 You who live by mighty waters,

  rich in treasures,
    your end has come,
      the thread of your life is cut.
14 The LORD of hosts has sworn
      by himself:
  Surely I will fill you with troops
      like a swarm of locusts,
  and they shall raise a shout of
      victory over you.

15 It is he who made the earth by
      his power,
    who established the world by
      his wisdom,
  and by his understanding stretched
      out the heavens.
16 When he utters his voice there is a
      tumult of waters in the
        heavens,
    and he makes the mist rise from
      the ends of the earth.
  He makes lightnings for the rain,
    and he brings out the wind
      from his storehouses.
17 Everyone is stupid and without
      knowledge;
    goldsmiths are all put to shame
      by their idols;
  for their images are false,
    and there is no breath in them.
18 They are worthless, a work of
      delusion;
    at the time of their punishment
      they shall perish.
19 Not like these is the LORD,ᵉ the
      portion of Jacob,
    for he is the one who formed
      all things,
  and Israel is the tribe of his
      inheritance;
    the LORD of hosts is his name.

20 You are my war club, my weapon
      of battle:

ᵉ Heb lacks *the* LORD

51.11–19: Terse military commands precede the attack by the *Medes;* either a reference
to Media which, lying northeast of Babylonia, participated in the overthrow of Assyria
and was a threat to Babylonia during this
period, or an oblique reference to Persia. The
phrase *mighty waters* refers to the Euphrates
and the network of canals around Babylon. In
connection with God's judgment on Babylon, the writer appends the oracle against the
idols from 10.12–16 (50.38).
  51.20–23: This oracle of the hammer ad-

with you I smash nations;
with you I destroy kingdoms;
21 with you I smash the horse and
its rider;
with you I smash the chariot
and the charioteer;
22 with you I smash man and
woman;
with you I smash the old man
and the boy;
with you I smash the young man
and the girl;
23 with you I smash shepherds and
their flocks;
with you I smash farmers and
their teams;
with you I smash governors and
deputies.

24 I will repay Babylon and all the
inhabitants of Chaldea before your very
eyes for all the wrong that they have
done in Zion, says the LORD.

25 I am against you, O destroying
mountain,
says the LORD,
that destroys the whole earth;
I will stretch out my hand
against you,
and roll you down from the
crags,
and make you a burned-out
mountain.
26 No stone shall be taken from you
for a corner
and no stone for a foundation,
but you shall be a perpetual waste,
says the LORD.

27 Raise a standard in the land,
blow the trumpet among the
nations;
prepare the nations for war
against her,
summon against her the
kingdoms,
Ararat, Minni, and Ashkenaz;
appoint a marshal against her,
bring up horses like bristling
locusts.
28 Prepare the nations for war
against her,
the kings of the Medes, with
their governors and
deputies,
and every land under their
dominion.
29 The land trembles and writhes,
for the LORD's purposes against
Babylon stand,
to make the land of Babylon a
desolation,
without inhabitant.
30 The warriors of Babylon have
given up fighting,
they remain in their
strongholds;
their strength has failed,
they have become women;
her buildings are set on fire,
her bars are broken.
31 One runner runs to meet another,
and one messenger to meet
another,
to tell the king of Babylon
that his city is taken from end
to end:
32 the fords have been seized,
the marshes have been burned
with fire,
and the soldiers are in panic.
33 For thus says the LORD of hosts,
the God of Israel:
Daughter Babylon is like a
threshing floor

---

dresses the destroying enemy of Babylon.
24–26: As did Assyria (Isa 10.5, 15), so too
Babylon will fall. *Destroying mountain* may
reflect the great ziggurat or temple tower of
Babylon, thrusting its shrine into the heavens, and here symbolizing Babylon itself.
Burned-out mountain, perhaps "mountain of
burnt bricks" (Gen 11.3).
51.27–33: As Babylon subdued the nations (25.15–26) so the nations will gather
against Babylon. *Ararat,* ancient Urartu,
modern Armenia, north of Lake Van. *Minni,*
Mannaeans living south of Lake Urmia. *Ashkenaz,* the Scythians—all defeated by the
Medes in the early sixth century. Desolate
Babylon (4.6–7) is as barren as a threshing
floor thoroughly cleaned before the harvest.
51.34–40: (*e*) Again Jerusalem's deliver-

at the time when it is trodden;
yet a little while
and the time of her harvest
will come.

34 "King Nebuchadrezzar of Babylon
has devoured me,
he has crushed me;
he has made me an empty vessel,
he has swallowed me like
a monster;
he has filled his belly with my
delicacies,
he has spewed me out.
35 May my torn flesh be avenged on
Babylon,"
the inhabitants of Zion shall say.
"May my blood be avenged on
the inhabitants of Chaldea,"
Jerusalem shall say.
36 Therefore thus says the LORD:
I am going to defend your cause
and take vengeance for you.
I will dry up her sea
and make her fountain dry;
37 and Babylon shall become a heap
of ruins,
a den of jackals,
an object of horror and of hissing,
without inhabitant.

38 Like lions they shall roar together;
they shall growl like lions'
whelps.
39 When they are inflamed, I will set
out their drink
and make them drunk, until
they become merry
and then sleep a perpetual sleep
and never wake, says the LORD.
40 I will bring them down like lambs
to the slaughter,
like rams and goats.

41 How Sheshach*f* is taken,
the pride of the whole earth
seized!

How Babylon has become
an object of horror among the
nations!
42 The sea has risen over Babylon;
she has been covered by its
tumultuous waves.
43 Her cities have become an object
of horror,
a land of drought and a desert,
a land in which no one lives,
and through which no mortal
passes.
44 I will punish Bel in Babylon,
and make him disgorge what he
has swallowed.
The nations shall no longer stream
to him;
the wall of Babylon has fallen.

45 Come out of her, my people!
Save your lives, each of you,
from the fierce anger of the
LORD!
46 Do not be fainthearted or fearful
at the rumors heard in
the land—
one year one rumor comes,
the next year another,
rumors of violence in the land
and of ruler against ruler.

47 Assuredly, the days are coming
when I will punish the images
of Babylon;
her whole land shall be put to
shame,
and all her slain shall fall in
her midst.
48 Then the heavens and the earth,
and all that is in them,
shall shout for joy over Babylon;
for the destroyers shall come
against them out of
the north,
says the LORD.

f Sheshach is a cryptogram for *Babel,* Babylon

---

ance is viewed as a mighty act of God (50.34);
he will make Babylonia a desert (9.10); the
once mighty lion will become drunk (25.15–
16) and fall into a helpless, permanent sleep.
Hissing, see Zeph 2.15 n.
51.41–43: Babylon (written *Sheshach,* an
atbash cipher; see 25.26 n.), will be inundated
by the waves of her attackers (46.7–8; Isa
8.7–8); when the flood recedes, she will be a
trackless wasteland. **44–49:** In antiquity, the
fall of a land was viewed as the defeat of its
god(s) (Isa 37.12).

49 Babylon must fall for the slain of
　　Israel,
　　　as the slain of all the earth have
　　　　fallen because of Babylon.

50 You survivors of the sword,
　　go, do not linger!
　　Remember the LORD in a distant
　　　land,
　　　and let Jerusalem come into
　　　　your mind:
51 We are put to shame, for we have
　　heard insults;
　　　dishonor has covered our face,
　　for aliens have come
　　　into the holy places of the
　　　　LORD's house.

52 Therefore the time is surely
　　　coming, says the LORD,
　　when I will punish her idols,
　　and through all her land
　　　the wounded shall groan.
53 Though Babylon should mount up
　　　to heaven,
　　　and though she should fortify
　　　　her strong height,
　　from me destroyers would come
　　　upon her,
　　says the LORD.

54 Listen!—a cry from Babylon!
　　A great crashing from the land
　　　of the Chaldeans!
55 For the LORD is laying Babylon
　　waste,
　　　and stilling her loud clamor.
　　Their waves roar like mighty
　　　waters,
　　　the sound of their clamor
　　　　resounds;
56 for a destroyer has come
　　　against her,
　　against Babylon;

her warriors are taken,
　　their bows are broken;
for the LORD is a God of
　　recompense,
　　he will repay in full.
57 I will make her officials and her
　　sages drunk,
　　　also her governors, her deputies,
　　　　and her warriors;
　　they shall sleep a perpetual sleep
　　　and never wake,
　　says the King, whose name is
　　　the LORD of hosts.

58 Thus says the LORD of hosts:
　　The broad wall of Babylon
　　　shall be leveled to the ground,
　　and her high gates
　　　shall be burned with fire.
　　The peoples exhaust themselves
　　　for nothing,
　　and the nations weary
　　　themselves only for fire. *g*

59 The word that the prophet Jeremiah commanded Seraiah son of Neriah son of Mahseiah, when he went with King Zedekiah of Judah to Babylon, in the fourth year of his reign. Seraiah was the quartermaster. 60 Jeremiah wrote in a *h* scroll all the disasters that would come on Babylon, all these words that are written concerning Babylon. 61 And Jeremiah said to Seraiah: "When you come to Babylon, see that you read all these words, 62 and say, 'O LORD, you yourself threatened to destroy this place so that neither human beings nor animals shall live in it, and it shall be desolate forever.' 63 When you finish reading this scroll, tie a stone to it, and throw it into

g Gk Syr Compare Hab 2.13: Heb *and the nations for fire, and they are weary*　　*h* Or *one*

51.50–58: (*f*) The doubts raised by the destruction of the temple (a problem which Ezekiel also considers) should be allayed by God's assurance of certain punishment for Babylon which desecrated the holy places. **51.59–64:** The oracles are written in a book and taken to Babylon. Though we know of no journey of Zedekiah, he may have gone to renew his fealty after the abortive plot of 594 B.C. (chs 27–28) became known to Nebuchadrezzar. *Seraiah*, brother of Baruch (32.12). **63:** By a symbolic act the oracle against Babylon is emphasized; this oracle may have been the causative factor for the composition of 50.1–51.58. Though for a time God endowed Babylon with great pow-

the middle of the Euphrates, [64]and say, 'Thus shall Babylon sink, to rise no more, because of the disasters that I am bringing on her.' "[i]

Thus far are the words of Jeremiah.

52 Zedekiah was twenty-one years old when he began to reign; he reigned eleven years in Jerusalem. His mother's name was Hamutal daughter of Jeremiah of Libnah. [2]He did what was evil in the sight of the LORD, just as Jehoiakim had done. [3]Indeed, Jerusalem and Judah so angered the LORD that he expelled them from his presence.

Zedekiah rebelled against the king of Babylon. [4]And in the ninth year of his reign, in the tenth month, on the tenth day of the month, King Nebuchadrezzar of Babylon came with all his army against Jerusalem, and they laid siege to it; they built siegeworks against it all around. [5]So the city was besieged until the eleventh year of King Zedekiah. [6]On the ninth day of the fourth month the famine became so severe in the city that there was no food for the people of the land. [7]Then a breach was made in the city wall;[j] and all the soldiers fled and went out from the city by night by the way of the gate between the two walls, by the king's garden, though the Chaldeans were all around the city. They went in the direction of the Arabah. [8]But the army of the Chaldeans pursued the king, and overtook Zedekiah in the plains of Jericho; and all his army was scattered, deserting him. [9]Then they captured the king, and brought him up to the king of

Babylon at Riblah in the land of Hamath, and he passed sentence on him. [10]The king of Babylon killed the sons of Zedekiah before his eyes, and also killed all the officers of Judah at Riblah. [11]He put out the eyes of Zedekiah, and bound him in fetters, and the king of Babylon took him to Babylon, and put him in prison until the day of his death.

12 In the fifth month, on the tenth day of the month—which was the nineteenth year of King Nebuchadrezzar, king of Babylon—Nebuzaradan the captain of the bodyguard who served the king of Babylon, entered Jerusalem. [13]He burned the house of the LORD, the king's house, and all the houses of Jerusalem; every great house he burned down. [14]All the army of the Chaldeans, who were with the captain of the guard, broke down all the walls around Jerusalem. [15]Nebuzaradan the captain of the guard carried into exile some of the poorest of the people and the rest of the people who were left in the city and the deserters who had defected to the king of Babylon, together with the rest of the artisans. [16]But Nebuzaradan the captain of the guard left some of the poorest people of the land to be vinedressers and tillers of the soil.

17 The pillars of bronze that were in the house of the LORD, and the stands and the bronze sea that were in the house of the LORD, the Chaldeans broke in pieces, and carried all the bronze to Babylon. [18]They took away the pots, the shovels,

i Gk: Heb *on her. And they shall weary themselves*
j Heb lacks *wall*

---

er, he did so that it might accomplish his purposes, not that it should be established as a great power forever.

52.1–34: Historical appendix. Though largely a duplicate of 2 Kings 24.18–25.30, this historical section, together with 39.1–10 and 40.7–43.7, provides much important complementary information (for similar historical additions, see Isa chs 36–39). **1–3:** Reign of Zedekiah (2 Kings 24.18–20; 597–587 B.C.). For details on the contemporary religious situation, see Ezek ch 8. **4–27:** Siege and fall of Jerusalem (39.1–10; 2 Kings

24.20b–25.26). **4–11:** January 588 B.C. to August 587. The last scene enacted before Zedekiah, the rebellious vassal (37.1; Ezek 17.18–21), at Nebuchadrezzar's headquarters at *Riblah* (in the central valley northeast of Byblos) was the slaughter of his sons and court officials. Then he was blinded and taken to Babylon to die in prison.

52.12–16: August 587 B.C. The reasons for the actions of Nebuzaradan, Nebuchadrezzar's field general, are unknown.

52.17–23: A detailed description of the booty taken from the temple.

the snuffers, the basins, the ladles, and all the vessels of bronze used in the temple service. ¹⁹ The captain of the guard took away the small bowls also, the firepans, the basins, the pots, the lampstands, the ladles, and the bowls for libation, both those of gold and those of silver. ²⁰ As for the two pillars, the one sea, the twelve bronze bulls that were under the sea, and the stands,* which King Solomon had made for the house of the LORD, the bronze of all these vessels was beyond weighing. ²¹ As for the pillars, the height of the one pillar was eighteen cubits, its circumference was twelve cubits; it was hollow and its thickness was four fingers. ²² Upon it was a capital of bronze; the height of the one capital was five cubits; latticework and pomegranates, all of bronze, encircled the top of the capital. And the second pillar had the same, with pomegranates. ²³ There were ninety-six pomegranates on the sides; all the pomegranates encircling the latticework numbered one hundred.

24 The captain of the guard took the chief priest Seraiah, the second priest Zephaniah, and the three guardians of the threshold; ²⁵ and from the city he took an officer who had been in command of the soldiers, and seven men of the king's council who were found in the city; the secretary of the commander of the army who mustered the people of the land; and sixty men of the people of the land who were found inside the city. ²⁶ Then Nebuzaradan the captain of the guard took them, and brought them to the king of Babylon at Riblah. ²⁷ And the king of Babylon struck them down, and put them to death at Riblah in the land of Hamath. So Judah went into exile out of its land.

28 This is the number of the people whom Nebuchadrezzar took into exile: in the seventh year, three thousand twenty-three Judeans; ²⁹ in the eighteenth year of Nebuchadrezzar he took into exile from Jerusalem eight hundred thirty-two persons; ³⁰ in the twenty-third year of Nebuchadrezzar, Nebuzaradan the captain of the guard took into exile of the Judeans seven hundred forty-five persons; all the persons were four thousand six hundred.

31 In the thirty-seventh year of the exile of King Jehoiachin of Judah, in the twelfth month, on the twenty-fifth day of the month, King Evil-merodach of Babylon, in the year he began to reign, showed favor to King Jehoiachin of Judah and brought him out of prison; ³² he spoke kindly to him, and gave him a seat above the seats of the other kings who were with him in Babylon. ³³ So Jehoiachin put aside his prison clothes, and every day of his life he dined regularly at the king's table. ³⁴ For his allowance, a regular daily allowance was given him by the king of Babylon, as long as he lived, up to the day of his death.

k Cn: Heb *that were under the stands*

52.24–27: 2 Kings 25.18–21. *Seraiah,* perhaps the same as in 36.26; otherwise unknown. *Zephaniah,* 21.1; 29.29. **28–30**: The first two deportations listed here coincide with the surrender of Jehoiachin (597 B.C.; 2 Kings 24.12–16) and the suppression of Zedekiah's revolt (587 B.C.). The occasion for the third deportation (582 B.C.) is not known. **31–34**: 2 Kings 25.27–30. The presence of this material proves that the editing of Jeremiah's material was later than 560 B.C. Jehoiachin's restoration may well have been viewed by his contemporaries as the beginning of Judah's restoration (23.5–6).

Lamentations

Lamentations is a small psalter of communal laments over Jerusalem, following its destruction by the Babylonians in 587 (586) B.C. These psalms have been traditionally ascribed to Jeremiah because of 2 Chr 35.25, but those laments were for the death of Josiah, not for the desolation of Jerusalem. The thought and diction are sufficiently unlike Jeremiah's to make his authorship unlikely. The first four chapters are alphabetic acrostics (with a stanza for each of the twenty-two letters of the Hebrew alphabet), and the fifth has the same number of verses as the alphabet. All were composed or adapted for public recitation on days of fasting and mourning (Joel 2.15–17; Zech 7.2–3), notably that of the ninth of Ab (August), which commemorated the disaster of 587 (586).

Chapters 1, 2, and 4 are in form dirges over the dead city. The elegy's limping 3:2 meter (three beats followed by two beats) is recognizable even in translation, together with the exclamatory "How . . ." in the opening line as an expression of grief (2 Sam 1.25, 27; Isa 1.21). In ch 3 the sadness of the desolate people and reflection upon the meaning of the disaster are voiced by an individual. Chapter 5 in its form and language recalls the liturgies for use in time of national trouble, such as Pss 74 and 79. The common theme of all the poems is the agony of the people, the apparent desertion of Zion by God, and the hope that God will yet restore a humbled and repentant Israel.

Lamentations may be the work of one or several authors, speaking out of the dreadful situation of the inhabitants of Jerusalem following the overthrow of the city.

1 How lonely sits the city
   that once was full of people!
How like a widow she has
      become,
   she that was great among the
      nations!
She that was a princess among the
      provinces
   has become a vassal.

2 She weeps bitterly in the night,
   with tears on her cheeks;
among all her lovers
   she has no one to comfort her;
all her friends have dealt
      treacherously with her,
   they have become her enemies.

3 Judah has gone into exile with
      suffering
   and hard servitude;
she lives now among the nations,
   and finds no resting place;
her pursuers have all overtaken her
   in the midst of her distress.

4 The roads to Zion mourn,
   for no one comes to the
      festivals;
all her gates are desolate,
   her priests groan;
her young girls grieve,*a*
   and her lot is bitter.

5 Her foes have become the masters,
   her enemies prosper,
because the LORD has made
      her suffer
   for the multitude of her
      transgressions;
her children have gone away,
   captives before the foe.

6 From daughter Zion has departed
   all her majesty.
Her princes have become like stags
   that find no pasture;
they fled without strength
   before the pursuer.

7 Jerusalem remembers,
   in the days of her affliction and
      wandering,
all the precious things
   that were hers in days of old.
When her people fell into the hand
   of the foe,
   and there was no one to help
      her,
the foe looked on mocking
   over her downfall.

8 Jerusalem sinned grievously,
   so she has become a mockery;
all who honored her despise her,
   for they have seen her
      nakedness;
she herself groans,
   and turns her face away.

9 Her uncleanness was in her skirts;
   she took no thought of her
      future;
her downfall was appalling,
   with none to comfort her.
"O LORD, look at my affliction,
   for the enemy has triumphed!"

10 Enemies have stretched out
      their hands
   over all her precious things;
she has even seen the nations
   invade her sanctuary,
those whom you forbade
   to enter your congregation.

11 All her people groan
   as they search for bread;
they trade their treasures for food
   to revive their strength.
Look, O LORD, and see
   how worthless I have become.

12 Is it nothing to you,*a* all you who
      pass by?
   Look and see
if there is any sorrow like my
      sorrow,

a Meaning of Heb uncertain

1.1–11a: A lament over Zion. 6: *Daughter
Zion,* the city is personified as a maiden.

1.11b–22: A lament uttered by Zion. 17:
The lament is briefly interrupted.

which was brought upon me,
which the LORD inflicted
on the day of his fierce anger.

13 From on high he sent fire;
it went deep into my bones;
he spread a net for my feet;
he turned me back;
he has left me stunned,
faint all day long.

14 My transgressions were bound[b]
into a yoke;
by his hand they were fastened
together;
they weigh on my neck,
sapping my strength;
the Lord handed me over
to those whom I cannot
withstand.

15 The LORD has rejected
all my warriors in the midst
of me;
he proclaimed a time against me
to crush my young men;
the Lord has trodden as in a
wine press
the virgin daughter Judah.

16 For these things I weep;
my eyes flow with tears;
for a comforter is far from me,
one to revive my courage;
my children are desolate,
for the enemy has prevailed.

17 Zion stretches out her hands,
but there is no one to
comfort her;
the LORD has commanded against
Jacob
that his neighbors should
become his foes;
Jerusalem has become
a filthy thing among them.

18 The LORD is in the right,
for I have rebelled against
his word;
but hear, all you peoples,
and behold my suffering;
my young women and young men
have gone into captivity.

19 I called to my lovers
but they deceived me;
my priests and elders
perished in the city
while seeking food
to revive their strength.

20 See, O LORD, how distressed I am;
my stomach churns,
my heart is wrung within me,
because I have been very
rebellious.
In the street the sword bereaves;
in the house it is like death.

21 They heard how I was groaning,
with no one to comfort me.
All my enemies heard of my
trouble;
they are glad that you have
done it.
Bring on the day you have
announced,
and let them be as I am.

22 Let all their evil doing come
before you;
and deal with them
as you have dealt with me
because of all my transgressions;
for my groans are many
and my heart is faint.

2 How the Lord in his anger
has humiliated[b] daughter Zion!
He has thrown down from heaven
to earth

b Meaning of Heb uncertain

1.18–22: Zion's confession of sin is followed by an appeal to *bring on the day* when the enemies will be punished (see 3.64–66).

2.1–22: The people's agony and their cry to God for mercy. 1: *Footstool,* i.e. the temple. **6:** *Booth,* "tabernacle," i.e. the tem-

the splendor of Israel;
he has not remembered his
footstool
in the day of his anger.

2 The Lord has destroyed without
mercy
all the dwellings of Jacob;
in his wrath he has broken down
the strongholds of daughter
Judah;
he has brought down to the
ground in dishonor
the kingdom and its rulers.

3 He has cut down in fierce anger
all the might of Israel;
he has withdrawn his right hand
from them
in the face of the enemy;
he has burned like a flaming fire
in Jacob,
consuming all around.

4 He has bent his bow like an
enemy,
with his right hand set like a
foe;
he has killed all in whom we
took pride
in the tent of daughter Zion;
he has poured out his fury like
fire.

5 The Lord has become like an
enemy;
he has destroyed Israel;
He has destroyed all its palaces,
laid in ruins its strongholds,
and multiplied in daughter Judah
mourning and lamentation.

6 He has broken down his booth
like a garden,
he has destroyed his tabernacle;
the LORD has abolished in Zion
festival and sabbath,

and in his fierce indignation has
spurned
king and priest.

7 The Lord has scorned his altar,
disowned his sanctuary;
he has delivered into the hand of
the enemy
the walls of her palaces;
a clamor was raised in the house
of the LORD
as on a day of festival.

8 The LORD determined to lay
in ruins
the wall of daughter Zion;
he stretched the line;
he did not withhold his hand
from destroying;
he caused rampart and wall
to lament;
they languish together.

9 Her gates have sunk into
the ground;
he has ruined and broken
her bars;
her king and princes are among
the nations;
guidance is no more,
and her prophets obtain
no vision from the LORD.

10 The elders of daughter Zion
sit on the ground in silence;
they have thrown dust on
their heads
and put on sackcloth;
the young girls of Jerusalem
have bowed their heads to
the ground.

11 My eyes are spent with weeping;
my stomach churns;
my bile is poured out on the
ground
because of the destruction of
my people,

---

ple. **7:** *Clamor* of battle in place of the shout
of acclamation in worship.
2.8: *Line,* a measuring line (Job 38.5). **9:**
Among the nations, i.e. in exile. *Guidance* or
teaching given by priests (Jer 18.18; Mal
2.5–8).

because infants and babes faint
    in the streets of the city.

12 They cry to their mothers,
    "Where is bread and wine?"
as they faint like the wounded
    in the streets of the city,
as their life is poured out
    on their mothers' bosom.

13 What can I say for you, to what
        compare you,
    O daughter Jerusalem?
To what can I liken you, that I
        may comfort you,
    O virgin daughter Zion?
For vast as the sea is your ruin;
    who can heal you?

14 Your prophets have seen for you
        false and deceptive visions;
they have not exposed your
        iniquity
    to restore your fortunes,
but have seen oracles for you
    that are false and misleading.

15 All who pass along the way
    clap their hands at you;
they hiss and wag their heads
    at daughter Jerusalem;
"Is this the city that was called
    the perfection of beauty,
    the joy of all the earth?"

16 All your enemies
    open their mouths against you;
they hiss, they gnash their teeth,
    they cry: "We have devoured
        her!
Ah, this is the day we longed for;
    at last we have seen it!"

17 The LORD has done what he
        purposed,
    he has carried out his threat;
as he ordained long ago,
    he has demolished without pity;

he has made the enemy rejoice
        over you,
    and exalted the might of
        your foes.

18 Cry aloud[c] to the Lord!
    O wall of daughter Zion!
Let tears stream down like a
        torrent
    day and night!
Give yourself no rest,
    your eyes no respite!

19 Arise, cry out in the night,
    at the beginning of the watches!
Pour out your heart like water
    before the presence of the Lord!
Lift your hands to him
    for the lives of your children,
who faint for hunger
    at the head of every street.

20 Look, O LORD, and consider!
    To whom have you done this?
Should women eat their offspring,
    the children they have borne?
Should priest and prophet be
        killed
    in the sanctuary of the Lord?

21 The young and the old are lying
    on the ground in the streets;
my young women and my
        young men
    have fallen by the sword;
in the day of your anger you have
        killed them,
    slaughtering without mercy.

22 You invited my enemies from
        all around
    as if for a day of festival;
and on the day of the anger of
        the LORD
    no one escaped or survived;
those whom I bore and reared
    my enemy has destroyed.

c Cn: Heb *Their heart cried*

2.14: *Deceptive visions* of victory (Jer 23.25–27). 15: *Clap their hands,* or "slap their hands," in derision. 17: *His threat* to punish disobedi- ence (see 1 Kings 9.6–9). 20: 2 Kings 6.28–29. 22: Am 5.18–20.

3 I am one who has seen affliction
under the rod of God's [d] wrath;
2 he has driven and brought me
into darkness without any light;
3 against me alone he turns
his hand,
again and again, all day long.

4 He has made my flesh and my
skin waste away,
and broken my bones;
5 he has besieged and enveloped me
with bitterness and tribulation;
6 he has made me sit in darkness
like the dead of long ago.

7 He has walled me about so that I
cannot escape;
he has put heavy chains on me;
8 though I call and cry for help,
he shuts out my prayer;
9 he has blocked my ways with
hewn stones,
he has made my paths crooked.

10 He is a bear lying in wait for me,
a lion in hiding;
11 he led me off my way and tore
me to pieces;
he has made me desolate;
12 he bent his bow and set me
as a mark for his arrow.

13 He shot into my vitals
the arrows of his quiver;
14 I have become the laughingstock
of all my people,
the object of their taunt-songs
all day long.
15 He has filled me with bitterness,
he has sated me with
wormwood.

16 He has made my teeth grind
on gravel,
and made me cower in ashes;
17 my soul is bereft of peace;
I have forgotten what
happiness is;
18 so I say, "Gone is my glory,
and all that I had hoped for
from the LORD."

19 The thought of my affliction and
my homelessness
is wormwood and gall!
20 My soul continually thinks of it
and is bowed down within me.
21 But this I call to mind,
and therefore I have hope:

22 The steadfast love of the LORD
never ceases, [e]
his mercies never come to an
end;
23 they are new every morning;
great is your faithfulness.
24 "The LORD is my portion," says
my soul,
"therefore I will hope in him."

25 The LORD is good to those who
wait for him,
to the soul that seeks him.
26 It is good that one should wait
quietly
for the salvation of the LORD.
27 It is good for one to bear
the yoke in youth,
28 to sit alone in silence
when the Lord has imposed it,
29 to put one's mouth to the dust
(there may yet be hope),

[d] Heb *his*   [e] Syr Tg: Heb LORD, *we are not
cut off*

3.1–66: An acrostic in three parts, with
three verses to each letter of the alphabet.
**3.1–18: A psalm of personal distress and
trust in God** (compare Ps 56). The distress is
expressed in terms recalling Job's complaints
against God; v. 1, compare Job 9.34; v. 2,
compare Job 19.8; v. 3, compare Job 7.18;
v. 4, compare Job 7.5; 30.30; v. 5, compare
Job 19.6, 12; v. 6, compare Job 23.16–17;
v. 7, compare Job 19.8; v. 8, compare Job
30.20; v. 9, compare Job 19.8; vv. 10–11,
compare Job 16.9; vv. 12–13, compare Job
16.12–13; v. 14, compare Job 30.9; v. 15,
compare Job 9.18; vv. 16–18, compare Job
19.10; 30.19.
**3.19–51: A sage counsels submission
and penitence in acknowledgment of
God's righteousness and mercy.**
3.29: *One's mouth in the dust,* i.e. in self-
abasement. **33:** *Willingly,* lit. "from his

30 to give one's cheek to the smiter,
and be filled with insults.

31 For the Lord will not
reject forever.
32 Although he causes grief, he will
have compassion
according to the abundance of
his steadfast love;
33 for he does not willingly afflict
or grieve anyone.

34 When all the prisoners of the land
are crushed under foot,
35 when human rights are perverted
in the presence of the Most
High,
36 when one's case is subverted
—does the Lord not see it?

37 Who can command and have
it done,
if the Lord has not ordained it?
38 Is it not from the mouth of the
Most High
that good and bad come?
39 Why should any who draw breath
complain
about the punishment of
their sins?

40 Let us test and examine our ways,
and return to the LORD.
41 Let us lift up our hearts as well as
our hands
to God in heaven.
42 We have transgressed and rebelled,
and you have not forgiven.

43 You have wrapped yourself with
anger and pursued us,
killing without pity;
44 you have wrapped yourself with
a cloud
so that no prayer can pass
through.

45 You have made us filth and
rubbish
among the peoples.

46 All our enemies
have opened their mouths
against us;
47 panic and pitfall have come
upon us,
devastation and destruction.
48 My eyes flow with rivers of tears
because of the destruction of
my people.

49 My eyes will flow without
ceasing,
without respite,
50 until the LORD from heaven
looks down and sees.
51 My eyes cause me grief
at the fate of all the young
women in my city.

52 Those who were my enemies
without cause
have hunted me like a bird;
53 they flung me alive into a pit
and hurled stones on me;
54 water closed over my head;
I said, "I am lost."

55 I called on your name, O LORD,
from the depths of the pit;
56 you heard my plea, "Do not close
your ear
to my cry for help, but give
me relief!"
57 You came near when I called
on you;
you said, "Do not fear!"

58 You have taken up my cause,
O Lord,
you have redeemed my life.
59 You have seen the wrong done to
me, O LORD;

---

heart." **38:** *Good and bad,* compare Isa 45.7;
Am 3.6.

3.40: *Us,* i.e. the nation Israel. **48:** See 2.11;
Jer 9.1.

**3.52–66: An individual psalm pleading
God's past mercies and praying for vindi-**
cation and the requiting of the enemy
(compare 5.1–22). The sentiment of v. 59 is
in contrast to that of v. 39 in the previous
section. **54:** Symbol of drowning, see Jon
2.3–6.

judge my cause.

60 You have seen all their malice,
  all their plots against me.

61 You have heard their taunts,
    O LORD,
  all their plots against me.
62 The whispers and murmurs of my
    assailants
  are against me all day long.
63 Whether they sit or rise—see,
    I am the object of their
    taunt-songs.

64 Pay them back for their deeds,
    O LORD,
  according to the work of their
    hands!
65 Give them anguish of heart;
  your curse be on them!
66 Pursue them in anger and
    destroy them
  from under the LORD's heavens.

4 How the gold has grown dim,
  how the pure gold is changed!
The sacred stones lie scattered
  at the head of every street.

2 The precious children of Zion,
  worth their weight in fine
    gold—
how they are reckoned as
    earthen pots,
  the work of a potter's hands!

3 Even the jackals offer the breast
  and nurse their young,
but my people has become cruel,
  like the ostriches in the
    wilderness.

4 The tongue of the infant sticks
  to the roof of its mouth for
    thirst;

the children beg for food,
  but no one gives them anything.

5 Those who feasted on delicacies
  perish in the streets;
those who were brought up
    in purple
  cling to ash heaps.

6 For the chastisement*f* of my
    people has been greater
  than the punishment*g* of
    Sodom,
which was overthrown in a
    moment,
  though no hand was laid on it. *h*

7 Her princes were purer than snow,
  whiter than milk;
their bodies were more ruddy
    than coral,
  their hair*h* like sapphire. *i*

8 Now their visage is blacker
    than soot;
  they are not recognized in
    the streets.
Their skin has shriveled on
    their bones;
  it has become as dry as wood.

9 Happier were those pierced by
    the sword
  than those pierced by hunger,
whose life drains away, deprived
  of the produce of the field.

10 The hands of compassionate
    women
  have boiled their own children;
they became their food
  in the destruction of my people.

f Or *iniquity*   *g* Or *sin*   *h* Meaning of Heb
uncertain   *i* Or *lapis lazuli*

3.64–66: These verses recall the cries against the enemies of the persecuted righteous in the Psalms; see Pss 3.7; 17.13, 14; 35.26; 59.11–13; compare Jer 11.20–23; 18.21, 22.
  4.1–22: The horrors of the siege and sack of Jerusalem. 1–2: *Gold . . . sacred stones,* the temple treasures; but the people are more precious to God. **3:** Jerusalem can no longer care for the children.
  4.4–5, 9–10: See 2 Kings 25.3. **12:** See Isa chs 36–37. **13–16:** Judah's religious leaders had been moral lepers.

11 The LORD gave full vent to
    his wrath;
  he poured out his hot anger,
and kindled a fire in Zion
    that consumed its foundations.

12 The kings of the earth did
    not believe,
  nor did any of the inhabitants
    of the world,
that foe or enemy could enter
    the gates of Jerusalem.

13 It was for the sins of her prophets
  and the iniquities of her priests,
who shed the blood of the
    righteous
in the midst of her.

14 Blindly they wandered through
    the streets,
  so defiled with blood
that no one was able
    to touch their garments.

15 "Away! Unclean!" people shouted
    at them;
  "Away! Away! Do not touch!"
So they became fugitives and
    wanderers;
  it was said among the nations,
  "They shall stay here no
    longer."

16 The LORD himself has scattered
    them,
  he will regard them no more;
no honor was shown to the
    priests,
  no favor to the elders.

17 Our eyes failed, ever watching
  vainly for help;
we were watching eagerly
  for a nation that could not save.

18 They dogged our steps
  so that we could not walk in
    our streets;
our end drew near; our days were
    numbered;
  for our end had come.

19 Our pursuers were swifter
  than the eagles in the heavens;
they chased us on the mountains,
  they lay in wait for us in the
    wilderness.

20 The LORD's anointed, the breath of
    our life,
  was taken in their pits—
the one of whom we said, "Under
    his shadow
  we shall live among the
    nations."

21 Rejoice and be glad, O daughter
    Edom,
  you that live in the land of Uz;
but to you also the cup shall pass;
  you shall become drunk and
    strip yourself bare.

22 The punishment of your iniquity,
    O daughter Zion, is
    accomplished,
  he will keep you in exile
    no longer;
but your iniquity, O daughter
    Edom, he will punish,
  he will uncover your sins.

5 Remember, O LORD, what has
    befallen us;
  look, and see our disgrace!
2 Our inheritance has been turned
    over to strangers,
  our homes to aliens.
3 We have become orphans,
    fatherless;

4.17–20: Jerusalem is again personified;
compare 1.12–16. **17**: *A nation that could not
save,* Egypt (Jer 37.5–10). **20**: *The LORD's
anointed,* king Zedekiah (2 Kings 25.4–6).
Pits, hunters' traps.
4.21–22: Unlike chs 1–3, which end with
a prayer for mercy and deliverance, ch 4 ends
with a vow or prophecy that *Edom* too will
suffer for treachery to Israel (see Ob 8–14).
5.1–22: **A community psalm of lament
and petition for restoration** (like Pss 44; 74;
79; 80). **1–14**: The misery of Judah's people

our mothers are like widows.

4 We must pay for the water
we drink;
the wood we get must be
bought.

5 With a yoke[j] on our necks we are
hard driven;
we are weary, we are given
no rest.

6 We have made a pact with[k] Egypt
and Assyria,
to get enough bread.

7 Our ancestors sinned; they are
no more,
and we bear their iniquities.

8 Slaves rule over us;
there is no one to deliver us
from their hand.

9 We get our bread at the peril of
our lives,
because of the sword in the
wilderness.

10 Our skin is black as an oven
from the scorching heat
of famine.

11 Women are raped in Zion,
virgins in the towns of Judah.

12 Princes are hung up by
their hands;
no respect is shown to the
elders.

13 Young men are compelled to
grind,

and boys stagger under loads
of wood.

14 The old men have left the city
gate,
the young men their music.

15 The joy of our hearts has ceased;
our dancing has been turned
to mourning.

16 The crown has fallen from
our head;
woe to us, for we have sinned!

17 Because of this our hearts are sick,
because of these things our eyes
have grown dim:

18 because of Mount Zion, which
lies desolate;
jackals prowl over it.

19 But you, O Lord, reign forever;
your throne endures to all
generations.

20 Why have you forgotten us
completely?
Why have you forsaken us these
many days?

21 Restore us to yourself, O Lord,
that we may be restored;
renew our days as of old—

22 unless you have utterly rejected
us,
and are angry with us beyond
measure.

j Symmachus: Heb lacks *With a yoke*
k Heb *have given the hand to*

---

under the heel of Babylon. **8**: *Slaves rule,* important posts were sometimes given to slaves of the king. **10**: *Black as an oven,* or "burned as in an oven."

5.15–18: The Davidic monarchy is no more, and the temple site is desolate. **16**: *Crown,* a double reference to the garland of the dancers and to the king's crown. **19–22**: A plea for divine remembrance and mercy (compare Pss 74.1–2; 79.5–8; 80.1–7).

Ezekiel

Ezekiel was a priest whose ministry to his fellow exiles extended from 593 (1.2) to perhaps 563 B.C., if the enigmatic "thirtieth year" in 1.1 is taken as the thirtieth year after his call and as the date of the initial compilation of the book of Ezekiel by the prophet himself. The latest dated oracle included in the book is of the year 571 B.C. (29.17). The book of Ezekiel has the most thoroughgoing chronological notations of any of the books of the prophets, with only three dates out of order (26.1; 29.17; 33.21). The dates do not necessarily apply to all the oracles following a given date, and the chronological position of undated oracles may be ascertained from the nature of their contents. The capture and destruction of Jerusalem in 587 B.C. was a decisive factor in Ezekiel's ministry. The oracles of warning (chs 1–24) are to be dated before the fall of Jerusalem. The oracles of hope (chs 33–48) belong after the fall of Jerusalem. The oracles against the foreign nations (chs 25–32) belong to the middle period of Ezekiel's ministry (587 to 585 B.C.; but 29.17–21 dates from 571 B.C.).

The original collection was rewritten and expanded by an editor, but Ezekiel's distinctive prose and poetry may be recognized throughout the book. The text has suffered much in transmission; as a consequence, the interpretation is frequently uncertain.

There was no other prophet in Israel whose words and actions expressed a personality so distinctive as Ezekiel's. In his vision reports, such as that of the throne chariot of the LORD (ch 1) or of the abominations in the temple of Jerusalem (ch 8), his use of language and symbolism is peculiar. His symbolic actions, such as his eating a scroll (2.8–3.3) or his forgoing the customary rites of mourning at the death of his wife (24.15–27), seem bizarre. His long allegories of erring Israel and Judah (chs 16 and 23) are repellent, at least to our taste. But there is no way to know whether the strange modes of presentation that he employed were due to his double role as both priest and prophet, or whether Ezekiel's personality was confounded by the disorienting experience of being exiled to Babylon. In the innovativeness with which he employed the language of vision, he laid the groundwork for the symbolic universe of apocalypticism. But he made it clear that no matter how specific his description of godly matters was, there was no human language that could do them justice ("This was the appearance of the likeness of the glory of the LORD," 1.28).

As a prophet to the exiles, Ezekiel assured his hearers of the abiding presence of God among them. He constantly emphasized the LORD's role in the events of the day, so that Israel and the nations "will know that I am the LORD" (a refrain that occurs many times throughout the book). He underscored the integrity of the individual and each one's personal responsibility to God (see e.g. ch 18). To a helpless and hopeless people he brought hope of restoration to homeland and temple by their just and holy God. In Ezekiel we have an unparalleled synthesis of the terrestrial and celestial in Israel's religion, truly fitting for one whose ministry marks the transition from pre-exilic Israelite religion to post-exilic Judaism.

1 In the thirtieth year, in the fourth month, on the fifth day of the month, as I was among the exiles by the river Chebar, the heavens were opened, and I saw visions of God. 2 On the fifth day of the month (it was the fifth year of the exile of King Jehoiachin), 3 the word of the LORD came to the priest Ezekiel son of Buzi, in the land of the Chaldeans by the river Chebar; and the hand of the LORD was on him there.

4 As I looked, a stormy wind came out of the north: a great cloud with brightness around it and fire flashing forth continually, and in the middle of the fire, something like gleaming amber. 5 In the middle of it was something like four living creatures. This was their appearance: they were of human form. 6 Each had four faces, and each of them had four wings. 7 Their legs were straight, and the soles of their feet were like the sole of a calf's foot; and they sparkled like burnished bronze. 8 Under their wings on their four sides they had human hands. And the four had their faces and their wings thus: 9 their wings touched one another; each of them moved straight ahead, without turning as they moved. 10 As for the appearance of their faces: the four had the face of a human being, the face of a lion on the right side, the face of an ox on the left side, and the face of an eagle; 11 such were their faces. Their wings were spread out above; each creature had two wings, each of which touched the wing of another, while two covered their bodies. 12 Each moved straight ahead; wherever the spirit would go, they went, without turning as they went. 13 In the middle of*a* the living creatures there was something that looked like burning coals of fire, like torches moving to and fro among the living creatures; the fire was bright, and lightning issued from the fire. 14 The living creatures darted to and fro, like a flash of lightning.

15 As I looked at the living creatures, I saw a wheel on the earth beside the living creatures, one for each of the four of them. *b* 16 As for the appearance of the wheels and their construction: their appearance was like the gleaming of beryl; and the four had the same form, their construction being something like a wheel within a wheel. 17 When they moved, they moved in any of the four directions without veering as they moved. 18 Their rims were tall and awesome, for the rims of all four were full of eyes all around. 19 When the living creatures moved, the wheels moved beside them; and when the living creatures rose from the earth, the wheels rose. 20 Wherever the spirit would go, they went, and the wheels rose along with them; for the spirit of the living creatures was in the wheels. 21 When they moved, the others

a Gk OL: Heb *And the appearance of*   *b* Heb *of their faces*

1.1–3.27: The call of Ezekiel. 1.1–3: Superscription. *The thirtieth year,* perhaps the thirtieth year after Ezekiel's call, and if so, the date of the initial composition of the book, 563 B.C. (compare Jer 36.1–3). *Fifth day* of the *fourth month . . . , fifth year of the exile* would be July 31, 593 B.C. This is reckoned from a lunar calendar, with the year beginning in the spring. The name *Ezekiel* means "God strengthens." *Chebar,* a canal which is mentioned also in the Babylonian records, flowing southeast from its fork above Babylon, through Nippur, and rejoining the Euphrates near Erech. *Hand of the* LORD expresses Ezekiel's sense of divine compulsion (3.14, 22; 8.1; 33.22; 37.1; 40.1). **1.4–28a: The throne chariot vision.**

Compare the imagery in 1 Kings 22.19–22; Isa 6.1–9. **4:** *Out of the north,* a literary figure drawn from Canaanite mythology, according to which the gods lived in the north. *Stormy wind* (1 Kings 19.11), *cloud* (Ex 19.16), and *fire* (1 Kings 19.11–12) are all elements in the theophany (manifestation) of God. **5:** The *living creatures* (Rev 4.7) are cherubim, guardians of God's throne (see Ex 25.10–22; 1 Kings 6.23–28), namely winged human-headed lions or oxen, symbolizing mobility, intelligence, and strength. **1.15–21:** *The four wheels* (compare the four faces of the creatures) symbolize omnidirectional mobility. **18:** *Full of eyes,* symbolic of the all-seeing nature of God (Rev 4.6, 8). **1.22:** In ancient cosmology, the firmament

moved; when they stopped, the others stopped; and when they rose from the earth, the wheels rose along with them; for the spirit of the living creatures was in the wheels.

22 Over the heads of the living creatures there was something like a dome, shining like crystal,ᶜ spread out above their heads. 23 Under the dome their wings were stretched out straight, one toward another; and each of the creatures had two wings covering its body. 24 When they moved, I heard the sound of their wings like the sound of mighty waters, like the thunder of the Almighty,ᵈ a sound of tumult like the sound of an army; when they stopped, they let down their wings. 25 And there came a voice from above the dome over their heads; when they stopped, they let down their wings.

26 And above the dome over their heads there was something like a throne, in appearance like sapphire;ᵉ and seated above the likeness of a throne was something that seemed like a human form. 27 Upward from what appeared like the loins I saw something like gleaming amber, something that looked like fire enclosed all around; and downward from what looked like the loins I saw something that looked like fire, and there was a splendor all around. 28 Like the bow in a cloud on a rainy day, such was the appearance of the splendor all around. This was the appearance of the likeness of the glory of the Lᴏʀᴅ.

When I saw it, I fell on my face, and I heard the voice of someone speaking.

2 He said to me: O mortal,ᶠ stand up on your feet, and I will speak with you. 2 And when he spoke to me, a spirit entered into me and set me on my feet; and I heard him speaking to me. 3 He said to me, Mortal, I am sending you to the people of Israel, to a nationᵍ of rebels who have rebelled against me; they and their ancestors have transgressed against me to this very day. 4 The descendants are impudent and stubborn. I am sending you to them, and you shall say to them, "Thus says the Lᴏʀᴅ Gᴏᴅ." 5 Whether they hear or refuse to hear (for they are a rebellious house), they shall know that there has been a prophet among them. 6 And you, O mortal, do not be afraid of them, and do not be afraid of their words, though briers and thorns surround you and you live among scorpions; do not be afraid of their words, and do not be dismayed at their looks, for they are a rebellious house. 7 You shall speak my words to them, whether they hear or refuse to hear; for they are a rebellious house.

8 But you, mortal, hear what I say to you; do not be rebellious like that rebellious house; open your mouth and eat what I give you. 9 I looked, and a hand was stretched out to me, and a written scroll was in it. 10 He spread it before me; it had writing on the front and on the back, and written on it were words of lamentation and mourning and woe.

3 He said to me, O mortal, eat what is offered to you; eat this scroll, and go, speak to the house of Israel. 2 So I

ᶜ Gk: Heb *like the awesome crystal*
ᵈ Traditional rendering of Heb *Shaddai*
ᵉ Or *lapis lazuli*   ᶠ Or *son of man*; Heb *ben adam* (and so throughout the book when Ezekiel is addressed)   ᵍ Syr: Heb *to nations*

---

separated the waters above the earth from the earth (Gen 1.6–8).

1.26–28: Thus the Lᴏʀᴅ was enthroned above his creatures; compare the Lᴏʀᴅ enthroned above the cherubim in Ex 37.9 (on the ark); 1 Sam 4.4.

1.28b–3.27: The five commissions. 1.28b–2.8a: The first commission. The nonmessianic term *mortal* (literally "son of man") occurs ninety-three times in Ezekiel; it emphasizes the prophet's finite dependence and insignificance before God's infinite power and glory (Ps 8.4). **2.2:** The *Spirit* is identified with "Spirit of the Lᴏʀᴅ" (11.5; 37.1; Isa 61.1). **5:** *Rebellious house*, a designation of Judah, whose apostasy was the cause of the exile (Jer 2.29; 3.13). **6–7:** Compare the Lᴏʀᴅ's encouragement of the prophet in Jer 1.6–8, 16–19.

2.8b–3.3: So that he will speak only what the Lᴏʀᴅ has written, Ezekiel is told to eat a papyrus *scroll* filled with words of woe (com-

opened my mouth, and he gave me the scroll to eat. [3]He said to me, Mortal, eat this scroll that I give you and fill your stomach with it. Then I ate it; and in my mouth it was as sweet as honey.

4 He said to me: Mortal, go to the house of Israel and speak my very words to them. [5]For you are not sent to a people of obscure speech and difficult language, but to the house of Israel— [6]not to many peoples of obscure speech and difficult language, whose words you cannot understand. Surely, if I sent you to them, they would listen to you. [7]But the house of Israel will not listen to you, for they are not willing to listen to me; because all the house of Israel have a hard forehead and a stubborn heart. [8]See, I have made your face hard against their faces, and your forehead hard against their foreheads. [9]Like the hardest stone, harder than flint, I have made your forehead; do not fear them or be dismayed at their looks, for they are a rebellious house. [10]He said to me: Mortal, all my words that I shall speak to you receive in your heart and hear with your ears; [11]then go to the exiles, to your people, and speak to them. Say to them, "Thus says the Lord GOD"; whether they hear or refuse to hear.

12 Then the spirit lifted me up, and as the glory of the LORD rose[h] from its place, I heard behind me the sound of loud rumbling; [13]it was the sound of the wings of the living creatures brushing against one another, and the sound of the wheels beside them, that sounded like a loud rumbling. [14]The spirit lifted me up and bore me away; I went in bitterness in the heat of my spirit, the hand of the LORD being strong upon me. [15]I came to the exiles at Tel-abib, who lived by the river Chebar.[i] And I sat there among them, stunned, for seven days.

16 At the end of seven days, the word of the LORD came to me: [17]Mortal, I have made you a sentinel for the house of Israel; whenever you hear a word from my mouth, you shall give them warning from me. [18]If I say to the wicked, "You shall surely die," and you give them no warning, or speak to warn the wicked from their wicked way, in order to save their life, those wicked persons shall die for their iniquity; but their blood I will require at your hand. [19]But if you warn the wicked, and they do not turn from their wickedness, or from their wicked way, they shall die for their iniquity; but you will have saved your life. [20]Again, if the righteous turn from their righteousness and commit iniquity, and I lay a stumbling block before them, they shall die; because you have not warned them, they shall die for their sin, and their righteous deeds that they have done shall not be remembered; but their blood I will require at your hand. [21]If, however, you warn the righteous not to sin, and they do not sin, they shall surely live, because they took warning; and you will have saved your life.

22 Then the hand of the LORD was upon me there; and he said to me, Rise up, go out into the valley, and there I will speak with you. [23]So I rose up and went out into the valley; and the glory of the LORD stood there, like the glory that I had seen by the river Chebar; and I fell on my face. [24]The spirit entered into me, and set me on my feet; and he spoke with

h Cn: Heb *and blessed be the glory of the* LORD
i Two Mss Syr: Heb *Chebar, and to where they lived.* Another reading is *Chebar, and I sat where they sat*

---

pare Jer 15.16; Zech 5.1–4; Rev 10.8–11). *Sweet*, Ps 19.10.
3.4–9: Second commission. Ezekiel's determination to prophesy (Jer 1.18) must be stronger than Israel's refusal to hear (Am 7.10–17; Jer 20.7–18). **10–15:** Third commission. An editorial reduplication of the preceding materials, emphasizing the mission to the exiles (2.4–5). *Tel-abib* (derived from Bab-

ylonian "til abubi," "mound of the deluge"), Jewish settlement near Nippur, not far from the Chebar canal.
3.16–21: Fourth commission. This *sentinel* passage (Jer 6.17; Hos 9.8) is an application of Ezekiel's doctrine of personal responsibility (18.1–32) to the prophetic office (33.7–16). **22–27:** Fifth commission. *The valley* is the southern Tigris-Euphrates valley (37.1; Gen

me and said to me: Go, shut yourself inside your house. 25 As for you, mortal, cords shall be placed on you, and you shall be bound with them, so that you cannot go out among the people; 26 and I will make your tongue cling to the roof of your mouth, so that you shall be speechless and unable to reprove them; for they are a rebellious house. 27 But when I speak with you, I will open your mouth, and you shall say to them, "Thus says the Lord GOD"; let those who will hear, hear; and let those who refuse to hear, refuse; for they are a rebellious house.

4 And you, O mortal, take a brick and set it before you. On it portray a city, Jerusalem; 2 and put siegeworks against it, and build a siege wall against it, and cast up a ramp against it; set camps also against it, and plant battering rams against it all around. 3 Then take an iron plate and place it as an iron wall between you and the city; set your face toward it, and let it be in a state of siege, and press the siege against it. This is a sign for the house of Israel.

4 Then lie on your left side, and place the punishment of the house of Israel upon it; you shall bear their punishment for the number of the days that you lie there. 5 For I assign to you a number of days, three hundred ninety days, equal to the number of the years of their punishment; and so you shall bear the punishment of the house of Israel. 6 When you have completed these, you shall lie down

a second time, but on your right side, and bear the punishment of the house of Judah; forty days I assign you, one day for each year. 7 You shall set your face toward the siege of Jerusalem, and with your arm bared you shall prophesy against it. 8 See, I am putting cords on you so that you cannot turn from one side to the other until you have completed the days of your siege.

9 And you, take wheat and barley, beans and lentils, millet and spelt; put them into one vessel, and make bread for yourself. During the number of days that you lie on your side, three hundred ninety days, you shall eat it. 10 The food that you eat shall be twenty shekels a day by weight; at fixed times you shall eat it. 11 And you shall drink water by measure, one-sixth of a hin; at fixed times you shall drink. 12 You shall eat it as a barley-cake, baking it in their sight on human dung. 13 The LORD said, "Thus shall the people of Israel eat their bread, unclean, among the nations to which I will drive them." 14 Then I said, "Ah Lord GOD! I have never defiled myself; from my youth up until now I have never eaten what died of itself or was torn by animals, nor has carrion flesh come into my mouth." 15 Then he said to me, "See, I will let you have cow's dung instead of human dung, on which you may prepare your bread."

16 Then he said to me, Mortal, I am going to break the staff of bread in Jerusalem; they shall eat bread by weight and

11.2). Ezekiel's *speechless* state may refer to his apparent inability to speak of anything but the doom of Judah and Jerusalem for the following seven and one-half years (24.26–27; 33.21–22).

4.1–5.17: Symbolic actions describing the coming siege of Jerusalem. 4.1–3: A sun-dried *brick* (common in Babylonia) with a relief drawing of Jerusalem under siege, and *an iron plate,* a griddle, symbolizing God's role in Jerusalem's fall (Jer 21.5).

4.4–8: As a symbol of the years of the punishment of Israel (exact significance unknown), the prophet is directed to lie on his left side *three-hundred ninety days,* and as a symbol of the years of the punishment of

Judah to lie on his right side *forty days* (compare the forty years in the wilderness, Num 14.33; Jer 25.11–12). The actual performance of this command seems most unlikely.

4.9–17: Unclean food and rationing reflect the rigors of the siege during which the people even practiced cannibalism (Jer 19.9; Lam 4.10). **9:** Mixing the grains indicates scarcity of foodstuffs, not uncleanness. **12:** *Human dung* was considered unclean (Deut 23.12–14). Dried *cow dung* (v. 15) was and is common fuel in the East. **14:** Compare Lev 17.10–16. **16:** *Staff of bread,* 5.16. Jerusalem's water during the siege came from cisterns and two springs, Gihon in the Kidron Valley and En-rogel to the south (2 Kings 20.20).

with fearfulness; and they shall drink water by measure and in dismay. [17]Lacking bread and water, they will look at one another in dismay, and waste away under their punishment.

5 And you, O mortal, take a sharp sword; use it as a barber's razor and run it over your head and your beard; then take balances for weighing, and divide the hair. [2]One third of the hair you shall burn in the fire inside the city, when the days of the siege are completed; one third you shall take and strike with the sword all around the city;[j] and one third you shall scatter to the wind, and I will unsheathe the sword after them. [3]Then you shall take from these a small number, and bind them in the skirts of your robe. [4]From these, again, you shall take some, throw them into the fire and burn them up; from there a fire will come out against all the house of Israel.

5 Thus says the Lord God: This is Jerusalem; I have set her in the center of the nations, with countries all around her. [6]But she has rebelled against my ordinances and my statutes, becoming more wicked than the nations and the countries all around her, rejecting my ordinances and not following my statutes. [7]Therefore thus says the Lord God: Because you are more turbulent than the nations that are all around you, and have not followed my statutes or kept my ordinances, but have acted according to the ordinances of the nations that are all around you; [8]therefore thus says the Lord God: I, I myself, am coming against you; I will execute judgments among you in the sight of the nations. [9]And because of all your abominations, I will do to you what I have never yet done, and the like of which I will never do again. [10]Surely, parents shall eat their children in your midst, and children shall eat their parents; I will execute judgments on you, and any of you who survive I will scatter to every wind. [11]Therefore, as I live, says the Lord God, surely, because you have defiled my sanctuary with all your detestable things and with all your abominations— therefore I will cut you down;[k] my eye will not spare, and I will have no pity. [12]One third of you shall die of pestilence or be consumed by famine among you; one third shall fall by the sword around you; and one third I will scatter to every wind and will unsheathe the sword after them.

13 My anger shall spend itself, and I will vent my fury on them and satisfy myself; and they shall know that I, the Lord, have spoken in my jealousy, when I spend my fury on them. [14]Moreover I will make you a desolation and an object of mocking among the nations around you, in the sight of all that pass by. [15]You shall be[l] a mockery and a taunt, a warning and a horror, to the nations around you, when I execute judgments on you in anger and fury, and with furious punishments—I, the Lord, have spoken— [16]when I loose against you[m] my deadly arrows of famine, arrows for destruction, which I will let loose to destroy you, and when I bring more and more famine upon you, and break your staff of bread. [17]I will send famine and wild animals against you, and they will rob you of your children; pestilence and bloodshed shall pass through you; and I will bring the sword upon you. I, the Lord, have spoken.

6 The word of the Lord came to me: [2]O mortal, set your face toward the mountains of Israel, and prophesy

---

j Heb *it*   k Another reading is *I will withdraw*
l Gk Syr Vg Tg: Heb *It shall be*   m Heb *them*

5.1–17: Shorn hair as a symbol of the fate awaiting the people of Jerusalem. The *sword* is appropriately used as a *razor* to convey the idea of military defeat.
5.5: *Center of the nations*, Jerusalem, the holy city, was designated as the geographical center of the earth (see 38.12, "center [literally, navel] of the earth"), as later Rome was looked upon as the navel of the earth.
5.14: 36.34; compare Jer 24.9–10. **17**: 14.21.
6.1–14: **Oracle against the mountains.**

against them, [3]and say, You mountains of Israel, hear the word of the Lord GOD! Thus says the Lord GOD to the mountains and the hills, to the ravines and the valleys: I, I myself will bring a sword upon you, and I will destroy your high places. [4]Your altars shall become desolate, and your incense stands shall be broken; and I will throw down your slain in front of your idols. [5]I will lay the corpses of the people of Israel in front of their idols; and I will scatter your bones around your altars. [6]Wherever you live, your towns shall be waste and your high places ruined, so that your altars will be waste and ruined, [n] your idols broken and destroyed, your incense stands cut down, and your works wiped out. [7]The slain shall fall in your midst; then you shall know that I am the LORD.

[8] But I will spare some. Some of you shall escape the sword among the nations and be scattered through the countries. [9]Those of you who escape shall remember me among the nations where they are carried captive, how I was crushed by their wanton heart that turned away from me, and their wanton eyes that turned after their idols. Then they will be loathsome in their own sight for the evils that they have committed, for all their abominations. [10]And they shall know that I am the LORD; I did not threaten in vain to bring this disaster upon them.

[11] Thus says the Lord GOD: Clap your hands and stamp your foot, and say, Alas for all the vile abominations of the house of Israel! For they shall fall by the sword, by famine, and by pestilence. [12]Those far off shall die of pestilence; those nearby shall fall by the sword; and

any who are left and are spared shall die of famine. Thus I will spend my fury upon them. [13]And you shall know that I am the LORD, when their slain lie among their idols around their altars, on every high hill, on all the mountain tops, under every green tree, and under every leafy oak, wherever they offered pleasing odor to all their idols. [14]I will stretch out my hand against them, and make the land desolate and waste, throughout all their settlements, from the wilderness to Riblah. [o] Then they shall know that I am the LORD.

7 The word of the LORD came to me: [2]You, O mortal, thus says the Lord GOD to the land of Israel:

An end! The end has come
    upon the four corners of the
    land.
[3] Now the end is upon you,
    I will let loose my anger
        upon you;
  I will judge you according to
        your ways,
    I will punish you for all your
        abominations.
[4] My eye will not spare you, I will
        have no pity.
  I will punish you for your
        ways,
    while your abominations are
        among you.
Then you shall know that I am the LORD.
[5] Thus says the Lord GOD:
  Disaster after disaster! See,
        it comes.

n Syr Vg Tg: Heb *and be made guilty*
o Another reading is *Diblah*

Compare 36.1–15. The address to the mountains is figurative for the *high places,* funerary cairns used as open-air pagan sanctuaries. The cultic equipment at these sanctuaries included *incense stands* and statues or other representations (such as the oak or other trees; v. 13; Hos 4.13) of fertility goddesses such as Asherah and Anath. *Idols* (v. 4) translates Ezekiel's characteristic term "gillulim" (literally, "dung balls"), found thirty-nine times in Ezekiel, compared with nine times in the rest of the Old Testament. *Works* (v. 6) may refer

to the sacred pillars or standing stones, which may have been commemorative monuments (Ex 23.24; Deut 7.5). **14**: *From the wilderness to Riblah* (the wilderness of south Judah and Riblah in central Syria) was the maximum extent of Israelite territory (47.16; 1 Kings 8.65).

7.1–27: Oracles on the approaching judgment. 1–4: The theme, *the end,* is reminiscent of Am 8.2; it is the sequel to the judgments of chs 4–6. **3**: Compare Ps 78.49.

6 An end has come, the end
   has come.
It has awakened against you; see,
   it comes!
7 Your doom[p] has come to you,
   O inhabitant of the land.
The time has come, the day
   is near—
of tumult, not of reveling on
   the mountains.
8 Soon now I will pour out my
   wrath upon you;
I will spend my anger
   against you.
I will judge you according to
   your ways,
   and punish you for all your
   abominations.
9 My eye will not spare; I will have
   no pity.
I will punish you according to
   your ways,
   while your abominations are
   among you.
Then you shall know that it is I the Lord
who strike.
10 See, the day! See, it comes!
   Your doom[p] has gone out.
The rod has blossomed, pride has
   budded.
11 Violence has grown into a rod
   of wickedness.
None of them shall remain,
   not their abundance, not their
   wealth;
   no pre-eminence among them. [p]
12 The time has come, the day
   draws near;
let not the buyer rejoice, nor the
   seller mourn,
   for wrath is upon all their
   multitude.
13 For the sellers shall not return to what
has been sold as long as they remain
alive. For the vision concerns all their
multitude; it shall not be revoked. Because of their iniquity, they cannot maintain their lives. [p]
14 They have blown the horn and
   made everything ready;
but no one goes to battle,
   for my wrath is upon all their
   multitude.
15 The sword is outside, pestilence
   and famine are inside;
those in the field die by
   the sword;
those in the city—famine and
   pestilence devour them.
16 If any survivors escape,
   they shall be found on the
   mountains
   like doves of the valleys,
all of them moaning over
   their iniquity.
17 All hands shall grow feeble,
   all knees turn to water.
18 They shall put on sackcloth,
   horror shall cover them.
Shame shall be on all faces,
   baldness on all their heads.
19 They shall fling their silver into
   the streets,
   their gold shall be treated as
   unclean.
Their silver and gold cannot save them
on the day of the wrath of the Lord.
They shall not satisfy their hunger or fill
their stomachs with it. For it was the
stumbling block of their iniquity.
20 From their[q] beautiful ornament, in
which they took pride, they made their
abominable images, their detestable
things; therefore I will make of it an unclean thing to them.
21 I will hand it over to strangers
   as booty,
   to the wicked of the earth
   as plunder;
   they shall profane it.

p Meaning of Heb uncertain
q Syr Symmachus: Heb *its*

7.7: *The day* (Joel 1.15; Mal 4.1; Heb 10.25)
is "the day of the Lord" (Am 5.18–20; Isa
2.11, 12–17).
7.10–23a: As Jerusalem's end approaches,
people will sell at a loss to recover what they
can, and the buyers will have little prospect of
retaining their purchases. The trumpet calls
to the defense, but instead of battle-dress one
sees only signs of mourning (Jer 16.6–9; Isa
15.2). The gold and silver they had cast into
idols cannot nourish the body, just as the
idols themselves are useless (Jer 2.26–28;

22 I will avert my face from them,
   so that they may profane my
      treasured[r] place;
   the violent shall enter it,
      they shall profane it.
23 Make a chain![s]
   For the land is full of bloody
      crimes;
   the city is full of violence.
24 I will bring the worst of the
      nations
   to take possession of their
      houses.
   I will put an end to the arrogance
      of the strong,
   and their holy places shall be
      profaned.
25 When anguish comes, they will
      seek peace,
   but there shall be none.
26 Disaster comes upon disaster,
      rumor follows rumor;
   they shall keep seeking a vision
      from the prophet;
   instruction shall perish from
      the priest,
   and counsel from the elders.
27 The king shall mourn,
   the prince shall be wrapped
      in despair,
   and the hands of the people of
      the land shall tremble.
   According to their way I will deal
      with them;
   according to their own
      judgments I will judge
      them.
And they shall know that I am the LORD.

8 In the sixth year, in the sixth month,
on the fifth day of the month, as I sat
in my house, with the elders of Judah
sitting before me, the hand of the Lord
GOD fell upon me there. 2 I looked, and
there was a figure that looked like a human being;[t] below what appeared to be
its loins it was fire, and above the loins
it was like the appearance of brightness,
like gleaming amber. 3 It stretched out
the form of a hand, and took me by a
lock of my head; and the spirit lifted me
up between earth and heaven, and
brought me in visions of God to Jerusalem, to the entrance of the gateway of the
inner court that faces north, to the seat of
the image of jealousy, which provokes to
jealousy. 4 And the glory of the God of
Israel was there, like the vision that I had
seen in the valley.

5 Then God[u] said to me, "O mortal,
lift up your eyes now in the direction of
the north." So I lifted up my eyes toward
the north, and there, north of the altar
gate, in the entrance, was this image of
jealousy. 6 He said to me, "Mortal, do
you see what they are doing, the great
abominations that the house of Israel are
committing here, to drive me far from
my sanctuary? Yet you will see still
greater abominations."

7 And he brought me to the entrance
of the court; I looked, and there was a
hole in the wall. 8 Then he said to me,
"Mortal, dig through the wall"; and
when I dug through the wall, there was
an entrance. 9 He said to me, "Go in, and
see the vile abominations that they are
committing here." 10 So I went in and
looked; there, portrayed on the wall all
around, were all kinds of creeping
things, and loathsome animals, and all
the idols of the house of Israel. 11 Before
them stood seventy of the elders of the
house of Israel, with Jaazaniah son of
Shaphan standing among them. Each

---

r Or *secret*   s Meaning of Heb uncertain
t Gk: Heb *like fire*   u Heb *he*

10.1–16; Zeph 1.18) except as booty for the
conqueror. **22**: *My treasured place,* the temple.
23b–27: The confusion and brutality of the
invasion and siege. *Peace* recalls Jer 6.14; see
also Ezek 22.28; the distraught leadership is
reminiscent of Jer 4.9–10; 13.13.
   8.1–11.25: The temple visions.
   8.1–18: The vision of idolatry (September

17, 592 B.C.). **1–4**: The gateway to the inner
court was the third gate leading north from
the palace complex into the temple precincts
(1 Kings 7.12; 2 Kings 20.4). The *seat . . .* was
perhaps a niche for a figured slab.
   8.7–13: The description may suggest the
worship of Osiris, since the wall-paintings
recall those of the Egyptian Book of the

had his censer in his hand, and the fragrant cloud of incense was ascending. 12 Then he said to me, "Mortal, have you seen what the elders of the house of Israel are doing in the dark, each in his room of images? For they say, 'The LORD does not see us, the LORD has forsaken the land.'" 13 He said also to me, "You will see still greater abominations that they are committing."

14 Then he brought me to the entrance of the north gate of the house of the LORD; women were sitting there weeping for Tammuz. 15 Then he said to me, "Have you seen this, O mortal? You will see still greater abominations than these."

16 And he brought me into the inner court of the house of the LORD; there, at the entrance of the temple of the LORD, between the porch and the altar, were about twenty-five men, with their backs to the temple of the LORD, and their faces toward the east, prostrating themselves to the sun toward the east. 17 Then he said to me, "Have you seen this, O mortal? Is it not bad enough that the house of Judah commits the abominations done here? Must they fill the land with violence, and provoke my anger still further? See, they are putting the branch to their nose! 18 Therefore I will act in wrath; my eye will not spare, nor will I have pity; and though they cry in my hearing with a loud voice, I will not listen to them."

9 Then he cried in my hearing with a loud voice, saying, "Draw near, you executioners of the city, each with his destroying weapon in his hand." 2 And

six men came from the direction of the upper gate, which faces north, each with his weapon for slaughter in his hand; among them was a man clothed in linen, with a writing case at his side. They went in and stood beside the bronze altar.

3 Now the glory of the God of Israel had gone up from the cherub on which it rested to the threshold of the house. The LORD called to the man clothed in linen, who had the writing case at his side; 4 and said to him, "Go through the city, through Jerusalem, and put a mark on the foreheads of those who sigh and groan over all the abominations that are committed in it." 5 To the others he said in my hearing, "Pass through the city after him, and kill; your eye shall not spare, and you shall show no pity. 6 Cut down old men, young men and young women, little children and women, but touch no one who has the mark. And begin at my sanctuary." So they began with the elders who were in front of the house. 7 Then he said to them, "Defile the house, and fill the courts with the slain. Go!" So they went out and killed in the city. 8 While they were killing, and I was left alone, I fell prostrate on my face and cried out, "Ah Lord GOD! will you destroy all who remain of Israel as you pour out your wrath upon Jerusalem?" 9 He said to me, "The guilt of the house of Israel and Judah is exceedingly great; the land is full of bloodshed and the city full of perversity; for they say, 'The LORD has forsaken the land, and the LORD does not see.' 10 As for me, my eye will not spare, nor will I have pity, but

---

Dead, but the prophet may have in mind a variety of Babylonian worship.

8.14–15: *Tammuz,* the Sumero-Accadian vegetation god; the weeping was for his descent into the underworld, coinciding with the annual decline of vegetation.

8.16–18: This may reflect the Egyptian worship of the sun-god, who was thought to bring forth all vegetation, or the worship of Tammuz-Adonis (compare Isa 17.10). The *branch,* or vine-sprout, may be a symbol in the cult rites, or it may be a reference to an obscene gesture.

9.1–11: The punishment of the guilty. 1–2: From the *north* (see 1.4 n.) came the divine *executioners* (compare the temple guards, or "watchmen," in 2 Kings 11.18, designated by the same Hebrew word) and *a man clothed in linen,* who functioned as the LORD's scribe, as did Nabu in the Babylonian pantheon. *Linen,* a ritually clean fabric, was worn by priests (Lev 6.10) and angels (Dan 10.5). **4:** The *mark* was the Hebrew letter "tau," made like an X (compare Rev 7.3–4). **6:** The *elders* are those of 8.16. **8:** *All who remain,* those remaining in Palestine after 597 B.C.

I will bring down their deeds upon their heads."

11 Then the man clothed in linen, with the writing case at his side, brought back word, saying, "I have done as you commanded me."

10 Then I looked, and above the dome that was over the heads of the cherubim there appeared above them something like a sapphire,ᵛ in form resembling a throne. ²He said to the man clothed in linen, "Go within the wheelwork underneath the cherubim; fill your hands with burning coals from among the cherubim, and scatter them over the city." He went in as I looked on. ³Now the cherubim were standing on the south side of the house when the man went in; and a cloud filled the inner court. ⁴Then the glory of the LORD rose up from the cherub to the threshold of the house; the house was filled with the cloud, and the court was full of the brightness of the glory of the LORD. ⁵The sound of the wings of the cherubim was heard as far as the outer court, like the voice of God Almightyʷ when he speaks.

6 When he commanded the man clothed in linen, "Take fire from within the wheelwork, from among the cherubim," he went in and stood beside a wheel. ⁷And a cherub stretched out his hand from among the cherubim to the fire that was among the cherubim, took some of it and put it into the hands of the man clothed in linen, who took it and went out. ⁸The cherubim appeared to have the form of a human hand under their wings.

9 I looked, and there were four wheels beside the cherubim, one beside each cherub; and the appearance of the wheels was like gleaming beryl. ¹⁰And as for their appearance, the four looked alike, something like a wheel within a wheel. ¹¹When they moved, they moved in any of the four directions without veering as they moved; but in whatever direction the front wheel faced, the others followed without veering as they moved. ¹²Their entire body, their rims, their spokes, their wings, and the wheels—the wheels of the four of them—were full of eyes all around. ¹³As for the wheels, they were called in my hearing "the wheelwork." ¹⁴Each one had four faces: the first face was that of the cherub, the second face was that of a human being, the third that of a lion, and the fourth that of an eagle.

15 The cherubim rose up. These were the living creatures that I saw by the river Chebar. ¹⁶When the cherubim moved, the wheels moved beside them; and when the cherubim lifted up their wings to rise up from the earth, the wheels at their side did not veer. ¹⁷When they stopped, the others stopped, and when they rose up, the others rose up with them; for the spirit of the living creatures was in them.

18 Then the glory of the LORD went out from the threshold of the house and stopped above the cherubim. ¹⁹The cherubim lifted up their wings and rose up from the earth in my sight as they went out with the wheels beside them. They stopped at the entrance of the east gate of the house of the LORD; and the glory of the God of Israel was above them.

20 These were the living creatures

ᵛ Or *lapis lazuli*   ʷ Traditional rendering of Heb *El Shaddai*

10.1–22 and 11.22–25: The LORD leaves his temple, and instructs the scribe to get coals from the fire between the cherubim (1.13) and to scatter them over the city (compare Gen 19.1–29; Rev 8.5). **1:** *Cherubim,* see 1.5 n. **3–4:** The *cloud of the glory of the LORD* recalls Ex 16.10; Num 10.34. The expression *the glory of the LORD* appears nineteen times in Ezekiel in the sense of the objective overpowering majesty of God, not so much an attribute as an expression for God himself (Lev 9.23; Num 20.6). **10.9–17:** A description of the throne-chariot of God (see 1.5–21). **19:** *East gate,* the main processional gate into the temple precinct (Pss 118.19–20; 24.7, 9). Here God in his throne chariot paused a moment and then forsook his sanctuary, desecrated by pagan rites and superficial worship. The departure continues in 11.22–25.

that I saw underneath the God of Israel by the river Chebar; and I knew that they were cherubim. ²¹ Each had four faces, each four wings, and underneath their wings something like human hands. ²² As for what their faces were like, they were the same faces whose appearance I had seen by the river Chebar. Each one moved straight ahead.

11 The spirit lifted me up and brought me to the east gate of the house of the LORD, which faces east. There, at the entrance of the gateway, were twenty-five men; among them I saw Jaazaniah son of Azzur, and Pelatiah son of Benaiah, officials of the people. ² He said to me, "Mortal, these are the men who devise iniquity and who give wicked counsel in this city; ³ they say, 'The time is not near to build houses; this city is the pot, and we are the meat.' ⁴ Therefore prophesy against them; prophesy, O mortal."

5 Then the spirit of the LORD fell upon me, and he said to me, "Say, Thus says the LORD: This is what you think, O house of Israel; I know the things that come into your mind. ⁶ You have killed many in this city, and have filled its streets with the slain. ⁷ Therefore thus says the Lord GOD: The slain whom you have placed within it are the meat, and this city is the pot; but you shall be taken out of it. ⁸ You have feared the sword; and I will bring the sword upon you, says the Lord GOD. ⁹ I will take you out of it and give you over to the hands of foreigners, and execute judgments upon you. ¹⁰ You shall fall by the sword; I will judge you at the border of Israel. And

you shall know that I am the LORD. ¹¹ This city shall not be your pot, and you shall not be the meat inside it; I will judge you at the border of Israel. ¹² Then you shall know that I am the LORD, whose statutes you have not followed, and whose ordinances you have not kept, but you have acted according to the ordinances of the nations that are around you."

13 Now, while I was prophesying, Pelatiah son of Benaiah died. Then I fell down on my face, cried with a loud voice, and said, "Ah Lord GOD! will you make a full end of the remnant of Israel?"

14 Then the word of the LORD came to me: ¹⁵ Mortal, your kinsfolk, your own kin, your fellow exiles, ˣ the whole house of Israel, all of them, are those of whom the inhabitants of Jerusalem have said, "They have gone far from the LORD; to us this land is given for a possession." ¹⁶ Therefore say: Thus says the Lord GOD: Though I removed them far away among the nations, and though I scattered them among the countries, yet I have been a sanctuary to them for a little while ʸ in the countries where they have gone. ¹⁷ Therefore say: Thus says the Lord GOD: I will gather you from the peoples, and assemble you out of the countries where you have been scattered, and I will give you the land of Israel. ¹⁸ When they come there, they will remove from it all its detestable things and all its abominations. ¹⁹ I will give them

x Gk Syr: Heb *people of your kindred*   y Or *to some extent*

11.1–25: Judgment and promise. 1–13: *Jaazaniah* and *Pelatiah* are otherwise unknown. *Wicked counsel* perhaps refers to the plot between Egypt and Zedekiah's pro-Egyptian counselors against Nebuchadrezzar (Jer 27.1–3; 37.5, 7, 11). Encouraged by the negotiations, they assure the populace of the city's security and urge them to continue domestic construction. Accusing the leaders of gross violence (ch 22; Jer 34.8–16), Ezekiel tells them that the city's walls (the *pot;* 24.1–14) will not protect them, but they will be

taken to *the border of Israel,* there to be judged (at Riblah?; Jer 52.24–27). **11.13:** Ezekiel added this note when these oracles were transcribed (1.1). **11.14–21:** This passage condemns the attitude which maintained that the exiles had borne God's punishment and their property now belonged to those who remained. Ezekiel warns them that God is with his exiled people and will restore them, while they, the presumptuous idolators, will be punished (Jer 24.1–10). New *heart,* Jer 32.37–41.

one[z] heart, and put a new spirit within them; I will remove the heart of stone from their flesh and give them a heart of flesh, [20]so that they may follow my statutes and keep my ordinances and obey them. Then they shall be my people, and I will be their God. [21]But as for those whose heart goes after their detestable things and their abominations,[a] I will bring their deeds upon their own heads, says the Lord God.

[22] Then the cherubim lifted up their wings, with the wheels beside them; and the glory of the God of Israel was above them. [23]And the glory of the Lord ascended from the middle of the city, and stopped on the mountain east of the city. [24]The spirit lifted me up and brought me in a vision by the spirit of God into Chaldea, to the exiles. Then the vision that I had seen left me. [25]And I told the exiles all the things that the Lord had shown me.

12 The word of the Lord came to me: [2]Mortal, you are living in the midst of a rebellious house, who have eyes to see but do not see, who have ears to hear but do not hear; [3]for they are a rebellious house. Therefore, mortal, prepare for yourself an exile's baggage, and go into exile by day in their sight; you shall go like an exile from your place to another place in their sight. Perhaps they will understand, though they are a rebellious house. [4]You shall bring out your baggage by day in their sight, as baggage for exile; and you shall go out yourself at evening in their sight, as those do who go into exile. [5]Dig through the wall in their sight, and carry the baggage through it. [6]In their sight you shall lift the baggage on your shoulder, and carry

it out in the dark; you shall cover your face, so that you may not see the land; for I have made you a sign for the house of Israel.

[7] I did just as I was commanded. I brought out my baggage by day, as baggage for exile, and in the evening I dug through the wall with my own hands; I brought it out in the dark, carrying it on my shoulder in their sight.

[8] In the morning the word of the Lord came to me: [9]Mortal, has not the house of Israel, the rebellious house, said to you, "What are you doing?" [10]Say to them, "Thus says the Lord God: This oracle concerns the prince in Jerusalem and all the house of Israel in it." [11]Say, "I am a sign for you: as I have done, so shall it be done to them; they shall go into exile, into captivity." [12]And the prince who is among them shall lift his baggage on his shoulder in the dark, and shall go out; he[b] shall dig through the wall and carry it through; he shall cover his face, so that he may not see the land with his eyes. [13]I will spread my net over him, and he shall be caught in my snare; and I will bring him to Babylon, the land of the Chaldeans, yet he shall not see it; and he shall die there. [14]I will scatter to every wind all who are around him, his helpers and all his troops; and I will unsheathe the sword behind them. [15]And they shall know that I am the Lord, when I disperse them among the nations and scatter them through the countries. [16]But I will let a few of them escape from the sword, from famine and pestilence, so that they may tell of all their abominations among

z Another reading is *a new* a Cn: Heb *And to the heart of their detestable things and their abominations their heart goes* b Gk Syr: Heb *they*

11.22–25: See 10.1–22. **23:** *The mountain east of the city* is the Mount of Olives.
12.1–20: Symbols of the Exile. 1–16: Ezekiel is commanded to symbolize by appropriate actions the collecting of whatever goods the exiles could carry (Jer 10.17) and the leaving of the city in the cool of the evening through its breached wall. *Dig through the wall* indicates a Babylonian locale in which

mud-brick ("adobe") houses were common. Many scholars conclude this oracle has been revised to refer to Zedekiah (the *prince*) who left Jerusalem by night, was captured (17.20), taken to Riblah and blinded (*may not see,* vv. 6, 12; Jer 39.1–10).
12.14: Seems to reflect 5.2, 10, 12. **15–16:** 14.21–23.

the nations where they go; then they shall know that I am the LORD.

17 The word of the LORD came to me: [18] Mortal, eat your bread with quaking, and drink your water with trembling and with fearfulness; [19] and say to the people of the land, Thus says the Lord GOD concerning the inhabitants of Jerusalem in the land of Israel: They shall eat their bread with fearfulness, and drink their water in dismay, because their land shall be stripped of all it contains, on account of the violence of all those who live in it. [20] The inhabited cities shall be laid waste, and the land shall become a desolation; and you shall know that I am the LORD.

21 The word of the LORD came to me: [22] Mortal, what is this proverb of yours about the land of Israel, which says, "The days are prolonged, and every vision comes to nothing"? [23] Tell them therefore, "Thus says the Lord GOD: I will put an end to this proverb, and they shall use it no more as a proverb in Israel." But say to them, The days are near, and the fulfillment of every vision. [24] For there shall no longer be any false vision or flattering divination within the house of Israel. [25] But I the LORD will speak the word that I speak, and it will be fulfilled. It will no longer be delayed; but in your days, O rebellious house, I will speak the word and fulfill it, says the Lord GOD.

26 The word of the LORD came to me: [27] Mortal, the house of Israel is saying, "The vision that he sees is for many years ahead; he prophesies for distant times." [28] Therefore say to them, Thus says the Lord GOD: None of my words will be delayed any longer, but the word that I speak will be fulfilled, says the Lord GOD.

13 The word of the LORD came to me: [2] Mortal, prophesy against the prophets of Israel who are prophesying; say to those who prophesy out of their own imagination: "Hear the word of the LORD!" [3] Thus says the Lord GOD, Alas for the senseless prophets who follow their own spirit, and have seen nothing! [4] Your prophets have been like jackals among ruins, O Israel. [5] You have not gone up into the breaches, or repaired a wall for the house of Israel, so that it might stand in battle on the day of the LORD. [6] They have envisioned falsehood and lying divination; they say, "Says the LORD," when the LORD has not sent them, and yet they wait for the fulfillment of their word! [7] Have you not seen a false vision or uttered a lying divination, when you have said, "Says the LORD," even though I did not speak?

8 Therefore thus says the Lord GOD: Because you have uttered falsehood and envisioned lies, I am against you, says the Lord GOD. [9] My hand will be against the prophets who see false visions and utter lying divinations; they shall not be in the council of my people, nor be enrolled in the register of the house of Israel, nor shall they enter the land of Israel; and you shall know that I am the Lord GOD. [10] Because, in truth, because they have misled my people, saying, "Peace," when there is no peace; and because, when the people build a wall, these prophets*c* smear whitewash on it. [11] Say

c Heb *they*

12.17–20: The people's terror at the approaching invasion (4.9–11, 16–17; Jer 4.19–21).

12.21–14.23: **Of prophets and people.** 12.21–28: A condemnation of the popular attitude that prophetic visions could be safely ignored (Hos 12.10; Jer 14.14–15; 23.28–29), and a correction of the impression that the fulfillment of prophecies of destruction such as those of Jeremiah was distant (Jer 5.12–13; 17.15). 27: Compare Isa 22.13; 1 Cor 15.32.

13.1–16: Against false prophets. The absence of objective criteria (Jer 28.8–9) resulted in the perennial problem of identifying the true prophet (1 Kings ch 22; Mic 3.5; Isa 9.15; Jer chs 14–15; with this passage, compare Jer 23.9–32). *Divination* (Ex 28.30; 1 Sam 28.6) played practically no role among the great prophets of Israel, but was common in contemporary non-Israelite prophetism, and apparently also in Israel. The false prophets' message (Jer chs 28–29) was like whitewash on a mud-brick wall (12.5) which provided no protection against the storm. 9: *Register,* Ezra ch 2; Ex 32.32–33.

to those who smear whitewash on it that it shall fall. There will be a deluge of rain,^d great hailstones will fall, and a stormy wind will break out. 12 When the wall falls, will it not be said to you, "Where is the whitewash you smeared on it?" 13 Therefore thus says the Lord GOD: In my wrath I will make a stormy wind break out, and in my anger there shall be a deluge of rain, and hailstones in wrath to destroy it. 14 I will break down the wall that you have smeared with whitewash, and bring it to the ground, so that its foundation will be laid bare; when it falls, you shall perish within it; and you shall know that I am the LORD. 15 Thus I will spend my wrath upon the wall, and upon those who have smeared it with whitewash; and I will say to you, The wall is no more, nor those who smeared it— 16 the prophets of Israel who prophesied concerning Jerusalem and saw visions of peace for it, when there was no peace, says the Lord GOD.

17 As for you, mortal, set your face against the daughters of your people, who prophesy out of their own imagination; prophesy against them 18 and say, Thus says the Lord GOD: Woe to the women who sew bands on all wrists, and make veils for the heads of persons of every height, in the hunt for human lives! Will you hunt down lives among my people, and maintain your own lives? 19 You have profaned me among my people for handfuls of barley and for pieces of bread, putting to death persons who should not die and keeping alive persons who should not live, by your lies to my people, who listen to lies.

20 Therefore thus says the Lord GOD: I am against your bands with which you hunt lives;^e I will tear them from your arms, and let the lives go free, the lives

that you hunt down like birds. 21 I will tear off your veils, and save my people from your hands; they shall no longer be prey in your hands; and you shall know that I am the LORD. 22 Because you have disheartened the righteous falsely, although I have not disheartened them, and you have encouraged the wicked not to turn from their wicked way and save their lives; 23 therefore you shall no longer see false visions or practice divination; I will save my people from your hand. Then you will know that I am the LORD.

14 Certain elders of Israel came to me and sat down before me. 2 And the word of the LORD came to me: 3 Mortal, these men have taken their idols into their hearts, and placed their iniquity as a stumbling block before them; shall I let myself be consulted by them? 4 Therefore speak to them, and say to them, Thus says the Lord GOD: Any of those of the house of Israel who take their idols into their hearts and place their iniquity as a stumbling block before them, and yet come to the prophet—I the LORD will answer those who come with the multitude of their idols, 5 in order that I may take hold of the hearts of the house of Israel, all of whom are estranged from me through their idols.

6 Therefore say to the house of Israel, Thus says the Lord GOD: Repent and turn away from your idols; and turn away your faces from all your abominations. 7 For any of those of the house of Israel, or of the aliens who reside in Israel, who separate themselves from me, taking their idols into their hearts and placing their iniquity as a stumbling

d Heb rain and you   *e Gk Syr: Heb* lives for birds

13.17–23: Against prophetesses. The reference here is to sorceresses and mediums (vv. 18, 20; 1 Sam 28.7–25), who were probably outside the cult of Yahweh, which provided few opportunities for participation of women (Miriam, Ex 15.20; Deborah, Judg 4.4; Huldah, 2 Kings 22.14 are widely scattered Old Testament examples; compare

1 Cor 14.34). The significance of *bands* and *veils* is unknown.

14.1–11: Against idolators. Religious duplicity was so reprehensible to God that God himself would punish anyone guilty of it. *Idols,* see 6.14 n. *Aliens* probably refers to proselytes who were considered equal with Israelites (47.22; Lev 17.8).

block before them, and yet come to a prophet to inquire of me by him, I the LORD will answer them myself. ⁸I will set my face against them; I will make them a sign and a byword and cut them off from the midst of my people; and you shall know that I am the LORD.

9 If a prophet is deceived and speaks a word, I, the LORD, have deceived that prophet, and I will stretch out my hand against him, and will destroy him from the midst of my people Israel. ¹⁰And they shall bear their punishment—the punishment of the inquirer and the punishment of the prophet shall be the same— ¹¹so that the house of Israel may no longer go astray from me, nor defile themselves any more with all their transgressions. Then they shall be my people, and I will be their God, says the Lord GOD.

12 The word of the LORD came to me: ¹³Mortal, when a land sins against me by acting faithlessly, and I stretch out my hand against it, and break its staff of bread and send famine upon it, and cut off from it human beings and animals, ¹⁴even if Noah, Daniel,ᶠ and Job, these three, were in it, they would save only their own lives by their righteousness, says the Lord GOD. ¹⁵If I send wild animals through the land to ravage it, so that it is made desolate, and no one may pass through because of the animals; ¹⁶even if these three men were in it, as I live, says the Lord GOD, they would save neither sons nor daughters; they alone would be saved, but the land would be desolate. ¹⁷Or if I bring a sword upon that land and say, 'Let a sword pass through the land,' and I cut off human beings and animals from it; ¹⁸though these three men were in it, as I live, says the Lord GOD, they would save neither

sons nor daughters, but they alone would be saved. ¹⁹Or if I send a pestilence into that land, and pour out my wrath upon it with blood, to cut off humans and animals from it; ²⁰even if Noah, Daniel,ᶠ and Job were in it, as I live, says the Lord GOD, they would save neither son nor daughter; they would save only their own lives by their righteousness.

21 For thus says the Lord GOD: How much more when I send upon Jerusalem my four deadly acts of judgment, sword, famine, wild animals, and pestilence, to cut off humans and animals from it! ²²Yet, survivors shall be left in it, sons and daughters who will be brought out; they will come out to you. When you see their ways and their deeds, you will be consoled for the evil that I have brought upon Jerusalem, for all that I have brought upon it. ²³They shall console you, when you see their ways and their deeds; and you shall know that it was not without cause that I did all that I have done in it, says the Lord GOD.

15 The word of the LORD came to me:
2 O mortal, how does the wood of
    the vine surpass all
        other wood—
    the vine branch that is among
        the trees of the forest?
3 Is wood taken from it to make
        anything?
    Does one take a peg from it on
        which to hang any object?
4 It is put in the fire for fuel;
    when the fire has consumed
        both ends of it
    and the middle of it is charred,
    is it useful for anything?

ᶠ Or, as otherwise read, *Danel*

14.12–23: **Against false hopes.** One would be saved only if righteous; righteousness is non-transferable and non-cumulative (33.12). Noah and Job are known in the Bible for their righteousness, as is Daniel. Ezekiel's reference to Daniel (see 28.3 n.), however, suggests the Canaanite Danel (so spelled in Ezekiel also) rather than the biblical Daniel.

There were undoubtedly cycles of literature associated with these patriarchs of which we now know very little.

15.1–8: **Allegory of the vine.** Vines and vineyards are common figures (Judg 9.8–15; Isa 5.1–7; Jer 2.21) but the reference to the *wood* is unique. The wood of the vine is good only when it produces satisfactorily; other-

5 When it was whole it was used
for nothing;
how much less—when the fire
has consumed it,
and it is charred—
can it ever be used for anything!

6 Therefore thus says the Lord GOD: Like the wood of the vine among the trees of the forest, which I have given to the fire for fuel, so I will give up the inhabitants of Jerusalem. 7I will set my face against them; although they escape from the fire, the fire shall still consume them; and you shall know that I am the LORD, when I set my face against them. 8And I will make the land desolate, because they have acted faithlessly, says the Lord GOD.

16 The word of the LORD came to me: 2Mortal, make known to Jerusalem her abominations, 3and say, Thus says the Lord GOD to Jerusalem: Your origin and your birth were in the land of the Canaanites; your father was an Amorite, and your mother a Hittite. 4As for your birth, on the day you were born your navel cord was not cut, nor were you washed with water to cleanse you, nor rubbed with salt, nor wrapped in cloths. 5No eye pitied you, to do any of these things for you out of compassion for you; but you were thrown out in the open field, for you were abhorred on the day you were born.

6 I passed by you, and saw you flailing about in your blood. As you lay in your blood, I said to you, "Live! 7and grow up*g* like a plant of the field." You

grew up and became tall and arrived at full womanhood;*h* your breasts were formed, and your hair had grown; yet you were naked and bare.

8 I passed by you again and looked on you; you were at the age for love. I spread the edge of my cloak over you, and covered your nakedness: I pledged myself to you and entered into a covenant with you, says the Lord GOD, and you became mine. 9Then I bathed you with water and washed off the blood from you, and anointed you with oil. 10I clothed you with embroidered cloth and with sandals of fine leather; I bound you in fine linen and covered you with rich fabric.*i* 11I adorned you with ornaments: I put bracelets on your arms, a chain on your neck, 12a ring on your nose, earrings in your ears, and a beautiful crown upon your head. 13You were adorned with gold and silver, while your clothing was of fine linen, rich fabric,*i* and embroidered cloth. You had choice flour and honey and oil for food. You grew exceedingly beautiful, fit to be a queen. 14Your fame spread among the nations on account of your beauty, for it was perfect because of my splendor that I had bestowed on you, says the Lord GOD.

15 But you trusted in your beauty, and played the whore because of your fame, and lavished your whorings on

g Gk Syr: Heb *Live! I made you a myriad*
h Cn: Heb *ornament of ornaments*   *i* Meaning of Heb uncertain

---

wise, even as fuel, it is practically useless. This is a figure for Judah; unproductive Judah must be destroyed (compare Jn 15.1–11).

16.1–63: The allegory of the unfaithful wife. 1–7: Jerusalem, the foundling. Ezekiel uses a folk-tale as an allegory. Jerusalem's ancestry was pagan and not related to the covenant. The *Canaanites* were the Semitic-speaking residents in Palestine before the Israelite invasion in the thirteenth century B.C., probably largely a part of the Amorite irruption into the Fertile Crescent in the early second millennium B.C. The *Hittites* were an Armenoid people living in Palestine with the Canaanites (Gen ch 23; Josh 3.10; compare

2 Sam 11.3). Unwanted and denied the common Palestinian amenities at birth, she, like female children in pagan antiquity, constituted a financial liability and was abandoned to die. By God's help, however, she grew into full maidenhood.

16.8–14: The maiden. She was adopted (by marriage) into God's covenant (compare 2 Sam 5.6–10) and became queen, receiving lavish adornment and generous nourishment (Jerusalem in Israel's golden age under Solomon).

16.15–22: The degenerate. The word *whorings* is used here in a double sense, referring both to the actual practice of cult prosti-

any passer-by.*j* 16 You took some of your garments, and made for yourself colorful shrines, and on them played the whore; nothing like this has ever been or ever shall be.*k* 17 You also took your beautiful jewels of my gold and my silver that I had given you, and made for yourself male images, and with them played the whore; 18 and you took your embroidered garments to cover them, and set my oil and my incense before them. 19 Also my bread that I gave you—I fed you with choice flour and oil and honey—you set it before them as a pleasing odor; and so it was, says the Lord God. 20 You took your sons and your daughters, whom you had borne to me, and these you sacrificed to them to be devoured. As if your whorings were not enough! 21 You slaughtered my children and delivered them up as an offering to them. 22 And in all your abominations and your whorings you did not remember the days of your youth, when you were naked and bare, flailing about in your blood.

23 After all your wickedness (woe, woe to you! says the Lord God), 24 you built yourself a platform and made yourself a lofty place in every square; 25 at the head of every street you built your lofty place and prostituted your beauty, offering yourself to every passer-by, and multiplying your whoring. 26 You played the whore with the Egyptians, your lustful neighbors, multiplying your whoring, to provoke me to anger. 27 Therefore I stretched out my hand against you, re-duced your rations, and gave you up to the will of your enemies, the daughters of the Philistines, who were ashamed of your lewd behavior. 28 You played the whore with the Assyrians, because you were insatiable; you played the whore with them, and still you were not satisfied. 29 You multiplied your whoring with Chaldea, the land of merchants; and even with this you were not satisfied.

30 How sick is your heart, says the Lord God, that you did all these things, the deeds of a brazen whore; 31 building your platform at the head of every street, and making your lofty place in every square! Yet you were not like a whore, because you scorned payment. 32 Adulterous wife, who receives strangers instead of her husband! 33 Gifts are given to all whores; but you gave your gifts to all your lovers, bribing them to come to you from all around for your whorings. 34 So you were different from other women in your whorings: no one solicited you to play the whore; and you gave payment, while no payment was given to you; you were different.

35 Therefore, O whore, hear the word of the Lord: 36 Thus says the Lord God, Because your lust was poured out and your nakedness uncovered in your whoring with your lovers, and because of all your abominable idols, and because of the blood of your children that you gave to them, 37 therefore, I will gather

j Heb adds *let it be his*   *k* Meaning of Heb uncertain

tution and to Jerusalem's unfaithfulness (Hos 4.13–14), which became widespread in the days of Manasseh (687–642 B.C.; 2 Kings 21.1–18) and Zedekiah (ch 8). She used her *garments* to make shrines and on them indulged in her promiscuity (Am 2.7–8). She fashioned her jewelry into idols (Judg 8.24–27) and amulets. God's gifts were used as offerings to other gods. **16.23–34**: Jerusalem is condemned for her religious infidelity and her proclivity toward foreign alliances—a major cause of her degeneration. But she was worse than a common prostitute who is paid for her services; Jerusalem invited her lovers and paid them (Isa 30.6; Hos ch 2; 8.9; Jer chs 2–3). **16.35–43**: Her lovers will turn against her and strip her. God himself will divorce her and expose her to be stoned (Deut 22.21, 24). Her beauty will be destroyed and she will again find herself ignoble and ignored. **44–52**: This section expands on the preceding theme, showing Jerusalem to be so much worse than her "elder sister" *Samaria* and the "younger sister" *Sodom* (Jer 3.6–11), both of whom were destroyed, that they *appear righteous* by comparison.

all your lovers, with whom you took pleasure, all those you loved and all those you hated; I will gather them against you from all around, and will uncover your nakedness to them, so that they may see all your nakedness. [38] I will judge you as women who commit adultery and shed blood are judged, and bring blood upon you in wrath and jealousy. [39] I will deliver you into their hands, and they shall throw down your platform and break down your lofty places; they shall strip you of your clothes and take your beautiful objects and leave you naked and bare. [40] They shall bring up a mob against you, and they shall stone you and cut you to pieces with their swords. [41] They shall burn your houses and execute judgments on you in the sight of many women; I will stop you from playing the whore, and you shall also make no more payments. [42] So I will satisfy my fury on you, and my jealousy shall turn away from you; I will be calm, and will be angry no longer. [43] Because you have not remembered the days of your youth, but have enraged me with all these things; therefore, I have returned your deeds upon your head, says the Lord GOD.

Have you not committed lewdness beyond all your abominations? [44] See, everyone who uses proverbs will use this proverb about you, "Like mother, like daughter." [45] You are the daughter of your mother, who loathed her husband and her children; and you are the sister of your sisters, who loathed their husbands and their children. Your mother was a Hittite and your father an Amorite. [46] Your elder sister is Samaria, who lived with her daughters to the north of you; and your younger sister, who lived to the south of you, is Sodom with her daughters. [47] You not only followed their ways, and acted according to their abominations; within a very little time you were more corrupt than they in all your ways. [48] As I live, says the Lord GOD,

your sister Sodom and her daughters have not done as you and your daughters have done. [49] This was the guilt of your sister Sodom: she and her daughters had pride, excess of food, and prosperous ease, but did not aid the poor and needy. [50] They were haughty, and did abominable things before me; therefore I removed them when I saw it. [51] Samaria has not committed half your sins; you have committed more abominations than they, and have made your sisters appear righteous by all the abominations that you have committed. [52] Bear your disgrace, you also, for you have brought about for your sisters a more favorable judgment; because of your sins in which you acted more abominably than they, they are more in the right than you. So be ashamed, you also, and bear your disgrace, for you have made your sisters appear righteous.

[53] I will restore their fortunes, the fortunes of Sodom and her daughters and the fortunes of Samaria and her daughters, and I will restore your own fortunes along with theirs, [54] in order that you may bear your disgrace and be ashamed of all that you have done, becoming a consolation to them. [55] As for your sisters, Sodom and her daughters shall return to their former state, Samaria and her daughters shall return to their former state, and you and your daughters shall return to your former state. [56] Was not your sister Sodom a byword in your mouth in the day of your pride, [57] before your wickedness was uncovered? Now you are a mockery to the daughters of Aram[1] and all her neighbors, and to the daughters of the Philistines, those all around who despise you. [58] You must bear the penalty of your lewdness and your abominations, says the LORD.

[59] Yes, thus says the Lord GOD: I will deal with you as you have done, you

1 Another reading is *Edom*

16.53–63: *Aram* (v. 57) should probably be read as "Edom"; the Edomites occupied Judahite territory after 587 B.C. (Lam 4.21–22). This reference is evidence for the revision of the oracle. All three "sisters" will be restored (Jer 12.14–17) and a new covenant es-

who have despised the oath, breaking the covenant; 60 yet I will remember my covenant with you in the days of your youth, and I will establish with you an everlasting covenant. 61 Then you will remember your ways, and be ashamed when I*m* take your sisters, both your elder and your younger, and give them to you as daughters, but not on account of my*n* covenant with you. 62 I will establish my covenant with you, and you shall know that I am the Lord, 63 in order that you may remember and be confounded, and never open your mouth again because of your shame, when I forgive you all that you have done, says the Lord God.

17 The word of the Lord came to me: 2 O mortal, propound a riddle, and speak an allegory to the house of Israel. 3 Say: Thus says the Lord God:

A great eagle, with great wings
and long pinions,
rich in plumage of many colors,
came to the Lebanon.
He took the top of the cedar,
4   broke off its topmost shoot;
He carried it to a land of trade,
set it in a city of merchants.
5   Then he took a seed from the
land,
placed it in fertile soil;
A plant*o* by abundant waters,
he set it like a willow twig.
6   It sprouted and became a vine
spreading out, but low;
Its branches turned toward him,
its roots remained where it
stood.
So it became a vine;
it brought forth branches,
put forth foliage.

7   There was another great eagle,
with great wings and much
plumage.
And see! This vine stretched out
its roots toward him;
It shot out its branches toward
him,
so that he might water it.
From the bed where it was planted
8   it was transplanted
to good soil by abundant waters,
so that it might produce
branches
and bear fruit
and become a noble vine.
9 Say: Thus says the Lord God:
Will it prosper?
Will he not pull up its roots,
cause its fruit to rot*o* and
wither,
its fresh sprouting leaves to
fade?
No strong arm or mighty army
will be needed
to pull it from its roots.
10 When it is transplanted, will
it thrive?
When the east wind strikes it,
will it not utterly wither,
wither on the bed where
it grew?

11 Then the word of the Lord came to me: 12 Say now to the rebellious house: Do you not know what these things mean? Tell them: The king of Babylon came to Jerusalem, took its king and its officials, and brought them back with him to Babylon. 13 He took one of the royal offspring and made a covenant with him, putting him under oath (he

m Syr: Heb *you*   *n* Heb lacks *my*   *o* Meaning of Heb uncertain

---

tablished (Jer 31.31–34). **62**: *I will establish my covenant* (see also v. 60), this phrase is characteristic of the "P" source in the Pentateuch (see Gen 6.18; 9.9, 11; 17.7, 19, etc.).
  17.1–21: **The allegory of the eagles.** Dramatis personae: *great eagle*, Nebuchadrezzar; *the top of the cedar*, house of David (Jer 22.5–6, 23); *topmost shoot*, Jehoiachin; *land of*

trade, Babylonia; *city of merchants*, Babylon; *seed from the land*, Zedekiah; *another great eagle*, Psammetichus II (594–588 B.C.), who engaged Zedekiah and other western states in anti-Babylonian intrigue (Jer ch 27). **5**: *Planted it*, i.e. made him king.
  17.9–10: Zedekiah will be unable long to resist Nebuchadrezzar (*the east wind*), though

had taken away the chief men of the land), ¹⁴so that the kingdom might be humble and not lift itself up, and that by keeping his covenant it might stand. ¹⁵But he rebelled against him by sending ambassadors to Egypt, in order that they might give him horses and a large army. Will he succeed? Can one escape who does such things? Can he break the covenant and yet escape? ¹⁶As I live, says the Lord GOD, surely in the place where the king resides who made him king, whose oath he despised, and whose covenant with him he broke—in Babylon he shall die. ¹⁷Pharaoh with his mighty army and great company will not help him in war, when ramps are cast up and siege walls built to cut off many lives. ¹⁸Because he despised the oath and broke the covenant, because he gave his hand and yet did all these things, he shall not escape. ¹⁹Therefore thus says the Lord GOD: As I live, I will surely return upon his head my oath that he despised, and my covenant that he broke. ²⁰I will spread my net over him, and he shall be caught in my snare; I will bring him to Babylon and enter into judgment with him there for the treason he has committed against me. ²¹All the pick *p* of his troops shall fall by the sword, and the survivors shall be scattered to every wind; and you shall know that I, the LORD, have spoken.

22 Thus says the Lord GOD:
I myself will take a sprig
from the lofty top of a cedar;
I will set it out.
I will break off a tender one
from the topmost of its
young twigs;

I myself will plant it
on a high and lofty mountain.
23 On the mountain height of Israel
I will plant it,
in order that it may produce
boughs and bear fruit,
and become a noble cedar.
Under it every kind of bird
will live;
in the shade of its branches
will nest
winged creatures of every kind.
24 All the trees of the field shall
know
that I am the LORD.
I bring low the high tree,
I make high the low tree;
I dry up the green tree
and make the dry tree flourish.
I the LORD have spoken;
I will accomplish it.

18 The word of the LORD came to me: ²What do you mean by repeating this proverb concerning the land of Israel, "The parents have eaten sour grapes, and the children's teeth are set on edge"? ³As I live, says the Lord GOD, this proverb shall no more be used by you in Israel. ⁴Know that all lives are mine; the life of the parent as well as the life of the child is mine: it is only the person who sins that shall die.

5 If a man is righteous and does what is lawful and right— ⁶if he does not eat upon the mountains or lift up his eyes to the idols of the house of Israel, does not defile his neighbor's wife or approach a woman during her menstrual period,

p Another reading is *fugitives*

---

the siege lasted nineteen months (Jer ch 52). **17**: *Pharaoh* has been added through editorial revision; read, "and not with a mighty army . . . shall he [Nebuchadrezzar] deal with him . . ." (Jer 37.3–11). Both Jeremiah and Ezekiel felt that Zedekiah should honor his oath of fealty to Nebuchadrezzar; his revolt was seen as rebellion against God's design (Jer 27.6–7). **17.22–24**: **Allegory of the cedar,** a messianic allegory. For similar imagery, see 31.1–9; for the Messiah as a *branch* compare Jer 23.5–6; Zech 3.8; *lofty mountain,* Mount Zion (Mic 4.1).

18.1–32: **Individual responsibility. 1–4**: It is human to blame someone else for one's plight; the exiles did this, blaming their ancestors for their misfortunes (Jer 31.27–30), presumably based on the covenant provision in Ex 20.5. Ezekiel points out that the blame rests directly upon themselves. This discussion parallels the provision regarding three generations in Ex 20.5. **18.5–9**: **First generation.** *Eat upon the mountains,* sacred meals in pagan high places (6.1–14). The list includes moral and religious provisions; note the strong legalistic

7 does not oppress anyone, but restores to the debtor his pledge, commits no robbery, gives his bread to the hungry and covers the naked with a garment, 8 does not take advance or accrued interest, withholds his hand from iniquity, executes true justice between contending parties, 9 follows my statutes, and is careful to observe my ordinances, acting faithfully—such a one is righteous; he shall surely live, says the Lord GOD.

10 If he has a son who is violent, a shedder of blood, 11 who does any of these things (though his father*q* does none of them), who eats upon the mountains, defiles his neighbor's wife, 12 oppresses the poor and needy, commits robbery, does not restore the pledge, lifts up his eyes to the idols, commits abomination, 13 takes advance or accrued interest; shall he then live? He shall not. He has done all these abominable things; he shall surely die; his blood shall be upon himself.

14 But if this man has a son who sees all the sins that his father has done, considers, and does not do likewise, 15 who does not eat upon the mountains or lift up his eyes to the idols of the house of Israel, does not defile his neighbor's wife, 16 does not wrong anyone, exacts no pledge, commits no robbery, but gives his bread to the hungry and covers the naked with a garment, 17 withholds his hand from iniquity,*r* takes no advance or accrued interest, observes my ordinances, and follows my statutes; he shall not die for his father's iniquity; he shall surely live. 18 As for his father, because he practiced extortion, robbed his brother, and did what is not good among his people, he dies for his iniquity.

19 Yet you say, "Why should not the son suffer for the iniquity of the father?" When the son has done what is lawful and right, and has been careful to observe all my statutes, he shall surely live. 20 The person who sins shall die. A child shall not suffer for the iniquity of a parent, nor a parent suffer for the iniquity of a child; the righteousness of the righteous shall be his own, and the wickedness of the wicked shall be his own.

21 But if the wicked turn away from all their sins that they have committed and keep all my statutes and do what is lawful and right, they shall surely live; they shall not die. 22 None of the transgressions that they have committed shall be remembered against them; for the righteousness that they have done they shall live. 23 Have I any pleasure in the death of the wicked, says the Lord GOD, and not rather that they should turn from their ways and live? 24 But when the righteous turn away from their righteousness and commit iniquity and do the same abominable things that the wicked do, shall they live? None of the righteous deeds that they have done shall be remembered; for the treachery of which they are guilty and the sin they have committed, they shall die.

25 Yet you say, "The way of the Lord is unfair." Hear now, O house of Israel: Is my way unfair? Is it not your ways that are unfair? 26 When the righteous turn away from their righteousness and commit iniquity, they shall die for it; for the iniquity that they have committed they shall die. 27 Again, when the wicked turn away from the wickedness they have committed and do what is lawful and right, they shall save their life. 28 Because they considered and turned away from all the transgressions that they had committed, they shall surely live; they shall not die. 29 Yet the house of Israel says,

q Heb *he*   *r* Gk: Heb *the poor*

---

emphasis: if anyone is "careful to observe my ordinances such a one is righteous." **10–13:** Second generation. *Shedder of blood,* a murderer. A life opposite to that of the father is represented. **18.14–18:** Third generation. Again a reversal from the father's life. **19–20:** Summation: neither the righteousness nor the iniquities of a previous generation are transferable to the next.
18.21–24: Within one's life the same principle of non-extension pertains.
18.25–29: Objection to this principle is a misunderstanding of God's justice. **30–32:**

"The way of the Lord is unfair."
O house of Israel, are my ways unfair? Is it not your ways that are unfair?

30 Therefore I will judge you, O house of Israel, all of you according to your ways, says the Lord God. Repent and turn from all your transgressions; otherwise iniquity will be your ruin. *s* ³¹ Cast away from you all the transgressions that you have committed against me, and get yourselves a new heart and a new spirit! Why will you die, O house of Israel? ³² For I have no pleasure in the death of anyone, says the Lord God. Turn, then, and live.

19 As for you, raise up a lamentation for the princes of Israel, ² and say:
What a lioness was your mother
  among lions!
She lay down among young lions,
  rearing her cubs.
³ She raised up one of her cubs;
  he became a young lion,
and he learned to catch prey;
  he devoured humans.
⁴ The nations sounded an alarm
  against him;
he was caught in their pit;
and they brought him with hooks
  to the land of Egypt.
⁵ When she saw that she was
  thwarted,
  that her hope was lost,
she took another of her cubs
  and made him a young lion.
⁶ He prowled among the lions;
  he became a young lion,
and he learned to catch prey;
  he devoured people.
⁷ And he ravaged their
  strongholds, *t*
  and laid waste their towns;

the land was appalled, and all
  in it,
  at the sound of his roaring.
⁸ The nations set upon him
  from the provinces all around;
they spread their net over him;
  he was caught in their pit.
⁹ With hooks they put him in a
  cage,
  and brought him to the king
  of Babylon;
they brought him into custody,
so that his voice should be heard
  no more
  on the mountains of Israel.
¹⁰ Your mother was like a vine in
  a vineyard *u*
  transplanted by the water,
fruitful and full of branches
  from abundant water.
¹¹ Its strongest stem became
  a ruler's scepter; *v*
  it towered aloft
  among the thick boughs;
  it stood out in its height
  with its mass of branches.
¹² But it was plucked up in fury,
  cast down to the ground;
the east wind dried it up;
  its fruit was stripped off,
its strong stem was withered;
  the fire consumed it.
¹³ Now it is transplanted into
  the wilderness,
  into a dry and thirsty land.
¹⁴ And fire has gone out from
  its stem,

s Or so that they shall not be a stumbling block of
iniquity to you   t Heb his widows
u Cn: Heb in your blood   v Heb Its strongest
stems became rulers' scepters

Because God is just, Israel's only hope is to repent and renew their covenant with him (v. 23; 36.24–32; Lam 3.33).
19.1–14: Two laments. 1–9: The *lioness* is Judah (Gen 49.9; symbol of Judah in 1 Kings 10.18–20 and found on Israelite seals). The first whelp is Jehoahaz, who was taken to Egypt (Jer 22.10–12; 2 Kings 23.30–34). Although some identify the second whelp as Jehoiachin (Jer 22.24–30; 2 Kings 24.8–16), it is more likely to be Zedekiah, likewise exiled to Babylon (Jer 39.7); both Jehoahaz and Zedekiah had the same mother, Hamutal (2 Kings 23.31; 24.18).
19.10–14: The *vine* is Judah (Isa 5.1–7; Jer 2.21). The *strongest stem* is Zedekiah (17.13) who was stripped (Jer 6.9) by *the east wind* (Nebuchadrezzar) and taken to Babylon (*transplanted;* Jer 52.1–11).

has consumed its branches
and fruit,
so that there remains in it no
strong stem,
no scepter for ruling.

This is a lamentation, and it is used as a lamentation.

20 In the seventh year, in the fifth month, on the tenth day of the month, certain elders of Israel came to consult the LORD, and sat down before me. 2 And the word of the LORD came to me: 3 Mortal, speak to the elders of Israel, and say to them: Thus says the Lord GOD: Why are you coming? To consult me? As I live, says the Lord GOD, I will not be consulted by you. 4 Will you judge them, mortal, will you judge them? Then let them know the abominations of their ancestors, 5 and say to them: Thus says the Lord GOD: On the day when I chose Israel, I swore to the offspring of the house of Jacob—making myself known to them in the land of Egypt—I swore to them, saying, I am the LORD your God. 6 On that day I swore to them that I would bring them out of the land of Egypt into a land that I had searched out for them, a land flowing with milk and honey, the most glorious of all lands. 7 And I said to them, Cast away the detestable things your eyes feast on, every one of you, and do not defile yourselves with the idols of Egypt; I am the LORD your God. 8 But they rebelled against me and would not listen to me; not one of them cast away the detestable things their eyes feasted on, nor did they forsake the idols of Egypt.

Then I thought I would pour out my wrath upon them and spend my anger against them in the midst of the land of Egypt. 9 But I acted for the sake of my name, that it should not be profaned in the sight of the nations among whom they lived, in whose sight I made myself known to them in bringing them out of the land of Egypt. 10 So I led them out of the land of Egypt and brought them into the wilderness. 11 I gave them my statutes and showed them my ordinances, by whose observance everyone shall live. 12 Moreover I gave them my sabbaths, as a sign between me and them, so that they might know that I the LORD sanctify them. 13 But the house of Israel rebelled against me in the wilderness; they did not observe my statutes but rejected my ordinances, by whose observance everyone shall live; and my sabbaths they greatly profaned.

Then I thought I would pour out my wrath upon them in the wilderness, to make an end of them. 14 But I acted for the sake of my name, so that it should not be profaned in the sight of the nations, in whose sight I had brought them out. 15 Moreover I swore to them in the wilderness that I would not bring them into the land that I had given them, a land flowing with milk and honey, the most glorious of all lands, 16 because they rejected my ordinances and did not observe my statutes, and profaned my sabbaths; for their heart went after their idols. 17 Nevertheless my eye spared them, and I did not destroy them or make an end of them in the wilderness.

18 I said to their children in the wil-

20.1–44: The fall and rise of Israel (compare Ps 106). **1–4:** Setting: August 14, 591 B.C. *Elders* of the Exile, 14.1–11. **5–8:** Apostasy in Egypt, where Israel served idols (Josh 24.14). There is some chronological confusion: God's promise to Israel was traditionally in a Palestinian setting (Gen 28.13–15); his revelation, *I am the LORD your God,* in the wilderness (Ex 20.2). **20.9–26:** Apostasy in the wilderness. *For the sake of my name* (36.22; Jer 14.7; Ps 106.8) expresses the important concept that Israel's

delivery from Egypt, the wilderness, and eventually from the Exile, was not accomplished because of Israel's intrinsic worth, but to demonstrate to all who would see that God is faithful and he alone is God (v. 44; Num 14.13–19). **12–13:** Of interest is the picture here of the institution of the *sabbaths* in the wilderness period (Ex 31.13); compare the emphasis on the divine institution of the sabbath at creation in the "P" source in Genesis (Gen 2.1–3). The proper observance of the sabbath (see Jer 17.19–27 n.) became increas-

derness, Do not follow the statutes of your parents, nor observe their ordinances, nor defile yourselves with their idols. ¹⁹I the LORD am your God; follow my statutes, and be careful to observe my ordinances, ²⁰and hallow my sabbaths that they may be a sign between me and you, so that you may know that I the LORD am your God. ²¹But the children rebelled against me; they did not follow my statutes, and were not careful to observe my ordinances, by whose observance everyone shall live; they profaned my sabbaths.

Then I thought I would pour out my wrath upon them and spend my anger against them in the wilderness. ²²But I withheld my hand, and acted for the sake of my name, so that it should not be profaned in the sight of the nations, in whose sight I had brought them out. ²³Moreover I swore to them in the wilderness that I would scatter them among the nations and disperse them through the countries, ²⁴because they had not executed my ordinances, but had rejected my statutes and profaned my sabbaths, and their eyes were set on their ancestors' idols. ²⁵Moreover I gave them statutes that were not good and ordinances by which they could not live. ²⁶I defiled them through their very gifts, in their offering up all their firstborn, in order that I might horrify them, so that they might know that I am the LORD.

27 Therefore, mortal, speak to the house of Israel and say to them, Thus says the Lord GOD: In this again your ancestors blasphemed me, by dealing treacherously with me. ²⁸For when I had brought them into the land that I swore to give them, then wherever they saw any high hill or any leafy tree, there they offered their sacrifices and presented the provocation of their offering; there they sent up their pleasing odors, and there they poured out their drink offerings. ²⁹(I said to them, What is the high place to which you go? So it is called Bamah^w to this day.) ³⁰Therefore say to the house of Israel, Thus says the Lord GOD: Will you defile yourselves after the manner of your ancestors and go astray after their detestable things? ³¹When you offer your gifts and make your children pass through the fire, you defile yourselves with all your idols to this day. And shall I be consulted by you, O house of Israel? As I live, says the Lord GOD, I will not be consulted by you.

32 What is in your mind shall never happen—the thought, "Let us be like the nations, like the tribes of the countries, and worship wood and stone."

33 As I live, says the Lord GOD, surely with a mighty hand and an outstretched arm, and with wrath poured out, I will be king over you. ³⁴I will bring you out from the peoples and gather you out of the countries where you are scattered, with a mighty hand and an outstretched arm, and with wrath poured out; ³⁵and I will bring you into the wilderness of the peoples, and there I will enter into judgment with you face to face. ³⁶As I entered into judgment with your ancestors in the wilderness of the land of Egypt, so I will enter into judgment with you, says the Lord GOD. ³⁷I will make you pass under the staff, and will bring you within the bond of the covenant. ³⁸I will purge out the rebels among you, and those who transgress against me; I will bring them out of the

w That is *High Place*

---

ingly important in post-exilic Judaism (Mt 12.1–8; Jn 9.13–16). **25–26**: This seems to contradict Jer 7.31, Lev 18.21, and the nature of God himself. One may interpret it as God's allowing his people to degenerate to this level so that they might conclusively learn his superiority to any deity made by human hands (Jer 19.4–6); these events, however, post-date the wilderness period.

20.27–29: Apostasy in Canaan. The emphasis is on the fertility cult associated with the high places (6.1–7; 16.15–22). **30–31**: In view of their constant faithlessness, they should not presume to approach God now.
20.32–39: As in the Sinai wilderness (Num 14.13–25), the unfaithful will be purged in the Syrian *wilderness*. But Israel will be preserved as God's people.

land where they reside as aliens, but they shall not enter the land of Israel. Then you shall know that I am the LORD.

39 As for you, O house of Israel, thus says the Lord GOD: Go serve your idols, everyone of you now and hereafter, if you will not listen to me; but my holy name you shall no more profane with your gifts and your idols.

40 For on my holy mountain, the mountain height of Israel, says the Lord GOD, there all the house of Israel, all of them, shall serve me in the land; there I will accept them, and there I will require your contributions and the choicest of your gifts, with all your sacred things. ⁴¹ As a pleasing odor I will accept you, when I bring you out from the peoples, and gather you out of the countries where you have been scattered; and I will manifest my holiness among you in the sight of the nations. ⁴² You shall know that I am the LORD, when I bring you into the land of Israel, the country that I swore to give to your ancestors. ⁴³ There you shall remember your ways and all the deeds by which you have polluted yourselves; and you shall loathe yourselves for all the evils that you have committed. ⁴⁴ And you shall know that I am the LORD, when I deal with you for my name's sake, not according to your evil ways, or corrupt deeds, O house of Israel, says the Lord GOD.

45 ˣ The word of the LORD came to me: ⁴⁶ Mortal, set your face toward the south, preach against the south, and prophesy against the forest land in the Negeb; ⁴⁷ say to the forest of the Negeb, Hear the word of the LORD: Thus says the Lord GOD, I will kindle a fire in you, and it shall devour every green tree in you and every dry tree; the blazing flame shall not be quenched, and all faces from south to north shall be scorched by it. ⁴⁸ All flesh shall see that I the LORD have kindled it; it shall not be quenched. ⁴⁹ Then I said, "Ah Lord GOD! they are saying of me, 'Is he not a maker of allegories?' "

21 ʸ The word of the LORD came to me: ²Mortal, set your face toward Jerusalem and preach against the sanctuaries; prophesy against the land of Israel ³ and say to the land of Israel, Thus says the LORD: I am coming against you, and will draw my sword out of its sheath, and will cut off from you both righteous and wicked. ⁴ Because I will cut off from you both righteous and wicked, therefore my sword shall go out of its sheath against all flesh from south to north; ⁵ and all flesh shall know that I the LORD have drawn my sword out of its sheath; it shall not be sheathed again. ⁶ Moan therefore, mortal; moan with breaking heart and bitter grief before their eyes. ⁷ And when they say to you, "Why do you moan?" you shall say, "Because of the news that has come. Every heart will melt and all hands will be feeble, every spirit will faint and all knees will turn to water. See, it comes and it will be fulfilled," says the Lord GOD.

8 And the word of the LORD came to me: ⁹ Mortal, prophesy and say: Thus says the Lord; Say:

A sword, a sword is sharpened,
  it is also polished;
10 It is sharpened for slaughter,
  honed to flash like lightning!

---

x Ch 21.1 in Heb   *y* Ch 21.6 in Heb

20.40–44: After the new Exodus (Jer 23.7–8) God will restore his people to Zion (17.22–24) and their sacrifices will again be acceptable (Ps 51.15–19).
20.45–49: Oracle against the south (i.e. Judah), which will be consumed by the invader from the north (Jer 5.14–17). One should read "south," not *Negeb,* in vv. 46–47.
21.1–32: Oracles on the sword, one of four conventional instruments of God's judgment (14.21; Isa 34.5; Rev 6.8; compare 6.11 and Jeremiah's three, Jer 14.12; compare Jer 5.6). **1–7:** God draws his sword. Because of its heterodoxy (*sanctuaries;* ch 20), Judah (*all flesh,* v. 4) will be cut down; everyone (*all flesh,* v. 5) will see the terrifying judgment (Jer 4.9).
21.8–17: Song of the sword (Jer 50.35–37). God's judgment is irrevocable. His flashing sword reaps its grim toll across the land (6.3). **12:** *Strike the thigh,* a sign of mourning (Jer 31.19).

How can we make merry?
  You have despised the rod,
  and all discipline. *z*
11 The sword*a* is given to be
    polished,
  to be grasped in the hand;
It is sharpened, the sword is
    polished,
  to be placed in the slayer's hand.
12 Cry and wail, O mortal,
  for it is against my people;
it is against all Israel's princes;
  they are thrown to the sword,
  together with my people.
  Ah! Strike the thigh!
13 For consider: What! If you despise the rod, will it not happen? *z* says the Lord GOD.
14 And you, mortal, prophesy;
  Strike hand to hand.
Let the sword fall twice, thrice;
  it is a sword for killing.
A sword for great slaughter—
  it surrounds them;
15 therefore hearts melt
  and many stumble.
At all their gates I have set
  the point *z* of the sword.
Ah! It is made for flashing,
  it is polished *b* for slaughter.
16 Attack to the right!
  Engage to the left!
  Wherever your edge is directed.
17 I too will strike hand to hand,
  I will satisfy my fury;
  I the LORD have spoken.

18 The word of the LORD came to me:
19 Mortal, mark out two roads for the sword of the king of Babylon to come; both of them shall issue from the same land. And make a signpost, make it for a fork in the road leading to a city;
20 mark out the road for the sword to come to Rabbah of the Ammonites or to Judah and to *c* Jerusalem the fortified.
21 For the king of Babylon stands at the parting of the way, at the fork in the two roads, to use divination; he shakes the arrows, he consults the teraphim, *d* he inspects the liver. 22 Into his right hand comes the lot for Jerusalem, to set battering rams, to call out for slaughter, for raising the battle cry, to set battering rams against the gates, to cast up ramps, to build siege towers. 23 But to them it will seem like a false divination; they have sworn solemn oaths; but he brings their guilt to remembrance, bringing about their capture.

24 Therefore thus says the Lord GOD: Because you have brought your guilt to remembrance, in that your transgressions are uncovered, so that in all your deeds your sins appear—because you have come to remembrance, you shall be taken in hand. *e*
25 As for you, vile, wicked prince
      of Israel,
  you whose day has come,
  the time of final punishment,
26 thus says the Lord GOD:
Remove the turban, take off
      the crown;
  things shall not remain as
      they are.
Exalt that which is low,
  abase that which is high.
27 A ruin, a ruin, a ruin—
  I will make it!
  (Such has never occurred.)
Until he comes whose right it is;
  to him I will give it.

28 As for you, mortal, prophesy, and say, Thus says the Lord GOD concerning

z Meaning of Heb uncertain   *a* Heb *It*
b Tg: Heb *wrapped up*   *c* Gk Syr: Heb *Judah in*   *d* Or *the household gods*   *e* Or *be taken captive*

21.18–24: The sword of Nebuchadrezzar. Perhaps from Riblah (compare 2 Kings 25.6), Nebuchadrezzar determines which rebel to attack first by using divination: belomancy (i.e. casting of arrows with names of projected victims on the heads); consultation of teraphim, an oracular device the use of which is unclear (Hos 3.4); and hepatoscopy (analyti-cal observation of the configurations and markings of sheep livers). The sword is about to fall—on Jerusalem! *Rabbah,* Ammonite capital (Jer 49.3).
    21.25–27: The sword will cut down Zedekiah, Judah's king (Jer 21.7).
    21.28–32: The sword against Ammon (v. 20). With familiar phraseology (compare

the Ammonites, and concerning their reproach; say:

A sword, a sword! Drawn for
slaughter
Polished to consume,*f* to flash
like lightning.

29 Offering false visions for you,
divining lies for you,
they place you over the necks
of the vile, wicked ones—
those whose day has come,
the time of final punishment.

30 Return it to its sheath!
In the place where you were
created,
in the land of your origin,
I will judge you.

31 I will pour out my indignation
upon you,
with the fire of my wrath
I will blow upon you.
I will deliver you into brutish
hands,
those skillful to destroy.

32 You shall be fuel for the fire,
your blood shall enter the earth;
You shall be remembered no
more,
for I the LORD have spoken.

22 The word of the LORD came to
me: 2 You, mortal, will you
judge, will you judge the bloody city?
Then declare to it all its abominable
deeds. 3 You shall say, Thus says the
Lord GOD: A city! Shedding blood within itself; its time has come; making its
idols, defiling itself. 4 You have become
guilty by the blood that you have shed,
and defiled by the idols that you have
made; you have brought your day near,
the appointed time of your years has
come. Therefore I have made you a disgrace before the nations, and a mockery

to all the countries. 5 Those who are near
and those who are far from you will
mock you, you infamous one, full of tumult.

6 The princes of Israel in you, everyone according to his power, have been
bent on shedding blood. 7 Father and
mother are treated with contempt in you;
the alien residing within you suffers extortion; the orphan and the widow are
wronged in you. 8 You have despised my
holy things, and profaned my sabbaths.
9 In you are those who slander to shed
blood, those in you who eat upon the
mountains, who commit lewdness in
your midst. 10 In you they uncover their
fathers' nakedness; in you they violate
women in their menstrual periods.
11 One commits abomination with his
neighbor's wife; another lewdly defiles
his daughter-in-law; another in you defiles his sister, his father's daughter. 12 In
you, they take bribes to shed blood; you
take both advance interest and accrued
interest, and make gain of your neighbors by extortion; and you have forgotten me, says the Lord GOD.

13 See, I strike my hands together at
the dishonest gain you have made, and at
the blood that has been shed within you.
14 Can your courage endure, or can your
hands remain strong in the days when I
shall deal with you? I the LORD have spoken, and I will do it. 15 I will scatter you
among the nations and disperse you
through the countries, and I will purge
your filthiness out of you. 16 And I*g* shall
be profaned through you in the sight of
the nations; and you shall know that I am
the LORD.

f Cn: Heb *to contain*   *g* Gk Syr Vg: Heb *you*

---

v. 28 with v. 8), Judah's co-conspirators, the
Ammonites, will also succumb to the sword
of God's wrath (25.1–7). This oracle, as also
vv. 18–24, may come from the time of the
assassination of Gedaliah (Jer 40.13–41.18).
22.1–31: Oracles of indictment. 1–16:
This writ of indictment contains a catalogue
of sins (18.5–18) including idolatry (6.2–14;

14.3–5), injustice (18.12), violence (7.23),
slander (Jer 6.28), adultery and fornication
(18.6; Jer 3.1–4), and extortion—a list reminiscent of the regulations in the Holiness
Code, Lev chs 17–26. In scorn and anger
(*strike my hands together*, 6.11; 21.14, 17) God
will wreak punishment upon this *bloody city*
(Nah 3.1, there referring to Nineveh).

17 The word of the LORD came to me: 18 Mortal, the house of Israel has become dross to me; all of them, silver, *h* bronze, tin, iron, and lead. In the smelter they have become dross. 19 Therefore thus says the Lord GOD: Because you have all become dross, I will gather you into the midst of Jerusalem. 20 As one gathers silver, bronze, iron, lead, and tin into a smelter, to blow the fire upon them in order to melt them; so I will gather you in my anger and in my wrath, and I will put you in and melt you. 21 I will gather you and blow upon you with the fire of my wrath, and you shall be melted within it. 22 As silver is melted in a smelter, so you shall be melted in it; and you shall know that I the LORD have poured out my wrath upon you.

23 The word of the LORD came to me: 24 Mortal, say to it: You are a land that is not cleansed, not rained upon in the day of indignation. 25 Its princes *i* within it are like a roaring lion tearing the prey; they have devoured human lives; they have taken treasure and precious things; they have made many widows within it. 26 Its priests have done violence to my teaching and have profaned my holy things; they have made no distinction between the holy and the common, neither have they taught the difference between the unclean and the clean, and they have disregarded my sabbaths, so that I am profaned among them. 27 Its officials within it are like wolves tearing the prey,

shedding blood, destroying lives to get dishonest gain. 28 Its prophets have smeared whitewash on their behalf, seeing false visions and divining lies for them, saying, "Thus says the Lord GOD," when the LORD has not spoken. 29 The people of the land have practiced extortion and committed robbery; they have oppressed the poor and needy, and have extorted from the alien without redress. 30 And I sought for anyone among them who would repair the wall and stand in the breach before me on behalf of the land, so that I would not destroy it; but I found no one. 31 Therefore I have poured out my indignation upon them; I have consumed them with the fire of my wrath; I have returned their conduct upon their heads, says the Lord GOD.

23 The word of the LORD came to me: 2 Mortal, there were two women, the daughters of one mother; 3 they played the whore in Egypt; they played the whore in their youth; their breasts were caressed there, and their virgin bosoms were fondled. 4 Oholah was the name of the elder and Oholibah the name of her sister. They became mine, and they bore sons and daughters. As for their names, Oholah is Samaria, and Oholibah is Jerusalem.

5 Oholah played the whore while she was mine; she lusted after her lovers the

h Transposed from the end of the verse; compare verse 20   *i* Gk: Heb *indignation.* 25 A *conspiracy of its prophets*

22.17–22: The judgment will be like a smelter in which base metals are removed; so Judah must endure the rigorous refining process and be purified of her baseness (Isa 1.22–25; Jer 6.27–30).
22.23–31: This oracle seems to come after Jerusalem's fall (587 B.C.; v. 31) and describes Judah's sinfulness in retrospect. All classes of Judean society were corrupt (Jer 8.8–10): princes, priests (Jer 2.8; Zeph 3.4), nobles (Mic 7.3), prophets (13.10–16), and people (12.19); and all must be punished (Jer 6.27).
23.1–49: **The allegory of the sisters, Oholah and Oholibah** (compare ch 16). **1–4:** Introduction. Israel's apostasy began in Egypt (20.5–9). The word-play *Oholah*, "her [own] tent," (i.e. Samaria), and *Oholibah*,

"my tent [is] in her" (i.e. Jerusalem), suggests that though Samaria had a sanctuary (tent), THE sanctuary was in Jerusalem, thus emphasizing the enormity of Judah's apostasy. On the marrying of sisters, see Gen 31.41; Lev 18.18.
23.5–10: Oholah. Like Hosea (8.9–10), Isaiah (7.1–9), and Jeremiah (4.30; same word for "lovers" in Ezek 23.9), Ezekiel viewed foreign alliances as disloyalty to God, though alternatives were not always politically available. Jehu (842–815 B.C.) surrendered to Shalmaneser III of Assyria; Jehoahaz (815–801) paid tribute to Adad-Nirari III, as did Menahem (745–738) to Tiglath-Pileser III (2 Kings 15.19–29) and Hoshea (732–724) to Shalmaneser V (2 Kings 17.1–14).

Assyrians, warriors[j] 6clothed in blue, governors and commanders, all of them handsome young men, mounted horsemen. 7She bestowed her favors upon them, the choicest men of Assyria all of them; and she defiled herself with all the idols of everyone for whom she lusted. 8She did not give up her whorings that she had practiced since Egypt; for in her youth men had lain with her and fondled her virgin bosom and poured out their lust upon her. 9Therefore I delivered her into the hands of her lovers, into the hands of the Assyrians, for whom she lusted. 10These uncovered her nakedness; they seized her sons and her daughters; and they killed her with the sword. Judgment was executed upon her, and she became a byword among women.

11 Her sister Oholibah saw this, yet she was more corrupt than she in her lusting and in her whorings, which were worse than those of her sister. 12She lusted after the Assyrians, governors and commanders, warriors[j] clothed in full armor, mounted horsemen, all of them handsome young men. 13And I saw that she was defiled; they both took the same way. 14But she carried her whorings further; she saw male figures carved on the wall, images of the Chaldeans portrayed in vermilion, 15with belts around their waists, with flowing turbans on their heads, all of them looking like officers— a picture of Babylonians whose native land was Chaldea. 16When she saw them she lusted after them, and sent messengers to them in Chaldea. 17And the Babylonians came to her into the bed of love,

and they defiled her with their lust; and after she defiled herself with them, she turned from them in disgust. 18When she carried on her whorings so openly and flaunted her nakedness, I turned in disgust from her, as I had turned from her sister. 19Yet she increased her whorings, remembering the days of her youth, when she played the whore in the land of Egypt 20and lusted after her paramours there, whose members were like those of donkeys, and whose emission was like that of stallions. 21Thus you longed for the lewdness of your youth, when the Egyptians[k] fondled your bosom and caressed[l] your young breasts.

22 Therefore, O Oholibah, thus says the Lord GOD: I will rouse against you your lovers from whom you turned in disgust, and I will bring them against you from every side: 23the Babylonians and all the Chaldeans, Pekod and Shoa and Koa, and all the Assyrians with them, handsome young men, governors and commanders all of them, officers and warriors,[m] all of them riding on horses. 24They shall come against you from the north[n] with chariots and wagons and a host of peoples; they shall set themselves against you on every side with buckler, shield, and helmet, and I will commit the judgment to them, and they shall judge you according to their ordinances. 25I will direct my indignation against you,

j Meaning of Heb uncertain   k Two Mss: MT *from Egypt*   l Cn: Heb *for the sake of*   m Compare verses 6 and 12: Heb *officers and called ones*   n Gk: Meaning of Heb uncertain

23.11–21: Oholibah. Judah, like Samaria, was tributary to Assyria, Ahaz (735–715 B.C.) to Tiglath-Pileser III (2 Kings 16.7–9), Hezekiah (715–687) to Sennacherib (2 Kings 18.1–36), and Manasseh (687–642) to Esarhaddon. Judah made alliances with Babylon; Hezekiah with Merodach-Baladan (2 Kings 20.12–21), Jehoiakim and Zedekiah with Nebuchadrezzar (2 Kings 24.1; compare Jer 22.18–23; 2 Kings 24.17; compare Jer 27.1–22) as well as with Egypt (2 Kings 23.35), and in both instances probably others not known to us (Hos 7.11). The prophets' objections were

based on the inherent, and demonstrated, dangers of syncretism and apostasy (2 Kings 16.7–19).

23.22–35: Such faithlessness, religious and political, will be punished. *The Babylonians,* their Aramean mercenaries (*Pekod,* Jer 50.21; *Shoa* and *Koa*), and Assyrian auxiliaries will devastate the land (*strip you; leave you naked*). *From the north,* the usual invasion route from Mesopotamia into Palestine (Jer 4.6; 6.1; 25.9). The oracle of the cup of wrath (vv. 32–34; Jer 25.15–29; Hab 2.16) assigns Samaria's fate to Jerusalem.

in order that they may deal with you in fury. They shall cut off your nose and your ears, and your survivors shall fall by the sword. They shall seize your sons and your daughters, and your survivors shall be devoured by fire. 26 They shall also strip you of your clothes and take away your fine jewels. 27 So I will put an end to your lewdness and your whoring brought from the land of Egypt; you shall not long for them, or remember Egypt any more. 28 For thus says the Lord GOD: I will deliver you into the hands of those whom you hate, into the hands of those from whom you turned in disgust; 29 and they shall deal with you in hatred, and take away all the fruit of your labor, and leave you naked and bare, and the nakedness of your whorings shall be exposed. Your lewdness and your whorings 30 have brought this upon you, because you played the whore with the nations, and polluted yourself with their idols. 31 You have gone the way of your sister; therefore I will give her cup into your hand. 32 Thus says the Lord GOD:

You shall drink your sister's cup,
    deep and wide;
you shall be scorned and derided,
    it holds so much.
33 You shall be filled with
    drunkenness and sorrow.
A cup of horror and desolation
    is the cup of your sister
        Samaria;
34 you shall drink it and drain it out,
    and gnaw its sherds,
    and tear out your breasts;

for I have spoken, says the Lord GOD. 35 Therefore thus says the Lord GOD: Because you have forgotten me and cast me behind your back, therefore bear the consequences of your lewdness and whorings.

36 The LORD said to me: Mortal, will you judge Oholah and Oholibah? Then declare to them their abominable deeds. 37 For they have committed adultery, and

blood is on their hands; with their idols they have committed adultery; and they have even offered up to them for food the children whom they had borne to me. 38 Moreover this they have done to me: they have defiled my sanctuary on the same day and profaned my sabbaths. 39 For when they had slaughtered their children for their idols, on the same day they came into my sanctuary to profane it. This is what they did in my house.

40 They even sent for men to come from far away, to whom a messenger was sent, and they came. For them you bathed yourself, painted your eyes, and decked yourself with ornaments; 41 you sat on a stately couch, with a table spread before it on which you had placed my incense and my oil. 42 The sound of a raucous multitude was around her, with many of the rabble brought in drunken from the wilderness; and they put bracelets on the arms ° of the women, and beautiful crowns upon their heads.

43 Then I said, Ah, she is worn out with adulteries, but they carry on their sexual acts with her. 44 For they have gone in to her, as one goes in to a whore. Thus they went in to Oholah and to Oholibah, wanton women. 45 But righteous judges shall declare them guilty of adultery and of bloodshed; because they are adulteresses and blood is on their hands.

46 For thus says the Lord GOD: Bring up an assembly against them, and make them an object of terror and of plunder. 47 The assembly shall stone them and with their swords they shall cut them down; they shall kill their sons and their daughters, and burn up their houses. 48 Thus will I put an end to lewdness in the land, so that all women may take warning and not commit lewdness as you have done. 49 They shall repay you for your lewdness, and you shall bear the

o Heb *hands*

23.36–49: As the adulteress is stoned (Lev 20.10), so the adultery (i.e. human sacrifice, pagan worship, profanation of the sabbath) of Samaria and Judah (Jer 4.30) has been and will be punished—with death.

penalty for your sinful idolatry; and you shall know that I am the Lord GOD.

24 In the ninth year, in the tenth month, on the tenth day of the month, the word of the LORD came to me: 2Mortal, write down the name of this day, this very day. The king of Babylon has laid siege to Jerusalem this very day. 3And utter an allegory to the rebellious house and say to them, Thus says the Lord GOD:

Set on the pot, set it on,
  pour in water also;
4 put in it the pieces,
  all the good pieces, the thigh
    and the shoulder;
  fill it with choice bones.
5 Take the choicest one of the flock,
  pile the logs*p* under it;
boil its pieces,*q*
  seethe*r* also its bones in it.

6 Therefore thus says the Lord GOD:
Woe to the bloody city,
  the pot whose rust is in it,
  whose rust has not gone out of
    it!
Empty it piece by piece,
  making no choice at all.*s*
7 For the blood she shed is inside it;
  she placed it on a bare rock;
  she did not pour it out on the
    ground,
  to cover it with earth.
8 To rouse my wrath, to take
    vengeance,
  I have placed the blood she shed
    on a bare rock,
  so that it may not be covered.
9Therefore thus says the Lord GOD:

Woe to the bloody city!
  I will even make the pile great.
10 Heap up the logs, kindle the fire;
  boil the meat well, mix in
    the spices,
  let the bones be burned.
11 Stand it empty upon the coals,
  so that it may become hot, its
    copper glow,
  its filth melt in it, its rust
    be consumed.
12 In vain I have wearied myself;*t*
  its thick rust does not depart.
  To the fire with its rust!*u*
13 Yet, when I cleansed you in your
    filthy lewdness,
  you did not become clean from
    your filth;
  you shall not again be cleansed
  until I have satisfied my fury
    upon you.
14I the LORD have spoken; the time is coming, I will act. I will not refrain, I will not spare, I will not relent. According to your ways and your doings I will judge you, says the Lord GOD.

15 The word of the LORD came to me: 16Mortal, with one blow I am about to take away from you the delight of your eyes; yet you shall not mourn or weep, nor shall your tears run down. 17Sigh, but not aloud; make no mourning for the dead. Bind on your turban, and put your sandals on your feet; do not cover your upper lip or eat the bread of mourners.*v*

---

p Compare verse 10: Heb *the bones*   *q* Two Mss: Heb *its boilings*   *r* Cn: Heb *its bones seethe* *s* Heb *piece, no lot has fallen on it* *t* Cn: Meaning of Heb uncertain   *u* Meaning of Heb uncertain   *v* Vg Tg: Heb *of men*

24.1–27: The beginning of the end.
24:1–14: The allegory of the pot (Jer 1.13–19) combines two themes, a *pot* boiling meat until it is completely destroyed and a pot full of *rust* heated until all the impurity is burned off. In the *pot* (i.e. Jerusalem, 11.3–12) everyone, the good and bad (21.4; Mic 3.2–3), will be boiled as the besiegers heap wood (siege equipment) around it. The contents will be thoroughly boiled, the pot emptied (i.e. after the siege), and the bones burned (a reference to the sacking of the city). Verses 6 and 11 introduce the theme of corrosion, referring to Jerusalem's bloody past (22.2–12; Gen 4.10–11). Recalling the theme of refining (22.17–22), this corrosion must be burned out 1: The date of the oracle is January 15, 588 B.C.

24.15–27: Oracle at the death of Ezekiel's wife. Ezekiel was instructed to forego the customary rites of mourning (Jer 16.5–9; Mic 1.8) as a sign to the people that the loss of cherished persons and things would bring a numbing, inexpressible grief (compare Jer 16.1–4).

¹⁸So I spoke to the people in the morning, and at evening my wife died. And on the next morning I did as I was commanded.

19 Then the people said to me, "Will you not tell us what these things mean for us, that you are acting this way?" ²⁰Then I said to them: The word of the LORD came to me: ²¹Say to the house of Israel, Thus says the Lord GOD: I will profane my sanctuary, the pride of your power, the delight of your eyes, and your heart's desire; and your sons and your daughters whom you left behind shall fall by the sword. ²²And you shall do as I have done; you shall not cover your upper lip or eat the bread of mourners. *w* ²³Your turbans shall be on your heads and your sandals on your feet; you shall not mourn or weep, but you shall pine away in your iniquities and groan to one another. ²⁴Thus Ezekiel shall be a sign to you; you shall do just as he has done. When this comes, then you shall know that I am the Lord GOD.

25 And you, mortal, on the day when I take from them their stronghold, their joy and glory, the delight of their eyes and their heart's affection, and also*x* their sons and their daughters, ²⁶on that day, one who has escaped will come to you to report to you the news. ²⁷On that day your mouth shall be opened to the one who has escaped, and you shall speak and no longer be silent. So you shall be a sign to them; and they shall know that I am the LORD.

25 The word of the LORD came to me: ²Mortal, set your face to-ward the Ammonites and prophesy against them. ³Say to the Ammonites, Hear the word of the Lord GOD: Thus says the Lord GOD, Because you said, "Aha!" over my sanctuary when it was profaned, and over the land of Israel when it was made desolate, and over the house of Judah when it went into exile; ⁴therefore I am handing you over to the people of the east for a possession. They shall set their encampments among you and pitch their tents in your midst; they shall eat your fruit, and they shall drink your milk. ⁵I will make Rabbah a pasture for camels and Ammon a fold for flocks. Then you shall know that I am the LORD. ⁶For thus says the Lord GOD: Because you have clapped your hands and stamped your feet and rejoiced with all the malice within you against the land of Israel, ⁷therefore I have stretched out my hand against you, and will hand you over as plunder to the nations. I will cut you off from the peoples and will make you perish out of the countries; I will destroy you. Then you shall know that I am the LORD.

8 Thus says the Lord GOD: Because Moab*y* said, The house of Judah is like all the other nations, ⁹therefore I will lay open the flank of Moab from the towns*z* on its frontier, the glory of the country, Beth-jeshimoth, Baal-meon, and Kiriathaim. ¹⁰I will give it along with Ammon to the people of the east as a possession. Thus Ammon shall be remembered

w Vg Tg: Heb *of men*   *x* Heb lacks *and also*
y Gk Old Latin: Heb *Moab and Seir*
z Heb *towns from its towns*

24.25–27: When word of the fall of Jerusalem reached Ezekiel (33.21–22), his tongue would be loosed (3.24–27), and he would proclaim a new message.

25.1–32.32: **Oracles against the nations.** Compare Isa chs 13–23; Jer chs 46–51. Seven nations (Ammon, Moab, Edom, Philistia, Tyre, Sidon, and Egypt; compare Deut 7.1) will be punished before Israel is restored (36.5–7).

25.1–17: **Oracles against Ammon,** **Moab, Edom, and Philistia. 1–7:** Against Ammon (21.28–32; Am 1.13–15; Jer 49.1–6). Ammon, which had seized one-time Israelite territory (Jer 49.1), is to be occupied by *the people of the east* (Isa 11.14), i.e. Arab tribes, whose expansion during this period pushed the Edomites into southern Judah (vv. 12–14). Centuries later this Arab expansion produced the Nabatean empire (compare 2 Cor 11.32).

25.8–11: Against Moab. Compare Jer

no more among the nations, [11]and I will execute judgments upon Moab. Then they shall know that I am the LORD.

12 Thus says the Lord GOD: Because Edom acted revengefully against the house of Judah and has grievously offended in taking vengeance upon them, [13]therefore thus says the Lord GOD, I will stretch out my hand against Edom, and cut off from it humans and animals, and I will make it desolate; from Teman even to Dedan they shall fall by the sword. [14]I will lay my vengeance upon Edom by the hand of my people Israel; and they shall act in Edom according to my anger and according to my wrath; and they shall know my vengeance, says the Lord GOD.

15 Thus says the Lord GOD: Because with unending hostilities the Philistines acted in vengeance, and with malice of heart took revenge in destruction; [16]therefore thus says the Lord GOD, I will stretch out my hand against the Philistines, cut off the Cherethites, and destroy the rest of the seacoast. [17]I will execute great vengeance on them with wrathful punishments. Then they shall know that I am the LORD, when I lay my vengeance on them.

26 In the eleventh year, on the first day of the month, the word of the LORD came to me: [2]Mortal, because Tyre said concerning Jerusalem,

"Aha, broken is the gateway of
    the peoples;
  it has swung open to me;
I shall be replenished,
  now that it is wasted."

[3]Therefore, thus says the Lord GOD:
  See, I am against you, O Tyre!
  I will hurl many nations
    against you,
  as the sea hurls its waves.
[4] They shall destroy the walls
    of Tyre
  and break down its towers.
  I will scrape its soil from it
  and make it a bare rock.
[5] It shall become, in the midst of
    the sea,
  a place for spreading nets.
I have spoken, says the Lord GOD.
  It shall become plunder for
    the nations,
[6]   and its daughter-towns in
    the country
  shall be killed by the sword.
Then they shall know that I am the LORD.

7 For thus says the Lord GOD: I will bring against Tyre from the north King Nebuchadrezzar of Babylon, king of kings, together with horses, chariots, cavalry, and a great and powerful army.
[8] Your daughter-towns in the
    country
  he shall put to the sword.
He shall set up a siege wall
    against you,
  cast up a ramp against you,
  and raise a roof of shields
    against you.
[9] He shall direct the shock of his
    battering rams against
    your walls
  and break down your towers
    with his axes.

48.1–47. The Arab expansion would also envelop Moab. **12–14:** Against Edom. Compare Isa ch 34; Jer 49.7–22.

25.15–17: Against Philistia. Compare Jer ch 47. *The Cherethites,* who lived between Gerar and Sharuhen (1 Sam 30.14), were perhaps originally Cretans (Jer 47.4).

26.1–28.19: Oracles against Tyre.

26.1–21: Tyre is to be destroyed by Nebuchadrezzar. Note the four divisions of the oracle, each with the introductory "Thus says the Lord GOD" (vv. 3, 7, 15, 19). **1–6:** Announcement of the judgment. For its failure to aid its ally Jerusalem (Jer 27.3), and because of its inordinate pride (28.2–10), Tyre is to fall. **4:** *Rock,* in Hebrew a play on the word "Tyre," which could also be read as "rock."

26.7–14: With mighty siege works, Nebuchadrezzar will besiege the city, which lay a half mile off-shore on an island, and he will occupy its suburbs (*daughter-towns*) on the mainland. Nebuchadrezzar's thirteen-year siege of Tyre apparently began shortly after Jerusalem's fall (*the eleventh year* of v. 1 should perhaps be read "the twelfth year" with the

10 His horses shall be so many
    that their dust shall cover you.
At the noise of cavalry, wheels,
        and chariots
    your very walls shall shake,
    when he enters your gates
        like those entering a
            breached city.
11 With the hoofs of his horses
    he shall trample all your streets.
He shall put your people to
        the sword,
    and your strong pillars shall fall
        to the ground.
12 They will plunder your riches
    and loot your merchandise;
they shall break down your walls
    and destroy your fine houses.
Your stones and timber and soil
    they shall cast into the water.
13 I will silence the music of
        your songs;
    the sound of your lyres shall be
        heard no more.
14 I will make you a bare rock;
    you shall be a place for
        spreading nets.
You shall never again be rebuilt,
    for I the Lord have spoken,
        says the Lord God.

15 Thus says the Lord God to Tyre:
Shall not the coastlands shake at the
sound of your fall, when the wounded
groan, when slaughter goes on within
you? 16 Then all the princes of the sea
shall step down from their thrones; they
shall remove their robes and strip off
their embroidered garments. They shall
clothe themselves with trembling, and
shall sit on the ground; they shall tremble
every moment, and be appalled at you.

17 And they shall raise a lamentation over
you, and say to you:

How you have vanished[a] from
        the seas,
    O city renowned,
once mighty on the sea,
    you and your inhabitants,[b]
who imposed your[c] terror
    on all the mainland![d]
18 Now the coastlands tremble
    on the day of your fall;
the coastlands by the sea
    are dismayed at your passing.

19 For thus says the Lord God: When
I make you a city laid waste, like cities
that are not inhabited, when I bring up
the deep over you, and the great waters
cover you, 20 then I will thrust you down
with those who descend into the Pit, to
the people of long ago, and I will make
you live in the world below, among primeval ruins, with those who go down to
the Pit, so that you will not be inhabited
or have a place[e] in the land of the living.
21 I will bring you to a dreadful end, and
you shall be no more; though sought for,
you will never be found again, says the
Lord God.

27 The word of the Lord came to
me: 2 Now you, mortal, raise a
lamentation over Tyre, 3 and say to Tyre,
which sits at the entrance to the sea, merchant of the peoples on many coastlands,
Thus says the Lord God:

O Tyre, you have said,
    "I am perfect in beauty."

---

a Gk OL Aquila: Heb *have vanished, O inhabited
one,* *b* Heb *it and its inhabitants* *c* Heb *their*
d Cn: Heb *its inhabitants* *e* Gk: Heb *I will give
beauty*

Septuagint; namely 586 B.C.), and ended in a
negotiated settlement (29.18). Tyre finally
fell to Alexander the Great in 332 B.C. **14:**
Compare vv. 4–5.
26.15–18: Lamentation of *the princes of the
sea*, perhaps cities in trade alliance with Tyre.
26.19–21: The waters of the primordial
deep (Gen 1.2) will cover the wasted city,
which will *descend into the Pit* (Sheol, see
31.15–18; Isa 14.15), the abode of the dead.
27.1–36: Lamentation over Tyre. The

good ship Tyre was constructed of the best
materials (*I am perfect;* compare 28.2–10).
Royal *purple* dye was the chief export product
of Phoenicia, the word "Phoenicia" comes
from the Greek word for "purple" (the word
"Canaan" may also mean "purple"). *Senir* is
Mount Hermon (Deut 3.9). *Bashan* is east of
the Sea of Galilee. *Elishah* is probably Cyprus. *Arvad*, like Tyre, was an island city,
two miles off-shore, 120 miles north of Tyre.
Gebal, later known as Byblos, was 70 miles

4 Your borders are in the heart of
    the seas;
    your builders made perfect
        your beauty.
5 They made all your planks
    of fir trees from Senir;
    they took a cedar from Lebanon
        to make a mast for you.
6 From oaks of Bashan
    they made your oars;
    they made your deck of pines*f*
        from the coasts of Cyprus,
        inlaid with ivory.
7 Of fine embroidered linen
    from Egypt
    was your sail,
        serving as your ensign;
    blue and purple from the coasts
        of Elishah
    was your awning.
8 The inhabitants of Sidon and
    Arvad
    were your rowers;
    skilled men of Zemer*g* were
        within you,
    they were your pilots.
9 The elders of Gebal and its artisans
        were within you,
    caulking your seams;
    all the ships of the sea with their
        mariners were within you,
    to barter for your wares.
10 Paras*h* and Lud and Put
    were in your army,
    your mighty warriors;
    they hung shield and helmet
        in you;
    they gave you splendor.

11 Men of Arvad and Helech*i*
    were on your walls all around;
    men of Gamad were at
        your towers.
    They hung their quivers all around
        your walls;
    they made perfect your beauty.
12 Tarshish did business with you out
of the abundance of your great wealth;
silver, iron, tin, and lead they exchanged
for your wares. 13 Javan, Tubal, and Meshech traded with you; they exchanged
human beings and vessels of bronze for
your merchandise. 14 Beth-togarmah exchanged for your wares horses, war
horses, and mules. 15 The Rhodians*j*
traded with you; many coastlands were
your own special markets; they brought
you in payment ivory tusks and ebony.
16 Edom*k* did business with you because
of your abundant goods; they exchanged
for your wares turquoise, purple, embroidered work, fine linen, coral, and rubies. 17 Judah and the land of Israel traded
with you; they exchanged for your merchandise wheat from Minnith, millet,*l*
honey, oil, and balm. 18 Damascus traded
with you for your abundant goods—
because of your great wealth of every
kind—wine of Helbon, and white wool.
19 Vedan and Javan from Uzal*l* entered
into trade for your wares; wrought iron,
cassia, and sweet cane were bartered for

f Or *boxwood*   *g* Cn Compare Gen 10.18: Heb
your skilled men, O Tyre   *h* Or *Persia*
i Or *and your army*   *j* Gk: Heb *The Dedanites*
k Another reading is *Aram*   *l* Meaning of
Heb uncertain

north of Tyre. **10–25a**: A (mostly prose) intrusion into the lament over Tyre. **10–11**:
Tyre's mercenaries included those from *Lud*,
identified with Lydia in Asia Minor (Gen
10.13; Jer 46.9 n.). *Put*, to be identified with
Libya (30.5). *Helech*, possibly Cilicia. The
identity of *Gamad* is uncertain; perhaps the
word should be read Gomerim, i.e. the Cimmerians in Cappadocia.
27.12–15a: Tyre's commercial empire is
described generally from west to east (*Tarshish*, Sardinia, or Tartessus in southern
Spain; *Javan*, Ionians, i.e. Greeks, Gen 10.2;
Tubal, Assyrian "Tabal," a people settled
in what is now southern Turkey, east of

the Anti-Taurus mountains; *Meshech*, Assyrian "Mushku," west of the Anti-Taurus
mountains; *Beth-togarmah*, Assyrian "Til-
garimmu," in what is now central Turkey,
east of the southernmost Halys River, east of
Tubal), from south to north (*Helbon*, famed
wine center thirteen miles north of Damascus), from southwest to northeast (*Uzal*,
modern Sana in Yemen; *Dedan*, in west central Arabia; *Sheba*, in southwest Arabia; *Haran*, on the Balikh River in Mesopotamia,
Gen 11.31–32; *Eden*, in Assyrian records Bit-
Adini, the Beth-eden of Am 1.5, and *Canneh*
southeast of Haran; *Asshur*, south of Nineveh;
Chilmad, an unidentified Mesopotamian

your merchandise. ²⁰ Dedan traded with
you in saddlecloths for riding. ²¹ Arabia
and all the princes of Kedar were your
favored dealers in lambs, rams, and
goats; in these they did business with
you. ²² The merchants of Sheba and Raamah traded with you; they exchanged for
your wares the best of all kinds of spices,
and all precious stones, and gold. ²³ Haran, Canneh, Eden, the merchants of
Sheba, Asshur, and Chilmad traded with
you. ²⁴ These traded with you in choice
garments, in clothes of blue and embroidered work, and in carpets of colored
material, bound with cords and made secure; in these they traded with you. ^m
²⁵ The ships of Tarshish traveled for you
in your trade.

So you were filled and heavily
laden
in the heart of the seas.
²⁶ Your rowers have brought you
into the high seas.
The east wind has wrecked you
in the heart of the seas.
²⁷ Your riches, your wares, your
merchandise,
your mariners and your pilots,
your caulkers, your dealers in
merchandise,
and all your warriors within
you,
with all the company
that is with you,
sink into the heart of the seas
on the day of your ruin.
²⁸ At the sound of the cry of
your pilots
the countryside shakes,
²⁹ and down from their ships
come all that handle the oar.
The mariners and all the pilots
of the sea
stand on the shore

³⁰ and wail aloud over you,
and cry bitterly.
They throw dust on their heads
and wallow in ashes;
³¹ they make themselves bald for
you,
and put on sackcloth,
and they weep over you in
bitterness of soul,
with bitter mourning.
³² In their wailing they raise a
lamentation for you,
and lament over you:
"Who was ever destroyed ⁿ
like Tyre
in the midst of the sea?
³³ When your wares came from
the seas,
you satisfied many peoples;
with your abundant wealth and
merchandise
you enriched the kings of
the earth.
³⁴ Now you are wrecked by the seas,
in the depths of the waters;
your merchandise and all your
crew
have sunk with you.
³⁵ All the inhabitants of the
coastlands
are appalled at you;
and their kings are horribly afraid,
their faces are convulsed.
³⁶ The merchants among the peoples
hiss at you;
you have come to a dreadful
end
and shall be no more forever."

28 The word of the Lord came to
me: ² Mortal, say to the prince of
Tyre, Thus says the Lord God:

---

m Cn: Heb *in your market*   *n* Tg Vg: Heb *like
silence*

---

city). *Sweet cane* (Jer 6.20) was used for oil and
sacrifice.
27.25b–36: This section continues vv. 1–
9. The sinking of the ship by the *east wind*
(19.12; Jer 18.17). The sudden end of the great
commercial city brings fear and lamentation
to seamen, merchants, and inhabitants of the
coastlands.

28.1–10. Oracle against Tyre, apparently using mythological themes, including the
Canaanite story of Danel (see 14.12–23 n.) in
which Danel is the wise judge of widows and
orphans. Pride leads to self-deification. Thus
the prince of Tyre sat *in the seat of the gods*
(compare Isa 14.13–14); but he will be slain
by *the most terrible of the nations,* Babylonia

Because your heart is proud
  and you have said, "I am a god;
I sit in the seat of the gods,
  in the heart of the seas,"
yet you are but a mortal, and
    no god,
  though you compare your mind
  with the mind of a god.
3 You are indeed wiser than
    Daniel; *o*
  no secret is hidden from you;
4 by your wisdom and your
    understanding
  you have amassed wealth
    for yourself,
and have gathered gold and silver
  into your treasuries.
5 By your great wisdom in trade
  you have increased your wealth,
  and your heart has become
    proud in your wealth.
6 Therefore thus says the Lord God:
Because you compare your mind
  with the mind of a god,
7 therefore, I will bring strangers
    against you,
  the most terrible of the nations;
they shall draw their swords
  against the beauty of
    your wisdom
  and defile your splendor.
8 They shall thrust you down to
    the Pit,
  and you shall die a violent death
  in the heart of the seas.
9 Will you still say, "I am a god,"
  in the presence of those who
    kill you,
though you are but a mortal, and
    no god,
  in the hands of those who
    wound you?
10 You shall die the death of the
    uncircumcised
  by the hand of foreigners;

for I have spoken, says the
  Lord God.
11 Moreover the word of the Lord
came to me: 12 Mortal, raise a lamentation over the king of Tyre, and say to
him, Thus says the Lord God:
You were the signet of
    perfection, *p*
  full of wisdom and perfect in
    beauty.
13 You were in Eden, the garden
    of God;
  every precious stone was your
    covering,
carnelian, chrysolite, and
    moonstone,
  beryl, onyx, and jasper,
sapphire, *q* turquoise, and emerald;
  and worked in gold were your
    settings
  and your engravings. *p*
On the day that you were created
  they were prepared.
14 With an anointed cherub as
    guardian I placed you; *p*
  you were on the holy mountain
    of God;
  you walked among the stones
    of fire.
15 You were blameless in your ways
  from the day that you were
    created,
  until iniquity was found in you.
16 In the abundance of your trade
  you were filled with violence,
    and you sinned;
so I cast you as a profane thing
    from the mountain of God,
and the guardian cherub drove
    you out
from among the stones of fire.

o Or, as otherwise read, *Danel*   *p* Meaning of
Heb uncertain   *q* Or *lapis lazuli*

(30.10–11), and come to an ignominious end
(*the death of the uncircumcised*) in Sheol (31.14–
18). **28.11–19: Lamentation over the king of
Tyre,** based on a variant version of the Eden
story. The primal human being, created perfect, dwelt in Eden, with a covering of pre-
cious stones (compare the twelve precious stones of the ephod in Ex 28.17–20 and the
description of the heavenly Jerusalem in Rev
4.1–6; 21.15–21). But pride and idolatry led
to banishment by the guardian cherub (Gen
3.24). **16:** *The mountain of God* in Canaanite
myth was Mount Sapon, modern Jebel Aqra',
north of Ugarit.

17 Your heart was proud because of
          your beauty;
     you corrupted your wisdom for
          the sake of your splendor.
   I cast you to the ground;
     I exposed you before kings,
        to feast their eyes on you.
18 By the multitude of your
          iniquities,
     in the unrighteousness of
          your trade,
     you profaned your sanctuaries.
   So I brought out fire from
          within you;
     it consumed you,
   and I turned you to ashes on
          the earth
     in the sight of all who saw you.
19 All who know you among the
          peoples
     are appalled at you;
   you have come to a dreadful end
     and shall be no more forever.

20 The word of the Lord came to me:
21 Mortal, set your face toward Sidon,
and prophesy against it, 22 and say, Thus
says the Lord God:
   I am against you, O Sidon,
     and I will gain glory in your
          midst.
   They shall know that I am
          the Lord
     when I execute judgments in it,
     and manifest my holiness in it;
23 for I will send pestilence into it,
     and bloodshed into its streets;
   and the dead shall fall in its midst,
     by the sword that is against it
          on every side.
   And they shall know that I am
          the Lord.

24 The house of Israel shall no longer
find a pricking brier or a piercing thorn

among all their neighbors who have
treated them with contempt. And they
shall know that I am the Lord God.

25 Thus says the Lord God: When I
gather the house of Israel from the peoples among whom they are scattered,
and manifest my holiness in them in the
sight of the nations, then they shall settle
on their own soil that I gave to my servant Jacob. 26 They shall live in safety in
it, and shall build houses and plant vineyards. They shall live in safety, when I
execute judgments upon all their neighbors who have treated them with contempt. And they shall know that I am the
Lord their God.

29 In the tenth year, in the tenth
month, on the twelfth day of the
month, the word of the Lord came to
me: 2 Mortal, set your face against Pharaoh king of Egypt, and prophesy against
him and against all Egypt; 3 speak, and
say, Thus says the Lord God:
   I am against you,
     Pharaoh king of Egypt,
   the great dragon sprawling
     in the midst of its channels,
     saying, "My Nile is my own;
       I made it for myself."
4 I will put hooks in your jaws,
     and make the fish of your
          channels stick to your
          scales.
   I will draw you up from your
          channels,
     with all the fish of your
          channels
     sticking to your scales.
5 I will fling you into the
          wilderness,
     you and all the fish of your
          channels;
   you shall fall in the open field,

28.20–23: Oracle against Sidon, north
of Tyre, an ally of Jerusalem against Nebuchadrezzar (Jer 27.3). **22–23:** Compare 20.41;
36.23.
28.24–26: Restoration of Israel. An editorial addition, concluding the first section of
the oracles concerning foreign nations. **24:**
Compare Num 33.55. **25–26:** Compare
34.28; Jer 23.6; Lev 25.19.

29.1–32.32: Against Egypt.
29.1–16: Against Pharaoh (January 7,
587 B.C.). Hophra's attack against Nebuchadrezzar in the spring of 588 had failed to relieve Jerusalem (vv. 6–9; Jer 37.1–10). Hophra is depicted as the great sea dragon (Hebrew "tannin"; Isa 27.1; Job ch 41) whom God will catch
and let his body become carrion (32.1–8). *His
channels,* the Nile delta and canals. *Fish of your*

and not be gathered and buried.
To the animals of the earth and to
the birds of the air
I have given you as food.
6 Then all the inhabitants of Egypt
shall know
that I am the LORD
because you' were a staff of reed
to the house of Israel;
7 when they grasped you with the
hand, you broke,
and tore all their shoulders;
and when they leaned on you,
you broke,
and made all their legs
unsteady.*

8 Therefore, thus says the Lord GOD:
I will bring a sword upon you, and will
cut off from you human being and animal; 9 and the land of Egypt shall be a
desolation and a waste. Then they shall
know that I am the LORD.

Because you' said, "The Nile is mine,
and I made it," 10 therefore, I am against
you, and against your channels, and I
will make the land of Egypt an utter
waste and desolation, from Migdol to
Syene, as far as the border of Ethiopia.*
11 No human foot shall pass through it,
and no animal foot shall pass through it;
it shall be uninhabited forty years. 12 I
will make the land of Egypt a desolation
among desolated countries; and her cities
shall be a desolation forty years among
cities that are laid waste. I will scatter the
Egyptians among the nations, and disperse them among the countries.

13 Further, thus says the Lord GOD:
At the end of forty years I will gather the
Egyptians from the peoples among
whom they were scattered; 14 and I will
restore the fortunes of Egypt, and bring

them back to the land of Pathros, the
land of their origin; and there they shall
be a lowly kingdom. 15 It shall be the
most lowly of the kingdoms, and never
again exalt itself above the nations; and I
will make them so small that they will
never again rule over the nations. 16 The
Egyptians' shall never again be the reliance of the house of Israel; they will recall
their iniquity, when they turned to them
for aid. Then they shall know that I am
the Lord GOD.

17 In the twenty-seventh year, in the
first month, on the first day of the
month, the word of the LORD came to
me: 18 Mortal, King Nebuchadrezzar of
Babylon made his army labor hard
against Tyre; every head was made bald
and every shoulder was rubbed bare; yet
neither he nor his army got anything
from Tyre to pay for the labor that he
had expended against it. 19 Therefore
thus says the Lord GOD: I will give the
land of Egypt to King Nebuchadrezzar
of Babylon; and he shall carry off its
wealth and despoil it and plunder it; and
it shall be the wages for his army. 20 I
have given him the land of Egypt as his
payment for which he labored, because
they worked for me, says the Lord GOD.

21 On that day I will cause a horn to
sprout up for the house of Israel, and I
will open your lips among them. Then
they shall know that I am the LORD.

30 The word of the LORD came to
me: 2 Mortal, prophesy, and say,
Thus says the Lord GOD:
Wail, "Alas for the day!"
3   For a day is near,

r Gk Syr Vg: Heb *they*   s Syr: Heb *stand*
t Gk Syr Vg: Heb *he*   u Or *Nubia*; Heb *Cush*
v Heb *It*

channels, the Egyptians and their mercenaries.
10: *Migdol* (southwest of Pelusium, 30.15; Jer
44.1) and *Syene* (Aswan, at the First Cataract
of the Nile) are the northern and southern
limits of Egypt. **11:** *Forty years,* 4.6; Num
14.33. **14:** *Pathros,* see Jer 44.1 n.
**29.17–21: Egypt as "wages" for Nebu-
chadrezzar.** This is Ezekiel's latest dated oracle, April 26, 571 B.C., shortly after Ahmo-

sis II forced Hophra to make him co-regent.
Nebuchadrezzar will get sufficient booty
from Egypt to compensate for his lack of
booty from the siege of Tyre (26.7; Jer 46.2–
26). *Horn* may refer to the re-establishment of
the Davidic line in Israel (Ps 132.17).
30.1–26: The doom of Egypt. 1–5: The
day of the LORD since the time of Amos (Am
5.18–20) was God's judgment day (15.5; Isa

the day of the LORD is near;
it will be a day of clouds,
a time of doom[w] for the
nations.
4 A sword shall come upon Egypt,
and anguish shall be in
Ethiopia,[x]
when the slain fall in Egypt,
and its wealth is carried away,
and its foundations are
torn down.
5 Ethiopia,[x] and Put, and Lud, and all
Arabia, and Libya,[y] and the people of the
allied land[z] shall fall with them by the
sword.

6 Thus says the LORD:
Those who support Egypt shall
fall,
and its proud might shall
come down;
from Migdol to Syene
they shall fall within it by
the sword,
says the Lord GOD.
7 They shall be desolated among
other desolated countries,
and their cities shall lie among
cities laid waste.
8 Then they shall know that I am
the LORD,
when I have set fire to Egypt,
and all who help it are broken.
9 On that day, messengers shall go
out from me in ships to terrify the unsuspecting Ethiopians;[a] and anguish shall
come upon them on the day of Egypt's
doom;[b] for it is coming!

10 Thus says the Lord GOD:
I will put an end to the hordes
of Egypt,
by the hand of King
Nebuchadrezzar of
Babylon.

11 He and his people with him, the
most terrible of the nations,
shall be brought in to destroy
the land;
and they shall draw their swords
against Egypt,
and fill the land with the slain.
12 I will dry up the channels,
and will sell the land into the
hand of evildoers;
I will bring desolation upon the
land and everything in it
by the hand of foreigners;
I the LORD have spoken.

13 Thus says the Lord GOD:
I will destroy the idols
and put an end to the images
in Memphis;
there shall no longer be a prince in
the land of Egypt;
so I will put fear in the land
of Egypt.
14 I will make Pathros a desolation,
and will set fire to Zoan,
and will execute acts of
judgment on Thebes.
15 I will pour my wrath upon
Pelusium,
the stronghold of Egypt,
and cut off the hordes of
Thebes.
16 I will set fire to Egypt;
Pelusium shall be in great
agony;
Thebes shall be breached,
and Memphis face adversaries
by day.
17 The young men of On and of
Pi-beseth shall fall by
the sword;

w Heb lacks *of doom*   x Or Nubia; Heb *Cush*
y Compare Gk Syr Vg: Heb *Cub*   z Meaning
of Heb uncertain   a Or *Nubians*; Heb *Cush*
b Heb *the day of Egypt*

2.12; Jer 30.7; Zeph 1.14–18); later it became
the day of Israel's restoration and remained
doomsday for the Gentiles. *Sword*, 21.3–17.
Put and Lud, see 27.10–11 n.
    30.6–9: Egypt's mercenaries (27.10–11)
will collapse before the onslaught; Egypt will
be reduced to ashes (28.18). *Migdol to Syene*,

see 29.10 n. **10–12:** Nebuchadrezzar (26.7),
king of *the most terrible of the nations* (28.7), is
God's instrument (Jer 27.6).
    30.13–19: All Egypt will be destroyed.
Memphis, the ancient capital of lower Egypt.
Pathros, see Jer 44.1 n. *Zoan*, in the Greek
period Tanis, in the northeast delta region.

and the cities themselves[c] shall
    go into captivity.
18 At Tehaphnehes the day shall
    be dark,
when I break there the
    dominion of Egypt,
and its proud might shall come to
    an end;
the city[d] shall be covered by
    a cloud,
and its daughter-towns shall go
    into captivity.
19 Thus I will execute acts of
    judgment on Egypt.
Then they shall know that I am
    the LORD.

20 In the eleventh year, in the first
month, on the seventh day of the month,
the word of the LORD came to me:
21 Mortal, I have broken the arm of Pharaoh king of Egypt; it has not been bound
up for healing or wrapped with a bandage, so that it may become strong to
wield the sword. 22 Therefore thus says
the Lord GOD: I am against Pharaoh king
of Egypt, and will break his arms, both
the strong arm and the one that was broken; and I will make the sword fall from
his hand. 23 I will scatter the Egyptians
among the nations, and disperse them
throughout the lands. 24 I will strengthen
the arms of the king of Babylon, and put
my sword in his hand; but I will break
the arms of Pharaoh, and he will groan
before him with the groans of one mortally wounded. 25 I will strengthen the
arms of the king of Babylon, but the
arms of Pharaoh shall fall. And they shall
know that I am the LORD, when I put my
sword into the hand of the king of Babylon. He shall stretch it out against the
land of Egypt, 26 and I will scatter the
Egyptians among the nations and disperse them throughout the countries.
Then they shall know that I am the
LORD.

31 In the eleventh year, in the third
month, on the first day of the
month, the word of the LORD came to
me: 2 Mortal, say to Pharaoh king of
Egypt and to his hordes:
    Whom are you like in your
        greatness?
3   Consider Assyria, a cedar
        of Lebanon,
with fair branches and forest
        shade,
and of great height,
    its top among the clouds.[e]
4 The waters nourished it,
    the deep made it grow tall,
making its rivers flow[f]
    around the place it was planted,
sending forth its streams
    to all the trees of the field.
5 So it towered high
    above all the trees of the field;
its boughs grew large
    and its branches long,
from abundant water in
    its shoots.
6 All the birds of the air
    made their nests in its boughs;
under its branches all the animals
    of the field
gave birth to their young;
    and in its shade
all great nations lived.
7 It was beautiful in its greatness,
    in the length of its branches;
for its roots went down
    to abundant water.

---

c Heb *and they*   d Heb *she*   e Gk: Heb *thick
boughs*   f Gk: Heb *rivers going*

Thebes is present-day Karnak (Jer 46.25). *Pelusium,* east of Zoan. *On* is Heliopolis (see Jer
43.13 n.). *Pi-beseth,* Bubastis. *Tehaphnehes,*
Tahpanhes (see Jer 2.14–19 n.).
    30.20–26: April 29, 587 B.C. Nebuchadrezzar had broken one arm of Hophra the year
before (see 29.1–16 n.); the next time, he will
break both arms.
    31.1–18: Allegory of the cedar (compare

ch 17). Date, June 21, 587 B.C. Ezekiel uses an
ancient Babylonian myth to emphasize that,
as with Tyre (28.1–5), the cause for Egypt's
fall was pride (and political unreliability,
29.6–9). **4:** *The deep* (Hebrew "tehom;" compare Tiamat, the Babylonian dragon of the
watery chaos; see Isa 51.9–11 n.), ancient
mythological opponent of the gods, nourished the tree so that it reached into the heav-

8 The cedars in the garden of God
    could not rival it,
  nor the fir trees equal its
    boughs;
  the plane trees were as nothing
    compared with its branches;
  no tree in the garden of God
    was like it in beauty.
9 I made it beautiful
    with its mass of branches,
  the envy of all the trees of Eden
    that were in the garden of God.
10 Therefore thus says the Lord God:
Because it*g* towered high and set its top
among the clouds,*h* and its heart was
proud of its height, 11 I gave it into the
hand of the prince of the nations; he has
dealt with it as its wickedness deserves.
I have cast it out. 12 Foreigners from the
most terrible of the nations have cut it
down and left it. On the mountains and
in all the valleys its branches have fallen,
and its boughs lie broken in all the watercourses of the land; and all the peoples of
the earth went away from its shade and
left it.
13 On its fallen trunk settle
    all the birds of the air,
  and among its boughs lodge
    all the wild animals.
14 All this is in order that no trees by the
waters may grow to lofty height or set
their tops among the clouds,*h* and that
no trees that drink water may reach up to
them in height.
  For all of them are handed over
    to death,
  to the world below;
  along with all mortals,
    with those who go down to
      the Pit.

15 Thus says the Lord God: On the
day it went down to Sheol I closed the
deep over it and covered it; I restrained
its rivers, and its mighty waters were
checked. I clothed Lebanon in gloom for
it, and all the trees of the field fainted
because of it. 16 I made the nations quake
at the sound of its fall, when I cast it
down to Sheol with those who go down
to the Pit; and all the trees of Eden, the
choice and best of Lebanon, all that were
well watered, were consoled in the world
below. 17 They also went down to Sheol
with it, to those killed by the sword,
along with its allies,*i* those who lived in
its shade among the nations.
18 Which among the trees of Eden
was like you in glory and in greatness?
Now you shall be brought down with
the trees of Eden to the world below; you
shall lie among the uncircumcised, with
those who are killed by the sword. This
is Pharaoh and all his horde, says the
Lord God.

32 In the twelfth year, in the twelfth
month, on the first day of the
month, the word of the Lord came to
me: 2 Mortal, raise a lamentation over
Pharaoh king of Egypt, and say to him:
  You consider yourself a lion
    among the nations,
  but you are like a dragon in
    the seas;
  you thrash about in your streams,
  trouble the water with your
    feet,
  and foul your*j* streams.
3 Thus says the Lord God:

g Syr Vg: Heb *you*   *h* Gk: Heb *thick boughs*
i Heb *its arms*   *j* Heb *their*

---

ens and sheltered all life. **7–9:** The splendor
of the great cedar, Egypt, was incomparable
(28.11–19), surpassing those *in the garden of
God* (thus rivaling God, Gen 11.1–9).
  31.10–18: But God will cut it down; the life
it sheltered will be dispersed, and it will go
down to Sheol (28.8–10) where it will lie not
with the honored dead, *the trees of Eden,* but
with those who died untimely, violent, or
dishonorable deaths. Later the great world
tree, or tree of life, became a motif in Near
Eastern (e.g. Persian) apocalypticism.

32.1–16: **Lament over Pharaoh.** The
date is March 3, 585 B.C. Though Pharaoh
thought himself *a lion* (symbol of royal power, see 19.1–9 n.; compare the lion-bodied
sphinx), he is only a sea monster whom God
will capture with a net (12.13), as Marduk
captured Tiamat (see 31.4 n.), and will expose on land for carrion (29.1–16). Verses
7–8 recall the day of the Lord (see 30.1–5 n.;
Isa 13.10; Joel 2.2) in which God overcomes
all that oppose him. The recurring *sword* of
God (vv. 10–12; 21.1–32; 30.25; compare

In an assembly of many peoples
I will throw my net over you;
and I[k] will haul you up in
my dragnet.

4 I will throw you on the ground,
on the open field I will fling
you,
and will cause all the birds of the
air to settle on you,
and I will let the wild animals
of the whole earth gorge
themselves with you.

5 I will strew your flesh on the
mountains,
and fill the valleys with your
carcass.[l]

6 I will drench the land with your
flowing blood
up to the mountains,
and the watercourses will be
filled with you.

7 When I blot you out, I will cover
the heavens,
and make their stars dark;
I will cover the sun with a cloud,
and the moon shall not give
its light.

8 All the shining lights of the
heavens
I will darken above you,
and put darkness on your land,
says the Lord GOD.

9 I will trouble the hearts of many
peoples,
as I carry you captive[m] among
the nations,
into countries you have not
known.

10 I will make many peoples appalled
at you;
their kings shall shudder because
of you.
When I brandish my sword before
them,
they shall tremble every
moment
for their lives, each one of them,
on the day of your downfall.

11 For thus says the Lord GOD:
The sword of the king of Babylon
shall come against you.

12 I will cause your hordes to fall
by the swords of mighty ones,
all of them most terrible among
the nations.
They shall bring to ruin the pride
of Egypt,
and all its hordes shall perish.

13 I will destroy all its livestock
from beside abundant waters;
and no human foot shall trouble
them any more,
nor shall the hoofs of cattle
trouble them.

14 Then I will make their waters
clear,
and cause their streams to run
like oil, says the Lord GOD.

15 When I make the land of Egypt
desolate
and when the land is stripped of
all that fills it,
when I strike down all who live in
it,
then they shall know that I am
the LORD.

16 This is a lamentation; it shall
be chanted.
The women of the nations shall
chant it.
Over Egypt and all its hordes they
shall chant it,
says the Lord GOD.

17 In the twelfth year, in the first
month,[n] on the fifteenth day of the
month, the word of the LORD came to
me:

18 Mortal, wail over the hordes
of Egypt,
and send them down,

k Gk Vg: Heb *they*   l Symmachus Syr Vg:
Heb *your height*   m Gk: Heb *bring your
destruction*   n Gk: Heb lacks *in the first month*

Lev 26.33; Isa 34.5–6; Jer 12.12) recalls the
sword-wielding Baal depicted in statues
found by archaeologists. Egypt will become
a lifeless wasteland (Jer 4.23–28), mourned by
professional mourning women (Jer 9.17–18).
32.17–32: Egypt in the underworld

with Egypt° and the daughters of
majestic nations,
to the world below,
with those who go down to
the Pit.
19 "Whom do you surpass in beauty?
Go down! Be laid to rest with
the uncircumcised!"
20 They shall fall among those who are
killed by the sword. Egypt^p has been
handed over to the sword; carry away
both it and its hordes. 21 The mighty
chiefs shall speak of them, with their
helpers, out of the midst of Sheol: "They
have come down, they lie still, the uncircumcised, killed by the sword."

22 Assyria is there, and all its company, their graves all around it, all of them
killed, fallen by the sword. 23 Their
graves are set in the uttermost parts of
the Pit. Its company is all around its
grave, all of them killed, fallen by the
sword, who spread terror in the land of
the living.

24 Elam is there, and all its hordes
around its grave; all of them killed, fallen
by the sword, who went down uncircumcised into the world below, who
spread terror in the land of the living.
They bear their shame with those who
go down to the Pit. 25 They have made
Elam° a bed among the slain with all its
hordes, their graves all around it, all of
them uncircumcised, killed by the
sword; for terror of them was spread in
the land of the living, and they bear their
shame with those who go down to the
Pit; they are placed among the slain.

26 Meshech and Tubal are there, and
all their multitude, their graves all
around them, all of them uncircumcised,
killed by the sword; for they spread terror in the land of the living. 27 And they
do not lie with the fallen warriors of long
ago^q who went down to Sheol with their
weapons of war, whose swords were laid
under their heads, and whose shields^r
are upon their bones; for the terror of the
warriors was in the land of the living.
28 So you shall be broken and lie among
the uncircumcised, with those who are
killed by the sword.

29 Edom is there, its kings and all its
princes, who for all their might are laid
with those who are killed by the sword;
they lie with the uncircumcised, with
those who go down to the Pit.

30 The princes of the north are there,
all of them, and all the Sidonians, who
have gone down in shame with the slain,
for all the terror that they caused by their
might; they lie uncircumcised with those
who are killed by the sword, and bear
their shame with those who go down to
the Pit.

31 When Pharaoh sees them, he will
be consoled for all his hordes—Pharaoh
and all his army, killed by the sword,
says the Lord GOD. 32 For he^s spread terror in the land of the living; therefore he
shall be laid to rest among the uncircumcised, with those who are slain by the
sword—Pharaoh and all his multitude,
says the Lord GOD.

33 The word of the LORD came to
me: 2 O Mortal, speak to your
people and say to them, If I bring the
sword upon a land, and the people of the
land take one of their number as their
sentinel; 3 and if the sentinel sees the
sword coming upon the land and blows
the trumpet and warns the people; 4 then

o Heb *it* p Heb *It* q Gk Old Latin: Heb *of
the uncircumcised* r Cn: Heb *iniquities*
s Cn: Heb *I*

(April 27, 586 B.C.). Like Tyre (28.10), Egypt
will join those in that section of the underworld reserved for the uncircumcised and
those who are executed or who die violent or
untimely deaths. They do not enjoy the status
of the honored war-dead, who were properly
buried (v. 27). In its dishonorable demise
Egypt will join others who were objects of
God's wrath (Isa 14.9–11), such as *Assyria*
(Nah chs 1–3); *Elam* (Jer 49.34–39); *Meshech
and Tubal* (see 27.12–25 n.; Gen 10.2 n.);
Edom (25.12–14); *Sidon* (28.20–23); *princes of
the north* (v. 30), unnamed rulers in Phoenicia
and Syria.

**33.1–39.29: Oracles of restoration.
33.1–20: Responsibility. 1–9:** The prophet
as sentinel (3.16–21; Isa 21.6; Jer 6.17) applies Ezekiel's doctrine of individual respon-

if any who hear the sound of the trumpet do not take warning, and the sword comes and takes them away, their blood shall be upon their own heads. 5 They heard the sound of the trumpet and did not take warning; their blood shall be upon themselves. But if they had taken warning, they would have saved their lives. 6 But if the sentinel sees the sword coming and does not blow the trumpet, so that the people are not warned, and the sword comes and takes any of them, they are taken away in their iniquity, but their blood I will require at the sentinel's hand.

7 So you, mortal, I have made a sentinel for the house of Israel; whenever you hear a word from my mouth, you shall give them warning from me. 8 If I say to the wicked, "O wicked ones, you shall surely die," and you do not speak to warn the wicked to turn from their ways, the wicked shall die in their iniquity, but their blood I will require at your hand. 9 But if you warn the wicked to turn from their ways, and they do not turn from their ways, the wicked shall die in their iniquity, but you will have saved your life.

10 Now you, mortal, say to the house of Israel, Thus you have said: "Our transgressions and our sins weigh upon us, and we waste away because of them; how then can we live?" 11 Say to them, As I live, says the Lord GOD, I have no pleasure in the death of the wicked, but that the wicked turn from their ways and live; turn back, turn back from your evil ways; for why will you die, O house of Israel? 12 And you, mortal, say to your people, The righteousness of the righteous shall not save them when they transgress; and as for the wickedness of the wicked, it shall not make them stumble when they turn from their wicked-

ness; and the righteous shall not be able to live by their righteousness *t* when they sin. 13 Though I say to the righteous that they shall surely live, yet if they trust in their righteousness and commit iniquity, none of their righteous deeds shall be remembered; but in the iniquity that they have committed they shall die. 14 Again, though I say to the wicked, "You shall surely die," yet if they turn from their sin and do what is lawful and right— 15 if the wicked restore the pledge, give back what they have taken by robbery, and walk in the statutes of life, committing no iniquity—they shall surely live, they shall not die. 16 None of the sins that they have committed shall be remembered against them; they have done what is lawful and right, they shall surely live.

17 Yet your people say, "The way of the Lord is not just," when it is their own way that is not just. 18 When the righteous turn from their righteousness, and commit iniquity, they shall die for it. *u* 19 And when the wicked turn from their wickedness, and do what is lawful and right, they shall live by it. *u* 20 Yet you say, "The way of the Lord is not just." O house of Israel, I will judge all of you according to your ways!

21 In the twelfth year of our exile, in the tenth month, on the fifth day of the month, someone who had escaped from Jerusalem came to me and said, "The city has fallen." 22 Now the hand of the LORD had been upon me the evening before the fugitive came; but he had opened my mouth by the time the fugitive came to me in the morning; so my mouth was opened, and I was no longer unable to speak.

23 The word of the LORD came to me: 24 Mortal, the inhabitants of these waste

t Heb *by it*   *u* Heb *them*

sibility. Death (vv. 8–9) may refer to premature death as contrasted with blessed longevity.

33.10–20: On individual responsibility. This oracle re-emphasizes 14.12–23 and 18.5–32.

33.21–33: Miscellanea. 21–22: January

19, 586 B.C. (read "eleventh" for *twelfth*). A few months after the fall of Jerusalem, word reached Ezekiel and his tongue was loosed (3.24–27). Perhaps these two verses, which are unrelated to this context, should follow 24.27.

places in the land of Israel keep saying, "Abraham was only one man, yet he got possession of the land; but we are many; the land is surely given us to possess." 25 Therefore say to them, Thus says the Lord GOD: You eat flesh with the blood, and lift up your eyes to your idols, and shed blood; shall you then possess the land? 26 You depend on your swords, you commit abominations, and each of you defiles his neighbor's wife; shall you then possess the land? 27 Say this to them, Thus says the Lord GOD: As I live, surely those who are in the waste places shall fall by the sword; and those who are in the open field I will give to the wild animals to be devoured; and those who are in strongholds and in caves shall die by pestilence. 28 I will make the land a desolation and a waste, and its proud might shall come to an end; and the mountains of Israel shall be so desolate that no one will pass through. 29 Then they shall know that I am the LORD, when I have made the land a desolation and a waste because of all their abominations that they have committed.

30 As for you, mortal, your people who talk together about you by the walls, and at the doors of the houses, say to one another, each to a neighbor, "Come and hear what the word is that comes from the LORD." 31 They come to you as people come, and they sit before you as my people, and they hear your words, but they will not obey them. For flattery is on their lips, but their heart is set on their gain. 32 To them you are like a singer of love songs, *v* one who has a beautiful voice and plays well on an instrument; they hear what you say, but they will not do it. 33 When this comes— and come it will!—then they shall know that a prophet has been among them.

34 The word of the LORD came to me: 2 Mortal, prophesy against the shepherds of Israel: prophesy, and say to them—to the shepherds: Thus says the Lord GOD: Ah, you shepherds of Israel who have been feeding yourselves! Should not shepherds feed the sheep? 3 You eat the fat, you clothe yourselves with the wool, you slaughter the fatlings; but you do not feed the sheep. 4 You have not strengthened the weak, you have not healed the sick, you have not bound up the injured, you have not brought back the strayed, you have not sought the lost, but with force and harshness you have ruled them. 5 So they were scattered, because there was no shepherd; and scattered, they became food for all the wild animals. 6 My sheep were scattered, they wandered over all the mountains and on every high hill; my sheep were scattered over all the face of the earth, with no one to search or seek for them.

7 Therefore, you shepherds, hear the word of the LORD: 8 As I live, says the Lord GOD, because my sheep have become a prey, and my sheep have become food for all the wild animals, since there was no shepherd; and because my shepherds have not searched for my sheep, but the shepherds have fed themselves, and have not fed my sheep; 9 therefore, you shepherds, hear the word of the LORD: 10 Thus says the Lord GOD, I am against the shepherds; and I will demand my sheep at their hand, and put a stop to their feeding the sheep; no longer shall the shepherds feed themselves. I will rescue my sheep from their mouths, so that they may not be food for them.

11 For thus says the Lord GOD: I myself will search for my sheep, and will

v Cn: Heb *like a love song*

33.23–29: Expansion on the problem of possession of Palestinian property formerly belonging to deportees. Note the three scourges in v. 27 (14.21; 21.3–5).
33.30–33: Consistent with human nature, the people heard what they wanted to hear. But the day approaches when they will be reminded that with the words of God's love

were also the words of God's justice (Jer 5.12–13). *Song,* oracles may have been chanted (2 Kings 3.15).
34.1–31: **The shepherds of Israel. 1–10:** The shepherds (i.e. kings) had misused their people (Jer 23.13–17) and scattered them (Jer 10.21; 23.1–4). This oracle applies the doctrine of individual responsibility (18.5–32) to

seek them out. ¹²As shepherds seek out their flocks when they are among their scattered sheep, so I will seek out my sheep. I will rescue them from all the places to which they have been scattered on a day of clouds and thick darkness. ¹³I will bring them out from the peoples and gather them from the countries, and will bring them into their own land; and I will feed them on the mountains of Israel, by the watercourses, and in all the inhabited parts of the land. ¹⁴I will feed them with good pasture, and the mountain heights of Israel shall be their pasture; there they shall lie down in good grazing land, and they shall feed on rich pasture on the mountains of Israel. ¹⁵I myself will be the shepherd of my sheep, and I will make them lie down, says the Lord GOD. ¹⁶I will seek the lost, and I will bring back the strayed, and I will bind up the injured, and I will strengthen the weak, but the fat and the strong I will destroy. I will feed them with justice.

17 As for you, my flock, thus says the Lord GOD: I shall judge between sheep and sheep, between rams and goats: ¹⁸Is it not enough for you to feed on the good pasture, but you must tread down with your feet the rest of your pasture? When you drink of clear water, must you foul the rest with your feet? ¹⁹And must my sheep eat what you have trodden with your feet, and drink what you have fouled with your feet?

20 Therefore, thus says the Lord GOD to them: I myself will judge between the fat sheep and the lean sheep. ²¹Because you pushed with flank and shoulder, and butted at all the weak animals with your horns until you scattered them far and wide, ²²I will save my flock, and they shall no longer be ravaged; and I will judge between sheep and sheep.

23 I will set up over them one shepherd, my servant David, and he shall feed them: he shall feed them and be their shepherd. ²⁴And I, the LORD, will be their God, and my servant David shall be prince among them; I, the LORD, have spoken.

25 I will make with them a covenant of peace and banish wild animals from the land, so that they may live in the wild and sleep in the woods securely. ²⁶I will make them and the region around my hill a blessing; and I will send down the showers in their season; they shall be showers of blessing. ²⁷The trees of the field shall yield their fruit, and the earth shall yield its increase. They shall be secure on their soil; and they shall know that I am the LORD, when I break the bars of their yoke, and save them from the hands of those who enslaved them. ²⁸They shall no more be plunder for the nations, nor shall the animals of the land devour them; they shall live in safety, and no one shall make them afraid. ²⁹I will provide for them a splendid vegetation so that they shall no more be consumed with hunger in the land, and no longer suffer the insults of the nations. ³⁰They shall know that I, the LORD their God, am with them, and that they, the house of Israel, are my people, says the

---

the rulers, who are also subject to God's law (2 Sam 12.1–15). *Wild animals,* Judah's attackers, especially Babylonia.

34.11–16: God is the Good Shepherd (Isa 40.11; Jer 31.10) who will gather the dispersed and injured flock. This passage suggests a return to theocracy (Hos 8.4; 1 Sam 8.7).

34.17–22: Sheep, good and bad, are found in the flock; the bad must be separated out and punished (Mt 25.31–32). The figure may have a double meaning and refer also to the nations which oppress Israel.

34.23–24: God will place his *servant David*

(2 Sam 3.18), i.e. a restored monarchy, over his people (37.22–25; Jer 23.5–6). *One shepherd,* see Hos 1.11; Jn 10.16.

34.25–31: Using the oldest term for covenant making (see Jer 31.31–34), the prophet affirms that God will make a *covenant of peace* (37.26; Heb 13.20). God, again resident on Mount Zion (*my hill*), will preserve the proper sequence of seasons (Gen 8.21–22), assuring his people of continuous prosperity (Am 9.13–14), free from fear of destruction within (*wild animals,* Lev 26.6) and without (*plunder for the nations*).

Lord God. [31] You are my sheep, the sheep of my pasture[w] and I am your God, says the Lord God.

35 The word of the Lord came to me: [2] Mortal, set your face against Mount Seir, and prophesy against it, [3] and say to it, Thus says the Lord God:

I am against you, Mount Seir;
  I stretch out my hand
    against you
  to make you a desolation and
    a waste.
[4] I lay your towns in ruins;
  you shall become a desolation,
  and you shall know that I am
    the Lord.

[5] Because you cherished an ancient enmity, and gave over the people of Israel to the power of the sword at the time of their calamity, at the time of their final punishment; [6] therefore, as I live, says the Lord God, I will prepare you for blood, and blood shall pursue you; since you did not hate bloodshed, bloodshed shall pursue you. [7] I will make Mount Seir a waste and a desolation; and I will cut off from it all who come and go. [8] I will fill its mountains with the slain; on your hills and in your valleys and in all your watercourses those killed with the sword shall fall. [9] I will make you a perpetual desolation, and your cities shall never be inhabited. Then you shall know that I am the Lord.

[10] Because you said, "These two nations and these two countries shall be mine, and we will take possession of them,"—although the Lord was there— [11] therefore, as I live, says the Lord God, I will deal with you according to the anger and envy that you showed because of your hatred against them; and I will make myself known among you,[x] when I judge you. [12] You shall know that I, the Lord, have heard all the abusive speech that you uttered against the mountains of Israel, saying, "They are laid desolate, they are given us to devour." [13] And you magnified yourselves against me with your mouth, and multiplied your words against me; I heard it. [14] Thus says the Lord God: As the whole earth rejoices, I will make you desolate. [15] As you rejoiced over the inheritance of the house of Israel, because it was desolate, so I will deal with you; you shall be desolate, Mount Seir, and all Edom, all of it. Then they shall know that I am the Lord.

36 And you, mortal, prophesy to the mountains of Israel, and say: O mountains of Israel, hear the word of the Lord. [2] Thus says the Lord God: Because the enemy said of you, "Aha!" and, "The ancient heights have become our possession," [3] therefore prophesy, and say: Thus says the Lord God: Because they made you desolate indeed, and crushed you from all sides, so that you became the possession of the rest of the nations, and you became an object of gossip and slander among the people; [4] therefore, O mountains of Israel, hear the word of the Lord God: Thus says the Lord God to the mountains and the hills, the watercourses and the valleys, the desolate wastes and the deserted towns, which have become a source of plunder and an object of derision to the rest of the

w Gk OL: Heb *pasture, you are people*
x Gk: Heb *them*

35.1–15: The oracle against Edom is an editorial expansion of an oracle of Ezekiel, using Ezekiel's phraseology (compare vv. 1–3 with 6.1–3). Properly belonging with 25.12–14, it was placed here to be contrasted with ch 36 (especially 36.5), forming the prelude to Israel's restoration. It reflects the growing intensity of Jewish hatred for Edom subsequent to Edom's occupation of southern Judah (Jer 49.7–22). In contrast with Egypt (29.13–16), Edom will remain forever desolate. *Mount Seir* is the plateau rising east of the Arabah in which Sela, the Edomite capital, was located.

36.1–39.29: The new Israel.

36.1–38: Prophecy to the mountains of Israel; the restoration of Israel. 1–7: Mountains of Israel, the highlands, represent all Israel (Deut 3.25). This oracle reverses ch 6; though dispossessed by *Edom* (35.1–15) and surrounding nations (Neh 2.19), Israel will be restored to her heritage (Mal 1.2–5).

nations all around; [5]therefore thus says the Lord GOD: I am speaking in my hot jealousy against the rest of the nations, and against all Edom, who, with whole-hearted joy and utter contempt, took my land as their possession, because of its pasture, to plunder it. [6]Therefore prophesy concerning the land of Israel, and say to the mountains and hills, to the watercourses and valleys, Thus says the Lord GOD: I am speaking in my jealous wrath, because you have suffered the insults of the nations; [7]therefore thus says the Lord GOD: I swear that the nations that are all around you shall themselves suffer insults.

8 But you, O mountains of Israel, shall shoot out your branches, and yield your fruit to my people Israel; for they shall soon come home. [9]See now, I am for you; I will turn to you, and you shall be tilled and sown; [10]and I will multiply your population, the whole house of Israel, all of it; the towns shall be inhabited and the waste places rebuilt; [11]and I will multiply human beings and animals upon you. They shall increase and be fruitful; and I will cause you to be inhabited as in your former times, and will do more good to you than ever before. Then you shall know that I am the LORD. [12]I will lead people upon you—my people Israel—and they shall possess you, and you shall be their inheritance. No longer shall you bereave them of children.

13 Thus says the Lord GOD: Because they say to you, "You devour people, and you bereave your nation of children," [14]therefore you shall no longer devour people and no longer bereave your nation of children, says the Lord GOD; [15]and no longer will I let you hear the insults of the nations, no longer shall you bear the disgrace of the peoples; and no longer shall you cause your nation to stumble, says the Lord GOD.

16 The word of the LORD came to me: [17]Mortal, when the house of Israel lived on their own soil, they defiled it with their ways and their deeds; their conduct in my sight was like the uncleanness of a woman in her menstrual period. [18]So I poured out my wrath upon them for the blood that they had shed upon the land, and for the idols with which they had defiled it. [19]I scattered them among the nations, and they were dispersed through the countries; in accordance with their conduct and their deeds I judged them. [20]But when they came to the nations, wherever they came, they profaned my holy name, in that it was said of them, "These are the people of the LORD, and yet they had to go out of his land." [21]But I had concern for my holy name, which the house of Israel had profaned among the nations to which they came.

22 Therefore say to the house of Israel, Thus says the Lord GOD: It is not for your sake, O house of Israel, that I am about to act, but for the sake of my holy name, which you have profaned among the nations to which you came. [23]I will sanctify my great name, which has been profaned among the nations, and which you have profaned among them; and the nations shall know that I am the LORD, says the Lord GOD, when through you I display my holiness before their eyes. [24]I will take you from the nations, and gather you from all the countries, and bring you into your own land. [25]I will sprinkle clean water upon you, and you shall be clean from all your uncleannesses, and

---

36.8–15: The land will surpass its pristine productivity (*in your former times,* in the years just after the Exodus; Hos 11.1–4; Jer 2.1–3). 14: *No longer devour people,* the hilltops will not serve again as fertility cult sanctuaries (6.1–14; Deut 12.1–3, 29–31), perhaps involving human sacrifices. 36.16–21: The land was defiled by Israel's offering of pagan sacrifices, thereby profan-ing God's name (Lev 18.21; 20.3). 17: *Uncleanness,* see Lev 15.19–30. 36.22–32: As in ch 20, God reminds Israel that he will restore them, his people (Ex 6.7; Lev 20.24), for the sake of his name; like Ezekiel (12.6; 24.27) they will be a sign for all nations. 25–26: God will *cleanse* them (compare Ex 30.17–21), inasmuch as human beings cannot make themselves clean. Again,

from all your idols I will cleanse you. 26 A new heart I will give you, and a new spirit I will put within you; and I will remove from your body the heart of stone and give you a heart of flesh. 27 I will put my spirit within you, and make you follow my statutes and be careful to observe my ordinances. 28 Then you shall live in the land that I gave to your ancestors; and you shall be my people, and I will be your God. 29 I will save you from all your uncleannesses, and I will summon the grain and make it abundant and lay no famine upon you. 30 I will make the fruit of the tree and the produce of the field abundant, so that you may never again suffer the disgrace of famine among the nations. 31 Then you shall remember your evil ways, and your dealings that were not good; and you shall loathe yourselves for your iniquities and your abominable deeds. 32 It is not for your sake that I will act, says the Lord GOD; let that be known to you. Be ashamed and dismayed for your ways, O house of Israel.

33 Thus says the Lord GOD: On the day that I cleanse you from all your iniquities, I will cause the towns to be inhabited, and the waste places shall be rebuilt. 34 The land that was desolate shall be tilled, instead of being the desolation that it was in the sight of all who passed by. 35 And they will say, "This land that was desolate has become like the garden of Eden; and the waste and desolate and ruined towns are now inhabited and fortified." 36 Then the nations that are left all around you shall know that I, the LORD, have rebuilt the ruined places, and replanted that which was desolate; I, the LORD, have spoken, and I will do it.

37 Thus says the Lord GOD: I will also let the house of Israel ask me to do this for them: to increase their population like a flock. 38 Like the flock for sacrifices, *y*

like the flock at Jerusalem during her appointed festivals, so shall the ruined towns be filled with flocks of people. Then they shall know that I am the LORD.

37 The hand of the LORD came upon me, and he brought me out by the spirit of the LORD and set me down in the middle of a valley; it was full of bones. 2 He led me all around them; there were very many lying in the valley, and they were very dry. 3 He said to me, "Mortal, can these bones live?" I answered, "O Lord GOD, you know." 4 Then he said to me, "Prophesy to these bones, and say to them: O dry bones, hear the word of the LORD. 5 Thus says the Lord GOD to these bones: I will cause breath *z* to enter you, and you shall live. 6 I will lay sinews on you, and will cause flesh to come upon you, and cover you with skin, and put breath *z* in you, and you shall live; and you shall know that I am the LORD."

7 So I prophesied as I had been commanded; and as I prophesied, suddenly there was a noise, a rattling, and the bones came together, bone to its bone. 8 I looked, and there were sinews on them, and flesh had come upon them, and skin had covered them; but there was no breath in them. 9 Then he said to me, "Prophesy to the breath, prophesy, mortal, and say to the breath: *a* Thus says the Lord GOD: Come from the four winds, O breath, *a* and breathe upon these slain, that they may live." 10 I prophesied as he commanded me, and the breath came into them, and they lived, and stood on their feet, a vast multitude.

11 Then he said to me, "Mortal, these bones are the whole house of Israel. They say, 'Our bones are dried up, and our

y Heb *flock of holy things*   *z* Or *spirit*
a Or *wind* or *spirit*

---

the *new heart* will be created by God's spirit (ch 37; Isa 44.3; Rom 8.3–6).
36.33–36: The skeptical nations (v. 20) will recognize Israel's restoration (Jer 31.23–28) as God's act. **37–38:** *Like a flock,* see ch 34.
37.1–14: Vision of the valley of dry

bones. 1: *Valley,* "plain" in 3.22; 8.4. The *bones* are the exiles, who have no more hope of resuscitating the kingdom of Israel than of putting flesh on a skeleton and calling it to life. **9:** Hebrew "rûah" means "spirit, breath, wind"; thus there is a constant word-play

hope is lost; we are cut off completely.' ¹²Therefore prophesy, and say to them, Thus says the Lord GOD: I am going to open your graves, and bring you up from your graves, O my people; and I will bring you back to the land of Israel. ¹³And you shall know that I am the LORD, when I open your graves, and bring you up from your graves, O my people. ¹⁴I will put my spirit within you, and you shall live, and I will place you on your own soil; then you shall know that I, the LORD, have spoken and will act," says the LORD.

15 The word of the LORD came to me: ¹⁶Mortal, take a stick and write on it, "For Judah, and the Israelites associated with it"; then take another stick and write on it, "For Joseph (the stick of Ephraim) and all the house of Israel associated with it"; ¹⁷and join them together into one stick, so that they may become one in your hand. ¹⁸And when your people say to you, "Will you not show us what you mean by these?" ¹⁹say to them, Thus says the Lord GOD: I am about to take the stick of Joseph (which is in the hand of Ephraim) and the tribes of Israel associated with it; and I will put the stick of Judah upon it, *ᵇ* and make them one stick, in order that they may be one in my hand. ²⁰When the sticks on which you write are in your hand before their eyes, ²¹then say to them, Thus says the Lord GOD: I will take the people of Israel from the nations among which they have gone, and will gather them from every quarter, and bring them to their own land. ²²I will make them one nation in the land, on the mountains of Israel; and

one king shall be king over them all. Never again shall they be two nations, and never again shall they be divided into two kingdoms. ²³They shall never again defile themselves with their idols and their detestable things, or with any of their transgressions. I will save them from all the apostasies into which they have fallen, *ᶜ* and will cleanse them. Then they shall be my people, and I will be their GOD.

24 My servant David shall be king over them; and they shall all have one shepherd. They shall follow my ordinances and be careful to observe my statutes. ²⁵They shall live in the land that I gave to my servant Jacob, in which your ancestors lived; they and their children and their children's children shall live there forever; and my servant David shall be their prince forever. ²⁶I will make a covenant of peace with them; it shall be an everlasting covenant with them; and I will bless *ᵈ* them and multiply them, and will set my sanctuary among them forevermore. ²⁷My dwelling place shall be with them; and I will be their God, and they shall be my people. ²⁸Then the nations shall know that I the LORD sanctify Israel, when my sanctuary is among them forevermore.

38 The word of the LORD came to me: ²Mortal, set your face toward Gog, of the land of Magog, the chief prince of Meshech and Tubal. Prophesy against him ³and say: Thus says the Lord GOD: I am against you,

ᵇ Heb *I will put them upon it*   *ᶜ* Another reading is *from all the settlements in which they have sinned*   *ᵈ* Tg: Heb *give*

---

here. *Four winds* may refer to God's omnipresence (1.17). **14:** This vision, which refers to the re-establishment of Israel, is indirectly an anticipation of the doctrine of resurrection. **37.15–28: Oracle of the two sticks** (Zech 11.7–14) envisions the re-unification of the long-divided land and the establishment of a united Israel, ruled by one king; then the situation earlier predicted will prevail (34.28): law-abiding living (11.20) in the promised land (28.25) under the Davidic king (34.23–

24); the covenant of peace (34.25); and the re-established central sanctuary (45.1–8). **25:** *Prince,* the king (compare 12.10; 34.24).

38.1–39.29: The Gog and Magog oracles describe in apocalyptic language the coming of the foe from the north (38.15; Jer 6.22) against God's people living peacefully in the promised land (38.8, 11–12). After a cataclysmic battle the aggressor forces will be completely defeated, and God will be acknowledged by all nations as the undisputed

O Gog, chief prince of Meshech and Tubal; [4]I will turn you around and put hooks into your jaws, and I will lead you out with all your army, horses and horsemen, all of them clothed in full armor, a great company, all of them with shield and buckler, wielding swords. [5]Persia, Ethiopia,[e] and Put are with them, all of them with buckler and helmet; [6]Gomer and all its troops; Beth-togarmah from the remotest parts of the north with all its troops—many peoples are with you.

7 Be ready and keep ready, you and all the companies that are assembled around you, and hold yourselves in reserve for them. [8]After many days you shall be mustered; in the latter years you shall go against a land restored from war, a land where people were gathered from many nations on the mountains of Israel, which had long lain waste; its people were brought out from the nations and now are living in safety, all of them. [9]You shall advance, coming on like a storm; you shall be like a cloud covering the land, you and all your troops, and many peoples with you.

10 Thus says the Lord God: On that day thoughts will come into your mind, and you will devise an evil scheme. [11]You will say, "I will go up against the land of unwalled villages; I will fall upon the quiet people who live in safety, all of them living without walls, and having no bars or gates"; [12]to seize spoil and carry off plunder; to assail the waste places that are now inhabited, and the

people who were gathered from the nations, who are acquiring cattle and goods, who live at the center[f] of the earth. [13]Sheba and Dedan and the merchants of Tarshish and all its young warriors[g] will say to you, "Have you come to seize spoil? Have you assembled your horde to carry off plunder, to carry away silver and gold, to take away cattle and goods, to seize a great amount of booty?"

14 Therefore, mortal, prophesy, and say to Gog: Thus says the Lord God: On that day when my people Israel are living securely, you will rouse yourself[h] [15]and come from your place out of the remotest parts of the north, you and many peoples with you, all of them riding on horses, a great horde, a mighty army; [16]you will come up against my people Israel, like a cloud covering the earth. In the latter days I will bring you against my land, so that the nations may know me, when through you, O Gog, I display my holiness before their eyes.

17 Thus says the Lord God: Are you he of whom I spoke in former days by my servants the prophets of Israel, who in those days prophesied for years that I would bring you against them? [18]On that day, when Gog comes against the land of Israel, says the Lord God, my wrath shall be aroused. [19]For in my jealousy and in my blazing wrath I declare:

e Or Nubia; Heb *Cush*   *f* Heb *navel*
g Heb *young lions*   *h* Gk: Heb *will you not know?*

victor (38.23; 39.21–29; 36.22–23). Since the foe from the north in Jeremiah (Jer 25.9) and Ezekiel (Ezek 26.7) was Babylon, it is probable that the foe here described is a grandiose surrogate for Babylon. The conflict is that preceding the fall of Babylon, and the victory includes Israel's restoration to its land (34.11–16; 36.8–38; Jer 31.23–28). The limited scope and goal of these oracles has been expanded to cosmic proportions by the writer of Rev 20.7–10.
38.1–9: *Gog,* king of *Magog,* both unidentified, though general location is in the north. *Meshech, Tubal, Put,* see 27.10–25a n. *Cush,* Ethiopia. *Gomer,* Assyrian "Gimirrai," Cim-

merians in central Asia Minor (Gen 10.2–3). *Beth-togarmah,* see 27.10–25a n. Though people and places in apocalyptic literature can often be identified, they are part of the literary equipment and should rarely be taken literally. *In the latter years* corresponds to "in the latter days" (Hos 3.5; Jer 30.24) before the re-establishment of David's line (34.23–24; Jer 23.5–6).
38.10–23: The plot against God's defenseless people is doomed to defeat. God will marshal all the forces of nature against Gog (Ps 18.7–15; Isa 24.18–20; 30.27–33; Joel 2.28–32). **12:** *Center,* see 5.5 n. **13:** *Sheba,* 27.22. *Dedan,* 25.13. *Tarshish,* see Jer 10.9 n.

On that day there shall be a great shaking in the land of Israel; 20 the fish of the sea, and the birds of the air, and the animals of the field, and all creeping things that creep on the ground, and all human beings that are on the face of the earth, shall quake at my presence, and the mountains shall be thrown down, and the cliffs shall fall, and every wall shall tumble to the ground. 21 I will summon the sword against Gog[i] in[j] all my mountains, says the Lord GOD; the swords of all will be against their comrades. 22 With pestilence and bloodshed I will enter into judgment with him; and I will pour down torrential rains and hailstones, fire and sulfur, upon him and his troops and the many peoples that are with him. 23 So I will display my greatness and my holiness and make myself known in the eyes of many nations. Then they shall know that I am the LORD.

39 And you, mortal, prophesy against Gog, and say: Thus says the Lord GOD: I am against you, O Gog, chief prince of Meshech and Tubal! 2 I will turn you around and drive you forward, and bring you up from the remotest parts of the north, and lead you against the mountains of Israel. 3 I will strike your bow from your left hand, and will make your arrows drop out of your right hand. 4 You shall fall upon the mountains of Israel, you and all your troops and the peoples that are with you; I will give you to birds of prey of every kind and to the wild animals to be devoured. 5 You shall fall in the open field; for I have spoken, says the Lord GOD. 6 I will send fire on Magog and on those who live securely in the coastlands; and they shall know that I am the LORD.

7 My holy name I will make known among my people Israel; and I will not let my holy name be profaned any more; and the nations shall know that I am the LORD, the Holy One in Israel. 8 It has come! It has happened, says the Lord GOD. This is the day of which I have spoken.

9 Then those who live in the towns of Israel will go out and make fires of the weapons and burn them—bucklers and shields, bows and arrows, handpikes and spears—and they will make fires of them for seven years. 10 They will not need to take wood out of the field or cut down any trees in the forests, for they will make their fires of the weapons; they will despoil those who despoiled them, and plunder those who plundered them, says the Lord GOD.

11 On that day I will give to Gog a place for burial in Israel, the Valley of the Travelers[k] east of the sea; it shall block the path of the travelers, for there Gog and all his horde will be buried; it shall be called the Valley of Hamon-gog.[l] 12 Seven months the house of Israel shall spend burying them, in order to cleanse the land. 13 All the people of the land shall bury them; and it will bring them honor on the day that I show my glory, says the Lord GOD. 14 They will set apart men to pass through the land regularly and bury any invaders[m] who remain on the face of the land, so as to cleanse it; for seven months they shall make their search. 15 As the searchers[m] pass through the land, anyone who sees a human bone shall set up a sign by it, until the buriers have buried it in the Valley of Hamon-gog.[l] 16 (A city Hamonah[n] is there also.) Thus they shall cleanse the land.

17 As for you, mortal, thus says the Lord GOD: Speak to the birds of every kind and to all the wild animals: Assemble and come, gather from all around to

---

i Heb *him* j Heb *to or for* k Or *of the Abarim* l That is, *the Horde of Gog* m Heb *travelers* n That is *The Horde*

39.1–20: Gog's defeat. The magnitude of the opposing forces, and thus of God's victory, is emphasized by the huge quantities of war material (enough *wood* for fuel for *seven years*) and by the *seven months* required to re-move the enemy dead to the region east of the Dead Sea. The slaughter of those who dared oppose God and oppress his people is regarded as a sacrifice to God's glory.

the sacrificial feast that I am preparing for you, a great sacrificial feast on the mountains of Israel, and you shall eat flesh and drink blood. 18 You shall eat the flesh of the mighty, and drink the blood of the princes of the earth—of rams, of lambs, and of goats, of bulls, all of them fatlings of Bashan. 19 You shall eat fat until you are filled, and drink blood until you are drunk, at the sacrificial feast that I am preparing for you. 20 And you shall be filled at my table with horses and charioteers,⁰ with warriors and all kinds of soldiers, says the Lord GOD.

21 I will display my glory among the nations; and all the nations shall see my judgment that I have executed, and my hand that I have laid on them. 22 The house of Israel shall know that I am the LORD their God, from that day forward. 23 And the nations shall know that the house of Israel went into captivity for their iniquity, because they dealt treacherously with me. So I hid my face from them and gave them into the hand of their adversaries, and they all fell by the sword. 24 I dealt with them according to their uncleanness and their transgressions, and hid my face from them.

25 Therefore thus says the Lord GOD: Now I will restore the fortunes of Jacob, and have mercy on the whole house of Israel; and I will be jealous for my holy name. 26 They shall forgetᵖ their shame, and all the treachery they have practiced against me, when they live securely in their land with no one to make them afraid, 27 when I have brought them back from the peoples and gathered them from their enemies' lands, and through them have displayed my holiness in the sight of many nations. 28 Then they shall know that I am the LORD their God because I sent them into exile among the nations, and then gathered them into their own land. I will leave none of them behind; 29 and I will never again hide my face from them, when I pour out my spirit upon the house of Israel, says the Lord GOD.

40 In the twenty-fifth year of our exile, at the beginning of the year, on the tenth day of the month, in the fourteenth year after the city was struck down, on that very day, the hand of the LORD was upon me, and he brought me there. 2 He brought me, in visions of God, to the land of Israel, and set me down upon a very high mountain, on which was a structure like a city to the south. 3 When he brought me there, a man was there, whose appearance shone like bronze, with a linen cord and a measuring reed in his hand; and he was standing in the gateway. 4 The man said to me, "Mortal, look closely and listen attentively, and set your mind upon all that I shall show you, for you were brought here in order that I might show it to you; declare all that you see to the house of Israel."

5 Now there was a wall all around the outside of the temple area. The length of the measuring reed in the man's hand was six long cubits, each being a cubit and a handbreadth in length; so he mea-

o Heb *chariots*   p Another reading is *They
shall bear*

39.21–29: Conclusion of the oracles—the victory and restoration told in familiar terms (5.8; 28.26; 34.30).

40.1–48.35: **The vision of the restored temple and land.** This section is really a continuation of the theme of chs 33–39, but because of its unified description of Israel's religious and political restoration it may be separated from the less homogeneous preceding section. The temple, though somewhat idealized, is the Solomonic temple, in which Ezekiel probably served as priest before his exile (1.3) and which was destroyed in 587 B.C.

40.1–47: **The temple area, gates, outer and inner courts. 1–5:** April 28, 573 B.C. On the twenty-fifth anniversary of his exile, Ezekiel is transported in a vision (compare 8.2–3) to the temple mountain (*very high mountain,* 17.22; Mic 4.1). **3:** The *reed* was about ten feet, four inches long.

40.5: The *long cubit* was about 20.68 inches; the ordinary cubit was about 17.5 inches. The equal height and width of the outer retaining

sured the thickness of the wall, one reed; and the height, one reed. 6 Then he went into the gateway facing east, going up its steps, and measured the threshold of the gate, one reed deep. *q* There were 7 recesses, and each recess was one reed wide and one reed deep; and the space between the recesses, five cubits; and the threshold of the gate by the vestibule of the gate at the inner end was one reed deep. 8 Then he measured the inner vestibule of the gateway, one cubit. 9 Then he measured the vestibule of the gateway, eight cubits; and its pilasters, two cubits; and the vestibule of the gate was at the inner end. 10 There were three recesses on either side of the east gate; the three were of the same size; and the pilasters on either side were of the same size. 11 Then he measured the width of the opening of the gateway, ten cubits; and the width of the gateway, thirteen cubits. 12 There was a barrier before the recesses, one cubit on either side; and the recesses were six cubits on either side. 13 Then he measured the gate from the back *r* of the one recess to the back *r* of the other, a width of twenty-five cubits, from wall to wall. *s* 14 He measured *t* also the vestibule, twenty cubits; and the gate next to the pilaster on every side of the court. *u* 15 From the front of the gate at the entrance to the end of the inner vestibule of the gate was fifty cubits. 16 The recesses and their pilasters had windows, with shutters *u* on the inside of the gateway all around, and the vestibules also had windows on the inside all around; and on the pilasters were palm trees.

17 Then he brought me into the outer court; there were chambers there, and a pavement, all around the court; thirty chambers fronted on the pavement. 18 The pavement ran along the side of the gates, corresponding to the length of the gates; this was the lower pavement. 19 Then he measured the distance from the inner front of *v* the lower gate to the outer front of the inner court, one hundred cubits. *w*

20 Then he measured the gate of the outer court that faced north—its depth and width. 21 Its recesses, three on either side, and its pilasters and its vestibule were of the same size as those of the first gate; its depth was fifty cubits, and its width twenty-five cubits. 22 Its windows, its vestibule, and its palm trees were of the same size as those of the gate that faced toward the east. Seven steps led up to it; and its vestibule was on the inside. *x* 23 Opposite the gate on the north, as on the east, was a gate to the inner court; he measured from gate to gate, one hundred cubits.

24 Then he led me toward the south, and there was a gate on the south; and he measured its pilasters and its vestibule; they had the same dimensions as the others. 25 There were windows all around in it and in its vestibule, like the windows of the others; its depth was fifty cubits, and its width twenty-five cubits. 26 There were seven steps leading up to it; its ves-

q Heb *deep, and one threshold, one reed deep*
r Gk: Heb *roof*   *s* Heb *opening facing opening*
t Heb *made*   *u* Meaning of Heb uncertain
v Compare Gk: Heb *from before*   *w* Heb adds *the east and the north*   *x* Gk: Heb *before them*

wall corresponds to the symmetry of the enclosure itself (42.16–20). **6–16:** The detailed description of the east or processional gate (10.19) makes it possible to compare it with Solomonic gateways found at Megiddo, Gezer, and Hazor, all having the same design and essentially the same dimensions. After ascending seven steps (vv. 22, 26) one came to the temple gate which had a double entrance vestibule, the second narrower than the first, and closed off from the interior by doors. The wall joined the gate at the entrance vestibules; the remainder of the gate extended into the outer court. The interior had three chambers, or guardrooms, on either side separated by heavy piers. Passing the last pier, one entered the outer court, or possibly first into an inner vestibule and then the outer court. The side chambers had back doors opening into the outer court. The windows, as the temple itself, were of Phoenician, or Tyrian, design, narrowing inwards. Palm trees were a common Near Eastern decorative motif. **17–19:** The thirty chambers were probably for the use of people and Levites worshiping in the outer court.

tibule was on the inside.*y* It had palm trees on its pilasters, one on either side. 27 There was a gate on the south of the inner court; and he measured from gate to gate toward the south, one hundred cubits.

28 Then he brought me to the inner court by the south gate, and he measured the south gate; it was of the same dimensions as the others. 29 Its recesses, its pilasters, and its vestibule were of the same size as the others; and there were windows all around in it and in its vestibule; its depth was fifty cubits, and its width twenty-five cubits. 30 There were vestibules all around, twenty-five cubits deep and five cubits wide. 31 Its vestibule faced the outer court, and palm trees were on its pilasters, and its stairway had eight steps.

32 Then he brought me to the inner court on the east side, and he measured the gate; it was of the same size as the others. 33 Its recesses, its pilasters, and its vestibule were of the same dimensions as the others; and there were windows all around in it and in its vestibule; its depth was fifty cubits, and its width twenty-five cubits. 34 Its vestibule faced the outer court, and it had palm trees on its pilasters, on either side; and its stairway had eight steps.

35 Then he brought me to the north gate, and he measured it; it had the same dimensions as the others. 36 Its recesses, its pilasters, and its vestibule were of the same size as the others;*z* and it had windows all around. Its depth was fifty cubits, and its width twenty-five cubits. 37 Its vestibule*a* faced the outer court, and it had palm trees on its pilasters, on either side; and its stairway had eight steps.

38 There was a chamber with its door in the vestibule of the gate,*b* where the burnt offering was to be washed. 39 And in the vestibule of the gate were two tables on either side, on which the burnt offering and the sin offering and the guilt offering were to be slaughtered. 40 On the outside of the vestibule*c* at the entrance of the north gate were two tables; and on the other side of the vestibule of the gate were two tables. 41 Four tables were on the inside, and four tables on the outside of the side of the gate, eight tables, on which the sacrifices were to be slaughtered. 42 There were also four tables of hewn stone for the burnt offering, a cubit and a half long, and one cubit and a half wide, and one cubit high, on which the instruments were to be laid with which the burnt offerings and the sacrifices were slaughtered. 43 There were pegs, one handbreadth long, fastened all around the inside. And on the tables the flesh of the offering was to be laid.

44 On the outside of the inner gateway there were chambers for the singers in the inner court, one*d* at the side of the north gate facing south, the other at the side of the east gate facing north. 45 He said to me, "This chamber that faces south is for the priests who have charge of the temple, 46 and the chamber that faces north is for the priests who have charge of the altar; these are the descendants of Zadok, who alone among the descendants of Levi may come near to the Lord to minister to him." 47 He measured the court, one hundred cubits deep, and one hundred cubits wide, a square; and the altar was in front of the temple.

48 Then he brought me to the vesti-

y Gk: Heb *before them*　　*z* One Ms: Compare verses 29 and 33: MT lacks *were of the same size as the others*　　*a* Gk Vg Compare verses 26, 31, 34: Heb *pilasters*　　*b* Cn: Heb *at the pilasters of the gates*　　*c* Cn: Heb *to him who goes up*　*d* Heb lacks *one*

40.20–27: Northern and southern gates were similar to the east gate.

40.28–37: Eight steps above the outer court were the gates leading to the inner court corresponding to those leading to the outer court. The temple was built on a succession of terraces.

40.38–43: At the east gate leading to the inner court were the facilities for the preparation of the sacrifices (Lev 1.1–7.38).

40.44–47: On the north and south sides of the square inner court were buildings for the use of the Zadokite priests (43.19; 44.15–16).

40.48–41.26: **The temple arrange-**

bule of the temple and measured the pilasters of the vestibule, five cubits on either side; and the width of the gate was fourteen cubits; and the sidewalls of the gate were three cubits *e* on either side. *49* The depth of the vestibule was twenty cubits, and the width twelve *f* cubits; ten steps led up *g* to it; and there were pillars beside the pilasters on either side.

41 Then he brought me to the nave, and measured the pilasters; on each side six cubits was the width of the pilasters. *h* *2* The width of the entrance was ten cubits; and the sidewalls of the entrance were five cubits on either side. He measured the length of the nave, forty cubits, and its width, twenty cubits. *3* Then he went into the inner room and measured the pilasters of the entrance, two cubits; and the width of the entrance, six cubits; and the sidewalls *i* of the entrance, seven cubits. *4* He measured the depth of the room, twenty cubits, and its width, twenty cubits, beyond the nave. And he said to me, This is the most holy place.

5 Then he measured the wall of the temple, six cubits thick; and the width of the side chambers, four cubits, all around the temple. *6* The side chambers were in three stories, one over another, thirty in each story. There were offsets *j* all around the wall of the temple to serve as supports for the side chambers, so that they should not be supported by the wall of the temple. *7* The passageway *k* of the side chambers widened from story to story; for the structure was supplied with a stairway all around the temple. For this reason the structure became wider from story to story. One ascended from the bottom story to the uppermost story by way of the middle one. *8* I saw also that the temple had a raised platform all around; the foundations of the side chambers measured a full reed of six long cubits. *9* The thickness of the outer wall of the side chambers was five cubits; and the free space between the side chambers of the temple *10* and the chambers of the court was a width of twenty cubits all around the temple on every side. *11* The side chambers opened onto the area left free, one door toward the north, and another door toward the south; and the width of the part that was left free was five cubits all around.

12 The building that was facing the temple yard on the west side was seventy cubits wide; and the wall of the building was five cubits thick all around, and its depth ninety cubits.

13 Then he measured the temple, one hundred cubits deep; and the yard and the building with its walls, one hundred cubits deep; *14* also the width of the east front of the temple and the yard, one hundred cubits.

15 Then he measured the depth of the building facing the yard at the west, together with its galleries *l* on either side, one hundred cubits.

e Gk: Heb *and the width of the gate was three cubits*
f Gk: Heb *eleven*   *g* Gk: Heb *and by steps that went up*   *h* Compare Gk: Heb *tent*
i Gk: Heb *width*   *j* Gk Compare 1 Kings 6.6: Heb *they entered*   *k* Cn: Heb *it was surrounded*
l Cn: Meaning of Heb uncertain

ments. 40.48–49: The temple was ten steps (about ten feet; see 41.8) above the level of the inner court. In front of the vestibule were Solomon's free-standing pillars (1 Kings 7.15–22).

41.1–4: The tripartite temple division is known also from the thirteenth century Canaanite temple at Hazor and the eighth century temple at Tell Tainat (Hattina) in northern Syria. *The vestibule* (40.48–49) was 35½ by 20½ feet; *the nave* (41.1–2), 35½ by 71 feet; and *the inner room* (41.3–4), or *most holy place,* which Ezekiel did not enter (Lev 16.1–34), 35½ by 35½ feet (1 Kings 6.1–8.66; Ex 26.31–37). **5–11:** The three tiers of thirty chambers per tier on the sides of the temple (1 Kings 6.5–10) were probably for equipment for the temple services, storage, and the temple treasures (1 Kings 14.26; 2 Kings 14.14).

41.12–15a: The purpose of the auxiliary building is unknown; earlier it may have served as a stable for sacred horses, now removed (2 Kings 23.11).

41.15b–26: The windows were like those in the gates (40.16); for the decorative motifs, see 1 Kings 6.29–30; on the *cherubim,* see 1.5 n. **22:** The *table* (perhaps for the bread

The nave of the temple and the inner room and the outer*m* vestibule [16]were paneled,*n* and, all around, all three had windows with recessed*o* frames. Facing the threshold the temple was paneled with wood all around, from the floor up to the windows (now the windows were covered), [17]to the space above the door, even to the inner room, and on the outside. And on all the walls all around in the inner room and the nave there was a pattern.*p* [18]It was formed of cherubim and palm trees, a palm tree between cherub and cherub. Each cherub had two faces: [19]a human face turned toward the palm tree on the one side, and the face of a young lion turned toward the palm tree on the other side. They were carved on the whole temple all around; [20]from the floor to the area above the door, cherubim and palm trees were carved on the wall.*q*

21 The doorposts of the nave were square. In front of the holy place was something resembling [22]an altar of wood, three cubits high, two cubits long, and two cubits wide;*r* its corners, its base,*s* and its walls were of wood. He said to me, "This is the table that stands before the LORD." [23]The nave and the holy place had each a double door. [24]The doors had two leaves apiece, two swinging leaves for each door. [25]On the doors of the nave were carved cherubim and palm trees, such as were carved on the walls; and there was a canopy of wood in front of the vestibule outside. [26]And there were recessed windows and palm trees on either side, on the sidewalls of the vestibule.*t*

42 Then he led me out into the outer court, toward the north, and he brought me to the chambers that were opposite the temple yard and opposite

the building on the north. [2]The length of the building that was on the north side*u* was*v* one hundred cubits, and the width fifty cubits. [3]Across the twenty cubits that belonged to the inner court, and facing the pavement that belonged to the outer court, the chambers rose*w* gallery*x* by gallery*x* in three stories. [4]In front of the chambers was a passage on the inner side, ten cubits wide and one hundred cubits deep,*y* and its*z* entrances were on the north. [5]Now the upper chambers were narrower, for the galleries*x* took more away from them than from the lower and middle chambers in the building. [6]For they were in three stories, and they had no pillars like the pillars of the outer*a* court; for this reason the upper chambers were set back from the ground more than the lower and the middle ones. [7]There was a wall outside parallel to the chambers, toward the outer court, opposite the chambers, fifty cubits long. [8]For the chambers on the outer court were fifty cubits long, while those opposite the temple were one hundred cubits long. [9]At the foot of these chambers ran a passage that one entered from the east in order to enter them from the outer court. [10]The width of the passage*b* is fixed by the wall of the court.

On the south*c* also, opposite the va-

m Gk: Heb *of the court*   *n* Gk: Heb *the thresholds*
o Cn Compare Gk 1 Kings 6.4: Meaning of Heb uncertain   *p* Heb *measures*   *q* Cn Compare verse 25: Heb *and the wall*   *r* Gk: Heb lacks *two cubits wide*   *s* Gk: Heb *length*
t Cn: Heb *vestibule. And the side chambers of the temple and the canopies.*   *u* Gk: Heb *door*
v Gk: Heb *before the length*   *w* Heb lacks *the chambers rose*   *x* Meaning of Heb uncertain
y Gk Syr: Heb *a way of one cubit*   *z* Heb *their*
a Gk: Heb lacks *outer*   *b* Heb lacks *of the passage*   *c* Gk: Heb *east*

---

of the Presence), see 1 Kings 6.20–22; Lev 24.5–9.

42.1–14: The priests' chambers. The description is unclear. Perhaps the three stories of chambers were arranged terrace-fashion against the north and south terrace-walls retaining the inner court, with staircases leading to the upper stories and halls on each level

into which the chamber doors opened. A second, smaller, apartment-building was opposite, on the outer court side. In them the priests were to store their share of the sacrifices (44.28–31; Lev 2.1–10; 7.7–10), eat their meals, and leave the garments they wore for the services before entering the outer court.

cant area and opposite the building, there were chambers [11] with a passage in front of them; they were similar to the chambers on the north, of the same length and width, with the same exits[d] and arrangements and doors. [12] So the entrances of the chambers to the south were entered through the entrance at the head of the corresponding passage, from the east, along the matching wall. [e]

13 Then he said to me, "The north chambers and the south chambers opposite the vacant area are the holy chambers, where the priests who approach the LORD shall eat the most holy offerings; there they shall deposit the most holy offerings—the grain offering, the sin offering, and the guilt offering, for the place is holy. [14] When the priests enter the holy place, they shall not go out of it into the outer court without laying there the vestments in which they minister, for these are holy; they shall put on other garments before they go near to the area open to the people."

15 When he had finished measuring the interior of the temple area, he led me out by the gate that faces east, and measured the temple area all around. [16] He measured the east side with the measuring reed, five hundred cubits by the measuring reed. [17] Then he turned and measured[f] the north side, five hundred cubits by the measuring reed. [18] Then he turned and measured[f] the south side, five hundred cubits by the measuring reed. [19] Then he turned to the west side and measured, five hundred cubits by the measuring reed. [20] He measured it on the four sides. It had a wall around it, five hundred cubits long and five hundred cubits wide, to make a separation between the holy and the common.

43 Then he brought me to the gate, the gate facing east. [2] And there, the glory of the God of Israel was coming from the east; the sound was like the sound of mighty waters; and the earth shone with his glory. [3] The[g] vision I saw was like the vision that I had seen when he came to destroy the city, and[h] like the vision that I had seen by the river Chebar; and I fell upon my face. [4] As the glory of the LORD entered the temple by the gate facing east, [5] the spirit lifted me up, and brought me into the inner court; and the glory of the LORD filled the temple.

6 While the man was standing beside me, I heard someone speaking to me out of the temple. [7] He said to me: Mortal, this is the place of my throne and the place for the soles of my feet, where I will reside among the people of Israel forever. The house of Israel shall no more defile my holy name, neither they nor their kings, by their whoring, and by the corpses of their kings at their death. [i] [8] When they placed their threshold by my threshold and their doorposts beside my doorposts, with only a wall between me and them, they were defiling my holy name by their abominations that they committed; therefore I have consumed them in my anger. [9] Now let them put away their idolatry and the corpses of their kings far from me, and I will reside among them forever.

10 As for you, mortal, describe the temple to the house of Israel, and let them measure the pattern; and let them

d Heb *and all their exits*   e Meaning of Heb uncertain   f Gk: Heb *measuring reed all around. He measured*   g Gk: Heb *Like the vision* h Syr: Heb *and the visions*   i Or *on their high places*

42.15–20: The total temple area was five hundred cubits square (861.63 feet square; see 40.5 n.).

43.1–12: Return of the glory of God. As God on his throne-chariot had forsaken the temple by the east gate (10.18–19; 11.22–23), so he returns from the same direction and re-consecrates the purified temple by his Presence (Ex 40.34–38; 1 Kings 8.10–11).

Sound of many waters, 1.24; Rev 14.2; 19.6. Like the brightness of the sun, *the earth shone with his glory* (Isa 60.1–3).

43.6–12: The restored temple is to be restricted to purely religious uses, as commanded by God; especially discouraged is any influence from the crown (ch 8; 1 Kings 7.1–12; 11.33; Am 7.13).

be ashamed of their iniquities. [11] When they are ashamed of all that they have done, make known to them the plan of the temple, its arrangement, its exits and its entrances, and its whole form—all its ordinances and its entire plan and all its laws; and write it down in their sight, so that they may observe and follow the entire plan and all its ordinances. [12] This is the law of the temple: the whole territory on the top of the mountain all around shall be most holy. This is the law of the temple.

13 These are the dimensions of the altar by cubits (the cubit being one cubit and a handbreadth): its base shall be one cubit high,[j] and one cubit wide, with a rim of one span around its edge. This shall be the height of the altar: [14] From the base on the ground to the lower ledge, two cubits, with a width of one cubit; and from the smaller ledge to the larger ledge, four cubits, with a width of one cubit; [15] and the altar hearth, four cubits; and from the altar hearth projecting upward, four horns. [16] The altar hearth shall be square, twelve cubits long by twelve wide. [17] The ledge also shall be square, fourteen cubits long by fourteen wide, with a rim around it half a cubit wide, and its surrounding base, one cubit. Its steps shall face east.

18 Then he said to me: Mortal, thus says the Lord God: These are the ordinances for the altar: On the day when it is erected for offering burnt offerings upon it and for dashing blood against it, [19] you shall give to the levitical priests of the family of Zadok, who draw near to me to minister to me, says the Lord God, a bull for a sin offering. [20] And you shall take some of its blood, and put it on the four horns of the altar, and on the four corners of the ledge, and upon the rim all around; thus you shall purify it and make atonement for it. [21] You shall also take the bull of the sin offering, and it shall be burnt in the appointed place belonging to the temple, outside the sacred area.

22 On the second day you shall offer a male goat without blemish for a sin offering; and the altar shall be purified, as it was purified with the bull. [23] When you have finished purifying it, you shall offer a bull without blemish and a ram from the flock without blemish. [24] You shall present them before the Lord, and the priests shall throw salt on them and offer them up as a burnt offering to the Lord. [25] For seven days you shall provide daily a goat for a sin offering; also a bull and a ram from the flock, without blemish, shall be provided. [26] Seven days shall they make atonement for the altar and cleanse it, and so consecrate it. [27] When these days are over, then from the eighth day onward the priests shall offer upon the altar your burnt offerings and your offerings of well-being; and I will accept you, says the Lord God.

44 Then he brought me back to the outer gate of the sanctuary, which faces east; and it was shut. [2] The Lord said to me: This gate shall remain shut; it shall not be opened, and no one shall enter by it; for the Lord, the God of Israel, has entered by it; therefore it shall remain shut. [3] Only the prince, because he is a prince, may sit in it to eat food before the Lord; he shall enter by

j Gk: Heb lacks *high*

43.13–27: The altar of burnt offering. 13–17: The altar was built in three superimposed squares of sixteen, fourteen, and twelve cubits respectively, resting on a foundation platform, "the bosom of the earth," with steps on the east leading to the *altar hearth,* "the mountain of God," like the Mesopotamian ziggurats (see Gen 11.4 n.). The total height of the altar is twelve cubits (20.68 feet). **43.18–26:** On the analogy of earlier consecratory rites (Ex 29.36–37; 40.1–38; Lev 8.14–15), the priests of Zadok (44.5–31) are to dedicate the altar.

44.1–31: Temple ordinances. 1–5: The outer east gate should remain closed, perhaps symbolizing its consecration by God's entrance and God's continued presence (43.7). The prince (see 37.15–28 n.) apparently entered from the temple side for the ceremonial meal (Lev 7.15; Deut 12.7), and thus the gate remained closed (compare 46.1).

way of the vestibule of the gate, and shall go out by the same way.

4 Then he brought me by way of the north gate to the front of the temple; and I looked, and lo! the glory of the LORD filled the temple of the LORD; and I fell upon my face. 5 The LORD said to me: Mortal, mark well, look closely, and listen attentively to all that I shall tell you concerning all the ordinances of the temple of the LORD and all its laws; and mark well those who may be admitted to[k] the temple and all those who are to be excluded from the sanctuary. 6 Say to the rebellious house,[l] to the house of Israel, Thus says the Lord GOD: O house of Israel, let there be an end to all your abominations 7 in admitting foreigners, uncircumcised in heart and flesh, to be in my sanctuary, profaning my temple when you offer to me my food, the fat and the blood. You[m] have broken my covenant with all your abominations. 8 And you have not kept charge of my sacred offerings; but you have appointed foreigners[n] to act for you in keeping my charge in my sanctuary.

9 Thus says the Lord GOD: No foreigner, uncircumcised in heart and flesh, of all the foreigners who are among the people of Israel, shall enter my sanctuary. 10 But the Levites who went far from me, going astray from me after their idols when Israel went astray, shall bear their punishment. 11 They shall be ministers in my sanctuary, having oversight at the gates of the temple, and serving in the temple; they shall slaughter the burnt offering and the sacrifice for the people, and they shall attend on them and serve them. 12 Because they ministered to them before their idols and made the house of Israel stumble into iniquity, therefore I have sworn concerning them, says the Lord GOD, that they shall bear their punishment. 13 They shall not come near to me, to serve me as priest, nor come near any of my sacred offerings, the things that are most sacred; but they shall bear their shame, and the consequences of the abominations that they have committed. 14 Yet I will appoint them to keep charge of the temple, to do all its chores, all that is to be done in it.

15 But the levitical priests, the descendants of Zadok, who kept the charge of my sanctuary when the people of Israel went astray from me, shall come near to me to minister to me; and they shall attend me to offer me the fat and the blood, says the Lord GOD. 16 It is they who shall enter my sanctuary, it is they who shall approach my table, to minister to me, and they shall keep my charge. 17 When they enter the gates of the inner court, they shall wear linen vestments; they shall have nothing of wool on them, while they minister at the gates of the inner court, and within. 18 They shall have linen turbans on their heads, and linen undergarments on their loins; they shall not bind themselves with anything that causes sweat. 19 When they go out into the outer court to the people, they shall remove the vestments in which they have been ministering, and lay them in

k Cn: Heb *the entrance of* l Gk: Heb lacks
house m Gk Syr Vg: Heb *They*
n Heb lacks *foreigners*

44.6–9: No more foreigners may serve in the temple (Josh 9.23; Num 31.30, 47), *uncircumcised in heart and flesh* (compare Deut 30.6; Lev 26.41; Jer 4.4). **10–14:** Because they had been party to Israel's apostasy, the Levites are demoted from being priests to temple servants (compare Deut 17.18–18.8). **44.15–31:** The priesthood. Solomon elevated *Zadok* to be chief priest after Abiathar's defection and banishment (2 Sam 15.24–29; 1 Kings 1.7–8; 2.26–27). 1 Chr 6.50–53 and 24.31 trace the lineage back to Eleazar, Aaron's son. Their present status is perhaps derived from Hilkiah's role in Josiah's reform (2 Kings 22.11–14; 23.4, 24). **16:** *My table,* the table in the nave (41.22) or the altar. **17:** *Linen vestments,* see 9.2 n. **19:** The care of the vestments, compare 42.14; Hag 2.10–12. **20:** The care of the *hair,* Lev 21.5; Deut 14.1–2. **21:** No *wine* before service, Lev 10.9. **22:** Proper marriage, Lev 21.7, 13–15. **23:** Teaching of the people, Lev 10.11. **24:** *Judges,* Deut 21.1–5. **25:** Defilement by a dead person, Lev 21.1–3.

the holy chambers; and they shall put on other garments, so that they may not communicate holiness to the people with their vestments. 20 They shall not shave their heads or let their locks grow long; they shall only trim the hair of their heads. 21 No priest shall drink wine when he enters the inner court. 22 They shall not marry a widow, or a divorced woman, but only a virgin of the stock of the house of Israel, or a widow who is the widow of a priest. 23 They shall teach my people the difference between the holy and the common, and show them how to distinguish between the unclean and the clean. 24 In a controversy they shall act as judges, and they shall decide it according to my judgments. They shall keep my laws and my statutes regarding all my appointed festivals, and they shall keep my sabbaths holy. 25 They shall not defile themselves by going near to a dead person; for father or mother, however, and for son or daughter, and for brother or unmarried sister they may defile themselves. 26 After he has become clean, they shall count seven days for him. 27 On the day that he goes into the holy place, into the inner court, to minister in the holy place, he shall offer his sin offering, says the Lord God.

28 This shall be their inheritance: I am their inheritance; and you shall give them no holding in Israel; I am their holding. 29 They shall eat the grain offering, the sin offering, and the guilt offering; and every devoted thing in Israel shall be theirs. 30 The first of all the first fruits of all kinds, and every offering of all kinds from all your offerings, shall belong to the priests; you shall also give to the priests the first of your dough, in order that a blessing may rest on your house.

31 The priests shall not eat of anything, whether bird or animal, that died of itself or was torn by animals.

45 When you allot the land as an inheritance, you shall set aside for the LORD a portion of the land as a holy district, twenty-five thousand cubits long and twenty*o* thousand cubits wide; it shall be holy throughout its entire extent. 2 Of this, a square plot of five hundred by five hundred cubits shall be for the sanctuary, with fifty cubits for an open space around it. 3 In the holy district you shall measure off a section twenty-five thousand cubits long and ten thousand wide, in which shall be the sanctuary, the most holy place. 4 It shall be a holy portion of the land; it shall be for the priests, who minister in the sanctuary and approach the LORD to minister to him; and it shall be both a place for their houses and a holy place for the sanctuary. 5 Another section, twenty-five thousand cubits long and ten thousand cubits wide, shall be for the Levites who minister at the temple, as their holding for cities to live in.*p*

6 Alongside the portion set apart as the holy district you shall assign as a holding for the city an area five thousand cubits wide, and twenty-five thousand cubits long; it shall belong to the whole house of Israel.

7 And to the prince shall belong the land on both sides of the holy district and the holding of the city, alongside the holy district and the holding of the city, on the west and on the east, corresponding in length to one of the tribal portions, and extending from the western to the

o Gk: Heb *ten*   *p* Gk: Heb *as their holding, twenty chambers*

44.28: No *inheritance,* Josh 13.14; Num 18.20–32. 29–30: Eating sacrifices and *first fruits,* Lev 2.3–10; 6.14–18; Deut 18.3–5. 31: Eating of unslaughtered meat, Lev 7.24.

45.1–9: The distribution of land (continued in 47.13–48.35) is completely idealistic. The *holy district* is divided into two sections, each 25,000 by 10,000 cubits; the northern section is for the Levites, the south-

ern for the Zadokite priests. The latter included the five hundred cubit square section for the temple area plus an enclosure-space fifty cubits deep not mentioned earlier (42.20). The 25,000 by 5,000 cubit property of Jerusalem south of the holy district (v. 6) made with the holy district (v. 1) an area 25,000 cubits (8.3 miles) square.

eastern boundary [8]of the land. It is to be his property in Israel. And my princes shall no longer oppress my people; but they shall let the house of Israel have the land according to their tribes.

9 Thus says the Lord GOD: Enough, O princes of Israel! Put away violence and oppression, and do what is just and right. Cease your evictions of my people, says the Lord GOD.

10 You shall have honest balances, an honest ephah, and an honest bath. [q] [11]The ephah and the bath shall be of the same measure, the bath containing one-tenth of a homer, and the ephah one-tenth of a homer; the homer shall be the standard measure. [12]The shekel shall be twenty gerahs. Twenty shekels, twenty-five shekels, and fifteen shekels shall make a mina for you.

13 This is the offering that you shall make: one-sixth of an ephah from each homer of wheat, and one-sixth of an ephah from each homer of barley, [14]and as the fixed portion of oil, [r] one-tenth of a bath from each cor (the cor, [s] like the homer, contains ten baths); [15]and one sheep from every flock of two hundred, from the pastures of Israel. This is the offering for grain offerings, burnt offerings, and offerings of well-being, to make atonement for them, says the Lord GOD. [16]All the people of the land shall join with the prince in Israel in making this offering. [17]But this shall be the obligation of the prince regarding the burnt offerings, grain offerings, and drink offerings, at the festivals, the new moons, and the sabbaths, all the appointed festivals of the house of Israel: he shall provide the sin offerings, grain offerings, the burnt offerings, and the offerings of well-being, to make atonement for the house of Israel.

18 Thus says the Lord GOD: In the first month, on the first day of the month, you shall take a young bull without blemish, and purify the sanctuary. [19]The priest shall take some of the blood of the sin offering and put it on the doorposts of the temple, the four corners of the ledge of the altar, and the posts of the gate of the inner court. [20]You shall do the same on the seventh day of the month for anyone who has sinned through error or ignorance; so you shall make atonement for the temple.

21 In the first month, on the fourteenth day of the month, you shall celebrate the festival of the passover, and for seven days unleavened bread shall be eaten. [22]On that day the prince shall provide for himself and all the people of the land a young bull for a sin offering. [23]And during the seven days of the festival he shall provide as a burnt offering to the LORD seven young bulls and seven rams without blemish, on each of the seven days; and a male goat daily for a sin offering. [24]He shall provide as a grain offering an ephah for each bull, an ephah for each ram, and a hin of oil to each ephah. [25]In the seventh month, on the fifteenth day of the month and for the seven days of the festival, he shall make the same provision for sin offerings, burnt offerings, and grain offerings, and for the oil.

46 Thus says the Lord GOD: The gate of the inner court that faces east shall remain closed on the six working days; but on the sabbath day it shall be opened and on the day of the new moon it shall be opened. [2]The prince shall enter by the vestibule of the gate from outside, and shall take his stand by

---

[q] A Heb measure of volume   [r] Cn: Heb *oil, the bath the oil*   [s] Vg: Heb *homer*

45.10–17: Weights and measures. For present-day equivalents of *ephah, bath, homer, shekel, gerah,* and *mina,* see pp. 424–425. **13–17:** The people will bring token gifts to the prince, who, as their representative, will offer them to God. **45.18–25: Festival regulations:** *passover* (Ex 23.15; Deut 16.1–8; Lev 23.4–8) and

booths (the reference in v. 25; see Ex 23.16; Deut 16.13–15; Lev 23.33–36); omitted are the festival of first fruits (Ex 23.16; Deut 16.9–12; Lev 23.15–21) and the day of atonement (Lev 23.26–32). **24:** *A hin,* see p. 425. **46.1–18: Regulations regarding the prince. 1–8:** The prince (see 37.15–28 n.) shall bring his offerings (45.13–16) through

the post of the gate. The priests shall offer his burnt offering and his offerings of well-being, and he shall bow down at the threshold of the gate. Then he shall go out, but the gate shall not be closed until evening. ³The people of the land shall bow down at the entrance of that gate before the LORD on the sabbaths and on the new moons. ⁴The burnt offering that the prince offers to the LORD on the sabbath day shall be six lambs without blemish and a ram without blemish; ⁵and the grain offering with the ram shall be an ephah, and the grain offering with the lambs shall be as much as he wishes to give, together with a hin of oil to each ephah. ⁶On the day of the new moon he shall offer a young bull without blemish, and six lambs and a ram, which shall be without blemish; ⁷as a grain offering he shall provide an ephah with the bull and an ephah with the ram, and with the lambs as much as he wishes, together with a hin of oil to each ephah. ⁸When the prince enters, he shall come in by the vestibule of the gate, and he shall go out by the same way.

9 When the people of the land come before the LORD at the appointed festivals, whoever enters by the north gate to worship shall go out by the south gate; and whoever enters by the south gate shall go out by the north gate: they shall not return by way of the gate by which they entered, but shall go out straight ahead. ¹⁰When they come in, the prince shall come in with them; and when they go out, he shall go out.

11 At the festivals and the appointed seasons the grain offering with a young bull shall be an ephah, and with a ram an ephah, and with the lambs as much as one wishes to give, together with a hin of oil to an ephah. ¹²When the prince provides a freewill offering, either a burnt offering or offerings of well-being as a freewill offering to the LORD, the gate facing east shall be opened for him; and he shall offer his burnt offering or his offerings of well-being as he does on the sabbath day. Then he shall go out, and after he has gone out the gate shall be closed.

13 He shall provide a lamb, a yearling, without blemish, for a burnt offering to the LORD daily; morning by morning he shall provide it. ¹⁴And he shall provide a grain offering with it morning by morning regularly, one-sixth of an ephah, and one-third of a hin of oil to moisten the choice flour, as a grain offering to the LORD; this is the ordinance for all time. ¹⁵Thus the lamb and the grain offering and the oil shall be provided, morning by morning, as a regular burnt offering.

16 Thus says the Lord GOD: If the prince makes a gift to any of his sons out of his inheritance,ᵗ it shall belong to his sons, it is their holding by inheritance. ¹⁷But if he makes a gift out of his inheritance to one of his servants, it shall be his to the year of liberty; then it shall revert to the prince; only his sons may keep a gift from his inheritance. ¹⁸The prince shall not take any of the inheritance of the people, thrusting them out of their holding; he shall give his sons their inheritance out of his own holding, so that none of my people shall be dispossessed of their holding.

19 Then he brought me through the entrance, which was at the side of the gate, to the north row of the holy cham-

ᵗ Gk: Heb *it is his inheritance*

---

the *gate . . . that faces east* of the *inner court* to its inner vestibule where he will stand (2 Kings 11.14) by the ritually cleansed doorway (45.18–19), while the priests offer the sacrifices on the altar directly ahead. The people stand in the outer court. For the sacrifices, see Ex 29.38–42; Num 28.3–15. *New moons,* first day of the month; note the increasingly strong emphasis on the sabbath.

46.9–10: The great festival crowds require regulations for controlled egress.
46.11–15: When the prince makes a *freewill offering* (Lev 22.18–23), the east gate (v. 1) will also be opened. The prince must also provide for daily sacrifices (Ex 29.38–42; 1 Kings 18.29; 2 Kings 16.15).
46.16–18: Crown property could not be disposed of permanently. When given to a

bers for the priests; and there I saw a place at the extreme western end of them. 20 He said to me, "This is the place where the priests shall boil the guilt offering and the sin offering, and where they shall bake the grain offering, in order not to bring them out into the outer court and so communicate holiness to the people."

21 Then he brought me out to the outer court, and led me past the four corners of the court; and in each corner of the court there was a court— 22 in the four corners of the court were small*u* courts, forty cubits long and thirty wide; the four were of the same size. 23 On the inside, around each of the four courts*v* was a row of masonry, with hearths made at the bottom of the rows all around. 24 Then he said to me, "These are the kitchens where those who serve at the temple shall boil the sacrifices of the people."

47 Then he brought me back to the entrance of the temple; there, water was flowing from below the threshold of the temple toward the east (for the temple faced east); and the water was flowing down from below the south end of the threshold of the temple, south of the altar. 2 Then he brought me out by way of the north gate, and led me around on the outside to the outer gate that faces toward the east;*w* and the water was coming out on the south side.

3 Going on eastward with a cord in his hand, the man measured one thousand cubits, and then led me through the water; and it was ankle-deep. 4 Again he measured one thousand, and led me through the water; and it was knee-deep.

Again he measured one thousand, and led me through the water; and it was up to the waist. 5 Again he measured one thousand, and it was a river that I could not cross, for the water had risen; it was deep enough to swim in, a river that could not be crossed. 6 He said to me, "Mortal, have you seen this?"

Then he led me back along the bank of the river. 7 As I came back, I saw on the bank of the river a great many trees on the one side and on the other. 8 He said to me, "This water flows toward the eastern region and goes down into the Arabah; and when it enters the sea, the sea of stagnant waters, the water will become fresh. 9 Wherever the river goes,*x* every living creature that swarms will live, and there will be very many fish, once these waters reach there. It will become fresh; and everything will live where the river goes. 10 People will stand fishing beside the sea*y* from En-gedi to En-eglaim; it will be a place for the spreading of nets; its fish will be of a great many kinds, like the fish of the Great Sea. 11 But its swamps and marshes will not become fresh; they are to be left for salt. 12 On the banks, on both sides of the river, there will grow all kinds of trees for food. Their leaves will not wither nor their fruit fail, but they will bear fresh fruit every month, because the water for them flows from the sanctuary. Their fruit will be for food, and their leaves for healing."

13 Thus says the Lord GOD: These are the boundaries by which you shall divide

u Gk Syr Vg: Meaning of Heb uncertain
v Heb *the four of them*   w Meaning of Heb uncertain   x Gk Syr Vg Tg: Heb *the two rivers go*   y Heb *it*

---

non-relative, it had to be returned on the *year of liberty* (jubilee year, Lev 25.8–17).

46.19–20: The priests' quarters, see 42.1–14. *Communicate holiness,* see 42.14; 44.19. **21–24:** The temple kitchens, for the common-meal sacrifices.

47.1–12: The sacred river. This figure, known from Ugaritic (Canaanite) and Mesopotamian sources, is also found in Joel 3.18; Zech 14.8; and Rev ch 22. From the throne of God (the temple; 43.7) issue the waters of life

by which the saline waters of the Dead Sea become fresh, and in the desolate wilderness of Judah fruit trees flourish, making a new Paradise, as it were, and the inhabitants of the land will benefit. **8:** *Arabah,* the valley south of the Dead Sea. **10:** *En-gedi to En-eglaim* ('Ain Feshkha, about one and one-half miles south of Khirbet Qumran), northwest coast of the Dead Sea.

47.13–20: Israel's boundaries (Num 34.1–12): north, apparently that of David's

the land for inheritance among the twelve tribes of Israel. Joseph shall have two portions. [14] You shall divide it equally; I swore to give it to your ancestors, and this land shall fall to you as your inheritance.

15 This shall be the boundary of the land: On the north side, from the Great Sea by way of Hethlon to Lebo-hamath, and on to Zedad, [z] [16] Berothah, Sibraim (which lies between the border of Damascus and the border of Hamath), as far as Hazer-hatticon, which is on the border of Hauran. [17] So the boundary shall run from the sea to Hazar-enon, which is north of the border of Damascus, with the border of Hamath to the north. [a] This shall be the north side.

18 On the east side, between Hauran and Damascus; along the Jordan between Gilead and the land of Israel; to the eastern sea and as far as Tamar. [b] This shall be the east side.

19 On the south side, it shall run from Tamar as far as the waters of Meribath-kadesh, from there along the Wadi of Egypt [c] to the Great Sea. This shall be the south side.

20 On the west side, the Great Sea shall be the boundary to a point opposite Lebo-hamath. This shall be the west side.

21 So you shall divide this land among you according to the tribes of Israel. [22] You shall allot it as an inheritance for yourselves and for the aliens who reside among you and have begotten children among you. They shall be to you as citizens of Israel; with you they shall be allotted an inheritance among the tribes of Israel. [23] In whatever tribe aliens reside, there you shall assign them their inheritance, says the Lord GOD.

48 These are the names of the tribes: Beginning at the northern border, on the Hethlon road, [d] from Lebo-hamath, as far as Hazar-enon (which is on the border of Damascus, with Hamath to the north), and [e] extending from the east side to the west, [f] Dan, one portion. [2] Adjoining the territory of Dan, from the east side to the west, Asher, one portion. [3] Adjoining the territory of Asher, from the east side to the west, Naphtali, one portion. [4] Adjoining the territory of Naphtali, from the east side to the west, Manasseh, one portion. [5] Adjoining the territory of Manasseh, from the east side to the west, Ephraim, one portion. [6] Adjoining the territory of Ephraim, from the east side to the west, Reuben, one portion. [7] Adjoining the territory of Reuben, from the east side to the west, Judah, one portion.

8 Adjoining the territory of Judah, from the east side to the west, shall be the portion that you shall set apart, twenty-five thousand cubits in width, and in length equal to one of the tribal portions, from the east side to the west, with the sanctuary in the middle of it. [9] The portion that you shall set apart for the LORD

---

z Gk: Heb *Lebo-zedad,* [16] *Hamath*   a Meaning of Heb uncertain   b Compare Syr: Heb *you shall measure*   c Heb lacks *of Egypt*   d Compare 47.15: Heb *by the side of the way*   e Cn: Heb *and they shall be his*   f Gk Compare verses 2–8: Heb *the east side the west*

---

empire in north Syria (2 Sam 8.5–12; Num 34.7–9); east, *Hazar-enon,* between Damascus and Palmyra, to the Dead Sea (Num 34.10–12); south, along the southern Negeb to *the Wadi of Egypt* (Num 34.3–5); west, the Mediterranean Sea (Num 34.6). *Sibraim,* the same as *Sepharvaim* (see Isa 36.19 n.).

47.21–23: Equal allotment. Proselytes (*aliens*) and Jews should be treated alike (Num 15.29; Lev 19.33–34).

48.1–29: The allotment of the land. All tribes receive equal allotments west of the Jordan, with Ephraim and Manasseh each given a portion and Levi, as the priestly tribe (see Josh 14.3–4), omitted. This division of the country ignores geographical reality. **1–7:** Seven tribes north of the holy district, with Judah immediately contiguous to it. **1:** *Hethlon,* Heitala, east of Tripoli (47.15). *Lebo-hamath,* territory between Riblah and Kadesh on the Orontes River.

48.8–22: The properties of the Zadokite priests and the Levites, like that of the prince on either side, were to be held in perpetuity (45.1–9). The size of Jerusalem, with the surrounding open area, was exactly ten times

shall be twenty-five thousand cubits in length, and twenty[g] thousand in width. [10] These shall be the allotments of the holy portion: the priests shall have an allotment measuring twenty-five thousand cubits on the northern side, ten thousand cubits in width on the western side, ten thousand in width on the eastern side, and twenty-five thousand in length on the southern side, with the sanctuary of the LORD in the middle of it. [11] This shall be for the consecrated priests, the descendants[h] of Zadok, who kept my charge, who did not go astray when the people of Israel went astray, as the Levites did. [12] It shall belong to them as a special portion from the holy portion of the land, a most holy place, adjoining the territory of the Levites. [13] Alongside the territory of the priests, the Levites shall have an allotment twenty-five thousand cubits in length and ten thousand in width. The whole length shall be twenty-five thousand cubits and the width twenty[i] thousand. [14] They shall not sell or exchange any of it; they shall not transfer this choice portion of the land, for it is holy to the LORD.

15 The remainder, five thousand cubits in width and twenty-five thousand in length, shall be for ordinary use for the city, for dwellings and for open country. In the middle of it shall be the city; [16] and these shall be its dimensions: the north side four thousand five hundred cubits, the south side four thousand five hundred, the east side four thousand five hundred, and the west side four thousand and five hundred. [17] The city shall have open land: on the north two hundred fifty cubits, on the south two hundred fifty, on the east two hundred fifty, on the west two hundred fifty. [18] The remainder of the length alongside the holy portion shall be ten thousand cubits to the east, and ten thousand to the west, and it shall be alongside the holy portion. Its pro-

duce shall be food for the workers of the city. [19] The workers of the city, from all the tribes of Israel, shall cultivate it. [20] The whole portion that you shall set apart shall be twenty-five thousand cubits square, that is, the holy portion together with the property of the city.

21 What remains on both sides of the holy portion and of the property of the city shall belong to the prince. Extending from the twenty-five thousand cubits of the holy portion to the east border, and westward from the twenty-five thousand cubits to the west border, parallel to the tribal portions, it shall belong to the prince. The holy portion with the sanctuary of the temple in the middle of it, [22] and the property of the Levites and of the city, shall be in the middle of that which belongs to the prince. The portion of the prince shall lie between the territory of Judah and the territory of Benjamin.

23 As for the rest of the tribes: from the east side to the west, Benjamin, one portion. [24] Adjoining the territory of Benjamin, from the east side to the west, Simeon, one portion. [25] Adjoining the territory of Simeon, from the east side to the west, Issachar, one portion. [26] Adjoining the territory of Issachar, from the east side to the west, Zebulun, one portion. [27] Adjoining the territory of Zebulun, from the east side to the west, Gad, one portion. [28] And adjoining the territory of Gad to the south, the boundary shall run from Tamar to the waters of Meribath-kadesh, from there along the Wadi of Egypt[j] to the Great Sea. [29] This is the land that you shall allot as an inheritance among the tribes of Israel, and these are their portions, says the Lord GOD.

g Compare 45.1: Heb *ten*   *h* One Ms Gk: Heb *of the descendants*   *i* Gk: Heb *ten*
j Heb lacks *of Egypt*

---

that of the temple (42.20), i.e. 5,000 cubits square (1.6 miles). **15**: *Ordinary use,* see 22.26; 42.20; 44.23.

48.23–29: Five tribes south of Jerusalem

and the holy district beginning with Benjamin.

48.30–35: **The new Jerusalem.** Three gates on each of the city's four sides, with

30 These shall be the exits of the city: On the north side, which is to be four thousand five hundred cubits by measure, ³¹ three gates, the gate of Reuben, the gate of Judah, and the gate of Levi, the gates of the city being named after the tribes of Israel. ³² On the east side, which is to be four thousand five hundred cubits, three gates, the gate of Joseph, the gate of Benjamin, and the gate of Dan. ³³ On the south side, which is to be four thousand five hundred cubits by measure, three gates, the gate of Simeon, the gate of Issachar, and the gate of Zebulun. ³⁴ On the west side, which is to be four thousand five hundred cubits, three gates, [k] the gate of Gad, the gate of Asher, and the gate of Naphtali. ³⁵ The circumference of the city shall be eighteen thousand cubits. And the name of the city from that time on shall be, The Lord is There.

k One Ms Gk Syr: MT *their gates three*

each gate named after a tribe, including Levi, and with Ephraim and Manasseh combined in Joseph (compare Rev 21.12–14). **35**: The new city receives a new name (Isa 62.2), "The Lord is there" (Yahweh-shammah); for other names see Jer 3.17; Zech 8.3; Isa 1.26 n.; 60.14 (compare Jer 23.6).

Daniel

The six stories and four dream-visions of the book of Daniel make up the only apocalyptic book in the Hebrew Scriptures. Other examples of the genre are 1 Enoch, Syriac Baruch, and the New Testament book of Revelation (see "Apocalyptic Literature," pp. 362–363 NT). These apocalypses come from times of national or community tribulation, and are not actual history, but, through symbols and signs, are interpretations of current history with its background and predictions of a future where tribulations and sorrows will give place to triumph and peace. The Apocalyptists usually set forth their messages under the name of some ancient worthy, e.g. Adam, Enoch, Noah, Abraham, or some other figure of note.

This book appears under the name of Daniel, or Danel, a worthy twice referred to in Ezekiel (Ezek 14.14; 28.3), and whose name appears also in the North Canaanite clay-tablet texts found at Ras Shamra. The author was a pious Jew living under the persecution of Antiochus Epiphanes, 167–164 B.C. (see "Survey of . . . Bible Lands," §19). To encourage his suffering fellow-believers he tells six stories, set in earlier days in Babylon just before and just after the Persian conquest, which illustrate how faithful Jews, loyally practicing their religion, were enabled by divine aid to triumph over their enemies. These were traditional tales, which were already written down and collected in the late third or early second century B.C. Then in four visions he ventures to interpret current history and predict the coming consummation when the faithful Jews will have ultimate victory. The section from 2.4b to 7.28 is written in Aramaic, though the remainder is in Hebrew.

1 In the third year of the reign of King Jehoiakim of Judah, King Nebuchadnezzar of Babylon came to Jerusalem and besieged it. 2 The Lord let King Jehoiakim of Judah fall into his power, as well as some of the vessels of the house of God. These he brought to the land of Shinar,*a* and placed the vessels in the treasury of his gods.

3 Then the king commanded his palace master Ashpenaz to bring some of the Israelites of the royal family and of the nobility, 4 young men without physical defect and handsome, versed in every branch of wisdom, endowed with knowledge and insight, and competent to serve in the king's palace; they were to be taught the literature and language of the Chaldeans. 5 The king assigned them a daily portion of the royal rations of food and wine. They were to be educated for three years, so that at the end of that time they could be stationed in the king's court. 6 Among them were Daniel, Hananiah, Mishael, and Azariah, from the tribe of Judah. 7 The palace master gave them other names: Daniel he called Belteshazzar, Hananiah he called Shadrach, Mishael he called Meshach, and Azariah he called Abednego.

8 But Daniel resolved that he would not defile himself with the royal rations of food and wine; so he asked the palace master to allow him not to defile himself. 9 Now God allowed Daniel to receive favor and compassion from the palace master. 10 The palace master said to Daniel, "I am afraid of my lord the king; he has appointed your food and your drink. If he should see you in poorer condition

than the other young men of your own age, you would endanger my head with the king." 11 Then Daniel asked the guard whom the palace master had appointed over Daniel, Hananiah, Mishael, and Azariah: 12 "Please test your servants for ten days. Let us be given vegetables to eat and water to drink. 13 You can then compare our appearance with the appearance of the young men who eat the royal rations, and deal with your servants according to what you observe." 14 So he agreed to this proposal and tested them for ten days. 15 At the end of ten days it was observed that they appeared better and fatter than all the young men who had been eating the royal rations. 16 So the guard continued to withdraw their royal rations and the wine they were to drink, and gave them vegetables. 17 To these four young men God gave knowledge and skill in every aspect of literature and wisdom; Daniel also had insight into all visions and dreams.

18 At the end of the time that the king had set for them to be brought in, the palace master brought them into the presence of Nebuchadnezzar, 19 and the king spoke with them. And among them all, no one was found to compare with Daniel, Hananiah, Mishael, and Azariah; therefore they were stationed in the king's court. 20 In every matter of wisdom and understanding concerning which the king inquired of them, he found them ten times better than all the magicians and enchanters in his whole

a Gk Theodotion: Heb adds *to the house of his own gods*

1.1–21: Daniel and his friends. A story to teach how faithful observance of the law is rewarded. **1:** The *third year* of Jehoiakim was 606 B.C. There is no other record of a siege of Jerusalem at this time. *Nebuchadnezzar* is a Jewish form of the name Nabuchadrezzar, who in 597 B.C. carried away temple treasure and captives to Babylon (2 Kings 24.10–15), in v. 2 called by the ancient name *Shinar* (Gen 10.10; Isa 11.11). **2:** *His gods* at Babylon included Marduk and Nebo. **3–5:** Promising young men among the captives were trained for *three years* to be royal pages. **6:** As such

they were given Babylonian names and provided with food and drink from the royal table. **1.8:** *Defile himself*, i.e. by eating food not permitted by Jewish dietary laws. **10:** *Poorer condition*, lit. "your faces looking sorrowful." **1.20:** *Magicians and enchanters* here probably stand merely as names for court sages whose learning and wisdom were surpassed by that divinely given to Daniel and his friends (v. 17). **21:** The *first year* of *King Cyrus* at Babylon was 538 B.C. (Ezra 1.1), almost seventy years after Daniel's coming to Babylon.

kingdom. 21 And Daniel continued there until the first year of King Cyrus.

2 In the second year of Nebuchadnezzar's reign, Nebuchadnezzar dreamed such dreams that his spirit was troubled and his sleep left him. 2 So the king commanded that the magicians, the enchanters, the sorcerers, and the Chaldeans be summoned to tell the king his dreams. When they came in and stood before the king, 3 he said to them, "I have had such a dream that my spirit is troubled by the desire to understand it." 4 The Chaldeans said to the king (in Aramaic),*b* "O king, live forever! Tell your servants the dream, and we will reveal the interpretation." 5 The king answered the Chaldeans, "This is a public decree: if you do not tell me both the dream and its interpretation, you shall be torn limb from limb, and your houses shall be laid in ruins. 6 But if you do tell me the dream and its interpretation, you shall receive from me gifts and rewards and great honor. Therefore tell me the dream and its interpretation." 7 They answered a second time, "Let the king first tell his servants the dream, then we can give its interpretation." 8 The king answered, "I know with certainty that you are trying to gain time, because you see I have firmly decreed: 9 if you do not tell me the dream, there is but one verdict for you. You have agreed to speak lying and misleading words to me until things take a turn. Therefore, tell me the dream, and I shall know that you can give me its interpretation." 10 The Chaldeans answered the king, "There is no one on earth who can reveal what the king de-

mands! In fact no king, however great and powerful, has ever asked such a thing of any magician or enchanter or Chaldean. 11 The thing that the king is asking is too difficult, and no one can reveal it to the king except the gods, whose dwelling is not with mortals."

12 Because of this the king flew into a violent rage and commanded that all the wise men of Babylon be destroyed. 13 The decree was issued, and the wise men were about to be executed; and they looked for Daniel and his companions, to execute them. 14 Then Daniel responded with prudence and discretion to Arioch, the king's chief executioner, who had gone out to execute the wise men of Babylon; 15 he asked Arioch, the royal official, "Why is the decree of the king so urgent?" Arioch then explained the matter to Daniel. 16 So Daniel went in and requested that the king give him time and he would tell the king the interpretation.

17 Then Daniel went to his home and informed his companions, Hananiah, Mishael, and Azariah, 18 and told them to seek mercy from the God of heaven concerning this mystery, so that Daniel and his companions with the rest of the wise men of Babylon might not perish. 19 Then the mystery was revealed to Daniel in a vision of the night, and Daniel blessed the God of heaven.
20 Daniel said:

"Blessed be the name of God from
age to age,

b The text from this point to the end of chapter 7 is in Aramaic

---

2.1–49: Nebuchadnezzar's dream. A story to teach the feebleness of human wisdom compared with that conferred by God. **1:** *Second year* is a slip (Daniel has already been at court for three years, 1.5, 18). **2:** *Chaldeans* here means not an ethnic group but a caste of wise men. In ancient as in modern times it was believed that important matters were revealed in *dreams* (Gen 20.3; 41.1–32; Mt 27.19). **2.4–11:** Cuneiform tablets on divination through dreams are still extant. If the king

would *tell* the dream they could consult dream-books and find the interpretation, but asking them to *tell* the dream itself was unheard of and *difficult* beyond human skill. **11:** The word *gods* is used, since these Chaldeans were polytheists.
2.14: On the ground of 1.20, Daniel and his friends were counted among the court's *wise men.* In what follows, Daniel represents the ideal sage. **18–19:** This is a *mystery,* but through prayer even mysteries may be *revealed* by God's wisdom (v. 30).

for wisdom and power are his.
21 He changes times and seasons,
deposes kings and sets up kings;
he gives wisdom to the wise
and knowledge to those who
have understanding.
22 He reveals deep and hidden things;
he knows what is in the
darkness,
and light dwells with him.
23 To you, O God of my ancestors,
I give thanks and praise,
for you have given me wisdom
and power,
and have now revealed to me
what we asked of you,
for you have revealed to us
what the king ordered."

24 Therefore Daniel went to Arioch, whom the king had appointed to destroy the wise men of Babylon, and said to him, "Do not destroy the wise men of Babylon; bring me in before the king, and I will give the king the interpretation." 25 Then Arioch quickly brought Daniel before the king and said to him: "I have found among the exiles from Judah a man who can tell the king the interpretation." 26 The king said to Daniel, whose name was Belteshazzar, "Are you able to tell me the dream that I have seen and its interpretation?" 27 Daniel answered the king, "No wise men, enchanters, magicians, or diviners can show to the king the mystery that the king is asking, 28 but there is a God in heaven who reveals mysteries, and he has disclosed to King Nebuchadnezzar what will happen at the end of days. Your dream and the visions of your head as you lay in bed were these: 29 To you, O king, as you lay in bed, came thoughts of what would be hereafter, and the revealer of mysteries disclosed to you what is to be. 30 But as for me, this mystery has not been revealed to me because of any wisdom that I have more than any other living being, but in order that the interpretation may be known to the king and that you may understand the thoughts of your mind.

31 "You were looking, O king, and lo! there was a great statue. This statue was huge, its brilliance extraordinary; it was standing before you, and its appearance was frightening. 32 The head of that statue was of fine gold, its chest and arms of silver, its middle and thighs of bronze, 33 its legs of iron, its feet partly of iron and partly of clay. 34 As you looked on, a stone was cut out, not by human hands, and it struck the statue on its feet of iron and clay and broke them in pieces. 35 Then the iron, the clay, the bronze, the silver, and the gold, were all broken in pieces and became like the chaff of the summer threshing floors; and the wind carried them away, so that not a trace of them could be found. But the stone that struck the statue became a great mountain and filled the whole earth.

36 "This was the dream; now we will tell the king its interpretation. 37 You, O king, the king of kings—to whom the God of heaven has given the kingdom, the power, the might, and the glory, 38 into whose hand he has given human beings, wherever they live, the wild animals of the field, and the birds of the air, and whom he has established as ruler over them all—you are the head of gold. 39 After you shall arise another kingdom inferior to yours, and yet a third kingdom of bronze, which shall rule over the whole earth. 40 And there shall be a fourth kingdom, strong as iron; just as iron crushes and smashes everything, *c* it shall crush and shatter all these. 41 As you

c Gk Theodotion Syr Vg: Aram adds *and like
iron that crushes*

2.26: *Belteshazzar*, as 1.7; 4.8, 9. **28**: *End of days* is an idiomatic expression meaning "a future time."
2.32–33: *Legs*, i.e. the lower legs, the upper legs being the *thighs*. **34–35**: The image was of human construction, but the *stone* was quarried supernaturally. This could become a *mountain* filling the *whole earth* because the earth was pictured as a disk beneath the heavenly vault.
2.36–45: For this writer the five kingdoms are the Babylonian, Median, Persian,

saw the feet and toes partly of potter's clay and partly of iron, it shall be a divided kingdom; but some of the strength of iron shall be in it, as you saw the iron mixed with the clay. 42 As the toes of the feet were part iron and part clay, so the kingdom shall be partly strong and partly brittle. 43 As you saw the iron mixed with clay, so will they mix with one another in marriage,*d* but they will not hold together, just as iron does not mix with clay. 44 And in the days of those kings the God of heaven will set up a kingdom that shall never be destroyed, nor shall this kingdom be left to another people. It shall crush all these kingdoms and bring them to an end, and it shall stand forever; 45 just as you saw that a stone was cut from the mountain not by hands, and that it crushed the iron, the bronze, the clay, the silver, and the gold. The great God has informed the king what shall be hereafter. The dream is certain, and its interpretation trustworthy."

46 Then King Nebuchadnezzar fell on his face, worshiped Daniel, and commanded that a grain offering and incense be offered to him. 47 The king said to Daniel, "Truly, your God is God of gods and Lord of kings and a revealer of mysteries, for you have been able to reveal this mystery!" 48 Then the king promoted Daniel, gave him many great gifts, and made him ruler over the whole province of Babylon and chief prefect over all the wise men of Babylon. 49 Daniel made a request of the king, and he appointed Shadrach, Meshach, and Abednego over the affairs of the province of Babylon. But Daniel remained at the king's court.

3 King Nebuchadnezzar made a golden statue whose height was sixty cubits and whose width was six cubits; he set it up on the plain of Dura in the province of Babylon. 2 Then King Nebuchadnezzar sent for the satraps, the prefects, and the governors, the counselors, the treasurers, the justices, the magistrates, and all the officials of the provinces to assemble and come to the dedication of the statue that King Nebuchadnezzar had set up. 3 So the satraps, the prefects, and the governors, the counselors, the treasurers, the justices, the magistrates, and all the officials of the provinces, assembled for the dedication of the statue that King Nebuchadnezzar had set up. When they were standing before the statue that Nebuchadnezzar had set up, 4 the herald proclaimed aloud, "You are commanded, O peoples, nations, and languages, 5 that when you hear the sound of the horn, pipe, lyre, trigon, harp, drum, and entire musical ensemble, you are to fall down and worship the golden statue that King Nebuchadnezzar has set up. 6 Whoever does not fall down and worship shall immediately be thrown into a furnace of blazing fire." 7 Therefore, as soon as all the peoples heard the sound of the horn, pipe, lyre, trigon, harp, drum, and entire musical ensemble, all the peoples, nations, and languages fell down and worshiped the golden statue that King Nebuchadnezzar had set up.

8 Accordingly, at this time certain Chaldeans came forward and denounced the Jews. 9 They said to King Nebuchadnezzar, "O king, live forever! 10 You, O king, have made a decree, that everyone who hears the sound of the horn, pipe, lyre, trigon, harp, drum, and entire

d Aram *by human seed*

Greek, and the coming universal kingdom of God. **41:** The declining strength of the fourth kingdom means the divided kingdoms of the Seleucids and Ptolemies (see "Survey of . . . Bible Lands," §19) whose rulers, though they intermarried, did not hold together.

2.46–49: That the king *worshiped* Daniel, recognized his God, and *promoted* him and his friends anticipates the future triumph of the God of the Jews.

3.1–30: The three youths in the fiery furnace. A story to show how martyrdom is preferable to apostasy. **1:** Huge *statues* of deities were common in ancient times. This one was gold-plated. *Dura* is unidentifiable. **2–4:** Ceremonies of *dedication* are well attested (1 Kings 8.63; 2 Chr 7.9; Neh 12.27; and title of Ps 30), at which *officials* were expected to appear, for they represented the various *peoples, nations, languages* of the kingdom. **5:** The

musical ensemble, shall fall down and worship the golden statue, [11] and whoever does not fall down and worship shall be thrown into a furnace of blazing fire. [12] There are certain Jews whom you have appointed over the affairs of the province of Babylon: Shadrach, Meshach, and Abednego. These pay no heed to you, O King. They do not serve your gods and they do not worship the golden statue that you have set up."

13 Then Nebuchadnezzar in furious rage commanded that Shadrach, Meshach, and Abednego be brought in; so they brought those men before the king. [14] Nebuchadnezzar said to them, "Is it true, O Shadrach, Meshach, and Abednego, that you do not serve my gods and you do not worship the golden statue that I have set up? [15] Now if you are ready when you hear the sound of the horn, pipe, lyre, trigon, harp, drum, and entire musical ensemble to fall down and worship the statue that I have made, well and good. [e] But if you do not worship, you shall immediately be thrown into a furnace of blazing fire, and who is the god that will deliver you out of my hands?"

16 Shadrach, Meshach, and Abednego answered the king, "O Nebuchadnezzar, we have no need to present a defense to you in this matter. [17] If our God whom we serve is able to deliver us from the furnace of blazing fire and out of your hand, O king, let him deliver us. [f] [18] But if not, be it known to you, O king, that we will not serve your gods and we will not worship the golden statue that you have set up."

19 Then Nebuchadnezzar was so filled with rage against Shadrach, Meshach, and Abednego that his face was distorted. He ordered the furnace heated up seven times more than was custom-ary, [20] and ordered some of the strongest guards in his army to bind Shadrach, Meshach, and Abednego and to throw them into the furnace of blazing fire. [21] So the men were bound, still wearing their tunics, [g] their trousers, [g] their hats, and their other garments, and they were thrown into the furnace of blazing fire. [22] Because the king's command was urgent and the furnace was so overheated, the raging flames killed the men who lifted Shadrach, Meshach, and Abednego. [23] But the three men, Shadrach, Meshach, and Abednego, fell down, bound, into the furnace of blazing fire.

24 Then King Nebuchadnezzar was astonished and rose up quickly. He said to his counselors, "Was it not three men that we threw bound into the fire?" They answered the king, "True, O king." [25] He replied, "But I see four men unbound, walking in the middle of the fire, and they are not hurt; and the fourth has the appearance of a god." [h] [26] Nebuchadnezzar then approached the door of the furnace of blazing fire and said, "Shadrach, Meshach, and Abednego, servants of the Most High God, come out! Come here!" So Shadrach, Meshach, and Abednego came out from the fire. [27] And the satraps, the prefects, the governors, and the king's counselors gathered together and saw that the fire had not had any power over the bodies of those men; the hair of their heads was not singed, their tunics [g] were not harmed, and not even the smell of fire came from them. [28] Nebuchadnezzar said, "Blessed be the God of Shadrach, Meshach, and Abed-

e Aram lacks *well and good*   f Or *If our God
whom we serve is able to deliver us, he will deliver
us from the furnace of blazing fire and out of your
hand, O king.*   g Meaning of Aram word
uncertain   h Aram *a son of the gods*

---

trigon was a stringed instrument. **6**: Punishment by burning alive was not uncommon (Gen 38.24; Josh 7.15). The *furnace* was the local kiln-type oven, with openings at top and at ground level.
3.19: The *seven times* is merely rhetorical. **21**: Usually victims were stripped; the binding and clothing here heightens the miracle of their deliverance. **22**: They were cast in at the top opening, flames from which *killed* the soldiers. The king was looking into the ground level opening and saw where they fell.
3.25: *The appearance of a god*, i.e. a celestial being (v. 28). **26**: They emerged from the ground level opening. **29**: This dismember-

nego, who has sent his angel and delivered his servants who trusted in him. They disobeyed the king's command and yielded up their bodies rather than serve and worship any god except their own God. 29 Therefore I make a decree: Any people, nation, or language that utters blasphemy against the God of Shadrach, Meshach, and Abednego shall be torn limb from limb, and their houses laid in ruins; for there is no other god who is able to deliver in this way." 30 Then the king promoted Shadrach, Meshach, and Abednego in the province of Babylon.

4 *i* King Nebuchadnezzar to all peoples, nations, and languages that live throughout the earth: May you have abundant prosperity! 2 The signs and wonders that the Most High God has worked for me I am pleased to recount.
3 How great are his signs,
    how mighty his wonders!
  His kingdom is an everlasting
      kingdom,
    and his sovereignty is from
        generation to generation.
4 *j* I, Nebuchadnezzar, was living at ease in my home and prospering in my palace. 5 I saw a dream that frightened me; my fantasies in bed and the visions of my head terrified me. 6 So I made a decree that all the wise men of Babylon should be brought before me, in order that they might tell me the interpretation of the dream. 7 Then the magicians, the enchanters, the Chaldeans, and the diviners came in, and I told them the dream, but they could not tell me its interpretation. 8 At last Daniel came in before me—he who was named Belteshazzar after the name of my god, and who is endowed with a spirit of the holy

gods *k*—and I told him the dream: 9 "O Belteshazzar, chief of the magicians, I know that you are endowed with a spirit of the holy gods *k* and that no mystery is too difficult for you. Hear *l* the dream that I saw; tell me its interpretation.
10 *m* Upon my bed this is what I saw;
      there was a tree at the center of
          the earth,
        and its height was great.
11 The tree grew great and strong,
      its top reached to heaven,
        and it was visible to the ends of
            the whole earth.
12 Its foliage was beautiful,
      its fruit abundant,
        and it provided food for all.
    The animals of the field found
        shade under it,
      the birds of the air nested in
          its branches,
        and from it all living beings
            were fed.

13 I continued looking, in the visions of my head as I lay in bed, and there was a holy watcher, coming down from heaven. 14 He cried aloud and said:
  'Cut down the tree and chop off
      its branches,
    strip off its foliage and scatter
        its fruit.
  Let the animals flee from
      beneath it
    and the birds from its branches.
15 But leave its stump and roots in
        the ground,

i Ch 3.31 in Aram   *j* Ch 4.1 in Aram
k Or *a holy, divine spirit*   *l* Theodotion: Aram
The visions of   *m* Theodotion Syr Compare
Gk: Aram adds *The visions of my head*

---

ment of person and property was not uncommon punishment (2.5).
  4.1–37: Nebuchadnezzar's madness. A story to show how helpless is the greatest heathen power against Israel's God. **1:** Many ancient monarchs imagined they ruled *throughout the earth.*
  4.6: As in 2.2 court sages are expected to be skilled interpreters of dreams. **9:** For Daniel's office compare 2.48; 5.11. What the *spirit of*

the holy gods (v. 18; 5.11) means is quite obscure. **10:** *Tree,* see Ezek 31.3–14. It is the world-tree at the center of the earth disk, whose branches touch the heavenly vault and stretch in all directions to where this vault touches the rim of the disk (see 2.34 n.).
  4.13, 17, 23: The *watcher* was a celestial being (see, in the pseudepigraphic literature, 1 Enoch 12.2, 3; Jubilees 4.15), who *cried aloud* to his attendants (v. 14); such celestial

with a band of iron and bronze,
  in the tender grass of the field.
Let him be bathed with the dew
    of heaven.
  and let his lot be with the
    animals of the field
  in the grass of the earth.
16 Let his mind be changed from that
    of a human,
  and let the mind of an animal be
    given to him.
  And let seven times pass
    over him.
17 The sentence is rendered by decree
    of the watchers,
  the decision is given by order of
    the holy ones,
  in order that all who live
    may know
  that the Most High is sovereign
    over the kingdom of
    mortals;
  he gives it to whom he will
    and sets over it the lowliest of
    human beings.'

18 This is the dream that I, King Nebuchadnezzar, saw. Now you, Belteshazzar, declare the interpretation, since all the wise men of my kingdom are unable to tell me the interpretation. You are able, however, for you are endowed with a spirit of the holy gods."*n*

19 Then Daniel, who was called Belteshazzar, was severely distressed for a while. His thoughts terrified him. The king said, "Belteshazzar, do not let the dream or the interpretation terrify you." Belteshazzar answered, "My lord, may the dream be for those who hate you, and its interpretation for your enemies! 20 The

tree that you saw, which grew great and strong, so that its top reached to heaven and was visible to the end of the whole earth, 21 whose foliage was beautiful and its fruit abundant, and which provided food for all, under which animals of the field lived, and in whose branches the birds of the air had nests— 22 it is you, O king! You have grown great and strong. Your greatness has increased and reaches to heaven, and your sovereignty to the ends of the earth. 23 And whereas the king saw a holy watcher coming down from heaven and saying, 'Cut down the tree and destroy it, but leave its stump and roots in the ground, with a band of iron and bronze, in the grass of the field; and let him be bathed with the dew of heaven, and let his lot be with the animals of the field, until seven times pass over him'— 24 this is the interpretation, O king, and it is a decree of the Most High that has come upon my lord the king. 25 You shall be driven away from human society, and your dwelling shall be with the wild animals. You shall be made to eat grass like oxen, you shall be bathed with the dew of heaven, and seven times shall pass over you, until you have learned that the Most High has sovereignty over the kingdom of mortals, and gives it to whom he will. 26 As it was commanded to leave the stump and roots of the tree, your kingdom shall be re-established for you from the time that you learn that Heaven is sovereign. 27 Therefore, O king, may my counsel be acceptable to you: atone for*o* your sins with righteousness, and your iniquities with

n Or a holy, divine spirit   o Aram break off

beings, as a kind of heavenly council, execute the *sentence* given by God (v. 17). **16:** The *seven times* means a conventional number of years. **19:** *Distressed,* i.e. embarrassed and perplexed, so he uses a stereotyped formula of aversion. **22:** The tree is *you,* for a tree not uncommonly symbolized a man (Pss 1.3; 37.35; Jer 17.8).
4.25: The king will suffer temporarily from "insania zoanthropia," a form of insanity in which a man acts like a beast (v. 33).

History knows of no such affliction affecting Nebuchadnezzar. One of Nebuchadnezzar's successors, Nabonidus, was absent from Babylon for several years and neglected his duties. A Jewish document found among the Dead Sea Scrolls ("The Prayer of Nabonidus") says that Nabonidus was smitten for seven years, but was restored with the help of a Jewish diviner. The story of Nebuchadnezzar's madness may have arisen from this tradition.

mercy to the oppressed, so that your prosperity may be prolonged."

28 All this came upon King Nebuchadnezzar. 29 At the end of twelve months he was walking on the roof of the royal palace of Babylon, 30 and the king said, "Is this not magnificent Babylon, which I have built as a royal capital by my mighty power and for my glorious majesty?" 31 While the words were still in the king's mouth, a voice came from heaven: "O King Nebuchadnezzar, to you it is declared: The kingdom has departed from you! 32 You shall be driven away from human society, and your dwelling shall be with the animals of the field. You shall be made to eat grass like oxen, and seven times shall pass over you, until you have learned that the Most High has sovereignty over the kingdom of mortals and gives it to whom he will." 33 Immediately the sentence was fulfilled against Nebuchadnezzar. He was driven away from human society, ate grass like oxen, and his body was bathed with the dew of heaven, until his hair grew as long as eagles' feathers and his nails became like birds' claws.

34 When that period was over, I, Nebuchadnezzar, lifted my eyes to heaven, and my reason returned to me.
I blessed the Most High,
 and praised and honored the one
  who lives forever.
For his sovereignty is an
  everlasting sovereignty,
 and his kingdom endures from
  generation to generation.
35 All the inhabitants of the earth are
  accounted as nothing,
 and he does what he wills with
  the host of heaven

and the inhabitants of the earth.
There is no one who can stay
  his hand
 or say to him, "What are you
  doing?"

36 At that time my reason returned to me; and my majesty and splendor were restored to me for the glory of my kingdom. My counselors and my lords sought me out, I was re-established over my kingdom, and still more greatness was added to me. 37 Now I, Nebuchadnezzar, praise and extol and honor the King of heaven,
 for all his works are truth,
  and his ways are justice;
 and he is able to bring low
  those who walk in pride.

5 King Belshazzar made a great festival for a thousand of his lords, and he was drinking wine in the presence of the thousand.

2 Under the influence of the wine, Belshazzar commanded that they bring in the vessels of gold and silver that his father Nebuchadnezzar had taken out of the temple in Jerusalem, so that the king and his lords, his wives, and his concubines might drink from them. 3 So they brought in the vessels of gold and silver[p] that had been taken out of the temple, the house of God in Jerusalem, and the king and his lords, his wives, and his concubines drank from them. 4 They drank the wine and praised the gods of gold and silver, bronze, iron, wood, and stone.

5 Immediately the fingers of a human

p Theodotion Vg: Aram lacks *and silver*

4.30: Nebuchadnezzar had a reputation as a builder. 36: His *counselors* were his ministers of state; his *lords* his courtiers.
5.1–31: **Belshazzar's festival.** A story to show that divine punishment is visited on sacrilege. 1: *King Belshazzar*, son of Nabonidus, the last Neo-Babylonian ruler, was only viceroy during his father's absences. *Great festival* probably means a state banquet. 2: For the *vessels* see 1.2; Ezra 1.7–11. *His father*, i.e. his predecessor, for three kings ruled between

Nebuchadnezzar and Nabonidus. From vv. 11, 13, 18, 22, however, we see that this writer thought Nebuchadnezzar was his real father. 4: During the drinking, libations were poured out to heathen deities.
5.5: This sacrilege brought forth *the fingers of a human hand* writing a message of doom. 7: The *writing*, we learn from v. 25, was Aramaic, but the message was cryptic and needed *interpretation*.

hand appeared and began writing on the plaster of the wall of the royal palace, next to the lampstand. The king was watching the hand as it wrote. 6 Then the king's face turned pale, and his thoughts terrified him. His limbs gave way, and his knees knocked together. 7 The king cried aloud to bring in the enchanters, the Chaldeans, and the diviners; and the king said to the wise men of Babylon, "Whoever can read this writing and tell me its interpretation shall be clothed in purple, have a chain of gold around his neck, and rank third in the kingdom." 8 Then all the king's wise men came in, but they could not read the writing or tell the king the interpretation. 9 Then King Belshazzar became greatly terrified and his face turned pale, and his lords were perplexed.

10 The queen, when she heard the discussion of the king and his lords, came into the banqueting hall. The queen said, "O king, live forever! Do not let your thoughts terrify you or your face grow pale. 11 There is a man in your kingdom who is endowed with a spirit of the holy gods. *q* In the days of your father he was found to have enlightenment, understanding, and wisdom like the wisdom of the gods. Your father, King Nebuchadnezzar, made him chief of the magicians, enchanters, Chaldeans, and diviners, *r* 12 because an excellent spirit, knowledge, and understanding to interpret dreams, explain riddles, and solve problems were found in this Daniel, whom the king named Belteshazzar. Now let Daniel be called, and he will give the interpretation."

13 Then Daniel was brought in before the king. The king said to Daniel, "So you are Daniel, one of the exiles of Judah, whom my father the king brought from Judah? 14 I have heard of you that a spirit of the gods *s* is in you, and that enlightenment, understanding, and excellent wisdom are found in you. 15 Now the wise men, the enchanters, have been brought in before me to read this writing and tell me its interpretation, but they were not able to give the interpretation of the matter. 16 But I have heard that you can give interpretations and solve problems. Now if you are able to read the writing and tell me its interpretation, you shall be clothed in purple, have a chain of gold around your neck, and rank third in the kingdom."

17 Then Daniel answered in the presence of the king, "Let your gifts be for yourself, or give your rewards to someone else! Nevertheless I will read the writing to the king and let him know the interpretation. 18 O king, the Most High God gave your father Nebuchadnezzar kingship, greatness, glory, and majesty. 19 And because of the greatness that he gave him, all peoples, nations, and languages trembled and feared before him. He killed those he wanted to kill, kept alive those he wanted to keep alive, honored those he wanted to honor, and degraded those he wanted to degrade. 20 But when his heart was lifted up and his spirit was hardened so that he acted proudly, he was deposed from his kingly throne, and his glory was stripped from him. 21 He was driven from human society, and his mind was made like that of an animal. His dwelling was with the wild asses, he was fed grass like oxen, and his body was bathed with the dew of heaven, until he learned that the Most High God has sovereignty over the kingdom of mortals, and sets over it whomever he will. 22 And you, Belshazzar his son, have not humbled your heart, even though you knew all this! 23 You have exalted yourself against the Lord of heaven! The vessels of his temple have been brought in before you, and you and your lords, your wives and your concubines have been drinking wine from them. You have praised the gods of silver and gold, of bronze, iron, wood, and stone, which do not see or hear or know; but

q Or *a holy, divine spirit*   *r* Aram adds *the king your father*   *s* Or *a divine spirit*

5.10–11: The *queen* was the queen-mother.

the God in whose power is your very breath, and to whom belong all your ways, you have not honored.

24 "So from his presence the hand was sent and this writing was inscribed. 25 And this is the writing that was inscribed: MENE, MENE, TEKEL, and PARSIN. 26 This is the interpretation of the matter: MENE, God has numbered the days of[t] your kingdom and brought it to an end; 27 TEKEL, you have been weighed on the scales and found wanting; 28 PERES,[u] your kingdom is divided and given to the Medes and Persians."

29 Then Belshazzar gave the command, and Daniel was clothed in purple, a chain of gold was put around his neck, and a proclamation was made concerning him that he should rank third in the kingdom.

30 That very night Belshazzar, the Chaldean king, was killed. 31[v] And Darius the Mede received the kingdom, being about sixty-two years old.

6 It pleased Darius to set over the kingdom one hundred twenty satraps, stationed throughout the whole kingdom, 2 and over them three presidents, including Daniel; to these the satraps gave account, so that the king might suffer no loss. 3 Soon Daniel distinguished himself above all the other presidents and satraps because an excellent spirit was in him, and the king planned to appoint him over the whole kingdom. 4 So the presidents and the satraps tried to find grounds for complaint against Daniel in connection with the kingdom. But they could find no grounds for complaint or any corruption, because he was faithful, and no negligence or corruption could be found in him. 5 The men said, "We shall not find any ground for complaint against this Daniel unless we find it in connection with the law of his God."

6 So the presidents and satraps conspired and came to the king and said to him, "O King Darius, live forever! 7 All the presidents of the kingdom, the prefects and the satraps, the counselors and the governors are agreed that the king should establish an ordinance and enforce an interdict, that whoever prays to anyone, divine or human, for thirty days, except to you, O king, shall be thrown into a den of lions. 8 Now, O king, establish the interdict and sign the document, so that it cannot be changed, according to the law of the Medes and the Persians, which cannot be revoked." 9 Therefore King Darius signed the document and interdict.

10 Although Daniel knew that the document had been signed, he continued

t Aram lacks *the days of*   *u* The singular of *Parsin*   *v* Ch 6.1 in Aram

---

5.25–28: All three words represent weights, *mene* a mina, *tekel* a shekel, *parsin* two half-minas, but they may also be read as the verbs "to number," "to weigh," and "to divide." So the meaning may be that Evilmerodach, who succeeded Nebuchadnezzar, weighs a mina, his successor Neriglissar also a mina, his successor Labashi-Marduk only a shekel, Nabonidus and Belshazzar half a mina each, but that now by the judgment of numbering, weighing, and dividing the Neo-Babylonian kingdom is to be conquered by the Medo-Persians. **31**: No such person as *Darius the Mede* is known to history, but there were three Persian kings named Darius. It was Cyrus who overthrew the Neo-Babylonian power. Cyrus's general, Gobyras, was about *sixty-two* when he occupied Babylon.

6.1–28: **Daniel in the lions' den.** A story to show how God will deliver his faithful servants. **1**: Darius I of Persia set up the satrapies (i.e. provinces, each governed by a satrap), but *Darius* here means the Mede of 5.31. Jewish tradition increased the twenty-odd satrapies into over one hundred and twenty (Esth 1.1; 8.9). **6.2–5**: Each satrap had a military chief and a civil secretary, these being the *three presidents*. The king, however, planned to appoint Daniel as grand vizier. From this pinnacle jealous officials seek to topple him, using his religion as their lever. **7**: Their request is that for a month the king should be treated as divine. From vv. 17, 23, 24 it appears that the *den* was thought of as a pit. **8**: For the unchangeableness of Medo-Persian law see Esth 1.19; 8.8.

6.10: Daniel had a roof-chamber with windows allowing him to face Jerusalem in

to go to his house, which had windows in its upper room open toward Jerusalem, and to get down on his knees three times a day to pray to his God and praise him, just as he had done previously. [11] The conspirators came and found Daniel praying and seeking mercy before his God. [12] Then they approached the king and said concerning the interdict, "O king! Did you not sign an interdict, that anyone who prays to anyone, divine or human, within thirty days except to you, O king, shall be thrown into a den of lions?" The king answered, "The thing stands fast, according to the law of the Medes and Persians, which cannot be revoked." [13] Then they responded to the king, "Daniel, one of the exiles from Judah, pays no attention to you, O king, or to the interdict you have signed, but he is saying his prayers three times a day."

14 When the king heard the charge, he was very much distressed. He was determined to save Daniel, and until the sun went down he made every effort to rescue him. [15] Then the conspirators came to the king and said to him, "Know, O king, that it is a law of the Medes and Persians that no interdict or ordinance that the king establishes can be changed."

16 Then the king gave the command, and Daniel was brought and thrown into the den of lions. The king said to Daniel, "May your God, whom you faithfully serve, deliver you!" [17] A stone was brought and laid on the mouth of the den, and the king sealed it with his own signet and with the signet of his lords, so that nothing might be changed concerning Daniel. [18] Then the king went to his palace and spent the night fasting; no food was brought to him, and sleep fled from him.

19 Then, at break of day, the king got up and hurried to the den of lions. [20] When he came near the den where Daniel was, he cried out anxiously to Daniel, "O Daniel, servant of the living God, has your God whom you faithfully serve been able to deliver you from the lions?" [21] Daniel then said to the king, "O king, live forever! [22] My God sent his angel and shut the lions' mouths so that they would not hurt me, because I was found blameless before him; and also before you, O king, I have done no wrong." [23] Then the king was exceedingly glad and commanded that Daniel be taken up out of the den. So Daniel was taken up out of the den, and no kind of harm was found on him, because he had trusted in his God. [24] The king gave a command, and those who had accused Daniel were brought and thrown into the den of lions—they, their children, and their wives. Before they reached the bottom of the den the lions overpowered them and broke all their bones in pieces.

25 Then King Darius wrote to all peoples and nations of every language throughout the whole world: "May you have abundant prosperity! [26] I make a decree, that in all my royal dominion people should tremble and fear before the God of Daniel:

For he is the living God,
  enduring forever.
His kingdom shall never be
    destroyed,
  and his dominion has no end.
[27] He delivers and rescues,
  he works signs and wonders in
    heaven and on earth;
  for he has saved Daniel
  from the power of the lions."
[28] So this Daniel prospered during the reign of Darius and the reign of Cyrus the Persian.

---

prayer (1 Kings 8.44, 48). For the three daily prayers see Ps 55.17, and for kneeling 2 Chr 6.13; Ezra 9.5; Acts 20.36. **14:** The monarch sees through the conspiracy but is helpless.
  6.22: For shutting the lions' mouths see 1 Macc 2.60; Heb 11.33. **24:** Well recognized family solidarity is involved here, as punishment falls also on the families of the accusers (compare Num 16.25–33; Josh 7.24; 2 Sam 21.6, 9; Esth 9.13). **28:** Cyrus was the first foreign king to rule over Babylon, beginning in 538 B.C., but this writer obviously thinks of his *Darius* the Mede as preceding him.

7 In the first year of King Belshazzar of Babylon, Daniel had a dream and visions of his head as he lay in bed. Then he wrote down the dream:[w] 2 I,[x] Daniel, saw in my vision by night the four winds of heaven stirring up the great sea, 3 and four great beasts came up out of the sea, different from one another. 4 The first was like a lion and had eagles' wings. Then, as I watched, its wings were plucked off, and it was lifted up from the ground and made to stand on two feet like a human being; and a human mind was given to it. 5 Another beast appeared, a second one, that looked like a bear. It was raised up on one side, had three tusks[y] in its mouth among its teeth and was told, "Arise, devour many bodies!" 6 After this, as I watched, another appeared, like a leopard. The beast had four wings of a bird on its back and four heads; and dominion was given to it. 7 After this I saw in the visions by night a fourth beast, terrifying and dreadful and exceedingly strong. It had great iron teeth and was devouring, breaking in pieces, and stamping what was left with its feet. It was different from all the beasts that preceded it, and it had ten horns. 8 I was considering the horns, when another horn appeared, a little one coming up among them; to make room for it, three of the earlier horns were plucked up by the roots. There were eyes like human eyes in this horn, and a mouth speaking arrogantly.

9 As I watched,

thrones were set in place,
and an Ancient One[z] took
his throne,
his clothing was white as snow,
and the hair of his head like
pure wool;
his throne was fiery flames,
and its wheels were burning
fire.
10 A stream of fire issued
and flowed out from his
presence.
A thousand thousands served him,
and ten thousand times ten
thousand stood
attending him.
The court sat in judgment,
and the books were opened.

11 I watched then because of the noise of the arrogant words that the horn was speaking. And as I watched, the beast was put to death, and its body destroyed and given over to be burned with fire. 12 As for the rest of the beasts, their dominion was taken away, but their lives were prolonged for a season and a time. 13 As I watched in the night visions,

I saw one like a human being[a]
coming with the clouds
of heaven.
And he came to the Ancient One[b]
and was presented before him.

w Q Ms Theodotion: MT adds *the beginning of the words; he said*   x Theodotion: Aram *Daniel answered and said, "I*   y Or *ribs*   z Aram *an Ancient of Days*   a Aram *one like a son of man* b Aram *the Ancient of Days*

7.1–28: The vision of the four beasts. A vision of the passing of kingdoms to make way for the kingdom of God. **1:** *The first year,* i.e. 554 B.C., when he began to act as viceroy for his father. **2:** The *great sea* is a traditional symbol of chaos, often associated with dragons and monsters (Isa 51.9–10; Isa 27.1; Job 26.12; Rev 13.1). In accord with the view that there were four kingdoms (see ch 2), *four winds* appropriately introduce the beasts symbolizing these kingdoms. **7.4–8:** The winged lion represents the Babylonian empire, the bear the Medes, the four-headed winged leopard the Persians, the dragon-like beast the Greeks, whose ten horns represent the ten rulers who succeeded

Alexander. The *little* horn (compare 8.9) is Antiochus Epiphanes, who gained his throne by uprooting others. **7.9–14:** The divine judgment. God, the *Ancient One,* appears on his fiery throne surrounded by his court (1 Kings 22.19); the record books are examined and judgment is given. **11–12:** The Greek empire is singled out for destruction. **13–14:** God will then give the *one like a human being* a universal and *everlasting dominion.* This figure represents the faithful Jews, and should probably be identified as the archangel Michael (see 10.13, 21; 12.1), although he was traditionally identified as the messiah.

14 To him was given dominion
  and glory and kingship,
  that all peoples, nations, and
    languages
  should serve him.
  His dominion is an everlasting
    dominion
  that shall not pass away,
  and his kingship is one
  that shall never be destroyed.

15 As for me, Daniel, my spirit was troubled within me,*c* and the visions of my head terrified me. 16 I approached one of the attendants to ask him the truth concerning all this. So he said that he would disclose to me the interpretation of the matter: 17 "As for these four great beasts, four kings shall arise out of the earth. 18 But the holy ones of the Most High shall receive the kingdom and possess the kingdom forever—forever and ever."

19 Then I desired to know the truth concerning the fourth beast, which was different from all the rest, exceedingly terrifying, with its teeth of iron and claws of bronze, and which devoured and broke in pieces, and stamped what was left with its feet; 20 and concerning the ten horns that were on its head, and concerning the other horn, which came up and to make room for which three of them fell out—the horn that had eyes and a mouth that spoke arrogantly, and that seemed greater than the others. 21 As I looked, this horn made war with the holy ones and was prevailing over them, 22 until the Ancient One*d* came; then judgment was given for the holy ones of the Most High, and the time arrived when the holy ones gained possession of the kingdom.

23 This is what he said: "As for the fourth beast,
  there shall be a fourth kingdom
    on earth

  that shall be different from all
    the other kingdoms;
  it shall devour the whole earth,
  and trample it down, and break
    it to pieces.
24 As for the ten horns,
  out of this kingdom ten kings
    shall arise,
  and another shall arise after
    them.
  This one shall be different from
    the former ones,
  and shall put down three kings.
25 He shall speak words against the
    Most High,
  shall wear out the holy ones of
    the Most High,
  and shall attempt to change the
    sacred seasons and the law;
and they shall be given into
    his power
  for a time, two times,*e* and half
    a time.
26 Then the court shall sit in
    judgment,
  and his dominion shall be
    taken away,
  to be consumed and totally
    destroyed.
27 The kingship and dominion
  and the greatness of the
    kingdoms under the
    whole heaven
  shall be given to the people of
    the holy ones of the
    Most High;
their kingdom shall be an
    everlasting kingdom,
  and all dominions shall serve
    and obey them."

28 Here the account ends. As for me, Daniel, my thoughts greatly terrified

---

c Aram *troubled in its sheath*   *d* Aram *the Ancient of Days*   *e* Aram *a time, times*

7.16: At the seer's request an angel gives the *interpretation*. **21**: "Holy ones" are angels (compare 8.13). Here they represent the righteous Jews persecuted by Antiochus Epiphanes.
7.25–28: Catalogue of Antiochus' enormities with prediction of their end after three and a half years (i.e. *a time, two times, and half a time;* compare 8.14; 9.27; 12.7, 11, 12). After his end the expected kingdom will be established.

me, and my face turned pale; but I kept the matter in my mind.

8 In the third year of the reign of King Belshazzar a vision appeared to me, Daniel, after the one that had appeared to me at first. ²In the vision I was looking and saw myself in Susa the capital, in the province of Elam,*f* and I was by the river Ulai.*g* ³I looked up and saw a ram standing beside the river.*h* It had two horns. Both horns were long, but one was longer than the other, and the longer one came up second. ⁴I saw the ram charging westward and northward and southward. All beasts were powerless to withstand it, and no one could rescue from its power; it did as it pleased and became strong.

5 As I was watching, a male goat appeared from the west, coming across the face of the whole earth without touching the ground. The goat had a horn*i* between its eyes. ⁶It came toward the ram with the two horns that I had seen standing beside the river,*h* and it ran at it with savage force. ⁷I saw it approaching the ram. It was enraged against it and struck the ram, breaking its two horns. The ram did not have power to withstand it; it threw the ram down to the ground and trampled upon it, and there was no one who could rescue the ram from its power. ⁸Then the male goat grew exceedingly great; but at the height of its power, the great horn was broken, and in its place there came up four prominent horns toward the four winds of heaven.

9 Out of one of them came another*j*

horn, a little one, which grew exceedingly great toward the south, toward the east, and toward the beautiful land. ¹⁰It grew as high as the host of heaven. It threw down to the earth some of the host and some of the stars, and trampled on them. ¹¹Even against the prince of the host it acted arrogantly; it took the regular burnt offering away from him and overthrew the place of his sanctuary. ¹²Because of wickedness, the host was given over to it together with the regular burnt offering;*k* it cast truth to the ground, and kept prospering in what it did. ¹³Then I heard a holy one speaking, and another holy one said to the one that spoke, "For how long is this vision concerning the regular burnt offering, the transgression that makes desolate, and the giving over of the sanctuary and host to be trampled?"*k* ¹⁴And he answered him,*l* "For two thousand three hundred evenings and mornings; then the sanctuary shall be restored to its rightful state."

15 When I, Daniel, had seen the vision, I tried to understand it. Then someone appeared standing before me, having the appearance of a man, ¹⁶and I heard a human voice by the Ulai, calling, "Gabriel, help this man understand the vision." ¹⁷So he came near where I stood; and when he came, I became frightened

---

f Gk Theodotion: MT Q Ms repeat *in the vision
I was looking*   *g* Or *the Ulai Gate*   *h* Or *gate*
i Theodotion: Gk *one horn*; Heb *a horn of vision*
j Cn Compare 7.8: Heb *one*   *k* Meaning of
Heb uncertain   *l* Gk Theodotion Syr Vg:
Heb *me*

8.1–27: The vision of the ram and the male goat. 1–2: The vision is dated two years later than the previous one (7.1), placing the seer in *Susa,* the winter capital of the Persian kings. The *Ulai* is the Eulaeus. **3–4:** The two-horned ram is the Medo-Persian empire (v. 20), the advance of the Persians being irresistible. **8.5–7:** The male goat from the west is Alexander the Great (v. 21), who overthrew the Persian empire. **8:** Alexander's empire was divided, the four *prominent* leaders being Cassander, Lysimachus, Seleucus, and Ptolemy. **8.9–14:** From the Seleucids sprang Antiochus Epiphanes who in 167 B.C. conquered

Palestine, violated the *sanctuary,* and prohibited worship there (vv. 23–25; see "Survey of . . . Bible Lands," §19). The angels comment and calculate that this will continue about three and a half years (*two thousand three hundred evenings and mornings* are one thousand, one hundred and fifty days; see 7.25–28 n.). *The sanctuary* was *restored* on 25th Chislev (December 14th) 164 B.C., only three years after it was desecrated. **8.16:** *A human voice* i.e. a celestial being speaking human language. **17:** *The end,* compare v. 19; 11.35; 12.4, 9, 13. **23–24:** The shameless insolence, "double talk," and ruthlessness of Antiochus were notorious. *The*

and fell prostrate. But he said to me, "Understand, O mortal, *m* that the vision is for the time of the end.".

18 As he was speaking to me, I fell into a trance, face to the ground; then he touched me and set me on my feet. 19 He said, "Listen, and I will tell you what will take place later in the period of wrath; for it refers to the appointed time of the end. 20 As for the ram that you saw with the two horns, these are the kings of Media and Persia. 21 The male goat *n* is the king of Greece, and the great horn between its eyes is the first king. 22 As for the horn that was broken, in place of which four others arose, four kingdoms shall arise from his *o* nation, but not with his power.

23 At the end of their rule,
 when the transgressions have
  reached their full measure,
 a king of bold countenance
  shall arise,
 skilled in intrigue.
24 He shall grow strong in power, *p*
 shall cause fearful destruction,
 and shall succeed in what
  he does.
He shall destroy the powerful
 and the people of the holy ones.
25 By his cunning
 he shall make deceit prosper
  under his hand,
 and in his own mind he shall
  be great.
Without warning he shall
  destroy many
 and shall even rise up against
  the Prince of princes.
But he shall be broken, and not
  by human hands.
26 The vision of the evenings and the mornings that has been told is true. As for you, seal up the vision, for it refers to many days from now."

27 So I, Daniel, was overcome and lay sick for some days; then I arose and went about the king's business. But I was dismayed by the vision and did not understand it.

9 In the first year of Darius son of Ahasuerus, by birth a Mede, who became king over the realm of the Chaldeans— 2 in the first year of his reign, I, Daniel, perceived in the books the number of years that, according to the word of the LORD to the prophet Jeremiah, must be fulfilled for the devastation of Jerusalem, namely, seventy years.

3 Then I turned to the Lord God, to seek an answer by prayer and supplication with fasting and sackcloth and ashes. 4 I prayed to the LORD my God and made confession, saying,

"Ah, Lord, great and awesome God, keeping covenant and steadfast love with those who love you and keep your commandments, 5 we have sinned and done wrong, acted wickedly and rebelled, turning aside from your commandments and ordinances. 6 We have not listened to your servants the prophets, who spoke in your name to our kings, our princes, and our ancestors, and to all the people of the land.

7 "Righteousness is on your side, O Lord, but open shame, as at this day, falls on us, the people of Judah, the inhabitants of Jerusalem, and all Israel, those who are near and those who are far

---

m Heb *son of man*   *n* Or *shaggy male goat*
o Gk Theodotion Vg: Heb *the*   *p* Theodotion
and one Gk Ms: Heb repeats (from 8.22) *but not*
with his power

---

people of the holy ones are the godly Jews (7.25, 27). **25**: *The Prince of princes* is God, to whom Antiochus' self-deification (11.36) was an affront, and by whom he was *broken* (2 Macc 9.5). **8.26–27**: Since Daniel is pictured as being in Susa in 552 B.C., and the vision concerns events of 164, the account is to be preserved for that time, for naturally Daniel did *not understand it.*

9.1–27: **The prophecy of the seventy weeks,** expounding a prophecy of Jeremiah. **1**: According to this writer's chronology the *first year* of *Darius* the Mede was 538 B.C. *Ahasuerus* means Xerxes, a fictitious parent for a fictitious Darius. **2**: The *seventy years* are referred to in Jer 25.11, 12; 29.10. **4–19**: With Daniel's prayer of confession compare Neh chs 1 and 9.

away, in all the lands to which you have driven them, because of the treachery that they have committed against you. 8 Open shame, O LORD, falls on us, our kings, our officials, and our ancestors, because we have sinned against you. 9 To the Lord our God belong mercy and forgiveness, for we have rebelled against him, 10 and have not obeyed the voice of the LORD our God by following his laws, which he set before us by his servants the prophets.

11 "All Israel has transgressed your law and turned aside, refusing to obey your voice. So the curse and the oath written in the law of Moses, the servant of God, have been poured out upon us, because we have sinned against you. 12 He has confirmed his words, which he spoke against us and against our rulers, by bringing upon us a calamity so great that what has been done against Jerusalem has never before been done under the whole heaven. 13 Just as it is written in the law of Moses, all this calamity has come upon us. We did not entreat the favor of the LORD our God, turning from our iniquities and reflecting on his*q* fidelity. 14 So the LORD kept watch over this calamity until he brought it upon us. Indeed, the LORD our God is right in all that he has done; for we have disobeyed his voice.

15 "And now, O Lord our God, who brought your people out of the land of Egypt with a mighty hand and made your name renowned even to this day— we have sinned, we have done wickedly. 16 O Lord, in view of all your righteous acts, let your anger and wrath, we pray, turn away from your city Jerusalem, your holy mountain; because of our sins and the iniquities of our ancestors, Jerusalem and your people have become a disgrace among all our neighbors. 17 Now therefore, O our God, listen to the prayer of your servant and to his supplication, and for your own sake, Lord,*r* let your face shine upon your desolated sanctuary. 18 Incline your ear, O my God, and hear. Open your eyes and look at our desolation and the city that bears your name. We do not present our supplication before you on the ground of our righteousness, but on the ground of your great mercies. 19 O Lord, hear; O Lord, forgive; O Lord, listen and act and do not delay! For your own sake, O my God, because your city and your people bear your name!"

20 While I was speaking, and was praying and confessing my sin and the sin of my people Israel, and presenting my supplication before the LORD my God on behalf of the holy mountain of my God— 21 while I was speaking in prayer, the man Gabriel, whom I had seen before in a vision, came to me in swift flight at the time of the evening sacrifice. 22 He came*s* and said to me, "Daniel, I have now come out to give you wisdom and understanding. 23 At the beginning of your supplications a word went out, and I have come to declare it, for you are greatly beloved. So consider the word and understand the vision:

24 "Seventy weeks are decreed for your people and your holy city: to finish the transgression, to put an end to sin, and to atone for iniquity, to bring in everlasting righteousness, to seal both vision and prophet, and to anoint a most holy place.*t* 25 Know therefore and understand: from the time that the word went out to restore and rebuild Jerusalem until the time of an anointed prince, there shall be seven weeks; and for sixty-two weeks it shall be built again with streets

q Heb *your*   r Theodotion Vg Compare Syr: Heb *for the Lord's sake*   s Gk Syr: Heb *He made to understand*   t Or *thing* or *one*

9.11: For this *curse* see Deut 28.15–45. 13: For what was *written* see Lev 26.14–22. 17: The *desolated* sanctuary is a hint of Antiochus' "abomination that desolates" (v. 27; 8.13; see 11.29–31 n.).

9.21: Gabriel is called *the man* because he appeared in human form (8.15). 24: Jeremiah's seventy years (see v. 2 n.) mean *seventy weeks*, or 490 years (i.e. 70 × 7 years), after which the final kingdom will come, fulfilling both *vision and prophet*.

9.25–27: The *anointed prince* is either

and moat, but in a troubled time. ²⁶ After the sixty-two weeks, an anointed one shall be cut off and shall have nothing, and the troops of the prince who is to come shall destroy the city and the sanctuary. Its*u* end shall come with a flood, and to the end there shall be war. Desolations are decreed. ²⁷ He shall make a strong covenant with many for one week, and for half of the week he shall make sacrifice and offering cease; and in their place*v* shall be an abomination that desolates, until the decreed end is poured out upon the desolator."

10 In the third year of King Cyrus of Persia a word was revealed to Daniel, who was named Belteshazzar. The word was true, and it concerned a great conflict. He understood the word, having received understanding in the vision.

2 At that time I, Daniel, had been mourning for three weeks. ³ I had eaten no rich food, no meat or wine had entered my mouth, and I had not anointed myself at all, for the full three weeks. ⁴ On the twenty-fourth day of the first month, as I was standing on the bank of the great river (that is, the Tigris), ⁵ I looked up and saw a man clothed in linen, with a belt of gold from Uphaz around his waist. ⁶ His body was like beryl, his face like lightning, his eyes like flaming torches, his arms and legs like the gleam of burnished bronze, and the sound of his words like the roar of a multitude. ⁷ I, Daniel, alone saw the vision; the people who were with me did not see the vision, though a great trembling fell upon them, and they fled and hid themselves. ⁸ So I was left alone to see this great vision. My strength left me, and my complexion grew deathly pale, and I retained no strength. ⁹ Then I heard the sound of his words; and when I heard the sound of his words, I fell into a trance, face to the ground.

10 But then a hand touched me and roused me to my hands and knees. ¹¹ He said to me, "Daniel, greatly beloved, pay attention to the words that I am going to speak to you. Stand on your feet, for I have now been sent to you." So while he was speaking this word to me, I stood up trembling. ¹² He said to me, "Do not fear, Daniel, for from the first day that you set your mind to gain understanding and to humble yourself before your God, your words have been heard, and I have come because of your words. ¹³ But the prince of the kingdom of Persia opposed me twenty-one days. So Michael, one of the chief princes, came to help me, and I left him there with the prince of the kingdom of Persia, *w* ¹⁴ and have come to help you understand what is to happen to your people at the end of days. For there is a further vision for those days."

15 While he was speaking these words to me, I turned my face toward

u Or *His*   v Cn: Meaning of Heb uncertain
w Gk Theodotion: Heb *I was left there with the kings of Persia*

Zerubbabel or Joshua (Ezra 3.2). The one cut off (v. 26) is Onias III (2 Macc 4.34, Dan 11.22). The *prince who is to come* is doubtless Antiochus, who wrought such *desolations,* made *a covenant* with the Hellenizing Jews, and desecrated the temple. *Seven weeks,* i.e. forty-nine years. *Sixty-two weeks,* i.e. 434 years. *One week,* i.e. seven years. *Half of the week,* i.e. three and one-half years; see 7.25–28 n.
10.1–21: **A vision of the last days** (which continues through 12.13, ch 10 being prologue, ch 11 the vision, ch 12 the epilogue). **1:** *The third year of King Cyrus,* 535 B.C. For the first *conflict* see v. 13. **4:** *Tigris* is a gloss on *great river,* for Babylon is on the Euphrates.

10.5: The *man* was a celestial being, probably Gabriel, in human form, but having angelic splendor. *Uphaz,* see Jer 10.9. **7:** His companions did not see the vision but were affected by the supernatural presence, as in Acts 9.7.
10.12–14: *Michael . . . came to help,* as in 9.21–23 Gabriel is God's response to supplication. Conflict in heaven among the angelic patrons of the nations had delayed his coming three weeks, and he got away because *Michael,* the Jews' patron angel (v. 21), had helped him against the patron angel of Persia. In vv. 20–21 he says he must return, when he expects to be involved with the patron angel of Greece also.

the ground and was speechless. 16 Then one in human form touched my lips, and I opened my mouth to speak, and said to the one who stood before me, "My lord, because of the vision such pains have come upon me that I retain no strength. 17 How can my lord's servant talk with my lord? For I am shaking,* no strength remains in me, and no breath is left in me."

18 Again one in human form touched me and strengthened me. 19 He said, "Do not fear, greatly beloved, you are safe. Be strong and courageous!" When he spoke to me, I was strengthened and said, "Let my lord speak, for you have strengthened me." 20 Then he said, "Do you know why I have come to you? Now I must return to fight against the prince of Persia, and when I am through with him, the prince of Greece will come. 21 But I am to tell you what is inscribed in the book of truth. There is no one with me who contends against these princes except Michael, your prince. 11 1 As for me, in the first year of Darius the Mede, I stood up to support and strengthen him.

2 "Now I will announce the truth to you. Three more kings shall arise in Persia. The fourth shall be far richer than all of them, and when he has become strong through his riches, he shall stir up all against the kingdom of Greece. 3 Then a warrior king shall arise, who shall rule with great dominion and take action as he pleases. 4 And while still rising in power, his kingdom shall be broken and divided toward the four winds of heaven, but not to his posterity, nor accord-

ing to the dominion with which he ruled; for his kingdom shall be uprooted and go to others besides these.

5 "Then the king of the south shall grow strong, but one of his officers shall grow stronger than he and shall rule a realm greater than his own realm. 6 After some years they shall make an alliance, and the daughter of the king of the south shall come to the king of the north to ratify the agreement. But she shall not retain her power, and his offspring shall not endure. She shall be given up, she and her attendants and her child and the one who supported her.

"In those times 7 a branch from her roots shall rise up in his place. He shall come against the army and enter the fortress of the king of the north, and he shall take action against them and prevail. 8 Even their gods, with their idols and with their precious vessels of silver and gold, he shall carry off to Egypt as spoils of war. For some years he shall refrain from attacking the king of the north; 9 then the latter shall invade the realm of the king of the south, but will return to his own land.

10 "His sons shall wage war and assemble a multitude of great forces, which shall advance like a flood and pass through, and again shall carry the war as far as his fortress. 11 Moved with rage, the king of the south shall go out and do battle against the king of the north, who shall muster a great multitude, which shall, however, be defeated by his enemy. 12 When the multitude has been car-

x Gk: Heb *from now*

11.1–12.13: Interpretation of the vision of history unfolding. 11.1: The angel proceeds to unfold history. **2:** If the *three more* are those after Cyrus they are probably Cambyses, Darius I, and Xerxes I, who warred with the Greeks. Some suggest that Darius III, the last king of Persia, is the *fourth;* others believe Artaxerxes is meant. **3:** The *warrior king* is Alexander the Great. **4:** For his kingdom broken into four, see 8.8 n. None of his successors were of *his posterity.*

11.5–6: The kings of *the south* are the Ptolemies, those of *the north* the Seleucids. Here the

king is Ptolemy I, and the officer who was *stronger* is Seleucus I. The *alliance* was that of about 250 B.C., when Ptolemy II gave his daughter Bernice to Antiochus II; but Bernice, her *attendants,* her *child,* and her husband fell, owing to the plotting of Laodice, mother of Seleucus II. **7–8:** The *branch* is Ptolemy III, who captured the *fortress* of Seleucia and brought back immense booty. **9:** A reference to the campaign of Seleucus II against Egypt in 242 B.C., which came to disaster.

11.10: *His sons* were Seleucus III and Antiochus III, the latter of whom attacked Egypt.

ried off, his heart shall be exalted, and he shall overthrow tens of thousands, but he shall not prevail. 13 For the king of the north shall again raise a multitude, larger than the former, and after some years[y] he shall advance with a great army and abundant supplies.

14 "In those times many shall rise against the king of the south. The lawless among your own people shall lift themselves up in order to fulfill the vision, but they shall fail. 15 Then the king of the north shall come and throw up siege-works, and take a well-fortified city. And the forces of the south shall not stand, not even his picked troops, for there shall be no strength to resist. 16 But he who comes against him shall take the actions he pleases, and no one shall withstand him. He shall take a position in the beautiful land, and all of it shall be in his power. 17 He shall set his mind to come with the strength of his whole kingdom, and he shall bring terms of peace[z] and perform them. In order to destroy the kingdom,[a] he shall give him a woman in marriage; but it shall not succeed or be to his advantage. 18 Afterward he shall turn to the coastlands, and shall capture many. But a commander shall put an end to his insolence; indeed,[b] he shall turn his insolence back upon him. 19 Then he shall turn back toward the fortresses of his own land, but he shall stumble and fall, and shall not be found.

20 "Then shall arise in his place one who shall send an official for the glory of the kingdom; but within a few days he shall be broken, though not in anger or in battle. 21 In his place shall arise a contemptible person on whom royal majesty had not been conferred; he shall come in without warning and obtain the kingdom through intrigue. 22 Armies shall be utterly swept away and broken before him, and the prince of the covenant as well. 23 And after an alliance is made with him, he shall act deceitfully and become strong with a small party. 24 Without warning he shall come into the richest parts[c] of the province and do what none of his predecessors had ever done, lavishing plunder, spoil, and wealth on them. He shall devise plans against strongholds, but only for a time. 25 He shall stir up his power and determination against the king of the south with a great army, and the king of the south shall wage war with a much greater and stronger army. But he shall not succeed, for plots shall be devised against him 26 by those who eat of the royal rations. They shall break him, his army shall be swept away, and many shall fall slain. 27 The two kings, their minds bent on evil, shall sit at one table and exchange lies. But it shall not succeed, for there remains an end at the time appointed. 28 He shall return to his land with great wealth, but his heart shall be set against the holy covenant. He shall work his will, and return to his own land.

y Heb *and at the end of the times years*
z Gk: Heb *kingdom, and upright ones with him*
a Heb *it*   b Meaning of Heb uncertain
c Or *among the richest men*

11–13: Ptolemy IV sent armies through Palestine and defeated Antiochus at Raphia, but Antiochus in turn crushed the Egyptians at Banias.

11.14: A reference to the Egyptian insurrections under the child king, Ptolemy V. **15–17**: Antiochus III campaigned against Egypt, taking possession of Palestine, then made peace with Egypt, sealing it by marrying his daughter to the youthful Ptolemy. **18–19**: Antiochus undertook a campaign to capture the *coastlands* of Asia Minor, but, checked by the Roman *commander,* he started plundering on his return journey and died at Elymais.

11.20: Seleucus IV succeeded him and, to help pay the Roman indemnity, sent to Jerusalem an *official,* Heliodorus, to seize the temple treasure. He failed and died ingloriously. **21–45**: These verses concern Antiochus IV Epiphanes, the *contemptible person* who attained power by guile. **22–24**: His southern campaign, during which he deposed *the prince,* i.e. the high-priest Onias III, by *alliance* appointed Jason as high-priest, and let his troops plunder Palestine. **25–28**: His Egyptian campaign. In 169 he invaded Egypt and captured Ptolemy VI. Troubles at home forced him to leave Egypt, and on his way back with great booty, he sacked Jerusalem and plundered the treasury.

29 "At the time appointed he shall return and come into the south, but this time it shall not be as it was before. 30 For ships of Kittim shall come against him, and he shall lose heart and withdraw. He shall be enraged and take action against the holy covenant. He shall turn back and pay heed to those who forsake the holy covenant. 31 Forces sent by him shall occupy and profane the temple and fortress. They shall abolish the regular burnt offering and set up the abomination that makes desolate. 32 He shall seduce with intrigue those who violate the covenant; but the people who are loyal to their God shall stand firm and take action. 33 The wise among the people shall give understanding to many; for some days, however, they shall fall by sword and flame, and suffer captivity and plunder. 34 When they fall victim, they shall receive a little help, and many shall join them insincerely. 35 Some of the wise shall fall, so that they may be refined, purified, and cleansed, *d* until the time of the end, for there is still an interval until the time appointed.

36 "The king shall act as he pleases. He shall exalt himself and consider himself greater than any god, and shall speak horrendous things against the God of gods. He shall prosper until the period of wrath is completed, for what is determined shall be done. 37 He shall pay no respect to the gods of his ancestors, or to the one beloved by women; he shall pay no respect to any other god, for he shall consider himself greater than all. 38 He shall honor the god of fortresses instead of these; a god whom his ancestors did not know he shall honor with gold and silver, with precious stones and costly gifts. 39 He shall deal with the strongest fortresses by the help of a foreign god. Those who acknowledge him he shall make more wealthy, and shall appoint them as rulers over many, and shall distribute the land for a price.

40 "At the time of the end the king of the south shall attack him. But the king of the north shall rush upon him like a whirlwind, with chariots and horsemen, and with many ships. He shall advance against countries and pass through like a flood. 41 He shall come into the beautiful land, and tens of thousands shall fall victim, but Edom and Moab and the main part of the Ammonites shall escape from his power. 42 He shall stretch out his hand against the countries, and the land of Egypt shall not escape. 43 He shall become ruler of the treasures of gold and of silver, and all the riches of Egypt; and the Libyans and the Ethiopians *e* shall follow in his train. 44 But reports from the east and the north shall alarm him, and he shall go out with great fury to bring ruin and complete destruction to many. 45 He shall pitch his palatial tents between the sea and the beautiful holy mountain. Yet he shall come to his end, with no one to help him.

12 "At that time Michael, the great prince, the protector of your people, shall arise. There shall be a time of

d Heb *made them white*   *e* Or *Nubians;* Heb *Cushites*

11.29–31: The second campaign against Egypt, when Romans (*Kittim:* the name comes from "Citium," Cyprus, and is applied to any westerners) forced him to withdraw. Being *enraged* he attacked Jerusalem again in 167 B.C., setting up the *abomination that makes desolate,* i.e. a heathen altar in the temple (see "Survey of . . . Bible Lands," §19). **11.32–35**: The resistance movement. Some Hellenizing Jews sided with Antiochus, forsaking the *covenant,* but the *wise,* though persecuted, maintained resistance. The *little help* may refer to the Maccabees, Mattathias and his son, Judas Maccabeus (1 Macc ch 2).

36–39: Antiochus' march towards his doom. He abandoned *the gods of his ancestors* and the Tammuz-Adonis cult, being interested in Zeus Olympius and claiming divine honors for himself. **11.40–45**: Predictions that Ptolemy will provoke another war with disastrous results, so that Antiochus will conquer Libya to the west of Egypt and Ethiopia to the south, but on his way back will perish somewhere along the coastal route. None of these predictions was fulfilled. **12.1–13: The final consummation. 1:** These campaigns of Antiochus only intro-

anguish, such as has never occurred since nations first came into existence. But at that time your people shall be delivered, everyone who is found written in the book. ²Many of those who sleep in the dust of the earth*ᶠ* shall awake, some to everlasting life, and some to shame and everlasting contempt. ³Those who are wise shall shine like the brightness of the sky,*ᵍ* and those who lead many to righteousness, like the stars forever and ever. ⁴But you, Daniel, keep the words secret and the book sealed until the time of the end. Many shall be running back and forth, and evil*ʰ* shall increase."

5 Then I, Daniel, looked, and two others appeared, one standing on this bank of the stream and one on the other. ⁶One of them said to the man clothed in linen, who was upstream, "How long shall it be until the end of these wonders?" ⁷The man clothed in linen, who was upstream, raised his right hand and his left hand toward heaven. And I heard him swear by the one who lives forever that it would be for a time, two times,

and half a time,*ⁱ* and that when the shattering of the power of the holy people comes to an end, all these things would be accomplished. ⁸I heard but could not understand; so I said, "My lord, what shall be the outcome of these things?" ⁹He said, "Go your way, Daniel, for the words are to remain secret and sealed until the time of the end. ¹⁰Many shall be purified, cleansed, and refined, but the wicked shall continue to act wickedly. None of the wicked shall understand, but those who are wise shall understand. ¹¹From the time that the regular burnt offering is taken away and the abomination that desolates is set up, there shall be one thousand two hundred ninety days. ¹²Happy are those who persevere and attain the thousand three hundred thirty-five days. ¹³But you, go your way,*ʲ* and rest; you shall rise for your reward at the end of the days."

f Or *the land of dust*   *g* Or *dome*
h Cn Compare Gk: Heb *knowledge*   *i* Heb *a time, times, and a half*   *j* Gk Theodotion: Heb adds *to the end*

---

duce the great tribulation which precedes the end of the age. *Michael,* see 10.12–13 n. **2:** The first clear reference to resurrection in the Bible. Not all are raised, but those who are raised will find everlasting reward or punishment. **4:** Knowledge of these matters is to be kept secret *until the end.*

12.6–7: When will the end be? One angelic

response to Daniel's question: it will be in three and a half years (see 7.25–28 n.). **8–12:** Two new and variant calculations setting the end somewhat later, i.e. 1,290 days and 1,335 days, perhaps added after the three and one-half years (1,150 days) had passed. **13:** The promise that Daniel will have a place in that final consummation.

Hosea

The book of Hosea stands first in that part of the Latter Prophets called the Book of the Twelve, also known as the Minor Prophets because of the brevity of the books in comparison with Isaiah, Jeremiah, and Ezekiel (see Introduction to the Old Testament, p. xxxi OT). By the second century B.C. these twelve constituted a unit (see Sir 49.10). See xxi.

Hosea's ministry to the northern kingdom followed closely upon that of Amos (see Introduction to Amos). While Amos had spoken as a southerner to a prosperous Israel enjoying an era of peace, Hosea spoke as a native to his own people who were suffering from war with Assyria and in virtual anarchy. Four Israelite kings had been assassinated within fourteen years after the death of Jeroboam II. After the Assyrian conquest of 733–732, which resulted in the fall of Damascus, Samaria itself soon fell to the Assyrians (721).

Not all aspects of Hosea's life are clear. It can best be reconstructed, however, from the first three chapters by observing that Hosea deals with Gomer as the LORD deals with Israel; the prophet's personal life is an embodiment of God's redeeming love. Accordingly the sensitive prophet, obedient to his call to take "a wife of whoredom," married the prostitute Gomer. She bore three children, of whom Hosea was presumably not the father (2.4–5), and then left him. But Hosea brought her back publicly (3.1–5) and took her again to himself.

Hosea's ministry dramatizes his message. Although the book is divided into two uneven parts (chs 1–3; 4–14) and presents serious textual difficulties, it is thoroughly unified by the dominant theme of divine compassion and the love that will not let Israel go. At the heart of Hosea's preaching is a gospel of redeeming love.

1 The word of the LORD that came to Hosea son of Beeri, in the days of Kings Uzziah, Jotham, Ahaz, and Hezekiah of Judah, and in the days of King Jeroboam son of Joash of Israel.

2 When the LORD first spoke through Hosea, the LORD said to Hosea, "Go, take for yourself a wife of whoredom and have children of whoredom, for the land commits great whoredom by forsaking the LORD." 3 So he went and took Gomer daughter of Diblaim, and she conceived and bore him a son.

4 And the LORD said to him, "Name him Jezreel; a for in a little while I will punish the house of Jehu for the blood of Jezreel, and I will put an end to the kingdom of the house of Israel. 5 On that day I will break the bow of Israel in the valley of Jezreel."

6 She conceived again and bore a daughter. Then the LORD said to him, "Name her Lo-ruhamah, b for I will no longer have pity on the house of Israel or forgive them. 7 But I will have pity on the house of Judah, and I will save them by the LORD their God; I will not save them by bow, or by sword, or by war, or by horses, or by horsemen."

8 When she had weaned Lo-ruhamah, she conceived and bore a son. 9 Then the LORD said, "Name him Lo-ammi, c for you are not my people and I am not your God." d

10 e Yet the number of the people of Israel shall be like the sand of the sea, which can be neither measured nor numbered; and in the place where it was said to them, "You are not my people," it shall be said to them, "Children of the living God." 11 The people of Judah and the people of Israel shall be gathered together, and they shall appoint for themselves one head; and they shall take possession of f the land, for great shall be the day of Jezreel.

2 g Say to your brother, h Ammi, i and to your sister, j Ruhamah. k

2 Plead with your mother, plead—
  for she is not my wife,
  and I am not her husband—

a That is God sows   b That is Not pitied
c That is Not my people   d Heb I am not yours
e Ch 2.1 in Heb   f Heb rise up from
g Ch 2.3 in Heb   h Gk: Heb brothers
i That is My People   j Gk Vg: Heb sisters
k That is Pitied

1.1: Superscription. The conviction that *the word of the LORD* comes to a prophet (Joel 1.1; Mic 1.1; Zeph 1.1; Hag 1.1; Zech 1.1) is fundamental to Hebrew prophecy; it asserts that the prophet's inspiration and authority are not self-generated, but come from God, whose will is disclosed through the prophet (Ezek 2.3–5; 3.10–11; Am 3.7; Zech 1.6), whose personal agent the prophet is (Ex 4.15–16; Isa 6.8), and whom alone the prophet must obey (1 Kings ch 13; Am 7.14–17; Acts 4.18–20). *Hosea* means "salvation" or "deliverance." *Kings . . . of Judah*, see Isa 1.1 n., and Mic 1.1 n. Hosea prophesied not only during the reign of Jeroboam II (786–746 B.C.; see Introduction to Amos), but also after his death.
1.2–3.5: Hosea's disciplinary actions against his unfaithful wife and her children and his redemptive love for them exemplify God's dealings with harlotrous Israel. They form a living basis for his preaching during the reign of Jeroboam II.
1.2–9: Hosea marries the prostitute Gomer and gives her children prophetically significant names (Isa 7.3; 8.3) at the LORD's command. **2:** *For the land . . .* expresses the essence of Israel's sin and its need of redemption. **4–5:** *Jezreel,* meaning "God sows," points backward to the sin of the house of *Jehu* (1 Kings 19.15–17; 2 Kings chs 9–10) and forward to Israel's restoration (Hos 2.21–23). Jeroboam II belonged to the dynasty of Jehu.
1.6: Because of Israel's sin the LORD will *no longer have pity,* and the name of the second child (Hebrew *Lo-ruhamah,* see note b) will be a living reminder of this. **7:** This verse, which exempts *Judah,* is a later addition. **8–9:** The name of the third child (Hebrew *Lo-ammi,* see note c) signifies the breaking of the covenant relationship between the LORD and Israel (Ex 6.7; 19.5; compare Isa 40.1; Jer 31.31–34).
1.10–2.1: Israel's punishment is not final; afterwards it *shall be like the sand of the sea* (compare Gen 22.17) and again be God's *children* (compare Rom 9.25–26). Though these words may have been uttered at a different time, the thought is genuinely Hosean (see 14.4–7).
2.2–13: Israel will suffer public shame and personal privation like a harlot, be-

that she put away her whoring
    from her face,
and her adultery from between
    her breasts,
3 or I will strip her naked
    and expose her as in the day she
        was born,
and make her like a wilderness,
    and turn her into a parched
        land,
and kill her with thirst.
4 Upon her children also I will
    have no pity,
because they are children of
    whoredom.
5 For their mother has played
    the whore;
she who conceived them has
    acted shamefully.
For she said, "I will go after
    my lovers;
they give me my bread and
    my water,
my wool and my flax, my oil
    and my drink."
6 Therefore I will hedge up her[1]
    way with thorns;
and I will build a wall
    against her,
so that she cannot find her
    paths.
7 She shall pursue her lovers,
    but not overtake them;
and she shall seek them,
    but shall not find them.
Then she shall say, "I will go
    and return to my first husband,
    for it was better with me then
        than now."
8 She did not know
    that it was I who gave her
    the grain, the wine, and the oil,

and who lavished upon her silver
    and gold that they used for
        Baal.
9 Therefore I will take back
    my grain in its time,
and my wine in its season;
and I will take away my wool and
    my flax,
which were to cover her
    nakedness.
10 Now I will uncover her shame
    in the sight of her lovers,
and no one shall rescue her out
    of my hand.
11 I will put an end to all her mirth,
    her festivals, her new moons,
    her sabbaths,
and all her appointed festivals.
12 I will lay waste her vines and her
    fig trees,
of which she said,
"These are my pay,
    which my lovers have given
    me."
I will make them a forest,
    and the wild animals shall
    devour them.
13 I will punish her for the festival
    days of the Baals,
when she offered incense to
    them
and decked herself with her ring
    and jewelry,
and went after her lovers,
    and forgot me, says the LORD.

14 Therefore, I will now allure her,
    and bring her into the
    wilderness,
and speak tenderly to her.

l Gk Syr: Heb *your*

---

cause it has adulterated the worship of the
LORD with Canaanite Baalism. **3:** Isa 47.2–3;
Ezek 16.37–39; Rev 17.16. **5:** Jer 2.23–25;
3.1–2. Israel's *lovers* and their gifts refer to the
immoral fertility rites of Canaanite religion.
    2.7: Hosea takes the metaphor of marriage
from the cult of Baal, and boldly calls the
LORD *husband*. **8:** The LORD, the universal
Creator, is the giver of all good gifts, though
Israel may fail to acknowledge this (Gen

1.29–30; Deut 7.13; Jas 1.17). **9:** Am 4.6–8.
11: Isa 1.12–17; Am 5.21–24. **13:** *Baals*, see
v. 16 n. *Incense* was peculiarly associated with
pagan worship (Jer 44.8, 17). *Decked herself*,
compare Isa 3.16–22.
    **2.14–23: The LORD will allure Israel
back,** renew his covenant with her, and be-
troth her to himself forever. **14:** The *wilderness*
signifies Israel's early years after the Exodus
when it was faithful to the covenant (Ex

15 From there I will give her her
> vineyards,
> and make the Valley of Achor a
> door of hope.
> There she shall respond as in the
> days of her youth,
> as at the time when she came
> out of the land of Egypt.

16 On that day, says the LORD, you will call me, "My husband," and no longer will you call me, "My Baal."[m] 17 For I will remove the names of the Baals from her mouth, and they shall be mentioned by name no more. 18 I will make for you[n] a covenant on that day with the wild animals, the birds of the air, and the creeping things of the ground; and I will abolish[o] the bow, the sword, and war from the land; and I will make you lie down in safety. 19 And I will take you for my wife forever; I will take you for my wife in righteousness and in justice, in steadfast love, and in mercy. 20 I will take you for my wife in faithfulness; and you shall know the LORD.

21 On that day I will answer, says
> the LORD,
> I will answer the heavens
> and they shall answer the earth;
22 and the earth shall answer the
> grain, the wine, and the oil,
> and they shall answer Jezreel;[p]
23 and I will sow him[q] for myself

in the land.
> And I will have pity on
> Lo-ruhamah,[r]
> and I will say to Lo-ammi,[s]
> "You are my people";
> and he shall say, "You are my
> God."

3 The LORD said to me again, "Go, love a woman who has a lover and is an adulteress, just as the LORD loves the people of Israel, though they turn to other gods and love raisin cakes." 2 So I bought her for fifteen shekels of silver and a homer of barley and a measure of wine.[t] 3 And I said to her, "You must remain as mine for many days; you shall not play the whore, you shall not have intercourse with a man, nor I with you." 4 For the Israelites shall remain many days without king or prince, without sacrifice or pillar, without ephod or teraphim. 5 Afterward the Israelites shall return and seek the LORD their God, and David their king; they shall come in awe to the LORD and to his goodness in the latter days.

4 Hear the word of the LORD,
> O people of Israel;

m That is, *"My master"*   n Heb *them*
o Heb *break*   p That is *God sows*
q Cn: Heb *her*   r That is *Not pitied*   s That is *Not my people*   t Gk: Heb *a homer of barley and a lethech of barley*

---

chs 19–24; Jer 2.2–3; compare Ezek 20.33–38). **15:** When entering the Promised Land, Israel sinned at *the Valley of Achor* (Josh 7.20–26; Isa 65.10). **16:** *Baal,* often used as a proper name of the leading Canaanite deity, means "master," "lord."

2.17–23: The conclusion of the chapter promises the removal of the *Baals,* the establishment of a universal *covenant* (Lev 26.6; Job 5.23; Isa 11.6–9; Ezek 34.25–31), the abolition of *war* (Ps 46.9; Isa 2.4), and betrothal to the LORD in *steadfast love* and *faithfulness.*

3.1–5: The restoration of Gomer. Hosea buys back his adulterous wife, disciplines her, and affirms his devotion. In a like manner the LORD will restore Israel. **1:** *Just as the LORD loves the people of Israel, though they turn to other gods;* these words disclose the central message of this book: divine love necessitates both Israel's temporary chastisement and its ultimate redemption. *Raisin cakes* were used

in pagan festivals (Isa 16.7; Jer 7.18). **2:** *Shekel,* about eleven grams; *homer,* about six and one-half bushels. From whom Hosea *bought her* is not indicated. True redemption is costly (Ps 49.7–8; 1 Cor 6.19–20; 7.23; Gal 4.4–5); Hosea redeemed his wife by love and for a price. **4:** During the period of corrective punishment Israel will be deprived of civil and ceremonial institutions. *Pillar,* Gen 31.45. *Ephod,* see Judg 8.24–28 n.; 1 Sam 2.18 n. *Teraphim,* household gods (Gen 31.19). **5:** The phrase *and David their king* is probably a later Judean addition.

4.1–14.9: Because of constant rebellion, the judgment of the LORD is upon Israel; yet it will eventually be restored. These themes are elaborated in a series of discourses written after the death of Jeroboam II. The lack of clear order may in part reflect the style of the prophet and may in part be due to later editorial rearrangement.

for the LORD has an indictment
    against the inhabitants
    of the land.
There is no faithfulness or loyalty,
    and no knowledge of God in
    the land.
2 Swearing, lying, and murder,
    and stealing and adultery
    break out;
    bloodshed follows bloodshed.
3 Therefore the land mourns,
    and all who live in it languish;
together with the wild animals
    and the birds of the air,
    even the fish of the sea are
    perishing.

4 Yet let no one contend,
    and let none accuse,
    for with you is my contention,
    O priest. *u*
5 You shall stumble by day;
    the prophet also shall stumble
    with you by night,
    and I will destroy your mother.
6 My people are destroyed for lack
    of knowledge;
    because you have rejected
    knowledge,
    I reject you from being a priest
    to me.
And since you have forgotten the
    law of your God,
    I also will forget your children.

7 The more they increased,
    the more they sinned against
    me;
    they changed *v* their glory
    into shame.
8 They feed on the sin of my
    people;

they are greedy for their
    iniquity.
9 And it shall be like people,
    like priest;
    I will punish them for their
    ways,
    and repay them for their deeds.
10 They shall eat, but not be satisfied;
    they shall play the whore, but
    not multiply;
    because they have forsaken
    the LORD
    to devote themselves to
    11 whoredom.

Wine and new wine
    take away the understanding.
12 My people consult a piece of
    wood,
    and their divining rod gives
    them oracles.
For a spirit of whoredom has led
    them astray,
    and they have played the whore,
    forsaking their God.
13 They sacrifice on the tops of the
    mountains,
    and make offerings upon
    the hills,
under oak, poplar, and terebinth,
    because their shade is good.

Therefore your daughters play
    the whore,
    and your daughters-in-law
    commit adultery.
14 I will not punish your daughters
    when they play the whore,
    nor your daughters-in-law when
    they commit adultery;

u Cn: Meaning of Heb uncertain   *v* Ancient
Heb tradition: MT *I will change*

**4.1–8.14: All Israel has forgotten the
LORD and has sought help from other
gods and other nations. 4.1–3:** *The* LORD *has
an indictment against* Israel because they love
neither God nor their fellow Israelites (Mic
6.2). *Faithfulness, loyalty* (steadfast love), and
knowledge of God are major theological terms
in Hosea (2.19–20; 4.6; 5.4, 7; 6.3, 6; 10.12;
11.3–4, 12b; 12.6). **4–6:** The controversy is

first of all with the priest and prophet; for,
having rejected the knowledge and teaching
of God, they have destroyed the people.
    4.7–14: Sin has infected them all, *like peo-
ple, like priest.* **13:** A reference to high-place
sanctuaries and their sacred trees (Jer 2.20). **14:**
Temple prostitutes, see Gen 38.15 n.; Deut
23.17 n. **15:** *Beth-aven* is Bethel (Am 5.5).

for the men themselves go aside
  with whores,
and sacrifice with temple
  prostitutes;
thus a people without
  understanding comes to
  ruin.

15 Though you play the whore,
    O Israel,
  do not let Judah become guilty.
  Do not enter into Gilgal,
  or go up to Beth-aven,
  and do not swear, "As the
    LORD lives."
16 Like a stubborn heifer,
    Israel is stubborn;
  can the LORD now feed them
    like a lamb in a broad pasture?

17 Ephraim is joined to idols—
    let him alone.
18 When their drinking is ended, they
    indulge in sexual orgies;
  they love lewdness more than
    their glory. *w*
19 A wind has wrapped them *x* in
    its wings,
  and they shall be ashamed
    because of their altars. *y*

5 Hear this, O priests!
    Give heed, O house of Israel!
  Listen, O house of the king!
    For the judgment pertains to
      you;
  for you have been a snare at
    Mizpah,
  and a net spread upon Tabor,
2 and a pit dug deep in Shittim; *z*
    but I will punish all of them.

3 I know Ephraim,
    and Israel is not hidden from
      me;

for now, O Ephraim, you have
  played the whore;
  Israel is defiled.
4 Their deeds do not permit them
    to return to their God.
  For the spirit of whoredom is
    within them,
  and they do not know the
    LORD.

5 Israel's pride testifies against him;
    Ephraim *a* stumbles in his guilt;
  Judah also stumbles with them.
6 With their flocks and herds they
    shall go
  to seek the LORD,
  but they will not find him;
    he has withdrawn from them.
7 They have dealt faithlessly with
    the LORD;
  for they have borne illegitimate
    children.
  Now the new moon shall
    devour them along with
    their fields.

8 Blow the horn in Gibeah,
    the trumpet in Ramah.
  Sound the alarm at Beth-aven;
    look behind you, Benjamin!
9 Ephraim shall become a desolation
    in the day of punishment;
  among the tribes of Israel
    I declare what is sure.
10 The princes of Judah have become
    like those who remove the
      landmark;
  on them I will pour out
    my wrath like water.
11 Ephraim is oppressed, crushed in
    judgment,

w Cn Compare Gk: Meaning of Heb uncertain
x Heb *her*   *y* Gk Syr: Heb *sacrifices*
z Cn: Meaning of Heb uncertain   *a* Heb *Israel*
and Ephraim

5.1–2: Israel's leaders have been a snare, so God will chastise all of them. *Tabor,* Mount Tabor, where there seems to have been a sanctuary (compare Deut 33.19; Judg 4.6). *Mizpah,* either the city in Transjordan (Judg 10.17) or north of Jerusalem (1 Sam 7.5). **3–4:** The spirit of whoredom is so confirmed within them that they cannot return to their God.
5.5b: 1.7. **6:** Contrast Jer 29.13. **7:** *Illegitimate children,* offspring faithless to the LORD. *New moon,* perhaps referring to the new moon festivals. **8–14:** In the day of punishment the LORD will become the enemy of

because he was determined to
go after vanity. *b*

12 Therefore I am like maggots to
Ephraim,
and like rottenness to the house
of Judah.

13 When Ephraim saw his sickness,
and Judah his wound,
then Ephraim went to Assyria,
and sent to the great king. *c*
But he is not able to cure you
or heal your wound.

14 For I will be like a lion to
Ephraim,
and like a young lion to the
house of Judah.
I myself will tear and go away;
I will carry off, and no one
shall rescue.

15 I will return again to my place
until they acknowledge their
guilt and seek my face.
In their distress they will beg
my favor:

6 "Come, let us return to the LORD;
for it is he who has torn, and he
will heal us;
he has struck down, and he will
bind us up.

2 After two days he will revive us;
on the third day he will raise
us up,
that we may live before him.

3 Let us know, let us press on to
know the LORD;
his appearing is as sure as
the dawn;
he will come to us like the
showers,
like the spring rains that water
the earth."

4 What shall I do with you,
O Ephraim?

What shall I do with you,
O Judah?
Your love is like a morning cloud,
like the dew that goes
away early.

5 Therefore I have hewn them by
the prophets,
I have killed them by the words
of my mouth,
and my *d* judgment goes forth as
the light.

6 For I desire steadfast love and not
sacrifice,
the knowledge of God rather
than burnt offerings.

7 But at *e* Adam they transgressed
the covenant;
there they dealt faithlessly
with me.

8 Gilead is a city of evildoers,
tracked with blood.

9 As robbers lie in wait *f* for
someone,
so the priests are banded
together; *g*
they murder on the road to
Shechem,
they commit a monstrous
crime.

10 In the house of Israel I have seen a
horrible thing;
Ephraim's whoredom is there,
Israel is defiled.

11 For you also, O Judah, a harvest
is appointed.

When I would restore the fortunes
of my people,

b Gk: Meaning of Heb uncertain   *c* Cn: Heb
to a king who will contend   *d* Gk Syr: Heb *your*
e Cn: Heb *like*   *f* Cn: Meaning of Heb
uncertain   *g* Syr: Heb *are a company*

Israel (Am 9.2–4). This section refers to the
Syro-Ephraimitic War (2 Kings 15.27–30).
10: *Remove the landmark,* Deut 19.14; Prov
22.28.
   5.15–6.3: If Israel will but return to the
LORD, God will heal and revive Israel. **6.3:**
Spring rains, the later rains, Deut 11.14. **4–6:**
What the LORD really desires is *steadfast love*

and *knowledge* of God (4.1–3). **6:** This verse
epitomizes much of Hosea's message (2.19–
20; compare Mt 9.13; 12.7). *Love and not sacrifice,* compare Am 5.23, 24; Mic 6.6–8.
   6.7–11: By its harlotry Ephraim has transgressed the Sinaitic covenant (Ex 24.3–8;
Deut 5.1–3). **7:** *Adam,* geographical allusion
uncertain; compare Josh 3.16.

7 ¹ when I would heal Israel,
the corruption of Ephraim is
revealed,
and the wicked deeds of
Samaria;
for they deal falsely,
the thief breaks in,
and the bandits raid outside.
² But they do not consider
that I remember all their
wickedness.
Now their deeds surround them,
they are before my face.
³ By their wickedness they make the
king glad,
and the officials by their
treachery.
⁴ They are all adulterers;
they are like a heated oven,
whose baker does not need to stir
the fire,
from the kneading of the dough
until it is leavened.
⁵ On the day of our king the
officials
became sick with the heat
of wine;
he stretched out his hand with
mockers.
⁶ For they are kindled *h* like an
oven, their heart burns
within them;
all night their anger smolders;
in the morning it blazes like a
flaming fire.
⁷ All of them are hot as an oven,
and they devour their rulers.
All their kings have fallen;
none of them calls upon me.

⁸ Ephraim mixes himself with
the peoples;
Ephraim is a cake not turned.
⁹ Foreigners devour his strength,

but he does not know it;
gray hairs are sprinkled upon him,
but he does not know it.
¹⁰ Israel's pride testifies against *i* him;
yet they do not return to the
Lord their God,
or seek him, for all this.

¹¹ Ephraim has become like a dove,
silly and without sense;
they call upon Egypt, they go
to Assyria.
¹² As they go, I will cast my net
over them;
I will bring them down like
birds of the air;
I will discipline them according
to the report made to their
assembly. *j*
¹³ Woe to them, for they have
strayed from me!
Destruction to them, for they
have rebelled against me!
I would redeem them,
but they speak lies against me.

¹⁴ They do not cry to me from
the heart,
but they wail upon their beds;
they gash themselves for grain
and wine;
they rebel against me.
¹⁵ It was I who trained and
strengthened their arms,
yet they plot evil against me.
¹⁶ They turn to that which does not
profit; *k*
they have become like a
defective bow;
their officials shall fall by
the sword

h Gk Syr: Heb *brought near*　　*i* Or *humbles*
j Meaning of Heb uncertain　　*k* Cn: Meaning
of Heb uncertain

7.1–7: Because of its wicked deeds, Samaria will also be punished. **4:** The time between the kneading of the dough and its fer-
mentation is apparently when the oven was
the hottest. **5:** *The day of our king,* perhaps at
the celebration of enthronement. **8–10:**
Ephraim is a half-baked cake. **11–13:** Ephraim
is like a silly dove. *Call upon Egypt . . . ,* seek
alliances.
7.14–16: Because they turn to Baal the Israelites shall bear the consequences. *Gash
themselves,* Deut 14.1; 1 Kings 18.28. *Babbling,* i.e. the negotiations conducted in the
Egyptian language.

because of the rage of their
    tongue.
So much for their babbling in the
    land of Egypt.

8 Set the trumpet to your lips!
    One like a vulture[l] is over the
        house of the LORD,
because they have broken my
    covenant,
    and transgressed my law.
2 Israel cries to me,
    "My God, we—Israel—know
        you!"
3 Israel has spurned the good;
    the enemy shall pursue him.

4 They made kings, but not
        through me;
    they set up princes, but without
        my knowledge.
With their silver and gold they
        made idols
    for their own destruction.
5 Your calf is rejected, O Samaria.
    My anger burns against them.
How long will they be incapable
    of innocence?
6   For it is from Israel,
an artisan made it;
    it is not God.
The calf of Samaria
    shall be broken to pieces.[m]

7 For they sow the wind,
    and they shall reap the
        whirlwind.
The standing grain has no heads,
    it shall yield no meal;
if it were to yield,
    foreigners would devour it.
8 Israel is swallowed up;
    now they are among the nations
        as a useless vessel.

9 For they have gone up to Assyria,
    a wild ass wandering alone;
    Ephraim has bargained
        for lovers.
10 Though they bargain with
        the nations,
    I will now gather them up.
They shall soon writhe
    under the burden of kings
        and princes.

11 When Ephraim multiplied altars
        to expiate sin,
    they became to him altars for
        sinning.
12 Though I write for him the
        multitude of my
        instructions,
    they are regarded as a strange
        thing.
13 Though they offer choice
        sacrifices,[n]
    though they eat flesh,
    the LORD does not accept them.
Now he will remember their
        iniquity,
    and punish their sins;
    they shall return to Egypt.
14 Israel has forgotten his Maker,
        and built palaces;
and Judah has multiplied fortified
        cities;
    but I will send a fire upon
        his cities,
    and it shall devour his
        strongholds.

9 Do not rejoice, O Israel!
    Do not exult[o] as other nations
        do;

l Meaning of Heb uncertain   m Or *shall go up
in flames*   n Cn: Meaning of Heb uncertain
o Gk: Heb *To exultation*

8.1: *The trumpet* warns of the approach of
the enemy (compare Jer 6.1; Neh 4.18–20;
Joel 2.1). *Vulture,* an eagle, the Assyrian army
(Jer 49.22). **4:** Probably an allusion to the rapid succession of kings after Jeroboam II (see
Introduction). **5:** *Your calf,* see 1 Kings 12.28,
29. *An artisan made it,* compare Isa 40.19, 20.
8.7–10: As punishment for political defec-

tion, Israel shall be without king and princes.
11–14: As punishment for religious defection,
Israel *shall return to Egypt* (compare 9.3, 6). **14:**
Reliance on palaces and walled cities alone is
sin. *Fire,* Am 1.4, 7.
**9.1–11.12: Israel has rejected the LORD
and must undergo punishment that will
bring loss of king, children, places of**

for you have played the whore,
    departing from your God.
You have loved a prostitute's
    pay
on all threshing floors.
2 Threshing floor and wine vat shall
    not feed them,
and the new wine shall fail
    them.
3 They shall not remain in the land
    of the LORD;
but Ephraim shall return to
    Egypt,
and in Assyria they shall eat
    unclean food.

4 They shall not pour drink
    offerings of wine to the
    LORD,
and their sacrifices shall not
    please him.
Such sacrifices shall be like
    mourners' bread;
all who eat of it shall be defiled;
for their bread shall be for their
    hunger only;
it shall not come to the house of
    the LORD.

5 What will you do on the day of
    appointed festival,
and on the day of the festival of
    the LORD?
6 For even if they escape
    destruction,
Egypt shall gather them,
Memphis shall bury them.
Nettles shall possess their precious
    things of silver;*p*
thorns shall be in their tents.

7 The days of punishment have
    come,
the days of recompense
    have come;

Israel cries,*q*
"The prophet is a fool,
    the man of the spirit is mad!"
Because of your great iniquity,
    your hostility is great.
8 The prophet is a sentinel for my
    God over Ephraim,
yet a fowler's snare is on all
    his ways,
and hostility in the house of
    his God.
9 They have deeply corrupted
    themselves
as in the days of Gibeah;
he will remember their iniquity,
    he will punish their sins.

10 Like grapes in the wilderness,
    I found Israel.
Like the first fruit on the fig tree,
    in its first season,
    I saw your ancestors.
But they came to Baal-peor,
    and consecrated themselves to a
    thing of shame,
and became detestable like the
    thing they loved.
11 Ephraim's glory shall fly away like
    a bird—
no birth, no pregnancy, no
    conception!
12 Even if they bring up children,
    I will bereave them until no one
    is left.
Woe to them indeed
    when I depart from them!
13 Once I saw Ephraim as a young
    palm planted in a lovely
    meadow,*p*
but now Ephraim must lead out
    his children for slaughter.
14 Give them, O LORD—

p Meaning of Heb uncertain   *q* Cn Compare
Gk: Heb *shall know*

---

worship, and country. 9.1–17: Israel is destined for prey and slaughter and to be deprived of land and cultus (2.11; 8.11–14; 11.5). 1: *Prostitute's pay,* Gen 38.17; Mic 1.7. 4: *Mourners' bread,* unclean because of association with the dead.
9.6: *Memphis,* the capital of Lower Egypt.

7: *Man of the spirit,* that is, the prophet (see Mic 3.8 n.). 8: *Sentinel,* compare Ezek 33.6–8. 9: 1 Sam ch 10.
9.10–17: *In the wilderness* Israel covenanted with the LORD, but in Canaan it *consecrated* itself to the fertility god Baal; therefore the LORD will make it barren. 10: *Baal-peor,* Num

what will you give?
Give them a miscarrying womb
and dry breasts.

15 Every evil of theirs began at
Gilgal;
there I came to hate them.
Because of the wickedness of
their deeds
I will drive them out of
my house.
I will love them no more;
all their officials are rebels.

16 Ephraim is stricken,
their root is dried up,
they shall bear no fruit.
Even though they give birth,
I will kill the cherished offspring
of their womb.

17 Because they have not listened
to him,
my God will reject them;
they shall become wanderers
among the nations.

10 Israel is a luxuriant vine
that yields its fruit.
The more his fruit increased
the more altars he built;
as his country improved,
he improved his pillars.

2 Their heart is false;
now they must bear their guilt.
The LORD *r* will break down
their altars,
and destroy their pillars.

3 For now they will say:
"We have no king,
for we do not fear the LORD,
and a king—what could he do
for us?"

4 They utter mere words;
with empty oaths they make
covenants;

so litigation springs up like
poisonous weeds
in the furrows of the field.

5 The inhabitants of Samaria
tremble
for the calf *s* of Beth-aven.
Its people shall mourn for it,
and its idolatrous priests shall
wail *t* over it,
over its glory that has departed
from it.

6 The thing itself shall be carried to
Assyria
as tribute to the great king. *u*
Ephraim shall be put to shame,
and Israel shall be ashamed of
his idol. *v*

7 Samaria's king shall perish
like a chip on the face of
the waters.

8 The high places of Aven, the sin
of Israel,
shall be destroyed.
Thorn and thistle shall grow up
on their altars.
They shall say to the mountains,
Cover us,
and to the hills, Fall on us.

9 Since the days of Gibeah you have
sinned, O Israel;
there they have continued.
Shall not war overtake them in
Gibeah?

10 I will come *w* against the wayward
people to punish them;
and nations shall be gathered
against them
when they are punished *x* for
their double iniquity.

r Heb *he*   *s* Gk Syr: Heb *calves*   *t* Cn: Heb
exult   *u* Cn: Heb *to a king who will contend*
v Cn: Heb *counsel*   *w* Cn Compare Gk: Heb
In my desire   *x* Gk: Heb *bound*

25.1–18. **15**: *Gilgal,* 4.15. *Drive them out of my
house,* as a husband his faithless wife.
   10.1–2: In Canaan Israel *increased* in sin as
it increased in prosperity. *Pillars,* Ex 23.24.
   10.3–10: Samaria's *king shall perish,* for it
sinned by establishing the monarchy in
the days of Saul. **5**: *Calf,* 8.5. **8**: Am 9.1;
Lk 23.30; Rev 6.16. *Aven* means "Wickedness," a reference to Bethel. **9**: *Gibeah,* see
Judg ch 19.

11 Ephraim was a trained heifer
    that loved to thresh,
    and I spared her fair neck;
  but I will make Ephraim break the
    ground;
  Judah must plow;
  Jacob must harrow for himself.
12 Sow for yourselves righteousness;
    reap steadfast love;
  break up your fallow ground;
  for it is time to seek the LORD,
    that he may come and rain
    righteousness upon you.

13 You have plowed wickedness,
    you have reaped injustice,
    you have eaten the fruit of lies.
  Because you have trusted in
    your power
  and in the multitude of your
    warriors,
14 therefore the tumult of war shall
    rise against your people,
  and all your fortresses shall
    be destroyed,
  as Shalman destroyed Beth-arbel
    on the day of battle
  when mothers were dashed in
    pieces with their children.
15 Thus it shall be done to you,
    O Bethel,
  because of your great
    wickedness.
  At dawn the king of Israel
    shall be utterly cut off.

11 When Israel was a child, I
    loved him,
  and out of Egypt I called my
    son.
2 The more I[y] called them,
    the more they went from me;[z]
  they kept sacrificing to the Baals,
    and offering incense to idols.

3 Yet it was I who taught Ephraim
    to walk,
  I took them up in my[a] arms;
  but they did not know that I
    healed them.
4 I led them with cords of human
    kindness,
  with bands of love.
  I was to them like those
    who lift infants to their cheeks.[b]
  I bent down to them and
    fed them.

5 They shall return to the land of
    Egypt,
  and Assyria shall be their king,
  because they have refused to
    return to me.
6 The sword rages in their cities,
    it consumes their oracle-priests,
  and devours because of their
    schemes.
7 My people are bent on turning
    away from me.
  To the Most High they call,
  but he does not raise them up
    at all.[c]

8 How can I give you up, Ephraim?
  How can I hand you over,
    O Israel?
  How can I make you like Admah?
  How can I treat you like
    Zeboiim?
  My heart recoils within me;
    my compassion grows warm
    and tender.
9 I will not execute my fierce anger;
  I will not again destroy
    Ephraim;

y Gk: Heb *they*   *z* Gk: Heb *them*   *a* Gk Syr
Vg: Heb *his*   *b* Or *who ease the yoke on their
jaws*   *c* Meaning of Heb uncertain

10.11–12: Repentance will bring *righteous-ness.* **12:** Compare 6.3; Isa 44.3; 45.8; 55.10; Jer 4.3; 2 Cor 9.10). **13–15:** Rebellious Israel shall hear *the tumult of war.* **14:** *Shalman . . . Beth-arbel,* the allusion is unknown and the text uncertain.
11.1–7: The LORD as a loving and patient

parent, must now chastise disobedient and wilful children with exile (v. 5). **1:** Ex 4.22; Mt 2.15. **3:** Jer 31.1–3.

11.8–9: In this striking soliloquy *compas-sion* restrains the divine anger; such is the nature of superhuman love. *Admah, Zeboiim,* cities destroyed along with Sodom and Go-

for I am God and no mortal,
the Holy One in your midst,
and I will not come in wrath. *d*

10 They shall go after the LORD,
who roars like a lion;
when he roars,
his children shall come
trembling from the west.
11 They shall come trembling like
birds from Egypt,
and like doves from the land of
Assyria;
and I will return them to their
homes, says the LORD.

12*e* Ephraim has surrounded me
with lies,
and the house of Israel with
deceit;
but Judah still walks*f* with God,
and is faithful to the Holy One.

12 Ephraim herds the wind,
and pursues the east wind all
day long;
they multiply falsehood and
violence;
they make a treaty with Assyria,
and oil is carried to Egypt.

2 The LORD has an indictment
against Judah,
and will punish Jacob according
to his ways,
and repay him according to
his deeds.
3 In the womb he tried to supplant
his brother,
and in his manhood he strove
with God.
4 He strove with the angel and
prevailed,

he wept and sought his favor;
he met him at Bethel,
and there he spoke with him. *g*
5 The LORD the God of hosts,
the LORD is his name!
6 But as for you, return to your
God,
hold fast to love and justice,
and wait continually for
your God.

7 A trader, in whose hands are false
balances,
he loves to oppress.
8 Ephraim has said, "Ah, I am rich,
I have gained wealth for myself;
in all of my gain
no offense has been found in me
that would be sin." *d*
9 I am the LORD your God
from the land of Egypt;
I will make you live in tents
again,
as in the days of the appointed
festival.

10 I spoke to the prophets;
it was I who multiplied visions,
and through the prophets I will
bring destruction.
11 In Gilead*h* there is iniquity,
they shall surely come to
nothing.
In Gilgal they sacrifice bulls,
so their altars shall be like
stone heaps
on the furrows of the field.
12 Jacob fled to the land of Aram,

d Meaning of Heb uncertain   *e* Ch 12.1 in
Heb   *f* Heb *roams* or *rules*   *g* Gk Syr: Heb
us   *h* Compare Syr: Heb *Gilead*

morrah (Gen ch 19; Deut 29.23). **10:** Contrast
5.14. **11:** *Doves,* contrast 7.11. *Egypt* and *Assyria,* contrast 9.3. **12:** *But Judah still . . . ,*
some regard the last part of this verse as a
Judean gloss (compare 1.7).
   12.1–14.9: Rebellion and restoration.
Although the LORD led Israel by the prophets,
it turned to Baal and died. Yet the LORD is the
savior. When Israel acknowledges God, Israel
will be restored to abundant life because of
God's love.

12.2–6: Israel should hold fast to God's
love and justice, as Jacob strove with the angel at the Jabbok River (Gen 32.24). **3:** *Tried
to supplant his brother,* Gen 25.26. **4:** *Strove with
the angel,* Gen 32.22–30. *He met him at Bethel,*
Gen 28.11–17; 35.5–8. **9:** As a punishment,
the LORD will again make Israel *live in tents,*
as in the wilderness during the Exodus.
   12.10–14: Because Israel spurned the
prophets, it will be held responsible for its
bitter offense (Am 3.2; compare Gal 6.7). **12:**

there Israel served for a wife,
and for a wife he guarded
sheep. [i]
13 By a prophet the LORD brought
Israel up from Egypt,
and by a prophet he was
guarded.
14 Ephraim has given bitter offense,
so his Lord will bring his crimes
down on him
and pay him back for his
insults.

13 When Ephraim spoke, there
was trembling;
he was exalted in Israel;
but he incurred guilt through
Baal and died.
2 And now they keep on sinning
and make a cast image for
themselves,
idols of silver made according to
their understanding,
all of them the work of artisans.
"Sacrifice to these," they say. [j]
People are kissing calves!
3 Therefore they shall be like the
morning mist
or like the dew that goes
away early,
like chaff that swirls from the
threshing floor
or like smoke from a window.

4 Yet I have been the LORD your
God
ever since the land of Egypt;
you know no God but me,
and besides me there is no
savior.
5 It was I who fed [k] you in the
wilderness,
in the land of drought.
6 When I fed [l] them, they were
satisfied;

they were satisfied, and their
heart was proud;
therefore they forgot me.
7 So I will become like a lion
to them,
like a leopard I will lurk beside
the way.
8 I will fall upon them like a bear
robbed of her cubs,
and will tear open the covering
of their heart;
there I will devour them like
a lion,
as a wild animal would
mangle them.

9 I will destroy you, O Israel;
who can help you? [m]
10 Where now is [n] your king, that he
may save you?
Where in all your cities are
your rulers,
of whom you said,
"Give me a king and rulers"?
11 I gave you a king in my anger,
and I took him away in
my wrath.

12 Ephraim's iniquity is bound up;
his sin is kept in store.
13 The pangs of childbirth come
for him,
but he is an unwise son;
for at the proper time he does not
present himself
at the mouth of the womb.

14 Shall I ransom them from the
power of Sheol?

i Heb lacks *sheep*   j Cn Compare Gk: Heb *To
these they say sacrifices of people*   k Gk Syr: Heb
knew   l Cn: Heb *according to their pasture*
m Gk Syr: Heb *for in me is your help*
n Gk Syr Vg: Heb *I will be*

Perhaps by a Judean editor; i.e. the cults of
Gilead and Gilgal are as worthless as that of
Bethel.
    13.1–3: Through the sin of Canaanite idol-
atry, Israel died (2.13; compare Rom 6.23;
8.13). **2:** *People are kissing calves,* at Dan and
Bethel. *Morning mist,* compare 6.4.

13.4–13: Because they forgot their savior,
the LORD will rend and destroy the Israelites,
and neither king nor princes will save them.
4: Ex 20.2; Isa 45.21. **14–16:** The LORD will
not redeem this rebellious people from Death;
Samaria shall bear its guilt. **14:** Quoted in
1 Cor 15.55.

Shall I redeem them from
Death?
O Death, where are° your
plagues?
O Sheol, where is° your
destruction?
Compassion is hidden from
my eyes.

15 Although he may flourish among
rushes, *p*
the east wind shall come, a blast
from the LORD,
rising from the wilderness;
and his fountain shall dry up,
his spring shall be parched.
It shall strip his treasury
of every precious thing.
16*q* Samaria shall bear her guilt,
because she has rebelled against
her God;
they shall fall by the sword,
their little ones shall be dashed
in pieces,
and their pregnant women
ripped open.

14 Return, O Israel, to the LORD
your God,
for you have stumbled because
of your iniquity.
2 Take words with you
and return to the LORD;
say to him,
"Take away all guilt;
accept that which is good,
and we will offer
the fruit*r* of our lips.
3 Assyria shall not save us;
we will not ride upon horses;
we will say no more, 'Our God,'
to the work of our hands.
In you the orphan finds mercy."

4 I will heal their disloyalty;
I will love them freely,
for my anger has turned
from them.
5 I will be like the dew to Israel;
he shall blossom like the lily,
he shall strike root like the
forests of Lebanon. *s*
6 His shoots shall spread out;
his beauty shall be like the
olive tree,
and his fragrance like that of
Lebanon.
7 They shall again live beneath
my*t* shadow,
they shall flourish as a garden;*u*
they shall blossom like the vine,
their fragrance shall be like the
wine of Lebanon.

8 O Ephraim, what have I*v* to do
with idols?
It is I who answer and look
after you. *w*
I am like an evergreen cypress;
your faithfulness*x* comes
from me.
9 Those who are wise understand
these things;
those who are discerning
know them.
For the ways of the LORD are
right,
and the upright walk in them,
but transgressors stumble
in them.

o Gk Syr: Heb *I will be*   p Or *among brothers*
q Ch 14.1 in Heb   r Gk Syr: Heb *bulls*
s Cn: Heb *like Lebanon*   t Heb *his*
u Cn: Heb *they shall grow grain*   v Or *What
more has Ephraim*   w Heb *him*   x Heb *your
fruit*

14.1–3: *Israel* may still return to the LORD,
for with God there is mercy (Ps 130.7–8; Isa
55.6–9). 3: The futility of foreign alliances
and idols. *Horses,* i.e. we will no longer depend on an alliance with Egypt (2 Kings
18.24; Isa 36.9).

14.4–7: And because God loves freely, Israel will be healed and brought back to fruitful life (Isa 54.5–8). 8: The LORD alone guides
and sustains Israel (2.8). 9: A later postscript,
in the style of wisdom literature (compare
Prov 4.11–12; see also Eccl 12.9–14 n.).

Joel

Of Joel himself nothing is known except that he was the son of Pethuel (1.1). Like Obadiah, even his name ("the LORD is God") is not unique, for he shares it with a dozen other Old Testament figures. From his book it appears that he lived in Judah during the Persian period of Jewish history (539–331 B.C.). He was not only well acquainted with the temple at Jerusalem, but was so much interested in its priesthood and services that, like Haggai and Zechariah, he can be considered a "cultic prophet," that is, a prophet who could exercise his ministry within the life of the temple, even using liturgical forms, and whose message may have been transmitted through priestly circles. As such, Joel helps to mark a notable change in Old Testament prophecy. Taking the characteristic forms of classical prophecy, he expands their apocalyptic and liturgical dimensions.

The dates of Joel's ministry cannot be determined with precision. The majority of historical references in his book, the absence of any mention of the Assyrians or Babylonians, and the heavy borrowing from earlier prophets point to the period from about 400 to 350 B.C.

Joel viewed a locust plague that ravished the country as God's judgment on the people and called them to repentance (1.2–2.27). Using this catastrophe as a dire warning, Joel went on to depict the advent of the day of the LORD and its final judgments and blessings (2.28–3.21).

1 The word of the LORD that came to Joel son of Pethuel:

2 Hear this, O elders,
    give ear, all inhabitants of
        the land!
Has such a thing happened in
        your days,
    or in the days of your ancestors?
3 Tell your children of it,
    and let your children tell their
        children,
    and their children another
        generation.

4 What the cutting locust left,
    the swarming locust has eaten.
What the swarming locust left,
    the hopping locust has eaten,
and what the hopping locust left,
    the destroying locust has eaten.

5 Wake up, you drunkards, and
        weep;
    and wail, all you wine-drinkers,
over the sweet wine,
    for it is cut off from your
        mouth.
6 For a nation has invaded my land,
        powerful and innumerable;
    its teeth are lions' teeth,
    and it has the fangs of a lioness.
7 It has laid waste my vines,
    and splintered my fig trees;
it has stripped off their bark and
        thrown it down;
    their branches have turned
        white.

8 Lament like a virgin dressed in
        sackcloth
    for the husband of her youth.
9 The grain offering and the drink
        offering are cut off
    from the house of the LORD.

The priests mourn,
    the ministers of the LORD.
10 The fields are devastated,
    the ground mourns;
for the grain is destroyed,
    the wine dries up,
    the oil fails.

11 Be dismayed, you farmers,
    wail, you vinedressers,
over the wheat and the barley;
    for the crops of the field
    are ruined.
12 The vine withers,
    the fig tree droops.
Pomegranate, palm, and apple—
    all the trees of the field are
        dried up;
surely, joy withers away
    among the people.

13 Put on sackcloth and lament,
        you priests;
    wail, you ministers of the altar.
Come, pass the night in sackcloth,
    you ministers of my God!
Grain offering and drink offering
    are withheld from the house of
        your God.

14 Sanctify a fast,
    call a solemn assembly.
Gather the elders
    and all the inhabitants of
        the land
to the house of the LORD your
        God,
    and cry out to the LORD.

15 Alas for the day!
For the day of the LORD is near,
    and as destruction from the
        Almighty[a] it comes.

a Traditional rendering of Heb *Shaddai*

1.1–2.27: The locust plague.
1.1: Superscription. *The word of the
LORD . . . ,* see Hos 1.1 n. *Pethuel* does not
occur elsewhere. **2–7:** The land is devastated
by a swarm of locusts, described as a nation
with lions' teeth (v. 6; compare 2.4–9; Rev
9.7–8).

1.8–14: The whole community is called
upon to lament and return to the LORD. **13–14:**
Ministers of the altar of burnt offering, who
serve in the inner court (see Ezek 43.18–27).
A fast, 2 Kings 18.6; Jer 36.6; Jon 3.5.
1.15–20: The lament. **15:** This calamitous
day portends the final day of the LORD which

16 Is not the food cut off
  before our eyes,
joy and gladness
  from the house of our God?

17 The seed shrivels under the
    clods, *b*
  the storehouses are desolate;
the granaries are ruined
  because the grain has failed.
18 How the animals groan!
  The herds of cattle wander
    about
because there is no pasture for
    them;
  even the flocks of sheep are
    dazed. *c*

19 To you, O LORD, I cry.
  For fire has devoured
    the pastures of the wilderness,
and flames have burned
  all the trees of the field.
20 Even the wild animals cry to you
  because the watercourses are
    dried up,
and fire has devoured
  the pastures of the wilderness.

2 Blow the trumpet in Zion;
  sound the alarm on my holy
    mountain!
Let all the inhabitants of the land
    tremble,
  for the day of the LORD is
    coming, it is near—
2 a day of darkness and gloom,
  a day of clouds and thick
    darkness!
Like blackness spread upon the
    mountains
  a great and powerful army
    comes;
their like has never been from
    of old,

nor will be again after them
  in ages to come.

3 Fire devours in front of them,
  and behind them a flame burns.
Before them the land is like the
    garden of Eden,
  but after them a desolate
    wilderness,
and nothing escapes them.

4 They have the appearance of
    horses,
  and like war-horses they charge.
5 As with the rumbling of chariots,
  they leap on the tops of the
    mountains,
like the crackling of a flame of fire
  devouring the stubble,
like a powerful army
  drawn up for battle.

6 Before them peoples are in
    anguish,
  all faces grow pale. *b*
7 Like warriors they charge,
  like soldiers they scale the wall.
Each keeps to its own course,
  they do not swerve from *d*
    their paths.
8 They do not jostle one another,
  each keeps to its own track;
they burst through the weapons
  and are not halted.
9 They leap upon the city,
  they run upon the walls;
they climb up into the houses,
  they enter through the windows
    like a thief.

10 The earth quakes before them,
  the heavens tremble.

b Meaning of Heb uncertain   *c* Compare Gk
Syr Vg: Meaning of Heb uncertain   *d* Gk Syr
Vg: Heb *they do not take a pledge along*

---

comes as destruction from the Almighty
(Zeph 1.14–18). **19–20**: *Fire* and *flames* epito-
mize the most severe destruction.
  2.1–11: The cry of alarm at the approaching catastrophe. **1**: The priests blow the ram's

horn ("shofar") to warn of the imminent danger (Hos 5.8; Am 3.6; Zeph 1.16; Rev 8.6–
13).
  2.4–9: The locusts, which symbolize that
dreadful day, approach with the relentless and

The sun and the moon are
darkened,
and the stars withdraw their
shining.
11 The LORD utters his voice
at the head of his army;
how vast is his host!
Numberless are those who obey
his command.
Truly the day of the LORD is great;
terrible indeed—who can
endure it?

12 Yet even now, says the LORD,
return to me with all your
heart,
with fasting, with weeping, and
with mourning;
13   rend your hearts and not
your clothing.
Return to the LORD, your God,
for he is gracious and merciful,
slow to anger, and abounding in
steadfast love,
and relents from punishing.
14 Who knows whether he will not
turn and relent,
and leave a blessing behind him,
a grain offering and a drink
offering
for the LORD, your God?

15 Blow the trumpet in Zion;
sanctify a fast;
call a solemn assembly;
16   gather the people.
Sanctify the congregation;
assemble the aged;
gather the children,
even infants at the breast.

Let the bridegroom leave his
room,
and the bride her canopy.
17 Between the vestibule and the altar
let the priests, the ministers of
the LORD, weep.
Let them say, "Spare your people,
O LORD,
and do not make your heritage a
mockery,
a byword among the nations.
Why should it be said among the
peoples,
'Where is their God?' "

18 Then the LORD became jealous for
his land,
and had pity on his people.
19 In response to his people the
LORD said:
I am sending you
grain, wine, and oil,
and you will be satisfied;
and I will no more make you
a mockery among the nations.

20 I will remove the northern army
far from you,
and drive it into a parched and
desolate land,
its front into the eastern sea,
and its rear into the western sea;
its stench and foul smell will
rise up.
Surely he has done great things!

21 Do not fear, O soil;
be glad and rejoice,
for the LORD has done great
things!

devastating force of a powerful *army* (2.25;
Rev 9.7–10). **6**: Nah 2.10. **10**: At the time of
the divine visitation the sun, moon, and stars
refuse to shine (Am 8.9). **11**: Rev 6.17.
2.12–17: A call to repentance, by which
total calamity may yet be averted. **13**: The
pleas to *rend your hearts* and *return to the LORD*
express the preaching of the classical prophets, placed here in a cultic context. *Gracious
and merciful . . .* is a frequent designation of
the LORD, rooted in Israel's ancient formula-
tions of faith (Ex 34.6; Neh 9.17, 31; Ps

86.15). **14**: Characteristically Joel regards
temple offerings as *a blessing*. **15**: 1.14. **17**:
Between the vestibule (1 Kings 6.3) *and the altar*
of burnt offering (2 Chr 4.1); this was in the
inner court of the priests.
2.18–27: God's gracious answer promises
the remission of the plague (vv. 20, 25), the
return of fertility (vv. 19, 21–24), and the restoration of the covenant (vv. 26–27). **18**: *Jealous* includes the meaning zealous (Deut 4.24).
20: *The northern army* refers to the destructive
agent, probably the locusts. *The eastern sea,*

22 Do not fear, you animals of
the field,
for the pastures of the
wilderness are green;
the tree bears its fruit,
the fig tree and vine give their
full yield.

23 O children of Zion, be glad
and rejoice in the LORD your
God;
for he has given the early rain*
for your vindication,
he has poured down for you
abundant rain,
the early and the later rain,
as before.

24 The threshing floors shall be full
of grain,
the vats shall overflow with
wine and oil.

25 I will repay you for the years
that the swarming locust has
eaten,
the hopper, the destroyer, and the
cutter,
my great army, which I sent
against you.

26 You shall eat in plenty and be
satisfied,
and praise the name of the LORD
your God,
who has dealt wondrously
with you.
And my people shall never again
be put to shame.

27 You shall know that I am in the
midst of Israel,

and that I, the LORD, am your
God and there is no other.
And my people shall never again
be put to shame.

28*f* Then afterward
I will pour out my spirit on
all flesh;
your sons and your daughters shall
prophesy,
your old men shall dream
dreams,
and your young men shall
see visions.

29 Even on the male and female
slaves,
in those days, I will pour out
my spirit.

30 I will show portents in the heavens
and on the earth, blood and fire and columns of smoke. 31 The sun shall be
turned to darkness, and the moon to
blood, before the great and terrible day
of the LORD comes. 32 Then everyone
who calls on the name of the LORD shall
be saved; for in Mount Zion and in Jerusalem there shall be those who escape, as
the LORD has said, and among the survivors shall be those whom the LORD calls.

3 *g* For then, in those days and at that
time, when I restore the fortunes of
Judah and Jerusalem, 2 I will gather all the
nations and bring them down to the valley of Jehoshaphat, and I will enter into
judgment with them there, on account of
my people and my heritage Israel, be-

e Meaning of Heb uncertain   *f* Ch 3.1 in Heb
g Ch 4.1 in Heb

---

the Dead Sea. *The western sea*, the Mediterranean Sea. **27**: By God's gifts of abundance the
LORD's people will know that the LORD alone
is their God (Isa 45.5, 6, 18; Ezek 36.11;
39.28) and dwells in their midst (3.17, 21).
**2.28–3.21: The day of the LORD. 2.28–
32**: The outpouring of the spirit and the signs
foretelling the great day. **28–29**: For Joel *all
flesh* meant primarily the Jews (3.2, 17, 19–
21; Ezek 39.29); for Peter at Pentecost it included all nations (Acts 2.17). **30–31**: *The
heavens* and *the earth* will warn of *the great and*

terrible day (Mk 13.24; Rev 6.12). **31**: Here
blood indicates color. **32**: Those who worship
the LORD (Gen 4.26; 12.8; Ps 116.13) will be
delivered (Acts 2.21; Rom 10.13). **32b**: Ob
17.
3.1–3: The day of judgment on all the nations. The trial will be held in *the valley of
Jehoshaphat* ("the LORD judges"), called the
valley of decision in v. 14 (Jer 25.31). Joel probably had no exact spot in mind, although the
context (2.32; 3.1, 16, 17, 21) points to the
general area of Jerusalem, and tradition lo-

cause they have scattered them among the nations. They have divided my land, 3 and cast lots for my people, and traded boys for prostitutes, and sold girls for wine, and drunk it down.

4 What are you to me, O Tyre and Sidon, and all the regions of Philistia? Are you paying me back for something? If you are paying me back, I will turn your deeds back upon your own heads swiftly and speedily. 5 For you have taken my silver and my gold, and have carried my rich treasures into your temples. *h* 6 You have sold the people of Judah and Jerusalem to the Greeks, removing them far from their own border. 7 But now I will rouse them to leave the places to which you have sold them, and I will turn your deeds back upon your own heads. 8 I will sell your sons and your daughters into the hand of the people of Judah, and they will sell them to the Sabeans, to a nation far away; for the LORD has spoken.

9 Proclaim this among the nations:
    Prepare war, *i*
        stir up the warriors.
    Let all the soldiers draw near,
        let them come up.
10 Beat your plowshares into swords,
        and your pruning hooks into
            spears;
    let the weakling say, "I am a
        warrior."

11 Come quickly, *j*
    all you nations all around,
        gather yourselves there.
    Bring down your warriors,
        O LORD.

12 Let the nations rouse themselves,
        and come up to the valley of
            Jehoshaphat;
    for there I will sit to judge
        all the neighboring nations.

13 Put in the sickle,
        for the harvest is ripe.
    Go in, tread,
        for the wine press is full.
    The vats overflow,
        for their wickedness is great.

14 Multitudes, multitudes,
        in the valley of decision!
    For the day of the LORD is near
        in the valley of decision.
15 The sun and the moon are
            darkened,
        and the stars withdraw their
            shining.

16 The LORD roars from Zion,
        and utters his voice from
            Jerusalem,
    and the heavens and the earth
        shake.
    But the LORD is a refuge for his
        people,
    a stronghold for the people of
        Israel.

17 So you shall know that I, the
        LORD your God,
    dwell in Zion, my holy
        mountain.
    And Jerusalem shall be holy,

h Or *palaces*   *i* Heb *sanctify war*   *j* Meaning of Heb uncertain

---

cates it in the Kidron valley. **3a**: Ob 11. **4–8**: *Tyre* (Isa ch 23; Ezek 26.1–28.19; Am 1.9–10; Zech 9.3–4), *Sidon* (Ezek 28.20–26; Zech 9.2), *and Philistia* (Isa 14.29–31; Jer ch 47; Ezek 25.15–17; Am 1.6–8; Zeph 2.4–7; Zech 9.5–7) will receive their recompense first (this prose paragraph may be a later addition).
    3.6–8: Because the Philistines and Phoenicians sold Israelites as slaves to Gentiles (Ezek 27.13; Am 1.6, 9), their children will be sold to the *Sabeans,* famous traders from southwest Arabia.

3.9–12: The holy war between the LORD's warriors and all the nations round about (compare Ezek chs 38–39). **10**: The picture in Isa 2.4 and Mic 4.3 is reversed.
    3.13–17: *The day of the LORD is near.* **13**: The inordinate wickedness of the Gentiles is compared to the harvest which is ripe and vats which overflow (Isa 63.1–6; Mt 3.10–12; Mk 4.29; Rev 14.15–20). **16**: Am 1.2. **17**: Jerusalem shall now be holy, as the LORD's earthly tabernacle (Isa 1.24–28; 52.1; Zech 8.3; 14.21; Rev 21.2–3, 27).

and strangers shall never again
pass through it.

18 In that day
the mountains shall drip sweet
wine,
the hills shall flow with milk,
and all the stream beds of Judah
shall flow with water;
a fountain shall come forth from
the house of the LORD
and water the Wadi Shittim.

19 Egypt shall become a desolation

and Edom a desolate wilderness,
because of the violence done to the
people of Judah,
in whose land they have shed
innocent blood.
20 But Judah shall be inhabited
forever,
and Jerusalem to all generations.
21 I will avenge their blood, and I
will not clear the guilty, *k*
for the LORD dwells in Zion.

k Gk Syr: Heb *I will hold innocent their blood that
I have not held innocent*

3.18–21: The divine blessings bestowed *in
that day* include the paradisaical fertility of the
land, the destruction of Israel's ancient enemies, and the restoration of Judah. **18:** *Moun-*
tains, Am 9.13; *fountain,* Ezek 47.1–12 n. **19:**
Edom, Ob 11 n. **20–21:** Isa 12.6; Zech 2.10–
12.

Amos

During the long and peaceful reign of Jeroboam II (786–746 B.C.) Israel attained a height of territorial expansion and national prosperity never again reached. The military security and economic affluence which characterized this age were taken by many Israelites as signs of the LORD's special favor that they felt they deserved because of their extravagant support of the official shrines.

Into this scene stepped the prophet Amos, probably sometime during the decade 760–750 B.C. A native of the small Judean village of Tekoa, Amos was called by God from a shepherd's task (7.14–15) to the difficult mission of preaching harsh words in a smooth season. He denounced Israel, as well as its neighbors, for reliance upon military might, and for grave injustice in social dealings, abhorrent immorality, and shallow, meaningless piety. Amos's forceful, uncompromising preaching brought him into conflict with the religious authorities of his day. His personal confrontation with the priest Amaziah (7.10–17) remains one of the unforgettable scenes in Hebrew prophecy.

Expelled from the royal sanctuary at Bethel and commanded not to prophesy there again, Amos perhaps returned to Judah and wrote down the essence of his public preaching in substantially its present form. The book falls into three parts: chs 1–2, oracles against Israel's neighbors; chs 3–6, indictment of Israel itself for sin and injustice; and chs 7–9, visions of Israel's coming doom. Amos became the first in a brilliant succession of prophets whose words, now preserved in written form, have left their indelible stamp on later thought about God and human history.

1 The words of Amos, who was among the shepherds of Tekoa, which he saw concerning Israel in the days of King Uzziah of Judah and in the days of King Jeroboam son of Joash of Israel, two years[a] before the earthquake. 2 And he said:

The LORD roars from Zion,
  and utters his voice from
    Jerusalem;
the pastures of the shepherds
    wither,
  and the top of Carmel dries up.

3 Thus says the LORD:
For three transgressions of
    Damascus,
  and for four, I will not revoke
    the punishment;[b]
because they have threshed Gilead
  with threshing sledges of iron.
4 So I will send a fire on the house
    of Hazael,
  and it shall devour the
    strongholds of Ben-hadad.
5 I will break the gate bars of
    Damascus,
  and cut off the inhabitants from
    the Valley of Aven,
  and the one who holds the scepter
    from Beth-eden;
  and the people of Aram shall go
    into exile to Kir,
          says the LORD.

6 Thus says the LORD:
For three transgressions of Gaza,
  and for four, I will not revoke
    the punishment;[b]
because they carried into exile
    entire communities,
  to hand them over to Edom.
7 So I will send a fire on the wall
    of Gaza,
  fire that shall devour its
    strongholds.
8 I will cut off the inhabitants from
    Ashdod,
  and the one who holds the
    scepter from Ashkelon;
I will turn my hand against
    Ekron,
  and the remnant of the
    Philistines shall perish,
          says the Lord GOD.

9 Thus says the LORD:
For three transgressions of Tyre,
  and for four, I will not revoke
    the punishment;[b]
because they delivered entire
    communities over to Edom,
  and did not remember the
    covenant of kinship.
10 So I will send a fire on the wall
    of Tyre,
  fire that shall devour its
    strongholds.

a Or *during two years*   b Heb *cause it to return*

1.1: Superscription. *He saw,* see Nah 1.1 n. *King Uzziah* reigned over the southern kingdom from 783 to 742 B.C. The *earthquake,* mentioned again in Zech 14.5, cannot be precisely dated.
1.2–2.16: Indictment of neighboring peoples, Israel, and Judah. Amos applies the same demanding standards to Israel's neighbors and to Israel. **1.2:** This verse occurs in similar form in Joel 3.16 (compare Jer 25.30). Some scholars consider it an editorial addition. It sounds the solemn theme of the whole section. **3, 6, 9, etc.:** *Thus says the LORD* and the conclusion *says the LORD* (Hebrew, "the oracle of Yahweh") are standard formulas identifying prophetic oracles. The expression *for three transgressions . . . and for four* indicates "more than enough" (Job 33.14; Prov 30.18). Here *transgression* means rebellion. **3–5:** Against Damascus (compare Isa 17.1–3; Jer 49.23–27; Zech 9.1–4). *Damascus* was the capital of Syria (*Aram*); *Hazael* and *Ben-hadad* III were rulers (2 Kings 13.3); and *Kir* is called the place of Syrian origins (Am 9.7) and exile (2 Kings 16.9). *The Valley of Aven,* perhaps to be read "the Valley of On," may be between the Lebanon and Anti-Lebanon ranges. *Beth-eden* is the Bit-idini of the Assyrian records, by the Euphrates. The Syrian conquest of *Gilead,* 2 Kings 10.32–33. **1.6–8:** Against Philistia. Four Philistine cities are condemned because of their slave traffic with Edom (2 Chr 21.16–17; Joel 3.4–8; Zeph 2.4 7). **9 10:** Against Tyre (see Joel 3.4–8). *Covenant of kinship,* compare 1 Kings 9.13.

11 Thus says the LORD:
For three transgressions of Edom,
   and for four, I will not revoke
      the punishment; *c*
because he pursued his brother
      with the sword
   and cast off all pity;
he maintained his anger
      perpetually, *d*
   and kept his wrath *e* forever.
12 So I will send a fire on Teman,
   and it shall devour the
      strongholds of Bozrah.

13 Thus says the LORD:
For three transgressions of the
      Ammonites,
   and for four, I will not revoke
      the punishment; *c*
because they have ripped open
      pregnant women in Gilead
   in order to enlarge their
      territory.
14 So I will kindle a fire against the
      wall of Rabbah,
   fire that shall devour its
      strongholds,
with shouting on the day of battle,
   with a storm on the day of the
      whirlwind;
15 then their king shall go into exile,
   he and his officials together,
                  says the LORD.

2 Thus says the LORD:
For three transgressions of Moab,
   and for four, I will not revoke
      the punishment; *c*
because he burned to lime
   the bones of the king of Edom.
2 So I will send a fire on Moab,

and it shall devour the
      strongholds of Kerioth,
   and Moab shall die amid uproar,
      amid shouting and the sound of
         the trumpet;
3 I will cut off the ruler from its
         midst,
   and will kill all its officials
         with him,
                  says the LORD.

4 Thus says the LORD:
For three transgressions of Judah,
   and for four, I will not revoke
      the punishment; *c*
because they have rejected the law
      of the LORD,
   and have not kept his statutes,
but they have been led astray by
      the same lies
   after which their ancestors
         walked.
5 So I will send a fire on Judah,
   and it shall devour the
      strongholds of Jerusalem.

6 Thus says the LORD:
For three transgressions of Israel,
   and for four, I will not revoke
      the punishment; *c*
because they sell the righteous
      for silver,
   and the needy for a pair of
         sandals—
7 they who trample the head of the
      poor into the dust of
         the earth,

c Heb *cause it to return*   *d* Syr Vg: Heb *and his*
anger tore perpetually   *e* Gk Syr Vg: Heb *and*
his wrath kept

1.11–12: Against Edom. This oracle, recalling later oracles against Edom (see Introduction to Obadiah), is regarded by many as secondary. *Brother,* that is, Judah, Mal 1.2. **13–15**: Against the Ammonites (Zeph 2.8–11), who are to be punished because of inhuman atrocities against Israelites in Transjordan in *Gilead.*
   2.1–3: Against Moab, Israel's ancient enemy to the south-east of the Jordan (Isa 15.1–

16.14). *Burned to lime,* perhaps involving a desecration (compare 2 Kings 23.16–20; Isa 33.12). **4–5**: Against Judah. In part because of its language, many consider this oracle a later addition, but it is consistent with Amos's view that all nations stand accountable before God.
   2.6–16: Against Israel. Using the same literary form, Amos shows that Israel and her neighbors are judged by the same standards.

and push the afflicted out of
the way;
father and son go in to the
same girl,
so that my holy name is
profaned;
8 they lay themselves down beside
every altar
on garments taken in pledge;
and in the house of their God
they drink
wine bought with fines they
imposed.

9 Yet I destroyed the Amorite
before them,
whose height was like the
height of cedars,
and who was as strong as oaks;
I destroyed his fruit above,
and his roots beneath.
10 Also I brought you up out of the
land of Egypt,
and led you forty years in
the wilderness,
to possess the land of the
Amorite.
11 And I raised up some of your
children to be prophets
and some of your youths to be
nazirites.*f*
Is it not indeed so, O people
of Israel?
says the LORD.

12 But you made the nazirites*f*
drink wine,
and commanded the prophets,
saying, "You shall not
prophesy."

13 So, I will press you down in
your place,
just as a cart presses down

when it is full of sheaves.*g*
14 Flight shall perish from the swift,
and the strong shall not retain
their strength,
nor shall the mighty save
their lives;
15 those who handle the bow shall
not stand,
and those who are swift of foot
shall not save themselves,
nor shall those who ride horses
save their lives;
16 and those who are stout of heart
among the mighty
shall flee away naked in
that day,
says the LORD.

3 Hear this word that the LORD has
spoken against you, O people of Israel, against the whole family that I
brought up out of the land of Egypt:
2 You only have I known
of all the families of the earth;
therefore I will punish you
for all your iniquities.

3 Do two walk together
unless they have made an
appointment?
4 Does a lion roar in the forest,
when it has no prey?
Does a young lion cry out from
its den,
if it has caught nothing?
5 Does a bird fall into a snare on
the earth,
when there is no trap for it?
Does a snare spring up from
the ground,
when it has taken nothing?

---

f That is, *those separated* or *those consecrated*
g Meaning of Heb uncertain

7–8: *Same girl,* cult prostitutes. **9:** *Amorite,* the pre-Israelite inhabitants of Canaan (see Ezek 16.3). **11:** *Nazirites,* Num 6.2 n.; Judg 13.5 n. **3.1–6.14: Israel's sinfulness and God's punishment.** Three sections, each with the same opening exhortation (3.1; 4.1; 5.1). **3.1–15: The privileges of election create a greater responsibility** (Lk 12.48b).

Because Israel had been favored above *all the families of the earth* (Ex 19.4–6; Deut 7.6), the neighboring nations are assembled to witness her chastisement. **3–8:** The call to prophesy cannot be resisted. **6:** Israel's God is operative in history, sending calamity (*disaster*) as corrective discipline (Job 2.10; Isa 45.7).

6 Is a trumpet blown in a city,
and the people are not afraid?
Does disaster befall a city,
unless the LORD has done it?
7 Surely the Lord GOD does
nothing,
without revealing his secret
to his servants the prophets.
8 The lion has roared;
who will not fear?
The Lord GOD has spoken;
who can but prophesy?

9 Proclaim to the strongholds
in Ashdod,
and to the strongholds in the
land of Egypt,
and say, "Assemble yourselves on
Mount*h* Samaria,
and see what great tumults are
within it,
and what oppressions are in
its midst."
10 They do not know how to do
right, says the LORD,
those who store up violence and
robbery in their
strongholds.
11 Therefore thus says the Lord GOD:
An adversary shall surround
the land,
and strip you of your defense;
and your strongholds shall be
plundered.

12 Thus says the LORD: As the shepherd rescues from the mouth of the lion two legs, or a piece of an ear, so shall the people of Israel who live in Samaria be rescued, with the corner of a couch and part*i* of a bed.

13 Hear, and testify against the house
of Jacob,
says the Lord GOD, the God
of hosts:
14 On the day I punish Israel for its
transgressions,
I will punish the altars of
Bethel,
and the horns of the altar shall
be cut off
and fall to the ground.
15 I will tear down the winter house
as well as the summer
house;
and the houses of ivory shall
perish,
and the great houses*j* shall come
to an end,
says the LORD.

4 Hear this word, you cows
of Bashan
who are on Mount Samaria,
who oppress the poor, who crush
the needy,
who say to their husbands,
"Bring something to
drink!"
2 The Lord GOD has sworn by
his holiness:
The time is surely coming
upon you,
when they shall take you away
with hooks,
even the last of you with
fishhooks.
3 Through breaches in the wall you
shall leave,

h Gk Syr: Heb *the mountains of*   *i* Meaning of
Heb uncertain   *j* Or *many houses*

3.7: The Old Testament concept of revelation is grounded in the conviction that the God who is sovereign over creation and history discloses a transcendent purpose to *his servants the prophets* (2.11; Gen 18.17–19; Ex 4.15–16; Jer 7.25; Dan 9.10; Lk 1.70). 12: Although a few may be rescued from the catastrophe (5.3), the burden of Amos's message is the thoroughness of the divine punishment (9.1). 15: Houses of ivory, see 1 Kings 22.39 n.

4.1–13: **Israel's luxurious excesses and vain piety.** Israel has not heeded the LORD's repeated warnings. 1–3: The wealthy and greedy women of *Samaria,* being as guilty as their husbands, will also be punished (compare Isa 3.16–26). Cows of Bashan, Ps 22.12. 4–5: Israel's love of manifold public rites at the chief sanctuaries is satirized.

4.6–12: Having ignored the LORD's repeated warnings through nature and history, Israel must now prepare to meet its God, who

each one straight ahead;
and you shall be flung out into
Harmon,*k*
                    says the LORD.

4 Come to Bethel—and transgress;
to Gilgal—and multiply
transgression;
bring your sacrifices every
morning,
your tithes every three days;
5 bring a thank offering of leavened
bread,
and proclaim freewill offerings,
publish them;
for so you love to do, O people
of Israel!
                    says the Lord GOD.

6 I gave you cleanness of teeth in all
your cities,
and lack of bread in all your
places,
yet you did not return to me,
                    says the LORD.

7 And I also withheld the rain
from you
when there were still three
months to the harvest;
I would send rain on one city,
and send no rain on another
city;
one field would be rained upon,
and the field on which it did not
rain withered;
8 so two or three towns wandered
to one town
to drink water, and were not
satisfied;
yet you did not return to me,
                    says the LORD.

9 I struck you with blight and
mildew;
I laid waste*l* your gardens and
your vineyards;

the locust devoured your fig
trees and your olive trees;
yet you did not return to me,
                    says the LORD.

10 I sent among you a pestilence after
the manner of Egypt;
I killed your young men with
the sword;
I carried away your horses;*m*
and I made the stench of your
camp go up into your
nostrils;
yet you did not return to me,
                    says the LORD.

11 I overthrew some of you,
as when God overthrew Sodom
and Gomorrah,
and you were like a brand
snatched from the fire;
yet you did not return to me,
                    says the LORD.

12 Therefore thus I will do to you,
O Israel;
because I will do this to you,
prepare to meet your God,
O Israel!

13 For lo, the one who forms the
mountains, creates the
wind,
reveals his thoughts to mortals,
makes the morning darkness,
and treads on the heights of
the earth—
the LORD, the God of hosts, is
his name!

5 Hear this word that I take up over
you in lamentation, O house of
Israel:

k Meaning of Heb uncertain   *l* Cn: Heb *the
multitude of*   *m* Heb *with the captivity of your
horses*

---

is characterized by patient love and inexorable
justice. **6**: *Cleanness of teeth,* i.e. a famine. **13**:
This doxology, like those in 5.8–9 and 9.5–6,
with which it shares the refrain *the LORD . . .
is his name,* may be a later addition.

**5.1–6.14: The horror and finality of Is-
rael's deserved punishment. 5.1–3**: A lament for the fallen and forsaken nation.
Marched out a thousand, a reference to military
forces.

2 Fallen, no more to rise,
    is maiden Israel;
forsaken on her land,
    with no one to raise her up.

3 For thus says the Lord GOD:
The city that marched out
    a thousand
shall have a hundred left,
and that which marched out
    a hundred
shall have ten left. *n*

4 For thus says the LORD to the
    house of Israel:
Seek me and live;
5   but do not seek Bethel,
and do not enter into Gilgal
    or cross over to Beer-sheba;
for Gilgal shall surely go into
    exile,
    and Bethel shall come to
    nothing.

6 Seek the LORD and live,
    or he will break out against the
    house of Joseph like fire,
    and it will devour Bethel, with
    no one to quench it.
7 Ah, you that turn justice to
    wormwood,
    and bring righteousness to
    the ground!

8 The one who made the Pleiades
    and Orion,
and turns deep darkness into
    the morning,
and darkens the day into night,
who calls for the waters of the
    sea,
    and pours them out on the
    surface of the earth,
the LORD is his name,
9 who makes destruction flash out
    against the strong,
    so that destruction comes upon
    the fortress.

10 They hate the one who reproves
    in the gate,
    and they abhor the one who
    speaks the truth.
11 Therefore because you trample
    on the poor
    and take from them levies
    of grain,
you have built houses of
    hewn stone,
    but you shall not live in them;
you have planted pleasant
    vineyards,
    but you shall not drink
    their wine.
12 For I know how many are your
    transgressions,
    and how great are your sins—
you who afflict the righteous, who
    take a bribe,
    and push aside the needy in
    the gate.
13 Therefore the prudent will keep
    silent in such a time;
    for it is an evil time.

14 Seek good and not evil,
    that you may live;
and so the LORD, the God of
    hosts, will be with you,
    just as you have said.
15 Hate evil and love good,
    and establish justice in the gate;
it may be that the LORD, the God
    of hosts,
    will be gracious to the remnant
    of Joseph.

16 Therefore thus says the LORD, the
    God of hosts, the Lord:
In all the squares there shall be
    wailing;
    and in all the streets they shall
    say, "Alas! alas!"
They shall call the farmers to
    mourning,

n Heb adds *to the house of Israel*

5.4–7, 14–15: There is yet time to seek the
LORD and live. 8–9: The second doxology
(see 4.13 n.).

5.10–13, 16–17: Amos warns Israel again
about the nature and effect of its transgressions. 18–20: *The day of the LORD*, in which

and those skilled in lamentation,
to wailing;
17 in all the vineyards there shall
be wailing,
for I will pass through the midst
of you,
says the LORD.

18 Alas for you who desire the day
of the LORD!
Why do you want the day of
the LORD?
It is darkness, not light;
19 as if someone fled from a lion,
and was met by a bear;
or went into the house and rested
a hand against the wall,
and was bitten by a snake.
20 Is not the day of the LORD
darkness, not light,
and gloom with no brightness
in it?

21 I hate, I despise your festivals,
and I take no delight in your
solemn assemblies.
22 Even though you offer me your
burnt offerings and grain
offerings,
I will not accept them;
and the offerings of well-being of
your fatted animals
I will not look upon.
23 Take away from me the noise of
your songs;
I will not listen to the melody
of your harps.
24 But let justice roll down like
waters,
and righteousness like an
ever-flowing stream.

25 Did you bring to me sacrifices and
offerings the forty years in the wilderness, O house of Israel? 26 You shall take
up Sakkuth your king, and Kaiwan your
star-god, your images,[o] which you
made for yourselves; 27 therefore I will
take you into exile beyond Damascus,
says the LORD, whose name is the God of
hosts.

6 Alas for those who are at ease
in Zion,
and for those who feel secure on
Mount Samaria,
the notables of the first of the
nations,
to whom the house of Israel
resorts!
2 Cross over to Calneh, and see;
from there go to Hamath
the great;
then go down to Gath of the
Philistines.
Are you better[p] than these
kingdoms?
Or is your[q] territory greater
than their[r] territory,
3 O you that put far away the
evil day,
and bring near a reign of
violence?

4 Alas for those who lie on beds
of ivory,
and lounge on their couches,
and eat lambs from the flock,
and calves from the stall;
5 who sing idle songs to the sound
of the harp,

o Heb *your images, your star-god* *p* Or *Are they better* *q* Heb *their* *r* Heb *your*

Israelites piously expected to be vindicated against their enemies, will be darkness and gloom (Am 8.9–14; see Zeph 1.14–18 n.). Amos's profound reinterpretation of this popular concept is among his most significant contributions.

5.21–27: The LORD delights not in an abundance of festivals and sacrifices but in *justice* and *righteousness*. Verse 24 expresses the heart of Amos's preaching. **25**: In the desert Israel had a pure, direct relation with God which rendered sacrifices unnecessary (Jer 2.2–3; Hos 2.14–20; 9.10). **26**: *Sakkuth* and *Kaiwan* designate known Assyrian deities (Acts 7.42–43).

6.1–7: The Israelites, who feel themselves secure in their false confidence and lie upon beds of ivory in luxurious self-indulgence, will be the first of those whom God will send into exile. **2**: Compare Isa 10.9–11; 2 Chr

and like David improvise on
instruments of music;
6 who drink wine from bowls,
and anoint themselves with the
finest oils,
but are not grieved over the
ruin of Joseph!
7 Therefore they shall now be the
first to go into exile,
and the revelry of the loungers
shall pass away.

8 The Lord GOD has sworn by
himself
(says the LORD, the God of hosts):
I abhor the pride of Jacob
and hate his strongholds;
and I will deliver up the city
and all that is in it.

9 If ten people remain in one house,
they shall die. 10 And if a relative, one
who burns the dead,^{s} shall take up the
body to bring it out of the house, and
shall say to someone in the innermost
parts of the house, "Is anyone else with
you?" the answer will come, "No."
Then the relative^t shall say, "Hush! We
must not mention the name of the
LORD."

11 See, the LORD commands,
and the great house shall be
shattered to bits,
and the little house to pieces.
12 Do horses run on rocks?
Does one plow the sea with
oxen?^{u}
But you have turned justice
into poison
and the fruit of righteousness
into wormwood—

13 you who rejoice in Lo-debar,^{v}
who say, "Have we not by our
own strength
taken Karnaim^w for ourselves?"
14 Indeed, I am raising up against
you a nation,
O house of Israel, says the
LORD, the God of hosts,
and they shall oppress you from
Lebo-hamath
to the Wadi Arabah.

7 This is what the Lord GOD showed
me: he was forming locusts at the
time the latter growth began to sprout (it
was the latter growth after the king's
mowings). 2 When they had finished eating the grass of the land, I said,
"O Lord GOD, forgive, I beg you!
How can Jacob stand?
He is so small!"
3 The LORD relented concerning
this;
"It shall not be," said the LORD.

4 This is what the Lord GOD showed
me: the Lord GOD was calling for a
shower of fire,^{x} and it devoured the
great deep and was eating up the land.
5 Then I said,
"O Lord GOD, cease, I beg you!
How can Jacob stand?
He is so small!"
6 The LORD relented concerning
this;
"This also shall not be," said the
Lord GOD.

s Or *who makes a burning for him*   t Heb *he*
u Or *Does one plow them with oxen*   v Or *in a*
thing of nothingness   w Or *horns*   x Or *for a*
judgment by fire

26.6. *Calneh* and *Hamath,* important commercial centers in Syria, represent wealth and
security. **5:** *David,* 1 Chr 23.5; Neh 12.36.
6.8–14: Because Israel has turned faith into
pride (Isa 28.1; Hos 5.5; Am 8.7) and justice
into poison, it will be thoroughly punished.
9–10: The sequel to siege or pestilence. *Not
mention the name of the LORD,* perhaps out of
fear of further judgment. **13:** *Lo-debar . . . Karnaim,* two towns east of the Jordan recovered

for Israel by Jeroboam II (2 Kings 14.25). **14:**
From *Lebo-hamath* to the *Wadi Arabah,* the
farthest extent of Israelite territory (see
2 Kings 14.25).
**7.1–9.15: Five visions of God's judg-
ment, and a prophecy of restoration. 7.1–
9:** The first three visions. **1–3:** Judgment by
locusts (Joel 1.2–7 n.). **4–6:** Judgment by fire
(see 1.4). The judgments are halted when
Amos pleads. **7–9:** *Plumb line* (2 Kings 21.13–

7 This is what he showed me: the Lord was standing beside a wall built with a plumb line, with a plumb line in his hand. 8 And the LORD said to me, "Amos, what do you see?" And I said, "A plumb line." Then the Lord said,

"See, I am setting a plumb line
in the midst of my people Israel;
I will never again pass them by;
9 the high places of Isaac shall be
made desolate,
and the sanctuaries of Israel shall
be laid waste,
and I will rise against the house
of Jeroboam with the
sword."

10 Then Amaziah, the priest of Bethel, sent to King Jeroboam of Israel, saying, "Amos has conspired against you in the very center of the house of Israel; the land is not able to bear all his words. 11 For thus Amos has said,

'Jeroboam shall die by the sword,
and Israel must go into exile
away from his land.' "

12 And Amaziah said to Amos, "O seer, go, flee away to the land of Judah, earn your bread there, and prophesy there; 13 but never again prophesy at Bethel, for it is the king's sanctuary, and it is a temple of the kingdom."

14 Then Amos answered Amaziah, "I am[y] no prophet, nor a prophet's son; but I am[y] a herdsman, and a dresser of sycamore trees, 15 and the LORD took me from following the flock, and the LORD said to me, 'Go, prophesy to my people Israel.'
16 "Now therefore hear the word of the LORD.

You say, 'Do not prophesy against
Israel,
and do not preach against the
house of Isaac.'
17 Therefore thus says the LORD:
'Your wife shall become a
prostitute in the city,
and your sons and your
daughters shall fall by
the sword,
and your land shall be parceled
out by line;
you yourself shall die in an
unclean land,
and Israel shall surely go into
exile away from its land.' "

8 This is what the Lord GOD showed me—a basket of summer fruit.[z] 2 He said, "Amos, what do you see?" And I said, "A basket of summer fruit."[z] Then the LORD said to me,

"The end[a] has come upon my
people Israel;
I will never again pass them by.
3 The songs of the temple[b] shall
become wailings in
that day,"
says the Lord GOD;
"the dead bodies shall be many,
cast out in every place. Be
silent!"

4 Hear this, you that trample on
the needy,
and bring to ruin the poor of
the land,

y Or *was*   z Heb *qayits*   a Heb *qets*
b Or *palace*

---

15). The people are found warped beyond correction; God decrees an irrevocable sentence of destruction.

7.10–17: Amos and Amaziah. A prose biographical incident, introduced here perhaps because of the threat against the house of Jeroboam in v. 9. Amaziah was the official priest of the royal sanctuary at Bethel. **12:** *Earn your bread,* literally "eat bread." **14:** Amos asserts that he is not a professional prophet (1 Sam 9.6–10; Mic 3.5–8, 11) or a

member of a prophetic guild (2 Kings 2.3 n.; 1 Sam 10.5; 1 Kings 22.6), but simply one whom the LORD took and sent to prophesy to the people (3.3–8; 2 Sam 7.8).

8.1–3: Fourth vision. *A basket of ripe summer fruit* symbolizes the immediacy of Israel's *end.*

8.4–14: The indictment of Israel and the coming of the day of mourning (see 5.18–20 n.). **5–6:** The merchants are impatient for the holy days to pass so they can resume their

5 saying, "When will the new moon
be over
so that we may sell grain;
and the sabbath,
so that we may offer wheat
for sale?
We will make the ephah small and
the shekel great,
and practice deceit with false
balances,
6 buying the poor for silver
and the needy for a pair of
sandals,
and selling the sweepings of
the wheat."

7 The LORD has sworn by the pride
of Jacob:
Surely I will never forget any of
their deeds.
8 Shall not the land tremble on
this account,
and everyone mourn who lives
in it,
and all of it rise like the Nile,
and be tossed about and sink
again, like the Nile of
Egypt?

9 On that day, says the Lord GOD,
I will make the sun go down
at noon,
and darken the earth in
broad daylight.
10 I will turn your feasts into
mourning,
and all your songs into
lamentation;
I will bring sackcloth on all loins,
and baldness on every head;
I will make it like the mourning
for an only son,
and the end of it like a
bitter day.

11 The time is surely coming, says
the Lord GOD,

when I will send a famine on
the land;
not a famine of bread, or a thirst
for water,
but of hearing the words of
the LORD.
12 They shall wander from sea to sea,
and from north to east;
they shall run to and fro, seeking
the word of the LORD,
but they shall not find it.

13 In that day the beautiful young
women and the young men
shall faint for thirst.
14 Those who swear by Ashimah
of Samaria,
and say, "As your god lives,
O Dan,"
and, "As the way of Beer-sheba
lives"—
they shall fall, and never
rise again.

9 I saw the LORD standing beside[c] the
altar, and he said:
Strike the capitals until the
thresholds shake,
and shatter them on the heads of
all the people;[d]
and those who are left I will kill
with the sword;
not one of them shall flee away,
not one of them shall escape.

2 Though they dig into Sheol,
from there shall my hand
take them;
though they climb up to heaven,
from there I will bring
them down.
3 Though they hide themselves on
the top of Carmel,

c Or *on*   d Heb *all of them*

fraudulent business (Isa 1.13–17; Lev 19.35–
36; Deut 25.13–16). **10:** *It,* i.e. the earth (see
vv. 8 and 9c). **13–14:** The patron deities of
pagan shrines, from farthest north (*Dan*) to

farthest south (*Beer-sheba*), will be of no help
in that day.
9.1–4: Fifth vision. The last pronouncement is the most dreadful. **1:** *Thresholds shake,*

from there I will search out and
      take them;
and though they hide from my
      sight at the bottom of
      the sea,
there I will command the
      sea-serpent, and it
      shall bite them.
4 And though they go into captivity
      in front of their enemies,
there I will command the
      sword, and it shall kill
      them;
and I will fix my eyes on them
      for harm and not for good.

5 The Lord, GOD of hosts,
he who touches the earth and
      it melts,
      and all who live in it mourn,
and all of it rises like the Nile,
      and sinks again, like the Nile
      of Egypt;
6 who builds his upper chambers in
      the heavens,
      and founds his vault upon
      the earth;
who calls for the waters of the
      sea,
      and pours them out upon the
      surface of the earth—
the LORD is his name.

7 Are you not like the Ethiopians*e*
      to me,
      O people of Israel? says the
      LORD.
Did I not bring Israel up from the
      land of Egypt,

and the Philistines from Caphtor
      and the Arameans from
      Kir?
8 The eyes of the Lord GOD are
      upon the sinful kingdom,
      and I will destroy it from the
      face of the earth
—except that I will not utterly
      destroy the house of Jacob,
            says the LORD.

9 For lo, I will command,
      and shake the house of Israel
            among all the nations
as one shakes with a sieve,
      but no pebble shall fall to
      the ground.
10 All the sinners of my people shall
      die by the sword,
who say, "Evil shall not
      overtake or meet us."

11 On that day I will raise up
      the booth of David that is
      fallen,
and repair its*f* breaches,
      and raise up its*g* ruins,
      and rebuild it as in the days of
      old;
12 in order that they may possess the
      remnant of Edom
      and all the nations who are
      called by my name,
says the LORD who does this.

---

e Or *Nubians*; Heb *Cushites*   *f* Gk: Heb *their*
g Gk: Heb *his*

---

see Isa 6.4. **2:** *Sheol,* the place of the dead (Job
10.19–22; Isa 14.11, 15), which offers no hiding place from God (Ps 139.7–12). **3:** *Sea-
serpent,* mythological dragon inhabiting the
deep sea (Job 41.1–34 n.) **9.5–6: The third doxology,** see 4.13 n.
The LORD of nature and history (9.1–4) is also
the Creator.
**9.7–10: Israel has no claim to special
privilege in the moral realm,** for the LORD
will destroy every sinful kingdom. Here
Amos's universalism (see 1.2–2.16 n.) complements the doctrine of election (3.2). **7:**
Caphtor, see Judg 13.1 n. *Kir,* see 1.5 n.

**9.11–15: Prophecies of the restoration
of the Davidic dynasty** (vv. 11–12) **and the
glorious age to come** (vv. 13–15). This section, so affirmative in its emphasis, is generally considered a later addition. It expands,
however, genuine elements in Amos's own
thought (compare 3.12; 5.3, 4, 6, 11–15) and
also evidences a basic connection: the full prophetic word contains both judgment and salvation (see Introduction to Micah). **11–12:**
Note the use of this passage by James at the
meeting of the apostles and the elders of the
church at Jerusalem in Acts 15.16–17. **13:** See
Joel 3.18–21 n. *Shall overtake,* Lev 26.5.

13 The time is surely coming, says
        the LORD,
    when the one who plows shall
        overtake the one who reaps,
    and the treader of grapes the
        one who sows the seed;
the mountains shall drip
        sweet wine,
    and all the hills shall flow with
        it.
14 I will restore the fortunes of my
        people Israel,

and they shall rebuild the ruined
        cities and inhabit them;
they shall plant vineyards and
        drink their wine,
    and they shall make gardens and
        eat their fruit.
15 I will plant them upon their land,
    and they shall never again be
        plucked up
out of the land that I have
        given them,
        says the LORD your God.

Obadiah

Nothing is known of the person of the prophet Obadiah; even his name is not distinctive (see v. 1 n.). The problems of date and composition of his brief work, the shortest book in the Old Testament, are not easily solved. Some of the oracles in the book were spoken, or written, soon after Jerusalem fell to the Babylonians in 587 (586) B.C. Obadiah may not himself have written all of the present book, but may have drawn upon collections of sayings transmitted orally among prophetic circles. This could account for the similarity between vv. 1–9 and Jer 49.7–22. But the work is dominated by the theme of justice and judgment, and may well be the product of one voice.

Verses 1–14 indict the Edomites for outrageous and hostile actions when their Israelite kinfolk were in peril. Starting from these recent events in the experience of his people, Obadiah, like Joel (see Introduction to Joel), moves on to portray their future consummation. Verses 15–18 announce the day of the LORD's recompense upon the nations for their shameful behavior. A final section (vv. 19–21) proclaims the return of Israel's exiles to the Promised Land, their dominion over Edom, and the LORD's universal sovereignty.

1 The vision of Obadiah.

Thus says the Lord GOD
concerning Edom:
We have heard a report from
the LORD,
and a messenger has been sent
among the nations:
"Rise up! Let us rise against it
for battle!"
2 I will surely make you least
among the nations;
you shall be utterly despised.
3 Your proud heart has deceived
you,
you that live in the clefts of
the rock, *a*
whose dwelling is in the
heights.
You say in your heart,
"Who will bring me down to
the ground?"
4 Though you soar aloft like
the eagle,
though your nest is set among
the stars,
from there I will bring you
down,
says the LORD.

5 If thieves came to you,
if plunderers by night
—how you have been
destroyed!—
would they not steal only what
they wanted?
If grape-gatherers came to you,
would they not leave gleanings?
6 How Esau has been pillaged,
his treasures searched out!
7 All your allies have deceived you,
they have driven you to
the border;
your confederates have prevailed
against you;
those who ate *b* your bread have
set a trap for you—
there is no understanding of it.
8 On that day, says the LORD,
I will destroy the wise out
of Edom,
and understanding out of
Mount Esau.
9 Your warriors shall be shattered,
O Teman,
so that everyone from Mount
Esau will be cut off.
10 For the slaughter and violence
done to your brother Jacob,
shame shall cover you,
and you shall be cut off forever.
11 On the day that you stood aside,
on the day that strangers carried
off his wealth,
and foreigners entered his gates
and cast lots for Jerusalem,
you too were like one of them.
12 But you should not have gloated *c*
over *d* your brother
on the day of his misfortune;
you should not have rejoiced over
the people of Judah
on the day of their ruin;
you should not have boasted

a Or *clefts of Sela*   *b* Cn: Heb lacks *those who
ate*   *c* Heb *But do not gloat* (and similarly
through verse 14)   *d* Heb *on the day of*

1: Introduction. *Vision,* see Nah 1.1 n.
Obadiah, meaning "Servant of the LORD," is
the name of twelve different persons in the
Old Testament. *Edom,* southeast of Palestine.
2–4: The humbling of Edom. The non-
Israelite *nations,* the Gentiles, will share in the
spectacle. **3:** The same Hebrew word means
rock and Sela, an Edomite fortress also called
Teman (v. 9) and Petra. **4:** *Says the LORD,* see
Am 1.3 n.
5–7: The pillaging and betrayal of Edom.
6: *Esau,* Jacob's twin, was the father of the
Edomites (Gen 25.30; 36.1). **8–11:** The destruction of Edom. **8:** Compare Zeph 1.9, 10;
3.16; Hag 2.23. Edom was renowned for its
wisdom (Jer 49.7). **10–11:** *Jacob* was the father
of the Israelites (Gen 49.2; Isa 43.1). By failing
to help his *brother* in his adversity, Edom became as guilty as the *foreigners* who *entered his
gates.* Obadiah shares this bitterness against
Edom with other writers of the sixth and fifth
centuries B.C. (Ps 137.7; Isa 34.5–7; 63.1–6;
Lam 4.21; Ezek 25.12–14; Mal 1.2–5; see
also, from an earlier period, Am 1.11–12).
12–14: Eight-fold indictment of Edom, cast in
a very forceful, repetitive style.

on the day of distress.

13 You should not have entered the
gate of my people
on the day of their calamity;
you should not have joined in the
gloating over Judah's*e*
disaster
on the day of his calamity;
you should not have looted
his goods
on the day of his calamity.

14 You should not have stood at
the crossings
to cut off his fugitives;
you should not have handed over
his survivors
on the day of distress.

15 For the day of the LORD is near
against all the nations.
As you have done, it shall be done
to you;
your deeds shall return on your
own head.

16 For as you have drunk on my
holy mountain,
all the nations around you
shall drink;
they shall drink and gulp down,*f*
and shall be as though they had
never been.

17 But on Mount Zion there shall be
those that escape,
and it shall be holy;
and the house of Jacob shall take

possession of those who
dispossessed them.

18 The house of Jacob shall be a fire,
the house of Joseph a flame,
and the house of Esau stubble;
they shall burn them and
consume them,
and there shall be no survivor of
the house of Esau;
for the LORD has spoken.

19 Those of the Negeb shall possess
Mount Esau,
and those of the Shephelah the
land of the Philistines;
they shall possess the land of
Ephraim and the land
of Samaria,
and Benjamin shall possess
Gilead.

20 The exiles of the Israelites who are
in Halah*g*
shall possess*h* Phoenicia as far
as Zarephath;
and the exiles of Jerusalem who
are in Sepharad
shall possess the towns of
the Negeb.

21 Those who have been saved*i* shall
go up to Mount Zion
to rule Mount Esau;
and the kingdom shall be
the LORD's.

e Heb *his*   *f* Meaning of Heb uncertain
g Cn: Heb *in this army*   *h* Cn: Meaning of
Heb uncertain   *i* Or *Saviors*

15–18: The day of the LORD's judgment
against all the nations, especially Edom. But
Israel shall be saved. **15:** *The day of the LORD,*
see Am 5.18 n. **16:** *You,* i.e. the Israelites.
Compare Jer 25.15–29. **17:** Joel 2.32.
**19–21: The division of the land, and the
LORD's kingship. 19:** *Negeb* is the arid south;
Shephelah, the western foothills. *Gilead* is in
Transjordan. **20:** *Halah* (the Hebrew text is

uncertain) is in northern Mesopotamia
(2 Kings 17.6); *Zarephath,* a town in southern
Phoenicia (1 Kings 17.9). *Sepharad,* perhaps
in Asia Minor. **21:** The immediate reference of
this verse is political: Israel will dominate
Edom. But it carries also a broader theological reference: power belongs to the LORD,
who shall reign over all (Pss 22.28; 47; 99.1–
2).

Jonah

The book of Jonah is unique among the prophetic books. It contains no collection of oracles in verse against Israel and foreign nations, but presents a prose narrative about the prophet himself. Instead of portraying a prophet who is an obedient servant of the LORD, calling his people to repentance, it features a recalcitrant prophet who flees from his mission and sulks when his hearers repent.

The book is a didactic narrative that has taken older material from the realm of popular legend and put it to new, more consequential use. Its two parts, chs 1–2 and 3–4, are now united by having in common a central character (Jonah), a similar plot (the ironical conversion of the heathen), and an identical theme (the breadth of God's saving love).

The principal figure of this artful story is an obscure Galilean prophet from Gath-hepher who counseled Jeroboam II (786–746 B.C.) in a successful conflict with the Syrians (2 Kings 14.25) and with whom some of the earlier traditional material was probably associated. Its author, however, probably lived in the post-exilic period because he shows the influence of Jeremiah and Second Isaiah, and opposes the type of a narrow sectarianism and exclusivism. Although the linguistic evidence is indecisive, a date sometime in the fifth or fourth century B.C. seems indicated.

With skill and finesse this little book calls Israel to repentance and reminds it of its mission to preach to all the nations the wideness of God's mercy and forgiveness (Gen 12.1–3; Isa 42.6–7; 49.6). In spirit, therefore, the book remains truly prophetic and justifies its place in the Book of the Twelve Prophets.

1 Now the word of the LORD came to Jonah son of Amittai, saying, 2 "Go at once to Nineveh, that great city, and cry out against it; for their wickedness has come up before me." 3 But Jonah set out to flee to Tarshish from the presence of the LORD. He went down to Joppa and found a ship going to Tarshish; so he paid his fare and went on board, to go with them to Tarshish, away from the presence of the LORD.

4 But the LORD hurled a great wind upon the sea, and such a mighty storm came upon the sea that the ship threatened to break up. 5 Then the mariners were afraid, and each cried to his god. They threw the cargo that was in the ship into the sea, to lighten it for them. Jonah, meanwhile, had gone down into the hold of the ship and had lain down, and was fast asleep. 6 The captain came and said to him, "What are you doing sound asleep? Get up, call on your god! Perhaps the god will spare us a thought so that we do not perish."

7 The sailors*a* said to one another, "Come, let us cast lots, so that we may know on whose account this calamity has come upon us." So they cast lots, and the lot fell on Jonah. 8 Then they said to him, "Tell us why this calamity has come upon us. What is your occupation? Where do you come from? What is your country? And of what people are you?" 9 "I am a Hebrew," he replied. "I worship the LORD, the God of heaven, who made the sea and the dry land." 10 Then the men were even more afraid, and said to

him, "What is this that you have done!" For the men knew that he was fleeing from the presence of the LORD, because he had told them so.

11 Then they said to him, "What shall we do to you, that the sea may quiet down for us?" For the sea was growing more and more tempestuous. 12 He said to them, "Pick me up and throw me into the sea; then the sea will quiet down for you; for I know it is because of me that this great storm has come upon you." 13 Nevertheless the men rowed hard to bring the ship back to land, but they could not, for the sea grew more and more stormy against them. 14 Then they cried out to the LORD, "Please, O LORD, we pray, do not let us perish on account of this man's life. Do not make us guilty of innocent blood; for you, O LORD, have done as it pleased you." 15 So they picked Jonah up and threw him into the sea; and the sea ceased from its raging. 16 Then the men feared the LORD even more, and they offered a sacrifice to the LORD and made vows.

17*b* But the LORD provided a large fish to swallow up Jonah; and Jonah was in the belly of the fish three days and three nights.

2 Then Jonah prayed to the LORD his God from the belly of the fish, 2 saying,

"I called to the LORD out of my
        distress,
    and he answered me;

a Heb *They*   *b* Ch 2.1 in Heb

1.1–16: Jonah's first call to preach to Nineveh (the capital of the Assyrians, who destroyed Samaria in 722–721), although unheeded, results in the conversion of the heathen sailors. **1–3:** Jonah rebels against the divine commission and attempts to flee from God. **1:** *The word of the LORD,* Hos 1.1 n. The word Jonah means "dove." **2:** For the *wickedness* of the Assyrian capital, *Nineveh,* see Nah ch 3. **3:** *Tarshish,* probably in southern Spain (Isa 23.1–12; Ezek 27.12, 25), represents the farthest point to which Jonah could sail. **1.4–16:** The LORD procures Jonah's recall. God causes a great tempest; for *the LORD, the God of heaven,* rules *the sea and the dry land* (Pss

65.5–7; 107.23–32; 139.7–12; Mk 4.35–41; Acts ch 27). **16:** Jonah becomes a missionary of his God in spite of himself.
1.17–2.10: Jonah is miraculously saved and returned to *dry land.* The fish merely serves as a sea-going vehicle to deposit Jonah back where he started (see 2.10). Like the tempest, the east wind, the plant, and the worm, it is an obedient agent of God's purpose. **17:** Mt 12.38–41.
2.1–10: Jonah's prayer and its answer. This psalm of thanksgiving (instead of an expected petition for help) may originally have been independent of the prose narrative. But it now serves to express Jonah's thanks for his

out of the belly of Sheol I cried,
and you heard my voice.
3 You cast me into the deep,
into the heart of the seas,
and the flood surrounded me;
all your waves and your billows
passed over me.
4 Then I said, 'I am driven away
from your sight;
how[c] shall I look again
upon your holy temple?'
5 The waters closed in over me;
the deep surrounded me;
weeds were wrapped around
my head
6 at the roots of the mountains.
I went down to the land
whose bars closed upon me
forever;
yet you brought up my life from
the Pit,
O LORD my God.
7 As my life was ebbing away,
I remembered the LORD;
and my prayer came to you,
into your holy temple.
8 Those who worship vain idols
forsake their true loyalty.
9 But I with the voice of
thanksgiving
will sacrifice to you;
what I have vowed I will pay.
Deliverance belongs to the
LORD!"
10 Then the LORD spoke to the fish, and it
spewed Jonah out upon the dry land.

3 The word of the LORD came to Jonah a second time, saying, 2 "Get up, go to Nineveh, that great city, and pro-

claim to it the message that I tell you." 3 So Jonah set out and went to Nineveh, according to the word of the LORD. Now Nineveh was an exceedingly large city, a three days' walk across. 4 Jonah began to go into the city, going a day's walk. And he cried out, "Forty days more, and Nineveh shall be overthrown!" 5 And the people of Nineveh believed God; they proclaimed a fast, and everyone, great and small, put on sackcloth.

6 When the news reached the king of Nineveh, he rose from his throne, removed his robe, covered himself with sackcloth, and sat in ashes. 7 Then he had a proclamation made in Nineveh: "By the decree of the king and his nobles: No human being or animal, no herd or flock, shall taste anything. They shall not feed, nor shall they drink water. 8 Human beings and animals shall be covered with sackcloth, and they shall cry mightily to God. All shall turn from their evil ways and from the violence that is in their hands. 9 Who knows? God may relent and change his mind; he may turn from his fierce anger, so that we do not perish."

10 When God saw what they did, how they turned from their evil ways, God changed his mind about the calamity that he had said he would bring upon them; and he did not do it.

4 But this was very displeasing to Jonah, and he became angry. 2 He prayed to the LORD and said, "O LORD! Is not this what I said while I was still in my own country? That is why I fled to

c Theodotion: Heb *surely*

---

deliverance (v. 9). **2**: Pss 18.6; 120.1. *Sheol,* or *Pit* (v. 6), is the region of darkness and death (Ps 88.3– 12). **3a**: Ps 88.6. **3b**: Ps 42.7. **6**: *Land,* of the departed, Sheol.
3.1–4.11: Jonah's second call to preach to Nineveh, although reluctantly and grudgingly obeyed, results in the wholesale conversion of the heathen city. **3**: *Exceedingly large city;* excavations have revealed a city about three miles in length and somewhat less than one and one-half miles wide. The message of the story, not the size of the city, is of primary

import. **5–10**: Again Jonah is a successful missionary in spite of himself.
3.6: *Sackcloth* and *ashes,* traditional signs of mourning and repentance (2 Sam 3.31; Job 42.6; Dan 9.3; Mt 11.21). **7–9**: The reaction of *the king of Nineveh* is modelled on Jer 18.8– 9. He sets a better example than Jonah (4.1). **10**: Repentance and deliverance are themes dominating the story of Jonah and its use in the New Testament (Mt 12.38–41; Lk 11.29– 32).
4.1–4: Jonah is not only disobedient but

Tarshish at the beginning; for I knew that you are a gracious God and merciful, slow to anger, and abounding in steadfast love, and ready to relent from punishing. ³ And now, O LORD, please take my life from me, for it is better for me to die than to live." ⁴ And the LORD said, "Is it right for you to be angry?" ⁵ Then Jonah went out of the city and sat down east of the city, and made a booth for himself there. He sat under it in the shade, waiting to see what would become of the city.

6 The LORD God appointed a bush, *d* and made it come up over Jonah, to give shade over his head, to save him from his discomfort; so Jonah was very happy about the bush. ⁷ But when dawn came up the next day, God appointed a worm that attacked the bush, so that it with-ered. ⁸ When the sun rose, God prepared a sultry east wind, and the sun beat down on the head of Jonah so that he was faint and asked that he might die. He said, "It is better for me to die than to live."

9 But God said to Jonah, "Is it right for you to be angry about the bush?" And he said, "Yes, angry enough to die." ¹⁰ Then the LORD said, "You are concerned about the bush, for which you did not labor and which you did not grow; it came into being in a night and perished in a night. ¹¹ And should I not be concerned about Nineveh, that great city, in which there are more than a hundred and twenty thousand persons who do not know their right hand from their left, and also many animals?"

d Heb *qiqayon*, possibly *the castor bean plant*

---

bigoted; the reason for his flight was that he did not want to see the Assyrians spared by *a gracious God* (compare Ex 34.6–7). Literary artistry is apparent in the death wish and question of the Lord in vv. 3–4 and 9. **11**: God has concern for every creature in the universe, even *animals*. The divine love extends beyond any covenant.

Micah

The superscription to this book indicates that Micah was a younger contemporary of Isaiah. The characteristics of the era in which Micah spoke were similar, therefore, to those seen in the ministries of Hosea and Isaiah. The prosperous half-century of peace enjoyed by the northern kingdom of Israel was ended by the death of Jeroboam II and the westward advance of the Assyrians. In 731 B.C. the fall of Damascus marked the end of Syria's freedom; a brief decade later brought the same fate to Israel and its capital at Samaria. Although the kingdom of Judah was spared at that time from the Assyrian yoke, the anti-Assyrian policies of Hezekiah brought Sennacherib of Assyria to Judah in 701 B.C., and Judah was left a weak vassal state.

But unlike Isaiah, Micah was neither of noble descent nor a native of the capital city. He came, rather, from the common people, being a citizen of the small village of Moresheth in the Judean foothills southwest of Jerusalem. Therefore Micah looked upon the corruptions and pretensions of the capital through different eyes. Perhaps it was this difference that accounts for Micah's prophecy of the fall of Jerusalem (3.9–12), a major note distinguishing him from his contemporaries and for which he was long remembered (Jer 26.18).

Micah stands solidly with Amos, Hosea, and Isaiah as a fierce champion of pure worship of the LORD and of social justice, and he shares with them both the word of judgment against God's own people (1.2–3.12; 6.1–7.7) and the promise of divine forgiveness and hope in a future restoration. The latter theme of the prophet comes to us in an expanded and edited post-exilic form in 4.1–5.15. The picture of the restored Jerusalem in 7.8–20 comes from the post-exilic period.

1 The word of the L ORD that came to Micah of Moresheth in the days of Kings Jotham, Ahaz, and Hezekiah of Judah, which he saw concerning Samaria and Jerusalem.

2 Hear, you peoples, all of you;
　listen, O earth, and all that is
　　in it;
　and let the Lord G OD be a witness
　　against you,
　the Lord from his holy temple.
3 For lo, the L ORD is coming out of
　　his place,
　and will come down and tread
　　upon the high places of
　　the earth.
4 Then the mountains will melt
　　under him
　and the valleys will burst open,
　like wax near the fire,
　　like waters poured down a
　　steep place.
5 All this is for the transgression
　　of Jacob
　and for the sins of the house
　　of Israel.
　What is the transgression of Jacob?
　Is it not Samaria?
　And what is the high place*a*
　　of Judah?
　Is it not Jerusalem?
6 Therefore I will make Samaria a
　　heap in the open country,
　a place for planting vineyards.
　I will pour down her stones into
　　the valley,
　and uncover her foundations.
7 All her images shall be beaten
　　to pieces,
　all her wages shall be burned
　　with fire,

and all her idols I will lay waste;
　for as the wages of a prostitute she
　　gathered them,
　and as the wages of a prostitute
　　they shall again be used.

8 For this I will lament and wail;
　I will go barefoot and naked;
　I will make lamentation like
　　the jackals,
　and mourning like the ostriches.
9 For her wound*b* is incurable.
　It has come to Judah;
　it has reached to the gate of
　　my people,
　to Jerusalem.

10 Tell it not in Gath,
　　weep not at all;
　in Beth-leaphrah
　　roll yourselves in the dust.
11 Pass on your way,
　　inhabitants of Shaphir,
　　in nakedness and shame;
　the inhabitants of Zaanan
　　do not come forth;
　Beth-ezel is wailing
　　and shall remove its support
　　from you.
12 For the inhabitants of Maroth
　　wait anxiously for good,
　yet disaster has come down from
　　the L ORD
　　to the gate of Jerusalem.
13 Harness the steeds to the chariots,
　　inhabitants of Lachish;
　it was the beginning of sin
　　to daughter Zion,
　for in you were found

a Heb *what are the high places*　　*b* Gk Syr Vg:
Heb *wounds*

1.1: **Superscription.** *The word of the* L ORD *. . . which he saw,* see Hos 1.1 n. and Nah 1.1 n. *Jotham* reigned over Judah from 750 to 735 B.C.; *Ahaz* from 735 to 715; and *Hezekiah* from 715 to 687.
1.2–3.12: **Threats directed against Samaria and Jerusalem because of the corruption of their religious and political leaders** form the first major section of Micah's book. 1.2–7: Samaria will be destroyed

when the L ORD comes in judgment to punish *the house of Israel* for its *transgression.* This oracle is to be dated before 721 B.C. **2:** Compare Isa 1.2; Hab 2.20.
1.8–16: In a series of forceful literary puns on the names of the areas to be devastated, the prophet laments that the *wound* of his people *is incurable* and that God *will again bring a conqueror* even *to the gate of Jerusalem.* This *lament,* perhaps acted out by the prophet him-

the transgressions of Israel.
14 Therefore you shall give
parting gifts
to Moresheth-gath;
the houses of Achzib shall be
a deception
to the kings of Israel.
15 I will again bring a conqueror
upon you,
inhabitants of Mareshah;
the glory of Israel
shall come to Adullam.
16 Make yourselves bald and cut off
your hair
for your pampered children;
make yourselves as bald as
the eagle,
for they have gone from you
into exile.

2 Alas for those who devise
wickedness
and evil deeds*c* on their beds!
When the morning dawns, they
perform it,
because it is in their power.
2 They covet fields, and seize them;
houses, and take them away;
they oppress householder
and house,
people and their inheritance.
3 Therefore thus says the LORD:
Now, I am devising against this
family an evil
from which you cannot remove
your necks;
and you shall not walk haughtily,
for it will be an evil time.
4 On that day they shall take up a
taunt song against you,
and wail with bitter
lamentation,
and say, "We are utterly ruined;
the LORD *d* alters the inheritance
of my people;

how he removes it from me!
Among our captors*e* he parcels
out our fields."
5 Therefore you will have no one to
cast the line by lot
in the assembly of the LORD.

6 "Do not preach"—thus they
preach—
"one should not preach of
such things;
disgrace will not overtake us."
7 Should this be said, O house
of Jacob?
Is the LORD's patience
exhausted?
Are these his doings?
Do not my words do good
to one who walks uprightly?
8 But you rise up against my
people*f* as an enemy;
you strip the robe from the
peaceful,*g*
from those who pass by trustingly
with no thought of war.
9 The women of my people you
drive out
from their pleasant houses;
from their young children you
take away
my glory forever.
10 Arise and go;
for this is no place to rest,
because of uncleanness that
destroys
with a grievous destruction.*h*
11 If someone were to go about
uttering empty falsehoods,
saying, "I will preach to you of
wine and strong drink,"

c Cn: Heb *work evil*   *d* Heb *he*   *e* Cn: Heb
the rebellious   *f* Cn: Heb *But yesterday my*
people rose   *g* Cn: Heb *from before a garment*
h Meaning of Heb uncertain

self, *barefoot* and *naked* (v. 8), may refer to the
campaign of 701 B.C. **16:** *Cut off your hair,* a
common rite of mourning (Jer 7.29).
2.1–11: Because of their uncleanness,
which Micah depicts with clarity (v. 2) and
bitter realism (v. 11), the Israelites must experience a grievous destruction, for the LORD
will hold the nation responsible for its social

and moral abuses. **1–2:** Ps 36.4; Isa 5.8–12;
32.7; Am 8.4. **3:** *An evil time,* Jer 18.11; compare Am 5.13. **4:** *A taunt song,* compare Hab
2.6. **5:** *Cast the line by lot,* on a piece of land,
thus acquiring title to it (Josh 18.6; Ps 16.6).
6: Micah quotes the popular preaching of the
false prophets (Isa 30.10; Jer 5.31; Am 2.12).

such a one would be the
    preacher for this people!

12 I will surely gather all of you,
        O Jacob,
    I will gather the survivors
        of Israel;
    I will set them together
        like sheep in a fold,
    like a flock in its pasture;
        it will resound with people.
13 The one who breaks out will go
        up before them;
    they will break through and
        pass the gate,
    going out by it.
    Their king will pass on before
        them,
    the LORD at their head.

3 And I said:
    Listen, you heads of Jacob
        and rulers of the house of Israel!
    Should you not know justice?—
2   you who hate the good and love
        the evil,
    who tear the skin off my people, *i*
        and the flesh off their bones;
3   who eat the flesh of my people,
        flay their skin off them,
    break their bones in pieces,
        and chop them up like meat *j* in
        a kettle,
    like flesh in a caldron.

4 Then they will cry to the LORD,
        but he will not answer them;
    he will hide his face from them at
        that time,

because they have acted
    wickedly.

5 Thus says the LORD concerning
        the prophets
    who lead my people astray,
    who cry "Peace"
        when they have something to
        eat,
    but declare war against those
        who put nothing into their
        mouths.
6 Therefore it shall be night to you,
        without vision,
    and darkness to you, without
        revelation.
    The sun shall go down upon
        the prophets,
    and the day shall be black
        over them;
7 the seers shall be disgraced,
        and the diviners put to shame;
    they shall all cover their lips,
        for there is no answer from
        God.
8 But as for me, I am filled
        with power,
    with the spirit of the LORD,
        and with justice and might,
    to declare to Jacob his
        transgression
    and to Israel his sin.

9 Hear this, you rulers of the house
        of Jacob
    and chiefs of the house of Israel,
    who abhor justice

i Heb *from them*   *j* Gk: Heb *as*

2.12–13: This picture of the restored remnant of all Israel, which presupposes the capture of the country and its destruction, was
probably contributed by an exilic or post-
exilic editor.
   3.1–4: The avaricious and irresponsible
deeds of Israel's *rulers*. 2–3: Compare Isa
5.20. Micah here uses the figures of a butcher
and wild beast. 4: Futile *cry to the LORD,* compare Isa 1.15.
   3.5–8: The mercenary prophets and seers,
who prostitute their solemn office, shall be
cut off from God, the source of true revelation and well-being. 7: *Seers,* see 1 Sam 9.9.

Shall . . . cover their lips, as a sign of mourning
(Lev 13.45; Ezek 27.17, 22). 8: In contrast
with the false officials, Micah's call has im-
bued him with power, justice, and might,
because it filled him with *the spirit of the LORD*
(Isa 11.2; 61.1; Ezek 2.2).
   3.9–12: The first section reaches its climax
with Micah's bold assertion that, because of
corruption which permeates all of its rulers,
Jerusalem and the temple itself shall be destroyed. 9: This recalls Am 5.6–7. 10: Compare Hab 2.12. 12: This was quoted at the trial
of Jeremiah (Jer 26.18).

and pervert all equity,

10 who build Zion with blood
and Jerusalem with wrong!

11 Its rulers give judgment for a
bribe,
its priests teach for a price,
its prophets give oracles
for money;
yet they lean upon the LORD
and say,
"Surely the LORD is with us!
No harm shall come upon us."

12 Therefore because of you
Zion shall be plowed as a field;
Jerusalem shall become a heap
of ruins,
and the mountain of the house a
wooded height.

4 In days to come
the mountain of the LORD's
house
shall be established as the highest
of the mountains,
and shall be raised up above
the hills.
Peoples shall stream to it,

2 and many nations shall come
and say:
"Come, let us go up to the
mountain of the LORD,
to the house of the God of
Jacob;
that he may teach us his ways
and that we may walk in
his paths."
For out of Zion shall go forth
instruction,
and the word of the LORD
from Jerusalem.

3 He shall judge between many
peoples,

and shall arbitrate between
strong nations far away;
they shall beat their swords into
plowshares,
and their spears into pruning
hooks;
nation shall not lift up sword
against nation,
neither shall they learn war
any more;

4 but they shall all sit under their
own vines and under their
own fig trees,
and no one shall make
them afraid;
for the mouth of the LORD of
hosts has spoken.

5 For all the peoples walk,
each in the name of its god,
but we will walk in the name of
the LORD our God
forever and ever.

6 In that day, says the LORD,
I will assemble the lame
and gather those who have been
driven away,
and those whom I have
afflicted.

7 The lame I will make the remnant,
and those who were cast off, a
strong nation;
and the LORD will reign over them
in Mount Zion
now and forevermore.

8 And you, O tower of the flock,
hill of daughter Zion,
to you it shall come,
the former dominion shall
come,

**4.1–5.15: Prophecies of Israel's glorious
future and the restoration of the Davidic
kingdom** constitute the second major section of the book. Though many scholars date
the origin of these prophecies in the post-
exilic period, they are quite possibly based on
genuine oracles from Micah, edited in their
present form after the Exile. **4.1–5:** The exaltation of Jerusalem as a center of worship by the
nations; the new age of peace. Verses 1–3 are
duplicated in Isa 2.2–4. Here, as sometimes

elsewhere in the prophetic books, the author
of a passage may be unknown, as with Isa
chs 40–66. **1:** *The mountain of the LORD's house,*
Mount Zion. *Highest of the mountains,* compare Ezek 40.2. **4:** Compare Zech 3.10. **5:**
Compare Isa 2.5.

4.6–8: The LORD's gracious exaltation of
the lame and rejected, for whom he will restore his kingdom (Isa 40.9–11; Ezek 34.11–
16; 37.24–28; Zeph 3.19; Jn 10.7–16).

the sovereignty of daughter
Jerusalem.

9 Now why do you cry aloud?
Is there no king in you?
Has your counselor perished,
that pangs have seized you like a
woman in labor?
10 Writhe and groan,*k* O daughter
Zion,
like a woman in labor;
for now you shall go forth from
the city
and camp in the open country;
you shall go to Babylon.
There you shall be rescued,
there the LORD will redeem you
from the hands of your enemies.

11 Now many nations
are assembled against you,
saying, "Let her be profaned,
and let our eyes gaze upon
Zion."
12 But they do not know
the thoughts of the LORD;
they do not understand his plan,
that he has gathered them as
sheaves to the threshing
floor.
13 Arise and thresh,
O daughter Zion,
for I will make your horn iron
and your hoofs bronze;
you shall beat in pieces many
peoples,
and shall*l* devote their gain to
the LORD,
their wealth to the Lord of the
whole earth.

5 *m* Now you are walled around with
a wall;*n*

siege is laid against us;
with a rod they strike the ruler
of Israel
upon the cheek.

2*o* But you, O Bethlehem of
Ephrathah,
who are one of the little clans
of Judah,
from you shall come forth for me
one who is to rule in Israel,
whose origin is from of old,
from ancient days.
3 Therefore he shall give them up
until the time
when she who is in labor has
brought forth;
then the rest of his kindred shall
return
to the people of Israel.
4 And he shall stand and feed his
flock in the strength of
the LORD,
in the majesty of the name of
the LORD his God.
And they shall live secure, for
now he shall be great
to the ends of the earth;
5 and he shall be the one of peace.

If the Assyrians come into
our land
and tread upon our soil,*p*
we will raise against them seven
shepherds
and eight installed as rulers.
6 They shall rule the land of Assyria
with the sword,

k Meaning of Heb uncertain　　*l* Gk Syr Tg:
Heb *and I will*　　*m* Ch 4.14 in Heb
n Cn Compare Gk: Meaning of Heb uncertain
o Ch 5.1 in Heb　　*p* Gk: Heb *in our palaces*

4.9–5.1: In three poetic units the humiliation and travail that Israel must experience
in the near future (see "now" in 4.9, 11, and
5.1) are contrasted with the final triumph. **4.9:**
Compare 5.3 and Jer 8.19. **10:** The reference
to exile in Babylon is a later addition (Isa
48.20; 52.9–12).
5.2–6: The shepherd king who is to be
ruler of Israel will be born not in Jerusalem,
but, like David, in Bethlehem, among the
insignificant clans of Judah. **2:** *Bethlehem,* Gen
35.19; Ruth 4.11; 1 Sam 17.12. *Me,* refers to
God. *From of old* could mean from the days of
David, rather than pre-existence from the beginning of time. The New Testament interprets this statement as referring to the birth of
Jesus (Mt 2.6; compare also Jn 7.40–43).

and the land of Nimrod with
the drawn sword;*q*
they*r* shall rescue us from the
Assyrians
if they come into our land
or tread within our border.

7 Then the remnant of Jacob,
surrounded by many peoples,
shall be like dew from the LORD,
like showers on the grass,
which do not depend upon people
or wait for any mortal.
8 And among the nations the
remnant of Jacob,
surrounded by many peoples,
shall be like a lion among the
animals of the forest,
like a young lion among the
flocks of sheep,
which, when it goes through,
treads down
and tears in pieces, with no one
to deliver.
9 Your hand shall be lifted up over
your adversaries,
and all your enemies shall be
cut off.

10 In that day, says the LORD,
I will cut off your horses from
among you
and will destroy your chariots;
11 and I will cut off the cities of
your land
and throw down all your
strongholds;
12 and I will cut off sorceries from
your hand,
and you shall have no more
soothsayers;

13 and I will cut off your images
and your pillars from among
you,
and you shall bow down no more
to the work of your hands;
14 and I will uproot your sacred
poles*s* from among you
and destroy your towns.
15 And in anger and wrath I will
execute vengeance
on the nations that did not
obey.

6 Hear what the LORD says:
Rise, plead your case before
the mountains,
and let the hills hear your voice.
2 Hear, you mountains, the
controversy of the LORD,
and you enduring foundations
of the earth;
for the LORD has a controversy
with his people,
and he will contend with Israel.

3 "O my people, what have I done
to you?
In what have I wearied you?
Answer me!
4 For I brought you up from the
land of Egypt,
and redeemed you from the
house of slavery;
and I sent before you Moses,
Aaron, and Miriam.
5 O my people, remember now
what King Balak of Moab
devised,

q Cn: Heb *in its entrances*   *r* Heb *he*
s Heb *Asherim*

5.6: *The land of Nimrod,* synonymous with
Assyria (Gen 10.9–11).
5.7–9: Then *the remnant* shall be a blessing
upon those who receive them and a curse
upon those who reject them (Gen 12.2–3; Gal
3.8).
5.10–15: *And in that day* the LORD will
abolish war in Israel (4.3), purify its worship,
and punish disobedient nations. **10:** Zech
9.10. **13–14:** *Pillars* and *sacred poles,* pagan
cult-objects (see Ex 34.13 n.).

**6.1–7.7: A series of laments, threats,
and denunciations, directed against all
classes of Israelites,** forms the third major
section. As extensions of themes sounded in
the first section (1.2–3.12), these oracles, if
not Micah's, are consonant with his thought.
6.1–8: The LORD has a *controversy* with this
people because they have forgotten the saving
acts of old and what it means *to walk humbly
with your God* (Isa 3.13; Hos 4.1–3; 12.2). **5:**
Num 22.1–6; 25.1–5; Josh 4.19–24. **8:** In this

what Balaam son of Beor
      answered him,
and what happened from Shittim
      to Gilgal,
   that you may know the saving
      acts of the LORD."

6 "With what shall I come before
      the LORD,
   and bow myself before God
      on high?
   Shall I come before him with
      burnt offerings,
   with calves a year old?
7 Will the LORD be pleased with
      thousands of rams,
   with ten thousands of rivers
      of oil?
   Shall I give my firstborn for my
      transgression,
   the fruit of my body for the sin
      of my soul?"
8 He has told you, O mortal, what
      is good;
   and what does the LORD require
      of you
   but to do justice, and to love
      kindness,
   and to walk humbly with
      your God?

9 The voice of the LORD cries to
      the city
   (it is sound wisdom to fear
      your name):
   Hear, O tribe and assembly of
      the city!*t*
10   Can I forget*u* the treasures of
      wickedness in the house of
      the wicked,
   and the scant measure that is
      accursed?
11 Can I tolerate wicked scales
   and a bag of dishonest weights?
12 Your*v* wealthy are full of
      violence;

your*w* inhabitants speak lies,
   with tongues of deceit in
      their mouths.
13 Therefore I have begun*x* to strike
      you down,
   making you desolate because of
      your sins.
14 You shall eat, but not be satisfied,
   and there shall be a gnawing
      hunger within you;
   you shall put away, but not save,
   and what you save, I will hand
      over to the sword.
15 You shall sow, but not reap;
   you shall tread olives, but not
      anoint yourselves with oil;
   you shall tread grapes, but not
      drink wine.
16 For you have kept the statutes
      of Omri*y*
   and all the works of the house
      of Ahab,
   and you have followed their
      counsels.
   Therefore I will make you a
      desolation, and your*z*
      inhabitants an object
      of hissing;
   so you shall bear the scorn of
      my people.

7 Woe is me! For I have become
      like one who,
   after the summer fruit has
      been gathered,
   after the vintage has been
      gleaned,
   finds no cluster to eat;
   there is no first-ripe fig for
      which I hunger.

t Cn Compare Gk: Heb *tribe, and who has
appointed it yet?*   *u* Cn: Meaning of Heb
uncertain   *v* Heb *Whose*   *w* Heb *whose*
x Gk Syr Vg: Heb *have made sick*   *y* Gk Syr
Vg Tg: Heb *the statutes of Omri are kept*
z Heb *its*

single sentence the prophet sums up the legal,
ethical, and covenantal requirements of religion, and sounds major notes of Amos (Am
5.24), Hosea (Hos 2.19–20; 6.6), and Isaiah
(Isa 7.9; 30.15).

6.9–16: Jerusalem, which is as wicked as
was Samaria (see v. 16), must be destroyed.
16: *Object of hissing,* see Zeph 2.15 n.
7.1–7: Not only the prince and the judge,
but the best and most upright of Jerusalem's

2 The faithful have disappeared from
the land,
and there is no one left who
is upright;
they all lie in wait for blood,
and they hunt each other
with nets.
3 Their hands are skilled to do evil;
the official and the judge ask for
a bribe,
and the powerful dictate what
they desire;
thus they pervert justice. *a*
4 The best of them is like a brier,
the most upright of them a
thorn hedge.
The day of their *b* sentinels, of
their *b* punishment,
has come;
now their confusion is at hand.
5 Put no trust in a friend,
have no confidence in a
loved one;
guard the doors of your mouth
from her who lies in your
embrace;
6 for the son treats the father with
contempt,
the daughter rises up against
her mother,
the daughter-in-law against her
mother-in-law;
your enemies are members of
your own household.
7 But as for me, I will look to
the LORD,
I will wait for the God of my
salvation;
my God will hear me.

8 Do not rejoice over me, O my
enemy;
when I fall, I shall rise;
when I sit in darkness,
the LORD will be a light to me.

9 I must bear the indignation of
the LORD,
because I have sinned
against him,
until he takes my side
and executes judgment for me.
He will bring me out to the light;
I shall see his vindication.
10 Then my enemy will see,
and shame will cover her who
said to me,
"Where is the LORD your God?"
My eyes will see her downfall; *c*
now she will be trodden down
like the mire of the streets.

11 A day for the building of your
walls!
In that day the boundary shall
be far extended.
12 In that day they will come to you
from Assyria to *d* Egypt,
and from Egypt to the River,
from sea to sea and from
mountain to mountain.
13 But the earth will be desolate
because of its inhabitants,
for the fruit of their doings.

14 Shepherd your people with
your staff,
the flock that belongs to you,
which lives alone in a forest
in the midst of a garden land;
let them feed in Bashan and Gilead
as in the days of old.
15 As in the days when you came out
of the land of Egypt,
show us *e* marvelous things.
16 The nations shall see and be
ashamed

a Cn: Heb *they weave it*   *b* Heb *your*
c Heb lacks *downfall*   *d* One Ms: MT *Assyria
and cities of*   *e* Cn: Heb *I will show him*

---

inhabitants, were corrupt, and the enemies of
the godly were neighbors and kinfolk close at
hand. There was no basis for mutual confidence. **1**: *Woe is me!* The speaker is feminine,
probably Samaria personified.
**7.8–20: God will show his steadfast
love to Israel, and shame will cover her**
enemies. This final section was probably
written in the early post-exilic period; compare Ps 137; Isa chs 33; 40–66. **8–10**: The
prophet speaks as Israel. **9**: Jer 10.19. **10**: *My
enemy,* Heb. feminine, apparently referring to
Damascus. *Where is . . . your God?* Ps 79.10;
Joel 2.17. **11–13**: Addressed to Jerusalem.

of all their might;
they shall lay their hands on
their mouths;
their ears shall be deaf;
17 they shall lick dust like a snake,
like the crawling things of
the earth;
they shall come trembling out of
their fortresses;
they shall turn in dread to the
LORD our God,
and they shall stand in fear
of you.

18 Who is a God like you, pardoning
iniquity
and passing over the
transgression
of the remnant of your*f*
possession?

He does not retain his anger
forever,
because he delights in showing
clemency.
19 He will again have compassion
upon us;
he will tread our iniquities
under foot.
You will cast all our*g* sins
into the depths of the sea.
20 You will show faithfulness to
Jacob
and unswerving loyalty to
Abraham,
as you have sworn to our
ancestors
from the days of old.

f Heb *his*   *g* Gk Syr Vg Tg: Heb *their*

7.14: A prayer to God. Compare Jer 50.19.
15: The LORD speaks. **16–17**: The nations will
be dumb, deaf, and humbled. Israel speaks.

7.18–20: Compare Ex 34.6–7; Ps 103.8–
10. **20**: Gen 12.1–3; 17.6–8; Lk 1.55.

Nahum

As is typical of Hebrew prophecy, Nahum's words were prompted by the dramatic events of international history. The proud Assyrian empire, whose power had for centuries been felt and feared from Mesopotamia to the Mediterranean, crumbled quickly after the death of Asshurbanipal (about 630 B.C.). Under the combined assaults of the vigorous Medes from north of Persia and the Chaldeans from southern Babylonia, the ancient capital city of Asshur fell in 614 B.C. When the renowned Nineveh was destroyed in 612, Assyrian domination of the Near East was ended.

The fervent reaction to the overthrow of Assyria, expressed by the peoples long subjected to its yoke, is nowhere more clearly seen than in the book of Nahum. The core of the book is a superb, vivid poem extolling Nineveh's destruction, which Nahum felt to be inevitable. The prophet spells out the reason for the Assyrian downfall in unequivocal terms: it is the LORD's judgment upon an unscrupulous, defiant nation.

This basic theme makes clear that Nahum's thought is passionately partisan. It asserts boldly that the LORD is the avenger of cruelty and immorality. But it fails, perhaps only because of a resolute singleness of purpose, to indicate the consequences of this divine justice for Israel itself.

The date of Nahum's triumphal ode lies close to the event it foretells, probably between 626 and 612 B.C. Its author is identified only by his name; even the location of Elkosh (1.1) is uncertain.

1 An oracle concerning Nineveh. The
book of the vision of Nahum of Elkosh.

2 A jealous and avenging God is
the LORD,
the LORD is avenging and
wrathful;
the LORD takes vengeance on his
adversaries
and rages against his enemies.
3 The LORD is slow to anger but
great in power,
and the LORD will by no means
clear the guilty.

His way is in whirlwind and
storm,
and the clouds are the dust of
his feet.
4 He rebukes the sea and makes
it dry,
and he dries up all the rivers;
Bashan and Carmel wither,
and the bloom of Lebanon
fades.
5 The mountains quake before him,
and the hills melt;
the earth heaves before him,
the world and all who live in it.

6 Who can stand before his
indignation?
Who can endure the heat of
his anger?
His wrath is poured out like fire,
and by him the rocks are broken
in pieces.
7 The LORD is good,

a stronghold in a day of trouble;
he protects those who take refuge
in him,
8 even in a rushing flood.
He will make a full end of his
adversaries, *a*
and will pursue his enemies into
darkness.
9 Why do you plot against the
LORD?
He will make an end;
no adversary will rise up twice.
10 Like thorns they are entangled,
like drunkards they are drunk;
they are consumed like dry
straw.
11 From you one has gone out
who plots evil against the LORD,
who counsels wickedness.

12 Thus says the LORD,
"Though they are at full strength
and many, *b*
they will be cut off and
pass away.
Though I have afflicted you,
I will afflict you no more.
13 And now I will break off his yoke
from you
and snap the bonds that
bind you."

14 The LORD has commanded
concerning you:
"Your name shall be
perpetuated no longer;

a Gk: Heb *of her place*   *b* Meaning of Heb
uncertain

1.1: **Title.** *Oracle,* literally "burden," is a
technical term describing the prophetic word
(Isa 13.1; Jer 23.33–40 n.; Hab 1.1; Zech 9.1;
Mal 1.1). The prophet "saw" this word (Am
1.1; Mic 1.1; Hab 1.1) as a *vision* (Ob 1).
Elkosh may be a village in Galilee, known to
Jerome as Elkesi, but more probably was located in SW Judah.
1.2–14: **The coming of the avenging
LORD.** An incomplete acrostic psalm, probably inserted either by Nahum or an editor to
introduce the prophet's own poem on Nineveh's fall (mainly 1.15–3.19). Now in some
disorder, it runs at least through v. 9, but its

precise relation to the main poem is unclear.
2: *Jealous,* see Ex 34.14 n. **3b:** Isa 29.6.
1.4: *Bashan, Carmel,* and *Lebanon* were regions famous for their fertility. **9:** *You,* perhaps the Israelites. **11:** *Counsels wickedness;* alternate translation, "a wicked counselor,"
which may refer to a specific person (the same
word, "wicked," later transliterated "Belial,"
is used in v. 15), such as Sennacherib. The *you*
refers to Israel's enemy.
1.12–13: The LORD assures Israel of deliverance. **12:** *They,* i.e. the days of your affliction. **14:** The announcement of Nineveh's
end.

from the house of your gods I will
　　cut off
　　the carved image and the
　　　cast image.
I will make your grave, for you
　　are worthless."

15*c* Look! On the mountains the feet
　　of one
　　who brings good tidings,
　　who proclaims peace!
Celebrate your festivals, O Judah,
　　fulfill your vows,
for never again shall the wicked
　　invade you;
　　they are utterly cut off.

2 A shatterer*d* has come up
　　against you.
　　Guard the ramparts;
　　watch the road;
gird your loins;
　　collect all your strength.

2 (For the LORD is restoring the
　　majesty of Jacob,
　　as well as the majesty of Israel,
though ravagers have ravaged
　　them
　　and ruined their branches.)

3 The shields of his warriors are red;
　　his soldiers are clothed in
　　　crimson.
The metal on the chariots flashes
　　on the day when he musters
　　　them;
　　the chargers*e* prance.
4 The chariots race madly through
　　the streets,
　　they rush to and fro through
　　the squares;

their appearance is like torches,
　　they dart like lightning.
5 He calls his officers;
　　they stumble as they come
　　　forward;
they hasten to the wall,
　　and the mantelet*f* is set up.
6 The river gates are opened,
　　the palace trembles.
7 It is decreed*f* that the city*g*
　　be exiled,
　　its slave women led away,
moaning like doves
　　and beating their breasts.
8 Nineveh is like a pool
　　whose waters*h* run away.
"Halt! Halt!"—
　　but no one turns back.
9 "Plunder the silver,
　　plunder the gold!
There is no end of treasure!
　　An abundance of every precious
　　　thing!"

10 Devastation, desolation, and
　　destruction!
Hearts faint and knees tremble,
　　all loins quake,
　　all faces grow pale!
11 What became of the lions' den,
　　the cave*i* of the young lions,
　　where the lion goes,
　　and the lion's cubs, with no one
　　　to disturb them?
12 The lion has torn enough for
　　his whelps

c Ch 2.1 in Heb　　*d* Cn: Heb *scatterer*
e Cn Compare Gk Syr: Heb *cypresses*
f Meaning of Heb uncertain　　*g* Heb *it*
h Cn Compare Gk: Heb *a pool, from the days that
she has become, and they*　　*i* Cn: Heb *pasture*

1.15–3.19: The sack of Nineveh. The
LORD's punishment of the defiant Assyrians
and Israel's proclamation of the good news.
1.15: Isa 40.9; 52.7; Rom 10.15. Read here 2.2.
　2.1–13: The assault upon Nineveh. The
plundering of its treasures and the terror of its
inhabitants will come because the LORD of
hosts is against that rapacious city (Isa 5.26–
30; Jer 5.15–17). **1:** Nahum is so thoroughly
convinced of the LORD's destruction of Nineveh that he can say that the enemy *has* already

come up. **2:** *Jacob* refers to the northern kingdom (see Am 6.8; 8.7); *Israel* refers to the
southern kingdom, regarded as the remnant
of Israel after the fall of the northern kingdom
(see Mic 1.13–15).
　2.7: The poem anticipates that the women
of Nineveh will be led away as humiliated
slaves (see also Isa 47.2–3). **11:** *Lion* often symbolizes the destroyer (Isa 5.29; Jer 4.7; Hos
5.14; Mic 5.8).

and strangled prey for his
lionesses;
he has filled his caves with prey
and his dens with torn flesh.

13 See, I am against you, says the
LORD of hosts, and I will burn your[j]
chariots in smoke, and the sword shall
devour your young lions; I will cut off
your prey from the earth, and the voice
of your messengers shall be heard no
more.

3 Ah! City of bloodshed,
utterly deceitful, full of booty—
no end to the plunder!
2 The crack of whip and rumble
of wheel,
galloping horse and bounding
chariot!
3 Horsemen charging,
flashing sword and glittering
spear,
piles of dead,
heaps of corpses,
dead bodies without end—
they stumble over the bodies!
4 Because of the countless
debaucheries of
the prostitute,
gracefully alluring, mistress
of sorcery,
who enslaves[k] nations through her
debaucheries,
and peoples through her
sorcery,
5 I am against you,
says the LORD of hosts,
and will lift up your skirts over
your face;
and I will let nations look on your
nakedness
and kingdoms on your shame.
6 I will throw filth at you
and treat you with contempt,
and make you a spectacle.

7 Then all who see you will shrink
from you and say,
"Nineveh is devastated; who will
bemoan her?"
Where shall I seek comforters
for you?

8 Are you better than Thebes[l]
that sat by the Nile,
with water around her,
her rampart a sea,
water her wall?
9 Ethiopia[m] was her strength,
Egypt too, and that without
limit;
Put and the Libyans were her[n]
helpers.

10 Yet she became an exile,
she went into captivity;
even her infants were dashed
in pieces
at the head of every street;
lots were cast for her nobles,
all her dignitaries were bound
in fetters.
11 You also will be drunken,
you will go into hiding;[o]
you will seek
a refuge from the enemy.
12 All your fortresses are like fig
trees
with first-ripe figs—
if shaken they fall
into the mouth of the eater.
13 Look at your troops:
they are women in your midst.
The gates of your land
are wide open to your foes;
fire has devoured the bars of
your gates.

j Heb *her*   *k* Heb *sells*   *l* Heb *No-amon*
m Or *Nubia*; Heb *Cush*   *n* Gk: Heb *your*
o Meaning of Heb uncertain

3.1–19: The arrogant Assyrian empire will
pass away and its capital, a city of rapine and
blood, will fall because the LORD is against it.
4: The LORD condemns Nineveh for treacherous and deceitful dealings with other nations.
Compare Jeremiah's indictment of Jerusalem
(5.1–3, 26–31) and the judgment against Babylon (Rome) in Rev 17.1–6; 18.1–24. **5–7:**
Such immorality is an offense to God and to
the nations and will not go unpunished.
3.8: The Egyptian capital *Thebes* (Hebrew,
No-amon) had been captured by the Assyrians themselves in 663 B.C. **10:** Ps 137.8–9.

14 Draw water for the siege,
strengthen your forts;
  trample the clay,
  tread the mortar,
  take hold of the brick mold!
15 There the fire will devour you,
the sword will cut you off.
  It will devour you like the
    locust.

  Multiply yourselves like the
    locust,
  multiply like the grasshopper!
16 You increased your merchants
more than the stars of the
    heavens.
  The locust sheds its skin and
    flies away.
17 Your guards are like grasshoppers,
your scribes like swarms*p*
  of locusts

settling on the fences
  on a cold day—
when the sun rises, they fly away;
no one knows where they
  have gone.

18 Your shepherds are asleep,
O king of Assyria;
  your nobles slumber.
Your people are scattered on the
    mountains
  with no one to gather them.
19 There is no assuaging your hurt,
your wound is mortal.
All who hear the news about you
  clap their hands over you.
For who has ever escaped
  your endless cruelty?

p Meaning of Heb uncertain

3.14–15: Nahum's warnings are satirical, for Nineveh is already doomed. **15:** *Locust* indicates both vast multitudes and destructive swarms. **18:** *Shepherds,* ruling officials (com-pare Zech 11.4–17 n.). **19:** *Clap their hands,* in derision (Lam 2.15; Job 27.23). *Who has ever escaped,* compare Isa 37.10–13.

Habakkuk

In the present book of Habakkuk at least three distinct literary forms can be recognized. The section 1.2–2.5 is constructed as a dialogue between the prophet and God; the next section (2.6–20), consisting of five woes against a wicked nation, is cast in classical prophetic style; and ch 3 is a lengthy poem, similar in structure to the Psalms and in its final form obviously meant for liturgical use. Moreover, various historical allusions discernible in the three sections point to different periods. These materials, therefore, may not originally have been a unit. But they are connected by the common theme of theodicy (justification of the ways of God) and now appear as the work of a Hebrew prophet who lived during the height of Babylonian power, most likely in the decade 608–598 B.C.

Although a certain Habakkuk appears in the apocryphal story of Bel and the Dragon, nothing is known about the life of the Old Testament prophet, not even his father's name. His thought, however, is laid bare by the searching questions he asks.

The author is confronting honestly the profoundly disturbing problem of why a just God is "silent when the wicked swallow those more righteous than they" (1.13). To this perennial question the prophet receives an answer that is eternally valid: God is still sovereign, and in God's own way and at the proper time will deal with the wicked; "but the righteous shall live by their faith" (2.4).

1 The oracle that the prophet Habakkuk saw.

2 O LORD, how long shall I cry
  for help,
  and you will not listen?
  Or cry to you "Violence!"
  and you will not save?
3 Why do you make me see
  wrongdoing
  and look at trouble?
  Destruction and violence are
  before me;
  strife and contention arise.
4 So the law becomes slack
  and justice never prevails.
  The wicked surround the
  righteous—
  therefore judgment comes forth
  perverted.

5 Look at the nations, and see!
  Be astonished! Be astounded!
  For a work is being done in
  your days
  that you would not believe if
  you were told.
6 For I am rousing the Chaldeans,
  that fierce and impetuous nation,
  who march through the breadth of
  the earth
  to seize dwellings not their
  own.
7 Dread and fearsome are they;
  their justice and dignity proceed
  from themselves.

8 Their horses are swifter than
  leopards,
  more menacing than wolves
  at dusk;
  their horses charge.
  Their horsemen come from
  far away;
  they fly like an eagle swift
  to devour.
9 They all come for violence,
  with faces pressing*a* forward;
  they gather captives like sand.
10 At kings they scoff,
  and of rulers they make sport.
  They laugh at every fortress,
  and heap up earth to take it.
11 Then they sweep by like the wind;
  they transgress and become
  guilty;
  their own might is their god!

12 Are you not from of old,
  O LORD my God, my Holy
  One?
  You*b* shall not die.
  O LORD, you have marked them
  for judgment;
  and you, O Rock, have
  established them for
  punishment.
13 Your eyes are too pure to
  behold evil,

a Meaning of Heb uncertain   *b* Ancient Heb
tradition: MT *We*

1.1: Title. *Oracle,* see Nah 1.1 n.
1.2–2.5: A dialogue in two cycles (1.2–
11; 1.12–2.5), between the prophet and God,
raising the question of the LORD's just government of the world.
1.2–11: The first cycle. 2–4: Habakkuk
protests that God neither hears nor acts
(compare Ps 22.1–2) and thereby negates *law*
and *justice.* This section may originally have
been concerned with unrighteous members
of the Israelite community; in its present
context it is directed against the Chaldeans
(vv. 6–11, 15–17). **2:** Ps 13.1–2; Jer 14.9; Rev
6.9–10. **3:** Jer 20.8, 10. **4:** Ps 119.126; Isa
59.12–14; Jer 12.1–4.
1.5–11: The LORD replies that the Chaldeans, wicked as they are, are the instrument of
God's own choosing (Isa 10.5–27; 41.2–3;
42.24; 44.28; 45.1–6; Jer 5.14–19; 27.6–7;
ch 51). **5:** Quoted in Acts 13.41. **6:** The *Chaldeans,* or neo-Babylonians, ruled the ancient
Near East from 612 to 539 B.C. (see Introduction to Nahum). **7–11:** This tyrannical nation
administers a *justice* all its own and worships
its own *might.*
1.12–2.5: The second cycle. Habakkuk
asks how long the Holy One will look on
while the faithless persecute those more righteous than they; he then takes his *stand* to
receive the LORD's answer. **12:** *Rock,* a metaphor frequently applied to God (Deut 32.4,
18, 30, 31; 2 Sam 23.3; Pss 18.2, 31; 92.15;
95.1; Isa 30.29; and elsewhere). **13:** The heart
of Habakkuk's problem (see Ps 5.4–6).

and you cannot look on
wrongdoing;
why do you look on the
treacherous,
and are silent when the wicked
swallow
those more righteous than they?
14 You have made people like the
fish of the sea,
like crawling things that have
no ruler.

15 The enemy^c brings all of them up
with a hook;
he drags them out with his net,
he gathers them in his seine;
so he rejoices and exults.
16 Therefore he sacrifices to his net
and makes offerings to his seine;
for by them his portion is lavish,
and his food is rich.
17 Is he then to keep on emptying
his net,
and destroying nations without
mercy?

2 I will stand at my watchpost,
and station myself on the
rampart;
I will keep watch to see what he
will say to me,
and what he^d will answer
concerning my complaint.
2 Then the LORD answered me
and said:
Write the vision;
make it plain on tablets,
so that a runner may read it.
3 For there is still a vision for the
appointed time;

it speaks of the end, and does
not lie.
If it seems to tarry, wait for it;
it will surely come, it will
not delay.
4 Look at the proud!
Their spirit is not right in them,
but the righteous live by
their faith. ^e
5 Moreover, wealth^f is treacherous;
the arrogant do not endure.
They open their throats wide
as Sheol;
like Death they never have
enough.
They gather all nations for
themselves,
and collect all peoples as
their own.

6 Shall not everyone taunt such people and, with mocking riddles, say about
them,
"Alas for you who heap up what
is not your own!"
How long will you load
yourselves with goods
taken in pledge?
7 Will not your own creditors
suddenly rise,
and those who make you
tremble wake up?
Then you will be booty for
them.
8 Because you have plundered many
nations,

c Heb *He*   d Syr: Heb *I*   e Or *faithfulness*
f Other Heb Mss read *wine*

2.1: The *watchpost* indicates both the
prophet's eager desire for, and confidence in,
the LORD's reply (Pss 5.3; 130.5–6; Isa 21.8;
Hos 9.8). **2–5:** The LORD responds with the
assurance that although the prophet may not
see its final issue, the divine justice is inexorable and will come in due time; in the meanwhile the *righteous* must *live* faithfully. **2:** The
LORD's answer is to be as *plain* as a highway
sign (Isa 8.1; Rev 1.19). **3:** Num 23.19; Dan
8.19; 2 Pet 3.8–10; Heb 10.37. **4:** The heart of
the matter is that the *righteous* who are faithful

to God have power to *live* but the *proud* (i.e.
unrighteous) *do not endure.* Here the contrast
is primarily between Israelites and Chaldeans;
but the verse has, properly, received wider
application (Rom 1.17; Gal 3.11; Heb 10.38–
39).
2.6–20: The five woes. These are directed against a nation that plunders *peoples,* obtains *gain* by violence, builds towns *with
blood,* shamelessly degrades its *neighbors,* and
trusts in *idols.* Applied originally to the Assyrians, Babylonians, or Macedonians, the woes

all that survive of the peoples
shall plunder you—
because of human bloodshed, and
violence to the earth,
to cities and all who live in
them.

9 "Alas for you who get evil gain
for your houses,
setting your nest on high
to be safe from the reach of
harm!"

10 You have devised shame for
your house
by cutting off many peoples;
you have forfeited your life.

11 The very stones will cry out from
the wall,
and the plaster*g* will respond
from the woodwork.

12 "Alas for you who build a town
by bloodshed,
and found a city on iniquity!"

13 Is it not from the LORD of hosts
that peoples labor only to feed
the flames,
and nations weary themselves
for nothing?

14 But the earth will be filled
with the knowledge of the glory
of the LORD,
as the waters cover the sea.

15 "Alas for you who make your
neighbors drink,
pouring out your wrath*h* until
they are drunk,
in order to gaze on their
nakedness!"

16 You will be sated with contempt
instead of glory.
Drink, you yourself, and
stagger!*i*
The cup in the LORD's right hand

will come around to you,
and shame will come upon
your glory!

17 For the violence done to Lebanon
will overwhelm you;
the destruction of the animals
will terrify you—*j*
because of human bloodshed and
violence to the earth,
to cities and all who live in
them.

18 What use is an idol
once its maker has shaped it—
a cast image, a teacher of lies?
For its maker trusts in what has
been made,
though the product is only an
idol that cannot speak!

19 Alas for you who say to the
wood, "Wake up!"
to silent stone, "Rouse
yourself!"
Can it teach?
See, it is gold and silver plated,
and there is no breath in it at
all.

20 But the LORD is in his holy
temple;
let all the earth keep silence
before him!

3 A prayer of the prophet Habakkuk
according to Shigionoth.

2 O LORD, I have heard of your
renown,
and I stand in awe, O LORD, of
your work.
In our own time revive it;

---

g Or *beam*   *h* Or *poison*   *i* Q Ms Gk: MT
be uncircumcised   *j* Gk Syr: Meaning of Heb
uncertain

---

have universal reference, indicting all human
tyranny. **14**: Isa 11.9. **20**: Ps 11.4; Zeph 1.7;
Zech 2.13.
3.1–19: Habakkuk's prayer. This is really a hymn, extolling the marching forth of the

LORD in victory *to save* his *people* (v. 13). This
magnificent poem exhibits the characteristics
of a psalm, including liturgical directions, and
was probably added later, possibly from the
circles of cultic prophecy (see Introduction to

in our own time make it
   known;
in wrath may you remember
   mercy.
3 God came from Teman,
   the Holy One from Mount
      Paran.   *Selah*
His glory covered the heavens,
   and the earth was full of
      his praise.
4 The brightness was like the
      sun;
   rays came forth from his hand,
   where his power lay hidden.
5 Before him went pestilence,
   and plague followed close
      behind.
6 He stopped and shook the earth;
   he looked and made the
      nations tremble.
The eternal mountains were
      shattered;
   along his ancient pathways
the everlasting hills sank low.
7 I saw the tents of Cushan under
      affliction;
   the tent-curtains of the land of
      Midian trembled.
8 Was your wrath against the
      rivers,*k* O Lord?
   Or your anger against the
      rivers,*k*
or your rage against the sea,*l*
when you drove your horses,
   your chariots to victory?
9 You brandished your naked bow,
   sated*m* were the arrows at your
      command.*n*   *Selah*
You split the earth with rivers.
10 The mountains saw you,
      and writhed;
   a torrent of water swept by;
the deep gave forth its voice.
The sun*o* raised high its hands;
11 the moon*p* stood still in its exalted
      place,

at the light of your arrows
      speeding by,
at the gleam of your flashing
      spear.
12 In fury you trod the earth,
   in anger you trampled nations.
13 You came forth to save your
      people,
   to save your anointed.
You crushed the head of the
      wicked house,
   laying it bare from foundation
      to roof.*n*   *Selah*
14 You pierced with his own arrows
      the head*q* of his warriors,*r*
   who came like a whirlwind to
      scatter us,*s*
   gloating as if ready to devour
      the poor who were in
      hiding.
15 You trampled the sea with
      your horses,
   churning the mighty waters.

16 I hear, and I tremble within;
   my lips quiver at the sound.
Rottenness enters into my
      bones,
   and my steps tremble*t*
      beneath me.
I wait quietly for the day of
      calamity
   to come upon the people who
      attack us.

17 Though the fig tree does not
      blossom,
   and no fruit is on the vines;
though the produce of the
      olive fails

---

k Or *against River*   *l* Or *against Sea*
m Cn: Heb *oaths*   *n* Meaning of Heb
uncertain   *o* Heb *It*   *p* Heb *sun, moon*
q Or *leader*   *r* Vg Compare Gk Syr: Meaning
of Heb uncertain   *s* Heb *me*
t Cn Compare Gk: Meaning of Heb uncertain

Joel). *Shigionoth, Selah,* and *choirmaster* are
technical terms in the Psalter (Pss 7, title; 4,
title).
   3.3, 7: God appears from the region of Sinai and marches toward Edom, as at the Exodus (Deut 33.2; Judg 5.4). **8b:** Deut 33.26.
   3.13: *Anointed* indicates either the king or
the nation. **13:** *Your anointed,* i.e. the king of

and the fields yield no food;
though the flock is cut off from
the fold
and there is no herd in the
stalls,
18 yet I will rejoice in the LORD;
I will exult in the God of my
salvation.
19 GOD, the Lord, is my strength;

he makes my feet like the feet
of a deer,
and makes me tread upon the
heights. *u*

To the leader: with stringed *v*
instruments.

u Heb *my heights*   *v* Heb *my stringed*

Judah. **18:** The psalm reaches its height by rejoicing in the LORD as the saving God (Pss 25.5; 27.1; 68.19–20) and thus exhibits a clear

thematic connection with chs 1–2 (see also 3.12–14, 16).

Zephaniah

The superscription to Zephaniah's book traces his ancestry back to Hezekiah and dates his ministry in the reign of Josiah (640–609 B.C.). Since the name Hezekiah is uncommon in the Old Testament, here it perhaps refers to the famous Judean king (715–687 B.C.), who was favorably influenced by the preaching of Isaiah and Micah. Zephaniah's intimate knowledge of Jerusalem and its court circles, his failure to denounce the king personally, and the absence of any concern with the poor of the land, may support the inference that he was of royal descent.

His condemnation of the corrupt practices and religious perversions (1.4–6, 8, 9, 12; 3.1–3, 7), officially legislated against by Deuteronomy, suggests that Zephaniah prophesied before Josiah's reforms of 621 B.C. (2 Kings ch 23). The allusion to imminent threat from the north (see 1.10 n.), perhaps the barbaric Scythian hordes, further suggests the decade about 630 for Zephaniah's public ministry.

The book can be divided into three sections: Chapters 1.2–2.3 proclaim doom on Judah for its religious syncretism, in the form of the destructive day of the LORD, which is "near and hastening fast." Chapter 2 (vv. 4–15) extends the divine judgment to other nations (especially Israel's ancient enemies), which are also guilty; however, a humble seeking after righteousness may mitigate the wrath of that day. Chapter 3, after condemning Jerusalem (vv. 1–7), promises comfort and consolation to those who wait patiently for the LORD and serve God "with one accord." The inhabitants of Jerusalem shall rejoice that the LORD their King is in their midst to save them and gather them home (3.20).

1 The word of the LORD that came to
Zephaniah son of Cushi son of Gedaliah son of Amariah son of Hezekiah, in
the days of King Josiah son of Amon of
Judah.

2 I will utterly sweep away
    everything
  from the face of the earth, says
    the LORD.
3 I will sweep away humans and
    animals;
  I will sweep away the birds of
    the air
  and the fish of the sea.
  I will make the wicked stumble. *a*
  I will cut off humanity
  from the face of the earth, says
    the LORD.
4 I will stretch out my hand against
    Judah,
  and against all the inhabitants of
    Jerusalem;
  and I will cut off from this place
    every remnant of Baal
  and the name of the idolatrous
    priests; *b*
5 those who bow down on the roofs
    to the host of the heavens;
  those who bow down and swear
    to the LORD,
  but also swear by Milcom; *c*
6 those who have turned back from
    following the LORD,
  who have not sought the LORD
    or inquired of him.

7 Be silent before the Lord GOD!
  For the day of the LORD is
    at hand;
  the LORD has prepared a sacrifice,
    he has consecrated his guests.
8 And on the day of the LORD's
    sacrifice
  I will punish the officials and the
    king's sons
  and all who dress themselves in
    foreign attire.
9 On that day I will punish
  all who leap over the threshold,
  who fill their master's house
    with violence and fraud.

10 On that day, says the LORD,
   a cry will be heard from the
     Fish Gate,
   a wail from the Second Quarter,
   a loud crash from the hills.
11 The inhabitants of the Mortar
     wail,
   for all the traders have perished;
   all who weigh out silver are
     cut off.
12 At that time I will search
     Jerusalem with lamps,
   and I will punish the people
   who rest complacently *d* on
     their dregs,
   those who say in their hearts,

a Cn: Heb *sea, and those who cause the wicked to
stumble*   b Compare Gk: Heb *the idolatrous
priests with the priests*   c Gk Mss Syr Vg: Heb
Malcam (or, *their king*)   d Heb *who thicken*

1.1: Superscription: *The word of the LORD,*
see Hos 1.1 n. Three others in the Old Testament are named *Zephaniah* (2 Kings 25.18;
1 Chr 6.36; Zech 6.10, 14). Elsewhere *Cushi*
means "Ethiopian" or "Cushite."
   **1.2–6: The threat of universal destruc-
tion** caused by Judah's religious syncretism.
Zephaniah vigorously condemns the adulter-
ation of the pure worship of the LORD with
elements of Canaanite (*Baal*), Ammonite
(*Milcom*), and Assyrian (*host of the heavens*)
religions. **2–3:** *Humans, humanity,* and *earth*
are alliterative in Hebrew, as in Gen ch 2 (see
Gen 2.7 n.), suggesting the negation of creation. *Says the LORD,* see Am 1.3 n.
   **1.7–9: Court officials and royal family
will be punished** *on the day of the LORD's*

sacrifice. **7:** *Prepared a sacrifice,* i.e. a slaughter
of sinners. **9:** *Leap over the threshold,* a pagan
religious practice (1 Sam 5.5 n.).
   **1.10–13: Merchants and traders will be
cut off** and the religiously indifferent will be
desolate *on that day.* **10:** The enemy will approach Jerusalem from the north, first
through *the Fish Gate* (Neh 3.1–6; 12.39;
compare Jer 1.13–16). *Second Quarter,* a section of Jerusalem (2 Kings 22.14). **11:** *Traders,*
literally, "people of Canaan." **12:** *Rest complacently* (literally "thicken," note *d*) *on their dregs,*
a figure drawn from wine-making to portray
indolence. If the wine is not stirred up while
it is fermenting, it becomes thick and lacks
strength (see also Jer 48.11).

"The LORD will not do good,
  nor will he do harm."
13 Their wealth shall be plundered,
  and their houses laid waste.
Though they build houses,
  they shall not inhabit them;
though they plant vineyards,
  they shall not drink wine
    from them.

14 The great day of the LORD is near,
  near and hastening fast;
the sound of the day of the LORD
  is bitter,
  the warrior cries aloud there.
15 That day will be a day of wrath,
  a day of distress and anguish,
a day of ruin and devastation,
  a day of darkness and gloom,
a day of clouds and thick
    darkness,
16   a day of trumpet blast and
      battle cry
against the fortified cities
  and against the lofty
    battlements.

17 I will bring such distress upon
    people
  that they shall walk like
    the blind;
because they have sinned against
    the LORD,
their blood shall be poured out
    like dust,
  and their flesh like dung.
18 Neither their silver nor their gold
  will be able to save them
  on the day of the LORD's wrath;
in the fire of his passion
  the whole earth shall be
    consumed;
for a full, a terrible end

he will make of all the
  inhabitants of the earth.

2 Gather together, gather,
  O shameless nation,
2 before you are driven away
  like the drifting chaff, *e*
before there comes upon you
  the fierce anger of the LORD,
before there comes upon you
  the day of the LORD's wrath.
3 Seek the LORD, all you humble of
    the land,
  who do his commands;
seek righteousness, seek humility;
  perhaps you may be hidden
  on the day of the LORD's wrath.
4 For Gaza shall be deserted,
  and Ashkelon shall become a
    desolation;
Ashdod's people shall be driven
    out at noon,
  and Ekron shall be uprooted.

5 Ah, inhabitants of the seacoast,
  you nation of the Cherethites!
The word of the LORD is
    against you,
O Canaan, land of the
    Philistines;
  and I will destroy you until no
    inhabitant is left.
6 And you, O seacoast, shall be
    pastures,
  meadows for shepherds
  and folds for flocks.
7 The seacoast shall become the
    possession
  of the remnant of the house
    of Judah,

e Cn Compare Gk Syr: Heb *before a decree is
born; like chaff a day has passed away*

1.14–18: The day of the LORD is at hand,
a day of *wrath* and *ruin*, of *distress* and *devastation*. On this day neither *silver nor gold* shall
save sinners from the *fire* of the LORD's zeal.
Zephaniah elaborates the preaching of Amos
(Am 5.18–20; 8.9–14) that the day of the
LORD will be darkness and not light, woe and
not weal (Isa 45.7) upon Israel as well as the
Gentiles (see also Isa 13.9–16; Ezek 7.19; Joel
1.15; 2.1–2). **18:** *End,* compare Ezek 7.2–7.
  **2.1–3: The humble of the land who
seek the LORD in righteousness** *may be hidden* from God's *fierce anger.* **3:** *Seek,* Isa 55.6–
9; Am 5.6–7. *Humble,* compare 3.12–13.
  2.4–15: Against the nations. 4–7: The
Philistines shall be desolated because *the word*

on which they shall pasture,
and in the houses of Ashkelon
they shall lie down at evening.
For the LORD their God will be
mindful of them
and restore their fortunes.

8 I have heard the taunts of Moab
and the revilings of the
Ammonites,
how they have taunted my people
and made boasts against
their territory.
9 Therefore, as I live, says the LORD
of hosts,
the God of Israel,
Moab shall become like Sodom
and the Ammonites like
Gomorrah,
a land possessed by nettles and
salt pits,
and a waste forever.
The remnant of my people shall
plunder them,
and the survivors of my nation
shall possess them.
10 This shall be their lot in return for
their pride,
because they scoffed and boasted
against the people of the LORD
of hosts.
11 The LORD will be terrible
against them;
he will shrivel all the gods of
the earth,
and to him shall bow down,
each in its place,
all the coasts and islands of
the nations.

12 You also, O Ethiopians,*f*
shall be killed by my sword.

13 And he will stretch out his hand
against the north,

and destroy Assyria;
and he will make Nineveh a
desolation,
a dry waste like the desert.
14 Herds shall lie down in it,
every wild animal;*g*
the desert owl*h* and the
screech owl*h*
shall lodge on its capitals;
the owl*i* shall hoot at the
window,
the raven*j* croak on the
threshold;
for its cedar work will be
laid bare.
15 Is this the exultant city
that lived secure,
that said to itself,
"I am, and there is no one else"?
What a desolation it has become,
a lair for wild animals!
Everyone who passes by it
hisses and shakes the fist.

3 Ah, soiled, defiled,
oppressing city!
2 It has listened to no voice;
it has accepted no correction.
It has not trusted in the LORD;
it has not drawn near to its
God.

3 The officials within it
are roaring lions;
its judges are evening wolves
that leave nothing until
the morning.
4 Its prophets are reckless,
faithless persons;
its priests have profaned what
is sacred,

f Or *Nubians*; Heb *Cushites*   *g* Tg Compare
Gk: Heb *nation*   *h* Meaning of Heb uncertain
i Cn: Heb *a voice*   *j* Gk Vg: Heb *desolation*

---

of the LORD is against them (see Joel 3.4–8).
8–11: The Moabites and Ammonites, traditional enemies of Israel who lived in Transjordan, shall be annihilated, becoming *like
Sodom* and *Gomorrah* (Isa chs 15–16; 25.10–
12; Jer 48.1–49.6; Ezek 25.8–11; Am 1.13–
2.3). **9:** Gen 19.24–28, 30–38.

2.12–15: The Ethiopians shall be slain, and
the Assyrians humiliated because of their arrogance (Isa 10.5–34). **12:** Isa ch 18. **14:** The
Hebrew text is uncertain. **15:** See Nah 3.4 n.
Hisses, to ward off a similar fate (Jer 18.16).
   3.1–7: **Woe to Jerusalem,** because the *officials, judges, prophets,* and *priests are corrupt*

they have done violence to
the law.
5 The LORD within it is righteous;
he does no wrong.
Every morning he renders his
judgment,
each dawn without fail;
but the unjust knows no shame.

6 I have cut off nations;
their battlements are in ruins;
I have laid waste their streets
so that no one walks in them;
their cities have been made
desolate,
without people, without
inhabitants.
7 I said, "Surely the city[k] will
fear me,
it will accept correction;
it will not lose sight[l]
of all that I have brought
upon it."
But they were the more eager
to make all their deeds corrupt.

8 Therefore wait for me, says
the LORD,
for the day when I arise as
a witness.
For my decision is to gather
nations,
to assemble kingdoms,
to pour out upon them my
indignation,
all the heat of my anger;
for in the fire of my passion
all the earth shall be consumed.

9 At that time I will change the
speech of the peoples
to a pure speech,
that all of them may call on the
name of the LORD

and serve him with one accord.
10 From beyond the rivers of
Ethiopia[m]
my suppliants, my scattered
ones,
shall bring my offering.

11 On that day you shall not be put
to shame
because of all the deeds by
which you have rebelled
against me;
for then I will remove from
your midst
your proudly exultant ones,
and you shall no longer be
haughty
in my holy mountain.
12 For I will leave in the midst of
you
a people humble and lowly.
They shall seek refuge in the name
of the LORD—
13 the remnant of Israel;
they shall do no wrong
and utter no lies,
nor shall a deceitful tongue
be found in their mouths.
Then they will pasture and
lie down,
and no one shall make them
afraid.

14 Sing aloud, O daughter Zion;
shout, O Israel!
Rejoice and exult with all
your heart,
O daughter Jerusalem!
15 The LORD has taken away the
judgments against you,

k Heb *it*   l Gk Syr: Heb *its dwelling will not be
cut off*   m Or *Nubia*; Heb *Cush*

and do not fear the *righteous* LORD, though
God has continually warned them (compare
Am 4.6–12).
**3.8–13: The nations will be converted
and a righteous remnant will be left in
Israel. 9:** The gift of *a pure speech* symbolizes
fidelity (Isa 6.5–7), removes the curse of Babel (Gen 11.1–9), and, for Christians, anticipates Pentecost (Acts 2.1–11). **13:** Ezek
34.13–16; Zech 8.3, 16.
**3.14–20: The glorious gospel of salva-
tion,** promising the joy of restoration to Jerusalem. This passage is generally held to be a
later addition. **14–15:** These verses resemble a
psalm of the enthronement of the LORD (compare Pss 47; 97). **14:** Zech 9.9. **15:** Isa 12.6;

he has turned away your
enemies.
The king of Israel, the LORD, is in
your midst;
you shall fear disaster no more.
16 On that day it shall be said to
Jerusalem:
Do not fear, O Zion;
do not let your hands grow
weak.
17 The LORD, your God, is in
your midst,
a warrior who gives victory;
he will rejoice over you with
gladness,
he will renew you[n] in his love;
he will exult over you with
loud singing
18 as on a day of festival.[o]
I will remove disaster from you,[p]
so that you will not bear
reproach for it.

19 I will deal with all your oppressors
at that time.
And I will save the lame
and gather the outcast,
and I will change their shame
into praise
and renown in all the earth.
20 At that time I will bring you
home,
at the time when I gather you;
for I will make you renowned
and praised
among all the peoples of
the earth,
when I restore your fortunes
before your eyes, says the LORD.

n Gk Syr: Heb *he will be silent*   o Gk Syr:
Meaning of Heb uncertain   p Cn: Heb *I will
remove from you; they were*

40.2; 41.10. **17**: Ex 15.3; Isa 12.2; 62.5. **19–20**:
The major elements of post-exilic eschatology are found here: destruction of the enemy (Ob 15, 16; Mic 5.9; Zech 12.9), ingathering of the exiles (Mic 4.6–7; Zech 10.8–12), and return to the Holy Land (Isa 62.1–5; Zech 8.7–8). By making Israel *renowned* (literally "a name") *and praised among all the peoples of the earth,* the LORD fulfills the longstanding promise to the patriarchs (Gen 12.2–3).

Haggai

When Cyrus conquered Babylon, he not only published a decree (538 B.C.) allowing the captive Jews to return to Palestine but also encouraged them to rebuild the temple at Jerusalem (Ezra 1.1–4). It is possible that under Sheshbazzar's leadership rebuilding was immediately attempted. By 520 B.C., however, no significant progress was evident. A successful effort was then begun, and the new temple was completed in the spring of 515 B.C.

Assisted by Zechariah (see Introduction to Zechariah), the man principally responsible for this major accomplishment was the prophet Haggai. In five addresses, dating from the sixth through the ninth months of 520 B.C., Haggai exhorted Zerubbabel the governor and Joshua the high priest, the joint leaders of the Judean community, to assume official leadership in the reconstruction of the temple, and urged the priests to purify the cultic worship. These twin projects were, first of all, urgent practical steps toward unifying the disrupted religious life of the community. But Haggai saw them also as necessary preparations for the messianic age. Upon the completion of these enterprises, the wonderful era foreseen by earlier prophets would come; for God would bless this people with fruitfulness and prosperity, overthrow the Gentiles, and establish Zerubbabel as the messianic king on the throne of David.

1 In the second year of King Darius, in the sixth month, on the first day of the month, the word of the LORD came by the prophet Haggai to Zerubbabel son of Shealtiel, governor of Judah, and to Joshua son of Jehozadak, the high priest: [2]Thus says the LORD of hosts: These people say the time has not yet come to rebuild the LORD's house. [3]Then the word of the LORD came by the prophet Haggai, saying: [4]Is it a time for you yourselves to live in your paneled houses, while this house lies in ruins? [5]Now therefore thus says the LORD of hosts: Consider how you have fared. [6]You have sown much, and harvested little; you eat, but you never have enough; you drink, but you never have your fill; you clothe yourselves, but no one is warm; and you that earn wages earn wages to put them into a bag with holes.

[7] Thus says the LORD of hosts: Consider how you have fared. [8]Go up to the hills and bring wood and build the house, so that I may take pleasure in it and be honored, says the LORD. [9]You have looked for much, and, lo, it came to little; and when you brought it home, I blew it away. Why? says the LORD of hosts. Because my house lies in ruins, while all of you hurry off to your own houses. [10]Therefore the heavens above you have withheld the dew, and the earth has withheld its produce. [11]And I have called for a drought on the land and the hills, on the grain, the new wine, the oil, on what the soil produces, on human beings and animals, and on all their labors.

[12] Then Zerubbabel son of Shealtiel, and Joshua son of Jehozadak, the high priest, with all the remnant of the people, obeyed the voice of the LORD their God, and the words of the prophet Haggai, as the LORD their God had sent him; and the people feared the LORD. [13]Then Haggai, the messenger of the LORD, spoke to the people with the LORD's message, saying, I am with you, says the LORD. [14]And the LORD stirred up the spirit of Zerubbabel son of Shealtiel, governor of Judah, and the spirit of Joshua son of Jehozadak, the high priest, and the spirit of all the remnant of the people; and they came and worked on the house of the LORD of hosts, their God, [15]on the twenty-fourth day of the month, in the sixth month.

2 In the second year of King Darius, [1]in the seventh month, on the twenty-first day of the month, the word of the LORD came by the prophet Haggai, saying: [2]Speak now to Zerubbabel son of Shealtiel, governor of Judah, and to Joshua son of Jehozadak, the high priest, and to the remnant of the people, and say, [3]Who is left among you that saw this house in its former glory? How does it look to you now? Is it not in your sight as nothing? [4]Yet now take courage, O Zerubbabel, says the LORD; take courage, O Joshua, son of Jehozadak, the high priest; take courage, all you people of the land, says the LORD; work, for I am with you, says the LORD of hosts, [5]according to the promise that I made you when you came out of Egypt. My spirit abides among you; do not fear. [6]For thus says the LORD of hosts: Once again, in a

---

1.1–15a: Because the people have neglected the temple, God has punished them. 1: *Darius* was king of the Persian empire from 521 to 485 B.C. *In the second year . . . in the sixth month,* mid-August to mid-September 520 B.C. *Haggai* means "festal." *Zerubbabel,* the grandson of Jehoiachin (2 Kings 24.8–17; 1 Chr 3.17), was a royal descent from David; as the Persian-appointed governor of Judah, he shared authority with *Joshua,* the high priest, who also had come from Babylon. **4:** The temple had been plundered and burned in 587 B.C. (2 Kings 25.8–

17). **5–11:** The poor conditions in Judah were God's punishment for the neglect of the Temple. **15a:** It is best to regard this as the date for the oracle in 2.15–19, which originally may have stood at this point.

1.15b–2.9: The new temple to be more splendid than the old. 2.1: *The seventh month,* mid-September to mid-October. **3:** A few of the people then present may have seen Solomon's temple before its destruction. **5:** Haggai sees God's *spirit* standing in Israel's midst and is reminded of the Exodus (Ex 13.21–22; 14.19–20). **6:** Referred to and par-

little while, I will shake the heavens and the earth and the sea and the dry land; 7 and I will shake all the nations, so that the treasure of all nations shall come, and I will fill this house with splendor, says the LORD of hosts. 8 The silver is mine, and the gold is mine, says the LORD of hosts. 9 The latter splendor of this house shall be greater than the former, says the LORD of hosts; and in this place I will give prosperity, says the LORD of hosts.

10 On the twenty-fourth day of the ninth month, in the second year of Darius, the word of the LORD came by the prophet Haggai, saying: 11 Thus says the LORD of hosts: Ask the priests for a ruling: 12 If one carries consecrated meat in the fold of one's garment, and with the fold touches bread, or stew, or wine, or oil, or any kind of food, does it become holy? The priests answered, "No." 13 Then Haggai said, "If one who is unclean by contact with a dead body touches any of these, does it become unclean?" The priests answered, "Yes, it becomes unclean." 14 Haggai then said, So is it with this people, and with this nation before me, says the LORD; and so with every work of their hands; and what they offer there is unclean. 15 But now, consider what will come to pass from this day on. Before a stone was placed upon a stone in the LORD's tem-

ple, 16 how did you fare?*a* When one came to a heap of twenty measures, there were but ten; when one came to the winevat to draw fifty measures, there were but twenty. 17 I struck you and all the products of your toil with blight and mildew and hail; yet you did not return to me, says the LORD. 18 Consider from this day on, from the twenty-fourth day of the ninth month. Since the day that the foundation of the LORD's temple was laid, consider: 19 Is there any seed left in the barn? Do the vine, the fig tree, the pomegranate, and the olive tree still yield nothing? From this day on I will bless you.

20 The word of the LORD came a second time to Haggai on the twenty-fourth day of the month: 21 Speak to Zerubbabel, governor of Judah, saying, I am about to shake the heavens and the earth, 22 and to overthrow the throne of kingdoms; I am about to destroy the strength of the kingdoms of the nations, and overthrow the chariots and their riders; and the horses and their riders shall fall, every one by the sword of a comrade. 23 On that day, says the LORD of hosts, I will take you, O Zerubbabel my servant, son of Shealtiel, says the LORD, and make you like a signet ring; for I have chosen you, says the LORD of hosts.

a Gk: Heb *since they were*

tially quoted in Heb 12.26. **7**: *Treasures of all nations;* compare Isa 40.14; 60.6; 61.6.

2.10–14: **The offering made by a defiled people is unclean. 10**: *The ninth month,* mid-November to mid-December. **14**: The people condemned here may be the Samaritans, who opposed the work of the post-exilic community (Ezra 4.1–5; Neh 4.7–8).

2.15–19: **When the foundation of the temple is laid, God will bless this needy, obedient people.** This section may have stood after 1.1–14. *The twenty-fourth day* (v. 18; see 1.15a n.) probably marks the laying of the foundation stone. **12–13**: From a

ritual point of view, the *unclean* is more "contagious" than the consecrated, or clean. **17**: See Am 4.9. **18**: *Ninth,* probably error for "sixth" or a gloss. **19**: Good harvests were signs of God's favor; see Ps 128.

2.20–23: **When the LORD establishes the kingdom, Zerubbabel will be the messiah. 21**: Compare 2.6; Joel 3.16. **22**: *Overthrow the chariots,* God will again redeem this people, as at the Exodus (Ex 15.1). Haggai links Israel's earlier traditions with the coming messianic age. **23**: *My servant,* Num 12.7–8; 2 Sam 3.18; Isa 42.1; Jer 27.6; see Zech 3.8 n. *Signet ring,* compare Jer 22.24.

Zechariah

Zechariah, whose prophecies date from 520 to 518 B.C. and are found in chapters 1–8, was contemporary with Haggai (Ezra 5.1; 6.14). He shared Haggai's zeal for a rebuilt temple, a purified community, and the coming of the messianic age (see Introduction to Haggai). Like Haggai also, Zechariah forms a link between earlier prophecy (especially Ezekiel) and mature apocalyptic thought (Dan chs 7–12). But Zechariah differs from his contemporary in the form and presentation of his message, employing the literary style of night visions and dialogues between God, seer, and interpreting angel. With him, therefore, both the form and imagery of Jewish apocalyptic thought are significantly developed.

Chapters 9–14, which nowhere claim to be from Zechariah, portray nothing of the early Persian period but speak rather of the Greeks (9.13). Instead of Joshua and Zerubbabel, unnamed shepherds lead the community. Instead of peace and rebuilding, there are expectations of universal warfare and the siege of Jerusalem. Style, vocabulary, and theological ideas differentiate these chapters from Zechariah's work. Although they may contain some earlier bits, they were written during the Greek period, principally in the fourth and third centuries B.C., by unknown authors. Since the eschatological and messianic themes found in the first section are here further elaborated, the authors are spiritual disciples of Zechariah. The pictures of the messianic Prince of Peace and the Good Shepherd smitten for the flock are used in the New Testament in order to describe the person and work of Jesus Christ.

1 In the eighth month, in the second year of Darius, the word of the LORD came to the prophet Zechariah son of Berechiah son of Iddo, saying: ²The LORD was very angry with your ancestors. ³Therefore say to them, Thus says the LORD of hosts: Return to me, says the LORD of hosts, and I will return to you, says the LORD of hosts. ⁴Do not be like your ancestors, to whom the former prophets proclaimed, "Thus says the LORD of hosts, Return from your evil ways and from your evil deeds." But they did not hear or heed me, says the LORD. ⁵Your ancestors, where are they? And the prophets, do they live forever? ⁶But my words and my statutes, which I commanded my servants the prophets, did they not overtake your ancestors? So they repented and said, "The LORD of hosts has dealt with us according to our ways and deeds, just as he planned to do."

7 On the twenty-fourth day of the eleventh month, the month of Shebat, in the second year of Darius, the word of the LORD came to the prophet Zechariah son of Berechiah son of Iddo; and Zechariah*ᵃ* said, ⁸In the night I saw a man riding on a red horse! He was standing among the myrtle trees in the glen; and behind him were red, sorrel, and white horses. ⁹Then I said, "What are these, my lord?" The angel who talked with me said to me, "I will show you what they are." ¹⁰So the man who was standing among the myrtle trees answered, "They are those whom the LORD has sent to patrol the earth." ¹¹Then they spoke to the angel of the LORD who was standing among the myrtle trees, "We have patrolled the earth, and lo, the whole earth remains at peace." ¹²Then the angel of the LORD said, "O LORD of hosts, how long will you withhold mercy from Jerusalem and the cities of Judah, with which you have been angry these seventy years?" ¹³Then the LORD replied with gracious and comforting words to the angel who talked with me. ¹⁴So the angel who talked with me said to me, Proclaim this message: Thus says the LORD of hosts; I am very jealous for Jerusalem and for Zion. ¹⁵And I am extremely angry with the nations that are at ease; for while I was only a little angry, they made the disaster worse. ¹⁶Therefore, thus says the LORD, I have returned to Jerusalem with compassion; my house shall be built in it, says the LORD of hosts, and the measuring line shall be stretched out over Jerusalem. ¹⁷Proclaim further: Thus says the LORD of hosts: My cities shall again overflow with prosperity; the LORD will again comfort Zion and again choose Jerusalem.

18*ᵇ* And I looked up and saw four horns. ¹⁹I asked the angel who talked

a Heb *and he*   *b* Ch 2.1 in Heb

1.1–6: Introduction: a call to repentance. 1: The first of three dates (see also 1.7; 6.16): mid–October to mid–November 520 B.C. *Zechariah,* meaning "The LORD is renowned," was the *son of* the priest *Iddo* (Ezra 5.1; 6.14; Neh 12.16). The presence of the phrase *son of Berechiah* is due probably to a scribal confusion with the son of Jeberechiah in Isa 8.2, with whom Zechariah the son of Jehoiada (2 Chr 24.20–22) is also confused in Mt 23.35. Zechariah had prophetic and priestly interests (see Introduction to Joel). **4:** The *prophets* had continually invited repentance (Isa 1.16–20; 30.15; 55.6–9; Jer 3.12, 22; 4.3–4; Hos 10.12; 14.1–7; Am 5.4, 6, 14). **6:** Both law (*statutes*) and prophecy (*words*) continue to be valid. *So they repented . . . ,* perhaps an editorial addition (contrast 7.11).

1.7–6.15: The word of the LORD to Zechariah in a series of eight "night visions." There is a standard pattern: (*a*) vision, (*b*) question, (*c*) answer. **1.7:** Mid–January to mid–February, 519 B.C.

1.8–17: The first vision: divine horsemen patrolling the earth. 9: *My lord,* i.e. *the angel.* **10:** Job 1.6–7; 2.1–2. **11:** *Remains at peace,* the catastrophic Day of Judgment has not yet appeared (compare v. 15; Hag 2.6). **12:** The *seventy years* refer without precision to the period from 587 to 519 B.C. (Jer 25.11; 29.10). **15:** Ps 123.4; Am 6.1. **17:** 8.3; Isa 44.26; 51.3; 54.8–10; Jer 31.38–39.

1.18–21: The second vision: four horns and four smiths. The *four horns* symbolize the powerful nations of the world (compare Ps 75.4, 5; Dan 7.19–27) which the *blacksmiths*

with me, "What are these?" And he answered me, "These are the horns that have scattered Judah, Israel, and Jerusalem." [20] Then the LORD showed me four blacksmiths. [21] And I asked, "What are they coming to do?" He answered, "These are the horns that scattered Judah, so that no head could be raised; but these have come to terrify them, to strike down the horns of the nations that lifted up their horns against the land of Judah to scatter its people."[c]

2 [d] I looked up and saw a man with a measuring line in his hand. [2] Then I asked, "Where are you going?" He answered me, "To measure Jerusalem, to see what is its width and what is its length." [3] Then the angel who talked with me came forward, and another angel came forward to meet him, [4] and said to him, "Run, say to that young man: Jerusalem shall be inhabited like villages without walls, because of the multitude of people and animals in it. [5] For I will be a wall of fire all around it, says the LORD, and I will be the glory within it."

6 Up, up! Flee from the land of the north, says the LORD; for I have spread you abroad like the four winds of heaven, says the LORD. [7] Up! Escape to Zion, you that live with daughter Babylon. [8] For thus said the LORD of hosts (after his glory[e] sent me) regarding the nations that plundered you: Truly, one who touches you touches the apple of my eye.[f] [9] See now, I am going to raise[g] my

hand against them, and they shall become plunder for their own slaves. Then you will know that the LORD of hosts has sent me. [10] Sing and rejoice, O daughter Zion! For lo, I will come and dwell in your midst, says the LORD. [11] Many nations shall join themselves to the LORD on that day, and shall be my people; and I will dwell in your midst. And you shall know that the LORD of hosts has sent me to you. [12] The LORD will inherit Judah as his portion in the holy land, and will again choose Jerusalem.

13 Be silent, all people, before the LORD; for he has roused himself from his holy dwelling.

3 Then he showed me the high priest Joshua standing before the angel of the LORD, and Satan[h] standing at his right hand to accuse him. [2] And the LORD said to Satan,[h] "The LORD rebuke you, O Satan![h] The LORD who has chosen Jerusalem rebuke you! Is not this man a brand plucked from the fire?" [3] Now Joshua was dressed with filthy clothes as he stood before the angel. [4] The angel said to those who were standing before him, "Take off his filthy clothes." And to him he said, "See, I have taken your guilt away from you, and I will clothe you with festal apparel." [5] And I said, "Let them put a clean turban on his

---

c Heb *it*   *d* Ch 2.5 in Heb   *e* Cn: Heb *after glory he*   *f* Heb *his eye*   *g* Or *wave*
h Or *the Accuser;* Heb *the Adversary*

(the LORD's agents of destruction) will scatter (Isa 54.16–17; Hag 2.21–22).

2.1–5: The third vision: a man going to measure Jerusalem. 1: *Measuring line,* see Ezek 40.3–4; Rev 11.1; 21.15–17. This measurement, as in Ezek 41.13, is part of the restoration, but the presence of the LORD, as at the Exodus, is what will afford true protection. **5:** God is Jerusalem's guard and glory (Isa 4.5 n.; Ezek 43.2–5).

2.6–13: An appeal to the exiles. 6: Babylonia is the *land of the north* (6.6, 8; Jer 3.18; 46.20; compare Jer 51.45–49). **8:** *Apple of my eye* indicates special favor and affection (Deut 32.10; see Ps 17.8 n.). **10:** Isa 54.1–3; 65.18–19; Zeph 3.14; Ezek 43.6–9. **11:** 8.20–23. Jerusalem will be the religious center of the

world. **12:** This is the only Old Testament reference to Palestine as *the holy land.* The concept reappears in the Apocrypha (Wisdom of Solomon 12.3; 2 Macc 1.7). **13:** Hab 2.20; Zeph 1.7.

3.1–10: The fourth vision: Joshua and Satan. 1: *Joshua,* see Hag 1.1 n. In the Old Testament *Satan* (literally, "the Adversary") is not the incarnation of evil but a functionary of the heavenly court who accuses mortals of wrong (1 Chr 21.1; see Job 1.6–8 n.). God then acquits ("justifies") or condemns the defendant. **3–4:** *Filthy clothes* represent the sin of both priest and people. **5:** The *clean turban* (Ex 28.4) and *apparel* (Lev 8.1–9) symbolize ritual purity, in preparation for the advent of the Messiah. **7:** *Access,* to the heavenly courts;

head." So they put a clean turban on his head and clothed him with the apparel; and the angel of the LORD was standing by.

6 Then the angel of the LORD assured Joshua, saying [7]"Thus says the LORD of hosts: If you will walk in my ways and keep my requirements, then you shall rule my house and have charge of my courts, and I will give you the right of access among those who are standing here. [8]Now listen, Joshua, high priest, you and your colleagues who sit before you! For they are an omen of things to come: I am going to bring my servant the Branch. [9]For on the stone that I have set before Joshua, on a single stone with seven facets, I will engrave its inscription, says the LORD of hosts, and I will remove the guilt of this land in a single day. [10]On that day, says the LORD of hosts, you shall invite each other to come under your vine and fig tree."

4 The angel who talked with me came again, and wakened me, as one is wakened from sleep. [2]He said to me, "What do you see?" And I said, "I see a lampstand all of gold, with a bowl on the top of it; there are seven lamps on it, with seven lips on each of the lamps that are on the top of it. [3]And by it there are two olive trees, one on the right of the bowl and the other on its left." [4]I said to the angel who talked with me, "What are these, my lord?" [5]Then the angel who talked with me answered me, "Do you not know what these are?" I said, "No, my lord." [6]He said to me, "This is the

word of the LORD to Zerubbabel: Not by might, nor by power, but by my spirit, says the LORD of hosts. [7]What are you, O great mountain? Before Zerubbabel you shall become a plain; and he shall bring out the top stone amid shouts of 'Grace, grace to it!' "

8 Moreover the word of the LORD came to me, saying, [9]"The hands of Zerubbabel have laid the foundation of this house; his hands shall also complete it. Then you will know that the LORD of hosts has sent me to you. [10]For whoever has despised the day of small things shall rejoice, and shall see the plummet in the hand of Zerubbabel.

"These seven are the eyes of the LORD, which range through the whole earth." [11]Then I said to him, "What are these two olive trees on the right and the left of the lampstand?" [12]And a second time I said to him, "What are these two branches of the olive trees, which pour out the oil[i] through the two golden pipes?" [13]He said to me, "Do you not know what these are?" I said, "No, my lord." [14]Then he said, "These are the two anointed ones who stand by the Lord of the whole earth."

5 Again I looked up and saw a flying scroll. [2]And he said to me, "What do you see?" I answered, "I see a flying scroll; its length is twenty cubits, and its width ten cubits." [3]Then he said to me, "This is the curse that goes out over the face of the whole land; for everyone who

i Cn: Heb *gold*

Joshua will present Jerusalem's prayers to heaven. **8**: *The Branch,* a Davidic figure who is to usher in the messianic age (compare Ps 132.17; Isa 4.2; 11.1; Jer 23.5; 33.15), here refers to Zerubbabel (see 6.9–15 n.). **9**: *Stone* may symbolize Joshua's high-priesthood (Ex 28.9–12; see Zech 6.9–15 n.). **10**: Mic 4.4.
4.1–14: **The fifth vision: a golden lampstand and two olive trees.** This vision emphasizes the important positions held by Joshua and Zerubbabel (v. 14) in the restored Jewish community. **2**: Contrast Ex 25.31–40. The stand with seven lamps signifies God's presence in *the whole earth* (v. 10b; compare Rev 5.6).

4.3, 11–14: Rev 11.4. The *two olive trees* may provide oil for the lamps (v. 12), or merely flank them. **6–10a**: Through God's spirit, Zerubbabel will complete the temple. **12**: Perhaps a gloss; no pipes have been previously mentioned. **14**: *Anointed,* literally, "sons of oil," is not the usual designation for "messiah."
5.1–4: **The sixth vision: a flying scroll. 1**: The *scroll* represents the word of God materialized (Ezek 2.9–3.3; Rev 10.8–11; see Jer 36.1–3). **2**: A cubit, about eighteen inches. **3**: The *curse* is to purify both civil and cultic life (compare Deut 27.14–26).

steals shall be cut off according to the writing on one side, and everyone who swears falsely[j] shall be cut off according to the writing on the other side. [4]I have sent it out, says the LORD of hosts, and it shall enter the house of the thief, and the house of anyone who swears falsely by my name; and it shall abide in that house and consume it, both timber and stones."

5 Then the angel who talked with me came forward and said to me, "Look up and see what this is that is coming out." [6]I said, "What is it?" He said, "This is a basket[k] coming out." And he said, "This is their iniquity[l] in all the land." [7]Then a leaden cover was lifted, and there was a woman sitting in the basket![k] [8]And he said, "This is Wickedness." So he thrust her back into the basket,[k] and pressed the leaden weight down on its mouth. [9]Then I looked up and saw two women coming forward. The wind was in their wings; they had wings like the wings of a stork, and they lifted up the basket[k] between earth and sky. [10]Then I said to the angel who talked with me, "Where are they taking the basket?"[k] [11]He said to me, "To the land of Shinar, to build a house for it; and when this is prepared, they will set the basket[k] down there on its base."

6 And again I looked up and saw four chariots coming out from between two mountains—mountains of bronze. [2]The first chariot had red horses, the second chariot black horses, [3]the third chariot white horses, and the fourth chariot dappled gray[m] horses. [4]Then I said to the angel who talked with me, "What are these, my lord?" [5]The angel answered me, "These are the four winds[n] of heav-

en going out, after presenting themselves before the LORD of all the earth. [6]The chariot with the black horses goes toward the north country, the white ones go toward the west country,[o] and the dappled ones go toward the south country." [7]When the steeds came out, they were impatient to get off and patrol the earth. And he said, "Go, patrol the earth." So they patrolled the earth. [8]Then he cried out to me, "Lo, those who go toward the north country have set my spirit at rest in the north country."

9 The word of the LORD came to me: [10]Collect silver and gold[p] from the exiles—from Heldai, Tobijah, and Jedaiah—who have arrived from Babylon; and go the same day to the house of Josiah son of Zephaniah. [11]Take the silver and gold and make a crown,[q] and set it on the head of the high priest Joshua son of Jehozadak; [12]say to him: Thus says the LORD of hosts: Here is a man whose name is Branch: for he shall branch out in his place, and he shall build the temple of the LORD. [13]It is he that shall build the temple of the LORD; he shall bear royal honor, and shall sit and rule on his throne. There shall be a priest by his throne, with peaceful understanding between the two of them. [14]And the crown[r] shall be in the care of Heldai,[s]

---

[j] The word *falsely* added from verse 4
[k] Heb *ephah*   [l] Gk Compare Syr: Heb *their eye*   [m] Compare Gk: Meaning of Heb uncertain   [n] Or *spirits*   [o] Cn: Heb *go after them*   [p] Cn Compare verse 11: Heb lacks *silver and gold*   [q] Gk Mss Syr Tg: Heb *crowns*   [r] Gk Syr: Heb *crowns*   [s] Syr Compare verse 10: Heb *Helem*

5.5–11: The seventh vision: a woman in a basket. Judah will be purified by sending its sin, personified as a woman in a basket (literally, "ephah"), to Babylon, where the basket will be worshiped. **6:** *Basket,* here a container with the capacity of an ephah, about six gallons. **11:** *Shinar,* Babylonia (Gen 10.10; 11.2, 9; Dan 1.2). *House* here means temple. *On its base,* as though the basket were an image.
6.1–8: The eighth vision: four chariots. Although details are not clear, this vision

probably refers to the inauguration of the messianic age with its judgment upon all the earth. **1–3:** See Rev 6.2–8. **8:** *North country,* Babylonia, see 2.6 n.
6.9–15: The crowning of the messianic leader. This section abounds with difficulties. Originally it probably directed the crowning of Zerubbabel as messianic king, but was revised to refer to Joshua. **11:** Zerubbabel should be read for, or perhaps along with, *Joshua.* **12:** *Branch* is a messianic designa-

Tobijah, Jedaiah, and Josiah[t] son of Zephaniah, as a memorial in the temple of the LORD.

15 Those who are far off shall come and help to build the temple of the LORD; and you shall know that the LORD of hosts has sent me to you. This will happen if you diligently obey the voice of the LORD your God.

7 In the fourth year of King Darius, the word of the LORD came to Zechariah on the fourth day of the ninth month, which is Chislev. 2 Now the people of Bethel had sent Sharezer and Regem-melech and their men, to entreat the favor of the LORD, 3 and to ask the priests of the house of the LORD of hosts and the prophets, "Should I mourn and practice abstinence in the fifth month, as I have done for so many years?" 4 Then the word of the LORD of hosts came to me: 5 Say to all the people of the land and the priests: When you fasted and lamented in the fifth month and in the seventh, for these seventy years, was it for me that you fasted? 6 And when you eat and when you drink, do you not eat and drink only for yourselves? 7 Were not these the words that the LORD proclaimed by the former prophets, when Jerusalem was inhabited and in prosperity, along with the towns around it, and when the Negeb and the Shephelah were inhabited?

8 The word of the LORD came to Zechariah, saying: 9 Thus says the LORD of hosts: Render true judgments, show kindness and mercy to one another; 10 do not oppress the widow, the orphan, the alien, or the poor; and do not devise evil in your hearts against one another. 11 But they refused to listen, and turned a stubborn shoulder, and stopped their ears in order not to hear. 12 They made their hearts adamant in order not to hear the law and the words that the LORD of hosts had sent by his spirit through the former prophets. Therefore great wrath came from the LORD of hosts. 13 Just as, when I[u] called, they would not hear, so, when they called, I would not hear, says the LORD of hosts, 14 and I scattered them with a whirlwind among all the nations that they had not known. Thus the land they left was desolate, so that no one went to and fro, and a pleasant land was made desolate.

8 The word of the LORD of hosts came to me, saying: 2 Thus says the LORD of hosts: I am jealous for Zion with great jealousy, and I am jealous for her with great wrath. 3 Thus says the LORD: I will return to Zion, and will dwell in the midst of Jerusalem; Jerusalem shall be called the faithful city, and the mountain of the LORD of hosts shall be called the holy mountain. 4 Thus says the LORD of hosts: Old men and old women shall again sit in the streets of Jerusalem, each with staff in hand because of their great age. 5 And the streets of the city shall be full of boys and girls playing in its streets. 6 Thus says the LORD of hosts: Even though it seems impossible to the remnant of this people in these days, should it also seem impossible to me,

t Syr Compare verse 10: Heb *Hen*   *u* Heb *he*

---

tion (see 3.8 n.; Hag 2.20–23). **15:** The new age will see the return of the exiles, conversion of the Gentiles, and completion of the temple (8.20–23; Ob 19–21; Mic 2.12; 4.1–5, 6–8; Zeph 3.9–10, 20; Mal 1.11). This verse is probably an addition.

7.1–14: An answer to an inquiry about fasting. 1: Mid-November to mid-December 518 B.C. **5:** The temple at Jerusalem was burned *in the fifth month* and Gedaliah the governor was murdered *in the seventh* (2 Kings 25.8–9, 25). *These seventy years,* see 1.12 n. **9–10:** This is the essence of prophetic moral teaching (Isa 1.16–17; 55.6–9; 58.6–12; Jer 7.5–7; Am 5.14–15, 21–24; Mic 6.8). **12:** Neh 9.3; Jer 5.3; 7.25–26; 11.10; Dan 9.11–14. **13:** Isa 1.15, Jer 7.13–15.

8.1–23: The LORD will return to Zion and do good to Jerusalem and Judah. Ten sections introduced by the words, *Thus says the LORD of hosts* (except in v. 3 where the words "of hosts" are not present). All of them deal with the messianic era. **2:** Joel 2.18 n. **3:** Isa 1.26; 11.9; 62.11–12; Jer 31.23. **4–5:** Isa 65.19–20. **7:** *East country* and *west country,* i.e. from all the lands of the dispersion. **8:** The

says the LORD of hosts? [7]Thus says the LORD of hosts: I will save my people from the east country and from the west country; [8]and I will bring them to live in Jerusalem. They shall be my people and I will be their God, in faithfulness and in righteousness.

9 Thus says the LORD of hosts: Let your hands be strong—you that have recently been hearing these words from the mouths of the prophets who were present when the foundation was laid for the rebuilding of the temple, the house of the LORD of hosts. [10]For before those days there were no wages for people or for animals, nor was there any safety from the foe for those who went out or came in, and I set them all against one other. [11]But now I will not deal with the remnant of this people as in the former days, says the LORD of hosts. [12]For there shall be a sowing of peace; the vine shall yield its fruit, the ground shall give its produce, and the skies shall give their dew; and I will cause the remnant of this people to possess all these things. [13]Just as you have been a cursing among the nations, O house of Judah and house of Israel, so I will save you and you shall be a blessing. Do not be afraid, but let your hands be strong.

14 For thus says the LORD of hosts: Just as I purposed to bring disaster upon you, when your ancestors provoked me to wrath, and I did not relent, says the LORD of hosts, [15]so again I have purposed in these days to do good to Jerusalem and to the house of Judah; do not be afraid. [16]These are the things that you shall do: Speak the truth to one another, render in your gates judgments that are true and make for peace, [17]do not devise evil in your hearts against one another, and love no false oath; for all these are things that I hate, says the LORD.

18 The word of the LORD of hosts came to me, saying: [19]Thus says the LORD of hosts: The fast of the fourth month, and the fast of the fifth, and the fast of the seventh, and the fast of the tenth, shall be seasons of joy and gladness, and cheerful festivals for the house of Judah: therefore love truth and peace.

20 Thus says the LORD of hosts: Peoples shall yet come, the inhabitants of many cities; [21]the inhabitants of one city shall go to another, saying, "Come, let us go to entreat the favor of the LORD, and to seek the LORD of hosts; I myself am going." [22]Many peoples and strong nations shall come to seek the LORD of hosts in Jerusalem, and to entreat the favor of the LORD. [23]Thus says the LORD of hosts: In those days ten men from nations of every language shall take hold of a Jew, grasping his garment and saying, "Let us go with you, for we have heard that God is with you."

9   An Oracle.

The word of the LORD is against
    the land of Hadrach
    and will rest upon Damascus.
For to the LORD belongs the
    capital[v] of Aram,[w]
    as do all the tribes of Israel;
[2] Hamath also, which borders on it,
    Tyre and Sidon, though they
        are very wise.

---

[v] Heb *eye*   [w] Cn: Heb *of Adam* (or *of humankind*)

---

covenant shall be reaffirmed (Ex 6.7; Jer 31.33). **9:** Hag 1.6–11; 2.4–9, 15–19. **13:** Jer 29.18; Gen 12.2–3. **14–17:** God still demands right living (see 7.9–10 n.; Eph 4.25–32; 1 Thess 5.12–22).

8.18–19: These fasts commemorated the fall and humiliation of Jerusalem. In *the fourth month* the Babylonians breached the walls of Jerusalem, and in *the tenth* month they had begun the siege (2 Kings 25.1–4; Jer 39.2; 52.4–7); for the other months, see 7.5 n. **20–**

23: The prophecy of Zechariah ends on a note of universalism (compare 2.11; Isa 2.3; 45.14, 24).

9.1–11.17: The restoration of Israel; the day of the LORD.

9.1–8: The shattering of Israel's enemies foreshadows the messianic era (see Zeph 3.14–20 n.); here may be a reflection of Alexander the Great's conquests after 333 B.C. **1–6:** These cities of Syria (*Aram*) and *Philistia* fall within the ideal limits of the

3 Tyre has built itself a rampart,
and heaped up silver like dust,
and gold like the dirt of the
streets.
4 But now, the Lord will strip it of
its possessions
and hurl its wealth into the sea,
and it shall be devoured by fire.

5 Ashkelon shall see it and be afraid;
Gaza too, and shall writhe in
anguish;
Ekron also, because its hopes
are withered.
The king shall perish from Gaza;
Ashkelon shall be uninhabited;
6 a mongrel people shall settle
in Ashdod,
and I will make an end of the
pride of Philistia.
7 I will take away its blood from
its mouth,
and its abominations from
between its teeth;
it too shall be a remnant for
our God;
it shall be like a clan in Judah,
and Ekron shall be like the
Jebusites.
8 Then I will encamp at my house
as a guard,
so that no one shall march to
and fro;
no oppressor shall again overrun
them,
for now I have seen with my
own eyes.

9 Rejoice greatly, O daughter Zion!
Shout aloud, O daughter
Jerusalem!

Lo, your king comes to you;
triumphant and victorious is he,
humble and riding on a donkey,
on a colt, the foal of a donkey.
10 He*x* will cut off the chariot
from Ephraim
and the war horse from
Jerusalem;
and the battle bow shall be cut
off,
and he shall command peace to
the nations;
his dominion shall be from sea
to sea,
and from the River to the ends
of the earth.

11 As for you also, because of the
blood of my covenant
with you,
I will set your prisoners free
from the waterless pit.
12 Return to your stronghold,
O prisoners of hope;
today I declare that I will restore
to you double.
13 For I have bent Judah as my bow;
I have made Ephraim its arrow.
I will arouse your sons, O Zion,
against your sons, O Greece,
and wield you like a
warrior's sword.

14 Then the LORD will appear
over them,
and his arrow go forth like
lightning;
the Lord GOD will sound the
trumpet

x Gk: Heb I

---

promised land (Gen 15.18–21; Ex 23.31). **7:**
When converted, the Philistines will observe
Jewish dietary laws (compare Gen 9.4; Lev
11.2–47; Deut 14.3–21). *Jebusites,* 2 Sam 5.6–
9 n.
9.9–10: The Prince of Peace. 9: Not two
animals (Mt 21.5) but one (Jn 12.14–15)
young animal is meant. The *triumphant* king
comes as a *humble* and peaceful monarch. **10:**
Pss 46.8–10; 72.8; Isa 11.6–9; 57.19; Hos
2.18; Mic 4.1–4; compare Eph 2.14–18.

**9.11–17: The ingathering of dispersed
Israelites. 11:** *Blood of my covenant,* Ex 24.8;
Mk 14.24; Heb 9.20–22. *Waterless pit,* a dungeon. **12:** Isa 40.1–2, 9–10; 61.7. **13:** *Judah* and
Ephraim, the southern and northern king-
doms. *Greece,* Hebrew "yawan" (Gen 10.2–
4), may be taken literally or as a symbol of
foreigners in the eschatological battle. **14:**
Them, i.e. Judah. **16–17:** Ezek 34.25–31 n.; Jer
31.12–14.

and march forth in the
whirlwinds of the south.
15 The LORD of hosts will protect
them,
and they shall devour and tread
down the slingers;[y]
they shall drink their blood[z]
like wine,
and be full like a bowl,
drenched like the corners of
the altar.

16 On that day the LORD their God
will save them
for they are the flock of
his people;
for like the jewels of a crown
they shall shine on his land.
17 For what goodness and beauty
are his!
Grain shall make the young
men flourish,
and new wine the young
women.

10 Ask rain from the LORD
in the season of the
spring rain,
from the LORD who makes the
storm clouds,
who gives showers of rain
to you,[a]
the vegetation in the field
to everyone.
2 For the teraphim[b] utter nonsense,
and the diviners see lies;
the dreamers tell false dreams,
and give empty consolation.
Therefore the people wander
like sheep;
they suffer for lack of a
shepherd.

3 My anger is hot against the
shepherds,
and I will punish the leaders;[c]

for the LORD of hosts cares for his
flock, the house of Judah,
and will make them like his
proud war horse.
4 Out of them shall come the
cornerstone,
out of them the tent peg,
out of them the battle bow,
out of them every commander.
5 Together they shall be like
warriors in battle,
trampling the foe in the mud of
the streets;
they shall fight, for the LORD is
with them,
and they shall put to shame the
riders on horses.

6 I will strengthen the house
of Judah,
and I will save the house
of Joseph.
I will bring them back because I
have compassion on them,
and they shall be as though I
had not rejected them;
for I am the LORD their God
and I will answer them.
7 Then the people of Ephraim shall
become like warriors,
and their hearts shall be glad as
with wine.
Their children shall see it and
rejoice,
their hearts shall exult in
the LORD.

8 I will signal for them and gather
them in,
for I have redeemed them,
and they shall be as numerous as
they were before.

y Cn: Heb *the slingstones*   z Gk: Heb *shall
drink*   a Heb *them*   b Or *household gods*
c Or *male goats*

10.1–2: **The LORD alone controls na-
ture,** and gives rain. **1:** 14.17; Joel 2.18–27;
Am 4.7–8. **2:** *Teraphim*, see Ezek 21.18–24 n.
10.3–12: **The LORD alone controls history,** and will gather in the redeemed. **6–7:**

Both Israel and Judah will be saved. **6:** 8.7–8;
Isa 41.17–20; 54.8. **8–12:** The restoration. **8:**
Isa 43.1–7, 14–21; Jer 23.3. **10:** Deut 30.1–5;
Hos 11.11. **12:** 14.9; Mic 4.5. *Them,* i.e. Judah
and Ephraim.

9 Though I scattered them among
the nations,
yet in far countries they shall
remember me,
and they shall rear their children
and return.
10 I will bring them home from the
land of Egypt,
and gather them from Assyria;
I will bring them to the land of
Gilead and to Lebanon,
until there is no room for them.
11 They*d* shall pass through the sea
of distress,
and the waves of the sea shall be
struck down,
and all the depths of the Nile
dried up.
The pride of Assyria shall be
laid low,
and the scepter of Egypt
shall depart.
12 I will make them strong in
the LORD,
and they shall walk in his name,
says the LORD.

11 Open your doors, O Lebanon,
so that fire may devour
your cedars!
2 Wail, O cypress, for the cedar
has fallen,
for the glorious trees are ruined!
Wail, oaks of Bashan,
for the thick forest has
been felled!
3 Listen, the wail of the shepherds,
for their glory is despoiled!
Listen, the roar of the lions,
for the thickets of the Jordan
are destroyed!

4 Thus said the LORD my God: Be a
shepherd of the flock doomed to slaughter. 5 Those who buy them kill them and
go unpunished; and those who sell them
say, "Blessed be the LORD, for I have
become rich"; and their own shepherds
have no pity on them. 6 For I will no
longer have pity on the inhabitants of the
earth, says the LORD. I will cause them,
every one, to fall each into the hand of a
neighbor, and each into the hand of the
king; and they shall devastate the earth,
and I will deliver no one from their hand.
7 So, on behalf of the sheep merchants, I became the shepherd of the
flock doomed to slaughter. I took two
staffs; one I named Favor, the other I
named Unity, and I tended the sheep. 8 In
one month I disposed of the three shepherds, for I had become impatient with
them, and they also detested me. 9 So I
said, "I will not be your shepherd. What
is to die, let it die; what is to be destroyed, let it be destroyed; and let those
that are left devour the flesh of one another!" 10 I took my staff Favor and broke
it, annulling the covenant that I had
made with all the peoples. 11 So it was
annulled on that day, and the sheep merchants, who were watching me, knew
that it was the word of the LORD. 12 I then
said to them, "If it seems right to you,
give me my wages; but if not, keep
them." So they weighed out as my
wages thirty shekels of silver. 13 Then the
LORD said to me, "Throw it into the
treasury"*e*—this lordly price at which I
was valued by them. So I took the thirty
shekels of silver and threw them into the
treasury in the house of the LORD.
14 Then I broke my second staff Unity,
annulling the family ties between Judah
and Israel.
15 Then the LORD said to me: Take

d Gk: Heb *He*   *e* Syr: Heb *it to the potter*

11.1–3: The fall of the tyrants. *Cedars,
shepherds,* and *lions* refer to the rulers. **1:** Lebanon's cedars were proverbial symbols of
strength (Ezek 31.2–9). **2:** *Oaks of Bashan,* Isa
2.13; Ezek 27.6; compare Ps 22.12; Am 4.1.
3: *Thickets of the Jordan,* Jer 12.5 n.
11.4–17: The two shepherds. Because of
their sins, God allows the Israelites to be
abused by their rulers and breaks the covenant with them. **4–14:** The prophet portrays
a good shepherd rejected by his sheep. **5:**
Those who buy and *sell* are the Ptolemaic over-
lords; *their own shepherds* are native appointees. **8:** *The three shepherds,* probably contemporary officials, cannot now be identified.
12–13: *Thirty shekels,* the price of a slave (Ex

once more the implements of a worthless shepherd. 16 For I am now raising up in the land a shepherd who does not care for the perishing, or seek the wandering,*f* or heal the maimed, or nourish the healthy,*g* but devours the flesh of the fat ones, tearing off even their hoofs.
17 Oh, my worthless shepherd,
    who deserts the flock!
May the sword strike his arm
    and his right eye!
Let his arm be completely
        withered,
    his right eye utterly blinded!

12   An Oracle.

The word of the LORD concerning Israel: Thus says the LORD, who stretched out the heavens and founded the earth and formed the human spirit within: 2 See, I am about to make Jerusalem a cup of reeling for all the surrounding peoples; it will be against Judah also in the siege against Jerusalem. 3 On that day I will make Jerusalem a heavy stone for all the peoples; all who lift it shall grievously hurt themselves. And all the nations of the earth shall come together against it. 4 On that day, says the LORD, I will strike every horse with panic, and its rider with madness. But on the house of Judah I will keep a watchful eye, when I strike every horse of the peoples with blindness. 5 Then the clans of Judah shall say to themselves, "The inhabitants of Jerusalem have strength through the LORD of hosts, their God."
6 On that day I will make the clans of Judah like a blazing pot on a pile of wood, like a flaming torch among sheaves; and they shall devour to the right and to the left all the surrounding peoples, while Jerusalem shall again be inhabited in its place, in Jerusalem.

7 And the LORD will give victory to the tents of Judah first, that the glory of the house of David and the glory of the inhabitants of Jerusalem may not be exalted over that of Judah. 8 On that day the LORD will shield the inhabitants of Jerusalem so that the feeblest among them on that day shall be like David, and the house of David shall be like God, like the angel of the LORD, at their head. 9 And on that day I will seek to destroy all the nations that come against Jerusalem.

10 And I will pour out a spirit of compassion and supplication on the house of David and the inhabitants of Jerusalem, so that, when they look on the one*h* whom they have pierced, they shall mourn for him, as one mourns for an only child, and weep bitterly over him, as one weeps over a firstborn. 11 On that day the mourning in Jerusalem will be as great as the mourning for Hadadrimmon in the plain of Megiddo. 12 The land shall mourn, each family by itself; the family of the house of David by itself, and their wives by themselves; the family of the house of Nathan by itself, and their wives by themselves; 13 the family of the house of Levi by itself, and their wives by themselves; the family of the Shimeites by itself, and their wives by them-

f Syr Compare Gk Vg: Heb *the youth*
g Meaning of Heb uncertain   *h* Heb *on me*

21.32; compare Mt 26.15; 27.9). *Lordly price* is ironic. **14**: Perhaps a reference to the division between Jews and Samaritans. **15–17**: The prophet portrays a worthless shepherd who exploits the sheep (Ezek 34.2–10; Mic 3.1–7; Jn 10.1, 8–13).
  12.1–14.21: **The coming great day of the LORD,** when Jerusalem shall be cleansed from sin, the covenant re-established, and God will reign over all the earth.
  12.1–13.6: **Jerusalem shall lament its sin and be purged of idolatry. 12.1**: Isa 42.5. **2**: Ps 75.8; Isa 51.17–23. **7–8**: The ideal king will be from *the house of David.* **10–14**: With the picture of Jerusalem mourning over a prophet or king whom it has martyred compare Isa 52.13–53.12; Mt 23.37; Jn 19.34–37; Rev 1.7. **10**: *Only child* and *firstborn,* see Lk 2.7; Jn 1.14, 18. **11**: *Hadad-rimmon,* a popular fertility god whose seasonal death was widely mourned (compare Ezek 8.14; Hos 7.14). **12**: *Nathan,* David's son (2 Sam 5.14). **13**: Compare Num 3.21.

selves; [14] and all the families that are left, each by itself, and their wives by themselves.

13 On that day a fountain shall be opened for the house of David and the inhabitants of Jerusalem, to cleanse them from sin and impurity.

2 On that day, says the LORD of hosts, I will cut off the names of the idols from the land, so that they shall be remembered no more; and also I will remove from the land the prophets and the unclean spirit. [3] And if any prophets appear again, their fathers and mothers who bore them will say to them, "You shall not live, for you speak lies in the name of the LORD"; and their fathers and their mothers who bore them shall pierce them through when they prophesy. [4] On that day the prophets will be ashamed, every one, of their visions when they prophesy; they will not put on a hairy mantle in order to deceive, [5] but each of them will say, "I am no prophet, I am a tiller of the soil; for the land has been my possession[i] since my youth." [6] And if anyone asks them, "What are these wounds on your chest?"[j] the answer will be "The wounds I received in the house of my friends."

[7] "Awake, O sword, against my
    shepherd,
  against the man who is my
    associate,"
            says the LORD of hosts.
Strike the shepherd, that the sheep
    may be scattered;
  I will turn my hand against
    the little ones.

[8] In the whole land, says the LORD,
  two-thirds shall be cut off
    and perish,
  and one-third shall be left alive.
[9] And I will put this third into
    the fire,
  refine them as one refines silver,
    and test them as gold is tested.
They will call on my name,
  and I will answer them.
I will say, "They are my people";
  and they will say, "The LORD is
    our God."

14 See, a day is coming for the LORD, when the plunder taken from you will be divided in your midst. [2] For I will gather all the nations against Jerusalem to battle, and the city shall be taken and the houses looted and the women raped; half the city shall go into exile, but the rest of the people shall not be cut off from the city. [3] Then the LORD will go forth and fight against those nations as when he fights on a day of battle. [4] On that day his feet shall stand on the Mount of Olives, which lies before Jerusalem on the east; and the Mount of Olives shall be split in two from east to west by a very wide valley; so that one half of the Mount shall withdraw northward, and the other half southward. [5] And you shall flee by the valley of the LORD's mountain,[k] for the valley between the mountains shall reach to Azal;[l] and you shall flee as you fled from the earthquake

---

i Cn: Heb *for humankind has caused me to possess*
j Heb *wounds between your hands*   k Heb *my mountains*   l Meaning of Heb uncertain

---

13.1: Ps 46.4; Ezek 47.1–12; Joel 3.18; Jn 4.10–14; 7.38; Rev 21.6; 22.1–2. **2–6:** Professional prophecy, fallen into disrepute, will cease. **2:** Hos 2.17; Zeph 1.4; 3.4. **3:** Jer 23.9–22. **4:** 2 Kings 1.8. **5:** Am 7.14. **6:** Compare 1 Kings 18.28.
13.7–9: God's shepherd, smitten for the sheep. A separate messianic oracle (Mt 26.31; Mk 14.27), closely linked with chs 9–11. After the death of the shepherd, a remnant of his flock will be purified and saved. **7:** *Man,* here the shepherd-king. **8:** Isa 6.13. **9:** *Refines,*

Isa 1.25–26; 48.10. *My people . . . our God,* signifies the re-establishing of the covenant (Hos 2.21–23).
14.1–21: The final warfare and the final victory (Isa 66.15–23; Ezek chs 38–39; Joel 3.9–21; Mk 13.7–27; Rev chs 20–22). An eschatological description in which Jerusalem suffers (v. 2), but will be defended by the LORD, who will become *king over all the earth* (v. 9), worshiped in *Jerusalem* where *the nations* will come in pilgrimage (v. 16). **5:** See Am 1.1 n.

in the days of King Uzziah of Judah. Then the LORD my God will come, and all the holy ones with him.

6 On that day there shall not be[m] either cold or frost.[n] 7 And there shall be continuous day (it is known to the LORD), not day and not night, for at evening time there shall be light.

8 On that day living waters shall flow out from Jerusalem, half of them to the eastern sea and half of them to the western sea; it shall continue in summer as in winter.

9 And the LORD will become king over all the earth; on that day the LORD will be one and his name one.

10 The whole land shall be turned into a plain from Geba to Rimmon south of Jerusalem. But Jerusalem shall remain aloft on its site from the Gate of Benjamin to the place of the former gate, to the Corner Gate, and from the Tower of Hananel to the king's wine presses. 11 And it shall be inhabited, for never again shall it be doomed to destruction; Jerusalem shall abide in security.

12 This shall be the plague with which the LORD will strike all the peoples that wage war against Jerusalem: their flesh shall rot while they are still on their feet; their eyes shall rot in their sockets, and their tongues shall rot in their mouths. 13 On that day a great panic from the LORD shall fall on them, so that each will seize the hand of a neighbor, and the hand of the one will be raised against the hand of the other; 14 even Judah will fight at Jerusalem. And the wealth of all the surrounding nations shall be collected—gold, silver, and garments in great abundance. 15 And a plague like this plague shall fall on the horses, the mules, the camels, the donkeys, and whatever animals may be in those camps.

16 Then all who survive of the nations that have come against Jerusalem shall go up year after year to worship the King, the LORD of hosts, and to keep the festival of booths.[o] 17 If any of the families of the earth do not go up to Jerusalem to worship the King, the LORD of hosts, there will be no rain upon them. 18 And if the family of Egypt do not go up and present themselves, then on them shall[p] come the plague that the LORD inflicts on the nations that do not go up to keep the festival of booths.[o] 19 Such shall be the punishment of Egypt and the punishment of all the nations that do not go up to keep the festival of booths.[o]

20 On that day there shall be inscribed on the bells of the horses, "Holy to the LORD." And the cooking pots in the house of the LORD shall be as holy as[q] the bowls in front of the altar; 21 and every cooking pot in Jerusalem and Judah shall be sacred to the LORD of hosts, so that all who sacrifice may come and use them to boil the flesh of the sacrifice. And there shall no longer be traders[r] in the house of the LORD of hosts on that day.

m Cn: Heb *there shall not be light*   n Compare Gk Syr Vg Tg: Meaning of Heb uncertain
o Or *tabernacles*; Heb *succoth*   p Gk Syr: Heb *shall not*   q Heb *shall be like*
r Or *Canaanites*

14.6–7: Isa 24.23; Rev 22.5. **8**: 13.1. *Eastern sea,* the Dead Sea. *Western sea,* the Mediterranean Sea (Joel 2.20). **9**: Deut 6.4; Ps 99.1–5; Mal 1.11; Rev 11.15–18; 15.3–4. **10–11**: Compare 2 Kings 23.8; Jer 31.38–40. **12–15**: 14.3; Ex 7.4–5.

14.16–19: *Booths,* Lev 23.33–36 n. Originally a nature festival (note *rain*), in the postexilic period this feast was probably associated with renewing the covenant and celebrating the rule of God. **17**: Ex 34.21–24; Jn 7.2. **18–19**: Compare Isa ch 19; Jer ch 46; Ezek chs 29–32. **20–21**: Lev 27.30–33; Joel 3.17; Rev 21.27. **21b**: The meaning of this sentence is not clear. Either no *traders* will be needed because everything will be holy, or nothing will be permitted that defiles pure worship (Jn 2.16).

Malachi

Nothing is known about the person of Malachi. Even his name, which means "My messenger," may be only an appellation, based on 3.1 (compare 2.7). The book, however, presents a substantial amount of information about its author's viewpoints. Living probably in the period 500–450 B.C., this prophetic voice was devoted to the temple and held a high view of the priesthood and its responsibilities. He speaks frequently of the covenant (2.4, 5, 8, 10, 14; 3.1) and shows great respect for the priestly "instruction" (*torah,* 2.6–9). Instead of adopting the oracular style normally used by earlier prophets, Malachi employs a distinctive question-and-answer method of stating his argument. Nevertheless, his emphases upon sin, judgment, and repentance and upon the advent of the day of the LORD (3.1–5, 7; 4.1–3, 6) mark him as a prophet, and he may best be understood as a "cultic prophet" (see Introduction to Joel).

One central theme dominates Malachi's thought: fidelity to the LORD's covenant and its teachings. From this standpoint he both condemns the priests for corrupting worship and misleading the people (1.6–2.9; 3.3–4) and charges his brethren to remain faithful to their wives from the community of covenant and thus have "godly offspring" (2.13–16). In exhorting Israel to faithfulness, Malachi gives also striking descriptions of the worship of the LORD by the Gentiles (1.11, 14b), the ideal priest (2.5–7), and the blessing of obedience (3.10–12, 16–17; 4.2–3). He asserts, further, that all members of the community are begotten by the same God (2.10) and adds to the concept of the day of the LORD (Zeph 1.14–18) the figure of the appointed forerunner (3.1–4; compare 4.5–6). The messenger who "prepares the way" (3.1) and the sending of Elijah (4.5) suggested to New Testament writers (Mt 11.7–15; Lk 1.16–17) a connection with the coming of the Messiah.

1 An oracle. The word of the Lord to Israel by Malachi. *a*

2 I have loved you, says the Lord. But you say, "How have you loved us?" Is not Esau Jacob's brother? says the Lord. Yet I have loved Jacob 3 but I have hated Esau; I have made his hill country a desolation and his heritage a desert for jackals. 4 If Edom says, "We are shattered but we will rebuild the ruins," the Lord of hosts says: They may build, but I will tear down, until they are called the wicked country, the people with whom the Lord is angry forever. 5 Your own eyes shall see this, and you shall say, "Great is the Lord beyond the borders of Israel!"

6 A son honors his father, and servants their master. If then I am a father, where is the honor due me? And if I am a master, where is the respect due me? says the Lord of hosts to you, O priests, who despise my name. You say, "How have we despised your name?" 7 By offering polluted food on my altar. And you say, "How have we polluted it?" *b* By thinking that the Lord's table may be despised. 8 When you offer blind animals in sacrifice, is that not wrong? And when you offer those that are lame or sick, is that not wrong? Try presenting that to your governor; will he be pleased with you or show you favor? says the Lord of hosts. 9 And now implore the favor of God, that he may be gracious to us. The fault is yours. Will he show favor to any of you? says the Lord of hosts. 10 Oh, that someone among you would shut the temple *c* doors, so that you would not kindle fire on my altar in vain! I have no pleasure in you, says the Lord of hosts,

and I will not accept an offering from your hands. 11 For from the rising of the sun to its setting my name is great among the nations, and in every place incense is offered to my name, and a pure offering; for my name is great among the nations, says the Lord of hosts. 12 But you profane it when you say that the Lord's table is polluted, and the food for it *d* may be despised. 13 "What a weariness this is," you say, and you sniff at me, *e* says the Lord of hosts. You bring what has been taken by violence or is lame or sick, and this you bring as your offering! Shall I accept that from your hand? says the Lord. 14 Cursed be the cheat who has a male in the flock and vows to give it, and yet sacrifices to the Lord what is blemished; for I am a great King, says the Lord of hosts, and my name is reverenced among the nations.

2 And now, O priests, this command is for you. 2 If you will not listen, if you will not lay it to heart to give glory to my name, says the Lord of hosts, then I will send the curse on you and I will curse your blessings; indeed I have already cursed them, *f* because you do not lay it to heart. 3 I will rebuke your offspring, and spread dung on your faces, the dung of your offerings, and I will put you out of my presence. *g*

4 Know, then, that I have sent this command to you, that my covenant with

a Or *by my messenger*   *b* Gk: Heb *you*
c Heb lacks *temple*   *d* Compare Syr Tg: Heb
its fruit, its food   *e* Another reading is *at it*
f Heb *it*   *g* Cn Compare Gk Syr: Heb *and he
shall bear you to it*

1.1: Superscription: Compare Zech 9.1; 12.1. *Oracle,* see Nah 1.1 n. *The word of the Lord,* see Hos 1.1 n.
1.2–5: God loves Israel. 2: *Esau, Jacob's* twin *brother,* was the ancestor of the Edomites (Gen 25.24–26, 30; 36.1). **2–3:** Rom 9.13. In biblical idiom, love and hatred here designate preference, not emotional outburst. **2–5:** Isa ch 34; 63.1–6; Jer 49.7–22; Ezek 25.12–14; Am 1.11–12; see Introduction to Obadiah.
1.6–2.9: The priests have despised their God and their solemn vocation. 1.6: Ex

20.12; Prov 30.11. **8:** *Blind* or *lame* animals were unacceptable for sacrifice (Lev 22.17–25; Deut 15.21). **10:** *Shut the temple doors* to halt vain worship (Isa 1.13; Am 5.21–24). **11:** In contrast with Judah's present corrupt practices, the Gentiles render a *pure offering* to the Lord (compare v. 14b; Ps 102.15; Joel 2.28; Zeph 3.9–10).
2.3: Ex 29.14; Nah 3.6. **4:** *Levi* was the priestly tribe (Num 3.45; 18.21–24; Deut 33.8–11). **5:** Num 25.12–13. **7:** Lev 10.11; Deut 21.5. **8–9:** Mic 3.11; Mt 23.1–36.

Levi may hold, says the LORD of hosts.
⁵My covenant with him was a covenant
of life and well-being, which I gave him;
this called for reverence, and he revered
me and stood in awe of my name. ⁶True
instruction was in his mouth, and no
wrong was found on his lips. He walked
with me in integrity and uprightness,
and he turned many from iniquity. ⁷For
the lips of a priest should guard knowledge, and people should seek instruction
from his mouth, for he is the messenger
of the LORD of hosts. ⁸But you have
turned aside from the way; you have
caused many to stumble by your instruction; you have corrupted the covenant of
Levi, says the LORD of hosts, ⁹and so I
make you despised and abased before all
the people, inasmuch as you have not
kept my ways but have shown partiality
in your instruction.

10 Have we not all one father? Has
not one God created us? Why then are we
faithless to one another, profaning the
covenant of our ancestors? ¹¹Judah has
been faithless, and abomination has been
committed in Israel and in Jerusalem; for
Judah has profaned the sanctuary of the
LORD, which he loves, and has married
the daughter of a foreign god. ¹²May the
LORD cut off from the tents of Jacob anyone who does this—any to witness^h or
answer, or to bring an offering to the
LORD of hosts.

13 And this you do as well: You cover
the LORD's altar with tears, with weeping
and groaning because he no longer regards the offering or accepts it with favor
at your hand. ¹⁴You ask, "Why does he
not?" Because the LORD was a witness

between you and the wife of your youth,
to whom you have been faithless, though
she is your companion and your wife by
covenant. ¹⁵Did not one God make her?ⁱ
Both flesh and spirit are his.^j And what
does the one God^k desire? Godly offspring. So look to yourselves, and do not
let anyone be faithless to the wife of his
youth. ¹⁶For I hate^l divorce, says the
LORD, the God of Israel, and covering
one's garment with violence, says the
LORD of hosts. So take heed to yourselves and do not be faithless.

17 You have wearied the LORD with
your words. Yet you say, "How have we
wearied him?" By saying, "All who do
evil are good in the sight of the LORD,
and he delights in them." Or by asking,
"Where is the God of justice?"

3 See, I am sending my messenger to
prepare the way before me, and the
Lord whom you seek will suddenly
come to his temple. The messenger of
the covenant, in whom you delight—
indeed, he is coming, says the LORD of
hosts. ²But who can endure the day of
his coming, and who can stand when he
appears?

For he is like a refiner's fire and like
fullers' soap; ³he will sit as a refiner and
purifier of silver, and he will purify
the descendants of Levi and refine them
like gold and silver, until they present
offerings to the LORD in righteous-
ness.^m ⁴Then the offering of Judah and

h Cn Compare Gk: Heb *arouse*   *i* Or *Has he
not made one?*   *j* Cn: Heb *and a remnant of spirit
was his*   *k* Heb *he*   *l* Cn: Heb *he hates*
m Or *right offerings to the LORD*

2.10–16: **God hates divorce and demands marital fidelity. 10:** Common derivation from and loyalty to the same God bespeaks unity and solidarity in the community
of covenant (v. 15; compare Deut 32.6, 18; Isa
63.16; 64.8; Jer 31.1–3; Hos 11.1; Eph 4.6). **11:**
The daughter of a foreign god refers to heathen
women. **14–16:** Marriage is a sacred *covenant*,
honored by members of the community of
covenant (Gen 2.24; Ezek 16.8; Hos 2.19; Mk
10.2–9; Eph 5.21–33).
2.17–3.5: **The LORD will send a messen-**

ger to prepare for the coming of the day
of judgment. **2.17:** Job 21.7–16; Hab 1.2–
4, 13.
3.1–4: The *messenger* or angel (Gen 16.7;
22.11; Ex 3.2; Isa 63.9) will prepare for *the day
of* God's *coming* (Isa 40.3; Mt 11.10; Mk 1.2;
Lk 1.17, 76; 7.27) first by purifying the priesthood. **2:** Mt 3.10–12. **5:** God will appear for
judgment against the wicked and godless
(Zeph 1.14–18; 3.1–8; Mk 13.14–37; 2 Thess
2.1–12).

Jerusalem will be pleasing to the LORD as in the days of old and as in former years.

5 Then I will draw near to you for judgment; I will be swift to bear witness against the sorcerers, against the adulterers, against those who swear falsely, against those who oppress the hired workers in their wages, the widow and the orphan, against those who thrust aside the alien, and do not fear me, says the LORD of hosts.

6 For I the LORD do not change; therefore you, O children of Jacob, have not perished. 7 Ever since the days of your ancestors you have turned aside from my statutes and have not kept them. Return to me, and I will return to you, says the LORD of hosts. But you say, "How shall we return?"

8 Will anyone rob God? Yet you are robbing me! But you say, "How are we robbing you?" In your tithes and offerings! 9 You are cursed with a curse, for you are robbing me—the whole nation of you! 10 Bring the full tithe into the storehouse, so that there may be food in my house, and thus put me to the test, says the LORD of hosts; see if I will not open the windows of heaven for you and pour down for you an overflowing blessing. 11 I will rebuke the locust[n] for you, so that it will not destroy the produce of your soil; and your vine in the field shall not be barren, says the LORD of hosts. 12 Then all nations will count you happy, for you will be a land of delight, says the LORD of hosts.

13 You have spoken harsh words against me, says the LORD. Yet you say, "How have we spoken against you?" 14 You have said, "It is vain to serve God. What do we profit by keeping his command or by going about as mourners before the LORD of hosts? 15 Now we count the arrogant happy; evildoers not only prosper, but when they put God to the test they escape."

16 Then those who revered the LORD spoke with one another. The LORD took note and listened, and a book of remembrance was written before him of those who revered the LORD and thought on his name. 17 They shall be mine, says the LORD of hosts, my special possession on the day when I act, and I will spare them as parents spare their children who serve them. 18 Then once more you shall see the difference between the righteous and the wicked, between one who serves God and one who does not serve him.

4 [o] See, the day is coming, burning like an oven, when all the arrogant and all evildoers will be stubble; the day that comes shall burn them up, says the LORD of hosts, so that it will leave them neither root nor branch. 2 But for you who revere my name the sun of righteousness shall rise, with healing in its wings. You shall go out leaping like calves from the stall. 3 And you shall tread down the wicked, for they will be ashes under the soles of your feet, on the day when I act, says the LORD of hosts.

n Heb *devourer*   o Ch 4.1-6 are Ch 3.19-24 in Heb

3.6–12: **If the people will return to God with a full measure of devotion, God will bless them. 6**: Num 23.19; Heb 13.8; Jas 1.17. **8–9**: The *tithes* required by the law (Lev 27.30; Num 18.21–24) were being withheld; therefore the *curse* of crop failure resulted (v. 11). **10**: Deut 28.2–12; Ezek 34.25–31. *Storehouse,* i.e. the public storehouse for offerings of grain, wine, oil, etc. (Neh 13.10–13). **12**: Isa 61.6–9; 62.4.

3.13–4.3: **When the day of judgment comes, the true worshipers will be spared. 3.14**: Job 21.15; Isa 58.3. **16**: *A book of*

remembrance, compare Ex 32.32–34; Ps 69.28; Isa 4.3; 65.6; Dan 7.10; 12.1; Rev 20.12; 21.27.

4.1: Compare 3.2–5. **2**: *The sun of* God's *righteousness* symbolizes health and vindication (2 Sam 23.4; Ps 84.11; Isa 60.1).

4.4–6: **Two appendices:** the first (v. 4) exhorts obedience to the teaching of Moses; the second (vv. 5–6) identifies the forerunner with *Elijah.* Standing at the very end of the prophetic corpus, these two appendices serve to link *the teaching* and *the prophet.* **4**: *Horeb,* Mount Sinai (Deut 5.1–3), is also the place

4 Remember the teaching of my servant Moses, the statutes and ordinances that I commanded him at Horeb for all Israel.

5 Lo, I will send you the prophet Elijah before the great and terrible day of the LORD comes. 6He will turn the hearts of parents to their children and the hearts of children to their parents, so that I will not come and strike the land with a curse.*p*

p Or a ban of utter destruction

where God was revealed to Elijah (1 Kings 19.1–18). **5**: A commentary on 3.1. For an expansion of this tradition, see Mt 11.7–15; 17.10–13; Mk 6.14–16; compare Ecclesiasticus 18.10 **6**: *Turn the hearts* . . . applied specifically to John the Baptist in Lk 1.17.

NOTES

NOTES

NOTES

NOTES

NOTES

THE APOCRYPHAL/
DEUTEROCANONICAL BOOKS
OF THE
OLD TESTAMENT

New Revised Standard Version

Introduction to the Apocryphal/Deuterocanonical Books

MEANINGS AND USAGE OF THE TERMS "APOCRYPHAL" AND "DEUTEROCANONICAL"

The word "apocrypha" is used in a variety of ways that can be confusing to the general reader. Confusion arises partly from the ambiguity of the ancient usage of the word, and partly from the modern application of the term to different groups of books. Etymologically the word means "things that are hidden," but why it was chosen to describe certain books is not clear. Some have suggested that the books were "hidden" or withdrawn from common use because they were deemed to contain mysterious or esoteric lore, too profound to be communicated to any except the initiated (compare 2 Esd 14.45–46). Others have suggested that the term was employed by those who held that such books deserved to be "hidden" because they were spurious or heretical. Thus it appears that in antiquity the term had an honorable significance as well as a derogatory one, depending upon the point of view of those who made use of the word.

According to traditional usage "Apocrypha" has been the designation applied to the fifteen books, or portions of books, listed below. (In many earlier editions of the Apocrypha, the Letter of Jeremiah is incorporated as the final chapter of the Book of Baruch; hence in these editions there are fourteen books.)

Tobit	The Prayer of Azariah and the Song
Judith	of the Three Jews
The Additions to the Book of Esther	Susanna
(contained in the Greek version of Esther)	Bel and the Dragon
The Wisdom of Solomon	1 Maccabees
Ecclesiasticus,	2 Maccabees
or the Wisdom of Jesus son of Sirach	1 Esdras
Baruch	The Prayer of Manasseh
The Letter of Jeremiah	2 Esdras

In addition, the present expanded edition includes the following three texts that are of special interest to Eastern Orthodox readers (see p. iv):

3 Maccabees   4 Maccabees   Psalm 151

None of these books is included in the Hebrew canon of Holy Scripture. All of them, however, with the exception of 2 Esdras, are present in copies of the Greek version of the Old Testament known as the Septuagint. The Old Latin translations of the Old Testament, made from the Septuagint, also include them, along with 2 Esdras. As a consequence, many of the early Church Fathers quoted most of these books as authoritative Scripture (see p. vi AP).

At the end of the fourth century Pope Damasus commissioned Jerome, the most learned biblical scholar of his day, to prepare a standard Latin version of the Scriptures (the Latin Vulgate). In the Old Testament Jerome followed the Hebrew canon and by means of prefaces called the reader's attention to the separate category of the apocryphal books. Subsequent copyists of the Latin Bible, however, were not always careful to transmit Jerome's prefaces, and during the medieval period the Western Church generally regarded these books as part of the holy Scriptures. In 1546 the Council of Trent decreed that the canon of the Old Testament includes them (except the Prayer of Manasseh and 1 and 2 Esdras). Subsequent editions of the Latin Vulgate text, officially approved by the Roman Catholic Church, contain these books incorporated within the sequence of the Old Testament books. Thus Tobit and Judith stand after Nehemiah; the Wisdom of Solomon and Ecclesiasticus stand after the Song of Solomon; Baruch (with the Letter of Jeremiah as chapter 6) stands after Lamentations; and 1 and 2 Maccabees conclude the books of the Old Testament. An appendix after the New Testament contains the Prayer of Manasseh and 1 and 2 Esdras, without implying canonical status.

Editions of the Bible prepared by Protestants have followed the Hebrew canon. The disputed books have generally been placed in a separate section, usually bound between the Old and New Testaments, but occasionally placed after the close of the New Testament.

Modern Roman Catholic scholars commonly employ a distinction introduced by Sixtus of Sienna in 1566 to designate the two groups of books. The terms "protocanonical" and "deuterocanonical" are used to signify respectively those books of Scripture that were received by the entire Church from the beginning as inspired, and those whose inspiration came to be recognized later, after the matter had been disputed by certain Fathers and local churches. Thus Roman Catholics accept as fully canonical those books and parts of books that Protestants call the Apocrypha (except the Prayer of Manasseh and 1 and 2 Esdras, which both groups regard as apocryphal). In short, as a popular Roman Catholic Catechism puts it, "*Deuterocanonical* does not mean *Apocryphal,* but simply 'later added to the canon.' "

The Eastern Orthodox Churches recognize several other books as authoritative. Editions of the Old Testament approved by the Holy Synod of the Greek Orthodox Church contain, besides the Deuterocanonical books, 1 Esdras, Psalm 151, the Prayer of Manasseh, and 3 Maccabees, while 4 Maccabees stands in an appendix. Slavonic Bibles approved by the Russian Orthodox Church contain, besides the Deuterocanonical books, 1 and 2 Esdras (called 2 and 3 Esdras), Psalm 151, and 3 Maccabees.

Besides the books that are included in the present edition, many other Jewish and Jewish-Christian works have survived from the period between about 200 B.C. to about A.D. 200. Since most of these profess to have been written by ancient worthies of Israel, who lived long before the books were actually composed, they are generally called "pseudepigrapha," meaning writings "falsely ascribed." (For a description of several of the more noteworthy pseudepigrapha, see pp. xi–xii AP.)

KINDS OF LITERATURE

The apocryphal/deuterocanonical books represent several different literary genres, including the historical, novelistic, didactic, devotional, epistolary, and apocalyptic types. Though several of the books combine material belonging to more than one of these genres, most of the books can be classified as predominantly of one type or another. Thus 1 Esdras, 1 Maccabees, and, in a certain sense, 2 Maccabees belong to the genre of historical writing. Second Maccabees, which is characterized by bombastic rhetoric, fiery arguments, exaggerated numbers, and superabundant use of invectives against the enemies of Jewish orthodoxy, falls more precisely into the category, then so popular in the Hellenistic world, known as "pathetic history"—a type of literature that uses all possible means to strike the imagination and move the emotions of the reader.

Ostensibly historical but actually quite imaginative are the books of Tobit, Judith, Susanna, and Bel and the Dragon, which may be called moralistic novels. In fact, the last two are noteworthy as ancient examples of the detective story.

Of a serious and didactic nature are the two treatises on wisdom, the Wisdom of Solomon and Ecclesiasticus, or the Wisdom of Jesus son of Sirach. The latter shows particularly close connections with the style and content of the Old Testament book of Proverbs, from which it is a natural development.

The Prayer of Manasseh takes its place with devotional literature of a relatively high order. The psalmody of the Prayer of Azariah and the Song of the Three Jews is of a decidedly liturgical cast.

The Old Testament contains no books that are in the form of a letter, but twenty-one of the twenty-seven books of the New Testament are in epistolary form. The Letter of Jeremiah, which dates from inter-testamental times, may have provided later writers with an example of how this literary form could be used for religious purposes, a form that offers the possibility of combining profound theological content with a direct personal approach to the reader.

Finally, 2 Esdras, a book that purports to reveal the future, is a specimen of the type of literature called apocalyptic (see "Apocalyptic Literature," pp. 362–363 NT). An apocalypse is literally "an unveiling." Like the last six chapters of Daniel in the Old Testament and the book of Revelation in the New Testament, which also are apocalypses, 2 Esdras includes many symbols involving mysterious numbers, strange beasts, and the disclosure of hitherto hidden truths through angelic visitants.

Despite the diversities of literary form, most of which are parallel to, or developments from, similar genres in the Old Testament, the attentive reader of the Apocrypha will be struck by the absence of the prophetic element. From first to last these books bear testimony to the assertion of the Jewish historian Josephus (*Against Apion,* i.8), that "the exact succession of the prophets" had been broken after the close of the Hebrew canon of the Old Testament. Sometimes there is a direct confession that the gift of prophecy had departed (1 Macc 9.27); at other times a hope is expressed that it might one day return (1 Macc 4.46; 14.41). When a writer imitates the prophetic character, as in the book of Baruch, he repeats with slight modifications the language of the older prophets. But the introductory phrase, "Thus says the LORD," which occurs so frequently in the Old Testament, is conspicuous by its absence from the apocryphal/deuterocanonical books.

DIVERGENT ATTITUDES IN THE CHRISTIAN CHURCH

Ecclesiastical opinions concerning the nature and worth of the books of the Apocrypha have varied with age and place.

None of the authors of the books of the New Testament makes a direct quotation from any of the fifteen books of the Apocrypha, though frequent quotations occur from most of the thirty-nine books of the Hebrew canon of the Old Testament. On the other hand, several New Testament writers make occasional allusions to one or more apocryphal books. For example, what seem to be literary echoes from the Wisdom of Solomon are present in Paul's Letter to the Romans (compare Rom 1.20–29 with Wis 13.5,8; 14.24,27; and Rom 9.20–23 with Wis 12.12,20; 15.7) and in his correspondence with the Corinthians (compare 2 Cor 5.1,4 with Wis 9.15). The short Letter of James, a typical bit of "wisdom literature" in the New Testament, contains allusions not only to the Old Testament book of Proverbs but to gnomic sayings in Sirach as well (compare Jas 1.19 with Sir 5.11; and Jas 1.13 with Sir 15.11–12).

During the early Christian centuries most Greek and Latin Church Fathers, such as Irenaeus, Tertullian, Clement of Alexandria, and Cyprian (none of whom knew any Hebrew), quoted passages from the Greek text of apocryphal/deuterocanonical books as "Scripture," "divine Scripture," "inspired," and the like. In this period only an occasional Father made an effort to learn the limits of the Palestinian Jewish canon (as Melito of Sardis) or to distinguish between the Hebrew text of Daniel and the addition of the story of Susanna in the Greek version (as Africanus).

In the fourth century many Greek Fathers (including Eusebius, Athanasius, Cyril of Jerusalem, Gregory of Nazianzus, Amphilochius, and Epiphanius) came to recognize a distinction between the books in the Hebrew canon and the rest, though the latter were still customarily cited as Scripture. During the following centuries usage fluctuated in the East, but at the important Synod of Jerusalem in 1672 the books of Tobit, Judith, Ecclesiasticus, and Wisdom were expressly designated as canonical.

In the Latin Church, on the other hand, though opinion has not been unanimous, a generally high regard for these books has prevailed. More than one local synodical council (e.g. Hippo, A.D. 393, and Carthage, 397 and 419) justified and authorized their use as Scripture. The so-called *Decretum Gelasianum,* a Latin document handed down most frequently under the name of Pope Gelasius (A.D. 492–496), but in some manuscripts as the work of Damasus (366–384) or Hormisdas (514–523), contains, among other material, lists of the books to be read as divine Scripture and of books to be avoided as apocryphal. The former list, which is not present in all the manuscripts, includes among the biblical books Tobit, Judith, Wisdom, Ecclesiasticus, and 1 and 2 Maccabees. Irrespective of the problem of its authorship (many scholars today believe it to be the work of a cleric who lived in south Gaul), the list without doubt reflects the views of the Roman Church at the beginning of the sixth century.

There were, however, occasional voices raised to question the legitimacy of regarding the disputed books as Scripture. At the close of the fourth century, Jerome spoke out decidedly for the Hebrew canon, declaring unreservedly that books that were outside that canon should be classed as apocryphal. When he prepared his celebrated revision of the Latin Bible, the Vulgate, he scrupulously separated the apocryphal Additions to Daniel and Esther, marking them with prefatory notes as absent from the original Hebrew. But, as was remarked above, subsequent scribes were not always careful to transmit Jerome's explanatory material, and during the Middle Ages most readers of the Latin Bible made no distinction between the two classes of books. It is noteworthy, however, that throughout these centuries more than one highly respected ecclesiastical writer (such as Gregory the Great, Walafrid Strabo, Hugh of St. Victor, Hugh of St. Cher, and Nicholas of Lyra), being influenced by the great authority of Jerome, raised theoretical doubts about the disputed books.

Toward the close of the fourteenth century John Wyclif ("the father of English prose") and his disciples, Nicholas of Hereford and John Purvey, produced the first English version of the Bible (see "English Versions of the Bible," pp. 400–406). This translation, having been rendered from the Latin Vulgate, included all of the disputed books, with the exception of 2 Esdras. In the Prologue to the Old Testament, however, a distinction is made between the books of the Hebrew canon, which are thereupon enumerated, and the others which, the writer says, "shal be set among apocrifa, that is, with outen autorite of bileue." In the case of the books of Esther and Daniel, the translators included a rendering of Jerome's notes calling the reader's attention to the additions.

In the controversies that arose at the time of the Reformation, Protestant leaders soon recognized the need to distinguish between books that were authoritative for the establishment of doctrine and those that were not. Thus, disputes over the doctrines of Purgatory and of the efficacy of prayers and Masses for the dead inevitably involved discussion concerning the authority of 2 Maccabees, which contains what was held to be scriptural warrant for them (12.43–45).

The first extensive discussion of the canon from the Protestant point of view was a treatise in Latin, *De Canonicis Scripturis Libellus,* published at Wittenberg in 1520 by Andreas Bodenstein, who is commonly known as Carlstadt, the name of his birthplace. Besides distinguishing the canonical books of the Hebrew Old Testament from the books of the Apocrypha, Carlstadt classified the latter into two divisions. Of one group, containing Wisdom, Ecclesiasticus, Judith, Tobit, and 1 and 2 Maccabees, he says, "These are Apocrypha, that is, are outside the Hebrew canon; yet they are holy writings" (sect. 114). In explaining his view of the status and worth of such books as Tobit, Wisdom, and Ecclesiasticus, he writes:

> What they contain is not to be despised at once; still it is not right that Christians should relieve, much less slake, their thirst with them. . . . Before all things the best books must be read, that is, those that are canonical beyond all controversy; afterwards, if one has the time, it is allowed to peruse the controverted books, provided that you have the set purpose of comparing and collating the non-canonical books with those which are truly canonical (sect. 118).

The second group of apocryphal books, namely 1 and 2 Esdras, Baruch, Prayer of Manasseh, and the Additions to Daniel, Carlstadt declared to be filled with ridiculous puerilities worthy of the censor's ban, and therefore to be contemptuously discarded.

The first Bible in a modern vernacular to segregate the apocryphal books from the others was the Dutch Bible published by Jacob van Liesveldt in 1526 at Antwerp. After Malachi there follows a section embodying the Apocrypha, which is entitled, "The books which are not in the canon, that is to say, which one does not find among the Jews in the Hebrew."

The first edition of the Swiss-German Bible, prepared by ministers of the Church in Zurich, was published in six volumes (Zurich, 1527–29), the fifth of which contains the Apocrypha. The title page of this volume states, "These are the books which are not reckoned as biblical by the ancients, nor are found among the Hebrews." A one-volume edition of the Zurich Bible, which appeared in 1530, contains the apocryphal books grouped together after the New Testament. In commenting on the attitude of Protestants respecting the disputed books, Œcolampadius, perhaps on the whole the best representative of the Swiss Reformers, declared in a formal statement issued in 1530: "We do not despise Judith, Tobit, Ecclesiasticus, Baruch, the last two books of Esdras, the three books of Maccabees, the Additions to Daniel; but we do not allow them divine authority with the others."

In reaction to Protestant criticism of the disputed books, on April 8, 1546, the Council of Trent gave what is regarded by Roman Catholics as the first infallible and effectually promulgated declaration on the canon of the Holy Scriptures. After enumerating the books, which in the Old Testament include Tobit, Judith, Wisdom, Ecclesiasticus, Baruch, and the two books of Maccabees, the decree pronounces an anathema upon anyone who "does not accept as sacred and canonical the aforesaid books in their entirety and with all their parts, as they have been accustomed to be read in the Catholic Church and as they are contained in the old Latin Vulgate Edition" (trans. by Father H. J. Schroeder). The reference to "books in their entirety and with all their parts" is intended to cover the Letter of Jeremiah as chapter 6 of Baruch, the Additions to Esther, and the chapters in Daniel concerning the Song of the Three Jews, Susanna, and Bel and the Dragon. It is noteworthy, however, that the Prayer of Manasseh and 1 and 2 Esdras, though included in some manuscripts of the Latin Vulgate, were denied canonical status by the Council. In the official edition of the Vulgate, published in 1592, these three are printed as an appendix after the New Testament, "lest they should perish altogether."

In England, though Protestants were unanimous in declaring that the apocryphal books were not to be used to establish any doctrine, differences arose as to the proper use and place of non-canonical books. The milder view prevailed in the Church of England, and the lectionary attached to the Book of Common Prayer, from 1549 onward, has always contained prescribed lessons from the Apocrypha. In reply to those who urged the discontinuance of reading lessons from apocryphal books, as being inconsistent with the sufficiency of Scripture, the bishops at the Savoy Conference, held in 1661, replied that the same objection could be raised against the preaching of sermons, and that it was much to be desired that all sermons should give as useful instruction as did the chapters selected from the Apocrypha.

A more strict point of view was taken by the Puritans, who felt uneasy that there should be any books included within the covers of the Bible besides those that they regarded as authoritative. In time this aversion to associating merely human books with those acknowledged as the only sacred and canonical ones found a natural expression in the publication of editions of the Bible from which the section devoted to the Apocrypha was omitted. The earliest copies of the English Bible that excluded the Apocrypha are certain Geneva Bibles printed in 1599 mainly in the Low Countries. The omission of the sheets containing the Apocrypha was presumably due to those responsible for binding the copies, for the titles of the apocryphal books occur in the table of contents at the beginning of the edition.

It would seem that the practice of issuing copies of the Bible without the Apocrypha continued, for in 1615 George Abbot, Archbishop of Canterbury, who had been one of the translators of the King James Version of 1611, directed public notices to be given that no Bibles were to be bound up and sold without the Apocrypha on pain of a whole year's imprisonment. Despite the severe penalty, however, not a few printings of the King James Version appeared in London and Cambridge without the Apocrypha; copies lacking the disputed books are dated 1616, 1618, 1620, 1622, 1626, 1627, 1629, 1630, and 1633. Like the copies of the Geneva Bible of 1599, these seem to have been the work of publishers who wished to satisfy a growing demand for less bulky and less expensive editions of the Bible.

During subsequent centuries the editions of Bibles that lacked the books of the Apocrypha came to outnumber by far those that included them, and soon it became difficult to obtain ordinary editions of the King James Version containing the Apocrypha.

THE PERVASIVE INFLUENCE OF THESE BOOKS

Most readers will probably be surprised to learn how pervasive the influence of the Apocrypha has been over the centuries. Not only have these books inspired homilies, meditations, and liturgical forms, but poets, dramatists, composers, and artists have drawn freely upon them for subject matter. Common proverbs and familiar names are derived from their pages. Even the discovery of the New World was due in part to the influence of a passage in 2 Esdras upon Christopher Columbus. In what follows the reader will find a representative selection of such examples, most of them chosen from *An Introduction to the Apocrypha* by B. M. Metzger (Oxford University Press), and arranged under the headings of (*a*) English Literature, (*b*) Music, (*c*) Art, and (*d*) Miscellaneous.

(*a*) English Literature. Sometime during the ninth or the tenth century an unknown poet, using the West-Saxon dialect, turned the story of Judith into an Old English epic of twelve cantos, transforming at the same time the heroine into a Christian. It is thought that the poem was written to celebrate the prowess of Æthelflæd, "The Lady of the Mercians," who, like the indomitable Judith, delivered her people from the fury of invaders, the heathen Northmen.

During the fourteenth and fifteenth centuries a poem called "The Pistill [i.e. Epistle] of Swete Susan" circulated in Scotland. Written in stanzas of thirteen lines and characterized by an unusual combination of alliteration and rhyme, the ancient apocryphal story was adorned with many imaginative details by the author, thought to have been a certain Huchown (Hugh) of Ayrshire in western Scotland.

How conversant Shakespeare was with the contents of the Bible is a question that, like many another concerning the bard of Avon, has been keenly debated. In any case, it is a fact that two of the poet's daughters bore the names of two of the chief heroines of the Apocrypha—Susanna and Judith—and, what is of greater significance, allusions to about eighty passages from eleven books of the Apocrypha have been identified in his plays.

Noteworthy among American writers who have drawn upon the Apocrypha for themes as well as subject matter is Henry Wadsworth Longfellow. His *New England Tragedies* contains references to 1 and 2 Maccabees, and the chief episodes of the courageous Maccabean uprising are included in his poetic dramatization, *Judas Maccabaeus*.

(*b*) Music. More than one hymn writer has drawn inspiration, as well as, in some cases, the words themselves, from the Apocrypha. For example, the exalted hymn of thanksgiving, "Nun danket alle Gott," written by Pastor Martin Rinkart about 1636 when the devastating Thirty Years War was nearing its end, is dependent upon Luther's translation of Sir 50.22–24. Two stanzas of the hymn, as translated by Catherine Winkworth, will show the amount of borrowing (here printed in italics):

Now thank we all our God	O may this bounteous God
With heart and hands and voices,	*Through all our life be near us,*
Who wondrous things hath done,	With ever joyful hearts
In whom His world rejoices;	And blessed peace to cheer us;
Who, from our mother's arms,	*And keep us in His grace,*
Hath blessed us on our way	And guide us when perplexed,
With countless gifts of love,	*And free us from all ills*
And still is ours today.	In this world and the next.

Strange though it may seem, ideas included in the Christmas hymn, "It Came upon the Midnight Clear," are traceable to the Old Testament Apocrypha. In the New Testament accounts of the Nativity, nothing is said of the exact time of Jesus' birth. The subsequent identification of the hour of his birth as midnight is doubtless due to the influence of a remarkable passage in the Wisdom of Solomon. At an early century in the Christian era the imagination of more than one Church Father was caught by pseudo-Solomon's vivid reference to the time when

God's "all-powerful word [the Logos] leaped from heaven, from the royal throne," namely when "night in its swift course was now half gone" (Wis 18.14–15). Despite the context of the passage, which speaks of the destruction of the first-born Egyptians at the time of the Exodus, the words were interpreted as referring to the Incarnation of the eternal Word of God, Jesus Christ. Thus by a curious, not to say ironical, twist of fortune, a passage that tells of a stern warrior with a sharp sword filling a doomed land with death has had a share in fixing popular traditions concerning the time and circumstances of the birth of the Prince of Peace.

The influence of the Apocrypha can also be traced in many an anthem, cantata, oratorio, and opera. Handel's oratorios *Susanna* and *Judas Maccabaeus,* as well as his *Alexander Balus,* an historical sequel to the latter, will occur at once to music lovers. At an early date in operatic history the stirring story of Judith was found to lend itself admirably to dramatic presentation. Italian and German operas on this theme were written by Andrea Salvadori, Marco da Gagliano, Martin Opitz, and Joachim Beccau. In the nineteenth century the noted Russian pianist and composer, Anton Rubenstein, published *The Maccabees,* an opera of monumental proportions, the libretto of which was written by one of his collaborators, Dr. H. S. von Mosenthal.

(*c*) Art. During the Renaissance and later, many painters chose subjects from the deuterocanonical books. Almost every large gallery in Europe and America has one or more works of the old masters depicting Judith, Tobit, or Susanna, who were the three most popular subjects from these books.

Besides paintings, down through the ages artists in almost every other medium have chosen themes from the Apocrypha. Were space available here for an inventory, examples could be cited from such divergent types of *objets d'art* as mosaics, frescoes, gems, ivories, sarcophagi, enameled plaques, terra cottas, stained glass, manuscript illumination, sculpture, and tapestries.

(*d*) Miscellaneous. The influence of the Apocrypha in everyday life can be observed in the currency of such names as Edna, Susanna (or one of its many derivatives, such as Susan, Suzanne, and Sue), Judith (or Judy), Raphael, and Tobias (or Toby).

The word "macabre," according to the opinion of several lexicographers, may be derived ultimately from "Maccabee," alluding to the grisly and gruesome tortures inflicted upon the Jewish martyrs.

Some of the most common expressions and proverbs have come from the Apocrypha. The sententious sayings, "A good name endures forever" and "You can't touch pitch without being defiled," are derived from Sir 41.13 and 13.1. The noble affirmation in 1 Esd 4.41, "Great is Truth, and mighty above all things" (King James Version), or its Latin form, *Magna est veritas et praevalet,* has been used frequently as a motto or maxim in a wide variety of contexts.

A passage from the Apocrypha encouraged Christopher Columbus in the enterprise that resulted in his discovery of the New World. To be sure, the verse in 2 Esdras is an erroneous comment upon the Genesis narrative of creation, and Columbus was mistaken in attributing its authority to the "prophet Ezra" of the Old Testament, but—for all that—it played a significant part in pushing back the earth's horizons, both figuratively and literally. The words of 2 Esd 6.42 concerning God's work of creation ("On the third day you commanded the waters to be gathered together in a seventh part of the earth; six parts you dried up and kept so that some of them might be planted and cultivated and be of service before you") led Columbus to reason that, if only one-seventh of the earth's surface is covered with water, the ocean between the west coast of Europe and the east coast of Asia could be no great width and might be navigated in a few days with a fair wind. It was partly by quoting this verse from what was regarded as an authoritative book that Columbus managed to persuade Ferdinand and Isabella of Spain to provide the necessary financial support for his voyage.

OTHER APOCRYPHAL AND PSEUDEPIGRAPHICAL LITERATURE

Besides the fifteen books or parts of books that are traditionally called the Apocrypha, there are many other Jewish or Jewish-Christian works, dating from the centuries immediately before and after the beginning of the Christian era, which for a time were popular among certain groups of Jews and in early Eastern Churches. It is customary (though not entirely appropriate) to classify these writings as Palestinian pseudepigrapha (those composed in Hebrew or Aramaic) and Alexandrian pseudepigrapha (those composed in Greek). Of the scores of such documents that are known to have circulated more or less widely, the following have been chosen as representative examples. (For a definition of pseudepigrapha, see p. iv AP.)

(*a*) Palestinian pseudepigrapha. The Book of Jubilees is a legendary expansion of Gen 1.1–Ex 12.47, written in Hebrew not long before 100 B.C. by an unknown author of nationalist and rigoristic outlook, who deplored contemporary laxity. It attempts to show that the Mosaic law, with its prescriptions about festivals, the Sabbath offerings, abstinence from blood and from fornication (which for the writer includes intermarriage with Gentiles), was promulgated in patriarchal times, and indeed existed eternally with God in heaven. Events recorded in Genesis are dated exactly (but fictitiously) according to the jubilee (every forty-nine years) and its subdivisions. The book has been transmitted in its entirety in an Ethiopic translation, and portions of the text survive in Greek, Latin, and Syriac versions. At about the middle of the present century five fragmentary manuscripts of Jubilees, written in a good style of Hebrew, were discovered at Qumran by the Dead Sea. These manuscripts, which preserve portions of fifteen of the fifty chapters of the book, show that the Latin and Ethiopic versions are faithful translations of the original.

In addition to the 150 psalms comprising the Book of Psalms in the Hebrew Bible, during the intertestamental period other psalms were composed in Hebrew and in other languages. One of these, which celebrates the prowess of young David in slaying Goliath, is appended (as Ps 151) to the Psalter in Greek manuscripts. Part of this psalm (see pp. 283–284 AP) came to be incorporated in the Ethiopian coronation ritual.

The Psalms of Solomon is a collection of eighteen songs of generally exalted sentiments, composed in Hebrew during the last century B.C. They are extant today in Greek and Syriac. The author, who is usually thought to reflect Pharisaic polemic against Sadducean dominance in the religious ceremonial of his day, looked forward to the time when the Messiah would reign as king at Jerusalem. According to an extended description of the coming Messiah (chs 17–18), he is to be sinless, strong through the spirit of holiness, gaining his wisdom from God, shepherding the flock of the Lord with fidelity and righteousness, and conquering the entire heathen world without warfare, "by the word of his mouth."

The book of Enoch, also called 1 Enoch or Ethiopic Enoch, is a heterogeneous collection of apocalypses and other material written by several authors in Aramaic (or Hebrew) during the last two centuries B.C. It embodies a series of revelations, of which Enoch is the professed recipient, on such matters as the origin of evil, the angels and their destinies, the nature of Gehenna and Paradise, and the pre-existent Messiah. Interspersed throughout the lengthy and rambling work are sections that have been called "the book of celestial physics." These sections, which are one of the curiosities of ancient pseudo-scientific literature, set forth contemporary speculations concerning such meteorological and astronomical phenomena as lightning, hail, snow, the twelve winds, the heavenly luminaries, and the like. The entire work is preserved in an Ethiopic translation, which includes what have been thought to be Christian interpolations in chs 37–71, where the Messiah is called the Son of Man. Portions of the book are extant in Greek and Latin; recently eight manuscripts of part of the work (but not chs 37–71) have turned up in Aramaic at Qumran. It is of interest that a quotation from the book of Enoch (1.9) occurs in the New Testament letter of Jude (vv. 14–15).

(b) Alexandrian pseudepigrapha. Third Maccabees is a religious novel written in Greek by an Alexandrian Jew sometime between 100 B.C. and A.D. 70. The title is a misnomer, for the book has nothing to do with the Maccabees. With many legendary embellishments the author recounts three stories of conflict between Ptolemy IV (221–203 B.C.) and the Jews of Egypt. The most dramatic section describes how the Jews were herded into the hippodrome near Alexandria, to be trampled under the feet of intoxicated elephants. After the king's purpose had been several times providentially delayed, it was finally foiled by a vision of angels which turned the elephants upon the persecutors.

Fourth Maccabees is a Greek philosophical treatise addressed to Jews on the supremacy of devout reason over the passions of body and soul. In the form of a Stoic diatribe, or popular address, the author begins with a philosophical exposition of his theme, which he then illustrates with examples drawn from 2 Maccabees. He describes at length the gruesome tortures that tested the fortitude of Eleazar, the seven brothers, and their mother, all of whom preferred death to committing apostasy. The book was probably written by a Hellenistic Jew of Alexandria at some time later than 2 Maccabees and before A.D. 70. In early Christianity the Maccabean martyrs were eventually canonized and accorded a yearly festival in the ecclesiastical calendar (August 1).

From what has been said above the reader will be able to form some opinion of the importance of apocryphal and pseudepigraphical literature, both for its own sake as well as for the information it supplies concerning the development of Jewish life and thought just prior to the beginning of the Christian era. The stirring political fortunes of the Jews in the time of the Maccabees; the rise of what has been called normative Judaism, and the emergence of the sects of the Pharisees and the Sadducees; the lush growth of popular belief in the activities of angels and demons, and the use of apotropaic magic to avert the malevolent influence of the latter; the growing preoccupation concerning original sin and its relation to the "evil inclination" present in every person; the blossoming of apocalyptic hopes relating to the coming Messiah, the resurrection of the body, and the vindication of the righteous—all these and many other topics receive welcome light from the apocryphal/deuterocanonical books.

Titles Given to Books
Associated with Ezra and Nehemiah
in Selected Versions

Version \\ Document	Old Testament book of Ezra	Old Testament book of Nehemiah	Paraphrase of 2 Chronicles chs 35—36; the whole book of Ezra; Nehemiah 7.38—8.12; plus a tale about Darius' bodyguards	A Latin Apocalypse
Greek Bible (Septuagint)	2 Esdras		1 Esdras	
Latin Vulgate Bible	1 Esdras	2 Esdras	3 Esdras	4 Esdras
Many later Latin Manuscripts	1 Esdras		3 Esdras	2 Esdras = chs 1—2 4 Esdras = chs 3—14 5 Esdras = chs 15—16
Douay English Version (1609–1610)	1 Esdras	2 Esdras	3 Esdras	4 Esdras
Russian Bible, Moscow Patriarchate (1956)	1 Esdras	Nehemiah	2 Esdras	3 Esdras
King James and New Revised Standard Versions	Ezra	Nehemiah	1 Esdras	2 Esdras

Chronological Tables
of Rulers

THE SELEUCID DYNASTY

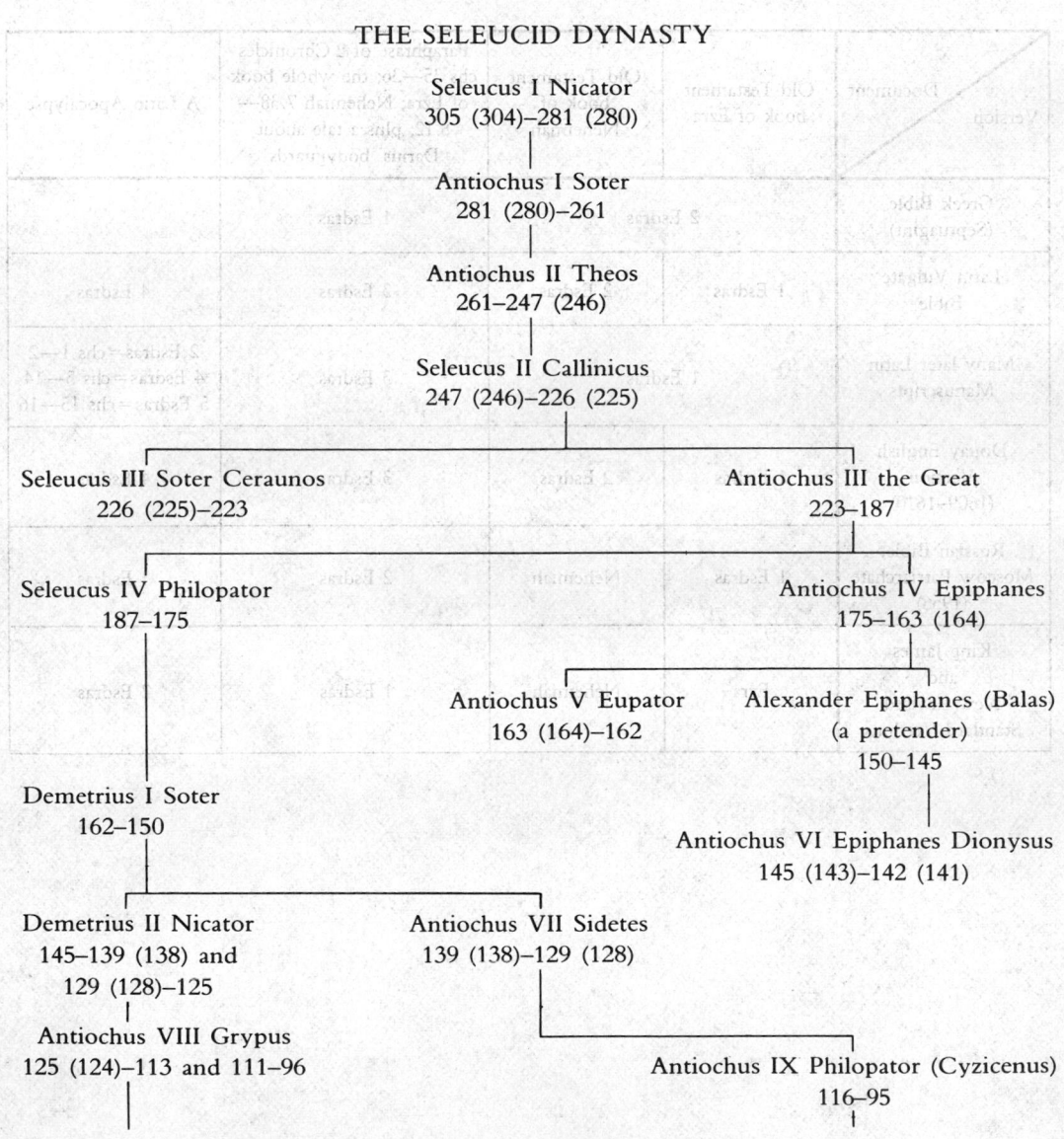

Seleucus I Nicator
305 (304)–281 (280)

Antiochus I Soter
281 (280)–261

Antiochus II Theos
261–247 (246)

Seleucus II Callinicus
247 (246)–226 (225)

Seleucus III Soter Ceraunos
226 (225)–223

Antiochus III the Great
223–187

Seleucus IV Philopator
187–175

Antiochus IV Epiphanes
175–163 (164)

Antiochus V Eupator
163 (164)–162

Alexander Epiphanes (Balas)
(a pretender)
150–145

Demetrius I Soter
162–150

Antiochus VI Epiphanes Dionysus
145 (143)–142 (141)

Demetrius II Nicator
145–139 (138) and
129 (128)–125

Antiochus VII Sidetes
139 (138)–129 (128)

Antiochus VIII Grypus
125 (124)–113 and 111–96

Antiochus IX Philopator (Cyzicenus)
116–95

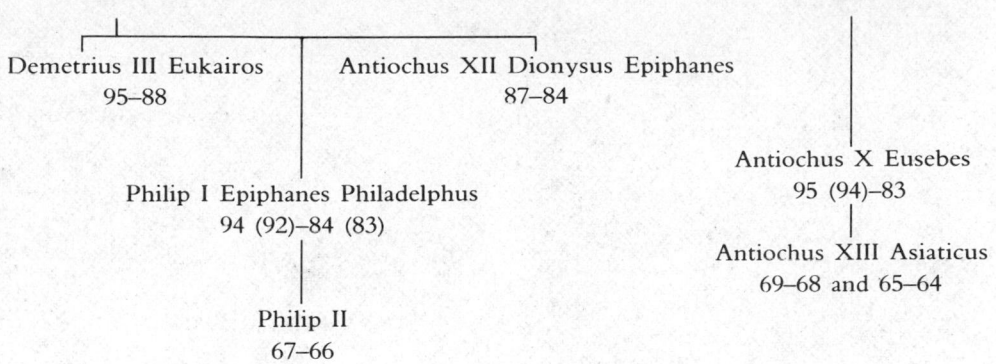

Demetrius III Eukairos   Antiochus XII Dionysus Epiphanes
95–88   87–84

                                                              Antiochus X Eusebes
                                                                  95 (94)–83

Philip I Epiphanes Philadelphus
94 (92)–84 (83)   Antiochus XIII Asiaticus
                                                        69–68 and 65–64

Philip II
67–66

THE HOUSE OF THE MACCABEES (HASMONEANS)

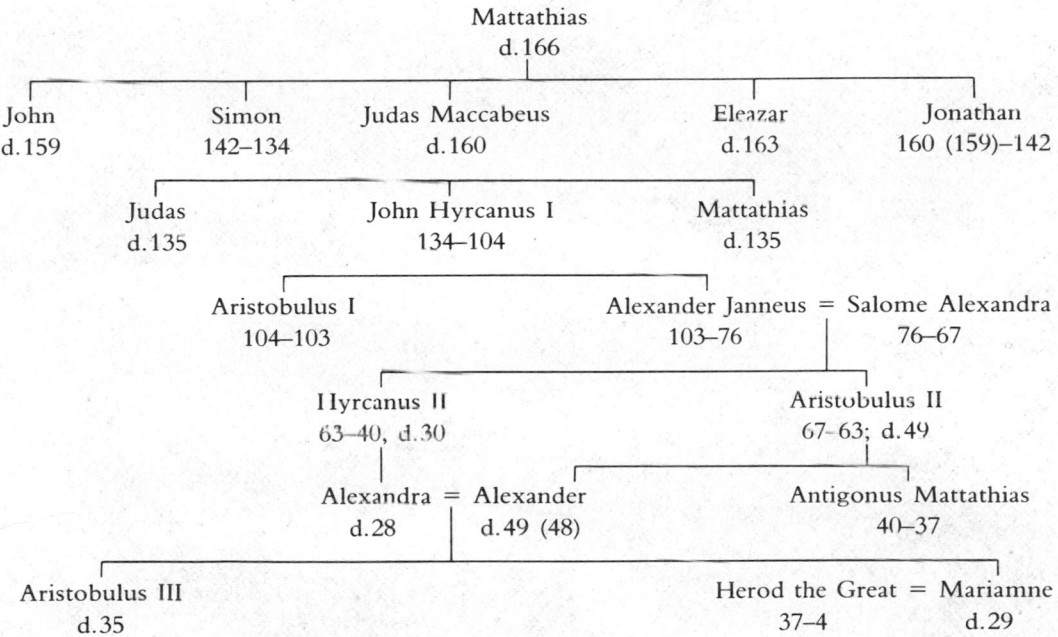

Mattathias
d. 166

John   Simon   Judas Maccabeus   Eleazar   Jonathan
d. 159   142–134   d. 160   d. 163   160 (159)–142

Judas   John Hyrcanus I   Mattathias
d. 135   134–104   d. 135

Aristobulus I   Alexander Janneus = Salome Alexandra
104–103   103–76   76–67

Hyrcanus II   Aristobulus II
63–40, d. 30   67–63; d. 49

Alexandra = Alexander   Antigonus Mattathias
d. 28   d. 49 (48)   40–37

Aristobulus III   Herod the Great = Mariamne
d. 35   37–4   d. 29

These two family trees include the names of the more important rulers. The presence of short parallel lines between two names indicates marriage. The date of an individual's death or the dates of his rule are in some cases disputed, and alternative possibilities are given within parentheses. All the dates are B.C. Besides the standard works on the chronology of the inter-testamental period, special mention should be made of F. M. Heichelheim's "Chronological Table from 323–30 B.C." in *Proceedings of the IX International Congress of Papyrology*, Oslo, 1958, pp. 163–182 (also published separately by the Norwegian Universities Press). Seleucus I became king in the seventh year of the Seleucid era; see A. J. Sachs and D. J. Wiseman, "A Babylonian King List of the Hellenistic Period," *Iraq*, xvi (1954), p. 205. Syria became a Roman province in 64 B.C., and Pompey conquered Jerusalem in 63 B.C.

Tobit

The book of Tobit is named after a generous and God-fearing Jew whose blindness and poverty in Nineveh are the direct result of his performing one of his most characteristic good deeds, namely, burying an executed fellow-Jew. Thanks to the courageous efforts of his devoted son Tobias, who is assisted by the angel Raphael disguised as Azariah, Tobit not only recovers his sight and fortune but also gains a pious daughter-in-law, Sarah. From her, Tobias exorcises Asmodeus, the demon who had claimed the lives of her seven previous husbands on their wedding night. On his deathbed, Tobit has Tobias promise to move the family from Nineveh to Ecbatana, where Tobias lives to a rich old age.

Despite all its trappings as an historical account, Tobit is best understood as a romance of Diaspora Judaism, relating the outcome of a successful quest. The tale is intended to entertain as well as to inspire faith in God and human effort. Without Tobias' own devotion and courage, neither Tobit nor Sarah would have been delivered, the help of Raphael notwithstanding.

The author created his narrative out of three well-known secular folktales: (1) the tale of the Grateful Dead (the story about a man who is impoverished but ultimately rewarded for burying an abused corpse); (2) the tale of the Monster in the Bridal Chamber (the story of a demon who kills the bride's husbands on their wedding night); and (3) the tale of Ahiqar (the account of a wise courtier who, though falsely incriminated by his adopted son, is vindicated).

The distinctive structure, images, and spirit of the book come, however, from the Hebrew Bible. Although only Amos and Nahum are actually mentioned, other biblical books are far more influential for the story, notably Genesis (with its stories of the betrothals of Isaac and Jacob), Job, and Isaiah (especially with regard to the Exile and the Return). Deuteronomy's doctrine of retribution (i.e. ultimately the righteous are rewarded and the wicked punished) provides the basic theology for Tobit.

Even though there is a strong cultic concern in chs 1–3 and a vision of a new and greater Jerusalem and temple in chs 13–14, the book's primary emphasis is on the everyday practical, moral, and sapiental aspects of being and doing good. The entire book—its plot, characters, and message—is ironic, since the reader's knowledge of the situation of the characters is greater than that of the characters themselves. Tobit's name ("God is my good") and those of the other major human characters are ironic, especially at the height of their personal anguish. The book's irony is purchased at the cost of a loss of suspense, the reader assured by ch 3 that Tobit and Sarah will be delivered; and by ch 6, how.

The author effectively uses a wide variety of literary techniques to express his religious ideas: monologue and dialogue, prayers and hymns, demonology and angelology, wisdom sayings, and death-bed testimony. The personalities of the main characters are clearly and individually developed. Despite their individual shortcomings, they are good but quite ordinary people, with none of the heroic stature of such other exilic characters as Daniel, Esther, or Judith.

The author of Tobit was a Jew, writing originally in Hebrew or Aramaic (copies of it in those languages have been found at Qumran), probably somewhere between 225–175 B.C., and, possibly, in Palestine.

Tobit is represented by three major Greek recensions and two Latin translations. Unlike the RSV, the NRSV of Tobit is based upon the Sinaiticus family as supplemented by the Old Latin. There are also some late Hebrew translations, which are based upon a Greek text, as are the older Syriac, Ethiopic, and Sahidic versions.

1 This book tells the story of Tobit son of Tobiel son of Hananiel son of Aduel son of Gabael son of Raphael*ᵃ* of the descendants of Asiel, of the tribe of Naphtali, ²who in the days of King Shalmaneser*ᵇ* of the Assyrians was taken into captivity from Thisbe, which is to the south of Kedesh Naphtali in Upper Galilee, above Asher toward the west, and north of Phogor.

3 I, Tobit, walked in the ways of truth and righteousness all the days of my life. I performed many acts of charity for my kindred and my people who had gone with me in exile to Nineveh in the land of the Assyrians. ⁴When I was in my own country, in the land of Israel, while I was still a young man, the whole tribe of my ancestor Naphtali deserted the house of David and Jerusalem. This city had been chosen from among all the tribes of Israel, where all the tribes of Israel should offer sacrifice and where the temple, the dwelling of God, had been consecrated and established for all generations forever.

5 All my kindred and our ancestral house of Naphtali sacrificed to the calf*ᶜ* that King Jeroboam of Israel had erected in Dan and on all the mountains of Gali-lee. ⁶But I alone went often to Jerusalem for the festivals, as it is prescribed for all Israel by an everlasting decree. I would hurry off to Jerusalem with the first fruits of the crops and the firstlings of the flock, the tithes of the cattle, and the first shearings of the sheep. ⁷I would give these to the priests, the sons of Aaron, at the altar; likewise the tenth of the grain, wine, olive oil, pomegranates, figs, and the rest of the fruits to the sons of Levi who ministered at Jerusalem. Also for six years I would save up a second tenth in money and go and distribute it in Jerusalem. ⁸A third tenth*ᵈ* I would give to the orphans and widows and to the converts who had attached themselves to Israel. I would bring it and give it to them in the third year, and we would eat it according to the ordinance decreed concerning it in the law of Moses and according to the instructions of Deborah, the mother of my father Tobiel,*ᵉ* for my father had died and left me an orphan. ⁹When I be-

a Other ancient authorities lack *son of Raphael son of Raguel*   *b* Gk *Enemessaros*   *c* Other ancient authorities read *heifer*   *d* A third tenth added from other ancient authorities   *e* Lat: Gk *Hananiel*

1.1–2: Title. 2: *Shalmaneser* (or rather Sargon; see v. 15 n.) took Samaria, the capital of Israel, in 722 B.C. and transported a large part of the population to Assyria (2 Kings 17.1–6). *Thisbe* is unidentified. *Kedesh Naphtali,* 2 Kings 15.29. *Asher* is probably Hazor; *Phogor* is unknown. **1.3–3.6: Tobit's own account of his virtuous life and unhappy fate. 1.3–22: Tobit's piety brings him into conflict with the king. 3:** *Nineveh* was the capital of Assyria. **4:** Since the rebellion of the northern tribes against Jerusalem (1 Kings 12.19–20) occurred about 922 B.C., Tobit could not have been *still a young man,* or even born, when it happened. Such chronological, topographical, and other historical difficulties make it clear that the story is fiction (compare v. 15 n., 6.2 n., 9.2 n.). **5:** *Calf,* 1 Kings 12.28–29. **6–8:** During the apostasy, Tobit alone remains loyal to the divinely-appointed temple in Jerusalem. **6:** *An everlasting decree,* Deut 12.11, 13–14. **9:** The name *Tobias* means "God is my good." **1.10–12:** Even in captivity among Gentiles Tobit refuses to violate the dietary laws. **14:** *Media* is the northern part of modern Iran, east of Assyria. *Ten talents,* while impossible

came a man I married a woman,*f* a member of our own family, and by her I became the father of a son whom I named Tobias.

10 After I was carried away captive to Assyria and came as a captive to Nineveh, everyone of my kindred and my people ate the food of the Gentiles, 11 but I kept myself from eating the food of the Gentiles. 12 Because I was mindful of God with all my heart, 13 the Most High gave me favor and good standing with Shalmaneser,*g* and I used to buy everything he needed. 14 Until his death I used to go into Media, and buy for him there. While in the country of Media I left bags of silver worth ten talents in trust with Gabael, the brother of Gabri. 15 But when Shalmaneser*g* died, and his son Sennacherib reigned in his place, the highways into Media became unsafe and I could no longer go there.

16 In the days of Shalmaneser*g* I performed many acts of charity to my kindred, those of my tribe. 17 I would give my food to the hungry and my clothing to the naked; and if I saw the dead body of any of my people thrown out behind the wall of Nineveh, I would bury it. 18 I also buried any whom King Sennacherib put to death when he came fleeing from Judea in those days of judgment that the king of heaven executed upon him because of his blasphemies. For in his anger he put to death many Israelites; but I would secretly remove the bodies and bury them. So when Sennacherib looked for them he could not find them. 19 Then one of the Ninevites went and informed the king about me, that I was burying them; so I hid myself. But when I realized that the king knew about me and that I was being searched for to be put to death, I was afraid and ran away. 20 Then all my property was confiscated; nothing was left to me that was not taken into the royal treasury except my wife Anna and my son Tobias.

21 But not forty*h* days passed before two of Sennacherib's*i* sons killed him, and they fled to the mountains of Ararat, and his son Esar-haddon*j* reigned after him. He appointed Ahikar, the son of my brother Hanael*k* over all the accounts of his kingdom, and he had authority over the entire administration. 22 Ahikar interceded for me, and I returned to Nineveh. Now Ahikar was chief cupbearer, keeper of the signet, and in charge of administrations of the accounts under King Sennacherib of Assyria; so Esar-haddon*j* reappointed him. He was my nephew and so a close relative.

2 Then during the reign of Esar-haddon*j* I returned home, and my wife Anna and my son Tobias were restored to me. At our festival of Pentecost, which is the sacred festival of weeks, a good dinner was prepared for me and I reclined to eat. 2 When the table was set for me and an abundance of food placed before me, I said to my son Tobias, "Go, my child, and bring whatever poor person you may find of our people among the exiles in Nineveh, who is wholeheartedly mindful of God,*l* and he

---

f Other ancient authorities add *Anna*
g Gk *Enemessaros*   *h* Other ancient authorities read either *forty-five* or *fifty*   *i* Gk *his*
j Gk *Sacherdonos*   *k* Other authorities read *Hananael*   *l* Lat: Gk *wholeheartedly mindful*

---

to equate to present-day purchasing power, it was a substantial sum of money. **1.15–20:** Tobit arouses Sennacherib's wrath and flees the country. **15:** *Shalmaneser* actually died before the fall of Samaria, which was taken by Sargon II. *Sennacherib* succeeded his father Sargon II in 705 B.C. **17:** It was for the Jews a great calamity that a dead body should lie unburied. **21–22:** Under a new king, *Esar-haddon* (681–669 B.C.), Tobit is able to return. **21:** *Ahikar* is patterned after Ahiqar, a legendary ancient wise man whose story survives in several Near-Eastern or Semitic languages. An Aramaic version of his adventures, dating from the fifth century B.C., was found among the Jewish papyri at Elephantine in upper Egypt (see also 14.10 n.).

2.1–14: Another act of charity results in Tobit's blindness and impoverishment. 1: The name *Anna* means "grace." *Pentecost* is approximately *seven weeks* after passover (Lev 23.15–21; Deut 16.9–11). **2:** Generosity toward the poor is one of the virtues taught by this book (4.7–11, 16). **3:** *Strangled,* presum-

shall eat together with me. I will wait for
you, until you come back." 3 So Tobias
went to look for some poor person of our
people. When he had returned he said,
"Father!" And I replied, "Here I am, my
child." Then he went on to say, "Look,
father, one of our own people has been
murdered and thrown into the market
place, and now he lies there strangled."
4 Then I sprang up, left the dinner before
even tasting it, and removed the body[m]
from the square[n] and laid it[m] in one of
the rooms until sunset when I might
bury it.[m] 5 When I returned, I washed
myself and ate my food in sorrow.
6 Then I remembered the prophecy of
Amos, how he said against Bethel,[o]

"Your festivals shall be turned
into mourning,
and all your songs into
lamentation."

And I wept.

7 When the sun had set, I went and
dug a grave and buried him. 8 And my
neighbors laughed and said, "Is he still
not afraid? He has already been hunted
down to be put to death for doing this,
and he ran away; yet here he is again
burying the dead!" 9 That same night I
washed myself and went into my courtyard and slept by the wall of the courtyard; and my face was uncovered because of the heat. 10 I did not know that
there were sparrows on the wall; their
fresh droppings fell into my eyes and
produced white films. I went to physicians to be healed, but the more they
treated me with ointments the more my
vision was obscured by the white films,
until I became completely blind. For four
years I remained unable to see. All my
kindred were sorry for me, and Ahikar
took care of me for two years before he
went to Elymais.

11 At that time, also, my wife Anna
earned money at women's work. 12 She
used to send what she made to the owners and they would pay wages to her.
One day, the seventh of Dystrus, when
she cut off a piece she had woven and sent
it to the owners, they paid her full wages
and also gave her a young goat for a
meal. 13 When she returned to me, the
goat began to bleat. So I called her and
said, "Where did you get this goat? It is
surely not stolen, is it? Return it to the
owners; for we have no right to eat anything stolen." 14 But she said to me, "It
was given to me as a gift in addition to
my wages." But I did not believe her,
and told her to return it to the owners. I
became flushed with anger against her
over this. Then she replied to me,
"Where are your acts of charity? Where
are your righteous deeds? These things
are known about you!"[p]

3 Then with much grief and anguish
of heart I wept, and with groaning
began to pray:

2 "You are righteous, O Lord,
and all your deeds are just;
all your ways are mercy and truth;
you judge the world.[q]
3 And now, O Lord, remember me
and look favorably upon me.
Do not punish me for my sins
and for my unwitting offenses
and those that my ancestors
committed before you.
They sinned against you,
4 and disobeyed your
commandments.
So you gave us over to plunder,
exile, and death,
to become the talk, the byword,
and an object of reproach
among all the nations among

m Gk *him* *n* Other ancient authorities lack *from the square* *o* Other ancient authorities read *against Bethlehem* *p* Or *to you*; Gk *with you* *q* Other ancient authorities read *you render true and righteous judgment forever*

---

ably executed (compare 1.18). Leaving the body unburied was intended as additional punishment, so Tobit's act of charity was an act of defiance toward the king. **5:** *Washed myself,* ceremonially, after touching a corpse (Num 19.11–13).

2.6: *Prophecy of Amos,* see Am 8.10. **10:** *Elymais,* a city, or possibly a region, in Persia (1 Macc 6.1). **12:** *Dystrus,* the Greek name for the Semitic month of Adar (February–March).

3.1–6: Tobit's prayer.

whom you have dispersed
us.
5 And now your many judgments
are true
in exacting penalty from me for
my sins.
For we have not kept your
commandments
and have not walked in
accordance with truth
before you.
6 So now deal with me as you will;
command my spirit to be taken
from me,
so that I may be released from
the face of the earth and
become dust.
For it is better for me to die than
to live,
because I have had to listen to
undeserved insults,
and great is the sorrow within
me.
Command, O Lord, that I be
released from this distress;
release me to go to the eternal
home,
and do not, O Lord, turn your
face away from me.
For it is better for me to die
than to see so much distress in
my life
and to listen to insults."

7 On the same day, at Ecbatana in
Media, it also happened that Sarah, the
daughter of Raguel, was reproached by
one of her father's maids. [8] For she had
been married to seven husbands, and the
wicked demon Asmodeus had killed
each of them before they had been with
her as is customary for wives. So the
maid said to her, "You are the one who
kills[r] your husbands! See, you have already been married to seven husbands
and have not borne the name of[s] a single

one of them. [9] Why do you beat us? Because your husbands are dead? Go with
them! May we never see a son or daughter of yours!"

10 On that day she was grieved in
spirit and wept. When she had gone up
to her father's upper room, she intended
to hang herself. But she thought it over
and said, "Never shall they reproach my
father, saying to him, 'You had only one
beloved daughter but she hanged herself
because of her distress.' And I shall bring
my father in his old age down in sorrow
to Hades. It is better for me not to hang
myself, but to pray the Lord that I may
die and not listen to these reproaches
anymore." [11] At that same time, with
hands outstretched toward the window,
she prayed and said,

"Blessed are you, merciful God!
Blessed is your name forever;
let all your works praise you
forever.
12 And now, Lord,[t] I turn my face
to you,
and raise my eyes toward you.
13 Command that I be released from
the earth
and not listen to such reproaches
any more.
14 You know, O Master, that I am
innocent
of any defilement with a man,
15 and that I have not disgraced my
name
or the name of my father in the
land of my exile.
I am my father's only child;
he has no other child to be his
heir;
and he has no close relative or
other kindred

r Other ancient authorities read *strangles*
s Other ancient authorities read *have had no
benefit from*   t Other ancient authorities lack
Lord

**3.7–17: God hears the prayer of Tobit,
and also of Sarah, plagued by a demon-
lover.** From this point on the story is told in
the third person, thereby enabling the narrator to know even the thoughts of the characters in the books. **7–10:** Sarah ("Mistress")

contemplates suicide. **7:** *Ecbatana,* capital of
Media, in Persia. **8:** The name *Asmodeus*
means "destroyer" in Hebrew, but he may
represent the Persian demon Aeshma Daeva.
3.11–15: Sarah's prayer. **11:** *Blessed are you*
is the traditional beginning of a Jewish prayer

for whom I should keep myself
as wife.
Already seven husbands of mine
have died.
Why should I still live?
But if it is not pleasing to you,
O Lord, to take my life,
hear me in my disgrace."
16 At that very moment, the prayers of both of them were heard in the glorious presence of God. 17 So Raphael was sent to heal both of them: Tobit, by removing the white films from his eyes, so that he might see God's light with his eyes; and Sarah, daughter of Raguel, by giving her in marriage to Tobias son of Tobit, and by setting her free from the wicked demon Asmodeus. For Tobias was entitled to have her before all others who had desired to marry her. At the same time that Tobit returned from the courtyard into his house, Sarah daughter of Raguel came down from her upper room.

4 That same day Tobit remembered the money that he had left in trust with Gabael at Rages in Media, 2 and he said to himself, "Now I have asked for death. Why do I not call my son Tobias and explain to him about the money before I die?" 3 Then he called his son Tobias, and when he came to him he said, "My son, when I die," give me a proper burial. Honor your mother and do not abandon her all the days of her life. Do whatever pleases her, and do not grieve her in anything. 4 Remember her, my son, because she faced many dangers for you while you were in her womb. And when she dies, bury her beside me in the same grave.

5 "Revere the Lord all your days, my son, and refuse to sin or to transgress his commandments. Live uprightly all the days of your life, and do not walk in the ways of wrongdoing; 6 for those who act in accordance with truth will prosper in all their activities. To all those who practice righteousness *v* 7 give alms from your possessions, and do not let your eye begrudge the gift when you make it. Do not turn your face away from anyone who is poor, and the face of God will not be turned away from you. 8 If you have many possessions, make your gift from them in proportion; if few, do not be afraid to give according to the little you have. 9 So you will be laying up a good treasure for yourself against the day of necessity. 10 For almsgiving delivers from death and keeps you from going into the Darkness. 11 Indeed, almsgiving, for all who practice it, is an excellent offering in the presence of the Most High.

12 "Beware, my son, of every kind of fornication. First of all, marry a woman

u Lat *v* The text of codex Sinaiticus goes directly from verse 6 to verse 19, reading *To those who practice righteousness* 19*the Lord will give good counsel.* In order to fill the lacuna verses 7 to 18 are derived from other ancient authorities

---

(compare 8.5, 15 and see Jdt 13.17). **16–17:** The angel Raphael is sent in answer to both the prayers (see 12.12–15 n.). **17:** The name *Raphael* means "God heals." *Entitled to have her,* 6.12. The phrase *at the same time* is a dramatic device which heightens the interest of the story.
4.1–21: Preparing to send his son for the trust-money, Tobit imparts his philosophy of life. 1: *The money . . . at Rages, Rages* was an important city whose ruins are located about five miles southeast of modern Teheran. **5–19:** This section of general ethical counsel epitomizes the moral teaching of the book. There are many close parallels with other biblical books of wisdom, such as Proverbs and Sirach, as well as with wisdom literature of other nations and peoples of the an-

cient Near East. **6:** Morality guarantees prosperity; a dogma of orthodox Hebrew wisdom (Ps 1.1–3; Prov 10.27–30). **7–11:** The value of almsgiving; the emphasis is typical of the books, as well as of the period (12.8–9; 14.10–11; Sir 3.30; 35.2; Mt 6.2–4).
4.12–13: One should marry within his own family group; this is a keynote of the book (1.9; 3.15; 6.11–12). **13:** *Pride,* Prov 16.18. *Idleness,* Prov 19.15; Sir 22.1–2. **14:** Lev 19.13. **15:** *What you hate, do not do,* the Golden Rule (Mt 7.12) in negative form, which was enunciated also by the great Jewish teacher, Hillel (flourished in the time of Herod the Great, 37–4 B.C.). *Wine,* Prov 23.29–35; Sir 31.29–31. **16:** Compare vv. 7–11. **17:** Placing food on graves was a pagan practice, forbidden in the Hebrew Bible (Deut 26.14) and

from among the descendants of your ancestors; do not marry a foreign woman, who is not of your father's tribe; for we are the descendants of the prophets. Remember, my son, that Noah, Abraham, Isaac, and Jacob, our ancestors of old, all took wives from among their kindred. They were blessed in their children, and their posterity will inherit the land. 13 So now, my son, love your kindred, and in your heart do not disdain your kindred, the sons and daughters of your people, by refusing to take a wife for yourself from among them. For in pride there is ruin and great confusion. And in idleness there is loss and dire poverty, because idleness is the mother of famine.

14 "Do not keep over until the next day the wages of those who work for you, but pay them at once. If you serve God you will receive payment. "Watch yourself, my son, in everything you do, and discipline yourself in all your conduct. 15 And what you hate, do not do to anyone. Do not drink wine to excess or let drunkenness go with you on your way. 16 Give some of your food to the hungry, and some of your clothing to the naked. Give all your surplus as alms, and do not let your eye begrudge your giving of alms. 17 Place your bread on the grave of the righteous, but give none to sinners. 18 Seek advice from every wise person and do not despise any useful counsel. 19 At all times bless the Lord God, and ask him that your ways may be made straight and that all your paths and plans may prosper. For none of the nations has understanding, but the Lord himself will give them good counsel; but if he chooses otherwise, he casts down to deepest Hades. So now, my child, remember these commandments, and do not let them be erased from your heart.

20 "And now, my son, let me explain to you that I left ten talents of silver in trust with Gabael son of Gabrias, at Rages in Media. 21 Do not be afraid, my son, because we have become poor. You have great wealth if you fear God and flee from every sin and do what is good in the sight of the Lord your God."

5 Then Tobias answered his father Tobit, "I will do everything that you have commanded me, father; 2 but how can I obtain the money *w* from him, since he does not know me and I do not know him? What evidence *x* am I to give him so that he will recognize and trust me, and give me the money? Also, I do not know the roads to Media, or how to get there." 3 Then Tobit answered his son Tobias, "He gave me his bond and I gave him my bond. I *y* divided his in two; we each took one part, and I put one with the money. And now twenty years have passed since I left this money in trust. So now, my son, find yourself a trustworthy man to go with you, and we will pay him wages until you return. But get back the money from Gabael." *z*

4 So Tobias went out to look for a man to go with him to Media, someone who was acquainted with the way. He went out and found the angel Raphael standing in front of him; but he did not perceive that he was an angel of God. 5 Tobias *a* said to him, "Where do you come from, young man?" "From your kindred, the Israelites," he replied, "and I have come here to work." Then Tobias *b* said to him, "Do you know the way to go to Media?" 6 "Yes," he replied, "I have been there many times; I am acquainted with it and know all the roads.

w Gk it   *x* Gk *sign*   *y* Other authorities
read *He*   *z* Gk *from him*   *a* Gk *He*
b Gk *he*

deprecated by many Jews (compare Sir 30.18). Some interpret the verse as a reference to the meals provided the mourners at funerals (compare Jer 16.7; Ezek 24.17); more likely, since this practice is also found in the proverbs of the Aramaic Ahiqar, the passage here is an echo of that proverb. **19**: A sound moral life needs to be sustained by prayer.

5.1–22: Raphael, in the disguise of Azariah, is employed as Tobias' guide. 4: Tobias *did not perceive that he was an angel*, as frequently in folklore, where angels (or gods) traveling in disguise are a favorite theme (compare Gen ch 18; Heb 13.2). **6**: *A journey of two days*, the distance between Rages and Ecbatana is actually 185 miles.

I have often traveled to Media, and would stay with our kinsman Gabael who lives in Rages of Media. It is a journey of two days from Ecbatana to Rages; for it lies in a mountainous area, while Ecbatana is in the middle of the plain." [7] Then Tobias said to him, "Wait for me, young man, until I go in and tell my father; for I do need you to travel with me, and I will pay you your wages." [8] He replied, "All right, I will wait; but do not take too long."

9 So Tobias[c] went in to tell his father Tobit and said to him, "I have just found a man who is one of our own Israelite kindred!" He replied, "Call the man in, my son, so that I may learn about his family and to what tribe he belongs, and whether he is trustworthy enough to go with you."

10 Then Tobias went out and called him, and said, "Young man, my father is calling for you." So he went in to him, and Tobit greeted him first. He replied, "Joyous greetings to you!" But Tobit retorted, "What joy is left for me any more? I am a man without eyesight; I cannot see the light of heaven, but I lie in darkness like the dead who no longer see the light. Although still alive, I am among the dead. I hear people but I cannot see them." But the young man[c] said, "Take courage; the time is near for God to heal you; take courage." Then Tobit said to him, "My son Tobias wishes to go to Media. Can you accompany him and guide him? I will pay your wages, brother." He answered, "I can go with him and I know all the roads, for I have often gone to Media and have crossed all its plains, and I am familiar with its mountains and all of its roads."

11 Then Tobit[c] said to him, "Brother, of what family are you and from what tribe? Tell me, brother." [12] He replied, "Why do you need to know my tribe?" But Tobit[c] said, "I want to be sure, brother, whose son you are and what

your name is." [13] He replied, "I am Azariah, the son of the great Hananiah, one of your relatives." [14] Then Tobit said to him, "Welcome! God save you, brother. Do not feel bitter toward me, brother, because I wanted to be sure about your ancestry. It turns out that you are a kinsman, and of good and noble lineage. For I knew Hananiah and Nathan,[d] the two sons of Shemeliah,[e] and they used to go with me to Jerusalem and worshiped with me there, and were not led astray. Your kindred are good people; you come of good stock. Hearty welcome!"

15 Then he added, "I will pay you a drachma a day as wages, as well as expenses for yourself and my son. So go with my son, [16] and[f] I will add something to your wages." Raphael[g] answered, "I will go with him; so do not fear. We shall leave in good health and return to you in good health, because the way is safe." [17] So Tobit[c] said to him, "Blessings be upon you, brother."

Then he called his son and said to him, "Son, prepare supplies for the journey and set out with your brother. May God in heaven bring you safely there and return you in good health to me; and may his angel, my son, accompany you both for your safety."

Before he went out to start his journey, he kissed his father and mother. Tobit then said to him, "Have a safe journey."

18 But his mother[h] began to weep, and said to Tobit, "Why is it that you have sent my child away? Is he not the staff of our hand as he goes in and out before us? [19] Do not heap money upon money, but let it be a ransom for our child. [20] For the life that is given to us by the Lord is enough for us." [21] Tobit[g] said

c Gk *he*   *d* Other ancient authorities read *Jathan* or *Nathamiah*   *e* Other ancient authorities read *Shemaiah*   *f* Other ancient authorities add *when you return safely*   *g* Gk *He*   *h* Other ancient authorities add *Anna*

5.13: In Hebrew *Azariah* means "God helps." **15:** Evidently a *drachma* was the normal day's wage for an artisan.

5.22: Pleasing irony here (and in v. 17); Tobit does not know that Raphael is the *good angel*.

to her, "Do not worry; our child will leave in good health and return to us in good health. Your eyes will see him on the day when he returns to you in good health. Say no more! Do not fear for them, my sister. ²²For a good angel will accompany him; his journey will be successful, and he will come back in good health." ¹So she stopped weeping.

6 The young man went out and the angel went with him; ²and the dog came out with him and went along with them. So they both journeyed along, and when the first night overtook them they camped by the Tigris river. ³Then the young man went down to wash his feet in the Tigris river. Suddenly a large fish leaped up from the water and tried to swallow the young man's foot, and he cried out. ⁴But the angel said to the young man, "Catch hold of the fish and hang on to it!" So the young man grasped the fish and drew it up on the land. ⁵Then the angel said to him, "Cut open the fish and take out its gall, heart, and liver. Keep them with you, but throw away the intestines. For its gall, heart, and liver are useful as medicine." ⁶So after cutting open the fish the young man gathered together the gall, heart, and liver; then he roasted and ate some of the fish, and kept some to be salted.

The two continued on their way together until they were near Media. ʲ ⁷Then the young man questioned the angel and said to him, "Brother Azariah, what medicinal value is there in the fish's heart and liver, and in the gall?" ⁸He replied, "As for the fish's heart and liver, you must burn them to make a smoke in the presence of a man or woman afflicted by a demon or evil spirit, and every affliction will flee away and never remain

with that person any longer. ⁹And as for the gall, anoint a person's eyes where white films have appeared on them; blow upon them, upon the white films, and the eyesʲ will be healed."

10 When he entered Media and already was approaching Ecbatana, ᵏ ¹¹Raphael said to the young man, "Brother Tobias." "Here I am," he answered. Then Raphaelˡ said to him, "We must stay this night in the home of Raguel. He is your relative, and he has a daughter named Sarah. ¹²He has no male heir and no daughter except Sarah only, and you, as next of kin to her, have before all other men a hereditary claim on her. Also it is right for you to inherit her father's possessions. Moreover, the girl is sensible, brave, and very beautiful, and her father is a good man." ¹³He continued, "You have every right to take her in marriage. So listen to me, brother; tonight I will speak to her father about the girl, so that we may take her to be your bride. When we return from Rages we will celebrate her marriage. For I know that Raguel can by no means keep her from you or promise her to another man without incurring the penalty of death according to the decree of the book of Moses. Indeed he knows that you, rather than any other man, are entitled to marry his daughter. So now listen to me, brother, and tonight we shall speak concerning the girl and arrange her engagement to you. And when we return from Rages we will take her and bring her back with us to your house."

14 Then Tobias said in answer to Raphael, "Brother Azariah, I have heard

---

ⁱ Other ancient authorities read *Ecbatana*
ʲ Gk *they*   ᵏ Other ancient authorities read *Rages*   ˡ Gk *he*

6.1–9: On the journey, Raphael instructs Tobias in obtaining magical medicines from a fish. 2: *The dog,* which in some versions is mentioned in 5.17 as Tobias' pet, is evidently a part of the folktale in an earlier form (see 11.4 n.). *The Tigris* is actually west of Nineveh, so they would not have crossed

it going to Persia (see 1.4 n.). **6–9:** Belief in the healing properties of the fish's organs is typical of folklore.

6.10–18: Raphael prepares Tobias to seek the hand of Sarah. 13: *According to the decree of the book of Moses,* presumably Num 36.6–8, although there is no mention of a

that she already has been married to seven husbands and that they died in the bridal chamber. On the night when they went in to her, they would die. I have heard people saying that it was a demon that killed them. [15] It does not harm her, but it kills anyone who desires to approach her. So now, since I am the only son my father has, I am afraid that I may die and bring my father's and mother's life down to their grave, grieving for me—and they have no other son to bury them."

16 But Raphael[m] said to him, "Do you not remember your father's orders when he commanded you to take a wife from your father's house? Now listen to me, brother, and say no more about this demon. Take her. I know that this very night she will be given to you in marriage. [17] When you enter the bridal chamber, take some of the fish's liver and heart, and put them on the embers of the incense. An odor will be given off; [18] the demon will smell it and flee, and will never be seen near her any more. Now when you are about to go to bed with her, both of you must first stand up and pray, imploring the Lord of heaven that mercy and safety may be granted to you. Do not be afraid, for she was set apart for you before the world was made. You will save her, and she will go with you. I presume that you will have children by her, and they will be as brothers to you. Now say no more!" When Tobias heard the words of Raphael and learned that she was his kinswoman,[n] related through his father's lineage, he loved her very much, and his heart was drawn to her.

7 Now when they[o] entered Ecbatana, Tobias[m] said to him, "Brother Azariah, take me straight to our brother Raguel." So he took him to Raguel's house, where they found him sitting beside the courtyard door. They greeted him first, and he replied, "Joyous greetings, brothers; welcome and good health!" Then he brought them into his house. [2] He said to his wife Edna, "How much the young man resembles my kinsman Tobit!" [3] Then Edna questioned them, saying, "Where are you from, brothers?" They answered, "We belong to the descendants of Naphtali who are exiles in Nineveh." [4] She said to them, "Do you know our kinsman Tobit?" And they replied, "Yes, we know him." Then she asked them, "Is he[p] in good health?" [5] They replied, "He is alive and in good health." And Tobias added, "He is my father!" [6] At that Raguel jumped up and kissed him and wept. [7] He also spoke to him as follows, "Blessings on you, my child, son of a good and noble father!"[q] "O most miserable of calamities that such an upright and beneficent man has become blind!" He then embraced his kinsman Tobias and wept. [8] His wife Edna also wept for him, and their daughter Sarah likewise wept. [9] Then Raguel[m] slaughtered a ram from the flock and received them very warmly.

When they had bathed and washed themselves and had reclined to dine, Tobias said to Raphael, "Brother Azariah,

m Gk *he*   *n* Gk *sister*   *o* Other ancient authorities read *he*   *p* Other ancient authorities add *alive and*   *q* Other ancient authorities add *When he heard that Tobit had lost his sight, he was stricken with grief and wept. Then he said,*

---

death penalty. **15**: *To bury them,* one of the chief concerns of this book (see 1.17 n.; 4.3–4; 14.10). **18**: *Pray, imploring the Lord,* magic is not enough; prayer is necessary too. *Before the world was made,* "marriages are made in heaven" (compare Gen 24.14).

7.1–9a: Tobias and Raphael arrive at the home of Sarah's father. The following conversation implies that Sarah and her family had not previously been aware of Tobias' existence. **2**: *Edna* ("Pleasure"), again,

an ironic name. **4–5**: The Syriac and Latin Vulgate omit the conversation about Tobit health, possibly on account of his blindness.

7.9b–16: Tobias proposes and the wedding takes place. 12: *Took her by the hand,* like a modern parent "giving away the bride," he marries her to Tobias. **13**: Signing the *contract* was the only other ceremony required, although there is no such reference to it in the Hebrew Bible.

8.1–9a: On the wedding night, Tobias

ask Raguel to give me my kinswoman[r] Sarah." 10 But Raguel overheard it and said to the lad, "Eat and drink, and be merry tonight. For no one except you, brother, has the right to marry my daughter Sarah. Likewise I am not at liberty to give her to any other man than yourself, because you are my nearest relative. But let me explain to you the true situation more fully, my child. 11 I have given her to seven men of our kinsmen, and all died on the night when they went in to her. But now, my child, eat and drink, and the Lord will act on behalf of you both." But Tobias said, "I will neither eat nor drink anything until you settle the things that pertain to me." So Raguel said, "I will do so. She is given to you in accordance with the decree in the book of Moses, and it has been decreed from heaven that she be given to you. Take your kinswoman;[r] from now on you are her brother and she is your sister. She is given to you from today and forever. May the Lord of heaven, my child, guide and prosper you both this night and grant you mercy and peace." 12 Then Raguel summoned his daughter Sarah. When she came to him he took her by the hand and gave her to Tobias,[s] saying, "Take her to be your wife in accordance with the law and decree written in the book of Moses. Take her and bring her safely to your father. And may the God of heaven prosper your journey with his peace." 13 Then he called her mother and told her to bring writing material; and he wrote out a copy of a marriage contract, to the effect that he gave her to him as wife according to the decree of the law of Moses. 14 Then they began to eat and drink.

15 Raguel called his wife Edna and said to her, "Sister, get the other room ready, and take her there." 16 So she went and made the bed in the room as he had told her, and brought Sarah[t] there. She

wept for her daughter.[t] Then, wiping away the tears,[u] she said to her, "Take courage, my daughter; the Lord of heaven grant you joy[v] in place of your sorrow. Take courage, my daughter." Then she went out.

8 When they had finished eating and drinking they wanted to retire; so they took the young man and brought him into the bedroom. 2 Then Tobias remembered the words of Raphael, and he took the fish's liver and heart out of the bag where he had them and put them on the embers of the incense. 3 The odor of the fish so repelled the demon that he fled to the remotest parts[w] of Egypt. But Raphael followed him, and at once bound him there hand and foot.

4 When the parents[x] had gone out and shut the door of the room, Tobias got out of bed and said to Sarah,[t] "Sister, get up, and let us pray and implore our Lord that he grant us mercy and safety." 5 So she got up, and they began to pray and implore that they might be kept safe. Tobias[y] began by saying,

"Blessed are you, O God of our
    ancestors,
  and blessed is your name in all
    generations forever.
Let the heavens and the whole
    creation bless you forever.
6 You made Adam, and for him
    you made his wife Eve
  as a helper and support.
From the two of them the
    human race has sprung.
You said, 'It is not good that the
    man should be alone;
  let us make a helper for him like
    himself.'

---

r Gk *sister*   s Gk *him*   t Gk *her*   u Other
ancient authorities read *the tears of her daughter*
v Other ancient authorities read *favor*
w Or *fled through the air to the parts*   x Gk *they*
y Gk *He*

---

routs the demon. 3: *Egypt* was the traditional home of magic and witchcraft (compare Ex 7.11). **4:** See 6.18 n. **5–8:** Tobias and Sarah join in prayer before consummating the marriage, although according to the Vulgate they waited until the end of the third night to consummate their marriage. **5:** *Blessed are you,* see 3.11 n.

7 I now am taking this kinswoman
of mine,
not because of lust,
but with sincerity.
Grant that she and I may find
mercy
and that we may grow old
together."

8 And they both said, "Amen, Amen."
9 Then they went to sleep for the night.

But Raguel arose and called his servants to him, and they went and dug a
grave, 10 for he said, "It is possible that he
will die and we will become an object of
ridicule and derision." 11 When they had
finished digging the grave, Raguel went
into his house and called his wife, 12 saying, "Send one of the maids and have her
go in to see if he is alive. But if he is dead,
let us bury him without anyone knowing
it." 13 So they sent the maid, lit a lamp,
and opened the door; and she went in and
found them sound asleep together.
14 Then the maid came out and informed
them that he was alive and that nothing
was wrong. 15 So they blessed the God of
heaven, and Raguel*z* said,

"Blessed are you, O God, with
every pure blessing;
let all your chosen ones bless
you.*a*
Let them bless you forever.
16 Blessed are you because you have
made me glad.
It has not turned out as I
expected,
but you have dealt with us
according to your great
mercy.

17 Blessed are you because you had
compassion
on two only children.
Be merciful to them, O Master,
and keep them safe;
bring their lives to fulfillment
in happiness and mercy."
18 Then he ordered his servants to fill in
the grave before daybreak.

19 After this he asked his wife to bake
many loaves of bread; and he went out to
the herd and brought two steers and four
rams and ordered them to be slaughtered. So they began to make preparations. 20 Then he called for Tobias and
swore on oath to him in these words:*b*
"You shall not leave here for fourteen
days, but shall stay here eating and
drinking with me; and you shall cheer up
my daughter, who has been depressed.
21 Take at once half of what I own and
return in safety to your father; the other
half will be yours when my wife and I
die. Take courage, my child. I am your
father and Edna is your mother, and we
belong to you as well as to your wife*c*
now and forever. Take courage, my
child."

9 Then Tobias called Raphael and said
to him, 2 "Brother Azariah, take
four servants and two camels with you
and travel to Rages. Go to the home of
Gabael, give him the bond, get the money, and then bring him with you to the
wedding celebration. 4 For you know
that my father must be counting the

z Gk *they*   *a* Other ancient authorities lack
this line   *b* Other ancient authorities read
Tobias and said to him   *c* Gk *sister*

**8.9b–21: Raguel's fears are happily dis-
appointed and he provides an extended
wedding feast. 9b:** *Dug a grave,* because he
did not, of course, know that Tobias was
provided with an effective means to drive
away the demon. **20:** For joy Raguel doubles
the usual length of a wedding feast (11.18;
Judg 14.12). The oath complicates Tobias'
affairs and makes necessary Raphael's solitary
mission in 9.3–5, although Tobias' sending
Raphael for the money shows how much Tobias trusts him.

**9.1–6: Raphael goes to Rages and ob-
tains the money from Gabael. 2:** From Ecbatana to *Rages* was a journey which, according to the ancient historian Arrian
(*Anabasis,* III. 19–20), took Alexander's army
eleven days of forced marches; the author evidently supposed it to be much shorter (see
1.4 n.). **3:** *Has sworn,* 8.20. **4:** Tobias' tender
concern for his father is typical of the spirit of
the book. The son's unwillingness to prolong
his visit is thoroughly justified by the touching description of his parent's uneasiness in

days, and if I delay even one day I will upset him very much. ³You are witness to the oath Raguel has sworn, and I cannot violate his oath."*d* ⁵So Raphael with the four servants and two camels went to Rages in Media and stayed with Gabael. Raphael*e* gave him the bond and informed him that Tobit's son Tobias had married and was inviting him to the wedding celebration. So Gabael*f* got up and counted out to him the money bags, with their seals intact; then they loaded them on the camels.*g* ⁶In the morning they both got up early and went to the wedding celebration. When they came into Raguel's house they found Tobias reclining at table. He sprang up and greeted Gabael,*h* who wept and blessed him with the words, "Good and noble son of a father good and noble, upright and generous! May the Lord grant the blessing of heaven to you and your wife, and to your wife's father and mother. Blessed be God, for I see in Tobias the very image of my cousin Tobit."

10 Now, day by day, Tobit kept counting how many days Tobias*f* would need for going and for returning. And when the days had passed and his son did not appear, ²he said, "Is it possible that he has been detained? Or that Gabael has died, and there is no one to give him the money?" ³And he began to worry. ⁴His wife Anna said, "My child has perished and is no longer among the living." And she began to weep and mourn for her son, saying, ⁵"Woe to me, my child, the light of my eyes, that I let you make the journey." ⁶But Tobit kept saying to her, "Be quiet and stop worrying, my dear;*i* he is all right. Probably something unexpected has happened there. The man who went

with him is trustworthy and is one of our own kin. Do not grieve for him, my dear;*i* he will soon be here." ⁷She answered him, "Be quiet yourself! Stop trying to deceive me! My child has perished." She would rush out every day and watch the road her son had taken, and would heed no one.*j* When the sun had set she would go in and mourn and weep all night long, getting no sleep at all.

Now when the fourteen days of the wedding celebration had ended that Raguel had sworn to observe for his daughter, Tobias came to him and said, "Send me back, for I know that my father and mother do not believe that they will see me again. So I beg of you, father, to let me go so that I may return to my own father. I have already explained to you how I left him." ⁸But Raguel said to Tobias, "Stay, my child, stay with me; I will send messengers to your father Tobit and they will inform him about you." ⁹But he said, "No! I beg you to send me back to my father." ¹⁰So Raguel promptly gave Tobias his wife Sarah, as well as half of all his property: male and female slaves, oxen and sheep, donkeys and camels, clothing, money, and household goods. ¹¹Then he saw them safely off; he embraced Tobias*h* and said, "Farewell, my child; have a safe journey. The Lord of heaven prosper you and your wife Sarah, and may I see children of yours before I die." ¹²Then he kissed his daughter Sarah and said to her, "My daughter, honor your father-in-law and

d In other ancient authorities verse 3 precedes verse 4   *e* Gk *He*   *f* Gk *he*   *g* Other ancient authorities lack *on the camels*   *h* Gk *him*   *i* Gk *sister*   *j* Other ancient authorities read *and she would eat nothing*

10.1–7. **5–6:** Like all the other characters in the book (except the demon), Gabael shows himself to be a loving and trustworthy person.

10.1–7a: Tobias' father and mother grow anxious at their son's absence. 1: In *counting . . . days,* Tobit had naturally made no allowance for a two-week wedding celebration. **4:** *My child has perished,* Anna's ten-

dency to suspect the worst and her husband's courageous attempts to console her illustrate the author's fine sensitivity to a broad spectrum of human reactions to the same events.

10.7b–13: Tobias and Sarah start for home. 11: *The Lord of heaven* was a favorite name for Israel's God in the Persian period and later (Jdt 5.8; Ezra 1.2).

your mother-in-law, [k] since from now on they are as much your parents as those who gave you birth. Go in peace, daughter, and may I hear a good report about you as long as I live." Then he bade them farewell and let them go. Then Edna said to Tobias, "My child and dear brother, the Lord of heaven bring you back safely, and may I live long enough to see children of you and of my daughter Sarah before I die. In the sight of the Lord I entrust my daughter to you; do nothing to grieve her all the days of your life. Go in peace, my child. From now on I am your mother and Sarah is your beloved wife. [l] May we all prosper together all the days of our lives." Then she kissed them both and saw them safely off. [13] Tobias parted from Raguel with happiness and joy, praising the Lord of heaven and earth, King over all, because he had made his journey a success. Finally, he blessed Raguel and his wife Edna, and said, "I have been commanded by the Lord to honor you all the days of my life." [m]

11 When they came near to Kaserin, which is opposite Nineveh, Raphael said, [2] "You are aware of how we left your father. [3] Let us run ahead of your wife and prepare the house while they are still on the way." [4] As they went on together Raphael [n] said to him, "Have the gall ready." And the dog [o] went along behind them.

[5] Meanwhile Anna sat looking intently down the road by which her son would come. [6] When she caught sight of him coming, she said to his father, "Look, your son is coming, and the man who went with him!"

[7] Raphael said to Tobias, before he had approached his father, "I know that his eyes will be opened. [8] Smear the gall of the fish on his eyes; the medicine will make the white films shrink and peel off from his eyes, and your father will regain his sight and see the light."

[9] Then Anna ran up to her son and threw her arms around him, saying, "Now that I have seen you, my child, I am ready to die." And she wept. [10] Then Tobit got up and came stumbling out through the courtyard door. Tobias went up to him, [11] with the gall of the fish in his hand, and holding him firmly, he blew into his eyes, saying, "Take courage, father." With this he applied the medicine on his eyes, [12] and it made them smart. [m] [13] Next, with both his hands he peeled off the white films from the corners of his eyes. Then Tobit [n] saw his son and [p] threw his arms around him, [14] and he wept and said to him, "I see you, my son, the light of my eyes!" Then he said,

"Blessed be God,
　　and blessed be his great name,
　　and blessed be all his holy
　　　　angels.
May his holy name be blessed [q]
　　throughout all the ages.
[15] Though he afflicted me,
　　he has had mercy upon me. [r]
　　Now I see my son Tobias!"

So Tobit went in rejoicing and praising God at the top of his voice. Tobias reported to his father that his journey had been successful, that he had brought the money, that he had married Raguel's daughter Sarah, and that she was, indeed, on her way there, very near to the gate of Nineveh.

[16] Then Tobit, rejoicing and praising God, went out to meet his daughter-in-law at the gate of Nineveh. When the

k Other ancient authorities lack parts of *Then . . . mother-in-law*　l Gk *sister*　m Lat: Meaning of Gk uncertain　n Gk *he*　o Codex Sinaiticus reads *And the Lord*　p Other ancient authorities lack *saw his son and*　q Codex Sinaiticus reads *May his great name be upon us and blessed be all the angels*　r Lat: Gk lacks this line

---

11.1–15: Tobias and Raphael precede Sarah into the city and heal Tobit's blindness. 4: *The dog* appears again for the first time since 6.2; perhaps his presence in the story is a survival from an older folktale, in which he had a real function.

11.16–18: Tobit meets his daughter-in-law and celebrates the marriage. 18: *Ahi-*

people of Nineveh saw him coming, walking along in full vigor and with no one leading him, they were amazed. [17] Before them all, Tobit acknowledged that God had been merciful to him and had restored his sight. When Tobit met Sarah the wife of his son Tobias, he blessed her saying, "Come in, my daughter, and welcome. Blessed be your God who has brought you to us, my daughter. Blessed be your father and your mother, blessed be my son Tobias, and blessed be you, my daughter. Come in now to your home, and welcome, with blessing and joy. Come in, my daughter." So on that day there was rejoicing among all the Jews who were in Nineveh. [18] Ahikar and his nephew Nadab were also present to share Tobit's joy. With merriment they celebrated Tobias's wedding feast for seven days, and many gifts were given to him.[s]

12 When the wedding celebration was ended, Tobit called his son Tobias and said to him, "My child, see to paying the wages of the man who went with you, and give him a bonus as well." [2] He replied, "Father, how much shall I pay him? It would do no harm to give him half of the possessions brought back with me. [3] For he has led me back to you safely, he cured my wife, he brought the money back with me, and he healed you. How much extra shall I give him as a bonus?" [4] Tobit said, "He deserves, my child, to receive half of all that he brought back." [5] So Tobias[t] called him and said, "Take for your wages half of all that you brought back, and farewell."

[6] Then Raphael[t] called the two of them privately and said to them, "Bless God and acknowledge him in the presence of all the living for the good things he has done for you. Bless and sing praise to his name. With fitting honor declare to all people the deeds[u] of God. Do not be slow to acknowledge him. [7] It is good to conceal the secret of a king, but to acknowledge and reveal the works of God, and with fitting honor to acknowledge him. Do good and evil will not overtake you. [8] Prayer with fasting[v] is good, but better than both is almsgiving with righteousness. A little with righteousness is better than wealth with wrongdoing.[w] It is better to give alms than to lay up gold. [9] For almsgiving saves from death and purges away every sin. Those who give alms will enjoy a full life, [10] but those who commit sin and do wrong are their own worst enemies.

[11] "I will now declare the whole truth to you and will conceal nothing from you. Already I have declared it to you when I said, 'It is good to conceal the secret of a king, but to reveal with due honor the works of God.' [12] So now when you and Sarah prayed, it was I who brought and read[x] the record of your prayer before the glory of the Lord, and likewise whenever you would bury the dead. [13] And that time when you did not hesitate to get up and leave your dinner to go and bury the dead, [14] I was sent to

s Other ancient authorities lack parts of this sentence   t Gk *he*   u Gk *words*; other ancient authorities read *words of the deeds*
v Codex Sinaiticus *with sincerity*   w Lat
x Lat: Gk lacks *and read*

---

kar . . . Nadab, see 14.10 n. *Seven days,* apparently the normal period of a wedding celebration (see 8.20 n.).
12.1–22: Raphael, being offered his wages, gives good advice and discloses his true identity. 1–5: Tobias generously wishes to reward Raphael far beyond the amount agreed upon (5.15). **6–10:** In the style of a Jewish teacher of wisdom, Raphael delivers a brief exhortation on the good life, similar to that of Tobit in ch 4. **8:** *Prayer . . . fasting . . . almsgiving . . .* and *righteousness* ("piety")

are mentioned together also in Mt 6.1–18 (on almsgiving, compare 4.7–11 and see Sir 3.30 n.).
12.11: Verse 7. **12–16:** Raphael reveals himself as an angelic intercessor who brings the prayers of mortals into the presence of God. From v. 15 we learn that there are six others. "Uriel" is named in 2 Esd 4.1; "Gabriel" and "Michael," respectively, in Dan 9.21 and 10.13. The growth of angelology was characteristic of the Judaism of the period; this was partly due to an increasing sense of God's

you to test you. And at the same time God sent me to heal you and Sarah your daughter-in-law. 15 I am Raphael, one of the seven angels who stand ready and enter before the glory of the Lord."

16 The two of them were shaken; they fell face down, for they were afraid. 17 But he said to them, "Do not be afraid; peace be with you. Bless God forevermore. 18 As for me, when I was with you, I was not acting on my own will, but by the will of God. Bless him each and every day; sing his praises. 19 Although you were watching me, I really did not eat or drink anything—but what you saw was a vision. 20 So now get up from the ground,[y] and acknowledge God. See, I am ascending to him who sent me. Write down all these things that have happened to you." And he ascended. 21 Then they stood up, and could see him no more. 22 They kept blessing God and singing his praises, and they acknowledged God for these marvelous deeds of his, when an angel of God had appeared to them.

13 Then Tobit[z] said:
"Blessed be God who lives forever,
because his kingdom[a] lasts throughout all ages.
2 For he afflicts, and he shows mercy;
he leads down to Hades in the lowest regions of the earth,
and he brings up from the great abyss,[b]
and there is nothing that can escape his hand.
3 Acknowledge him before the nations, O children of Israel;

for he has scattered you among them.
4 He has shown you his greatness even there.
Exalt him in the presence of every living being,
because he is our Lord and he is our God;
he is our Father and he is God forever.
5 He will afflict[c] you for your iniquities,
but he will again show mercy on all of you.
He will gather you from all the nations
among whom you have been scattered.
6 If you turn to him with all your heart and with all your soul,
to do what is true before him,
then he will turn to you
and will no longer hide his face from you.
So now see what he has done for you;
acknowledge him at the top of your voice.
Bless the Lord of righteousness,
and exalt the King of the ages.[d]
In the land of my exile I acknowledge him,
and show his power and majesty to a nation of sinners:

y Other ancient authorities read *now bless the Lord on earth*　z Gk *he*　a Other ancient authorities read *forever, and his kingdom*　b Gk *from destruction*　c Other ancient authorities read *He afflicted*　d The lacuna in codex Sinaiticus, verses 6b to 10a, is filled in from other ancient authorities

---

transcendence and partly, perhaps, to Persian influences. **17:** *Do not be afraid,* compare Mt 28.5, 10.
13.1–17: Tobit's hymn of praise. Some scholars believe that chs 13 and 14 were added to the book much later in order to give substance to the words of 12.22a and to round out the account of Tobit's life. The hymn contains numerous echoes of Old Testament passages, especially Isa 40–55, and has no particular appropriateness to Tobit's personal situation. On the other hand, while Tobit and his family have been delivered, the other exiles have not; therefore chs 13–14 can be seen as completing the story of Israel's deliverance. **1–6:** Exhortation to the exiles. **2:** 1 Sam 2.6–8; Lk 1.52–53. **4:** *Our Father,* Isa 63.16; 64.8; Sir 23.1, 4; Mt 6.9.

'Turn back, you sinners, and do
what is right before him;
perhaps he may look with favor
upon you and show you
mercy.'

7 As for me, I exalt my God,
and my soul rejoices in the King
of heaven.

8 Let all people speak of his majesty,
and acknowledge him in
Jerusalem.

9 O Jerusalem, the holy city,
he afflicted*e* you for the deeds
of your hands,*f*
but will again have mercy on
the children of the
righteous.

10 Acknowledge the Lord, for he is
good,*g*
and bless the King of the ages,
so that his tent*h* may be rebuilt
in you in joy.
May he cheer all those within you
who are captives,
and love all those within you
who are distressed,
to all generations forever.

11 A bright light will shine to all the
ends of the earth;
many nations will come to you
from far away,
the inhabitants of the remotest
parts of the earth to your
holy name,
bearing gifts in their hands for
the King of heaven.
Generation after generation will
give joyful praise in you,
the name of the chosen city will
endure forever.

12 Cursed are all who speak a harsh
word against you;
cursed are all who conquer you
and pull down your walls,
all who overthrow your towers
and set your homes on fire.
But blessed forever will be all
who revere you. *i*

13 Go, then, and rejoice over the
children of the righteous,
for they will be gathered
together
and will praise the Lord of the
ages.

14 Happy are those who love you,
and happy are those who rejoice
in your prosperity.
Happy also are all people who
grieve with you
because of your afflictions;
for they will rejoice with you
and witness all your glory
forever.

15 My soul blesses*j* the Lord, the
great King!

16 For Jerusalem will be built*k* as
his house for all ages.
How happy I will be if a remnant
of my descendants should
survive
to see your glory and
acknowledge the King of
heaven.
The gates of Jerusalem will be
built with sapphire and
emerald,
and all your walls with precious
stones.
The towers of Jerusalem will be
built with gold,
and their battlements with pure
gold.
The streets of Jerusalem will be
paved
with ruby and with stones of
Ophir.

17 The gates of Jerusalem will sing
hymns of joy,
and all her houses will cry,
'Hallelujah!
Blessed be the God of Israel!'

e Other ancient authorities read *will afflict*
f Other ancient authorities read *your children*
g Other ancient authorities read *Lord worthily*
h Or *tabernacle*   i Other ancient authorities
read *who build you up*   j Or *O my soul, bless*
k Other ancient authorities add *for a city*

13.8–17: God's favor to Jerusalem. **10:**
Tent, temple. **16–17:** That *Jerusalem will be*
built with precious stones is an echo of Isa
54.11–12 (compare Rev 21.18–21).

and the blessed will bless the
holy name forever and
ever."

14 So ended Tobit's words of praise.
2 Tobit[l] died in peace when he
was one hundred twelve years old, and
was buried with great honor in Nineveh.
He was sixty-two[m] years old when he
lost his eyesight, and after regaining it he
lived in prosperity, giving alms and continually blessing God and acknowledging God's majesty.

3 When he was about to die, he called
his son Tobias and the seven sons of Tobias[n] and gave this command: "My son,
take your children 4 and hurry off to Media, for I believe the word of God that
Nahum spoke about Nineveh, that all
these things will take place and overtake
Assyria and Nineveh. Indeed, everything that was spoken by the prophets of
Israel, whom God sent, will occur. None
of all their words will fail, but all will
come true at their appointed times. So it
will be safer in Media than in Assyria and
Babylon. For I know and believe that
whatever God has said will be fulfilled
and will come true; not a single word of
the prophecies will fail. All of our kindred, inhabitants of the land of Israel,
will be scattered and taken as captives
from the good land; and the whole land
of Israel will be desolate, even Samaria
and Jerusalem will be desolate. And the
temple of God in it will be burned to the
ground, and it will be desolate for a
while.[o]

5 "But God will again have mercy on
them, and God will bring them back into
the land of Israel; and they will rebuild
the temple of God, but not like the first
one until the period when the times of
fulfillment shall come. After this they all
will return from their exile and will rebuild Jerusalem in splendor; and in it the

temple of God will be rebuilt, just as the
prophets of Israel have said concerning
it. 6 Then the nations in the whole world
will all be converted and worship God in
truth. They will all abandon their idols,
which deceitfully have led them into
their error; 7 and in righteousness they
will praise the eternal God. All the Israelites who are saved in those days and are
truly mindful of God will be gathered
together; they will go to Jerusalem and
live in safety forever in the land of Abraham, and it will be given over to them.
Those who sincerely love God will rejoice, but those who commit sin and injustice will vanish from all the earth.
8,9 So now, my children, I command
you, serve God faithfully and do what is
pleasing in his sight. Your children are
also to be commanded to do what is right
and to give alms, and to be mindful of
God and to bless his name at all times
with sincerity and with all their strength.
So now, my son, leave Nineveh; do not
remain here. 10 On whatever day you
bury your mother beside me, do not stay
overnight within the confines of the city.
For I see that there is much wickedness
within it, and that much deceit is practiced within it, while the people are without shame. See, my son, what Nadab did
to Ahikar who had reared him. Was he
not, while still alive, brought down into
the earth? For God repaid him to his face
for this shameful treatment. Ahikar came
out into the light, but Nadab went into
the eternal darkness, because he tried to
kill Ahikar. Because he gave alms, Ahikar[p] escaped the fatal trap that Nadab
had set for him, but Nadab fell into it

l Gk *He*   *m* Other ancient authorities read
fifty-eight   *n* Lat: Gk lacks *and the seven sons of
Tobias*   *o* Lat: Other ancient authorities read
*of God will be in distress and will be burned for a
while*   *p* Gk *he*; other ancient authorities read
Manasses

**14.1–15: Tobit's final counsel and
death. 3–9:** He advises his son to leave Nineveh, which is to be destroyed, and predicts
the future course of Israel's history. **4:** *The
word . . . Nahum spoke,* see Nahum 1.1; 2.8–
10, 13; 3.18–19. **6:** That *the nations . . . will all*

be converted to Judaism was a characteristic
belief of the post-exilic age (e.g. Zech 8.20–
23). **10:** *Nadab* (also Nasbas or Nadin) is the
villain of the *Ahikar* story (see 1.21 n.). *Gave
alms,* 4.7–11.

himself, and was destroyed. [11] So now, my children, see what almsgiving accomplishes, and what injustice does—it brings death! But now my breath fails me."

Then they laid him on his bed, and he died; and he received an honorable funeral. [12] When Tobias's mother died, he buried her beside his father. Then he and his wife and children*q* returned to Media and settled in Ecbatana with Raguel his father-in-law. [13] He treated his parents-in-law*r* with great respect in their old age, and buried them in Ecbatana of Media. He inherited both the property of Raguel and that of his father Tobit. [14] He died highly respected at the age of one hundred seventeen*s* years. [15] Before he died he heard*t* of the destruction of Nineveh, and he saw its prisoners being led into Media, those whom King Cyaxares*u* of Media had taken captive. Tobias*v* praised God for all he had done to the people of Nineveh and Assyria; before he died he rejoiced over Nineveh, and he blessed the Lord God forever and ever. Amen.*w*

q Codex Sinaiticus lacks *and children*
r Gk *them*   *s* Other authorities read other numbers   *t* Codex Sinaiticus reads *saw and heard*   *u* Cn: Codex Sinaiticus *Ahikar*; other ancient authorities read *Nebuchadnezzar and Ahasuerus*   *v* Gk *He*   *w* Other ancient authorities lack *Amen*

Judith

No other biblical book, in either its parts or its totality, is as quintessentially ironic as Judith. All its important scenes, as well as its major and minor characters, are ironic. Able to defeat mighty nations in both the East and the West (chs 1–7), the Assyrian army of Nebuchadnezzar, under the command of Holofernes, is routed by Bethulia, a small Samaritan town blocking the army's route to Jerusalem and the temple (chs 8–16). Even more ironic, Holofernes, the "invincible" head of the Assyrian army, is beheaded by Judith, a beautiful and wealthy Israelite widow who is known for her piety and self-denial. The book of Judith, then, is about a saint who risks her life to slay the enemy of her people.

Often characterized as a novel, Judith is best understood as a folktale about a pious widow who, strengthened by her faith in the God of Israel, takes matters into her own hands and so saves her people and Jerusalem. The story's characters are vividly drawn and take on a life of their own. Their speeches, conversations, and prayers, as well as the story's plot, clearly and effectively express the storyteller's theology and ethics.

As for the book's religious ideas, neither God's titles nor attributes are in any way unusual, let alone objectionable. Moreover, the importance of Jerusalem, the efficacy of prayer, fasting and the wearing of sackcloth, the importance of observing dietary laws—all are unquestioned. With the exception of almsgiving and the baptizing of Gentile converts, virtually all the traditional practices of Maccabean Pharisaism are mentioned. God's covenant with Israel is interpreted largely in Deuteronomistic terms. Finally, the atmosphere of the tale is entirely realistic, with no aura of the miraculous.

And yet, the book bristles with problems, as its struggles for canonicity so clearly attest. Although the book purports to be an historical account, it abounds in serious problems concerning both history and geography (chs 1–2). Despite the wealth of geographical and topographical clues throughout the story, the location of Bethulia, the principal scene of the action, is unknown.

Moreover, many readers, past and present, have censured Judith's character and conduct. For though the reader is assured by the narrator that Judith was diligent in prayer and fasting, strict in observing dietary laws, forever celibate after her husband's death, ever fearing the Lord and always honored by all, the discerning reader also recognizes that in Judith's dealings with Holofernes she showed herself to be a shameless flatterer, a bold-faced liar, and a ruthless assassin.

The basis of the Greek version of Judith was certainly Hebrew. The author of the Semitic version was probably a Palestinian Jew.

Despite the story's post-exilic setting and a significant number of Persian nouns and names, it also has unmistakable Hellenistic features, as well as distinctively Maccabean/Hasmonean elements, notably the worshiping of a king as god, the sweeping political and military powers of the high priest, and the supremacy of the Jerusalem council.

Other elements in the story are reminiscent of the general circumstances, terminology, spirit, and tradition of the days of Judas Maccabeus (167–161 B.C.) and his defeat of Nicanor, the general under the infamous Antiochus IV Epiphanes (175–163 B.C.), as narrated in 1 Macc 7.43–50. The book was probably composed sometime during the reign of John Hyrcanus I (135–105 B.C.).

The story is extant in four slightly different Greek versions, two Latin ones, a Syriac one, as well as several later Hebrew recensions. The tale has inspired numerous works of painting, sculpture, and literature, including an Anglo-Saxon epic (see p. ix AP).

1 It was the twelfth year of the reign of Nebuchadnezzar, who ruled over the Assyrians in the great city of Nineveh. In those days Arphaxad ruled over the Medes in Ecbatana. ²He built walls around Ecbatana with hewn stones three cubits thick and six cubits long; he made the walls seventy cubits high and fifty cubits wide. ³At its gates he raised towers one hundred cubits high and sixty cubits wide at the foundations. ⁴He made its gates seventy cubits high and forty cubits wide to allow his armies to march out in force and his infantry to form their ranks. ⁵Then King Nebuchadnezzar made war against King Arphaxad in the great plain that is on the borders of Ragau. ⁶There rallied to him all the people of the hill country and all those who lived along the Euphrates, the Tigris, and the Hydaspes, and, on the plain, Arioch, king of the Elymeans. Thus, many nations joined the forces of the Chaldeans. ᵃ

7 Then Nebuchadnezzar, king of the Assyrians, sent messengers to all who lived in Persia and to all who lived in the west, those who lived in Cilicia and Damascus, Lebanon and Antilebanon, and all who lived along the seacoast, ⁸and those among the nations of Carmel and Gilead, and Upper Galilee and the great plain of Esdraelon, ⁹and all who were in Samaria and its towns, and beyond the Jordan as far as Jerusalem and Bethany and Chelous and Kadesh and the river of Egypt, and Tahpanhes and Raamses and the whole land of Goshen, ¹⁰even beyond Tanis and Memphis, and all who lived in Egypt as far as the borders of Ethiopia. ¹¹But all who lived in the whole region disregarded the summons of Nebuchadnezzar, king of the Assyrians, and refused to join him in the war; for they were not afraid of him, but regarded him as only one man. ᵇ So they sent back his messengers empty-handed and in disgrace.

12 Then Nebuchadnezzar became very angry with this whole region, and swore by his throne and kingdom that he would take revenge on the whole territory of Cilicia and Damascus and Syria, that he would kill with his sword also all the inhabitants of the land of Moab, and the people of Ammon, and all Judea, and every one in Egypt, as far as the coasts of the two seas.

13 In the seventeenth year he led his forces against King Arphaxad and defeated him in battle, overthrowing the whole army of Arphaxad and all his cavalry and all his chariots. ¹⁴Thus he took possession of his towns and came to Ecbatana, captured its towers, plundered its markets, and turned its glory into disgrace. ¹⁵He captured Arphaxad in the

a Syr: Gk *Cheleoudites*   *b* Or *a man*

1.1–6: Nebuchadnezzar declares war on Arphaxad, king of Media. 1: *Nebuchadnezzar* (605 [or 604]–562 B.C.) was second ruler over the Neo-Babylonian Empire (not *over the Assyrians*). It was he who destroyed Jerusalem in 587–86 B.C. and carried the Jews off into their Babylonian Exile (2 Kings 24.1–25.26). The author of the book of Judith, in complete disregard of history, represents him as flourishing after the Exile (4.3; 5.19). Some scholars believe that the historical confusion of the book, of which this is but one example, is deliberate, intended to stamp the work unmistakably as fiction. *Arphaxad* is unknown. *The Medes* inhabited the northern part of modern Iran and had their capital at *Ecbatana* (Tob 3.7). **5:** *Ragau,* the Median city where Arphaxad was later slain (v. 15), located 200 miles northeast of Ecbatana and six miles southwest of modern Teheran. **6:** *Euphrates . . . Tigris,* the principal rivers of Mesopotamia (modern Iraq). *Hydaspes,* a river in India, but here evidently placed in Mesopotamia. *Arioch,* unknown. *Elymeans,* possibly the inhabitants of Elyma, a Persian district (Polybius v. 44.9). *Chaldeans,* the Neo-Babylonians (see v. 1 n.).

1.7–11: The Persians and the western nations refuse Nebuchadnezzar's plea for help. 7–10: The nations enumerated correspond to modern Syria, Lebanon, Palestine, and Egypt. **7:** *Persia,* the southern part of modern Iran.

1.12–16: Angry at the western nations, Nebuchadnezzar defeats Arphaxad without their assistance. 12: *The two seas,* unclear, possibly the Red and the Mediterranean.

mountains of Ragau and struck him down with his spears, thus destroying him once and for all. 16 Then he returned to Nineveh, he and all his combined forces, a vast body of troops; and there he and his forces rested and feasted for one hundred twenty days.

2 In the eighteenth year, on the twenty-second day of the first month, there was talk in the palace of Nebuchadnezzar, king of the Assyrians, about carrying out his revenge on the whole region, just as he had said. 2 He summoned all his ministers and all his nobles and set before them his secret plan and recounted fully, with his own lips, all the wickedness of the region. *c* 3 They decided that every one who had not obeyed his command should be destroyed.

4 When he had completed his plan, Nebuchadnezzar, king of the Assyrians, called Holofernes, the chief general of his army, second only to himself, and said to him, 5 "Thus says the Great King, the lord of the whole earth: Leave my presence and take with you men confident in their strength, one hundred twenty thousand foot soldiers and twelve thousand cavalry. 6 March out against all the land to the west, because they disobeyed my orders. 7 Tell them to prepare earth and water, for I am coming against them in my anger, and will cover the whole face of the earth with the feet of my troops, to whom I will hand them over to be plundered. 8 Their wounded shall fill their ravines and gullies, and the swelling river shall be filled with their dead. 9 I will lead them away captive to the ends of the whole earth. 10 You shall go and seize all their territory for me in advance. They must yield themselves to you, and you shall hold them for me until the day of their punishment. 11 But to those who resist show no mercy, but hand them over to slaughter and plunder throughout your whole region. 12 For as I live, and by the power of my kingdom, what I have spoken I will accomplish by my own hand. 13 And you—take care not to transgress any of your lord's commands, but carry them out exactly as I have ordered you; do it without delay."

14 So Holofernes left the presence of his lord, and summoned all the commanders, generals, and officers of the Assyrian army. 15 He mustered the picked troops by divisions as his lord had ordered him to do, one hundred twenty thousand of them, together with twelve thousand archers on horseback, 16 and he organized them as a great army is marshaled for a campaign. 17 He took along a vast number of camels and donkeys and mules for transport, and innumerable sheep and oxen and goats for food; 18 also ample rations for everyone, and a huge amount of gold and silver from the royal palace.

19 Then he set out with his whole army, to go ahead of King Nebuchadnezzar and to cover the whole face of the earth to the west with their chariots and cavalry and picked foot soldiers. 20 Along with them went a mixed crowd like a swarm of locusts, like the dust *d* of the earth—a multitude that could not be counted.

c Meaning of Gk uncertain   *d* Gk sand

2.1–13: Nebuchadnezzar orders Holofernes to lead a punitive expedition against the West. 4: Next to Judith, *Holofernes* is the principal character of the book. So far as is known, Nebuchadnezzar had no such general, but the name is found in classical authors of a much later period. His career may have been suggested by memories of a Persian general of similar name who was a leader in the expedition which invaded the West under Artaxerxes III about 350 B.C. (Diodorus Siculus, *Hist.* XXXI. 19; see 12.11 n.).
2.7: *Earth and water* were characteristic signs of submission demanded by Persian (not Assyrian or Babylonian) kings (Herodotus, *Hist.* VI. 48).
2.14–27: Holofernes brings an enormous army to Damascus. 21: The *three days* march is impossible, since *Nineveh,* the capital of Assyria, is at least three hundred miles from *Bectileth* (an unidentified site), which is described as *north of Upper Cilicia.* Cilicia is in southeastern Asia Minor.

21 They marched for three days from Nineveh to the plain of Bectileth, and camped opposite Bectileth near the mountain that is to the north of Upper Cilicia. 22 From there Holofernes[e] took his whole army, the infantry, cavalry, and chariots, and went up into the hill country. 23 He ravaged Put and Lud, and plundered all the Rassisites and the Ishmaelites on the border of the desert, south of the country of the Chelleans. 24 Then he followed[f] the Euphrates and passed through Mesopotamia and destroyed all the fortified towns along the brook Abron, as far as the sea. 25 He also seized the territory of Cilicia, and killed everyone who resisted him. Then he came to the southern borders of Japheth, facing Arabia. 26 He surrounded all the Midianites, and burned their tents and plundered their sheepfolds. 27 Then he went down into the plain of Damascus during the wheat harvest, and burned all their fields and destroyed their flocks and herds and sacked their towns and ravaged their lands and put all their young men to the sword.

28 So fear and dread of him fell upon all the people who lived along the seacoast, at Sidon and Tyre, and those who lived in Sur and Ocina and all who lived in Jamnia. Those who lived in Azotus and Ascalon feared him greatly.

3 They therefore sent messengers to him to sue for peace in these words: 2 "We, the servants of Nebuchadnezzar, the Great King, lie prostrate before you. Do with us whatever you will. 3 See, our buildings and all our land and all our wheat fields and our flocks and herds and all our encampments[g] lie before you; do with them as you please. 4 Our towns and their inhabitants are also your slaves; come and deal with them as you see fit."

5 The men came to Holofernes and told him all this. 6 Then he went down to the seacoast with his army and stationed garrisons in the fortified towns and took picked men from them as auxiliaries. 7 These people and all in the countryside welcomed him with garlands and dances and tambourines. 8 Yet he demolished all their shrines[h] and cut down their sacred groves; for he had been commissioned to destroy all the gods of the land, so that all nations should worship Nebuchadnezzar alone, and that all their dialects and tribes should call upon him as a god.

e Gk he   f Or crossed   g Gk all the sheepfolds of our tents   h Syr: Gk borders

---

2.23: *Put and Lud,* probably in Asia Minor. *Rassisites,* unknown. *Ishmaelites,* Arabs (Gen 16.11–12). *Chelleans,* unknown. **24:** The geography here is confused. *Abron,* unknown. The natural line of march would be directly south from Cilicia to Damascus. **26:** *Midianites,* archaic for Arabs (Judg 6.1–6). **27:** *Damascus,* the ancient and beautiful capital of Syria, was noted for its fertile surroundings. **2.28–3.8: Through fear the people of the seacoast submit voluntarily. 2.28:** *Sidon and Tyre,* on the Phoenician coast, west of Damascus. *Sur and Ocina,* unknown. *Jamnia,* just north of *Azotus* (Ashdod) *and Ascalon* (Ashkelon), important (and at one time Philistine) cities in southwest Palestine. **3.8:** Holofernes seems here to go beyond his original commission, which was merely punitive; now he seeks to impose religious unity by forcing all peoples to worship Nebuchadnezzar alone. The language is suggested directly by such passages as Dan chs 3 and 6, and indirectly by the persecutions of Antiochus Epiphanes as related in 1 Macc 1.10–2.26 and 2 Macc chs 6–7. Holofernes' invasion thus becomes a threat to Israel's religious integrity as well as the people's national security.

3.9–4.15: When Holofernes reaches the soil of Palestine, the Jews prepare to resist. 3.9: *Esdraelon* is the great plain which cuts across Palestine just north of Mt. Carmel. *Dothan* is a short distance south of the plain. *Scythopolis* is identical with Beth-shan, located at the junction of Esdraelon and the Jordan Valley. **4.3:** The statement that *they had only recently returned from exile* is flagrantly anachronistic, since it was Nebuchadnezzar who had begun the captivity (compare 1.1 n.), and the return took place nearly fifty years later under the Persian Empire, which had succeeded the Babylonian (Ezra 1.1–3). **4:** The inclusion of *Samaria* is puzzling, since the people of Judea (v. 1) and the Samaritans were separate and increasingly hostile communities in the post-exilic era. **6:** The name of the high priest, *Joakim,* is probably derived from Neh 12.26. The identification of *Bethulia,* the center of the story's action, is one of the major problems of the book; the most

9 Then he came toward Esdraelon, near Dothan, facing the great ridge of Judea; ¹⁰he camped between Geba and Scythopolis, and remained for a whole month in order to collect all the supplies for his army.

4 When the Israelites living in Judea heard of everything that Holofernes, the general of Nebuchadnezzar, the king of the Assyrians, had done to the nations, and how he had plundered and destroyed all their temples, ²they were therefore greatly terrified at his approach; they were alarmed both for Jerusalem and for the temple of the Lord their God. ³For they had only recently returned from exile, and all the people of Judea had just now gathered together, and the sacred vessels and the altar and the temple had been consecrated after their profanation. ⁴So they sent word to every district of Samaria, and to Kona, Beth-horon, Belmain, and Jericho, and to Choba and Aesora, and the valley of Salem. ⁵They immediately seized all the high hilltops and fortified the villages on them and stored up food in preparation for war—since their fields had recently been harvested.

6 The high priest, Joakim, who was in Jerusalem at the time, wrote to the people of Bethulia and Betomesthaim, which faces Esdraelon opposite the plain near Dothan, ⁷ordering them to seize the mountain passes, since by them Judea could be invaded; and it would be easy to stop any who tried to enter, for the approach was narrow, wide enough for only two at a time to pass.

8 So the Israelites did as they had been ordered by the high priest Joakim and the senate of the whole people of Israel, in session at Jerusalem. ⁹And every man of

Israel cried out to God with great fervor, and they humbled themselves with much fasting. ¹⁰They and their wives and their children and their cattle and every resident alien and hired laborer and purchased slave—they all put sackcloth around their waists. ¹¹And all the Israelite men, women, and children living at Jerusalem prostrated themselves before the temple and put ashes on their heads and spread out their sackcloth before the Lord. ¹²They even draped the altar with sackcloth and cried out in unison, praying fervently to the God of Israel not to allow their infants to be carried off and their wives to be taken as booty, and the towns they had inherited to be destroyed, and the sanctuary to be profaned and desecrated to the malicious joy of the Gentiles.

13 The Lord heard their prayers and had regard for their distress; for the people fasted many days throughout Judea and in Jerusalem before the sanctuary of the Lord Almighty. ¹⁴The high priest Joakim and all the priests who stood before the Lord and ministered to the Lord, with sackcloth around their loins, offered the daily burnt offerings, the votive offerings, and freewill offerings of the people. ¹⁵With ashes on their turbans, they cried out to the Lord with all their might to look with favor on the whole house of Israel.

5 It was reported to Holofernes, the general of the Assyrian army, that the people of Israel had prepared for war and had closed the mountain passes and fortified all the high hilltops and set up barricades in the plains. ²In great anger he called together all the princes of Moab and the commanders of Ammon and all the governors of the coastland, ³and said

---

probable suggestion is that it is a pseudonym for Shechem (see 6.11 n.; 7.18 n.; and 10.10 n.).
4.8: *Senate,* an anachronism of the author (see 1.1 n. and 1 Macc 12.6 n.). **14:** *The daily burnt offerings* were the prescribed sacrifices (Ex 29.38–42); the *freewill offerings* were presented voluntarily as occasion required (Lev 22.18–30).

5.1–24: Holofernes is advised by Achior that the Jews are invincible as long as they keep God's law. 2: *Moab* was the region directly east of the Dead Sea, while *Ammon* lay north and east of it. Both nations were traditional enemies of the Jews (Judg 3.12–30; 2 Sam 10–12; 2 Kings 3.4–27; 24.2). **3:** The circumstance that Israel inhabited the rugged *hill country* had often helped her to

to them, "Tell me, you Canaanites, what people is this that lives in the hill country? What towns do they inhabit? How large is their army, and in what does their power and strength consist? Who rules over them as king and leads their army? ⁴And why have they alone, of all who live in the west, refused to come out and meet me?"

5 Then Achior, the leader of all the Ammonites, said to him, "May my lord please listen to a report from the mouth of your servant, and I will tell you the truth about this people that lives in the mountain district near you. No falsehood shall come from your servant's mouth. ⁶These people are descended from the Chaldeans. ⁷At one time they lived in Mesopotamia, because they did not wish to follow the gods of their ancestors who were in Chaldea. ⁸Since they had abandoned the ways of their ancestors, and worshiped the God of heaven, the God they had come to know, their ancestors[i] drove them out from the presence of their gods. So they fled to Mesopotamia, and lived there for a long time. ⁹Then their God commanded them to leave the place where they were living and go to the land of Canaan. There they settled, and grew very prosperous in gold and silver and very much livestock. ¹⁰When a famine spread over the land of Canaan they went down to Egypt and lived there as long as they had food. There they became so great a multitude that their race could not be counted. ¹¹So the king of Egypt became hostile to them; he exploited them and forced them to make bricks. ¹²They cried out to their

God, and he afflicted the whole land of Egypt with incurable plagues. So the Egyptians drove them out of their sight. ¹³Then God dried up the Red Sea before them, ¹⁴and he led them by the way of Sinai and Kadesh-barnea. They drove out all the people of the desert, ¹⁵and took up residence in the land of the Amorites, and by their might destroyed all the inhabitants of Heshbon; and crossing over the Jordan they took possession of all the hill country. ¹⁶They drove out before them the Canaanites, the Perizzites, the Jebusites, the Shechemites, and all the Gergesites, and lived there a long time.

17 "As long as they did not sin against their God they prospered, for the God who hates iniquity is with them. ¹⁸But when they departed from the way he had prescribed for them, they were utterly defeated in many battles and were led away captive to a foreign land. The temple of their God was razed to the ground, and their towns were occupied by their enemies. ¹⁹But now they have returned to their God, and have come back from the places where they were scattered, and have occupied Jerusalem, where their sanctuary is, and have settled in the hill country, because it was uninhabited.

20 "So now, my master and lord, if there is any oversight in this people and they sin against their God and we find out their offense, then we can go up and defeat them. ²¹But if they are not a guilty nation, then let my lord pass them by; for their Lord and God will defend them,

i Gk they

---

preserve her independence in the past. **5–21:** To explain the character of the Israelites, Achior summarizes their entire history from Abraham to the return from exile. **6–9:** The migration of Abraham and the prosperity of his descendants (Gen 11.27–37.1). **6:** *Chaldeans,* Abraham came from Ur of the Chaldees (Gen 11.27–31). **7:** *Mesopotamia* here is the region in the north around Haran (Gen 11.31). Late Jewish tradition ascribed the migration of Abraham's family to a desire to escape the influence of polytheism.

5.8: *The God (or Lord) of heaven* was a favorite name for Israel's God during the Persian period and later (Tob 10.11; Ezra 1.2). **10–13:** The descent into Egypt and the Exodus (Gen 37.2–Ex 18.27). **14:** *Sinai,* Ex 19.1–Num 10.10. *Kadesh-barnea,* Num 20.1. **15–16:** The conquests in Transjordan and Canaan (Num 20.14–Josh 11.23). **15:** *Amorites . . . Heshbon,* Num 21.21–32.

5.17: *They prospered,* the philosophy of Deuteronomy (28.1–14). **18:** The exile (see 1.1 n.). **19:** The return from exile (Ezra chs 1–

and we shall become the laughingstock of the whole world."

22 When Achior had finished saying these things, all the people standing around the tent began to complain; Holofernes' officers and all the inhabitants of the seacoast and Moab insisted that he should be cut to pieces. 23 They said, "We are not afraid of the Israelites; they are a people with no strength or power for making war. 24 Therefore let us go ahead, Lord Holofernes, and your vast army will swallow them up."

6 When the disturbance made by the people outside the council had died down, Holofernes, the commander of the Assyrian army, said to Achior*j* in the presence of all the foreign contingents: 2 "Who are you, Achior and you mercenaries of Ephraim, to prophesy among us as you have done today and tell us not to make war against the people of Israel because their God will defend them? What god is there except Nebuchadnezzar? He will send his forces and destroy them from the face of the earth. Their God will not save them; 3 we the king's*k* servants will destroy them as one man. They cannot resist the might of our cavalry. 4 We will overwhelm them;*l* their mountains will be drunk with their blood, and their fields will be full of their dead. Not even their footprints will survive our attack; they will utterly perish. So says King Nebuchadnezzar, lord of the whole earth. For he has spoken; none of his words shall be in vain.

5 "As for you, Achior, you Ammonite mercenary, you have said these words in a moment of perversity; you shall not see my face again from this day until I take revenge on this race that came out of Egypt. 6 Then at my return the sword of my army and the spear*m* of my servants shall pierce your sides, and you shall fall among their wounded. 7 Now my slaves

are going to take you back into the hill country and put you in one of the towns beside the passes. 8 You will not die until you perish along with them. 9 If you really hope in your heart that they will not be taken, then do not look downcast! I have spoken, and none of my words shall fail to come true."

10 Then Holofernes ordered his slaves, who waited on him in his tent, to seize Achior and take him away to Bethulia and hand him over to the Israelites. 11 So the slaves took him and led him out of the camp into the plain, and from the plain they went up into the hill country and came to the springs below Bethulia. 12 When the men of the town saw them,*n* they seized their weapons and ran out of the town to the top of the hill, and all the slingers kept them from coming up by throwing stones at them. 13 So having taken shelter below the hill, they bound Achior and left him lying at the foot of the hill, and returned to their master.

14 Then the Israelites came down from their town and found him; they untied him and brought him into Bethulia and placed him before the magistrates of their town, 15 who in those days were Uzziah son of Micah, of the tribe of Simeon, and Chabris son of Gothoniel, and Charmis son of Melchiel. 16 They called together all the elders of the town, and all their young men and women ran to the assembly. They set Achior in the midst of all their people, and Uzziah questioned him about what had happened. 17 He answered and told them what had taken place at the council of Holofernes, and all that he had said in the presence of the Assyrian leaders, and all that Holo-

---

j Other ancient authorities add *and to all the Moabites*   *k* Gk *his*   *l* Other ancient authorities add *with it*   *m* Lat Syr: Gk *people*   *n* Other ancient authorities add *on the top of the hill*

---

3); see 4.3 n. **22–24:** Holofernes' other advisers oppose the view of Achior.

6.1–21: For his presumption, Achior is handed over to the Jews to perish with them in the fall of Bethulia. 1–9: Holofernes' denunciation of Achior. **10–13:** Achior is

left bound in sight of the inhabitants of Bethulia. **11:** With whatever place the mysterious Bethulia may be identified (see 4.6 n.), it is clearly pictured as a city on a hill with *springs* below it.

6.14–21: The Israelites bring Achior inside

fernes had boasted he would do against the house of Israel. ¹⁸ Then the people fell down and worshiped God, and cried out:

19 "O Lord God of heaven, see their arrogance, and have pity on our people in their humiliation, and look kindly today on the faces of those who are consecrated to you."

20 Then they reassured Achior, and praised him highly. ²¹ Uzziah took him from the assembly to his own house and gave a banquet for the elders; and all that night they called on the God of Israel for help.

7 The next day Holofernes ordered his whole army, and all the allies who had joined him, to break camp and move against Bethulia, and to seize the passes up into the hill country and make war on the Israelites. ²So all their warriors marched off that day; their fighting forces numbered one hundred seventy thousand infantry and twelve thousand cavalry, not counting the baggage and the foot soldiers handling it, a very great multitude. ³They encamped in the valley near Bethulia, beside the spring, and they spread out in breadth over Dothan as far as Balbaim and in length from Bethulia to Cyamon, which faces Esdraelon.

4 When the Israelites saw their vast numbers, they were greatly terrified and said to one another, "They will now strip clean the whole land; neither the high mountains nor the valleys nor the hills will bear their weight." ⁵Yet they all seized their weapons, and when they had kindled fires on their towers, they remained on guard all that night.

6 On the second day Holofernes led out all his cavalry in full view of the Israelites in Bethulia. ⁷He reconnoitered the approaches to their town, and visited the springs that supplied their water; he seized them and set guards of soldiers over them, and then returned to his army.

8 Then all the chieftains of the Edomites and all the leaders of the Moabites and the commanders of the coastland came to him and said, ⁹"Listen to what we have to say, my lord, and your army will suffer no losses. ¹⁰This people, the Israelites, do not rely on their spears but on the height of the mountains where they live, for it is not easy to reach the tops of their mountains. ¹¹Therefore, my lord, do not fight against them in regular formation, and not a man of your army will fall. ¹²Remain in your camp, and keep all the men in your forces with you; let your servants take possession of the spring of water that flows from the foot of the mountain, ¹³for this is where all the people of Bethulia get their water. So thirst will destroy them, and they will surrender their town. Meanwhile, we and our people will go up to the tops of the nearby mountains and camp there to keep watch to see that no one gets out of the town. ¹⁴They and their wives and children will waste away with famine, and before the sword reaches them they will be strewn about in the streets where they live. ¹⁵Thus you will pay them back with evil, because they rebelled and did not receive you peaceably."

16 These words pleased Holofernes and all his attendants, and he gave orders to do as they had said. ¹⁷So the army of the Ammonites moved forward, together with five thousand Assyrians, and they encamped in the valley and seized the water supply and the springs of the Israelites. ¹⁸And the Edomites and Ammonites went up and encamped in the

---

the city, which then hears the story of Holofernes' arrogance.

7.1–7: Holofernes and his allies advance to Bethulia and survey the situation. 1–3: The line of march is south from Esdraelon. **3:** *Beside the spring,* see 6.11 n. *Dothan,* see 3.9 n. *Balbaim* and *Cyamon* are unknown. *Esdraelon,* see 3.9 n.

7.8–18: They cut off the city's water supply by seizing the springs. 8: The

Edomites, not previously mentioned in the narrative, were located southeast of the Dead Sea; they were traditional enemies of the Israelites (see e.g. Ob 18 and 1 Macc 5.1–5). **18:** The localities of *Egrebeh* and *Chusi* and the valley of *Mochmur* have all been identified (with varying degrees of probability) with sites in the neighborhood of Shechem (see 4.6 n.).

hill country opposite Dothan; and they sent some of their men toward the south and the east, toward Egrebeh, which is near Chusi beside the Wadi Mochmur. The rest of the Assyrian army encamped in the plain, and covered the whole face of the land. Their tents and supply trains spread out in great number, and they formed a vast multitude.

19 The Israelites then cried out to the Lord their God, for their courage failed, because all their enemies had surrounded them, and there was no way of escape from them. 20 The whole Assyrian army, their infantry, chariots, and cavalry, surrounded them for thirty-four days, until all the water containers of every inhabitant of Bethulia were empty; 21 their cisterns were going dry, and on no day did they have enough water to drink, for their drinking water was rationed. 22 Their children were listless, and the women and young men fainted from thirst and were collapsing in the streets of the town and in the gateways; they no longer had any strength.

23 Then all the people, the young men, the women, and the children, gathered around Uzziah and the rulers of the town and cried out with a loud voice, and said before all the elders, 24 "Let God judge between you and us! You have done us a great injury in not making peace with the Assyrians. 25 For now we have no one to help us; God has sold us into their hands, to be strewn before them in thirst and exhaustion. 26 Now summon them and surrender the whole town as booty to the army of Holofernes and to all his forces. 27 For it would be better for us to be captured by them. *o* We shall indeed become slaves, but our lives will be spared, and we shall not witness our little ones dying before our eyes, and our wives and children drawing their last breath. 28 We call to witness

against you heaven and earth and our God, the Lord of our ancestors, who punishes us for our sins and the sins of our ancestors; do today the things that we have described!"

29 Then great and general lamentation arose throughout the assembly, and they cried out to the Lord God with a loud voice. 30 But Uzziah said to them, "Courage, my brothers and sisters!*p* Let us hold out for five days more; by that time the Lord our God will turn his mercy to us again, for he will not forsake us utterly. 31 But if these days pass by, and no help comes for us, I will do as you say."

32 Then he dismissed the people to their various posts, and they went up on the walls and towers of their town. The women and children he sent home. In the town they were in great misery.

8 Now in those days Judith heard about these things: she was the daughter of Merari son of Ox son of Joseph son of Oziel son of Elkiah son of Ananias son of Gideon son of Raphain son of Ahitub son of Elijah son of Hilkiah son of Eliab son of Nathanael son of Salamiel son of Sarasadai son of Israel. 2 Her husband Manasseh, who belonged to her tribe and family, had died during the barley harvest. 3 For as he stood overseeing those who were binding sheaves in the field, he was overcome by the burning heat, and took to his bed and died in his town Bethulia. So they buried him with his ancestors in the field between Dothan and Balamon. 4 Judith remained as a widow for three years and four months 5 at home where she set up a tent for herself on the roof of her house. She put sackcloth around her waist and dressed in widow's clothing. 6 She fasted

---

o Other ancient authorities add *than to die of thirst*   *p* Gk *Courage, brothers*

7.19–28: **Driven to desperation, the citizens of Bethulia urge their leaders to capitulate.** 23: *Uzziah,* 6.15.

7.29–32: **Uzziah advises a delay of five days.**

8.1–8: **The character of Judith.** 1: The

name *Judith* means "Jewess." Her ancestors cannot be identified: some of the names are unparalleled in the Old Testament. 2–3: Her husband had died of sunstroke (compare 2 Kings 4.18–20). 6: Judith's rigorous fasting, which the author obviously approves of, is in

all the days of her widowhood, except the day before the sabbath and the sabbath itself, the day before the new moon and the day of the new moon, and the festivals and days of rejoicing of the house of Israel. 7 She was beautiful in appearance, and was very lovely to behold. Her husband Manasseh had left her gold and silver, men and women slaves, livestock, and fields; and she maintained this estate. 8 No one spoke ill of her, for she feared God with great devotion.

9 When Judith heard the harsh words spoken by the people against the ruler, because they were faint for lack of water, and when she heard all that Uzziah said to them, and how he promised them under oath to surrender the town to the Assyrians after five days, 10 she sent her maid, who was in charge of all she possessed, to summon Uzziah and*q* Chabris and Charmis, the elders of her town. 11 They came to her, and she said to them,

"Listen to me, rulers of the people of Bethulia! What you have said to the people today is not right; you have even sworn and pronounced this oath between God and you, promising to surrender the town to our enemies unless the Lord turns and helps us within so many days. 12 Who are you to put God to the test today, and to set yourselves up in the place of*r* God in human affairs? 13 You are putting the Lord Almighty to the test, but you will never learn anything! 14 You cannot plumb the depths of the human heart or understand the workings of the human mind; how do you expect to search out God, who made all these things, and find out his mind or comprehend his thought? No, my brothers, do not anger the Lord our God. 15 For if he does not choose to help us

within these five days, he has power to protect us within any time he pleases, or even to destroy us in the presence of our enemies. 16 Do not try to bind the purposes of the Lord our God; for God is not like a human being, to be threatened, or like a mere mortal, to be won over by pleading. 17 Therefore, while we wait for his deliverance, let us call upon him to help us, and he will hear our voice, if it pleases him.

18 "For never in our generation, nor in these present days, has there been any tribe or family or people or town of ours that worships gods made with hands, as was done in days gone by. 19 That was why our ancestors were handed over to the sword and to pillage, and so they suffered a great catastrophe before our enemies. 20 But we know no other god but him, and so we hope that he will not disdain us or any of our nation. 21 For if we are captured, all Judea will be captured and our sanctuary will be plundered; and he will make us pay for its desecration with our blood. 22 The slaughter of our kindred and the captivity of the land and the desolation of our inheritance—all this he will bring on our heads among the Gentiles, wherever we serve as slaves; and we shall be an offense and a disgrace in the eyes of those who acquire us. 23 For our slavery will not bring us into favor, but the Lord our God will turn it to dishonor.

24 "Therefore, my brothers, let us set an example for our kindred, for their lives depend upon us, and the sanctuary—both the temple and the altar—rests upon us. 25 In spite of everything let us give thanks to the Lord our

---

q Other ancient authorities lack Uzziah and *(see verses 28 and 35)   r Or* above

---

accord with the practices of the Pharisees. She omits fasting only on days when it is forbidden. **7**: In the eyes of some biblical narrators, beauty and wealth are normal elements in the character of a romantic heroine.
8.9–27: **Judith presents sound theological arguments against Uzziah's proposal. 12**: Note the neat balance of her antitheses:

it is not for us to *put God to the test* (Deut 6.16); he is (v. 25) putting us to the test. **16**: *God is not like a human being,* Num 23.19; 1 Sam 15.29.
8.18: The people are safe because they have been entirely loyal to God (compare 5.21). Judith's optimistic judgment of her contemporaries stands in sharp contrast to the atti-

God, who is putting us to the test as he did our ancestors. 26 Remember what he did with Abraham, and how he tested Isaac, and what happened to Jacob in Syrian Mesopotamia, while he was tending the sheep of Laban, his mother's brother. 27 For he has not tried us with fire, as he did them, to search their hearts, nor has he taken vengeance on us; but the Lord scourges those who are close to him in order to admonish them."

28 Then Uzziah said to her, "All that you have said was spoken out of a true heart, and there is no one who can deny your words. 29 Today is not the first time your wisdom has been shown, but from the beginning of your life all the people have recognized your understanding, for your heart's disposition is right. 30 But the people were so thirsty that they compelled us to do for them what we have promised, and made us take an oath that we cannot break. 31 Now since you are a God-fearing woman, pray for us, so that the Lord may send us rain to fill our cisterns. Then we will no longer feel faint from thirst."

32 Then Judith said to them, "Listen to me. I am about to do something that will go down through all generations of our descendants. 33 Stand at the town gate tonight so that I may go out with my maid; and within the days after which you have promised to surrender the town to our enemies, the Lord will deliver Israel by my hand. 34 Only, do not try to find out what I am doing; for I will not tell you until I have finished what I am about to do."

35 Uzziah and the rulers said to her,

"Go in peace, and may the Lord God go before you, to take vengeance on our enemies." 36 So they returned from the tent and went to their posts.

9 Then Judith prostrated herself, put ashes on her head, and uncovered the sackcloth she was wearing. At the very time when the evening incense was being offered in the house of God in Jerusalem, Judith cried out to the Lord with a loud voice, and said,

2 "O Lord God of my ancestor Simeon, to whom you gave a sword to take revenge on those strangers who had torn off a virgin's clothing[s] to defile her, and exposed her thighs to put her to shame, and polluted her womb to disgrace her; for you said, 'It shall not be done'—yet they did it. 3 So you gave up their rulers to be killed, and their bed, which was ashamed of the deceit they had practiced, was stained with blood, and you struck down slaves along with princes, and princes on their thrones. 4 You gave up their wives for booty and their daughters to captivity, and all their booty to be divided among your beloved children who burned with zeal for you and abhorred the pollution of their blood and called on you for help—O God, my God, hear me also—a widow.

5 "For you have done these things and those that went before and those that followed. You have designed the things that are now, and those that are to come. What you had in mind has happened; 6 the things you decided on presented themselves and said, 'Here we are!' For all your ways are prepared in advance,

s Cn: Gk *loosed her womb*

---

tude of the great prophets, but is similar to that of the author of Ps 44.17–18. **27**: Their present sufferings are not punitive, but educative.

8.28–35: Answering Uzziah's plea, she personally pledges to deliver the city. 30–31: The vow, though wrong, as Judith said, had been made and its consequences could be avoided only by an act of God.

8.33: Judith promises that God will act through her.

9.1–14: Judith's prayer. 1: *Evening incense,* Ex 30.8; Ps 141.2. It may have become customary to offer prayer regularly at this time of day. **2**: *My ancestor Simeon,* the patriarch, son of Jacob, who, with Levi, avenged their sister Dinah (Gen 34.25–26). **4**: God's special care for the *widow* was an article of Israel's faith (Deut 10.18; Ps 146.9).

9.5–6: God's absolute foreknowledge and control of history, both past and future (Isa 44.6–8). **10**: The author of the book apparent-

and your judgment is with foreknowledge.

7 "Here now are the Assyrians, a greatly increased force, priding themselves in their horses and riders, boasting in the strength of their foot soldiers, and trusting in shield and spear, in bow and sling. They do not know that you are the Lord who crushes wars; the Lord is your name. 8 Break their strength by your might, and bring down their power in your anger; for they intend to defile your sanctuary, and to pollute the tabernacle where your glorious name resides, and to break off the horns[t] of your altar with the sword. 9 Look at their pride, and send your wrath upon their heads. Give to me, a widow, the strong hand to do what I plan. 10 By the deceit of my lips strike down the slave with the prince and the prince with his servant; crush their arrogance by the hand of a woman.

11 "For your strength does not depend on numbers, nor your might on the powerful. But you are the God of the lowly, helper of the oppressed, upholder of the weak, protector of the forsaken, savior of those without hope. 12 Please, please, God of my father, God of the heritage of Israel, Lord of heaven and earth, Creator of the waters, King of all your creation, hear my prayer! 13 Make my deceitful words bring wound and bruise on those who have planned cruel things against your covenant, and against your sacred house, and against Mount Zion, and against the house your children possess. 14 Let your whole nation and every tribe know and understand that you are God, the God of all power and might, and that there is no other who protects the people of Israel but you alone!"

10 When Judith[u] had stopped crying out to the God of Israel, and had ended all these words, 2 she rose from where she lay prostrate. She called her maid and went down into the house where she lived on sabbaths and on her festal days. 3 She removed the sackcloth she had been wearing, took off her widow's garments, bathed her body with water, and anointed herself with precious ointment. She combed her hair, put on a tiara, and dressed herself in the festive attire that she used to wear while her husband Manasseh was living. 4 She put sandals on her feet, and put on her anklets, bracelets, rings, earrings, and all her other jewelry. Thus she made herself very beautiful, to entice the eyes of all the men who might see her. 5 She gave her maid a skin of wine and a flask of oil, and filled a bag with roasted grain, dried fig cakes, and fine bread;[v] then she wrapped up all her dishes and gave them to her to carry.

6 Then they went out to the town gate of Bethulia and found Uzziah standing there with the elders of the town, Chabris and Charmis. 7 When they saw her transformed in appearance and dressed differently, they were very

t Syr: Gk *horn*   *u* Gk *she*   *v* Other ancient authorities add *and cheese*

---

ly feels no moral inconsistency in having Judith pray for divine help in practicing *deceit* (see v. 13 n.). That an enemy should be destroyed *by the hand of a woman* would be not only remarkable, but, in the thinking of the time, particularly ignominious (Judg 9.54). **11:** *Your strength does not depend on numbers,* Judg 7.2; 1 Sam 14.6.

9.12: *King of all your creation,* a favorite late Jewish title for God. **13:** This prayer for God's blessing on a lie has been frequently criticized as a serious, and basic, moral blemish in the book. While Judith's conduct may, in the abstract, seem indefensible, it has often been imitated by both Jews and Christians in critical or life-threatening circumstances. The moral question is complex and the practical solution sometimes agonizing.

10.1–5: **Judith beautifies herself. 2:** *Went down into the house,* from the roof where she apparently lived (presumably in a tent) except on the sabbath and festivals. **4:** The Latin version states that God gave her a supernatural beauty because her motive in adorning herself was virtuous, not sensual. **5:** *Dishes,* for cooking her food in accordance with Jewish dietary laws.

10.6–10: **She leaves the city with the elders' blessing. 8:** *She bowed down to God,* prayer and acts of piety accompany every one

greatly astounded at her beauty and said to her, ⁸"May the God of our ancestors grant you favor and fulfill your plans, so that the people of Israel may glory and Jerusalem may be exalted." She bowed down to God.

9 Then she said to them, "Order the gate of the town to be opened for me so that I may go out and accomplish the things you have just said to me." So they ordered the young men to open the gate for her, as she requested. ¹⁰When they had done this, Judith went out, accompanied by her maid. The men of the town watched her until she had gone down the mountain and passed through the valley, where they lost sight of her.

11 As the women*ʷ* were going straight on through the valley, an Assyrian patrol met her ¹²and took her into custody. They asked her, "To what people do you belong, and where are you coming from, and where are you going?" She replied, "I am a daughter of the Hebrews, but I am fleeing from them, for they are about to be handed over to you to be devoured. ¹³I am on my way to see Holofernes the commander of your army, to give him a true report; I will show him a way by which he can go and capture all the hill country without losing one of his men, captured or slain."

14 When the men heard her words, and observed her face—she was in their eyes marvelously beautiful—they said to her, ¹⁵"You have saved your life by hurrying down to see our lord. Go at once to his tent; some of us will escort you and hand you over to him. ¹⁶When you stand before him, have no fear in your heart, but tell him what you have just said, and he will treat you well."

17 They chose from their number a hundred men to accompany her and her maid, and they brought them to the tent of Holofernes. ¹⁸There was great excitement in the whole camp, for her arrival was reported from tent to tent. They came and gathered around her as she stood outside the tent of Holofernes, waiting until they told him about her. ¹⁹They marveled at her beauty and admired the Israelites, judging them by her. They said to one another, "Who can despise these people, who have women like this among them? It is not wise to leave one of their men alive, for if we let them go they will be able to beguile the whole world!"

20 Then the guards of Holofernes and all his servants came out and led her into the tent. ²¹Holofernes was resting on his bed under a canopy that was woven with purple and gold, emeralds and other precious stones. ²²When they told him of her, he came to the front of the tent, with silver lamps carried before him. ²³When Judith came into the presence of Holofernes*ˣ* and his servants, they all marveled at the beauty of her face. She prostrated herself and did obeisance to him, but his slaves raised her up.

11 Then Holofernes said to her, "Take courage, woman, and do not be afraid in your heart, for I have never hurt anyone who chose to serve Nebuchadnezzar, king of all the earth. ²Even now, if your people who live in the hill country had not slighted me, I would never have lifted my spear against them. They have brought this on themselves. ³But now tell me why you have

w Gk *they*   *x* Gk *him*

---

of her decisive acts. **10:** *Down the mountain . . . through the valley,* the geographical location of Bethulia, on the mountain with the "Assyrian" camp at the entrance of the valley leading up to it, could hardly be described more clearly.
10.11–23: Arriving at the enemy lines, she is brought into Holofernes' presence.
13: *I will show him a way,* the first of her "deceitful words" (see 9.13 n.).
10.21: The *canopy* was some kind of deco-

rated mosquito net. The Anglo-Saxon poem of Judith describes it thus: "There was hung/All golden a fair fly-net round the bed/Of the folk-leader, that the baleful one,/The chief of warriors, might look through on each/Child of the brave who came therein, and none/Might look on him. . . ."
11.1–4: Holofernes graciously receives her.

fled from them and have come over to us. In any event, you have come to safety. Take courage! You will live tonight and ever after. 4No one will hurt you. Rather, all will treat you well, as they do the servants of my lord King Nebuchadnezzar."

5 Judith answered him, "Accept the words of your slave, and let your servant speak in your presence. I will say nothing false to my lord this night. 6If you follow out the words of your servant, God will accomplish something through you, and my lord will not fail to achieve his purposes. 7By the life of Nebuchadnezzar, king of the whole earth, and by the power of him who has sent you to direct every living being! Not only do human beings serve him because of you, but also the animals of the field and the cattle and the birds of the air will live, because of your power, under Nebuchadnezzar and all his house. 8For we have heard of your wisdom and skill, and it is reported throughout the whole world that you alone are the best in the whole kingdom, the most informed and the most astounding in military strategy.

9 "Now as for Achior's speech in your council, we have heard his words, for the people of Bethulia spared him and he told them all he had said to you. 10Therefore, lord and master, do not disregard what he said, but keep it in your mind, for it is true. Indeed our nation cannot be punished, nor can the sword prevail against them, unless they sin against their God.

11 "But now, in order that my lord may not be defeated and his purpose frustrated, death will fall upon them, for a sin has overtaken them by which they are about to provoke their God to anger when they do what is wrong. 12Since their food supply is exhausted and their water has almost given out, they have planned to kill their livestock and have determined to use all that God by his laws has forbidden them to eat. 13They have decided to consume the first fruits of the grain and the tithes of the wine and oil, which they had consecrated and set aside for the priests who minister in the presence of our God in Jerusalem— things it is not lawful for any of the people even to touch with their hands. 14Since even the people in Jerusalem have been doing this, they have sent messengers there in order to bring back permission from the council of the elders. 15When the response reaches them and they act upon it, on that very day they will be handed over to you to be destroyed.

16 "So when I, your slave, learned all this, I fled from them. God has sent me to accomplish with you things that will astonish the whole world wherever people shall hear about them. 17Your servant is indeed God-fearing and serves the God of heaven night and day. So, my lord, I will remain with you; but every night your servant will go out into the valley and pray to God. He will tell me when they have committed their sins. 18Then I will come and tell you, so that you may go out with your whole army, and not one of them will be able to withstand you. 19Then I will lead you through Judea, until you come to Jerusalem; there I

11.5–23: Judith's explanation of her flight commends her to the general and his advisers. With the exception of v. 6 (which the narrator intends to be ironic) and vv. 9–10, every verse contains an equivocation, half-truth, or misrepresentation. **6:** The idea that God can work through heathen kings and their armies is fairly common in the Old Testament (Isa 10.5; 44.28; Jer 25.9). **7:** This verse is as obscure in the Greek as in the English. **11.9:** *Achior's speech,* compare 5.5–21. **11–**

15: It is not true to say that the imminent fall of Bethulia is due to the circumstance that its citizens are about to appropriate to common use food that the Mosaic law assigned to God and the temple. **13:** *First fruits,* Ex 23.19. *Tithes,* Lev 27.30. **14:** *Council of the elders,* the Sanhedrin, the supreme religious authority of later Judaism. **17:** *God of heaven,* see 5.8 n. By telling him that *every night* she *will go out into the valley* Judith is preparing a ruse for her eventual escape (13.10).

will set your throne. *y* You will drive them like sheep that have no shepherd, and no dog will so much as growl at you. For this was told me to give me foreknowledge; it was announced to me, and I was sent to tell you."

20 Her words pleased Holofernes and all his servants. They marveled at her wisdom and said, 21 "No other woman from one end of the earth to the other looks so beautiful or speaks so wisely!" 22 Then Holofernes said to her, "God has done well to send you ahead of the people, to strengthen our hands and bring destruction on those who have despised my lord. 23 You are not only beautiful in appearance, but wise in speech. If you do as you have said, your God shall be my God, and you shall live in the palace of King Nebuchadnezzar and be renowned throughout the whole world."

12 Then he commanded them to bring her in where his silver dinnerware was kept, and ordered them to set a table for her with some of his own delicacies, and with some of his own wine to drink. 2 But Judith said, "I cannot partake of them, or it will be an offense; but I will have enough with the things I brought with me." 3 Holofernes said to her, "If your supply runs out, where can we get you more of the same? For none of your people are here with us." 4 Judith replied, "As surely as you live, my lord, your servant will not use up the supplies I have with me before the Lord carries out by my hand what he has determined."

5 Then the servants of Holofernes brought her into the tent, and she slept until midnight. Toward the morning watch she got up 6 and sent this message to Holofernes: "Let my lord now give orders to allow your servant to go out and pray." 7 So Holofernes commanded his guards not to hinder her. She remained in the camp three days. She went out each night to the valley of Bethulia, and bathed at the spring in the camp. *z* 8 After bathing, she prayed the Lord God of Israel to direct her way for the triumph of his *a* people. 9 Then she returned purified and stayed in the tent until she ate her food toward evening.

10 On the fourth day Holofernes held a banquet for his personal attendants only, and did not invite any of his officers. 11 He said to Bagoas, the eunuch who had charge of his personal affairs, "Go and persuade the Hebrew woman who is in your care to join us and to eat and drink with us. 12 For it would be a disgrace if we let such a woman go without having intercourse with her. If we do not seduce her, she will laugh at us."

13 So Bagoas left the presence of Holofernes, and approached her and said, "Let this pretty girl not hesitate to come to my lord to be honored in his presence, and to enjoy drinking wine with us, and to become today like one of the Assyrian women who serve in the palace of Nebuchadnezzar." 14 Judith replied, "Who am I to refuse my lord? Whatever pleases him I will do at once, and it will be a joy

y Or *chariot*   *z* Other ancient authorities lack *in the camp*   *a* Other ancient authorities read *her*

12.1–9: For three days Judith remains and establishes a pattern of conduct. 2: Once again the narrative stresses Judith's meticulous observance of the Jewish dietary laws, even in the presence of Gentiles. Her behavior contrasts markedly with that of Esther (Esth 2.20; compare Dan 1.8). **4:** *Before the Lord carries out by my hand* is a skillful foreshadowing of the murder. **6–7:** Since she had gone out every night, her departure after Holofernes' assassination would occasion no surprise. **8:** Her prayer was genuine enough, but not for the purpose previously announced (11.17).

12.10–20: Judith invited to Holofernes' banquet. 11: *Bagoas* is a well-known Persian name, spelled "Bigvai" in Ezra 2.2 and the Elephantine papyri. Diodorus Siculus (*Hist.* XVI. 47) mentions an officer Bagoas in the army of Artaxerxes III (see 2.4 n.). In oriental kingdoms the officer in charge of the women was normally a *eunuch;* eunuchs often attained also to positions of considerable responsibility in the state (Dan 1.3; Acts 8.27).

to me until the day of my death." ¹⁵So she proceeded to dress herself in all her woman's finery. Her maid went ahead and spread for her on the ground before Holofernes the lambskins she had received from Bagoas for her daily use in reclining.

16 Then Judith came in and lay down. Holofernes' heart was ravished with her and his passion was aroused, for he had been waiting for an opportunity to seduce her from the day he first saw her. ¹⁷So Holofernes said to her, "Have a drink and be merry with us!" ¹⁸Judith said, "I will gladly drink, my lord, because today is the greatest day in my whole life." ¹⁹Then she took what her maid had prepared and ate and drank before him. ²⁰Holofernes was greatly pleased with her, and drank a great quantity of wine, much more than he had ever drunk in any one day since he was born.

13 When evening came, his slaves quickly withdrew. Bagoas closed the tent from outside and shut out the attendants from his master's presence. They went to bed, for they all were weary because the banquet had lasted so long. ²But Judith was left alone in the tent, with Holofernes stretched out on his bed, for he was dead drunk.

3 Now Judith had told her maid to stand outside the bedchamber and to wait for her to come out, as she did on the other days; for she said she would be going out for her prayers. She had said the same thing to Bagoas. ⁴So everyone went out, and no one, either small or great, was left in the bedchamber. Then Judith, standing beside his bed, said in her heart, "O Lord God of all might, look in this hour on the work of my hands for the exaltation of Jerusalem. ⁵Now indeed is the time to help your heritage and to carry out my design to destroy the enemies who have risen up against us."

6 She went up to the bedpost near Holofernes' head, and took down his sword that hung there. ⁷She came close to his bed, took hold of the hair of his head, and said, "Give me strength today, O Lord God of Israel!" ⁸Then she struck his neck twice with all her might, and cut off his head. ⁹Next she rolled his body off the bed and pulled down the canopy from the posts. Soon afterward she went out and gave Holofernes' head to her maid, ¹⁰who placed it in her food bag.

Then the two of them went out together, as they were accustomed to do for prayer. They passed through the camp, circled around the valley, and went up the mountain to Bethulia, and came to its gates. ¹¹From a distance Judith called out to the sentries at the gates, "Open, open the gate! God, our God, is with us, still showing his power in Israel and his strength against our enemies, as he has done today!"

12 When the people of her town heard her voice, they hurried down to the town gate and summoned the elders of the town. ¹³They all ran together, both small and great, for it seemed unbelievable that she had returned. They opened the gate and welcomed them. Then they lit a fire to give light, and gathered around them. ¹⁴Then she said to them with a loud voice, "Praise God, O praise him! Praise God, who has not withdrawn his mercy from the house of Israel, but has destroyed our enemies by my hand this very night!"

12.14: Judith's words *Who am I to refuse my lord? Whatever pleases him I will do at once* are a masterpiece of irony; for Holofernes thinks that he is her lord, but the reader knows otherwise.
13.1–10a: Judith beheads Holofernes.
4–7: The irony is that Holofernes is mastered by the very person he thought he had mastered—and with his own sword. Judith's prayer before she decapitates the enemy of her people is the most dramatic, horrifying, and ironic moment in the story, for the pious woman prays for strength to slay the tyrant. **9**: *Pulled down the canopy* (10.21) and carried it off as a trophy (v. 15). **10a**: Even the bag in which she brought her food (10.5) is now seen to have been part of a well-laid plan.
13.10b–20: She escapes and returns to her own people. 10b: Her established habit of leaving the camp each night for prayer

15 Then she pulled the head out of the bag and showed it to them, and said, "See here, the head of Holofernes, the commander of the Assyrian army, and here is the canopy beneath which he lay in his drunken stupor. The Lord has struck him down by the hand of a woman. 16 As the Lord lives, who has protected me in the way I went, I swear that it was my face that seduced him to his destruction, and that he committed no sin with me, to defile and shame me."

17 All the people were greatly astonished. They bowed down and worshiped God, and said with one accord, "Blessed are you our God, who have this day humiliated the enemies of your people."

18 Then Uzziah said to her, "O daughter, you are blessed by the Most High God above all other women on earth; and blessed be the Lord God, who created the heavens and the earth, who has guided you to cut off the head of the leader of our enemies. 19 Your praise[b] will never depart from the hearts of those who remember the power of God. 20 May God grant this to be a perpetual honor to you, and may he reward you with blessings, because you risked your own life when our nation was brought low, and you averted our ruin, walking in the straight path before our God." And all the people said, "Amen. Amen."

14 Then Judith said to them, "Listen to me, my friends. Take this head and hang it upon the parapet of your wall. 2 As soon as day breaks and the sun rises on the earth, each of you take up your weapons, and let every able-bodied man go out of the town; set a captain over them, as if you were going down to the plain against the Assyrian outpost; only do not go down. 3 Then they will seize their arms and go into the camp and rouse the officers of the Assyrian army. They will rush into the tent of Holofernes and will not find him. Then panic will come over them, and they will flee before you. 4 Then you and all who live within the borders of Israel will pursue them and cut them down in their tracks. 5 But before you do all this, bring Achior the Ammonite to me so that he may see and recognize the man who despised the house of Israel and sent him to us as if to his death."

6 So they summoned Achior from the house of Uzziah. When he came and saw the head of Holofernes in the hand of one of the men in the assembly of the people, he fell down on his face in a faint. 7 When they raised him up he threw himself at Judith's feet, and did obeisance to her, and said, "Blessed are you in every tent of Judah! In every nation those who hear your name will be alarmed. 8 Now tell me what you have done during these days."

So Judith told him in the presence of the people all that she had done, from the day she left until the moment she began speaking to them. 9 When she had finished, the people raised a great shout and made a joyful noise in their town. 10 When Achior saw all that the God of Israel had done, he believed firmly in God. So he was circumcised, and joined the house of Israel, remaining so to this day.

b Other ancient authorities read *hope*

(11.17; 12.7) permits her an easy escape. **17:** *Blessed are you our God,* is the most common formula of late Jewish prayers (see Tob 3.11 n.). **18:** Uzziah's words are reminiscent of those spoken concerning Jael under similar circumstances (Judg 5.24); also of Melchizedek's greeting to Abraham (Gen 14.19–20).

14.1–10: Achior identifies the head of Holofernes and is converted to Judaism. 1–4: Judith suggests tactics that will put the enemy to flight with minimum effort by the Jewish forces. In the Latin version, v. 5 is omitted and vv. 6–7, more logically, precede vv. 1–4. **5:** *Achior,* the Ammonite leader whom Holofernes had left to be destroyed with the Jews (5.5–6.21). The full irony of the situation is evident when one remembers that Achior had been told he would see that face again only on the day of Holofernes' vengeance (6.5). **10:** The fact that the conversion of an Ammonite to Judaism is strictly forbidden by the law (Deut 23.3) may explain why

11 As soon as it was dawn they hung the head of Holofernes on the wall. Then they all took their weapons, and they went out in companies to the mountain passes. 12When the Assyrians saw them they sent word to their commanders, who then went to the generals and the captains and to all their other officers. 13They came to Holofernes' tent and said to the steward in charge of all his personal affairs, "Wake up our lord, for the slaves have been so bold as to come down against us to give battle, to their utter destruction."

14 So Bagoas went in and knocked at the entry of the tent, for he supposed that he was sleeping with Judith. 15But when no one answered, he opened it and went into the bedchamber and found him sprawled on the floor dead, with his head missing. 16He cried out with a loud voice and wept and groaned and shouted, and tore his clothes. 17Then he went to the tent where Judith had stayed, and when he did not find her, he rushed out to the people and shouted, 18"The slaves have tricked us! One Hebrew woman has brought disgrace on the house of King Nebuchadnezzar. Look, Holofernes is lying on the ground, and his head is missing!"

19 When the leaders of the Assyrian army heard this, they tore their tunics and were greatly dismayed, and their loud cries and shouts rose up throughout the camp.

15 When the men in the tents heard it, they were amazed at what had happened. 2Overcome with fear and trembling, they did not wait for one another, but with one impulse all rushed out and fled by every path across the plain and through the hill country. 3Those who had camped in the hills around Bethulia also took to flight. Then the Israelites, everyone that was a soldier, rushed out upon them. 4Uzziah sent men to Betomasthaim*c* and Choba and Kola, and to all the frontiers of Israel, to tell what had taken place and to urge all to rush out upon the enemy to destroy them. 5When the Israelites heard it, with one accord they fell upon the enemy,*d* and cut them down as far as Choba. Those in Jerusalem and all the hill country also came, for they were told what had happened in the camp of the enemy. The men in Gilead and in Galilee outflanked them with great slaughter, even beyond Damascus and its borders. 6The rest of the people of Bethulia fell upon the Assyrian camp and plundered it, acquiring great riches. 7And the Israelites, when they returned from the slaughter, took possession of what remained. Even the villages and towns in the hill country and in the plain got a great amount of booty, since there was a vast quantity of it.

8 Then the high priest Joakim and the elders of the Israelites who lived in Jerusalem came to witness the good things that the Lord had done for Israel, and to see Judith and to wish her well. 9When they met her, they all blessed her with one accord and said to her, "You are the glory of Jerusalem, you are the great boast of Israel, you are the great pride of our nation! 10You have done all this with your own hand; you have done great good to Israel, and God is well pleased

---

c Other ancient authorities add *and Bebai*
d Gk *them*

---

the book of Judith was never canonized by the Jews.

14.11–19: The enemy discovers Holofernes' death. 11–13: As Judith had planned, the threatening movements of the Jews lead to the discovery of Holofernes' body.

15.1–7: The Assyrians, fleeing in panic, are slaughtered and despoiled by the Jews. 3: *Those who had camped in the hills* were the Edomites and Ammonites (7.18). **4:** As with so many place names in this book, *Betomasthaim and Choba and Kola* have never been satisfactorily identified. **5:** *Gilead,* which was situated in the northern part of Transjordan, and *Galilee,* in the north of Palestine proper, lay on either flank of the enemy's northeastward flight through *Damascus* and back toward Assyria.

15.8–13: Judith is led in triumph to Jerusalem. 11: The *thirty days* of plundering is

with it. May the Almighty Lord bless you forever!" And all the people said, "Amen."

11 All the people plundered the camp for thirty days. They gave Judith the tent of Holofernes and all his silver dinnerware, his beds, his bowls, and all his furniture. She took them and loaded her mules and hitched up her carts and piled the things on them.

12 All the women of Israel gathered to see her, and blessed her, and some of them performed a dance in her honor. She took ivy-wreathed wands in her hands and distributed them to the women who were with her; 13 and she and those who were with her crowned themselves with olive wreaths. She went before all the people in the dance, leading all the women, while all the men of Israel followed, bearing their arms and wearing garlands and singing hymns.

14 Judith began this thanksgiving before all Israel, and all the people loudly sang this song of praise. 1 And Judith said,

16

Begin a song to my God with
    tambourines,
    sing to my Lord with cymbals.
Raise to him a new psalm; *e*
    exalt him, and call upon his
    name.
2 For the Lord is a God who
    crushes wars;
    he sets up his camp among his
    people;
    he delivered me from the hands
    of my pursuers.

3 The Assyrian came down from the
    mountains of the north;
    he came with myriads of his
    warriors;
    their numbers blocked up the
    wadis,
    and their cavalry covered the
    hills.
4 He boasted that he would burn up
    my territory,
    and kill my young men with the
    sword,
    and dash my infants to the
    ground,
    and seize my children as booty,
    and take my virgins as spoil.

5 But the Lord Almighty has foiled
    them
    by the hand of a woman. *f*
6 For their mighty one did not fall
    by the hands of the young
    men,
    nor did the sons of the Titans
    strike him down,
    nor did tall giants set upon him;
    but Judith daughter of Merari
    with the beauty of her
    countenance undid him.

7 For she put away her widow's
    clothing
    to exalt the oppressed in Israel.

e Other ancient authorities read *a psalm and
praise*   *f* Other ancient authorities add *he has
confounded them*

obviously unrealistic. While the people apparently intend that Judith herself shall have Holofernes' treasure, she accepts the gifts only with the intention of dedicating them to God (16.19). **12:** *A dance,* compare 1 Sam 18.6; Ps 149.3–9. *Ivy-wreathed wands,* Ps 118.27; 1 Macc 13.51; 2 Macc 10.7. **13:** The reference to *olive wreaths* seems to indicate a late date for the book, for the custom is Greek, not Jewish. The destination of the procession is the temple in Jerusalem (16.18). **16.1–17:** **Judith's thanksgiving psalm.** In the book of Tobit (ch 13) a prayer composed by Tobit occupies a similar position at the end of the story. It is Israel, personified as a woman, who sings the hymn; Judith herself is referred to only in the third person (v. 7). **1–2:** A call to praise (compare Ex 15.21; Judg 5.2). **1:** The phrase "a new" psalm (compare v. 13) is a cliché drawn from the Psalter (e.g. Pss 96.1; 98.1). **3–12:** Description of the victory. **6:** *Sons of the Titans* is Greek; perhaps the Semitic original had "sons of Rephaim" (compare Deut 3.11). **13–16:** A general hymn of praise. **13:** *I will sing . . . a new song,* Ps 144.9. The hymn begins in the next line. **14:** Pss 33.6; 104.30. **16:** *Every sacrifice . . . is a small thing;* to compare sacrifice unfavorably with moral obedience ("to fear the Lord") is a commonplace of Old Testament religion,

She anointed her face with
    perfume;
8 she fastened her hair with a tiara
    and put on a linen gown to
    beguile him.
9 Her sandal ravished his eyes,
    her beauty captivated his mind,
    and the sword severed his neck!
10 The Persians trembled at her
    boldness,
    the Medes were daunted at her
    daring.

11 Then my oppressed people
    shouted;
    my weak people cried out, *g* and
    the enemy *h* trembled;
    they lifted up their voices, and
    the enemy *h* were turned
    back.
12 Sons of slave-girls pierced them
    through
    and wounded them like the
    children of fugitives;
    they perished before the army of
    my Lord.

13 I will sing to my God a new song:
    O Lord, you are great and
    glorious,
    wonderful in strength,
    invincible.
14 Let all your creatures serve you,
    for you spoke, and they were
    made.
    You sent forth your spirit, *i* and it
    formed them; *j*
    there is none that can resist your
    voice.
15 For the mountains shall be shaken
    to their foundations with
    the waters;
    before your glance the rocks
    shall melt like wax.

But to those who fear you
    you show mercy.
16 For every sacrifice as a fragrant
    offering is a small thing,
    and the fat of all whole burnt
    offerings to you is a very
    little thing;
    but whoever fears the Lord is
    great forever.

17 Woe to the nations that rise up
    against my people!
    The Lord Almighty will take
    vengeance on them in the
    day of judgment;
    he will send fire and worms into
    their flesh;
    they shall weep in pain forever.

18 When they arrived at Jerusalem, they worshiped God. As soon as the people were purified, they offered their burnt offerings, their freewill offerings, and their gifts. 19 Judith also dedicated to God all the possessions of Holofernes, which the people had given her; and the canopy that she had taken for herself from his bedchamber she gave as a votive offering. 20 For three months the people continued feasting in Jerusalem before the sanctuary, and Judith remained with them.

21 After this they all returned home to their own inheritances. Judith went to Bethulia, and remained on her estate. For the rest of her life she was honored throughout the whole country. 22 Many desired to marry her, but she gave herself to no man all the days of her life after her husband Manasseh died and was gath-

g Other ancient authorities read *feared*
h Gk *they*   *i* Or *breath*   *j* Other ancient
authorities read *they were created*

---

especially in later times (1 Sam 15.22; Pss 40.6–8; 50.8–15; 51.16–17; Hos 6.6; compare Sir 34.18–19). **17:** Concluding anathema on Israel's enemies (compare Judg 5.31). *Fire and worms,* Isa 66.24; Sir 7.17. As in Dan 12.2, the punishment of the wicked is eternal.
16.18–20: A victory celebration in Je-

rusalem. **20:** *Three months,* Syriac, "a month of days."
16.21–25: Judith remains a widow and dies at a ripe old age. 24: She *distributed her property,* according to the Mosaic law (Num 27.11).

ered to his people. 23 She became more and more famous, and grew old in her husband's house, reaching the age of one hundred five. She set her maid free. She died in Bethulia, and they buried her in the cave of her husband Manasseh; 24 and the house of Israel mourned her for seven days. Before she died she distributed her property to all those who were next of kin to her husband Manasseh, and to her own nearest kindred. 25 No one ever again spread terror among the Israelites during the lifetime of Judith, or for a long time after her death.

NOTE. The deuterocanonical portions of the Book of Esther are several additional passages found in the Greek translation of the Hebrew Book of Esther, a translation that differs also in other respects from the Hebrew text (the latter is translated in the NRSV Old Testament). The disordered chapter numbers come from the displacement of the additions to the end of the canonical Book of Esther by Jerome in his Latin translation and from the subsequent division of the Bible into chapters by Stephen Langton, who numbered the additions consecutively as though they formed a direct continuation of the Hebrew text. So that the additions may be read in their proper context, the whole of the Greek version is here translated, though certain familiar names are given according to their Hebrew rather than their Greek form; for example, Mordecai and Vashti instead of Mardocheus and Astin. The order followed is that of the Greek text, but the chapter and verse numbers conform to those of the King James or Authorized Version. The additions, conveniently indicated by the letters A–F, are located as follows: A, before 1.1; B, after 3.13; C and D, after 4.17; E, after 8.12; F, after 10.3.

Esther

(The Greek Version Containing
the Additional Chapters)

The Additions to the book of Esther comprise 107 verses. Their contents are as follows: Addition A: Mordecai's dream (11.2–12) and his discovery of a plot against the king (12.1–6); Addition B: The royal edict dictated by Haman, announcing a pogrom against the Jews (13.1–7); Addition C: The prayers of Mordecai (13.8–18) and Esther (14.1–19); Addition D: Esther's appearing, unsummoned, before the king (15.4–19); and Addition E: The royal edict dictated by Mordecai, counteracting the edict sent by Haman (16.1–24); and Addition F: The interpretation of Mordecai's dream (10.4–13) and the colophon (an inscription at the end of a manuscript) to the Greek version (11.1).

There is no mention of God in the Hebrew narrative, but in the Greek Additions the words "Lord" or "God" appear more than fifty times. In fact, occasionally God is mentioned also in the portions of the Greek version that correspond to the canonical Hebrew text, as when Mordecai instructs Esther prior to her becoming queen that she should "fear God and keep his laws" (2.20) and then later urges her to "call upon the Lord" (4.8) before appearing, unsummoned and unannounced, before the king. We also read in the Greek text, "That night the Lord took sleep from the king" (compare the Hebrew at 6.1). Likewise, Haman's wife and his friends caution Haman that if Mordecai be Jewish, then "the living God is with him" (compare the Hebrew at 6.13).

The additions are clearly intrusive and secondary, for they contradict the Hebrew at a number of points. They sometimes make the characters and events more vivid or dramatic, and they always give the narrative an explicitly religious character that is lacking in the Hebrew.

Moreover, the additions provide their authors with an opportunity to express their own particular theological views. Additions A and F especially emphasize God's providential care for the people Israel in a universally hostile world. Addition C attests to the efficacy of prayer and well expresses Queen Esther's abhorrence at being married to a Gentile, her loathing of all things worldly and courtly, and her strict observance of dietary laws—none of which are so much as hinted at in the Hebrew. Thanks in part to Addition D, the climax of the Greek version is reached when God miraculously changes to gentleness the king's "fierce anger" at Esther's unannounced entrance. This is lacking in the Hebrew. Taken together, the six additions de-emphasize the establishment of Purim and express a strong anti-Gentile spirit.

Originally, A, C, D, and F were probably composed in either Hebrew or Aramaic and, if so, were already part of that particular Semitic text used by the Greek translator. B and E are so florid and rhetorical that they must have been originally composed in Greek.

The additions were not composed at the same time. The latest possible date for B, C, D, and E is A.D. 93. The colophon's location (11.1) immediately after F suggests that A as well as F were part of the Semitic text at the time that Lysimachus of Jerusalem made his Greek translation of it, about 114 B.C. It is impossible to say who composed the Semitic or Greek Additions.

Additions B and E may have been composed in a sophisticated Greek Jewish center, such as Alexandria, but a Palestinian provenance for the others is likely.

ADDITION A

11 *a* 2 In the second year of the reign of Artaxerxes the Great, on the first day of Nisan, Mordecai son of Jair son of Shimei *b* son of Kish, of the tribe of Benjamin, had a dream. 3 He was a Jew living in the city of Susa, a great man, serving in the court of the king. 4 He was one of the captives whom King Nebuchadnezzar of Babylon had brought from Jerusalem with King Jeconiah of Judea. And this was his dream: 5 Noises *c* and confusion, thunders and earthquake, tumult on the earth! 6 Then two great dragons came forward, both ready to fight, and they roared terribly. 7 At their roaring every nation prepared for war, to fight against the righteous nation. 8 It was a day of darkness and gloom, of tribulation and distress, affliction and great tumult on the earth! 9 And the whole righteous nation was troubled; they feared the evils that threatened them, *d* and were ready to perish. 10 Then they cried out to God; and at their outcry, as though from a tiny spring, there came a great river, with abundant water; 11 light came, and the sun rose, and the lowly were exalted and devoured those held in honor.

12 Mordecai saw in this dream what God had determined to do, and after he awoke he had it on his mind, seeking all day to understand it in every detail.

12 Now Mordecai took his rest in the courtyard with Gabatha and Tharra, the two eunuchs of the king who kept watch in the courtyard. 2 He overheard their conversation and inquired

a Chapters 11.2—12.6 correspond to chapter A 1-17 in some translations. b Gk *Semeios*
c Or *Voices* d Gk *their own evils*

11.2–12: Mordecai's dream of impending conflict between two *dragons* (Mordecai and Haman; compare 10.7) pictures how the *righteous nation* Israel, threatened with annihilation, is delivered by God. **2:** *Artaxerxes,* the Greek name used throughout the Greek version of Esther and to be identified with Heb Ahasuerus, which is Xerxes I (486 [or 485]–465 B.C.). His *second year* was 485 [or 484] B.C. As the first month of the new year, Nisan (Babylonian name for March-April) is the appropriate time for a new and significant dream for Mordecai. All the month names in the book are of Babylonian derivation. **4:** This verse dates Mordecai's captivity in 597 B.C. (2 Kings 24.15); v. 2 dates his dream 112 years later. **7:** *The righteous nation,* that is, the Jews. **10:** The expressions *tiny spring* and *great river* refer to Esther (10.6). **12:** Mordecai knows that the *dream* forecasts God's action, but does not yet grasp the interpretation given in 10.6–12.

12.1–6: Mordecai saves the king's life when two eunuchs posted to protect the king plot instead to kill him. It is unclear whether this plot is the same as the one in Esth 2.19–23 of the Hebrew text or is an earlier one. **5:** The king rewards Mordecai (in Esth 6.3, however, it is said later that "You have not done anything for him"). **6:** It is implied that Haman shared in the plot and so resented Mordecai's action. *Bougean* represents a term of reproach, which in Esth 3.1 translates the Hebrew word Agagite (compare 1 Sam 15.8).

into their purposes, and learned that they were preparing to lay hands on King Artaxerxes; and he informed the king concerning them. ³Then the king examined the two eunuchs, and after they had confessed it, they were led away to execution. ⁴The king made a permanent record of these things, and Mordecai wrote an account of them. ⁵And the king ordered Mordecai to serve in the court, and rewarded him for these things. ⁶But Haman son of Hammedatha, a Bougean, who was in great honor with the king, determined to injure Mordecai and his people because of the two eunuchs of the king.

END OF ADDITION A

---

1 It was after this that the following things happened in the days of Artaxerxes, the same Artaxerxes who ruled over one hundred twenty-seven provinces from India to Ethiopia. ᵉ ²In those days, when King Artaxerxes was enthroned in the city of Susa, ³in the third year of his reign, he gave a banquet for his Friends and other persons of various nations, the Persians and Median nobles, and the governors of the provinces. ⁴After this, when he had displayed to them the riches of his kingdom and the splendor of his bountiful celebration during the course of one hundred eighty days, ⁵at the end of the festivityᶠ the king gave a drinking party for the people of various nations who lived in the city. This was held for six days in the courtyard of the royal palace, ⁶which was adorned with curtains of fine linen and cotton, held by

cords of purple linen attached to gold and silver blocks on pillars of marble and other stones. Gold and silver couches were placed on a mosaic floor of emerald, mother-of-pearl, and marble. There were coverings of gauze, embroidered in various colors, with roses arranged around them. ⁷The cups were of gold and silver, and a miniature cup was displayed, made of ruby, worth thirty thousand talents. There was abundant sweet wine, such as the king himself drank. ⁸The drinking was not according to a fixed rule; but the king wished to have it so, and he commanded his stewards to comply with his pleasure and with that of the guests.

9 Meanwhile, Queen Vashtiᵍ gave a drinking party for the women in the palace where King Artaxerxes was.

10 On the seventh day, when the king was in good humor, he told Haman, Bazan, Tharra, Boraze, Zatholtha, Abataza, and Tharaba, the seven eunuchs who served King Artaxerxes, ¹¹to escort the queen to him in order to proclaim her as queen and to place the diadem on her head, and to have her display her beauty to all the governors and the people of various nations, for she was indeed a beautiful woman. ¹²But Queen Vashtiᵍ refused to obey him and would not come with the eunuchs. This offended the king and he became furious. ¹³He said to his Friends, "This is how Vashtiᵍ has answered me. ʰ Give therefore your ruling and judgment on this matter." ¹⁴Arkезасиз, Sарзатhacuз, and Malезсar, then

---

e Other ancient authorities lack *to Ethiopia*
f Gk *marriage feast*   g Gk *Astin*   h Gk *Astin has said thus and so*

1.1–9: Artaxerxes' banquet. The king gives a lavish seven-day drinking party to celebrate his marriage. **2:** That the king *was enthroned* underscores that he now rules securely, having successfully put down earlier uprisings throughout his empire. **5:** *The festivity* (literally, "the marriage feast"), in the Hebrew text there is no mention of this being a marriage feast (but see 1.11 of the Greek). Here, the Greek speaks of *six days* of partying; but it has "seven" in 1.10 (so also 1.5, 10 of the Hebrew). **7:** The Hebrew text makes no

mention of the fabulously expensive *miniature cup.*

1.10–22: The dismissal of Queen Vashti, the result of her refusing to appear before the king and his guests. **11:** *To proclaim her as queen,* here, Vashti's coronation is the purpose of her appearance and not, as in the Hebrew, merely for the king to display her fabled beauty to his drunken friends. **14:** In place of the seven privy counselors named in the Hebrew text, three *governors* are named here: *Arkesaeus, Sarsathaeus, and Malesear.*

the governors of the Persians and Medes who were closest to the king—Arkesaeus, Sarsathaeus, and Malesear, who sat beside him in the chief seats—came to him [15] and told him what must be done to Queen Vashti[i] for not obeying the order that the king had sent her by the eunuchs. [16] Then Muchaeus said to the king and the governors, "Queen Vashti[i] has insulted not only the king but also all the king's governors and officials" [17] (for he had reported to them what the queen had said and how she had defied the king). "And just as she defied King Artaxerxes, [18] so now the other ladies who are wives of the Persian and Median governors, on hearing what she has said to the king, will likewise dare to insult their husbands. [19] If therefore it pleases the king, let him issue a royal decree, inscribed in accordance with the laws of the Medes and Persians so that it may not be altered, that the queen may no longer come into his presence; but let the king give her royal rank to a woman better than she. [20] Let whatever law the king enacts be proclaimed in his kingdom, and thus all women will give honor to their husbands, rich and poor alike." [21] This speech pleased the king and the governors, and the king did as Muchaeus had recommended. [22] The king sent the decree into all his kingdom, to every province in its own language, so that in every house respect would be shown to every husband.

2 After these things, the king's anger abated, and he no longer was concerned about Vashti[i] or remembered what he had said and how he had condemned her. [2] Then the king's servants said, "Let beautiful and virtuous girls be sought out for the king. [3] The king shall appoint officers in all the provinces of his kingdom, and they shall select beautiful young virgins to be brought to the harem in Susa, the capital. Let them be en-

trusted to the king's eunuch who is in charge of the women, and let ointments and whatever else they need be given them. [4] And the woman who pleases the king shall be queen instead of Vashti.[i] This pleased the king, and he did so.

[5] Now there was a Jew in Susa the capital whose name was Mordecai son of Jair son of Shimei[j] son of Kish, of the tribe of Benjamin; [6] he had been taken captive from Jerusalem among those whom King Nebuchadnezzar of Babylon had captured. [7] And he had a foster child, the daughter of his father's brother, Aminadab, and her name was Esther. When her parents died, he brought her up to womanhood as his own. The girl was beautiful in appearance. [8] So, when the decree of the king was proclaimed, and many girls were gathered in Susa the capital in custody of Gai, Esther also was brought to Gai, who had custody of the women. [9] The girl pleased him and won his favor, and he quickly provided her with ointments and her portion of food,[k] as well as seven maids chosen from the palace; he treated her and her maids with special favor in the harem. [10] Now Esther had not disclosed her people or country, for Mordecai had commanded her not to make it known. [11] And every day Mordecai walked in the courtyard of the harem, to see what would happen to Esther.

[12] Now the period after which a girl was to go to the king was twelve months. During this time the days of beautification are completed—six months while they are anointing themselves with oil of myrrh, and six months with spices and ointments for women. [13] Then she goes in to the king; she is handed to the person appointed, and goes with him from the harem to the

i Gk *Astin*   j Gk *Semeios*   k Gk lacks *of food*

2.1–18: Esther becomes the new queen by following carefully the instructions of her foster father Mordecai and the harem etiquette as offered by Gai, the eunuch in charge

of the harem. **7:** *Esther,* whose Hebrew name ("Hadassah") is lacking here, is the daughter of *Aminadab* ("Abihail" in Hebrew at 2.15). **2.16:** Esther went in to the king *in the twelfth*

king's palace. [14] In the evening she enters and in the morning she departs to the second harem, where Gai the king's eunuch is in charge of the women; and she does not go in to the king again unless she is summoned by name.

15 When the time was fulfilled for Esther daughter of Aminadab, the brother of Mordecai's father, to go in to the king, she neglected none of the things that Gai, the eunuch in charge of the women, had commanded. Now Esther found favor in the eyes of all who saw her. [16] So Esther went in to King Artaxerxes in the twelfth month, which is Adar, in the seventh year of his reign. [17] And the king loved Esther and she found favor beyond all the other virgins, so he put on her the queen's diadem. [18] Then the king gave a banquet lasting seven days for all his Friends and the officers to celebrate his marriage to Esther; and he granted a remission of taxes to those who were under his rule.

19 Meanwhile Mordecai was serving in the courtyard. [20] Esther had not disclosed her country—such were the instructions of Mordecai; but she was to fear God and keep his laws, just as she had done when she was with him. So Esther did not change her mode of life.

21 Now the king's eunuchs, who were chief bodyguards, were angry because of Mordecai's advancement, and they plotted to kill King Artaxerxes. [22] The matter became known to Mordecai, and he warned Esther, who in turn revealed the plot to the king. [23] He investigated the two eunuchs and hanged them. Then the king ordered a memorandum to be deposited in the royal library in praise of the goodwill shown by Mordecai.

3 After these events King Artaxerxes promoted Haman son of Hammedatha, a Bougean, advancing him and granting him precedence over all the king's[l] Friends. [2] So all who were at court used to do obeisance to Haman,[m] for so the king had commanded to be done. Mordecai, however, did not do obeisance. [3] Then the king's courtiers said to Mordecai, "Mordecai, why do you disobey the king's command?" [4] Day after day they spoke to him, but he would not listen to them. Then they informed Haman that Mordecai was resisting the king's command. Mordecai had told them that he was a Jew. [5] So when Haman learned that Mordecai was not doing obeisance to him, he became furiously angry, [6] and plotted to destroy all the Jews under Artaxerxes' rule.

7 In the twelfth year of King Artaxerxes Haman[n] came to a decision by casting lots, taking the days and the months one by one, to fix on one day to destroy the whole race of Mordecai. The lot fell on the fourteenth[o] day of the month of Adar.

8 Then Haman[n] said to King Artax-

---

l Gk *all his*   *m* Gk *him*   *n* Gk *he*
o Other ancient witnesses read *thirteenth*; see 8.12

---

month . . . *Adar* (March–April), but the Hebrew at 2.16 has "tenth month . . . Tebeth" (December–January). **17**: *The queen's diadem,* the actual queen of Xerxes was Amestris (Herodotus, *Hist.* VII, 61), a Persian woman. **18**: *Granted a remission of taxes,* Hebrew "granted a holiday."
2.19–23: A plot to kill King Artaxerxes. According to the Greek text this is a second plot discovered and reported by Mordecai (see Addition A [12.1–6]). **20**: *She was to fear God and keep his laws* is lacking in the Hebrew. *Esther did not change her mode of life,* there is no hint of this in the Hebrew, where she is a Jew more by ethnicity than religious faith or observance. **21**: *Angry because of Mor-* *decai's advancement,* the Hebrew offers no motive for the bodyguards' anger against the king.
3.1–7: Mordecai refuses to do obeisance to Haman, thereby earning the latter's determination to destroy Mordecai and his people. **7**: In keeping with the Greek's playing down the establishing of the festival of Purim, here (in contrast to 3.7 of the Hebrew) no clue is given until 9.26 that this *casting lots* will provide the name for the festival of Purim ("Lots"). Instead of *fourteenth day,* 8.12 of the Greek and the Hebrew rightly have "the thirteenth."
3.8–13: The royal decree against the Jews, instigated by Haman, is empire-wide

erxes, "There is a certain nation scattered among the other nations in all your kingdom; their laws are different from those of every other nation, and they do not keep the laws of the king. It is not expedient for the king to tolerate them. ⁹If it pleases the king, let it be decreed that they are to be destroyed, and I will pay ten thousand talents of silver into the king's treasury." ¹⁰So the king took off his signet ring and gave it to Haman to seal the decree*ᵖ* that was to be written against the Jews. ¹¹The king told Haman, "Keep the money, and do whatever you want with that nation."

12 So on the thirteenth day of the first month the king's secretaries were summoned, and in accordance with Haman's instructions they wrote in the name of King Artaxerxes to the magistrates and the governors in every province from India to Ethiopia. There were one hundred twenty-seven provinces in all, and the governors were addressed each in his own language. ¹³Instructions were sent by couriers throughout all the empire of Artaxerxes to destroy the Jewish people on a given day of the twelfth month, which is Adar, and to plunder their goods.

ADDITION B

13 ᵠ This is a copy of the letter: "The Great King, Artaxerxes, writes the following to the governors of the hundred twenty-seven provinces from India to Ethiopia and to the officials under them:

2 "Having become ruler of many na-

tions and master of the whole world (not elated with presumption of authority but always acting reasonably and with kindness), I have determined to settle the lives of my subjects in lasting tranquility and, in order to make my kingdom peaceable and open to travel throughout all its extent, to restore the peace desired by all people.

3 "When I asked my counselors how this might be accomplished, Haman—who excels among us in sound judgment, and is distinguished for his unchanging goodwill and steadfast fidelity, and has attained the second place in the kingdom— ⁴pointed out to us that among all the nations in the world there is scattered a certain hostile people, who have laws contrary to those of every nation and continually disregard the ordinances of kings, so that the unifying of the kingdom that we honorably intend cannot be brought about. ⁵We understand that this people, and it alone, stands constantly in opposition to every nation, perversely following a strange manner of life and laws, and is ill-disposed to our government, doing all the harm they can so that our kingdom may not attain stability.

6 "Therefore we have decreed that those indicated to you in the letters written by Haman, who is in charge of affairs and is our second father, shall all—wives and children included—be utterly destroyed by the swords of their enemies, without pity or restraint, on the four-

---

p Gk lacks *the decree*   *q* Chapter 13.1–7 corresponds to chapter B 1–7 in some translations.

---

and fixed for the thirteenth of Adar. **8:** *Their laws are different,* i.e. the Mosaic law. **13:** *Instructions were sent,* the Greek omits the Hebrew's hyperbolic claim "to every province in its own script and every people in its own language" (3.12). Aramaic was normally used for such official correspondence. *To destroy the Jewish people* is a tame rendering of the Hebrew's far more explicitly brutal "to destroy, to kill, and to annihilate all Jews, young and old, women and children." *On a*

given day, i.e. the thirteenth (so 8.12 of the Greek and of the Hebrew). **13.1–7: The king's letter ordering the massacre of the Jews.** This addition follows Esth 3.1–13, in which Haman has induced Artaxerxes to send a letter to all provinces of his kingdom, ordering complete annihilation of *a certain hostile people,* the Jews, because they observe *a strange manner of life and laws;* they observe the Mosaic law. **6:** *Our second father* implies that Haman ranked second only

teenth day of the twelfth month, Adar, of this present year, 7so that those who have long been hostile and remain so may in a single day go down in violence to Hades, and leave our government completely secure and untroubled hereafter."

END OF ADDITION B

3 14Copies of the document were posted in every province, and all the nations were ordered to be prepared for that day. 15The matter was expedited also in Susa. And while the king and Haman caroused together, the city of Susa*r* was thrown into confusion.

4 When Mordecai learned of all that had been done, he tore his clothes, put on sackcloth, and sprinkled himself with ashes; then he rushed through the street of the city, shouting loudly: "An innocent nation is being destroyed!" 2He got as far as the king's gate, and there he stopped, because no one was allowed to enter the courtyard clothed in sackcloth and ashes. 3And in every province where the king's proclamation had been posted there was a loud cry of mourning and lamentation among the Jews, and they put on sackcloth and ashes. 4When the queen's*s* maids and eunuchs came and told her, she was deeply troubled by what she heard had happened, and sent some clothes to Mordecai to put on instead of sackcloth; but he would not consent. 5Then Esther summoned Hachratheus, the eunuch who attended her, and ordered him to get accurate information for her from Mordecai.*t*

7 So Mordecai told him what had happened and how Haman had promised to pay ten thousand talents into the royal treasury to bring about the destruction of the Jews. 8He also gave him a copy of what had been posted in Susa for their destruction, to show to Esther; and he told him to charge her to go in to the king and plead for his favor in behalf of the people. "Remember," he said, "the days when you were an ordinary person, being brought up under my care—for Haman, who stands next to the king, has spoken against us and demands our death. Call upon the Lord; then speak to the king in our behalf, and save us from death."

9 Hachratheus went in and told Esther all these things. 10And she said to him, "Go to Mordecai and say, 11'All nations of the empire know that if any man or woman goes to the king inside the inner court without being called, there is no escape for that person. Only the one to whom the king stretches out the golden scepter is safe—and it is now thirty days since I was called to go to the king.'"

12 When Hachratheus delivered her entire message to Mordecai, 13Mordecai told him to go back and say to her, "Esther, do not say to yourself that you alone among all the Jews will escape alive. 14For if you keep quiet at such a time as this, help and protection will come to the Jews from another quarter, but you and your father's family will perish. Yet, who knows whether it was not for such a time as this that you were made queen?" 15Then Esther gave the messenger this answer to take back to Mordecai: 16"Go and gather all the Jews who are in Susa and fast on my behalf;

r Gk *the city*   *s* Gk *When her*   *t* Other ancient witnesses add *6So Hachratheus went out to Mordecai in the street of the city opposite the city gate.*

---

to the king. *Fourteenth day,* according to 16.20 (and Esth 3.13; 8.12; 9.1) it was the thirteenth day. *Adar,* February–March.

3.14–15: The letter is delivered and panic ensues.

4.1–17: Mordecai seeks Esther's aid. He orders her to risk her life by going, unsummoned, before the king to ask him to spare her people. **8:** *Call upon the Lord,* Mordecai makes no such request in the Hebrew.

4.11: Although *without being called, there is no escape* is simply a security precaution, it effectively sets the stage for Esther's dramatic appearance in Addition D. **14:** *Help . . . from another quarter,* in contrast to the Hebrew, here there can be no doubt that divine help is

for three days and nights do not eat or drink, and my maids and I will also go without food. After that I will go to the king, contrary to the law, even if I must die." [17] So Mordecai went away and did what Esther had told him to do.

ADDITION C

13 [8] "Then Mordecai[v] prayed to the Lord, calling to remembrance all the works of the Lord.

9 He said, "O Lord, Lord, you rule as King over all things, for the universe is in your power and there is no one who can oppose you when it is your will to save Israel, [10] for you have made heaven and earth and every wonderful thing under heaven. [11] You are Lord of all, and there is no one who can resist you, the Lord. [12] You know all things; you know, O Lord, that it was not in insolence or pride or for any love of glory that I did this, and refused to bow down to this proud Haman; [13] for I would have been willing to kiss the soles of his feet to save Israel! [14] But I did this so that I might not set human glory above the glory of God, and I will not bow down to anyone but you, who are my Lord; and I will not do these things in pride. [15] And now, O Lord God and King, God of Abraham, spare your people; for the eyes of our foes are upon us[w] to annihilate us, and they desire to destroy the inheritance that has been yours from the beginning. [16] Do not neglect your portion, which you redeemed for yourself out of the land of Egypt. [17] Hear my prayer, and have mercy upon your inheritance; turn our mourning into feasting that we may

live and sing praise to your name, O Lord; do not destroy the lips[x] of those who praise you."

18 And all Israel cried out mightily, for their death was before their eyes.

14 Then Queen Esther, seized with deadly anxiety, fled to the Lord. [2] She took off her splendid apparel and put on the garments of distress and mourning, and instead of costly perfumes she covered her head with ashes and dung, and she utterly humbled her body; every part that she loved to adorn she covered with her tangled hair. [3] She prayed to the Lord God of Israel, and said: "O my Lord, you only are our king; help me, who am alone and have no helper but you, [4] for my danger is in my hand. [5] Ever since I was born I have heard in the tribe of my family that you, O Lord, took Israel out of all the nations, and our ancestors from among all their forebears, for an everlasting inheritance, and that you did for them all that you promised. [6] And now we have sinned before you, and you have handed us over to our enemies [7] because we glorified their gods. You are righteous, O Lord! [8] And now they are not satisfied that we are in bitter slavery, but they have covenanted with their idols [9] to abolish what your mouth has ordained, and to destroy your inheritance, to stop the mouths of those who praise you and to quench your altar and the glory of your house, [10] to open the mouths of the nations for the

---

u Chapters 13.8—15.16 correspond to chapters C 1-30 and D 1-16 in some translations.
v Gk *he*   *w* Gk *for they are eying us*
x Gk *mouth*

---

meant. **16:** *Fast on my behalf,* the act of fasting is also here in the Hebrew; it is the only religious activity explicitly mentioned in the Hebrew text (see 4.3, 14; 9.31).
13.8–14.19: The prayers of Mordecai and Esther. The canonical book of Esther never mentions God or prayer or calls Israel God's chosen people. These prayers are added to give the book an explicitly religious tone. They call God *Lord* and *King* and *Lord God of Abraham.* He is righteous, has created and rules all things, has redeemed his people Israel

from Egypt, answers prayer, and can save them now.
13.12–14: Mordecai did not *bow down* to Haman, as the king's other servants did (Esth 3.2), because such homage is due only to God. **18:** All Israel echoed Mordecai's prayer.
14.1–19: Esther joins in her people's prayer for deliverance. 1–2: She discards every trace of queenly attire and elegance and prays as an unworthy member of Israel. **6–7:** Her people's captivity is due to their sinfulness and idolatry while living in Palestine

praise of vain idols, and to magnify forever a mortal king.

11 "O Lord, do not surrender your scepter to what has no being; and do not let them laugh at our downfall; but turn their plan against them, and make an example of him who began this against us. 12Remember, O Lord; make yourself known in this time of our affliction, and give me courage, O King of the gods and Master of all dominion! 13Put eloquent speech in my mouth before the lion, and turn his heart to hate the man who is fighting against us, so that there may be an end of him and those who agree with him. 14But save us by your hand, and help me, who am alone and have no helper but you, O Lord. 15You have knowledge of all things, and you know that I hate the splendor of the wicked and abhor the bed of the uncircumcised and of any alien. 16You know my necessity—that I abhor the sign of my proud position, which is upon my head on days when I appear in public. I abhor it like a filthy rag, and I do not wear it on the days when I am at leisure. 17And your servant has not eaten at Haman's table, and I have not honored the king's feast or drunk the wine of libations. 18Your servant has had no joy since the day that I was brought here until now, except in you, O Lord God of Abraham. 19O God, whose might is over all, hear the voice of the despairing, and save us from the hands of evildoers. And save me from my fear!"

END OF ADDITION C

ADDITION D

15 On the third day, when she ended her prayer, she took off the garments in which she had worshiped, and arrayed herself in splendid attire. 2Then, majestically adorned, after invoking the aid of the all-seeing God and Savior, she took two maids with her; 3on one she leaned gently for support, 4while the other followed, carrying her train. 5She was radiant with perfect beauty, and she looked happy, as if beloved, but her heart was frozen with fear. 6When she had gone through all the doors, she stood before the king. He was seated on his royal throne, clothed in the full array of his majesty, all covered with gold and precious stones. He was most terrifying.

7 Lifting his face, flushed with splendor, he looked at her in fierce anger. The queen faltered, and turned pale and faint, and collapsed on the head of the maid who went in front of her. 8Then God changed the spirit of the king to gentleness, and in alarm he sprang from his throne and took her in his arms until she came to herself. He comforted her with soothing words, and said to her, 9"What is it, Esther? I am your husband. *y* Take courage; 10You shall not die, for our law applies only to our subjects. *z* Come near."

11 Then he raised the golden scepter and touched her neck with it; 12he embraced her, and said, "Speak to me." 13She said to him, "I saw you, my lord, like an angel of God, and my heart was

y Gk *brother*   *z* Meaning of Gk uncertain

(Dan 9.16); Israel does not deserve to be saved.

14.11: Esther begs God not to allow pagan idols, which have no real existence, to rule the world. **16–18:** Esther says that she hates her position as queen and wife of the king, and has never *eaten at Haman's table* or *honored the king's feast* (but compare Esth 2.18, as well as Esth 5.5 and 7.1).

15.1–16: Esther risks her life to appeal to the king. This passage expands and exaggerates Esth 5.1–2, bringing out religious ele-

ments absent from the Hebrew text. **1–2:** *She had worshiped . . . invoking the aid of the all-seeing God and Savior* is lacking in the Hebrew. Anyone who entered the king's presence without his summons or permission was put to death, unless the king forgave the intrusion (Esth 4.11). To avoid personal danger and to induce the king to reverse his decree, Esther enhanced her charm by splendid attire; then with fearful heart she entered the king's presence.

15.8: That *God changed the spirit of the king*

shaken with fear at your glory. 14For you are wonderful, my lord, and your countenance is full of grace." 15And while she was speaking, she fainted and fell. 16Then the king was agitated, and all his servants tried to comfort her.

END OF ADDITION D

5 *a* 3The king said to her, "What do you wish, Esther? What is your request? It shall be given you, even to half of my kingdom." 4And Esther said, "Today is a special day for me. If it pleases the king, let him and Haman come to the dinner that I shall prepare today." 5Then the king said, "Bring Haman quickly, so that we may do as Esther desires." So they both came to the dinner that Esther had spoken about. 6While they were drinking wine, the king said to Esther, "What is it, Queen Esther? It shall be granted you." 7She said, "My petition and request is: 8if I have found favor in the sight of the king, let the king and Haman come to the dinner that I shall prepare them, and tomorrow I will do as I have done today."

9 So Haman went out from the king joyful and glad of heart. But when he saw Mordecai the Jew in the courtyard, he was filled with anger. 10Nevertheless, he went home and summoned his friends and his wife Zosara. 11And he told them about his riches and the honor that the king had bestowed on him, and how he had advanced him to be the first in the kingdom. 12And Haman said, "The queen did not invite anyone to the dinner with the king except me; and I am invited again tomorrow. 13But these things give

me no pleasure as long as I see Mordecai the Jew in the courtyard." 14His wife Zosara and his friends said to him, "Let a gallows be made, fifty cubits high, and in the morning tell the king to have Mordecai hanged on it. Then, go merrily with the king to the dinner." This advice pleased Haman, and so the gallows was prepared.

6 That night the Lord took sleep from the king, so he gave orders to his secretary to bring the book of daily records, and to read to him. 2He found the words written about Mordecai, how he had told the king about the two royal eunuchs who were on guard and sought to lay hands on King Artaxerxes. 3The king said, "What honor or dignity did we bestow on Mordecai?" The king's servants said, "You have not done anything for him." 4While the king was inquiring about the goodwill shown by Mordecai, Haman was in the courtyard. The king asked, "Who is in the courtyard?" Now Haman had come to speak to the king about hanging Mordecai on the gallows that he had prepared. 5The servants of the king answered, "Haman is standing in the courtyard." And the king said, "Summon him." 6Then the king said to Haman, "What shall I do for the person whom I wish to honor?" And Haman said to himself, "Whom would the king wish to honor more than me?" 7So he said to the king, "For a person whom the king wishes to honor, 8let the king's servants bring out the fine linen robe that the king has worn, and the horse on which the king rides, 9and let

a In Greek, Chapter D replaces verses 1 and 2 in Hebrew.

---

is in no way indicated in the Hebrew text. **13**: *You . . . like an angel of God,* lacking in Hebrew, is pure flattery.

5.3–8: Esther before the king. 3: *To half of my kingdom* (also 7.2) is a customary hyperbole (compare Mk 6.23). **4, 8**: No subtle reason need be sought for the first and second *dinner,* nor for Haman's being the only guest; these are but literary embellishments needful for the story.

5.9–14: Haman's plot against Mordecai will, temporarily, assuage his anger against Mordecai. **14**: *Gallows . . . fifty cubits high* (more than 75 feet high) is a hyperbole for literary effect.

6.1–13: Mordecai's generous reward from the king is, ironically, proposed by Haman himself. **1**: *The Lord took sleep from the king,* Heb "The king could not sleep." **3**: *You have not done anything for him,* in contrast to the

both be given to one of the king's honored Friends, and let him robe the person whom the king loves and mount him on the horse, and let it be proclaimed through the open square of the city, saying, 'Thus shall it be done to everyone whom the king honors.' " [10]Then the king said to Haman, "You have made an excellent suggestion! Do just as you have said for Mordecai the Jew, who is on duty in the courtyard. And let nothing be omitted from what you have proposed." [11]So Haman got the robe and the horse; he put the robe on Mordecai and made him ride through the open square of the city, proclaiming, "Thus shall it be done to everyone whom the king wishes to honor." [12]Then Mordecai returned to the courtyard, and Haman hurried back to his house, mourning and with his head covered. [13]Haman told his wife Zosara and his friends what had befallen him. His friends and his wife said to him, "If Mordecai is of the Jewish people, and you have begun to be humiliated before him, you will surely fall. You will not be able to defend yourself, because the living God is with him."

14 While they were still talking, the eunuchs arrived and hurriedly brought Haman to the banquet that Esther had prepared. [1]So the king and Haman went in to drink with the queen. [2]And the second day, as they were drinking wine, the king said, "What is it, Queen Esther? What is your petition and what is your request? It shall be granted to you, even to half of my kingdom." [3]She answered and said, "If I have found favor with the king, let my life be granted me at my petition, and my people at

7

my request. [4]For we have been sold, I and my people, to be destroyed, plundered, and made slaves—we and our children—male and female slaves. This has come to my knowledge. Our antagonist brings shame on[b] the king's court." [5]Then the king said, "Who is the person that would dare to do this thing?" [6]Esther said, "Our enemy is this evil man Haman!" At this, Haman was terrified in the presence of the king and queen.

7 The king rose from the banquet and went into the garden, and Haman began to beg for his life from the queen, for he saw that he was in serious trouble. [8]When the king returned from the garden, Haman had thrown himself on the couch, pleading with the queen. The king said, "Will he dare even assault my wife in my own house?" Haman, when he heard, turned away his face. [9]Then Bugathan, one of the eunuchs, said to the king, "Look, Haman has even prepared a gallows for Mordecai, who gave information of concern to the king; it is standing at Haman's house, a gallows fifty cubits high." So the king said, "Let Haman be hanged on that." [10]So Haman was hanged on the gallows he had prepared for Mordecai. With that the anger of the king abated.

8 On that very day King Artaxerxes granted to Esther all the property of the persecutor[c] Haman. Mordecai was summoned by the king, for Esther had told the king[d] that he was related to her. [2]The king took the ring that had been taken from Haman, and gave it to Mor-

b Gk *is not worthy of*   *c* Gk *slanderer*
d Gk *him*

---

king's rewarding Mordecai for his information concerning the first plot (see Addition A [12.5]). **13**: *Because the living God is with him* foreshadows for the reader Haman's sure and certain demise.

6.14–7.6: At Esther's second banquet Haman is caught unawares by her and is revealed for the evil counselor he is.

7.7–10: Haman's punishment of being hanged on the gallows he had erected for Mordecai is poetic justice. **8**: *Even assault my*

wife, by even touching Esther's couch to beg for his life Haman is violating harem law and committing a capital offense. **9**: *Bugathan,* Heb "Harbona."

8.1–12: The king's favor toward the Jews is proved by Mordecai's being invested with all the powers of Haman and by having Haman's royal edict against the Jews neutralized by a new royal edict. **2**: *The king took the ring . . . from Haman,* this transference of the signet ring to Mordecai indicates his promo-

decai; and Esther set Mordecai over everything that had been Haman's.

3 Then she spoke once again to the king and, falling at his feet, she asked him to avert all the evil that Haman had planned against the Jews. 4 The king extended his golden scepter to Esther, and she rose and stood before the king. 5 Esther said, "If it pleases you, and if I have found favor, let an order be sent rescinding the letters that Haman wrote and sent to destroy the Jews in your kingdom. 6 How can I look on the ruin of my people? How can I be safe if my ancestral nation*e* is destroyed?" 7 The king said to Esther, "Now that I*f* have granted all of Haman's property to you and have hanged him on a tree because he acted against the Jews, what else do you request? 8 Write in my name what you think best and seal it with my ring; for whatever is written at the king's command and sealed with my ring cannot be contravened."

9 The secretaries were summoned on the twenty-third day of the first month, that is, Nisan, in the same year; and all that he commanded with respect to the Jews was given in writing to the administrators and governors of the provinces from India to Ethiopia, one hundred twenty-seven provinces, to each province in its own language. 10 The edict was written*g* with the king's authority and sealed with his ring, and sent out by couriers. 11 He ordered the Jews in every city to observe their own laws, to defend themselves, and to act as they wished against their opponents and enemies 12 on a certain day, the thirteenth of the twelfth month, which is Adar, throughout all the kingdom of Artaxerxes.

ADDITION E

16 *h* The following is a copy of this letter:

"The Great King, Artaxerxes, to the governors of the provinces from India to Ethiopia, one hundred twenty-seven provinces, and to those who are loyal to our government, greetings.

2 "Many people, the more they are honored with the most generous kindness of their benefactors, the more proud do they become, 3 and not only seek to injure our subjects, but in their inability to stand prosperity, they even undertake to scheme against their own benefactors. 4 They not only take away thankfulness from others, but, carried away by the boasts of those who know nothing of goodness, they even assume that they will escape the evil-hating justice of God, who always sees everything. 5 And often many of those who are set in places of authority have been made in part responsible for the shedding of innocent blood, and have been involved in irremediable calamities, by the persuasion of friends who have been entrusted with the administration of public affairs, 6 when these persons by the false trickery of their evil natures beguile the sincere goodwill of their sovereigns.

7 "What has been wickedly accomplished through the pestilent behavior of those who exercise authority unworthily can be seen, not so much from the more ancient records that we hand on, as from investigation of matters close at hand.*i* 8 In the future we will take care to render our kingdom quiet and peaceable for all, 9 by changing our methods and always judging what comes before our eyes with more equitable consideration. 10 For Haman son of Hammedatha, a Macedonian (really an alien to the Persian blood, and quite devoid of our kindliness), having become our guest, 11 enjoyed so fully the

e Gk *country* *f* Gk *If I*
g Gk *It was written* *h* Chapter 16.1-24 corresponds to chapter E 1-24 in some translations. *i* Gk *matters beside* (your) *feet*

---

tion to the rank of grand vizier. **3:** *Spoke once again to the king* but with none of the "weeping and pleading" in 8.3 of the Hebrew. **8:** *Whatever is written . . . cannot be contravened* (see

1.19), this belief in the immutability of Persian law, a literary invention for effect, also appears in Dan 6.8.

8.9: *First month . . . Nisan,* Hebrew "third

goodwill that we have for every nation that he was called our father and was continually bowed down to by all as the person second to the royal throne. 12 But, unable to restrain his arrogance, he undertook to deprive us of our kingdom and our life,*j* 13 and with intricate craft and deceit asked for the destruction of Mordecai, our savior and perpetual benefactor, and of Esther, the blameless partner of our kingdom, together with their whole nation. 14 He thought that by these methods he would catch us undefended and would transfer the kingdom of the Persians to the Macedonians.

15 "But we find that the Jews, who were consigned to annihilation by this thrice-accursed man, are not evildoers, but are governed by most righteous laws 16 and are children of the living God, most high, most mighty,*k* who has directed the kingdom both for us and for our ancestors in the most excellent order.

17 "You will therefore do well not to put in execution the letters sent by Haman son of Hammedatha, 18 since he, the one who did these things, has been hanged at the gate of Susa with all his household—for God, who rules over all things, has speedily inflicted on him the punishment that he deserved.

19 "Therefore post a copy of this letter publicly in every place, and permit the Jews to live under their own laws. 20 And give them reinforcements, so that on the thirteenth day of the twelfth month, Adar, on that very day, they may defend themselves against those who attack them at the time of oppression. 21 For God, who rules over all things, has made this day to be a joy for his chosen people instead of a day of destruction for them.

22 "Therefore you shall observe this with all good cheer as a notable day among your commemorative festivals, 23 so that both now and hereafter it may represent deliverance for you*l* and the loyal Persians, but that it may be a reminder of destruction for those who plot against us.

24 "Every city and country, without exception, that does not act accordingly shall be destroyed in wrath with spear and fire. It shall be made not only impassable for human beings, but also most hateful to wild animals and birds for all time.

END OF ADDITION E

8 13 "Let copies of the decree be posted conspicuously in all the kingdom, and let all the Jews be ready on that day to fight against their enemies."

14 So the messengers on horseback set out with all speed to perform what the king had commanded; and the decree was published also in Susa. 15 Mordecai went out dressed in the royal robe and wearing a gold crown and a turban of purple linen. The people in Susa rejoiced on seeing him. 16 And the Jews had light and gladness 17 in every city and province

j Gk *our spirit*   *k* Gk *greatest*   *l* Other ancient authorities read *for us*

---

month . . . Sivan [May-June]." **11:** This explicit permission for the *Jews . . . to observe their own laws,* which is a distinctly religious concern, is lacking in the Hebrew.

16.1–24: The king's second letter, denouncing Haman and directing his subjects to help the Jews.

16.10–14: Haman, already executed (Esth 7.6–10), is (falsely) called a *Macedonian* trying to overthrow Persian rule.

16.15–16: The king commends the Mosaic *laws,* praises the Jews, and recognizes God's rule. All Persian subjects are commanded to ignore Haman's letter (according to 13.1–6 it was really the king's letter) and help the Jews defend themselves on the thirteenth day of Adar, the prescribed day of annihilation. **18:** *Hanged . . . with all his household* contradicts 7.10 of the Hebrew, where only Haman is hanged (see also 9.6–10, 14 of the Greek text). **24:** Destruction faces every place that does not defend and respect the Jews.

8.13–17: Dispatch and posting of the king's decree. 15: Mordecai appears in his state robes. **17:** *A banquet and a holiday* contrast with the mourning and lamentation of 4.3. *Gentiles were circumcised,* the Hebrew says merely "professed to be Jews."

wherever the decree was published; wherever the proclamation was made, the Jews had joy and gladness, a banquet and a holiday. And many of the Gentiles were circumcised and became Jews out of fear of the Jews.

9 Now on the thirteenth day of the twelfth month, which is Adar, the decree written by the king arrived. 2 On that same day the enemies of the Jews perished; no one resisted, because they feared them. 3 The chief provincial governors, the princes, and the royal secretaries were paying honor to the Jews, because fear of Mordecai weighed upon them. 4 The king's decree required that Mordecai's name be held in honor throughout the kingdom. *m* 6 Now in the city of Susa the Jews killed five hundred people, 7 including Pharsannestain, Delphon, Phasga, 8 Pharadatha, Barea, Sarbacha, 9 Marmasima, Aruphaeus, Arsaeus, Zabutheus, 10 the ten sons of Haman son of Hammedatha, the Bougean, the enemy of the Jews—and they indulged*n* themselves in plunder.

11 That very day the number of those killed in Susa was reported to the king. 12 The king said to Esther, "In Susa, the capital, the Jews have destroyed five hundred people. What do you suppose they have done in the surrounding countryside? Whatever more you ask will be done for you." 13 And Esther said to the king, "Let the Jews be allowed to do the same tomorrow. Also, hang up the bodies of Haman's ten sons." 14 So he permitted this to be done, and handed over to the Jews of the city the bodies of Haman's sons to hang up. 15 The Jews who

were in Susa gathered on the fourteenth and killed three hundred people, but took no plunder.

16 Now the other Jews in the kingdom gathered to defend themselves, and got relief from their enemies. They destroyed fifteen thousand of them, but did not engage in plunder. 17 On the fourteenth day they rested and made that same day a day of rest, celebrating it with joy and gladness. 18 The Jews who were in Susa, the capital, came together also on the fourteenth, but did not rest. They celebrated the fifteenth with joy and gladness. 19 On this account then the Jews who are scattered around the country outside Susa keep the fourteenth of Adar as a joyful holiday, and send presents of food to one another, while those who live in the large cities keep the fifteenth day of Adar as their joyful holiday, also sending presents to one another.

20 Mordecai recorded these things in a book, and sent it to the Jews in the kingdom of Artaxerxes both near and far, 21 telling them that they should keep the fourteenth and fifteenth days of Adar, 22 for on these days the Jews got relief from their enemies. The whole month (namely, Adar), in which their condition had been changed from sorrow into gladness and from a time of distress to a holiday, was to be celebrated as a time for feasting*o* and gladness and

m Meaning of Gk uncertain. Some ancient authorities add verse 5, *So the Jews struck down all their enemies with the sword, killing and destroying them, and they did as they pleased to those who hated them.* *n* Other ancient authorities read *did not indulge* *o* Gk *of weddings*

9.1–19: Victory of the Jews over their enemies on the thirteenth of Adar is total throughout the empire, with the exception of Susa, which has complete success the next day (v. 15). Therefore, Jews outside of Susa celebrate their victory on the fourteenth of Adar; those in Susa on the fifteenth. **2:** That *no one resisted,* which contradicts the Hebrew ("no one could withstand them"), is patently false; for thousands did (see 9.11, 16). **9.10:** That *the ten sons of Haman* were killed

on the thirteenth of Adar agrees with the Hebrew but contradicts Addition E (16.18). *Indulged themselves in plunder* contradicts the Hebrew. **16:** *Destroyed fifteen thousand,* Hebrew "seventy-five thousand."

9.20–10.3: The festival of Purim, ordered first by Mordecai and then later confirmed by Queen Esther, is to be celebrated by all Jews on the fourteenth and fifteenth of Adar.

for sending presents of food to their friends and to the poor.

23 So the Jews accepted what Mordecai had written to them 24—how Haman son of Hammedatha, the Macedonian, *p* fought against them, how he made a decree and cast lots*q* to destroy them, 25 and how he went in to the king, telling him to hang Mordecai; but the wicked plot he had devised against the Jews came back upon himself, and he and his sons were hanged. 26 Therefore these days were called "Purim," because of the lots (for in their language this is the word that means "lots"). And so, because of what was written in this letter, and because of what they had experienced in this affair and what had befallen them, Mordecai established this festival,*r* 27 and the Jews took upon themselves, upon their descendants, and upon all who would join them, to observe it without fail.*s* These days of Purim should be a memorial and kept from generation to generation, in every city, family, and country. 28 These days of Purim were to be observed for all time, and the commemoration of them was never to cease among their descendants.

29 Then Queen Esther daughter of Aminadab along with Mordecai the Jew wrote down what they had done, and gave full authority to the letter about Purim.*t* 31 And Mordecai and Queen Esther established this decision on their own responsibility, pledging their own well-being to the plan.*s* 32 Esther established it by a decree forever, and it was written for a memorial.

10 The king levied a tax upon his kingdom both by land and sea. 2 And as for his power and bravery, and the wealth and glory of his kingdom, they were recorded in the annals of the kings of the Persians and the Medes.

3 Mordecai acted with authority on behalf of King Artaxerxes and was great in the kingdom, as well as honored by the Jews. His way of life was such as to make him beloved to his whole nation.

ADDITION F

4*u* And Mordecai said, "These things have come from God; 5 for I remember the dream that I had concerning these matters, and none of them has failed to be fulfilled. 6 There was the little spring that became a river, and there was light and sun and abundant water—the river is Esther, whom the king married and made queen. 7 The two dragons are Haman and myself. 8 The nations are those that gathered to destroy the name of the Jews. 9 And my nation, this is Israel, who cried out to God and were saved. The Lord has saved his people; the Lord has rescued us from all these evils; God has done great signs and wonders, wonders that have never happened among the nations. 10 For this purpose he made two lots, one for the people of God and one for all the nations, 11 and these two lots came to the hour and moment and day of decision before God and among all the nations. 12 And God remembered his people and vindicated his inheritance. 13 So they will observe these days in the month of Adar, on the fourteenth and fifteenth*v* of that month, with an assembly and joy and gladness before God,

p Other ancient witnesses read *the Bougean*
q Gk *a lot*　　*r* Gk *he established* (it)
s Meaning of Gk uncertain　　*t* Verse 30 in Heb is lacking in Gk: *Letters were sent to all the Jews, to the one hundred twenty-seven provinces of the kingdom of Ahasuerus, in words of peace and truth.*
u Chapter 10.4–13 and 11.1 correspond to chapter F1–11 in some translations.　　*v* Other ancient authorities lack *and fifteenth*

9.23: Here, *the Macedonian* ("the Agagite" in 9.24 of the Hebrew) is a pejorative epithet. **26**: *In their language,* "Purim" is the Hebrew cognate of the Akkadian word for "lots." **10.4–13**: **Mordecai's dream interpreted.** This passage interprets the dream of 11.5–11. The threat to Israel is here ascribed not to one man, Haman, as in the canonical Esther, but to a general anti-Semitic hostility against the Jews. **11**: *Two lots,* not mentioned in Addition A, but mentioned here because of the word Purim ("lots"). **13**: As in Esth 9.20–22, Israel is to celebrate this deliverance annually.

from generation to generation forever among his people Israel." **1** In the fourth year of the reign of Ptolemy and Cleopatra, Dositheus, who said that he was a priest and a Levite, *w* and his son Ptolemy brought to Egypt *x* the preceding Letter about Purim, which they said was authentic and had been translat-

11

ed by Lysimachus son of Ptolemy, one of the residents of Jerusalem.

END OF ADDITION F

w Or *priest, and Levitas*   *x* Cn: Gk *brought in*

11.1: The book of Esther attested as genuine. 1: *Fourth year,* that is 114–113 B.C. *The preceding Letter,* a reference not merely to Mordecai's letter (Esth 9.20–22) but to the entire book of Esther, including, probably,

some of the Additions. *Lysimachus* is a Greek name; on his work of translating the Hebrew book of Esther into Greek, see the Introduction.

The Wisdom of Solomon

This book is attributed to Solomon, though his name never appears (even in 9.8). But 7.1–14 and 8.17–9.18 deliberately reflect his prayer for wisdom as found in 1 Kings 3.6–9 and 2 Chr 1.8–10. The work was composed in Greek by an unknown Hellenistic Jew, probably at Alexandria, the largest Jewish center in the Diaspora (the "dispersion," the name for the Jewish communities outside the land of Israel). The date is uncertain, probably the latter part of the first century B.C. The literary genre is generally described as protreptic, a form of didactic exhortation. It is clearly a product of Hellenistic culture, as indicated by the mention of the four cardinal virtues in 8.7 and the philosophical treatment of the knowledge of God in 13.1–9. There is also a wide range of Greek vocabulary in the work. At the same time, it is intensely Jewish. The author has borrowed phraseology from the Septuagint, the Greek translation that would have been the Bible of his Jewish compatriots, and it is written in the poetic parallelism characteristic of the Hebrew Bible.

Chapters 1–5 deal with the gift of immortality, and this constitutes a breakthrough in biblical thought (see also Dan 12.2–3); righteousness is undying (1.15; see also 3.1–5). When the wicked behold the just person before the throne of God, they realize their mistake; the just one is numbered among the children of God and given the portion of the holy ones in the heavenly court (5.5). In chs 6–9 pseudo-Solomon speaks to the "judges" of the earth (in reality to the Jews) concerning Wisdom (a spirit and personified as in Prov 8; Sir 24) and her attributes. Unlike other wisdom books, which usually leave the sacred history unmentioned, this work portrays wisdom as a savior of Israel's ancestors (10.1–21). In chs 11–19 there is presented an elaborate system of contrasts between God's treatment of the Israelites and of the Egyptians at the time of the plagues. The theme is spelled out in 11.5: Israel was benefited through the very things that punished their enemies (thus, water from the rock, instead of the plague of the Nile, for instance); in the same way, "one is punished by the very things by which one sins" (11.16). Interspersed in this meditation on the Egyptian plagues are several digressions about God's power and mercy (11.17–12.22), false worship (13.1–15.17), and other matters.

To the Jews of the dispersion the Wisdom of Solomon offered strength and consolation. Theirs was a true wisdom, which surpassed even that of the Greeks. Theirs was immortality (a gift of God to the righteous, not the result of having an immortal or spiritual "soul"). Their Lord was the "author of beauty" (13.3); to know him was complete righteousness and to know his power was the root of immortality (15.3), for his immortal spirit was in all things (12.1).

1 Love righteousness, you rulers of the earth,
think of the Lord in goodness
and seek him with sincerity of heart;

2 because he is found by those who do not put him to the test,
and manifests himself to those who do not distrust him.

3 For perverse thoughts separate people from God,
and when his power is tested, it exposes the foolish;

4 because wisdom will not enter a deceitful soul,
or dwell in a body enslaved to sin.

5 For a holy and disciplined spirit will flee from deceit,
and will leave foolish thoughts behind,
and will be ashamed at the approach of unrighteousness.

6 For wisdom is a kindly spirit,
but will not free blasphemers from the guilt of their words;
because God is witness of their inmost feelings,
and a true observer of their hearts,
and a hearer of their tongues.

7 Because the spirit of the Lord has filled the world,
and that which holds all things together knows what is said,

8 therefore those who utter unrighteous things will not escape notice,
and justice, when it punishes, will not pass them by.

9 For inquiry will be made into the counsels of the ungodly,
and a report of their words will come to the Lord,
to convict them of their lawless deeds;

10 because a jealous ear hears all things,
and the sound of grumbling does not go unheard.

11 Beware then of useless grumbling,
and keep your tongue from slander;
because no secret word is without result, *a*
and a lying mouth destroys the soul.

12 Do not invite death by the error of your life,
or bring on destruction by the works of your hands;

13 because God did not make death,
and he does not delight in the death of the living.

14 For he created all things so that they might exist;
the generative forces *b* of the world are wholesome,
and there is no destructive poison in them,
and the dominion *c* of Hades is not on earth.

15 For righteousness is immortal.

16 But the ungodly by their words and deeds summoned death; *d*

a Or *will go unpunished*   *b* Or *the creatures*
c Or *palace*   *d* Gk *him*

1.1–6.25: Commendation of wisdom as guide to happiness and immortality. 1.1–5: *Rulers* (compare 6.1) are urged to *love righteousness* and seek God. *Wisdom* dwells only in a sincere, holy, and disciplined soul.
1.6–11: The ungodly will not escape punishment; God knows their unrighteous thoughts, words, and deeds and will judge them. **6:** *A kindly spirit,* literally "a philanthropic spirit," i.e. loving humankind; wisdom's concern for human welfare will not tolerate blasphemous or unrighteous words.
1.12–15: God has made human beings for immortality. 15: *Immortal* is literally "undying." Immortality is seen as a gift of God; it is not due to the nature of the soul, but to *righteousness,* the vital relationship of persons to God.
1.16–2.24: The reasoning of the materialist or sensualist. 2.1–5: The ungodly say that life is *short,* birth the result of *mere chance,* life without real meaning, and physical death

considering him a friend, they
    pined away
and made a covenant with him,
because they are fit to belong to
    his company.

2 For they reasoned unsoundly,
    saying to themselves,
"Short and sorrowful is our life,
and there is no remedy when a life
    comes to its end,
and no one has been known to
    return from Hades.

2 For we were born by mere
    chance,
and hereafter we shall be as
    though we had never been,
for the breath in our nostrils is
    smoke,
and reason is a spark kindled by
    the beating of our hearts;

3 when it is extinguished, the body
    will turn to ashes,
and the spirit will dissolve like
    empty air.

4 Our name will be forgotten in
    time,
and no one will remember our
    works;
our life will pass away like the
    traces of a cloud,
and be scattered like mist
that is chased by the rays of the
    sun
and overcome by its heat.

5 For our allotted time is the passing
    of a shadow,
and there is no return from our
    death,
because it is sealed up and no one
    turns back.

6 "Come, therefore, let us enjoy the
    good things that exist,
and make use of the creation to
    the full as in youth.

7 Let us take our fill of costly wine
    and perfumes,

and let no flower of spring pass us
    by.

8 Let us crown ourselves with
    rosebuds before they
    wither.

9 Let none of us fail to share in our
    revelry;
everywhere let us leave signs of
    enjoyment,
because this is our portion, and
    this our lot.

10 Let us oppress the righteous poor
    man;
let us not spare the widow
or regard the gray hairs of the
    aged.

11 But let our might be our law of
    right,
for what is weak proves itself to
    be useless.

12 "Let us lie in wait for the
    righteous man,
because he is inconvenient to us
    and opposes our actions;
he reproaches us for sins against
    the law,
and accuses us of sins against our
    training.

13 He professes to have knowledge of
    God,
and calls himself a child*e* of the
    Lord.

14 He became to us a reproof of our
    thoughts;

15 the very sight of him is a burden
    to us,
because his manner of life is unlike
    that of others,
and his ways are strange.

16 We are considered by him as
    something base,
and he avoids our ways as
    unclean;

e Or servant

---

the end of existence. **6–9**: They encourage
one another to live lives of sensual satisfaction
before death snuffs out existence.
   2.10–20: They urge one another to oppress *righteous*, helpless folk, whose godly
words and lives reproach them, and to persecute, torture, and kill those who call *God* their
father.

he calls the last end of the
righteous happy,
and boasts that God is his father.

17 Let us see if his words are true,
and let us test what will happen at
the end of his life;

18 for if the righteous man is God's
child, he will help him,
and will deliver him from the
hand of his adversaries.

19 Let us test him with insult and
torture,
so that we may find out how
gentle he is,
and make trial of his forbearance.

20 Let us condemn him to a shameful
death,
for, according to what he says, he
will be protected."

21 Thus they reasoned, but they were
led astray,
for their wickedness blinded them,

22 and they did not know the secret
purposes of God,
nor hoped for the wages of
holiness,
nor discerned the prize for
blameless souls;

23 for God created us for
incorruption,
and made us in the image of his
own eternity,*f*

24 but through the devil's envy death
entered the world,
and those who belong to his
company experience it.

3 But the souls of the righteous are
in the hand of God,
and no torment will ever touch
them.

2 In the eyes of the foolish they
seemed to have died,
and their departure was thought to
be a disaster,

3 and their going from us to be
their destruction;
but they are at peace.

4 For though in the sight of others
they were punished,
their hope is full of immortality.

5 Having been disciplined a little,
they will receive great
good,
because God tested them and
found them worthy of
himself;

6 like gold in the furnace he tried
them,
and like a sacrificial burnt offering
he accepted them.

7 In the time of their visitation they
will shine forth,
and will run like sparks through
the stubble.

8 They will govern nations and rule
over peoples,
and the Lord will reign over them
forever.

9 Those who trust in him will
understand truth,
and the faithful will abide with
him in love,
because grace and mercy are upon
his holy ones,
and he watches over his elect.*g*

10 But the ungodly will be punished
as their reasoning deserves,

f Other ancient authorities read *nature*   *g* Text
of this line uncertain; omitted by some ancient
authorities. Compare 4.15

2.21–24: Such false reasoning arises from
wickedness and consequent failure to know
God. Human beings were made in the divine
image for immortality, but *the devil's envy*
(compare 1.16) brought *death* into *the world*.
**3.1–9: The blessed estate of the righ-
teous.** Though affliction, suffering, and the
early death of *the righteous* may seem to be
divine punishment, after death they are forever safe and at *peace* with God; they enjoy sure

immortality. **7–9:** Their discipline and testing
will be followed by a divine *visitation;* God
will vindicate them, let them share in his rule
over all peoples, and give them understanding of his ways.
**3.10–4.6: The punishment of the un-
godly.** The *ungodly* will meet a sad end because, disregarding what *the righteous* could
teach them, they *rebelled against the Lord*
(v. 10). In Israel the possession of many chil-

those who disregarded the
    righteous[h]
and rebelled against the Lord;

11 for those who despise wisdom and
    instruction are miserable.
Their hope is vain, their labors are
    unprofitable,
and their works are useless.

12 Their wives are foolish, and their
    children evil;

13 their offspring are accursed.
For blessed is the barren woman
    who is undefiled,
who has not entered into a sinful
    union;
she will have fruit when God
    examines souls.

14 Blessed also is the eunuch whose
    hands have done no lawless
    deed,
and who has not devised wicked
    things against the Lord;
for special favor will be shown
    him for his faithfulness,
and a place of great delight in the
    temple of the Lord.

15 For the fruit of good labors is
    renowned,
and the root of understanding does
    not fail.

16 But children of adulterers will not
    come to maturity,
and the offspring of an unlawful
    union will perish.

17 Even if they live long they will be
    held of no account,
and finally their old age will be
    without honor.

18 If they die young, they will have
    no hope
and no consolation on the day of
    judgment.

19 For the end of an unrighteous
    generation is grievous.

4 Better than this is childlessness
    with virtue,
for in the memory of virtue[i] is
    immortality,
because it is known both by God
    and by mortals.

2 When it is present, people imitate[j]
    it,
and they long for it when it has
    gone;
throughout all time it marches,
    crowned in triumph,
victor in the contest for prizes that
    are undefiled.

3 But the prolific brood of the
    ungodly will be of no use,
and none of their illegitimate
    seedlings will strike a deep
    root
or take a firm hold.

4 For even if they put forth boughs
    for a while,
standing insecurely they will be
    shaken by the wind,
and by the violence of the winds
    they will be uprooted.

5 The branches will be broken off
    before they come to
    maturity,
and their fruit will be useless,
not ripe enough to eat, and good
    for nothing.

6 For children born of unlawful
    unions
are witnesses of evil against their
    parents when God examines
    them.[k]

7 But the righteous, though they die
    early, will be at rest.

8 For old age is not honored for
    length of time,

---

h Or *what is right*   i Gk *it*   j Other ancient
authorities read *honor*   k Gk *at their examination*

---

dren and a long life were often considered
proof of God's favor. But the *barren woman*
and the *eunuch* are also blessed by God
(vv. 13–14); a virtuous life is what counts, for
it alone leads to *immortality*. The wicked, even
if they *live long* (v. 17) and have many children (4.3), have *no* justified *hope* for the future; their children usually die early, are *of no*
account, and attest their parents' wickedness
(3.16–18).

**4.7–19: The blessedness of the righ-
teous despite premature death.** The righteous who die young *will be at rest* with God
(v. 7). *A blameless life* means more than long
years; and early death ends the danger of falling into evil.

or measured by number of years;
9 but understanding is gray hair for
anyone,
and a blameless life is ripe old age.

10 There were some who pleased
God and were loved by
him,
and while living among sinners
were taken up.
11 They were caught up so that evil
might not change their
understanding
or guile deceive their souls.
12 For the fascination of wickedness
obscures what is good,
and roving desire perverts the
innocent mind.
13 Being perfected in a short time,
they fulfilled long years;
14 for their souls were pleasing to the
Lord,
therefore he took them quickly
from the midst of
wickedness.
15 Yet the peoples saw and did not
understand,
or take such a thing to heart,
that God's grace and mercy are
with his elect,
and that he watches over his holy
ones.

16 The righteous who have died will
condemn the ungodly who
are living,
and youth that is quickly
perfected*l* will condemn
the prolonged old age of
the unrighteous.
17 For they will see the end of the
wise,
and will not understand what the
Lord purposed for them,
and for what he kept them safe.
18 The unrighteous*m* will see, and
will have contempt for
them,

but the Lord will laugh them to
scorn.
After this they will become
dishonored corpses,
and an outrage among the dead
forever;
19 because he will dash them
speechless to the ground,
and shake them from the
foundations;
they will be left utterly dry and
barren,
and they will suffer anguish,
and the memory of them will
perish.

20 They will come with dread when
their sins are reckoned up,
and their lawless deeds will
convict them to their face.

5 Then the righteous will stand with
great confidence
in the presence of those who have
oppressed them
and those who make light of their
labors.
2 When the unrighteous*n* see them,
they will be shaken with
dreadful fear,
and they will be amazed at the
unexpected salvation of the
righteous.
3 They will speak to one another in
repentance,
and in anguish of spirit they will
groan, and say,
4 "These are persons whom we once
held in derision
and made a byword of
reproach—fools that we
were!
We thought that their lives were
madness
and that their end was without
honor.

l Or ended   m Gk They   n Gk they

4.10–15: Enoch is an example of a good
man *perfected in a short time;* he lived 365 years,
several hundred years less than any other listed in Gen ch 5. **14–15:** *The peoples . . . did not*
understand, they fail to recognize God's providence in an early death (compare v. 17).
  4.20–5.23: The judgment. In contrast to
their arrogant speech in 1.16–2.24, the un-

5 Why have they been numbered
  among the children of God?
And why is their lot among the
  saints?
6 So it was we who strayed from
  the way of truth,
and the light of righteousness did
  not shine on us,
and the sun did not rise upon us.
7 We took our fill of the paths of
  lawlessness and destruction,
and we journeyed through
  trackless deserts,
but the way of the Lord we have
  not known.
8 What has our arrogance profited
  us?
And what good has our boasted
  wealth brought us?

9 "All those things have vanished
  like a shadow,
and like a rumor that passes by;
10 like a ship that sails through the
  billowy water,
and when it has passed no trace
  can be found,
no track of its keel in the waves;
11 or as, when a bird flies through
  the air,
no evidence of its passage is
  found;
the light air, lashed by the beat of
  its pinions
and pierced by the force of its
  rushing flight,
is traversed by the movement of
  its wings,
and afterward no sign of its
  coming is found there;
12 or as, when an arrow is shot at a
  target,
the air, thus divided, comes
  together at once,
so that no one knows its pathway.
13 So we also, as soon as we were

born, ceased to be,
and we had no sign of virtue to
  show,
but were consumed in our
  wickedness."
14 Because the hope of the ungodly is
  like thistledown[o] carried by
  the wind,
and like a light frost[p] driven away
  by a storm;
it is dispersed like smoke before
  the wind,
and it passes like the remembrance
  of a guest who stays but a
  day.

15 But the righteous live forever,
and their reward is with the Lord;
the Most High takes care of them.
16 Therefore they will receive a
  glorious crown
and a beautiful diadem from the
  hand of the Lord,
because with his right hand he will
  cover them,
and with his arm he will shield
  them.
17 The Lord[q] will take his zeal as his
  whole armor,
and will arm all creation to repel[r]
  his enemies;
18 he will put on righteousness as a
  breastplate,
and wear impartial justice as a
  helmet;
19 he will take holiness as an
  invincible shield,
20 and sharpen stern wrath for a
  sword,
and creation will join with him to
  fight against his frenzied
  foes.

o Other ancient authorities read *dust*   p Other
ancient authorities read *spider's web*   q Gk He
r Or *punish*

---

godly are now filled with remorse at their
own folly. **5.5:** *Among the children of God . . .
among the saints:* that is, the righteous are to be
counted in the heavenly court, with God's
family (angels). **6–7:** The *way of the Lord* is
contrasted with the *paths of lawlessness*. **9–14:**

The examples of *ship, bird,* and *arrow,* which
leave *no trace,* underscore the folly of the unrighteous. **5.15–16:** The *reward* of the righteous is a
glorious crown and divine protection. **17–22:**
The punishment of the Lord's enemies is

21 Shafts of lightning will fly with
    true aim,
  and will leap from the clouds to
    the target, as from a
    well-drawn bow,
22 and hailstones full of wrath will be
    hurled as from a catapult;
  the water of the sea will rage
    against them,
  and rivers will relentlessly
    overwhelm them;
23 a mighty wind will rise against
    them,
  and like a tempest it will winnow
    them away.
  Lawlessness will lay waste the
    whole earth,
  and evildoing will overturn the
    thrones of rulers.

6 Listen therefore, O kings, and
    understand;
  learn, O judges of the ends of the
    earth.
2 Give ear, you that rule over
    multitudes,
  and boast of many nations.
3 For your dominion was given you
    from the Lord,
  and your sovereignty from the
    Most High;
  he will search out your works and
    inquire into your plans.
4 Because as servants of his
    kingdom you did not rule
    rightly,
  or keep the law,
  or walk according to the purpose
    of God,
5 he will come upon you terribly
    and swiftly,
  because severe judgment falls on
    those in high places.

6 For the lowliest may be pardoned
    in mercy,
  but the mighty will be mightily
    tested.
7 For the Lord of all will not stand
    in awe of anyone,
  or show deference to greatness;
  because he himself made both
    small and great,
  and he takes thought for all alike.
8 But a strict inquiry is in store for
    the mighty.
9 To you then, O monarchs, my
    words are directed,
  so that you may learn wisdom and
    not transgress.
10 For they will be made holy who
    observe holy things in
    holiness,
  and those who have been taught
    them will find a defense.
11 Therefore set your desire on my
    words;
  long for them, and you will be
    instructed.

12 Wisdom is radiant and unfading,
  and she is easily discerned by
    those who love her,
  and is found by those who seek
    her.
13 She hastens to make herself
    known to those who desire
    her.
14 One who rises early to seek her
    will have no difficulty,
  for she will be found sitting at the
    gate.
15 To fix one's thought on her is
    perfect understanding,
  and one who is vigilant on her
    account will soon be free
    from care,

---

reminiscent of Isa 59.16–19; see also Eph
6.14–17. **23**: The statement that *evildoing*
overthrows *thrones* recalls 1.1 and prepares for
6.1–11.
  6.1–25: Exhortation to seek wisdom.
1–11: Further admonition to rulers (see 1.1),
who receive their authority from God and
must answer for lawless, godless acts; God

treats alike lowly and *mighty*, . . . *small and
great* (vv. 6–7). *Monarchs* need *wisdom* and will
receive it if they desire it (vv. 9–11).
  6.12–16: Wisdom is personified as a woman, as in Prov 8. She is easily found, for she
seeks out those who *desire her*, and pseudo-
Solomon will seek her out as a bride (8.2).

16 because she goes about seeking
  those worthy of her,
and she graciously appears to them
  in their paths,
and meets them in every thought.

17 The beginning of wisdom*s* is the
  most sincere desire for
  instruction,
and concern for instruction is love
  of her,
18 and love of her is the keeping of
  her laws,
and giving heed to her laws is
  assurance of immortality,
19 and immortality brings one near
  to God;
20 so the desire for wisdom leads to a
  kingdom.

21 Therefore if you delight in thrones
  and scepters, O monarchs
  over the peoples,
honor wisdom, so that you may
  reign forever.
22 I will tell you what wisdom is and
  how she came to be,
and I will hide no secrets from
  you,
but I will trace her course from
  the beginning of creation,
and make knowledge of her clear,
and I will not pass by the truth;
23 nor will I travel in the company of
  sickly envy,
for envy*t* does not associate with
  wisdom.
24 The multitude of the wise is the
  salvation of the world,
and a sensible king is the stability
  of any people.

25 Therefore be instructed by my
  words, and you will profit.

7 I also am mortal, like everyone
  else,
a descendant of the first-formed
  child of earth;
and in the womb of a mother I
  was molded into flesh,
2 within the period of ten months,
  compacted with blood,
from the seed of a man and the
  pleasure of marriage.
3 And when I was born, I began to
  breathe the common air,
and fell upon the kindred earth;
my first sound was a cry, as is
  true of all.
4 I was nursed with care in
  swaddling cloths.
5 For no king has had a different
  beginning of existence;
6 there is for all one entrance into
  life, and one way out.
7 Therefore I prayed, and
  understanding was given
  me;
I called on God, and the spirit of
  wisdom came to me.
8 I preferred her to scepters and
  thrones,
and I accounted wealth as nothing
  in comparison with her.
9 Neither did I liken to her any
  priceless gem,
because all gold is but a little sand
  in her sight,
and silver will be accounted as
  clay before her.

s Gk *Her beginning*   *t* Gk *this*

6.17–21: The steps from the love of wisdom to immortality; a sorites (a form of logic much used by the Stoics), tracing the path from *desire* for wisdom to its results in *immortality* and being *near to God*. 20: A *kingdom*, God's eternal kingdom, in which rulers and others who truly desire wisdom participate. 21–25: An invitation to learn of wisdom from Solomon. As in 1.1 and 6.1, *kings* are addressed, but the real audience is the Jewish community.

7.1–8.21: **A description of wisdom and Solomon's quest for her.** 7.1–22a: Recognizing his need for wisdom, Solomon *prayed* for and received wisdom, valuing it above every other gift (1 Kings 3.5–15). With the gift of wisdom he received *all good things, friendship with God*, and *unerring knowledge* of the world, the heavenly bodies, and plant and animal life. This is the encyclopedic knowledge that was particularly cultivated in the Hellenistic world.

10 I loved her more than health and
   beauty,
and I chose to have her rather than
   light,
because her radiance never ceases.
11 All good things came to me along
   with her,
and in her hands uncounted
   wealth.
12 I rejoiced in them all, because
   wisdom leads them;
but I did not know that she was
   their mother.
13 I learned without guile and I
   impart without grudging;
I do not hide her wealth,
14 for it is an unfailing treasure for
   mortals;
those who get it obtain friendship
   with God,
commended for the gifts that
   come from instruction.

15 May God grant me to speak with
   judgment,
and to have thoughts worthy of
   what I have received;
for he is the guide even of
   wisdom
and the corrector of the wise.
16 For both we and our words are in
   his hand,
as are all understanding and skill
   in crafts.
17 For it is he who gave me unerring
   knowledge of what exists,
to know the structure of the world
   and the activity of the
   elements;
18 the beginning and end and middle
   of times,
the alternations of the solstices and
   the changes of the seasons,
19 the cycles of the year and the
   constellations of the stars,
20 the natures of animals and the
   tempers of wild animals,

the powers of spirits[u] and the
   thoughts of human beings,
the varieties of plants and the
   virtues of roots;
21 I learned both what is secret and
   what is manifest,
22 for wisdom, the fashioner of all
   things, taught me.

There is in her a spirit that is
   intelligent, holy,
unique, manifold, subtle,
mobile, clear, unpolluted,
distinct, invulnerable, loving the
   good, keen,
irresistible, 23 beneficent, humane,
steadfast, sure, free from anxiety,
all-powerful, overseeing all,
and penetrating through all spirits
that are intelligent, pure, and
   altogether subtle.
24 For wisdom is more mobile than
   any motion;
because of her pureness she
   pervades and penetrates all
   things.
25 For she is a breath of the power of
   God,
and a pure emanation of the glory
   of the Almighty;
therefore nothing defiled gains
   entrance into her.
26 For she is a reflection of eternal
   light,
a spotless mirror of the working
   of God,
and an image of his goodness.
27 Although she is but one, she can
   do all things,
and while remaining in herself, she
   renews all things;
in every generation she passes into
   holy souls
and makes them friends of God,
   and prophets;

u Or *winds*

7.22b–8.1: **The nature and beneficial
works of wisdom. 22b–23:** Wisdom's
twenty-one (3 × 7) attributes. **24:** She is a
pure, freely moving, and all penetrating *spirit*

(see 1.6; 9.12). **25–26:** She emanates from
God; her *power, glory,* purity, *light,* and *goodness* are expressed through her (see also Jn
1.1–14; Heb 1.1–3).

28 for God loves nothing so much as
the person who lives with
wisdom.
29 She is more beautiful than the sun,
and excels every constellation of
the stars.
Compared with the light she is
found to be superior,
30 for it is succeeded by the night,
but against wisdom evil does not
prevail.

8 She reaches mightily from one end
of the earth to the other,
and she orders all things well.
2 I loved her and sought her from
my youth;
I desired to take her for my bride,
and became enamored of her
beauty.
3 She glorifies her noble birth by
living with God,
and the Lord of all loves her.
4 For she is an initiate in the
knowledge of God,
and an associate in his works.
5 If riches are a desirable possession
in life,
what is richer than wisdom, the
active cause of all things?
6 And if understanding is effective,
who more than she is fashioner of
what exists?
7 And if anyone loves righteousness,
her labors are virtues;
for she teaches self-control and
prudence,
justice and courage;
nothing in life is more profitable
for mortals than these.
8 And if anyone longs for wide
experience,
she knows the things of old, and
infers the things to come;
she understands turns of speech
and the solutions of riddles;

she has foreknowledge of signs
and wonders
and of the outcome of seasons and
times.
9 Therefore I determined to take her
to live with me,
knowing that she would give me
good counsel
and encouragement in cares and
grief.
10 Because of her I shall have glory
among the multitudes
and honor in the presence of the
elders, though I am young.
11 I shall be found keen in judgment,
and in the sight of rulers I shall be
admired.
12 When I am silent they will wait
for me,
and when I speak they will give
heed;
if I speak at greater length,
they will put their hands on their
mouths.
13 Because of her I shall have
immortality,
and leave an everlasting
remembrance to those who
come after me.
14 I shall govern peoples,
and nations will be subject to me;
15 dread monarchs will be afraid of
me when they hear of me;
among the people I shall show
myself capable, and
courageous in war.
16 When I enter my house, I shall
find rest with her;
for companionship with her has
no bitterness,
and life with her has no pain, but
gladness and joy.
17 When I considered these things
inwardly,
and pondered in my heart

7.27–8.1: She is everywhere, *orders all
things all well*, and *can do all things*. She enters
holy souls and makes them *friends of God; evil*
cannot defeat her.
8.2–21: **Solomon desires wisdom as his
bride and teacher. 4–6**: She is God's *associate in his works*, and his agent in making all

things (Prov 8.22–30; see also Jn 1.3; Col
1.16; Heb 1.2). **7**: She *teaches self-control, prudence, justice,* and *courage,* which (according to
Plato and the Stoics) are the four cardinal
virtues. **8**: She knows the past and the future,
and so can give good counsel.
8.10–16: *Because of her* Solomon will re-

that in kinship with wisdom there
is immortality,

18 and in friendship with her, pure
delight,
and in the labors of her hands,
unfailing wealth,
and in the experience of her
company, understanding,
and renown in sharing her words,
I went about seeking how to get
her for myself.

19 As a child I was naturally gifted,
and a good soul fell to my lot;

20 or rather, being good, I entered an
undefiled body.

21 But I perceived that I would not
possess wisdom unless God
gave her to me—
and it was a mark of insight to
know whose gift she was—
so I appealed to the Lord and
implored him,
and with my whole heart I said:

9 "O God of my ancestors and Lord
of mercy,
who have made all things by your
word,

2 and by your wisdom have formed
humankind
to have dominion over the
creatures you have made,

3 and rule the world in holiness and
righteousness,
and pronounce judgment in
uprightness of soul,

4 give me the wisdom that sits by
your throne,
and do not reject me from among
your servants.

5 For I am your servant[v] the son of
your serving girl,
a man who is weak and
short-lived,
with little understanding of
judgment and laws;

6 for even one who is perfect among
human beings

will be regarded as nothing
without the wisdom that
comes from you.

7 You have chosen me to be king of
your people
and to be judge over your sons
and daughters.

8 You have given command to build
a temple on your holy
mountain,
and an altar in the city of your
habitation,
a copy of the holy tent that you
prepared from the
beginning.

9 With you is wisdom, she who
knows your works
and was present when you made
the world;
she understands what is pleasing in
your sight
and what is right according to
your commandments.

10 Send her forth from the holy
heavens,
and from the throne of your glory
send her,
that she may labor at my side,
and that I may learn what is
pleasing to you.

11 For she knows and understands all
things,
and she will guide me wisely in
my actions
and guard me with her glory.

12 Then my works will be
acceptable,
and I shall judge your people
justly,
and shall be worthy of the throne[w]
of my father.

13 For who can learn the counsel of
God?
Or who can discern what the Lord
wills?

v Gk *slave*   w Gk *thrones*

---

ceive *honor* and respect, *immortality,* a great
empire, *rest* and *joy.* **19–20:** The Platonic view
of the soul as pre-existent seems to be reflected here, but unlike Plato's view, there is
union with an *undefiled body.*

9.1–18: Solomon's prayer for wisdom
(1 Kings 3.6–9). **1–2:** God *made all* things by
word and *wisdom* (Jn 1.1–3); here "word" and
"wisdom" are synonymous. **9:** Compare 8.3–
4 and Prov 8.27–30.

14 For the reasoning of mortals is
  worthless,
 and our designs are likely to fail;
15 for a perishable body weighs
  down the soul,
 and this earthy tent burdens the
  thoughtful[x] mind.
16 We can hardly guess at what is on
  earth,
 and what is at hand we find with
  labor;
 but who has traced out what is in
  the heavens?
17 Who has learned your counsel,
 unless you have given wisdom
 and sent your holy spirit from on
  high?
18 And thus the paths of those on
  earth were set right,
 and people were taught what
  pleases you,
 and were saved by wisdom."

10 Wisdom[y] protected the
   first-formed father of the
   world, when he alone had
   been created;
 she delivered him from his
  transgression,
2 and gave him strength to rule all
  things.
3 But when an unrighteous man
  departed from her in his
  anger,
 he perished because in rage he
  killed his brother.
4 When the earth was flooded
  because of him, wisdom
  again saved it,
 steering the righteous man by a
  paltry piece of wood.

5 Wisdom[y] also, when the nations
  in wicked agreement had
  been put to confusion,
 recognized the righteous man and
  preserved him blameless
  before God,
 and kept him strong in the face of
  his compassion for his
  child.

6 Wisdom[y] rescued a righteous man
  when the ungodly were
  perishing;
 he escaped the fire that descended
  on the Five Cities.[z]
7 Evidence of their wickedness still
  remains:
 a continually smoking wasteland,
 plants bearing fruit that does not
  ripen,
 and a pillar of salt standing as a
  monument to an
  unbelieving soul.

8 For because they passed wisdom
  by,
 they not only were hindered from
  recognizing the good,
 but also left for humankind a
  reminder of their folly,
 so that their failures could never
  go unnoticed.

9 Wisdom rescued from troubles
  those who served her.
10 When a righteous man fled from
  his brother's wrath,
 she guided him on straight paths;
 she showed him the kingdom of
  God,

x Or *anxious*   y Gk *She*   z Or *on Pentapolis*

9.13–15: The *perishable body,* though not
called evil (8.20), *burdens* and hampers the
mind. 16–17: Even earthly things we know
imperfectly; only through *wisdom,* also called
the holy spirit, can we learn heavenly things
(1 Cor 2.7–12). 18: *Wisdom* guides, teaches,
and saves her followers. This is illustrated in
ch 10 in the lives of the righteous ancestors.
10.1–11.4: **Seven historical illustrations**
of the saving power of wisdom. 10.1–2:
Adam (Gen 1.26–5.5). 3: Cain (Gen 4.1–6).
Perished, spiritual death. 4: Noah (Gen 5.28–
9.29). 5: Abraham (Gen 11.26–15.10); he is
linked with the *nations* of Gen 11.1–9.
10.6–8: *A righteous man,* Lot (Gen 19.23–
26); throughout this chapter personal names
are deliberately replaced by the category of
"righteous." 9–12: Jacob (Gen 25.19–49.33,

and gave him knowledge of holy
things;
she prospered him in his labors,
and increased the fruit of his toil.
11 When his oppressors were
covetous,
she stood by him and made him
rich.
12 She protected him from his
enemies,
and kept him safe from those who
lay in wait for him;
in his arduous contest she gave
him the victory,
so that he might learn that
godliness is more powerful
than anything else.

13 When a righteous man was sold,
wisdom*a* did not desert
him,
but delivered him from sin.
She descended with him into the
dungeon,
14 and when he was in prison she did
not leave him,
until she brought him the scepter
of a kingdom
and authority over his masters.
Those who accused him she
showed to be false,
and she gave him everlasting
honor.

15 A holy people and blameless race
wisdom delivered from a nation of
oppressors.
16 She entered the soul of a servant
of the Lord,
and withstood dread kings with
wonders and signs.
17 She gave to holy people the
reward of their labors;
she guided them along a
marvelous way,
and became a shelter to them by
day,

and a starry flame through the
night.
18 She brought them over the Red
Sea,
and led them through deep waters;
19 but she drowned their enemies,
and cast them up from the depth
of the sea.
20 Therefore the righteous plundered
the ungodly;
they sang hymns, O Lord, to your
holy name,
and praised with one accord your
defending hand;
21 for wisdom opened the mouths of
those who were mute,
and made the tongues of infants
speak clearly.

11 Wisdom*b* prospered their
works by the hand of a
holy prophet.
2 They journeyed through an
uninhabited wilderness,
and pitched their tents in
untrodden places.
3 They withstood their enemies and
fought off their foes.
4 When they were thirsty, they
called upon you,
and water was given them out of
flinty rock,
and from hard stone a remedy for
their thirst.
5 For through the very things by
which their enemies were
punished,
they themselves received benefit in
their need.
6 Instead of the fountain of an
ever-flowing river,
stirred up and defiled with blood
7 in rebuke for the decree to kill the
infants,

a Gk *she*   *b* Gk *She*

---

especially ch 28). **13–14:** Joseph (Gen 39;
41.39–47).
   10.15–11.4: Through Moses, *a holy prophet*
(11.1), wisdom delivered Israel from Egypt
(Ex 1.1–15.21, especially chs 14–15), leading

the people *through an uninhabited wilderness* (Ex
15.22–17.6).
   11.5–19.22: The writer composes an historical meditation, consisting of comparisons between Israel and Egypt. **11.5:** The prin-

you gave them abundant water
     unexpectedly,
8 showing by their thirst at that
          time
how you punished their enemies.
9 For when they were tried, though
          they were being disciplined
          in mercy,
they learned how the ungodly
          were tormented when
          judged in wrath.
10 For you tested them as a parent[c]
          does in warning,
but you examined the ungodly[d] as
          a stern king does in
          condemnation.
11 Whether absent or present, they
          were equally distressed,
12 for a twofold grief possessed
          them,
and a groaning at the memory of
          what had occurred.
13 For when they heard that through
          their own punishments
the righteous[e] had received
          benefit, they perceived it
          was the Lord's doing.
14 For though they had mockingly
          rejected him who long
          before had been cast out
          and exposed,
at the end of the events they
          marveled at him,
when they felt thirst in a different
          way from the righteous.

15 In return for their foolish and
          wicked thoughts,
which led them astray to worship
          irrational serpents and
          worthless animals,
you sent upon them a multitude of
          irrational creatures to
          punish them,

16 so that they might learn that one
          is punished by the very
          things by which one sins.
17 For your all-powerful hand,
          which created the world out of
          formless matter,
did not lack the means to send
          upon them a multitude of
          bears, or bold lions,
18 or newly-created unknown beasts
          full of rage,
or such as breathe out fiery breath,
or belch forth a thick pall of
          smoke,
or flash terrible sparks from their
          eyes;
19 not only could the harm they did
          destroy people,[f]
but the mere sight of them could
          kill by fright.
20 Even apart from these, people[e]
          could fall at a single breath
when pursued by justice
and scattered by the breath of
          your power.
But you have arranged all things
          by measure and number
          and weight.

21 For it is always in your power to
          show great strength,
and who can withstand the might
          of your arm?
22 Because the whole world before
          you is like a speck that tips
          the scales,
and like a drop of morning dew
          that falls on the ground.
23 But you are merciful to all, for
          you can do all things,
and you overlook people's sins, so
          that they may repent.

c Gk *a father*   *d* Gk *those*   *e* Gk *they*
f Gk *them*

---

ciple behind all the comparisons: *through the
very things by which* Egypt was *punished,* Israel
received benefit.
11.6–17: Instead of the Nile plague, Israel
received water from the rock (vv. 4, 7; Ex
17.1–7), but Egypt was afflicted with the
small animals. **7:** *You,* refers to the Lord, in
the direct address which continues to 19.22.

15: *Irrational creatures,* such as the frogs and
flies, etc. (Ex chs 8 and 10), in accordance
with the principle enunciated in v. 16 (see also
Ps 7.15–16).
     **11.17–12.2: God's mercy and love for
all. 11.17:** *Formless matter,* a concept derived
from Greek philosophy, is used to describe the
chaos of Gen 1.2.

24 For you love all things that exist,
and detest none of the things that
you have made,
for you would not have made
anything if you had hated
it.
25 How would anything have
endured if you had not
willed it?
Or how would anything not called
forth by you have been
preserved?
26 You spare all things, for they are
yours, O Lord, you who
love the living.

12 For your immortal spirit is in
all things.
2 Therefore you correct little by
little those who trespass,
and you remind and warn them of
the things through which
they sin,
so that they may be freed from
wickedness and put their
trust in you, O Lord.

3 Those who lived long ago in your
holy land
4 you hated for their detestable
practices,
their works of sorcery and unholy
rites,
5 their merciless slaughter*g* of
children,
and their sacrificial feasting on
human flesh and blood.
These initiates from the midst of a
heathen cult,*h*
6 these parents who murder helpless
lives,
you willed to destroy by the hands
of our ancestors,
7 so that the land most precious of
all to you
might receive a worthy colony of
the servants*i* of God.

8 But even these you spared, since
they were but mortals,
and sent wasps*j* as forerunners of
your army
to destroy them little by little,
9 though you were not unable to
give the ungodly into the
hands of the righteous in
battle,
or to destroy them at one blow by
dread wild animals or your
stern word.
10 But judging them little by little
you gave them an
opportunity to repent,
though you were not unaware that
their origin*k* was evil
and their wickedness inborn,
and that their way of thinking
would never change.
11 For they were an accursed race
from the beginning,
and it was not through fear of
anyone that you left them
unpunished for their sins.

12 For who will say, "What have you
done?"
Or will resist your judgment?
Who will accuse you for the
destruction of nations that
you made?
Or who will come before you to
plead as an advocate for the
unrighteous?
13 For neither is there any god
besides you, whose care is
for all people,*l*
to whom you should prove that
you have not judged
unjustly;

g Gk *slaughterers*   *h* Meaning of Gk uncertain
i Or *children*   *j* Or *hornets*   *k* Or *nature*
l Or *all things*

12.3–11: Even the Canaanites were objects of divine leniency. Despite their evil *practices,* God judged them *little by little* (v. 8; see Ex 23.29–30), giving them *an opportunity to repent* (v. 10; compare 2 Esd 9.11; Heb 12.17).

12.12–18: God's supreme power delights in benevolence. God, *sovereign,* all-powerful, is answerable to no one; he is *righteous,* condemns no one unjustly, and judges with *mildness* and *forbearance.* He cares for all people (v. 13) and rules all things (v. 15).

14 nor can any king or monarch
confront you about those
whom you have punished.
15 You are righteous and you rule all
things righteously,
deeming it alien to your power
to condemn anyone who does not
deserve to be punished.
16 For your strength is the source of
righteousness,
and your sovereignty over all
causes you to spare all.
17 For you show your strength when
people doubt the
completeness of your
power,
and you rebuke any insolence
among those who know
it.*m*
18 Although you are sovereign in
strength, you judge with
mildness,
and with great forbearance you
govern us;
for you have power to act
whenever you choose.

19 Through such works you have
taught your people
that the righteous must be kind,
and you have filled your children
with good hope,
because you give repentance for
sins.
20 For if you punished with such
great care and indulgence*n*
the enemies of your servants*o* and
those deserving of death,
granting them time and
opportunity to give up their
wickedness,
21 with what strictness you have
judged your children,

to whose ancestors you gave oaths
and covenants full of good
promises!
22 So while chastening us you
scourge our enemies ten
thousand times more,
so that, when we judge, we may
meditate upon your
goodness,
and when we are judged, we may
expect mercy.

23 Therefore those who lived
unrighteously, in a life of
folly,
you tormented through their own
abominations.
24 For they went far astray on the
paths of error,
accepting as gods those animals
that even their enemies*p*
despised;
they were deceived like foolish
infants.
25 Therefore, as though to children
who cannot reason,
you sent your judgment to mock
them.
26 But those who have not heeded
the warning of mild
rebukes
will experience the deserved
judgment of God.
27 For when in their suffering they
became incensed
at those creatures that they had
thought to be gods, being
punished by means of
them,

m Meaning of Gk uncertain   *n* Other ancient
authorities lack *and indulgence*; others read *and
entreaty*   *o* Or *children*   *p* Gk *they*

**12.19–22: God's mercy is an example to
Israel.** Israel may think that God has *judged*
them strictly (v. 21), but he *scourges* Israel's
enemies far *more;* God chastens Israel in *mercy*
and for their own good.
**12.23–27: Further comments on the
punishment of the Egyptians** (in accord
with the principle set forth in 11.16). The
Egyptians were tormented by the *animals* that
they worshiped. **27:** Though the Egyptians
recognized Israel's God as *the true God,* they
refused to let Israel go. *The utmost condemnation,* i.e. the death of their firstborn sons and
the disaster at the Red Sea.

they saw and recognized as the
true God the one whom
they had before refused to
know.
Therefore the utmost
condemnation came upon
them.

13 For all people who were
ignorant of God were
foolish by nature;
and they were unable from the
good things that are seen to
know the one who exists,
nor did they recognize the artisan
while paying heed to his
works;

2 but they supposed that either fire
or wind or swift air,
or the circle of the stars, or
turbulent water,
or the luminaries of heaven were
the gods that rule the
world.

3 If through delight in the beauty of
these things people assumed
them to be gods,
let them know how much better
than these is their Lord,
for the author of beauty created
them.

4 And if people*q* were amazed at
their power and working,
let them perceive from them
how much more powerful is the
one who formed them.

5 For from the greatness and beauty
of created things
comes a corresponding perception
of their Creator.

6 Yet these people are little to be
blamed,

for perhaps they go astray
while seeking God and desiring to
find him.

7 For while they live among his
works, they keep searching,
and they trust in what they see,
because the things that are
seen are beautiful.

8 Yet again, not even they are to be
excused;

9 for if they had the power to know
so much
that they could investigate the
world,
how did they fail to find sooner
the Lord of these things?

10 But miserable, with their hopes set
on dead things, are those
who give the name "gods" to the
works of human hands,
gold and silver fashioned with
skill,
and likenesses of animals,
or a useless stone, the work of an
ancient hand.

11 A skilled woodcutter may saw
down a tree easy to handle
and skillfully strip off all its bark,
and then with pleasing
workmanship
make a useful vessel that serves
life's needs,

12 and burn the cast-off pieces of his
work
to prepare his food, and eat his
fill.

13 But a cast-off piece from among
them, useful for nothing,
a stick crooked and full of knots,

q Gk *they*

**13.1–15.17: A digression on false wor-
ship. 13.1–9:** Nature worship is the least cul-
pable form of false worship. The implication
is that God can be known apart from revelation *from the good things that are seen.* He is *the
author of beauty,* and from *his works* he is
known by analogy (*a corresponding perception
of their Creator*). **6–7:** The writer partly excuses such idolatry as arising from an honest
search for God. **8–9:** But such people are not
to be excused; they should have discerned *the
Lord* of created *things.*

13.10–19: The folly of image-worship.
The polemic against idols is frequent in the
Bible; see Isa 44.9–20; Jer 10.1–16; Ps 135.15–
18. This is a satirical description of how a
woodcutter fells a *tree,* makes of some knotty
cast-off piece a *likeness* of a *human being* or
animal, and calls it his god, though it cannot
stand or act and has no life.

he takes and carves with care in
his leisure,
and shapes it with skill gained in
idleness;[r]
he forms it in the likeness of a
human being,
14 or makes it like some worthless
animal,
giving it a coat of red paint and
coloring its surface red
and covering every blemish in it
with paint;
15 then he makes a suitable niche for
it,
and sets it in the wall, and fastens
it there with iron.
16 He takes thought for it, so that it
may not fall,
because he knows that it cannot
help itself,
for it is only an image and has
need of help.
17 When he prays about possessions
and his marriage and
children,
he is not ashamed to address a
lifeless thing.
18 For health he appeals to a thing
that is weak;
for life he prays to a thing that is
dead;
for aid he entreats a thing that is
utterly inexperienced;
for a prosperous journey, a thing
that cannot take a step;
19 for money-making and work and
success with his hands
he asks strength of a thing whose
hands have no strength.

14 Again, one preparing to sail
and about to voyage over
raging waves
calls upon a piece of wood more
fragile than the ship that
carries him.

2 For it was desire for gain that
planned that vessel,
and wisdom was the artisan who
built it;
3 but it is your providence,
O Father, that steers its
course,
because you have given it a path
in the sea,
and a safe way through the waves,
4 showing that you can save from
every danger,
so that even a person who lacks
skill may put to sea.
5 It is your will that works of your
wisdom should not be
without effect;
therefore people trust their lives
even to the smallest piece of
wood,
and passing through the billows
on a raft they come safely
to land.
6 For even in the beginning, when
arrogant giants were
perishing,
the hope of the world took refuge
on a raft,
and guided by your hand left to
the world the seed of a new
generation.
7 For blessed is the wood by which
righteousness comes.

8 But the idol made with hands is
accursed, and so is the one
who made it—
he for having made it, and the
perishable thing because it
was named a god.
9 For equally hateful to God are the
ungodly and their
ungodliness;

r Other ancient authorities read *with intelligent
skill*

**14.1–14: The folly of the seafarer who
trusts in the wooden image on the ship's
prow.** It is the *providence* of the *Father* that is
responsible for a safe voyage.
   14.6–7: The reference is to Noah, *the hope*

of the world. The *wood* refers not (as some have
thought) to the cross of Christ, but to Noah's
ark, which carried forward God's righteous
will. In contrast to the *blessed wood* is the *accursed* idolator.

10 for what was done will be
punished together with the
one who did it.
11 Therefore there will be a visitation
also upon the heathen idols,
because, though part of what God
created, they became an
abomination,
snares for human souls
and a trap for the feet of the
foolish.

12 For the idea of making idols was
the beginning of
fornication,
and the invention of them was the
corruption of life;
13 for they did not exist from the
beginning,
nor will they last forever.
14 For through human vanity they
entered the world,
and therefore their speedy end has
been planned.

15 For a father, consumed with grief
at an untimely
bereavement,
made an image of his child, who
had been suddenly taken
from him;
he now honored as a god what
was once a dead human
being,
and handed on to his dependents
secret rites and initiations.
16 Then the ungodly custom, grown
strong with time, was kept
as a law,
and at the command of monarchs
carved images were
worshiped.
17 When people could not honor
monarchs^s in their
presence, since they lived at
a distance,

they imagined their appearance far
away,
and made a visible image of the
king whom they honored,
so that by their zeal they might
flatter the absent one as
though present.
18 Then the ambition of the artisan
impelled
even those who did not know the
king to intensify their
worship.
19 For he, perhaps wishing to please
his ruler,
skillfully forced the likeness to
take more beautiful form,
20 and the multitude, attracted by the
charm of his work,
now regarded as an object of
worship the one whom
shortly before they had
honored as a human being.
21 And this became a hidden trap for
humankind,
because people, in bondage to
misfortune or to royal
authority,
bestowed on objects of stone or
wood the name that ought
not to be shared.

22 Then it was not enough for them
to err about the knowledge
of God,
but though living in great strife
due to ignorance,
they call such great evils peace.
23 For whether they kill children in
their initiations, or celebrate
secret mysteries,
or hold frenzied revels with
strange customs,

s Gk *them*

14.15–21: The origins of idolatry. The
cult of the dead is traced to a grief-stricken
father who made and *worshiped* an *image* of his
dead child (vv. 15–16). Ruler worship is explained by the subjects who lived at a distance, and worshiped a *visible image* of their
king (v. 17); moreover skilled craftsmen
made the image *more beautiful* than the king
(vv. 18–20). (Euhemerus, about 300 B.C.,
taught that all gods were deified rulers).
14.22–31: The evil results of idolatry
(Rom 1.24–32). Ignorance of God, *strife,*

24 they no longer keep either their
    lives or their marriages
    pure,
but they either treacherously kill
    one another, or grieve one
    another by adultery,
25 and all is a raging riot of blood
    and murder, theft and
    deceit, corruption,
    faithlessness, tumult,
    perjury,
26 confusion over what is good,
    forgetfulness of favors,
defiling of souls, sexual
    perversion,
disorder in marriages, adultery,
    and debauchery.
27 For the worship of idols not to be
    named
is the beginning and cause and end
    of every evil.
28 For their worshipers *t* either rave
    in exultation,
or prophesy lies, or live
    unrighteously, or readily
    commit perjury;
29 for because they trust in lifeless
    idols
they swear wicked oaths and
    expect to suffer no harm.
30 But just penalties will overtake
    them on two counts:
because they thought wrongly
    about God in devoting
    themselves to idols,
and because in deceit they swore
    unrighteously through
    contempt for holiness.
31 For it is not the power of the
    things by which people
    swear, *u*
but the just penalty for those who
    sin,
that always pursues the
    transgression of the
    unrighteous.

15 But you, our God, are kind
    and true,
patient, and ruling all things *v* in
    mercy.
2 For even if we sin we are yours,
    knowing your power;
but we will not sin, because we
    know that you
    acknowledge us as yours.
3 For to know you is complete
    righteousness,
and to know your power is the
    root of immortality.
4 For neither has the evil intent of
    human art misled us,
nor the fruitless toil of painters,
a figure stained with varied colors,
5 whose appearance arouses yearning
    in fools,
so that they desire *w* the lifeless
    form of a dead image.
6 Lovers of evil things and fit for
    such objects of hope *x*
are those who either make or
    desire or worship them.

7 A potter kneads the soft earth
and laboriously molds each vessel
    for our service,
fashioning out of the same clay
both the vessels that serve clean
    uses
and those for contrary uses,
    making all alike;
but which shall be the use of each
    of them
the worker in clay decides.
8 With misspent toil, these workers
    form a futile god from the
    same clay—
these mortals who were made of
    earth a short time before

t Gk *they*   *u* Or *of the oaths people swear*
v Or *ruling the universe*   *w* Gk *and he desires*
x Gk *such hopes*

---

moral wrong, and perversion (v. 26) will
bring sure punishment.
    **15.1–17: The contrast between the wor-
shipers of the true God and idolaters. 1–5:**
Speaking to God (vv. 1–3; see also 11.26–
12.2), the writer notes the purifying influence
of the worship of the true God on the life of
Israel. **3:** *To know* God is *righteousness* and
eternal life (Jn 17.3).
    15.6–17: The folly and wickedness of
making and worshiping clay idols. For financial profit *a potter* molds from one mass of *clay*
both useful *vessels* and *counterfeit gods;* cheap
imitations of gold, silver, and copper *images.*

and after a little while go to the
earth from which all
mortals are taken,
when the time comes to return the
souls that were borrowed.
9 But the workers are not concerned
that mortals are destined to
die
or that their life is brief,
but they compete with workers in
gold and silver,
and imitate workers in copper;
and they count it a glorious thing
to mold counterfeit gods.
10 Their heart is ashes, their hope is
cheaper than dirt,
and their lives are of less worth
than clay,
11 because they failed to know the
one who formed them
and inspired them with active
souls
and breathed a living spirit into
them.
12 But they considered our existence
an idle game,
and life a festival held for profit,
for they say one must get money
however one can, even by
base means.
13 For these persons, more than all
others, know that they sin
when they make from earthy
matter fragile vessels and
carved images.

14 But most foolish, and more
miserable than an infant,
are all the enemies who oppressed
your people.
15 For they thought that all their
heathen idols were gods,
though these have neither the use
of their eyes to see with,
nor nostrils with which to draw
breath,
nor ears with which to hear,

nor fingers to feel with,
and their feet are of no use for
walking.
16 For a human being made them,
and one whose spirit is borrowed
formed them;
for none can form gods that are
like themselves.
17 People are mortal, and what they
make with lawless hands is
dead;
for they are better than the objects
they worship,
since*y* they have life, but the
idols*z* never had.

18 Moreover, they worship even the
most hateful animals,
which are worse than all others
when judged by their lack
of intelligence;
19 and even as animals they are not
so beautiful in appearance
that one would desire them,
but they have escaped both the
praise of God and his
blessing.

16 Therefore those people*a* were
deservedly punished
through such creatures,
and were tormented by a
multitude of animals.
2 Instead of this punishment you
showed kindness to your
people,
and you prepared quails to eat,
a delicacy to satisfy the desire of
appetite;
3 in order that those people, when
they desired food,
might lose the least remnant of
appetite*b*
because of the odious creatures
sent to them,

y Other ancient authorities read *of which*
z Gk *but they*   a Gk *they*   b Gk *loathed
the necessary appetite*

14–17: The stupidity of Israel's *enemies,* who
worship useless, lifeless images.
15.18–16.4: **Resumption of the topic of
the plague of animals** (11.15). 16.1–4: The

Egyptians, *tormented* by the *animals* they wor-
shiped (Ex chs 8 and 10), lost all appetite; the
Israelites, after brief hunger, enjoyed quails
(Num ch 11).

while your people,[c] after suffering
    want a short time,
might partake of delicacies.
4 For it was necessary that upon
    those oppressors inescapable
    want should come,
while to these others it was merely
    shown how their enemies
    were being tormented.

5 For when the terrible rage of wild
    animals came upon your
    people[d]
and they were being destroyed by
    the bites of writhing
    serpents,
your wrath did not continue to the
    end;
6 they were troubled for a little
    while as a warning,
and received a symbol of
    deliverance to remind them
    of your law's command.

7 For the one who turned toward it
    was saved, not by the thing
    that was beheld,
but by you, the Savior of all.
8 And by this also you convinced
    our enemies
that it is you who deliver from
    every evil.
9 For they were killed by the bites
    of locusts and flies,
and no healing was found for
    them,
because they deserved to be
    punished by such things.
10 But your children were not
    conquered even by the
    fangs of venomous
    serpents,
for your mercy came to their help
    and healed them.

11 To remind them of your oracles
    they were bitten,
and then were quickly delivered,
so that they would not fall into
    deep forgetfulness
and become unresponsive[e] to
    your kindness.
12 For neither herb nor poultice
    cured them,
but it was your word, O Lord,
    that heals all people.
13 For you have power over life and
    death;
you lead mortals down to the
    gates of Hades and back
    again.
14 A person in wickedness kills
    another,
but cannot bring back the departed
    spirit,
or set free the imprisoned soul.

15 To escape from your hand is
    impossible;
16 for the ungodly, refusing to know
    you,
were flogged by the strength of
    your arm,
pursued by unusual rains and hail
    and relentless storms,
and utterly consumed by fire.
17 For—most incredible of all—in
    water, which quenches all
    things,
the fire had still greater effect,
for the universe defends the
    righteous.
18 At one time the flame was
    restrained,
so that it might not consume the
    creatures sent against the
    ungodly,

c Gk *they*   d Gk *them*   e Meaning of Gk
uncertain

**16.5–14: Another contrast between
Egyptians and Israelites.** The Egyptians
were killed by bites of locusts and flies; Israel,
bitten by serpents (Num 21.6–9), suffered
briefly as a *warning* (vv. 6, 11) but were quickly healed by God's word (vv. 11–12), not by
the bronze serpent.
    16.13–14: God has *power* to bestow *life* or

lead to *death,* in contrast to the wicked, who
cannot free the soul *imprisoned* by the gates of
Hades, or Sheol (1 Sam 2.6; Ps 9.13).
    **16.15–29: The elements combine to
punish the Egyptians and to serve Israel.**
The food of angels is manna, suited to the
taste of each Israelite (vv. 20–21, 25).

but that seeing this they might know
that they were being pursued by the judgment of God;

19 and at another time even in the midst of water it burned more intensely than fire,
to destroy the crops of the unrighteous land.

20 Instead of these things you gave your people food of angels,
and without their toil you supplied them from heaven with bread ready to eat,
providing every pleasure and suited to every taste.

21 For your sustenance manifested your sweetness toward your children;
and the bread, ministering*f* to the desire of the one who took it,
was changed to suit everyone's liking.

22 Snow and ice withstood fire without melting,
so that they might know that the crops of their enemies
were being destroyed by the fire that blazed in the hail
and flashed in the showers of rain;

23 whereas the fire,*g* in order that the righteous might be fed,
even forgot its native power.

24 For creation, serving you who made it,
exerts itself to punish the unrighteous,
and in kindness relaxes on behalf of those who trust in you.

25 Therefore at that time also, changed into all forms,
it served your all-nourishing bounty,
according to the desire of those who had need,*h*

26 so that your children, whom you loved, O Lord, might learn
that it is not the production of crops that feeds humankind
but that your word sustains those who trust in you.

27 For what was not destroyed by fire
was melted when simply warmed by a fleeting ray of the sun,

28 to make it known that one must rise before the sun to give you thanks,
and must pray to you at the dawning of the light;

29 for the hope of an ungrateful person will melt like wintry frost,
and flow away like waste water.

17 Great are your judgments and hard to describe;
therefore uninstructed souls have gone astray.

2 For when lawless people supposed that they held the holy nation in their power,
they themselves lay as captives of darkness and prisoners of long night,
shut in under their roofs, exiles from eternal providence.

3 For thinking that in their secret sins they were unobserved
behind a dark curtain of forgetfulness,
they were scattered, terribly*i* alarmed,
and appalled by specters.

f Gk *and it, ministering*   *g* Gk *this*
h Or *who made supplication*   *i* Other ancient authorities read *unobserved, they were darkened behind a dark curtain of forgetfulness, terribly*

16.22–27: The *fire* God sent executed his judgment; now he kept fire from melting *snow and ice* (a poetical expression for the manna; 19.21); now *a fleeting ray of the sun* melted it (Ex 16.21). The fire and manna show how *creation* serves God (5.17), so that Israel might learn that it is sustained by God's word (v. 26). **28–29:** One should *rise* at dawn to thank God, for God's blessing escapes the *ungrateful person* as fast as the rising sun melts *wintry frost*.
17.1–18.4: The contrast between the plague of darkness and the pillar of fire. 1–7: God's judgment made the *lawless* Egyp-

4 For not even the inner chamber
  that held them protected
  them from fear,
but terrifying sounds rang out
  around them,
and dismal phantoms with gloomy
  faces appeared.
5 And no power of fire was able to
  give light,
nor did the brilliant flames of the
  stars
avail to illumine that hateful night.
6 Nothing was shining through to
  them
except a dreadful, self-kindled fire,
and in terror they deemed the
  things that they saw
to be worse than that unseen
  appearance.
7 The delusions of their magic art
  lay humbled,
and their boasted wisdom was
  scornfully rebuked.
8 For those who promised to drive
  off the fears and disorders
  of a sick soul
were sick themselves with
  ridiculous fear.
9 For even if nothing disturbing
  frightened them,
yet, scared by the passing of wild
  animals and the hissing of
  snakes
10 they perished in trembling fear,
refusing to look even at the air,
  though it nowhere could be
  avoided.
11 For wickedness is a cowardly
  thing, condemned by its
  own testimony;[j]
distressed by conscience, it has
  always exaggerated[k] the
  difficulties.
12 For fear is nothing but a giving up
  of the helps that come from
  reason;
13 and hope, defeated by this inward
  weakness,

prefers ignorance of what causes
  the torment.
14 But throughout the night, which
  was really powerless
and which came upon them from
  the recesses of powerless
  Hades,
they all slept the same sleep,
15 and now were driven by
  monstrous specters,
and now were paralyzed by their
  souls' surrender;
for sudden and unexpected fear
  overwhelmed them.
16 And whoever was there fell down,
and thus was kept shut up in a
  prison not made of iron;
17 for whether they were farmers or
  shepherds
or workers who toiled in the
  wilderness,
they were seized, and endured the
  inescapable fate;
for with one chain of darkness
  they all were bound.
18 Whether there came a whistling
  wind,
or a melodious sound of birds in
  wide-spreading branches,
or the rhythm of violently rushing
  water,
19 or the harsh crash of rocks hurled
  down,
or the unseen running of leaping
  animals,
or the sound of the most savage
  roaring beasts,
or an echo thrown back from a
  hollow of the mountains,
it paralyzed them with terror.
20 For the whole world was
  illumined with brilliant
  light,
and went about its work
  unhindered,

j Meaning of Gk uncertain
k Other ancient authorities read *anticipated*

---

tians the terrified *captives of darkness,* unaided
by their *magic art* or *boasted wisdom* (Ex 1.21–
23).
  17.11: *Conscience,* in the moral sense, occurs

only here in the Greek Old Testament. **12:** A
remarkable description of fear. **18–19:** The
terror of the Egyptians is illustrated by seven
incidents.

21 while over those people alone
  heavy night was spread,
an image of the darkness that was
  destined to receive them;
but still heavier than darkness
  were they to themselves.

18 But for your holy ones there
  was very great light.
Their enemies[1] heard their voices
  but did not see their forms,
and counted them happy for not
  having suffered,
2 and were thankful that your holy
  ones,[m] though previously
  wronged, were doing them
  no injury;
and they begged their pardon for
  having been at variance
  with them.[m]
3 Therefore you provided a flaming
  pillar of fire
as a guide for your people's[n]
  unknown journey,
and a harmless sun for their
  glorious wandering.
4 For their enemies[o] deserved to be
  deprived of light and
  imprisoned in darkness,
those who had kept your children
  imprisoned,
through whom the imperishable
  light of the law was to be
  given to the world.

5 When they had resolved to kill the
  infants of your holy ones,
and one child had been abandoned
  and rescued,
you in punishment took away a
  multitude of their children;
and you destroyed them all
  together by a mighty flood.
6 That night was made known
  beforehand to our
  ancestors,
so that they might rejoice in sure

knowledge of the oaths in
  which they trusted.
7 The deliverance of the righteous
  and the destruction of their
  enemies
were expected by your people.
8 For by the same means by which
  you punished our enemies
you called us to yourself and
  glorified us.
9 For in secret the holy children of
  good people offered
  sacrifices,
and with one accord agreed to the
  divine law,
so that the saints would share alike
  the same things,
both blessings and dangers;
and already they were singing the
  praises of the ancestors.[p]
10 But the discordant cry of their
  enemies echoed back,
and their piteous lament for their
  children was spread abroad.
11 The slave was punished with the
  same penalty as the master,
and the commoner suffered the
  same loss as the king;
12 and they all together, by the one
  form[q] of death,
had corpses too many to count.
For the living were not sufficient
  even to bury them,
since in one instant their most
  valued children had been
  destroyed.
13 For though they had disbelieved
  everything because of their
  magic arts,
yet, when their firstborn were
  destroyed, they
  acknowledged your people
  to be God's child.

l Gk *They*   *m* Meaning of Gk uncertain
n Gk *their*   *o* Gk *those persons*   *p* Other
ancient authorities read *dangers, the ancestors
already leading the songs of praise*   *q* Gk *name*

18.1: *Your holy ones:* the Israelites. **3–4:** *Pillar of fire . . . a harmless sun,* Ex 13.21–22. *Light of the law,* in contrast to the *light* and *darkness* of their enemies.
18.5–25: The contrast between the tenth plague and the fate of the Israelites. **5:** *Resolved,* Ex 1.16. *Abandoned and rescued,* Ex 2.1–10. *A multitude of their children,* the Egyptian firstborn (Ex 12.29). *That night,* the first passover.

14 For while gentle silence enveloped
      all things,
  and night in its swift course was
      now half gone,
15 your all-powerful word leaped
      from heaven, from the
      royal throne,
  into the midst of the land that was
      doomed,
  a stern warrior
16 carrying the sharp sword of your
      authentic command,
  and stood and filled all things with
      death,
  and touched heaven while standing
      on the earth.
17 Then at once apparitions in
      dreadful dreams greatly
      troubled them,
  and unexpected fears assailed
      them;
18 and one here and another there,
      hurled down half dead,
  made known why they were
      dying;
19 for the dreams that disturbed them
      forewarned them of this,
  so that they might not perish
      without knowing why they
      suffered.

20 The experience of death touched
      also the righteous,
  and a plague came upon the
      multitude in the desert,
  but the wrath did not long
      continue.
21 For a blameless man was quick to
      act as their champion;
  he brought forward the shield of
      his ministry,
  prayer and propitiation by incense;
  he withstood the anger and put an
      end to the disaster,
  showing that he was your servant.
22 He conquered the wrath[r] not by
      strength of body,

not by force of arms,
  but by his word he subdued the
      avenger,
  appealing to the oaths and
      covenants given to our
      ancestors.
23 For when the dead had already
      fallen on one another in
      heaps,
  he intervened and held back the
      wrath,
  and cut off its way to the living.
24 For on his long robe the whole
      world was depicted,
  and the glories of the ancestors
      were engraved on the four
      rows of stones,
  and your majesty was on the
      diadem upon his head.
25 To these the destroyer yielded,
      these he[s] feared;
  for merely to test the wrath was
      enough.

19 But the ungodly were assailed
      to the end by pitiless anger,
  for God[t] knew in advance even
      their future actions:
2 how, though they themselves had
      permitted[u] your people to
      depart
  and hastily sent them out,
  they would change their minds
      and pursue them.
3 For while they were still engaged
      in mourning,
  and were lamenting at the graves
      of their dead,
  they reached another foolish
      decision,
  and pursued as fugitives those
      whom they had begged and
      compelled to leave.

r Cn: Gk *multitude*   s Other ancient authorities
read *they*   t Gk *he*   u Other ancient
authorities read *had changed their minds to permit*

18.15: God's *all-powerful word* (Greek "logos"), as a *stern warrior, leaped from heaven* and carried out God's judgment (this recalls Rev 19.13 rather than Jn 1.1–18). **20–25**: When *a plague* struck Israel *in the desert*, Aaron stopped the destroying angel from inflicting further death (Num 16.41–50). **24**: From Ex 28.15–21, and Jewish tradition. **19.1–22**: **God judged the Egyptians and delivered Israel at the Red Sea. 1–5**: The

4 For the fate they deserved drew
      them on to this end,
and made them forget what had
      happened,
in order that they might fill up the
      punishment that their
      torments still lacked,
5 and that your people might
      experience*v* an incredible
      journey,
but they themselves might meet a
      strange death.

6 For the whole creation in its
      nature was fashioned anew,
complying with your commands,
so that your children*w* might be
      kept unharmed.
7 The cloud was seen
      overshadowing the camp,
and dry land emerging where
      water had stood before,
an unhindered way out of the Red
      Sea,
and a grassy plain out of the
      raging waves,
8 where those protected by your
      hand passed through as one
      nation,
after gazing on marvelous
      wonders.
9 For they ranged like horses,
and leaped like lambs,
praising you, O Lord, who
      delivered them.
10 For they still recalled the events of
      their sojourn,
how instead of producing animals
      the earth brought forth
      gnats,

and instead of fish the river
      spewed out vast numbers of
      frogs.
11 Afterward they saw also a new
      kind*x* of birds,
when desire led them to ask for
      luxurious food;
12 for, to give them relief, quails
      came up from the sea.

13 The punishments did not come
      upon the sinners
without prior signs in the violence
      of thunder,
for they justly suffered because of
      their wicked acts;
for they practiced a more bitter
      hatred of strangers.
14 Others had refused to receive
      strangers when they came
      to them,
but these made slaves of guests
      who were their benefactors.
15 And not only so—but, while
      punishment of some sort
      will come upon the former
for having received strangers with
      hostility,
16 the latter, having first received
      them with festal
      celebrations,
afterward afflicted with terrible
      sufferings
those who had already shared the
      same rights.
17 They were stricken also with loss
      of sight—

v Other ancient authorities read *accomplish*
w Or *servants*   *x* Or *production*

Egyptians' foolish decision to pursue Israel and enslave them again. **6–12:** As before (16.24), creation serves God's purposes. **7:** *Cloud,* Ex 13.21–22. *Dry land,* Ex 14.21–22. **10:** Ex 8.1–24. **11–12:** Ex 16.13 and Num 11.4, 20, 31–32.
    19.13–17: The Egyptians treated *strangers* worse than did the inhabitants of Sodom, and so deserved greater *punishment.* **14:** *Others,* those of Sodom (Gen 19.1–11). *These,* the Egyptians. *Guests,* Israel had been invited to

Egypt (Gen 45.16–20). **17:** Gen 19.11; Ex 10.21–23. **18–21:** In the plagues and at the Red Sea nature and *animals* changed their customary action to effect God's redemptive purpose (16.24). **19:** *Land animals,* apparently a reference to Israel and their cattle crossing the Red Sea. *Creatures that swim,* frogs (Ex 8.1–7). **20–21:** 16.17. *Heavenly food,* the manna. **22:** This abrupt ending gives the lesson of the historical survey: God's *help* for this people.

just as were those at the door of
the righteous man—
when, surrounded by yawning
darkness,
all of them tried to find the way
through their own doors.

18 For the elements changed*y* places
with one another,
as on a harp the notes vary the
nature of the rhythm,
while each note remains the
same. *z*
This may be clearly inferred from
the sight of what took
place.
19 For land animals were transformed
into water creatures,
and creatures that swim moved
over to the land.

20 Fire even in water retained its
normal power,
and water forgot its fire-quenching
nature.
21 Flames, on the contrary, failed to
consume
the flesh of perishable creatures
that walked among them,
nor did they melt*a* the crystalline,
quick-melting kind of
heavenly food.

22 For in everything, O Lord, you
have exalted and glorified
your people,
and you have not neglected to
help them at all times and
in all places.

y Gk *changing*   *z* Meaning of Gk uncertain
a Cn: Gk *nor could be melted*

Ecclesiasticus,
or the Wisdom of Jesus Son of
Sirach

This work is known as Ecclesiasticus, or the Wisdom of Ben Sira, or simply Sirach (the Greek spelling of Sira). "Ecclesiasticus" suggests its use as a "church book" in the early Christian community, which accepted it into its canon. The Jewish community, followed by the Reformers, excluded it from the canon. This explains the extraordinary history of its text. The original Hebrew text was lost to the western world from about 400 to 1900; the book survived in Greek, Latin, Syriac, and other translations. Since about 1900, extensive fragments of copies of the Hebrew have been discovered in various places: Cairo, Qumran, and Masada, so that now two-thirds of the Hebrew text exists. The translation here is one of a critically established text, using both Hebrew and other witnesses to the original. The reader will occasionally find a slightly different verse numbering from that in traditional renderings, since the NRSV follows the numbering of the critical Greek text edited by J. Ziegler.

Ben Sira signs his name (50.27), describes his profession (39.1–11), and invites students to his school (51.23). Sometime before 180 B.C. and the ensuing Maccabean revolt, he committed his wisdom to writing, probably in Jerusalem (see his description in 50.1–24 of Simeon II, who was high priest from 219–196). Sometime after 132 (see the Prologue) his grandson translated the original Hebrew into Greek. The grandson rightly stresses the profound knowledge that Sirach had of Hebrew traditions ("the Law and the Prophets and the other books"—already the three-fold division of the Hebrew Bible was in the process of formation).

Although Sirach lacks any clear structure, it resembles the book of Proverbs in many ways. It stresses characteristic wisdom teachings: proper speech, riches and poverty, honesty, diligence, choice of friends, sin and death, retribution, and wisdom itself. Unlike Proverbs 10ff., individual proverbs are not set apart, but are incorporated into smooth-flowing poems of some length (often in accordance with the number of letters in the Hebrew alphabet). The doctrine is surprisingly traditional, almost as if Job and Ecclesiastes had never been written. Sirach is not unaware of the problem of suffering (2.1–6; 11.4; 40.1–10), but he is a firm believer in the justice of divine retribution. God will reward everyone according to one's deserts (15.11–16.23). There is no intimation of a future life with God in the Hebrew text; rather, all go to Sheol (14.12–19; 38.16–23). This is the usual Hebrew teaching, which understood immortality only in terms of one's progeny and good name (44.13–15). Traditionally, wisdom literature never appealed to Israel's sacred history or covenant. Sirach is an outstanding exception, in view of his "Praise of the Ancestors" (chs 44–50) and his identification of personified wisdom with the Torah or Law (24.23). At the same time, his book belongs definitely to the genre of wisdom literature, with its stress on the lessons of experience and on the "fear of the Lord" (1.11–30; 25.10–11; 40.25–27).

THE PROLOGUE

Many great teachings have been given to us through the Law and the Prophets and the others*a* that followed them, and for these we should praise Israel for instruction and wisdom. Now, those who read the scriptures must not only themselves understand them, but must also as lovers of learning be able through the spoken and written word to help the outsiders. So my grandfather Jesus, who had devoted himself especially to the reading of the Law and the Prophets and the other books of our ancestors, and had acquired considerable proficiency in them, was himself also led to write something pertaining to instruction and wisdom, so that by becoming familiar also with his book*b* those who love learning might make even greater progress in living according to the law.

You are invited therefore to read it with goodwill and attention, and to be indulgent in cases where, despite our diligent labor in translating, we may seem to have rendered some phrases imperfectly. For what was originally expressed in Hebrew does not have exactly the same sense when translated into another language. Not only this book, but even the Law itself, the Prophecies, and the rest of the books differ not a little when read in the original.

When I came to Egypt in the thirty-eighth year of the reign of Euergetes and stayed for some time, I found opportunity for no little instruction.*c* It seemed highly necessary that I should myself devote some diligence and labor to the translation of this book. During that time I have applied my skill day and night to complete and publish the book for those living abroad who wished to gain learning and are disposed to live according to the law.

1 All wisdom is from the Lord,
    and with him it remains forever.
2 The sand of the sea, the drops of
        rain,
    and the days of eternity—who
        can count them?
3 The height of heaven, the breadth
        of the earth,
    the abyss, and wisdom*d*—who
        can search them out?
4 Wisdom was created before all
        other things,
    and prudent understanding from
        eternity.*e*
6 The root of wisdom—to whom
        has it been revealed?
    Her subtleties—who knows
        them?*f*
8 There is but one who is wise,
        greatly to be feared,
    seated upon his throne—the
        Lord.
9 It is he who created her;
    he saw her and took her
        measure;
    he poured her out upon all his
        works,
10 upon all the living according to his
        gift;
    he lavished her upon those who
        love him.*g*

11 The fear of the Lord is glory and
        exultation,

a Or *other books*   *b* Gk *with these things*
c Other ancient authorities read *I found a copy
affording no little instruction*   *d* Other ancient
authorities read *the depth of the abyss*   *e* Other
ancient authorities add as verse 5, *The source of
wisdom is God's word in the highest heaven, and her
ways are the eternal commandments.*   *f* Other
ancient authorities add as verse 7, *The knowledge
of wisdom—to whom was it manifested? And her
abundant experience—who has understood it?*
g Other ancient authorities add *Love of the Lord
is glorious wisdom; to those to whom he appears he
apportions her, that they may see him.*

1.1–10: The personification of Wisdom
(see also ch 24; Job 28; Prov 8). **1:** *From the
Lord*, a divine origin, as in 24.3; Prov 8.22–
25. **2–3:** Illustrations of the impossibility of
fathoming the depths of divine wisdom
(18.4–7; Rom 11.33). **4:** Prov 8.22–31. **6–10:**
Only with God is Wisdom to be found (Job
28); yet God has *lavished* her upon *all the living,* i.e. Jews and Gentiles, but especially
upon *those who love him* (Israel).
  **1.11–30: A poem that identifies wisdom
and fear of the Lord** (vv. 11–14, 16, 18,

and gladness and a crown of
rejoicing.

12 The fear of the Lord delights the
heart,
and gives gladness and joy and
long life. *h*

13 Those who fear the Lord will have
a happy end;
on the day of their death they
will be blessed.

14 To fear the Lord is the beginning
of wisdom;
she is created with the faithful in
the womb.

15 She made *i* among human beings
an eternal foundation,
and among their descendants she
will abide faithfully.

16 To fear the Lord is fullness of
wisdom;
she inebriates mortals with her
fruits;

17 she fills their *j* whole house with
desirable goods,
and their *j* storehouses with her
produce.

18 The fear of the Lord is the crown
of wisdom,
making peace and perfect health
to flourish. *k*

19 She rained down knowledge and
discerning comprehension,
and she heightened the glory of
those who held her fast.

20 To fear the Lord is the root of
wisdom,
and her branches are long life. *l*

22 Unjust anger cannot be justified,
for anger tips the scale to one's
ruin.

23 Those who are patient stay calm
until the right moment,
and then cheerfulness comes
back to them.

24 They hold back their words until
the right moment;

then the lips of many tell of
their good sense.

25 In the treasuries of wisdom are
wise sayings,
but godliness is an abomination
to a sinner.

26 If you desire wisdom, keep the
commandments,
and the Lord will lavish her
upon you.

27 For the fear of the Lord is wisdom
and discipline,
fidelity and humility are his
delight.

28 Do not disobey the fear of the
Lord;
do not approach him with a
divided mind.

29 Do not be a hypocrite before
others,
and keep watch over your lips.

30 Do not exalt yourself, or you may
fall
and bring dishonor upon
yourself.
The Lord will reveal your secrets
and overthrow you before the
whole congregation,
because you did not come in the
fear of the Lord,
and your heart was full of
deceit.

2 My child, when you come to
serve the Lord,
prepare yourself for testing. *m*

h Other ancient authorities add *The fear of the
Lord is a gift from the Lord; also for love he makes
firm paths.* *i* Gk *made as a nest* *j* Other
ancient authorities read *her* *k* Other ancient
authorities add *Both are gifts of God for peace; glory
opens out for those who love him. He saw her and
took her measure.* *l* Other ancient authorities
add as verse 21, *The fear of the Lord drives away
sins; and where it abides, it will turn away all anger.*
m Or *trials*

20–21, 25, 27–28, 30). **12:** *Long life,* typical of
wisdom teaching; see v. 20; Prov 3.16. **14:**
Prov 1.7; 9.10; Job 28.28; Ps 111.10.

2.1–17: A poem on trusting God. 1–6:
The testing of faith is a common biblical
theme (Gen 22; Job 1–2; Jas 1.2–4; Luke

2 Set your heart right and be
steadfast,
and do not be impetuous in
time of calamity.
3 Cling to him and do not depart,
so that your last days may be
prosperous.
4 Accept whatever befalls you,
and in times of humiliation be
patient.
5 For gold is tested in the fire,
and those found acceptable, in
the furnace of humiliation. *n*
6 Trust in him, and he will help
you;
make your ways straight, and
hope in him.

7 You who fear the Lord, wait for
his mercy;
do not stray, or else you may
fall.
8 You who fear the Lord, trust in
him,
and your reward will not be
lost.
9 You who fear the Lord, hope for
good things,
for lasting joy and mercy. *o*
10 Consider the generations of old
and see:
has anyone trusted in the Lord
and been disappointed?
Or has anyone persevered in the
fear of the Lord *p* and been
forsaken?
Or has anyone called upon him
and been neglected?
11 For the Lord is compassionate and
merciful;
he forgives sins and saves in
time of distress.

12 Woe to timid hearts and to slack
hands,
and to the sinner who walks a
double path!

13 Woe to the fainthearted who have
no trust!
Therefore they will have no
shelter.
14 Woe to you who have lost your
nerve!
What will you do when the
Lord's reckoning comes?

15 Those who fear the Lord do not
disobey his words,
and those who love him keep
his ways.
16 Those who fear the Lord seek to
please him,
and those who love him are
filled with his law.
17 Those who fear the Lord prepare
their hearts,
and humble themselves before
him.
18 Let us fall into the hands of the
Lord,
but not into the hands of
mortals;
for equal to his majesty is his
mercy,
and equal to his name are his
works. *q*

3 Listen to me your father,
O children;
act accordingly, that you may
be kept in safety.
2 For the Lord honors a father
above his children,
and he confirms a mother's
right over her children.
3 Those who honor their father
atone for sins,

n Other ancient authorities add *in sickness
and poverty put your trust in him*
o Other ancient authorities add *For his reward is
an everlasting gift with joy.*   p Gk *of him*
q Syr: Gk lacks this line

11.4). **7–9:** Vv. 15–17. **9:** *Joy,* in this life, not
in the next.
   2.10–11: Israel's history shows the efficacy
of trust in God, who is *compassionate and merci-*
ful (Ex 34.6–7; Ps 103.8–9; Jon 4.2). **17:** 2 Sam
24.14; 1 Chr 21.13.
   3.1–16: Filial duty and its reward (Ex
20.12; Deut 5.16; Eph 6.1–3). In accord with

4 and those who respect their
mother are like those who
lay up treasure.

5 Those who honor their father will
have joy in their own
children,
and when they pray they will be
heard.

6 Those who respect their father
will have long life,
and those who honor' their
mother obey the Lord;

7 they will serve their parents as
their masters.*

8 Honor your father by word and
deed,
that his blessing may come
upon you.

9 For a father's blessing strengthens
the houses of the children,
but a mother's curse uproots
their foundations.

10 Do not glorify yourself by
dishonoring your father,
for your father's dishonor is no
glory to you.

11 The glory of one's father is one's
own glory,
and it is a disgrace for children
not to respect their mother.

12 My child, help your father in his
old age,
and do not grieve him as long
as he lives;

13 even if his mind fails, be patient
with him;
because you have all your
faculties do not despise
him.

14 For kindness to a father will not
be forgotten,
and will be credited to you
against your sins;

15 in the day of your distress it will
be remembered in your
favor;
like frost in fair weather, your
sins will melt away.

16 Whoever forsakes a father is like a
blasphemer,
and whoever angers a mother is
cursed by the Lord.

17 My child, perform your tasks with
humility;*
then you will be loved by those
whom God accepts.

18 The greater you are, the more you
must humble yourself;
so you will find favor in the
sight of the Lord. *

20 For great is the might of the Lord;
but by the humble he is
glorified.

21 Neither seek what is too difficult
for you,
nor investigate what is beyond
your power.

22 Reflect upon what you have been
commanded,
for what is hidden is not your
concern.

23 Do not meddle in matters that are
beyond you,
for more than you can
understand has been shown
you.

24 For their conceit has led many
astray,
and wrong opinion has impaired
their judgment.

r Heb: Other ancient authorities read *comfort*
s In other ancient authorities this line is preceded
by *Those who fear the Lord honor their father,*
t Heb: Gk *meekness*   u Other ancient
authorities add as verse 19, *Many are lofty and
renowned, but to the humble he reveals his secrets.*

the Jewish doctrine that the observance of the
Mosaic law is meritorious, Sirach teaches that
the keeping of the commandment to honor
one's parents (Ex 20.12) *atones for sins* (vv. 3,
14–15).
   3.14: *Kindness,* literally "righteousness,"
which came to be centered in almsgiving as

the characteristic sign (3.10–4.6; 7.10; 29.8–
13; Tob 14.10–11).
   3.17–24: On humility. A frequent topic
in wisdom instruction (Prov 11.2; 15.33;
22.4; and Sir 7.16–17; 10.28). **21–24:** Perhaps
Sirach is warning against Greek learning. **21:**
Ps 131; Eccl 7.24.

25 Without eyes there is no light;
   without knowledge there is no
      wisdom. *v*
26 A stubborn mind will fare badly at
      the end,
   and whoever loves danger will
      perish in it.
27 A stubborn mind will be burdened
      by troubles,
   and the sinner adds sin to sins.
28 When calamity befalls the proud,
      there is no healing,
   for an evil plant has taken root
      in him.
29 The mind of the intelligent
      appreciates proverbs,
   and an attentive ear is the desire
      of the wise.

30 As water extinguishes a blazing
      fire,
   so almsgiving atones for sin.
31 Those who repay favors give
      thought to the future;
   when they fall they will find
      support.

4 My child, do not cheat the poor
      of their living,
   and do not keep needy eyes
      waiting.
2 Do not grieve the hungry,
   or anger one in need.
3 Do not add to the troubles of the
      desperate,
   or delay giving to the needy.
4 Do not reject a suppliant in
      distress,
   or turn your face away from the
      poor.
5 Do not avert your eye from the
      needy,
   and give no one reason to curse
      you;
6 for if in bitterness of soul some
      should curse you,
   their Creator will hear their
      prayer.

7 Endear yourself to the
      congregation;
   bow your head low to the great.
8 Give a hearing to the poor,
   and return their greeting
      politely.
9 Rescue the oppressed from the
      oppressor;
   and do not be hesitant in giving
      a verdict.
10 Be a father to orphans,
   and be like a husband to their
      mother;
   you will then be like a son of the
      Most High,
   and he will love you more than
      does your mother.

11 Wisdom teaches *w* her children
   and gives help to those who
      seek her.
12 Whoever loves her loves life,
   and those who seek her from
      early morning are filled
      with joy.
13 Whoever holds her fast inherits
      glory,
   and the Lord blesses the place
      she *x* enters.
14 Those who serve her minister to
      the Holy One;
   the Lord loves those who love
      her.
15 Those who obey her will judge
      the nations,
   and all who listen to her will
      live secure.
16 If they remain faithful, they will
      inherit her;
   their descendants will also
      obtain her.
17 For at first she will walk with
      them on tortuous paths;

v Heb: Other ancient authorities lack verse 25
w Heb Syr: Gk *exalts*   *x* Or *he*

3.25–29: On docility. The *stubborn* (literally "heavy heart") refuse the lessons of wisdom.
3.30–4.10: On almsgiving. See 3.14 n.
4.6: Ex 22.22. **9–10:** 34.21–27; see 35.14–

15 n. Orphans and widows were particularly
indigent (Ex 22.21–23).
4.11–19: The rewards of Wisdom. Personified Wisdom (1.10; ch 24) loves her followers, but also tests them. **14:** Service of

she will bring fear and dread
upon them,
and will torment them by her
discipline
until she trusts them, *y*
and she will test them with her
ordinances.
18 Then she will come straight back
to them again and gladden
them,
and will reveal her secrets to
them.
19 If they go astray she will forsake
them,
and hand them over to their
ruin.

20 Watch for the opportune time, and
beware of evil,
and do not be ashamed to be
yourself.
21 For there is a shame that leads to
sin,
and there is a shame that is
glory and favor.
22 Do not show partiality, to your
own harm,
or deference, to your downfall.
23 Do not refrain from speaking at
the proper moment, *z*
and do not hide your wisdom. *a*
24 For wisdom becomes known
through speech,
and education through the
words of the tongue.
25 Never speak against the truth,
but be ashamed of your
ignorance.
26 Do not be ashamed to confess
your sins,
and do not try to stop the
current of a river.
27 Do not subject yourself to a fool,
or show partiality to a ruler.
28 Fight to the death for truth,

and the Lord God will fight for
you.
29 Do not be reckless in your speech,
or sluggish and remiss in your
deeds.
30 Do not be like a lion in your
home,
or suspicious of your servants.
31 Do not let your hand be stretched
out to receive
and closed when it is time to
give.

5 Do not rely on your wealth,
or say, "I have enough."
2 Do not follow your inclination
and strength
in pursuing the desires of your
heart.
3 Do not say, "Who can have power
over me?"
for the Lord will surely punish
you.
4 Do not say, "I sinned, yet what
has happened to me?"
for the Lord is slow to anger.
5 Do not be so confident of
forgiveness *b*
that you add sin to sin.
6 Do not say, "His mercy is great,
he will forgive *c* the multitude
of my sins,"
for both mercy and wrath are with
him,
and his anger will rest on
sinners.
7 Do not delay to turn back to the
Lord,

y Or *until they remain faithful in their heart*
z Heb: Gk *at a time of salvation*   *a* So some
Gk Mss and Heb Syr Lat: Other Gk Mss lack
and do not hide your wisdom   *b* Heb: Gk
atonement   *c* Heb: Gk *he* (or *it*) *will atone for*

wisdom is service of God, and God will love
those who love her (Prov 8.17). **17–18:** Prov
3.11–12.
4.20–5.8: Precepts for everyday life.
20–28: True and false shame (20.22–23;
41.14–42.8). **26:** It is as futile to hide one's
sins from God as to try to stop a river from
flowing. **30:** *A lion,* wild, relentless, destructive. **31:** *Hand,* Deut 15.7–8; Acts 20.35.
5.1–8: A poem against presumption. **5:**
Sin to sin, presumption is the added sin. **6:**
16.11–12. The Lord is a God of both *mercy and
wrath.*

and do not postpone it from day
to day;
for suddenly the wrath of the Lord
will come upon you,
and at the time of punishment
you will perish.

8 Do not depend on dishonest
wealth,
for it will not benefit you on
the day of calamity.

9 Do not winnow in every wind,
or follow every path. *d*

10 Stand firm for what you know,
and let your speech be
consistent.

11 Be quick to hear,
but deliberate in answering.

12 If you know what to say, answer
your neighbor;
but if not, put your hand over
your mouth.

13 Honor and dishonor come from
speaking,
and the tongue of mortals may
be their downfall.

14 Do not be called double-tongued*e*
and do not lay traps with your
tongue;
for shame comes to the thief,
and severe condemnation to the
double-tongued.

15 In great and small matters cause
no harm,*f*

6 1 and do not become an enemy
instead of a friend;
for a bad name incurs shame and
reproach;
so it is with the double-tongued
sinner.

2 Do not fall into the grip of
passion,*g*

or you may be torn apart as by
a bull. *h*

3 Your leaves will be devoured and
your fruit destroyed,
and you will be left like a
withered tree.

4 Evil passion destroys those who
have it,
and makes them the
laughingstock of their
enemies.

5 Pleasant speech multiplies friends,
and a gracious tongue multiplies
courtesies.

6 Let those who are friendly with
you be many,
but let your advisers be one in a
thousand.

7 When you gain friends, gain them
through testing,
and do not trust them hastily.

8 For there are friends who are such
when it suits them,
but they will not stand by you
in time of trouble.

9 And there are friends who change
into enemies,
and tell of the quarrel to your
disgrace.

10 And there are friends who sit at
your table,
but they will not stand by you
in time of trouble.

11 When you are prosperous, they
become your second self,
and lord it over your servants;

12 but if you are brought low, they
turn against you,
and hide themselves from you.

d Gk adds *so it is with the double-tongued sinner*
(see 6.1)   *e* Heb: Gk *a slanderer*   *f* Heb Syr:
Gk *be ignorant*   *g* Heb: Meaning of Gk
uncertain   *h* Meaning of Gk uncertain

5.9–6.1: **Honesty and sincerity. 9:** A
condemnation of duplicity. **12:** *Hand . . .
mouth,* to keep silence, since one has no
competency to speak. **13:** The topic of the
tongue (use and abuse) is taken up frequently;
see 19.6–17; 20.16–20; 22.27–23.15; 28.12–
26.
6.2–4: Self-control; a warning against
lustful passions (18.30–19.3).
6.5–17: **True and false friendship.** A
frequent topic (see 11.29–12.18; 22.19–26;
37.1–6). Friends are to be tested, lest they
turn out to be "fair weather" friends (vv. 5–
12). True friends are a *treasure,* a gift from
God (vv. 13–17).

13 Keep away from your enemies,
   and be on guard with your
      friends.

14 Faithful friends are a sturdy
      shelter:
   whoever finds one has found a
      treasure.
15 Faithful friends are beyond price;
   no amount can balance their
      worth.
16 Faithful friends are life-saving
      medicine;
   and those who fear the Lord
      will find them.
17 Those who fear the Lord direct
      their friendship aright,
   for as they are, so are their
      neighbors also.

18 My child, from your youth choose
      discipline,
   and when you have gray hair
      you will still find wisdom.
19 Come to her like one who plows
      and sows,
   and wait for her good harvest.
   For when you cultivate her you
      will toil but little,
   and soon you will eat of her
      produce.
20 She seems very harsh to the
      undisciplined;
   fools cannot remain with her.
21 She will be like a heavy stone to
      test them,
   and they will not delay in
      casting her aside.
22 For wisdom is like her name;
   she is not readily perceived by
      many.

23 Listen, my child, and accept my
      judgment;
   do not reject my counsel.

24 Put your feet into her fetters,
   and your neck into her collar.
25 Bend your shoulders and carry
      her,
   and do not fret under her
      bonds.
26 Come to her with all your soul,
   and keep her ways with all your
      might.
27 Search out and seek, and she will
      become known to you;
   and when you get hold of her,
      do not let her go.
28 For at last you will find the rest
      she gives,
   and she will be changed into joy
      for you.
29 Then her fetters will become for
      you a strong defense,
   and her collar a glorious robe.
30 Her yoke*i* is a golden ornament,
   and her bonds a purple cord.
31 You will wear her like a glorious
      robe,
   and put her on like a splendid
      crown.*j*

32 If you are willing, my child, you
      can be disciplined,
   and if you apply yourself you
      will become clever.
33 If you love to listen you will gain
      knowledge,
   and if you pay attention you
      will become wise.
34 Stand in the company of the
      elders.
   Who is wise? Attach yourself to
      such a one.
35 Be ready to listen to every godly
      discourse,

i Heb: Gk *Upon her*   *j* Heb: Gk *crown of
gladness*

6.18–37: Encouragement to seek Wisdom. *Discipline* (Hebrew "musar"; see 1.27; 4.17; 22.6; 23.2; 26.14; 32.14) is needed (vv. 18–22). There is a wordplay on "musar," which can also mean "withdrawn," and thus *not readily perceived by many* (v. 22). **6.23–31:** Ben Sira often speaks to *my child,* in the style of the sages (Prov 4.10). One is to submit to the *yoke* of Wisdom, which is in reality a *golden ornament.* **32–37:** One way of becoming wise is to seek out the intelligent and frequent their company. Attaining wisdom means fidelity to the Law (v. 37; see also 15.1; 19.20; 23.27; 24.23).

and let no wise proverbs escape
  you.
36 If you see an intelligent person,
    rise early to visit him;
  let your foot wear out his
    doorstep.
37 Reflect on the statutes of the Lord,
    and meditate at all times on his
    commandments.
  It is he who will give insight to[k]
    your mind,
  and your desire for wisdom will
    be granted.

7 Do no evil, and evil will never
    overtake you.
2 Stay away from wrong, and it will
    turn away from you.
3 Do[l] not sow in the furrows of
    injustice,
  and you will not reap a
    sevenfold crop.

4 Do not seek from the Lord high
    office,
  or the seat of honor from the
    king.
5 Do not assert your righteousness
    before the Lord,
  or display your wisdom before
    the king.
6 Do not seek to become a judge,
    or you may be unable to root
    out injustice;
  you may be partial to the
    powerful,
  and so mar your integrity.
7 Commit no offense against the
    public,
  and do not disgrace yourself
    among the people.

8 Do not commit a sin twice;
  not even for one will you go
    unpunished.

9 Do not say, "He will consider the
    great number of my gifts,
  and when I make an offering to
    the Most High God, he will
    accept it."
10 Do not grow weary when you
    pray;
  do not neglect to give alms.
11 Do not ridicule a person who is
    embittered in spirit,
  for there is One who humbles
    and exalts.
12 Do not devise[m] a lie against your
    brother,
  or do the same to a friend.
13 Refuse to utter any lie,
  for it is a habit that results in no
    good.
14 Do not babble in the assembly of
    the elders,
  and do not repeat yourself when
    you pray.

15 Do not hate hard labor
  or farm work, which was
    created by the Most High.
16 Do not enroll in the ranks of
    sinners;
  remember that retribution does
    not delay.
17 Humble yourself to the utmost,
  for the punishment of the
    ungodly is fire and
    worms.[n]

18 Do not exchange a friend for
    money,
  or a real brother for the gold of
    Ophir.
19 Do not dismiss[o] a wise and good
    wife,

k Heb: Gk *will confirm*   l Gk *My child, do*
m Heb: Gk *plow*   n Heb *for the expectation of*
mortals is worms   o Heb: Gk *deprive yourself of*

7.1–17: **Various ethical maxims. 1–3:**
Avoidance of sin. **4–7:** Proper conduct in
public life. **9:** One cannot buy off God. **11:**
11.4. The role of God is emphasized also in
1 Sam 2.8. **14:** Eccl 5.1 [2]; Mt 6.8. **16:** In the
context of v. 15, this means to consider oneself better than others. **17:***Worms,* or corruption in the grave, awaits all (not "fire and
worms" as the Greek has it).
  7.18–36: Various duties. 18: *Ophir,* on
the coast of southern Arabia or eastern Africa,
was famous for its gold (1 Kings 9.28). **19:**

for her charm is worth more
than gold.
20 Do not abuse slaves who work
faithfully,
or hired laborers who devote
themselves to their task.
21 Let your soul love intelligent
slaves;*p*
do not withhold from them
their freedom.

22 Do you have cattle? Look after
them;
if they are profitable to you,
keep them.
23 Do you have children? Discipline
them,
and make them obedient*q* from
their youth.
24 Do you have daughters? Be
concerned for their
chastity,*r*
and do not show yourself too
indulgent with them.
25 Give a daughter in marriage, and
you complete a great task;
but give her to a sensible man.
26 Do you have a wife who pleases
you?*s* Do not divorce her;
but do not trust yourself to one
whom you detest.

27 With all your heart honor your
father,
and do not forget the birth
pangs of your mother.
28 Remember that it was of your
parents*t* you were born;
how can you repay what they
have given to you?

29 With all your soul fear the Lord,
and revere his priests.

30 With all your might love your
Maker,
and do not neglect his ministers.
31 Fear the Lord and honor the
priest,
and give him his portion, as you
have been commanded:
the first fruits, the guilt offering,
the gift of the shoulders,
the sacrifice of sanctification,
and the first fruits of the
holy things.

32 Stretch out your hand to the poor,
so that your blessing may be
complete.
33 Give graciously to all the living;
do not withhold kindness even
from the dead.
34 Do not avoid those who weep,
but mourn with those who
mourn.
35 Do not hesitate to visit the sick,
because for such deeds you will
be loved.
36 In all you do, remember the end
of your life,
and then you will never sin.

8 Do not contend with the
powerful,
or you may fall into their hands.
2 Do not quarrel with the rich,
in case their resources outweigh
yours;
for gold has ruined many,
and has perverted the minds of
kings.
3 Do not argue with the loud of
mouth,

p Heb *Love a wise slave as yourself*
q Gk *bend their necks*   *r* Gk *body*   *s* Heb Syr
lack who pleases you   *t* Gk *them*

Such a *wife* is not to be divorced; on wives,
see 9.1; 26.1–18; 36.27–31. **21**: After six years
of service a Hebrew slave was entitled to *freedom* (Ex 21.2; Lev 25.9–43).
   7.24–25: On *daughters*, see 42.9–14.
   7.29–31: In supporting and showing hon-
or to the priests one indirectly honors God
himself; for the priestly portion, see Num
18.9–20. **31**: *As . . . commanded*, Ex 29.27; Lev

7.31–34; Num 18.8–20; Deut 18.3. **33**: *Kindness*, probably provision for an honorable
burial (30.18; Tob 1.17). **34**: Rom 12.15. **36**:
The realization that the end of a sinner is
bound to be a sad one should serve as a deter-
rence from sin.
   8.1–19: **Advice about various types of
people. 1–2**: The *gold* of the *rich* can function
as a bribe. **5–7**: Have respect for the penitent,

and do not heap wood on their
    fire.

4 Do not make fun of one who is
    ill-bred,
  or your ancestors may be
    insulted.
5 Do not reproach one who is
    turning away from sin;
  remember that we all deserve
    punishment.
6 Do not disdain one who is old,
  for some of us are also growing
    old.
7 Do not rejoice over any one's
    death;
  remember that we must all die.

8 Do not slight the discourse of the
    sages,
  but busy yourself with their
    maxims;
  because from them you will learn
    discipline
  and how to serve princes.
9 Do not ignore the discourse of the
    aged,
  for they themselves learned
    from their parents;*u*
  from them you learn how to
    understand
  and to give an answer when the
    need arises.

10 Do not kindle the coals of sinners,
  or you may be burned in their
    flaming fire.
11 Do not let the insolent bring you
    to your feet,
  or they may lie in ambush
    against your words.
12 Do not lend to one who is
    stronger than you;
  but if you do lend anything,
    count it as a loss.
13 Do not give surety beyond your
    means;

but if you give surety, be
    prepared to pay.

14 Do not go to law against a judge,
  for the decision will favor him
    because of his standing.
15 Do not go traveling with the
    reckless,
  or they will be burdensome to
    you;
  for they will act as they please,
  and through their folly you will
    perish with them.
16 Do not pick a fight with the
    quick-tempered,
  and do not journey with them
    through lonely country,
  because bloodshed means nothing
    to them,
  and where no help is at hand,
    they will strike you down.
17 Do not consult with fools,
  for they cannot keep a secret.
18 In the presence of strangers do
    nothing that is to be kept
    secret,
  for you do not know what they
    will divulge.*v*
19 Do not reveal your thoughts to
    anyone,
  or you may drive away your
    happiness.*w*

9 Do not be jealous of the wife of
    your bosom,
  or you will teach her an evil
    lesson to your own hurt.
2 Do not give yourself to a woman
  and let her trample down your
    strength.
3 Do not go near a loose woman,
  or you will fall into her snares.
4 Do not dally with a singing girl,
  or you will be caught by her
    tricks.

u Or *ancestors*   *v* Or *it will bring forth*
w Heb: Gk *and let him not return a favor to you*

---

the aged, and the departed. **8–9**: Learn from
people with experience. **10**: Lighting *coals* is
an image for arousing passion. **12–13**: Sirach
is cautious about loans and surety, but not
opposed to them (29.1–7), in contrast to Prov
6.1–5; 11.15. **17–19**: The dangers of confiding
in people.
  9.1–9: **Conduct with women. 1–2**: Do
not be jealous (v.1), but be firm (v.2). **3**: *Loose*,
on the "strange" woman, see Prov 2.16; 5.3,

5 Do not look intently at a virgin,
  or you may stumble and incur
    penalties for her.
6 Do not give yourself to
    prostitutes,
  or you may lose your
    inheritance.
7 Do not look around in the streets
    of a city,
  or wander about in its deserted
    sections.
8 Turn away your eyes from a
    shapely woman,
  and do not gaze at beauty
    belonging to another;
many have been seduced by a
    woman's beauty,
  and by it passion is kindled like
    a fire.
9 Never dine with another man's
    wife,
  or revel with her at wine;
  or your heart may turn aside to
    her,
  and in blood[x] you may be
    plunged into destruction.

10 Do not abandon old friends,
  for new ones cannot equal
    them.
A new friend is like new wine;
  when it has aged, you can drink
    it with pleasure.

11 Do not envy the success of
    sinners,
  for you do not know what their
    end will be like.
12 Do not delight in what pleases the
    ungodly;
  remember that they will not be
    held guiltless all their lives.

13 Keep far from those who have
    power to kill,
  and you will not be haunted by
    the fear of death.

But if you approach them, make
    no misstep,
  or they may rob you of your
    life.
Know that you are stepping
    among snares,
  and that you are walking on the
    city battlements.

14 As much as you can, aim to know
    your neighbors,
  and consult with the wise.
15 Let your conversation be with
    intelligent people,
  and let all your discussion be
    about the law of the Most
    High.
16 Let the righteous be your dinner
    companions,
  and let your glory be in the fear
    of the Lord.

17 A work is praised for the skill of
    the artisan;
  so a people's leader is proved
    wise by his words.
18 The loud of mouth are feared in
    their city,
  and the one who is reckless in
    speech is hated.

10 A wise magistrate educates his
    people,
  and the rule of an intelligent
    person is well ordered.
2 As the people's judge is, so are his
    officials;
  as the ruler of the city is, so are
    all its inhabitants.
3 An undisciplined king ruins his
    people,
  but a city becomes fit to live in
    through the understanding
    of its rulers.
4 The government of the earth is in
    the hand of the Lord,

x Heb: Gk *by your spirit*

20; 7.5. **4–9:** The following prohibitions have various motive clauses, mostly practical.
  9.10–16: Precepts about various types of persons.

9.17–10.5: Concerning rulers. 10.4: *For the* right *time*.
  10.6–18: Concerning arrogance and pride, especially in rulers (Prov 16, 18). Si-

and over it he will raise up the
    right leader for the time.
5 Human success is in the hand of
    the Lord,
  and it is he who confers honor
    upon the lawgiver.ʸ

6 Do not get angry with your
    neighbor for every injury,
  and do not resort to acts of
    insolence.
7 Arrogance is hateful to the Lord
    and to mortals,
  and injustice is outrageous to
    both.
8 Sovereignty passes from nation to
    nation
  on account of injustice and
    insolence and wealth.ᶻ
9 How can dust and ashes be proud?
  Even in life the human body
    decays.ᵃ
10 A long illness baffles the
    physician;ᵇ
  the king of today will die
    tomorrow.
11 For when one is dead
  he inherits maggots and
    verminᶜ and worms.
12 The beginning of human pride is
    to forsake the Lord;
  the heart has withdrawn from
    its Maker.
13 For the beginning of pride is sin,
  and the one who clings to it
    pours out abominations.
  Therefore the Lord brings upon
    them unheard-of calamities,
  and destroys them completely.
14 The Lord overthrows the thrones
    of rulers,
  and enthrones the lowly in their
    place.
15 The Lord plucks up the roots of
    the nations,ᵈ
  and plants the humble in their
    place.

16 The Lord lays waste the lands of
    the nations,
  and destroys them to the
    foundations of the earth.
17 He removes some of them and
    destroys them,
  and erases the memory of them
    from the earth.
18 Pride was not created for human
    beings,
  or violent anger for those born
    of women.

19 Whose offspring are worthy of
    honor?
  Human offspring.
  Whose offspring are worthy of
    honor?
  Those who fear the Lord.
  Whose offspring are unworthy of
    honor?
  Human offspring.
  Whose offspring are unworthy of
    honor?
  Those who break the
    commandments.
20 Among family members their
    leader is worthy of honor,
  but those who fear the Lord are
    worthy of honor in his
    eyes.ᵉ
22 The rich, and the eminent, and the
    poor—
  their glory is the fear of the
    Lord.
23 It is not right to despise one who
    is intelligent but poor,
  and it is not proper to honor
    one who is sinful.

y Heb: Gk *scribe*   z Other ancient authorities
add here or after verse 9a, *Nothing is more wicked
than one who loves money, for such a person puts his
own soul up for sale*.   a Heb: Meaning of Gk
uncertain   b Heb Lat: Meaning of Gk
uncertain   c Heb: Gk *wild animals*   d Other
ancient authorities read *proud nations*
e Other ancient authorities add as verse 21, *The
fear of the Lord is the beginning of acceptance;
obduracy and pride are the beginning of rejection.*

---

rach may have in mind certain Ptolemaic or
Seleucid kings, but several verses apply to
ordinary people (vv. 12, 18).

**10.19–11.6: True honor is fear of the
Lord. 19:** This key verse is developed in what
follows. **28:** See 3.17 n. Humility is not false

24 The prince and the judge and the
   ruler are honored,
   but none of them is greater than
      the one who fears the Lord.
25 Free citizens will serve a wise
      servant,
   and an intelligent person will
      not complain.

26 Do not make a display of your
      wisdom when you do your
      work,
   and do not boast when you are
      in need.
27 Better is the worker who has
      goods in plenty
   than the boaster who lacks
      bread.

28 My child, honor yourself with
      humility,
   and give yourself the esteem
      you deserve.
29 Who will acquit those who
      condemn*f* themselves?
   And who will honor those who
      dishonor themselves?*g*
30 The poor are honored for their
      knowledge,
   while the rich are honored for
      their wealth.
31 One who is honored in poverty,
      how much more in wealth!
   And one dishonored in wealth,
      how much more in
      poverty!

11 The wisdom of the humble lifts
      their heads high,
   and seats them among the great.
2 Do not praise individuals for their
      good looks,
   or loathe anyone because of
      appearance alone.

3 The bee is small among flying
      creatures,
   but what it produces is the best
      of sweet things.
4 Do not boast about wearing fine
      clothes,
   and do not exalt yourself when
      you are honored;
for the works of the Lord are
      wonderful,
   and his works are concealed
      from humankind.
5 Many kings have had to sit on the
      ground,
   but one who was never thought
      of has worn a crown.
6 Many rulers have been utterly
      disgraced,
   and the honored have been
      handed over to others.

7 Do not find fault before you
      investigate;
   examine first, and then criticize.
8 Do not answer before you listen,
   and do not interrupt when
      another is speaking.
9 Do not argue about a matter that
      does not concern you,
   and do not sit with sinners
      when they judge a case.

10 My child, do not busy yourself
      with many matters;
   if you multiply activities, you
      will not be held blameless.
If you pursue, you will not
      overtake,
   and by fleeing you will not
      escape.
11 There are those who work and
      struggle and hurry,

f Heb: Gk *sin against*   *g* Heb Lat: Gk *their own
life*

---

grovelling. **30–31:** *Wealth* does make a difference, even for the wise, *honored for their knowledge.*
   11.2–4: Do not judge by appearances,
whether *looks,* or size, or *clothes.*
   **11.7–9: Six prohibitions against hasty
and rash actions.**
   11.10–13: Avoidance of anxiety.
   11.14–19: All things come from the

Lord, both good and evil (Isa 45.7; Job
1.21). **19:** Lk 12.19.
   11.20–28: Retribution. 20–26: God's reward or punishment occurs in this life, down
to the very end. **27–28:** The *close of one's life*
revives memories of the past, and is a test of
one's life; see 7.36 n.
   **11.29–12.18: Choosing companions. 29–
34:** Sirach advises extreme care.

but are so much the more in
    want.
12 There are others who are slow and
    need help,
  who lack strength and abound
    in poverty;
but the eyes of the Lord look
    kindly upon them;
  he lifts them out of their lowly
    condition
13 and raises up their heads
  to the amazement of the many.

14 Good things and bad, life and
    death,
  poverty and wealth, come from
    the Lord.*h*
17 The Lord's gift remains with the
    devout,
  and his favor brings lasting
    success.
18 One becomes rich through
    diligence and self-denial,
  and the reward allotted to him
    is this:
19 when he says, "I have found rest,
  and now I shall feast on my
    goods!"
he does not know how long it
    will be
  until he leaves them to others
    and dies.

20 Stand by your agreement and
    attend to it,
  and grow old in your work.
21 Do not wonder at the works of a
    sinner,
  but trust in the Lord and keep
    at your job;
for it is easy in the sight of the
    Lord
  to make the poor rich suddenly,
    in an instant.
22 The blessing of the Lord is*i* the
    reward of the pious,
  and quickly God causes his
    blessing to flourish.
23 Do not say, "What do I need,
  and what further benefit can be
    mine?"
24 Do not say, "I have enough,

and what harm can come to me
    now?"
25 In the day of prosperity, adversity
    is forgotten,
  and in the day of adversity,
    prosperity is not
    remembered.
26 For it is easy for the Lord on the
    day of death
  to reward individuals according
    to their conduct.
27 An hour's misery makes one
    forget past delights,
  and at the close of one's life
    one's deeds are revealed.
28 Call no one happy before his
    death;
  by how he ends, a person
    becomes known.*j*

29 Do not invite everyone into your
    home,
  for many are the tricks of the
    crafty.
30 Like a decoy partridge in a cage,
  so is the mind of the proud,
  and like spies they observe your
    weakness;*k*
31 for they lie in wait, turning good
    into evil,
  and to worthy actions they
    attach blame.
32 From a spark many coals are
    kindled,
  and a sinner lies in wait to shed
    blood.
33 Beware of scoundrels, for they
    devise evil,
  and they may ruin your
    reputation forever.
34 Receive strangers into your home
    and they will stir up trouble
    for you,
  and will make you a stranger to
    your own family.

*h Other ancient authorities add as verses 15 and
16, 15Wisdom, understanding, and knowledge of the
law come from the Lord; affection and the ways of
good works come from him. 16Error and darkness were
created with sinners; evil grows old with those who
take pride in malice.*   *i Heb: Gk is in*   *j Heb:
Gk and through his children a person becomes known*
k Heb: Gk downfall

12 If you do good, know to whom you do it,
and you will be thanked for your good deeds.

2 Do good to the devout, and you will be repaid—
if not by them, certainly by the Most High.

3 No good comes to one who persists in evil
or to one who does not give alms.

4 Give to the devout, but do not help the sinner.

5 Do good to the humble, but do not give to the ungodly;
hold back their bread, and do not give it to them,
for by means of it they might subdue you;
then you will receive twice as much evil
for all the good you have done to them.

6 For the Most High also hates sinners
and will inflict punishment on the ungodly.[l]

7 Give to the one who is good, but do not help the sinner.

8 A friend is not known[m] in prosperity,
nor is an enemy hidden in adversity.

9 One's enemies are friendly[n] when one prospers,
but in adversity even one's friend disappears.

10 Never trust your enemy,
for like corrosion in copper, so is his wickedness.

11 Even if he humbles himself and walks bowed down,
take care to be on your guard against him.
Be to him like one who polishes a mirror,
to be sure it does not become completely tarnished.

12 Do not put him next to you,
or he may overthrow you and take your place.
Do not let him sit at your right hand,
or else he may try to take your own seat,
and at last you will realize the truth of my words,
and be stung by what I have said.

13 Who pities a snake charmer when he is bitten,
or all those who go near wild animals?

14 So no one pities a person who associates with a sinner
and becomes involved in the other's sins.

15 He stands by you for a while,
but if you falter, he will not be there.

16 An enemy speaks sweetly with his lips,
but in his heart he plans to throw you into a pit;
an enemy may have tears in his eyes,
but if he finds an opportunity he will never have enough of your blood.

17 If evil comes upon you, you will find him there ahead of you;
pretending to help, he will trip you up.

18 Then he will shake his head, and clap his hands,
and whisper much, and show his true face.

l Other ancient authorities add *and he is keeping them for the day of their punishment*
m Other ancient authorities read *punished*
n Heb: Gk *grieved*

12.1–7: While Sirach's advice is positive, it is also self-serving. The contrast between the *sinners* and the *devout* is sharply drawn. As sinner, the *ungodly* one deserves rejection by God (v. 6, "hate" is not a mere emotion). **8–18:** A warning against false friends (see 6.5–17). **13.1–14.2: Warnings about associates. 1:**

13 Whoever touches pitch gets dirty,
and whoever associates with a proud person becomes like him.

2 Do not lift a weight too heavy for you,
or associate with one mightier and richer than you.
How can the clay pot associate with the iron kettle?
The pot will strike against it and be smashed.

3 A rich person does wrong, and even adds insults;
a poor person suffers wrong, and must add apologies.

4 A rich person*o* will exploit you if you can be of use to him,
but if you are in need he will abandon you.

5 If you own something, he will live with you;
he will drain your resources without a qualm.

6 When he needs you he will deceive you,
and will smile at you and encourage you;
he will speak to you kindly and say, "What do you need?"

7 He will embarrass you with his delicacies,
until he has drained you two or three times,
and finally he will laugh at you.
Should he see you afterwards, he will pass you by
and shake his head at you.

8 Take care not to be led astray
and humiliated when you are enjoying yourself.*p*

9 When an influential person invites you, be reserved,
and he will invite you more insistently.

10 Do not be forward, or you may be rebuffed;
do not stand aloof, or you will be forgotten.

11 Do not try to treat him as an equal,
or trust his lengthy conversations;
for he will test you by prolonged talk,
and while he smiles he will be examining you.

12 Cruel are those who do not keep your secrets;
they will not spare you harm or imprisonment.

13 Be on your guard and very careful,
for you are walking about with your own downfall.*q*

15 Every creature loves its like,
and every person the neighbor.

16 All living beings associate with their own kind,
and people stick close to those like themselves.

17 What does a wolf have in common with a lamb?
No more has a sinner with the devout.

18 What peace is there between a hyena and a dog?
And what peace between the rich and the poor?

19 Wild asses in the wilderness are the prey of lions;
likewise the poor are feeding grounds for the rich.

20 Humility is an abomination to the proud;
likewise the poor are an abomination to the rich.

o Gk *He*   *p* Other ancient authorities read *in your folly*   *q* Other ancient authorities add as verse 14, *When you hear these things in your sleep, wake up! During all your life love the Lord, and call on him for your salvation.*

This maxim is developed in the following sayings. **3–7**: Wealth makes a difference; rich and poor are not treated alike (see 10.31). **13.9–13**: Caution in dealing with the powerful.

13.13–23: "Like loves like," and inequalities between the rich and the poor (vv. 20–23) are obvious (see vv. 3–7). **24**: 11.10; 31.8. **25**: The exterior appearance reflects what is within.

21 When the rich person totters, he is
    supported by friends,
but when the humble[r] falls, he
    is pushed away even by
    friends.
22 If the rich person slips, many
    come to the rescue;
he speaks unseemly words, but
    they justify him.
If the humble person slips, they
    even criticize him;
he talks sense, but is not given a
    hearing.
23 The rich person speaks and all are
    silent;
they extol to the clouds what he
    says.
The poor person speaks and they
    say, "Who is this fellow?"
And should he stumble, they
    even push him down.
24 Riches are good if they are free
    from sin;
poverty is evil only in the
    opinion of the ungodly.

25 The heart changes the
    countenance,
either for good or for evil.[s]
26 The sign of a happy heart is a
    cheerful face,
but to devise proverbs requires
    painful thinking.

14 Happy are those who do not
    blunder with their lips,
and need not suffer remorse for
    sin.
2 Happy are those whose hearts do
    not condemn them,
and who have not given up
    their hope.

3 Riches are inappropriate for a
    small-minded person;
and of what use is wealth to a
    miser?
4 What he denies himself he collects
    for others;

and others will live in luxury on
    his goods.
5 If one is mean to himself, to
    whom will he be generous?
He will not enjoy his own
    riches.
6 No one is worse than one who is
    grudging to himself;
this is the punishment for his
    meanness.
7 If ever he does good, it is by
    mistake;
and in the end he reveals his
    meanness.
8 The miser is an evil person;
he turns away and disregards
    people.
9 The eye of the greedy person is
    not satisfied with his share;
greedy injustice withers the
    soul.
10 A miser begrudges bread,
and it is lacking at his table.

11 My child, treat yourself well,
    according to your means,
and present worthy offerings to
    the Lord.
12 Remember that death does not
    tarry,
and the decree[t] of Hades has
    not been shown to you.
13 Do good to friends before you
    die,
and reach out and give to them
    as much as you can.
14 Do not deprive yourself of a day's
    enjoyment;
do not let your share of desired
    good pass by you.
15 Will you not leave the fruit of
    your labors to another,
and what you acquired by toil
    to be divided by lot?

r Other ancient authorities read *poor*
s Other ancient authorities add *and a glad heart makes
a cheerful countenance*   t Heb Syr: Gk *covenant*

**14.3–19: The proper use of wealth. 3–
10:** A description of the *miser* and his fate.
14.11–19: In contrast to vv. 3–10, riches are
to be used generously because there will be no
opportunity after death (vv. 12, 16), which is
inevitable (vv. 17–19).

16 Give, and take, and indulge
yourself,
because in Hades one cannot
look for luxury.
17 All living beings become old like a
garment,
for the decree*u* from of old is,
"You must die!"
18 Like abundant leaves on a
spreading tree
that sheds some and puts forth
others,
so are the generations of flesh and
blood:
one dies and another is born.
19 Every work decays and ceases to
exist,
and the one who made it will
pass away with it.

20 Happy is the person who
meditates on*v* wisdom
and reasons intelligently,
21 who*w* reflects in his heart on her
ways
and ponders her secrets,
22 pursuing her like a hunter,
and lying in wait on her paths;
23 who peers through her windows
and listens at her doors;
24 who camps near her house
and fastens his tent peg to her
walls;
25 who pitches his tent near her,
and so occupies an excellent
lodging place;
26 who places his children under her
shelter,
and lodges under her boughs;
27 who is sheltered by her from the
heat,
and dwells in the midst of her
glory.

15 Whoever fears the Lord will do
this,

and whoever holds to the law
will obtain wisdom.*x*
2 She will come to meet him like a
mother,
and like a young bride she will
welcome him.
3 She will feed him with the bread
of learning,
and give him the water of
wisdom to drink.
4 He will lean on her and not fall,
and he will rely on her and not
be put to shame.
5 She will exalt him above his
neighbors,
and will open his mouth in the
midst of the assembly.
6 He will find gladness and a crown
of rejoicing,
and will inherit an everlasting
name.
7 The foolish will not obtain her,
and sinners will not see her.
8 She is far from arrogance,
and liars will never think of her.
9 Praise is unseemly on the lips of a
sinner,
for it has not been sent from the
Lord.
10 For in wisdom must praise be
uttered,
and the Lord will make it
prosper.

11 Do not say, "It was the Lord's
doing that I fell away";
for he does not do*y* what he
hates.
12 Do not say, "It was he who led
me astray";
for he has no need of the sinful.
13 The Lord hates all abominations;

u Heb: Gk *covenant*   *v* Other ancient
authorities read *dies in*   *w* The structure adopted
in verses 21–27 follows the Heb   *x* Gk *her*
y Heb: Gk *you ought not do*

14.20–15.10: **The search for Wisdom
and her blessings. 14.20–27:** One should
pursue Wisdom as a beloved, with intensity
and resolution.

15.1–10: Wisdom's response to those who
seek her, and hold *to the law.*
15.11–20: **Free will. 11–13:** God cannot be
blamed for human sinfulness. **14–17:** Al-

such things are not loved by
those who fear him.

14 It was he who created humankind
in the beginning,
and he left them in the power of
their own free choice.

15 If you choose, you can keep the
commandments,
and to act faithfully is a matter
of your own choice.

16 He has placed before you fire and
water;
stretch out your hand for
whichever you choose.

17 Before each person are life and
death,
and whichever one chooses will
be given.

18 For great is the wisdom of the
Lord;
he is mighty in power and sees
everything;

19 his eyes are on those who fear
him,
and he knows every human
action.

20 He has not commanded anyone to
be wicked,
and he has not given anyone
permission to sin.

16 Do not desire a multitude of
worthless *z* children,
and do not rejoice in ungodly
offspring.

2 If they multiply, do not rejoice in
them,
unless the fear of the Lord is in
them.

3 Do not trust in their survival,
or rely on their numbers; *a*

for one can be better than a
thousand,
and to die childless better than
to have ungodly children.

4 For through one intelligent person
a city can be filled with
people,
but through a clan of outlaws it
becomes desolate.

5 Many such things my eye has
seen,
and my ear has heard things
more striking than these.

6 In an assembly of sinners a fire is
kindled,
and in a disobedient nation
wrath blazes up.

7 He did not forgive the ancient
giants
who revolted in their might.

8 He did not spare the neighbors of
Lot,
whom he loathed on account of
their arrogance.

9 He showed no pity on the
doomed nation,
on those dispossessed because of
their sins; *b*

10 or on the six hundred thousand
foot soldiers
who assembled in their
stubbornness. *c*

z Heb: Gk *unprofitable*   *a* Other ancient
authorities add *For you will groan in untimely
mourning, and will know of their sudden end.*
b Other ancient authorities add *All these things he
did to the hard-hearted nations, and by the multitude
of his holy ones he was not appeased.*   *c* Other
ancient authorities add *Chastising, showing mercy,
striking, healing, the Lord persisted in mercy and
discipline.*

---

though the Bible affirms human responsibility, it does not speak of "free will," which is
the meaning of "yeṣer" in v. 14 (*free choice*).
The Hebrew word is a technical term, sometimes used in a good sense (Isa 26.3; 1 Chr
29.18); but usually it refers to an evil tendency
or inclination toward sin (Gen 6.5; 8.21; compare 2 Esd 4.30–31). In post-biblical times the
doctrine arose of a good and an evil "yeṣer"
that every person possesses. **16:** *Fire and water,*
representing opposite extremes. *Fire* here has

no eschatological significance. **17:** Deut 30.19;
Jer 21.8. **18–20:** The divine attributes mentioned here reaffirm the point of vv. 11–13.
**16.1–4: The misfortune of having un-
godly children** (40.15–16; 41.5–13).
**16.5–23: The certainty of punishment
for sin. 6–8:** Sirach refers to the fate of Korah, Dathan, and Abiram (v. 6; Num 16), *the
ancient giants* of Gen 6.4, and the people of
Sodom (v. 8; Ezek 16.49–50; Gen 19).
16.9: The Canaanites are meant. **10:** The

11 Even if there were only one
        stiff-necked person,
    it would be a wonder if he
        remained unpunished.
    For mercy and wrath are with the
        Lord;*d*
    he is mighty to forgive—but he
        also pours out wrath.
12 Great as his mercy, so also is his
        chastisement;
    he judges a person according to
        one's deeds.
13 The sinner will not escape with
        plunder,
    and the patience of the godly
        will not be frustrated.
14 He makes room for every act of
        mercy;
    everyone receives in accordance
        with one's deeds.*e*

17 Do not say, "I am hidden from
        the Lord,
    and who from on high has me
        in mind?
    Among so many people I am
        unknown,
    for what am I in a boundless
        creation?
18 Lo, heaven and the highest
        heaven,
    the abyss and the earth, tremble
        at his visitation!*f*
19 The very mountains and the
        foundations of the earth
    quiver and quake when he looks
        upon them.
20 But no human mind can grasp
        this,
    and who can comprehend his
        ways?
21 Like a tempest that no one can
        see,

so most of his works are
        concealed.*g*
22 Who is to announce his acts of
        justice?
    Or who can await them? For his
        decree*h* is far off."*i*
23 Such are the thoughts of one
        devoid of understanding;
    a senseless and misguided
        person thinks foolishly.

24 Listen to me, my child, and
        acquire knowledge,
    and pay close attention to my
        words.
25 I will impart discipline precisely*j*
    and declare knowledge
        accurately.

26 When the Lord created*k* his works
        from the beginning,
    and, in making them,
        determined their
        boundaries,
27 he arranged his works in an
        eternal order,
    and their dominion*l* for all
        generations.
    They neither hunger nor grow
        weary,

---

d Gk *him*   e Other ancient authorities
add *15The Lord hardened Pharaoh so that
he did not recognize him, in order that his works
might be known under heaven. 16His mercy is
manifest to the whole of creation, and he divided his
light and darkness with a plumb line.*   f Other
ancient authorities add *The whole world past and
present is in his will.*   g Meaning of Gk
uncertain: Heb Syr *If I sin, no eye can see me, and
if I am disloyal all in secret, who is to know?*
h Heb *the decree*: Gk *the covenant*   i Other
ancient authorities add *and a scrutiny for all comes
at the end*   j Gk *by weight*   k Heb: Gk *judged*
l Or *elements*

600,000 are the Israelites in the desert (see
46.8; Num 14). **11**: *Mercy and wrath*, 5.6. **12–
14**: Good works are efficacious, and God
judges accordingly, as Sirach frequently affirms (3.3, 14–16; 11.26; 15.19; 29.11–12).
  16.17–23: Let no one think it is possible to
escape God's attention in the immense and
mysterious realm of creation! In v. 23 Sirach

expresses his judgment on the ideas expressed
in vv. 17–22.
  **16.24–17.14: Divine wisdom seen in
creation. 26–30**: The order and harmony of
created things. God provides for all *living beings*, and eventually they return to the earth
(17.1; 40.11; Gen 3.19).

and they do not abandon their
tasks.

28 They do not crowd one another,
and they never disobey his
word.

29 Then the Lord looked upon the
earth,
and filled it with his good
things.

30 With all kinds of living beings he
covered its surface,
and into it they must return.

17 The Lord created human beings
out of earth,
and makes them return to it
again.

2 He gave them a fixed number of
days,
but granted them authority over
everything on the earth. *m*

3 He endowed them with strength
like his own, *n*
and made them in his own
image.

4 He put the fear of them *o* in all
living beings,
and gave them dominion over
beasts and birds. *p*

6 Discretion and tongue and eyes,
ears and a mind for thinking he
gave them.

7 He filled them with knowledge
and understanding,
and showed them good and
evil.

8 He put the fear of him into *q* their
hearts
to show them the majesty of his
works. *r*

10 And they will praise his holy
name,

9 to proclaim the grandeur of his
works.

11 He bestowed knowledge upon
them,

and allotted to them the law of
life. *s*

12 He established with them an
eternal covenant,
and revealed to them his
decrees.

13 Their eyes saw his glorious
majesty,
and their ears heard the glory of
his voice.

14 He said to them, "Beware of all
evil."
And he gave commandment to
each of them concerning
the neighbor.

15 Their ways are always known to
him;
they will not be hid from his
eyes. *t*

17 He appointed a ruler for every
nation,
but Israel is the Lord's own
portion. *u*

19 All their works are as clear as the
sun before him,
and his eyes are ever upon their
ways.

20 Their iniquities are not hidden
from him,

m Lat: Gk *it*   *n* Lat: Gk *proper to them*
o Syr: Gk *him*   *p* Other ancient authorities add
as verse 5, *They obtained the use of the five faculties
of the Lord; as sixth he distributed to them the gift of mind,
and as seventh, reason, the interpreter of one's faculties.*
q Other ancient authorities read *He set his eye upon*
r Other ancient authorities add *and he gave them
to boast of his marvels forever*   *s* Other ancient
authorities add *so that they may know that they who
are alive now are mortal*
t Other ancient authorities add *16Their ways from
youth tend toward evil, and they are unable to make
for themselves hearts of flesh in place of their stony
hearts. 17For in the division of the nations of the
whole earth, he appointed*   *u* Other ancient
authorities add as verse 18, *whom, being his
firstborn, he brings up with discipline, and allotting to
him the light of his love, he does not neglect him.*

17.1–14: The creation of *human beings* is a
reason for praise, because of God's gracious
gifts, especially the *law of life*, the *covenant*,
and *commandment* (3–14).
17.15–24: **The divine judge.** God knows

the good (especially *almsgiving*, v. 22) and the
sinful actions of humans, and will requite accordingly. **17:** See Deut 32.8–9, an allusion to
Israel's special position. **24:** *A return*, to God's
favor.

and all their sins are before the
Lord. *v*

22 One's almsgiving is like a
signet ring with the Lord, *w*
and he will keep a person's
kindness like the apple of
his eye. *x*

23 Afterward he will rise up and
repay them,
and he will bring their
recompense on their heads.

24 Yet to those who repent he grants
a return,
and he encourages those who
are losing hope.

25 Turn back to the Lord and forsake
your sins;
pray in his presence and lessen
your offense.

26 Return to the Most High and turn
away from iniquity, *y*
and hate intensely what he
abhors.

27 Who will sing praises to the Most
High in Hades
in place of the living who give
thanks?

28 From the dead, as from one who
does not exist, thanksgiving
has ceased;
those who are alive and well
sing the Lord's praises.

29 How great is the mercy of the
Lord,
and his forgiveness for those
who return to him!

30 For not everything is within
human capability,
since human beings are not
immortal.

31 What is brighter than the sun? Yet
it can be eclipsed.
So flesh and blood devise evil.

32 He marshals the host of the height
of heaven;

but all human beings are dust
and ashes.

18 He who lives forever created
the whole universe;
2 the Lord alone is just. *z*
4 To none has he given power to
proclaim his works;
and who can search out his
mighty deeds?
5 Who can measure his majestic
power?
And who can fully recount his
mercies?
6 It is not possible to diminish or
increase them,
nor is it possible to fathom the
wonders of the Lord.
7 When human beings have finished,
they are just beginning,
and when they stop, they are
still perplexed.
8 What are human beings, and of
what use are they?
What is good in them, and what
is evil?
9 The number of days in their life is
great if they reach one
hundred years. *a*
10 Like a drop of water from the sea
and a grain of sand,
so are a few years among the
days of eternity.

v Other ancient authorities add as verse 21, *But
the Lord, who is gracious and knows how they are
formed, has neither left them nor abandoned them, but
has spared them.* *w* Gk *him* *x* Other
ancient authorities add *apportioning repentance to
his sons and daughters* *y* Other ancient
authorities add *for he will lead you out of darkness
to the light of health.* *z* Other ancient
authorities add *and there is no other beside him;* *3he
steers the world with the span of his hand, and all
things obey his will; for he is king of all things by his
power, separating among them the holy things from
the profane.* *a* Other ancient authorities add *but
the death of each one is beyond the calculation of all*

**17.25–32: Exhortation to turn to God.
27–28:** One reason for repentance is that
there is no loving contact with God in *Hades*
(Sheol); see Pss 6.5; 30.9; 88.5; 115.17–18; Isa
38.18–19; Bar 2.17. By repentance, one can
be *alive*, and *sing the Lord's praises.* **31:** If the *sun*
can suffer an eclipse, a mere human can also
fail.

18.1–14: A hymn to God's majesty. Although the Lord is transcendent (vv. 1–7), and
humans are weak and short-lived (vv. 8–10), God
is *patient* and merciful to them (vv. 11–14).

11 That is why the Lord is patient
   with them
   and pours out his mercy upon
   them.
12 He sees and recognizes that their
   end is miserable;
   therefore he grants them
   forgiveness all the more.
13 The compassion of human beings
   is for their neighbors,
   but the compassion of the Lord
   is for every living thing.
   He rebukes and trains and teaches
   them,
   and turns them back, as a
   shepherd his flock.
14 He has compassion on those who
   accept his discipline
   and who are eager for his
   precepts.

15 My child, do not mix reproach
   with your good deeds,
   or spoil your gift by harsh
   words.
16 Does not the dew give relief from
   the scorching heat?
   So a word is better than a gift.
17 Indeed, does not a word surpass a
   good gift?
   Both are to be found in a
   gracious person.
18 A fool is ungracious and abusive,
   and the gift of a grudging giver
   makes the eyes dim.

19 Before you speak, learn;
   and before you fall ill, take care
   of your health.
20 Before judgment comes, examine
   yourself;
   and at the time of scrutiny you
   will find forgiveness.
21 Before falling ill, humble yourself;
   and when you have sinned,
   repent.
22 Let nothing hinder you from
   paying a vow promptly,
   and do not wait until death to
   be released from it.
23 Before making a vow, prepare
   yourself;
   do not be like one who puts the
   Lord to the test.
24 Think of his wrath on the day of
   death,
   and of the moment of
   vengeance when he turns
   away his face.
25 In the time of plenty think of the
   time of hunger;
   in days of wealth think of
   poverty and need.
26 From morning to evening
   conditions change;
   all things move swiftly before
   the Lord.

27 One who is wise is cautious in
   everything;
   when sin is all around, one
   guards against wrongdoing.
28 Every intelligent person knows
   wisdom,
   and praises the one who finds
   her.
29 Those who are skilled in words
   become wise themselves,
   and pour forth apt proverbs. *b*

SELF–CONTROL *c*

30 Do not follow your base desires,
   but restrain your appetites.
31 If you allow your soul to take
   pleasure in base desire,
   it will make you the
   laughingstock of your
   enemies.
32 Do not revel in great luxury,
   or you may become
   impoverished by its
   expense.

b Other ancient authorities add *Better is confidence
in the one Lord than clinging with a dead heart to a
dead one.*   *c* This heading is included in the Gk
text.

**18.15–19.19: Various warnings and ob-
servations. 18.15–18:** Do not humiliate the re-
ceiver of charity (v. 16; see also 43.22); be a
gracious giver. **19–21:** The need of foresight.

22–26: Be careful in making vows. **27–29:**
Characteristics of the *wise.*

18.30–19.4: An exhortation to self-
control. **19.5–12:** Warnings against gossip. **8:**

33 Do not become a beggar by
    feasting with borrowed
    money,
  when you have nothing in your
    purse. *d*

19 The one who does this*e* will
    not become rich;
  one who despises small things
    will fail little by little.
2 Wine and women lead intelligent
    men astray,
  and the man who consorts with
    prostitutes is reckless.
3 Decay and worms will take
    possession of him,
  and the reckless person will be
    snatched away.

4 One who trusts others too quickly
    has a shallow mind,
  and one who sins does wrong to
    oneself.
5 One who rejoices in wickedness*f*
    will be condemned,*g*
6   but one who hates gossip has
    less evil.
7 Never repeat a conversation,
    and you will lose nothing at all.
8 With friend or foe do not report
    it,
  and unless it would be a sin for
    you, do not reveal it;
9 for someone may have heard you
    and watched you,
  and in time will hate you.
10 Have you heard something? Let it
    die with you,
  Be brave, it will not make you
    burst!
11 Having heard something, the fool
    suffers birth pangs
  like a woman in labor with a
    child.
12 Like an arrow stuck in a person's
    thigh,
  so is gossip inside a fool.

13 Question a friend; perhaps he did
    not do it;

or if he did, so that he may not
    do it again.
14 Question a neighbor; perhaps he
    did not say it;
  or if he said it, so that he may
    not repeat it.
15 Question a friend, for often it is
    slander;
  so do not believe everything
    you hear.
16 A person may make a slip without
    intending it.
  Who has not sinned with his
    tongue?
17 Question your neighbor before
    you threaten him;
  and let the law of the Most
    High take its course. *h*

20 The whole of wisdom is fear of
    the Lord,
  and in all wisdom there is the
    fulfillment of the law. *i*
22 The knowledge of wickedness is
    not wisdom,
  nor is there prudence in the
    counsel of sinners.
23 There is a cleverness that is
    detestable,
  and there is a fool who merely
    lacks wisdom.
24 Better are the God-fearing who
    lack understanding
  than the highly intelligent who
    transgress the law.

d Other ancient authorities add *for you will be
plotting against your own life*   *e* Heb: Gk
A worker who is a drunkard   *f* Other ancient
authorities read *heart*   *g* Other ancient
authorities add *but one who withstands pleasures
crowns his life. 6One who controls the tongue will live
without strife,*   *h* Other ancient authorities
add *and do not be angry. 18The fear of the Lord
is the beginning of acceptance, and wisdom obtains
his love. 19The knowledge of the Lord's commandments
is life-giving discipline; and those who do what is pleasing
to him enjoy the fruit of the tree of immortality.*
i Other ancient authorities add *and the knowledge of
his omnipotence. 21When a slave says to his master, "I
will not act as you wish," even if later he does it, he
angers the one who supports him.*

Unless by keeping silent *it would be a sin.* **16–
18**: On administering reproof. Note that v. 18
is in note *h.*

**19.20–30: Wisdom and craftiness con-
trasted. 23**: *Who merely lacks wisdom,* and is
guileless.

25 There is a cleverness that is exact
but unjust,
and there are people who abuse
favors to gain a verdict.
26 There is the villain bowed down
in mourning,
but inwardly he is full of deceit.
27 He hides his face and pretends not
to hear,
but when no one notices, he
will take advantage of you.
28 Even if lack of strength keeps him
from sinning,
he will nevertheless do evil
when he finds the
opportunity.
29 A person is known by his
appearance,
and a sensible person is known
when first met, face to face.
30 A person's attire and hearty
laughter,
and the way he walks, show
what he is.

20 There is a rebuke that is
untimely,
and there is the person who is
wise enough to keep silent.
2 How much better it is to rebuke
than to fume!
3 And the one who admits his fault
will be kept from failure.
4 Like a eunuch lusting to violate a
girl
is the person who does right
under compulsion.
5 Some people keep silent and are
thought to be wise,
while others are detested for
being talkative.
6 Some people keep silent because
they have nothing to say,
while others keep silent because
they know when to speak.
7 The wise remain silent until the
right moment,

but a boasting fool misses the
right moment.
8 Whoever talks too much is
detested,
and whoever pretends to
authority is hated.*j*

9 There may be good fortune for a
person in adversity,
and a windfall may result in a
loss.
10 There is the gift that profits you
nothing,
and the gift to be paid back
double.
11 There are losses for the sake of
glory,
and there are some who have
raised their heads from
humble circumstances.
12 Some buy much for little,
but pay for it seven times over.
13 The wise make themselves
beloved by only few
words,*k*
but the courtesies of fools are
wasted.
14 A fool's gift will profit you
nothing,*l*
for he looks for recompense
sevenfold.*m*
15 He gives little and upbraids much;
he opens his mouth like a town
crier.
Today he lends and tomorrow he
asks it back;
such a one is hateful to God and
humans.*n*
16 The fool says, "I have no friends,

*j Other ancient authorities add How good it is to show
repentance when you are reproved, for so you will
escape deliberate sin! k Heb: Gk by words
l Other ancient authorities add so it is with the
envious who give under compulsion m Syr: Gk he
has many eyes instead of one n Other ancient
authorities lack to God and humans*

19.26–30: Appearances can be both deceiving and revealing.
20.1–8: Times for speech and silence. 4:
The point of the comparison is that one cannot be compelled to do what is *right.* **5–6:**

Prov 17.27–28. **7:** *The right moment,* Prov
15.23; 25.11.
**20.9–17: Paradoxes in the appearances
of things.**

and I get no thanks for my
good deeds.
Those who eat my bread are
evil-tongued."
17 How many will ridicule him, and
how often![o]
18 A slip on the pavement is better
than a slip of the tongue;
the downfall of the wicked will
occur just as speedily.
19 A coarse person is like an
inappropriate story,
continually on the lips of the
ignorant.
20 A proverb from a fool's lips will
be rejected,
for he does not tell it at the
proper time.
21 One may be prevented from
sinning by poverty;
so when he rests he feels no
remorse.
22 One may lose his life through
shame,
or lose it because of human
respect.[p]
23 Another out of shame makes
promises to a friend,
and so makes an enemy for
nothing.

24 A lie is an ugly blot on a person;
it is continually on the lips of
the ignorant.
25 A thief is preferable to a habitual
liar,
but the lot of both is ruin.
26 A liar's way leads to disgrace,
and his shame is ever with him.

PROVERBIAL SAYINGS[q]

27 The wise person advances himself
by his words,
and one who is sensible pleases
the great.
28 Those who cultivate the soil heap
up their harvest,
and those who please the great
atone for injustice.
29 Favors and gifts blind the eyes of
the wise;
like a muzzle on the mouth they
stop reproofs.
30 Hidden wisdom and unseen
treasure,
of what value is either?
31 Better are those who hide their
folly
than those who hide their
wisdom.[r]

21 Have you sinned, my child? Do
so no more,
but ask forgiveness for your
past sins.
2 Flee from sin as from a snake;
for if you approach sin, it will
bite you.
Its teeth are lion's teeth,
and can destroy human lives.
3 All lawlessness is like a two-edged
sword;
there is no healing for the
wound it inflicts.

4 Panic and insolence will waste
away riches;
thus the house of the proud will
be laid waste.[s]

o Other ancient authorities add *for he has not honestly received what he has, and what he does not have is unimportant to him*
p Other ancient authorities read *his foolish look*
q This heading is included in the Gk text.
r Other ancient authorities add [32]*Unwearied endurance in seeking the Lord is better than a masterless charioteer of one's own life.*
s Other ancient authorities read *uprooted*

20.18–20: Inappropriate speech. 19: Syriac: "As the fat tail of a sheep [Ex 29.22], eaten without salt, so is a word spoken out of season."
20.21–23: Reflections on poverty and shame. 23: The *promises* are either insincere or impossible.
20.24–26: Lying (7.13; 25.2; Prov 6.16–19; 12.22).
20.27–31: Recommendations for wisdom. 29: The danger of *gifts,* or bribes.
21.1–10: Warnings against sin. 2–3: The characteristics of sin are subtlety, strength, and deadliness. **5:** 35.17–21.

5 The prayer of the poor goes from
    their lips to the ears of
    God, *t*
    and his judgment comes
    speedily.
6 Those who hate reproof walk in
    the sinner's steps,
    but those who fear the Lord
    repent in their heart.
7 The mighty in speech are widely
    known;
    when they slip, the sensible
    person knows it.

8 Whoever builds his house with
    other people's money
    is like one who gathers stones
    for his burial mound. *u*
9 An assembly of the wicked is like
    a bundle of tow,
    and their end is a blazing fire.
10 The way of sinners is paved with
    smooth stones,
    but at its end is the pit of
    Hades.

11 Whoever keeps the law controls
    his thoughts,
    and the fulfillment of the fear of
    the Lord is wisdom.
12 The one who is not clever cannot
    be taught,
    but there is a cleverness that
    increases bitterness.
13 The knowledge of the wise will
    increase like a flood,
    and their counsel like a
    life-giving spring.
14 The mind *v* of a fool is like a
    broken jar;
    it can hold no knowledge.

15 When an intelligent person hears a
    wise saying,

he praises it and adds to it;
    when a fool *w* hears it, he laughs
    at *x* it
    and throws it behind his back.
16 A fool's chatter is like a burden on
    a journey,
    but delight is found in the
    speech of the intelligent.
17 The utterance of a sensible person
    is sought in the assembly,
    and they ponder his words in
    their minds.

18 Like a house in ruins is wisdom to
    a fool,
    and to the ignorant, knowledge
    is talk that has no meaning.
19 To a senseless person education is
    fetters on his feet,
    and like manacles on his right
    hand.
20 A fool raises his voice when he
    laughs,
    but the wise *y* smile quietly.
21 To the sensible person education is
    like a golden ornament,
    and like a bracelet on the right
    arm.

22 The foot of a fool rushes into a
    house,
    but an experienced person waits
    respectfully outside.
23 A boor peers into the house from
    the door,
    but a cultivated person remains
    outside.
24 It is ill-mannered for a person to
    listen at a door;

t Gk *his ears*   *u* Other ancient authorities
read *for the winter*   *v* Syr Lat: Gk *entrails*
w Syr: Gk *reveler*   *x* Syr: Gk *dislikes*
y Syr Lat: Gk *clever*

21.7b: An ironic comment on the *mighty* talker, who is unaware of his errors. **8:** That is, by oppression of others, he prepares for his own destruction. **9:** The image of *fire* is meant to suggest punishment in this life (see v. 10). **10:** *Hades,* Sheol; see 17.27 n.
21.11–28: A series of contrasts between the wise and fools. 13–14: The contrast is between a *spring* and a *broken jar* (or cistern, Jer 2.13), which holds no water. All the comparisons that follow are in favor of the wise: increasing wisdom (v. 15), giving *delight* (v. 16), a *golden ornament* (v. 21), courtesy (vv. 22–24), careful *words* (vv. 25–26). **27:** *Curses himself,* in the sense that the curse recoils upon him.

the discreet would be grieved by
    the disgrace.

25 The lips of babblers speak of what
    is not their concern. *z*
    but the words of the prudent are
    weighed in the balance.
26 The mind of fools is in their
    mouth,
    but the mouth of the wise is in *a*
    their mind.
27 When an ungodly person curses an
    adversary, *b*
    he curses himself.
28 A whisperer degrades himself
    and is hated in his
    neighborhood.

22 The idler is like a filthy stone,
    and every one hisses at his
    disgrace.
2 The idler is like the filth of
    dunghills;
    anyone that picks it up will
    shake it off his hand.

3 It is a disgrace to be the father of
    an undisciplined son,
    and the birth of a daughter is a
    loss.
4 A sensible daughter obtains a
    husband of her own,
    but one who acts shamefully is a
    grief to her father.
5 An impudent daughter disgraces
    father and husband,
    and is despised by both.
6 Like music in time of mourning is
    ill-timed conversation,
    but a thrashing and discipline
    are at all times wisdom. *c*

9 Whoever teaches a fool is like one
    who glues potsherds
    together,

or who rouses a sleeper from
    deep slumber.
10 Whoever tells a story to a fool
    tells it to a drowsy man;
    and at the end he will say,
    "What is it?"
11 Weep for the dead, for he has left
    the light behind;
    and weep for the fool, for he
    has left intelligence behind.
    Weep less bitterly for the dead, for
    he is at rest;
    but the life of the fool is worse
    than death.
12 Mourning for the dead lasts seven
    days,
    but for the foolish or the
    ungodly it lasts all the days
    of their lives.

13 Do not talk much with a senseless
    person
    or visit an unintelligent person. *d*
    Stay clear of him, or you may
    have trouble,
    and be spattered when he shakes
    himself off.
    Avoid him and you will find rest,
    and you will never be wearied
    by his lack of sense.
14 What is heavier than lead?
    And what is its name except
    "Fool"?
15 Sand, salt, and a piece of iron
    are easier to bear than a stupid
    person.

z Other ancient authorities read *of strangers speak of
these things*   *a* Other ancient authorities omit *in*
b Or *curses Satan*   *c* Other ancient authorities
add *7Children who are brought up in a good life,
conceal the lowly birth of their parents. 8Children who
are disdainfully and boorishly haughty stain the
nobility of their kindred.*
d Other ancient authorities add *For being without
sense he will despise everything about you*

22.1–18: Concerning laziness and foolishness. 1: *Filthy stone,* a stone used in lieu of toilet paper; see v. 2. **3–6:** These verses deal with wicked offspring. **3b:** This reflects the misogyny of the age.
22.7: A thankless task. **11–12:** One can *weep for a fool* more than for the *dead; seven days* was the usual period of mourning (Gen 50.10).
22.13–18: Avoid fools. **13:** *Shakes himself off,* as would an animal, to remove filth. **14–15:** 21.16; the fool is difficult *to bear.* The *mind* of the wise is contrasted with the *resolve* of the fool.

16 A wooden beam firmly bonded
　　into a building
　is not loosened by an
　　earthquake;
　so the mind firmly resolved after
　　due reflection
　will not be afraid in a crisis.
17 A mind settled on an intelligent
　　thought
　is like stucco decoration that
　　makes a wall smooth.
18 Fences*e* set on a high place
　will not stand firm against the
　　wind;
　so a timid mind with a fool's
　　resolve
　will not stand firm against any
　　fear.

19 One who pricks the eye brings
　　tears,
　and one who pricks the heart
　　makes clear its feelings.
20 One who throws a stone at birds
　　scares them away,
　and one who reviles a friend
　　destroys a friendship.
21 Even if you draw your sword
　　against a friend,
　do not despair, for there is a
　　way back.
22 If you open your mouth against
　　your friend,
　do not worry, for reconciliation
　　is possible.
　But as for reviling, arrogance,
　　disclosure of secrets, or a
　　treacherous blow—
　in these cases any friend will
　　take to flight.

23 Gain the trust of your neighbor in
　　his poverty,
　so that you may rejoice with
　　him in his prosperity.
　Stand by him in time of distress,

　so that you may share with him
　　in his inheritance.*f*
24 The vapor and smoke of the
　　furnace precede the fire;
　so insults precede bloodshed.
25 I am not ashamed to shelter a
　　friend,
　and I will not hide from him.
26 But if harm should come to me
　　because of him,
　whoever hears of it will beware
　　of him.

27 Who will set a guard over my
　　mouth,
　and an effective seal upon my
　　lips,
　so that I may not fall because of
　　them,
　and my tongue may not destroy
　　me?

23 O Lord, Father and Master of
　　my life,
　do not abandon me to their
　　designs,
　and do not let me fall because of
　　them!
2 Who will set whips over my
　　thoughts,
　and the discipline of wisdom
　　over my mind,
　so as not to spare me in my
　　errors,
　and not overlook my*g* sins?
3 Otherwise my mistakes may be
　　multiplied,
　and my sins may abound,
　and I may fall before my
　　adversaries,
　and my enemy may rejoice over
　　me.*h*

e Other ancient authorities read *Pebbles*
f Other ancient authorities add *For one should not
always despise restricted circumstances, or admire a rich
person who is stupid.*　　*g* Gk *their*
h Other ancient authorities add *From them
the hope of your mercy is remote*

**22.19–26: The preservation of friend-
ship. 19:** As the *eye* can weep, the *heart* can
show *feelings*. **20–22:** The things that destroy
friendship. **23–26:** Fidelity in friendship.

**22.27–23.27: A prayer for self-control,
with reflections on adultery. 22.27–23.1:**
Sins of the *tongue,* which is also the topic of
23.7–15.

4 O Lord, Father and God of my
life,
do not give me haughty eyes,
5 and remove evil desire from me.
6 Let neither gluttony nor lust
overcome me,
and do not give me over to
shameless passion.

DISCIPLINE OF THE TONGUE [i]
7 Listen, my children, to instruction
concerning the mouth;
the one who observes it will
never be caught.
8 Sinners are overtaken through
their lips;
by them the reviler and the
arrogant are tripped up.
9 Do not accustom your mouth to
oaths,
nor habitually utter the name of
the Holy One;
10 for as a servant who is constantly
under scrutiny
will not lack bruises,
so also the person who always
swears and utters the Name
will never be cleansed[j] from
sin.
11 The one who swears many oaths
is full of iniquity,
and the scourge will not leave
his house.
If he swears in error, his sin
remains on him,
and if he disregards it, he sins
doubly;
if he swears a false oath, he will
not be justified,
for his house will be filled with
calamities.

12 There is a manner of speaking
comparable to death;[k]
may it never be found in the
inheritance of Jacob!

Such conduct will be far from the
godly,
and they will not wallow in
sins.
13 Do not accustom your mouth to
coarse, foul language,
for it involves sinful speech.
14 Remember your father and mother
when you sit among the great,
or you may forget yourself in
their presence,
and behave like a fool through
bad habit;
then you will wish that you had
never been born,
and you will curse the day of
your birth.
15 Those who are accustomed to
using abusive language
will never become disciplined as
long as they live.

16 Two kinds of individuals multiply
sins,
and a third incurs wrath.
Hot passion that blazes like a fire
will not be quenched until it
burns itself out;
one who commits fornication with
his near of kin
will never cease until the fire
burns him up.
17 To a fornicator all bread is sweet;
he will never weary until he
dies.
18 The one who sins against his
marriage bed
says to himself, "Who can see
me?
Darkness surrounds me, the walls
hide me,

i This heading is included in the Gk text.
j Syr *be free*   k Other ancient authorities read
clothed about with death

23.2–6: Sins of passion, especially lust,
which is the topic of 23.16–26. **11:** *Doubly,*
because he not only swears rashly, but *disregards* the oath by failing to keep it.
23.12: *Comparable to death,* a reference to
blasphemy (Lev 24.11–16).

23.16–17: This numerical saying singles
out three kinds of lechers: the dissolute, the
incestuous, and the adulterer. **17:** *All bread is
sweet,* a euphemism for the adulterer's search
for a partner.
23.18–21: Reflections about the adulterer.

and no one sees me. Why
should I worry?
The Most High will not
remember sins."
19 His fear is confined to human eyes
and he does not realize that the
eyes of the Lord
are ten thousand times brighter
than the sun;
they look upon every aspect of
human behavior
and see into hidden corners.
20 Before the universe was created, it
was known to him,
and so it is since its completion.
21 This man will be punished in the
streets of the city,
and where he least suspects it,
he will be seized.

22 So it is with a woman who leaves
her husband
and presents him with an heir
by another man.
23 For first of all, she has disobeyed
the law of the Most High;
second, she has committed an
offense against her husband;
and third, through her fornication
she has committed adultery
and brought forth children by
another man.
24 She herself will be brought before
the assembly,
and her punishment will extend
to her children.
25 Her children will not take root,
and her branches will not bear
fruit.
26 She will leave behind an accursed
memory
and her disgrace will never be
blotted out.
27 Those who survive her will
recognize

that nothing is better than the
fear of the Lord,
and nothing sweeter than to heed
the commandments of the
Lord.[l]

THE PRAISE OF WISDOM [m]

24 Wisdom praises herself,
and tells of her glory in the
midst of her people.
2 In the assembly of the Most High
she opens her mouth,
and in the presence of his hosts
she tells of her glory:
3 "I came forth from the mouth of
the Most High,
and covered the earth like a
mist.
4 I dwelt in the highest heavens,
and my throne was in a pillar of
cloud.
5 Alone I compassed the vault of
heaven
and traversed the depths of the
abyss.
6 Over waves of the sea, over all
the earth,
and over every people and
nation I have held sway.[n]
7 Among all these I sought a resting
place;
in whose territory should I
abide?

8 "Then the Creator of all things
gave me a command,
and my Creator chose the place
for my tent.
He said, 'Make your dwelling in
Jacob,

l Other ancient authorities add as verse 28, *It is a
great honor to follow God, and to be received by him
is long life.*   m This heading is included in the
Gk text   n Other ancient authorities read
I have acquired a possession

20: God knows things even before creation.
22–26: Reflections about the adulteress. 27:
A conclusion.
   24.1–34: **Praise of Lady Wisdom** (Prov
8). 1–2: Introduction of Wisdom, speaking to
the people and to members of the heavenly

court (*assembly of the Most High*). 3–7: Wisdom's speech describes her divine origin and
her search for a resting place, though she rules
both in *heavens* and on *the earth*.
   24.8–12: She obeys the divine command
to dwell *in Jacob*, and there eternal Wisdom

and in Israel receive your
inheritance.'
9 Before the ages, in the beginning,
he created me,
and for all the ages I shall not
cease to be.
10 In the holy tent I ministered
before him,
and so I was established in
Zion.
11 Thus in the beloved city he gave
me a resting place,
and in Jerusalem was my
domain.
12 I took root in an honored people,
in the portion of the Lord, his
heritage.

13 "I grew tall like a cedar in
Lebanon,
and like a cypress on the heights
of Hermon.
14 I grew tall like a palm tree in
En-gedi,[o]
and like rosebushes in Jericho;
like a fair olive tree in the field,
and like a plane tree beside
water[p] I grew tall.
15 Like cassia and camel's thorn I
gave forth perfume,
and like choice myrrh I spread
my fragrance,
like galbanum, onycha, and stacte,
and like the odor of incense in
the tent.
16 Like a terebinth I spread out my
branches,
and my branches are glorious
and graceful.
17 Like the vine I bud forth delights,
and my blossoms become
glorious and abundant
fruit.[q]

19 "Come to me, you who desire
me,
and eat your fill of my fruits.
20 For the memory of me is sweeter
than honey,
and the possession of me
sweeter than the
honeycomb.
21 Those who eat of me will hunger
for more,
and those who drink of me will
thirst for more.
22 Whoever obeys me will not be put
to shame,
and those who work with me
will not sin."

23 All this is the book of the
covenant of the Most High
God,
the law that Moses commanded
us
as an inheritance for the
congregations of Jacob.[r]
25 It overflows, like the Pishon, with
wisdom,
and like the Tigris at the time of
the first fruits.
26 It runs over, like the Euphrates,
with understanding,
and like the Jordan at harvest
time.
27 It pours forth instruction like the
Nile,[s]
like the Gihon at the time of
vintage.
28 The first man did not know

o Other ancient authorities read *on the beaches*
p Other ancient authorities omit *beside water*
q Other ancient authorities add as verse 18, *I am the mother of beautiful love, of fear, of knowledge, and of holy hope; being eternal, I am given to all my children, to those who are named by him.*
r Other ancient authorities add as verse 24, *"Do not cease to be strong in the Lord, cling to him so that he may strengthen you; the Lord Almighty alone is God, and besides him there is no savior."*
s Syr: Gk *It makes instruction shine forth like light*

(vv. 9–10) ministers to the Lord in the Jerusalem temple. **24.13–17**: Her majesty is compared to trees, plants, and exotic aromatics. **19–22**: She concludes her speech with an invitation to her unusual meal, which increases one's hunger, and keeps from sin; see Prov 9.5; 8.35. **24.23–29**: Sirach identifies Wisdom with the Torah that *overflows . . . with wisdom*

wisdom[t] fully,
nor will the last one fathom her.
29 For her thoughts are more
abundant than the sea,
and her counsel deeper than the
great abyss.

30 As for me, I was like a canal from
a river,
like a water channel into a
garden.
31 I said, "I will water my garden
and drench my flower-beds."
And lo, my canal became a river,
and my river a sea.
32 I will again make instruction shine
forth like the dawn,
and I will make it clear from far
away.
33 I will again pour out teaching like
prophecy,
and leave it to all future
generations.
34 Observe that I have not labored
for myself alone,
but for all who seek wisdom.[t]

25 I take pleasure in three things,
and they are beautiful in the
sight of God and of
mortals:[u]
agreement among brothers and
sisters, friendship among
neighbors,
and a wife and a husband who
live in harmony.
2 I hate three kinds of people,
and I loathe their manner of life:
a pauper who boasts, a rich person
who lies,
and an old fool who commits
adultery.

3 If you gathered nothing in your
youth,
how can you find anything in
your old age?
4 How attractive is sound judgment
in the gray-haired,
and for the aged to possess good
counsel!
5 How attractive is wisdom in the
aged,
and understanding and counsel
in the venerable!
6 Rich experience is the crown of
the aged,
and their boast is the fear of the
Lord.

7 I can think of nine whom I would
call blessed,
and a tenth my tongue
proclaims:
a man who can rejoice in his
children;
a man who lives to see the
downfall of his foes.
8 Happy the man who lives with a
sensible wife,
and the one who does not plow
with ox and ass together.[v]
Happy is the one who does not
sin with the tongue,
and the one who has not served
an inferior.
9 Happy is the one who finds a
friend,[w]
and the one who speaks to
attentive listeners.

t Gk *her*   *u* Syr Lat: Gk *In three things I was*
beautiful and I stood in beauty before the Lord and
mortals.   *v* Heb Syr: Gk lacks *and the one who does*
not plow with ox and ass together   *w* Lat Syr: Gk
good sense

(vv. 23–24). **28**: Just as the *first man* (Adam)
did not know her (since the Torah was given
first to Moses), so the *last one* will never exhaust such profound wisdom.
24.30–34: Sirach describes how in channeling this wisdom in his book, he has become indeed a *river,* a veritable *sea* whose
instruction (comparable to *prophecy*) reaches
all future generations.
25.1–2: **A numerical saying about**
three beautiful and three hateful things.
25.3–6: **The attractiveness of wisdom**
in the aged.
25.7–11: **A numerical saying of ten**
blessings.
25.13–26.27: **Wicked and virtuous**
women. 13–15: The wickedness of the *wife* is
heightened by the perils inherent in polygamy (*vengeance, wrath*).

10 How great is the one who finds
    wisdom!
    But none is superior to the one
        who fears the Lord.
11 Fear of the Lord surpasses
        everything;
    to whom can we compare the
        one who has it?*

13 Any wound, but not a wound of
        the heart!
    Any wickedness, but not the
        wickedness of a woman!
14 Any suffering, but not suffering
        from those who hate!
    And any vengeance, but not the
        vengeance of enemies!
15 There is no venom* worse than a
        snake's venom, *
    and no anger worse than a
        woman's* wrath.

16 I would rather live with a lion and
        a dragon
    than live with an evil woman.
17 A woman's wickedness changes
        her appearance,
    and darkens her face like that of
        a bear.
18 Her husband sits* among the
        neighbors,
    and he cannot help sighing*
        bitterly.
19 Any iniquity is small compared to
        a woman's iniquity;
    may a sinner's lot befall her!
20 A sandy ascent for the feet of the
        aged—
    such is a garrulous wife to a
        quiet husband.
21 Do not be ensnared by a woman's
        beauty,
    and do not desire a woman for
        her possessions.*
22 There is wrath and impudence and
        great disgrace

when a wife supports her
        husband.
23 Dejected mind, gloomy face,
    and wounded heart come from
        an evil wife.
    Drooping hands and weak knees
        come from the wife who does
        not make her husband
        happy.
24 From a woman sin had its
        beginning,
    and because of her we all die.
25 Allow no outlet to water,
    and no boldness of speech to an
        evil wife.
26 If she does not go as you direct,
    separate her from yourself.

26 Happy is the husband of a
    good wife;
    the number of his days will be
        doubled.
2 A loyal wife brings joy to her
        husband,
    and he will complete his years
        in peace.
3 A good wife is a great blessing;
    she will be granted among the
        blessings of the man who
        fears the Lord.
4 Whether rich or poor, his heart is
        content,
    and at all times his face is
        cheerful.
5 Of three things my heart is
        frightened,
    and of a fourth I am in great
        fear:*

x Other ancient authorities add as verse 12,
*The fear of the Lord is the beginning of love
for him, and faith is the beginning of clinging to him.*
y Syr: Gk *head*   *z* Other ancient
authorities read *an enemy's*   *a* Heb Syr: Gk
loses heart   *b* Other ancient authorities read *and
listening he sighs*   *c* Heb Syr: Other Gk
authorities read *for her beauty*
d Syr: Meaning of Gk uncertain

25.19: *Sinner's lot,* by her being married to
a sinner. **24:** Chronologically, the *woman* in
Eden sinned first.
25.26b: Literally, "cut her off from your
flesh," that is, divorce her (Deut 24.1); hith-

erto they had been "one flesh" (Gen 2.24).
    26.5–6: A numerical saying that illustrates
the problems of polygamy (v. 5, the culminating evil, *worse than death*).

Slander in the city, the gathering
of a mob,
and false accusation—all these
are worse than death.
6 But it is heartache and sorrow
when a wife is jealous of a
rival,
and a tongue-lashing makes it
known to all.
7 A bad wife is a chafing yoke;
taking hold of her is like
grasping a scorpion.
8 A drunken wife arouses great
anger;
she cannot hide her shame.
9 The haughty stare betrays an
unchaste wife;
her eyelids give her away.

10 Keep strict watch over a
headstrong daughter,
or else, when she finds liberty,
she will make use of it.
11 Be on guard against her impudent
eye,
and do not be surprised if she
sins against you.
12 As a thirsty traveler opens his
mouth
and drinks from any water near
him,
so she will sit in front of every
tent peg
and open her quiver to the
arrow.

13 A wife's charm delights her
husband,
and her skill puts flesh on his
bones.
14 A silent wife is a gift from the
Lord,
and nothing is so precious as her
self-discipline.
15 A modest wife adds charm to
charm,
and no scales can weigh the
value of her chastity.
16 Like the sun rising in the heights
of the Lord,
so is the beauty of a good wife
in her well-ordered home.

17 Like the shining lamp on the holy
lampstand,
so is a beautiful face on a stately
figure.
18 Like golden pillars on silver bases,
so are shapely legs and steadfast
feet.

———————

Other ancient authorities add
verses 19–27:

19 *My child, keep sound the bloom of
your youth,
and do not give your strength to
strangers.*
20 *Seek a fertile field within the whole
plain,
and sow it with your own seed,
trusting in your fine stock.*
21 *So your offspring will prosper,
and, having confidence in their good
descent, will grow great.*
22 *A prostitute is regarded as spittle,
and a married woman as a tower of
death to her lovers.*
23 *A godless wife is given as a portion to
a lawless man,
but a pious wife is given to the man
who fears the Lord.*
24 *A shameless woman constantly acts
disgracefully,
but a modest daughter will even be
embarrassed before her
husband.*
25 *A headstrong wife is regarded as a
dog,
but one who has a sense of shame
will fear the Lord.*
26 *A wife honoring her husband will
seem wise to all,
but if she dishonors him in her
pride she will be known to all
as ungodly.
Happy is the husband of a good
wife;
for the number of his years will be
doubled.*
27 *A loud-voiced and garrulous wife is
like a trumpet sounding the
charge,
and every person like this lives in
the anarchy of war.*

———————

28 At two things my heart is grieved,
and because of a third anger
comes over me:
a warrior in want through
poverty,
intelligent men who are treated
contemptuously,
and a man who turns back from
righteousness to sin—
the Lord will prepare him for
the sword!

29 A merchant can hardly keep from
wrongdoing,
nor is a tradesman innocent of
sin.

27 Many have committed sin for
gain,*
and those who seek to get rich
will avert their eyes.

2 As a stake is driven firmly into a
fissure between stones,
so sin is wedged in between
selling and buying.

3 If a person is not steadfast in the
fear of the Lord,
his house will be quickly
overthrown.

4 When a sieve is shaken, the refuse
appears;
so do a person's faults when he
speaks.

5 The kiln tests the potter's vessels;
so the test of a person is in his
conversation.

6 Its fruit discloses the cultivation of
a tree;
so a person's speech discloses
the cultivation of his mind.

7 Do not praise anyone before he
speaks,
for this is the way people are
tested.

8 If you pursue justice, you will
attain it

and wear it like a glorious robe.

9 Birds roost with their own kind,
so honesty comes home to those
who practice it.

10 A lion lies in wait for prey;
so does sin for evildoers.

11 The conversation of the godly is
always wise,
but the fool changes like the
moon.

12 Among stupid people limit your
time,
but among thoughtful people
linger on.

13 The talk of fools is offensive,
and their laughter is wantonly
sinful.

14 Their cursing and swearing make
one's hair stand on end,
and their quarrels make others
stop their ears.

15 The strife of the proud leads to
bloodshed,
and their abuse is grievous to
hear.

16 Whoever betrays secrets destroys
confidence,
and will never find a congenial
friend.

17 Love your friend and keep faith
with him;
but if you betray his secrets, do
not follow after him.

18 For as a person destroys his
enemy,
so you have destroyed the
friendship of your
neighbor.

19 And as you allow a bird to escape
from your hand,
so you have let your neighbor
go, and will not catch him
again.

e Other ancient authorities read *a trifle*

26.28–28.26: Miscellaneous observations. 26.28: *A warrior,* Syriac, "a wealthy man." **26.29–27.3:** The moral dangers in commercial activities. **27.1:** Prov 28.21. **4–7:** Tests in life.

27.8–10: Reward and retribution. **11–15:** Kinds of talk.
27.16–21: Against disclosing secrets (Prov 20.19; 25.8–10). **21:** *Without hope,* of reconciliation (22.22).

20 Do not go after him, for he is too
   far off,
   and has escaped like a gazelle
   from a snare.
21 For a wound may be bandaged,
   and there is reconciliation after
   abuse,
   but whoever has betrayed
   secrets is without hope.

22 Whoever winks the eye plots
   mischief,
   and those who know him will
   keep their distance.
23 In your presence his mouth is all
   sweetness,
   and he admires your words;
   but later he will twist his speech
   and with your own words he
   will trip you up.
24 I have hated many things, but him
   above all;
   even the Lord hates him.
25 Whoever throws a stone straight
   up throws it on his own
   head,
   and a treacherous blow opens
   up many wounds.
26 Whoever digs a pit will fall into it,
   and whoever sets a snare will be
   caught in it.
27 If a person does evil, it will roll
   back upon him,
   and he will not know where it
   came from.
28 Mockery and abuse issue from the
   proud,
   but vengeance lies in wait for
   them like a lion.
29 Those who rejoice in the fall of
   the godly will be caught in
   a snare,
   and pain will consume them
   before their death.

30 Anger and wrath, these also are
   abominations,

yet a sinner holds on to them.
28 The vengeful will face the
   Lord's vengeance,
   for he keeps a strict account of[f]
   their sins.
2 Forgive your neighbor the wrong
   he has done,
   and then your sins will be
   pardoned when you pray.
3 Does anyone harbor anger against
   another,
   and expect healing from the
   Lord?
4 If one has no mercy toward
   another like himself,
   can he then seek pardon for his
   own sins?
5 If a mere mortal harbors wrath,
   who will make an atoning
   sacrifice for his sins?
6 Remember the end of your life,
   and set enmity aside;
   remember corruption and death,
   and be true to the
   commandments.
7 Remember the commandments,
   and do not be angry with
   your neighbor;
   remember the covenant of the
   Most High, and overlook
   faults.

8 Refrain from strife, and your sins
   will be fewer;
   for the hot-tempered kindle
   strife,
9 and the sinner disrupts friendships
   and sows discord among those
   who are at peace.
10 In proportion to the fuel, so will
   the fire burn,
   and in proportion to the
   obstinacy, so will strife
   increase;[g]

f Other ancient authorities read *for he firmly
establishes*   g Other ancient authorities read
burn

27.22–27: Hypocrisy and punishment.
22: *Winks,* with insincerity (Prov 6.13;
10.10). 25–27: Evil actions beget evil results
(Ps 7.14–16; 9.15–16; Prov 26.27; Eccl 10.8).
   27.28–28.1: *Vengeance* comes upon the

evil, ultimately from the Lord. 28.2–7: Forgiveness for oneself is contingent upon showing forgiveness to others (Mt 6.12, 14; Mk
11.25).
   28.8–11: Quarreling. 12–26: Evils of the

in proportion to a person's
strength will be his anger,
and in proportion to his wealth
he will increase his wrath.
11 A hasty quarrel kindles a fire,
and a hasty dispute sheds blood.
12 If you blow on a spark, it will
glow;
if you spit on it, it will be put
out;
yet both come out of your
mouth.

13 Curse the gossips and the
double-tongued,
for they destroy the peace of
many.
14 Slander[h] has shaken many,
and scattered them from nation
to nation;
it has destroyed strong cities,
and overturned the houses of
the great.
15 Slander[i] has driven virtuous
women from their homes,
and deprived them of the fruit
of their toil.
16 Those who pay heed to slander[j]
will not find rest,
nor will they settle down in
peace.
17 The blow of a whip raises a welt,
but a blow of the tongue
crushes the bones.
18 Many have fallen by the edge of
the sword,
but not as many as have fallen
because of the tongue.
19 Happy is the one who is protected
from it,
who has not been exposed to its
anger,
who has not borne its yoke,
and has not been bound with its
fetters.

20 For its yoke is a yoke of iron,
and its fetters are fetters of
bronze;
21 its death is an evil death,
and Hades is preferable to it.
22 It has no power over the godly;
they will not be burned in its
flame.
23 Those who forsake the Lord will
fall into its power;
it will burn among them and
will not be put out.
It will be sent out against them
like a lion;
like a leopard it will mangle
them.
24a As you fence in your property
with thorns,
25b so make a door and a bolt for
your mouth.
24b As you lock up your silver and
gold,
25a so make balances and scales for
your words.
26 Take care not to err with your
tongue,[k]
and fall victim to one lying in
wait.

29 The merciful lend to their
neighbors;
by holding out a helping hand
they keep the
commandments.
2 Lend to your neighbor in his time
of need;
repay your neighbor when a
loan falls due.
3 Keep your promise and be honest
with him,
and on every occasion you will
find what you need.

h *Gk A third tongue*   i *Gk a third tongue*
j *Gk it*   k *Gk with it*

tongue. **12**: *Mouth,* see 5.13; Prov 15.1; Jas
3.9–10.
28.17: Prov 25.15. **21**: *Hades* (Sheol) is *preferable* to the living *death* inflicted by the *evil*
tongue. **24–25**: Just as one protects valuables,
one should be careful about one's *words.*
29:1–20: **Loans, Alms, Surety. 1–7**: Ex

22.25; loans were given without charging interest (Ex 22.24), but interest was permitted
if one loaned to a Gentile (Deut 23.20–21). **3**:
If a loan is repaid on time, one can count on
receiving a future loan. **4–7**: Warnings about
abuses in borrowing.

4 Many regard a loan as a windfall,
    and cause trouble to those who
      help them.
5 One kisses another's hands until
    he gets a loan,
    and is deferential in speaking of
      his neighbor's money;
    but at the time for repayment he
      delays,
    and pays back with empty
      promises,
    and finds fault with the time.
6 If he can pay, his creditor[l] will
    hardly get back half,
    and will regard that as a
      windfall.
    If he cannot pay, the borrower[l]
      has robbed the other of his
      money,
    and he has needlessly made him
      an enemy;
    he will repay him with curses and
      reproaches,
    and instead of glory will repay
      him with dishonor.
7 Many refuse to lend, not because
    of meanness,
    but from fear[m] of being
      defrauded needlessly.

8 Nevertheless, be patient with
    someone in humble
      circumstances,
    and do not keep him waiting for
      your alms.
9 Help the poor for the
    commandment's sake,
    and in their need do not send
      them away empty-handed.
10 Lose your silver for the sake of a
    brother or a friend,
    and do not let it rust under a
      stone and be lost.
11 Lay up your treasure according to
    the commandments of the
      Most High,
    and it will profit you more than
      gold.

12 Store up almsgiving in your
    treasury,
    and it will rescue you from
      every disaster;
13 better than a stout shield and a
    sturdy spear,
    it will fight for you against the
      enemy.

14 A good person will be surety for
    his neighbor,
    but the one who has lost all
      sense of shame will fail
      him.
15 Do not forget the kindness of
    your guarantor,
    for he has given his life for you.
16 A sinner wastes the property of
    his guarantor,
17   and the ungrateful person
    abandons his rescuer.
18 Being surety has ruined many
    who were prosperous,
    and has tossed them about like
      waves of the sea;
    it has driven the influential into
      exile,
    and they have wandered among
      foreign nations.
19 The sinner comes to grief through
    surety;
    his pursuit of gain involves him
      in lawsuits.
20 Assist your neighbor to the best of
    your ability,
    but be careful not to fall
      yourself.

21 The necessities of life are water,
    bread, and clothing,
    and also a house to assure
      privacy.
22 Better is the life of the poor under
    their own crude roof

l Gk *he*   *m* Other ancient authorities
read *many refuse to lend, therefore, because of
such meanness; they are afraid*

29.8–13: As regards the poor, almsgiving
is a duty (see 40.1–6). **9:** *For the commandment's
sake,* Deut 15.7–11.

29.14–20: The dangers in guaranteeing a
loan for another (see 8.12–13 n.); Sirach advises caution.

than sumptuous food in the
house of others.
23 Be content with little or much,
and you will hear no reproach
for being a guest.[n]
24 It is a miserable life to go from
house to house;
as a guest you should not open
your mouth;
25 you will play the host and provide
drink without being
thanked,
and besides this you will hear
rude words like these:
26 "Come here, stranger, prepare the
table;
let me eat what you have
there."
27 "Be off, stranger, for an honored
guest is here;
my brother has come for a visit,
and I need the
guest-room."
28 It is hard for a sensible person to
bear
scolding about lodging[o] and the
insults of the moneylender.

CONCERNING CHILDREN[p]

30 He who loves his son will whip
him often,
so that he may rejoice at the
way he turns out.
2 He who disciplines his son will
profit by him,
and will boast of him among
acquaintances.
3 He who teaches his son will make
his enemies envious,
and will glory in him among his
friends.
4 When the father dies he will not
seem to be dead,
for he has left behind him one
like himself,
5 whom in his life he looked upon
with joy
and at death, without grief.
6 He has left behind him an avenger
against his enemies,
and one to repay the kindness of
his friends.

7 Whoever spoils his son will bind
up his wounds,
and will suffer heartache at
every cry.
8 An unbroken horse turns out
stubborn,
and an unchecked son turns out
headstrong.
9 Pamper a child, and he will
terrorize you;
play with him, and he will
grieve you.
10 Do not laugh with him, or you
will have sorrow with him,
and in the end you will gnash
your teeth.
11 Give him no freedom in his
youth,
and do not ignore his errors.
12 Bow down his neck in his youth,[q]
and beat his sides while he is
young,
or else he will become stubborn
and disobey you,
and you will have sorrow of
soul from him.[r]
13 Discipline your son and make his
yoke heavy,[s]
so that you may not be offended
by his shamelessness.

n Lat: Gk *reproach from your family;* other ancient authorities lack this line *o* Or *scolding from the household* *p* This heading is included in the Gk text. *q* Other ancient authorities lack this line and the preceding line *r* Other ancient authorities lack this line *s* Heb: Gk *take pains with him*

29.21–28: Frugality. 21: The basic necessities are expanded in 39.26. **24–28:** Do not live beyond your means so that you do not become dependent upon others.

30.1–13: Training of children. 1: Physical punishment is taken for granted (22.6; Prov 13.24; 19.18; 23.13–14; 29.15). **4:** The child is the image of the *father;* see Tob 9.6. **6:** *An avenger,* as in Ps 127.5.

30.7–13: Dangers that result if discipline is not applied.

14 Better off poor, healthy, and fit
     than rich and afflicted in body.
15 Health and fitness are better than
     any gold,
   and a robust body than
     countless riches.
16 There is no wealth better than
     health of body,
   and no gladness above joy of
     heart.
17 Death is better than a life of
     misery,
   and eternal sleep*t* than chronic
     sickness.

CONCERNING FOODS *u*

18 Good things poured out upon a
     mouth that is closed
   are like offerings of food placed
     upon a grave.
19 Of what use to an idol is a
     sacrifice?
   For it can neither eat nor smell.
   So is the one punished by the
     Lord;
20 he sees with his eyes and groans
   as a eunuch groans when
     embracing a girl. *v*

21 Do not give yourself over to
     sorrow,
   and do not distress yourself
     deliberately.
22 A joyful heart is life itself,
   and rejoicing lengthens one's life
     span.
23 Indulge yourself*w* and take
     comfort,
   and remove sorrow far from
     you,
   for sorrow has destroyed many,
   and no advantage ever comes
     from it.
24 Jealousy and anger shorten life,
   and anxiety brings on premature
     old age.

25 Those who are cheerful and merry
     at table
   will benefit from their food.

31 Wakefulness over wealth wastes
     away one's flesh,
   and anxiety about it drives away
     sleep.
2 Wakeful anxiety prevents slumber,
   and a severe illness carries off
     sleep. *x*
3 The rich person toils to amass a
     fortune,
   and when he rests he fills
     himself with his dainties.
4 The poor person toils to make a
     meager living,
   and if ever he rests he becomes
     needy.
5 One who loves gold will not be
     justified;
   one who pursues money will be
     led astray*y* by it.
6 Many have come to ruin because
     of gold,
   and their destruction has met
     them face to face.
7 It is a stumbling block to those
     who are avid for it,
   and every fool will be taken
     captive by it.
8 Blessed is the rich person who is
     found blameless,
   and who does not go after gold.
9 Who is he, that we may praise
     him?
   For he has done wonders among
     his people.

t Other ancient authorities lack *eternal sleep*
u This heading is included in the Gk text; other
ancient authorities place the heading before
verse 16   *v* Other ancient authorities add *So
is the person who does right under compulsion*
w Other ancient authorities read *Beguile yourself*
x Other ancient authorities read *sleep carries off
a severe illness*   *y* Heb Syr: Gk *pursues destruction
will be filled*

30.14–20: **Concerning health. 17**: Job
3.11, 13, 17; Tob 3.6, 10, 13. **18**: Let Jer 27.28;
Bel 3 (compare Ps 115.5–7).
   30.21–25: Cheerfulness.

31.1–11: **Right attitude toward wealth.**
Sirach has strict views concerning the rich
(vv. 5–7; 11.11), and the implication of verses
8–11 is that few rich people are also just.

10 Who has been tested by it and
    been found perfect?
  Let it be for him a ground for
    boasting.
  Who has had the power to
    transgress and did not
    transgress,
  and to do evil and did not do it?
11 His prosperity will be
    established, z
  and the assembly will proclaim
    his acts of charity.

12 Are you seated at the table of the
    great? a
  Do not be greedy at it,
  and do not say, "How much
    food there is here!"
13 Remember that a greedy eye is a
    bad thing.
  What has been created more
    greedy than the eye?
  Therefore it sheds tears for any
    reason.
14 Do not reach out your hand for
    everything you see,
  and do not crowd your
    neighbor b at the dish.
15 Judge your neighbor's feelings by
    your own,
  and in every matter be
    thoughtful.
16 Eat what is set before you like a
    well brought-up person, c
  and do not chew greedily, or
    you will give offense.
17 Be the first to stop, as befits good
    manners,
  and do not be insatiable, or you
    will give offense.
18 If you are seated among many
    persons,
  do not help yourself d before
    they do.

19 How ample a little is for a
    well-disciplined person!

  He does not breathe heavily
    when in bed.
20 Healthy sleep depends on
    moderate eating;
  he rises early, and feels fit.
  The distress of sleeplessness and of
    nausea
  and colic are with the glutton.
21 If you are overstuffed with food,
  get up to vomit, and you will
    have relief.
22 Listen to me, my child, and do
    not disregard me,
  and in the end you will
    appreciate my words.
  In everything you do be
    moderate, e
  and no sickness will overtake
    you.
23 People bless the one who is liberal
    with food,
  and their testimony to his
    generosity is trustworthy.
24 The city complains of the one
    who is stingy with food,
  and their testimony to his
    stinginess is accurate.

25 Do not try to prove your strength
    by wine-drinking,
  for wine has destroyed many.
26 As the furnace tests the work of
    the smith, f
  so wine tests hearts when the
    insolent quarrel.
27 Wine is very life to human beings
    if taken in moderation.
  What is life to one who is without
    wine?
  It has been created to make
    people happy.

z Other ancient authorities add *because of this*
a Heb Syr; Gk *at a great table*   b Gk *him*
c Heb: Gk *like a human being*   d Gk *reach out
your hand*   e Heb Syr: Gk *industrious*   f Heb:
Gk *tests the hardening of steel by dipping*

**31.12–31: Temperance in food and
drink.** One should be particularly aware of
the need of politeness and kindness when eating with others. **13:** *Greedy eye,* as in 14.10.
  31.25–26: Excessive drinking is not a
proof of a person's worth. As metal is tested
by *fire, so hearts* are tested by *wine;* see vv. 29–
30. **27–28:** Wine is a staple food in Mediterranean countries (Ps 104.15).

28 Wine drunk at the proper time and
in moderation
  is rejoicing of heart and gladness
  of soul.
29 Wine drunk to excess leads to
bitterness of spirit,
  to quarrels and stumbling.
30 Drunkenness increases the anger of
a fool to his own hurt,
  reducing his strength and adding
  wounds.
31 Do not reprove your neighbor at a
banquet of wine,
  and do not despise him in his
  merrymaking;
  speak no word of reproach to
  him,
  and do not distress him by
  making demands of him.

32 If they make you master of the
feast, do not exalt yourself;
  be among them as one of their
  number.
  Take care of them first and then
  sit down;
2   when you have fulfilled all your
  duties, take your place,
  so that you may be merry along
  with them
  and receive a wreath for your
  excellent leadership.

3 Speak, you who are older, for it is
your right,
  but with accurate knowledge,
  and do not interrupt the
  music.
4 Where there is entertainment, do
not pour out talk;
  do not display your cleverness
  at the wrong time.
5 A ruby seal in a setting of gold
  is a concert of music at a
  banquet of wine.
6 A seal of emerald in a rich setting
of gold

is the melody of music with
  good wine.
7 Speak, you who are young, if you
are obliged to,
  but no more than twice, and
  only if asked.
8 Be brief; say much in few words;
  be as one who knows and can
  still hold his tongue.
9 Among the great do not act as
their equal;
  and when another is speaking,
  do not babble.

10 Lightning travels ahead of the
thunder,
  and approval goes before one
  who is modest.
11 Leave in good time and do not be
the last;
  go home quickly and do not
  linger.
12 Amuse yourself there to your
heart's content,
  but do not sin through proud
  speech.
13 But above all bless your Maker,
  who fills you with his good
  gifts.

14 The one who seeks God*g* will
accept his discipline,
  and those who rise early to seek
  him*h* will find favor.
15 The one who seeks the law will be
filled with it,
  but the hypocrite will stumble
  at it.
16 Those who fear the Lord will
form true judgments,
  and they will kindle righteous
  deeds like a light.
17 The sinner will shun reproof,

g Heb: Gk *who fears the Lord*   *h* Other ancient
authorities lack *to seek him*

32.1–13: Proper behavior at a banquet.
1–2: The *master of the feast* had to care for the
needs of those at the table. **3–10:** Directions
to the *older* people and the *young.* **11–13:** These
words are meant for all.

**32.14–33.6: The God-fearing man con-
trasted with the sinner. 32.14:** *Discipline,*
Hebrew "musar," is a frequent topic (see
6.18 n.). **17:** *A decision,* i.e. an interpretation
(of the law).

and will find a decision
according to his liking.

18 A sensible person will not
overlook a thoughtful
suggestion;
an insolent[i] and proud person
will not be deterred by
fear.[j]

19 Do nothing without deliberation,
but when you have acted, do
not regret it.

20 Do not go on a path full of
hazards,
and do not stumble at an
obstacle twice.[k]

21 Do not be overconfident on a
smooth[l] road,

22 and give good heed to your
paths.[m]

23 Guard[n] yourself in every act,
for this is the keeping of the
commandments.

24 The one who keeps the law
preserves himself,[o]
and the one who trusts the Lord
will not suffer loss.

33 No evil will befall the one who
fears the Lord,
but in trials such a one will be
rescued again and again.

2 The wise will not hate the law,
but the one who is hypocritical
about it is like a boat in a
storm.

3 The sensible person will trust in
the law;
for such a one the law is as
dependable as a divine
oracle.

4 Prepare what to say, and then you
will be listened to;
draw upon your training, and
give your answer.

5 The heart of a fool is like a cart
wheel,
and his thoughts like a turning
axle.

6 A mocking friend is like a stallion
that neighs no matter who the
rider is.

7 Why is one day more important
than another,
when all the daylight in the year
is from the sun?

8 By the Lord's wisdom they were
distinguished,
and he appointed the different
seasons and festivals.

9 Some days he exalted and
hallowed,
and some he made ordinary
days.

10 All human beings come from the
ground,
and humankind[p] was created
out of the dust.

11 In the fullness of his knowledge
the Lord distinguished them
and appointed their different
ways.

12 Some he blessed and exalted,
and some he made holy and
brought near to himself;
but some he cursed and brought
low,
and turned them out of their
place.

13 Like clay in the hand of the
potter,
to be molded as he pleases,

---

i Heb: Gk *alien*   j Meaning of Gk uncertain.
Other ancient authorities add *and after acting,
with him, without deliberation*   k Heb: Gk *stumble
on stony ground*   l Or *an unexplored*   m Heb
Syr: Gk *and beware of your children*   n Heb Syr:
Gk *Trust*   o Heb: Gk *who believes the law heeds
the commandments*   p Heb: Gk *Adam*

33.2: Hebrew, "He that hates the law is not
wise, and is tossed about like a boat in a
storm." **3**: *Urim,* 45.10; Ex 28.30; Num
27.21; 1 Sam 14.41–42.
33.7–15: **Divinely ordained opposites
in creation.** Corresponding to wise (righ-

teous) and foolish (sinner) are the polarities to
be found in life, such as holy days and *ordinary
days* (v. 9), the blessed and the cursed (v. 12).
13: *Clay . . . potter,* a common biblical image
(Isa 29.16; 45.8; 64.8; Jer 18.4, 6; Wis 15.7–8;
Rom 9.21) that underscores divine determi-

so all are in the hand of their
Maker,
to be given whatever he decides.

14 Good is the opposite of evil,
and life the opposite of death;
so the sinner is the opposite of
the godly.
15 Look at all the works of the Most
High;
they come in pairs, one the
opposite of the other.

16 Now I was the last to keep vigil;
I was like a gleaner following
the grape-pickers;
17 by the blessing of the Lord I
arrived first,
and like a grape-picker I filled
my wine press.
18 Consider that I have not labored
for myself alone,
but for all who seek instruction.
19 Hear me, you who are great
among the people,
and you leaders of the
congregation, pay heed!

20 To son or wife, to brother or
friend,
do not give power over
yourself, as long as you
live;
and do not give your property to
another,
in case you change your mind
and must ask for it.
21 While you are still alive and have
breath in you,
do not let anyone take your
place.
22 For it is better that your children
should ask from you
than that you should look to the
hand of your children.
23 Excel in all that you do;

bring no stain upon your honor.
24 At the time when you end the
days of your life,
in the hour of death, distribute
your inheritance.

25 Fodder and a stick and burdens for
a donkey;
bread and discipline and work
for a slave.
26 Set your slave to work, and you
will find rest;
leave his hands idle, and he will
seek liberty.
27 Yoke and thong will bow the
neck,
and for a wicked slave there are
racks and tortures.
28 Put him to work, in order that he
may not be idle,
29 for idleness teaches much evil.
30 Set him to work, as is fitting for
him,
and if he does not obey, make
his fetters heavy.
Do not be overbearing toward
anyone,
and do nothing unjust.

31 If you have but one slave, treat
him like yourself,
because you have bought him
with blood.
If you have but one slave, treat
him like a brother,
for you will need him as you
need your life.
32 If you ill-treat him, and he leaves
you and runs away,
33 which way will you go to seek
him?

34 The senseless have vain and
false hopes,
and dreams give wings to fools.

---

nism at the same time that human free will is
affirmed (15.11–20). **14–15:** A summary
statement of the theme.
 **33.16–19: The author's qualifications as
a teacher.** See also 24.30–34; 39.12–13;
51.13–28.

**33.20–24: On preserving financial in-
dependence.**
33.25–33: On the treatment of slaves.
Slavery is taken for granted as a feature of
society. Although Sirach's words are strong,
especially for the lazy (vv. 25–29), he urges

2 As one who catches at a shadow
and pursues the wind,
so is anyone who believes in^q
dreams.

3 What is seen in dreams is but a
reflection,
the likeness of a face looking at
itself.

4 From an unclean thing what can
be clean?
And from something false what
can be true?

5 Divinations and omens and
dreams are unreal,
and like a woman in labor, the
mind has fantasies.

6 Unless they are sent by
intervention from the Most
High,
pay no attention to them.

7 For dreams have deceived many,
and those who put their hope in
them have perished.

8 Without such deceptions the law
will be fulfilled,
and wisdom is complete in the
mouth of the faithful.

9 An educated^r person knows many
things,
and one with much experience
knows what he is talking
about.

10 An inexperienced person knows
few things,

11 but he that has traveled acquires
much cleverness.

12 I have seen many things in my
travels,
and I understand more than I
can express.

13 I have often been in danger of
death,
but have escaped because of
these experiences.

14 The spirit of those who fear the
Lord will live,

15 for their hope is in him who
saves them.

16 Those who fear the Lord will not
be timid,
or play the coward, for he is
their hope.

17 Happy is the soul that fears the
Lord!

18 To whom does he look? And
who is his support?

19 The eyes of the Lord are on those
who love him,
a mighty shield and strong
support,
a shelter from scorching wind and
a shade from noonday sun,
a guard against stumbling and a
help against falling.

20 He lifts up the soul and makes the
eyes sparkle;
he gives health and life and
blessing.

21 If one sacrifices ill-gotten goods,
the offering is blemished;^s

22 the gifts^t of the lawless are not
acceptable.

23 The Most High is not pleased
with the offerings of the
ungodly,

q Syr: Gk *pays heed to*   r Other ancient
authorities read *A traveled*   s Other ancient
authorities read *is made in mockery*   t Other
ancient authorities read *mockeries*

justice (v.30). The rights of a slave are defined in the law codes (Ex 21.2–6, 20–21,
26–27; Lev 25.46; Deut 15.12–18).
**34.1–8: The vanity of dreams and
omens** (Deut 13.2–5; 18.9–14; Eccl 5.7; Jer
29.8). **2:** *Pursues the wind,* the image as in Hos
12.1; Eccl 1.14; and often. **3:** *Dreams* have no
reality; they are a reflection of one's concerns.
4: Job 14.4. **6:** Allowance is made for God-
given dreams, (Gen 37.5ff.; Judg 7.13ff.; Job
33.15–18).

**34.9–13: Learning from experience
and travel.**
**34.14–20: The blessings of those who
fear God. 19:** *Shelter,* Ps 61.2–4; 91.1–4; Isa
25.4; the *scorching wind* is the well known
sirocco in spring and autumn.
34.21–31: Unacceptable sacrifices (1 Sam
15.22; Ps 51.16–19; Prov 15.8; 21.3; Hos
6.6; Am 5.21–24; Mt 23.23). Sirach strikes
out against abuses and against *offerings of the
ungodly* (v. 22).

nor for a multitude of sacrifices
   does he forgive sins.
24 Like one who kills a son before
   his father's eyes
  is the person who offers a
   sacrifice from the property
   of the poor.
25 The bread of the needy is the life
   of the poor;
  whoever deprives them of it is a
   murderer.
26 To take away a neighbor's living
   is to commit murder;
27   to deprive an employee of
   wages is to shed blood.

28 When one builds and another tears
   down,
  what do they gain but hard
   work?
29 When one prays and another
   curses,
  to whose voice will the Lord
   listen?
30 If one washes after touching a
   corpse, and touches it
   again,
  what has been gained by
   washing?
31 So if one fasts for his sins,
  and goes again and does the
   same things,
  who will listen to his prayer?
  And what has he gained by
   humbling himself?

35 The one who keeps the law
   makes many offerings;
2   one who heeds the commandments
  makes an offering of
   well-being.
3 The one who returns a kindness
  offers choice flour,

4 and one who gives alms
   sacrifices a thank offering.
5 To keep from wickedness is
   pleasing to the Lord,
  and to forsake unrighteousness
   is an atonement.
6 Do not appear before the Lord
   empty-handed,
7 for all that you offer is in
   fulfillment of the
   commandment.
8 The offering of the righteous
   enriches the altar,
  and its pleasing odor rises before
   the Most High.
9 The sacrifice of the righteous is
   acceptable,
  and it will never be forgotten.
10 Be generous when you worship
   the Lord,
  and do not stint the first fruits
   of your hands.
11 With every gift show a cheerful
   face,
  and dedicate your tithe with
   gladness.
12 Give to the Most High as he has
   given to you,
  and as generously as you can
   afford.
13 For the Lord is the one who
   repays,
  and he will repay you sevenfold.
14 Do not offer him a bribe, for he
   will not accept it;
15   and do not rely on a dishonest
   sacrifice;
  for the Lord is the judge,
  and with him there is no
   partiality.
16 He will not show partiality to the
   poor;
  but he will listen to the prayer
   of one who is wronged.

34.28: *One builds,* i.e. the poor person victimized by the rich man, who *tears down;* both of them lose. **29**: The rich man *prays* hypocritically, and the poor *curses* him, but it is clear that the Lord will listen to the *voice* of the poor (35.20–21). **30–31**: The futility of purificatory acts by an unrepentant sinner.
35.1–13: **Acceptable sacrifices** (Isa 1.11–18; Mic 6.6–8; Mk 12.33). **1–2**: Keeping the law (especially almsgiving) is equivalent to *many offerings.*
35.6–13: The spirit with which one should offer sacrifices.
35.14–26: **A warning against exploitation of the poor. 14**: *Dishonest sacrifice,* i.e. one made possible by injustice (34.21). *No*

17 He will not ignore the supplication
of the orphan,
or the widow when she pours
out her complaint.
18 Do not the tears of the widow run
down her cheek
19 as she cries out against the one
who causes them to fall?
20 The one whose service is pleasing
to the Lord will be
accepted,
and his prayer will reach to the
clouds.
21 The prayer of the humble pierces
the clouds,
and it will not rest until it
reaches its goal;
it will not desist until the Most
High responds
22 and does justice for the
righteous, and executes
judgment.
Indeed, the Lord will not delay,
and like a warrior^u will not be
patient
until he crushes the loins of the
unmerciful
23 and repays vengeance on the
nations;
until he destroys the multitude of
the insolent,
and breaks the scepters of the
unrighteous;
24 until he repays mortals according
to their deeds,
and the works of all according
to their thoughts;
25 until he judges the case of his
people
and makes them rejoice in his
mercy.
26 His mercy is as welcome in time
of distress

as clouds of rain in time of
drought.

36 Have mercy upon us, O God^v
of all,
2 and put all the nations in fear of
you.
3 Lift up your hand against foreign
nations
and let them see your might.
4 As you have used us to show your
holiness to them,
so use them to show your glory
to us.
5 Then they will know,^w as we have
known
that there is no God but you,
O Lord.
6 Give new signs, and work other
wonders;
7 make your hand and right arm
glorious.
8 Rouse your anger and pour out
your wrath;
9 destroy the adversary and wipe
out the enemy.
10 Hasten the day, and remember the
appointed time,^x
and let people recount your
mighty deeds.
11 Let survivors be consumed in the
fiery wrath,
and may those who harm your
people meet destruction.
12 Crush the heads of hostile rulers
who say, "There is no one but
ourselves."

u Heb: Gk *and with them*   *v* Heb: Gk *O Master,
the God*   *w* Heb: Gk *And let them know
you*   *x* Other ancient authorities read *remember
your oath*

partiality, Deut 10.17; Job 34.19; Wis 6.7; Acts
10.34; Gal 2.6. **17–18:** The orphan and the
widow in particular need help since there is
no one to plead their cause (except God).
35.22–26: This serves as an introduction
to the prayer in 36.1–19; Sirach has in mind
the oppression of the chosen people by pagan
oppressors.
36.1–22: **A prayer for the deliverance**

and restoration of Israel. 1: *God of all,* not
just of Israel. **2:** *All the nations,* the Seleucids
in particular are meant; they had power over
Palestine after 198 B.C. **4:** As God showed his
holiness by punishing Israel (Ezek 20.41;
28.25), so now he shows his *glory* by punishing the Gentiles. **10:** *The appointed time,* the
coming of the Messianic era.

13 Gather all the tribes of Jacob, *y*
16 and give them their inheritance,
as at the beginning.
17 Have mercy, O Lord, on the
people called by your
name,
on Israel, whom you have
named *z* your firstborn,
18 Have pity on the city of your
sanctuary, *a*
Jerusalem, the place of your
dwelling. *b*
19 Fill Zion with your majesty, *c*
and your temple *d* with your
glory.
20 Bear witness to those whom you
created in the beginning,
and fulfill the prophecies spoken
in your name.
21 Reward those who wait for you
and let your prophets be found
trustworthy.
22 Hear, O Lord, the prayer of your
servants, according to your
goodwill toward *e* your
people,
and all who are on the earth will
know
that you are the Lord, the God
of the ages.

23 The stomach will take any food,
yet one food is better than
another.
24 As the palate tastes the kinds of
game,
so an intelligent mind detects
false words.
25 A perverse mind will cause grief,
but a person with experience
will pay him back.
26 A woman will accept any man as
a husband,
but one girl is preferable to
another.

27 A woman's beauty lights up a
man's face,
and there is nothing he desires
more.
28 If kindness and humility mark her
speech,
her husband is more fortunate
than other men.
29 He who acquires a wife gets his
best possession, *f*
a helper fit for him and a pillar
of support. *g*
30 Where there is no fence, the
property will be plundered;
and where there is no wife, a
man will become a fugitive
and a wanderer. *h*
31 For who will trust a nimble
robber
that skips from city to city?
So who will trust a man that has
no nest,
but lodges wherever night
overtakes him?

37 Every friend says, "I too am a
friend";
but some friends are friends
only in name.
2 Is it not a sorrow like that for
death itself
when a dear friend turns into an
enemy?
3 O inclination to evil, why were
you formed

y Owing to a dislocation in the Greek Mss of
Sirach, the verse numbers 14 and 15 are not
used in chapter 36, though no text is missing.
z Other ancient authorities read *you have
likened to*   *a* Or *on your holy city*   *b* Heb:
Gk *your rest*   *c* Heb Syr: Gk *the celebration of
your wondrous deeds*   *d* Heb Syr: Gk Lat *people*
e Heb and two Gk witnesses: Lat and most Gk
witnesses read *according to the blessing of Aaron for*
f Heb: Gk *enters upon a possession*   *g* Heb: Gk
rest   *h* Heb: Gk *wander about and sigh*

36.23–31: **Concerning discrimination**
(vv. 23–25, in general; vv. 26–31, in choosing a wife). **26:** *Will accept,* because the marriage was arranged by her father (see 7.25).
29: *A helper fit for him,* Gen 2.18.
37.1–6: **False friends** (6.7–13). **3:** *Inclination to evil,* the evil "yeṣer" (see 15.14 n.). **4:**
Happiness, arising from feasting (see v. 5).
37.7–15: Concerning counselors. 14:
Sentinels, probably astrologers.
37.16–26: True and false wisdom.
37.27–31: Temperance (see 31.12–
32.13).

to cover the land with deceit?

4 Some companions rejoice in the
happiness of a friend,
but in time of trouble they are
against him.

5 Some companions help a friend
for their stomachs' sake,
yet in battle they will carry his
shield.

6 Do not forget a friend during the
battle,[i]
and do not be unmindful of him
when you distribute your
spoils.[j]

7 All counselors praise the counsel
they give,
but some give counsel in their
own interest.

8 Be wary of a counselor,
and learn first what is his interest,
for he will take thought for
himself.
He may cast the lot against you

9 and tell you, "Your way is
good,"
and then stand aside to see what
happens to you.

10 Do not consult the one who
regards you with suspicion;
hide your intentions from those
who are jealous of you.

11 Do not consult with a woman
about her rival
or with a coward about war,
with a merchant about business
or with a buyer about selling,
with a miser about generosity[k]
or with the merciless about
kindness,
with an idler about any work
or with a seasonal laborer about
completing his work,
with a lazy servant about a big
task—
pay no attention to any advice
they give.

12 But associate with a godly person
whom you know to be a keeper
of the commandments,
who is like-minded with yourself,
and who will grieve with you if
you fail.

13 And heed[l] the counsel of your
own heart,
for no one is more faithful to
you than it is.

14 For our own mind sometimes
keeps us better informed
than seven sentinels sitting high
on a watchtower.

15 But above all pray to the Most
High
that he may direct your way in
truth.

16 Discussion is the beginning of
every work,
and counsel precedes every
undertaking.

17 The mind is the root of all
conduct;

18 it sprouts four branches,[m]
good and evil, life and death;
and it is the tongue that
continually rules them.

19 Some people may be clever
enough to teach many,
and yet be useless to themselves.

20 A skillful speaker may be hated;
he will be destitute of all food,

21 for the Lord has withheld the gift
of charm,
since he is lacking in all
wisdom.

22 If a person is wise to his own
advantage,
the fruits of his good sense will
be praiseworthy.[n]

23 A wise person instructs his own
people,
and the fruits of his good sense
will endure.

24 A wise person will have praise
heaped upon him,
and all who see him will call
him happy.

25 The days of a person's life are
numbered,
but the days of Israel are
without number.

---

i Heb: Gk *in your heart*   *j* Heb: Gk *him in
your wealth*   *k* Heb: Gk *gratitude*   *l* Heb: Gk
establish   *m* Heb: Gk *As a clue to changes of
heart four kinds of destiny appear*   *n* Other
ancient witnesses read *trustworthy*

26 One who is wise among his
 people will inherit honor, *o*
 and his name will live forever.

27 My child, test yourself while you
 live;
 see what is bad for you and do
 not give in to it.
28 For not everything is good for
 everyone,
 and no one enjoys everything.
29 Do not be greedy for every
 delicacy,
 and do not eat without restraint;
30 for overeating brings sickness,
 and gluttony leads to nausea.
31 Many have died of gluttony,
 but the one who guards against
 it prolongs his life.

38 Honor physicians for their
 services,
 for the Lord created them;
2 for their gift of healing comes
 from the Most High,
 and they are rewarded by the
 king.
3 The skill of physicians makes
 them distinguished,
 and in the presence of the great
 they are admired.
4 The Lord created medicines out of
 the earth,
 and the sensible will not despise
 them.
5 Was not water made sweet with a
 tree
 in order that its*p* power might
 be known?
6 And he gave skill to human beings
 that he*q* might be glorified in
 his marvelous works.
7 By them the physician*r* heals and
 takes away pain;
8 the pharmacist makes a mixture
 from them.
 God's*s* works will never be
 finished;

 and from him health*t* spreads
 over all the earth.

9 My child, when you are ill, do
 not delay,
 but pray to the Lord, and he
 will heal you.
10 Give up your faults and direct
 your hands rightly,
 and cleanse your heart from all
 sin.
11 Offer a sweet-smelling sacrifice,
 and a memorial portion of
 choice flour,
 and pour oil on your offering,
 as much as you can
 afford. *u*
12 Then give the physician his place,
 for the Lord created him;
 do not let him leave you, for
 you need him.
13 There may come a time when
 recovery lies in the hands of
 physicians, *v*
14 for they too pray to the Lord
 that he grant them success in
 diagnosis*w*
 and in healing, for the sake of
 preserving life.
15 He who sins against his Maker,
 will be defiant toward the
 physician. *x*

16 My child, let your tears fall for
 the dead,
 and as one in great pain begin
 the lament.
 Lay out the body with due
 ceremony,
 and do not neglect the burial.
17 Let your weeping be bitter and
 your wailing fervent;

o Other ancient authorities read *confidence*
p Or *his*   *q* Or *they*   *r* Heb: Gk *he*
s Gk *His*   *t* Or *peace*   *u* Heb: Lat lacks *as
much as you can afford*; Meaning of Gk uncertain
v Gk *in their hands*   *w* Heb: Gk *rest*
x Heb: Gk *may he fall into the hands of the
physician*

38.1–15: Concerning physicians. Their
work is a continuation of God's creative work
(v. 8).

38.16–23: On mourning for the dead
(22.11–12).

make your mourning worthy of
the departed,
for one day, or two, to avoid
criticism;
then be comforted for your
grief.

18 For grief may result in death,
and a sorrowful heart saps one's
strength.

19 When a person is taken away,
sorrow is over;
but the life of the poor weighs
down the heart.

20 Do not give your heart to grief;
drive it away, and remember
your own end.

21 Do not forget, there is no coming
back;
you do the dead*y* no good, and
you injure yourself.

22 Remember his*z* fate, for yours is
like it;
yesterday it was his,*a* and today
it is yours.

23 When the dead is at rest, let his
remembrance rest too,
and be comforted for him when
his spirit has departed.

24 The wisdom of the scribe depends
on the opportunity of
leisure;
only the one who has little
business can become wise.

25 How can one become wise who
handles the plow,
and who glories in the shaft of a
goad,
who drives oxen and is occupied
with their work,
and whose talk is about bulls?

26 He sets his heart on plowing
furrows,
and he is careful about fodder
for the heifers.

27 So too is every artisan and master
artisan
who labors by night as well as
by day;
those who cut the signets of seals,
each is diligent in making a
great variety;
they set their heart on painting a
lifelike image,
and they are careful to finish
their work.

28 So too is the smith, sitting by the
anvil,
intent on his iron-work;
the breath of the fire melts his
flesh,
and he struggles with the heat
of the furnace;
the sound of the hammer deafens
his ears,*b*
and his eyes are on the pattern
of the object.
He sets his heart on finishing his
handiwork,
and he is careful to complete its
decoration.

29 So too is the potter sitting at his
work
and turning the wheel with his
feet;
he is always deeply concerned
over his products,
and he produces them in
quantity.

30 He molds the clay with his arm
and makes it pliable with his
feet;
he sets his heart to finish the
glazing,
and he takes care in firing*c* the
kiln.

y Gk *him*   *z* Heb: Gk *my*   *a* Heb: Gk
mine   *b* Cn: Gk *renews his ear*   *c* Cn: Gk
cleaning

38.24–39.11: Various craftsmen contrasted with the scribe, a student of divine
wisdom. Sirach shows high regard for the
farmer and seal-maker, the *smith* and the *potter,* and he describes their work with enthusiasm. But the highest vocation is that of the
scribe (v. 24) who is concerned with the *law,
wisdom* (i.e. the Writings), and *prophecies;* this
reflects the threefold division of the Hebrew
Bible (39.1–3; see also the prologue by his
grandson).

31 All these rely on their hands,
and all are skillful in their own
work.

32 Without them no city can be
inhabited,
and wherever they live, they
will not go hungry. *d*
Yet they are not sought out for
the council of the people, *e*

33 nor do they attain eminence in
the public assembly.
They do not sit in the judge's seat,
nor do they understand the
decisions of the courts;
they cannot expound discipline or
judgment,
and they are not found among
the rulers. *f*

34 But they maintain the fabric of the
world,
and their concern is for *g* the
exercise of their trade.

How different the one who
devotes himself
to the study of the law of the
Most High!

39 He seeks out the wisdom of all
the ancients,
and is concerned with
prophecies;

2 he preserves the sayings of the
famous
and penetrates the subtleties of
parables;

3 he seeks out the hidden meanings
of proverbs
and is at home with the
obscurities of parables.

4 He serves among the great
and appears before rulers;
he travels in foreign lands
and learns what is good and evil
in the human lot.

5 He sets his heart to rise early
to seek the Lord who made
him,
and to petition the Most High;

he opens his mouth in prayer
and asks pardon for his sins.

6 If the great Lord is willing,
he will be filled with the spirit
of understanding;
he will pour forth words of
wisdom of his own
and give thanks to the Lord in
prayer.

7 The Lord *h* will direct his counsel
and knowledge,
as he meditates on his mysteries.

8 He will show the wisdom of what
he has learned,
and will glory in the law of the
Lord's covenant.

9 Many will praise his
understanding;
it will never be blotted out.
His memory will not disappear,
and his name will live through
all generations.

10 Nations will speak of his wisdom,
and the congregation will
proclaim his praise.

11 If he lives long, he will leave a
name greater than a
thousand,
and if he goes to rest, it is
enough *i* for him.

12 I have more on my mind to
express;
I am full like the full moon.

13 Listen to me, my faithful children,
and blossom
like a rose growing by a stream
of water.

14 Send out fragrance like incense,
and put forth blossoms like a
lily.

d Syr: Gk *and people can neither live nor walk there*
e Most ancient authorities lack this line
f Cn: Gk *among parables* *g* Syr: Gk *prayer is in*
h Gk *He himself* *i* Cn: Meaning of Gk
uncertain

39.12–35: **A hymn of praise** about the
goodness and purposefulness of creation (see
vv. 16, 21, 33–35). Both *good* and *bad* things
are used by God for the divine purpose
(vv. 25–31). See 33.7–15.

Scatter the fragrance, and sing a
hymn of praise;
bless the Lord for all his works.
15 Ascribe majesty to his name
and give thanks to him with
praise,
with songs on your lips, and with
harps;
this is what you shall say in
thanksgiving:

16 "All the works of the Lord are
very good,
and whatever he commands will
be done at the appointed
time.
17 No one can say, 'What is this?' or
'Why is that?'—
for at the appointed time all
such questions will be
answered.
At his word the waters stood in a
heap,
and the reservoirs of water at
the word of his mouth.
18 When he commands, his every
purpose is fulfilled,
and none can limit his saving
power.
19 The works of all are before him,
and nothing can be hidden from
his eyes.
20 From the beginning to the end of
time he can see everything,
and nothing is too marvelous
for him.
21 No one can say, 'What is this?' or
'Why is that?'—
for everything has been created
for its own purpose.

22 "His blessing covers the dry land
like a river,
and drenches it like a flood.
23 But his wrath drives out the
nations,
as when he turned a watered
land into salt.
24 To the faithful his ways are
straight,
but full of pitfalls for the
wicked.

25 From the beginning good things
were created for the good,
but for sinners good things and
bad.*j*
26 The basic necessities of human life
are water and fire and iron and
salt
and wheat flour and milk and
honey,
the blood of the grape and oil
and clothing.
27 All these are good for the godly,
but for sinners they turn into
evils.

28 "There are winds created for
vengeance,
and in their anger they can
dislodge mountains;*k*
on the day of reckoning they will
pour out their strength
and calm the anger of their
Maker.
29 Fire and hail and famine and
pestilence,
all these have been created for
vengeance;
30 the fangs of wild animals and
scorpions and vipers,
and the sword that punishes the
ungodly with destruction.
31 They take delight in doing his
bidding,
always ready for his service on
earth;
and when their time comes they
never disobey his
command."

32 So from the beginning I have been
convinced of all this
and have thought it out and left
it in writing:
33 All the works of the Lord are
good,
and he will supply every need in
its time.
34 No one can say, "This is not as
good as that,"

j Heb Lat: Gk *sinners bad things*   *k* Heb Syr:
Gk *can scourge mightily*

for everything proves good in
  its appointed time.
35 So now sing praise with all your
  heart and voice,
  and bless the name of the Lord.

40 Hard work was created for
  everyone,
  and a heavy yoke is laid on the
  children of Adam,
from the day they come forth
  from their mother's womb
until the day they return to[l] the
  mother of all the living.[m]
2 Perplexities and fear of heart are
  theirs,
  and anxious thought of the day
  of their death.
3 From the one who sits on a
  splendid throne
  to the one who grovels in dust
  and ashes,
4 from the one who wears purple
  and a crown
  to the one who is clothed in
  burlap,
5 there is anger and envy and
  trouble and unrest,
  and fear of death, and fury and
  strife.
And when one rests upon his bed,
  his sleep at night confuses his
  mind.
6 He gets little or no rest;
  he struggles in his sleep as he
  did by day.[n]
He is troubled by the visions of
  his mind
  like one who has escaped from
  the battlefield.
7 At the moment he reaches safety
  he wakes up,
  astonished that his fears were
  groundless.
8 To all creatures, human and
  animal,
  but to sinners seven times more,

9 come death and bloodshed and
  strife and sword,
  calamities and famine and ruin
  and plague.
10 All these were created for the
  wicked,
  and on their account the flood
  came.
11 All that is of earth returns to
  earth,
  and what is from above returns
  above.[o]

12 All bribery and injustice will be
  blotted out,
  but good faith will last forever.
13 The wealth of the unjust will dry
  up like a river,
  and crash like a loud clap of
  thunder in a storm.
14 As a generous person has cause to
  rejoice,
  so lawbreakers will utterly fail.
15 The children of the ungodly put
  out few branches;
  they are unhealthy roots on
  sheer rock.
16 The reeds by any water or river
  bank
  are plucked up before any grass;
17 but kindness is like a garden of
  blessings,
  and almsgiving endures forever.

18 Wealth and wages make life
  sweet,[p]
  but better than either is finding
  a treasure.
19 Children and the building of a city
  establish one's name,
  but better than either is the one
  who finds wisdom.

l Other Gk and Lat authorities read *are
buried in*   *m* Heb: Gk *of all*   *n* Arm: Meaning
of Gk uncertain   *o* Heb Syr: Gk Lat *from the
waters returns to the sea*   *p* Heb: Gk *Life is sweet
for the self-reliant worker*

**40.1–41.13: Reflections on the human
condition,** its miseries (40.1–17; 40.28–
41.13) and its joys (40.18–27). **40.10:** A typical
expression of the traditional view of retribution (39.29–31).

40.11: *What is from above,* the life-breath of
God (Gen 2.7; Eccl 12.7). **27:** The climax of
the ten "better"-sayings is *fear of the Lord* (see
1.11–30).

Cattle and orchards make one
    prosperous;*q*
  but a blameless wife is
    accounted better than
    either.
20 Wine and music gladden the heart,
  but the love of friends*r* is better
    than either.
21 The flute and the harp make sweet
    melody,
  but a pleasant voice is better
    than either.
22 The eye desires grace and beauty,
  but the green shoots of grain
    more than either.
23 A friend or companion is always
    welcome,
  but a sensible wife*s* is better
    than either.
24 Kindred and helpers are for a time
    of trouble,
  but almsgiving rescues better
    than either.
25 Gold and silver make one stand
    firm,
  but good counsel is esteemed
    more than either.
26 Riches and strength build up
    confidence,
  but the fear of the Lord is better
    than either.
  There is no want in the fear of the
    Lord,
  and with it there is no need to
    seek for help.
27 The fear of the Lord is like a
    garden of blessing,
  and covers a person better than
    any glory.
28 My child, do not lead the life of a
    beggar;
  it is better to die than to beg.
29 When one looks to the table of
    another,
  one's way of life cannot be
    considered a life.

One loses self-respect with another
    person's food,
  but one who is intelligent and
    well instructed guards
    against that.
30 In the mouth of the shameless
    begging is sweet,
  but it kindles a fire inside him.

41 O death, how bitter is the
    thought of you
  to the one at peace among
    possessions,
  who has nothing to worry about
    and is prosperous in
    everything,
  and still is vigorous enough to
    enjoy food!
2 O death, how welcome is your
    sentence
  to one who is needy and failing
    in strength,
  worn down by age and anxious
    about everything;
  to one who is contrary, and has
    lost all patience!
3 Do not fear death's decree for you;
  remember those who went
    before you and those who
    will come after.
4 This is the Lord's decree for all
    flesh;
  why then should you reject the
    will of the Most High?
  Whether life lasts for ten years or
    a hundred or a thousand,
  there are no questions asked in
    Hades.

5 The children of sinners are
    abominable children,
  and they frequent the haunts of
    the ungodly.

q Heb Syr: Gk lacks *but better .. prosperous*
r Heb: Gk *wisdom*   *s* Heb Compare Syr: Gk
wife with her husband

41.3–4: Resignation to death is characteristic of Old Testament thought (see 14.11–19;
17.25–28; 38.16–23). **5–13:** A contrast between the fate of the virtuous and the wicked.

Verses 8–10 show that Sirach has in mind
those who abandon the Law for another, perhaps Greek, way of life.

6 The inheritance of the children of
      sinners will perish,
   and on their offspring will be a
      perpetual disgrace.
7 Children will blame an ungodly
      father,
   for they suffer disgrace because
      of him.
8 Woe to you, the ungodly,
   who have forsaken the law of
      the Most High God!
9 If you have children, calamity will
      be theirs;
   you will beget them only for
      groaning.
   When you stumble, there is lasting
      joy;*t*
   and when you die, a curse is
      your lot.
10 Whatever comes from earth
      returns to earth;
   so the ungodly go from curse to
      destruction.

11 The human body is a fleeting
      thing,
   but a virtuous name will never
      be blotted out.*u*
12 Have regard for your name, since
      it will outlive you
   longer than a thousand hoards
      of gold.
13 The days of a good life are
      numbered,
   but a good name lasts forever.

14 My children, be true to your
      training and be at peace;
   hidden wisdom and unseen
      treasure—
   of what value is either?
15 Better are those who hide their
      folly
   than those who hide their
      wisdom.
16 Therefore show respect for my
      words;

for it is not good to feel shame in
      every circumstance,
   nor is every kind of abashment
      to be approved.*v*

17 Be ashamed of sexual immorality,
      before your father or
      mother;
   and of a lie, before a prince or a
      ruler;
18 of a crime, before a judge or
      magistrate;
   and of a breach of the law,
      before the congregation and
      the people;
   of unjust dealing, before your
      partner or your friend;
19 and of theft, in the place where
      you live.
   Be ashamed of breaking an oath or
      agreement,*w*
   and of leaning on your elbow at
      meals;
   of surliness in receiving or giving,
20   and of silence, before those who
      greet you;
   of looking at a prostitute,
21   and of rejecting the appeal of a
      relative;
   of taking away someone's portion
      or gift,
   and of gazing at another man's
      wife;
22 of meddling with his servant-
      girl—
   and do not approach her bed;
   of abusive words, before friends—
   and do not be insulting after
      making a gift.

42 Be ashamed of repeating what
      you hear,

t Heb: Meaning of Gk uncertain   *u* Heb: Gk
*People grieve over the death of the body, but the bad
name of sinners will be blotted out*   *v* Heb: Gk *and
not everything is confidently esteemed by everyone*
w Heb: Gk *before the truth of God and the covenant*

41.14–42.8: A poem on true shame
(41.16–42.1) **and false shame** (42.2–8).
**42.9–14: A father's concern for his
daughter** (7.24–25; 25.13–26.18).

**42.15–43.33: In praise of God, the om-
nipotent and omniscient Creator. 42.15:**
The *works of the Lord* in creation are meant.

and of betraying secrets.
Then you will show proper shame,
and will find favor with
everyone.

Of the following things do not be
ashamed,
and do not sin to save face:

2 Do not be ashamed of the law of
the Most High and his
covenant,
and of rendering judgment to
acquit the ungodly;

3 of keeping accounts with a partner
or with traveling
companions,
and of dividing the inheritance
of friends;

4 of accuracy with scales and
weights,
and of acquiring much or little;

5 of profit from dealing with
merchants,
and of frequent disciplining of
children,
and of drawing blood from the
back of a wicked slave.

6 Where there is an untrustworthy
wife, a seal is a good thing;
and where there are many
hands, lock things up.

7 When you make a deposit, be sure
it is counted and weighed,
and when you give or receive,
put it all in writing.

8 Do not be ashamed to correct the
stupid or foolish
or the aged who are guilty of
sexual immorality.
Then you will show your sound
training,
and will be approved by all.

9 A daughter is a secret anxiety to
her father,
and worry over her robs him of
sleep;
when she is young, for fear she
may not marry,
or if married, for fear she may
be disliked;

10 while a virgin, for fear she may be
seduced
and become pregnant in her
father's house;
or having a husband, for fear she
may go astray,
or, though married, for fear she
may be barren.

11 Keep strict watch over a
headstrong daughter,
or she may make you a
laughingstock to your
enemies,
a byword in the city and the
assembly of[x] the people,
and put you to shame in public
gatherings.[y]
See that there is no lattice in her
room,
no spot that overlooks the
approaches to the house.[z]

12 Do not let her parade her beauty
before any man,
or spend her time among
married women;[a]

13 for from garments comes the moth,
and from a woman comes
woman's wickedness.

14 Better is the wickedness of a man
than a woman who does
good;
it is woman who brings shame
and disgrace.

15 I will now call to mind the works
of the Lord,
and will declare what I have
seen.
By the word of the Lord his
works are made;
and all his creatures do his
will.[b]

16 The sun looks down on
everything with its light,
and the work of the Lord is full
of his glory.

x Heb: Meaning of Gk uncertain   y Heb: Gk
to shame before the great multitude   z Heb: Gk
lacks *See . . . house*   a Heb: Meaning of Gk
uncertain   b Syr Compare Heb: most Gk
witnesses lack *and all . . . will*

17 The Lord has not empowered
even his holy ones
to recount all his marvelous
works,
which the Lord the Almighty has
established
so that the universe may stand
firm in his glory.
18 He searches out the abyss and the
human heart;
he understands their innermost
secrets.
For the Most High knows all that
may be known;
he sees from of old the things
that are to come. *c*
19 He discloses what has been and
what is to be,
and he reveals the traces of
hidden things.
20 No thought escapes him,
and nothing is hidden from
him.
21 He has set in order the splendors
of his wisdom;
he is from all eternity one and
the same.
Nothing can be added or taken
away,
and he needs no one to be his
counselor.
22 How desirable are all his works,
and how sparkling they are to
see! *d*
23 All these things live and remain
forever;
each creature is preserved to
meet a particular need. *e*
24 All things come in pairs, one
opposite the other,
and he has made nothing
incomplete.
25 Each supplements the virtues of
the other.
Who could ever tire of seeing
his glory?

43 The pride of the higher realms
is the clear vault of the sky,
as glorious to behold as the
sight of the heavens.
2 The sun, when it appears,
proclaims as it rises
what a marvelous instrument it
is, the work of the Most
High.
3 At noon it parches the land,
and who can withstand its
burning heat?
4 A man tending *f* a furnace works
in burning heat,
but three times as hot is the sun
scorching the mountains;
it breathes out fiery vapors,
and its bright rays blind the
eyes.
5 Great is the Lord who made it;
at his orders it hurries on its
course.
6 It is the moon that marks the
changing seasons, *g*
governing the times, their
everlasting sign.
7 From the moon comes the sign for
festal days,
a light that wanes when it
completes its course.
8 The new moon, as its name
suggests, renews itself; *h*
how marvelous it is in this
change,
a beacon to the hosts on high,
shining in the vault of the
heavens!

9 The glory of the stars is the
beauty of heaven,

c Heb: Gk *he sees the sign(s) of the age*   *d* Meaning
of Gk uncertain   *e* Heb: Gk *forever for every need,
and all are obedient*   *f* Other ancient authorities
read *blowing upon*   *g* Heb: Meaning of Gk
uncertain   *h* Heb: Gk *The month is named after
the moon*

42.17: *Holy ones,* the members of the heavenly court, or angels, as in Job 5.1; 15.15; Ps
89.8. Even the angels are unable adequately to
declare God's *marvelous works.*
42.18–21: The divine omniscience extends
to the *abyss* and its chaotic forces, and to the
human heart (see Prov 15.11), to past and
future (v. 19), because God is *everlasting*
(v. 21). **22–25:** The beauty and splendor of
his *works* are increased by their purposefulness (v. 23), and their polarity (v. 24; see
33.15).

a glittering array in the heights
of the Lord.
10 On the orders of the Holy One
they stand in their
appointed places;
they never relax in their
watches.
11 Look at the rainbow, and praise
him who made it;
it is exceedingly beautiful in its
brightness.
12 It encircles the sky with its
glorious arc;
the hands of the Most High
have stretched it out.

13 By his command he sends the
driving snow
and speeds the lightnings of his
judgment.
14 Therefore the storehouses are
opened,
and the clouds fly out like birds.
15 In his majesty he gives the clouds
their strength,
and the hailstones are broken in
pieces.
17a The voice of his thunder rebukes
the earth;
16 when he appears, the mountains
shake.
At his will the south wind blows;
17b so do the storm from the north
and the whirlwind.
He scatters the snow like birds
flying down,
and its descent is like locusts
alighting.
18 The eye is dazzled by the beauty
of its whiteness,
and the mind is amazed as it
falls.
19 He pours frost over the earth like
salt,
and icicles form like pointed
thorns.

20 The cold north wind blows,
and ice freezes on the water;
it settles on every pool of water,
and the water puts it on like a
breastplate.
21 He consumes the mountains and
burns up the wilderness,
and withers the tender grass like
fire.
22 A mist quickly heals all things;
the falling dew gives
refreshment from the heat.

23 By his plan he stilled the deep
and planted islands in it.
24 Those who sail the sea tell of its
dangers,
and we marvel at what we hear.
25 In it are strange and marvelous
creatures,
all kinds of living things, and
huge sea-monsters.
26 Because of him each of his
messengers succeeds,
and by his word all things hold
together.

27 We could say more but could
never say enough;
let the final word be: "He is the
all."
28 Where can we find the strength to
praise him?
For he is greater than all his
works.
29 Awesome is the Lord and very
great,
and marvelous is his power.
30 Glorify the Lord and exalt him as
much as you can,
for he surpasses even that.
When you exalt him, summon all
your strength,
and do not grow weary, for you
cannot praise him enough.

43.1–12: The splendor of heavenly phenomena: *sky, sun, moon* (the Jews followed
a lunar calendar), *stars,* and *rainbow.*
43.13–26: A list of various things in nature
that fulfill the divine will. **23–25:** The *deep*
or *sea* occurs frequently in the Bible as the
personified power of chaos with which Rahab
or Leviathan are associated (Ps 104.24–26).
27: *He is the all,* in the sense that all creatures
reveal the divine presence.

31 Who has seen him and can
   describe him?
   Or who can extol him as he is?
32 Many things greater than these lie
   hidden,
   for I[i] have seen but few of his
   works.
33 For the Lord has made all things,
   and to the godly he has given
   wisdom.

HYMN IN HONOR OF OUR ANCESTORS[j]

44 Let us now sing the praises of
   famous men,
   our ancestors in their
   generations.
2 The Lord apportioned to them[k]
   great glory,
   his majesty from the beginning.
3 There were those who ruled in
   their kingdoms,
   and made a name for themselves
   by their valor;
   those who gave counsel because
   they were intelligent;
   those who spoke in prophetic
   oracles;
4 those who led the people by their
   counsels
   and by their knowledge of the
   people's lore;
   they were wise in their words of
   instruction;
5 those who composed musical
   tunes,
   or put verses in writing;
6 rich men endowed with resources,
   living peacefully in their
   homes—
7 all these were honored in their
   generations,
   and were the pride of their
   times.
8 Some of them have left behind a
   name,

so that others declare their
   praise.
9 But of others there is no memory;
   they have perished as though
   they had never existed;
   they have become as though they
   had never been born,
   they and their children after
   them.
10 But these also were godly men,
   whose righteous deeds have not
   been forgotten;
11 their wealth will remain with their
   descendants,
   and their inheritance with their
   children's children.[l]
12 Their descendants stand by the
   covenants;
   their children also, for their
   sake.
13 Their offspring will continue
   forever,
   and their glory will never be
   blotted out.
14 Their bodies are buried in peace,
   but their name lives on
   generation after generation.
15 The assembly declares[m] their
   wisdom,
   and the congregation proclaims
   their praise.

16 Enoch pleased the Lord and was
   taken up,
   an example of repentance to all
   generations.

17 Noah was found perfect and
   righteous;
   in the time of wrath he kept the
   race alive;[n]

---

i Heb: Gk *we*   j This title is included in the
Gk text.   k Heb: Gk *created*   l Heb Compare
Lat Syr: Meaning of Gk uncertain   m Heb: Gk
Peoples declare   n Heb: Gk *was taken in exchange*

44.1–50.24: In praise of famous men.
Sirach celebrates the covenant with the patriarchs and Israel by commenting on the heroes of Israel's history.
44.1–15: Introduction. **1:** *Famous men,* Hebrew and Syriac, "men of piety." **3–7:** Sirach

lists twelve classes of heroes who were *the pride of their times,* and will be exemplified in the names to follow. **9–10:** Some good people have left no memorial, but they will not be forgotten (by God).
44.16–18: Enoch. The popularity of spec-

therefore a remnant was left on
the earth
when the flood came.
18 Everlasting covenants were made
with him
that all flesh should never again
be blotted out by a flood.

19 Abraham was the great father of a
multitude of nations,
and no one has been found like
him in glory.
20 He kept the law of the Most
High,
and entered into a covenant
with him;
he certified the covenant in his
flesh,
and when he was tested he
proved faithful.
21 Therefore the Lord[o] assured him
with an oath
that the nations would be
blessed through his
offspring;
that he would make him as
numerous as the dust of the
earth,
and exalt his offspring like the
stars,
and give them an inheritance from
sea to sea
and from the Euphrates[p] to the
ends of the earth.
22 To Isaac also he gave the same
assurance
for the sake of his father
Abraham.
The blessing of all people and the
covenant
23 he made to rest on the head of
Jacob;
he acknowledged him with his
blessings,

and gave him his inheritance;
he divided his portions,
and distributed them among
twelve tribes.

From his descendants the Lord[o]
brought forth a godly man,
who found favor in the sight of
all

45 [1] and was beloved by God and
people,
Moses, whose memory is
blessed.
2 He made him equal in glory to the
holy ones,
and made him great, to the
terror of his enemies.
3 By his words he performed swift
miracles;[q]
the Lord[o] glorified him in the
presence of kings.
He gave him commandments for
his people,
and revealed to him his glory.
4 For his faithfulness and meekness
he consecrated him,
choosing him out of all
humankind.
5 He allowed him to hear his voice,
and led him into the dark cloud,
and gave him the commandments
face to face,
the law of life and knowledge,
so that he might teach Jacob the
covenant,
and Israel his decrees.

6 He exalted Aaron, a holy man like
Moses[r]
who was his brother, of the
tribe of Levi.

o Gk *he*   p Syr: Heb Gk *River*   q Heb: Gk
caused signs to cease   r Gk *him*

ulation about *Enoch* in Sirach's time may be
the reason he heads the list (although the
name is lacking in several ancient witnesses),
and closes it in 49.14. *Noah* deserves early
mention as the second founder of the human
race after the flood.
**44.19–47.11: Others with whom God
has made a covenant:** Abraham, Isaac, Ja-
cob/Israel, Moses, Aaron, Phinehas, and David. The glorification of Aaron reflects the
ascendancy of the high priesthood in Sirach's
day. Much more space is given to him and
Phinehas (and then to Simeon the high priest
in ch 50) than to Moses; note the prayer for
high priests in 45.26.

7 He made an everlasting covenant
    with him,
    and gave him the priesthood of
        the people.
He blessed him with stateliness,
    and put a glorious robe on him.
8 He clothed him in perfect splendor,
    and strengthened him with the
        symbols of authority,
    the linen undergarments, the
        long robe, and the ephod.
9 And he encircled him with
        pomegranates,
    with many golden bells all
        around,
to send forth a sound as he walked,
    to make their ringing heard in
        the temple
    as a reminder to his people;
10 with the sacred vestment, of gold
        and violet
    and purple, the work of an
        embroiderer;
with the oracle of judgment, Urim
        and Thummim;
11   with twisted crimson, the work
        of an artisan;
with precious stones engraved like
        seals,
    in a setting of gold, the work of
        a jeweler,
    to commemorate in engraved
        letters
    each of the tribes of Israel;
12 with a gold crown upon his
        turban,
    inscribed like a seal with
        "Holiness,"
    a distinction to be prized, the
        work of an expert,
    a delight to the eyes, richly
        adorned.
13 Before him such beautiful things
        did not exist.
    No outsider ever put them on,
but only his sons
    and his descendants in
        perpetuity.
14 His sacrifices shall be wholly
        burned
    twice every day continually.
15 Moses ordained him,
    and anointed him with holy oil;

it was an everlasting covenant for
        him
    and for his descendants as long
        as the heavens endure,
to minister to the Lord[s] and serve
        as priest
    and bless his people in his name.
16 He chose him out of all the living
    to offer sacrifice to the Lord,
incense and a pleasing odor as a
        memorial portion,
    to make atonement for the[t]
        people.
17 In his commandments he gave him
        authority and statutes and[u]
        judgments,
    to teach Jacob the testimonies,
    and to enlighten Israel with his
        law.
18 Outsiders conspired against him,
    and envied him in the
        wilderness,
Dathan and Abiram and their
        followers
    and the company of Korah, in
        wrath and anger.
19 The Lord saw it and was not
        pleased,
    and in the heat of his anger they
        were destroyed;
he performed wonders against them
    to consume them in flaming fire.
20 He added glory to Aaron
    and gave him a heritage;
he allotted to him the best of the
        first fruits,
    and prepared bread of first fruits
        in abundance;
21 for they eat the sacrifices of the
        Lord,
    which he gave to him and his
        descendants.
22 But in the land of the people he
        has no inheritance,
    and he has no portion among
        the people;
    for the Lord[v] himself is his[w]
        portion and inheritance.

---

s Gk *him*   t Other ancient authorities read *his*
or *your*   u Heb: Gk *authority in covenants of*
v Gk *he*   w Other ancient authorities read
your

23 Phinehas son of Eleazar ranks
  third in glory
  for being zealous in the fear of
    the Lord,
  and standing firm, when the
    people turned away,
  in the noble courage of his soul;
  and he made atonement for
    Israel.
24 Therefore a covenant of friendship
    was established with him,
  that he should be leader of the
    sanctuary and of his people,
  that he and his descendants should
    have
  the dignity of the priesthood
    forever.
25 Just as a covenant was established
    with David
  son of Jesse of the tribe of
    Judah,
  that the king's heritage passes only
    from son to son,
  so the heritage of Aaron is for
    his descendants alone.

26 And now bless the Lord
  who has crowned you with
    glory. ˣ
  May the Lord ʸ grant you wisdom
    of mind
  to judge his people with
    justice,
  so that their prosperity may not
    vanish,
  and that their glory may endure
    through all their generations.

46 Joshua son of Nun was mighty
    in war,
  and was the successor of Moses
    in the prophetic office.
  He became, as his name implies,
  a great savior of God's ᶻ elect,
  to take vengeance on the enemies
    that rose against them,

  so that he might give Israel
    its inheritance.
2 How glorious he was when he
    lifted his hands
  and brandished his sword
    against the cities!
3 Who before him ever stood so
    firm?
  For he waged the wars of the
    Lord.
4 Was it not through him that the
    sun stood still
  and one day became as long as
    two?
5 He called upon the Most High,
    the Mighty One,
  when enemies pressed him on
    every side,
  and the great Lord answered him
    with hailstones of mighty
    power.
6 He overwhelmed that nation in
    battle,
  and on the slope he destroyed
    his opponents,
  so that the nations might know his
    armament,
  that he was fighting in the sight
    of the Lord;
  for he was a devoted follower of
    the Mighty One.
7 And in the days of Moses he
    proved his loyalty,
  he and Caleb son of Jephunneh:
  they opposed the congregation, ᵃ
  restrained the people from sin,
  and stilled their wicked
    grumbling.
8 And these two alone were spared
    out of six hundred thousand
    infantry,
  to lead the people ᵇ into their
    inheritance,

x Heb. Gk lacks *who . . . glory*   y Gk *he*
z Gk *his*   a Other ancient authorities read
the enemy   b Gk *them*

46.1–20: Joshua (vv. 1–7), **Caleb** (vv. 7–
10), **the judges** (vv. 11–12), **and Samuel** (vv. 13–20). **1**: *His name* means "The
Lord is salvation." **2–8**: Josh chs 6–11. **4**: Josh
10.12–14. **6**: Josh 10.11. **7**: Num 14.6–
10; 1 Macc 2.55–56. **8**: Num 11.21; 14.38;
26.65. **9**: Josh 14.6–11.
  46.12: *Bones send forth new life* (49.10), the
meaning is to be interpreted in the light of the
last line of the verse. **13**: *Anointed rulers,* 1 Sam
10.1; 16.13. **15**: 1 Sam 3.19–20. **16–18**: 1 Sam
7.9–11. **19**: 1 Sam 12.3. **20**: 1 Sam 28.18–19.

the land flowing with milk and
honey.

9 The Lord gave Caleb strength,
which remained with him in his
old age,
so that he went up to the hill
country,
and his children obtained it for
an inheritance,

10 so that all the Israelites might see
how good it is to follow the
Lord.

11 The judges also, with their
respective names,
whose hearts did not fall into
idolatry
and who did not turn away from
the Lord—
may their memory be blessed!

12 May their bones send forth new
life from where they lie,
and may the names of those
who have been honored
live again in their children!

13 Samuel was beloved by his Lord;
a prophet of the Lord, he
established the kingdom
and anointed rulers over his
people.

14 By the law of the Lord he judged
the congregation,
and the Lord watched over
Jacob.

15 By his faithfulness he was proved
to be a prophet,
and by his words he became
known as a trustworthy
seer.

16 He called upon the Lord, the
Mighty One,
when his enemies pressed him
on every side,
and he offered in sacrifice a
suckling lamb.

17 Then the Lord thundered from
heaven,
and made his voice heard with a
mighty sound;

18 he subdued the leaders of the
enemy*c*
and all the rulers of the
Philistines.

19 Before the time of his eternal
sleep,
Samuel*d* bore witness before the
Lord and his anointed:
"No property, not so much as a
pair of shoes,
have I taken from anyone!"
And no one accused him.

20 Even after he had fallen asleep, he
prophesied
and made known to the king his
death,
and lifted up his voice from the
ground
in prophecy, to blot out the
wickedness of the people.

47 After him Nathan rose up
to prophesy in the days of
David.

2 As the fat is set apart from the
offering of well-being,
so David was set apart from the
Israelites.

3 He played with lions as though
they were young goats,
and with bears as though they
were lambs of the flock.

4 In his youth did he not kill a
giant,
and take away the people's
disgrace,
when he whirled the stone in the
sling
and struck down the boasting
Goliath?

5 For he called on the Lord, the
Most High,

c Heb: Gk *leaders of the people of Tyre*
d Gk *he*

47.1–22: Nathan (v.1), **David** (vv.2–11),
and Solomon (vv.12–22). **1:** 2 Sam 7.2–3;
12.1; 1 Chr 17.1. **2:** *The fat,* the portion reserved for sacrifice (Lev 3.3–5). **3:** 1 Sam 17.34.
4: 1 Sam 17.49–51.

and he gave strength to his right
arm
to strike down a mighty warrior,
and to exalt the power*ᵉ* of his
people.
6 So they glorified him for the tens
of thousands he conquered,
and praised him for the
blessings bestowed by the
Lord,
when the glorious diadem was
given to him.
7 For he wiped out his enemies on
every side,
and annihilated his adversaries
the Philistines;
he crushed their power*ᵉ* to our
own day.
8 In all that he did he gave thanks
to the Holy One, the Most
High, proclaiming his
glory;
he sang praise with all his heart,
and he loved his Maker.
9 He placed singers before the altar,
to make sweet melody with
their voices.*ᶠ*
10 He gave beauty to the festivals,
and arranged their times
throughout the year,*ᵍ*
while they praised God's*ʰ* holy
name,
and the sanctuary resounded
from early morning.
11 The Lord took away his sins,
and exalted his power*ᵉ* forever;
he gave him a covenant of
kingship
and a glorious throne in Israel.

12 After him a wise son rose up
who because of him lived in
security:*ⁱ*
13 Solomon reigned in an age of
peace,
because God made all his
borders tranquil,

so that he might build a house in
his name
and provide a sanctuary to stand
forever.
14 How wise you were when you
were young!
You overflowed like the Nile*ʲ*
with understanding.
15 Your influence spread throughout
the earth,
and you filled it with proverbs
having deep meaning.
16 Your fame reached to far-off
islands,
and you were loved for your
peaceful reign.
17 Your songs, proverbs, and
parables,
and the answers you gave
astounded the nations.
18 In the name of the Lord God,
who is called the God of Israel,
you gathered gold like tin
and amassed silver like lead.
19 But you brought in women to lie
at your side,
and through your body you
were brought into
subjection.
20 You stained your honor,
and defiled your family line,
so that you brought wrath upon
your children,
and they were grieved*ᵏ* at your
folly,
21 because the sovereignty was
divided
and a rebel kingdom arose out
of Ephraim.
22 But the Lord will never give up
his mercy,
or cause any of his works to
perish;

---

e Gk *horn*   *f* Other ancient authorities add
and daily they sing his praises   *g* Gk *to completion*
h Gk *his*   *i* Heb: Gk *in a broad place*   *j* Heb:
Gk *a river*   *k* Other ancient authorities read *I
was grieved*

47.6: 1 Sam 18.7. **7:** 2 Sam 5.7; 8.1. **9:**
1 Chr 16.4. **11:** 2 Sam 12.13. **13–17:** 1 Kings
4.21–32.

47.14: Compare the address to Elijah in
48.4–11. **18:** 1 Kings 10.21, 27. **19:** 1 Kings
11.1. **21:** 1 Kings 12.15–20. **22:** 2 Sam 7.15; Ps

he will never blot out the
    descendants of his chosen
    one,
or destroy the family line of
    him who loved him.
So he gave a remnant to Jacob,
    and to David a root from his
    own family.

23 Solomon rested with his ancestors,
    and left behind him one of his
    sons,
broad in*l* folly and lacking in
    sense,
Rehoboam, whose policy drove
    the people to revolt.
Then Jeroboam son of Nebat led
    Israel into sin
and started Ephraim on its sinful
    ways.
24 Their sins increased more and
    more,
until they were exiled from their
    land.
25 For they sought out every kind of
    wickedness,
until vengeance came upon
    them.

48 Then Elijah arose, a prophet
    like fire,
    and his word burned like a
    torch.
2 He brought a famine upon them,
    and by his zeal he made them
    few in number.
3 By the word of the Lord he shut
    up the heavens,
    and also three times brought
    down fire.
4 How glorious you were, Elijah, in
    your wondrous deeds!
    Whose glory is equal to yours?

5 You raised a corpse from death
    and from Hades, by the word of
    the Most High.
6 You sent kings down to
    destruction,
    and famous men, from their
    sickbeds.
7 You heard rebuke at Sinai
    and judgments of vengeance at
    Horeb.
8 You anointed kings to inflict
    retribution,
    and prophets to succeed you. *m*
9 You were taken up by a
    whirlwind of fire,
    in a chariot with horses of fire.
10 At the appointed time, it is
    written, you are destined*n*
to calm the wrath of God before
    it breaks out in fury,
to turn the hearts of parents to
    their children,
    and to restore the tribes of
    Jacob.
11 Happy are those who saw you
    and were adorned*o* with your
    love!
    For we also shall surely live. *p*

12 When Elijah was enveloped in the
    whirlwind,
Elisha was filled with his spirit.
He performed twice as many
    signs,
    and marvels with every
    utterance of his mouth. *q*
Never in his lifetime did he
    tremble before any ruler,

l Heb (with a play on the name Rehoboam)
Syr: Gk *the people's*   *m* Heb: Gk *him*   *n* Heb:
Gk *are for reproofs*   *o* Other ancient authorities
read *and have died*   *p* Text and meaning of
Gk uncertain   *q* Heb: Gk lacks *He performed . . .
mouth*

89.33. Or . . . *perish,* Hebrew "He will let
none of his words fall to the ground."
    47.23–48.14: Kings and prophets. Solo-
mon's son, Rehoboam, and Jeroboam in the
northern kingdom are mentioned briefly, and
the deeds of Elijah and Elisha are described in
detail. **23:** 1 Kings 11.43; 12.10–14; for Jeroboam, 1 Kings 12.28–30. **24:** 2 Kings 17.6,
18.

48.1: 1 Kings 17.1. *Torch,* Hebrew "furnace." **2:** Jas 5.17. **3:** 1 Kings 18.38; 2 Kings
1.10–12. **5:** 1 Kings 17.21–22. **6:** 2 Kings
1.16.
    48.7: 1 Kings 19.8. **8:** 1 Kings 19.15–16. **9:**
2 Kings 2.11. **10:** Mal 4.5–6. **12:** 2 Kings 2.9,
13. **13:** 2 Kings 13.20–21. **15:** 2 Kings 18.11–
12.
    48.15–49.13: In the kingdom of Judah,

nor could anyone intimidate
him at all.
13 Nothing was too hard for him,
and when he was dead, his body
prophesied.
14 In his life he did wonders,
and in death his deeds were
marvelous.

15 Despite all this the people did not
repent,
nor did they forsake their sins,
until they were carried off as
plunder from their land,
and were scattered over all the
earth.
The people were left very few in
number,
but with a ruler from the house
of David.
16 Some of them did what was right,
but others sinned more and
more.

17 Hezekiah fortified his city,
and brought water into its
midst;
he tunneled the rock with iron
tools,
and built cisterns for the water.
18 In his days Sennacherib invaded
the country;
he sent his commander* and
departed;
he shook his fist against Zion,
and made great boasts in his
arrogance.
19 Then their hearts were shaken and
their hands trembled,
and they were in anguish, like
women in labor.
20 But they called upon the Lord
who is merciful,
spreading out their hands
toward him.

The Holy One quickly heard them
from heaven,
and delivered them through
Isaiah.
21 The Lord* struck down the camp
of the Assyrians,
and his angel wiped them out.
22 For Hezekiah did what was
pleasing to the Lord,
and he kept firmly to the ways
of his ancestor David,
as he was commanded by the
prophet Isaiah,
who was great and trustworthy
in his visions.
23 In Isaiah's* days the sun went
backward,
and he prolonged the life of the
king.
24 By his dauntless spirit he saw the
future,
and comforted the mourners in
Zion.
25 He revealed what was to occur to
the end of time,
and the hidden things before
they happened.

49 The name* of Josiah is like
blended incense
prepared by the skill of the
perfumer;
his memory* is as sweet as honey
to every mouth,
and like music at a banquet of
wine.
2 He did what was right by
reforming the people,
and removing the wicked
abominations.

r Other ancient authorities add *from Lachish*
s Gk *He*   t Gk *his*   u Heb: Gk *memory*
v Heb: Gk *it*

the heroes are Hezekiah and Isaiah, followed
by Jeremiah, Ezekiel (and Job; see Ezek 14.14),
and the twelve prophets, which probably
constituted a corpus of writings by the time
of Sirach. The heroes of the restoration are
Zerubbabel, Jeshua, and Nehemiah, with
Ezra being notably absent. **48.17:** 2 Kings
20.20. **18:** 2 Kings 18.13, 17; Isa 36.1. *Made
. . . boasts,* Hebrew and Syriac read "blasphemed God." **20:** 2 Kings 19.15–20.
48.21: 2 Kings 19.35; Isa 37.36; 1 Macc
7.41. **22:** 2 Kings 18.3. **23:** 2 Kings 20.10–11;
Isa 38.8. **24–25:** Isa 40.1; 42.9.
49.2a: Hebrew, "For he was grieved over

3 He kept his heart fixed on the
   Lord;
   in lawless times he made
   godliness prevail.

4 Except for David and Hezekiah
   and Josiah,
   all of them were great sinners,
   for they abandoned the law of the
   Most High;
   the kings of Judah came to an
   end.

5 They*w* gave their power to others,
   and their glory to a foreign
   nation,

6 who set fire to the chosen city of
   the sanctuary,
   and made its streets desolate,
   as Jeremiah had foretold. *x*

7 For they had mistreated him,
   who even in the womb had
   been consecrated a prophet,
   to pluck up and ruin and destroy,
   and likewise to build and to
   plant.

8 It was Ezekiel who saw the vision
   of glory,
   which God*y* showed him above
   the chariot of the cherubim.

9 For God*z* also mentioned Job
   who held fast to all the ways of
   justice. *a*

10 May the bones of the Twelve
   Prophets
   send forth new life from where
   they lie,
   for they comforted the people of
   Jacob
   and delivered them with
   confident hope.

11 How shall we magnify
   Zerubbabel?
   He was like a signet ring on the
   right hand,

12 and so was Jeshua son of
   Jozadak;

in their days they built the house
   and raised a temple*b* holy to the
   Lord,
   destined for everlasting glory.

13 The memory of Nehemiah also is
   lasting;
   he raised our fallen walls,
   and set up gates and bars,
   and rebuilt our ruined houses.

14 Few have*c* ever been created on
   earth like Enoch,
   for he was taken up from the
   earth.

15 Nor was anyone ever born like
   Joseph;*d*
   even his bones were cared for.

16 Shem and Seth and Enosh were
   honored,*e*
   but above every other created
   living being was Adam.

50 The leader of his brothers and
   the pride of his people*f*
   was the high priest, Simon son
   of Onias,
   who in his life repaired the house,
   and in his time fortified the
   temple.

2 He laid the foundations for the
   high double walls,
   the high retaining walls for the
   temple enclosure.

3 In his days a water cistern was
   dug,*g*
   a reservoir like the sea in
   circumference.

4 He considered how to save his
   people from ruin,

w Heb *He*   x Gk *by the hand of Jeremiah*
y Gk *He*   z Gk *he*   a Heb Compare Syr:
Meaning of Gk uncertain   b Other ancient
authorities read *people*   c Heb Syr: Gk *No one
has*   d Heb Syr: Gk adds *the leader of his
brothers, the support of the people*   e Heb: Gk
Shem and Seth were honored by people
f Heb Syr: Gk lacks this line. Compare 49.15
g Heb: Meaning of Gk uncertain

our backslidings" (2 Kings 22.11–13). **3:**
2 Kings 23.3, 25. **5–7:** 2 Chr 36.17–19. *Jeremiah,* Jer 1.5–10; 39.8.
   49.8–9: Ezek 1.3–15; 13.11; 38.9, 16, 22.

10: *Bones,* see 46.12 n. **11:** Ezra 3.2; Hag 2.23.
12: Ezra 3.2; Hag 1.12; 2.2; Zech 3.1. **13:** Neh
7.1.
   49.14–16: Conclusion. The mention of

and fortified the city against
siege.
5 How glorious he was, surrounded
by the people,
as he came out of the house of
the curtain.
6 Like the morning star among the
clouds,
like the full moon at the festal
season;[h]
7 like the sun shining on the temple
of the Most High,
like the rainbow gleaming in
splendid clouds;
8 like roses in the days of first
fruits,
like lilies by a spring of water,
like a green shoot on Lebanon
on a summer day;
9 like fire and incense in the censer,
like a vessel of hammered gold
studded with all kinds of
precious stones;
10 like an olive tree laden with fruit,
and like a cypress towering in
the clouds.
11 When he put on his glorious robe
and clothed himself in perfect
splendor,
when he went up to the holy
altar,
he made the court of the
sanctuary glorious.

12 When he received the portions
from the hands of the
priests,
as he stood by the hearth of the
altar
with a garland of brothers around
him,
he was like a young cedar on
Lebanon
surrounded by the trunks of
palm trees.
13 All the sons of Aaron in their
splendor

held the Lord's offering in their
hands
before the whole congregation
of Israel.
14 Finishing the service at the altars,[i]
and arranging the offering to the
Most High, the Almighty,
15 he held out his hand for the cup
and poured a drink offering of
the blood of the grape;
he poured it out at the foot of the
altar,
a pleasing odor to the Most
High, the king of all.
16 Then the sons of Aaron shouted;
they blew their trumpets of
hammered metal;
they sounded a mighty fanfare
as a reminder before the Most
High.
17 Then all the people together
quickly
fell to the ground on their faces
to worship their Lord,
the Almighty, God Most High.

18 Then the singers praised him with
their voices
in sweet and full-toned
melody.[j]
19 And the people of the Lord Most
High offered
their prayers before the Merciful
One,
until the order of worship of the
Lord was ended,
and they completed his ritual.
20 Then Simon[k] came down and
raised his hands
over the whole congregation of
Israelites,
to pronounce the blessing of the
Lord with his lips,

h Heb: Meaning of Gk uncertain   i Other
ancient authorities read *altar*   j Other ancient
authorities read *in sweet melody throughout
the house*   k Gk *he*

Enoch (see also 44.16) and Joseph along with
Shem, Seth, and Enosh leads back to Adam.
14: Gen 5.24; Heb 11.5. **15**: Gen 39.1ff.;
50.25–26. **16**: Gen 5.3, 32.

50.1–24: Simon, son of Onias. 1: Simon II was high priest about 219–196 (Josephus, *Antiquities,* XII. iv. 10). *Onias* is the
Greek form of Johanan. *The house,* of God.

and to glory in his name;

21 and they bowed down in worship
a second time,
to receive the blessing from the
Most High.

22 And now bless the God of all,
who everywhere works great
wonders,
who fosters our growth from
birth,
and deals with us according to
his mercy.

23 May he give us[l] gladness of heart,
and may there be peace in our[m]
days
in Israel, as in the days of old.

24 May he entrust to us his mercy,
and may he deliver us in our[n]
days!

25 Two nations my soul detests,
and the third is not even a
people:

26 Those who live in Seir,[o] and the
Philistines,
and the foolish people that live
in Shechem.

27 Instruction in understanding and
knowledge
I have written in this book,
Jesus son of Eleazar son of Sirach[p]
of Jerusalem,
whose mind poured forth
wisdom.

28 Happy are those who concern
themselves with these
things,
and those who lay them to heart
will become wise.

29 For if they put them into practice,
they will be equal to
anything,
for the fear[q] of the Lord is their
path.

PRAYER OF JESUS SON OF SIRACH[r]

51 I give you thanks, O Lord and
King,
and praise you, O God my
Savior.
I give thanks to your name,
2 for you have been my protector
and helper
and have delivered me from
destruction
and from the trap laid by a
slanderous tongue,
from lips that fabricate lies.
In the face of my adversaries
you have been my helper 3 and
delivered me,
in the greatness of your mercy
and of your name,
from grinding teeth about to
devour me,
from the hand of those seeking
my life,
from the many troubles I
endured,
4 from choking fire on every side,
and from the midst of fire that I
had not kindled,
5 from the deep belly of Hades,
from an unclean tongue and
lying words—
6 the slander of an unrighteous
tongue to the king.
My soul drew near to death,
and my life was on the brink of
Hades below.
7 They surrounded me on every
side,
and there was no one to help
me;

l Other ancient authorities read *you*   *m* Other
ancient authorities read *your*   *n* Other ancient
authorities read *his*   *o* Heb Compare Lat: Gk *on
the mountain of Samaria*   *p* Heb: Meaning of Gk
uncertain   *q* Heb: Other ancient authorities
read *light*   *r* This title is included in the Gk text.

50.16: *Trumpets,* Num 10.2; 31.6. **20:** *The
blessing,* namely Num 6.24–27. *His name,*
only the high priest (and only once a year, on
the Day of Atonement) could utter the ineffable
name, YHWH.
50.22–24: Doxology. **24:** Hebrew, "May
his love abide upon Simeon, and may he keep

in him the covenant of Phinehas; may one
never be cut off from him; and as for his
offspring, (may it be) as the days of heaven."
50.25–26: An invective against Edomites
(*Seir*), pagans or Hellenists (*Philistines*),
and Samaritans (*Shechem*).
50.27–29: A postscript, probably the

I looked for human assistance,
   and there was none.
8 Then I remembered your mercy,
      O Lord,
   and your kindness[s] from of old,
   for you rescue those who wait for
      you
   and save them from the hand of
      their enemies.
9 And I sent up my prayer from the
      earth,
   and begged for rescue from
      death.
10 I cried out, "Lord, you are my
      Father;[t]
   do not forsake me in the days of
      trouble,
   when there is no help against
      the proud.
11 I will praise your name
      continually,
   and will sing hymns of
      thanksgiving."
   My prayer was heard,
12   for you saved me from
         destruction
   and rescued me in time of
      trouble.
   For this reason I thank you and
      praise you,
   and I bless the name of the
      Lord.

Heb adds:
Give thanks to the LORD, for he is
   good,
   for his mercy endures forever;

Give thanks to the God of praises,
   for his mercy endures forever;

Give thanks to the guardian of Israel,
   for his mercy endures forever;

Give thanks to him who formed all
   things,
   for his mercy endures forever;

Give thanks to the redeemer of Israel,
   for his mercy endures forever;

Give thanks to him who gathers the
   dispersed of Israel,
   for his mercy endures forever;

Give thanks to him who rebuilt his
   city and his sanctuary,
   for his mercy endures forever;

Give thanks to him who makes a
   horn to sprout for the house of
   David,
   for his mercy endures forever;

Give thanks to him who has chosen
   the sons of Zadok to be priests,
   for his mercy endures forever;

Give thanks to the shield of Abraham,
   for his mercy endures forever;

Give thanks to the rock of Isaac,
   for his mercy endures forever;

Give thanks to the mighty one of
   Jacob,
   for his mercy endures forever;

Give thanks to him who has chosen
   Zion,
   for his mercy endures forever;

Give thanks to the King of the kings
   of kings,
   for his mercy endures forever;

He has raised up a horn for his
   people,
   praise for all his loyal ones.

For the children of Israel, the people
   close to him.
   Praise the LORD!

---

s Other ancient authorities read *work*   t Heb: Gk
the Father of my lord

---

original ending of the author, who gives his
full name and recommends *this book.*
   51.1–30: Two appendices: A psalm of
thanksgiving (vv.1–12) and a 23-line acrostic poem about Sirach's love for Wisdom
(vv.13–30). Verses 13–20a have been preserved in a psalms scroll from Qumran
Cave II.

13 While I was still young, before I
    went on my travels,
    I sought wisdom openly in my
        prayer.
14 Before the temple I asked for her,
    and I will search for her until
        the end.

15 From the first blossom to the
        ripening grape
    my heart delighted in her;
    my foot walked on the straight
        path;
    from my youth I followed her
        steps.

16 I inclined my ear a little and
        received her,
    and I found for myself much
        instruction.
17 I made progress in her;
    to him who gives wisdom I will
        give glory.

18 For I resolved to live according to
        wisdom,ᵘ
    and I was zealous for the good,
    and I shall never be
        disappointed.
19 My soul grappled with wisdom,ᵘ
    and in my conduct I was strict;ᵛ

    I spread out my hands to the
        heavens,
    and lamented my ignorance of
        her.
20 I directed my soul to her,
    and in purity I found her.

    With her I gained understanding
        from the first;
    therefore I will never be
        forsaken.
21 My heart was stirred to seek her;

therefore I have gained a prize
        possession.
22 The Lord gave me my tongue as a
        reward,
    and I will praise him with it.

23 Draw near to me, you who are
        uneducated,
    and lodge in the house of
        instruction.
24 Why do you say you are lacking
        in these things,ʷ
    and why do you endure such
        great thirst?
25 I opened my mouth and said,
    Acquire wisdomˣ for yourselves
        without money.

26 Put your neck under herʸ yoke,
    and let your souls receive
        instruction;
    it is to be found close by.

27 See with your own eyes that I
        have labored but little
    and found for myself much
        serenity.
28 Hear but a little of my instruction,
    and through me you will
        acquire silver and gold.ᶻ

29 May your soul rejoice in God'sᵃ
        mercy,
    and may you never be ashamed
        to praise him.
30 Do your work in good time,
    and in his own time Godᵇ will
        give you your reward.

u Gk *her*   v Meaning of Gk uncertain
w Cn Compare Heb Syr: Meaning of Gk
uncertain   x Heb: Gk lacks *wisdom*   y Heb:
other ancient authorities read *the*
z Syr Compare Heb: Gk *Get instruction with a
large sum of silver, and you will gain by it much
gold.*   a Gk *his*   b Gk *he*

Baruch

The book of Baruch was probably written sometime between 200 and 60 B.C.; it is set, however, during the Babylonian exile of the early sixth-century B.C., and attributed to Jeremiah's friend and secretary, Baruch son of Neriah (Jer 32.12; 36.4; 43.3; 45.1). Although Jeremiah and Baruch both are reported (Jer 43.1–7) to have been taken to Egypt in 582 B.C., a tradition developed later, which is reflected in these works, that Baruch went to Babylonia. If the book of Daniel was composed in the first half of the second century B.C., then Baruch would have been written after about 150 B.C., because Bar 1.15–2.19 is largely Dan 9.4–19 rewritten. Most of Baruch is made up of pastiches of biblical passages copied or paraphrased (from, e.g., Dan 9, Job 28, and Isa 40–66). Certainly Baruch himself would not have made the numerous mistakes contained in Bar 1.1–14.

Baruch falls into two main sections, each of which is made up of two parts. The first section, in prose, includes an introduction (1.1–14) and a corporate confession of sin (1.15–3.8) for Jews in Jerusalem to recite at the altar there, along with appropriate sacrifices, on various festival days and seasons. The idea of a letter or scroll written in Babylon to be read aloud in Jerusalem is derived from the exchange of letters recorded in Jer 29 and the scroll of Jeremiah's oracles penned by Baruch and read before King Jehoiakim in 605 B.C. (Jer 36). The corporate confession is modeled on Dan 9.4–19 (compare Ezra 9.6–15; Neh 9.6–37).

The second section is made up of two poems. The first (3.9–4.4) is a paean of praise of Wisdom, which though elusive is largely identified as Torah, God's precious gift to Israel. The second comprises an address by Jerusalem to the people of Israel (4.5–29), and by a rhetorical apostrophe to Jerusalem (4.30–5.9), inspired no doubt by Isa 51.17–52.10; 54; and 60–62, where the speaker is also not always clearly defined.

The basic text of the following translation is the Greek Septuagint; there are also ancient Syriac, Latin, Coptic, Ethiopic, Arabic, and Armenian versions based on the Greek. The prose section (1.1–3.8) has long been viewed as translated from a lost Hebrew original; recent research indicates that the poetic sections also derive from Hebrew originals.

There may have been as many as four different authors contributing to what is now called Baruch, one for each of the three major sections and a final redactor. Different names for God are used in the confession and in the poems. The theologies of the confession and of the second poem are quite compatible, but are both so clearly drawn from biblical passages paraphrased or rewritten that it is difficult to discern the particular thinking of the author(s). The first poem, on Wisdom, is similarly drawn from the Bible (Job 28 largely), and thus conceals the mind of its composer; it has, however, a different tone and style from the second. An argument could be made that there were only two basic contributors, and that one of these was the redactor. As a whole literary piece, Baruch would have well served Jewish communities in Judah and the Dispersion during the Seleucid and later eras of suffering and repression.

1 These are the words of the book that Baruch son of Neriah son of Mahseiah son of Zedekiah son of Hasadiah son of Hilkiah wrote in Babylon, 2in the fifth year, on the seventh day of the month, at the time when the Chaldeans took Jerusalem and burned it with fire.

3 Baruch read the words of this book to Jeconiah son of Jehoiakim, king of Judah, and to all the people who came to hear the book, 4and to the nobles and the princes, and to the elders, and to all the people, small and great, all who lived in Babylon by the river Sud.

5 Then they wept, and fasted, and prayed before the Lord; 6they collected as much money as each could give, 7and sent it to Jerusalem to the high priest*a* Jehoiakim son of Hilkiah son of Shallum, and to the priests, and to all the people who were present with him in Jerusalem. 8At the same time, on the tenth day of Sivan, Baruch*b* took the vessels of the house of the Lord, which had been carried away from the temple, to return them to the land of Judah—the silver vessels that Zedekiah son of Josiah, king of Judah, had made, 9after King Nebuchadnezzar of Babylon had carried away from Jerusalem Jeconiah and the princes and the prisoners and the nobles and the people of the land, and brought them to Babylon.

10 They said: Here we send you money; so buy with the money burnt offerings and sin offerings and incense, and prepare a grain offering, and offer them on the altar of the Lord our God; 11and pray for the life of King Nebuchadnezzar of Babylon, and for the life of his son

Belshazzar, so that their days on earth may be like the days of heaven. 12The Lord will give us strength, and light to our eyes; we shall live under the protection*c* of King Nebuchadnezzar of Babylon, and under the protection of his son Belshazzar, and we shall serve them many days and find favor in their sight. 13Pray also for us to the Lord our God, for we have sinned against the Lord our God, and to this day the anger of the Lord and his wrath have not turned away from us. 14And you shall read aloud this scroll that we are sending you, to make your confession in the house of the Lord on the days of the festivals and at appointed seasons.

15 And you shall say: The Lord our God is in the right, but there is open shame on us today, on the people of Judah, on the inhabitants of Jerusalem, 16and on our kings, our rulers, our priests, our prophets, and our ancestors, 17because we have sinned before the Lord. 18We have disobeyed him, and have not heeded the voice of the Lord our God, to walk in the statutes of the Lord that he set before us. 19From the time when the Lord brought our ancestors out of the land of Egypt until today, we have been disobedient to the Lord our God, and we have been negligent, in not heeding his voice. 20So to this day there have clung to us the calamities and the curse that the Lord declared through his servant Moses at the time when he brought our ancestors out of the land of Egypt to give to us a land flowing with milk and

a Gk *the priest*   *b* Gk *he*   *c* Gk *in the shadow*

1.1–14: Historical introduction. 1–2: Authorship and date. **1:** *Baruch,* Jeremiah's secretary (Jer 32.12). **2:** *Fifth year,* after the fall of Jerusalem in 587/6 B.C. **3–4:** The book is read before the exiles. **3:** *Jeconiah,* also called Jehoiachin (2 Kings 24.15; Jer 24.1). **4:** *Sud,* unknown. **1.5–14:** A gift of money, the temple vessels, and the book are sent to Jerusalem. **5:** The word *Lord* occurs only in the first part of the book (1.1–3.8). **7:** *The high priest Jehoia-* kim, otherwise unknown. **8:** *Sivan,* the third month (May–June). For the return of gold and *silver vessels,* see Ezra 1.7–11. **9:** Jer 24.1; 2 Kings 24.10–16. **11:** Jer 29.7. *Belshazzar* was actually the son of Nabonidus. *Like the days of heaven,* without end (Deut 11.21).

1.15–3.8: Confession of sin, for the Jerusalem community (1.15–2.5), and for the exiles (2.6–3.8); compare 1.15–16 and 2.6. **1.15–2.5:** Disobedience brought the judgment of exile.

honey. 21 We did not listen to the voice of the Lord our God in all the words of the prophets whom he sent to us, 22 but all of us followed the intent of our own wicked hearts by serving other gods and doing what is evil in the sight of the Lord our God.

2 So the Lord carried out the threat he spoke against us: against our judges who ruled Israel, and against our kings and our rulers and the people of Israel and Judah. 2 Under the whole heaven there has not been done the like of what he has done in Jerusalem, in accordance with the threats that were[d] written in the law of Moses. 3 Some of us ate the flesh of their sons and others the flesh of their daughters. 4 He made them subject to all the kingdoms around us, to be an object of scorn and a desolation among all the surrounding peoples, where the Lord has scattered them. 5 They were brought down and not raised up, because our nation[e] sinned against the Lord our God, in not heeding his voice.

6 The Lord our God is in the right, but there is open shame on us and our ancestors this very day. 7 All those calamities with which the Lord threatened us have come upon us. 8 Yet we have not entreated the favor of the Lord by turning away, each of us, from the thoughts of our wicked hearts. 9 And the Lord has kept the calamities ready, and the Lord has brought them upon us, for the Lord is just in all the works that he has commanded us to do. 10 Yet we have not obeyed his voice, to walk in the statutes of the Lord that he set before us.

11 And now, O Lord God of Israel, who brought your people out of the land of Egypt with a mighty hand and with signs and wonders and with great power and outstretched arm, and made yourself a name that continues to this day, 12 we have sinned, we have been ungodly, we have done wrong, O Lord our God, against all your ordinances. 13 Let your anger turn away from us, for we are left, few in number, among the nations where you have scattered us. 14 Hear, O Lord, our prayer and our supplication, and for your own sake deliver us, and grant us favor in the sight of those who have carried us into exile; 15 so that all the earth may know that you are the Lord our God, for Israel and his descendants are called by your name.

16 O Lord, look down from your holy dwelling, and consider us. Incline your ear, O Lord, and hear; 17 open your eyes, O Lord, and see, for the dead who are in Hades, whose spirit has been taken from their bodies, will not ascribe glory or justice to the Lord; 18 but the person who is deeply grieved, who walks bowed and feeble, with failing eyes and famished soul, will declare your glory and righteousness, O Lord.

19 For it is not because of any righteous deeds of our ancestors or our kings that we bring before you our prayer for mercy, O Lord our God. 20 For you have sent your anger and your wrath upon us, as you declared by your servants the prophets, saying: 21 Thus says the Lord: Bend your shoulders and serve the king of Babylon, and you will remain in the land that I gave to your ancestors. 22 But if you will not obey the voice of the Lord and will not serve the king of Babylon, 23 I will make to cease from the towns of Judah and from the region around Jerusalem the voice of mirth and the voice of gladness, the voice of the bridegroom and the voice of the bride, and the whole land will be a desolation without inhabitants.

24 But we did not obey your voice, to

d Gk *in accordance with what is*
e Gk *because we*

1.15–18: Based on Dan 9.7–10. **15:** Ezra 9.7. **20:** Deut ch 28; Jer 11.3–5. **21:** Jer 7.25–26; Dan 9.5.

2.1–2: Dan 9.12–13. **3:** Lev 26.29; Deut 28.53; Jer 19.9; Lam 4.10. **5:** Deut 28.13.

2.6–10: Confession of guilt. **8:** Dan 9.13.

9: Dan 9.14.

2.11–26: Supplication and confession. **11–14;** Dan 9.15–17. **13:** Deut 4.27; Jer 42.2. **16:** Deut 26.15. **17:** Pss 6.5; 30.9; Isa 38.18; Sir 17.27–28. *Hades,* Sheol. **21:** Jer 27.11–12. **23:** Jer 7.34. **25:** Jer 36.30. **26:** Jer 7.14.

serve the king of Babylon; and you have carried out your threats, which you spoke by your servants the prophets, that the bones of our kings and the bones of our ancestors would be brought out of their resting place; 25 and indeed they have been thrown out to the heat of day and the frost of night. They perished in great misery, by famine and sword and pestilence. 26 And the house that is called by your name you have made as it is today, because of the wickedness of the house of Israel and the house of Judah.

27 Yet you have dealt with us, O Lord our God, in all your kindness and in all your great compassion, 28 as you spoke by your servant Moses on the day when you commanded him to write your law in the presence of the people of Israel, saying, 29 "If you will not obey my voice, this very great multitude will surely turn into a small number among the nations, where I will scatter them. 30 For I know that they will not obey me, for they are a stiff-necked people. But in the land of their exile they will come to themselves 31 and know that I am the Lord their God. I will give them a heart that obeys and ears that hear; 32 they will praise me in the land of their exile, and will remember my name 33 and turn from their stubbornness and their wicked deeds; for they will remember the ways of their ancestors, who sinned before the Lord. 34 I will bring them again into the land that I swore to give to their ancestors, to Abraham, Isaac, and Jacob, and they will rule over it; and I will increase them, and they will not be diminished. 35 I will make an everlasting covenant with them to be their God and they shall be my people; and I will never again remove my people Israel from the land that I have given them."

3 O Lord Almighty, God of Israel, the soul in anguish and the wearied spirit cry out to you. 2 Hear, O Lord, and have mercy, for we have sinned before you. 3 For you are enthroned forever, and we are perishing forever. 4 O Lord Almighty, God of Israel, hear now the prayer of the people*f* of Israel, the children of those who sinned before you, who did not heed the voice of the Lord their God, so that calamities have clung to us. 5 Do not remember the iniquities of our ancestors, but in this crisis remember your power and your name. 6 For you are the Lord our God, and it is you, O Lord, whom we will praise. 7 For you have put the fear of you in our hearts so that we would call upon your name; and we will praise you in our exile, for we have put away from our hearts all the iniquity of our ancestors who sinned against you. 8 See, we are today in our exile where you have scattered us, to be reproached and cursed and punished for all the iniquities of our ancestors, who forsook the Lord our God.

9 Hear the commandments of life,
        O Israel;
    give ear, and learn wisdom!
10 Why is it, O Israel, why is it that
        you are in the land of your
        enemies,
    that you are growing old in a
        foreign country,
    that you are defiled with the dead,
11   that you are counted among
        those in Hades?
12 You have forsaken the fountain of
        wisdom.
13 If you had walked in the way of
        God,

f Gk *dead*

2.27–35: Repentance and restoration under an everlasting covenant. 28–29: Deut 28.58, 62. 30: 1 Kings 8.47. 31: Jer 24.7. 33: Deut 9.6. 34: Lev 26.42; Deut 6.10; Jer 32.37. 35: Jer 32.38–40; Ezek 36.26–29; Am 9.15.
3.1–8: Impassioned plea of repentant exiles ("though penitent we are still in exile!"). 4: The Israelites in exile were as though "dead" (vv. 10–11; Isa 59.10b; Lam 3.6). 7: Jer 32.40b. 8: The ancestors' sins (2.33; 3.4–5) are visited on later generations (Ex 34.7; Lam 5.7; contrast Jer 31.29 and Ezek 18.2–32).
3.9–4.4: **Wisdom, found by God, was given to Israel as Torah.**
3.9–14: Introduction to the poem. 10: *Growing old,* the exile has been long (contrast

you would be living in peace
    forever.
14 Learn where there is wisdom,
    where there is strength,
    where there is understanding,
  so that you may at the same time
      discern
    where there is length of days,
      and life,
    where there is light for the eyes,
      and peace.

15 Who has found her place?
    And who has entered her
      storehouses?
16 Where are the rulers of the
      nations,
    and those who lorded it over
      the animals on earth;
17 those who made sport of the birds
      of the air,
    and who hoarded up silver and
      gold
  in which people trust,
    and there is no end to their
      getting;
18 those who schemed to get silver,
      and were anxious,
    but there is no trace of their
      works?
19 They have vanished and gone
      down to Hades,
    and others have arisen in their
      place.

20 Later generations have seen the
      light of day,
    and have lived upon the earth;
  but they have not learned the way
      to knowledge,
    nor understood her paths,
    nor laid hold of her.
21 Their descendants have strayed far
    from her*g* way.

22 She has not been heard of in
    Canaan,
    or seen in Teman;
23 the descendants of Hagar, who
      seek for understanding on
      the earth,
    the merchants of Merran and
      Teman,
    the story-tellers and the seekers
      for understanding,
  have not learned the way to
      wisdom,
    or given thought to her paths.

24 O Israel, how great is the house of
      God,
    how vast the territory that he
      possesses!
25 It is great and has no bounds;
    it is high and immeasurable.
26 The giants were born there, who
      were famous of old,
    great in stature, expert in war.
27 God did not choose them,
    or give them the way to
      knowledge;
28 so they perished because they had
      no wisdom,
    they perished through their
      folly.

29 Who has gone up into heaven, and
      taken her,
    and brought her down from the
      clouds?
30 Who has gone over the sea, and
      found her,
    and will buy her for pure gold?
31 No one knows the way to her,
    or is concerned about the path
      to her.

g Other ancient authorities read *their*

1.2). **11**: Pss 28.1; 88.4; Isa 53.12. **12**: Prov
18.4; Jer 2.13. **14**: Prov 3.16; 8.14.
   3.15–28: The rulers of the world and the
mighty have not found wisdom. **15**: Job
28.12. **16b–17a**: Jer 27.6; Dan 2.38; Jdt 11.7.
22: *Canaan*, Ezek 28.3–5 associates Tyre (in
Canaan) with wisdom. *Teman*, in Edom, was
reputed for its wisdom (Jer 49.7; Ob 8–9). **23**:

Merran, perhaps a corruption that arose in the
Hebrew text for "Midian," a son of Keturah
(Gen 25.2). **24**: *House of God,* the created
world. **26**: Gen 6.4; Wis 14.6; compare the
book of Enoch 7.1–6.
   3.29–37: God found wisdom and gave her
to Israel (Sir 24.1–12). **29–30**: Deut 30.12–
13; Job 28.13–14.

32 But the one who knows all things
knows her,
he found her by his
understanding.
The one who prepared the earth
for all time
filled it with four-footed
creatures;
33 the one who sends forth the light,
and it goes;
he called it, and it obeyed him,
trembling;
34 the stars shone in their watches,
and were glad;
he called them, and they said,
"Here we are!"
They shone with gladness for
him who made them.
35 This is our God;
no other can be compared to
him.
36 He found the whole way to
knowledge,
and gave her to his servant
Jacob
and to Israel, whom he loved.
37 Afterward she appeared on earth
and lived with humankind.

4 She is the book of the
commandments of God,
the law that endures forever.
All who hold her fast will live,
and those who forsake her will
die.
2 Turn, O Jacob, and take her;
walk toward the shining of her
light.
3 Do not give your glory to
another,
or your advantages to an alien
people.
4 Happy are we, O Israel,

for we know what is pleasing to
God.
5 Take courage, my people,
who perpetuate Israel's name!
6 It was not for destruction
that you were sold to the
nations,
but you were handed over to your
enemies
because you angered God.
7 For you provoked the one who
made you
by sacrificing to demons and
not to God.
8 You forgot the everlasting God,
who brought you up,
and you grieved Jerusalem, who
reared you.
9 For she saw the wrath that came
upon you from God,
and she said:
Listen, you neighbors of Zion,
God has brought great sorrow
upon me;
10 for I have seen the exile of my
sons and daughters,
which the Everlasting brought
upon them.
11 With joy I nurtured them,
but I sent them away with
weeping and sorrow.
12 Let no one rejoice over me, a
widow
and bereaved of many;
I was left desolate because of the
sins of my children,
because they turned away from
the law of God.
13 They had no regard for his
statutes;
they did not walk in the ways
of God's commandments,

3.32–34: Job 28.23–26; Prov 8.22–31. **33:**
Light, Gen 1.3. **34:** *Stars . . . were glad,* Job 38.7.
37: Many early Christian commentators took
this as an allusion to the Incarnation.
4.1–3: Wisdom is Torah. **1:** Sir 24.23. **2:** Isa
60.3.
4.5–5.9: Poem of comfort and restoration.

4.5–20: Israel provoked God, and Zion
now mourns for her captive children. **5:** *Take
courage, my people,* inspired by Isa 40.1 (compare Deut 31.6). **7:**
Demons, Deut 32.16–17; Ps 106.37; 1 Cor
10.20. **9b–16:** Jerusalem speaks to her *neighbors* (i.e. neighboring cities). **12:** *Widow,* Lam
1.1. **15:** Deut 28.49–50; Jer 6.15.

or tread the paths his
  righteousness showed them.
14 Let the neighbors of Zion come;
  remember the capture of my
    sons and daughters,
  which the Everlasting brought
    upon them.
15 For he brought a distant nation
    against them,
  a nation ruthless and of a
    strange language,
  which had no respect for the aged
    and no pity for a child.
16 They led away the widow's
    beloved sons,
  and bereaved the lonely woman
    of her daughters.

17 But I, how can I help you?
18 For he who brought these
    calamities upon you
  will deliver you from the hand
    of your enemies.
19 Go, my children, go;
  for I have been left desolate.
20 I have taken off the robe of peace
  and put on sackcloth for my
    supplication;
  I will cry to the Everlasting all
    my days.

21 Take courage, my children, cry to
    God,
  and he will deliver you from the
    power and hand of the
    enemy.
22 For I have put my hope in the
    Everlasting to save you,
  and joy has come to me from
    the Holy One,
  because of the mercy that will
    soon come to you
  from your everlasting savior. *h*
23 For I sent you out with sorrow
    and weeping,

but God will give you back to
  me with joy and gladness
  forever.
24 For as the neighbors of Zion have
    now seen your capture,
  so they soon will see your
    salvation by God,
  which will come to you with great
    glory
  and with the splendor of the
    Everlasting.
25 My children, endure with patience
    the wrath that has come
    upon you from God.
  Your enemy has overtaken you,
  but you will soon see their
    destruction
  and will tread upon their necks.
26 My pampered children have
    traveled rough roads;
  they were taken away like a
    flock carried off by the
    enemy.

27 Take courage, my children, and
    cry to God,
  for you will be remembered by
    the one who brought this
    upon you.
28 For just as you were disposed to
    go astray from God,
  return with tenfold zeal to seek
    him.
29 For the one who brought these
    calamities upon you
  will bring you everlasting joy
    with your salvation.

30 Take courage, O Jerusalem,
  for the one who named you will
    comfort you.
31 Wretched will be those who
    mistreated you

h Or from the Everlasting, your savior

4.17–29: Jerusalem encourages her exiled children. **17–18:** Only God can help. **20:** *Robe of peace,* garment worn in time of prosperity. *Sackcloth for my supplication,* garment worn by a suppliant. **23:** Ps 126.6; Jer 31.12–13. **24:** Isa 60.1–3. **25:** *The wrath* is only temporary (Isa 54.7–8).

4.30–5.9: Jerusalem encouraged with promises concerning the destruction of her enemy and the return of her children. **4.30:** *The one who named you,* see 5.4 n. **31–35:** Contrast the attitude toward Babylon in 1.11–12.

and who rejoiced at your fall.
32 Wretched will be the cities that
your children served as
slaves;
wretched will be the city that
received your offspring.
33 For just as she rejoiced at your fall
and was glad for your ruin,
so she will be grieved at her
own desolation.
34 I will take away her pride in her
great population,
and her insolence will be turned
to grief.
35 For fire will come upon her from
the Everlasting for many
days,
and for a long time she will be
inhabited by demons.

36 Look toward the east,
O Jerusalem,
and see the joy that is coming
to you from God.
37 Look, your children are coming,
whom you sent away;
they are coming, gathered from
east and west,
at the word of the Holy One,
rejoicing in the glory of God.

5 Take off the garment of your
sorrow and affliction,
O Jerusalem,
and put on forever the beauty of
the glory from God.
2 Put on the robe of the
righteousness that comes
from God;

put on your head the diadem of
the glory of the Everlasting;
3 for God will show your splendor
everywhere under heaven.
4 For God will give you evermore
the name,
"Righteous Peace, Godly
Glory."

5 Arise, O Jerusalem, stand upon
the height;
look toward the east,
and see your children gathered
from west and east
at the word of the Holy One,
rejoicing that God has
remembered them.
6 For they went out from you on
foot,
led away by their enemies;
but God will bring them back to
you,
carried in glory, as on a royal
throne.
7 For God has ordered that every
high mountain and the
everlasting hills be made
low
and the valleys filled up, to
make level ground,
so that Israel may walk safely in
the glory of God.
8 The woods and every fragrant tree
have shaded Israel at God's
command.
9 For God will lead Israel with joy,
in the light of his glory,
with the mercy and
righteousness that come
from him.

4.35: *Fire,* Jer 51.58. *Demons,* Isa 13.21. **36**:
Isa 40.9–11. **37**: Isa 43.5.
5.1–9: Glorification of Jerusalem and return of the exiles. **1–2**: Isa 61.3, 10. **4**: Isa

60.14; 62.4; Jer 33.16; Ezek 48.35. *Righteous
Peace,* Isa 32.17. **5**: Isa 49.18; 60.4. **6**: Isa 49.22;
66.20. **7**: Isa 42.16–17.

The Letter of Jeremiah

These seventy-three verses purport to be a letter that Jeremiah composed for those about to be taken into exile from Judah to Babylonia in 597 (or 587) B.C. by Nebuchadnezzar's forces. It was undoubtedly inspired by Jeremiah's letter (Jer 29.1–23) to those who had been taken hostage in 597, a decade before the final defeat of Judah and the destruction of Jerusalem. It is an impassioned sermon against idol worship and polytheism based on Jer 10, and particularly Jer 10.11: "May the gods, who did not make heaven and earth, perish from the earth and from beneath the heavens." This verse is the only one in the book of Jeremiah in Aramaic, and because it is so central to the Letter it has been suggested that the latter too may originally have been composed in Aramaic. The Letter is also influenced, however, by other biblical polemics against idol worship (Pss 115.4–8; 135.15–18; Isa 40.18–20; 41.6–7; 46.1–7; etc.) and may well have been written in Hebrew. A few scholars have suggested that it was originally composed in Greek. The various parts of the sermon cohere as a series of warnings to Jews, who might be attracted to idol worship, to recognize and be wary of the idolatry of their time; each part ends on a common refrain, with variations, insisting that idols are not gods nor to be confused with the one, true God (vv. 16, 23, 29, 40, 44, 52, 56, 65, 69, 72).

Most scholars date the Letter in the hellenistic period. The reference in v. 3 (so some have argued) would yield a date around 317 B.C.; others date the Letter still later. The allusion to the Letter in 2 Macc 2.1–3 would indicate a date at least in the second century B.C., perhaps the third.

The Letter is found at different locations in various manuscripts and versions. It stands as a discrete work between Lamentations and Ezekiel in two major Greek Septuagint manuscripts (fourth-century Vaticanus and fifth-century Alexandrinus), in the Milan Syriac Hexaplar, and in Arabic. In other Greek and Syriac manuscripts, and in the Latin version, it appears as the sixth chapter of Baruch. Since it is, however, clearly independent of Baruch, the New Revised Standard Version includes it as a separate book.

6 *a* A copy of a letter that Jeremiah sent to those who were to be taken to Babylon as exiles by the king of the Babylonians, to give them the message that God had commanded him.

2 Because of the sins that you have committed before God, you will be taken to Babylon as exiles by Nebuchadnezzar, king of the Babylonians. ³ Therefore when you have come to Babylon you will remain there for many years, for a long time, up to seven generations; after

a The King James Version (like the Latin Vulgate) prints The Letter of Jeremiah as Chapter 6 of the Book of Baruch, and the chapter and verse numbers are here retained. In the Greek Septuagint, the Letter is separated from Baruch by the Book of Lamentations.

that I will bring you away from there in peace. 4 Now in Babylon you will see gods made of silver and gold and wood, which people carry on their shoulders, and which cause the heathen to fear. 5 So beware of becoming at all like the foreigners or of letting fear for these gods[b] possess you 6 when you see the multitude before and behind them worshiping them. But say in your heart, "It is you, O Lord, whom we must worship." 7 For my angel is with you, and he is watching over your lives.

8 Their tongues are smoothed by the carpenter, and they themselves are overlaid with gold and silver; but they are false and cannot speak. 9 People[c] take gold and make crowns for the heads of their gods, as they might for a girl who loves ornaments. 10 Sometimes the priests secretly take gold and silver from their gods and spend it on themselves, 11 or even give some of it to the prostitutes on the terrace. They deck their gods[d] out with garments like human beings—these gods of silver and gold and wood 12 that cannot save themselves from rust and corrosion. When they have been dressed in purple robes, 13 their faces are wiped because of the dust from the temple, which is thick upon them. 14 One of them holds a scepter, like a district judge, but is unable to destroy anyone who offends it. 15 Another has a dagger in its right hand, and an ax, but cannot defend itself from war and rob-

bers. 16 From this it is evident that they are not gods; so do not fear them.

17 For just as someone's dish is useless when it is broken, 18 so are their gods when they have been set up in the temples. Their eyes are full of the dust raised by the feet of those who enter. And just as the gates are shut on every side against anyone who has offended a king, as though under sentence of death, so the priests make their temples secure with doors and locks and bars, in order that they may not be plundered by robbers. 19 They light more lamps for them than they light for themselves, though their gods[e] can see none of them. 20 They are[f] just like a beam of the temple, but their hearts, it is said, are eaten away when crawling creatures from the earth devour them and their robes. They do not notice 21 when their faces have been blackened by the smoke of the temple. 22 Bats, swallows, and birds alight on their bodies and heads; and so do cats. 23 From this you will know that they are not gods; so do not fear them.

24 As for the gold that they wear for beauty—it[g] will not shine unless someone wipes off the tarnish; for even when they were being cast, they did not feel it. 25 They are bought without regard to cost, but there is no breath in them. 26 Having no feet, they are carried on the shoulders of others, revealing to human-

b Gk *for them*   *c* Gk *They*   *d* Gk *them*
e Gk *they*   *f* Gk *It is*   *g* Lat Syr: Gk *they*

6.1–7: **Historical introduction. 1:** The exile of 597 B.C. (2 Kings 24.10–17). *Letter,* according to Jer 29.1 a letter is sent to Babylon. *King of the Babylonians,* but "king of Babylon" in the book of Jeremiah (Jer 20.4; 21.2; etc.). **2:** Jer 10–13; etc. **3:** *Seven generations,* contrast seventy years in Jer 29.10, forty years in Ezek 4.6, seventy weeks of years in Dan 9.24. **4:** *Silver and gold,* overlaid on wood (v. 55; Isa 40.19; Jer 10.3–4). *Which people carry on their shoulders,* perhaps an allusion to the Babylonian New Year procession, or a reflection of Isa 46.7; Jer 10.5. **5:** *Like the foreigners,* Jer 10.2. **7:** *My angel,* Ex 23.23; 32.34; Ps 91.11–12; etc.

8–73: **Condemnation of idolatry.**

8–16: **Idols are decked out like people. 8:** *Carpenter,* Isa 40.20; 44.13; Jer 10.3–4; etc. **11:** *Prostitutes,* probably cult prostitutes. **12:** *Purple robes,* Jer 10.9. **14:** *Scepter,* Esth 5.2. *Destroy,* put to death. **15:** *Dagger,* the Hebrew word behind the Greek could also mean "sword." Archaeologists have found representations of deities bearing scepters, swords, daggers, and battle-axes.

17–23: **Uselessness and helplessness of idols. 17:** *Dish . . . broken,* Isa 30.14; Jer 19.11; 22.28. **18:** *Gates* of the palace or doors of courtyard prison (Jer 32.2). **19:** Ps 115.5. *Lamps* have been found in excavated temples. **22:** This is the earliest Jewish reference to *cats,* which were first domesticated in Egypt.

kind their worthlessness. And those who serve them are put to shame [27] because, if any of these gods falls[h] to the ground, they themselves must pick it up. If anyone sets it upright, it cannot move itself; and if it is tipped over, it cannot straighten itself. Gifts are placed before them just as before the dead. [28] The priests sell the sacrifices that are offered to these gods[i] and use the money themselves. Likewise their wives preserve some of the meat[j] with salt, but give none to the poor or helpless. [29] Sacrifices to them may even be touched by women in their periods or at childbirth. Since you know by these things that they are not gods, do not fear them.

30 For how can they be called gods? Women serve meals for gods of silver and gold and wood; [31] and in their temples the priests sit with their clothes torn, their heads and beards shaved, and their heads uncovered. [32] They howl and shout before their gods as some do at a funeral banquet. [33] The priests take some of the clothing of their gods[k] to clothe their wives and children. [34] Whether one does evil to them or good, they will not be able to repay it. They cannot set up a king or depose one. [35] Likewise they are not able to give either wealth or money; if one makes a vow to them and does not keep it, they will not require it. [36] They cannot save anyone from death or rescue the weak from the strong. [37] They cannot restore sight to the blind; they cannot rescue one who is in distress. [38] They cannot take pity on a widow or do good to an orphan. [39] These things that are made of wood and overlaid with gold and sil-

ver are like stones from the mountain, and those who serve them will be put to shame. [40] Why then must anyone think that they are gods, or call them gods?

Besides, even the Chaldeans themselves dishonor them; for when they see someone who cannot speak, they bring Bel and pray that the mute may speak, as though Bel[l] were able to understand! [41] Yet they themselves cannot perceive this and abandon them, for they have no sense. [42] And the women, with cords around them, sit along the passageways, burning bran for incense. [43] When one of them is led off by one of the passers-by and is taken to bed by him, she derides the woman next to her, because she was not as attractive as herself and her cord was not broken. [44] Whatever is done for these idols[m] is false. Why then must anyone think that they are gods, or call them gods?

45 They are made by carpenters and goldsmiths; they can be nothing but what the artisans wish them to be. [46] Those who make them will certainly not live very long themselves; [47] how then can the things that are made by them be gods? They have left only lies and reproach for those who come after. [48] For when war or calamity comes upon them, the priests consult together as to where they can hide themselves and their gods.[n] [49] How then can one fail to see that these are not gods, for they cannot save themselves from war or calamity?

---

h Gk *if they fall*   *i* Gk *to them*   *j* Gk *of them*
k Gk *some of their clothing*   *l* Gk *he*
m Gk *them*   *n* Gk *them*

24–29: Idols are unable to feel or move: Isa 40.20; 44.9; 46.2; Jer 10.4–5. **25:** *Without regard to cost,* great cost *No breath,* Ps 135.17; Jer 10.14; Hab 2.19. **26:** Isa 46.1, 7. **27:** *Cannot move,* Isa 46.7; Jer 10.4. *Gifts for the dead,* Ps 106.28; Sir 30.18–19. **29:** Lev 12.1–8.

30–40a: Idols cannot repay good or evil, or help worshipers. 30: *Women,* there were only male ministrants in the Jewish temple. **31–32:** Ritual lamentations for dying gods (such as Tammuz, Ezek 8.14; compare Lev 21.5, 10; Ezek 24.17). **34b:** Job 12.18;

Dan 2.21. **35b:** Deut 23.21. **36:** On the contrary, the Lord can do this (Deut 32.39; Ps 49.15). **37:** Ps 146.8. **38:** Deut 10.18; Ps 146.9; Jer 7.6. **39:** Hab 2.19.

40b–44: The Chaldeans dishonor their own idols. 40b: *Bel,* Marduk (Isa 46.1). **43:** A similar Babylonian practice is described by Herodotus (*Hist.* I. 199), according to which cult prostitutes sat among roped-off passageways.

45–52: Idols are but the work of human hands. 45: Ps 115.4; Isa 40.19; Jer 10.9. **47:** Idolaters bequeath lies and reproach, not

50 Since they are made of wood and overlaid with gold and silver, it will afterward be known that they are false. 51 It will be manifest to all the nations and kings that they are not gods but the work of human hands, and that there is no work of God in them. 52 Who then can fail to know that they are not gods?*o*

53 For they cannot set up a king over a country or give rain to people. 54 They cannot judge their own cause or deliver one who is wronged, for they have no power; 55 they are like crows between heaven and earth. When fire breaks out in a temple of wooden gods overlaid with gold or silver, their priests will flee and escape, but the gods*p* will be burned up like timbers. 56 Besides, they can offer no resistance to king or enemy. Why then must anyone admit or think that they are gods?

57 Gods made of wood and overlaid with silver and gold are unable to save themselves from thieves or robbers. 58 Anyone who can will strip them of their gold and silver and of the robes they wear, and go off with this booty, and they will not be able to help themselves. 59 So it is better to be a king who shows his courage, or a household utensil that serves its owner's need, than to be these false gods; better even the door of a house that protects its contents, than these false gods; better also a wooden pillar in a palace, than these false gods.

60 For sun and moon and stars are bright, and when sent to do a service, they are obedient. 61 So also the lightning, when it flashes, is widely seen; and the wind likewise blows in every land.

62 When God commands the clouds to go over the whole world, they carry out his command. 63 And the fire sent from above to consume mountains and woods does what it is ordered. But these idols*q* are not to be compared with them in appearance or power. 64 Therefore one must not think that they are gods, nor call them gods, for they are not able either to decide a case or to do good to anyone. 65 Since you know then that they are not gods, do not fear them.

66 They can neither curse nor bless kings; 67 they cannot show signs in the heavens for the nations, or shine like the sun or give light like the moon. 68 The wild animals are better than they are, for they can flee to shelter and help themselves. 69 So we have no evidence whatever that they are gods; therefore do not fear them.

70 Like a scarecrow in a cucumber bed, which guards nothing, so are their gods of wood, overlaid with gold and silver. 71 In the same way, their gods of wood, overlaid with gold and silver, are like a thornbush in a garden on which every bird perches; or like a corpse thrown out in the darkness. 72 From the purple and linen*r* that rot upon them you will know that they are not gods; and they will finally be consumed themselves, and be a reproach in the land. 73 Better, therefore, is someone upright who has no idols; such a person will be far above reproach.

o Meaning of Gk uncertain   *p* Gk *they*
q Gk *these things*   *r* Cn: Gk *marble,* Syr *silk*

---

real gods, to posterity. **50**: *Afterward,* when the veneer has been exposed for what it is. *False,* a fraud.

53–56: The impotence of idols. 53: See v. 34b. *Give rain,* Deut 11.14; 28.12; Ps 147.8.

57–65: Idols are helpless, useless, and not to be compared with celestial phenomena. 60: Gen 1.14–18. **61–62:** Job 38.24–27; Ps 97.4. **63:** *Fire,* lightning. **64:** *Decide a case,* the true God does this (Ex 18.19; Ps 43.1; Isa 41.21).

66–69: The helplessness of idols. 67: *Signs,* portents (Jer 10.2; Joel 2.30; Mt 16.1).

70–73: Idols are compared with a scarecrow, thornbush, and corpse. 70: *Scarecrow,* Jer 10.5. **71:** *Thornbush,* an ordinary, useless shrub (compare Judg 9.14–15). **72:** The Greek text ("marble," see note *r*) is a misinterpretation of the Hebrew word "shesh," which means both "linen" and "marble" ("alabaster"). **73:** The conclusion of the matter.

The Additions to the Greek Book of Daniel

As in the case of the Book of Esther, the ancient Greek version of the Book of Daniel is considerably longer than the surviving Hebrew text. The Greek Daniel has, apart from numerous textual differences throughout the book, three extended passages that are lacking in the Masoretic Hebrew text. All Greek witnesses place the Prayer of Azarias and the Song of the Three Jews in Dan ch 3. The Septuagint places the prose stories of Bel and the Dragon and of Susanna, in that order, at the end of Daniel after 12.13.

The Greek translation made by Theodotion, though corrected to the Hebrew and therefore shorter than the Septuagint text of Daniel, nonetheless includes all the outstanding passages in the Greek Daniel as integral parts of the book, with Susanna at the beginning of the book and Bel and the Dragon at the end of Dan 6; in this manner chronology is observed in a semibiographical story starting with the young Daniel's early detective work concerning Susanna, and continuing with further sleuthing triumphs over Bel and his priests and over the Dragon, as happening in the court of Cyrus. The Old Latin, Coptic, and Arabic versions follow Theodotion.

In all the witnesses Azariah's prayer and the song of the three Jews appear quite logically after Dan 3.23 (Theodotion; 3.24 in Septuagint), where we find the three gifted and handsome youths (Dan 1.4) in the burning furnace. In the Hebrew and in Theodotion they are called by the Babylonian names given them by Nebuchadnezzar's chief eunuch (Shadrach, Meshach, and Abednego, Dan 1.7); while in the Septuagint they retain their Hebrew names as pronounced in Greek, Ananias, Azarias, and Misael (see 2.17).

Although Theodotion "corrected" in many particulars the Greek text of Daniel to the emerging proto-Masoretic text of the beginning of the second century B.C., he nonetheless retained all the so-called additions of the Septuagint, albeit in a more suitable order for biographic purposes. Jerome's Latin Vulgate followed Theodotion basically, but made the story of Susanna ch 13, and the account of Bel and the Dragon ch 14, thus inverting the Greek order. As is mentioned in the Preface, "To the Reader" (pp. ix–xiv), the New RSV gives a translation of Theodotion's Greek text of Daniel.

The Prayer of Azariah
and the Song of the Three Jews

The passage that follows Dan 3.23(24) in all the previously mentioned witnesses (see p. 173 AP) except the Hebrew, consists of three parts: the prayer of Azariah, in vv. 1–22; a short report in prose on the welfare of the three Jews in the furnace, in vv. 23–27; and a long hymn sung by the three, untouched while the flames danced around them, in vv. 28–68. Both the prayer and the song are included as numbers 7 and 8 of the fifteen "Odes" appended to the Psalter in a few Septuagint manuscripts (see the Introduction to Ps 151).

We learn in 3.24–25 that Nebuchadnezzar was himself witness to the punishment he had decreed against the three Jews for refusing to worship his gods, including the golden statue he had erected (Dan 3.1) in a plain of the province of Babylon. What he saw in the furnace were four men walking about unharmed (Dan 3.25). The short prose report supplies the necessary information that the fourth was an angel of the Lord (v. 26; see Dan 3.28) who made the inside of the furnace quite bearable for the three. Such a miracle deserved liturgical material as well, and so we find in the Greek and Latin traditions the requisite prayer for deliverance before the report (vv. 3–22), and a song of praise after it (vv. 29–68), as indicated by the biblical pattern in such cases (Ex 15; 1 Sam 2; 2 Sam 23). The song is in two parts: the first a song of thanksgiving for deliverance, and the second a litany exhorting all creation to praise God, in the manner of Pss 136 and 148.

Early Jews composed many such poems, as shown in the abundant non-Masoretic and non-biblical psalmody in the Dead Sea Scrolls. Whether these parts of the Greek Daniel were first composed in Hebrew or in a highly semitized Greek is debated; the balance is in favor of Hebrew originals. When they were composed is also debated, but the second century B.C. is indicated. The interesting question is whether they were a part of a pre-Masoretic Hebrew Daniel, which is usually dated in the first half of the second century B.C.; each, the Septuagint and the Hebrew text of Daniel, has its own integrity, especially if read through from beginning to end. Whether that integrity was in the mind of an original author, early redactor, or late redactor, in Hebrew or Greek, will probably not be determined without further discovery of early manuscript evidence.

(Additions to Daniel, inserted
between 3.23 and 3.24)

1 They*a* walked around in the midst
of the flames, singing hymns to God and
blessing the Lord. 2 Then Azariah stood
still in the fire and prayed aloud:

3 "Blessed are you, O Lord, God of
   our ancestors, and worthy
   of praise;
   and glorious is your name
   forever!

4 For you are just in all you have
   done;
   all your works are true and your
   ways right,
   and all your judgments are true.

5 You have executed true judgments
   in all you have brought
   upon us
   and upon Jerusalem, the holy
   city of our ancestors;
   by a true judgment you have
   brought all this upon us
   because of our sins.

6 For we have sinned and broken
   your law in turning away
   from you;
   in all matters we have sinned
   grievously.

7 We have not obeyed your
   commandments,
   we have not kept them or done
   what you have commanded
   us for our own good.

8 So all that you have brought upon
   us,
   and all that you have done to
   us,
   you have done by a true
   judgment.

9 You have handed us over to our
   enemies, lawless and hateful
   rebels,
   and to an unjust king, the most
   wicked in all the world.

10 And now we cannot open our
    mouths;
    we, your servants who worship
    you, have become a shame
    and a reproach.

11 For your name's sake do not give
    us up forever,
    and do not annul your
    covenant.

12 Do not withdraw your mercy
    from us,
   for the sake of Abraham your
    beloved
    and for the sake of your servant
    Isaac
    and Israel your holy one,

13 to whom you promised
    to multiply their descendants
    like the stars of heaven
    and like the sand on the shore
    of the sea.

14 For we, O Lord, have become
    fewer than any other
    nation,
    and are brought low this day in
    all the world because of our
    sins.

15 In our day we have no ruler, or
    prophet, or leader,
    no burnt offering, or sacrifice,
    or oblation, or incense,
    no place to make an offering
    before you and to find
    mercy.

16 Yet with a contrite heart and a
    humble spirit may we be
    accepted,

17 as though it were with burnt
    offerings of rams and bulls,
    or with tens of thousands of fat
    lambs;

a That is, Hananiah, Mishael, and Azariah (Dan
2.17), the original names of Shadrach, Meshach,
and Abednego (Dan 1.6-7)

1–22: The prayer of Azariah. 1: *They,*
the three men mentioned in Dan 1.6–7; 3.22–
23. **3:** 1 Chr 29.10, 20. **4:** Neh 9.33; Rev 16.7;
19.2. **6–7:** Isa 59.12–13; Dan 9.5–8; Bar
1.17–18. **8–10:** Lev 26.14, 38; Deut 28.15,
63–64; 30.1–3.

12: *Abraham your beloved,* 2 Chr 20.7; Isa
41.8; Jas 2.23. **13:** Gen 15.5; 22.17. **14:** Deut
7.7; Jer 42.2; Bar 2.13. **15:** Lam 2.9; Hos 3.4;
2 Esd 10.21–22. **16:** Ps 51.16–17; Hos 6.6. **19:**
Ps 25.3. **21:** Ps 35.26. **22:** Ps 83.18.

such may our sacrifice be in
your sight today,
and may we unreservedly
follow you, *b*
for no shame will come to those
who trust in you.
18 And now with all our heart we
follow you;
we fear you and seek your
presence.
19 Do not put us to shame,
but deal with us in your
patience
and in your abundant mercy.
20 Deliver us in accordance with
your marvelous works,
and bring glory to your name,
O Lord.
21 Let all who do harm to your
servants be put to shame;
let them be disgraced and
deprived of all power,
and let their strength be broken.
22 Let them know that you alone are
the Lord God,
glorious over the whole world."

23 Now the king's servants who
threw them in kept stoking the furnace
with naphtha, pitch, tow, and brushwood. 24 And the flames poured out
above the furnace forty-nine cubits,
25 and spread out and burned those Chaldeans who were caught near the furnace.
26 But the angel of the Lord came down
into the furnace to be with Azariah and
his companions, and drove the fiery
flame out of the furnace, 27 and made the
inside of the furnace as though a moist
wind were whistling through it. The fire
did not touch them at all and caused them
no pain or distress.
28 Then the three with one voice
praised and glorified and blessed God in
the furnace:
29 "Blessed are you, O Lord, God of
our ancestors,

and to be praised and highly
exalted forever;
30 And blessed is your glorious, holy
name,
and to be highly praised and
highly exalted forever.
31 Blessed are you in the temple of
your holy glory,
and to be extolled and highly
glorified forever.
32 Blessed are you who look into the
depths from your throne on
the cherubim,
and to be praised and highly
exalted forever.
33 Blessed are you on the throne of
your kingdom,
and to be extolled and highly
exalted forever.
34 Blessed are you in the firmament
of heaven,
and to be sung and glorified
forever.
35 "Bless the Lord, all you works of
the Lord;
sing praise to him and highly
exalt him forever.
36 Bless the Lord, you heavens;
sing praise to him and highly
exalt him forever.
37 Bless the Lord, you angels of the
Lord;
sing praise to him and highly
exalt him forever.
38 Bless the Lord, all you waters
above the heavens;
sing praise to him and highly
exalt him forever.
39 Bless the Lord, all you powers of
the Lord;
sing praise to him and highly
exalt him forever.
40 Bless the Lord, sun and moon;
sing praise to him and highly
exalt him forever.

b Meaning of Gk uncertain

**23–27: The continued stoking of the
furnace, and the descent of the angel of
the Lord. 23:** *Naphtha,* a natural petroleum.
28–68: Song of the three young men.

32–37: Ps 148. **35:** Pss 103.22; 145.10. **37:** Pss
103.20; 148.2. **38:** Ps 148.4. **39:** *All you powers,* i.e. heavenly bodies or angels. **40:** Ps
148.3.

41 Bless the Lord, stars of heaven;
  sing praise to him and highly
    exalt him forever.

42 "Bless the Lord, all rain and dew;
  sing praise to him and highly
    exalt him forever.
43 Bless the Lord, all you winds;
  sing praise to him and highly
    exalt him forever.
44 Bless the Lord, fire and heat;
  sing praise to him and highly
    exalt him forever.
45 Bless the Lord, winter cold and
    summer heat;
  sing praise to him and highly
    exalt him forever.
46 Bless the Lord, dews and falling
    snow;
  sing praise to him and highly
    exalt him forever.
47 Bless the Lord, nights and days;
  sing praise to him and highly
    exalt him forever.
48 Bless the Lord, light and darkness;
  sing praise to him and highly
    exalt him forever.
49 Bless the Lord, ice and cold;
  sing praise to him and highly
    exalt him forever.
50 Bless the Lord, frosts and snows;
  sing praise to him and highly
    exalt him forever.
51 Bless the Lord, lightnings and
    clouds;
  sing praise to him and highly
    exalt him forever.

52 "Let the earth bless the Lord;
  let it sing praise to him and
    highly exalt him forever.
53 Bless the Lord, mountains and
    hills;
  sing praise to him and highly
    exalt him forever.
54 Bless the Lord, all that grows in
    the ground;

  sing praise to him and highly
    exalt him forever.
55 Bless the Lord, seas and rivers;
  sing praise to him and highly
    exalt him forever.
56 Bless the Lord, you springs;
  sing praise to him and highly
    exalt him forever.
57 Bless the Lord, you whales and all
    that swim in the waters;
  sing praise to him and highly
    exalt him forever.
58 Bless the Lord, all birds of
    the air;
  sing praise to him and highly
    exalt him forever.
59 Bless the Lord, all wild animals
    and cattle;
  sing praise to him and highly
    exalt him forever.

60 "Bless the Lord, all people on
    earth;
  sing praise to him and highly
    exalt him forever.
61 Bless the Lord, O Israel;
  sing praise to him and highly
    exalt him forever.
62 Bless the Lord, you priests of the
    Lord;
  sing praise to him and highly
    exalt him forever.
63 Bless the Lord, you servants of the
    Lord;
  sing praise to him and highly
    exalt him forever.
64 Bless the Lord, spirits and souls of
    the righteous;
  sing praise to him and highly
    exalt him forever.
65 Bless the Lord, you who are holy
    and humble in heart;
  sing praise to him and highly
    exalt him forever.

66 "Bless the Lord, Hananiah,
    Azariah, and Mishael;

44: Ps 148.8. **53**: Ps 148.9. **58–59**: Ps 148.10. **61–62**: Ps 135.19. **63**: Ps 134.1. **65**: *Holy and humble in heart,* Pss 18.25, 27; 86.1–2; Zeph 2.3. **67–68**: Pss 106.1; 136.1–2.

sing praise to him and highly
exalt him forever.
For he has rescued us from Hades
and saved us from the
power of death,
and delivered us from the midst
of the burning fiery
furnace;
from the midst of the fire he has
delivered us.

67 Give thanks to the Lord, for he is
good,
for his mercy endures forever.
68 All who worship the Lord, bless
the God of gods,
sing praise to him and give
thanks to him,
for his mercy endures forever."

c Gk hand

Susanna
(Chapter 13 of the Greek
Version of Daniel)

The story of Susanna and the young Daniel in the Septuagint is strikingly different from the same in Theodotion. In Theodotion it reads like a delightful yarn about two wicked elders who falsely accuse the virtuous young Susanna and are exposed by the youthful and sagacious Daniel. In the Septuagint the story reads like a story mirroring all virtuous Jewish youth who are the beloved of Jacob and whose qualities should be sponsored in all Jewish communities. The Septuagint prefers general titles whereas Theodotion provides specific names; the trial takes place in a synagogue in the Septuagint, but in Susanna's home in Theodotion. There is actually little verbatim agreement between the two accounts, though the story is clearly the same. The perverted elders who maligned the young woman, the Septuagint makes clear from the start of its account, were judges corrupted by foreign influence.

Although the Theodotionic account reads better, perhaps, than that in the Septuagint, the story itself is one of the most engaging short stories in world literature. It is a celebration of the triumph of virtue over villainy, and of the innocence of youth over jaded elders corrupted by power and the anxiety of aging. While innocence and vulnerability may be generic to youth itself, it is God (Theodotion) or an angel (Septuagint) who inspires youth with a spirit of insight and understanding, integrity and courage. While the characters, protagonists and antagonists, are Jews, the story is about human strengths and weaknesses, and hence is universal in its appeal.

It is difficult to determine if the account concerning Susanna was composed in Greek or rests upon a Semitic original. Puns based on the Greek in vv. 54–55 and 58–59 might indicate a Greek original, though good translators often try to match puns from original compositions in the receptor language. The Greek text, like other purely Greek portions of Daniel, is full of semitisms; most scholars have thought in terms of a Semitic original, a few thinking of two separate ones lying behind the Theodotionic and the Septuagint versions. The date of composition would have been sometime in the second century B.C.

1 There was a man living in Babylon whose name was Joakim. [2]He married the daughter of Hilkiah, named Susanna, a very beautiful woman and one who feared the Lord. [3]Her parents were righteous, and had trained their daughter according to the law of Moses. [4]Joakim was very rich, and had a fine garden adjoining his house; the Jews used to come to him because he was the most honored of them all.

5 That year two elders from the people were appointed as judges. Concerning them the Lord had said: "Wickedness came forth from Babylon, from elders who were judges, who were supposed to govern the people." [6]These men were frequently at Joakim's house, and all who had a case to be tried came to them there.

7 When the people left at noon, Susanna would go into her husband's garden to walk. [8]Every day the two elders used to see her, going in and walking about, and they began to lust for her. [9]They suppressed their consciences and turned away their eyes from looking to Heaven or remembering their duty to administer justice. [10]Both were overwhelmed with passion for her, but they did not tell each other of their distress, [11]for they were ashamed to disclose their lustful desire to seduce her. [12]Day after day they watched eagerly to see her.

13 One day they said to each other, "Let us go home, for it is time for lunch." So they both left and parted from each other. [14]But turning back, they met again; and when each pressed the other for the reason, they confessed their lust. Then together they arranged for a time when they could find her alone.

15 Once, while they were watching for an opportune day, she went in as before with only two maids, and wished to bathe in the garden, for it was a hot day. [16]No one was there except the two elders, who had hidden themselves and were watching her. [17]She said to her maids, "Bring me olive oil and ointments, and shut the garden doors so that I can bathe." [18]They did as she told them: they shut the doors of the garden and went out by the side doors to bring what they had been commanded; they did not see the elders, because they were hiding.

19 When the maids had gone out, the two elders got up and ran to her. [20]They said, "Look, the garden doors are shut, and no one can see us. We are burning with desire for you; so give your consent, and lie with us. [21]If you refuse, we will testify against you that a young man was with you, and this was why you sent your maids away."

22 Susanna groaned and said, "I am completely trapped. For if I do this, it will mean death for me; if I do not, I cannot escape your hands. [23]I choose not to do it; I will fall into your hands, rather than sin in the sight of the Lord."

24 Then Susanna cried out with a loud voice, and the two elders shouted against her. [25]And one of them ran and opened the garden doors. [26]When the people in the house heard the shouting in the garden, they rushed in at the side door to see what had happened to her. [27]And when the elders told their story,

---

1–4: **Introduction.** The setting of the story is Babylon during the exile. 1: The name *Joakim* means "the Lord will establish." 2: The names *Susanna* and *Hilkiah* mean respectively "lily" and "the Lord is my portion." 4: Some Jews prospered during the exile (Jer 29.5).

5–14: **The two lustful elders.** 5: *That year,* apparently the year of Joakim's marriage (v. 2). The *two elders* are identified by Jewish tradition to be the two false prophets mentioned in Jer 29.21–23. The quotation (*"Wickedness . . . people"*) is an unknown prophetic saying probably based on Jer 23.14–15. 9: *Heaven,* a metonym, a word associated with another word and used in its place, for God (see 1 Macc 3.18 n. and compare Lk 15.18). Such metonymy was a way of avoiding the pronunciation of God's name.

15–27: **The attempted seduction.** 17: *Oil and* (perfumed) *ointments* were used after bathing. 22: The Mosaic law prescribed *death* as punishment for an unfaithful wife (Lev 20.10; Deut 22.22). 23: See Joseph's reply to his tempter (Gen 39.9).

the servants felt very much ashamed, for nothing like this had ever been said about Susanna.

28 The next day, when the people gathered at the house of her husband Joakim, the two elders came, full of their wicked plot to have Susanna put to death. In the presence of the people they said, 29 "Send for Susanna daughter of Hilkiah, the wife of Joakim." 30 So they sent for her. And she came with her parents, her children, and all her relatives. 31 Now Susanna was a woman of great refinement and beautiful in appearance. 32 As she was veiled, the scoundrels ordered her to be unveiled, so that they might feast their eyes on her beauty. 33 Those who were with her and all who saw her were weeping.

34 Then the two elders stood up before the people and laid their hands on her head. 35 Through her tears she looked up toward Heaven, for her heart trusted in the Lord. 36 The elders said, "While we were walking in the garden alone, this woman came in with two maids, shut the garden doors, and dismissed the maids. 37 Then a young man, who was hiding there, came to her and lay with her. 38 We were in a corner of the garden, and when we saw this wickedness we ran to them. 39 Although we saw them embracing, we could not hold the man, because he was stronger than we, and he opened the doors and got away. 40 We did, however, seize this woman and asked who the young man was, 41 but she would not tell us. These things we testify."

Because they were elders of the people and judges, the assembly believed them and condemned her to death.

42 Then Susanna cried out with a loud voice, and said, "O eternal God, you know what is secret and are aware of all things before they come to be; 43 you know that these men have given false evidence against me. And now I am to die, though I have done none of the wicked things that they have charged against me!"

44 The Lord heard her cry. 45 Just as she was being led off to execution, God stirred up the holy spirit of a young lad named Daniel, 46 and he shouted with a loud voice, "I want no part in shedding this woman's blood!"

47 All the people turned to him and asked, "What is this you are saying?" 48 Taking his stand among them he said, "Are you such fools, O Israelites, as to condemn a daughter of Israel without examination and without learning the facts? 49 Return to court, for these men have given false evidence against her."

50 So all the people hurried back. And the rest of the *a* elders said to him, "Come, sit among us and inform us, for God has given you the standing of an elder." 51 Daniel said to them, "Separate them far from each other, and I will examine them."

52 When they were separated from each other, he summoned one of them and said to him, "You old relic of wicked days, your sins have now come home, which you have committed in the past, 53 pronouncing unjust judgments, condemning the innocent and acquitting the guilty, though the Lord said, 'You shall not put an innocent and righteous person to death.' 54 Now then, if you really saw this woman, tell me this: Under what tree did you see them being intimate with each other?" He answered, "Under

a Gk lacks *rest of the*

28–43: Susanna falsely accused and condemned to death. 34: The judges play the part of witnesses by laying their hands on the head of the accused (Lev 24.14). **35:** *She looked up toward Heaven,* appealing her cause to a higher tribunal (vv. 42–43). **41:** Since, according to Jewish law, a witness could not be the judge, the sentence of death is passed by the credulous *assembly*. **42:** *God, you know what is secret,* 1 Sam 16.7; Jer 11.20; 20.12; Ps 33.13–15; Prov 15.11; Heb 4.13.

44–59: Susanna rescued and acquitted. 50: Here *the elders* are obviously not the two who had testified, but their colleagues on the bench. **53:** Ex 23.7. **54–59:** The wordplay of the original (see notes *b* and *c*) may be also

a mastic tree." [b] 55 And Daniel said, "Very well! This lie has cost you your head, for the angel of God has received the sentence from God and will immediately cut [b] you in two."

56 Then, putting him to one side, he ordered them to bring the other. And he said to him, "You offspring of Canaan and not of Judah, beauty has beguiled you and lust has perverted your heart. 57 This is how you have been treating the daughters of Israel, and they were intimate with you through fear; but a daughter of Judah would not tolerate your wickedness. 58 Now then, tell me: Under what tree did you catch them being intimate with each other?" He answered, "Under an evergreen oak." [c] 59 Daniel said to him, "Very well! This lie has cost you also your head, for the angel of God is waiting with his sword to split [c] you in two, so as to destroy you both."

60 Then the whole assembly raised a great shout and blessed God, who saves those who hope in him. 61 And they took action against the two elders, because out of their own mouths Daniel had convicted them of bearing false witness; they did to them as they had wickedly planned to do to their neighbor. 62 Acting in accordance with the law of Moses, they put them to death. Thus innocent blood was spared that day.

63 Hilkiah and his wife praised God for their daughter Susanna, and so did her husband Joakim and all her relatives, because she was found innocent of a shameful deed. 64 And from that day onward Daniel had a great reputation among the people.

b The Greek words for *mastic tree* and *cut* are similar, thus forming an ironic wordplay
c The Greek words for *evergreen oak* and *split* are similar, thus forming an ironic wordplay

represented in English by the paraphrase, "Under a *clove* tree . . . the angel will *cleave* you"; "under a *yew* tree . . . the angel will *hew* you asunder."

60–62: **The two elders condemned to death. 62**: *The law of Moses,* concerning false witnesses (Deut 19.16–21).

Bel and the Dragon
(Chapter 14 of the Greek Version of Daniel)

The Septuagint and Theodotionic versions of this story cohere quite well and differ only in details, except that, as noted earlier (p. 173 AP), the version in Theodotion is integrated into the larger book of Daniel in such a fashion that something like a folk-biography is presented. The Septuagint story begins as though the reader knows nothing of Daniel at all. True also to the style of the Theodotionic Daniel, Habakkuk in v. 33 is identified as the biblical prophet, whereas in the Septuagint he is simply a man with the same name as the prophet. Theodotion has in several ways integrated the story into the larger biblical framework.

The story has three distinct episodes: the story of Daniel's exposing the fraud of the priests of Bel; the story of Daniel's proving the vulnerability of the so-called dragon, or snake; and the closing story of Daniel's surviving once more a sojourn in the lions' den, as in Dan 6, but this time for six days (v. 31). After Daniel has proved to the king the fraudulent character of the king's religion, and the king once more is favorably impressed with Daniel and his faith, the Babylonians conspire against the king, even threatening to kill him if he does not hand Daniel over to them. God summons Habakkuk from Judea and again an angel intervenes (as in Dan 6.22) to save him by transporting Habakkuk with the food to the lions' den in Babylon. One is left to assume that the angel had closed the lions' mouths, as in ch 6.

The whole account revolves around the motif of eating—the priests eating the food that the king set out for Bel, the dragon eating the concoction Daniel stirred up that killed the beast, and the focus in the final episode on Daniel's need to eat, which Habakkuk satisfies. All this is apparently set over against the lions' not eating Daniel. Another motif is set in the claim that the king's gods, in contrast to the God of Daniel, were not living gods. The complete conversion of the king occurs between his confession of the greatness of Bel in v. 18 and his confession of the greatness of the God of Daniel in v. 41.

The two genres, one of the favor of the wise courtier in the eyes of a foreign king and the other of court conflict, which are kept distinct in Dan 1–6, are combined in this story. The story reaches beyond the court, however, to the differences between those who, like the king, are open to worshiping the living God and any others who are not. The author was ridiculing, in typical early Jewish literary fashion, the two main characters of the Babylonian creation myth, the *Enuma Elish,* Bel or Marduk, and Tiamat, a sea serpent or monster whom Marduk slew as the major act of creation. By stark contrast, the God of Daniel was "the living God who created heaven and earth and has dominion over all living creatures" (v. 5).

Oppressed folk often ridicule what their malefactors hold dear, and satire is an antidote for the suffering endured. The account of Bel and the Dragon would have had such universal appeal (indeed, still does) that it could have been written anywhere at any time that oppression was severe. It was clearly intended to encourage Jews to remain faithful; it could have been refuted by Babylonian apologists. The original may have been in either Hebrew or Aramaic. Although there are few clues as to when Bel and the Dragon should be dated, the account, like the other portions of the Greek Daniel not in the Hebrew, surely originates in the second century B.C.

1 When King Astyages was laid to rest with his ancestors, Cyrus the Persian succeeded to his kingdom. 2 Daniel was a companion of the king, and was the most honored of all his friends.

3 Now the Babylonians had an idol called Bel, and every day they provided for it twelve bushels of choice flour and forty sheep and six measures*a* of wine. 4 The king revered it and went every day to worship it. But Daniel worshiped his own God.

So the king said to him, "Why do you not worship Bel?" 5 He answered, "Because I do not revere idols made with hands, but the living God, who created heaven and earth and has dominion over all living creatures."

6 The king said to him, "Do you not think that Bel is a living god? Do you not see how much he eats and drinks every day?" 7 And Daniel laughed, and said, "Do not be deceived, O king, for this thing is only clay inside and bronze outside, and it never ate or drank anything."

8 Then the king was angry and called the priests of Bel*b* and said to them, "If you do not tell me who is eating these provisions, you shall die. 9 But if you prove that Bel is eating them, Daniel shall die, because he has spoken blasphemy against Bel." Daniel said to the king, "Let it be done as you have said."

10 Now there were seventy priests of Bel, besides their wives and children. So the king went with Daniel into the temple of Bel. 11 The priests of Bel said, "See, we are now going outside; you yourself, O king, set out the food and prepare the wine, and shut the door and seal it with your signet. 12 When you return in the morning, if you do not find that Bel has eaten it all, we will die; otherwise Daniel will, who is telling lies about us." 13 They

were unconcerned, for beneath the table they had made a hidden entrance, through which they used to go in regularly and consume the provisions. 14 After they had gone out, the king set out the food for Bel. Then Daniel ordered his servants to bring ashes, and they scattered them throughout the whole temple in the presence of the king alone. Then they went out, shut the door and sealed it with the king's signet, and departed. 15 During the night the priests came as usual, with their wives and children, and they ate and drank everything.

16 Early in the morning the king rose and came, and Daniel with him. 17 The king said, "Are the seals unbroken, Daniel?" He answered, "They are unbroken, O king." 18 As soon as the doors were opened, the king looked at the table, and shouted in a loud voice, "You are great, O Bel, and in you there is no deceit at all!"

19 But Daniel laughed and restrained the king from going in. "Look at the floor," he said, "and notice whose footprints these are." 20 The king said, "I see the footprints of men and women and children."

21 Then the king was enraged, and he arrested the priests and their wives and children. They showed him the secret doors through which they used to enter to consume what was on the table. 22 Therefore the king put them to death, and gave Bel over to Daniel, who destroyed it and its temple.

23 Now in that place*c* there was a great dragon, which the Babylonians revered. 24 The king said to Daniel, "You cannot deny that this is a living god; so

a A little more than fifty gallons   *b* Gk *his priests*   *c* Other ancient authorities lack *in that place*

1–2: **Introduction. 1:** *Cyrus the Persian* (Dan 6.28) became conquering king of Babylon in 538 B.C.

3–22: **The story of Bel. 3:** *Bel*, or Bel-Marduk (compare Merodach, Jer 50.2; see Isa 46.1 and Letter of Jeremiah 6.41), was the chief god in the Babylonian pantheon. Sever-

al ancient sources testify to the enormous quantities of sacrifices presented to Marduk in the daily ritual. **7:** Daniel ridicules the king's argument: clay and bronze do not eat (Sir 30.19). **11:** Dan 6.17. Archaeologists have found many Babylonian signets.

16–22: **The fraud detected** (compare

worship him." 25 Daniel said, "I worship the Lord my God, for he is the living God. 26 But give me permission, O king, and I will kill the dragon without sword or club." The king said, "I give you permission."

27 Then Daniel took pitch, fat, and hair, and boiled them together and made cakes, which he fed to the dragon. The dragon ate them, and burst open. Then Daniel said, "See what you have been worshiping!"

28 When the Babylonians heard about it, they were very indignant and conspired against the king, saying, "The king has become a Jew; he has destroyed Bel, and killed the dragon, and slaughtered the priests." 29 Going to the king, they said, "Hand Daniel over to us, or else we will kill you and your household." 30 The king saw that they were pressing him hard, and under compulsion he handed Daniel over to them.

31 They threw Daniel into the lions' den, and he was there for six days. 32 There were seven lions in the den, and every day they had been given two human bodies and two sheep; but now they were given nothing, so that they would devour Daniel.

33 Now the prophet Habakkuk was in Judea; he had made a stew and had broken bread into a bowl, and was going into the field to take it to the reapers. 34 But the angel of the Lord said to Habakkuk, "Take the food that you have to Babylon, to Daniel, in the lions' den." 35 Habakkuk said, "Sir, I have never seen Babylon, and I know nothing about the den." 36 Then the angel of the Lord took him by the crown of his head and carried him by his hair; with the speed of the wind*d* he set him down in Babylon, right over the den.

37 Then Habakkuk shouted, "Daniel, Daniel! Take the food that God has sent you." 38 Daniel said, "You have remembered me, O God, and have not forsaken those who love you." 39 So Daniel got up and ate. And the angel of God immediately returned Habakkuk to his own place.

40 On the seventh day the king came to mourn for Daniel. When he came to the den he looked in, and there sat Daniel! 41 The king shouted with a loud voice, "You are great, O Lord, the God of Daniel, and there is no other besides you!" 42 Then he pulled Daniel*e* out, and threw into the den those who had attempted his destruction, and they were instantly eaten before his eyes.

d Or *by the power of his spirit*   *e* Gk *him*

Dan 2.12; 6.24). **22:** According to ancient historians it was Xerxes who destroyed Bel's temple.
23–42: The story of the dragon. 23: *A great dragon,* that is, a live serpent worshiped as a god (compare Num 21.8–9; 2 Kings 18.4). **26:** *Permission* was granted because the king believed in the immortality of the serpent god. **31–32:** The second time that Daniel is put in *the lions' den* (Dan 6.16–24).

33–39: The intervention of Habakkuk. 33: The author in the Greek translation made by Theodotion intends to identify this Habakkuk with the Minor Prophet of that name; chronologically, however, such an identification is impossible, nor is it reflected in the Septuagint version. **36:** *Hair,* Ezek 8.3. **37:** 1 Kings 17.4.
40–42: Daniel's liberation. 41: Compare Dan 6.26–27.

1 Maccabees

First Maccabees recounts the origins of the Hasmonean dynasty. This begins with the first appearance of Mattathias (see 2.1n.) and his five sons as leaders of the resistance against Seleucid oppression. It leads to the eventual enthronement of the dynasty at the head of the independent state of Judea. The book begins with the death of Alexander the Great and the rise to power of the Seleucid king, Antiochus IV Epiphanes. When Antiochus attacked Jerusalem and desecrated the temple, Judas Maccabeus (see 2.2–5n.) rallied the Jews to recover and purify it, thereby establishing himself as a powerful leader. The Hasmoneans were as skilled in diplomacy as in warfare, and were able to benefit from the instability of the Seleucid throne. After Judas' death his brother Jonathan was chosen leader of the Jews and was appointed high priest by the Seleucid king Demetrius, who needed help in opposing his rival Alexander Epiphanes. On Jonathan's death his brother Simon was installed as leader and high priest, and Judea was granted independence by Demetrius II. With the accession of Simon's son, John Hyrcanus, the Hasmonean dynasty was in place.

The complex problem of cultural assimilation pervades the book. Since the conquest of Alexander, Greek and Semitic culture had mingled together, and the Hellenistic culture that resulted dominated all the countries of the Mediterranean basin. The books of Sirach and the Wisdom of Solomon, as well as the translation of the Hebrew Scriptures into Greek (the Septuagint), testify to the influence of Greek ideas on Judaism. The balance between adapting to the dominant culture and remaining faithful to the teachings of the Torah was delicate, and with the harsh rule of the Seleucids that delicate balance was tipped.

Before the time of the Maccabees, Jews had usually cooperated peacefully with their foreign rulers, for since the time of Jeremiah they had believed that such rulers were part of God's plan to chastise and redeem Israel. (Paul instructs the early Christians in this theology when he exhorts them in Rom 13.1–7 to support the state.) Not all Jews, therefore, supported the Maccabees in their armed revolt against the Seleucid government. Many tried to adapt to Seleucid demands without abandoning the Torah (1.11–15, 52); others preferred martyrdom (1.62–63; 2.28–38). The author of 1 Maccabees believed that the Hasmonean rebellion and subsequent rule was in accord with the divine will.

Although the order of events sometimes differs from the parallel accounts in 2 Maccabees and the Jewish historian Josephus, 1 Maccabees is generally regarded as an accurate historical source, often corroborated by Polybius and other Greek historians. The diplomatic correspondence and royal pronouncements occurring throughout the book (8.23–32; 9.18–20, 25–45; 11.30–37; 12.5–18, 19–23; 13.36–40; 14.20–23, 27–45; 15.9; 15.16–21) appear to be authentic.

The style echoes the biblical books of Samuel and Kings, giving it the aura of an official history of Israel. Commentary is provided in the form of poetic fragments, whose biblical language provides a theological framework for the narrated events. Frequent allusions to biblical prophecies also help guide the reader to see the Hasmonean dynasty as the fulfillment of the divine will.

All extant manuscripts of 1 Maccabees are in Greek or Latin, the original Hebrew having been lost at an early time. The book was probably written shortly after the death of the Hasmonean king John Hyrcanus I in 104 B.C. (see 16.23–24).

1 After Alexander son of Philip, the Macedonian, who came from the land of Kittim, had defeated*a* King Darius of the Persians and the Medes, he succeeded him as king. (He had previously become king of Greece.) ²He fought many battles, conquered strongholds, and put to death the kings of the earth. ³He advanced to the ends of the earth, and plundered many nations. When the earth became quiet before him, he was exalted, and his heart was lifted up. ⁴He gathered a very strong army and ruled over countries, nations, and princes, and they became tributary to him.

5 After this he fell sick and perceived that he was dying. ⁶So he summoned his most honored officers, who had been brought up with him from youth, and divided his kingdom among them while he was still alive. ⁷And after Alexander had reigned twelve years, he died.

8 Then his officers began to rule, each in his own place. ⁹They all put on crowns after his death, and so did their descendants after them for many years; and they caused many evils on the earth.

10 From them came forth a sinful root, Antiochus Epiphanes, son of King Antiochus; he had been a hostage in Rome. He began to reign in the one hundred thirty-seventh year of the kingdom of the Greeks. *b*

11 In those days certain renegades came out from Israel and misled many, saying, "Let us go and make a covenant with the Gentiles around us, for since we separated from them many disasters have come upon us." ¹²This proposal pleased them, ¹³and some of the people eagerly went to the king, who authorized them to observe the ordinances of the Gentiles. ¹⁴So they built a gymnasium in Jerusalem, according to Gentile custom, ¹⁵and removed the marks of circumcision, and abandoned the holy covenant. They joined with the Gentiles and sold themselves to do evil.

16 When Antiochus saw that his kingdom was established, he determined to become king of the land of Egypt, in order that he might reign over both kingdoms. ¹⁷So he invaded Egypt with a strong force, with chariots and elephants

a Gk adds *and he defeated*   b 175 B.C.

1.1–10: Introduction. A summary of history from Alexander to Antiochus IV. **1**: *Alexander* the Great (356–323 B.C.), son of Philip of Macedon, who had conquered *Kittim* (Greece), swept through Asia Minor, and *defeated King Darius* III at Issus (333 B.C.) and at Gaugamela (331 B.C.). **3**: After taking Egypt, Mesopotamia, and Persia he advanced to the *ends of the earth* (to Bactria and India). *He was exalted,* i.e. he was deified. Such pride in rulers is castigated in Isa 2.5–22. **4**: He planned a universal empire dominated by Greek culture. **5**: *He fell sick* in Babylon. **6**: A complex history of power struggles lies behind this statement. **8–9**: By 275 B.C. three dynasties were established, the Antigonids of Macedonia, the Ptolemies of Egypt, and the Seleucids of Syria. **9**. *Crowns,* lit. "diadems"; these were a strip of white cloth decorated on the edges. **10**: *Sinful root,* Isa 11.10; Dan 11.7. *Antiochus* IV, who took the name *Epiphanes* ("god manifest"), reigned 175–164 B.C.; he was son *of King Antiochus* III the Great (223–187 B.C.), who had wrested Palestine from Egypt at the battle of Paneas in 198 B.C. but lost most of Asia Minor to Rome at Magnesia in 190 B.C. (compare Dan 11.18). Because of this defeat the

son *had been a hostage in Rome. One hundred thirty-seventh year* of the Seleucid era; reckoning of this era varied in different places; dates given in the notes (*b, c,* etc.) are approximate. **1.11–15: The paganizing program.** Greek culture had penetrated Palestine peacefully, but now enthusiasts introduced Greek religion (2 Macc 4,11–17). **11**: *Certain renegades,* led by Jason, whom Antiochus appointed in place of his brother Onias III (2 Macc 4.7). *Renegades,* lit. "lawless ones," those who compromised the Law of Moses. The term is used throughout 1 Maccabees to describe Jews who did not support the Hasmoneans. *Covenant,* prohibition that was a cornerstone of the Law, because of the danger of idolatry (Ex 34.15–16, Deut 7.1–6). See Jer 44.15–23. *Disasters,* loss of business and prestige because relations with Syria had deteriorated. **14**: *A gymnasium,* see 2 Macc 4.9–10 n. **1.16–40: Antiochus invades Egypt and Palestine.** Invasion of Egypt is followed by plundering of the temple in Jerusalem (2 Macc 5.1, 11–26). **17**: The Syrian army had *elephants,* though the treaty of Apamea with

and cavalry and with a large fleet. [18] He engaged King Ptolemy of Egypt in battle, and Ptolemy turned and fled before him, and many were wounded and fell. [19] They captured the fortified cities in the land of Egypt, and he plundered the land of Egypt.

20 After subduing Egypt, Antiochus returned in the one hundred forty-third year.[c] He went up against Israel and came to Jerusalem with a strong force. [21] He arrogantly entered the sanctuary and took the golden altar, the lampstand for the light, and all its utensils. [22] He took also the table for the bread of the Presence, the cups for drink offerings, the bowls, the golden censers, the curtain, the crowns, and the gold decoration on the front of the temple; he stripped it all off. [23] He took the silver and the gold, and the costly vessels; he took also the hidden treasures that he found. [24] Taking them all, he went into his own land.

He shed much blood,
and spoke with great arrogance.
25 Israel mourned deeply in every
community,
26   rulers and elders groaned,
young women and young men
became faint,
the beauty of the women faded.
27 Every bridegroom took up the
lament;
she who sat in the bridal
chamber was mourning.
28 Even the land trembled for its
inhabitants,
and all the house of Jacob was
clothed with shame.

29 Two years later the king sent to the cities of Judah a chief collector of tribute,

and he came to Jerusalem with a large force. [30] Deceitfully he spoke peaceable words to them, and they believed him; but he suddenly fell upon the city, dealt it a severe blow, and destroyed many people of Israel. [31] He plundered the city, burned it with fire, and tore down its houses and its surrounding walls. [32] They took captive the women and children, and seized the livestock. [33] Then they fortified the city of David with a great strong wall and strong towers, and it became their citadel. [34] They stationed there a sinful people, men who were renegades. These strengthened their position; [35] they stored up arms and food, and collecting the spoils of Jerusalem they stored them there, and became a great menace,
36 for the citadel[d] became an ambush
against the sanctuary,
an evil adversary of Israel at all
times.
37 On every side of the sanctuary
they shed innocent blood;
they even defiled the sanctuary.
38 Because of them the residents of
Jerusalem fled;
she became a dwelling of
strangers;
she became strange to her
offspring,
and her children forsook her.
39 Her sanctuary became desolate like
a desert;
her feasts were turned into
mourning,
her sabbaths into a reproach,
her honor into contempt.
40 Her dishonor now grew as great
as her glory;

c 169 B.C.   d Gk *it*

Rome (188 B.C.) had forbidden this. **18:** *Ptolemy VI Philometor* reigned 180–145 B.C. **20:** *Antiochus returned* because the Roman envoy, Popilius Laenas, threatened him with war if he annexed Egypt; also, news of internal strife in Jerusalem had reached him, and he feared a revolt (2 Macc ch 5). **21:** See Isa 10.5–11.
1.24–28: Fragment of a contemporary poem. **28:** *The house of Jacob,* Israel, the Jewish people. **33:** *City of David,* a term with several

different meanings in the Hebrew Scriptures (compare Isa 22.9 and 1 Kings 11.27 with 2 Sam 5.7, 9); the precise location of this fortification within Jerusalem is unknown. **34:** *Sinful people,* like "lawless ones," a biblical term used primarily for evil Israelites (Isa 1.4); it here connotes Jews who supported the Seleucids. **36–40:** Poetic fragment (compare Pss 74; 79).
1.41–64: Desecration of the temple.

her exaltation was turned into mourning.

41 Then the king wrote to his whole kingdom that all should be one people, 42 and that all should give up their particular customs. 43 All the Gentiles accepted the command of the king. Many even from Israel gladly adopted his religion; they sacrificed to idols and profaned the sabbath. 44 And the king sent letters by messengers to Jerusalem and the towns of Judah; he directed them to follow customs strange to the land, 45 to forbid burnt offerings and sacrifices and drink offerings in the sanctuary, to profane sabbaths and festivals, 46 to defile the sanctuary and the priests, 47 to build altars and sacred precincts and shrines for idols, to sacrifice swine and other unclean animals, 48 and to leave their sons uncircumcised. They were to make themselves abominable by everything unclean and profane, 49 so that they would forget the law and change all the ordinances. 50 He added, *e* "And whoever does not obey the command of the king shall die."

51 In such words he wrote to his whole kingdom. He appointed inspectors over all the people and commanded the towns of Judah to offer sacrifice, town by town. 52 Many of the people, everyone who forsook the law, joined them, and they did evil in the land; 53 they drove Israel into hiding in every place of refuge they had.

54 Now on the fifteenth day of Chis-lev, in the one hundred forty-fifth year, *f* they erected a desolating sacrilege on the altar of burnt offering. They also built altars in the surrounding towns of Judah, 55 and offered incense at the doors of the houses and in the streets. 56 The books of the law that they found they tore to pieces and burned with fire. 57 Anyone found possessing the book of the covenant, or anyone who adhered to the law, was condemned to death by decree of the king. 58 They kept using violence against Israel, against those who were found month after month in the towns. 59 On the twenty-fifth day of the month they offered sacrifice on the altar that was on top of the altar of burnt offering. 60 According to the decree, they put to death the women who had their children circumcised, 61 and their families and those who circumcised them; and they hung the infants from their mothers' necks.

62 But many in Israel stood firm and were resolved in their hearts not to eat unclean food. 63 They chose to die rather than to be defiled by food or to profane the holy covenant; and they did die. 64 Very great wrath came upon Israel.

2 In those days Mattathias son of John son of Simeon, a priest of the family of Joarib, moved from Jerusalem and settled in Modein. 2 He had five sons, John surnamed Gaddi, 3 Simon called Thassi, 4 Judas called Maccabeus, 5 Eleazar called Avaran, and Jonathan called Apphus.

e Gk lacks *He added*   *f* 167 B.C.

The first outright religious persecution of the Jews, which is also reflected in Dan 11.29–39 (compare 2 Macc 6.1–11). **41–42:** *His whole kingdom,* Syria, Palestine, Mesopotamia, Persia, and parts of Asia Minor. *One people,* unified in language, religion, culture, and even dress; Judaism, with its revealed law and rejection of other gods, opposed this. **47:** *Unclean animals* were not dirty but ritually impure and unacceptable for sacrifice (Lev 22.17–30). **1.54:** *Chislev,* approximately December. The *desolating sacrilege* (Dan 11.31; 12.11; 2 Macc 6.2) was an altar to Olympian Zeus and perhaps a statue of him. **59:** *Offered sacrifice,* probably of swine (2 Macc 6.4–5). **60–**64: 2 Macc chs 6–7, and 4 Maccabees contain stories of martyrdoms. *Chose to die rather than to be defiled by food,* compare Dan 3.8–18. *Wrath came upon Israel,* as a punishment for sin (2 Macc 6.12–16).

2.1–48: Revolt of Mattathias. 1: The family of *Mattathias* is known as Hasmoneans (see p. xiv AP), from a traditional ancestor Hashmonia, not mentioned in 1 Maccabees but named in Josephus (*Antiquities,* XII. vi. 1). *Joarib* was first in the list of divisions of priests (1 Chr 24.7; Neh 11.10). *Modein,* in the mountains on the road to Beth-horon, about seventeen miles northwest of Jerusalem. **2–5:** *Simon,* third of the family to rule (chs 13–16). *Maccabeus,* probably from a Hebrew word

6 He saw the blasphemies being committed in Judah and Jerusalem, 7 and said,
"Alas! Why was I born to see this,
the ruin of my people, the ruin
of the holy city,
and to live there when it was
given over to the enemy,
the sanctuary given over to
aliens?
8 Her temple has become like a
person without honor;[g]
9 her glorious vessels have been
carried into exile.
Her infants have been killed in her
streets,
her youths by the sword of the
foe.
10 What nation has not inherited her
palaces[h]
and has not seized her spoils?
11 All her adornment has been taken
away;
no longer free, she has become a
slave.
12 And see, our holy place, our
beauty,
and our glory have been laid
waste;
the Gentiles have profaned them.
13 Why should we live any
longer?"

14 Then Mattathias and his sons tore their clothes, put on sackcloth, and mourned greatly.

15 The king's officers who were enforcing the apostasy came to the town of Modein to make them offer sacrifice. 16 Many from Israel came to them; and Mattathias and his sons were assembled. 17 Then the king's officers spoke to Mattathias as follows: "You are a leader, honored and great in this town, and supported by sons and brothers. 18 Now be the first to come and do what the king commands, as all the Gentiles and the people of Judah and those that are left in Jerusalem have done. Then you and your sons will be numbered among the Friends of the king, and you and your sons will be honored with silver and gold and many gifts."

19 But Mattathias answered and said in a loud voice: "Even if all the nations that live under the rule of the king obey him, and have chosen to obey his commandments, everyone of them abandoning the religion of their ancestors, 20 I and my sons and my brothers will continue to live by the covenant of our ancestors. 21 Far be it from us to desert the law and the ordinances. 22 We will not obey the king's words by turning aside from our religion to the right hand or to the left."

23 When he had finished speaking these words, a Jew came forward in the sight of all to offer sacrifice on the altar in Modein, according to the king's command. 24 When Mattathias saw it, he burned with zeal and his heart was stirred. He gave vent to righteous anger; he ran and killed him on the altar. 25 At the same time he killed the king's officer who was forcing them to sacrifice, and he tore down the altar. 26 Thus he burned with zeal for the law, just as Phinehas did against Zimri son of Salu.

27 Then Mattathias cried out in the town with a loud voice, saying: "Let everyone who is zealous for the law and supports the covenant come out with me!" 28 Then he and his sons fled to the hills and left all that they had in the town.

29 At that time many who were seeking righteousness and justice went down to the wilderness to live there, 30 they,

---

g Meaning of Gk uncertain   *h* Other ancient authorities read *has not had a part in her kingdom*

---

meaning "hammer." The other surnames are of uncertain derivation. *Jonathan,* successor of Judas (chs 9–12). **7–13:** Poetic fragment; compare Pss 44; 74; 79; and the book of Lamentations. **2.14:** *Tore their clothes, put on sackcloth,* signs of mourning (Gen 37.34). **18:** *The Friends of the king* were a special class of potentates and courtiers who wore distinctive purple dress and insignia. **2.23:** Elsewhere in chs 1–13, "Israelite" is used instead of the term *Jew,* which here perhaps means "Judean." **24:** *His heart,* literally "his kidneys," which were considered the seat of emotion. **26:** *As Phinehas did,* the whole episode is written to echo Num 25.6–15. **28:**

their sons, their wives, and their livestock, because troubles pressed heavily upon them. 31 And it was reported to the king's officers, and to the troops in Jerusalem the city of David, that those who had rejected the king's command had gone down to the hiding places in the wilderness. 32 Many pursued them, and overtook them; they encamped opposite them and prepared for battle against them on the sabbath day. 33 They said to them, "Enough of this! Come out and do what the king commands, and you will live." 34 But they said, "We will not come out, nor will we do what the king commands and so profane the sabbath day." 35 Then the enemy[i] quickly attacked them. 36 But they did not answer them or hurl a stone at them or block up their hiding places, 37 for they said, "Let us all die in our innocence; heaven and earth testify for us that you are killing us unjustly." 38 So they attacked them on the sabbath, and they died, with their wives and children and livestock, to the number of a thousand persons.

39 When Mattathias and his friends learned of it, they mourned for them deeply. 40 And all said to their neighbors: "If we all do as our kindred have done and refuse to fight with the Gentiles for our lives and for our ordinances, they will quickly destroy us from the earth." 41 So they made this decision that day: "Let us fight against anyone who comes to attack us on the sabbath day; let us not all die as our kindred died in their hiding places."

42 Then there united with them a company of Hasideans, mighty warriors of Israel, all who offered themselves willingly for the law. 43 And all who became fugitives to escape their troubles joined them and reinforced them. 44 They organized an army, and struck down sinners in their anger and renegades in their wrath; the survivors fled to the Gentiles for safety. 45 And Mattathias and his friends went around and tore down the altars; 46 they forcibly circumcised all the uncircumcised boys that they found within the borders of Israel. 47 They hunted down the arrogant, and the work prospered in their hands. 48 They rescued the law out of the hands of the Gentiles and kings, and they never let the sinner gain the upper hand.

49 Now the days drew near for Mattathias to die, and he said to his sons: "Arrogance and scorn have now become strong; it is a time of ruin and furious anger. 50 Now, my children, show zeal for the law, and give your lives for the covenant of our ancestors.

51 "Remember the deeds of the ancestors, which they did in their generations; and you will receive great honor and an everlasting name. 52 Was not Abraham found faithful when tested, and it was reckoned to him as righteousness? 53 Joseph in the time of his distress kept the commandment, and became lord of Egypt. 54 Phinehas our ancestor, because he was deeply zealous, received the covenant of everlasting priesthood. 55 Joshua, because he fulfilled the command, became a judge in Israel. 56 Caleb, because he testified in the assembly, received an inheritance in the land. 57 David, because he was merciful, inherited the throne of the kingdom forever. 58 Elijah, because of great zeal for the law, was taken up into heaven. 59 Hananiah, Azariah, and Mishael believed and were

i Gk *they*

---

2 Macc 5.27. **29–31**: In *the wilderness* of Judea they found *hiding places* in grottoes and caves (Judg 20.47). **37**: 1.63.

2.41: The earliest statement of the principle that one may profane one sabbath in order to keep all the others. **42**: *Hasideans,* "the pious," a group not concerned for Jewish nationalism but only for the religious law. At first they resisted passively (1.62–63; 2.37), but now turned to violent action.

2.49–70: **Death of Mattathias.** Mattathias is portrayed like Jacob in Gen ch 49. **52**: *Faithful when tested,* Gen 22.1–18. *Reckoned to him,* Gen 15.6; Rom 4.3. **53**: *Joseph,* Gen chs 39–45. **54**: *Phinehas,* v. 26. **55–56**: *Joshua . . . Caleb,* Num 13.1–14.12; 26.65; Josh 1.1–9.

2.57: *Merciful,* or perhaps "loyal" (2 Sam 7.16; Pss 89.35–37; 132.11–12). **58**: 2 Kings 2.9–12. **59–60**: Dan 3.8–30; 6.1–24. **63**: Ps

saved from the flame. [60]Daniel, because of his innocence, was delivered from the mouth of the lions.

61 "And so observe, from generation to generation, that none of those who put their trust in him will lack strength. [62]Do not fear the words of sinners, for their splendor will turn into dung and worms. [63]Today they will be exalted, but tomorrow they will not be found, because they will have returned to the dust, and their plans will have perished. [64]My children, be courageous and grow strong in the law, for by it you will gain honor.

65 "Here is your brother Simeon who, I know, is wise in counsel; always listen to him; he shall be your father. [66]Judas Maccabeus has been a mighty warrior from his youth; he shall command the army for you and fight the battle against the peoples.*j* [67]You shall rally around you all who observe the law, and avenge the wrong done to your people. [68]Pay back the Gentiles in full, and obey the commands of the law."

69 Then he blessed them, and was gathered to his ancestors. [70]He died in the one hundred forty-sixth year*k* and was buried in the tomb of his ancestors at Modein. And all Israel mourned for him with great lamentation.

3 Then his son Judas, who was called Maccabeus, took command in his place. [2]All his brothers and all who had joined his father helped him; they gladly fought for Israel.

[3] He extended the glory of his
    people.
Like a giant he put on his
    breastplate;
he bound on his armor of war and
    waged battles,
protecting the camp by his
    sword.
[4] He was like a lion in his deeds,

like a lion's cub roaring for
    prey.
[5] He searched out and pursued those
    who broke the law;
he burned those who troubled
    his people.
[6] Lawbreakers shrank back for fear
    of him;
all the evildoers were
    confounded;
and deliverance prospered by his
    hand.
[7] He embittered many kings,
but he made Jacob glad by his
    deeds,
and his memory is blessed
    forever.
[8] He went through the cities of
    Judah;
he destroyed the ungodly out of
    the land;*l*
thus he turned away wrath from
    Israel.
[9] He was renowned to the ends of
    the earth;
he gathered in those who were
    perishing.

10 Apollonius now gathered together Gentiles and a large force from Samaria to fight against Israel. [11]When Judas learned of it, he went out to meet him, and he defeated and killed him. Many were wounded and fell, and the rest fled. [12]Then they seized their spoils; and Judas took the sword of Apollonius, and used it in battle the rest of his life.

13 When Seron, the commander of the Syrian army, heard that Judas had gathered a large company, including a body of faithful soldiers who stayed with him and went out to battle, [14]he said, "I will make a name for myself and win honor in the kingdom. I will make war on Judas and his companions, who scorn

j Or *of the people*   *k* 166 B.C.   *l* Gk *it*

37.10, 35–36. **69:** *Gathered to his ancestors,* buried with his ancestors (Judg 2.10).
3.1–12: Defeat of Apollonius. 3–9: From a contemporary poem. **4:** *Like a lion,* Hos 5.14. **8:** *He turned away wrath,* i.e. God's punishment, through his exploits (2 Macc

7.38). **10:** *Apollonius,* according to Josephus (*Antiquities,* XII. v. 5; vii.1), was governor of Samaria.
3.13–26: Battle of Beth-horon. This was Judas' first great victory. **16:** *The ascent of Beth-horon* was a route from the coastal plain

the king's command." [15] Once again a strong army of godless men went up with him to help him, to take vengeance on the Israelites.

16 When he approached the ascent of Beth-horon, Judas went out to meet him with a small company. [17] But when they saw the army coming to meet them, they said to Judas, "How can we, few as we are, fight against so great and so strong a multitude? And we are faint, for we have eaten nothing today." [18] Judas replied, "It is easy for many to be hemmed in by few, for in the sight of Heaven there is no difference between saving by many or by few. [19] It is not on the size of the army that victory in battle depends, but strength comes from Heaven. [20] They come against us in great insolence and lawlessness to destroy us and our wives and our children, and to despoil us; [21] but we fight for our lives and our laws. [22] He himself will crush them before us; as for you, do not be afraid of them."

23 When he finished speaking, he rushed suddenly against Seron and his army, and they were crushed before him. [24] They pursued them*m* down the descent of Beth-horon to the plain; eight hundred of them fell, and the rest fled into the land of the Philistines. [25] Then Judas and his brothers began to be feared, and terror fell on the Gentiles all around them. [26] His fame reached the king, and the Gentiles talked of the battles of Judas.

27 When King Antiochus heard these reports, he was greatly angered; and he sent and gathered all the forces of his kingdom, a very strong army. [28] He opened his coffers and gave a year's pay to his forces, and ordered them to be ready for any need. [29] Then he saw that the money in the treasury was exhausted, and that the revenues from the country were small because of the dissension and disaster that he had caused in the land by abolishing the laws that had existed from the earliest days. [30] He feared that he might not have such funds as he had before for his expenses and for the gifts that he used to give more lavishly than preceding kings. [31] He was greatly perplexed in mind; then he determined to go to Persia and collect the revenues from those regions and raise a large fund.

32 He left Lysias, a distinguished man of royal lineage, in charge of the king's affairs from the river Euphrates to the borders of Egypt. [33] Lysias was also to take care of his son Antiochus until he returned. [34] And he turned over to Lysias*n* half of his forces and the elephants, and gave him orders about all that he wanted done. As for the residents of Judea and Jerusalem, [35] Lysias was to send a force against them to wipe out and destroy the strength of Israel and the remnant of Jerusalem; he was to banish the memory of them from the place, [36] settle aliens in all their territory, and distribute their land by lot. [37] Then the king took the remaining half of his forces and left Antioch his capital in the one hundred and forty-seventh year.*o* He crossed the Euphrates river and went through the upper provinces.

38 Lysias chose Ptolemy son of Dorymenes, and Nicanor and Gorgias, able men among the Friends of the king, [39] and sent with them forty thousand in-

m Other ancient authorities read *him*
n Gk *him*   *o* 165 B.C.

---

to the Judean highlands. The town is about twelve miles northwest of Jerusalem. **18**: The word *Heaven* was used to avoid pronouncing God's name (compare "he himself," v. 22, and see Sus 9 n.). *By many or by few,* 1 Sam 14.6. **24**: *Land of the Philistines,* the coastal plain.
3.27–4.35: **Campaigns of Lysias.** Antiochus IV goes to Persia; Judas defeats Lysias at Emmaus and Beth-zur.

3.28: *Any need* implies that Seleucid power was beginning to decline. **30**: Antiochus was noted for his extravagance (see 2 Macc 4.30 n.). **33**: *Antiochus* V Eupator, *his son,* was only nine years old; he reigned 164–162 B.C. **36**: *Settle aliens,* as the Assyrians had done (2 Kings 17.24). **37**: *Antioch,* modern Antakya, was built by Seleucus I in 300 B.C. and expanded by Antiochus IV. *Upper provinces,* Persia.

fantry and seven thousand cavalry to go into the land of Judah and destroy it, as the king had commanded. 40 So they set out with their entire force, and when they arrived they encamped near Emmaus in the plain. 41 When the traders of the region heard what was said to them, they took silver and gold in immense amounts, and fetters,*p* and went to the camp to get the Israelites for slaves. And forces from Syria and the land of the Philistines joined with them.

42 Now Judas and his brothers saw that misfortunes had increased and that the forces were encamped in their territory. They also learned what the king had commanded to do to the people to cause their final destruction. 43 But they said to one another, "Let us restore the ruins of our people, and fight for our people and the sanctuary." 44 So the congregation assembled to be ready for battle, and to pray and ask for mercy and compassion.
45 Jerusalem was uninhabited like a
    wilderness;
  not one of her children went in
    or out.
  The sanctuary was trampled
    down,
  and aliens held the citadel;
  it was a lodging place for the
    Gentiles.
  Joy was taken from Jacob;
  the flute and the harp ceased to
    play.

46 Then they gathered together and went to Mizpah, opposite Jerusalem, because Israel formerly had a place of prayer in Mizpah. 47 They fasted that day, put on sackcloth and sprinkled ashes on their heads, and tore their clothes. 48 And they opened the book of the law to inquire into those matters about which the Gentiles consulted the likenesses of their gods. 49 They also brought the vestments of the priesthood and the first fruits and the tithes, and they stirred up the nazirites*q* who had completed their days; 50 and they cried aloud to Heaven, saying,

  "What shall we do with these?
    Where shall we take them?
51 Your sanctuary is trampled down
    and profaned,
  and your priests mourn in
    humiliation.
52 Here the Gentiles are assembled
    against us to destroy us;
  you know what they plot
    against us.
53 How will we be able to withstand
    them,
  if you do not help us?"

54 Then they sounded the trumpets and gave a loud shout. 55 After this Judas appointed leaders of the people, in charge of thousands and hundreds and fifties and tens. 56 Those who were building houses, or were about to be married, or were planting a vineyard, or were fainthearted, he told to go home again, according to the law. 57 Then the army marched out and encamped to the south of Emmaus. 58 And Judas said, "Arm yourselves and be courageous. Be ready early in the morning to fight with these Gentiles who have assembled against us to destroy us and our sanctuary. 59 It is better for us to die in battle than to see the misfortunes of our nation and of the sanctuary. 60 But as his will in heaven may be, so shall he do."

p Syr: Gk Mss, Vg *slaves*   *q* That is *those
separated* or *those consecrated*

3.38: *Ptolemy,* known as Macron (2 Macc 10.12). *Nicanor,* 2 Macc 8.9. *Gorgias,* 2 Macc 10.14. **40:** *Emmaus* (not the Emmaus of Lk 24.13), was about twenty-five miles west of Jerusalem. **41:** Some pro-Syrian Jews joined Antiochus' army. **45:** Compare Ps 74; Isa 24.8.
3.46: *Mizpah,* perhaps en-Nebi Samwil, seven miles northwest of Jerusalem, but sometimes identified with Tell en-Nasbeh, nine miles north of the city. See 1 Sam 7.5–11. **48:** They expected guidance from *the book of the law,* the Pentateuch, while the Greeks sought oracles from *the likenesses of their gods.* **49:** *Tithes* were brought to Jerusalem and there distributed (Neh 10.35–38). *Nazirites,* Num 6.1–21. **50–53:** Verse 45. **54:** *Trumpets,* to summon the army (Num 10.1–10). **55:** In Moses' day such *leaders* assisted in civic administration (Ex 18.25); here, as in the Essene *War Scroll* from Qumran, they have a military

4 Now Gorgias took five thousand infantry and one thousand picked cavalry, and this division moved out by night ²to fall upon the camp of the Jews and attack them suddenly. Men from the citadel were his guides. ³But Judas heard of it, and he and his warriors moved out to attack the king's force in Emmaus ⁴while the division was still absent from the camp. ⁵When Gorgias entered the camp of Judas by night, he found no one there, so he looked for them in the hills, because he said, "These men are running away from us."

6 At daybreak Judas appeared in the plain with three thousand men, but they did not have armor and swords such as they desired. ⁷And they saw the camp of the Gentiles, strong and fortified, with cavalry all around it; and these men were trained in war. ⁸But Judas said to those who were with him, "Do not fear their numbers or be afraid when they charge. ⁹Remember how our ancestors were saved at the Red Sea, when Pharaoh with his forces pursued them. ¹⁰And now, let us cry to Heaven, to see whether he will favor us and remember his covenant with our ancestors and crush this army before us today. ¹¹Then all the Gentiles will know that there is one who redeems and saves Israel."

12 When the foreigners looked up and saw them coming against them, ¹³they went out from their camp to battle. Then the men with Judas blew their trumpets ¹⁴and engaged in battle. The Gentiles were crushed, and fled into the plain, ¹⁵and all those in the rear fell by the sword. They pursued them to Gazara, and to the plains of Idumea, and to Azotus and Jamnia; and three thousand of them fell. ¹⁶Then Judas and his force turned back from pursuing them, ¹⁷and he said to the people, "Do not be greedy for plunder, for there is a battle before us; ¹⁸Gorgias and his force are near us in the hills. But stand now against our enemies and fight them, and afterward seize the plunder boldly."

19 Just as Judas was finishing this speech, a detachment appeared, coming out of the hills. ²⁰They saw that their army[r] had been put to flight, and that the Jews[r] were burning the camp, for the smoke that was seen showed what had happened. ²¹When they perceived this, they were greatly frightened, and when they also saw the army of Judas drawn up in the plain for battle, ²²they all fled into the land of the Philistines. ²³Then Judas returned to plunder the camp, and they seized a great amount of gold and silver, and cloth dyed blue and sea purple, and great riches. ²⁴On their return they sang hymns and praises to Heaven—"For he is good, for his mercy endures forever." ²⁵Thus Israel had a great deliverance that day.

26 Those of the foreigners who escaped went and reported to Lysias all that had happened. ²⁷When he heard it, he was perplexed and discouraged, for things had not happened to Israel as he had intended, nor had they turned out as the king had ordered. ²⁸But the next year he mustered sixty thousand picked infantry and five thousand cavalry to subdue them. ²⁹They came into Idumea and encamped at Beth-zur, and Judas met them with ten thousand men.

30 When he saw that their army was strong, he prayed, saying, "Blessed are

r Gk *they*

---

function (2 Macc 8.22–23). **56:** Deut 20.5–8.
4.2: *Men from the citadel,* Jewish refugees opposed to Judas. **9:** Ex 14.21–29. **15:** The pursuit went in all directions. *Gazara,* or Gezer (Josh 21.21; 1 Kings 9.17), was five miles northwest of Emmaus. *Idumea* was far to the south. *Azotus,* or Ashdod, and *Jamnia,* lay west and southwest. **17–18:** Judas maintained discipline (2 Macc 8.26).
4.19: *The hills,* the Judean highland. **24:**

Heaven, see 3.18 n. See Ps 118.1; 136.1. **26–35:** The account in 2 Macc 11.1–12 puts the rout of Lysias after the death of Timothy. **28:** *The next year,* perhaps as late as autumn, 164 B.C. **29:** *Beth-zur,* about twenty miles south of Jerusalem on the road to Hebron. Lysias decided to attack Jerusalem from the south.
4.30–35: The account in 2 Macc 11.6–15

you, O Savior of Israel, who crushed the attack of the mighty warrior by the hand of your servant David, and gave the camp of the Philistines into the hands of Jonathan son of Saul, and of the man who carried his armor. ³¹ Hem in this army by the hand of your people Israel, and let them be ashamed of their troops and their cavalry. ³² Fill them with cowardice; melt the boldness of their strength; let them tremble in their destruction. ³³ Strike them down with the sword of those who love you, and let all who know your name praise you with hymns."

34 Then both sides attacked, and there fell of the army of Lysias five thousand men; they fell in action.ˢ ³⁵ When Lysias saw the rout of his troops and observed the boldness that inspired those of Judas, and how ready they were either to live or to die nobly, he withdrew to Antioch and enlisted mercenaries in order to invade Judea again with an even larger army.

36 Then Judas and his brothers said, "See, our enemies are crushed; let us go up to cleanse the sanctuary and dedicate it." ³⁷ So all the army assembled and went up to Mount Zion. ³⁸ There they saw the sanctuary desolate, the altar profaned, and the gates burned. In the courts they saw bushes sprung up as in a thicket, or as on one of the mountains. They saw also the chambers of the priests in ruins. ³⁹ Then they tore their clothes and mourned with great lamentation; they sprinkled themselves with ashes ⁴⁰ and fell face down on the ground. And when the signal was given with the trumpets, they cried out to Heaven.

41 Then Judas detailed men to fight against those in the citadel until he had cleansed the sanctuary. ⁴² He chose blameless priests devoted to the law, ⁴³ and they cleansed the sanctuary and removed the defiled stones to an unclean place. ⁴⁴ They deliberated what to do about the altar of burnt offering, which had been profaned. ⁴⁵ And they thought it best to tear it down, so that it would not be a lasting shame to them that the Gentiles had defiled it. So they tore down the altar, ⁴⁶ and stored the stones in a convenient place on the temple hill until a prophet should come to tell what to do with them. ⁴⁷ Then they took unhewnᵗ stones, as the law directs, and built a new altar like the former one. ⁴⁸ They also rebuilt the sanctuary and the interior of the temple, and consecrated the courts. ⁴⁹ They made new holy vessels, and brought the lampstand, the altar of incense, and the table into the temple. ⁵⁰ Then they offered incense on the altar and lit the lamps on the lampstand, and these gave light in the temple. ⁵¹ They placed the bread on the table and hung up the curtains. Thus they finished all the work they had undertaken.

52 Early in the morning on the twenty-fifth day of the ninth month, which is the month of Chislev, in the one hundred forty-eighth year,ᵘ ⁵³ they rose and offered sacrifice, as the law directs, on the new altar of burnt offering that they had built. ⁵⁴ At the very season and on the very day that the Gentiles had profaned it, it was dedicated with songs and harps and lutes and cymbals. ⁵⁵ All the people fell on their faces and wor-

---

ˢ Or *and some fell on the opposite side*
ᵗ Gk *whole* ᵘ 164 B.C.

---

agrees that Judas won the battle, but states that there was a negotiated peace.

4.36–61: Rededication of the temple. **38**: *Chambers of the priests* perhaps surrounded the sanctuary on three sides. **41**: *The citadel* (1.33–35) was occupied by a Syrian garrison until the time of Simon (13.49–52). **46**: Malachi was regarded as the last *prophet;* though such men as John Hyrcanus I and John the Baptist were thought to have prophetic gifts, this was not universally recognized. **47**: Ex 20.25; Deut 27.5–6. **50**: *Offered incense . . . lit the lamps,* Ex 30.7–8. **51**: *The bread,* of the Presence (Ex 25.30).

4.52–59: Judas set the rededication of the temple exactly three years after its pollution (1.54) and three and a half years after Antiochus' capture of Jerusalem (Dan 7.25; but see 2 Macc 10.3). The Hanukkah festival, celebrated *for eight days* like Hezekiah's reconsecration (2 Chr 29.17), commemorates this event.

shiped and blessed Heaven, who had prospered them. ⁵⁶ So they celebrated the dedication of the altar for eight days, and joyfully offered burnt offerings; they offered a sacrifice of well-being and a thanksgiving offering. ⁵⁷ They decorated the front of the temple with golden crowns and small shields; they restored the gates and the chambers for the priests, and fitted them with doors. ⁵⁸ There was very great joy among the people, and the disgrace brought by the Gentiles was removed.

59 Then Judas and his brothers and all the assembly of Israel determined that every year at that season the days of dedication of the altar should be observed with joy and gladness for eight days, beginning with the twenty-fifth day of the month of Chislev.

60 At that time they fortified Mount Zion with high walls and strong towers all around, to keep the Gentiles from coming and trampling them down as they had done before. ⁶¹ Judas*ᵛ* stationed a garrison there to guard it; he also fortified Beth-zur to guard it, so that the people might have a stronghold that faced Idumea.

5 When the Gentiles all around heard that the altar had been rebuilt and the sanctuary dedicated as it was before, they became very angry, ²and they determined to destroy the descendants of Jacob who lived among them. So they began to kill and destroy among the people. ³ But Judas made war on the descendants of Esau in Idumea, at Akrabattene, because they kept lying in wait for Israel. He dealt them a heavy blow and humbled them and despoiled them. ⁴ He also

remembered the wickedness of the sons of Baean, who were a trap and a snare to the people and ambushed them on the highways. ⁵ They were shut up by him in their*ʷ* towers; and he encamped against them, vowed their complete destruction, and burned with fire their towers and all who were in them. ⁶ Then he crossed over to attack the Ammonites, where he found a strong band and many people, with Timothy as their leader. ⁷ He engaged in many battles with them, and they were crushed before him; he struck them down. ⁸ He also took Jazer and its villages; then he returned to Judea.

9 Now the Gentiles in Gilead gathered together against the Israelites who lived in their territory, and planned to destroy them. But they fled to the stronghold of Dathema, ¹⁰ and sent to Judas and his brothers a letter that said, "The Gentiles around us have gathered together to destroy us. ¹¹ They are preparing to come and capture the stronghold to which we have fled, and Timothy is leading their forces. ¹² Now then, come and rescue us from their hands, for many of us have fallen, ¹³ and all our kindred who were in the land of Tob have been killed; the enemy*ˣ* have captured their wives and children and goods, and have destroyed about a thousand persons there."

14 While the letter was still being read, other messengers, with their garments torn, came from Galilee and made a similar report; ¹⁵ they said that the people of Ptolemais and Tyre and Sidon, and

---

v Gk *He*   *w* Gk *her*   *x* Gk *they*

5.1–68: Campaigns in all directions. Judas now attacked Idumea in the south (vv. 3–5, 65), Ammon and Gilead east of the Jordan (vv. 6–13, 24–51), Galilee in the north (vv. 21–23), and the coastal plain. These events may have occurred after the death of Antiochus IV (6.16). **2:** *Descendants of Jacob,* Israelites or Jews. **3:** *Descendants of Esau,* Edomites or Idumeans, south of the Dead Sea. *Akrabattene,* perhaps on the border between Idumea and Judea. **4:** *Baean,* probably in Transjordan (Num 32.3). **6:** *Ammonites,* a

Semitic people east of the Jordan near the present Amman. **8:** *Jazer,* west of Amman, fifteen miles north of Heshbon (Num 32.3). **5.9:** *Gilead,* east of the Jordan between the Yarmuk and the Arnon (Josh 22.9). *Dathema,* possibly el-Hosn, in Gilead opposite Beisan; or Ramtha, now near the Syrian border. **13:** *Land of Tob,* possibly Hippos, twelve miles southeast of the Sea of Galilee (Judg 11.3; 2 Macc 12.17). **15:** *Ptolemais,* or Acco (Judg 1.31), now Acre, north of Haifa on the coast. *Tyre and Sidon,* farther north in Lebanon. As

197
AP

all Galilee of the Gentiles,*y* had gathered together against them "to annihilate us." [16]When Judas and the people heard these messages, a great assembly was called to determine what they should do for their kindred who were in distress and were being attacked by enemies. *z* [17]Then Judas said to his brother Simon, "Choose your men and go and rescue your kindred in Galilee; Jonathan my brother and I will go to Gilead." [18]But he left Joseph, son of Zechariah, and Azariah, a leader of the people, with the rest of the forces, in Judea to guard it; [19]and he gave them this command, "Take charge of this people, but do not engage in battle with the Gentiles until we return." [20]Then three thousand men were assigned to Simon to go to Galilee, and eight thousand to Judas for Gilead.

21 So Simon went to Galilee and fought many battles against the Gentiles, and the Gentiles were crushed before him. [22]He pursued them to the gate of Ptolemais; as many as three thousand of the Gentiles fell, and he despoiled them. [23]Then he took the Jews*a* of Galilee and Arbatta, with their wives and children, and all they possessed, and led them to Judea with great rejoicing.

24 Judas Maccabeus and his brother Jonathan crossed the Jordan and made three days' journey into the wilderness. [25]They encountered the Nabateans, who met them peaceably and told them all that had happened to their kindred in Gilead: [26]"Many of them have been shut up in Bozrah and Bosor, in Alema and Chaspho, Maked and Carnaim"—all these towns were strong and large— [27]"and some have been shut up in the other towns of Gilead; the enemy*b* are getting ready to attack the strongholds tomorrow and capture and destroy all these people in a single day."

28 Then Judas and his army quickly turned back by the wilderness road to Bozrah; and he took the town, and killed every male by the edge of the sword; then he seized all its spoils and burned it with fire. [29]He left the place at night, and they went all the way to the stronghold of Dathema. *c* [30]At dawn they looked out and saw a large company, which could not be counted, carrying ladders and engines of war to capture the stronghold, and attacking the Jews within. *d* [31]So Judas saw that the battle had begun and that the cry of the town went up to Heaven, with trumpets and loud shouts, [32]and he said to the men of his forces, "Fight today for your kindred!" [33]Then he came up behind them in three companies, who sounded their trumpets and cried aloud in prayer. [34]And when the army of Timothy realized that it was Maccabeus, they fled before him, and he dealt them a heavy blow. As many as eight thousand of them fell that day.

35 Next he turned aside to Maapha, *e* and fought against it and took it; and he killed every male in it, plundered it, and burned it with fire. [36]From there he marched on and took Chaspho, Maked, and Bosor, and the other towns of Gilead.

37 After these things Timothy gathered another army and encamped opposite Raphon, on the other side of the stream. [38]Judas sent men to spy out the camp, and they reported to him, "All the Gentiles around us have gathered to him; it is a very large force. [39]They also have hired Arabs to help them, and they are encamped across the stream, ready

---

y Gk *aliens*   *z* Gk *them*   *a* Gk *those*
b Gk *they*   *c* Gk lacks *of Dathema*. See verse 9
d Gk *and they were attacking them*   *e* Other ancient authorities read *Alema*

---

yet few Jews lived in *Galilee of the Gentiles* (Isa 9.1; Mt 4.15). **23:** *Arbatta*, either near the Sea of Galilee, or the Arabah depression south of the Dead Sea (Deut 1.7; Josh 11.16).
5.25: *Nabateans*, or Nebaioth (Gen 25.13), an Arab people living as nomads in the desert east of Palestine as far north as Palmyra. **26:**

Bozrah, southeast of Dera'a (Isa 63.1; Jer 48.24). *Bosor*, Bezer in the desert (Deut 4.43). Alema, unidentified. *Chaspho* and *Maked*, cities of Gilead (v. 36). *Carnaim*, Gen 14.5; Am 6.13; 2 Macc 12.21, 26. **28:** *Killed every male*, Gen 34.25.
5.37: *The stream*, a tributary of the Yar-

to come and fight against you." And Judas went to meet them.

40 Now as Judas and his army drew near to the stream of water, Timothy said to the officers of his forces, "If he crosses over to us first, we will not be able to resist him, for he will surely defeat us. 41 But if he shows fear and camps on the other side of the river, we will cross over to him and defeat him." 42 When Judas approached the stream of water, he stationed the officers[f] of the army at the stream and gave them this command, "Permit no one to encamp, but make them all enter the battle." 43 Then he crossed over against them first, and the whole army followed him. All the Gentiles were defeated before him, and they threw away their arms and fled into the sacred precincts at Carnaim. 44 But he took the town and burned the sacred precincts with fire, together with all who were in them. Thus Carnaim was conquered; they could stand before Judas no longer.

45 Then Judas gathered together all the Israelites in Gilead, the small and the great, with their wives and children and goods, a very large company, to go to the land of Judah. 46 So they came to Ephron. This was a large and very strong town on the road, and they could not go around it to the right or to the left; they had to go through it. 47 But the people of the town shut them out and blocked up the gates with stones.

48 Judas sent them this friendly message, "Let us pass through your land to get to our land. No one will do you harm; we will simply pass by on foot." But they refused to open to him. 49 Then Judas ordered proclamation to be made to the army that all should encamp where they were. 50 So the men of the forces encamped, and he fought against the town all that day and all the night, and the town was delivered into his hands. 51 He destroyed every male by the edge of the sword, and razed and plundered the town. Then he passed through the town over the bodies of the dead.

52 Then they crossed the Jordan into the large plain before Beth-shan. 53 Judas kept rallying the laggards and encouraging the people all the way until he came to the land of Judah. 54 So they went up to Mount Zion with joy and gladness, and offered burnt offerings, because they had returned in safety; not one of them had fallen.

55 Now while Judas and Jonathan were in Gilead and their[g] brother Simon was in Galilee before Ptolemais, 56 Joseph son of Zechariah, and Azariah, the commanders of the forces, heard of their brave deeds and of the heroic war they had fought. 57 So they said, "Let us also make a name for ourselves; let us go and make war on the Gentiles around us." 58 So they issued orders to the men of the forces that were with them and marched against Jamnia. 59 Gorgias and his men came out of the town to meet them in battle. 60 Then Joseph and Azariah were routed, and were pursued to the borders of Judea; as many as two thousand of the people of Israel fell that day. 61 Thus the people suffered a great rout because, thinking to do a brave deed, they did not listen to Judas and his brothers. 62 But they did not belong to the family of those men through whom deliverance was given to Israel.

63 The man Judas and his brothers were greatly honored in all Israel and among all the Gentiles, wherever their

f Or *scribes*　　*g* Gk *his*

---

muk. **39:** *Arabs* were not usually hostile to the Jews but could be *hired* as mercenaries. **40–41:** Judas heard Timothy's order or decided to make a surprise attack (compare 1 Sam 14.7–10). **43:** *Sacred precincts*, of Atargatis, the Syrian fish goddess (2 Macc 12.26). **46:** *Ephron*, eight miles east of the Jordan, opposite Bethshan (v. 52), and west of Irbid (Arbela).

5.48–51: Num 21.21–24. **52:** *The large plain*, between the Jordan and Mt. Gilboa. *Beth-shan*, Beisan, about eighteen miles south of the Sea of Galilee (Judg 1.27; 1 Kings 4.12). **58:** *Jamnia*, 4.15. **62:** Only *the family* of the Hasmoneans is regarded as divinely chosen to save Israel.

name was heard. 64 People gathered to them and praised them.

65 Then Judas and his brothers went out and fought the descendants of Esau in the land to the south. He struck Hebron and its villages and tore down its strongholds and burned its towers on all sides. 66 Then he marched off to go into the land of the Philistines, and passed through Marisa. *h* 67 On that day some priests, who wished to do a brave deed, fell in battle, for they went out to battle unwisely. 68 But Judas turned aside to Azotus in the land of the Philistines; he tore down their altars, and the carved images of their gods he burned with fire; he plundered the towns and returned to the land of Judah.

6 King Antiochus was going through the upper provinces when he heard that Elymais in Persia was a city famed for its wealth in silver and gold. 2 Its temple was very rich, containing golden shields, breastplates, and weapons left there by Alexander son of Philip, the Macedonian king who first reigned over the Greeks. 3 So he came and tried to take the city and plunder it, but he could not because his plan had become known to the citizens 4 and they withstood him in battle. So he fled and in great disappointment left there to return to Babylon.

5 Then someone came to him in Persia and reported that the armies that had gone into the land of Judah had been routed; 6 that Lysias had gone first with a strong force, but had turned and fled before the Jews; *i* that the Jews *j* had grown strong from the arms, supplies, and abundant spoils that they had taken from the armies they had cut down; 7 that they had torn down the abomination that he had erected on the altar in Jerusalem; and that they had surrounded the sanctuary with high walls as before, and also Bethzur, his town.

8 When the king heard this news, he was astounded and badly shaken. He took to his bed and became sick from disappointment, because things had not turned out for him as he had planned. 9 He lay there for many days, because deep disappointment continually gripped him, and he realized that he was dying. 10 So he called all his Friends and said to them, "Sleep has departed from my eyes and I am downhearted with worry. 11 I said to myself, 'To what distress I have come! And into what a great flood I now am plunged! For I was kind and beloved in my power.' 12 But now I remember the wrong I did in Jerusalem. I seized all its vessels of silver and gold, and I sent to destroy the inhabitants of Judah without good reason. 13 I know that it is because of this that these misfortunes have come upon me; here I am, perishing of bitter disappointment in a strange land."

14 Then he called for Philip, one of his Friends, and made him ruler over all his kingdom. 15 He gave him the crown and his robe and the signet, so that he

---

h Other ancient authorities read *Samaria*
i Gk *them* *j* Gk *they*

5.65: *Descendants of Esau*, Edomites. *Hebron*, the old capital of David, twenty miles south of Jerusalem (Gen 23.2; 2 Sam 2.11). **66:** *Marisa*, or Mareshah (Josh 15.44), near Beit-Jibrin.
6.1–17: Death of Antiochus IV and accession of Antiochus V (2 Macc ch 9). **1:** *The upper provinces*, Persia and Mesopotamia (3.31–37). *Elymais*, biblical Elam or Susiana; but according to 2 Macc 9.2 the incident occurred in Persepolis. **2:** *Its temple* was that of Nanea (2 Macc 1.13–16), or Anahita, identified with Artemis. **5:** *In Persia*, perhaps at Ecbatana (2 Macc 9.3). According to Polybius (*History*, xxxi. 11) the king took sick and died at Tabae (perhaps Gabae, modern Isfahan). **7:** *The abomination*, statue of a pagan god.
6.8–9: *Deep disappointment*, perhaps insanity; according to 2 Macc 9.5–12 he was stricken with a loathsome physical malady. **12:** *Its vessels*, Dan 5.2. **14–15:** *Philip . . . ruler*, Lysias, satrap in the west, had previously been given this commission (3.32). *The signet*, a symbol of transfer of authority to the regent. **16:** 2 Macc 11.33 implies that Antiochus IV had died before the restoration of the temple at Jerusalem. **17:** The word *Eupator* means "of a good father."

might guide his son Antiochus and bring him up to be king. 16 Thus King Antiochus died there in the one hundred forty-ninth year.[k] 17 When Lysias learned that the king was dead, he set up Antiochus the king's[l] son to reign. Lysias[m] had brought him up from boyhood; he named him Eupator.

18 Meanwhile the garrison in the citadel kept hemming Israel in around the sanctuary. They were trying in every way to harm them and strengthen the Gentiles. 19 Judas therefore resolved to destroy them, and assembled all the people to besiege them. 20 They gathered together and besieged the citadel[n] in the one hundred fiftieth year;[o] and he built siege towers and other engines of war. 21 But some of the garrison escaped from the siege and some of the ungodly Israelites joined them. 22 They went to the king and said, "How long will you fail to do justice and to avenge our kindred? 23 We were happy to serve your father, to live by what he said, and to follow his commands. 24 For this reason the sons of our people besieged the citadel[p] and became hostile to us; moreover, they have put to death as many of us as they have caught, and they have seized our inheritances. 25 It is not against us alone that they have stretched out their hands; they have also attacked all the lands on their borders. 26 And see, today they have encamped against the citadel in Jerusalem to take it; they have fortified both the sanctuary and Beth-zur; 27 unless you quickly prevent them, they will do still greater things, and you will not be able to stop them."

28 The king was enraged when he heard this. He assembled all his Friends, the commanders of his forces and those in authority.[q] 29 Mercenary forces also came to him from other kingdoms and from islands of the seas. 30 The number of his forces was one hundred thousand foot soldiers, twenty thousand horsemen, and thirty-two elephants accustomed to war. 31 They came through Idumea and encamped against Beth-zur, and for many days they fought and built engines of war; but the Jews[r] sallied out and burned these with fire, and fought courageously.

32 Then Judas marched away from the citadel and encamped at Beth-zechariah, opposite the camp of the king. 33 Early in the morning the king set out and took his army by a forced march along the road to Beth-zechariah, and his troops made ready for battle and sounded their trumpets. 34 They offered the elephants the juice of grapes and mulberries, to arouse them for battle. 35 They distributed the animals among the phalanxes; with each elephant they stationed a thousand men armed with coats of mail, and with brass helmets on their heads; and five hundred picked horsemen were assigned to each beast. 36 These took their position beforehand wherever the animal was; wherever it went, they went with it, and they never left it. 37 On the elephants[s] were wooden towers, strong and covered; they were fastened on each animal by special harness, and on each were four[t] armed men who fought from there, and also its Indian driver. 38 The rest of the cavalry were stationed on either side, on the two flanks of the

k 163 B.C. *l* Gk *his* *m* Gk *He* *n* Gk *it*
o 162 B.C. *p* Meaning of Gk uncertain
q Gk *those over the reins* *r* Gk *they*
s Gk *them* *t* Cn: Some authorities read *thirty;* others *thirty-two*

6.18–54: Attack on the citadel and second battle at Beth-zur. The citadel was equally important to the Syrians and to Judas, for without it the Seleucid monarchy could not maintain sovereignty in Palestine. **21:** *Ungodly,* i.e. pro-Greek.

6.28–30: Lysias' second campaign is here dated 162 B.C. (v. 20), but in 2 Macc 13.1 a year earlier. **31:** Judas had won the first battle at *Beth-zur* (4.29–34) and had fortified it (4.61). **32:** *Beth-zechariah* was six miles from Beth-zur and ten miles southwest of Jerusalem. **34–35:** *The juice* may have been to simulate blood; but *elephants* were sometimes given wine to madden them. Here the animals were used to force an opening in the ranks. *Phalanxes,* the Greek infantry formation, eight to eighteen men deep, highly disciplined and mobile. The Seleucids could muster twenty thousand of such infantry.

army, to harass the enemy while being themselves protected by the phalanxes. 39 When the sun shone on the shields of gold and brass, the hills were ablaze with them and gleamed like flaming torches.

40 Now a part of the king's army was spread out on the high hills, and some troops were on the plain, and they advanced steadily and in good order. 41 All who heard the noise made by their multitude, by the marching of the multitude and the clanking of their arms, trembled, for the army was very large and strong. 42 But Judas and his army advanced to the battle, and six hundred of the king's army fell. 43 Now Eleazar, called Avaran, saw that one of the animals was equipped with royal armor. It was taller than all the others, and he supposed that the king was on it. 44 So he gave his life to save his people and to win for himself an everlasting name. 45 He courageously ran into the midst of the phalanx to reach it; he killed men right and left, and they parted before him on both sides. 46 He got under the elephant, stabbed it from beneath, and killed it; but it fell to the ground upon him and he died. 47 When the Jews*u* saw the royal might and the fierce attack of the forces, they turned away in flight.

48 The soldiers of the king's army went up to Jerusalem against them, and the king encamped in Judea and at Mount Zion. 49 He made peace with the people of Beth-zur, and they evacuated the town because they had no provisions there to withstand a siege, since it was a sabbatical year for the land. 50 So the king took Beth-zur and stationed a guard there to hold it. 51 Then he encamped before the sanctuary for many days. He set up siege towers, engines of war to throw fire and stones, machines to shoot arrows, and catapults. 52 The Jews*u* also made engines of war to match theirs, and fought for many days. 53 But they had no food in storage, *v* because it was the seventh year; those who had found safety in Judea from the Gentiles had consumed the last of the stores. 54 Only a few men were left in the sanctuary; the rest scattered to their own homes, for the famine proved too much for them.

55 Then Lysias heard that Philip, whom King Antiochus while still living had appointed to bring up his son Antiochus to be king, 56 had returned from Persia and Media with the forces that had gone with the king, and that he was trying to seize control of the government. 57 So he quickly gave orders to withdraw, and said to the king, to the commanders of the forces, and to the troops, men, "Daily we grow weaker, our food supply is scant, the place against which we are fighting is strong, and the affairs of the kingdom press urgently on us. 58 Now then let us come to terms with these people, and make peace with them and with all their nation. 59 Let us agree to let them live by their laws as they did before; for it was on account of their laws that we abolished that they became angry and did all these things."

60 The speech pleased the king and the commanders, and he sent to the Jews*w* an offer of peace, and they accepted it. 61 So the king and the commanders gave them their oath. On these conditions the Jews*u* evacuated the stronghold. 62 But when the king entered Mount Zion and saw what a strong fortress the place was, he broke the oath he

u Gk *they*   *v* Other ancient authorities read *in the sanctuary*   *w* Gk *them*

6.43: *Eleazar,* brother of Judas (2.5). **48:** *Mount Zion,* south of the temple (1.33). **49:** The garrison was promised immunity if it surrendered. Every seventh year the land had to lie fallow (Ex 23.11; Lev 25.3–7). This *sabbatical year* was apparently 162 B.C. (v. 20) or possibly a year earlier.
6.55–63: Lysias makes peace. The return of Philip caused a diversion; Lysias abandoned the siege and restored Jewish religious rights. **55:** *Philip* had received the symbols of sovereignty (v. 15), though Antiochus IV had previously appointed Lysias. **59:** Judea had generally accepted Seleucid rule until Antiochus IV began his program of hellenization, though there was always a faction engaged in intrigue with Egypt. **62:** Judas had also built a citadel on *Mount Zion;* its wall was now destroyed.

had sworn and gave orders to tear down the wall all around. [63] Then he set off in haste and returned to Antioch. He found Philip in control of the city, but he fought against him, and took the city by force.

7 In the one hundred fifty-first year[x] Demetrius son of Seleucus set out from Rome, sailed with a few men to a town by the sea, and there began to reign. [2] As he was entering the royal palace of his ancestors, the army seized Antiochus and Lysias to bring them to him. [3] But when this act became known to him, he said, "Do not let me see their faces!" [4] So the army killed them, and Demetrius took his seat on the throne of his kingdom.

[5] Then there came to him all the renegade and godless men of Israel; they were led by Alcimus, who wanted to be high priest. [6] They brought to the king this accusation against the people: "Judas and his brothers have destroyed all your Friends, and have driven us out of our land. [7] Now then send a man whom you trust; let him go and see all the ruin that Judas[y] has brought on us and on the land of the king, and let him punish them and all who help them."

[8] So the king chose Bacchides, one of the king's Friends, governor of the province Beyond the River; he was a great man in the kingdom and was faithful to the king. [9] He sent him, and with him he sent the ungodly Alcimus, whom he made high priest; and he commanded him to take vengeance on the Israelites. [10] So they marched away and came with a large force into the land of Judah; and he sent messengers to Judas and his brothers with peaceable but treacherous words. [11] But they paid no attention to their words, for they saw that they had come with a large force.

[12] Then a group of scribes appeared in a body before Alcimus and Bacchides to ask for just terms. [13] The Hasideans were first among the Israelites to seek peace from them, [14] for they said, "A priest of the line of Aaron has come with the army, and he will not harm us." [15] Alcimus[z] spoke peaceable words to them and swore this oath to them, "We will not seek to injure you or your Friends." [16] So they trusted him; but he seized sixty of them and killed them in one day, in accordance with the word that was written,

[17] "The flesh of your faithful ones
    and their blood
  they poured out all around
      Jerusalem,
    and there was no one to bury
      them."

[18] Then the fear and dread of them fell on all the people, for they said, "There is no truth or justice in them, for they have violated the agreement and the oath that they swore."

x 161 B.C.   y Gk *he*   z Gk *He*

7.1–4: Demetrius I becomes king (2 Macc 14.1–2). **1:** *Demetrius* I Soter (reigned 162–150 B.C.) was *son of Seleucus* IV Philopator, elder brother of Antiochus IV. When Rome demanded hostages (1.10), Antiochus IV was sent to Rome; later when Seleucus became king, his son Demetrius replaced Antiochus. On the latter's death, he vainly petitioned the senate to be released. Subsequently he escaped from Rome with a small group of men and landed in Tripolis, *a town by the sea* (2 Macc 14.1). **2:** *Antiochus,* that is, Antiochus V Eupator (6.17). **3:** *Do not let me see their faces,* a signal for the murder.

7.5–25: Alcimus as high priest. Legitimate high priests were descended from a particular family. Antiochus IV appointed Jason in place of his brother Onias III (2 Macc 4.7); Jason was in turn supplanted by Menelaus (2 Macc 4.23–26), who was put to death about 162 B.C., after having officiated for ten years (2 Macc 13.1–8). Either Onias III or his son Onias IV, last legitimate claimant, fled to Egypt and established a temple at Heliopolis (Cairo). *Alcimus,* or Jakim (2 Macc 14.3), was not a member of the high-priestly family; he belonged to the hellenizing faction and was willing to further Demetrius' plans. **8:** *Beyond the River,* the province west of the Euphrates (Ezra 4.11).

7.12–14: *The Hasideans* (2.42), probably the same as the *group of scribes,* had no political ambitions and were content to live under Syrian rule if they were permitted to keep the Mosaic law. **17:** Ps 79.2–3. All ancients regarded an unburied dead body with horror,

19 Then Bacchides withdrew from Jerusalem and encamped in Beth-zaith. And he sent and seized many of the men who had deserted to him,*ᵃ* and some of the people, and killed them and threw them into a great pit. ²⁰He placed Alcimus in charge of the country and left with him a force to help him; then Bacchides went back to the king.

21 Alcimus struggled to maintain his high priesthood, ²²and all who were troubling their people joined him. They gained control of the land of Judah and did great damage in Israel. ²³And Judas saw all the wrongs that Alcimus and those with him had done among the Israelites; it was more than the Gentiles had done. ²⁴So Judas*ᵇ* went out into all the surrounding parts of Judea, taking vengeance on those who had deserted and preventing those in the city*ᶜ* from going out into the country. ²⁵When Alcimus saw that Judas and those with him had grown strong, and realized that he could not withstand them, he returned to the king and brought malicious charges against them.

26 Then the king sent Nicanor, one of his honored princes, who hated and detested Israel, and he commanded him to destroy the people. ²⁷So Nicanor came to Jerusalem with a large force, and treacherously sent to Judas and his brothers this peaceable message, ²⁸"Let there be no fighting between you and me; I shall come with a few men to see you face to face in peace."

29 So he came to Judas, and they greeted one another peaceably; but the enemy were preparing to kidnap Judas. ³⁰It became known to Judas that Nicanor*ᵇ* had come to him with treacherous intent, and he was afraid of him and

would not meet him again. ³¹When Nicanor learned that his plan had been disclosed, he went out to meet Judas in battle near Caphar-salama. ³²About five hundred of the army of Nicanor fell, and the rest*ᵈ* fled into the city of David.

33 After these events Nicanor went up to Mount Zion. Some of the priests from the sanctuary and some of the elders of the people came out to greet him peaceably and to show him the burnt offering that was being offered for the king. ³⁴But he mocked them and derided them and defiled them and spoke arrogantly, ³⁵and in anger he swore this oath, "Unless Judas and his army are delivered into my hands this time, then if I return safely I will burn up this house." And he went out in great anger. ³⁶At this the priests went in and stood before the altar and the temple; they wept and said,

37 "You chose this house to be called
    by your name,
  and to be for your people a
    house of prayer and
    supplication.
38 Take vengeance on this man and
    on his army,
  and let them fall by the sword;
  remember their blasphemies,
  and let them live no longer."

39 Now Nicanor went out from Jerusalem and encamped in Beth-horon, and the Syrian army joined him. ⁴⁰Judas encamped in Adasa with three thousand men. Then Judas prayed and said, ⁴¹"When the messengers from the king spoke blasphemy, your angel went out and struck down one hundred eighty-

a Or *many of his men who had deserted*
b Gk *he*   *c* Gk *and they were prevented*
d Gk *they*

---

and to leave foes unburied was the ultimate outrage. **19:** *Beth-zaith,* perhaps three miles north of Beth-zur; or Bezetha, north of the temple area in Jerusalem.
  7.26–50: Defeat of Nicanor. The last of Judas' great victories (2 Macc 14.12–15.36). **26:** According to Josephus (*Antiquities,* xii. x. 4), *Nicanor* was one of the men who had escaped from Rome with Demetrius (see

7.1 n.). **30:** Judas never trusted the Syrians (vv. 10–11). **31:** *Capharsalama,* perhaps Khirbet Deir Sellam, about five miles northeast of Jerusalem. **33:** The Jews customarily offered sacrifices to God for the welfare of their rulers. **37–38:** Compare 1 Kings 8.29, 43; Pss 68.16; 87.1–2.
  7.40: *Adasa,* about seven miles from Bethhoron on the road to Jerusalem. **41:** 2 Kings

five thousand of the Assyrians.*e* 42So also crush this army before us today; let the rest learn that Nicanor*f* has spoken wickedly against the sanctuary, and judge him according to this wickedness."

43 So the armies met in battle on the thirteenth day of the month of Adar. The army of Nicanor was crushed, and he himself was the first to fall in the battle. 44When his army saw that Nicanor had fallen, they threw down their arms and fled. 45The Jews*g* pursued them a day's journey, from Adasa as far as Gazara, and as they followed they kept sounding the battle call on the trumpets. 46People came out of all the surrounding villages of Judea, and they outflanked the enemy*h* and drove them back to their pursuers,*i* so that they all fell by the sword; not even one of them was left. 47Then the Jews*g* seized the spoils and the plunder; they cut off Nicanor's head and the right hand that he had so arrogantly stretched out, and brought them and displayed them just outside Jerusalem. 48The people rejoiced greatly and celebrated that day as a day of great gladness. 49They decreed that this day should be celebrated each year on the thirteenth day of Adar. 50So the land of Judah had rest for a few days.

8 Now Judas heard of the fame of the Romans, that they were very strong and were well-disposed toward all who made an alliance with them, that they pledged friendship to those who came to them, 2and that they were very strong. He had been told of their wars and of the brave deeds that they were doing among the Gauls, how they had defeated them and forced them to pay tribute, 3and what they had done in the land of Spain to get control of the silver and gold mines there, 4and how they had gained control of the whole region by their planning and patience, even though the place was far distant from them. They also subdued the kings who came against them from the ends of the earth, until they crushed them and inflicted great disaster on them; the rest paid them tribute every year. 5They had crushed in battle and conquered Philip, and King Perseus of the Macedonians,*j* and the others who rose up against them. 6They also had defeated Antiochus the Great, king of Asia, who went to fight against them with one hundred twenty elephants and with cavalry and chariots and a very large army. He was crushed by them; 7they took him alive and decreed that he and those who would reign after him should pay a heavy tribute and give hostages and surrender some of their best provinces, 8the countries of India, Media, and Lydia. These they took from him and gave to King Eumenes. 9The Greeks planned to come and destroy them, 10but this became known to them, and they sent a general against the

e Gk of them   f Gk he   g Gk they
h Gk them   i Gk there   j Or Kittim

19.35. **43**: *Adar,* roughly March, in 161 B.C. (see 2 Macc 15.36 n.). **45**: *Gazara,* Gezer (4.15). **47**: *Head . . . right hand,* in punishment for blasphemy and for raising his hand against the temple (2 Macc 15.32). See 1 Sam 17.54. **49**: 2 Macc 15.36. This festival, which came to be called Nicanor Day, was one of the days on which the Jews prohibited mourning (compare 13.52). It is no longer observed. **8.1–32**: **Treaty with Rome.** The author of 1 Maccabees emphasizes friendly relations between Rome and the Jews because the Romans checked the ambitions of the Seleucids. After 190 B.C. Rome steadily increased her influence in the Near East and Syrian power declined. **1**: *The Romans had made an alliance* with kings in Asia Minor and Egypt. **2**: Two nations of *Gauls* were *defeated,* those of upper Italy in 190 B.C., and the Galatians of Asia Minor in 189 B.C. **3–4**: Rome conquered the Carthaginian colonies of *Spain,* not *the whole region,* in the Second Punic War. **5**: *Philip,* defeated at Cynoscephalae in 197 B.C.; *Perseus,* his son, last Macedonian king, beaten at Pydna in 168 B.C. **6–8**: *Antiochus* was not captured, but had to *give hostages* (7.1). *India* was not part of his domain; he kept *Media,* but he surrendered *Lydia* and other parts of Asia Minor. *Eumenes* II of Pergamum was given much of Seleucid Asia Minor. **9**: *The Greeks,* possibly the Macedonians (v. 5), or the Achaean league somewhat later. **12**: This is the

Greeks*k* and attacked them. Many of them were wounded and fell, and the Romans*l* took captive their wives and children; they plundered them, conquered the land, tore down their strongholds, and enslaved them to this day. 11 The remaining kingdoms and islands, as many as ever opposed them, they destroyed and enslaved; 12 but with their friends and those who rely on them they have kept friendship. They have subdued kings far and near, and as many as have heard of their fame have feared them. 13 Those whom they wish to help and to make kings, they make kings, and those whom they wish they depose; and they have been greatly exalted. 14 Yet for all this not one of them has put on a crown or worn purple as a mark of pride, 15 but they have built for themselves a senate chamber, and every day three hundred twenty senators constantly deliberate concerning the people, to govern them well. 16 They trust one man each year to rule over them and to control all their land; they all heed the one man, and there is no envy or jealousy among them.

17 So Judas chose Eupolemus son of John son of Accos, and Jason son of Eleazar, and sent them to Rome to establish friendship and alliance, 18 and to free themselves from the yoke; for they saw that the kingdom of the Greeks was enslaving Israel completely. 19 They went to Rome, a very long journey; and they entered the senate chamber and spoke as follows: 20 "Judas, who is also called Maccabeus, and his brothers and the people of the Jews have sent us to you to establish alliance and peace with you, so that we may be enrolled as your allies and friends." 21 The proposal pleased them, 22 and this is a copy of the letter that they wrote in reply, on bronze tablets, and sent to Jerusalem to remain with them there as a memorial of peace and alliance:

23 "May all go well with the Romans and with the nation of the Jews at sea and on land forever, and may sword and enemy be far from them. 24 If war comes first to Rome or to any of their allies in all their dominion, 25 the nation of the Jews shall act as their allies wholeheartedly, as the occasion may indicate to them. 26 To the enemy that makes war they shall not give or supply grain, arms, money, or ships, just as Rome has decided; and they shall keep their obligations without receiving any return. 27 In the same way, if war comes first to the nation of the Jews, the Romans shall willingly act as their allies, as the occasion may indicate to them. 28 And to their enemies there shall not be given grain, arms, money, or ships, just as Rome has decided; and they shall keep these obligations and do so without deceit. 29 Thus on these terms the Romans make a treaty with the Jewish people. 30 If after these terms are in effect both parties shall determine to add or delete anything, they shall do so at their discretion, and any addition or deletion that they may make shall be valid.

31 "Concerning the wrongs that King Demetrius is doing to them, we have written to him as follows, 'Why have you made your yoke heavy on our friends and allies the Jews? 32 If now they

k Gk *them*   *l* Gk *they*

---

estimate of a partisan; it is not true that they always *kept friendship.* 14: *A crown,* Rome wanted no king, but magistrates, senators, and knights wore *purple* borders on their garments. 15: *Every day,* actually senate meetings were held three times a month, and on the festivals. 16: *One man,* in reality there were two consuls, and *envy* and *jealousy* were constant. The author of 1 Maccabees idealizes the Romans because their republican institutions were congenial to the Jews.

8.17: *Eupolemus,* 2 Macc 4.11. *Accos,* a priestly family (Ezra 2.61). **19:** *A very long*

journey, emphasizes that this alliance was in accord with the law of Moses, which forbade only covenants with foreigners surrounding Israel. **22:** Important documents were often inscribed *on bronze tablets.* **23–30:** The treaty letter begins with the conventional formula and is drawn up as though the two parties were equals, and Judea a sovereign state. **31–32:** This postscript was not part of the treaty, and is correctly omitted from it by Josephus (*Antiquities,* xii. x. 6); there is no evidence that the Romans helped Judas against Demetrius.

appeal again for help against you, we will defend their rights and fight you on sea and on land.' "

9 When Demetrius heard that Nicanor and his army had fallen in battle, he sent Bacchides and Alcimus into the land of Judah a second time, and with them the right wing of the army. [2] They went by the road that leads to Gilgal and encamped against Mesaloth in Arbela, and they took it and killed many people. [3] In the first month of the one hundred fifty-second year[m] they encamped against Jerusalem; [4] then they marched off and went to Berea with twenty thousand foot soldiers and two thousand cavalry.

5 Now Judas was encamped in Elasa, and with him were three thousand picked men. [6] When they saw the huge number of the enemy forces, they were greatly frightened, and many slipped away from the camp, until no more than eight hundred of them were left.

7 When Judas saw that his army had slipped away and the battle was imminent, he was crushed in spirit, for he had no time to assemble them. [8] He became faint, but he said to those who were left, "Let us get up and go against our enemies. We may have the strength to fight them." [9] But they tried to dissuade him, saying, "We do not have the strength. Let us rather save our own lives now, and let us come back with our kindred and fight them; we are too few." [10] But Judas said, "Far be it from us to do such a thing as to flee from them. If our time has come, let us die bravely for our kindred, and leave no cause to question our honor."

11 Then the army of Bacchides[n]

marched out from the camp and took its stand for the encounter. The cavalry was divided into two companies, and the slingers and the archers went ahead of the army, as did all the chief warriors. [12] Bacchides was on the right wing. Flanked by the two companies, the phalanx advanced to the sound of the trumpets; and the men with Judas also blew their trumpets. [13] The earth was shaken by the noise of the armies, and the battle raged from morning until evening.

14 Judas saw that Bacchides and the strength of his army were on the right; then all the stouthearted men went with him, [15] and they crushed the right wing, and he pursued them as far as Mount Azotus. [16] When those on the left wing saw that the right wing was crushed, they turned and followed close behind Judas and his men. [17] The battle became desperate, and many on both sides were wounded and fell. [18] Judas also fell, and the rest fled.

19 Then Jonathan and Simon took their brother Judas and buried him in the tomb of their ancestors at Modein, [20] and wept for him. All Israel made great lamentation for him; they mourned many days and said,

[21] "How is the mighty fallen,
   the savior of Israel!"

[22] Now the rest of the acts of Judas, and his wars and the brave deeds that he did, and his greatness, have not been recorded, but they were very many.

23 After the death of Judas, the renegades emerged in all parts of Israel; all the wrongdoers reappeared. [24] In those days a very great famine occurred, and the

m 160 B.C.   *n* Gk *the army*

9.1–22: Death of Judas. Continual Syrian pressure weakened Judas' forces, which were a guerrilla band facing a highly organized army. Judas decided that it was better to fall in battle than to withdraw. **1–2:** *Bacchides,* 7.8. *Alcimus,* see 7.5 n. Josephus (*Antiquities,* XII. xi. 1) says that they started from Antioch, camped at Arbela in Galilee (not *Gilgal*), besieged refugees in the caves, and went on toward Jerusalem. *Mesaloth,* perhaps the Hebrew word for "steps," i.e. ascents to *Arbela.*

4: *Berea,* perhaps el-Bireh, opposite Ramallah, ten miles north of Jerusalem. **5:** *Elasa* has not been identified.

9.15: *Mount Azotus,* perhaps el'Asur, six miles northeast of el-Bireh. **19:** *Modein,* 2.1, 70; 13.27–30. **21:** 2 Sam 1.19. **22:** The expression *Now the rest of the acts* imitates the style of Hebrew chronicles (1 Kings 11.41).

9.23–73: Jonathan becomes leader and defeats Bacchides. 23: *The renegades,* i.e. the pro-Syrian element. **24:** *The country,* the ma-

country went over to their side. 25 Bacchides chose the godless and put them in charge of the country. 26 They made inquiry and searched for the friends of Judas, and brought them to Bacchides, who took vengeance on them and made sport of them. 27 So there was great distress in Israel, such as had not been since the time that prophets ceased to appear among them.

28 Then all the friends of Judas assembled and said to Jonathan, 29 "Since the death of your brother Judas there has been no one like him to go against our enemies and Bacchides, and to deal with those of our nation who hate us. 30 Now therefore we have chosen you today to take his place as our ruler and leader, to fight our battle." 31 So Jonathan accepted the leadership at that time in place of his brother Judas.

32 When Bacchides learned of this, he tried to kill him. 33 But Jonathan and his brother Simon and all who were with him heard of it, and they fled into the wilderness of Tekoa and camped by the water of the pool of Asphar. 34 Bacchides found this out on the sabbath day, and he with all his army crossed the Jordan.

35 So Jonathan*o* sent his brother as leader of the multitude and begged the Nabateans, who were his friends, for permission to store with them the great amount of baggage that they had. 36 But the family of Jambri from Medeba came out and seized John and all that he had, and left with it.

37 After these things it was reported to Jonathan and his brother Simon, "The family of Jambri are celebrating a great wedding, and are conducting the bride, a daughter of one of the great nobles of Canaan, from Nadabath with a large escort." 38 Remembering how their brother John had been killed, they went up and hid under cover of the mountain. 39 They looked out and saw a tumultuous procession with a great amount of baggage; and the bridegroom came out with his friends and his brothers to meet them with tambourines and musicians and many weapons. 40 Then they rushed on them from the ambush and began killing them. Many were wounded and fell, and the rest fled to the mountain; and the Jews*p* took all their goods. 41 So the wedding was turned into mourning and the voice of their musicians into a funeral dirge. 42 After they had fully avenged the blood of their brother, they returned to the marshes of the Jordan.

43 When Bacchides heard of this, he came with a large force on the sabbath day to the banks of the Jordan. 44 And Jonathan said to those with him, "Let us get up now and fight for our lives, for today things are not as they were before. 45 For look! the battle is in front of us and behind us; the water of the Jordan is on this side and on that, with marsh and thicket; there is no place to turn. 46 Cry out now to Heaven that you may be delivered from the hands of our enemies." 47 So the battle began, and Jonathan stretched out his hand to strike Bacchides, but he eluded him and went to the

o Gk he   p Gk they

jority of the nation, or perhaps the rural population; resistance now seemed futile. **25**: *The godless,* i.e. hellenized Jews. **27**: *Prophets ceased,* see 4.46 n. **30–31**: *We have chosen you,* Judas was self-appointed, but *Jonathan* was elected by his peers; he became leader about 160 or 159 B.C. and high priest in 152 (10.21). **33**: *Tekoa,* five miles southeast of Bethlehem (Am 1.1); *the wilderness* reached from here to the Dead Sea. *The pool of Asphar* may be three miles south of Tekoa. **34**: *Bacchides* thought the Jews might be surprised *on the sabbath day.* He apparently came from Jerusalem, *crossed the Jordan,* and camped on the east side.

9.35: *His brother,* John (v. 36; 2.2). *The Nabateans,* see 5.25 n. **36**: *Family of Jambri,* evidently a Nabatean tribe. *Medeba* or Madeba, twelve miles southeast of the north end of the Dead Sea. **37**: *Nadabath,* perhaps Nebo, a little north of Medeba (Num 33.47; Deut 32.49). **38**: *John . . . killed,* the sequel to v. 36 was the murder of John and all his companions (Josephus, *Antiquities,* XIII. i. 2). **45**: The Jews were apparently on the east side of the Jordan, between the river and the Syrian forces.

9.50–53: Bacchides established forces and garrisons to prevent guerrilla operations. *Em-*

rear. [48] Then Jonathan and the men with him leaped into the Jordan and swam across to the other side, and the enemy[q] did not cross the Jordan to attack them. [49] And about one thousand of Bacchides' men fell that day.

50 Then Bacchides[r] returned to Jerusalem and built strong cities in Judea: the fortress in Jericho, and Emmaus, and Beth-horon, and Bethel, and Timnath, and[s] Pharathon, and Tephon, with high walls and gates and bars. [51] And he placed garrisons in them to harass Israel. [52] He also fortified the town of Beth-zur, and Gazara, and the citadel, and in them he put troops and stores of food. [53] And he took the sons of the leading men of the land as hostages and put them under guard in the citadel at Jerusalem.

54 In the one hundred and fifty-third year,[t] in the second month, Alcimus gave orders to tear down the wall of the inner court of the sanctuary. He tore down the work of the prophets! [55] But he only began to tear it down, for at that time Alcimus was stricken and his work was hindered; his mouth was stopped and he was paralyzed, so that he could no longer say a word or give commands concerning his house. [56] And Alcimus died at that time in great agony. [57] When Bacchides saw that Alcimus was dead, he returned to the king, and the land of Judah had rest for two years.

58 Then all the lawless plotted and said, "See! Jonathan and his men are living in quiet and confidence. So now let us bring Bacchides back, and he will capture them all in one night." [59] And they went and consulted with him. [60] He started to come with a large force, and secretly sent letters to all his allies in Judea, telling them to seize Jonathan and his men; but they were unable to do it, because their plan became known. [61] And Jonathan's men[q] seized about fifty of the men of the country who were leaders in this treachery, and killed them.

62 Then Jonathan with his men, and Simon, withdrew to Bethbasi in the wilderness; he rebuilt the parts of it that had been demolished, and they fortified it. [63] When Bacchides learned of this, he assembled all his forces, and sent orders to the men of Judea. [64] Then he came and encamped against Bethbasi; he fought against it for many days and made machines of war.

65 But Jonathan left his brother Simon in the town, while he went out into the country; and he went with only a few men. [66] He struck down Odomera and his kindred and the people of Phasiron in their tents. [67] Then he[u] began to attack and went into battle with his forces; and Simon and his men sallied out from the town and set fire to the machines of war. [68] They fought with Bacchides, and he was crushed by them. They pressed him very hard, for his plan and his expedition had been in vain. [69] So he was very angry at the renegades who had counseled him to come into the country, and he killed many of them. Then he decided to go back to his own land.

70 When Jonathan learned of this, he sent ambassadors to him to make peace with him and obtain release of the cap-

q Gk they   r Gk he   s Some authorities omit and   t 159 B.C.   u Other ancient authorities read they

maus, see 3.40 n. *Beth-horon*, see 3.16 n. *Bethel*, now Beitin, about twelve miles north of Jerusalem. *Timnath*, perhaps twelve miles northwest of Bethel. *Pharathon*, six miles southwest of Shechem or Nablus. *Tephon*, Tappuah, twenty-five miles north of Jerusalem. *Beth-zur*, see 4.29 n. *Gazara*, see 4.15 n. **54:** *The wall of the inner court* separated this part from the rest of the temple mount, which was open to Gentiles. Pagans were now to have access to *the sanctuary*. *The prophets*, Haggai and Zechariah had built the second temple. **56:** *Alcimus* had been high priest for about two years (7.5). **57:** *He returned*, believing that with the fortresses garrisoned the situation was stable; his departure gave Jonathan a free hand. **61:** *Fifty . . . leaders*, probably the hellenizers (see v. 25 n.).
9.62: *Bethbasi*, perhaps Khirbet Beit-Bassa, about three miles northeast of Tekoa. **66:** *Odomera . . . people of Phasiron*, probably bedouin sheikhs. **68:** *He was crushed*, for Simon attacked by surprise and *Bacchides* had depended heavily on the war machines

tives. [71] He agreed, and did as he said; and he swore to Jonathan[v] that he would not try to harm him as long as he lived. [72] He restored to him the captives whom he had taken previously from the land of Judah; then he turned and went back to his own land, and did not come again into their territory. [73] Thus the sword ceased from Israel. Jonathan settled in Michmash and began to judge the people; and he destroyed the godless out of Israel.

10 In the one hundred sixtieth year[w] Alexander Epiphanes, son of Antiochus, landed and occupied Ptolemais. They welcomed him, and there he began to reign. [2] When King Demetrius heard of it, he assembled a very large army and marched out to meet him in battle. [3] Demetrius sent Jonathan a letter in peaceable words to honor him; [4] for he said to himself, "Let us act first to make peace with him[x] before he makes peace with Alexander against us, [5] for he will remember all the wrongs that we did to him and to his brothers and his nation." [6] So Demetrius[y] gave him authority to recruit troops, to equip them with arms, and to become his ally; and he commanded that the hostages in the citadel should be released to him.

[7] Then Jonathan came to Jerusalem and read the letter in the hearing of all the people and of those in the citadel. [8] They were greatly alarmed when they heard that the king had given him authority to recruit troops. [9] But those in the citadel released the hostages to Jonathan, and he returned them to their parents.

[10] And Jonathan took up residence in Jerusalem and began to rebuild and restore the city. [11] He directed those who were doing the work to build the walls and encircle Mount Zion with squared stones, for better fortification; and they did so.

[12] Then the foreigners who were in the strongholds that Bacchides had built fled; [13] all of them left their places and went back to their own lands. [14] Only in Beth-zur did some remain who had forsaken the law and the commandments, for it served as a place of refuge.

[15] Now King Alexander heard of all the promises that Demetrius had sent to Jonathan, and he heard of the battles that Jonathan[y] and his brothers had fought, of the brave deeds that they had done, and of the troubles that they had endured. [16] So he said, "Shall we find another such man? Come now, we will make him our friend and ally." [17] And he wrote a letter and sent it to him, in the following words:

[18] "King Alexander to his brother Jonathan, greetings. [19] We have heard about you, that you are a mighty warrior and worthy to be our friend. [20] And so we have appointed you today to be the high priest of your nation; you are to be called the king's Friend and you are to

---

v Gk *him*   w 152 B.C.   x Gk *them*
y Gk *he*

(v. 67). **72:** Josephus says that there was an exchange of prisoners (*Antiquities*, XIII. i. 6). **73:** *Thus the sword ceased,* for about seven years, until the events of 10.69. *Michmash,* now Mukhmas, eight miles northeast of Jerusalem (1 Sam 14.5–23). *To judge the people,* as a natural leader, like Samuel and those in the book of Judges.

10.1–21: Alexander Balas appoints Jonathan high priest. 1: *Alexander* I *Epiphanes,* who came from Ephesus and whose given name was Balas (or Ba'al), posed as the *son of Antiochus* IV. He claimed the kingship from 150 B.C. onward, and reigned until about 145. Attalus II of Pergamum and Ptolemy VI persuaded the Roman senate to recognize him. **6:**

Gave him authority, as a local prince or governor, but not independence. *The hostages,* 9.53. The Syrians held the citadel at Jerusalem.

10.10: *Jonathan* now left Michmash. **11:** Lysias had ordered the wall of the Jewish fortress torn down (6.62). **14:** *Some,* i.e. hellenized Jews opposed to the Hasmoneans. **18–20:** *Alexander,* hearing of Demetrius' letter, decided to outbid him. Until the time of Antiochus IV, the hereditary high priest had been confirmed but not appointed by the ruler; now Alexander appointed Jonathan and made him one of his *Friends* (see 2.18 n.); the Jews had not elected him. **21:** *The sacred vestments,* Ex 28.1–39; 39.1–26. *Booths,* a seven-day festival held in September (Lev 23.33–43), had

take our side and keep friendship with us." He also sent him a purple robe and a golden crown.

21 So Jonathan put on the sacred vestments in the seventh month of the one hundred sixtieth year,*z* at the festival of booths,*a* and he recruited troops and equipped them with arms in abundance. 22 When Demetrius heard of these things he was distressed and said, 23 "What is this that we have done? Alexander has gotten ahead of us in forming a friendship with the Jews to strengthen himself. 24 I also will write them words of encouragement and promise them honor and gifts, so that I may have their help." 25 So he sent a message to them in the following words:

"King Demetrius to the nation of the Jews, greetings. 26 Since you have kept your agreement with us and have continued your friendship with us, and have not sided with our enemies, we have heard of it and rejoiced. 27 Now continue still to keep faith with us, and we will repay you with good for what you do for us. 28 We will grant you many immunities and give you gifts.

29 "I now free you and exempt all the Jews from payment of tribute and salt tax and crown levies, 30 and instead of collecting the third of the grain and the half of the fruit of the trees that I should receive, I release them from this day and henceforth. I will not collect them from the land of Judah or from the three districts added to it from Samaria and Galilee, from this day and for all time. 31 Jerusalem and its environs, its tithes and its revenues, shall be holy and free from tax. 32 I release also my control of the citadel in Jerusalem and give it to the high priest,

so that he may station in it men of his own choice to guard it. 33 And everyone of the Jews taken as a captive from the land of Judah into any part of my kingdom, I set free without payment; and let all officials cancel also the taxes on their livestock.

34 "All the festivals and sabbaths and new moons and appointed days, and the three days before a festival and the three after a festival—let them all be days of immunity and release for all the Jews who are in my kingdom. 35 No one shall have authority to exact anything from them or annoy any of them about any matter.

36 "Let Jews be enrolled in the king's forces to the number of thirty thousand men, and let the maintenance be given them that is due to all the forces of the king. 37 Let some of them be stationed in the great strongholds of the king, and let some of them be put in positions of trust in the kingdom. Let their officers and leaders be of their own number, and let them live by their own laws, just as the king has commanded in the land of Judah.

38 "As for the three districts that have been added to Judea from the country of Samaria, let them be annexed to Judea so that they may be considered to be under one ruler and obey no other authority than the high priest. 39 Ptolemais and the land adjoining it I have given as a gift to the sanctuary in Jerusalem, to meet the necessary expenses of the sanctuary. 40 I also grant fifteen thousand shekels of silver yearly out of the king's revenues from appropriate places. 41 And all the

---

z 152 B.C.   a Or *tabernacles*

---

come to be associated with the hope of victory over the Gentiles (Zech 14.16–19).

10.22–50: Demetrius's offer to the Jews; his defeat. 25: The letter was addressed *to the nation,* ignoring Jonathan. Demetrius thought that he could drive a wedge between leader and people. **29–30:** *All the Jews* in the Seleucid realm, not merely in Judea. *Tribute,* direct taxes proportionate to individual wealth. *Salt tax,* on salt from the marshes and the Dead Sea. *Crown levies,* fixed amounts of

money. *The three districts* (11.34) that Alexander the Great had transferred from Samaria to Judea and Antiochus IV had reassigned to Samaria were now restored (v. 38). **32:** *Release* of *control of the citadel* would free Jerusalem from military domination.

10.34: *Appointed days,* other public festivals. **36–37:** Opening the army and the civil service to Jews might strengthen their loyalty to the crown. **39:** *Ptolemais* was in the hands of Alexander (v. 1). This was an invitation to

additional funds that the government officials have not paid as they did in the first years,[b] they shall give from now on for the service of the temple.[c] 42 Moreover, the five thousand shekels of silver that my officials[d] have received every year from the income of the services of the temple, this too is canceled, because it belongs to the priests who minister there. 43 And all who take refuge at the temple in Jerusalem, or in any of its precincts, because they owe money to the king or are in debt, let them be released and receive back all their property in my kingdom.

44 "Let the cost of rebuilding and restoring the structures of the sanctuary be paid from the revenues of the king. 45 And let the cost of rebuilding the walls of Jerusalem and fortifying it all around, and the cost of rebuilding the walls in Judea, also be paid from the revenues of the king."

46 When Jonathan and the people heard these words, they did not believe or accept them, because they remembered the great wrongs that Demetrius[e] had done in Israel and how much he had oppressed them. 47 They favored Alexander, because he had been the first to speak peaceable words to them, and they remained his allies all his days.

48 Now King Alexander assembled large forces and encamped opposite Demetrius. 49 The two kings met in battle, and the army of Demetrius fled, and Alexander[f] pursued him and defeated them. 50 He pressed the battle strongly until the sun set, and on that day Demetrius fell.

51 Then Alexander sent ambassadors to Ptolemy king of Egypt with the following message: 52 "Since I have returned to my kingdom and have taken my seat on the throne of my ancestors, and established my rule—for I crushed Demetrius and gained control of our country; 53 I met him in battle, and he and his army were crushed by us, and we have taken our seat on the throne of his kingdom—54 now therefore let us establish friendship with one another; give me now your daughter as my wife, and I will become your son-in-law, and will make gifts to you and to her in keeping with your position."

55 Ptolemy the king replied and said, "Happy was the day on which you returned to the land of your ancestors and took your seat on the throne of their kingdom. 56 And now I will do for you as you wrote, but meet me at Ptolemais, so that we may see one another, and I will become your father-in-law, as you have said."

57 So Ptolemy set out from Egypt, he and his daughter Cleopatra, and came to Ptolemais in the one hundred sixty-second year.[g] 58 King Alexander met him, and Ptolemy[e] gave him his daughter Cleopatra in marriage, and celebrated her wedding at Ptolemais with great pomp, as kings do.

59 Then King Alexander wrote to Jonathan to come and meet him. 60 So he went with pomp to Ptolemais and met the two kings; he gave them and their Friends silver and gold and many gifts, and found favor with them. 61 A group of malcontents from Israel, renegades, gathered together against him to accuse him; but the king paid no attention to them. 62 The king gave orders to take off Jonathan's garments and to clothe him in purple, and they did so. 63 The king also seated him at his side; and he said to his

b Meaning of Gk uncertain   c Gk *house*
d Gk *they*   e Gk *he*   f Other ancient authorities read *Alexander fled, and Demetrius*
g 150 B.C.

the Jews to help Demetrius recapture it. **41**: *The additional funds* were grants once made to the temple by Ptolemaic and Seleucid kings, but *not paid* since the time of Antiochus IV. **10.44–45**: Here Demetrius followed the custom of Persian kings (Ezra 6.8; 7.20). **47**: *Alexander* was also recognized as king by the Jews' allies, the Romans. **50**: *Demetrius fell* probably in 150 B.C. (v. 57).

10.51–66: **Alexander's relations with Egypt and Judea. 51**: *Ptolemy* VI Philometor (1.18). **55**: *Ptolemy* recognized Alexander as legitimate (see 10.1–21 n.).

officers, "Go out with him into the middle of the city and proclaim that no one is to bring charges against him about any matter, and let no one annoy him for any reason." 64 When his accusers saw the honor that was paid him, in accord with the proclamation, and saw him clothed in purple, they all fled. 65 Thus the king honored him and enrolled him among his chief[h] Friends, and made him general and governor of the province. 66 And Jonathan returned to Jerusalem in peace and gladness.

67 In the one hundred sixty-fifth year[i] Demetrius son of Demetrius came from Crete to the land of his ancestors. 68 When King Alexander heard of it, he was greatly distressed and returned to Antioch. 69 And Demetrius appointed Apollonius the governor of Coelesyria, and he assembled a large force and encamped against Jamnia. Then he sent the following message to the high priest Jonathan:

70 "You are the only one to rise up against us, and I have fallen into ridicule and disgrace because of you. Why do you assume authority against us in the hill country? 71 If you now have confidence in your forces, come down to the plain to meet us, and let us match strength with each other there, for I have with me the power of the cities. 72 Ask and learn who I am and who the others are that are helping us. People will tell you that you cannot stand before us, for your ancestors were twice put to flight in their own land. 73 And now you will not be able to withstand my cavalry and such an army in the plain, where there is no stone or pebble, or place to flee."

74 When Jonathan heard the words of Apollonius, his spirit was aroused. He chose ten thousand men and set out from Jerusalem, and his brother Simon met him to help him. 75 He encamped before Joppa, but the people of the city closed its gates, for Apollonius had a garrison in Joppa. 76 So they fought against it, and the people of the city became afraid and opened the gates, and Jonathan gained possession of Joppa.

77 When Apollonius heard of it, he mustered three thousand cavalry and a large army, and went to Azotus as though he were going farther. At the same time he advanced into the plain, for he had a large troop of cavalry and put confidence in it. 78 Jonathan[j] pursued him to Azotus, and the armies engaged in battle. 79 Now Apollonius had secretly left a thousand cavalry behind them. 80 Jonathan learned that there was an ambush behind him, for they surrounded his army and shot arrows at his men from early morning until late afternoon. 81 But his men stood fast, as Jonathan had commanded, and the enemy's[k] horses grew tired.

82 Then Simon brought forward his force and engaged the phalanx in battle (for the cavalry was exhausted); they were overwhelmed by him and fled, 83 and the cavalry was dispersed in the plain. They fled to Azotus and entered Beth-dagon, the temple of their idol, for safety. 84 But Jonathan burned Azotus and the surrounding towns and plundered them; and the temple of Dagon,

h Gk *first*   *i* 147 B.C.   *j* Gk *he*   *k* Gk *their*

10.57: *Cleopatra* III, who later married her uncle, Ptolemy VIII. 62: A change of *garments* often signified honor or dishonor (Zech 3.3–5; Gen 41.42). 63: See Esth 6.6–9. 65: *Chief Friends*, see 2.18 n. *General and governor*, with military and civil authority.

10.67–89: **Victories of Jonathan.** 67: *Demetrius* II *son of Demetrius* I disputed the throne with Alexander and later with Tryphon and Antiochus VI; in 138 B.C. he was taken captive by the Parthians. He reigned again from 129 till his death in 125 B.C. 69:

Coelesyria, meaning "hollow Syria," originally designated the country between the Lebanon and anti-Lebanon mountains; here it is Palestine and Transjordan, including the coast. *Jamnia,* 4.15. 72: *Twice put to flight,* 6.54; 9.18. 74: Jonathan now had forces for more than guerrilla engagements (v. 65); he had troops organized as phalanxes (v. 82). 75: *Joppa,* now Jaffa, a seaport near Jamnia and forty miles from Jerusalem.

10.82: *His force* had been held in reserve and was fresh. 83: *Beth-dagon,* house of Dagon,

and those who had taken refuge in it, he burned with fire. 85 The number of those who fell by the sword, with those burned alive, came to eight thousand.

86 Then Jonathan left there and encamped against Askalon, and the people of the city came out to meet him with great pomp.

87 He and those with him then returned to Jerusalem with a large amount of booty. 88 When King Alexander heard of these things, he honored Jonathan still more; 89 and he sent to him a golden buckle, such as it is the custom to give to the King's Kinsmen. He also gave him Ekron and all its environs as his possession.

11 Then the king of Egypt gathered great forces, like the sand by the seashore, and many ships; and he tried to get possession of Alexander's kingdom by trickery and add it to his own kingdom. 2 He set out for Syria with peaceable words, and the people of the towns opened their gates to him and went to meet him, for King Alexander had commanded them to meet him, since he was Alexander's*l* father-in-law. 3 But when Ptolemy entered the towns he stationed forces as a garrison in each town.

4 When he*m* approached Azotus, they showed him the burnt-out temple of Dagon, and Azotus and its suburbs destroyed, and the corpses lying about, and the charred bodies of those whom Jonathan*n* had burned in the war, for they had piled them in heaps along his route. 5 They also told the king what Jonathan

had done, to throw blame on him; but the king kept silent. 6 Jonathan met the king at Joppa with pomp, and they greeted one another and spent the night there. 7 And Jonathan went with the king as far as the river called Eleutherus; then he returned to Jerusalem.

8 So King Ptolemy gained control of the coastal cities as far as Seleucia by the sea, and he kept devising wicked designs against Alexander. 9 He sent envoys to King Demetrius, saying, "Come, let us make a covenant with each other, and I will give you in marriage my daughter who was Alexander's wife, and you shall reign over your father's kingdom. 10 I now regret that I gave him my daughter, for he has tried to kill me." 11 He threw blame on Alexander*o* because he coveted his kingdom. 12 So he took his daughter away from him and gave her to Demetrius. He was estranged from Alexander, and their enmity became manifest.

13 Then Ptolemy entered Antioch and put on the crown of Asia. Thus he put two crowns on his head, the crown of Egypt and that of Asia. 14 Now King Alexander was in Cilicia at that time, because the people of that region were in revolt. 15 When Alexander heard of it, he came against him in battle. Ptolemy marched out and met him with a strong force, and put him to flight. 16 So Alexander fled into Arabia to find protection there, and King Ptolemy was trium-

l Gk *his* *m* Other ancient authorities read *they* *n* Gk *he* *o* Gk *him*

---

the Philistine grain god (Judg 16.23). **86:** *Askalon,* about twelve miles north of Gaza. **89:** *A golden buckle,* with which the most select of the Friends of the king fastened their purple robes. *Ekron,* northernmost of the Philistine cities, was given to Jonathan as a personal possession, and its taxes were assigned to him. See 1 Sam 27.6.
11.1–19: **Invasion of Ptolemy VI and victory of Demetrius II. 1:** Josephus says that Ptolemy came to aid Alexander, his son-in-law, but that the latter plotted against Ptolemy's life (*Antiquities,* XIII. iv. 5–6). **5:** *Kept silent,* he had not yet broken with Alexander, and was not ready to commit himself. **7:** *Eleu-*

therus, now Nahr el-Kebir, north of Tripolis. **11.8:** *Seleucia* in Pieria, the main port for Antioch, near the mouth of the Orontes. **9:** *My daughter,* Cleopatra III (10.57). **13:** According to Josephus, the army proclaimed *Ptolemy* as king, but he persuaded the people of Antioch to support Demetrius. **14:** *Cilicia,* on the south coast of Turkey, always closely related to Syria, and the only section of Asia Minor then part of the Seleucid Empire. **15:** *Ptolemy marched out,* according to Josephus (*Antiquities,* XIII. iv. 8) with his son-in-law, Demetrius, who had already married Cleopatra. **16:** *Arabia* here includes the country east of Aleppo and Damascus. **18:** *Ptolemy died* of

phant. [17]Zabdiel the Arab cut off the head of Alexander and sent it to Ptolemy. [18]But King Ptolemy died three days later, and his troops in the strongholds were killed by the inhabitants of the strongholds. [19]So Demetrius became king in the one hundred sixty-seventh year.[p]

20 In those days Jonathan assembled the Judeans to attack the citadel in Jerusalem, and he built many engines of war to use against it. [21]But certain renegades who hated their nation went to the king and reported to him that Jonathan was besieging the citadel. [22]When he heard this he was angry, and as soon as he heard it he set out and came to Ptolemais; and he wrote Jonathan not to continue the siege, but to meet him for a conference at Ptolemais as quickly as possible.

23 When Jonathan heard this, he gave orders to continue the siege. He chose some of the elders of Israel and some of the priests, and put himself in danger, [24]for he went to the king at Ptolemais, taking silver and gold and clothing and numerous other gifts. And he won his favor. [25]Although certain renegades of his nation kept making complaints against him, [26]the king treated him as his predecessors had treated him; he exalted him in the presence of all his Friends. [27]He confirmed him in the high priesthood and in as many other honors as he had formerly had, and caused him to be reckoned among his chief[q] Friends. [28]Then Jonathan asked the king to free Judea and the three districts of Samaria[r] from tribute, and promised him three hundred talents. [29]The king consented, and wrote a letter to Jonathan about all these things; its contents were as follows:

30 "King Demetrius to his brother Jonathan and to the nation of the Jews, greetings. [31]This copy of the letter that we wrote concerning you to our kinsman Lasthenes we have written to you also, so that you may know what it says. [32]'King Demetrius to his father Lasthenes, greetings. [33]We have determined to do good to the nation of the Jews, who are our friends and fulfill their obligations to us, because of the goodwill they show toward us. [34]We have confirmed as their possession both the territory of Judea and the three districts of Aphairema and Lydda and Rathamin; the latter, with all the region bordering them, were added to Judea from Samaria. To all those who offer sacrifice in Jerusalem we have granted release from[s] the royal taxes that the king formerly received from them each year, from the crops of the land and the fruit of the trees. [35]And the other payments henceforth due to us of the tithes, and the taxes due to us, and the salt pits and the crown taxes due to us—from all these we shall grant them release. [36]And not one of these grants shall be canceled from this time on forever. [37]Now therefore take care to make a copy of this, and let it be given to Jonathan and put up in a conspicuous place on the holy mountain.' "

38 When King Demetrius saw that the land was quiet before him and that

p 145 B.C. *q* Gk *first* *r* Cn: Gk *the three districts and Samaria*
s Or *Samaria, for all those who offer sacrifice in Jerusalem, in place of*

---

wounds suffered in the victory over Alexander. **19**: *Demetrius* had claimed to be *king* since 150 B.C. (see 10.67 n.).

11.20–37: Agreement between Demetrius II and Jonathan. 20: Demetrius I had once promised to turn *the citadel* over to the high priest (10.32), but he refused to recognize Jonathan as such because the latter had supported Alexander. Now Jonathan resolved to attack the citadel and make Judea independent. **23–24**: Jonathan decided to *continue the siege* so as to negotiate from strength, but his dangerous visit and gifts showed that he was willing to make terms. **27**: 10.20.

11.30–37: The letter repeats earlier promises (10.25–45), but says nothing of the citadel. **31**: *Lasthenes,* probably governor of Coelesyria. **34**: *The three districts,* 10.30, 38. *Aphairema,* probably et-Taiyibeh, four miles northeast of Bethel, the Ephraim of Jn 11.54. *Lydda,* or Lod, east of Jaffa. *Rathamin,* perhaps Ramathaim-zophim (1 Sam 1.1), the Arimathea of Mk 15.43.

11.38–52: Jonathan's aid to Demetrius. 38: *Demetrius* in overconfidence reduced his army, probably to save money. This made

there was no opposition to him, he dismissed all his troops, all of them to their own homes, except the foreign troops that he had recruited from the islands of the nations. So all the troops who had served under his predecessors hated him. 39 A certain Trypho had formerly been one of Alexander's supporters; he saw that all the troops were grumbling against Demetrius. So he went to Imalkue the Arab, who was bringing up Antiochus, the young son of Alexander, 40 and insistently urged him to hand Antiochus *t* over to him, to become king in place of his father. He also reported to Imalkue *t* what Demetrius had done and told of the hatred that the troops of Demetrius *u* had for him; and he stayed there many days.

41 Now Jonathan sent to King Demetrius the request that he remove the troops of the citadel from Jerusalem, and the troops in the strongholds; for they kept fighting against Israel. 42 And Demetrius sent this message back to Jonathan: "Not only will I do these things for you and your nation, but I will confer great honor on you and your nation, if I find an opportunity. 43 Now then you will do well to send me men who will help me, for all my troops have revolted." 44 So Jonathan sent three thousand stalwart men to him at Antioch, and when they came to the king, the king rejoiced at their arrival.

45 Then the people of the city assembled within the city, to the number of a hundred and twenty thousand, and they wanted to kill the king. 46 But the king fled into the palace. Then the people of the city seized the main streets of the city and began to fight. 47 So the king called the Jews to his aid, and they all rallied around him and then spread out through the city; and they killed on that day about one hundred thousand. 48 They set fire to the city and seized a large amount of spoil on that day, and saved the king. 49 When the people of the city saw that the Jews had gained control of the city as they pleased, their courage failed and they cried out to the king with this entreaty: 50 "Grant us peace, and make the Jews stop fighting against us and our city." 51 And they threw down their arms and made peace. So the Jews gained glory in the sight of the king and of all the people in his kingdom, and they returned to Jerusalem with a large amount of spoil.

52 So King Demetrius sat on the throne of his kingdom, and the land was quiet before him. 53 But he broke his word about all that he had promised; he became estranged from Jonathan and did not repay the favors that Jonathan *v* had done him, but treated him very harshly.

54 After this Trypho returned, and with him the young boy Antiochus who began to reign and put on the crown. 55 All the troops that Demetrius had discharged gathered around him; they fought against Demetrius, *t* and he fled and was routed. 56 Trypho captured the elephants *w* and gained control of Antioch. 57 Then the young Antiochus wrote to Jonathan, saying, "I confirm you in the high priesthood and set you over the four districts and make you one of the king's Friends." 58 He also sent him gold plate and a table service, and granted him the right to drink from gold cups and dress in purple and wear a gold buckle. 59 He appointed Jonathan's *x* brother Si-

t Gk *him*　　*u* Gk *his troops*
v Gk *he*　　*w* Gk *animals*　　*x* Gk *his*

---

him unpopular. **42–43**: At last Demetrius seemed to agree (compare v. 53) to evacuate the citadel and other fortresses. **45–47**: *People of the city,* a mob, not an army. *Jews,* Judeans (see 2.23 n.).

11.53–74: Estrangement of Demetrius and Jonathan. 53: Josephus says that Demetrius now demanded tribute as before (*Antiquities,* XIII. v. 3). **54**: *Antiochus* VI Epiphanes, son of Alexander Balas, reigned nominally from about 145 to 142 B.C. **55**: *All the troops,* v. 38. **57**: *The four districts,* the three of v. 34 and probably Ekron (10.89). **58**: *Gold cups,* Esth 1.7. *Gold buckle,* sign of being a *Friend* of the king (see 2.18 n.; 10.89 n.). **59**: *The Ladder of Tyre,* the coastline between Ptolemais and Tyre. *The borders of Egypt,* probably Wadi el-Arish.

mon governor from the Ladder of Tyre to the borders of Egypt.

60 Then Jonathan set out and traveled beyond the river and among the towns, and all the army of Syria gathered to him as allies. When he came to Askalon, the people of the city met him and paid him honor. 61 From there he went to Gaza, but the people of Gaza shut him out. So he besieged it and burned its suburbs with fire and plundered them. 62 Then the people of Gaza pleaded with Jonathan, and he made peace with them, and took the sons of their rulers as hostages and sent them to Jerusalem. And he passed through the country as far as Damascus.

63 Then Jonathan heard that the officers of Demetrius had come to Kadesh in Galilee with a large army, intending to remove him from office. 64 He went to meet them, but left his brother Simon in the country. 65 Simon encamped before Beth-zur and fought against it for many days and hemmed it in. 66 Then they asked him to grant them terms of peace, and he did so. He removed them from there, took possession of the town, and set a garrison over it.

67 Jonathan and his army encamped by the waters of Gennesaret. Early in the morning they marched to the plain of Hazor, 68 and there in the plain the army of the foreigners met him; they had set an ambush against him in the mountains, but they themselves met him face to face. 69 Then the men in ambush emerged from their places and joined battle. 70 All the men with Jonathan fled; not one of them was left except Mattathias son of Absalom and Judas son of Chalphi, commanders of the forces of the army. 71 Jonathan tore his clothes, put dust on his head, and prayed. 72 Then he turned back to the battle against the enemy[y] and routed them, and they fled. 73 When his men who were fleeing saw this, they returned to him and joined him in the pursuit as far as Kadesh, to their camp, and there they encamped. 74 As many as three thousand of the foreigners fell that day. And Jonathan returned to Jerusalem.

12 Now when Jonathan saw that the time was favorable for him, he chose men and sent them to Rome to confirm and renew the friendship with them. 2 He also sent letters to the same effect to the Spartans and to other places. 3 So they went to Rome and entered the senate chamber and said, "The high priest Jonathan and the Jewish nation have sent us to renew the former friendship and alliance with them." 4 And the Romans[z] gave them letters to the people in every place, asking them to provide for the envoys[y] safe conduct to the land of Judah.

5 This is a copy of the letter that Jonathan wrote to the Spartans: 6 "The high priest Jonathan, the senate of the nation, the priests, and the rest of the Jewish people to their brothers the Spartans, greetings. 7 Already in time past a letter was sent to the high priest Onias from Arius,[a] who was king among you, stating that you are our brothers, as the appended copy shows. 8 Onias welcomed the envoy with honor, and received the letter, which contained a clear declaration of alliance and friendship. 9 Therefore, though we have no need of these things, since we have as encouragement the holy books that are in our hands,

y Gk them   z Gk they   a Vg Compare
verse 20: Gk Darius

11.60: *The river*, Jordan. 61: *Gaza*, southernmost of the Philistine cities, near the Egyptian border (Judg 16.1). 62: *Damascus*, outside Jonathan's control, but his influence extended nearly that far. 63: *Kadesh*, northwest of Lake Huleh or Merom (Judg 4.9). 67: *The waters of Gennesaret*, the Sea of Galilee. *Hazor*, southwest of Lake Huleh (Josh 11.1).
12.1–23: Alliances with the Romans and Spartans (compare ch 8). 2: *The Spartans* had not joined the Achaean league against Rome. 4: *The Romans* continued the old alliance in order to keep Syria weak. 6: *The senate*, over which the high priest presided; it corresponds to the later council or Sanhedrin (Mk 14.55). 7: *Onias* I, high priest 320–290 B.C. *Arius*, king of Sparta 309–265 B.C.

10 we have undertaken to send to renew our family ties and friendship with you, so that we may not become estranged from you, for considerable time has passed since you sent your letter to us. 11 We therefore remember you constantly on every occasion, both at our festivals and on other appropriate days, at the sacrifices that we offer and in our prayers, as it is right and proper to remember brothers. 12 And we rejoice in your glory. 13 But as for ourselves, many trials and many wars have encircled us; the kings around us have waged war against us. 14 We were unwilling to annoy you and our other allies and friends with these wars, 15 for we have the help that comes from Heaven for our aid, and so we were delivered from our enemies, and our enemies were humbled. 16 We therefore have chosen Numenius son of Antiochus and Antipater son of Jason, and have sent them to Rome to renew our former friendship and alliance with them. 17 We have commanded them to go also to you and greet you and deliver to you this letter from us concerning the renewal of our family ties. 18 And now please send us a reply to this."

19 This is a copy of the letter that they sent to Onias: 20 "King Arius of the Spartans, to the high priest Onias, greetings. 21 It has been found in writing concerning the Spartans and the Jews that they are brothers and are of the family of Abraham. 22 And now that we have learned this, please write us concerning your welfare; 23 we on our part write to you that your livestock and your property belong to us, and ours belong to you. We therefore command that our envoys[b] report to you accordingly."

24 Now Jonathan heard that the commanders of Demetrius had returned, with a larger force than before, to wage war against him. 25 So he marched away from Jerusalem and met them in the region of Hamath, for he gave them no opportunity to invade his own country. 26 He sent spies to their camp, and they returned and reported to him that the enemy[b] were being drawn up in formation to attack the Jews[c] by night. 27 So when the sun had set, Jonathan commanded his troops to be alert and to keep their arms at hand so as to be ready all night for battle, and he stationed outposts around the camp. 28 When the enemy heard that Jonathan and his troops were prepared for battle, they were afraid and were terrified at heart; so they kindled fires in their camp and withdrew.[d] 29 But Jonathan and his troops did not know it until morning, for they saw the fires burning. 30 Then Jonathan pursued them, but he did not overtake them, for they had crossed the Eleutherus river. 31 So Jonathan turned aside against the Arabs who are called Zabadeans, and he crushed them and plundered them. 32 Then he broke camp and went to Damascus, and marched through all that region.

33 Simon also went out and marched through the country as far as Askalon and the neighboring strongholds. He turned aside to Joppa and took it by surprise, 34 for he had heard that they were ready to hand over the stronghold to those whom Demetrius had sent. And he stationed a garrison there to guard it.

35 When Jonathan returned he convened the elders of the people and planned with them to build strongholds in Judea, 36 to build the walls of Jerusalem still higher, and to erect a high barrier between the citadel and the city to separate it from the city, in order to isolate it so that its garrison[b] could neither buy

---

b Gk *they*   c Gk *them*   d Other ancient authorities omit *and withdrew*

12.21: *Brothers . . . of the family of Abraham,* compare v. 7; such a tradition was evidently current in the East.
12.24–53: Jonathan captured by Trypho. 24–25: Jonathan met the Syrians at the border of Judea to prevent an invasion. *Ha-* math, on the Orontes, modern Hama in Syria. **28:** *Kindled fires,* so that Jonathan would think they were still in camp. **30:** *The Eleutherus* is too far north (see 11.7 n.); perhaps the Orontes. **31:** *Zabadeans,* perhaps people northwest of Damascus.

nor sell. 37 So they gathered together to rebuild the city; part of the wall on the valley to the east had fallen, and he repaired the section called Chaphenatha. 38 Simon also built Adida in the Shephelah; he fortified it and installed gates with bolts.

39 Then Trypho attempted to become king in Asia and put on the crown, and to raise his hand against King Antiochus. 40 He feared that Jonathan might not permit him to do so, but might make war on him, so he kept seeking to seize and kill him, and he marched out and came to Beth-shan. 41 Jonathan went out to meet him with forty thousand picked warriors, and he came to Beth-shan. 42 When Trypho saw that he had come with a large army, he was afraid to raise his hand against him. 43 So he received him with honor and commended him to all his Friends, and he gave him gifts and commanded his Friends and his troops to obey him as they would himself. 44 Then he said to Jonathan, "Why have you put all these people to so much trouble when we are not at war? 45 Dismiss them now to their homes and choose for yourself a few men to stay with you, and come with me to Ptolemais. I will hand it over to you as well as the other strongholds and the remaining troops and all the officials, and will turn around and go home. For that is why I am here."

46 Jonathan[e] trusted him and did as he said; he sent away the troops, and they returned to the land of Judah. 47 He kept with himself three thousand men, two thousand of whom he left in Galilee, while one thousand accompanied him. 48 But when Jonathan entered Ptolemais, the people of Ptolemais closed the gates and seized him, and they killed with the sword all who had entered with him.

49 Then Trypho sent troops and cavalry into Galilee and the Great Plain to destroy all Jonathan's soldiers. 50 But they realized that Jonathan had been seized and had perished along with his men, and they encouraged one another and kept marching in close formation, ready for battle. 51 When their pursuers saw that they would fight for their lives, they turned back. 52 So they all reached the land of Judah safely, and they mourned for Jonathan and his companions and were in great fear; and all Israel mourned deeply. 53 All the nations around them tried to destroy them, for they said, "They have no leader or helper. Now therefore let us make war on them and blot out the memory of them from humankind."

13 Simon heard that Trypho had assembled a large army to invade the land of Judah and destroy it, 2 and he saw that the people were trembling with fear. So he went up to Jerusalem, and gathering the people together 3 he encouraged them, saying to them, "You yourselves know what great things my brothers and I and the house of my father have done for the laws and the sanctuary; you know also the wars and the difficulties that my brothers and I have seen. 4 By reason of this all my brothers have perished for the sake of Israel, and I alone am left. 5 And now, far be it from me to spare my life in any time of distress, for I am not better than my brothers. 6 But I will avenge my nation and the sanctuary and your wives and children, for all the nations have gathered together out of hatred to destroy us."

7 The spirit of the people was rekin-

e Gk *he*

12.36: The purpose was to starve out the garrison. **37**: *The valley to the east,* the Kidron (1 Kings 2.37; Jn 18.1); here the slope was sometimes steep. The location of *Chaphenatha* is unknown. **38**: *Adida,* about three miles east of Lydda (Ezra 2.33; Neh 7.37). *Shephelah,* the foothill country between the coastal plain and the central highlands. **40**: *Beth-shan,* see 5.52 n. **45**: *The other strongholds,* probably along the coast. **52**: *They mourned,* because of the supposition that Jonathan had been slain; but see 13.23.

13.1–30: **Simon becomes leader; death of Jonathan. 1**: *Simon* (2.3) was governor of the coastal area (11.59). **4**: Eleazar, Judas, and John had died (6.46; 9.18, 42), and Simon supposed that Jonathan had been slain. **7–8**: Jonathan had been chosen by his friends

dled when they heard these words, [8] and they answered in a loud voice, "You are our leader in place of Judas and your brother Jonathan. [9] Fight our battles, and all that you say to us we will do." [10] So he assembled all the warriors and hurried to complete the walls of Jerusalem, and he fortified it on every side. [11] He sent Jonathan son of Absalom to Joppa, and with him a considerable army; he drove out its occupants and remained there.

12 Then Trypho left Ptolemais with a large army to invade the land of Judah, and Jonathan was with him under guard. [13] Simon encamped in Adida, facing the plain. [14] Trypho learned that Simon had risen up in place of his brother Jonathan, and that he was about to join battle with him, so he sent envoys to him and said, [15] "It is for the money that your brother Jonathan owed the royal treasury, in connection with the offices he held, that we are detaining him. [16] Send now one hundred talents of silver and two of his sons as hostages, so that when released he will not revolt against us, and we will release him."

17 Simon knew that they were speaking deceitfully to him, but he sent to get the money and the sons, so that he would not arouse great hostility among the people, who might say, [18] "It was because Simon[f] did not send him the money and the sons, that Jonathan[g] perished." [19] So he sent the sons and the hundred talents, but Trypho[g] broke his word and did not release Jonathan.

20 After this Trypho came to invade the country and destroy it, and he circled around by the way to Adora. But Simon and his army kept marching along opposite him to every place he went. [21] Now the men in the citadel kept sending envoys to Trypho urging him to come to them by way of the wilderness and to send them food. [22] So Trypho got all his cavalry ready to go, but that night a very heavy snow fell, and he did not go because of the snow. He marched off and went into the land of Gilead. [23] When he approached Baskama, he killed Jonathan, and he was buried there. [24] Then Trypho turned and went back to his own land.

25 Simon sent and took the bones of his brother Jonathan, and buried him in Modein, the city of his ancestors. [26] All Israel bewailed him with great lamentation, and mourned for him many days. [27] And Simon built a monument over the tomb of his father and his brothers; he made it high so that it might be seen, with polished stone at the front and back. [28] He also erected seven pyramids, opposite one another, for his father and mother and four brothers. [29] For the pyramids[h] he devised an elaborate setting, erecting about them great columns, and on the columns he put suits of armor for a permanent memorial, and beside the suits of armor he carved ships, so that they could be seen by all who sail the sea. [30] This is the tomb that he built in Modein; it remains to this day.

31 Trypho dealt treacherously with the young King Antiochus; he killed him [32] and became king in his place, putting on the crown of Asia; and he brought

f Gk I   g Gk he   h Gk For these

(9.28–30); *the people* now elected Simon *leader,* but not yet high priest (compare 14.41). **11:** *Jonathan son of Absalom,* perhaps a brother of Mattathias (11.70). **15–16:** He regarded *Jonathan* as a vassal of Syria who had to pay for *the offices he held* (11.57).

13.20: *Adora,* or Adoraim, now Dura, five miles southwest of Hebron (2 Chr 11.9). **22–23:** *He marched off,* south of the Dead Sea. *Baskama,* possibly northeast of the Sea of Galilee. *He killed Jonathan,* late in 143 or early in 142 B.C. **25:** *Modein,* 2.1; 9.19. **28:** *Pyramids,* combining the Greek custom of building a monument with a suit of armor at the victory site with the Hebrew custom of burial at the ancestral home. **29:** *Carved ships,* symbols claiming domination of the sea, found also on coins of Herod and Archelaus. They were not visible from the sea; *so that they could be seen by all who sail the sea,* they were intended to serve as a warning to potential naval invaders. The Hasmoneans held the seaport of Joppa (13.11; 14.5, 34).

13.31–53: Simon makes Judea independent. 31–32: *He killed him,* probably in 142 B.C. *Antiochus* VI was about seven years

great calamity on the land. [33] But Simon built up the strongholds of Judea and walled them all around, with high towers and great walls and gates and bolts, and he stored food in the strongholds. [34] Simon also chose emissaries and sent them to King Demetrius with a request to grant relief to the country, for all that Trypho did was to plunder. [35] King Demetrius sent him a favorable reply to this request, and wrote him a letter as follows, [36] "King Demetrius to Simon, the high priest and friend of kings, and to the elders and nation of the Jews, greetings. [37] We have received the gold crown and the palm branch that you[i] sent, and we are ready to make a general peace with you and to write to our officials to grant you release from tribute. [38] All the grants that we have made to you remain valid, and let the strongholds that you have built be your possession. [39] We pardon any errors and offenses committed to this day, and cancel the crown tax that you owe; and whatever other tax has been collected in Jerusalem shall be collected no longer. [40] And if any of you are qualified to be enrolled in our bodyguard,[j] let them be enrolled, and let there be peace between us."

[41] In the one hundred seventieth year[k] the yoke of the Gentiles was removed from Israel, [42] and the people began to write in their documents and contracts, "In the first year of Simon the great high priest and commander and leader of the Jews."

[43] In those days Simon[l] encamped against Gazara[m] and surrounded it with troops. He made a siege engine, brought it up to the city, and battered and captured one tower. [44] The men in the siege engine leaped out into the city, and a great tumult arose in the city. [45] The men in the city, with their wives and children, went up on the wall with their clothes torn, and they cried out with a loud voice, asking Simon to make peace with them; [46] they said, "Do not treat us according to our wicked acts but according to your mercy." [47] So Simon reached an agreement with them and stopped fighting against them. But he expelled them from the city and cleansed the houses in which the idols were located, and then entered it with hymns and praise. [48] He removed all uncleanness from it, and settled in it those who observed the law. He also strengthened its fortifications and built in it a house for himself.

[49] Those who were in the citadel at Jerusalem were prevented from going in and out to buy and sell in the country. So they were very hungry, and many of them perished from famine. [50] Then they cried to Simon to make peace with them, and he did so. But he expelled them from there and cleansed the citadel from its pollutions. [51] On the twenty-third day of the second month, in the one hundred seventy-first year,[n] the Jews[o] entered it with praise and palm branches, and with harps and cymbals and stringed instruments, and with hymns and songs, because a great enemy had been crushed and removed from Israel. [52] Simon[p] decreed that every year they should celebrate this day with rejoicing. He strengthened the fortifications of the temple hill alongside the citadel, and he

---

i The word *you* in verses 37–40 is plural
j Or *court* k 142 B.C. l Gk *he*
m Cn: Gk *Gaza* n 141 B.C. o Gk *they*
p Gk *He*

---

old and had reigned since 145 (11.54). *Became king,* about 142 or 141 B.C. **34:** *Demetrius* II (see 10.67 n.) now disputed the throne with Trypho. **36–40:** The letter, addressed to *the elders and nation* and to *Simon* as head of a priestly state, recognizes sovereignty (compare v. 42). The weakness of Demetrius II made possible a great diplomatic victory. **42:** The new era, replacing the Seleucid era, is a mark of sovereignty. It is debated whether Simon was the first of the Hasmoneans to strike coins. **43:** A Greek inscription hostile to Simon has been found at *Gazara. Siege engine,* a tower on wheels, in which men with catapults and battering rams could breach fortified walls. **47–48:** The later Hasmoneans continued the policy of settling Jews in strategic places. **51:** *The second month,* Iyyar or May. *Palm branches* symbolized victory (2 Macc 10.7). **53:** *John* Hyrcanus reigned as high priest 134–104 B.C.

and his men lived there. ⁵³Simon saw that his son John had reached manhood, and so he made him commander of all the forces; and he lived at Gazara.

14 In the one hundred seventy-second year*q* King Demetrius assembled his forces and marched into Media to obtain help, so that he could make war against Trypho. ²When King Arsaces of Persia and Media heard that Demetrius had invaded his territory, he sent one of his generals to take him alive. ³The general*r* went and defeated the army of Demetrius, and seized him and took him to Arsaces, who put him under guard.

⁴ The land*s* had rest all the days of
    Simon.
  He sought the good of his
    nation;
his rule was pleasing to them,
  as was the honor shown him, all
    his days.
⁵ To crown all his honors he took
    Joppa for a harbor,
  and opened a way to the isles of
    the sea.
⁶ He extended the borders of his
    nation,
  and gained full control of the
    country.
⁷ He gathered a host of captives;
  he ruled over Gazara and
    Beth-zur and the citadel,
  and he removed its uncleanness
    from it;
  and there was none to oppose
    him.
⁸ They tilled their land in peace;
  the ground gave its increase,
  and the trees of the plains their
    fruit.

⁹ Old men sat in the streets;
  they all talked together of good
    things,
  and the youths put on splendid
    military attire.
¹⁰ He supplied the towns with food,
  and furnished them with the
    means of defense,
  until his renown spread to the
    ends of the earth.
¹¹ He established peace in the land,
  and Israel rejoiced with great
    joy.
¹² All the people sat under their own
    vines and fig trees,
  and there was none to make
    them afraid.
¹³ No one was left in the land to
    fight them,
  and the kings were crushed in
    those days.
¹⁴ He gave help to all the humble
    among his people;
  he sought out the law,
  and did away with all the
    renegades and outlaws.
¹⁵ He made the sanctuary glorious,
  and added to the vessels of the
    sanctuary.

16 It was heard in Rome, and as far away as Sparta, that Jonathan had died, and they were deeply grieved. ¹⁷When they heard that his brother Simon had become high priest in his stead, and that he was ruling over the country and the towns in it, ¹⁸they wrote to him on bronze tablets to renew with him the friendship and alliance that they had established with his brothers Judas and

q 140 B.C.   *r* Gk *He*   *s* Other ancient authorities add *of Judah*

14.1–15: Capture of Demetrius II. 1: Other historians date this invasion in 138 B.C., the year in which he was captured. *Media,* lying west of Teheran, was still claimed by the Seleucids. **2–3:** *Arsaces* VI Mithradates I (171–138 B.C.), founder of the Parthian Empire, treated Demetrius kindly and later married him to his sister.
14.4–15: A contemporary poem of rejoicing, describing the reign of Simon as the fulfillment of biblical prophecies. **4:** 1 Kings 5.5.

5–7: 13.41–53. *A harbor,* important for trade connections with the sea (13.11). *The isles of the sea,* Cyprus, Rhodes, and Crete. **8:** Zech 8.12. **9:** Zech 8.4. **10:** 12.38; 13.33, 52. **12:** 1 Kings 4.25; Mic 4.4; Zech 3.10. **13–14:** Isa 11.3–4. **14:** *Sought out,* he studied Torah to learn God's will, as the kings of Israel had consulted prophets. See Ezra 7.10.
14.16–24: Alliances with Rome and Sparta. Simon was perhaps the first high priest recognized by the Roman senate as rul-

Jonathan. [19] And these were read before the assembly in Jerusalem.

20 This is a copy of the letter that the Spartans sent:

"The rulers and the city of the Spartans to the high priest Simon and to the elders and the priests and the rest of the Jewish people, our brothers, greetings. [21] The envoys who were sent to our people have told us about your glory and honor, and we rejoiced at their coming. [22] We have recorded what they said in our public decrees, as follows, 'Numenius son of Antiochus and Antipater son of Jason, envoys of the Jews, have come to us to renew their friendship with us. [23] It has pleased our people to receive these men with honor and to put a copy of their words in the public archives, so that the people of the Spartans may have a record of them. And they have sent a copy of this to the high priest Simon.' "

24 After this Simon sent Numenius to Rome with a large gold shield weighing one thousand minas, to confirm the alliance with the Romans. [t]

25 When the people heard these things they said, "How shall we thank Simon and his sons? [26] For he and his brothers and the house of his father have stood firm; they have fought and repulsed Israel's enemies and established its freedom." [27] So they made a record on bronze tablets and put it on pillars on Mount Zion.

This is a copy of what they wrote: "On the eighteenth day of Elul, in the one hundred seventy-second year, [u] which is the third year of the great high priest Simon, [28] in Asaramel, [v] in the great assembly of the priests and the people and the rulers of the nation and the elders of the country, the following was proclaimed to us:

29 "Since wars often occurred in the country, Simon son of Mattathias, a priest of the sons [w] of Joarib, and his brothers, exposed themselves to danger and resisted the enemies of their nation, in order that their sanctuary and the law might be preserved; and they brought great glory to their nation. [30] Jonathan rallied the [x] nation, became their high priest, and was gathered to his people. [31] When their enemies decided to invade their country and lay hands on their sanctuary, [32] then Simon rose up and fought for his nation. He spent great sums of his own money; he armed the soldiers of his nation and paid them wages. [33] He fortified the towns of Judea, and Beth-zur on the borders of Judea, where formerly the arms of the enemy had been stored, and he placed there a garrison of Jews. [34] He also fortified Joppa, which is by the sea, and Gazara, which is on the borders of Azotus, where the enemy formerly lived. He settled Jews there, and provided in those towns [t] whatever was necessary for their restoration.

35 "The people saw Simon's faithfulness [y] and the glory that he had resolved to win for his nation, and they made him their leader and high priest, because he had done all these things and because of

---

t Gk *them*   *u* 140 B.C.   *v* This word resembles the Hebrew words for *the court of the people of God* or *the prince of the people of God*   *w* Meaning of Gk uncertain   *x* Gk *their*   *y* Other ancient authorities read *conduct*

---

er of the Jews. **19:** *The assembly,* the people as a whole. **22:** *Numenius* and *Antipater,* 12.16. **24:** *Weighing one thousand minas,* an obvious exaggeration; a Greek mina is over 15 ounces (troy).

14.25–49: **Simon elected high priest, military commander, and ruler.** The formal document of vv. 27–49 served as a constitution for the new state of Judea. **27–28:** *Bronze tablets,* see 8.22 n. *Mount Zion,* 1.33; 4.37. *Elul,* August-September. *Third year,* see 13.42 n. *The great assembly* represented all "states" or classes. In theory the high priest

held his office by divine appointment, indicated by descent from a particular family. Since there was no legitimate claimant, Simon was legitimized by a process known in ancient Israel. See Ex ch 19; 2 Kings ch 23; Ezra ch 10; Neh ch 9. **29:** *Sons of Joarib,* see 2.1 n. **30:** The decree recognizes in retrospect the office of *Jonathan,* first Hasmonean *high priest.* **32:** Use of *his own money* had not been previously mentioned. **33–34:** 13.43–48; 14.3–7.

14.35: The high priest must have moral responsibility along with his powers. **36–37:**

the justice and loyalty that he had maintained toward his nation. He sought in every way to exalt his people. [36] In his days things prospered in his hands, so that the Gentiles were put out of the[z] country, as were also those in the city of David in Jerusalem, who had built themselves a citadel from which they used to sally forth and defile the environs of the sanctuary, doing great damage to its purity. [37] He settled Jews in it and fortified it for the safety of the country and of the city, and built the walls of Jerusalem higher.

[38] "In view of these things King Demetrius confirmed him in the high priesthood, [39] made him one of his Friends, and paid him high honors. [40] For he had heard that the Jews were addressed by the Romans as friends and allies and brothers, and that the Romans[a] had received the envoys of Simon with honor.

[41] "The Jews and their priests have resolved that Simon should be their leader and high priest forever, until a trustworthy prophet should arise, [42] and that he should be governor over them and that he should take charge of the sanctuary and appoint officials over its tasks and over the country and the weapons and the strongholds, and that he should take charge of the sanctuary, [43] and that he should be obeyed by all, and that all contracts in the country should be written in his name, and that he should be clothed in purple and wear gold.

[44] "None of the people or priests shall be permitted to nullify any of these decisions or to oppose what he says, or to convene an assembly in the country without his permission, or to be clothed in purple or put on a gold buckle. [45] Whoever acts contrary to these decisions or rejects any of them shall be liable to punishment."

[46] All the people agreed to grant Simon the right to act in accordance with these decisions. [47] So Simon accepted and agreed to be high priest, to be commander and ethnarch of the Jews and priests, and to be protector of them all. [b] [48] And they gave orders to inscribe this decree on bronze tablets, to put them up in a conspicuous place in the precincts of the sanctuary, [49] and to deposit copies of them in the treasury, so that Simon and his sons might have them.

15 Antiochus, son of King Demetrius, sent a letter from the islands of the sea to Simon, the priest and ethnarch of the Jews, and to all the nation; [2] its contents were as follows: "King Antiochus to Simon the high priest and ethnarch and to the nation of the Jews, greetings. [3] Whereas certain scoundrels have gained control of the kingdom of our ancestors, and I intend to lay claim to the kingdom so that I may restore it as it formerly was, and have recruited a host of mercenary troops and have equipped warships, [4] and intend to make a landing in the country so that I may proceed against those who have destroyed our country and those who have devastated many cities in my kingdom, [5] now therefore I confirm to you all the tax remis-

z Meaning of Gk uncertain   a Gk they
b Or to preside over them all

1.34; 4.41, 60; 6.18; 13.49–52. **41–43:** The office was to be hereditary in Simon's family, but since this was an act of the nation rather than of God, a *trustworthy prophet* might annul or confirm the decision (see 4.46 n.). *Clothed in purple and wear gold,* borrowing the dress of the Seleucid kings and courtiers. From Alexander Janneus onwards (103–76 B.C.), the Hasmoneans assumed the political title of king. Little is known about the political role of high priest during the period of the Second Temple. A letter written in 408 B.C. from the Jews at Elephantine requesting help from the Judean government is addressed to the high priest and the nobles. The power vested in Simon is probably not entirely new. **46–47:** In this social contract, both the people and Simon accept the conditions. *Ethnarch,* civil magistrate. **49:** The *treasury* in the temple served as a national archive.

15.1–14: Arrival of Antiochus VII. 1: *Antiochus* VII (known as Sidetes because reared at Side in Pamphylia), younger brother of Demetrius II, reigned 138–129 B.C. After his brother's capture he married Cleopatra III (10.57–58; 11.12). **3:** *Scoundrels,* Trypho and

sions that the kings before me have granted you, and a release from all the other payments from which they have released you. [6]I permit you to mint your own coinage as money for your country, [7]and I grant freedom to Jerusalem and the sanctuary. All the weapons that you have prepared and the strongholds that you have built and now hold shall remain yours. [8]Every debt you owe to the royal treasury and any such future debts shall be canceled for you from henceforth and for all time. [9]When we gain control of our kingdom, we will bestow great honor on you and your nation and the temple, so that your glory will become manifest in all the earth."

10 In the one hundred seventy-fourth year[c] Antiochus set out and invaded the land of his ancestors. All the troops rallied to him, so that there were only a few with Trypho. [11]Antiochus pursued him, and Trypho[d] came in his flight to Dor, which is by the sea; [12]for he knew that troubles had converged on him, and his troops had deserted him. [13]So Antiochus encamped against Dor, and with him were one hundred twenty thousand warriors and eight thousand cavalry. [14]He surrounded the town, and the ships joined battle from the sea; he pressed the town hard from land and sea, and permitted no one to leave or enter it.

15 Then Numenius and his companions arrived from Rome, with letters to the kings and countries, in which the following was written: [16]"Lucius, consul of the Romans, to King Ptolemy, greetings. [17]The envoys of the Jews have come to us as our friends and allies to renew our ancient friendship and alliance. They had been sent by the high priest Simon and by the Jewish people [18]and have brought a gold shield weighing one thousand minas. [19]We therefore have decided to write to the kings and countries that they should not seek their harm or make war against them and their cities and their country, or make alliance with those who war against them. [20]And it has seemed good to us to accept the shield from them. [21]Therefore if any scoundrels have fled to you from their country, hand them over to the high priest Simon, so that he may punish them according to their law."

22 The consul[e] wrote the same thing to King Demetrius and to Attalus and Ariarathes and Arsaces, [23]and to all the countries, and to Sampsames,[f] and to the Spartans, and to Delos, and to Myndos, and to Sicyon, and to Caria, and to Samos, and to Pamphylia, and to Lycia, and to Halicarnassus, and to Rhodes, and to Phaselis, and to Cos, and to Side, and to Aradus and Gortyna and Cnidus and Cyprus and Cyrene. [24]They also sent a copy of these things to the high priest Simon.

25 King Antiochus besieged Dor for the second time, continually throwing his forces against it and making engines of war; and he shut Trypho up and kept him from going out or in. [26]And Simon sent to Antiochus[g] two thousand picked troops, to fight for him, and silver and gold and a large amount of military equipment. [27]But he refused to receive them, and broke all the agreements he

---

c 138 B.C.   *d* Gk *he*   *e* Gk *He*
f The name is uncertain   *g* Gk *him*

---

his faction. **5**: He reaffirms his brother's grants (13.39). **6**: *To mint your own coinage* was legal recognition of independence. When the Seleucids permitted subject cities to coin money, it bore the king's name. **7–8**: 13.38–39. **10**: He first landed in Seleucia (11.8), where Cleopatra was living. **11**: *Dor,* about nine miles north of Caesarea (Judg 1.27). **13**: The numbers are probably exaggerated.
15.15–24: **Renewal of alliance with Rome.** The letter follows logically after 14.24. **16**: If the letter is genuine, this is *Lucius*

Calpurnius Piso, consul 140–139 B.C. *Ptolemy* VII Physcon reigned 145–116 B.C. **18**: *Shield,* see 14.24 n. **22–23**: *Demetrius* II was still a prisoner in Parthia; the Romans had not recognized Antiochus VII. *Attalus* II, king of Pergamum 159–138 B.C.; *Ariarathes* V, king of Cappadocia 162–130 B.C. *Delos* in the Cyclades and the other localities were free states in Greece, the Greek islands and Asia Minor. *Cyrene,* capital of Libya.
15.25–16.10: **War with Antiochus VII.** **15.27**: Josephus says that Antiochus ac-

formerly had made with Simon, and became estranged from him. [28] He sent to him Athenobius, one of his Friends, to confer with him, saying, "You hold control of Joppa and Gazara and the citadel in Jerusalem; they are cities of my kingdom. [29] You have devastated their territory, you have done great damage in the land, and you have taken possession of many places in my kingdom. [30] Now then, hand over the cities that you have seized and the tribute money of the places that you have conquered outside the borders of Judea; [31] or else pay me five hundred talents of silver for the destruction that you have caused and five hundred talents more for the tribute money of the cities. Otherwise we will come and make war on you."

[32] So Athenobius, the king's Friend, came to Jerusalem, and when he saw the splendor of Simon, and the sideboard with its gold and silver plate, and his great magnificence, he was amazed. When he reported to him the king's message, [33] Simon said to him in reply: "We have neither taken foreign land nor seized foreign property, but only the inheritance of our ancestors, which at one time had been unjustly taken by our enemies. [34] Now that we have the opportunity, we are firmly holding the inheritance of our ancestors. [35] As for Joppa and Gazara, which you demand, they were causing great damage among the people and to our land; for them we will give you one hundred talents."

Athenobius[h] did not answer him a word, [36] but returned in wrath to the king and reported to him these words, and also the splendor of Simon and all that he had seen. And the king was very angry.

[37] Meanwhile Trypho embarked on a ship and escaped to Orthosia. [38] Then the king made Cendebeus commander-in-chief of the coastal country, and gave him troops of infantry and cavalry. [39] He commanded him to encamp against Judea, to build up Kedron and fortify its gates, and to make war on the people; but the king pursued Trypho. [40] So Cendebeus came to Jamnia and began to provoke the people and invade Judea and take the people captive and kill them. [41] He built up Kedron and stationed horsemen and troops there, so that they might go out and make raids along the highways of Judea, as the king had ordered him.

16 John went up from Gazara and reported to his father Simon what Cendebeus had done. [2] And Simon called in his two eldest sons Judas and John, and said to them: "My brothers and I and my father's house have fought the wars of Israel from our youth until this day, and things have prospered in our hands so that we have delivered Israel many times. [3] But now I have grown old, and you by Heaven's[i] mercy are mature in years. Take my place and my brother's, and go out and fight for our nation, and may the help that comes from Heaven be with you."

[4] So John[j] chose out of the country twenty thousand warriors and cavalry, and they marched against Cendebeus and camped for the night in Modein. [5] Early in the morning they started out and marched into the plain, where a large force of infantry and cavalry was coming to meet them; and a stream lay between them. [6] Then he and his army lined up against them. He saw that the soldiers

---

h Gk *He*   i Gk *his*   j Other ancient authorities read *he*

---

cepted this aid (*Antiquities,* XIII. vii. 2). **30:** *Outside . . . Judea,* perhaps the districts of 11.34; but compare v. 8. **33:** The Hasmoneans claimed that all Palestine had always belonged by right to the Jews. Greek law recognized the right to reclaim ancestral property that had been seized. **35:** *Joppa and Gazara,* 12.33; 13.43–48. **37:** *Orthosia,* a few miles north of Tripolis; from there Trypho went to Apamea, where he was besieged and slain. **39–40:** *Kedron,* perhaps Gedereth, southwest of Ekron (Josh 15.41). The plan was to control the coastal plain and recover Gazara and Joppa.

16.1: *John* Hyrcanus I commanded Gazara (13.53). **4:** *Cavalry* are now for the first time

were afraid to cross the stream, so he crossed over first; and when his troops saw him, they crossed over after him. [7]Then he divided the army and placed the cavalry in the center of the infantry, for the cavalry of the enemy were very numerous. [8]They sounded the trumpets, and Cendebeus and his army were put to flight; many of them fell wounded and the rest fled into the stronghold. [9]At that time Judas the brother of John was wounded, but John pursued them until Cendebeus[k] reached Kedron, which he had built. [10]They also fled into the towers that were in the fields of Azotus, and John[k] burned it with fire, and about two thousand of them fell. He then returned to Judea safely.

[11] Now Ptolemy son of Abubus had been appointed governor over the plain of Jericho; he had a large store of silver and gold, [12]for he was son-in-law of the high priest. [13]His heart was lifted up; he determined to get control of the country, and made treacherous plans against Simon and his sons, to do away with them. [14]Now Simon was visiting the towns of the country and attending to their needs, and he went down to Jericho with his sons Mattathias and Judas, in the one hundred seventy-seventh year,[l] in the eleventh month, which is the month of Shebat. [15]The son of Abubus received them treacherously in the little stronghold called Dok, which he had built; he gave them a great banquet, and hid men

there. [16]When Simon and his sons were drunk, Ptolemy and his men rose up, took their weapons, rushed in against Simon in the banquet hall and killed him and his two sons, as well as some of his servants. [17]So he committed an act of great treachery and returned evil for good.

[18] Then Ptolemy wrote a report about these things and sent it to the king, asking him to send troops to aid him and to turn over to him the towns and the country. [19]He sent other troops to Gazara to do away with John; he sent letters to the captains asking them to come to him so that he might give them silver and gold and gifts; [20]and he sent other troops to take possession of Jerusalem and the temple hill. [21]But someone ran ahead and reported to John at Gazara that his father and brothers had perished, and that "he has sent men to kill you also." [22]When he heard this, he was greatly shocked; he seized the men who came to destroy him and killed them, for he had found out that they were seeking to destroy him.

[23] The rest of the acts of John and his wars and the brave deeds that he did, and the building of the walls that he completed, and his achievements, [24]are written in the annals of his high priesthood, from the time that he became high priest after his father.

k Gk *he*   *l* 134 B.C.

---

part of the Judean army. **7**: The meaning is unclear. The author may refer to the prevailing Roman tactic of providing each unit of infantry with a unit of cavalry. **8**: *The stronghold,* Kedron (15.39). **10**: *Azotus,* destroyed by Jonathan (10.84).

16.11–24: Death of Simon and accession of John Hyrcanus I. 11: *Plain of Jericho,* the fertile region north of the Dead Sea. **12**: *The high priest,* Simon. **14**: *Shebat,* February–March. **15**: *Dok,* 'Ain Duq, three miles northwest of Jericho. **18**: *The king,* Antiochus VII. **23**: *John* was high priest 134–104 B.C. When

Antiochus later besieged Jerusalem, John was defeated but made peace and accompanied the king on an expedition to Parthia, where Antiochus was killed. Afterward he gained control of most of Palestine, and forced the Idumeans to adopt Judaism. Late in his reign the Pharisees turned against him and demanded that he give up the high priesthood. **24**: *The annals* have been lost. The book is concluded in a manner similar to the biblical accounts of the kings of Israel (1 Kings 11.41; 2 Kings 10.34; 12.19; 20.20, etc.).

2 Maccabees

Second Maccabees narrates the events of Jewish history during the persecutions of three Seleucid kings: Seleucus IV, Antiochus IV (Epiphanes), and Antiochus Eupator (from about 180–161 B.C.; see Introduction to First Maccabees). The narrative begins with the author's preface (2.19–32), explaining that the work is an abridgment presenting highlights of a five-volume work by Jason of Cyrene. No evidence of Jason's work has survived. The stories in the first half tell of persecution and martyrdom, and in the second half of the victories won by Judas Maccabeus over the Seleucid oppressors. An eloquent epilogue addresses the reader in a style similar to the preface. Two letters to the Jewish communities in Egypt have been added to the beginning of the book (1.1–2.18) by an unknown author.

Although 2 Maccabees parallels chapters 1–8 of 1 Maccabees, it is distinguished by its literary style and theological point of view. The author of 2 Maccabees addresses the reader directly, in the manner of Greek historians, providing theological guidance. The unforgettable stories are told with dramatic artistry and vivid detail. Like the Greek historians the author guides the reader's interpretation by means of elaborate speeches in the mouths of central characters (6.24–28; 7.27–29, 30–38; 15.22–23).

The stories are unified by the idea that history reflects the divine plan, which can be affected by the prayers and deeds of the faithful. Like the tyrants of Egypt, Assyria, and Babylonia, who oppressed Israel in biblical times, the Seleucid kings are seen as unwitting tools of God, aiding in the chastisement of Israel. The author firmly believed in the sanctity of the Temple, but also held that abuses would bring divine judgment, as the desecrations of Antiochus had proved. The author also believed, however, that the suffering of the martyrs and the prayers of the courageous Judas Maccabeus moved God to intervene and end the time of divine wrath.

Several important theological ideas not found in the Hebrew Scriptures but important in Judaism and Christianity appear in 2 Maccabees. These include resurrection of the dead (hinted at in Dan 12.2, but stated clearly in 2 Macc 7–8); the doctrine that the world was created out of nothing (*creatio ex nihilo,* 7.28); and the efficacy of praying for the dead (12.39–45). The beginning of the later rabbinic principle of measure for measure, which teaches that divine retribution is perfectly just, is graphically portrayed in 3.27–28; 9.28; 13.8 and 15.32–33. The stories of the aged scribe Eleazar going willingly to his death rather than eat food forbidden by the Torah, and of the mother encouraging her seven sons to die for their faith in certain hope of resurrection, became models for later Jewish and Christian martyrologies.

Second Maccabees was written in Greek and was translated into Latin, Syriac and Armenian in antiquity. It was written between 104 and 63 B.C. (see 15.37).

1

The Jews in Jerusalem and those in the land of Judea,

To their Jewish kindred in Egypt,

Greetings and true peace.

2 May God do good to you, and may he remember his covenant with Abraham and Isaac and Jacob, his faithful servants. ³May he give you all a heart to worship him and to do his will with a strong heart and a willing spirit. ⁴May he open your heart to his law and his commandments, and may he bring peace. ⁵May he hear your prayers and be reconciled to you, and may he not forsake you in time of evil. ⁶We are now praying for you here.

7 In the reign of Demetrius, in the one hundred sixty-ninth year,ᵃ we Jews wrote to you, in the critical distress that came upon us in those years after Jason and his company revolted from the holy land and the kingdom ⁸and burned the gate and shed innocent blood. We prayed to the Lord and were heard, and we offered sacrifice and grain offering, and we lit the lamps and set out the loaves. ⁹And now see that you keep the festival of booths in the month of Chislev, in the one hundred eighty-eighth year.ᵇ

10 The people of Jerusalem and of Judea and the senate and Judas,

To Aristobulus, who is of the family of the anointed priests, teacher of King Ptolemy, and to the Jews in Egypt,

Greetings and good health.

11 Having been saved by God out of grave dangers we thank him greatly for taking our side against the king,ᶜ ¹²for he drove out those who fought against the holy city. ¹³When the leader reached Persia with a force that seemed irresistible, they were cut to pieces in the temple of Nanea by a deception employed by the priests of the goddessᵈ Nanea. ¹⁴On the pretext of intending to marry her, Antiochus came to the place together with his Friends, to secure most of its treasures as a dowry. ¹⁵When the priests of the temple of Nanea had set out the treasures and Antiochus had come with a few men inside the wall of the sacred precinct, they closed the temple as soon as he entered it. ¹⁶Opening a secret door in the ceiling, they threw stones and

---

ᵃ 143 B.C.   ᵇ 124 B.C.   ᶜ Cn: Gk *as those who array themselves against a king*   ᵈ Gk lacks *the goddess*

1.1–9: Letter to the Jews in Egypt. The authorship of these letters is uncertain. The first is addressed to the large Jewish community that had lived in Egypt since Alexander the Great (1 Macc 1.1). **1:** Greek letters usually began with the word *greetings*, and Jewish ones with *peace* (Rom 1.7). **2:** Gen 15.18; 26.3; 35.12; Lev 26.27–45. **5:** To live outside Judea was thought of as divine punishment. **7:** The previous letter was in the reign of Demetrius II (see 1 Macc 10.67 n.). *The critical distress* was the capture and murder of the high priest Jonathan (1 Macc 12.48; see 13.23 n.). *Jason and his company,* 4.7–22. *The kingdom,* rule of the legitimate high priests. **8:** *Burned the gate,* 1 Macc 4.38. *Shed innocent blood,* 1 Macc 1.60–61. *We . . . were heard,* i.e. by God, when Simon made Judea independent (1 Macc 13.1–42). **9:** *The festival of booths* would properly be kept in September (Lev 23.33–43). This refers to Hanukkah, celebrated on the 25th of *Chislev* (November–December), commemorating Judas Maccabeus' restoration of the temple (10.1–8; 1 Macc 4.59). Palestinian Jews now wished the Egyptian Jews to observe the feast in 124 B.C., when they wrote.

1.10–2.18: Letter to Aristobulus. The purpose of the letter is to show why the new eight-day festival should be kept, though it had not been prescribed by the Mosaic law. Keeping the festival implied accepting that Judas, like Nehemiah, was doing God's will. Nehemiah's rededication of the temple was a precedent (1.18–36).

1.10: *The senate,* see 1 Macc 12.6 n. *The anointed priests,* descendants of Zadok (2 Chr 31.10), from whom high priests were chosen. One branch of these came to Egypt with Ptolemy I. The *king,* Ptolemy VII Physcon, who reigned 145–116 B.C. **11:** *Grave dangers,* in the time of king Antiochus IV (4.7). **13:** *The leader,* Antiochus IV, died later; his forces *were cut to pieces* (9.1–4; 1 Macc 6.1–4). *Nanea,* a Syrian goddess equated with Artemis or Aphrodite and the Persian Anahita. **14:** *Marry her,* the goddess, so as to obtain a large *dowry* from the treasures at her temple (compare 9.2; 1 Macc 6.1–4). **18:** *The festival of booths,* compare 1 Macc 10.21; 1 Kings 8.2; Neh 8.13–18.

struck down the leader and his men; they dismembered them and cut off their heads and threw them to the people outside. [17] Blessed in every way be our God, who has brought judgment on those who have behaved impiously.

18 Since on the twenty-fifth day of Chislev we shall celebrate the purification of the temple, we thought it necessary to notify you, in order that you also may celebrate the festival of booths and the festival of the fire given when Nehemiah, who built the temple and the altar, offered sacrifices.

19 For when our ancestors were being led captive to Persia, the pious priests of that time took some of the fire of the altar and secretly hid it in the hollow of a dry cistern, where they took such precautions that the place was unknown to anyone. [20] But after many years had passed, when it pleased God, Nehemiah, having been commissioned by the king of Persia, sent the descendants of the priests who had hidden the fire to get it. And when they reported to us that they had not found fire but only a thick liquid, he ordered them to dip it out and bring it. [21] When the materials for the sacrifices were presented, Nehemiah ordered the priests to sprinkle the liquid on the wood and on the things laid upon it. [22] When this had been done and some time had passed, and when the sun, which had been clouded over, shone out, a great fire blazed up, so that all marveled. [23] And while the sacrifice was being consumed, the priests offered prayer—the priests and everyone. Jonathan led, and the rest responded, as did Nehemiah. [24] The prayer was to this effect:

"O Lord, Lord God, Creator of all things, you are awe-inspiring and strong and just and merciful, you alone are king and are kind, [25] you alone are bountiful, you alone are just and almighty and eternal. You rescue Israel from every evil; you chose the ancestors and consecrated them. [26] Accept this sacrifice on behalf of all your people Israel and preserve your portion and make it holy. [27] Gather together our scattered people, set free those who are slaves among the Gentiles, look on those who are rejected and despised, and let the Gentiles know that you are our God. [28] Punish those who oppress and are insolent with pride. [29] Plant your people in your holy place, as Moses promised."

30 Then the priests sang the hymns. [31] After the materials of the sacrifice had been consumed, Nehemiah ordered that the liquid that was left should be poured on large stones. [32] When this was done, a flame blazed up; but when the light from the altar shone back, it went out. [33] When this matter became known, and it was reported to the king of the Persians that, in the place where the exiled priests had hidden the fire, the liquid had appeared with which Nehemiah and his associates had burned the materials of the sacrifice, [34] the king investigated the matter, and enclosed the place and made it sacred. [35] And with those persons whom the king favored he exchanged many excellent gifts. [36] Nehemiah and his associates called this "nephthar," which means purification, but by most people it is called naphtha. *e*

e Gk *nephthai*

The festival of the fire, vv. 19–36. Fire and light are associated with Hanukkah, which is celebrated with a nine-branched candlestick. A Talmudic tradition tells of a small amount of oil that burned miraculously for a long time till new oil could be consecrated.

1.19: *Persia,* actually Babylonia (2 Kings 24.14), later part of the Persian empire. **20:** *Nehemiah . . . commissioned,* Neh 2.7–8; his book does not contain the legend of the fire. *Thick liquid,* naphtha or petroleum (v. 36).

22: 1 Kings 18.33–38. **25:** *You chose,* Gen 12.1–3; 22.15–18; Deut 14.2; Mal 1.2. **26:** *Your portion,* Israel (Deut 32.9). *Holy,* Lev 19.2. **27:** *Gather together,* Ps 147.2; Jer 23.8; Sir 36.11; Bar 5.6. **28:** *Punish . . . insolent,* 1 Sam 2.3–4; Lk 1.51–52. **29:** *As Moses promised,* Deut 30.5. **34:** Localities where miracles occurred were *enclosed* as *sacred.* The Persians considered fire holy. **36:** *Nephthar* is an otherwise unknown word.

2 One finds in the records that the prophet Jeremiah ordered those who were being deported to take some of the fire, as has been mentioned, ²and that the prophet, after giving them the law, instructed those who were being deported not to forget the commandments of the Lord, or to be led astray in their thoughts on seeing the gold and silver statues and their adornment. ³And with other similar words he exhorted them that the law should not depart from their hearts.

4 It was also in the same document that the prophet, having received an oracle, ordered that the tent and the ark should follow with him, and that he went out to the mountain where Moses had gone up and had seen the inheritance of God. ⁵Jeremiah came and found a cave-dwelling, and he brought there the tent and the ark and the altar of incense; then he sealed up the entrance. ⁶Some of those who followed him came up intending to mark the way, but could not find it. ⁷When Jeremiah learned of it, he rebuked them and declared: "The place shall remain unknown until God gathers his people together again and shows his mercy. ⁸Then the Lord will disclose these things, and the glory of the Lord and the cloud will appear, as they were shown in the case of Moses, and as Solomon asked that the place should be specially consecrated."

9 It was also made clear that being possessed of wisdom Solomon*ᶠ* offered sacrifice for the dedication and completion of the temple. ¹⁰Just as Moses prayed to the Lord, and fire came down from heaven and consumed the sacrifices, so also Solomon prayed, and the fire came down and consumed the whole burnt offerings. ¹¹And Moses said, "They were consumed because the sin offering had not been eaten." ¹²Likewise Solomon also kept the eight days.

13 The same things are reported in the records and in the memoirs of Nehemiah, and also that he founded a library and collected the books about the kings and prophets, and the writings of David, and letters of kings about votive offerings. ¹⁴In the same way Judas also collected all the books that had been lost on account of the war that had come upon us, and they are in our possession. ¹⁵So if you have need of them, send people to get them for you.

16 Since, therefore, we are about to celebrate the purification, we write to you. Will you therefore please keep the days? ¹⁷It is God who has saved all his people, and has returned the inheritance to all, and the kingship and the priesthood and the consecration, ¹⁸as he promised through the law. We have hope in God that he will soon have mercy on us and will gather us from everywhere under heaven into his holy place, for he has

f Gk *he*

2.1: No such *records* are known. *Jeremiah* remained in Judea after the exile (Jer 29.1–23; 40.1–42.7). In the apocryphal Epistle of Jeremiah the writer similarly exhorts the exiles. **4:** Solomon brought the *tent* to Jerusalem with the ark (1 Kings 8.4). There is no further record in the Old Testament of the tent, but the ark was kept in the first temple; according to Alexander Polyhistor (first century B.C.), perhaps from the historian Eupolemus, Jeremiah concealed the ark after the temple was destroyed in 587–6 B.C. *The mountain*, Nebo (Deut 32.49). **8:** *The glory* and *the cloud* indicate God's direct presence (Ex 16.10; Mk 9.2–8). *Solomon*, 1 Kings 8.11. **2.9:** Solomon's *wisdom*, 1 Kings 3.3–28; 4.29–34. *Offered sacrifice*, 1 Kings 8.62–64. **10:** *Moses prayed*, Lev 9.24. *Solomon*, 2 Chr 7.1. **11:** The meaning is obscure, but see Lev 10.16–19. **12:** *Eight days*, 1 Kings 8.65; 2 Chr 7.9. **13:** *The memoirs of Nehemiah*, the biblical book of Nehemiah does not contain these references. See Ezra 3.14 and 1 Esdras 5.46–50. There is no record that *he founded a library*, but the Pentateuch was canonized in his time, and he may have collected the books of Kings. *Votive offerings*, made to the temple (Ezra 7.15–20). **14:** *Judas* Maccabeus may have *collected all the books* remaining after the destruction in the time of Antiochus IV (1 Macc 1.56–57). **16:** 1.18. **17:** *The kingship*, independence; the Hasmoneans were not yet called kings. **18:** 1.27; Deut 30.3.

rescued us from great evils and has purified the place.

19 The story of Judas Maccabeus and his brothers, and the purification of the great temple, and the dedication of the altar, 20 and further the wars against Antiochus Epiphanes and his son Eupator, 21 and the appearances that came from heaven to those who fought bravely for Judaism, so that though few in number they seized the whole land and pursued the barbarian hordes, 22 and regained possession of the temple famous throughout the world, and liberated the city, and re-established the laws that were about to be abolished, while the Lord with great kindness became gracious to them— 23 all this, which has been set forth by Jason of Cyrene in five volumes, we shall attempt to condense into a single book. 24 For considering the flood of statistics involved and the difficulty there is for those who wish to enter upon the narratives of history because of the mass of material, 25 we have aimed to please those who wish to read, to make it easy for those who are inclined to memorize, and to profit all readers. 26 For us who have undertaken the toil of abbreviating, it is no light matter but calls for sweat and loss of sleep, 27 just as it is not easy for one who prepares a banquet and seeks the benefit of others. Nevertheless, to secure the gratitude of many we will gladly endure the uncomfortable

toil, 28 leaving the responsibility for exact details to the compiler, while devoting our effort to arriving at the outlines of the condensation. 29 For as the master builder of a new house must be concerned with the whole construction, while the one who undertakes its painting and decoration has to consider only what is suitable for its adornment, such in my judgment is the case with us. 30 It is the duty of the original historian to occupy the ground, to discuss matters from every side, and to take trouble with details, 31 but the one who recasts the narrative should be allowed to strive for brevity of expression and to forego exhaustive treatment. 32 At this point therefore let us begin our narrative, without adding any more to what has already been said; for it would be foolish to lengthen the preface while cutting short the history itself.

3 While the holy city was inhabited in unbroken peace and the laws were strictly observed because of the piety of the high priest Onias and his hatred of wickedness, 2 it came about that the kings themselves honored the place and glorified the temple with the finest presents, 3 even to the extent that King Seleucus of Asia defrayed from his own revenues all the expenses connected with the service of the sacrifices.

4 But a man named Simon, of the tribe of Benjamin, who had been made

---

2.19–32: The epitomist's preface. Following the custom of Greek histories, the Preface presents highlights and entices readers (see Thucydides i 23.1–3 and Tacitus *Histories* i 2–3). He summarizes parts of the book. **20–21:** *Appearances* (Gr. "epiphaneiai"), true divine manifestations, in contrast to Antiochus' boastful title *Epiphanes*, "god manifest." *Judaism,* first known use of this term for the religion, in contrast to Hellenism (4.13). *Barbarian,* used by the Greeks for all non-Greeks, connotes "one who speaks a foreign language." It can also mean "savage," as in 4.25; 5.22; 10.4; 15.2. **23:** *Jason of Cyrene* is not mentioned in ancient sources, and no work of his survives. Resting one's work on the authority of a purported ancient source is a literary tradition of

antiquity; it is not certain whether the author of 2 Maccabees had an actual source or is using the literary conceit that he had such a document. **3.1–4.6: Simon's plot against Onias.** **3.1:** Jerusalem was not *in unbroken peace,* though quieter than in later years. The high priest Onias III, son of Simon the Just (Sir 50.1–21), ruled before 175 B.C. He turned against Syria and collaborated with Egypt, while his cousins, the family of Tobias, to which Simon (v. 4) belonged, were proSyrian. **2:** *The kings,* i.e. the Ptolemies of Egypt and Antiochus III the Great (reigned 233–187 B.C.). **3:** *Seleucus* IV Philopator, son of Antiochus III, reigned 187–175 B.C. The events of 3.1–4.6 were in his reign. He was

captain of the temple, had a disagreement with the high priest about the administration of the city market. 5 Since he could not prevail over Onias, he went to Apollonius of Tarsus,[g] who at that time was governor of Coelesyria and Phoenicia, 6 and reported to him that the treasury in Jerusalem was full of untold sums of money, so that the amount of the funds could not be reckoned, and that they did not belong to the account of the sacrifices, but that it was possible for them to fall under the control of the king. 7 When Apollonius met the king, he told him of the money about which he had been informed. The king[h] chose Heliodorus, who was in charge of his affairs, and sent him with commands to effect the removal of the reported wealth. 8 Heliodorus at once set out on his journey, ostensibly to make a tour of inspection of the cities of Coelesyria and Phoenicia, but in fact to carry out the king's purpose.

9 When he had arrived at Jerusalem and had been kindly welcomed by the high priest of[i] the city, he told about the disclosure that had been made and stated why he had come, and he inquired whether this really was the situation. 10 The high priest explained that there were some deposits belonging to widows and orphans, 11 and also some money of Hyrcanus son of Tobias, a man of very prominent position, and that it totaled in all four hundred talents of silver and two hundred of gold. To such an extent the impious Simon had misrepresented the facts. 12 And he said that it was utterly impossible that wrong should be done to those people who had trusted in the holiness of the place and in the sanctity and inviolability of the temple that is honored throughout the whole world.

13 But Heliodorus, because of the orders he had from the king, said that this money must in any case be confiscated for the king's treasury. 14 So he set a day and went in to direct the inspection of these funds.

There was no little distress throughout the whole city. 15 The priests prostrated themselves before the altar in their priestly vestments and called toward heaven upon him who had given the law about deposits, that he should keep them safe for those who had deposited them. 16 To see the appearance of the high priest was to be wounded at heart, for his face and the change in his color disclosed the anguish of his soul. 17 For terror and bodily trembling had come over the man, which plainly showed to those who looked at him the pain lodged in his heart. 18 People also hurried out of their houses in crowds to make a general supplication because the holy place was about to be brought into dishonor. 19 Women, girded with sackcloth under their breasts, thronged the streets. Some of the young women who were kept indoors ran together to the gates, and some to the walls, while others peered out of the windows. 20 And holding up their hands to heaven, they all made supplication. 21 There was something pitiable in the prostration of the whole populace

g Gk *Apollonius son of Tharseas*   h Gk *He*
i Other ancient authorities read *and*

---

assassinated by Heliodorus (v. 7). **4:** *Simon* was a grandson of Tobias, who married a sister of Onias II. When Onias II refused to pay tribute to Egypt, Ptolemy III took away his civil authority and appointed Joseph, son of Tobias, *captain of the temple.* His son Simon succeeded him. **5:** *Tarsus,* capital of Cilicia (Acts 9.11), then part of the Seleucid empire. *Coelesyria,* see 1 Macc 10.69 n. *Apollonius* was removed from office at the death of Seleucus IV in 175 B.C. **7:** *Heliodorus,* see v. 3 n.

3.9: *The high priest,* Onias III. **11:** *Hyrcanus,* actually son of Joseph and half-brother of Si-

mon (v. 4), was pro-Egyptian. He fled east of the Jordan after 198 B.C. and built the fortress of 'Araq el-Emir. He committed suicide on the accession of Antiochus IV in 175 B.C. *Simon had misrepresented the facts* only in part; Onias and Hyrcanus probably withheld tribute. **18:** Temples, whether pagan or Jewish, were considered inviolate. **19:** *Sackcloth,* robes of goat hair, a sign of mourning and penitence. Some young women were *kept indoors* until marriage (Sir 42.9–12).

3.20: *Holding up their hands,* the ancient gesture of prayer (1 Kings 8.54; 1 Tim 2.8).

and the anxiety of the high priest in his great anguish.

22 While they were calling upon the Almighty Lord that he would keep what had been entrusted safe and secure for those who had entrusted it, 23 Heliodorus went on with what had been decided. 24 But when he arrived at the treasury with his bodyguard, then and there the Sovereign of spirits and of all authority caused so great a manifestation that all who had been so bold as to accompany him were astounded by the power of God, and became faint with terror. 25 For there appeared to them a magnificently caparisoned horse, with a rider of frightening mien; it rushed furiously at Heliodorus and struck at him with its front hoofs. Its rider was seen to have armor and weapons of gold. 26 Two young men also appeared to him, remarkably strong, gloriously beautiful and splendidly dressed, who stood on either side of him and flogged him continuously, inflicting many blows on him. 27 When he suddenly fell to the ground and deep darkness came over him, his men took him up, put him on a stretcher, 28 and carried him away—this man who had just entered the aforesaid treasury with a great retinue and all his bodyguard but was now unable to help himself. They recognized clearly the sovereign power of God.

29 While he lay prostrate, speechless because of the divine intervention and deprived of any hope of recovery, 30 they praised the Lord who had acted marvelously for his own place. And the temple, which a little while before was full of fear and disturbance, was filled with joy and gladness, now that the Almighty Lord had appeared. 31 Some of Heliodorus's friends quickly begged Onias to call upon the Most High to grant life to one who was lying quite at his last breath. 32 So the high priest, fearing that the king might

get the notion that some foul play had been perpetrated by the Jews with regard to Heliodorus, offered sacrifice for the man's recovery. 33 While the high priest was making an atonement, the same young men appeared again to Heliodorus dressed in the same clothing, and they stood and said, "Be very grateful to the high priest Onias, since for his sake the Lord has granted you your life. 34 And see that you, who have been flogged by heaven, report to all people the majestic power of God." Having said this they vanished.

35 Then Heliodorus offered sacrifice to the Lord and made very great vows to the Savior of his life, and having bidden Onias farewell, he marched off with his forces to the king. 36 He bore testimony to all concerning the deeds of the supreme God, which he had seen with his own eyes. 37 When the king asked Heliodorus what sort of person would be suitable to send on another mission to Jerusalem, he replied, 38 "If you have any enemy or plotter against your government, send him there, for you will get him back thoroughly flogged, if he survives at all; for there is certainly some power of God about the place. 39 For he who has his dwelling in heaven watches over that place himself and brings it aid, and he strikes and destroys those who come to do it injury." 40 This was the outcome of the episode of Heliodorus and the protection of the treasury.

4 The previously mentioned Simon, who had informed about the money against[j] his own country, slandered Onias, saying that it was he who had incited Heliodorus and had been the real cause of the misfortune. 2 He dared to designate as a plotter against the government the man who was the benefactor of the city, the protector of his compatriots, and a zealot for the laws. 3 When his ha-

j Gk *and*

24: *Manifestation,* see 2.21 n. 29: *Speechless,* Lk 1.20. 31: *The Most High* (Gen 14.18), a title often used by non-Jews (Dan 3.26; Mk 5.7).
4.1–6: Intrigues concerning the high

priesthood. *Simon* (see 3.4 n.) was disturbed because *Onias* and *Heliodorus* were now friends. The latter may already have planned to kill Seleucus IV and wanted the high

tred progressed to such a degree that even murders were committed by one of Simon's approved agents, [4]Onias recognized that the rivalry was serious and that Apollonius son of Menesthcus,[k] and governor of Coelesyria and Phoenicia, was intensifying the malice of Simon. [5]So he appealed to the king, not accusing his compatriots but having in view the welfare, both public and private, of all the people. [6]For he saw that without the king's attention public affairs could not again reach a peaceful settlement, and that Simon would not stop his folly.

7 When Seleucus died and Antiochus, who was called Epiphanes, succeeded to the kingdom, Jason the brother of Onias obtained the high priesthood by corruption, [8]promising the king at an interview[l] three hundred sixty talents of silver, and from another source of revenue eighty talents. [9]In addition to this he promised to pay one hundred fifty more if permission were given to establish by his authority a gymnasium and a body of youth for it, and to enroll the people of Jerusalem as citizens of Antioch. [10]When the king assented and Jason[m] came to office, he at once shifted his compatriots over to the Greek way of life.

11 He set aside the existing royal concessions to the Jews, secured through John the father of Eupolemus, who went on the mission to establish friendship and

alliance with the Romans; and he destroyed the lawful ways of living and introduced new customs contrary to the law. [12]He took delight in establishing a gymnasium right under the citadel, and he induced the noblest of the young men to wear the Greek hat. [13]There was such an extreme of Hellenization and increase in the adoption of foreign ways because of the surpassing wickedness of Jason, who was ungodly and no true[n] high priest, [14]that the priests were no longer intent upon their service at the altar. Despising the sanctuary and neglecting the sacrifices, they hurried to take part in the unlawful proceedings in the wrestling arena after the signal for the discus-throwing, [15]disdaining the honors prized by their ancestors and putting the highest value upon Greek forms of prestige. [16]For this reason heavy disaster overtook them, and those whose ways of living they admired and wished to imitate completely became their enemies and punished them. [17]It is no light thing to show irreverence to the divine laws—a fact that later events will make clear.

18 When the quadrennial games were being held at Tyre and the king was present, [19]the vile Jason sent envoys,

k Vg Compare verse 21: Meaning of Gk uncertain   l Or *by a petition*   m Gk *he*
n Gk lacks *true*

---

priest's good will. *Apollonius,* in favor with Seleucus, continued to support Simon. **5:** Before Onias arrived in Antioch, Seleucus had already been assassinated by Heliodorus (175 B.C.).

4.7–22: Jason as high priest. 7: *Antiochus IV Epiphanes,* "god manifest," called Epimanes, "madman," by his enemies, was brother of Seleucus IV, and *succeeded to the kingdom* despite Heliodorus' attempt at revolution. He reigned 175–164 B.C. He had great ability but intense passion and pride (1 Macc 1.1–10). *Jason the brother of Onias* III (3.1), originally named Joshua, took a Greek name. **9–10:** Like Alexander the Great and his successors, Antiochus promoted *the Greek way of life* in order to strengthen his kingdom through cultural unity; this involved worship of other gods. *A gymnasium and a body of youth for it* were necessary *to enroll the people of Jeru-*

salem as citizens of Antioch, so that the city could coin money and have honors and commercial advantages. The gymnasium was the center of political and cultural education, as well as sports.

4.11: *Royal concessions,* granted by Antiochus III (3.2). The mission of *Eupolemus* (1 Macc 8.17) was later. *Destroyed,* 1 Macc 1.15, 44–50. **12:** The broad-brimmed *Greek hat* was worn by the god Hermes; headgear has usually had national or religious significance in the East. **13:** *Hellenization,* Greek religion and culture (see vv. 9–10 n.). *No true high priest,* because he got the office by bribery and did not keep the Mosaic law. **16–17:** For the interpretation of disaster as the result of forsaking Torah see 1 Kings 17.5–18; 2 Chr 36.11–21; Neh ch 9. **18:** *Tyre,* an important port north of Palestine (Josh 19.29; 1 Kings 7.13); *quadrennial games* had been held

chosen as being Antiochian citizens from Jerusalem, to carry three hundred silver drachmas for the sacrifice to Hercules. Those who carried the money, however, thought best not to use it for sacrifice, because that was inappropriate, but to expend it for another purpose. 20 So this money was intended by the sender for the sacrifice to Hercules, but by the decision of its carriers it was applied to the construction of triremes.

21 When Apollonius son of Menestheus was sent to Egypt for the coronation*o* of Philomotor as king, Antiochus learned that Philomotor*p* had become hostile to his government, and he took measures for his own security. Therefore upon arriving at Joppa he proceeded to Jerusalem. 22 He was welcomed magnificently by Jason and the city, and ushered in with a blaze of torches and with shouts. Then he marched his army into Phoenicia.

23 After a period of three years Jason sent Menelaus, the brother of the previously mentioned Simon, to carry the money to the king and to complete the records of essential business. 24 But he, when presented to the king, extolled him with an air of authority, and secured the high priesthood for himself, outbidding Jason by three hundred talents of silver. 25 After receiving the king's orders he returned, possessing no qualification for the high priesthood, but having the hot temper of a cruel tyrant and the rage of a savage wild beast. 26 So Jason, who after supplanting his own brother was supplanted by another man, was driven as a

fugitive into the land of Ammon. 27 Although Menelaus continued to hold the office, he did not pay regularly any of the money promised to the king. 28 When Sostratus the captain of the citadel kept requesting payment—for the collection of the revenue was his responsibility—the two of them were summoned by the king on account of this issue. 29 Menelaus left his own brother Lysimachus as deputy in the high priesthood, while Sostratus left Crates, the commander of the Cyprian troops.

30 While such was the state of affairs, it happened that the people of Tarsus and of Mallus revolted because their cities had been given as a present to Antiochis, the king's concubine. 31 So the king went hurriedly to settle the trouble, leaving Andronicus, a man of high rank, to act as his deputy. 32 But Menelaus, thinking he had obtained a suitable opportunity, stole some of the gold vessels of the temple and gave them to Andronicus; other vessels, as it happened, he had sold to Tyre and the neighboring cities. 33 When Onias became fully aware of these acts, he publicly exposed them, having first withdrawn to a place of sanctuary at Daphne near Antioch. 34 Therefore Menelaus, taking Andronicus aside, urged him to kill Onias. Andronicus*q* came to Onias, and resorting to treachery, offered him sworn pledges and gave him his right hand; he persuaded him, though still suspicious, to come out from

o Meaning of Gk uncertain   *p* Gk *he*
q Gk *He*

---

there as early as the time of Alexander the Great. **19**: *Hercules,* the Greek name of the god Melkart of Tyre. **20**: *Triremes,* war vessels manned by three benches of rowers. **21**: *Apollonius, v. 4. The coronation* of Ptolemy VI *Philometor* occurred about 172 B.C., some time after the death of his mother, Cleopatra I, and he ruled until 146 or 145 B.C. His advisers abandoned Cleopatra's policy, became *hostile* to Syria, and claimed Palestine. *Joppa,* the port forty miles from Jerusalem. **22**: *Phoenicia,* the coastal plain.

4.23–50: Menelaus as high priest. 23: *Menelaus* reigned from about 172 to 162 B.C.,

when he was executed (13.3–8) and replaced by Alcimus (14.3–14). **26**: *Land of Ammon,* east of the Jordan, near the present Amman. **29**: *The Cyprian troops* were mercenaries. **30**: *Mallus* was on the Pyramus river east of *Tarsus* (3.5). Hellenistic kings often provided a wife or *concubine* with a regular income by giving her a city. Antiochus, being extravagant (see 1 Macc 3.30 n.), was often in need of money. **32**: *Gave them,* either to pay tribute or as a bribe. **33**: *Daphne,* about five miles from *Antioch,* had a *place of sanctuary* to Apollo and Artemis.

the place of sanctuary; then, with no regard for justice, he immediately put him out of the way.

35 For this reason not only Jews, but many also of other nations, were grieved and displeased at the unjust murder of the man. [36] When the king returned from the region of Cilicia, the Jews in the city[r] appealed to him with regard to the unreasonable murder of Onias, and the Greeks shared their hatred of the crime. [37] Therefore Antiochus was grieved at heart and filled with pity, and wept because of the moderation and good conduct of the deceased. [38] Inflamed with anger, he immediately stripped off the purple robe from Andronicus, tore off his clothes, and led him around the whole city to that very place where he had committed the outrage against Onias, and there he dispatched the bloodthirsty fellow. The Lord thus repaid him with the punishment he deserved.

39 When many acts of sacrilege had been committed in the city by Lysimachus with the connivance of Menelaus, and when report of them had spread abroad, the populace gathered against Lysimachus, because many of the gold vessels had already been stolen. [40] Since the crowds were becoming aroused and filled with anger, Lysimachus armed about three thousand men and launched an unjust attack, under the leadership of a certain Auranus, a man advanced in years and no less advanced in folly. [41] But when the Jews[s] became aware that Lysimachus was attacking them, some picked up stones, some blocks of wood, and others took handfuls of the ashes that were lying around, and threw them in wild confusion at Lysimachus and his men. [42] As a result, they wounded many of them, and killed some, and put all the rest to flight; the temple robber himself they killed close by the treasury.

43 Charges were brought against Menelaus about this incident. [44] When the king came to Tyre, three men sent by the senate presented the case before him. [45] But Menelaus, already as good as beaten, promised a substantial bribe to Ptolemy son of Dorymenes to win over the king. [46] Therefore Ptolemy, taking the king aside into a colonnade as if for refreshment, induced the king to change his mind. [47] Menelaus, the cause of all the trouble, he acquitted of the charges against him, while he sentenced to death those unfortunate men, who would have been freed uncondemned if they had pleaded even before Scythians. [48] And so those who had spoken for the city and the villages[t] and the holy vessels quickly suffered the unjust penalty. [49] Therefore even the Tyrians, showing their hatred of the crime, provided magnificently for their funeral. [50] But Menelaus, because of the greed of those in power, remained in office, growing in wickedness, having become the chief plotter against his compatriots.

5 About this time Antiochus made his second invasion of Egypt. [2] And it happened that, for almost forty days, there appeared over all the city golden-clad cavalry charging through the air, in companies fully armed with lances and drawn swords— [3] troops of cavalry drawn up, attacks and counterattacks

r Or *in each city*   s Gk *they*   t Other ancient authorities read *the people*

4.35: *Unjust murder,* he had been lured from a place protected by the gods. **38:** *Stripped off the purple robe,* degrading him before execution. **39:** *The city,* Jerusalem. *Menelaus* was still in Antioch. **42:** *The temple robber,* Lysimachus. **44:** *The senate,* see 1 Macc 12.6 n. **45:** *Dorymenes* had fought for Ptolemy IV against Antiochus III; his son *Ptolemy* had been governor of Cyprus and deserted to Antiochus IV (see 10.12–13 n.). **47:** The

Scythians (Col 3.11) lived in what is now southern Russia and were proverbial for their brutality.

5.1–27: Antiochus IV desecrates the temple. 1: *Second invasion,* in 169 B.C.; perhaps the writer regards the coming of the Seleucid army into Palestine in 171 B.C. (4.21–22) as the first invasion. We would speak of them as the first and second phases of the invasion (compare 1 Macc 1.16–19).

made on this side and on that, brandishing of shields, massing of spears, hurling of missiles, the flash of golden trappings, and armor of all kinds. 4 Therefore everyone prayed that the apparition might prove to have been a good omen.

5 When a false rumor arose that Antiochus was dead, Jason took no fewer than a thousand men and suddenly made an assault on the city. When the troops on the wall had been forced back and at last the city was being taken, Menelaus took refuge in the citadel. 6 But Jason kept relentlessly slaughtering his compatriots, not realizing that success at the cost of one's kindred is the greatest misfortune, but imagining that he was setting up trophies of victory over enemies and not over compatriots. 7 He did not, however, gain control of the government; in the end he got only disgrace from his conspiracy, and fled again into the country of the Ammonites. 8 Finally he met a miserable end. Accused*u* before Aretas the ruler of the Arabs, fleeing from city to city, pursued by everyone, hated as a rebel against the laws, and abhorred as the executioner of his country and his compatriots, he was cast ashore in Egypt. 9 There he who had driven many from their own country into exile died in exile, having embarked to go to the Lacedaemonians in hope of finding protection because of their kinship. 10 He who had cast out many to lie unburied had no one to mourn for him; he had no funeral of any sort and no place in the tomb of his ancestors.

11 When news of what had happened reached the king, he took it to mean that Judea was in revolt. So, raging inwardly, he left Egypt and took the city by storm.

12 He commanded his soldiers to cut down relentlessly everyone they met and to kill those who went into their houses. 13 Then there was massacre of young and old, destruction of boys, women, and children, and slaughter of young girls and infants. 14 Within the total of three days eighty thousand were destroyed, forty thousand in hand-to-hand fighting, and as many were sold into slavery as were killed.

15 Not content with this, Antiochus*v* dared to enter the most holy temple in all the world, guided by Menelaus, who had become a traitor both to the laws and to his country. 16 He took the holy vessels with his polluted hands, and swept away with profane hands the votive offerings that other kings had made to enhance the glory and honor of the place. 17 Antiochus was elated in spirit, and did not perceive that the Lord was angered for a little while because of the sins of those who lived in the city, and that this was the reason he was disregarding the holy place. 18 But if it had not happened that they were involved in many sins, this man would have been flogged and turned back from his rash act as soon as he came forward, just as Heliodorus had been, whom King Seleucus sent to inspect the treasury. 19 But the Lord did not choose the nation for the sake of the holy place, but the place for the sake of the nation. 20 Therefore the place itself shared in the misfortunes that befell the nation and afterward participated in its benefits; and what was forsaken in the wrath of the Almighty was restored again in all its glory when the great Lord became reconciled.

u Cn: Gk *Imprisoned*   *v* Gk *he*

2–4: 3.25–26. *The city,* Jerusalem. 5–8: *Jason* was an Oniad and pro-Egyptian (see 3.1 n.). Thinking *that Antiochus was dead,* he planned, with Egyptian help, to recover the high priesthood. He was opposed by *Menelaus* the Tobiad (4.23) and also by the Jews loyal to Judaism; he massacred people of both factions. *Ammonites,* 4.26. *Aretas,* king of Nabatean Arabia, south and east of Palestine; his capital was at Petra. 9–10: Rejected in Egypt,

Jason fled to Sparta (1 Macc 12.7). *Unburied,* see 1 Macc 7.17 n.; 1 Kings 13.22.

5.11–14: So confused was the situation that Antiochus thought all *Judea was in revolt.* He was *raging inwardly* because the Romans had forced him out of Egypt (see 1 Macc 1.20 n.); both his foreign and his domestic programs were collapsing. 11: *The city,* Jerusalem.

5.15–23a: The temple had been pillaged after the first Egyptian invasion (1 Macc

21 So Antiochus carried off eighteen hundred talents from the temple, and hurried away to Antioch, thinking in his arrogance that he could sail on the land and walk on the sea, because his mind was elated. 22 He left governors to oppress the people: at Jerusalem, Philip, by birth a Phrygian and in character more barbarous than the man who appointed him; 23 and at Gerizim, Andronicus; and besides these Menelaus, who lorded it over his compatriots worse than the others did. In his malice toward the Jewish citizens, *w* 24 Antiochus *x* sent Apollonius, the captain of the Mysians, with an army of twenty-two thousand, and commanded him to kill all the grown men and to sell the women and boys as slaves. 25 When this man arrived in Jerusalem, he pretended to be peaceably disposed and waited until the holy sabbath day; then, finding the Jews not at work, he ordered his troops to parade under arms. 26 He put to the sword all those who came out to see them, then rushed into the city with his armed warriors and killed great numbers of people.

27 But Judas Maccabeus, with about nine others, got away to the wilderness, and kept himself and his companions alive in the mountains as wild animals do; they continued to live on what grew wild, so that they might not share in the defilement.

6 Not long after this, the king sent an Athenian *y* senator *z* to compel the Jews to forsake the laws of their ancestors and no longer to live by the laws of God; 2 also to pollute the temple in Jerusalem and to call it the temple of Olympian Zeus, and to call the one in Gerizim the temple of Zeus-the-Friend-of-Strangers, as did the people who lived in that place.

3 Harsh and utterly grievous was the onslaught of evil. 4 For the temple was filled with debauchery and reveling by the Gentiles, who dallied with prostitutes and had intercourse with women within the sacred precincts, and besides brought in things for sacrifice that were unfit. 5 The altar was covered with abominable offerings that were forbidden by the laws. 6 People could neither keep the sabbath, nor observe the festivals of their ancestors, nor so much as confess themselves to be Jews.

7 On the monthly celebration of the king's birthday, the Jews *a* were taken, under bitter constraint, to partake of the sacrifices; and when a festival of Dionysus was celebrated, they were compelled to wear wreathes of ivy and to walk in the procession in honor of Dionysus. 8 At the suggestion of the people of Ptolemais *b* a decree was issued to the neighboring Greek cities that they should adopt the same policy toward the Jews and

w Or *worse than the others did in his malice toward the Jewish citizens*   *x* Gk *he*   *y* Other ancient authorities read *Antiochian*
z Or *Geron an Athenian*   *a* Gk *they*
b Cn: Gk *suggestion of the Ptolemies* (or *of Ptolemy*)

1.21–28). *Angered for a little while,* not permanently (compare 6.12–16). **21**: *His arrogance* was that of a god manifest (see 4.7 n.). **22–23**: *Philip,* probably not the later regent (9.29). *Andronicus* (4.31) was now made governor of Samaria. **24–26**: *Antiochus sent Apollonius* about two years after the events of vv. 15–23 (see 1 Macc 1.29). Loyal Jews did not yet fight on the *sabbath* (1 Macc 2.32–41). **27**: *Judas Maccabeus,* the third son of Mattathias, of the Hasmonean family (1 Macc 2.1–28). *The defilement,* 4.11; 1 Macc 1.48, 63.

6.1–6: Campaign against Judaism. What had been voluntary (4.9–17) was now enforced (see 1 Macc 1.41–64 n.). **2**: *Olympian Zeus* was now identified with the God of Israel and probably with Antiochus. *To pollute the temple,* they set up a statue or pagan altar (1 Macc 1.54). The Samaritans, descendants of the ten northern tribes and Assyrian settlers (2 Kings 17.6, 24), had built the temple on Mount *Gerizim.* **4**: *Intercourse . . . sacred precincts,* as in Syrian fertility cults (see Let Jer 6.11 n. and 6.43 n.). *Things . . . unfit,* swine (Lev 11.7; 1 Macc 1.47). **6**: 1 Macc 1.45–51. *Jews,* originally "Judeans"; here "those loyal to the religion" (Judaism, 2.21).

6.7–17: The first martyrdoms. Chs 6–7 are the earliest martyrologies, a type of writing popular subsequently in Christianity, designed to encourage the faithful when persecuted. **7**: *Dionysus,* god of wine and the grape

make them partake of the sacrifices, 9 and should kill those who did not choose to change over to Greek customs. One could see, therefore, the misery that had come upon them. 10 For example, two women were brought in for having circumcised their children. They publicly paraded them around the city, with their babies hanging at their breasts, and then hurled them down headlong from the wall. 11 Others who had assembled in the caves nearby, in order to observe the seventh day secretly, were betrayed to Philip and were all burned together, because their piety kept them from defending themselves, in view of their regard for that most holy day.

12 Now I urge those who read this book not to be depressed by such calamities, but to recognize that these punishments were designed not to destroy but to discipline our people. 13 In fact, it is a sign of great kindness not to let the impious alone for long, but to punish them immediately. 14 For in the case of the other nations the Lord waits patiently to punish them until they have reached the full measure of their sins; but he does not deal in this way with us, 15 in order that he may not take vengeance on us afterward when our sins have reached their height. 16 Therefore he never withdraws his mercy from us. Although he disciplines us with calamities, he does not forsake his own people. 17 Let what we have said serve as a reminder; we must go on briefly with the story.

18 Eleazar, one of the scribes in high position, a man now advanced in age and of noble presence, was being forced to open his mouth to eat swine's flesh. 19 But he, welcoming death with honor rather than life with pollution, went up to the rack of his own accord, spitting out the flesh, 20 as all ought to go who have the courage to refuse things that it is not right to taste, even for the natural love of life.

21 Those who were in charge of that unlawful sacrifice took the man aside because of their long acquaintance with him, and privately urged him to bring meat of his own providing, proper for him to use, and to pretend that he was eating the flesh of the sacrificial meal that had been commanded by the king, 22 so that by doing this he might be saved from death, and be treated kindly on account of his old friendship with them. 23 But making a high resolve, worthy of his years and the dignity of his old age and the gray hairs that he had reached with distinction and his excellent life even from childhood, and moreover according to the holy God-given law, he declared himself quickly, telling them to send him to Hades.

24 "Such pretense is not worthy of our time of life," he said, "for many of the young might suppose that Eleazar in his ninetieth year had gone over to an alien religion, 25 and through my pretense, for the sake of living a brief moment longer, they would be led astray because of me, while I defile and disgrace my old age. 26 Even if for the present I would avoid the punishment of mortals, yet whether I live or die I shall not escape the hands of the Almighty. 27 Therefore, by bravely giving up my life now, I will show myself worthy of my old age 28 and leave to the young a noble example of how to die a good death willingly and nobly for the revered and holy laws."

When he had said this, he went*c* at once to the rack. 29 Those who a little

c Other ancient authorities read *was dragged*

---

harvest; *ivy* was one of his symbols. **8**: *Ptolemais,* formerly Accho, or Acco, modern Acre, a coastal city eight miles north of Mt. Carmel. If the correct reading is *Ptolemies* or *Ptolemy* (as in note *b*), see 4.45 n. *The same policy toward the Jews* living as citizens of the Antiochene republic. **10**: See 1 Macc 1.60–61. **11**: A different interpretation is given in 1 Macc 2.29–41. **12–17**: The victories of Israel's enemies are explained as God's corrective punishment, always followed by mercy (compare Isa 54.7–8; Ps 94.12–15).

6.18–31: Martyrdom of Eleazar. The story is told more elaborately in 4 Maccabees. **18**: *Scribes,* scholars learned in the Mosaic law, not necessarily priests. **24–28**: Eleazar's

before had acted toward him with good-will now changed to ill will, because the words he had uttered were in their opinion sheer madness. *d* 30 When he was about to die under the blows, he groaned aloud and said: "It is clear to the Lord in his holy knowledge that, though I might have been saved from death, I am enduring terrible sufferings in my body under this beating, but in my soul I am glad to suffer these things because I fear him."

31 So in this way he died, leaving in his death an example of nobility and a memorial of courage, not only to the young but to the great body of his nation.

7 It happened also that seven brothers and their mother were arrested and were being compelled by the king, under torture with whips and thongs, to partake of unlawful swine's flesh. 2 One of them, acting as their spokesman, said, "What do you intend to ask and learn from us? For we are ready to die rather than transgress the laws of our ancestors."

3 The king fell into a rage, and gave orders to have pans and caldrons heated. 4 These were heated immediately, and he commanded that the tongue of their spokesman be cut out and that they scalp him and cut off his hands and feet, while the rest of the brothers and the mother looked on. 5 When he was utterly helpless, the king*e* ordered them to take him to the fire, still breathing, and to fry him in a pan. The smoke from the pan spread widely, but the brothers*f* and their mother encouraged one another to die nobly, saying, 6 "The Lord God is watching over us and in truth has compassion on us, as Moses declared in his song that bore witness against the people to their faces, when he said, 'And he will have compassion on his servants.' "*g*

7 After the first brother had died in this way, they brought forward the second for their sport. They tore off the skin of his head with the hair, and asked him, "Will you eat rather than have your body punished limb by limb?" 8 He replied in the language of his ancestors and said to them, "No." Therefore he in turn underwent tortures as the first brother had done. 9 And when he was at his last breath, he said, "You accursed wretch, you dismiss us from this present life, but the King of the universe will raise us up to an everlasting renewal of life, because we have died for his laws."

10 After him, the third was the victim of their sport. When it was demanded, he quickly put out his tongue and courageously stretched forth his hands, 11 and said nobly, "I got these from Heaven, and because of his laws I disdain them, and from him I hope to get them back again." 12 As a result the king himself and those with him were astonished at the young man's spirit, for he regarded his sufferings as nothing.

13 After he too had died, they maltreated and tortured the fourth in the same way. 14 When he was near death, he said, "One cannot but choose to die at the hands of mortals and to cherish the hope God gives of being raised again by him. But for you there will be no resurrection to life!"

15 Next they brought forward the fifth and maltreated him. 16 But he looked at the king,*h* and said, "Because you have authority among mortals, though you also are mortal, you do what you please. But do not think that God has forsaken our people. 17 Keep on, and see how his mighty power will torture you and your descendants!"

d Meaning of Gk uncertain   *e* Gk *he*
f Gk *they*   *g* Gk *slaves*   *h* Gk *at him*

---

speech resembles the last speech of Socrates in the *Apology.* **30:** *Fear,* revere (Job 28.28; Ps 19.9).

7.1–42: Martyrdom of seven brothers and their mother. This story is the principal subject of 4 Maccabees. **2:** Dan 3.16–18. **6:** Deut 32.36. **9:** God is often addressed in later Jewish prayer as *King of the universe.* The idea of resurrection is now clearly stated (Dan 12.2).

7.11: *Heaven,* a circumlocution for God; used also in v. 34. **17:** Antiochus IV died in

18 After him they brought forward the sixth. And when he was about to die, he said, "Do not deceive yourself in vain. For we are suffering these things on our own account, because of our sins against our own God. Therefore[i] astounding things have happened. 19 But do not think that you will go unpunished for having tried to fight against God!"

20 The mother was especially admirable and worthy of honorable memory. Although she saw her seven sons perish within a single day, she bore it with good courage because of her hope in the Lord. 21 She encouraged each of them in the language of their ancestors. Filled with a noble spirit, she reinforced her woman's reasoning with a man's courage, and said to them, 22 "I do not know how you came into being in my womb. It was not I who gave you life and breath, nor I who set in order the elements within each of you. 23 Therefore the Creator of the world, who shaped the beginning of humankind and devised the origin of all things, will in his mercy give life and breath back to you again, since you now forget yourselves for the sake of his laws."

24 Antiochus felt that he was being treated with contempt, and he was suspicious of her reproachful tone. The youngest brother being still alive, Antiochus[j] not only appealed to him in words, but promised with oaths that he would make him rich and enviable if he would turn from the ways of his ancestors, and that he would take him for his Friend and entrust him with public affairs. 25 Since the young man would not listen to him at all, the king called the mother to him and urged her to advise the youth to save himself. 26 After much urging on his part, she undertook to persuade her son. 27 But, leaning close to him, she spoke in their native language as follows, deriding the cruel tyrant: "My son, have pity on me. I carried you nine months in my womb, and nursed you for three years, and have reared you and brought you up to this point in your life, and have taken care of you.[k] 28 I beg you, my child, to look at the heaven and the earth and see everything that is in them, and recognize that God did not make them out of things that existed.[l] And in the same way the human race came into being. 29 Do not fear this butcher, but prove worthy of your brothers. Accept death, so that in God's mercy I may get you back again along with your brothers."

30 While she was still speaking, the young man said, "What are you[m] waiting for? I will not obey the king's command, but I obey the command of the law that was given to our ancestors through Moses. 31 But you,[n] who have contrived all sorts of evil against the Hebrews, will certainly not escape the hands of God. 32 For we are suffering because of our own sins. 33 And if our living Lord is angry for a little while, to rebuke and discipline us, he will again be reconciled with his own servants.[o] 34 But you, unholy wretch, you most defiled of all mortals, do not be elated in vain and puffed up by uncertain hopes, when you raise your hand against the children of heaven. 35 You have not yet escaped the judgment of the almighty, all-seeing God. 36 For our brothers after enduring a brief suffering have drunk[p] of ever-flowing life, under God's covenant; but you, by the judgment of God, will receive just punishment for your arrogance. 37 I, like my brothers, give up body and life for the laws of our ancestors, appealing to God to show mercy soon to our nation and by trials and plagues to make you confess that he alone is God, 38 and

i Lat: Other ancient authorities lack *Therefore*
j Gk *he*   k Or *have borne the burden of your education*   l Or *God made them out of things that did not exist*   m The Gk here for *you* is plural
n The Gk here for *you* is singular   o Gk *slaves*
p Cn: Gk *fallen*

misery and his son was murdered (9.5–28).
18–19: 6.12–16.
   7.28: This is the first appearance in Jewish Scriptures of the idea, borrowed from the Greek philosophers, that God created the universe out of nothing. **33:** 5.17; 6.12–16. **38:** *To bring*

through me and my brothers to bring to an end the wrath of the Almighty that has justly fallen on our whole nation."

39 The king fell into a rage, and handled him worse than the others, being exasperated at his scorn. 40 So he died in his integrity, putting his whole trust in the Lord.

41 Last of all, the mother died, after her sons.

42 Let this be enough, then, about the eating of sacrifices and the extreme tortures.

8 Meanwhile Judas, who was also called Maccabeus, and his companions secretly entered the villages and summoned their kindred and enlisted those who had continued in the Jewish faith, and so they gathered about six thousand. 2 They implored the Lord to look upon the people who were oppressed by all; and to have pity on the temple that had been profaned by the godless; 3 to have mercy on the city that was being destroyed and about to be leveled to the ground; to hearken to the blood that cried out to him; 4 to remember also the lawless destruction of the innocent babies and the blasphemies committed against his name; and to show his hatred of evil.

5 As soon as Maccabeus got his army organized, the Gentiles could not withstand him, for the wrath of the Lord had turned to mercy. 6 Coming without warning, he would set fire to towns and villages. He captured strategic positions and put to flight not a few of the enemy. 7 He found the nights most advantageous

for such attacks. And talk of his valor spread everywhere.

8 When Philip saw that the man was gaining ground little by little, and that he was pushing ahead with more frequent successes, he wrote to Ptolemy, the governor of Coelesyria and Phoenicia, to come to the aid of the king's government. 9 Then Ptolemy*q* promptly appointed Nicanor son of Patroclus, one of the king's chief*r* Friends, and sent him, in command of no fewer than twenty thousand Gentiles of all nations, to wipe out the whole race of Judea. He associated with him Gorgias, a general and a man of experience in military service. 10 Nicanor determined to make up for the king the tribute due to the Romans, two thousand talents, by selling the captured Jews into slavery. 11 So he immediately sent to the towns on the seacoast, inviting them to buy Jewish slaves and promising to hand over ninety slaves for a talent, not expecting the judgment from the Almighty that was about to overtake him.

12 Word came to Judas concerning Nicanor's invasion; and when he told his companions of the arrival of the army, 13 those who were cowardly and distrustful of God's justice ran off and got away. 14 Others sold all their remaining property, and at the same time implored the Lord to rescue those who had been sold by the ungodly Nicanor before he ever met them, 15 if not for their own sake, then for the sake of the covenants made with their ancestors, and because he had called them by his holy and glorious name. 16 But Maccabeus gathered his

q Gk he   *r* Gk one of the first

---

to an end the wrath of the Almighty, not by atoning for Israel's sins through their deaths (as in 4 Macc 1.11; 17.20–22), but by increasing the suffering of Israel to such a degree that God would be moved to intervene for them. See Deut 32.36; Judg 2.18.

8.1–7: Judas begins the revolt.

8.8–29: First victory over Nicanor. Judas assembled his forces at Mizpah and attacked Gorgias' army at Emmaus (see 1 Macc 3.40 n.). **8:** *Philip,* see 5.22 n. *Ptolemy* (see 4.45 n.), appointed by Lysias after Antiochus

had left for Persia (1 Macc 3.38). **9:** *Gorgias,* not Nicanor, is the principal figure in 1 Macc 3.38–4.25. **10:** Since the battle of Magnesia (see 1 Macc 1.10 n.) the Seleucids had been forced to pay *tribute;* perhaps the *two thousand talents* represented the last installment. **11:** Slave traders accompanied the expedition (compare 8.34 and 1 Macc 3.41). **13:** Compare 1 Macc 3.56. **15:** The idea expressed by these words, echoing Dan 9.19, occurs often in later Jewish prayers. *Covenants,* with the patriarchs and at Sinai (see 1.24–29 n.; Ex 19.5–6). *Called*

forces together, to the number six thousand, and exhorted them not to be frightened by the enemy and not to fear the great multitude of Gentiles who were wickedly coming against them, but to fight nobly, [17] keeping before their eyes the lawless outrage that the Gentiles[s] had committed against the holy place, and the torture of the derided city, and besides, the overthrow of their ancestral way of life. [18] "For they trust to arms and acts of daring," he said, "but we trust in the Almighty God, who is able with a single nod to strike down those who are coming against us, and even, if necessary, the whole world."

19 Moreover, he told them of the occasions when help came to their ancestors; how, in the time of Sennacherib, when one hundred eighty-five thousand perished, [20] and the time of the battle against the Galatians that took place in Babylonia, when eight thousand Jews[t] fought along with four thousand Macedonians; yet when the Macedonians were hard pressed, the eight thousand, by the help that came to them from heaven, destroyed one hundred twenty thousand Galatians[u] and took a great amount of booty.

21 With these words he filled them with courage and made them ready to die for their laws and their country; then he divided his army into four parts. [22] He appointed his brothers also, Simon and Joseph and Jonathan, each to command a division, putting fifteen hundred men under each. [23] Besides, he appointed Eleazar to read aloud[v] from the holy book, and gave the watchword, "The help of God"; then, leading the first division himself, he joined battle with Nicanor.

24 With the Almighty as their ally, they killed more than nine thousand of the enemy, and wounded and disabled most of Nicanor's army, and forced them all to flee. [25] They captured the money of those who had come to buy them as slaves. After pursuing them for some distance, they were obliged to return because the hour was late. [26] It was the day before the sabbath, and for that reason they did not continue their pursuit. [27] When they had collected the arms of the enemy and stripped them of their spoils, they kept the sabbath, giving great praise and thanks to the Lord, who had preserved them for that day and allotted it to them as the beginning of mercy. [28] After the sabbath they gave some of the spoils to those who had been tortured and to the widows and orphans, and distributed the rest among themselves and their children. [29] When they had done this, they made common supplication and implored the merciful Lord to be wholly reconciled with his servants.[w]

30 In encounters with the forces of Timothy and Bacchides they killed more than twenty thousand of them and got possession of some exceedingly high strongholds, and they divided a very large amount of plunder, giving to those who had been tortured and to the orphans and widows, and also to the aged, shares equal to their own. [31] They collected the arms of the enemy,[x] and carefully stored all of them in strategic places;

s Gk *they*   *t* Gk lacks *Jews*   *u* Gk lacks
Galatians   *v* Meaning of Gk uncertain
w Gk *slaves*   *x* Gk *their arms*

---

them by his . . . *name,* as God's people (Deut 28.10). **17:** *Lawless outrage,* 5.15–16. *Ancestral way of life,* the Torah.

8.19: 2 Kings 19.35. **20:** *The Galatians* from Asia Minor often served as mercenaries. Jewish forces evidently aided Antiochus III and *the Macedonians.* **22:** *Simon,* high priest 142–134 B.C., and *Jonathan,* from 160 to 143 or 142 B.C. *Joseph,* called John in 1 Macc 2.2; 9.36. **23:** *Eleazar,* another brother, was killed at Beth-zechariah (1 Macc 2.5; 6.43–46). The

motto *"The help of God"* is prescribed by the Qumran *War Scroll* for one of the banners of the army returning from battle. **25:** *Slaves,* vv. 11, 34. **26:** Gorgias and his army were in the hills (1 Macc 4.16–18). **27–29:** The victory was a sign of God's favor, but the campaign had not yet been won (6.12–16; 1 Macc 4.19–25).

8.30–36: **Other victories** (1 Macc 5.37–44 tells of a battle against *Timothy* at Raphon). **30:** *Bacchides,* 1 Macc 7.8. **33:** *City of their an-*

the rest of the spoils they carried to Jerusalem. 32 They killed the commander of Timothy's forces, a most wicked man, and one who had greatly troubled the Jews. 33 While they were celebrating the victory in the city of their ancestors, they burned those who had set fire to the sacred gates, Callisthenes and some others, who had fled into one little house; so these received the proper reward for their impiety. y

34 The thrice-accursed Nicanor, who had brought the thousand merchants to buy the Jews, 35 having been humbled with the help of the Lord by opponents whom he regarded as of the least account, took off his splendid uniform and made his way alone like a runaway slave across the country until he reached Antioch, having succeeded chiefly in the destruction of his own army! 36 So he who had undertaken to secure tribute for the Romans by the capture of the people of Jerusalem proclaimed that the Jews had a Defender, and that therefore the Jews were invulnerable, because they followed the laws ordained by him.

9 About that time, as it happened, Antiochus had retreated in disorder from the region of Persia. 2 He had entered the city called Persepolis and attempted to rob the temples and control the city. Therefore the people rushed to the rescue with arms, and Antiochus and his army were defeated, z with the result that Antiochus was put to flight by the inhabitants and beat a shameful retreat. 3 While he was in Ecbatana, news came to him of what had happened to Nicanor and the forces of Timothy. 4 Transported with rage, he conceived the idea of turning upon the Jews the injury done by those who had put him to flight; so he ordered his charioteer to drive without stopping until he completed the journey. But the judgment of heaven rode with him! For in his arrogance he said, "When I get there I will make Jerusalem a cemetery of Jews."

5 But the all-seeing Lord, the God of Israel, struck him with an incurable and invisible blow. As soon as he stopped speaking he was seized with a pain in his bowels, for which there was no relief, and with sharp internal tortures— 6 and that very justly, for he had tortured the bowels of others with many and strange inflictions. 7 Yet he did not in any way stop his insolence, but was even more filled with arrogance, breathing fire in his rage against the Jews, and giving orders to drive even faster. And so it came about that he fell out of his chariot as it was rushing along, and the fall was so hard as to torture every limb of his body. 8 Thus he who only a little while before had thought in his superhuman arrogance that he could command the waves of the sea, and had imagined that he could weigh the high mountains in a balance, was brought down to earth and carried in a litter, making the power of God manifest to all. 9 And so the ungodly man's body swarmed with worms, and while he was still living in anguish and pain, his flesh rotted away, and because of the stench the whole army felt revulsion at his decay. 10 Because of his intolerable stench no one was able to carry the man who a little while before had thought that he could touch the stars of

y Meaning of Gk uncertain   z Gk they were defeated

cestors, Jerusalem, with its *sacred gates,* Judas' ancestral home (1 Macc 2.1). **34:** *Thrice-accursed,* 15.3; Add Est 16.15.

9.1–12: Antiochus' illness (1 Macc 6.1–16). Here this story is placed before the purification of the temple (10.1–8; 1 Macc 4.36–61), Judas' southern campaigns (10.14–38; 1 Macc ch 5), and Lysias' first expedition (11.1–15; 1 Macc 4.26–35). **1:** *Antiochus* went to *Persia* to strengthen his authority there and to get funds. **2:** *Persepolis,* near Shiraz, the capital of Persia, founded by Darius I. **3:** Antiochus was on his way to Babylon (1 Macc 6.4) but went north by way of *Ecbatana,* Hamadan. **4:** 5.11; 7.3. **5:** See 1 Macc 6.9 n. **8:** *Command the waves,* like Xerxes invading Greece. *Weigh the high mountains,* like God (see 5.21 n.; Isa 40.12). **9:** *Worms,* Acts 12.23. **10:** Isa 14.12–19.

heaven. 11 Then it was that, broken in spirit, he began to lose much of his arrogance and to come to his senses under the scourge of God, for he was tortured with pain every moment. 12 And when he could not endure his own stench, he uttered these words, "It is right to be subject to God; mortals should not think that they are equal to God."*a*

13 Then the abominable fellow made a vow to the Lord, who would no longer have mercy on him, stating 14 that the holy city, which he was hurrying to level to the ground and to make a cemetery, he was now declaring to be free; 15 and the Jews, whom he had not considered worth burying but had planned to throw out with their children for the wild animals and for the birds to eat, he would make, all of them, equal to citizens of Athens; 16 and the holy sanctuary, which he had formerly plundered, he would adorn with the finest offerings; and all the holy vessels he would give back, many times over; and the expenses incurred for the sacrifices he would provide from his own revenues; 17 and in addition to all this he also would become a Jew and would visit every inhabited place to proclaim the power of God. 18 But when his sufferings did not in any way abate, for the judgment of God had justly come upon him, he gave up all hope for himself and wrote to the Jews the following letter, in the form of a supplication. This was its content:

19 "To his worthy Jewish citizens, Antiochus their king and general sends hearty greetings and good wishes for their health and prosperity. 20 If you and your children are well and your affairs are as you wish, I am glad. As my hope is in heaven, 21 I remember with affection your esteem and goodwill. On my way back from the region of Persia I suffered an annoying illness, and I have deemed it necessary to take thought for the general security of all. 22 I do not despair of my condition, for I have good hope of recovering from my illness, 23 but I observed that my father, on the occasions when he made expeditions into the upper country, appointed his successor, 24 so that, if anything unexpected happened or any unwelcome news came, the people throughout the realm would not be troubled, for they would know to whom the government was left. 25 Moreover, I understand how the princes along the borders and the neighbors of my kingdom keep watching for opportunities and waiting to see what will happen. So I have appointed my son Antiochus to be king, whom I have often entrusted and commended to most of you when I hurried off to the upper provinces; and I have written to him what is written here. 26 I therefore urge and beg you to remember the public and private services rendered to you and to maintain your present goodwill, each of you, toward me and my son. 27 For I am sure that he will follow my policy and will treat you with moderation and kindness."

28 So the murderer and blasphemer, having endured the more intense suffering, such as he had inflicted on others, came to the end of his life by a most pitiable fate, among the mountains in a strange land. 29 And Philip, one of his courtiers, took his body home; then, fearing the son of Antiochus, he withdrew to Ptolemy Philometor in Egypt.

a Or *not think thoughts proper only to God*

9.13–29: Repentance and death of Antiochus. 15: *Citizens of Athens* were proud of their heritage, though the city no longer had actual power. **16**: 5.16. **17**: 7.37; Dan 4.31–35. **9.19–27**: The letter is no supplication (v. 18); it is addressed to Jews loyal to the king and bids them support his *son Antiochus* V (vv. 25–27). **23**: *My father,* Antiochus III (see 3.3 n.), who *appointed* Seleucus IV as *his successor. The upper country,* Babylonia and Persia (1 Macc 3.37). **28**: Antiochus IV died *among the mountains,* perhaps at Gabae or Isfahan (see 1 Macc 6.5 n.). **29**: *Philip* was perhaps Antiochus V's guardian (see 1 Macc 6.14–15 n.). *Fearing* Lysias, viceroy in the west, rather than *the son of Antiochus,* who was a child, he went over to Syria's enemy, *Ptolemy* VI (see 4.21 n.). Josephus says that Philip took over the Seleucid government and was later killed (*Antiquities,* XII. ix. 7).

10 Now Maccabeus and his followers, the Lord leading them on, recovered the temple and the city; ²they tore down the altars that had been built in the public square by the foreigners, and also destroyed the sacred precincts. ³They purified the sanctuary, and made another altar of sacrifice; then, striking fire out of flint, they offered sacrifices, after a lapse of two years, and they offered incense and lighted lamps and set out the bread of the Presence. ⁴When they had done this, they fell prostrate and implored the Lord that they might never again fall into such misfortunes, but that, if they should ever sin, they might be disciplined by him with forbearance and not be handed over to blasphemous and barbarous nations. ⁵It happened that on the same day on which the sanctuary had been profaned by the foreigners, the purification of the sanctuary took place, that is, on the twenty-fifth day of the same month, which was Chislev. ⁶They celebrated it for eight days with rejoicing, in the manner of the festival of booths, remembering how not long before, during the festival of booths, they had been wandering in the mountains and caves like wild animals. ⁷Therefore, carrying ivy-wreathed wands and beautiful branches and also fronds of palm, they offered hymns of thanksgiving to him who had given success to the purifying of his own holy place. ⁸They decreed by public edict, ratified by vote, that the whole nation of the Jews should observe these days every year.

⁹ Such then was the end of Antiochus, who was called Epiphanes.

¹⁰ Now we will tell what took place under Antiochus Eupator, who was the son of that ungodly man, and will give a brief summary of the principal calamities of the wars. ¹¹This man, when he succeeded to the kingdom, appointed one Lysias to have charge of the government and to be chief governor of Coelesyria and Phoenicia. ¹²Ptolemy, who was called Macron, took the lead in showing justice to the Jews because of the wrong that had been done to them, and attempted to maintain peaceful relations with them. ¹³As a result he was accused before Eupator by the king's Friends. He heard himself called a traitor at every turn, because he had abandoned Cyprus, which Philometor had entrusted to him, and had gone over to Antiochus Epiphanes. Unable to command the respect due his office, ᵇ he took poison and ended his life.

¹⁴ When Gorgias became governor of the region, he maintained a force of mercenaries, and at every turn kept attacking the Jews. ¹⁵Besides this, the Idumeans, who had control of important strongholds, were harassing the Jews; they received those who were banished

ᵇ Cn: Meaning of Gk uncertain

10.1–9: Purification of the temple (compare 1 Macc 4.36–61). **1:** They *recovered the temple*, desecrated by Antiochus (6.2–4; 1 Macc 1.54), *and the city*, except for the citadel (1 Macc 4.60; 6.18). **2:** *The altars* had been used for pagan worship. **3:** They *purified the sanctuary* by removing the desecrated stones (1 Macc 1.44–46). The reference to *striking fire out of flint* ignores the legends of 1.19–2.1. *Two years*, according to 1 Macc 1.54 and 4.52 it was three years. The *incense, lamps,* and *bread of the Presence*, prescribed by Ex 30.7–8; 25.30. **4:** 5.17–20; 6.12–16. **5–6:** *Chislev*, December, 164 B.C. (see 1 Macc 4.52–59 n.). At the normal time of *the festival of booths*, in September, *they had been wandering* like their ancestors (Lev 23.43) and could not celebrate it. **7:** *Ivy-wreathed wands*, here in honor of God (compare 6.7). *Branches* were carried in procession at the festival of booths. *Fronds of palm* symbolize victory (1 Macc 13.51; Jn 12.13).

10.10–13: Antiochus V and Ptolemy Macron. 10–11: *Antiochus V Eupator* (9.25), son of Antiochus IV, reigned from 164 to 162 B.C., when he was murdered by order of Demetrius I. He was about nine years old; his father had *appointed . . . . Lysias* as regent (1 Macc 3.32–33). **12–13:** *Ptolemy* had changed allegiance from Egypt to Syria (see 4.45 n.; 6.8); now he was friendly *to the Jews*.

10.14–23: Attacks on the Idumeans (1 Macc 5.1–3, 9–54). **14:** *Gorgias* succeeded Ptolemy. **15:** *Idumeans*, or Edomites (see 1 Macc 5.3 n.); John Hyrcanus later forced

from Jerusalem, and endeavored to keep up the war. ¹⁶But Maccabeus and his forces, after making solemn supplication and imploring God to fight on their side, rushed to the strongholds of the Idumeans. ¹⁷Attacking them vigorously, they gained possession of the places, and beat off all who fought upon the wall, and slaughtered those whom they encountered, killing no fewer than twenty thousand.

18 When at least nine thousand took refuge in two very strong towers well equipped to withstand a siege, ¹⁹Maccabeus left Simon and Joseph, and also Zacchaeus and his troops, a force sufficient to besiege them; and he himself set off for places where he was more urgently needed. ²⁰But those with Simon, who were money-hungry, were bribed by some of those who were in the towers, and on receiving seventy thousand drachmas let some of them slip away. ²¹When word of what had happened came to Maccabeus, he gathered the leaders of the people, and accused these men of having sold their kindred for money by setting their enemies free to fight against them. ²²Then he killed these men who had turned traitor, and immediately captured the two towers. ²³Having success at arms in everything he undertook, he destroyed more than twenty thousand in the two strongholds.

24 Now Timothy, who had been defeated by the Jews before, gathered a tremendous force of mercenaries and collected the cavalry from Asia in no small number. He came on, intending to take Judea by storm. ²⁵As he drew near, Maccabeus and his men sprinkled dust on their heads and girded their loins with sackcloth, in supplication to God. ²⁶Falling upon the steps before the altar, they

implored him to be gracious to them and to be an enemy to their enemies and an adversary to their adversaries, as the law declares. ²⁷And rising from their prayer they took up their arms and advanced a considerable distance from the city; and when they came near the enemy they halted. ²⁸Just as dawn was breaking, the two armies joined battle, the one having as pledge of success and victory not only their valor but also their reliance on the Lord, while the other made rage their leader in the fight.

29 When the battle became fierce, there appeared to the enemy from heaven five resplendent men on horses with golden bridles, and they were leading the Jews. ³⁰Two of them took Maccabeus between them, and shielding him with their own armor and weapons, they kept him from being wounded. They showered arrows and thunderbolts on the enemy, so that, confused and blinded, they were thrown into disorder and cut to pieces. ³¹Twenty thousand five hundred were slaughtered, besides six hundred cavalry.

32 Timothy himself fled to a stronghold called Gazara, especially well garrisoned, where Chaereas was commander. ³³Then Maccabeus and his men were glad, and they besieged the fort for four days. ³⁴The men within, relying on the strength of the place, kept blaspheming terribly and uttering wicked words. ³⁵But at dawn of the fifth day, twenty young men in the army of Maccabeus, fired with anger because of the blasphemies, bravely stormed the wall and with savage fury cut down everyone they met. ³⁶Others who came up in the same way wheeled around against the defenders and set fire to the towers; they kindled fires and burned the blasphemers

---

them to adopt Judaism. *Those . . . banished,* supporters of Menelaus. **19:** *Urgently needed,* perhaps in Ammon and Gilead (1 Macc 5.6–13).

10.24–38: Victory over Timothy. 24: They met *Timothy* (8.30) at *dawn* (v. 28) at Dathema east of the Jordan (1 Macc 5.28–34). **26:** Ex 23.22. **29:** 3.24–26. **31:** *Twenty thou-* sand *five hundred,* compare the number of fatalities mentioned in 8.30; 10.17, 23. **32–38:** The fort of *Gazara* (1 Macc 4.15; 7.45) was well garrisoned; Simon captured it much later (1 Macc 13.43–48). **37:** *They killed Timothy;* but a Timothy reappears in 12.2, 18–25 (compare 1 Macc 5.11–40).

alive. Others broke open the gates and let in the rest of the force, and they occupied the city. 37 They killed Timothy, who was hiding in a cistern, and his brother Chaereas, and Apollophanes. 38 When they had accomplished these things, with hymns and thanksgivings they blessed the Lord who shows great kindness to Israel and gives them the victory.

11 Very soon after this, Lysias, the king's guardian and kinsman, who was in charge of the government, being vexed at what had happened, 2 gathered about eighty thousand infantry and all his cavalry and came against the Jews. He intended to make the city a home for Greeks, 3 and to levy tribute on the temple as he did on the sacred places of the other nations, and to put up the high priesthood for sale every year. 4 He took no account whatever of the power of God, but was elated with his ten thousands of infantry, and his thousands of cavalry, and his eighty elephants. 5 Invading Judea, he approached Beth-zur, which was a fortified place about five stadia ᶜ from Jerusalem, and pressed it hard.

6 When Maccabeus and his men got word that Lysias ᵈ was besieging the strongholds, they and all the people, with lamentations and tears, prayed the Lord to send a good angel to save Israel. 7 Maccabeus himself was the first to take up arms, and he urged the others to risk their lives with him to aid their kindred. Then they eagerly rushed off together. 8 And there, while they were still near Jerusalem, a horseman appeared at their head, clothed in white and brandishing weapons of gold. 9 And together they all praised the merciful God, and were strengthened in heart, ready to assail not only humans but the wildest animals or walls of iron. 10 They advanced in battle order, having their heavenly ally, for the Lord had mercy on them. 11 They hurled themselves like lions against the enemy, and laid low eleven thousand of them and sixteen hundred cavalry, and forced all the rest to flee. 12 Most of them got away stripped and wounded, and Lysias himself escaped by disgraceful flight.

13 As he was not without intelligence, he pondered over the defeat that had befallen him, and realized that the Hebrews were invincible because the mighty God fought on their side. So he sent to them 14 and persuaded them to settle everything on just terms, promising that he would persuade the king, constraining him to be their friend. ᶜ 15 Maccabeus, having regard for the common good, agreed to all that Lysias urged. For the king granted every request in behalf of the Jews which Maccabeus delivered to Lysias in writing.

16 The letter written to the Jews by Lysias was to this effect:

"Lysias to the people of the Jews, greetings. 17 John and Absalom, who were sent by you, have delivered your signed communication and have asked about the matters indicated in it. 18 I have informed the king of everything that needed to be brought before him, and he has agreed to what was possible. 19 If you will maintain your goodwill toward the government, I will endeavor in the future to help promote your welfare.

c Meaning of Gk uncertain   *d* Gk *he*

11.1–15: Victory over Lysias at Beth-zur. This probably occurred before the dedication of the temple (1 Macc 4.26–35). **1:** *Lysias,* see 10.10–13 n. **3:** In many Greek cults the priesthood was *for sale every year;* Antiochus IV had twice disposed of the Jewish high priesthood (4.7, 24). **4:** *Elephants,* see 1 Macc 1.17 n.; 6.34–35 n. **5:** *Beth-zur,* about twenty miles south of Jerusalem on the road to Hebron. **6:** *Good angel,* 15.23; Ex 23.20; Josh 5.13–15; Judg 6.11; 2 Kings 19.35. **13–15:** According to 1 Macc 4.35 no peace was made, but Lysias returned to Antioch for reinforcements. He may have heard of Antiochus' death and hastened home to take control.

11.16–38: Letters of Lysias, Antiochus V, and the Romans. If Lysias heard of Philip's plot (see 9.29 n.), he may have wished to gain time through friendly gestures to the Jews. **16:** He wrote *to the people;* he did not recognize Judas' authority. **19:** Part of the Jews had *goodwill toward the government.* See Introduction to 1 Maccabees. **21:** The date is early December, 164 B.C., before Judas

20 And concerning such matters and their details, I have ordered these men and my representatives to confer with you. 21 Farewell. The one hundred forty-eighth year,ᵉ Dioscorinthius twenty-fourth."

22 The king's letter ran thus:

"King Antiochus to his brother Lysias, greetings. 23 Now that our father has gone on to the gods, we desire that the subjects of the kingdom be undisturbed in caring for their own affairs. 24 We have heard that the Jews do not consent to our father's change to Greek customs, but prefer their own way of living and ask that their own customs be allowed them. 25 Accordingly, since we choose that this nation also should be free from disturbance, our decision is that their temple be restored to them and that they shall live according to the customs of their ancestors. 26 You will do well, therefore, to send word to them and give them pledges of friendship, so that they may know our policy and be of good cheer and go on happily in the conduct of their own affairs."

27 To the nation the king's letter was as follows:

"King Antiochus to the senate of the Jews and to the other Jews, greetings. 28 If you are well, it is as we desire. We also are in good health. 29 Menelaus has informed us that you wish to return home and look after your own affairs. 30 Therefore those who go home by the thirtieth of Xanthicus will have our pledge of friendship and full permission 31 for the Jews to enjoy their own food and laws, just as formerly, and none of them shall be molested in any way for what may have been done in ignorance. 32 And I have also sent Menelaus to encourage you. 33 Farewell. The one hundred forty-eighth year,ᵉ Xanthicus fifteenth."

34 The Romans also sent them a letter, which read thus:

"Quintus Memmius and Titus Manius, envoys of the Romans, to the people of the Jews, greetings. 35 With regard to what Lysias the kinsman of the king has granted you, we also give consent. 36 But as to the matters that he decided are to be referred to the king, as soon as you have considered them, send some one promptly so that we may make proposals appropriate for you. For we are on our way to Antioch. 37 Therefore make haste and send messengers so that we may have your judgment. 38 Farewell. The one hundred forty-eighth year,ᵉ Xanthicus fifteenth."

12 When this agreement had been reached, Lysias returned to the king, and the Jews went about their farming.

2 But some of the governors in various places, Timothy and Apollonius son of Gennaeus, as well as Hieronymus and Demophon, and in addition to these Nicanor the governor of Cyprus, would not let them live quietly and in peace. 3 And the people of Joppa did so ungodly a deed as this: they invited the Jews who lived among them to embark, with their wives and children, on boats that they had provided, as though there were no ill will to the Jews;ᶠ 4 and this was done by public vote of the city. When they ac-

ᵉ 164 B.C.   ᶠ Gk *to them*

---

rededicated the temple (1 Macc 4.52). **23**: *Our father*, Antiochus IV, *has gone on to the gods;* in his lifetime he had been worshiped. **25**: 1 Macc 4.36–61 says nothing of this, but Lysias may have instructed the citadel garrison not to interfere with the temple.

11.27: The letter *to the senate* (1.10) and people ignores Judas (compare vv. 16–21). **29**: *Menelaus* had gone to Antioch and advised the king to let the Jews *return* to Jerusalem. He was now sent back (v. 32), hoping to regain the high priesthood. **30**: *Xanthicus*, March–

April. **31**: *Their own food and laws,* 1 Macc 1.47–49. The words *in ignorance* imply that the king still maintained his claims and merely granted pardon (1 Macc 13.39). **34–37**: The *envoys* acted as intermediaries in *matters . . . referred to the king* that were still under negotiation.

12.1–16: **Attacks on Joppa, Jamnia, and Caspin. 2**: *Timothy,* 8.30–33; 10.24–37. *Apollonius*, not the Apollonius of 4.21. *Nicanor* is called *governor of Cyprus;* it was under Egypt's rule till 58 B.C., but Syria may have claimed

cepted, because they wished to live peaceably and suspected nothing, the people of Joppa*g* took them out to sea and drowned them, at least two hundred. 5 When Judas heard of the cruelty visited on his compatriots, he gave orders to his men 6 and, calling upon God, the righteous judge, attacked the murderers of his kindred. He set fire to the harbor by night, burned the boats, and massacred those who had taken refuge there. 7 Then, because the city's gates were closed, he withdrew, intending to come again and root out the whole community of Joppa. 8 But learning that the people in Jamnia meant in the same way to wipe out the Jews who were living among them, 9 he attacked the Jamnites by night and set fire to the harbor and the fleet, so that the glow of the light was seen in Jerusalem, thirty miles*h* distant.

10 When they had gone more than a mile*i* from there, on their march against Timothy, at least five thousand Arabs with five hundred cavalry attacked them. 11 After a hard fight, Judas and his companions, with God's help, were victorious. The defeated nomads begged Judas to grant them pledges of friendship, promising to give him livestock and to help his people*j* in all other ways. 12 Judas, realizing that they might indeed be useful in many ways, agreed to make peace with them; and after receiving his pledges they went back to their tents.

13 He also attacked a certain town that was strongly fortified with earthworks*k* and walls, and inhabited by all sorts of Gentiles. Its name was Caspin. 14 Those who were within, relying on the strength of the walls and on their supply of provisions, behaved most insolently toward Judas and his men, railing at them and even blaspheming and saying unholy things. 15 But Judas and his men, calling against the great Sovereign of the world, who without battering-rams or engines of war overthrew Jericho in the days of Joshua, rushed furiously upon the walls. 16 They took the town by the will of God, and slaughtered untold numbers, so that the adjoining lake, a quarter of a mile*l* wide, appeared to be running over with blood.

17 When they had gone ninety-five miles*m* from there, they came to Charax, to the Jews who are called Toubiani. 18 They did not find Timothy in that region, for he had by then left there without accomplishing anything, though in one place he had left a very strong garrison. 19 Dositheus and Sosipater, who were captains under Maccabeus, marched out and destroyed those whom Timothy had left in the stronghold, more than ten thousand men. 20 But Maccabeus arranged his army in divisions, set men*j* in command of the divisions, and hurried after Timothy, who had with him one hundred twenty thousand infantry and two thousand five hundred cavalry. 21 When Timothy learned of the approach of Judas, he sent off the women and the children and also the baggage to a place called Carnaim; for that place was hard to besiege and difficult of access because of the narrowness of all the approaches. 22 But when Judas's first division appeared, terror and fear came over the enemy at the manifestation to them of him who sees all things. In their flight they rushed headlong in every direction, so that often they were injured by their own men and pierced by the points of their own swords. 23 Judas pressed the pursuit with the utmost vigor, putting the sinners to the sword, and destroyed as many as thirty thousand.

g Gk *they*   *h* Gk *two hundred forty stadia*
i Gk *nine stadia*   *j* Gk *them*
k Meaning of Gk uncertain   *l* Gk *two stadia*
m Gk *seven hundred fifty stadia*

it after the defection of Ptolemy Macron (10.13). There may have been two Nicanors (see 14.12 n.). **8:** *Jamnia,* about twelve miles south of Joppa. **13:** *Caspin,* perhaps Chaspho (1 Macc 5.36). **15:** Josh 6.1–21.
12.17–31: Battles in the northeast (the account supplements 1 Macc 5.9–32). **17:** *Toubiani,* perhaps people of Tob (1 Macc 5.13). **18:** *One place,* perhaps Bozrah, southeast of Tob (1 Macc 5.28). **21:** *Carnaim,* a little north of Dera'a in Syria (Gen 14.5; 1 Macc 5.26).

24 Timothy himself fell into the hands of Dositheus and Sosipater and their men. With great guile he begged them to let him go in safety, because he held the parents of most of them, and the brothers of some, to whom no consideration would be shown. 25 And when with many words he had confirmed his solemn promise to restore them unharmed, they let him go, for the sake of saving their kindred.

26 Then Judas[n] marched against Carnaim and the temple of Atargatis, and slaughtered twenty-five thousand people. 27 After the rout and destruction of these, he marched also against Ephron, a fortified town where Lysias lived with multitudes of people of all nationalities.[o] Stalwart young men took their stand before the walls and made a vigorous defense; and great stores of war engines and missiles were there. 28 But the Jews[p] called upon the Sovereign who with power shatters the might of his enemies, and they got the town into their hands, and killed as many as twenty-five thousand of those who were in it.

29 Setting out from there, they hastened to Scythopolis, which is seventy-five miles[q] from Jerusalem. 30 But when the Jews who lived there bore witness to the goodwill that the people of Scythopolis had shown them and their kind treatment of them in times of misfortune, 31 they thanked them and exhorted them to be well disposed to their race in the future also. Then they went up to Jerusalem, as the festival of weeks was close at hand.

32 After the festival called Pentecost, they hurried against Gorgias, the governor of Idumea, 33 who came out with three thousand infantry and four hundred cavalry. 34 When they joined battle, it happened that a few of the Jews fell. 35 But a certain Dositheus, one of Bacenor's men, who was on horseback and was a strong man, caught hold of Gorgias, and grasping his cloak was dragging him off by main strength, wishing to take the accursed man alive, when one of the Thracian cavalry bore down on him and cut off his arm; so Gorgias escaped and reached Marisa.

36 As Esdris and his men had been fighting for a long time and were weary, Judas called upon the Lord to show himself their ally and leader in the battle. 37 In the language of their ancestors he raised the battle cry, with hymns; then he charged against Gorgias's troops when they were not expecting it, and put them to flight.

38 Then Judas assembled his army and went to the city of Adullam. As the seventh day was coming on, they purified themselves according to the custom, and kept the sabbath there.

39 On the next day, as had now become necessary, Judas and his men went to take up the bodies of the fallen and to bring them back to lie with their kindred in the sepulchres of their ancestors. 40 Then under the tunic of each one of the dead they found sacred tokens of the idols of Jamnia, which the law forbids the Jews to wear. And it became clear to all that this was the reason these men had fallen. 41 So they all blessed the ways of

n Gk *he*   *o* Meaning of Gk uncertain
p Gk *they*   *q* Gk *six hundred stadia*

12.26: *Atargatis,* the Syrian goddess to whom fish were sacred. 27: He *marched* south *against Ephron,* eight miles east of the Jordan, opposite Scythopolis (v. 29; 1 Macc 5.46–51). *War engines* were large catapults. 29: *Scythopolis,* ancient Beth-shan, then and later an important city (see 1 Macc 5.52 n.). 31: The *festival of weeks* or Pentecost was at the time of the wheat harvest, seven weeks after Passover, and was celebrated in Jerusalem (Ex 34.22–24; Deut 16.9–12).
12.32–38: **Battle with Gorgias.** 35:

Marisa, in the foothills southwest of Jerusalem near Beit-Jibrin (see 1 Macc 5.66 n.). 36: *Esdris,* evidently a division leader (v. 20); the author of 2 Maccabees has abbreviated his source. 38: *City of Adullam,* northeast of Marisa (Josh 12.15; 15.35). They *kept the sabbath,* when it was not necessary to fight (8.27; see 1 Macc 2.41 n.).
12.39–45: **Burial of the dead.** The author believed that many had been killed because they wore *sacred tokens* of pagan gods *which the law forbids* (v. 40; Deut 7.25–26), but

the Lord, the righteous judge, who reveals the things that are hidden; [42] and they turned to supplication, praying that the sin that had been committed might be wholly blotted out. The noble Judas exhorted the people to keep themselves free from sin, for they had seen with their own eyes what had happened as the result of the sin of those who had fallen. [43] He also took up a collection, man by man, to the amount of two thousand drachmas of silver, and sent it to Jerusalem to provide for a sin offering. In doing this he acted very well and honorably, taking account of the resurrection. [44] For if he were not expecting that those who had fallen would rise again, it would have been superfluous and foolish to pray for the dead. [45] But if he was looking to the splendid reward that is laid up for those who fall asleep in godliness, it was a holy and pious thought. Therefore he made atonement for the dead, so that they might be delivered from their sin.

13 In the one hundred forty-ninth year[r] word came to Judas and his men that Antiochus Eupator was coming with a great army against Judea, [2] and with him Lysias, his guardian, who had charge of the government. Each of them had a Greek force of one hundred ten thousand infantry, five thousand three hundred cavalry, twenty-two elephants, and three hundred chariots armed with scythes.

[3] Menelaus also joined them and with utter hypocrisy urged Antiochus on, not for the sake of his country's welfare, but because he thought that he would be established in office. [4] But the King of kings aroused the anger of Antiochus against the scoundrel; and when Lysias informed him that this man was to blame for all the trouble, he ordered them to take him to Beroea and to put him to death by the method that is customary in that place. [5] For there is a tower there, fifty cubits high, full of ashes, and it has a rim running around it that on all sides inclines precipitously into the ashes. [6] There they all push to destruction anyone guilty of sacrilege or notorious for other crimes. [7] By such a fate it came about that Menelaus the lawbreaker died, without even burial in the earth. [8] And this was eminently just; because he had committed many sins against the altar whose fire and ashes were holy, he met his death in ashes.

[9] The king with barbarous arrogance was coming to show the Jews things far worse than those that had been done[s] in his father's time. [10] But when Judas heard of this, he ordered the people to call upon the Lord day and night, now if ever to help those who were on the point of being deprived of the law and their country and the holy temple, [11] and not to let the people who had just begun to revive fall into the hands of the blasphemous Gentiles. [12] When they had all joined in the same petition and had implored the merciful Lord with weeping and fasting and lying prostrate for three days without ceasing, Judas exhorted them and ordered them to stand ready.

[13] After consulting privately with the elders, he determined to march out and decide the matter by the help of God before the king's army could enter Judea

---

r 163 B.C.   *s Or the worst of the things
that had been done*

Josephus says (*Antiquities*, XII. viii. 6) this reverse befell them because they had disobeyed Judas' instructions not to join battle before his arrival. See also 1 Macc 5.67. This is the first known statement of the doctrine that a *sin offering* and prayer make *atonement* for the sins of *the dead* (v. 45), and it is justified by the hope that *those who had fallen would rise again* (vv. 43–44; 7.11; 14.46). *Fall asleep*, die (1 Cor 15.20).

13.1–8: Death of Menelaus. 1–2: *Antio-* chus and *Lysias*, see 10.10–11 n. *Chariots armed with scythes* to cut down foot soldiers had been used since the days of the Persian Empire. **4:** *The King of kings*, God (Deut 10.17; Ps 136.3; Rev 19.16). What *aroused* his *anger* is not known (but see 4.27). *Beroea*, now Aleppo in northern Syria. The *method* of execution (vv. 5–6) was Persian.

13.9–17: Preliminary skirmish. 12: Jews employed such acts of penitence particularly when there was danger of sacrilege (3.15;

and get possession of the city. ¹⁴ So, committing the decision to the Creator of the world and exhorting his troops to fight bravely to the death for the laws, temple, city, country, and commonwealth, he pitched his camp near Modein. ¹⁵ He gave his troops the watchword, "God's victory," and with a picked force of the bravest young men, he attacked the king's pavilion at night and killed as many as two thousand men in the camp. He stabbed* the leading elephant and its rider. ¹⁶ In the end they filled the camp with terror and confusion and withdrew in triumph. ¹⁷ This happened, just as day was dawning, because the Lord's help protected him.

18 The king, having had a taste of the daring of the Jews, tried strategy in attacking their positions. ¹⁹ He advanced against Beth-zur, a strong fortress of the Jews, was turned back, attacked again,ᵘ and was defeated. ²⁰ Judas sent in to the garrison whatever was necessary. ²¹ But Rhodocus, a man from the ranks of the Jews, gave secret information to the enemy; he was sought for, caught, and put in prison. ²² The king negotiated a second time with the people in Beth-zur, gave pledges, received theirs, withdrew, attacked Judas and his men, was defeated; ²³ he got word that Philip, who had been left in charge of the government, had revolted in Antioch; he was dismayed, called in the Jews, yielded and swore to observe all their rights, settled with them and offered sacrifice, honored the sanctuary and showed generosity to the holy place. ²⁴ He received Maccabeus, left

Hegemonides as governor from Ptolemais to Gerar, ²⁵ and went to Ptolemais. The people of Ptolemais were indignant over the treaty; in fact they were so angry that they wanted to annul its terms.ᵗ ²⁶ Lysias took the public platform, made the best possible defense, convinced them, appeased them, gained their goodwill, and set out for Antioch. This is how the king's attack and withdrawal turned out.

14 Three years later, word came to Judas and his men that Demetrius son of Seleucus had sailed into the harbor of Tripolis with a strong army and a fleet, ² and had taken possession of the country, having made away with Antiochus and his guardian Lysias.

3 Now a certain Alcimus, who had formerly been high priest but had willfully defiled himself in the times of separation,ᵛ realized that there was no way for him to be safe or to have access again to the holy altar, ⁴ and went to King Demetrius in about the one hundred fifty-first year,ʷ presenting to him a crown of gold and a palm, and besides these some of the customary olive branches from the temple. During that day he kept quiet. ⁵ But he found an opportunity that furthered his mad purpose when he was invited by Demetrius to a meeting of the council and was asked about the attitude and intentions of the Jews. He answered:

6 "Those of the Jews who are called

t Meaning of Gk uncertain   *u* Or *faltered*
v Other ancient authorities read *of mixing*
w 161 B.C.

10.4; 1 Macc 4.40). **14**: The Syrian army had invaded Judea from the south, through Idumea (1 Macc 6.31). Judas first *pitched his camp near Modein* to watch the Syrian line along the coast. The first battle occurred at Beth-zechariah (1 Macc 6.32–47). **15**: "*God's victory*," see 8.23 n. Eleazar *stabbed the leading elephant* (1 Macc 6.43–46). **16**: According to 1 Macc 6.47 the Jews fled. **17**: Ps 46.6.

13.18–26: **Attack on Beth-zur. 19**: The Syrians were defeated in the first attempt (1 Macc 6.31). **21–22**: The garrison surrendered because of lack of food (1 Macc 6.49); possibly this was the *secret information*. **23**:

Philip, see 9.29 n.; 1 Macc 6.14–15, 55–56. **24**: *Gerar*, south of Gaza on the coastal plain. **26**: 1 Macc 6.63.

14.1–10: **Accession of Demetrius I** (1 Macc 7.1–7). **1**: *Three years later*, about 161 B.C. *Demetrius* I Soter, *son of Seleucus IV*, reigned 162–150. *Tripolis*, see 1 Macc 7.1 n. **2**: *Antiochus . . . Lysias*, 1 Macc 7.3–4. **3**: *Alcimus* may not have *been high priest. Defiled*, 4.11–15. **4**: 1 Macc 7.5–7 may record an earlier visit. *Crown*, emblem of sovereignty; the *palm*, victory. **6**: *Hasideans*, see 1 Macc 2.42 n. **7**: *Ancestral glory*, he claimed legitimate succession.

Hasideans, whose leader is Judas Macca-
bous, are keeping up war and stirring up
sedition, and will not let the kingdom
attain tranquility. [7] Therefore I have laid
aside my ancestral glory—I mean the
high priesthood—and have now come
here, [8] first because I am genuinely concerned for the interests of the king, and
second because I have regard also for my
compatriots. For through the folly of
those whom I have mentioned our whole
nation is now in no small misfortune.
[9] Since you are acquainted, O king, with
the details of this matter, may it please
you to take thought for our country and
our hard-pressed nation with the gracious kindness that you show to all.
[10] For as long as Judas lives, it is impossible for the government to find peace."
[11] When he had said this, the rest of the
king's Friends, [x] who were hostile to Judas, quickly inflamed Demetrius still
more. [12] He immediately chose Nicanor,
who had been in command of the elephants, appointed him governor of Judea, and sent him off [13] with orders to kill
Judas and scatter his troops, and to install
Alcimus as high priest of the great [y] temple. [14] And the Gentiles throughout Judea, who had fled before [z] Judas, flocked
to join Nicanor, thinking that the misfortunes and calamities of the Jews
would mean prosperity for themselves.
[15] When the Jews [a] heard of Nicanor's coming and the gathering of the
Gentiles, they sprinkled dust on their
heads and prayed to him who established
his own people forever and always upholds his own heritage by manifesting
himself. [16] At the command of the leader,
they [b] set out from there immediately
and engaged them in battle at a village
called Dessau. [z] [17] Simon, the brother of

Judas, had encountered Nicanor, but had
been temporarily [c] checked because of
the sudden consternation created by the
enemy.
[18] Nevertheless Nicanor, hearing of
the valor of Judas and his troops and their
courage in battle for their country,
shrank from deciding the issue by bloodshed. [19] Therefore he sent Posidonius,
Theodotus, and Mattathias to give and
receive pledges of friendship. [20] When the
terms had been fully considered, and the
leader had informed the people, and it
had appeared that they were of one
mind, they agreed to the covenant.
[21] The leaders [d] set a day on which to
meet by themselves. A chariot came forward from each army; seats of honor
were set in place; [22] Judas posted armed
men in readiness at key places to prevent
sudden treachery on the part of the enemy; so they duly held the consultation.
[23] Nicanor stayed on in Jerusalem
and did nothing out of the way, but dismissed the flocks of people that had gathered. [24] And he kept Judas always in his
presence; he was warmly attached to the
man. [25] He urged him to marry and have
children; so Judas [b] married, settled
down, and shared the common life.
[26] But when Alcimus noticed their
goodwill for one another, he took the
covenant that had been made and went to
Demetrius. He told him that Nicanor
was disloyal to the government, since he
had appointed that conspirator against
the kingdom, Judas, to be his successor.
[27] The king became excited and, provoked by the false accusations of that de-

x Gk of the Friends
y Gk greatest   z Meaning of Gk uncertain
a Gk they   b Gk he   c Other ancient
authorities read slowly   d Gk They

14.11–14: **Appointment of Nicanor and
Alcimus.** This story omits the expedition of
Bacchides (1 Macc 7.8–25). Josephus says
that *Nicanor* had escaped from Rome with
Demetrius (*Antiquities*, XII. x. 4); if he is the
person in 8.9–36 he must have gone from
Syria to Rome to assist the escape.
14.15–36: **Nicanor seeks friendship
with Judas. 15:** *Sprinkled dust,* Josh 7.6. **16:**

The leader, Judas, or possibly Nicanor. *Dessau,* perhaps Adasa (1 Macc 7.40–45). **20–21:**
The leader, Nicanor. *The people,* his army.
Afterward *the leaders,* Nicanor and Judas,
met. **22:** 1 Macc 7.12–18. **24:** *Warmly attached*
only so long as things went well (compare
vv. 31–33).
14.26: *Alcimus* failed to get civil power and
feared that *Judas* would be made his *successor*

praved man, wrote to Nicanor, stating that he was displeased with the covenant and commanding him to send Maccabeus to Antioch as a prisoner without delay.

28 When this message came to Nicanor, he was troubled and grieved that he had to annul their agreement when the man had done no wrong. 29 Since it was not possible to oppose the king, he watched for an opportunity to accomplish this by a stratagem. 30 But Maccabeus, noticing that Nicanor was more austere in his dealings with him and was meeting him more rudely than had been his custom, concluded that this austerity did not spring from the best motives. So he gathered not a few of his men, and went into hiding from Nicanor. 31 When the latter became aware that he had been cleverly outwitted by the man, he went to the great*e* and holy temple while the priests were offering the customary sacrifices, and commanded them to hand the man over. 32 When they declared on oath that they did not know where the man was whom he wanted, 33 he stretched out his right hand toward the sanctuary, and swore this oath: "If you do not hand Judas over to me as a prisoner, I will level this shrine of God to the ground and tear down the altar, and build here a splendid temple to Dionysus."

34 Having said this, he went away. Then the priests stretched out their hands toward heaven and called upon the constant Defender of our nation, in these words: 35 "O Lord of all, though you have need of nothing, you were pleased that there should be a temple for your habitation among us; 36 so now, O holy One, Lord of all holiness, keep undefiled forever this house that has been so recently purified."

37 A certain Razis, one of the elders of Jerusalem, was denounced to Nicanor as a man who loved his compatriots and was very well thought of and for his goodwill was called father of the Jews. 38 In former times, when there was no mingling with the Gentiles, he had been accused of Judaism, and he had most zealously risked body and life for Judaism. 39 Nicanor, wishing to exhibit the enmity that he had for the Jews, sent more than five hundred soldiers to arrest him; 40 for he thought that by arresting*f* him he would do them an injury. 41 When the troops were about to capture the tower and were forcing the door of the courtyard, they ordered that fire be brought and the doors burned. Being surrounded, Razis*g* fell upon his own sword, 42 preferring to die nobly rather than to fall into the hands of sinners and suffer outrages unworthy of his noble birth. 43 But in the heat of the struggle he did not hit exactly, and the crowd was now rushing in through the doors. He courageously ran up on the wall, and bravely threw himself down into the crowd. 44 But as they quickly drew back, a space opened and he fell in the middle of the empty space. 45 Still alive and aflame with anger, he rose, and though his blood gushed forth and his wounds were severe he ran through the crowd; and standing upon a steep rock, 46 with his blood now completely drained from him, he tore out his entrails, took them in both hands and hurled them at the crowd, calling upon the Lord of life and spirit to give them back to him again. This was the manner of his death.

15 When Nicanor heard that Judas and his troops were in the region of Samaria, he made plans to attack them with complete safety on the day of rest. 2 When the Jews who were compelled to follow him said, "Do not destroy so sav-

---

e Gk *greatest* *f* Meaning of Gk uncertain *g* Gk *he*

---

as high priest. **33**: *Stretched out his right hand . . . and swore*, 15.32–33. *Dionysus*, 6.7 n. **35–36**: 1 Kings 8.27–30. *Purified*, 10.1–8.

14.37–46: Death of Razis. A martyrology in the style of 6.18–7.42. **37**: *Elders*, 13.13.

42: See 1 Sam 31.4. **46**: He expected his body to be restored in the resurrection (7.11).

15.1–36: Death of Nicanor (1 Macc 7.39–50). **1**: *Nicanor camped at Beth-horon, and Judas was at Adasa, between Beth-horon*

agely and barbarously, but show respect for the day that he who sees all things has honored and hallowed above other days," ³ the thrice-accursed wretch asked if there were a sovereign in heaven who had commanded the keeping of the sabbath day. ⁴ When they declared, "It is the living Lord himself, the Sovereign in heaven, who ordered us to observe the seventh day," ⁵ he replied, "But I am a sovereign also, on earth, and I command you to take up arms and finish the king's business." Nevertheless, he did not succeed in carrying out his abominable design.

6 This Nicanor in his utter boastfulness and arrogance had determined to erect a public monument of victory over Judas and his forces. ⁷ But Maccabeus did not cease to trust with all confidence that he would get help from the Lord. ⁸ He exhorted his troops not to fear the attack of the Gentiles, but to keep in mind the former times when help had come to them from heaven, and so to look for the victory that the Almighty would give them. ⁹ Encouraging them from the law and the prophets, and reminding them also of the struggles they had won, he made them the more eager. ¹⁰ When he had aroused their courage, he issued his orders, at the same time pointing out the perfidy of the Gentiles and their violation of oaths. ¹¹ He armed each of them not so much with confidence in shields and spears as with the inspiration of brave words, and he cheered them all by relating a dream, a sort of vision, ʰ which was worthy of belief.

12 What he saw was this: Onias, who had been high priest, a noble and good man, of modest bearing and gentle manner, one who spoke fittingly and had been trained from childhood in all that belongs to excellence, was praying with outstretched hands for the whole body of the Jews. ¹³ Then in the same fashion another appeared, distinguished by his gray hair and dignity, and of marvelous majesty and authority. ¹⁴ And Onias spoke, saying, "This is a man who loves the family of Israel and prays much for the people and the holy city—Jeremiah, the prophet of God." ¹⁵ Jeremiah stretched out his right hand and gave to Judas a golden sword, and as he gave it he addressed him thus: ¹⁶ "Take this holy sword, a gift from God, with which you will strike down your adversaries."

17 Encouraged by the words of Judas, so noble and so effective in arousing valor and awaking courage in the souls of the young, they determined not to carry on a campaignⁱ but to attack bravely, and to decide the matter by fighting hand to hand with all courage, because the city and the sanctuary and the temple were in danger. ¹⁸ Their concern for wives and children, and also for brothers and sistersʲ and relatives, lay upon them less heavily; their greatest and first fear was for the consecrated sanctuary. ¹⁹ And those who had to remain in the city were in no little distress, being anxious over the encounter in the open country.

20 When all were now looking forward to the coming issue, and the enemy was already close at hand with their army drawn up for battle, the elephantsᵏ strategically stationed and the cavalry deployed on the flanks, ²¹ Maccabeus, observing the masses that were in front of him and the varied supply of arms and the savagery of the elephants, stretched out his hands toward heaven and called

h Meaning of Gk uncertain   i Or *to remain in camp*   j Gk *for brothers*   k Gk *animals*

---

and Jerusalem. 2: *The Jews* in Nicanor's army wished to honor the sabbath. 3: *Thrice-accursed*, 8.34. 4–5: Ex 20.8–11; Dan 3.16–18. 8: 1 Macc 7.41. 9: *The law and the prophets* were now regarded as scripture (compare the Prologue to Sirach); not all the other books had been collected. 10: *Violation of oaths*, 11.27–32; 14.20–28.

15.12: *Onias*, 3.1–40. 15–16: The *golden sword* was a sign that God approved the Jews' self-defense on the sabbath. 18: *First fear*, compare 14.33.
15.20: *Elephants*, to break through the Jewish infantry; the *cavalry* protected the *flanks* of the Syrian infantry. 22–23: See 11.6 n.; 2 Kings 19.35.

upon the Lord who works wonders; for he knew that it is not by arms, but as the Lord[1] decides, that he gains the victory for those who deserve it. 22 He called upon him in these words: "O Lord, you sent your angel in the time of King Hezekiah of Judea, and he killed fully one hundred eighty-five thousand in the camp of Sennacherib. 23 So now, O Sovereign of the heavens, send a good angel to spread terror and trembling before us. 24 By the might of your arm may these blasphemers who come against your holy people be struck down." With these words he ended his prayer.

25 Nicanor and his troops advanced with trumpets and battle songs, 26 but Judas and his troops met the enemy in battle with invocations to God and prayers. 27 So, fighting with their hands and praying to God in their hearts, they laid low at least thirty-five thousand, and were greatly gladdened by God's manifestation.

28 When the action was over and they were returning with joy, they recognized Nicanor, lying dead, in full armor. 29 Then there was shouting and tumult, and they blessed the Sovereign Lord in the language of their ancestors. 30 Then the man who was ever in body and soul the defender of his people, the man who maintained his youthful goodwill toward his compatriots, ordered them to cut off Nicanor's head and arm and carry them to Jerusalem. 31 When he arrived there and had called his compatriots together and stationed the priests before the altar, he sent for those who were in the citadel. 32 He showed them the vile Nicanor's head and that profane man's arm, which had been boastfully stretched out against the holy house of the Almighty. 33 He cut out the tongue of the ungodly Nicanor and said that he would feed it piecemeal to the birds and would hang up these rewards of his folly opposite the sanctuary. 34 And they all, looking to heaven, blessed the Lord who had manifested himself, saying, "Blessed is he who has kept his own place undefiled!" 35 Judas[m] hung Nicanor's head from the citadel, a clear and conspicuous sign to everyone of the help of the Lord. 36 And they all decreed by public vote never to let this day go unobserved, but to celebrate the thirteenth day of the twelfth month—which is called Adar in the Aramaic language—the day before Mordecai's day.

37 This, then, is how matters turned out with Nicanor, and from that time the city has been in the possession of the Hebrews. So I will here end my story.

38 If it is well told and to the point, that is what I myself desired; if it is poorly done and mediocre, that was the best I could do. 39 For just as it is harmful to drink wine alone, or, again, to drink water alone, while wine mixed with water is sweet and delicious and enhances one's enjoyment, so also the style of the story delights the ears of those who read the work. And here will be the end.

l Gk *he*   *m* Gk *He*

15.29: *Language of their ancestors,* Hebrew. Palestinian Jews spoke Aramaic, but formal prayer, using the language of Scripture, was often in Hebrew. **30**: 1 Sam 17.54. **31**: *The citadel* on the Ophel hill was held by Syrians (1 Macc 1.33; 6.18); but the Jews had built another fort (1 Macc 4.60). **32**: *Head . . . arm,* 14.33. **35**: 1 Sam 31.9; Jdt 14.1; 1 Macc 7.47. **36**: The *twelfth month* (February-March) was *called Adar in the Aramaic language* and also in Hebrew. If there was but one month of Adar in this year (probably 161 B.C.), it was *the day*

before Mordecai's day, but in some years a second month of Adar was intercalated to harmonize the calendar. Nicanor's day was observed up to A.D. 70.
15.37–39: **Conclusion.** The epitomist wrote some time before the Jewish war (A.D. 66-70), when Jerusalem was still in Jewish hands. **39**: The strong wine of Greece was usually tempered with water. *The style* will delight *the ears of those who read,* because in antiquity it was the custom to read literary works aloud, even to oneself.

(b) The books from 1 Esdras through 3 Maccabees are recognized as Deuterocanonical Scripture by the Greek and the Russian Churches. They are not so recognized by the Roman Catholic Church, but 1 Esdras and the Prayer of Manasseh (together with 2 Esdras) are placed in an appendix to the Latin Vulgate Bible.

1 Esdras

The book that is known in the Apocrypha as 1 Esdras is called 3 Esdras in the Latin Vulgate Bible, where it is now placed (since the Council of Trent) in an appendix after the New Testament. None of the other apocryphal books is so intimately connected with the Old Testament. Beginning somewhat abruptly with a description of the great passover held by King Josiah in Jerusalem (about 621 B.C.), the book reproduces the substance of 2 Chr 35.1–36.23, the whole of Ezra, and Neh 7.38–8.12, breaking off in the middle of a sentence after an account of Ezra's reforms (about 458 B.C.). There are numerous minor discrepancies between the apocryphal and canonical accounts, including a rearrangement of the materials, and the story of the three young men in the court of Darius (3.1–5.6) has no parallel in the Old Testament.

The origin of the work is debated. Is it an earlier form of the Greek translation of biblical Ezra, with the Ezra materials found in Nehemiah (7.38–9.38) partially included? Some interpreters believe that to be the case. Several biblical books are known to have had more than a single Hebrew edition in post-exilic Israel, and 1 Esdras may be a translation of an alternative collection of Ezra memoirs, plus lists and other materials, and including the story of the three young men at the court of Darius.

Or is the book an apocryphal work, a translation of a later Hebrew/Aramaic version of the Ezra story, belonging to the late second century B.C., designed to stress the importance of Josiah, Zerubbabel, and Ezra in the establishment of temple worship and fidelity to the Torah? Both views have their supporters, but the former view is the more probable. The date of 1 Esdras in this Greek translation is probably not later than 100 B.C., since the work in its Greek form was used by Josephus in his *Antiquities of the Jewish People* (A.D. 93–94). See the Introductions to 1 Chronicles and Ezra for the origin and purpose of Ezra.

1 Josiah kept the passover to his Lord in Jerusalem; he killed the passover lamb on the fourteenth day of the first month, [2]having placed the priests according to their divisions, arrayed in their vestments, in the temple of the Lord. [3]He told the Levites, the temple servants of Israel, that they should sanctify themselves to the Lord and put the holy ark of the Lord in the house that King Solomon, son of David, had built; [4]and he said, "You need no longer carry it on your shoulders. Now worship the Lord your God and serve his people Israel; prepare yourselves by your families and kindred, [5]in accordance with the directions of King David of Israel and the magnificence of his son Solomon. Stand in order in the temple according to the groupings of the ancestral houses of you Levites, who minister before your kindred the people of Israel, [6]and kill the passover lamb and prepare the sacrifices for your kindred, and keep the passover according to the commandment of the Lord that was given to Moses."

[7] To the people who were present Josiah gave thirty thousand lambs and kids, and three thousand calves; these were given from the king's possessions, as he promised, to the people and the priests and Levites. [8]Hilkiah, Zechariah, and Jehiel,*a* the chief officers of the temple, gave to the priests for the passover two thousand six hundred sheep and three hundred calves. [9]And Jeconiah and Shemaiah and his brother Nethanel, and Hashabiah and Ochiel and Joram, captains over thousands, gave the Levites for the passover five thousand sheep and seven hundred calves.

[10] This is what took place. The priests and the Levites, having the unleavened bread, stood in proper order according to kindred [11]and the grouping of the ancestral houses, before the people, to make the offering to the Lord as it is written in the book of Moses; this they did in the morning. [12]They roasted the passover lamb with fire, as required; and they boiled the sacrifices in bronze pots and caldrons, with a pleasing odor, [13]and carried them to all the people. Afterward they prepared the passover for themselves and for their kindred the priests, the sons of Aaron, [14]because the priests were offering the fat until nightfall; so the Levites prepared it for themselves and for their kindred the priests, the sons of Aaron. [15]The temple singers, the sons of Asaph, were in their place according to the arrangement made by David, and also Asaph, Zechariah, and Eddinus, who represented the king. [16]The gatekeepers were at each gate; no one needed to interrupt his daily duties, for their kindred the Levites prepared the passover for them.

[17] So the things that had to do with the sacrifices to the Lord were accomplished that day: the passover was kept [18]and the sacrifices were offered on the altar of the Lord, according to the command of King Josiah. [19]And the people of Israel who were present at that time kept the passover and the festival of unleavened bread seven days. [20]No passover like it had been kept in Israel since the times of the prophet Samuel; [21]none of the kings of Israel had kept such a passover as was kept by Josiah and the priests and Levites and the people of Judah and all of Israel who were living in Jerusalem. [22]In the eighteenth year of the reign of Josiah this passover was kept.

[23] And the deeds of Josiah were up-

a Gk *Esyelus*

1.1–33: Josiah's passover; his effort to intercept the Egyptians at Megiddo, and his death (2 Chr 35.1–27). According to 2 Kings 23.21–23 the passover celebration concluded Josiah's religious reform, the account of which is omitted in 1 Esdras. **7–9:** The list of offerings differs slightly from that in 2 Chr 35.7–9.

1.15: *Zechariah* and *Eddinus* appear in 2 Chr 35.15 as Heman and Jeduthun. **1.17–33:** 1 Esdras follows the account in 2 Chr 35.16–27 faithfully, apart from the omission of the Pharaoh's name, Josiah's disguising himself, and his being struck by an arrow.

right in the sight of the Lord, for his heart was full of godliness. 24 In ancient times the events of his reign have been recorded—concerning those who sinned and acted wickedly toward the Lord beyond any other people or kingdom, and how they grieved the Lord*b* deeply, so that the words of the Lord fell upon Israel.

25 After all these acts of Josiah, it happened that Pharaoh, king of Egypt, went to make war at Carchemish on the Euphrates, and Josiah went out against him. 26 And the king of Egypt sent word to him saying, "What have we to do with each other, O king of Judea? 27 I was not sent against you by the Lord God, for my war is at the Euphrates. And now the Lord is with me! The Lord is with me, urging me on! Stand aside, and do not oppose the Lord."

28 Josiah, however, did not turn back to his chariot, but tried to fight with him, and did not heed the words of the prophet Jeremiah from the mouth of the Lord. 29 He joined battle with him in the plain of Megiddo, and the commanders came down against King Josiah. 30 The king said to his servants, "Take me away from the battle, for I am very weak." And immediately his servants took him out of the line of battle. 31 He got into his second chariot; and after he was brought back to Jerusalem he died, and was buried in the tomb of his ancestors.

32 In all Judea they mourned for Josiah. The prophet Jeremiah lamented for Josiah, and the principal men, with the women,*c* have made lamentation for him to this day; it was ordained that this should always be done throughout the whole nation of Israel. 33 These things are written in the book of the histories of the kings of Judea; and every one of the acts of Josiah, and his splendor, and his understanding of the law of the Lord, and the things that he had done before, and these that are now told, are recorded in the book of the kings of Israel and Judah.

34 The men of the nation took Jeconiah*d* son of Josiah, who was twenty-three years old, and made him king in succession to his father Josiah. 35 He reigned three months in Judah and Jerusalem. Then the king of Egypt deposed him from reigning in Jerusalem, 36 and fined the nation one hundred talents of silver and one talent of gold. 37 The king of Egypt made his brother Jehoiakim king of Judea and Jerusalem. 38 Jehoiakim put the nobles in prison, and seized his brother Zarius and brought him back from Egypt.

39 Jehoiakim was twenty-five years old when he began to reign in Judea and Jerusalem; he did what was evil in the sight of the Lord. 40 King Nebuchadnezzar of Babylon came up against him; he bound him with a chain of bronze and took him away to Babylon. 41 Nebuchadnezzar also took some holy vessels of the Lord, and carried them away, and stored them in his temple in Babylon. 42 But the things that are reported about Jehoiakim,*b* and his uncleanness and impiety, are written in the annals of the kings.

43 His son Jehoiachin*e* became king in his place; when he was made king he

b Gk *him*   *c* Or *their wives*   *d* 2 Kings
23.30; 2 Chr 36.1 *Jehoahaz*   *e* Gk *Jehoiakim*

1.34–58: The last kings of Judah; Jerusalem's fall to the Babylonians (2 Chr 36.1–21). **34:** The expression *men of the nation* corresponds to the "people of the land" of 2 Chr 36.1; in pre-exilic times these were conservative landowners who often came to the support of reforming kings or who themselves instituted reforms (2 Kings 12.18, 20; 21.24; 23.30). *Jeconiah,* also called Jehoahaz (2 Kings 23.30–31; 2 Chr 36.1–2) and Shallum (Jer 22.11). **38:** The author has misunderstood 2 Chr 36.4. Neco of Egypt removed

Jehoahaz from the throne and installed Josiah's elder son Eliakim as king, changing his name to Jehoiakim. Jehoahaz was taken to Egypt where presumably he died (Jer 22.10–12). 1 Esdras has *Jehoiakim* bring up his brother *Zarius* from Egypt; the name *Zarius* is apparently an orthographic corruption (through confusion of the Hebrew letters *d* and *r*) of Zedekiah, who was a brother of Jehoiakim (2 Kings 24.17). **39:** 1 Esdras omits the length of Jehoiakim's reign, which was eleven years (2 Chr 36.5).

was eighteen years old, 44 and he reigned three months and ten days in Jerusalem. He did what was evil in the sight of the Lord. 45 A year later Nebuchadnezzar sent and removed him to Babylon, with the holy vessels of the Lord, 46 and made Zedekiah king of Judea and Jerusalem.

Zedekiah was twenty-one years old, and he reigned eleven years. 47 He also did what was evil in the sight of the Lord, and did not heed the words that were spoken by the prophet Jeremiah from the mouth of the Lord. 48 Although King Nebuchadnezzar had made him swear by the name of the Lord, he broke his oath and rebelled; he stiffened his neck and hardened his heart and transgressed the laws of the Lord, the God of Israel. 49 Even the leaders of the people and of the priests committed many acts of sacrilege and lawlessness beyond all the unclean deeds of all the nations, and polluted the temple of the Lord in Jerusalem—the temple that God had made holy. 50 The God of their ancestors sent his messenger to call them back, because he would have spared them and his dwelling place. 51 But they mocked his messengers, and whenever the Lord spoke, they scoffed at his prophets, 52 until in his anger against his people because of their ungodly acts he gave command to bring against them the kings of the Chaldeans. 53 These killed their young men with the sword around their holy temple, and did not spare young man or young woman,*f* old man or child, for he gave them all into their hands. 54 They took all the holy vessels of the Lord, great and small, the treasure chests of the Lord, and the royal stores, and carried them away to Babylon. 55 They burned the house of the Lord, broke down the

walls of Jerusalem, burned their towers with fire, 56 and utterly destroyed all its glorious things. The survivors he led away to Babylon with the sword, 57 and they were servants to him and to his sons until the Persians began to reign, in fulfillment of the word of the Lord by the mouth of Jeremiah, 58 saying, "Until the land has enjoyed its sabbaths, it shall keep sabbath all the time of its desolation until the completion of seventy years."

2 In the first year of Cyrus as king of the Persians, so that the word of the Lord by the mouth of Jeremiah might be accomplished— 2 the Lord stirred up the spirit of King Cyrus of the Persians, and he made a proclamation throughout all his kingdom and also put it in writing:

3 "Thus says Cyrus king of the Persians: The Lord of Israel, the Lord Most High, has made me king of the world, 4 and he has commanded me to build him a house at Jerusalem, which is in Judea. 5 If any of you, therefore, are of his people, may your Lord be with you; go up to Jerusalem, which is in Judea, and build the house of the Lord of Israel—he is the Lord who dwells in Jerusalem, 6 and let each of you, wherever you may live, be helped by the people of your place with gold and silver, 7 with gifts and with horses and cattle, besides the other things added as votive offerings for the temple of the Lord that is in Jerusalem."

8 Then arose the heads of families of the tribes of Judah and Benjamin, and the priests and the Levites, and all whose spirit the Lord had stirred to go up to build the house in Jerusalem for the Lord; 9 their neighbors helped them with everything, with silver and gold, with

f Gk virgin

1.43: 1 Esdras mistakenly gives Jehoiakim (see note *e*) as the name of that king's son and successor, rather than Jehoiachin; but the king's age at the beginning of his reign is correctly given (*eighteen years;* not eight, as in 2 Chr 36.9).
1.58: To *keep sabbath* means that the land is to continue in a state of "sabbath" rest (i.e. to lie untended as in the seventh or sabbatical

years) until the exiles return (Jer 25.11–12; 29.10; compare Lev 25.1–7; 26.27–39).
2.1–15: Cyrus of Persia permits the exiles to return (Ezra 1.1–11). The text is virtually identical with that in Ezra, although the inventory of the sacred vessels appears to be better preserved in 1 Esdras than in Ezra. **1:** *First year of Cyrus,* i.e. 538 B.C.

horses and cattle, and with a very great number of votive offerings from many whose hearts were stirred.

10 King Cyrus also brought out the holy vessels of the Lord that Nebuchadnezzar had carried away from Jerusalem and stored in his temple of idols. 11 When King Cyrus of the Persians brought these out, he gave them to Mithridates, his treasurer, 12 and by him they were given to Sheshbazzar,*g* the governor of Judea. 13 The number of these was: one thousand gold cups, one thousand silver cups, twenty-nine silver censers, thirty gold bowls, two thousand four hundred ten silver bowls, and one thousand other vessels. 14 All the vessels were handed over, gold and silver, five thousand four hundred sixty-nine, 15 and they were carried back by Sheshbazzar with the returning exiles from Babylon to Jerusalem.

16 In the time of King Artaxerxes of the Persians, Bishlam, Mithridates, Tabeel, Rehum, Beltethmus, the scribe Shimshai, and the rest of their associates, living in Samaria and other places, wrote him the following letter, against those who were living in Judea and Jerusalem:

17 "To King Artaxerxes our lord, your servants the recorder Rehum and the scribe Shimshai and the other members of their council, and the judges in Coelesyria and Phoenicia: 18 Let it now be known to our lord the king that the Jews who came up from you to us have gone to Jerusalem and are building that rebellious and wicked city, repairing its

market places and walls and laying the foundations for a temple. 19 Now if this city is built and the walls finished, they will not only refuse to pay tribute but will even resist kings. 20 Since the building of the temple is now going on, we think it best not to neglect such a matter, 21 but to speak to our lord the king, in order that, if it seems good to you, search may be made in the records of your ancestors. 22 You will find in the annals what has been written about them, and will learn that this city was rebellious, troubling both kings and other cities, 23 and that the Jews were rebels and kept setting up blockades in it from of old. That is why this city was laid waste. 24 Therefore we now make known to you, O lord and king, that if this city is built and its walls finished, you will no longer have access to Coelesyria and Phoenicia."

25 Then the king, in reply to the recorder Rehum, Beltethmus, the scribe Shimshai, and the others associated with them and living in Samaria and Syria and Phoenicia, wrote as follows:

26 "I have read the letter that you sent me. So I ordered search to be made, and it has been found that this city from of old has fought against kings, 27 that the people in it were given to rebellion and war, and that mighty and cruel kings ruled in Jerusalem and exacted tribute from Coelesyria and Phoenicia. 28 Therefore I have now issued orders to prevent

g Gk *Sanabassaros*

2.16–30: **Opposition to the rebuilding of the temple and the city walls** (Ezra 4.7–24). A misplaced account of opposition to rebuilding the walls of Jerusalem in the time of Artaxerxes I (464–424 B.C.). Cyrus was succeeded by Cambyses (529–521), who was followed by Darius I (521–485). Josephus (*Ant.* XI. ii.1–3) substitutes Cambyses for Artaxerxes, thus providing the correct sequence of Persian kings. The original location of the passage was probably between Ezra ch 10 and Neh ch 1. **16:** The name *Beltethmus* is a Greek transliteration of the Aramaic title of the office held by *Rehum;* the same mistake occurs in v. 25. **17:** The persons named are

officials of the Persian province called "Beyond the River" (Ezra 4.10), which included the lands of Syria, Phoenicia, and Palestine. *Rehum* is designated "the commander" in Ezra 4.8f.; the translation *recorder* is supported by Josephus.

2.20: The account differs considerably from that in Ezra 4.14, which contains no reference to the rebuilding of the *temple* at this point (compare Ezra 4.24, however, where work on the *temple* is said to have stopped).

2.25: *Rehum,* the governor is again identified as the *recorder* rather than as the commander of Persian forces in Samaria (Ezra 4.17). As in v. 16 the name *Beltethmus* is a

these people from building the city and to take care that nothing more be done 29 and that such wicked proceedings go no further to the annoyance of kings."

30 Then, when the letter from King Artaxerxes was read, Rehum and the scribe Shimshai and their associates went quickly to Jerusalem, with cavalry and a large number of armed troops, and began to hinder the builders. And the building of the temple in Jerusalem stopped until the second year of the reign of King Darius of the Persians.

3 Now King Darius gave a great banquet for all that were under him, all that were born in his house, and all the nobles of Media and Persia, 2 and all the satraps and generals and governors that were under him in the hundred twenty-seven satrapies from India to Ethiopia. 3 They ate and drank, and when they were satisfied they went away, and King Darius went to his bedroom; he went to sleep, but woke up again.

4 Then the three young men of the bodyguard, who kept guard over the person of the king, said to one another, 5 "Let each of us state what one thing is strongest; and to the one whose statement seems wisest, King Darius will give rich gifts and great honors of victo-

ry. 6 He shall be clothed in purple, and drink from gold cups, and sleep on a gold bed, *h* and have a chariot with gold bridles, and a turban of fine linen, and a necklace around his neck; 7 and because of his wisdom he shall sit next to Darius and shall be called Kinsman of Darius."

8 Then each wrote his own statement, and they sealed them and put them under the pillow of King Darius, 9 and said, "When the king wakes, they will give him the writing; and to the one whose statement the king and the three nobles of Persia judge to be wisest the victory shall be given according to what is written." 10 The first wrote, "Wine is strongest." 11 The second wrote, "The king is strongest." 12 The third wrote, "Women are strongest, but above all things truth is victor." *i*

13 When the king awoke, they took the writing and gave it to him, and he read it. 14 Then he sent and summoned all the nobles of Persia and Media and the satraps and generals and governors and prefects, 15 and he took his seat in the council chamber, and the writing was read in their presence. 16 He said, "Call

h Gk *on gold*   *i* Or *but truth is victor over all things*

---

transliteration of the Aramaic title of *Rehum* and is not the name of a third addressee. **30:** An erroneous reference (as also in Ezra 4.24) to the halting of work on the *temple*.

3.1–5.6: The three young bodyguards in the court of Darius. This famous story, found only in 1 Esdras among the several works attributed to Ezra, provides sufficient reason for the preservation of the book throughout the centuries. The story probably originated outside the Jewish community as a popular tale praising the relative strength of wine, kings, and women (the original order was perhaps kings, wine, and women). The praise of the strength of truth (4.33–41; compare 3.12) was added later in the transmission of the story, perhaps by a Greek-speaking editor (this part of the story has close parallels to Greek thought and literature). The author of 1 Esdras, adopting the story, needed only to identify the third youth with Zerubbabel (4.13) and to add a sequel to the tale, relating how Darius rewarded Zerubbabel by sup-

porting the rebuilding of Jerusalem and its temple (4.42–5.6). The version of the story found in Josephus (*Ant.* xi. iii.2–9) differs from the one given here in several particulars.

3.1–17a: The contest planned. 1–3: Apparently Darius's banquet was held at Susa, though the location is not explicitly mentioned. **2:** During Darius's reign (521–485 B.C.) there were actually only about twenty provinces (*satrapies*); this number was increased during Seleucid times (after 312 B.C.), and the total one *hundred twenty-seven* became conventional in later literature (Esth 1.1; Josephus, *Ant.* xi. iii.2). **4–12:** The three bodyguards decide upon a form of entertainment for the king that would bring riches and honor to one of them. According to Josephus (*Ant.* xi. iii.2) it was the king who proposed the contest.

3.13–17a: The entire court is assembled to hear the guardsmen defend their respective answers; such a scene is entirely consonant with court practices in the ancient world.

the young men, and they shall explain their statements." So they were summoned, and came in. 17They said to them, "Explain to us what you have written."

Then the first, who had spoken of the strength of wine, began and said: 18"Gentlemen, how is wine the strongest? It leads astray the minds of all who drink it. 19It makes equal the mind of the king and the orphan, of the slave and the free, of the poor and the rich. 20It turns every thought to feasting and mirth, and forgets all sorrow and debt. 21It makes all hearts feel rich, forgets kings and satraps, and makes everyone talk in millions.*j* 22When people drink they forget to be friendly with friends and kindred, and before long they draw their swords. 23And when they recover from the wine, they do not remember what they have done. 24Gentlemen, is not wine the strongest, since it forces people to do these things?" When he had said this, he stopped speaking.

4 Then the second, who had spoken of the strength of the king, began to speak: 2"Gentlemen, are not men strongest, who rule over land and sea and all that is in them? 3But the king is stronger; he is their lord and master, and whatever he says to them they obey. 4If he tells them to make war on one another, they do it; and if he sends them out against the enemy, they go, and conquer mountains, walls, and towers. 5They kill and are killed, and do not disobey the king's command; if they win the victory, they bring everything to the king—whatever spoil they take and everything else. 6Likewise those who do not serve in the army or make war but till the soil; when-

ever they sow and reap, and bring some to the king; and they compel one another to pay taxes to the king. 7And yet he is only one man! If he tells them to kill, they kill; if he tells them to release, they release; 8if he tells them to attack, they attack; if he tells them to lay waste, they lay waste; if he tells them to build, they build; 9if he tells them to cut down, they cut down; if he tells them to plant, they plant. 10All his people and his armies obey him. Furthermore, he reclines, he eats and drinks and sleeps, 11but they keep watch around him, and no one may go away to attend to his own affairs, nor do they disobey him. 12Gentlemen, why is not the king the strongest, since he is to be obeyed in this fashion?" And he stopped speaking.

13 Then the third, who had spoken of women and truth (and this was Zerubbabel), began to speak: 14"Gentlemen, is not the king great, and are not men many, and is not wine strong? Who is it, then, that rules them, or has the mastery over them? Is it not women? 15Women gave birth to the king and to every people that rules over sea and land. 16From women they came; and women brought up the very men who plant the vineyards from which comes wine. 17Women make men's clothes; they bring men glory; men cannot exist without women. 18If men gather gold and silver or any other beautiful thing, and then see a woman lovely in appearance and beauty, 19they let all those things go, and gape at her, and with open mouths stare at her, and all prefer her to gold or silver or any other beautiful thing. 20A man leaves his

j Gk talents

3.17b–24: **In praise of the strength of wine.** *Wine* is the great leveler in society; it takes away one's capacity for discernment and remembrance, overpowering king and commoner alike.

4.1–12: **In praise of the strength of kings.** The arbitrary power of oriental kings here portrayed is quite true to the actual situation in the ancient world. No polemic against kingship need be seen in the passage.

4.13–32: **In praise of the strength of women.** The third youth, identified for the first time as *Zerubbabel* (v. 13), depicts the strength of *women,* who give birth to kings, who receive from men the treasures won in warfare and heroic deeds, who can humiliate their masters, including kings, and yet are sought after and fawned upon by those whom they humiliate.

own father, who brought him up, and his own country, and clings to his wife. 21 With his wife he ends his days, with no thought of his father or his mother or his country. 22 Therefore you must realize that women rule over you!

"Do you not labor and toil, and bring everything and give it to women? 23 A man takes his sword, and goes out to travel and rob and steal and to sail the sea and rivers; 24 he faces lions, and he walks in darkness, and when he steals and robs and plunders, he brings it back to the woman he loves. 25 A man loves his wife more than his father or his mother. 26 Many men have lost their minds because of women, and have become slaves because of them. 27 Many have perished, or stumbled, or sinned because of women. 28 And now do you not believe me?

"Is not the king great in his power? Do not all lands fear to touch him? 29 Yet I have seen him with Apame, the king's concubine, the daughter of the illustrious Bartacus; she would sit at the king's right hand 30 and take the crown from the king's head and put it on her own, and slap the king with her left hand. 31 At this the king would gaze at her with mouth agape. If she smiles at him, he laughs; if she loses her temper with him, he flatters her, so that she may be reconciled to him. 32 Gentlemen, why are not women strong, since they do such things?"

33 Then the king and the nobles looked at one another; and he began to speak about truth: 34 "Gentlemen, are not

women strong? The earth is vast, and heaven is high, and the sun is swift in its course, for it makes the circuit of the heavens and returns to its place in one day. 35 Is not the one who does these things great? But truth is great, and stronger than all things. 36 The whole earth calls upon truth, and heaven blesses her. All God's works[k] quake and tremble, and with him there is nothing unrighteous. 37 Wine is unrighteous, the king is unrighteous, women are unrighteous, all human beings are unrighteous, all their works are unrighteous, and all such things. There is no truth in them and in their unrighteousness they will perish. 38 But truth endures and is strong forever, and lives and prevails forever and ever. 39 With it there is no partiality or preference, but it does what is righteous instead of anything that is unrighteous or wicked. Everyone approves its deeds, 40 and there is nothing unrighteous in its judgment. To it belongs the strength and the kingship and the power and the majesty of all the ages. Blessed be the God of truth!" 41 When he stopped speaking, all the people shouted and said, "Great is truth, and strongest of all!"

42 Then the king said to him, "Ask what you wish, even beyond what is written, and we will give it to you, for you have been found to be the wisest. You shall sit next to me, and be called my Kinsman." 43 Then he said to the king,

k Gk *All the works*

4.29: The king's concubine *Apame,* daughter of Bartacus, cannot be identified.
4.33–41: **In praise of the strength of truth.** The strength of *truth,* an addition to the original story probably made prior to the story's adaptation to the Jewish author's purpose, is portrayed in imagery akin to the depiction of truth in Greek literature. The Jewish adapter of the story may have modified the original somewhat to make truth more nearly akin to Hebraic ideas of truth (firmness, reliability). The closing references to truth suggest that it is virtually equivalent to the will of God: "Blessed be the God of truth!" (v. 40). The audience responds (v. 41) with the declaration "Great is truth, and strongest of all!" The Latin proverb "Magna est

veritas et praevalet" ("Great is truth, and it prevails") is the most famous line from the (Clementine) Vulgate text of 1 Esdras.
4.42–57: **Zerubbabel's reward.** Darius authorizes Zerubbabel to return to Jerusalem and rebuild the temple, with generous support from the Persian treasury. **43**: The historically improbable vow of Darius to rebuild Jerusalem and its temple upon his accession to the kingship is not otherwise attested; indeed, the author has already recounted Cyrus' proclamation authorizing the return of the exiles and the restoration of the temple vessels (2.1–15). **45**: *The Edomites* are credited with having burned the temple, contrary to 1.55 (see Ob 11–14).

"Remember the vow that you made on the day when you became king, to build Jerusalem, 44 and to send back all the vessels that were taken from Jerusalem, which Cyrus set apart when he began[l] to destroy Babylon, and vowed to send them back there. 45 You also vowed to build the temple, which the Edomites burned when Judea was laid waste by the Chaldeans. 46 And now, O lord the king, this is what I ask and request of you, and this befits your greatness. I pray therefore that you fulfill the vow whose fulfillment you vowed to the King of heaven with your own lips."

47 Then King Darius got up and kissed him, and wrote letters for him to all the treasurers and governors and generals and satraps, that they should give safe conduct to him and to all who were going up with him to build Jerusalem. 48 And he wrote letters to all the governors in Coelesyria and Phoenicia and to those in Lebanon, to bring cedar timber from Lebanon to Jerusalem, and to help him build the city. 49 He wrote in behalf of all the Jews who were going up from his kingdom to Judea, in the interest of their freedom, that no officer or satrap or governor or treasurer should forcibly enter their doors; 50 that all the country that they would occupy should be theirs without tribute; that the Idumeans should give up the villages of the Jews that they held; 51 that twenty talents a year should be given for the building of the temple until it was completed, 52 and an additional ten talents a year for burnt offerings to be offered on the altar every day, in accordance with the commandment to make seventeen offerings; 53 and that all who came from Babylonia to build the city should have their freedom, they and their children and all the priests who came. 54 He wrote also concerning their support and the priests' vestments in which[m] they were to minister. 55 He wrote that the support for the Levites should be provided until the day when the temple would be finished and Jerusalem built. 56 He wrote that land and wages should be provided for all who guarded the city. 57 And he sent back from Babylon all the vessels that Cyrus had set apart; everything that Cyrus had ordered to be done, he also commanded to be done and to be sent to Jerusalem.

58 When the young man went out, he lifted up his face to heaven toward Jerusalem, and praised the King of heaven, saying, 59 "From you comes the victory; from you comes wisdom, and yours is the glory. I am your servant. 60 Blessed are you, who have given me wisdom; I give you thanks, O Lord of our ancestors."

61 So he took the letters, and went to Babylon and told this to all his kindred. 62 And they praised the God of their ancestors, because he had given them release and permission 63 to go up and build Jerusalem and the temple that is called by his name; and they feasted, with music and rejoicing, for seven days.

l Cn: Gk *vowed*   *m* Gk *in what priestly vestments*

4.48–57: Darius magnificently supports the program outlined by Zerubbabel. The historical background is reflected more accurately in the decree issued by Darius after the governor of Samaria had complained about the rebuilding of the temple (Ezra 6.1–13; 1 Esd 6.23–34). The decree of Cyrus allowing the exiles to return and restore the temple and its cult (2.1–15) is no doubt historical, and Darius confirmed this decree (6.23–34); but Zerubbabel's return was hardly supported by Darius in the manner here portrayed.
4.58–60: Zerubbabel's prayer. The language of this prayer is similar to a prayer of Daniel (Dan 2.20–23) and may be dependent upon it.
4.61–5.6: Preparations for the return. Zerubbabel journeys (perhaps from Susa) to Babylon and there recruits leaders for the returning exiles (5.4–6). The list of the leaders is hopelessly confused. *Jeshua* (5.5) is clearly the leading priest, and *Zerubbabel* is the hero of the story, not his son *Joakim*. In Neh 12.10, 26, Joakim appears as the son of Jeshua; 1 Chr 3.17–24 gives a different genealogy for Zerubbabel (where he is said to be a grandson of Jehoiachin).

5 After this the heads of ancestral houses were chosen to go up, according to their tribes, with their wives and sons and daughters, and their male and female servants, and their livestock. 2 And Darius sent with them a thousand cavalry to take them back to Jerusalem in safety, with the music of drums and flutes; 3 all their kindred were making merry. And he made them go up with them.

4 These are the names of the men who went up, according to their ancestral houses in the tribes, over their groups: 5 the priests, the descendants of Phinehas son of Aaron; Jeshua son of Jozadak son of Seraiah and Joakim son of Zerubbabel son of Shealtiel, of the house of David, of the lineage of Phares, of the tribe of Judah, 6 who spoke wise words before King Darius of the Persians, in the second year of his reign, in the month of Nisan, the first month.

7 These are the Judeans who came up out of their sojourn in exile, whom King Nebuchadnezzar of Babylon had carried away to Babylon 8 and who returned to Jerusalem and the rest of Judea, each to his own town. They came with Zerubbabel and Jeshua, Nehemiah, Seraiah, Resaiah, Eneneus, Mordecai, Beelsarus, Aspharasus, Reeliah, Rehum, and Baanah, their leaders.

9 The number of those of the nation and their leaders: the descendants of Parosh, two thousand one hundred seventy-two. The descendants of Shephatiah, four hundred seventy-two. 10 The descendants of Arah, seven hundred fifty-six. 11 The descendants of Pahath-moab, of the descendants of Jeshua and Joab, two thousand eight hundred twelve. 12 The descendants of Elam, one thousand two hundred fifty-four. The descendants of Zattu, nine hundred

forty-five. The descendants of Chorbe, seven hundred five. The descendants of Bani, six hundred forty-eight. 13 The descendants of Bebai, six hundred twenty-three. The descendants of Azgad, one thousand three hundred twenty-two. 14 The descendants of Adonikam, six hundred sixty-seven. The descendants of Bigvai, two thousand sixty-six. The descendants of Adin, four hundred fifty-four. 15 The descendants of Ater, namely of Hezekiah, ninety-two. The descendants of Kilan and Azetas, sixty-seven. The descendants of Azaru, four hundred thirty-two. 16 The descendants of Annias, one hundred one. The descendants of Arom. The descendants of Bezai, three hundred twenty-three. The descendants of Arsiphurith, one hundred twelve. 17 The descendants of Baiterus, three thousand five. The descendants of Bethlomon, one hundred twenty-three. 18 Those from Netophah, fifty-five. Those from Anathoth, one hundred fifty-eight. Those from Bethasmoth, forty-two. 19 Those from Kiriatharim, twenty-five. Those from Chephirah and Beeroth, seven hundred forty-three. 20 The Chadiasans and Ammidians, four hundred twenty-two. Those from Kirama and Geba, six hundred twenty-one. 21 Those from Macalon, one hundred twenty-two. Those from Betolio, fifty-two. The descendants of Niphish, one hundred fifty-six. 22 The descendants of the other Calamolalus and Ono, seven hundred twenty-five. The descendants of Jerechus, three hundred forty-five. 23 The descendants of Senaah, three thousand three hundred thirty.

24 The priests: the descendants of Jedaiah son of Jeshua, of the descendants of Anasib, nine hundred seventy-two. The descendants of Immer, one thousand and fifty-two. 25 The descendants of Pashhur,

---

5.7–46: A list of the returning exiles (Ezra 2.1–70 and Neh 7.6–73a). The list in 1 Esdras differs from that in Ezra at many points, both as to names and numbers. The totals, however, are almost identical. The numbers of the priests and Levites are almost identical in the three lists, an indication that priestly and levitical genealogies were more carefully preserved than the other lists. **5.24–25:** Only four divisions of priests are given, while in 1 Chr ch 24 twenty-four divisions appear.

one thousand two hundred forty-seven. The descendants of Charme, one thousand seventeen.

26 The Levites: the descendants of Jeshua and Kadmiel and Bannas and Sudias, seventy-four. 27 The temple singers: the descendants of Asaph, one hundred twenty-eight. 28 The gatekeepers: the descendants of Shallum, the descendants of Ater, the descendants of Talmon, the descendants of Akkub, the descendants of Hatita, the descendants of Shobai, in all one hundred thirty-nine.

29 The temple servants: the descendants of Esau, the descendants of Hasupha, the descendants of Tabbaoth, the descendants of Keros, the descendants of Sua, the descendants of Padon, the descendants of Lebanah, the descendants of Hagabah, 30 the descendants of Akkub, the descendants of Uthai, the descendants of Ketab, the descendants of Hagab, the descendants of Subai, the descendants of Hana, the descendants of Cathua, the descendants of Geddur, 31 the descendants of Jairus, the descendants of Daisan, the descendants of Noeba, the descendants of Chezib, the descendants of Gazera, the descendants of Uzza, the descendants of Phinoe, the descendants of Hasrah, the descendants of Basthai, the descendants of Asnah, the descendants of Maani, the descendants of Nephisim, the descendants of Acuph,[n] the descendants of Hakupha, the descendants of Asur, the descendants of Pharakim, the descendants of Bazluth, 32 the descendants of Mehida, the descendants of Cutha, the descendants of Charea, the descendants of Barkos, the descendants of Serar, the descendants of Temah, the descendants of Neziah, the descendants of Hatipha.

33 The descendants of Solomon's servants: the descendants of Assaphioth, the descendants of Peruda, the descendants of Jaalah, the descendants of Lozon, the descendants of Isdael, the descendants of Shephatiah, 34 the descendants of Agia, the descendants of Pocherethhazzebaim, the descendants of Sarothie, the descendants of Masiah, the descendants of Gas, the descendants of Addus, the descendants of Subas, the descendants of Apherra, the descendants of Barodis, the descendants of Shaphat, the descendants of Allon.

35 All the temple servants and the descendants of Solomon's servants were three hundred seventy-two.

36 The following are those who came up from Tel-melah and Tel-harsha, under the leadership of Cherub, Addan, and Immer, 37 though they could not prove by their ancestral houses or lineage that they belonged to Israel: the descendants of Delaiah son of Tobiah, and the descendants of Nekoda, six hundred fifty-two.

38 Of the priests the following had assumed the priesthood but were not found registered: the descendants of Habaiah, the descendants of Hakkoz, and the descendants of Jaddus who had married Agia, one of the daughters of Barzillai, and was called by his name. 39 When a search was made in the register and the genealogy of these men was not found, they were excluded from serving as priests. 40 And Nehemiah and Attharias[o] told them not to share in the holy things until a high priest should appear wearing Urim and Thummim.[p]

41 All those of Israel, twelve or more years of age, besides male and female ser-

n Other ancient authorities read *Acub* or *Acum*
o Or *the governor*   p Gk *Manifestation and Truth*

5.40: The name *Nehemiah* is not found in the lists in Ezra and Nehemiah; "the governor" (see note *o*) orders the community to await the appearance of a high priest before participating in the holy things. The name is an addition to the text, arising from the circumstance that Nehemiah served as governor of Judah under appointment by Artaxerxes I

(Neh 5.14). *Urim and Thummim* are the sacred lots used by the priests to receive oracular decisions (Ex 28.30; Lev 8.8; Deut 33.8; 1 Sam 14.41). **41:** The total exceeds the sum of the several groups listed, it being assumed that others were present who are not specifically mentioned in the list.

vants, were forty-two thousand three hundred sixty; [42] their male and female servants were seven thousand three hundred thirty-seven; there were two hundred forty-five musicians and singers. [43] There were four hundred thirty-five camels, and seven thousand thirty-six horses, two hundred forty-five mules, and five thousand five hundred twenty-five donkeys.

44 Some of the heads of families, when they came to the temple of God that is in Jerusalem, vowed that, to the best of their ability, they would erect the house on its site, [45] and that they would give to the sacred treasury for the work a thousand minas of gold, five thousand minas of silver, and one hundred priests' vestments.

46 The priests, the Levites, and some of the people[q] settled in Jerusalem and its vicinity; and the temple singers, the gate-keepers, and all Israel in their towns.

47 When the seventh month came, and the Israelites were all in their own homes, they gathered with a single purpose in the square before the first gate toward the east. [48] Then Jeshua son of Jozadak, with his fellow priests, and Zerubbabel son of Shealtiel, with his kinsmen, took their places and prepared the altar of the God of Israel, [49] to offer burnt offerings upon it, in accordance with the directions in the book of Moses the man of God. [50] And some joined them from the other peoples of the land. And they erected the altar in its place, for all the peoples of the land were hostile to them and were stronger than they; and they offered sacrifices at the proper times and burnt offerings to the Lord morning and evening. [51] They kept the festival of booths, as it is commanded in the law, and offered the proper sacrifices every day, [52] and thereafter the regular offerings and sacrifices on sabbaths and at new moons and at all the consecrated feasts. [53] And all who had made any vow to God began to offer sacrifices to God, from the new moon of the seventh month, though the temple of God was not yet built. [54] They gave money to the masons and the carpenters, and food and drink [55] and carts[r] to the Sidonians and the Tyrians, to bring cedar logs from Lebanon and convey them in rafts to the harbor of Joppa, according to the decree that they had in writing from King Cyrus of the Persians.

56 In the second year after their coming to the temple of God in Jerusalem, in the second month, Zerubbabel son of Shealtiel and Jeshua son of Jozadak made a beginning, together with their kindred and the levitical priests and all who had come back to Jerusalem from exile; [57] and they laid the foundation of the temple of God on the new moon of the second

q Or *those who were of the people*
r Meaning of Gk uncertain

5.47–73: Work on the temple commences and is interrupted (Ezra 3.1–4.5; compare Josephus, *Ant.* XI. iv.1–3). This section is confused because the building of the temple is placed both in the reign of Cyrus (538–529 B.C.) and that of Darius (521–485 B.C.). A first return of exiles under Sheshbazzar and a second return under Zerubbabel and Jeshua have been merged. The true sequence of events is that Sheshbazzar returned to Judah shortly after 538 B.C., restored the sacrificial altar, resumed the cultic services, and laid the foundation of the temple. The work was halted until the return of additional exiles under Zerubbabel and Jeshua; when the work was resumed, opposition from Samaria, the capital of the province to which Judah belonged, quickly developed. Haggai and Zechariah encouraged the community to complete the temple, and work was begun once more; the temple was finally dedicated in 516 B.C. **47–55:** It is highly doubtful that Zerubbabel and Jeshua were involved in the initial work under Sheshbazzar; the events recorded here belong to the period of rebuilding begun in the second year of Darius, not Cyrus. **51:** The *festival of booths,* a harvest celebration, is observed for one week beginning on the fifteenth day of the seventh month (Lev 23.39). **54:** Minted *money* was in use in the Persian period. **56:** Apparently the *second year* of Cyrus is intended, but the second year of Darius is the correct date.

month in the second year after they came to Judea and Jerusalem. 58 They appointed the Levites who were twenty or more years of age to have charge of the work of the Lord. And Jeshua arose, and his sons and kindred and his brother Kadmiel and the sons of Jeshua Emadabun and the sons of Joda son of Iliadun, with their sons and kindred, all the Levites, pressing forward the work on the house of God with a single purpose.

So the builders built the temple of the Lord. 59 And the priests stood arrayed in their vestments, with musical instruments and trumpets, and the Levites, the sons of Asaph, with cymbals, 60 praising the Lord and blessing him, according to the directions of King David of Israel; 61 they sang hymns, giving thanks to the Lord, "For his goodness and his glory are forever upon all Israel." 62 And all the people sounded trumpets and shouted with a great shout, praising the Lord for the erection of the house of the Lord. 63 Some of the levitical priests and heads of ancestral houses, old men who had seen the former house, came to the building of this one with outcries and loud weeping, 64 while many came with trumpets and a joyful noise, 65 so that the people could not hear the trumpets because of the weeping of the people.

For the multitude sounded the trumpets loudly, so that the sound was heard far away; 66 and when the enemies of the tribe of Judah and Benjamin heard it, they came to find out what the sound of the trumpets meant. 67 They learned that those who had returned from exile were building the temple for the Lord God of Israel. 68 So they approached Zerubbabel and Jeshua and the heads of the ancestral houses and said to them, "We will build with you. 69 For we obey your Lord just as you do and we have been sacrificing to him ever since the days of King Esarhaddonˢ of the Assyrians, who brought us here." 70 But Zerubbabel and Jeshua and the heads of the ancestral houses in Israel said to them, "You have nothing to do with us in building the house for the Lord our God, 71 for we alone will build it for the Lord of Israel, as Cyrus, the king of the Persians, has commanded us." 72 But the peoples of the land pressed hardᵗ upon those in Judea, cut off their supplies, and hindered their building; 73 and by plots and demagoguery and uprisings they prevented the completion of the building as long as King Cyrus lived. They were kept from building for two years, until the reign of Darius.

6 Now in the second year of the reign of Darius, the prophets Haggai and Zechariah son of Iddo prophesied to the Jews who were in Judea and Jerusalem; they prophesied to them in the name of the Lord God of Israel. 2 Then Zerubbabel son of Shealtiel and Jeshua son of Jozadak began to build the house of the Lord that is in Jerusalem, with the help of the prophets of the Lord who were with them.

3 At the same time Sisinnes the governor of Syria and Phoenicia and Sathrabuzanes and their associates came to them and said, 4 "By whose order are you building this house and this roof and finishing all the other things? And who are the builders that are finishing these

---

s Gk *Asbasareth*   *t* Meaning of Gk uncertain

5.59–65: The author erroneously speaks of the temple's being built at this time; the ceremony described in Ezra 3.10–13 occurred when the foundation of the temple was laid. **66–73:** Enemies interrupt the work; they hear the sound of celebration, a detail not found in Ezra. **5.69:** Instead of *Esar-haddon,* Josephus (*Ant.* xi. iv.3) reads Shalmaneser (as in 2 Kings ch 17). **73:** The *two years* from the reign of Cyrus to that of Darius (compare 2.30) is a mistake; Ezra lacks this detail, although in Ezra 4.24 the cessation of work until the time of Darius introduces a similar confusion.

6.1–7.15: The temple completed (Ezra 4.24–6.22). Haggai and Zechariah encourage the resumption of work on the temple and succeed in gaining support for Zerubbabel and Jeshua (= Joshua; compare Hag 1.1–4; 2.1–4; Zech 4.9; 6.15). **6.3:** *Sisinnes* is Tattenai, governor of the province "Beyond the River"; *Sathrabuzanes* is Shetherbozenai (Ezra 5.3).

things?" 5 Yet the elders of the Jews were dealt with kindly, for the providence of the Lord was over the captives; 6 they were not prevented from building until word could be sent to Darius concerning them and a report made.

7 A copy of the letter that Sisinnes the governor of Syria and Phoenicia, and Sathrabuzanes, and their associates the local rulers in Syria and Phoenicia, wrote and sent to Darius:

8 "To King Darius, greetings. Let it be fully known to our lord the king that, when we went to the country of Judea and entered the city of Jerusalem, we found the elders of the Jews, who had been in exile, 9 building in the city of Jerusalem a great new house for the Lord, of hewn stone, with costly timber laid in the walls. 10 These operations are going on rapidly, and the work is prospering in their hands and being completed with all splendor and care. 11 Then we asked these elders, 'At whose command are you building this house and laying the foundations of this structure?' 12 In order that we might inform you in writing who the leaders are, we questioned them and asked them for a list of the names of those who are at their head. 13 They answered us, 'We are the servants of the Lord who created the heaven and the earth. 14 The house was built many years ago by a king of Israel who was great and strong, and it was finished. 15 But when our ancestors sinned against the Lord of Israel who is in heaven, and provoked him, he gave them over into the hands of King Nebuchadnezzar of Babylon, king of the Chaldeans; 16 and they pulled down the house, and burned it, and carried the people away captive to Babylon. 17 But in the first year that Cyrus reigned over the country of Babylonia, King Cyrus wrote that this house should be rebuilt. 18 And the holy vessels of gold and of silver, which Nebuchadnezzar had taken out of the house in Jeru-

salem and stored in his own temple, these King Cyrus took out again from the temple in Babylon, and they were delivered to Zerubbabel and Sheshbazzar*u* the governor 19 with the command that he should take all these vessels back and put them in the temple at Jerusalem, and that this temple of the Lord should be rebuilt on its site. 20 Then this Sheshbazzar, after coming here, laid the foundations of the house of the Lord that is in Jerusalem. Although it has been in process of construction from that time until now, it has not yet reached completion.' 21 Now therefore, O king, if it seems wise to do so, let search be made in the royal archives of our lord*v* the king that are in Babylon; 22 if it is found that the building of the house of the Lord in Jerusalem was done with the consent of King Cyrus, and if it is approved by our lord the king, let him send us directions concerning these things."

23 Then Darius commanded that search be made in the royal archives that were deposited in Babylon. And in Ecbatana, the fortress that is in the country of Media, a scroll*w* was found in which this was recorded: 24 "In the first year of the reign of King Cyrus, he ordered the building of the house of the Lord in Jerusalem, where they sacrifice with perpetual fire; 25 its height to be sixty cubits and its width sixty cubits, with three courses of hewn stone and one course of new native timber; the cost to be paid from the treasury of King Cyrus; 26 and that the holy vessels of the house of the Lord, both of gold and of silver, which Nebuchadnezzar took out of the house in Jerusalem and carried away to Babylon, should be restored to the house in Jerusalem, to be placed where they had been."

27 So Darius*x* commanded Sisinnes

---

u Gk *Sanabassarus* *v* Other ancient authorities read *of Cyrus* *w* Other authorities read *passage* *x* Gk *he*

6.14: *A king of Israel,* namely Solomon. 18: *Zerubbabel* is an addition; only Sheshbazzar is mentioned in Ezra 5.14 and in Josephus (*Ant.*

xi. iv.4). 23: *Ecbatana* was the summer residence of Darius.

6.32: Ezra 6.11 prescribes that violators of

the governor of Syria and Phoenicia, and Sathrabuzanes, and their associates, and those who were appointed as local rulers in Syria and Phoenicia, to keep away from the place, and to permit Zerubbabel, the servant of the Lord and governor of Judea, and the elders of the Jews to build this house of the Lord on its site. 28 "And I command that it be built completely, and that full effort be made to help those who have returned from the exile of Judea, until the house of the Lord is finished; 29 and that out of the tribute of Coelesyria and Phoenicia a portion be scrupulously given to these men, that is, to Zerubbabel the governor, for sacrifices to the Lord, for bulls and rams and lambs, 30 and likewise wheat and salt and wine and oil, regularly every year, without quibbling, for daily use as the priests in Jerusalem may indicate, 31 in order that libations may be made to the Most High God for the king and his children, and prayers be offered for their lives."

32 He commanded that if anyone should transgress or nullify any of the things herein written, *y* a beam should be taken out of the house of the perpetrator, who then shall be impaled upon it, and all property forfeited to the king.

33 "Therefore may the Lord, whose name is there called upon, destroy every king and nation that shall stretch out their hands to hinder or damage that house of the Lord in Jerusalem.

34 "I, King Darius, have decreed that it be done with all diligence as here prescribed."

7 Then Sisinnes the governor of Coelesyria and Phoenicia, and Sathrabuzanes, and their associates, following the orders of King Darius, 2 supervised the holy work with very great care, assisting the elders of the Jews and the chief officers of the temple. 3 The holy work prospered, while the prophets Haggai and Zechariah prophesied; 4 and they completed it by the command of the Lord God of Israel. So with the consent of Cyrus and Darius and Artaxerxes, kings of the Persians, 5 the holy house was finished by the twenty-third day of the month of Adar, in the sixth year of King Darius. 6 And the people of Israel, the priests, the Levites, and the rest of those who returned from exile who joined them, did according to what was written in the book of Moses. 7 They offered at the dedication of the temple of the Lord one hundred bulls, two hundred rams, four hundred lambs, 8 and twelve male goats for the sin of all Israel, according to the number of the twelve leaders of the tribes of Israel; 9 and the priests and the Levites stood arrayed in their vestments, according to kindred, for the services of the Lord God of Israel in accordance with the book of Moses; and the gatekeepers were at each gate.

10 The people of Israel who came from exile kept the passover on the fourteenth day of the first month, after the priests and the Levites were purified together. 11 Not all of the returned captives were purified, but the Levites were all purified together, *z* 12 and they sacrificed the passover lamb for all the returned captives and for their kindred the priests and for themselves. 13 The people of Israel who had returned from exile ate it, all those who had separated themselves from the abominations of the peoples of the land and sought the Lord. 14 They also kept the festival of unleavened bread seven days, rejoicing before the Lord, 15 because he had changed the will of the king of the Assyrians concerning them,

y Other authorities read *stated above* or *added in writing*   *z* Meaning of Gk uncertain

the decree be impaled and their house be made a dunghill (2 Kings 10.27; Dan 2.5).

7.4: *Artaxerxes* (see 8.1–9.36 n.) is erroneously named here (as also in Ezra 6.14); the name is omitted by Josephus because of the anachronism. **5:** The date intended is February–March, 516 B.C. **7–8:** Compare the account of the dedication of the first temple (1 Kings 8.5, 63). **9:** Compare Ezra 6.18.

7.13: Contrary to Ezra 6.21 the account here seems to suggest that only the returned exiles participated in the passover. **15:** The expression *king of the Assyrians* may be used because the Persian empire comprised the for-

to strengthen their hands for the service of the Lord God of Israel.

8 After these things, when Artaxerxes, the king of the Persians, was reigning, Ezra came, the son of Seraiah son of Azariah son of Hilkiah son of Shallum ²son of Zadok son of Ahitub son of Amariah son of Uzzi son of Bukki son of Abishua son of Phineas son of Eleazar son of Aaron the high*ᵃ* priest. ³This Ezra came up from Babylon as a scribe skilled in the law of Moses, which was given by the God of Israel; ⁴and the king showed him honor, for he found favor before the king*ᵇ* in all his requests. ⁵There came up with him to Jerusalem some of the people of Israel and some of the priests and Levites and temple singers and gatekeepers and temple servants, ⁶in the seventh year of the reign of Artaxerxes, in the fifth month (this was the king's seventh year); for they left Babylon on the new moon of the first month and arrived in Jerusalem on the new moon of the fifth month, by the prosperous journey that the Lord gave them.*ᶜ* ⁷For Ezra possessed great knowledge, so that he omitted nothing from the law of the Lord or the commandments, but taught all Israel all the ordinances and judgments.

8 The following is a copy of the written commission from King Artaxerxes that was delivered to Ezra the priest and reader of the law of the Lord:

9 "King Artaxerxes to Ezra the priest and reader of the law of the Lord, greeting. ¹⁰In accordance with my gracious decision, I have given orders that those of the Jewish nation and of the priests and Levites and others in our realm, those who freely choose to do so, may go with you to Jerusalem. ¹¹Let as many as are so disposed, therefore, leave with you, just as I and the seven Friends who are my counselors have decided, ¹²in order to look into matters in Judea and Jerusalem, in accordance with what is in the law of the Lord, ¹³and to carry to Jerusalem the gifts for the Lord of Israel that I and my Friends have vowed, and to collect for the Lord in Jerusalem all the gold and silver that may be found in the country of Babylonia, ¹⁴together with what is given by the nation for the temple of their Lord that is in Jerusalem, both gold and silver for bulls and rams and lambs and what goes with them, ¹⁵so as to offer sacrifices on the altar of their Lord that is in Jerusalem. ¹⁶Whatever you and your kindred are minded to do with the gold and silver, perform it in accordance with the will of your God; ¹⁷deliver the holy vessels of the Lord that are given you for the use of the temple of your God that is in Jerusalem. ¹⁸And whatever else occurs to you as necessary for the temple of your God, you may provide out of the royal treasury.

19 "I, King Artaxerxes, have commanded the treasurers of Syria and Phoenicia that whatever Ezra the priest and reader of the law of the Most High God

a Gk *the first*   *b* Gk *him*   *c* Other authorities add *for him* or *upon him*

mer empire of Assyria. Josephus refers to the Persian king (*Ant.* XI. iv.8).

8.1–9.55: The history of Ezra (Ezra 7.1–10.44 and Neh 7.73–8.12). Ezra, whose name appears as author or central personality in 1 Esdras, is first introduced at this point in the document.

8.1–9.36: Ezra leads a group of exiles from Babylonia. The author ignores the work of Nehemiah (Neh chs 1–7), as Sirach ignores the work of Ezra (Sir 49.13). It is possible that Ezra came to Judea under Artaxerxes II (404–358 B.C.) rather than under Artaxerxes I (464–423 B.C.), but the sequence

Ezra (458–7 B.C.), Nehemiah (445–4 B.C.) is more probable. This sequence is presupposed in Ezra and Nehemiah.

8.1–7: Ezra identified. 1–2: The genealogy is briefer than that in Ezra 7.1–5. **6:** *The seventh year* of Artaxerxes I was 458 or 457 B.C. (If Ezra came in the seventh year of Artaxerxes II, the date would be 398 or 397 B.C.) **7:** Ezra comes specifically as a teacher of *the law of the Lord.*

8.8–24: The letter of Artaxerxes to Ezra (Ezra 7.12–26). **11:** *The seven Friends* or counselors of the king are referred to in Esth 1.14 and Herodotus, *Hist.* III. 84.

sends for, they shall take care to give him, [20] up to a hundred talents of silver, and likewise up to a hundred cors of wheat, a hundred baths of wine, and salt in abundance. [21] Let all things prescribed in the law of God be scrupulously fulfilled for the Most High God, so that wrath may not come upon the kingdom of the king and his sons. [22] You are also informed that no tribute or any other tax is to be laid on any of the priests or Levites or temple singers or gatekeepers or temple servants or persons employed in this temple, and that no one has authority to impose any tax on them.

[23] "And you, Ezra, according to the wisdom of God, appoint judges and justices to judge all those who know the law of your God, throughout all Syria and Phoenicia; and you shall teach it to those who do not know it. [24] All who transgress the law of your God or the law of the kingdom shall be strictly punished, whether by death or some other punishment, either fine or imprisonment."

[25] Then Ezra the scribe said,[d] "Blessed be the Lord alone, who put this into the heart of the king, to glorify his house that is in Jerusalem, [26] and who honored me in the sight of the king and his counselors and all his Friends and nobles. [27] I was encouraged by the help of the Lord my God, and I gathered men from Israel to go up with me."

[28] These are the leaders, according to their ancestral houses and their groups, who went up with me from Babylon, in the reign of King Artaxerxes: [29] Of the descendants of Phineas, Gershom. Of the descendants of Ithamar, Gamael. Of the descendants of David, Hattush son of Shecaniah. [30] Of the descendants of Parosh, Zechariah, and with him a hundred fifty men enrolled. [31] Of the descendants

of Pahath-moab, Eliehoenai son of Zerahiah, and with him two hundred men. [32] Of the descendants of Zattu, Shecaniah son of Jahaziel, and with him three hundred men. Of the descendants of Adin, Obed son of Jonathan, and with him two hundred fifty men. [33] Of the descendants of Elam, Jeshaiah son of Gotholiah, and with him seventy men. [34] Of the descendants of Shephatiah, Zeraiah son of Michael, and with him seventy men. [35] Of the descendants of Joab, Obadiah son of Jehiel, and with him two hundred twelve men. [36] Of the descendants of Bani, Shelomith son of Josiphiah, and with him a hundred sixty men. [37] Of the descendants of Bebai, Zechariah son of Bebai, and with him twenty-eight men. [38] Of the descendants of Azgad, Johanan son of Hakkatan, and with him a hundred ten men. [39] Of the descendants of Adonikam, the last ones, their names being Eliphelet, Jeuel, and Shemaiah, and with them seventy men. [40] Of the descendants of Bigvai, Uthai son of Istalcurus, and with him seventy men.

[41] I assembled them at the river called Theras, and we encamped there three days, and I inspected them. [42] When I found there none of the descendants of the priests or of the Levites, [43] I sent word to Eliezar, Iduel, Maasmas, [44] Elnathan, Shemaiah, Jarib, Nathan, Elnathan, Zechariah, and Meshullam, who were leaders and men of understanding; [45] I told them to go to Iddo, who was the leading man at the place of the treasury, [46] and ordered them to tell Iddo and his kindred and the treasurers at that place to send us men to serve as priests in the house of our Lord. [47] And by the mighty

---

[d] Other ancient authorities lack *Then Ezra the scribe said*

8.20: The *talent* was 75.5 U.S. pounds; the *cor* 6.5 bushels; and the *bath* about 6 gallons. **22**: Temple personnel are exempt from all taxes. **23–24**: Ezra is given authority to appoint judges throughout the entire province in order to maintain the Jewish law.

8.25–60: **Ezra leads the exiles to Jerusalem** (Ezra 7.27–8.30). **28–40**: The list of

those who returned differs in a few particulars from that found in Ezra 8.1–14.

8.41: *The river . . . Theras* (Ahava in Ezra 8.21) is probably a tributary of the Euphrates. **42–49**: Because neither priests nor Levites were among the group first assembled by Ezra, special measures had to be taken to secure the required number of both.

hand of our Lord they brought us competent men of the descendants of Mahli son of Levi, son of Israel, namely Sherebiah[e] with his descendants and kinsmen, eighteen; [48]also Hashabiah and Annunus and his brother Jeshaiah, of the descendants of Hananiah, and their descendants, twenty men; [49]and of the temple servants, whom David and the leaders had given for the service of the Levites, two hundred twenty temple servants; the list of all their names was reported.

50 There I proclaimed a fast for the young men before our Lord, to seek from him a prosperous journey for ourselves and for our children and the livestock that were with us. [51]For I was ashamed to ask the king for foot soldiers and cavalry and an escort to keep us safe from our adversaries; [52]for we had said to the king, "The power of our Lord will be with those who seek him, and will support them in every way." [53]And again we prayed to our Lord about these things, and we found him very merciful.

54 Then I set apart twelve of the leaders of the priests, Sherebiah and Hashabiah, and ten of their kinsmen with them; [55]and I weighed out to them the silver and the gold and the holy vessels of the house of our Lord, which the king himself and his counselors and the nobles and all Israel had given. [56]I weighed and gave to them six hundred fifty talents of silver, and silver vessels worth a hundred talents, and a hundred talents of gold, [57]and twenty golden bowls, and twelve bronze vessels of fine bronze that glittered like gold. [58]And I said to them, "You are holy to the Lord, and the vessels are holy, and the silver and the gold are vowed to the Lord, the Lord of our ancestors. [59]Be watchful and on guard until you deliver them to the leaders of the priests and the Levites, and to the heads of the ancestral houses of Israel, in Jerusalem, in the chambers of the house of our Lord." [60]So the priests and the Levites who took the silver and the gold and the vessels that had been in Jerusalem carried them to the temple of the Lord.

61 We left the river Theras on the twelfth day of the first month; and we arrived in Jerusalem by the mighty hand of our Lord, which was upon us; he delivered us from every enemy on the way, and so we came to Jerusalem. [62]When we had been there three days, the silver and the gold were weighed and delivered in the house of our Lord to the priest Meremoth son of Uriah; [63]with him was Eleazar son of Phinehas, and with them were Jozabad son of Jeshua and Moeth son of Binnui,[f] the Levites. [64]The whole was counted and weighed, and the weight of everything was recorded at that very time. [65]And those who had returned from exile offered sacrifices to the Lord, the God of Israel, twelve bulls for all Israel, ninety-six rams, [66]seventy-two lambs, and as a thank offering twelve male goats—all as a sacrifice to the Lord. [67]They delivered the king's orders to the royal stewards and to the governors of Coelesyria and Phoenicia; and these officials[g] honored the people and the temple of the Lord.

68 After these things had been done, the leaders came to me and said, [69]"The people of Israel and the rulers and the priests and the Levites have not put away

e Gk *Asbebias*   f Gk *Sabannus*   g Gk *they*

8.50: Fasting prior to an important undertaking was common (2 Chr 20.3; Esth 4.16; Jer 36.9). **58**: *Holy* objects could be entrusted only to those who were *holy* themselves. **8.61–67**: **Arrival in Jerusalem** (Ezra 8.31–36). The treasures are placed in the temple storehouses (*chambers*, v. 59), sacrifices are offered to God, and the king's orders delivered to the provincial officers; the latter have no choice but to obey.

8.68–9.36: **Mixed marriages in Judah** (Ezra 9.1–10.44). No sooner does Ezra arrive than he is presented with evidence that the community has been corrupted by mixed marriages. The older legislation had warned against marriage with the population of Canaan upon entrance into the land (Deut 7.3) but had not expressly forbidden mixed marriages. Strong warnings had been issued, however, against Israel's adopting the abomi-

from themselves the alien peoples of the land and their pollutions, the Canaanites, the Hittites, the Perizzites, the Jebusites, the Moabites, the Egyptians, and the Edomites. ⁷⁰For they and their descendants have married the daughters of these people,ʰ and the holy race has been mixed with the alien peoples of the land; and from the beginning of this matter the leaders and the nobles have been sharing in this iniquity."

71 As soon as I heard these things I tore my garments and my holy mantle, and pulled out hair from my head and beard, and sat down in anxiety and grief. ⁷²And all who were ever moved atⁱ the word of the Lord of Israel gathered around me, as I mourned over this iniquity, and I sat grief-stricken until the evening sacrifice. ⁷³Then I rose from my fast, with my garments and my holy mantle torn, and kneeling down and stretching out my hands to the Lord ⁷⁴I said,

"O Lord, I am ashamed and confused before your face. ⁷⁵For our sins have risen higher than our heads, and our mistakes have mounted up to heaven ⁷⁶from the times of our ancestors, and we are in great sin to this day. ⁷⁷Because of our sins and the sins of our ancestors, we with our kindred and our kings and our priests were given over to the kings of the earth, to the sword and exile and plundering, in shame until this day. ⁷⁸And now in some measure mercy has come to us from you, O Lord, to leave to us a root and a name in your holy place, ⁷⁹and to uncover a light for us in the house of the Lord our God, and to give us food in the time of our servitude. ⁸⁰Even in our bondage we were not forsaken by our Lord, but he brought us into favor with the kings of the Persians, so that they have given us food ⁸¹and glorified the temple of our Lord, and raised Zion from desolation, to give us a stronghold in Judea and Jerusalem.

82 "And now, O Lord, what shall we say, when we have these things? For we have transgressed your commandments, which you gave by your servants the prophets, saying, ⁸³'The land that you are entering to take possession of is a land polluted with the pollution of the aliens of the land, and they have filled it with their uncleanness. ⁸⁴Therefore do not give your daughters in marriage to their descendants, and do not take their daughters for your descendants; ⁸⁵do not seek ever to have peace with them, so that you may be strong and eat the good things of the land and leave it for an inheritance to your children forever.' ⁸⁶And all that has happened to us has come about because of our evil deeds and our great sins. For you, O Lord, lifted the burden of our sins ⁸⁷and gave us such a root as this; but we turned back again to transgress your law by mixing with the uncleanness of the peoples of the land. ⁸⁸Were you not angry enough with us to destroy us without leaving a root or seed or name? ⁸⁹O Lord of Israel, you are faithful; for we are left as a root to this day. ⁹⁰See, we are now before you in our iniquities; for we can no longer stand in your presence because of these things."

91 While Ezra was praying and making his confession, weeping and lying on the ground before the temple, there gathered around him a very great crowd of men and women and youths from Jerusalem; for there was great weeping among the multitude. ⁹²Then Shecaniah son of Jehiel, one of the men of Israel,

---

h Gk *their daughters*   *i* Or *zealous for*

---

nable practices of the surrounding nations (Lev 18.24–30). During the exile Israel had been able to survive only on the basis of maintaining a relatively high level of racial integrity. The strict separation carried out by Ezra is therefore understandable; in the exile standards were probably higher on this issue than they were in Judah. (Nehemiah also faced the same problem; see Neh 10.28–30; 13.3, 23–30.)

8.74–90: Ezra's prayer (Ezra 9.6–15). Ezra speaks for the entire community, acknowledging the sin of all and the justice of their punishment by God. **82–85:** The prophetic books contain no such statement; the author may have in mind Lev 18.19–30.

called out, and said to Ezra, "We have sinned against the Lord, and have married foreign women from the peoples of the land; but even now there is hope for Israel. [93] Let us take an oath to the Lord about this, that we will put away all our foreign wives, with their children, [94] as seems good to you and to all who obey the law of the Lord. [95] Rise up[j] and take action, for it is your task, and we are with you to take strong measures." [96] Then Ezra rose up and made the leaders of the priests and Levites of all Israel swear that they would do this. And they swore to it.

9 Then Ezra set out and went from the court of the temple to the chamber of Jehohanan son of Eliashib, [2] and spent the night there; and he did not eat bread or drink water, for he was mourning over the great iniquities of the multitude. [3] And a proclamation was made throughout Judea and Jerusalem to all who had returned from exile that they should assemble at Jerusalem, [4] and that if any did not meet there within two or three days, in accordance with the decision of the ruling elders, their livestock would be seized for sacrifice and the men themselves[k] expelled from the multitude of those who had returned from the captivity.

[5] Then the men of the tribe of Judah and Benjamin assembled at Jerusalem within three days; this was the ninth month, on the twentieth day of the month. [6] All the multitude sat in the open square before the temple, shivering because of the bad weather that prevailed. [7] Then Ezra stood up and said to them, "You have broken the law and married foreign women, and so have increased the sin of Israel. [8] Now then make confession and give glory to the Lord the God of our ancestors, [9] and do his will; separate yourselves from the peoples of the land and from your foreign wives."

[10] Then all the multitude shouted and said with a loud voice, "We will do as you have said. [11] But the multitude is great and it is winter, and we are not able to stand in the open air. This is not a work we can do in one day or two, for we have sinned too much in these things. [12] So let the leaders of the multitude stay, and let all those in our settlements who have foreign wives come at the time appointed, [13] with the elders and judges of each place, until we are freed from the wrath of the Lord over this matter."

[14] Jonathan son of Asahel and Jahzeiah son of Tikvah[l] undertook the matter on these terms, and Meshullam and Levi and Shabbethai served with them as judges. [15] And those who had returned from exile acted in accordance with all this.

[16] Ezra the priest chose for himself the leading men of their ancestral houses, all of them by name; and on the new moon of the tenth month they began their sessions to investigate the matter. [17] And the cases of the men who had foreign wives were brought to an end by the new moon of the first month.

[18] Of the priests, those who were brought in and found to have foreign wives were: [19] of the descendants of Jeshua son of Jozadak and his kindred, Maaseiah, Eliezar, Jarib, and Jodan. [20] They

j Other ancient authorities read *as seems good to you." And all who obeyed the law of the Lord rose and said to Ezra, "Rise up* **k** Gk *he himself* *l* Gk *Thocanos*

8.91–9.36: The people repent and dismiss their foreign wives (Ezra 10.1–44). **9.4:** *The ruling elders* issue orders for the entire community to assemble within two or three days; Ezra is the religious, not the political, authority in the land. **7:** *The law,* i.e. Deut 7.3. **8:** To *give glory to the Lord* is to acknowledge themselves to be in the wrong (compare Josh 7.19). **9.11–13:** Because of the severe winter weather, it is agreed that the separation should take place in the local districts, and the multitude is dismissed. **16–17:** Three months are required to settle the cases, from the first of the *tenth month* (Tebet = December-January) to the first of the *first month* (Nisan = March-April). **18–36:** The list of those who put away foreign wives, including priests, Levites, and the laity. The list was probably preserved in the temple archives.

pledged themselves to put away their wives, and to offer rams in expiation of their error. 21 Of the descendants of Immer: Hanani and Zebadiah and Maaseiah and Shemaiah and Jehiel and Azariah. 22 Of the descendants of Pashhur: Elioenai, Maaseiah, Ishmael, and Nathanael, and Gedaliah, and Salthas.

23 And of the Levites: Jozabad and Shimei and Kelaiah, who was Kelita, and Pethahiah and Judah and Jonah. 24 Of the temple singers: Eliashib and Zaccur.*m* 25 Of the gatekeepers: Shallum and Telem.*n*

26 Of Israel: of the descendants of Parosh: Ramiah, Izziah, Malchijah, Mijamin, and Eleazar, and Asibias, and Benaiah. 27 Of the descendants of Elam: Mattaniah and Zechariah, Jezrielus and Abdi, and Jeremoth and Elijah. 28 Of the descendants of Zamoth: Eliadas, Eliashib, Othoniah, Jeremoth, and Zabad and Zerdaiah. 29 Of the descendants of Bebai: Jehohanan and Hananiah and Zabbai and Emathis. 30 Of the descendants of Mani: Olamus, Mamuchus, Adaiah, Jashub, and Sheal and Jeremoth. 31 Of the descendants of Addi: Naathus and Moossias, Laccunus and Naidus, and Bescaspasmys and Sesthel, and Belnuus and Manasseas. 32 Of the descendants of Annan, Elionas and Asaias and Melchias and Sabbaias and Simon Chosamaeus. 33 Of the descendants of Hashum: Mattenai and Mattattah and Zabad and Eliphelet and Manasseh and Shimei. 34 Of the descendants of Bani: Jeremai, Momdius, Maerus, Joel, Mamdai and Bedeiah and Vaniah, Carabasion and Eliashib and Mamitanemus, Eliasis, Binnui, Elialis, Shimei, Shelemiah, Nethaniah. Of the descendants of Ezora: Shashai, Azarel,

Azael, Samatus, Zambris, Joseph. 35 Of the descendants of Nooma: Mazitias, Zabad, Iddo, Joel, Benaiah. 36 All these had married foreign women, and they put them away together with their children.

37 The priests and the Levites and the Israelites settled in Jerusalem and in the country. On the new moon of the seventh month, when the people of Israel were in their settlements, 38 the whole multitude gathered with one accord in the open square before the east gate of the temple; 39 they told Ezra the chief priest and reader to bring the law of Moses that had been given by the Lord God of Israel. 40 So Ezra the chief priest brought the law, for all the multitude, men and women, and all the priests to hear the law, on the new moon of the seventh month. 41 He read aloud in the open square before the gate of the temple from early morning until midday, in the presence of both men and women; and all the multitude gave attention to the law. 42 Ezra the priest and reader of the law stood on the wooden platform that had been prepared; 43 and beside him stood Mattathiah, Shema, Ananias, Azariah, Uriah, Hezekiah, and Baalsamus on his right, 44 and on his left Pedaiah, Mishael, Malchijah, Lothasubus, Nabariah, and Zechariah. 45 Then Ezra took up the book of the law in the sight of the multitude, for he had the place of honor in the presence of all. 46 When he opened the law, they all stood erect. And Ezra blessed the Lord God Most High, the God of hosts, the Almighty, 47 and the multitude answered, "Amen." They lifted up their hands, and fell to the ground

m Gk *Bacchurus*   *n* Gk *Tolbanes*

9.20: An offering of a ram as a guilt offering (Ezra 10.19) was made *in expiation* of the sin. 22: *Gedaliah,* Greek "Ocidelos" (the parallel in Ezra 10.22 reads "Jozabad"; compare Ezra 10.18).

9.37–55: **Ezra's public reading of the law** (Neh 7.73–8.12). 37: *The new moon* or first day *of the seventh month* was a day of holy convocation (Lev 23.23–24; Num 29.1), the day of the New Year. 39: Ezra is not identi-

fied elsewhere as the *chief priest* (in Neh 8.2 he is called the priest). *The law of Moses* is either the present Pentateuch or (more probably) the major legal portions of it. 42: The *platform* erected for Ezra probably continued the tradition whereby kings would appear before the people to reaffirm the covenant law on the festal occasion at the turn of the year (compare 2 Chr 20.5; 23.13; 29.4).

and worshiped the Lord. [48]Jeshua and Anniuth and Sherebiah, Jadinus, Akkub, Shabbethai, Hodiah, Maiannas and Kelita, Azariah and Jozabad, Hanan, Pelaiah, the Levites, taught the law of the Lord, [o] at the same time explaining what was read.

49 Then Attharates[p] said to Ezra the chief priest and reader, and to the Levites who were teaching the multitude, and to all, [50]"This day is holy to the Lord"— now they were all weeping as they heard the law— [51]"so go your way, eat the fat and drink the sweet, and send portions to those who have none; [52]for the day is holy to the Lord; and do not be sorrowful, for the Lord will exalt you." [53]The Levites commanded all the people, saying, "This day is holy; do not be sorrowful." [54]Then they all went their way, to eat and drink and enjoy themselves, and to give portions to those who had none, and to make great rejoicing; [55]because they were inspired by the words which they had been taught. And they came together. [q]

o Other ancient authorities add *and read the law of the Lord to the multitude*   *p* Or *the governor*
q The Greek text ends abruptly: compare Neh 8.13

9.48: The Levites explained the law to the people, perhaps translating it (or its difficult portions) into Aramaic for those who may not have been familiar with Hebrew. **49**: *Attharates* is a corruption of "tirshatha," *governor,* in Neh 8.9. 1 Esdras does not intend to indicate that Nehemiah, whom some have identified with Attharates the governor, was a participant in the festivity (in Neh 8.9 the name of Nehemiah is an intrusion). **50–55**:

The people are to rejoice even though the words of the law cause them to recognize their sin. **55**: The book ends abruptly; originally it may have continued with the story of the great celebration of the festival of booths (Neh 8.13–18). This would have been a fitting conclusion to the work, since it begins with the account of Josiah's great passover celebration.

The Prayer of Manasseh

While the Prayer of Manasseh is considered deuterocanonical by Eastern Orthodox communions, it is not so considered by Jews, Protestants, or Roman Catholics. Since the Council of Trent it has been included in an appendix to the Latin Vulgate along with 3 and 4 Ezra (1 and 2 Esdras). In some manuscripts of the Greek Septuagint it stands immediately following the Psalter among a group of fifteen psalms or songs, under the heading "Odes," which, while not in the Psalter itself, are, with the exception of our Prayer and one other, found in some other book in the Old or New Testament (Ex 15, Deut 32, Luke 1, etc.). Most scholars think that it was originally composed in a semitizing Greek probably late in the first century B.C.

King Manasseh is presented in 2 Kings (21.1–18) as the worst possible sinner and the basic cause of the downfall of Judah. In 2 Chronicles, while his wicked deeds are not in any way denied, Manasseh is pictured as praying earnestly and humbly to God during exile. His prayer was heard, and God restored him to the throne in Jerusalem where he instigated a reform program of true worship of God (2 Chr 33.10–17). The account in 2 Chronicles ends with the comment that the whole account of his sins, his penitence, his prayer, God's answer, and his restoration had been written in the records of the seers (33.19). But 2 Chronicles fails to record the prayer. This lack would have been noted by many ancient readers of the biblical Chronicles account; our author filled the obvious gap.

The prayer is a classic of penitential devotion and even has literary value. The theology is consonant with that of early Judaism: God, though clearly the God of justice, is also the God of mercy and repentance. He is the God of those who repent (v. 13). This theme was important to the emerging theology of exilic and post-exilic Judaism, whose existence was always tenuous, living under one repressive régime after another in various parts of the world.

Manasseh, who as king had the power to do all the evil things that brought about Judah's fall, was in the Jewish mind sorely in need of redemption. If he was irredeemable there might be doubt about who could repent and be heard, in other words, doubt about God's measure of mercy being as great as God's measure of justice. God is not only Creator and Sustainer, but also Redeemer—compassionate, long-suffering and very merciful (v. 7). God, therefore, appointed repentance not for the righteous, but ordained it for sinners, even for Manasseh (v. 8).

1 O Lord Almighty,
  God of our ancestors,
  of Abraham and Isaac and Jacob
  and of their righteous offspring;
2 you who made heaven and
    earth
  with all their order;
3 who shackled the sea by your
    word of command,
  who confined the deep
  and sealed it with your terrible
    and glorious name;
4 at whom all things shudder,
  and tremble before your power,

5 for your glorious splendor cannot
    be borne,
and the wrath of your threat to
    sinners is unendurable;
6 yet immeasurable and unsearchable
    is your promised mercy,
7 for you are the Lord Most High,
    of great compassion,
        long-suffering, and very
        merciful,
and you relent at human suffering.
O Lord, according to your great
    goodness
you have promised repentance and
    forgiveness
to those who have sinned against
    you,
and in the multitude of your
    mercies
you have appointed repentance for
    sinners,
so that they may be saved. *a*
8 Therefore you, O Lord, God of
    the righteous,
have not appointed repentance for
    the righteous,
for Abraham and Isaac and Jacob,
    who did not sin against
    you,
but you have appointed repentance
    for me, who am a sinner.
9 For the sins I have committed are
    more in number than the
    sand of the sea;
my transgressions are multiplied,
    O Lord, they are
    multiplied!
I am not worthy to look up and
    see the height of heaven
because of the multitude of my
    iniquities.

10 I am weighted down with many
    an iron fetter,
so that I am rejected *b* because of
    my sins,
and I have no relief;
for I have provoked your wrath
and have done what is evil in your
    sight,
setting up abominations and
    multiplying offenses.
11 And now I bend the knee of my
    heart,
imploring you for your kindness.
12 I have sinned, O Lord, I have
    sinned,
and I acknowledge my
    transgressions.
13 I earnestly implore you,
forgive me, O Lord, forgive me!
Do not destroy me with my
    transgressions!
Do not be angry with me forever
    or store up evil for me;
do not condemn me to the depths
    of the earth.
For you, O Lord, are the God of
    those who repent,
14 and in me you will manifest your
    goodness;
for, unworthy as I am, you will
    save me according to your
    great mercy,
15 and I will praise you continually
    all the days of my life.
For all the host of heaven sings
    your praise,
and yours is the glory forever.
    Amen.

a Other ancient authorities lack *O Lord, according
. . . be saved*   *b* Other ancient authorities read
so that I cannot lift up my head

**1–8: Invocation and ascription of
praise to God,** whose majesty is displayed in
creation (vv. 1–4), and whose mercy grants
repentance to sinners (vv. 6–8). **1:** *God of our
ancestors,* Ex 3.15–16; Dan 2.23; Acts 3.13. **2:**
All their order, splendor and orderly array. **3:**
Shackled the sea, Job 38.8–11. **7:** The second
part of this verse (*O Lord . . . may be saved*) is
preserved in the later Greek manuscripts and
in the Latin and Syriac versions. **8:** *Not . . . for
the righteous,* Mk 2.17; Lk 5.32; 1 Tim 1.15.
For me . . . a sinner, Lk 15.7; 18.13.

9–10: Personal confession of sin. For
the background see 2 Kings 21.1–18; 2 Chr
33.1–9.

11–15a: Supplication for pardon. 11:
Knee of my heart, an expression indicating special depth of feeling. See Joel 2.13. **12:** *I acknowledge my transgressions,* compare Ps 19.12.
13: *The depths of the earth,* probably Sheol or
Hades is meant (Ps 63.9; 88.5–6).

15b: Concluding doxology. *Host of
heaven,* multitude of angelic beings (2 Chr
18.18; Lk 2.13).

Psalm 151

Some manuscripts of the Masoretic text of the Hebrew Bible number the psalms in the Psalter sequentially up to 150. The oldest complete Masoretic manuscript of the Bible (Leningradensis), which forms the basis for all current translations into modern languages, numbers them only to 149, combining the "traditional" Ps 115 with 114. Those two psalms are also combined as Ps 113 in Greek Septuagint manuscripts (= "traditional" Ps 114), whereas "traditional" Ps 116 is divided into Pss 114 and 115 in Septuagint manuscripts.

In addition to these rather slight variations in the shape of the "traditional" Psalter, most Greek manuscripts of the Psalter have 151 psalms, the last not appearing in any Hebrew manuscript of the Psalter until the discovery in 1956 of a Psalter manuscript among the Dead Sea Scrolls, from Qumran Cave XI, designated 11QPsᵃ. The latter contains forty canonical psalms from the last third of the Masoretic or traditional Hebrew Psalter (not all in the traditional sequence) along with eight compositions not in other known Psalters, except for Ps 151 in the Septuagint, and Pss 151, 154 and 155, which are included in a Syriac Psalter. The Hebrew of Ps 151 is subject to more than one interpretation; this is especially the case in vv. 2b–4.

In codex Alexandrinus and most other Septuagint manuscripts Ps 151 is presented as an appendix to the Psalter with a superscription designating it "outside the number" (supposedly of the "traditional" 150 psalms); but it is attached to the Psalter itself and not included among the fifteen "Odes" that in some Septuagint manuscripts (notably codex Alexandrinus) are appended to the Greek Psalter after Ps 151. In codex Sinaiticus, one of the oldest and in some ways the most reliable and almost complete manuscript of the Septuagint, Ps 151 is included as an integral part of the Psalter, for which there is a subscription that reads "The 151 Psalms of David." A translation of the Hebrew text, made by J. A. Sanders, the original editor of the scroll, in *Discoveries in the Judaean Desert of Jordan,* Vol. IV (Oxford, 1965), is given here (with two slight modifications) for purposes of comparison with the Septuagint form of the psalm. It should be mentioned that the Hebrew script presents several palaeographical and philological uncertainties.

A Hallelujah of David the Son of Jesse

¹Smaller was I than my brothers
  and the youngest of the sons of my father,
Yet he made me shepherd of his flock
  and ruler over his kids.

²My hands have made an instrument
  and my fingers a lyre;
And [so] have I rendered glory to the Lord,
  thought I, within my soul.

³The mountains do not witness to him,
  nor do the hills proclaim;
The trees have cherished my words
  and the flock my works.

⁴For who can proclaim and who can bespeak
  and who can recount the deeds of the Lord?
Everything has God seen,
  everything has he heard and he has heeded.

[5]He sent his prophet to anoint me,
  Samuel to make me great;
My brothers went out to meet him,
  handsome of figure and appearance.

[6]Though they were tall of stature
  and handsome by their hair,

The Lord God chose
  them not.

[7]But he sent and took me from behind the flock
  and anointed me with holy oil,
And he made me leader of his people
  and ruler over the people of his covenant.

In the line following the Hebrew text of this psalm another psalm (with its heading) begins, of which only two poorly preserved lines remain. Apparently they celebrate David's victory over Goliath; Sanders's translation is as follows:

> At the beginning of David's power after
> the prophet of God had anointed him.

> [1]Then I [saw] a Philistine
> uttering defiances from the r[anks of the
> Philistines].

It thus appears that the Greek text of Ps 151 is a condensed recension of what was originally two separate psalms in Hebrew.

This psalm is ascribed to David as his own composition (though it is outside the number[a]), after he had fought in single combat with Goliath.

1 I was small among my brothers,
  and the youngest in my father's
  house;
I tended my father's sheep.

2 My hands made a harp;
  my fingers fashioned a lyre.

3 And who will tell my Lord?
  The Lord himself; it is he who
  hears. [b]

4 It was he who sent his messenger[c]
  and took me from my father's
  sheep,

and anointed me with his
  anointing oil.

5 My brothers were handsome and
  tall,
but the Lord was not pleased
  with them.

6 I went out to meet the Philistine,[d]
  and he cursed me by his idols.

7 But I drew his own sword;
  I beheaded him, and took away
  disgrace from the people of
  Israel.

a Other ancient authorities add *of the one hundred fifty* (psalms) b Other ancient authorities add *everything*; others add *me*; others read *who will hear me* c Or *angel* d Or *foreigner*

1: *Small . . . youngest,* 1 Sam 16.7 and 11. *Tended . . . sheep,* 1 Sam 16.11. One form of the Syriac version continues, "and I met a lion and also a wolf, and I killed them and tore them in pieces" (compare 1 Sam 17.34–36). **2:** *My hands made a harp,* 2 Chr 29.26. **4:** *Anointed me,* 1 Sam 16.13; Ps 89.20. *Took me from my father's sheep,* Ps 78.70. **5:** *The Lord was not pleased with them,* 1 Sam 16.7–10. **6:** *He cursed* . . . *by his idols,* 1 Sam 17.43. Certain manuscripts of the Old Latin, Arabic, and Ethiopic versions continue, with minor deviations: "And I slung three stones at him in the middle of his forehead, and laid him low by the might of the Lord." According to 1 Sam 17.49–50 David felled Goliath with only one stone. **7:** *His own sword,* 1 Sam 17.51.

3 Maccabees

The title of the book known as 3 Maccabees is a misnomer, for the contents deal not with the exploits of the Maccabean heroes, but with the struggles of Egyptian Jews who suffered under Ptolemy IV Philopator (221–203 B.C.), half a century prior to the Maccabean period with its persecution of Palestinian Jewry under Antiochus IV Epiphanes (175–164 B.C.). The book has been transmitted in manuscripts of the Greek Septuagint and the Syriac Peshitta, as well as in most manuscripts of the Armenian Bible. It is not, however, included in the Latin Vulgate. This may explain why the book has been accorded deuterocanonical status in Eastern Christendom (i.e. of lesser authority than canonical works), while both the Roman and Reformed Churches of the West regard it as apocryphal.

After a brief introduction, 3 Maccabees begins by describing the attempt of King Ptolemy of Egypt to enter the holy of holies in the Jerusalem temple. The desecration is averted by divine intervention in response to the prayer of the high priest Simon (1.1–2.24). Upon his return to Egypt, the king determines to wreak vengeance on the Jews for his humiliation in Jerusalem. He alters their civic status and attempts to impose upon them by force the pagan cult of Dionysus, promising to those who comply equal citizenship with the Alexandrians (2.25–33). The vast majority of Jews resist, and with great cruelty they are herded together to be registered, tortured, and put to death. Again through divine intervention, after forty days the registration remains incomplete because writing materials have been exhausted (3.1–4.21). A third miracle occurs in answer to the prayer of the aged priest Eleazar, and the Jews are spared from being trampled by a great herd of elephants (5.1–6.21). The king, struck with fear as the elephants turn back upon his own forces, repents and becomes the patron of the Jews. He addresses a letter of protection on their behalf to his provincial governors, and the Jews return to their homes in safety and with rejoicing (6.22–7.23).

The book was obviously written to console, exhort, and teach Egyptian Jews, who during the first century B.C. were frequently threatened by the efforts of Roman administrators to alter their civic status. The author intends to inspire faith in the providence of God (4.21) by recounting how Jews had been delivered from similar tribulations in the past.

The work was originally drawn up in Greek by an unknown Alexandrian Jew, sometime between the Battle of Raphia in 217 B.C. and the fall of the Jerusalem temple in A.D. 70. A likely date, considering its literary affinities with 2 Maccabees and the Letter of Aristeas, is the early first century B.C. The author has composed his work by using the familiar principles of concentric parallelism, and focusing on the miracle surrounding the registration in 4.14–21.

    A: Ptolemy threatens to desecrate the temple [1.1–29]
      B: Simon's prayer of intercession (divine intervention) [2.1–24]
        C: Ptolemy's cruel treatment of the Jews [2.25–4.13]
          D: Thwarting the registration (divine intervention) [4.14–21]
        C': Ptolemy's cruel treatment of the Jews [5.1–51]
      B': Eleazar's prayer of intercession (divine intervention) [6.1–29]
    A': Ptolemy delivers and defends the Jews [6.30–7.23]

The style of the book is pseudoclassical, utilizing many uncommon and poetical words. Sentences are awkwardly constructed and frequently repetitious. The author's exaggerated rhetoric aims to drive home his message that God remains faithful to his chosen people, to bless and preserve them throughout the vicissitudes of their experiences.

1 When Philopator learned from those who returned that the regions that he had controlled had been seized by Antiochus, he gave orders to all his forces, both infantry and cavalry, took with him his sister Arsinoë, and marched out to the region near Raphia, where the army of Antiochus was encamped. 2 But a certain Theodotus, determined to carry out the plot he had devised, took with him the best of the Ptolemaic arms that had been previously issued to him,*a* and crossed over by night to the tent of Ptolemy, intending single-handed to kill him and thereby end the war. 3 But Dositheus, known as the son of Drimylus, a Jew by birth who later changed his religion and apostatized from the ancestral traditions, had led the king away and arranged that a certain insignificant man should sleep in the tent; and so it turned out that this man incurred the vengeance meant for the king.*b* 4 When a bitter fight resulted, and matters were turning out rather in favor of Antiochus, Arsinoë went to the troops with wailing and tears, her locks all disheveled, and exhorted them to defend themselves and their children and wives bravely, promising to give them each two minas of gold if they won the battle. 5 And so it came about that the enemy was routed in the action, and many captives also were taken. 6 Now that he had foiled the plot, Ptolemy*c* decided to visit the neighboring cities and encourage them. 7 By doing this, and by endowing their sacred enclosures with gifts, he strengthened the morale of his subjects.

8 Since the Jews had sent some of their council and elders to greet him, to bring him gifts of welcome, and to congratulate him on what had happened, he was all the more eager to visit them as soon as possible. 9 After he had arrived in Jerusalem, he offered sacrifice to the supreme God*d* and made thank offerings and did what was fitting for the holy place.*e* Then, upon entering the place and being impressed by its excellence and its beauty, 10 he marveled at the good order of the temple, and conceived a desire to enter the sanctuary. 11 When they said that this was not permitted, because not even members of their own nation were allowed to enter, not even all of the priests, but only the high priest who was pre-eminent over all—and he only once a year—the king was by no means persuaded. 12 Even after the law had been read to him, he did not cease to maintain that he ought to enter, saying, "Even if those men are deprived of this honor, I ought not to be." 13 And he inquired why, when he entered every other temple,*f* no one there had stopped him.

a Or the best of the Ptolemaic soldiers previously put under his command   b Gk that one   c Gk he   d Gk the greatest God   e Gk the place   f Or entered the temple precincts

1.1–7: The battle of Raphia (217 B.C.). The abruptness with which the book opens and the use of the Greek conjunctive particle "de" indicate that the introduction to 3 Maccabees has not survived (see also 2.25 n.). **1:** Ptolemy IV *Philopator* was king of Egypt 221–203 B.C. *From those who returned,* fugitives who had escaped. *Antiochus* III, later called the Great, was king of Syria 223–187 B.C. *Raphia,* a city of Palestine, three miles from Gaza and not far from the Egyptian frontier. *Arsinoë,* Ptolemy's sister, who became his wife, was later put to death at the instigation of her husband. **2:** *Theodotus* had been chief commander of the Egyptian forces in Syria, but subsequently became disaffected and deserted to Antiochus III (so Polybius, v. 40). **3:** A *Dositheus* is mentioned in Hibeh papyrus 90 as priest of Alexander in 222 B.C. **5:** According to Polybius (v. 86.5–6) Antiochus lost nearly 10,000 infantry, 300 cavalry, and 4,000 prisoners; Ptolemy 1,500 infantry and 700 cavalry.

1.8–15: Ptolemy attempts to enter the sanctuary at Jerusalem. 9: Reference to *the supreme God,* Greek "megistos theos," occurs frequently in 3 Maccabees (1.9, 16; 3.11; 4.16; 5.25; 7.22) as well as in 2 Maccabees (3.36). That the pagan Ptolemy should have offered *sacrifice* to the God of the Jews was not an unusual practice in an age of religious syncretism. *The holy place,* a surrogate for "the temple" in 3 Maccabees and other Jewish literature. **10:** Ptolemy's eagerness to inspect the interior of the temple may have been motivated by his curiosity concerning its architecture, for he considered himself a con-

14 And someone answered thoughtlessly that it was wrong to take that as a portent.[g] 15 "But since this has happened," the king[h] said, "why should not I at least enter, whether they wish it or not?"

16 Then the priests in all their vestments prostrated themselves and entreated the supreme God[i] to aid in the present situation and to avert the violence of this evil design, and they filled the temple with cries and tears; 17 those who remained behind in the city were agitated and hurried out, supposing that something mysterious was occurring. 18 Young women who had been secluded in their chambers rushed out with their mothers, sprinkled their hair with dust,[i] and filled the streets with groans and lamentations. 19 Those women who had recently been arrayed for marriage abandoned the bridal chambers[k] prepared for wedded union, and, neglecting proper modesty, in a disorderly rush flocked together in the city. 20 Mothers and nurses abandoned even newborn children here and there, some in houses and some in the streets, and without a backward look they crowded together at the most high temple. 21 Various were the supplications of those gathered there because of what the king was profanely plotting. 22 In addition, the bolder of the citizens would not tolerate the completion of his plans or the fulfillment of his intended purpose. 23 They shouted to their compatriots to take arms and die courageously for the ancestral law, and created a considerable disturbance in the holy place;[l] and being barely restrained by the old men and the elders,[m] they resorted to the same

posture of supplication as the others. 24 Meanwhile the crowd, as before, was engaged in prayer, 25 while the elders near the king tried in various ways to change his arrogant mind from the plan that he had conceived. 26 But he, in his arrogance, took heed of nothing, and began now to approach, determined to bring the aforesaid plan to a conclusion. 27 When those who were around him observed this, they turned, together with our people, to call upon him who has all power to defend them in the present trouble and not to overlook this unlawful and haughty deed. 28 The continuous, vehement, and concerted cry of the crowds[n] resulted in an immense uproar; 29 for it seemed that not only the people but also the walls and the whole earth around echoed, because indeed all at that time[o] preferred death to the profanation of the place.

2 Then the high priest Simon, facing the sanctuary, bending his knees and extending his hands with calm dignity, prayed as follows:[p] 2 "Lord, Lord, king of the heavens, and sovereign of all creation, holy among the holy ones, the only ruler, almighty, give attention to us who are suffering grievously from an impious and profane man, puffed up in his audacity and power. 3 For you, the

g Or to boast of this   h Gk he   i Gk the greatest God   j Other ancient authorities add and ashes   k Or the canopies   l Gk the place   m Other ancient authorities read priests   n Other ancient authorities read vehement cry of the assembled crowds   o Other ancient authorities lack at that time   p Other ancient authorities lack verse 1

noisseur of the arts. **11:** *High priest . . . once a year,* Ex 30.10; Lev 16.2, 11–12, 15, 34; Heb 9.7; Josephus, *Ant.* XII.iii.3.

1.16–29: Jewish reaction to Ptolemy's determination to enter the sanctuary. It is possible that because of religious superstition many regarded Ptolemy's desire as an evil omen of something calamitous that would befall them. **18:** *Young women,* 2 Macc 3.19. **19:** *Bridal chambers,* Joel 2.16; 2 Esd 16.33, 34; Bar 2.23. **21:** *Various* with respect to the persons offering prayer. **23:** *Die courageously for*

the ancestral law, 1 Macc 2.40; 3.21; 13.3–4; 2 Macc 8.21.

2.1–20: The prayer of Simon, the high priest. This was probably Simon II, son of Onias II and high priest about 219–196 B.C. (see Sir 50.1 n.). The prayer is in a classic Jewish form that, like Eleazar's prayer in 6.1–15, follows the pattern of Pss 105 and 106 in addressing God in terms of his power, glory, and great works reflected in historical deliverances of Israel. **2:** *Holy among the holy ones,* Isa 57.15 LXX. **3:** Ex 18.11; Ps 31.23. **4:** *Giants,*

creator of all things and the governor of all, are a just Ruler, and you judge those who have done anything in insolence and arrogance. 4 You destroyed those who in the past committed injustice, among whom were even giants who trusted in their strength and boldness, whom you destroyed by bringing on them a boundless flood. 5 You consumed with fire and sulfur the people of Sodom who acted arrogantly, who were notorious for their vices;*q* and you made them an example to those who should come afterward. 6 You made known your mighty power by inflicting many and varied punishments on the audacious Pharaoh who had enslaved your holy people Israel. 7 And when he pursued them with chariots and a mass of troops, you overwhelmed him in the depths of the sea, but carried through safely those who had put their confidence in you, the Ruler over the whole creation. 8 And when they had seen works of your hands, they praised you, the Almighty. 9 You, O King, when you had created the boundless and immeasurable earth, chose this city and sanctified this place for your name, though you have no need of anything; and when you had glorified it by your magnificent manifestation,*r* you made it a firm foundation for the glory of your great and honored name. 10 And because you love the house of Israel, you promised that if we should have reverses and tribulation should overtake us, you would listen to our petition when we come to this place and pray. 11 And indeed you are faithful and true.

12 And because oftentimes when our fathers were oppressed you helped them in their humiliation, and rescued them from great evils, 13 see now, O holy King, that because of our many and great sins we are crushed with suffering, subjected to our enemies, and overtaken by helplessness. 14 In our downfall this audacious and profane man undertakes to violate the holy place on earth dedicated to your glorious name. 15 For your dwelling is the heaven of heavens, unapproachable by human beings. 16 But because you graciously bestowed your glory on your people Israel, you sanctified this place. 17 Do not punish us for the defilement committed by these men, or call us to account for this profanation, otherwise the transgressors will boast in their wrath and exult in the arrogance of their tongue, saying, 18 'We have trampled down the house of the sanctuary as the houses of the abominations are trampled down.' 19 Wipe away our sins and disperse our errors, and reveal your mercy at this hour. 20 Speedily let your mercies overtake us, and put praises in the mouth of those who are downcast and broken in spirit, and give us peace."

21 Thereupon God, who oversees all things, the first Father of all, holy among the holy ones, having heard the lawful supplication, scourged him who had exalted himself in insolence and audacity. 22 He shook him on this side and that as a reed is shaken by the wind, so that he lay helpless on the ground and, besides being paralyzed in his limbs, was unable

q Other ancient authorities read secret in their vices   r Or epiphany

---

Jdt 16.7; Wis 14.6; Sir 16.7; Bar 3.26; 1 Enoch 7.2; 15.8. **5:** *Sodom,* Gen 19.24; Deut 29.23; Wis 10.7. *An example,* 2 Pet 2.6. **2.6:** *Power,* Ex 9.16; Rom 9.17. *Varied punishments,* Ex chs 5–12. *Your holy people Israel,* Ex 19.6; 1 Pet 2.9. **7:** *Overwhelmed him,* Ex 14.21–28. **8:** *They praised,* Ex 15.1–21; Wis 19.8–9; 1 Macc 4.9. **9:** *Boundless and immeasurable earth,* Bar 3.24–25. *This city,* Jerusalem. *Place,* see 1.9 n.; 1 Kings 9.3. *No need of anything,* Acts 17.25; 2 Macc 14.35–36. **10:** *Promised . . . you would listen,* Deut 4.30; 30.1–6; 1 Kings 8.33–34, 48–50. **12:** *Our fathers . . .*

you . . . rescued, 1 Sam 12.10–11; Ps 22.4–5; 106.43; Neh 9.28. **2.14:** *Glorious name,* Jdt 9.8. **15:** *The heaven of heavens,* 1 Kings 8.27; Isa 66.1. *Unapproachable by human beings,* Prov 30.4; Isa 57.15; Bar 3.29. **18:** *Trampled down,* Isa 10.10–11; Dan 8.13. **19:** *Wipe away our sins,* Ps 51.2, 9. **20:** *Mercies,* Ps 79.8, 13. **2.21–24:** **The punishment of Ptolemy.** **21:** *Holy ones,* see Isa 57.15 LXX. **22:** *Paralyzed,* compare the punishment of Heliodorus (2 Macc 3.22–30) and of Antiochus (2 Macc 9.4–7). **23:** *Friends,* the higher officers and

even to speak, since he was smitten[s] by a righteous judgment. 23 Then both friends and bodyguards, seeing the severe punishment that had overtaken him, and fearing that he would lose his life, quickly dragged him out, panic-stricken in their exceedingly great fear. 24 After a while he recovered, and though he had been punished, he by no means repented, but went away uttering bitter threats.

25 When he arrived in Egypt, he increased in his deeds of malice, abetted by the previously mentioned drinking companions and comrades, who were strangers to everything just. 26 He was not content with his uncounted licentious deeds, but even continued with such audacity that he framed evil reports in the various localities; and many of his friends, intently observing the king's purpose, themselves also followed his will. 27 He proposed to inflict public disgrace on the Jewish community,[t] and he set up a stone[u] on the tower in the courtyard with this inscription: 28 "None of those who do not sacrifice shall enter their sanctuaries, and all Jews shall be subjected to a registration involving poll tax and to the status of slaves. Those who object to this are to be taken by force and put to death; 29 those who are registered are also to be branded on their bodies by fire with the ivy-leaf symbol of Dionysus, and they shall also be reduced to their former limited status." 30 In order that he might not appear to be an enemy of all, he inscribed below: "But if any of them prefer to join those who have been initiated into the mysteries, they shall have equal citizenship with the Alexandrians."

31 Now some, however, with an obvious abhorrence of the price to be exacted for maintaining the religion of their city,[v] readily gave themselves up, since they expected to enhance their reputation by their future association with the king. 32 But the majority acted firmly with a courageous spirit and did not abandon their religion; and by paying money in exchange for life they confidently attempted to save themselves from the registration. 33 They remained resolutely hopeful of obtaining help, and they abhorred those who separated themselves from them, considering them to be enemies of the Jewish nation,[t] and depriving them of companionship and mutual help.

3 When the impious king comprehended this situation, he became so infuriated that not only was he enraged against those Jews who lived in Alexandria, but was still more bitterly hostile toward those in the countryside; and he ordered that all should promptly be gathered into one place, and put to death by the most cruel means. 2 While these matters were being arranged, a hostile rumor was circulated against the Jewish nation by some who conspired to do them ill, a pretext being given by a report that they hindered others[w] from the observance of their customs. 3 The Jews, however, con-

---

s Other ancient authorities read *pierced*
t Gk *the nation*   u Gk *stele*   v Meaning of Gk uncertain   w Gk *them*

---

courtiers of the king (see 4 Macc 8.5 n.). *Lose his life,* 2 Macc 3.31. **24**: *By no means repented,* 2 Macc 9.7.

2.25–33: Hostile measures against the Jews of Alexandria. 25: Inasmuch as these *companions* have not, in fact, been *previously mentioned* in the present text of 3 Maccabees, we have additional evidence that the opening section has been lost (see 1.1–7 n.). **26**: *Friends,* see v. 23 n. **28**: *Registration,* a rare Greek word ("laographia"), which has been found in Greek papyri from Egypt, refers to a list of all people of the lower classes and of the slaves. **29**: Such branding in honor of a deity was not uncommon in ancient times (compare Rev 7.3; 13.16–17). According to 2 Macc 6.7 Antiochus introduced the worship of Dionysus into Jerusalem. **32**: *Paying money,* as bribes. **33**: *Depriving them of companionship,* 2 Jn 10–11.

3.1–10: The Jews and their neighbors. 1: The distinction between Jews in *Alexandria* and those in the *countryside* is made also in 4.11–12. **2**: The *rumor* (Esth 3.8) maliciously represents the Jews as hostile to the best interests of the state. **4**: *Separateness with respect to*

tinued to maintain goodwill and unswerving loyalty toward the dynasty; [4] but because they worshiped God and conducted themselves by his law, they kept their separateness with respect to foods. For this reason they appeared hateful to some; [5] but since they adorned their style of life with the good deeds of upright people, they were established in good repute with everyone. [6] Nevertheless those of other races paid no heed to their good service to their nation, which was common talk among all; [7] instead they gossiped about the differences in worship and foods, alleging that these people were loyal neither to the king nor to his authorities, but were hostile and greatly opposed to his government. So they attached no ordinary reproach to them.

[8] The Greeks in the city, though wronged in no way, when they saw an unexpected tumult around these people and the crowds that suddenly were forming, were not strong enough to help them, for they lived under tyranny. They did try to console them, being grieved at the situation, and expected that matters would change; [9] for such a great community ought not be left to its fate when it had committed no offense. [10] And already some of their neighbors and friends and business associates had taken some of them aside privately and were pledging to protect them and to exert more earnest efforts for their assistance.

[11] Then the king, boastful of his present good fortune, and not considering the might of the supreme God,[x] but assuming that he would persevere constantly in his same purpose, wrote this letter against them:

[12] "King Ptolemy Philopator to his generals and soldiers in Egypt and all its districts, greetings and good health:

[13] "I myself and our government are faring well. [14] When our expedition took place in Asia, as you yourselves know, it was brought to conclusion, according to plan, by the gods' deliberate alliance with us in battle, [15] and we considered that we should not rule the nations inhabiting Coelesyria and Phoenicia by the power of the spear, but should cherish them with clemency and great benevolence, gladly treating them well. [16] And when we had granted very great revenues to the temples in the cities, we came on to Jerusalem also, and went up to honor the temple of those wicked people, who never cease from their folly. [17] They accepted our presence by word, but insincerely by deed, because when we proposed to enter their inner temple and honor it with magnificent and most beautiful offerings, [18] they were carried away by their traditional arrogance, and excluded us from entering; but they were spared the exercise of our power because of the benevolence that we have toward all. [19] By maintaining their manifest ill-will toward us, they become the only people among all nations who hold their heads high in defiance of kings and their own benefactors, and are unwilling to regard any action as sincere.

[20] "But we, when we arrived in Egypt victorious, accommodated ourselves to their folly and did as was proper, since we treat all nations with benevolence. [21] Among other things, we made known to all our amnesty toward their compatriots here, both because of their

---

[x] Gk *the greatest God*

---

foods, for a defense of the observance of Jewish dietary rules, see the *Letter of Aristeas,* §§128–166. **5**: Deut 4.5–6; Col 4.5; 1 Thess 4.12. **7**: For similar charges see Esth 3.8; Add Esth 13.4–5. **8**: *The Greeks,* the nobler, cultivated class, in distinction from *those of other races* (v. 6).

3.11–30: Ptolemy orders the arrest of all Jews in his kingdom. 11: *Assuming that he*

would persevere, a reference to the calamity that came upon him by which he forgot his own previous commands (5.27–28).

3.15: *Benevolence,* Greek "philanthropia" (see vv. 18, 20), was regarded as a major political virtue during the Hellenistic (and Byzantine) period. **18**: *Excluded,* 1.10–12. **21**: For the confidence placed in Jews, see 6.25 and Josephus, *Ant.* XIX.v.2.

alliance with us and the myriad affairs liberally entrusted to them from the beginning; and we ventured to make a change, by deciding both to deem them worthy of Alexandrian citizenship and to make them participants in our regular religious rites. [y] 22 But in their innate malice they took this in a contrary spirit, and disdained what is good. Since they incline constantly to evil, 23 they not only spurn the priceless citizenship, but also both by speech and by silence they abominate those few among them who are sincerely disposed toward us; in every situation, in accordance with their infamous way of life, they secretly suspect that we may soon alter our policy. 24 Therefore, fully convinced by these indications that they are ill-disposed toward us in every way, we have taken precautions so that, if a sudden disorder later arises against us, we shall not have these impious people behind our backs as traitors and barbarous enemies. 25 Therefore we have given orders that, as soon as this letter arrives, you are to send to us those who live among you, together with their wives and children, with insulting and harsh treatment, and bound securely with iron fetters, to suffer the sure and shameful death that befits enemies. 26 For when all of these have been punished, we are sure that for the remaining time the government will be established for ourselves in good order and in the best state. 27 But those who shelter any of the Jews, whether old people or children or even infants, will be tortured to death with the most hateful torments, together with their families. 28 Any who are willing to give information will receive the property of those who incur the punishment, and also two thousand drachmas from the royal treasury, and will be awarded their freedom. [z] 29 Every place detected sheltering a Jew is to be made unapproachable and burned with fire, and shall become useless for all time to any mortal creature." 30 The letter was written in the above form.

4 In every place, then, where this decree arrived, a feast at public expense was arranged for the Gentiles with shouts and gladness, for the inveterate enmity that had long ago been in their minds was now made evident and outspoken. 2 But among the Jews there was incessant mourning, lamentation, and tearful cries; everywhere their hearts were burning, and they groaned because of the unexpected destruction that had suddenly been decreed for them. 3 What district or city, or what habitable place at all, or what streets were not filled with mourning and wailing for them? 4 For with such a harsh and ruthless spirit were they being sent off, all together, by the generals in the several cities, that at the sight of their unusual punishments, even some of their enemies, perceiving the common object of pity before their eyes, reflected on the uncertainty of life and shed tears at the most miserable expulsion of these people. 5 For a multitude of gray-headed old men, sluggish and bent with age, was being led away, forced to march at a swift pace by the violence with which they were driven in such a shameful manner. 6 And young women who had just entered the bridal chamber [a] to share married life exchanged joy for wailing, their myrrh-perfumed hair sprinkled with ashes, and were carried away unveiled, all together raising a lament instead of a wedding song, as they were torn by the harsh treatment of the heathen. [b] 7 In bonds and in public view they were violently dragged along as far

[y] Other ancient authorities read *partners of our regular priests*   [z] Gk *crowned with freedom*   [a] Or *the canopy*   [b] Other ancient authorities read *as though torn by heathen whelps*

3.24: *Behind our backs*, Ex 1.10. **28**: *Awarded their freedom*, another rendering is "crowned at the Eleutheria" (a festival of Dionysus; see 2.29 n.). **29**: *Useless for all time*, Add Esth 16.24.
4.1–21: **The Jews brought to Alexandria and imprisoned. 1**: *Enmity* on the part of native-born Egyptians for the Jews is assumed in 3 Maccabees; contrast 3.8. **2**: *Mourning, lamentation, and tearful cries*, Esth 4.3. **4**: *Unusual punishments*, 2 Macc 9.6. **6**: *Young women*, 1 Macc 1.26–27.

as the place of embarkation. [8] Their husbands, in the prime of youth, their necks encircled with ropes instead of garlands, spent the remaining days of their marriage festival in lamentations instead of good cheer and youthful revelry, seeing death immediately before them. [c] [9] They were brought on board like wild animals, driven under the constraint of iron bonds; some were fastened by the neck to the benches of the boats, others had their feet secured by unbreakable fetters, [10] and in addition they were confined under a solid deck, so that, with their eyes in total darkness, they would undergo treatment befitting traitors during the whole voyage.

[11] When these people had been brought to the place called Schedia, and the voyage was concluded as the king had decreed, he commanded that they should be enclosed in the hippodrome that had been built with a monstrous perimeter wall in front of the city, and that was well suited to make them an obvious spectacle to all coming back into the city and to those from the city[d] going out into the country, so that they could neither communicate with the king's forces nor in any way claim to be inside the circuit of the city. [e] [12] And when this had happened, the king, hearing that the Jews' compatriots from the city frequently went out in secret to lament bitterly the ignoble misfortune of their kindred, [13] ordered in his rage that these people be dealt with in precisely the same fashion as the others, not omitting any detail of their punishment. [14] The entire race was to be registered individually, not for the hard labor that has been briefly mentioned before, but to be tortured

with the outrages that he had ordered, and at the end to be destroyed in the space of a single day. [15] The registration of these people was therefore conducted with bitter haste and zealous intensity from the rising of the sun until its setting, coming to an end after forty days but still uncompleted.

[16] The king was greatly and continually filled with joy, organizing feasts in honor of all his idols, with a mind alienated from truth and with a profane mouth, praising speechless things that are not able even to communicate or to come to one's help, and uttering improper words against the supreme God.[f] [17] But after the previously mentioned interval of time the scribes declared to the king that they were no longer able to take the census of the Jews because of their immense number, [18] though most of them were still in the country, some still residing in their homes, and some at the place;[g] the task was impossible for all the generals in Egypt. [19] After he had threatened them severely, charging that they had been bribed to contrive a means of escape, he was clearly convinced about the matter [20] when they said and proved that both the paper[h] and the pens they used for writing had already given out. [21] But this was an act of the invincible providence of him who was aiding the Jews from heaven.

5 Then the king, completely inflexible, was filled with overpowering anger and wrath; so he summoned Her-

c Gk *seeing Hades already lying at their feet*
d Gk *those of them*   e Or *claim protection of the walls*; meaning of Gk uncertain   f Gk *the greatest God*   g Other ancient authorities read *on the way*   h Or *paper factory*

4.11: *Schedia,* a promontory about three miles from Alexandria. *The hippodrome* was situated at the east or Canobic gate of Alexandria; according to Strabo (XVII.1.10, 16) a canal joined Schedia and the Canobic gate. *Inside . . . the city,* implies that Jews living in Alexandria (see v. 12, *compatriots from the city*) had been thus far unmolested (yet compare 3.1). **14:** *Mentioned before,* see 2.28. The registration for poll tax and slave status is transformed into an instrument to serve the execu-

tion of *the entire race* of Jews. **15:** *Forty days,* see 2 Macc 5.2. **16:** *Feasts,* Dan 5.4. *Supreme God,* see 1.9 n. **4.17:** *Their immense number,* an obvious hyperbole. **19:** *Bribed,* see 2.32. **21:** *An act of the invincible providence* prevents the registration and serves as the focal point of the entire narrative. As in the past, the Jews will be saved from destruction by *him* who aids them *from heaven.*

5.1–51: **Ptolemy orders the execution**

mon, keeper of the elephants, [2]and ordered him on the following day to drug all the elephants—five hundred in number—with large handfuls of frankincense and plenty of unmixed wine, and to drive them in, maddened by the lavish abundance of drink, so that the Jews might meet their doom. [3]When he had given these orders he returned to his feasting, together with those of his Friends and of the army who were especially hostile toward the Jews. [4]And Hermon, keeper of the elephants, proceeded faithfully to carry out the orders. [5]The servants in charge of the Jews[i] went out in the evening and bound the hands of the wretched people and arranged for their continued custody through the night, convinced that the whole nation would experience its final destruction. [6]For to the Gentiles it appeared that the Jews were left without any aid, [7]because in their bonds they were forcibly confined on every side. But with tears and a voice hard to silence they all called upon the Almighty Lord and Ruler of all power, their merciful God and Father, praying [8]that he avert with vengeance the evil plot against them and in a glorious manifestation rescue them from the fate now prepared for them. [9]So their entreaty ascended fervently to heaven.

10 Hermon, however, when he had drugged the pitiless elephants until they had been filled with a great abundance of wine and satiated with frankincense, presented himself at the courtyard early in the morning to report to the king about these preparations. [11]But the Lord[j] sent upon the king a portion of sleep, that beneficence that from the beginning, night and day, is bestowed by him who

grants it to whomever he wishes. [12]And by the action of the Lord he was overcome by so pleasant and deep a sleep[k] that he quite failed in his lawless purpose and was completely frustrated in his inflexible plan. [13]Then the Jews, since they had escaped the appointed hour, praised their holy God and again implored him who is easily reconciled to show the might of his all-powerful hand to the arrogant Gentiles.

14 But now, since it was nearly the middle of the tenth hour, the person who was in charge of the invitations, seeing that the guests were assembled, approached the king and nudged him. [15]And when he had with difficulty roused him, he pointed out that the hour of the banquet was already slipping by, and he gave him an account of the situation. [16]The king, after considering this, returned to his drinking, and ordered those present for the banquet to recline opposite him. [17]When this was done he urged them to give themselves over to revelry and to make the present[l] portion of the banquet joyful by celebrating all the more. [18]After the party had been going on for some time, the king summoned Hermon and with sharp threats demanded to know why the Jews had been allowed to remain alive through the present day. [19]But when he, with the corroboration of his Friends, pointed out that while it was still night he had carried out completely the order given him, [20]the king,[j] possessed by a savagery worse than that of Phalaris, said that the

i Gk *them*   j Gk *he*   k Other ancient authorities add *from evening until the ninth hour*   l Other ancient authorities read *delayed* (Gk *untimely*)

---

of the Jews, but is twice thwarted (vv. 12, 27–28). **2:** *Five hundred elephants* is an exaggeration; Ptolemy had seventy-three elephants at the battle of Raphia. **3:** *Returned to his feasting,* 4.16. **5:** *Bound the hands,* according to 3.25 they had already been bound securely, but perhaps their hands had been loosened when the Jews were enclosed within the hippodrome (see 5.49 n.). **6:** *It appeared,* although the reader understands the Jews are constantly

aided by divine Providence. **7:** The title *Father* is also given to God in Tob 13.4; Wis 11.10. **5.8:** *Glorious manifestation,* 2 Macc 2.21. **11:** The divine gift of *sleep* is extolled in the Latin poets (Seneca, *Hercules Furens* 1066 ff.; Statius, *Silvae* v. 4); compare Ps 127.2. **14:** *Middle of the tenth hour,* 3:30 P.M. **19:** *Friends,* see 2.23 n. **20:** *Phalaris,* tyrant of Agrigentum (c. 570–554 B.C.) whose cruelty was proverbial (Polybius XII.25).

Jews[m] were benefited by today's sleep, "but," he added, "tomorrow without delay prepare the elephants in the same way for the destruction of the lawless Jews!" 21 When the king had spoken, all those present readily and joyfully with one accord gave their approval, and all went to their own homes. 22 But they did not so much employ the duration of the night in sleep as in devising all sorts of insults for those they thought to be doomed.

23 Then, as soon as the cock had crowed in the early morning, Hermon, having equipped[n] the animals, began to move them along in the great colonnade. 24 The crowds of the city had been assembled for this most pitiful spectacle and they were eagerly waiting for daybreak. 25 But the Jews, at their last gasp—since the time had run out—stretched their hands toward heaven and with most tearful supplication and mournful dirges implored the supreme God[o] to help them again at once. 26 The rays of the sun were not yet shed abroad, and while the king was receiving his Friends, Hermon arrived and invited him to come out, indicating that what the king desired was ready for action. 27 But he, on receiving the report and being struck by the unusual invitation to come out—since he had been completely overcome by incomprehension—inquired what the matter was for which this had been so zealously completed for him. 28 This was the act of God who rules over all things, for he had implanted in the king's mind a forgetfulness of the things he had previously devised. 29 Then Hermon and all the king's Friends[p] pointed out that the animals and the armed forces were ready, "O king, according to your eager purpose."[q] 30 But at these words he was filled with an overpowering wrath, because by the providence of God his whole mind had been deranged concern-

ing these matters; and with a threatening look he said, 31 "If your parents or children were present, I would have prepared them to be a rich feast for the savage animals instead of the Jews, who give me no ground for complaint and have exhibited to an extraordinary degree a full and firm loyalty to my ancestors. 32 In fact you would have been deprived of life instead of these, if it were not for an affection arising from our nurture in common and your usefulness." 33 So Hermon suffered an unexpected and dangerous threat, and his eyes wavered and his face fell. 34 The king's Friends one by one sullenly slipped away and dismissed[r] the assembled people to their own occupations. 35 Then the Jews, on hearing what the king had said, praised the manifest Lord God, King of kings, since this also was his aid that they had received.

36 The king, however, reconvened the party in the same manner and urged the guests to return to their celebrating. 37 After summoning Hermon he said in a threatening tone, "How many times, you poor wretch, must I give you orders about these things? 38 Equip[s] the elephants now once more for the destruction of the Jews tomorrow!" 39 But the officials who were at table with him, wondering at his instability of mind, remonstrated as follows: 40 "O king, how long will you put us to the test, as though we are idiots, ordering now for a third time that they be destroyed, and again revoking your decree in the matter?[t] 41 As a result the city is in a tumult because of its expectation; it is crowded

m Gk *they* n Or *armed* o Gk *the greatest God* p Gk *all the Friends* q Other ancient authorities read *pointed to the beasts and the armed forces, saying, "They are ready, O king, according to your eager purpose."* r Other ancient authorities read *he dismissed* s Or *Arm* t Other ancient authorities read *when the matter is in hand*

5.23: *The great colonnade* was no doubt some well-known place in Alexandria. 28: *The act of God,* Prov 21.1. 29: An interpolation in several Greek manuscripts indicates that though Ptolemy was moved by compas-

sion and determined to release the Jews, Hermon influenced him to proceed with his plans to destroy them.

5.41: The revolutionary character of the Alexandrians in ancient times is well known.

with masses of people, and also in constant danger of being plundered."

42 At this the king, a Phalaris in everything and filled with madness, took no account of the changes of mind that had come about within him for the protection of the Jews, and he firmly swore an irrevocable oath that he would send them to death[u] without delay, mangled by the knees and feet of the animals, [43] and would also march against Judea and rapidly level it to the ground with fire and spear, and by burning to the ground the temple inaccessible to him[v] would quickly render it forever empty of those who offered sacrifices there. [44] Then the Friends and officers departed with great joy, and they confidently posted the armed forces at the places in the city most favorable for keeping guard.

45 Now when the animals had been brought virtually to a state of madness, so to speak, by the very fragrant draughts of wine mixed with frankincense and had been equipped with frightful devices, the elephant keeper [46] entered at about dawn into the courtyard—the city now being filled with countless masses of people crowding their way into the hippodrome—and urged the king on to the matter at hand. [47] So he, when he had filled his impious mind with a deep rage, rushed out in full force along with the animals, wishing to witness, with invulnerable heart and with his own eyes, the grievous and pitiful destruction of the aforementioned people.

48 When the Jews saw the dust raised by the elephants going out at the gate and by the following armed forces, as well as by the trampling of the crowd, and heard the loud and tumultuous noise, [49] they thought that this was their last moment of life, the end of their most miserable suspense, and giving way to lamentation and groans they kissed each other, embracing relatives and falling into one another's arms[w]—parents and children, mothers and daughters, and others with babies at their breasts who were drawing their last milk. [50] Not only this, but when they considered the help that they had received before from heaven, they prostrated themselves with one accord on the ground, removing the babies from their breasts, [51] and cried out in a very loud voice, imploring the Ruler over every power to manifest himself and be merciful to them, as they stood now at the gates of death.[u]

6 Then a certain Eleazar, famous among the priests of the country, who had attained a ripe old age and throughout his life had been adorned with every virtue, directed the elders around him to stop calling upon the holy God, and he prayed as follows: [2] "King of great power, Almighty God Most High, governing all creation with mercy, [3] look upon the descendants of Abraham, O Father, upon the children of the sainted Jacob, a people of your consecrated portion who are perishing as foreigners in a foreign land. [4] Pharaoh with his abundance of chariots, the former ruler of this Egypt, exalted with lawless insolence and boastful tongue, you destroyed

u Gk *Hades*   v Gk *us*   w Gk *falling upon*
their necks

42: *Phalaris,* see v. 20 n. 45: The *frightful devices* were probably scythes, knives, and other military equipment attached to the different parts of the bodies of the elephants. 49: *Embracing,* but according to 6.27 they are still bound.
5.50–51: The author continues to build up his thesis that God was the only resort for the captives, and that their hope would not be frustrated. 51: *Gates of death,* Pss 9.13; 107.18.
6.1–15: **The prayer of Eleazar.** Like Si-

mon's prayer in 2.1–20, it is plainly Jewish in form and style, containing doxology, thanksgiving for God's earlier interventions in Israel's history, and petition for a new miracle. The emphasis is on the exclusiveness and separate standing of Israel before God (v. 3). Eleazar expects God's intervention, not on account of Israel's virtues or merits, but because of divine mercy. 1: *Eleazar,* a favorite name for a Jewish hero (2 Macc 6.18). *Priests,* perhaps those of the Jewish temple at Leonto-

together with his arrogant army by drowning them in the sea, manifesting the light of your mercy on the nation of Israel. 5 Sennacherib exulting in his countless forces, oppressive king of the Assyrians, who had already gained control of the whole world by the spear and was lifted up against your holy city, speaking grievous words with boasting and insolence, you, O Lord, broke in pieces, showing your power to many nations. 6 The three companions in Babylon who had voluntarily surrendered their lives to the flames so as not to serve vain things, you rescued unharmed, even to a hair, moistening the fiery furnace with dew and turning the flame against all their enemies. 7 Daniel, who through envious slanders was thrown down into the ground to lions as food for wild animals, you brought up to the light unharmed. 8 And Jonah, wasting away in the belly of a huge, sea-born monster, you, Father, watched over and restored[x] unharmed to all his family. 9 And now, you who hate insolence, all-merciful and protector of all, reveal yourself quickly to those of the nation of Israel[y]—who are being outrageously treated by the abominable and lawless Gentiles.

10 "Even if our lives have become entangled in impieties in our exile, rescue us from the hand of the enemy, and destroy us, Lord, by whatever fate you choose. 11 Let not the vain-minded praise their vanities[z] at the destruction of your beloved people, saying, 'Not even their god has rescued them.' 12 But you, O Eternal One, who have all might and all power, watch over us now and have mercy on us who by the senseless insolence of the lawless are being deprived of life in the manner of traitors. 13 And let

the Gentiles cower today in fear of your invincible might, O honored One, who have power to save the nation of Jacob. 14 The whole throng of infants and their parents entreat you with tears. 15 Let it be shown to all the Gentiles that you are with us, O Lord, and have not turned your face from us; but just as you have said, 'Not even when they were in the land of their enemies did I neglect them,' so accomplish it, O Lord."

16 Just as Eleazar was ending his prayer, the king arrived at the hippodrome with the animals and all the arrogance of his forces. 17 And when the Jews observed this they raised great cries to heaven so that even the nearby valleys resounded with them and brought an uncontrollable terror upon the army. 18 Then the most glorious, almighty, and true God revealed his holy face and opened the heavenly gates, from which two glorious angels of fearful aspect descended, visible to all but the Jews. 19 They opposed the forces of the enemy and filled them with confusion and terror, binding them with immovable shackles. 20 Even the king began to shudder bodily, and he forgot his sullen insolence. 21 The animals turned back upon the armed forces following them and began trampling and destroying them.

22 Then the king's anger was turned to pity and tears because of the things that he had devised beforehand. 23 For when he heard the shouting and saw them all fallen headlong to destruction, he wept and angrily threatened his Friends, saying, 24 "You are committing

x Other ancient authorities read *rescued and restored*; others, *mercifully restored* y Other ancient authorities read *to the saints of Israel* z Or *bless their vain gods*

---

polis in Egypt. 4: *Pharaoh,* Ex 14.28. 5: *Sennacherib,* 2 Kings 18.13; 19.35–37. 6: *Three companions in Babylon,* Dan 3.22, 27; Song of Thr 22–27. 7: *Daniel,* Dan 6.22.

6.8: *Jonah,* Jon 2.10. 11: *Ps 115.2.* 15: *As you have said,* Lev 26.44.

6.16–29: **The Jews are delivered, and the king now favors them.** 18: *Most glorious,* Greek "megalodoxos," compare 1 Enoch

14.20 and Testament of Levi 3.4, where God is called "the Great Glory." *Angels of fearful aspect,* for similar terror-inspiring apparitions, see Wis 17.3, 15; 18.17; 2 Macc 3.25–29; 10.29. *Visible to all but the Jews,* Dan 10.7; Acts 9.7; 22.6–9.

6.21: *The animals turned* upon the king's own *forces,* a detail found also in Josephus's account (*Against Apion,* ii.5), compare Pss

treason and surpassing tyrants in cruelty; and even me, your benefactor, you are now attempting to deprive of dominion and life by secretly devising acts of no advantage to the kingdom. 25 Who has driven from their homes those who faithfully kept our country's fortresses, and foolishly gathered every one of them here? 26 Who is it that has so lawlessly encompassed with outrageous treatment those who from the beginning differed from*a* all nations in their goodwill toward us and often have accepted willingly the worst of human dangers? 27 Loose and untie their unjust bonds! Send them back to their homes in peace, begging pardon for your former actions!*b* 28 Release the children of the almighty and living God of heaven, who from the time of our ancestors until now has granted an unimpeded and notable stability to our government." 29 These then were the things he said; and the Jews, immediately released, praised their holy God and Savior, since they now had escaped death.

30 Then the king, when he had returned to the city, summoned the official in charge of the revenues and ordered him to provide to the Jews both wines and everything else needed for a festival of seven days, deciding that they should celebrate their rescue with all joyfulness in that same place in which they had expected to meet their destruction. 31 Accordingly those disgracefully treated and near to death,*c* or rather, who stood at its gates, arranged for a banquet of deliverance instead of a bitter and lamentable death, and full of joy they apportioned to celebrants the place that had been prepared for their destruction and burial. 32 They stopped their chanting of dirges and took up the song of their ancestors, praising God, their Savior and worker of wonders.*d* Putting an end to all mourn-

ing and wailing, they formed choruses*e* as a sign of peaceful joy. 33 Likewise also the king, after convening a great banquet to celebrate these events, gave thanks to heaven unceasingly and lavishly for the unexpected rescue that he*f* had experienced. 34 Those who had previously believed that the Jews would be destroyed and become food for birds, and had joyfully registered them, groaned as they themselves were overcome by disgrace, and their fire-breathing boldness was ignominiously*g* quenched.

35 The Jews, as we have said before, arranged the aforementioned choral group*h* and passed the time in feasting to the accompaniment of joyous thanksgiving and psalms. 36 And when they had ordained a public rite for these things in their whole community and for their descendants, they instituted the observance of the aforesaid days as a festival, not for drinking and gluttony, but because of the deliverance that had come to them through God. 37 Then they petitioned the king, asking for dismissal to their homes. 38 So their registration was carried out from the twenty-fifth of Pachon to the fourth of Epeiph,*i* for forty days; and their destruction was set for the fifth to the seventh of Epeiph,*j* the three days 39 on which the Lord of all most gloriously revealed his mercy and rescued them all together and unharmed. 40 Then they feasted, being provided with everything by the king, until the fourteenth day,*k* on which also they made the petition for

a Or *excelled above*   *b* Other ancient authorities read *revoking your former commands*   *c* Gk *Hades*   *d* Other ancient authorities read *praising Israel and the wonder-working God*; or *praising Israel's Savior, the wonder-working God*   *e* Or *dances*   *f* Other ancient authorities read *they*   *g* Other ancient authorities read *completely*   *h* Or *dance*   *i* July 7–August 15   *j* August 16–18   *k* August 25

7.15–16; 9.15–16; 35.8; 57.6. **25:** *Faithfully*, contrast the king's language in 3.24. **28:** *Children of . . . God*, Wis 18.13.
    6.30–41: The Jews celebrate their deliverance. 32: *The song of their ancestors*, perhaps Ps 136, which was used earlier as a hymn of thanksgiving (1 Chr 16.41; 2 Chr 5.13; 7.3;

Ezra 3.11). **34:** *Food for birds*, Gen 40.19; Ezek 39.4; 2 Macc 9.15.
    6.36: The institution of Jewish festivals is a common feature at this period (compare Esth 9.15; 1 Macc 4.56; 7.49; 13.51; 2 Macc 10.6; 15.36). **38:** *Epeiph*, or Epiphi.

their dismissal. [41] The king granted their request at once and wrote the following letter for them to the generals in the cities, magnanimously expressing his concern:

7 "King Ptolemy Philopator to the generals in Egypt and all in authority in his government, greetings and good health:

2 "We ourselves and our children are faring well, the great God guiding our affairs according to our desire. [3] Certain of our friends, frequently urging us with malicious intent, persuaded us to gather together the Jews of the kingdom in a body and to punish them with barbarous penalties as traitors; [4] for they declared that our government would never be firmly established until this was accomplished, because of the ill-will that these people had toward all nations. [5] They also led them out with harsh treatment as slaves, or rather as traitors, and, girding themselves with a cruelty more savage than that of Scythian custom, they tried without any inquiry or examination to put them to death. [6] But we very severely threatened them for these acts, and in accordance with the clemency that we have toward all people we barely spared their lives. Since we have come to realize that the God of heaven surely defends the Jews, always taking their part as a father does for his children, [7] and since we have taken into account the friendly and firm goodwill that they had toward us and our ancestors, we justly have acquitted them of every charge of whatever kind. [8] We also have ordered all people to return to their own homes, with no one in any place[l] doing them harm at all or reproaching them for the irrational things that have happened. [9] For you should know that if we devise any evil against them or cause them any grief at all, we always shall have not a mortal but the Ruler over every power, the Most High God, in everything and inescapably as an antagonist to avenge such acts. Farewell."

10 On receiving this letter the Jews[m] did not immediately hurry to make their departure, but they requested of the king that at their own hands those of the Jewish nation who had willfully transgressed against the holy God and the law of God should receive the punishment they deserved. [11] They declared that those who for the belly's sake had transgressed the divine commandments would never be favorably disposed toward the king's government. [12] The king[n] then, admitting and approving the truth of what they said, granted them a general license so that freely, and without royal authority or supervision, they might destroy those everywhere in his kingdom who had transgressed the law of God. [13] When they had applauded him in fitting manner, their priests and the whole multitude shouted the Hallelujah and joyfully departed. [14] And so on their way they punished and put to a public and shameful death any whom they met of their compatriots who had become defiled. [15] In that day they put to death more than three hundred men; and they kept the day as a joyful festival, since they had destroyed the profaners. [16] But those

l Other ancient authorities read *way*
m Gk *they* n Gk *He*

7.1–9: **Ptolemy's letter on behalf of the Jews.** 2: *Children,* Philopator had only one legitimate son, born in 209–8, who reigned later as Ptolemy V Epiphanes (203–181 B.C.). Either the author had no knowledge of Philopator's family life, or the king is referring in general terms to include all members of his court. 3: The king seeks to exonerate himself, blaming others. 4: *Ill-will,* 3.2, 7; Esth 3.8; Add Esth 13.4–5. 5: *Scythian custom,* see 2 Macc 4.47 n.; 4 Macc 10.7 n. 6: *Threatened them,* that is, the enemies of the Jews. *Father,* 5.7 n.; Ps 103.13. 8: *In any place* through which the Jews might pass on their return.

7.10–23: **The Jews punish the renegades and return home.** 10: *They requested of the king,* in the later periods of their history the Jews were obliged to seek permission from their foreign rulers to execute their own laws pertaining to capital punishment (Deut 13.6–18; Esth 8.8–11; Jn 18.31). 13: *Hallelujah,* Tob 13.18.

7.16: Bar 5.6, 8. 17: This *Ptolemais* was probably not the city of this name near

who had held fast to God even to death and had received the full enjoyment of deliverance began their departure from the city, crowned with all sorts of very fragrant flowers, joyfully and loudly giving thanks to the one God of their ancestors, the eternal Savior*o* of Israel, in words of praise and all kinds of melodious songs.

17 When they had arrived at Ptolemais, called "rose-bearing" because of a characteristic of the place, the fleet waited for them, in accordance with the common desire, for seven days. 18 There they celebrated their deliverance,*p* for the king had generously provided all things to them for their journey until all of them arrived at their own houses. 19 And when they had all landed in peace with appropriate thanksgiving, there too in like manner they decided to observe these days as a joyous festival during the time of their stay. 20 Then, after inscribing them as holy on a pillar and dedicating a place of prayer at the site of the festival, they departed unharmed, free, and overjoyed, since at the king's command they had all of them been brought safely by land and sea and river to their own homes. 21 They also possessed greater prestige among their enemies, being held in honor and awe; and they were not subject at all to confiscation of their belongings by any one. 22 Besides, they all recovered all of their property, in accordance with the registration, so that those who held any of it restored it to them with extreme fear.*q* So the supreme God perfectly performed great deeds for their deliverance. 23 Blessed be the Deliverer of Israel through all times! Amen.

o Other ancient authorities read *the holy Savior*; others, *the holy one*   p Gk *they made a cup of deliverance*   q Other ancient authorities read *with a very large supplement*

Thebes in Upper Egypt, but "Ptolemais at the harbor" in the Arsinoite nome (province), about twelve miles from present-day Cairo. *Rose-bearing* is not elsewhere applied to Ptolemais. 20: *And sea,* there was no sea to cross in Egypt.

7.22–23: The book closes with a benediction to the *Supreme God . . . Deliverer;* this suggests that it may have been read liturgically to celebrate the Alexandrian festival described in 6.30–36.

2 Esdras

The book commonly known as 2 Esdras differs from the other fourteen books of the Apocrypha in being an apocalypse (for the characteristics of apocalyptic literature, see p. 362 NT). The main part of 2 Esdras is a series of seven revelations (3.1–5.20; 5.21–6.34; 6.35–9.25; 9.38–10.59; 11.1–12.51; 13.1–58; 14.1–48), in which the seer is instructed by the angel Uriel concerning some of the great mysteries of the moral world.

The problems concerning the composition and transmission of 2 Esdras are extremely complicated. The author of the central portion (chs 3–14) was an unknown Palestinian Jew who probably wrote in Hebrew or Aramaic near the close of the first century A.D. Subsequently the book was translated into Greek. About the middle of the next century an unknown Christian editor added in Greek an introductory section, which now comprises chs 1–2. Nearly a century later another unknown Christian appended chs 15–16, also in Greek.

The Semitic original and almost all of the Greek text have been lost (only 15.57–59 survives on a scrap of Greek papyrus). Before the text of the central section (chs 3–14) perished, however, translations were made into several other languages, namely Syriac, Coptic, Ethiopic, Arabic (two independent versions), Armenian, and Georgian. In the West the entire book (chs 1–16) circulated in several Old Latin versions. A later form of the Latin text is printed, since the Council of Trent, as an appendix to the New Testament in the Roman Catholic Vulgate Bible, where it is called the Fourth Book of Esdras.

The purpose of the original author of 2 Esdras was manifold. The apocalypses that are found in chs 11–12 and 13 are concerned chiefly with denunciation of the wickedness of Rome (under the image of "Babylon"). Chapter 14 contains a most important legend about the preservation of the Scriptures and about the authority of the "hidden" books of the Bible—the Apocrypha and Pseudepigrapha of the Hebrew Scriptures. Chapters 3–10 contain the author's wrestling with some of the central questions confronting the Jewish people in the first century A.D. These include in particular the question of how to affirm God's justice, wisdom, power, and goodness, given the many evils and trials that beset the human community, and the people of Israel in particular. These issues are addressed in a series of questions put to an angel by Ezra, illuminated by visions that the seer Ezra is shown, and followed by interpretations of the visions by the angel. Ezra's laments and questions pose the author's own concerns and hint at his own answers. The answers given by the angel are more conventional responses to the questions than they are definitive teachings of the author of 2 Esdras. Readers of 2 Esdras should therefore give special attention to the content and imagery of the complaints and the visions in order to discern the author's own "answers" to haunting questions that the human community continues to face.

Comprising what is sometimes
called 5 Ezra (chapters 1–2),
4 Ezra (chapters 3–14),
and 6 Ezra (chapters 15–16)

1 The book[a] of the prophet Ezra son
of Seraiah son of Azariah son of Hilkiah son of Shallum son of Zadok son of
Ahitub 2son of Ahijah son of Phinehas
son of Eli son of Amariah son of Azariah
son of Meraimoth son of Arna son of
Uzzi son of Borith son of Abishua son of
Phinehas son of Eleazar 3son of Aaron,
of the tribe of Levi, who was a captive in
the country of the Medes in the reign of
Artaxerxes, king of the Persians. [b]

4 The word of the Lord came to me,
saying, 5"Go, declare to my people their
evil deeds, and to their children the iniquities that they have committed against
me, so that they may tell[c] their children's children 6that the sins of their parents have increased in them, for they
have forgotten me and have offered sacrifices to strange gods 7Was it not I who
brought them out of the land of Egypt,
out of the house of bondage? But they
have angered me and despised my counsels. 8Now you, pull out the hair of your
head and hurl[d] all evils upon them, for
they have not obeyed my law—they are
a rebellious people. 9How long shall I
endure them, on whom I have bestowed
such great benefits? 10For their sake I
have overthrown many kings; I struck
down Pharaoh with his servants and all
his army. 11I destroyed all nations before
them, and scattered in the east the peoples of two provinces,[e] Tyre and Sidon;
I killed all their enemies.

12 "But speak to them and say, Thus
says the Lord: 13Surely it was I who

brought you through the sea, and made
safe highways for you where there was
no road; I gave you Moses as leader and
Aaron as priest; 14I provided light for
you from a pillar of fire, and did great
wonders among you. Yet you have forgotten me, says the Lord.

15 "Thus says the Lord Almighty:[f]
The quails were a sign to you; I gave you
camps for your protection, and in them
you complained. 16You have not exulted
in my name at the destruction of your
enemies, but to this day you still complain.[g] 17Where are the benefits that I
bestowed on you? When you were hungry and thirsty in the wilderness, did you
not cry out to me, 18saying, 'Why have
you led us into this wilderness to kill us?
It would have been better for us to serve
the Egyptians than to die in this wilderness.' 19I pitied your groanings and gave
you manna for food; you ate the bread of
angels. 20When you were thirsty, did I
not split the rock so that waters flowed
in abundance? Because of the heat I
clothed you with the leaves of trees.[h] 21I
divided fertile lands among you; I drove

a Other ancient authorities read *The second book*
b Other ancient authorities, which place
chapters 1 and 2 after 16.78, lack verses 1–3 and
begin the chapter: *The word of the Lord that came
to Ezra son of Chusi in the days of King
Nebuchadnezzar, saying,* "Go,   c Other ancient
authorities read *nourish*   d Other ancient
authorities read *and shake out*   e Other ancient
authorities read *Did I not destroy the city of
Bethsaida because of you, and to the south burn two
cities . . . ?*   f Other ancient authorities lack
Almighty   g Other ancient authorities read
verse 16, *Your pursuer with his army I sank in the
sea, but still the people complain also concerning their
own destruction.*   h Other ancient authorities
read *I made for you trees with leaves*

1.1–2.48: **Ezra is commanded to reprove the Jewish people. 1.1–3: The genealogy of Ezra,** who is of priestly descent
(compare the somewhat different genealogies
in Ezra 7.1–5 and 1 Esd 8.1–2). **3:** Either *Artaxerxes* I, who reigned 464–424 B.C., or *Artaxerxes* II, who reigned 404–358 B.C.; the
former is more probable.
1.4–11: Ezra receives a prophetic call. 4:
The expression, *the word of the Lord came . . .*,
so typical of prophetic authorization, is absent from the canonical book of Ezra. **5:** Isa

58.1. **8:** The command to *pull out his hair* is to
be connected with Ezra's denunciation (*hurl
all evils*) of his people. **10:** Ex 14.28. **11:** The
author is confused; *Tyre and Sidon,* which
were cities, not *provinces,* lay to the west of
the land of the Medes (v. 3).
**1.12–23: Summary of God's mercies to
Israel. 13:** Ex 14.29. **14:** Ex 13.21. **15:** Ex
16.13; Ps 105.40. **17–18:** Num 14.3. **19:** *The
bread of angels,* Ps 78.25; Wis 16.20. **20:** Num
20.11; Wis 11.4. **22–23:** Ex 15.22–25.

out the Canaanites, the Perizzites, and the Philistines[i] before you. What more can I do for you? says the Lord. [22]Thus says the Lord Almighty:[j] When you were in the wilderness, at the bitter stream, thirsty and blaspheming my name, [23]I did not send fire on you for your blasphemies, but threw a tree into the water and made the stream sweet.

24 "What shall I do to you, O Jacob? You, Judah, would not obey me. I will turn to other nations and will give them my name, so that they may keep my statutes. [25]Because you have forsaken me, I also will forsake you. When you beg mercy of me, I will show you no mercy. [26]When you call to me, I will not listen to you; for you have defiled your hands with blood, and your feet are swift to commit murder. [27]It is not as though you had forsaken me; you have forsaken yourselves, says the Lord.

28 "Thus says the Lord Almighty: Have I not entreated you as a father entreats his sons or a mother her daughters or a nurse her children, [29]so that you should be my people and I should be your God, and that you should be my children and I should be your father? [30]I gathered you as a hen gathers her chicks under her wings. But now, what shall I do to you? I will cast you out from my presence. [31]When you offer oblations to me, I will turn my face from you; for I have rejected your[k] festal days, and new moons, and circumcisions of the flesh.[l] [32]I sent you my servants the prophets, but you have taken and killed them and torn their bodies[m] in pieces; I will require their blood of you, says the Lord.[n]

33 "Thus says the Lord Almighty: Your house is desolate; I will drive you out as the wind drives straw; [34]and your sons will have no children, because with you[o] they have neglected my commandment and have done what is evil in my

sight. [35]I will give your houses to a people that will come, who without having heard me will believe. Those to whom I have shown no signs will do what I have commanded. [36]They have seen no prophets, yet will recall their former state.[p] [37]I call to witness the gratitude of the people that is to come, whose children rejoice with gladness;[q] though they do not see me with bodily eyes, yet with the spirit they will believe the things I have said.

38 "And now, father,[r] look with pride and see the people coming from the east; [39]to them I will give as leaders Abraham, Isaac, and Jacob, and Hosea and Amos and Micah and Joel and Obadiah and Jonah [40]and Nahum and Habakkuk, Zephaniah, Haggai, Zechariah and Malachi, who is also called the messenger of the Lord.[s]

2 "Thus says the Lord: I brought this people out of bondage, and I gave

i Other ancient authorities read *Perizzites and their children*   j Other ancient authorities lack *Almighty*   k Other ancient authorities read *I have not commanded for you*   l Other ancient authorities lack *of the flesh*   m Other ancient authorities read *the bodies of the apostles*   n Other ancient authorities add *Thus says the Lord Almighty: Recently you also laid hands on me, crying out before the judge's seat for him to deliver me to you. You took me as a sinner, not as a father who freed you from slavery, and you delivered me to death by hanging me on the tree; these are the things you have done. Therefore, says the Lord, let my Father and his angels return and judge between you and me; if I have not kept the commandment of the Father, if I have not nourished you, if I have not done the things my Father commanded, I will contend in judgment with you, says the Lord.*   o Other ancient authorities lack *with you*   p Other ancient authorities read *their iniquities*   q Other ancient authorities read *The apostles bear witness to the coming people with joy*   r Other ancient authorities read *brother*   s Other ancient authorities read *and Jacob, Elijah and Enoch, Zechariah and Hosea, Amos, Joel, Micah, Obadiah, Zephaniah, [40]Nahum, Jonah, Mattia (or Mattathias), Habakkuk, and twelve angels with flowers*

1.24–32: The casting off of Israel. 26: Isa 1.15; 59.7. **29**: Jer 24.7; Heb 8.10. **30**: The similarity with Mt 23.37 and Lk 13.34 suggests that the author of this part of 2 Esdras was a Jewish Christian. **31**: The rejection of circumcision also reveals the Christian identi-

ty of the author. **32**: Compare Mt 23.34–35. **1.33–40: God will give Israel's houses to another people. 35–36**: Gentile Christians are meant (compare Rom 10.14–20). **37**: *With bodily eyes,* Jn 20.29. **38**: God is represented as addressing Ezra as *father* of the na-

them commandments through my servants the prophets; but they would not listen to them, and made my counsels void. ²The mother who bore them[t] says to them, 'Go, my children, because I am a widow and forsaken. ³I brought you up with gladness; but with mourning and sorrow I have lost you, because you have sinned before the Lord God and have done what is evil in my sight.[u] ⁴But now what can I do for you? For I am a widow and forsaken. Go, my children, and ask for mercy from the Lord.' ⁵Now I call upon you, father, as a witness in addition to the mother of the children, because they would not keep my covenant, ⁶so that you may bring confusion on them and bring their mother to ruin, so that they may have no offspring. ⁷Let them be scattered among the nations; let their names be blotted out from the earth, because they have despised my covenant.

8 "Woe to you, Assyria, who conceal the unrighteous within you! O wicked nation, remember what I did to Sodom and Gomorrah, ⁹whose land lies in lumps of pitch and heaps of ashes.[v] That is what I will do to those who have not listened to me, says the Lord Almighty."

10 Thus says the Lord to Ezra: "Tell my people that I will give them the kingdom of Jerusalem, which I was going to give to Israel. ¹¹Moreover, I will take back to myself their glory, and will give to these others the everlasting habitations, which I had prepared for Israel.[w] ¹²The tree of life shall give them fragrant perfume, and they shall neither toil nor become weary. ¹³Go[x] and you will re-

ceive; pray that your days may be few, that they may be shortened. The kingdom is already prepared for you; be on the watch! ¹⁴Call, O call heaven and earth to witness: I set aside evil and created good; for I am the Living One, says the Lord.

15 "Mother, embrace your children; bring them up with gladness, as does a dove; strengthen their feet, because I have chosen you, says the Lord. ¹⁶And I will raise up the dead from their places, and bring them out from their tombs, because I recognize my name in them. ¹⁷Do not fear, mother of children, for I have chosen you, says the Lord. ¹⁸I will send you help, my servants Isaiah and Jeremiah. According to their counsel I have consecrated and prepared for you twelve trees loaded with various fruits, ¹⁹and the same number of springs flowing with milk and honey, and seven mighty mountains on which roses and lilies grow; by these I will fill your children with joy.

20 "Guard the rights of the widow, secure justice for the ward, give to the needy, defend the orphan, clothe the naked, ²¹care for the injured and the weak, do not ridicule the lame, protect the maimed, and let the blind have a vision of my splendor. ²²Protect the old and the young within your walls; ²³When you find any who are dead, commit them to

t Other ancient authorities read *They begat for themselves a mother who*   u Other ancient authorities read *in his sight*   v Other ancient authorities read *Gomorrah, whose land descends to hell*   w Lat *for those*   x Other ancient authorities read *Seek*

tion. **39–40**: The three patriarchs and the twelve minor prophets, arranged in the order of the Septuagint.

**2.1–9: The Lord's anger against Israel.
1**: *My servants the prophets,* Zech 1.6. **2**: *The mother who bore them,* Jerusalem (Isa 54.1; Gal 4.26–27). *Go . . . ,* Bar 4.19. **3**: Bar 4.11. **5**: Again the author has God address Ezra as *father.* **6**: *To ruin,* in the fall of Jerusalem, A.D. 70. **8**: By the name *Assyria,* Israel's ancient foe, the author refers cryptically to Rome. *Sodom and Gomorrah,* Gen 19.24.
**2.10–14: Israel's habitation to be given

to others. 10**: *My people,* i.e. the Christians (compare Hos 2.23). **11**: *Everlasting habitations,* Lk 16.9. **12**: Rev 2.7; 22.2, 14. **13**: Mt 7.7–8; Lk 11.9–10; Mt 25.34. **14**: Isa 1.2.

**2.15–32: Exhortation to good works.
15**: *Mother,* probably a reference to the church.
2.18: *Twelve trees,* Rev 22.2. **19**: *Milk and honey,* Deut 31.20. *Seven mighty mountains,* from 1 Enoch 24.2, showing the author's familiarity with this section of the apocalypse of Enoch. **23**: *Find any . . . dead,* compare Tob 1.17–19. **26**: Jn 17.12.

the grave and mark it,[y] and I will give you the first place in my resurrection. 24 Pause and be quiet, my people, because your rest will come.

25 "Good nurse, nourish your children; strengthen their feet. 26 Not one of the servants[z] whom I have given you will perish, for I will require them from among your number. 27 Do not be anxious, for when the day of tribulation and anguish comes, others shall weep and be sorrowful, but you shall rejoice and have abundance. 28 The nations shall envy you, but they shall not be able to do anything against you, says the Lord. 29 My power will protect[a] you, so that your children may not see hell.[b]

30 "Rejoice, O mother, with your children, because I will deliver you, says the Lord. 31 Remember your children that sleep, because I will bring them out of the hiding places of the earth, and will show mercy to them; for I am merciful, says the Lord Almighty. 32 Embrace your children until I come, and proclaim mercy to them; because my springs run over, and my grace will not fail."

33 I, Ezra, received a command from the Lord on Mount Horeb to go to Israel. When I came to them they rejected me and refused the Lord's commandment. 34 Therefore I say to you, O nations that hear and understand, "Wait for your shepherd; he will give you everlasting rest, because he who will come at the end of the age is close at hand. 35 Be ready for the rewards of the kingdom, because perpetual light will shine on you forevermore. 36 Flee from the shadow of this age, receive the joy of your glory; I publicly call on my savior to witness.[c] 37 Receive what the Lord has entrusted to you and be joyful, giving thanks to him who has called you to the celestial kingdoms. 38 Rise, stand erect and see the number of those who have been sealed at the feast of the Lord. 39 Those who have departed from the shadow of this age have received glorious garments from the Lord. 40 Take again your full number, O Zion, and close the list of your people who are clothed in white, who have fulfilled the law of the Lord. 41 The number of your children, whom you desired, is now complete; implore the Lord's authority that your people, who have been called from the beginning, may be made holy."

42 I, Ezra, saw on Mount Zion a great multitude that I could not number, and they all were praising the Lord with songs. 43 In their midst was a young man of great stature, taller than any of the others, and on the head of each of them he placed a crown, but he was more exalted than they. And I was held spellbound. 44 Then I asked an angel, "Who are these, my lord?" 45 He answered and said to me, "These are they who have put off mortal clothing and have put on the immortal, and have confessed the name of God. Now they are being crowned, and receive palms." 46 Then I said to the angel, "Who is that young man who is placing crowns on them and putting palms in their hands?" 47 He answered and said to me, "He is the Son of God, whom they confessed in the world." So I began to praise those who had stood valiantly for the name of the Lord.[d] 48 Then the angel said to me, "Go, tell my people how great and how many are the wonders of the Lord God that you have seen."

3 In the thirtieth year after the destruction of the city, I was in Babylon— I, Salathiel, who am also called Ezra. I was troubled as I lay on my bed, and my

[y] Or *seal it*; or *mark them and commit them to the grave*   [z] Or *slaves*   [a] Lat *hands will cover*   [b] Lat *Gehenna*   [c] Other ancient authorities read *I testify that my savior has been commissioned by the Lord*   [d] Other ancient authorities read *to praise and glorify the Lord*

2.33–41: Rejected by Israel, Ezra turns to the Gentiles. 33: *On Mount Horeb,* like a second Moses (Ex 3.1; 2 Chr 5.10). **35:** *Perpetual light,* Isa 60.20; Rev 21.23; 22.5. **40:** *Zion,* the church is identified as Zion, Heb 12.22–23. *Clothed in white,* Rev 3.4; 6.11; 7.14. **41:** *The number . . . is complete,* see 4.36–37 n.; Rev 6.11. *Called,* Rom 8.29–30.

thoughts welled up in my heart, ²because I saw the desolation of Zion and the wealth of those who lived in Babylon. ³My spirit was greatly agitated, and I began to speak anxious words to the Most High, and said, ⁴"O sovereign Lord, did you not speak at the beginning when you planted*ᵉ* the earth—and that without help—and commanded the dust*ᶠ* ⁵and it gave you Adam, a lifeless body? Yet he was the creation of your hands, and you breathed into him the breath of life, and he was made alive in your presence. ⁶And you led him into the garden that your right hand had planted before the earth appeared. ⁷And you laid upon him one commandment of yours; but he transgressed it, and immediately you appointed death for him and for his descendants. From him there sprang nations and tribes, peoples and clans without number. ⁸And every nation walked after its own will; they did ungodly things in your sight and rejected your commands, and you did not hinder them. ⁹But again, in its time you brought the flood upon the inhabitants of the world and destroyed them. ¹⁰And the same fate befell all of them: just as death came upon Adam, so the flood upon them. ¹¹But you left one of them, Noah with his household, and all the righteous who have descended from him.

12 "When those who lived on earth began to multiply, they produced children and peoples and many nations, and again they began to be more ungodly than were their ancestors. ¹³And when they were committing iniquity in your sight, you chose for yourself one of them, whose name was Abraham; ¹⁴you loved him, and to him alone you revealed the end of the times, secretly by night. ¹⁵You made an everlasting covenant with him, and promised him that you would never forsake his descendants; and you gave him Isaac, and to Isaac you gave Jacob and Esau. ¹⁶You set apart Jacob for yourself, but Esau you rejected; and Jacob became a great multitude. ¹⁷And when you led his descendants out of Egypt, you brought them to Mount Sinai. ¹⁸You bent down the heavens and shook*ᵍ* the earth, and moved the world, and caused the depths to tremble, and troubled the times. ¹⁹Your glory passed through the four gates of fire and earthquake and wind and ice, to give the law to the descendants of Jacob, and your commandment to the posterity of Israel.

20 "Yet you did not take away their evil heart from them, so that your law might produce fruit in them. ²¹For the first Adam, burdened with an evil heart, transgressed and was overcome, as were also all who were descended from him. ²²Thus the disease became permanent; the law was in the hearts of the people along with the evil root; but what was

e Other ancient authorities read *formed*
f Syr Ethiop: Lat *people* or *world*
g Syr Ethiop Arab 1 Georg: Lat *set fast*

2.42–48: Ezra's vision of a great multitude. 42: Rev 7.9. **43**: *A young man,* compare v. 47 and 1 Enoch 46.1.

3.1–5.20: The first vision. 3.1–3: Introduction. 1: *The thirtieth year after the destruction* of Jerusalem by Nebuchadnezzar in 587/6 B.C. (2 Kings 25.1ff.) would be 557/6 B.C. The date specified may imply that the author was writing about A.D. 100 (i.e. thirty years after the fall of Jerusalem in A.D. 70). *Salathiel* is the Greek form of Shealtiel (Ezra 3.2; 5.2; Neh 12.1). The words *who am also called Ezra* are an anachronistic gloss; Ezra lived almost a century later. **3.4–36: The author raises perplexing questions.** Whence comes sin with its consequent misery? How can Israel's continuing affliction be reconciled with God's justice? **4–5**: The creation of Adam. **6–8**: Adam's sin brings death on all mortals. **7**: The words *immediately you appointed death* imply that Adam was not originally intended to be mortal (see Gen 3.22 and compare Wis 1.13–14; 2.23–24). **8**: Gen 6.12. **9–11**: The flood (Gen 6.11ff.).

3.12–16: The choice of Abraham (Gen 12.1; 17.5). **14**: *By night,* Gen 15.5, 12, 17. **3.17–19**: The Exodus and the giving of the law. **18**: Compare Ex 19.16–18; Ps 68.7–8. **20–27**: The tendency to sin is universal and permanent. **20**: *Evil heart,* the evil "yeser" (see Sir 15.14–17 n.).

good departed, and the evil remained. 23 So the times passed and the years were completed, and you raised up for yourself a servant, named David. 24 You commanded him to build a city for your name, and there to offer you oblations from what is yours. 25 This was done for many years; but the inhabitants of the city transgressed, 26 in everything doing just as Adam and all his descendants had done, for they also had the evil heart. 27 So you handed over your city to your enemies.

28 "Then I said in my heart, Are the deeds of those who inhabit Babylon any better? Is that why it has gained dominion over Zion? 29 For when I came here I saw ungodly deeds without number, and my soul has seen many sinners during these thirty years. *h* And my heart failed me, 30 because I have seen how you endure those who sin, and have spared those who act wickedly, and have destroyed your people, and protected your enemies, 31 and have not shown to anyone how your way may be comprehended. *i* Are the deeds of Babylon better than those of Zion? 32 Or has another nation known you besides Israel? Or what tribes have so believed the covenants as these tribes of Jacob? 33 Yet their reward has not appeared and their labor has borne no fruit. For I have traveled widely among the nations and have seen that they abound in wealth, though they are unmindful of your commandments. 34 Now therefore weigh in a balance our iniquities and those of the inhabitants of the world; and it will be found which way the turn of the scale will incline. 35 When have the inhabitants of the earth not sinned in your sight? Or what nation

has kept your commandments so well? 36 You may indeed find individuals who have kept your commandments, but nations you will not find."

4 Then the angel that had been sent to me, whose name was Uriel, answered 2 and said to me, "Your understanding has utterly failed regarding this world, and do you think you can comprehend the way of the Most High?" 3 Then I said, "Yes, my lord." And he replied to me, "I have been sent to show you three ways, and to put before you three problems. 4 If you can solve one of them for me, then I will show you the way you desire to see, and will teach you why the heart is evil."

5 I said, "Speak, my lord."

And he said to me, "Go, weigh for me the weight of fire, or measure for me a blast *j* of wind, or call back for me the day that is past."

6 I answered and said, "Who of those that have been born can do that, that you should ask me about such things?"

7 And he said to me, "If I had asked you, 'How many dwellings are in the heart of the sea, or how many streams are at the source of the deep, or how many streams are above the firmament, or which are the exits of Hades, or which are the entrances *k* of paradise?' 8 perhaps you would have said to me, 'I never went down into the deep, nor as yet into Hades, neither did I ever ascend into heaven.' 9 But now I have asked you only

h Ethiop Arab 1 Arm: Lat Syr *in this thirtieth year* *i* Syr; compare Ethiop: Lat *how this way should be forsaken* *j* Syr Ethiop Arab 1 Arab 2 Georg *a measure* *k* Syr Compare Ethiop Arab 2 Arm: Lat lacks *of Hades, or which are the entrances*

3.27: 2 Kings 25.1–21. **28–36**: The deeds of Babylon compared with those of Israel. **28**: *Babylon,* i.e. Rome (Rev 14.8). **29**: *Came here,* to Rome. The first *thirty years* of the Babylonian exile are meant.

3.30–31: Here the author expresses the essence of the problem: God permits evildoers to continue in their wickedness, does not spare the suffering people of God, and does not let anyone understand why this should be

so. **34**: For God's balance, compare Job 31.6; Ps 62.9; Prov 16.2; Dan 5.27; Enoch 41.1; 61.8. **36**: *Individuals* among the Gentiles.

4.1–5.19: **The reply: God's ways are beyond human comprehension.**

4.1–12: **The limitations of the human mind** (Wis 9.16). **1**: The name *Uriel* in Hebrew means "the fire of God." According to Enoch 20.2 Uriel is a watcher over the world and over Tartarus, the lowest part of hell

about fire and wind and the day—things that you have experienced and from which you cannot be separated, and you have given me no answer about them." [10]He said to me, "You cannot understand the things with which you have grown up; [11]how then can your mind comprehend the way of the Most High? And how can one who is already worn out[*l*] by the corrupt world understand incorruption?"[*m*] When I heard this, I fell on my face[*n*] [12]and said to him, "It would have been better for us not to be here than to come here and live in ungodliness, and to suffer and not understand why."

13 He answered me and said, "I went into a forest of trees of the plain, and they made a plan [14]and said, 'Come, let us go and make war against the sea, so that it may recede before us and so that we may make for ourselves more forests.' [15]In like manner the waves of the sea also made a plan and said, 'Come, let us go up and subdue the forest of the plain so that there also we may gain more territory for ourselves.' [16]But the plan of the forest was in vain, for the fire came and consumed it; [17]likewise also the plan of the waves of the sea was in vain,[*o*] for the sand stood firm and blocked it. [18]If now you were a judge between them, which would you undertake to justify, and which to condemn?"

19 I answered and said, "Each made a foolish plan, for the land has been assigned to the forest, and the locale of the sea a place to carry its waves."

20 He answered me and said, "You have judged rightly, but why have you not judged so in your own case? [21]For as the land has been assigned to the forest and the sea to its waves, so also those who inhabit the earth can understand only what is on the earth, and he who is[*p*]

above the heavens can understand what is above the height of the heavens."

22 Then I answered and said, "I implore you, my lord, why[*q*] have I been endowed with the power of understanding? [23]For I did not wish to inquire about the ways above, but about those things that we daily experience: why Israel has been given over to the Gentiles in disgrace; why the people whom you loved has been given over to godless tribes, and the law of our ancestors has been brought to destruction and the written covenants no longer exist. [24]We pass from the world like locusts, and our life is like a mist,[*r*] and we are not worthy to obtain mercy. [25]But what will he do for his[*s*] name that is invoked over us? It is about these things that I have asked."

26 He answered me and said, "If you are alive, you will see, and if you live long,[*t*] you will often marvel, because the age is hurrying swiftly to its end. [27]It will not be able to bring the things that have been promised to the righteous in their appointed times, because this age is full of sadness and infirmities. [28]For the evil about which[*u*] you ask me has been sown, but the harvest of it has not yet come. [29]If therefore that which has been sown is not reaped, and if the place where the evil has been sown does not pass away, the field where the good has been sown will not come. [30]For a grain of evil seed was sown in Adam's heart from the beginning, and how much ungodliness it has produced until now—and will produce until the time of thresh-

l Meaning of Lat uncertain *m* Syr Ethiop
the way of the incorruptible? *n* Syr Ethiop Arab 1:
Meaning of Lat uncertain *o* Lat lacks *was in vain*
p Or *those who are* *q* Syr Ethiop Arm:
Meaning of Lat uncertain *r* Syr Ethiop Arab
Georg: Lat *a trembling* *s* Ethiop adds *holy*
t Syr: Lat *live* *u* Syr Ethiop: Meaning of
Lat uncertain

(compare 2 Pet 2.4 note *l*). **12**: For the seer, to live without understanding of life's meaning is intolerable.
4.13–21: **Parable of the conflict between the forest and the sea. 21**: Isa 55.8–9; Ps 104.5–9; Jn 3.31; 1 Cor 2.14.
4.22–32: **Additional questions. 26–32**:

Ezra protests that he is inquiring only about the meaning of earthly, historical happenings, not about cosmic events. The angel answers that the new age, soon to dawn, will solve all problems; but first the evil that is sown must be reaped. **30**: *A grain of evil seed,* the evil "yeṣer" (see Sir 15.14–17 n.).

ing comes! 31 Consider now for yourself how much fruit of ungodliness a grain of evil seed has produced. 32 When heads of grain without number are sown, how great a threshing floor they will fill!"

33 Then I answered and said, "How long?ᵛ When will these things be? Why are our years few and evil?" 34 He answered me and said, "Do not be in a greater hurry than the Most High. You, indeed, are in a hurry for yourself,ʷ but the Highest is in a hurry on behalf of many. 35 Did not the souls of the righteous in their chambers ask about these matters, saying, 'How long are we to remain here?ˣ And when will the harvest of our reward come? 36 And the archangel Jeremiel answered and said, 'When the number of those like yourselves is completed;ʸ for he has weighed the age in the balance, 37 and measured the times by measure, and numbered the times by number; and he will not move or arouse them until that measure is fulfilled.' "

38 Then I answered and said, "But, O sovereign Lord, all of us also are full of ungodliness. 39 It is perhaps on account of us that the time of threshing is delayed for the righteous—on account of the sins of those who inhabit the earth."

40 He answered me and said, "Go and ask a pregnant woman whether, when her nine months have been completed, her womb can keep the fetus within her any longer."

41 And I said, "No, lord, it cannot."

He said to me, "In Hades the chambers of the souls are like the womb. 42 For just as a woman who is in labor makes haste to escape the pangs of birth, so also do these places hasten to give back those things that were committed to them from the beginning. 43 Then the things that you desire to see will be disclosed to you."

44 I answered and said, "If I have found favor in your sight, and if it is possible, and if I am worthy, 45 show me this also: whether more time is to come than has passed, or whether for us the greater part has gone by. 46 For I know what has gone by, but I do not know what is to come."

47 And he said to me, "Stand at my right side, and I will show you the interpretation of a parable."

48 So I stood and looked, and lo, a flaming furnace passed by before me, and when the flame had gone by I looked, and lo, the smoke remained. 49 And after this a cloud full of water passed before me and poured down a heavy and violent rain, and when the violent rainstorm had passed, drops still remained in the cloud.ᶻ

50 He said to me, "Consider it for yourself; for just as the rain is more than the drops, and the fire is greater than the smoke, so the quantity that passed was far greater; but drops and smoke remained."

51 Then I prayed and said, "Do you think that I shall live until those days? Or who will be alive in those days?"

52 He answered me and said, "Concerning the signs about which you ask me, I can tell you in part; but I was not sent to tell you concerning your life, for I do not know.

5 "Now concerning the signs: lo, the days are coming when those who inhabit the earth shall be seized with great terror,ᵛ and the way of truth shall be hidden, and the land shall be barren of

ᵛ Syr Ethiop: Meaning of Lat uncertain
ʷ Syr Ethiop Arab Arm: Meaning of Lat uncertain   ˣ Syr Ethiop Arab 2 Georg: Lat *How long do I hope thus?*   ʸ Syr Ethiop Arab 2: Lat *number of seeds is completed for you*
ᶻ Lat *in it*

4.33–43: The seer asks when the new age will come; he is told that first the predetermined number of the righteous must be completed. **35:** *The righteous,* i.e. the righteous dead. *Chambers,* literally "storehouses" or "garners"; according to rabbinical teaching the souls of the righteous dead are beneath the

throne of God (compare Rev 6.9f.). **36:** *Jeremiel,* probably the same as Remiel, the seventh of seven archangels mentioned in 1 Enoch 20.1–8. *Completed,* 2.41; Rev 6.11. **36–37:** *Weighed . . . measured . . . numbered,* God has determined the times and periods of history (see Sir 36.10 n.). There may be an

faith. ²Unrighteousness shall be increased beyond what you yourself see, and beyond what you heard of formerly. ³And the land that you now see ruling shall be a trackless waste, and people shall see it desolate. ⁴But if the Most High grants that you live, you shall see it thrown into confusion after the third period;*a*

and the sun shall suddenly begin
to shine at night,
and the moon during the day.
5　Blood shall drip from wood,
and the stone shall utter its
voice;
the peoples shall be troubled,
and the stars shall fall. *b*

⁶And one shall reign whom those who inhabit the earth do not expect, and the birds shall fly away together; ⁷and the Dead Sea*c* shall cast up fish; and one whom the many do not know shall make his voice heard by night, and all shall hear his voice. *d* ⁸There shall be chaos also in many places, fire shall often break out, the wild animals shall roam beyond their haunts, and menstruous women shall bring forth monsters. ⁹Salt waters shall be found in the sweet, and all friends shall conquer one another; then shall reason hide itself, and wisdom shall withdraw into its chamber, ¹⁰and it shall be sought by many but shall not be found, and unrighteousness and unrestraint shall increase on earth. ¹¹One country shall ask its neighbor, 'Has righteousness, or anyone who does right, passed through you?' And it will answer, 'No.' ¹²At that time people shall hope

but not obtain; they shall labor, but their ways shall not prosper. ¹³These are the signs that I am permitted to tell you, and if you pray again, and weep as you do now, and fast for seven days, you shall hear yet greater things than these."

14　Then I woke up, and my body shuddered violently, and my soul was so troubled that it fainted. ¹⁵But the angel who had come and talked with me held me and strengthened me and set me on my feet.

16　Now on the second night Phaltiel, a chief of the people, came to me and said, "Where have you been? And why is your face sad? ¹⁷Or do you not know that Israel has been entrusted to you in the land of their exile? ¹⁸Rise therefore and eat some bread, and do not forsake us, like a shepherd who leaves the flock in the power of savage wolves."

19　Then I said to him, "Go away from me and do not come near me for seven days; then you may come to me." He heard what I said and left me. ²⁰So I fasted seven days, mourning and weeping, as the angel Uriel had commanded me.

21　After seven days the thoughts of my heart were very grievous to me again. ²²Then my soul recovered the spirit of understanding, and I began once

---

a Literally *after the third,* Ethiop *after three months;* Arm *after the third vision;* Georg *after the third day*
b Ethiop Compare Syr and Arab: Meaning of Lat uncertain　　*c* Lat *Sea of Sodom*
d Cn: Lat *fish; and it shall make its voice heard by night, which the many have not known, but all shall hear its voice.*

---

allusion to Dan 5.24–28, the handwriting on the wall. **41**: *Chambers,* see v. 35 n.
4.44–50: The seer asks what proportion of time remains; he is told by a parable that the end is near.
4.51–5.13: The seer asks whether the end will come during his own lifetime; he is given a description of the signs that will precede the end (compare Mt 24.4–31; Mk 13.5–27; Lk 21.8–28). **5.2**: Mt 24.12. **3**: *The land that you now see ruling,* i.e. the Roman Empire. **4**: The reference to *the third period* is cryptic (compare 14.11–13). **5**: *The stone shall utter its voice,* Hab 2.11; Lk 19.40. **6**: *The birds,*

foreseeing impending disasters, *shall fly away.*
5.8a: Syriac, "a fissure shall arise over wide regions" (compare Zech 14.4). *Shall often break out,* or "shall burst forth for a long period." **10–11**: Isa 59.14–15. **13**: For the author, fasting prepared one to receive a divine revelation; he refers to three fasts each of seven days (5.20; 6.35; 12.51).
5.14–20: Conclusion of the vision. 14: *Then I woke up,* from the dream-vision. *My soul . . . fainted,* Pss 84.2; 107.5; Jon 2.7; compare Isa 6.5; Dan 10.17b. **16**: *Phaltiel,* the historical reference is uncertain; compare Paltiel in Num 34.26; 2 Sam 3.15.

more to speak words in the presence of the Most High. 23 I said, "O sovereign Lord, from every forest of the earth and from all its trees you have chosen one vine, 24 and from all the lands of the world you have chosen for yourself one region, *e* and from all the flowers of the world you have chosen for yourself one lily, 25 and from all the depths of the sea you have filled for yourself one river, and from all the cities that have been built you have consecrated Zion for yourself, 26 and from all the birds that have been created you have named for yourself one dove, and from all the flocks that have been made you have provided for yourself one sheep, 27 and from all the multitude of peoples you have gotten for yourself one people; and to this people, whom you have loved, you have given the law that is approved by all. 28 And now, O Lord, why have you handed the one over to the many, and dishonored *f* the one root beyond the others, and scattered your only one among the many? 29 And those who opposed your promises have trampled on those who believed your covenants. 30 If you really hate your people, they should be punished at your own hands."

31 When I had spoken these words, the angel who had come to me on a previous night was sent to me. 32 He said to me, "Listen to me, and I will instruct you; pay attention to me, and I will tell you more."

33 Then I said, "Speak, my lord." And he said to me, "Are you greatly disturbed in mind over Israel? Or do you love him more than his Maker does?"

34 I said, "No, my lord, but because of my grief I have spoken; for every hour I suffer agonies of heart, while I strive to understand the way of the Most High and to search out some part of his judgment."

35 He said to me, "You cannot." And I said, "Why not, my lord? Why then was I born? Or why did not my mother's womb become my grave, so that I would not see the travail of Jacob and the exhaustion of the people of Israel?"

36 He said to me, "Count up for me those who have not yet come, and gather for me the scattered raindrops, and make the withered flowers bloom again for me; 37 open for me the closed chambers, and bring out for me the winds shut up in them, or show me the picture of a voice; and then I will explain to you the travail that you ask to understand." *g*

38 I said, "O sovereign Lord, who is able to know these things except he whose dwelling is not with mortals? 39 As for me, I am without wisdom, and how can I speak concerning the things that you have asked me?"

40 He said to me, "Just as you cannot do one of the things that were mentioned, so you cannot discover my judgment, or the goal of the love that I have promised to my people."

41 I said, "Yet, O Lord, you have charge of those who are alive at the end, but what will those do who lived before me, or we, ourselves, or those who come after us?"

42 He said to me, "I shall liken my judgment to a circle; *h* just as for those who are last there is no slowness, so for those who are first there is no haste."

e Ethiop: Lat *pit*   *f* Syr Ethiop Arab: Lat *prepared*   *g* Lat *see*   *h* Or *crown*

5.21–6.34: **The second vision. 5.21–30: The seer reiterates his complaints of divine inequity in dealing with Israel. 23–28:** Most of the figures representing Israel have been drawn from the Old Testament: the *vine* (v. 23), Ps 80.8–15; the *lily* (v. 24), Song 2.2 (interpreted allegorically); Hos 14.5; the *river* (v. 25), Isa 8.6; the city of *Zion* (v. 25), Ps 132.13; the *dove* (v. 26), Ps 74.19; the *sheep* (v. 26), Ps 79.13; Isa 53.7; the *root* (v. 28), 1 Enoch 93.8, compare Rom 11.17–18. **33:** It is unthinkable that human beings should love Israel more than God their *Maker does* (8.47). **35:** Job 3.11; 10.18–19. The seer insists that mortals must be able to comprehend some portion of God's ways; if not, life has no purpose at all. **36–40:** If the seer cannot understand the things of earth, how can he expect to fathom the judgments and purpose of God? **40:** *My . . . I,* the angel speaks in God's name.

43 Then I answered and said, "Could you not have created at one time those who have been and those who are and those who will be, so that you might show your judgment the sooner?"

44 He replied to me and said, "The creation cannot move faster than the Creator, nor can the world hold at one time those who have been created in it."

45 I said, "How have you said to your servant that you[i] will certainly give life at one time to your creation? If therefore all creatures will live at one time[j] and the creation will sustain them, it might even now be able to support all of them present at one time."

46 He said to me, "Ask a woman's womb, and say to it, 'If you bear ten[k] children, why one after another?' Request it therefore to produce ten at one time."

47 I said, "Of course it cannot, but only each in its own time."

48 He said to me, "Even so I have given the womb of the earth to those who from time to time are sown in it. 49 For as an infant does not bring forth, and a woman who has become old does not bring forth any longer, so I have made the same rule for the world that I created."

50 Then I inquired and said, "Since you have now given me the opportunity, let me speak before you. Is our mother, of whom you have told me, still young? Or is she now approaching old age?"

51 He replied to me, "Ask a woman who bears children, and she will tell you. 52 Say to her, 'Why are those whom you have borne recently not like those whom you bore before, but smaller in stature?' 53 And she herself will answer you,

'Those born in the strength of youth are different from those born during the time of old age, when the womb is failing.' 54 Therefore you also should consider that you and your contemporaries are smaller in stature than those who were before you, 55 and those who come after you will be smaller than you, as born of a creation that already is aging and passing the strength of youth."

56 I said, "I implore you, O Lord, if I have found favor in your sight, show your servant through whom you will visit your creation."

6 He said to me, "At the beginning of the circle of the earth, before[l] the portals of the world were in place, and before the assembled winds blew, 2 and before the rumblings of thunder sounded, and before the flashes of lightning shone, and before the foundations of paradise were laid, 3 and before the beautiful flowers were seen, and before the powers of movements[m] were established, and before the innumerable hosts of angels were gathered together, 4 and before the heights of the air were lifted up, and before the measures of the firmaments were named, and before the footstool of Zion was established, 5 and before the present years were reckoned and before the imaginations of those who now sin were estranged, and before those who stored

i Syr Ethiop Arab 1: Meaning of Lat uncertain
j Lat lacks *If . . . one time*   k Syr Ethiop
Arab 2 Arm: Meaning of Lat uncertain
l Meaning of Lat uncertain: Compare Syr *The
beginning by the hand of humankind, but the end by
my own hands. For as before the land of the world
existed there, and before;* Ethiop: *At first by the Son
of Man, and afterwards I myself. For before the earth
and the lands were created, and before*
m Or *earthquakes*

5.41–55: The place of successive generations in the divine plan for the world. The seer inquires about the status of those who have died before the messianic age shall begin (v. 41); he is told, in effect, that the last shall be as the first, and the first as the last (v. 42). He inquires why all human generations could not have lived at the same time, namely at the beginning of the messianic age (v. 43); the reply is that generations must follow one another (vv. 44–49). *Mother* earth has become old and the last generations are inferior to the early ones (vv. 50 55).

5.52: *Smaller in stature*, compare Gen 6.4 (the Nephilim, "giants"); Num 13.33.

5.56–6.6: The end of the age. As God alone created the world (without an intermediate agency), so he will bring about its end by himself alone (7.39–44). The use of "before . . ." (6.1) is characteristic of creation stories. See the Babylonian creation story "Enuma Elish" and Gen 2.4b–5.

up treasures of faith were sealed— [6] then I planned these things, and they were made through me alone and not through another; just as the end shall come through me alone and not through another."

7 I answered and said, "What will be the dividing of the times? Or when will be the end of the first age and the beginning of the age that follows?"

8 He said to me, "From Abraham to Isaac,[n] because from him were born Jacob and Esau, for Jacob's hand held Esau's heel from the beginning. [9] Now Esau is the end of this age, and Jacob is the beginning of the age that follows. [10] The beginning of a person is the hand, and the end of a person is the heel;[o] seek for nothing else, Ezra, between the heel and the hand, Ezra!"

11 I answered and said, "O sovereign Lord, if I have found favor in your sight, [12] show your servant the last of your signs of which you showed me a part on a previous night."

13 He answered and said to me, "Rise to your feet and you will hear a full, resounding voice. [14] And if the place where you are standing is greatly shaken [15] while the voice is speaking, do not be terrified; because the word concerns the end, and the foundations of the earth will understand [16] that the speech concerns them. They will tremble and be shaken, for they know that their end must be changed."

17 When I heard this, I got to my feet and listened; a voice was speaking, and its sound was like the sound of mighty[p] waters. [18] It said, "The days are coming when I draw near to visit the inhabitants of the earth, [19] and when I require from the doers of iniquity the penalty of their iniquity, and when the humiliation of Zion is complete. [20] When the seal is placed upon the age that is about to pass away, then I will show these signs: the books shall be opened before the face of the firmament, and all shall see my judgment[q] together. [21] Children a year old shall speak with their voices, and pregnant women shall give birth to premature children at three and four months, and these shall live and leap about. [22] Sown places shall suddenly appear unsown, and full storehouses shall suddenly be found to be empty; [23] the trumpet shall sound aloud, and when all hear it, they shall suddenly be terrified. [24] At that time friends shall make war on friends like enemies, the earth and those who inhabit it shall be terrified, and the springs of the fountains shall stand still, so that for three hours they shall not flow.

25 "It shall be that whoever remains after all that I have foretold to you shall be saved and shall see my salvation and the end of my world. [26] And they shall see those who were taken up, who from their birth have not tasted death; and the heart of the earth's[r] inhabitants shall be changed and converted to a different spirit. [27] For evil shall be blotted out, and deceit shall be quenched; [28] faithfulness shall flourish, and corruption shall be overcome, and the truth, which has been so long without fruit, shall be revealed."

29 While he spoke to me, little by lit-

---

n Other ancient authorities read *to Abraham*
o Syr: Meaning of Lat uncertain   *p* Lat *many*   *q* Syr: Lat lacks *my judgment*
r Syr Compare Ethiop Arab 1 Arm: Lat lacks *earth's*

6.7–10: The dividing of the times. In allegorical language the seer is told that the present corrupt age (symbolized by *Esau*) will be followed immediately, without a break, by the glorious age to come (symbolized by *Jacob*).

6.11–28: The signs of the end of the age. **12:** *Of which you showed me a part,* 4.51–5.13. **17:** *I got to my feet,* presumably the author had previously been lying down, experiencing a dream-vision. *Mighty waters,* Rev 1.15; 14.2;

19.6. **20:** *The books shall be opened,* i.e. the celestial books in which are written the deeds of humankind (Dan 7.10; 12.1; Mal 3.16; Rev 20.12; compare Ex 32.32; Ps 69.28; Lk 10.20; Heb 12.23). **6.23:** *The trumpet,* 1 Cor 15.52; 1 Thess 4.16. **26:** *Those who were taken up,* such as Enoch (Gen 5.24; Sir 44.16) and Elijah (2 Kings 2.11–12); compare also 14.9. *Shall be . . . converted,* by the preaching of Elijah (Mal 4.6).

tle the place where I was standing began to rock to and fro. *s* 30 And he said to me, "I have come to show you these things this night. *t* 31 If therefore you will pray again and fast again for seven days, I will again declare to you greater things than these, *u* 32 because your voice has surely been heard by the Most High; for the Mighty One has seen your uprightness and has also observed the purity that you have maintained from your youth. 33 Therefore he sent me to show you all these things, and to say to you: 'Believe and do not be afraid! 34 Do not be quick to think vain thoughts concerning the former times; then you will not act hastily in the last times.'"

35 Now after this I wept again and fasted seven days in the same way as before, in order to complete the three weeks that had been prescribed for me. 36 Then on the eighth night my heart was troubled within me again, and I began to speak in the presence of the Most High. 37 My spirit was greatly aroused, and my soul was in distress.

38 I said, "O Lord, you spoke at the beginning of creation, and said on the first day, 'Let heaven and earth be made,' and your word accomplished the work. 39 Then the spirit was blowing, and darkness and silence embraced everything; the sound of human voices was not yet there. *v* 40 Then you commanded a ray of light to be brought out from your storechambers, so that your works could be seen.

41 "Again, on the second day, you created the spirit of the firmament, and commanded it to divide and separate the waters, so that one part might move upward and the other part remain beneath.

42 "On the third day you commanded the waters to be gathered together in a seventh part of the earth; six parts you dried up and kept so that some of them might be planted and cultivated and be of service before you. 43 For your word went forth, and at once the work was done. 44 Immediately fruit came forth in endless abundance and of varied appeal to the taste, and flowers of inimitable color, and odors of inexpressible fragrance. These were made on the third day.

45 "On the fourth day you commanded the brightness of the sun, the light of the moon, and the arrangement of the stars to come into being; 46 and you commanded them to serve humankind, about to be formed.

47 "On the fifth day you commanded the seventh part, where the water had been gathered together, to bring forth living creatures, birds, and fishes; and so it was done. 48 The dumb and lifeless water produced living creatures, as it was commanded, so that therefore the nations might declare your wondrous works.

49 "Then you kept in existence two living creatures; *w* the one you called Be-

s Syr Ethiop Compare Arab Arm: Meaning of Lat uncertain   *t* Syr Compare Ethiop: Meaning of Lat uncertain   *u* Syr Ethiop Arab 1 Arm: Lat adds *by day*   *v* Syr Ethiop: Lat *was not yet from you*   *w* Syr Ethiop: Lat *two souls*

6.29–34: Conclusion of the vision. 34: The seer is cautioned against being oversolicitous.
6.35–9.25: The third vision. 6.35–37: Introduction. 35: *I . . . fasted seven days,* see 5.13 n. *The three weeks* (compare Dan 10.2–3), so far only two fasts of seven days have been mentioned (here and at 5.20); presumably the author is thinking also of another fast prior to the first vision (3.1–5.20), not mentioned in the present form of the book.
6.38–59: The seer recounts God's work in creation. If the world was created for Israel (v. 55), why has the nation not pos-

sessed its inheritance? **38–54**: Gen ch 1. **38**: *Your word accomplished the work,* Ps 33.6; Heb 11.3; 2 Pet 3.5. **40**: God's *store-chambers* are in heaven. **41**: *The spirit of the firmament* is an angel (compare the angel with power over fire, Rev 14.18, and the angel of water, Rev 16.5). **46**: Ps 8.6–8.
6.49–52: *Behemoth* and *Leviathan* are two primeval monsters (compare Job 7.12; 26.12–13; Ps 74.12–15; 89.10–11; Isa 30.7; 51.9–10). **55**: The idea that the world was created for the sake of Israel (7.11) is not found in the Old Testament, but was deduced by Jewish rabbis from such passages as Ex 4.22; Deut

hemoth[x] and the name of the other Leviathan. 50 And you separated one from the other, for the seventh part where the water had been gathered together could not hold them both. 51 And you gave Behemoth[x] one of the parts that had been dried up on the third day, to live in it, where there are a thousand mountains; 52 but to Leviathan you gave the seventh part, the watery part; and you have kept them to be eaten by whom you wish, and when you wish.

53 "On the sixth day you commanded the earth to bring forth before you cattle, wild animals, and creeping things; 54 and over these you placed Adam, as ruler over all the works that you had made; and from him we have all come, the people whom you have chosen.

55 "All this I have spoken before you, O Lord, because you have said that it was for us that you created this world.[y] 56 As for the other nations that have descended from Adam, you have said that they are nothing, and that they are like spittle, and you have compared their abundance to a drop from a bucket. 57 And now, O Lord, these nations, which are reputed to be as nothing, domineer over us and devour us. 58 But we your people, whom you have called your firstborn, only begotten, zealous for you,[z] and most dear, have been given into their hands. 59 If the world has indeed been created for us, why do we not possess our world as an inheritance? How long will this be so?"

7 When I had finished speaking these words, the angel who had been sent to me on the former nights was sent to me again. 2 He said to me, "Rise, Ezra, and listen to the words that I have come to speak to you."

3 I said, "Speak, my lord." And he said to me, "There is a sea set in a wide expanse so that it is deep and vast, 4 but it has an entrance set in a narrow place, so that it is like a river. 5 If there are those who wish to reach the sea, to look at it or to navigate it, how can they come to the broad part unless they pass through the narrow part? 6 Another example: There is a city built and set on a plain, and it is full of all good things; 7 but the entrance to it is narrow and set in a precipitous place, so that there is fire on the right hand and deep water on the left. 8 There is only one path lying between them, that is, between the fire and the water, so that only one person can walk on the path. 9 If now the city is given to someone as an inheritance, how will the heir receive the inheritance unless by passing through the appointed danger?"

10 I said, "That is right, lord." He said to me, "So also is Israel's portion. 11 For I made the world for their sake, and when Adam transgressed my statutes, what had been made was judged. 12 And so the entrances of this world were made narrow and sorrowful and toilsome; they are few and evil, full of dangers and involved in great hardships. 13 But the entrances of the greater world are broad and safe, and yield the fruit of immortality. 14 Therefore unless the living pass through the difficult and futile experiences, they can never receive those things that have been reserved for them. 15 Now therefore why are you disturbed, seeing that you are to perish? Why are you moved, seeing that you are mortal? 16 Why have you not considered in your mind what is to come, rather than what is now present?"

17 Then I answered and said,

x Other Lat authorities read *Enoch*
y Syr Ethiop Arab 2: Lat *the firstborn world*
Compare Arab 1 *first world*   z Meaning of Lat uncertain

10.15; 14.2. **56:** *A drop from a bucket,* Isa 40.15.
7.1–25: The angel instructs the seer. The wickedness of this world makes the path to the next world narrow and dangerous. **1:** *The former nights,* at the beginning of each vision. **11:** Though *the world* was created for Israel's *sake,* that inheritance was spoiled *when*

Adam transgressed (compare Rom 5.18–20). **12 and 13:** *The entrances,* Ethiopic, "the ways," i.e. the paths of life here on earth, and in the world of immortality. **13:** *The greater world,* Syriac, "the future world."
7.14: *Things . . . reserved for them,* 1 Cor 2.9.
15–16: The seer should not brood over diffi-

"O sovereign Lord, you have ordained in your law that the righteous shall inherit these things, but that the ungodly shall perish. 18 The righteous, therefore, can endure difficult circumstances while hoping for easier ones; but those who have done wickedly have suffered the difficult circumstances and will never see the easier ones."

19 He said to me, "You are not a better judge than the Lord, *a* or wiser than the Most High! 20 Let many perish who are now living, rather than that the law of God that is set before them be disregarded! 21 For the Lord *b* strictly commanded those who came into the world, when they came, what they should do to live, and what they should observe to avoid punishment. 22 Nevertheless they were not obedient, and spoke against him;

they devised for themselves vain
        thoughts,
23   and proposed to themselves
        wicked frauds;
they even declared that the Most
        High does not exist,
and they ignored his ways.
24   They scorned his law,
        and denied his covenants;
they have been unfaithful to his
        statutes,
and have not performed his
        works.

25 "That is the reason, Ezra, that empty things are for the empty, and full things are for the full.

26 "For indeed the time will come, when the signs that I have foretold to you will come to pass, that the city that now is not seen shall appear, *c* and the land that now is hidden shall be disclosed. 27 Everyone who has been delivered from the evils that I have foretold shall see my wonders. 28 For my son the Messiah *d* shall be revealed with those who are with him, and those who remain shall rejoice four hundred years. 29 After those years my son the Messiah shall die, and all who draw human breath. *e* 30 Then the world shall be turned back to primeval silence for seven days, as it was at the first beginnings, so that no one shall be left. 31 After seven days the world that is not yet awake shall be roused, and that which is corruptible shall perish. 32 The earth shall give up those who are asleep in it, and the dust those who rest there in silence; and the chambers shall give up the souls that have been committed to them. 33 The Most High shall be revealed on the seat of judgment, and compassion shall pass away, and patience shall be withdrawn. *f* 34 Only judgment shall remain, truth shall stand, and faithfulness shall grow strong. 35 Recompense shall follow, and the reward shall be manifested; righteous deeds shall awake, and unrighteous

a Other ancient authorities read *God*; Ethiop Georg *the only One* *b* Other ancient authorities read *God* *c* Arm: Lat Syr *that the bride shall appear, even the city appearing* *d* Syr Arab 1: Ethiop *my Messiah*; Arab 2 *the Messiah*; Arm *the Messiah of God*; Lat *my son Jesus* *e* Arm *all who have continued in faith and in patience* *f* Lat *shall gather together*

culties and death; though inevitable, they are but preliminary to something better (2 Cor 4.18). **17–18:** The seer inquires whether the future bliss is only for the righteous. They, after all, have hope of bliss with God beyond this life. But what of the wicked, who suffer now and are doomed to suffer more in the judgment? **19–25:** The angel replies that those who disregard God's law will be punished; they had fair warning. **25:** Mt 13.12.
7.26–44: The messianic kingdom and the end of the world. 26: *The signs . . . foretold,* 6.20–24. *The city,* the Jerusalem that is to come; see 10.25–54. *The land,* the paradise that is to come. **28:** *Those who remain,* after the tribulations that will precede the inauguration of the messianic kingdom. *Four hundred years,* so the Latin and Arabic 1; Syriac, "thirty years"; Arabic 2, "one thousand years"; Ethiopic and Armenian omit. **31:** *The world . . . not yet awake,* i.e. the world to come. **32:** Dan 12.2. *Chambers,* see 4.35 n. **33:** *Judgment,* Syriac adds, "and then comes the end." *Away,* Syriac adds, "and pity shall be far off." The final judgment will be conducted in strict accord with justice and truth. **34:** *Grow strong,* i.e. triumph. **35:** *Righteous deeds shall awake,* acts of charity hitherto concealed shall be disclosed (compare Mt 25.35–46).

deeds shall not sleep.*g* *36* The pit*h* of torment shall appear, and opposite it shall be the place of rest; and the furnace of hell*i* shall be disclosed, and opposite it the paradise of delight. *37* Then the Most High will say to the nations that have been raised from the dead, 'Look now, and understand whom you have denied, whom you have not served, whose commandments you have despised. *38* Look on this side and on that; here are delight and rest, and there are fire and torments.' Thus he will*j* speak to them on the day of judgment— *39* a day that has no sun or moon or stars, *40* or cloud or thunder or lightning, or wind or water or air, or darkness or evening or morning, *41* or summer or spring or heat or winter*k* or frost or cold, or hail or rain or dew, *42* or noon or night, or dawn or shining or brightness or light, but only the splendor of the glory of the Most High, by which all shall see what has been destined. *43* It will last as though for a week of years. *44* This is my judgment and its prescribed order; and to you alone I have shown these things."

45 I answered and said, "O sovereign Lord, I said then and*l* I say now: Blessed are those who are alive and keep your commandments! *46* But what of those for whom I prayed? For who among the living is there that has not sinned, or who is there among mortals that has not transgressed your covenant? *47* And now I see that the world to come will bring delight to few, but torments to many. *48* For an evil heart has grown up in us, which has alienated us from God,*m* and has brought us into corruption and the ways of death, and has shown us the paths of perdition and removed us far

from life—and that not merely for a few but for almost all who have been created."

49 He answered me and said, "Listen to me, Ezra,*n* and I will instruct you, and will admonish you once more. *50* For this reason the Most High has made not one world but two. *51* Inasmuch as you have said that the righteous are not many but few, while the ungodly abound, hear the explanation for this.

52 "If you have just a few precious stones, will you add to them lead and clay?"*o* *53* I said, "Lord, how could that be?" *54* And he said to me, "Not only that, but ask the earth and she will tell you; defer to her, and she will declare it to you. *55* Say to her, 'You produce gold and silver and bronze, and also iron and lead and clay; *56* but silver is more abundant than gold, and bronze than silver, and iron than bronze, and lead than iron, and clay than lead.' *57* Judge therefore which things are precious and desirable, those that are abundant or those that are rare?"

58 I said, "O sovereign Lord, what is plentiful is of less worth, for what is more rare is more precious."

59 He answered me and said, "Consider within yourself*p* what you have thought, for the person who has what is hard to get rejoices more than the person who has what is plentiful. *60* So also will

---

g The passage from verse 36 to verse 105, formerly missing, has been restored to the text h Syr Ethiop: Lat *place* i Lat Syr Ethiop *Gehenna* j Syr Ethiop Arab 1: Lat *you shall* k Or *storm* l Syr: Lat *And I answered, "I said then, O Lord, and* m Cn: Lat Syr Ethiop *from these* n Syr Arab 1 Georg: Lat Ethiop lack *Ezra* o Arab 1: Meaning of Lat Syr Ethiop uncertain p Syr Ethiop Arab 1: Meaning of Lat uncertain

---

7.36–105: These verses are lacking from the standard editions of the Latin Vulgate and from the King James Version. They are present in the Syriac, Ethiopic, Arabic, and Armenian versions, and in two Latin manuscripts. The section was probably deliberately cut out of an ancestor of most extant Latin manuscripts because of dogmatic reasons, for the passage contains an emphatic denial of the value of prayers for the dead (v. *105*). **36:** Pit, Rev 9.2. *Opposite*, Lk 16.23–24. **37:** Mt

25.31ff. **39–43:** Description of the day of judgment. **42:** *Only the uncreated light of the Most High* will serve to illuminate the judgment scene (compare Isa 60.19–20; Rev 21.23). **43:** *A week of years*, seven years. **7.45–61:** **The small number of the saved** (contrast Lk 13.23–30). **45–46:** The seer's chief concern has only been deepened. What hope is there for sinners? And are not all, or almost all (7.48) sinners? **48:** *An evil heart*, see 3.20 n. **49:** 5.32.

be the judgment[q] that I have promised; for I will rejoice over the few who shall be saved, because it is they who have made my glory to prevail now, and through them my name has now been honored. [61] I will not grieve over the great number of those who perish; for it is they who are now like a mist, and are similar to a flame and smoke—they are set on fire and burn hotly, and are extinguished."

[62] I replied and said, "O earth, what have you brought forth, if the mind is made out of the dust like the other created things? [63] For it would have been better if the dust itself had not been born, so that the mind might not have been made from it. [64] But now the mind grows with us, and therefore we are tormented, because we perish and we know it. [65] Let the human race lament, but let the wild animals of the field be glad; let all who have been born lament, but let the cattle and the flocks rejoice. [66] It is much better with them than with us; for they do not look for a judgment, and they do not know of any torment or salvation promised to them after death. [67] What does it profit us that we shall be preserved alive but cruelly tormented? [68] For all who have been born are entangled in[r] iniquities, and are full of sins and burdened with transgressions. [69] And if after death we were not to come into judgment, perhaps it would have been better for us."

[70] He answered me and said, "When the Most High made the world and Adam and all who have come from him, he first prepared the judgment and the things that pertain to the judgment. [71] But now, understand from your own words—for you have said that the mind

grows with us. [72] For this reason, therefore, those who live on earth shall be tormented, because though they had understanding, they committed iniquity; and though they received the commandments, they did not keep them; and though they obtained the law, they dealt unfaithfully with what they received. [73] What, then, will they have to say in the judgment, or how will they answer in the last times? [74] How long the Most High has been patient with those who inhabit the world!—and not for their sake, but because of the times that he has foreordained."

[75] I answered and said, "If I have found favor in your sight, O Lord, show this also to your servant: whether after death, as soon as everyone of us yields up the soul, we shall be kept in rest until those times come when you will renew the creation, or whether we shall be tormented at once?"

[76] He answered me and said, "I will show you that also, but do not include yourself with those who have shown scorn, or number yourself among those who are tormented. [77] For you have a treasure of works stored up with the Most High, but it will not be shown to you until the last times. [78] Now concerning death, the teaching is: When the decisive decree has gone out from the Most High that a person shall die, as the spirit leaves the body to return again to him who gave it, first of all it adores the glory of the Most High. [79] If it is one of those who have shown scorn and have not kept the way of the Most High, who have

q Syr Arab 1: Lat *creation*   r Syr *defiled with*

7.52: The question implies that the number of the elect cannot be increased by adding base elements. **61:** The angel will not grieve over the sinners who perish, but Ezra will! (7.62–69).
7.62–74: The seer's lament over the human race. 63: 4.12. **64:** The possession of reasoning powers intensifies sufferings. **67:** Ezra, like the whole human race, is one of the sinners, and thus is worse off than the animals, who cannot know and meditate on their fate. Contrast 6.32–34. **70:** *Things that*

pertain to the judgment, according to rabbinical teaching, before the beginning of the world God created Paradise and Gehenna.
7.75–101: The state of the departed after death and before the judgment. 76: Ezra is not to include himself among the wicked, says the angel. But the seer continues to do so (8.47). **77:** *A treasure of works,* 8.33, 36. *Will not be shown,* see v. 35 n. **78:** Eccl 12.7. The first act of the departed spirit (whether righteous or wicked) is to adore God.

despised his law and hated those who fear God— [80] such spirits shall not enter into habitations, but shall immediately wander about in torments, always grieving and sad, in seven ways. [81] The first way, because they have scorned the law of the Most High. [82] The second way, because they cannot now make a good repentance so that they may live. [83] The third way, they shall see the reward laid up for those who have trusted the covenants of the Most High. [84] The fourth way, they shall consider the torment laid up for themselves in the last days. [85] The fifth way, they shall see how the habitations of the others are guarded by angels in profound quiet. [86] The sixth way, they shall see how some of them will cross over[s] into torments. [87] The seventh way, which is worse[t] than all the ways that have been mentioned, because they shall utterly waste away in confusion and be consumed with shame,[u] and shall wither with fear at seeing the glory of the Most High in whose presence they sinned while they were alive, and in whose presence they are to be judged in the last times.

[88] "Now this is the order of those who have kept the ways of the Most High, when they shall be separated from their mortal body.[v] [89] During the time that they lived in it,[w] they laboriously served the Most High, and withstood danger every hour so that they might keep the law of the Lawgiver perfectly. [90] Therefore this is the teaching concerning them: [91] First of all, they shall see with great joy the glory of him who receives them, for they shall have rest in seven orders. [92] The first order, because they have striven with great effort to overcome the evil thought that was formed with them, so that it might not lead them astray from life into death. [93] The second order, because they see the perplexity in which

the souls of the ungodly wander and the punishment that awaits them. [94] The third order, they see the witness that he who formed them bears concerning them, that throughout their life they kept the law with which they were entrusted. [95] The fourth order, they understand the rest that they now enjoy, being gathered into their chambers and guarded by angels in profound quiet, and the glory waiting for them in the last days. [96] The fifth order, they rejoice that they have now escaped what is corruptible and shall inherit what is to come; and besides they see the straits and toil[x] from which they have been delivered, and the spacious liberty that they are to receive and enjoy in immortality. [97] The sixth order, when it is shown them how their face is to shine like the sun, and how they are to be made like the light of the stars, being incorruptible from then on. [98] The seventh order, which is greater than all that have been mentioned, because they shall rejoice with boldness, and shall be confident without confusion, and shall be glad without fear, for they press forward to see the face of him whom they served in life and from whom they are to receive their reward when glorified. [99] This is the order of the souls of the righteous, as henceforth is announced;[y] and the previously mentioned are the ways of torment that those who would not give heed shall suffer hereafter."

[100] Then I answered and said, "Will time therefore be given to the souls, after they have been separated from the bodies, to see what you have described to me?"

s Cn: Meaning of Lat uncertain   t Lat Syr Ethiop *greater*   u Syr Ethiop: Meaning of Lat uncertain   v Lat *the corruptible vessel*
w Syr: Ethiop: Meaning of Lat uncertain
x Syr Ethiop: Lat *fullness*   y Syr: Meaning of Lat uncertain

---

7.80–87: Seven kinds of torment for the wicked. **80:** *Habitations,* Lk 16.9; elsewhere called "chambers," see 4.35 n. **83:** Compare Lk 16.23. **85:** *The others,* i.e. the righteous.
7.88–99: Seven kinds of joyous rest for the righteous. The author implies that the

mortal body has been merely a prison-house for the spirit (contrast 1 Cor 15.53; 2 Cor 5.2–4). **92:** *The evil thought,* the evil *yeṣer* (see Sir 15.14–17 n.). **95:** *Chambers,* see 4.35 n. *In the last days,* better, "at their latter end." **97:** *Shine,* v. 125; Dan 12.3; Mt 13.43. **98:** *To see*

101 He said to me, "They shall have freedom for seven days, so that during these seven days they may see the things of which you have been told, and afterwards they shall be gathered in their habitations."

102 I answered and said, "If I have found favor in your sight, show further to me, your servant, whether on the day of judgment the righteous will be able to intercede for the ungodly or to entreat the Most High for them— *103* fathers for sons or sons for parents, brothers for brothers, relatives for their kindred, or friends for those who are most dear."

104 He answered me and said, "Since you have found favor in my sight, I will show you this also. The day of judgment is decisive^z and displays to all the seal of truth. Just as now a father does not send his son, or a son his father, or a master his servant, or a friend his dearest friend, to be ill^a or sleep or eat or be healed in his place, *105* so no one shall ever pray for another on that day, neither shall anyone lay a burden on another;^b for then all shall bear their own righteousness and unrighteousness."

36 *106* I answered and said, "How then do we find that first Abraham prayed for the people of Sodom, and Moses for our ancestors who sinned in the desert, 37 *107* and Joshua after him for Israel in the days of Achan, 38 *108* and Samuel in the days of Saul,^c and David for the plague, and Solomon for those at the dedication, 39 *109* and Elijah for those who received the rain, and for the one who was dead, that he might live, 40 *110* and Hezekiah for the people in the days of Sennacherib, and many others prayed for many? 41 *111* So if now, when

corruption has increased and unrighteousness has multiplied, the righteous have prayed for the ungodly, why will it not be so then as well?"

42 *112* He answered me and said, "This present world is not the end; the full glory does not^d remain in it;^e therefore those who were strong prayed for the weak. 43 *113* But the day of judgment will be the end of this age and the beginning^f of the immortal age to come, in which corruption has passed away, 44 *114* sinful indulgence has come to an end, unbelief has been cut off, and righteousness has increased and truth has appeared. 45 *115* Therefore no one will then be able to have mercy on someone who has been condemned in the judgment, or to harm^g someone who is victorious."

46 *116* I answered and said, "This is my first and last comment: it would have been better if the earth had not produced Adam, or else, when it had produced him, had restrained him from sinning. 47 *117* For what good is it to all that they live in sorrow now and expect punishment after death? 48 *118* O Adam, what have you done? For though it was you who sinned, the fall was not yours alone, but ours also who are your descendants. 49 *119* For what good is it to us, if an immortal time has been promised to us, but we have done deeds that bring death? 50 *120* And what good is it that an everlasting hope has been promised to us, but we

z Lat *bold*   *a* Syr Ethiop
Arm: Lat *to understand*   *b* Syr Ethiop: Lat
lacks *on that . . . another*   *c* Syr Ethiop Arab 1:
Lat Arab 2 Arm lack *in the days of Saul*
d Lat lacks *not*   *e* Or *the glory does not*
continuously abide in it   *f* Syr Ethiop: Lat lacks
the beginning   *g* Syr Ethiop: Lat *overwhelm*

the face of God (Mt 5.8; Heb 12.14; 1 Jn 3.2; Rev 22.4). *Reward,* 1 Cor 3.14; Rev 22.12. **101**: *Habitations,* see v. *80* n. **7.102–115:** **No intercession for the wicked on the day of judgment** (compare Deut 24.16; Jer 31.30; Ezek 18.1–32). **106:** At verse *106* we come to the continuation of chapter 7 as preserved in the standard editions of the Latin Vulgate; NRSV resumes the Latin numbering here, designating verses *106–140* as 36–*70*, but with the numbers *106–140* added as well. Gen 18.23; Ex 32.11. **107:** Josh

7.6–7. **108:** *Samuel,* 1 Sam 7.9; 12.23. *David,* 2 Sam 24.17. *Solomon,* 1 Kings 8.22–23, 30. **109:** 1 Kings 18.42, 45; 17.20–21. **110:** 2 Kings 19.15–19.

7.112–115: During the present order, intercession *for the weak* is possible, but the day of judgment means the closing of all accounts on the basis of strict justice (see v. 33 n.).

7.116–131: **The seer laments the fate of the mass of humanity. 116:** *My first . . . comment,* 3.5ff. **118:** 4.30–31.

have miserably failed? 51 *121* Or that safe and healthful habitations have been reserved for us, but we have lived wickedly? 52 *122* Or that the glory of the Most High will defend those who have led a pure life, but we have walked in the most wicked ways? 53 *123* Or that a paradise shall be revealed, whose fruit remains unspoiled and in which are abundance and healing, but we shall not enter it 54 *124* because we have lived in perverse ways?ʰ 55 *125* Or that the faces of those who practiced self-control shall shine more than the stars, but our faces shall be blacker than darkness? 56 *126* For while we lived and committed iniquity we did not consider what we should suffer after death."

57 *127* He answered and said, "This is the significance of the contest that all who are born on earth shall wage: 58 *128* if they are defeated they shall suffer what you have said, but if they are victorious they shall receive what I have said. ⁱ 59 *129* For this is the way of which Moses, while he was alive, spoke to the people, saying, 'Choose life for yourself, so that you may live!' 60 *130* But they did not believe him or the prophets after him, or even myself who have spoken to them. 61 *131* Therefore there shall not beʲ grief at their destruction, so much as joy over those to whom salvation is assured."

62 *132* I answered and said, "I know, O Lord, that the Most High is now called merciful, because he has mercy on those who have not yet come into the world; 63 *133* and gracious, because he is gracious to those who turn in repentance to his law; 64 *134* and patient, because he shows patience toward those who have sinned, since they are his own creatures; 65 *135* and bountiful, because he would rather give than take away;ᵏ 66 *136* and abundant in compassion, because he makes his compassions abound more and more to those now living and to those who are gone and to those yet to come— 67 *137* for if he did not make them abound, the world with those who inhabit it would not have life— 68 *138* and he is called the giver, because if he did not give out of his goodness so that those who have committed iniquities might be relieved of them, not one ten-thousandth of humankind could have life; 69 *139* and the judge, because if he did not pardon those who were created by his word and blot out the multitude of their sins, ˡ 70 *140* there would probably be left only very few of the innumerable multitude."

8 He answered me and said, "The Most High made this world for the sake of many, but the world to come for the sake of only a few. 2 But I tell you a parable, Ezra. Just as, when you ask the earth, it will tell you that it provides a large amount of clay from which earthenware is made, but only a little dust from which gold comes, so is the course of the present world. 3 Many have been created, but only a few shall be saved."

4 I answered and said, "Then drink your fill of understanding,ᵐ O my soul, and drink wisdom, O my heart. 5 For

h Cn: Lat Syr *places*   *i* Syr Ethiop Arab 1: Lat *what I say*   *j* Syr: Lat *there was not*   *k* Or *he is ready to give according to requests*   *l* Lat *contempts*   *m* Syr: Lat *Then release understanding*

7.123: *Fruit,* compare Ezek 47.12; Rev 22.2. 125: *Shine more than the stars,* Dan 12.3; compare Mt 13.43. *Darkness,* Mt 8.12; 22.13; Jude 13. 127–129: Human beings are responsible for their choices (Deut 30.19; Ezek 18.1–32).
7.132–8.3: **The seer acknowledges (and implicitly appeals to) God's mercy.** Will a merciful God permit so many to perish? Ezra is told that nothing can alter their doom, for *many have been created, but only a few shall be saved* (8.3). 132–139: For the sevenfold attributes of God, compare Ex 34.6–7. 132:

O Lord, better, "sir." 135: Acts 20.35. 138: *Life,* i.e. eternal life.
8.2: *A parable,* an analogous illustration (as in 7.54–57). 3: Mt 22.14.
8.4–36: **The seer implores God to show mercy upon his creation.** 4–19a: Why should God wonderfully fashion and sustain all humankind, only to destroy a great majority? 4–5: The pre-existence of the soul is implied here (Wis 8.19). 7: Isa 44.6; 45.11; 60.21. 14: *Was fashioned by your command,* Ps 139.14–15. 15–16: The seer leaves the fate of *humankind* in God's hands, and speaks partic-

not of your own will did you come into the world, *n* and against your will you depart, for you have been given only a short time to live. 6 O Lord above us, grant to your servant that we may pray before you, and give us a seed for our heart and cultivation of our understanding so that fruit may be produced, by which every mortal who bears the likeness *o* of a human being may be able to live. 7 For you alone exist, and we are a work of your hands, as you have declared. 8 And because you give life to the body that is now fashioned in the womb, and furnish it with members, what you have created is preserved amid fire and water, and for nine months the womb *p* endures your creature that has been created in it. 9 But that which keeps and that which is kept shall both be kept by your keeping. *n* And when the womb gives up again what has been created in it, 10 you have commanded that from the members themselves (that is, from the breasts) milk, the fruit of the breasts, should be supplied, 11 so that what has been fashioned may be nourished for a time; and afterwards you will still guide it in your mercy. 12 You have nurtured it in your righteousness, and instructed it in your law, and reproved it in your wisdom. 13 You put it to death as your creation, and make it live as your work. 14 If then you will suddenly and quickly *q* destroy what with so great labor was fashioned by your command, to what purpose was it made? 15 And now I will speak out: About all humankind you know best; but I will speak about your people, for whom I am grieved, 16 and about your inheritance, for whom I lament, and about Israel, for whom I am sad, and about the seed of Jacob, for whom I am

troubled. 17 Therefore I will pray before you for myself and for them, for I see the failings of us who inhabit the earth; 18 and now also *r* I have heard of the swiftness of the judgment that is to come. 19 Therefore hear my voice and understand my words, and I will speak before you."

The beginning of the words of Ezra's prayer, *s* before he was taken up. He said: 20 "O Lord, you who inhabit eternity, *t* whose eyes are exalted *u* and whose upper chambers are in the air, 21 whose throne is beyond measure and whose glory is beyond comprehension, before whom the hosts of angels stand trembling 22 and at whose command they are changed to wind and fire, *v* whose word is sure and whose utterances are certain, whose command is strong and whose ordinance is terrible, 23 whose look dries up the depths and whose indignation makes the mountains melt away, and whose truth is established *w* forever— 24 hear, O Lord, the prayer of your servant, and give ear to the petition of your creature; attend to my words. 25 For as long as I live I will speak, and as long as I have understanding I will answer. 26 O do not look on the sins of your people, but on those who serve you in truth. 27 Do not take note of the endeavors of those who act wickedly, but of the endeavors of those who have kept your covenants amid afflictions. 28 Do not think of those who have lived wickedly in your sight,

n Syr: Meaning of Lat uncertain   *o* Syr: Lat place   *p* Lat *what you have formed*
q Syr: Lat *will with a light command*   *r* Syr: Lat but   *s* Syr Ethiop; Lat *beginning of Ezra's words*
t Or *you who abide forever*   *u* Another Lat text reads *whose are the highest heavens*
v Syr: Lat *they whose service takes the form of wind and fire*   *w* Arab 2: Other authorities read *truth bears witness*

---

ularly about Israel, God's *inheritance* (Ps 28.9).

8.19b–36: A beautiful and liturgically structured prayer (invocation to God, whose attributes are recalled, vv. 20–23; petitions, interspersed with confession and intercessions, vv. 24–35; concluding ascription of praise, v. 36). This prayer also occurs separately, with the title "Confessio Esdrae," in the section of canticles and hymns contained in many manuscripts of the Latin Vulgate

Bible. This circumstance accounts for the presence (in v. 19b) of a superscription in the third person. **19b:** The words, *before he was taken up,* indicate that the belief was current that Ezra, like Enoch and Elijah, was translated to heaven without dying.

8.22: *Wind and fire,* Ps 104.4; Heb 1.7. **23:** *Dries up,* Isa 50.2; 51.10. *Mountains melt,* Mic 1.4; Sir 16.18–19. **32:** Rom 3.19–26. **33:** 7.77.

but remember those who have willingly acknowledged that you are to be feared. [29]Do not will the destruction of those who have the ways of cattle, but regard those who have gloriously taught your law.[x] [30]Do not be angry with those who are deemed worse than wild animals, but love those who have always put their trust in your glory. [31]For we and our ancestors have passed our lives in ways that bring death;[y] but it is because of us sinners that you are called merciful. [32]For if you have desired to have pity on us, who have no works of righteousness, then you will be called merciful. [33]For the righteous, who have many works laid up with you, shall receive their reward in consequence of their own deeds. [34]But what are mortals, that you are angry with them; or what is a corruptible race, that you are so bitter against it? [35]For in truth there is no one among those who have been born who has not acted wickedly; among those who have existed[z] there is no one who has not done wrong. [36]For in this, O Lord, your righteousness and goodness will be declared, when you are merciful to those who have no store of good works."

[37] He answered me and said, "Some things you have spoken rightly, and it will turn out according to your words. [38]For indeed I will not concern myself about the fashioning of those who have sinned, or about their death, their judgment, or their destruction; [39]but I will rejoice over the creation of the righteous,

over their pilgrimage also, and their salvation, and their receiving their reward. [40]As I have spoken, therefore, so it shall be.

[41] "For just as the farmer sows many seeds in the ground and plants a multitude of seedlings, and yet not all that have been sown will come up[a] in due season, and not all that were planted will take root; so also those who have been sown in the world will not all be saved."

[42] I answered and said, "If I have found favor in your sight, let me speak. [43]If the farmer's seed does not come up, because it has not received your rain in due season, or if it has been ruined by too much rain, it perishes.[b] [44]But people, who have been formed by your hands and are called your own image because they are made like you, and for whose sake you have formed all things—have you also made them like the farmer's seed? [45]Surely not, O Lord[c] above! But spare your people and have mercy on your inheritance, for you have mercy on your own creation."

[46] He answered me and said, "Things that are present are for those who live now, and things that are future are for those who will live hereafter. [47]For you

---

x Syr *have received the brightness of your law*
y Syr Ethiop: Meaning of Lat uncertain
z Syr: Meaning of Lat uncertain   a Syr Ethiop *will live*; Lat *will be saved*   b Cn: Compare Syr Arab 1 Arm Georg 2: Meaning of Lat uncertain
c Ethiop Arab Compare Syr: Lat lacks *O Lord*

8.37–40: The divine reply to the seer's prayer: God will rejoice in the righteous and forget the sinners (the central petition of the prayer—mercy on the wicked—is ignored). **39:** *Their pilgrimage,* i.e. their return home to God (compare 2 Cor 5.6–8). **40:** Instead of *I have spoken,* the reading of the Ethiopic, "you have spoken," is to be preferred in view of v. 37 and the irony of the divine reply: "it is to be as you have *spoken,* but not as you had intended" (in vv. 26–36 the seer prayed God to ignore the wicked and their doings and pay attention to the righteous only; this, the Almighty replies, he will do, but in the sense of being unconcerned about the *destruction* of the wicked, v. 38).

8.41–45: Humankind is like the farmer's seed; only a few individuals will escape destruction. **45:** An anguished entreaty: *spare your people,* Joel 2.17.

8.46–62a: The final divine reply: The seer is assured that his lot is with the blessed, and is advised to think no more about sinners, who deserve their doom because they have *despised the Most High* (v. 56). **46–47:** The seer's objection (v. 44) is invalid, for the simile of the seeds suits the *present* corruptible order; the *future* has standards of its own. Moreover, God's love for his *creation* far exceeds human love (see 5.33 n.). **8.47–50:** The angel is almost annoyed with Ezra for insisting upon placing himself

come far short of being able to love my creation more than I love it. But you have often compared yourself[d] to the unrighteous. Never do so! 48 But even in this respect you will be praiseworthy before the Most High, 49 because you have humbled yourself, as is becoming for you, and have not considered yourself to be among the righteous. You will receive the greatest glory, 50 for many miseries will affect those who inhabit the world in the last times, because they have walked in great pride. 51 But think of your own case, and inquire concerning the glory of those who are like yourself, 52 because it is for you that paradise is opened, the tree of life is planted, the age to come is prepared, plenty is provided, a city is built, rest is appointed,[e] goodness is established and wisdom perfected beforehand. 53 The root of evil[f] is sealed up from you, illness is banished from you, and death[g] is hidden; Hades has fled and corruption has been forgotten;[h] 54 sorrows have passed away, and in the end the treasure of immortality is made manifest. 55 Therefore do not ask any more questions about the great number of those who perish. 56 For when they had opportunity to choose, they despised the Most High, and were contemptuous of his law, and abandoned his ways. 57 Moreover, they have even trampled on his righteous ones, 58 and said in their hearts that there is no God—though they knew well that they must die. 59 For just as the things that I have predicted await[i] you, so the thirst and torment that are prepared await them. For the Most High did not intend that anyone should be destroyed; 60 but those who were created

have themselves defiled the name of him who made them, and have been ungrateful to him who prepared life for them now. 61 Therefore my judgment is now drawing near; 62 I have not shown this to all people, but only to you and a few like you."

Then I answered and said, 63 "O Lord, you have already shown me a great number of the signs that you will do in the last times, but you have not shown me when you will do them."

9 He answered me and said, "Measure carefully in your mind, and when you see that some of the predicted signs have occurred, 2 then you will know that it is the very time when the Most High is about to visit the world that he has made. 3 So when there shall appear in the world earthquakes, tumult of peoples, intrigues of nations, wavering of leaders, confusion of princes, 4 then you will know that it was of these that the Most High spoke from the days that were of old, from the beginning. 5 For just as with everything that has occurred in the world, the beginning is evident,[j] and the end manifest; 6 so also are the times of the Most High: the beginnings are manifest in wonders and mighty works, and the end in penalties[k] and in signs.

7 It shall be that all who will be saved and will be able to escape on account of their works, or on account of the faith by

d Syr Ethiop: Lat *brought yourself near*
e Syr Ethiop: Lat *allowed*   f Lat lacks *of evil*
g Syr Ethiop Arm: Lat lacks *death*   h Syr: Lat *Hades and corruption have fled into oblivion*; or *corruption has fled into Hades to be forgotten*
i Syr: Lat *will receive*   j Syr: Ethiop *is in the word*; Meaning of Lat uncertain   k Syr: Lat Ethiop *in effects*

---

with the wicked. Humility is fine, for pride is dangerous indeed. But Ezra has no need to fear the judgment. The fact that the angel *cannot seem even to understand* that Ezra's concern is not for himself but for the myriads of sinners doomed to perish is a good indication that the author of 2 Esd 3–10 thinks as Ezra does, not as the angel does. **48:** *In this respect,* i.e. the seer's humility (compare Lk 18.13–14). **52:** The future joys of heaven are already in existence and may be contemplated now (1 Pet 1.4). *Tree of life,* 7.123; Rev 2.7; 22.2.

53: *Hades* is personified (as in Rev 6.8). **56:** *Opportunity to choose,* i.e. free will. **58:** Ps 14.1; 53.1. **59:** *The things . . . predicted,* in vv. 52–54. *Thirst,* in the fire of hell (Lk 16.24). *The Most High did not intend* humankind's destruction (Mt 18.14; 1 Tim 2.4). **62:** *A few like you,* i.e. prophets (apocalyptists) like the seer.

8.62b–9.13: **The end, and the signs that will precede it** (4.51–5.13; 6.11–24). **8.63:** *When,* 4.33; contrast Acts 1.7. **9.3:** The messianic woes on earth. **7:** De-

which they have believed, [8] will survive the dangers that have been predicted, and will see my salvation in my land and within my borders, which I have sanctified for myself from the beginning. [9] Then those who have now abused my ways shall be amazed, and those who have rejected them with contempt shall live in torments. [10] For as many as did not acknowledge me in their lifetime, though they received my benefits, [11] and as many as scorned my law while they still had freedom, and did not understand but despised it[l] while an opportunity of repentance was still open to them, [12] these must in torment acknowledge it[l] after death. [13] Therefore, do not continue to be curious about how the ungodly will be punished; but inquire how the righteous will be saved, those to whom the age belongs and for whose sake the age was made."[m]

14 I answered and said, [15] "I said before, and I say now, and will say it again: there are more who perish than those who will be saved, [16] as a wave is greater than a drop of water."

17 He answered me and said, "As is the field, so is the seed; and as are the flowers, so are the colors; and as is the work, so is the product; and as is the farmer, so is the threshing floor. [18] For there was a time in this age when I was preparing for those who now exist, before the world was made for them to live in, and no one opposed me then, for no one existed; [19] but now those who have been created in this world, which is supplied both with an unfailing table and an inexhaustible pasture,[n] have become corrupt in their ways. [20] So I considered my world, and saw that it was lost. I saw that my earth was in peril because of the devices of those who[o] had come into it. [21] And I saw and spared some[p] with great difficulty, and saved for myself one grape out of a cluster, and one plant out of a great forest.[q] [22] So let the multitude perish that has been born in vain, but let my grape and my plant be saved, because with much labor I have perfected them.

23 "Now, if you will let seven days more pass—do not, however, fast during them, [24] but go into a field of flowers where no house has been built, and eat only of the flowers of the field, and taste no meat and drink no wine, but eat only flowers, [25] and pray to the Most High continually. Then I will come and talk with you."

26 So I went, as he directed me, into the field that is called Ardat;[r] there I sat among the flowers and ate of the plants of the field, and the nourishment they afforded satisfied me. [27] After seven days, while I lay on the grass, my heart was troubled again as it was before. [28] Then my mouth was opened, and I began to speak before the Most High, and said, [29] "O Lord, you showed yourself among us, to our ancestors in the wilderness when they came out from Egypt and when they came into the untrodden and unfruitful wilderness; [30] and you said, 'Hear me, O Israel, and give heed to my

l Or *me*   *m* Syr: Lat *saved, and whose is the age and for whose sake the age was made and when*
n Cn: Lat *law*   *o* Cn: Lat *devices that*
p Lat *them*   *q* Syr Ethiop Arab 1: Lat *tribe*
r Syr Ethiop *Arpad*; Arm *Ardab*

---

liverance may come on the basis of good deeds or of belief in God (compare Jas 2.14–26 and Gal 2.11–21).

9.9–12: The state of the wicked immediately after death. **11:** Opportunity of repentance (Wis 12.10, 20; Heb 12.17). **12:** *Acknowledge,* their earlier opportunity of repentance; or the word may be translated "be brought to know."

9.14–25: Recapitulation: The seer again deplores the fate of the wicked, and the small number of the saved is explained a last time. **19:** Restore "law" (see note *n*) to the text: the meaning is that despite God's gracious provision of earthly sustenance and divine law, mortals *have become corrupt.*

9.21–22: The preservation of a small remnant is the result of God's grace. **24:** Likewise Daniel and his companions ate only vegetables (Dan 1.8–16; compare 2 Macc 5.27).

9.26–10.59: The fourth vision. 9.26–28: Introduction. 26: *Ardat,* an unknown location, probably of symbolical or mystic significance.

9.29–37: The abiding glory of the Mosaic law, contrasted with Israel. **29:** Ex 19.9;

words, O descendants of Jacob. ³¹For I sow my law in you, and it shall bring forth fruit in you, and you shall be glorified through it forever.' ³²But though our ancestors received the law, they did not keep it and did not observe the*ˢ* statutes; yet the fruit of the law did not perish—for it could not, because it was yours. ³³Yet those who received it perished, because they did not keep what had been sown in them. ³⁴Now this is the general rule that, when the ground has received seed, or the sea a ship, or any dish food or drink, and when it comes about that what was sown or what was launched or what was put in is destroyed, ³⁵they are destroyed, but the things that held them remain; yet with us it has not been so. ³⁶For we who have received the law and sinned will perish, as well as our hearts that received it; ³⁷the law, however, does not perish but survives in its glory."

38 When I said these things in my heart, I looked around,*ᵗ* and on my right I saw a woman; she was mourning and weeping with a loud voice, and was deeply grieved at heart; her clothes were torn, and there were ashes on her head. ³⁹Then I dismissed the thoughts with which I had been engaged, and turned to her ⁴⁰and said to her, "Why are you weeping, and why are you grieved at heart?"

41 She said to me, "Let me alone, my lord, so that I may weep for myself and continue to mourn, for I am greatly embittered in spirit and deeply distressed."

42 I said to her, "What has happened to you? Tell me."

43 And she said to me, "Your servant was barren and had no child, though I lived with my husband for thirty years.

⁴⁴Every hour and every day during those thirty years I prayed to the Most High, night and day. ⁴⁵And after thirty years God heard your servant, and looked upon my low estate, and considered my distress, and gave me a son. I rejoiced greatly over him, I and my husband and all my neighbors;*ᵘ* and we gave great glory to the Mighty One. ⁴⁶And I brought him up with much care. ⁴⁷So when he grew up and I came to take a wife for him, I set a day for the marriage feast.

10 "But it happened that when my son entered his wedding chamber, he fell down and died. ²So all of us put out our lamps, and all my neighbors*ᵛ* attempted to console me; I remained quiet until the evening of the second day. ³But when all of them had stopped consoling me, encouraging me to be quiet, I got up in the night and fled, and I came to this field, as you see. ⁴And now I intend not to return to the town, but to stay here; I will neither eat nor drink, but will mourn and fast continually until I die."

5 Then I broke off the reflections with which I was still engaged, and answered her in anger and said, ⁶"You most foolish of women, do you not see our mourning, and what has happened to us? ⁷For Zion, the mother of us all, is in deep grief and great distress. ⁸It is most appropriate to mourn now, because we are all mourning, and to be sorrowful, because we are all sorrowing; you are sorrowing for one son, but we, the whole world, for our mother.*ʷ* ⁹Now ask the earth,

s Lat *my*   *t* Syr Arab Arm: Lat *I looked about
me with my eyes*   *u* Literally *all my citizens*
v Literally *all my citizens*   *w* Compare Syr:
Meaning of Lat uncertain

24.10; Deut 4.12. **30–37**: The seer draws the contrast between the vessel that contains a precious object and the object it contains. The latter may be used up or destroyed, but normally the former survives. With us it is just the opposite; God keeps alive the divine law, but Israel, its receptacle, is destroyed. Could God not care a bit more for the receptacle? **9.38–10.24**: **The seer speaks with a dis-**

consolate woman. **38**: *Ashes on her head,* a sign of mourning. **47**: It was customary for the father to arrange for the wedding (see Sir 7.25 n.).

10.2: *Lamps,* because weddings took place at night. **2**: *I remained quiet,* shows the depth of her grief, for ordinarily there was loud lamentation. **7**: *Zion, the mother of us all,* Gal 4.26, compare Lam 2.20–22.

and she will tell you that it is she who ought to mourn over so many who have come into being upon her. 10 From the beginning all have been born of her, and others will come; and, lo, almost all go*x* to perdition, and a multitude of them will come to doom. 11 Who then ought to mourn the more, she who lost so great a multitude, or you who are grieving for one alone? 12 But if you say to me, 'My lamentation is not like the earth's, for I have lost the fruit of my womb, which I brought forth in pain and bore in sorrow; 13 but it is with the earth according to the way of the earth—the multitude that is now in it goes as it came'; 14 then I say to you, 'Just as you brought forth in sorrow, so the earth also has from the beginning given her fruit, that is, humankind, to him who made her.' 15 Now, therefore, keep your sorrow to yourself, and bear bravely the troubles that have come upon you. 16 For if you acknowledge the decree of God to be just, you will receive your son back in due time, and will be praised among women. 17 Therefore go into the town to your husband."

18 She said to me, "I will not do so; I will not go into the city, but I will die here."

19 So I spoke again to her, and said, 20 "Do not do that, but let yourself be persuaded—for how many are the adversities of Zion?—and be consoled because of the sorrow of Jerusalem. 21 For you see how our sanctuary has been laid waste, our altar thrown down, our temple destroyed; 22 our harp has been laid low, our song has been silenced, and our rejoicing has been ended; the light of our lampstand has been put out, the ark of our covenant has been plundered, our holy things have been polluted, and the name by which we are called has been almost profaned; our children*y* have suffered abuse, our priests have been burned to death, our Levites have gone into exile, our virgins have been defiled, and our wives have been ravished; our righteous men*z* have been carried off, our little ones have been cast out, our young men have been enslaved and our strong men made powerless. 23 And, worst of all, the seal of Zion has been deprived of its glory, and given over into the hands of those that hate us. 24 Therefore shake off your great sadness and lay aside your many sorrows, so that the Mighty One may be merciful to you again, and the Most High may give you rest, a respite from your troubles."

25 While I was talking to her, her face suddenly began to shine exceedingly; her countenance flashed like lightning, so that I was too frightened to approach her, and my heart was terrified. While*a* I was wondering what this meant, 26 she suddenly uttered a loud and fearful cry, so that the earth shook at the sound. 27 When I looked up, the woman was no longer visible to me, but a city was being built,*b* and a place of huge foundations showed itself. I was afraid, and cried with a loud voice and said, 28 "Where is

x Literally *walk*   y Ethiop *free men*
z Syr *our seers*   a Syr Ethiop Arab 1: Lat lacks
I was too . . . terrified. While   b Lat: Syr
Ethiop Arab 1 Arab 2 Arm *but there was an
established city*

10.16: To *acknowledge* the justice of God's *decree* is equivalent to pious submission to his will. *You will receive your son back in due time,* i.e. in the birth of another son, after returning to her husband (v. 17). **21–23**: A pathetic account of the utter ruin of Israel. **22**: *Harp* symbolizes the service of praise. The extinction of the perpetually burning lamp marked the cessation of temple services. *Our holy things* are enumerated in 1 Macc 4.49–51. *The name* Israel was bestowed by God (Gen 32.28). **23**: *The seal* of a nation is symbolic of its independence.

10.25–28: **A vision of the transformed Jerusalem. 27**: *A city was being built,* Zech 2.1–5; Heb 11.10; Rev 21.9–21. The new or transformed Zion is under construction, but its extent is known only to God. The author of 2 Esd 3–10 here offers hope for the wicked that a place will exist in Zion for them. Mother Zion will not only be restored but will become the city transformed to accommodate all whom God is pleased to redeem. The Latin text of 10.27, 42 is to be preferred, as in NRSV. **28**: *At first,* 4.1.

10.29–59: **Interpretation of the vision.**

the angel Uriel, who came to me at first? For it was he who brought me into this overpowering bewilderment; my end has become corruption, and my prayer a reproach."

29 While I was speaking these words, the angel who had come to me at first came to me, and when he saw me ³⁰lying there like a corpse, deprived of my understanding, he grasped my right hand and strengthened me and set me on my feet, and said to me, ³¹"What is the matter with you? And why are you troubled? And why are your understanding and the thoughts of your mind troubled?"

32 I said, "It was because you abandoned me. I did as you directed, and went out into the field, and lo, what I have seen I saw, and can still see, I am unable to explain."

33 He said to me, "Stand up like a man, and I will instruct you."

34 I said, "Speak, my lord; only do not forsake me, so that I may not die before my time.ᶜ ³⁵For I have seen what I did not know, and I hearᵈ what I do not understand ³⁶—or is my mind deceived, and my soul dreaming? ³⁷Now therefore I beg you to give your servant an explanation of this bewildering vision."

38 He answered me and said, "Listen to me, and I will teach you, and tell you about the things that you fear, for the Most High has revealed many secrets to you. ³⁹He has seen your righteous conduct, and that you have sorrowed continually for your people and mourned greatly over Zion. ⁴⁰This therefore is the meaning of the vision. ⁴¹The woman who appeared to you a little while ago, whom you saw mourning and whom you began to console ⁴²(you do not now see the form of a woman, but there appeared to you a city being built)ᵉ ⁴³and who told you about the misfortune of her son—this is the interpretation: ⁴⁴The woman whom you saw is Zion, which you now behold as a city being built.ᶠ ⁴⁵And as for her telling you that she was barren for thirty years, the reason is that there were three thousandᵍ years in the world before any offering was offered in it.ʰ ⁴⁶And after three thousandⁱ years Solomon built the city, and offered offerings; then it was that the barren woman bore a son. ⁴⁷And as for her telling you that she brought him up with much care, that was the period of residence in Jerusalem. ⁴⁸And as for her saying to you, 'My son died as he entered his wedding chamber,' and that misfortune had overtaken her,ʲ this was the destruction that befell Jerusalem. ⁴⁹So you saw her likeness, how she mourned for her son, and you began to console her for what had happened.ᵏ ⁵⁰For now the Most High, seeing that you are sincerely grieved and profoundly distressed for her, has shown you the brilliance of her glory, and the loveliness of her beauty. ⁵¹Therefore I told you to remain in the field where no house had been built, ⁵²for I knew that the Most High would reveal these things to you. ⁵³Therefore I told you to go into the field where there was no foundation of any building, ⁵⁴because no work of human construction could endure in a place where the city of the Most High was to be revealed.

55 "Therefore do not be afraid, and do not let your heart be terrified; but go in and see the splendor orˡ the vastness

c Syr Ethiop Arab: Lat *die to no purpose*
d Other ancient authorities read *have heard*
e Lat: Syr Ethiop Arab 1 Arab 2 Arm *an established city*   f Cn: Lat *an established city*
g Most Lat Mss read *three*   h Cn: Lat Syr Arab Arm *her*   i Syr Ethiop Arab Arm: Lat *three*   j Or *him*   k Most Lat Mss and Arab 1 add *these were the things to be opened to you*
l Other ancient authorities read *and*

30: *Like a corpse,* Rev 1.17; compare Dan 8.18; 10.9. 32: *And can still see,* the vision is still before the seer's eyes. *Unable to explain,* compare 2 Cor 12.3–4 (also of an ecstatic experience). 33: *Stand up,* 5.15; 6.13, 17. 44: *Zion,* i.e. the transformed Jerusalem. 45: *In it,* in the world. 46: *A son,* i.e. the historical Jeru-salem. 49: The *likeness,* or model, of the earthly city is the heavenly Zion, who *mourned for her son* (the ruined earthly Jerusalem). For the idea of a heavenly counterpart or model, compare Ex 25.9, 40; Heb 8.5. 10.55–56: 1 Cor 2.9; 2 Cor 12.4. *Go in and see,* the city is conceived as still present to

of the building, as far as it is possible for your eyes to see it, 56 and afterward you will hear as much as your ears can hear. 57 For you are more blessed than many, and you have been called to be with*m* the Most High as few have been. 58 But tomorrow night you shall remain here, 59 and the Most High will show you in those dream visions what the Most High will do to those who inhabit the earth in the last days."

So I slept that night and the following one, as he had told me.

11 On the second night I had a dream: I saw rising from the sea an eagle that had twelve feathered wings and three heads. 2 I saw it spread its wings over*n* the whole earth, and all the winds of heaven blew upon it, and the clouds were gathered around it. *o* 3 I saw that out of its wings there grew opposing wings; but they became little, puny wings. 4 But its heads were at rest; the middle head was larger than the other heads, but it too was at rest with them. 5 Then I saw that the eagle flew with its wings, and it reigned over the earth and over those who inhabit it. 6 And I saw how all things under heaven were subjected to it, and no one spoke against it—not a single creature that was on the earth. 7 Then I saw the eagle rise upon its talons, and it uttered a cry to its wings, saying, 8 "Do not all watch at the same time; let each sleep in its own place, and watch in its turn; 9 but let the heads be reserved for the last."

10 I looked again and saw that the voice did not come from its heads, but from the middle of its body. 11 I counted its rival wings, and there were eight of them. 12 As I watched, one wing on the right side rose up, and it reigned over all the earth. 13 And after a time its reign came to an end, and it disappeared, so that even its place was no longer visible. Then the next wing rose up and reigned, and it continued to reign a long time. 14 While it was reigning its end came also, so that it disappeared like the first. 15 And a voice sounded, saying to it, 16 "Listen to me, you who have ruled the earth all this time; I announce this to you before you disappear. 17 After you no one shall rule as long as you have ruled, not even half as long."

18 Then the third wing raised itself up, and held the rule as the earlier ones had done, and it also disappeared. 19 And so it went with all the wings; they wielded power one after another and then were never seen again. 20 I kept looking, and in due time the wings that followed*p* also rose up on the right*q* side, in order to rule. There were some of them that ruled, yet disappeared suddenly; 21 and others of them rose up, but did not hold the rule.

22 And after this I looked and saw that the twelve wings and the two little wings had disappeared, 23 and nothing remained on the eagle's body except the three heads that were at rest and six little wings.

24 As I kept looking I saw that two little wings separated from the six and remained under the head that was on the right side; but four remained in their place. 25 Then I saw that these little

---

m Or been named by   n Arab 2 Arm: Lat Syr Ethiop *in   o* Syr: Compare Ethiop Arab: Lat lacks *the clouds* and *around it   p* Syr Arab 2 *the little wings   q* Some Ethiop Mss read *left*

Ezra. **57:** *You have been called to be with the Most High,* Arabic 1, "your name is known [or recognized] before the Most High," i.e. God has singled you out for honor (Isa 45.3–4).

11.1–12.51: The fifth vision (the eagle vision). 11.1: *From the sea,* Dan 7.3; Rev 13.1. *An eagle,* symbol of the Roman Empire. **2:** *Spread its wings,* asserted its dominion. *The winds,* 13.2; Dan 7.2. **3:** *Opposing wings,* symbolizing usurpers who revolted against the Roman emperors. *But they became little,* i.e. they were subdued. **4:** *Were at rest,* i.e. were not troubled by the opposing wings.

11.13: *Its reign came to an end,* i.e. the ruler perished. **36:** *Look in front of you,* the seer is alerted to the special importance of what follows. **43:** Dan 5.20.

12.3b–39: The interpretation of the vision. 3b–9: The seer awakes and asks for an interpretation of the vision.

wings[r] planned to set themselves up and hold the rule. 26 As I kept looking, one was set up, but suddenly disappeared; 27 a second also, and this disappeared more quickly than the first. 28 While I continued to look the two that remained were planning between themselves to reign together; 29 and while they were planning, one of the heads that were at rest (the one that was in the middle) suddenly awoke; it was greater than the other two heads. 30 And I saw how it allied the two heads with itself, 31 and how the head turned with those that were with it and devoured the two little wings[r] that were planning to reign. 32 Moreover this head gained control of the whole earth, and with much oppression dominated its inhabitants; it had greater power over the world than all the wings that had gone before.

33 After this I looked again and saw the head in the middle suddenly disappear, just as the wings had done. 34 But the two heads remained, which also in like manner ruled over the earth and its inhabitants. 35 And while I looked, I saw the head on the right side devour the one on the left.

36 Then I heard a voice saying to me, "Look in front of you and consider what you see." 37 When I looked, I saw what seemed to be a lion roused from the forest, roaring; and I heard how it uttered a human voice to the eagle, and spoke, saying, 38 "Listen and I will speak to you. The Most High says to you, 39 'Are you not the one that remains of the four beasts that I had made to reign in my world, so that the end of my times might come through them? 40 You, the fourth that has come, have conquered all the beasts that have gone before; and you have held sway over the world with great terror, and over all the earth with grievous oppression; and for so long you have lived on the earth with deceit.[s] 41 You have judged the earth, but not with truth, 42 for you have oppressed the meek and injured the peaceable; you have hated those who tell the truth, and have loved liars; you have destroyed the homes of those who brought forth fruit,

and have laid low the walls of those who did you no harm. 43 Your insolence has come up before the Most High, and your pride to the Mighty One. 44 The Most High has looked at his times; now they have ended, and his ages have reached completion. 45 Therefore you, eagle, will surely disappear, you and your terrifying wings, your most evil little wings, your malicious heads, your most evil talons, and your whole worthless body, 46 so that the whole earth, freed from your violence, may be refreshed and relieved, and may hope for the judgment and mercy of him who made it.'"

12 While the lion was saying these words to the eagle, I looked 2 and saw that the remaining head had disappeared. The two wings that had gone over to it rose up and[t] set themselves up to reign, and their reign was brief and full of tumult. 3 When I looked again, they were already vanishing. The whole body of the eagle was burned, and the earth was exceedingly terrified.

Then I woke up in great perplexity of mind and great fear, and I said to my spirit, 4 "You have brought this upon me, because you search out the ways of the Most High. 5 I am still weary in mind and very weak in my spirit, and not even a little strength is left in me, because of the great fear with which I have been terrified tonight. 6 Therefore I will now entreat the Most High that he may strengthen me to the end."

7 Then I said, "O sovereign Lord, if I have found favor in your sight, and if I have been accounted righteous before you beyond many others, and if my prayer has indeed come up before your face, 8 strengthen me and show me, your servant, the interpretation and meaning of this terrifying vision so that you may fully comfort my soul. 9 For you have judged me worthy to be shown the end of the times and the last events of the times."

10 He said to me, "This is the inter-

r Syr: Lat *underwings*   s Syr Arab Arm: Lat Ethiop *The fourth came, however, and conquered . . . and held sway . . . and for so long lived*   t Ethiop: Lat lacks *rose up and*

pretation of this vision that you have seen: [11] The eagle that you saw coming up from the sea is the fourth kingdom that appeared in a vision to your brother Daniel. [12] But it was not explained to him as I now explain to you or have explained it. [13] The days are coming when a kingdom shall rise on earth, and it shall be more terrifying than all the kingdoms that have been before it. [14] And twelve kings shall reign in it, one after another. [15] But the second that is to reign shall hold sway for a longer time than any other one of the twelve. [16] This is the interpretation of the twelve wings that you saw.

17 "As for your hearing a voice that spoke, coming not from the eagle's[u] heads but from the midst of its body, this is the interpretation: [18] In the midst of[v] the time of that kingdom great struggles shall arise, and it shall be in danger of falling; nevertheless it shall not fall then, but shall regain its former power.[w] [19] As for your seeing eight little wings[x] clinging to its wings, this is the interpretation: [20] Eight kings shall arise in it, whose times shall be short and their years swift; [21] two of them shall perish when the middle of its time draws near; and four shall be kept for the time when its end approaches, but two shall be kept until the end.

22 "As for your seeing three heads at rest, this is the interpretation: [23] In its last days the Most High will raise up three kings,[y] and they[z] shall renew many things in it, and shall rule the earth [24] and its inhabitants more oppressively than all who were before them. Therefore they are called the heads of the eagle, [25] be-cause it is they who shall sum up his wickedness and perform his last actions. [26] As for your seeing that the large head disappeared, one of the kings[a] shall die in his bed, but in agonies. [27] But as for the two who remained, the sword shall devour them. [28] For the sword of one shall devour him who was with him; but he also shall fall by the sword in the last days.

29 As for your seeing two little wings[b] passing over to[c] the head which was on the right side, [30] this is the interpretation: It is these whom the Most High has kept for the eagle's[d] end; this was the reign which was brief and full of tumult, as you have seen.

31 "And as for the lion whom you saw rousing up out of the forest and roaring and speaking to the eagle and reproving him for his unrighteousness, and as for all his words that you have heard, [32] this is the Messiah[e] whom the Most High has kept until the end of days, who will arise from the offspring of David, and will come and speak[f] with them. He will denounce them for their ungodliness and for their wickedness, and will display before them their contemptuous dealings. [33] For first he will bring them alive before his judgment seat, and when he has reproved them, then he will destroy them. [34] But in mercy he will set free the remnant of my people, those

u Lat *his*   v Syr Arm: Lat *After* Arab 1 Arm: Lat Syr *its beginning*   w Ethiop *underwings*   x Syr: Lat *kingdoms*   y Syr Ethiop Arab Arm: Lat *them*   z Syr Ethiop Arm: Lat *he*   a Lat *them*   b Arab 1: Lat *underwings*   c Syr Ethiop: Lat lacks *to*   d Lat *his*   e Literally *anointed one*   f Syr: Lat lacks *of days . . . and speak*

12.11: *The fourth kingdom* in Daniel's vision (Dan 7.7) symbolized the Greek or Macedonian Empire; here, however, it is reinterpreted (compare v. 12) to refer to the Roman Empire (see 11.1 n.). **13:** *The days are coming*, the seer is represented as prophesying during the exile. **17:** 11.10; compare 11.15. **18:** There is nothing in the vision that corresponds to what is said in this verse. The author probably refers to *the time* of *great struggles* for power that followed the death of Nero A.D. 68. **19:** *Little wings*, 11.3, 11. *Clinging to its wings*, Armenian, "sprouting out around its great wings." **20:** *In it*, within the Roman Empire. **21:** *Its time*, the time of the kingdom. *Its end*, the end of the kingdom. **23:** *Its last days*, the last days of the kingdom. **23–24:** 11.30–32.

12.28: *But . . . days*, there is nothing corresponding to this in the vision. **30:** *As you have seen*, v. 3. **31:** *The lion*, 11.37ff. **32:** *Whom the Most High has kept until the end of days*, the hidden Messiah preserved by God for the last days (Dan 7.13–14; Enoch 48.6; 62.7). *He will denounce them*, 13.37. **34:** *He will make them*

who have been saved throughout my borders, and he will make them joyful until the end comes, the day of judgment, of which I spoke to you at the beginning. 35 This is the dream that you saw, and this is its interpretation. 36 And you alone were worthy to learn this secret of the Most High. 37 Therefore write all these things that you have seen in a book, put it*g* in a hidden place; 38 and you shall teach them to the wise among your people, whose hearts you know are able to comprehend and keep these secrets. 39 But as for you, wait here seven days more, so that you may be shown whatever it pleases the Most High to show you." Then he left me.

40 When all the people heard that the seven days were past and I had not returned to the city, they all gathered together, from the least to the greatest, and came to me and spoke to me, saying, 41 "How have we offended you, and what harm have we done you, that you have forsaken us and sit in this place? 42 For of all the prophets you alone are left to us, like a cluster of grapes from the vintage, and like a lamp in a dark place, and like a haven for a ship saved from a storm. 43 Are not the disasters that have befallen us enough? 44 Therefore if you forsake us, how much better it would have been for us if we also had been consumed in the burning of Zion. 45 For we are no better than those who died there." And they wept with a loud voice.

Then I answered them and said, 46 "Take courage, O Israel; and do not be sorrowful, O house of Jacob; 47 for the Most High has you in remembrance, and the Mighty One has not forgotten you in your struggle. 48 As for me, I have neither forsaken you nor withdrawn from you; but I have come to this place to pray on account of the desolation of Zion, and to seek mercy on account of the humiliation of our*h* sanctuary. 49 Now go to your homes, every one of you, and after these days I will come to you." 50 So the people went into the city, as I told them to do. 51 But I sat in the field seven days, as the angel*i* had commanded me; and I ate only of the flowers of the field, and my food was of plants during those days.

13 After seven days I dreamed a dream in the night. 2 And lo, a wind arose from the sea and stirred up*j* all its waves. 3 As I kept looking the wind made something like the figure of a man come up out of the heart of the sea. And I saw*k* that this man flew*l* with the clouds of heaven; and wherever he turned his face to look, everything under his gaze trembled, 4 and whenever his voice issued from his mouth, all who heard his voice melted as wax melts*m* when it feels the fire.

5 After this I looked and saw that an innumerable multitude of people were gathered together from the four winds of heaven to make war against the man who came up out of the sea. 6 And I looked and saw that he carved out for himself a great mountain, and flew up on to it. 7 And I tried to see the region or place from which the mountain was carved, but I could not.

8 After this I looked and saw that all who had gathered together against him,

g Ethiop Arab 1 Arab 2 Arm: Lat Syr *them*
h Syr Ethiop: Lat *your*   i Literally *he*
j Other ancient authorities read *I saw a wind
arise from the sea and stir up*   k Syr: Lat lacks *the
wind . . . I saw*   l Syr Ethiop Arab Arm: Lat
grew strong   m Syr: Lat *burned as the earth rests*

---

joyful, 7.28. **35**: *The dream,* 11.1. **37**: The seer is bidden to compose an esoteric book. **37–38**: To *put the book in a hidden place* suggests that it is an apocryphal book, which only the elect (*the wise*) can *comprehend.* **39**: *He,* the angel Uriel (see 4.1 n.).
12.40–51: **The seer comforts those who were grieved because of his absence. 40**: *The seven days,* 9.23, 27. **42**: *A lamp,* 2 Pet 1.19. **49**: *These days,* v. 39. **51**: 9.24–26.
13.1–58: **The sixth vision (the man from the sea). 3**: *Something like the figure of a man,* the Messiah (Dan 7.13); compare v. 32 "my Son," i.e. the Son of God. *Flew with the clouds,* Isa 19.1; Dan 7.13; Rev 1.7. **4**: *As wax melts,* Mic 1.4; Jdt 16.15. **6**: *Carved out,* Dan 2.45.

to wage war with him, were filled with fear, and yet they dared to fight. [9] When he saw the onrush of the approaching multitude, he neither lifted his hand nor held a spear or any weapon of war; [10] but I saw only how he sent forth from his mouth something like a stream of fire, and from his lips a flaming breath, and from his tongue he shot forth a storm of sparks. [n] [11] All these were mingled together, the stream of fire and the flaming breath and the great storm, and fell on the onrushing multitude that was prepared to fight, and burned up all of them, so that suddenly nothing was seen of the innumerable multitude but only the dust of ashes and the smell of smoke. When I saw it, I was amazed.

12 After this I saw the same man come down from the mountain and call to himself another multitude that was peaceable. [13] Then many people[o] came to him, some of whom were joyful and some sorrowful; some of them were bound, and some were bringing others as offerings.

Then I woke up in great terror, and prayed to the Most High, and said, [14] "From the beginning you have shown your servant these wonders, and have deemed me worthy to have my prayer heard by you; [15] now show me the interpretation of this dream also. [16] For as I consider it in my mind, alas for those who will be left in those days! And still more, alas for those who are not left! [17] For those who are not left will be sad [18] because they understand the things that are reserved for the last days, but cannot attain them. [19] But alas for those also who are left, and for that very reason! For they shall see great dangers and much distress, as these dreams show. [20] Yet it is better[p]

to come into these things,[q] though incurring peril, than to pass from the world like a cloud, and not to see what will happen in the last days."

He answered me and said, [21] "I will tell you the interpretation of the vision, and I will also explain to you the things that you have mentioned. [22] As for what you said about those who survive, and concerning those who do not survive,[r] this is the interpretation: [23] The one who brings the peril at that time will protect those who fall into peril, who have works and faith toward the Almighty. [24] Understand therefore that those who are left are more blessed than those who have died.

25 "This is the interpretation of the vision: As for your seeing a man come up from the heart of the sea, [26] this is he whom the Most High has been keeping for many ages, who will himself deliver his creation; and he will direct those who are left. [27] And as for your seeing wind and fire and a storm coming out of his mouth, [28] and as for his not holding a spear or weapon of war, yet destroying the onrushing multitude that came to conquer him, this is the interpretation: [29] The days are coming when the Most High will deliver those who are on the earth. [30] And bewilderment of mind shall come over those who inhabit the earth. [31] They shall plan to make war against one another, city against city, place against place, people against people, and kingdom against kingdom. [32] When these things take place and the signs occur that I showed you before, then my

n Meaning of Lat uncertain   o Lat Syr
Arab 2 literally *the faces of many people*   p Ethiop
Compare Arab 2: Lat *easier*   q Syr: Lat *this*
r Syr Arab 1: Lat lacks *and . . . not survive*

13.10: Isa 11.4. **13a**: *Some . . . were bound,* Jews who came from captivity. *Others as offerings,* Isa 66.20.
13.13b–20: **The seer prays that God will interpret the vision to him. 14**: *From the beginning,* when the seer first began to have the visions. *My prayer,* 9.25ff. **19**: *For that very reason,* better, "for this reason—" (the reason follows).

13.21–58: **The interpretation of the vision. 21**: *Things . . . mentioned,* in vv. 16–20. **23**: *The one who brings the peril,* the Messiah, whose advent is preceded by the messianic woes. **26**: *He whom the Most High has been keeping for many ages,* the hidden Messiah (v. 52; 12.32). **27–28**: Verses 9–11. **13.31**: Isa 19.2; Mt 24.7. **32**: *Then my Son will be revealed,* 7.28; Mt 24.30; Mk 13.26. **34**:

Son will be revealed, whom you saw as a man coming up from the sea. [s]

33 "Then, when all the nations hear his voice, all the nations shall leave their own lands and the warfare that they have against one another; [34] and an innumerable multitude shall be gathered together, as you saw, wishing to come and conquer him. [35] But he shall stand on the top of Mount Zion. [36] And Zion shall come and be made manifest to all people, prepared and built, as you saw the mountain carved out without hands. [37] Then he, my Son, will reprove the assembled nations for their ungodliness (this was symbolized by the storm), [38] and will reproach them to their face with their evil thoughts and the torments with which they are to be tortured (which were symbolized by the flames), and will destroy them without effort by means of the law [t] (which was symbolized by the fire).

39 "And as for your seeing him gather to himself another multitude that was peaceable, [40] these are the nine [u] tribes that were taken away from their own land into exile in the days of King Hoshea, whom Shalmaneser, king of the Assyrians, made captives; he took them across the river, and they were taken into another land. [41] But they formed this plan for themselves, that they would leave the multitude of the nations and go to a more distant region, where no human beings had ever lived, [42] so that there at least they might keep their statutes that they had not kept in their own land. [43] And they went in by the narrow passages of the Euphrates river. [44] For at that time the Most High performed signs for them, and stopped the channels of the river until they had crossed over. [45] Through that region there was a long

way to go, a journey of a year and a half; and that country is called Arzareth. [v]

46 "Then they lived there until the last times; and now, when they are about to come again, [47] the Most High will stop [w] the channels of the river again, so that they may be able to cross over. Therefore you saw the multitude gathered together in peace. [48] But those who are left of your people, who are found within my holy borders, shall be saved. [x] [49] Therefore when he destroys the multitude of the nations that are gathered together, he will defend the people who remain. [50] And then he will show them very many wonders."

51 I said, "O sovereign Lord, explain this to me: Why did I see the man coming up from the heart of the sea?"

52 He said to me, "Just as no one can explore or know what is in the depths of the sea, so no one on earth can see my Son or those who are with him, except in the time of his day. [y] [53] This is the interpretation of the dream that you saw. And you alone have been enlightened about this, [54] because you have forsaken your own ways and have applied yourself to mine, and have searched out my law; [55] for you have devoted your life to wisdom, and called understanding your mother. [56] Therefore I have shown you these things; for there is a reward laid up with the Most High. For it will be that after three more days I will tell you other things, and explain weighty and wondrous matters to you."

s Syr and most Lat Mss lack *from the sea*
t Syr: Lat *effort and the law*   u Other Lat Mss
ten; Syr Ethiop Arab 1 Arm *nine and a half*
v That is *Another Land*   w Syr: Lat *stops*
x Syr: Lat lacks *shall be saved*   y Syr: Ethiop
except when his time and his day have come. Lat
lacks *his*

Rev 16.16; 19.19. **36**: *Zion,* the heavenly Jerusalem (7.26; Rev 21.2, 9f.). *Without hands,* Dan 2.34, 45. **37**: 12.32. **40**: 2 Kings 17.1–6. *The nine tribes,* the Northern Kingdom (usually "ten tribes"—see note *u*). *The river,* the Euphrates.
13.44: *Stopped . . . the river,* Josh 3.14–16. **45**: *Arzareth,* Hebrew for "Another Land"

(see note *v*; compare Deut 29.28). **47**: *Will stop . . . the river,* Isa 11.15–16. **49**: *The people who remain,* presumably Israel, including the ten tribes who have returned to Palestine (v. 48). **50**: *Then,* in the messianic age. **52**: *Those . . . with him,* perhaps angels (Mt 24.31; 25.31). *Except . . . day,* until the day on which the Messiah appears.

57 Then I got up and walked in the field, giving great glory and praise to the Most High for the wonders that he does[z] from time to time, 58 and because he governs the times and whatever things come to pass in their seasons. And I stayed there three days.

14 On the third day, while I was sitting under an oak, suddenly a voice came out of a bush opposite me and said, "Ezra, Ezra!" 2 And I answered, "Here I am, Lord," and I rose to my feet. 3 Then he said to me, "I revealed myself in a bush and spoke to Moses when my people were in bondage in Egypt; 4 and I sent him and led[a] my people out of Egypt; and I led him up on Mount Sinai, where I kept him with me many days. 5 I told him many wondrous things, and showed him the secrets of the times and declared to him[b] the end of the times. Then I commanded him, saying, 6 'These words you shall publish openly, and these you shall keep secret.' 7 And now I say to you: 8 Lay up in your heart the signs that I have shown you, the dreams that you have seen, and the interpretations that you have heard; 9 for you shall be taken up from among humankind, and henceforth you shall live with my Son and with those who are like you, until the times are ended. 10 The age has lost its youth, and the times begin to grow old. 11 For the age is divided into twelve parts, and nine[c] of its parts have already passed, 12 as well as half of the tenth part; so two of its parts remain, besides half of the tenth part.[d] 13 Now therefore, set your house in order, and reprove your people; comfort the lowly among them, and instruct those that are wise.[e] And now renounce the life that is corruptible, 14 and put away from you mortal thoughts; cast away from you the burdens of humankind, and divest yourself now of your weak nature; 15 lay to one side the thoughts that are most grievous to you, and hurry to escape from these times. 16 For evils worse than those that you have now seen happen shall take place hereafter. 17 For the weaker the world becomes through old age, the more shall evils be increased upon its inhabitants. 18 Truth shall go farther away, and falsehood shall come near. For the eagle[f] that you saw in the vision is already hurrying to come."

19 Then I answered and said, "Let me speak[g] in your presence, Lord. 20 For I will go, as you have commanded me, and I will reprove the people who are now living; but who will warn those who will be born hereafter? For the world lies in darkness, and its inhabitants are without light. 21 For your law has been burned, and so no one knows the things which have been done or will be done by you. 22 If then I have found favor with you, send the holy spirit into me, and I will write everything that has happened in the world from the beginning, the things that were written in your law, so that people may be able to find the path, and that those who want to live in the last days may do so."

23 He answered me and said, "Go and gather the people, and tell them not to seek you for forty days. 24 But prepare for yourself many writing tablets, and take with you Sarea, Dabria, Selemia, Ethanus, and Asiel—these five, who are trained to write rapidly; 25 and you shall come here, and I will light in your heart

z Lat *did* a Syr Arab 1 Arab 2 *he led* b Syr Ethiop Arab Arm: Lat lacks *declared to him* c Cn: Lat Ethiop *ten* d Syr lacks verses 11, 12: Ethiop *For the world is divided into ten parts, and has come to the tenth, and half of the tenth remains. Now . . .* e Lat lacks *and . . . wise* f Syr Ethiop Arab Arm: Meaning of Lat uncertain g Most Lat Mss lack *Let me speak*

14.1–48: **The seventh vision (the legend of Ezra and the holy Scriptures). 1–18:** God speaks to Ezra. **1:** *A bush,* compare Ex 3.4. **4:** *Many days,* forty days (Ex 34.28). **9:** *My Son,* the hidden Messiah (7.28; 13.32, 52). **14.10:** 5.50–55. **13:** *House,* of Israel. **14:** 2 Cor 5.4. **16:** Mt 24.8. **18:** *The eagle,* ch 11.

14.19–26: **Ezra's prayer for inspiration to restore the holy Scriptures. 20:** *Without light,* without the light of God's law (Ps 19.8b). **21:** 4.23. **22:** *The holy spirit* will guide Ezra in rewriting the law, which has been burned (v. 21). **23:** *Forty days,* Ex 24.18; 34.28; Deut 9.9, 18. **24:** *Many,* compare v. 44.

the lamp of understanding, which shall not be put out until what you are about to write is finished. 26 And when you have finished, some things you shall make public, and some you shall deliver in secret to the wise; tomorrow at this hour you shall begin to write."

27 Then I went as he commanded me, and I gathered all the people together, and said, 28 "Hear these words, O Israel. 29 At first our ancestors lived as aliens in Egypt, and they were liberated from there 30 and received the law of life, which they did not keep, which you also have transgressed after them. 31 Then land was given to you for a possession in the land of Zion; but you and your ancestors committed iniquity and did not keep the ways that the Most High commanded you. 32 And since he is a righteous judge, in due time he took from you what he had given. 33 And now you are here, and your people*h* are farther in the interior.*i* 34 If you, then, will rule over your minds and discipline your hearts, you shall be kept alive, and after death you shall obtain mercy. 35 For after death the judgment will come, when we shall live again; and then the names of the righteous shall become manifest, and the deeds of the ungodly shall be disclosed. 36 But let no one come to me now, and let no one seek me for forty days."

37 So I took the five men, as he commanded me, and we proceeded to the field, and remained there. 38 And on the next day a voice called me, saying, "Ezra, open your mouth and drink what I give you to drink." 39 So I opened my mouth, and a full cup was offered to me;

it was full of something like water, but its color was like fire. 40 I took it and drank; and when I had drunk it, my heart poured forth understanding, and wisdom increased in my breast, for my spirit retained its memory, 41 and my mouth was opened and was no longer closed. 42 Moreover, the Most High gave understanding to the five men, and by turns they wrote what was dictated, using characters that they did not know.*j* They sat forty days; they wrote during the daytime, and ate their bread at night. 43 But as for me, I spoke in the daytime and was not silent at night. 44 So during the forty days, ninety-four*k* books were written. 45 And when the forty days were ended, the Most High spoke to me, saying, "Make public the twenty-four*l* books that you wrote first, and let the worthy and the unworthy read them; 46 but keep the seventy that were written last, in order to give them to the wise among your people. 47 For in them is the spring of understanding, the fountain of wisdom, and the river of knowledge." 48 And I did so.*m*

---

h Lat *brothers*   i Syr Ethiop Arm: Lat *are among you*   j Syr Compare Ethiop Arab 2 Arm: Meaning of Lat uncertain   k Syr Ethiop Arab 1 Arm: Meaning of Lat uncertain
l Syr Arab 1: Lat lacks *twenty-four*   m Syr adds *in the seventh year of the sixth week, five thousand years and three months and twelve days after creation. At that time Ezra was caught up, and taken to the place of those who are like him, after he had written all these things. And he was called the scribe of the knowledge of the Most High for ever and ever.* Ethiop Arab 1 Arm have a similar ending

---

26: *Some things . . . make public,* namely, the rewritten books of the Old Testament. *Some . . . deliver in secret,* namely, the apocalypses (see 12.37–38 n.).
14.27–36: **The last words of Ezra. 29**: *Aliens,* Gen 47.4. **30**: *The law,* which, if observed, would confer *life*. **33**: *Farther,* 13.45. **34**: *Kept alive,* i.e. spiritually alive. **36**: *Forty days,* v. 23.
14.37–48: **The restoration of the holy Scriptures. 37**: *The five men,* v. 24. **39**: *A full cup* of inspiration, containing the fire of the spirit (v. 22). **41**: *Was opened,* in fluent speech.

42: *Using characters that they did not know,* in a new Hebrew script, the (modern) square characters. **43**: Apparently Ezra dictated constantly, repeating what the scribes did not hear during their times of rest and sleep. **45**: *The twenty-four books* of the Hebrew canon comprise the five books of the Law (Gen, Ex, Lev, Num, Deut), eight books of the Prophets (the former prophets, Josh, Judg, 1 and 2 Sam [as one book], 1 and 2 Kings [as one book]; the latter prophets, Isa, Jer, Ezek, and the Twelve [counted as one book]), and eleven books of the Writings (Ps, Prov, Job,

15 *[n]* Speak in the ears of my people the words of the prophecy that I will put in your mouth, says the Lord, 2and cause them to be written on paper; for they are trustworthy and true. 3Do not fear the plots against you, and do not be troubled by the unbelief of those who oppose you. 4For all unbelievers shall die in their unbelief. *[o]*

5 Beware, says the Lord, I am bringing evils upon the world, the sword and famine, death and destruction, 6because iniquity has spread throughout every land, and their harmful doings have reached their limit. 7Therefore, says the Lord, 8I will be silent no longer concerning their ungodly acts that they impiously commit, neither will I tolerate their wicked practices. Innocent and righteous blood cries out to me, and the souls of the righteous cry out continually. 9I will surely avenge them, says the Lord, and will receive to myself all the innocent blood from among them. 10See, my people are being led like a flock to the slaughter; I will not allow them to live any longer in the land of Egypt, 11but I will bring them out with a mighty hand and with an uplifted arm, and will strike Egypt with plagues, as before, and will destroy all its land.

12 Let Egypt mourn, and its foundations, because of the plague of chastisement and castigation that the Lord will bring upon it. 13Let the farmers that till the ground mourn, because their seed shall fail to grow *[p]* and their trees shall be ruined by blight and hail and by a terrible tempest. 14Alas for the world and for those who live in it! 15For the sword and misery draw near them, and nation shall rise up to fight against nation, with swords in their hands. 16For there shall be unrest among people; growing strong against one another, they shall in their might have no respect for their king or the chief of their leaders. 17For a person will desire to go into a city, and shall not be able to do so. 18Because of their pride the cities shall be in confusion, the houses shall be destroyed, and people shall be afraid. 19People shall have no pity for their neighbors, but shall make an assault upon *[q]* their houses with the sword, and plunder their goods, because of hunger for bread and because of great tribulation.

20 See how I am calling together all the kings of the earth to turn to me, says God, from the rising sun and from the south, from the east and from Lebanon; to turn and repay what they have given them. 21Just as they have done to my elect until this day, so I will do, and will repay into their bosom. Thus says the Lord God: 22My right hand will not spare the sinners, and my sword will not cease from those who shed innocent blood on earth. 23And a fire went forth from his wrath, and consumed the foundations of the earth and the sinners, like burnt straw. 24Alas for those who sin and do not observe my commandments, says the Lord; *[r]* 25I will not spare them. Depart, you faithless children! Do not pollute my sanctuary. 26For God *[s]* knows all who sin against him; therefore he will hand them over to death and slaughter. 27Already calamities have come upon the whole earth, and you shall remain in them; God *[s]* will not deliver you, because you have sinned against him.

28 What a terrifying sight, appearing

---

n Chapters 15 and 16 (except 15.57-59, which has been found in Greek) are extant only in Lat *o* Other ancient authorities add *and all who believe shall be saved by their faith* *p* Lat lacks *to grow* *q* Cn: Lat *shall empty* *r* Other ancient authorities read *God* *s* Other ancient authorities read *the Lord*

Song, Ruth, Lam, Eccl, Esth, Dan, Ezra-Neh [as one book], 1 and 2 Chr [as one book]). **46**: *The seventy* are esoteric, apocalyptic books (see 12.37-38 n.).

15.1-16.78: An appendix. 15.1-4: The certainty of this prophecy. **1**: *That I will put in your mouth,* Isa 51.16; Jer 1.9.

15.5-11: God will take vengeance upon the wicked. 9: *All the innocent blood,* i.e. all the souls of the righteous (compare Rev 6.10; 19.2). **10**: Ps 44.22; Isa 53.7. **11**: *Will strike Egypt . . . as before,* perhaps an allusion to the occurrence during the reign of Gallienus (A.D. 260-268) of a terrible famine, followed

from the east! 29 The nations of the dragons of Arabia shall come out with many chariots, and from the day that they set out, their hissing shall spread over the earth, so that all who hear them will fear and tremble. 30 Also the Carmonians, raging in wrath, shall go forth like wild boars[t] from the forest, and with great power they shall come and engage them in battle, and with their tusks they shall devastate a portion of the land of the Assyrians with their teeth. 31 And then the dragons,[u] remembering their origin, shall become still stronger; and if they combine in great power and turn to pursue them, 32 then these shall be disorganized and silenced by their power, and shall turn and flee.[v] 33 And from the land of the Assyrians an enemy in ambush shall attack them and destroy one of them, and fear and trembling shall come upon their army, and indecision upon their kings.

34 See the clouds from the east, and from the north to the south! Their appearance is exceedingly threatening, full of wrath and storm. 35 They shall clash against one another and shall pour out a heavy tempest on the earth, and their own tempest;[w] and there shall be blood from the sword as high as a horse's belly 36 and a man's thigh and a camel's hock. 37 And there shall be fear and great trembling on the earth; those who see that wrath shall be horror-stricken, and they shall be seized with trembling. 38 After that, heavy storm clouds shall be stirred up from the south, and from the north, and another part from the west. 39 But the winds from the east shall prevail over the cloud that was[x] raised in wrath, and shall dispel it; and the tempest[y] that was to cause destruction by the east wind shall be driven violently toward the south and west. 40 Great and mighty clouds, full of wrath and tempest, shall rise and destroy all the earth and its inhabitants, and shall pour out upon every high and lofty place[z] a terrible tempest, 41 fire and hail and flying swords and floods of water, so that all the fields and all the streams shall be filled with the abundance of those waters. 42 They shall destroy cities and walls, mountains and hills, trees of the forests, and grass of the meadows, and their grain. 43 They shall go on steadily to Babylon and blot it out. 44 They shall come to it and surround it; they shall pour out on it the tempest[y] and all its fury;[a] then the dust and smoke shall reach the sky, and all who are around it shall mourn for it. 45 And those who survive shall serve those who have destroyed it.

46 And you, Asia, who share in the splendor of Babylon and the glory of her person— 47 woe to you, miserable wretch! For you have made yourself like her; you have decked out your daughters for prostitution to please and glory in your lovers, who have always lusted after you. 48 You have imitated that hateful one in all her deeds and devices.[b] Therefore God[c] says, 49 I will send evils upon you: widowhood, poverty, famine, sword, and pestilence, bringing ruin to your houses, bringing destruction and death. 50 And the glory of your strength

t Other ancient authorities lack *like wild boars*
u Cn: Lat *dragon*   v Other ancient authorities read *turn their face to the north*   w Meaning of Lat uncertain   x Literally *that he*
y Meaning of Lat uncertain   z Or *eminent person*   a Other ancient authorities add *until they destroy it to its foundations*   b Other ancient authorities add *you have followed after that one about to gratify her magnates and leaders so that you may be made proud and be pleased by her fornications*
c Other ancient authorities read *the Lord*

by a plague, which killed two-thirds of the population of Alexandria.

15.12–27: The signs of the end. 15: *Nation . . . against nation,* Mt 24.7; Mk 13.8; Lk 21.10. **18:** Lk 21.26.

15.28–63: A vision of warfare. This section is thought to reflect events of the third century A.D., including the attack of King Sapor I of Persia (A.D. 240–273) upon the Roman province of Syria. **30:** *The Carmonians,* from Carmania (Kirman), the southern province of the Parthian empire. *Like wild boars,* Ps 80.13. **35:** *As high as . . . ,* Rev 14.20.

15.43: *Babylon,* i.e. Rome. **47–48:** Rev 14.8; 17.4–5. **49:** Rev 18.7–8.

shall wither like a flower when the heat shall rise that is sent upon you. [51] You shall be weakened like a wretched woman who is beaten and wounded, so that you cannot receive your mighty lovers. [52] Would I have dealt with you so violently, says the Lord, [53] if you had not killed my chosen people continually, exulting and clapping your hands and talking about their death when you were drunk?

54 Beautify your face! [55] The reward of a prostitute is in your lap; therefore you shall receive your recompense. [56] As you will do to my chosen people, says the Lord, so God will do to you, and will hand you over to adversities. [57] Your children shall die of hunger, and you shall fall by the sword; your cities shall be wiped out, and all your people who are in the open country shall fall by the sword. [58] Those who are in the mountains and highlands[d] shall perish of hunger, and they shall eat their own flesh in hunger for bread and drink their own blood in thirst for water. [59] Unhappy above all others, you shall come and suffer fresh miseries. [60] As they pass by they shall crush the hateful[e] city, and shall destroy a part of your land and abolish a portion of your glory, when they return from devastated Babylon. [61] You shall be broken down by them like stubble,[f] and they shall be like fire to you. [62] They shall devour you and your cities, your land and your mountains; they shall burn with fire all your forests and your fruitful trees. [63] They shall carry your children away captive, plunder your wealth, and mar the glory of your countenance.

16 Woe to you, Babylon and Asia! Woe to you, Egypt and Syria! [2] Bind on sackcloth and cloth of goats' hair,[g] and wail for your children, and lament for them; for your destruction is at hand. [3] The sword has been sent upon you, and who is there to turn it back? [4] A fire has been sent upon you, and who is there to quench it? [5] Calamities have been sent upon you, and who is there to drive them away? [6] Can one drive off a hungry lion in the forest, or quench a fire in the stubble once it has started to burn?[h] [7] Can one turn back an arrow shot by a strong archer? [8] The Lord God sends calamities, and who will drive them away? [9] Fire will go forth from his wrath, and who is there to quench it? [10] He will flash lightning, and who will not be afraid? He will thunder, and who will not be terrified? [11] The Lord will threaten, and who will not be utterly shattered at his presence? [12] The earth and its foundations quake, the sea is churned up from the depths, and its waves and the fish with them shall be troubled at the presence of the Lord and the glory of his power. [13] For his right hand that bends the bow is strong, and his arrows that he shoots are sharp and when they are shot to the ends of the world will not miss once. [14] Calamities are sent forth and shall not return until they come over the earth. [15] The fire is kindled, and shall not be put out until it consumes the foundations of the earth. [16] Just as an arrow shot by a mighty archer does not return, so the calamities that are sent upon the earth shall not return. [17] Alas for me! Alas for me! Who will deliver me in those days?

18 The beginning of sorrows, when there shall be much lamentation; the beginning of famine, when many shall perish; the beginning of wars, when the powers shall be terrified; the beginning of calamities, when all shall tremble.

d Gk: Lat omits *and highlands*   e Another reading is *idle* or *unprofitable*   f Other ancient authorities read *like dry straw*   g Other ancient authorities lack *cloth of goats' hair*   h Other ancient authorities read *fire when dry straw has been set on fire*

16.1–34: **Denunciation of Babylon, Asia, Egypt, and Syria.** 1: *Babylon,* i.e. Rome. 2: *Sackcloth and cloth of goats' hair,* signs of mourning.
16.12: Ps 18.15. 15: *Until it consumes . . . the earth,* an apocalyptic idea from Persian eschatology (compare 2 Pet 3.10). 29: Isa 17.6.
16.35–50: **God's people are warned of impending disasters. 38**: 4.40. **41**: 1 Cor 7.29–31.
16.51–67: **The impossibility of hiding sin from God.**

What shall they do, when the calamities come? ¹⁹Famine and plague, tribulation and anguish are sent as scourges for the correction of humankind. ²⁰Yet for all this they will not turn from their iniquities, or ever be mindful of the scourges. ²¹Indeed, provisions will be so cheap upon earth that people will imagine that peace is assured for them, and then calamities shall spring up on the earth—the sword, famine, and great confusion. ²²For many of those who live on the earth shall perish by famine; and those who survive the famine shall die by the sword. ²³And the dead shall be thrown out like dung, and there shall be no one to console them; for the earth shall be left desolate, and its cities shall be demolished. ²⁴No one shall be left to cultivate the earth or to sow it. ²⁵The trees shall bear fruit, but who will gather it? ²⁶The grapes shall ripen, but who will tread them? For in all places there shall be great solitude; ²⁷a person will long to see another human being, or even to hear a human voice. ²⁸For ten shall be left out of a city; and two, out of the field, those who have hidden themselves in thick groves and clefts in the rocks. ²⁹Just as in an olive orchard three or four olives may be left on every tree, ³⁰or just as when a vineyard is gathered, some clusters may be left¹ by those who search carefully through the vineyard, ³¹so in those days three or four shall be left by those who search their houses with the sword. ³²The earth shall be left desolate, and its fields shall be plowed up,ʲ and its roads and all its paths shall bring forth thorns, because no sheep will go along them. ³³Virgins shall mourn because they have no bridegrooms; women shall mourn because they have no husbands; their daughters shall mourn, because they have no help. ³⁴Their bridegrooms shall be killed in war, and their husbands shall perish of famine.

35 Listen now to these things, and understand them, you who are servants of the Lord. ³⁶This is word of the Lord; receive it and do not disbelieve what the Lord says.ᵏ ³⁷The calamities draw near, and are not delayed. ³⁸Just as a pregnant woman, in the ninth month when the time of her delivery draws near, has great pains around her womb for two or three hours beforehand, but when the child comes forth from the womb, there will not be a moment's delay, ³⁹so the calamities will not delay in coming upon the earth, and the world will groan, and pains will seize it on every side.

40 Hear my words, O my people; prepare for battle, and in the midst of the calamities be like strangers on the earth. ⁴¹Let the one who sells be like one who will flee; let the one who buys be like one who will lose; ⁴²let the one who does business be like one who will not make a profit; and let the one who builds a house be like one who will not live in it; ⁴³let the one who sows be like one who will not reap; so also the one who prunes the vines, like one who will not gather the grapes; ⁴⁴those who marry, like those who will have no children; and those who do not marry, like those who are widowed. ⁴⁵Because of this those who labor, labor in vain; ⁴⁶for strangers shall gather their fruits, and plunder their goods, overthrow their houses, and take their children captive; for in captivity and famine they will produce their children.ˡ ⁴⁷Those who conduct business, do so only to have it plundered; the more they adorn their cities, their houses and possessions, and their persons, ⁴⁸the more angry I will be with them for their sins, says the Lord. ⁴⁹Just as a respectable and virtuous woman abhors a prostitute, ⁵⁰so righteousness shall abhor iniquity, when she decks herself out, and shall accuse her to her face when he comes who will defend the one who searches out every sin on earth.

51 Therefore do not be like her or her works. ⁵²For in a very short time iniquity will be removed from the earth, and righteousness will reign over us. ⁵³Sinners must not say that they have not

i Other ancient authorities read *a cluster may remain exposed*   *j* Other ancient authorities read *be for briers*   *k* Cn: Lat *do not believe the gods of whom the Lord speaks*   *l* Other ancient authorities read *therefore those who are married may know that they will produce children for captivity and famine*

sinned;[m] for God[n] will burn coals of fire on the head of everyone who says, "I have not sinned before God and his glory." 54 The Lord[o] certainly knows everything that people do; he knows their imaginations and their thoughts and their hearts. 55 He said, "Let the earth be made," and it was made, and "Let the heaven be made," and it was made. 56 At his word the stars were fixed in their places, and he knows the number of the stars. 57 He searches the abyss and its treasures; he has measured the sea and its contents; 58 he has confined the sea in the midst of the waters;[p] and by his word he has suspended the earth over the water. 59 He has spread out the heaven like a dome and made it secure upon the waters; 60 he has put springs of water in the desert, and pools on the tops of the mountains, so as to send rivers from the heights to water the earth. 61 He formed human beings and put a heart in the midst of each body, and gave each person breath and life and understanding 62 and the spirit[q] of Almighty God,[r] who surely made all things and searches out hidden things in hidden places. 63 He knows your imaginations and what you think in your hearts! Woe to those who sin and want to hide their sins! 64 The Lord will strictly examine all their works, and will make a public spectacle of all of you. 65 You shall be put to shame when your sins come out before others, and your own iniquities shall stand as your accusers on that day. 66 What will you do? Or how will you hide your sins before the Lord and his glory? 67 Indeed, God[s] is the judge; fear him! Cease from your sins, and forget your iniquities, never to commit them again; so God[s] will lead you forth and deliver you from all tribulation.

68 The burning wrath of a great multitude is kindled over you; they shall drag some of you away and force you to eat what was sacrificed to idols. 69 And those who consent to eat shall be held in derision and contempt, and shall be trampled under foot. 70 For in many places[t] and in neighboring cities there shall be a great uprising against those who fear the Lord. 71 They shall[u] be like maniacs, sparing no one, but plundering and destroying those who continue to fear the Lord. [v] 72 For they shall destroy and plunder their goods, and drive them out of house and home. 73 Then the tested quality of my elect shall be manifest, like gold that is tested by fire.

74 Listen, my elect ones, says the Lord; the days of tribulation are at hand, but I will deliver you from them. 75 Do not fear or doubt, for God[w] is your guide. 76 You who keep my commandments and precepts, says the Lord God, must not let your sins weigh you down, or your iniquities prevail over you. 77 Woe to those who are choked by their sins and overwhelmed by their iniquities! They are like a field choked with underbrush and its path[x] overwhelmed with thorns, so that no one can pass through. 78 It is shut off and given up to be consumed by fire.

---

m Other ancient authorities add *or the unjust done injustice*   *n* Lat *for he*   *o* Other ancient authorities read *Lord God*   *p* Other ancient authorities read *confined the world between the waters and the waters*   *q* Or *breath*   *r* Other ancient authorities read *of the Lord Almighty*   *s* Other ancient authorities read *the Lord*   *t* Meaning of Lat uncertain   *u* Other ancient authorities read *For people, because of their misfortunes, shall*   *v* Other ancient authorities read *fear God*   *w* Other ancient authorities read *the Lord*   *x* Other ancient authorities read *seed*

16.68–78: Though persecuted, God's elect will be delivered. 73: Zech 13.9; 1 Pet 1.7.

4 Maccabees

The book known as 4 Maccabees is included in important manuscripts of the Greek Bible, and was early translated into Syriac. Although never canonized, it has deeply influenced the preaching and piety of the Eastern Churches.

At one time 4 Maccabees was attributed to Josephus and given the title *On the Supremacy of Reason*. This describes it well, for it is a diatribe or lecture, or perhaps a panegyric, on the mastery of the passions by religious reason, as exemplified by the story of the martyrdoms of Eleazar, the seven brothers, and their mother. Its traditional title was no doubt adopted because the account is an expansion of 2 Macc 6.12–7.42, and the story belongs to the Maccabean period.

The book is a classic example of the interpretation of Judaism in terms of Greek philosophy. The ideas are Stoic (with some significant differences), and so is the terminology. The numerous Old Testament quotations are taken exclusively from the Greek Septuagint. The treatise was written originally in Greek, and in the florid Asiatic style. Possibly it was first delivered as an oration at a festival commemorating the Maccabean martyrs or at the Feast of Dedication (1.10; 3.19; 14.9; compare Jn 10.22).

The author's theology, with its emphasis on the absolute sovereignty of the Law, is genuinely Jewish but with two special characteristics. The martyrdoms are a substitutionary atonement that expiates the nation's sin and purifies the land (1.11; 17.21; 18.4). The martyrs are immediately immortal, received by the patriarchs and living in God (7.19; 16.25). Whereas 2 Maccabees reflects Persian influence with its emphasis on resurrection of the body, 4 Maccabees echoes the Greek idea of the immortality of the soul (14.5–6; 16.13; 17.12; 18.23; see Lk 16.22).

The book has frequently been assigned to the period A.D. 20–54, when Cilicia was joined to Syria and Phoenicia as a single province (4.2), and it is tempting to date it to the reign of Caligula (A.D. 37–41), who proposed to violate the Jerusalem temple (compare 4.5–14). The fact, however, that its concern is with a philosophical question, rather than with persecution *per se*, makes any such hypothesis conjectural. The most that can be said with certainty is that it was written sometime between the end of the Hasmonean dynasty in 63 B.C. and the destruction of the Jerusalem temple in A.D. 70.

Alexandria has been proposed as the place of composition, and Jerusalem cannot be excluded. Antioch, however, has the best claim, for the martyrs might have been brought to the royal capital (5.1), and in Antioch the Jews were called "Hebrews," as in this book.

1 The subject that I am about to discuss is most philosophical, that is, whether devout reason is sovereign over the emotions. So it is right for me to advise you to pay earnest attention to philosophy. 2For the subject is essential to everyone who is seeking knowledge, and in addition it includes the praise of the highest virtue—I mean, of course, rational judgment. 3If, then, it is evident that reason rules over those emotions that hinder self-control, namely, gluttony and lust, 4it is also clear that it masters the emotions that hinder one from justice, such as malice, and those that stand in the way of courage, namely anger, fear, and pain. 5Some might perhaps ask, "If reason rules the emotions, why is it not sovereign over forgetfulness and ignorance?" Their attempt at argument is ridiculous!*a* 6For reason does not rule its own emotions, but those that are opposed to justice, courage, and self-control;*b* and it is not for the purpose of destroying them, but so that one may not give way to them.

7 I could prove to you from many and various examples that reason*c* is dominant over the emotions, 8but I can demonstrate it best from the noble bravery of those who died for the sake of virtue, Eleazar and the seven brothers and their mother. 9All of these, by despising sufferings that bring death, demonstrated that reason controls the emotions. 10On this anniversary*d* it is fitting for me to praise for their virtues those who, with their mother, died for the sake of nobility and goodness, but I would also call them blessed for the honor in which they are held. 11All people, even their torturers, marveled at their courage and endurance,

and they became the cause of the downfall of tyranny over their nation. By their endurance they conquered the tyrant, and thus their native land was purified through them. 12I shall shortly have an opportunity to speak of this; but, as my custom is, I shall begin by stating my main principle, and then I shall turn to their story, giving glory to the all-wise God.

13 Our inquiry, accordingly, is whether reason is sovereign over the emotions. 14We shall decide just what reason is and what emotion is, how many kinds of emotions there are, and whether reason rules over all these. 15Now reason is the mind that with sound logic prefers the life of wisdom. 16Wisdom, next, is the knowledge of divine and human matters and the causes of these. 17This, in turn, is education in the law, by which we learn divine matters reverently and human affairs to our advantage. 18Now the kinds of wisdom are rational judgment, justice, courage, and self-control. 19Rational judgment is supreme over all of these, since by means of it reason rules over the emotions. 20The two most comprehensive types*e* of the emotions are pleasure and pain; and each of these is by nature concerned with both body and soul. 21The emotions of both pleasure and pain have many consequences. 22Thus desire precedes pleasure and delight follows it. 23Fear precedes pain and sorrow comes after. 24Anger, as a person will see by

a Or *They are attempting to make my argument ridiculous!*   *b* Other ancient authorities add *and rational judgment*   *c* Other ancient authorities read *devout reason*   *d* Gk *At this time*   *e* Or *sources*

1.1–3.18: Philosophical introduction. The principal thesis is stated in 1.3–12, which concludes with a short doxology, and is developed further in 1.13–3.18. **1:** *Devout,* Greek "eusebēs," religious or pious; compare 5.7, 31. **2–4:** *Rational judgment . . . self-control . . . justice . . . courage,* the four cardinal virtues of the Platonic and Stoic traditions. **5:** The objection is dealt with in 2.24–3.1. **8:** *Eleazar,* 2 Macc 6.18; 3 Macc 6.1. **10:** Compare 3.19 and see Introduction.

1.11: *Tyranny,* the attempt of Antiochus IV (4.15) to impose pagan worship on the Jewish nation; see 5.1 n. *Purified,* the idea of expiation is developed further in 6.28–29; 17.21. **17:** *Education in the law,* Jews regarded the Mosaic law as philosophical and the highest form of education (see 11.21 n.; 18.6–19 n.). *We learn . . . to our advantage,* compare the Stoic definition of wisdom in Cicero's *Tusculan Disputations* iv.25.57. **18:** The four *kinds* are found also in Wis 8.7.

reflecting on this experience, is an emotion embracing pleasure and pain. 25 In pleasure there exists even a malevolent tendency, which is the most complex of all the emotions. 26 In the soul it is boastfulness, covetousness, thirst for honor, rivalry, and malice; 27 in the body, indiscriminate eating, gluttony, and solitary gormandizing.

28 Just as pleasure and pain are two plants growing from the body and the soul, so there are many offshoots of these plants,*f* 29 each of which the master cultivator, reason, weeds and prunes and ties up and waters and thoroughly irrigates, and so tames the jungle of habits and emotions. 30 For reason is the guide of the virtues, but over the emotions it is sovereign.

Observe now, first of all, that rational judgment is sovereign over the emotions by virtue of the restraining power of self-control. 31 Self-control, then, is dominance over the desires. 32 Some desires are mental, others are physical, and reason obviously rules over both. 33 Otherwise, how is it that when we are attracted to forbidden foods we abstain from the pleasure to be had from them? Is it not because reason is able to rule over appetites? I for one think so. 34 Therefore when we crave seafood and fowl and animals and all sorts of foods that are forbidden to us by the law, we abstain because of domination by reason. 35 For the emotions of the appetites are restrained, checked by the temperate mind, and all the impulses of the body are bridled by reason.

2 And why is it amazing that the desires of the mind for the enjoyment of beauty are rendered powerless? 2 It is for this reason, certainly, that the temperate Joseph is praised, because by men-

tal effort*g* he overcame sexual desire. 3 For when he was young and in his prime for intercourse, by his reason he nullified the frenzy*h* of the passions. 4 Not only is reason proved to rule over the frenzied urge of sexual desire, but also over every desire.*i* 5 Thus the law says, "You shall not covet your neighbor's wife or anything that is your neighbor's." 6 In fact, since the law has told us not to covet, I could prove to you all the more that reason is able to control desires.

Just so it is with the emotions that hinder one from justice. 7 Otherwise how could it be that someone who is habitually a solitary gormandizer, a glutton, or even a drunkard can learn a better way, unless reason is clearly lord of the emotions? 8 Thus, as soon as one adopts a way of life in accordance with the law, even though a lover of money, one is forced to act contrary to natural ways and to lend without interest to the needy and to cancel the debt when the seventh year arrives. 9 If one is greedy, one is ruled by the law through reason so that one neither gleans the harvest nor gathers the last grapes from the vineyard.

In all other matters we can recognize that reason rules the emotions. 10 For the law prevails even over affection for parents, so that virtue is not abandoned for their sakes. 11 It is superior to love for one's wife, so that one rebukes her when she breaks the law. 12 It takes precedence over love for children, so that one punishes them for misdeeds. 13 It is sovereign over the relationship of friends, so that one rebukes friends when they act wick-

f Other ancient authorities read *these emotions*
g Other ancient authorities add *in reasoning*
h Or *gadfly*   *i* Or *all covetousness*

1:24: *As a person . . . experience,* the Greek is obscure. 27: Job 31.17. 33: *Reason is able to rule over appetites,* in Judaism, desires are not to be extirpated, as Stoics taught, but are to be controlled; compare v. 6; Mishnah *P. Aboth* 4.1. 34: *Seafood,* Lev 11.1–31; Deut 14.3–21; Acts 10.10–14. 35: *Emotions . . . restrained . . . impulses bridled;* reason, informed

by the law, dominates the passions of both mind and body.

2.1: *Enjoyment of beauty* refers to sexual desire but also suggests the concept of "eros" in Plato's *Symposium.* 2: *Joseph,* Gen 39.7–12. 8: *Lend without interest,* to other Jews, Ex 22.25. *Seventh year,* Deut 15.1–3. 9: Lev 19.9–10; Deut 20.19–20. 11–12: Mt 10.37; Lk 14.26.

edly. [14]Do not consider it paradoxical when reason, through the law, can prevail even over enmity. The fruit trees of the enemy are not cut down, but one preserves the property of enemies from marauders and helps raise up what has fallen.[j]

15 It is evident that reason rules even[k] the more violent emotions: lust for power, vainglory, boasting, arrogance, and malice. [16]For the temperate mind repels all these malicious emotions, just as it repels anger—for it is sovereign over even this. [17]When Moses was angry with Dathan and Abiram, he did nothing against them in anger, but controlled his anger by reason. [18]For, as I have said, the temperate mind is able to get the better of the emotions, to correct some, and to render others powerless. [19]Why else did Jacob, our most wise father, censure the households of Simeon and Levi for their irrational slaughter of the entire tribe of the Shechemites, saying, "Cursed be their anger"? [20]For if reason could not control anger, he would not have spoken thus. [21]Now when God fashioned human beings, he planted in them emotions and inclinations, [22]but at the same time he enthroned the mind among the senses as a sacred governor over them all. [23]To the mind he gave the law; and one who lives subject to this will rule a kingdom that is temperate, just, good, and courageous.

24 How is it then, one might say, that if reason is master of the emotions, it does not control forgetfulness and ignorance? [1]But this argument is entirely ridiculous; for it is evident that reason rules not over its own emotions, but

3

over those of the body. [2]No one of us[l] can eradicate that kind of desire, but reason can provide a way for us not to be enslaved by desire. [3]No one of us can eradicate anger from the mind, but reason can help to deal with anger. [4]No one of us can eradicate malice, but reason can fight at our side so that we are not overcome by malice. [5]For reason does not uproot the emotions but is their antagonist.

6 Now this can be explained more clearly by the story of King David's thirst. [7]David had been attacking the Philistines all day long, and together with the soldiers of his nation had killed many of them. [8]Then when evening fell, he[m] came, sweating and quite exhausted, to the royal tent, around which the whole army of our ancestors had encamped. [9]Now all the rest were at supper, [10]but the king was extremely thirsty, and though springs were plentiful there, he could not satisfy his thirst from them. [11]But a certain irrational desire for the water in the enemy's territory tormented and inflamed him, undid and consumed him. [12]When his guards complained bitterly because of the king's craving, two staunch young soldiers, respecting[n] the king's desire, armed themselves fully, and taking a pitcher climbed over the enemy's ramparts. [13]Eluding the sentinels at the gates, they went searching throughout the enemy camp [14]and found the spring, and from it boldly brought the king a drink. [15]But Da-

j Or the beasts that have fallen   k Other ancient authorities read through   l Gk you   m Other ancient authorities read he hurried and   n Or embarrassed because of

---

2.14: Deut 20.19–20; Ex 23.4–5; Josephus, *Against Apion* ii.211–212. **17:** *Dathan and Abiram,* Num 16.1–35; Sir 45.18. **19:** Gen 49.7. **21:** According to rabbinic Judaism, God *planted* the good and the evil *inclinations in human beings;* the latter is to be controlled, and in itself is not essentially evil. **23:** *Will rule a kingdom,* according to the Stoics and Philo, the wise man is a king. Compare the different concept of the reign of the saints in 1 Cor 4.8; 6.2–3; 1 Pet 2.9.
3.1: *Those of the body,* but the emotions of

vv. 2–4 are those of the mind; thus 1.6 would fit better here.
3.6–18: King David's thirst. Some details are different in 2 Sam 23.13–17; 1 Chr 11.15–19. **7:** *Philistines,* literally "foreigners"; in the Greek Bible this word usually translates the Hebrew word "Philistines." **15:** *Equivalent to blood,* 2 Sam 23.17, "Can I drink the blood . . . ?" **17:** *Frenzied desires,* in Greek mythology the "oistros" was the gadfly that tormented Io, and it became a symbol of uncontrolled sexual desire.

vid,[o] though he was burning with thirst, considered it an altogether fearful danger to his soul to drink what was regarded as equivalent to blood. [16]Therefore, opposing reason to desire, he poured out the drink as an offering to God. [17]For the temperate mind can conquer the drives of the emotions and quench the flames of frenzied desires; [18]it can overthrow bodily agonies even when they are extreme, and by nobility of reason spurn all domination by the emotions.

[19] The present occasion now invites us to a narrative demonstration of temperate reason.

[20] At a time when our ancestors were enjoying profound peace because of their observance of the law and were prospering, so that even Seleucus Nicanor, king of Asia, had both appropriated money to them for the temple service and recognized their commonwealth— [21]just at that time certain persons attempted a revolution against the public harmony and caused many and various disasters.

4 Now there was a certain Simon, a political opponent of the noble and good man, Onias, who then held the high priesthood for life. When despite all manner of slander he was unable to injure Onias in the eyes of the nation, he fled the country with the purpose of betraying it. [2]So he came to Apollonius, governor of Syria, Phoenicia, and Cilicia, and said, [3]"I have come here because I am loyal to the king's government, to report that in the Jerusalem treasuries there are deposited tens of thousands in private funds, which are not the property of the temple but belong to King Seleucus." [4]When Apollonius learned the de-

tails of these things, he praised Simon for his service to the king and went up to Seleucus to inform him of the rich treasure. [5]On receiving authority to deal with this matter, he proceeded quickly to our country accompanied by the accursed Simon and a very strong military force. [6]He said that he had come with the king's authority to seize the private funds in the treasury. [7]The people indignantly protested his words, considering it outrageous that those who had committed deposits to the sacred treasury should be deprived of them, and did all that they could to prevent it. [8]But, uttering threats, Apollonius went on to the temple. [9]While the priests together with women and children were imploring God in the temple to shield the holy place that was being treated so contemptuously, [10]and while Apollonius was going up with his armed forces to seize the money, angels on horseback with lightning flashing from their weapons appeared from heaven, instilling in them great fear and trembling. [11]Then Apollonius fell down half dead in the temple area that was open to all, stretched out his hands toward heaven, and with tears begged the Hebrews to pray for him and propitiate the wrath of the heavenly army. [12]For he said that he had committed a sin deserving of death, and that if he were spared he would praise the blessedness of the holy place before all people. [13]Moved by these words, the high priest Onias, although otherwise he had scruples about doing so, prayed for him so that King Seleucus would not suppose that Apollo-

[o] Gk *he*

3.19–4.14: Attempt on the temple treasury. Compare 2 Macc 3.1–40. **19:** *The present occasion,* perhaps when 4 Maccabees was first read publicly (see Introduction). **20:** *Profound peace,* see 2 Macc 3.1 n. *Seleucus Nicanor,* the author is confused. Seleucus I Nicator ruled 305/304–281/280 B.C.; the king who is meant here is Seleucus IV Philopator, 187–175 B.C. (see 2 Macc 3.3 n.).

4.1: *Simon,* 2 Macc 3.4 n. *Onias* III, 2 Macc 3.1 n. Life tenure of *the high priesthood* was the regular rule until the first century A.D. when

the Roman procurators disregarded it, Jn 18.13 n.; Josephus, *Ant.*XVIII.ii.2; xx.10. **2:** *Cilicia* was joined to Phoenicia as one province only in A.D. 20–54; 2 Macc 3.5 is more accurate. **3:** *Private funds* were often deposited in temples, as in a bank. **5:** *Authority,* but according to 2 Macc 3.7–8, Heliodorus was put in command. **6:** *Private funds* in the Jerusalem temple, see Josephus, *B.J.*I.v.2. **10:** Compare 3 Macc 2.21–24; 6.18.

Antiochus's persecution
of the Jews

nius had been overcome by human treachery and not by divine justice. 14 So Apollonius,ᵖ having been saved beyond all expectations, went away to report to the king what had happened to him.

15 When King Seleucus died, his son Antiochus Epiphanes succeeded to the throne, an arrogant and terrible man, 16 who removed Onias from the priesthood and appointed Onias'sᑫ brother Jason as high priest. 17 Jasonʳ agreed that if the office were conferred on him he would pay the king three thousand six hundred sixty talents annually. 18 So the king appointed him high priest and ruler of the nation. 19 Jasonʳ changed the nation's way of life and altered its form of government in complete violation of the law, 20 so that not only was a gymnasium constructed at the very citadelˢ of our native land, but also the temple service was abolished. 21 The divine justice was angered by these acts and caused Antiochus himself to make war on them. 22 For when he was warring against Ptolemy in Egypt, he heard that a rumor of his death had spread and that the people of Jerusalem had rejoiced greatly. He speedily marched against them, 23 and after he had plundered them he issued a decree that if any of them were found observing the ancestral law they should die. 24 When, by means of his decrees, he had not been able in any way to put an end to the people's observance of the law, but saw

that all his threats and punishments were being disregarded 25 —even to the extent that women, because they had circumcised their sons, were thrown headlong from heights along with their infants, though they had known beforehand that they would suffer this— 26 when, I say, his decrees were despised by the people, he himself tried through torture to compel everyone in the nation to eat defiling foods and to renounce Judaism.

5 The tyrant Antiochus, sitting in state with his counselors on a certain high place, and with his armed soldiers standing around him, 2 ordered the guards to seize each and every Hebrew and to compel them to eat pork and food sacrificed to idols. 3 If any were not willing to eat defiling food, they were to be broken on the wheel and killed. 4 When many persons had been rounded up, one man, Eleazar by name, leader of the flock, was broughtᵗ before the king. He was a man of priestly family, learned in the law, advanced in age, and known to many in the tyrant's court because of his philosophy.ᵘ

5 When Antiochus saw him he said, 6 "Before I begin to torture you, old man, I would advise you to save yourself by eating pork, 7 for I respect your age and

p Gk he   q Gk his   r Gk He   s Or high
place   t Or was the first of the flock to be brought
u Other ancient authorities read his advanced age

4.15–26: Antiochus's persecution of the Jews. Compare 1 Macc 1.20–64; 2 Macc 5.11–6.11. **15:** *Antiochus* IV *Epiphanes* was the brother of *Seleucus* IV, and son of Antiochus III; see 1 Macc 1.10 n. **16:** *Jason*, 2 Macc 4.7 n. **4.20:** *At the very citadel,* more probably "under the citadel," as in 2 Macc 4.12. **21:** *The divine justice,* a theological interpretation of 2 Macc 4.16–17. **22:** *Ptolemy* VI Philometor (180–145 B.C.), 1 Macc 1.16–19.

5.1–7.23: Martyrdom of Eleazar. Compare 2 Macc 6.18–31. **5.1:** *Tyrant,* in Greek usually with a bad connotation, denoting not a legitimate monarch but one who rules by force. *Sitting in state,* perhaps in Jerusalem; but early Christian tradition located this in Antioch, and a church was erected there in honor of the martyrs. **2:** Jews regarded the eating of *pork and food sacrificed to idols* as idolatry and profanation of the divine name because it was a public defiance of God's law; compare 1 Cor 10.6–22. **3:** *Defiling,* the Greek word, peculiar to 4 Maccabees, implies that forbidden foods were polluted and particularly odious. **4:** *Eleazar* may mean "God has helped"; it is the same name as Lazarus (Lk 16.20; Jn 11.1) and, as a male name, serves as a symbol for a man of great piety; compare 3 Macc 6.1; 2 Macc 8.23 n. **7–8:** A Stoic *philosopher* regarded the distinctions of national religions and laws, such as those of Judaism, as unimportant, whereas Jews considered the Mosaic law to be the highest philosophy; compare 1.17 n. Stoics also taught that one should live according to *nature.* **7:** *Religion,* Greek "thrēskeia," religious practice or cult; compare v. 31.

your gray hairs. Although you have had them for so long a time, it does not seem to me that you are a philosopher when you observe the religion of the Jews. [8]When nature has granted it to us, why should you abhor eating the very excellent meat of this animal? [9]It is senseless not to enjoy delicious things that are not shameful, and wrong to spurn the gifts of nature. [10]It seems to me that you will do something even more senseless if, by holding a vain opinion concerning the truth, you continue to despise me to your own hurt. [11]Will you not awaken from your foolish philosophy, dispel your futile reasonings, adopt a mind appropriate to your years, philosophize according to the truth of what is beneficial, [12]and have compassion on your old age by honoring my humane advice? [13]For consider this: if there is some power watching over this religion of yours, it will excuse you from any transgression that arises out of compulsion."

14 When the tyrant urged him in this fashion to eat meat unlawfully, Eleazar asked to have a word. [15]When he had received permission to speak, he began to address the people as follows: [16]"We, O Antiochus, who have been persuaded to govern our lives by the divine law, think that there is no compulsion more powerful than our obedience to the law. [17]Therefore we consider that we should not transgress it in any respect. [18]Even if, as you suppose, our law were not truly divine and we had wrongly held it to be divine, not even so would it be right for us to invalidate our reputation for piety. [19]Therefore do not suppose that it would be a petty sin if we were to eat defiling food; [20]to transgress the law in matters either small or great is of equal seriousness, [21]for in either case the law is equally despised. [22]You scoff at our philosophy as though living by it were irrational, [23]but it teaches us self-control, so that we master all pleasures and desires, and it also trains us in courage, so that we endure any suffering willingly; [24]it instructs us in justice, so that in all our dealings we act impartially,[v] and it teaches us piety, so that with proper reverence we worship the only living God.

25 "Therefore we do not eat defiling food; for since we believe that the law was established by God, we know that in the nature of things the Creator of the world in giving us the law has shown sympathy toward us. [26]He has permitted us to eat what will be most suitable for our lives,[w] but he has forbidden us to eat meats that would be contrary to this. [27]It would be tyrannical for you to compel us not only to transgress the law, but also to eat in such a way that you may deride us for eating defiling foods, which are most hateful to us. [28]But you shall have no such occasion to laugh at me, [29]nor will I transgress the sacred oaths of my ancestors concerning the keeping of the law, [30]not even if you gouge out my eyes and burn my entrails. [31]I am not so old and cowardly as not to be young in reason on behalf of piety. [32]Therefore get your torture wheels ready and fan the fire more vehemently! [33]I do not so pity my old age as to break the ancestral law by my own act. [34]I will not play false to you, O law that trained me, nor will I renounce you, beloved self-control. [35]I will not put you to shame, philosophical reason, nor will I reject you, honored priesthood and knowledge of the law. [36]You, O king,[x] shall not defile the honorable mouth of my old age, nor my

v Or *so that we hold in balance all our habitual
inclinations*   w Or *souls*   x Gk lacks *O king*

5.13: *Some power watching over,* a Greek philosophical expression; compare 2 Macc 7.35; 9.5; 3 Macc 2.21. 23–24: *Self-control . . . courage . . . justice,* cardinal virtues (see 1.2–4 n.). In Xenophon's *Memorabilia,* and sometimes in Philo, *piety* or religion is the fourth virtue.

5.27: *Deride us,* because this would bring discredit on the Jewish people, the Mosaic law, and the God who gave it. God's name must be hallowed (Mt 6.9). 31: *Piety,* Greek "eusebeia," proper reverence toward God (see v. 7 n.; 1.1 n.). No single Greek word corresponds to the English word "religion."

long life lived lawfully. [37] My ancestors will receive me as pure, as one who does not fear your violence even to death. [38] You may tyrannize the ungodly, but you shall not dominate my religious principles, either by words or through deeds."

6 When Eleazar in this manner had made eloquent response to the exhortations of the tyrant, the guards who were standing by dragged him violently to the instruments of torture. [2] First they stripped the old man, though he remained adorned with the gracefulness of his piety. [3] After they had tied his arms on each side they flogged him, [4] while a herald who faced him cried out, "Obey the king's commands!" [5] But the courageous and noble man, like a true Eleazar, was unmoved, as though being tortured in a dream; [6] yet while the old man's eyes were raised to heaven, his flesh was being torn by scourges, his blood flowing, and his sides were being cut to pieces. [7] Although he fell to the ground because his body could not endure the agonies, he kept his reason upright and unswerving. [8] One of the cruel guards rushed at him and began to kick him in the side to make him get up again after he fell. [9] But he bore the pains and scorned the punishment and endured the tortures. [10] Like a noble athlete the old man, while being beaten, was victorious over his torturers; [11] in fact, with his face bathed in sweat, and gasping heavily for breath, he amazed even his torturers by his courageous spirit.

12 At that point, partly out of pity for his old age, [13] partly out of sympathy from their acquaintance with him, partly out of admiration for his endurance, some of the king's retinue came to him and said, [14] "Eleazar, why are you so irra-tionally destroying yourself through these evil things? [15] We will set before you some cooked meat; save yourself by pretending to eat pork."

16 But Eleazar, as though more bitterly tormented by this counsel, cried out: [17] "Never may we, the children of Abraham,y think so basely that out of cowardice we feign a role unbecoming to us! [18] For it would be irrational if having lived in accordance with truth up to old age and having maintained in accordance with law the reputation of such a life, we should now change our course [19] and ourselves become a pattern of impiety to the young by setting them an example in the eating of defiling food. [20] It would be shameful if we should survive for a little while and during that time be a laughing-stock to all for our cowardice, [21] and be despised by the tyrant as unmanly by not contending even to death for our divine law. [22] Therefore, O children of Abraham, die nobly for your religion! [23] And you, guards of the tyrant, why do you delay?"

24 When they saw that he was so courageous in the face of the afflictions, and that he had not been changed by their compassion, the guards brought him to the fire. [25] There they burned him with maliciously contrived instruments, threw him down, and poured stinking liquids into his nostrils. [26] When he was now burned to his very bones and about to expire, he lifted up his eyes to God and said, [27] "You know, O God, that though I might have saved myself, I am dying in burning torments for the sake of the law. [28] Be merciful to your people, and let our punishment suffice for them. [29] Make my blood their purification, and take my life

y Or O *children of Abraham*

37: 13.17; 17.12; Mk 12.26–27. Immortality of the martyrs is implied in 9.22; 2 Macc 7.36. **6.5:** *True Eleazar,* see 5.4 n. **6:** *Eyes . . . to heaven,* a natural gesture in prayer, particularly that of a martyr. Compare v. 26 and Stephen's supplication in Acts 7.55. **10:** *Noble athlete,* 1 Cor 9.24–27; Heb 12.1.

6.17–21: See 5.27 n. **23:** *Why do you delay?* Compare 9.1; 2 Macc 7.30. **29:** *In exchange,* Greek "antipsychon," a word used by the martyr Ignatius of Antioch in his letters. Compare 1.11; 9.24; 12.17; 17.21–22; 18.4; Mk 10.45. The idea of expiation derives ultimately from Isa 53.5–12 and is also found in

in exchange for theirs." 30 After he said this, the holy man died nobly in his tortures; even in the tortures of death he resisted, by virtue of reason, for the sake of the law.

31 Admittedly, then, devout reason is sovereign over the emotions. 32 For if the emotions had prevailed over reason, we would have testified to their domination. 33 But now that reason has conquered the emotions, we properly attribute to it the power to govern. 34 It is right for us to acknowledge the dominance of reason when it masters even external agonies. It would be ridiculous to deny it. ᶻ 35 I have proved not only that reason has mastered agonies, but also that it masters pleasures and in no respect yields to them.

7 For like a most skillful pilot, the reason of our father Eleazar steered the ship of religion over the sea of the emotions, 2 and though buffeted by the stormings of the tyrant and overwhelmed by the mighty waves of tortures, 3 in no way did he turn the rudder of religion until he sailed into the haven of immortal victory. 4 No city besieged with many ingenious war machines has ever held out as did that most holy man. Although his sacred life was consumed by tortures and racks, he conquered the besiegers with the shield of his devout reason. 5 For in setting his mind firm like a jutting cliff, our father Eleazar broke the maddening waves of the emotions. 6 O priest, worthy of the priesthood, you neither defiled your sacred teeth nor profaned your stomach, which had room only for reverence and purity, by eating defiling foods. 7 O man in harmony with the law and philosopher of divine life! 8 Such should be those who are administ-

trators of the law, shielding it with their own blood and noble sweat in sufferings even to death. 9 You, father, strengthened our loyalty to the law through your glorious endurance, and you did not abandon the holiness that you praised, but by your deeds you made your words of divineᵃ philosophy credible. 10 O aged man, more powerful than tortures; O elder, fiercer than fire; O supreme king over the passions, Eleazar! 11 For just as our father Aaron, armed with the censer, ran through the multitude of the people and conquered the fieryᵇ angel, 12 so the descendant of Aaron, Eleazar, though being consumed by the fire, remained unmoved in his reason. 13 Most amazing, indeed, though he was an old man, his body no longer tense and firm, ᶜ his muscles flabby, his sinews feeble, he became young again 14 in spirit through reason; and by reason like that of Isaac he rendered the many-headed rack ineffective. 15 O man of blessed age and of venerable gray hair and of law-abiding life, whom the faithful seal of death has perfected!

16 If, therefore, because of piety an aged man despised tortures even to death, most certainly devout reason is governor of the emotions. 17 Some perhaps might say, "Not all have full command of their emotions, because not all have prudent reason." 18 But as many as attend to religion with a whole heart, these alone are able to control the passions of the flesh, 19 since they believe that they, like our patriarchs Abraham

z Syr: Meaning of Gk uncertain   a Other ancient authorities lack *divine*   b Other ancient authorities lack *fiery*   c Gk *the tautness of the body already loosed*

---

the Qumran *Manual of Discipline.* 31: The transition from religious language in vv. 27–29 to the philosophical note of *devout reason* is abrupt; but for this author the two are one.

7.1–3: The metaphor of the *pilot* and *the ship*, common in Greek literature, recurs in 13.6–7 and in 15.31–32 as a reference to Noah's ark; compare 1 Pet 3.20. 6: *Defiled . . . profaned,* the Jewish concept was realistic, as though a physical infection were incurred (see 2 Macc 6.20); contrast Mk 7.15; Acts 10.13–

15. 8: *Administrators,* literally "those who make (or create) something"; the Greek is obscure. A rabbi or priest was responsible for maintenance of the law in the community (Mal 2.7). 9: *Credible,* in both Judaism and Stoicism, the ultimate test is the conformity of one's deeds to one's profession. 10: *Eleazar,* see 5.4 n.

7.11: Num 16.46–50. 14: Compare 2 Cor 4.7–18. *Isaac,* Gen 22.1–14. 19: *Abraham and Isaac and Jacob* are living (Mk 12.26). *Live to*

and Isaac and Jacob, do not die to God, but live to God. 20 No contradiction therefore arises when some persons appear to be dominated by their emotions because of the weakness of their reason. 21 What person who lives as a philosopher by the whole rule of philosophy, and trusts in God, 22 and knows that it is blessed to endure any suffering for the sake of virtue, would not be able to overcome the emotions through godliness? 23 For only the wise and courageous are masters of their emotions.

8 For this is why even the very young, by following a philosophy in accordance with devout reason, have prevailed over the most painful instruments of torture. 2 For when the tyrant was conspicuously defeated in his first attempt, being unable to compel an aged man to eat defiling foods, then in violent rage he commanded that others of the Hebrew captives be brought, and that any who ate defiling food would be freed after eating, but if any were to refuse, they would be tortured even more cruelly.

3 When the tyrant had given these orders, seven brothers—handsome, modest, noble, and accomplished in every way—were brought before him along with their aged mother. 4 When the tyrant saw them, grouped about their mother as though a chorus, he was pleased with them. And struck by their appearance and nobility, he smiled at them, and summoned them nearer and said, 5 "Young men, with favorable feelings I admire each and every one of you, and greatly respect the beauty and the number of such brothers. Not only do I advise you not to display the same madness as that of the old man who has just been tortured, but I also exhort you to yield to me and enjoy my friendship. 6 Just as I am able to punish those who disobey my orders, so I can be a benefactor to those who obey me. 7 Trust me, then, and you will have positions of authority in my government if you will renounce the ancestral tradition of your national life. 8 Enjoy your youth by adopting the Greek way of life and by changing your manner of living. 9 But if by disobedience you rouse my anger, you will compel me to destroy each and every one of you with dreadful punishments through tortures. 10 Therefore take pity on yourselves. Even I, your enemy, have compassion for your youth and handsome appearance. 11 Will you not consider this, that if you disobey, nothing remains for you but to die on the rack?"

12 When he had said these things, he ordered the instruments of torture to be brought forward so as to persuade them out of fear to eat the defiling food. 13 When the guards had placed before them wheels and joint-dislocators, rack and hooks*d* and catapults*e* and caldrons, braziers and thumbscrews and iron claws and wedges and bellows, the tyrant resumed speaking: 14 "Be afraid, young fellows; whatever justice you revere will be merciful to you when you transgress under compulsion."

d Meaning of Gk uncertain   *e* Here and elsewhere in 4 Macc an instrument of torture

God, a similar expression is found in Lk 20.38; Rom 6.10; 14.8; Gal 2.19. Compare 16.25.
8.1–9.9: The seven brothers defy the tyrant. This account is an amplification of 2 Macc 7.1–2. **3**: *Accomplished,* or graceful, the Greek ideal of physical beauty joined to perfect education. **4**: A Greek *chorus* was a company of dancers, who often moved in a circle and spoke lines in unison; compare 14.7. **5**: *Beauty,* see 2.1 n. *Friendship,* almost a technical term, because the "friends" of a Hellenistic king were employed in the government; see v. 7 and 3 Macc 2.23 n. **6**: *Benefac-*tor, a title often adopted by Hellenistic monarchs; in Lk 22.25 it seems ironical, as here. **8**: *Adopting the Greek way of life,* Antiochus could have believed sincerely that this was the highest civilization and that Judaism was "superstition." Thus it is a tragic conflict between two points of view.
8.13: *Hooks,* these and some of the other instruments of torture cannot be described precisely; compare 11.10. **14**: *Justice,* a philosophical way of speaking of God. **15**: *Nullified his tyranny,* Epictetus the Stoic taught that, while the tyrant might chain a person's leg or

15 But when they had heard the inducements and saw the dreadful devices, not only were they not afraid, but they also opposed the tyrant with their own philosophy, and by their right reasoning nullified his tyranny. 16 Let us consider, on the other hand, what arguments might have been used if some of them had been cowardly and unmanly. Would they not have been the following? 17 "O wretches that we are and so senseless! Since the king has summoned and exhorted us to accept kind treatment if we obey him, 18 why do we take pleasure in vain resolves and venture upon a disobedience that brings death? 19 O men and brothers, should we not fear the instruments of torture and consider the threats of torments, and give up this vain opinion and this arrogance that threatens to destroy us? 20 Let us take pity on our youth and have compassion on our mother's age; 21 and let us seriously consider that if we disobey we are dead! 22 Also, divine justice will excuse us for fearing the king when we are under compulsion. 23 Why do we banish ourselves from this most pleasant life and deprive ourselves of this delightful world? 24 Let us not struggle against compulsion *f* or take hollow pride in being put to the rack. 25 Not even the law itself would arbitrarily put us to death for fearing the instruments of torture. 26 Why does such contentiousness excite us and such a fatal stubbornness please us, when we can live in peace if we obey the king?"

27 But the youths, though about to be tortured, neither said any of these things nor even seriously considered them. 28 For they were contemptuous of the emotions and sovereign over agonies, 29 so that as soon as the tyrant had

ceased counseling them to eat defiling food, all with one voice together, as from one mind, said:

9 "Why do you delay, O tyrant? For we are ready to die rather than transgress our ancestral commandments; 2 we are obviously putting our forebears to shame unless we should practice ready obedience to the law and to Moses *g* our counselor. 3 Tyrant and counselor of lawlessness, in your hatred for us do not pity us more than we pity ourselves. *h* 4 For we consider this pity of yours, which insures our safety through transgression of the law, to be more grievous than death itself. 5 You are trying to terrify us by threatening us with death by torture, as though a short time ago you learned nothing from Eleazar. 6 And if the aged men of the Hebrews because of their religion lived piously *i* while enduring torture, it would be even more fitting that we young men should die despising your coercive tortures, which our aged instructor also overcame. 7 Therefore, tyrant, put us to the test; and if you take our lives because of our religion, do not suppose that you can injure us by torturing us. 8 For we, through this severe suffering and endurance, shall have the prize of virtue and shall be with God, on whose account we suffer; 9 but you, because of your bloodthirstiness toward us, will deservedly undergo from the divine justice eternal torment by fire."

10 When they had said these things, the tyrant was not only indignant, as at those who are disobedient, but also infu-

---

f Or *fate*   *g* Other ancient authorities read
knowledge   *h* Meaning of Gk uncertain
i Other ancient authorities read *died*

---

cut off one's head, he could neither chain nor cut off one's moral purpose (*Discourses* i.18.17).

8.25: *The law* would not condemn them for *fearing;* but a Jew could not be excused for committing idolatry, even under duress (see 5.2 n.). **29**: *With one voice,* as if they were a chorus (see 8.4 n.).

9.6: *Aged men,* the plural may refer to He-

brew prophets such as Isaiah, who, according to Jewish tradition, were also martyred; compare Heb 11.35–37. **7**: *Do not suppose that you can injure us,* a Stoic principle; suffering cannot affect the essential nature of those who are wise. **8**: *Prize of virtue,* an athletic metaphor; compare 6.10; Wis 10.12; 1 Cor 9.24.

9.10–25: **Martyrdom of the eldest.** The details do not agree with the earlier account

riated, as at those who are ungrateful. [11] Then at his command the guards brought forward the eldest, and having torn off his tunic, they bound his hands and arms with thongs on each side. [12] When they had worn themselves out beating him with scourges, without accomplishing anything, they placed him upon the wheel. [13] When the noble youth was stretched out around this, his limbs were dislocated, [14] and with every member disjointed he denounced the tyrant, saying, [15] "Most abominable tyrant, enemy of heavenly justice, savage of mind, you are mangling me in this manner, not because I am a murderer, or as one who acts impiously, but because I protect the divine law." [16] And when the guards said, "Agree to eat so that you may be released from the tortures," [17] he replied, "You abominable lackeys, your wheel is not so powerful as to strangle my reason. Cut my limbs, burn my flesh, and twist my joints; [18] through all these tortures I will convince you that children of the Hebrews alone are invincible where virtue is concerned." [19] While he was saying these things, they spread fire under him, and while fanning the flames[j] they tightened the wheel further. [20] The wheel was completely smeared with blood, and the heap of coals was being quenched by the drippings of gore, and pieces of flesh were falling off the axles of the machine. [21] Although the ligaments joining his bones were already severed, the courageous youth, worthy of Abraham, did not groan, [22] but as though transformed by fire into immortality, he nobly endured the rackings. [23] "Imitate me, brothers," he said. "Do not leave your post in my struggle[k] or renounce our courageous family ties. [24] Fight the sacred and noble battle for religion. Thereby the just Providence of our ancestors may become merciful to our nation and take vengeance on the accursed tyrant." [25] When he had said this, the saintly youth broke the thread of life.

[26] While all were marveling at his courageous spirit, the guards brought in the next eldest, and after fitting themselves with iron gauntlets having sharp hooks, they bound him to the torture machine and catapult. [27] Before torturing him, they inquired if he were willing to eat, and they heard his noble decision.[l] [28] These leopard-like beasts tore out his sinews with the iron hands, flayed all his flesh up to his chin, and tore away his scalp. But he steadfastly endured this agony and said, [29] "How sweet is any kind of death for the religion of our ancestors!" [30] To the tyrant he said, "Do you not think, you most savage tyrant, that you are being tortured more than I, as you see the arrogant design of your tyranny being defeated by our endurance for the sake of religion? [31] I lighten my pain by the joys that come from virtue, [32] but you suffer torture by the threats that come from impiety. You will not escape, you most abominable tyrant, the judgments of the divine wrath."

---

j Meaning of Gk uncertain   *k* Other ancient authorities read *post forever*   *l* Other ancient authorities read *having heard his noble decision, they tore him to shreds*

---

in 2 Macc 7.3–6; here the author allows himself the freedom of an historical novelist. **17:** See v. 7 n.

9.22: *Transformed,* the same Greek verb is used in Phil 3.21; synonymous verbs in 1 Cor 15.51–52; 2 Cor 3.18. *Immortality,* literally "incorruption," "that which is imperishable," as in 17.12; 1 Cor 15.53. Here the author comes close to the Greek doctrine that the soul is by nature immortal, but puts the emphasis on reward and punishment after this life (compare Lk 16.19–31; 23.43). This is in contrast to the doctrine of the resurrection expressed in 2 Macc 12.44–45. **23:** *Fami-*ly *ties,* referring to the immediate family, but kinship with the whole Jewish nation may be implied. **24:** *Fight,* compare 2 Tim 4.7. *Accursed,* Greek "alastōr," a word from the Greek tragedies.

9.26–12.19: **Martyrdom of the other brothers.** These stories follow 2 Macc 7.7–40 in general, but are made more vivid and sensational. **28:** *Beasts,* 1 Cor 15.32. **29:** *How sweet,* compare 2 Macc 6.30. One may compare the letter to the Romans by the Christian martyr Ignatius and also the Latin saying, "Dulce et decorum est pro patria mori."

10 When he too had endured a glorious death, the third was led in, and many repeatedly urged him to save himself by tasting the meat. 2 But he shouted, "Do you not know that the same father begot me as well as those who died, and the same mother bore me, and that I was brought up on the same teachings? 3 I do not renounce the noble kinship that binds me to my brothers."*m* 5 Enraged by the man's boldness, they disjointed his hands and feet with their instruments, dismembering him by prying his limbs from their sockets, 6 and breaking his fingers and arms and legs and elbows. 7 Since they were not able in any way to break his spirit,*n* they abandoned the instruments*o* and scalped him with their fingernails in a Scythian fashion. 8 They immediately brought him to the wheel, and while his vertebrae were being dislocated by this, he saw his own flesh torn all around and drops of blood flowing from his entrails. 9 When he was about to die, he said, 10 "We, most abominable tyrant, are suffering because of our godly training and virtue, 11 but you, because of your impiety and bloodthirstiness, will undergo unceasing torments."

12 When he too had died in a manner worthy of his brothers, they dragged in the fourth, saying, 13 "As for you, do not give way to the same insanity as your brothers, but obey the king and save yourself." 14 But he said to them, "You do not have a fire hot enough to make me play the coward. 15 No—by the blessed death of my brothers, by the eternal destruction of the tyrant, and by the everlasting life of the pious, I will not renounce our noble family ties. 16 Contrive tortures, tyrant, so that you may learn

from them that I am a brother to those who have just now been tortured." 17 When he heard this, the bloodthirsty, murderous, and utterly abominable Antiochus gave orders to cut out his tongue. 18 But he said, "Even if you remove my organ of speech, God hears also those who are mute. 19 See, here is my tongue; cut it off, for in spite of this you will not make our reason speechless. 20 Gladly, for the sake of God, we let our bodily members be mutilated. 21 God will visit you swiftly, for you are cutting out a tongue that has been melodious with divine hymns."

11 When he too died, after being cruelly tortured, the fifth leaped up, saying, 2 "I will not refuse, tyrant, to be tortured for the sake of virtue. 3 I have come of my own accord, so that by murdering me you will incur punishment from the heavenly justice for even more crimes. 4 Hater of virtue, hater of humankind, for what act of ours are you destroying us in this way? 5 Is it because*p* we revere the Creator of all things and live according to his virtuous law? 6 But these deeds deserve honors, not tortures."*q* 9 While he was saying these things, the guards bound him and dragged him to the catapult; 10 they tied him to it on his knees, and fitting iron

---

m Other ancient authorities add verse 4 *So if you have any instrument of torture, apply it to my body; for you cannot touch my soul, even if you wish."*
n Gk *to strangle him*   *o* Other ancient authorities read *they tore off his skin*   *p* Other ancient authorities read *Or does it seem evil to you that*   *q* Other authorities add verses 7 and 8, 7 *If you but understood human feelings and had hope of salvation from God—* 8 *but, as it is, you are a stranger to God and persecute those who serve him."*

10.4: *You cannot touch my soul;* this verse, which does not occur in certain manuscripts, may be a later interpolation. See 9.7 n. and compare Mt 10.28; Lk 12.4–5. **5:** *Boldness,* Acts 4.13; 2 Cor 3.12. Freedom of speech was one of the ideals of Greek democracy. **7:** *Scythian fashion,* the Scythians were notorious for their barbarous cruelty (2 Macc 4.47; 3 Macc 7.5).
10.15: *Everlasting life,* or glorious life; see

9.22 n. *Family ties,* see 9.23; the seven brothers and their mother represent the entire Jewish nation. **19:** 2 Macc 7.10.
11.3: A new idea; he welcomes torture so that the tyrant may be punished the more. **7:** *Human feelings,* ironical because the compassion expressed in 12.2 was genuine but did not go far enough. Verses 7–8 are almost certainly interpolations. **10:** *Wedge on the wheel,* it is not certain how the wheel was constructed.

clamps on them, they twisted his back[r] around the wedge on the wheel,[s] so that he was completely curled back like a scorpion, and all his members were disjointed. [11]In this condition, gasping for breath and in anguish of body, [12]he said, "Tyrant, they are splendid favors that you grant us against your will, because through these noble sufferings you give us an opportunity to show our endurance for the law."

13 When he too had died, the sixth, a mere boy, was led in. When the tyrant inquired whether he was willing to eat and be released, he said, [14]"I am younger in age than my brothers, but I am their equal in mind. [15]Since to this end we were born and bred, we ought likewise to die for the same principles. [16]So if you intend to torture me for not eating defiling foods, go on torturing!" [17]When he had said this, they led him to the wheel. [18]He was carefully stretched tight upon it, his back was broken, and he was roasted[t] from underneath. [19]To his back they applied sharp spits that had been heated in the fire, and pierced his ribs so that his entrails were burned through. [20]While being tortured he said, "O contest befitting holiness, in which so many of us brothers have been summoned to an arena of sufferings for religion, and in which we have not been defeated! [21]For religious knowledge, O tyrant, is invincible. [22]I also, equipped with nobility, will die with my brothers, [23]and I myself will bring a great avenger upon you, you inventor of tortures and enemy of those who are truly devout. [24]We six boys have paralyzed your tyranny. [25]Since you have not been able to persuade us to change our mind or to force us to eat defiling foods, is not this your downfall? [26]Your fire is cold to us, and the catapults painless, and your violence powerless. [27]For it is not the guards of the tyrant but those of the divine law that are set over us; therefore, unconquered, we hold fast to reason."

12 When he too, thrown into the caldron, had died a blessed death, the seventh and youngest of all came forward. [2]Even though the tyrant had been vehemently reproached by the brothers, he felt strong compassion for this child when he saw that he was already in fetters. He summoned him to come nearer and tried to persuade him, saying, [3]"You see the result of your brothers' stupidity, for they died in torments because of their disobedience. [4]You too, if you do not obey, will be miserably tortured and die before your time, [5]but if you yield to persuasion you will be my friend and a leader in the government of the kingdom." [6]When he had thus appealed to him, he sent for the boy's mother to show compassion on her who had been bereaved of so many sons and to influence her to persuade the surviving son to obey and save himself. [7]But when his mother had exhorted him in the Hebrew language, as we shall tell a little later, [8]he said, "Let me loose, let me speak to the king and to all his friends that are with him." [9]Extremely pleased by the boy's declaration, they freed him at once. [10]Running to the nearest of the braziers, [11]he said, "You profane tyrant, most impious of all the wicked, since you have received good things and also your kingdom from God, were you not ashamed to murder his servants and torture on the

r Gk *loins*   s Meaning of Gk uncertain
t Other ancient authorities add *by fire*

11.13–27: 2 Macc 7.18–19. 20: *Contest,* compare 6.10. *Arena,* literally "gymnasium." 21: *Religious knowledge,* or science. In Greek thought true knowledge almost always leads to virtue; in Judaism, knowledge of the Mosaic law at least predisposes one toward it; see 1.17 n.; 18.6–19 n. 25: *Downfall,* see 8.15 n. 26: *Painless,* compare Heb 12.2, "despising the shame."

12.2: *The tyrant* could feel *strong compassion;* see 8.10. 5: *Friend,* see 8.5 n. 7: *Hebrew language,* or perhaps Aramaic, as in Acts 21.40; compare 2 Macc 7.21, 27. Many Palestinians, and certainly the author's first readers, spoke Greek. Her use of Hebrew indicates her devotion to the sacred tongue. *A little later,* for dramatic effect, the author postpones the speech to 16.16–23. 11: *Those who practice,*

wheel those who practice religion? [12] Because of this, justice has laid up for you intense and eternal fire and tortures, and these throughout all time[u] will never let you go. [13] As a man, were you not ashamed, you most savage beast, to cut out the tongues of men who have feelings like yours and are made of the same elements as you, and to maltreat and torture them in this way? [14] Surely they by dying nobly fulfilled their service to God, but you will wail bitterly for having killed without cause the contestants for virtue." [15] Then because he too was about to die, he said, [16] "I do not desert the excellent example[v] of my brothers, [17] and I call on the God of our ancestors to be merciful to our nation;[w] [18] but on you he will take vengeance both in this present life and when you are dead." [19] After he had uttered these imprecations, he flung himself into the braziers and so ended his life.[x]

13 Since, then, the seven brothers despised sufferings even unto death, everyone must concede that devout reason is sovereign over the emotions. [2] For if they had been slaves to their emotions and had eaten defiling food, we would say that they had been conquered by these emotions. [3] But in fact it was not so. Instead, by reason, which is praised before God, they prevailed over their emotions. [4] The supremacy of the mind over these cannot be overlooked, for the brothers[y] mastered both emotions and pains. [5] How then can one fail to confess the sovereignty of right reason over emotion in those who were not turned back by fiery agonies? [6] For just as towers jutting out over harbors hold back the threatening waves and make it calm for those who sail into the inner basin, [7] so the seven-towered right reason of the youths, by fortifying the harbor of religion, conquered the tempest of the emotions. [8] For they constituted a holy chorus of religion and encouraged one another, saying, [9] "Brothers, let us die like brothers for the sake of the law; let us imitate the three youths in Assyria who despised the same ordeal of the furnace. [10] Let us not be cowardly in the demonstration of our piety." [11] While one said, "Courage, brother," another said, "Bear up nobly," [12] and another reminded them, "Remember whence you came, and the father by whose hand Isaac would have submitted to being slain for the sake of religion." [13] Each of them and all of them together looking at one another, cheerful and undaunted, said, "Let us with all our hearts consecrate ourselves to God, who gave us our lives,[z] and let us use our bodies as a bulwark for the law. [14] Let us not fear him who thinks he is killing us, [15] for great is the struggle of the soul and the danger of eternal torment lying before those who transgress the commandment of God. [16] Therefore let us put on the full armor of self-control, which is divine reason. [17] For if we so die,[a] Abraham and Isaac and Jacob will welcome us, and all the fathers will praise us." [18] Those who were left behind said to each of the brothers who were

u Gk throughout the whole age   *v Other ancient authorities read the witness*   *w Other ancient authorities read my race*   *x Gk and so gave up; other ancient authorities read gave up his spirit or his soul*   *y Gk they*   *z Or souls*   *a Other ancient authorities read suffer*

Greek "askētas," almost "the athletes of religion"; compare v. 14 and Philo, *On Dreams* i.59.

12.13: *Feelings like yours,* a Stoic idea, found also in Wis 7.1–6; Acts 14.15. **14:** *Contestants,* see 11.20 n. **19:** *Flung himself,* as the mother does in 17.1. The remaining defenders of Masada at the end of the Jewish War of A.D. 66–73 killed one another. Jews, like Stoics, approved of suicide in certain circumstances.

13.1–14.10: Philosophical interpretation. The martyrdoms attest the supremacy of pious reason. Compare 6.31–35. **13.8:** *Chorus,* see 8.4 n.; 14.7. **9:** Dan ch 3. **12:** *Remember whence you came,* Isa 51.1–2. *The father* and *Isaac,* 15.28; Gen 22.1–19; Wis 10.5. Their story became a favorite theme for Christians (Heb 11.17–19).

13.13: *Each of them and all of them together,* as in a Greek chorus (see 8.4 n.). **14:** Mt 10.28; Lk 12.4. **17:** 5.37 n.; compare Lk 16.22. **19:**

being dragged away, "Do not put us to shame, brother, or betray the brothers who have died before us."

19 You are not ignorant of the affection of family ties, which the divine and all-wise Providence has bequeathed through the fathers to their descendants and which was implanted in the mother's womb. 20 There each of the brothers spent the same length of time and was shaped during the same period of time; and growing from the same blood and through the same life, they were brought to the light of day. 21 When they were born after an equal time of gestation, they drank milk from the same fountains. From such embraces brotherly-loving souls are nourished; 22 and they grow stronger from this common nurture and daily companionship, and from both general education and our discipline in the law of God.

23 Therefore, when sympathy and brotherly affection had been so established, the brothers were the more sympathetic to one another. 24 Since they had been educated by the same law and trained in the same virtues and brought up in right living, they loved one another all the more. 25 A common zeal for nobility strengthened their goodwill toward one another, and their concord, 26 because they could make their brotherly love more fervent with the aid of their religion. 27 But although nature and companionship and virtuous habits had augmented the affection of family ties, those who were left endured for the sake of religion, while watching their brothers being maltreated and tortured to death.

14 Furthermore, they encouraged them to face the torture, so that they not only despised their agonies, but also mastered the emotions of brotherly love.

2 O reason,[b] more royal than kings

and freer than the free! 3 O sacred and harmonious concord of the seven brothers on behalf of religion! 4 None of the seven youths proved coward or shrank from death, 5 but all of them, as though running the course toward immortality, hastened to death by torture. 6 Just as the hands and feet are moved in harmony with the guidance of the mind, so those holy youths, as though moved by an immortal spirit of devotion, agreed to go to death for its sake. 7 O most holy seven, brothers in harmony! For just as the seven days of creation move in choral dance around religion, 8 so these youths, forming a chorus, encircled the sevenfold fear of tortures and dissolved it. 9 Even now, we ourselves shudder as we hear of the suffering of these young men; they not only saw what was happening, not only heard the direct word of threat, but also bore the sufferings patiently, and in agonies of fire at that. 10 What could be more excruciatingly painful than this? For the power of fire is intense and swift, and it consumed their bodies quickly.

11 Do not consider it amazing that reason had full command over these men in their tortures, since the mind of woman despised even more diverse agonies, 12 for the mother of the seven young men bore up under the rackings of each one of her children.

13 Observe how complex is a mother's love for her children, which draws everything toward an emotion felt in her inmost parts. 14 Even unreasoning animals, as well as human beings, have a sympathy and parental love for their offspring. 15 For example, among birds, the ones that are tame protect their young by building on the housetops, 16 and the others, by building in precipitous chasms

b Or *O minds*

Affection, Greek "philtra," a magical charm believed to produce love. *Divine and all-wise Providence*, a Stoic concept (9.24; 17.22). **14.5**: 9.8 n.; Heb 12.1. **8**: *Sevenfold fear*, the Greek here is obscure. **9**: *Even now*, see 3.19 n.

14.11–17.1: **The mother of the seven.** Her death is merely mentioned in 2 Macc 7.41; here it is made the climax of the oration. **14.14**: The analogy between *unreasoning animals* and *human beings* (vv. 14–19) was a theme of popular Greek philosophy. **20**: *Did not sway, the*

and in holes and tops of trees, hatch the nestlings and ward off the intruder. [17]If they are not able to keep the intruder[c] away, they do what they can to help their young by flying in circles around them in the anguish of love, warning them with their own calls. [18]And why is it necessary to demonstrate sympathy for children by the example of unreasoning animals, [19]since even bees at the time for making honeycombs defend themselves against intruders and, as though with an iron dart, sting those who approach their hive and defend it even to the death? [20]But sympathy for her children did not sway the mother of the young men; she was of the same mind as Abraham.

15 O reason of the children, tyrant over the emotions! O religion, more desirable to the mother than her children! [2]Two courses were open to this mother, that of religion, and that of preserving her seven sons for a time, as the tyrant had promised. [3]She loved religion more, the religion that preserves them for eternal life according to God's promise.[d] [4]In what manner might I express the emotions of parents who love their children? We impress upon the character of a small child a wondrous likeness both of mind and of form. Especially is this true of mothers, who because of their birth pangs have a deeper sympathy toward their offspring than do the fathers. [5]Considering that mothers are the weaker sex and give birth to many, they are more devoted to their children.[e] [6]The mother of the seven boys, more than any other mother, loved her children. In seven pregnancies she had implanted in herself tender love toward them, [7]and because of the many pains she suffered with each of them she had sympathy for them; [8]yet because of the fear of God she disdained the temporary safety of her children. [9]Not only so, but also because of

the nobility of her sons and their ready obedience to the law, she felt a greater tenderness toward them. [10]For they were righteous and self-controlled and brave and magnanimous, and loved their brothers and their mother, so that they obeyed her even to death in keeping the ordinances.

11 Nevertheless, though so many factors influenced the mother to suffer with them out of love for her children, in the case of none of them were the various tortures strong enough to pervert her reason. [12]But each child separately and all of them together the mother urged on to death for religion's sake. [13]O sacred nature and affection of parental love, yearning of parents toward offspring, nurture and indomitable suffering by mothers! [14]This mother, who saw them tortured and burned one by one, because of religion did not change her attitude. [15]She watched the flesh of her children being consumed by fire, their toes and fingers scattered[f] on the ground, and the flesh of the head to the chin exposed like masks.

16 O mother, tried now by more bitter pains than even the birth pangs you suffered for them! [17]O woman, who alone gave birth to such complete devotion! [18]When the firstborn breathed his last, it did not turn you aside, nor when the second in torments looked at you piteously nor when the third expired; [19]nor did you weep when you looked at the eyes of each one in his tortures gazing boldly at the same agonies, and saw in their nostrils the signs of the approach of death. [20]When you saw the flesh of children burned upon the flesh of other children, severed hands upon hands, scalped

*c Gk it   d Gk according to God   e Or For to
the degree that mothers are weaker and the more
children they bear, the more they are devoted to their
children.   f Or quivering*

---

mother's constancy is the supreme proof of the dominance of religious reason. *Abraham* had offered Isaac; see 13.12 n.

15.2: *Two courses,* the two ways of Jer 21.8 became a pattern of Jewish thought. *Religion,*

5.31 n. **4:** *Of mind and of form,* a Stoic idea. **13:** *Indomitable suffering,* actually it is the mother who is *indomitable* in spite of her *love* and *suffering.*

15.17: *Gave birth,* includes the idea of spiri-

heads upon heads, and corpses fallen on other corpses, and when you saw the place filled with many spectators of the torturings, you did not shed tears. [21] Neither the melodies of sirens nor the songs of swans attract the attention of their hearers as did the voices of the children in torture calling to their mother. [22] How great and how many torments the mother then suffered as her sons were tortured on the wheel and with the hot irons! [23] But devout reason, giving her heart a man's courage in the very midst of her emotions, strengthened her to disregard, for the time, her parental love.

24 Although she witnessed the destruction of seven children and the ingenious and various rackings, this noble mother disregarded all these[g] because of faith in God. [25] For as in the council chamber of her own soul she saw mighty advocates—nature, family, parental love, and the rackings of her children— [26] this mother held two ballots, one bearing death and the other deliverance for her children. [27] She did not approve the deliverance that would preserve the seven sons for a short time, [28] but as the daughter of God-fearing Abraham she remembered his fortitude.

29 O mother of the nation, vindicator of the law and champion of religion, who carried away the prize of the contest in your heart! [30] O more noble than males in steadfastness, and more courageous than men in endurance! [31] Just as Noah's ark, carrying the world in the universal flood, stoutly endured the waves, [32] so you, O guardian of the law, overwhelmed from every side by the flood of your emotions and the violent winds, the torture of your sons, endured nobly and withstood the wintry storms that assail religion.

16 If, then, a woman, advanced in years and mother of seven sons, endured seeing her children tortured to death, it must be admitted that devout reason is sovereign over the emotions. [2] Thus I have demonstrated not only that men have ruled over the emotions, but also that a woman has despised the fiercest tortures. [3] The lions surrounding Daniel were not so savage, nor was the raging fiery furnace of Mishael so intensely hot, as was her innate parental love, inflamed as she saw her seven sons tortured in such varied ways. [4] But the mother quenched so many and such great emotions by devout reason.

5 Consider this also: If this woman, though a mother, had been fainthearted, she would have mourned over them and perhaps spoken as follows: [6] "O how wretched am I and many times unhappy! After bearing seven children, I am now the mother of none! [7] O seven childbirths all in vain, seven profitless pregnancies, fruitless nurturings and wretched nursings! [8] In vain, my sons, I endured many birth pangs for you, and the more grievous anxieties of your upbringing. [9] Alas for my children, some unmarried, others married and without offspring.[h] I shall not see your children or have the happiness of being called grandmother. [10] Alas, I who had so many and beautiful children am a widow and alone, with many sorrows.[i] [11] And when I die, I shall have none of my sons to bury me."

12 Yet that holy and God-fearing mother did not wail with such a lament for any of them, nor did she dissuade any of them from dying, nor did she grieve as they were dying. [13] On the contrary, as though having a mind like adamant and giving rebirth for immortality to the whole number of her sons, she implored them and urged them on to death for the sake of religion. [14] O mother, soldier of

g Other ancient authorities read *having bidden them farewell, surrendered them*   h Gk *without benefit*   i Or *much to be pitied*

---

tual birth; compare 16.13 n.; 17.6; Gal 4.19. **25–26:** *Ballots,* as though she were in the *council chamber* of a Greek city. **28:** See 13.12 n. **31:** *Noah's ark,* see 7.1–3 n.; also Wis 14.6.
  16.3: *Daniel,* Dan 6.1–24. *Mishael,* Dan 1.7; 3.19–30. **11:** *None of my sons to bury me,* for both Jews and Greeks a supreme calamity. **13:** *Giving rebirth for immortality,* see 15.17 n.; Jn 3.5. **15:** *Hebrew language,* see 12.7 n.

God in the cause of religion, elder and woman! By steadfastness you have conquered even a tyrant, and in word and deed you have proved more powerful than a man. 15 For when you and your sons were arrested together, you stood and watched Eleazar being tortured, and said to your sons in the Hebrew language, 16 "My sons, noble is the contest to which you are called to bear witness for the nation. Fight zealously for our ancestral law. 17 For it would be shameful if, while an aged man endures such agonies for the sake of religion, you young men were to be terrified by tortures. 18 Remember that it is through God that you have had a share in the world and have enjoyed life, 19 and therefore you ought to endure any suffering for the sake of God. 20 For his sake also our father Abraham was zealous to sacrifice his son Isaac, the ancestor of our nation; and when Isaac saw his father's hand wielding a knife*j* and descending upon him, he did not cower. 21 Daniel the righteous was thrown to the lions, and Hananiah, Azariah, and Mishael were hurled into the fiery furnace and endured it for the sake of God. 22 You too must have the same faith in God and not be grieved. 23 It is unreasonable for people who have religious knowledge not to withstand pain."

24 By these words the mother of the seven encouraged and persuaded each of her sons to die rather than violate God's commandment. 25 They knew also that those who die for the sake of God live to God, as do Abraham and Isaac and Jacob and all the patriarchs.

17 Some of the guards said that when she also was about to be seized and put to death she threw herself into the flames so that no one might touch her body.

2 O mother, who with your seven sons nullified the violence of the tyrant, frustrated his evil designs, and showed the courage of your faith! 3 Nobly set like a roof on the pillars of your sons, you held firm and unswerving against the earthquake of the tortures. 4 Take courage, therefore, O holy-minded mother, maintaining firm an enduring hope in God. 5 The moon in heaven, with the stars, does not stand so august as you, who, after lighting the way of your starlike seven sons to piety, stand in honor before God and are firmly set in heaven with them. 6 For your children were true descendants of father Abraham. *k*

7 If it were possible for us to paint the history of your religion as an artist might, would not those who first beheld it have shuddered as they saw the mother of the seven children enduring their varied tortures to death for the sake of religion? 8 Indeed it would be proper to inscribe on their tomb these words as a reminder to the people of our nation: *l*

9 "Here lie buried an aged priest and an aged woman and seven sons, because of the violence of the tyrant who wished to destroy the way of life of the Hebrews. 10 They vindicated their nation,

j Gk *sword* *k* Gk *For your childbearing was from Abraham the father*; other ancient authorities read *For . . . Abraham the servant*
l Or *as a memorial to the heroes of our people*

16.16–23: Compare this more rhetorical speech with 2 Macc 7.27–29. **16:** *Contest*, 6.10; 11.20. **20:** *Isaac*, see 13.12 n. **21:** *Daniel*, v. 3. *Hananiah, Azariah, and Mishael*, Dan ch 3. **25:** *Live to God*, see 7.19 n.
17.1: *Threw herself*, compare 12.19; 2 Macc 7.41. *Touch her body*, this would be a violation of her chastity.
17.2–18.5: Panegyric on the mother. The author has already pronounced encomiums on Eleazar (7.1–15), the brothers (14.2–10), and the mother (ch 15), sometimes addressing them directly. The oration now comes to its climax. **3:** *Roof . . . pillars . . .*

earthquake, metaphors appropriate to Antioch, where there were frequent earthquakes, but applicable also to Palestine. **5:** Stoics regarded *the stars* as living beings; for Jews the language was metaphorical. *Lighting the way*, mystical language; compare Jn 12.35–36, 46. **6:** *True descendants*, compare 15.28; 13.12 n.; the giving of new birth is like Isaac's return from impending death; compare Heb 11. 17–19. **7:** *Possible*, or "permitted." At this time the Jews may have taken the commandment of Ex 20.4 strictly. **9:** *Way of life*, Greek "politeia," "commonwealth."

looking to God and enduring torture even to death."

11 Truly the contest in which they were engaged was divine, 12 for on that day virtue gave the awards and tested them for their endurance. The prize was immortality in endless life. 13 Eleazar was the first contestant, the mother of the seven sons entered the competition, and the brothers contended. 14 The tyrant was the antagonist, and the world and the human race were the spectators. 15 Reverence for God was victor and gave the crown to its own athletes. 16 Who did not admire the athletes of the divine*m* legislation? Who were not amazed?

17 The tyrant himself and all his council marveled at their*n* endurance, 18 because of which they now stand before the divine throne and live the life of eternal blessedness. 19 For Moses says, "All who are consecrated are under your hands." 20 These, then, who have been consecrated for the sake of God,*o* are honored, not only with this honor, but also by the fact that because of them our enemies did not rule over our nation, 21 the tyrant was punished, and the homeland purified—they having become, as it were, a ransom for the sin of our nation. 22 And through the blood of those devout ones and their death as an atoning sacrifice, divine Providence preserved Israel that previously had been mistreated.

23 For the tyrant Antiochus, when he saw the courage of their virtue and their endurance under the tortures, proclaimed them to his soldiers as an example for their own endurance, 24 and this made them brave and courageous for infantry battle and siege, and he ravaged and conquered all his enemies.

18 O Israelite children, offspring of the seed of Abraham, obey this law and exercise piety in every way, 2 knowing that devout reason is master of all emotions, not only of sufferings from within, but also of those from without.

3 Therefore those who gave over their bodies in suffering for the sake of religion were not only admired by mortals, but also were deemed worthy to share in a divine inheritance. 4 Because of them the nation gained peace, and by reviving observance of the law in the homeland they ravaged the enemy. 5 The tyrant Antiochus was both punished on earth and is being chastised after his death. Since in no way whatever was he able to compel the Israelites to become pagans and to abandon their ancestral customs, he left Jerusalem and marched against the Persians.

6 The mother of seven sons expressed also these principles to her children: 7 "I was a pure virgin and did not go outside my father's house; but I guarded the rib from which woman was made.*p* 8 No seducer corrupted me on a desert plain, nor did the destroyer, the deceitful serpent, defile the purity of my virginity. 9 In the time of my maturity I remained with my husband, and when these sons had grown up their father died. A happy man was he, who lived out his life with

m Other ancient authorities read *true*
n Other ancient authorities add *virtue and*
o Other ancient authorities lack *for the sake of God*   *p* Gk *the rib that was built*

17.11–12: *Contest . . . awards . . . prize,* 6.10; 11.20. **21:** *Ransom,* see 6.29 n. **22:** *Through the blood,* Rom 3.25; *atoning sacrifice,* Greek "hilastērion," as in Rom 3.25; compare Heb 9.11–15; 1 Pet 1.19; 1 Jn 1.7. **24:** *Ravaged and conquered,* but Antiochus was not successful, and died in Babylon (1 Macc 6.1–16). **18.1–5:** The exhortation seems repetitious after 17.7–24, but it is the author's method to employ recurrent themes. **5:** *Chastised after his death,* balances the immediate immortality bestowed upon the righteous martyrs. See 18.22–23. *Marched against the Persians,*

17.24 n. The second sentence in this verse does not fit well with the first. **18.6–19:** **The mother's last words.** Compare 2 Macc 7.22–29. The mother is the supreme heroine of the story. This is a quiet passage after the highly emotional parts, designed to move the reader to reflection. Jewish education began in the home, and the mother's influence was always important. **7:** *Rib,* Gen 2.22. **8:** *Desert plain,* Deut 22.25–27. In such places women were in danger from men and also from demons, who were believed to inhabit the wilderness. **11:** *Abel,* Gen

good children, and did not have the grief of bereavement. 10While he was still with you, he taught you the law and the prophets. 11He read to you about Abel slain by Cain, and Isaac who was offered as a burnt offering, and about Joseph in prison. 12He told you of the zeal of Phinehas, and he taught you about Hananiah, Azariah, and Mishael in the fire. 13He praised Daniel in the den of the lions and blessed him. 14He reminded you of the scripture of Isaiah, which says, 'Even though you go through the fire, the flame shall not consume you.' 15He sang to you songs of the psalmist David, who said, 'Many are the afflictions of the righteous.' 16He recounted to you Solomon's proverb, 'There is a tree of life for those who do his will.' 17He confirmed the query of Ezekiel, 'Shall these dry bones live?' 18For he did not forget to teach you the song that Moses taught, which says,

19'I kill and I make alive: this is your life and the length of your days.' "

20 O bitter was that day—and yet not bitter—when that bitter tyrant of the Greeks quenched fire with fire in his cruel caldrons, and in his burning rage brought those seven sons of the daughter of Abraham to the catapult and back again to more*q* tortures, 21pierced the pupils of their eyes and cut out their tongues, and put them to death with various tortures. 22For these crimes divine justice pursued and will pursue the accursed tyrant. 23But the sons of Abraham with their victorious mother are gathered together into the chorus of the fathers, and have received pure and immortal*r* souls from God, 24to whom be glory forever and ever. Amen.

q Other ancient authorities read *to all his*
r Other ancient authorities read *victorious*

4.2–15. *Isaac,* Gen 22.1–19. *Joseph,* Gen 39.1–23. **12:** *Phinehas,* Num 25.1–9. *Hananiah,* 16.21. **13:** *Daniel,* Dan 6.1–24. **14:** Isa 43.2. **15:** Ps 34.19. **16:** Prov 3.18, modified slightly. **17:** Ezek 37.2–3. **19:** Deut 32.39; 30.20.

18.20–24: Conclusion. This peroration sums up many previous themes set forth in 4 Maccabees **23:** *Abraham,* 13.12 n. *Chorus,* see 8.4 n. *Immortal,* see 9.22 n. **24:** Compare Rom 11.36; 16.27; 2 Tim 4.18; Heb 13.21.

INDEX TO THE ANNOTATIONS
APOCRYPHAL/DEUTEROCANONICAL BOOKS

The following index lists important persons, places, and ideas that are mentioned in the general introduction and the annotations. In order to gain the fullest information, the verses of the passage of the Apocryphal/Deuterocanonical Books, as well as the annotation itself, should be read, and all cross references should be consulted.

NOTES

THE NEW COVENANT
COMMONLY CALLED
THE NEW TESTAMENT
OF OUR LORD AND SAVIOR
JESUS CHRIST

New Revised Standard Version

Introduction to the New Testament

The New Testament comprises the twenty-seven books that constitute the second of the two portions into which the Bible is naturally divided. A more appropriate word than "testament" to designate the character of these books is "covenant." This is the word that the Bible uses in referring to the relationship that God established with his people. In a sublime passage written more than five hundred years before Christ, the prophet Jeremiah predicted that the covenant relation of God with his people, instituted through Moses on Mount Sinai, would give place in the future to a more inward and personal one (Jer 31. 31–34). In accord with this prophecy the Apostle Paul regarded the Christian dispensation as based on a new covenant, which he contrasted with the old covenant enshrined in the books of Moses (2 Cor 3.6–15). Another early Christian writer, doubtless thinking of the words of Jesus when he instituted the Last Supper (Mt 26.28; Mk 14.24; 1 Cor 11.25), declared that by his sacrificial death Christ became the mediator of a new covenant (Heb 9.15–20; compare 10.16). It is quite understandable, therefore, that the two collections of documents that belong respectively to the two dispensations of God's dealings with his people have been brought together in one volume as "The Holy Bible, containing the Old and New Testaments."

Unlike the books of the Old Testament, which originated during a period extending over many centuries, the books of the New Testament were written within a period of somewhat less than one hundred years. These books fall into four different literary forms. Four of them are "Gospels" because they tell the "gospel" (a word derived from the Anglo-Saxon *gōd-spell*, meaning "good tidings") of Jesus Christ, that is, his birth, baptism, ministry of teaching and healing, death, and resurrection. Church history is represented in the Acts of the Apostles, which is an account of the spread of the Christian faith during the first thirty or so years after the death and resurrection of Jesus Christ. Twenty-one of the books of the New Testament are in the form of letters. The last book of the New Testament is an apocalypse (see pp. 362–363 NT), that is, a revelation (Greek *apokalypsis*) or disclosure of God's will for the future.

Jesus himself left no literary remains; information regarding his words and works comes from his immediate followers (the apostles) and their disciples. At first this information was circulated orally. As far as we know today, the first attempt to produce a written Gospel was made by John Mark, who according to tradition was a disciple of the Apostle Peter. This Gospel, along with a collection of sayings of Jesus and several other special sources, formed the basis of the Gospels attributed to Matthew and Luke. The reasons for regarding the Gospel of Mark as one of the sources of Matthew and Luke include the following facts. (1) Apart from details Mark contains very little that is not in Matthew or in Luke. (2) When Mark and Matthew differ as to sequence of matter, Luke agrees with Mark, and when Mark and Luke differ as to sequence, Matthew agrees with Mark. (3) Matthew and Luke never agree as to sequence against Mark.

Because these three Gospels have so much in common, they are generally called the Synoptic Gospels (from the Greek word *synoptikos,* viewing together).

In contrast to the Synoptics, which narrate mainly Jesus' public teaching and ministry to the crowds in Galilee, the Gospel according to John contains information regarding Jesus' early Judean ministry as well as extensive discourses to the disciples concerning the union of the Christian with Christ (chs 14–17). John presents no parables in the Synoptic sense, and he frequently weaves the words of Jesus so closely with his own interpretation of them that it is difficult to find the break between the two. (For further information concerning the genre of the gospels, see pp. viii–x NT.)

The letters in the New Testament were written by various teachers in the primitive church to congregations and to individuals in order to provide further instruction in the Christian faith and to correct certain abuses and disorders that had arisen. The letter-form happily combines the advantages inherent in both conversation and treatise. By their nature letters permit the writer to communicate with the recipients in a personal and indeed affectionate manner; at the same time, they also allow the writer to treat abstract subjects with accuracy and fullness.

The New Testament letters fall into two main groups, those attributed to Paul and those attributed to other writers. Within each group the present sequence seems to have been determined in accordance with the length of the letters, the longest standing first. The present titles of the letters were not part of the original documents, and would not have been needed until the individual letters had been gathered into a collection. (For further information on ancient letter-writing, see pp. 204–205 NT.)

The language in which the books of the New Testament were written was the *koine* or common Greek of the time. This form of Greek, which lacks many of the subtle refinements of classical Greek of an earlier period, was known and used by most of the peoples of the Roman Empire to whom the first Christian missionaries carried the gospel. Noticeably different grades of koine Greek are found in the several New Testament documents. The most highly literary as regards sentence structure and vocabulary are the Letter to the Hebrews and the two books written by Luke (the Gospel and the Acts). Those that are the furthest removed from classical Greek standards and closest to colloquial Greek are the Gospel of Mark and the book of Revelation. Furthermore, since all the authors represented in the New Testament appear to have been either Jews or Jewish proselytes before becoming Christians, it is natural that their use of koine Greek was colored by their familiarity with the special characteristics of the Greek translation of the Hebrew Old Testament (the Septuagint). Here and there the Gospels and the first half of Acts preserve in Greek certain turns of expression that reflect an underlying Aramaic idiom, which was the mother tongue of Jesus and his disciples. Aramaic is a Semitic language related to Hebrew.

The original manuscripts of the several books of the New Testament have long since disappeared. Three sources of information exist today for our knowledge of the text of the New Testament. They are Greek manuscripts, early translations into other languages (primarily Syriac, Latin, and Coptic), and quotations from the New Testament made by early ecclesiastical writers. The total number of Greek manuscripts of all or part of the New Testament is close to five thousand. Not quite one-half of this number are manuscripts that contain only the four Gospels. There are fifty-nine manuscripts that contain all twenty-seven books of the New Testament. Of the five thousand, the most important are, in general, the oldest; more than three hundred, written on papyrus or parchment, date from the second to the eighth century. In evaluating the significance of this rich store of manuscripts, it should be recalled that the writings of many ancient classical authors have survived in only a few copies (or even in only one), and that not infrequently these copies date from the late Middle Ages, separated from the time of the composition of the originals by more than a thousand years.

Before the invention of printing with movable type (about A.D. 1450–1456) all literary works had to be copied by hand. Owing to the rapid expansion of the early church and the growing demand for additional copies of the Christian Scriptures, sometimes speed in the preparation of manuscripts seemed to be of greater importance than strict accuracy of detail. Consequently unintentional errors, inevitable in all transcription, were multiplied in what was copied. At other times alterations were deliberately made; for example, scribes would occasionally improve the grammar and style, correct real or imagined errors in history and geography, adjust quotations from the Old Testament in accordance with current copies of the Greek Septuagint translation, and harmonize differing accounts in the Synoptic Gospels. Thus, not only inadvertence but also well-intentioned efforts resulted in the creation of thousands of divergencies among the manuscripts of the New Testament. It should be added, however, that the great majority of these variant readings involve inconsequential details, such as alternative spellings, differing order of words, and interchange of synonyms. Among the relatively few variants that involve the essential meaning of the text, modern scholars are usually able to determine with more or less probability what the original text was. In deciding among the variant readings scholars usually give preference to those that are preserved in the older manuscripts, as well as those supported by witnesses that are widely separated geographically. Preference is also often given to the shorter reading, for scribes were much more prone to make additions than deliberately to omit. Another scribal tendency was the harmonization of divergent accounts. In general, the reading is preferred that best explains the rise of the other reading(s). (For examples of various kinds of additions to the text, see Mt 6.13 note *j*; 23.13 note *q*; Mk 9.43 note *u*; 16.8 note *r*; Lk 17.35 note *l*; 24.12 note *b*; Jn 5.3 note *i*; 8.11 note *k*; Acts 8.36 note *n*.)

As long as copies of the books of the New Testament were written on scrolls, practical difficulties of length prevented the collection of many of the books of the New Testament into one scroll. Early in the second century, however, the adoption by many Christians of the codex or leaf-form of manuscript permitted the gathering, for example, of all of the Gospels or all of Paul's letters into one volume. This process of assembling various books in a single codex was accompanied by the development of the idea of the canon of the New Testament.

The word "canon" is a Greek word borrowed from a Semitic root, meaning a stalk or a reed, which could be used for measuring. By extension the word came to denote a rule, norm, or list (of books). Why, how, and when the present books of the New Testament were finally gathered into one collection are questions difficult to answer because of the lack of explicit information. It is possible, however, to reconstruct some of the influences that must have contributed to the emergence of the canon of the New Testament.

The Bible of the earliest Christians was the Old Testament (2 Tim 3.15–17). Of equal authority to these writings were the remembered words of Jesus (Acts 20.35; 1 Cor 7.10, 12; 9.14; 1 Tim 5.18). Parallel with the oral circulation of Jesus' teaching were apostolic interpretations of his person and significance for the life of the church. It is natural that when the gospels and the apostolic letters were written, incorporating these two kinds of authoritative materials, they would be treasured, circulated, and read in services of worship (Col 4.16; 1 Thess 5.27; Rev 1.3).

At first a local church would have only a few apostolic letters and perhaps one or two gospels. During the course of the second century most churches came to acknowledge a canon that included the present four Gospels, the Acts, thirteen letters of Paul, 1 Peter, and 1 John. Seven books still lacked general recognition: Hebrews, James, 2 Peter, 2 and 3 John, Jude, and Revelation. On the other hand, certain writings, such as the Letter of Barnabas or the *Shepherd* of Hermas, were accepted as Scripture by several ecclesiastical writers, though rejected by the majority. During the third century there was a sifting of the disputed books; certain of them came to be acknowledged as canonical and others were rejected as apocryphal. The fourth century was marked by authoritative pronouncements first by bishops of provincial churches and later by synods or councils. St. Athanasius in his Festal Letter for A.D. 367 was the first to name the twenty-seven books of the New Testament as exclusively canonical. Not all Christian communities, however, were willing at first to follow the Athanasian list. In the East, for example, the Armenian Church accepted the Third Letter of Paul to the Corinthians, and in the West many Latin manuscripts contained the spurious Letter to the Laodiceans. (Strangely enough, this brief letter follows Galatians in all eighteen German Bibles printed prior to Luther's translation of the Greek New Testament.)

Various external circumstances assisted in the process of canonization of the New Testament books. The emergence of heretical sects with their own sacred books made it imperative for the church to determine the limits of the canon. Likewise, when Christians were persecuted for their faith it became a matter of the utmost importance to know which books could be renounced and which were essential to the faith.

The tests of canonicity that were most frequently applied seem to have been apostolic authorship, or at least apostolic content, and general harmony with the Old Testament and the rest of the New Testament. Prior, however, to the issuing of pronouncements by church councils was the intuitive insight of individual Christians who had discerned the inherent significance of the canonical books. In the most basic sense neither individuals nor councils created the canon, but only came to recognize and acknowledge the self-authenticating quality of these writings, which imposed themselves as canonical upon the church. (For further discussion reference may be made to B. M. Metzger, *The Canon of the New Testament: Its Origin, Development, and Significance* [Oxford University Press, 1987].)

Chronological Tables of Rulers during New Testament Times

ROMAN EMPERORS

27 B.C.–A.D. 14	Augustus
A.D. 14–37	Tiberius
A.D. 37–41	Caligula
A.D. 41–54	Claudius
A.D. 54–68	Nero
A.D. 68–69	Galba; Otho; Vitellius
A.D. 69–79	Vespasian
A.D. 79–81	Titus
A.D. 81–96	Domitian

HERODIAN RULERS

37–4 B.C.	Herod the Great, king of the Jews
4 B.C.–A.D. 6	Archelaus, ethnarch of Judea
4 B.C.–A.D. 39	Herod Antipas, tetrarch of Galilee and Perea
4 B.C.–A.D. 34	Philip, tetrarch of Ituraea, Trachonitis, etc.
A.D. 37–44	Herod Agrippa I, from 37 to 44 king over the former tetrarchy of Philip, and from 41 to 44 over Judea, Galilee, and Perea
A.D. 53–about 100	Herod Agrippa II, king over the former tetrarchy of Philip and Lysanias, and from 56 (or 61) over parts of Galilee and Perea

PROCURATORS OF JUDEA AFTER THE REIGN OF ARCHELAUS TO THE REIGN OF HEROD AGRIPPA I

A.D. 6–8	Coponius
A.D. 9–12	M. Ambivius
A.D. 12–15	Annius Rufus
A.D. 15–26	Valerius Gratus
A.D. 26–36	Pontius Pilate
A.D. 37	Marullus
A.D. 37–41	Herennius Capito

PROCURATORS OF PALESTINE FROM THE REIGN OF HEROD AGRIPPA I TO THE JEWISH REVOLT

A.D. 44–about 46	Cuspius Fadus
A.D. about 46–48	Tiberius Alexander
A.D. 48–52	Ventidius Cumanus
A.D. 52–60	M. Antonius Felix
A.D. 60–62	Porcius Festus
A.D. 62–64	Clodius Albinus
A.D. 64–66	Gessius Florus

The Narrative Books—
Gospels and Acts

THE GOSPELS

1. *Their Literary Genre*

What is a gospel? To what literary genre do the four gospels of the New Testament belong? Of course there are differences among the four, particularly between the three Synoptic Gospels on the one hand and the Gospel according to John on the other, but to what type of ancient literature would a second-century librarian in Alexandria have assigned a gospel if a manuscript copy had been presented to a library?

Scholars have given various answers to such questions, depending upon the kinds and styles of contemporary Jewish and Hellenistic literature with which the gospels are compared. Some have pointed out similarities between the Christian gospels and Graeco-Roman biographies, such as those that report the lives of Epictetus, Apollonius, and Socrates. Seeking to refine still further the broad category of biography, other scholars have argued that the gospels belong to a certain species of biography called aretalogy. An aretalogy, as defined by the dictionary, is "a narrative of the miraculous deeds of a god or hero." According to this understanding, the gospels are held to advance Jesus as a wonder-working divine man. It must be acknowledged, however, that since we have no complete text surviving from the past that is specifically labeled aretalogy, to attempt to clarify the gospels by grouping them with a type of writing that has no known surviving examples is far from illuminating.

Still others have regarded the Synoptic Gospels as examples of ancient laudatory (or encomium) biography, of which many examples survive, such as Xenophon's *Agesilaus,* Isocrates' *Evagoras,* Tacitus' *Agricola,* and Lucian's *Life of Demonax.* The purpose of this kind of literature is to praise the greatness and merit of the person who is the subject of the document.

Scholars who do not wish to go so far afield to seek for examples have preferred to look to a Jewish milieu for clues to the genre of the canonical gospels. Certainly some of the gospel pericopes—the individual incidents of which the gospels are composed—bear a closer stylistic resemblance to the rabbinic stories preserved in ancient Jewish sources than to any other body of literature. Furthermore, the all-inclusive role of Moses as he is presented in the book of Exodus, culminating as the mediator of the covenant, or as he is presented in Philo's *Life of Moses,* obviously presents certain similarities with the life of Jesus. Still others have found an Old Testament model in the cycle of Elisha stories.

It must, however, be concluded that none of these proposed models—nor even a combination of models, such as aretalogical biography—provides a suitable literary category for classifying the gospel genre. The canonical gospels are not romances or folk-tales; they purport to retell actual events. They are not biographies; they concentrate on the public career of Jesus with little or no attention given to his environment, training, and development of character. They are not simply memoirs of a teacher, philosopher, or wise man; the ministry of Jesus embraced not merely word and example but action. And as regards this action, the gospels do not give a neutral account of what happened; rather they tell of the work of God in the career of Jesus, and they present their story as an offer of salvation for all who will believe. In short, the gospels represent a genre all their own because they present the tradition of Jesus from the viewpoint of faith in him as redeemer. Hence it was the intention of the four evangelists that their gospels be understood not only as narrative, but at the same time and especially as proclamation.

2. *The Origin of the Gospels*

The gospels, like the other New Testament books and most of the rest of the Bible, are the literary productions of a believing community. They are not telling a story for its entertainment value, though they will use literary techniques such as suspense, metaphor, characterization, and so on. Nor are they simply stenographic reports of words and deeds. They are written with the aim of changing the reader or of building up the community's faith. The story they have to tell is intended to make a difference in the reader's life.

That story, most scholars now agree, was put together over a period of several decades and in several stages. (1) During the time of oral transmission, the teachings of Jesus, and individual stories about him, were communicated among his followers by word of mouth. Of these stories, the most lengthy and important would have been the so-called "Passion Narratives" or accounts covering the week or so leading up to his death, and the subsequent accounts of his resurrection; there would have been other stories as well, such as those of healings and confrontations with the authorities. It was also during this time that some of the materials were re-told in Greek, the common language outside of Palestine, instead of the Aramaic that Jesus and his first followers spoke. (2) Subsequently written collections were assembled, consisting of selections of these teachings and stories. Included among such accounts would have been a complete Passion Narrative, or a collection of short sayings, or some parables. (3) Finally, the four evangelists collected these materials, both oral and written, and combined them into the gospels that we have today.

The literary relationship among the first three gospels—Matthew, Mark, and Luke—is also of interest. These three are called the "Synoptic Gospels" because they "view together" (*synoptikos* in Greek) the events of Jesus' life. Many of the sayings, parables, and incidents appear in more than one gospel, often in similar or virtually identical words. Among various explanations for this, the most widely held view holds (*a*) that Matthew and Luke depended on Mark for the general outline and many of the incidents of Jesus' ministry; (*b*) that they also had a separate list of sayings of Jesus, which scholars have called "Q" (for the German *Quelle,* "source"); and (*c*) that they each had additional materials (for instance, the story of the wise men from the East in Matthew ch 2, or the parable of the Good Samaritan in Luke ch 10). It is generally agreed that John derives largely from a different tradition, since most of the incidents and their treatment in his gospel differ from those in the Synoptics.

Each of the four gospel writers had his own point of view and special reading public, and these would have guided the selection, arrangement, and synthesis of the many units handed down concerning the life, deeds, sayings, death, and resurrection of Jesus the Messiah. Yet all four gospels show the same basic intention: they are addressed to readers in order to minister to their faith. In fact, two of the gospels contain expressions of their author's intentions. The prologue of Luke's gospel states that it was written so that Theophilus would "know the truth concerning the things about which you have been instructed" (Luke 1.4). A similar expression of intention stands near the close of the Fourth Gospel: These things "are written so that you may come to believe that Jesus is the Messiah, the Son of God, and that through believing you may have life in his name" (John 20.31).

Thus, even if in literary form or shape the gospels are not in all respects a novelty, the distinctive purpose of the four authors accounts for the special nature of their literary work. They were not memorializing the deeds and words of a revered but deceased master; they were explaining the universal significance of the life and death of someone whom they proclaimed to be still alive and knowable. In a word, they were evangelizing—and consequently, for good reason, they came to be known in the church as the four Evangelists.

THE ACTS OF THE APOSTLES

Inasmuch as the book of Acts forms the second part of a two-volume work, of which the Gospel according to Luke is the first part, it is natural to look for literary correspondences between the two. Apparently the authors of Matthew, Mark, and John saw no need for a second volume, one that contained an account of the growth of the church in order to complement their gospel. Luke-Acts reflects the conviction that both the story of Jesus and the story of the apostolic church are incomplete without the other as complement.

The thread of the narrative in the Gospel according to Luke runs from the birth of John the Baptist, soon followed by the birth of Jesus, through the events of the latter's ministry of healing and teaching to the climax of his arrest, trial, crucifixion. The gospel narrative concludes with the resurrection and ascension of the Lord. It is at this point that the book of Acts picks up the narrative by recounting in greater detail the ascension of Jesus, with the subsequent coming of the Holy Spirit upon the waiting church. There follows an account of the planting and extension of the church by the establishment of radiating centers at certain salient points throughout a significant part of the Roman Empire, beginning at Jerusalem and ending at Rome. Thus it was not the purpose of the writer to produce full-scale biographies of Peter or Paul or other apostles; these characters are described only insofar as necessary in showing how the church was formed, how it broadened to receive Gentiles, and how it extended from Jerusalem to Rome. So, too, it was evidently not Luke's aim to write all that he knew of the history of any local church, at Jerusalem or Antioch or Philippi, but only to show how the witnessing of Christian messengers resulted in the establishment of such societies, and how they aided in fulfilling the task laid upon them to be witnesses "in Jerusalem, in all Judea and Samaria, and to the ends of the earth" (Acts 1.8).

Thus Luke's historical art and technique produced two volumes that present a unified narrative because the major human characters—John the Baptist, Jesus, the apostles, and Paul—share in a mission that expresses a single controlling purpose: fulfilling the purpose of God. Since God's purpose repeatedly encounters rejection, the narrative is loaded with tension. When the book of Acts is compared with other ancient historical narratives (or even with historical novels), Luke's masterful command of the "dramatic episode style" is obvious. His artistic technique holds the reader's attention, focused on the sequence of individual events, some of which involved threats and dangers plotted against the messengers of the gospel. As a result the narrative throbs with life, and the reader is enabled to discern a unified action with a specific goal—the proclamation of the gospel to Jew and Gentile alike. (See also the Introduction to the Acts of the Apostles, p. 160 NT.)

The Gospel According to
Matthew

The Gospel according to Matthew is a manual of Christian teaching in which Jesus Christ, Lord of the new-yet-old community, the church, is described particularly as the fulfiller and the fulfillment of God's will disclosed in the Old Testament. Jesus is set forth as Israel's royal Messiah (1.1; 19.28) in whom God's purpose culminates and by whose words and life his followers, the true Israel (25.34), may gain divine forgiveness and fellowship.

The accounts of Jesus' deeds and words, drawn from Christian sources both oral and written, are arranged in a generally biographical order: chs 1–2, Birth of Jesus; 3.1–12, Activity of John the Baptist; 3.13–4.11, Baptism and temptation of Jesus; 4.12–18.35, Jesus' preaching and teaching in Galilee; chs 19–20, Journey to Jerusalem; chs 21–27, The last week, concluding with Jesus' crucifixion and burial; ch 28, The resurrection; Jesus' commission to his disciples.

Within this natural framework the accounts of what Jesus said or did are grouped by common subject matter. The five discourses of Jesus, a noteworthy feature of this Gospel (see 7.28 n.), are Matthew's collections of teachings on specific themes: chs 5–7, The Sermon on the Mount; ch 10, Instructions for missionary disciples; ch 13, The parables of the kingdom of Heaven; ch 18, On sincere discipleship; chs 24–25, On the end of the present age. The author seems to have deliberately built his gospel around these five great discourses as though his object was especially to show the fullness of Jesus' teaching.

Much of the material peculiar to this Gospel is concerned with the Jews or with the fulfillment of Old Testament prophecies. All of the Gospels mention Jesus' Davidic lineage, but Matthew emphasizes this relationship by referring to it much more often than does any of the other Evangelists. In addition to what he has in common with one or more of the other Gospels, he includes also the testimony of the two blind men (9.27), the multitude (12.23), the Canaanite woman (15.22), the crowds at the triumphant entry into Jerusalem (21.9), and the children in the temple (21.15). In short, the special aim of Matthew is to show that Jesus is the legitimate heir to the royal house of David.

In this connection the author's frequent appeal to the fulfillment of prophecy is a noteworthy feature of this Gospel. All four evangelists cite Old Testament prophecies that they regard as having been fulfilled in the person and work of Jesus, but Matthew includes nine additional such prophetic proof-texts, all of which are characterized by a certain verbal literalism that would make special appeal to readers having a Jewish background (1.22–23; 2.15; 2.17–18; 2.23; 4.14–16; 8.17; 12.17–21; 13.35; 21.4–5; 27.9–10).

This Gospel is anonymous. The unknown Christian teacher who prepared it during the last third of the first century may have used as one of his sources a collection of Jesus' sayings that the apostle Matthew, according to second-century writers, is said to have drawn up. In time a title containing Matthew's name, and signifying apostolic authority, came to identify the whole.

For the literary genre of the gospels, see pp. viii–x NT.

1
NT

1 An account of the genealogy^a of Jesus the Messiah,^b the son of David, the son of Abraham.

2 Abraham was the father of Isaac, and Isaac the father of Jacob, and Jacob the father of Judah and his brothers, 3 and Judah the father of Perez and Zerah by Tamar, and Perez the father of Hezron, and Hezron the father of Aram, 4 and Aram the father of Aminadab, and Aminadab the father of Nahshon, and Nahshon the father of Salmon, 5 and Salmon the father of Boaz by Rahab, and Boaz the father of Obed by Ruth, and Obed the father of Jesse, 6 and Jesse the father of King David.

And David was the father of Solomon by the wife of Uriah, 7 and Solomon the father of Rehoboam, and Rehoboam the father of Abijah, and Abijah the father of Asaph,^c 8 and Asaph^c the father of Jehoshaphat, and Jehoshaphat the father of Joram, and Joram the father of Uzziah, 9 and Uzziah the father of Jotham, and Jotham the father of Ahaz, and Ahaz the father of Hezekiah, 10 and Hezekiah the father of Manasseh, and Manasseh the father of Amos,^d and Amos^d the father of Josiah, 11 and Josiah the father of Jechoniah and his brothers, at the time of the deportation to Babylon.

12 And after the deportation to Babylon: Jechoniah was the father of Salathiel, and Salathiel the father of Zerubbabel, 13 and Zerubbabel the father of Abiud, and Abiud the father of Eliakim, and Eliakim the father of Azor, 14 and Azor the father of Zadok, and Zadok the father of Achim, and Achim the father of Eliud, 15 and Eliud the father of Eleazar, and Eleazar the father of Matthan, and Matthan the father of Jacob, 16 and Jacob the father of Joseph the husband of Mary, of whom Jesus was born, who is called the Messiah.^c

17 So all the generations from Abraham to David are fourteen generations; and from David to the deportation to Babylon, fourteen generations; and from the deportation to Babylon to the Messiah,^e fourteen generations.

18 Now the birth of Jesus the Messiah^b took place in this way. When his mother Mary had been engaged to Joseph, but before they lived together, she was found to be with child from the Holy Spirit. 19 Her husband Joseph, being a righteous man and unwilling to expose her to public disgrace, planned to dismiss her quietly. 20 But just when he had resolved to do this, an angel of the Lord appeared to him in a dream and said, "Joseph, son of David, do not be afraid to take Mary as your wife, for the child conceived in her is from the Holy Spirit. 21 She will bear a son, and you are to name him Jesus, for he will save his people from their sins." 22 All this took place to fulfill what had been spoken by the Lord through the prophet:

23 "Look, the virgin shall conceive
and bear a son,

a Or *birth*   b Or *Jesus Christ*   c Other ancient authorities read *Asa*   d Other ancient authorities read *Amon*   e Or *the Christ*

1.1–17: Jesus' royal descent (Lk 3.23–38) is traced through *King David* (22.41–45; Rom 1.3) back to Abraham the patriarch (Gal 3.16). **1**: *Messiah,* Hebrew word meaning "anointed one," translated into Greek as "Christos." **3–6**: Ruth 4.18–22; 1 Chr 2.1–15. **8**: For the sake of the pattern (v. 17) the names of Ahaziah, Joash, and Amaziah (1 Chr 3.11–12) have been omitted; such omission was quite consistent with Jewish practice in forming genealogies. **11**: *The deportation,* 2 Kings 24.8–16; Jer 27.20. **12**: *Jechoniah,* or *Jehoiachin* (2 Kings 24.6; 1 Chr 3.16). *Salathiel* apparently transmitted the line of legal descent from *Jechoniah* to *Zerubbabel* (Ezra 3.2; Hag 2.2; Lk 3.27), although the Chronicler traces it through Pedaiah (1 Chr 3.16–19). **13–16**: The persons from *Abiud* to *Jacob* are otherwise unknown. **17**: The device of making three groups of names is an aid to memory. *Fourteen* is the sum of the numerical value of the three letters in the name David in Hebrew (DWD).

1.18–2.23: Jesus' birth and infancy (Lk 1.26–2.40). **20**: *Angel,* see Heb 1.14 n. **21**: The Hebrew and Aramaic forms of *Jesus* and *he will save* are similar. The point could be suggested by translating, "You shall call his name 'Savior' because he will save." **22–23**: See Isa 7.14 n.

and they shall name him
Emmanuel,"
which means, "God is with us." 24 When
Joseph awoke from sleep, he did as the
angel of the Lord commanded him; he
took her as his wife, 25 but had no marital
relations with her until she had borne a
son;*f* and he named him Jesus.

2 In the time of King Herod, after
Jesus was born in Bethlehem of Judea, wise men*g* from the East came to
Jerusalem, 2 asking, "Where is the child
who has been born king of the Jews? For
we observed his star at its rising,*h* and
have come to pay him homage." 3 When
King Herod heard this, he was frightened, and all Jerusalem with him; 4 and
calling together all the chief priests and
scribes of the people, he inquired of them
where the Messiah*i* was to be born.
5 They told him, "In Bethlehem of Judea;
for so it has been written by the prophet:
6 'And you, Bethlehem, in the land
    of Judah,
  are by no means least among
    the rulers of Judah;
  for from you shall come a ruler
  who is to shepherd*j* my
    people Israel.' "
7 Then Herod secretly called for the
wise men*g* and learned from them the
exact time when the star had appeared.
8 Then he sent them to Bethlehem, saying, "Go and search diligently for the
child; and when you have found him,
bring me word so that I may also go and
pay him homage." 9 When they had
heard the king, they set out; and there,
ahead of them, went the star that they
had seen at its rising,*h* until it stopped
over the place where the child was.
10 When they saw that the star had

stopped,*k* they were overwhelmed with
joy. 11 On entering the house, they saw
the child with Mary his mother; and they
knelt down and paid him homage. Then,
opening their treasure chests, they offered him gifts of gold, frankincense, and
myrrh. 12 And having been warned in a
dream not to return to Herod, they left
for their own country by another road.

13 Now after they had left, an angel
of the Lord appeared to Joseph in a dream
and said, "Get up, take the child and his
mother, and flee to Egypt, and remain
there until I tell you; for Herod is about
to search for the child, to destroy him."
14 Then Joseph*l* got up, took the child
and his mother by night, and went to
Egypt, 15 and remained there until the
death of Herod. This was to fulfill what
had been spoken by the Lord through the
prophet, "Out of Egypt I have called
my son."

16 When Herod saw that he had been
tricked by the wise men,*g* he was infuriated, and he sent and killed all the children in and around Bethlehem who were
two years old or under, according to
the time that he had learned from the
wise men.*g* 17 Then was fulfilled what
had been spoken through the prophet
Jeremiah:
18 "A voice was heard in Ramah,
    wailing and loud lamentation,
  Rachel weeping for her children;
    she refused to be consoled,
      because they are no more."
19 When Herod died, an angel of the
Lord suddenly appeared in a dream to
Joseph in Egypt and said, 20 "Get up, take

f Other ancient authorities read *her firstborn son*
g Or *astrologers*; Gk *magi*   *h* Or *in the East*
i Or *the Christ*   *j* Or *rule*   *k* Gk *saw the star*
l Gk *he*

2.1–12: The wise men (Magi). **1:** The *wise
men,* a learned class in ancient Persia. **2:** Jer
23.5; Num 24.17. **3:** Herod's fears were
aroused that his own children might be excluded from the throne. **5:** Jn 7.42. **6:** Mic 5.2.
11: *Frankincense and myrrh,* aromatic gum re-
sins obtained from shrubs found in tropical
countries of the East.
2.13–23: Escape to Egypt and return.

15: *Out of Egypt . . . ,* a quotation from Hos
11.1, where the reference is to Israel (compare
Ex 4.22). **18:** Quoted from Jer 31.15. *Rachel,*
wife of Jacob, died in childbirth and according to Gen 35.16–20 was buried near Bethlehem. *Ramah,* north of Jerusalem, was the
scene of national grief (Jer 40.1) inflicted by
an enemy.
2.19: *Herod* the Great *died* early in 4 B.C. **22:**

the child and his mother, and go to the land of Israel, for those who were seeking the child's life are dead." 21 Then Joseph*m* got up, took the child and his mother, and went to the land of Israel. 22 But when he heard that Archelaus was ruling over Judea in place of his father Herod, he was afraid to go there. And after being warned in a dream, he went away to the district of Galilee. 23 There he made his home in a town called Nazareth, so that what had been spoken through the prophets might be fulfilled, "He will be called a Nazorean."

3 In those days John the Baptist appeared in the wilderness of Judea, proclaiming, 2 "Repent, for the kingdom of heaven has come near."*n* 3 This is the one of whom the prophet Isaiah spoke when he said,

"The voice of one crying out in
the wilderness:
'Prepare the way of the Lord,
make his paths straight.' "

4 Now John wore clothing of camel's hair with a leather belt around his waist, and his food was locusts and wild honey. 5 Then the people of Jerusalem and all Judea were going out to him, and all the region along the Jordan, 6 and they were baptized by him in the river Jordan, confessing their sins.

7 But when he saw many Pharisees and Sadducees coming for baptism, he said to them, "You brood of vipers! Who warned you to flee from the wrath to come? 8 Bear fruit worthy of repentance. 9 Do not presume to say to yourselves, 'We have Abraham as our ancestor'; for I tell you, God is able from these stones to raise up children to Abraham. 10 Even now the ax is lying at the root of the trees; every tree therefore that does not bear good fruit is cut down and thrown into the fire.

11 "I baptize you with*o* water for repentance, but one who is more powerful than I is coming after me; I am not worthy to carry his sandals. He will baptize you with*o* the Holy Spirit and fire. 12 His winnowing fork is in his hand, and he will clear his threshing floor and will gather his wheat into the granary; but the chaff he will burn with unquenchable fire."

13 Then Jesus came from Galilee to John at the Jordan, to be baptized by

m Gk *he*   *n* Or *is at hand*   *o* Or *in*

Archelaus, who was almost as cruel as his father Herod, reigned from 4 B.C. to A.D. 6 and was replaced by a Roman procurator. **23:** There is a similarity in sound and possibly in meaning between the Aramaic word for *Nazareth* and the Hebrew word translated *branch* (Isa 11.1). *Nazareth* was situated in a secluded valley in lower Galilee, a little north of the Esdraelon plain.

3.1–12: Activity of John the Baptist (Mk 1.1–8; Lk 3.1–18; Jn 1.6–8, 19–28). **1:** *John* resembled the Old Testament prophets (compare v. 4 with 2 Kings 1.8; Zech 13.4). Christian faith understood him to fulfill Isa 40.3; Mal 3.1; 4.5 (see 3.3; 17.10–12). His influence outside Christianity is attested by Acts 18.25; 19.1–7. *Those days,* namely, when Jesus began his public life. *The wilderness of Judea* lay east and southeast of Jerusalem. **2:** *Repent,* literally "return," meant to come back to the way of life charted by the covenant between God and Israel (Ex 19.3–6; 24.3–8; Jer 31.31–34). *The kingdom,* see 4.17 n. **3:** Isa 40.3. **6:** See Mk 1.4 n.

3.7: *Pharisees* and *Sadducees* formed two major divisions among the Jews (for differences between them, see 22.23 n. and Acts 23.6–10). A third Jewish sect in Palestine was the Essenes (see Josephus, *B.J.* II. viii. 2–13); their beliefs and practices are reflected in the Dead Sea Scrolls found at Qumran. *The wrath to come,* God's judgment (1 Thess 1.10). **8–10:** See Lk 3.7–9 n.; Jn 8.33. **11–12:** See Lk 12.49 n.; Acts 2.17–21; 19.1–7; 18.24–26.

3.13–17: Jesus' baptism (Mk 1.9–11; Lk 3.21–22; Jn 1.31–34). **13–15:** Jesus recognized John's authority and identified himself with those who responded in faith to John's call. **16–17:** A description of the surge of certainty and self-understanding that came to Jesus at his baptism. The language, akin to Old Testament speech, portrays a spiritual experience which words cannot adequately describe. *My Son, the Beloved,* see Mk 1.11 n.

4.1–11: Jesus' temptation (Mk 1.12–13; Lk 4.1–13; Heb 2.18; 4.15). The accounts illustrate Jesus' habitual refusal to allow his

him. 14John would have prevented him, saying, "I need to be baptized by you, and do you come to me?" 15But Jesus answered him, "Let it be so now; for it is proper for us in this way to fulfill all righteousness." Then he consented. 16And when Jesus had been baptized, just as he came up from the water, suddenly the heavens were opened to him and he saw the Spirit of God descending like a dove and alighting on him. 17And a voice from heaven said, "This is my Son, the Beloved,p with whom I am well pleased."

4 Then Jesus was led up by the Spirit into the wilderness to be tempted by the devil. 2He fasted forty days and forty nights, and afterwards he was famished. 3The tempter came and said to him, "If you are the Son of God, command these stones to become loaves of bread." 4But he answered, "It is written,

'One does not live by bread alone,
    but by every word that comes
        from the mouth of God.' "

5 Then the devil took him to the holy city and placed him on the pinnacle of the temple, 6saying to him, "If you are the Son of God, throw yourself down; for it is written,

'He will command his angels
        concerning you,'
    and 'On their hands they will
        bear you up,
so that you will not dash your
        foot against a stone.' "

7Jesus said to him, "Again it is written,

'Do not put the Lord your God to the test.' "

8 Again, the devil took him to a very high mountain and showed him all the kingdoms of the world and their splendor; 9and he said to him, "All these I will give you, if you will fall down and worship me." 10Jesus said to him, "Away with you, Satan! for it is written,

'Worship the Lord your God,
    and serve only him.' "

11Then the devil left him, and suddenly angels came and waited on him.

12 Now when Jesusq heard that John had been arrested, he withdrew to Galilee. 13He left Nazareth and made his home in Capernaum by the sea, in the territory of Zebulun and Naphtali, 14so that what had been spoken through the prophet Isaiah might be fulfilled:

15 "Land of Zebulun, land of
        Naphtali,
    on the road by the sea, across
        the Jordan, Galilee of the
        Gentiles—
16 the people who sat in darkness
        have seen a great light,
    and for those who sat in the
            region and shadow of death
        light has dawned."

17From that time Jesus began to proclaim, "Repent, for the kingdom of heaven has come near."r

p Or *my beloved Son*   q Gk *he*   r Or *is at hand*

---

sense of mission to be influenced by concern for his safety or for merely practical interests. **1:** *The devil, tempter* (v. 3), and *Satan* (v. 10) are names for evil conceived as a personal will actively hostile to God (see Lk 13.11, 16 n.). **2:** *Forty,* compare Ex 34.28; 1 Kings 19.8. **3:** *If you are the Son of God;* but see the declaration in 3.17. **4:** Deut 8.3.

4.5: *The holy city,* Jerusalem. The *pinnacle* mentioned in the next clause most likely overlooked the temple courts and the deep valley of the Kidron below. **6:** Ps 91.11–12. **7:** Deut 6.16. **10:** Deut 6.13.

4.12–25: Beginnings of Jesus' activity in Galilee. 12–17: Mk 1.14–15; Lk 4.14–15.

13: *Capernaum,* on the northwest coast of the Sea of Galilee. **15–16:** Isa 9.1–2. **16:** *The people who sat in darkness,* those who suffered most from the Assyrian invasions. **17:** *From that time,* the arrest of John (v. 12). *The kingdom of heaven* is Matthew's usual way of expressing the equivalent phrase, "the kingdom of God," found in parallel accounts in the other gospels. In asserting that God's *kingdom has come near* Jesus meant that all God's past dealings with his creation were coming to climax and fruition. Jesus taught both the present reality of God's rule (Lk 10.18; 11.20; 17.21) and its future realization (Mt 6.10). See Mk 1.15 n.

18 As he walked by the Sea of Galilee, he saw two brothers, Simon, who is called Peter, and Andrew his brother, casting a net into the sea—for they were fishermen. 19 And he said to them, "Follow me, and I will make you fish for people." 20 Immediately they left their nets and followed him. 21 As he went from there, he saw two other brothers, James son of Zebedee and his brother John, in the boat with their father Zebedee, mending their nets, and he called them. 22 Immediately they left the boat and their father, and followed him.

23 Jesus^s went throughout Galilee, teaching in their synagogues and proclaiming the good news^t of the kingdom and curing every disease and every sickness among the people. 24 So his fame spread throughout all Syria, and they brought to him all the sick, those who were afflicted with various diseases and pains, demoniacs, epileptics, and paralytics, and he cured them. 25 And great crowds followed him from Galilee, the Decapolis, Jerusalem, Judea, and from beyond the Jordan.

5 When Jesus^u saw the crowds, he went up the mountain; and after he sat down, his disciples came to him. 2 Then he began to speak, and taught them, saying:

3 "Blessed are the poor in spirit, for theirs is the kingdom of heaven.

4 "Blessed are those who mourn, for they will be comforted.

5 "Blessed are the meek, for they will inherit the earth.

6 "Blessed are those who hunger and thirst for righteousness, for they will be filled.

7 "Blessed are the merciful, for they will receive mercy.

8 "Blessed are the pure in heart, for they will see God.

9 "Blessed are the peacemakers, for they will be called children of God.

10 "Blessed are those who are persecuted for righteousness' sake, for theirs is the kingdom of heaven.

11 "Blessed are you when people revile you and persecute you and utter all kinds of evil against you falsely^v on my account. 12 Rejoice and be glad, for your reward is great in heaven, for in the same way they persecuted the prophets who were before you.

13 "You are the salt of the earth; but if salt has lost its taste, how can its saltiness be restored? It is no longer good for anything, but is thrown out and trampled under foot.

14 "You are the light of the world. A city built on a hill cannot be hid. 15 No one after lighting a lamp puts it under the bushel basket, but on the lampstand, and

s Gk *He*   t Gk *gospel*   u Gk *he*   v Other ancient authorities lack *falsely*

4.18–22: Mk 1.16–20; Lk 5.1–11; Jn 1.35–42. **24**: *Demoniacs,* persons controlled in body or will, or in both, by evil forces (Mt 8.16, 28; 9.32; 15.22; Mk 5.15; see Lk 13.11, 16 n.). *Demons,* see Lk 4.33 n. **25**: *Decapolis,* see Mk 5.20 n.
5.1–7.27: **The Sermon on the Mount** sounds the keynote of the new age that Jesus came to introduce. Internal analysis and comparison with Luke's Gospel suggest that the Evangelist (in accord with his habit of synthesis) has inserted into this account of the Sermon portions of Jesus' teaching given on other occasions. **1**: *He sat down,* the usual position of Jewish rabbis while teaching (compare Lk 4.20–21).
5.3–12: **The Beatitudes** (Lk 6.17, 20–23) proclaim God's favor toward those who aspire to live under his rule. **3**: *Poor in spirit,* those who feel a deep sense of spiritual poverty (Isa 66.2). **4**: *Comforted,* the word implies strengthening as well as consolation. **5.5**: Ps 37.11. **6**: Isa 55.1–2; Jn 4.14; 6.48–51. **7**: *Will receive mercy,* on the day of judgment. **8**: Purity of *heart* is single-mindedness or sincerity, freedom from mixed motives; it is not synonymous with chastity, but includes it (Ps 24.4; Heb 12.14). *See God,* 1 Cor 13.12; 1 Jn 3.2; Rev 22.4.
5.9: *Peacemakers* are not merely "peaceable," but those who work earnestly to "make" peace. *Will be called children of God,* will be acknowledged as such by God. **10**: 1 Pet 3.14; 4.14. **12**: 2 Chr 36.15–16; Mt 23.37; Acts 7.52.
5.13–16: **The witness of the disciples.** **13**: Mk 9.49–50; Lk 14.34–35. **14**: Phil 2.15; Jn 8.12. **15**: See Mk 4.21 n. **16**: 1 Pet 2.12.

it gives light to all in the house. 16 In the same way, let your light shine before others, so that they may see your good works and give glory to your Father in heaven.

17 "Do not think that I have come to abolish the law or the prophets; I have come not to abolish but to fulfill. 18 For truly I tell you, until heaven and earth pass away, not one letter,*w* not one stroke of a letter, will pass from the law until all is accomplished. 19 Therefore, whoever breaks*x* one of the least of these commandments, and teaches others to do the same, will be called least in the kingdom of heaven; but whoever does them and teaches them will be called great in the kingdom of heaven. 20 For I tell you, unless your righteousness exceeds that of the scribes and Pharisees, you will never enter the kingdom of heaven.

21 "You have heard that it was said to those of ancient times, 'You shall not murder'; and 'whoever murders shall be liable to judgment.' 22 But I say to you that if you are angry with a brother or sister,*y* you will be liable to judgment; and if you insult*z* a brother or sister,*a* you will be liable to the council; and if you say, 'You fool,' you will be liable to the hell*b* of fire. 23 So when you are offering your gift at the altar, if you remember that your brother or sister*c* has something against you, 24 leave your gift there before the altar and go; first be reconciled to your brother or sister,*c* and then come and offer your gift. 25 Come to

terms quickly with your accuser while you are on the way to court*d* with him, or your accuser may hand you over to the judge, and the judge to the guard, and you will be thrown into prison. 26 Truly I tell you, you will never get out until you have paid the last penny.

27 "You have heard that it was said, 'You shall not commit adultery.' 28 But I say to you that everyone who looks at a woman with lust has already committed adultery with her in his heart. 29 If your right eye causes you to sin, tear it out and throw it away; it is better for you to lose one of your members than for your whole body to be thrown into hell.*b* 30 And if your right hand causes you to sin, cut it off and throw it away; it is better for you to lose one of your members than for your whole body to go into hell.*b*

31 "It was also said, 'Whoever divorces his wife, let him give her a certificate of divorce.' 32 But I say to you that anyone who divorces his wife, except on the ground of unchastity, causes her to commit adultery; and whoever marries a divorced woman commits adultery.

33 "Again, you have heard that it was said to those of ancient times, 'You shall not swear falsely, but carry out the vows you have made to the Lord.' 34 But I say to you, Do not swear at all, either by

w Gk *one iota*   *x* Or *annuls*   *y* Gk *a brother*; other ancient authorities add *without cause*
z Gk *say Raca to* (an obscure term of abuse)
a Gk *a brother*   *b* Gk *Gehenna*   *c* Gk *your brother*   *d* Gk lacks *to court*

5.17–20: The relation of Jesus' message to the Jewish law was a great concern to followers with a Jewish background. **17:** *The prophets* in the Hebrew Scriptures comprise the books of Joshua, Judges, Samuel, Kings, Isaiah, Jeremiah, Ezekiel, and the twelve minor prophets (see Lk 24.27 n., 44 n.). Many Jews esteemed the prophets less than the law; hence the word *or* here. **18:** Mk 13.31; Lk 16.17. **19:** *Breaks,* or "sets aside." *Teaches,* Jas 3.1. **20:** *Righteousness,* one's acceptance of God's requirements and one's being accepted by God (Lk 18.10–14).
5.21–48: Illustrations of the true understanding of the Law. 21: The *judgment,*

a local Jewish court established in every town or city in accordance with the command in Deut 16.18. **22:** *Council,* the Sanhedrin, comprising seventy members. **25–26:** Lk 12.57–59. **26:** *Penny,* see Lk 12.59 n.
5.27: Ex 20.14; Deut 5.18. *Adultery* carried the death penalty (Lev 20.10; Deut 22.22). **29–30:** Mk 9.43–48; Mt 18.8–9. **31:** *It was also said,* Deut 24.1–4. **32:** The expression *except . . . unchastity* occurs also in 19.9; it is absent from the accounts in Mk 10.11–12 and Lk 16.18 (compare also Rom 7.2–3; 1 Cor 7.10–11).
5.33–37: Lev 19.12; Num 30.2; Deut 23.21; Mt 23.16–22; Jas 5.12. **35:** Isa 66.1.

heaven, for it is the throne of God, 35 or by the earth, for it is his footstool, or by Jerusalem, for it is the city of the great King. 36 And do not swear by your head, for you cannot make one hair white or black. 37 Let your word be 'Yes, Yes' or 'No, No'; anything more than this comes from the evil one. *e*

38 "You have heard that it was said, 'An eye for an eye and a tooth for a tooth.' 39 But I say to you, Do not resist an evildoer. But if anyone strikes you on the right cheek, turn the other also; 40 and if anyone wants to sue you and take your coat, give your cloak as well; 41 and if anyone forces you to go one mile, go also the second mile. 42 Give to everyone who begs from you, and do not refuse anyone who wants to borrow from you.

43 "You have heard that it was said, 'You shall love your neighbor and hate your enemy.' 44 But I say to you, Love your enemies and pray for those who persecute you, 45 so that you may be children of your Father in heaven; for he makes his sun rise on the evil and on the good, and sends rain on the righteous and on the unrighteous. 46 For if you love those who love you, what reward do you have? Do not even the tax collectors do the same? 47 And if you greet only your brothers and sisters, *f* what more are you doing than others? Do not even the Gentiles do the same? 48 Be perfect, therefore, as your heavenly Father is perfect.

6 "Beware of practicing your piety before others in order to be seen by them; for then you have no reward from your Father in heaven.

2 "So whenever you give alms, do not sound a trumpet before you, as the hypocrites do in the synagogues and in the streets, so that they may be praised by others. Truly I tell you, they have received their reward. 3 But when you give alms, do not let your left hand know what your right hand is doing, 4 so that your alms may be done in secret; and your Father who sees in secret will reward you. *g*

5 "And whenever you pray, do not be like the hypocrites; for they love to stand and pray in the synagogues and at the street corners, so that they may be seen by others. Truly I tell you, they have received their reward. 6 But whenever you pray, go into your room and shut the door and pray to your Father who is in secret; and your Father who sees in secret will reward you. *g*

7 "When you are praying, do not heap up empty phrases as the Gentiles do; for they think that they will be heard because of their many words. 8 Do not be like them, for your Father knows what you need before you ask him.

9 "Pray then in this way:

e Or *evil*   *f* Gk *your brothers*   *g* Other ancient authorities add *openly*

5.38: Ex 21.23–24; Lev 24.19–20; Deut 19.21. Though this principle controlled retaliation in primitive society, it did not justify it. **39–42:** Lk 6.29–30; Rom 12.17; 1 Cor 6.7; 1 Pet 2.19; 3.9. **40:** To give the *cloak,* a long outer garment, was a proof of greater self-denial than to give the *coat* (inner tunic; see 10.10 n.). **41:** Soldiers could compel civilians to carry their baggage; to go a *second mile* would relieve another from the burden. **44–48:** Lk 6.27–28, 32–36. **45:** To be *children of* God is to pattern attitudes after God's. The words *children of* commonly mean persons who show the quality named or trait of character implied (see 23.31 n.; Lk 6.35; 10.6; Jn 8.39–47). **48:** *Be perfect,* in love to all (Col 3.14; 1 Jn 4.19).

6.1–34: **Teachings in practical piety;**

Jesus emphasizes a sincere response to God that identifies oneself with his purposes. **1:** 23.5. **2:** Giving *alms* was considered by Judaism to be the foremost act of piety. *Have received their reward,* i.e. the praise of others.

6.5: *Have received their reward,* i.e. the acts done to win the applause of others have no reward beyond that. Lk 18.10–14.

6.9–13: The Lord's Prayer (compare Lk 11.2–4) falls into two parts: after the opening invocation, there are three petitions concerning God's glory, followed by those concerning our needs. The phrase, *on earth as it is in heaven* (v. 10), belongs to each of the first three petitions. On the basis of David's prayer (1 Chr 29.11–13) the early church added an appropriate concluding doxology (see note *j*). **9:** Isa 63.16; 64.8. **12:** *As we also,* i.e.

Our Father in heaven,
   hallowed be your name.
10   Your kingdom come.
Your will be done,
   on earth as it is in heaven.
11   Give us this day our daily
     bread. *h*
12   And forgive us our debts,
   as we also have forgiven
   our debtors.
13   And do not bring us to the time
   of trial, *i*
but rescue us from the
   evil one. *j*

14 For if you forgive others their trespasses, your heavenly Father will also forgive you; 15 but if you do not forgive others, neither will your Father forgive your trespasses.

16 "And whenever you fast, do not look dismal, like the hypocrites, for they disfigure their faces so as to show others that they are fasting. Truly I tell you, they have received their reward. 17 But when you fast, put oil on your head and wash your face, 18 so that your fasting may be seen not by others but by your Father who is in secret; and your Father who sees in secret will reward you. *k*

19 "Do not store up for yourselves treasures on earth, where moth and rust *l* consume and where thieves break in and steal; 20 but store up for yourselves treasures in heaven, where neither moth nor rust *l* consumes and where thieves do not break in and steal. 21 For where your treasure is, there your heart will be also.

22 "The eye is the lamp of the body. So, if your eye is healthy, your whole body will be full of light; 23 but if your eye is unhealthy, your whole body will be full of darkness. If then the light in you is darkness, how great is the darkness!

24 "No one can serve two masters; for a slave will either hate the one and love the other, or be devoted to the one and despise the other. You cannot serve God and wealth. *m*

25 "Therefore I tell you, do not worry about your life, what you will eat or what you will drink, *n* or about your body, what you will wear. Is not life more than food, and the body more than clothing? 26 Look at the birds of the air; they neither sow nor reap nor gather into barns, and yet your heavenly Father feeds them. Are you not of more value than they? 27 And can any of you by worrying add a single hour to your span of life? *o* 28 And why do you worry about clothing? Consider the lilies of the field, how they grow; they neither toil nor spin, 29 yet I tell you, even Solomon in all his glory was not clothed like one of these. 30 But if God so clothes the grass of the field, which is alive today and tomorrow is thrown into the oven, will he not much more clothe you—you of little faith? 31 Therefore do not worry, saying, 'What will we eat?' or 'What will we drink?' or 'What will we wear?' 32 For it is the Gentiles who strive for all these things; and indeed your heavenly Father knows that you need all these things. 33 But strive first for the kingdom of God *p* and his *q* righteousness, and all these things will be given to you as well.

34 "So do not worry about tomorrow, for tomorrow will bring worries of

h Or *our bread for tomorrow*   *i* Or *us into temptation*   *j* Or *from evil.* Other ancient authorities add, in some form, *For the kingdom and the power and the glory are yours forever. Amen.* *k* Other ancient authorities add *openly* *l* Gk *eating*   *m* Gk *mammon*   *n* Other ancient authorities lack *or what you will drink* *o* Or *add one cubit to your height*   *p* Other ancient authorities lack *of God*   *q* Or *its*

---

we cannot ask for ourselves what we deny to others. **13:** 2 Thess 3.3; Jas 1.13. **14–15:** 18.35; Mk 11.25–26; Eph 4.32; Col 3.13.

   6.16–18: Acceptable fasting (Isa 58.5). Especially pious Jews used to fast twice a week (see Lk 18.12 n.). **19–21:** The uselessness of trusting in worldly goods (Jas 5.2–3). *Moth,* in antiquity a large part of riches often con-

sisted of costly garments especially liable to destruction by moths.
   6.22–23: Lk 11.34–36. **24:** Lk 16.13.
   6.25–33: Lk 12.22–31. **25:** Lk 10.41; 12.11; Phil 4.6. **29:** 1 Kings 10.4–7. **30:** *You of little faith* are unwilling to rest in the assurance that God cares about your lives (8.26; 14.31; 16.8). **33:** Mk 10.29–30; Lk 18.29–30.

its own. Today's trouble is enough for today.

7 "Do not judge, so that you may not be judged. ²For with the judgment you make you will be judged, and the measure you give will be the measure you get. ³Why do you see the speck in your neighbor'sʳ eye, but do not notice the log in your own eye? ⁴Or how can you say to your neighbor,ˢ 'Let me take the speck out of your eye,' while the log is in your own eye? ⁵You hypocrite, first take the log out of your own eye, and then you will see clearly to take the speck out of your neighbor'sʳ eye.

6 "Do not give what is holy to dogs; and do not throw your pearls before swine, or they will trample them under foot and turn and maul you.

7 "Ask, and it will be given you; search, and you will find; knock, and the door will be opened for you. ⁸For everyone who asks receives, and everyone who searches finds, and for everyone who knocks, the door will be opened. ⁹Is there anyone among you who, if your child asks for bread, will give a stone? ¹⁰Or if the child asks for a fish, will give a snake? ¹¹If you then, who are evil, know how to give good gifts to your children, how much more will your Father in heaven give good things to those who ask him!

12 "In everything do to others as you would have them do to you; for this is the law and the prophets.

13 "Enter through the narrow gate; for the gate is wide and the road is easyᵗ that leads to destruction, and there are many who take it. ¹⁴For the gate is narrow and the road is hard that leads to life, and there are few who find it.

15 "Beware of false prophets, who come to you in sheep's clothing but inwardly are ravenous wolves. ¹⁶You will know them by their fruits. Are grapes gathered from thorns, or figs from thistles? ¹⁷In the same way, every good tree bears good fruit, but the bad tree bears bad fruit. ¹⁸A good tree cannot bear bad fruit, nor can a bad tree bear good fruit. ¹⁹Every tree that does not bear good fruit is cut down and thrown into the fire. ²⁰Thus you will know them by their fruits.

21 "Not everyone who says to me, 'Lord, Lord,' will enter the kingdom of heaven, but only the one who does the will of my Father in heaven. ²²On that day many will say to me, 'Lord, Lord, did we not prophesy in your name, and cast out demons in your name, and do many deeds of power in your name?' ²³Then I will declare to them, 'I never knew you; go away from me, you evildoers.'

24 "Everyone then who hears these words of mine and acts on them will be like a wise man who built his house on rock. ²⁵The rain fell, the floods came, and the winds blew and beat on that

r Gk *brother's*   s Gk *brother*   t Other ancient authorities read *for the road is wide and easy*

7.1–29: Illustrations of the practical meaning of Jesus' message. 1–5: Judgment of others (Lk 6.37–38, 41–42; Mk 4.24; Rom 2.1; 14.10). **6**: Concerning discrimination in judging. *What is holy,* i.e. the flesh of Jewish sacrifices. **7–11**: Encouragement to prayer (6.8; Mk 11.23–24; Jn 15.7; 1 Jn 3.22; 5.14). **9–10**: That is, a round *stone* like a loaf of *bread,* and a dried *fish* resembling a *snake.*
7.12: Others had formulated a negative Golden Rule that counsels inaction ("Do not do to others . . ."); the positive form here requires active contribution to the welfare and happiness of others. Lk 6.31; Mt 22.39–40; Rom 13.8–10. **13–14**: Lk 13.23–24; Jer 21.8; Ps 1; Deut 30.19; Jn 10.7; 14.6.

7.15–20: Lk 6.43–45. **15**: 24.11, 24; Ezek 22.27; 1 Jn 4.1; Jn 10.12. *Sheep* often symbolize a group of followers in a religious sense (Ezek 34.1–24; Lk 12.32). **16**: 3.8; 12.33–35; Lk 6.43–45. **19**: 3.10; Lk 13.6–9; Jas 3.10–12. **22**: *That day,* the day of judgment. Jesus speaks as the divine judge.
7.24–27: Lk 6.47–49; Jas 1.22–25. **28**: *When Jesus had finished saying these things,* this (or a similar) formula marks the conclusion of each of the five main discourses in the gospel (see Introduction and 11.1; 13.53; 19.1; 26.1). **29**: *Not as their scribes,* Jesus speaks on his own responsibility without appeal to traditional authority (Mk 1.22; 11.18; Lk 4.32).

house, but it did not fall, because it had been founded on rock. 26 And everyone who hears these words of mine and does not act on them will be like a foolish man who built his house on sand. 27 The rain fell, and the floods came, and the winds blew and beat against that house, and it fell—and great was its fall!"

28 Now when Jesus had finished saying these things, the crowds were astounded at his teaching, 29 for he taught them as one having authority, and not as their scribes.

8 When Jesus" had come down from the mountain, great crowds followed him; 2 and there was a leper" who came to him and knelt before him, saying, "Lord, if you choose, you can make me clean." 3 He stretched out his hand and touched him, saying, "I do choose. Be made clean!" Immediately his leprosy" was cleansed. 4 Then Jesus said to him, "See that you say nothing to anyone; but go, show yourself to the priest, and offer the gift that Moses commanded, as a testimony to them."

5 When he entered Capernaum, a centurion came to him, appealing to him 6 and saying, "Lord, my servant is lying at home paralyzed, in terrible distress." 7 And he said to him, "I will come and cure him." 8 The centurion answered, "Lord, I am not worthy to have you come under my roof; but only speak the word, and my servant will be healed. 9 For I also am a man under authority, with soldiers under me; and I say to one, 'Go,' and he goes, and to another, 'Come,' and he comes, and to my slave,

'Do this,' and the slave does it." 10 When Jesus heard him, he was amazed and said to those who followed him, "Truly I tell you, in no one" in Israel have I found such faith. 11 I tell you, many will come from east and west and will eat with Abraham and Isaac and Jacob in the kingdom of heaven, 12 while the heirs of the kingdom will be thrown into the outer darkness, where there will be weeping and gnashing of teeth." 13 And to the centurion Jesus said, "Go; let it be done for you according to your faith." And the servant was healed in that hour.

14 When Jesus entered Peter's house, he saw his mother-in-law lying in bed with a fever; 15 he touched her hand, and the fever left her, and she got up and began to serve him. 16 That evening they brought to him many who were possessed with demons; and he cast out the spirits with a word, and cured all who were sick. 17 This was to fulfill what had been spoken through the prophet Isaiah, "He took our infirmities and bore our diseases."

18 Now when Jesus saw great crowds around him, he gave orders to go over to the other side. 19 A scribe then approached and said, "Teacher, I will follow you wherever you go." 20 And Jesus said to him, "Foxes have holes, and birds of the air have nests; but the Son of Man has nowhere to lay his head." 21 Another of his disciples said to him, "Lord, first let me go and bury my father." 22 But

---

u Gk *he*   *v* The terms *leper* and *leprosy* can refer to several diseases   *w* Other ancient authorities read *Truly I tell you, not even*

8.1–9.38: Events in Galilee. 8.2–4: Mk 1.40–44; Lk 5.12–14. *Leprosy,* a skin disorder of an uncertain nature. Several diseases were possibly referred to by this name (see Lev 13.1–59 n.; Num 5.1–4). Its presence excluded the sufferer from associating with others. *Make me clean,* the leper seeks not merely healing but the freedom to rejoin the Jewish community. **4:** Lev 14.2–32. **5–13:** Lk 7.1–10; Jn 4.46–53. The *centurion,* a non-Jewish military officer in command of from fifty to one hundred soldiers, is convinced that diseases are as obedient to Jesus as soldiers are to

him. **10:** *Faith* refers to the centurion's trust and recognition of Jesus' power (v. 13; Mk 11.23 n., 24 n.). **11–12:** See Lk 14.15 n.; Isa 49.12; 59.19; Mt 13.42, 50; 22.13; 24.51; 25.30.

8.14–17: Mk 1.29–34; Lk 4.38–41. **16:** *Demons,* see 4.24 n.; 12.22 n; Lk 4.33 n.; 7.33 n.; 13.16 n. **17:** Isa 53.4.

8.18–22: Mk 4.35; Lk 8.22; 9.57–60. **18:** *The other side,* the eastern shore of the Sea of Galilee. **20:** *Son of Man,* see Mk 2.10 n. **22:** *Follow me,* Jesus implies that obedience to his call must take precedence over every other

Jesus said to him, "Follow me, and let the dead bury their own dead."

23 And when he got into the boat, his disciples followed him. 24 A windstorm arose on the sea, so great that the boat was being swamped by the waves; but he was asleep. 25 And they went and woke him up, saying, "Lord, save us! We are perishing!" 26 And he said to them, "Why are you afraid, you of little faith?" Then he got up and rebuked the winds and the sea; and there was a dead calm. 27 They were amazed, saying, "What sort of man is this, that even the winds and the sea obey him?"

28 When he came to the other side, to the country of the Gadarenes, *x* two demoniacs coming out of the tombs met him. They were so fierce that no one could pass that way. 29 Suddenly they shouted, "What have you to do with us, Son of God? Have you come here to torment us before the time?" 30 Now a large herd of swine was feeding at some distance from them. 31 The demons begged him, "If you cast us out, send us into the herd of swine." 32 And he said to them, "Go!" So they came out and entered the swine; and suddenly, the whole herd rushed down the steep bank into the sea and perished in the water. 33 The swineherds ran off, and on going into the town, they told the whole story about what had happened to the demoniacs. 34 Then the whole town came out to meet Jesus; and when they saw him, they begged him to leave their neighborhood.

9 ¹ And after getting into a boat he crossed the sea and came to his own town.

2 And just then some people were carrying a paralyzed man lying on a bed.

When Jesus saw their faith, he said to the paralytic, "Take heart, son; your sins are forgiven." ³ Then some of the scribes said to themselves, "This man is blaspheming." 4 But Jesus, perceiving their thoughts, said, "Why do you think evil in your hearts? 5 For which is easier, to say, 'Your sins are forgiven,' or to say, 'Stand up and walk'? 6 But so that you may know that the Son of Man has authority on earth to forgive sins"—he then said to the paralytic—"Stand up, take your bed and go to your home." 7 And he stood up and went to his home. 8 When the crowds saw it, they were filled with awe, and they glorified God, who had given such authority to human beings.

9 As Jesus was walking along, he saw a man called Matthew sitting at the tax booth; and he said to him, "Follow me." And he got up and followed him.

10 And as he sat at dinner *y* in the house, many tax collectors and sinners came and were sitting *z* with him and his disciples. 11 When the Pharisees saw this, they said to his disciples, "Why does your teacher eat with tax collectors and sinners?" 12 But when he heard this, he said, "Those who are well have no need of a physician, but those who are sick. 13 Go and learn what this means, 'I desire mercy, not sacrifice.' For I have come to call not the righteous but sinners."

14 Then the disciples of John came to him, saying, "Why do we and the Pharisees fast often, *a* but your disciples do not

---

x Other ancient authorities read *Gergesenes*; others, *Gerasenes*   *y* Gk *reclined*   *z* Gk *were reclining*   *a* Other ancient authorities lack *often*

---

duty or love (compare 10.37). *Let the dead,* i.e. the spiritually dead, who are not alive to the greater demands of the kingdom of God.
8.23–27: Mk 4.36–41; Lk 8.22–24. **25:** See Lk 8.24 n.
8.28–34: Mk 5.1–20; Lk 8.26–39. *Gadarenes,* the inhabitants of the city, or of the surrounding district, of Gadara, the capital of Perea. **31:** See v. 16 n. and Mk 5.13 n.
9.1–8: Healing a paralytic (Mk 2.1–12;

Lk 5.17–26). **1:** *His own town,* Capernaum. **8:** 7.28–29.
   9.9–13: Mk 2.13–17; Lk 5.27–32. **10:** *The house,* presumably Matthew's house (see v. 9). Lk 7.34; 15.1–2. **13:** Hos 6.6; Mt 12.7; 15.2–6. Jesus uses a biblical quotation to challenge a conventional religious idea (see Lk 5.32 n.).
   9.14–17: Mk 2.18–22; Lk 5.33–39. **15:** Jesus recognizes the principle of fasting, but

fast?" 15 And Jesus said to them, "The wedding guests cannot mourn as long as the bridegroom is with them, can they? The days will come when the bridegroom is taken away from them, and then they will fast. 16 No one sews a piece of unshrunk cloth on an old cloak, for the patch pulls away from the cloak, and a worse tear is made. 17 Neither is new wine put into old wineskins; otherwise, the skins burst, and the wine is spilled, and the skins are destroyed; but new wine is put into fresh wineskins, and so both are preserved."

18 While he was saying these things to them, suddenly a leader of the synagogue*b* came in and knelt before him, saying, "My daughter has just died; but come and lay your hand on her, and she will live." 19 And Jesus got up and followed him, with his disciples. 20 Then suddenly a woman who had been suffering from hemorrhages for twelve years came up behind him and touched the fringe of his cloak, 21 for she said to herself, "If I only touch his cloak, I will be made well." 22 Jesus turned, and seeing her he said, "Take heart, daughter; your faith has made you well." And instantly the woman was made well. 23 When Jesus came to the leader's house and saw the flute players and the crowd making a commotion, 24 he said, "Go away; for the girl is not dead but sleeping." And they laughed at him. 25 But when the crowd had been put outside, he went in and took her by the hand, and the girl got up. 26 And the report of this spread throughout that district.

27 As Jesus went on from there, two blind men followed him, crying loudly, "Have mercy on us, Son of David!" 28 When he entered the house, the blind men came to him; and Jesus said to them, "Do you believe that I am able to do this?" They said to him, "Yes, Lord." 29 Then he touched their eyes and said, "According to your faith let it be done to you." 30 And their eyes were opened. Then Jesus sternly ordered them, "See that no one knows of this." 31 But they went away and spread the news about him throughout that district.

32 After they had gone away, a demoniac who was mute was brought to him. 33 And when the demon had been cast out, the one who had been mute spoke; and the crowds were amazed and said, "Never has anything like this been seen in Israel." 34 But the Pharisees said, "By the ruler of the demons he casts out the demons."*c*

35 Then Jesus went about all the cities and villages, teaching in their synagogues, and proclaiming the good news of the kingdom, and curing every disease and every sickness. 36 When he saw the crowds, he had compassion for them, because they were harassed and helpless, like sheep without a shepherd. 37 Then he said to his disciples, "The harvest is plentiful, but the laborers are few; 38 therefore ask the Lord of the harvest to send out laborers into his harvest."

10 Then Jesus*d* summoned his twelve disciples and gave them

b Gk lacks *of the synagogue*   *c* Other ancient authorities lack this verse   *d* Gk *he*

---

denies that it fits the circumstances of his life. **16–17:** The two pictorial sayings defend the practices of John's disciples and the practices of his own disciples; Jesus insists that the two ways should not be joined.

9.18–26: Mk 5.21–43; Lk 8.40–56. **21:** The Greek word here translated *be made well* (also v. 22; Mk 5.23, 28, 34; 10.52; Lk 8.36, 48, 50; 17.19; 18.42) carries with it the idea of rescue from impending destruction or from a superior power. **22:** Mk 11.23 n., 24 n. **23:** Jer 9.17–18. *Flute players,* hired mourners. **24:** Jesus speaks in the perspective of the kingdom of God in which physical death is not finally destructive of a person's existence but is a temporary cessation of personal activity (and analogous to sleeping). Verse 18 and the crowd's attitude clearly assert the fact of physical death.

9.27–31: 20.29–34. **29:** 9.22 n. **30:** 8.4. **32–34:** 12.22–24; Lk 11.14–15. **34:** See 12.24 n.; Mk 3.22 n.; Jn 7.20.

9.35–38: 4.23–25. **36:** Mk 6.34; Mt 14.14; 15.32; Num 27.17; Ezek 34.1–6; Zech 10.2.

10.1–11.1: Commissioning and instruction of the Twelve. 10.1–4: Mk 6.7; 3.13–

authority over unclean spirits, to cast them out, and to cure every disease and every sickness. 2These are the names of the twelve apostles: first, Simon, also known as Peter, and his brother Andrew; James son of Zebedee, and his brother John; 3Philip and Bartholomew; Thomas and Matthew the tax collector; James son of Alphaeus, and Thaddaeus;*e* 4Simon the Cananaean, and Judas Iscariot, the one who betrayed him.

5 These twelve Jesus sent out with the following instructions: "Go nowhere among the Gentiles, and enter no town of the Samaritans, 6but go rather to the lost sheep of the house of Israel. 7As you go, proclaim the good news, 'The kingdom of heaven has come near.'*f* 8Cure the sick, raise the dead, cleanse the lepers,*g* cast out demons. You received without payment; give without payment. 9Take no gold, or silver, or copper in your belts, 10no bag for your journey, or two tunics, or sandals, or a staff; for laborers deserve their food. 11Whatever town or village you enter, find out who in it is worthy, and stay there until you leave. 12As you enter the house, greet it. 13If the house is worthy, let your peace come upon it; but if it is not worthy, let your peace return to you. 14If anyone will not welcome you or listen to your words, shake off the dust from your feet as you leave that house or town. 15Truly I tell you, it will be more tolerable for the land of Sodom and Gomorrah on the day of judgment than for that town.

16 "See, I am sending you out like sheep into the midst of wolves; so be wise as serpents and innocent as doves.

17Beware of them, for they will hand you over to councils and flog you in their synagogues; 18and you will be dragged before governors and kings because of me, as a testimony to them and the Gentiles. 19When they hand you over, do not worry about how you are to speak or what you are to say; for what you are to say will be given to you at that time; 20for it is not you who speak, but the Spirit of your Father speaking through you. 21Brother will betray brother to death, and a father his child, and children will rise against parents and have them put to death; 22and you will be hated by all because of my name. But the one who endures to the end will be saved. 23When they persecute you in one town, flee to the next; for truly I tell you, you will not have gone through all the towns of Israel before the Son of Man comes.

24 "A disciple is not above the teacher, nor a slave above the master; 25it is enough for the disciple to be like the teacher, and the slave like the master. If they have called the master of the house Beelzebul, how much more will they malign those of his household!

26 "So have no fear of them; for nothing is covered up that will not be uncovered, and nothing secret that will not become known. 27What I say to you in the dark, tell in the light; and what you hear whispered, proclaim from the housetops. 28Do not fear those who kill the body but cannot kill the soul; rather fear

e Other ancient authorities read *Lebbaeus,* or *Lebbaeus called Thaddaeus* *f* Or *is at hand* *g* The terms *leper* and *leprosy* can refer to several diseases

---

19; Lk 9.1; 6.12–16. **1**: *Unclean spirits,* see Mk 1.23 n.

10.5–15: Mk 6.8–11; Lk 9.2–5; 10.3–12. **5**: 15.21–28; Lk 9.52; Jn 4.9. **6**: 15.24. **7**: The primary message. Through acceptance, or at least openness to this message and its bearer, healing would follow (see 4.17 n.; 4.23; 9.21, 35). **9**: Lk 22.35–36. **10**: *Tunic,* a short-sleeved garment of knee-length, held in at the waist by a girdle (Mk 1.6). *Deserve,* 1 Cor 9.14. **12**: *Greet it,* the usual form was, "Peace be to this house." **15**: Life and death depend on one's response to God's kingdom. *Sodom and Go-*

morrah illustrate God's judgment on wickedness (Gen 18.16–33; ch 19).

10.16–25: 24.9, 13; Mk 13.9–13; Lk 21.12–17, 19. **20**: Jn 16.7–11. **21**: 10.35–36; Lk 12.52–53. **22**: *Because of my name,* "because of me and my cause." **23**: The words stress the urgency of the disciples' task. **25**: Lk 6.40; Jn 13.16; 15.20; Mt 9.34; 12.24; Mk 3.22.

10.26–33: Lk 12.2–9. **27**: *Proclaim from the housetops,* unlike the Essenes, Jesus was opposed to secret doctrines revealed only to initiates. **28**: Heb 10.31. **29–33**: 6.26–33. **29**: See Lk 12.6 n. **31**: 12.12.

him who can destroy both soul and body in hell. *h* 29 Are not two sparrows sold for a penny? Yet not one of them will fall to the ground apart from your Father. 30 And even the hairs of your head are all counted. 31 So do not be afraid; you are of more value than many sparrows.

32 "Everyone therefore who acknowledges me before others, I also will acknowledge before my Father in heaven; 33 but whoever denies me before others, I also will deny before my Father in heaven.

34 "Do not think that I have come to bring peace to the earth; I have not come to bring peace, but a sword.

35 For I have come to set a man
against his father,
and a daughter against her mother,
and a daughter-in-law against her
mother-in-law;
36 and one's foes will be members of
one's own household.

37 Whoever loves father or mother more than me is not worthy of me; and whoever loves son or daughter more than me is not worthy of me; 38 and whoever does not take up the cross and follow me is not worthy of me. 39 Those who find their life will lose it, and those who lose their life for my sake will find it.

40 "Whoever welcomes you welcomes me, and whoever welcomes me welcomes the one who sent me. 41 Whoever welcomes a prophet in the name of a prophet will receive a prophet's reward; and whoever welcomes a righteous person in the name of a righteous person will receive the reward of the righteous; 42 and whoever gives even a cup of cold water to one of these little ones in the name of a disciple—truly I tell you, none of these will lose their reward."

11 Now when Jesus had finished instructing his twelve disciples, he went on from there to teach and proclaim his message in their cities.

2 When John heard in prison what the Messiah *i* was doing, he sent word by his *j* disciples 3 and said to him, "Are you the one who is to come, or are we to wait for another?" 4 Jesus answered them, "Go and tell John what you hear and see: 5 the blind receive their sight, the lame walk, the lepers *k* are cleansed, the deaf hear, the dead are raised, and the poor have good news brought to them. 6 And blessed is anyone who takes no offense at me."

7 As they went away, Jesus began to speak to the crowds about John: "What did you go out into the wilderness to look at? A reed shaken by the wind? 8 What then did you go out to see? Someone *l* dressed in soft robes? Look, those who wear soft robes are in royal palaces. 9 What then did you go out to see? A

h Gk *Gehenna*   i Or *the Christ*   j Other ancient authorities read *two of his*   k The terms *leper* and *leprosy* can refer to several diseases   l Or *Why then did you go out? To see someone*

10.32–33: Jesus claims to mediate God's will; a favorable response to him is a response to God (compare vv. 40–42).
10.34–36: Lk 12.51–53. 35: Mic 7.6. 37–39: 16.24–25; Mk 8.34–35; Lk 9.23–24; 14.26–27; 17.33. 37: Compare the stronger form of expression in Lk 14.26. 38: A *cross*, a Roman means of execution, was carried by the condemned to the scene of death. Jesus sees that the acceptance of his message with its promise also brings seeming destruction (v. 34). Only those who in faith accept the threat of destruction will find life (v. 39; 5.11–12; 16.24; Mk 8.34–35; 10.29–31; Lk 9.24–25, 14.27; 17.33; Jn 12.25). 41: In the *name of a prophet*, out of respect for the office and work of a prophet. 42: *Little ones*, see 18.6 n. 11.1: *Finished*, see 7.28 n.
11.2–12.50: **Narratives illustrating the authority claimed by Jesus. 11.2–19**: Jesus and John (Lk 7.18–35; 16.16). 2: *In prison*, at Machaerus, a fortified place about five miles east of the Dead Sea. 4–5: Jesus performs the works of the predicted Messiah (Isa 29.18–19; 35.5–6; 61.1; compare Lk 4.18–19). 6: Jesus invites John to answer his own question, basing his decision on what he hears of Jesus' activities interpreted in comparison with Isaiah's words (compare Lk 4.17–21).
11.7–15: John was important because he

prophet?*m* Yes, I tell you, and more than a prophet. [10]This is the one about whom it is written,

'See, I am sending my messenger
　　ahead of you,
　who will prepare your way
　　before you.'

[11]Truly I tell you, among those born of women no one has arisen greater than John the Baptist; yet the least in the kingdom of heaven is greater than he. [12]From the days of John the Baptist until now the kingdom of heaven has suffered violence,*n* and the violent take it by force. [13]For all the prophets and the law prophesied until John came; [14]and if you are willing to accept it, he is Elijah who is to come. [15]Let anyone with ears*o* listen!

16 "But to what will I compare this generation? It is like children sitting in the marketplaces and calling to one another,

[17] 'We played the flute for you, and
　　you did not dance;
　we wailed, and you did not
　　mourn.'

[18]For John came neither eating nor drinking, and they say, 'He has a demon'; [19]the Son of Man came eating and drinking, and they say, 'Look, a glutton and a drunkard, a friend of tax collectors and sinners!' Yet wisdom is vindicated by her deeds."*p*

20 Then he began to reproach the cities in which most of his deeds of power had been done, because they did not repent. [21]"Woe to you, Chorazin! Woe to you, Bethsaida! For if the deeds of power done in you had been done in Tyre and Sidon, they would have repented long ago in sackcloth and ashes. [22]But I tell you, on the day of judgment it will be more tolerable for Tyre and Sidon than for you. [23]And you, Capernaum,
　will you be exalted to heaven?
　　No, you will be brought down
　　　to Hades.
For if the deeds of power done in you had been done in Sodom, it would have remained until this day. [24]But I tell you that on the day of judgment it will be more tolerable for the land of Sodom than for you."

25 At that time Jesus said, "I thank*q* you, Father, Lord of heaven and earth, because you have hidden these things from the wise and the intelligent and have revealed them to infants; [26]yes, Father, for such was your gracious will.*r* [27]All things have been handed over to me by my Father; and no one knows the Son except the Father, and no one knows the Father except the Son and anyone to whom the Son chooses to reveal him.

28 "Come to me, all you that are weary and are carrying heavy burdens, and I will give you rest. [29]Take my yoke upon you, and learn from me; for I am gentle and humble in heart, and you will find rest for your souls. [30]For my yoke is easy, and my burden is light."

12 At that time Jesus went through the grainfields on the sabbath; his disciples were hungry, and they began to pluck heads of grain and to eat. [2]When the Pharisees saw it, they said to him,

m Other ancient authorities read *Why then did you go out? To see a prophet?*　*n* Or *has been coming violently*　*o* Other ancient authorities add *to hear*　*p* Other ancient authorities read *children*　*q* Or *praise*　*r* Or *for so it was well-pleasing in your sight*

---

introduced the new manifestation (or "coming") of God's kingdom. **10:** From Mal 3.1; compare Mk 1.2. **12:** *The violent* are the eager, ardent multitudes. **14:** Mal 4.5; Lk 1.17; Mk 9.11–13. Biblical prophecy depends on human acceptance of God's terms for fulfillment. If John's message were accepted, his activity would become that foretold in Elijah's name. Jesus seems not to have expected the literal return of *Elijah* (17.10–13; Mk 9.9–13).

11.18: See Lk 7.33 n. **19:** That is, divine *wisdom* is proved right by its results. **21:** *Chorazin,* two and a half miles north of Capernaum, on the coast of the Sea of Galilee. *Bethsaida,* near the northern extremity of the Sea of Galilee. *Tyre and Sidon,* Phoenician cities on the coast of the Mediterranean Sea. **23:** Isa 14.13, 15. **25–30:** Lk 10.21–22.
　11.25: 9.13; 10.42; see 16.17 n; Lk 10.21–22; 24.16. **27:** Jesus claimed a special relation to God which he could share with others (Jn

"Look, your disciples are doing what is not lawful to do on the sabbath." ³He said to them, "Have you not read what David did when he and his companions were hungry? ⁴He entered the house of God and ate the bread of the Presence, which it was not lawful for him or his companions to eat, but only for the priests. ⁵Or have you not read in the law that on the sabbath the priests in the temple break the sabbath and yet are guiltless? ⁶I tell you, something greater than the temple is here. ⁷But if you had known what this means, 'I desire mercy and not sacrifice,' you would not have condemned the guiltless. ⁸For the Son of Man is lord of the sabbath."

9 He left that place and entered their synagogue; ¹⁰a man was there with a withered hand, and they asked him, "Is it lawful to cure on the sabbath?" so that they might accuse him. ¹¹He said to them, "Suppose one of you has only one sheep and it falls into a pit on the sabbath; will you not lay hold of it and lift it out? ¹²How much more valuable is a human being than a sheep! So it is lawful to do good on the sabbath." ¹³Then he said to the man, "Stretch out your hand." He stretched it out, and it was restored, as sound as the other. ¹⁴But the Pharisees went out and conspired against him, how to destroy him.

15 When Jesus became aware of this, he departed. Many crowds⁵ followed him, and he cured all of them, ¹⁶and he ordered them not to make him known. ¹⁷This was to fulfill what had been spoken through the prophet Isaiah:

18 "Here is my servant, whom I
    have chosen,
  my beloved, with whom my
    soul is well pleased.
I will put my Spirit upon him,
  and he will proclaim justice
    to the Gentiles.
19 He will not wrangle or cry aloud,
  nor will anyone hear his voice
    in the streets.
20 He will not break a bruised reed
  or quench a smoldering wick
    until he brings justice to victory.
21   And in his name the Gentiles
    will hope."

22 Then they brought to him a demoniac who was blind and mute; and he cured him, so that the one who had been mute could speak and see. ²³All the crowds were amazed and said, "Can this be the Son of David?" ²⁴But when the Pharisees heard it, they said, "It is only by Beelzebul, the ruler of the demons, that this fellow casts out the demons."

s Other ancient authorities lack *crowds*

3.35; 13.3) **29**: The rabbis spoke of the *yoke* of the Law. Jesus regarded his claim as more demanding and more rewarding (5.17–20). **12.1–14**: **Jesus and sabbath laws** (Mk 2.23–3.6; Lk 6.1–11). **1**: Deut 23.25. **2**: The objection rested on the traditional interpretation that plucking grain by hand was an activity forbidden by Ex 20.8–11. **3–4**: 1 Sam 21.1–6; Lev 24.5–9. **4**: *Bread of the Presence*, twelve loaves of bread continually kept on a table in the holy place of the temple. They were a symbol of communion with God. **5**: Num 28.9–10. **6**: Since no penalty was exacted from those who set aside provisions of the Law for the sake of some human need or some more significant service to God, Jesus' disciples eat because of their need and serve him who is greater than the institutions of the Law (see vv. 41–42). **7**: Hos 6.6; Mt 9.13. **8**: Jesus claims, by virtue of his mission as the Messiah, authority over another's obedience to God (11.27; Jn 5.1–18). **11–12**: The rabbis agreed with the principle of attending to accidental injury and danger on the sabbath, but they thought that chronic conditions should wait (Lk 13.14). For Jesus it was important to restore a person to useful life. **12**: 10.31. **12.15–21**: **Work of healing** (Mk 3.7–12; Lk 6.17–19; 4.40). **17–21**: Isa 42.1–4. **20**: *Smoldering wick*, a lamp wick whose flame has almost gone out. **12.22–37**: **Sources of Jesus' power** (Mk 3.20–30; Lk 11.14–23; 12.10). **22–24**: The dumbness here said to be caused by demonic possession is said in Lk 11.14 to describe the demon itself. The biblical writers speak either of *curing* the victim or casting out the demon (v. 24; 9.32–33; Lk 11.14–15). **23**: *Son of David*, a title of the Messiah (21.9). **24**: The issue is how to account for Jesus' manifest power. The Pharisees attribute it to evil forces hostile to humankind (see Lk 7.33 n.). *Beelzebul*, see

²⁵He knew what they were thinking and said to them, "Every kingdom divided against itself is laid waste, and no city or house divided against itself will stand. ²⁶If Satan casts out Satan, he is divided against himself; how then will his kingdom stand? ²⁷If I cast out demons by Beelzebul, by whom do your own exorcists' cast them out? Therefore they will be your judges. ²⁸But if it is by the Spirit of God that I cast out demons, then the kingdom of God has come to you. ²⁹Or how can one enter a strong man's house and plunder his property, without first tying up the strong man? Then indeed the house can be plundered. ³⁰Whoever is not with me is against me, and whoever does not gather with me scatters. ³¹Therefore I tell you, people will be forgiven for every sin and blasphemy, but blasphemy against the Spirit will not be forgiven. ³²Whoever speaks a word against the Son of Man will be forgiven, but whoever speaks against the Holy Spirit will not be forgiven, either in this age or in the age to come.

33 "Either make the tree good, and its fruit good; or make the tree bad, and its fruit bad; for the tree is known by its fruit. ³⁴You brood of vipers! How can you speak good things, when/ you are evil? For out of the abundance of the heart the mouth speaks. ³⁵The good person brings good things out of a good treasure, and the evil person brings evil things out of an evil treasure. ³⁶I tell you, on the day of judgment you will have to give an account for every careless word you utter; ³⁷for by your words you will be justified, and by your words you will be condemned."

38 Then some of the scribes and Pharisees said to him, "Teacher, we wish to see a sign from you." ³⁹But he answered them, "An evil and adulterous generation asks for a sign, but no sign will be given to it except the sign of the prophet Jonah. ⁴⁰For just as Jonah was three days and three nights in the belly of the sea monster, so for three days and three nights the Son of Man will be in the heart of the earth. ⁴¹The people of Nineveh will rise up at the judgment with this generation and condemn it, because they repented at the proclamation of Jonah, and see, something greater than Jonah is here! ⁴²The queen of the South will rise up at the judgment with this generation and condemn it, because she came from the ends of the earth to listen to the wisdom of Solomon, and see, something greater than Solomon is here!

43 "When the unclean spirit has gone out of a person, it wanders through waterless regions looking for a resting place, but it finds none. ⁴⁴Then it says, 'I will return to my house from which I came.' When it comes, it finds it empty, swept, and put in order. ⁴⁵Then it goes and brings along seven other spirits more evil than itself, and they enter and live there; and the last state of that person is worse than the first. So will it be also with this evil generation."

46 While he was still speaking to the crowds, his mother and his brothers

t Gk *sons*

---

2 Kings 1.2 n.; Mk 3.22 n. **27**: *Your own exorcists,* your disciples (compare 1 Pet 5.13). Exorcising demons was not limited to Jesus and his followers (7.22–23; Mk 9.38; Acts 19.13–19). **28**: Lk 4.18–20. **31–32**: The unforgivable sin is the utter rebellion against God that denies him as the doer of his own acts (Lk 12.10). **32**; Mk 3.28–30.
  12.33–36: 7.16–20; Mk 7.14–23; Lk 6.43–45. **33**: *Make,* recognize that fruit and tree will be alike (Jas 3.11–12). **36**: *Careless,* useless; *barren* in Jas 2.20. **37**: Compare Rom 2.6.
  12.38–42: **Request for a sign** (Lk 11.16, 29–32). **39**: *Adulterous* was used by Old Testament prophets to describe Israel's turning away from God (Jer 3.8; Ezek 23.37; Hos 2.2–10). *Sign,* compare v. 40. **41**: Jon 3.5; Mt 11.20–24; 12.6. **42**: *The queen of the South,* the queen of Sheba (1 Kings 10.1–10; 2 Chr 9.1–9).
  12.43–45: **The return of the unclean spirit** (Lk 11.24–26; see Mk 1.23 n.). **43**: Waterless places, or deserts, supposed to be the favorite abode of demons (compare Isa 13.21–22; 34.14). **44**: *My house,* the person previously possessed by the demon. *Empty,* though evil has been temporarily expelled, nothing good has been put in its place.

were standing outside, wanting to speak to him. 47 Someone told him, "Look, your mother and your brothers are standing outside, wanting to speak to you." *u* 48 But to the one who had told him this, Jesus*v* replied, "Who is my mother, and who are my brothers?" 49 And pointing to his disciples, he said, "Here are my mother and my brothers! 50 For whoever does the will of my Father in heaven is my brother and sister and mother."

13 That same day Jesus went out of the house and sat beside the sea. 2 Such great crowds gathered around him that he got into a boat and sat there, while the whole crowd stood on the beach. 3 And he told them many things in parables, saying: "Listen! A sower went out to sow. 4 And as he sowed, some seeds fell on the path, and the birds came and ate them up. 5 Other seeds fell on rocky ground, where they did not have much soil, and they sprang up quickly, since they had no depth of soil. 6 But when the sun rose, they were scorched; and since they had no root, they withered away. 7 Other seeds fell among thorns, and the thorns grew up and choked them. 8 Other seeds fell on good soil and brought forth grain, some a hundredfold, some sixty, some thirty. 9 Let anyone with ears*w* listen!"

10 Then the disciples came and asked him, "Why do you speak to them in parables?" 11 He answered, "To you it has been given to know the secrets*x* of the kingdom of heaven, but to them it has not been given. 12 For to those who have, more will be given, and they will have an abundance; but from those who have nothing, even what they have will be taken away. 13 The reason I speak to them in parables is that 'seeing they do not perceive, and hearing they do not listen, nor do they understand.' 14 With them indeed is fulfilled the prophecy of Isaiah that says:

'You will indeed listen, but never
        understand,
    and you will indeed look, but
        never perceive.
15 For this people's heart has
        grown dull,
    and their ears are hard of
        hearing,
    and they have shut their eyes;
        so that they might not look
        with their eyes,
    and listen with their ears,
    and understand with their heart
        and turn—
    and I would heal them.'
16 But blessed are your eyes, for they see, and your ears, for they hear. 17 Truly I tell you, many prophets and righteous people longed to see what you see, but did not see it, and to hear what you hear, but did not hear it.

u Other ancient authorities lack verse 47
v Gk *he*   *w* Other ancient authorities add *to
hear*   *x* Or *mysteries*

12.46–50: Jesus' true family (Mk 3.31–35; Lk 8.19–21).

13.1–52: Teaching in parables (Mk 4.1–34; Lk 8.4–18; 13.18–21). **1:** *The sea,* of Galilee. **3:** *Parables* are stories describing situations in everyday life which, as Jesus used them, convey a spiritual meaning. In general the teaching of each parable relates to a single point, and apart from this the details may, or may not, have a particular meaning. Jesus used this method of teaching because: (*a*) it gave vivid, memorable expression to his teachings; (*b*) it led those who heard to reflect on his words and bear responsibility for their decision to accept or oppose his claim; (*c*) it probably reduced specific grounds for contention by hostile listeners. **3b–8: The sower,** explained in vv. 18–23 (see Mk 4.1–9). **11:** The disciples heard and accepted the message about God's kingdom and by their faith had access to deeper understanding (see Mk 4.11 n.). **12:** 25.29; Mk 4.24–25; Lk 8.18; 19.26. **13:** The parables do not obscure truth but present it; the hearers receive the message through their physical senses but do not comprehend (see 11.25 n.). **14–15:** Isa 6.9–10; Mk 8.18; see Acts 28.26 n. **16–17:** See Lk 10.23–24 n. **17:** *See . . . hear,* Jesus' message about God's kingdom.

18 "Hear then the parable of the sower. 19 When anyone hears the word of the kingdom and does not understand it, the evil one comes and snatches away what is sown in the heart; this is what was sown on the path. 20 As for what was sown on rocky ground, this is the one who hears the word and immediately receives it with joy; 21 yet such a person has no root, but endures only for a while, and when trouble or persecution arises on account of the word, that person immediately falls away. *y* 22 As for what was sown among thorns, this is the one who hears the word, but the cares of the world and the lure of wealth choke the word, and it yields nothing. 23 But as for what was sown on good soil, this is the one who hears the word and understands it, who indeed bears fruit and yields, in one case a hundredfold, in another sixty, and in another thirty."

24 He put before them another parable: "The kingdom of heaven may be compared to someone who sowed good seed in his field; 25 but while everybody was asleep, an enemy came and sowed weeds among the wheat, and then went away. 26 So when the plants came up and bore grain, then the weeds appeared as well. 27 And the slaves of the householder came and said to him, 'Master, did you not sow good seed in your field? Where, then, did these weeds come from?' 28 He answered, 'An enemy has done this.' The slaves said to him, 'Then do you want us to go and gather them?' 29 But he replied, 'No; for in gathering the weeds you would uproot the wheat along with them. 30 Let both of them grow together until the harvest; and at harvest time I will tell the reapers, Collect the weeds first and bind them in bundles to be burned, but gather the wheat into my barn.' "

31 He put before them another parable: "The kingdom of heaven is like a mustard seed that someone took and sowed in his field; 32 it is the smallest of all the seeds, but when it has grown it is the greatest of shrubs and becomes a tree, so that the birds of the air come and make nests in its branches."

33 He told them another parable: "The kingdom of heaven is like yeast that a woman took and mixed in with *z* three measures of flour until all of it was leavened."

34 Jesus told the crowds all these things in parables; without a parable he told them nothing. 35 This was to fulfill what had been spoken through the prophet: *a*

"I will open my mouth to speak
in parables;
I will proclaim what has been
hidden from the foundation
of the world." *b*

36 Then he left the crowds and went into the house. And his disciples approached him, saying, "Explain to us the parable of the weeds of the field." 37 He answered, "The one who sows the good seed is the Son of Man; 38 the field is the world, and the good seed are the children of the kingdom; the weeds are the children of the evil one, 39 and the enemy who sowed them is the devil; the harvest is the end of the age, and the reapers are angels. 40 Just as the weeds are collected and burned up with fire, so will it be at the end of the age. 41 The Son of Man will send his angels, and they will collect out

y Gk *stumbles* *z* Gk *hid in* *a* Other
ancient authorities read *the prophet Isaiah*
b Other ancient authorities lack *of the world*

13.18–23: Response to Jesus' message affected by the circumstances of human life. 22: 19.23.
13.24–30: **Weeds in the wheat.** God allows good and evil to exist together until the close of human history (vv. 36–43).
13.31–32: **The mustard seed** (Lk 13.18–19). The beginnings of God's kingdom are small, but it has an inherent nature that will grow to its intended end, startlingly different in size from its beginning. 32: Dan 4.12.
13.33–43: **Yeast** (Lk 13.20–21). God's rule, like *yeast* working in a hidden way, will pervade one's life, giving it a new quality. 35: *The prophet*, i.e. Asaph the seer (2 Chr 29.30), named as the author of Ps 78, from which (v. 2) the quotation is taken. 42: See Lk 12.49 n. 43: Dan 12.3.

of his kingdom all causes of sin and all evildoers, 42 and they will throw them into the furnace of fire, where there will be weeping and gnashing of teeth. 43 Then the righteous will shine like the sun in the kingdom of their Father. Let anyone with ears^c listen!

44 "The kingdom of heaven is like treasure hidden in a field, which someone found and hid; then in his joy he goes and sells all that he has and buys that field.

45 "Again, the kingdom of heaven is like a merchant in search of fine pearls; 46 on finding one pearl of great value, he went and sold all that he had and bought it.

47 "Again, the kingdom of heaven is like a net that was thrown into the sea and caught fish of every kind; 48 when it was full, they drew it ashore, sat down, and put the good into baskets but threw out the bad. 49 So it will be at the end of the age. The angels will come out and separate the evil from the righteous 50 and throw them into the furnace of fire, where there will be weeping and gnashing of teeth.

51 "Have you understood all this?" They answered, "Yes." 52 And he said to them, "Therefore every scribe who has been trained for the kingdom of heaven is like the master of a household who brings out of his treasure what is new and what is old." 53 When Jesus had finished these parables, he left that place.

54 He came to his hometown and began to teach the people^d in their synagogue, so that they were astounded and said, "Where did this man get this wisdom and these deeds of power? 55 Is not this the carpenter's son? Is not his mother called Mary? And are not his brothers James and Joseph and Simon and Judas? 56 And are not all his sisters with us? Where then did this man get all this?" 57 And they took offense at him. But Jesus said to them, "Prophets are not without honor except in their own country and in their own house." 58 And he did not do many deeds of power there, because of their unbelief.

14 At that time Herod the ruler^e heard reports about Jesus; 2 and he said to his servants, "This is John the Baptist; he has been raised from the dead, and for this reason these powers are at work in him." 3 For Herod had arrested John, bound him, and put him in prison on account of Herodias, his brother Philip's wife,^f 4 because John had been telling him, "It is not lawful for you to have her." 5 Though Herod^g wanted to put him to death, he feared the crowd, because they regarded him as a prophet. 6 But when Herod's birthday came, the daughter of Herodias danced before the company, and she pleased Herod 7 so much that he promised on oath to grant her whatever she might ask. 8 Prompted by her mother, she said, "Give me the head of John the Baptist here on a platter." 9 The king was grieved, yet out of regard for his oaths and for the guests, he commanded it to be given; 10 he sent and

c Other ancient authorities add *to hear*
d Gk *them*   e Gk *tetrarch*   f Other ancient authorities read *his brother's wife*   g Gk *he*

13.44–46: **Hidden treasure and the pearl of great value. 44:** Some people respond in whole-hearted dedication to Jesus' message without any other thought than to have what it yields. **45–46:** Some people dedicate themselves to God's kingdom because, being able to judge the value of other claims being made on them, they value it more.
13.47–52: **The dragnet. 52:** *Scribe,* an expert in the Mosaic law, having become a disciple of Jesus, is able to preserve past insights and enlarge them.
13.53–17.27: **Events of decisive acceptance or rejection of Jesus. 13.53–58:** Re-

jection at home. **53:** *Finished,* see 7.28 n. **54:** *His hometown,* Nazareth (Lk 4.16, 23). **55–56:** *Mother . . . brothers . . . sisters,* the two latter terms may possibly refer to relatives other than siblings; see also 12.46–50; Mk 3.31–32; 6.3; Lk 8.19–20; Jn 2.12; 7.3, 5; Acts 1.14; 1 Cor 9.5; Gal 1.19. **58:** See Mk 6.5–6 n.
14.1–12: **Death of John** (Mk 6.14–29; Lk 9.7–9). **1:** *Herod* Antipas, son of Herod the Great. **3:** *Philip,* not the ruler mentioned in Lk 3.1, but a half-brother of *Herod* Antipas. **4:** Lev 18.16; 20.21. **6:** *The daughter* was Salome (Josephus, *Ant.*x.viii.5.4). **10:** *The prison,* see 11.2 n.

had John beheaded in the prison. 11 The head was brought on a platter and given to the girl, who brought it to her mother. 12 His disciples came and took the body and buried it; then they went and told Jesus.

13 Now when Jesus heard this, he withdrew from there in a boat to a deserted place by himself. But when the crowds heard it, they followed him on foot from the towns. 14 When he went ashore, he saw a great crowd; and he had compassion for them and cured their sick. 15 When it was evening, the disciples came to him and said, "This is a deserted place, and the hour is now late; send the crowds away so that they may go into the villages and buy food for themselves." 16 Jesus said to them, "They need not go away; you give them something to eat." 17 They replied, "We have nothing here but five loaves and two fish." 18 And he said, "Bring them here to me." 19 Then he ordered the crowds to sit down on the grass. Taking the five loaves and the two fish, he looked up to heaven, and blessed and broke the loaves, and gave them to the disciples, and the disciples gave them to the crowds. 20 And all ate and were filled; and they took up what was left over of the broken pieces, twelve baskets full. 21 And those who ate were about five thousand men, besides women and children.

22 Immediately he made the disciples get into the boat and go on ahead to the other side, while he dismissed the crowds. 23 And after he had dismissed the crowds, he went up the mountain by himself to pray. When evening came, he was there alone, 24 but by this time the boat, battered by the waves, was far from the land, *h* for the wind was against them. 25 And early in the morning he came walking toward them on the sea. 26 But when the disciples saw him walking on the sea, they were terrified, saying, "It is a ghost!" And they cried out in fear. 27 But immediately Jesus spoke to them and said, "Take heart, it is I; do not be afraid."

28 Peter answered him, "Lord, if it is you, command me to come to you on the water." 29 He said, "Come." So Peter got out of the boat, started walking on the water, and came toward Jesus. 30 But when he noticed the strong wind, *i* he became frightened, and beginning to sink, he cried out, "Lord, save me!" 31 Jesus immediately reached out his hand and caught him, saying to him, "You of little faith, why did you doubt?" 32 When they got into the boat, the wind ceased. 33 And those in the boat worshiped him, saying, "Truly you are the Son of God."

34 When they had crossed over, they came to land at Gennesaret. 35 After the people of that place recognized him, they sent word throughout the region and brought all who were sick to him, 36 and begged him that they might touch even the fringe of his cloak; and all who touched it were healed.

15 Then Pharisees and scribes came to Jesus from Jerusalem and said, 2 "Why do your disciples break the tradition of the elders? For they do not wash their hands before they eat." 3 He answered them, "And why do you break the commandment of God for the sake of

---

h Other ancient authorities read *was out on the sea*   *i* Other ancient authorities read *the wind*

14.13–21: Five thousand fed (Mk 6.30–44; Lk 9.10–17; Jn 6.1–13). **13:** After John's death Jesus faced a new stage in his life (compare his reaction to John's imprisonment, Mk 1.14–15). **14:** 20.25–28. **21:** According to custom the *women and children* would stand or sit separate from the men.
14.22–36: Jesus walks on water (Mk 6.45–52; Jn 6.15–21). **24:** *Was far from the land,* Greek literally "was many stadia from the land"; a stadion was about one-eighth of a mile. **25:** *Early in the morning,* Greek literally "in the fourth watch of the night" (the fourth watch was from 3 to 6 A.M.). **26:** Lk 24.37. **33:** Mk 6.51–52. **34:** *Gennesaret,* a district on the northwestern shore of the Sea of Galilee, which was also called the Lake of Gennesaret.
15.1–20: Tradition of the elders (Mk 7.1–23). **2:** *The tradition of the elders,* the rabbinical exposition of the Law of Moses. *They*

your tradition? [4]For God said,[j] 'Honor your father and your mother,' and, 'Whoever speaks evil of father or mother must surely die.' [5]But you say that whoever tells father or mother, 'Whatever support you might have had from me is given to God,'[k] then that person need not honor the father.[l] [6]So, for the sake of your tradition, you make void the word[m] of God. [7]You hypocrites! Isaiah prophesied rightly about you when he said:

[8] 'This people honors me with
their lips,
but their hearts are far from me;
[9] in vain do they worship me,
teaching human precepts as
doctrines.' "

10 Then he called the crowd to him and said to them, "Listen and understand: [11]it is not what goes into the mouth that defiles a person, but it is what comes out of the mouth that defiles." [12]Then the disciples approached and said to him, "Do you know that the Pharisees took offense when they heard what you said?" [13]He answered, "Every plant that my heavenly Father has not planted will be uprooted. [14]Let them alone; they are blind guides of the blind.[n] And if one blind person guides another, both will fall into a pit." [15]But Peter said to him, "Explain this parable to us." [16]Then he said, "Are you also still without understanding? [17]Do you not see that whatever goes into the mouth enters the stomach, and goes out into the sewer? [18]But what comes out of the mouth proceeds from the heart, and this is what defiles.

[19]For out of the heart come evil intentions, murder, adultery, fornication, theft, false witness, slander. [20]These are what defile a person, but to eat with unwashed hands does not defile."

21 Jesus left that place and went away to the district of Tyre and Sidon. [22]Just then a Canaanite woman from that region came out and started shouting, "Have mercy on me, Lord, Son of David; my daughter is tormented by a demon." [23]But he did not answer her at all. And his disciples came and urged him, saying, "Send her away, for she keeps shouting after us." [24]He answered, "I was sent only to the lost sheep of the house of Israel." [25]But she came and knelt before him, saying, "Lord, help me." [26]He answered, "It is not fair to take the children's food and throw it to the dogs." [27]She said, "Yes, Lord, yet even the dogs eat the crumbs that fall from their masters' table." [28]Then Jesus answered her, "Woman, great is your faith! Let it be done for you as you wish." And her daughter was healed instantly.

29 After Jesus had left that place, he passed along the Sea of Galilee, and he went up the mountain, where he sat down. [30]Great crowds came to him, bringing with them the lame, the maimed, the blind, the mute, and many others. They put them at his feet, and he cured them, [31]so that the crowd was

[j] Other ancient authorities read *commanded, saying* [k] Or *is an offering* [l] Other ancient authorities add *or the mother* [m] Other ancient authorities read *law*; others, *commandment* [n] Other ancient authorities lack *of the blind*

do not wash, see Lk 11.38 n. **4:** Ex 20.12; Deut 5.16; Ex 21.17; Lev 20.9. **7–9:** Isa 29.13 (see Mk 7.6–7 n.).

15.10–20: The teaching here depends on the principle in the Law that certain physical conditions can and do render an individual unfit to share in the worship of the community. **11.** *Defiles,* renders unfit to share in public ritual (Acts 10.14–15; 1 Tim 4.3). **13:** Isa 60.21. **14:** Lk 6.39; Mt 23.16, 24. **19–20:** Violations of the rights and interests of another hinder worship (5.23–24).

15.21–28: The Canaanite woman (Mk 7.24–30). **21:** *Jesus ... went* northwesterly

from upper Galilee into Phoenicia, *the district of Tyre and Sidon.* **22:** The woman, though a Gentile, speaks to Jesus as the Jewish Messiah. **24:** 10.6, 23. Jesus consistently said that his primary mission was to call Jews back to God. The Gentile woman's claim must be based on her own personal acceptance of his message. The distinction is between his mission and his willingness to respond to faith wherever found. **26–27:** See Mk 7.27 n. **27:** The woman accepts Jesus' mission and as a Gentile asks his help.

15.29–31: Healings (Mk 7.31–37).

amazed when they saw the mute speaking, the maimed whole, the lame walking, and the blind seeing. And they praised the God of Israel.

32 Then Jesus called his disciples to him and said, "I have compassion for the crowd, because they have been with me now for three days and have nothing to eat; and I do not want to send them away hungry, for they might faint on the way." 33 The disciples said to him, "Where are we to get enough bread in the desert to feed so great a crowd?" 34 Jesus asked them, "How many loaves have you?" They said, "Seven, and a few small fish." 35 Then ordering the crowd to sit down on the ground, 36 he took the seven loaves and the fish; and after giving thanks he broke them and gave them to the disciples, and the disciples gave them to the crowds. 37 And all of them ate and were filled; and they took up the broken pieces left over, seven baskets full. 38 Those who had eaten were four thousand men, besides women and children. 39 After sending away the crowds, he got into the boat and went to the region of Magadan. *o*

16 The Pharisees and Sadducees came, and to test Jesus *p* they asked him to show them a sign from heaven. 2 He answered them, "When it is evening, you say, 'It will be fair weather, for the sky is red.' 3 And in the morning, 'It will be stormy today, for the sky is red and threatening.' You know how to interpret the appearance of the sky, but you cannot interpret the signs of the times. *q* 4 An evil and adulterous generation asks for a sign, but no sign will be given to it except the sign of Jonah." Then he left them and went away.

5 When the disciples reached the other side, they had forgotten to bring any bread. 6 Jesus said to them, "Watch out, and beware of the yeast of the Pharisees and Sadducees." 7 They said to one another, "It is because we have brought no bread." 8 And becoming aware of it, Jesus said, "You of little faith, why are you talking about having no bread? 9 Do you still not perceive? Do you not remember the five loaves for the five thousand, and how many baskets you gathered? 10 Or the seven loaves for the four thousand, and how many baskets you gathered? 11 How could you fail to perceive that I was not speaking about bread? Beware of the yeast of the Pharisees and Sadducees!" 12 Then they understood that he had not told them to beware of the yeast of bread, but of the teaching of the Pharisees and Sadducees.

13 Now when Jesus came into the district of Caesarea Philippi, he asked his disciples, "Who do people say that the Son of Man is?" 14 And they said, "Some say John the Baptist, but others Elijah, and still others Jeremiah or one of the

o Other ancient authorities read Magdala *or* Magdalan *p* Gk *him* *q Other ancient authorities lack* 2When it is . . . of the times

---

15.32–39: Four thousand fed (see Mk 8.1–10 n.). **39:** *Magadan* was apparently on the west side of the Sea of Galilee.

16.1–4: Demand for signs (Mk 8.11–13; Lk 11.16, 29; 12.54–56). **3:** *The signs of the times* may refer to 15.29–31; compare 11.2–6. **4:** See 12.39 n., 40 n.; Jon 3.4–5.

16.5–12: Yeast of the Pharisees (Mk 8.14–21; Lk 12.1). **5:** *The other side,* the eastern shore of the Sea of Galilee. **6:** *Yeast,* see Mk 8.15 n. **9:** 14.17–21. **10:** 15.34–38.

16.13–23: Peter's confession (Mk 8.27–33; Lk 9.18–22). **13:** See Mk 8.27 n. *Son of Man* here is equivalent to "I." **14:** *Elijah,* believed by Jews to be a forerunner of the Messiah. **16:** Peter asserts that Jesus is the Messiah, not merely one of the prophets (v. 14). He identifies Jesus with the figure of Mal 3.1–4 (compare Mk 1.2; Mt 1.16; Jn 1.49; 11.27). **17:** *Simon* was Peter's personal name. *Flesh and blood,* human beings (1 Cor 15.50; Gal 1.16; Eph 6.12). *Revealed,* understanding spiritual realities involves God's disclosure (see 11.25 n.; Lk 24.16; 1 Cor 1.18–25; 2.6–16). **18:** The Greek text involves a play on two words, "Petros" ("Peter") and "petra" ("rock"). Palestinian Aramaic, which Jesus usually spoke, used the same word for both proper name and common noun: "You are 'Kepha' [Cephas; compare 1 Cor 15.5; Gal 2.9], and on this 'kepha' [rock] I will build . . ." For the view that all the apostles also form the foun-

prophets." 15 He said to them, "But who do you say that I am?" 16 Simon Peter answered, "You are the Messiah,^r the Son of the living God." 17 And Jesus answered him, "Blessed are you, Simon son of Jonah! For flesh and blood has not revealed this to you, but my Father in heaven. 18 And I tell you, you are Peter,^s and on this rock^t I will build my church, and the gates of Hades will not prevail against it. 19 I will give you the keys of the kingdom of heaven, and whatever you bind on earth will be bound in heaven, and whatever you loose on earth will be loosed in heaven." 20 Then he sternly ordered the disciples not to tell anyone that he was^u the Messiah.^r

21 From that time on, Jesus began to show his disciples that he must go to Jerusalem and undergo great suffering at the hands of the elders and chief priests and scribes, and be killed, and on the third day be raised. 22 And Peter took him aside and began to rebuke him, saying, "God forbid it, Lord! This must never happen to you." 23 But he turned and said to Peter, "Get behind me, Satan! You are a stumbling block to me; for you are setting your mind not on divine things but on human things."

24 Then Jesus told his disciples, "If any want to become my followers, let them deny themselves and take up their cross and follow me. 25 For those who want to save their life will lose it, and those who lose their life for my sake will find it. 26 For what will it profit them if they gain the whole world but forfeit their life? Or what will they give in return for their life?

27 "For the Son of Man is to come with his angels in the glory of his Father, and then he will repay everyone for what has been done. 28 Truly I tell you, there are some standing here who will not taste death before they see the Son of Man coming in his kingdom."

17 Six days later, Jesus took with him Peter and James and his brother John and led them up a high mountain, by themselves. 2 And he was transfigured before them, and his face shone like the sun, and his clothes became dazzling white. 3 Suddenly there appeared to them Moses and Elijah, talking with him. 4 Then Peter said to Jesus, "Lord, it is good for us to be here; if you wish, I^v will make three dwellings^w here, one for you, one for Moses, and one for Elijah." 5 While he was still speaking, suddenly a bright cloud overshadowed them, and from the cloud a voice said, "This is my Son, the Beloved;^x with him I am well pleased; listen to him!" 6 When the disciples heard this, they fell to the ground and were overcome by fear. 7 But Jesus came and touched them, saying, "Get up and do not be afraid." 8 And when they looked up, they saw no one except Jesus himself alone.

9 As they were coming down the mountain, Jesus ordered them, "Tell no one about the vision until after the Son of Man has been raised from the dead."

r Or *the Christ*   s Gk *Petros*   t Gk *petra*
u Other ancient authorities add *Jesus*   v Other ancient authorities read *we*   w Or *tents*
x Or *my beloved Son*

---

dation of the church, see Eph 2.20; Rev 21.14. *Church,* see Gal 1.13 n. **19**: *The keys of the kingdom* are a symbol of Peter's power as the leader of the church. *Bind* and *loose* are technical rabbinic terms meaning "forbid" and "permit" some action about which a question has arisen. Later the authority of binding and loosing was also conferred upon all the apostles (18.18). **20**: See Mk 8.30 n. **21**: See Lk 9.22 n. **22–23**: See Mk 8.32 n., 33 n.

16.24–28: **On discipleship** (Mk 8.34–9.1; Lk 9.23–27). **24**: See 10.38 n. **25**: See Mk 8.35 n. **26**: Here *life* is not merely physical

existence, but the higher or spiritual life, the real self (compare Lk 9.25; 12.15). **27**: Ps 62.12; Mt 10.33; Lk 12.8–9; Rom 2.6; 1 Jn 2.28; Rev 22.12. **28**: See Mk 9.1 n.; 1 Cor 16.22; 1 Thess 4.15–18; Jas 5.7; Rev 1.7.

17.1–8: **The transfiguration.** See notes on the parallel passages, Mk 9.2–8; Lk 9.28–36. **1**: *Six days later,* after Peter's confession (16.16). *A high mountain,* probably Mount Hermon, near Caesarea Philippi. It is 9,000 feet high.

17.9–13: **Prophecies about Elijah** (Mk 9.9–13). **9**: See Mk 8.30 n.

10 And the disciples asked him, "Why, then, do the scribes say that Elijah must come first?" 11 He replied, "Elijah is indeed coming and will restore all things; 12 but I tell you that Elijah has already come, and they did not recognize him, but they did to him whatever they pleased. So also the Son of Man is about to suffer at their hands." 13 Then the disciples understood that he was speaking to them about John the Baptist.

14 When they came to the crowd, a man came to him, knelt before him, 15 and said, "Lord, have mercy on my son, for he is an epileptic and he suffers terribly; he often falls into the fire and often into the water. 16 And I brought him to your disciples, but they could not cure him." 17 Jesus answered, "You faithless and perverse generation, how much longer must I be with you? How much longer must I put up with you? Bring him here to me." 18 And Jesus rebuked the demon,*y* and it*z* came out of him, and the boy was cured instantly. 19 Then the disciples came to Jesus privately and said, "Why could we not cast it out?" 20 He said to them, "Because of your little faith. For truly I tell you, if you have faith the size of a*a* mustard seed, you will say to this mountain, 'Move from here to there,' and it will move; and nothing will be impossible for you."*b*

22 As they were gathering*c* in Galilee, Jesus said to them, "The Son of Man is going to be betrayed into human hands, 23 and they will kill him, and on the third day he will be raised." And they were greatly distressed.

24 When they reached Capernaum, the collectors of the temple tax*d* came to Peter and said, "Does your teacher not pay the temple tax?"*d* 25 He said, "Yes, he does." And when he came home, Jesus spoke of it first, asking, "What do you think, Simon? From whom do kings of the earth take toll or tribute? From their children or from others?" 26 When Peter*e* said, "From others," Jesus said to him, "Then the children are free. 27 However, so that we do not give offense to them, go to the sea and cast a hook; take the first fish that comes up; and when you open its mouth, you will find a coin;*f* take that and give it to them for you and me."

18 At that time the disciples came to Jesus and asked, "Who is the greatest in the kingdom of heaven?" 2 He called a child, whom he put among them, 3 and said, "Truly I tell you, unless you change and become like children, you will never enter the kingdom of heaven. 4 Whoever becomes humble like this child is the greatest in the kingdom of heaven. 5 Whoever welcomes one such child in my name welcomes me.

6 "If any of you put a stumbling block before one of these little ones who believe in me, it would be better for you if a great millstone were fastened around your neck and you were drowned in the

y Gk *it* or *him*   *z* Gk *the demon*   *a* Gk *faith
as a grain of*   *b* Other ancient authorities add
verse 21, *But this kind does not come out except by
prayer and fasting*   *c* Other ancient authorities
read *living*   *d* Gk *didrachma*   *e* Gk *he*
f Gk *stater*; the stater was worth two didrachmas

17.10: See 11.14 n. **12**: *Elijah has already come,* in the person of John the Baptist.
17.14–21: **An epileptic child healed** (Mk 9.14–29; Lk 9.37–42). **15**: To be *epileptic* was attributed to the baleful influences of the moon, a demonic force (compare Ps 121.6). **20**: *Little faith* as distinguished from unbelief (13.58). Jesus' saying is in figurative language; faith is concerned with God's will, not with moving mountains (compare 21.21–22; Mk 11.22–23; Lk 17.6; 1 Cor 13.2; Jas 1.6).
17.22–23: **The Passion foretold a second time** (Mk 9.30–32; Lk 9.43–45). Compare 16.21; 20.17–19. **22**: *Were gathering,* for the pilgrimage to Jerusalem for the Passover.
17.24–27: **Money for the temple tax. 24**: The half-shekel tax was paid by Jewish males annually in March for the upkeep of the temple. On the value see 26.15 n. (Ex 30.13; 38.26). **27**: The *coin* (Greek "stater") was exactly enough (two didrachmas) to pay for both.
18.1–35: **Sayings on humility and forgiveness. 1–5**: True greatness (Mk 9.33–37; Lk 9.46–48). **3**: *Change and become like children,* turn away from self-chosen goals and

depth of the sea. 7 Woe to the world because of stumbling blocks! Occasions for stumbling are bound to come, but woe to the one by whom the stumbling block comes!

8 "If your hand or your foot causes you to stumble, cut it off and throw it away; it is better for you to enter life maimed or lame than to have two hands or two feet and to be thrown into the eternal fire. 9 And if your eye causes you to stumble, tear it out and throw it away; it is better for you to enter life with one eye than to have two eyes and to be thrown into the hell*g* of fire.

10 "Take care that you do not despise one of these little ones; for, I tell you, in heaven their angels continually see the face of my Father in heaven.*h* 12 What do you think? If a shepherd has a hundred sheep, and one of them has gone astray, does he not leave the ninety-nine on the mountains and go in search of the one that went astray? 13 And if he finds it, truly I tell you, he rejoices over it more than over the ninety-nine that never went astray. 14 So it is not the will of your*i* Father in heaven that one of these little ones should be lost.

15 "If another member of the church*j* sins against you,*k* go and point out the fault when the two of you are alone. If the member listens to you, you have regained that one.*l* 16 But if you are not listened to, take one or two others along with you, so that every word may be confirmed by the evidence of two or three witnesses. 17 If the member refuses to listen to them, tell it to the church; and if the offender refuses to listen even to the church, let such a one be to you as a Gentile and a tax collector. 18 Truly I tell you, whatever you bind on earth will be bound in heaven, and whatever you loose on earth will be loosed in heaven. 19 Again, truly I tell you, if two of you agree on earth about anything you ask, it will be done for you by my Father in heaven. 20 For where two or three are gathered in my name, I am there among them."

21 Then Peter came and said to him, "Lord, if another member of the church*m* sins against me, how often should I forgive? As many as seven times?" 22 Jesus said to him, "Not seven times, but, I tell you, seventy-seven*n* times.

23 "For this reason the kingdom of heaven may be compared to a king who wished to settle accounts with his slaves. 24 When he began the reckoning, one who owed him ten thousand talents*o* was brought to him; 25 and, as he could not pay, his lord ordered him to be sold, together with his wife and children and all his possessions, and payment to be made. 26 So the slave fell on his knees before him, saying, 'Have patience with me, and I will pay you everything.' 27 And out of pity for him, the lord of that slave released him and forgave him the debt. 28 But that same slave, as he went out, came upon one of his fellow slaves who owed him a hundred denarii;*p* and seizing him by the throat, he

g Gk *Gehenna*   *h* Other ancient authorities add verse 11, *For the Son of Man came to save the lost*   *i* Other ancient authorities read *my*   *j* Gk *If your brother*   *k* Other ancient authorities lack *against you*   *l* Gk *the brother*   *m* Gk *if my brother*   *n* Or *seventy times seven*   *o* A talent was worth more than fifteen years' wages of a laborer   *p* The denarius was the usual day's wage for a laborer

relate oneself to God as to a father. Childlike relations to a parent, not childish behavior, are in view. Mk 10.15; Lk 18.17; 1 Pet 2.2. **6:** *Little ones,* disciples of Jesus, whom he calls "children" (Mk 10.24; compare Mt 11.25).
18.7–9: Warnings of hell (Mk 9.42–48; Lk 17.1–2). **8–9:** In vivid language Jesus speaks of the terrible danger in yielding to temptation (5.29–30).
18.10–14: The lost sheep (Lk 15.3–7). **10:**

Little ones, see v. 6 n. *Angels,* see Acts 12.15 n.
18.15–20: Discipline among followers (Lk 17.3). 1 Cor 6.1–6; Gal 6.1; Jas 5.19–20; Lev 19.17. **15:** *Alone,* solitary reproof is more gracious than reproof given in public. **16:** Deut 19.15. **18:** See 16.19 n.; Jn 20.21–23 n.
18.21–35: Forgiveness. 21–22: Lk 17.4. Forgiveness is beyond calculating. **23:** 25.19. **25:** *To be sold,* this was permitted by the law of Moses (Lev 25.39; 2 Kings 4.1). **26:** 8.2;

said, 'Pay what you owe.' 29 Then his fellow slave fell down and pleaded with him, 'Have patience with me, and I will pay you.' 30 But he refused; then he went and threw him into prison until he would pay the debt. 31 When his fellow slaves saw what had happened, they were greatly distressed, and they went and reported to their lord all that had taken place. 32 Then his lord summoned him and said to him, 'You wicked slave! I forgave you all that debt because you pleaded with me. 33 Should you not have had mercy on your fellow slave, as I had mercy on you?' 34 And in anger his lord handed him over to be tortured until he would pay his entire debt. 35 So my heavenly Father will also do to every one of you, if you do not forgive your brother or sister*q* from your heart."

19 When Jesus had finished saying these things, he left Galilee and went to the region of Judea beyond the Jordan. 2 Large crowds followed him, and he cured them there.

3 Some Pharisees came to him, and to test him they asked, "Is it lawful for a man to divorce his wife for any cause?" 4 He answered, "Have you not read that the one who made them at the beginning 'made them male and female,' 5 and said, 'For this reason a man shall leave his father and mother and be joined to his wife, and the two shall become one flesh'? 6 So they are no longer two, but one flesh. Therefore what God has joined together, let no one separate." 7 They said to him, "Why then did Moses command us to give a certificate of dismissal and to divorce her?" 8 He said to them, "It was because you were so hardhearted that Moses allowed you to divorce your wives, but from the beginning it was not so. 9 And I say to you, whoever divorces his wife, except for unchastity, and marries another commits adultery."*r*

10 His disciples said to him, "If such is the case of a man with his wife, it is better not to marry." 11 But he said to them, "Not everyone can accept this teaching, but only those to whom it is given. 12 For there are eunuchs who have been so from birth, and there are eunuchs who have been made eunuchs by others, and there are eunuchs who have made themselves eunuchs for the sake of the kingdom of heaven. Let anyone accept this who can."

13 Then little children were being brought to him in order that he might lay his hands on them and pray. The disciples spoke sternly to those who brought them; 14 but Jesus said, "Let the little children come to me, and do not stop them; for it is to such as these that the kingdom of heaven belongs." 15 And he laid his hands on them and went on his way.

16 Then someone came to him and said, "Teacher, what good deed must I do to have eternal life?" 17 And he said to him, "Why do you ask me about what is good? There is only one who is good. If you wish to enter into life, keep the com-

---

q Gk *brother*   *r* Other ancient authorities read *except on the ground of unchastity, causes her to commit adultery*; others add at the end of the verse *and he who marries a divorced woman commits adultery*

17.14. **32–33**: Lk 7.41–43. **34**: *To be tortured,* in order to discover whether the debtor was concealing any money or other valuables.

19.1–20.34: From Galilee to Jerusalem (Mk 10.1–52; Lk 18.15–19.27). **1**: Jesus took the Perean route from Galilee to Jerusalem, east of the Jordan, thus avoiding Samaria, whose hostile inhabitants sometimes attacked Jewish pilgrim bands.

19.1–12: Marriage and divorce (Mk 10.1–12). **1**: *Finished,* see 7.28 n. **3**: The Mosaic law gives no answer to this question and the rabbis differed in their opinions. **4–6**: Gen 1.27; 2.24. Jesus appeals to God's purpose of unity in marriage as shown in the account of creation. **7**: Deut 24.1–4. **8**: See Mk 10.5 n. **9**: See 5.32 n.; Lk 16.18; 1 Cor 7.10–13. **11–12**: Jesus recognizes a place for voluntary celibacy in the service of God's kingdom (compare 1 Cor 7.1–9).

19.13–15: Blessing the children (Mk 10.13–16; Lk 18.15–17). **14**: See Mk 10.15 n.; compare Mt 18.2–4; 1 Cor 14.20.

19.16–30: The rich young man (Mk

mandments." 18 He said to him, "Which ones?" And Jesus said, "You shall not murder; You shall not commit adultery; You shall not steal; You shall not bear false witness; 19 Honor your father and mother; also, You shall love your neighbor as yourself." 20 The young man said to him, "I have kept all these;*s* what do I still lack?" 21 Jesus said to him, "If you wish to be perfect, go, sell your possessions, and give the money*t* to the poor, and you will have treasure in heaven; then come, follow me." 22 When the young man heard this word, he went away grieving, for he had many possessions.

23 Then Jesus said to his disciples, "Truly I tell you, it will be hard for a rich person to enter the kingdom of heaven. 24 Again I tell you, it is easier for a camel to go through the eye of a needle than for someone who is rich to enter the kingdom of God." 25 When the disciples heard this, they were greatly astounded and said, "Then who can be saved?" 26 But Jesus looked at them and said, "For mortals it is impossible, but for God all things are possible."

27 Then Peter said in reply, "Look, we have left everything and followed you. What then will we have?" 28 Jesus said to them, "Truly I tell you, at the renewal of all things, when the Son of Man is seated on the throne of his glory, you who have followed me will also sit on twelve thrones, judging the twelve tribes of Israel. 29 And everyone who has left houses or brothers or sisters or father or mother or children or fields, for my name's sake, will receive a hundredfold,*u* and will inherit eternal life. 30 But many who are first will be last, and the last will be first.

20 "For the kingdom of heaven is like a landowner who went out early in the morning to hire laborers for his vineyard. 2 After agreeing with the laborers for the usual daily wage,*v* he sent them into his vineyard. 3 When he went out about nine o'clock, he saw others standing idle in the marketplace; 4 and he said to them, 'You also go into the vineyard, and I will pay you whatever is right.' So they went. 5 When he went out again about noon and about three o'clock, he did the same. 6 And about five o'clock he went out and found others standing around; and he said to them, 'Why are you standing here idle all day?' 7 They said to him, 'Because no one has hired us.' He said to them, 'You also go into the vineyard.' 8 When evening came, the owner of the vineyard said to his manager, 'Call the laborers and give them their pay, beginning with the last and then going to the first.' 9 When those

s Other ancient authorities add *from my youth*
t Gk lacks *the money*   *u* Other ancient
authorities read *manifold*   *v* Gk *a denarius*

10.17–31; Lk 18.18–30). **16**: Lk 10.25; Lev 18.5. The question concerns the way of life which Jesus will guarantee as satisfying God (see Lk 18.26 n.). **17**: Jesus replies that the good way of life is obedience to God's will (15.2–3, 6). *Keep the commandments,* the tense in Greek implies not a single action but a continued process. **16**: Ex 20.12–16; Deut 5.16–20; Rom 13.9; Jas 2.11. **19**: Lev 19.18; Mt 22.39; Rom 13.8; Jas 2.8–9. **21**: Jesus consistently turned people's attention from concern over their own religious standing, calling them to involve themselves in the basic, vital interests of others. Neither wealth, poverty, nor formal piety was so important as sharing in the working out of God's life-giving design for all people (5.23–24, 43–48; 6.33). Eternal life will be found through utter dependence on God, not through a ritual that wealth makes possible (see Lk 12.33 n.; Acts 2.44–45; 4.34, 35). **24**: See Mk 10.25 n. **26**: *For God all things are possible,* Gen 18.14; Jer 32.17.

19.27: *We,* not like that rich man. **28**: *The renewal of all things* refers to the consummation of God's purpose (compare Rom 8.18–25). **29**: *Inherit eternal life* means *enter the kingdom of God* (vv. 23, 24), and *inherit the kingdom* (25.34). **30**: 20.16; Mk 10.31; Lk 13.30.

20.1–16: **Laborers in the vineyard. 1**: *Early,* approximately six a.m. **3**: *Others,* who had not been there earlier. **8**: Lev 19.13; Deut 24.14–15. **9**: *The usual daily wage,* Greek literally "a denarius." Smaller coins existed (see Lk 12.59 n.); therefore payment could have been made on an hourly basis.

hired about five o'clock came, each of them received the usual daily wage. *w* 10 Now when the first came, they thought they would receive more; but each of them also received the usual daily wage. *w* 11 And when they received it, they grumbled against the landowner, 12 saying, 'These last worked only one hour, and you have made them equal to us who have borne the burden of the day and the scorching heat.' 13 But he replied to one of them, 'Friend, I am doing you no wrong; did you not agree with me for the usual daily wage? *w* 14 Take what belongs to you and go; I choose to give to this last the same as I give to you. 15 Am I not allowed to do what I choose with what belongs to me? Or are you envious because I am generous?' *x* 16 So the last will be first, and the first will be last." *y*

17 While Jesus was going up to Jerusalem, he took the twelve disciples aside by themselves, and said to them on the way, 18 "See, we are going up to Jerusalem, and the Son of Man will be handed over to the chief priests and scribes, and they will condemn him to death; 19 then they will hand him over to the Gentiles to be mocked and flogged and crucified; and on the third day he will be raised."

20 Then the mother of the sons of Zebedee came to him with her sons, and kneeling before him, she asked a favor of him. 21 And he said to her, "What do you want?" She said to him, "Declare that these two sons of mine will sit, one at your right hand and one at your left, in your kingdom." 22 But Jesus answered,

"You do not know what you are asking. Are you able to drink the cup that I am about to drink?" *z* They said to him, "We are able." 23 He said to them, "You will indeed drink my cup, but to sit at my right hand and at my left, this is not mine to grant, but it is for those for whom it has been prepared by my Father."

24 When the ten heard it, they were angry with the two brothers. 25 But Jesus called them to him and said, "You know that the rulers of the Gentiles lord it over them, and their great ones are tyrants over them. 26 It will not be so among you; but whoever wishes to be great among you must be your servant, 27 and whoever wishes to be first among you must be your slave; 28 just as the Son of Man came not to be served but to serve, and to give his life a ransom for many."

29 As they were leaving Jericho, a large crowd followed him. 30 There were two blind men sitting by the roadside. When they heard that Jesus was passing by, they shouted, "Lord, *a* have mercy on us, Son of David!" 31 The crowd sternly ordered them to be quiet; but they shouted even more loudly, "Have mercy on us, Lord, Son of David!" 32 Jesus stood still and called them, saying, "What do you want me to do for you?" 33 They said to him, "Lord, let our eyes be opened." 34 Moved with compas-

---

w Gk *a denarius* *x* Gk *is your eye evil because I am good?* *y* Other ancient authorities add *for many are called but few are chosen* *z* Other ancient authorities add *or to be baptized with the baptism that I am baptized with?* *a* Other ancient authorities lack *Lord*

20.14: The point of the parable is the willingness of the owner to exceed conventional practices, and his freedom to do so within the limits of agreement. **15**: The first sentence is not a statement of economic theory except as it claims the right to enter into differing contracts. The second sentence expresses the sense of the Greek text, which is literally translated in note *x*. **16**: Compare 19.30.

20.17–19: **The Passion foretold a third time** (Mk 10.32–34; Lk 18.31–34); a prophecy remarkable for its detailed character; compare 16.21; 17.22.

20.20–28: **James and John seek honor**

(Mk 10.35–45; Lk 22.24–27). **20**: *The mother,* Salome (see 27.56 and Mk 15.40). **22**: *Cup,* see Lk 22.42 n. **23**: Acts 12.2; Rev 1.9; Mt 13.11. **24**: *They were angry,* being afraid of losing something themselves. **26**: See Mk 9.35. **28**: 26.39; 1 Tim 2.5–6; Jn 13.15–16; Titus 2.14; 1 Pet 1.18. The thought seems to be based on Isa ch 53.

20.29–34: **Two blind men of Jericho** (Mk 10.46–52; Lk 18.35–43). Jesus responds not to the Messianic title *Son of David* (v. 30) but to the cry of need (v. 34; compare 15.22–28).

sion, Jesus touched their eyes. Immediately they regained their sight and followed him.

21 When they had come near Jerusalem and had reached Bethphage, at the Mount of Olives, Jesus sent two disciples, ²saying to them, "Go into the village ahead of you, and immediately you will find a donkey tied, and a colt with her; untie them and bring them to me. ³If anyone says anything to you, just say this, 'The Lord needs them.' And he will send them immediately.*ᵇ* ⁴This took place to fulfill what had been spoken through the prophet, saying,

5 "Tell the daughter of Zion,
Look, your king is coming to
        you,
    humble, and mounted on
        a donkey,
    and on a colt, the foal of
        a donkey."

⁶The disciples went and did as Jesus had directed them; ⁷they brought the donkey and the colt, and put their cloaks on them, and he sat on them. ⁸A very large crowdᶜ spread their cloaks on the road, and others cut branches from the trees and spread them on the road. ⁹The crowds that went ahead of him and that followed were shouting,

"Hosanna to the Son of David!
    Blessed is the one who comes in
        the name of the Lord!
Hosanna in the highest heaven!"

¹⁰When he entered Jerusalem, the whole city was in turmoil, asking, "Who is this?" ¹¹The crowds were saying, "This is the prophet Jesus from Nazareth in Galilee."

12 Then Jesus entered the temple*ᵈ* and drove out all who were selling and buying in the temple, and he overturned the tables of the money changers and the seats of those who sold doves. ¹³He said to them, "It is written,

'My house shall be called a house
    of prayer';
but you are making it a den
    of robbers."

14 The blind and the lame came to him in the temple, and he cured them. ¹⁵But when the chief priests and the scribes saw the amazing things that he did, and heardᵉ the children crying out in the temple, "Hosanna to the Son of David," they became angry ¹⁶and said to him, "Do you hear what these are saying?" Jesus said to them, "Yes; have you never read,

'Out of the mouths of infants and
    nursing babies
you have prepared praise for
    yourself'?"

¹⁷He left them, went out of the city to Bethany, and spent the night there.

18 In the morning, when he returned to the city, he was hungry. ¹⁹And seeing a fig tree by the side of the road, he went to it and found nothing at all on it but leaves. Then he said to it, "May no fruit

b Or 'The Lord needs them and will send them back immediately.'   *c* Or Most of the crowd
d Other ancient authorities add of God
e Gk lacks heard

21.1–27.66: The last week (Mk 11.1–15.47; Lk 19.28–23.56).
21.1–11: Palm Sunday (Mk 11.1–10; Lk 19.28–38; Jn 12.12–18). **1:** See Mk 11.1 n. **3:** *And he will send them,* most probably the owner was a secret disciple of Jesus. **5:** Isa 62.11; Zech 9.9. The Hebrew text refers not to two animals but to one. The reference to the two in v. 7 may have arisen through misunderstanding the form of Hebrew poetic expression in Zech 9.9. **8:** Tokens of honor (2 Kings 9.13). **9:** Ps 118.26. *Hosanna,* originally a Hebrew invocation addressed to God, meaning, "O save!"; later it was used as a cry of joyous

acclamation. **11:** The identification reflects an unchanged attitude toward Jesus. His parable (see Mk 11.1 n.) is seen and not understood (Jn 6.14; 7.40; Acts 3.22; Mk 6.15; Lk 13.33).
21.12–17: Cleansing the temple (Mk 11.11, 15–19; Lk 19.45–48; Jn 2.13–17). **12:** The animals for sale were acceptable for sacrifice; the money changers converted Gentile coins into Jewish money that could properly be presented in the temple (Ex 30.13; Lev 1.14). **13:** Isa 56.7; Jer 7.11. **15:** Lk 19.39; Mt 21.9. *Hosanna,* see v. 9 n. **16:** Ps 8.2.
21.18–22: Fig tree cursed (Mk 11.12–14, 20–25). See Mk 11.13 n. **19:** The leaves of the

ever come from you again!" And the fig tree withered at once. 20 When the disciples saw it, they were amazed, saying, "How did the fig tree wither at once?" 21 Jesus answered them, "Truly I tell you, if you have faith and do not doubt, not only will you do what has been done to the fig tree, but even if you say to this mountain, 'Be lifted up and thrown into the sea,' it will be done. 22 Whatever you ask for in prayer with faith, you will receive."

23 When he entered the temple, the chief priests and the elders of the people came to him as he was teaching, and said, "By what authority are you doing these things, and who gave you this authority?" 24 Jesus said to them, "I will also ask you one question; if you tell me the answer, then I will also tell you by what authority I do these things. 25 Did the baptism of John come from heaven, or was it of human origin?" And they argued with one another, "If we say, 'From heaven,' he will say to us, 'Why then did you not believe him?' 26 But if we say, 'Of human origin,' we are afraid of the crowd; for all regard John as a prophet." 27 So they answered Jesus, "We do not know." And he said to them, "Neither will I tell you by what authority I am doing these things.

28 "What do you think? A man had two sons; he went to the first and said, 'Son, go and work in the vineyard today.' 29 He answered, 'I will not'; but later he changed his mind and went. 30 The father*f* went to the second and said the same; and he answered, 'I go, sir'; but he did not go. 31 Which of the two did the will of his father?" They said, "The first." Jesus said to them, "Truly I tell

you, the tax collectors and the prostitutes are going into the kingdom of God ahead of you. 32 For John came to you in the way of righteousness and you did not believe him, but the tax collectors and the prostitutes believed him; and even after you saw it, you did not change your minds and believe him.

33 "Listen to another parable. There was a landowner who planted a vineyard, put a fence around it, dug a wine press in it, and built a watchtower. Then he leased it to tenants and went to another country. 34 When the harvest time had come, he sent his slaves to the tenants to collect his produce. 35 But the tenants seized his slaves and beat one, killed another, and stoned another. 36 Again he sent other slaves, more than the first; and they treated them in the same way. 37 Finally he sent his son to them, saying, 'They will respect my son.' 38 But when the tenants saw the son, they said to themselves, 'This is the heir; come, let us kill him and get his inheritance.' 39 So they seized him, threw him out of the vineyard, and killed him. 40 Now when the owner of the vineyard comes, what will he do to those tenants?" 41 They said to him, "He will put those wretches to a miserable death, and lease the vineyard to other tenants who will give him the produce at the harvest time."

42 Jesus said to them, "Have you never read in the scriptures:

'The stone that the builders
rejected
has become the cornerstone;*g*
this was the Lord's doing,
and it is amazing in our eyes'?

f Gk *He*   *g* Or *keystone*

---

fig tree normally appear after the fruit. **21:** See 17.20 n.

21.23–32: Jesus' authority (Mk 11.27–33; Lk 20.1–8). Jn 2.18–22. **23:** *By what authority?* Jesus had not been ordained as a rabbi. **26:** 11.9; 14.5; Lk 1.76. **27:** Jesus declined to answer because his listeners declined to heed. **28–32:** 20.1; 21.33; Lk 15.11–32. **32:** Lk

7.29–30. *The way of righteousness* led to reconciliation with God by faith.

21.33–46: Parable of the vineyard (Mk 12.1–12; Lk 20.9–19). **33:** Compare Isa 5.1–7, which forms the background of Jesus' parable. **34:** 22.3. **41:** 8.11; Acts 13.46; 18.6; 28.28.

21.42: Jesus agrees with his listeners' an-

43 Therefore I tell you, the kingdom of God will be taken away from you and given to a people that produces the fruits of the kingdom. *h* 44 The one who falls on this stone will be broken to pieces; and it will crush anyone on whom it falls." *i*

45 When the chief priests and the Pharisees heard his parables, they realized that he was speaking about them. 46 They wanted to arrest him, but they feared the crowds, because they regarded him as a prophet.

22 Once more Jesus spoke to them in parables, saying: 2 "The kingdom of heaven may be compared to a king who gave a wedding banquet for his son. 3 He sent his slaves to call those who had been invited to the wedding banquet, but they would not come. 4 Again he sent other slaves, saying, 'Tell those who have been invited: Look, I have prepared my dinner, my oxen and my fat calves have been slaughtered, and everything is ready; come to the wedding banquet.' 5 But they made light of it and went away, one to his farm, another to his business, 6 while the rest seized his slaves, mistreated them, and killed them. 7 The king was enraged. He sent his troops, destroyed those murderers, and burned their city. 8 Then he said to his slaves, 'The wedding is ready, but those invited were not worthy. 9 Go therefore into the main streets, and invite everyone you find to the wedding banquet.' 10 Those slaves went out into the streets and gathered all whom they found, both good and bad; so the wedding hall was filled with guests.

11 "But when the king came in to see the guests, he noticed a man there who was not wearing a wedding robe, 12 and he said to him, 'Friend, how did you get in here without a wedding robe?' And he was speechless. 13 Then the king said to the attendants, 'Bind him hand and foot, and throw him into the outer darkness, where there will be weeping and gnashing of teeth.' 14 For many are called, but few are chosen."

15 Then the Pharisees went and plotted to entrap him in what he said. 16 So they sent their disciples to him, along with the Herodians, saying, "Teacher, we know that you are sincere, and teach the way of God in accordance with truth, and show deference to no one; for you do not regard people with partiality. 17 Tell us, then, what you think. Is it lawful to pay taxes to the emperor, or not?" 18 But Jesus, aware of their malice, said, "Why are you putting me to the test, you hypocrites? 19 Show me the coin used for the tax." And they brought him a denarius. 20 Then he said to them, "Whose head is this, and whose title?" 21 They answered, "The emperor's." Then he said to them, "Give therefore to the emperor the things that are the emperor's, and to God the things that are God's." 22 When they heard this, they were amazed; and they left him and went away.

23 The same day some Sadducees came to him, saying there is no resurrection; *j* and they asked him a question, saying, 24 "Teacher, Moses said, 'If a man dies childless, his brother shall marry the

h Gk *the fruits of it*   *i* Other ancient authorities lack verse 44   *j* Other ancient authorities read *who say that there is no resurrection*

---

swer (v. 41) and quotes Ps 118.22–23 to support his teaching (Acts 4.11; 1 Pet 2.7).
22.1–14: The marriage feast (Lk 14.16–24). **3**: 21.34. **10**: 13.47. **13**: 8.12.
22.15–22: Paying taxes to Caesar (Mk 12.13–17; Lk 20.20–26). **15**: Mk 3.6; 8.15. **16**: *Herodians,* Mk 3.6 n. In asking Jesus for a pronouncement affecting all Jews, his enemies thought to bring him into conflict with sectarian views. **17**: If Jesus approved paying taxes he would offend the nationalistic parties; if he disapproved payment he could be reported as disloyal to the empire. **21**: Rom 13.7; 1 Pet 2.17.
22.23–33: Question about the resurrection (Mk 12.18–27; Lk 20.27–40). **23**: Belief in the *resurrection* was held by the Pharisees in Jesus' day, but rejected by the Sadducees (Acts 4.1–2, 23.6–10). **24**: Deut 25.5.

widow, and raise up children for his brother.' 25 Now there were seven brothers among us; the first married, and died childless, leaving the widow to his brother. 26 The second did the same, so also the third, down to the seventh. 27 Last of all, the woman herself died. 28 In the resurrection, then, whose wife of the seven will she be? For all of them had married her."

29 Jesus answered them, "You are wrong, because you know neither the scriptures nor the power of God. 30 For in the resurrection they neither marry nor are given in marriage, but are like angels*k* in heaven. 31 And as for the resurrection of the dead, have you not read what was said to you by God, 32 'I am the God of Abraham, the God of Isaac, and the God of Jacob'? He is God not of the dead, but of the living." 33 And when the crowd heard it, they were astounded at his teaching.

34 When the Pharisees heard that he had silenced the Sadducees, they gathered together, 35 and one of them, a lawyer, asked him a question to test him. 36 "Teacher, which commandment in the law is the greatest?" 37 He said to him, " 'You shall love the Lord your God with all your heart, and with all your soul, and with all your mind.' 38 This is the greatest and first commandment. 39 And a second is like it: 'You shall love your neighbor as yourself.' 40 On these two commandments hang all the law and the prophets."

41 Now while the Pharisees were gathered together, Jesus asked them this question: 42 "What do you think of the Messiah?*l* Whose son is he?" They said to him, "The son of David." 43 He said to them, "How is it then that David by the Spirit*m* calls him Lord, saying,

44 'The Lord said to my Lord,
 "Sit at my right hand,
  until I put your enemies under
   your feet" '?
45 If David thus calls him Lord, how can he be his son?" 46 No one was able to give him an answer, nor from that day did anyone dare to ask him any more questions.

23 Then Jesus said to the crowds and to his disciples, 2 "The scribes and the Pharisees sit on Moses' seat; 3 therefore, do whatever they teach you and follow it; but do not do as they do, for they do not practice what they teach. 4 They tie up heavy burdens, hard to bear,*n* and lay them on the shoulders of others; but they themselves are unwilling to lift a finger to move them. 5 They do all their deeds to be seen by others; for they make their phylacteries broad and their fringes long. 6 They love to have the place of honor at banquets and the best seats in the synagogues, 7 and to be greeted with respect in the marketplaces, and to have people call them rabbi. 8 But you are not to be called rabbi, for you have one teacher, and you are all students.*o* 9 And call no one your father on earth, for

k Other ancient authorities add *of God*
l Or *Christ*   *m* Gk *in spirit*   *n* Other ancient authorities lack *hard to bear*   *o* Gk *brothers*

22.29: The Sadducees fail to see God's purpose and do not trust his *power*. 31–32: Ex 3.6. The idea here is that those who are related to God in faith have life even though physically dead; resurrection is the divine act by which they will achieve the fullness of life intended in creation and lost through sin and death (see Lk 20.34–36 n.). 22.34–40: **The great commandment** (Mk 12.28–34; Lk 10.25–28). 37: Deut 6.5. 39: Lev 19.18; compare Mt 19.19; Rom 13.9; Gal 5.14; Jas 2.8. 40: The Law contains many ways of applying to life the principle of love.

22.41–46: **David's son** (Mk 12.35–37; Lk 20.41–44). 44: The first *Lord* refers to God, the second *Lord* is taken here to refer to the Messiah (see Ps 110.1; Acts 2.34–35; Heb 1.13; 10.12–13). 23.1–36: **Woe to scribes and Pharisees.** 4: *Heavy burdens*, minute and perplexing interpretations of the law. Lk 11.46; Mt 11.28–30; Acts 15.10. 5: 6.1; 5.16; Ex 13.9; Deut 6.8. *Phylacteries*, little leather boxes worn on the left arm and forehead, containing strips of parchment bearing the text of Ex 13.9, 16; Deut 6.4–9; 11.18–20. *Fringes*, see Mk 6.56 n.

you have one Father—the one in heaven. [10]Nor are you to be called instructors, for you have one instructor, the Messiah. [p] [11]The greatest among you will be your servant. [12]All who exalt themselves will be humbled, and all who humble themselves will be exalted.

13 "But woe to you, scribes and Pharisees, hypocrites! For you lock people out of the kingdom of heaven. For you do not go in yourselves, and when others are going in, you stop them. [q] [15]Woe to you, scribes and Pharisees, hypocrites! For you cross sea and land to make a single convert, and you make the new convert twice as much a child of hell[r] as yourselves.

16 "Woe to you, blind guides, who say, 'Whoever swears by the sanctuary is bound by nothing, but whoever swears by the gold of the sanctuary is bound by the oath.' [17]You blind fools! For which is greater, the gold or the sanctuary that has made the gold sacred? [18]And you say, 'Whoever swears by the altar is bound by nothing, but whoever swears by the gift that is on the altar is bound by the oath.' [19]How blind you are! For which is greater, the gift or the altar that makes the gift sacred? [20]So whoever swears by the altar, swears by it and by everything on it; [21]and whoever swears by the sanctuary, swears by it and by the one who dwells in it; [22]and whoever swears by heaven, swears by the throne of God and by the one who is seated upon it.

23 "Woe to you, scribes and Pharisees, hypocrites! For you tithe mint, dill, and cummin, and have neglected the weightier matters of the law: justice and mercy and faith. It is these you ought to have practiced without neglecting the others. [24]You blind guides! You strain out a gnat but swallow a camel!

25 "Woe to you, scribes and Pharisees, hypocrites! For you clean the outside of the cup and of the plate, but inside they are full of greed and self-indulgence. [26]You blind Pharisee! First clean the inside of the cup, [s] so that the outside also may become clean.

27 "Woe to you, scribes and Pharisees, hypocrites! For you are like whitewashed tombs, which on the outside look beautiful, but inside they are full of the bones of the dead and of all kinds of filth. [28]So you also on the outside look righteous to others, but inside you are full of hypocrisy and lawlessness.

29 "Woe to you, scribes and Pharisees, hypocrites! For you build the tombs of the prophets and decorate the graves of the righteous, [30]and you say, 'If we had lived in the days of our ancestors, we would not have taken part with them in shedding the blood of the prophets.' [31]Thus you testify against yourselves that you are descendants of those who murdered the prophets. [32]Fill up, then, the measure of your ancestors. [33]You snakes, you brood of vipers! How can

p Or *the Christ*   q Other authorities add here (or after verse 12) verse 14, *Woe to you, scribes and Pharisees, hypocrites! For you devour widows' houses and for the sake of appearance you make long prayers; therefore you will receive the greater condemnation* r Gk *Gehenna*   s Other ancient authorities add *and of the plate*

6–7: Mk 12.38–39; Lk 11.43; 14.7–11; 20.46. 8: Jas 3.1. 12: Lk 14.11; 18.14; Mt 18.4; 1 Pet 5.6.

23.13: Seven "woes" follow; the denunciations are an indictment of some, not all, Pharisees. Lk 11.52. 15: Acts 2.10; 6.5; 13.43.

23.16: 5.33–37; 15.14. 17: Ex 30.29. 21: 1 Kings 8.13; Ps 26.8. 23–24: Lk 11.42; Lev 27.30; Mic 6.8. *Strain out a gnat* which had fallen into the wine that was to be drunk. 23.25–26: Lk 11.39–41; Mk 7.4. 27–28: Tombs were *whitewashed* before Passover so

that Jewish travelers might not unwittingly touch them and become ceremonially unclean. Lk 11.44; Acts 23.3; Ps 5.9. 28: See Lk 20.20 n. 29–32: Lk 11.47–48; Acts 7.51–53. 30: *Blood of the prophets,* only one such murder is mentioned in the Hebrew Scriptures (2 Chr 24.20–22), but Jewish legend had added others to the list of national martyrs. 31: The scribes and Pharisees would admit to being *descendants of those who murdered the prophets;* Jesus insists that their attitudes are also similar (v. 28). 33: 3.7; Lk 3.7.

you escape being sentenced to hell?[t] [34]Therefore I send you prophets, sages, and scribes, some of whom you will kill and crucify, and some you will flog in your synagogues and pursue from town to town, [35]so that upon you may come all the righteous blood shed on earth, from the blood of righteous Abel to the blood of Zechariah son of Barachiah, whom you murdered between the sanctuary and the altar. [36]Truly I tell you, all this will come upon this generation.

[37] "Jerusalem, Jerusalem, the city that kills the prophets and stones those who are sent to it! How often have I desired to gather your children together as a hen gathers her brood under her wings, and you were not willing! [38]See, your house is left to you, desolate.[u] [39]For I tell you, you will not see me again until you say, 'Blessed is the one who comes in the name of the Lord.' "

24 As Jesus came out of the temple and was going away, his disciples came to point out to him the buildings of the temple. [2]Then he asked them, "You see all these, do you not? Truly I tell you, not one stone will be left here upon another; all will be thrown down."

[3] When he was sitting on the Mount of Olives, the disciples came to him privately, saying, "Tell us, when will this be, and what will be the sign of your coming and of the end of the age?" [4]Jesus answered them, "Beware that no one leads you astray. [5]For many will come in my name, saying, 'I am the Messiah!'[v] and they will lead many astray. [6]And you will hear of wars and rumors of wars; see that you are not alarmed; for this must take place, but the end is not yet. [7]For nation will rise against nation, and kingdom against kingdom, and there will be famines[w] and earthquakes in various places: [8]all this is but the beginning of the birth pangs.

[9] "Then they will hand you over to be tortured and will put you to death, and you will be hated by all nations because of my name. [10]Then many will fall away,[x] and they will betray one another and hate one another. [11]And many false prophets will arise and lead many astray. [12]And because of the increase of lawlessness, the love of many will grow cold. [13]But the one who endures to the end will be saved. [14]And this good news[y] of the kingdom will be proclaimed throughout the world, as a testimony to all the nations; and then the end will come.

[t] Gk *Gehenna*   [u] Other ancient authorities lack *desolate*   [v] Or *the Christ*   [w] Other ancient authorities add *and pestilences*   [x] Or *stumble*   [y] Or *gospel*

23.34–36: Lk 11.49–51. 34: See Lk 11.49 n.; Mt 10.17, 23; 2 Chr 36.15–16. *Prophets, sages, and scribes* are terms of Jewish origin applied here to Christian missionaries. 35: Gen 4.8; Heb 11.4; 2 Chr 24.20–22; Zech 1.1. The identifying words *son of Barachiah* (not in Lk 11.51) probably were mistakenly added to the text of Matthew at an early date because of confusion over which *Zechariah* was meant. The meaning of the sentence is to indicate the sweep of time from the first to the last victim of murder mentioned in the Old Testament (2 Chronicles stands last in the order of books in the Hebrew Bible). 23.37–39: **Lament over Jerusalem** (Lk 13.34–35). 37: The words *how often* suggest repeated efforts, made perhaps during an earlier Judean ministry (see Lk 4.44 n.). 38: *Your house,* the city itself. 1 Kings 9.7; Jer 12.7; 22.5. 39: 21.9; Ps 118.26.

24.1–3: **Destruction of the temple foretold** (Mk 13.1–2; Lk 21.5–7). 1: These verses, together with the discourse that follows, seem to merge teachings about an immediate destruction of Jerusalem with details associated in Scripture with the end of human history. These teachings were set down by the Evangelist in the light of events between A.D. 30 and 70. It is difficult to be certain what the original form of Jesus' words was. 3: Lk 17.20–21; Mt 13.39, 40, 49; 16.27.
24.4–51: **On the end of the age** (Mk 13.3–37; Lk 21.8–36). 5: 1 Jn 2.18. 6–7: Rev 6.3–8, 12–17. 8: *The birth pangs* signal the imminence of the new age, which was announced at the beginning of Jesus' public ministry as *come near* (4.17), but is to be realized only after a period of witness to Jesus' message (v. 14). Verses 5–14 seem to include a larger community of followers than the

had directed them. 17When they saw him, they worshiped him; but some doubted. 18And Jesus came and said to them, "All authority in heaven and on earth has been given to me. 19Go therefore and make disciples of all nations, baptizing them in the name of the Father and of the Son and of the Holy Spirit, 20and teaching them to obey everything that I have commanded you. And remember, I am with you always, to the end of the age."*o*

o Other ancient authorities add *Amen*

24.11. *Worshiped,* Greek literally "prostrated themselves in worship"; they had not done this before the crucifixion. **18**: 11.27; Lk 10.22; Phil 2.9; Eph 1.20–22. *All authority,* compare Dan 7.14. **19**: *All nations,* contrast 10.5, and compare Mk 16.15; Lk 24.47; Acts 1.8. According to Hebrew usage *in the name of* means in the possession and protection of (Ps 124.8). **20**: *I am with you,* 18.20; Acts 18.10.

15 "So when you see the desolating sacrilege standing in the holy place, as was spoken of by the prophet Daniel (let the reader understand), 16 then those in Judea must flee to the mountains; 17 the one on the housetop must not go down to take what is in the house; 18 the one in the field must not turn back to get a coat. 19 Woe to those who are pregnant and to those who are nursing infants in those days! 20 Pray that your flight may not be in winter or on a sabbath. 21 For at that time there will be great suffering, such as has not been from the beginning of the world until now, no, and never will be. 22 And if those days had not been cut short, no one would be saved; but for the sake of the elect those days will be cut short. 23 Then if anyone says to you, 'Look! Here is the Messiah!' z or 'There he is!'—do not believe it. 24 For false messiahs a and false prophets will appear and produce great signs and omens, to lead astray, if possible, even the elect. 25 Take note, I have told you beforehand. 26 So, if they say to you, 'Look! He is in the wilderness,' do not go out. If they say, 'Look! He is in the inner rooms,' do not believe it. 27 For as the lightning comes from the east and flashes as far as the west, so will be the coming of the Son of Man. 28 Wherever the corpse is, there the vultures will gather.

29 "Immediately after the suffering of those days

the sun will be darkened,
    and the moon will not give
        its light;
the stars will fall from heaven,
    and the powers of heaven will
        be shaken.

30 Then the sign of the Son of Man will appear in heaven, and then all the tribes of the earth will mourn, and they will see 'the Son of Man coming on the clouds of heaven' with power and great glory. 31 And he will send out his angels with a loud trumpet call, and they will gather his elect from the four winds, from one end of heaven to the other.

32 "From the fig tree learn its lesson: as soon as its branch becomes tender and puts forth its leaves, you know that summer is near. 33 So also, when you see all these things, you know that he b is near, at the very gates. 34 Truly I tell you, this generation will not pass away until all these things have taken place. 35 Heaven and earth will pass away, but my words will not pass away.

36 "But about that day and hour no one knows, neither the angels of heaven, nor the Son, c but only the Father. 37 For as the days of Noah were, so will be the coming of the Son of Man. 38 For as in those days before the flood they were eating and drinking, marrying and giving in marriage, until the day Noah entered the ark, 39 and they knew nothing until the flood came and swept them all away, so too will be the coming of the Son of Man. 40 Then two will be in the field; one will be taken and one will be left. 41 Two women will be grinding meal together; one will be taken and one will be left. 42 Keep awake therefore, for you do not know on what day d your Lord is coming. 43 But understand this: if

z Or *the Christ*   a Or *christs*   b Or *it*
c Other ancient authorities lack *nor the Son*
d Other ancient authorities read *at what hour*

---

original disciples. **9**: 10.17–18, 22; Jn 15.18; 16.2. **13**: 10.22; Rev 2.7. **14**: 28.19; Rom 10.18.

24.15: Dan 9.27; 11.31; 12.11; see Mk 13.14 n. **17–18**: Lk 17.31. **21**: Dan 12.1; Joel 2.2. **28**: See Lk 17.37 n.; Job 39.30.

24.29–31: The language here is drawn from the Old Testament; God's victory over sin is to be established by the Son of Man whom he sends (Rev 8.12; Isa 13.10; 34.4; Ezek 32.7; Joel 2.10–11; Zeph 1.15). **30**: 16.27; Dan 7.13; Rev 1.7. **31**: 1 Cor 15.52; 1 Thess 4.16; Isa 27.13; Zech 2.10; 9.14. **34**: 10.23; 16.28. The normal meaning of *this generation* would be "people of our time," and the words would refer to a period of 20–30 years. What Jesus meant, however, is uncertain. **35**: 5.18; Lk 16.17.

24.36: Acts 1.6–7. **37–39**: Lk 17.26–27; Gen 6.5–8; 7.6–24. **40–41**: Lk 17.34–35. **42**: Mk 13.35; Lk 12.40; 21.34–46; Mt 25.13. **43–51**: Lk 12.39–46. **43**: 1 Thess 5.2; Rev 3.3.

the owner of the house had known in what part of the night the thief was coming, he would have stayed awake and would not have let his house be broken into. 44 Therefore you also must be ready, for the Son of Man is coming at an unexpected hour.

45 "Who then is the faithful and wise slave, whom his master has put in charge of his household, to give the other slaves*e* their allowance of food at the proper time? 46 Blessed is that slave whom his master will find at work when he arrives. 47 Truly I tell you, he will put that one in charge of all his possessions. 48 But if that wicked slave says to himself, 'My master is delayed,' 49 and he begins to beat his fellow slaves, and eats and drinks with drunkards, 50 the master of that slave will come on a day when he does not expect him and at an hour that he does not know. 51 He will cut him in pieces*f* and put him with the hypocrites, where there will be weeping and gnashing of teeth.

25 "Then the kingdom of heaven will be like this. Ten bridesmaids*g* took their lamps and went to meet the bridegroom.*h* 2 Five of them were foolish, and five were wise. 3 When the foolish took their lamps, they took no oil with them; 4 but the wise took flasks of oil with their lamps. 5 As the bridegroom was delayed, all of them became drowsy and slept. 6 But at midnight there was a shout, 'Look! Here is the bridegroom! Come out to meet him.' 7 Then all those bridesmaids*g* got up and trimmed their lamps. 8 The foolish said to the wise, 'Give us some of your oil, for our lamps are going out.' 9 But the wise replied, 'No! there will not be enough for you and for us; you had better go to the dealers and buy some for yourselves.'

10 And while they went to buy it, the bridegroom came, and those who were ready went with him into the wedding banquet; and the door was shut. 11 Later the other bridesmaids*g* came also, saying, 'Lord, lord, open to us.' 12 But he replied, 'Truly I tell you, I do not know you.' 13 Keep awake therefore, for you know neither the day nor the hour.*i*

14 "For it is as if a man, going on a journey, summoned his slaves and entrusted his property to them; 15 to one he gave five talents,*j* to another two, to another one, to each according to his ability. Then he went away. 16 The one who had received the five talents went off at once and traded with them, and made five more talents. 17 In the same way, the one who had the two talents made two more talents. 18 But the one who had received the one talent went off and dug a hole in the ground and hid his master's money. 19 After a long time the master of those slaves came and settled accounts with them. 20 Then the one who had received the five talents came forward, bringing five more talents, saying, 'Master, you handed over to me five talents; see, I have made five more talents.' 21 His master said to him, 'Well done, good and trustworthy slave; you have been trustworthy in a few things, I will put you in charge of many things; enter into the joy of your master.' 22 And the one with the two talents also came forward, saying, 'Master, you handed over to me two talents; see, I have made two more talents.' 23 His master said to him, 'Well done, good and trustworthy slave;

e Gk *to give them*   *f* Or *cut him off*
g Gk *virgins*   *h* Other ancient authorities add
and the bride   *i* Other ancient authorities add *in
which the Son of Man is coming*   *j* A talent was
worth more than fifteen years' wages of a laborer

25.1–46: **Teachings on the coming of the kingdom. 1–13: The parable of the wise and foolish bridesmaids** is based on the Palestinian custom that *the bridegroom* fetched his bride from her parents' home to his own. **1**: Lk 12.35–38; Mk 13.34. **2**: 7.24–27. **10**: Rev 19.9. **11–12**: Lk 13.25; Mt 7.21–23. **13**: 24.42; Mk 13.35; Lk 12.40.

25.14–30: **Parable of the talents.** Lk 19.12–27. **15**: On the value of this *talent* see note *j*. **21**: Lk 16.10. **29**: The statement, *From those who have nothing . . . taken away,* illustrates Jesus' way of speaking in two settings at once: as the master's servant had his original talent, yet had earned nothing by it, so individuals can have their earthly existence

you have been trustworthy in a few things, I will put you in charge of many things; enter into the joy of your master.' 24 Then the one who had received the one talent also came forward, saying, 'Master, I knew that you were a harsh man, reaping where you did not sow, and gathering where you did not scatter seed; 25 so I was afraid, and I went and hid your talent in the ground. Here you have what is yours.' 26 But his master replied, 'You wicked and lazy slave! You knew, did you, that I reap where I did not sow, and gather where I did not scatter? 27 Then you ought to have invested my money with the bankers, and on my return I would have received what was my own with interest. 28 So take the talent from him, and give it to the one with the ten talents. 29 For to all those who have, more will be given, and they will have an abundance; but from those who have nothing, even what they have will be taken away. 30 As for this worthless slave, throw him into the outer darkness, where there will be weeping and gnashing of teeth.'

31 "When the Son of Man comes in his glory, and all the angels with him, then he will sit on the throne of his glory. 32 All the nations will be gathered before him, and he will separate people one from another as a shepherd separates the sheep from the goats, 33 and he will put the sheep at his right hand and the goats at the left. 34 Then the king will say to those at his right hand, 'Come, you that are blessed by my Father, inherit the kingdom prepared for you from the foundation of the world; 35 for I was hungry and you gave me food, I was thirsty and you gave me something to drink, I was a stranger and you welcomed me, 36 I

was naked and you gave me clothing, I was sick and you took care of me, I was in prison and you visited me.' 37 Then the righteous will answer him, 'Lord, when was it that we saw you hungry and gave you food, or thirsty and gave you something to drink? 38 And when was it that we saw you a stranger and welcomed you, or naked and gave you clothing? 39 And when was it that we saw you sick or in prison and visited you?' 40 And the king will answer them, 'Truly I tell you, just as you did it to one of the least of these who are members of my family, k you did it to me.' 41 Then he will say to those at his left hand, 'You that are accursed, depart from me into the eternal fire prepared for the devil and his angels; 42 for I was hungry and you gave me no food, I was thirsty and you gave me nothing to drink, 43 I was a stranger and you did not welcome me, naked and you did not give me clothing, sick and in prison and you did not visit me.' 44 Then they also will answer, 'Lord, when was it that we saw you hungry or thirsty or a stranger or naked or sick or in prison, and did not take care of you?' 45 Then he will answer them, 'Truly I tell you, just as you did not do it to one of the least of these, you did not do it to me.' 46 And these will go away into eternal punishment, but the righteous into eternal life."

26 When Jesus had finished saying all these things, he said to his disciples, 2 "You know that after two days the Passover is coming, and the Son of Man will be handed over to be crucified."

3 Then the chief priests and the elders

k Gk *these my brothers*

---

and all that derives from it, yet lack merit in the final judgment (v. 30). **30:** *Worthless,* without value to his master.

25.31–46: The Great Judgment. 31: 16.27; 19.28. **32:** Ezek 34.17. *The nations,* probably those who do not know the God of Israel (compare Rom 2.13–16). **34:** Lk 12.32; Mt 5.3; Rev 13.8; 17.8. **35–36:** Isa 58.7; Jas 1.27; 2.15–16; Heb 13.2; 2 Tim 1.16. **40:** 10.42; Mk 9.41; Heb 6.10; Prov 19.17. **41:** Mk 9.48, Rev 20.10. **46:** Dan 12.2, Jn 5.29. *Go away . . . into eternal life* expresses the same idea as *inherit the kingdom* (v. 34).

26.1–27.66: Jesus' death (Mk 14.1–15.47; Lk 22.1–23.56; Jn 13.1–19.42). **26.1:** *Finished,* see 7.28 n. **2–5:** Mk 14.1–2; Lk 22.1–2; Jn 11.47–53. **2:** *The Passover* commemorated the escape from Egypt under Moses (Ex 12.1–

of the people gathered in the palace of the high priest, who was called Caiaphas, [4] and they conspired to arrest Jesus by stealth and kill him. [5] But they said, "Not during the festival, or there may be a riot among the people."

6 Now while Jesus was at Bethany in the house of Simon the leper,[l] [7] a woman came to him with an alabaster jar of very costly ointment, and she poured it on his head as he sat at the table. [8] But when the disciples saw it, they were angry and said, "Why this waste? [9] For this ointment could have been sold for a large sum, and the money given to the poor." [10] But Jesus, aware of this, said to them, "Why do you trouble the woman? She has performed a good service for me. [11] For you always have the poor with you, but you will not always have me. [12] By pouring this ointment on my body she has prepared me for burial. [13] Truly I tell you, wherever this good news[m] is proclaimed in the whole world, what she has done will be told in remembrance of her."

14 Then one of the twelve, who was called Judas Iscariot, went to the chief priests [15] and said, "What will you give me if I betray him to you?" They paid him thirty pieces of silver. [16] And from that moment he began to look for an opportunity to betray him.

17 On the first day of Unleavened Bread the disciples came to Jesus, saying, "Where do you want us to make the preparations for you to eat the Passover?" [18] He said, "Go into the city to a certain man, and say to him, 'The Teacher says, My time is near; I will keep the Passover at your house with my disciples.'" [19] So the disciples did as Jesus had directed them, and they prepared the Passover meal.

20 When it was evening, he took his place with the twelve;[n] [21] and while they were eating, he said, "Truly I tell you, one of you will betray me." [22] And they became greatly distressed and began to say to him one after another, "Surely not I, Lord?" [23] He answered, "The one who has dipped his hand into the bowl with me will betray me. [24] The Son of Man goes as it is written of him, but woe to that one by whom the Son of Man is betrayed! It would have been better for that one not to have been born." [25] Judas, who betrayed him, said, "Surely not I, Rabbi?" He replied, "You have said so."

26 While they were eating, Jesus took a loaf of bread, and after blessing it he broke it, gave it to the disciples, and said, "Take, eat; this is my body." [27] Then he took a cup, and after giving thanks he gave it to them, saying, "Drink from it,

l The terms *leper* and *leprosy* can refer to several diseases   m Or *gospel*   n Other ancient authorities add *disciples*

---

20). **3**: Joseph *Caiaphas,* son-in-law of Annas, was appointed high priest in A.D. 26 by the Roman procurator Valerius Gratus.

26.6–13: Mk 14.3–9; Jn 12.1–8. A similar event is reported in Lk 7.36–50. **6**: The identity of this Simon is unknown. **7**: Jn 12.3; see Lk 7.37 n., 46. **10**: The *good service* is what is good and fitting under the circumstances of impending death. The same Greek words are translated "good works" in 5.16. **12**: Jn 19.40. The woman's act won higher praise from Jesus than any other mentioned in the New Testament.

26.14–16: Mk 14.10–11; Lk 22.3–6. **14**: See Mk 14.10 n. **15**: Ex 21.32; Zech 11.12. The value of the *thirty pieces of silver* is uncertain. Matthew's quotation refers to silver shekels; at four denarii to the shekel this was one hundred and twenty days' wages (20.2).

26.17–29: **The Last Supper. 17–19**: Mk 14.12–16; Lk 22.7–13. **17**: *Eat the Passover,* i.e. the paschal lamb (Ex 12.18–27; Deut 16.5–8). **18**: Lk 22.10 n., 11 n.; Jn 7.6; 12.23; 13.1; 17.1. **19**: 21.6; Deut 16.5–8. **20–25**: Mk 14.17–21; Lk 22.14, 21–23; Jn 13.21–30. **24**: Ps 41.9; Lk 24.25; 1 Cor 15.3; Acts 17.2–3; Mt 18.7. **25**: Judas' question is phrased to imply that the answer will be in the negative. *You have said so,* a common form of assent in Palestine.

26.26–29: Mk 14.22–25; Lk 22.15–20; 1 Cor 10.16; 11.23–26; Mt 14.19; 15.36; see Lk 22.17 n. **28**: Heb 9.20; Mt 20.28; Mk 1.4; Ex 24.6–8; see Mk 14.24 n. In the background of Jesus' words are several important ideas of Jewish religion: one's sins lead to death; God has rescued his people, as from Egypt, and may be trusted to deliver from death itself; in mercy God forgives those who

all of you; [28] for this is my blood of the[o] covenant, which is poured out for many for the forgiveness of sins. [29] I tell you, I will never again drink of this fruit of the vine until that day when I drink it new with you in my Father's kingdom."

30 When they had sung the hymn, they went out to the Mount of Olives.

31 Then Jesus said to them, "You will all become deserters because of me this night; for it is written,

'I will strike the shepherd,
    and the sheep of the flock will
        be scattered.'

[32] But after I am raised up, I will go ahead of you to Galilee." [33] Peter said to him, "Though all become deserters because of you, I will never desert you." [34] Jesus said to him, "Truly I tell you, this very night, before the cock crows, you will deny me three times." [35] Peter said to him, "Even though I must die with you, I will not deny you." And so said all the disciples.

36 Then Jesus went with them to a place called Gethsemane; and he said to his disciples, "Sit here while I go over there and pray." [37] He took with him Peter and the two sons of Zebedee, and began to be grieved and agitated. [38] Then he said to them, "I am deeply grieved, even to death; remain here, and stay awake with me." [39] And going a little farther, he threw himself on the ground and prayed, "My Father, if it is possible, let this cup pass from me; yet not what I want but what you want." [40] Then he came to the disciples and found them sleeping; and he said to Peter, "So, could you not stay awake with me one hour?

[41] Stay awake and pray that you may not come into the time of trial;[p] the spirit indeed is willing, but the flesh is weak." [42] Again he went away for the second time and prayed, "My Father, if this cannot pass unless I drink it, your will be done." [43] Again he came and found them sleeping, for their eyes were heavy. [44] So leaving them again, he went away and prayed for the third time, saying the same words. [45] Then he came to the disciples and said to them, "Are you still sleeping and taking your rest? See, the hour is at hand, and the Son of Man is betrayed into the hands of sinners. [46] Get up, let us be going. See, my betrayer is at hand."

47 While he was still speaking, Judas, one of the twelve, arrived; with him was a large crowd with swords and clubs, from the chief priests and the elders of the people. [48] Now the betrayer had given them a sign, saying, "The one I will kiss is the man; arrest him." [49] At once he came up to Jesus and said, "Greetings, Rabbi!" and kissed him. [50] Jesus said to him, "Friend, do what you are here to do." Then they came and laid hands on Jesus and arrested him. [51] Suddenly, one of those with Jesus put his hand on his sword, drew it, and struck the slave of the high priest, cutting off his ear. [52] Then Jesus said to him, "Put your sword back into its place; for all who take the sword will perish by the sword. [53] Do you think that I cannot appeal to my Fa-

o Other ancient authorities add *new*
p Or *into temptation*

---

obey him; God will make a new covenant (Jer 31.31–34). **29**: See Lk 14.15; 22.18, 30; Rev 19.9.

26.30–56: Gethsemane. 30–35: Mk 14.26–31; Lk 22.31–34, 39; Jn 14.31; 18.1; 13.36–38. **30**: *The hymn* would be Pss 115–118, the second part of the Hallel Psalms. *To the Mount of Olives*, on the east of Jerusalem. **31**: Zech 13.7; Jn 16.32. **32**: 28.7, 10, 16. **34**: *Before the cock crows*, before dawn.

26.36–46: Mk 14.32–42; Lk 22.40–46. **38**: Jn 12.27; Heb 5.7–8; Ps 42.6. *I am*, Greek literally "My soul is." **39**: Ezek 23.31–34; Jn 18.11; Mt 20.22. Jesus does not desire death but accepts God's will even including death. *Cup*, see Lk 22.42 n. **41**: 6.13; Lk 11.4. *The time of trial*, when one's best intentions may give way. **42**: Jn 4.34; 5.30; 6.38. **45**: 26.18 n.; Jn 12.23; 13.1; 17.1. **47–56**: Mk 14.43–52; Lk 22.47–53; Jn 18.2–11.

26.47: *A large crowd*, evidently considerable resistance was expected. **50**: *Friend*, "comrade." The synoptic gospels do not report Judas' movements on this night (compare Jn 13.30; 18.3). **51**: Jn 18.10. **52**: Gen 9.6; Rev 13.10. **53**: *Twelve legions*, 72,000. **54**:

ther, and he will at once send me more than twelve legions of angels? 54 But how then would the scriptures be fulfilled, which say it must happen in this way?" 55 At that hour Jesus said to the crowds, "Have you come out with swords and clubs to arrest me as though I were a bandit? Day after day I sat in the temple teaching, and you did not arrest me. 56 But all this has taken place, so that the scriptures of the prophets may be fulfilled." Then all the disciples deserted him and fled.

57 Those who had arrested Jesus took him to Caiaphas the high priest, in whose house the scribes and the elders had gathered. 58 But Peter was following him at a distance, as far as the courtyard of the high priest; and going inside, he sat with the guards in order to see how this would end. 59 Now the chief priests and the whole council were looking for false testimony against Jesus so that they might put him to death, 60 but they found none, though many false witnesses came forward. At last two came forward 61 and said, "This fellow said, 'I am able to destroy the temple of God and to build it in three days.' " 62 The high priest stood up and said, "Have you no answer? What is it that they testify against you?" 63 But Jesus was silent. Then the high priest said to him, "I put you under oath before the living God, tell us if you are the Messiah,*q* the Son of God." 64 Jesus said to him, "You have said so. But I tell you,

From now on you will see the
   Son of Man
seated at the right hand of
   Power
and coming on the clouds
   of heaven."

65 Then the high priest tore his clothes and said, "He has blasphemed! Why do we still need witnesses? You have now heard his blasphemy. 66 What is your verdict?" They answered, "He deserves death." 67 Then they spat in his face and struck him; and some slapped him, 68 saying, "Prophesy to us, you Messiah!*q* Who is it that struck you?"

69 Now Peter was sitting outside in the courtyard. A servant-girl came to him and said, "You also were with Jesus the Galilean." 70 But he denied it before all of them, saying, "I do not know what you are talking about." 71 When he went out to the porch, another servant-girl saw him, and she said to the bystanders, "This man was with Jesus of Nazareth."*r* 72 Again he denied it with an oath, "I do not know the man." 73 After a little while the bystanders came up and said to Peter, "Certainly you are also one of them, for your accent betrays you." 74 Then he began to curse, and he swore an oath, "I do not know the man!" At that moment the cock crowed. 75 Then Peter remembered what Jesus had said: "Before the cock crows, you will deny me three times." And he went out and wept bitterly.

27 When morning came, all the chief priests and the elders of the people conferred together against Jesus in order to bring about his death. 2 They bound him, led him away, and handed him over to Pilate the governor.

3 When Judas, his betrayer, saw that Jesus*s* was condemned, he repented and brought back the thirty pieces of silver to the chief priests and the elders. 4 He said,

---

q Or *Christ*   *r* Gk *the Nazorean*   *s* Gk *he*

Faith in God cannot claim his promise (4.6) so as to counteract his purpose. **55**: Lk 19.47; Jn 18.19–21.

26.57–75: **Jesus before Caiaphas. 57**: The reference is to the Jewish supreme court (the Sanhedrin; see Jn 11.47 n.). **59**: See Jn 14.55 n. **61**: 24.2; 27.40; Acts 6.14; Jn 2.19. **63**: 27.11; Jn 18.33. **64**: *You have said so,* see 26.25 n. *The Son of Man,* 16.28; Dan 7.13; Ps 110.1.

26.65: Num 14.6; Acts 14.14; Lev 24.16. **66**: Lev 24.16. **68**: *Prophesy,* at this moment Jesus was blindfolded (Lk 22.64). **73**: Peter spoke with a Galilean accent differing from the Judean; see Acts 2.7. **75**: Compare v. 34. **27.1–26**: **Jesus before Pilate. 1–2**: Mk 15.1; Lk 23.1; Jn 18.28–32. Jewish law required that the Sanhedrin take formal action by daylight. Apparently 26.57–68 describes a pre-dawn hearing. **3–10**: Acts 1.16–20. The

"I have sinned by betraying innocent[t] blood." But they said, "What is that to us? See to it yourself." 5 Throwing down the pieces of silver in the temple, he departed; and he went and hanged himself. 6 But the chief priests, taking the pieces of silver, said, "It is not lawful to put them into the treasury, since they are blood money." 7 After conferring together, they used them to buy the potter's field as a place to bury foreigners. 8 For this reason that field has been called the Field of Blood to this day. 9 Then was fulfilled what had been spoken through the prophet Jeremiah,[u] "And they took[v] the thirty pieces of silver, the price of the one on whom a price had been set,[w] on whom some of the people of Israel had set a price, 10 and they gave[x] them for the potter's field, as the Lord commanded me."

11 Now Jesus stood before the governor; and the governor asked him, "Are you the King of the Jews?" Jesus said, "You say so." 12 But when he was accused by the chief priests and elders, he did not answer. 13 Then Pilate said to him, "Do you not hear how many accusations they make against you?" 14 But he gave him no answer, not even to a single charge, so that the governor was greatly amazed.

15 Now at the festival the governor was accustomed to release a prisoner for the crowd, anyone whom they wanted. 16 At that time they had a notorious prisoner, called Jesus[y] Barabbas. 17 So after they had gathered, Pilate said to them, "Whom do you want me to release for you, Jesus[y] Barabbas or Jesus who is called the Messiah?"[z] 18 For he realized that it was out of jealousy that they had handed him over. 19 While he was sitting on the judgment seat, his wife sent word to him, "Have nothing to do with that innocent man, for today I have suffered a great deal because of a dream about him." 20 Now the chief priests and the elders persuaded the crowds to ask for Barabbas and to have Jesus killed. 21 The governor again said to them, "Which of the two do you want me to release for you?" And they said, "Barabbas." 22 Pilate said to them, "Then what should I do with Jesus who is called the Messiah?"[z] All of them said, "Let him be crucified!" 23 Then he asked, "Why, what evil has he done?" But they shouted all the more, "Let him be crucified!"

24 So when Pilate saw that he could do nothing, but rather that a riot was beginning, he took some water and washed his hands before the crowd, saying, "I am innocent of this man's blood;[a] see to it yourselves." 25 Then the people as a whole answered, "His blood be on us and on our children!" 26 So he released Barabbas for them; and after flogging Jesus, he handed him over to be crucified.

27 Then the soldiers of the governor took Jesus into the governor's headquarters,[b] and they gathered the whole cohort around him. 28 They stripped him and put a scarlet robe on him, 29 and after twisting some thorns into a crown, they put it on his head. They put a reed in his right hand and knelt before him and mocked him, saying, "Hail, King of the Jews!" 30 They spat on him, and took the

t Other ancient authorities read *righteous*
u Other ancient authorities read *Zechariah* or *Isaiah*　v Or *I took*　w Or *the price of the precious One*　x Other ancient authorities read *I gave*　y Other ancient authorities lack *Jesus*　z Or *the Christ*　a Other ancient authorities read *this righteous blood*, or *this righteous man's blood*　b Gk *the praetorium*

details of Judas' end are obscure. Each account connects him in death with a cemetery for foreigners in Jerusalem. **6:** See Deut 23.18 for scruples about ill-gotten gains for sacred purposes. **9–10:** Zech 11.12–13; Jer 18.1–3; 32.6–15.

27.11–14: Mk 15.2–5; Lk 23.2–5; Jn 18.29–19.16. **14:** Lk 23.9; Mt 26.62; Mk 14.60; 1 Tim 6.13. **15–26:** Mk 15.6–15; Lk 23.18–25; Jn 18.38–40; 19.4–16. **19:** Lk 23.4. **21:** Acts 3.13–14. **24:** Deut 21.6–9; Ps 26.6. **25:** Acts 5.28; Josh 2.19. **26:** Scourging with a multi-thonged whip ordinarily preceded execution.

27.27–44: The crucifixion. 27–31: Mk 15.16–20; Jn 19.1–3. **27:** *The cohort* at full strength numbered about five hundred men.

reed and struck him on the head. 31 After mocking him, they stripped him of the robe and put his own clothes on him. Then they led him away to crucify him.

32 As they went out, they came upon a man from Cyrene named Simon; they compelled this man to carry his cross. 33 And when they came to a place called Golgotha (which means Place of a Skull), 34 they offered him wine to drink, mixed with gall; but when he tasted it, he would not drink it. 35 And when they had crucified him, they divided his clothes among themselves by casting lots;*c* 36 then they sat down there and kept watch over him. 37 Over his head they put the charge against him, which read, "This is Jesus, the King of the Jews."

38 Then two bandits were crucified with him, one on his right and one on his left. 39 Those who passed by derided*d* him, shaking their heads 40 and saying, "You who would destroy the temple and build it in three days, save yourself! If you are the Son of God, come down from the cross." 41 In the same way the chief priests also, along with the scribes and elders, were mocking him, saying, 42 "He saved others; he cannot save himself.*e* He is the King of Israel; let him come down from the cross now, and we will believe in him. 43 He trusts in God; let God deliver him now, if he wants to; for he said, 'I am God's Son.'" 44 The bandits who were crucified with him also taunted him in the same way.

45 From noon on, darkness came over the whole land*f* until three in the afternoon. 46 And about three o'clock

Jesus cried with a loud voice, "Eli, Eli, lema sabachthani?" that is, "My God, my God, why have you forsaken me?" 47 When some of the bystanders heard it, they said, "This man is calling for Elijah." 48 At once one of them ran and got a sponge, filled it with sour wine, put it on a stick, and gave it to him to drink. 49 But the others said, "Wait, let us see whether Elijah will come to save him."*g* 50 Then Jesus cried again with a loud voice and breathed his last.*h* 51 At that moment the curtain of the temple was torn in two, from top to bottom. The earth shook, and the rocks were split. 52 The tombs also were opened, and many bodies of the saints who had fallen asleep were raised. 53 After his resurrection they came out of the tombs and entered the holy city and appeared to many. 54 Now when the centurion and those with him, who were keeping watch over Jesus, saw the earthquake and what took place, they were terrified and said, "Truly this man was God's Son!"*i*

55 Many women were also there, looking on from a distance; they had followed Jesus from Galilee and had provided for him. 56 Among them were Mary Magdalene, and Mary the mother of

c Other ancient authorities add in order that what had been spoken through the prophet might be fulfilled, "They divided my clothes among themselves, and for my clothing they cast lots."
d Or blasphemed   e Or is he unable to save himself?   f Or earth   g Other ancient authorities add And another took a spear and pierced his side, and out came water and blood   h Or gave up his spirit   i Or a son of God

---

32–44: Mk 15.21–32; Lk 23.26, 33–43; Jn 19.17–24. 32: The procession included Jesus, two other prisoners, a centurion, and a few soldiers. *Simon,* see Mk 15.21 n. 34: *Gall,* any bitter liquid, possibly the myrrh of Mk 15.23. 35: *They divided his clothes,* see Jn 19.23 n.; Ps 22.18. 37: Indication of the offense was customary. Since the Romans recognized the ruling Herods, it seems implied that Jesus was alleged to be a pretender and revolutionary. 39: Ps 22.7–8; 109.25. 40: 26.61; Acts 6.14; Jn 2.19. 42–43: The taunts stress religious aspects of Jesus' works and words. *Israel* (rather than *the Jews,* v. 37) refers to the religious community rather than the political state. 43: Ps 22.8.

27.45–66: **The death of Jesus.** 45–56: Mk 15.33–41; Lk 23.44–49; Jn 19.28–37. 46: *Eli . . . sabachthani,* quoted from Ps 22.1. 47: *Elijah* (similar in sound to *Eli*) was expected to usher in the final period (Mal 4.5–6; Mt 27.49). 48: Ps 69.21. The motive in offering the *sour wine* may have been to revive him and hence prolong the ordeal. 51: Heb 9.8; 10.19; Ex 26.31–35; Mt 28.2; see Mk 15.38 n. 56: *James,* possibly the James of 10.3; Lk 24.10; Acts 1.13.

James and Joseph, and the mother of the sons of Zebedee.

57 When it was evening, there came a rich man from Arimathea, named Joseph, who was also a disciple of Jesus. [58] He went to Pilate and asked for the body of Jesus; then Pilate ordered it to be given to him. [59] So Joseph took the body and wrapped it in a clean linen cloth [60] and laid it in his own new tomb, which he had hewn in the rock. He then rolled a great stone to the door of the tomb and went away. [61] Mary Magdalene and the other Mary were there, sitting opposite the tomb.

62 The next day, that is, after the day of Preparation, the chief priests and the Pharisees gathered before Pilate [63] and said, "Sir, we remember what that impostor said while he was still alive, 'After three days I will rise again.' [64] Therefore command the tomb to be made secure until the third day; otherwise his disciples may go and steal him away, and tell the people, 'He has been raised from the dead,' and the last deception would be worse than the first." [65] Pilate said to them, "You have a guard[j] of soldiers; go, make it as secure as you can."[k] [66] So they went with the guard and made the tomb secure by sealing the stone.

28 After the sabbath, as the first day of the week was dawning, Mary Magdalene and the other Mary went to see the tomb. [2] And suddenly there was a great earthquake; for an angel of the Lord, descending from heaven, came and rolled back the stone and sat on it. [3] His appearance was like lightning, and his clothing white as snow. [4] For fear of him the guards shook and became like dead men. [5] But the angel said to the women, "Do not be afraid; I know that you are looking for Jesus who was crucified. [6] He is not here; for he has been raised, as he said. Come, see the place where he[l] lay. [7] Then go quickly and tell his disciples, 'He has been raised from the dead,[m] and indeed he is going ahead of you to Galilee; there you will see him.' This is my message for you." [8] So they left the tomb quickly with fear and great joy, and ran to tell his disciples. [9] Suddenly Jesus met them and said, "Greetings!" And they came to him, took hold of his feet, and worshiped him. [10] Then Jesus said to them, "Do not be afraid; go and tell my brothers to go to Galilee; there they will see me."

11 While they were going, some of the guard went into the city and told the chief priests everything that had happened. [12] After the priests[n] had assembled with the elders, they devised a plan to give a large sum of money to the soldiers, [13] telling them, "You must say, 'His disciples came by night and stole him away while we were asleep.' [14] If this comes to the governor's ears, we will satisfy him and keep you out of trouble." [15] So they took the money and did as they were directed. And this story is still told among the Jews to this day.

16 Now the eleven disciples went to Galilee, to the mountain to which Jesus

j Or *Take a guard* *k* Gk *you know how*
l Other ancient authorities read *the Lord*
m Other ancient authorities lack *from the dead*
n Gk *they*

27.57–61: Mk 15.42–47; Lk 23.50–56; Jn 19.38–42; Acts 13.29. **58:** Bodies of the executed were normally denied burial. **60:** See Mk 16.3–5 n.; Acts 13.29. **61:** 27.56.
27.62: *Next day,* the sabbath (Mk 15.42). **65:** *You have a guard,* permission to use the Temple police, under the authority of the Sanhedrin; or, *Take a guard,* i.e. granting them a squad of Roman soldiers.
28.1–10: The first Easter (Mk 16.1–8; Lk 24.1–11; Jn 20.1–10). **3:** *Lightning,* Dan 10.6. **4:** *The guards,* 27.62–66. **7:** 26.32; 28.16; Jn 21.1–23; 1 Cor 15.3–4, 12, 20. **8:** Compare Lk 24.9, 22–23; the sequence of events cannot be worked out. Each account is a separate summary of early Christian testimony to the fact of Jesus' resurrection. **9:** Jn 20.14–18.
28.11–15: Bribing the guard. 11: 27.62–66. **12:** *Large* enough to persuade *the soldiers* to invent a story contrary to the truth, at their own peril. **14:** *We will satisfy him,* in the same way they had persuaded the soldiers—by bribes. **15:** *This day,* i.e. the time when the Gospel according to Matthew was written.
28.16–20: Jesus' commission to his disciples. 17: 1 Cor 15.5–6; Jn 21.1–23; Lk

had directed them. 17When they saw
him, they worshiped him; but some
doubted. 18And Jesus came and said to
them, "All authority in heaven and on
earth has been given to me. 19Go therefore and make disciples of all nations,
baptizing them in the name of the Father

and of the Son and of the Holy Spirit,
20and teaching them to obey everything
that I have commanded you. And remember, I am with you always, to the
end of the age."*o*

o Other ancient authorities add *Amen*

24.11. *Worshiped,* Greek literally "prostrated
themselves in worship"; they had not done
this before the crucifixion. **18**: 11.27; Lk
10.22; Phil 2.9; Eph 1.20–22. *All authority,*
compare Dan 7.14. **19**: *All nations,* contrast

10.5, and compare Mk 16.15; Lk 24.47; Acts
1.8. According to Hebrew usage *in the name
of* means in the possession and protection of
(Ps 124.8). **20**: *I am with you,* 18.20; Acts
18.10.

The Gospel According to
Mark

The Gospel according to Mark is generally recognized as the earliest attempt to reduce the apostolic tradition concerning Jesus the Messiah to written form (see pp. ix–x NT). The Evangelist presents Jesus as the Son of God (1.1, 11; 5.7; 9.7; 14.61–62; 15.39) whose ministry was characterized by a succession of mighty works that, to those who had eyes to see, were signs of the presence of God's power and kingdom.

This Gospel, the shortest of the four, is largely a collection of narratives that depict Jesus as being almost constantly active; a favorite word in Mark is the Greek word meaning *immediately* or *at once* or *then,* which occurs about forty times in sixteen chapters. On the other hand, Mark records fewer words of Jesus than does any of the other Gospels; it contains one collection of sayings in the form of a discourse (ch 13) and a few parables (e.g. ch 4).

The main divisions are the following: 1.1–13, Opening events of Jesus' public life (John the Baptist; baptism and temptation of Jesus); 1.14–9.50, Jesus' preaching, teaching, and healing ministry in Galilee; ch 10, Journey to Jerusalem; chs 11–15, The last week, concluding with Jesus' crucifixion and burial; 16.1–8, The resurrection.

The style of the Marcan narrative is vivid and concrete, with an obvious concern for detail. The Evangelist notes, for example, the stunned reaction of the crowds (1.27f.; 2.12) and the fear and amazement of the disciples (9.6; 10.24, 32), as well as the pity, anger mingled with grief, and godly sorrow (or exasperation) experienced by Jesus (1.41; 3.5; 8.12).

In the earliest Greek manuscripts and versions (Latin, Syriac, Coptic, Armenian) the author's account breaks off suddenly with the words "for they were afraid" (16.8). Later manuscripts provide as a more suitable close for the book either a shorter or a longer ending, or sometimes both (see note *r* at 16.8). Whether Mark was prevented by death from completing his Gospel, or whether the original copy was accidentally mutilated, losing a portion at the close, no one can say.

Mark's ordering of the story of Jesus, beginning with the preaching of John the Baptist, preserves to some extent the form of the first Christian missionary message. The early disciples reinforced their claim that Jesus was the divinely appointed Messiah (8.29) by retelling significant moments from his days in Galilee and the last week of his life (compare Peter's account in Acts 10.34–43).

Although the Gospel is anonymous, an ancient tradition may well be correct in ascribing it to John Mark (mentioned in Acts 12.12; 15.37), who is said to have composed it at Rome as a summary of Peter's preaching (compare 1 Pet 5.13). The presence in this Gospel of several Latinisms, as well as Aramaic words and phrases (3.17; 5.41; 7.34; 14.36; 15.34) that are translated into Greek, points to a Gentile circle of readers living perhaps in Italy.

The date of writing, though uncertain, was probably prior to the fall of Jerusalem in A.D. 70. The persecution of Christians by Nero following the disastrous fire that destroyed half the city of Rome during the summer of A.D. 64 may account for the addition of the phrase "with persecutions" in 10.30 (a phrase not found in the Matthean and Lucan parallels).

For the literary genre of the gospels, see pp. viii–x NT.

1 The beginning of the good news *a* of Jesus Christ, the Son of God. *b*

2 As it is written in the prophet Isaiah, *c*

"See, I am sending my messenger
    ahead of you, *d*
who will prepare your way;
3 the voice of one crying out in the
    wilderness:
'Prepare the way of the Lord,
    make his paths straight,'"

4 John the baptizer appeared *e* in the wilderness, proclaiming a baptism of repentance for the forgiveness of sins. 5 And people from the whole Judean countryside and all the people of Jerusalem were going out to him, and were baptized by him in the river Jordan, confessing their sins. 6 Now John was clothed with camel's hair, with a leather belt around his waist, and he ate locusts and wild honey. 7 He proclaimed, "The one who is more powerful than I is coming after me; I am not worthy to stoop down and untie the thong of his sandals. 8 I have baptized you with *f* water; but he will baptize you with *f* the Holy Spirit."

9 In those days Jesus came from Nazareth of Galilee and was baptized by John in the Jordan. 10 And just as he was coming up out of the water, he saw the heavens torn apart and the Spirit descending like a dove on him. 11 And a voice came from heaven, "You are my Son, the Beloved; *g* with you I am well pleased."

12 And the Spirit immediately drove him out into the wilderness. 13 He was in the wilderness forty days, tempted by Satan; and he was with the wild beasts; and the angels waited on him.

14 Now after John was arrested, Jesus came to Galilee, proclaiming the good news *a* of God, *h* 15 and saying, "The time is fulfilled, and the kingdom of God has come near; *i* repent, and believe in the good news." *a*

16 As Jesus passed along the Sea of Galilee, he saw Simon and his brother Andrew casting a net into the sea—for they were fishermen. 17 And Jesus said to them, "Follow me and I will make you fish for people." 18 And immediately they left their nets and followed him. 19 As he went a little farther, he saw James son of Zebedee and his brother John, who were in their boat mending the nets. 20 Immediately he called them; and they left their father Zebedee in the boat with the hired men, and followed him.

21 They went to Capernaum; and when the sabbath came, he entered the

a Or *gospel* *b* Other ancient authorities lack *the Son of God* *c* Other ancient authorities read *in the prophets* *d* Gk *before your face* *e* Other ancient authorities read *John was baptizing* *f* Or *in* *g* Or *my beloved Son* *h* Other ancient authorities read *of the kingdom* *i* Or *is at hand*

1.1–8: Activity of John the Baptist (Mt 3.1–12; Lk 3.1–20; Jn 1.6, 15, 19–28). **1:** *The good news* begins with John's call to repentance. **2:** Mal 3.1 (God is the speaker); Mt 11.10; Lk 7.27. **3:** Isa 40.3. **4:** Acts 13.24. John called people to baptism *with water,* thereby symbolizing recognition and confession of sin together with acceptance of God's judgment and forgiveness. **8:** Baptism *with the Holy Spirit* would draw people into spiritual communion with God (Acts 2.17–21; Joel 2.28–29).

1.9–11: Jesus' baptism (Mt 3.13–17; Lk 3.21–22; Jn 1.29–34). **11:** *Beloved,* similar in meaning to *chosen* (Isa 42.1), refers to an act of will rather than of feeling (Ps 2.7; Lk 9.35; 2 Pet 1.17).

1.12–13: Jesus' temptation (Mt 4.1–11; Lk 4.1–13).

1.14–39: Beginnings of Jesus' activity in Galilee. 14–15: Mt 4.12–17; Lk 4.14–15. **15:** Jesus' message summarized. The whole of Mark is an expansion of this verse. *Repent,* return to God's way; *believe the good news,* accept the message (see Mt 4.17 n.). **16–20:** Mt 4.18–22; Lk 5.1–11; Jn 1.35–42. **16:** *Sea of Galilee,* other names are Sea of Tiberias and Lake of Gennesaret. It is 12¾ miles long by 7½ miles wide.

1.21–22: Mt 7.28–29; Lk 4.31–32. **22:** *One having authority . . . not as the scribes,* Jesus spoke without citing various spiritual authorities in support of his teaching. **23–28:** Lk 4.33–37. **23:** The *spirit,* or demon, was called *unclean* because the effect of the condition was to separate people from the worship of God. **27:** See Mt 7.29 n.

synagogue and taught. 22They were astounded at his teaching, for he taught them as one having authority, and not as the scribes. 23Just then there was in their synagogue a man with an unclean spirit, 24and he cried out, "What have you to do with us, Jesus of Nazareth? Have you come to destroy us? I know who you are, the Holy One of God." 25But Jesus rebuked him, saying, "Be silent, and come out of him!" 26And the unclean spirit, convulsing him and crying with a loud voice, came out of him. 27They were all amazed, and they kept on asking one another, "What is this? A new teaching—with authority! He*j* commands even the unclean spirits, and they obey him." 28At once his fame began to spread throughout the surrounding region of Galilee.

29 As soon as they*k* left the synagogue, they entered the house of Simon and Andrew, with James and John. 30Now Simon's mother-in-law was in bed with a fever, and they told him about her at once. 31He came and took her by the hand and lifted her up. Then the fever left her, and she began to serve them.

32 That evening, at sundown, they brought to him all who were sick or possessed with demons. 33And the whole city was gathered around the door. 34And he cured many who were sick with various diseases, and cast out many demons; and he would not permit the demons to speak, because they knew him.

35 In the morning, while it was still very dark, he got up and went out to a deserted place, and there he prayed. 36And Simon and his companions hunted for him. 37When they found him, they said to him, "Everyone is searching for you." 38He answered, "Let us go on to the neighboring towns, so that I may proclaim the message there also; for that is what I came out to do." 39And he went throughout Galilee, proclaiming the message in their synagogues and casting out demons.

40 A leper*l* came to him begging him, and kneeling*m* he said to him, "If you choose, you can make me clean." 41Moved with pity,*n* Jesus*o* stretched out his hand and touched him, and said to him, "I do choose. Be made clean!" 42Immediately the leprosy*l* left him, and he was made clean. 43After sternly warning him he sent him away at once, 44saying to him, "See that you say nothing to anyone; but go, show yourself to the priest, and offer for your cleansing what Moses commanded, as a testimony to them." 45But he went out and began to proclaim it freely, and to spread the word, so that Jesus*o* could no longer go into a town openly, but stayed out in the country; and people came to him from every quarter.

2 When he returned to Capernaum after some days, it was reported that he was at home. 2So many gathered around that there was no longer room for them, not even in front of the door; and he was speaking the word to them.

j Or *A new teaching! With authority he*
k Other ancient authorities read *he*
l The terms *leper* and *leprosy* can refer to several diseases   *m* Other ancient authorities lack *kneeling*   *n* Other ancient authorities read *anger*   *o* Gk *he*

1.29–34: Mt 8.14–17; Lk 4.38–41. **32:** The sabbath ended at sunset. **33:** *The door,* of Peter's house. **34:** See vv. 43–44 n.
1.35–39: Mt 4.23–25; Lk 4.42–44. **35:** See Lk 3.21 n. **38:** He *came out* from Capernaum (2.1).
1.40–9.50: Ministry and controversy, chiefly in Galilee. 1.40–45: Mt 8.2–4; Lk 5.12–16. **43–44:** Jesus wished the healing to carry with it a spiritual obligation. He apparently feared that rumors of miracles would gather the curious and foster cries for help only in physical terms, thus hindering his message (v. 45).
2.1–12: Healing a paralytic (Mt 9.1–8; Lk 5.17–26). **2:** *The word* was everything he had to say to people about God's purposes. **4:** Dwelling houses in Palestine usually had a flight of stone steps built on the outside and leading to *the roof,* which was flat and made probably of sticks and packed earth. **7:** The Jewish leaders sense in Jesus' words (v. 5) either some claim of his own or the assertion of a way of forgiveness different from that

3 Then some people*p* came, bringing to him a paralyzed man, carried by four of them. 4 And when they could not bring him to Jesus because of the crowd, they removed the roof above him; and after having dug through it, they let down the mat on which the paralytic lay. 5 When Jesus saw their faith, he said to the paralytic, "Son, your sins are forgiven." 6 Now some of the scribes were sitting there, questioning in their hearts, 7 "Why does this fellow speak in this way? It is blasphemy! Who can forgive sins but God alone?" 8 At once Jesus perceived in his spirit that they were discussing these questions among themselves; and he said to them, "Why do you raise such questions in your hearts? 9 Which is easier, to say to the paralytic, 'Your sins are forgiven,' or to say, 'Stand up and take your mat and walk'? 10 But so that you may know that the Son of Man has authority on earth to forgive sins"—he said to the paralytic— 11 "I say to you, stand up, take your mat and go to your home." 12 And he stood up, and immediately took the mat and went out before all of them; so that they were all amazed and glorified God, saying, "We have never seen anything like this!"

13 Jesus*q* went out again beside the sea; the whole crowd gathered around him, and he taught them. 14 As he was walking along, he saw Levi son of Alphaeus sitting at the tax booth, and he said to him, "Follow me." And he got up and followed him.

15 And as he sat at dinner*r* in Levi's*s* house, many tax collectors and sinners were also sitting*t* with Jesus and his disciples—for there were many who followed him. 16 When the scribes of*u* the Pharisees saw that he was eating with sinners and tax collectors, they said to his disciples, "Why does he eat*v* with tax collectors and sinners?" 17 When Jesus heard this, he said to them, "Those who are well have no need of a physician, but those who are sick; I have come to call not the righteous but sinners."

18 Now John's disciples and the Pharisees were fasting; and people*p* came and said to him, "Why do John's disciples and the disciples of the Pharisees fast, but your disciples do not fast?" 19 Jesus said to them, "The wedding guests cannot fast while the bridegroom is with them, can they? As long as they have the bridegroom with them, they cannot fast. 20 The days will come when the bridegroom is taken away from them, and then they will fast on that day.

21 "No one sews a piece of unshrunk cloth on an old cloak; otherwise, the patch pulls away from it, the new from the old, and a worse tear is made. 22 And no one puts new wine into old wine-

p Gk *they*   *q* Gk *He*   *r* Gk *reclined*
s Gk *his*   *t* Gk *reclining*   *u* Other ancient
authorities read *and*   *v* Other ancient authorities
add *and drink*

taught by the scribes. **10:** *Son of Man,* a title which Jesus used of himself, probably seemed to his listeners to carry either of two meanings: (*a*) that Jesus called himself a typical human being in accordance with the common meaning of *son of* (see Mt 5.45 n.); or (*b*) that Jesus (contrary to the humble conditions of his daily life) linked himself to the prophesied figure of Dan 7.13–14 who was popularly regarded as the coming Messiah (see Acts 7.56 n.). Jesus nowhere fully discloses his own understanding of the term (but see 8.32 n.). However, each meaning by itself, as well as both together (see Mt 25.29 n.), could have appealed to him. It was also characteristic of him to speak in such a way as to oblige his hearers to determine their own personal attitudes toward him as part of the process of understanding his words (see Mt 13.3 n.).

2.13–17: **The call of Levi** (Mt 9.9–13; Lk 5.27–32). **13:** *The sea,* of Galilee. **14:** *Levi,* Matthew. This experience was not necessarily Matthew's first contact with Jesus (v. 15). **16:** *Sinners* here means Jews who do not observe the dietary and other laws. **17:** See Lk 5.32 n.

2.18–22: **Fasting** (Mt 9.14–17; Lk 5.33–39). **18:** *John's,* i.e. the Baptist's *disciples.* The Pharisees fasted on Mondays and Thursdays. **19–20:** Isa 62.5; Lk 17.22. The implication is that Jesus comes as a *bridegroom* for his followers (the bride). Fasting was inappropriate to a wedding, and to joyous association with himself (see Jn 3.27–29 n.).

skins; otherwise, the wine will burst the skins, and the wine is lost, and so are the skins; but one puts new wine into fresh wineskins."*w*

23 One sabbath he was going through the grainfields; and as they made their way his disciples began to pluck heads of grain. 24 The Pharisees said to him, "Look, why are they doing what is not lawful on the sabbath?" 25 And he said to them, "Have you never read what David did when he and his companions were hungry and in need of food? 26 He entered the house of God, when Abiathar was high priest, and ate the bread of the Presence, which it is not lawful for any but the priests to eat, and he gave some to his companions." 27 Then he said to them, "The sabbath was made for humankind, and not humankind for the sabbath; 28 so the Son of Man is lord even of the sabbath."

3 Again he entered the synagogue, and a man was there who had a withered hand. 2 They watched him to see whether he would cure him on the sabbath, so that they might accuse him. 3 And he said to the man who had the withered hand, "Come forward." 4 Then he said to them, "Is it lawful to do good or to do harm on the sabbath, to save life or to kill?" But they were silent. 5 He looked around at them with anger; he was grieved at their hardness of heart and said to the man, "Stretch out your hand." He stretched it out, and his hand was restored. 6 The Pharisees went out and immediately conspired with the Herodians against him, how to destroy him.

7 Jesus departed with his disciples to the sea, and a great multitude from Galilee followed him; 8 hearing all that he was doing, they came to him in great numbers from Judea, Jerusalem, Idumea, beyond the Jordan, and the region around Tyre and Sidon. 9 He told his disciples to have a boat ready for him because of the crowd, so that they would not crush him; 10 for he had cured many, so that all who had diseases pressed upon him to touch him. 11 Whenever the unclean spirits saw him, they fell down before him and shouted, "You are the Son of God!" 12 But he sternly ordered them not to make him known.

13 He went up the mountain and called to him those whom he wanted, and they came to him. 14 And he appointed twelve, whom he also named apostles,*x* to be with him, and to be sent out to proclaim the message, 15 and to have authority to cast out demons. 16 So he appointed the twelve:*y* Simon (to whom he gave the name Peter); 17 James son of Zebedee and John the brother of James (to whom he gave the name Boanerges, that is, Sons of Thunder); 18 and Andrew, and Philip, and Bartholomew, and Matthew, and Thomas, and James son of Al-

w Other ancient authorities lack *but one puts new wine into fresh wineskins* *x* Other ancient authorities lack *whom he also named apostles* *y* Other ancient authorities lack *So he appointed the twelve*

2.23–3.6: Jesus and sabbath laws (Mt 12.1–14; Lk 6.1–11). **2.24:** The Pharisees attack Jesus by attacking his disciples (vv. 16, 18). **26:** *Abiathar* was *high priest* during David's reign (2 Sam 15.35); his father Ahimelech was priest at the time David ate the consecrated bread (1 Sam 21.1–6). **27:** Ex 23.12; Deut 5.14. **3.3–4:** Jesus, acting by the principle stated in 2.27, equated acts to meet human need with acts *lawful . . . on the sabbath.* **6:** *Herodians,* apparently a group supporting the royal family. Nothing definite is known about them, but probably their interests were secular. The Pharisees sought allies wherever they might be found (12.13).

3.7–12: Work of healing (Mt 4.24–25; 12.15–21; Lk 6.17–19; 4.41). **7–8:** 1.28, 38, 45. Jesus' fame has spread. **10:** 5.29, 34; 6.54–56. **12:** 1.43–44.

3.13–19a: The Twelve chosen (Mt 10.1–4; Lk 6.12–16). Jesus invited the Twelve to live intimately with him, adopting his way of life as well as his message. Spiritual strength came through the community he established (see 6.7–13 n.). **13:** Lk 6.12. **18:** This *Alphaeus* was apparently not identical with Levi's father (2.14). *Simon the Cananaean* belonged to a Jewish patriotic group also called Zealots (Lk 6.15; Acts 1.13).

phaeus, and Thaddaeus, and Simon the Cananaean, [19]and Judas Iscariot, who betrayed him.

Then he went home; [20]and the crowd came together again, so that they could not even eat. [21]When his family heard it, they went out to restrain him, for people were saying, "He has gone out of his mind." [22]And the scribes who came down from Jerusalem said, "He has Beelzebul, and by the ruler of the demons he casts out demons." [23]And he called them to him, and spoke to them in parables, "How can Satan cast out Satan? [24]If a kingdom is divided against itself, that kingdom cannot stand. [25]And if a house is divided against itself, that house will not be able to stand. [26]And if Satan has risen up against himself and is divided, he cannot stand, but his end has come. [27]But no one can enter a strong man's house and plunder his property without first tying up the strong man; then indeed the house can be plundered.

[28] "Truly I tell you, people will be forgiven for their sins and whatever blasphemies they utter; [29]but whoever blasphemes against the Holy Spirit can never have forgiveness, but is guilty of an eternal sin"— [30]for they had said, "He has an unclean spirit."

[31] Then his mother and his brothers came; and standing outside, they sent to him and called him. [32]A crowd was sitting around him; and they said to him, "Your mother and your brothers and sisters[z] are outside, asking for you." [33]And he replied, "Who are my mother and my brothers?" [34]And looking at those who sat around him, he said, "Here are my mother and my brothers! [35]Whoever does the will of God is my brother and sister and mother."

[4] Again he began to teach beside the sea. Such a very large crowd gathered around him that he got into a boat on the sea and sat there, while the whole crowd was beside the sea on the land. [2]He began to teach them many things in parables, and in his teaching he said to them: [3]"Listen! A sower went out to sow. [4]And as he sowed, some seed fell on the path, and the birds came and ate it up. [5]Other seed fell on rocky ground, where it did not have much soil, and it sprang up quickly, since it had no depth of soil. [6]And when the sun rose, it was scorched; and since it had no root, it withered away. [7]Other seed fell among thorns, and the thorns grew up and choked it, and it yielded no grain. [8]Other seed fell into good soil and brought forth grain, growing up and increasing and yielding thirty and sixty and a hundredfold." [9]And he said, "Let anyone with ears to hear listen!"

[10] When he was alone, those who were around him along with the twelve asked him about the parables. [11]And he said to them, "To you has been given the secret[a] of the kingdom of God, but for those outside, everything comes in parables; [12]in order that

'they may indeed look, but not
    perceive,
and may indeed listen, but not
    understand;
so that they may not turn again
    and be forgiven.' "

z Other ancient authorities lack *and sisters*
a Or *mystery*

3.19b–35: Questions about Jesus' power (Mt 12.22–37; Lk 11.14–23; 12.10; 6.43–45). **21:** *His family* (perhaps his mother and his brothers, v. 31) were concerned both for his safety, amidst the intense emotions rising round him, and his sanity. **22:** The Pharisees attribute his acts to demonic power (see Lk 7.33 n.; Jn 10.20). *Beelzebul,* a pagan god (see 2 Kings 1.2 n.) identified here with Satan. **24–27:** Inner division is destructive. **29:** Mt 12.31–32 n. **31–35:** *Brothers,* see Mt 13.55 n.

4.1–34: Teaching in parables (Mt ch 13; Lk 8.4–18; 13.18–21). On Jesus' use of parables, see Mt 13.3 n. **3:** The *sower* scattered seed broadcast. **5:** The rock lay just below the surface of the ground. **11:** *Has been given,* i.e. by God. Jesus distinguished between his followers' spiritual opportunities and those of others who were *outside,* i.e. those who do not belong to the elect (Lk 10.23–24). **12:** An adaptation of Isa 6.9–10.

4.13–20: See Mt 13.18–23 n. **21:** *A bushel,*

13 And he said to them, "Do you not understand this parable? Then how will you understand all the parables? 14 The sower sows the word. 15 These are the ones on the path where the word is sown: when they hear, Satan immediately comes and takes away the word that is sown in them. 16 And these are the ones sown on rocky ground: when they hear the word, they immediately receive it with joy. 17 But they have no root, and endure only for a while; then, when trouble or persecution arises on account of the word, immediately they fall away.*b* 18 And others are those sown among the thorns: these are the ones who hear the word, 19 but the cares of the world, and the lure of wealth, and the desire for other things come in and choke the word, and it yields nothing. 20 And these are the ones sown on the good soil: they hear the word and accept it and bear fruit, thirty and sixty and a hundredfold."

21 He said to them, "Is a lamp brought in to be put under the bushel basket, or under the bed, and not on the lampstand? 22 For there is nothing hidden, except to be disclosed; nor is anything secret, except to come to light. 23 Let anyone with ears to hear listen!" 24 And he said to them, "Pay attention to what you hear; the measure you give will be the measure you get, and still more will be given you. 25 For to those who have, more will be given; and from those who have nothing, even what they have will be taken away."

26 He also said, "The kingdom of God is as if someone would scatter seed on the ground, 27 and would sleep and rise night and day, and the seed would sprout and grow, he does not know how. 28 The earth produces of itself, first the stalk, then the head, then the full grain in the head. 29 But when the grain is ripe, at once he goes in with his sickle, because the harvest has come."

30 He also said, "With what can we compare the kingdom of God, or what parable will we use for it? 31 It is like a mustard seed, which, when sown upon the ground, is the smallest of all the seeds on earth; 32 yet when it is sown it grows up and becomes the greatest of all shrubs, and puts forth large branches, so that the birds of the air can make nests in its shade."

33 With many such parables he spoke the word to them, as they were able to hear it; 34 he did not speak to them except in parables, but he explained everything in private to his disciples.

35 On that day, when evening had come, he said to them, "Let us go across to the other side." 36 And leaving the crowd behind, they took him with them in the boat, just as he was. Other boats were with him. 37 A great windstorm arose, and the waves beat into the boat, so that the boat was already being swamped. 38 But he was in the stern, asleep on the cushion; and they woke him up and said to him, "Teacher, do you not care that we are perishing?" 39 He woke up and rebuked the wind, and said to the sea, "Peace! Be still!" Then the wind ceased, and there was a dead calm. 40 He said to them, "Why are you afraid? Have you still no faith?" 41 And they were filled with great awe and said to one

b Or *stumble*

(Greek "modios"), a pan or container holding about eight quarts (Mt 5.15; Lk 8.16; 11.33). **22**: Mt 10.26; Lk 8.17; 12.2. **24–25**: Mt 7.2; 13.12; Lk 6.38.
4.26–29: The seed growing secretly (Mt 13.24–30). The growth of God's kingdom in the world is beyond human understanding or control. Yet people may recognize its progress and play a part in it. **29**: Joel 3.13.

4.30–34: The mustard seed (see Mt 13.31–32 n.; Lk 13.18–19). **32**: Dan 4.12, 21; Ezek 17.23; 31.6. **33–34**: Mt 13.34–35.
4.35–41: Wind and sea calmed (Mt 8.18, 23–27; Lk 8.22–25). Jesus' authority is shown to extend to the world of natural forces. **38**: *Teacher,* see Lk 8.24 n. **40**: *Faith* trusts God to achieve his purpose, even through apparent destruction (Mt 6.25–33; 10.38 n.; Mk 14.35–36).

another, "Who then is this, that even the wind and the sea obey him?"

5 They came to the other side of the sea, to the country of the Gerasenes. *c* 2 And when he had stepped out of the boat, immediately a man out of the tombs with an unclean spirit met him. 3 He lived among the tombs; and no one could restrain him any more, even with a chain; 4 for he had often been restrained with shackles and chains, but the chains he wrenched apart, and the shackles he broke in pieces; and no one had the strength to subdue him. 5 Night and day among the tombs and on the mountains he was always howling and bruising himself with stones. 6 When he saw Jesus from a distance, he ran and bowed down before him; 7 and he shouted at the top of his voice, "What have you to do with me, Jesus, Son of the Most High God? I adjure you by God, do not torment me." 8 For he had said to him, "Come out of the man, you unclean spirit!" 9 Then Jesus *d* asked him, "What is your name?" He replied, "My name is Legion; for we are many." 10 He begged him earnestly not to send them out of the country. 11 Now there on the hillside a great herd of swine was feeding; 12 and the unclean spirits *e* begged him, "Send us into the swine; let us enter them." 13 So he gave them permission. And the unclean spirits came out and entered the swine; and the herd, numbering about two thousand, rushed down the steep bank into the sea, and were drowned in the sea.

14 The swineherds ran off and told it in the city and in the country. Then people came to see what it was that had happened. 15 They came to Jesus and saw the demoniac sitting there, clothed and in his right mind, the very man who had had the legion; and they were afraid. 16 Those who had seen what had happened to the demoniac and to the swine reported it. 17 Then they began to beg Jesus *f* to leave their neighborhood. 18 As he was getting into the boat, the man who had been possessed by demons begged him that he might be with him. 19 But Jesus *d* refused, and said to him, "Go home to your friends, and tell them how much the Lord has done for you, and what mercy he has shown you." 20 And he went away and began to proclaim in the Decapolis how much Jesus had done for him; and everyone was amazed.

21 When Jesus had crossed again in the boat *g* to the other side, a great crowd gathered around him; and he was by the sea. 22 Then one of the leaders of the synagogue named Jairus came and, when he saw him, fell at his feet 23 and begged him repeatedly, "My little daughter is at the point of death. Come and lay your hands on her, so that she may be made well, and live." 24 So he went with him.

And a large crowd followed him and pressed in on him. 25 Now there was a woman who had been suffering from hemorrhages for twelve years. 26 She had endured much under many physicians, and had spent all that she had; and she was no better, but rather grew worse. 27 She had heard about Jesus, and came up behind him in the crowd and touched his cloak, 28 for she said, "If I but touch his clothes, I will be made well." 29 Immedi-

c Other ancient authorities read *Gergesenes;* others, *Gadarenes* *d* Gk *he* *e* Gk *they* *f* Gk *him* *g* Other ancient authorities lack *in the boat*

5.1–43: Preaching tour. 1–20: The Gerasene demoniac (Mt 8.28–34; Lk 8.26–39). **7**: 1.24. **9**: A legion, a major unit in the Roman army, consisted of four thousand to six thousand men. Jesus would have the man witness in an area where he was known. **13**: Since *swine* were unclean animals, it was considered fitting for *unclean spirits* to enter them. **20**: *Decapolis,* a federation of about ten cities in eastern Palestine.

5.21–43: Jairus' daughter raised (Mt 9.18–26; Lk 8.40–56). **23**: See Mt 9.21 n. **25**: Lev 15.25–30. **28**: 5.23. **30**: Lk 5.17. **34**: *Has made you well,* see Mt 9.21 n. **36**: 4.40. **39**: *Them,* i.e. the professional mourners who were present on such occasions. *Not dead but sleeping,* see Mt 9.24 n. **41**: The phrase *talitha cum* preserves the actual (as distinct from translated) Aramaic words of Jesus (see 2 Kings 18.26 n.).

ately her hemorrhage stopped; and she felt in her body that she was healed of her disease. 30 Immediately aware that power had gone forth from him, Jesus turned about in the crowd and said, "Who touched my clothes?" 31 And his disciples said to him, "You see the crowd pressing in on you; how can you say, 'Who touched me?' " 32 He looked all around to see who had done it. 33 But the woman, knowing what had happened to her, came in fear and trembling, fell down before him, and told him the whole truth. 34 He said to her, "Daughter, your faith has made you well; go in peace, and be healed of your disease."

35 While he was still speaking, some people came from the leader's house to say, "Your daughter is dead. Why trouble the teacher any further?" 36 But overhearing*h* what they said, Jesus said to the leader of the synagogue, "Do not fear, only believe." 37 He allowed no one to follow him except Peter, James, and John, the brother of James. 38 When they came to the house of the leader of the synagogue, he saw a commotion, people weeping and wailing loudly. 39 When he had entered, he said to them, "Why do you make a commotion and weep? The child is not dead but sleeping." 40 And they laughed at him. Then he put them all outside, and took the child's father and mother and those who were with him, and went in where the child was. 41 He took her by the hand and said to her, "Talitha cum," which means, "Little girl, get up!" 42 And immediately the girl got up and began to walk about (she was twelve years of age). At this they were overcome with amazement. 43 He strictly ordered them that no one should know this, and told them to give her something to eat.

6 He left that place and came to his hometown, and his disciples followed him. 2 On the sabbath he began to teach in the synagogue, and many who heard him were astounded. They said, "Where did this man get all this? What is this wisdom that has been given to him? What deeds of power are being done by his hands! 3 Is not this the carpenter, the son of Mary*i* and brother of James and Joses and Judas and Simon, and are not his sisters here with us?" And they took offense*j* at him. 4 Then Jesus said to them, "Prophets are not without honor, except in their hometown, and among their own kin, and in their own house." 5 And he could do no deed of power there, except that he laid his hands on a few sick people and cured them. 6 And he was amazed at their unbelief.

Then he went about among the villages teaching. 7 He called the twelve and began to send them out two by two, and gave them authority over the unclean spirits. 8 He ordered them to take nothing for their journey except a staff; no bread, no bag, no money in their belts; 9 but to wear sandals and not to put on two tunics. 10 He said to them, "Wherever you enter a house, stay there until you leave the place. 11 If any place will not welcome you and they refuse to hear you, as you leave, shake off the dust that is on your feet as a testimony against them." 12 So they went out and proclaimed that all should repent. 13 They cast out many demons, and anointed with oil many who were sick and cured them.

14 King Herod heard of it, for Jesus'*k*

h Or *ignoring*; other ancient authorities read *hearing*   *i* Other ancient authorities read *son of the carpenter and of Mary*   *j* Or *stumbled*   *k* Gk *his*

6.1–6: **Rejection at home** (Mt 13.53–58; Lk 4.16–30). **3**: See Mt 13.55 n. **5–6**: Jesus required faith on the part of those who sought healing for themselves or for others (occasional apparent exceptions occur; e.g. Jn 5.13).
6.7–13: **Commissioning and instruction of the Twelve** (Mt 10.1, 9–11, 14; Lk 9.1–6). See 3.13–19 n. **7**: The disciples share Jesus' *authority* over malign, destructive forces (vv. 7–9); his refusal to engage in self-seeking (v. 10), or violence (v. 11); his message (v. 12; 1.14–15); and his sympathy for human suffering (v. 13). **9**: *Tunics,* see Mt 10.10 n. **12**: 1.14–15. **13**: Jas 5.14; Isa 1.6; Lk 10.34.

name had become known. Some were[l] saying, "John the baptizer has been raised from the dead; and for this reason these powers are at work in him." 15 But others said, "It is Elijah." And others said, "It is a prophet, like one of the prophets of old." 16 But when Herod heard of it, he said, "John, whom I beheaded, has been raised."

17 For Herod himself had sent men who arrested John, bound him, and put him in prison on account of Herodias, his brother Philip's wife, because Herod[m] had married her. 18 For John had been telling Herod, "It is not lawful for you to have your brother's wife." 19 And Herodias had a grudge against him, and wanted to kill him. But she could not, 20 for Herod feared John, knowing that he was a righteous and holy man, and he protected him. When he heard him, he was greatly perplexed;[n] and yet he liked to listen to him. 21 But an opportunity came when Herod on his birthday gave a banquet for his courtiers and officers and for the leaders of Galilee. 22 When his daughter Herodias[o] came in and danced, she pleased Herod and his guests; and the king said to the girl, "Ask me for whatever you wish, and I will give it." 23 And he solemnly swore to her, "Whatever you ask me, I will give you, even half of my kingdom." 24 She went out and said to her mother, "What should I ask for?" She replied, "The head of John the baptizer." 25 Immediately she rushed back to the king and requested, "I want you to give me at once the head of John the Baptist on a platter." 26 The king was deeply grieved; yet out of regard for his oaths and for the guests, he did not want to refuse her. 27 Immediately the king sent a soldier of the guard with orders to bring John's[p] head. He went and beheaded him in the prison, 28 brought his head on a platter, and gave it to the girl. Then the girl gave it to her mother. 29 When his disciples heard about it, they came and took his body, and laid it in a tomb.

30 The apostles gathered around Jesus, and told him all that they had done and taught. 31 He said to them, "Come away to a deserted place all by yourselves and rest a while." For many were coming and going, and they had no leisure even to eat. 32 And they went away in the boat to a deserted place by themselves. 33 Now many saw them going and recognized them, and they hurried there on foot from all the towns and arrived ahead of them. 34 As he went ashore, he saw a great crowd; and he had compassion for them, because they were like sheep without a shepherd; and he began to teach them many things. 35 When it grew late, his disciples came to him and said, "This is a deserted place, and the hour is now very late; 36 send them away so that they may go into the surrounding country and villages and buy something for themselves to eat." 37 But he answered them, "You give them something to eat." They said to him, "Are we to go and buy two hundred denarii[q] worth of bread, and give it to them to eat?" 38 And he said to them, "How many loaves have you? Go and see." When they had found out, they said, "Five, and two fish." 39 Then he ordered them to get all the people to sit down in groups on the green grass. 40 So they sat down in groups of hundreds and of fifties. 41 Taking the five

l Other ancient authorities read *He was*
m Gk *he*   n Other ancient authorities read *he did many things*   o Other ancient authorities read *the daughter of Herodias herself*   p Gk *his*
q The denarius was the usual day's wage for a laborer

6.14–29: Death of John (Mt 14.1–12; Lk 9.7–9). **14:** *Herod* Antipas, son of Herod the Great, was tetrarch of Galilee, and only by courtesy could he be called *King.*
6.17: *In prison,* at Machaerus, east of the Dead Sea, on the Nabataean border (see Josephus, *Ant.* XVIII.109–119). **20:** Mt 21.26.

22: *His daughter Herodias,* the manuscripts here differ; see note *o.* Contrary to Josephus (see Mt 14.6 n.) the daughter of Herodias may also have been named Herodias. **27:** *In the prison,* at Machaerus, where a fort and prison had been built. It was about five miles east of the Dead Sea.

loaves and the two fish, he looked up to heaven, and blessed and broke the loaves, and gave them to his disciples to set before the people; and he divided the two fish among them all. ⁴²And all ate and were filled; ⁴³and they took up twelve baskets full of broken pieces and of the fish. ⁴⁴Those who had eaten the loaves numbered five thousand men.

45 Immediately he made his disciples get into the boat and go on ahead to the other side, to Bethsaida, while he dismissed the crowd. ⁴⁶After saying farewell to them, he went up on the mountain to pray.

47 When evening came, the boat was out on the sea, and he was alone on the land. ⁴⁸When he saw that they were straining at the oars against an adverse wind, he came towards them early in the morning, walking on the sea. He intended to pass them by. ⁴⁹But when they saw him walking on the sea, they thought it was a ghost and cried out; ⁵⁰for they all saw him and were terrified. But immediately he spoke to them and said, "Take heart, it is I; do not be afraid." ⁵¹Then he got into the boat with them and the wind ceased. And they were utterly astounded, ⁵²for they did not understand about the loaves, but their hearts were hardened.

53 When they had crossed over, they came to land at Gennesaret and moored the boat. ⁵⁴When they got out of the boat, people at once recognized him, ⁵⁵and rushed about that whole region and began to bring the sick on mats to wherever they heard he was. ⁵⁶And wherever he went, into villages or cities or farms, they laid the sick in the marketplaces, and begged him that they might touch even the fringe of his cloak; and all who touched it were healed.

7 Now when the Pharisees and some of the scribes who had come from Jerusalem gathered around him, ²they noticed that some of his disciples were eating with defiled hands, that is, without washing them. ³(For the Pharisees, and all the Jews, do not eat unless they thoroughly wash their hands,ʳ thus observing the tradition of the elders; ⁴and they do not eat anything from the market unless they wash it;ˢ and there are also many other traditions that they observe, the washing of cups, pots, and bronze kettles.ᵗ) ⁵So the Pharisees and the scribes asked him, "Why do your disciples not liveᵘ according to the tradition of the elders, but eat with defiled hands?" ⁶He said to them, "Isaiah prophesied rightly about you hypocrites, as it is written,

'This people honors me with
        their lips,
    but their hearts are far from me;
⁷ in vain do they worship me,
        teaching human precepts as
        doctrines.'

r Meaning of Gk uncertain   s Other ancient authorities read *and when they come from the marketplace, they do not eat unless they purify themselves*   t Other ancient authorities add *and beds*   u Gk *walk*

6.30–44: **Five thousand fed** (Mt 14.13–21; Lk 9.10–17; Jn 6.1–13; compare Mk 8.1–10). **34**: *Sheep without a shepherd,* a familiar picture of aimlessness (Num 27.17; 1 Kings 22.17; Ezek 34.5). **44**: *Men,* see Mt 14.21 n.

6.45–52: **Jesus walks on water** (Mt 14.22–33; Jn 6.15–21). **45**: *Bethsaida,* a fishing village at the north end of the Sea of Galilee. **48**: *Early in the morning,* Greek literally "about the fourth watch of the night" (the fourth watch was from 3 to 6 A.M.). *He intended to pass them by* describes the way Jesus appeared to his disciples. **52**: The disciples miss the true import of Jesus' acts through lack of faith (Mk 3.5; 8.17; Jn 12.40; Rom 11.7, 25; 2 Cor 3.14; Eph 4.18; but compare Mt 14.33).

6.53–56: **Belief in Jesus' power to heal** (Mt 14.34–36). Compare Mt 4.24; Mk 1.32–34; 3.10; Lk 4.40–41; 6.18, 19. **56**: *Fringe,* the blue twisted threads at the four corners of male garments, as a reminder to obey God's commandments (Num 15.38–40; Deut 22.12). That Jesus wore this fringe indicates his observance of the Law.

7.1–23: **Tradition of the elders** (Mt 15.1–20). Whereas the common people were moved by elemental needs, and Jesus was aroused to compassion for human suffering (6.53–56), the religious leaders were concerned with details of ritual. **3**: See Lk 11.38 n. **4**: Mt 23.25; Lk 11.39. **5**: Gal 1.14. **6–7**: Isa 29.13, according to the Septuagint.

8 You abandon the commandment of God and hold to human tradition."

9 Then he said to them, "You have a fine way of rejecting the commandment of God in order to keep your tradition! 10 For Moses said, 'Honor your father and your mother'; and, 'Whoever speaks evil of father or mother must surely die.' 11 But you say that if anyone tells father or mother, 'Whatever support you might have had from me is Corban' (that is, an offering to God*v*)— 12 then you no longer permit doing anything for a father or mother, 13 thus making void the word of God through your tradition that you have handed on. And you do many things like this."

14 Then he called the crowd again and said to them, "Listen to me, all of you, and understand: 15 there is nothing outside a person that by going in can defile, but the things that come out are what defile."*w*

17 When he had left the crowd and entered the house, his disciples asked him about the parable. 18 He said to them, "Then do you also fail to understand? Do you not see that whatever goes into a person from outside cannot defile, 19 since it enters, not the heart but the stomach, and goes out into the sewer?" (Thus he declared all foods clean.) 20 And he said, "It is what comes out of a person that defiles. 21 For it is from within, from the human heart, that evil intentions come: fornication, theft, murder, 22 adultery, avarice, wickedness, deceit, licentiousness, envy, slander, pride, folly. 23 All these evil things come from within, and they defile a person."

24 From there he set out and went away to the region of Tyre.*x* He entered a house and did not want anyone to know he was there. Yet he could not escape notice, 25 but a woman whose little daughter had an unclean spirit immediately heard about him, and she came and bowed down at his feet. 26 Now the woman was a Gentile, of Syrophoenician origin. She begged him to cast the demon out of her daughter. 27 He said to her, "Let the children be fed first, for it is not fair to take the children's food and throw it to the dogs." 28 But she answered him, "Sir,*y* even the dogs under the table eat the children's crumbs." 29 Then he said to her, "For saying that, you may go—the demon has left your daughter." 30 So she went home, found the child lying on the bed, and the demon gone.

31 Then he returned from the region of Tyre, and went by way of Sidon towards the Sea of Galilee, in the region of the Decapolis. 32 They brought to him a deaf man who had an impediment in his speech; and they begged him to lay his hand on him. 33 He took him aside in private, away from the crowd, and put his fingers into his ears, and he spat and touched his tongue. 34 Then looking up to heaven, he sighed and said to him, "Ephphatha," that is, "Be opened." 35 And immediately his ears were opened, his tongue was released, and he spoke plainly. 36 Then Jesus*z* ordered them to tell no one; but the more he ordered them, the more zealously they proclaimed it. 37 They were astounded beyond measure, saying, "He has done everything well; he even makes the deaf to hear and the mute to speak."

8 In those days when there was again a great crowd without anything to eat, he called his disciples and said to

---

v Gk lacks *to God*   *w* Other ancient authorities add verse 16, *"Let anyone with ears to hear listen"*   *x* Other ancient authorities add *and Sidon*   *y* Or *Lord*; other ancient authorities prefix *Yes*   *z* Gk *he*

11: *An offering to God* verbally, but in fact retained for private use. 15: See Mt 15.10–20 n. 19: *Clean,* i.e. ritually. 21–23: Gal 5.19–21.
7.24–30: **The Syrophoenician woman** (Mt 15.21–28). 27: In Jesus' reply, the Jews are represented as *children,* the Gentiles as *dogs,* and the benefits of Jesus' ministry as *the* children's *food.* See Mt 15.24 n. 28: See Mt 15.27 n.
7.31–37: **Healings** (Mt 15.29–31). 31: *Decapolis,* see 5.20 n. 33: *Touched his tongue,* i.e. with the saliva. 34: Jesus looked *up to heaven,*

them, 2 "I have compassion for the crowd, because they have been with me now for three days and have nothing to eat. 3 If I send them away hungry to their homes, they will faint on the way—and some of them have come from a great distance." 4 His disciples replied, "How can one feed these people with bread here in the desert?" 5 He asked them, "How many loaves do you have?" They said, "Seven." 6 Then he ordered the crowd to sit down on the ground; and he took the seven loaves, and after giving thanks he broke them and gave them to his disciples to distribute; and they distributed them to the crowd. 7 They had also a few small fish; and after blessing them, he ordered that these too should be distributed. 8 They ate and were filled; and they took up the broken pieces left over, seven baskets full. 9 Now there were about four thousand people. And he sent them away. 10 And immediately he got into the boat with his disciples and went to the district of Dalmanutha.*a*

11 The Pharisees came and began to argue with him, asking him for a sign from heaven, to test him. 12 And he sighed deeply in his spirit and said, "Why does this generation ask for a sign? Truly I tell you, no sign will be given to this generation." 13 And he left them, and getting into the boat again, he went across to the other side.

14 Now the disciples*b* had forgotten to bring any bread; and they had only one loaf with them in the boat. 15 And he cautioned them, saying, "Watch out—beware of the yeast of the Pharisees and the yeast of Herod."*c* 16 They said to one another, "It is because we have no bread." 17 And becoming aware of it, Jesus said to them, "Why are you talking about having no bread? Do you still not perceive or understand? Are your hearts hardened? 18 Do you have eyes, and fail to see? Do you have ears, and fail to hear? And do you not remember? 19 When I broke the five loaves for the five thousand, how many baskets full of broken pieces did you collect?" They said to him, "Twelve." 20 "And the seven for the four thousand, how many baskets full of broken pieces did you collect?" And they said to him, "Seven." 21 Then he said to them, "Do you not yet understand?"

22 They came to Bethsaida. Some people*d* brought a blind man to him and begged him to touch him. 23 He took the blind man by the hand and led him out of the village; and when he had put saliva on his eyes and laid his hands on him, he asked him, "Can you see anything?" 24 And the man*e* looked up and said, "I can see people, but they look like trees, walking." 25 Then Jesus*e* laid his hands on his eyes again; and he looked intently and his sight was restored, and he saw everything clearly. 26 Then he sent him away to his home, saying, "Do not even go into the village."*f*

27 Jesus went on with his disciples to the villages of Caesarea Philippi; and on

a Other ancient authorities read *Mageda* or *Magdala*   *b* Gk *they*   *c* Other ancient authorities read *the Herodians*   *d* Gk *They*   *e* Gk *he*   *f* Other ancient authorities add *or tell anyone in the village*

---

thus indicating to the deaf man that he was praying for him. *Ephphatha,* an Aramaic word (see 5.41 n.).

8.1–10: Four thousand fed (Mt 15.32–39). This narrative repeats the thought of 6.30–44. Some regard these passages as alternate ways of describing one original event, the details of which we can no longer determine. **10:** *Dalmanutha,* the location is unknown (see note *a* and Mt 15.39).

8.11–13: Sayings on signs (Mt 16.1–4; 12.38–39; Lk 11.29; 11.16; 12.54–56). **13:** *The other side,* the eastern side of the Sea of Galilee.

8.14–21: Yeast of the Pharisees (Mt 16.5–12; Lk 12.1). **14–15:** Jesus joins two ideas. **15:** *Yeast* here seems to refer to settled conviction which affects all of life as yeast raises dough. *Yeast of the Pharisees* is hypocrisy (Lk 12.1), which spreads its influence by means of their teaching (Mt 16.12). *Yeast of Herod* is worldliness and irreligion. **17–18:** Isa 6.9–10; Jer 5.21; Ezek 12.2; Mt 13.10–15; Mk 6.52; Jn 12.36–41. **19:** 6.41–44. **20:** 8.1–10.

8.22–26: A blind man healed. 10.46–52; Jn 9.1–7. **22:** *Bethsaida,* 6.45; Lk 9.10.

8.27–33: Peter's confession (Mt 16.13–

the way he asked his disciples, "Who do people say that I am?" 28 And they answered him, "John the Baptist; and others, Elijah; and still others, one of the prophets." 29 He asked them, "But who do you say that I am?" Peter answered him, "You are the Messiah."*g* 30 And he sternly ordered them not to tell anyone about him.

31 Then he began to teach them that the Son of Man must undergo great suffering, and be rejected by the elders, the chief priests, and the scribes, and be killed, and after three days rise again. 32 He said all this quite openly. And Peter took him aside and began to rebuke him. 33 But turning and looking at his disciples, he rebuked Peter and said, "Get behind me, Satan! For you are setting your mind not on divine things but on human things."

34 He called the crowd with his disciples, and said to them, "If any want to become my followers, let them deny themselves and take up their cross and follow me. 35 For those who want to save their life will lose it, and those who lose their life for my sake, and for the sake of the gospel,*h* will save it. 36 For what will it profit them to gain the whole world and forfeit their life? 37 Indeed, what can they give in return for their life? 38 Those who are ashamed of me and of my words*i* in this adulterous and sinful generation, of them the Son of Man will also be ashamed when he comes in the glory of his Father with the holy angels."

9 1 And he said to them, "Truly I tell you, there are some standing here who will not taste death until they see that the kingdom of God has come with*j* power."

2 Six days later, Jesus took with him Peter and James and John, and led them up a high mountain apart, by themselves. And he was transfigured before them, 3 and his clothes became dazzling white, such as no one*k* on earth could bleach them. 4 And there appeared to them Elijah with Moses, who were talking with Jesus. 5 Then Peter said to Jesus, "Rabbi, it is good for us to be here; let us make three dwellings,*l* one for you, one for Moses, and one for Elijah." 6 He did not know what to say, for they were terrified. 7 Then a cloud overshadowed them, and from the cloud there came a voice, "This is my Son, the Beloved;*m* listen to him!" 8 Suddenly when they looked around, they saw no one with them any more, but only Jesus.

9 As they were coming down the

---

g Or *the Christ*   *h* Other ancient authorities read *lose their life for the sake of the gospel*
i Other ancient authorities read *and of mine*
j Or *in*   *k* Gk *no fuller*   *l* Or *tents*
m Or *my beloved Son*

23; Lk 9.18–22). **27:** *Caesarea Philippi* was a site of pagan worship. **28:** 6.14–16. **29:** Jn 6.66–69. **30:** Jesus consistently sought to repress sensational reports (see 1.43–44 n.). **31:** First prediction of the Passion (compare 9.30–32; 10.33–34). **32:** The idea that the *Son of Man* (the Messiah) was to suffer was in complete contrast to Jewish expectation (see 2.10 n.; 9.10 n.; Mt 16.22). **33:** Jesus saw in Peter's words a continuation of Satan's temptation (Mt 4.10; Lk 4.8).

8.34–9.1: On discipleship (Mt 16.24–28; Lk 9.23–27). **8.34:** *Deny themselves,* i.e. renounce self-centeredness; see Mt 10.38 n. **35:** Mt 10.39; Lk 17.33; Jn 12.25. The apparent contradiction here is overcome because everyone's existence depends on God. **38:** See Mt 12.39 n. **9.1:** *Taste,* become personally acquainted with.

9.2–8: The transfiguration (Mt 17.1–8; Lk 9.28–36). By this narrative the author means to describe a vision of Jesus in heavenly glory as the Messiah (see Lk 9.28–36 n.). **2:** *Six days later,* after Peter's confession of Jesus as the Messiah (8.29). *Transfigured,* having a non-earthly appearance. **4:** The prophet *Elijah* was expected to appear on earth before the Messiah appeared (Mal 4.5–6). *Moses,* the lawgiver, the traditional author of the first five books of the Bible, which formed the basic authority in Jewish religion. **5:** *Three dwellings,* booth-like, temporary shelters. **7:** Mt 3.17; Jn 12.28–29; 2 Pet 1.17–18.

9.9–13: Prophecies about Elijah (Mt 17.9–13). **10:** The disciples could not associate belief in resurrection with the Son of Man (v. 9; see 2.10 n.). **11:** See Mt 11.14 n. **13:** *Elijah has come* in the person of John the Baptist

mountain, he ordered them to tell no one about what they had seen, until after the Son of Man had risen from the dead. [10] So they kept the matter to themselves, questioning what this rising from the dead could mean. [11] Then they asked him, "Why do the scribes say that Elijah must come first?" [12] He said to them, "Elijah is indeed coming first to restore all things. How then is it written about the Son of Man, that he is to go through many sufferings and be treated with contempt? [13] But I tell you that Elijah has come, and they did to him whatever they pleased, as it is written about him."

14 When they came to the disciples, they saw a great crowd around them, and some scribes arguing with them. [15] When the whole crowd saw him, they were immediately overcome with awe, and they ran forward to greet him. [16] He asked them, "What are you arguing about with them?" [17] Someone from the crowd answered him, "Teacher, I brought you my son; he has a spirit that makes him unable to speak; [18] and whenever it seizes him, it dashes him down; and he foams and grinds his teeth and becomes rigid; and I asked your disciples to cast it out, but they could not do so." [19] He answered them, "You faithless generation, how much longer must I be among you? How much longer must I put up with you? Bring him to me." [20] And they brought the boy[n] to him. When the spirit saw him, immediately it convulsed the boy,[n] and he fell on the ground and rolled about, foaming at the mouth. [21] Jesus[o] asked the father, "How long has this been happening to him?"

And he said, "From childhood. [22] It has often cast him into the fire and into the water, to destroy him; but if you are able to do anything, have pity on us and help us." [23] Jesus said to him, "If you are able!—All things can be done for the one who believes." [24] Immediately the father of the child cried out,[p] "I believe; help my unbelief!" [25] When Jesus saw that a crowd came running together, he rebuked the unclean spirit, saying to it, "You spirit that keeps this boy from speaking and hearing, I command you, come out of him, and never enter him again!" [26] After crying out and convulsing him terribly, it came out, and the boy was like a corpse, so that most of them said, "He is dead." [27] But Jesus took him by the hand and lifted him up, and he was able to stand. [28] When he had entered the house, his disciples asked him privately, "Why could we not cast it out?" [29] He said to them, "This kind can come out only through prayer."[q]

30 They went on from there and passed through Galilee. He did not want anyone to know it; [31] for he was teaching his disciples, saying to them, "The Son of Man is to be betrayed into human hands, and they will kill him, and three days after being killed, he will rise again." [32] But they did not understand what he was saying and were afraid to ask him.

33 Then they came to Capernaum; and when he was in the house he asked

n Gk *him* *o* Gk *He* *p* Other ancient authorities add *with tears* *q* Other ancient authorities add *and fasting*

(Mt 11.14; Lk 1.17, 76), and John was treated as Elijah had been treated (1 Kings 19.2, 10). **9.14–29: Epileptic child healed** (Mt 17.14–21; Lk 9.37–42). **15:** Perhaps the crowd was *overcome with awe* by the coincidence of the disciples' failure (v. 18) and Jesus' unexpected return. **18:** In modern diagnosis the symptoms point to an epileptic seizure. **19:** Failure is attributed to wrong attitude (v. 29); the disciple must speak from faith not argument (see Lk 10.20 n.; Mk 11.23 n., 24 n.). **28:** A reasonable question in the light of 6.7,

13 where spiritual power accompanied the message. **29:** Prayer to God is faith in God, and contrasts with the argumentative attitude in v. 14. The potency in faith rests with God and is not under the believer's control.
9.30–32: The Passion foretold again (Mt 17.22–23; Lk 9.43–45); compare 8.31; 10.33. **31:** The burden of Jesus' teaching was on his coming violent death and resurrection (8.31; 10.33–34; see Lk 9.22 n.). **32:** See 9.10 n.; Jn 12.16.
9.33–37: True greatness (Mt 18.1–5; Lk

them, "What were you arguing about on the way?" 34 But they were silent, for on the way they had argued with one another who was the greatest. 35 He sat down, called the twelve, and said to them, "Whoever wants to be first must be last of all and servant of all." 36 Then he took a little child and put it among them; and taking it in his arms, he said to them, 37 "Whoever welcomes one such child in my name welcomes me, and whoever welcomes me welcomes not me but the one who sent me."

38 John said to him, "Teacher, we saw someone* casting out demons in your name, and we tried to stop him, because he was not following us." 39 But Jesus said, "Do not stop him; for no one who does a deed of power in my name will be able soon afterward to speak evil of me. 40 Whoever is not against us is for us. 41 For truly I tell you, whoever gives you a cup of water to drink because you bear the name of Christ will by no means lose the reward.

42 "If any of you put a stumbling block before one of these little ones who believe in me,* it would be better for you if a great millstone were hung around your neck and you were thrown into the sea. 43 If your hand causes you to stumble, cut it off; it is better for you to enter life maimed than to have two hands and to go to hell,* to the unquenchable fire.* 45 And if your foot causes you to stumble, cut it off; it is better for you to enter life lame than to have two feet and to be thrown into hell.*.* 47 And if your

eye causes you to stumble, tear it out; it is better for you to enter the kingdom of God with one eye than to have two eyes and to be thrown into hell,* 48 where their worm never dies, and the fire is never quenched.

49 "For everyone will be salted with fire.* 50 Salt is good; but if salt has lost its saltiness, how can you season it?* Have salt in yourselves, and be at peace with one another."

10 He left that place and went to the region of Judea and* beyond the Jordan. And crowds again gathered around him; and, as was his custom, he again taught them.

2 Some Pharisees came, and to test him they asked, "Is it lawful for a man to divorce his wife?" 3 He answered them, "What did Moses command you?" 4 They said, "Moses allowed a man to write a certificate of dismissal and to divorce her." 5 But Jesus said to them, "Because of your hardness of heart he wrote this commandment for you. 6 But from the beginning of creation, 'God made them male and female.' 7 'For this reason a man shall leave his father and mother and be joined to his wife,* 8 and the two

r Other ancient authorities add *who does not follow us* *s* Other ancient authorities lack *in me* *t* Gk *Gehenna* *u* Verses 44 and 46 (which are identical with verse 48) are lacking in the best ancient authorities *v* Other ancient authorities either add or substitute *and every sacrifice will be salted with salt* *w* Or *how can you restore its saltiness?* *x* Other ancient authorities lack *and* *y* Other ancient authorities lack *and be joined to his wife*

9.46–48). **34:** Lk 22.24. **35:** 10.43–44; Mt 20.26–27; 23.11; Lk 9.48; 22.26. **36:** 10.16. **37:** *In my name* means "because of regard for who and what I am."

9.38–41: The unknown exorcist (Lk 9.49–50). **39–40:** Mt 12.30; Lk 11.23. Each saying arose in a situation which gave it specific meaning. Compare the account of Eldad and Medad (Num 11.26–29) where Moses rebukes Joshua for the same jealous attitude. **41:** Mt 10.42; Mk 9.37.

9.42–48: Warnings of hell (Mt 18.6–9; 5.29–30; Lk 17.1–2). **42:** *Little ones,* followers (see Mt 18.6 n.). The *great millstone* was turned by a donkey. **48:** Isa 66.24.

9.49–50: Salty disciples. Mt 5.13; Lk 14.34–35. **50a:** Mt 5.13. **50b:** Perhaps the meaning is: "Maintain peacefully your own distinct character and service."

10.1–52: From Galilee to Jerusalem (Mt 19.1–20.34; Lk 18.15–19.27). **1–12: On marriage and divorce** (Mt 19.1–12). **1:** *Beyond the Jordan,* Perea. Lk 9.51; Jn 10.40; 11.7. **2:** See Mt 19.3 n. **3–4:** See Deut 24.1–4 n. **5:** Jesus is saying that the Law was shaped to the character of those for whom it was written. **6:** Gen 1.27; 5.2. **7–8:** Gen 2.24. **11:** See Mt 5.31–32 n. **12:** This provision was not applicable to Palestine, where women could not sue for divorce.

shall become one flesh.' So they are no longer two, but one flesh. ⁹Therefore what God has joined together, let no one separate."

10 Then in the house the disciples asked him again about this matter. ¹¹He said to them, "Whoever divorces his wife and marries another commits adultery against her; ¹²and if she divorces her husband and marries another, she commits adultery."

13 People were bringing little children to him in order that he might touch them; and the disciples spoke sternly to them. ¹⁴But when Jesus saw this, he was indignant and said to them, "Let the little children come to me; do not stop them; for it is to such as these that the kingdom of God belongs. ¹⁵Truly I tell you, whoever does not receive the kingdom of God as a little child will never enter it." ¹⁶And he took them up in his arms, laid his hands on them, and blessed them.

17 As he was setting out on a journey, a man ran up and knelt before him, and asked him, "Good Teacher, what must I do to inherit eternal life?" ¹⁸Jesus said to him, "Why do you call me good? No one is good but God alone. ¹⁹You know the commandments: 'You shall not murder; You shall not commit adultery; You shall not steal; You shall not bear false witness; You shall not defraud; Honor your father and mother.'" ²⁰He said to him, "Teacher, I have kept all these since my youth." ²¹Jesus, looking at him, loved him and said, "You lack one thing; go, sell what you own, and give the money ᶻ to the poor, and you will have treasure in heaven; then come, follow me." ²²When he heard this, he was shocked and went away grieving, for he had many possessions.

23 Then Jesus looked around and said to his disciples, "How hard it will be for those who have wealth to enter the kingdom of God!" ²⁴And the disciples were perplexed at these words. But Jesus said to them again, "Children, how hard it is ᵃ to enter the kingdom of God! ²⁵It is easier for a camel to go through the eye of a needle than for someone who is rich to enter the kingdom of God." ²⁶They were greatly astounded and said to one another, ᵇ "Then who can be saved?" ²⁷Jesus looked at them and said, "For mortals it is impossible, but not for God; for God all things are possible."

28 Peter began to say to him, "Look, we have left everything and followed you." ²⁹Jesus said, "Truly I tell you, there is no one who has left house or brothers or sisters or mother or father or children or fields, for my sake and for the sake of the good news, ᶜ ³⁰who will not receive a hundredfold now in this age—houses, brothers and sisters, mothers and children, and fields with persecutions—and in the age to come eternal life. ³¹But many who are first will be last, and the last will be first."

32 They were on the road, going up to Jerusalem, and Jesus was walking ahead of them; they were amazed, and those who followed were afraid. He took

z Gk lacks *the money*   *a* Other ancient authorities add *for those who trust in riches*   *b* Other ancient authorities read *to him*   *c* Or *gospel*

10.13–16: Blessing the children (Mt 19.13–15; Lk 18.15–17). **14:** *Was indignant,* because the disciples interfered with the manifestation of his love. **15:** To receive the kingdom as a child is to depend in trustful simplicity on what God offers.
10.17–31: The rich man (Mt 19.16–30; Lk 18.18–30). **17:** Lk 10.25; Mk 1.40. **19:** Ex 20.12–16; Deut 5.16–20. **21:** Jesus' invitation to *sell . . . give . . . come* is a drastic test of the questioner's concern for spiritual satisfaction (see Lk 12.33–34 n.). **24:** It was supposed that wealth made possible the performance of religious duties. Jesus' point is that by nature people do not submit to God's rule (compare v. 15), but sincere submission is essential to salvation. **25:** A proverbial expression denoting a human impossibility (compare v. 27). **26:** To *be saved* is to enter the kingdom (v. 25). **28:** 1.16–20.
10.32–34: The Passion foretold a third time (Mt 20.17–19; Lk 18.31–34); compare 8.31; 9.31. **32:** *Walking ahead of them,* a vivid picture of Jesus' resolute demeanor (compare

the twelve aside again and began to tell them what was to happen to him, ³³saying, "See, we are going up to Jerusalem, and the Son of Man will be handed over to the chief priests and the scribes; then they will condemn him to death; then they will hand him over to the Gentiles; ³⁴they will mock him, and spit upon him, and flog him, and kill him; and after three days he will rise again."

35 James and John, the sons of Zebedee, came forward to him and said to him, "Teacher, we want you to do for us whatever we ask of you." ³⁶And he said to them, "What is it you want me to do for you?" ³⁷And they said to him, "Grant us to sit, one at your right hand and one at your left, in your glory." ³⁸But Jesus said to them, "You do not know what you are asking. Are you able to drink the cup that I drink, or be baptized with the baptism that I am baptized with?" ³⁹They replied, "We are able." Then Jesus said to them, "The cup that I drink you will drink; and with the baptism with which I am baptized, you will be baptized; ⁴⁰but to sit at my right hand or at my left is not mine to grant, but it is for those for whom it has been prepared."

41 When the ten heard this, they began to be angry with James and John. ⁴²So Jesus called them and said to them, "You know that among the Gentiles those whom they recognize as their rulers lord it over them, and their great ones are tyrants over them. ⁴³But it is not so among you; but whoever wishes to become great among you must be your servant, ⁴⁴and whoever wishes to be first among you must be slave of all. ⁴⁵For the Son of Man came not to be served but to serve, and to give his life a ransom for many."

46 They came to Jericho. As he and his disciples and a large crowd were leaving Jericho, Bartimaeus son of Timaeus, a blind beggar, was sitting by the roadside. ⁴⁷When he heard that it was Jesus of Nazareth, he began to shout out and say, "Jesus, Son of David, have mercy on me!" ⁴⁸Many sternly ordered him to be quiet, but he cried out even more loudly, "Son of David, have mercy on me!" ⁴⁹Jesus stood still and said, "Call him here." And they called the blind man, saying to him, "Take heart; get up, he is calling you." ⁵⁰So throwing off his cloak, he sprang up and came to Jesus. ⁵¹Then Jesus said to him, "What do you want me to do for you?" The blind man said to him, "My teacher,ᵈ let me see again." ⁵²Jesus said to him, "Go; your faith has made you well." Immediately he regained his sight and followed him on the way.

11 When they were approaching Jerusalem, at Bethphage and Bethany, near the Mount of Olives, he sent two of his disciples ²and said to them, "Go into the village ahead of you, and immediately as you enter it, you will find

ᵈ Aramaic *Rabbouni*

Lk 9.51) despite the sufferings which awaited him at Jerusalem (see 8.31 n.). **34:** See Mt 10.38 n.; Mk 14.65; 15.19, 26–32.
10.35–45: James and John seek honor (Mt 20.20–28; Lk 22.24–27). **37:** See Mt 19.28 n. The seats symbolize positions of special dignity. **38:** Lk 12.50; Jn 18.11; Mk 14.36. *Cup,* see Lk 22.42 n. Acceptance of *baptism* symbolizes acceptance of God's way (see 1.4 n.). **40:** *Has been prepared,* in Mt 20.23 the phrase is made more precise by adding *by my Father.* **45:** *A ransom,* that which is given to gain release. Jesus speaks of his life and death as achieving freedom for *many,* i.e. with no specified limit to "a few," but does not spell out details (14.24; Lk 4.18; 1 Tim 2.5–6).
10.46–52: Blind Bartimaeus (Mt 20.29–34; Lk 18.35–43). **46:** *Bartimaeus* means "son of Timaeus" in Aramaic. **47:** By addressing Jesus as *Son of David,* the beggar was publicly identifying him as King of the Jews, the Messiah, a dangerous thing politically. **50:** *Cloak,* the outer garment. **52:** See Mt 9.21 n.; Mk 11.23 n., 24 n.
11.1–15.47: The last week (Mt 21.1–27.66; Lk 19.28–23.56).
11.1–11: Palm Sunday (Mt 21.1–9; Lk 19.28–38). **1:** Jesus dramatized his offer of himself as the Messiah, putting his emphasis on humility. Like a parable, his action had to

tied there a colt that has never been ridden; untie it and bring it. ³If anyone says to you, 'Why are you doing this?' just say this, 'The Lord needs it and will send it back here immediately.'" ⁴They went away and found a colt tied near a door, outside in the street. As they were untying it, ⁵some of the bystanders said to them, "What are you doing, untying the colt?" ⁶They told them what Jesus had said; and they allowed them to take it. ⁷Then they brought the colt to Jesus and threw their cloaks on it; and he sat on it. ⁸Many people spread their cloaks on the road, and others spread leafy branches that they had cut in the fields. ⁹Then those who went ahead and those who followed were shouting,

"Hosanna!
 Blessed is the one who comes in
  the name of the Lord!
10 Blessed is the coming kingdom
  of our ancestor David!
 Hosanna in the highest heaven!"

11 Then he entered Jerusalem and went into the temple; and when he had looked around at everything, as it was already late, he went out to Bethany with the twelve.

12 On the following day, when they came from Bethany, he was hungry. ¹³Seeing in the distance a fig tree in leaf, he went to see whether perhaps he would find anything on it. When he came to it, he found nothing but leaves, for it was not the season for figs. ¹⁴He said to it, "May no one ever eat fruit from you again." And his disciples heard it.

15 Then they came to Jerusalem. And he entered the temple and began to drive out those who were selling and those who were buying in the temple, and he overturned the tables of the money changers and the seats of those who sold doves; ¹⁶and he would not allow anyone to carry anything through the temple. ¹⁷He was teaching and saying, "Is it not written,

'My house shall be called a house
  of prayer for all the
  nations'?
 But you have made it a den
  of robbers.'"

18And when the chief priests and the scribes heard it, they kept looking for a way to kill him; for they were afraid of him, because the whole crowd was spellbound by his teaching. ¹⁹And when evening came, Jesus and his disciples*ᵉ* went out of the city.

20 In the morning as they passed by, they saw the fig tree withered away to its roots. ²¹Then Peter remembered and said to him, "Rabbi, look! The fig tree that you cursed has withered." ²²Jesus answered them, "Have*ᶠ* faith in God. ²³Truly I tell you, if you say to this mountain, 'Be taken up and thrown into the sea,' and if you do not doubt in your heart, but believe that what you say will come to pass, it will be done for you. ²⁴So I tell you, whatever you ask for in prayer, believe that you have received*ᵍ* it, and it will be yours.

25 "Whenever you stand praying, forgive, if you have anything against

e Gk *they*: other ancient authorities read *he*
f Other ancient authorities read *"If you have*
g Other ancient authorities read *are receiving*

---

be understood and accepted. **7–10:** Jn 12.12–15. **9:** Ps 118.26; Mt 21.15; 23.39; Zech 9.9. *Hosanna,* see Mt 21.9 n. **11:** Mt 21.10–11, 17.
 11.12–14: Fig tree cursed (Mt 21.18–19; compare Lk 13.6–9). **12:** Monday. **13:** The leaves showed the possibility of green fruit. Jesus' meaning, probably symbolic, is not clear.
 11.15–19: Cleansing the temple (Mt 21.12–13; Lk 19.45–48; Jn 2.13–17). **15:** See Mt 21.12 n. **16:** The outer precinct of the temple, known as the Court of the Gentiles, was used for the sale of sacrificial animals. **17:** Isa 56.7; Jer 7.11. **19:** Lk 21.37–38.
 11.20–25: The meaning of the withered fig tree (Mt 21.18–22). **23:** See Mt 17.20 n. Jesus emphasizes not power in faith but the power of God, his illustration being figurative. Faith will command only according to God's will (Mt 4.3–4; Mk 14.35–36). **24:** See Lk 11.9 n. What God wills is possible both to himself and to the person who shares his will. **25:** Mt 6.14–15; 18.35.

anyone; so that your Father in heaven may also forgive you your trespasses."[h]

27 Again they came to Jerusalem. As he was walking in the temple, the chief priests, the scribes, and the elders came to him 28 and said, "By what authority are you doing these things? Who gave you this authority to do them?" 29 Jesus said to them, "I will ask you one question; answer me, and I will tell you by what authority I do these things. 30 Did the baptism of John come from heaven, or was it of human origin? Answer me." 31 They argued with one another, "If we say, 'From heaven,' he will say, 'Why then did you not believe him?' 32 But shall we say, 'Of human origin'?"—they were afraid of the crowd, for all regarded John as truly a prophet. 33 So they answered Jesus, "We do not know." And Jesus said to them, "Neither will I tell you by what authority I am doing these things."

12 Then he began to speak to them in parables. "A man planted a vineyard, put a fence around it, dug a pit for the wine press, and built a watchtower; then he leased it to tenants and went to another country. 2 When the season came, he sent a slave to the tenants to collect from them his share of the produce of the vineyard. 3 But they seized him, and beat him, and sent him away empty-handed. 4 And again he sent another slave to them; this one they beat over the head and insulted. 5 Then he sent another, and that one they killed. And so it was with many others; some they beat, and others they killed. 6 He had still one other, a beloved son. Finally he sent him to them, saying, 'They will respect my son.' 7 But those tenants said to one another, 'This is the heir; come, let us kill him, and the inheritance will be ours.'

8 So they seized him, killed him, and threw him out of the vineyard. 9 What then will the owner of the vineyard do? He will come and destroy the tenants and give the vineyard to others. 10 Have you not read this scripture:

'The stone that the builders
  rejected
    has become the cornerstone;[i]
11 this was the Lord's doing,
  and it is amazing in our eyes'?"

12 When they realized that he had told this parable against them, they wanted to arrest him, but they feared the crowd. So they left him and went away.

13 Then they sent to him some Pharisees and some Herodians to trap him in what he said. 14 And they came and said to him, "Teacher, we know that you are sincere, and show deference to no one; for you do not regard people with partiality, but teach the way of God in accordance with truth. Is it lawful to pay taxes to the emperor, or not? 15 Should we pay them, or should we not?" But knowing their hypocrisy, he said to them, "Why are you putting me to the test? Bring me a denarius and let me see it." 16 And they brought one. Then he said to them, "Whose head is this, and whose title?" They answered, "The emperor's." 17 Jesus said to them, "Give to the emperor the things that are the emperor's, and to God the things that are God's." And they were utterly amazed at him.

18 Some Sadducees, who say there is no resurrection, came to him and asked him a question, saying, 19 "Teacher, Moses wrote for us that 'if a man's brother dies, leaving a wife but no child, the

h Other ancient authorities add verse 26, *"But if you do not forgive, neither will your Father in heaven forgive your trespasses."*   i Or *keystone*

11.27–33: On Jesus' authority (Mt 21.23–27; Lk 20.1–8; Jn 2.18–22). See Mt 21.27 n.
12.1–12: Parable of the vineyard (Mt 21.33–46; Lk 20.9–19). Isa 5.1–7. **10:** Ps 118.22–23.
12.13–17: Paying taxes to Caesar (Mt 22.15–22; Lk 20.20–26). **13:** 3.6 n.; Lk 11.53–54. **14:** See Mt 22.16 n. **17:** Rom 13.7.
12.18–27: Question about the resurrection (Mt 22.23–33; Lk 20.27–40). **18:** See Mt 22.23 n. **19:** Deut 25.5. **24:** See Mt 22.29 n. **26:** Ex 3.6; see Mt 22.31–32 n.; Lk 20.34–36 n.

man[j] shall marry the widow and raise up children for his brother.' 20 There were seven brothers; the first married and, when he died, left no children; 21 and the second married her and died, leaving no children; and the third likewise; 22 none of the seven left children. Last of all the woman herself died. 23 In the resurrection[k] whose wife will she be? For the seven had married her."

24 Jesus said to them, "Is not this the reason you are wrong, that you know neither the scriptures nor the power of God? 25 For when they rise from the dead, they neither marry nor are given in marriage, but are like angels in heaven. 26 And as for the dead being raised, have you not read in the book of Moses, in the story about the bush, how God said to him, 'I am the God of Abraham, the God of Isaac, and the God of Jacob'? 27 He is God not of the dead, but of the living; you are quite wrong."

28 One of the scribes came near and heard them disputing with one another, and seeing that he answered them well, he asked him, "Which commandment is the first of all?" 29 Jesus answered, "The first is, 'Hear, O Israel: the Lord our God, the Lord is one; 30 you shall love the Lord your God with all your heart, and with all your soul, and with all your mind, and with all your strength.' 31 The second is this, 'You shall love your neighbor as yourself.' There is no other commandment greater than these." 32 Then the scribe said to him, "You are right, Teacher; you have truly said that 'he is one, and besides him there is no other'; 33 and 'to love him with all the

heart, and with all the understanding, and with all the strength,' and 'to love one's neighbor as oneself,'—this is much more important than all whole burnt offerings and sacrifices." 34 When Jesus saw that he answered wisely, he said to him, "You are not far from the kingdom of God." After that no one dared to ask him any question.

35 While Jesus was teaching in the temple, he said, "How can the scribes say that the Messiah[l] is the son of David? 36 David himself, by the Holy Spirit, declared,

'The Lord said to my Lord,
"Sit at my right hand,
until I put your enemies under
your feet."'

37 David himself calls him Lord; so how can he be his son?" And the large crowd was listening to him with delight.

38 As he taught, he said, "Beware of the scribes, who like to walk around in long robes, and to be greeted with respect in the marketplaces, 39 and to have the best seats in the synagogues and places of honor at banquets! 40 They devour widows' houses and for the sake of appearance say long prayers. They will receive the greater condemnation."

41 He sat down opposite the treasury, and watched the crowd putting money into the treasury. Many rich people put in large sums. 42 A poor widow came and put in two small copper coins, which are worth a penny. 43 Then he called his disciples and said to them, "Truly I tell you, this poor widow has put in more than all

---

j Gk *his brother*   *k* Other ancient authorities add *when they rise*   *l* Or *the Christ*

12.28–34: The great commandment (Mt 22.34–40; Lk 10.25–28). **29:** The words of Deut 6.4, which are both preface to, and part of, the first commandment, define qualitatively the wholeness of the love that God requires. **33:** 1 Sam 15.22; Hos 6.6; Mic 6.6–8; Mt 9.13. Sacrifices were offered only at the temple in Jerusalem.
12.35–37: David's son (Mt 22.41–46; Lk 20.41–44). **36:** *By the Holy Spirit,* i.e. by inspiration. *Lord,* see Mt 22.44 n.
12.38–40: Sayings on pride and humil-

ity. **38:** Mt 23.1; Lk 20.45. **39:** Mt 23.6; Lk 20.46. The *best seats* were at the front, facing the congregation (Jas 2.2, 3). The *places of honor* were the couches at the host's table (Lk 11.43; 14.7–11). **40:** Lk 20.47.
12.41–44: The widow's offering (Lk 21.1–4). **41:** *Into the treasury,* i.e. into the donation chests. Thirteen of these were in the temple court, each one labeled telling the purpose for which the money would be used; see Lk 21.1 n. **42:** See Lk 12.59 n. **43:** See Lk 21.2 n.

those who are contributing to the treasury. 44 For all of them have contributed out of their abundance; but she out of her poverty has put in everything she had, all she had to live on."

13 As he came out of the temple, one of his disciples said to him, "Look, Teacher, what large stones and what large buildings!" 2 Then Jesus asked him, "Do you see these great buildings? Not one stone will be left here upon another; all will be thrown down."

3 When he was sitting on the Mount of Olives opposite the temple, Peter, James, John, and Andrew asked him privately, 4 "Tell us, when will this be, and what will be the sign that all these things are about to be accomplished?" 5 Then Jesus began to say to them, "Beware that no one leads you astray. 6 Many will come in my name and say, 'I am he!'*m* and they will lead many astray. 7 When you hear of wars and rumors of wars, do not be alarmed; this must take place, but the end is still to come. 8 For nation will rise against nation, and kingdom against kingdom; there will be earthquakes in various places; there will be famines. This is but the beginning of the birth pangs.

9 "As for yourselves, beware; for they will hand you over to councils; and you will be beaten in synagogues; and you will stand before governors and kings because of me, as a testimony to them. 10 And the good news*n* must first be proclaimed to all nations. 11 When they bring you to trial and hand you over, do not worry beforehand about what you are to say; but say whatever is given you at that time, for it is not you who speak, but the Holy Spirit. 12 Brother will betray brother to death, and a

father his child, and children will rise against parents and have them put to death; 13 and you will be hated by all because of my name. But the one who endures to the end will be saved.

14 "But when you see the desolating sacrilege set up where it ought not to be (let the reader understand), then those in Judea must flee to the mountains; 15 the one on the housetop must not go down or enter the house to take anything away; 16 the one in the field must not turn back to get a coat. 17 Woe to those who are pregnant and to those who are nursing infants in those days! 18 Pray that it may not be in winter. 19 For in those days there will be suffering, such as has not been from the beginning of the creation that God created until now, no, and never will be. 20 And if the Lord had not cut short those days, no one would be saved; but for the sake of the elect, whom he chose, he has cut short those days. 21 And if anyone says to you at that time, 'Look! Here is the Messiah!'*o* or 'Look! There he is!'—do not believe it. 22 False messiahs*p* and false prophets will appear and produce signs and omens, to lead astray, if possible, the elect. 23 But be alert; I have already told you everything.

24 "But in those days, after that suffering,
 the sun will be darkened,
 and the moon will not give
 its light,
25 and the stars will be falling
 from heaven,
 and the powers in the heavens
 will be shaken.
26 Then they will see 'the Son of Man coming in clouds' with great power and

m Gk *I am*   *n* Gk *gospel*   *o* Or *the Christ*
p Or *christs*

13.1–2: Destruction of Jerusalem foretold (Mt 24.1–3; Lk 21.5–7). See Mt 24.1 n. **1:** *The temple,* begun by Herod the Great, was as yet unfinished. *What large stones,* most of them were 37½ feet long, 18 feet wide, and 12 feet thick. **2:** Lk 19.43–44; Mk 14.58; 15.29; Jn 2.19; Acts 6.14. This temple was destroyed A.D. 70.

13.3–37: On the end of the age (Mt 24.4–36; Lk 21.8–36). **4:** Lk 17.20. **6:** Jn 8.24; 1 Jn 2.18. **8:** *Birth pangs,* see Mt 24.8 n. **9–13:** Mt 10.17–22. **11:** Jn 14.26; 16.7–11; Lk 12.11–12. **13:** Jn 15.18–21.
13.14: Dan 9.27; 11.31; 12.11. *The desolating sacrilege,* the intrusion of Gentile practices into the temple. **17:** Lk 23.29. **22:** Mt 7.15; Jn

glory. 27 Then he will send out the angels, and gather his elect from the four winds, from the ends of the earth to the ends of heaven.

28 "From the fig tree learn its lesson: as soon as its branch becomes tender and puts forth its leaves, you know that summer is near. 29 So also, when you see these things taking place, you know that he *q* is near, at the very gates. 30 Truly I tell you, this generation will not pass away until all these things have taken place. 31 Heaven and earth will pass away, but my words will not pass away.

32 "But about that day or hour no one knows, neither the angels in heaven, nor the Son, but only the Father. 33 Beware, keep alert; *r* for you do not know when the time will come. 34 It is like a man going on a journey, when he leaves home and puts his slaves in charge, each with his work, and commands the doorkeeper to be on the watch. 35 Therefore, keep awake—for you do not know when the master of the house will come, in the evening, or at midnight, or at cockcrow, or at dawn, 36 or else he may find you asleep when he comes suddenly. 37 And what I say to you I say to all: Keep awake."

14 It was two days before the Passover and the festival of Unleavened Bread. The chief priests and the scribes were looking for a way to arrest Jesus *s* by stealth and kill him; 2 for they said, "Not during the festival, or there may be a riot among the people."

3 While he was at Bethany in the house of Simon the leper, *t* as he sat at the table, a woman came with an alabaster jar of very costly ointment of nard, and she broke open the jar and poured the ointment on his head. 4 But some were there who said to one another in anger, "Why was the ointment wasted in this way? 5 For this ointment could have been sold for more than three hundred denarii, *u* and the money given to the poor." And they scolded her. 6 But Jesus said, "Let her alone; why do you trouble her? She has performed a good service for me. 7 For you always have the poor with you, and you can show kindness to them whenever you wish; but you will not always have me. 8 She has done what she could; she has anointed my body beforehand for its burial. 9 Truly I tell you, wherever the good news *v* is proclaimed in the whole world, what she has done will be told in remembrance of her."

10 Then Judas Iscariot, who was one of the twelve, went to the chief priests in order to betray him to them. 11 When they heard it, they were greatly pleased, and promised to give him money. So he began to look for an opportunity to betray him.

12 On the first day of Unleavened Bread, when the Passover lamb is sacrificed, his disciples said to him, "Where do you want us to go and make the preparations for you to eat the Passover?" 13 So he sent two of his disciples, saying to them, "Go into the city, and a man

q Or it   *r* Other ancient authorities add *and pray*   *s* Gk *him*   *t* The terms *leper* and *leprosy* can refer to several diseases   *u* The denarius was the usual day's wage for a laborer   *v* Or *gospel*

4.48. **26:** 8.38; Mt 10.23; Dan 7.13; 1 Thess 4.13–18. **30:** See Mt 24.34 n.; Mk 9.1. **31:** Mt 5.18; Lk 16.17.

13.32: Acts 1.7. **33:** Eph 6.18; Col 4.2. **34:** Mt 25.14. **35:** Lk 12.35–40. Domestic division of the night into four parts, namely 9 P.M., 12 P.M., 3 A.M., and 6 A.M.

14.1–15.47: Jesus' death (Mt 26.1–27.66; Lk 22.1–23.56; Jn 13.1–19.42). **14.1:** *The festival of Unleavened Bread* was part of the commemoration of the escape from Egypt under Moses (Ex 12.1–20). **3–9:** See Mt 26.6 n.

Nard was imported from India. **5:** On the value of a denarius, see note *u*. **6:** See Mt 26.10 n. **7:** Deut 15.11. **8:** The woman has shown personal regard for Jesus within her ability and opportunity. Jn 19.40.

14.10–11: Mt 26.14–16; Lk 22.3–6. **10:** *One of the twelve;* the words do not so much identify *Judas* as intensify the horror of the betrayal. Judas makes possible a change in the priests' plans (vv. 1–2).

14.12–25: The Last Supper. 12–16: Mt 26.17–19; Lk 22.7–13; see 22.10 n. **13:** *Two of*

carrying a jar of water will meet you; follow him, 14 and wherever he enters, say to the owner of the house, 'The Teacher asks, Where is my guest room where I may eat the Passover with my disciples?' 15 He will show you a large room upstairs, furnished and ready. Make preparations for us there." 16 So the disciples set out and went to the city, and found everything as he had told them; and they prepared the Passover meal.

17 When it was evening, he came with the twelve. 18 And when they had taken their places and were eating, Jesus said, "Truly I tell you, one of you will betray me, one who is eating with me." 19 They began to be distressed and to say to him one after another, "Surely, not I?" 20 He said to them, "It is one of the twelve, one who is dipping bread*w* into the bowl*x* with me. 21 For the Son of Man goes as it is written of him, but woe to that one by whom the Son of Man is betrayed! It would have been better for that one not to have been born."

22 While they were eating, he took a loaf of bread, and after blessing it he broke it, gave it to them, and said, "Take; this is my body." 23 Then he took a cup, and after giving thanks he gave it to them, and all of them drank from it. 24 He said to them, "This is my blood of the*y* covenant, which is poured out for many. 25 Truly I tell you, I will never again drink of the fruit of the vine until that day when I drink it new in the kingdom of God."

26 When they had sung the hymn, they went out to the Mount of Olives. 27 And Jesus said to them, "You will all become deserters; for it is written,

'I will strike the shepherd,
    and the sheep will be scattered.'
28 But after I am raised up, I will go before you to Galilee." 29 Peter said to him, "Even though all become deserters, I will not." 30 Jesus said to him, "Truly I tell you, this day, this very night, before the cock crows twice, you will deny me three times." 31 But he said vehemently, "Even though I must die with you, I will not deny you." And all of them said the same.

32 They went to a place called Gethsemane; and he said to his disciples, "Sit here while I pray." 33 He took with him Peter and James and John, and began to be distressed and agitated. 34 And he said to them, "I am deeply grieved, even to death; remain here, and keep awake." 35 And going a little farther, he threw himself on the ground and prayed that, if it were possible, the hour might pass from him. 36 He said, "Abba, *z* Father, for you all things are possible; remove this cup from me; yet, not what I want, but what you want." 37 He came and found them sleeping; and he said to Peter, "Simon, are you asleep? Could you not keep awake one hour? 38 Keep awake and pray that you may not come into the time of trial;*a* the spirit indeed is willing, but the flesh is weak." 39 And again he went away and prayed, saying the same words. 40 And once more he came and found them sleeping, for their eyes were very heavy; and they did not know what

---

w Gk lacks *bread*   *x* Other ancient authorities read *same bowl*   *y* Other ancient authorities add *new*   *z* Aramaic for *Father*   *a* Or *into temptation*

---

his disciples, according to Luke (22.8) they were Peter and John. **14:** See Lk 22.12 n.
14.17–21: Mt 26.20–25; Lk 22.14, 21–23; Jn 13.21–30. **18:** Ps 41.9. **19:** The question was so worded as to imply that the answer would be negative. **22–25:** Mt 26.26–29; Lk 22.15–20; 1 Cor 11.23–26. **23:** 1 Cor 10.16. **24:** See Mt 26.28 n. Jesus speaks of his blood as being the mediating reality in a new relationship between God and humankind (see 10.45 n.). **25:** Lk 13.29; see 22.16 n.
14.26–52: Gethsemane. 26–31: Mt 26.30–

35; Lk 22.39, 31–34. **26:** The Passover meal ended with the singing of Pss 115–118, the second part of the Hallel psalms. *They went out,* Jn 18.1–2. **27:** Zech 13.7; Jn 16.32. **28:** 16.7. **30:** 14.66–72; Jn 13.36–38; 18.17–18, 25–27. **32–42:** Mt 26.36–46; Lk 22.40–46. **32:** Jn 18.1; Heb 5.7–8. **34:** Jn 12.27. **35–36:** Jesus would not accept for himself the possibility of anything contrary to God's will (see 11.23–24 n.). *Abba,* "father" in Aramaic (see Rom 8.15 n.; Gal 4.6). **36:** *Cup,* see Lk 22.42 n. **38:** Mt 6.13; Lk 11.4.

to say to him. 41 He came a third time and said to them, "Are you still sleeping and taking your rest? Enough! The hour has come; the Son of Man is betrayed into the hands of sinners. 42 Get up, let us be going. See, my betrayer is at hand."

43 Immediately, while he was still speaking, Judas, one of the twelve, arrived; and with him there was a crowd with swords and clubs, from the chief priests, the scribes, and the elders. 44 Now the betrayer had given them a sign, saying, "The one I will kiss is the man; arrest him and lead him away under guard." 45 So when he came, he went up to him at once and said, "Rabbi!" and kissed him. 46 Then they laid hands on him and arrested him. 47 But one of those who stood near drew his sword and struck the slave of the high priest, cutting off his ear. 48 Then Jesus said to them, "Have you come out with swords and clubs to arrest me as though I were a bandit? 49 Day after day I was with you in the temple teaching, and you did not arrest me. But let the scriptures be fulfilled." 50 All of them deserted him and fled.

51 A certain young man was following him, wearing nothing but a linen cloth. They caught hold of him, 52 but he left the linen cloth and ran off naked.

53 They took Jesus to the high priest; and all the chief priests, the elders, and the scribes were assembled. 54 Peter had followed him at a distance, right into the courtyard of the high priest; and he was sitting with the guards, warming himself at the fire. 55 Now the chief priests and the whole council were looking for testimony against Jesus to put him to death; but they found none. 56 For many gave false testimony against him, and their testimony did not agree. 57 Some stood up and gave false testimony against him, saying, 58 "We heard him say, 'I will destroy this temple that is made with hands, and in three days I will build another, not made with hands.'" 59 But even on this point their testimony did not agree. 60 Then the high priest stood up before them and asked Jesus, "Have you no answer? What is it that they testify against you?" 61 But he was silent and did not answer. Again the high priest asked him, "Are you the Messiah, *b* the Son of the Blessed One?" 62 Jesus said, "I am; and

'you will see the Son of Man
seated at the right hand of
the Power,'
and 'coming with the clouds
of heaven.'"

63 Then the high priest tore his clothes and said, "Why do we still need witnesses? 64 You have heard his blasphemy! What is your decision?" All of them condemned him as deserving death. 65 Some began to spit on him, to blindfold him, and to strike him, saying to him, "Prophesy!" The guards also took him over and beat him.

66 While Peter was below in the courtyard, one of the servant-girls of the high priest came by. 67 When she saw Peter warming himself, she stared at him

b Or the Christ

14.43–52: Mt 26.47–56; Lk 22.47–53; Jn 18.2–11. **43:** See Mt 26.50 n. **47:** *One . . . who stood near,* according to Jn 18.10 it was Simon Peter. **49:** Lk 19.47; Jn 18.19–21. **51:** The young man's identity is not disclosed. Perhaps he was sleeping in the house where Jesus ate the Last Supper and rose hastily from bed to follow Jesus to Gethsemane. If the house was that of Mary, the mother of John Mark (where the disciples met at a later date; Acts 12.12), it is possible that the *young man* was the Evangelist himself. **14.53–72: Jesus before Caiaphas. 55:** At least two witnesses who agreed were required by Num 35.30 and Deut 19.15 (compare Mt 18.16). **58:** *Another, not made with hands,* an accusation that Jesus practices wizardry, which according to Lev 20.27 was a capital crime; see Jn 2.19–21. **61–62:** *The Blessed One . . . Power,* Jewish ways of referring to God. **62:** Dan 7.13 combined with the thought of Ps 110.1. **63:** *Tore his clothes,* an action expressing grief. Acts 14.14; Joel 2.12–13. **64:** Lev 24.16. *All of them,* but according to Lk 23.51 Joseph of Arimathea, a member of the council, had not assented. **66:** 14.30.

and said, "You also were with Jesus, the man from Nazareth." 68 But he denied it, saying, "I do not know or understand what you are talking about." And he went out into the forecourt.*ᶜ* Then the cock crowed.*ᵈ* 69 And the servant-girl, on seeing him, began again to say to the bystanders, "This man is one of them." 70 But again he denied it. Then after a little while the bystanders again said to Peter, "Certainly you are one of them; for you are a Galilean." 71 But he began to curse, and he swore an oath, "I do not know this man you are talking about." 72 At that moment the cock crowed for the second time. Then Peter remembered that Jesus had said to him, "Before the cock crows twice, you will deny me three times." And he broke down and wept.

15 As soon as it was morning, the chief priests held a consultation with the elders and scribes and the whole council. They bound Jesus, led him away, and handed him over to Pilate. 2 Pilate asked him, "Are you the King of the Jews?" He answered him, "You say so." 3 Then the chief priests accused him of many things. 4 Pilate asked him again, "Have you no answer? See how many charges they bring against you." 5 But Jesus made no further reply, so that Pilate was amazed.

6 Now at the festival he used to release a prisoner for them, anyone for whom they asked. 7 Now a man called Barabbas was in prison with the rebels who had committed murder during the insurrection. 8 So the crowd came and began to ask Pilate to do for them according to his custom. 9 Then he answered them, "Do you want me to release for you the King of the Jews?" 10 For he realized that it was out of jealousy that the chief priests had handed him over. 11 But the chief priests stirred up the crowd to have him release Barabbas for them instead. 12 Pilate spoke to them again, "Then what do you wish me to do*ᵉ* with the man you call*ᶠ* the King of the Jews?" 13 They shouted back, "Crucify him!" 14 Pilate asked them, "Why, what evil has he done?" But they shouted all the more, "Crucify him!" 15 So Pilate, wishing to satisfy the crowd, released Barabbas for them; and after flogging Jesus, he handed him over to be crucified.

16 Then the soldiers led him into the courtyard of the palace (that is, the governor's headquarters*ᵍ*); and they called together the whole cohort. 17 And they clothed him in a purple cloak; and after twisting some thorns into a crown, they put it on him. 18 And they began saluting him, "Hail, King of the Jews!" 19 They struck his head with a reed, spat upon him, and knelt down in homage to him. 20 After mocking him, they stripped him of the purple cloak and put his own clothes on him. Then they led him out to crucify him.

21 They compelled a passer-by, who was coming in from the country, to carry his cross; it was Simon of Cyrene, the

c Or *gateway*   *d* Other ancient authorities lack *Then the cock crowed*   *e* Other ancient authorities read *what should I do*   *f* Other ancient authorities lack *the man you call*   *g* Gk *the praetorium*

14.70: See Mt 26.73 n. **72:** Before dawn, compare 13.35. *He broke down,* the meaning of the Greek is uncertain.
15.1–15: Jesus before Pilate. 1: Pontius *Pilate,* Roman governor of Judea, Samaria, and Idumaea (A.D. 26–36). See Mt 27.1–2 n.; Lk 23.1; Jn 18.28–32. **2–5:** Mt 27.11–14; Lk 23.2–5; Jn 18.29–38.
15.6–15: Mt 27.15–26; Lk 23.18–25; Jn 18.38–40; 19.4–16. **7:** The *insurrection* is unrecorded; the crime was more serious than that of brigandage (15.27; Jn 18.40). **11:** Acts 3.14. **13:** According to Jewish law a religious curse was implicit in crucifixion (Deut 21.23). **15:** See Mt 27.26 n.
15.16–47: The crucifixion. 16–20: Mt 27.27–31; Lk 23.11; Jn 19.1–3. **16:** See Mt 27.27 n. *Of Cyrene,* capital city of the north African district of Cyrenaica, which had a large Jewish community.
15.21–32: Mt 27.32–44; Lk 23.33–43; Jn 19.17–24. **21:** The men named were probably known to the Christians who first read Mark. A connection with the *Rufus* mentioned in

father of Alexander and Rufus. 22 Then they brought Jesus*h* to the place called Golgotha (which means the place of a skull). 23 And they offered him wine mixed with myrrh; but he did not take it. 24 And they crucified him, and divided his clothes among them, casting lots to decide what each should take.

25 It was nine o'clock in the morning when they crucified him. 26 The inscription of the charge against him read, "The King of the Jews." 27 And with him they crucified two bandits, one on his right and one on his left. *i* 29 Those who passed by derided*j* him, shaking their heads and saying, "Aha! You who would destroy the temple and build it in three days, 30 save yourself, and come down from the cross!" 31 In the same way the chief priests, along with the scribes, were also mocking him among themselves and saying, "He saved others; he cannot save himself. 32 Let the Messiah,*k* the King of Israel, come down from the cross now, so that we may see and believe." Those who were crucified with him also taunted him.

33 When it was noon, darkness came over the whole land*l* until three in the afternoon. 34 At three o'clock Jesus cried out with a loud voice, "Eloi, Eloi, lema sabachthani?" which means, "My God, my God, why have you forsaken me?"*m* 35 When some of the bystanders heard it, they said, "Listen, he is calling for Elijah." 36 And someone ran, filled a sponge with sour wine, put it on a stick, and gave it to him to drink, saying, "Wait, let us see whether Elijah will come to take him down." 37 Then Jesus gave a loud cry and breathed his last. 38 And the curtain of the temple was torn in two, from top to bottom. 39 Now when the centurion, who stood facing him, saw that in this way he*n* breathed his last, he said, "Truly this man was God's Son!"*o*

40 There were also women looking on from a distance; among them were Mary Magdalene, and Mary the mother of James the younger and of Joses, and Salome. 41 These used to follow him and provided for him when he was in Galilee; and there were many other women who had come up with him to Jerusalem.

42 When evening had come, and since it was the day of Preparation, that is, the day before the sabbath, 43 Joseph of Arimathea, a respected member of the council, who was also himself waiting expectantly for the kingdom of God, went boldly to Pilate and asked for the body of Jesus. 44 Then Pilate wondered if he were already dead; and summoning the centurion, he asked him whether he had been dead for some time. 45 When he learned from the centurion that he was dead, he granted the body to Joseph. 46 Then Joseph*p* bought a linen cloth, and taking down the body,*q* wrapped it in the linen cloth, and laid it in a tomb that had been hewn out of the rock. He then rolled a stone against the door of the tomb. 47 Mary Magdalene and Mary the mother of Joses saw where the body*q* was laid.

16 When the sabbath was over, Mary Magdalene, and Mary the

h Gk him　　*i* Other ancient authorities add verse 28, *And the scripture was fulfilled that says,* "*And he was counted among the lawless.*"
j Or blasphemed　　*k* Or the Christ　　*l* Or earth
m Other ancient authorities read *made me a reproach*　　*n* Other ancient authorities add *cried out and*　　*o* Or a son of God　　*p* Gk he
q Gk it

Rom 16.13 is possible but not established. **23:** *Wine . . . with myrrh* was a sedative. **24:** Ps 22.18. **29:** 13.2; 14.58; Jn 2.19. **31:** Ps 22.7–8. **15.33:** See Mt 27.45 n. **34:** See Mt 27.46 n. **36:** Ps 69.21; see Mt 27.48 n. **38:** The *curtain* closed off the Holy of Holies (Heb 9.3), the inner sanctuary which represented God's presence with his people (compare 2 Kings 19.14–15; 2 Chr 6.1–2, 18–21). The damage to the curtain, whatever the underlying event may have been, symbolized for Christian faith the unhindered access to God achieved for all by Jesus' death (Heb 10.19–20). **41:** Lk 8.1–3.
15.42–47: Mt 27.57–61; Lk 23.50–56; Jn 19.38–42. **42:** Late on Friday afternoon. **46:** Acts 13.29.
16.1–8: The first Easter (Mt 28.1–10; Lk

mother of James, and Salome bought spices, so that they might go and anoint him. 2 And very early on the first day of the week, when the sun had risen, they went to the tomb. 3 They had been saying to one another, "Who will roll away the stone for us from the entrance to the tomb?" 4 When they looked up, they saw that the stone, which was very large, had already been rolled back. 5 As they entered the tomb, they saw a young man, dressed in a white robe, sitting on the right side; and they were alarmed. 6 But he said to them, "Do not be alarmed; you are looking for Jesus of Nazareth, who was crucified. He has been raised; he is not here. Look, there is the place they laid him. 7 But go, tell his disciples and Peter that he is going ahead of you to Galilee; there you will see him, just as he told you." 8 So they went out and fled from the tomb, for terror and amazement had seized them; and they said nothing to anyone, for they were afraid. *r*

THE SHORTER ENDING OF MARK

⟦And all that had been commanded them they told briefly to those around Peter. And afterward Jesus himself sent out through them, from east to west, the sacred and imperishable proclamation of eternal salvation. *s*⟧

THE LONGER ENDING OF MARK

9 ⟦Now after he rose early on the first day of the week, he appeared first to

r Some of the most ancient authorities bring the book to a close at the end of verse 8. One authority concludes the book with the shorter ending; others include the shorter ending and then continue with verses 9–20. In most authorities verses 9–20 follow immediately after verse 8, though in some of these authorities the passage is marked as being doubtful.   *s* Other ancient authorities add *Amen*

24.1–11; Jn 20.1–10). **1**: Lk 23.56; Jn 19.39. *Sabbath* ended at sundown on Saturday. The women came to complete the rites of burial. **3–4**: The disk-shaped *stone* rolled edgeways in a gutter to close the opening of the tomb. **5**: The main chamber of a tomb was normally furnished with niches to receive the bodies. The young man's clothes indicate him to be a heavenly messenger. **7**: 14.28; Jn 21.1–23; Mt 28.7. **8**: *For they were afraid.* The Greek expression is unusual in style and abrupt in effect, especially if, as is possible, it originally ended the Gospel. In contrast with Mt 28.8–10, where fear is part of an emotional state that includes joy (v. 8) and is controlled by worship (v. 9) and acceptance of mission (v. 10), fear here (probably in the sense of overwhelming awe) is the pervasive consequence of amazement (v. 5) and of trembling and astonishment (v. 8) that resulted in flight and silence (v. 8). On the silence of the women compare Mt 28.8 n.; Lk 24.9–11, 22–24, and vv. 9–10 below.
16.9–20: **The traditional close of the Gospel of Mark.** Nothing is certainly known either about how this Gospel originally ended or about the origin of vv. 9–20, which, because of the textual evidence as well as stylistic differences from the rest of the Gospel, cannot have been part of the original text of Mark. Certain important witnesses to the text, including some ancient ones, end the Gospel with v. 8. Though it is possible that the compiler of the Gospel intended this abrupt ending, one can find hints that he intended to describe events after the resurrection: for example, Mk 14.28 looks forward to an account of at least one experience of the disciples with Jesus in Galilee after the resurrection, while the friendly reference to Peter (16.7) may anticipate the recounting of the otherwise unrecorded moment of reconciliation between Peter and his Lord (compare Lk 24.34; 1 Cor 15.5). If accounts such as these were originally part of Mark's Gospel, the loss of them took place very shortly after the Gospel was written, under circumstances beyond present knowledge. Many witnesses, some ancient, end the Gospel with vv. 9–20, thus showing that from early Christian times these verses have been accepted traditionally and generally as part of the canonical Gospel of Mark. A variety of other manuscripts conclude the Gospel with the shorter ending, either alone or followed by verses 9–20, thus indicating that different attempts were made to provide a suitable ending for the Gospel. The longer ending may have been compiled early in the second century as a didactic summary of grounds for belief in Jesus' resurrection, being appended to the Gospel by the middle of the second century. On the Christian belief in continuing unrecorded memories about Jesus in the first century see Lk

Mary Magdalene, from whom he had cast out seven demons. ¹⁰ She went out and told those who had been with him, while they were mourning and weeping. ¹¹ But when they heard that he was alive and had been seen by her, they would not believe it.

12 After this he appeared in another form to two of them, as they were walking into the country. ¹³ And they went back and told the rest, but they did not believe them.

14 Later he appeared to the eleven themselves as they were sitting at the table; and he upbraided them for their lack of faith and stubbornness, because they had not believed those who saw him after he had risen. *ᵗ* ¹⁵ And he said to them, "Go into all the world and proclaim the good news *ᵘ* to the whole creation. ¹⁶ The one who believes and is baptized will be saved; but the one who does not believe will be condemned. ¹⁷ And these signs will accompany those who believe: by using my name they will cast out demons; they will speak in new tongues;

¹⁸ they will pick up snakes in their hands, *ᵛ* and if they drink any deadly thing, it will not hurt them; they will lay their hands on the sick, and they will recover."

19 So then the Lord Jesus, after he had spoken to them, was taken up into heaven and sat down at the right hand of God. ²⁰ And they went out and proclaimed the good news everywhere, while the Lord worked with them and confirmed the message by the signs that accompanied it. *ʷ* ⟧

t Other ancient authorities add, in whole or in part, *And they excused themselves, saying, "This age of lawlessness and unbelief is under Satan, who does not allow the truth and power of God to prevail over the unclean things of the spirits. Therefore reveal your righteousness now"—thus they spoke to Christ. And Christ replied to them, "The term of years of Satan's power has been fulfilled, but other terrible things draw near. And for those who have sinned I was handed over to death, that they may return to the truth and sin no more, that they may inherit the spiritual and imperishable glory of righteousness that is in heaven."* *u* Or *gospel* *v* Other ancient authorities lack *in their hands* *w* Other ancient authorities add *Amen*

1.1–2; Jn 20.30; 21.25; Acts 20.35 n.; 1 Cor 15.3; also compare Mt 28.20; Jn 16.12–33; Rev 1.12–16 n.; 2.18.
16.9–18: Post-resurrection appearances of Jesus. 9–10: Mary is associated with other women in vv. 1, 7–8 and parallels; she is apparently alone in Jn 20.1–2, 11–19. *Seven demons,* Lk 8.2. **11:** Lk 24.11, 22–25; Jn 20.19–29; 1 Cor 15.5. Here, as in Jn 20.19–29, the disciples are convinced of the truth of Jesus' resurrection by their own immediate experience with him, though they should have heeded the witness of others as later generations must do (Jn 20.29). **12–13:** Lk 24.12–35. **13:** Compare Lk 24.34.
16.14–18: Mt 28.19; Lk 24.47. **16:** Acts 2.37–42; 10.47–48; Rom 10.9. **17–18:** The reality of faith in believers' lives as they respond to the apostolic witness is signified by events that both correspond with biblically recorded happenings in the lives of the apostles and

conform to apostolic statements about the gifts of the Spirit (for example, 1 Cor 12.8–11, 28; 14.2–5; Heb 2.3–4): exorcism (Acts 8.6–7, 16.18, 19.11–20); new tongues (see Acts 2.4–11 n.; 10.46; 19.6; 1 Cor 12.10, 28; 14.2–33); healing (Acts 28.8; 1 Cor 12.9; Jas 5.13–16). Instances of picking up snakes and drinking poison, without injury to the believer in either case, lack New Testament parallels. However, the former resembles the harmless accidental attack upon Paul in Acts 28.3–6, and the latter appears occasionally in Christian literature from the second century onward.
16.19–20: Jesus' exaltation. 19: For the concept of Jesus' exaltation, Phil 2.9–11; Heb 1.3; for the language *was taken up,* Acts 1.2, 11, 22; 1 Tim 3.16 (seemingly a Christian hymn); for the image of the *right hand of God,* Ps 110.1 n.; Acts 7.55; Heb 1.3. **20:** Verses 17–18; Heb 2.3–4.

The Gospel According to

Luke

The Gospel according to Luke sets forth the words and works of Jesus as the divine-human Savior, whose compassion and tenderness extended to all who were needy. The universal mission of Jesus is emphasized (*a*) by tracing his genealogy back to Adam (3.38; contrast Mt 1.1–2); (*b*) by including references that commend members of a despised race, the Samaritans (10.30–37; 17.11–19; see Acts 8.5 n.); (*c*) by indicating that women have a new place of importance among the followers of Jesus (7.36–50; 8.3; 10.38–42); and (*d*) by promising that Gentiles would have an opportunity to accept the gospel (2.32; 3.6; 24.27; compare 15.4 n.).

In addition to presenting the story of Jesus' work in Galilee and his last week at Jerusalem, Luke includes more episodes of Jesus' final journey to Jerusalem than do any of the other Evangelists. This special section (9.51–18.14) also preserves many of the most beloved of Jesus' parables—such as the Good Samaritan (10.25–37), the Prodigal Son (15.11–32), the Unjust Judge (18.1–8), and the Pharisee and the Tax Collector (18.9–14).

It is obvious from a number of features that the Evangelist envisages a Gentile rather than a Jewish reading public. Thus, he makes comparatively few quotations from the Old Testament, which would have been a strange and almost unknown book to most non-Jews. For the same reason Luke seldom appeals to the argument from prophecy. Furthermore, instead of using the Jewish word "rabbi," Luke is the only New Testament author who employs the classical Greek equivalent, a word meaning *master* (5.5; 8.24, 45; 9.33, 49; 17.13).

The major divisions are: chs 1–2, Births of John and Jesus; the boy Jesus in the temple; 3.1–22, Activity of John the Baptist; baptism of Jesus; 3.23–38, Genealogy of Jesus; 4.1–13, Temptation of Jesus; 4.14–9.50, Jesus' activity, chiefly in Galilee; 9.51–19.27, Journey to Jerusalem; 19.28–23.56, The last week, concluding with Jesus' crucifixion and burial; ch 24, The resurrection; commissioning the disciples.

Although the Gospel is anonymous and the evidence pertaining to its author is inconclusive, many considerations support the early Christian tradition that the author was the physician Luke, a Gentile convert and friend of the apostle Paul (Col 4.14; compare 2 Tim 4.11; Philem 24). The Gospel appears to have been written, perhaps at Antioch, during the last third of the first century, though the precise date is unknown. Luke, who was not an eyewitness of the life of Jesus, tells us that he used great care in collecting information for his book (1.1–4). He dedicated the book, along with the Acts of the Apostles (1.1), to a certain Theophilus, who was probably a Roman of high rank.

Finally, mention must be made of the high quality of Luke's literary style (see also Introduction to the Acts of the Apostles). Of all four Evangelists, he is preeminently a person of broad culture, capable of adapting his Greek diction to different occasions, writing sometimes formal, classical prose, sometimes a racy narrative style in the vernacular of his own day, and sometimes a Semitic "Bible Greek" in which the Septuagint was written. As a gifted literary artist he produced what has justly been described as "the most beautiful book in the world."

For the literary genre of the gospels, see pp. viii–x NT.

1 Since many have undertaken to set down an orderly account of the events that have been fulfilled among us, ²just as they were handed on to us by those who from the beginning were eye-witnesses and servants of the word, ³I too decided, after investigating everything carefully from the very first,ᵃ to write an orderly account for you, most excellent Theophilus, ⁴so that you may know the truth concerning the things about which you have been instructed.

5 In the days of King Herod of Judea, there was a priest named Zechariah, who belonged to the priestly order of Abijah. His wife was a descendant of Aaron, and her name was Elizabeth. ⁶Both of them were righteous before God, living blamelessly according to all the commandments and regulations of the Lord. ⁷But they had no children, because Elizabeth was barren, and both were getting on in years.

8 Once when he was serving as priest before God and his section was on duty, ⁹he was chosen by lot, according to the custom of the priesthood, to enter the sanctuary of the Lord and offer incense. ¹⁰Now at the time of the incense offering, the whole assembly of the people was praying outside. ¹¹Then there appeared to him an angel of the Lord, standing at the right side of the altar of incense. ¹²When Zechariah saw him, he was terrified; and fear overwhelmed him. ¹³But the angel said to him, "Do not be afraid, Zechariah, for your prayer has been heard. Your wife Elizabeth will bear you a son, and you will name him John. ¹⁴You will have joy and gladness, and many will rejoice at his birth, ¹⁵for he will be great in the sight of the Lord. He must never drink wine or strong drink; even before his birth he will be filled with the Holy Spirit. ¹⁶He will turn many of the people of Israel to the Lord their God. ¹⁷With the spirit and power of Elijah he will go before him, to turn the hearts of parents to their children, and the disobedient to the wisdom of the righteous, to make ready a people prepared for the Lord." ¹⁸Zechariah said to the angel, "How will I know that this is so? For I am an old man, and my wife is getting on in years." ¹⁹The angel replied, "I am Gabriel. I stand in the presence of God, and I have been sent to speak to you and to bring you this good news. ²⁰But now, because you did not believe my words, which will be fulfilled in their time, you will become mute, unable to speak, until the day these things occur."

21 Meanwhile the people were waiting for Zechariah, and wondered at his delay in the sanctuary. ²²When he did come out, he could not speak to them, and they realized that he had seen a vision in the sanctuary. He kept motioning to them and remained unable to speak. ²³When his time of service was ended, he went to his home.

24 After those days his wife Elizabeth conceived, and for five months she remained in seclusion. She said, ²⁵"This is what the Lord has done for me when he looked favorably on me and took away the disgrace I have endured among my people."

ᵃ Or *for a long time*

1.1–4: **Introduction.** 1: Jn 20.30–31; 21.25. The writers of the gospels used sources of information now lost (see 6.17–49 n.). 2: Acts 1.21; 10.39; Heb 2.3; 1 Jn 1.1. 3: See Acts 1.1 n. *Theophilus,* an unknown Christian, perhaps of social prominence (*most excellent,* see Acts 23.26 n.). 4: Jn 20.31.
1.5–2.40: **The births of John and Jesus** (Mt 1.18–2.23). 1.5: *Herod* the Great reigned 37–4 B.C. The date intended here is approximately 7–6 B.C. *The priestly order of Abijah,* there were twenty-four priestly orders, of which Abijah's was the eighth (1 Chr 24.10).
1.8–9: 2 Chr 31.2; Ex 30.1, 6–8. **14–17:** Canticle in honor of John. **15:** Num 6.1–4; Lk 7.33. 17: *He will go before him,* that is, he will be the forerunner of the Messiah (Mal 4.5–6; Mt 11.14). **19:** Dan 8.16; 9.21. **22:** *He could not speak,* he was unable to pronounce the priestly blessing for which the people were waiting. **25:** Among Jews barrenness was regarded as a sign of divine disfavor and therefore a *disgrace* (see Gen 16.2 n.; 30.23; 1 Sam 1.1–18; Ps 128.3).

26 In the sixth month the angel Gabriel was sent by God to a town in Galilee called Nazareth, 27 to a virgin engaged to a man whose name was Joseph, of the house of David. The virgin's name was Mary. 28 And he came to her and said, "Greetings, favored one! The Lord is with you."*b* 29 But she was much perplexed by his words and pondered what sort of greeting this might be. 30 The angel said to her, "Do not be afraid, Mary, for you have found favor with God. 31 And now, you will conceive in your womb and bear a son, and you will name him Jesus. 32 He will be great, and will be called the Son of the Most High, and the Lord God will give to him the throne of his ancestor David. 33 He will reign over the house of Jacob forever, and of his kingdom there will be no end." 34 Mary said to the angel, "How can this be, since I am a virgin?"*c* 35 The angel said to her, "The Holy Spirit will come upon you, and the power of the Most High will overshadow you; therefore the child to be born*d* will be holy; he will be called Son of God. 36 And now, your relative Elizabeth in her old age has also conceived a son; and this is the sixth month for her who was said to be barren. 37 For nothing will be impossible with God." 38 Then Mary said, "Here am I, the servant of the Lord; let it be with me according to your word." Then the angel departed from her.

39 In those days Mary set out and went with haste to a Judean town in the hill country, 40 where she entered the house of Zechariah and greeted Elizabeth. 41 When Elizabeth heard Mary's greeting, the child leaped in her womb. And Elizabeth was filled with the Holy Spirit 42 and exclaimed with a loud cry, "Blessed are you among women, and blessed is the fruit of your womb. 43 And why has this happened to me, that the mother of my Lord comes to me? 44 For as soon as I heard the sound of your greeting, the child in my womb leaped for joy. 45 And blessed is she who believed that there would be*e* a fulfillment of what was spoken to her by the Lord."

46 And Mary*f* said,
"My soul magnifies the Lord,
47 and my spirit rejoices in God
    my Savior,
48 for he has looked with favor on
    the lowliness of his servant.
Surely, from now on all
    generations will call
    me blessed;
49 for the Mighty One has done
    great things for me,
and holy is his name.
50 His mercy is for those who fear
    him
from generation to generation.
51 He has shown strength with
    his arm;
he has scattered the proud in the
    thoughts of their hearts.
52 He has brought down the
    powerful from their
    thrones,
and lifted up the lowly;
53 he has filled the hungry with
    good things,
and sent the rich away empty.
54 He has helped his servant Israel,
    in remembrance of his mercy,
55 according to the promise he made
    to our ancestors,
to Abraham and to his
    descendants forever."

56 And Mary remained with her about three months and then returned to her home.

---

b Other ancient authorities add *Blessed are you among women*   c Gk *I do not know a man*
d Other ancient authorities add *of you*
e Or *believed, for there will be*   f Other ancient authorities read *Elizabeth*

1.26–38: The birth of Jesus is foretold (the Annunciation). **26**: *In the sixth month,* after the conception of John the Baptist. **31**: *Jesus,* the Greek form of the Hebrew Joshua (see Mt 1.21 n.). **33**: Mt 28.18; Dan 2.44. **42**: 11.27–28.

1.46–55: The "Magnificat" (so called from the first word of the Latin translation) is based largely on Hannah's prayer in 1 Sam 2.1–10. *Magnifies,* i.e. declares the greatness of. **47**: 1 Tim 2.3; Titus 3.4; Jude 25. **55**: Gen 17.7; 18.18; 22.17; Mic 7.20.

57 Now the time came for Elizabeth to give birth, and she bore a son. 58 Her neighbors and relatives heard that the Lord had shown his great mercy to her, and they rejoiced with her.

59 On the eighth day they came to circumcise the child, and they were going to name him Zechariah after his father. 60 But his mother said, "No; he is to be called John." 61 They said to her, "None of your relatives has this name." 62 Then they began motioning to his father to find out what name he wanted to give him. 63 He asked for a writing tablet and wrote, "His name is John." And all of them were amazed. 64 Immediately his mouth was opened and his tongue freed, and he began to speak, praising God. 65 Fear came over all their neighbors, and all these things were talked about throughout the entire hill country of Judea. 66 All who heard them pondered them and said, "What then will this child become?" For, indeed, the hand of the Lord was with him.

67 Then his father Zechariah was filled with the Holy Spirit and spoke this prophecy:

68 "Blessed be the Lord God of
    Israel,
  for he has looked favorably on
    his people and redeemed
    them.
69 He has raised up a mighty savior*g*
    for us
  in the house of his servant
    David,
70 as he spoke through the mouth of
    his holy prophets from of old,

71 that we would be saved from
    our enemies and from the
    hand of all who hate us.
72 Thus he has shown the mercy
    promised to our ancestors,
  and has remembered his
    holy covenant,
73 the oath that he swore to our
    ancestor Abraham,
  to grant us 74 that we, being
    rescued from the hands of
    our enemies,
  might serve him without fear, 75 in
    holiness and righteousness
  before him all our days.
76 And you, child, will be called the
    prophet of the Most High;
  for you will go before the Lord
    to prepare his ways,
77 to give knowledge of salvation to
    his people
  by the forgiveness of their sins.
78 By the tender mercy of our God,
    the dawn from on high will
    break upon*h* us,
79 to give light to those who sit in
    darkness and in the shadow
    of death,
  to guide our feet into the way
    of peace."

80 The child grew and became strong in spirit, and he was in the wilderness until the day he appeared publicly to Israel.

2 In those days a decree went out from Emperor Augustus that all the

g Gk *a horn of salvation*   *h* Other ancient authorities read *has broken upon*

1.57–66: Birth of John the Baptist. **59:** Lev 12.3; Gen 17.12; Lk 2.21. **63:** See v. 13. **65:** *Fear* (rendered "awe" in 5.26) indicates recognition of the limits of human understanding and power before God (2.9; 7.16; Acts 2.43, 46–47; 5.5, 11; 19.17).
1.67–79: The "Benedictus," so called from the first word in the Latin translation. **69:** *A mighty savior,* one who will bring salvation; see Ps 18.1–3; 92.10–11; 132.17–18. **76:** Mal 4.5; Lk 7.26. **77:** Mk 1.4. **78:** Mal 4.2; Eph 5.14. *The dawn* will be when God fulfills his purpose to bless humankind. **79:** Isa 9.2; Mt 4.16; Lk 4.18. **80:** These words cover a period of approximately thirty years. *The day he appeared publicly,* 3.2, 3.
2.1–40: **The birth of Jesus** (Mt 1.18–2.23). **1:** About 6–5 B.C. (see 1.5 n.). The Emperor *Augustus* reigned from 27 B.C. to A.D. 14. *World* here refers to the Roman Empire. **2:** *Quirinius* was at this time a special legate or commissioner of Augustus to carry

world should be registered. 2 This was the first registration and was taken while Quirinius was governor of Syria. 3 All went to their own towns to be registered. 4 Joseph also went from the town of Nazareth in Galilee to Judea, to the city of David called Bethlehem, because he was descended from the house and family of David. 5 He went to be registered with Mary, to whom he was engaged and who was expecting a child. 6 While they were there, the time came for her to deliver her child. 7 And she gave birth to her firstborn son and wrapped him in bands of cloth, and laid him in a manger, because there was no place for them in the inn.

8 In that region there were shepherds living in the fields, keeping watch over their flock by night. 9 Then an angel of the Lord stood before them, and the glory of the Lord shone around them, and they were terrified. 10 But the angel said to them, "Do not be afraid; for see—I am bringing you good news of great joy for all the people: 11 to you is born this day in the city of David a Savior, who is the Messiah, *i* the Lord. 12 This will be a sign for you: you will find a child wrapped in bands of cloth and lying in a manger." 13 And suddenly there was with the angel a multitude of the heavenly host, *j* praising God and saying,

14 "Glory to God in the highest
   heaven,

and on earth peace among those
   whom he favors!" *k*

15 When the angels had left them and gone into heaven, the shepherds said to one another, "Let us go now to Bethlehem and see this thing that has taken place, which the Lord has made known to us." 16 So they went with haste and found Mary and Joseph, and the child lying in the manger. 17 When they saw this, they made known what had been told them about this child; 18 and all who heard it were amazed at what the shepherds told them. 19 But Mary treasured all these words and pondered them in her heart. 20 The shepherds returned, glorifying and praising God for all they had heard and seen, as it had been told them.

21 After eight days had passed, it was time to circumcise the child; and he was called Jesus, the name given by the angel before he was conceived in the womb.

22 When the time came for their purification according to the law of Moses, they brought him up to Jerusalem to present him to the Lord 23 (as it is written in the law of the Lord, "Every firstborn male shall be designated as holy to the Lord"), 24 and they offered a sacrifice according to what is stated in the law of the Lord, "a pair of turtledoves or two young pigeons."

25 Now there was a man in Jerusalem

---

i Or *the Christ*   *j* Gk *army*   *k* Other ancient authorities read *peace, goodwill among people*

---

on a war against a rebellious tribe, the Homonadenses. As such he was military governor of Syria, while the civil administration was in the hands of Varus. **7**: According to Catholic teaching, the expression *firstborn son* is used here simply as a Semitic legal term and does not necessarily imply subsequent births. *Bands of cloth* were customarily wrapped around a newly born infant. **9**: *Terrified*, see 1.65 n. **11**: *The city of David*, Bethlehem. Three great Christian claims about Jesus are that he is *Savior, Messiah*, and *Lord* (see Mt 1.21 n.; 16.16 n.; Jn 4.42; Acts 2.36; 5.31; Phil 2.11). **14**: 3.22; 19.38. *Peace . . .* , the lack of one letter in the later Greek manuscripts accounts for the rendering in note *k*. *Those whom he favors* means "those whom God has

chosen in accord with his good pleasure." **2.21**: See Mt 1.21 n. **22–24**: Lev 12.2–8. **23**: Ex 13.2, 12.

2.25–38: *Simeon* and *Anna*, not otherwise known, express faith in Jesus as Savior, Messiah, and universal Lord (see v. 11 n.). **25**: *The consolation of Israel* was the salvation which the Messiah was to bring (vv. 26, 38; 23.51). **26**: *The Lord's Messiah*, the Christ of God (9.20). **29–32**: The "Nunc Dimittis," so called from the first words of the Latin translation. **29**: *You are dismissing your servant*, the figure is taken from the manumission of a slave. *In peace*, i.e. in a state of peace with God. **30**: 3.6; Isa 52.10. **32**: Isa 42.6; 49.6; Acts 13.47; 26.23.

whose name was Simeon;[l] this man was righteous and devout, looking forward to the consolation of Israel, and the Holy Spirit rested on him. 26 It had been revealed to him by the Holy Spirit that he would not see death before he had seen the Lord's Messiah.[m] 27 Guided by the Spirit, Simeon[n] came into the temple; and when the parents brought in the child Jesus, to do for him what was customary under the law, 28 Simeon[o] took him in his arms and praised God, saying,

29 "Master, now you are dismissing
   your servant[p] in peace,
   according to your word;
30 for my eyes have seen your
   salvation,
31 which you have prepared in the
   presence of all peoples,
32 a light for revelation to the
   Gentiles
   and for glory to your people
   Israel."

33 And the child's father and mother were amazed at what was being said about him. 34 Then Simeon[l] blessed them and said to his mother Mary, "This child is destined for the falling and the rising of many in Israel, and to be a sign that will be opposed 35 so that the inner thoughts of many will be revealed—and a sword will pierce your own soul too."

36 There was also a prophet, Anna[q] the daughter of Phanuel, of the tribe of Asher. She was of a great age, having lived with her husband seven years after her marriage, 37 then as a widow to the age of eighty-four. She never left the temple but worshiped there with fasting and prayer night and day. 38 At that moment she came, and began to praise God and to speak about the child[r] to all who were looking for the redemption of Jerusalem.

39 When they had finished everything required by the law of the Lord, they returned to Galilee, to their own town of Nazareth. 40 The child grew and became strong, filled with wisdom; and the favor of God was upon him.

41 Now every year his parents went to Jerusalem for the festival of the Passover. 42 And when he was twelve years old, they went up as usual for the festival. 43 When the festival was ended and they started to return, the boy Jesus stayed behind in Jerusalem, but his parents did not know it. 44 Assuming that he was in the group of travelers, they went a day's journey. Then they started to look for him among their relatives and friends. 45 When they did not find him, they returned to Jerusalem to search for him. 46 After three days they found him in the temple, sitting among the teachers, listening to them and asking them questions. 47 And all who heard him were amazed at his understanding and his answers. 48 When his parents[s] saw him they were astonished; and his mother said to him, "Child, why have you treated us like this? Look, your father and I have been searching for you in great anxiety." 49 He said to them, "Why were you searching for me? Did you not know that I must be in my Father's house?"[t] 50 But they did not understand what he said to them. 51 Then he went down with them and came to Nazareth, and was obedient to them. His mother treasured all these things in her heart.

52 And Jesus increased in wisdom and in years,[u] and in divine and human favor.

l Gk Symeon   m Or the Lord's Christ
n Gk In the Spirit, he   o Gk he   p Gk slave
q Gk Hanna   r Gk him   s Gk they
t Or be about my Father's interests?   u Or in stature

2.33: Luke could call Joseph the *father* of Jesus notwithstanding 1.34–35 since he was Mary's husband and Jesus' legal father (compare Mt 13.55; Lk 3.23). 36: Josh 19.24.
2.41–52: **The boy Jesus at Jerusalem.** This is the only information in the Bible on Jesus' maturing. 41: Ex 23.15; Deut 16.1–8. Mary was not legally obligated to go up to Passover festivals, but did so perhaps from pious motives. 46: *The teachers* were the experts in the Jewish religion. 48: Mk 3.31–35. 50–51: 2.19. 52: 1 Sam 2.26; Lk 1.80; 2.40.

3 In the fifteenth year of the reign of Emperor Tiberius, when Pontius Pilate was governor of Judea, and Herod was ruler*v* of Galilee, and his brother Philip ruler*v* of the region of Ituraea and Trachonitis, and Lysanias ruler*v* of Abilene, 2during the high priesthood of Annas and Caiaphas, the word of God came to John son of Zechariah in the wilderness. 3He went into all the region around the Jordan, proclaiming a baptism of repentance for the forgiveness of sins, 4as it is written in the book of the words of the prophet Isaiah,

"The voice of one crying out in
 the wilderness:
'Prepare the way of the Lord,
 make his paths straight.
5 Every valley shall be filled,
 and every mountain and hill
  shall be made low,
and the crooked shall be made
  straight,
 and the rough ways made
  smooth;
6 and all flesh shall see the salvation
 of God.' "

7 John said to the crowds that came out to be baptized by him, "You brood of vipers! Who warned you to flee from the wrath to come? 8Bear fruits worthy of repentance. Do not begin to say to yourselves, 'We have Abraham as our ancestor'; for I tell you, God is able from these stones to raise up children to Abraham. 9Even now the ax is lying at the root of the trees; every tree therefore that does not bear good fruit is cut down and thrown into the fire."

10 And the crowds asked him, "What then should we do?" 11In reply he said to them, "Whoever has two coats must share with anyone who has none; and whoever has food must do likewise." 12Even tax collectors came to be baptized, and they asked him, "Teacher, what should we do?" 13He said to them, "Collect no more than the amount prescribed for you." 14Soldiers also asked him, "And we, what should we do?" He said to them, "Do not extort money from anyone by threats or false accusation, and be satisfied with your wages."

15 As the people were filled with expectation, and all were questioning in their hearts concerning John, whether he might be the Messiah,*w* 16John answered all of them by saying, "I baptize you with water; but one who is more powerful than I is coming; I am not worthy to untie the thong of his sandals. He will baptize you with*x* the Holy Spirit and fire. 17His winnowing fork is in his hand, to clear his threshing floor and to gather the wheat into his granary; but the chaff he will burn with unquenchable fire."

18 So, with many other exhortations, he proclaimed the good news to the people. 19But Herod the ruler,*v* who had been rebuked by him because of Herodias, his brother's wife, and because of all the evil things that Herod had done,

v Gk *tetrarch* *w* Or *the Christ* *x* Or *in*

3.1–20: Activity of John the Baptist (Mt 3.1–12; Mk 1.1–8). **1:** The year A.D. 26 or 27 is meant. *Pilate,* a Roman procurator, had final authority in *Judea* (23.1). The remainder of the kingdom of Herod the Great was divided between his sons *Herod* Antipas (9.7; 23.6, 7) and *Philip. Abilene,* north of Philip's rule, was closely associated with it during the first century. **2:** Annas and his son-in-law Caiaphas (Jn 18.13) controlled the Jewish temple and priests. Caiaphas was the high priest (Mt 26.3; Jn 11.49); Annas, though retired, retained his prestige (Acts 4.6). *John,* see Mt 3.1 n. **3:** See Mk 1.4 n. **4–6:** Isa 40.3–5. **5:** The language expresses the idea of moral and spiritual renewal, but the coming of God's salvation also meant judgment (v. 7; Am 5.18–20). **3.7:** *Vipers,* poisonous reptiles that frequented the area (Isa 30.6; 59.5; Mt 12.34; 23.33). *Wrath,* see Mt 3.7 n. **8:** John demands right living based on a sincere search for God's will (Mt 7.15–20; Gal 5.22–23) and suited to the protestations of repentance (see Mt 3.2 n.). The claim to have *Abraham as . . . ancestor* was a claim to privileged standing with God through natural birth (Jn 8.33, 39; Rom 2.28, 29). **9:** *Fire,* a symbol of judgment (Mt 7.19; 13.40–42; Heb 6.7–8). **10–11:** 6.29; Acts 2.44–45; 4.32–35. **12–13:** 19.2, 8. **15:** Acts 13.25; Jn 1.19–22; Lk 7.19. **16:** Acts 1.5; 11.16; 19.4. **18:** *He proclaimed the good news,* or

20 added to them all by shutting up John in prison.

21 Now when all the people were baptized, and when Jesus also had been baptized and was praying, the heaven was opened, 22 and the Holy Spirit descended upon him in bodily form like a dove. And a voice came from heaven, "You are my Son, the Beloved;*y* with you I am well pleased."*z*

23 Jesus was about thirty years old when he began his work. He was the son (as was thought) of Joseph son of Heli, 24 son of Matthat, son of Levi, son of Melchi, son of Jannai, son of Joseph, 25 son of Mattathias, son of Amos, son of Nahum, son of Esli, son of Naggai, 26 son of Maath, son of Mattathias, son of Semein, son of Josech, son of Joda, 27 son of Joanan, son of Rhesa, son of Zerubbabel, son of Shealtiel, *a* son of Neri, 28 son of Melchi, son of Addi, son of Cosam, son of Elmadam, son of Er, 29 son of Joshua, son of Eliezer, son of Jorim, son of Matthat, son of Levi, 30 son of Simeon, son of Judah, son of Joseph, son of Jonam, son of Eliakim, 31 son of Melea, son of Menna, son of Mattatha, son of Nathan, son of David, 32 son of Jesse, son of Obed, son of Boaz, son of Sala, *b* son of Nahshon, 33 son of Amminadab, son of Admin, son of Arni, *c* son of Hezron son of Perez, son of Judah, 34 son of Jacob, son of Isaac, son of Abraham, son of Terah, son of Nahor, 35 son of Serug,

son of Reu, son of Peleg, son of Eber, son of Shelah, 36 son of Cainan, son of Arphaxad, son of Shem, son of Noah, son of Lamech, 37 son of Methuselah, son of Enoch, son of Jared, son of Mahalaleel, son of Cainan, 38 son of Enos, son of Seth, son of Adam, son of God.

4 Jesus, full of the Holy Spirit, returned from the Jordan and was led by the Spirit in the wilderness, 2 where for forty days he was tempted by the devil. He ate nothing at all during those days, and when they were over, he was famished. 3 The devil said to him, "If you are the Son of God, command this stone to become a loaf of bread." 4 Jesus answered him, "It is written, 'One does not live by bread alone.'"

5 Then the devil*d* led him up and showed him in an instant all the kingdoms of the world. 6 And the devil*d* said to him, "To you I will give their glory and all this authority; for it has been given over to me, and I give it to anyone I please. 7 If you, then, will worship me, it will all be yours." 8 Jesus answered him, "It is written,

'Worship the Lord your God,
    and serve only him.'"

y Or *my beloved Son*   *z* Other ancient authorities read *You are my Son, today I have begotten you*   *a* Gk *Salathiel*   *b* Other ancient authorities read *Salmon*   *c* Other ancient authorities read *Amminadab, son of Aram*; others vary widely   *d* Gk *he*

"preached the gospel," refers to the message of forgiveness (v. 3) and the advent of a new relationship between individuals and God (vv. 15–17). **19–20:** Mt 14.3–4; Mk 6.17–18. **3.21–22: Jesus' baptism** (Mt 3.13–17; Mk 1.9–11; Jn 1.29–34). See Mt 3.16–17 n. **21:** Prayer was part of many recorded momentous events in Jesus' life (e.g. Mk 1.35; Lk 5.16; 6.12; 9.18, 28; 11.1; 22.41–46). **22:** *Beloved,* see Mk 1.11 n.; Ps 2.7; Isa 42.1; Lk 9.35. **3.23–38: The genealogy of Jesus** is traced through Old Testament lines back to Adam, the first man (compare Mt 1.1–17). By linking Jesus' line with God's original creation Luke shows Jesus' common humanity,

as contrasted with Matthew's emphasis on Jesus' Jewish lineage (see Mt 1.1 n.). **23:** Jesus' age is approximate (Jn 8.57). The persons named from *Heli* to *Zerubbabel* (v. 27) are otherwise unknown. For the rest of the genealogy, compare Gen 5.3–32; 11.10–26; Ruth 4.18–22; 1 Chr 1.1–4, 24–28; 2.1–15. **4.1–13: Jesus' temptation** (Mt 4.1–11; Mk 1.12–13). The order of temptations differs from Matthew's but the testing remains the same. **1:** *Full of the Holy Spirit* is a Christian phrase (Acts 2.4; 6.3, 5; 7.55; 11.24); the work of Jesus and the church begins as God acts through individuals. **2:** Deut 9.9; 1 Kings 19.8; see Mt 4.2 n. **3–4:** See Mt 4.3–4 n.; Deut 8.3. **5–8:** Deut 6.13; 1 Jn 5.19.

9 Then the devil[e] took him to Jerusalem, and placed him on the pinnacle of the temple, saying to him, "If you are the Son of God, throw yourself down from here, 10for it is written,

'He will command his angels
        concerning you,
    to protect you,'

11and

'On their hands they will bear
        you up,
    so that you will not dash your
        foot against a stone.' "

12Jesus answered him, "It is said, 'Do not put the Lord your God to the test.' " 13When the devil had finished every test, he departed from him until an opportune time.

14 Then Jesus, filled with the power of the Spirit, returned to Galilee, and a report about him spread through all the surrounding country. 15He began to teach in their synagogues and was praised by everyone.

16 When he came to Nazareth, where he had been brought up, he went to the synagogue on the sabbath day, as was his custom. He stood up to read, 17and the scroll of the prophet Isaiah was given to him. He unrolled the scroll and found the place where it was written:

18 "The Spirit of the Lord is upon
        me,
    because he has anointed me
    to bring good news to the
        poor.
    He has sent me to proclaim release
        to the captives
    and recovery of sight to the
        blind,

to let the oppressed go free,
19 to proclaim the year of the Lord's
        favor."

20And he rolled up the scroll, gave it back to the attendant, and sat down. The eyes of all in the synagogue were fixed on him. 21Then he began to say to them, "Today this scripture has been fulfilled in your hearing." 22All spoke well of him and were amazed at the gracious words that came from his mouth. They said, "Is not this Joseph's son?" 23He said to them, "Doubtless you will quote to me this proverb, 'Doctor, cure yourself!' And you will say, 'Do here also in your hometown the things that we have heard you did at Capernaum.' " 24And he said, "Truly I tell you, no prophet is accepted in the prophet's hometown. 25But the truth is, there were many widows in Israel in the time of Elijah, when the heaven was shut up three years and six months, and there was a severe famine over all the land; 26yet Elijah was sent to none of them except to a widow at Zarephath in Sidon. 27There were also many lepers[f] in Israel in the time of the prophet Elisha, and none of them was cleansed except Naaman the Syrian." 28When they heard this, all in the synagogue were filled with rage. 29They got up, drove him out of the town, and led him to the brow of the hill on which their town was built, so that they might hurl him off the cliff. 30But he passed through the midst of them and went on his way.

31 He went down to Capernaum, a city in Galilee, and was teaching them on

---

e Gk *he*   *f* The terms *leper* and *leprosy* can refer to several diseases

---

4.9–12: See Mt 4.5–7; Ps 91.11–12; Deut 6.16. **13:** *Until an opportune time,* to renew temptation (compare Mk 8.33; Lk 22.28).
4.14–9.50: Events and teachings in Galilee. 4.14–15: Jesus' return to Galilee (Mt 4.12–17; Mk 1.14–15). **14:** 5.17.
4.16–30: In the synagogue at Nazareth (Mt 13.53–58; Mk 6.1–6). **16:** Jesus' *custom* was not merely to worship in the synagogue but to present his message there (v. 15; Mt 4.23; 9.35). Details of synagogue worship are

mentioned here and in vv. 17, 20 (compare Acts 13.15). **17:** The *scroll of Isaiah was given to* Jesus by the chazzan or attendant of the synagogue (v. 20). **18–19:** Isa 61.1, 2; 58.6; see Mt 3.1 n.; Lk 13.11, 16 n. **20:** Having read while standing, Jesus *sat down* according to custom to preach the sermon. **21:** *Fulfilled,* see Mt 11.4–6 n. **23:** The popular reaction became increasingly hostile as the magnitude of Jesus' message sank home. **24–27:** According to the Old Testament, foreigners sometimes

the sabbath. 32 They were astounded at his teaching, because he spoke with authority. 33 In the synagogue there was a man who had the spirit of an unclean demon, and he cried out with a loud voice, 34 "Let us alone! What have you to do with us, Jesus of Nazareth? Have you come to destroy us? I know who you are, the Holy One of God." 35 But Jesus rebuked him, saying, "Be silent, and come out of him!" When the demon had thrown him down before them, he came out of him without having done him any harm. 36 They were all amazed and kept saying to one another, "What kind of utterance is this? For with authority and power he commands the unclean spirits, and out they come!" 37 And a report about him began to reach every place in the region.

38 After leaving the synagogue he entered Simon's house. Now Simon's mother-in-law was suffering from a high fever, and they asked him about her. 39 Then he stood over her and rebuked the fever, and it left her. Immediately she got up and began to serve them.

40 As the sun was setting, all those who had any who were sick with various kinds of diseases brought them to him; and he laid his hands on each of them and cured them. 41 Demons also came out of many, shouting, "You are the Son of God!" But he rebuked them and would not allow them to speak, because they knew that he was the Messiah. g

42 At daybreak he departed and went into a deserted place. And the crowds were looking for him; and when they reached him, they wanted to prevent him from leaving them. 43 But he said to them, "I must proclaim the good news of the kingdom of God to the other cities also; for I was sent for this purpose." 44 So he continued proclaiming the message in the synagogues of Judea. h

5 Once while Jesus i was standing beside the lake of Gennesaret, and the crowd was pressing in on him to hear the word of God, 2 he saw two boats there at the shore of the lake; the fishermen had gone out of them and were washing their nets. 3 He got into one of the boats, the one belonging to Simon, and asked him to put out a little way from the shore. Then he sat down and taught the crowds from the boat. 4 When he had finished speaking, he said to Simon, "Put out into the deep water and let down your nets for a catch." 5 Simon answered, "Master, we have worked all night long but have caught nothing. Yet if you say so, I will let down the nets." 6 When they had done this, they caught so many fish that their nets were beginning to break. 7 So they signaled their partners in the other boat to come and help them. And they came and filled both boats, so that they began to sink. 8 But when Simon Peter saw it,

g Or *the Christ*   h Other ancient authorities read *Galilee*   i Gk *he*

knew God's help when Israel did not (1 Kings 17.1, 8–16; 18.1; 2 Kings 5.1–14). **30**: It is not known that he ever visited Nazareth again.

4.31–37: The synagogue at Capernaum (Mt 7.28–29; Mk 1.21–28). **32**: *Teaching* here means the substance of what Jesus taught. **33**: See Mk 1.23 n. Demons were thought of in Jesus' day as non-material existences of a personal sort, hostile to human welfare and rebellious against God. The gospels reflect widespread dread of demons and a general sense of helplessness before demonic activity. Jesus is portrayed here and elsewhere (compare 11.20–22) as one who can deliver from demonic oppression and from Satan himself, evil's supreme embodiment (see Mt 4.24 n.; 12.22 n.; Lk 7.33 n.; 13.16 n.).

4.38–44: Healing and preaching (Mt 8.14–17; Mk 1.29–39). **40**: *Each of them*, Jesus gives attention to individuals, illustrating vv. 18–19. **41**: *Demons*, see v. 33 n. **44**: This is the only express mention outside the Fourth Gospel of Jesus' early Judean ministry (but compare the implications of Mt 23.37 and Lk 13.34). The original text, however, is uncertain, for *Galilee* appears in some ancient witnesses here as well as in the parallel accounts (Mt 4.23; Mk 1.39).

5.1–11: The unexpected catch (Mt 4.18–22; Mk 1.16–20). **1**: *Lake of Gennesaret*, Sea of Galilee. **8**: Although the great catch is not described as a miracle, Peter sees in Jesus' guidance a more-than-human power, and he responds by personal self-judgment.

he fell down at Jesus' knees, saying, "Go away from me, Lord, for I am a sinful man!" 9For he and all who were with him were amazed at the catch of fish that they had taken; 10and so also were James and John, sons of Zebedee, who were partners with Simon. Then Jesus said to Simon, "Do not be afraid; from now on you will be catching people." 11When they had brought their boats to shore, they left everything and followed him.

12 Once, when he was in one of the cities, there was a man covered with leprosy.*j* When he saw Jesus, he bowed with his face to the ground and begged him, "Lord, if you choose, you can make me clean." 13Then Jesus*k* stretched out his hand, touched him, and said, "I do choose. Be made clean." Immediately the leprosy*j* left him. 14And he ordered him to tell no one. "Go," he said, "and show yourself to the priest, and, as Moses commanded, make an offering for your cleansing, for a testimony to them." 15But now more than ever the word about Jesus*l* spread abroad; many crowds would gather to hear him and to be cured of their diseases. 16But he would withdraw to deserted places and pray.

17 One day, while he was teaching, Pharisees and teachers of the law were sitting near by (they had come from every village of Galilee and Judea and from Jerusalem); and the power of the Lord was with him to heal.*m* 18Just then some men came, carrying a paralyzed man on a bed. They were trying to bring him in and lay him before Jesus;*l* 19but finding no way to bring him in because of the crowd, they went up on the roof and let him down with his bed through the tiles into the middle of the crowd*n* in front of Jesus. 20When he saw their faith, he said,

"Friend,*o* your sins are forgiven you." 21Then the scribes and the Pharisees began to question, "Who is this who is speaking blasphemies? Who can forgive sins but God alone?" 22When Jesus perceived their questionings, he answered them, "Why do you raise such questions in your hearts? 23Which is easier, to say, 'Your sins are forgiven you,' or to say, 'Stand up and walk'? 24But so that you may know that the Son of Man has authority on earth to forgive sins"—he said to the one who was paralyzed—"I say to you, stand up and take your bed and go to your home." 25Immediately he stood up before them, took what he had been lying on, and went to his home, glorifying God. 26Amazement seized all of them, and they glorified God and were filled with awe, saying, "We have seen strange things today."

27 After this he went out and saw a tax collector named Levi, sitting at the tax booth; and he said to him, "Follow me." 28And he got up, left everything, and followed him.

29 Then Levi gave a great banquet for him in his house; and there was a large crowd of tax collectors and others sitting at the table*p* with them. 30The Pharisees and their scribes were complaining to his disciples, saying, "Why do you eat and drink with tax collectors and sinners?" 31Jesus answered, "Those who are well have no need of a physician, but those who are sick; 32I have come to call not the righteous but sinners to repentance."

33 Then they said to him, "John's disciples, like the disciples of the Pharisees,

j The terms *leper* and *leprosy* can refer to several diseases *k* Gk *he* *l* Gk *him* *m* Other ancient authorities read *was present to heal them* *n* Gk *into the midst* *o* Gk *Man* *p* Gk *reclining*

5.12–16: A leper healed (Mt 8.1–4; Mk 1.40–45). **12:** *Leprosy*, see Mt 8.2–4 n. **14:** Lev 13.2–3; 14.2–32. **16:** See 3.21 n.
5.17–26: Forgiveness of sins (Mt 9.1–8; Mk 2.1–12). **17:** *Power*, 4.14. **24:** *Son of Man*, see Mk 2.10 n.
5.27–32: Call of Levi (Matthew) (Mt 9.9–13; Mk 2.13–17). **30:** *Sinners*, see Mk

2.16 n. **32:** Jesus sought to draw in (i.e. *call*) outcasts whom the Pharisees excluded from society. *Call*, however, also meant "invite" (as to a banquet); Jesus' invitation to the Messianic banquet (see 12.37 n.; Rev 3.20 n.) was extended not to the self-judged righteous but to *sinners* (Mt 8.11–12; compare Lk 14.15–24).

frequently fast and pray, but your disciples eat and drink. 34Jesus said to them, "You cannot make wedding guests fast while the bridegroom is with them, can you? 35The days will come when the bridegroom will be taken away from them, and then they will fast in those days." 36He also told them a parable: "No one tears a piece from a new garment and sews it on an old garment; otherwise the new will be torn, and the piece from the new will not match the old. 37And no one puts new wine into old wineskins; otherwise the new wine will burst the skins and will be spilled, and the skins will be destroyed. 38But new wine must be put into fresh wineskins. 39And no one after drinking old wine desires new wine, but says, 'The old is good.' "q

6 One sabbathr while Jesuss was going through the grainfields, his disciples plucked some heads of grain, rubbed them in their hands, and ate them. 2But some of the Pharisees said, "Why are you doing what is not lawfult on the sabbath?" 3Jesus answered, "Have you not read what David did when he and his companions were hungry? 4He entered the house of God and took and ate the bread of the Presence, which it is not lawful for any but the priests to eat, and gave some to his companions?" 5Then he said to them, "The Son of Man is lord of the sabbath."

6 On another sabbath he entered the synagogue and taught, and there was a man there whose right hand was withered. 7The scribes and the Pharisees watched him to see whether he would cure on the sabbath, so that they might find an accusation against him. 8Even though he knew what they were thinking, he said to the man who had the withered hand, "Come and stand here." He got up and stood there. 9Then Jesus said to them, "I ask you, is it lawful to do good or to do harm on the sabbath, to save life or to destroy it?" 10After looking around at all of them, he said to him, "Stretch out your hand." He did so, and his hand was restored. 11But they were filled with fury and discussed with one another what they might do to Jesus.

12 Now during those days he went out to the mountain to pray; and he spent the night in prayer to God. 13And when day came, he called his disciples and chose twelve of them, whom he also named apostles: 14Simon, whom he named Peter, and his brother Andrew, and James, and John, and Philip, and Bartholomew, 15and Matthew, and Thomas, and James son of Alphaeus, and Simon, who was called the Zealot, 16and Judas son of James, and Judas Iscariot, who became a traitor.

17 He came down with them and

q Other ancient authorities read *better*; others lack verse 39   r Other ancient authorities read *On the second first sabbath*   s Gk *he*   t Other ancient authorities add *to do*

5.33–39: On fasting (Mt 9.14–17; Mk 2.18–22). **34–35:** See Mk 2.19–20 n. **36–37:** See Mt 9.16–17 n. **39:** This sentence explains why the disciples of John and of the Pharisees (v. 33) continued their religious forms notwithstanding Jesus' message; pious conservatism prevented their accepting the new revelation.

6.1–11: Jesus and sabbath laws (see Mt 12.1–14 n.; Mk 2.23–3.6). **1:** Deut 23.25. **2:** Ex 20.10; 23.12; Deut 5.14. **3:** 1 Sam 21.1–6. **4:** Lev 24.5–9. **11:** *Filled with fury*, at their being thwarted (v. 7).

6.12–16: Choosing the Twelve (Mt 10.1–4; Mk 3.13–19). **12:** See 3.21 n. **13:** The *disciples* were a larger group of followers from whom Jesus chose twelve for close compan-

ionship (Acts 1.21–22). *Apostle* means "appointed representative" and was not always limited to the Twelve (Rom 16.7; Acts 14.14).

6.17–49: The Sermon on the Plain. Luke gives here a number of Jesus' sayings found in the Sermon on the Mount (Mt chs 5–7). He has fewer teachings than appear there; he gives others found elsewhere in Matthew. Verses 24–26 are without parallel. The "sermons" in the gospels were formed from collections of memorized words of Jesus to instruct Christian converts (compare Jn 20.30–31; 21.25).

6.17–19: Mt 4.24–25; 12.15–21; Mk 3.7–12. **17:** The setting for the discourse in vv. 20–49 (compare Mt 5.1–2). Each gospel distin-

stood on a level place, with a great crowd of his disciples and a great multitude of people from all Judea, Jerusalem, and the coast of Tyre and Sidon. 18 They had come to hear him and to be healed of their diseases; and those who were troubled with unclean spirits were cured. 19 And all in the crowd were trying to touch him, for power came out from him and healed all of them.

20 Then he looked up at his disciples and said:

"Blessed are you who are poor,
for yours is the kingdom of
God.
21 "Blessed are you who are
hungry now,
for you will be filled.

"Blessed are you who weep now,
for you will laugh.

22 "Blessed are you when people hate you, and when they exclude you, revile you, and defame you *u* on account of the Son of Man. 23 Rejoice in that day and leap for joy, for surely your reward is great in heaven; for that is what their ancestors did to the prophets.
24 "But woe to you who are rich,
for you have received your
consolation.
25 "Woe to you who are full now,
for you will be hungry.

"Woe to you who are laughing
now,
for you will mourn and weep.

26 "Woe to you when all speak well of you, for that is what their ancestors did to the false prophets.

27 "But I say to you that listen, Love your enemies, do good to those who hate you, 28 bless those who curse you, pray for those who abuse you. 29 If anyone

strikes you on the cheek, offer the other also; and from anyone who takes away your coat do not withhold even your shirt. 30 Give to everyone who begs from you; and if anyone takes away your goods, do not ask for them again. 31 Do to others as you would have them do to you.

32 "If you love those who love you, what credit is that to you? For even sinners love those who love them. 33 If you do good to those who do good to you, what credit is that to you? For even sinners do the same. 34 If you lend to those from whom you hope to receive, what credit is that to you? Even sinners lend to sinners, to receive as much again. 35 But love your enemies, do good, and lend, expecting nothing in return. *v* Your reward will be great, and you will be children of the Most High; for he is kind to the ungrateful and the wicked. 36 Be merciful, just as your Father is merciful.

37 "Do not judge, and you will not be judged; do not condemn, and you will not be condemned. Forgive, and you will be forgiven; 38 give, and it will be given to you. A good measure, pressed down, shaken together, running over, will be put into your lap; for the measure you give will be the measure you get back."

39 He also told them a parable: "Can a blind person guide a blind person? Will not both fall into a pit? 40 A disciple is not above the teacher, but everyone who is fully qualified will be like the teacher. 41 Why do you see the speck in your neighbor's *w* eye, but do not notice the

u Gk *cast out your name as evil*   *v* Other ancient authorities read *despairing of no one*
w Gk *brother's*

---

guishes between the crowds in general and Jesus' followers. The teachings are primarily for the latter, though not exclusively so (7.1).
18: *Unclean spirits,* see Mk 1.23 n. **20–23**: Mt 5.3–12; Lk 4.18–19.
6.22: *Exclude you,* this refers to social ostracism and exclusion from the synagogue and temple. **24–26**: Material satisfactions will not last (11.38–52; 17.1; 21.23; 22.22). **25**: 12.19–20; 16.25; Jas 5.1–5.
6.27–30: Mt 5.39–42; Rom 12.17; 13.8–

10. **28**: Luke's Jesus gives the greatest example of this on the cross, if the verse is authentic; see 23.34 and note *o.* **29**: The reference is to a robber who grabs the outer garment (*coat*). He is not to be restrained from taking the *shirt* also. **31**: Mt 7.12. **32–36**: Mt 5.44–48. **35**: *The Most High,* a common expression referring to God.
6.37–42: Mt 7.1–5. **39**: Mt 15.14. **40**: Mt 10.24–25; Jn 13.16. **41–42**: Mt 7.3–5.

log in your own eye? 42 Or how can you say to your neighbor, *x* 'Friend, *x* let me take out the speck in your eye,' when you yourself do not see the log in your own eye? You hypocrite, first take the log out of your own eye, and then you will see clearly to take the speck out of your neighbor's *y* eye.

43 "No good tree bears bad fruit, nor again does a bad tree bear good fruit; 44 for each tree is known by its own fruit. Figs are not gathered from thorns, nor are grapes picked from a bramble bush. 45 The good person out of the good treasure of the heart produces good, and the evil person out of evil treasure produces evil; for it is out of the abundance of the heart that the mouth speaks.

46 "Why do you call me 'Lord, Lord,' and do not do what I tell you? 47 I will show you what someone is like who comes to me, hears my words, and acts on them. 48 That one is like a man building a house, who dug deeply and laid the foundation on rock; when a flood arose, the river burst against that house but could not shake it, because it had been well built. *z* 49 But the one who hears and does not act is like a man who built a house on the ground without a foundation. When the river burst against it, immediately it fell, and great was the ruin of that house."

7 After Jesus *a* had finished all his sayings in the hearing of the people, he entered Capernaum. 2 A centurion there had a slave whom he valued highly, and who was ill and close to death. 3 When he heard about Jesus, he sent some Jewish elders to him, asking him to come and heal his slave. 4 When they came to Jesus, they appealed to him earnestly, saying, "He is worthy of having you do this for him, 5 for he loves our people, and it is he who built our synagogue for us." 6 And Jesus went with them, but when he was not far from the house, the centurion sent friends to say to him, "Lord, do not trouble yourself, for I am not worthy to have you come under my roof; 7 therefore I did not presume to come to you. But only speak the word, and let my servant be healed. 8 For I also am a man set under authority, with soldiers under me; and I say to one, 'Go,' and he goes, and to another, 'Come,' and he comes, and to my slave, 'Do this,' and the slave does it." 9 When Jesus heard this he was amazed at him, and turning to the crowd that followed him, he said, "I tell you, not even in Israel have I found such faith." 10 When those who had been sent returned to the house, they found the slave in good health.

11 Soon afterwards *b* he went to a town called Nain, and his disciples and a large crowd went with him. 12 As he approached the gate of the town, a man who had died was being carried out. He was his mother's only son, and she was a widow; and with her was a large crowd from the town. 13 When the Lord saw her, he had compassion for her and said to her, "Do not weep." 14 Then he came forward and touched the bier, and the bearers stood still. And he said, "Young man, I say to you, rise!" 15 The dead man

x Gk *brother*   *y* Gk *brother's*   *z* Other ancient authorities read *founded upon the rock* *a* Gk *he*   *b* Other ancient authorities read *Next day*

6.43–45: Mt 7.16–21; 12.33–35; Jas 3.11–12. **45:** Mk 7.14–23. **46–49:** Mt 7.24–27; Jas 1.22–25. The differences between Matthew and Luke probably derive from different collections of words of Jesus.
7.1–10: The centurion's slave. This narrative appears to be another version of the story recounted in Mt 8.5–13 (compare Jn 4.46–53). **3:** *Elders,* leaders in the Jewish community. **5:** Acts 10.2. **9:** Though Luke lacks the climactic utterance of Mt 8.13, his intent is the same: the faith of a Gentile is acceptable to Jesus (4.27; 5.32).
7.11–17: The widow's son at Nain. Mk 5.21–24, 35–43; Jn 11.1–44; 1 Kings 17.17–24; 2 Kings 4.32–37; Lk 4.25–26. **11:** *Nain,* about twenty-five miles southwest of Capernaum. **12:** No burial was allowed within the walls of a Jewish city or town. In this case the funeral procession was passing through *the gate of the town.* The *large crowd* is evidence of deep sympathy for the loss of her only son. **13:** *Lord* is used frequently in Luke as a title for Jesus. **16:** *Fear,* see 1.65 n.

sat up and began to speak, and Jesus*c*
gave him to his mother. 16 Fear seized all
of them; and they glorified God, saying,
"A great prophet has risen among us!"
and "God has looked favorably on his
people!" 17 This word about him spread
throughout Judea and all the surrounding country.

18 The disciples of John reported all
these things to him. So John summoned
two of his disciples 19 and sent them to
the Lord to ask, "Are you the one who
is to come, or are we to wait for another?" 20 When the men had come to him,
they said, "John the Baptist has sent us to
you to ask, 'Are you the one who is to
come, or are we to wait for another?' "
21 Jesus*d* had just then cured many people
of diseases, plagues, and evil spirits, and
had given sight to many who were blind.
22 And he answered them, "Go and tell
John what you have seen and heard: the
blind receive their sight, the lame walk,
the lepers*e* are cleansed, the deaf hear,
the dead are raised, the poor have good
news brought to them. 23 And blessed is
anyone who takes no offense at me."

24 When John's messengers had
gone, Jesus*c* began to speak to the
crowds about John:*f* "What did you go
out into the wilderness to look at? A reed
shaken by the wind? 25 What then did you
go out to see? Someone*g* dressed in soft
robes? Look, those who put on fine
clothing and live in luxury are in royal
palaces. 26 What then did you go out to
see? A prophet? Yes, I tell you, and more
than a prophet. 27 This is the one about
whom it is written,

'See, I am sending my messenger
ahead of you,

who will prepare your way
before you.'
28 I tell you, among those born of women
no one is greater than John; yet the least
in the kingdom of God is greater than
he." 29 (And all the people who heard
this, including the tax collectors, acknowledged the justice of God, *h* because
they had been baptized with John's baptism. 30 But by refusing to be baptized by
him, the Pharisees and the lawyers rejected God's purpose for themselves.)

31 "To what then will I compare the
people of this generation, and what are
they like? 32 They are like children sitting
in the marketplace and calling to one another,

'We played the flute for you, and
you did not dance;
we wailed, and you did not
weep.'
33 For John the Baptist has come eating
no bread and drinking no wine, and you
say, 'He has a demon'; 34 the Son of Man
has come eating and drinking, and you
say, 'Look, a glutton and a drunkard, a
friend of tax collectors and sinners!'
35 Nevertheless, wisdom is vindicated by
all her children."

36 One of the Pharisees asked Jesus*f*
to eat with him, and he went into the
Pharisee's house and took his place at the
table. 37 And a woman in the city, who
was a sinner, having learned that he was
eating in the Pharisee's house, brought
an alabaster jar of ointment. 38 She stood
behind him at his feet, weeping, and be-

c Gk *he*   *d* Gk *He*   *e* The terms *leper* and
leprosy can refer to several diseases   *f* Gk *him*
g Or *Why then did you go out? To see someone*
h Or *praised God*

7.18–35: Jesus and John (Mt 11.2–19).
18: *John the Baptist* was at this time in prison
at Machaerus. **19:** *Lord,* see v. 13 n. *The one
who is to come,* the Messiah (see Mt 11.2–3 n.).
21–22: Isa 29.18–19; 35.5–6; 61.1; Lk 4.18–
19. An appeal to John to believe because of the
evidence that God's purposes were being realized. **27:** Mal 3.1; Mk 1.2. **29:** *Acknowledged
the justice of God,* in the sense that they accepted Jesus' word about John and God's purpose
through him, in contrast with the thought of

v. 30. **30:** Not all Pharisees are referred to
here (see Mt 3.7 n.). The *lawyers,* also called
scribes, were experts in Jewish law. **33:** *Demon* possession is appealed to in order to explain unconventional behavior (Mt 11.18; Mk
3.22; Lk 8.27) or attitudes considered basically false (Jn 7.20; 10.20; 1 Tim 4.1). See 4.33 n.
7.36–50: The woman who was a sinner
(compare Mt 26.6–13; Mk 14.3–9; Jn 12.1–
8). **36:** 11.37; 14.1. **37:** Mt 9.10; Lk 5.29–30;
7.34. Houses seem to have been somewhat

gan to bathe his feet with her tears and to dry them with her hair. Then she continued kissing his feet and anointing them with the ointment. 39 Now when the Pharisee who had invited him saw it, he said to himself, "If this man were a prophet, he would have known who and what kind of woman this is who is touching him—that she is a sinner." 40 Jesus spoke up and said to him, "Simon, I have something to say to you." "Teacher," he replied, "Speak." 41 "A certain creditor had two debtors; one owed five hundred denarii, *i* and the other fifty. 42 When they could not pay, he canceled the debts for both of them. Now which of them will love him more?" 43 Simon answered, "I suppose the one for whom he canceled the greater debt." And Jesus *j* said to him, "You have judged rightly." 44 Then turning toward the woman, he said to Simon, "Do you see this woman? I entered your house; you gave me no water for my feet, but she has bathed my feet with her tears and dried them with her hair. 45 You gave me no kiss, but from the time I came in she has not stopped kissing my feet. 46 You did not anoint my head with oil, but she has anointed my feet with ointment. 47 Therefore, I tell you, her sins, which were many, have been forgiven; hence she has shown great love. But the one to whom little is forgiven, loves little." 48 Then he said to her, "Your sins are forgiven." 49 But those who were at the table with him began to say among themselves, "Who is this who

even forgives sins?" 50 And he said to the woman, "Your faith has saved you; go in peace."

8 Soon afterwards he went on through cities and villages, proclaiming and bringing the good news of the kingdom of God. The twelve were with him, 2 as well as some women who had been cured of evil spirits and infirmities: Mary, called Magdalene, from whom seven demons had gone out, 3 and Joanna, the wife of Herod's steward Chuza, and Susanna, and many others, who provided for them *k* out of their resources.

4 When a great crowd gathered and people from town after town came to him, he said in a parable: 5 "A sower went out to sow his seed; and as he sowed, some fell on the path and was trampled on, and the birds of the air ate it up. 6 Some fell on the rock; and as it grew up, it withered for lack of moisture. 7 Some fell among thorns, and the thorns grew with it and choked it. 8 Some fell into good soil, and when it grew, it produced a hundredfold." As he said this, he called out, "Let anyone with ears to hear listen!"

9 Then his disciples asked him what this parable meant. 10 He said, "To you it has been given to know the secrets *l* of the kingdom of God; but to others I speak *m* in parables, so that

i The denarius was the usual day's wage for a laborer   *j* Gk *he*   *k* Other ancient authorities read *him*   *l* Or *mysteries*   *m* Gk lacks *I speak*

---

open to intrusion of this sort (Mk 1.33; 2.2). The woman may have intended to anoint Jesus' head, a sign of regard (v. 46) as well as of personal grooming (Mt 6.17), but was overcome by a sense of humility and gratitude for his message (5.32). **38:** Jesus was reclining at the table and his feet were stretched out on the couch behind him; hence the woman could easily approach and anoint his feet. **39:** Jesus does not share the Pharisee's concern (compare Mk 1.41 with Mk 7.3–4). **41:** *Denarius,* see 10.35 note *r.* **42:** Mt 18.25. **44–46:** Jesus notes contrasts in responses to himself. **47:** The conclusion: the woman knows the meaning of forgiveness, the host does not.

Hence she has shown great love, her great love proves that her many sins have been forgiven. **48:** Mt 9.2; Mk 2.5; 11.23 n., 24 n.; Lk 5.20.
8.1–3: On tour. 1: Mt 4.23; 9.35; Mk 3.14; Lk 23.49. **2:** Mt 27.55–56; Mk 15.40–41. *Mary, called Magdalene,* apparently came from Magdala on the Sea of Galilee. There is no evidence to identify her with the woman in 7.36–50. **3:** Herod's *steward* was probably a domestic administrator. *Others,* i.e. other women.
8.4–15: Parable of the sower (Mt 13.1–23; Mk 4.1–20). **4:** *Parable,* see Mt 13.3 n. **5:** See Mk 4.3 n. **6:** See Mk 4.5 n. **10:** See Mk 4.11 n.; Mt 13.11 n.; Isa 6.9–10; Jer 5.21;

'looking they may not perceive,
and listening they may not
understand.'
11 "Now the parable is this: The seed
is the word of God. 12 The ones on the
path are those who have heard; then the
devil comes and takes away the word
from their hearts, so that they may not
believe and be saved. 13 The ones on the
rock are those who, when they hear the
word, receive it with joy. But these have
no root; they believe only for a while and
in a time of testing fall away. 14 As for
what fell among the thorns, these are the
ones who hear; but as they go on their
way, they are choked by the cares and
riches and pleasures of life, and their fruit
does not mature. 15 But as for that in the
good soil, these are the ones who, when
they hear the word, hold it fast in an
honest and good heart, and bear fruit
with patient endurance.

16 "No one after lighting a lamp
hides it under a jar, or puts it under a bed,
but puts it on a lampstand, so that those
who enter may see the light. 17 For nothing is hidden that will not be disclosed,
nor is anything secret that will not become known and come to light. 18 Then
pay attention to how you listen; for to
those who have, more will be given; and
from those who do not have, even what
they seem to have will be taken away."

19 Then his mother and his brothers
came to him, but they could not reach
him because of the crowd. 20 And he was
told, "Your mother and your brothers
are standing outside, wanting to see
you." 21 But he said to them, "My mother and my brothers are those who hear
the word of God and do it."

22 One day he got into a boat with his
disciples, and he said to them, "Let us go
across to the other side of the lake." So
they put out, 23 and while they were sailing he fell asleep. A windstorm swept
down on the lake, and the boat was filling with water, and they were in danger.
24 They went to him and woke him up,
shouting, "Master, Master, we are perishing!" And he woke up and rebuked
the wind and the raging waves; they
ceased, and there was a calm. 25 He said
to them, "Where is your faith?" They
were afraid and amazed, and said to one
another, "Who then is this, that he commands even the winds and the water, and
they obey him?"

26 Then they arrived at the country of
the Gerasenes,[n] which is opposite Galilee. 27 As he stepped out on land, a man
of the city who had demons met him.
For a long time he had worn[o] no clothes,
and he did not live in a house but in the
tombs. 28 When he saw Jesus, he fell
down before him and shouted at the top
of his voice, "What have you to do with
me, Jesus, Son of the Most High God? I
beg you, do not torment me"— 29 for
Jesus[p] had commanded the unclean spirit
to come out of the man. (For many times

n Other ancient authorities read *Gadarenes;*
others, *Gergesenes*   *o* Other ancient authorities
read *a man of the city who had had demons for a long
time met him. He wore*   *p* Gk *he*

Ezek 12.2. **11:** 1 Thess 2.13; 1 Pet 1.23. **15:** The
words *honest and good,* here spoken of the
heart (compare Mk 7.21–23), echo the classical Greek description of the true gentleman.
  8.16–18: On obedient listening (Mk
4.21–25). **16:** See Mk 4.21 n. **17:** See Mk
4.22 n. **18:** See Mt 13.12 n.
  8.19–21: Jesus' true family (Mt 12.46–
50; Mk 3.31–35). **19:** *Brothers,* see Mt 13.55 n.
21: 11.28.
  8.22–25: Wind and sea calmed (Mt
8.18, 23–27; Mk 4.35–41). See notes on the
passage in Mark. **24:** *Master* and the titles for

Jesus in the parallels express aspects of the
disciples' attitude toward Jesus (Mt 17.4; Mk
9.5; 11.21; 14.45; Lk 17.13; Jn 1.38).
  8.26–39: The Gerasene demoniac (Mt
8.28–34; Mk 5.1–20). **27:** *Demons,* see 4.33 n.
31: *Abyss,* a place of confinement for demonic
forces which, though hostile to God, are ultimately under his control (Rev 9.1–11; 11.7;
17.8; 20.1–3). The words attribute to Jesus a
judicial authority (as Mt 7.21–23; 11.20–24).
36: *Had been healed,* see Mt 9.21 n. **37:** *Leave,*
5.8.

it had seized him; he was kept under guard and bound with chains and shackles, but he would break the bonds and be driven by the demon into the wilds.) 30 Jesus then asked him, "What is your name?" He said, "Legion"; for many demons had entered him. 31 They begged him not to order them to go back into the abyss.

32 Now there on the hillside a large herd of swine was feeding; and the demons*q* begged Jesus*r* to let them enter these. So he gave them permission. 33 Then the demons came out of the man and entered the swine, and the herd rushed down the steep bank into the lake and was drowned.

34 When the swineherds saw what had happened, they ran off and told it in the city and in the country. 35 Then people came out to see what had happened, and when they came to Jesus, they found the man from whom the demons had gone sitting at the feet of Jesus, clothed and in his right mind. And they were afraid. 36 Those who had seen it told them how the one who had been possessed by demons had been healed. 37 Then all the people of the surrounding country of the Gerasenes*s* asked Jesus*r* to leave them; for they were seized with great fear. So he got into the boat and returned. 38 The man from whom the demons had gone begged that he might be with him; but Jesus*t* sent him away, saying, 39 "Return to your home, and declare how much God has done for you." So he went away, proclaiming throughout the city how much Jesus had done for him.

40 Now when Jesus returned, the crowd welcomed him, for they were all waiting for him. 41 Just then there came a man named Jairus, a leader of the synagogue. He fell at Jesus' feet and begged him to come to his house, 42 for he had an only daughter, about twelve years old, who was dying.

As he went, the crowds pressed in on him. 43 Now there was a woman who had been suffering from hemorrhages for twelve years; and though she had spent all she had on physicians,*u* no one could cure her. 44 She came up behind him and touched the fringe of his clothes, and immediately her hemorrhage stopped. 45 Then Jesus asked, "Who touched me?" When all denied it, Peter*v* said, "Master, the crowds surround you and press in on you." 46 But Jesus said, "Someone touched me; for I noticed that power had gone out from me." 47 When the woman saw that she could not remain hidden, she came trembling; and falling down before him, she declared in the presence of all the people why she had touched him, and how she had been immediately healed. 48 He said to her, "Daughter, your faith has made you well; go in peace."

49 While he was still speaking, someone came from the leader's house to say, "Your daughter is dead; do not trouble the teacher any longer." 50 When Jesus heard this, he replied, "Do not fear. Only believe, and she will be saved." 51 When he came to the house, he did not allow anyone to enter with him, except Peter, John, and James, and the child's father and mother. 52 They were all weeping and wailing for her; but he said, "Do not weep; for she is not dead but sleeping." 53 And they laughed at him, knowing that she was dead. 54 But he took her by the hand and called out, "Child, get up!" 55 Her spirit returned, and she got up at once. Then he directed them to give her something to eat. 56 Her parents were astounded; but he ordered them to tell no one what had happened.

q Gk *they*   *r* Gk *him*   *s* Other ancient authorities read *Gadarenes*; others, *Gergesenes*   *t* Gk *he*   *u* Other ancient authorities lack *and had spent all she had on physicians*   *v* Other ancient authorities add *and those who were with him*

8.40–56: **Jairus' daughter raised** (Mt 9.18–26; Mk 5.21–43). See notes on the parallel passages. 43: Lev 15.25–30. 46: 5.17. 48: See Mt 9.21 n., 22; Lk 7.50; 17.19; 18.42. 50: See Mt 9.21 n.

9 Then Jesus[w] called the twelve together and gave them power and authority over all demons and to cure diseases, 2and he sent them out to proclaim the kingdom of God and to heal. 3He said to them, "Take nothing for your journey, no staff, nor bag, nor bread, nor money—not even an extra tunic. 4Whatever house you enter, stay there, and leave from there. 5Wherever they do not welcome you, as you are leaving that town shake the dust off your feet as a testimony against them." 6They departed and went through the villages, bringing the good news and curing diseases everywhere.

7 Now Herod the ruler[x] heard about all that had taken place, and he was perplexed, because it was said by some that John had been raised from the dead, 8by some that Elijah had appeared, and by others that one of the ancient prophets had arisen. 9Herod said, "John I beheaded; but who is this about whom I hear such things?" And he tried to see him.

10 On their return the apostles told Jesus[y] all they had done. He took them with him and withdrew privately to a city called Bethsaida. 11When the crowds found out about it, they followed him; and he welcomed them, and spoke to them about the kingdom of God, and healed those who needed to be cured.

12 The day was drawing to a close, and the twelve came to him and said, "Send the crowd away, so that they may go into the surrounding villages and countryside, to lodge and get provisions; for we are here in a deserted place." 13But he said to them, "You give them something to eat." They said, "We have no more than five loaves and two fish—unless we are to go and buy food for all these people." 14For there were about five thousand men. And he said to his disciples, "Make them sit down in groups of about fifty each." 15They did so and made them all sit down. 16And taking the five loaves and the two fish, he looked up to heaven, and blessed and broke them, and gave them to the disciples to set before the crowd. 17And all ate and were filled. What was left over was gathered up, twelve baskets of broken pieces.

18 Once when Jesus[w] was praying alone, with only the disciples near him, he asked them, "Who do the crowds say that I am?" 19They answered, "John the Baptist; but others, Elijah; and still others, that one of the ancient prophets has arisen." 20He said to them, "But who do you say that I am?" Peter answered, "The Messiah[z] of God."

21 He sternly ordered and commanded them not to tell anyone, 22saying, "The Son of Man must undergo great suffering, and be rejected by the elders, chief priests, and scribes, and be killed, and on the third day be raised."

23 Then he said to them all, "If any want to become my followers, let them deny themselves and take up their cross daily and follow me. 24For those who want to save their life will lose it, and those who lose their life for my sake will

w Gk he   x Gk tetrarch   y Gk him
z Or The Christ

9.1–6: **Commissioning and instruction of the Twelve** (Mt 9.35; 10.1, 9–11, 14; Mk 6.7–13; Lk 10.4–11). See notes on the passages in Matthew and Mark. **3:** *Tunic,* see Mt 10.10 n.
9.7–9: **Herod asks about Jesus** (Mt 14.1–2; Mk 6.14–16). *Herod* Antipas, son of Herod the Great. See notes on the parallel passages; compare Mt 16.14; Lk 9.18–19.
9.10–17: **Five thousand fed** (Mt 14.13–21; Mk 6.30–44; Jn 6.1–14). See notes on the parallel passages. **13:** 2 Kings 4.42–44. **16:** 22.19; 24.30–31; Acts 2.42; 20.11; 27.35.

9.18–22: **Peter's confession** (Mt 16.13–23; Mk 8.27–33). See notes on the parallel passages. **18:** See 3.21 n. **19:** 9.7; Mk 9.11–13; see Mt 14.2. **22:** 9.43–45; 18.31–34; 17.25; see Mk 9.31 n. Jesus accepted rejection because he insisted that he himself be freely followed.
9.23–27: **On discipleship** (Mt 16.24–28; Mk 8.34–9.1). See notes on the parallel passages. **23:** The language suggests that Jesus frequently spoke in this way (see Mt 10.38 n.). **26:** Mt 10.33; Lk 12.9; 1 Jn 2.28. **27:** *Taste death,* die (compare Jn 8.52; Heb 2.9).

save it. 25 What does it profit them if they gain the whole world, but lose or forfeit themselves? 26 Those who are ashamed of me and of my words, of them the Son of Man will be ashamed, when he comes in his glory and the glory of the Father and of the holy angels. 27 But truly I tell you, there are some standing here who will not taste death before they see the kingdom of God."

28 Now about eight days after these sayings Jesus*a* took with him Peter and John and James, and went up on the mountain to pray. 29 And while he was praying, the appearance of his face changed, and his clothes became dazzling white. 30 Suddenly they saw two men, Moses and Elijah, talking to him. 31 They appeared in glory and were speaking of his departure, which he was about to accomplish at Jerusalem. 32 Now Peter and his companions were weighed down with sleep; but since they had stayed awake,*b* they saw his glory and the two men who stood with him. 33 Just as they were leaving him, Peter said to Jesus, "Master, it is good for us to be here; let us make three dwellings,*c* one for you, one for Moses, and one for Elijah"—not knowing what he said. 34 While he was saying this, a cloud came and overshadowed them; and they were terrified as they entered the cloud. 35 Then from the cloud came a voice that said, "This is my Son, my Chosen;*d* listen to him!" 36 When the voice had spoken, Jesus was found alone. And they kept silent and in those days told no one any of the things they had seen.

37 On the next day, when they had come down from the mountain, a great crowd met him. 38 Just then a man from the crowd shouted, "Teacher, I beg you to look at my son; he is my only child. 39 Suddenly a spirit seizes him, and all at once he*e* shrieks. It convulses him until he foams at the mouth; it mauls him and will scarcely leave him. 40 I begged your disciples to cast it out, but they could not." 41 Jesus answered, "You faithless and perverse generation, how much longer must I be with you and bear with you? Bring your son here." 42 While he was coming, the demon dashed him to the ground in convulsions. But Jesus rebuked the unclean spirit, healed the boy, and gave him back to his father. 43 And all were astounded at the greatness of God.

While everyone was amazed at all that he was doing, he said to his disciples, 44 "Let these words sink into your ears: The Son of Man is going to be betrayed into human hands." 45 But they did not understand this saying; its meaning was concealed from them, so that they could not perceive it. And they were afraid to ask him about this saying.

46 An argument arose among them as to which one of them was the greatest. 47 But Jesus, aware of their inner thoughts, took a little child and put it by his side, 48 and said to them, "Whoever welcomes this child in my name welcomes me, and whoever welcomes me welcomes the one who sent me; for the least among all of you is the greatest."

a Gk *he*   *b* Or *but when they were fully awake*   *c* Or *tents*   *d* Other ancient authorities read *my Beloved*   *e* Or *it*

9.28–36: The transfiguration (Mt 17.1–8; Mk 9.2–8). This event began as prayer (v. 29; see 3.21 n.) and grew into an intense religious experience, the exact nature of which is uncertain (Mt 17.9 uses the word *vision*). The aura of unnatural brilliance is associated with mystical experiences elsewhere (Ex 34.29–35; Acts 9.3). **31:** *Departure,* death. **32:** Apparently the experience took place at night. **35:** *Chosen* has about the same meaning as *beloved* in the parallel passages (see Mk 1.11 n.; Lk 3.22; Jn 12.28–30). **9.37–43a: Epileptic child healed** (Mt 17.14–21; Mk 9.14–29). See notes on the parallel passages. **43:** Acts 2.22.

9.43b–45: The Passion foretold again (Mt 17.22–23; Mk 9.30–32); compare 9.22; 18.31–33. **44:** 9.22; 18.31–34; 17.25. **45:** The comment is based on the point of view that (a) the Messiah's death was not part of the disciples' Jewish faith, and (b) spiritual truth must be revealed (see 24.16 n.; Mt 16.17 n.; 1 Cor 2.14).

9.46–48: True greatness (Mt 18.1–5; Mk 9.33–37). **47:** 18.17. **48:** See 10.16 n.; Mk 9.35 n.

49 John answered, "Master, we saw someone casting out demons in your name, and we tried to stop him, because he does not follow with us." 50 But Jesus said to him, "Do not stop him; for whoever is not against you is for you."

51 When the days drew near for him to be taken up, he set his face to go to Jerusalem. 52 And he sent messengers ahead of him. On their way they entered a village of the Samaritans to make ready for him; 53 but they did not receive him, because his face was set toward Jerusalem. 54 When his disciples James and John saw it, they said, "Lord, do you want us to command fire to come down from heaven and consume them?"*f* 55 But he turned and rebuked them. 56 Then*g* they went on to another village.

57 As they were going along the road, someone said to him, "I will follow you wherever you go." 58 And Jesus said to him, "Foxes have holes, and birds of the air have nests; but the Son of Man has nowhere to lay his head." 59 To another he said, "Follow me." But he said, "Lord, first let me go and bury my father." 60 But Jesus*h* said to him, "Let the dead bury their own dead; but as for you, go and proclaim the kingdom of God." 61 Another said, "I will follow you, Lord; but let me first say farewell to those at my home." 62 Jesus said to him, "No one who puts a hand to the plow and looks back is fit for the kingdom of God."

10 After this the Lord appointed seventy*i* others and sent them on ahead of him in pairs to every town and place where he himself intended to go. 2 He said to them, "The harvest is plentiful, but the laborers are few; therefore ask the Lord of the harvest to send out laborers into his harvest. 3 Go on your way. See, I am sending you out like lambs into the midst of wolves. 4 Carry no purse, no bag, no sandals; and greet no one on the road. 5 Whatever house you enter, first say, 'Peace to this house!' 6 And if anyone is there who shares in peace, your peace will rest on that person; but if not, it will return to you. 7 Remain in the same house, eating and drinking whatever they provide, for the laborer deserves to be paid. Do not move about from house to house. 8 Whenever you enter a town and its people welcome you, eat what is set before you; 9 cure the sick who are there, and say to them, 'The kingdom of God has come near to you.'*j* 10 But whenever you enter a town and they do not welcome you, go out into its streets and say, 11 'Even the dust of your town that clings to our feet, we wipe off in protest against you. Yet know this: the kingdom of God has come near.'*k* 12 I tell you, on that day it will be more tolerable for Sodom than for that town.

13 "Woe to you, Chorazin! Woe to you, Bethsaida! For if the deeds of power done in you had been done in Tyre and

f Other ancient authorities add *as Elijah did*
g Other ancient authorities read *rebuked them, and said, "You do not know what spirit you are of, 56for the Son of Man has not come to destroy the lives of human beings but to save them." Then* h Gk he
i Other ancient authorities read *seventy-two*
j Or *is at hand for you* k Or *is at hand*

9.49–50: The unknown exorcist (Mk 9.38–41). **49:** 11.19. **50:** 11.23; see Mk 9.39–40 n.
9.51–18.14: Events and teachings on the way to Jerusalem. This section, which is peculiar to Luke, reports what is generally called the Perean ministry of Jesus, though the scene of that ministry was partly in Judea. **9.51–56: The hostile Samaritans** (compare Jn 4.9). **52:** *Messengers,* whose errand was to arrange for lodging and food (Mt 10.5 was at an earlier stage of time and purpose). **53:** The Samaritans *did not* help pilgrims going to keep a feast at what they regarded as the wrong sanctuary (compare Jn 4.20). **54:** Mk 3.17; 2 Kings 1.9–16.
9.57–62: Claims of discipleship (Mt 8.19–22). **60:** See Mt 8.22 n. **61:** Phil 3.13; Heb 6.4–6.
10.1–16: Mission of the Seventy (Mt 9.37–38; 10.7–16). **1:** 9.1–5, 51–52; Mk 6.7–11. **2:** Mt 9.37–38; Jn 4.35. **4:** *Greet no one,* which might cause delay. **5:** 1 Sam 25.6. **6:** *Anyone . . . who shares in peace,* a peaceful person (compare Mt 5.45 n.). **7:** 1 Cor 9.4–14; 1 Tim 5.18; Deut 24.15. **9:** 11.20. **11:** Acts 13.51. **12:** Mt 11.24; Gen 19.24–28; Jude 7.

Sidon, they would have repented long ago, sitting in sackcloth and ashes. [14] But at the judgment it will be more tolerable for Tyre and Sidon than for you. [15] And you, Capernaum,

> will you be exalted to heaven?
> No, you will be brought down
>   to Hades.

16 "Whoever listens to you listens to me, and whoever rejects you rejects me, and whoever rejects me rejects the one who sent me."

17 The seventy[l] returned with joy, saying, "Lord, in your name even the demons submit to us!" [18] He said to them, "I watched Satan fall from heaven like a flash of lightning. [19] See, I have given you authority to tread on snakes and scorpions, and over all the power of the enemy; and nothing will hurt you. [20] Nevertheless, do not rejoice at this, that the spirits submit to you, but rejoice that your names are written in heaven."

21 At that same hour Jesus[m] rejoiced in the Holy Spirit[n] and said, "I thank[o] you, Father, Lord of heaven and earth, because you have hidden these things from the wise and the intelligent and have revealed them to infants; yes, Father, for such was your gracious will. [p] [22] All things have been handed over to me by my Father; and no one knows who the Son is except the Father, or who the Father is except the Son and anyone to whom the Son chooses to reveal him."

23 Then turning to the disciples, Jesus[m] said to them privately, "Blessed are the eyes that see what you see! [24] For I tell you that many prophets and kings desired to see what you see, but did not see it, and to hear what you hear, but did not hear it."

25 Just then a lawyer stood up to test Jesus. [q] "Teacher," he said, "what must I do to inherit eternal life?" [26] He said to him, "What is written in the law? What do you read there?" [27] He answered, "You shall love the Lord your God with all your heart, and with all your soul, and with all your strength, and with all your mind; and your neighbor as yourself." [28] And he said to him, "You have given the right answer; do this, and you will live."

29 But wanting to justify himself, he asked Jesus, "And who is my neighbor?" [30] Jesus replied, "A man was going down from Jerusalem to Jericho, and fell into the hands of robbers, who stripped him, beat him, and went away, leaving him half dead. [31] Now by chance a priest was going down that road; and when he saw

l Other ancient authorities read *seventy-two*
m Gk *he*   *n* Other authorities read *in the spirit*
o Or *praise*   *p* Or *for so it was well-pleasing in
your sight*   *q* Gk *him*

10.13–15: Mt 11.21–23; Lk 6.24–26. **15:** The words echo Isa 14.13–15 (see also v. 18). **16:** Mt 10.40; 18.5; Mk 9.37; Lk 9.48; Jn 13.20; 12.48; Gal 4.14.
10.17–20: Return of the Seventy. 17: *Even the demons submit to us,* unlike the Twelve (9.1), the Seventy had not been promised this power (see 13.16 n.). **18:** *I watched Satan fall from heaven,* alluding to the victory of the Seventy over the demons; compare Jn 12.31; Rev 12.7–12. **19:** *Authority,* Mk 6.7; Lk 22.29. *The enemy,* Satan (Mt 13.39). **20:** Jesus regarded exorcism as not in itself a sign of God's kingdom (11.19). *Written in heaven,* Dan 12.1; Ps 69.28; Ex 32.32; Phil 4.3; Heb 12.23; Rev 3.5; 13.8; 21.27.
10.21–22: Prayer of Jesus (see 3.21 n. and Mt 11.25–27 n.). **21:** 1 Cor 1.26–29. *In the Holy Spirit,* in spiritual ecstasy. **22:** Mt 28.18; Jn 3.35; 13.3; 10.15; 17.25.
10.23–24: Mt 13.16–17; Jn 8.56; Heb 11.13; 1 Pet 1.10–12. Jesus speaks both of spiritual perception guided by faith (Mk 4.9; Lk 8.10), and of the fulfillment of God's purpose (2.26–32).
10.25–28: A lawyer's question (Mt 22.23–40; Mk 12.28–31). **25:** Mt 19.16 n.; Mk 10.17; Lk 18.18. *Inherit,* see Mt 19.29 n. **27:** See Mk 12.29 n., 31 n. Rom 13.8–10; Gal 5.14; and Jas 2.8 implicitly link Deut 6.4–5 and Lev 19.18, stressing as here (vv. 29–37) the principle that acts of love are the final requirement of the law (compare Deut 6.4ff. and notes; Lev 19.17–18 n.). **28:** Mk 12.34; Lev 18.1–5.
10.29–37: The Good Samaritan. 29: The questioner intended to prove his right to eternal life by defining the limits of his duty, and showing how he had fulfilled it. *Justify himself* means to show himself to be righteous, acceptable to God (18.9–14). **31–33:** The *priest* represented the highest religious leadership among the Jews; the *Levite* (v. 32)

him, he passed by on the other side. 32 So likewise a Levite, when he came to the place and saw him, passed by on the other side. 33 But a Samaritan while traveling came near him; and when he saw him, he was moved with pity. 34 He went to him and bandaged his wounds, having poured oil and wine on them. Then he put him on his own animal, brought him to an inn, and took care of him. 35 The next day he took out two denarii, *r* gave them to the innkeeper, and said, 'Take care of him; and when I come back, I will repay you whatever more you spend.' 36 Which of these three, do you think, was a neighbor to the man who fell into the hands of the robbers?" 37 He said, "The one who showed him mercy." Jesus said to him, "Go and do likewise."

38 Now as they went on their way, he entered a certain village, where a woman named Martha welcomed him into her home. 39 She had a sister named Mary, who sat at the Lord's feet and listened to what he was saying. 40 But Martha was distracted by her many tasks; so she came to him and asked, "Lord, do you not care that my sister has left me to do all the work by myself? Tell her then to help me." 41 But the Lord answered her, "Martha, Martha, you are worried and distracted by many things; 42 there is need of only one thing. *s* Mary has chosen the better part, which will not be taken away from her."

11 He was praying in a certain place, and after he had finished, one of his disciples said to him, "Lord, teach us to pray, as John taught his disciples." 2 He said to them, "When you pray, say:

Father, *t* hallowed be your name.
Your kingdom come. *u*
3 Give us each day our daily
bread. *v*
4 And forgive us our sins,
for we ourselves forgive
everyone indebted to us.
And do not bring us to the time
of trial." *w*

5 And he said to them, "Suppose one of you has a friend, and you go to him at midnight and say to him, 'Friend, lend me three loaves of bread; 6 for a friend of mine has arrived, and I have nothing to set before him.' 7 And he answers from within, 'Do not bother me; the door has already been locked, and my children are with me in bed; I cannot get up and give you anything.' 8 I tell you, even though he will not get up and give him anything because he is his friend, at least because of his persistence he will get up and give him whatever he needs.

9 "So I say to you, Ask, and it will be given you; search, and you will find; knock, and the door will be opened for you. 10 For everyone who asks receives, and everyone who searches finds, and for everyone who knocks, the door will be opened. 11 Is there anyone among you

r The denarius was the usual day's wage for a laborer   s Other ancient authorities read *few things are necessary, or only one*   t Other ancient authorities read *Our Father in heaven*   u A few ancient authorities read *Your Holy Spirit come upon us and cleanse us*. Other ancient authorities add *Your will be done, on earth as in heaven*   v Or *our bread for tomorrow*   w Or *us into temptation*. Other ancient authorities add *but rescue us from the evil one* (or *from evil*)

---

was the designated lay-associate of the priest. In contrast it was a *Samaritan,* a foreigner not expected to show sympathy to Jews (see Jn 4.9 n.; Acts 8.5 n.), who *was moved with pity.* **34:** *Oil and wine,* ancient medication.

10.38–42: Martha and Mary, compare Jn 11.1 where they are introduced as well-known persons living at Bethany. **41:** *By many things,* which were not important enough to call for excessive attention or worry. **42:** With delicate ambiguity Jesus rebuked Martha's choice of values; a simple meal (one dish) is

sufficient for hospitality. Jesus approved Mary's preference for listening to his teaching (thereby accepting a woman as a disciple) as contrasted with Martha's unneeded acts of hospitality (the more usual woman's role).

11.1–13: Sayings on prayer. 1–4: See Mt 6.9–13 n.; Lk 3.21 n. **4:** Mk 11.25; Mt 18.35. **5–8:** Lk 18.1–5.

11.9–13: Mt 7.7–11. **9:** Mt 18.19; 21.22; Mk 11.24; Jas 1.5–8; 1 Jn 5.14–15; Jn 14.13; 15.7; 17.23–24.

11.14–28: Sources of Jesus' power (Mt

who, if your child asks for[x] a fish, will give a snake instead of a fish? 12 Or if the child asks for an egg, will give a scorpion? 13 If you then, who are evil, know how to give good gifts to your children, how much more will the heavenly Father give the Holy Spirit[y] to those who ask him!"

14 Now he was casting out a demon that was mute; when the demon had gone out, the one who had been mute spoke, and the crowds were amazed. 15 But some of them said, "He casts out demons by Beelzebul, the ruler of the demons." 16 Others, to test him, kept demanding from him a sign from heaven. 17 But he knew what they were thinking and said to them, "Every kingdom divided against itself becomes a desert, and house falls on house. 18 If Satan also is divided against himself, how will his kingdom stand? —for you say that I cast out the demons by Beelzebul. 19 Now if I cast out the demons by Beelzebul, by whom do your exorcists[z] cast them out? Therefore they will be your judges. 20 But if it is by the finger of God that I cast out the demons, then the kingdom of God has come to you. 21 When a strong man, fully armed, guards his castle, his property is safe. 22 But when one stronger than he attacks him and overpowers him, he takes away his armor in which he trusted and divides his plunder. 23 Whoever is not with me is against me, and whoever does not gather with me scatters.

24 "When the unclean spirit has gone out of a person, it wanders through waterless regions looking for a resting place, but not finding any, it says, 'I will return to my house from which I came.' 25 When it comes, it finds it swept and put in order. 26 Then it goes and brings seven other spirits more evil than itself, and they enter and live there; and the last state of that person is worse than the first."

27 While he was saying this, a woman in the crowd raised her voice and said to him, "Blessed is the womb that bore you and the breasts that nursed you!" 28 But he said, "Blessed rather are those who hear the word of God and obey it!"

29 When the crowds were increasing, he began to say, "This generation is an evil generation; it asks for a sign, but no sign will be given to it except the sign of Jonah. 30 For just as Jonah became a sign to the people of Nineveh, so the Son of Man will be to this generation. 31 The queen of the South will rise at the judgment with the people of this generation and condemn them, because she came from the ends of the earth to listen to the wisdom of Solomon, and see, something greater than Solomon is here! 32 The people of Nineveh will rise up at the judgment with this generation and condemn it, because they repented at the proclamation of Jonah, and see, something greater than Jonah is here!

33 "No one after lighting a lamp puts it in a cellar,[a] but on the lampstand so that those who enter may see the light. 34 Your eye is the lamp of your body. If your eye is healthy, your whole body is full of light; but if it is not healthy, your body is full of darkness. 35 Therefore consider whether the light in you is not darkness. 36 If then your whole body is full of light, with no part of it in darkness, it will be as full of light as when a lamp gives you light with its rays."

x Other ancient authorities add *bread, will give a stone; or if your child asks for*   y Other ancient authorities read *the Father give the Holy Spirit from heaven*   z Gk *sons*   a Other ancient authorities add *or under the bushel basket*

12.22–30; Mk 3.22–27). **14**: Mt 9.32–34; see 12.22–24 n. **15**: See Mk 3.22 n. **16**: Mt 12.38; 16.1–4; Mk 8.11–12; Jn 2.18; 6.30; 1 Cor 1.22. **19**: See Mt 12.27 n. **20**: Ex 8.19. *Finger* stands for God's power. **23**: Mt 12.30. Jesus used similar language for a different point in Mk 9.38–40; Lk 9.49–50.

11.24–26: See Mt 12.43–45 n. **28**: 8.21. **11.29–32**: **Request for a sign. 29**: Mt 12.39. **30**: Jon 3.4. **31**: 1 Kings 10.1–10; 2 Chr 9.1–9. **32**: Mt 12.6. **11.33–36**: **Concerning light** (Mt 5.15; 6.22–23). **33**: See Mk 4.21 n.

Jesus denounces
Pharisees and lawyers

37 While he was speaking, a Pharisee invited him to dine with him; so he went in and took his place at the table. 38 The Pharisee was amazed to see that he did not first wash before dinner. 39 Then the Lord said to him, "Now you Pharisees clean the outside of the cup and of the dish, but inside you are full of greed and wickedness. 40 You fools! Did not the one who made the outside make the inside also? 41 So give for alms those things that are within; and see, everything will be clean for you.

42 "But woe to you Pharisees! For you tithe mint and rue and herbs of all kinds, and neglect justice and the love of God; it is these you ought to have practiced, without neglecting the others. 43 Woe to you Pharisees! For you love to have the seat of honor in the synagogues and to be greeted with respect in the marketplaces. 44 Woe to you! For you are like unmarked graves, and people walk over them without realizing it."

45 One of the lawyers answered him, "Teacher, when you say these things, you insult us too." 46 And he said, "Woe also to you lawyers! For you load people with burdens hard to bear, and you yourselves do not lift a finger to ease them. 47 Woe to you! For you build the tombs of the prophets whom your ancestors killed. 48 So you are witnesses and approve of the deeds of your ancestors;

for they killed them, and you build their tombs. 49 Therefore also the Wisdom of God said, 'I will send them prophets and apostles, some of whom they will kill and persecute,' 50 so that this generation may be charged with the blood of all the prophets shed since the foundation of the world, 51 from the blood of Abel to the blood of Zechariah, who perished between the altar and the sanctuary. Yes, I tell you, it will be charged against this generation. 52 Woe to you lawyers! For you have taken away the key of knowledge; you did not enter yourselves, and you hindered those who were entering."

53 When he went outside, the scribes and the Pharisees began to be very hostile toward him and to cross-examine him about many things, 54 lying in wait for him, to catch him in something he might say.

12 Meanwhile, when the crowd gathered by the thousands, so that they trampled on one another, he began to speak first to his disciples, "Beware of the yeast of the Pharisees, that is, their hypocrisy. 2 Nothing is covered up that will not be uncovered, and nothing secret that will not become known. 3 Therefore whatever you have said in the dark will be heard in the light, and what you have whispered behind closed doors will be proclaimed from the housetops.

4 "I tell you, my friends, do not fear

11.37–54: Against Pharisees and lawyers (Mt 23.1–36, in different order). **37:** 7.36; 14.1. **38:** *Wash before dinner,* ceremonial washing (literally "baptize before dinner"; compare Mk 7.1–5). **39–41:** Mt 23.25–26. Jesus turns back the Pharisees' criticism that the outside is unwashed by insisting that the inner life is equal in importance to the outer (v. 40); indeed, that it exercises a cleansing or corrupting power over the outward (v. 41; Mk 7.23). **41:** Titus 1.15.
11.42: *Rue,* a medicinal herb. Mt 23.23; Lev 27.30; Mic 6.8. **43:** See Mk 12.38–39 n. **44:** Mt 23.27. **45:** *Lawyer,* a teacher of the Jewish law. **46:** Mt 23.4. **47–48:** Mt 23.29–32; Acts 7.51–53. **49–51:** Mt 23.34–36. **49:** 1 Cor 1.24; Col 2.3. *The Wisdom of God* is either the title of an otherwise unknown book, or (more probably) a reference to the divine decrees as

interpreted by Jesus (compare Jesus' personification of wisdom in 7.35); in Mt 23.34 the words are attributed to Jesus himself. **51:** See Mt 23.35 n. **52:** Mt 23.13. *The key of knowledge,* the clue to proper understanding of God's purpose. The lawyers have not *entered,* i.e. have rejected, God's rule, and keep others from it.
12.1–12: Encouragement of disciples (Mt 10.26–33). **1:** Mt 16.6, 12; see Mk 8.15 n. **2:** See Mk 4.22 n. **3:** Similar to but different from Mt 10.27. **5:** Heb 10.31. **6:** Mt 10.29. The *penny* (assarion) was one-sixteenth of a denarius (see 12.59 n.). **7:** 21.18; Acts 27.34; Mt 12.12. **9:** Mk 8.38; Lk 9.26; 2 Tim 2.12. **10:** See Mt 12.31 n.; Mk 3.28–29. **11:** Mt 10.19; Mk 13.11; Lk 21.14–15. **12:** 2 Tim 4.17.

100
NT

those who kill the body, and after that can do nothing more. [5]But I will warn you whom to fear: fear him who, after he has killed, has authority[b] to cast into hell.[c] Yes, I tell you, fear him! [6]Are not five sparrows sold for two pennies? Yet not one of them is forgotten in God's sight. [7]But even the hairs of your head are all counted. Do not be afraid; you are of more value than many sparrows.

[8] "And I tell you, everyone who acknowledges me before others, the Son of Man also will acknowledge before the angels of God; [9]but whoever denies me before others will be denied before the angels of God. [10]And everyone who speaks a word against the Son of Man will be forgiven; but whoever blasphemes against the Holy Spirit will not be forgiven. [11]When they bring you before the synagogues, the rulers, and the authorities, do not worry about how[d] you are to defend yourselves or what you are to say; [12]for the Holy Spirit will teach you at that very hour what you ought to say."

[13] Someone in the crowd said to him, "Teacher, tell my brother to divide the family inheritance with me." [14]But he said to him, "Friend, who set me to be a judge or arbitrator over you?" [15]And he said to them, "Take care! Be on your guard against all kinds of greed; for one's life does not consist in the abundance of possessions." [16]Then he told them a parable: "The land of a rich man produced abundantly. [17]And he thought to himself, 'What should I do, for I have no place to store my crops?' [18]Then he said, 'I will do this: I will pull down my barns and build larger ones, and there I will store all my grain and my goods. [19]And I will say to my soul, 'Soul, you have ample goods laid up for many years; relax, eat, drink, be merry.' [20]But God said to him, 'You fool! This very night

your life is being demanded of you. And the things you have prepared, whose will they be?' [21]So it is with those who store up treasures for themselves but are not rich toward God."

[22] He said to his disciples, "Therefore I tell you, do not worry about your life, what you will eat, or about your body, what you will wear. [23]For life is more than food, and the body more than clothing. [24]Consider the ravens: they neither sow nor reap, they have neither storehouse nor barn, and yet God feeds them. Of how much more value are you than the birds! [25]And can any of you by worrying add a single hour to your span of life?[e] [26]If then you are not able to do so small a thing as that, why do you worry about the rest? [27]Consider the lilies, how they grow: they neither toil nor spin;[f] yet I tell you, even Solomon in all his glory was not clothed like one of these. [28]But if God so clothes the grass of the field, which is alive today and tomorrow is thrown into the oven, how much more will he clothe you—you of little faith! [29]And do not keep striving for what you are to eat and what you are to drink, and do not keep worrying. [30]For it is the nations of the world that strive after all these things, and your Father knows that you need them. [31]Instead, strive for his[g] kingdom, and these things will be given to you as well.

[32] "Do not be afraid, little flock, for it is your Father's good pleasure to give you the kingdom. [33]Sell your possessions, and give alms. Make purses for yourselves that do not wear out, an unfailing treasure in heaven, where no thief comes near and no moth destroys. [34]For

b Or *power*   c Gk *Gehenna*   d Other ancient authorities add *or what*   e Or *add a cubit to your stature*   f Other ancient authorities read *Consider the lilies; they neither spin nor weave*   g Other ancient authorities read *God's*

12.13–21: Parable of the rich fool. 13: According to Deut 21.17 the elder received double the younger's share. **15:** *One's life,* i.e. full human experience (1 Tim 6.6–10). **20:** Jer 17.11; Job 27.8; Ps 39.6; Lk 12.33–34.

12.22–34: On anxiety (Mt 6.25–33, 19–21). **24:** 12.6–7. **25:** See Mt 6.27 n. **27:** 1 Kings 10.1–10. **30:** Mt 6.8.
12.32: *Flock* refers to the Messiah's people (Ezek ch 34). **33–34:** Compare Mt 6.19–21;

where your treasure is, there your heart will be also.

35 "Be dressed for action and have your lamps lit; 36 be like those who are waiting for their master to return from the wedding banquet, so that they may open the door for him as soon as he comes and knocks. 37 Blessed are those slaves whom the master finds alert when he comes; truly I tell you, he will fasten his belt and have them sit down to eat, and he will come and serve them. 38 If he comes during the middle of the night, or near dawn, and finds them so, blessed are those slaves.

39 "But know this: if the owner of the house had known at what hour the thief was coming, he*h* would not have let his house be broken into. 40 You also must be ready, for the Son of Man is coming at an unexpected hour."

41 Peter said, "Lord, are you telling this parable for us or for everyone?" 42 And the Lord said, "Who then is the faithful and prudent manager whom his master will put in charge of his slaves, to give them their allowance of food at the proper time? 43 Blessed is that slave whom his master will find at work when he arrives. 44 Truly I tell you, he will put that one in charge of all his possessions. 45 But if that slave says to himself, 'My master is delayed in coming,' and if he begins to beat the other slaves, men and women, and to eat and drink and get drunk, 46 the master of that slave will come on a day when he does not expect him and at an hour that he does not know, and will cut him in pieces,*i* and put him with the unfaithful. 47 That slave who knew what his master wanted, but did not prepare himself or do what was wanted, will receive a severe beating. 48 But the one who did not know and did what deserved a beating will receive a light beating. From everyone to whom much has been given, much will be required; and from the one to whom much has been entrusted, even more will be demanded.

49 "I came to bring fire to the earth, and how I wish it were already kindled! 50 I have a baptism with which to be baptized, and what stress I am under until it is completed! 51 Do you think that I have come to bring peace to the earth? No, I tell you, but rather division! 52 From now on five in one household will be divided, three against two and two against three; 53 they will be divided:

father against son
  and son against father,
mother against daughter
  and daughter against mother,
mother-in-law against her
    daughter-in-law
  and daughter-in-law against
    mother-in-law."

54 He also said to the crowds, "When you see a cloud rising in the west, you immediately say, 'It is going to rain'; and so it happens. 55 And when you see the south wind blowing, you say, 'There will be scorching heat'; and it happens. 56 You hypocrites! You know how to interpret the appearance of earth and sky, but why do you not know how to interpret the present time?

h Other ancient authorities add *would have
watched and*   *i* Or *cut him off*

Mk 10.21; Lk 18.22; Acts 2.45; 4.32–35. Jesus spoke against abuse, not possession, of property (v. 15). **12.35–48: On watchfulness** (Mt 24.43–51). **35**: Eph 6.14; Mt 25.1–13; Mk 13.33–37. **37**: The language suggests that Jesus' mind moved to the Messianic banquet (13.29; 22.16), to which a marriage feast served as an analogy. **38**: The time, between 9 P.M. and 3 A.M. **12.39–40**: Mt 24.43–44; 1 Thess 5.2; Rev 3.3; 16.15; 2 Pet 3.10. **42**: Mt 24.45–51. **47–**

48: Deut 25.2–3; Num 15.29–30; Lk 8.18; 19.26. **12.49–56: On the end of the age. 49**: *Fire,* a symbol of judgment (Mt 3.11; 7.19; Mk 9.48; Lk 3.16). **50**: Mk 10.38–39; Jn 12.27. **51–53**: Mt 10.34–36; Lk 21.16; Mic 7.6. **12.54–56**: Mt 16.2–3. Winds from the west blew off the Mediterranean; those from the south off the desert. Jesus says there are many signs of spiritual crisis that human beings neglect.

57 "And why do you not judge for yourselves what is right? 58 Thus, when you go with your accuser before a magistrate, on the way make an effort to settle the case,ʲ or you may be dragged before the judge, and the judge hand you over to the officer, and the officer throw you in prison. 59 I tell you, you will never get out until you have paid the very last penny."

13 At that very time there were some present who told him about the Galileans whose blood Pilate had mingled with their sacrifices. 2 He asked them, "Do you think that because these Galileans suffered in this way they were worse sinners than all other Galileans? 3 No, I tell you; but unless you repent, you will all perish as they did. 4 Or those eighteen who were killed when the tower of Siloam fell on them—do you think that they were worse offenders than all the others living in Jerusalem? 5 No, I tell you; but unless you repent, you will all perish just as they did."

6 Then he told this parable: "A man had a fig tree planted in his vineyard; and he came looking for fruit on it and found none. 7 So he said to the gardener, 'See here! For three years I have come looking for fruit on this fig tree, and still I find none. Cut it down! Why should it be wasting the soil?' 8 He replied, 'Sir, let it alone for one more year, until I dig around it and put manure on it. 9 If it

bears fruit next year, well and good; but if not, you can cut it down.' "

10 Now he was teaching in one of the synagogues on the sabbath. 11 And just then there appeared a woman with a spirit that had crippled her for eighteen years. She was bent over and was quite unable to stand up straight. 12 When Jesus saw her, he called her over and said, "Woman, you are set free from your ailment." 13 When he laid his hands on her, immediately she stood up straight and began praising God. 14 But the leader of the synagogue, indignant because Jesus had cured on the sabbath, kept saying to the crowd, "There are six days on which work ought to be done; come on those days and be cured, and not on the sabbath day." 15 But the Lord answered him and said, "You hypocrites! Does not each of you on the sabbath untie his ox or his donkey from the manger, and lead it away to give it water? 16 And ought not this woman, a daughter of Abraham whom Satan bound for eighteen long years, be set free from this bondage on the sabbath day?" 17 When he said this, all his opponents were put to shame; and the entire crowd was rejoicing at all the wonderful things that he was doing.

18 He said therefore, "What is the kingdom of God like? And to what should I compare it? 19 It is like a mustard

ʲ Gk *settle with him*

12.57–59: Mt 5.25–26. **59:** *Penny,* Greek "lepton," the smallest Greek coin in circulation. There were two lepta to a quadrans ("penny" in Mt 5.26; Mk 12.42), eight to an assarion ("penny" in Lk 12.6) and one hundred and twenty-eight to a denarius, the daily wage in Mt 20.2.
13.1–9: On repentance. 1: They had been slain by Pilate's order while sacrificing in the temple at Jerusalem. **2:** The reports carried with them questions of a deeper meaning. According to common Jewish belief, painful experiences were signs of God's judgment (Jn 9.2–3). Jesus does not argue here (as in Mt 5.45) for a disconnection between natural and moral good and evil. Here suffering represents God's judgment and is a call to repen-

tance lest spiritual catastrophe overtake his hearers. **4:** *Siloam,* a section of Jerusalem.
13.6–9: Mt 21.18–20; Mk 11.12–14, 20–21. **7:** Mt 3.10; 7.19; Lk 3.9.
13.10–17: A crippled woman healed. This miraculous cure was unasked for by the woman or by anyone in her behalf. **14:** Ex 20.9–10; Mt 12.11–12; Lk 6.6–11; 14.1–6; Jn 5.1–18. **16:** Jesus attributes physical (and psychical) disorders to the work of *Satan* (see Mt 4.1 n.; 12.24 n.). They are in conflict with God's purpose of salvation in his covenant with Abraham, and are the concern of his saving activity (4.18; Mt 8.14–17).
13.18–21: Parables of mustard seed and yeast (see Mt 13.31–33 n.; Mk 4.30–32).

seed that someone took and sowed in the garden; it grew and became a tree, and the birds of the air made nests in its branches."

20 And again he said, "To what should I compare the kingdom of God? 21 It is like yeast that a woman took and mixed in with[k] three measures of flour until all of it was leavened."

22 Jesus[l] went through one town and village after another, teaching as he made his way to Jerusalem. 23 Someone asked him, "Lord, will only a few be saved?" He said to them, 24 "Strive to enter through the narrow door; for many, I tell you, will try to enter and will not be able. 25 When once the owner of the house has got up and shut the door, and you begin to stand outside and to knock at the door, saying, 'Lord, open to us,' then in reply he will say to you, 'I do not know where you come from.' 26 Then you will begin to say, 'We ate and drank with you, and you taught in our streets.' 27 But he will say, 'I do not know where you come from; go away from me, all you evildoers!' 28 There will be weeping and gnashing of teeth when you see Abraham and Isaac and Jacob and all the prophets in the kingdom of God, and you yourselves thrown out. 29 Then people will come from east and west, from north and south, and will eat in the kingdom of God. 30 Indeed, some are last who will be first, and some are first who will be last."

31 At that very hour some Pharisees came and said to him, "Get away from here, for Herod wants to kill you." 32 He said to them, "Go and tell that fox for me,[m] 'Listen, I am casting out demons and performing cures today and tomorrow, and on the third day I finish my work. 33 Yet today, tomorrow, and the next day I must be on my way, because it is impossible for a prophet to be killed outside of Jerusalem.' 34 Jerusalem, Jerusalem, the city that kills the prophets and stones those who are sent to it! How often have I desired to gather your children together as a hen gathers her brood under her wings, and you were not willing! 35 See, your house is left to you. And I tell you, you will not see me until the time comes when[n] you say, 'Blessed is the one who comes in the name of the Lord.'"

14 On one occasion when Jesus[o] was going to the house of a leader of the Pharisees to eat a meal on the sabbath, they were watching him closely. 2 Just then, in front of him, there was a man who had dropsy. 3 And Jesus asked the lawyers and Pharisees, "Is it lawful to cure people on the sabbath, or not?" 4 But they were silent. So Jesus[o] took him and healed him, and sent him away. 5 Then he said to them, "If one of you has a child[p] or an ox that has fallen into a well, will you not immediately pull it out on a sabbath day?" 6 And they could not reply to this.

7 When he noticed how the guests chose the places of honor, he told them a parable. 8 "When you are invited by someone to a wedding banquet, do not sit down at the place of honor, in case

k Gk *hid in*   l Gk *He*   m Gk lacks *for me*
n Other ancient authorities lack *the time comes
when*   o Gk *he*   p Other ancient authorities
read *a donkey*

13.22–30: On the end of the age. 22–24: Mt 7.13–14; Jn 10.7; Lk 9.51 n. **25:** Mt 25.10–12. **26–29:** The host clearly is meant to be the Messiah (compare 14.15–24).

13.31–33: Words to Herod Antipas. 31: *Here,* Herod's domain (Galilee and Perea). **32:** *Listen . . . ,* no cunning threats of *that fox* can shorten *my work. Third day,* not literally; a short and limited time is meant. **33:** *It is impossible . . . ,* bitter irony.

13.34–35: Lament over Jerusalem (Mt 23.37–39). **34:** *How often,* see Mt 23.37 n. **35:** Jer 22.5; Ps 118.26.

14.1–6: Healing the man with dropsy. Mt 12.9–14; Mk 3.1–6; Lk 6.6–11; 13.10–17. **2:** *Dropsy,* swelling from abnormal fluid retention.

14.7–14: On humility. 8: Prov 25.6–7; Lk 11.43; 20.46. **12:** Jas 2.2–4; Mt 5.43–48. **14:** Jesus appeals not to a spirit of material gain, i.e. a hope of reward in the judgment, but rather to the faith that the principle of love will be vindicated (Col 3.23–24).

14.15–24: The great dinner (Mt 22.1–10; compare also Mt 8.11; 26.29; see Lk

someone more distinguished than you has been invited by your host; 9 and the host who invited both of you may come and say to you, 'Give this person your place,' and then in disgrace you would start to take the lowest place. 10 But when you are invited, go and sit down at the lowest place, so that when your host comes, he may say to you, 'Friend, move up higher'; then you will be honored in the presence of all who sit at the table with you. 11 For all who exalt themselves will be humbled, and those who humble themselves will be exalted."

12 He said also to the one who had invited him, "When you give a luncheon or a dinner, do not invite your friends or your brothers or your relatives or rich neighbors, in case they may invite you in return, and you would be repaid. 13 But when you give a banquet, invite the poor, the crippled, the lame, and the blind. 14 And you will be blessed, because they cannot repay you, for you will be repaid at the resurrection of the righteous."

15 One of the dinner guests, on hearing this, said to him, "Blessed is anyone who will eat bread in the kingdom of God!" 16 Then Jesus*q* said to him, "Someone gave a great dinner and invited many. 17 At the time for the dinner he sent his slave to say to those who had been invited, 'Come; for everything is ready now.' 18 But they all alike began to make excuses. The first said to him, 'I have bought a piece of land, and I must go out and see it; please accept my regrets.' 19 Another said, 'I have bought five yoke of oxen, and I am going to try them out; please accept my regrets.' 20 Another said, 'I have just been married, and therefore I cannot come.' 21 So the slave returned and reported this to his master. Then the owner of the house became angry and said to his slave, 'Go out at once into the streets and lanes of the town and bring in the poor, the crippled, the blind, and the lame.' 22 And the slave said, 'Sir, what you ordered has been done, and there is still room.' 23 Then the master said to the slave, 'Go out into the roads and lanes, and compel people to come in, so that my house may be filled. 24 For I tell you,*r* none of those who were invited will taste my dinner.' "

25 Now large crowds were traveling with him; and he turned and said to them, 26 "Whoever comes to me and does not hate father and mother, wife and children, brothers and sisters, yes, and even life itself, cannot be my disciple. 27 Whoever does not carry the cross and follow me cannot be my disciple. 28 For which of you, intending to build a tower, does not first sit down and estimate the cost, to see whether he has enough to complete it? 29 Otherwise, when he has laid a foundation and is not able to finish, all who see it will begin to ridicule him, 30 saying, 'This fellow began to build and was not able to finish.' 31 Or what king, going out to wage war against another king, will not sit down first and consider whether he is able with ten thousand to oppose the one who comes against him with twenty thousand? 32 If he cannot, then, while the other is still far away, he sends a delegation and asks for the terms of peace. 33 So therefore, none of you can become my disciple if you do not give up all your possessions.

34 "Salt is good; but if salt has lost its taste, how can its saltiness be restored?*s* 35 It is fit neither for the soil nor for the

q Gk *he*   *r* The Greek word for *you* here is plural   *s* Or *how can it be used for seasoning?*

5.32 n.; 13.29). The point is that though some people imagine they highly prize the thought of sharing God's kingdom, they may in fact be rejecting appeals to act so that they may enter it. **15**: Lk 22.16, 18, 28–30; Rev 19.9. **20**: Deut 24.5; 1 Cor 7.33. **24**: *You* is plural here; Jesus drops the parabolic form and speaks to the guests (v. 15) in his own person.

14.25–35: **Conditions of discipleship. 26–27**: Mt 10.37–38. **26**: Jn 12.25. *Hate* is used in vigorous, vivid hyperbole; the parallel passage in Mt 10.37 reflects Jesus' meaning. **27**: See Mt 10.38 n. **31–32**: Possibly Jesus alludes to some contemporary event. **33**: 9.57–62; 12.33; 18.29–30; Phil 3.7. **34–35**: Mt 5.13; Mk 9.49–50.

manure pile; they throw it away. Let anyone with ears to hear listen!"

15 Now all the tax collectors and sinners were coming near to listen to him. ² And the Pharisees and the scribes were grumbling and saying, "This fellow welcomes sinners and eats with them."

3 So he told them this parable: ⁴"Which one of you, having a hundred sheep and losing one of them, does not leave the ninety-nine in the wilderness and go after the one that is lost until he finds it? ⁵When he has found it, he lays it on his shoulders and rejoices. ⁶And when he comes home, he calls together his friends and neighbors, saying to them, 'Rejoice with me, for I have found my sheep that was lost.' ⁷Just so, I tell you, there will be more joy in heaven over one sinner who repents than over ninety-nine righteous persons who need no repentance.

8 "Or what woman having ten silver coins,ᵗ if she loses one of them, does not light a lamp, sweep the house, and search carefully until she finds it? ⁹When she has found it, she calls together her friends and neighbors, saying, 'Rejoice with me, for I have found the coin that I had lost.' ¹⁰Just so, I tell you, there is joy in the presence of the angels of God over one sinner who repents."

11 Then Jesusᵘ said, "There was a man who had two sons. ¹²The younger of them said to his father, 'Father, give me the share of the property that will belong to me.' So he divided his property between them. ¹³A few days later the younger son gathered all he had and traveled to a distant country, and there he squandered his property in dissolute liv-

ing. ¹⁴When he had spent everything, a severe famine took place throughout that country, and he began to be in need. ¹⁵So he went and hired himself out to one of the citizens of that country, who sent him to his fields to feed the pigs. ¹⁶He would gladly have filled himself withᵛ the pods that the pigs were eating; and no one gave him anything. ¹⁷But when he came to himself he said, 'How many of my father's hired hands have bread enough and to spare, but here I am dying of hunger! ¹⁸I will get up and go to my father, and I will say to him, "Father, I have sinned against heaven and before you; ¹⁹I am no longer worthy to be called your son; treat me like one of your hired hands."' ²⁰So he set off and went to his father. But while he was still far off, his father saw him and was filled with compassion; he ran and put his arms around him and kissed him. ²¹Then the son said to him, 'Father, I have sinned against heaven and before you; I am no longer worthy to be called your son.'ʷ ²²But the father said to his slaves, 'Quickly, bring out a robe—the best one—and put it on him; put a ring on his finger and sandals on his feet. ²³And get the fatted calf and kill it, and let us eat and celebrate; ²⁴for this son of mine was dead and is alive again; he was lost and is found!' And they began to celebrate.

25 "Now his elder son was in the field; and when he came and approached the house, he heard music and dancing. ²⁶He called one of the slaves and asked

---

t Gk drachmas, each worth about a day's wage for a laborer   *u Gk he*   *v Other ancient authorities read filled his stomach with*   *w Other ancient authorities add treat me as one of your hired servants*

15.1–32: Parables about the lost. 3–7: The lost sheep. Mt 18.12–14. The parable illustrates God's concern for those who lack ability to find him; he seeks them. **4:** The phrase *until he finds it* is in harmony with Luke's universalism (see Introduction and contrast "if" in Mt 18.13).

15.8–10: The lost coin. This parable intensifies the picture of human helplessness and divine concern. The *coin* was approximately equivalent to the denarius (see note *t*).

15.11–32: The lost son. 12: See 12.13 n. **15:** *Pigs,* the culminating indignity for a Jew. **22–24:** His place as son is freely restored. The parable illustrates God's acceptance of those who rebel and return. **22:** *Robe,* a festal garment (not to be worn while working); *ring,* symbol of authority; *sandals,* slaves would have been unshod. Gen 41.42; Zech 3.4. **24:** 1 Tim 5.6; Eph 2.1; Lk 9.60.

15.25–32: Jesus' aim was to portray the difference between God's loving forgiveness

what was going on. 27 He replied, 'Your brother has come, and your father has killed the fatted calf, because he has got him back safe and sound.' 28 Then he became angry and refused to go in. His father came out and began to plead with him. 29 But he answered his father, 'Listen! For all these years I have been working like a slave for you, and I have never disobeyed your command; yet you have never given me even a young goat so that I might celebrate with my friends. 30 But when this son of yours came back, who has devoured your property with prostitutes, you killed the fatted calf for him!' 31 Then the father *x* said to him, 'Son, you are always with me, and all that is mine is yours. 32 But we had to celebrate and rejoice, because this brother of yours was dead and has come to life; he was lost and has been found.' "

16 Then Jesus *x* said to the disciples, "There was a rich man who had a manager, and charges were brought to him that this man was squandering his property. 2 So he summoned him and said to him, 'What is this that I hear about you? Give me an accounting of your management, because you cannot be my manager any longer.' 3 Then the manager said to himself, 'What will I do, now that my master is taking the position away from me? I am not strong enough to dig, and I am ashamed to beg. 4 I have decided what to do so that, when I am dismissed as manager, people may welcome me into their homes.' 5 So, summoning his master's debtors one by one, he asked the first, 'How much do you owe my master?' 6 He answered, 'A hundred jugs of olive oil.' He said to him, 'Take your bill, sit down quickly, and make it fifty.'

7 Then he asked another, 'And how much do you owe?' He replied, 'A hundred containers of wheat.' He said to him, 'Take your bill and make it eighty.' 8 And his master commended the dishonest manager because he had acted shrewdly; for the children of this age are more shrewd in dealing with their own generation than are the children of light. 9 And I tell you, make friends for yourselves by means of dishonest wealth *y* so that when it is gone, they may welcome you into the eternal homes. *z*

10 "Whoever is faithful in a very little is faithful also in much; and whoever is dishonest in a very little is dishonest also in much. 11 If then you have not been faithful with the dishonest wealth, *y* who will entrust to you the true riches? 12 And if you have not been faithful with what belongs to another, who will give you what is your own? 13 No slave can serve two masters; for a slave will either hate the one and love the other, or be devoted to the one and despise the other. You cannot serve God and wealth." *y*

14 The Pharisees, who were lovers of money, heard all this, and they ridiculed him. 15 So he said to them, "You are those who justify yourselves in the sight of others; but God knows your hearts; for what is prized by human beings is an abomination in the sight of God.

16 "The law and the prophets were in effect until John came; since then the good news of the kingdom of God is proclaimed, and everyone tries to enter it by force. *a* 17 But it is easier for heaven and earth to pass away, than for one

x Gk *he*   *y* Gk *mammon*   *z* Gk *tents*
a Or *everyone is strongly urged to enter it*

and the self-centered complacency that not only denies love, but cannot understand it.
16.1–13: The dishonest manager. The point is in v. 8, the application in v. 9: the dishonest manager was prudent in using the things of this life to ensure the future; believers should do the same. **8:** *His master* is the rich man of v. 1. *Children of light,* those who are spiritually enlightened. The phrase appears in Jn 12.36; Eph 5.8; 1 Thess 5.5, as well as in the Dead Sea Scrolls, where it is contrasted with the children of darkness. **10:** Mt 25.21; Lk 19.17. **13:** See Mt 6.24 n.
16.14–18: Teaching about the law. 14– 15: Mt 19.16–30; Lk 18.9–14. *What is prized,* that is, what is given a regard due only to God. **17:** Mt 5.17–18; Lk 21.33. **18:** Mt 5.31– 32; 19.9; Mk 10.11–12; 1 Cor 7.10–11.

stroke of a letter in the law to be dropped.

18 "Anyone who divorces his wife and marries another commits adultery, and whoever marries a woman divorced from her husband commits adultery.

19 "There was a rich man who was dressed in purple and fine linen and who feasted sumptuously every day. 20 And at his gate lay a poor man named Lazarus, covered with sores, 21 who longed to satisfy his hunger with what fell from the rich man's table; even the dogs would come and lick his sores. 22 The poor man died and was carried away by the angels to be with Abraham. *b* The rich man also died and was buried. 23 In Hades, where he was being tormented, he looked up and saw Abraham far away with Lazarus by his side. *c* 24 He called out, 'Father Abraham, have mercy on me, and send Lazarus to dip the tip of his finger in water and cool my tongue; for I am in agony in these flames.' 25 But Abraham said, 'Child, remember that during your lifetime you received your good things, and Lazarus in like manner evil things; but now he is comforted here, and you are in agony. 26 Besides all this, between you and us a great chasm has been fixed, so that those who might want to pass from here to you cannot do so, and no one can cross from there to us.' 27 He said, 'Then, father, I beg you to send him to my father's house— 28 for I have five brothers—that he may warn them, so that they will not also come into this place of torment.' 29 Abraham replied, 'They have Moses and the prophets; they

should listen to them.' 30 He said, 'No, father Abraham; but if someone goes to them from the dead, they will repent.' 31 He said to him, 'If they do not listen to Moses and the prophets, neither will they be convinced even if someone rises from the dead.' "

17 Jesus*d* said to his disciples, "Occasions for stumbling are bound to come, but woe to anyone by whom they come! 2 It would be better for you if a millstone were hung around your neck and you were thrown into the sea than for you to cause one of these little ones to stumble. 3 Be on your guard! If another disciple*e* sins, you must rebuke the offender, and if there is repentance, you must forgive. 4 And if the same person sins against you seven times a day, and turns back to you seven times and says, 'I repent,' you must forgive."

5 The apostles said to the Lord, "Increase our faith!" 6 The Lord replied, "If you had faith the size of a*f* mustard seed, you could say to this mulberry tree, 'Be uprooted and planted in the sea,' and it would obey you.

7 "Who among you would say to your slave who has just come in from plowing or tending sheep in the field, 'Come here at once and take your place at the table'? 8 Would you not rather say to him, 'Prepare supper for me, put on your apron and serve me while I eat and drink; later you may eat and drink'? 9 Do you thank the slave for doing what was

*b Gk to Abraham's bosom   c Gk in his bosom
d Gk He   e Gk your brother   f Gk faith as a
grain of*

16.19–31: The rich man and Lazarus. It is not known whether Jesus alluded here to a contemporary incident. The main point (vv. 27–31) is that the Old Testament speaks an urgent and sufficient call to repentance (v. 17). As a whole the story seems to illustrate vv. 10–15. Jesus draws details from contemporary ways of expressing Jewish faith, including items not found in the Old Testament. **19**: The *rich man,* though unnamed, is commonly called "Dives" (Latin for "rich man"). *Purple* was a rich cloth dyed with the liquid obtained from a species of shellfish. **20**: The person named here is not to be identified with the Lazarus of Jn 11.1–44; 12.1, 9. **21–22**: The moral quality of Lazarus is passed over to illustrate the fatal deficiency in the life of the other and the impossibility of changing his condemnation. **22**: See 23.42 n. **29**: Jn 5.45–47; Acts 15.21. **30**: 3.8; 19.9.

17.1–2: Mt 18.6, 7; Mk 9.42; 1 Cor 8.12. **2**: *Little ones,* disciples (see Mt 18.6 n.). **5**: Mk 11.23 n., 24 n.

17.7–10: One's relation to God makes obedience to God a duty to be fulfilled and not an occasion for reward.

commanded? 10So you also, when you have done all that you were ordered to do, say, 'We are worthless slaves; we have done only what we ought to have done!' "

11 On the way to Jerusalem Jesus*g* was going through the region between Samaria and Galilee. 12As he entered a village, ten lepers*h* approached him. Keeping their distance, 13they called out, saying, "Jesus, Master, have mercy on us!" 14When he saw them, he said to them, "Go and show yourselves to the priests." And as they went, they were made clean. 15Then one of them, when he saw that he was healed, turned back, praising God with a loud voice. 16He prostrated himself at Jesus'*i* feet and thanked him. And he was a Samaritan. 17Then Jesus asked, "Were not ten made clean? But the other nine, where are they? 18Was none of them found to return and give praise to God except this foreigner?" 19Then he said to him, "Get up and go on your way; your faith has made you well."

20 Once Jesus*g* was asked by the Pharisees when the kingdom of God was coming, and he answered, "The kingdom of God is not coming with things that can be observed; 21nor will they say, 'Look, here it is!' or 'There it is!' For, in fact, the kingdom of God is among*j* you."

22 Then he said to the disciples, "The days are coming when you will long to see one of the days of the Son of Man, and you will not see it. 23They will say

to you, 'Look there!' or 'Look here!' Do not go, do not set off in pursuit. 24For as the lightning flashes and lights up the sky from one side to the other, so will the Son of Man be in his day.*k* 25But first he must endure much suffering and be rejected by this generation. 26Just as it was in the days of Noah, so too it will be in the days of the Son of Man. 27They were eating and drinking, and marrying and being given in marriage, until the day Noah entered the ark, and the flood came and destroyed all of them. 28Likewise, just as it was in the days of Lot: they were eating and drinking, buying and selling, planting and building, 29but on the day that Lot left Sodom, it rained fire and sulfur from heaven and destroyed all of them 30—it will be like that on the day that the Son of Man is revealed. 31On that day, anyone on the housetop who has belongings in the house must not come down to take them away; and likewise anyone in the field must not turn back. 32Remember Lot's wife. 33Those who try to make their life secure will lose it, but those who lose their life will keep it. 34I tell you, on that night there will be two in one bed; one will be taken and the other left. 35There will be two women grinding meal together; one will be taken and the other left."*l* 37Then they asked him, "Where, Lord?" He said to them,

g Gk *he*   h The terms *leper* and *leprosy* can refer to several diseases   i Gk *his*
j Or *within*   k Other ancient authorities lack *in his day*   l Other ancient authorities add verse 36, *"Two will be in the field; one will be taken and the other left."*

17.11–19: Ten lepers cleansed. 12: Lev 13.45–46; see Mt 8.2 n. **14:** *Priest,* Lev 13.2–3; 14.2–32. **17:** *The other nine* were, presumably, Jews. **18:** 7.9. **19:** Mt 9.22; Mk 5.34; Lk 8.48; 18.42. *Made you well,* see Mt 9.21 n.; Mk 11.23 n., 24 n.
17.20–21: The kingdom is among you, in the person of Jesus. **20:** 19.11; 21.7; Acts 1.6. **21:** The reality of God's *kingdom* is present and available. The questioners had in mind a kingdom bringing material benefits.
17.22–37: The end of the age. Mt ch 24 has similar teachings, in a different order and setting. **22:** Mt 9.15; Mk 2.20; Lk 5.35. **23–**

24: Mt 24.23, 26, 27; Mk 13.21; Rev 1.7. **23:** *Look, there* is the Son of Man! etc. **24:** The coming will be sudden and visible to all. **25:** 9.22. **26–27:** Mt 24.37–39; Gen 6.5–8; 7.6–24. **28–30:** Gen 18.16–19.28. **31:** Mt 24.17–18; Mk 13.15–16; Lk 21.21. **32:** Gen 19.26. **33:** See Mt 10.38 n., 39. **34–35:** Mt 24.40–41. **37:** Mt 24.28. Jesus' answer is a significant appeal to faith. The questioners wish to know *where* the Messiah and his people will be located. Instead of answering them directly, Jesus warns: As surely as vultures find the carcass, so surely will divine judgment come; therefore always be ready!

"Where the corpse is, there the vultures will gather."

18 Then Jesus[m] told them a parable about their need to pray always and not to lose heart. 2 He said, "In a certain city there was a judge who neither feared God nor had respect for people. 3 In that city there was a widow who kept coming to him and saying, 'Grant me justice against my opponent.' 4 For a while he refused; but later he said to himself, 'Though I have no fear of God and no respect for anyone, 5 yet because this widow keeps bothering me, I will grant her justice, so that she may not wear me out by continually coming.' "[n] 6 And the Lord said, "Listen to what the unjust judge says. 7 And will not God grant justice to his chosen ones who cry to him day and night? Will he delay long in helping them? 8 I tell you, he will quickly grant justice to them. And yet, when the Son of Man comes, will he find faith on earth?"

9 He also told this parable to some who trusted in themselves that they were righteous and regarded others with contempt: 10 "Two men went up to the temple to pray, one a Pharisee and the other a tax collector. 11 The Pharisee, standing by himself, was praying thus, 'God, I thank you that I am not like other people: thieves, rogues, adulterers, or even like this tax collector. 12 I fast twice a week; I give a tenth of all my income.' 13 But the tax collector, standing far off, would not even look up to heaven, but was beating his breast and saying, 'God, be merciful to me, a sinner!' 14 I tell you, this man went down to his home justified rather than the other; for all who exalt themselves will be humbled, but all who humble themselves will be exalted."

15 People were bringing even infants to him that he might touch them; and when the disciples saw it, they sternly ordered them not to do it. 16 But Jesus called for them and said, "Let the little children come to me, and do not stop them; for it is to such as these that the kingdom of God belongs. 17 Truly I tell you, whoever does not receive the kingdom of God as a little child will never enter it."

18 A certain ruler asked him, "Good Teacher, what must I do to inherit eternal life?" 19 Jesus said to him, "Why do you call me good? No one is good but God alone. 20 You know the commandments: 'You shall not commit adultery; You shall not murder; You shall not steal; You shall not bear false witness; Honor your father and mother.' " 21 He replied, "I have kept all these since my youth." 22 When Jesus heard this, he said to him, "There is still one thing lacking. Sell all that you own and distribute the money[o] to the poor, and you will have treasure in heaven; then come, follow me." 23 But when he heard this, he became sad; for he was very rich. 24 Jesus looked at him and said, "How hard it is for those who have wealth to enter the kingdom of God! 25 Indeed, it is easier for a camel to go through the eye of a needle

m Gk *he*   *n* Or *so that she may not finally come and slap me in the face*   *o* Gk lacks *the money*

18.1–8: **The unjust judge. 1:** The point is carefully stated (compare 11.5–8), perhaps because the details are incongruous (as in 16.1–9). **7:** Rev 6.10; Mt 24.22; Rom 8.33; Col 3.12; 2 Tim 2.10. **8:** *Comes,* from heaven, in judgment. *Faith,* a requisite for this persistent prayer (v. 1).
18.9–14: **Pharisee and tax collector. 9:** *Righteous,* that is, acceptable to God because of their ritual observance (vv. 11–12; see Mt 5.20 n.). **11:** Mt 6.5; Mk 11.25. **12:** *Twice a week,* Mondays and Thursdays. **14:** *Justified* means "accepted by God," "right with God."

God receives those who in contrition implore his mercy rather than those who parade their supposed virtues (15.7).
18.15–19.27: **From Galilee to Jerusalem** (Mt 19.1–20.34; Mk 10.1–52).
18.15–17: **Blessing the children** (Mt 19.13–15; 18.3; Mk 10.13–16). **16–17:** God's kingdom is shared by those who depend in trustful simplicity on God. **17:** *As a little child,* with teachable humility.
18.18–30: **The rich ruler** (Mt 19.16–30; Mk 10.17–31). **18:** 10.25. **20:** See Mt 19.18 n. The order of the commandments, varying

than for someone who is rich to enter the kingdom of God."

26 Those who heard it said, "Then who can be saved?" 27 He replied, "What is impossible for mortals is possible for God."

28 Then Peter said, "Look, we have left our homes and followed you." 29 And he said to them, "Truly I tell you, there is no one who has left house or wife or brothers or parents or children, for the sake of the kingdom of God, 30 who will not get back very much more in this age, and in the age to come eternal life."

31 Then he took the twelve aside and said to them, "See, we are going up to Jerusalem, and everything that is written about the Son of Man by the prophets will be accomplished. 32 For he will be handed over to the Gentiles; and he will be mocked and insulted and spat upon. 33 After they have flogged him, they will kill him, and on the third day he will rise again." 34 But they understood nothing about all these things; in fact, what he said was hidden from them, and they did not grasp what was said.

35 As he approached Jericho, a blind man was sitting by the roadside begging. 36 When he heard a crowd going by, he asked what was happening. 37 They told him, "Jesus of Nazareth*p* is passing by." 38 Then he shouted, "Jesus, Son of David, have mercy on me!" 39 Those who were in front sternly ordered him to be quiet; but he shouted even more loudly,

"Son of David, have mercy on me!" 40 Jesus stood still and ordered the man to be brought to him; and when he came near, he asked him, 41 "What do you want me to do for you?" He said, "Lord, let me see again." 42 Jesus said to him, "Receive your sight; your faith has saved you." 43 Immediately he regained his sight and followed him, glorifying God; and all the people, when they saw it, praised God.

19 He entered Jericho and was passing through it. 2 A man was there named Zacchaeus; he was a chief tax collector and was rich. 3 He was trying to see who Jesus was, but on account of the crowd he could not, because he was short in stature. 4 So he ran ahead and climbed a sycamore tree to see him, because he was going to pass that way. 5 When Jesus came to the place, he looked up and said to him, "Zacchaeus, hurry and come down; for I must stay at your house today." 6 So he hurried down and was happy to welcome him. 7 All who saw it began to grumble and said, "He has gone to be the guest of one who is a sinner." 8 Zacchaeus stood there and said to the Lord, "Look, half of my possessions, Lord, I will give to the poor; and if I have defrauded anyone of anything, I will pay back four times as much." 9 Then Jesus said to him, "Today salvation has come to this house, because he too is a son of Abraham. 10 For the Son

p Gk *the Nazorean*

---

from the Hebrew Scripture, follows the ancient Greek translation of the Old Testament. **22:** See 12.33 n. **25:** See Mk 10.25 n.

18.26: To *be saved* refers to the same spiritual experience as to *inherit eternal life* (v. 18), and to *enter the kingdom of God* (v. 25). The heart of this story lies in the questioner's sense of personal lack notwithstanding his opportunity (because of his wealth) to fulfill all ritual requirements. **27:** Gen 18.14; Job 42.2; Jer 32.17; Lk 1.37. **28:** 5.1–11.

18.31–34: **The Passion foretold again** (Mt 20.17–19; Mk 10.32–34); compare 9.22, 44–45; 17.25.

18.35–43: **A blind man healed** (Mt 20.29–34; Mk 10.46–52). Mt 9.27–31; Mk

8.22; Jn 9.1–3. **42:** See Mt 9.21 n.; Mk 11.23 n., 24 n.

19.1–10: **Zacchaeus. 1:** *Jericho* was on a main trade route, and was an important customs center. **2:** As *chief tax collector* Zacchaeus had contracted for the right to collect revenues in the district. His neighbors despised him for thus sharing in the Roman domination (v. 7). **7:** 5.29–30; 15.1–2. **8:** *I will give,* henceforward, a vow. *Four times as much,* Ex 22.1; Lev 6.5; Num 5.6–7. **9:** *Salvation,* or the kingdom of God, *has come* in Jesus' message and Zacchaeus' response (17.20–21; see 18.26 n.). *A son of Abraham,* in spirit as well as by descent.

of Man came to seek out and to save the lost."

11 As they were listening to this, he went on to tell a parable, because he was near Jerusalem, and because they supposed that the kingdom of God was to appear immediately. 12 So he said, "A nobleman went to a distant country to get royal power for himself and then return. 13 He summoned ten of his slaves, and gave them ten pounds,*q* and said to them, 'Do business with these until I come back.' 14 But the citizens of his country hated him and sent a delegation after him, saying, 'We do not want this man to rule over us.' 15 When he returned, having received royal power, he ordered these slaves, to whom he had given the money, to be summoned so that he might find out what they had gained by trading. 16 The first came forward and said, 'Lord, your pound has made ten more pounds.' 17 He said to him, 'Well done, good slave! Because you have been trustworthy in a very small thing, take charge of ten cities.' 18 Then the second came, saying, 'Lord, your pound has made five pounds.' 19 He said to him, 'And you, rule over five cities.' 20 Then the other came, saying, 'Lord, here is your pound. I wrapped it up in a piece of cloth, 21 for I was afraid of you, because you are a harsh man; you take what you did not deposit, and reap what you did not sow.' 22 He said to him, 'I will judge you by your own words, you wicked slave! You knew, did you, that I was a harsh man, taking what I did not deposit and reaping what I did not sow? 23 Why then did you not put my money into the bank? Then when I returned, I could have collected it with interest.' 24 He said to the bystanders,

'Take the pound from him and give it to the one who has ten pounds.' 25 (And they said to him, 'Lord, he has ten pounds!') 26 'I tell you, to all those who have, more will be given; but from those who have nothing, even what they have will be taken away. 27 But as for these enemies of mine who did not want me to be king over them—bring them here and slaughter them in my presence.' "

28 After he had said this, he went on ahead, going up to Jerusalem.

29 When he had come near Bethphage and Bethany, at the place called the Mount of Olives, he sent two of the disciples, 30 saying, "Go into the village ahead of you, and as you enter it you will find tied there a colt that has never been ridden. Untie it and bring it here. 31 If anyone asks you, 'Why are you untying it?' just say this, 'The Lord needs it.' " 32 So those who were sent departed and found it as he had told them. 33 As they were untying the colt, its owners asked them, "Why are you untying the colt?" 34 They said, "The Lord needs it." 35 Then they brought it to Jesus; and after throwing their cloaks on the colt, they set Jesus on it. 36 As he rode along, people kept spreading their cloaks on the road. 37 As he was now approaching the path down from the Mount of Olives, the whole multitude of the disciples began to praise God joyfully with a loud voice for all the deeds of power that they had seen, 38 saying,

"Blessed is the king
who comes in the name of
the Lord!
Peace in heaven,

q The mina, rendered here by *pound,* was about three months' wages for a laborer

---

19.11–27: Parable of the pounds. Compare the somewhat similar parable of the talents (Mt 25.14–30). **11:** 9.51 n.; 13.22; 17.11; 18.31. **12:** The details here may reflect contemporary events. **13:** *Ten,* only three are mentioned later. **17:** 16.10. **21:** *You take . . . deposit,* probably a current proverbial expression for a grasping person. **26:** See Mt 13.12 n. **27:** Though all are judged, only the hostile are punished.

19.28–23.56: The last week (Mt 21.1–27.66; Mk 11.1–15.47). **19.28–44: Palm Sunday** (Mt 21.1–9; Mk 11.1–10). Jn 12.12–18. **32:** 22.13. **35:** See Mk 11.1 n. **36:** 2 Kings 9.13. **37:** The road traversed a ridge into the valley of the Kidron. **38:** Ps 118.26; Zech 9.9; Lk 13.35. **39–40:** Mt 21.15–16; Hab 2.11.

and glory in the highest
heaven!"
39 Some of the Pharisees in the crowd said to him, "Teacher, order your disciples to stop." 40 He answered, "I tell you, if these were silent, the stones would shout out."

41 As he came near and saw the city, he wept over it, 42 saying, "If you, even you, had only recognized on this day the things that make for peace! But now they are hidden from your eyes. 43 Indeed, the days will come upon you, when your enemies will set up ramparts around you and surround you, and hem you in on every side. 44 They will crush you to the ground, you and your children within you, and they will not leave within you one stone upon another; because you did not recognize the time of your visitation from God."r

45 Then he entered the temple and began to drive out those who were selling things there; 46 and he said, "It is written,

'My house shall be a house
    of prayer';
    but you have made it a den
        of robbers."

47 Every day he was teaching in the temple. The chief priests, the scribes, and the leaders of the people kept looking for a way to kill him; 48 but they did not find anything they could do, for all the people were spellbound by what they heard.

20 One day, as he was teaching the people in the temple and telling the good news, the chief priests and the scribes came with the elders 2 and said to him, "Tell us, by what authority are you doing these things? Who is it who gave you this authority?" 3 He answered them,

"I will also ask you a question, and you tell me: 4 Did the baptism of John come from heaven, or was it of human origin?" 5 They discussed it with one another, saying, "If we say, 'From heaven,' he will say, 'Why did you not believe him?' 6 But if we say, 'Of human origin,' all the people will stone us; for they are convinced that John was a prophet." 7 So they answered that they did not know where it came from. 8 Then Jesus said to them, "Neither will I tell you by what authority I am doing these things."

9 He began to tell the people this parable: "A man planted a vineyard, and leased it to tenants, and went to another country for a long time. 10 When the season came, he sent a slave to the tenants in order that they might give him his share of the produce of the vineyard; but the tenants beat him and sent him away empty-handed. 11 Next he sent another slave; that one also they beat and insulted and sent away empty-handed. 12 And he sent still a third; this one also they wounded and threw out. 13 Then the owner of the vineyard said, 'What shall I do? I will send my beloved son; perhaps they will respect him.' 14 But when the tenants saw him, they discussed it among themselves and said, 'This is the heir; let us kill him so that the inheritance may be ours.' 15 So they threw him out of the vineyard and killed him. What then will the owner of the vineyard do to them? 16 He will come and destroy those tenants and give the vineyard to others." When they heard this, they said, "Heaven forbid!" 17 But he looked at them and said, "What then does this text mean:

r Gk lacks *from God*

19.41: 13.33–34. 43: 21.20–24; 21.6; Isa 29.3; Jer 6.6; Ezek 4.2. *Your enemies,* the Roman armies. *Ramparts,* a palisade, which would keep out all supplies of food. 44: Ps 137.8–9; Hos 10.14–15; 13.16; see 1 Pet 2.12 n. *Your children,* the city's inhabitants. *The time of your visitation,* the time of Christ's ministry.
19.45–46: **Cleansing the temple** (Mt 21.12–13; Mk 11.15–19; Jn 2.13–17). 45: See Mt 21.12 n.; Ex 30.13; Lev 1.14. 46: Isa 56.7; Jer 7.11.
20.1–8: **On Jesus' authority** (Mt 21.23–27; Mk 11.27–33; Jn 2.18–22).
20.9–19: **Parable of the vineyard** (Mt 21.33–46; Mk 12.1–12). 9: Isa 5.1–7; Mt 25.14. 13: The use of *beloved* (not present in Matthew and Mark) identifies the *son* with Jesus. 16: Acts 13.46; 18.6; 28.28. 17: Ps 118.22–23; Acts 4.11; 1 Pet 2.7.

'The stone that the builders
    rejected
  has become the cornerstone'?^s
18 Everyone who falls on that stone will
be broken to pieces; and it will crush
anyone on whom it falls." 19 When the
scribes and chief priests realized that he
had told this parable against them, they
wanted to lay hands on him at that very
hour, but they feared the people.

20 So they watched him and sent
spies who pretended to be honest, in order to trap him by what he said, so as to
hand him over to the jurisdiction and
authority of the governor. 21 So they
asked him, "Teacher, we know that you
are right in what you say and teach, and
you show deference to no one, but teach
the way of God in accordance with truth.
22 Is it lawful for us to pay taxes to the
emperor, or not?" 23 But he perceived
their craftiness and said to them,
24 "Show me a denarius. Whose head and
whose title does it bear?" They said,
"The emperor's." 25 He said to them,
"Then give to the emperor the things
that are the emperor's, and to God the
things that are God's." 26 And they were
not able in the presence of the people to
trap him by what he said; and being
amazed by his answer, they became silent.

27 Some Sadducees, those who say
there is no resurrection, came to him
28 and asked him a question, "Teacher,
Moses wrote for us that if a man's brother dies, leaving a wife but no children,
the man^t shall marry the widow and
raise up children for his brother. 29 Now
there were seven brothers; the first mar-

ried, and died childless; 30 then the second
31 and the third married her, and so in the
same way all seven died childless. 32 Finally the woman also died. 33 In the resurrection, therefore, whose wife will the
woman be? For the seven had married
her."

34 Jesus said to them, "Those who
belong to this age marry and are given in
marriage; 35 but those who are considered
worthy of a place in that age and in the
resurrection from the dead neither marry
nor are given in marriage. 36 Indeed they
cannot die anymore, because they are
like angels and are children of God, being
children of the resurrection. 37 And the
fact that the dead are raised Moses himself showed, in the story about the bush,
where he speaks of the Lord as the God
of Abraham, the God of Isaac, and the
God of Jacob. 38 Now he is God not of
the dead, but of the living; for to him all
of them are alive." 39 Then some of the
scribes answered, "Teacher, you have
spoken well." 40 For they no longer dared
to ask him another question.

41 Then he said to them, "How can
they say that the Messiah^u is David's
son? 42 For David himself says in the
book of Psalms,

'The Lord said to my Lord,
  "Sit at my right hand,
43   until I make your enemies your
      footstool." '
44 David thus calls him Lord; so how can
he be his son?"

45 In the hearing of all the people he

---

s Or keystone   t Gk his brother   u Or the
Christ

20.18: Isa 8.14–15. **19**: Lk 19.47.
20.20–26: Paying taxes to Caesar (Mt
22.15–22; Mk 12.13–17). **20**: *Honest* translates a Greek word which normally means
"correct according to the law," i.e. "righteous." It is used here in the same sense of
false pretense that it has in Mt 23.28. **25**: Rom
13.7; Lk 23.2.
**20.27–40: Question about the resur-
rection.** (Mt 22.23–33; Mk 12.18–27). **27**:
Acts 4.1–2; 23.6–10. **28**: Deut 25.5; Gen 38.8.
20.34–36: Luke makes the same point as
Matthew and Mark, but in somewhat differ-

ent language: human relations in the home do
not exist in the same way beyond death. Jesus
distinguishes two ages and kinds of existence.
Mortals are part of this age by the fact of
physical birth, and of the age to come by
resurrection (v. 36; Rom 1.4). **38**: God is not
frustrated by physical death. **39**: Mk 12.28.
40: Mk 12.34; Mt 22.46.
20.41–44: David's son (Mt 22.41–46; Mk
12.35–37). **42**: Ps 110.1; see Mt 22.44 n. **44**:
The question is: How can the Messiah be
David's descendant if David calls him *Lord*?

said to the^v disciples, ⁴⁶"Beware of the scribes, who like to walk around in long robes, and love to be greeted with respect in the marketplaces, and to have the best seats in the synagogues and places of honor at banquets. ⁴⁷They devour widows' houses and for the sake of appearance say long prayers. They will receive the greater condemnation."

21 He looked up and saw rich people putting their gifts into the treasury; ²he also saw a poor widow put in two small copper coins. ³He said, "Truly I tell you, this poor widow has put in more than all of them; ⁴for all of them have contributed out of their abundance, but she out of her poverty has put in all she had to live on."

5 When some were speaking about the temple, how it was adorned with beautiful stones and gifts dedicated to God, he said, ⁶"As for these things that you see, the days will come when not one stone will be left upon another; all will be thrown down."

7 They asked him, "Teacher, when will this be, and what will be the sign that this is about to take place?" ⁸And he said, "Beware that you are not led astray; for many will come in my name and say, 'I am he!'^w and, 'The time is near!'^x Do not go after them.

9 "When you hear of wars and insurrections, do not be terrified; for these things must take place first, but the end will not follow immediately." ¹⁰Then he said to them, "Nation will rise against nation, and kingdom against kingdom; ¹¹there will be great earthquakes, and in various places famines and plagues; and there will be dreadful portents and great signs from heaven.

12 "But before all this occurs, they will arrest you and persecute you; they will hand you over to synagogues and prisons, and you will be brought before kings and governors because of my name. ¹³This will give you an opportunity to testify. ¹⁴So make up your minds not to prepare your defense in advance; ¹⁵for I will give you words^y and a wisdom that none of your opponents will be able to withstand or contradict. ¹⁶You will be betrayed even by parents and brothers, by relatives and friends; and they will put some of you to death. ¹⁷You will be hated by all because of my name. ¹⁸But not a hair of your head will perish. ¹⁹By your endurance you will gain your souls.

20 "When you see Jerusalem surrounded by armies, then know that its desolation has come near.^z ²¹Then those in Judea must flee to the mountains, and those inside the city must leave it, and those out in the country must not enter it; ²²for these are days of vengeance, as a fulfillment of all that is written. ²³Woe to those who are pregnant and to those who are nursing infants in those days! For there will be great distress on the earth and wrath against this people; ²⁴they will fall by the edge of the sword and be taken away as captives among all nations; and Jerusalem will be trampled on by the

v Other ancient authorities read *his*   *w* Gk *I am*   *x* Or *at hand*   *y* Gk *a mouth*   *z* Or *is at hand*

20.45–47: On pride and humility. 45: Mt 23.1; Mk 12.37. **46:** Mt 23.6; see Mk 12.39 n.; Lk 11.43; 14.7–11.

21.1–4: The widow's offering (Mk 12.41–44). **1:** *The treasury* refers here to a container (shaped like an inverted trumpet for protection against theft; see Mk 12.41 n.) to receive offerings; in Jn 8.20 to a room in the temple. **2:** The *copper coin* (lepton) was of little monetary value (see 12.59 n.) but of great spiritual significance because of its cost to this giver.

21.5–7: Destruction of the temple foretold (Mt 24.1–3; Mk 13.1–2). **5:** See Mt 24.1 n. **6:** See Mk 13.2 n. **7:** 17.20; Acts 1.6.

21.8–36: On the end of the age (Mt 24.4–36; Mk 13.3–37). **8:** 17.23; Mk 13.21; 1 Jn 2.18. **10:** 2 Chr 15.6; Isa 19.2. **12–17:** Mt 10.17–22.

21.12: Acts 25.24; Jn 16.2. **13:** Phil 1.12. **14–15:** 12.11–12. **16:** 12.52–53. **17:** Mt 10.22; Jn 15.18–25. **18:** 12.7; Mt 10.30; Acts 27.34; 1 Sam 14.45. **19:** *Gain your souls,* Mk 13.13; Mt 10.22; Rev 2.7.

21.20–22: 19.41–44; 23.28–31; 17.31. **20:** *Armies,* the Roman legions. **24:** Rom 11.25; Isa 63.18; Dan 8.13; Rev 11.2. *The times of the Gentiles* represent spiritual opportunity that

Gentiles, until the times of the Gentiles are fulfilled.

25 "There will be signs in the sun, the moon, and the stars, and on the earth distress among nations confused by the roaring of the sea and the waves. 26 People will faint from fear and foreboding of what is coming upon the world, for the powers of the heavens will be shaken. 27 Then they will see 'the Son of Man coming in a cloud' with power and great glory. 28 Now when these things begin to take place, stand up and raise your heads, because your redemption is drawing near."

29 Then he told them a parable: "Look at the fig tree and all the trees; 30 as soon as they sprout leaves you can see for yourselves and know that summer is already near. 31 So also, when you see these things taking place, you know that the kingdom of God is near. 32 Truly I tell you, this generation will not pass away until all things have taken place. 33 Heaven and earth will pass away, but my words will not pass away.

34 "Be on guard so that your hearts are not weighed down with dissipation and drunkenness and the worries of this life, and that day catch you unexpectedly, 35 like a trap. For it will come upon all who live on the face of the whole earth. 36 Be alert at all times, praying that you may have the strength to escape all these things that will take place, and to stand before the Son of Man."

37 Every day he was teaching in the temple, and at night he would go out and spend the night on the Mount of Olives, as it was called. 38 And all the people would get up early in the morning to listen to him in the temple.

22 Now the festival of Unleavened Bread, which is called the Passover, was near. 2 The chief priests and the scribes were looking for a way to put Jesus*a* to death, for they were afraid of the people.

3 Then Satan entered into Judas called Iscariot, who was one of the twelve; 4 he went away and conferred with the chief priests and officers of the temple police about how he might betray him to them. 5 They were greatly pleased and agreed to give him money. 6 So he consented and began to look for an opportunity to betray him to them when no crowd was present.

7 Then came the day of Unleavened Bread, on which the Passover lamb had to be sacrificed. 8 So Jesus*b* sent Peter and John, saying, "Go and prepare the Passover meal for us that we may eat it." 9 They asked him, "Where do you want us to make preparations for it?" 10 "Listen," he said to them, "when you have entered the city, a man carrying a jar of water will meet you; follow him into the house he enters 11 and say to the owner of the house, 'The teacher asks you, "Where is the guest room, where I may eat the Passover with my disciples?"' 12 He will show you a large room upstairs, already furnished. Make preparations for us there." 13 So they went and

a Gk *him*   *b* Gk *he*

God previously had given Jews and now extends to non-Jews (20.16; Mk 13.10; Rom 11.25). **25:** Rev 6.12–13; Isa 13.10; Joel 2.10; Zeph 1.15. **27:** Dan 7.13–14. **28:** *Redemption,* 2.38; Eph 4.30. **33:** 16.17.

21.34: 8.14; 12.22, 45; Mk 4.19; 1 Thess 5.6–7. **36:** Mt 7.21–23; Mk 13.33; 2 Cor 5.10. **37:** Mt 21.17; Mk 11.19; Lk 19.47. **38:** The story of the woman caught in adultery appears after this verse in some ancient manuscripts; see Jn 8.11, note *k*.

22.1–23.56: Jesus' death (Mt 26.1–27.66; Mk 14.1–15.47; Jn 13.1–19.42). **22.1–2:** Mt 26.2–5; Mk 14.1–2; Jn 11.47–53. The word "called" is a concession to the Gentile readers for whom Luke wrote. **3–6:** Mt 26.14–16; Mk 14.10–11; Jn 13.2. **5:** See Mt 26.15 n.

22.7–38: The Last Supper. 7–13: Mt 26.17–19; Mk 14.12–16. **7:** Ex 12.18–20; Deut 16.5–8. **10:** The plans rest on some prearrangement, apparently; a man carrying a jar of water would be doing woman's work and would be readily noticeable. The procedure hid the intended place of the meal from Jesus' enemies. **11:** The identity of the householder is unknown (see Mk 14.51 n.). **12:** The *large room upstairs* was on the second floor, probably served by an outside staircase.

22.14–23: Mt 26.20–29; Mk 14.17–25; Jn 13.21–30. **14:** *The hour,* of the meal, after sun-

found everything as he had told them; and they prepared the Passover meal.

14 When the hour came, he took his place at the table, and the apostles with him. [15]He said to them, "I have eagerly desired to eat this Passover with you before I suffer; [16]for I tell you, I will not eat it*c* until it is fulfilled in the kingdom of God." [17]Then he took a cup, and after giving thanks he said, "Take this and divide it among yourselves; [18]for I tell you that from now on I will not drink of the fruit of the vine until the kingdom of God comes." [19]Then he took a loaf of bread, and when he had given thanks, he broke it and gave it to them, saying, "This is my body, which is given for you. Do this in remembrance of me." [20]And he did the same with the cup after supper, saying, "This cup that is poured out for you is the new covenant in my blood.*d* [21]But see, the one who betrays me is with me, and his hand is on the table. [22]For the Son of Man is going as it has been determined, but woe to that one by whom he is betrayed!" [23]Then they began to ask one another, which one of them it could be who would do this.

24 A dispute also arose among them as to which one of them was to be regarded as the greatest. [25]But he said to them, "The kings of the Gentiles lord it over them; and those in authority over them are called benefactors. [26]But not so with you; rather the greatest among you must become like the youngest, and the leader like one who serves. [27]For who is greater, the one who is at the table or the one who serves? Is it not the one at the table? But I am among you as one who serves.

28 "You are those who have stood by me in my trials; [29]and I confer on you, just as my Father has conferred on me, a kingdom, [30]so that you may eat and drink at my table in my kingdom, and you will sit on thrones judging the twelve tribes of Israel.

31 "Simon, Simon, listen! Satan has demanded*e* to sift all of you like wheat, [32]but I have prayed for you that your own faith may not fail; and you, when once you have turned back, strengthen your brothers." [33]And he said to him, "Lord, I am ready to go with you to prison and to death!" [34]Jesus*f* said, "I tell you, Peter, the cock will not crow this day, until you have denied three times that you know me."

35 He said to them, "When I sent you out without a purse, bag, or sandals, did you lack anything?" They said, "No, not a thing." [36]He said to them, "But now, the one who has a purse must take it, and likewise a bag. And the one who has no sword must sell his cloak and buy one.

c Other ancient authorities read *never eat it again*
d Other ancient authorities lack, in whole or in part, verses 19b-20 (*which is given . . . in my blood*)
e Or *has obtained permission*   *f* Gk *He*

down. **15**: 12.49–50. **16**: Jesus thinks of the meal as pointing forward to the meal celebrating the fulfilling of God's kingdom (13.28–29; 14.15; 22.28–30). **17**: Some Jewish meals included prayers over the cup of wine and several such prayers might be offered during the meal (see v. 20). Luke's order of events may be related to this fact, or to variations among early Christians in the way they observed the Lord's supper. Jesus transformed a Jewish devotional meal into a continuing expression of association with himself in death and victory. **19**: 1 Cor 11.23–26. **21**: Ps 41.9; Jn 13.21–30.
22.24–30: Mt 20.25–28; Mk 10.42–45. **24**: 9.46; Mk 9.34; Jn 13.3–16. **25**: *Benefactors*, a title bestowed on Hellenistic kings. **26**: See Mk 9.35 n. **28**: *My trials*, 4.13; Heb 2.18; 4.15.

29: Mk 14.25; Heb 9.20. **30**: Mk 10.37; Rev 3.21; 20.4.
22.31–34: Mt 26.33–35; Mk 14.29–31; Jn 13.37–38. **31**: Job 1.6–12; Am 9.9. **32**: *You* in this verse is singular number (contrast v. 31). **34**: Verses 54–62.
22.35: 10.4. **36**: An example of Jesus' fondness for striking metaphors (see Mt 23.24; Mk 10.25), but the disciples take it literally. The *sword* apparently meant to Jesus a preparation to live by one's own resources against hostility. The natural meaning of v. 38 is that the disciples supposed he spoke of an actual sword, only to learn that two swords were sufficient for the whole enterprise, i.e. were not to be used at all. **37**: Isa 53.12.

37 For I tell you, this scripture must be fulfilled in me, 'And he was counted among the lawless'; and indeed what is written about me is being fulfilled." 38 They said, "Lord, look, here are two swords." He replied, "It is enough."

39 He came out and went, as was his custom, to the Mount of Olives; and the disciples followed him. 40 When he reached the place, he said to them, "Pray that you may not come into the time of trial."*g* 41 Then he withdrew from them about a stone's throw, knelt down, and prayed, 42 "Father, if you are willing, remove this cup from me; yet, not my will but yours be done." ⟦43 Then an angel from heaven appeared to him and gave him strength. 44 In his anguish he prayed more earnestly, and his sweat became like great drops of blood falling down on the ground.⟧*h* 45 When he got up from prayer, he came to the disciples and found them sleeping because of grief, 46 and he said to them, "Why are you sleeping? Get up and pray that you may not come into the time of trial."*g*

47 While he was still speaking, suddenly a crowd came, and the one called Judas, one of the twelve, was leading them. He approached Jesus to kiss him; 48 but Jesus said to him, "Judas, is it with a kiss that you are betraying the Son of Man?" 49 When those who were around him saw what was coming, they asked, "Lord, should we strike with the sword?" 50 Then one of them struck the slave of the high priest and cut off his right ear. 51 But Jesus said, "No more of this!" And he touched his ear and healed

him. 52 Then Jesus said to the chief priests, the officers of the temple police, and the elders who had come for him, "Have you come out with swords and clubs as if I were a bandit? 53 When I was with you day after day in the temple, you did not lay hands on me. But this is your hour, and the power of darkness!"

54 Then they seized him and led him away, bringing him into the high priest's house. But Peter was following at a distance. 55 When they had kindled a fire in the middle of the courtyard and sat down together, Peter sat among them. 56 Then a servant-girl, seeing him in the firelight, stared at him and said, "This man also was with him." 57 But he denied it, saying, "Woman, I do not know him." 58 A little later someone else, on seeing him, said, "You also are one of them." But Peter said, "Man, I am not!" 59 Then about an hour later still another kept insisting, "Surely this man also was with him; for he is a Galilean." 60 But Peter said, "Man, I do not know what you are talking about!" At that moment, while he was still speaking, the cock crowed. 61 The Lord turned and looked at Peter. Then Peter remembered the word of the Lord, how he had said to him, "Before the cock crows today, you will deny me three times." 62 And he went out and wept bitterly.

63 Now the men who were holding Jesus began to mock him and beat him; 64 they also blindfolded him and kept ask-

g Or into temptation   h Other ancient authorities lack verses 43 and 44

22.39–53: **Gethsemane. 39–46:** Mt 26.30, 36–46; Mk 14.26, 32–42. **39:** Jn 18.1–2. **41:** Heb 5.7–8; Lk 11.4. **42:** *Cup,* metaphor for that which is allotted by God, whether blessing (Ps 16.5; 116.13) or judgment (Isa 51.17; Lam 4.21). It here refers to Jesus' suffering and death: see Mt 20.22; Mk 10.38. **43–44:** Although it is probable that these verses were not part of the original Gospel of Luke (since important early manuscripts lack them), they were known to Christian writers of the second century and reflect tradition from the first century concerning the suffer-

ings of Jesus. **43:** Mt 4.11; see Acts 12.15 n. **44:** 2 Cor 8.9; Phil 2.6–8; Heb 2.9, 17–18; 4.15; 5.8; 1 Pet 2.21–24; 4.1. **22.47–53:** Mt 26.47–56; Mk 14.43–52; Jn 18.3–11. **47:** See Mt 26.50 n. **52:** Only Luke says that the prospect of arresting Jesus had attracted the leaders of the Jewish religion (compare Mt 26.47, 57; Mk 14.43; Jn 18.3). **22.54–71:** **Jesus before Caiaphas. 54–55:** Jn 18.12–16. **56–62:** Jn 18.16–18, 25–27. **59:** See Mt 26.73 n. **61:** 7.13; 22.34. **22.63–65:** Jn 18.22–24. **66:** See Mt 27.1 n. **23.1–5, 13–25:** **Jesus before Pilate. 1:** Mt

ing him, "Prophesy! Who is it that struck you?" 65 They kept heaping many other insults on him.

66 When day came, the assembly of the elders of the people, both chief priests and scribes, gathered together, and they brought him to their council. 67 They said, "If you are the Messiah, *i* tell us." He replied, "If I tell you, you will not believe; 68 and if I question you, you will not answer. 69 But from now on the Son of Man will be seated at the right hand of the power of God." 70 All of them asked, "Are you, then, the Son of God?" He said to them, "You say that I am." 71 Then they said, "What further testimony do we need? We have heard it ourselves from his own lips!"

23 Then the assembly rose as a body and brought Jesus *j* before Pilate. 2 They began to accuse him, saying, "We found this man perverting our nation, forbidding us to pay taxes to the emperor, and saying that he himself is the Messiah, a king." *k* 3 Then Pilate asked him, "Are you the king of the Jews?" He answered, "You say so." 4 Then Pilate said to the chief priests and the crowds, "I find no basis for an accusation against this man." 5 But they were insistent and said, "He stirs up the people by teaching throughout all Judea, from Galilee where he began even to this place."

6 When Pilate heard this, he asked whether the man was a Galilean. 7 And when he learned that he was under Herod's jurisdiction, he sent him off to Herod, who was himself in Jerusalem at that time. 8 When Herod saw Jesus, he was very glad, for he had been wanting to see him for a long time, because he had heard about him and was hoping to see him perform some sign. 9 He questioned him at some length, but Jesus *l* gave him no answer. 10 The chief priests and the scribes stood by, vehemently accusing him. 11 Even Herod with his soldiers treated him with contempt and mocked him; then he put an elegant robe on him, and sent him back to Pilate. 12 That same day Herod and Pilate became friends with each other; before this they had been enemies.

13 Pilate then called together the chief priests, the leaders, and the people, 14 and said to them, "You brought me this man as one who was perverting the people; and here I have examined him in your presence and have not found this man guilty of any of your charges against him. 15 Neither has Herod, for he sent him back to us. Indeed, he has done nothing to deserve death. 16 I will therefore have him flogged and release him." *m*

18 Then they all shouted out together, "Away with this fellow! Release Barabbas for us!" 19 (This was a man who had been put in prison for an insurrection that had taken place in the city, and for murder.) 20 Pilate, wanting to release Jesus, addressed them again; 21 but they kept shouting, "Crucify, crucify him!" 22 A third time he said to them, "Why, what evil has he done? I have found in him no ground for the sentence of death; I will therefore have him flogged and then release him." 23 But they kept urgently demanding with loud shouts that he should be crucified; and their voices prevailed. 24 So Pilate gave his verdict that their demand should be granted. 25 He released the man they asked for, the

i Or *the Christ*   *j* Gk *him*   *k* Or *is an anointed king*   *l* Gk *he*   *m* Here, or after verse 19, other ancient authorities add verse 17, *Now he was obliged to release someone for them at the festival*

27.1–2; Mk 15.1; Jn 18.28. **2**: 20.25. The charge is phrased to sound like treason. **3**: Mt 27.11–12; Mk 15.2–3; Jn 18.29–38; Lk 22.70. **4**: 23.14, 22, 41; Mt 27.24; Jn 19.4, 6; Acts 13.28. Pilate refused to take religious ideas in a political sense. Here and in the following verses Luke seems anxious to show that Pilate sought to free Jesus, but yielded at length to pressures.

23.6–12: Jesus before Herod Antipas. This episode is reported only by Luke. *Herod* was a son of Herod the Great (see 3.1 n.). **8**: 9.9; Acts 4.27–28. **9**: Mk 15.5. **11**: Mk 15.17–19; Jn 19.2–3. **14**: Verses 4, 22, 41. **16**: Jn 19.12–14.

23.18–25: Mt 27.20–26; Mk 15.11–15; Jn 18.38–40; 19.14–16; Acts 3.13–14.

one who had been put in prison for insurrection and murder, and he handed Jesus over as they wished.

26 As they led him away, they seized a man, Simon of Cyrene, who was coming from the country, and they laid the cross on him, and made him carry it behind Jesus. 27 A great number of the people followed him, and among them were women who were beating their breasts and wailing for him. 28 But Jesus turned to them and said, "Daughters of Jerusalem, do not weep for me, but weep for yourselves and for your children. 29 For the days are surely coming when they will say, 'Blessed are the barren, and the wombs that never bore, and the breasts that never nursed.' 30 Then they will begin to say to the mountains, 'Fall on us'; and to the hills, 'Cover us.' 31 For if they do this when the wood is green, what will happen when it is dry?"

32 Two others also, who were criminals, were led away to be put to death with him. 33 When they came to the place that is called The Skull, they crucified Jesus*n* there with the criminals, one on his right and one on his left. ⟦34 Then Jesus said, "Father, forgive them; for they do not know what they are doing."⟧*o* And they cast lots to divide his clothing. 35 And the people stood by, watching; but the leaders scoffed at him, saying, "He saved others; let him save himself if he is the Messiah*p* of God, his chosen one!" 36 The soldiers also mocked him, coming up and offering him sour wine, 37 and saying, "If you are the King of the Jews, save yourself!" 38 There was

also an inscription over him,*q* "This is the King of the Jews."

39 One of the criminals who were hanged there kept deriding*r* him and saying, "Are you not the Messiah?*p* Save yourself and us!" 40 But the other rebuked him, saying, "Do you not fear God, since you are under the same sentence of condemnation? 41 And we indeed have been condemned justly, for we are getting what we deserve for our deeds, but this man has done nothing wrong." 42 Then he said, "Jesus, remember me when you come into*s* your kingdom." 43 He replied, "Truly I tell you, today you will be with me in Paradise."

44 It was now about noon, and darkness came over the whole land*t* until three in the afternoon, 45 while the sun's light failed;*u* and the curtain of the temple was torn in two. 46 Then Jesus, crying with a loud voice, said, "Father, into your hands I commend my spirit." Having said this, he breathed his last. 47 When the centurion saw what had taken place, he praised God and said, "Certainly this man was innocent."*v* 48 And when all the crowds who had gathered there for this spectacle saw what had taken place, they returned home, beating their breasts. 49 But all his acquaintances, including the

n Gk *him*   *o* Other ancient authorities lack the sentence *Then Jesus . . . what they are doing*   *p* Or *the Christ*   *q* Other ancient authorities add *written in Greek and Latin and Hebrew* (that is, *Aramaic*)   *r* Or *blaspheming*   *s* Other ancient authorities read *in*   *t* Or *earth*   *u* Or *the sun was eclipsed.* Other ancient authorities read *the sun was darkened*   *v* Or *righteous*

23.26–56: The crucifixion. 26: See Mt 27.32 n.; Mk 15.21 n.; Jn 19.17. **28–32:** 21.23–24; 19.41–44. **30:** Hos 10.8. **31:** A proverbial saying which, in this context, probably means: If the innocent Jesus meets such a fate, what will be the fate of the guilty Jerusalem (v. 28)? Compare 1 Pet 4.17–18. **33–43:** Mt 27.33–44; Mk 15.22–32; Jn 19.17–24. **34:** Num 15.27–31; Acts 7.60; Ps 22.18. **35:** 4.23. **36:** Ps 69.21; see Mt 27.48 n.; Mk 15.23. **41:** Verses 4, 14, 22. **42:** The robber's appeal may be based on the charge against Jesus (vv. 2, 3, 38); he thinks in terms of 21.27–28. Jesus promises him much more than he had

asked, intimating also that God's kingly power is a present reality, not merely future (Mt 6.10). **43:** *Paradise* (like "Abraham's bosom" in 16.22) was a contemporary Jewish term for the lodging place of the righteous dead prior to resurrection. **43:** 2 Cor 12.3; Rev 2.7.

23.44–49: Mt 27.45–56; Mk 15.33–41; Jn 19.25–30. **44:** See Mt 27.45 n. **45:** Ex 26.31–35; Heb 9.8; 10.19–20. *The sun's light failed,* the translation is uncertain; see note *u*. **46:** Ps 31.5. **48:** The cause of this popular agitation is not clear (Zech 12.10). **49:** 8.1–3; 23.55–56; 24.10; Ps 38.11. **50–56:** Mt 27.57–61; Mk 15.42–47; Jn 19.38–42; Acts 13.29. **51:** The

women who had followed him from Galilee, stood at a distance, watching these things.

50 Now there was a good and righteous man named Joseph, who, though a member of the council, [51] had not agreed to their plan and action. He came from the Jewish town of Arimathea, and he was waiting expectantly for the kingdom of God. [52] This man went to Pilate and asked for the body of Jesus. [53] Then he took it down, wrapped it in a linen cloth, and laid it in a rock-hewn tomb where no one had ever been laid. [54] It was the day of Preparation, and the sabbath was beginning. [w] [55] The women who had come with him from Galilee followed, and they saw the tomb and how his body was laid. [56] Then they returned, and prepared spices and ointments.

On the sabbath they rested according to the commandment.

24 But on the first day of the week, at early dawn, they came to the tomb, taking the spices that they had prepared. [2] They found the stone rolled away from the tomb, [3] but when they went in, they did not find the body. [x] [4] While they were perplexed about this, suddenly two men in dazzling clothes stood beside them. [5] The women [y] were terrified and bowed their faces to the ground, but the men [z] said to them, "Why do you look for the living among the dead? He is not here, but has risen. [a] [6] Remember how he told you, while he was still in Galilee, [7] that the Son of Man must be handed over to sinners, and be crucified, and on the third day rise again." [8] Then they remembered his words, [9] and returning from the tomb,

they told all this to the eleven and to all the rest. [10] Now it was Mary Magdalene, Joanna, Mary the mother of James, and the other women with them who told this to the apostles. [11] But these words seemed to them an idle tale, and they did not believe them. [12] But Peter got up and ran to the tomb; stooping and looking in, he saw the linen cloths by themselves; then he went home, amazed at what had happened. [b]

13 Now on that same day two of them were going to a village called Emmaus, about seven miles [c] from Jerusalem, [14] and talking with each other about all these things that had happened. [15] While they were talking and discussing, Jesus himself came near and went with them, [16] but their eyes were kept from recognizing him. [17] And he said to them, "What are you discussing with each other while you walk along?" They stood still, looking sad. [d] [18] Then one of them, whose name was Cleopas, answered him, "Are you the only stranger in Jerusalem who does not know the things that have taken place there in these days?" [19] He asked them, "What things?" They replied, "The things about Jesus of Nazareth, [e] who was a prophet mighty in deed and word before God and all the people, [20] and how our chief priests and leaders handed him over to be con-

w Gk *was dawning*   x Other ancient authorities add *of the Lord Jesus*   y Gk *They*   z Gk *but they*   a Other ancient authorities lack *He is not here, but has risen*   b Other ancient authorities lack verse 12   c Gk *sixty stadia*; other ancient authorities read *a hundred sixty stadia*   d Other ancient authorities read *walk along, looking sad?"*   e Other ancient authorities read *Jesus the Nazorean*

council was the Sanhedrin (see Jn 11.47 n.). **52:** Mt 27.58. **53:** Acts 13.29. **54:** *The sabbath began at sundown.* Luke, having non-Jewish readers in mind, wishes to indicate the urgency of the burial in Jewish custom. **56:** Mk 16.1; Ex 12.16; 20.10.

24.1–12: **The first Easter** (Mt 28.1–10; Mk 16.1–8; Jn 20.1, 11–18). **1:** See Mk 16.1 n. **2:** See Mk 16.3 n. **4:** See Mk 16.5 n. **6:** 9.22; 13.32–33. *You* here suggests that Jesus' disciples as a group often included others than

those of the inner circle. **9:** See Mt 28.8 n.; Mk 16.8. **10:** Mk 16.1; Lk 8.1–3; Jn 19.25; 20.2. **12:** This verse (see note *b*), though appearing in valuable ancient manuscripts, seems to be an addition to the original text of Luke based on Jn 20.3–10.

24.13–35: On the road to Emmaus. 16: See Mt 16.17 n.; Jn 20.14; 21.4. The distinction is between perception and recognition. **19:** Mt 21.11; Lk 7.16; 13.33; Acts 3.22; 10.38. **25:** Mk 12.24. **26:** *Necessary* because of the

demned to death and crucified him. [21] But we had hoped that he was the one to redeem Israel.[f] Yes, and besides all this, it is now the third day since these things took place. [22] Moreover, some women of our group astounded us. They were at the tomb early this morning, [23] and when they did not find his body there, they came back and told us that they had indeed seen a vision of angels who said that he was alive. [24] Some of those who were with us went to the tomb and found it just as the women had said; but they did not see him." [25] Then he said to them, "Oh, how foolish you are, and how slow of heart to believe all that the prophets have declared! [26] Was it not necessary that the Messiah[g] should suffer these things and then enter into his glory?" [27] Then beginning with Moses and all the prophets, he interpreted to them the things about himself in all the scriptures.

[28] As they came near the village to which they were going, he walked ahead as if he were going on. [29] But they urged him strongly, saying, "Stay with us, because it is almost evening and the day is now nearly over." So he went in to stay with them. [30] When he was at the table with them, he took bread, blessed and broke it, and gave it to them. [31] Then their eyes were opened, and they recognized him; and he vanished from their sight. [32] They said to each other, "Were not our hearts burning within us[h] while he was talking to us on the road, while he was opening the scriptures to us?" [33] That same hour they got up and re-turned to Jerusalem; and they found the eleven and their companions gathered together. [34] They were saying, "The Lord has risen indeed, and he has appeared to Simon!" [35] Then they told what had happened on the road, and how he had been made known to them in the breaking of the bread.

[36] While they were talking about this, Jesus himself stood among them and said to them, "Peace be with you."[i] [37] They were startled and terrified, and thought that they were seeing a ghost. [38] He said to them, "Why are you frightened, and why do doubts arise in your hearts? [39] Look at my hands and my feet; see that it is I myself. Touch me and see; for a ghost does not have flesh and bones as you see that I have." [40] And when he had said this, he showed them his hands and his feet.[j] [41] While in their joy they were disbelieving and still wondering, he said to them, "Have you anything here to eat?" [42] They gave him a piece of broiled fish, [43] and he took it and ate in their presence.

[44] Then he said to them, "These are my words that I spoke to you while I was still with you—that everything written about me in the law of Moses, the prophets, and the psalms must be fulfilled." [45] Then he opened their minds to understand the scriptures, [46] and he said to

*f Or to set Israel free g Or the Christ
h Other ancient authorities lack within us
i Other ancient authorities lack and said to them,
"Peace be with you." j Other ancient
authorities lack verse 40*

---

divine plan. **27:** *Moses,* the traditional author of the first five books of the Old Testament. *The prophets,* a major section of the Jewish Scriptures (see v. 44 n.; Mt 5.17 n.; Acts 28.23).

24.28: Mk 6.48. **30:** Mk 6.41; 14.22; Lk 9.16; 22.19. **34:** Mk 16.7; 1 Cor 15.5. Peter's experience, doubtless of tremendous effect for the beginnings of Christianity, is not described.

24.36–53: Commissioning of the disciples. 36–43: Jn 20.19–23; 1 Cor 15.5. The experience with Jesus, affirmed in v. 36, is tentatively interpreted in v. 37 as encounter with the dead, but this explanation is rejected in v. 39. **39:** *Touch me,* 1 Jn 1.1.

24.44: Verses 26–27; Acts 28.23. The Psalms formed the opening, and the longest, part of the third division of the Jewish Scriptures (see p. 623 OT). **45:** 24.32. **46:** Hos 6.2; 1 Cor 15.3–4. **47:** Acts 1.4–8; Mt 28.19. **48:** 1.2; Acts 1.8. **49:** Acts 2.1–4; Jn 14.26; 20.21–23. The words allude to the energy of the Holy Spirit referred to in Joel 2.28–32 (compare Acts 2.1–21). The new age has begun but its power is not yet freely felt. **51:** Acts 1.9–11. **52–53:** Acts 1.12–14.

them, "Thus it is written, that the Messiah[k] is to suffer and to rise from the dead on the third day, [47] and that repentance and forgiveness of sins is to be proclaimed in his name to all nations,[l] beginning from Jerusalem. [48] You are witnesses of these things. [49] And see, I am sending upon you what my Father promised; so stay here in the city until you have been clothed with power from on high."

50 Then he led them out as far as Bethany, and, lifting up his hands, he blessed them. [51] While he was blessing them, he withdrew from them and was carried up into heaven.[m] [52] And they worshiped him, and[n] returned to Jerusalem with great joy; [53] and they were continually in the temple blessing God.[o]

k Or *the Christ*   l Or *nations. Beginning from Jerusalem you are witnesses*   m Other ancient authorities lack *and was carried up into heaven*   n Other ancient authorities lack *worshiped him, and*   o Other ancient authorities add *Amen*

The Gospel According to
John

The fourth Gospel explains the mystery of the person of Jesus. Like others among his contemporaries, yet also unlike them, he stands above them in unique, solitary grandeur. Whence this uniqueness? The Evangelist takes us behind the scenes of Jesus' ministry, giving us a glimpse into his eternal origin and divine nature. He was unique because "he was in the beginning with God," active in creation, the source of light and life (1.2–4). Hence, when he became incarnate in human flesh, he made known the eternal God, whom "no one has ever seen" (1.14, 18).

As do the other Evangelists, the author records real events, but he goes beyond them in interpreting these events. He uses symbolically a number of terms drawn from common experience—bread, water, light, life, word, shepherd, door, way—to make the significance of Christ both clear and gripping. After a magnificent prologue (1.1–18) he sets forth Jesus Christ as the object of faith (1.19–4.54), depicts Christ's conflict with unbelievers (chs 5–12), his fellowship with believers (chs 13–17), his death and resurrection (chs 18–20), and concludes with an epilogue (ch 21). A large part of the Gospel consists of discourses of Jesus. These discourses are not individual sayings (as in the Synoptic Gospels), nor even collections of sayings (as in the Sermon on the Mount, Matthew 5–7); they develop a particular theme. Furthermore, it is characteristic of the Johannine discourses that Jesus is interrupted by questions or objections from the hearers—something that never happens in the other Gospels.

The first half of the fourth Gospel contains accounts of seven miracles of Jesus, though the author knows that Jesus had performed many others as well (20.30). John's word for these wondrous deeds is "signs," because they are here regarded as symbols of Jesus' teaching or as a revelation of his glory (2.11). Their purpose is to evoke faith on the part of those who witness them (2.23), beginning with the disciples (2.11).

The conflict between Jesus and the Pharisees reported in the Synoptic Gospels is given marked attention in John (for example, 8.31–59; 10.19–39), and the expression "the Jews" (which should not be understood as a condemnation of Jews in particular or in general) virtually becomes a technical term for those who reject Jesus. These features no doubt reflect the heightened antagonism that developed in the latter part of the first century between church and synagogue, with mutual recrimination arising.

While the Synoptic Gospels preserve the sayings of Jesus in words closer to their original form, the fourth Evangelist employs more freely his own modes of thought and language in reporting and interpreting the teaching of Jesus. The fact, however, that this Gospel was soon placed side by side with the Synoptics indicates that the early church realized that Jesus' promise, as reported by John (14.26), had been fulfilled, "The Holy Spirit . . . will teach you everything, and remind you of all that I have said to you."

Who wrote this Gospel? Tradition says it was the apostle John. Many scholars, however, think that it was composed by a disciple of John who recorded his preaching as Mark did that of Peter. In any case, when the Gospel was published near the close of the first century, the church accepted it as authentic and apostolic testimony to Jesus (21.24), written that readers might "come to believe that Jesus is the Messiah, the Son of God," and thus "have life in his name" (20.31). (For the literary genre of the gospels, see pp. viii–x NT).

1 In the beginning was the Word, and the Word was with God, and the Word was God. ²He was in the beginning with God. ³All things came into being through him, and without him not one thing came into being. What has come into being ⁴in him was life,ᵃ and the life was the light of all people. ⁵The light shines in the darkness, and the darkness did not overcome it.

6 There was a man sent from God, whose name was John. ⁷He came as a witness to testify to the light, so that all might believe through him. ⁸He himself was not the light, but he came to testify to the light. ⁹The true light, which enlightens everyone, was coming into the world.ᵇ

10 He was in the world, and the world came into being through him; yet the world did not know him. ¹¹He came to what was his own,ᶜ and his own people did not accept him. ¹²But to all who received him, who believed in his name, he gave power to become children of God, ¹³who were born, not of blood or of the will of the flesh or of the will of man, but of God.

14 And the Word became flesh and lived among us, and we have seen his glory, the glory as of a father's only son,ᵈ full of grace and truth. ¹⁵(John testified to him and cried out, "This was he of whom I said, 'He who comes after me

ranks ahead of me because he was before me.' ") ¹⁶From his fullness we have all received, grace upon grace. ¹⁷The law indeed was given through Moses; grace and truth came through Jesus Christ. ¹⁸No one has ever seen God. It is God the only Son,ᵉ who is close to the Father's heart,ᶠ who has made him known.

19 This is the testimony given by John when the Jews sent priests and Levites from Jerusalem to ask him, "Who are you?" ²⁰He confessed and did not deny it, but confessed, "I am not the Messiah."ᵍ ²¹And they asked him, "What then? Are you Elijah?" He said, "I am not." "Are you the prophet?" He answered, "No." ²²Then they said to him, "Who are you? Let us have an answer for those who sent us. What do you say about yourself?" ²³He said,

"I am the voice of one crying out
    in the wilderness,
'Make straight the way of
    the Lord,' "

as the prophet Isaiah said.

---

a Or *³through him. And without him not one thing came into being that has come into being. ⁴In him was life*   *b* Or *He was the true light that enlightens everyone coming into the world*   *c* Or *to his own home*   *d* Or *the Father's only Son*   *e* Other ancient authorities read *It is an only Son, God,* or *It is the only Son*   *f* Gk *bosom*   *g* Or *the Christ*

1.1–18: The Prologue. 1–2: The *Word* (Greek "logos") of God is more than speech; it is God in action, creating (Gen 1.3; Ps 33.6), revealing (Am 3.7–8), redeeming (Ps 107.19–20). Jesus is this *Word* (v. 14). He was eternal (*in the beginning;* compare Gen 1.1); personal (*with God*); divine (*was God*). *Was,* not "became" (contrast v. 14). **3:** He was sole agent of creation (Gen 1.1; Prov 8.27–30; Col 1.16–17; Heb 1.2). **4:** Apart from him both physical (Col 1.17) and spiritual life would recede into nothingness (5.39–40; 8.12). **5:** *Darkness* is total evil in conflict with God; it cannot *overcome.*

1.6–8: John (the Baptist), climaxing the Old Testament prophets, was *sent* (commissioned by God, Mal 3.1) to point to Jesus (vv. 19–34). **9:** *True light* is real, underived light, contrasted not with false light, but with

those such as John, who was but a lamp (5.35). **11:** *His own people,* the Jews.

1.14–17: God's *glory* dwelt ("tabernacled") in the *flesh* (human nature) of Jesus, as did his *grace* (redeeming love) and *truth* (faithfulness to his promises). These are available to *all,* exhaustless (*grace upon grace*), a fulfillment of the *law* of Moses. **18:** *Close to the Father's heart,* complete communion (vv. 1–2). On seeing and knowing God, see 14.9.

1.19–34: The testimony of John. 19: *Jews,* the religious authorities. **21:** *Elijah* (2 Kings 2.11) was expected to return to prepare the Messiah's way (Mal 4.5). John is unconscious of this role, but Jesus later ascribed it to him (see Mt 11.14 n.; Mk 9.13 n.). *The prophet* was likewise an expected Messianic forerunner (6.14; 7.40; see Deut 18.15). **23:** As a *voice* John fulfills a prophetic role

24 Now they had been sent from the Pharisees. 25 They asked him, "Why then are you baptizing if you are neither the Messiah, *h* nor Elijah, nor the prophet?" 26 John answered them, "I baptize with water. Among you stands one whom you do not know, 27 the one who is coming after me; I am not worthy to untie the thong of his sandal." 28 This took place in Bethany across the Jordan where John was baptizing.

29 The next day he saw Jesus coming toward him and declared, "Here is the Lamb of God who takes away the sin of the world! 30 This is he of whom I said, 'After me comes a man who ranks ahead of me because he was before me.' 31 I myself did not know him; but I came baptizing with water for this reason, that he might be revealed to Israel." 32 And John testified, "I saw the Spirit descending from heaven like a dove, and it remained on him. 33 I myself did not know him, but the one who sent me to baptize with water said to me, 'He on whom you see the Spirit descend and remain is the one who baptizes with the Holy Spirit.' 34 And I myself have seen and have testified that this is the Son of God." *i*

35 The next day John again was standing with two of his disciples, 36 and as he watched Jesus walk by, he exclaimed, "Look, here is the Lamb of God!" 37 The two disciples heard him say this, and they followed Jesus. 38 When Jesus turned and saw them following, he said to them, "What are you looking for?" They said to him, "Rabbi" (which translated means Teacher), "where are

you staying?" 39 He said to them, "Come and see." They came and saw where he was staying, and they remained with him that day. It was about four o'clock in the afternoon. 40 One of the two who heard John speak and followed him was Andrew, Simon Peter's brother. 41 He first found his brother Simon and said to him, "We have found the Messiah" (which is translated Anointed *j*). 42 He brought Simon *k* to Jesus, who looked at him and said, "You are Simon son of John. You are to be called Cephas" (which is translated Peter *l*).

43 The next day Jesus decided to go to Galilee. He found Philip and said to him, "Follow me." 44 Now Philip was from Bethsaida, the city of Andrew and Peter. 45 Philip found Nathanael and said to him, "We have found him about whom Moses in the law and also the prophets wrote, Jesus son of Joseph from Nazareth." 46 Nathanael said to him, "Can anything good come out of Nazareth?" Philip said to him, "Come and see." 47 When Jesus saw Nathanael coming toward him, he said of him, "Here is truly an Israelite in whom there is no deceit!" 48 Nathanael asked him, "Where did you get to know me?" Jesus answered, "I saw you under the fig tree before Philip called you." 49 Nathanael replied, "Rabbi, you are the Son of God! You are the King of Israel!" 50 Jesus answered, "Do you believe because I told you that I saw you

h Or *the Christ*   *i* Other ancient authorities read *is God's chosen one*   *j* Or *Christ*   *k* Gk *him*   *l* From the word for *rock* in Aramaic (*kepha*) and Greek (*petra*), respectively

announcing the Messiah's coming (Isa 40.3). **25**: *Why . . . are you baptizing*, performing an official rite, without official status? **27**: *Untie . . . sandal*, a slave's task.

1.29: *Lamb*, Ex ch 12; Isa 53.7. *Of God*, provided by God. **30**: *He outranks me, because he was* (existed) *before me.* **31–33**: John's knowledge of Jesus' significance was given him by God at the baptism. **34**: *Son of God*, the Messiah (v. 49; 11.27).

1.35–51: The testimony of Jesus' first disciples. 37: *They followed Jesus*, the first known disciples of Jesus; later both became

apostles. **39**: *Come and see*, a call to personal following (8.12). **42**: In Aramaic *Cephas* (Greek "Peter") means "rock". **44**: *Bethsaida*, on the western shore of the Sea of Galilee. **45**: *Moses . . . prophets*, the Old Testament points to Christ. **46**: *Nathanael*, probably the same person as Bartholomew (Mt 10.3; Mk 3.18; Lk 6.14), lived in Cana, near Nazareth (21.2). **47**: *No deceit*, no qualities of Jacob before he became Israel (Gen 27.35; 32.28). **51**: What Jacob saw in vision (Gen 28.12) is now a reality in Jesus. *Son of Man*, a messenger from heaven to make God known (3.13), and to be

under the fig tree? You will see greater things than these." [51] And he said to him, "Very truly, I tell you,[m] you will see heaven opened and the angels of God ascending and descending upon the Son of Man."

2 On the third day there was a wedding in Cana of Galilee, and the mother of Jesus was there. [2]Jesus and his disciples had also been invited to the wedding. [3]When the wine gave out, the mother of Jesus said to him, "They have no wine." [4]And Jesus said to her, "Woman, what concern is that to you and to me? My hour has not yet come." [5]His mother said to the servants, "Do whatever he tells you." [6]Now standing there were six stone water jars for the Jewish rites of purification, each holding twenty or thirty gallons. [7]Jesus said to them, "Fill the jars with water." And they filled them up to the brim. [8]He said to them, "Now draw some out, and take it to the chief steward." So they took it. [9]When the steward tasted the water that had become wine, and did not know where it came from (though the servants who had drawn the water knew), the steward called the bridegroom [10]and said to him, "Everyone serves the good wine first, and then the inferior wine after the guests have become drunk. But you have kept the good wine until now." [11]Jesus did this, the first of his signs, in Cana of Galilee, and revealed his glory; and his disciples believed in him.

[12] After this he went down to Capernaum with his mother, his brothers, and his disciples; and they remained there a few days.

[13] The Passover of the Jews was near, and Jesus went up to Jerusalem. [14]In the temple he found people selling cattle, sheep, and doves, and the money changers seated at their tables. [15]Making a whip of cords, he drove all of them out of the temple, both the sheep and the cattle. He also poured out the coins of the money changers and overturned their tables. [16]He told those who were selling the doves, "Take these things out of here! Stop making my Father's house a marketplace!" [17]His disciples remembered that it was written, "Zeal for your house will consume me." [18]The Jews then said to him, "What sign can you show us for doing this?" [19]Jesus answered them, "Destroy this temple, and in three days I will raise it up." [20]The Jews then said, "This temple has been under construction for forty-six years, and will you raise it up in three days?" [21]But he was speaking of the temple of his body. [22]After he was raised from the dead, his disciples remembered that he had said this; and they believed the scripture and the word that Jesus had spoken.

[23] When he was in Jerusalem during the Passover festival, many believed in his name because they saw the signs that he was doing. [24]But Jesus on his part

m Both instances of the Greek word for *you* in this verse are plural

---

the final judge (5.27; see Mk 2.10 n.).

2.1–12: The wedding at Cana. 1: *On the third day,* from the day when Philip was called (1.43). *Cana,* a small village nine miles northwest of Nazareth. **4:** *Woman,* a term of solemn and respectful address (compare 19.26). The *hour* of Jesus' self-disclosure was determined by God, not by Mary's desires. His final manifestation was at the cross (7.30; 8.20; 12.23, 27; 13.1; 17.1). **6:** *Rites of purification* were ceremonial, not hygienic. **8:** *Steward,* headwaiter or toastmaster. **11:** Jesus' miracles were not wonders to astound, but *signs* pointing to *his glory* (God's presence in him). *First,* for the second see 4.46–54. **12:** *Capernaum,* on the northern shore of the Sea of Galilee.

2.13–25: The cleansing of the temple (compare Mt 21.12–17; Mk 11.15–19; Lk 19.45–48). **14:** Animals were sold for sacrifice; Roman money was changed into Jewish money to pay the temple tax. **15–16:** Not an outburst of temper, but the energy of righteousness against religious leaders to whom religion had become a business. *My Father's house* is a claim to lordship. **17:** Ps 69.9. **20:** *Forty-six years,* this was the temple begun by Herod the Great in 20 B.C. (it was finished by Herod Agrippa in A.D. 64). **21:** The reply of Jesus (v. 19) was a prediction of his own death and resurrection. **23–25:** Faith which rests merely on *signs* and not on him to whom they point is shallow and unstable.

would not entrust himself to them, because he knew all people [25] and needed no one to testify about anyone; for he himself knew what was in everyone.

3 Now there was a Pharisee named Nicodemus, a leader of the Jews. [2] He came to Jesus[n] by night and said to him, "Rabbi, we know that you are a teacher who has come from God; for no one can do these signs that you do apart from the presence of God." [3] Jesus answered him, "Very truly, I tell you, no one can see the kingdom of God without being born from above."[o] [4] Nicodemus said to him, "How can anyone be born after having grown old? Can one enter a second time into the mother's womb and be born?" [5] Jesus answered, "Very truly, I tell you, no one can enter the kingdom of God without being born of water and Spirit. [6] What is born of the flesh is flesh, and what is born of the Spirit is spirit.[p] [7] Do not be astonished that I said to you, 'You[q] must be born from above.'[r] [8] The wind[p] blows where it chooses, and you hear the sound of it, but you do not know where it comes from or where it goes. So it is with everyone who is born of the Spirit." [9] Nicodemus said to him, "How can these things be?" [10] Jesus answered him, "Are you a teacher of Israel, and yet you do not understand these things?

11 "Very truly, I tell you, we speak of what we know and testify to what we have seen; yet you[s] do not receive our testimony. [12] If I have told you about earthly things and you do not believe, how can you believe if I tell you about heavenly things? [13] No one has ascended into heaven except the one who descended from heaven, the Son of Man.[t] [14] And just as Moses lifted up the serpent in the wilderness, so must the Son of Man be lifted up, [15] that whoever believes in him may have eternal life.[u]

16 "For God so loved the world that he gave his only Son, so that everyone who believes in him may not perish but may have eternal life.

17 "Indeed, God did not send the Son into the world to condemn the world, but in order that the world might be saved through him. [18] Those who believe in him are not condemned; but those who do not believe are condemned already, because they have not believed in the name of the only Son of God. [19] And this is the judgment, that the light has come into the world, and people loved darkness rather than light because their deeds were evil. [20] For all who do evil hate the light and do not come to the light, so that their deeds may not be exposed. [21] But those who do what is true come to the light, so that it may be clearly seen that their deeds have been done in God."[u]

22 After this Jesus and his disciples went into the Judean countryside, and he spent some time there with them and baptized. [23] John also was baptizing at Aenon near Salim because water was abundant there; and people kept coming

n Gk *him*   o Or *born anew*   p The same Greek word means both *wind* and *spirit* q The Greek word for *you* here is plural r Or *anew*   s The Greek word for *you* here and in verse 12 is plural   t Other ancient authorities add *who is in heaven*   u Some interpreters hold that the quotation concludes with verse 15

3.1–21: Jesus and official Judaism. 1: The Pharisees were the most devout of Jews. *A leader,* a member of the Sanhedrin (see 11.47 n.). **3:** *The kingdom of God* is entered, not by moral achievement, but by a transformation wrought by God. **5:** Birth into the new order is through *water* (referring to baptism; 1.33; Eph 5.26) and *Spirit* (Ezek 36.25–27; Titus 3.5). **6:** Like begets like. **8–9:** See note *p* and Ezek 37.5–10. **12:** *Earthly things,* such as the parable of the wind; *heavenly things,* supreme spiritual realities. **13–15:** Jesus *descended from heaven* to bring *eternal life* (participation in God's life), through being *lifted up* on the cross (Num 21.9).

3.16: Luther called this verse "the Gospel in miniature." **17–20:** God's purpose is to save; individuals judge themselves by hiding their *evil deeds* from the *light* of Christ's holiness.

3.22–36: Further testimony of John (compare 1.19–34). **23:** The exact location of Aenon and Salim is uncertain.

and were being baptized ²⁴—John, of course, had not yet been thrown into prison.

25 Now a discussion about purification arose between John's disciples and a Jew.ᵛ ²⁶They came to John and said to him, "Rabbi, the one who was with you across the Jordan, to whom you testified, here he is baptizing, and all are going to him." ²⁷John answered, "No one can receive anything except what has been given from heaven. ²⁸You yourselves are my witnesses that I said, 'I am not the Messiah,ʷ but I have been sent ahead of him.' ²⁹He who has the bride is the bridegroom. The friend of the bridegroom, who stands and hears him, rejoices greatly at the bridegroom's voice. For this reason my joy has been fulfilled. ³⁰He must increase, but I must decrease."ˣ

31 The one who comes from above is above all; the one who is of the earth belongs to the earth and speaks about earthly things. The one who comes from heaven is above all. ³²He testifies to what he has seen and heard, yet no one accepts his testimony. ³³Whoever has accepted his testimony has certifiedʸ this, that God is true. ³⁴He whom God has sent speaks the words of God, for he gives the Spirit without measure. ³⁵The Father loves the Son and has placed all things in his hands. ³⁶Whoever believes in the Son has eternal life; whoever disobeys the Son will not see life, but must endure God's wrath.

4 Now when Jesusᶻ learned that the Pharisees had heard, "Jesus is making and baptizing more disciples than John" ²—although it was not Jesus himself but his disciples who baptized— ³he left Judea and started back to Galilee. ⁴But he had to go through Samaria. ⁵So he came to a Samaritan city called Sychar, near the plot of ground that Jacob had given to his son Joseph. ⁶Jacob's well was there, and Jesus, tired out by his journey, was sitting by the well. It was about noon.

7 A Samaritan woman came to draw water, and Jesus said to her, "Give me a drink." ⁸(His disciples had gone to the city to buy food.) ⁹The Samaritan woman said to him, "How is it that you, a Jew, ask a drink of me, a woman of Samaria?" (Jews do not share things in common with Samaritans.)ᵃ ¹⁰Jesus answered her, "If you knew the gift of God, and who it is that is saying to you, 'Give me a drink,' you would have asked him, and he would have given you living water." ¹¹The woman said to him, "Sir, you have no bucket, and the well is deep. Where do you get that living water? ¹²Are you greater than our ancestor Jacob, who gave us the well, and with his sons and his flocks drank from it?" ¹³Jesus said to her, "Everyone who drinks of this water will be thirsty again, ¹⁴but those who drink of the water that I will give them will never be thirsty. The water that I will give will become in them a spring of water gushing up to eternal life." ¹⁵The woman said to him, "Sir, give me this water, so that I may

v Other ancient authorities read *the Jews*
w Or *the Christ*   *x* Some interpreters hold that the quotation continues through verse 36
y Gk *set a seal to*   *z* Other ancient authorities read *the Lord*   *a* Other ancient authorities lack this sentence

---

3.25: *Purification,* Jewish religious ceremonies. **27–29:** John was only the *friend of the bridegroom,* leading Israel, the bride, to Jesus, the bridegroom. He *rejoices* in their union (see Mk 2.19–20 n.). **32–35:** *No one,* a generalization about the Jews. The author and others do believe, and attest that Jesus authentically speaks *the words of God.* **36:** Unbelief is disobedience. *Wrath* is the consuming fire of God's holiness.
4.1–42: Jesus and the Samaritans. 1–3:

The Pharisees, hostile to John, now turn on Jesus. **4:** *Samaria,* between Judea and Galilee, with a mixed people (see Acts 8.5 n.). **5:** Gen 33.19; 48.22; Josh 24.32. **6:** *Tired out,* shows Jesus' humanity. **5:** Gen 33.19; 48.22; Josh 24.32. **9:** Rabbis avoided speaking to a *woman* in public (v. 27). *Jews* held *Samaritans* in contempt, as religious apostates (2 Kings 17.24–34).
4.10: *Living water,* Jer 2.13; 17.13. **14:** Jesus' gift is abundant life (10.10). **17:** *I have no hus-*

never be thirsty or have to keep coming here to draw water."

16 Jesus said to her, "Go, call your husband, and come back." 17 The woman answered him, "I have no husband." Jesus said to her, "You are right in saying, 'I have no husband'; 18 for you have had five husbands, and the one you have now is not your husband. What you have said is true!" 19 The woman said to him, "Sir, I see that you are a prophet. 20 Our ancestors worshiped on this mountain, but you*b* say that the place where people must worship is in Jerusalem." 21 Jesus said to her, "Woman, believe me, the hour is coming when you will worship the Father neither on this mountain nor in Jerusalem. 22 You worship what you do not know; we worship what we know, for salvation is from the Jews. 23 But the hour is coming, and is now here, when the true worshipers will worship the Father in spirit and truth, for the Father seeks such as these to worship him. 24 God is spirit, and those who worship him must worship in spirit and truth." 25 The woman said to him, "I know that Messiah is coming" (who is called Christ). "When he comes, he will proclaim all things to us." 26 Jesus said to her, "I am he,*c* the one who is speaking to you."

27 Just then his disciples came. They were astonished that he was speaking with a woman, but no one said, "What do you want?" or, "Why are you speaking with her?" 28 Then the woman left her water jar and went back to the city. She said to the people, 29 "Come and see a man who told me everything I have ever done! He cannot be the Messiah,*d*

can he?" 30 They left the city and were on their way to him.

31 Meanwhile the disciples were urging him, "Rabbi, eat something." 32 But he said to them, "I have food to eat that you do not know about." 33 So the disciples said to one another, "Surely no one has brought him something to eat?" 34 Jesus said to them, "My food is to do the will of him who sent me and to complete his work. 35 Do you not say, 'Four months more, then comes the harvest'? But I tell you, look around you, and see how the fields are ripe for harvesting. 36 The reaper is already receiving*e* wages and is gathering fruit for eternal life, so that sower and reaper may rejoice together. 37 For here the saying holds true, 'One sows and another reaps.' 38 I sent you to reap that for which you did not labor. Others have labored, and you have entered into their labor."

39 Many Samaritans from that city believed in him because of the woman's testimony, "He told me everything I have ever done." 40 So when the Samaritans came to him, they asked him to stay with them; and he stayed there two days. 41 And many more believed because of his word. 42 They said to the woman, "It is no longer because of what you said that we believe, for we have heard for ourselves, and we know that this is truly the Savior of the world."

43 When the two days were over, he went from that place to Galilee 44 (for Jesus himself had testified that a prophet has no honor in the prophet's own coun-

b The Greek word for *you* here and in verses 21 and 22 is plural   *c* Gk *I am*   *d* Or *the Christ* *e* Or *35. . . the fields are already ripe for harvesting.* *36 The reaper is receiving*

---

band, a true answer, literally taken, though given with an intention to deceive. **19–20:** A *prophet* should be able to settle rival religious claims. **21:** *This mountain,* i.e. Mount Gerizim, where the Samaritans had had a temple. Jesus means that the place of worship is not of primary importance. **24:** Worship *in spirit* is our response to God's gift of himself (*the Father seeks,* v. 23). *In . . . truth,* in accord with God's nature seen in Christ.

4.27: See v. 9 n. **35:** *Do you not say,* i.e. is it not a common fact? **36:** *Wages,* the reward of gathering believers. **37–38:** Jesus *sows* (vv. 7–26), the disciples *reap;* the harvest comes from the *labor* of Jesus' life, death, and resurrection (12.23–24).

4.39–42: Faith based on the testimony of another (*the woman*) is vindicated in personal experience.

4.43–54: Jesus and the Gentiles. Illus-

try). 45 When he came to Galilee, the Galileans welcomed him, since they had seen all that he had done in Jerusalem at the festival; for they too had gone to the festival.

46 Then he came again to Cana in Galilee where he had changed the water into wine. Now there was a royal official whose son lay ill in Capernaum. 47 When he heard that Jesus had come from Judea to Galilee, he went and begged him to come down and heal his son, for he was at the point of death. 48 Then Jesus said to him, "Unless you*f* see signs and wonders you will not believe." 49 The official said to him, "Sir, come down before my little boy dies." 50 Jesus said to him, "Go; your son will live." The man believed the word that Jesus spoke to him and started on his way. 51 As he was going down, his slaves met him and told him that his child was alive. 52 So he asked them the hour when he began to recover, and they said to him, "Yesterday at one in the afternoon the fever left him." 53 The father realized that this was the hour when Jesus had said to him, "Your son will live." So he himself believed, along with his whole household. 54 Now this was the second sign that Jesus did after coming from Judea to Galilee.

5 After this there was a festival of the Jews, and Jesus went up to Jerusalem.
2 Now in Jerusalem by the Sheep Gate there is a pool, called in Hebrew*g* Beth-zatha,*h* which has five porticoes. 3 In these lay many invalids—blind,

lame, and paralyzed.*i* 5 One man was there who had been ill for thirty-eight years. 6 When Jesus saw him lying there and knew that he had been there a long time, he said to him, "Do you want to be made well?" 7 The sick man answered him, "Sir, I have no one to put me into the pool when the water is stirred up; and while I am making my way, someone else steps down ahead of me." 8 Jesus said to him, "Stand up, take your mat and walk." 9 At once the man was made well, and he took up his mat and began to walk.

Now that day was a sabbath. 10 So the Jews said to the man who had been cured, "It is the sabbath; it is not lawful for you to carry your mat." 11 But he answered them, "The man who made me well said to me, 'Take up your mat and walk.'" 12 They asked him, "Who is the man who said to you, 'Take it up and walk'?" 13 Now the man who had been healed did not know who it was, for Jesus had disappeared in*j* the crowd that was there. 14 Later Jesus found him in the temple and said to him, "See, you have been made well! Do not sin any more, so that nothing worse happens to you." 15 The man went away and told the Jews that it was Jesus who had made him well.

f Both instances of the Greek word for *you* in this verse are plural   *g* That is, *Aramaic*
h Other ancient authorities read *Bethesda*, others *Bethsaida*   *i* Other ancient authorities add, wholly or in part, *waiting for the stirring of the water;* *4for an angel of the Lord went down at certain seasons into the pool, and stirred up the water; whoever stepped in first after the stirring of the water was made well from whatever disease that person had.*
j Or *had left because of*

---

trates v. 42, Jesus as *Savior of the world* (Jew, Samaritan, Gentile—everyone; compare Isa 43.3, 11; 45.22).

4.46: *A royal official,* a Gentile military officer. *In Capernaum,* about eighteen miles from Cana. **48**: *You* is plural here, addressed to all who base faith on mere signs (compare v. 45). **49**: He desires life for his child, not a display. **50**: The official *believed* that Jesus' *word* had effected the cure, and he did not return to his home (which was only about eighteen miles away) until the next day (v. 52). **53**: *Believed,*

in the deepest sense. **54**: *Second,* for the first see 2.1–11.

5.1–18: **Healing the lame man on the sabbath. 2**: Excavations at the Pool of St. Anna in the northeast of the city have revealed *five porticoes.* **3**: After the word *paralyzed* later manuscripts add an explanatory statement; see note *i.* **7**: *When the water is stirred up* is explained by the addition to v. 3. Movement caused by an intermittent spring was attributed to divine action. **13**: *Jesus had disappeared* to avoid publicity. **14**: There are

16 Therefore the Jews started persecuting Jesus, because he was doing such things on the sabbath. 17 But Jesus answered them, "My Father is still working, and I also am working." 18 For this reason the Jews were seeking all the more to kill him, because he was not only breaking the sabbath, but was also calling God his own Father, thereby making himself equal to God.

19 Jesus said to them, "Very truly, I tell you, the Son can do nothing on his own, but only what he sees the Father doing; for whatever the Father*k* does, the Son does likewise. 20 The Father loves the Son and shows him all that he himself is doing; and he will show him greater works than these, so that you will be astonished. 21 Indeed, just as the Father raises the dead and gives them life, so also the Son gives life to whomever he wishes. 22 The Father judges no one but has given all judgment to the Son, 23 so that all may honor the Son just as they honor the Father. Anyone who does not honor the Son does not honor the Father who sent him. 24 Very truly, I tell you, anyone who hears my word and believes him who sent me has eternal life, and does not come under judgment, but has passed from death to life.

25 "Very truly, I tell you, the hour is coming, and is now here, when the dead will hear the voice of the Son of God, and those who hear will live. 26 For just as the Father has life in himself, so he has granted the Son also to have life in himself; 27 and he has given him authority to exe-

cute judgment, because he is the Son of Man. 28 Do not be astonished at this; for the hour is coming when all who are in their graves will hear his voice 29 and will come out—those who have done good, to the resurrection of life, and those who have done evil, to the resurrection of condemnation.

30 "I can do nothing on my own. As I hear, I judge; and my judgment is just, because I seek to do not my own will but the will of him who sent me.

31 "If I testify about myself, my testimony is not true. 32 There is another who testifies on my behalf, and I know that his testimony to me is true. 33 You sent messengers to John, and he testified to the truth. 34 Not that I accept such human testimony, but I say these things so that you may be saved. 35 He was a burning and shining lamp, and you were willing to rejoice for a while in his light. 36 But I have a testimony greater than John's. The works that the Father has given me to complete, the very works that I am doing, testify on my behalf that the Father has sent me. 37 And the Father who sent me has himself testified on my behalf. You have never heard his voice or seen his form, 38 and you do not have his word abiding in you, because you do not believe him whom he has sent.

39 "You search the scriptures because you think that in them you have eternal life; and it is they that testify on my behalf. 40 Yet you refuse to come to me to

k Gk *that one*

---

worse things than illness. **16:** *The Jews,* the religious authorities, opposed Jesus for his break with their legalism. **17:** God continually gives life and judges evil, as does Jesus. **18:** *Equal,* see 10.30–33.

5.19–29: Jesus' relation to God. 19–20: Jesus' sonship involves the identity of his will and actions with the Father's. The *greater works* are giving life (v. 21) and judgment (v. 22). **24:** *Anyone* who *believes* on the basis of Jesus' word *has passed* into the realm where death does not reign.

5.25: The *coming* age is already present in Jesus. To *hear* with the comprehension of faith makes the spiritually *dead* live. **26–29:**

They will share in the final *resurrection of life.*

5.30–40: Evidence of Jesus' relation to God. 30: Jesus' judgment is that of God, and therefore *just,* without favoritism or error. **32:** *Another,* the Father. **33–40:** God witnesses to Jesus through the ministry of *John* the Baptist (vv. 33–35), through Jesus' *works* (v. 36), and through *the scriptures* (vv. 37–40).

5.39: *You search,* in Greek the verb may be either imperative or indicative, i.e. "Search" or "You search"; the latter gives the better sense.

5.41–47: Jesus rebukes those who refuse his offer. 41: No human standards apply to him. **42:** No *love of God,* no love of Jesus.

have life. [41]I do not accept glory from human beings. [42]But I know that you do not have the love of God in[l] you. [43]I have come in my Father's name, and you do not accept me; if another comes in his own name, you will accept him. [44]How can you believe when you accept glory from one another and do not seek the glory that comes from the one who alone is God? [45]Do not think that I will accuse you before the Father; your accuser is Moses, on whom you have set your hope. [46]If you believed Moses, you would believe me, for he wrote about me. [47]But if you do not believe what he wrote, how will you believe what I say?"

6 After this Jesus went to the other side of the Sea of Galilee, also called the Sea of Tiberias.[m] [2]A large crowd kept following him, because they saw the signs that he was doing for the sick. [3]Jesus went up the mountain and sat down there with his disciples. [4]Now the Passover, the festival of the Jews, was near. [5]When he looked up and saw a large crowd coming toward him, Jesus said to Philip, "Where are we to buy bread for these people to eat?" [6]He said this to test him, for he himself knew what he was going to do. [7]Philip answered him, "Six months' wages[n] would not buy enough bread for each of them to get a little." [8]One of his disciples, Andrew, Simon Peter's brother, said to him, [9]"There is a boy here who has five barley loaves and two fish. But what are they among so many people?" [10]Jesus said, "Make the people sit down." Now there was a great deal of grass in the place; so they[o] sat down, about five thousand in all. [11]Then Jesus took the loaves, and when he had given thanks, he distributed them to those who were seated; so also the fish, as much as they wanted. [12]When they were satisfied, he told his disciples, "Gather up the fragments left over, so that nothing may be lost." [13]So they gathered them up, and from the fragments of the five barley loaves, left by those who had eaten, they filled twelve baskets. [14]When the people saw the sign that he had done, they began to say, "This is indeed the prophet who is to come into the world."

15 When Jesus realized that they were about to come and take him by force to make him king, he withdrew again to the mountain by himself.

16 When evening came, his disciples went down to the sea, [17]got into a boat, and started across the sea to Capernaum. It was now dark, and Jesus had not yet come to them. [18]The sea became rough because a strong wind was blowing. [19]When they had rowed about three or four miles,[p] they saw Jesus walking on the sea and coming near the boat, and they were terrified. [20]But he said to them, "It is I;[q] do not be afraid." [21]Then they wanted to take him into the boat, and immediately the boat reached the land toward which they were going.

22 The next day the crowd that had stayed on the other side of the sea saw that there had been only one boat there.

l Or *among*   *m* Gk *of Galilee of Tiberius*
n Gk *Two hundred denarii*; the denarius was the usual day's wage for a laborer   *o* Gk *the men*
p Gk *about twenty-five or thirty stadia*   *q* Gk *I am*

43–44: Judgment based on human pride. 45: 9.28; Rom 2.17. 47: Lk 16.29, 31.

6.1–15: **Feeding the five thousand;** the only miracle recorded by all four gospels (Mt 14.13–21; Mk 6.32–44; Lk 9.10–17). 1: *The other side,* the eastern shore. *Tiberias,* so named after the city, founded by Herod Antipas about A.D. 20 in honor of Tiberias Caesar. 6: *To test* Philip's faith. 9: *Barley loaves,* food of the poor. 12: *Gather,* an act of reverential economy toward the gift of God. 13: *Twelve baskets,* one for each disciple.

6.15: *To make him king,* as a political Messiah opposing Rome; but Jesus would not accept this (18.36).

6.16–21: **Jesus walks on the sea** (Mt 14.22–27; Mk 6.45–51). Jesus is greater than a political ruler (v. 15); he is Lord of the elements (Ps 107.29–30). 17: *Not yet come,* probably they expected to meet Jesus along the shore. 20–21: Jesus' presence dispels fear.

6.22–71: **Jesus, the bread of life.** 22–25: Note the clamor for more bread. 22: *On the other side,* the eastern side.

They also saw that Jesus had not got into the boat with his disciples, but that his disciples had gone away alone. 23 Then some boats from Tiberias came near the place where they had eaten the bread after the Lord had given thanks.*r* 24 So when the crowd saw that neither Jesus nor his disciples were there, they themselves got into the boats and went to Capernaum looking for Jesus.

25 When they found him on the other side of the sea, they said to him, "Rabbi, when did you come here?" 26 Jesus answered them, "Very truly, I tell you, you are looking for me, not because you saw signs, but because you ate your fill of the loaves. 27 Do not work for the food that perishes, but for the food that endures for eternal life, which the Son of Man will give you. For it is on him that God the Father has set his seal." 28 Then they said to him, "What must we do to perform the works of God?" 29 Jesus answered them, "This is the work of God, that you believe in him whom he has sent." 30 So they said to him, "What sign are you going to give us then, so that we may see it and believe you? What work are you performing? 31 Our ancestors ate the manna in the wilderness; as it is written, 'He gave them bread from heaven to eat.'" 32 Then Jesus said to them, "Very truly, I tell you, it was not Moses who gave you the bread from heaven, but it is my Father who gives you the true bread from heaven. 33 For the bread of God is that which*s* comes down from heaven and gives life to the world." 34 They said to him, "Sir, give us this bread always."

35 Jesus said to them, "I am the bread of life. Whoever comes to me will never be hungry, and whoever believes in me will never be thirsty. 36 But I said to you that you have seen me and yet do not believe. 37 Everything that the Father gives me will come to me, and anyone who comes to me I will never drive away; 38 for I have come down from heaven, not to do my own will, but the will of him who sent me. 39 And this is the will of him who sent me, that I should lose nothing of all that he has given me, but raise it up on the last day. 40 This is indeed the will of my Father, that all who see the Son and believe in him may have eternal life; and I will raise them up on the last day."

41 Then the Jews began to complain about him because he said, "I am the bread that came down from heaven." 42 They were saying, "Is not this Jesus, the son of Joseph, whose father and mother we know? How can he now say, 'I have come down from heaven'?" 43 Jesus answered them, "Do not complain among yourselves. 44 No one can come to me unless drawn by the Father who sent me; and I will raise that person up on the last day. 45 It is written in the prophets, 'And they shall all be taught by God.' Everyone who has heard and learned from the Father comes to me. 46 Not that anyone has seen the Father except the one who is from God; he has seen the Father. 47 Very truly, I tell you, whoever believes has eternal life. 48 I am

r Other ancient authorities lack *after the Lord had given thanks*   *s* Or *he who*

6.25: *The other side,* the crowd had now crossed in boats (v. 24) and were on the western side. **26:** *Signs,* pointing to Jesus as food for the soul. **27:** *Son of Man,* see 1.51 n. *Seal,* God's authentication, perhaps at the baptism (1.32). **28:** *Works,* 3.21; Rev 2.26. **29:** *Work,* singular number; not many works (v. 28), but obedient trust (*believe*) is the one thing pleasing to God (1 Jn 3.23). *Him . . . sent,* Jesus who reveals God. **30:** *See,* as a proof; but faith cannot be proved. **31:** The Messiah was expected to reproduce the miracle of the giving of manna (Ex 16.4, 15; Num 11.8; Ps 78.24; 105.40).
6.36–40: Jesus himself is God's gift of sustenance for time and eternity. Belief or unbelief involves a mystery known only to God, but no one who *comes* is rejected (v. 37). Faith is God's gift, not a human achievement; it gives *eternal life* now and issues in resurrection *on the last day.* **44–45:** The drawing is not coercive or mechanical. *Prophets,* Isa 54.13; compare Joel 2.28–29. Had they *heard* and *learned* God's voice in their scriptures, they would have recognized its accents in him who alone has direct communion with God.

the bread of life. ⁴⁹ Your ancestors ate the manna in the wilderness, and they died. ⁵⁰ This is the bread that comes down from heaven, so that one may eat of it and not die. ⁵¹ I am the living bread that came down from heaven. Whoever eats of this bread will live forever; and the bread that I will give for the life of the world is my flesh."

52 The Jews then disputed among themselves, saying, "How can this man give us his flesh to eat?" ⁵³ So Jesus said to them, "Very truly, I tell you, unless you eat the flesh of the Son of Man and drink his blood, you have no life in you. ⁵⁴ Those who eat my flesh and drink my blood have eternal life, and I will raise them up on the last day; ⁵⁵ for my flesh is true food and my blood is true drink. ⁵⁶ Those who eat my flesh and drink my blood abide in me, and I in them. ⁵⁷ Just as the living Father sent me, and I live because of the Father, so whoever eats me will live because of me. ⁵⁸ This is the bread that came down from heaven, not like that which your ancestors ate, and they died. But the one who eats this bread will live forever." ⁵⁹ He said these things while he was teaching in the synagogue at Capernaum.

60 When many of his disciples heard it, they said, "This teaching is difficult; who can accept it?" ⁶¹ But Jesus, being aware that his disciples were complaining about it, said to them, "Does this offend you? ⁶² Then what if you were to see the Son of Man ascending to where

he was before? ⁶³ It is the spirit that gives life; the flesh is useless. The words that I have spoken to you are spirit and life. ⁶⁴ But among you there are some who do not believe." For Jesus knew from the first who were the ones that did not believe, and who was the one that would betray him. ⁶⁵ And he said, "For this reason I have told you that no one can come to me unless it is granted by the Father."

66 Because of this many of his disciples turned back and no longer went about with him. ⁶⁷ So Jesus asked the twelve, "Do you also wish to go away?" ⁶⁸ Simon Peter answered him, "Lord, to whom can we go? You have the words of eternal life. ⁶⁹ We have come to believe and know that you are the Holy One of God." *t* ⁷⁰ Jesus answered them, "Did I not choose you, the twelve? Yet one of you is a devil." ⁷¹ He was speaking of Judas son of Simon Iscariot, *u* for he, though one of the twelve, was going to betray him.

7 After this Jesus went about in Galilee. He did not wish *v* to go about in Judea because the Jews were looking for an opportunity to kill him. ² Now the Jewish festival of Booths *w* was near. ³ So his brothers said to him, "Leave here and

t Other ancient authorities read *the Christ, the Son of the living God*   *u* Other ancient authorities read *Judas Iscariot son of Simon;* others, *Judas son of Simon from Karyot* (Kerioth)
v Other ancient authorities read *was not at liberty*
w Or *Tabernacles*

6.51–58: In these verses the Fourth Evangelist (who does not include an account of the institution of the Lord's Supper) presents teachings that have unmistakable eucharistic echoes. **51:** *The living bread . . . is my flesh,* the One who became flesh (assumed complete human nature, 1.14) offered himself to God in death, thus releasing his life *for the life of the world.* **53:** The separation of the *blood* from the *flesh* emphasizes the reality of Jesus' death. **54:** To *eat* and *drink* is to believe (v. 47), to appropriate, assimilate, and *abide* in Christ (v. 56). **58:** Since Christ is *bread . . . from heaven* (compare vv. 32–35), to eat him is to *live forever.* **6.60:** *This teaching is difficult* means it is offensive but not obscure. **62–63:** The ascen-

sion by which Jesus will be taken away as regards the flesh, will indicate that he has been speaking of spiritual realities and not the actual eating of his flesh. **64–65:** These truths can be discerned only by faith, which is God's gift, not a human achievement (Eph 2.8).

6.66–71: To receive God's gift of faith is to *know* God in Christ; to refuse it is to become an ally of the *devil.* Faith and unbelief mark the great divisions among even the disciples. **71:** *Iscariot,* identifies *Judas* as a man from Kerioth, a town in southern Judea.

7.1–52: **Jesus, the water of life. 2:** *Booths,* a festival which began on the fifteenth day of the seventh month (Sept.–Oct.), commemorating the wilderness wanderings (Lev

go to Judea so that your disciples also may see the works you are doing; [4] for no one who wants[x] to be widely known acts in secret. If you do these things, show yourself to the world." [5] (For not even his brothers believed in him.) [6] Jesus said to them, "My time has not yet come, but your time is always here. [7] The world cannot hate you, but it hates me because I testify against it that its works are evil. [8] Go to the festival yourselves. I am not[y] going to this festival, for my time has not yet fully come." [9] After saying this, he remained in Galilee.

10 But after his brothers had gone to the festival, then he also went, not publicly but as it were[z] in secret. [11] The Jews were looking for him at the festival and saying, "Where is he?" [12] And there was considerable complaining about him among the crowds. While some were saying, "He is a good man," others were saying, "No, he is deceiving the crowd." [13] Yet no one would speak openly about him for fear of the Jews.

14 About the middle of the festival Jesus went up into the temple and began to teach. [15] The Jews were astonished at it, saying, "How does this man have such learning,[a] when he has never been taught?" [16] Then Jesus answered them, "My teaching is not mine but his who sent me. [17] Anyone who resolves to do the will of God will know whether the teaching is from God or whether I am speaking on my own. [18] Those who speak on their own seek their own glory; but the one who seeks the glory of him

who sent him is true, and there is nothing false in him.

19 "Did not Moses give you the law? Yet none of you keeps the law. Why are you looking for an opportunity to kill me?" [20] The crowd answered, "You have a demon! Who is trying to kill you?" [21] Jesus answered them, "I performed one work, and all of you are astonished. [22] Moses gave you circumcision (it is, of course, not from Moses, but from the patriarchs), and you circumcise a man on the sabbath. [23] If a man receives circumcision on the sabbath in order that the law of Moses may not be broken, are you angry with me because I healed a man's whole body on the sabbath? [24] Do not judge by appearances, but judge with right judgment."

25 Now some of the people of Jerusalem were saying, "Is not this the man whom they are trying to kill? [26] And here he is, speaking openly, but they say nothing to him! Can it be that the authorities really know that this is the Messiah?[b] [27] Yet we know where this man is from; but when the Messiah[b] comes, no one will know where he is from." [28] Then Jesus cried out as he was teaching in the temple, "You know me, and you know where I am from. I have not come on my own. But the one who sent me is true, and you do not know him. [29] I know him, because I am from him, and

x Other ancient authorities read *wants it*
y Other ancient authorities add *yet*
z Other ancient authorities lack *as it were*
a Or *this man know his letters* b Or *the Christ*

23.39–43; Deut 16.13–15). **3–5:** Jesus' *brothers* challenge him to declare himself *to the world* of pilgrims at the feast; his claims (chs 5 and 6) must be verified at Jerusalem. **6–8:** Jesus' *time,* or *hour* (v. 30; see 2.4 n.; 8.20; 12.23; 17.1), was the time for his self-manifestation on the cross. **10:** Compare Jesus' *secret* journey with the public one in 12.12–15, and the reasons for both in v. 8 and 12.23.
7.11–13: Subdued debate (*complaining*) over Jesus' character is prompted by *fear of the Jews* (see 5.16 n.). **15:** Wherein lay the authority of Jesus' teaching (Mk 1.22) without the accreditation of official rabbinic study? **16–18:**

Jesus' *teaching* came *from God,* not from himself, a fact that can be recognized by anyone who wills to obey God's *will.*
7.19–24: The law of Moses condemns their desire to *kill* Jesus for healing on the sabbath (5.18), for it enjoins circumcision even when the eighth day falls on a sabbath (Lev 12.3). If circumcision, why not healings?
7.25–31: Possibility of Jesus' messiahship is denied because his origin is known, whereas it was believed that the Messiah's origin would be mysterious. Jesus' human origin is apparent, but not his true origin in God, by whom he was *sent* and whose authority he bears.

he sent me." [30]Then they tried to arrest him, but no one laid hands on him, because his hour had not yet come. [31]Yet many in the crowd believed in him and were saying, "When the Messiah[c] comes, will he do more signs than this man has done?"[d]

32 The Pharisees heard the crowd muttering such things about him, and the chief priests and Pharisees sent temple police to arrest him. [33]Jesus then said, "I will be with you a little while longer, and then I am going to him who sent me. [34]You will search for me, but you will not find me; and where I am, you cannot come." [35]The Jews said to one another, "Where does this man intend to go that we will not find him? Does he intend to go to the Dispersion among the Greeks and teach the Greeks? [36]What does he mean by saying, 'You will search for me and you will not find me' and 'Where I am, you cannot come'?"

37 On the last day of the festival, the great day, while Jesus was standing there, he cried out, "Let anyone who is thirsty come to me, [38]and let the one who believes in me drink. As[e] the scripture has said, 'Out of the believer's heart[f] shall flow rivers of living water.' " [39]Now he said this about the Spirit, which believers in him were to receive; for as yet there was no Spirit,[g] because Jesus was not yet glorified.

40 When they heard these words, some in the crowd said, "This is really the prophet." [41]Others said, "This is the Messiah."[c] But some asked, "Surely the Messiah[c] does not come from Galilee, does he? [42]Has not the scripture said that the Messiah[c] is descended from David and comes from Bethlehem, the village where David lived?" [43]So there was a division in the crowd because of him. [44]Some of them wanted to arrest him, but no one laid hands on him.

45 Then the temple police went back to the chief priests and Pharisees, who asked them, "Why did you not arrest him?" [46]The police answered, "Never has anyone spoken like this!" [47]Then the Pharisees replied, "Surely you have not been deceived too, have you? [48]Has any one of the authorities or of the Pharisees believed in him? [49]But this crowd, which does not know the law—they are accursed." [50]Nicodemus, who had gone to Jesus[h] before, and who was one of them, asked, [51]"Our law does not judge people without first giving them a hearing to find out what they are doing, does it?" [52]They replied, "Surely you are not also from Galilee, are you? Search and you will see that no prophet is to arise from Galilee."

c Or *the Christ*   d Other ancient authorities read *is doing*   e Or *come to me and drink.* [38]*The one who believes in me, as*   f Gk *out of his belly*   g Other ancient authorities read *for as yet the Spirit* (others, *Holy Spirit*) *had not been given*   h Gk *him*

7.32–36: The *chief priests* (Sadducees) and *Pharisees*, inveterate enemies, unite in sending *temple police* to arrest Jesus. This leads Jesus to speak of his death (*I am going to him who sent me*). Later the Jews will seek deliverance in the Messiah, but *will not find* him, for he will have ascended to where they *cannot come* because they are unbelievers (contrast 8.21 with 12.26; 17.24). They miss the point, thinking that he is going to the Jews dispersed among the *Greeks* (Gentiles).
7.37–39: For seven days water was carried in a golden pitcher from the Pool of Siloam to the temple as a reminder of the water from the rock in the desert (Num 20.2–13), and as a symbol of hope for the coming Messianic deliverance (Isa 12.3). Jesus is the true water of life, who turns the symbol into reality (Isa 44.3; 55.1). Believers become channels of life to others, through Christ's *Spirit* given at Pentecost after he was *glorified* (crucified, risen, ascended). The gift of the *Spirit* is a mark of the Messianic age (Joel 2.28–29; Acts 2.14–21).
7.40–44: There is a division among the people over superficial matters. That he is from God is important. *The prophet . . . the Messiah*, see 1.21 n. **42**: *Descended from David*, 2 Sam 7.12–13; Ps 89.3–4; 132.11–12. *Bethlehem*, Mic 5.2. **49**: *Crowd*, the masses who are indifferent to scrupulous Pharisaic observances. **52**: Sarcasm, expressing the contempt of Jerusalem aristocrats for Galilean peasants.

8 [53 Then each of them went home, 1 while Jesus went to the Mount of Olives. 2 Early in the morning he came again to the temple. All the people came to him and he sat down and began to teach them. 3 The scribes and the Pharisees brought a woman who had been caught in adultery; and making her stand before all of them, 4 they said to him, "Teacher, this woman was caught in the very act of committing adultery. 5 Now in the law Moses commanded us to stone such women. Now what do you say?" 6 They said this to test him, so that they might have some charge to bring against him. Jesus bent down and wrote with his finger on the ground. 7 When they kept on questioning him, he straightened up and said to them, "Let anyone among you who is without sin be the first to throw a stone at her." 8 And once again he bent down and wrote on the ground. *i* 9 When they heard it, they went away, one by one, beginning with the elders; and Jesus was left alone with the woman standing before him. 10 Jesus straightened up and said to her, "Woman, where are they? Has no one condemned you?" 11 She said, "No one, sir." *j* And Jesus said, "Neither do I condemn you. Go your way, and from now on do not sin again."] *k*

12 Again Jesus spoke to them, saying, "I am the light of the world. Whoever follows me will never walk in darkness but will have the light of life." 13 Then the Pharisees said to him, "You are testifying on your own behalf; your testimony is not valid." 14 Jesus answered, "Even if I testify on my own behalf, my testimony is valid because I know where I have come from and where I am going, but you do not know where I come from or where I am going. 15 You judge by human standards; *l* I judge no one. 16 Yet even if I do judge, my judgment is valid; for it is not I alone who judge, but I and the Father *m* who sent me. 17 In your law it is written that the testimony of two witnesses is valid. 18 I testify on my own behalf, and the Father who sent me testifies on my behalf." 19 Then they said to him, "Where is your Father?" Jesus answered, "You know neither me nor my Father. If you knew me, you would know my Father also." 20 He spoke these words while he was teaching in the treasury of the temple, but no one arrested him, because his hour had not yet come.

21 Again he said to them, "I am going away, and you will search for me, but you will die in your sin. Where I am

i Other ancient authorities add *the sins of each of them*   *j* Or *Lord*   *k* The most ancient authorities lack 7.53—8.11; other authorities add the passage here or after 7.36 or after 21.25 or after Luke 21.38, with variations of text; some mark the passage as doubtful.   *l* Gk *according to the flesh*   *m* Other ancient authorities read *he*

7.53—8.11: The woman caught in adultery. This account, omitted in many ancient manuscripts, appears to be an authentic incident in Jesus' ministry, though not belonging originally to John's Gospel.
8.1: *Mount of Olives,* a prominent hill east of Jerusalem, and separated from the city by the valley of the Kidron. **2:** *Early,* Lk 21.38; Acts 5.21. *Sat down,* Mt 5.1; 23.2; Mk 9.35. **5:** *In the law,* Lev 20.10; Deut 22.23–24. **7:** *Without sin,* Mt 23.28; Rom 2.1. **8:** *Wrote,* according to several later manuscripts, Jesus *wrote . . . on the ground* "the sins of each of them" (compare Jer 17.13). **11:** *Do not sin again,* 5.14.
8.12–59: Jesus, the light of life. Great golden lamps in the temple court were lit during the *festival of Booths* (7.2): therefore the appropriateness of Jesus' claim in v. 12 (Isa 49.6; 60.1–3). **13–18:** Jesus answers the objection to his bearing self-witness: (*a*) he comes from the world above; hence, he alone understands who he is (Mt 11.27); (*b*) the joint witness of the Father and the Son fulfills the Jewish requirement of *two* witnesses (Deut 19.15).
8.19: The question reveals judgment *by human standards* (v. 15); those who question Jesus have no ears to hear God speaking in Jesus. **20:** *The treasury,* in the Women's Court were thirteen bronze chests for the reception of voluntary offerings and the temple dues. **22:** Suicide leads to hell, they suggest. **23–24:** Jesus reiterates his source *from above* (in God); to believe in him is the only escape from sin and death.

going, you cannot come." 22 Then the Jews said, "Is he going to kill himself? Is that what he means by saying, 'Where I am going, you cannot come'?" 23 He said to them, "You are from below, I am from above; you are of this world, I am not of this world. 24 I told you that you would die in your sins, for you will die in your sins unless you believe that I am he."[n] 25 They said to him, "Who are you?" Jesus said to them, "Why do I speak to you at all?[o] 26 I have much to say about you and much to condemn; but the one who sent me is true, and I declare to the world what I have heard from him." 27 They did not understand that he was speaking to them about the Father. 28 So Jesus said, "When you have lifted up the Son of Man, then you will realize that I am he,[n] and that I do nothing on my own, but I speak these things as the Father instructed me. 29 And the one who sent me is with me; he has not left me alone, for I always do what is pleasing to him." 30 As he was saying these things, many believed in him.

31 Then Jesus said to the Jews who had believed in him, "If you continue in my word, you are truly my disciples; 32 and you will know the truth, and the truth will make you free." 33 They answered him, "We are descendants of Abraham and have never been slaves to anyone. What do you mean by saying, 'You will be made free'?"

34 Jesus answered them, "Very truly, I tell you, everyone who commits sin is a slave to sin. 35 The slave does not have a permanent place in the household; the son has a place there forever. 36 So if the Son makes you free, you will be free indeed. 37 I know that you are descendants of Abraham; yet you look for an opportunity to kill me, because there is no place in you for my word. 38 I declare what I have seen in the Father's presence; as for you, you should do what you have heard from the Father."[p]

39 They answered him, "Abraham is our father." Jesus said to them, "If you were Abraham's children, you would be doing[q] what Abraham did, 40 but now you are trying to kill me, a man who has told you the truth that I heard from God. This is not what Abraham did. 41 You are indeed doing what your father does." They said to him, "We are not illegitimate children; we have one father, God himself." 42 Jesus said to them, "If God were your Father, you would love me, for I came from God and now I am here. I did not come on my own, but he sent me. 43 Why do you not understand what I say? It is because you cannot accept my word. 44 You are from your father the devil, and you choose to do your father's desires. He was a murderer from the beginning and does not stand in the truth, because there is no truth in him. When he lies, he speaks according to his own nature, for he is a liar and the father of lies. 45 But because I tell the truth, you do not believe me. 46 Which of you convicts me of sin? If I tell the truth, why do you not believe me? 47 Whoever is from God hears the words of God. The reason you do not hear them is that you are not from God."

48 The Jews answered him, "Are we

n Gk I am o Or What I have told you from the beginning p Other ancient authorities read you do what you have heard from your father
q Other ancient authorities read If you are Abraham's children, then do

---

8.25–27: An indirect claim to oneness with God, which they miss. **28–29:** The oneness is based on obedience, of which the cross is the final proof (12.32).
8.32: *Truth,* not knowledge in general, but saving truth (14.6). *Free,* from the power of sin and its slavery (v. 34). **33–38:** As *descendants of Abraham* the Jews had the truth of the Mosaic law; yet the law left them *slaves to sin.* True freedom comes only through *the Son.*

8.39–47: Their desire *to kill* Jesus forfeits their claim to be heirs of Abraham's faith and true children of God. They insist (v. 41) that *God* is their father. Their murderous intention and resistance to the truth belie this and brand them as children of *the devil* (v. 44). The fault is in them and not in Jesus (v. 46). **43:** *Word,* teaching.
8.48: *The Jews* turn to insult and calumny; unable to deny the miracles of Jesus, they

not right in saying that you are a Samaritan and have a demon?" 49 Jesus answered, "I do not have a demon; but I honor my Father, and you dishonor me. 50 Yet I do not seek my own glory; there is one who seeks it and he is the judge. 51 Very truly, I tell you, whoever keeps my word will never see death." 52 The Jews said to him, "Now we know that you have a demon. Abraham died, and so did the prophets; yet you say, 'Whoever keeps my word will never taste death.' 53 Are you greater than our father Abraham, who died? The prophets also died. Who do you claim to be?" 54 Jesus answered, "If I glorify myself, my glory is nothing. It is my Father who glorifies me, he of whom you say, 'He is our God,' 55 though you do not know him. But I know him; if I would say that I do not know him, I would be a liar like you. But I do know him and I keep his word. 56 Your ancestor Abraham rejoiced that he would see my day; he saw it and was glad." 57 Then the Jews said to him, "You are not yet fifty years old, and have you seen Abraham?"*r* 58 Jesus said to them, "Very truly, I tell you, before Abraham was, I am." 59 So they picked up stones to throw at him, but Jesus hid himself and went out of the temple.

9 As he walked along, he saw a man blind from birth. 2 His disciples asked him, "Rabbi, who sinned, this man or his parents, that he was born blind?" 3 Jesus answered, "Neither this man nor his parents sinned; he was born blind so that God's works might be revealed in him. 4 We*s* must work the works of him who sent me*t* while it is day; night is coming when no one can work. 5 As long as I am in the world, I am the light of the world." 6 When he had

said this, he spat on the ground and made mud with the saliva and spread the mud on the man's eyes, 7 saying to him, "Go, wash in the pool of Siloam" (which means Sent). Then he went and washed and came back able to see. 8 The neighbors and those who had seen him before as a beggar began to ask, "Is this not the man who used to sit and beg?" 9 Some were saying, "It is he." Others were saying, "No, but it is someone like him." He kept saying, "I am the man." 10 But they kept asking him, "Then how were your eyes opened?" 11 He answered, "The man called Jesus made mud, spread it on my eyes, and said to me, 'Go to Siloam and wash.' Then I went and washed and received my sight." 12 They said to him, "Where is he?" He said, "I do not know."

13 They brought to the Pharisees the man who had formerly been blind. 14 Now it was a sabbath day when Jesus made the mud and opened his eyes. 15 Then the Pharisees also began to ask him how he had received his sight. He said to them, "He put mud on my eyes. Then I washed, and now I see." 16 Some of the Pharisees said, "This man is not from God, for he does not observe the sabbath." But others said, "How can a man who is a sinner perform such signs?" And they were divided. 17 So they said again to the blind man, "What do you say about him? It was your eyes he opened." He said, "He is a prophet."

18 The Jews did not believe that he had been blind and had received his sight until they called the parents of the man

---

r Other ancient authorities read *has Abraham seen you?* *s* Other ancient authorities read *I* *t* Other ancient authorities read *us*

---

attribute them to the agency of *a demon.* **50–54:** God is judge, and will vindicate believers in Jesus with eternal life. **56:** Refusal to believe severs them from *Abraham,* who *rejoiced* in the hope of the coming of the Messiah (Heb 11.17). **58:** The *I am* is the divine name (Ex 3.14), a claim to pre-existence and oneness with God (10.30–33).

9.1–41: Jesus manifests himself as the

light of life. 2–3: Suffering was attributed to sin, either of the parents or of the man prenatally. Jesus denies this and shifts attention from cause to purpose; this is an opportunity for God to act. **6:** *Saliva* was considered to have medicinal value. **7:** *Sent,* a symbol of Jesus as "sent" from God to give light.

9.13–14: The *Pharisees* would consider making *mud* as work, a breach of sabbath

who had received his sight [19] and asked them, "Is this your son, who you say was born blind? How then does he now see?" [20] His parents answered, "We know that this is our son, and that he was born blind; [21] but we do not know how it is that now he sees, nor do we know who opened his eyes. Ask him; he is of age. He will speak for himself." [22] His parents said this because they were afraid of the Jews; for the Jews had already agreed that anyone who confessed Jesus[u] to be the Messiah[v] would be put out of the synagogue. [23] Therefore his parents said, "He is of age; ask him."

24 So for the second time they called the man who had been blind, and they said to him, "Give glory to God! We know that this man is a sinner." [25] He answered, "I do not know whether he is a sinner. One thing I do know, that though I was blind, now I see." [26] They said to him, "What did he do to you? How did he open your eyes?" [27] He answered them, "I have told you already, and you would not listen. Why do you want to hear it again? Do you also want to become his disciples?" [28] Then they reviled him, saying, "You are his disciple, but we are disciples of Moses. [29] We know that God has spoken to Moses, but as for this man, we do not know where he comes from." [30] The man answered, "Here is an astonishing thing! You do not know where he comes from, and yet he opened my eyes. [31] We know that God does not listen to sinners, but he does listen to one who worships him and obeys his will. [32] Never since the world began has it been heard that anyone opened the eyes of a person born blind. [33] If this man were not from God, he could do nothing." [34] They answered him, "You were born entirely in sins, and are you trying to teach us?" And they drove him out.

35 Jesus heard that they had driven him out, and when he found him, he said, "Do you believe in the Son of Man?"[w] [36] He answered, "And who is he, sir?[x] Tell me, so that I may believe in him." [37] Jesus said to him, "You have seen him, and the one speaking with you is he." [38] He said, "Lord,[x] I believe." And he worshiped him. [39] Jesus said, "I came into this world for judgment so that those who do not see may see, and those who do see may become blind." [40] Some of the Pharisees near him heard this and said to him, "Surely we are not blind, are we?" [41] Jesus said to them, "If you were blind, you would not have sin. But now that you say, 'We see,' your sin remains.

10 "Very truly, I tell you, anyone who does not enter the sheepfold by the gate but climbs in by another way is a thief and a bandit. [2] The one who enters by the gate is the shepherd of the sheep. [3] The gatekeeper opens the gate for him, and the sheep hear his voice. He calls his own sheep by name and leads them out. [4] When he has brought out all his own, he goes ahead of them, and the

---

u Gk *him*   *v* Or *the Christ*   *w* Other ancient authorities read *the Son of God*   *x* *Sir* and *Lord* translate the same Greek word

---

laws. *They,* the neighbors and friends. **17:** The miracle authenticates Jesus as a messenger from God.

9.18–23: The *parents* are interrogated to avoid the possibility of mistaken identity. **22:** *Put out,* excommunicated (v. 34).

9.24: Not "*Give glory to God* rather than to Jesus," but a technical phrase adjuring the man to tell the truth (Josh 7.19). One who broke the sabbath must be *a sinner.* **25:** Recovered sight was more important than Pharisaic tradition. **28:** *You are his disciple,* an expression of baffled rage; the Pharisees attempt to cover their own defeat by denouncing the one

who had been blind. **31:** *We know,* from Old Testament teachings (Ps 66.18; Prov 15.29). **34:** Anger usurps reason.

9.35–41: Judgment was not the purpose, but the result, of Jesus' coming (3.17). Belief opens the eyes of the spirit (*he worshiped him*), unbelief blinds *the Pharisees.* Proud refusal to admit spiritual blindness demonstrates their *sin* (v. 41).

10.1–42: Jesus, the shepherd who gives his life. 1–6: The details are strikingly true to life. **3:** *The gatekeeper* is apparently someone paid by a number of shepherds to guard their sheep in one fold, since Jesus refers to several

sheep follow him because they know his voice. ⁵They will not follow a stranger, but they will run from him because they do not know the voice of strangers." ⁶Jesus used this figure of speech with them, but they did not understand what he was saying to them.

7 So again Jesus said to them, "Very truly, I tell you, I am the gate for the sheep. ⁸All who came before me are thieves and bandits; but the sheep did not listen to them. ⁹I am the gate. Whoever enters by me will be saved, and will come in and go out and find pasture. ¹⁰The thief comes only to steal and kill and destroy. I came that they may have life, and have it abundantly.

11 "I am the good shepherd. The good shepherd lays down his life for the sheep. ¹²The hired hand, who is not the shepherd and does not own the sheep, sees the wolf coming and leaves the sheep and runs away—and the wolf snatches them and scatters them. ¹³The hired hand runs away because a hired hand does not care for the sheep. ¹⁴I am the good shepherd. I know my own and my own know me, ¹⁵just as the Father knows me and I know the Father. And I lay down my life for the sheep. ¹⁶I have other sheep that do not belong to this fold. I must bring them also, and they will listen to my voice. So there will be one flock, one shepherd. ¹⁷For this reason the Father loves me, because I lay down my life in order to take it up again. ¹⁸No one takesʸ it from me, but I lay it down of my own accord. I have power to lay it down, and I have power to take it up again. I have received this command from my Father."

19 Again the Jews were divided because of these words. ²⁰Many of them were saying, "He has a demon and is out of his mind. Why listen to him?" ²¹Others were saying, "These are not the words of one who has a demon. Can a demon open the eyes of the blind?"

22 At that time the festival of the Dedication took place in Jerusalem. It was winter, ²³and Jesus was walking in the temple, in the portico of Solomon. ²⁴So the Jews gathered around him and said to him, "How long will you keep us in suspense? If you are the Messiah, ᶻ tell us plainly." ²⁵Jesus answered, "I have told you, and you do not believe. The works that I do in my Father's name testify to me; ²⁶but you do not believe, because you do not belong to my sheep. ²⁷My sheep hear my voice. I know them, and they follow me. ²⁸I give them eternal life, and they will never perish. No one will snatch them out of my hand. ²⁹What my Father has given me is greater than all else, and no one can snatch it out of the Father's hand. ᵃ ³⁰The Father and I are one."

31 The Jews took up stones again to stone him. ³²Jesus replied, "I have shown you many good works from the Father. For which of these are you going to stone me?" ³³The Jews answered, "It is not for a good work that we are going to stone you, but for blasphemy, because you, though only a human being, are making yourself God." ³⁴Jesus answered, "Is it

y Other ancient authorities read *has taken*
z Or *the Christ*   a Other ancient authorities read *My Father who has given them to me is greater than all, and no one can snatch them out of the Father's hand*

---

shepherds by inference (v. 16) and repeats *his own sheep* several times (vv. 3, 4, 12, 14). **7:** Christ is the *gate* into God's fold. **8:** Refers to Messianic pretenders. **9:** Christ provides (*a*) escape from the perils of sin, (*b*) freedom, and (*c*) spiritual sustenance (the bread, water, and light of life, chs 6–9). **10:** *Life,* see 3.13–15 n. *Abundantly,* beyond measure (Ps 23.5).
10.11: Jesus fulfills Old Testament promises that God himself will come to shepherd his people (Isa 40.11; Jer 23.1–6; Ezek 34, esp.

v. 11). **16:** *Other sheep,* Gentiles. *One flock,* Eph 2.11–22.
10.19–21: 7.43; 8.48; 9.16. **22:** *Dedication,* commemorating the rededication of the temple in 164 B.C. after its desecration by Antiochus Epiphanes (1 Macc 4.52–59). **23:** *The portico of Solomon,* a cloister on the east side of the temple buildings. **24–30:** Evidence of Jesus' oneness with God demands faith for its interpretation (see 8.58 n.).
10.31–39: Two arguments: (*a*) In the light

not written in your law,[b] 'I said, you are gods'? [35] If those to whom the word of God came were called 'gods'—and the scripture cannot be annulled— [36] can you say that the one whom the Father has sanctified and sent into the world is blaspheming because I said, 'I am God's Son'? [37] If I am not doing the works of my Father, then do not believe me. [38] But if I do them, even though you do not believe me, believe the works, so that you may know and understand[c] that the Father is in me and I am in the Father." [39] Then they tried to arrest him again, but he escaped from their hands.

40 He went away again across the Jordan to the place where John had been baptizing earlier, and he remained there. [41] Many came to him, and they were saying, "John performed no sign, but everything that John said about this man was true." [42] And many believed in him there.

11 Now a certain man was ill, Lazarus of Bethany, the village of Mary and her sister Martha. [2] Mary was the one who anointed the Lord with perfume and wiped his feet with her hair; her brother Lazarus was ill. [3] So the sisters sent a message to Jesus,[d] "Lord, he whom you love is ill." [4] But when Jesus heard it, he said, "This illness does not lead to death; rather it is for God's glory, so that the Son of God may be glorified through it." [5] Accordingly, though Jesus loved Martha and her sister and Lazarus, [6] after having heard that Lazarus[e] was ill, he stayed two days longer in the place where he was.

7 Then after this he said to the disciples, "Let us go to Judea again." [8] The disciples said to him, "Rabbi, the Jews were just now trying to stone you, and are you going there again?" [9] Jesus answered, "Are there not twelve hours of daylight? Those who walk during the day do not stumble, because they see the light of this world. [10] But those who walk at night stumble, because the light is not in them." [11] After saying this, he told them, "Our friend Lazarus has fallen asleep, but I am going there to awaken him." [12] The disciples said to him, "Lord, if he has fallen asleep, he will be all right." [13] Jesus, however, had been speaking about his death, but they thought that he was referring merely to sleep. [14] Then Jesus told them plainly, "Lazarus is dead. [15] For your sake I am glad I was not there, so that you may believe. But let us go to him." [16] Thomas, who was called the Twin,[f] said to his fellow disciples, "Let us also go, that we may die with him."

17 When Jesus arrived, he found that Lazarus[e] had already been in the tomb four days. [18] Now Bethany was near Jerusalem, some two miles[g] away, [19] and many of the Jews had come to Martha and Mary to console them about their brother. [20] When Martha heard that Jesus was coming, she went and met him, while Mary stayed at home. [21] Martha said to Jesus, "Lord, if you had been here, my brother would not have died. [22] But even now I know that God will give you whatever you ask of him." [23] Jesus said to her, "Your brother will

b Other ancient authorities read *in the law*
c Other ancient authorities lack *and understand*; others read *and believe*   d Gk *him*   e Gk *he*
f Gk *Didymus*   g Gk *fifteen stadia*

---

of Ps 82.6, titles are less important than realities; (*b*) Jesus' works authenticate him, for they are the kind which God does. **40:** Jesus withdraws to Perea for safety. **41:** 1.26–36.

11.1–57: The raising of Lazarus, the crowning miracle or *sign* (12.17–18), revealing Jesus as the giver of life (5.25–29), and precipitating his death (11.53). **1:** *Bethany,* v. 18. **2:** 12.1–3. **4:** *Death,* i.e. final death. **6:** The action of Jesus is never hurried.

11.9: His life would end when God willed; his enemies could not shorten it. **11:** *Asleep,* a

common New Testament description of death (see Mt 9.24 n.; Mk 5.39; Acts 7.60; 1 Cor 15.6). **15:** The miracle will confirm their faith. **16:** Courageous loyalty, not cynicism.

11.19: *Many came to console them,* Jewish mourning ceremonies were elaborate and attended by many. At least ten persons were expected to take part in them; they were continued for about thirty days. **20:** As in Lk 10.38–42, Martha is active, Mary contemplative. **21–22:** Disappointment blends with

rise again." 24 Martha said to him, "I know that he will rise again in the resurrection on the last day." 25 Jesus said to her, "I am the resurrection and the life. *h* Those who believe in me, even though they die, will live, 26 and everyone who lives and believes in me will never die. Do you believe this?" 27 She said to him, "Yes, Lord, I believe that you are the Messiah, *i* the Son of God, the one coming into the world."

28 When she had said this, she went back and called her sister Mary, and told her privately, "The Teacher is here and is calling for you." 29 And when she heard it, she got up quickly and went to him. 30 Now Jesus had not yet come to the village, but was still at the place where Martha had met him. 31 The Jews who were with her in the house, consoling her, saw Mary get up quickly and go out. They followed her because they thought that she was going to the tomb to weep there. 32 When Mary came where Jesus was and saw him, she knelt at his feet and said to him, "Lord, if you had been here, my brother would not have died." 33 When Jesus saw her weeping, and the Jews who came with her also weeping, he was greatly disturbed in spirit and deeply moved. 34 He said, "Where have you laid him?" They said to him, "Lord, come and see." 35 Jesus began to weep. 36 So the Jews said, "See how he loved him!" 37 But some of them said, "Could not he who opened the eyes of the blind man have kept this man from dying?"

38 Then Jesus, again greatly disturbed, came to the tomb. It was a cave, and a stone was lying against it. 39 Jesus said, "Take away the stone." Martha, the sister of the dead man, said to him, "Lord, already there is a stench because he has been dead four days." 40 Jesus said to her, "Did I not tell you that if you believed, you would see the glory of God?" 41 So they took away the stone. And Jesus looked upward and said, "Father, I thank you for having heard me. 42 I knew that you always hear me, but I have said this for the sake of the crowd standing here, so that they may believe that you sent me." 43 When he had said this, he cried with a loud voice, "Lazarus, come out!" 44 The dead man came out, his hands and feet bound with strips of cloth, and his face wrapped in a cloth. Jesus said to them, "Unbind him, and let him go."

45 Many of the Jews therefore, who had come with Mary and had seen what Jesus did, believed in him. 46 But some of them went to the Pharisees and told them what he had done. 47 So the chief priests and the Pharisees called a meeting of the council, and said, "What are we to do? This man is performing many signs. 48 If we let him go on like this, everyone will believe in him, and the Romans will come and destroy both our holy place *j* and our nation." 49 But one of them, Caiaphas, who was high priest that year, said to them, "You know nothing at all! 50 You do not understand that it is better for you to have one man die for the people than to have the whole nation de-

h Other ancient authorities lack *and the life*
i Or *the Christ*   *j* Or *our temple*; Greek *our place*

---

faint hope, *even now*. **24:** Belief in *the resurrection on the last day* was widespread among pious Jews in Jesus' day.

11.25–26: Jesus is not only the agent of final resurrection, but gives eternal life now (Rom 6.4–5; Col 2.12; 3.1). The body dies, but the person *will never die*. **27:** Martha rests her perplexity on the mystery of Jesus' person as the Messiah. **32:** See v. 21. **33:** *Moved*, stirred with indignation, probably at the power of death (12.27). **34:** *Laid*, buried. **35:** Jesus' humanity was real.

11.39: *Dead four days*, popular belief imag-

ined that the soul lingered near the body for three days, then left. **40:** *Glory of God*, i.e. God acting to reveal his nature as lifegiver. **41–42:** God hears even Jesus' unspoken thoughts, but Jesus wishes the people to know that he is no magician, but one sent from God.

11.46: *Some . . . went to the Pharisees*, this is a crucial turning point in Jesus' ministry. **47:** *Council*, the Sanhedrin, the official Jewish court, made up of seventy priests, scribes, and elders, presided over by the high priest. **49–53:** An unconscious prophecy with deep

stroyed." 51 He did not say this on his own, but being high priest that year he prophesied that Jesus was about to die for the nation, 52 and not for the nation only, but to gather into one the dispersed children of God. 53 So from that day on they planned to put him to death.

54 Jesus therefore no longer walked about openly among the Jews, but went from there to a town called Ephraim in the region near the wilderness; and he remained there with the disciples. 55 Now the Passover of the Jews was near, and many went up from the country to Jerusalem before the Passover to purify themselves. 56 They were looking for Jesus and were asking one another as they stood in the temple, "What do you think? Surely he will not come to the festival, will he?" 57 Now the chief priests and the Pharisees had given orders that anyone who knew where Jesus[k] was should let them know, so that they might arrest him.

12 Six days before the Passover Jesus came to Bethany, the home of Lazarus, whom he had raised from the dead. 2 There they gave a dinner for him. Martha served, and Lazarus was one of those at the table with him. 3 Mary took a pound of costly perfume made of pure nard, anointed Jesus' feet, and wiped them[l] with her hair. The house was filled with the fragrance of the perfume. 4 But Judas Iscariot, one of his disciples (the one who was about to betray him), said, 5 "Why was this perfume not sold for three hundred denarii[m] and the money given to the poor?" 6 (He said this not because he cared about the poor, but be-

cause he was a thief; he kept the common purse and used to steal what was put into it.) 7 Jesus said, "Leave her alone. She bought it[n] so that she might keep it for the day of my burial. 8 You always have the poor with you, but you do not always have me."

9 When the great crowd of the Jews learned that he was there, they came not only because of Jesus but also to see Lazarus, whom he had raised from the dead. 10 So the chief priests planned to put Lazarus to death as well, 11 since it was on account of him that many of the Jews were deserting and were believing in Jesus.

12 The next day the great crowd that had come to the festival heard that Jesus was coming to Jerusalem. 13 So they took branches of palm trees and went out to meet him, shouting,

"Hosanna!
Blessed is the one who comes in
the name of the Lord—
the King of Israel!"

14 Jesus found a young donkey and sat on it; as it is written:

15 "Do not be afraid, daughter of
Zion.
Look, your king is coming,
sitting on a donkey's colt!"

16 His disciples did not understand these things at first; but when Jesus was glorified, then they remembered that these things had been written of him and had been done to him. 17 So the crowd that

k Gk *he*   l Gk *his feet*   m Three hundred denarii would be nearly a year's wages for a laborer   n Gk lacks *She bought it*

---

theological meaning—Jesus' death would redeem not only the Jews but believers in all nations.
11.54: *No longer . . . openly*, because the Sanhedrin had ordered his arrest (v. 57). *Ephraim*, about fifteen miles north of Jerusalem.
12.1–11: **The anointing at Bethany.** In the full knowledge of a plot against his life (11.53, 57), Jesus returns to the vicinity of Jerusalem. **1**: *Bethany*, 11.18. **3**: *A* (Roman) *pound*, twelve ounces. **4–5**: Judas' reaction is a sign of his defection. **5**: *Three hundred denar-*

ii, see note *m*. **7**: Jesus foresees his death; the anointing is the last rite in preparation for burial. **8**: Such spontaneous love will not neglect the poor.
12.12–19: **Palm Sunday.** Jesus dramatically manifests himself as the Messiah. **13**: *Hosanna*, see Mt 21.9 n. *In the name of*, with the authority of God. *King of Israel*, the Messiah. **14–15**: Zech 9.9. Warlike kings rode on horses and in chariots; the king of peace, *on a donkey's colt*. **16**: The cross, resurrection, ascension, and Pentecost clarified many Old Testament passages.

had been with him when he called Lazarus out of the tomb and raised him from the dead continued to testify. *o* 18 It was also because they heard that he had performed this sign that the crowd went to meet him. 19 The Pharisees then said to one another, "You see, you can do nothing. Look, the world has gone after him!"

20 Now among those who went up to worship at the festival were some Greeks. 21 They came to Philip, who was from Bethsaida in Galilee, and said to him, "Sir, we wish to see Jesus." 22 Philip went and told Andrew; then Andrew and Philip went and told Jesus. 23 Jesus answered them, "The hour has come for the Son of Man to be glorified. 24 Very truly, I tell you, unless a grain of wheat falls into the earth and dies, it remains just a single grain; but if it dies, it bears much fruit. 25 Those who love their life lose it, and those who hate their life in this world will keep it for eternal life. 26 Whoever serves me must follow me, and where I am, there will my servant be also. Whoever serves me, the Father will honor.

27 "Now my soul is troubled. And what should I say—'Father, save me from this hour'? No, it is for this reason that I have come to this hour. 28 Father, glorify your name." Then a voice came from heaven, "I have glorified it, and I will glorify it again." 29 The crowd standing there heard it and said that it was thunder. Others said, "An angel has spoken to him." 30 Jesus answered, "This voice has come for your sake, not for mine. 31 Now is the judgment of this world; now the ruler of this world will be driven out. 32 And I, when I am lifted up from the earth, will draw all people*p* to myself." 33 He said this to indicate the kind of death he was to die. 34 The crowd answered him, "We have heard from the law that the Messiah*q* remains forever. How can you say that the Son of Man must be lifted up? Who is this Son of Man?" 35 Jesus said to them, "The light is with you for a little longer. Walk while you have the light, so that the darkness may not overtake you. If you walk in the darkness, you do not know where you are going. 36 While you have the light, believe in the light, so that you may become children of light."

After Jesus had said this, he departed and hid from them. 37 Although he had performed so many signs in their presence, they did not believe in him. 38 This was to fulfill the word spoken by the prophet Isaiah:

"Lord, who has believed our
          message,
    and to whom has the arm of the
          Lord been revealed?"
39 And so they could not believe, because Isaiah also said,
40 "He has blinded their eyes
          and hardened their heart,
    so that they might not look with
          their eyes,
    and understand with their heart
          and turn—
    and I would heal them."
41 Isaiah said this because*r* he saw his glo-

o Other ancient authorities read *with him began to testify that he had called. . .from the dead*
p Other ancient authorities read *all things*
q Or *the Christ*   *r* Other ancient witnesses read *when*

12.20–50: Jesus' public ministry concludes. 20: *Greeks,* Gentiles. **21–22:** Both *Philip* ("lover of horses") and *Andrew* ("manly") were Greek names. Both were from *Bethsaida,* on the shore of the Sea of Galilee. **23–24:** *The hour,* see 2.4 n. **25–26:** Mt 10.39; Mk 8.35; Lk 9.24; 14.26. **12.27–30:** Mk 14.32–42. A *voice . . . from heaven* was a common sign of divine reassurance (Mk 1.11; 9.7; Acts 9.7). **30:** *For your sake,* that they might believe that Jesus came from God. **31–33:** Reality reverses appearances: Jesus' death judges the *world,* not him; defeats Satan, not Jesus; draws, not repels, *all people.* **34–36a:** All evidence is in; it is time to act! *Children of light,* see Lk 16.8 n. **12.36b–43:** Refusal to *believe* evidence (*so many signs*) induces spiritual blindness (Isa 6.9–10; Mt 13.14–15; see Acts 28.26 n.). **41:** *His glory,* the glory of Christ. **42:** *Many . . . authorities,* Nicodemus and Joseph of Arimathea are the only known ones.

ry and spoke about him. [42]Nevertheless many, even of the authorities, believed in him. But because of the Pharisees they did not confess it, for fear that they would be put out of the synagogue; [43]for they loved human glory more than the glory that comes from God.

44 Then Jesus cried aloud: "Whoever believes in me believes not in me but in him who sent me. [45]And whoever sees me sees him who sent me. [46]I have come as light into the world, so that everyone who believes in me should not remain in the darkness. [47]I do not judge anyone who hears my words and does not keep them, for I came not to judge the world, but to save the world. [48]The one who rejects me and does not receive my word has a judge; on the last day the word that I have spoken will serve as judge, [49]for I have not spoken on my own, but the Father who sent me has himself given me a commandment about what to say and what to speak. [50]And I know that his commandment is eternal life. What I speak, therefore, I speak just as the Father has told me."

13 Now before the festival of the Passover, Jesus knew that his hour had come to depart from this world and go to the Father. Having loved his own who were in the world, he loved them to the end. [2]The devil had already put it into the heart of Judas son of Simon Iscariot to betray him. And during supper [3]Jesus, knowing that the Father had given all things into his hands, and that he had come from God and was going to God, [4]got up from the table,[s] took off his outer robe, and tied a towel around himself. [5]Then he poured water into a basin and began to wash the disciples' feet and to wipe them with the towel that was tied around him. [6]He came to Simon Peter, who said to him, "Lord, are you

going to wash my feet?" [7]Jesus answered, "You do not know now what I am doing, but later you will understand." [8]Peter said to him, "You will never wash my feet." Jesus answered, "Unless I wash you, you have no share with me." [9]Simon Peter said to him, "Lord, not my feet only but also my hands and my head!" [10]Jesus said to him, "One who has bathed does not need to wash, except for the feet,[t] but is entirely clean. And you[u] are clean, though not all of you." [11]For he knew who was to betray him; for this reason he said, "Not all of you are clean."

12 After he had washed their feet, had put on his robe, and had returned to the table, he said to them, "Do you know what I have done to you? [13]You call me Teacher and Lord—and you are right, for that is what I am. [14]So if I, your Lord and Teacher, have washed your feet, you also ought to wash one another's feet. [15]For I have set you an example, that you also should do as I have done to you. [16]Very truly, I tell you, servants[v] are not greater than their master, nor are messengers greater than the one who sent them. [17]If you know these things, you are blessed if you do them. [18]I am not speaking of all of you; I know whom I have chosen. But it is to fulfill the scripture, 'The one who ate my bread[w] has lifted his heel against me.' [19]I tell you this now, before it occurs, so that when it does occur, you may believe that I am he.[x] [20]Very truly, I tell you, whoever receives one whom I send receives me; and whoever receives me receives him who sent me."

s *Gk from supper*   t Other ancient authorities lack *except for the feet*   u The Greek word for *you* here is plural   v *Gk slaves*   w Other ancient authorities read *ate bread with me*   x *Gk I am*

12.44–50: Summary of Jesus' teaching.
13.1–38: The Last Supper. Preparation for the teaching (chs 14–17) and events (chs 18–21) to follow. **1:** *To the end,* the utmost. **5:** Lk 22.27. **6–9:** Though it seems incongruous, Peter must let Jesus *wash* him, the

reason made plain *later* in the cross. **13.11:** Outward washing alone does not cleanse (Lk 11.39–41; Heb 10.22). **15:** 1 Pet 2.21. **16:** Mt 10.24; Lk 6.40. **17:** Lk 11.28; Jas 1.25. **18:** Ps 41.9.

21 After saying this Jesus was troubled in spirit, and declared, "Very truly, I tell you, one of you will betray me." 22 The disciples looked at one another, uncertain of whom he was speaking. 23 One of his disciples—the one whom Jesus loved—was reclining next to him; 24 Simon Peter therefore motioned to him to ask Jesus of whom he was speaking. 25 So while reclining next to Jesus, he asked him, "Lord, who is it?" 26 Jesus answered, "It is the one to whom I give this piece of bread when I have dipped it in the dish." *y* So when he had dipped the piece of bread, he gave it to Judas son of Simon Iscariot. *z* 27 After he received the piece of bread, *a* Satan entered into him. Jesus said to him, "Do quickly what you are going to do." 28 Now no one at the table knew why he said this to him. 29 Some thought that, because Judas had the common purse, Jesus was telling him, "Buy what we need for the festival"; or, that he should give something to the poor. 30 So, after receiving the piece of bread, he immediately went out. And it was night.

31 When he had gone out, Jesus said, "Now the Son of Man has been glorified, and God has been glorified in him. 32 If God has been glorified in him, *b* God will also glorify him in himself and will glorify him at once. 33 Little children, I am with you only a little longer. You will look for me; and as I said to the Jews so now I say to you, 'Where I am going, you cannot come.' 34 I give you a new commandment, that you love one another. Just as I have loved you, you also should love one another. 35 By this everyone will know that you are my disciples, if you have love for one another."

36 Simon Peter said to him, "Lord, where are you going?" Jesus answered, "Where I am going, you cannot follow me now; but you will follow afterward." 37 Peter said to him, "Lord, why can I not follow you now? I will lay down my life for you." 38 Jesus answered, "Will you lay down your life for me? Very truly, I tell you, before the cock crows, you will have denied me three times.

14 "Do not let your hearts be troubled. Believe *c* in God, believe also in me. 2 In my Father's house there are many dwelling places. If it were not so, would I have told you that I go to prepare a place for you? *d* 3 And if I go and prepare a place for you, I will come again and will take you to myself, so that where I am, there you may be also. 4 And you know the way to the place where I am going." *e* 5 Thomas said to him, "Lord, we do not know where you are going. How can we know the way?" 6 Jesus said to him, "I am the way, and the truth, and the life. No one comes to the Father except through me. 7 If you know

y Gk *dipped it*   z Other ancient authorities read *Judas Iscariot son of Simon*; others, *Judas son of Simon from Karyot* (Kerioth)   a Gk *After the piece of bread*   b Other ancient authorities lack *If God has been glorified in him*   c Or *You believe*   d Or *If it were not so, I would have told you; for I go to prepare a place for you*   e Other ancient authorities read *Where I am going you know, and the way you know*

13.21–30: The betrayer must be dissuaded, or dismissed. Jesus honors him by seating him next to himself, handing him a *piece of bread* (Ruth 2.14), concealing his treachery from all but the beloved disciple. **13.31–35:** The death that Judas has gone to arrange will *glorify* (reveal the essence of) both Father and Son as holy love. The disciples are now the organ of this love. **33:** *Little children,* an expression of endearment found only here in the Gospels; see also 1 Jn 2.1, 12, 28; 3.7, 18; 4.4; 5.21. **36–38:** Peter is not yet ready to *follow* Jesus to death; *afterward,* according to tradition, he was martyred.
14.1–17.26: Jesus' farewell discourse and prayer; an interpretation of Jesus' completed work on earth and relation to both believers and the world after his resurrection and ascension. It is a meditation, which—like a love-letter—is difficult to outline.
14.1–31: The believers' relation to the glorified Christ; no separation, but deepened fellowship. **1:** Belief in God has new meaning in Jesus. **2–3:** For him to *go,* through death and resurrection, to his *Father's house* (with *dwelling places* for all) was to *prepare a place* of permanent fellowship with him (13.33, 36). **4–7:** Access to God is solely through Jesus (Mt 11.27; Jn 1.18; 6.46; Acts 4.12).

me, you will know*f* my Father also. From now on you do know him and have seen him."

8 Philip said to him, "Lord, show us the Father, and we will be satisfied." 9 Jesus said to him, "Have I been with you all this time, Philip, and you still do not know me? Whoever has seen me has seen the Father. How can you say, 'Show us the Father'? 10 Do you not believe that I am in the Father and the Father is in me? The words that I say to you I do not speak on my own; but the Father who dwells in me does his works. 11 Believe me that I am in the Father and the Father is in me; but if you do not, then believe me because of the works themselves. 12 Very truly, I tell you, the one who believes in me will also do the works that I do and, in fact, will do greater works than these, because I am going to the Father. 13 I will do whatever you ask in my name, so that the Father may be glorified in the Son. 14 If in my name you ask me*g* for anything, I will do it.

15 "If you love me, you will keep*h* my commandments. 16 And I will ask the Father, and he will give you another Advocate,*i* to be with you forever. 17 This is the Spirit of truth, whom the world cannot receive, because it neither sees him nor knows him. You know him, because he abides with you, and he will be in*j* you.

18 "I will not leave you orphaned; I am coming to you. 19 In a little while the world will no longer see me, but you will see me; because I live, you also will live. 20 On that day you will know that I am in my Father, and you in me, and I in you. 21 They who have my commandments and keep them are those who love me; and those who love me will be loved

by my Father, and I will love them and reveal myself to them." 22 Judas (not Iscariot) said to him, "Lord, how is it that you will reveal yourself to us, and not to the world?" 23 Jesus answered him, "Those who love me will keep my word, and my Father will love them, and we will come to them and make our home with them. 24 Whoever does not love me does not keep my words; and the word that you hear is not mine, but is from the Father who sent me.

25 "I have said these things to you while I am still with you. 26 But the Advocate,*i* the Holy Spirit, whom the Father will send in my name, will teach you everything, and remind you of all that I have said to you. 27 Peace I leave with you; my peace I give to you. I do not give to you as the world gives. Do not let your hearts be troubled, and do not let them be afraid. 28 You heard me say to you, 'I am going away, and I am coming to you.' If you loved me, you would rejoice that I am going to the Father, because the Father is greater than I. 29 And now I have told you this before it occurs, so that when it does occur, you may believe. 30 I will no longer talk much with you, for the ruler of this world is coming. He has no power over me; 31 but I do as the Father has commanded me, so that the world may know that I love the Father. Rise, let us be on our way.

15 "I am the true vine, and my Father is the vinegrower. 2 He removes every branch in me that bears no fruit. Every branch that bears fruit he

f Other ancient authorities read *If you had known me, you would have known*   *g* Other ancient authorities lack *me*   *h* Other ancient authorities read *me, keep*   *i* Or *Helper*
j Or *among*

14.8–11: Knowledge of God is solely through the person, *words*, and *works* of Jesus. **12–17**: *Greater works* (of a more exalted nature because redemption is achieved) will be done by the believers through prayer (vv. 13, 15), obedience (v. 15), and the Holy Spirit (*Advocate*, vv. 16, 17).
14.18–20: The Spirit imparts Christ's life (Acts 2.33) and unites the believer to God.

21–24: Fellowship with Christ is dependent on *love* which issues in obedience.
14.25–27: The *Holy Spirit* interprets Christ's teachings (v. 26), and imparts his peace (v. 27). **28–31**: To go to the Father meant Jesus' self-chosen conflict with *the ruler of this world* (i.e. Satan), whose *power* would be broken by death and resurrection.
15.1–27: **The pattern of the Christian**

prunes*k* to make it bear more fruit. ³You have already been cleansed*k* by the word that I have spoken to you. ⁴Abide in me as I abide in you. Just as the branch cannot bear fruit by itself unless it abides in the vine, neither can you unless you abide in me. ⁵I am the vine, you are the branches. Those who abide in me and I in them bear much fruit, because apart from me you can do nothing. ⁶Whoever does not abide in me is thrown away like a branch and withers; such branches are gathered, thrown into the fire, and burned. ⁷If you abide in me, and my words abide in you, ask for whatever you wish, and it will be done for you. ⁸My Father is glorified by this, that you bear much fruit and become*l* my disciples. ⁹As the Father has loved me, so I have loved you; abide in my love. ¹⁰If you keep my commandments, you will abide in my love, just as I have kept my Father's commandments and abide in his love. ¹¹I have said these things to you so that my joy may be in you, and that your joy may be complete.

12 "This is my commandment, that you love one another as I have loved you. ¹³No one has greater love than this, to lay down one's life for one's friends. ¹⁴You are my friends if you do what I command you. ¹⁵I do not call you servants*m* any longer, because the servant*n* does not know what the master is doing; but I have called you friends, because I have made known to you everything that I have heard from my Father. ¹⁶You did not choose me but I chose you. And I appointed you to go and bear fruit, fruit that will last, so that the Father will give

you whatever you ask him in my name. ¹⁷I am giving you these commands so that you may love one another.

18 "If the world hates you, be aware that it hated me before it hated you. ¹⁹If you belonged to the world,*o* the world would love you as its own. Because you do not belong to the world, but I have chosen you out of the world—therefore the world hates you. ²⁰Remember the word that I said to you, 'Servants*p* are not greater than their master.' If they persecuted me, they will persecute you; if they kept my word, they will keep yours also. ²¹But they will do all these things to you on account of my name, because they do not know him who sent me. ²²If I had not come and spoken to them, they would not have sin; but now they have no excuse for their sin. ²³Whoever hates me hates my Father also. ²⁴If I had not done among them the works that no one else did, they would not have sin. But now they have seen and hated both me and my Father. ²⁵It was to fulfill the word that is written in their law, 'They hated me without a cause.'

26 "When the Advocate*q* comes, whom I will send to you from the Father, the Spirit of truth who comes from the Father, he will testify on my behalf. ²⁷You also are to testify because you have been with me from the beginning.

16 "I have said these things to you to keep you from stumbling. ²They will put you out of the synagogues. In-

---

k The same Greek root refers to pruning and cleansing   *l* Or *be*   *m* Gk *slaves*
n Gk *slave*   *o* Gk *were of the world*
p Gk *Slaves*   *q* Or *Helper*

---

believer's life. Three dimensions are set forth: (*a*) **1–11**: The believer's relation to Christ—*abide*. As *the true vine* Jesus was the true Israel, fulfilling the vocation in which the old Israel had failed (Isa 5.1–7; Jer 2.21; Ezek 19.10–14). The *fruit* bearing (Gal 5.22–23) of the new Israel (the church) springs from union (actual incorporation) with him (v. 5), through prayer (v. 7), and loving obedience (vv. 9, 10), issuing in *joy* (v. 11).

15.12–17: (*b*) The relation of believers to one another—*love*. The measure is determined

by Jesus' death (v. 13). Fellowship with Jesus (vv. 14, 15), *fruit* bearing, and prayer (v. 16), are all dependent on obeying his *commands* to *love* (v. 17).

15.18–27: (*c*) The believer's relation to the world—to be separate from it (the world hates the church because it hates Christ who has judged it, vv. 18–25; Ps 35.19; 69.4), and to *testify on . . . behalf* of Christ in the power of the Holy *Spirit* (vv. 26–27; Acts 1.21–22; 5.32).

deed, an hour is coming when those who kill you will think that by doing so they are offering worship to God. 3 And they will do this because they have not known the Father or me. 4 But I have said these things to you so that when their hour comes you may remember that I told you about them.

"I did not say these things to you from the beginning, because I was with you. 5 But now I am going to him who sent me; yet none of you asks me, 'Where are you going?' 6 But because I have said these things to you, sorrow has filled your hearts. 7 Nevertheless I tell you the truth: it is to your advantage that I go away, for if I do not go away, the Advocate^r will not come to you; but if I go, I will send him to you. 8 And when he comes, he will prove the world wrong about^s sin and righteousness and judgment: 9 about sin, because they do not believe in me; 10 about righteousness, because I am going to the Father and you will see me no longer; 11 about judgment, because the ruler of this world has been condemned.

12 "I still have many things to say to you, but you cannot bear them now. 13 When the Spirit of truth comes, he will guide you into all the truth; for he will not speak on his own, but will speak whatever he hears, and he will declare to you the things that are to come. 14 He will glorify me, because he will take what is mine and declare it to you. 15 All that the Father has is mine. For this reason I said that he will take what is mine and declare it to you.

16 "A little while, and you will no longer see me, and again a little while, and you will see me." 17 Then some of his disciples said to one another, "What does he mean by saying to us, 'A little while, and you will no longer see me, and again a little while, and you will see me'; and 'Because I am going to the Father'?" 18 They said, "What does he mean by this 'a little while'? We do not know what he is talking about." 19 Jesus knew that they wanted to ask him, so he said to them, "Are you discussing among yourselves what I meant when I said, 'A little while, and you will no longer see me, and again a little while, and you will see me'? 20 Very truly, I tell you, you will weep and mourn, but the world will rejoice; you will have pain, but your pain will turn into joy. 21 When a woman is in labor, she has pain, because her hour has come. But when her child is born, she no longer remembers the anguish because of the joy of having brought a human being into the world. 22 So you have pain now; but I will see you again, and your hearts will rejoice, and no one will take your joy from you. 23 On that day you will ask nothing of me.^t Very truly, I tell you, if you ask anything of the Father in my name, he will give it to you.^u 24 Until now you have not asked for anything in my name. Ask and you will receive, so that your joy may be complete.

25 "I have said these things to you in figures of speech. The hour is coming when I will no longer speak to you in

r Or *Helper*   s Or *convict the world of*
t Or *will ask me no question*   u Other ancient authorities read *Father, he will give it to you in my name*

16.1–33: The Christian's relation to the world. 1–4a: Forewarning of conflict. It is to be expected that the world, even the religious world, will persecute the followers of Christ (vv. 2–3; Acts 22.3–5; 26.9–11). **16.4b–11:** The work of the *Advocate* (Holy Spirit) through the church. *Sorrow* at Jesus' departure is transformed by *the truth* that his death and resurrection make possible the Spirit's work (vv. 6–7). The Spirit *will prove the world wrong about sin* (unbelief in Jesus, v. 9), *about righteousness* (revealed in the cross, v. 10), *about judgment* (triumph over evil, v. 11); see also 12.31; 14.30; 1 Cor 2.8; Col 2.15. **16.12–15:** The guidance of the *Spirit* into the full *truth* about the historic Jesus (*will glorify*, will reveal the essential nature of). **16.16–24:** Temporary *pain* over Jesus' death yields to *joy* over his resurrection and abiding presence (vv. 20–22). This is sustained by prayer in Jesus' *name* (nature, all that the cross and resurrection reveal him to be, vv. 23–24). **16.25–33:** The pledge of triumph. Jesus' *going to the Father* (v. 28) makes plain all his

figures, but will tell you plainly of the Father. 26On that day you will ask in my name. I do not say to you that I will ask the Father on your behalf; 27for the Father himself loves you, because you have loved me and have believed that I came from God.*v* 28I came from the Father and have come into the world; again, I am leaving the world and am going to the Father."

29 His disciples said, "Yes, now you are speaking plainly, not in any figure of speech! 30Now we know that you know all things, and do not need to have anyone question you; by this we believe that you came from God." 31Jesus answered them, "Do you now believe? 32The hour is coming, indeed it has come, when you will be scattered, each one to his home, and you will leave me alone. Yet I am not alone because the Father is with me. 33I have said this to you, so that in me you may have peace. In the world you face persecution. But take courage; I have conquered the world!"

17 After Jesus had spoken these words, he looked up to heaven and said, "Father, the hour has come; glorify your Son so that the Son may glorify you, 2since you have given him authority over all people,*w* to give eternal life to all whom you have given him. 3And this is eternal life, that they may know you, the only true God, and Jesus Christ whom you have sent. 4I glorified you on earth by finishing the work that you gave me to do. 5So now, Father, glorify me in your own presence with the glory that I had in your presence before the world existed.

6 "I have made your name known to those whom you gave me from the world. They were yours, and you gave them to me, and they have kept your word. 7Now they know that everything you have given me is from you; 8for the words that you gave to me I have given to them, and they have received them and know in truth that I came from you; and they have believed that you sent me. 9I am asking on their behalf; I am not asking on behalf of the world, but on behalf of those whom you gave me, because they are yours. 10All mine are yours, and yours are mine; and I have been glorified in them. 11And now I am no longer in the world, but they are in the world, and I am coming to you. Holy Father, protect them in your name that you have given me, so that they may be one, as we are one. 12While I was with them, I protected them in your name that*x* you have given me. I guarded them, and not one of them was lost except the one destined to be lost,*y* so that the scripture might be fulfilled. 13But now I am coming to you, and I speak these things in the world so that they may have my joy made complete in themselves.*z* 14I have given them your word, and the world has hated them because they do not belong to the world, just as I do not belong to the world. 15I am not asking you to take them out of the world, but I ask you to protect them from the evil one.*a* 16They do not belong to the world, just as I do not belong to the world. 17Sanctify them in the truth; your word is truth. 18As you have

v Other ancient authorities read *the Father*
w Gk *flesh*   *x* Other ancient authorities read *protected in your name those whom*   *y* Gk *except the son of destruction*   *z* Or *among themselves*
a Or *from evil*

---

teachings, reveals God's love (v. 27), empowers prayer (v. 26), and offers *peace* even in *persecution* (v. 33). **32**: Mt 26.31; Mk 14.27; Zech 13.7. **33**: 14.27; 15.18; Rom 8.37; 2 Cor 2.14; Rev 3.21.
17.1–26: Jesus' high priestly prayer. This falls naturally into three parts: (*a*) **1–5**: Jesus' prayer for himself. The *hour* (see 2.4 n.) of Jesus' perfect obedience unto death *has come,* securing *eternal life* for *all people* through

knowledge (personal acquaintance) of God and his Son. *Finishing* his *work* (19.30), Jesus awaits the restoration of his pre-incarnate *glory.* (*b*) **6–19**: Jesus' prayer for his disciples, left *in the world* after his ascension (v. 11), is that they may *be one* as are the Father and the Son (v. 11), have *joy* (v. 13), be victorious over the *evil one* (v. 15), and fulfill their mission of representing Christ to *the world* (vv. 16–19). (*c*) **20–26**: Jesus' prayer for the church uni-

sent me into the world, so I have sent them into the world. 19 And for their sakes I sanctify myself, so that they also may be sanctified in truth.

20 "I ask not only on behalf of these, but also on behalf of those who will believe in me through their word, 21 that they may all be one. As you, Father, are in me and I am in you, may they also be in us, *b* so that the world may believe that you have sent me. 22 The glory that you have given me I have given them, so that they may be one, as we are one, 23 I in them and you in me, that they may become completely one, so that the world may know that you have sent me and have loved them even as you have loved me. 24 Father, I desire that those also, whom you have given me, may be with me where I am, to see my glory, which you have given me because you loved me before the foundation of the world.

25 "Righteous Father, the world does not know you, but I know you; and these know that you have sent me. 26 I made your name known to them, and I will make it known, so that the love with which you have loved me may be in them, and I in them."

18 After Jesus had spoken these words, he went out with his disciples across the Kidron valley to a place where there was a garden, which he and his disciples entered. 2 Now Judas, who betrayed him, also knew the place, because Jesus often met there with his disciples. 3 So Judas brought a detachment of soldiers together with police from the chief priests and the Pharisees, and they came there with lanterns and torches and weapons. 4 Then Jesus, knowing all that was to happen to him, came forward and asked them, "Whom are you looking for?" 5 They answered, "Jesus of Nazareth." *c* Jesus replied, "I am he." *d* Judas, who betrayed him, was standing with them. 6 When Jesus *e* said to them, "I am he," *d* they stepped back and fell to the ground. 7 Again he asked them, "Whom are you looking for?" And they said, "Jesus of Nazareth." *c* 8 Jesus answered, "I told you that I am he. *d* So if you are looking for me, let these men go." 9 This was to fulfill the word that he had spoken, "I did not lose a single one of those whom you gave me." 10 Then Simon Peter, who had a sword, drew it, struck the high priest's slave, and cut off his right ear. The slave's name was Malchus. 11 Jesus said to Peter, "Put your sword back into its sheath. Am I not to drink the cup that the Father has given me?"

12 So the soldiers, their officer, and the Jewish police arrested Jesus and bound him. 13 First they took him to Annas, who was the father-in-law of Caiaphas, the high priest that year. 14 Caiaphas was the one who had advised the Jews that it was better to have one person die for the people.

15 Simon Peter and another disciple followed Jesus. Since that disciple was known to the high priest, he went with Jesus into the courtyard of the high priest, 16 but Peter was standing outside at the gate. So the other disciple, who was known to the high priest, went out, spoke to the woman who guarded the gate, and brought Peter in. 17 The woman said to Peter, "You are not also one of this man's disciples, are you?" He said, "I am not." 18 Now the slaves and the police had made a charcoal fire because it was cold, and they were standing around it

---

b Other ancient authorities read *be one in us*
c Gk *the Nazorean*   *d* Gk *I am*   *e* Gk *he*

---

versal is that believers may be indwelt by the Father and the Son and express their unity in *love*, thus fulfilling its mission of leading *the world* to *believe*.

18.1–19.42: Arrest, trial, crucifixion, and burial of Jesus. 18.1: *Kidron valley*, between Jerusalem and the Mount of Olives. *A garden*, Gethsemane. **3:** Both Roman *soldiers* and the Jewish temple *police* made the arrest. **4:** Jesus' fate is self-chosen; he, not Judas or the soldiers, determines his death.

18.9: 6.39; 10.28; 17.12. **11:** *Cup,* see Lk 22.42 n. **13:** *Annas,* though deposed by the Romans in A.D. 15, was still the leading influence among the Jews through his son-in-law and four sons who succeeded him.

and warming themselves. Peter also was standing with them and warming himself.

19 Then the high priest questioned Jesus about his disciples and about his teaching. 20 Jesus answered, "I have spoken openly to the world; I have always taught in synagogues and in the temple, where all the Jews come together. I have said nothing in secret. 21 Why do you ask me? Ask those who heard what I said to them; they know what I said." 22 When he had said this, one of the police standing nearby struck Jesus on the face, saying, "Is that how you answer the high priest?" 23 Jesus answered, "If I have spoken wrongly, testify to the wrong. But if I have spoken rightly, why do you strike me?" 24 Then Annas sent him bound to Caiaphas the high priest.

25 Now Simon Peter was standing and warming himself. They asked him, "You are not also one of his disciples, are you?" He denied it and said, "I am not." 26 One of the slaves of the high priest, a relative of the man whose ear Peter had cut off, asked, "Did I not see you in the garden with him?" 27 Again Peter denied it, and at that moment the cock crowed.

28 Then they took Jesus from Caiaphas to Pilate's headquarters.*f* It was early in the morning. They themselves did not enter the headquarters,*f* so as to avoid ritual defilement and to be able to eat the Passover. 29 So Pilate went out to them and said, "What accusation do you bring against this man?" 30 They answered, "If this man were not a criminal, we would not have handed him over to you." 31 Pilate said to them, "Take him yourselves and judge him according to your law." The Jews replied, "We are not permitted to put anyone to death."

32 (This was to fulfill what Jesus had said when he indicated the kind of death he was to die.)

33 Then Pilate entered the headquarters*f* again, summoned Jesus, and asked him, "Are you the King of the Jews?" 34 Jesus answered, "Do you ask this on your own, or did others tell you about me?" 35 Pilate replied, "I am not a Jew, am I? Your own nation and the chief priests have handed you over to me. What have you done?" 36 Jesus answered, "My kingdom is not from this world. If my kingdom were from this world, my followers would be fighting to keep me from being handed over to the Jews. But as it is, my kingdom is not from here." 37 Pilate asked him, "So you are a king?" Jesus answered, "You say that I am a king. For this I was born, and for this I came into the world, to testify to the truth. Everyone who belongs to the truth listens to my voice." 38 Pilate asked him, "What is truth?"

After he had said this, he went out to the Jews again and told them, "I find no case against him. 39 But you have a custom that I release someone for you at the Passover. Do you want me to release for you the King of the Jews?" 40 They shouted in reply, "Not this man, but Barabbas!" Now Barabbas was a bandit.

19 Then Pilate took Jesus and had him flogged. 2 And the soldiers wove a crown of thorns and put it on his head, and they dressed him in a purple robe. 3 They kept coming up to him, saying, "Hail, King of the Jews!" and striking him on the face. 4 Pilate went out again and said to them, "Look, I am bringing him out to you to let you know

f Gk *the praetorium*

18.19–21: An informal trial, designed to indict Jesus for training disciples secretly as revolutionaries. **24:** He is sent to *Caiaphas* for formal trial before the Sanhedrin (Mt 26.57–75; Mk 14.53–72; Lk 22.54–71). **27:** 13.38. **18.28:** *Defilement*, entering a Gentile's house would make them ceremonially unclean. **29–31:** The Jews tried religious cases, but could not administer the death penalty.

32: Crucifixion (3.14; 12.32), rather than the Jewish method of stoning. **18.33–37:** The Jews charged Jesus with political treason. Jesus is king of *truth,* from God's world. **19.1–5:** Though Pilate found (18.38b) Jesus innocent of political insurrection, he has him *flogged.* **7:** Lev 24.16; Mk 14.61–64; Jn 5.18; 10.33. **9:** Unable to understand the

that I find no case against him." 5 So Jesus came out, wearing the crown of thorns and the purple robe. Pilate said to them, "Here is the man!" 6 When the chief priests and the police saw him, they shouted, "Crucify him! Crucify him!" Pilate said to them, "Take him yourselves and crucify him; I find no case against him." 7 The Jews answered him, "We have a law, and according to that law he ought to die because he has claimed to be the Son of God."

8 Now when Pilate heard this, he was more afraid than ever. 9 He entered his headquarters*g* again and asked Jesus, "Where are you from?" But Jesus gave him no answer. 10 Pilate therefore said to him, "Do you refuse to speak to me? Do you not know that I have power to release you, and power to crucify you?" 11 Jesus answered him, "You would have no power over me unless it had been given you from above; therefore the one who handed me over to you is guilty of a greater sin." 12 From then on Pilate tried to release him, but the Jews cried out, "If you release this man, you are no friend of the emperor. Everyone who claims to be a king sets himself against the emperor."

13 When Pilate heard these words, he brought Jesus outside and sat*h* on the judge's bench at a place called The Stone Pavement, or in Hebrew*i* Gabbatha. 14 Now it was the day of Preparation for the Passover; and it was about noon. He said to the Jews, "Here is your King!" 15 They cried out, "Away with him! Away with him! Crucify him!" Pilate asked them, "Shall I crucify your King?" The chief priests answered, "We have no king but the emperor." 16 Then he handed him over to them to be crucified.

So they took Jesus; 17 and carrying the cross by himself, he went out to what is called The Place of the Skull, which in Hebrew*i* is called Golgotha. 18 There they crucified him, and with him two others, one on either side, with Jesus between them. 19 Pilate also had an inscription written and put on the cross. It read, "Jesus of Nazareth,*j* the King of the Jews." 20 Many of the Jews read this inscription, because the place where Jesus was crucified was near the city; and it was written in Hebrew, *i* in Latin, and in Greek. 21 Then the chief priests of the Jews said to Pilate, "Do not write, 'The King of the Jews,' but, 'This man said, I am King of the Jews.' " 22 Pilate answered, "What I have written I have written." 23 When the soldiers had crucified Jesus, they took his clothes and divided them into four parts, one for each soldier. They also took his tunic; now the tunic was seamless, woven in one piece from the top. 24 So they said to one another, "Let us not tear it, but cast lots for it to see who will get it." This was to fulfill what the scripture says,

"They divided my clothes among
     themselves,
  and for my clothing they
          cast lots."

25 And that is what the soldiers did.

Meanwhile, standing near the cross of Jesus were his mother, and his mother's sister, Mary the wife of Clopas, and Mary Magdalene. 26 When Jesus saw his mother and the disciple whom he loved standing beside her, he said to his mother, "Woman, here is your son." 27 Then he said to the disciple, "Here is your

g Gk *the praetorium*   h Or *seated him*   i That is, *Aramaic*   j Gk *the Nazorean*

charge, Pilate is superstitiously afraid (Mt 27.19).
19.11: God controls evil, without setting aside human responsibility. The high priest, who *handed* Jesus *over* to Pilate, bore the greater responsibility. **12**: A threat of blackmail. **17**: *By himself,* until relieved by Simon of Cyrene (Mt 27.32; Mk 15.21; Lk 23.26). *Skull,* a place of skull-like appearance just outside the city walls (v. 20).

19.19–22: The trilingual caption expressed Pilate's contempt (v. 14). **23**: A Roman custom. *His clothes,* namely, head-dress; cloak or outer garment; belt; shoes; tunic or inner garment. *Four parts,* one of the above, in the order mentioned, leaving the *tunic* to be won by lot. **24**: Providence controlled even the soldiers' behavior (Ps 22.18). **26–27**: Indicates Jesus' real humanity and concern for human values.

mother." And from that hour the disciple took her into his own home.

28 After this, when Jesus knew that all was now finished, he said (in order to fulfill the scripture), "I am thirsty." 29 A jar full of sour wine was standing there. So they put a sponge full of the wine on a branch of hyssop and held it to his mouth. 30 When Jesus had received the wine, he said, "It is finished." Then he bowed his head and gave up his spirit.

31 Since it was the day of Preparation, the Jews did not want the bodies left on the cross during the sabbath, especially because that sabbath was a day of great solemnity. So they asked Pilate to have the legs of the crucified men broken and the bodies removed. 32 Then the soldiers came and broke the legs of the first and of the other who had been crucified with him. 33 But when they came to Jesus and saw that he was already dead, they did not break his legs. 34 Instead, one of the soldiers pierced his side with a spear, and at once blood and water came out. 35 (He who saw this has testified so that you also may believe. His testimony is true, and he knows *k* that he tells the truth.) 36 These things occurred so that the scripture might be fulfilled, "None of his bones shall be broken." 37 And again another passage of scripture says, "They will look on the one whom they have pierced."

38 After these things, Joseph of Arimathea, who was a disciple of Jesus, though a secret one because of his fear of the Jews, asked Pilate to let him take away the body of Jesus. Pilate gave him permission; so he came and removed his body. 39 Nicodemus, who had at first come to Jesus by night, also came, bringing a mixture of myrrh and aloes, weighing about a hundred pounds. 40 They took the body of Jesus and wrapped it with the spices in linen cloths, according to the burial custom of the Jews. 41 Now there was a garden in the place where he was crucified, and in the garden there was a new tomb in which no one had ever been laid. 42 And so, because it was the Jewish day of Preparation, and the tomb was nearby, they laid Jesus there.

20 Early on the first day of the week, while it was still dark, Mary Magdalene came to the tomb and saw that the stone had been removed from the tomb. 2 So she ran and went to Simon Peter and the other disciple, the one whom Jesus loved, and said to them, "They have taken the Lord out of the tomb, and we do not know where they have laid him." 3 Then Peter and the other disciple set out and went toward the tomb. 4 The two were running together, but the other disciple outran Peter and reached the tomb first. 5 He bent down to look in and saw the linen wrappings lying there, but he did not go in. 6 Then Simon Peter came, following him, and went into the tomb. He saw the linen wrappings lying there, 7 and the cloth that had been on Jesus' head, not lying with the linen wrappings but rolled up in a place by itself. 8 Then the other disciple,

k Or *there is one who knows*

19.28: *I am thirsty,* Ps 69.21. 30: *Finished,* all that God has sent him to do for the redemption of the world (17.4). 31: *A day of great solemnity,* especially holy since it fell on the first day of the Festival of Unleavened Bread. 32: *Broke the legs,* with a heavy mallet, to hasten death. 34: *Blood and water* indicate the reality of Jesus' humanity, and perhaps also the new covenant and baptism (Mk 14.24; 1 Cor 10.16; Jn 3.5; 1 Jn 5.6–8). 19.36: Jesus fulfills the passover (Ex 12.46; 1 Cor 5.7). 37: Zech 12.10. 38: *Joseph of Arimathea,* Mt 27.57–60; Mk 15.43; Lk 23.50–53. 39: *Nicodemus,* 3.1–15; 7.50–52. *Myrrh,* a resinous gum, which, when mixed with crushed or pounded *aloes,* was used for embalming. *A hundred* (Roman) *pounds,* about seventy-five pounds, avoirdupois weight.

20.1–31: **The resurrection. 1:** *First day,* Sunday. **2–3:** The empty tomb indicates actual resurrection, not mere immortality. **4:** *The other disciple* was younger. **6:** Peter shows characteristic boldness. **7:** Jesus' body had escaped without *the linen wrappings* being unwound. *The cloth,* which had been wrapped about *Jesus' head* (compare 11.44), lay apart, still *rolled up.* **8:** *Believed,* faith grasped the evidence that Jesus had not been resuscitated

who reached the tomb first, also went in, and he saw and believed; 9for as yet they did not understand the scripture, that he must rise from the dead. 10Then the disciples returned to their homes.

11 But Mary stood weeping outside the tomb. As she wept, she bent over to look[l] into the tomb; 12and she saw two angels in white, sitting where the body of Jesus had been lying, one at the head and the other at the feet. 13They said to her, "Woman, why are you weeping?" She said to them, "They have taken away my Lord, and I do not know where they have laid him." 14When she had said this, she turned around and saw Jesus standing there, but she did not know that it was Jesus. 15Jesus said to her, "Woman, why are you weeping? Whom are you looking for?" Supposing him to be the gardener, she said to him, "Sir, if you have carried him away, tell me where you have laid him, and I will take him away." 16Jesus said to her, "Mary!" She turned and said to him in Hebrew,[m] "Rabbouni!" (which means Teacher). 17Jesus said to her, "Do not hold on to me, because I have not yet ascended to the Father. But go to my brothers and say to them, 'I am ascending to my Father and your Father, to my God and your God.'" 18Mary Magdalene went and announced to the disciples, "I have seen the Lord"; and she told them that he had said these things to her.

19 When it was evening on that day, the first day of the week, and the doors of the house where the disciples had met were locked for fear of the Jews, Jesus came and stood among them and said, "Peace be with you." 20After he said this,

he showed them his hands and his side. Then the disciples rejoiced when they saw the Lord. 21Jesus said to them again, "Peace be with you. As the Father has sent me, so I send you." 22When he had said this, he breathed on them and said to them, "Receive the Holy Spirit. 23If you forgive the sins of any, they are forgiven them; if you retain the sins of any, they are retained."

24 But Thomas (who was called the Twin[n]), one of the twelve, was not with them when Jesus came. 25So the other disciples told him, "We have seen the Lord." But he said to them, "Unless I see the mark of the nails in his hands, and put my finger in the mark of the nails and my hand in his side, I will not believe."

26 A week later his disciples were again in the house, and Thomas was with them. Although the doors were shut, Jesus came and stood among them and said, "Peace be with you." 27Then he said to Thomas, "Put your finger here and see my hands. Reach out your hand and put it in my side. Do not doubt but believe." 28Thomas answered him, "My Lord and my God!" 29Jesus said to him, "Have you believed because you have seen me? Blessed are those who have not seen and yet have come to believe."

30 Now Jesus did many other signs in the presence of his disciples, which are not written in this book. 31But these are written so that you may come to believe[o] that Jesus is the Messiah,[p] the Son

---

l Gk lacks *to look* *m* That is, *Aramaic*
n Gk *Didymus* *o* Other ancient authorities read *may continue to believe* *p* Or *the Christ*

---

from a swoon, or stolen; he had been transformed without corruption into his resurrection body (Acts 2.24–31).

20.9: *The scripture,* the Old Testament (compare Lk 24.27, 32, 44–46; Acts 2.24–28).

20.16–17: The old title, *Teacher,* and Mary's effort to *hold on* to (cling) to him, were to be abandoned for the new relation with him as the *ascended* Lord (compare chs 14–17). **20**: *Hands, side,* identifying marks; also signs of glory through suffering (Lk 24.25–26). **22**: *He breathed on them,* the same image

that was used to describe the communication of the natural life (Gen 2.7) is here used to express the communication of the new, spiritual, life of re-created humanity. **23**: The church, having received the Spirit, embodies Christ's mission of forgiveness (see Mt 16.19 n.).

20.24–27: Thomas wanted visible proof. **28**: Climax of the book. **29**: Faith now rests on the apostolic testimony. **30–31**: Purpose of the Gospel according to John.

of God, and that through believing you may have life in his name.

21 After these things Jesus showed himself again to the disciples by the Sea of Tiberias; and he showed himself in this way. [2]Gathered there together were Simon Peter, Thomas called the Twin,[q] Nathanael of Cana in Galilee, the sons of Zebedee, and two others of his disciples. [3]Simon Peter said to them, "I am going fishing." They said to him, "We will go with you." They went out and got into the boat, but that night they caught nothing.

4 Just after daybreak, Jesus stood on the beach; but the disciples did not know that it was Jesus. [5]Jesus said to them, "Children, you have no fish, have you?" They answered him, "No." [6]He said to them, "Cast the net to the right side of the boat, and you will find some." So they cast it, and now they were not able to haul it in because there were so many fish. [7]That disciple whom Jesus loved said to Peter, "It is the Lord!" When Simon Peter heard that it was the Lord, he put on some clothes, for he was naked, and jumped into the sea. [8]But the other disciples came in the boat, dragging the net full of fish, for they were not far from the land, only about a hundred yards[r] off.

9 When they had gone ashore, they saw a charcoal fire there, with fish on it, and bread. [10]Jesus said to them, "Bring some of the fish that you have just caught." [11]So Simon Peter went aboard and hauled the net ashore, full of large fish, a hundred fifty-three of them; and though there were so many, the net was not torn. [12]Jesus said to them, "Come and have breakfast." Now none of the disciples dared to ask him, "Who are you?" because they knew it was the Lord. [13]Jesus came and took the bread and gave it to them, and did the same with the fish. [14]This was now the third time that Jesus appeared to the disciples after he was raised from the dead.

15 When they had finished breakfast, Jesus said to Simon Peter, "Simon son of John, do you love me more than these?" He said to him, "Yes, Lord; you know that I love you." Jesus said to him, "Feed my lambs." [16]A second time he said to him, "Simon son of John, do you love me?" He said to him, "Yes, Lord; you know that I love you." Jesus said to him, "Tend my sheep." [17]He said to him the third time, "Simon son of John, do you love me?" Peter felt hurt because he said to him the third time, "Do you love me?" And he said to him, "Lord, you know everything; you know that I love you." Jesus said to him, "Feed my sheep. [18]Very truly, I tell you, when you were younger, you used to fasten your own belt and to go wherever you wished. But when you grow old, you will stretch out your hands, and someone else will fasten a belt around you and take you where you do not wish to go." [19](He said this to indicate the kind of death by which he would glorify God.) After this he said to him, "Follow me."

20 Peter turned and saw the disciple whom Jesus loved following them; he was the one who had reclined next to Jesus at the supper and had said, "Lord,

q Gk Didymus   r Gk two hundred cubits

21.1–25: Epilogue. A post-resurrection appearance in Galilee (*Tiberias,* see 6.1 n.). **2–3:** Indicates Peter's natural leadership. **4–6:** Obedience to Jesus' command is rewarded. **5:** *Children,* a colloquialism perhaps meaning "lads."

21.9–14: Jesus' feeding of the disciples is a prelude to his command to Peter to *feed* others (vv. 15, 17). **14:** *The third time,* for the two previous appearances, see 20.19–23 and 26–29. **15–17:** *These,* other disciples (Mk

14.29). The triple question is reminiscent of Peter's triple denial (18.17, 25–27). **18–19:** According to tradition Peter was martyred under Nero at Rome about A.D. 64–68.

21.20–22: Each is to *follow* his Lord, regardless of others. *Until I come,* the second coming. **23:** A rumor that the End would come before the beloved disciple's death was falsely based. **24:** *We know,* confirmation by the elders of the church where the Gospel of John was written (see also 3 Jn v. 12).

who is it that is going to betray you?"
²¹When Peter saw him, he said to Jesus,
"Lord, what about him?" ²²Jesus said to
him, "If it is my will that he remain until
I come, what is that to you? Follow me!"
²³So the rumor spread in the community*s* that this disciple would not die. Yet
Jesus did not say to him that he would
not die, but, "If it is my will that he
remain until I come, what is that to
you?"*t*

24 This is the disciple who is testifying to these things and has written them,
and we know that his testimony is true.
²⁵But there are also many other things
that Jesus did; if every one of them were
written down, I suppose that the world
itself could not contain the books that
would be written.

s Gk *among the brothers*   *t* Other ancient
authorities lack *what is that to you*

The Acts
of the Apostles

The book of Acts continues the narrative of the Gospel according to Luke by tracing the story of the Christian movement from the resurrection of Jesus to the time when the apostle Paul was in Rome, proclaiming the gospel "with all boldness and without hindrance" (28.31). Most of the first half of Acts is occupied with the Jerusalem church, its leaders and relationships, while the latter half is dominated by Paul and his three missionary journeys, climaxed by his arrest and voyage to Rome. The progress of the narrative is mainly geographical: from Jerusalem the word spreads to Samaria (8.5), the seacoast (8.40), Damascus (9.10), Antioch (where the disciples were first called "Christians"; 11.26), Asia Minor (13.13), Europe (16.11), and finally Rome itself.

Like the four Gospels, Acts is an anonymous book. The tradition attributing the third Gospel and Acts to Luke, the companion for a time of Paul (Col 4.14; compare 2 Tim 4.11; Philem 24), begins in the latter part of the second century and remains constant thereafter.

About one-fifth of the book of Acts comprises the report of speeches and missionary discourses; of the latter there are six to Jewish audiences (2.14–39; 3.12–26; 4.9–12; 5.29–32; 10.34–43; 13.16–41), two to Gentiles (14.15–17; 17.22–31). The archaic flavor and Semitic idiom of the speeches to the Jews testify to Luke's skill in conveying a variety of emphases and nuances appropriate to the several speakers and circumstances.

Four portions in the latter part of the book suddenly fall into the first person plural (16.10–17; 20.5–15; 21.1–18; 27.1–28.16). It is possible that these "we-passages" —all of which begin or end with a sea voyage—come from a travel diary that Luke, as Paul's companion on these occasions, had drawn up at the time. It is also possible that they, like the contents of the letter of Claudius Lysias to Felix (23.26), are Luke's own free composition, drawn up in the manner similar to that of other ancient historians, who shaped speeches by prominent persons to illustrate the meaning of historical events. (See also the article on "Narrative Books," pp. viii–x NT.)

The date of the composition of Acts is disputed. Because there is no mention of the outcome of Paul's arrest (the apostle is awaiting trial at the close of the book), some have thought that the book was published prior to Paul's martyrdom under Nero, about A.D. 65–67. On the other hand, internal qualities that hint at the author's considerable degree of historical maturity in assessing the significance of the first thirty years of the church's history suggests a later date, perhaps in the 80s. Luke's purpose in writing was to awaken faith by showing the triumphant progress of the Good News and to defend Christians against the charge that they were destructive of Jewish institutions and a troublesome element in the empire. None of the judges and other authorities who hear Paul and other Christians find them guilty of anything wrong.

Another of the author's special interests, controlling the selection and presentation of details, was to show the activity of the Holy Spirit in the founding and development of the church. In fact, the book might appropriately be entitled "The Acts of the Holy Spirit," for the dominating theme is the power of the Spirit manifested in and through the members of the early church. But Luke had also an interest in history for its own sake and in the men and women of the story, in the details of lodging, entertainment, and travel, and all that constitutes local color. As a result his work entertains and pleases the reader while dealing with the spread of Christianity in selected portions of the Mediterranean world. From every point of view, the New Testament would be infinitely poorer without this first book of church history.

1 In the first book, Theophilus, I wrote about all that Jesus did and taught from the beginning [2]until the day when he was taken up to heaven, after giving instructions through the Holy Spirit to the apostles whom he had chosen. [3]After his suffering he presented himself alive to them by many convincing proofs, appearing to them during forty days and speaking about the kingdom of God. [4]While staying*a* with them, he ordered them not to leave Jerusalem, but to wait there for the promise of the Father. "This," he said, "is what you have heard from me; [5]for John baptized with water, but you will be baptized with*b* the Holy Spirit not many days from now."

[6] So when they had come together, they asked him, "Lord, is this the time when you will restore the kingdom to Israel?" [7]He replied, "It is not for you to know the times or periods that the Father has set by his own authority. [8]But you will receive power when the Holy Spirit has come upon you; and you will be my witnesses in Jerusalem, in all Judea and Samaria, and to the ends of the earth." [9]When he had said this, as they were watching, he was lifted up, and a cloud took him out of their sight. [10]While he was going and they were gazing up toward heaven, suddenly two men in white robes stood by them. [11]They said, "Men of Galilee, why do you stand looking up toward heaven? This Jesus, who has been taken up from you into heaven, will come in the same way as you saw him go into heaven."

[12] Then they returned to Jerusalem from the mount called Olivet, which is near Jerusalem, a sabbath day's journey away. [13]When they had entered the city, they went to the room upstairs where they were staying, Peter, and John, and James, and Andrew, Philip and Thomas, Bartholomew and Matthew, James son of Alphaeus, and Simon the Zealot, and Judas son of*c* James. [14]All these were constantly devoting themselves to prayer, together with certain women, including Mary the mother of Jesus, as well as his brothers.

[15] In those days Peter stood up among the believers*d* (together the crowd numbered about one hundred twenty persons) and said, [16]"Friends, *e* the scripture had to be fulfilled, which the Holy Spirit through David foretold concerning Judas, who became a guide for those who arrested Jesus— [17]for he was numbered among us and was allotted his share in this ministry." [18](Now this man acquired a field with the reward of his wickedness; and falling headlong,*f* he burst open in the middle and all his bowels gushed out. [19]This became known to all the residents of Jerusalem, so that the field was called in their language Hakeldama, that is, Field of

a Or *eating*   *b* Or *by*   *c* Or *the brother of*
d Gk *brothers*   *e* Gk Men, brothers
f Or *swelling up*

1.1–5: Introduction; the risen Christ. The author links his volume with *the first book*, the Gospel of Luke. **1**: *Theophilus,* "lover of God"; perhaps a Roman official to whom the two books were addressed, or any reader who loves God (see Lk 1.3 n.). **3**: Some of the *many proofs* are given in Lk 24.13–53. **4**: The Greek words for *staying* and *eating* (see note *a*) are identical. **5**: John the Baptist had predicted that the Messiah would baptize his people *with the Holy Spirit* (Mk 1.8; Mt 3.11; Lk 3.16; Jn 1.33). The idea is found also in the Dead Sea scrolls.
1.6–11: The ascension. Compare Lk 24.50–51. **6**: Since he was promised "the throne of his ancestor David" (Lk 1.32), the disciples would expect Jesus to *restore the kingdom to Israel*. **8**: For apostles as *witnesses* see Lk 24.48; Acts 1.22; 2.32.
1.12–26: The gathering of the Twelve. The gospels agree that the Eleven remained together; now the sacred number, corresponding to the tribes of Israel, is restored, in anticipation of the coming age (Lk 22.29–30). **12**: *Olivet,* the Mount of Olives was east of Jerusalem. *A sabbath day's journey,* a little more than half a mile. **13**: The same list as in Lk 6.14–16, but in a different order. **14**: *Brothers,* see Mt 13.55.
1.17: *Us,* the apostles. **18**: *Falling headlong,* literally "flat" or "prone," but here the meaning is uncertain; according to Mt 27.5, Judas

Blood.) 20"For it is written in the book of Psalms,

'Let his homestead become
        desolate,
    and let there be no one to live
        in it';

and

'Let another take his position of
        overseer.'

21So one of the men who have accompanied us during all the time that the Lord Jesus went in and out among us, 22beginning from the baptism of John until the day when he was taken up from us—one of these must become a witness with us to his resurrection." 23So they proposed two, Joseph called Barsabbas, who was also known as Justus, and Matthias. 24Then they prayed and said, "Lord, you know everyone's heart. Show us which one of these two you have chosen 25to take the place*g* in this ministry and apostleship from which Judas turned aside to go to his own place." 26And they cast lots for them, and the lot fell on Matthias; and he was added to the eleven apostles.

2 When the day of Pentecost had come, they were all together in one place. 2And suddenly from heaven there came a sound like the rush of a violent wind, and it filled the entire house where they were sitting. 3Divided tongues, as of fire, appeared among them, and a tongue rested on each of them. 4All of them were filled with the Holy Spirit and began to speak in other languages, as the Spirit gave them ability.

5 Now there were devout Jews from every nation under heaven living in Jerusalem. 6And at this sound the crowd gathered and was bewildered, because each one heard them speaking in the native language of each. 7Amazed and astonished, they asked, "Are not all these who are speaking Galileans? 8And how is it that we hear, each of us, in our own native language? 9Parthians, Medes, Elamites, and residents of Mesopotamia, Judea and Cappadocia, Pontus and Asia, 10Phrygia and Pamphylia, Egypt and the parts of Libya belonging to Cyrene, and visitors from Rome, both Jews and proselytes, 11Cretans and Arabs—in our own languages we hear them speaking about God's deeds of power." 12All were amazed and perplexed, saying to one another, "What does this mean?" 13But others sneered and said, "They are filled with new wine."

14 But Peter, standing with the eleven, raised his voice and addressed them, "Men of Judea and all who live in Jerusalem, let this be known to you, and listen to what I say. 15Indeed, these are not drunk, as you suppose, for it is only nine o'clock in the morning. 16No, this is what was spoken through the prophet Joel:

17 'In the last days it will be, God
        declares,
    that I will pour out my Spirit
        upon all flesh,
    and your sons and your
        daughters shall prophesy,
    and your young men shall
        see visions,
    and your old men shall
        dream dreams.

---

g Other ancient authorities read *the share*

---

hanged himself. *Their language,* Aramaic, a dialect related to Hebrew. **22:** *The baptism of John* is regarded as the beginning of the gospel (10.37; Mk 1.1–4). Every apostle is *a witness to* Jesus' *resurrection;* this includes Paul (1 Cor 15.8–9).
  2.1–47: The day of Pentecost (Lev 23.15–21). Jewish tradition held that the Law was given on this day, seven weeks after Passover. **1–13:** The gift of the Holy Spirit. **3:** John had promised a baptism of the Holy Spirit and *fire* (Lk 3.16). **4–11:** The *other languages* in the Corinthian church were an incoherent form of speech (1 Cor 14.1–33); here Luke thinks of a gift of foreign languages, as though the story of the tower of Babel (Gen 11.1–9) had been reversed. **11:** *Arabs,* i.e. men of Jewish descent who came from Arabia. **13:** *Filled with new wine* suggests the ecstatic utterance of 1 Cor ch 14.
  2.14–36: Peter's sermon. **17:** The gift of the Spirit to *all flesh,* and not just to chosen individuals, is a mark of the Messianic age (Joel 2.28–32). Like Paul (1 Cor 12.13), Acts

18 Even upon my slaves, both men
and women,
in those days I will pour out
my Spirit;
and they shall prophesy.

19 And I will show portents in the
heaven above
and signs on the earth below,
blood, and fire, and
smoky mist.

20 The sun shall be turned to
darkness
and the moon to blood,
before the coming of the
Lord's great and glorious
day.

21 Then everyone who calls on the
name of the Lord shall
be saved.'

22 "You that are Israelites,[h] listen to
what I have to say: Jesus of Nazareth,[i] a
man attested to you by God with deeds
of power, wonders, and signs that God
did through him among you, as you
yourselves know— 23 this man, handed
over to you according to the definite plan
and foreknowledge of God, you crucified and killed by the hands of those outside the law. 24 But God raised him up,
having freed him from death,[j] because it
was impossible for him to be held in its
power. 25 For David says concerning
him,

'I saw the Lord always before me,
for he is at my right hand so
that I will not be shaken;

26 therefore my heart was glad, and
my tongue rejoiced;
moreover my flesh will live
in hope.

27 For you will not abandon my soul
to Hades,
or let your Holy One experience
corruption.

28 You have made known to me the
ways of life;

you will make me full of
gladness with your
presence.'

29 "Fellow Israelites,[k] I may say to
you confidently of our ancestor David
that he both died and was buried, and his
tomb is with us to this day. 30 Since he
was a prophet, he knew that God had
sworn with an oath to him that he would
put one of his descendants on his throne.
31 Foreseeing this, David[l] spoke of the
resurrection of the Messiah,[m] saying,

'He was not abandoned to Hades,
nor did his flesh experience
corruption.'

32 This Jesus God raised up, and of that all
of us are witnesses. 33 Being therefore exalted at[n] the right hand of God, and having received from the Father the promise
of the Holy Spirit, he has poured out this
that you both see and hear. 34 For David
did not ascend into the heavens, but he
himself says,

'The Lord said to my Lord,
"Sit at my right hand,

35 until I make your enemies your
footstool." '

36 Therefore let the entire house of Israel
know with certainty that God has made
him both Lord and Messiah,[o] this Jesus
whom you crucified."

37 Now when they heard this, they
were cut to the heart and said to Peter and
to the other apostles, "Brothers,[k] what
should we do?" 38 Peter said to them,
"Repent, and be baptized every one of
you in the name of Jesus Christ so that
your sins may be forgiven; and you will
receive the gift of the Holy Spirit. 39 For
the promise is for you, for your children,
and for all who are far away, everyone
whom the Lord our God calls to him."

h Gk Men, Israelites   i Gk the Nazorean
j Gk the pains of death   k Gk Men, brothers
l Gk he   m Or the Christ   n Or by
o Or Christ

usually assumes that all Christians receive the
Spirit (10.44–48).
2.23: *Definite plan,* see Lk 24.26 n. **24**: *God
raised him up,* see 1 Cor 15.4–8. **25**: Ps 16.8–
11. **29**: *His tomb,* the tombs of the House of
David were among the notable features of

interest in Jerusalem. **30**: Ps 132.11. **31**: Ps
16.10. **36**: Early Christians believed that Jesus
was not only the *Messiah* on earth, but also the
heavenly *Lord;* vv. 34–35 use Ps 110.1 to
prove this.
2.37–42: The call to repentance. **39**: Isa

40 And he testified with many other arguments and exhorted them, saying, "Save yourselves from this corrupt generation." 41 So those who welcomed his message were baptized, and that day about three thousand persons were added. 42 They devoted themselves to the apostles' teaching and fellowship, to the breaking of bread and the prayers.

43 Awe came upon everyone, because many wonders and signs were being done by the apostles. 44 All who believed were together and had all things in common; 45 they would sell their possessions and goods and distribute the proceeds*p* to all, as any had need. 46 Day by day, as they spent much time together in the temple, they broke bread at home*q* and ate their food with glad and generous*r* hearts, 47 praising God and having the goodwill of all the people. And day by day the Lord added to their number those who were being saved.

3 One day Peter and John were going up to the temple at the hour of prayer, at three o'clock in the afternoon. 2 And a man lame from birth was being carried in. People would lay him daily at the gate of the temple called the Beautiful Gate so that he could ask for alms from those entering the temple. 3 When he saw Peter and John about to go into the temple, he asked them for alms. 4 Peter looked intently at him, as did John, and said, "Look at us." 5 And he fixed his attention on them, expecting to receive something from them. 6 But Peter said, "I have no silver or gold, but what I have

I give you; in the name of Jesus Christ of Nazareth, *s* stand up and walk." 7 And he took him by the right hand and raised him up; and immediately his feet and ankles were made strong. 8 Jumping up, he stood and began to walk, and he entered the temple with them, walking and leaping and praising God. 9 All the people saw him walking and praising God, 10 and they recognized him as the one who used to sit and ask for alms at the Beautiful Gate of the temple; and they were filled with wonder and amazement at what had happened to him.

11 While he clung to Peter and John, all the people ran together to them in the portico called Solomon's Portico, utterly astonished. 12 When Peter saw it, he addressed the people, "You Israelites, *t* why do you wonder at this, or why do you stare at us, as though by our own power or piety we had made him walk? 13 The God of Abraham, the God of Isaac, and the God of Jacob, the God of our ancestors has glorified his servant*u* Jesus, whom you handed over and rejected in the presence of Pilate, though he had decided to release him. 14 But you rejected the Holy and Righteous One and asked to have a murderer given to you, 15 and you killed the Author of life, whom God raised from the dead. To this we are witnesses. 16 And by faith in his name, his name itself has made this man

p Gk *them*   *q* Or *from house to house*
r Or *sincere*   *s* Gk *the Nazorean*   *t* Gk *Men,*
Israelites   *u* Or *child*

57.19; Joel 2.32. **42:** *The breaking of bread,* apparently a common meal which included the Lord's supper (see 1 Cor 11.17–34). **43–47:** Description of the early church. **44:** Jerusalem Christians for a time, like the Essenes (see Mt 3.7 n.), had everything in common (4.32–35), but 5.4 suggests that this was not a universal rule.
3.1–10: Healing at the Beautiful Gate. Two of the chief disciples manifest "the signs of a true apostle" (2 Cor 12.12; Mt 10.8). **1:** *Three o'clock in the afternoon,* when sacrifice was offered with prayer (Ex 29.39; Lev 6.20; Josephus, *Ant.* xiv.4.3). **2:** *The gate . . . called Beautiful* was made of polished Corinthian

bronze and ornamented with silver and gold; it was probably on the east side of the temple. **6:** Christians and others healed and cast out demons *in the name of Jesus* (19.13; Mk 9.38; Mt 7.22).
3.11–26: Peter's preaching. In a second sermon Peter gives a fuller account of the Christian message. **11:** *The portico called Solomon's* was probably on the east side of the temple. **13:** *Servant,* the word can be translated "child" (see note *u*), but in the Greek version of Isa 52.13 it is used for the suffering Servant of the Lord; compare the prayers in Acts 4.25, 27, 30. **15:** *Author,* the word can mean "pioneer" or "founder" (of a new city).

strong, whom you see and know; and the faith that is through Jesus*ᵛ* has given him this perfect health in the presence of all of you.

17 "And now, friends,*ʷ* I know that you acted in ignorance, as did also your rulers. 18 In this way God fulfilled what he had foretold through all the prophets, that his Messiah*ˣ* would suffer. 19 Repent therefore, and turn to God so that your sins may be wiped out, 20 so that times of refreshing may come from the presence of the Lord, and that he may send the Messiah*ʸ* appointed for you, that is, Jesus, 21 who must remain in heaven until the time of universal restoration that God announced long ago through his holy prophets. 22 Moses said, 'The Lord your God will raise up for you from your own people*ʷ* a prophet like me. You must listen to whatever he tells you. 23 And it will be that everyone who does not listen to that prophet will be utterly rooted out of the people.' 24 And all the prophets, as many as have spoken, from Samuel and those after him, also predicted these days. 25 You are the descendants of the prophets and of the covenant that God gave to your ancestors, saying to Abraham, 'And in your descendants all the families of the earth shall be blessed.' 26 When God raised up his servant,*ᶻ* he sent him first to you, to bless you by turning each of you from your wicked ways."

4 While Peter and John*ᵃ* were speaking to the people, the priests, the captain of the temple, and the Sadducees came to them, 2 much annoyed because they were teaching the people and proclaiming that in Jesus there is the resurrection of the dead. 3 So they arrested them and put them in custody until the next day, for it was already evening. 4 But many of those who heard the word believed; and they numbered about five thousand.

5 The next day their rulers, elders, and scribes assembled in Jerusalem, 6 with Annas the high priest, Caiaphas, John,*ᵇ* and Alexander, and all who were of the high-priestly family. 7 When they had made the prisoners*ᶜ* stand in their midst, they inquired, "By what power or by what name did you do this?" 8 Then Peter, filled with the Holy Spirit, said to them, "Rulers of the people and elders, 9 if we are questioned today because of a good deed done to someone who was sick and are asked how this man has been healed, 10 let it be known to all of you, and to all the people of Israel, that this man is standing before you in good health by the name of Jesus Christ of Nazareth,*ᵈ* whom you crucified, whom God raised from the dead. 11 This Jesus*ᵉ* is

'the stone that was rejected by
    you, the builders;
  it has become the cornerstone.'*ᶠ*
12 There is salvation in no one else, for there is no other name under heaven given among mortals by which we must be saved."

13 Now when they saw the boldness of Peter and John and realized that they were uneducated and ordinary men, they

v Gk *him*   *w* Gk *brothers*   *x* Or *his Christ*
y Or *the Christ*   *z* Or *child*   *a* Gk *While*
they   *b* Other ancient authorities read *Jonathan*
c Gk *them*   *d* Gk *the Nazorean*   *e* Gk *This*
f Or *keystone*

3.22: Jesus is successor of Moses as well as David (Deut 18.15–16). **23:** Deut 18.19; Lev 23.29. **25:** Gen 22.18.

4.1–31: Arrest and release of Peter and John. The first of many incidents in which the apostles defend the faith before the authorities. **1–2:** *Captain of the temple,* chief officer of the temple guard of priests and Levites. **3:** It was not legal to hold a judicial inquiry at night. The *priests* in control of *the temple* were usually Sadducees, who denied the *resurrection of the dead* (23.6–8). **5–6:** Perhaps the *rulers* are priests; with the *elders and scribes* they made up the Sanhedrin, or council, of which *the high priest* was head (see Jn 11.47 n.). *Annas* held this office A.D. 6–14, and *Caiaphas* was his son-in-law (Jn 18.13); *John* may be Jonathan, who succeeded Caiaphas. **11:** The stone which was rejected (Ps 118.22) is also identified with Jesus in Mk 12.10; 1 Pet 2.7. Essenes (see Mt 3.7 n.) and Christians used this and other Old Testament "stone" passages in teaching. **13:** Like Jesus (Jn 7.15), the apostles were considered *uneducated* because they

were amazed and recognized them as companions of Jesus. 14 When they saw the man who had been cured standing beside them, they had nothing to say in opposition. 15 So they ordered them to leave the council while they discussed the matter with one another. 16 They said, "What will we do with them? For it is obvious to all who live in Jerusalem that a notable sign has been done through them; we cannot deny it. 17 But to keep it from spreading further among the people, let us warn them to speak no more to anyone in this name." 18 So they called them and ordered them not to speak or teach at all in the name of Jesus. 19 But Peter and John answered them, "Whether it is right in God's sight to listen to you rather than to God, you must judge; 20 for we cannot keep from speaking about what we have seen and heard." 21 After threatening them again, they let them go, finding no way to punish them because of the people, for all of them praised God for what had happened. 22 For the man on whom this sign of healing had been performed was more than forty years old.

23 After they were released, they went to their friends*g* and reported what the chief priests and the elders had said to them. 24 When they heard it, they raised their voices together to God and said, "Sovereign Lord, who made the heaven and the earth, the sea, and everything in them, 25 it is you who said by the Holy Spirit through our ancestor David, your servant:*h*

'Why did the Gentiles rage,
     and the peoples imagine vain
          things?
26   The kings of the earth took their
          stand,
     and the rulers have gathered
          together

against the Lord and against
     his Messiah.'*i*
27 For in this city, in fact, both Herod and Pontius Pilate, with the Gentiles and the peoples of Israel, gathered together against your holy servant*h* Jesus, whom you anointed, 28 to do whatever your hand and your plan had predestined to take place. 29 And now, Lord, look at their threats, and grant to your servants*j* to speak your word with all boldness, 30 while you stretch out your hand to heal, and signs and wonders are performed through the name of your holy servant*h* Jesus." 31 When they had prayed, the place in which they were gathered together was shaken; and they were all filled with the Holy Spirit and spoke the word of God with boldness.

32 Now the whole group of those who believed were of one heart and soul, and no one claimed private ownership of any possessions, but everything they owned was held in common. 33 With great power the apostles gave their testimony to the resurrection of the Lord Jesus, and great grace was upon them all. 34 There was not a needy person among them, for as many as owned lands or houses sold them and brought the proceeds of what was sold. 35 They laid it at the apostles' feet, and it was distributed to each as any had need. 36 There was a Levite, a native of Cyprus, Joseph, to whom the apostles gave the name Barnabas (which means "son of encouragement"). 37 He sold a field that belonged to him, then brought the money, and laid it at the apostles' feet.

5 But a man named Ananias, with the consent of his wife Sapphira, sold a piece of property; 2 with his wife's knowledge, he kept back some of the

---

g Gk *their own*   *h* Or *child*   *i* Or *his Christ*
j Gk *slaves*

---

lacked rabbinical training. **25–26**: Ps 2.1–2. **27**: *Servant,* see 3.13 n. **29**: *Look at their threats,* a prayer to be shielded from harm.
    4.32–5.11: The sharing of goods. Christians took care of their needy (Rom 12.8;

1 Cor 13.3; Heb 13.16), but it was only in Jerusalem that this type of communal living (similar to that of the Essenes) was practiced for a time.

proceeds, and brought only a part and laid it at the apostles' feet. [3] "Ananias," Peter asked, "why has Satan filled your heart to lie to the Holy Spirit and to keep back part of the proceeds of the land? [4] While it remained unsold, did it not remain your own? And after it was sold, were not the proceeds at your disposal? How is it that you have contrived this deed in your heart? You did not lie to us[k] but to God!" [5] Now when Ananias heard these words, he fell down and died. And great fear seized all who heard of it. [6] The young men came and wrapped up his body,[l] then carried him out and buried him.

[7] After an interval of about three hours his wife came in, not knowing what had happened. [8] Peter said to her, "Tell me whether you and your husband sold the land for such and such a price." And she said, "Yes, that was the price." [9] Then Peter said to her, "How is it that you have agreed together to put the Spirit of the Lord to the test? Look, the feet of those who have buried your husband are at the door, and they will carry you out." [10] Immediately she fell down at his feet and died. When the young men came in they found her dead, so they carried her out and buried her beside her husband [11] And great fear seized the whole church and all who heard of these things.

[12] Now many signs and wonders were done among the people through the apostles. And they were all together in Solomon's Portico. [13] None of the rest dared to join them, but the people held them in high esteem. [14] Yet more than ever believers were added to the Lord, great numbers of both men and women, [15] so that they even carried out the sick into the streets, and laid them on cots and mats, in order that Peter's shadow might fall on some of them as he came by. [16] A great number of people would also gather from the towns around Jerusalem, bringing the sick and those tormented by unclean spirits, and they were all cured.

[17] Then the high priest took action; he and all who were with him (that is, the sect of the Sadducees), being filled with jealousy, [18] arrested the apostles and put them in the public prison. [19] But during the night an angel of the Lord opened the prison doors, brought them out, and said, [20] "Go, stand in the temple and tell the people the whole message about this life." [21] When they heard this, they entered the temple at daybreak and went on with their teaching.

When the high priest and those with him arrived, they called together the council and the whole body of the elders of Israel, and sent to the prison to have them brought. [22] But when the temple police went there, they did not find them in the prison; so they returned and reported, [23] "We found the prison securely locked and the guards standing at the doors, but when we opened them, we found no one inside." [24] Now when the captain of the temple and the chief priests heard these words, they were perplexed about them, wondering what might be going on. [25] Then someone arrived and announced, "Look, the men whom you put in prison are standing in the temple and teaching the people!" [26] Then the captain went with the temple police and brought them, but without violence, for they were afraid of being stoned by the people.

[27] When they had brought them, they had them stand before the council. The high priest questioned them, [28] saying, "We gave you strict orders not to teach in this name,[m] yet here you have

k Gk *to men*   l Meaning of Gk uncertain
m Other ancient authorities read *Did we not give you strict orders not to teach in this name?*

5.3: The apostles, or perhaps the church, represent *the Holy Spirit*. **4:** The property was at Ananias' *disposal* until he pretended to dedicate all his goods.
5.12–42: Second arrest of the apostles. The motif of obeying God rather than human beings, and the divided opinion of the council, are repeated (compare ch 4). **12:** *Solomon's Portico*, see 3.11 n. **13:** *None of the rest*, i.e. the non-believers. **16:** Compare Mk 1.32–34.
5.19: *An angel . . . opened the prison doors*, as in 12.6–11; compare 16.25–26. **21:** *Body of the*

filled Jerusalem with your teaching and you are determined to bring this man's blood on us." 29 But Peter and the apostles answered, "We must obey God rather than any human authority. *n* 30 The God of our ancestors raised up Jesus, whom you had killed by hanging him on a tree. 31 God exalted him at his right hand as Leader and Savior that he might give repentance to Israel and forgiveness of sins. 32 And we are witnesses to these things, and so is the Holy Spirit whom God has given to those who obey him."

33 When they heard this, they were enraged and wanted to kill them. 34 But a Pharisee in the council named Gamaliel, a teacher of the law, respected by all the people, stood up and ordered the men to be put outside for a short time. 35 Then he said to them, "Fellow Israelites, *o* consider carefully what you propose to do to these men. 36 For some time ago Theudas rose up, claiming to be somebody, and a number of men, about four hundred, joined him; but he was killed, and all who followed him were dispersed and disappeared. 37 After him Judas the Galilean rose up at the time of the census and got people to follow him; he also perished, and all who followed him were scattered. 38 So in the present case, I tell you, keep away from these men and let them alone; because if this plan or this undertaking is of human origin, it will fail; 39 but if it is of God, you will not be able to overthrow them—in that case you may even be found fighting against God!"

They were convinced by him, 40 and when they had called in the apostles, they had them flogged. Then they ordered them not to speak in the name of Jesus, and let them go. 41 As they left the council, they rejoiced that they were considered worthy to suffer dishonor for the sake of the name. 42 And every day in the temple and at home *p* they did not cease to teach and proclaim Jesus as the Messiah. *q*

6 Now during those days, when the disciples were increasing in number, the Hellenists complained against the Hebrews because their widows were being neglected in the daily distribution of food. 2 And the twelve called together the whole community of the disciples and said, "It is not right that we should neglect the word of God in order to wait on tables. *r* 3 Therefore, friends, *s* select from among yourselves seven men of good standing, full of the Spirit and of wisdom, whom we may appoint to this task, 4 while we, for our part, will devote ourselves to prayer and to serving the word." 5 What they said pleased the whole community, and they chose Stephen, a man full of faith and the Holy Spirit, together with Philip, Prochorus, Nicanor, Timon, Parmenas, and Nicolaus, a proselyte of Antioch. 6 They had these men stand before the apostles, who prayed and laid their hands on them.

n Gk *than men*   *o* Gk *Men, Israelites*
p Or *from house to house*   *q* Or *the Christ*
r Or *keep accounts*   *s* Gk *brothers*

---

elders, another word for *council* or Sanhedrin (see Jn 11.47 n.). **30:** The cross was a pole or *tree* to which a crossbeam was fixed. **31:** *Leader,* the word is translated *Author* in 3.15. *Savior,* a name given by the ancients to one who saves a city, rescues, or heals; the New Testament uses it of Jesus as healer and deliverer from sin and death.
5.34–37: *Gamaliel,* a famous liberal rabbi; the speech here fits his tolerant attitude. *Theudas,* according to Josephus (*Ant.* xx.5.1), raised his revolt later than this. *Judas the Galilean,* a revolutionary leader who in A.D. 6 or 7 opposed the imposition of new taxes following *the census.* **40:** *Had them flogged,* for

their disobedience to the Sanhedrin's first command (4.18). The penalty was probably thirty-nine stripes (see 2 Cor 11.24).
6.1–7: Choice of the Seven. These are traditionally regarded as the first deacons (Phil 1.1; 1 Tim 3.8–13), but their functions are more like those of presbyters or bishops (20.17, 28), for Stephen teaches (vv. 9–10) and does not merely *wait on tables* (v. 2). **1:** *The Hellenists,* Greek-speaking Jews or Jews who have adopted Greek customs; the *Hebrews* probably spoke Aramaic and were more conservative (compare 21.20). **2:** *To wait on tables,* perhaps to serve meals (Jesus thought this compatible with *the word of God,* Lk

7 The word of God continued to spread; the number of the disciples increased greatly in Jerusalem, and a great many of the priests became obedient to the faith.

8 Stephen, full of grace and power, did great wonders and signs among the people. 9 Then some of those who belonged to the synagogue of the Freedmen (as it was called), Cyrenians, Alexandrians, and others of those from Cilicia and Asia, stood up and argued with Stephen. 10 But they could not withstand the wisdom and the Spirit[t] with which he spoke. 11 Then they secretly instigated some men to say, "We have heard him speak blasphemous words against Moses and God." 12 They stirred up the people as well as the elders and the scribes; then they suddenly confronted him, seized him, and brought him before the council. 13 They set up false witnesses who said, "This man never stops saying things against this holy place and the law; 14 for we have heard him say that this Jesus of Nazareth[u] will destroy this place and will change the customs that Moses handed on to us." 15 And all who sat in the council looked intently at him, and they saw that his face was like the face of an angel.

7 Then the high priest asked him, "Are these things so?" 2 And Stephen replied:

"Brothers[v] and fathers, listen to me. The God of glory appeared to our ancestor Abraham when he was in Mesopotamia, before he lived in Haran, 3 and said to him, 'Leave your country and your relatives and go to the land that I will show you.' 4 Then he left the country of the Chaldeans and settled in Haran. After his father died, God had him move from there to this country in which you are now living. 5 He did not give him any of it as a heritage, not even a foot's length, but promised to give it to him as his possession and to his descendants after him, even though he had no child. 6 And God spoke in these terms, that his descendants would be resident aliens in a country belonging to others, who would enslave them and mistreat them during four hundred years. 7 'But I will judge the nation that they serve,' said God, 'and after that they shall come out and worship me in this place.' 8 Then he gave him the covenant of circumcision. And so Abraham[w] became the father of Isaac and circumcised him on the eighth day; and Isaac became the father of Jacob, and Jacob of the twelve patriarchs.

9 "The patriarchs, jealous of Joseph, sold him into Egypt; but God was with him, 10 and rescued him from all his afflictions, and enabled him to win favor and to show wisdom when he stood before Pharaoh, king of Egypt, who appointed him ruler over Egypt and over all his household. 11 Now there came a famine throughout Egypt and Canaan, and great suffering, and our ancestors could find no food. 12 But when Jacob heard that there was grain in Egypt, he sent our ancestors there on their first vis-

---

t Or spirit   u Gk the Nazorean   v Gk Men, brothers   w Gk he

---

22.27); the phrase can also include financial administration. **5:** Stephen, Philip, and the other names are Greek. Proselyte, a Gentile convert to Judaism, prior to becoming a Christian. **6:** Laid . . . hands, see 1 Tim 4.14 n.

6.8–8.1a: Preaching and martyrdom of Stephen. Like Jesus, Stephen was charged with prophesying against the temple, and forgave his persecutors.

6.9: Freedmen, former slaves, either Jews or proselytes; an inscription found in Jerusalem is thought to refer to this synagogue. Cilicia, the southeastern portion of Asia Minor. Asia, the Roman province of that name in western Asia Minor. **14:** Jesus . . . will destroy this place, Mk 14.58; Jn 2.19. Stephen saw more clearly than others that Jesus' teaching would change the customs (Mk 7.18–19; Mt 23.25–26; Lk 11.39–41). Christianity now begins to emerge as a world religion.

7.2–50: Early Christian preaching often included the stories of Abraham (v. 2), Joseph (v. 9), Moses (v. 20), and others (compare 13.16–24; Heb 11.4–40). **6:** Four hundred years, so Gen 15.13; more precisely, 430 years (Ex 12.40; see Gal 3.17 n.).

it. [13] On the second visit Joseph made himself known to his brothers, and Joseph's family became known to Pharaoh. [14] Then Joseph sent and invited his father Jacob and all his relatives to come to him, seventy-five in all; [15] so Jacob went down to Egypt. He himself died there as well as our ancestors, [16] and their bodies [x] were brought back to Shechem and laid in the tomb that Abraham had bought for a sum of silver from the sons of Hamor in Shechem.

[17] "But as the time drew near for the fulfillment of the promise that God had made to Abraham, our people in Egypt increased and multiplied [18] until another king who had not known Joseph ruled over Egypt. [19] He dealt craftily with our race and forced our ancestors to abandon their infants so that they would die. [20] At this time Moses was born, and he was beautiful before God. For three months he was brought up in his father's house; [21] and when he was abandoned, Pharaoh's daughter adopted him and brought him up as her own son. [22] So Moses was instructed in all the wisdom of the Egyptians and was powerful in his words and deeds.

[23] "When he was forty years old, it came into his heart to visit his relatives, the Israelites. [y] [24] When he saw one of them being wronged, he defended the oppressed man and avenged him by striking down the Egyptian. [25] He supposed that his kinsfolk would understand that God through him was rescuing them, but they did not understand. [26] The next day he came to some of them as they were quarreling and tried to reconcile them, saying, 'Men, you are brothers; why do you wrong each other?' [27] But the man who was wronging his neighbor pushed Moses [z] aside, saying, 'Who made you a ruler and a judge over us? [28] Do you want to kill me as you killed the Egyptian yesterday?' [29] When he heard this, Moses fled and became a resident alien in the land of Midian. There he became the father of two sons.

[30] "Now when forty years had passed, an angel appeared to him in the wilderness of Mount Sinai, in the flame of a burning bush. [31] When Moses saw it, he was amazed at the sight; and as he approached to look, there came the voice of the Lord: [32] 'I am the God of your ancestors, the God of Abraham, Isaac, and Jacob.' Moses began to tremble and did not dare to look. [33] Then the Lord said to him, 'Take off the sandals from your feet, for the place where you are standing is holy ground. [34] I have surely seen the mistreatment of my people who are in Egypt and have heard their groaning, and I have come down to rescue them. Come now, I will send you to Egypt.'

[35] "It was this Moses whom they rejected when they said, 'Who made you a ruler and a judge?' and whom God now sent as both ruler and liberator through the angel who appeared to him in the bush. [36] He led them out, having performed wonders and signs in Egypt, at the Red Sea, and in the wilderness for forty years. [37] This is the Moses who said to the Israelites, 'God will raise up a prophet for you from your own people [a] as he raised me up.' [38] He is the one who was in the congregation in the wilderness with the angel who spoke to him at Mount Sinai, and with our ancestors; and he received living oracles to give to us. [39] Our ancestors were unwilling to obey him; instead, they pushed him aside, and in their hearts they turned back to Egypt, [40] saying to Aaron, 'Make gods for us who will lead the way for us; as for this Moses who led us out from the land of Egypt, we do not know what has hap-

x Gk *they*   y Gk *his brothers, the sons of Israel*
z Gk *him*   a Gk *your brothers*

7.14: *Seventy-five,* according to the Septuagint of Gen 46.27; Ex 1.5. **16:** *Shechem;* but according to Gen 50.13 Jacob was buried at Hebron. *Abraham;* but according to Gen 33.19 and Josh 24.32 it was Jacob who bought the tomb at Shechem.
7.29: *Midian,* northwest Arabia. **37:** *A prophet,* 3.22.

pened to him.' ⁴¹ At that time they made a calf, offered a sacrifice to the idol, and reveled in the works of their hands. ⁴² But God turned away from them and handed them over to worship the host of heaven, as it is written in the book of the prophets:

'Did you offer to me slain victims
    and sacrifices
forty years in the wilderness,
    O house of Israel?
⁴³ No; you took along the tent
    of Moloch,
and the star of your god
    Rephan,
the images that you made
    to worship;
so I will remove you beyond
    Babylon.'

⁴⁴ "Our ancestors had the tent of testimony in the wilderness, as God ᵇ directed when he spoke to Moses, ordering him to make it according to the pattern he had seen. ⁴⁵ Our ancestors in turn brought it in with Joshua when they dispossessed the nations that God drove out before our ancestors. And it was there until the time of David, ⁴⁶ who found favor with God and asked that he might find a dwelling place for the house of Jacob. ᶜ ⁴⁷ But it was Solomon who built a house for him. ⁴⁸ Yet the Most High does not dwell in houses made with human hands; ᵈ as the prophet says,

⁴⁹ 'Heaven is my throne,
    and the earth is my footstool.
What kind of house will you build
    for me, says the Lord,
    or what is the place of my rest?

⁵⁰ Did not my hand make all these
    things?'

⁵¹ "You stiff-necked people, uncircumcised in heart and ears, you are forever opposing the Holy Spirit, just as your ancestors used to do. ⁵² Which of the prophets did your ancestors not persecute? They killed those who foretold the coming of the Righteous One, and now you have become his betrayers and murderers. ⁵³ You are the ones that received the law as ordained by angels, and yet you have not kept it."

⁵⁴ When they heard these things, they became enraged and ground their teeth at Stephen. ᵉ ⁵⁵ But filled with the Holy Spirit, he gazed into heaven and saw the glory of God and Jesus standing at the right hand of God. ⁵⁶ "Look," he said, "I see the heavens opened and the Son of Man standing at the right hand of God!" ⁵⁷ But they covered their ears, and with a loud shout all rushed together against him. ⁵⁸ Then they dragged him out of the city and began to stone him; and the witnesses laid their coats at the feet of a young man named Saul. ⁵⁹ While they were stoning Stephen, he prayed, "Lord Jesus, receive my spirit." ⁶⁰ Then he knelt down and cried out in a loud voice, "Lord, do not hold this sin against them." When he had said this, he died. ᶠ

8 ¹ And Saul approved of their killing him.

That day a severe persecution began against the church in Jerusalem, and all

b Gk *he*   *c* Other ancient authorities read *for the God of Jacob*   *d* Gk *with hands*   *e* Gk *him*   *f* Gk *fell asleep*

7.42–43: Am 5.25–27 is quoted to suggest that the Hebrews had always been idolaters. The *book of the* (twelve minor) *prophets* was thought of as a unit. **42**: *The host of heaven,* the stars (2 Kings 17.16). **43**: *Moloch,* a god of the Ammonites; children were offered to it as sacrifices. *Rephan,* the planet Saturn. **44–50**: Up to v. 42 Stephen has been on common ground with his hearers; now he argues that God *does not dwell in houses made with human hands* (Isa 66.1–2) and that it was wrong for Solomon to build *a house for him.* The Letter to the Hebrews (chs 8 and 9) likewise regards

the tent (tabernacle) as the true type of worship, and ignores the temple.
7.51: *Uncircumcised in heart and ears,* Ex 33.3, 5; Jer 9.26; Rom 2.29. **53**: *The law,* being *ordained by angels,* is considered valid; but Paul used this Jewish tradition to argue that the law is secondary (Gal 3.19). **56**: *Son of Man* in the gospels usually denotes Jesus as the glorified heavenly judge (see Mk 2.10 n.); elsewhere in the New Testament the phrase is found only here and in Rev 1.13. **58**: *The witnesses,* who were legally required to cast the first stones at the offender (Deut 17.7).

except the apostles were scattered throughout the countryside of Judea and Samaria. ²Devout men buried Stephen and made loud lamentation over him. ³But Saul was ravaging the church by entering house after house; dragging off both men and women, he committed them to prison.

4 Now those who were scattered went from place to place, proclaiming the word. ⁵Philip went down to the city*ᵍ* of Samaria and proclaimed the Messiah*ʰ* to them. ⁶The crowds with one accord listened eagerly to what was said by Philip, hearing and seeing the signs that he did, ⁷for unclean spirits, crying with loud shrieks, came out of many who were possessed; and many others who were paralyzed or lame were cured. ⁸So there was great joy in that city.

9 Now a certain man named Simon had previously practiced magic in the city and amazed the people of Samaria, saying that he was someone great. ¹⁰All of them, from the least to the greatest, listened to him eagerly, saying, "This man is the power of God that is called Great." ¹¹And they listened eagerly to him because for a long time he had amazed them with his magic. ¹²But when they believed Philip, who was proclaiming the good news about the kingdom of God and the name of Jesus Christ, they were baptized, both men and women. ¹³Even Simon himself believed. After being baptized, he stayed constantly with Philip and was amazed when he saw the signs and great miracles that took place.

14 Now when the apostles at Jerusa-

lem heard that Samaria had accepted the word of God, they sent Peter and John to them. ¹⁵The two went down and prayed for them that they might receive the Holy Spirit ¹⁶(for as yet the Spirit had not come*ⁱ* upon any of them; they had only been baptized in the name of the Lord Jesus). ¹⁷Then Peter and John*ʲ* laid their hands on them, and they received the Holy Spirit. ¹⁸Now when Simon saw that the Spirit was given through the laying on of the apostles' hands, he offered them money, ¹⁹saying, "Give me also this power so that anyone on whom I lay my hands may receive the Holy Spirit." ²⁰But Peter said to him, "May your silver perish with you, because you thought you could obtain God's gift with money! ²¹You have no part or share in this, for your heart is not right before God. ²²Repent therefore of this wickedness of yours, and pray to the Lord that, if possible, the intent of your heart may be forgiven you. ²³For I see that you are in the gall of bitterness and the chains of wickedness." ²⁴Simon answered, "Pray for me to the Lord, that nothing of what you*ᵏ* have said may happen to me."

25 Now after Peter and John*ˡ* had testified and spoken the word of the Lord, they returned to Jerusalem, proclaiming the good news to many villages of the Samaritans.

26 Then an angel of the Lord said to Philip, "Get up and go toward the south*ᵐ* to the road that goes down from Jerusalem to Gaza." (This is a wilderness

g Other ancient authorities read *a city*
h Or *the Christ*   i Gk *fallen*   j Gk *they*
k The Greek word for *you* and the verb *pray* are plural   l Gk *after they*   m Or *go at noon*

8.1b–40: Spread of the gospel to Samaria and the seacoast. Christianity now first reaches non-Jewish regions. **3:** *Saul was ravaging the church,* Gal 1.13. **5:** *Philip,* not the apostle, but one of the seven mentioned in 6.5. He now became an evangelist (21.8). *Samaria* was inhabited by mixed remnants of the northern tribes who worshiped the Lord God and used the Pentateuch. Jews despised them. In one tradition the disciples are forbidden to visit their towns (Mt 10.5), but according to

others Jesus was friendly to Samaritans (Lk 10.30–37; 17.11–19; Jn 4.4–42).

8.14–17: In Acts believers usually receive the Holy Spirit at baptism (2.38; 19.5–6), or before baptism (10.44). Here the Samaritans *received the Holy Spirit* (v. 17) some time subsequent to baptism, and only after the apostles' visit. **17:** *Laid . . . hands,* see 1 Tim 4.14 n. **18:** The term "simony" (buying church offices) is derived from this account.

8.27: *Ethiopian* at this time meant "Nubi-

road.) 27 So he got up and went. Now there was an Ethiopian eunuch, a court official of the Candace, queen of the Ethiopians, in charge of her entire treasury. He had come to Jerusalem to worship 28 and was returning home; seated in his chariot, he was reading the prophet Isaiah. 29 Then the Spirit said to Philip, "Go over to this chariot and join it." 30 So Philip ran up to it and heard him reading the prophet Isaiah. He asked, "Do you understand what you are reading?" 31 He replied, "How can I, unless someone guides me?" And he invited Philip to get in and sit beside him. 32 Now the passage of the scripture that he was reading was this:

"Like a sheep he was led to the
     slaughter,
   and like a lamb silent before
     its shearer,
     so he does not open his
       mouth.
33   In his humiliation justice was
       denied him.
   Who can describe his
       generation?
     For his life is taken away
       from the earth."

34 The eunuch asked Philip, "About whom, may I ask you, does the prophet say this, about himself or about someone else?" 35 Then Philip began to speak, and starting with this scripture, he proclaimed to him the good news about Jesus. 36 As they were going along the road, they came to some water; and the eunuch said, "Look, here is water! What is to prevent me from being baptized?" *n* 38 He commanded the chariot to stop,

and both of them, Philip and the eunuch, went down into the water, and Philip *o* baptized him. 39 When they came up out of the water, the Spirit of the Lord snatched Philip away; the eunuch saw him no more, and went on his way rejoicing. 40 But Philip found himself at Azotus, and as he was passing through the region, he proclaimed the good news to all the towns until he came to Caesarea.

9 Meanwhile Saul, still breathing threats and murder against the disciples of the Lord, went to the high priest 2 and asked him for letters to the synagogues at Damascus, so that if he found any who belonged to the Way, men or women, he might bring them bound to Jerusalem. 3 Now as he was going along and approaching Damascus, suddenly a light from heaven flashed around him. 4 He fell to the ground and heard a voice saying to him, "Saul, Saul, why do you persecute me?" 5 He asked, "Who are you, Lord?" The reply came, "I am Jesus, whom you are persecuting. 6 But get up and enter the city, and you will be told what you are to do." 7 The men who were traveling with him stood speechless because they heard the voice but saw no one. 8 Saul got up from the ground, and though his eyes were open, he could see nothing; so they led him by the hand and brought him into Damascus. 9 For three days he was without sight, and neither ate nor drank.

n Other ancient authorities add all or most of verse 37, And Philip said, "If you believe with all your heart, you may." And he replied, "I believe that Jesus Christ is the Son of God." o Gk he

---

an." *The Candace* was the title of the *queen of the Ethiopians.* **28:** *He was reading* aloud to himself (as was customary in antiquity); hence Philip *heard him* (v. 30). **30–35:** In the book of Isaiah the early Christians found many prophecies of Christ; Isa 53.7–8 deals with the servant of the Lord (see Acts 3.13 n., and compare Mt 8.17). **40:** *Caesarea,* important Palestinian seaport where the Roman procurator had his headquarters.
   9.1–22: Conversion of Saul of Tarsus. Slightly different versions are found in 22.4–

16; 26.9–18; compare Paul's account, Gal 1.13–17. **2:** *The Way,* i.e. the true way of the Lord, was one of the earliest names for Christianity. Those *who belonged to* it *at Damascus* were probably from Jerusalem; the empire granted the Jews the right to extradite offenders. **3:** The glory of God (or Christ) is often described as light (2 Cor 3.18; 4.6). **4–5:** In persecuting the disciples, he persecuted *Jesus* (Mt 10.40; 25.40). **7:** The Greek suggests that his companions heard the sound of the voice but not the words spoken (see Jn 12.29).

10 Now there was a disciple in Damascus named Ananias. The Lord said to him in a vision, "Ananias." He answered, "Here I am, Lord." 11 The Lord said to him, "Get up and go to the street called Straight, and at the house of Judas look for a man of Tarsus named Saul. At this moment he is praying, 12 and he has seen in a vision*p* a man named Ananias come in and lay his hands on him so that he might regain his sight." 13 But Ananias answered, "Lord, I have heard from many about this man, how much evil he has done to your saints in Jerusalem; 14 and here he has authority from the chief priests to bind all who invoke your name." 15 But the Lord said to him, "Go, for he is an instrument whom I have chosen to bring my name before Gentiles and kings and before the people of Israel; 16 I myself will show him how much he must suffer for the sake of my name." 17 So Ananias went and entered the house. He laid his hands on Saul*q* and said, "Brother Saul, the Lord Jesus, who appeared to you on your way here, has sent me so that you may regain your sight and be filled with the Holy Spirit." 18 And immediately something like scales fell from his eyes, and his sight was restored. Then he got up and was baptized, 19 and after taking some food, he regained his strength.

For several days he was with the disciples in Damascus, 20 and immediately he began to proclaim Jesus in the synagogues, saying, "He is the Son of God." 21 All who heard him were amazed and said, "Is not this the man who made havoc in Jerusalem among those who invoked this name? And has he not come here for the purpose of bringing them bound before the chief priests?" 22 Saul became increasingly more powerful and confounded the Jews who lived in Damascus by proving that Jesus*r* was the Messiah.*s*

23 After some time had passed, the Jews plotted to kill him, 24 but their plot became known to Saul. They were watching the gates day and night so that they might kill him; 25 but his disciples took him by night and let him down through an opening in the wall,*t* lowering him in a basket.

26 When he had come to Jerusalem, he attempted to join the disciples; and they were all afraid of him, for they did not believe that he was a disciple. 27 But Barnabas took him, brought him to the apostles, and described for them how on the road he had seen the Lord, who had spoken to him, and how in Damascus he had spoken boldly in the name of Jesus. 28 So he went in and out among them in Jerusalem, speaking boldly in the name of the Lord. 29 He spoke and argued with the Hellenists; but they were attempting to kill him. 30 When the believers*u* learned of it, they brought him down to Caesarea and sent him off to Tarsus.

31 Meanwhile the church throughout Judea, Galilee, and Samaria had peace and was built up. Living in the fear of the Lord and in the comfort of the Holy Spirit, it increased in numbers.

32 Now as Peter went here and there among all the believers,*v* he came down also to the saints living in Lydda. 33 There

p Other ancient authorities lack *in a vision*
q Gk *him*   r Gk *that this*   s Or *the Christ*
t Gk *through the wall*   u Gk *brothers*
v Gk *all of them*

9.10: *Ananias,* evidently one of the leaders of the believers at Damascus (see 22.12). **11:** *The street called Straight,* perhaps Darb el-Mostakim, which runs through Damascus from east to west. **15:** Saul, like the prophets, was chosen for a special purpose (Jer 1.5; Gal 1.15); *instrument,* literally "vessel" (as in Rom 9.22–23; 2 Cor 4.7), emphasizes his domination by Christ.
9.23–31: Saul's first visit to Jerusalem.

In Gal 1.15–20 Paul implies that his first visit was three years after his conversion. **25:** *A basket,* of the large kind used for carrying merchandise overland (2 Cor 11.32–33). **29:** *Hellenists,* Greek-speaking Jews. **30:** *Tarsus,* Paul's home city (22.3).
9.32–43: Peter's journey to Lydda and Joppa. 32: *Lydda,* in the plain of Sharon (v. 35), about ten miles SE of Joppa (v. 36). **39:** *Widows,* those associated with Dorcas in

he found a man named Aeneas, who had been bedridden for eight years, for he was paralyzed. ³⁴Peter said to him, "Aeneas, Jesus Christ heals you; get up and make your bed!" And immediately he got up. ³⁵And all the residents of Lydda and Sharon saw him and turned to the Lord.

36 Now in Joppa there was a disciple whose name was Tabitha, which in Greek is Dorcas. ʷ She was devoted to good works and acts of charity. ³⁷At that time she became ill and died. When they had washed her, they laid her in a room upstairs. ³⁸Since Lydda was near Joppa, the disciples, who heard that Peter was there, sent two men to him with the request, "Please come to us without delay." ³⁹So Peter got up and went with them; and when he arrived, they took him to the room upstairs. All the widows stood beside him, weeping and showing tunics and other clothing that Dorcas had made while she was with them. ⁴⁰Peter put all of them outside, and then he knelt down and prayed. He turned to the body and said, "Tabitha, get up." Then she opened her eyes, and seeing Peter, she sat up. ⁴¹He gave her his hand and helped her up. Then calling the saints and widows, he showed her to be alive. ⁴²This became known throughout Joppa, and many believed in the Lord. ⁴³Meanwhile he stayed in Joppa for some time with a certain Simon, a tanner.

10 In Caesarea there was a man named Cornelius, a centurion of the Italian Cohort, as it was called. ²He was a devout man who feared God with all his household; he gave alms generously to the people and prayed constantly to God. ³One afternoon at about three o'clock he had a vision in which he clearly saw an angel of God coming in and saying to him, "Cornelius." ⁴He stared at him in terror and said, "What is it, Lord?" He answered, "Your prayers and your alms have ascended as a memorial before God. ⁵Now send men to Joppa for a certain Simon who is called Peter; ⁶he is lodging with Simon, a tanner, whose house is by the seaside." ⁷When the angel who spoke to him had left, he called two of his slaves and a devout soldier from the ranks of those who served him, ⁸and after telling them everything, he sent them to Joppa.

9 About noon the next day, as they were on their journey and approaching the city, Peter went up on the roof to pray. ¹⁰He became hungry and wanted something to eat; and while it was being prepared, he fell into a trance. ¹¹He saw the heaven opened and something like a large sheet coming down, being lowered to the ground by its four corners. ¹²In it were all kinds of four-footed creatures and reptiles and birds of the air. ¹³Then he heard a voice saying, "Get up, Peter; kill and eat." ¹⁴But Peter said, "By no means, Lord; for I have never eaten anything that is profane or unclean." ¹⁵The voice said to him again, a second time, "What God has made clean, you must not call profane." ¹⁶This happened three times, and the thing was suddenly taken up to heaven.

w The name Tabitha in Aramaic and the name Dorcas in Greek mean *a gazelle*

---

good works; see 1 Tim 5.3–16 n. **43**: *A tanner* was practically an outcast; Jewish law regarded the work as defiling, since it required working with animal carcasses, which were ritually unclean. That Peter *stayed* with Simon shows that he had begun to disregard Jewish practices.
10.1–48: The conversion of Cornelius. Luke ascribes to Peter the honor of converting the first Gentile; but see 11.19–21. **1**: *Italian Cohort,* probably the Cohors II. Italica Civium Romanorum. **2**: Cornelius *feared God,* i.e. worshiped him, but had not adopted

the Jewish religion. **3**: *About three o'clock,* the time of afternoon prayer.
10.9: *Noon,* the usual Roman time for luncheon. *The roof,* see Mk 2.4 n. **14**: *Profane,* ritually impure; only certain mammals, fish, and insects might be eaten (Lev 11). **15**: *God has made clean* all foods through Jesus' word (Mk 7.14–19), but Peter did not realize this until now; some Jewish Christians continued even later to insist on food laws (15.29; Col 2.21). **16**: *Three times,* for emphasis and warning.

17 Now while Peter was greatly puzzled about what to make of the vision that he had seen, suddenly the men sent by Cornelius appeared. They were asking for Simon's house and were standing by the gate. 18 They called out to ask whether Simon, who was called Peter, was staying there. 19 While Peter was still thinking about the vision, the Spirit said to him, "Look, three*x* men are searching for you. 20 Now get up, go down, and go with them without hesitation; for I have sent them." 21 So Peter went down to the men and said, "I am the one you are looking for; what is the reason for your coming?" 22 They answered, "Cornelius, a centurion, an upright and God-fearing man, who is well spoken of by the whole Jewish nation, was directed by a holy angel to send for you to come to his house and to hear what you have to say." 23 So Peter*y* invited them in and gave them lodging.

The next day he got up and went with them, and some of the believers*z* from Joppa accompanied him. 24 The following day they came to Caesarea. Cornelius was expecting them and had called together his relatives and close friends. 25 On Peter's arrival Cornelius met him, and falling at his feet, worshiped him. 26 But Peter made him get up, saying, "Stand up; I am only a mortal." 27 And as he talked with him, he went in and found that many had assembled; 28 and he said to them, "You yourselves know that it is unlawful for a Jew to associate with or to visit a Gentile; but God has shown me that I should not call anyone profane or unclean. 29 So when I was sent for, I came without objection. Now may I ask why you sent for me?"

30 Cornelius replied, "Four days ago

at this very hour, at three o'clock, I was praying in my house when suddenly a man in dazzling clothes stood before me. 31 He said, 'Cornelius, your prayer has been heard and your alms have been remembered before God. 32 Send therefore to Joppa and ask for Simon, who is called Peter; he is staying in the home of Simon, a tanner, by the sea.' 33 Therefore I sent for you immediately, and you have been kind enough to come. So now all of us are here in the presence of God to listen to all that the Lord has commanded you to say."

34 Then Peter began to speak to them: "I truly understand that God shows no partiality, 35 but in every nation anyone who fears him and does what is right is acceptable to him. 36 You know the message he sent to the people of Israel, preaching peace by Jesus Christ—he is Lord of all. 37 That message spread throughout Judea, beginning in Galilee after the baptism that John announced: 38 how God anointed Jesus of Nazareth with the Holy Spirit and with power; how he went about doing good and healing all who were oppressed by the devil, for God was with him. 39 We are witnesses to all that he did both in Judea and in Jerusalem. They put him to death by hanging him on a tree; 40 but God raised him on the third day and allowed him to appear, 41 not to all the people but to us who were chosen by God as witnesses, and who ate and drank with him after he rose from the dead. 42 He commanded us to preach to the people and to testify that he is the one ordained by God as judge of the living and the dead. 43 All the prophets testify about him that everyone who

x One ancient authority reads *two*; others lack the word *y* Gk *he* *z* Gk *brothers*

10.26: *I am only a mortal,* 14.15; Rev 19.10. **28**: Food laws made association with Gentiles difficult. *But God has shown me,* Peter now understands the meaning of the vision that had puzzled him earlier (v. 17). **30**: See v. 3 n. **34**: *No partiality,* literally "God accepts no one's face"; compare Rom 2.11 where Paul teaches that Jews have no special privilege.

10.36–43: This sums up the gospel, beginning with the baptism of John and ending with the statement that Christ is judge (3.20–21). **38**: *God anointed Jesus,* Lk 3.22; 4.14. **41**: Jesus was seen only by those *chosen* (Lk 24.48; Acts 1.8, 22). **42**: *Judge of the living and the dead,* 1 Pet 4.5; 2 Tim 4.1.

believes in him receives forgiveness of sins through his name."

44 While Peter was still speaking, the Holy Spirit fell upon all who heard the word. 45 The circumcised believers who had come with Peter were astounded that the gift of the Holy Spirit had been poured out even on the Gentiles, 46 for they heard them speaking in tongues and extolling God. Then Peter said, 47 "Can anyone withhold the water for baptizing these people who have received the Holy Spirit just as we have?" 48 So he ordered them to be baptized in the name of Jesus Christ. Then they invited him to stay for several days.

11 Now the apostles and the believers *a* who were in Judea heard that the Gentiles had also accepted the word of God. 2 So when Peter went up to Jerusalem, the circumcised believers *b* criticized him, 3 saying, "Why did you go to uncircumcised men and eat with them?" 4 Then Peter began to explain it to them, step by step, saying, 5 "I was in the city of Joppa praying, and in a trance I saw a vision. There was something like a large sheet coming down from heaven, being lowered by its four corners; and it came close to me. 6 As I looked at it closely I saw four-footed animals, beasts of prey, reptiles, and birds of the air. 7 I also heard a voice saying to me, 'Get up, Peter; kill and eat.' 8 But I replied, 'By no means, Lord; for nothing profane or unclean has ever entered my mouth.' 9 But a second time the voice answered from heaven, 'What God has made clean, you must not call profane.' 10 This happened three times; then everything was pulled up again to heaven. 11 At that very moment three men, sent to me from Caesarea, arrived at the house where we were. 12 The Spirit told me to go with them and

not to make a distinction between them and us. *c* These six brothers also accompanied me, and we entered the man's house. 13 He told us how he had seen the angel standing in his house and saying, 'Send to Joppa and bring Simon, who is called Peter; 14 he will give you a message by which you and your entire household will be saved.' 15 And as I began to speak, the Holy Spirit fell upon them just as it had upon us at the beginning. 16 And I remembered the word of the Lord, how he had said, 'John baptized with water, but you will be baptized with the Holy Spirit.' 17 If then God gave them the same gift that he gave us when we believed in the Lord Jesus Christ, who was I that I could hinder God?" 18 When they heard this, they were silenced. And they praised God, saying, "Then God has given even to the Gentiles the repentance that leads to life."

19 Now those who were scattered because of the persecution that took place over Stephen traveled as far as Phoenicia, Cyprus, and Antioch, and they spoke the word to no one except Jews. 20 But among them were some men of Cyprus and Cyrene who, on coming to Antioch, spoke to the Hellenists *d* also, proclaiming the Lord Jesus. 21 The hand of the Lord was with them, and a great number became believers and turned to the Lord. 22 News of this came to the ears of the church in Jerusalem, and they sent Barnabas to Antioch. 23 When he came and saw the grace of God, he rejoiced, and he exhorted them all to remain faithful to the Lord with steadfast devotion; 24 for

a Gk brothers   *b* Gk lacks *believers*   *c* Or *not to hesitate*   *d* Other ancient authorities read *Greeks*

10.44–48: They knew by the *speaking in tongues* (2.4–11) that *the Holy Spirit* fell before baptism (see 8.14–17 n.).

11.1–18: Peter's defense. The baptism of Gentiles requires explanation. **2:** *Circumcised believers* are conservative Jewish Christians, who object to Peter's associating with Gentiles (15.1–5; 21.20; Gal 2.12). **16:** See 1.5 n.

11.19–26: Mission to the Greeks in Antioch. Stephen's death and the ensuing *persecution* led to the Gentile mission (8.1b–4). **19:** *Antioch* on the Orontes, in Syria, the third largest city in the Roman Empire, ranking next to Alexandria. **20:** *Hellenists*, Greek-speaking Jews. **22:** *Barnabas* came from Cyprus (4.36) and there were Cypriots in Anti-

he was a good man, full of the Holy Spirit and of faith. And a great many people were brought to the Lord. 25 Then Barnabas went to Tarsus to look for Saul, 26 and when he had found him, he brought him to Antioch. So it was that for an entire year they met with*ᵉ* the church and taught a great many people, and it was in Antioch that the disciples were first called "Christians."

27 At that time prophets came down from Jerusalem to Antioch. 28 One of them named Agabus stood up and predicted by the Spirit that there would be a severe famine over all the world; and this took place during the reign of Claudius. 29 The disciples determined that according to their ability, each would send relief to the believers*ᶠ* living in Judea; 30 this they did, sending it to the elders by Barnabas and Saul.

12 About that time King Herod laid violent hands upon some who belonged to the church. 2 He had James, the brother of John, killed with the sword. 3 After he saw that it pleased the Jews, he proceeded to arrest Peter also. (This was during the festival of Unleavened Bread.) 4 When he had seized him, he put him in prison and handed him over to four squads of soldiers to guard him, intending to bring him out to the people after the Passover. 5 While Peter was kept in prison, the church prayed fervently to God for him.

6 The very night before Herod was going to bring him out, Peter, bound with two chains, was sleeping between two soldiers, while guards in front of the door were keeping watch over the prison. 7 Suddenly an angel of the Lord appeared and a light shone in the cell. He tapped Peter on the side and woke him, saying, "Get up quickly." And the chains fell off his wrists. 8 The angel said to him, "Fasten your belt and put on your sandals." He did so. Then he said to him, "Wrap your cloak around you and follow me." 9 Peter*ᵍ* went out and followed him; he did not realize that what was happening with the angel's help was real; he thought he was seeing a vision. 10 After they had passed the first and the second guard, they came before the iron gate leading into the city. It opened for them of its own accord, and they went outside and walked along a lane, when suddenly the angel left him. 11 Then Peter came to himself and said, "Now I am sure that the Lord has sent his angel and rescued me from the hands of Herod and from all that the Jewish people were expecting."

12 As soon as he realized this, he went to the house of Mary, the mother of John whose other name was Mark, where many had gathered and were praying. 13 When he knocked at the outer gate, a maid named Rhoda came to answer. 14 On recognizing Peter's voice, she was

e Or *were guests of*   *f* Gk *brothers*   *g* Gk *He*

---

och. **25–26:** *Saul* was in *Tarsus,* compare Gal 1.21. *Christians,* a Latin word meaning "partisans of Christ," perhaps at first a term of reproach.

11.27–30: Famine relief sent to Jerusalem through Barnabas and Paul. **27:** *Prophets* were numerous in the early church (see 13.1 n.; 1 Cor 12.28–29); their utterances, while including the foretelling of events, were mainly exhortations to righteousness. They were next in authority to the apostles (Eph 2.20; 4.11). **28:** *Agabus,* 21.10–11. The *famine . . . during the reign of Claudius* (A.D. 41–54) probably occurred in A.D. 46.

12.1–19: Herod Agrippa's persecution. James the son of Zebedee is martyred; Peter is arrested but escapes. **1:** *Herod* Agrippa I,

grandson of Herod the Great and the Maccabean Mariamne, was made king by Claudius A.D. 41. He was popular because he favored Pharisaism. *James,* one of the sons of Zebedee. **4:** *Four squads of soldiers,* an indication of the political importance of the prisoner. *Passover* season, i.e. the days of unleavened bread (v. 3).

12.12: *John . . . Mark* accompanied Paul and Barnabas as far as Perga (12.25–13.13; compare also Mk 14.51 n.). **15:** It was believed that a person's guardian *angel* represented the person in heaven (Gen 48.16; Mt 18.10). **17:** *Tell . . . James,* the brother of Jesus next in age, and now titular head of the church at Jerusalem.

so overjoyed that, instead of opening the gate, she ran in and announced that Peter was standing at the gate. 15 They said to her, "You are out of your mind!" But she insisted that it was so. They said, "It is his angel." 16 Meanwhile Peter continued knocking; and when they opened the gate, they saw him and were amazed. 17 He motioned to them with his hand to be silent, and described for them how the Lord had brought him out of the prison. And he added, "Tell this to James and to the believers."*h* Then he left and went to another place.

18 When morning came, there was no small commotion among the soldiers over what had become of Peter. 19 When Herod had searched for him and could not find him, he examined the guards and ordered them to be put to death. Then he went down from Judea to Caesarea and stayed there.

20 Now Herod*i* was angry with the people of Tyre and Sidon. So they came to him in a body; and after winning over Blastus, the king's chamberlain, they asked for a reconciliation, because their country depended on the king's country for food. 21 On an appointed day Herod put on his royal robes, took his seat on the platform, and delivered a public address to them. 22 The people kept shouting, "The voice of a god, and not of a mortal!" 23 And immediately, because he had not given the glory to God, an angel of the Lord struck him down, and he was eaten by worms and died.

24 But the word of God continued to advance and gain adherents. 25 Then after completing their mission Barnabas and Saul returned to*j* Jerusalem and brought with them John, whose other name was Mark.

13 Now in the church at Antioch there were prophets and teachers: Barnabas, Simeon who was called Niger, Lucius of Cyrene, Manaen a member of the court of Herod the ruler,*k* and Saul. 2 While they were worshiping the Lord and fasting, the Holy Spirit said, "Set apart for me Barnabas and Saul for the work to which I have called them." 3 Then after fasting and praying they laid their hands on them and sent them off.

4 So, being sent out by the Holy Spirit, they went down to Seleucia; and from there they sailed to Cyprus. 5 When they arrived at Salamis, they proclaimed the word of God in the synagogues of the Jews. And they had John also to assist them. 6 When they had gone through the whole island as far as Paphos, they met a certain magician, a Jewish false prophet, named Bar-Jesus. 7 He was with the proconsul, Sergius Paulus, an intelligent man, who summoned Barnabas and Saul and wanted to hear the word of God. 8 But the magician Elymas (for that is the translation of his name) opposed them and tried to turn the proconsul away from the faith. 9 But Saul, also known as Paul, filled with the Holy Spirit, looked

h Gk *brothers*   *i* Gk *he*   *j* Other ancient authorities read *from*   *k* Gk *tetrarch*

12.20–23: Death of Herod Agrippa. This was in the spring of A.D. 44. Josephus (*Ant.* xix.8.2) tells how he was stricken by a mortal illness immediately after the people hailed him as a god.
12.24–13.12: Barnabas and Saul in Cyprus. The beginning of Paul's first missionary journey.
13.1: *Prophets and teachers* were important in the early church (see 11.27 n.; 1 Cor 12.28; Eph 4.11). *Niger* means "black." *Cyrene* had a large Jewish colony. *Manaen,* Greek form of Hebrew Manahem. *Herod* Antipas, the son of Herod the Great. **2–3:** Paul declared that he received his apostleship not from human authorities but from divine commissioning (Gal 1.1). *The Holy Spirit* called them; when the others *laid their hands on them* it was not to make them apostles but to bless them for the new work (see 1 Tim 4.14 n.). **4:** *Seleucia* Pieria, Antioch's seaport, about 16 miles west at the mouth of the river Orontes. **5:** *Salamis,* largest city of Cyprus, north of modern Famagusta. *John,* that is John Mark, the cousin of Barnabas (12.12). **6:** *Paphos,* capital of the island, located in the extreme west. *Bar-Jesus,* "son of Jesus (or Joshua)." **7:** The island was a senatorial province ruled by a *proconsul.* **8:** *Elymas* does not mean "Bar-Jesus." Perhaps his name was Elymas, son of Je-

intently at him [10] and said, "You son of the devil, you enemy of all righteousness, full of all deceit and villainy, will you not stop making crooked the straight paths of the Lord? [11] And now listen—the hand of the Lord is against you, and you will be blind for a while, unable to see the sun." Immediately mist and darkness came over him, and he went about groping for someone to lead him by the hand. [12] When the proconsul saw what had happened, he believed, for he was astonished at the teaching about the Lord.

13 Then Paul and his companions set sail from Paphos and came to Perga in Pamphylia. John, however, left them and returned to Jerusalem; [14] but they went on from Perga and came to Antioch in Pisidia. And on the sabbath day they went into the synagogue and sat down. [15] After the reading of the law and the prophets, the officials of the synagogue sent them a message, saying, "Brothers, if you have any word of exhortation for the people, give it." [16] So Paul stood up and with a gesture began to speak:

"You Israelites,[l] and others who fear God, listen. [17] The God of this people Israel chose our ancestors and made the people great during their stay in the land of Egypt, and with uplifted arm he led them out of it. [18] For about forty years he put up with[m] them in the wilderness. [19] After he had destroyed seven nations in the land of Canaan, he gave them their land as an inheritance [20] for about four hundred fifty years. After that he gave them judges until the time of the prophet Samuel. [21] Then they asked for a king; and God gave them Saul son of Kish, a man of the tribe of Benjamin, who reigned for forty years. [22] When he had

removed him, he made David their king. In his testimony about him he said, 'I have found David, son of Jesse, to be a man after my heart, who will carry out all my wishes.' [23] Of this man's posterity God has brought to Israel a Savior, Jesus, as he promised; [24] before his coming John had already proclaimed a baptism of repentance to all the people of Israel. [25] And as John was finishing his work, he said, 'What do you suppose that I am? I am not he. No, but one is coming after me; I am not worthy to untie the thong of the sandals[n] on his feet.'

26 "My brothers, you descendants of Abraham's family, and others who fear God, to us[o] the message of this salvation has been sent. [27] Because the residents of Jerusalem and their leaders did not recognize him or understand the words of the prophets that are read every sabbath, they fulfilled those words by condemning him. [28] Even though they found no cause for a sentence of death, they asked Pilate to have him killed. [29] When they had carried out everything that was written about him, they took him down from the tree and laid him in a tomb. [30] But God raised him from the dead; [31] and for many days he appeared to those who came up with him from Galilee to Jerusalem, and they are now his witnesses to the people. [32] And we bring you the good news that what God promised to our ancestors [33] he has fulfilled for us, their children, by raising Jesus; as also it is written in the second psalm,

'You are my Son;
today I have begotten you.'

---

l Gk *Men, Israelites*   *m* Other ancient authorities read *cared for*   *n* Gk *untie the sandals*   *o* Other ancient authorities read *you*

---

sus (or Joshua). **9**: *Saul* was his Jewish name, *Paul* his Roman name. A new source, using the latter, may begin here.

13.13–52: Journey to Antioch of Pisidia and Iconium. Paul's first work in inner Asia Minor. **13**: *Perga* is inland from Attalia, main seaport of Pamphylia ("region of all tribes"), which lies south of the Taurus range, between Cilicia and Lycia. **14**: Strictly, *Antioch*

near *Pisidia,* near the modern Yalovach. **15**: One lesson each from *the law and the prophets* was customary. **19**: *Seven nations,* Deut 7.1; Josh 14.1. The rabbis reckoned almost *four hundred fifty years* from the entrance into Canaan to the building of the temple. **22**: Ps 89.20; 1 Sam 13.14. **23**: *Savior,* see 5.31 n. **25**: Mk 1.7; Lk 3.16; Mt 3.11; Jn 1.20. **13.26–41**: For argument and style compare

34 As to his raising him from the dead, no more to return to corruption, he has spoken in this way,

'I will give you the holy promises made to David.'

35 Therefore he has also said in another psalm,

'You will not let your Holy One experience corruption.'

36 For David, after he had served the purpose of God in his own generation, died,[p] was laid beside his ancestors, and experienced corruption; 37 but he whom God raised up experienced no corruption. 38 Let it be known to you therefore, my brothers, that through this man forgiveness of sins is proclaimed to you; 39 by this Jesus[q] everyone who believes is set free from all those sins[r] from which you could not be freed by the law of Moses. 40 Beware, therefore, that what the prophets said does not happen to you:

41 'Look, you scoffers!

Be amazed and perish,

for in your days I am doing a work,

a work that you will never believe, even if someone tells you.' "

42 As Paul and Barnabas[s] were going out, the people urged them to speak about these things again the next sabbath. 43 When the meeting of the synagogue broke up, many Jews and devout converts to Judaism followed Paul and Barnabas, who spoke to them and urged them to continue in the grace of God.

44 The next sabbath almost the whole city gathered to hear the word of the Lord.[t] 45 But when the Jews saw the crowds, they were filled with jealousy; and blaspheming, they contradicted what was spoken by Paul. 46 Then both Paul and Barnabas spoke out boldly, say-

ing, "It was necessary that the word of God should be spoken first to you. Since you reject it and judge yourselves to be unworthy of eternal life, we are now turning to the Gentiles. 47 For so the Lord has commanded us, saying,

'I have set you to be a light for the Gentiles,

so that you may bring salvation to the ends of the earth.' "

48 When the Gentiles heard this, they were glad and praised the word of the Lord; and as many as had been destined for eternal life became believers. 49 Thus the word of the Lord spread throughout the region. 50 But the Jews incited the devout women of high standing and the leading men of the city, and stirred up persecution against Paul and Barnabas, and drove them out of their region. 51 So they shook the dust off their feet in protest against them, and went to Iconium. 52 And the disciples were filled with joy and with the Holy Spirit.

14 The same thing occurred in Iconium, where Paul and Barnabas[s] went into the Jewish synagogue and spoke in such a way that a great number of both Jews and Greeks became believers. 2 But the unbelieving Jews stirred up the Gentiles and poisoned their minds against the brothers. 3 So they remained for a long time, speaking boldly for the Lord, who testified to the word of his grace by granting signs and wonders to be done through them. 4 But the residents of the city were divided; some sided with the Jews, and some with the apostles. 5 And when an attempt was made by both Gentiles and Jews, with their rulers, to mistreat them and to

---

p Gk fell asleep   q Gk this   r Gk all
s Gk they   t Other ancient authorities read God

2.14–36. 33: Ps 2.7. 34: Isa 55.3. 35: Ps 16.10. 38–39: Paul's letters usually speak of justification, not forgiveness of sins. Set free, Greek "justified" or "acquitted." 41: Hab 1.5.

13.46: In Corinth also (18.6) Paul went first to the Jews and then to the Gentiles. His churches usually included Jews. 47: Isa 49.6. 51: They shook the dust off (Lk 10.11) to show

that their responsibility ended (18.6). Iconium, modern Konya, an important road-junction in the central plain.

14.1–28: **Ministry in the Iconium region and return.** Conclusion of the first missionary journey. 2: 17.5, 13.

14.6: Lystra, a Roman colony, modern

stone them, 6the apostles*u* learned of it and fled to Lystra and Derbe, cities of Lycaonia, and to the surrounding country; 7and there they continued proclaiming the good news.

8 In Lystra there was a man sitting who could not use his feet and had never walked, for he had been crippled from birth. 9He listened to Paul as he was speaking. And Paul, looking at him intently and seeing that he had faith to be healed, 10said in a loud voice, "Stand upright on your feet." And the man*v* sprang up and began to walk. 11When the crowds saw what Paul had done, they shouted in the Lycaonian language, "The gods have come down to us in human form!" 12Barnabas they called Zeus, and Paul they called Hermes, because he was the chief speaker. 13The priest of Zeus, whose temple was just outside the city,*w* brought oxen and garlands to the gates; he and the crowds wanted to offer sacrifice. 14When the apostles Barnabas and Paul heard of it, they tore their clothes and rushed out into the crowd, shouting, 15"Friends,*x* why are you doing this? We are mortals just like you, and we bring you good news, that you should turn from these worthless things to the living God, who made the heaven and the earth and the sea and all that is in them. 16In past generations he allowed all the nations to follow their own ways; 17yet he has not left himself without a witness in doing good—giving you rains from heaven and fruitful seasons, and filling you with food and your hearts with joy." 18Even with these words, they scarcely restrained the crowds from offering sacrifice to them.

19 But Jews came there from Antioch and Iconium and won over the crowds. Then they stoned Paul and dragged him out of the city, supposing that he was dead. 20But when the disciples surrounded him, he got up and went into the city. The next day he went on with Barnabas to Derbe.

21 After they had proclaimed the good news to that city and had made many disciples, they returned to Lystra, then on to Iconium and Antioch. 22There they strengthened the souls of the disciples and encouraged them to continue in the faith, saying, "It is through many persecutions that we must enter the kingdom of God." 23And after they had appointed elders for them in each church, with prayer and fasting they entrusted them to the Lord in whom they had come to believe.

24 Then they passed through Pisidia and came to Pamphylia. 25When they had spoken the word in Perga, they went down to Attalia. 26From there they sailed back to Antioch, where they had been commended to the grace of God for the work*y* that they had completed. 27When they arrived, they called the church together and related all that God had done with them, and how he had opened a door of faith for the Gentiles. 28And they stayed there with the disciples for some time.

15 Then certain individuals came down from Judea and were teaching the brothers, "Unless you are

u Gk *they*   *v* Gk *he*   *w* Or *The priest of Zeus-Outside-the-City*   *x* Gk *Men*
y Or *committed in the grace of God to the work*

Hatun-Serai, 25 miles southwest of Konya. *Derbe,* probably Kerti, a mound near Beydilli. **11–12:** According to a myth of this region, *Zeus* and *Hermes* visited Baucis and Philemon *in human form,* and rewarded their hospitality. Paul, *the chief speaker,* was hailed as the messenger of the gods (compare Gal 4.14), not the chief god. **14:** *Tore their clothes,* a sign of horror and dismay at what they looked upon as blasphemy. **15–17:** Compare 17.22–31. Paul, like Peter (10.26), rejects worship of himself. **14.19:** Paul tells of receiving a stoning (2 Cor

11.25). **23:** In Acts, Paul's churches are ruled by *elders* (compare 20.17). The word is not used in letters attributed to Paul except in 1 Timothy and Titus. **25:** *Attalia,* the seaport of Perga. **26:** *Antioch,* their starting point, from which they had been absent about a year and a half.

15.1–35: Controversy over admission of Gentiles. The Jerusalem church guides developments on the mission field (8.14; 11.1–3). **1:** *Certain individuals,* not identified here or in Gal 2.4; 5.12. **2:** *Were appointed,*

circumcised according to the custom of Moses, you cannot be saved." [2] And after Paul and Barnabas had no small dissension and debate with them, Paul and Barnabas and some of the others were appointed to go up to Jerusalem to discuss this question with the apostles and the elders. [3] So they were sent on their way by the church, and as they passed through both Phoenicia and Samaria, they reported the conversion of the Gentiles, and brought great joy to all the believers. [z] [4] When they came to Jerusalem, they were welcomed by the church and the apostles and the elders, and they reported all that God had done with them. [5] But some believers who belonged to the sect of the Pharisees stood up and said, "It is necessary for them to be circumcised and ordered to keep the law of Moses."

[6] The apostles and the elders met together to consider this matter. [7] After there had been much debate, Peter stood up and said to them, "My brothers, [a] you know that in the early days God made a choice among you, that I should be the one through whom the Gentiles would hear the message of the good news and become believers. [8] And God, who knows the human heart, testified to them by giving them the Holy Spirit, just as he did to us; [9] and in cleansing their hearts by faith he has made no distinction between them and us. [10] Now therefore why are you putting God to the test by placing on the neck of the disciples a yoke that neither our ancestors nor we have been able to bear? [11] On the contrary, we believe that we will be saved

through the grace of the Lord Jesus, just as they will."

[12] The whole assembly kept silence, and listened to Barnabas and Paul as they told of all the signs and wonders that God had done through them among the Gentiles. [13] After they finished speaking, James replied, "My brothers, [a] listen to me. [14] Simeon has related how God first looked favorably on the Gentiles, to take from among them a people for his name. [15] This agrees with the words of the prophets, as it is written,

[16] 'After this I will return,
and I will rebuild the dwelling of
David, which has fallen;
from its ruins I will rebuild it,
and I will set it up,
[17] so that all other peoples may seek
the Lord—
even all the Gentiles over whom
my name has been called.
Thus says the Lord, who has
been making these things
[18] known from long ago.' [b]

[19] Therefore I have reached the decision that we should not trouble those Gentiles who are turning to God, [20] but we should write to them to abstain only from things polluted by idols and from fornication and from whatever has been strangled[c] and from blood. [21] For in every city, for generations past, Moses has had those who proclaim him, for he has been read aloud every sabbath in the synagogues."

[22] Then the apostles and the elders, with the consent of the whole church,

z Gk *brothers*   a Gk *Men, brothers*   b Other ancient authorities read *things.* [18]*Known to God from of old are all his works.'*   c Other ancient authorities lack *and from whatever has been strangled*

Paul says he went "by revelation" (Gal 2.2; see Acts 11.27–30 n.). *Apostles and . . . elders,* or presbyters, rule the Jerusalem church; see 6.1–7 n. **5:** The strict believers of v. 1 were probably *Pharisees* (see 11.2 n.). **7:** *In the early days,* compare 10.9–48. **10:** The *yoke* is that of the law (Mt 11.29–30; 23.4). **11:** *We will be saved through . . . grace,* Rom 3.24.

15.13: *James,* the brother of the Lord (see Gal 1.19 n.). **14:** *Simeon,* the Semitic form of Peter's given name. **16–18:** Am 9.11–12; Jer

12.15; Isa 45.21. **20:** *What has been strangled* may mean the same as *blood* (omitted by some manuscripts), i.e. meat not ritually butchered. *Polluted by idols,* i.e. food sacrificed to them. The rabbis taught that these, and *fornication,* had been forbidden to Noah's sons, therefore to the righteous of all nations.

15.22: *Paul* is told of the decree in 21.25 as though for the first time. He did not absolutely forbid food offered to idols (1 Cor 10.27–29) and he rejected other restrictions on food

decided to choose men from among their members[d] and to send them to Antioch with Paul and Barnabas. They sent Judas called Barsabbas, and Silas, leaders among the brothers, 23 with the following letter: "The brothers, both the apostles and the elders, to the believers[e] of Gentile origin in Antioch and Syria and Cilicia, greetings. 24 Since we have heard that certain persons who have gone out from us, though with no instructions from us, have said things to disturb you and have unsettled your minds,[f] 25 we have decided unanimously to choose representatives[g] and send them to you, along with our beloved Barnabas and Paul, 26 who have risked their lives for the sake of our Lord Jesus Christ. 27 We have therefore sent Judas and Silas, who themselves will tell you the same things by word of mouth. 28 For it has seemed good to the Holy Spirit and to us to impose on you no further burden than these essentials: 29 that you abstain from what has been sacrificed to idols and from blood and from what is strangled[h] and from fornication. If you keep yourselves from these, you will do well. Farewell."

30 So they were sent off and went down to Antioch. When they gathered the congregation together, they delivered the letter. 31 When its members[i] read it, they rejoiced at the exhortation. 32 Judas and Silas, who were themselves prophets, said much to encourage and strengthen the believers.[e] 33 After they had been there for some time, they were sent off in peace by the believers[e] to those who had sent them.[j] 35 But Paul and Barnabas remained in Antioch, and there, with many others, they taught and proclaimed the word of the Lord.

36 After some days Paul said to Barnabas, "Come, let us return and visit the believers[e] in every city where we proclaimed the word of the Lord and see how they are doing." 37 Barnabas wanted to take with them John called Mark. 38 But Paul decided not to take with them one who had deserted them in Pamphylia and had not accompanied them in the work. 39 The disagreement became so sharp that they parted company; Barnabas took Mark with him and sailed away to Cyprus. 40 But Paul chose Silas and set out, the believers[e] commending him to the grace of the Lord. 41 He went through Syria and Cilicia, strengthening the churches.

16 Paul[k] went on also to Derbe and to Lystra, where there was a disciple named Timothy, the son of a Jewish woman who was a believer; but his father was a Greek. 2 He was well spoken of by the believers[e] in Lystra and Iconium. 3 Paul wanted Timothy to accompany him; and he took him and had him circumcised because of the Jews who were in those places, for they all knew that his father was a Greek. 4 As they went from town to town, they delivered to them for observance the decisions that had been reached by the apostles and elders who were in Jerusalem. 5 So the churches were strengthened in the faith and increased in numbers daily.

6 They went through the region of Phrygia and Galatia, having been forbidden by the Holy Spirit to speak the word

---

d Gk *from among them*   e Gk *brothers*
f Other ancient authorities add *saying, 'You must be circumcised and keep the law,'*   g Gk *men*
h Other ancient authorities lack *and from what is strangled*   i Gk *When they*   j Other ancient authorities add verse 34, *But it seemed good to Silas to remain there*   k Gk *He*

(Gal 2.11–12; Col 2.21). *Silas* may be the Silvanus of 2 Cor 1.19; 1 Thess 1.1; 2 Thess 1.1.
15.36–41: Departure on second missionary journey. 37–38: *John called Mark . . . had deserted,* see 12.12 n.; 13.13. **39:** The *disagreement* may also have involved the eating together of Jews and Gentiles (see 10.28 n.; Gal 2.13).

16.1–5: Timothy joins Paul. If *Timothy,* being *the son of a Jewish woman,* was considered a Jew, Paul *had him circumcised* without inconsistency despite his rejection of the rite for Gentiles.
16.6–10: Through Asia Minor to Troas. Journey through the interior to the Aegean. **6:** *The region* is probably the country northwest of Iconium where both Phrygians and Gala-

in Asia. [7] When they had come opposite Mysia, they attempted to go into Bithynia, but the Spirit of Jesus did not allow them; [8] so, passing by Mysia, they went down to Troas. [9] During the night Paul had a vision: there stood a man of Macedonia pleading with him and saying, "Come over to Macedonia and help us." [10] When he had seen the vision, we immediately tried to cross over to Macedonia, being convinced that God had called us to proclaim the good news to them.

11 We set sail from Troas and took a straight course to Samothrace, the following day to Neapolis, [12] and from there to Philippi, which is a leading city of the district[l] of Macedonia and a Roman colony. We remained in this city for some days. [13] On the sabbath day we went outside the gate by the river, where we supposed there was a place of prayer; and we sat down and spoke to the women who had gathered there. [14] A certain woman named Lydia, a worshiper of God, was listening to us; she was from the city of Thyatira and a dealer in purple cloth. The Lord opened her heart to listen eagerly to what was said by Paul. [15] When she and her household were baptized, she urged us, saying, "If you have judged me to be faithful to the Lord, come and stay at my home." And she prevailed upon us.

16 One day, as we were going to the place of prayer, we met a slave-girl who had a spirit of divination and brought her owners a great deal of money by fortune-telling. [17] While she followed Paul and us, she would cry out, "These men are slaves of the Most High God, who proclaim to you[m] a way of salvation." [18] She kept doing this for many days. But Paul, very much annoyed, turned and said to the spirit, "I order you in the name of Jesus Christ to come out of her." And it came out that very hour.

19 But when her owners saw that their hope of making money was gone, they seized Paul and Silas and dragged them into the marketplace before the authorities. [20] When they had brought them before the magistrates, they said, "These men are disturbing our city; they are Jews [21] and are advocating customs that are not lawful for us as Romans to adopt or observe." [22] The crowd joined in attacking them, and the magistrates had them stripped of their clothing and ordered them to be beaten with rods. [23] After they had given them a severe flogging, they threw them into prison and ordered the jailer to keep them securely. [24] Following these instructions, he put them in the innermost cell and fastened their feet in the stocks.

25 About midnight Paul and Silas were praying and singing hymns to God, and the prisoners were listening to them. [26] Suddenly there was an earthquake, so violent that the foundations of the prison were shaken; and immediately all the doors were opened and everyone's

l Other authorities read *a city of the first district*
m Other ancient authorities read *to us*

---

tians lived. *Asia,* the Roman province of Asia (in Asia Minor) included western Phrygia. **7:** *Opposite Mysia* (the region east of Troas) would be near Nacoleia, modern Seyitgazi. *Bithynia* was north of here. **9:** *Macedonia,* a Roman province in Europe including Philippi, Thessalonica, and Beroea. **16.11–40: Paul and Silas in Philippi.** Paul enters Europe for the first time. *We,* here begins the first of the "we-passages" in Acts, which may suggest that the author joined the others at Troas. **11:** *Samothrace,* an island midway between Troas and *Neapolis,* the seaport of Philippi. **12:** *Philippi* was *a leading city,* but not the capital, *of Macedonia. A Roman colony*

was often founded to provide land for veteran soldiers; as such it had special civic rights. **13:** *A place of prayer,* probably a synagogue. **14:** *Thyatira* (Rev 2.18–29), a city of Lydia, a country of Asia Minor. **17:** Pagans sometimes spoke of the God of Israel or the highest god of their pantheon as *the Most High God* (Num 24.16; Isa 14.14; Dan 3.26). **16.20:** *Magistrates,* Greek "generals"; here probably praetors. **21:** *Not lawful,* because Jews were forbidden to make converts of *Romans.* **24:** *The stocks* tortured prisoners by forcing their legs apart. **27:** *Drew his sword,* a Roman jailor whose prisoner escaped was liable to forfeit his life.

chains were unfastened. 27 When the jailer woke up and saw the prison doors wide open, he drew his sword and was about to kill himself, since he supposed that the prisoners had escaped. 28 But Paul shouted in a loud voice, "Do not harm yourself, for we are all here." 29 The jailer *n* called for lights, and rushing in, he fell down trembling before Paul and Silas. 30 Then he brought them outside and said, "Sirs, what must I do to be saved?" 31 They answered, "Believe on the Lord Jesus, and you will be saved, you and your household." 32 They spoke the word of the Lord *o* to him and to all who were in his house. 33 At the same hour of the night he took them and washed their wounds; then he and his entire family were baptized without delay. 34 He brought them up into the house and set food before them; and he and his entire household rejoiced that he had become a believer in God.

35 When morning came, the magistrates sent the police, saying, "Let those men go." 36 And the jailer reported the message to Paul, saying, "The magistrates sent word to let you go; therefore come out now and go in peace." 37 But Paul replied, "They have beaten us in public, uncondemned, men who are Roman citizens, and have thrown us into prison; and now are they going to discharge us in secret? Certainly not! Let them come and take us out themselves." 38 The police reported these words to the magistrates, and they were afraid when they heard that they were Roman citizens; 39 so they came and apologized to them. And they took them out and asked them to leave the city. 40 After leaving the prison they went to Lydia's home; and when they had seen and encouraged the brothers and sisters *p* there, they departed.

17 After Paul and Silas *q* had passed through Amphipolis and Apollonia, they came to Thessalonica, where there was a synagogue of the Jews. 2 And Paul went in, as was his custom, and on three sabbath days argued with them from the scriptures, 3 explaining and proving that it was necessary for the Messiah *r* to suffer and to rise from the dead, and saying, "This is the Messiah, *r* Jesus whom I am proclaiming to you." 4 Some of them were persuaded and joined Paul and Silas, as did a great many of the devout Greeks and not a few of the leading women. 5 But the Jews became jealous, and with the help of some ruffians in the marketplaces they formed a mob and set the city in an uproar. While they were searching for Paul and Silas to bring them out to the assembly, they attacked Jason's house. 6 When they could not find them, they dragged Jason and some believers *p* before the city authorities, *s* shouting, "These people who have been turning the world upside down have come here also, 7 and Jason has entertained them as guests. They are all acting contrary to the decrees of the emperor, saying that there is another king named Jesus." 8 The people and the city officials were disturbed when they heard this, 9 and after they had taken bail from Jason and the others, they let them go.

10 That very night the believers *p* sent Paul and Silas off to Beroea; and when they arrived, they went to the Jewish synagogue. 11 These Jews were more receptive than those in Thessalonica, for they welcomed the message very eagerly and examined the scriptures every day to see whether these things were so.

n Gk He　　*o* Other ancient authorities read *word of God*　　*p* Gk *brothers*　　*q* Gk *they*　　*r* Or *the Christ*　　*s* Gk *politarchs*

16.35: *Police,* lictors, whose symbol of office was a bundle of rods, sometimes bound around an ax (the fasces). **37–38:** By law *Roman citizens* were protected against scourging (yet see 2 Cor 11.25).
17.1–15: From Thessalonica to Athens. Paul founds churches in Thessalonica and Be-roea. **1:** *Amphipolis and Apollonia* were on the Via Egnatia between Philippi and *Thessalonica,* capital of the province. **6–7:** *City authorities,* i.e. "politarchs," a Macedonian title. **9:** *Had taken bail from Jason,* i.e. had made him legally responsible for future good behavior.
17.10: *Beroea,* a city of Macedonia, fifty

12 Many of them therefore believed, including not a few Greek women and men of high standing. 13 But when the Jews of Thessalonica learned that the word of God had been proclaimed by Paul in Beroea as well, they came there too, to stir up and incite the crowds. 14 Then the believers¹ immediately sent Paul away to the coast, but Silas and Timothy remained behind. 15 Those who conducted Paul brought him as far as Athens; and after receiving instructions to have Silas and Timothy join him as soon as possible, they left him.

16 While Paul was waiting for them in Athens, he was deeply distressed to see that the city was full of idols. 17 So he argued in the synagogue with the Jews and the devout persons, and also in the marketplace" every day with those who happened to be there. 18 Also some Epicurean and Stoic philosophers debated with him. Some said, "What does this babbler want to say?" Others said, "He seems to be a proclaimer of foreign divinities." (This was because he was telling the good news about Jesus and the resurrection.) 19 So they took him and brought him to the Areopagus and asked him, "May we know what this new teaching is that you are presenting? 20 It sounds rather strange to us, so we would like to know what it means." 21 Now all the Athenians and the foreigners living there would spend their time in nothing but telling or hearing something new.

22 Then Paul stood in front of the Areopagus and said, "Athenians, I see how extremely religious you are in every way. 23 For as I went through the city and looked carefully at the objects of your worship, I found among them an altar with the inscription, 'To an unknown god.' What therefore you worship as unknown, this I proclaim to you. 24 The God who made the world and everything in it, he who is Lord of heaven and earth, does not live in shrines made by human hands, 25 nor is he served by human hands, as though he needed anything, since he himself gives to all mortals life and breath and all things. 26 From one ancestor ᵛ he made all nations to inhabit the whole earth, and he allotted the times of their existence and the boundaries of the places where they would live, 27 so that they would search for God ʷ and perhaps grope for him and find him— though indeed he is not far from each one of us. 28 For 'In him we live and move and have our being'; as even some of your own poets have said,

'For we too are his offspring.'

29 Since we are God's offspring, we ought not to think that the deity is like gold, or silver, or stone, an image formed by the art and imagination of mortals. 30 While God has overlooked the times of human ignorance, now he commands all people everywhere to repent, 31 because he has fixed a day on which he will have the world judged in righteousness by a man whom he has appointed, and of this he has given assurance to all by raising him from the dead."

32 When they heard of the resurrec-

t Gk brothers   u Or civic center; Gk agora
v Gk From one; other ancient authorities read
From one blood   w Other ancient authorities
read the Lord

---

miles southwest of Thessalonica. **14**: *To the coast,* where he would take a ship to Athens (v. 15). **14–15**: *Timothy* must soon have joined Paul at Athens (1 Thess 3.2).

17.16–34: Paul at Athens. The apostle is portrayed as the first Christian philosopher, using Stoic and Jewish arguments. **17**: *The marketplace* or *Agora* has now been excavated. **18**: *Babbler,* literally "cock-sparrow," one who picks up scraps of learning. *Jesus and Anastasis* (= *resurrection*) were mistaken for two *foreign divinities.* **19**: Either the council of the Areopagus, or the hill itself, west of the Acropolis. **21**: *The Athenians* at this time were famous for their curiosity.

17.22: *Religious,* i.e. addicted to worship of gods. **23**: One who did not know which god to thank or propitiate might set up an altar *to an unknown god.* **26**: *From one ancestor,* Adam. **28**: The first quotation is sometimes attributed to Epimenides; the second is from the opening lines of the *Phaenomena* by Aratus, a Greek poet of Cilicia. He was a Stoic. **31**: *By a man,* i.e. Jesus. **32**: *Some scoffed,* as in 1 Cor

tion of the dead, some scoffed; but others said, "We will hear you again about this." ³³At that point Paul left them. ³⁴But some of them joined him and became believers, including Dionysius the Areopagite and a woman named Damaris, and others with them.

18 After this Paul^x left Athens and went to Corinth. ²There he found a Jew named Aquila, a native of Pontus, who had recently come from Italy with his wife Priscilla, because Claudius had ordered all Jews to leave Rome. Paul^y went to see them, ³and, because he was of the same trade, he stayed with them, and they worked together—by trade they were tentmakers. ⁴Every sabbath he would argue in the synagogue and would try to convince Jews and Greeks.

5 When Silas and Timothy arrived from Macedonia, Paul was occupied with proclaiming the word,^z testifying to the Jews that the Messiah^a was Jesus. ⁶When they opposed and reviled him, in protest he shook the dust from his clothes^b and said to them, "Your blood be on your own heads! I am innocent. From now on I will go to the Gentiles." ⁷Then he left the synagogue^c and went to the house of a man named Titius^d Justus, a worshiper of God; his house was next door to the synagogue. ⁸Crispus, the official of the synagogue, became a believer in the Lord, together with all his household; and many of the Corinthians who heard Paul became believers and were baptized. ⁹One night the Lord said to Paul in a vision, "Do not be afraid, but speak and do not be silent; ¹⁰for I am with you, and no one will lay a hand on you to harm you, for there are many in this city who are my people." ¹¹He stayed there a year and six months, teaching the word of God among them.

12 But when Gallio was proconsul of Achaia, the Jews made a united attack on Paul and brought him before the tribunal. ¹³They said, "This man is persuading people to worship God in ways that are contrary to the law." ¹⁴Just as Paul was about to speak, Gallio said to the Jews, "If it were a matter of crime or serious villainy, I would be justified in accepting the complaint of you Jews; ¹⁵but since it is a matter of questions about words and names and your own law, see to it yourselves; I do not wish to be a judge of these matters." ¹⁶And he dismissed them from the tribunal. ¹⁷Then all of them^e seized Sosthenes, the official of the synagogue, and beat him in front of the tribunal. But Gallio paid no attention to any of these things.

18 After staying there for a considerable time, Paul said farewell to the believers^f and sailed for Syria, accompanied by Priscilla and Aquila. At Cenchreae he had his hair cut, for he was under a vow. ¹⁹When they reached Ephesus, he left them there, but first he himself went into the synagogue and had a discussion with the Jews. ²⁰When they asked him to stay longer, he declined;

x Gk *he*   y Gk *He*   z Gk *with the word*
a Or *the Christ*   b Gk *reviled him, he shook out
his clothes*   c Gk *left there*   d Other ancient
authorities read *Titus*   e Other ancient
authorities read *all the Greeks*   f Gk *brothers*

15.12, 35. **34**: *Dionysius the Areopagite,* a member of the Court of the Areopagus, the governing power of the city.
18.1–17: Founding of the church in Corinth. See 1 Cor chs 1–4 for Paul's own account. **2**: *Pontus,* a Roman province of Asia Minor. Claudius' edict expelling the Jews from Rome was issued probably in A.D. 49. *Priscilla,* the Prisca of 1 Cor 16.19. **3**: *Tentmakers,* the Greek word usually means "leather-workers." **8**: *Crispus,* one of the few Corinthians baptized personally by Paul (1 Cor 1.14).

18.12: L. Junius *Gallio,* brother of the philosopher Seneca, *proconsul of Achaia* about A.D. 51. The *tribunal* has been excavated. *The law,* the Jewish law. **17**: *Sosthenes* may be the one mentioned in 1 Cor 1.1.
18.18–23: End of the second missionary journey and beginning of the third. Paul returns to Antioch and goes back to Asia Minor. **18**: Paul *had his hair cut* as a temporary nazirite *vow* (21.24; Num 6.1–21). *Cenchreae,* the eastern port of Corinth. **22**: *Caesarea,* see 8.40 n. *Antioch,* see 11.19 n. **23**: *Galatia,* see 16.6 n. *The region of Galatia and Phrygia,*

21but on taking leave of them, he said, "I*g* will return to you, if God wills." Then he set sail from Ephesus.

22 When he had landed at Caesarea, he went up to Jerusalem*h* and greeted the church, and then went down to Antioch. 23After spending some time there he departed and went from place to place through the region of Galatia*i* and Phrygia, strengthening all the disciples.

24 Now there came to Ephesus a Jew named Apollos, a native of Alexandria. He was an eloquent man, well-versed in the scriptures. 25He had been instructed in the Way of the Lord; and he spoke with burning enthusiasm and taught accurately the things concerning Jesus, though he knew only the baptism of John. 26He began to speak boldly in the synagogue; but when Priscilla and Aquila heard him, they took him aside and explained the Way of God to him more accurately. 27And when he wished to cross over to Achaia, the believers*j* encouraged him and wrote to the disciples to welcome him. On his arrival he greatly helped those who through grace had become believers, 28for he powerfully refuted the Jews in public, showing by the scriptures that the Messiah*k* is Jesus.

19 While Apollos was in Corinth, Paul passed through the interior regions and came to Ephesus, where he found some disciples. 2He said to them, "Did you receive the Holy Spirit when you became believers?" They replied, "No, we have not even heard that there is a Holy Spirit." 3Then he said, "Into what then were you baptized?" They an-

swered, "Into John's baptism." 4Paul said, "John baptized with the baptism of repentance, telling the people to believe in the one who was to come after him, that is, in Jesus." 5On hearing this, they were baptized in the name of the Lord Jesus. 6When Paul had laid his hands on them, the Holy Spirit came upon them, and they spoke in tongues and prophesied— 7altogether there were about twelve of them.

8 He entered the synagogue and for three months spoke out boldly, and argued persuasively about the kingdom of God. 9When some stubbornly refused to believe and spoke evil of the Way before the congregation, he left them, taking the disciples with him, and argued daily in the lecture hall of Tyrannus.*l* 10This continued for two years, so that all the residents of Asia, both Jews and Greeks, heard the word of the Lord.

11 God did extraordinary miracles through Paul, 12so that when the handkerchiefs or aprons that had touched his skin were brought to the sick, their diseases left them, and the evil spirits came out of them. 13Then some itinerant Jewish exorcists tried to use the name of the Lord Jesus over those who had evil spirits, saying, "I adjure you by the Jesus whom Paul proclaims." 14Seven sons of a Jewish high priest named Sceva were

g Other ancient authorities read *I must at all costs keep the approaching festival in Jerusalem, but I* *h* Gk *went up*   *i* Gk *the Galatian region* *j* Gk *brothers*   *k* Or *the Christ*   *l* Other ancient authorities read *of a certain Tyrannus, from eleven o'clock in the morning to four in the afternoon*

---

see 16.6 n. Paul perhaps came near Pessinus and Gordium and then turned west.

18.24–28: Apollos in Ephesus. For his activity in Corinth see 1 Cor 1.12; 3.1–9, 21–23. **24:** An Alexandrian *well-versed in the scriptures* would probably interpret them allegorically. **25:** He knew something *concerning Jesus* but was ignorant of such Christian rites as baptism in Jesus' name (19.5).

19.1–41: Paul's long ministry in Ephesus, the capital of the Roman province of Asia. This period of more than two years (v. 10; compare 20.31) forms the background of 1 and 2 Corinthians. **1:** *The interior regions*

of Asia Minor (18.23). *Disciples* elsewhere in Acts means Christians; these perhaps belonged to John the Baptist. **2:** All who read the Old Testament would know of *a Holy Spirit*. The reference may be to outward signs of the Spirit's presence. **5–6:** See 8.14–17 n. **9:** *The Way,* see 9.2 n. *Tyrannus,* head of a school of philosophical disputants. The hours mentioned in note *l* were during the heat of the day, when Tyrannus would not be using the hall. **10:** *Two years,* during which the seven churches of Asia would have been founded (see Rev 1.4 n.).

19.12: 5.15. **13:** Jewish and pagan sources

doing this. [15] But the evil spirit said to them in reply, "Jesus I know, and Paul I know; but who are you?" [16] Then the man with the evil spirit leaped on them, mastered them all, and so overpowered them that they fled out of the house naked and wounded. [17] When this became known to all residents of Ephesus, both Jews and Greeks, everyone was awe-struck; and the name of the Lord Jesus was praised. [18] Also many of those who became believers confessed and disclosed their practices. [19] A number of those who practiced magic collected their books and burned them publicly; when the value of these books[m] was calculated, it was found to come to fifty thousand silver coins. [20] So the word of the Lord grew mightily and prevailed.

[21] Now after these things had been accomplished, Paul resolved in the Spirit to go through Macedonia and Achaia, and then to go on to Jerusalem. He said, "After I have gone there, I must also see Rome." [22] So he sent two of his helpers, Timothy and Erastus, to Macedonia, while he himself stayed for some time longer in Asia.

[23] About that time no little disturbance broke out concerning the Way. [24] A man named Demetrius, a silversmith who made silver shrines of Artemis, brought no little business to the artisans. [25] These he gathered together, with the workers of the same trade, and said, "Men, you know that we get our wealth from this business. [26] You also see and hear that not only in Ephesus but in almost the whole of Asia this Paul has persuaded and drawn away a considerable number of people by saying that gods made with hands are not gods. [27] And there is danger not only that this trade of ours may come into disrepute but also that the temple of the great goddess Artemis will be scorned, and she will be deprived of her majesty that brought all Asia and the world to worship her."

[28] When they heard this, they were enraged and shouted, "Great is Artemis of the Ephesians!" [29] The city was filled with the confusion; and people[n] rushed together to the theater, dragging with them Gaius and Aristarchus, Macedonians who were Paul's travel companions. [30] Paul wished to go into the crowd, but the disciples would not let him; [31] even some officials of the province of Asia,[o] who were friendly to him, sent him a message urging him not to venture into the theater. [32] Meanwhile, some were shouting one thing, some another; for the assembly was in confusion, and most of them did not know why they had come together. [33] Some of the crowd gave instructions to Alexander, whom the Jews had pushed forward. And Alexander motioned for silence and tried to make a defense before the people. [34] But when they recognized that he was a Jew, for about two hours all of them shouted in unison, "Great is Artemis of the Ephesians!" [35] But when the town clerk had quieted the crowd, he said, "Citizens of Ephesus, who is there that does not know that the city of the Ephesians is the temple keeper of the great Artemis and of the statue that fell from heaven?[p] [36] Since these things cannot be denied, you ought to be quiet and do nothing

m Gk them   n Gk they   o Gk some of the
Asiarchs   p Meaning of Gk uncertain

show that non-Christians used *the name of the Lord Jesus* in exorcism (Mk 9.38–41). **19:** Ephesus was such a noted center of *magic* that magical books were often called "Ephesian Scripts."

19.21: Rom 1.13–15; 15.22–25. **22:** *Erastus,* Rom 16.23; 2 Tim 4.20; an inscription from Corinth mentions a Roman official (an aedile) by this name. **24:** *Shrines,* perhaps miniature temples or statues. **27:** *Artemis,* Ephesus' chief divinity; her *temple* was one of the seven wonders of the ancient world.

19.31: *Asiarchs* (in note *o*), an honorific title given civic benefactors in the Roman province of Asia. **35:** Cities and peoples were sometimes honored with the title of *temple keeper* to a god. *The statue that fell from heaven,* apparently a meteoric stone having some resemblance to a woman.

rash. 37 You have brought these men here who are neither temple robbers nor blasphemers of our*q* goddess. 38 If therefore Demetrius and the artisans with him have a complaint against anyone, the courts are open, and there are proconsuls; let them bring charges there against one another. 39 If there is anything further* you want to know, it must be settled in the regular assembly. 40 For we are in danger of being charged with rioting today, since there is no cause that we can give to justify this commotion." 41 When he had said this, he dismissed the assembly.

20 After the uproar had ceased, Paul sent for the disciples; and after encouraging them and saying farewell, he left for Macedonia. 2 When he had gone through those regions and had given the believers*s* much encouragement, he came to Greece, 3 where he stayed for three months. He was about to set sail for Syria when a plot was made against him by the Jews, and so he decided to return through Macedonia. 4 He was accompanied by Sopater son of Pyrrhus from Beroea, by Aristarchus and Secundus from Thessalonica, by Gaius from Derbe, and by Timothy, as well as by Tychicus and Trophimus from Asia. 5 They went ahead and were waiting for us in Troas; 6 but we sailed from Philippi after the days of Unleavened Bread, and in five days we joined them in Troas, where we stayed for seven days.

7 On the first day of the week, when we met to break bread, Paul was holding a discussion with them; since he intended to leave the next day, he continued speaking until midnight. 8 There were many lamps in the room upstairs where we were meeting. 9 A young man named Eutychus, who was sitting in the window, began to sink off into a deep sleep while Paul talked still longer. Overcome by sleep, he fell to the ground three floors below and was picked up dead. 10 But Paul went down, and bending over him took him in his arms, and said, "Do not be alarmed, for his life is in him." 11 Then Paul went upstairs, and after he had broken bread and eaten, he continued to converse with them until dawn; then he left. 12 Meanwhile they had taken the boy away alive and were not a little comforted.

13 We went ahead to the ship and set sail for Assos, intending to take Paul on board there; for he had made this arrangement, intending to go by land himself. 14 When he met us in Assos, we took him on board and went to Mitylene. 15 We sailed from there, and on the following day we arrived opposite Chios. The next day we touched at Samos, and*t* the day after that we came to Miletus. 16 For Paul had decided to sail past Ephesus, so that he might not have to spend time in Asia; he was eager to be in Jerusalem, if possible, on the day of Pentecost.

17 From Miletus he sent a message to Ephesus, asking the elders of the church to meet him. 18 When they came to him, he said to them:

"You yourselves know how I lived among you the entire time from the first day that I set foot in Asia, 19 serving the

q Other ancient authorities read *your*   *r* Other ancient authorities read *about other matters*   *s* Gk *given them*   *t* Other ancient authorities add *after remaining at Trogyllium*

20.1–6: The last visit to Greece. 2: *Through those regions,* revisiting the churches of Philippi, Thessalonica, and Beroea. **3:** *Three months,* at Corinth. **4:** His companions may be carrying relief money to Jerusalem (24.17; 1 Cor 16.1–4; 2 Cor 8.23). *Sopater,* perhaps the Sosipater of Rom 16.21. **5:** The "we-passages" resume here at Philippi, the city where the previous passage ended (16.17). **6:** *Days of Unleavened Bread,* 12.3; see Mk 14.1 n.
20.7–21.14: Paul's return to Palestine.

This concludes the third missionary journey.
20.7: *The first day of the week,* Sunday. *Break bread,* see 2.42 n. **9:** *Sitting in the window,* because of the smoke and fumes of the *many lamps* (v. 8). Windows were unglazed. **14:** *Mitylene,* capital city of the island of Lesbos. **15:** *Chios,* an island south of Lesbos. *Samos,* an island southwest of Chios.
20.17: *Miletus,* modern Balat, then an important port at the mouth of the Maeander.

Lord with all humility and with tears, enduring the trials that came to me through the plots of the Jews. 20 I did not shrink from doing anything helpful, proclaiming the message to you and teaching you publicly and from house to house, 21 as I testified to both Jews and Greeks about repentance toward God and faith toward our Lord Jesus. 22 And now, as a captive to the Spirit, " I am on my way to Jerusalem, not knowing what will happen to me there, 23 except that the Holy Spirit testifies to me in every city that imprisonment and persecutions are waiting for me. 24 But I do not count my life of any value to myself, if only I may finish my course and the ministry that I received from the Lord Jesus, to testify to the good news of God's grace.

25 "And now I know that none of you, among whom I have gone about proclaiming the kingdom, will ever see my face again. 26 Therefore I declare to you this day that I am not responsible for the blood of any of you, 27 for I did not shrink from declaring to you the whole purpose of God. 28 Keep watch over yourselves and over all the flock, of which the Holy Spirit has made you overseers, to shepherd the church of God *v* that he obtained with the blood of his own Son. *w* 29 I know that after I have gone, savage wolves will come in among you, not sparing the flock. 30 Some even from your own group will come distorting the truth in order to entice the disciples to follow them. 31 Therefore be alert, remembering that for three years I did not cease night or day to warn everyone with tears. 32 And now I commend you to God and to the message of his grace, a message that is able to build you up and to give you the inheritance among all who are sanctified. 33 I coveted no one's silver or gold or clothing. 34 You know for yourselves that I worked with my own hands to support myself and my companions. 35 In all this I have given you an example that by such work we must support the weak, remembering the words of the Lord Jesus, for he himself said, 'It is more blessed to give than to receive.' "

36 When he had finished speaking, he knelt down with them all and prayed. 37 There was much weeping among them all; they embraced Paul and kissed him, 38 grieving especially because of what he had said, that they would not see him again. Then they brought him to the ship.

21 When we had parted from them and set sail, we came by a straight course to Cos, and the next day to Rhodes, and from there to Patara. *x* 2 When we found a ship bound for Phoenicia, we went on board and set sail. 3 We came in sight of Cyprus; and leaving it on our left, we sailed to Syria and landed at Tyre, because the ship was to unload its cargo there. 4 We looked up the disciples and stayed there for seven days. Through the Spirit they told Paul not to go on to Jerusalem. 5 When our days there were ended, we left and proceeded on our journey; and all of them, with wives and children, escorted us outside the city. There we knelt down on the beach and prayed 6 and said farewell to one another. Then we went on board the ship, and they returned home.

7 When we had finished *y* the voyage from Tyre, we arrived at Ptolemais; and we greeted the believers *z* and stayed with them for one day. 8 The next day we left and came to Caesarea; and we went into the house of Philip the evangelist, one of the seven, and stayed with him.

u Or *And now, bound in the spirit*   *v* Other ancient authorities read *of the Lord*   *w* Or *with his own blood*; Gk *with the blood of his Own*   *x* Other ancient authorities add *and Myra*   *y* Or *continued*   *z* Gk *brothers*

24–25: *Finish my course . . . testify,* a premonition of martyrdom (2 Tim 4.6).

20.28: *Overseers,* Greek "bishops" (compare Titus 1.5–7), the elders of v. 17. As in 1 Pet 2.25, the bishop is a shepherd. **29–**

30: Mt 7.15; Mk 13.22. **34**: 1 Cor 9.1–18; 2 Cor 11.7–11. **35**: A saying of Jesus not found in the gospels.

21.7: *Ptolemais,* modern Acre, near Haifa. **8**: *Caesarea,* see 8.40 n. *Philip the evangelist,*

[9] He had four unmarried daughters[a] who had the gift of prophecy. [10] While we were staying there for several days, a prophet named Agabus came down from Judea. [11] He came to us and took Paul's belt, bound his own feet and hands with it, and said, "Thus says the Holy Spirit, 'This is the way the Jews in Jerusalem will bind the man who owns this belt and will hand him over to the Gentiles.'" [12] When we heard this, we and the people there urged him not to go up to Jerusalem. [13] Then Paul answered, "What are you doing, weeping and breaking my heart? For I am ready not only to be bound but even to die in Jerusalem for the name of the Lord Jesus." [14] Since he would not be persuaded, we remained silent except to say, "The Lord's will be done."

15 After these days we got ready and started to go up to Jerusalem. [16] Some of the disciples from Caesarea also came along and brought us to the house of Mnason of Cyprus, an early disciple, with whom we were to stay.

17 When we arrived in Jerusalem, the brothers welcomed us warmly. [18] The next day Paul went with us to visit James; and all the elders were present. [19] After greeting them, he related one by one the things that God had done among the Gentiles through his ministry. [20] When they heard it, they praised God. Then they said to him, "You see, brother, how many thousands of believers there are among the Jews, and they are all zealous for the law. [21] They have been told about you that you teach all the Jews living among the Gentiles to forsake Moses, and that you tell them not to circumcise their children or observe the customs. [22] What then is to be done? They will certainly hear that you have come. [23] So do what we tell you. We have four men who are under a vow. [24] Join these men, go through the rite of purification with them, and pay for the shaving of their heads. Thus all will know that there is nothing in what they have been told about you, but that you yourself observe and guard the law. [25] But as for the Gentiles who have become believers, we have sent a letter with our judgment that they should abstain from what has been sacrificed to idols and from blood and from what is strangled[b] and from fornication." [26] Then Paul took the men, and the next day, having purified himself, he entered the temple with them, making public the completion of the days of purification when the sacrifice would be made for each of them.

27 When the seven days were almost completed, the Jews from Asia, who had seen him in the temple, stirred up the whole crowd. They seized him, [28] shouting, "Fellow Israelites, help! This is the man who is teaching everyone everywhere against our people, our law, and this place; more than that, he has actually brought Greeks into the temple and has defiled this holy place." [29] For they had previously seen Trophimus the Ephesian with him in the city, and they supposed that Paul had brought him into the temple. [30] Then all the city was aroused, and the people rushed together. They seized Paul and dragged him out of the temple, and immediately the doors were shut.

a Gk *four daughters, virgins,*   b Other ancient authorities lack *and from what is strangled*

---

though not an apostle, did similar work. *One of the seven,* see 6.1–7 n. **10–11:** *Agabus* (11.28), like Old Testament prophets (Isa 20.2–6), performed a symbolic act.

21.15–26: Paul's conformity to Judaism. He takes a vow in order to allay suspicions. **15:** *James,* see 15.11 n. **20:** On Jewish Christians who were *zealous for the law,* see 11.2; 15.5; and the Introduction to Galatians. **21:** The charge was untrue (16.3; 1 Cor 9.20;

10.32). **23–24:** The temporary nazirite *vow* was a later development out of Num 6.1–21. **25:** He is told of the decree as though he had not heard of it (see 15.22 n.).

21.27–22.29: Paul's arrest and defense. Seized after a disturbance in the temple, Paul tells of his conversion.

21.28: It was a capital offense to bring *Greeks into the temple;* an inscription stating this has been discovered. **29:** *Trophimus,* 20.4;

31 While they were trying to kill him, word came to the tribune of the cohort that all Jerusalem was in an uproar. 32 Immediately he took soldiers and centurions and ran down to them. When they saw the tribune and the soldiers, they stopped beating Paul. 33 Then the tribune came, arrested him, and ordered him to be bound with two chains; he inquired who he was and what he had done. 34 Some in the crowd shouted one thing, some another; and as he could not learn the facts because of the uproar, he ordered him to be brought into the barracks. 35 When Paul*c* came to the steps, the violence of the mob was so great that he had to be carried by the soldiers. 36 The crowd that followed kept shouting, "Away with him!"

37 Just as Paul was about to be brought into the barracks, he said to the tribune, "May I say something to you?" The tribune*d* replied, "Do you know Greek? 38 Then you are not the Egyptian who recently stirred up a revolt and led the four thousand assassins out into the wilderness?" 39 Paul replied, "I am a Jew, from Tarsus in Cilicia, a citizen of an important city; I beg you, let me speak to the people." 40 When he had given him permission, Paul stood on the steps and motioned to the people for silence; and when there was a great hush, he addressed them in the Hebrew*e* language, saying:

22 "Brothers and fathers, listen to the defense that I now make before you."

2 When they heard him addressing them in Hebrew,*e* they became even more quiet. Then he said:

3 "I am a Jew, born in Tarsus in Cilicia, but brought up in this city at the feet of Gamaliel, educated strictly according to our ancestral law, being zealous for God, just as all of you are today. 4 I persecuted this Way up to the point of death by binding both men and women and putting them in prison, 5 as the high priest and the whole council of elders can testify about me. From them I also received letters to the brothers in Damascus, and I went there in order to bind those who were there and to bring them back to Jerusalem for punishment.

6 "While I was on my way and approaching Damascus, about noon a great light from heaven suddenly shone about me. 7 I fell to the ground and heard a voice saying to me, 'Saul, Saul, why are you persecuting me?' 8 I answered, 'Who are you, Lord?' Then he said to me, 'I am Jesus of Nazareth*f* whom you are persecuting.' 9 Now those who were with me saw the light but did not hear the voice of the one who was speaking to me. 10 I asked, 'What am I to do, Lord?' The Lord said to me, 'Get up and go to Damascus; there you will be told everything that has been assigned to you to do.' 11 Since I could not see because of the brightness of that light, those who were with me took my hand and led me to Damascus.

12 "A certain Ananias, who was a devout man according to the law and well spoken of by all the Jews living there, 13 came to me; and standing beside me, he said, 'Brother Saul, regain your sight!' In that very hour I regained my sight and saw him. 14 Then he said, 'The God of our ancestors has chosen you to know his will, to see the Righteous One and to hear his own voice; 15 for you will be his witness to all the world of what you have seen and heard. 16 And now why do you delay? Get up, be baptized, and have

c Gk *he*   *d* Gk *He*   *e* That is, *Aramaic*
f Gk *the Nazorean*

2 Tim 4.20. **31**: A military *tribune* commanded a *cohort;* this would have been stationed at Fort Antonia, which had access to the temple courtyard. **33**: Paul's case came into the hands of Roman, not Jewish, authorities. **38**: Concerning *the Egyptian,* a pseudo-messiah who, with several thousand followers, had planned to enter Jerusalem and overpower the Roman garrison, see Josephus, *War* ii.13.5.

22.3: *Brought up,* educated. *Gamaliel,* see 5.34 n. **4–21**: In 9.7 the companions heard the voice but saw no one; otherwise the account is similar to 9.1–18.

your sins washed away, calling on his name.'

17 "After I had returned to Jerusalem and while I was praying in the temple, I fell into a trance [18] and saw Jesus[g] saying to me, 'Hurry and get out of Jerusalem quickly, because they will not accept your testimony about me.' [19] And I said, 'Lord, they themselves know that in every synagogue I imprisoned and beat those who believed in you. [20] And while the blood of your witness Stephen was shed, I myself was standing by, approving and keeping the coats of those who killed him.' [21] Then he said to me, 'Go, for I will send you far away to the Gentiles.' "

22 Up to this point they listened to him, but then they shouted, "Away with such a fellow from the earth! For he should not be allowed to live." [23] And while they were shouting, throwing off their cloaks, and tossing dust into the air, [24] the tribune directed that he was to be brought into the barracks, and ordered him to be examined by flogging, to find out the reason for this outcry against him. [25] But when they had tied him up with thongs,[h] Paul said to the centurion who was standing by, "Is it legal for you to flog a Roman citizen who is uncondemned?" [26] When the centurion heard that, he went to the tribune and said to him, "What are you about to do? This man is a Roman citizen." [27] The tribune came and asked Paul,[g] "Tell me, are you a Roman citizen?" And he said, "Yes." [28] The tribune answered, "It cost me a large sum of money to get my citizenship." Paul said, "But I was born a citizen." [29] Immediately those who were about to examine him drew back from

him; and the tribune also was afraid, for he realized that Paul was a Roman citizen and that he had bound him.

30 Since he wanted to find out what Paul[i] was being accused of by the Jews, the next day he released him and ordered the chief priests and the entire council to meet. He brought Paul down and had him stand before them.

23 While Paul was looking intently at the council he said, "Brothers,[j] up to this day I have lived my life with a clear conscience before God." [2] Then the high priest Ananias ordered those standing near him to strike him on the mouth. [3] At this Paul said to him, "God will strike you, you whitewashed wall! Are you sitting there to judge me according to the law, and yet in violation of the law you order me to be struck?" [4] Those standing nearby said, "Do you dare to insult God's high priest?" [5] And Paul said, "I did not realize, brothers, that he was high priest; for it is written, 'You shall not speak evil of a leader of your people.' "

6 When Paul noticed that some were Sadducees and others were Pharisees, he called out in the council, "Brothers, I am a Pharisee, a son of Pharisees. I am on trial concerning the hope of the resurrection[k] of the dead." [7] When he said this, a dissension began between the Pharisees and the Sadducees, and the assembly was divided. [8] (The Sadducees say that there is no resurrection, or angel, or spirit; but the Pharisees acknowledge all three.) [9] Then a great clamor arose, and certain

g Gk him   h Or up for the lashes   i Gk he
j Gk Men, brothers   k Gk concerning hope and
resurrection

22.24: The *flogging* was to extract evidence, not for punishment. **25:** *A Roman citizen,* still uncondemned, was protected from scourging; see 16.37–38 n. **28:** In Claudius' reign *citizenship* was often purchased for *a large sum.*
22.30–23.10: Paul before the Sanhedrin; the tribune brings him before it to get evidence, not to try him.
23.2: *Ananias, high priest* in the reigns of

Claudius and Nero, was assassinated about A.D. 66. **3:** *Whitewashed wall* is obscure; perhaps a tomb (compare Mt 23.27), which is unclean. *In violation to the law,* because Paul had not yet been allowed to speak in his own behalf (Jn 7.51). **5:** Quoted from Ex 22.28. **6:** *Paul* claims to be a *Pharisee* (26.5; Phil 3.5). For his *hope* of the *resurrection* see 1 Cor ch 15; Phil 3.11. **8:** *Sadducees* denied the *resurrection,* see Mt 22.23 n.

scribes of the Pharisees' group stood up and contended, "We find nothing wrong with this man. What if a spirit or an angel has spoken to him?" 10 When the dissension became violent, the tribune, fearing that they would tear Paul to pieces, ordered the soldiers to go down, take him by force, and bring him into the barracks.

11 That night the Lord stood near him and said, "Keep up your courage! For just as you have testified for me in Jerusalem, so you must bear witness also in Rome."

12 In the morning the Jews joined in a conspiracy and bound themselves by an oath neither to eat nor drink until they had killed Paul. 13 There were more than forty who joined in this conspiracy. 14 They went to the chief priests and elders and said, "We have strictly bound ourselves by an oath to taste no food until we have killed Paul. 15 Now then, you and the council must notify the tribune to bring him down to you, on the pretext that you want to make a more thorough examination of his case. And we are ready to do away with him before he arrives."

16 Now the son of Paul's sister heard about the ambush; so he went and gained entrance to the barracks and told Paul. 17 Paul called one of the centurions and said, "Take this young man to the tribune, for he has something to report to him." 18 So he took him, brought him to the tribune, and said, "The prisoner Paul called me and asked me to bring this young man to you; he has something to tell you." 19 The tribune took him by the hand, drew him aside privately, and asked, "What is it that you have to report to me?" 20 He answered, "The Jews have agreed to ask you to bring Paul down to the council tomorrow, as though they were going to inquire more thoroughly into his case. 21 But do not be persuaded by them, for more than forty of their men are lying in ambush for him. They have bound themselves by an oath neither to eat nor drink until they kill him. They are ready now and are waiting for your consent." 22 So the tribune dismissed the young man, ordering him, "Tell no one that you have informed me of this."

23 Then he summoned two of the centurions and said, "Get ready to leave by nine o'clock tonight for Caesarea with two hundred soldiers, seventy horsemen, and two hundred spearmen. 24 Also provide mounts for Paul to ride, and take him safely to Felix the governor." 25 He wrote a letter to this effect:

26 "Claudius Lysias to his Excellency the governor Felix, greetings. 27 This man was seized by the Jews and was about to be killed by them, but when I had learned that he was a Roman citizen, I came with the guard and rescued him. 28 Since I wanted to know the charge for which they accused him, I had him brought to their council. 29 I found that he was accused concerning questions of their law, but was charged with nothing deserving death or imprisonment. 30 When I was informed that there would be a plot against the man, I sent him to you at once, ordering his accusers also to state before you what they have against him.¹"

31 So the soldiers, according to their instructions, took Paul and brought him during the night to Antipatris. 32 The next day they let the horsemen go on with him, while they returned to the barracks. 33 When they came to Caesarea and delivered the letter to the governor, they presented Paul also before him. 34 On reading the letter, he asked what province he belonged to, and when he learned that he was from Cilicia, 35 he said, "I will

1 Other ancient authorities add Farewell

23.11–35: Paul is sent to Caesarea, the seat of the Roman governor.

23.23: The number of soldiers is surprising. The Greek word translated *spearmen* is obscure; it may mean "bowmen" or "bodyguards." **24:** Antonius *Felix,* procurator of Judea (A.D. 52–58), brother of a favorite freedman of Claudius. **26:** *His Excellency,* the

give you a hearing when your accusers arrive." Then he ordered that he be kept under guard in Herod's headquarters. *m*

24 Five days later the high priest Ananias came down with some elders and an attorney, a certain Tertullus, and they reported their case against Paul to the governor. ²When Paul*ⁿ* had been summoned, Tertullus began to accuse him, saying:

"Your Excellency,*ᵒ* because of you we have long enjoyed peace, and reforms have been made for this people because of your foresight. ³We welcome this in every way and everywhere with utmost gratitude. ⁴But, to detain you no further, I beg you to hear us briefly with your customary graciousness. ⁵We have, in fact, found this man a pestilent fellow, an agitator among all the Jews throughout the world, and a ringleader of the sect of the Nazarenes.*ᵖ* ⁶He even tried to profane the temple, and so we seized him. *q* ⁸By examining him yourself you will be able to learn from him concerning everything of which we accuse him."

9 The Jews also joined in the charge by asserting that all this was true.

10 When the governor motioned to him to speak, Paul replied:

"I cheerfully make my defense, knowing that for many years you have been a judge over this nation. ¹¹As you can find out, it is not more than twelve days since I went up to worship in Jerusalem. ¹²They did not find me disputing with anyone in the temple or stirring up a crowd either in the synagogues or throughout the city. ¹³Neither can they prove to you the charge that they now bring against me. ¹⁴But this I admit to you, that according to the Way, which they call a sect, I worship the God of our ancestors, believing everything laid down according to the law or written in the prophets. ¹⁵I have a hope in God—a hope that they themselves also accept—that there will be a resurrection of both*ʳ* the righteous and the unrighteous. ¹⁶Therefore I do my best always to have a clear conscience toward God and all people. ¹⁷Now after some years I came to bring alms to my nation and to offer sacrifices. ¹⁸While I was doing this, they found me in the temple, completing the rite of purification, without any crowd or disturbance. ¹⁹But there were some Jews from Asia—they ought to be here before you to make an accusation, if they have anything against me. ²⁰Or let these men here tell what crime they had found when I stood before the council, ²¹unless it was this one sentence that I called out while standing before them, 'It is about the resurrection of the dead that I am on trial before you today.' "

22 But Felix, who was rather well informed about the Way, adjourned the hearing with the comment, "When Lysias the tribune comes down, I will decide your case." ²³Then he ordered the centurion to keep him in custody, but to let him have some liberty and not to prevent any of his friends from taking care of his needs.

24 Some days later when Felix came with his wife Drusilla, who was Jewish, he sent for Paul and heard him speak concerning faith in Christ Jesus. ²⁵And as he discussed justice, self-control, and the

m Gk *praetorium*   *n* Gk *he*   *o* Gk lacks *Your Excellency*   *p* Gk *Nazoreans*   *q* Other ancient authorities add *and we would have judged him according to our law.* 7*But the chief captain Lysias came and with great violence took him out of our hands,* 8*commanding his accusers to come before you.*   *r* Other ancient authorities read *of the dead, both of*

---

word used in addressing Theophilus (Lk 1.3). **31:** *Antipatris,* a city on the main road to Caesarea.

24.1–27: Paul before Felix. The apostle denies the charges, and the governor postpones the case. **5:** *Sect,* the word is usually used in a bad sense (24.14; 28.22). *Nazarenes,* Christians as followers of Jesus of Nazareth (2.22); a meaning not found elsewhere in early Christian literature.

24.14: *The Way,* see 9.2 n. **17:** *To bring alms,* see 20.1–6 n. **24:** *Drusilla* was sister of Herod Agrippa II and Bernice (25.13); noted for her beauty, she had left her husband Azizus king of Emessa to marry Felix. **25–26:** The behavior of *Felix* is like that of Herod Antipas (Mk

coming judgment, Felix became frightened and said, "Go away for the present; when I have an opportunity, I will send for you." 26 At the same time he hoped that money would be given him by Paul, and for that reason he used to send for him very often and converse with him.

27 After two years had passed, Felix was succeeded by Porcius Festus; and since he wanted to grant the Jews a favor, Felix left Paul in prison.

25 Three days after Festus had arrived in the province, he went up from Caesarea to Jerusalem 2 where the chief priests and the leaders of the Jews gave him a report against Paul. They appealed to him 3 and requested, as a favor to them against Paul, *s* to have him transferred to Jerusalem. They were, in fact, planning an ambush to kill him along the way. 4 Festus replied that Paul was being kept at Caesarea, and that he himself intended to go there shortly. 5 "So," he said, "let those of you who have the authority come down with me, and if there is anything wrong about the man, let them accuse him."

6 After he had stayed among them not more than eight or ten days, he went down to Caesarea; the next day he took his seat on the tribunal and ordered Paul to be brought. 7 When he arrived, the Jews who had gone down from Jerusalem surrounded him, bringing many serious charges against him, which they could not prove. 8 Paul said in his defense, "I have in no way committed an offense against the law of the Jews, or against the temple, or against the emperor." 9 But Festus, wishing to do the Jews a favor, asked Paul, "Do you wish to go up to Jerusalem and be tried there before me on these charges?" 10 Paul said, "I am

appealing to the emperor's tribunal; this is where I should be tried. I have done no wrong to the Jews, as you very well know. 11 Now if I am in the wrong and have committed something for which I deserve to die, I am not trying to escape death; but if there is nothing to their charges against me, no one can turn me over to them. I appeal to the emperor." 12 Then Festus, after he had conferred with his council, replied, "You have appealed to the emperor; to the emperor you will go."

13 After several days had passed, King Agrippa and Bernice arrived at Caesarea to welcome Festus. 14 Since they were staying there several days, Festus laid Paul's case before the king, saying, "There is a man here who was left in prison by Felix. 15 When I was in Jerusalem, the chief priests and the elders of the Jews informed me about him and asked for a sentence against him. 16 I told them that it was not the custom of the Romans to hand over anyone before the accused had met the accusers face to face and had been given an opportunity to make a defense against the charge. 17 So when they met here, I lost no time, but on the next day took my seat on the tribunal and ordered the man to be brought. 18 When the accusers stood up, they did not charge him with any of the crimes *t* that I was expecting. 19 Instead they had certain points of disagreement with him about their own religion and about a certain Jesus, who had died, but whom Paul asserted to be alive. 20 Since I was at a loss how to investigate these questions, I asked whether he wished to go to Jerusalem and be tried there on

s Gk *him*   *t* Other ancient authorities read *with anything*

6.20). Compare 18.14–17. 27: *After two years had passed,* either from Felix's appointment or Paul's arrest. *Porcius Festus,* appointed procurator of Judea by Nero in A.D. 60.

25.1–12: Appeal to the emperor. Paul insists on a Roman trial and Festus sends him to Rome. **9–10:** Paul fears being turned over to a Jewish court; therefore he insists on *the*

emperor's tribunal, i.e. trial according to Roman law. **11–12:** Paul's safety was in his Roman citizenship (22.27) with its right of appeal.

25.13–26.32: Paul's defense before Agrippa. Paul's speech is intended to present a model defense of Christianity.

25.13: *King Agrippa,* Herod Agrippa II,

these charges." 21 But when Paul had appealed to be kept in custody for the decision of his Imperial Majesty, I ordered him to be held until I could send him to the emperor." 22 Agrippa said to Festus, "I would like to hear the man myself." "Tomorrow," he said, "you will hear him."

23 So on the next day Agrippa and Bernice came with great pomp, and they entered the audience hall with the military tribunes and the prominent men of the city. Then Festus gave the order and Paul was brought in. 24 And Festus said, "King Agrippa and all here present with us, you see this man about whom the whole Jewish community petitioned me, both in Jerusalem and here, shouting that he ought not to live any longer. 25 But I found that he had done nothing deserving death; and when he appealed to his Imperial Majesty, I decided to send him. 26 But I have nothing definite to write to our sovereign about him. Therefore I have brought him before all of you, and especially before you, King Agrippa, so that, after we have examined him, I may have something to write— 27 for it seems to me unreasonable to send a prisoner without indicating the charges against him."

26 Agrippa said to Paul, "You have permission to speak for yourself." Then Paul stretched out his hand and began to defend himself:

2 "I consider myself fortunate that it is before you, King Agrippa, I am to make my defense today against all the accusations of the Jews, 3 because you are especially familiar with all the customs and controversies of the Jews; therefore I beg of you to listen to me patiently. 4 "All the Jews know my way of life from my youth, a life spent from the beginning among my own people and in Jerusalem. 5 They have known for a long time, if they are willing to testify, that I have belonged to the strictest sect of our religion and lived as a Pharisee. 6 And now I stand here on trial on account of my hope in the promise made by God to our ancestors, 7 a promise that our twelve tribes hope to attain, as they earnestly worship day and night. It is for this hope, your Excellency,ᵛ that I am accused by Jews! 8 Why is it thought incredible by any of you that God raises the dead?

9 "Indeed, I myself was convinced that I ought to do many things against the name of Jesus of Nazareth.ʷ 10 And that is what I did in Jerusalem; with authority received from the chief priests, I not only locked up many of the saints in prison, but I also cast my vote against them when they were being condemned to death. 11 By punishing them often in all the synagogues I tried to force them to blaspheme; and since I was so furiously enraged at them, I pursued them even to foreign cities.

12 "With this in mind, I was traveling to Damascus with the authority and commission of the chief priests, 13 when at midday along the road, your Excellency,ᵛ I saw a light from heaven, brighter than the sun, shining around me and my companions. 14 When we had all fallen to the ground, I heard a voice saying to me in the Hebrewˣ language, 'Saul, Saul, why are you persecuting me? It hurts you to kick against the goads.' 15 I asked, 'Who are you, Lord?' The Lord answered, 'I am Jesus whom you are persecuting. 16 But get up and stand on your feet; for I have appeared to you for this

u Gk on them   v Gk O king   w Gk the
Nazorean   x That is, Aramaic

---

who ruled parts of Palestine. He and *Bernice* were children of Herod Agrippa I (12.1–23). **25:** *His Imperial Majesty,* literally Augustus, one of the imperial titles. **26.3:** Agrippa was perhaps well acquainted with Judaism, but was not a practicing Jew. **4:** *Among my own people,* i.e. Cilicia. **8:** *You,* i.e. the Jewish people.

26.10: *Cast my vote,* for the stoning of Stephen (8.1); but according to this verse he had also helped to put others to death for their faith in Jesus. **12–20:** A third account of Paul's conversion (9.1–8; 22.4–16). Here he adds that *we had all fallen to the ground.* **14:** A goad is a sharp pointed stick used to prod an ox or donkey.

purpose, to appoint you to serve and testify to the things in which you have seen me[y] and to those in which I will appear to you. [17]I will rescue you from your people and from the Gentiles—to whom I am sending you [18]to open their eyes so that they may turn from darkness to light and from the power of Satan to God, so that they may receive forgiveness of sins and a place among those who are sanctified by faith in me.'

19 "After that, King Agrippa, I was not disobedient to the heavenly vision, [20]but declared first to those in Damascus, then in Jerusalem and throughout the countryside of Judea, and also to the Gentiles, that they should repent and turn to God and do deeds consistent with repentance. [21]For this reason the Jews seized me in the temple and tried to kill me. [22]To this day I have had help from God, and so I stand here, testifying to both small and great, saying nothing but what the prophets and Moses said would take place: [23]that the Messiah[z] must suffer, and that, by being the first to rise from the dead, he would proclaim light both to our people and to the Gentiles."

24 While he was making this defense, Festus exclaimed, "You are out of your mind, Paul! Too much learning is driving you insane!" [25]But Paul said, "I am not out of my mind, most excellent Festus, but I am speaking the sober truth. [26]Indeed the king knows about these things, and to him I speak freely; for I am certain that none of these things has escaped his notice, for this was not done in a corner. [27]King Agrippa, do you believe the prophets? I know that you believe." [28]Agrippa said to Paul, "Are you so quickly persuading me to become a Christian?"[a] [29]Paul replied, "Whether quickly or not, I pray to God that not only you but also all who are listening to me today might become such as I am—except for these chains."

30 Then the king got up, and with him the governor and Bernice and those who had been seated with them; [31]and as they were leaving, they said to one another, "This man is doing nothing to deserve death or imprisonment." [32]Agrippa said to Festus, "This man could have been set free if he had not appealed to the emperor."

27 When it was decided that we were to sail for Italy, they transferred Paul and some other prisoners to a centurion of the Augustan Cohort, named Julius. [2]Embarking on a ship of Adramyttium that was about to set sail to the ports along the coast of Asia, we put to sea, accompanied by Aristarchus, a Macedonian from Thessalonica. [3]The next day we put in at Sidon; and Julius treated Paul kindly, and allowed him to go to his friends to be cared for. [4]Putting out to sea from there, we sailed under the lee of Cyprus, because the winds were against us. [5]After we had sailed across the sea that is off Cilicia and Pamphylia, we came to Myra in Lycia. [6]There the centurion found an Alexandrian ship bound for Italy and put us on board. [7]We sailed slowly for a number of days and arrived with difficulty off Cnidus, and as the wind was against us, we sailed under

y Other ancient authorities read *the things that you have seen*   z Or *the Christ*   a Or *Quickly you will persuade me to play the Christian*

26.17: *Your people,* i.e. the Jewish people. **23**: *That the Messiah must suffer,* 8.32–35; Lk 24.26. **28**: *Christian,* see 11.26 n. **32**: See 25.11–12 n.

27.1–44: **The voyage to Malta.** A dangerous winter voyage ends in shipwreck. **1**: The "we-passages" resume here. *The Augustan Cohort,* a unit of this name was stationed in Syria in the first century. **2**: *Adramyttium,* an important seaport of Mysia. *Aristarchus,* 19.29; 20.4. *Thessalonica,* see 17.1 n. **3**: *Sidon,* a seaport on the coast of Phoenicia, twenty miles north of Tyre. **4**: *Under the lee of Cyprus,* apparently east of the island. **7**: *The lee of Crete,* to its south; *Salmone* is at the eastern end. **8**: *Fair Havens,* a bay on the southern coast of Crete. *Lasea,* a city about four miles east of Fair Havens.

27.9: *The Fast,* the Day of Atonement, in September or October. **17**: *The Syrtis,* a dangerous shoal west of Cyrene.

27.27: *Sea of Adria* then included the cen-

the lee of Crete off Salmone. ⁸Sailing past it with difficulty, we came to a place called Fair Havens, near the city of Lasea.

9 Since much time had been lost and sailing was now dangerous, because even the Fast had already gone by, Paul advised them, ¹⁰saying, "Sirs, I can see that the voyage will be with danger and much heavy loss, not only of the cargo and the ship, but also of our lives." ¹¹But the centurion paid more attention to the pilot and to the owner of the ship than to what Paul said. ¹²Since the harbor was not suitable for spending the winter, the majority was in favor of putting to sea from there, on the chance that somehow they could reach Phoenix, where they could spend the winter. It was a harbor of Crete, facing southwest and northwest.

13 When a moderate south wind began to blow, they thought they could achieve their purpose; so they weighed anchor and began to sail past Crete, close to the shore. ¹⁴But soon a violent wind, called the northeaster, rushed down from Crete. *b* ¹⁵Since the ship was caught and could not be turned head-on into the wind, we gave way to it and were driven. ¹⁶By running under the lee of a small island called Cauda*c* we were scarcely able to get the ship's boat under control. ¹⁷After hoisting it up they took measures*d* to undergird the ship; then, fearing that they would run on the Syrtis, they lowered the sea anchor and so were driven. ¹⁸We were being pounded by the storm so violently that on the next day they began to throw the cargo overboard, ¹⁹and on the third day with their own hands they threw the ship's tackle overboard. ²⁰When neither sun nor stars appeared for many days, and no small tempest raged, all hope of our being saved was at last abandoned.

21 Since they had been without food for a long time, Paul then stood up among them and said, "Men, you should have listened to me and not have set sail from Crete and thereby avoided this damage and loss. ²²I urge you now to keep up your courage, for there will be no loss of life among you, but only of the ship. ²³For last night there stood by me an angel of the God to whom I belong and whom I worship, ²⁴and he said, 'Do not be afraid, Paul; you must stand before the emperor; and indeed, God has granted safety to all those who are sailing with you.' ²⁵So keep up your courage, men, for I have faith in God that it will be exactly as I have been told. ²⁶But we will have to run aground on some island."

27 When the fourteenth night had come, as we were drifting across the sea of Adria, about midnight the sailors suspected that they were nearing land. ²⁸So they took soundings and found twenty fathoms; a little farther on they took soundings again and found fifteen fathoms. ²⁹Fearing that we might run on the rocks, they let down four anchors from the stern and prayed for day to come. ³⁰But when the sailors tried to escape from the ship and had lowered the boat into the sea, on the pretext of putting out anchors from the bow, ³¹Paul said to the centurion and the soldiers, "Unless these men stay in the ship, you cannot be saved." ³²Then the soldiers cut away the ropes of the boat and set it adrift.

33 Just before daybreak, Paul urged all of them to take some food, saying, "Today is the fourteenth day that you have been in suspense and remaining without food, having eaten nothing. ³⁴Therefore I urge you to take some food, for it will help you survive; for none of you will lose a hair from your heads." ³⁵After he had said this, he took bread; and giving thanks to God in the presence of all, he broke it and began to eat. ³⁶Then all of them were encouraged and took food for themselves. ³⁷(We were in all two hundred seventy-six*e* persons in the ship.) ³⁸After they had satisfied their hunger, they lightened the ship by throwing the wheat into the sea.

39 In the morning they did not recognize the land, but they noticed a bay with a beach, on which they planned to run the ship ashore, if they could. ⁴⁰So they

b Gk *it*   *c* Other ancient authorities read *Clauda*   *d* Gk *helps*   *e* Other ancient authorities read *seventy-six*; others, *about seventy-six*

cast off the anchors and left them in the sea. At the same time they loosened the ropes that tied the steering-oars; then hoisting the foresail to the wind, they made for the beach. 41 But striking a reef,*f* they ran the ship aground; the bow stuck and remained immovable, but the stern was being broken up by the force of the waves. 42 The soldiers' plan was to kill the prisoners, so that none might swim away and escape; 43 but the centurion, wishing to save Paul, kept them from carrying out their plan. He ordered those who could swim to jump overboard first and make for the land, 44 and the rest to follow, some on planks and others on pieces of the ship. And so it was that all were brought safely to land.

28 After we had reached safety, we then learned that the island was called Malta. 2 The natives showed us unusual kindness. Since it had begun to rain and was cold, they kindled a fire and welcomed all of us around it. 3 Paul had gathered a bundle of brushwood and was putting it on the fire, when a viper, driven out by the heat, fastened itself on his hand. 4 When the natives saw the creature hanging from his hand, they said to one another, "This man must be a murderer; though he has escaped from the sea, justice has not allowed him to live." 5 He, however, shook off the creature into the fire and suffered no harm. 6 They were expecting him to swell up or drop dead, but after they had waited a long time and saw that nothing unusual had happened to him, they changed their minds and began to say that he was a god.

7 Now in the neighborhood of that place were lands belonging to the leading man of the island, named Publius, who received us and entertained us hospitably for three days. 8 It so happened that the father of Publius lay sick in bed with fever and dysentery. Paul visited him and cured him by praying and putting his hands on him. 9 After this happened, the rest of the people on the island who had diseases also came and were cured. 10 They bestowed many honors on us, and when we were about to sail, they put on board all the provisions we needed.

11 Three months later we set sail on a ship that had wintered at the island, an Alexandrian ship with the Twin Brothers as its figurehead. 12 We put in at Syracuse and stayed there for three days; 13 then we weighed anchor and came to Rhegium. After one day there a south wind sprang up, and on the second day we came to Puteoli. 14 There we found believers*g* and were invited to stay with them for seven days. And so we came to Rome. 15 The believers*g* from there, when they heard of us, came as far as the Forum of Appius and Three Taverns to meet us. On seeing them, Paul thanked God and took courage.

16 When we came into Rome, Paul was allowed to live by himself, with the soldier who was guarding him.

17 Three days later he called together the local leaders of the Jews. When they had assembled, he said to them, "Brothers, though I had done nothing against our people or the customs of our ancestors, yet I was arrested in Jerusalem and

f Gk *place of two seas*   *g* Gk *brothers*

tral Mediterranean. **41:** *A reef,* Greek "a place of two seas," probably the bay now named for St. Paul.

28.1–10: Paul in Malta. Unharmed though bitten by a viper, Paul heals the father of Publius. **2:** The *natives* were non-Greeks who spoke a Semitic language. **7:** *Leading man,* a Greek term for a high official in Malta.

28.11–16: The journey to Rome. Paul goes by sea to Syracuse, Rhegium, and Puteoli, then by land to the capital. **11:** *The Twin Brothers,* Castor and Pollux, were worshiped by sailors. **12:** *Syracuse,* capital city of the is-

land of Sicily, and a Roman colony (see 16.12 n.). **13:** *Rhegium,* modern Reggio Calabria. *Puteoli,* Pozzuoli, on the north side of the Bay of Naples. **15:** The *Forum of Appius* was forty-three miles from Rome, and the *Three Taverns* thirty-three miles, both on the Via Appia. **16:** The "we-passages" would seem to end here. *By himself,* apparently in his own quarters, under house arrest or with a light chain (v. 20). This exceptional treatment was due to the favorable report of Festus (see 26.31) and the goodwill of the centurion.

28.17–28: Paul and the Jews of Rome.

handed over to the Romans. [18] When they had examined me, the Romans[h] wanted to release me, because there was no reason for the death penalty in my case. [19] But when the Jews objected, I was compelled to appeal to the emperor—even though I had no charge to bring against my nation. [20] For this reason therefore I have asked to see you and speak with you,[i] since it is for the sake of the hope of Israel that I am bound with this chain." [21] They replied, "We have received no letters from Judea about you, and none of the brothers coming here has reported or spoken anything evil about you. [22] But we would like to hear from you what you think, for with regard to this sect we know that everywhere it is spoken against."

23 After they had set a day to meet with him, they came to him at his lodgings in great numbers. From morning until evening he explained the matter to them, testifying to the kingdom of God and trying to convince them about Jesus both from the law of Moses and from the prophets. [24] Some were convinced by what he had said, while others refused to believe. [25] So they disagreed with each other; and as they were leaving, Paul made one further statement: "The Holy Spirit was right in saying to your ancestors through the prophet Isaiah,

[26] 'Go to this people and say,
  You will indeed listen, but never
    understand,
  and you will indeed look, but
    never perceive.
[27] For this people's heart has
    grown dull,
  and their ears are hard of
    hearing,
  and they have shut their eyes;
    so that they might not look
      with their eyes,
  and listen with their ears,
  and understand with their heart
    and turn—
  and I would heal them.'

[28] Let it be known to you then that this salvation of God has been sent to the Gentiles; they will listen."[j]

30 He lived there two whole years at his own expense[k] and welcomed all who came to him, [31] proclaiming the kingdom of God and teaching about the Lord Jesus Christ with all boldness and without hindrance.

h Gk *they*   *i* Or *I have asked you to see me and speak with me*   *j* Other ancient authorities add verse 29, *And when he had said these words, the Jews departed, arguing vigorously among themselves*   *k* Or *in his own hired dwelling*

Rejected by the Jews, he again turns to the Gentiles (see 13.46 n.). **20:** The *chain* may be actual or a metaphor of his restriction. **23:** *The prophets,* see Mt 5.17 n. **26:** Christians often used Isa 6.9–10 to explain the Jews' rejection of the gospel (see Mt 13.14–15 n.; Jn 12.40; compare the ideas of Rom chs 9–11).

28.30–31: Conclusion. Though under house arrest (v. 16) Paul preached *without hindrance* for *two whole years.* Thus the author concludes his narrative of the spread of the gospel from Jerusalem (1.8) to Rome, the capital of the empire.

Letters/Epistles in the
New Testament

CLASSIFICATION OF NEW TESTAMENT LETTERS

The predominant literary form in the New Testament is the letter form. Of the twenty-seven documents comprising the New Testament, seven-ninths of them have the format of a letter or epistle. Of these twenty-one, fourteen have been traditionally attributed to Paul, and the other seven to a variety of authors. In the case of the Pauline letters, the titles now given to them involve the names of the recipients (To the Romans; To the Galatians, To Philemon, etc.), whereas in the case of the others, called the General or Catholic Letters, the titles (except "To the Hebrews") identify the author (Letter of James, Letter of Jude, etc.). These titles, it goes without saying, were added by scribes and editors after the several letters had been collected into a corpus.

In almost all the complete manuscripts of the Greek New Testament the General Letters follow Acts and precede the Pauline Letters, whereas in many Latin manuscripts, the Pauline Letters often precede the General Letters. This latter sequence is followed in English Bibles. The Pauline Letters are in two series, those addressed to churches and those addressed to individuals. The present order of the General Letters and of each series of the Pauline Letters is in terms of their length, going from the longer to the shorter letters.

The letters included in the New Testament (with the exception of those included in Acts 15.23–29; 23.26–30, and the letters to the seven churches of Asia in chapters 2 and 3 of the Book of Revelation) are in general longer than other ancient letters. The letters attributed to Paul (with the exception of Hebrews) average about 1300 words each; Romans has 7101, Titus (the second shortest) 658, Philemon 335. Only 2 and 3 John are shorter than this, with 245 and 219 words respectively. The author of the anonymous Letter to the Hebrews concludes by saying, "I have written to you briefly" (13.22); since this letter contains 4951 words, we must perhaps understand the word "briefly" by considering the extent of the subject and its ramifications.

In comparison with the New Testament letters the following statistics are of interest. Thousands of private letters on papyrus have been unearthed, chiefly in Egypt; these average (according to one scholar's calculations) eighty-seven words in length—the shortest has eighteen words, the longest two hundred and nine. The literary collections of epistles from antiquity (for the distinction between letters and epistles, see p. 206 NT) have a much greater average length; the 796 genuine letters and epistles of Cicero the Roman orator average 295 words each (the shortest has 22 words, the longest 2530), and the 124 epistles of the Stoic philosopher Seneca average 995 words (from 149 to 4134 words).

MANNER OF WRITING LETTERS

In antiquity, just as today, people sometimes wrote letters personally, and sometimes they dictated to a scribe or amanuensis. In the latter case, one might dictate word by word, with the amanuensis writing each out verbatim. Or one might dictate more rapidly with the message being taken down stenographically. In this case the material would, of course, need to be transcribed again. Besides dictating word by word, there was also the practice of giving the amanuensis exact instructions about the content of the letter and leaving the choice of wording to the amanuensis.

In the case of the Letter to the Romans, we know that Paul dictated it to a certain Tertius, who added his own greeting among those that Paul was sending (Rom 16.22). In three other letters the apostle mentions explicitly that he has written the final greeting with his own hand, presumably as a guarantee of genuineness (1 Cor 16.21; Col 4.18; and especially 2 Thess 3.17). Near the close of his Letter to the Galatians Paul says to his readers, "See what large letters I make when I am writing in my own hand!" (6.11). Since the subject changes at this point, it is probable that here he has taken over the stylus from the amanuensis. In Philemon (v. 19) Paul writes a personal I O U that is apparently different from the script in the rest of the letter. From these data we may conclude that Paul's usual practice was to dictate the bulk of a given letter to a scribe and often to add notes in his own handwriting, perhaps also correcting what the scribe had written.

How far the New Testament contains letters for which a secretary had some responsibility for the language and style continues to be debated. It is possible that this is how we should understand the comment made near the close of 1 Peter: "Through Silvanus, whom I consider a faithful brother, I have written this short letter to encourage you" (5.12). In the case of several other letters (both Pauline and General) it is now thought that a friend and follower of the apostle drew up the document in the name of the apostle, perhaps actually after the death of the putative author. (See the Introductions to Ephesians, 1 Timothy, and 2 Peter.)

In light of such practices in antiquity, several consequences appear to be more or less probable. For one thing, a discontinuity in thought or expression may be the result of resuming the dictation after a break of some hours (or even on a subsequent day), and in a different mood from that reflected at the outset of the letter. Thus, for example, we can explain the change of tone between the close of ch 2 and what follows 3.1 of Paul's Letter to the Philippians. Furthermore, if a secretary took part in the composition of a letter, and not simply its transcription, then the vocabulary and style are not decisive criteria for settling its authenticity. Finally, the similarity between Ephesians and Colossians could be accounted for on the assumption that Paul had kept a copy of the earlier one, which he read before dictating the other. (We know that Cicero made a practice of keeping copies of the letters that he had sent.) It is also possible that these two letters were written by others, or another, with access to Paul's writings (see the Introduction to Ephesians and Colossians).

FORM OF ANCIENT LETTERS

The form of an ancient Greco-Roman letter was conspicuously different from the modern letter. At its beginning one finds the sender's name, followed by that of the recipient, and a short salutation. The message in the body of the letter was often introduced by an appeal to the gods in terms of gratitude and thanksgiving or of a prayer for the welfare of the recipient. At the end of the letter came the conveyance of greetings.

The apostle Paul's letters, while they fall within the category of such Hellenistic letters, show some development of the letter pattern based on the apostle's creativity. Identifications of the writer (co-workers are sometimes named) and of the addressees are followed by expanded descriptions of both parties in terms of their standing in relation to God through Christ. Specifically, the stereotyped Hellenistic "Greetings!" (Greek *chairein*) is replaced by "Grace (Greek *charis*) to you and peace from God our Father and the Lord Jesus Christ," where the reference to peace (compare Hebrew *shalom*) may echo the close of the old priestly blessing of Num 6.24–26. The Pauline letters and those influenced by them generally expand the element of thanksgiving for the addressees (for example, 1 Cor 1.4–9; Eph 1.15–23; Phil 1.3–11; 2 Thess 1.3–12; 1 Tim 1.3–7); in Galatians, however, Paul replaces the thanksgiving with a rebuke, "I am astonished . . ." (1.6–9). The body of Pauline letters is usually divided into two parts, one doctrinal in content, and the other hortatory, giving instructions for Christian conduct. The conclusion of his letters often contains personal news and a final greeting. This is never the ordinary Greek "Farewell" (*erroso,* as in Acts 15.29), but a characteristic blessing, "The grace of our Lord Jesus Christ be with you" (Rom 16.20; 1 Cor 16.23; 2 Cor 13.13 (expanded); Gal 6.18; Phil 4.23; 1 Thess 5.28; Philem 25).

DISTINCTION BETWEEN LETTERS AND EPISTLES

In modern usage a distinction is often made between the terms "letter" and "epistle." Letters, as Adolf Deissmann repeatedly emphasized, are generally private missives and deal with circumstances of the passing moment. Confidential and personal in nature, they are intended only for the person or persons to whom they are addressed. Epistles, on the other hand, are on a more sophisticated level of literary effort, and are written with the intention of being both public and more or less permanent. In brief, the letter is, so to speak, a slice of life, while the epistle is a product of literary art. Now the examples in the New Testament combine features of both the letter and the epistle. The Letters of Paul are all real letters, arising out of real situations in his life or the lives of those to whom he writes. At the same time, they are quasi-official letters from an apostle rather than merely private letters, and their style is more formal than the ephemeral letters preserved among the Greek papyri.

Some of the other letters in the New Testament are more akin to the form and style of a theological treatise (such as the Letter to the Hebrews) or an edifying homily (1 Peter and 1 John). Even these, however, were not originally directed to the whole of Christendom, but were intended for particular—even if loosely defined—destinations, and were written in order to deal with urgent difficulties experienced by the recipients, such as persecution, heresy, and religious or moral laxity. Thus, the letters of the New Testament incorporate elements of the on-going life of the churches, such as liturgical materials, moral guidelines, and religious instruction applicable to many believers besides the immediate recipients of the document. Furthermore, it should be noted that the literary form of a letter/epistle combines the advantages of a conversation and a treatise; in such a format it is possible to communicate truth, not only abstractly, but in close relation to the personal circumstances of the recipients.

DISPATCH OF LETTERS

In antiquity there was no organized postal service available to the general populace. The system of letter carriers (*cursus publicus*) created by the Emperor Augustus was restricted to the dispatch of government communiqués from Rome to officials in the provinces. Private individuals, therefore, had to make their own arrangements for the carrying and delivery of letters. Wealthy families would keep a certain number of slaves to serve as couriers—*grammatophoroi* ("letter carriers") as they were called in Greek, or *tabellarii* ("tablet men" from the wooden tablets, *tabellae*) in Latin. Families and friends living near one another would pool their couriers in order to increase the opportunities to get off or receive a letter. But there were never enough carriers to meet the needs, and delays were inevitable. The opposite side of the coin was the need to dash off some lines at breakneck speed in order to take advantage of an available courier.

The actual transmission of private letters was in practice exposed to many uncertainties, delays, and, at times, almost insuperable difficulties. Travel whether by land or by sea could be hazardous. In all the countries referred to in the New Testament (except Egypt) the terrain is mountainous. One could travel on land, of course, only by daylight. The rigors of winter prevented travel by sea as well as by land. There was also the ever-present possibility of being waylaid by bandits or by pirates. Then too, illness or accident might incapacitate the one carrying the letter. In view of such eventualities, for really important letters it was advisable to send duplicate copies through two different messengers. In any case, a letter carrier had to be physically fit as well as loyal and intelligent. It is not surprising to find in antiquity repeated reference to letters written but never received. Others were lost for months, or even years, before finally being delivered. For example, through a series of mishaps a letter written by Augustine of Hippo in North Africa to Jerome in Bethlehem took nine years for delivery!

In the case of several letters in the New Testament internal evidence allows us to infer who it was that served as letter carrier. Obviously the apostle Paul utilized Onesimus, the converted runaway slave, to deliver the letter to his master Philemon. Titus and an unidentified believer "who is famous among all the churches for his proclaiming the good news," carried what we know as Paul's Second Letter to the Corinthians (2 Cor 8.16–18). Tychicus carried the Letter to the Ephesians (Eph 6.21) and the Letter to the Colossians (Col 4.7–8), while Phoebe, whom Paul commends to the Christians at Rome (Rom 16.1–2), probably carried his Letter to the Romans. Epaphroditus, sent by the church at Philippi with a gift to Paul (Phil 4.18), would have carried back with him the apostle's letter of thanks to that congregation (Phil 2.25–28). Three of the converts at Corinth (Stephanas, Fortunatus, and Achaicus) probably brought Paul the letter that is referred to in 1 Cor 7.1, as well as carrying back Paul's reply (1 Cor 16.15–18).

CONCLUSION

The letters in the New Testament are real letters, and the issues they deal with belong to a particular period in the history of the church. At the same time, the treatment of the several problems that elicited a response on the part of Paul, Peter, John, and other leaders in the early church often involved an exposition of fundamental teachings of the Christian faith. Thus it comes about that these letters—having been recopied and circulated to other congregations, and finally collected together—eventually became, and have justly remained, among the primary documents of the Christian religion (on canonization, see p. v NT). Like the genre of gospels, the New Testament letters represent something startlingly new and original in the literature of their time. Both literary epistles, such as those of Cicero or Seneca, and the incidental papyrus letters preserved in the dry climate of Egypt are well known, but never before had the world ever seen anything quite like these rather lengthy letters/epistles, almost wholly concerned not just with personal details but with matters of Christian doctrine and conduct. Even the preliminary greeting and the concluding farewell are new and distinctively Christian formulae. If we ask what it was that provided the new dynamic behind the writing of these apostolic letters, the answer must certainly include reference to the several writers' loyal commitment to one whom they called "the Lord Jesus Christ."

The Letter of Paul to the
Romans

This letter, the first of Paul's letters in the canonical collection, is also the longest, the weightiest, and subsequently the most influential of all his correspondence. Written at the height of the apostle's career (between A.D. 54 and 58), it conveys the full richness of his experience of Christ as well as the full maturity of his thought.

The gospel as God's power for salvation to all who believe is the theme of Romans (see 1.16–17 n.). It is expressed especially in terms of God's saving righteousness, or justification by faith (see 3.24 n.), and with a universal concern for both Jew and Gentile (note the frequent use of "all" and "every," 1.16; 3.9, 19, 23–24; 4.11, 16; 5.12, 18). Reconciliation (5.10–11), forgiveness (4.6–8), life in Christ (chs 5–8), the Spirit and our adoption by God—all these intertwine with justification (8.5–30). There is also development of God's plan of salvation, from Adam through Abraham and Moses to Christ (chs 4–5) and the future fulfillment (8.18–25). Israel and the church, comprising the community of those who hear the proclamation about Christ and believe, are discussed in chs 9–11.

For several years—years of intense missionary activity in the northeastern Mediterranean area—Paul had been engaged in collecting contributions in his predominantly Gentile churches of Greece and Asia Minor for the needy Jerusalem church. It was his hope that these gifts would allay certain suspicions about him and his work and bring the two wings of the church, Jewish Christian and Gentile, closer together. The collection was now complete, and Paul, apparently in Corinth (15.25–27; compare 1 Cor 16.3–5 and Acts 20.2–3), was awaiting an opportunity to go to Jerusalem with it. He intended afterwards to carry the gospel to Spain (15.28) and, on his way, to stop at Rome, where the church had already been established by others. He therefore writes to the Christians at Rome to announce this intention and to explain his understanding of the gospel (1.10–15), perhaps also with a view to securing the support of believers there for his work farther to the west.

Because the text of Romans has existed from antiquity in several forms (see 15–16 n.; 16.25–27 n.), some have suggested that Paul, while writing to Rome, might also have sent a copy to Ephesus, for 16.1–16 reads like a letter of recommendation for Phoebe; or that he left a copy in Corinth (for the "strong" and the "weak" of 14.1–15.13 may reflect 1 Cor 8.9–13; 10.23–11.1). Others propose that Paul is writing with one eye on the situation in Jerusalem, where the collection is to be presented.

After the customary salutation and thanksgiving, Paul describes the world's need of redemption (1.18–3.20). He then discusses God's saving act in Christ: its nature (3.21–4.25) and the new life that it has made available (5.1–8.39). After a section dealing with the role of the Jewish nation in God's plan (chs 9–11), the letter closes with ethical teaching and some personal remarks (chs 12–16).

1 Paul, a servant[a] of Jesus Christ, called to be an apostle, set apart for the gospel of God, [2]which he promised beforehand through his prophets in the holy scriptures, [3]the gospel concerning his Son, who was descended from David according to the flesh [4]and was declared to be Son of God with power according to the spirit[b] of holiness by resurrection from the dead, Jesus Christ our Lord, [5]through whom we have received grace and apostleship to bring about the obedience of faith among all the Gentiles for the sake of his name, [6]including yourselves who are called to belong to Jesus Christ,

7 To all God's beloved in Rome, who are called to be saints:

Grace to you and peace from God our Father and the Lord Jesus Christ.

8 First, I thank my God through Jesus Christ for all of you, because your faith is proclaimed throughout the world. [9]For God, whom I serve with my spirit by announcing the gospel[c] of his Son, is my witness that without ceasing I remember you always in my prayers, [10]asking that by God's will I may somehow at last succeed in coming to you. [11]For I am longing to see you so that I may share with you some spiritual gift to strengthen you— [12]or rather so that we may be mutually encouraged by each other's faith, both yours and mine. [13]I

want you to know, brothers and sisters, [d] that I have often intended to come to you (but thus far have been prevented), in order that I may reap some harvest among you as I have among the rest of the Gentiles. [14]I am a debtor both to Greeks and to barbarians, both to the wise and to the foolish [15]—hence my eagerness to proclaim the gospel to you also who are in Rome.

16 For I am not ashamed of the gospel; it is the power of God for salvation to everyone who has faith, to the Jew first and also to the Greek. [17]For in it the righteousness of God is revealed through faith for faith; as it is written, "The one who is righteous will live by faith."[e]

18 For the wrath of God is revealed from heaven against all ungodliness and wickedness of those who by their wickedness suppress the truth. [19]For what can be known about God is plain to them, because God has shown it to them. [20]Ever since the creation of the world his eternal power and divine nature, invisible though they are, have been understood and seen through the things he has made. So they are without excuse; [21]for though they knew God, they did not honor him as God or give thanks to him, but they became futile in their thinking,

a Gk *slave*   b Or *Spirit*   c Gk *my spirit in the gospel*   d Gk *brothers*   e Or *The one who is righteous through faith will live*

1.1–7: Salutation. Ancient Greek letters customarily began with the names of the sender and of the recipient and a short greeting (compare Acts 23.26). Paul expands this form to express his Christian faith as well. **3–4:** God's *Son*, who came into the world physically *descended from David*, was manifested and installed in his true status at the resurrection. *The spirit of holiness*, the Holy Spirit. **7:** *Saints*, "holy ones," who belong to God, consecrated to God's service. *Grace . . . and peace*, conventional Greek and Hebrew greetings, now expressing gifts received in Christ (5.1–2). **1.8–15: Thanksgiving.** After the salutation in ancient letters there usually came a short prayer of thanksgiving or of petition for those addressed. This element also Paul expands in a characteristically Christian way, reflecting his plans and mutual hopes (vv. 10–15).

1.16–17: The theme of the letter. In Christ God has acted powerfully to save *Jew* and *Gentile*, offering righteousness and life, to be received in faith. **17:** *The righteousness of God* originates in the divine nature (3.5) acting to effect pardon or acceptance with God, a relationship that is not a human achievement but God's gift. *Through faith for faith*, faith is the sole condition of salvation. *The one who . . . will live*, Hab 2.4 in a Greek form allowing also the translation in the note; compare Gal 3.11; Phil 3.9; Heb 10.38.

1.18–32: God's judgment upon sin. 18: *Wrath*, see Col 3.6 n. **19:** *What can be known*, i.e. apart from God's revelation to Israel and

and their senseless minds were darkened. 22Claiming to be wise, they became fools; 23and they exchanged the glory of the immortal God for images resembling a mortal human being or birds or four-footed animals or reptiles.

24 Therefore God gave them up in the lusts of their hearts to impurity, to the degrading of their bodies among themselves, 25because they exchanged the truth about God for a lie and worshiped and served the creature rather than the Creator, who is blessed forever! Amen.

26 For this reason God gave them up to degrading passions. Their women exchanged natural intercourse for unnatural, 27and in the same way also the men, giving up natural intercourse with women, were consumed with passion for one another. Men committed shameless acts with men and received in their own persons the due penalty for their error.

28 And since they did not see fit to acknowledge God, God gave them up to a debased mind and to things that should not be done. 29They were filled with every kind of wickedness, evil, covetousness, malice. Full of envy, murder, strife, deceit, craftiness, they are gossips, 30slanderers, God-haters,*f* insolent, haughty, boastful, inventors of evil, rebellious toward parents, 31foolish, faithless, heartless, ruthless. 32They know God's decree, that those who practice such things deserve to die—yet they not only do them but even applaud others who practice them.

2 Therefore you have no excuse, whoever you are, when you judge others; for in passing judgment on another you condemn yourself, because you, the judge, are doing the very same things. 2You say,*g* "We know that God's judgment on those who do such things is in accordance with truth." 3Do you imagine, whoever you are, that when you judge those who do such things and yet do them yourself, you will escape the judgment of God? 4Or do you despise the riches of his kindness and forbearance and patience? Do you not realize that God's kindness is meant to lead you to repentance? 5But by your hard and impenitent heart you are storing up wrath for yourself on the day of wrath, when God's righteous judgment will be revealed. 6For he will repay according to each one's deeds: 7to those who by patiently doing good seek for glory and honor and immortality, he will give eternal life; 8while for those who are self-seeking and who obey not the truth but wickedness, there will be wrath and fury. 9There will be anguish and distress for everyone who does evil, the Jew first and also the Greek, 10but glory and honor and peace for everyone who does good, the Jew first and also the Greek. 11For God shows no partiality.

12 All who have sinned apart from the law will also perish apart from the law, and all who have sinned under the law will be judged by the law. 13For it is not the hearers of the law who are righteous in God's sight, but the doers of the law who will be justified. 14When Gentiles, who do not possess the law, do instinctively what the law requires, these, though not having the law, are a law to themselves. 15They show that what the law requires is written on their

f Or *God-hated*   *g* Gk lacks *You say*

---

in Christ. **20–21:** People have denied the knowledge about God available in the created world.

1.24, 26, 28: *God gave them up,* because in turning from God they violated their true nature, becoming involved in terrible and destructive perversions; God has let the process of death work itself out. **26–27:** See 1 Cor 6.9 n. **29–31:** Gal 5.19–21.

2.1–11: **Jews are under judgment,** as well as Gentiles (1.18–32). **5:** *The day of wrath,* God's condemnation of sin, already manifest (1.18), will be consummated in the final judgment. **9–10:** The Jew *first* (compare 1.16), because privileges granted to the covenant people (3.1–8) increase responsibility (1.16; Am 3.2; Lk 12.48).

2.12–29: **Basis for judgment:** law of Moses for the Jew; law of conscience for the Gentile. **14–15:** Paul recognizes (despite 1.18–

hearts, to which their own conscience also bears witness; and their conflicting thoughts will accuse or perhaps excuse them 16 on the day when, according to my gospel, God, through Jesus Christ, will judge the secret thoughts of all.

17 But if you call yourself a Jew and rely on the law and boast of your relation to God 18 and know his will and determine what is best because you are instructed in the law, 19 and if you are sure that you are a guide to the blind, a light to those who are in darkness, 20 a corrector of the foolish, a teacher of children, having in the law the embodiment of knowledge and truth, 21 you, then, that teach others, will you not teach yourself? While you preach against stealing, do you steal? 22 You that forbid adultery, do you commit adultery? You that abhor idols, do you rob temples? 23 You that boast in the law, do you dishonor God by breaking the law? 24 For, as it is written, "The name of God is blasphemed among the Gentiles because of you."

25 Circumcision indeed is of value if you obey the law; but if you break the law, your circumcision has become uncircumcision. 26 So, if those who are uncircumcised keep the requirements of the law, will not their uncircumcision be regarded as circumcision? 27 Then those who are physically uncircumcised but keep the law will condemn you that have the written code and circumcision but break the law. 28 For a person is not a Jew who is one outwardly, nor is true circumcision something external and phys-ical. 29 Rather, a person is a Jew who is one inwardly, and real circumcision is a matter of the heart—it is spiritual and not literal. Such a person receives praise not from others but from God.

3 Then what advantage has the Jew? Or what is the value of circumcision? 2 Much, in every way. For in the first place the Jews *h* were entrusted with the oracles of God. 3 What if some were unfaithful? Will their faithlessness nullify the faithfulness of God? 4 By no means! Although everyone is a liar, let God be proved true, as it is written,

"So that you may be justified in
  your words,
  and prevail in your judging." *i*

5 But if our injustice serves to confirm the justice of God, what should we say? That God is unjust to inflict wrath on us? (I speak in a human way.) 6 By no means! For then how could God judge the world? 7 But if through my falsehood God's truthfulness abounds to his glory, why am I still being condemned as a sinner? 8 And why not say (as some people slander us by saying that we say), "Let us do evil so that good may come"? Their condemnation is deserved!

9 What then? Are we any better off? *j* No, not at all; for we have already charged that all, both Jews and Greeks, are under the power of sin, 10 as it is written:

---

h Gk *they*   *i* Gk *when you are being judged*
j Or *at any disadvantage?*

---

32) that there are morally sensitive and responsible Gentiles, however short they may fall of God's righteous demands.

2.17–24: Many Jews, though possessing God's written law, fall short of even ordinary morality. **23:** *Do you . . . the law?,* may also be translated, "You dishonor God by breaking the law." **24:** Isa 52.5.

2.25: *Circumcision has become uncircumcision,* in the sense that the Jewish violator of the law stands before God precisely where the Gentile violator stands (Jer 9.25–26). **28:** Mt 3.9; Jn 8.39; Gal 6.15. **29:** *Circumcision . . . of the heart,* Deut 10.16; Jer 4.4; 9.26; Ezek 44.9.

3.1–8: The advantage of the Jews as the covenant people cannot be denied. To them were given the *oracles,* i.e. the Scriptures, and particularly the promises they contain. God's *faithfulness* in making the promises is not invalidated by the failure of the Jews to keep their part of the covenant; nor can that failure be excused on the plea that, because of it, God's *faithfulness* and *truthfulness* will shine the more brightly when God acts for good (Paul will discuss this problem more fully in chs 9–11). **4:** Ps 51.4. **5:** *Justice of God,* same phrase as in 1.17, "righteousness of God."

3.9–20: All are guilty. Jew and Greek, despite the former's advantages, stand on the same ground, *under the power of sin.* **10–18:** Pss

"There is no one who is righteous,
　　not even one;
11　　there is no one who has
　　　understanding,
　　there is no one who seeks
　　　God.
12　All have turned aside, together
　　　they have become
　　　worthless;
　　there is no one who shows
　　　kindness,
　　there is not even one."
13　"Their throats are opened graves;
　　they use their tongues to
　　　deceive."
　　"The venom of vipers is under
　　　their lips."
14　　"Their mouths are full of
　　　cursing and bitterness."
15　"Their feet are swift to shed
　　　blood;
16　　ruin and misery are in their
　　　paths,
17　and the way of peace they have
　　　not known."
18　　"There is no fear of God before
　　　their eyes."

19　Now we know that whatever the law says, it speaks to those who are under the law, so that every mouth may be silenced, and the whole world may be held accountable to God. 20 For "no human being will be justified in his sight" by deeds prescribed by the law, for through the law comes the knowledge of sin.

21　But now, apart from law, the righteousness of God has been disclosed, and is attested by the law and the prophets, 22 the righteousness of God through faith in Jesus Christ *k* for all who believe. For there is no distinction, 23 since all have sinned and fall short of the glory of God; 24 they are now justified by his grace as a gift, through the redemption that is in Christ Jesus, 25 whom God put forward as a sacrifice of atonement *l* by his blood, effective through faith. He did this to show his righteousness, because in his divine forbearance he had passed over the sins previously committed; 26 it was to prove at the present time that he himself is righteous and that he justifies the one who has faith in Jesus. *m*

27　Then what becomes of boasting? It is excluded. By what law? By that of works? No, but by the law of faith. 28 For we hold that a person is justified by faith apart from works prescribed by the law. 29 Or is God the God of Jews only? Is he not the God of Gentiles also? Yes, of Gentiles also, 30 since God is one; and he will justify the circumcised on the

---

k Or *through the faith of Jesus Christ*　*l* Or *a place of atonement*　*m* Or *who has the faith of Jesus*

14.1–2; 53.1–2; 5.9; 140.3; 10.7; Isa 59.7–8; Ps 36.1. The law succeeds only in making people aware of their condition. That indeed was God's purpose in giving it (7.7; see Gal 3.19–29 n.).
3.21–26: The true righteousness, now revealed in Christ, rests not upon obedience to law, but on faith in God's act of *redemption . . . in Christ Jesus.* **21:** *The law and the prophets,* the Hebrew Scriptures. **22:** *Faith in Jesus Christ;* the alternate rendering *faith of Jesus Christ* (also at 3.26; Gal 2.16, 20; 3.22; Phil 3.9) assumes Christ's faithful obedience unto death (Rom 5.18–21; Phil 2.8) or in becoming incarnate (Phil 2.6–8), or the entire ministry of Jesus, possibly involving an interchange of faith with *all who believe* in God (4.24) and Christ (Gal 2.16). **24:** *Justified,* same term as

1.17; 3.5, 21, 25, 26, "righteousness," "justice." *Redemption* means a ransoming or "buying back" (as of a slave or captive), and therefore emancipation or deliverance. Slaves of sin are set free through God's act in Christ (Eph 1.7; Col 1.14; Heb 9.15). **25:** *Sacrifice of atonement by his blood,* a reference to the death of Christ as a sacrifice for sin (1 Jn 2.2; Ex 25.22; Lev 16.12–15), demonstrating the seriousness with which God regards sin (despite divine *forbearance*); it reveals how God's love (5.8) effects forgiveness. **26:** *Faith,* see 3.22 n.
3.27–31: Boasting is excluded. One's *works* or deeds might be ground for boasting, but if salvation is by faith, pride *is excluded.* **27:** *Law,* here "principle," as at 7.21–23 n. **30:** Since *God is one* (Deut 6.4), Jews and Gentiles will be treated on the same basis, faith.

ground of faith and the uncircumcised through that same faith. 31 Do we then overthrow the law by this faith? By no means! On the contrary, we uphold the law.

4 What then are we to say was gained by[n] Abraham, our ancestor according to the flesh? 2 For if Abraham was justified by works, he has something to boast about, but not before God. 3 For what does the scripture say? "Abraham believed God, and it was reckoned to him as righteousness." 4 Now to one who works, wages are not reckoned as a gift but as something due. 5 But to one who without works trusts him who justifies the ungodly, such faith is reckoned as righteousness. 6 So also David speaks of the blessedness of those to whom God reckons righteousness apart from works:

7 "Blessed are those whose iniquities are forgiven,
and whose sins are covered;
8 blessed is the one against whom the Lord will not reckon sin."

9 Is this blessedness, then, pronounced only on the circumcised, or also on the uncircumcised? We say, "Faith was reckoned to Abraham as righteousness." 10 How then was it reckoned to him? Was it before or after he had been circumcised? It was not after, but before he was circumcised. 11 He received the sign of circumcision as a seal of the righteousness that he had by faith while he was still uncircumcised. The purpose was to make him the ancestor of all who believe without being circumcised and who thus have righteousness reckoned to them, 12 and likewise the ancestor of the circumcised who are not only circumcised but who also follow the example of the faith that our ancestor Abraham had before he was circumcised.

13 For the promise that he would inherit the world did not come to Abraham or to his descendants through the law but through the righteousness of faith. 14 If it is the adherents of the law who are to be the heirs, faith is null and the promise is void. 15 For the law brings wrath; but where there is no law, neither is there violation.

16 For this reason it depends on faith, in order that the promise may rest on grace and be guaranteed to all his descendants, not only to the adherents of the law but also to those who share the faith of Abraham (for he is the father of all of us, 17 as it is written, "I have made you the father of many nations")—in the presence of the God in whom he believed, who gives life to the dead and calls into existence the things that do not exist. 18 Hoping against hope, he believed that he would become "the father of many nations," according to what was said, "So numerous shall your descendants be." 19 He did not weaken in faith when he considered his own body, which was already[o] as good as dead (for he was about a hundred years old), or when he considered the barrenness of Sarah's womb. 20 No distrust made him

n Other ancient authorities read *say about*
o Other ancient authorities lack *already*

4.1–8: **Abraham justified by faith,** not by works. 2: *But not before God;* Abraham may have had grounds for glorying, but, Paul adds, not in God's sight. 3: According to Paul's understanding of Gen 15.6, Abraham's faith in God was credited to him as righteousness. The commercial term *reckoned* reminds Paul of Ps 32.2 (v. 8). 6–8: God's blessing belongs not to those who perfectly obey the law (as though that were possible), but to those who in faith accept God's free gift of forgiveness (Ps 32.1–2).
4.9–12: This justification of Abraham oc-

curred *before he was circumcised* (Gen 17.24), and therefore cannot have been dependent upon circumcision; it depended only upon faith (15.6). 11: Gen 17.10. 12: *Follow the example,* i.e. rely only on faith, as Abraham did.
4.13–25: **The true descendants of Abraham** are those who have faith in Christ, whether Jews or Gentiles. To them the benefits promised to Abraham belong (Gen 17.4–6; 22.17–18; Gal 3.29). 17: Gen 17.5. 18: Gen 15.5. 19: Gen 17.17; 18.11; Heb 11.11–12 (Sarah also). 22–23: See v. 3.

waver concerning the promise of God, but he grew strong in his faith as he gave glory to God, 21 being fully convinced that God was able to do what he had promised. 22 Therefore his faith*p* "was reckoned to him as righteousness." 23 Now the words, "it was reckoned to him," were written not for his sake alone, 24 but for ours also. It will be reckoned to us who believe in him who raised Jesus our Lord from the dead, 25 who was handed over to death for our trespasses and was raised for our justification.

5 Therefore, since we are justified by faith, we*q* have peace with God through our Lord Jesus Christ, 2 through whom we have obtained access*r* to this grace in which we stand; and we*s* boast in our hope of sharing the glory of God. 3 And not only that, but we*s* also boast in our sufferings, knowing that suffering produces endurance, 4 and endurance produces character, and character produces hope, 5 and hope does not disappoint us, because God's love has been poured into our hearts through the Holy Spirit that has been given to us.

6 For while we were still weak, at the right time Christ died for the ungodly. 7 Indeed, rarely will anyone die for a righteous person—though perhaps for a good person someone might actually dare to die. 8 But God proves his love for us in that while we still were sinners Christ died for us. 9 Much more surely then, now that we have been justified by his blood, will we be saved through him from the wrath of God.*t* 10 For if while we were enemies, we were reconciled to

God through the death of his Son, much more surely, having been reconciled, will we be saved by his life. 11 But more than that, we even boast in God through our Lord Jesus Christ, through whom we have now received reconciliation.

12 Therefore, just as sin came into the world through one man, and death came through sin, and so death spread to all because all have sinned— 13 sin was indeed in the world before the law, but sin is not reckoned when there is no law. 14 Yet death exercised dominion from Adam to Moses, even over those whose sins were not like the transgression of Adam, who is a type of the one who was to come.

15 But the free gift is not like the trespass. For if the many died through the one man's trespass, much more surely have the grace of God and the free gift in the grace of the one man, Jesus Christ, abounded for the many. 16 And the free gift is not like the effect of the one man's sin. For the judgment following one trespass brought condemnation, but the free gift following many trespasses brings justification. 17 If, because of the one man's trespass, death exercised dominion through that one, much more surely will those who receive the abundance of grace and the free gift of righteousness exercise dominion in life through the one man, Jesus Christ.

18 Therefore just as one man's trespass led to condemnation for all, so one

p Gk *Therefore it*   *q* Other ancient authorities read *let us*   *r* Other ancient authorities add *by faith*   *s* Or *let us*   *t* Gk *the wrath*

5.1–11: Consequences of justification. **1–5:** When we rely utterly upon God's grace and not at all upon ourselves, *we have peace,* i.e. reconciliation (5.10), *with God,* and other blessings. *Hope of . . . the glory of God,* though we had fallen short of the glorious destiny God intended for us (3.23), we now find ourselves confidently expecting it. **5.6–11:** Christ in death has borne the consequences of our sin and thus has reconciled us to God. Paul never speaks of God's being reconciled to us; it is we who were estranged.

9–10: Being *now . . . justified* and *reconciled by* Christ's *blood* (*death*), we *will . . . be saved* in the final judgment *by* (*through*) Christ's cross and resurrection (4.24, 25), including participation in his present *life* as the risen Lord. **11:** *Now,* under the gospel (3.21). **5.12–21: Adam and Christ; analogy and contrast.** Sin and death for all followed upon Adam's disobedience (Gen 2.17; 3.17–19). **13–16:** 1 Cor 15.21–23, 45–49. **5.18:** *Justification and life for all* followed upon Christ's perfect obedience. **20:** *Law . . .*

man's act of righteousness leads to justification and life for all. ¹⁹For just as by the one man's disobedience the many were made sinners, so by the one man's obedience the many will be made righteous. ²⁰But law came in, with the result that the trespass multiplied; but where sin increased, grace abounded all the more, ²¹so that, just as sin exercised dominion in death, so grace might also exercise dominion through justification*u* leading to eternal life through Jesus Christ our Lord.

6 What then are we to say? Should we continue in sin in order that grace may abound? ²By no means! How can we who died to sin go on living in it? ³Do you not know that all of us who have been baptized into Christ Jesus were baptized into his death? ⁴Therefore we have been buried with him by baptism into death, so that, just as Christ was raised from the dead by the glory of the Father, so we too might walk in newness of life.

5 For if we have been united with him in a death like his, we will certainly be united with him in a resurrection like his. ⁶We know that our old self was crucified with him so that the body of sin might be destroyed, and we might no longer be enslaved to sin. ⁷For whoever has died is freed from sin. ⁸But if we have died with Christ, we believe that we will also live with him. ⁹We know that Christ, being raised from the dead, will never die again; death no longer has dominion over him. ¹⁰The death he died, he died to

sin, once for all; but the life he lives, he lives to God. ¹¹So you also must consider yourselves dead to sin and alive to God in Christ Jesus.

12 Therefore, do not let sin exercise dominion in your mortal bodies, to make you obey their passions. ¹³No longer present your members to sin as instruments*v* of wickedness, but present yourselves to God as those who have been brought from death to life, and present your members to God as instruments*v* of righteousness. ¹⁴For sin will have no dominion over you, since you are not under law but under grace.

15 What then? Should we sin because we are not under law but under grace? By no means! ¹⁶Do you not know that if you present yourselves to anyone as obedient slaves, you are slaves of the one whom you obey, either of sin, which leads to death, or of obedience, which leads to righteousness? ¹⁷But thanks be to God that you, having once been slaves of sin, have become obedient from the heart to the form of teaching to which you were entrusted, ¹⁸and that you, having been set free from sin, have become slaves of righteousness. ¹⁹I am speaking in human terms because of your natural limitations.*w* For just as you once presented your members as slaves to impurity and to greater and greater iniquity, so now present your members as slaves to righteousness for sanctification.

20 When you were slaves of sin, you

---

u Or *righteousness*   *v* Or *weapons*   *w* Gk *the weakness of your flesh*

---

the trespass multiplied, this is explained in 7.7–13.
6.1–14: Dying and rising with Christ. Paul's insistence that salvation is entirely a gracious and undeserved gift of God may seem to have laid him open to the charge of encouraging sin. This charge Paul vigorously rejects. When Christians are *baptized,* they are united with Christ. They share in Christ's death and in the *newness of life* (v. 4) which his resurrection has made possible for us. But this death is a *death . . . to sin,* and the new life is *life . . . to God* (v. 10). *How can we who died to sin go on living in it?* (v. 2).

6.6: *The body of sin,* not the physical body as such, but the sinful self. **13:** *Your members,* all the faculties and functions of the person.
6.15–23: The two slaveries. In rejecting again the same charge (see v. 1 n.), Paul draws an analogy from slavery. Sinners are sin's slaves; but if they become God's slaves, how can they any longer obey their former master? **19:** *Sanctification,* the result of righteousness/justification (v. 22) and ongoing process of living this out in consecration to God and holiness (see 1.7 *saints;* 1 Thess 4.3– 8).

were free in regard to righteousness. 21 So what advantage did you then get from the things of which you now are ashamed? The end of those things is death. 22 But now that you have been freed from sin and enslaved to God, the advantage you get is sanctification. The end is eternal life. 23 For the wages of sin is death, but the free gift of God is eternal life in Christ Jesus our Lord.

7 Do you not know, brothers and sisters *x*—for I am speaking to those who know the law—that the law is binding on a person only during that person's lifetime? 2 Thus a married woman is bound by the law to her husband as long as he lives; but if her husband dies, she is discharged from the law concerning the husband. 3 Accordingly, she will be called an adulteress if she lives with another man while her husband is alive. But if her husband dies, she is free from that law, and if she marries another man, she is not an adulteress.

4 In the same way, my friends, *x* you have died to the law through the body of Christ, so that you may belong to another, to him who has been raised from the dead in order that we may bear fruit for God. 5 While we were living in the flesh, our sinful passions, aroused by the law, were at work in our members to bear fruit for death. 6 But now we are discharged from the law, dead to that which held us captive, so that we are slaves not under the old written code but in the new life of the Spirit.

7 What then should we say? That the law is sin? By no means! Yet, if it had not been for the law, I would not have known sin. I would not have known what it is to covet if the law had not said, "You shall not covet." 8 But sin, seizing an opportunity in the commandment, produced in me all kinds of covetousness. Apart from the law sin lies dead. 9 I was once alive apart from the law, but when the commandment came, sin revived 10 and I died, and the very commandment that promised life proved to be death to me. 11 For sin, seizing an opportunity in the commandment, deceived me and through it killed me. 12 So the law is holy, and the commandment is holy and just and good.

13 Did what is good, then, bring death to me? By no means! It was sin, working death in me through what is good, in order that sin might be shown to be sin, and through the commandment might become sinful beyond measure.

14 For we know that the law is spiritual; but I am of the flesh, sold into slavery under sin. *y* 15 I do not understand my own actions. For I do not do what I want, but I do the very thing I hate. 16 Now if I do what I do not want, I agree

x Gk *brothers*   *y* Gk *sold under sin*

7.1–6: An analogy from marriage. Christians who have died to sin are no more bound to it than is a woman to her deceased husband. **1–2:** *Law* generally, but specifically Num 5.20, 24; 30.10–14. **4–6:** *The law* of Moses, as in 2.12–27; 3.19–21, 28–31; 4.13–16.
7.7–13: The law and sin. *I, me:* Variously referred to Paul (before or after faith), Adam, the Jewish people, the carefree childhood of a Jewish boy, or all humanity, vv. 7–13 represent a dramatic device by Paul to look with Christian eyes on the law and the human situation. **7:** Though the law is *holy . . . and good* (v. 12), it not only makes one conscious of sin (see Gal 3.19 n.), but also incites to sin (e.g. covetousness; compare Ex 20.17; Deut 5.21). **9:** A Jewish boy was not held responsible for his actions until after the age of twelve when he became "Barmitzvah" (Son of the Commandment). **10:** Lev 18.5. **13:** The real enemy is sin, which uses even *what is good* (the law) to make persons more sinful than they would otherwise be.
7.14–23: The inner conflict. Sin is personified as an evil power that enters one's life and brings the true self into slavery to its rule.
7.14: *The law is spiritual,* divine in origin and nature, and holy (v. 12). *I am of the flesh,* referring not merely to our physical nature, but to our whole self, ruled by sin and selfishness (*I, me*), though the *inmost self* (v. 22) or *mind* (vv. 23, 25) rejoices in God's *law.* **17:** In emphasizing sin's power, Paul seems to deny one's responsibility for sin (compare v. 20), but other passages prevent inferring that he

that the law is good. [17] But in fact it is no longer I that do it, but sin that dwells within me. [18] For I know that nothing good dwells within me, that is, in my flesh. I can will what is right, but I cannot do it. [19] For I do not do the good I want, but the evil I do not want is what I do. [20] Now if I do what I do not want, it is no longer I that do it, but sin that dwells within me.

21 So I find it to be a law that when I want to do what is good, evil lies close at hand. [22] For I delight in the law of God in my inmost self, [23] but I see in my members another law at war with the law of my mind, making me captive to the law of sin that dwells in my members. [24] Wretched man that I am! Who will rescue me from this body of death? [25] Thanks be to God through Jesus Christ our Lord!

So then, with my mind I am a slave to the law of God, but with my flesh I am a slave to the law of sin.

8 There is therefore now no condemnation for those who are in Christ Jesus. [2] For the law of the Spirit[z] of life in Christ Jesus has set you[a] free from the law of sin and of death. [3] For God has done what the law, weakened by the flesh, could not do: by sending his own Son in the likeness of sinful flesh, and to deal with sin,[b] he condemned sin in the flesh, [4] so that the just requirement of the law might be fulfilled in us, who walk not according to the flesh but according to the Spirit.[z] [5] For those who live according to the flesh set their minds on the things of the flesh, but those who live according to the Spirit[z] set their minds on the things of the Spirit.[z] [6] To set the mind on the flesh is death, but to set the mind on the Spirit[z] is life and peace. [7] For this reason the mind that is set on the flesh is hostile to God; it does not submit to God's law—indeed it cannot, [8] and those who are in the flesh cannot please God.

9 But you are not in the flesh; you are in the Spirit,[z] since the Spirit of God dwells in you. Anyone who does not have the Spirit of Christ does not belong to him. [10] But if Christ is in you, though the body is dead because of sin, the Spirit[z] is life because of righteousness. [11] If the Spirit of him who raised Jesus from the dead dwells in you, he who raised Christ[c] from the dead will give life to your mortal bodies also through[d] his Spirit that dwells in you.

12 So then, brothers and sisters,[e] we

z Or *spirit*   a Here the Greek word *you* is singular number; other ancient authorities read *me* or *us*   b Or *and as a sin offering*   c Other ancient authorities read *the Christ* or *Christ Jesus* or *Jesus Christ*   d Other ancient authorities read *on account of*   e Gk *brothers*

---

means this (e.g. Rom 1.31–2.5; 5.12, 14). **21–23**: *Law,* here "principle" or "pattern," as at 3.27.

7.24–25: Despair and release. Threatened by utter defeat in the struggle with our enemy entrenched in our own lives, we cast ourselves upon God's mercy in Christ; only then do we find freedom from both the guilt and the power of sin. **24**: *This body of death,* i.e. the body, which is the instrument of sin, is under the dominion of death.

8.1–4: God's saving act. 1: *Condemnation* means judgment, doom, and death (2 Cor 3.7, 9), but not for those in Christ, because God has sentenced sin to death (*condemned sin,* v. 3). **2**: *The* life-giving *Spirit* is the divine principle (*law*) in the new order which God has created through Christ. To be *in Christ* is to belong to this new order and thus to have the Spirit (v. 9), who is the actual presence of God in our midst and in our hearts. **3**: *Sin offering* (in footnote), Lev 6.25; 14.19. **4**: Only through the power of *the Spirit* can we hope for the righteousness which *the law* requires but cannot enable us in our weakness to attain.

8.5–11: Life in the flesh and in the Spirit. 5: To live *according to the flesh* (see 7.14 n.) is to be dominated by selfish passions; to *live according to* (or *in,* v. 9) *the Spirit* is to belong to the new community of faith where God dwells as the Spirit. **8.9–10**: Note the interchangeable use of *the Spirit, the Spirit of God,* and *the Spirit of Christ,* and the interrelation with *Christ*. **10**: Gal 2.20; Eph 3.17. **11**: Jn 5.21.

8.12–17: The Spirit and adoption. The Spirit does not make slaves of us, but *children*

are debtors, not to the flesh, to live according to the flesh— 13 for if you live according to the flesh, you will die; but if by the Spirit you put to death the deeds of the body, you will live. 14 For all who are led by the Spirit of God are children of God. 15 For you did not receive a spirit of slavery to fall back into fear, but you have received a spirit of adoption. When we cry, "Abba!*f* Father!" 16 it is that very Spirit bearing witness*g* with our spirit that we are children of God, 17 and if children, then heirs, heirs of God and joint heirs with Christ—if, in fact, we suffer with him so that we may also be glorified with him.

18 I consider that the sufferings of this present time are not worth comparing with the glory about to be revealed to us. 19 For the creation waits with eager longing for the revealing of the children of God; 20 for the creation was subjected to futility, not of its own will but by the will of the one who subjected it, in hope 21 that the creation itself will be set free from its bondage to decay and will obtain the freedom of the glory of the children of God. 22 We know that the whole creation has been groaning in labor pains until now; 23 and not only the creation, but we ourselves, who have the first fruits of the Spirit, groan inwardly while we wait for adoption, the redemption of

our bodies. 24 For in*h* hope we were saved. Now hope that is seen is not hope. For who hopes*i* for what is seen? 25 But if we hope for what we do not see, we wait for it with patience.

26 Likewise the Spirit helps us in our weakness; for we do not know how to pray as we ought, but that very Spirit intercedes*j* with sighs too deep for words. 27 And God,*k* who searches the heart, knows what is the mind of the Spirit, because the Spirit*l* intercedes for the saints according to the will of God.*m*

28 We know that all things work together for good*n* for those who love God, who are called according to his purpose. 29 For those whom he foreknew he also predestined to be conformed to the image of his Son, in order that he might be the firstborn within a large family.*o* 30 And those whom he predestined he also called; and those whom he called he also justified; and those whom he justified he also glorified.

f Aramaic for *Father*   *g* Or 15 *a spirit of adoption, by which we cry, "Abba! Father!"* 16 *The Spirit itself bears witness*   *h* Or *by*   *i* Other ancient authorities read *awaits*   *j* Other ancient authorities add *for us*   *k* Gk *the one* *l* Gk *he* or *it*   *m* Gk *according to God* *n* Other ancient authorities read *God makes all things work together for good,* or *in all things God works for good*   *o* Gk *among many brothers*

---

of God. **15:** *Abba,* the Aramaic word meaning "Father," which Jesus used in his own prayers (Mk 14.36) and which passed into the prayer life of the early church. **16–17:** The fact that the *Spirit* prompts this ecstatic prayer proves our adoption (Gal 4.5–7).

8.18–25: **The hope of fulfillment. 18:** The Christian life involves *sufferings* (especially in Paul's mission work), but the apostle rejoices in the sure hope of *glory* (5.2). **20:** *The one who subjected* creation: God (Gen 3.17). **21:** When human beings (in Christ) are finally restored to their true destiny, nature will also share in this release from *bondage to decay* and in the *freedom of glory.* **22–23:** Nature is thought of as sharing in the stress, anxiety, and pain which we ourselves feel as we wait for the promised *redemption. The first fruits of the Spirit,* the Spirit, already received, is an advanced installment of the full adoption we

are yet to receive. *Our bodies,* as often in Paul, our "selves," expressed bodily (2 Cor 5.2–4). **24–25:** 1 Cor 2.9; 2 Cor 5.7; Heb 11.1.

8.26–30: **Human weakness is sustained** by the Spirit's intercession and by the knowledge of God's loving purpose. **28:** *His purpose,* or plan of salvation, is set forth in vv. 29–30. **29:** *To be conformed to . . . his Son* is to share the resurrection life of Christ, to be a "joint heir" (compare v. 17), to be *glorified* (v. 30).

8.31–39: **Our confidence in God. 31:** Ps 118.6. **32:** 4.25; 5.8; Jn 3.16. **35:** To be a Christian in the first century was both difficult and dangerous (compare 2 Cor 11.23–27). **36:** Ps 44.22. **38:** *Neither death, nor life,* i.e. whether we live or die (14.8) we shall not be separated. *Angels . . . rulers . . . powers,* supernatural beings, whether evil or good, and of various ranks (see Eph 6.12 n.). **39:** *Height*

31 What then are we to say about these things? If God is for us, who is against us? 32He who did not withhold his own Son, but gave him up for all of us, will he not with him also give us everything else? 33Who will bring any charge against God's elect? It is God who justifies. 34Who is to condemn? It is Christ Jesus, who died, yes, who was raised, who is at the right hand of God, who indeed intercedes for us.*p* 35Who will separate us from the love of Christ? Will hardship, or distress, or persecution, or famine, or nakedness, or peril, or sword? 36As it is written,

"For your sake we are being killed
 all day long;
we are accounted as sheep to be
 slaughtered."

37No, in all these things we are more than conquerors through him who loved us. 38For I am convinced that neither death, nor life, nor angels, nor rulers, nor things present, nor things to come, nor powers, 39nor height, nor depth, nor anything else in all creation, will be able to separate us from the love of God in Christ Jesus our Lord.

9 I am speaking the truth in Christ—I am not lying; my conscience confirms it by the Holy Spirit— 2I have great sorrow and unceasing anguish in my heart. 3For I could wish that I myself were accursed and cut off from Christ for the sake of my own people,*q* my kindred according to the flesh. 4They are Israelites, and to them belong the adoption, the glory, the covenants, the giving of the law, the worship, and the promises; 5to them belong the patriarchs, and from

them, according to the flesh, comes the Messiah,*r* who is over all, God blessed forever.*s* Amen.

6 It is not as though the word of God had failed. For not all Israelites truly belong to Israel, 7and not all of Abraham's children are his true descendants; but "It is through Isaac that descendants shall be named for you." 8This means that it is not the children of the flesh who are the children of God, but the children of the promise are counted as descendants. 9For this is what the promise said, "About this time I will return and Sarah shall have a son." 10Nor is that all; something similar happened to Rebecca when she had conceived children by one husband, our ancestor Isaac. 11Even before they had been born or had done anything good or bad (so that God's purpose of election might continue, 12not by works but by his call) she was told, "The elder shall serve the younger." 13As it is written,

"I have loved Jacob,
 but I have hated Esau."

14 What then are we to say? Is there injustice on God's part? By no means! 15For he says to Moses,

"I will have mercy on whom I
 have mercy,
and I will have compassion on
 whom I have compassion."

16So it depends not on human will or exertion, but on God who shows mercy. 17For the scripture says to Pharaoh, "I have raised you up for the very purpose

---

p Or *Is it Christ Jesus . . . for us?* *q* Gk *my brothers* *r* Or *the Christ* *s* Or *Messiah, who is God over all, blessed forever,* or *Messiah. May he who is God over all be blessed forever*

---

and *depth,* compare Ps 139.8; perhaps astrological, the highest point to which the stars rise and the abyss out of which they were thought to ascend; i.e. no such power can separate us from Christ or defeat God's purpose for us.

9.1–5: The problem of Israel's unbelief. 3: Ex 32.32. **4:** *Adoption,* i.e. as God's children (Ex 4.22; Jer 31.9). *Glory,* God's presence (Ex 16.10; 24.16). *Covenants,* plural because several existed with humanity and with Israel (Gen 6.18; 9.9; 15.8; 17.2, 7, 9; Ex 2.24). *Giving of the law,* Ex 20.1–17; Deut

5.1–21. *Worship,* in tabernacle and temple. **5:** *Messiah . . . God:* Whether Christ is called God here depends on the punctuation inserted (see footnote).

9.6–13: God's promise to Israel has not failed, because the promise was not made to all of Abraham's physical descendants, but to those whom God chose, i.e. Sarah, not Hagar (Gal 4.21–31); Isaac, not Ishmael; Jacob, not Esau. **7:** Gen 21.12. **9:** Gen 18.10. **10–12:** Gen 25.21, 23. **13:** Mal 1.2–3.

9.14–29: God's right to choose. 15: Ex 33.19. **17:** Ex 9.16. **19–21:** Isa 29.16; 45.9;

of showing my power in you, so that my name may be proclaimed in all the earth." 18 So then he has mercy on whomever he chooses, and he hardens the heart of whomever he chooses.

19 You will say to me then, "Why then does he still find fault? For who can resist his will?" 20 But who indeed are you, a human being, to argue with God? Will what is molded say to the one who molds it, "Why have you made me like this?" 21 Has the potter no right over the clay, to make out of the same lump one object for special use and another for ordinary use? 22 What if God, desiring to show his wrath and to make known his power, has endured with much patience the objects of wrath that are made for destruction; 23 and what if he has done so in order to make known the riches of his glory for the objects of mercy, which he has prepared beforehand for glory— 24 including us whom he has called, not from the Jews only but also from the Gentiles? 25 As indeed he says in Hosea,

"Those who were not my people I
    will call 'my people,'
and her who was not beloved I
    will call 'beloved.' "
26 "And in the very place where it
        was said to them, 'You are
        not my people,'
    there they shall be called
        children of the living God."

27 And Isaiah cries out concerning Israel, "Though the number of the children of Israel were like the sand of the sea, only a remnant of them will be saved; 28 for the Lord will execute his sentence on the earth quickly and decisively." *t* 29 And as Isaiah predicted,

"If the Lord of hosts had not left
    survivors *u* to us,
we would have fared like
    Sodom
and been made like Gomorrah."

30 What then are we to say? Gentiles, who did not strive for righteousness, have attained it, that is, righteousness through faith; 31 but Israel, who did strive for the righteousness that is based on the law, did not succeed in fulfilling that law. 32 Why not? Because they did not strive for it on the basis of faith, but as if it were based on works. They have stumbled over the stumbling stone, 33 as it is written,

"See, I am laying in Zion a stone
    that will make people
    stumble, a rock that will
    make them fall,
and whoever believes in him *v*
    will not be put to shame."

10 Brothers and sisters, *w* my heart's desire and prayer to God for them is that they may be saved. 2 I can testify that they have a zeal for God, but it is not enlightened. 3 For, being ignorant of the righteousness that comes from God, and seeking to establish their own, they have not submitted to God's righteousness. 4 For Christ is the end of the law so that there may be righteousness for everyone who believes.

5 Moses writes concerning the righteousness that comes from the law, that

t Other ancient authorities read *for he will finish his work and cut it short in righteousness, because the Lord will make the sentence shortened on the earth*
u Or *descendants*; Gk *seed*   *v* Or *trusts in it*
w Gk *Brothers*

64.8; Jer 18.6. **19:** As at 6.1, 7.7, and elsewhere, an objection, to which Paul responds. **9.24:** God's choice or election is not limited to *the Jews* (compare 3.29). **25–26:** The passage *in Hosea* (Hos 2.23; 1.10) refers to God's reclaiming of Israel after the nation had forsaken God and lost covenant status; Paul (as also 1 Pet 2.10) applies the promise to the Gentiles. **27–29:** God's promises never included all Israelites (Isa 10.22; 1.9). *Sodom* and *Gomorrah*, Gen 19.24–25.

9.30–10.13: True righteousness is by faith. 9.30: 3.22; 10.6, 20; Gal 2.16; 3.24; Phil 3.9; Heb 11.7. **32b–33:** The *stone* (in Isa 8.14–15, God, a *rock* over which Israel stumbles; in Isa 28.16, a symbol of salvation) is christologically interpreted (compare Mt 21.42). Belief or trust in it (Christ) brings salvation (10.10–11). **10.4:** *The end of the law,* its termination (Gal 3.23–26; 4.2–6) and/or its goal (3.31; 9.30–32, *strive* toward). **10.5:** Lev 18.5; Gal 3.12. One must actually

"the person who does these things will live by them." [6]But the righteousness that comes from faith says, "Do not say in your heart, 'Who will ascend into heaven?'" (that is, to bring Christ down) [7]"or 'Who will descend into the abyss?'" (that is, to bring Christ up from the dead). [8]But what does it say?

"The word is near you,
    on your lips and in your heart"
(that is, the word of faith that we proclaim); [9]because[x] if you confess with your lips that Jesus is Lord and believe in your heart that God raised him from the dead, you will be saved. [10]For one believes with the heart and so is justified, and one confesses with the mouth and so is saved. [11]The scripture says, "No one who believes in him will be put to shame." [12]For there is no distinction between Jew and Greek; the same Lord is Lord of all and is generous to all who call on him. [13]For, "Everyone who calls on the name of the Lord shall be saved."

14 But how are they to call on one in whom they have not believed? And how are they to believe in one of whom they have never heard? And how are they to hear without someone to proclaim him? [15]And how are they to proclaim him unless they are sent? As it is written, "How beautiful are the feet of those who bring good news!" [16]But not all have obeyed the good news;[y] for Isaiah says, "Lord, who has believed our message?" [17]So faith comes from what is heard, and what is heard comes through the word of Christ.[z]

18 But I ask, have they not heard? Indeed they have; for

"Their voice has gone out to all
    the earth,
and their words to the ends of
    the world."
[19]Again I ask, did Israel not understand? First Moses says,
"I will make you jealous of those
    who are not a nation;
with a foolish nation I will
    make you angry."
[20]Then Isaiah is so bold as to say,
"I have been found by those who
    did not seek me;
I have shown myself to those
    who did not ask for me."
[21]But of Israel he says, "All day long I have held out my hands to a disobedient and contrary people."

11 I ask, then, has God rejected his people? By no means! I myself am an Israelite, a descendant of Abraham, a member of the tribe of Benjamin. [2]God has not rejected his people whom he foreknew. Do you not know what the scripture says of Elijah, how he pleads with God against Israel? [3]"Lord, they have killed your prophets, they have demolished your altars; I alone am left, and they are seeking my life." [4]But what is the divine reply to him? "I have kept for myself seven thousand who have not bowed the knee to Baal." [5]So too at the present time there is a remnant, chosen by grace. [6]But if it is by grace, it is no longer on the basis of works, otherwise grace would no longer be grace.[a]

x Or *namely, that*   *y* Or *gospel*   *z* Or *about Christ*; other ancient authorities read *of God*
a Other ancient authorities add *But if it is by works, it is no longer on the basis of grace, otherwise work would no longer be work*

---

practice the law if one is to find life through it; this Paul has already shown to be impossible (3.9–20). But one has only to accept the free gift of the salvation in Christ (vv. 6–9; compare Deut 30.11–14). **10**: *Justified,* same term as *righteousness* (9.30; 10.3–6). Both faith and confession of Jesus are essential for justification and salvation. **11**: Isa 28.16. **13**: Joel 2.32. The early Christians often applied to Jesus Old Testament references to the *Lord,* which in their original context refer to God. **10.14–21: Israel responsible for its fail-**ure. **14–18**: The nation cannot claim that it has not had the opportunity of hearing the gospel. **15**: Isa 52.7. **16**: Isa 53.1. **10.18**: Ps 19.4. **19–21**: Nor can Israel claim that it has not understood the gospel; even Gentiles have been able to understand it. **19**: Deut 32.21. **20–21**: Isa 65.1–2.

11.1–16: Israel's rejection not final. 1–6: As in Elijah's time (1 Kings 19.10, 18), there is a *remnant* of the faithful. Paul as a Jew is no more alone than Elijah was.

7 What then? Israel failed to obtain what it was seeking. The elect obtained it, but the rest were hardened, [8] as it is written,

"God gave them a sluggish spirit,
eyes that would not see
and ears that would not hear,
down to this very day."

[9] And David says,

"Let their table become a snare
and a trap,
a stumbling block and a
retribution for them;

[10] let their eyes be darkened so that
they cannot see,
and keep their backs forever
bent."

11 So I ask, have they stumbled so as to fall? By no means! But through their stumbling[b] salvation has come to the Gentiles, so as to make Israel[c] jealous. [12] Now if their stumbling[b] means riches for the world, and if their defeat means riches for Gentiles, how much more will their full inclusion mean!

13 Now I am speaking to you Gentiles. Inasmuch then as I am an apostle to the Gentiles, I glorify my ministry [14] in order to make my own people[d] jealous, and thus save some of them. [15] For if their rejection is the reconciliation of the world, what will their acceptance be but life from the dead! [16] If the part of the dough offered as first fruits is holy, then the whole batch is holy; and if the root is holy, then the branches also are holy.

17 But if some of the branches were broken off, and you, a wild olive shoot, were grafted in their place to share the rich root[e] of the olive tree, [18] do not boast over the branches. If you do boast, remember that it is not you that support the root, but the root that supports you. [19] You will say, "Branches were broken off so that I might be grafted in." [20] That is true. They were broken off because of their unbelief, but you stand only through faith. So do not become proud, but stand in awe. [21] For if God did not spare the natural branches, perhaps he will not spare you.[f] [22] Note then the kindness and the severity of God: severity toward those who have fallen, but God's kindness toward you, provided you continue in his kindness; otherwise you also will be cut off. [23] And even those of Israel,[g] if they do not persist in unbelief, will be grafted in, for God has the power to graft them in again. [24] For if you have been cut from what is by nature a wild olive tree and grafted, contrary to nature, into a cultivated olive tree, how much more will these natural branches be grafted back into their own olive tree.

25 So that you may not claim to be wiser than you are, brothers and sisters,[h] I want you to understand this mystery: a hardening has come upon part of Israel,

b Gk *transgression*   c Gk *them*   d Gk *my flesh*
e Other ancient authorities read *the richness*
f Other ancient authorities read *neither will he
spare you*   g Gk lacks *of Israel*   h Gk *brothers*

11.7–12: The resistance to the gospel on the part of many Jews is providential; God has *hardened* their hearts for a loving purpose, namely, that the Gentiles might have an opportunity to hear and receive the gospel. **8**: Isa 29.10. **9**: Ps 69.22–23.
11.13–16: The *reconciliation* of Gentiles will have the effect of making Israelites *jealous* and thus of drawing *some of them* to Christ. **16**: *The dough* (Num 15.18–21 Septuagint) likely stands for a converted remnant of Jewish Christians; *the root* (Jer 11.16–17), Israel's *ancestors* (v. 28), through whom all Israel has been consecrated.
11.17–24: **The metaphor of the olive tree.** The tree, including root and branches, is Israel. The branches broken off are the un-believing Jews; the branches grafted in are Gentiles who believe in Christ. **20–22**: Having been made a part of the tree only because of faith (not merit or works), Gentile believers have no reason for pride, else God who has grafted them into the tree may later cut them off. **24**: The restoration of Israel will be easier for God than the call of the Gentiles.
11.25–36: **All Israel will be saved. 25–26**: A *mystery*, a truth once hidden, but now revealed by God. The *full number of the Gentiles* may mean the elect from among the Gentiles; and *all Israel* may mean all the elect (Gentiles and Jews), the elect of Israel as a people, or Israel as a whole, but not every Israelite. **26–27**: Isa 59.20–21; 27.9. **28–32**: Although some Jews may be temporarily *ene-*

until the full number of the Gentiles has come in. 26 And so all Israel will be saved; as it is written,

"Out of Zion will come the
    Deliverer;
  he will banish ungodliness
    from Jacob."
27 "And this is my covenant
    with them,
  when I take away their sins."
28 As regards the gospel they are enemies of God[i] for your sake; but as regards election they are beloved, for the sake of their ancestors; 29 for the gifts and the calling of God are irrevocable. 30 Just as you were once disobedient to God but have now received mercy because of their disobedience, 31 so they have now been disobedient in order that, by the mercy shown to you, they too may now[j] receive mercy. 32 For God has imprisoned all in disobedience so that he may be merciful to all.

33 O the depth of the riches and wisdom and knowledge of God! How unsearchable are his judgments and how inscrutable his ways!
34 "For who has known the mind of
    the Lord?
  Or who has been his
    counselor?"
35 "Or who has given a gift to him,
  to receive a gift in return?"
36 For from him and through him and to him are all things. To him be the glory forever. Amen.

12 I appeal to you therefore, brothers and sisters,[k] by the mercies of God, to present your bodies as a living sacrifice, holy and acceptable to God, which is your spiritual[l] worship. 2 Do not be conformed to this world,[m] but be transformed by the renewing of your minds, so that you may discern what is the will of God—what is good and acceptable and perfect.[n]

3 For by the grace given to me I say to everyone among you not to think of yourself more highly than you ought to think, but to think with sober judgment, each according to the measure of faith that God has assigned. 4 For as in one body we have many members, and not all the members have the same function, 5 so we, who are many, are one body in Christ, and individually we are members one of another. 6 We have gifts that differ according to the grace given to us: prophecy, in proportion to faith; 7 ministry, in ministering; the teacher, in teaching; 8 the exhorter, in exhortation; the giver, in generosity; the leader, in diligence; the compassionate, in cheerfulness.

9 Let love be genuine; hate what is evil, hold fast to what is good; 10 love one another with mutual affection; outdo one another in showing honor. 11 Do not lag in zeal, be ardent in spirit, serve the Lord.[o] 12 Rejoice in hope, be patient in suffering, persevere in prayer. 13 Contribute to the needs of the saints; extend hospitality to strangers.

14 Bless those who persecute you;

i Gk lacks *of God*   *j* Other ancient authorities lack *now*   *k* Gk *brothers*   *l* Or *reasonable*   *m* Gk *age*   *n* Or *what is the good and acceptable and perfect will of God*   *o* Other ancient authorities read *serve the opportune time*

---

mies of the *gospel,* the *election* of the Jews is *irrevocable.*
11.33: The wonder of God's providence. **34**: Isa 40.13. **35**: Job 35.7; 41.11. **36**: 1 Cor 8.6; 11.12; Col 1.16; Heb 2.10.
12.1–8: **The consecrated life. 1**: *Therefore,* in view of the argument above, especially 3.21–8.39. *Bodies,* as often in Paul, means "selves." *Living sacrifice,* as contrasted to the sacrifice of a slain beast. **2**: Christians are to live as belonging to the coming age, not this present world (Eph 2.2; 1 Jn 2.15). **3**: *Measure* *of faith,* the gift of faith to work miracles (1 Cor 13.2) or of trusting obedience in Christ, with which to measure oneself; or the faith or gospel that Christians confess. **4–8**: Gifts are to be used for the community, the *body in Christ;* see 1 Cor 12.4–31 n. **8**: *The leader,* administrator (1 Thess 5.12), perhaps of charity; or patron, benefactor (16.2).
12.9–21: **Exhortations. 9–10**: *Love,* compare 1 Cor 13 and 1 Thess 4.9. **13**: *Hospitality,* see 16.1–2 n.; Heb 13.2 n.; 3 Jn 5–8 n.
12.14: Mt 5.44. **19**: Vindication is God's

bless and do not curse them. 15 Rejoice with those who rejoice, weep with those who weep. 16 Live in harmony with one another; do not be haughty, but associate with the lowly;*p* do not claim to be wiser than you are. 17 Do not repay anyone evil for evil, but take thought for what is noble in the sight of all. 18 If it is possible, so far as it depends on you, live peaceably with all. 19 Beloved, never avenge yourselves, but leave room for the wrath of God;*q* for it is written, "Vengeance is mine, I will repay, says the Lord." 20 No, "if your enemies are hungry, feed them; if they are thirsty, give them something to drink; for by doing this you will heap burning coals on their heads." 21 Do not be overcome by evil, but overcome evil with good.

13 Let every person be subject to the governing authorities; for there is no authority except from God, and those authorities that exist have been instituted by God. 2 Therefore whoever resists authority resists what God has appointed, and those who resist will incur judgment. 3 For rulers are not a terror to good conduct, but to bad. Do you wish to have no fear of the authority? Then do what is good, and you will receive its approval; 4 for it is God's servant for your good. But if you do what is wrong, you should be afraid, for the authority*r* does not bear the sword in vain! It is the servant of God to execute wrath on the wrongdoer. 5 Therefore one must be subject, not only because of wrath but also because of conscience. 6 For the same

reason you also pay taxes, for the authorities are God's servants, busy with this very thing. 7 Pay to all what is due them—taxes to whom taxes are due, revenue to whom revenue is due, respect to whom respect is due, honor to whom honor is due.

8 Owe no one anything, except to love one another; for the one who loves another has fulfilled the law. 9 The commandments, "You shall not commit adultery; You shall not murder; You shall not steal; You shall not covet"; and any other commandment, are summed up in this word, "Love your neighbor as yourself." 10 Love does no wrong to a neighbor; therefore, love is the fulfilling of the law.

11 Besides this, you know what time it is, how it is now the moment for you to wake from sleep. For salvation is nearer to us now than when we became believers; 12 the night is far gone, the day is near. Let us then lay aside the works of darkness and put on the armor of light; 13 let us live honorably as in the day, not in reveling and drunkenness, not in debauchery and licentiousness, not in quarreling and jealousy. 14 Instead, put on the Lord Jesus Christ, and make no provision for the flesh, to gratify its desires.

14 Welcome those who are weak in faith,*s* but not for the purpose of quarreling over opinions. 2 Some believe

p Or give yourselves to humble tasks   q Gk the wrath   r Gk it   s Or conviction

---

prerogative, not ours (Deut 32.35). We are to pursue *good* in the face of evil (v. 21). **20:** To *heap burning coals* . . . makes enemies feel ashamed (Prov 25.21–22) and perhaps repentant. **13.1–7: The Christian and the state.** Though the Christian has no right to punish (12.19–21), the state does have that right and the Christian must respect it. Paul's confidence that the Roman state (under Nero!) is, on the whole, just and beneficent reflects Jewish teaching (Wis 6.1–3) and is matched in 1 Pet 2.13–17; 3.13. **13.8–10: Love fulfills the law. 8a:** Pay every debt; do not stand under any obligation

except the obligation to love. **8b:** Mk 12.31; Jas 2.8. **9:** Ex 20.13–17; Deut 5.17–21 (in the order of the Septuagint). **13.11–14: The imminence of Christ's second coming** makes it the more urgent that Christians *live honorably.* **14:** To *put on the Lord Jesus Christ* is to enter fully into the new order of existence which God has created through Christ (see 6.1–14 n.). **14.1–23: Love respects the scruples of others.** Some (Jewish) Christians had scruples about eating meat (v. 2), drinking wine (v. 21), and not observing the sabbath (vv. 5–6). Paul regards these scruples as unnecessary (v. 14; compare Mt 15.11) and designates

in eating anything, while the weak eat only vegetables. 3 Those who eat must not despise those who abstain, and those who abstain must not pass judgment on those who eat; for God has welcomed them. 4 Who are you to pass judgment on servants of another? It is before their own lord that they stand or fall. And they will be upheld, for the Lord*t* is able to make them stand.

5 Some judge one day to be better than another, while others judge all days to be alike. Let all be fully convinced in their own minds. 6 Those who observe the day, observe it in honor of the Lord. Also those who eat, eat in honor of the Lord, since they give thanks to God; while those who abstain, abstain in honor of the Lord and give thanks to God.

7 We do not live to ourselves, and we do not die to ourselves. 8 If we live, we live to the Lord, and if we die, we die to the Lord; so then, whether we live or whether we die, we are the Lord's. 9 For to this end Christ died and lived again, so that he might be Lord of both the dead and the living.

10 Why do you pass judgment on your brother or sister?*u* Or you, why do you despise your brother or sister?*u* For we will all stand before the judgment seat of God.*v* 11 For it is written,

"As I live, says the Lord, every
    knee shall bow to me,
and every tongue shall give
    praise to*w* God."

12 So then, each of us will be accountable to God.*x*

13 Let us therefore no longer pass judgment on one another, but resolve instead never to put a stumbling block or hindrance in the way of another.*y* 14 I

know and am persuaded in the Lord Jesus that nothing is unclean in itself; but it is unclean for anyone who thinks it unclean. 15 If your brother or sister*u* is being injured by what you eat, you are no longer walking in love. Do not let what you eat cause the ruin of one for whom Christ died. 16 So do not let your good be spoken of as evil. 17 For the kingdom of God is not food and drink but righteousness and peace and joy in the Holy Spirit. 18 The one who thus serves Christ is acceptable to God and has human approval. 19 Let us then pursue what makes for peace and for mutual upbuilding. 20 Do not, for the sake of food, destroy the work of God. Everything is indeed clean, but it is wrong for you to make others fall by what you eat; 21 it is good not to eat meat or drink wine or do anything that makes your brother or sister*u* stumble.*z* 22 The faith that you have, have as your own conviction before God. Blessed are those who have no reason to condemn themselves because of what they approve. 23 But those who have doubts are condemned if they eat, because they do not act from faith;*a* for whatever does not proceed from faith*a* is sin.*b*

15 We who are strong ought to put up with the failings of the weak, and not to please ourselves. 2 Each of us must please our neighbor for the good purpose of building up the neighbor.

t Other ancient authorities read *for God*
u Gk *brother*   *v* Other ancient authorities read *of Christ*   *w* Or *confess*   *x* Other ancient authorities lack *to God*   *y* Gk *of a brother*
z Other ancient authorities add *or be upset or be weakened*   *a* Or *conviction*   *b* Other authorities, some ancient, add here 16.25–27

---

those who are troubled by them *weak* or *weak in faith* (v. 1). But the "strong" must not pass judgment on the *weak* (v. 10). **11:** Isa 45.23.

14.13–15: The "strong" must also restrict their own liberty if they find that their example is injuring a *brother or sister* who has such scruples (1 Cor 8.9–13; 10.23–29a).

14.17–21: More important than our right to eat and drink as we please is our obligation not to *destroy the work of God* by making our

brother or sister stumble (1 Cor 10.23–24). **22:** *Blessed are those* who are free from misgivings as to the rightness of their practices, whether of eating or of not eating. **23:** Whatever is done against one's conscience *is sin.*

Chs 15–16: Some ancient writers knew a text of Romans without these chapters.

15.1–13: The strong should bear patiently burdens laid on them by *the failings of the weak.* **3:** Ps 69.9 is related to the self-

3 For Christ did not please himself; but, as it is written, "The insults of those who insult you have fallen on me." 4 For whatever was written in former days was written for our instruction, so that by steadfastness and by the encouragement of the scriptures we might have hope. 5 May the God of steadfastness and encouragement grant you to live in harmony with one another, in accordance with Christ Jesus, 6 so that together you may with one voice glorify the God and Father of our Lord Jesus Christ.

7 Welcome one another, therefore, just as Christ has welcomed you, for the glory of God. 8 For I tell you that Christ has become a servant of the circumcised on behalf of the truth of God in order that he might confirm the promises given to the patriarchs, 9 and in order that the Gentiles might glorify God for his mercy. As it is written,

"Therefore I will confess c you
among the Gentiles,
and sing praises to your name";
10 and again he says,
"Rejoice, O Gentiles, with his
people";
11 and again,
"Praise the Lord, all you Gentiles,
and let all the peoples
praise him";
12 and again Isaiah says,
"The root of Jesse shall come,
the one who rises to rule
the Gentiles;
in him the Gentiles shall hope."
13 May the God of hope fill you with all joy and peace in believing, so that you may abound in hope by the power of the Holy Spirit.

14 I myself feel confident about you, my brothers and sisters, d that you yourselves are full of goodness, filled with all knowledge, and able to instruct one another. 15 Nevertheless on some points I have written to you rather boldly by way of reminder, because of the grace given me by God 16 to be a minister of Christ Jesus to the Gentiles in the priestly service of the gospel of God, so that the offering of the Gentiles may be acceptable, sanctified by the Holy Spirit. 17 In Christ Jesus, then, I have reason to boast of my work for God. 18 For I will not venture to speak of anything except what Christ has accomplished e through me to win obedience from the Gentiles, by word and deed, 19 by the power of signs and wonders, by the power of the Spirit of God, f so that from Jerusalem and as far around as Illyricum I have fully proclaimed the good news g of Christ. 20 Thus I make it my ambition to proclaim the good news, g not where Christ has already been named, so that I do not build on someone else's foundation, 21 but as it is written,

"Those who have never been told
of him shall see,
and those who have never heard
of him shall understand."

22 This is the reason that I have so often been hindered from coming to you. 23 But now, with no further place for me in these regions, I desire, as I have for many years, to come to you 24 when I go to Spain. For I do hope to see you

c Or *thank*   d Gk *brothers*   e Gk *speak of those things that Christ has not accomplished* f Other ancient authorities read *of the Spirit* or *of the Holy Spirit*   g Or *gospel*

---

abasement of the pre-existent Christ (2 Cor 8.9; Phil 2.5–8), or the passion (Mk 15.29–32, 36). **8–12**: Christ had to minister to (become *a servant of*) the Jews to prove God's *truth* or faithfulness to promises to the *patriarchs* (9.4); but the promised salvation was also for the Gentiles, as Paul emphasizes with quotations from Ps 18.49; Deut 32.43; Ps 117.1; and Isa 11.10.
15.14–23: **Personal notes. 14–16**: An apology for Paul's apparent boldness in writing so long a letter to a church with which he had had no earlier connections.

15.14: *Able to instruct one another*, suggests a certain maturity and that the Roman church was not under the oversight of another apostle (compare v. 20). **19**: *Signs and wonders*, a reference to apostolic miracles (1 Cor 12.10; 2 Cor 12.12; Gal 3.5). *Illyricum*, modern Albania.

15.21: Isa 52.15. **23**: *No further place*, Paul conceives of his work as primarily that of an

on my journey and to be sent on by you, once I have enjoyed your company for a little while. 25 At present, however, I am going to Jerusalem in a ministry to the saints; 26 for Macedonia and Achaia have been pleased to share their resources with the poor among the saints at Jerusalem. 27 They were pleased to do this, and indeed they owe it to them; for if the Gentiles have come to share in their spiritual blessings, they ought also to be of service to them in material things. 28 So, when I have completed this, and have delivered to them what has been collected,*h* I will set out by way of you to Spain; 29 and I know that when I come to you, I will come in the fullness of the blessing*i* of Christ.

30 I appeal to you, brothers and sisters,*j* by our Lord Jesus Christ and by the love of the Spirit, to join me in earnest prayer to God on my behalf, 31 that I may be rescued from the unbelievers in Judea, and that my ministry*k* to Jerusalem may be acceptable to the saints, 32 so that by God's will I may come to you with joy and be refreshed in your company. 33 The God of peace be with all of you.*l* Amen.

16 I commend to you our sister Phoebe, a deacon*m* of the church at Cenchreae, 2 so that you may welcome her in the Lord as is fitting for the saints, and help her in whatever she may require from you, for she has been a benefactor of many and of myself as well.

3 Greet Prisca and Aquila, who work

with me in Christ Jesus, 4 and who risked their necks for my life, to whom not only I give thanks, but also all the churches of the Gentiles. 5 Greet also the church in their house. Greet my beloved Epaenetus, who was the first convert*n* in Asia for Christ. 6 Greet Mary, who has worked very hard among you. 7 Greet Andronicus and Junia,*o* my relatives*p* who were in prison with me; they are prominent among the apostles, and they were in Christ before I was. 8 Greet Ampliatus, my beloved in the Lord. 9 Greet Urbanus, our co-worker in Christ, and my beloved Stachys. 10 Greet Apelles, who is approved in Christ. Greet those who belong to the family of Aristobulus. 11 Greet my relative*q* Herodion. Greet those in the Lord who belong to the family of Narcissus. 12 Greet those workers in the Lord, Tryphaena and Tryphosa. Greet the beloved Persis, who has worked hard in the Lord. 13 Greet Rufus, chosen in the Lord; and greet his mother—a mother to me also. 14 Greet Asyncritus, Phlegon, Hermes, Patrobas, Hermas, and the brothers and sisters*j* who are with them. 15 Greet Philologus, Julia, Nereus and his sister, and Olym-

h Gk *have sealed to them this fruit* *i* Other ancient authorities add *of the gospel* *j* Gk *brothers* *k* Other ancient authorities read *my bringing of a gift* *l* One ancient authority adds 16.25-27 here *m* Or *minister* *n* Gk *first fruits* *o* Or *Junias*; other ancient authorities read *Julia* *p* Or *compatriots* *q* Or *compatriot*

---

evangelist taking the gospel to new territories. **25**: *Ministry to the saints,* the collection mentioned in 1 Cor 16.1-4; 2 Cor 8-9; and probably Gal 2.10.
15.30-31: Paul anticipates that he may have trouble in Jerusalem (Acts 21.7-28.31). *My ministry to Jerusalem,* the offering he is bringing.
16.1-23: **Greetings. 1-2**: Because ancient inns and hotels were often infested with prostitutes and bandits, Christians who traveled usually depended upon the hospitality of other believers (see 12.13 n.). *Deacon* may denote an office or simply one who serves. *Cenchreae*

was Corinth's seaport to the east. *Benefactor,* see 12.8 n.
16.3-16: Except for *Prisca and Aquila* (Acts ch 18; 1 Cor 16.19; 2 Tim 4.19), nothing is known about these persons beyond what is contained here. **5**: *The church in their house,* most congregations in the earliest period met in the homes of members (see Philem 2 n.). **6**: See v. 12. **7**: *Junia,* a woman's name; *Junias,* the same Greek read as a man's name. *Relatives,* Jews like Paul (9.3). *Apostles,* a larger group than the twelve; see 1 Cor 15.5, 7; Phil 2.25 *messenger* (and footnote). **11**: *Relative,* see v. 7 n. **12**: *Persis,* the name of a woman. **13**:

pas, and all the saints who are with them. 16 Greet one another with a holy kiss. All the churches of Christ greet you.

17 I urge you, brothers and sisters,^r to keep an eye on those who cause dissensions and offenses, in opposition to the teaching that you have learned; avoid them. 18 For such people do not serve our Lord Christ, but their own appetites,^s and by smooth talk and flattery they deceive the hearts of the simple-minded. 19 For while your obedience is known to all, so that I rejoice over you, I want you to be wise in what is good and guileless in what is evil. 20 The God of peace will shortly crush Satan under your feet. The grace of our Lord Jesus Christ be with you.^t

21 Timothy, my co-worker, greets you; so do Lucius and Jason and Sosipater, my relatives.^u

22 I Tertius, the writer of this letter, greet you in the Lord.^v

23 Gaius, who is host to me and to the whole church, greets you. Erastus, the city treasurer, and our brother Quartus, greet you.^w

25 Now to God^x who is able to strengthen you according to my gospel and the proclamation of Jesus Christ, according to the revelation of the mystery that was kept secret for long ages 26 but is now disclosed, and through the prophetic writings is made known to all the Gentiles, according to the command of the eternal God, to bring about the obedience of faith— 27 to the only wise God, through Jesus Christ, to whom^y be the glory forever! Amen.^z

r Gk *brothers*   s Gk *their own belly*   t Other ancient authorities lack this sentence
u Or *compatriots*   v Or *I Tertius, writing this letter in the Lord, greet you*   w Other ancient authorities add verse 24, *The grace of our Lord Jesus Christ be with all of you. Amen.*   x Gk *the one*   y Other ancient authorities lack *to whom.* The verse then reads, *to the only wise God be the glory through Jesus Christ forever. Amen.*
z Other ancient authorities lack 16.25–27 or include it after 14.23 or 15.33; others put verse 24 after verse 27

Rufus, see Mk 15.21 n. *His mother . . . also,* Paul felt for her the affection of a son. **16:** The *holy kiss,* a symbol of communal love among Christians, became a regular part of worship in the church (1 Cor 16.20; 2 Cor 13.12; 1 Thess 5.26; 1 Pet 5.14).

16.17: Gal 1.8–9; 2 Thess 3.6, 14; 2 Jn 10. **20:** *Crush Satan under your feet,* see Gen 3.15. **22:** *Tertius* is the amanuensis, or secretary, who wrote down what Paul dictated (see Col 4.18 n. and compare 1 Pet 5.12).
16.25–27: The benediction. Because these verses occur at different points in ancient Greek manuscripts and versions (see textual notes on 14.23, 15.33, and 16.27) and reflect a liturgical style and themes (like *mystery*) not found elsewhere in Romans, some consider 25–27 a fragment inserted variously by later scribes and editors. **25:** *Proclamation,* another word for "gospel" (1 Cor 1.21–24). **26:** *The prophetic writings* (the Old Testament) held a *mystery* or secret (v. 25) which became known only when Christ appeared.

The First Letter of Paul to the
Corinthians

This is one of the most valuable of Paul's letters, not only for the light it throws upon the character and mind of the apostle (e.g. 2.16; ch 4) and for its vigorous presentation of the gospel (1.18–31, "Christ crucified"; 15.1–11, the "good news" of Christ's death and resurrection), but also for the vivid pictures it brings us of the actual life and problems of a particular local church at the middle of the first century.

The church at Corinth was situated near the center of the Roman province of Achaia in one of the most important cities of Greece. Paul had himself brought the Christian message to Corinth and it was through his work that the church had been established there (Acts 18.1–11). One gathers from the two letters to the Corinthians in the New Testament canon that Paul's subsequent relations with this church were disturbed from time to time by doubts and suspicions on both sides, but that, in spite of even a "painful visit" to Corinth, he rejoiced at his confidence in this richly gifted community (2 Cor 2.1; 7.16). It is clear that he is writing this letter from Ephesus (16.8; compare Acts 19.1–40), just across the Aegean Sea from Corinth. The allusion to the "collection for the saints" (16.1; compare 2 Cor 9.1–2) shows that the letter must be dated earlier than Romans, but hardly more than two or three years earlier (see Introduction to Romans).

After the usual epistolary address (1.1–9), Paul lays the groundwork in "the message of the cross" (1.17–25) for dealing with divisions and disorders in the Corinthian community (1.10–6.20). Then a series of questions, posed by the Corinthians, is discussed, concerning marriage and other aspects of status in society (ch 7); practices in a pagan environment (8–11.1); propriety and principles in communal life, especially worship, in a group with many different spiritual gifts (11.2–14); the resurrection of Christ and of Christians (ch 15); and specific plans for the Corinthians and Paul (ch 16).

The letter is thus concerned directly or indirectly with doctrinal and ethical problems that were disturbing the Corinthian church. In settling these local problems Paul has bequeathed to the Church universal some of the most exalted chapters in his correspondence, such as the hymn on Christian love (ch 13) and the teaching on Christ's cross (ch 1) and resurrection (ch 15).

1 Paul, called to be an apostle of Christ Jesus by the will of God, and our brother Sosthenes,

2 To the church of God that is in Corinth, to those who are sanctified in Christ Jesus, called to be saints, together with all those who in every place call on the name of our Lord Jesus Christ, both their Lord*a* and ours:

3 Grace to you and peace from God our Father and the Lord Jesus Christ.

4 I give thanks to my*b* God always for you because of the grace of God that has been given you in Christ Jesus, 5 for in every way you have been enriched in him, in speech and knowledge of every kind— 6 just as the testimony of*c* Christ has been strengthened among you— 7 so that you are not lacking in any spiritual gift as you wait for the revealing of our Lord Jesus Christ. 8 He will also strengthen you to the end, so that you may be blameless on the day of our Lord Jesus Christ. 9 God is faithful; by him you were called into the fellowship of his Son, Jesus Christ our Lord.

10 Now I appeal to you, brothers and sisters,*d* by the name of our Lord Jesus Christ, that all of you be in agreement and that there be no divisions among you, but that you be united in the same mind and the same purpose. 11 For it has been reported to me by Chloe's people that there are quarrels among you, my brothers and sisters.*e* 12 What I mean is that each of you says, "I belong to Paul," or "I belong to Apollos," or "I belong to Cephas," or "I belong to Christ." 13 Has Christ been divided? Was Paul crucified for you? Or were you baptized in the name of Paul? 14 I thank God*f* that I baptized none of you except Crispus and Gaius, 15 so that no one can say that you were baptized in my name. 16 (I did baptize also the household of Stephanas; beyond that, I do not know whether I baptized anyone else.) 17 For Christ did not send me to baptize but to proclaim the gospel, and not with eloquent wisdom, so that the cross of Christ might not be emptied of its power.

18 For the message about the cross is foolishness to those who are perishing, but to us who are being saved it is the power of God. 19 For it is written,

"I will destroy the wisdom of
     the wise,
   and the discernment of the
     discerning I will thwart."

20 Where is the one who is wise? Where is the scribe? Where is the debater of this age? Has not God made foolish the wisdom of the world? 21 For since, in the wisdom of God, the world did not know God through wisdom, God decided, through the foolishness of our proclamation, to save those who believe. 22 For Jews demand signs and Greeks desire wisdom, 23 but we proclaim Christ cruci-

a Gk *theirs*   *b* Other ancient authorities lack
my   *c* Or *to*   *d* Gk *brothers*   *e* Gk *my*
brothers   *f* Other ancient authorities read *I am*
thankful

1.1–3: Salutation, see Rom 1.1–7 n. *Sosthenes,* perhaps the person referred to in Acts 18.17. **3:** *Grace . . . peace,* see Rom 1.7 n.
1.4–9: Thanksgiving, see Rom 1.8–15 n. Paul touches felicitously upon themes which he will discuss more critically later in the letter, namely, his readers' advanced understanding (*knowledge*) of the gospel, their eloquence (*speech*), and the variety of their spiritual gifts. **7–8:** *The revealing . . .* and *the day . . .* are allusions to the final time of judgment and salvation.
1.10–17: Divisions at Corinth. 11: *Chloe's people,* members of or slaves in her household who have either written or visited Paul. **12:** *Apollos,* Acts 18.24–28. *Cephas,* Peter (see Mt 16.18 n.). Possibly four parties existed in the church. **14–17:** Not to be taken to mean that Paul regarded baptism as unimportant, but only that he had not himself baptized many persons at Corinth. *Crispus,* former official of the synagogue at Corinth, Acts 18.8. *Gaius,* Rom 16.23. **16:** *Stephanas,* 1 Cor 16.15. **17:** *Eloquent wisdom,* some at Corinth took pride in their ability to talk in a sophisticated, philosophical way; over against this stands the wisdom of the cross.
1.18–2.5: Christ crucified is *the power of God* to save, a message to be accepted and lived. **19:** Isa 29.14 Gk. **20–21:** The saved are not the *wise* of this world (or the *powerful* or *noble,* v. 26), but *those who believe.*

fied, a stumbling block to Jews and foolishness to Gentiles, 24 but to those who are the called, both Jews and Greeks, Christ the power of God and the wisdom of God. 25 For God's foolishness is wiser than human wisdom, and God's weakness is stronger than human strength.

26 Consider your own call, brothers and sisters:*g* not many of you were wise by human standards,*h* not many were powerful, not many were of noble birth. 27 But God chose what is foolish in the world to shame the wise; God chose what is weak in the world to shame the strong; 28 God chose what is low and despised in the world, things that are not, to reduce to nothing things that are, 29 so that no one*i* might boast in the presence of God. 30 He is the source of your life in Christ Jesus, who became for us wisdom from God, and righteousness and sanctification and redemption, 31 in order that, as it is written, "Let the one who boasts, boast in*j* the Lord."

2 When I came to you, brothers and sisters,*g* I did not come proclaiming the mystery*k* of God to you in lofty words or wisdom. 2 For I decided to know nothing among you except Jesus Christ, and him crucified. 3 And I came to you in weakness and in fear and in much trembling. 4 My speech and my proclamation were not with plausible words of wisdom,*l* but with a demonstration of the Spirit and of power, 5 so that your faith might rest not on human wisdom but on the power of God.

6 Yet among the mature we do speak wisdom, though it is not a wisdom of this age or of the rulers of this age, who

are doomed to perish. 7 But we speak God's wisdom, secret and hidden, which God decreed before the ages for our glory. 8 None of the rulers of this age understood this; for if they had, they would not have crucified the Lord of glory. 9 But, as it is written,

"What no eye has seen, nor
    ear heard,
  nor the human heart conceived,
  what God has prepared for those
    who love him"—

10 these things God has revealed to us through the Spirit; for the Spirit searches everything, even the depths of God. 11 For what human being knows what is truly human except the human spirit that is within? So also no one comprehends what is truly God's except the Spirit of God. 12 Now we have received not the spirit of the world, but the Spirit that is from God, so that we may understand the gifts bestowed on us by God. 13 And we speak of these things in words not taught by human wisdom but taught by the Spirit, interpreting spiritual things to those who are spiritual.*m*

14 Those who are unspiritual*n* do not receive the gifts of God's Spirit, for they are foolishness to them, and they are unable to understand them because they are spiritually discerned. 15 Those who are spiritual discern all things, and they

g Gk *brothers*   *h* Gk *according to the flesh*
i Gk *no flesh*   *j* Or *of*   *k* Other ancient
authorities read *testimony*   *l* Other ancient
authorities read *the persuasiveness of wisdom*
m Or *interpreting spiritual things in spiritual
language,* or *comparing spiritual things with spiritual*
n Or *natural*

1.23: *Christ crucified,* the Christ who died on the cross, now risen and living. **28:** *Things that are not:* God chose the "nothings" in order to nullify the "somebodies" by worldly standards, so that boasting might be *in the Lord,* not ourselves. **31:** See Jer 9.23–24; 2 Cor 10.17; Gal 6.14.
2.6–16: Spiritual wisdom. God's Spirit imparts a deeper wisdom than any human speculation can achieve, a wisdom which only the *mature* (those who have been *taught by the Spirit*) can understand. **7:** *Wisdom, secret*

and hidden, God's plan for salvation, redemption in Christ, now present through the Spirit and soon to be fully manifested. **8:** *The rulers of this age,* political leaders (Acts 4.25–28), or cosmic, demonic powers (Eph 1.20–21; 3.10; 6.12), or both. Not recognizing Christ for what he was, they crucified him; but his cross was the means of his victory over them (Col 2.14–15). **9:** Compare Isa 64.4.
2.10–12: *The Spirit* we have received is God's own Spirit who knows what is in God, as our own spirits know what is in us. **15:**

are themselves subject to no one else's scrutiny.

16 "For who has known the mind
   of the Lord
      so as to instruct him?"

But we have the mind of Christ.

3 And so, brothers and sisters,[o] I could not speak to you as spiritual people, but rather as people of the flesh, as infants in Christ. 2I fed you with milk, not solid food, for you were not ready for solid food. Even now you are still not ready, 3for you are still of the flesh. For as long as there is jealousy and quarreling among you, are you not of the flesh, and behaving according to human inclinations? 4For when one says, "I belong to Paul," and another, "I belong to Apollos," are you not merely human?

5 What then is Apollos? What is Paul? Servants through whom you came to believe, as the Lord assigned to each. 6I planted, Apollos watered, but God gave the growth. 7So neither the one who plants nor the one who waters is anything, but only God who gives the growth. 8The one who plants and the one who waters have a common purpose, and each will receive wages according to the labor of each. 9For we are God's servants, working together; you are God's field, God's building.

10 According to the grace of God given to me, like a skilled master builder I laid a foundation, and someone else is building on it. Each builder must choose with care how to build on it. 11For no one can lay any foundation other than the one that has been laid; that foundation is Jesus Christ. 12Now if anyone builds on the foundation with gold, silver, precious stones, wood, hay, straw— 13the work of each builder will become visible, for the Day will disclose it, because it will be revealed with fire, and the fire will test what sort of work each has done. 14If what has been built on the foundation survives, the builder will receive a reward. 15If the work is burned up, the builder will suffer loss; the builder will be saved, but only as through fire.

16 Do you not know that you are God's temple and that God's Spirit dwells in you?[p] 17If anyone destroys God's temple, God will destroy that person. For God's temple is holy, and you are that temple.

18 Do not deceive yourselves. If you think that you are wise in this age, you should become fools so that you may become wise. 19For the wisdom of this world is foolishness with God. For it is written,

   "He catches the wise in their
      craftiness,"

20and again,

   "The Lord knows the thoughts of
      the wise,
      that they are futile."

o Gk *brothers*   p In verses 16 and 17 the Greek word for *you* is plural

---

Perhaps Paul means that the spiritual are not to be judged by the unspiritual (4.3–4). **16:** Those with the Spirit have the *mind of Christ,* and no one is in a position to instruct *Christ. The Lord* (Isa 40.13 Gk, referring to God) is here applied to Christ.

3.1–9: Dissension over leaders. Paul has apparently been criticized at Corinth for preaching too simple a gospel (2.1–2). He explains that, though he has wisdom to impart (2.6), he could impart it only to *spiritual people,* the mature (2.6); but the Corinthians are *infants* and do not qualify. They lack the real community to which God's servants like Apollos and Paul each contribute, as Corinthian party slogans show (1.12–13; 3.4).

3.10–23: Teachers and church under God. 10–15: Paul turns to the work of missionary preachers who edify the church; at the Judgment Day (v. 13) God will test what each has built. **16–17:** Since the community (*you,* plural) is a *temple* where God is present (6.19), those who divide and destroy it will suffer condemnation.

3.18–20: Most divisive and harmful is the "wisdom" that finds its center in "this age" (2.6) and not in God's wisdom, Christ and the cross (1.18–25). Paul refutes such pretensions by quoting Job 5.13 and Ps 94.11. **21–23:** You do not belong to any human leader; all of these leaders belong to you, the community as a whole, as indeed does everything else. But this is true only because you belong to Christ and thus to God.

21 So let no one boast about human leaders. For all things are yours, 22 whether Paul or Apollos or Cephas or the world or life or death or the present or the future—all belong to you, 23 and you belong to Christ, and Christ belongs to God.

4 Think of us in this way, as servants of Christ and stewards of God's mysteries. 2 Moreover, it is required of stewards that they be found trustworthy. 3 But with me it is a very small thing that I should be judged by you or by any human court. I do not even judge myself. 4 I am not aware of anything against myself, but I am not thereby acquitted. It is the Lord who judges me. 5 Therefore do not pronounce judgment before the time, before the Lord comes, who will bring to light the things now hidden in darkness and will disclose the purposes of the heart. Then each one will receive commendation from God.

6 I have applied all this to Apollos and myself for your benefit, brothers and sisters, *q* so that you may learn through us the meaning of the saying, "Nothing beyond what is written," so that none of you will be puffed up in favor of one against another. 7 For who sees anything different in you? *r* What do you have that you did not receive? And if you received it, why do you boast as if it were not a gift?

8 Already you have all you want! Already you have become rich! Quite apart from us you have become kings! Indeed, I wish that you had become kings, so that we might be kings with you! 9 For I think that God has exhibited us apostles as last of all, as though sentenced to death, because we have become a spectacle to the world, to angels and to mortals. 10 We are fools for the sake of Christ, but you are wise in Christ. We are weak, but you are strong. You are held in honor, but we in disrepute. 11 To the present hour we are hungry and thirsty, we are poorly clothed and beaten and homeless, 12 and we grow weary from the work of our own hands. When reviled, we bless; when persecuted, we endure; 13 when slandered, we speak kindly. We have become like the rubbish of the world, the dregs of all things, to this very day.

14 I am not writing this to make you ashamed, but to admonish you as my beloved children. 15 For though you might have ten thousand guardians in Christ, you do not have many fathers. Indeed, in Christ Jesus I became your father through the gospel. 16 I appeal to you, then, be imitators of me. 17 For this reason I sent *s* you Timothy, who is my beloved and faithful child in the Lord, to remind you of my ways in Christ Jesus, as I teach them everywhere in every church. 18 But some of you, thinking that I am not coming to you, have become arrogant. 19 But I will come to you soon, if the Lord wills, and I will find out not the talk of these arrogant people but their power. 20 For the kingdom of God de-

q Gk *brothers* *r* Or *Who makes you different from another?* *s* Or *am sending*

4.1–13: Applications. 1–7: The teachers themselves, as *servants of Christ,* are answerable only to the Lord. **4–5:** Since only the Lord knows the heart, both boasting and blaming are ruled out.

4.6: If this is true for such teachers as Paul and Apollos, it is true also for church factions. *What is written,* in Scripture, thus avoiding speculations. **7:** See 3.21–23 n.

4.8–13: Corinthian pretensions are rebuked in ironical statements by which Paul hopes to shame his critics. Some in Corinth felt they were already in God's final kingdom, rich and full (v. 8). Contrast the lot of apostles like Paul. **9:** A reference to captives thrown to beasts in the arena or to triumphal processions where military conquerors displayed their captives. **10:** *Sentenced to death, . . . a spectacle,* like condemned criminals in the arena. **12:** Paul's *work,* Acts 18.2–3; 1 Cor 9.6, 18.

4.14–21: Fatherly admonition and warning. 15: The *guardian* (translated "disciplinarian" in Gal 3.24) was the slave who looked after a child's conduct when not in school. A wide gulf separated this slave and the father; Paul's relation to the Corinthians was unique. **16:** *Imitators,* see 11.1. **17:** *Timothy,* one of Paul's most faithful helpers (Acts 16.1–3). **19, 20:** *Power,* of the Spirit.

pends not on talk but on power. 21 What would you prefer? Am I to come to you with a stick, or with love in a spirit of gentleness?

5 It is actually reported that there is sexual immorality among you, and of a kind that is not found even among pagans; for a man is living with his father's wife. 2 And you are arrogant! Should you not rather have mourned, so that he who has done this would have been removed from among you?

3 For though absent in body, I am present in spirit; and as if present I have already pronounced judgment 4 in the name of the Lord Jesus on the man who has done such a thing.*t* When you are assembled, and my spirit is present with the power of our Lord Jesus, 5 you are to hand this man over to Satan for the destruction of the flesh, so that his spirit may be saved in the day of the Lord.*u*

6 Your boasting is not a good thing. Do you not know that a little yeast leavens the whole batch of dough? 7 Clean out the old yeast so that you may be a new batch, as you really are unleavened. For our paschal lamb, Christ, has been sacrificed. 8 Therefore, let us celebrate the festival, not with the old yeast, the yeast of malice and evil, but with the unleavened bread of sincerity and truth.

9 I wrote to you in my letter not to associate with sexually immoral persons— 10 not at all meaning the immoral of this world, or the greedy and robbers, or idolaters, since you would then need to go out of the world. 11 But now I am writing to you not to associate with anyone who bears the name of brother or sister*v* who is sexually immoral or greedy, or is an idolater, reviler, drunkard, or robber. Do not even eat with such a one. 12 For what have I to do with judging those outside? Is it not those who are inside that you are to judge? 13 God will judge those outside. "Drive out the wicked person from among you."

6 When any of you has a grievance against another, do you dare to take it to court before the unrighteous, instead of taking it before the saints? 2 Do you not know that the saints will judge the world? And if the world is to be judged by you, are you incompetent to try trivial cases? 3 Do you not know that we are to judge angels—to say nothing of ordinary matters? 4 If you have ordinary cases, then, do you appoint as judges those who have no standing in the church? 5 I say this to your shame. Can it be that there is no one among you wise enough to decide between one believer*v* and another, 6 but a believer*v* goes to court against a believer*v*—and before unbelievers at that?

t Or on the man who has done such a thing in the name of the Lord Jesus   u Other ancient authorities add Jesus   v Gk brother

---

Chs 5–6: Disorders in Corinth.
5.1–13: A case for church discipline. 1: *Reported,* possibly by Chloe's people (1.11). Both Roman and Jewish law (Deut 22.20; 27.20; Lev 18.7–8) forbade marriage between a man and his stepmother. **2:** Paul is as much shocked by the complacency of the congregation in tolerating this sin as by the sin itself. **5:** Membership in the church protects one from Satan's destructive power; this man does not deserve such protection. Once excluded, he will be subject to disease and death (*the destruction of the flesh*), but there is hope that his spirit may be saved in the final judgment (see 1 Tim 1.20 n.). **7:** *Yeast . . . unleavened,* Ex 12.15. *Paschal lamb,* Ex 12.3–8, 21; Mk 14.12; Lk 22.7. **8:** Our new life is ethically a continuing festival, with bread unleavened by *malice and evil.*
5.9: An earlier *letter,* which has been lost (unless possibly a fragment of it is now found in 2 Cor 6.14–7.1). **11:** Immoral people are bound to be met with in public social life, but they must not be tolerated as brothers or sisters (compare Deut 17.2–7).
6.1–8: Lawsuits in pagan courts. Christians ought to settle their differences outside of Roman courts, for they will participate with Christ in the final, eschatological judgment of pagan magistrates; how absurd then that they should now abide by the judgments of outsiders! **1:** *Unrighteous,* unbelievers (v. 6). **2:** *Will judge,* compare Mt 19.28. **7–8:** A rebuke to both sides of every lawsuit.

7 In fact, to have lawsuits at all with one another is already a defeat for you. Why not rather be wronged? Why not rather be defrauded? 8 But you yourselves wrong and defraud—and believers*w* at that.

9 Do you not know that wrongdoers will not inherit the kingdom of God? Do not be deceived! Fornicators, idolaters, adulterers, male prostitutes, sodomites, 10thieves, the greedy, drunkards, revilers, robbers—none of these will inherit the kingdom of God. 11 And this is what some of you used to be. But you were washed, you were sanctified, you were justified in the name of the Lord Jesus Christ and in the Spirit of our God.

12 "All things are lawful for me," but not all things are beneficial. "All things are lawful for me," but I will not be dominated by anything. 13 "Food is meant for the stomach and the stomach for food,"*x* and God will destroy both one and the other. The body is meant not for fornication but for the Lord, and the Lord for the body. 14 And God raised the Lord and will also raise us by his power. 15 Do you not know that your bodies are members of Christ? Should I therefore take the members of Christ and make them members of a prostitute? Never! 16 Do you not know that whoever is united to a prostitute becomes one body with her? For it is said, "The two shall be one flesh." 17 But anyone united to the Lord becomes one spirit with him. 18 Shun fornication! Every sin that a person commits is outside the body; but the fornicator sins against the body itself. 19 Or do you not know that your body is a temple*y* of the Holy Spirit within you, which you have from God, and that you are not your own? 20 For you were bought with a price; therefore glorify God in your body.

7 Now concerning the matters about which you wrote: "It is well for a man not to touch a woman." 2 But because of cases of sexual immorality, each man should have his own wife and each woman her own husband. 3 The husband should give to his wife her conjugal rights, and likewise the wife to her husband. 4 For the wife does not have authority over her own body, but the husband does; likewise the husband does not have authority over his own body, but the wife does. 5 Do not deprive one another except perhaps by agreement for a set time, to devote yourselves to prayer, and then come together again, so that Satan may not tempt you because of your lack of self-control. 6 This I say by way of concession, not of command. 7 I wish that all were as I myself am. But each has a particular gift from God, one having one kind and another a different kind.

8 To the unmarried and the widows I say that it is well for them to remain

---

w Gk *brothers*   *x* The quotation may extend to the word *other*   *y* Or *sanctuary*

6.9–20: **A warning against laxity.** Corinth was known in antiquity as a particularly licentious city. **9–10:** A traditional, probably catechetical, early Christian list of vices. **6.9:** *Male prostitutes,* young men or boys in a pederastic relationship; *sodomites,* the older homosexual; see Rom 1.26–27; 1 Tim 1.10; possibly 1 Cor 11.4–7 is pertinent. **11bc:** Baptismal tradition (*washed*). **6.12–13:** Slogans quoted from Paul's opponents. **13:** The libertines argued that satisfying sexual desire was like taking food to satisfy one's hunger. Paul rejects this analogy. **16:** Immorality involves the whole *body,* which for Paul means one's entire personal life (compare Gen 2.24; Mt 19.5; Eph 5.31).

19: *Your body,* of each Christian; contrast 3.16. **20:** We belong to Christ, who has *bought* us with his own blood (7.23).

Chs 7–14, 16: Discussion of questions from the Corinthians.

7.1–16: Directions about marriage. 1: The phrase *Now concerning* introduces the first of a series of questions raised in Corinth (7.25; 8.1; 12.1; 16.1, 12; see 15.1). "*It . . . woman,*" a quotation from the Corinthians (cf. 6.12–13 n.), favoring asceticism. Paul's reasons for favoring celibacy (v. 8) are given in vv. 26, 32–34. **3–4:** Equality of *conjugal rights* between husband and wife and the need for mutual consideration.

7.8: *As I am:* Paul was therefore single or

unmarried as I am. 9 But if they are not practicing self-control, they should marry. For it is better to marry than to be aflame with passion.

10 To the married I give this command—not I but the Lord—that the wife should not separate from her husband 11 (but if she does separate, let her remain unmarried or else be reconciled to her husband), and that the husband should not divorce his wife.

12 To the rest I say—I and not the Lord—that if any believer *z* has a wife who is an unbeliever, and she consents to live with him, he should not divorce her. 13 And if any woman has a husband who is an unbeliever, and he consents to live with her, she should not divorce him. 14 For the unbelieving husband is made holy through his wife, and the unbelieving wife is made holy through her husband. Otherwise, your children would be unclean, but as it is, they are holy. 15 But if the unbelieving partner separates, let it be so; in such a case the brother or sister is not bound. It is to peace that God has called you. *a* 16 Wife, for all you know, you might save your husband. Husband, for all you know, you might save your wife.

17 However that may be, let each of you lead the life that the Lord has assigned, to which God called you. This is my rule in all the churches. 18 Was anyone at the time of his call already circumcised? Let him not seek to remove the marks of circumcision. Was anyone at the time of his call uncircumcised? Let him not seek circumcision. 19 Circumci-sion is nothing, and uncircumcision is nothing; but obeying the commandments of God is everything. 20 Let each of you remain in the condition in which you were called.

21 Were you a slave when called? Do not be concerned about it. Even if you can gain your freedom, make use of your present condition now more than ever. *b* 22 For whoever was called in the Lord as a slave is a freed person belonging to the Lord, just as whoever was free when called is a slave of Christ. 23 You were bought with a price; do not become slaves of human masters. 24 In whatever condition you were called, brothers and sisters, *c* there remain with God.

25 Now concerning virgins, I have no command of the Lord, but I give my opinion as one who by the Lord's mercy is trustworthy. 26 I think that, in view of the impending *d* crisis, it is well for you to remain as you are. 27 Are you bound to a wife? Do not seek to be free. Are you free from a wife? Do not seek a wife. 28 But if you marry, you do not sin, and if a virgin marries, she does not sin. Yet those who marry will experience distress in this life, *e* and I would spare you that. 29 I mean, brothers and sisters, *c* the appointed time has grown short; from now on, let even those who have wives be as though they had none, 30 and those who mourn as though they were not mourning, and those who rejoice as though

---

z Gk *brother*   a Other ancient authorities read
us   b Or *avail yourself of the opportunity*
c Gk *brothers*   d Or *present*   e Gk *in the flesh*

---

a widower. **10–11:** *Not I but the Lord,* a possible reference to the teaching of Jesus found in Mk 10.2–9.

7.12: *I and not the Lord,* Paul can cite no remembered saying of Jesus as authority for what he now writes. **14:** *Is made holy,* is brought in some sense within the sphere of salvation. *They are holy,* i.e. regarded as Christian children. **15:** *The brother or sister,* the Christian partner.

7.17–40: Eschatology and changes in social and marital status.

7.17–24: Because the end of the world is fast approaching (vv. 26, 29–31), it is better for everyone to remain as is and not try to change his or her outward situation. But believers are free from bondage to "this world" and its environment.

7.25–35: Paul repeats his counsel about marriage, justifying his caution with a reference to the shortness of the time and the desirability of not being involved in the distracting obligations of family life. **25:** *Virgins:* This group, about which the Corinthians had asked (see 7.1 n.), may refer to unmarried but engaged couples or possibly to a couple married but ascetically committed not to have sexual relations; see vv. 28, 34, 36–38.

they were not rejoicing, and those who buy as though they had no possessions, 31 and those who deal with the world as though they had no dealings with it. For the present form of this world is passing away.

32 I want you to be free from anxieties. The unmarried man is anxious about the affairs of the Lord, how to please the Lord; 33 but the married man is anxious about the affairs of the world, how to please his wife, 34 and his interests are divided. And the unmarried woman and the virgin are anxious about the affairs of the Lord, so that they may be holy in body and spirit; but the married woman is anxious about the affairs of the world, how to please her husband. 35 I say this for your own benefit, not to put any restraint upon you, but to promote good order and unhindered devotion to the Lord.

36 If anyone thinks that he is not behaving properly toward his fiancée,*f* if his passions are strong, and so it has to be, let him marry as he wishes; it is no sin. Let them marry. 37 But if someone stands firm in his resolve, being under no necessity but having his own desire under control, and has determined in his own mind to keep her as his fiancée,*f* he will do well. 38 So then, he who marries his fiancée*f* does well; and he who refrains from marriage will do better.

39 A wife is bound as long as her husband lives. But if the husband dies,*g* she is free to marry anyone she wishes, only in the Lord. 40 But in my judgment she is more blessed if she remains as she is. And I think that I too have the Spirit of God.

8 Now concerning food sacrificed to idols: we know that "all of us possess knowledge." Knowledge puffs up, but love builds up. 2 Anyone who claims to know something does not yet have the necessary knowledge; 3 but anyone who loves God is known by him.

4 Hence, as to the eating of food offered to idols, we know that "no idol in the world really exists," and that "there is no God but one." 5 Indeed, even though there may be so-called gods in heaven or on earth—as in fact there are many gods and many lords— 6 yet for us there is one God, the Father, from whom are all things and for whom we exist, and one Lord, Jesus Christ, through whom are all things and through whom we exist.

7 It is not everyone, however, who has this knowledge. Since some have become so accustomed to idols until now, they still think of the food they eat as food offered to an idol; and their conscience, being weak, is defiled. 8 "Food will not bring us close to God."*h* We are no worse off if we do not eat, and no better off if we do. 9 But take care that this liberty of yours does not somehow become a stumbling block to the weak.

f Gk *virgin*   *g* Gk *falls asleep*
h The quotation may extend to the end of the verse

7.36–38: The translation above assumes an engaged couple; see v. 9. Others, less likely, have interpreted *he, his,* and *him* as referring to a father and his daughter (note *f* Gk *virgin*), or a master and his slave, and her suitor; the father "gives her in marriage," v. 38. A third possibility is a couple pledged to virginity in a "spiritual marriage," who now wish to enter into normal conjugal relationships (vv. 3–5). Paul's preference, in any case, is consistent with vv. 7–8, 24, 26–28. **8.1–13: May a Christian eat food consecrated to an idol?** Another question from Corinth (7.1 n.). Much of the meat sold in the market places had come from animals sacrificed in pagan temples. Some Christians, weak believers (v. 11), had scruples about eating such meat (Rom ch 14). Others, however, felt superior to such scruples and contemptuous toward those troubled by them. Their views are quoted in vv. 1, 4, and 8. These superior people with their *"knowledge"* Paul rebukes. His principles are *love* for others that *builds up* community (v. 1) and renunciation of one's rights for the sake of others (v. 13). **2–3:** The true blessedness consists, not in knowing, but in being known by God, and it is in love that one is thus known. **8.6:** Basic Old Testament (Deut 6.4) and early Christian confessions (Phil 2.11) are expanded (see Mal 2.10; Rom 11.36) and combined to speak of God and Christ, each with

10 For if others see you, who possess knowledge, eating in the temple of an idol, might they not, since their conscience is weak, be encouraged to the point of eating food sacrificed to idols? 11 So by your knowledge those weak believers for whom Christ died are destroyed.[i] 12 But when you thus sin against members of your family,[j] and wound their conscience when it is weak, you sin against Christ. 13 Therefore, if food is a cause of their falling,[k] I will never eat meat, so that I may not cause one of them[l] to fall.

9 Am I not free? Am I not an apostle? Have I not seen Jesus our Lord? Are you not my work in the Lord? 2 If I am not an apostle to others, at least I am to you; for you are the seal of my apostleship in the Lord.

3 This is my defense to those who would examine me. 4 Do we not have the right to our food and drink? 5 Do we not have the right to be accompanied by a believing wife,[m] as do the other apostles and the brothers of the Lord and Cephas? 6 Or is it only Barnabas and I who have no right to refrain from working for a living? 7 Who at any time pays the expenses for doing military service? Who plants a vineyard and does not eat any of its fruit? Or who tends a flock and does not get any of its milk?

8 Do I say this on human authority? Does not the law also say the same? 9 For it is written in the law of Moses, "You shall not muzzle an ox while it is treading out the grain." Is it for oxen that God is concerned? 10 Or does he not speak entirely for our sake? It was indeed written for our sake, for whoever plows should plow in hope and whoever threshes should thresh in hope of a share in the crop. 11 If we have sown spiritual good among you, is it too much if we reap your material benefits? 12 If others share this rightful claim on you, do not we still more?

Nevertheless, we have not made use of this right, but we endure anything rather than put an obstacle in the way of the gospel of Christ. 13 Do you not know that those who are employed in the temple service get their food from the temple, and those who serve at the altar share in what is sacrificed on the altar? 14 In the same way, the Lord commanded that those who proclaim the gospel should get their living by the gospel.

15 But I have made no use of any of these rights, nor am I writing this so that they may be applied in my case. Indeed, I would rather die than that—no one will deprive me of my ground for boasting! 16 If I proclaim the gospel, this gives me no ground for boasting, for an obligation is laid on me, and woe to me if I do not proclaim the gospel! 17 For if I do this of my own will, I have a reward; but if not

i Gk *the weak brother . . . is destroyed*
j Gk *against the brothers*   k Gk *my brother's falling*   l Gk *cause my brother*   m Gk *a sister as wife*

---

regard to creation (see 10.26). **10**: Social clubs and guilds held banquets in pagan temples; since no real acknowledgment or worship of the idol was involved for Christians, many of them felt that there was no objection to attending. But Paul warns of the influence by example upon a brother or sister for whom such indulgence would be a violation of conscience and therefore destructive.

9.1–14: Paul's rights as an apostle. Paul himself practiced the principle of self-denial (v. 12b) that he has just been preaching (8.13). For the good of the church, he has refrained from asserting his rights as an apostle (vv. 15, 18; 1 Thess 2.7). **1**: *Free,* a theme developed in 9.19–27; *an apostle,* developed in 9.1–8; see 15.8–9; Gal 1.1, 11–12; Acts 9.3–6, 17. **2**: The Corinthian church is itself proof of the effectiveness and authenticity of Paul's *apostleship.* **4**: That is, at the expense of the church.

9.5: *Brothers of the Lord,* see Mt 13.55. For Peter's wife, compare Mk 1.30. **6**: See 4.10 n. **9**: The ox has a right to eat the grain (Deut 25.4). **13**: Lev 7.28–36, as also in pagan cult. **14**: Lk 10.7; compare 1 Tim 5.18.

9.15–27: Paul is free to waive his apostolic rights. 15–18: He deserves no credit for preaching the gospel—this he cannot help—but he does take pride in doing it without compensation; and the ground of this pride he will not give up. See 2 Cor 11.7–12.

of my own will, I am entrusted with a commission. 18 What then is my reward? Just this: that in my proclamation I may make the gospel free of charge, so as not to make full use of my rights in the gospel.

19 For though I am free with respect to all, I have made myself a slave to all, so that I might win more of them. 20 To the Jews I became as a Jew, in order to win Jews. To those under the law I became as one under the law (though I myself am not under the law) so that I might win those under the law. 21 To those outside the law I became as one outside the law (though I am not free from God's law but am under Christ's law) so that I might win those outside the law. 22 To the weak I became weak, so that I might win the weak. I have become all things to all people, that I might by all means save some. 23 I do it all for the sake of the gospel, so that I may share in its blessings.

24 Do you not know that in a race the runners all compete, but only one receives the prize? Run in such a way that you may win it. 25 Athletes exercise self-control in all things; they do it to receive a perishable wreath, but we an imperishable one. 26 So I do not run aimlessly, nor do I box as though beating the air; 27 but I punish my body and enslave it, so that after proclaiming to others I myself should not be disqualified.

10 I do not want you to be unaware, brothers and sisters, *n* that our ancestors were all under the cloud, and all passed through the sea, 2 and all were baptized into Moses in the cloud and in the sea, 3 and all ate the same spiritual food, 4 and all drank the same spiritual drink. For they drank from the spiritual rock that followed them, and the rock was Christ. 5 Nevertheless, God was not pleased with most of them, and they were struck down in the wilderness.

6 Now these things occurred as examples for us, so that we might not desire evil as they did. 7 Do not become idolaters as some of them did; as it is written, "The people sat down to eat and drink, and they rose up to play." 8 We must not indulge in sexual immorality as some of them did, and twenty-three thousand fell in a single day. 9 We must not put Christ *o* to the test, as some of them did, and were destroyed by serpents. 10 And do not complain as some of them did, and were destroyed by the destroyer. 11 These things happened to them to serve as an example, and they were written down to instruct us, on whom the ends of the ages have come. 12 So if you think you are standing, watch out that you do not fall. 13 No testing has overtaken you that is not common to everyone. God is faithful, and he will not let you be tested beyond your strength, but with the testing he will also provide the way out so that you may be able to endure it.

14 Therefore, my dear friends, *p* flee from the worship of idols. 15 I speak as to

n Gk *brothers*   *o* Other ancient authorities read *the Lord*   *p* Gk *my beloved*

9.19–23: Paul returns to the general topic of ch 8; the Christian is free except from the obligation to love; therefore free to serve. If taken in context vv. 20–22 will not be misunderstood: here are consideration and tact, not cowardice and compromise. 24–27: Athletic metaphors on the importance of self-discipline out of consideration for others. 10.1–13: **A warning against overconfidence.** Baptism and partaking the Lord's supper do not guarantee salvation, any more than corresponding acts sufficed for the ancient Hebrews. 2: *Cloud, sea,* Ex 13.21; 14.22.

3–4: Ex 16.4–35; 17.6; Num 20.7–11. A later legend told of the rock following the people in the desert. Paul sees the *rock* as a symbol of Christ, perhaps as the work of the pre-existent Christ. 5: Num 14.29–30. 7: Ex 32.4, 6. 8: Num 25.1–9. 9: Num 21.5–6. 10: Num 16.13–14, 41–49.

10.14–22: **Application: sacrifices to idols, again.** The scriptural examples cited in vv. 1–13 move Paul to warn against participation in pagan worship. To eat at a friend's table, or even at a banquet in a temple, food consecrated to an idol is one thing; taking part

sensible people; judge for yourselves what I say. ¹⁶ The cup of blessing that we bless, is it not a sharing in the blood of Christ? The bread that we break, is it not a sharing in the body of Christ? ¹⁷ Because there is one bread, we who are many are one body, for we all partake of the one bread. ¹⁸ Consider the people of Israel;*q* are not those who eat the sacrifices partners in the altar? ¹⁹ What do I imply then? That food sacrificed to idols is anything, or that an idol is anything? ²⁰ No, I imply that what pagans sacrifice, they sacrifice to demons and not to God. I do not want you to be partners with demons. ²¹ You cannot drink the cup of the Lord and the cup of demons. You cannot partake of the table of the Lord and the table of demons. ²² Or are we provoking the Lord to jealousy? Are we stronger than he?

23 "All things are lawful," but not all things are beneficial. "All things are lawful," but not all things build up. ²⁴ Do not seek your own advantage, but that of the other. ²⁵ Eat whatever is sold in the meat market without raising any question on the ground of conscience, ²⁶ for "the earth and its fullness are the Lord's." ²⁷ If an unbeliever invites you to a meal and you are disposed to go, eat whatever is set before you without raising any question on the ground of conscience. ²⁸ But if someone says to you, "This has been offered in sacrifice," then do not eat it,

out of consideration for the one who informed you, and for the sake of conscience— ²⁹ I mean the other's conscience, not your own. For why should my liberty be subject to the judgment of someone else's conscience? ³⁰ If I partake with thankfulness, why should I be denounced because of that for which I give thanks?

31 So, whether you eat or drink, or whatever you do, do everything for the glory of God. ³² Give no offense to Jews or to Greeks or to the church of God, ³³ just as I try to please everyone in everything I do, not seeking my own advantage, but that of many, so that they may **11** be saved. ¹ Be imitators of me, as I am of Christ.

2 I commend you because you remember me in everything and maintain the traditions just as I handed them on to you. ³ But I want you to understand that Christ is the head of every man, and the husband*r* is the head of his wife,*s* and God is the head of Christ. ⁴ Any man who prays or prophesies with something on his head disgraces his head, ⁵ but any woman who prays or prophesies with her head unveiled disgraces her head—it is one and the same thing as having her

---

q Gk *Israel according to the flesh*   *r* The same Greek word means *man* or *husband*   *s* Or *head of the woman*

---

in a pagan religious rite (comparable to the Lord's supper in the church, vv. 16–17) is quite another. **19–20:** Idols have no real existence; but demons stand behind them (Deut 32.17) and use them to destroy those who worship them. **21:** 2 Cor 6.15–16. **10.23–11.1: Principles: our freedom and responsibility for others.** Chapter 8 sides with those willing to eat food that had been offered to an idol but advised restraint in this right out of conscience' sake for weaker Christians. In 10.1–22 Paul uses a sacramental argument based on the Old Testament to conclude, with the "weak" (8.7, 10; 9.22), against participation in such eating and drinking. Paul resolves the matter by affirming Christian freedom (10.25–27, 29b–30), yet with consideration for others (vv. 24, 28–

29a), all to God's glory (v. 31), as Paul himself does (10.33–11.1). **10.23:** Slogans in Corinth, as at 6.12–13. **26:** Pss 24.1; 50.12. **28:** 8.7, 10–12. **32:** 8.13. **11.2–14.40: Problems in community life and worship.** **11.2–16: Propriety in dress at public prayer.** **2:** *The traditions,* see 11.23; 15.3; 2 Thess 2.15 n. **3–5:** A play on the word *head:* v. 3, "source," see v. 8; 15.28, but in 4a (and 5–7) a physical sense, and 4b *his head* and *her head* may denote Christ. Reflecting first-century culture, a man dishonors Christ by worshiping with his head covered; a woman dishonors both her husband and Christ by worshiping otherwise. **4–7a:** Confusion of gender, see 6.9 n.

head shaved. [6]For if a woman will not veil herself, then she should cut off her hair; but if it is disgraceful for a woman to have her hair cut off or to be shaved, she should wear a veil. [7]For a man ought not to have his head veiled, since he is the image and reflection[t] of God; but woman is the reflection[t] of man. [8]Indeed, man was not made from woman, but woman from man. [9]Neither was man created for the sake of woman, but woman for the sake of man. [10]For this reason a woman ought to have a symbol of[u] authority on her head,[v] because of the angels. [11]Nevertheless, in the Lord woman is not independent of man or man independent of woman. [12]For just as woman came from man, so man comes through woman; but all things come from God. [13]Judge for yourselves: is it proper for a woman to pray to God with her head unveiled? [14]Does not nature itself teach you that if a man wears long hair, it is degrading to him, [15]but if a woman has long hair, it is her glory? For her hair is given to her for a covering. [16]But if anyone is disposed to be contentious—we have no such custom, nor do the churches of God.

[17] Now in the following instructions I do not commend you, because when you come together it is not for the better but for the worse. [18]For, to begin with, when you come together as a church, I hear that there are divisions among you; and to some extent I believe it. [19]Indeed, there have to be factions among you, for only so will it become clear who among you are genuine. [20]When you come together, it is not really to eat the Lord's supper. [21]For when the time comes to eat, each of you goes ahead with your own supper, and one goes hungry and another becomes drunk. [22]What! Do you not have homes to eat and drink in? Or do you show contempt for the church of God and humiliate those who have nothing? What should I say to you? Should I commend you? In this matter I do not commend you!

[23] For I received from the Lord what I also handed on to you, that the Lord Jesus on the night when he was betrayed took a loaf of bread, [24]and when he had given thanks, he broke it and said, "This is my body that is for[w] you. Do this in remembrance of me." [25]In the same way he took the cup also, after supper, saying, "This cup is the new covenant in my blood. Do this, as often as you drink it, in remembrance of me." [26]For as often as you eat this bread and drink the cup, you proclaim the Lord's death until he comes.

[27] Whoever, therefore, eats the bread or drinks the cup of the Lord in an unworthy manner will be answerable for the body and blood of the Lord. [28]Examine yourselves, and only then eat of the

---

t Or *glory*   *u* Gk lacks *a symbol of*
v Or *have freedom of choice regarding her head*
w Other ancient authorities read *is broken for*

11.7b–9: Reflects Gen 2.21–23, not 1.26–27. **10:** *Authority,* to be exercised likely by women in worship (v. 5), less likely by men over women. See note *v.* If *angels* are thought of as administering the divine order (see 1 Tim 5.21 n.), women worshiping with head bare show disrespect for this order. Or *the angels* are cosmic, demonic powers against whom protection is needed. Or these *angels* are messengers from other churches (11.16; Gal 4.14; Lk 7.24; 9.52) whom Corinthian women leaders ought not to offend. **11–12:** Paul guards against a wrong inference (subordination of women) from what he has said. **16:** Paul shuts off further discussion, appealing to the example of other churches (14.36).

11.17–34: Directions in the face of abuses at the Lord's supper. 18–19: *Factions,* cliques, perhaps corresponding to social classes (compare also 1.11–12). **20–21:** The Lord's supper (see 10.3–5, 11–12, 16–17) took place in connection with a common meal, which came to be known as the agape, or love feast (Jude 12); see vv. 24–25, bread, supper, cup. **23:** *Received . . . handed on,* technical terms for transmitting an oral tradition. **11.23–25:** Mt 26.26–29; Mk 14.22–25; and especially Lk 22.14–20. **25:** *New covenant,* Jer 31.31. *In my blood,* here Christ's death (v. 26); the Mosaic covenant was also confirmed with blood (Ex 24.8). **26:** *You proclaim,* the celebration is a sermon on "Christ crucified" (1.23).

bread and drink of the cup. 29For all who eat and drink*x* without discerning the body,*y* eat and drink judgment against themselves. 30For this reason many of you are weak and ill, and some have died.*z* 31But if we judged ourselves, we would not be judged. 32But when we are judged by the Lord, we are disciplined*a* so that we may not be condemned along with the world.

33 So then, my brothers and sisters,*b* when you come together to eat, wait for one another. 34If you are hungry, eat at home, so that when you come together, it will not be for your condemnation. About the other things I will give instructions when I come.

12 Now concerning spiritual gifts,*c* brothers and sisters,*b* I do not want you to be uninformed. 2You know that when you were pagans, you were enticed and led astray to idols that could not speak. 3Therefore I want you to understand that no one speaking by the Spirit of God ever says "Let Jesus be cursed!" and no one can say "Jesus is Lord" except by the Holy Spirit.

4 Now there are varieties of gifts, but the same Spirit; 5and there are varieties of services, but the same Lord; 6and there are varieties of activities, but it is the same God who activates all of them in everyone. 7To each is given the manifestation of the Spirit for the common

good. 8To one is given through the Spirit the utterance of wisdom, and to another the utterance of knowledge according to the same Spirit, 9to another faith by the same Spirit, to another gifts of healing by the one Spirit, 10to another the working of miracles, to another prophecy, to another the discernment of spirits, to another various kinds of tongues, to another the interpretation of tongues. 11All these are activated by one and the same Spirit, who allots to each one individually just as the Spirit chooses.

12 For just as the body is one and has many members, and all the members of the body, though many, are one body, so it is with Christ. 13For in the one Spirit we were all baptized into one body—Jews or Greeks, slaves or free—and we were all made to drink of one Spirit.

14 Indeed, the body does not consist of one member but of many. 15If the foot would say, "Because I am not a hand, I do not belong to the body," that would not make it any less a part of the body. 16And if the ear would say, "Because I am not an eye, I do not belong to the body," that would not make it any less a part of the body. 17If the whole body were an eye, where would the hearing

x Other ancient authorities add *in an unworthy manner,* *y* Other ancient authorities read *the Lord's body* *z* Gk *fallen asleep* *a* Or *When we are judged, we are being disciplined by the Lord* *b* Gk *brothers* *c* Or *spiritual persons*

29: *Without discerning the body,* the community, one's relation to other Christians (vv. 20–22, 33); in v. 27 *body and blood,* the whole person of Christ.

11.33–34: *Wait for one another,* apparently those who came early left too little food for others. *Eat at home,* Paul says, if you cannot control your hunger.

Chs 12–14: The gifts of the Spirit. 1: *Now concerning,* see 7.1 n.

12.1–11: Variety of spiritual gifts. 2: A reference to the highly emotional practices in certain pagan cults. Ecstasy does not prove that one is led by the Holy Spirit; confession of Jesus as Lord is the criterion.

12.4–11: The real test of gifts is whether they come from the one God and contribute to *the common good* (v. 7) and edify the com-

munity (8.1). Note the suggestion of the Trinity, *Spirit . . . Lord . . . God* (vv. 4–5). **9:** *Faith,* not saving faith (Rom 1.17 n.) but miracle-working faith (13.2). **10:** *Miracles,* see Gal 3.5 n. *Prophecy,* inspired preaching. *Discernment of spirits,* power to recognize whether a prophet is true or false (1 Jn 4.1). *Interpretation of tongues,* ability to put into intelligible words what the ecstatics were saying (14.9–19).

12.12–31: The body requires a variety of members. 12–13: *Many* persons are made *one body* by baptism.

12.14–26: This body (Rom 12.4–5; Eph 4.14–16) is Christ's (v. 12) and needs the gifts of each, even those who feel unimportant. To *care for* and empathize with others prevents dissension (vv. 25–26).

be? If the whole body were hearing, where would the sense of smell be? 18 But as it is, God arranged the members in the body, each one of them, as he chose. 19 If all were a single member, where would the body be? 20 As it is, there are many members, yet one body. 21 The eye cannot say to the hand, "I have no need of you," nor again the head to the feet, "I have no need of you." 22 On the contrary, the members of the body that seem to be weaker are indispensable, 23 and those members of the body that we think less honorable we clothe with greater honor, and our less respectable members are treated with greater respect; 24 whereas our more respectable members do not need this. But God has so arranged the body, giving the greater honor to the inferior member, 25 that there may be no dissension within the body, but the members may have the same care for one another. 26 If one member suffers, all suffer together with it; if one member is honored, all rejoice together with it.

27 Now you are the body of Christ and individually members of it. 28 And God has appointed in the church first apostles, second prophets, third teachers; then deeds of power, then gifts of healing, forms of assistance, forms of leadership, various kinds of tongues. 29 Are all apostles? Are all prophets? Are all teachers? Do all work miracles? 30 Do all possess gifts of healing? Do all speak in tongues? Do all interpret? 31 But strive

for the greater gifts. And I will show you a still more excellent way.

13 If I speak in the tongues of mortals and of angels, but do not have love, I am a noisy gong or a clanging cymbal. 2 And if I have prophetic powers, and understand all mysteries and all knowledge, and if I have all faith, so as to remove mountains, but do not have love, I am nothing. 3 If I give away all my possessions, and if I hand over my body so that I may boast,*d* but do not have love, I gain nothing.

4 Love is patient; love is kind; love is not envious or boastful or arrogant 5 or rude. It does not insist on its own way; it is not irritable or resentful; 6 it does not rejoice in wrongdoing, but rejoices in the truth. 7 It bears all things, believes all things, hopes all things, endures all things.

8 Love never ends. But as for prophecies, they will come to an end; as for tongues, they will cease; as for knowledge, it will come to an end. 9 For we know only in part, and we prophesy only in part; 10 but when the complete comes, the partial will come to an end. 11 When I was a child, I spoke like a child, I thought like a child, I reasoned like a child; when I became an adult, I put an end to childish ways. 12 For now we see in a mirror, dimly,*e* but then we will see

d Other ancient authorities read *body to be burned*
e Gk *in a riddle*

12.27–30: For the one *church* (*you* in Corinth and those elsewhere) God provides varied leadership. *Apostles* (4.9; 9.5), *prophets* (14.3), and teachers are preeminent. Among other functions and gifts, *assistance* and *leadership* suggest "helpful deeds" and "guidance"; only later were they linked to "deacons" and "bishops" (Phil 1.1). **31:** For all spiritual gifts *love* provides a better way (ch 13; see 14.1). **13.1–13:** **Love, the greatest gift and way.** This lyric chapter, powerful by itself, is even more meaningful in its context. Paul is still discussing spiritual gifts (12.1–11); the great gift of the Spirit is not tongues or even prophecy, but *love*. This *love* is not love in an ordinary or general sense, but the love for

others which is known within the church (8.1), inspired ultimately by the love of God in Christ for us through the Holy Spirit (Rom 5.5). **1:** *Gong . . . cymbal,* as in noisy accompaniments of pagan worship (12.2). **2:** *Faith,* see 12.9 n.; Mt 17.20. **3:** *Boast,* or glory, as at 9.15 and 2 Cor 1.14; note *d* suggests voluntary self-immolation, martyrdom, or branding as a slave. **13.4–7:** The reverse of the proud, contemptuous, divisive spirit manifested in the behavior of some at Corinth. **13.8–13:** *Knowledge* and ability to express it (whether in *prophecies* or *tongues*) are too faulty for one to take pride in them. **12:** *A mirror,* a polished metal surface, not yielding

face to face. Now I know only in part; then I will know fully, even as I have been fully known. 13 And now faith, hope, and love abide, these three; and the greatest of these is love.

14 Pursue love and strive for the spiritual gifts, and especially that you may prophesy. 2 For those who speak in a tongue do not speak to other people but to God; for nobody understands them, since they are speaking mysteries in the Spirit. 3 On the other hand, those who prophesy speak to other people for their upbuilding and encouragement and consolation. 4 Those who speak in a tongue build up themselves, but those who prophesy build up the church. 5 Now I would like all of you to speak in tongues, but even more to prophesy. One who prophesies is greater than one who speaks in tongues, unless someone interprets, so that the church may be built up.

6 Now, brothers and sisters,*f* if I come to you speaking in tongues, how will I benefit you unless I speak to you in some revelation or knowledge or prophecy or teaching? 7 It is the same way with lifeless instruments that produce sound, such as the flute or the harp. If they do not give distinct notes, how will anyone know what is being played? 8 And if the bugle gives an indistinct sound, who will get ready for battle? 9 So with yourselves; if in a tongue you utter speech that is not intelligible, how will anyone know what is being said? For you will be speaking into the air. 10 There are doubtless many different kinds of sounds in the world, and nothing is without sound. 11 If then I do not know the meaning of a sound, I will be a foreigner to the speaker and the speaker a foreigner to me. 12 So with

yourselves; since you are eager for spiritual gifts, strive to excel in them for building up the church.

13 Therefore, one who speaks in a tongue should pray for the power to interpret. 14 For if I pray in a tongue, my spirit prays but my mind is unproductive. 15 What should I do then? I will pray with the spirit, but I will pray with the mind also; I will sing praise with the spirit, but I will sing praise with the mind also. 16 Otherwise, if you say a blessing with the spirit, how can anyone in the position of an outsider say the "Amen" to your thanksgiving, since the outsider does not know what you are saying? 17 For you may give thanks well enough, but the other person is not built up. 18 I thank God that I speak in tongues more than all of you; 19 nevertheless, in church I would rather speak five words with my mind, in order to instruct others also, than ten thousand words in a tongue.

20 Brothers and sisters,*f* do not be children in your thinking; rather, be infants in evil, but in thinking be adults. 21 In the law it is written,

"By people of strange tongues
    and by the lips of foreigners
I will speak to this people;
    yet even then they will not
        listen to me,"

says the Lord. 22 Tongues, then, are a sign not for believers but for unbelievers, while prophecy is not for unbelievers but for believers. 23 If, therefore, the whole church comes together and all speak in tongues, and outsiders or unbelievers enter, will they not say that you are out of your mind? 24 But if all prophesy, an unbeliever or outsider who enters is re-

f Gk *brothers*

---

a clear image. **13**: *Love is greatest* because it endures when *faith* becomes sight (v. 12) and *hope* is fulfilled, with God who is love (1 Jn 4.8–10). The triad appears elsewhere in Paul's letters (Rom 5.1–5; Gal 5.5–6; Col 1.4–5; 1 Thess 1.3; 5.8).

14.1–40: Among gifts, prophecy outranks tongues (vv. 1, 5, 39). Besides love (v. 1), Paul stresses order (vv. 33, 40) and

building up the community (vv. 3, 5, 12, 17, 26) when members come together. On *prophecy* and *speaking in tongues,* see 12.10 n.

14.16: *Say the "Amen,"* Neh 8.6; 1 Chr 16.36. **18**: For other evidence that Paul himself had ecstatic experiences, see 2 Cor 12.1–4. **21**: *Law,* the Old Testament (Isa 28.11–12). **25**: *The secrets . . . are disclosed,* compare 4.5; Jn 4.16–19. *God . . . among you,* Isa 45.14; Zech

proved by all and called to account by all. 25 After the secrets of the unbeliever's heart are disclosed, that person will bow down before God and worship him, declaring, "God is really among you."

26 What should be done then, my friends?g When you come together, each one has a hymn, a lesson, a revelation, a tongue, or an interpretation. Let all things be done for building up. 27 If anyone speaks in a tongue, let there be only two or at most three, and each in turn; and let one interpret. 28 But if there is no one to interpret, let them be silent in church and speak to themselves and to God. 29 Let two or three prophets speak, and let the others weigh what is said. 30 If a revelation is made to someone else sitting nearby, let the first person be silent. 31 For you can all prophesy one by one, so that all may learn and all be encouraged. 32 And the spirits of prophets are subject to the prophets, 33 for God is a God not of disorder but of peace.

(As in all the churches of the saints, 34 women should be silent in the churches. For they are not permitted to speak, but should be subordinate, as the law also says. 35 If there is anything they desire to know, let them ask their husbands at home. For it is shameful for a woman to speak in church. h 36 Or did the word of God originate with you? Or are you the only ones it has reached?)

37 Anyone who claims to be a prophet, or to have spiritual powers, must ac-

knowledge that what I am writing to you is a command of the Lord. 38 Anyone who does not recognize this is not to be recognized. 39 So, my friends,i be eager to prophesy, and do not forbid speaking in tongues; 40 but all things should be done decently and in order.

15 Now I would remind you, brothers and sisters,g of the good newsj that I proclaimed to you, which you in turn received, in which also you stand, 2 through which also you are being saved, if you hold firmly to the message that I proclaimed to you—unless you have come to believe in vain.

3 For I handed on to you as of first importance what I in turn had received: that Christ died for our sins in accordance with the scriptures, 4 and that he was buried, and that he was raised on the third day in accordance with the scriptures, 5 and that he appeared to Cephas, then to the twelve. 6 Then he appeared to more than five hundred brothers and sistersg at one time, most of whom are still alive, though some have died. k 7 Then he appeared to James, then to all the apostles. 8 Last of all, as to one untimely born, he appeared also to me. 9 For I am the least of the apostles, unfit to be called an apostle, because I persecuted the church

g Gk *brothers*   h Other ancient authorities put
verses 34–35 after verse 40   i Gk *my brothers*
j Or *gospel*   k Gk *fallen asleep*

8.23 applied to the Christian community; 1 Cor 3.16–17 n.

14.26: Features of a non-eucharistic worship-service of the word. *Hymn,* Eph 5.19. *Lesson,* a teaching.

14.33b–36: Since 11.5 reports that women had a role in worship at Corinth, some refer *women* here only to those married (see v. 35), or distinguish a house church in ch 11 from the total community in Corinth (v. 26), or take these culturally conditioned verses as an editorial insertion; see 1 Tim 2.11–12.

15.1–58: The resurrection.

15.1–11: The gospel of Christ's death and resurrection. 1: *Now I would remind you,* not necessarily a question from Corinth (see 7.1 n.), but Paul has heard that some "spiritu-

al persons" there deny "resurrection" (15.12) of the body (v. 35, on grounds in Greek philosophy of an "immortal soul") or in the future (vv. 23, 49, 51–52, on the grounds that Christians are already raised, 4.8; 2 Tim 2.17b–18). As a basis for our resurrection Paul reiterates that of Christ.

15.3: *Handed on . . . received,* see 11.23 n. *Scriptures,* Isa 53.5–12. **4:** *Scriptures,* Hos 6.2; Ps 16.10 (compare Acts 2.31). **7:** *James,* "the Lord's brother" of Gal 1.19. **8:** *One untimely born,* the meaning is obscure; Paul perhaps is referring to the separation in time between his own experience and those of the others. For other accounts of Paul's encounter with the risen Christ, see 9.1; Gal 1.16; Acts 9.3–6.

of God. 10 But by the grace of God I am what I am, and his grace toward me has not been in vain. On the contrary, I worked harder than any of them— though it was not I, but the grace of God that is with me. 11 Whether then it was I or they, so we proclaim and so you have come to believe.

12 Now if Christ is proclaimed as raised from the dead, how can some of you say there is no resurrection of the dead? 13 If there is no resurrection of the dead, then Christ has not been raised; 14 and if Christ has not been raised, then our proclamation has been in vain and your faith has been in vain. 15 We are even found to be misrepresenting God, because we testified of God that he raised Christ—whom he did not raise if it is true that the dead are not raised. 16 For if the dead are not raised, then Christ has not been raised. 17 If Christ has not been raised, your faith is futile and you are still in your sins. 18 Then those also who have died*l* in Christ have perished. 19 If for this life only we have hoped in Christ, we are of all people most to be pitied.

20 But in fact Christ has been raised from the dead, the first fruits of those who have died. *l* 21 For since death came through a human being, the resurrection of the dead has also come through a human being; 22 for as all die in Adam, so all will be made alive in Christ. 23 But each in his own order: Christ the first fruits, then at his coming those who belong to Christ. 24 Then comes the end,*m* when he hands over the kingdom to God the Father, after he has destroyed every ruler and every authority and power. 25 For he must reign until he has put all his enemies under his feet. 26 The last enemy to be destroyed is death. 27 For "God*n*" has put all things in subjection under his feet." But when it says, "All things are put in subjection," it is plain that this does not include the one who put all things in subjection under him. 28 When all things are subjected to him, then the Son himself will also be subjected to the one who put all things in subjection under him, so that God may be all in all.

29 Otherwise, what will those people do who receive baptism on behalf of the dead? If the dead are not raised at all, why are people baptized on their behalf?

30 And why are we putting ourselves in danger every hour? 31 I die every day! That is as certain, brothers and sisters,*o* as my boasting of you—a boast that I make in Christ Jesus our Lord. 32 If with merely human hopes I fought with wild animals at Ephesus, what would I have gained by it? If the dead are not raised,

"Let us eat and drink,
  for tomorrow we die."

33 Do not be deceived:

"Bad company ruins good
    morals."

34 Come to a sober and right mind, and sin no more; for some people have no knowledge of God. I say this to your shame.

35 But someone will ask, "How are the dead raised? With what kind of body

---

l Gk *fallen asleep*   *m* Or *Then come the rest*
n Gk *he*   *o* Gk *brothers*

15.12–34: The significance for us of the resurrection. 18: 1 Thess 4.16. **21–22**: Rom 5.12–18. **23**: *Coming,* the glorious return of Christ at the end of the age (1 Thess 2.19; 4.13–17). **24–27**: *His enemies* are the demonic powers dominating the present age; one of these is *death.* **27**: Ps 8.6. *His feet,* Christ's. **15.29**: A practice otherwise unknown. Presumably Christians accepted baptism on behalf of loved ones who had died without being baptized, in order that the latter might share in the final resurrection. Without advocating this practice, Paul makes it a point in his argument. **31**: *I die,* i.e. I risk death, *every*

day. **32**: One cannot say with assurance whether the fighting *with wild animals* is to be taken literally or is merely a strong metaphor (compare 4.9). In any case, Paul had bitter and dangerous enemies. The quotation is from Isa 22.13. **33**: Paul quotes a Greek proverb (from the Attic poet Menander), warning the Corinthians not to associate with those who deny the resurrection.

15.35–58: The nature of the resurrection. 35–44: Greeks had no trouble in conceiving of the immortality of the soul, but the idea of the raised body was difficult. Paul's point is that there are many kinds of "bodies";

do they come?" 36 Fool! What you sow does not come to life unless it dies. 37 And as for what you sow, you do not sow the body that is to be, but a bare seed, perhaps of wheat or of some other grain. 38 But God gives it a body as he has chosen, and to each kind of seed its own body. 39 Not all flesh is alike, but there is one flesh for human beings, another for animals, another for birds, and another for fish. 40 There are both heavenly bodies and earthly bodies, but the glory of the heavenly is one thing, and that of the earthly is another. 41 There is one glory of the sun, and another glory of the moon, and another glory of the stars; indeed, star differs from star in glory.

42 So it is with the resurrection of the dead. What is sown is perishable, what is raised is imperishable. 43 It is sown in dishonor, it is raised in glory. It is sown in weakness, it is raised in power. 44 It is sown a physical body, it is raised a spiritual body. If there is a physical body, there is also a spiritual body. 45 Thus it is written, "The first man, Adam, became a living being"; the last Adam became a life-giving spirit. 46 But it is not the spiritual that is first, but the physical, and then the spiritual. 47 The first man was from the earth, a man of dust; the second man is*p* from heaven. 48 As was the man of dust, so are those who are of the dust; and as is the man of heaven, so are those who are of heaven. 49 Just as we have borne the image of the man of dust, we will*q* also bear the image of the man of heaven.

50 What I am saying, brothers and sisters,*r* is this: flesh and blood cannot inherit the kingdom of God, nor does the perishable inherit the imperishable. 51 Listen, I will tell you a mystery! We will not all die,*s* but we will all be changed, 52 in a moment, in the twinkling of an eye, at the last trumpet. For the trumpet will sound, and the dead will be raised imperishable, and we will be changed. 53 For this perishable body must put on imperishability, and this mortal body must put on immortality. 54 When this perishable body puts on imperishability, and this mortal body puts on immortality, then the saying that is written will be fulfilled:

"Death has been swallowed up
   in victory."
55 "Where, O death, is your victory?
   Where, O death, is your sting?"
56 The sting of death is sin, and the power of sin is the law. 57 But thanks be to God, who gives us the victory through our Lord Jesus Christ.

58 Therefore, my beloved,*t* be steadfast, immovable, always excelling in the work of the Lord, because you know that in the Lord your labor is not in vain.

16 Now concerning the collection for the saints: you should follow the directions I gave to the churches of Galatia. 2 On the first day of every week, each of you is to put aside and save whatever extra you earn, so that collections need not be taken when I come. 3 And when I arrive, I will send any whom you approve with letters to take your gift to Jerusalem. 4 If it seems advisable that I should go also, they will accompany me.

5 I will visit you after passing through Macedonia—for I intend to pass through Macedonia— 6 and perhaps I will stay with you or even spend the winter, so that you may send me on my way, wherever I go. 7 I do not want to see you now just in passing, for I hope to spend some

---

p Other ancient authorities add *the Lord*
q Other ancient authorities read *let us*
r Gk *brothers*   *s* Gk *fall asleep*   *t* Gk *beloved brothers*

---

the resurrection body will be a new body (not *perishable,* v. 42, or *physical,* v. 44), which God will provide. **45–47**: Gen 2.7.

15.50: Jn 3.6. **51–52**: *Mystery,* a secret made known in Christ. *We will not all die* before the Lord's coming (1 Thess 4.13–17). **54–55**: Isa 25.8 Gk; Hos 13.14. **57**: *God gives us the victory*

over sin now (Rom 8.1–2) and hereafter over death (Rom 8.11). **58**: *Immovable,* not shaken by false teaching.

16.1–24: **Final messages. 1**: *Now concerning,* see 7.1 n. Collection, Rom 15.25–29; 2 Cor chs 8–9; Gal 2.10. **5**: Acts 19.21. **8–9**: Acts 18.19–20; 19.8–10, 20, 23–27.

time with you, if the Lord permits. ⁸But I will stay in Ephesus until Pentecost, ⁹for a wide door for effective work has opened to me, and there are many adversaries.

10 If Timothy comes, see that he has nothing to fear among you, for he is doing the work of the Lord just as I am; ¹¹therefore let no one despise him. Send him on his way in peace, so that he may come to me; for I am expecting him with the brothers.

12 Now concerning our brother Apollos, I strongly urged him to visit you with the other brothers, but he was not at all willing ᵘ to come now. He will come when he has the opportunity.

13 Keep alert, stand firm in your faith, be courageous, be strong. ¹⁴Let all that you do be done in love.

15 Now, brothers and sisters, ᵛ you know that members of the household of Stephanas were the first converts in Achaia, and they have devoted themselves to the service of the saints; ¹⁶I urge you to put yourselves at the service of such people, and of everyone who works and toils with them. ¹⁷I rejoice at the coming of Stephanas and Fortunatus and Achaicus, because they have made up for your absence; ¹⁸for they refreshed my spirit as well as yours. So give recognition to such persons.

19 The churches of Asia send greetings. Aquila and Prisca, together with the church in their house, greet you warmly in the Lord. ²⁰All the brothers and sisters ᵛ send greetings. Greet one another with a holy kiss.

21 I, Paul, write this greeting with my own hand. ²²Let anyone be accursed who has no love for the Lord. Our Lord, come! ʷ ²³The grace of the Lord Jesus be with you. ²⁴My love be with all of you in Christ Jesus. ˣ

---

u Or *it was not at all God's will for him*
v Gk *brothers*   w Gk *Marana tha.* These Aramaic words can also be read *Maran atha,* meaning *Our Lord has come*   x Other ancient authorities add *Amen*

16.10: *Timothy,* see Introduction to 2 Timothy. **12:** *Apollos,* Acts 18.24–26.

16.15: *Stephanas,* 1.16. **17:** Perhaps these men brought the letter mentioned in 7.1 and are to carry back Paul's reply. **19:** *Asia,* a Roman province in western Asia Minor. *Aquila and Prisca* are mentioned in Acts 18.2; Rom 16.3; 2 Tim 4.19. *The church in their house,* see

Philem 1 n. **20:** *Holy kiss,* see Rom 16.16 n.

16.21: Paul adds his personal signature (2 Thess 3.17), after dictating the rest; perhaps Sosthenes (1.1) had served as his secretary (compare Rom 16.22). **22:** *Our Lord, come!* is the preferable rendering of the words *Marana tha* (transliterated from two Aramaic words); see Rev 22.20; see also note *w.*

The Second Letter of Paul to the
Corinthians

Relations between Paul and the Corinthian church had deteriorated during the period after 1 Corinthians was written. The visit announced at 1 Cor 16.5–7 did not come off as planned, and the Corinthians felt that Paul had vacillated in his plans (2 Cor 1.15–23). Much of 2 Corinthians therefore deals with a crisis in confidence between apostle and community (1.12– 7.16; 1.1–11 provides the usual Pauline introduction to a letter). Having made a "painful visit" to the church there (2.1), the apostle delayed from making another visit (12.14; 13.1) because he had reason to believe that it too would be painful. Instead, he had written a severe letter (possibly lost) "out of much distress and anguish of heart and with many tears" (2.4) and sent it to Corinth, presumably by Titus, one of his fellow-workers. So anxious was he about the effects of this letter that he found it impossible to wait for Titus' return. He left Ephesus, hoping to meet Titus in Troas. Disappointed there, he went on to Macedonia (2.12–13), where Titus rejoined him, bringing a most reassuring report on the attitude of the Corinthian church toward him (7.13–16). In relief and gratitude, Paul wrote the letter before us, or at least much of it on this occasion.

Chapters 8 and 9 go on to speak at length about the collection for the church at Jerusalem (8.1–9.15), which was now almost complete.

Chapters 10 to 13 constitute a vigorous defense of Paul and his work and are written in a tone so different from that of chapters 1 to 9 that many scholars believe they are a fragment of a letter written to Corinth at some other time. Possibly chs 10–13 are the "severe letter" mentioned above.

Because so much of 2 Corinthians is a response to the words and feelings of others (often opponents), which are not recorded in the letter, it is sometimes difficult to follow the apostle's argument. Its pages preserve much, however, that illuminates Paul's own life, ministry, and the concern that he felt for the churches he had established (4.5; 5.20; 6.1–10; 11.22–33; 12.2–10).

1 Paul, an apostle of Christ Jesus by the will of God, and Timothy our brother,

To the church of God that is in Corinth, including all the saints throughout Achaia:

2 Grace to you and peace from God our Father and the Lord Jesus Christ.

3 Blessed be the God and Father of our Lord Jesus Christ, the Father of mercies and the God of all consolation, 4 who consoles us in all our affliction, so that we may be able to console those who are in any affliction with the consolation with which we ourselves are consoled by God. 5 For just as the sufferings of Christ are abundant for us, so also our consolation is abundant through Christ. 6 If we are being afflicted, it is for your consolation and salvation; if we are being consoled, it is for your consolation, which you experience when you patiently endure the same sufferings that we are also suffering. 7 Our hope for you is unshaken; for we know that as you share in our sufferings, so also you share in our consolation.

8 We do not want you to be unaware, brothers and sisters,*a* of the affliction we experienced in Asia; for we were so utterly, unbearably crushed that we despaired of life itself. 9 Indeed, we felt that we had received the sentence of death so that we would rely not on ourselves but on God who raises the dead. 10 He who rescued us from so deadly a peril will continue to rescue us; on him we have set our hope that he will rescue us again, 11 as you also join in helping us by your prayers, so that many will give thanks on our*b* behalf for the blessing granted us through the prayers of many.

12 Indeed, this is our boast, the testimony of our conscience: we have behaved in the world with frankness*c* and godly sincerity, not by earthly wisdom but by the grace of God—and all the more toward you. 13 For we write you nothing other than what you can read and also understand; I hope you will understand until the end— 14 as you have already understood us in part—that on the day of the Lord Jesus we are your boast even as you are our boast.

15 Since I was sure of this, I wanted to come to you first, so that you might have a double favor;*d* 16 I wanted to visit you on my way to Macedonia, and to come back to you from Macedonia and have you send me on to Judea. 17 Was I vacillating when I wanted to do this? Do I make my plans according to ordinary human standards,*e* ready to say "Yes, yes" and "No, no" at the same time? 18 As surely as God is faithful, our word to you has not been "Yes and No." 19 For the Son of God, Jesus Christ, whom we proclaimed among you, Silvanus and Timothy and I, was not "Yes and No"; but in him it is always "Yes." 20 For in him every one of God's promises is a "Yes." For this reason it is through him that we say the "Amen," to the glory of God. 21 But

---

a Gk *brothers*   *b* Other ancient authorities read *your*   *c* Other ancient authorities read *holiness*
d Other ancient authorities read *pleasure*
e Gk *according to the flesh*

1.1–11: Salutation and thanksgiving, see Rom 1.1–7 n. and 1.8–15 n. The thanksgiving (vv. 3–11) reflects the relief which Paul felt as a result of the news that Titus brought him (see Introduction). **1:** *Timothy,* see Introduction to 2 Timothy. *Achaia,* the Roman province in which Corinth was located. **2:** *Grace . . . peace,* see Rom 1.7 n.; 2 Thess 1.2 n.
1.8–10: It is not known what this terrible trial was; some identify it with the experience mentioned in 1 Cor 15.32; see also 2 Cor 4.11; 11.23. **8:** *Asia,* a Roman province in western Asia Minor.

1.12–7.16: Paul's apostolic ministry and the crisis with Corinth.
1.12–2.13: Recent relations with the church. See Introduction. **12–14:** Appeal for a full and fair hearing, even if, while writing in the first person (*we, I*), Paul "boasts." *Our boast,* a key theme in 2 Cor, along with "confidence" (3.4; 8.22; 10.2 *boldness;* see 1.15 *sure*). Compare Rom 1.29–31.
1.15: See 1 Cor 16.5–7. **16:** Paul *wanted* to go to both Macedonia and Corinth in order (among other reasons) to receive their contributions to the collection, which he would then take *to Judea.* **17–20:** His critics accused

it is God who establishes us with you in Christ and has anointed us, 22 by putting his seal on us and giving us his Spirit in our hearts as a first installment.

23 But I call on God as witness against me: it was to spare you that I did not come again to Corinth. 24 I do not mean to imply that we lord it over your faith; rather, we are workers with you for your joy, because you stand firm in the faith.

2 1 So I made up my mind not to make you another painful visit. 2 For if I cause you pain, who is there to make me glad but the one whom I have pained? 3 And I wrote as I did, so that when I came, I might not suffer pain from those who should have made me rejoice; for I am confident about all of you, that my joy would be the joy of all of you. 4 For I wrote you out of much distress and anguish of heart and with many tears, not to cause you pain, but to let you know the abundant love that I have for you.

5 But if anyone has caused pain, he has caused it not to me, but to some extent—not to exaggerate it—to all of you. 6 This punishment by the majority is enough for such a person; 7 so now instead you should forgive and console him, so that he may not be overwhelmed by excessive sorrow. 8 So I urge you to reaffirm your love for him. 9 I wrote for this reason: to test you and to know whether you are obedient in everything. 10 Anyone whom you forgive, I also for-

give. What I have forgiven, if I have forgiven anything, has been for your sake in the presence of Christ. 11 And we do this so that we may not be outwitted by Satan; for we are not ignorant of his designs.

12 When I came to Troas to proclaim the good news of Christ, a door was opened for me in the Lord; 13 but my mind could not rest because I did not find my brother Titus there. So I said farewell to them and went on to Macedonia.

14 But thanks be to God, who in Christ always leads us in triumphal procession, and through us spreads in every place the fragrance that comes from knowing him. 15 For we are the aroma of Christ to God among those who are being saved and among those who are perishing; 16 to the one a fragrance from death to death, to the other a fragrance from life to life. Who is sufficient for these things? 17 For we are not peddlers of God's word like so many,*f* but in Christ we speak as persons of sincerity, as persons sent from God and standing in his presence.

3 Are we beginning to commend ourselves again? Surely we do not need, as some do, letters of recommendation to you or from you, do we? 2 You yourselves are our letter, written on our*g* hearts, to be known and read by all; 3 and

f Other ancient authorities read *like the others*
g Other ancient authorities read *your*

---

him of being *vacillating*. **22:** The *Spirit*, already given, is an advance *installment* of what is in store for Christians (5.5; Eph 1.13); God will finish what has been begun (Rom 8.16–17, 23; Eph 1.14; Phil 1.6).

2.1: *Another painful visit*, assuming between the plans announced in 1 Cor 16.5–7 and the writing of 2 Cor or any part of it a visit where Paul was wronged by someone in Corinth and not supported by the community there (see 2.5–11; 7.8–12), a visit not recorded in Acts but hinted at in 2 Cor 12.14; 13.1. **3–4:** Many identify the tearful but severe letter referred to with chs 10–13 (see Introduction); but 2.5 suggests that an individual caused the pain, whereas chs 10–13 deal with problems caused by *false apostles* (11.5, 13). Some therefore have identified what Paul *wrote . . . out of*

much distress with 1 Cor, especially ch 5. Or the severe letter may simply be lost.

2.12–13: *A door was opened*, there was a good opportunity for preaching Christ. *Titus*, like Timothy, was one of Paul's helpers (see Introduction to Titus).

2.14–3.6: Our ministry, *as persons sent from God* (2.17; 3.5). **14:** A reference to the triumphal processions of conquerors returning to their capitals. The *fragrance* is the odor of incense in connection with such processions or with sacrifice (Col 2.15). It is knowledge of God in Christ (see 4.6). **16:** The decisiveness of one's acceptance or rejection of the gospel. **17:** *Peddlers*, a reference presumably to the "false apostles" of 11.4, 13.

3.1: Apparently Paul had been accused of commending himself. **3:** The church at Cor-

you show that you are a letter of Christ, prepared by us, written not with ink but with the Spirit of the living God, not on tablets of stone but on tablets of human hearts.

4 Such is the confidence that we have through Christ toward God. 5 Not that we are competent of ourselves to claim anything as coming from us; our competence is from God, 6 who has made us competent to be ministers of a new covenant, not of letter but of spirit; for the letter kills, but the Spirit gives life.

7 Now if the ministry of death, chiseled in letters on stone tablets,*h* came in glory so that the people of Israel could not gaze at Moses' face because of the glory of his face, a glory now set aside, 8 how much more will the ministry of the Spirit come in glory? 9 For if there was glory in the ministry of condemnation, much more does the ministry of justification abound in glory! 10 Indeed, what once had glory has lost its glory because of the greater glory; 11 for if what was set aside came through glory, much more has the permanent come in glory!

12 Since, then, we have such a hope, we act with great boldness, 13 not like Moses, who put a veil over his face to keep the people of Israel from gazing at the end of the glory that*i* was being set aside. 14 But their minds were hardened. Indeed, to this very day, when they hear the reading of the old covenant, that same veil is still there, since only in Christ is it set aside. 15 Indeed, to this very day whenever Moses is read, a veil lies over their minds; 16 but when one turns to the Lord, the veil is removed. 17 Now the Lord is the Spirit, and where the Spirit of the Lord is, there is freedom. 18 And all of us, with unveiled faces, seeing the glory of the Lord as though reflected in a mirror, are being transformed into the same image from one degree of glory to another; for this comes from the Lord, the Spirit.

4 Therefore, since it is by God's mercy that we are engaged in this ministry, we do not lose heart. 2 We have renounced the shameful things that one hides; we refuse to practice cunning or to falsify God's word; but by the open statement of the truth we commend ourselves to the conscience of everyone in the sight of God. 3 And even if our gospel is veiled, it is veiled to those who are perishing. 4 In their case the god of this world has blinded the minds of the unbelievers, to keep them from seeing the light of the gospel of the glory of Christ, who is the image of God. 5 For we do not proclaim ourselves; we proclaim Jesus Christ as Lord and ourselves as your slaves for Jesus' sake. 6 For it is the God

h Gk *on stones*   *i* Gk *of what*

---

inth (Paul's *letter of recommendation*) was the work of God; Paul was only the instrument (v. 5). *Tablets of stone,* like the *letter* (v. 6), suggest the covenant with Moses (Ex 24.12); Christ has instituted a new covenant. 6: *Kills,* produces not life but despair and death. For Paul's view of the purpose and effect of the law, see Gal 3.19–29 n. and Rom 7.7–13 n. **3.7–18: The ministry of the new covenant** (see 1 Cor 11.25 n.). The contrast announced in v. 6 is developed on the basis of Ex 34.29–35. **7:** *Ministry of death,* existence under the Mosaic law. **13:** Paul interprets Moses' *veil* as his effort to hide from the people how temporary the old covenant was to be. **14:** *The old covenant,* the books of the law. **3.15–17:** In Christ one sees the transiency of the old and knows the *freedom* and *glory* of the new. Paul here seems to identify the Spirit, known within the church, with the risen Lord Jesus, but since v. 16 is a quotation of Ex 34.34, the reference may be to Moses and God. **17:** *Lord,* God (v. 3); others see a reference to Christ (Rom 8.9–10). **18:** *All of us* who are *being saved* (2.15), not just Paul and colleagues. *Unveiled faces,* in contrast to v. 15. *Seeing . . . reflected in a mirror:* combines two suggested senses of a single Greek phrase, "beholding" and "reflecting." *Transformed,* Rom 12.2; see Rom 8.17, 29. *Image,* of God, that is, Christ, 4.4. *Lord,* see v. 17.

4.1–18: True treasure, mortal ministers. 1: Such a ministry (3.6), from God, brings confidence; see 1.12–14 n. **2:** Probably an oblique reference to the methods of the "false apostles" (see 2.17 n.). **3–4:** Paul has apparently been accused of not making the gospel clear. **3:** *Veiled,* 3.15, 18; 2.15. **4:** *God of this world,* Satan or Beliar (6.15). **6:** Gen 1.3, applied to our *new creation* (5.17).

who said, "Let light shine out of darkness," who has shone in our hearts to give the light of the knowledge of the glory of God in the face of Jesus Christ. 7 But we have this treasure in clay jars, so that it may be made clear that this extraordinary power belongs to God and does not come from us. 8 We are afflicted in every way, but not crushed; perplexed, but not driven to despair; 9 persecuted, but not forsaken; struck down, but not destroyed; 10 always carrying in the body the death of Jesus, so that the life of Jesus may also be made visible in our bodies. 11 For while we live, we are always being given up to death for Jesus' sake, so that the life of Jesus may be made visible in our mortal flesh. 12 So death is at work in us, but life in you.

13 But just as we have the same spirit of faith that is in accordance with scripture—"I believed, and so I spoke"—we also believe, and so we speak, 14 because we know that the one who raised the Lord Jesus will raise us also with Jesus, and will bring us with you into his presence. 15 Yes, everything is for your sake, so that grace, as it extends to more and more people, may increase thanksgiving, to the glory of God.

16 So we do not lose heart. Even though our outer nature is wasting away, our inner nature is being renewed day by day. 17 For this slight momentary affliction is preparing us for an eternal weight of glory beyond all measure, 18 because we look not at what can be seen but at what cannot be seen; for what can be seen

is temporary, but what cannot be seen is eternal.

5 For we know that if the earthly tent we live in is destroyed, we have a building from God, a house not made with hands, eternal in the heavens. 2 For in this tent we groan, longing to be clothed with our heavenly dwelling— 3 if indeed, when we have taken it off[j] we will not be found naked. 4 For while we are still in this tent, we groan under our burden, because we wish not to be unclothed but to be further clothed, so that what is mortal may be swallowed up by life. 5 He who has prepared us for this very thing is God, who has given us the Spirit as a guarantee.

6 So we are always confident; even though we know that while we are at home in the body we are away from the Lord— 7 for we walk by faith, not by sight. 8 Yes, we do have confidence, and we would rather be away from the body and at home with the Lord. 9 So whether we are at home or away, we make it our aim to please him. 10 For all of us must appear before the judgment seat of Christ, so that each may receive recompense for what has been done in the body, whether good or evil.

11 Therefore, knowing the fear of the Lord, we try to persuade others; but we ourselves are well known to God, and I hope that we are also well known to your consciences. 12 We are not commending ourselves to you again, but giving you an

j Other ancient authorities read *put it on*

4.7: *Clay jars,* a reference to the weakness of the body (Gen 2.7) and indeed to all human limitations. 10–11: In his sufferings and perils Paul shares in Jesus' death; but it is given to him also to share in the life of the risen, victorious Christ. 12: The apostle "dies" (i.e. suffers) that the Corinthians and others may know the life in Christ (v. 15).
4.13: A reference to the faith of the psalmist in the midst of troubles (Ps 116.10); belief leads to confession of faith. 15: As the gospel of God's grace (6.1) advances, *more and more people* respond in faith with thankfulness. 16: See 3.18. 17: Paul considers all of his sufferings as a *slight momentary affliction.*

5.1–10: **Confidence in facing death.** Continuing to think of his sufferings and constant peril, Paul uses the figures of a *tent* and a *building from God* in speaking of death and resurrection. This *building* is the "spiritual body" of 1 Cor 15.44–50. 3: *Naked,* Paul hopes that the Lord will come and that he will receive his new body before he has had to put off the old one. 4: 1 Cor 15.51–54. *Burden,* of afflictions (1.6; 4.8, 17). 5: See 1.22 n.
5.6–8: Believers are never *away from the Lord* in the absolute sense; but as long as they are in the body, they are to a degree separated from him (1 Cor 13.12).
5.11–6.13: **Further defense of his minis-**

opportunity to boast about us, so that you may be able to answer those who boast in outward appearance and not in the heart. [13]For if we are beside ourselves, it is for God; if we are in our right mind, it is for you. [14]For the love of Christ urges us on, because we are convinced that one has died for all; therefore all have died. [15]And he died for all, so that those who live might live no longer for themselves, but for him who died and was raised for them.

16 From now on, therefore, we regard no one from a human point of view;[k] even though we once knew Christ from a human point of view,[k] we know him no longer in that way. [17]So if anyone is in Christ, there is a new creation: everything old has passed away; see, everything has become new! [18]All this is from God, who reconciled us to himself through Christ, and has given us the ministry of reconciliation; [19]that is, in Christ God was reconciling the world to himself,[l] not counting their trespasses against them, and entrusting the message of reconciliation to us. [20]So we are ambassadors for Christ, since God is making his appeal through us; we entreat you on behalf of Christ, be reconciled to God. [21]For our sake he made him to be sin who knew no sin, so that in him we might become the righteousness of God.

6 As we work together with him,[m] we urge you also not to accept the grace of God in vain. [2]For he says,

"At an acceptable time I have
    listened to you,
  and on a day of salvation I have
    helped you."

See, now is the acceptable time; see, now is the day of salvation! [3]We are putting no obstacle in anyone's way, so that no fault may be found with our ministry, [4]but as servants of God we have commended ourselves in every way: through great endurance, in afflictions, hardships, calamities, [5]beatings, imprisonments, riots, labors, sleepless nights, hunger; [6]by purity, knowledge, patience, kindness, holiness of spirit, genuine love, [7]truthful speech, and the power of God; with the weapons of righteousness for the right hand and for the left; [8]in honor and dishonor, in ill repute and good repute. We are treated as impostors, and yet are true; [9]as unknown, and yet are well known; as dying, and see— we are alive; as punished, and yet not killed; [10]as sorrowful, yet always rejoicing; as poor, yet making many rich; as having nothing, and yet possessing everything.

11 We have spoken frankly to you Corinthians; our heart is wide open to you. [12]There is no restriction in our affections, but only in yours. [13]In return— I speak as to children—open wide your hearts also.

k *Gk according to the flesh*   l *Or God was in Christ reconciling the world to himself*   m *Gk As we work together*

---

try of reconciliation. **5.13**: *Beside ourselves,* in contrast to *in our right mind,* refers to a charge by opponents against Paul, perhaps of religious ecstasy (12.1–7). **14**: *The love of Christ,* Christ's love for us, as seen in his cross. *All have died,* to "die" in this sense is to live no longer for oneself but for Christ (v. 15).

5.16–17: Once Paul *from a human point of view,* by worldly standards, thought of Christ as rightly put to death; now he knows the crucified Christ as risen Lord, head of a *new creation* into which the believer as a new creature is incorporated. **18**: *Reconciled,* Rom 5.10; Col 1.20; Heb 1.3. **20**: *Be reconciled,* accept God's forgiveness in Christ. **21**: Paul does not

say "made him a sinner"; the sinless Christ bore the burden of our sin that we might be acquitted (Gal 3.13). *Sin* may mean sin offering (see Rom 8.3 n.; Isa 53.10). *Righteousness of God,* Rom 1.17 n. As at Rom 5.10–11, reference to *reconciliation* intertwines with justification.

6.1–2: Quoting Isa 49.8 Paul urges his readers to respond faithfully to God's grace in Christ. *Now,* at this moment, before the Lord returns. **4–5**: 11.23–29. **6–7**: See Gal 5.22–23. **10**: *Rich,* with spiritual gifts.

6.11: I have been frank with you, *Corinthians,* for you are in my *heart.* **12**: Any restraint upon our relations has been owing to you, not to me; see 1.12–14.

14 Do not be mismatched with unbelievers. For what partnership is there between righteousness and lawlessness? Or
what fellowship is there between light
and darkness? 15 What agreement does
Christ have with Beliar? Or what does a
believer share with an unbeliever?
16 What agreement has the temple of God
with idols? For we[n] are the temple of the
living God; as God said,
"I will live in them and walk
among them,
and I will be their God,
and they shall be my people.
17 Therefore come out from them,
and be separate from them, says
the Lord,
and touch nothing unclean;
then I will welcome you,
18 and I will be your father,
and you shall be my sons and
daughters,
says the Lord Almighty."

7 Since we have these promises, beloved, let us cleanse ourselves from
every defilement of body and of spirit,
making holiness perfect in the fear of
God.

2 Make room in your hearts[o] for us;
we have wronged no one, we have corrupted no one, we have taken advantage
of no one. 3 I do not say this to condemn
you, for I said before that you are in our
hearts, to die together and to live together. 4 I often boast about you; I have great
pride in you; I am filled with consolation;
I am overjoyed in all our affliction.

5 For even when we came into Macedonia, our bodies had no rest, but we
were afflicted in every way—disputes
without and fears within. 6 But God,
who consoles the downcast, consoled us
by the arrival of Titus, 7 and not only by
his coming, but also by the consolation
with which he was consoled about you,
as he told us of your longing, your
mourning, your zeal for me, so that I
rejoiced still more. 8 For even if I made
you sorry with my letter, I do not regret
it (though I did regret it, for I see that I
grieved you with that letter, though only
briefly). 9 Now I rejoice, not because you
were grieved, but because your grief led
to repentance; for you felt a godly grief,
so that you were not harmed in any way
by us. 10 For godly grief produces a repentance that leads to salvation and
brings no regret, but worldly grief produces death. 11 For see what earnestness
this godly grief has produced in you,
what eagerness to clear yourselves, what
indignation, what alarm, what longing,
what zeal, what punishment! At every
point you have proved yourselves guiltless in the matter. 12 So although I wrote
to you, it was not on account of the one
who did the wrong, nor on account of
the one who was wronged, but in order
that your zeal for us might be made
known to you before God. 13 In this we
find comfort.

In addition to our own consolation,

n Other ancient authorities read you
o Gk lacks in your hearts

**6.14–7.1: A parenthesis on relations
with unbelievers.** This passage represents
an abrupt change of subject and an interruption, since 7.2 seems to follow directly upon
6.13. It has been suggested that a fragment of
some other letter to Corinth (possibly that
mentioned in 1 Cor 5.9–11) was inserted here
when the letters of Paul were first collected
and published. But the verses do continue the
urgent appeal of 6.1 with an exhortation to
holiness (7.1). **6.15:** *Beliar,* an evil spirit in the
intertestamental literature; under, or identified with, Satan. **16:** 1 Cor 3.16–17 n. **16–18:**
A number of Old Testament passages are
rather loosely quoted to urge separateness
from the pagan world: Lev 26.12; Ezek 37.27;
Isa 52.11; 2 Sam 7.14. **7.1:** *Making holiness perfect,* completely dedicated to God.

7.2–16: Paul's joy at the restoration of
good relations. **2–4:** After the "parenthesis"
of 6.14–7.1 this statement follows naturally
upon 6.13. **4:** *Boast,* see 1.12–14 n.

7.5–16: See Introduction and 2.13. **5:** *Bodies,* a way of referring to his suffering and
anxiety; see 2.13 mind. **6–7:** *Consolation,* 1.3–
11. **8:** 2.3–4 n. **11:** *Punishment,* a penalty of
some kind imposed by the community on an
unnamed opponent of the apostle (2.5–9). **12:**
The one who was wronged, probably Paul himself.

we rejoiced still more at the joy of Titus, because his mind has been set at rest by all of you. 14For if I have been somewhat boastful about you to him, I was not disgraced; but just as everything we said to you was true, so our boasting to Titus has proved true as well. 15And his heart goes out all the more to you, as he remembers the obedience of all of you, and how you welcomed him with fear and trembling. 16I rejoice, because I have complete confidence in you.

8 We want you to know, brothers and sisters,*p* about the grace of God that has been granted to the churches of Macedonia; 2for during a severe ordeal of affliction, their abundant joy and their extreme poverty have overflowed in a wealth of generosity on their part. 3For, as I can testify, they voluntarily gave according to their means, and even beyond their means, 4begging us earnestly for the privilege*q* of sharing in this ministry to the saints— 5and this, not merely as we expected; they gave themselves first to the Lord and, by the will of God, to us, 6so that we might urge Titus that, as he had already made a beginning, so he should also complete this generous undertaking*r* among you. 7Now as you excel in everything—in faith, in speech, in knowledge, in utmost eagerness, and in our love for you*s*—so we want you to excel also in this generous undertaking.*r*

8 I do not say this as a command, but I am testing the genuineness of your love against the earnestness of others. 9For you know the generous act*t* of our Lord Jesus Christ, that though he was rich, yet for your sakes he became poor, so that by his poverty you might become rich. 10And in this matter I am giving my advice: it is appropriate for you who began last year not only to do something but even to desire to do something— 11now finish doing it, so that your eagerness may be matched by completing it according to your means. 12For if the eagerness is there, the gift is acceptable according to what one has—not according to what one does not have. 13I do not mean that there should be relief for others and pressure on you, but it is a question of a fair balance between 14your present abundance and their need, so that their abundance may be for your need, in order that there may be a fair balance. 15As it is written,

"The one who had much did not
have too much,
and the one who had little did
not have too little."

16 But thanks be to God who put in the heart of Titus the same eagerness for you that I myself have. 17For he not only accepted our appeal, but since he is more eager than ever, he is going to you of his own accord. 18With him we are sending the brother who is famous among all the churches for his proclaiming the good news;*u* 19and not only that, but he has

p Gk *brothers*   *q* Gk *grace*   *r* Gk *this grace*
s Other ancient authorities read *your love for us*
t Gk *the grace*   *u* Or *the gospel*

8.1–9.15: The collection for the relief of the Jerusalem church (Gal 2.1–10; 1 Cor 16.1–4; Rom 15.25–27). These chapters have sometimes been taken as a separate stewardship appeal or even as two administrative letters, ch 8 to Corinth, ch 9 to Achaia (9.2; 1.1 n.).
8.1: *Churches of Macedonia,* in a province to the north where Philippi and Thessalonica were located (Acts 16.9–10, 12, 17). are cited as examples of generosity; see 9.2, however.
8.6: *Titus,* see 1.12–13 n.; 8.16–17. **9:** *Though he was rich,* a reference to Christ's pre-existence. **10:** The raising of funds at Corinth had perhaps been interrupted by the rift between Paul and the church, a rift now healed. **13–14:** *A fair balance* or equality is stressed. **14:** *Your abundance* suggests that the Corinthian church was in better economic condition than some of Paul's other churches (compare the Macedonians' extreme poverty, 8.1–2). **15:** Ex 16.18.
8.16–9.5: Administrative arrangements for the collection. **18:** *The brother,* like *our brother* in v. 22, is unidentified. One gathers from v. 23 that two men had been appointed by the churches to assist Paul in the collection and, no doubt at Paul's insistence, to "audit" the accounts (vv. 20–21).

also been appointed by the churches to travel with us while we are administering this generous undertaking*v* for the glory of the Lord himself*w* and to show our goodwill. 20 We intend that no one should blame us about this generous gift that we are administering, 21 for we intend to do what is right not only in the Lord's sight but also in the sight of others. 22 And with them we are sending our brother whom we have often tested and found eager in many matters, but who is now more eager than ever because of his great confidence in you. 23 As for Titus, he is my partner and co-worker in your service; as for our brothers, they are messengers*x* of the churches, the glory of Christ. 24 Therefore openly before the churches, show them the proof of your love and of our reason for boasting about you.

9 Now it is not necessary for me to write you about the ministry to the saints, 2 for I know your eagerness, which is the subject of my boasting about you to the people of Macedonia, saying that Achaia has been ready since last year; and your zeal has stirred up most of them. 3 But I am sending the brothers in order that our boasting about you may not prove to have been empty in this case, so that you may be ready, as I said you would be; 4 otherwise, if some Macedonians come with me and find that you are not ready, we would be humiliated—to say nothing of you—in this undertaking.*y* 5 So I thought it necessary to urge the brothers to go on ahead to you, and arrange in advance for this bountiful gift that you have promised, so that it may be ready as a voluntary gift and not as an extortion.

6 The point is this: the one who sows sparingly will also reap sparingly, and the one who sows bountifully will also reap bountifully. 7 Each of you must give as you have made up your mind, not reluctantly or under compulsion, for God loves a cheerful giver. 8 And God is able to provide you with every blessing in abundance, so that by always having enough of everything, you may share abundantly in every good work. 9 As it is written,

"He scatters abroad, he gives to
      the poor;
   his righteousness*z* endures
      forever."

10 He who supplies seed to the sower and bread for food will supply and multiply your seed for sowing and increase the harvest of your righteousness.*z* 11 You will be enriched in every way for your great generosity, which will produce thanksgiving to God through us; 12 for the rendering of this ministry not only supplies the needs of the saints but also overflows with many thanksgivings to God. 13 Through the testing of this ministry you glorify God by your obedience to the confession of the gospel of Christ and by the generosity of your sharing with them and with all others, 14 while they long for you and pray for you because of the surpassing grace of God that he has given you. 15 Thanks be to God for his indescribable gift!

10 I myself, Paul, appeal to you by the meekness and gentleness of Christ—I who am humble when face to face with you, but bold toward you

---

v Gk *this grace*   *w* Other ancient authorities lack *himself*   *x* Gk *apostles*   *y* Other ancient authorities add *of boasting*   *z* Or *benevolence*

9.2: *Achaia,* see 1.1 n. **3:** *The brothers,* perhaps the same ones spoken of in 8.18, 22–23. **7:** *God . . . giver,* see Prov 22.8a Gk. **9:** Ps 112.9.
9.10: A reminiscence of Isa 55.10. **11–12:** The gift *which . . . through us* is to be delivered at Jerusalem will be the occasion of *many thanksgivings to God* on the part of those who receive it; see 4.15 n. **15:** *Thanks . . . to God,*

not only for the generosity inspired, but also, and much more, for the *gift* of Christ.
Chs 10–13: Vindicating Paul's authority. On the problem which the abrupt change of tone in these chapters presents, see Introduction. Some identify 10.1–13.10 with the "painful" letter referred to in 2.3–9; 7.8–12. This identification is uncertain, but one can see grounds for holding that these chapters

when I am away!— 2I ask that when I am present I need not show boldness by daring to oppose those who think we are acting according to human standards.*a* 3Indeed, we live as human beings,*b* but we do not wage war according to human standards;*a* 4for the weapons of our warfare are not merely human,*c* but they have divine power to destroy strongholds. We destroy arguments 5and every proud obstacle raised up against the knowledge of God, and we take every thought captive to obey Christ. 6We are ready to punish every disobedience when your obedience is complete.

7 Look at what is before your eyes. If you are confident that you belong to Christ, remind yourself of this, that just as you belong to Christ, so also do we. 8Now, even if I boast a little too much of our authority, which the Lord gave for building you up and not for tearing you down, I will not be ashamed of it. 9I do not want to seem as though I am trying to frighten you with my letters. 10For they say, "His letters are weighty and strong, but his bodily presence is weak, and his speech contemptible." 11Let such people understand that what we say by letter when absent, we will also do when present.

12 We do not dare to classify or compare ourselves with some of those who commend themselves. But when they measure themselves by one another, and compare themselves with one another, they do not show good sense. 13We, however, will not boast beyond limits, but will keep within the field that God has assigned to us, to reach out even as far as you. 14For we were not overstepping our limits when we reached you; we were the first to come all the way to you with the good news*d* of Christ. 15We do not boast beyond limits, that is, in the labors of others; but our hope is that, as your faith increases, our sphere of action among you may be greatly enlarged, 16so that we may proclaim the good news*d* in lands beyond you, without boasting of work already done in someone else's sphere of action. 17"Let the one who boasts, boast in the Lord." 18For it is not those who commend themselves that are approved, but those whom the Lord commends.

11 I wish you would bear with me in a little foolishness. Do bear with me! 2I feel a divine jealousy for you, for I promised you in marriage to one husband, to present you as a chaste virgin to Christ. 3But I am afraid that as the serpent deceived Eve by its cunning, your thoughts will be led astray from a sincere and pure*e* devotion to Christ. 4For if someone comes and proclaims another

---

a Gk *according to the flesh*   *b* Gk *in the flesh*
c Gk *fleshly*   *d* Or *the gospel*
e Other ancient authorities lack *and pure*

---

were not originally a part of the same letter as chs 1–9.

10.1–18: An appeal for true apostleship, Paul's, in contrast to the "super-apostles" (11.5, 12–15; 12.11–12). **1:** *I who am humble,* Paul is ironically quoting his opponents, as he often does in these last four chapters. On the paradox of *meekness* and *gentleness* with *boldness* in Paul and Christ, compare that of *weakness* and *power* in Christ and the true apostle (13.3–4). **6:** The "disobedient" (11.4) ones will be punished when the church as a whole returns to its loyalty and *obedience* to Paul, as he is sure it will eventually do.

10.7–12: The critics are probably members of the Corinthian church as well as "false apostles" from outside. They have apparently accused him of weakness and empty *boasting* (1.12–14 n.). **10:** 2.3. **13–14:** Paul is not doing more than God commissioned him to do when he exercises his authority as an apostle in the church at Corinth. **15–16:** A hint perhaps that his opponents were exploiting *the labors of others,* intruding where others had built; for Paul's missionary policy, compare Rom 15.20. On Paul's hopes of working west of Achaia, see Acts 19.21; Rom 1.11–15; 15.23–28. **17:** See 1 Cor 1.31 n.

11.1–15: Paul's reply to opponents. Because those who are "super-apostles" (11.5; 12.1) in their own estimation have intruded and preach a *different gospel* and *another Jesus* and threaten to *lead astray* the Corinthians, Paul adopts their method of boasting (11.13) to meet this menace, but *in the Lord* and *not beyond limits* (10.13, 15, 17). **2:** Hos 2.19–20;

Jesus than the one we proclaimed, or if you receive a different spirit from the one you received, or a different gospel from the one you accepted, you submit to it readily enough. 5 I think that I am not in the least inferior to these super-apostles. 6 I may be untrained in speech, but not in knowledge; certainly in every way and in all things we have made this evident to you.

7 Did I commit a sin by humbling myself so that you might be exalted, because I proclaimed God's good news*f* to you free of charge? 8 I robbed other churches by accepting support from them in order to serve you. 9 And when I was with you and was in need, I did not burden anyone, for my needs were supplied by the friends*g* who came from Macedonia. So I refrained and will continue to refrain from burdening you in any way. 10 As the truth of Christ is in me, this boast of mine will not be silenced in the regions of Achaia. 11 And why? Because I do not love you? God knows I do!

12 And what I do I will also continue to do, in order to deny an opportunity to those who want an opportunity to be recognized as our equals in what they boast about. 13 For such boasters are false apostles, deceitful workers, disguising themselves as apostles of Christ. 14 And no wonder! Even Satan disguises himself as an angel of light. 15 So it is not strange if his ministers also disguise themselves as ministers of righteousness. Their end will match their deeds.

16 I repeat, let no one think that I am a fool; but if you do, then accept me as a fool, so that I too may boast a little. 17 What I am saying in regard to this boastful confidence, I am saying not with the Lord's authority, but as a fool; 18 since many boast according to human standards,*h* I will also boast. 19 For you gladly put up with fools, being wise yourselves! 20 For you put up with it when someone makes slaves of you, or preys upon you, or takes advantage of you, or puts on airs, or gives you a slap in the face. 21 To my shame, I must say, we were too weak for that!

But whatever anyone dares to boast of—I am speaking as a fool—I also dare to boast of that. 22 Are they Hebrews? So am I. Are they Israelites? So am I. Are they descendants of Abraham? So am I. 23 Are they ministers of Christ? I am talking like a madman—I am a better one: with far greater labors, far more imprisonments, with countless floggings, and often near death. 24 Five times I have re-

*f Gk the gospel of God   g Gk brothers
h Gk according to the flesh*

Eph 5.26–27. **3**: Gen 3.4. **5–6**: That Paul as a speaker was being compared to others to his disadvantage appears also in 10.10 and, less clearly, in 1 Cor 2.1–5.

11.7–11: Even Paul's honorable and generous determination (so as not to be a financial burden) to work with his own hands (1 Cor 4.12; 1 Thess 2.9) had been perversely twisted into a charge against him, probably that manual labor degrades one. On the matter of Paul's support, see Acts 18.1–4; 2 Thess 3.7–9; 1 Cor 9.3–7; and especially Phil 4.10–20. Why he accepted aid of this kind only from the Philippian church in Macedonia we are not told. **8**: *I robbed,* a natural exaggeration (compare v. 9). Those *disguising themselves as apostles of Christ* (v. 13) are in reality *ministers* of Satan; they have not manifested the same disinterestedness as Paul and are seeking to force him through their criticisms to surrender a practice which embarrasses them. **10**:

Achaia, see 1.1 n. **13–15**: Paul's judgment of the *false apostles.*

11.16–33: **Paul's boasting.** Paul indulges in the "little foolishness" spoken of in v. 1. He knows it is foolish to boast, and perhaps even wrong (12.1), but accepts the challenge (v. 18) of some at Corinth who have forced him to assert his claims (12.11). Some verses are written in irony (19, 20–21).

11.22: Apparently even Paul's Judaism (Rom 9.1–3; 11.1; Gal 1.13–14; Phil 3.4–6; compare Acts 22.3; 26.4–6) was called into question by the self-assertions of the "super-apostles." **23–27**: Very little is known about these hardships; the book of Acts refers to only a few of them. See also 2 Cor 4.8–9; 6.4–5; and 1 Cor 4.9–13 on the lifestyle of true apostles. **24**: *Forty lashes,* a Jewish punishment; only thirty-nine were administered lest by a miscount the number exceed the maximum prescribed in Deut 25.3.

ceived from the Jews the forty lashes minus one. 25 Three times I was beaten with rods. Once I received a stoning. Three times I was shipwrecked; for a night and a day I was adrift at sea; 26 on frequent journeys, in danger from rivers, danger from bandits, danger from my own people, danger from Gentiles, danger in the city, danger in the wilderness, danger at sea, danger from false brothers and sisters;[i] 27 in toil and hardship, through many a sleepless night, hungry and thirsty, often without food, cold and naked. 28 And, besides other things, I am under daily pressure because of my anxiety for all the churches. 29 Who is weak, and I am not weak? Who is made to stumble, and I am not indignant?

30 If I must boast, I will boast of the things that show my weakness. 31 The God and Father of the Lord Jesus (blessed be he forever!) knows that I do not lie. 32 In Damascus, the governor[j] under King Aretas guarded the city of Damascus in order to[k] seize me, 33 but I was let down in a basket through a window in the wall,[l] and escaped from his hands.

12 It is necessary to boast; nothing is to be gained by it, but I will go on to visions and revelations of the Lord. 2 I know a person in Christ who fourteen years ago was caught up to the third heaven—whether in the body or out of the body I do not know; God knows. 3 And I know that such a person— whether in the body or out of the body I do not know; God knows— 4 was caught up into Paradise and heard things that are not to be told, that no mortal is permitted to repeat. 5 On behalf of such a one I will boast, but on my own behalf I will not boast, except of my weaknesses. 6 But if I wish to boast, I will not be a fool, for I will be speaking the truth. But I refrain from it, so that no one may think better of me than what is seen in me or heard from me, 7 even considering the exceptional character of the revelations. Therefore, to keep[m] me from being too elated, a thorn was given me in the flesh, a messenger of Satan to torment me, to keep me from being too elated.[n] 8 Three times I appealed to the Lord about this, that it would leave me, 9 but he said to me, "My grace is sufficient for you, for power[o] is made perfect in weakness." So, I will boast all the more gladly of my weaknesses, so that the power of Christ may dwell in me. 10 Therefore I am content with weaknesses, insults, hardships, persecutions, and calamities for the sake of Christ; for whenever I am weak, then I am strong.

11 I have been a fool! You forced me to it. Indeed you should have been the ones commending me, for I am not at all inferior to these super-apostles, even though I am nothing. 12 The signs of a true apostle were performed among you with utmost patience, signs and wonders

i Gk *brothers*   j Gk *ethnarch*   k Other ancient authorities read *and wanted to*
l Gk *through the wall*   m Other ancient authorities read *To keep*   n Other ancient authorities lack *to keep me from being too elated*
o Other ancient authorities read *my power*

11.25: *Beaten with rods,* a Roman punishment (Acts 16.22–23). *Stoning,* Acts 14.19. **32–33**: This item would seem logically to belong after v. 27; perhaps it is an afterthought (see also Acts 9.23–25). *Aretas* IV was king of Nabataea, southeast of Palestine. Apparently Damascus was under his jurisdiction at the time of Paul's escape.
12.1–13: Further boastings: strength in weakness, a theme begun at 11.30. Though Paul speaks of *revelations* (vv. 1, 7), he refers specifically to only one of them. **2**: *I know a person,* an oblique reference to himself (see v. 7). *The third heaven,* the highest ecstasy. Nothing is otherwise known of this experi-

ence unless it is that referred to in Gal 1.16 and 1 Cor 15.8; but this is not probable. **4**: *Things that are not to be told,* because they are too sacred.
12.5: Paul deserves no credit for these *revelations;* therefore he will not speak of them as though they were his own. **7–9**: Some have thought that *the thorn . . . in the flesh* was illness or a physical disability; others have suggested that it was a specific opponent or the opposition of his fellow Jews. No firm decision is possible. **12**: A reference to apostolic miracles (Rom 15.19; Gal 3.5). **13**: See 11.7– 11 n. *Forgive me this wrong,* for the irony compare 11.21a.

and mighty works. 13 How have you been worse off than the other churches, except that I myself did not burden you? Forgive me this wrong!

14 Here I am, ready to come to you this third time. And I will not be a burden, because I do not want what is yours but you; for children ought not to lay up for their parents, but parents for their children. 15 I will most gladly spend and be spent for you. If I love you more, am I to be loved less? 16 Let it be assumed that I did not burden you. Nevertheless (you say) since I was crafty, I took you in by deceit. 17 Did I take advantage of you through any of those whom I sent to you? 18 I urged Titus to go, and sent the brother with him. Titus did not take advantage of you, did he? Did we not conduct ourselves with the same spirit? Did we not take the same steps?

19 Have you been thinking all along that we have been defending ourselves before you? We are speaking in Christ before God. Everything we do, beloved, is for the sake of building you up. 20 For I fear that when I come, I may find you not as I wish, and that you may find me not as you wish; I fear that there may perhaps be quarreling, jealousy, anger, selfishness, slander, gossip, conceit, and disorder. 21 I fear that when I come again, my God may humble me before you, and that I may have to mourn over many who previously sinned and have not repented of the impurity, sexual immorality, and licentiousness that they have practiced.

13 This is the third time I am coming to you. "Any charge must be sustained by the evidence of two or three witnesses." 2 I warned those who sinned previously and all the others, and I warn them now while absent, as I did when present on my second visit, that if I come again, I will not be lenient— 3 since you desire proof that Christ is speaking in me. He is not weak in dealing with you, but is powerful in you. 4 For he was crucified in weakness, but lives by the power of God. For we are weak in him, *p* but in dealing with you we will live with him by the power of God.

5 Examine yourselves to see whether you are living in the faith. Test yourselves. Do you not realize that Jesus Christ is in you?—unless, indeed, you fail to meet the test! 6 I hope you will find out that we have not failed. 7 But we pray to God that you may not do anything wrong—not that we may appear to have met the test, but that you may do what is right, though we may seem to have failed. 8 For we cannot do anything against the truth, but only for the truth. 9 For we rejoice when we are weak and you are strong. This is what we pray for, that you may become perfect. 10 So I write these things while I am away from you, so that when I come, I may not have to be severe in using the authority that the Lord has given me for building up and not for tearing down.

p Other ancient authorities read *with him*

12.14–13.10: Paul plans to visit Corinth again. This *third* visit cannot be identified with assurance. Some suppose it to be the visit referred to in retrospect in 2.1; others, that it is the visit in prospect when chs 1–9 were written. **12.14–18:** Once again (see 11.7–11) Paul discusses financial support. **17–18:** Is this the sending of *Titus* and *the brother* mentioned in 8.16–19? If so, chs 10–13 cannot have been written earlier than chs 1–9, unless ch 8 be regarded as a separate note about the collection written earlier still. See Introduction and 8.1–9.15 n.; 10–13 n. **12.19–13.10: A call for self-examination** (13.5) **and amendment** before Paul comes. For his readers' sake, as well as for his own, the apostle wants the coming visit to be happy and mutually rewarding. If chs 10–13 are the "painful" letter referred to in 2.3–4, we know from chs 1–9 that Paul's hopes were fulfilled. **13.1:** *Witnesses,* Deut 19.15. **2–4:** Paul will vigorously assert his apostolic authority in dealing with wrongdoing at Corinth, but it will be Christ's authority. **5–9:** Paul hopes that he will not need to exert his authority, but that the church will discipline itself. **10:** See 12.19.

11 Finally, brothers and sisters, *q* farewell.* Put things in order, listen to my appeal,* agree with one another, live in peace; and the God of love and peace will be with you. 12Greet one another with a holy kiss. All the saints greet you.

13 The grace of the Lord Jesus Christ, the love of God, and the communion of* the Holy Spirit be with all of you.

q Gk *brothers*   *r* Or *rejoice*   *s* Or *encourage one another*   *t* Or *and the sharing in*

13.11–14: Conclusion. 12: The liturgical *kiss,* see Rom 16.16 n. **13:** The fullest of Paul's benedictions at the end of his letters (for the Trinitarian form, compare 1 Cor 12.4–6). The order is significant; the *grace of Christ* expresses and leads one toward the *love of God,* and the love of God when actualized through the *Spirit,* produces *communion* with God and with one another.

The Letter of Paul to the
Galatians

Often called the Magna Charta of Christian liberty, the Letter to the Galatians deals with the question whether Gentiles must become Jews before they can become Christians. Certain Judaizing teachers had infiltrated the churches of Galatia in central Asia Minor, which Paul had previously founded (Acts 16.6), declaring that in addition to having faith in Jesus Christ a Christian was obligated to keep the Mosaic law. Paul insists, on the contrary, that a person becomes right with God only by faith in Christ and not by the performance of good works, ritual observances, and the like (2.16; 3.24–25; 5.1; 6.12–15).

So serious was the crisis in Galatia that Paul dispenses with his customary expression of thanksgiving and commendation, and plunges directly into a vigorous defense of his apostolic authority and the validity of his teaching (1.1–2.21). His gospel must be asserted against *certain people . . . from James* in Jerusalem (2.12) and even against *Cephas* (Peter, see 2.14). The central part of the letter is an exposition of the doctrine of justification by faith alone (3.1–4.31). Lest some should imagine that this doctrine leads to a life of indifference to the moral code, Paul concludes with certain practical applications of his teaching (5.1–6.18). Some think these ethical admonitions were necessitated not by the Judaizing teachers but a second group of opponents, more gnosticizing in nature, who abused Christian freedom.

The importance of this brief letter is hard to overestimate. Written perhaps about A.D. 55 or slightly earlier during Paul's third missionary journey, it gives many autobiographical details of the apostle's earlier life and evangelistic activity. Here are set forth, with impassioned eloquence, the true function of the Mosaic law and its relation to God's grace manifested in Christ. The declaration of the principles reiterated in these six chapters made Christianity a world religion instead of a Jewish sect.

1 Paul an apostle—sent neither by human commission nor from human authorities, but through Jesus Christ and God the Father, who raised him from the dead— ²and all the members of God's family*ᵃ* who are with me,

To the churches of Galatia:

3 Grace to you and peace from God our Father and the Lord Jesus Christ, ⁴who gave himself for our sins to set us free from the present evil age, according to the will of our God and Father, ⁵to whom be the glory forever and ever. Amen.

6 I am astonished that you are so quickly deserting the one who called you in the grace of Christ and are turning to a different gospel— ⁷not that there is another gospel, but there are some who are confusing you and want to pervert the gospel of Christ. ⁸But even if we or an angel*ᵇ* from heaven should proclaim to you a gospel contrary to what we proclaimed to you, let that one be accursed! ⁹As we have said before, so now I repeat, if anyone proclaims to you a gospel contrary to what you received, let that one be accursed!

10 Am I now seeking human approval, or God's approval? Or am I trying to please people? If I were still pleasing people, I would not be a servant*ᶜ* of Christ.

11 For I want you to know, brothers and sisters,*ᵈ* that the gospel that was proclaimed by me is not of human origin; ¹²for I did not receive it from a human source, nor was I taught it, but I received it through a revelation of Jesus Christ. 13 You have heard, no doubt, of my earlier life in Judaism. I was violently persecuting the church of God and was trying to destroy it. ¹⁴I advanced in Judaism beyond many among my people of the same age, for I was far more zealous for the traditions of my ancestors. ¹⁵But when God, who had set me apart before I was born and called me through his grace, was pleased ¹⁶to reveal his Son to me,*ᵉ* so that I might proclaim him among the Gentiles, I did not confer with

a Gk *all the brothers*   *b* Or *a messenger*
c Gk *slave*   *d* Gk *brothers*   *e* Gk *in me*

1.1–5: The Salutation (see Rom 1.1–7 n.) emphasizes both Paul's divinely given authority as an *apostle* and in his gospel the atoning death of *Jesus Christ, who gave himself for our sins* (Mk 10.45; 1 Tim 2.6). **2:** *Galatia,* probably the area in north central Asia Minor around Ancyra (modern Ankara, Turkey), not the Roman province southward; see Acts 16.6; 18.23. **3:** *Grace* and *peace,* see Rom 1.7 n. and 2 Thess 1.2 n. **4:** *The present evil age* (see 2 Cor 4.4) is in contrast to the coming age to be inaugurated at the return of Christ. Our deliverance through Christ's self-giving rests upon *the will of God.*

1.6–10: The Galatian apostasy from the true gospel. Instead of the customary thanksgiving (Rom 1.8–15 n.) Paul launches into the issue: *deserting* . . . *Christ* for what is not the "good news." **6:** *The one who called you,* God, through Christ (see 1.15). **8–9:** A solemn anathema; Rom 9.3; 1 Cor 12.3, "(ac)cursed." **10:** Paul denies an opponent's charge that he conciliates people to win converts; see 1 Thess 2.4.

1.11–2.21: Defense of Paul's gospel and apostolic authority. In a first-person-singular narrative (*I, me*) Paul sets forth his call from God to *proclaim* Christ *among the* Gentiles (1.15–16) and his experiences as an *apostle* (1.1, 17; 2.8). Then (2.15–21) in propositional and personal terms he outlines his gospel of justification by faith.

1.11–24: Paul's vindication of his apostleship. 11: Despite his severity, Paul addresses the Galatians as *brothers and sisters.* A message merely *of human origin* would be no *gospel of Christ* (v. 7) or *revelation* (v. 12). **12:** *Through a revelation,* at the time of Paul's conversion (Acts 9.3–6).

1.13: *I was . . . persecuting,* compare Acts 8.3; 9.21; 22.4. The *church* is the people of God, called into fellowship with the Lord through the redemptive work of Jesus Christ. The word may refer to the total number of believers throughout the world, to those in a particular region (1.2), or to those in one locality, whether gathered for worship and instruction, engaged in mission, or scattered by persecution. **14:** Phil 3.4–6; Acts 22.3. **15–17:** Paul's conversion and commission as an apostle were due to the sovereign plan of God, not to human teachers (1.1). **16:** *To me,* but also through Paul (note *e*) as a missionary. **17:** *Arabia,* that is, the Nabataean kingdom, of which the capital was Petra. *Damascus,* Acts 9.19–25; 2 Cor 11.32–33.

any human being, [17] nor did I go up to Jerusalem to those who were already apostles before me, but I went away at once into Arabia, and afterwards I returned to Damascus.

18 Then after three years I did go up to Jerusalem to visit Cephas and stayed with him fifteen days; [19] but I did not see any other apostle except James the Lord's brother. [20] In what I am writing to you, before God, I do not lie! [21] Then I went into the regions of Syria and Cilicia, [22] and I was still unknown by sight to the churches of Judea that are in Christ; [23] they only heard it said, "The one who formerly was persecuting us is now proclaiming the faith he once tried to destroy." [24] And they glorified God because of me.

2 Then after fourteen years I went up again to Jerusalem with Barnabas, taking Titus along with me. [2] I went up in response to a revelation. Then I laid before them (though only in a private meeting with the acknowledged leaders) the gospel that I proclaim among the Gentiles, in order to make sure that I was not running, or had not run, in vain. [3] But even Titus, who was with me, was not compelled to be circumcised, though he was a Greek. [4] But because of false believers[f] secretly brought in, who slipped in to spy on the freedom we have in Christ Jesus, so that they might en-

slave us— [5] we did not submit to them even for a moment, so that the truth of the gospel might always remain with you. [6] And from those who were supposed to be acknowledged leaders (what they actually were makes no difference to me; God shows no partiality)—those leaders contributed nothing to me. [7] On the contrary, when they saw that I had been entrusted with the gospel for the uncircumcised, just as Peter had been entrusted with the gospel for the circumcised [8] (for he who worked through Peter making him an apostle to the circumcised also worked through me in sending me to the Gentiles), [9] and when James and Cephas and John, who were acknowledged pillars, recognized the grace that had been given to me, they gave to Barnabas and me the right hand of fellowship, agreeing that we should go to the Gentiles and they to the circumcised. [10] They asked only one thing, that we remember the poor, which was actually what I was[g] eager to do.

11 But when Cephas came to Antioch, I opposed him to his face, because he stood self-condemned; [12] for until certain people came from James, he used to eat with the Gentiles. But after they came, he drew back and kept himself separate for fear of the circumcision fac-

f Gk false brothers g Or had been

1.18: *Cephas*, the Aramaic equivalent of "Peter" (see Mt 16.18 n.). 19: *The Lord's brother*, see Mt 13.55 n. 21: *Cilicia*, of which Tarsus (see Acts 9.30 n.) was the capital.

2.1–10: **Paul's apostleship recognized in Jerusalem.** This visit is probably the one mentioned in Acts 15.2. 1: *Barnabas*, see Acts 11.22 n., 25; 13.1–3; 15.2; 1 Cor 9.6. *Titus*, one of Paul's most trusted helpers (2 Cor 7.6; 8.6, 16–17). 2–3: Although the Judaizers had apparently demanded that *Titus, a Greek*, should be circumcised, the apostles at Jerusalem, by not enforcing that demand, approved of Paul's work among the Gentiles (v. 9; see Acts 16.1–5 n.). On *Titus*, see Introduction to Titus. 4: *False believers* (literally "brothers," note *f*). The New Testament never combines "false" with the root "believe" or "have faith"; though the word "brothers [and

sisters]" regularly in Acts denotes Christians, Paul uses "false brothers" only here and at 2 Cor 11.26, of opponents (cf. 11.3). On their identity here, see Acts 15.1, 5. 10: *The poor*, Jerusalem Christians or a group within the Palestinian church, Rom 15.26; on the collection for them, see Acts 11.29–30; 24.17; 1 Cor 16.1–3; 2 Cor 8.1–15.

2.11–14: **Paul rebukes Peter's inconsistency at Antioch** when pressured by *people . . . from James*, more conservative on the Jewish law than Peter. *Cephas*, see 1.18 n. After his vision (Acts 10.10–35) Peter had recognized that God makes no distinction between Jew and Gentile. But at Antioch, when criticized for table fellowship with converted Gentiles, Peter yielded to the narrow prejudices of the Judaizers (v. 12); his inconsistency was contagious (v. 13).

tion. [13] And the other Jews joined him in this hypocrisy, so that even Barnabas was led astray by their hypocrisy. [14] But when I saw that they were not acting consistently with the truth of the gospel, I said to Cephas before them all, "If you, though a Jew, live like a Gentile and not like a Jew, how can you compel the Gentiles to live like Jews?"[h]

[15] We ourselves are Jews by birth and not Gentile sinners; [16] yet we know that a person is justified[i] not by the works of the law but through faith in Jesus Christ.[j] And we have come to believe in Christ Jesus, so that we might be justified by faith in Christ,[k] and not by doing the works of the law, because no one will be justified by the works of the law. [17] But if, in our effort to be justified in Christ, we ourselves have been found to be sinners, is Christ then a servant of sin? Certainly not! [18] But if I build up again the very things that I once tore down, then I demonstrate that I am a transgressor. [19] For through the law I died to the law, so that I might live to God. I have been crucified with Christ; [20] and it is no longer I who live, but it is Christ who lives in me. And the life I now live in the flesh I live by faith in the Son of God,[l] who loved me and gave himself for me. [21] I do not nullify the grace of God; for if justification[m] comes through the law, then Christ died for nothing.

3 You foolish Galatians! Who has bewitched you? It was before your eyes that Jesus Christ was publicly exhibited as crucified! [2] The only thing I want to learn from you is this: Did you receive the Spirit by doing the works of the law or by believing what you heard? [3] Are you so foolish? Having started with the Spirit, are you now ending with the flesh? [4] Did you experience so much for nothing?—if it really was for nothing. [5] Well then, does God[n] supply you with the Spirit and work miracles among you by your doing the works of the law, or by your believing what you heard?

[6] Just as Abraham "believed God, and it was reckoned to him as righteousness," [7] so, you see, those who believe are the descendants of Abraham. [8] And the scripture, foreseeing that God would justify the Gentiles by faith, declared the gospel beforehand to Abraham, saying, "All the Gentiles shall be blessed in you." [9] For this reason, those who believe are blessed with Abraham who believed.

[10] For all who rely on the works of the law are under a curse; for it is written, "Cursed is everyone who does not ob-

h Some interpreters hold that the quotation extends into the following paragraph
i Or *reckoned as righteous;* and so elsewhere
j Or *the faith of Jesus Christ*   k Or *the faith of Christ*   l Or *by the faith of the Son of God*
m Or *righteousness*   n Gk *he*

2.15–21: A statement of principle on the fundamental difference between the law and the gospel. These verses could be taken in quotation marks as a continuation of what Paul said to Cephas at Antioch (note *h*), but they function in the letter as a propositional statement on what is at issue, using *we* in vv. 15–17 and concluding with a declaration of Paul's own living faith. **16:** *Faith in Jesus Christ,* see notes *j* and *k* and Rom 3.22 n. Here Christ is the object of faith, in whom *we have come to believe.* **20:** Paul's union with Christ does not destroy his own personality, but sustains and molds his Christian life (compare Jesus' reference to the vine and the branches, Jn 15.1–5). *Faith,* see v. 16.
3.1–4.31: Evidence confirming Paul's gospel and its freedom. For *the truth of the gospel* (2.5, 14), that justification/righteous-

ness comes by faith and not by law (2.16, 21), a series of proofs are presented.
3.1–5: An appeal to experience, the Galatians' receipt of the Spirit, to prove that justification is by faith, not works. The sequence is hearing and believing the gospel, receiving the Spirit, and hence experiencing divine wondrous deeds. **5:** *Miracles,* wrought *among* (lit. "in") the Galatians, attest the truth of the gospel (on apostolic miracles, compare Rom 15.19; 1 Cor 12.10; 2 Cor 12.12).
3.6–14: An appeal to Abraham's experience in scripture. 6–7: *Abraham* is typical of all *those who believe* (Gen 15.6; Rom 4.3, 16). **8:** Gen 12.3; compare 18.18; Acts 3.25. **10:** Deut 27.26. **11:** Hab 2.4 Gk. **12:** Lev 18.5; Rom 10.5. **13:** The *tree* in Deut 21.23 is referred to Jesus' cross.

serve and obey all the things written in the book of the law." 11 Now it is evident that no one is justified before God by the law; for "The one who is righteous will live by faith." *o* 12 But the law does not rest on faith; on the contrary, "Whoever does the works of the law*p* will live by them." 13 Christ redeemed us from the curse of the law by becoming a curse for us—for it is written, "Cursed is everyone who hangs on a tree"— 14 in order that in Christ Jesus the blessing of Abraham might come to the Gentiles, so that we might receive the promise of the Spirit through faith.

15 Brothers and sisters, *q* I give an example from daily life: once a person's will*r* has been ratified, no one adds to it or annuls it. 16 Now the promises were made to Abraham and to his offspring;*s* it does not say, "And to offsprings,"*t* as of many; but it says, "And to your offspring,"*s* that is, to one person, who is Christ. 17 My point is this: the law, which came four hundred thirty years later, does not annul a covenant previously ratified by God, so as to nullify the promise. 18 For if the inheritance comes from the law, it no longer comes from the promise; but God granted it to Abraham through the promise.

19 Why then the law? It was added because of transgressions, until the offspring*s* would come to whom the promise had been made; and it was ordained through angels by a mediator. 20 Now a mediator involves more than one party; but God is one.

21 Is the law then opposed to the promises of God? Certainly not! For if a law had been given that could make alive, then righteousness would indeed come through the law. 22 But the scripture has imprisoned all things under the power of sin, so that what was promised through faith in Jesus Christ*u* might be given to those who believe.

23 Now before faith came, we were imprisoned and guarded under the law until faith would be revealed. 24 Therefore the law was our disciplinarian until Christ came, so that we might be justified by faith. 25 But now that faith has come, we are no longer subject to a disciplinarian, 26 for in Christ Jesus you are all children of God through faith. 27 As many of you as were baptized into Christ have clothed yourselves with Christ. 28 There is no longer Jew or Greek, there is no longer slave or free, there is no longer male and female; for all of you are one in Christ Jesus. 29 And if you belong to Christ, then you are Abraham's offspring,*s* heirs according to the promise.

o Or *The one who is righteous through faith will live*   *p* Gk *does them*   *q* Gk *Brothers*
r Or *covenant* (as in verse 17)   *s* Gk *seed*
t Gk *seeds*   *u* Or *through the faith of Jesus Christ*

3.15–18: An example from the covenant with Abraham: God gives our inheritance by promise, not law. **15:** The same Greek word means last *will* or testament and *covenant* (note *r*). **16:** *Offspring,* the word used in Gen 12.7 and 22.17–18 is literally "seed," a collective singular. **17:** The faith-principle is older and more fundamental than the Mosaic law. Paul follows the chronology found in some manuscripts of the Septuagint of Ex 12.40, according to which the 430 years included the sojourn of Israel's forebears in Canaan and in Egypt; on the other hand the Hebrew text of Ex 12.40 refers the 430 years solely to the sojourn in Egypt (see also Acts 7.6 n.).

3.19–25: The true purpose of the Mosaic law. Though the law could not make people righteous (v. 21), it revealed God's will so that they might recognize their *transgressions* (vv. 19 and 22; Rom 3.20; 7.7). It is like a *disciplinarian* (see 1 Cor 4.15 n.) who has temporary charge of a child (vv. 24–25). It was not even given by God directly, but (according to later Jewish belief) through *angels* (v. 19; Deut 33.2, Septuagint; see Acts 7.38, 53 n.; Heb 2.2) and *a mediator,* Moses (see Lev 26.46).

3.19: *Because of,* or even to produce, *transgressions.* **22:** *Faith in Jesus Christ,* see Rom 3.22 n. **23:** *Faith,* the era of Christ, Christianity.

3.26–29: An appeal to baptismal equality. In the imagery and implications about those *baptized into* (Rom 6.3–11) and *clothed . . . with Christ* (Rom 13.14; Col 3.10), Paul is here interested especially in *Jew* and *Greek* all being *one in Christ* and thus *Abraham's offspring* (spiritual kinship), so fulfilling God's promise (3.6–7, 16 n., 17 n.).

4 My point is this: heirs, as long as they are minors, are no better than slaves, though they are the owners of all the property; 2 but they remain under guardians and trustees until the date set by the father. 3 So with us; while we were minors, we were enslaved to the elemental spirits *v* of the world. 4 But when the fullness of time had come, God sent his Son, born of a woman, born under the law, 5 in order to redeem those who were under the law, so that we might receive adoption as children. 6 And because you are children, God has sent the Spirit of his Son into our *w* hearts, crying, "Abba! *x* Father!" 7 So you are no longer a slave but a child, and if a child then also an heir, through God. *y*

8 Formerly, when you did not know God, you were enslaved to beings that by nature are not gods. 9 Now, however, that you have come to know God, or rather to be known by God, how can you turn back again to the weak and beggarly elemental spirits? *z* How can you want to be enslaved to them again? 10 You are observing special days, and months, and seasons, and years. 11 I am afraid that my work for you may have been wasted.

12 Friends, *a* I beg you, become as I am, for I also have become as you are. You have done me no wrong. 13 You know that it was because of a physical infirmity that I first announced the gospel to you; 14 though my condition put you to the test, you did not scorn or despise me, but welcomed me as an angel of God, as Christ Jesus. 15 What has become of the goodwill you felt? For I testify that, had it been possible, you would have torn out your eyes and given them to me. 16 Have I now become your enemy by telling you the truth? 17 They make much of you, but for no good purpose; they want to exclude you, so that you may make much of them. 18 It is good to be made much of for a good purpose at all times, and not only when I am present with you. 19 My little children, for whom I am again in the pain of childbirth until Christ is formed in you, 20 I wish I were present with you now and could change my tone, for I am perplexed about you.

21 Tell me, you who desire to be subject to the law, will you not listen to the law? 22 For it is written that Abraham had two sons, one by a slave woman and the other by a free woman. 23 One, the child of the slave, was born according to the flesh; the other, the child of the free woman, was born through the promise. 24 Now this is an allegory: these women are two covenants. One woman, in fact,

v Or *the rudiments*   *w* Other ancient authorities read *your*   *x* Aramaic for *Father*   *y* Other ancient authorities read *an heir of God through Christ*   *z* Or *beggarly rudiments*   *a* Gk *Brothers*

4.1–11: Enslavement under the law; freedom for God's children. 3: *Elemental spirits of the world,* cosmic powers controlling the universe (4.8); or (note *v*) *the rudiments* of the world (earth, air, fire, water), or rudimentary rules and religious observances (vv. 9–10; Col 2.8, 20). **4**: Christ was sent at a time determined by God in order to ransom those who were in bondage *under the law.* **4.6**: *Abba,* see note *x* and Rom 8.15 n. **9**: *Elemental spirits,* see v. 3 n. **10**: They still observe Jewish fast-*days,* new moons (*months;* Col 2.16), Passover *seasons,* and sabbatical *years.* **4.12–20: An appeal to the Galatians in their relationship to Paul.** While Paul reproves the Galatians for their spiritual immaturity, the apostle points to past examples of fervent friendship. **13**: *First,* either "formerly" on his prior visit, or on the first of two visits (Acts 16.6; 18.23). **13–14**: On his first visit through the region of Galatia (Acts 16.6) an illness (was it eye trouble? see v. 15) detained Paul; though he was a care to the Galatians, they treated him with special consideration. **17**: *They,* the Judaizing teachers. **19**: *My little children,* Paul addresses the Galatians as their father in Jesus Christ (1 Cor 4.15). **4.21–5.1: A final proof,** the allegory about *Hagar* and Sarah, showing that those who rely upon the law instead of having faith in God's promise are to be excluded from the inheritance. **22**: Gen 16.15; 21.2, 9. **25**: *Jerusalem* was stressed by Paul's opponents (2.12). **26**: The heavenly Jerusalem (see Ezek 40; Zech 2.6–9; Rev 21.2), not that in Palestine, is *our mother.*

is Hagar, from Mount Sinai, bearing children for slavery. 25 Now Hagar is Mount Sinai in Arabia *b* and corresponds to the present Jerusalem, for she is in slavery with her children. 26 But the other woman corresponds to the Jerusalem above; she is free, and she is our mother. 27 For it is written,

"Rejoice, you childless one, you
who bear no children,
burst into song and shout, you
who endure no birth pangs;
for the children of the desolate
woman are more numerous
than the children of the one
who is married."

28 Now you, *c* my friends, *d* are children of the promise, like Isaac. 29 But just as at that time the child who was born according to the flesh persecuted the child who was born according to the Spirit, so it is now also. 30 But what does the scripture say? "Drive out the slave and her child; for the child of the slave will not share the inheritance with the child of the free woman." 31 So then, friends, *d* we are children, not of the slave but of the free woman. 1 For freedom Christ has set us free. Stand firm, therefore, and do not submit again to a yoke of slavery.

2 Listen! I, Paul, am telling you that if you let yourselves be circumcised, Christ will be of no benefit to you. 3 Once again I testify to every man who lets himself be circumcised that he is obliged to obey the entire law. 4 You who want to be justified by the law have cut yourselves off from Christ; you have fallen away from grace. 5 For through the Spirit, by faith, we eagerly wait for the hope of righteousness. 6 For in Christ Jesus neither circumcision nor uncircumcision counts for anything; the only thing that counts is faith working *e* through love.

7 You were running well; who prevented you from obeying the truth? 8 Such persuasion does not come from the one who calls you. 9 A little yeast leavens the whole batch of dough. 10 I am confident about you in the Lord that you will not think otherwise. But whoever it is that is confusing you will pay the penalty. 11 But my friends, *d* why am I still being persecuted if I am still preaching circumcision? In that case the offense of the cross has been removed. 12 I wish those who unsettle you would castrate themselves!

13 For you were called to freedom, brothers and sisters; *d* only do not use your freedom as an opportunity for self-indulgence, *f* but through love become slaves to one another. 14 For the whole law is summed up in a single commandment, "You shall love your neighbor as yourself." 15 If, however, you bite and

b Other ancient authorities read *For Sinai is a mountain in Arabia*   *c* Other ancient authorities read *we*   *d* Gk *brothers*   *e* Or *made effective*
f Gk *the flesh*

4.27: Isa 54.1 Gk, with Sarah as the once *childless one.* **29–30**: Gen 21.9–12. **5.1**: Both applies the allegory and sets the theme for the next section (see 5.13).
5.2–6.10: **Ethical exhortations,** growing out of Paul's gospel and of various sorts.
5.2–26: The nature of Christian liberty. 2–12: To seek justification by legal works is futile; Christ and the Mosaic law of circumcision are mutually exclusive. Faith alone justifies, but the faith that justifies is not alone— it produces good works *through love* (v. 6). **3**: See 3.10–12.
5.6: The ethical result of the gospel is one's faith expressing itself in loving deeds. If *made effective* (note *e*) is read, the sense may be

"coming to effective expression in love" or "made effective by God's *love*." **11**: *Still preaching circumcision,* as Paul might have if once a missionary for Judaism; or a charge based on 1 Cor 9.20 or on the case of Timothy reported in Acts 16.1–3. **12**: A bitterly satirical wish; see Phil 3.2. **13–26**: Though free from the law, Christians must not abuse their liberty. Paul's emphasis on ethical responsibility may be intended to answer those concerned about libertine opponents (see Introduction), though his letters regularly include moral imperatives (Rom 12.1ff.).
5.13: 5 6; Mt 20.26. **14**: Lev 19.18; Mt 22.39; compare Rom 13.8–10. **16**: *Spirit, flesh:* 3.3; Rom 8.5–11 n. **17**: Rom 7.15–23.

devour one another, take care that you are not consumed by one another.

16 Live by the Spirit, I say, and do not gratify the desires of the flesh. [17] For what the flesh desires is opposed to the Spirit, and what the Spirit desires is opposed to the flesh; for these are opposed to each other, to prevent you from doing what you want. [18] But if you are led by the Spirit, you are not subject to the law. [19] Now the works of the flesh are obvious: fornication, impurity, licentiousness, [20] idolatry, sorcery, enmities, strife, jealousy, anger, quarrels, dissensions, factions, [21] envy,[g] drunkenness, carousing, and things like these. I am warning you, as I warned you before: those who do such things will not inherit the kingdom of God.

22 By contrast, the fruit of the Spirit is love, joy, peace, patience, kindness, generosity, faithfulness, [23] gentleness, and self-control. There is no law against such things. [24] And those who belong to Christ Jesus have crucified the flesh with its passions and desires. [25] If we live by the Spirit, let us also be guided by the Spirit. [26] Let us not become conceited, competing against one another, envying one another.

6 My friends,[h] if anyone is detected in a transgression, you who have received the Spirit should restore such a one in a spirit of gentleness. Take care that you yourselves are not tempted. [2] Bear one another's burdens, and in this way you will fulfill[i] the law of Christ. [3] For if those who are nothing think they are something, they deceive themselves. [4] All must test their own work; then that work, rather than their neighbor's work, will become a cause for pride. [5] For all must carry their own loads.

6 Those who are taught the word must share in all good things with their teacher.

7 Do not be deceived; God is not mocked, for you reap whatever you sow. [8] If you sow to your own flesh, you will reap corruption from the flesh; but if you sow to the Spirit, you will reap eternal life from the Spirit. [9] So let us not grow weary in doing what is right, for we will reap at harvest time, if we do not give up. [10] So then, whenever we have an opportunity, let us work for the good of all, and especially for those of the family of faith.

11 See what large letters I make when I am writing in my own hand! [12] It is those who want to make a good showing in the flesh that try to compel you to be circumcised—only that they may not be persecuted for the cross of Christ. [13] Even the circumcised do not themselves obey the law, but they want you to be circumcised so that they may boast about your flesh. [14] May I never boast of anything except the cross of our Lord Jesus Christ, by which[j] the world has been crucified to me, and I to the world. [15] For[k] neither circumcision nor uncircumcision is anything; but a new creation is everything! [16] As for those who will follow this rule—peace be upon them, and mercy, and upon the Israel of God.

g Other ancient authorities add *murder*
h Gk *Brothers*　　i Other ancient authorities read *in this way fulfill*　　j Or *through whom*
k Other ancient authorities add *in Christ Jesus*

5.19–21: Catalogues of vices were common in the Graeco-Roman world (cf. Rom 1.29–31; 1 Cor 6.9–10), as were lists of virtues (cf. 2 Cor 6.6–7), *fruit* that comes from the Holy Spirit (5.22–23). **6.1–10: Specifics in the use of Christian liberty. 2:** *The law of Christ* (the term "law" is used figuratively; Rom 8.2 n.; 1 Cor 9.21) is the principle or norm of love for one another; see 5.14 n.

6.6: The church is obligated to support its teachers. **10:** *For the good of all,* for Christian love is not limited. **6.11–18: Paul's autograph postscript,** see 1 Cor 16.21 n.; 2 Thess 3.17. **11:** Large letters, compared with those of a trained scribe. **15:** Final declaration of what really matters in Christianity (see 5.6; 5.2–12 n.; 1 Cor 7.19). *New creation,* see 2 Cor 5.17. **16:** *Rule,* vv. 14–15. The church as those holding to Christ's

17 From now on, let no one make trouble for me; for I carry the marks of Jesus branded on my body.

18 May the grace of our Lord Jesus Christ be with your spirit, brothers and sisters.[1] Amen.

l Gk brothers

---

cross and *new creation* is God's *Israel* (3.7, 9, 14, 29).

6.17: *Marks of Jesus,* scars from Paul's sufferings as a missionary for Christ, like brand marks on a slave (1.10 note *c*). **18:** The addition of the words *brothers and sisters* to the benediction softens the severity of the whole letter.

The Letter of Paul to the
Ephesians

In powerful, poetic language, which was probably drawn in part from Christian hymns and liturgies, the Letter to the Ephesians celebrates the life of the church, a unique community established by God through the work of Jesus Christ, who is its head, and also the head of the whole creation. The church was established by God's eternal purpose, and in it believers already live in a union with God through Christ and the Holy Spirit that anticipates the full union in the life to come.

Reconciliation to God through the death of Christ has broken the power of evil and specifically the long-standing separation between Jews and non-Jews. Paul's own dignity and vocation as apostle are emphasized; his apostleship includes both his mission to the Gentiles and his stewardship of the mystery of the gospel, which could not be known until its revelation through Jesus Christ. The concluding ethical exhortation stresses unity in the church, love as an imitation of God, and separation from impurity. Advice to family groups follows the tradition of the time in emphasizing the authority of the father. The well-known concluding passage about putting on the armor of God pictures the life of faith as a conflict with evil, a theme that also appears earlier in the letter.

Some early manuscripts of this letter and some commentators in the early church make no reference to Ephesus in 1.1 (see note *a*), nor does the author deal with the problems of a particular congregation. Hence Ephesians is widely regarded as a "circular letter" that was not written specifically for Ephesus, but was distributed to several churches in Asia Minor. Through the years it has been assumed that Ephesians was written by the apostle Paul, late in his career and from prison, probably in Rome. There are, however, important contrasts between Ephesians and the letters that we can confidently ascribe to Paul. Many of the words in Ephesians do not appear elsewhere in the apostle's correspondence, and some important terms have a different meaning here from their meaning in letters that are surely Paul's. The style, also, with its loose collection of phrases and clauses in long sentences (many of which have been divided in the NRSV), is not characteristic of Paul's letters. Many scholars hold that these differences simply express changes in Paul's thought and style. Many other scholars, however, hold that Ephesians was written by a follower of Paul who had at hand a collection of Paul's letters, and who interpreted the mind of Paul to the church of a slightly later day. That would explain the many parallels to Paul's other letters that are woven into the text, a striking feature of this letter. Parts of Ephesians are strongly parallel to Colossians, which also may or may not have been written by Paul (see the Introduction to Colossians). If Paul was the author of both letters, they were written at nearly the same time; if Ephesians comes from a disciple of Paul, Colossians was an important source.

1 Paul, an apostle of Christ Jesus by the will of God,

To the saints who are in Ephesus and are faithful[a] in Christ Jesus:

2 Grace to you and peace from God our Father and the Lord Jesus Christ.

3 Blessed be the God and Father of our Lord Jesus Christ, who has blessed us in Christ with every spiritual blessing in the heavenly places, 4just as he chose us in Christ[b] before the foundation of the world to be holy and blameless before him in love. 5He destined us for adoption as his children through Jesus Christ, according to the good pleasure of his will, 6to the praise of his glorious grace that he freely bestowed on us in the Beloved. 7In him we have redemption through his blood, the forgiveness of our trespasses, according to the riches of his grace 8that he lavished on us. With all wisdom and insight 9he has made known to us the mystery of his will, according to his good pleasure that he set forth in Christ, 10as a plan for the fullness of time, to gather up all things in him, things in heaven and things on earth. 11In Christ we have also obtained an inheritance,[c] having been destined according to the purpose of him who accomplishes all things according to his counsel and will, 12so that we, who were the first to set our hope on Christ, might live for the praise of his glory. 13In him you also, when you had heard the word of truth, the gospel of your salvation, and had believed in him, were marked with the seal of the promised Holy Spirit; 14this[d] is the pledge of our inheritance toward redemption as God's own people, to the praise of his glory.

15 I have heard of your faith in the Lord Jesus and your love[e] toward all the saints, and for this reason 16I do not cease to give thanks for you as I remember you in my prayers. 17I pray that the God of our Lord Jesus Christ, the Father of glory, may give you a spirit of wisdom and revelation as you come to know him, 18so that, with the eyes of your heart enlightened, you may know what is the hope to which he has called you, what are the riches of his glorious inheritance among the saints, 19and what is the immeasurable greatness of his power for us who believe, according to the working of his great power. 20God[f] put this power to work in Christ when he raised him from the dead and seated him at his right hand in the heavenly places, 21far above all rule and authority and power and dominion, and above every name that is named, not only in this age but also in the age to come. 22And he has put all things under his feet and has made him the head over all things for the church, 23which is his body, the fullness of him who fills all in all.

2 You were dead through the trespasses and sins 2in which you once lived, following the course of this world, following the ruler of the power of the

---

a Other ancient authorities lack *in Ephesus*, reading *saints who are also faithful*   b Gk *in him*
c Or *been made a heritage*   d Other ancient authorities read *who*   e Other ancient authorities lack *and your love*   f Gk *He*

1.1–2: Salutation, see Introduction and Rom 1.1–7 n. **2:** *Grace . . . peace,* see 2 Thess 1.2 n.

1.3–23: Thanksgiving for blessings and prayer for spiritual wisdom (see Rom 1.8–15 n.). **3:** *In the heavenly places,* an expression, found only in this letter (1.20; 2.6; 3.10; 6.12), referring to the unseen spiritual world behind and above the material universe. **4–10:** God *chose us . . .* and *destined us* to be *his children* in accord with his eternal *good pleasure . . . in Christ . . . to gather up* all things in heaven and earth in him. **6:** *To the praise of his* [the Father's] *glorious grace* (v. 6), a refrain repeated in briefer form referring to Christ (v. 12) and the Holy Spirit (vv. 13–14). **7:** *Redemption through his blood,* Mk 10.45; 14.24; see Heb 9.11 n. **9:** The word *mystery* everywhere in Ephesians (except 5.32) and Colossians refers to God's age-long purpose, now disclosed to his chosen, to call Gentiles as well as Jews to share in Christ's redemptive work (3.4–6).

1.10: *Fullness of time,* Gal 4.4. **13:** *You also,* "you Gentiles as well as we Jews." *Promised,* Lk 24.49; Jn 14.26; Acts 1.4; 2.33. **14:** *Pledge,* see 2 Cor 1.22 n. **22–23:** The church, as the *fullness of Christ,* is the complement of his mystic person; he is the *head,* the church *is his body.*

2.1–22: Christ's benefits, for both Gen-

air, the spirit that is now at work among those who are disobedient. 3 All of us once lived among them in the passions of our flesh, following the desires of flesh and senses, and we were by nature children of wrath, like everyone else. 4 But God, who is rich in mercy, out of the great love with which he loved us 5 even when we were dead through our trespasses, made us alive together with Christ*g*—by grace you have been saved— 6 and raised us up with him and seated us with him in the heavenly places in Christ Jesus, 7 so that in the ages to come he might show the immeasurable riches of his grace in kindness toward us in Christ Jesus. 8 For by grace you have been saved through faith, and this is not your own doing; it is the gift of God— 9 not the result of works, so that no one may boast. 10 For we are what he has made us, created in Christ Jesus for good works, which God prepared beforehand to be our way of life.

11 So then, remember that at one time you Gentiles by birth,*h* called "the uncircumcision" by those who are called "the circumcision"—a physical circumcision made in the flesh by human hands— 12 remember that you were at that time without Christ, being aliens from the commonwealth of Israel, and strangers to the covenants of promise, having no hope and without God in the world. 13 But now in Christ Jesus you who once were far off have been brought

near by the blood of Christ. 14 For he is our peace; in his flesh he has made both groups into one and has broken down the dividing wall, that is, the hostility between us. 15 He has abolished the law with its commandments and ordinances, that he might create in himself one new humanity in place of the two, thus making peace, 16 and might reconcile both groups to God in one body*i* through the cross, thus putting to death that hostility through it.*j* 17 So he came and proclaimed peace to you who were far off and peace to those who were near; 18 for through him both of us have access in one Spirit to the Father. 19 So then you are no longer strangers and aliens, but you are citizens with the saints and also members of the household of God, 20 built upon the foundation of the apostles and prophets, with Christ Jesus himself as the cornerstone.*k* 21 In him the whole structure is joined together and grows into a holy temple in the Lord; 22 in whom you also are built together spiritually*l* into a dwelling place for God.

3 This is the reason that I Paul am a prisoner for*m* Christ Jesus for the sake of you Gentiles— 2 for surely you have already heard of the commission of God's grace that was given me for you,

g Other ancient authorities read *in Christ*
h Gk *in the flesh*   i Or *reconcile both of us in one body for God*   j Or *in him, or in himself*
k Or *keystone*   l Gk *in the Spirit*   m Or *of*

---

tiles and Jews. **1:** *Dead,* v. 5; Col 2.13. **2:** *The ruler of the power of the air,* Satan (6.11–12; Col 1.13). **3:** *By nature,* the human state apart from God's grace in Christ. *Wrath,* see Col 3.6 n. **5:** *Grace,* God's unmerited favor shown to humankind in Christ. **8:** *Through faith,* as the channel; Paul never says, "saved because of faith." *This,* namely, your salvation. **10:** *Good works* are the result, not the cause, of salvation. *Beforehand* ascribes the whole matter to God.

2.11: *Called . . . called,* since the distinction between Jew and Gentile is removed in Christ (Col 3.11) the terms "circumcision" and "uncircumcision" are obsolete. **12:** The plight of the Gentiles apart from Christ. **13–22:** *But now* the reconciling work of Christ avails also for Gentiles, *who once were far off* from God

and his people. *By the blood of Christ,* by the new covenant (see 1.7 n.). **14–16:** Christ has not only *broken down the . . . hostility* between Jew and Gentile (*made both groups into one*), but has reconciled *both . . . to God in one body,* the church. **14:** *The dividing wall,* an allusion to the barrier in the temple at Jerusalem separating the court of Israel from the court of the Gentiles. Gentiles who trespassed the barrier did so on pain of death, as warning notices proclaimed. **16:** *Both,* Jew and Gentile. **17:** Isa 52.7; 57.19. **18:** The unity of Christians *in one body,* the church (v. 16), is based on their participation *in one Spirit.* **19:** *Strangers,* v. 12. **20–22:** *The foundation,* laid by *the apostles* and Christian *prophets* (3.5; 4.11; Acts 13.1). The *whole structure* depends upon Christ as the *cornerstone* (a messianic designation, compare Isa

3 and how the mystery was made known to me by revelation, as I wrote above in a few words, 4 a reading of which will enable you to perceive my understanding of the mystery of Christ. 5 In former generations this mystery[n] was not made known to humankind, as it has now been revealed to his holy apostles and prophets by the Spirit: 6 that is, the Gentiles have become fellow heirs, members of the same body, and sharers in the promise in Christ Jesus through the gospel.

7 Of this gospel I have become a servant according to the gift of God's grace that was given me by the working of his power. 8 Although I am the very least of all the saints, this grace was given to me to bring to the Gentiles the news of the boundless riches of Christ, 9 and to make everyone see[o] what is the plan of the mystery hidden for ages in[p] God who created all things; 10 so that through the church the wisdom of God in its rich variety might now be made known to the rulers and authorities in the heavenly places. 11 This was in accordance with the eternal purpose that he has carried out in Christ Jesus our Lord, 12 in whom we have access to God in boldness and confidence through faith in him.[q] 13 I pray therefore that you[r] may not lose heart over my sufferings for you; they are your glory.

14 For this reason I bow my knees before the Father,[s] 15 from whom every family[t] in heaven and on earth takes its name. 16 I pray that, according to the riches of his glory, he may grant that you may be strengthened in your inner being with power through his Spirit, 17 and that Christ may dwell in your hearts through faith, as you are being rooted and grounded in love. 18 I pray that you may have the power to comprehend, with all the saints, what is the breadth and length and height and depth, 19 and to know the love of Christ that surpasses knowledge, so that you may be filled with all the fullness of God.

20 Now to him who by the power at work within us is able to accomplish abundantly far more than all we can ask or imagine, 21 to him be glory in the church and in Christ Jesus to all generations, forever and ever. Amen.

4 I therefore, the prisoner in the Lord, beg you to lead a life worthy of the calling to which you have been called, 2 with all humility and gentleness, with patience, bearing with one another in love, 3 making every effort to maintain the unity of the Spirit in the bond of peace. 4 There is one body and one Spirit, just as you were called to the one hope of your calling, 5 one Lord, one faith, one baptism, 6 one God and Father of all,

---

n Gk it   o Other ancient authorities read *to bring
to light*   p Or *by*   q Or *the faith of him*
r Or *I*   s Other ancient authorities add *of our
Lord Jesus Christ*   t Gk *fatherhood*

28.16; Mt 21.42). Christians are built into a growing *temple* or *dwelling place for God* (1 Cor 3.16–17; 1 Pet 2.4–5).

3.1–21: A prayer for wisdom, interrupted by a parenthesis on Paul's mission to the Gentiles (vv. 2–13). **1:** Paul was *a prisoner* because he had aroused the hostility of the Jews by advocating the equality of the *Gentiles* in the church (Acts 21.21, 28; 22.21–22). **2:** Col 1.25. **3:** *Mystery,* see 1.9 n. *By revelation,* 1.17; Acts 9.3–4; Gal 1.12. *In a few words,* in 1.9–10.

3.5: Christian *prophets,* see 2.20. **6:** The content of the *mystery,* v. 3. **7:** *Servant,* Col 1.23, 25. **8:** *Least,* 1 Cor 15.9. *Boundless,* in their nature, extent, and application (v. 19; Rom 11.33). **9:** *The mystery,* or *secret, that*

Gentiles, by faith, were to form an integral part of the new Israel. **10:** By means of *the church* God's *wisdom* in the plan of redemption is displayed to angelic beings (1 Cor 4.9) *in its rich variety.* **12:** 2.18.

3.14–19: Paul's prayer. **14:** *For this reason,* resumes v. 1. **14–15:** *Father, family,* a play on words in Greek ("pater," "patria," see note *t*); God is the Author of all family relationships. **18:** *The breadth . . . and depth,* of Christ's love. **20–21:** The doxology celebrates God's boundless generosity and his *glory* both *in the church and in Christ Jesus.*

4.1–6.20: Ethical implications of the doctrinal teaching in chs 1–3. **4.1–16: An appeal to maintain the unity of the faith. 2:** Col 3.12–13. **4–6:** Seven elements of unity.

who is above all and through all and in all.

7 But each of us was given grace according to the measure of Christ's gift. 8 Therefore it is said,

"When he ascended on high he
    made captivity itself
        a captive;
    he gave gifts to his people."

9 (When it says, "He ascended," what does it mean but that he had also descended[u] into the lower parts of the earth? 10 He who descended is the same one who ascended far above all the heavens, so that he might fill all things.) 11 The gifts he gave were that some would be apostles, some prophets, some evangelists, some pastors and teachers, 12 to equip the saints for the work of ministry, for building up the body of Christ, 13 until all of us come to the unity of the faith and of the knowledge of the Son of God, to maturity, to the measure of the full stature of Christ. 14 We must no longer be children, tossed to and fro and blown about by every wind of doctrine, by people's trickery, by their craftiness in deceitful scheming. 15 But speaking the truth in love, we must grow up in every way into him who is the head, into Christ, 16 from whom the whole body, joined and knit together by every ligament with which it is equipped, as each part is working properly, promotes the body's growth in building itself up in love.

17 Now this I affirm and insist on in the Lord: you must no longer live as the Gentiles live, in the futility of their minds. 18 They are darkened in their understanding, alienated from the life of God because of their ignorance and hard-ness of heart. 19 They have lost all sensitivity and have abandoned themselves to licentiousness, greedy to practice every kind of impurity. 20 That is not the way you learned Christ! 21 For surely you have heard about him and were taught in him, as truth is in Jesus. 22 You were taught to put away your former way of life, your old self, corrupt and deluded by its lusts, 23 and to be renewed in the spirit of your minds, 24 and to clothe yourselves with the new self, created according to the likeness of God in true righteousness and holiness.

25 So then, putting away falsehood, let all of us speak the truth to our neighbors, for we are members of one another. 26 Be angry but do not sin; do not let the sun go down on your anger, 27 and do not make room for the devil. 28 Thieves must give up stealing; rather let them labor and work honestly with their own hands, so as to have something to share with the needy. 29 Let no evil talk come out of your mouths, but only what is useful for building up,[v] as there is need, so that your words may give grace to those who hear. 30 And do not grieve the Holy Spirit of God, with which you were marked with a seal for the day of redemption. 31 Put away from you all bitterness and wrath and anger and wrangling and slander, together with all malice, 32 and be kind to one another, tenderhearted, forgiving one another, as God in Christ has forgiven you.[w]

5 ¹ Therefore be imitators of God, as beloved children, 2 and live in love, as

---

u Other ancient authorities add *first*   v Other ancient authorities read *building up faith*
w Other ancient authorities read *us*

4.7–16: Christian unity amid diversity of spiritual gifts. **8:** Ps 68.18. The psalm is applied to Christ's victory over spiritual powers (1.21; Col 2.15), but here attention is focused on his ascent. **9:** *Lower parts,* the region of the grave. **10:** The same Christ, after his burial, was exalted that he might fill all things with himself. **11:** *Pastors and teachers,* two aspects of one ministry. **12:** *All Christians are to be equipped for the work of active spiritual ser-*vice. **15:** The verb means both *speaking* and *doing.*

4.17–5.20: An appeal to renounce pagan ways. 4.17–19: Rom 1.21–25. **22:** *Put away,* like filthy, worn-out clothes (Col 3.9; see Rom 6.6 n.). **25:** Zech 8.16; Rom 12.5. **26:** If angry, let it neither be in a sinful spirit nor prolonged (Ps 4.4; Jas 1.19–20). **28:** *Share,* not merely in restitution, but in liberality. **29:** Col 3.8. **32:** *As God . . . has forgiven,*

Christ loved us[x] and gave himself up for us, a fragrant offering and sacrifice to God.

3 But fornication and impurity of any kind, or greed, must not even be mentioned among you, as is proper among saints. [4]Entirely out of place is obscene, silly, and vulgar talk; but instead, let there be thanksgiving. [5]Be sure of this, that no fornicator or impure person, or one who is greedy (that is, an idolater), has any inheritance in the kingdom of Christ and of God.

6 Let no one deceive you with empty words, for because of these things the wrath of God comes on those who are disobedient. [7]Therefore do not be associated with them. [8]For once you were darkness, but now in the Lord you are light. Live as children of light— [9]for the fruit of the light is found in all that is good and right and true. [10]Try to find out what is pleasing to the Lord. [11]Take no part in the unfruitful works of darkness, but instead expose them. [12]For it is shameful even to mention what such people do secretly; [13]but everything exposed by the light becomes visible, [14]for everything that becomes visible is light. Therefore it says,

"Sleeper, awake!
  Rise from the dead,
  and Christ will shine on you."

15 Be careful then how you live, not as unwise people but as wise, [16]making the most of the time, because the days are evil. [17]So do not be foolish, but understand what the will of the Lord is. [18]Do not get drunk with wine, for that is debauchery; but be filled with the Spirit, [19]as you sing psalms and hymns and spiritual songs among yourselves, singing and making melody to the Lord in your hearts, [20]giving thanks to God the Father at all times and for everything in the name of our Lord Jesus Christ.

21 Be subject to one another out of reverence for Christ.

22 Wives, be subject to your husbands as you are to the Lord. [23]For the husband is the head of the wife just as Christ is the head of the church, the body of which he is the Savior. [24]Just as the church is subject to Christ, so also wives ought to be, in everything, to their husbands.

25 Husbands, love your wives, just as Christ loved the church and gave himself up for her, [26]in order to make her holy by cleansing her with the washing of water by the word, [27]so as to present the church to himself in splendor, without a spot or wrinkle or anything of the kind—yes, so that she may be holy and without blemish. [28]In the same way, husbands should love their wives as they do their own bodies. He who loves his wife loves himself. [29]For no one ever hates his own body, but he nourishes and tenderly cares for it, just as Christ does for the church, [30]because we are members of his body.[y] [31]"For this reason a man will leave his father and mother and be joined to his wife, and the two will become one flesh." [32]This is a great mys-

x Other ancient authorities read *you*
y Other ancient authorities add *of his flesh and of his bones*

Christian conduct is the corollary of Christian doctrine (Phil 2.5–8; Col 3.1–3).
5.1: *Imitators of God,* in forgiving (4.32) and in loving. **2:** *Gave himself up* to death as a sacrifice *for us.* **4:** *Vulgar talk,* of a licentious nature. **5:** *Greedy,* see Col 3.5 n.
5.6: *Wrath,* see Col 3.6 n. **9:** Gal 5.22–23. **11:** Gal 5.19–21. **14:** Quoted perhaps from an early Christian hymn based on Isa 60.1. **16:** Col 4.5. **19:** Col 3.16.
5.21–6.9: The Christian household (Col 3.18–4.1): husband and wife (5.21–33), children and parents (6.1–4), masters and slaves (6.5–9). **5.21:** The general principle is that of mutual subjection (Phil 2.3) (with a presup-posed subordination of wife, children, and slaves) on which the following applications rest. **23:** Being *the head of the wife* involves responsibility for cherishing and protecting her (vv. 25, 29).
5.25: *Gave,* see 5.2 n. **26:** Christian baptism, under the figure of the ritual bath of purification taken by Jewish women, here prior to the marriage ceremony. **27:** Song 4.7; 2 Cor 11.2; Rev 21.2. **31:** Gen 2.24. **32:** *This ... mystery,* namely Christ's spiritual union with our humanity, is of great significance in understanding the nature of human marriage. **33:** Here, implicitly, monogamy is taken for granted.

tery, and I am applying it to Christ and the church. ³³ Each of you, however, should love his wife as himself, and a wife should respect her husband.

6 Children, obey your parents in the Lord, ^z for this is right. ² "Honor your father and mother"—this is the first commandment with a promise: ³ "so that it may be well with you and you may live long on the earth."

4 And, fathers, do not provoke your children to anger, but bring them up in the discipline and instruction of the Lord.

5 Slaves, obey your earthly masters with fear and trembling, in singleness of heart, as you obey Christ; ⁶ not only while being watched, and in order to please them, but as slaves of Christ, doing the will of God from the heart. ⁷ Render service with enthusiasm, as to the Lord and not to men and women, ⁸ knowing that whatever good we do, we will receive the same again from the Lord, whether we are slaves or free.

9 And, masters, do the same to them. Stop threatening them, for you know that both of you have the same Master in heaven, and with him there is no partiality.

10 Finally, be strong in the Lord and in the strength of his power. ¹¹ Put on the whole armor of God, so that you may be able to stand against the wiles of the devil. ¹² For our ^a struggle is not against enemies of blood and flesh, but against the rulers, against the authorities, against the cosmic powers of this present darkness, against the spiritual forces of evil in the heavenly places. ¹³ Therefore take up the whole armor of God, so that you may be able to withstand on that evil day, and having done everything, to stand firm. ¹⁴ Stand therefore, and fasten the belt of truth around your waist, and put on the breastplate of righteousness. ¹⁵ As shoes for your feet put on whatever will make you ready to proclaim the gospel of peace. ¹⁶ With all of these, ^b take the shield of faith, with which you will be able to quench all the flaming arrows of the evil one. ¹⁷ Take the helmet of salvation, and the sword of the Spirit, which is the word of God.

18 Pray in the Spirit at all times in every prayer and supplication. To that end keep alert and always persevere in supplication for all the saints. ¹⁹ Pray also for me, so that when I speak, a message may be given to me to make known with boldness the mystery of the gospel, ^c ²⁰ for which I am an ambassador in chains. Pray that I may declare it boldly, as I must speak.

21 So that you also may know how I am and what I am doing, Tychicus will tell you everything. He is a dear brother and a faithful minister in the Lord. ²² I am sending him to you for this very purpose, to let you know how we are, and to encourage your hearts.

23 Peace be to the whole community, ^d and love with faith, from God the Father and the Lord Jesus Christ. ²⁴ Grace be with all who have an undying love for our Lord Jesus Christ. ^e

z Other ancient authorities lack *in the Lord*
a Other ancient authorities read *your*   b Or *In all circumstances*   c Other ancient authorities lack *of the gospel*   d Gk *to the brothers*   e Other ancient authorities add *Amen*

6.2: Ex 20.12. **3**: Deut 5.16. **4**: Col 3.21. **8**: Col 3.22; on slavery, see Introduction to Philemon. **9**: God shows *no partiality* to social status (Col 3.25–4.1).

6.10–20: God's armor and the Christian's warfare. 12: *Blood and flesh,* that is, mere mortal beings. *Rulers . . . forces of evil,* organized forces of malevolent spirit beings (see Rom 8.38 n.; Rev 12.7–9). **13–17**: The armor that God wears (Isa 11.5; 59.17) and supplies.

6.15: Isa 52.7. **17**: *The sword* (the one offensive weapon mentioned here) is the word that God speaks through his servants (Hos 6.5; Mt 10.19–20; compare Heb 4.12). **18–20**: Persevering prayer (Col 4.2–4) is an aid in standing (v. 14). **19**: *Mystery,* see 1.9 n. **20**: *In chains,* as a prisoner awaiting trial.

6.21–24: Personal matters and benediction. 24: *Grace . . . undying love,* a worthy conclusion, returning to the fundamental thought of 1.3–14.

The Letter of Paul to the
Philippians

Philippi was a city of Macedonia, a stop on one of the main roads between East and West in the Roman Empire. The Christian community in Philippi was the first church established by Paul on European soil (Acts 16.11–13). Paul seems to have had a close and happy relationship with this church in the years that followed (4.15–16).

From the beginning of Paul's work in Philippi there had been opposition (1 Thess 2.2; Acts 16.19–40), and evidently opposition to the church was very active at the time Paul wrote the Letter to the Philippians, for a main theme of this letter is persistence in faith in the face of opposition and even the threat of death. Paul offers himself as an example of steadfast courage and of joy in the midst of harsh circumstances and possible death; he turns attention away from preoccupation with one's own fate toward the proclamation of the gospel, which is the task to be accomplished in the face of suffering, and toward the joy, which is real even in such difficult times, while one follows the "mind" of Christ (1.12–25; 2.1–5; 3.8–16; 4.11–13).

At the center of this theme is the moving and powerful hymn-like passage that celebrates the story of Christ's self-emptying even to death, for which Christ was awarded by God the name "Lord," which is above every name (2.6–11). The hymn was probably quoted and adapted by Paul from earlier tradition. Through the years and today, the theme of self-emptying, or "kenosis," has been a key concept according to which Christians have reflected about the meaning of Christ.

Except that Paul was in prison at the time of writing and awaiting trial (1.12–26), it is impossible to speak confidently of the time and place of the writing of this letter. Because of references to the imperial guard or praetorium (1.13) and to the emperor's household (4.22) and also because the situation reflected in the letter bears some resemblance to that described at the very end of the book of Acts, many scholars put its composition during the period of Paul's imprisonment at Rome (about A.D. 61–63); but the indications are by no means conclusive, and it is perhaps more probable that the letter was written from Caesarea or Ephesus at an earlier stage in Paul's career. The terms referring to the emperor's establishment were used also for government functions in other parts of the empire than Rome.

The immediate occasion of Paul's writing was the return to Philippi of Epaphroditus (2.25–30), who had been sent by the Philippian church with gifts for Paul (4.18), and who had been seriously ill while staying with Paul. The apostle took this opportunity to thank them for their gifts, and to set their difficulties in a wider framework by describing his own situation and the relation of both his and their situations to the reality of Christ.

The letter follows the usual pattern of Paul's letters in a general way. But some abrupt changes of topic (especially between 3.1 and 3.2), in addition to the fact that an early Christian writer (Polycarp, *Philippians* 2.3), speaks of "letters" of Paul to the Philippians, has led some scholars to conclude that our present Philippians is composed of parts of two or three letters that Paul wrote to Philippi. Many others, however, find that Philippians is a coherent whole as it stands.

1 Paul and Timothy, servants*a* of Christ Jesus,

To all the saints in Christ Jesus who are in Philippi, with the bishops*b* and deacons:*c*

2 Grace to you and peace from God our Father and the Lord Jesus Christ.

3 I thank my God every time I remember you, 4constantly praying with joy in every one of my prayers for all of you, 5because of your sharing in the gospel from the first day until now. 6I am confident of this, that the one who began a good work among you will bring it to completion by the day of Jesus Christ. 7It is right for me to think this way about all of you, because you hold me in your heart,*d* for all of you share in God's grace*e* with me, both in my imprisonment and in the defense and confirmation of the gospel. 8For God is my witness, how I long for all of you with the compassion of Christ Jesus. 9And this is my prayer, that your love may overflow more and more with knowledge and full insight 10to help you to determine what is best, so that in the day of Christ you may be pure and blameless, 11having produced the harvest of righteousness that comes through Jesus Christ for the glory and praise of God.

12 I want you to know, beloved,*f* that what has happened to me has actually helped to spread the gospel, 13so that it has become known throughout the whole imperial guard*g* and to everyone else that my imprisonment is for Christ;

14and most of the brothers and sisters,*f* having been made confident in the Lord by my imprisonment, dare to speak the word*h* with greater boldness and without fear.

15 Some proclaim Christ from envy and rivalry, but others from goodwill. 16These proclaim Christ out of love, knowing that I have been put here for the defense of the gospel; 17the others proclaim Christ out of selfish ambition, not sincerely but intending to increase my suffering in my imprisonment. 18What does it matter? Just this, that Christ is proclaimed in every way, whether out of false motives or true; and in that I rejoice.

Yes, and I will continue to rejoice, 19for I know that through your prayers and the help of the Spirit of Jesus Christ this will turn out for my deliverance. 20It is my eager expectation and hope that I will not be put to shame in any way, but that by my speaking with all boldness, Christ will be exalted now as always in my body, whether by life or by death. 21For to me, living is Christ and dying is gain. 22If I am to live in the flesh, that means fruitful labor for me; and I do not know which I prefer. 23I am hard pressed between the two: my desire is to depart and be with Christ, for that is far better; 24but to remain in the flesh is more necessary for you. 25Since I am convinced of

a Gk *slaves*   *b* Or *overseers*   *c* Or *overseers
and helpers*   *d* Or *because I hold you in my heart*
e Gk *in grace*   *f* Gk *brothers*   *g* Gk *whole
praetorium*   *h* Other ancient authorities read
word of God

1.1–2: Salutation, see Rom 1.1–7 n. **1:** *Bishops and deacons,* i.e. "overseers" and "helpers" (compare 1 Cor 12.28). **2:** *Grace . . . peace,* see 2 Thess 1.2 n.

1.3–11: Thanksgiving, see Rom 1.8–15 n. **5:** *Sharing,* see Introduction. *From the first day,* when Paul first preached at Philippi (Acts 16.13). **6:** *The day of Jesus Christ,* the day when he will return and the present age will end (1 Cor 1.8; compare 2 Thess 2.3 and 2 Pet 3.10). **10:** Rom 2.18.

1.12–30. Paul's present circumstances. 12: See v. 7. **13:** Paul is under military guard; and now all the members of the local headquarters (compare note *g*) know why he is

there, and so have heard at least the name of Christ.

1.15–18: That some of Paul's fellow-Christians regarded him with suspicion and ill-will appears in Galatians and in 1 and 2 Corinthians (especially 2 Cor chs 10–13). **21–24:** His life is not his own but belongs utterly to Christ; therefore *to depart and be with Christ* is *gain;* but this would mean loss to his churches.

1.25–26: It is not known whether Paul's expectation was fulfilled or not. **28–30:** The *evidence* consists in the church's brave, united witness to Christ even at the cost of *struggle* and *suffering.*

this, I know that I will remain and continue with all of you for your progress and joy in faith, 26 so that I may share abundantly in your boasting in Christ Jesus when I come to you again.

27 Only, live your life in a manner worthy of the gospel of Christ, so that, whether I come and see you or am absent and hear about you, I will know that you are standing firm in one spirit, striving side by side with one mind for the faith of the gospel, 28 and are in no way intimidated by your opponents. For them this is evidence of their destruction, but of your salvation. And this is God's doing. 29 For he has graciously granted you the privilege not only of believing in Christ, but of suffering for him as well— 30 since you are having the same struggle that you saw I had and now hear that I still have.

2 If then there is any encouragement in Christ, any consolation from love, any sharing in the Spirit, any compassion and sympathy, 2 make my joy complete: be of the same mind, having the same love, being in full accord and of one mind. 3 Do nothing from selfish ambition or conceit, but in humility regard others as better than yourselves. 4 Let each of you look not to your own interests, but to the interests of others. 5 Let the same mind be in you that was *i* in Christ Jesus,

6 who, though he was in the form
of God,
did not regard equality with
God
as something to be exploited,
7 but emptied himself,
taking the form of a slave,

being born in human likeness.
And being found in human form,
8 he humbled himself
and became obedient to the
point of death—
even death on a cross.

9 Therefore God also highly
exalted him
and gave him the name
that is above every name,
10 so that at the name of Jesus
every knee should bend,
in heaven and on earth and
under the earth,
11 and every tongue should confess
that Jesus Christ is Lord,
to the glory of God the Father.

12 Therefore, my beloved, just as you have always obeyed me, not only in my presence, but much more now in my absence, work out your own salvation with fear and trembling; 13 for it is God who is at work in you, enabling you both to will and to work for his good pleasure.

14 Do all things without murmuring and arguing, 15 so that you may be blameless and innocent, children of God without blemish in the midst of a crooked and perverse generation, in which you shine like stars in the world. 16 It is by your holding fast to the word of life that I can boast on the day of Christ that I did not run in vain or labor in vain. 17 But even if I am being poured out as a libation over the sacrifice and the offering of your faith, I am glad and rejoice with all of you— 18 and in the same way you also must be glad and rejoice with me.

i Or that you have

2.1–18: **Humility and the example of Christ. 1–2:** The "if" is rhetorical; with Christ's *encouragement* and moved by God's *love* for them, they are to *make* Paul's *joy complete.* **3:** He has reason to fear that they are somewhat divided by petty jealousies (4.2). **2.6:** *In the form of God,* that is, pre-existent and divine (Jn 1.1–3; Col 1.15). *Something to be exploited,* and never relinquished. **7:** *But emptied himself,* the extreme limit of self-denial. *Slave,* perhaps an allusion to Isa 52.13–53.12. *Born,* Jn 1.14. **8:** Mt 26.39; Jn 10.18; Rom 5.19; Heb 5.8; 12.2. **9:** *Exalted,* at the resurrection. *The name* is "Lord" (see 1 Thess 1.1 n.). **10–11:** Compare Isa 45.23. **2.12:** *With fear and trembling,* humbly and with constant dependence on God's help. **14:** *Without murmuring,* compare 1 Cor 10.9–10. **15:** *Crooked and perverse,* Deut 32.5. *Shine like stars,* Dan 12.3. **16:** *Day of Christ,* see 1.6 n. **17:** A reminder that Paul stands in danger of condemnation and death.

19 I hope in the Lord Jesus to send
Timothy to you soon, so that I may be
cheered by news of you. 20 I have no one
like him who will be genuinely concerned for your welfare. 21 All of them
are seeking their own interests, not those
of Jesus Christ. 22 But Timothy's[j] worth
you know, how like a son with a father
he has served with me in the work of the
gospel. 23 I hope therefore to send him as
soon as I see how things go with me;
24 and I trust in the Lord that I will also
come soon.

25 Still, I think it necessary to send to
you Epaphroditus—my brother and co-
worker and fellow soldier, your messenger[k] and minister to my need; 26 for he
has been longing for[l] all of you, and has
been distressed because you heard that he
was ill. 27 He was indeed so ill that he
nearly died. But God had mercy on him,
and not only on him but on me also, so
that I would not have one sorrow after
another. 28 I am the more eager to send
him, therefore, in order that you may
rejoice at seeing him again, and that I
may be less anxious. 29 Welcome him
then in the Lord with all joy, and honor
such people, 30 because he came close to
death for the work of Christ,[m] risking his
life to make up for those services that you
could not give me.

3 Finally, my brothers and sisters,[n]
rejoice[o] in the Lord.

To write the same things to you is not
troublesome to me, and for you it is a
safeguard.

2 Beware of the dogs, beware of the
evil workers, beware of those who muti-
late the flesh![p] 3 For it is we who are the
circumcision, who worship in the Spirit
of God[q] and boast in Christ Jesus and
have no confidence in the flesh— 4 even
though I, too, have reason for confidence
in the flesh.

If anyone else has reason to be confident in the flesh, I have more: 5 circumcised on the eighth day, a member of the
people of Israel, of the tribe of Benjamin,
a Hebrew born of Hebrews; as to the
law, a Pharisee; 6 as to zeal, a persecutor
of the church; as to righteousness under
the law, blameless.

7 Yet whatever gains I had, these I
have come to regard as loss because of
Christ. 8 More than that, I regard everything as loss because of the surpassing
value of knowing Christ Jesus my Lord.
For his sake I have suffered the loss of all
things, and I regard them as rubbish, in
order that I may gain Christ 9 and be
found in him, not having a righteousness
of my own that comes from the law, but
one that comes through faith in Christ,[r]
the righteousness from God based on
faith. 10 I want to know Christ[s] and the
power of his resurrection and the sharing
of his sufferings by becoming like him in

j Gk *his* *k* Gk *apostle* *l* Other ancient authorities read *longing to see* *m* Other ancient authorities read *of the Lord* *n* Gk *my brothers* *o* Or *farewell* *p* Gk *the mutilation* *q* Other ancient authorities read *worship God in spirit* *r* Or *through the faith of Christ* *s* Gk *him*

2.19–30: Timothy and Epaphroditus. **19**: *Timothy,* Acts 16.1–3; 1 Cor 16.10–11; see Introduction to 2 Timothy. **23–24**: A reference to the uncertainty of how his impending trial will end (1.19, 26). **25**: *Epaphroditus,* see Introduction and 4.18.

3.1–11: A warning. **1**: *The same things,* he has given this warning before. **2**: A bitter and ironical reference to those who preach the necessity of circumcision (compare Gal 5.12). **3**: *Circumcision,* Jer 4.4; Rom 2.28–29; Gal 6.14–15; Col 2.11–13. *The flesh,* outward states or rites. **5**: *A Hebrew . . .,* though living in a Greek city (Tarsus) Paul's family spoke the language of Palestine (Acts 21.40; 22.2). *Pharisee,* one who carefully observed the Jewish law.

3.6: *A persecutor,* Acts 9.1–2; 1 Cor 15.9; Gal 1.13. *Under the law, blameless,* so far as its external requirements are concerned. No contradiction need be found between this statement and what Paul says in Rom 3.20 and 7.7–25. **9**: *In Christ,* see Rom 3.22 n. *The righteousness from God,* a free gift, dependent only upon a human being's willingness humbly to receive it (Rom 1.16–4.25).

3.10–11: Actually to know Christ as risen and living is to have *power* to suffer like him and for him, and to possess the sure hope of rising and living with him.

his death, [11] if somehow I may attain the resurrection from the dead.

12 Not that I have already obtained this or have already reached the goal;[t] but I press on to make it my own, because Christ Jesus has made me his own. [13] Beloved,[u] I do not consider that I have made it my own;[v] but this one thing I do: forgetting what lies behind and straining forward to what lies ahead, [14] I press on toward the goal for the prize of the heavenly[w] call of God in Christ Jesus. [15] Let those of us then who are mature be of the same mind; and if you think differently about anything, this too God will reveal to you. [16] Only let us hold fast to what we have attained.

17 Brothers and sisters,[u] join in imitating me, and observe those who live according to the example you have in us. [18] For many live as enemies of the cross of Christ; I have often told you of them, and now I tell you even with tears. [19] Their end is destruction; their god is the belly; and their glory is in their shame; their minds are set on earthly things. [20] But our citizenship[x] is in heaven, and it is from there that we are expecting a Savior, the Lord Jesus Christ. [21] He will transform the body of our humiliation[y] that it may be conformed to the body of his glory,[z] by the power that also enables him to make all things subject to himself. [1] Therefore, my 4 brothers and sisters,[a] whom I love and long for, my joy and crown, stand firm in the Lord in this way, my beloved.

2 I urge Euodia and I urge Syntyche to be of the same mind in the Lord. [3] Yes, and I ask you also, my loyal companion,[b] help these women, for they have struggled beside me in the work of the gospel, together with Clement and the rest of my co-workers, whose names are in the book of life.

4 Rejoice[c] in the Lord always; again I will say, Rejoice.[c] [5] Let your gentleness be known to everyone. The Lord is near. [6] Do not worry about anything, but in everything by prayer and supplication with thanksgiving let your requests be made known to God. [7] And the peace of God, which surpasses all understanding, will guard your hearts and your minds in Christ Jesus.

8 Finally, beloved,[d] whatever is true, whatever is honorable, whatever is just, whatever is pure, whatever is pleasing, whatever is commendable, if there is any excellence and if there is anything worthy of praise, think about[e] these things. [9] Keep on doing the things that you have learned and received and heard and seen in me, and the God of peace will be with you.

10 I rejoice[f] in the Lord greatly that now at last you have revived your concern for me; indeed, you were concerned for me, but had no opportunity to show it.[g] [11] Not that I am referring to being in need; for I have learned to be content with whatever I have. [12] I know what it

---

t Or *have already been made perfect*
u Gk *Brothers*   v Other ancient authorities read *my own yet*   w Gk *upward*
x Or *commonwealth*   y Or *our humble bodies*
z Or *his glorious body*   a Gk *my brothers*
b Or *loyal Syzygus*   c Or *Farewell*
d Gk *brothers*   e Gk *take account of*   f Gk *I rejoiced*   g Gk lacks *to show it*

3.12–21: Confession and exhortation. 12–14: Though righteousness is God's gift, Christians are not relieved of the obligation of serious effort. *The goal,* the allusion is to Greek foot races and their finishing post. *The prize* is God's *heavenly call* to share God's glory (Rom 5.2). **3.18–19:** *Many,* presumably professing Christians, but probably not the Judaizers of 3.2, whom the description scarcely fits. **20:** *Our citizenship,* our real homeland. Paul's illustration is drawn from the fact that the citizenship of the Philippian was in Rome. **21:** Rom 8.23; 1 Cor 15.47–57; 2 Cor 5.1–5; Col 3.1–4.

4.1–23: Final appeals. 1: *Joy and crown,* 1 Thess 2.19–20. **2:** *Euodia . . . Syntyche,* two women in the Philippian church who had been disagreeing. **3:** *My loyal companion,* probably a leader in the church at Philippi. The Greek word for *companion* may be understood as a proper name, Syzygus. *Book of life,* Ex 32.32. **5:** Ps 119.151.

4.10: A reference to the gift which Epaph-

is to have little, and I know what it is to have plenty. In any and all circumstances I have learned the secret of being well-fed and of going hungry, of having plenty and of being in need. [13] I can do all things through him who strengthens me. [14] In any case, it was kind of you to share my distress.

15 You Philippians indeed know that in the early days of the gospel, when I left Macedonia, no church shared with me in the matter of giving and receiving, except you alone. [16] For even when I was in Thessalonica, you sent me help for my needs more than once. [17] Not that I seek the gift, but I seek the profit that accumulates to your account. [18] I have been paid in full and have more than enough;

I am fully satisfied, now that I have received from Epaphroditus the gifts you sent, a fragrant offering, a sacrifice acceptable and pleasing to God. [19] And my God will fully satisfy every need of yours according to his riches in glory in Christ Jesus. [20] To our God and Father be glory forever and ever. Amen.

21 Greet every saint in Christ Jesus. The friends [h] who are with me greet you. [22] All the saints greet you, especially those of the emperor's household.

23 The grace of the Lord Jesus Christ be with your spirit. [i]

h Gk *brothers*   *i* Other ancient authorities add *Amen*

---

roditus brought; see Introduction and 2.25–30. **14**: *My distress,* the deprivations and disappointments attendant upon imprisonment.

4.15: *When I left Macedonia,* Acts 16.40. **16**: According to Acts 17.1, Paul had gone to Thessalonica immediately after his stay in

Philippi. **22**: *The emperor's household,* Paul had taken advantage of his imprisonment to evangelize the soldiers who were his jailers, and they joined him in sending greetings to the Philippians (see 1.13 n.). The emperor at that time was probably Nero.

The Letter of Paul to the
Colossians

Colossae was a town of Phrygia in Asia Minor, not far from Ephesus. The church in Colossae had been founded, not by Paul, but probably by Epaphras (1.7; 4.12). The letter is confident that the believers are basically faithful to Paul's message. But right faith is threatened by the wrong teaching of persons who are not clearly identified. Our only clue to their teaching is the letter itself. It opposes certain ascetic practices ("Do not handle, do not taste," 2.21; "severe treatment of the body," 2.23) and some ritual practices that may have come from the Jewish tradition (dietary regulations and "festivals, new moons, and sabbaths," 2.16), as well as "philosophy and empty deceit" (2.8), which seems to mean esoteric teaching, and which included relying on the "elemental spirits of the universe" (2.8, 20). A combination of Jewish ritual practices and gnostic or theosophical speculation seems to have been what threatened the church at Colossae. Such syncretism or mixing of religious traditions was common at the time. Colossians opposes such teachings because they distribute the powers on which human beings depend among a variety of sources.

Against any such division of loyalty, the letter presents a strong and poetic image of the sole lordship of Christ, both throughout the whole cosmos and in the church. Christ is the unified and unifying power on which alone they are to depend. There is no stronger affirmation of the lordship of Christ in the New Testament.

The letter follows the pattern of most Pauline letters: introduction (1.1–11); moving directly into the basic statement (1.12–2.23), which opens with the powerful celebration of the cosmic and ecclesiastical Christ, and closes by arguing from Christ's lordship against the practices that were to be rejected in Colossae; practical application (3.1–4.6), which stresses being risen with Christ, and includes specific, somewhat patriarchal, directives about household and religious practices; and conclusion (4.7–18).

Colossians is closely related to the letters to Philemon and to Ephesians; the latter has drawn from it (see Introductions to Philemon and Ephesians).

The language of Colossians is similar to that of letters that are certainly Paul's, but many terms that are central to Paul elsewhere do not appear, and Paul's usual argumentative presentation of his thought is replaced by a more liturgical, celebrative style. The greatest contrast with letters that are surely Paul's is the emphasis on the transformation of the present by faith, instead of the usual Pauline tension between the partly-fulfilled present and the future that is hoped for. Scholars are divided about how to interpret these differences. Some hold that they are strong enough to conclude that Colossians was not written by Paul, as it claims, but by a disciple of Paul shortly after his time, to give Paul's authority to the continuing tradition of his teaching. Others think that the letter was written by Paul, while in prison (4.4, 18), presumably at Rome; the particular situation and, perhaps, changes in Paul's thinking, account for the contrasts.

1 Paul, an apostle of Christ Jesus by the will of God, and Timothy our brother,

2 To the saints and faithful brothers and sisters*ᵃ* in Christ in Colossae:

Grace to you and peace from God our Father.

3 In our prayers for you we always thank God, the Father of our Lord Jesus Christ, ⁴for we have heard of your faith in Christ Jesus and of the love that you have for all the saints, ⁵because of the hope laid up for you in heaven. You have heard of this hope before in the word of the truth, the gospel ⁶that has come to you. Just as it is bearing fruit and growing in the whole world, so it has been bearing fruit among yourselves from the day you heard it and truly comprehended the grace of God. ⁷This you learned from Epaphras, our beloved fellow servant.*ᵇ* He is a faithful minister of Christ on your*ᶜ* behalf, ⁸and he has made known to us your love in the Spirit.

9 For this reason, since the day we heard it, we have not ceased praying for you and asking that you may be filled with the knowledge of God's*ᵈ* will in all spiritual wisdom and understanding, ¹⁰so that you may lead lives worthy of the Lord, fully pleasing to him, as you bear fruit in every good work and as you grow in the knowledge of God. ¹¹May you be made strong with all the strength that comes from his glorious power, and may you be prepared to endure everything with patience, while joyfully ¹²giving thanks to the Father, who has enabled*ᵉ* you*ᶠ* to share in the inheritance of the saints in the light. ¹³He has rescued us from the power of darkness and transferred us into the kingdom of his beloved Son, ¹⁴in whom we have redemption, the forgiveness of sins.*ᵍ*

15 He is the image of the invisible God, the firstborn of all creation; ¹⁶for in*ʰ* him all things in heaven and on earth were created, things visible and invisible, whether thrones or dominions or rulers or powers—all things have been created through him and for him. ¹⁷He himself is before all things, and in*ʰ* him all things hold together. ¹⁸He is the head of the body, the church; he is the beginning, the firstborn from the dead, so that he might come to have first place in everything. ¹⁹For in him all the fullness of God was pleased to dwell, ²⁰and through him God was pleased to reconcile to himself all things, whether on earth or in heaven, by making peace through the blood of his cross.

21 And you who were once estranged and hostile in mind, doing evil deeds, ²²he has now reconciled*ⁱ* in his fleshly

a Gk *brothers*   *b* Gk *slave*   *c* Other ancient authorities read *our*   *d* Gk *his*   *e* Other ancient authorities read *called*   *f* Other ancient authorities read *us*   *g* Other ancient authorities add *through his blood*   *h* Or *by*   *i* Other ancient authorities read *you have now been reconciled*

1.1–2: Salutation, see Rom 1.1–7 n. **2:** *Grace . . . peace,* see 2 Thess 1.2 n.

1.3–14: Thanksgiving and intercession, see Rom 1.8–15 n. **3:** Eph 1.16. **4:** Philem 5. **4–5:** *Faith . . . love . . . hope,* 1 Cor 13.13. **6:** *The whole world,* in every quarter of the Roman Empire (v. 23). **7:** *Epaphras,* the founder of the church at Colossae, is now with Paul (4.12; Philem 23). **8–9:** Eph 1.15–17. **1.9–11:** A petition for sensitivity to God's will, issuing in Christian conduct and sustained by divine strength. **13:** *Rescued from Satan's power* (Mt 6.13 note *h*; Acts 26.18). **1.15–23: The supremacy of Christ** in the universe and in the church. **15:** The *image* perfectly reveals *the invisible God* (Jn 1.18; 2 Cor 4.4; Heb 1.3). *Firstborn* expresses priority to and supremacy over *all creation* (v. 17). **16:** Every *created* thing had its origin *in him* and exists *for him. Thrones . . . , powers* refer to various ranks of angels (Eph 6.12). **17:** Christ holds together the universe (Heb 1.3). **18:** *Head of . . . the church,* Eph 1.22–23; 4.15. *The beginning,* the origin or source (Rev 3.14). *Firstborn from the dead,* Acts 26.23; Rom 14.9; Rev 1.5. **19:** *The fullness of God,* a technical term for the plenitude of deity (see 2.9 n.), The verse may also be translated, "For it pleased God that in him [i.e. the Son] all the fullness of deity should dwell." **20:** *Blood of his cross,* the sacrificial death of Christ. **22:** *To present you,* the purpose and effect of the reconciliation (Eph 5.27).

body[j] through death, so as to present you holy and blameless and irreproachable before him— [23] provided that you continue securely established and steadfast in the faith, without shifting from the hope promised by the gospel that you heard, which has been proclaimed to every creature under heaven. I, Paul, became a servant of this gospel.

24 I am now rejoicing in my sufferings for your sake, and in my flesh I am completing what is lacking in Christ's afflictions for the sake of his body, that is, the church. [25] I became its servant according to God's commission that was given to me for you, to make the word of God fully known, [26] the mystery that has been hidden throughout the ages and generations but has now been revealed to his saints. [27] To them God chose to make known how great among the Gentiles are the riches of the glory of this mystery, which is Christ in you, the hope of glory. [28] It is he whom we proclaim, warning everyone and teaching everyone in all wisdom, so that we may present everyone mature in Christ. [29] For this I toil and struggle with all the energy that he powerfully inspires within me.

2 For I want you to know how much I am struggling for you, and for those in Laodicea, and for all who have not seen me face to face. [2] I want their hearts to be encouraged and united in love, so that they may have all the riches of assured understanding and have the knowledge of God's mystery, that is, Christ himself, [k] [3] in whom are hidden all the treasures of wisdom and knowledge. [4] I am saying this so that no one may deceive you with plausible arguments.

[5] For though I am absent in body, yet I am with you in spirit, and I rejoice to see your morale and the firmness of your faith in Christ.

6 As you therefore have received Christ Jesus the Lord, continue to live your lives[l] in him, [7] rooted and built up in him and established in the faith, just as you were taught, abounding in thanksgiving.

8 See to it that no one takes you captive through philosophy and empty deceit, according to human tradition, according to the elemental spirits of the universe, [m] and not according to Christ. [9] For in him the whole fullness of deity dwells bodily, [10] and you have come to fullness in him, who is the head of every ruler and authority. [11] In him also you were circumcised with a spiritual circumcision, [n] by putting off the body of the flesh in the circumcision of Christ; [12] when you were buried with him in baptism, you were also raised with him through faith in the power of God, who raised him from the dead. [13] And when you were dead in trespasses and the uncircumcision of your flesh, God[o] made you[p] alive together with him, when he forgave us all our trespasses, [14] erasing the record that stood against us with its legal demands. He set this aside, nailing it to the cross. [15] He disarmed[q] the rulers

j Gk *in the body of his flesh*   k Other ancient authorities read *of the mystery of God, both of the Father and of Christ*   l Gk *to walk*   m Or *the rudiments of the world*   n Gk *a circumcision made without hands*   o Gk *he*   p Other ancient authorities read *made us*; others, *made*   q Or *divested himself of*

1.24–2.7: Paul's interest in the Colossians, justifying his intervention in the affairs of a church where he was personally unknown (2.1). **1.24:** Because of the mystical union of the believers with Christ, what Paul suffers *for the sake of . . . the church* can be called *Christ's afflictions* (2 Cor 1.5; 4.10). **26:** *Mystery,* 2.2; 4.3; see Eph 1.9 n. **2.1:** *Laodicea,* chief city of Phrygia in Asia Minor, near Colossae (Rev 3.14–22).

2.8–23: Warning against false teaching. 8: Here *philosophy* refers to "vain speculation." *Elemental spirits,* see Gal 4.3 n. **9:** *In him,* the exalted Christ. *The whole fullness of deity,* not merely the divine attributes but the divine nature. *Dwells,* eternally. **10:** Eph 1.21–22. **13:** Eph 2.1, 5. **14:** *The record* of our transgressions (Eph 2.15; 1 Pet 2.24). **15:** *Rulers and authorities,* see 1.16 n. *Made a public example,* as captives, stripped of armor, are displayed in proof of victory.

and authorities and made a public example of them, triumphing over them in it.

16 Therefore do not let anyone condemn you in matters of food and drink or of observing festivals, new moons, or sabbaths. 17 These are only a shadow of what is to come, but the substance belongs to Christ. 18 Do not let anyone disqualify you, insisting on self-abasement and worship of angels, dwelling*r* on visions,*s* puffed up without cause by a human way of thinking,*t* 19 and not holding fast to the head, from whom the whole body, nourished and held together by its ligaments and sinews, grows with a growth that is from God.

20 If with Christ you died to the elemental spirits of the universe,*u* why do you live as if you still belonged to the world? Why do you submit to regulations, 21 "Do not handle, Do not taste, Do not touch"? 22 All these regulations refer to things that perish with use; they are simply human commands and teachings. 23 These have indeed an appearance of wisdom in promoting self-imposed piety, humility, and severe treatment of the body, but they are of no value in checking self-indulgence.*v*

3 So if you have been raised with Christ, seek the things that are above, where Christ is, seated at the right hand of God. 2 Set your minds on things that are above, not on things that are on earth, 3 for you have died, and your life is hidden with Christ in God. 4 When Christ who is your*w* life is revealed, then you also will be revealed with him in glory.

5 Put to death, therefore, whatever in you is earthly: fornication, impurity, passion, evil desire, and greed (which is idolatry). 6 On account of these the wrath of God is coming on those who are disobedient.*x* 7 These are the ways you also once followed, when you were living that life.*y* 8 But now you must get rid of all such things—anger, wrath, malice, slander, and abusive*z* language from your mouth. 9 Do not lie to one another, seeing that you have stripped off the old self with its practices 10 and have clothed yourselves with the new self, which is being renewed in knowledge according to the image of its creator. 11 In that renewal*a* there is no longer Greek and Jew, circumcised and uncircumcised, barbarian, Scythian, slave and free; but Christ is all and in all!

12 As God's chosen ones, holy and beloved, clothe yourselves with compassion, kindness, humility, meekness, and patience. 13 Bear with one another and, if anyone has a complaint against another, forgive each other; just as the Lord*b* has forgiven you, so you also must forgive. 14 Above all, clothe yourselves with love, which binds everything together in perfect harmony. 15 And let the peace of Christ rule in your hearts, to which indeed you were called in the one body. And be thankful. 16 Let the word of

r Other ancient authorities read *not dwelling*
s Meaning of Gk uncertain   t Gk *by the mind of his flesh*   u Or *the rudiments of the world*
v Or *are of no value, serving only to indulge the flesh*   w Other authorities read *our*
x Other ancient authorities lack *on those who are disobedient* (Gk *the children of disobedience*)
y Or *living among such people*   z Or *filthy*
a Gk *its creator,* 11*where*   b Other ancient authorities read *just as Christ*

2.16–23: The Colossian error involved excessive ritualism, asceticism, and the worship of angels. 16: *Festivals, new moons, . . . sabbaths,* annual, monthly, and weekly observances. 19: *The head,* Christ (Eph 1.22; 4.15).
2.20: *You died,* compare Rom 6.6–11; Gal 2.19. *Elemental spirits,* see Gal 4.3 n. *Regulations,* such as those quoted in v. 21.
3.1–17: **The true Christian life. 1:** *Raised,* 2.12. *Seated,* see Heb 1.3. **3:** *You have died to* the world (2.20); *your* new *life is secure in God,*

hidden *with Christ* from the world. **4:** Jn 14.6; 1 Jn 2.28; 3.2.
3.5: *Greed,* the covetous person sets up another object of worship besides God (Eph 5.5). **6:** *The wrath of God,* God's steadfast and holy hatred of sin (Rom 1.18–32). *Is coming,* at the Day of Judgment. **10:** *Image,* Gen 1.26–27; Eph 2.10; 4.24. **11:** Gal 3.28. *Greek,* i.e. Gentile.
3.12: *Clothe yourselves,* v. 12 takes up the metaphor of v. 10; these virtues belong to the

Christ[c] dwell in you richly; teach and admonish one another in all wisdom; and with gratitude in your hearts sing psalms, hymns, and spiritual songs to God.[d] 17 And whatever you do, in word or deed, do everything in the name of the Lord Jesus, giving thanks to God the Father through him.

18 Wives, be subject to your husbands, as is fitting in the Lord. 19 Husbands, love your wives and never treat them harshly.

20 Children, obey your parents in everything, for this is your acceptable duty in the Lord. 21 Fathers, do not provoke your children, or they may lose heart. 22 Slaves, obey your earthly masters[e] in everything, not only while being watched and in order to please them, but wholeheartedly, fearing the Lord.[e] 23 Whatever your task, put yourselves into it, as done for the Lord and not for your masters,[f] 24 since you know that from the Lord you will receive the inheritance as your reward; you serve[g] the Lord Christ. 25 For the wrongdoer will be paid back for whatever wrong has been done, and there is no partiality.

4 1 Masters, treat your slaves justly and fairly, for you know that you also have a Master in heaven.

2 Devote yourselves to prayer, keeping alert in it with thanksgiving. 3 At the same time pray for us as well that God will open to us a door for the word, that we may declare the mystery of Christ, for which I am in prison, 4 so that I may reveal it clearly, as I should.

5 Conduct yourselves wisely toward outsiders, making the most of the time.[h] 6 Let your speech always be gracious, seasoned with salt, so that you may know how you ought to answer everyone.

7 Tychicus will tell you all the news about me; he is a beloved brother, a faithful minister, and a fellow servant[i] in the Lord. 8 I have sent him to you for this very purpose, so that you may know how we are[j] and that he may encourage your hearts; 9 he is coming with Onesimus, the faithful and beloved brother, who is one of you. They will tell you about everything here.

10 Aristarchus my fellow prisoner greets you, as does Mark the cousin of Barnabas, concerning whom you have received instructions—if he comes to you, welcome him. 11 And Jesus who is called Justus greets you. These are the only ones of the circumcision among my co-workers for the kingdom of God, and they have been a comfort to me. 12 Epaphras, who is one of you, a servant[i] of Christ Jesus, greets you. He is always wrestling in his prayers on your behalf,

c Other ancient authorities read *of God*, or *of the Lord*   d Other ancient authorities read *to the Lord*   e In Greek the same word is used for *master* and *Lord*   f Gk *not for men*   g Or *you are slaves of*, or *be slaves of*   h Or *opportunity*   i Gk *slave*   j Other authorities read *that I may know how you are*

---

new nature. **15**: *Rule,* literally "be umpire." **17**: *Do everything* for Jesus' sake, that it may be as though he were doing it (v. 11).
3.18–4.6: The Christian's duties as regards family (vv. 18–20), masters and slaves (3.22–4.1) prayer, and social intercourse (4.2–6), much of which presupposes the patriarchal ethics of the time.
3.18–4.1: Eph 5.22–6.9. **3.22**: *Slaves, obey;* the case of Onesimus was then engaging Paul's attention (see Introduction to Philemon). **4.3**: *Open . . . a door,* see Rev 3.8 n. **5**: Eph 5.16. **6**: *Gracious,* courteous. *With salt,* with spiritual understanding (Mk 9.50).
4.7–18: Epilogue, personal messages (vv. 7–9), greetings (vv. 10–15), and final instructions (vv. 16–18). **7**: *Tychicus,* Eph 6.21–22. **9**: *Onesimus,* Philem 10. *Everything here,* presumably at Rome.
4.10: Acts 19.29; 27.2; Philem 24. **11**: *Jesus,* the Greek form of a fairly common Jewish name. **12**: 1.7; Philem 23. **14**: 2 Tim 4.10–11; Philem 24. **15**: *Church in . . . house,* see Philem 2 n. **16**: *The letter* left at *Laodicea,* either the letter to the Ephesians or some other Pauline letter no longer extant. **18**: *I . . . with my own hand,* up to this point Paul had been dictating (see 2 Thess 3.17 n.). *Remember my chains,* i.e. in prayer.

so that you may stand mature and fully assured in everything that God wills. [13] For I testify for him that he has worked hard for you and for those in Laodicea and in Hierapolis. [14] Luke, the beloved physician, and Demas greet you. [15] Give my greetings to the brothers and sisters[k] in Laodicea, and to Nympha and the church in her house. [16] And when this letter has been read among you, have it read also in the church of the Laodiceans;

and see that you read also the letter from Laodicea. [17] And say to Archippus, "See that you complete the task that you have received in the Lord."

18 I, Paul, write this greeting with my own hand. Remember my chains. Grace be with you.[l]

k Gk *brothers*   l Other ancient authorities add *Amen*

The First Letter of Paul to the
Thessalonians

Thessalonica was the capital of the Roman province of Macedonia, important for its location on both sea and land routes of travel. Paul founded the church in Thessalonica shortly after he left Philippi (2.1–2; Acts 17.1–8). Although Acts tells us that his initial contact in Thessalonica was with the synagogue, 1 Thessalonians is addressed to Gentile believers (1.9).

The letter reflects the life of a congregation that was devoted to its faith and strongly aware of its separation from the society in which its members had until recently found their standards and values. At the same time it was also a community that was threatened by social pressures and at times outright persecution to turn back to the life from which they had come.

Paul wrote to encourage the church, stressing that opposition is simply something to be expected (3.3), and expressing appreciation for their steadfastness (1.6; 3.6–10); but he also wrote to defend himself against accusations about his character and motives (2.1–12). He forcefully asserts his straightforwardness (2.3–6) as well as his affectionate caring for them like a nurse or a parent (2.7, 11). Paul skillfully links his situation to theirs by telling of his sending Timothy to them and of his own response to Timothy's encouraging report (3.1–10). This is followed by ethical admonitions that emphasize holiness or belonging to God, as well as the need to act in such a way as to make an acceptable impression on outsiders (4.1–12).

A further main theme of 1 Thessalonians is a response to questions about eschatology that were troubling the church, that is, questions about the end. Some of the Thessalonians were perplexed by the deaths of believers, and wondered whether those who had died would be excluded from the life of the resurrection. Paul replies with an apocalyptic narrative, a story of the events of the end-time, which makes clear that it would not be an advantage to be still living when Christ returns; all believers "will be with the Lord forever" (4.13–18). This section shows that both Paul and his readers expected the end very soon. But Paul follows with another statement, also expressed in traditional apocalyptic language, which emphasizes that one cannot predict the time of the end (5.1–11). These apocalyptic sections are to be compared with Paul's most extensive statement about hope for the end, 1 Corinthians 15.

In one passage the difficulties of the Thessalonian church are compared to those of the churches of Judea (2.13–16). This passage, so critical of the Jews, is held by many scholars to be a later addition to the letter, not only because the anti-Jewish language is not like what Paul writes elsewhere, but especially because the passage introduces an atypical second thanksgiving into the pattern of the letter.

Although Paul had sent Timothy to Thessalonica from Athens (3.1–2), he was probably in Corinth when he wrote 1 Thessalonians, perhaps in the early 50s. This letter is probably the earliest of Paul's extant correspondence and therefore the earliest of all New Testament writings.

1 Paul, Silvanus, and Timothy,
To the church of the Thessalonians in God the Father and the Lord Jesus Christ:
Grace to you and peace.

2 We always give thanks to God for all of you and mention you in our prayers, constantly ³remembering before our God and Father your work of faith and labor of love and steadfastness of hope in our Lord Jesus Christ. ⁴For we know, brothers and sisters*ᵃ* beloved by God, that he has chosen you, ⁵because our message of the gospel came to you not in word only, but also in power and in the Holy Spirit and with full conviction; just as you know what kind of persons we proved to be among you for your sake. ⁶And you became imitators of us and of the Lord, for in spite of persecution you received the word with joy inspired by the Holy Spirit, ⁷so that you became an example to all the believers in Macedonia and in Achaia. ⁸For the word of the Lord has sounded forth from you not only in Macedonia and Achaia, but in every place your faith in God has become known, so that we have no need to speak about it. ⁹For the people of those regions*ᵇ* report about us what kind of welcome we had among you, and how you turned to God from idols, to serve a living and true God, ¹⁰and to wait for his Son from heaven, whom he raised from the dead—Jesus, who rescues us from the wrath that is coming.

2 You yourselves know, brothers and sisters,*ᵃ* that our coming to you was not in vain, ²but though we had already suffered and been shamefully mistreated at Philippi, as you know, we had courage in our God to declare to you the gospel of God in spite of great opposition. ³For our appeal does not spring from deceit or impure motives or trickery, ⁴but just as we have been approved by God to be entrusted with the message of the gospel, even so we speak, not to please mortals, but to please God who tests our hearts. ⁵As you know and as God is our witness, we never came with words of flattery or with a pretext for greed; ⁶nor did we seek praise from mortals, whether from you or from others, ⁷though we might have made demands as apostles of Christ. But we were gentle*ᶜ* among you, like a nurse tenderly caring for her own children. ⁸So deeply do we care for you that we are determined to share with you not only the gospel of God but also our own selves, because you have become very dear to us.

9 You remember our labor and toil, brothers and sisters;*ᵃ* we worked night and day, so that we might not burden any of you while we proclaimed to you the gospel of God. ¹⁰You are witnesses, and God also, how pure, upright, and blameless our conduct was toward you believers. ¹¹As you know, we dealt with

a Gk *brothers*　　　*b* Gk *For they*　　　*c* Other ancient authorities read *infants*

1.1: Salutation, see Rom 1.1–7 n. *Silvanus*, identical with Silas (Acts 15.22, 40; 16.19–25; 17.4). *Timothy*, see Introduction to 2 Timothy. *Church*, see Gal 1.13 n. *Lord*, the title of Israel's covenant God, is applied by Christians to the risen and glorified Jesus (Phil 2.9–11). *Grace . . . and peace*, see 2 Thess 1.2 n.
1.2–10: Thanksgiving, see Rom 1.8–15 n. Paul is grateful that the Thessalonians have persevered, though aware that much yet needs to be accomplished in them. **4:** Israel's privileges as God's *chosen* are transferred to the church. **5:** *Our . . . gospel* is not just words; the saving power of God is at work in it.
1.6: The first converts of Paul and Silas at Thessalonica were subjected to much *persecution* (Acts 17.5–9). **7–8:** *Macedonia*, of which Thessalonica was the capital. *Achaia*, Greece (where Paul was then). **10:** The return of the risen Christ is the hope of God's people. *The wrath*, i.e. of God, which is not anger or irritation, but justice (Rom 1.28, 32; 2.8, 9; Eph 2.3; Col 3.6).
2.1–16: Paul's life and work at Thessalonica. 2: *Philippi*, Acts 16.19–40. **3–8:** The opponents of the gospel accused Paul of heresy, immorality, trickery, and greed. He replies by reminding his readers of his conduct among them. **4:** *Tests our hearts*, Prov 17.3.
2.9: Paul supported himself by working at his trade (Acts 18.3) so as not to burden the

each one of you like a father with his children, 12 urging and encouraging you and pleading that you lead a life worthy of God, who calls you into his own kingdom and glory.

13 We also constantly give thanks to God for this, that when you received the word of God that you heard from us, you accepted it not as a human word but as what it really is, God's word, which is also at work in you believers. 14 For you, brothers and sisters, *d* became imitators of the churches of God in Christ Jesus that are in Judea, for you suffered the same things from your own compatriots as they did from the Jews, 15 who killed both the Lord Jesus and the prophets, *e* and drove us out; they displease God and oppose everyone 16 by hindering us from speaking to the Gentiles so that they may be saved. Thus they have constantly been filling up the measure of their sins; but God's wrath has overtaken them at last.*f*

17 As for us, brothers and sisters, *d* when, for a short time, we were made orphans by being separated from you— in person, not in heart—we longed with great eagerness to see you face to face. 18 For we wanted to come to you— certainly I, Paul, wanted to again and again—but Satan blocked our way. 19 For what is our hope or joy or crown of boasting before our Lord Jesus at his coming? Is it not you? 20 Yes, you are our glory and joy!

3 Therefore when we could bear it no longer, we decided to be left alone in Athens; 2 and we sent Timothy, our brother and co-worker for God in proclaiming*g* the gospel of Christ, to strengthen and encourage you for the sake of your faith, 3 so that no one would be shaken by these persecutions. Indeed, you yourselves know that this is what we are destined for. 4 In fact, when we were with you, we told you beforehand that we were to suffer persecution; so it turned out, as you know. 5 For this reason, when I could bear it no longer, I sent to find out about your faith; I was afraid that somehow the tempter had tempted you and that our labor had been in vain.

6 But Timothy has just now come to us from you, and has brought us the good news of your faith and love. He has told us also that you always remember us kindly and long to see us—just as we long to see you. 7 For this reason, brothers and sisters, *d* during all our distress and persecution we have been encouraged about you through your faith. 8 For we now live, if you continue to stand firm in the Lord. 9 How can we thank God enough for you in return for all the joy that we feel before our God because of you? 10 Night and day we pray most earnestly that we may see you face to face and restore whatever is lacking in your faith.

11 Now may our God and Father himself and our Lord Jesus direct our way to you. 12 And may the Lord make you increase and abound in love for one another and for all, just as we abound in love for you. 13 And may he so strengthen your hearts in holiness that you may be blameless before our God and Father at the coming of our Lord Jesus with all his saints.

d Gk *brothers*   *e* Other ancient authorities read *their own prophets*   *f* Or *completely* or *forever*   *g* Gk lacks *proclaiming*

---

church. **14–16:** The severe language reflects the strenuous struggle between Paul and the Jews (Acts 14.2, 5, 19; 17.5, 13; 21.21; 25.2, 7). It is possible that 2.13–16 was added later to the letter (see Introduction).
2.17–3.13: Paul's affection for the Thessalonians. 2.18: Behind obstacles Paul recognizes the activity of Satan, the adversary of God's kingly rule (see Mt 4.1 n.; Rom 16.20; 2 Cor 11.14; 1 Tim 1.20).

3.3: *Persecutions* are normal, not exceptional. **5:** *Tempter,* see 2.18 n.
3.6: *Timothy has . . . come,* see Introduction. **7:** *Distress and persecution,* Acts 18.6, 12. **10:** The prayer was answered some years later (Acts 20.1–2). **13:** The heart is not the organ of feeling as in modern speech, but the controlling center of personality. *Saints,* all who belong to God.

4 Finally, brothers and sisters,[h] we ask and urge you in the Lord Jesus that, as you learned from us how you ought to live and to please God (as, in fact, you are doing), you should do so more and more. [2]For you know what instructions we gave you through the Lord Jesus. [3]For this is the will of God, your sanctification: that you abstain from fornication; [4]that each one of you know how to control your own body[i] in holiness and honor, [5]not with lustful passion, like the Gentiles who do not know God; [6]that no one wrong or exploit a brother or sister[j] in this matter, because the Lord is an avenger in all these things, just as we have already told you beforehand and solemnly warned you. [7]For God did not call us to impurity but in holiness. [8]Therefore whoever rejects this rejects not human authority but God, who also gives his Holy Spirit to you.

9 Now concerning love of the brothers and sisters,[h] you do not need to have anyone write to you, for you yourselves have been taught by God to love one another; [10]and indeed you do love all the brothers and sisters[h] throughout Macedonia. But we urge you, beloved,[h] to do so more and more, [11]to aspire to live quietly, to mind your own affairs, and to work with your hands, as we directed you, [12]so that you may behave properly toward outsiders and be dependent on no one.

13 But we do not want you to be uninformed, brothers and sisters,[h] about those who have died,[k] so that you may not grieve as others do who have no hope. [14]For since we believe that Jesus died and rose again, even so, through Jesus, God will bring with him those who have died.[k] [15]For this we declare to you by the word of the Lord, that we who are alive, who are left until the coming of the Lord, will by no means precede those who have died.[k] [16]For the Lord himself, with a cry of command, with the archangel's call and with the sound of God's trumpet, will descend from heaven, and the dead in Christ will rise first. [17]Then we who are alive, who are left, will be caught up in the clouds together with them to meet the Lord in the air; and so we will be with the Lord forever. [18]Therefore encourage one another with these words.

5 Now concerning the times and the seasons, brothers and sisters,[h] you do not need to have anything written to you. [2]For you yourselves know very well that the day of the Lord will come like a thief in the night. [3]When they say, "There is peace and security," then sudden destruction will come upon them, as labor pains come upon a pregnant woman, and there will be no escape! [4]But you, beloved,[h] are not in darkness, for that day to surprise you like a thief; [5]for you are all children of light and children of the day; we are not of the night or of darkness. [6]So then let us not fall asleep as others do, but let us keep awake and be

---

h Gk *brothers*   i Or *how to take a wife for himself*   j Gk *brother*   k Gk *fallen asleep*

4.1–12: Exhortation to purity. 3: The Gentile world was notorious for sexual license (1 Cor 6.18). **5:** *Gentiles who do not know God,* Ps 79.6. **8:** Paul's supreme argument for purity of life—the indwelling Holy Spirit (1 Cor 3.16–17).

4.9: *Taught by God,* Isa 54.13. **11:** *Work with your hands,* see 2.9 n. **12:** *Outsiders,* not belonging to God's people. *Be dependent on no one,* 2 Thess 3.11–12.

4.13–5.11: Questions concerning the coming of the Lord. 4.13: See Mt 9.24 n.; Jn 11.11 n. **14:** The Christian is united with Christ in his death and resurrection (Rom 6.3). **15:** *The word of the Lord,* authoritative revelation, not private opinion. **16:** Mt 24.30–31; Mk 13.26–27; 1 Cor 15.52.

5.1: Some *brothers and sisters* desired an eschatological time-table (Acts 1.6–7). **2:** Paul reiterates Jesus' warning (Mt 24.43–44; Lk 12.39–40). **3:** *Pregnant woman,* a common prophetic expression for the suddenness of the day of the Lord (Isa 13.8; Jer 6.24; Hos 13.13). **5:** Lk 16.8. The Essenes contrasted the Children of Light with the Children of Darkness. **5.6:** *Let us not fall asleep,* in carelessness and sin. **8:** The Christian's armor (compare Rom

sober; [7]for those who sleep sleep at night, and those who are drunk get drunk at night. [8]But since we belong to the day, let us be sober, and put on the breastplate of faith and love, and for a helmet the hope of salvation. [9]For God has destined us not for wrath but for obtaining salvation through our Lord Jesus Christ, [10]who died for us, so that whether we are awake or asleep we may live with him. [11]Therefore encourage one another and build up each other, as indeed you are doing.

[12]But we appeal to you, brothers and sisters,[l] to respect those who labor among you, and have charge of you in the Lord and admonish you; [13]esteem them very highly in love because of their work. Be at peace among yourselves. [14]And we urge you, beloved,[l] to admonish the idlers, encourage the faint hearted, help the weak, be patient with all of them. [15]See that none of you repays evil for evil, but always seek to do good

to one another and to all. [16]Rejoice always, [17]pray without ceasing, [18]give thanks in all circumstances; for this is the will of God in Christ Jesus for you. [19]Do not quench the Spirit. [20]Do not despise the words of prophets,[m] [21]but test everything; hold fast to what is good; [22]abstain from every form of evil.

[23]May the God of peace himself sanctify you entirely; and may your spirit and soul and body be kept sound[n] and blameless at the coming of our Lord Jesus Christ. [24]The one who calls you is faithful, and he will do this.

[25]Beloved,[o] pray for us.

[26]Greet all the brothers and sisters[l] with a holy kiss. [27]I solemnly command you by the Lord that this letter be read to all of them.[p]

[28]The grace of our Lord Jesus Christ be with you.[q]

l Gk brothers   m Gk despise prophecies
n Or complete   o Gk Brothers   p Gk to all
the brothers   q Other ancient authorities add
Amen

13.12; 2 Cor 6.7; 10.4; Eph 6.13–17). **10:** *Whether we are awake or asleep,* whether we are alive or dead when Christ returns. **11:** Paul frequently pictures the church as a temple under construction. The builder is God, but Christians can take part in the work of building (1 Cor 3.9; Eph 2.20).
5.12–28: Concluding exhortation. 19: *Do not quench* the fire of *the Spirit* (compare Mt 3.11; Acts 2.3; 2 Tim 1.6). **20:** *Words of prophets,* not only prediction, but the inspired word

of preaching (see Acts 11.27 n.). **21:** *Test everything,* this injunction probably relates to preaching mentioned in the preceding verse; see 1 Jn 4.1–3.
5.23: *Spirit and soul and body,* Paul does not think of a person as having three parts, but as a unity which may be viewed from three different points of view: the relation to God, the personal vitality, and the physical body. **24:** *Will do this,* v. 23. **26:** *Holy kiss,* see Rom 16.16 n. **27:** Col 4.16.

The Second Letter of Paul to the
Thessalonians

Second Thessalonians deals with issues that are also central in 1 Thessalonians: how to understand the fulfillment of God's promises in the return of Christ, and how to act in the meantime. It appears to be a letter that Paul wrote shortly after 1 Thessalonians, to correct the impression that "the day of the Lord is already here" (2.2); the Greek of this verse can also be rendered, "the day of the Lord is impending." First Thessalonians assumes that Christ's coming again is near (1 Thess 4.13–18), but also emphasizes that the appearance of Christ will be a surprise; we cannot know the time (1 Thess 5.1–11). The thrust of 2 Thessalonians is in the opposite direction—if we cannot know the exact time, nonetheless we can know that the day of the Lord will not come at once; a dire struggle with evil must take place first, and even this is to be delayed for a time. The details of this apocalyptic story were known to those who received the letter. Thus, the "rebellion" and "the lawless one" (2.3–4, 8) and the "mystery of lawlessness" (2.7), as well as "what is now restraining [the lawless one]" (2.6) and "the one who restrains it" (2.7), are references to things that the recipients of the letter already know (2.6), though these terms are not clear to us. But it is clear that knowing about the delayed struggle with lawlessness assured one that the end was not to come just yet. The message of 2 Thessalonians prepares the church for a period of continued life in this world.

The implication for daily life was very clear: it is important to continue the pursuits of daily life in a regular and orderly way, especially the life of work, as Paul himself had worked when he was among them (3.6–13). "Anyone unwilling to work should not eat" (3.10) has been cited as a maxim of the orderly life ever since. This theme also appears in 1 Thessalonians; the emphasis on it is even stronger in the second letter.

Most scholars have explained the contrast between the two Thessalonian letters by placing the second letter shortly after the first, perhaps even before Paul had heard how the church responded to his first letter. When he learned how disruptive the expectation of the imminent end was, he wrote, using the theme of apocalyptic signs (see 2.3–4n.), to explain that the end was not to come for a time, and to reinforce his teaching about the proper conduct of life.

Other scholars find it difficult to think that Paul would shift the emphasis of his apocalyptic teaching so abruptly, and such scholars also stress stylistic features of the letter that point to a different author: the letter contains two thanksgivings (1.3–4; 2.13–15); only 1 Thessalonians among Paul's letters has this feature, and some scholars hold that the second one is a later addition (see the Introduction to 1 Thessalonians), and the second letter draws from the first in a way that seems more likely to come from a later writer than from Paul himself. This interpretation sees the letter as pseudonymous, written in Paul's name to clarify a dispute about eschatology among the followers of Paul at a somewhat later date. (See the Introduction to 2 Peter on pseudonymous authorship.)

1 Paul, Silvanus, and Timothy,
To the church of the Thessalonians in God our Father and the Lord Jesus Christ:
2 Grace to you and peace from God our*a* Father and the Lord Jesus Christ.
3 We must always give thanks to God for you, brothers and sisters,*b* as is right, because your faith is growing abundantly, and the love of everyone of you for one another is increasing. 4Therefore we ourselves boast of you among the churches of God for your steadfastness and faith during all your persecutions and the afflictions that you are enduring.
5 This is evidence of the righteous judgment of God, and is intended to make you worthy of the kingdom of God, for which you are also suffering. 6For it is indeed just of God to repay with affliction those who afflict you, 7and to give relief to the afflicted as well as to us, when the Lord Jesus is revealed from heaven with his mighty angels 8in flaming fire, inflicting vengeance on those who do not know God and on those who do not obey the gospel of our Lord Jesus. 9These will suffer the punishment of eternal destruction, separated from the presence of the Lord and from the glory of his might, 10when he comes to be glorified by his saints and to be marveled at on that day among all who have believed, because our testimony to you was

believed. 11To this end we always pray for you, asking that our God will make you worthy of his call and will fulfill by his power every good resolve and work of faith, 12so that the name of our Lord Jesus may be glorified in you, and you in him, according to the grace of our God and the Lord Jesus Christ.

2 As to the coming of our Lord Jesus Christ and our being gathered together to him, we beg you, brothers and sisters,*b* 2not to be quickly shaken in mind or alarmed, either by spirit or by word or by letter, as though from us, to the effect that the day of the Lord is already here. 3Let no one deceive you in any way; for that day will not come unless the rebellion comes first and the lawless one*c* is revealed, the one destined for destruction.*d* 4He opposes and exalts himself above every so-called god or object of worship, so that he takes his seat in the temple of God, declaring himself to be God. 5Do you not remember that I told you these things when I was still with you? 6And you know what is now restraining him, so that he may be revealed when his time comes. 7For the mystery of lawlessness is already at work, but only until the one who now

a Other ancient authorities read *the*
b Gk *brothers*   c Gk *the man of lawlessness;*
other ancient authorities read *the man of sin*
d Gk *the son of destruction*

1.1–2: Salutation, see Rom 1.1–7 n. *Silvanus, Timothy,* see 1 Thess 1.1 n. **2:** *Grace . . . and peace* combine the conventional Greek and Hebrew salutations; the apostle's greeting, however, is not merely his own good wishes but the grace and peace of God given in Jesus Christ.
1.3–4: Thanksgiving, see Rom 1.8–15 n.
1.5–12: The judgment of God. Reference to *afflictions* (v. 4) leads to a digression of remarkable intensity. **8:** *Vengeance,* the word is not used in the sense of revenge, but of just recompense.
1.9: *Eternal destruction,* not annihilation, but endless ruin in separation from Christ. **10:** *That day,* see 2 Tim 1.12 n. *Glorified by his saints,* what Christ has done for his people will be seen to redound to his glory. **12:** *The name,* the character and fame (Phil 2.9).

2.1–12: The day of the Lord. *The rebellion . . . and the lawless one* will precede the second coming of Christ. **2:** The new teaching originated in or was supported by the utterances of an ecstatic individual, or perhaps in a letter purporting to come from Paul. **3–4:** The day of the Lord (2 Pet 3.10) will be preceded by persecutions, false prophets, and a "desolating sacrilege" (Dan 9.20–27; 11.31; 12.11; Mt ch 24; Mk ch 13; Lk 21.5–36). **6–8:** *The lawless one* has not yet come because something (v. 7, someone) is *restraining him.* Paul assumes that his readers understand this reference, but we do not. There are three main conjectures, none of which is entirely satisfactory: (*a*) The Roman empire and emperor; (*b*) A supernatural power; (*c*) Satan himself. **7:** *The mystery of lawlessness,* the counterfeit and opponent of the mystery of

restrains it is removed. [8] And then the lawless one will be revealed, whom the Lord Jesus[e] will destroy[f] with the breath of his mouth, annihilating him by the manifestation of his coming. [9] The coming of the lawless one is apparent in the working of Satan, who uses all power, signs, lying wonders, [10] and every kind of wicked deception for those who are perishing, because they refused to love the truth and so be saved. [11] For this reason God sends them a powerful delusion, leading them to believe what is false, [12] so that all who have not believed the truth but took pleasure in unrighteousness will be condemned.

[13] But we must always give thanks to God for you, brothers and sisters[g] beloved by the Lord, because God chose you as the first fruits[h] for salvation through sanctification by the Spirit and through belief in the truth. [14] For this purpose he called you through our proclamation of the good news,[i] so that you may obtain the glory of our Lord Jesus Christ. [15] So then, brothers and sisters,[g] stand firm and hold fast to the traditions that you were taught by us, either by word of mouth or by our letter.

[16] Now may our Lord Jesus Christ himself and God our Father, who loved us and through grace gave us eternal comfort and good hope, [17] comfort your hearts and strengthen them in every good work and word.

3 Finally, brothers and sisters,[g] pray for us, so that the word of the Lord may spread rapidly and be glorified everywhere, just as it is among you, [2] and that we may be rescued from wicked and evil people; for not all have faith. [3] But the Lord is faithful; he will strengthen you and guard you from the evil one.[j] [4] And we have confidence in the Lord concerning you, that you are doing and will go on doing the things that we command. [5] May the Lord direct your hearts to the love of God and to the steadfastness of Christ.

[6] Now we command you, beloved,[g] in the name of our Lord Jesus Christ, to keep away from believers who are[k] living in idleness and not according to the tradition that they[l] received from us. [7] For you yourselves know how you ought to imitate us; we were not idle when we were with you, [8] and we did not eat anyone's bread without paying for it; but with toil and labor we worked night and day, so that we might not burden any of you. [9] This was not because we do not have that right, but in order to give you an example to imitate. [10] For even when we were with you, we gave you this command: Anyone unwilling to work should not eat. [11] For we hear that some of you are living in idleness, mere busybodies, not doing any work. [12] Now such persons we command and exhort in the Lord Jesus Christ to do their work quietly and to earn their own living. [13] Brothers and sisters,[m] do not be weary in doing what is right.

[14] Take note of those who do not obey what is say in this letter; have

---

e Other ancient authorities lack *Jesus*   *f* Other ancient authorities read *consume*   *g* Gk *brothers*   *h* Other ancient authorities read *from the beginning*   *i* Or *through our gospel*   *j* Or *from evil*   *k* Gk *from every brother who is*   *l* Other ancient authorities read *you*   *m* Gk *Brothers*

---

godliness (Col 1.26; 1 Tim 3.16). **11**: Rom 1.28.

2.13–17: **Thanksgiving and exhortation. 13**: 1.3; Eph 1.4; 1 Pet 1.2. **15**: *The traditions* would involve doctrinal, moral, and liturgical teaching, as 1 Corinthians shows. **16**: 1 Thess 3.11; 1 Pet 1.3.

3.1–18: **Closing appeals, rebukes, and prayer. 2**: Rom 15.31. **3**: *The evil one*, Mt 6.13. **5**: *Love of God*, love for God.

3.6–15: Paul rebukes those who, because of the supposed imminence of Christ's coming, had ceased to work and were living on the generosity of others. **8**: 1 Thess 2.9; Acts 18.3. **9**: *That right*, Lk 10.7; Gal 6.6. **11**: *Busybodies*, 1 Thess 4.11. **13**: Gal 6.9. **15**: A reminder that those who disregard Paul's instructions are nevertheless to be treated with love. **16**: *Give you peace*, see the Aaronic benediction, Num 6.26. **17**: Paul adds a concluding note in his own handwriting as a guarantee of genuineness (see 2.2 n.).

nothing to do with them, so that they may be ashamed. [15] Do not regard them as enemies, but warn them as believers. [n]

16 Now may the Lord of peace himself give you peace at all times in all ways. The Lord be with all of you.

17 I, Paul, write this greeting with my own hand. This is the mark in every letter of mine; it is the way I write. [18] The grace of our Lord Jesus Christ be with all of you. [o]

n Gk *a brother*   *o* Other ancient authorities add *Amen*

The First Letter of Paul to
Timothy

The two letters to Timothy and the one to Titus, commonly called the Pastorals, are similar in character and in the problems they raise concerning authorship. It is difficult to ascribe them in their present form to the apostle Paul. The vocabulary and style of these letters differ widely from the acknowledged letters of Paul; some of his leading theological themes are entirely absent (e.g. the union of the believer with Christ, the power and witness of the Spirit, freedom from the law), and some of the expressions bear a different meaning from that in his customary usage (e.g. "the faith" as a synonym for the Christian religion rather than the believer's relationship to Christ).

A few scholars, attempting to maintain Pauline authorship, account for the differences by assuming changes in his environment as well as modifications of his vocabulary, style, and thought. But in view of the widespread custom in antiquity of pseudonymous authorship (that is, the use of a respected name to give authority to a writing actually written by someone else), it is easier to assume that a loyal disciple of Paul composed these letters. The purpose was to present Paul's teaching as it was then understood in the church, using it as a bulwark against wrong teaching and practice. Some scholars believe that fragments of letters written by Paul are incorporated into these three letters, while others hold that the personal greetings are simply a framework that the author used to give Paul's authority to the teachings of the letters. (On pseudonymous authorship see the Introduction to 2 Peter).

The First Letter to Timothy has a double purpose: to provide guidance in the problems of church administration, and to oppose false teaching of a speculative and moralistic type. Thus, it offers suggestions for the regulation of worship (2.1–15), lays down the qualifications of "bishops" (3.1–7) and "deacons" (3.8–13) (neither bishop nor deacon was yet defined as they later came to be, and the letter does not clearly describe their functions), while it also gives instructions as to the attitude of church leaders toward an asceticism that devalued the creation (4.1–10) and toward individual members (5.1–22), particularly widows (5.3–16), elders or presbyters (5.17–20), and slaves (6.1–2). It also warns strongly against teachers who lack understanding, wander into vain discussions, and end by making shipwreck of their faith (1.3–7, 19–20; 6.3–10). The author is especially sharp in reproving those who seek to make profit out of religion (6.3–10). The letter attacks an asceticism that was related to gnosticism, namely the attitude of one who claims to possess true knowledge (*gnosis*) and is therefore superior to the claims of the body, renouncing marriage and abstaining from certain foods (4.3). Speculative mythology is also denounced as "profane myths" (4.7) and "what is falsely called knowledge" (6.20).

For information concerning the person called Timothy, see the Introduction to 2 Timothy.

1 Paul, an apostle of Christ Jesus by the command of God our Savior and of Christ Jesus our hope,
2 To Timothy, my loyal child in the faith:
Grace, mercy, and peace from God the Father and Christ Jesus our Lord.

3 I urge you, as I did when I was on my way to Macedonia, to remain in Ephesus so that you may instruct certain people not to teach any different doctrine, 4 and not to occupy themselves with myths and endless genealogies that promote speculations rather than the divine training*a* that is known by faith. 5 But the aim of such instruction is love that comes from a pure heart, a good conscience, and sincere faith. 6 Some people have deviated from these and turned to meaningless talk, 7 desiring to be teachers of the law, without understanding either what they are saying or the things about which they make assertions.

8 Now we know that the law is good, if one uses it legitimately. 9 This means understanding that the law is laid down not for the innocent but for the lawless and disobedient, for the godless and sinful, for the unholy and profane, for those who kill their father or mother, for murderers, 10 fornicators, sodomites, slave traders, liars, perjurers, and whatever else is contrary to the sound teaching 11 that conforms to the glorious gospel of the blessed God, which he entrusted to me.

12 I am grateful to Christ Jesus our Lord, who has strengthened me, because he judged me faithful and appointed me to his service, 13 even though I was formerly a blasphemer, a persecutor, and a man of violence. But I received mercy because I had acted ignorantly in unbelief, 14 and the grace of our Lord overflowed for me with the faith and love that are in Christ Jesus. 15 The saying is sure and worthy of full acceptance, that Christ Jesus came into the world to save sinners—of whom I am the foremost. 16 But for that very reason I received mercy, so that in me, as the foremost, Jesus Christ might display the utmost patience, making me an example to those who would come to believe in him for eternal life. 17 To the King of the ages, immortal, invisible, the only God, be honor and glory forever and ever.*b* Amen.

18 I am giving you these instructions, Timothy, my child, in accordance with the prophecies made earlier about you, so that by following them you may fight the good fight, 19 having faith and a good conscience. By rejecting conscience, certain persons have suffered shipwreck in the faith; 20 among them are Hymenaeus and Alexander, whom I have turned over to Satan, so that they may learn not to blaspheme.

2 First of all, then, I urge that supplications, prayers, intercessions, and

a Or *plan*   *b* Gk *to the ages of the ages*

1.1–2: Salutation, see Rom 1.1–7 n. **2:** *Timothy,* see Introduction to 2 Timothy. *Mercy* is added to *grace and peace* (compare 2 Tim 1.2 and see 2 Thess 1.2 n.). **1.3–20: The defense of the truth. 4:** *Myths and endless genealogies,* probably Jewish speculations (compare Titus 1.14; 3.9). *The divine training* pictures the Christian life as the discipline of servants in a large household (the term translated *training* may also mean the divine "plan" of salvation; see note *a*). **5:** The goal of preaching is love: not romantic sentiment, but sharing God's generosity with one's neighbor. **6:** The neglect of love leads to empty talk.

1.12–14: The Lord displayed his grace in making an apostle out of Saul the persecutor (Acts 9.4; 1 Cor 15.9; Gal 1.13; Phil 3.6). **15:** *The saying is sure,* a formula characteristic of the letters to Timothy and Titus (3.1; 4.9; 2 Tim 2.11; Titus 3.8). **17:** This is the language of Jewish congregational prayer and praise. **20:** *Hymenaeus,* 2 Tim 2.17. *Alexander,* perhaps the same as Alexander the coppersmith, 2 Tim 4.14. *Satan* was regarded as the source of suffering and disease as well as moral evil (Lk 13.16; 2 Cor 12.7). Under the power of Satan, the sufferer may be moved to repentance (see 1 Cor 5.5 n.). **2.1–15: The regulation of worship. 2:**

thanksgivings be made for everyone, [2]for kings and all who are in high positions, so that we may lead a quiet and peaceable life in all godliness and dignity. [3]This is right and is acceptable in the sight of God our Savior, [4]who desires everyone to be saved and to come to the knowledge of the truth. [5]For

> there is one God;
> there is also one mediator
> between God and
> humankind,
> Christ Jesus, himself human,
> [6]   who gave himself a ransom
> for all

—this was attested at the right time. [7]For this I was appointed a herald and an apostle (I am telling the truth,[c] I am not lying), a teacher of the Gentiles in faith and truth.

[8]I desire, then, that in every place the men should pray, lifting up holy hands without anger or argument; [9]also that the women should dress themselves modestly and decently in suitable clothing, not with their hair braided, or with gold, pearls, or expensive clothes, [10]but with good works, as is proper for women who profess reverence for God. [11]Let a woman[d] learn in silence with full submission. [12]I permit no woman[d] to teach or to have authority over a man;[e] she is to keep silent. [13]For Adam was formed first, then Eve; [14]and Adam was not deceived, but the woman was deceived and became a transgressor. [15]Yet she will be saved through childbearing, provided they continue in faith and love and holiness, with modesty.

3 The saying is sure:[f] whoever aspires to the office of bishop[g] desires a no-

ble task. [2]Now a bishop[h] must be above reproach, married only once,[i] temperate, sensible, respectable, hospitable, an apt teacher, [3]not a drunkard, not violent but gentle, not quarrelsome, and not a lover of money. [4]He must manage his own household well, keeping his children submissive and respectful in every way— [5]for if someone does not know how to manage his own household, how can he take care of God's church? [6]He must not be a recent convert, or he may be puffed up with conceit and fall into the condemnation of the devil. [7]Moreover, he must be well thought of by outsiders, so that he may not fall into disgrace and the snare of the devil.

[8]Deacons likewise must be serious, not double-tongued, not indulging in much wine, not greedy for money; [9]they must hold fast to the mystery of the faith with a clear conscience. [10]And let them first be tested; then, if they prove themselves blameless, let them serve as deacons. [11]Women[j] likewise must be serious, not slanderers, but temperate, faithful in all things. [12]Let deacons be married only once,[k] and let them manage their children and their households well; [13]for those who serve well as deacons gain a good standing for themselves and great boldness in the faith that is in Christ Jesus.

---

c Other ancient authorities add *in Christ*
d Or *wife*   e Or *her husband*   f Some interpreters place these words at the end of the previous paragraph. Other ancient authorities read *The saying is commonly accepted*
g Or *overseer*   h Or *an overseer*   i Gk the *husband of one wife*   j Or *Their wives*, or *Women deacons*   k Gk *be husbands of one wife*

The Christian prays even for bad rulers (Rom 13.1). *A quiet and peaceable life,* a phrase that expresses the ideal of good citizenship. **4:** One of the strongest affirmations of the universality of God's grace. **5:** *Mediator,* Heb 9.15; 12.24. **6:** *Ransom,* Mt 20.28; Mk 10.45. **2.8:** *Lifting up holy hands,* a common posture for prayer (Ps 141.2). **9:** *Suitable clothing,* 1 Pet 3.3–6. **11:** *In silence,* 1 Cor 14.34–35. **12:** 1 Cor 11.2–16; compare Eph 5.22–33. **13:** Gen 2.7, 21–22. **14:** Gen 3.1–6. **15:** This much

debated verse has also been translated (*a*) "she will be saved through the birth of the Child" [referring to Jesus Christ], or (*b*) "she will be brought safely through childbirth."

3.1–16: Problems of administration. 1–7: The qualifications of a bishop (literally "overseer," note *g*). **2:** *Hospitable,* see 3 Jn 5 n. **6:** The devil accuses believers before God's judgment seat (Job 1.6–11; Rev 12.10).

3.8–13: The qualifications of deacons (compare Acts 6.1–6). **11:** Women also shared

14 I hope to come to you soon, but I am writing these instructions to you so that, 15if I am delayed, you may know how one ought to behave in the household of God, which is the church of the living God, the pillar and bulwark of the truth. 16Without any doubt, the mystery of our religion is great:

He[l] was revealed in flesh,
vindicated[m] in spirit,[n]
seen by angels,
proclaimed among Gentiles,
believed in throughout the
world,
taken up in glory.

4 Now the Spirit expressly says that in later[o] times some will renounce the faith by paying attention to deceitful spirits and teachings of demons, 2through the hypocrisy of liars whose consciences are seared with a hot iron. 3They forbid marriage and demand abstinence from foods, which God created to be received with thanksgiving by those who believe and know the truth. 4For everything created by God is good, and nothing is to be rejected, provided it is received with thanksgiving; 5for it is sanctified by God's word and by prayer.

6 If you put these instructions before the brothers and sisters,[p] you will be a good servant[q] of Christ Jesus, nourished on the words of the faith and of the sound teaching that you have followed. 7Have nothing to do with profane myths and old wives' tales. Train yourself in godliness, 8for, while physical training is of some value, godliness is valuable in every way, holding promise for both the present life and the life to come. 9The saying is sure and worthy of full acceptance. 10For to this end we toil and struggle,[r] because we have our hope set on the living God, who is the Savior of all people, especially of those who believe.

11 These are the things you must insist on and teach. 12Let no one despise your youth, but set the believers an example in speech and conduct, in love, in faith, in purity. 13Until I arrive, give attention to the public reading of scripture,[s] to exhorting, to teaching. 14Do not neglect the gift that is in you, which was given to you through prophecy with the laying on of hands by the council of elders.[t] 15Put these things into practice, devote yourself to them, so that all may see your progress. 16Pay close attention to yourself and to your teaching; continue in these things, for in doing this you will save both yourself and your hearers.

5 Do not speak harshly to an older man,[u] but speak to him as to a fa-

---

l Gk *Who*; other ancient authorities read *God*; others, *Which*   m Or *justified*   n Or *by the Spirit*   o Or *the last*   p Gk *brothers*
q Or *deacon*   r Other ancient authorities read *suffer reproach*   s Gk *to the reading*   t Gk *by the presbytery*   u Or *an elder, or a presbyter*

---

in the work of the deacons (Rom 16.1). **14–16:** Practical problems in perspective; administration is not a necessary evil but a spiritual task in the household of God (1 Cor 12.28). **16:** Christ is the content of *the mystery.* What follows the colon may be a quotation from an early Christian hymn.

4.1–16: False teachers. 1: *The* Holy *Spirit,* speaking through New Testament prophets (compare Acts 20.23; 21.11). *Later times,* that is, either times later than Paul, or else (see note *o*) the period before the day of the Lord (Mt 24.24; 1 Jn 2.18). *Deceitful spirits,* see 1 Jn 4.1 n.; 2 Jn 7. *Teachings of demons,* Eph 6.12; Jas 3.15; Rev 16.13–14. **3–5:** Gnostic teachers forbade legitimate *marriage* and certain *foods* (Col 2.16). The author's reply stresses the goodness of God's creation (Titus 1.15); since, however, the creation can be misused, *everything* should be *sanctified by God's word and by prayer.*

4.7: *Myths and old wives' tales,* foolish stories and speculations (1.4; Titus 1.14). **13:** The church adopted many liturgical practices of the synagogue, including *public reading of scripture, exhorting, teaching.* **14:** The laying on of hands (see 5.22 n.) is not an empty gesture but expresses the donation and reception of a gift; e.g. Jesus blesses children (Mk 10.16), he heals with a touch (Mk 6.5); the Spirit is given to the baptized (Acts 8.17; 19.6); believers are set aside for special tasks in the church (Acts 6.6; 13.3).

ther, to younger men as brothers, ²to older women as mothers, to younger women as sisters—with absolute purity.

3 Honor widows who are really widows. ⁴If a widow has children or grandchildren, they should first learn their religious duty to their own family and make some repayment to their parents; for this is pleasing in God's sight. ⁵The real widow, left alone, has set her hope on God and continues in supplications and prayers night and day; ⁶but the widow*ᵛ* who lives for pleasure is dead even while she lives. ⁷Give these commands as well, so that they may be above reproach. ⁸And whoever does not provide for relatives, and especially for family members, has denied the faith and is worse than an unbeliever.

9 Let a widow be put on the list if she is not less than sixty years old and has been married only once;*ʷ* ¹⁰she must be well attested for her good works, as one who has brought up children, shown hospitality, washed the saints' feet, helped the afflicted, and devoted herself to doing good in every way. ¹¹But refuse to put younger widows on the list; for when their sensual desires alienate them from Christ, they want to marry, ¹²and so they incur condemnation for having violated their first pledge. ¹³Besides that, they learn to be idle, gadding about from house to house; and they are not merely idle, but also gossips and busybodies, saying what they should not say. ¹⁴So I would have younger widows marry, bear children, and manage their households, so as to give the adversary no occasion to revile us. ¹⁵For some have already turned away to follow Satan. ¹⁶If any believing woman*ˣ* has relatives who are really widows, let her assist them; let the church not be burdened, so that it can assist those who are real widows.

17 Let the elders who rule well be considered worthy of double honor,*ʸ* especially those who labor in preaching and teaching; ¹⁸for the scripture says, "You shall not muzzle an ox while it is treading out the grain," and, "The laborer deserves to be paid." ¹⁹Never accept any accusation against an elder except on the evidence of two or three witnesses. ²⁰As for those who persist in sin, rebuke them in the presence of all, so that the rest also may stand in fear. ²¹In the presence of God and of Christ Jesus and of the elect angels, I warn you to keep these instructions without prejudice, doing nothing on the basis of partiality. ²²Do not ordain*ᶻ* anyone hastily, and do not participate in the sins of others; keep yourself pure.

23 No longer drink only water, but take a little wine for the sake of your stomach and your frequent ailments.

24 The sins of some people are conspicuous and precede them to judgment, while the sins of others follow them there. ²⁵So also good works are conspicuous; and even when they are not, they cannot remain hidden.

6 Let all who are under the yoke of slavery regard their masters as worthy of all honor, so that the name of God

---

v Gk *she*   *w* Gk *the wife of one husband*
x Other ancient authorities read *believing man or woman*; others, *believing man*
y Or *compensation*   *z* Gk *Do not lay hands on*

5.1–6.2: The pastor and the flock. 5.3–16: Three classes of *widows* are mentioned: (a) *real* widows are older women who depend upon the church for support (Acts 6.1); (b) those widows *put on the list* are Christian workers whose qualifications are detailed; they pledge themselves to the service of Christ; (c) *younger* widows are encouraged to remarry. *Put on the list,* so as to receive communal assistance. **3:** *Widows who are really widows* are not to be treated as burdens but with honor.

5.17–22: *Elders,* see Titus 1.5–7 n. **18:** Deut 25.4; Mt 10.10; Lk 10.7; 1 Cor 9.9–14. **19:** *Two or three witnesses,* Deut 19.15; Mt 18.16. **21:** *Elect angels,* chosen by God as his ministers (see 1 Cor 11.10 n.; Heb 1.14 n.). **22:** *Ordain* is literally "lay hands on" (see 4.14). Here the phrase may refer to the restoration of penitents after discipline.

6.1–2: Slaves and masters (1 Cor 7.21–23; Eph 6.5–9), see Introduction to Philemon.

6.3–21: Final directions. 3–5: *Sound words,* the apostolic testimony. The false

and the teaching may not be blasphemed. ²Those who have believing masters must not be disrespectful to them on the ground that they are members of the church;ᵃ rather they must serve them all the more, since those who benefit by their service are believers and beloved.ᵇ

Teach and urge these duties. ³Whoever teaches otherwise and does not agree with the sound words of our Lord Jesus Christ and the teaching that is in accordance with godliness, ⁴is conceited, understanding nothing, and has a morbid craving for controversy and for disputes about words. From these come envy, dissension, slander, base suspicions, ⁵and wrangling among those who are depraved in mind and bereft of the truth, imagining that godliness is a means of gain.ᶜ ⁶Of course, there is great gain in godliness combined with contentment; ⁷for we brought nothing into the world, so thatᵈ we can take nothing out of it; ⁸but if we have food and clothing, we will be content with these. ⁹But those who want to be rich fall into temptation and are trapped by many senseless and harmful desires that plunge people into ruin and destruction. ¹⁰For the love of money is a root of all kinds of evil, and in their eagerness to be rich some have wandered away from the faith and pierced themselves with many pains.

11 But as for you, man of God, shun all this; pursue righteousness, godliness, faith, love, endurance, gentleness. ¹²Fight the good fight of the faith; take hold of the eternal life, to which you were called and for which you madeᵉ the good confession in the presence of many witnesses. ¹³In the presence of God, who gives life to all things, and of Christ Jesus, who in his testimony before Pontius Pilate made the good confession, I charge you ¹⁴to keep the commandment without spot or blame until the manifestation of our Lord Jesus Christ, ¹⁵which he will bring about at the right time—he who is the blessed and only Sovereign, the King of kings and Lord of lords. ¹⁶It is he alone who has immortality and dwells in unapproachable light, whom no one has ever seen or can see; to him be honor and eternal dominion. Amen.

17 As for those who in the present age are rich, command them not to be haughty, or to set their hopes on the uncertainty of riches, but rather on God who richly provides us with everything for our enjoyment. ¹⁸They are to do good, to be rich in good works, generous, and ready to share, ¹⁹thus storing up for themselves the treasure of a good foundation for the future, so that they may take hold of the life that really is life.

20 Timothy, guard what has been entrusted to you. Avoid the profane chatter and contradictions of what is falsely called knowledge; ²¹by professing it some have missed the mark as regards the faith.

Grace be with you.ᶠ

a Gk *are brothers*   b Or *since they are believers and beloved, who devote themselves to good deeds*   c Other ancient authorities add *Withdraw yourself from such people*   d Other ancient authorities read *world—it is certain that*   e Gk *confessed*   f The Greek word for *you* here is plural; in other ancient authorities it is singular. Other ancient authorities add *Amen*

---

teachers are conceited, contentious, and greedy. **6**: 2 Cor 9.8. **10**: Frequently misquoted: it is the *love of money* which is a *root of all kinds of evil*.

6.12: *Confession,* a word from the language of worship, signifying adoration and praise of God; in baptism, or before the Roman tribunal, the believer praises God by affirming that Jesus is Lord. **14**: *The commandment,* here probably synonymous with the Christian way of life. **16**: God *alone . . . has immortality,* 1.17; Jn 5.26. *Unapproachable light,* Ps 104.2. *No one . . . . can see* God (Jn 1.18; 1 Jn 4.12), but the Son can and will reveal him (Mt 11.27; 1 Jn 3.2; Rev 22.4).

6.17–19: Warnings to the wealthy. **20**: Gnostic pretensions are dealt a final blow in the reference to *what is falsely called knowledge.*

The Second Letter of Paul to
Timothy

From the Book of Acts we learn that Timothy was from Lystra in Asia Minor, and was the son of a Greek father and a Jewish mother who had become a Christian (Acts 16.1), and in Acts Timothy is mentioned as a companion of Paul in his travels. According to 2 Tim 1.5; 3.15, Timothy had known the Hebrew Scriptures from his childhood, and through his mother's and grandmother's influence he had become a Christian before Paul's arrival. In several of Paul's letters Timothy is mentioned as a trusted companion and fellow worker (Rom 16.21; 1 Cor 4.17, etc.) or he is joined with Paul in the greeting of the letter (2 Cor 1.1; Phil 1.1, etc.).

Second Timothy is the most personal of the Pastoral Letters; most of it is directed specifically to Timothy. By contrast, most of Titus and 1 Timothy consist of general teaching for the congregation. This letter, however, presents Paul speaking directly to Timothy, and thus of all the Pastorals it has the best claim to direct Pauline authorship or to having within it substantial fragments that Paul had written to Timothy. It must be recognized, furthermore, that in this period the form of the personal letter (along with more general forms of instruction) was used pseudonymously to give teaching that could be applied beyond the situation of a particular recipient. If it was written in the generation after Paul, it presents in Paul's name a set of instructions for church leaders. The personal cast of the letter may thus reflect either direct or indirect authorship by Paul, but in either case it is a presentation of Paul's teaching to the churches. See the Introduction to 1 Timothy for further discussion of date and authorship.

Second Timothy is an earnest pastoral letter from a veteran missionary to a younger colleague. Like the other Pastorals, it presents a teaching suitable for continuing life in the world, though it does refer to the "last days" (3.1). It pictures Timothy as responsible for a group of churches and for preserving them from destructive influences from without and from dissidents within. Endurance is a prime quality for such a leader. Timothy is urged to rekindle the gift of God within him (1.3–7); he is not to be ashamed of witnessing to the Lord (1.8–18); and is to take his share of suffering as a good soldier of Jesus Christ (2.1–13). As he encounters false teachers with their "profane chatter," he must endeavor to be a sound worker, handling the word rightly (2.14–19), and to purify himself from what is ignoble, to be a vessel fit for the Master's use (2.20–26). In this he can be helped by the example of Paul (3.10–17), who is now at the end of his career and awaits the crown of righteousness. The farewell words (4.6–8) are a moving testimony of Christian fortitude and hope in the face of certain martyrdom.

1 Paul, an apostle of Christ Jesus by the will of God, for the sake of the promise of life that is in Christ Jesus, 2 To Timothy, my beloved child:

Grace, mercy, and peace from God the Father and Christ Jesus our Lord.

3 I am grateful to God—whom I worship with a clear conscience, as my ancestors did—when I remember you constantly in my prayers night and day. 4 Recalling your tears, I long to see you so that I may be filled with joy. 5 I am reminded of your sincere faith, a faith that lived first in your grandmother Lois and your mother Eunice and now, I am sure, lives in you. 6 For this reason I remind you to rekindle the gift of God that is within you through the laying on of my hands; 7 for God did not give us a spirit of cowardice, but rather a spirit of power and of love and of self-discipline.

8 Do not be ashamed, then, of the testimony about our Lord or of me his prisoner, but join with me in suffering for the gospel, relying on the power of God, 9 who saved us and called us with a holy calling, not according to our works but according to his own purpose and grace. This grace was given to us in Christ Jesus before the ages began, 10 but it has now been revealed through the appearing of our Savior Christ Jesus, who abolished death and brought life and immortality to light through the gospel. 11 For this gospel I was appointed a herald and an apostle and a teacher, *a* 12 and for this reason I suffer as I do. But I am not ashamed, for I know the one in whom I have put my trust, and I am sure that he is able to guard until that day what I have entrusted to him. *b* 13 Hold to the standard of sound teaching that you have heard from me, in the faith and love that are in Christ Jesus. 14 Guard the good treasure entrusted to you, with the help of the Holy Spirit living in us.

15 You are aware that all who are in Asia have turned away from me, including Phygelus and Hermogenes. 16 May the Lord grant mercy to the household of Onesiphorus, because he often refreshed me and was not ashamed of my chain; 17 when he arrived in Rome, he eagerly *c* searched for me and found me 18—may the Lord grant that he will find mercy from the Lord on that day! And you know very well how much service he rendered in Ephesus.

2 You then, my child, be strong in the grace that is in Christ Jesus; 2 and what you have heard from me through many witnesses entrust to faithful people who will be able to teach others as well. 3 Share in suffering like a good soldier of Christ Jesus. 4 No one serving in the army gets entangled in everyday affairs; the soldier's aim is to please the enlisting officer. 5 And in the case of an athlete, no one is crowned without competing according to the rules. 6 It is the farmer who

a Other ancient authorities add *of the Gentiles*
b Or *what has been entrusted to me*
c Or *promptly*

1.1–2: Salutation, see Rom 1.1–7 n. *Grace . . .* see 2 Thess 1.2 n. and 1 Tim 1.2 n. **1.3–7: Thanksgiving and exhortation,** see Rom 1.8–15 n. **5:** Timothy's family, see Introduction. **6:** God does not regret or withdraw his gifts, but they can become ineffective through one's neglect of them; hence Timothy is encouraged to *rekindle the gift of God. Laying on . . . hands,* see 1 Tim 4.14 n. **7:** Throughout the letter Timothy is exhorted to courage and endurance, which are possibilities, not because of native human qualities, but through the gift of the Holy Spirit (Rom 8.15; Eph 1.17). **1.8–2.13: An appeal to show courage.**

Timothy was apparently overawed by his surroundings and did not make his witness boldly. **1.9:** In the gospel God invites men and women to become his own, not as a reward of their accomplishments, but because he is gracious (Eph 1.3–4). **10:** *The appearing,* the Incarnation of Christ. *Abolished death,* Rom 6.9; 8.2; Heb 2.14–15. *Life and immortality,* immortal life (Rom 8.11). **12:** *That day,* the second coming of Christ (v. 18; 2 Thess 1.10). **13:** *The standard of sound teaching,* the apostolic testimony to the gospel. **1.15:** *Asia,* the Roman province of that name in Asia Minor. **18:** *That day,* see v. 12 n. **2.3–7:** Three illustrations (*soldier, athlete,*

does the work who ought to have the first share of the crops. 7 Think over what I say, for the Lord will give you understanding in all things.

8 Remember Jesus Christ, raised from the dead, a descendant of David—that is my gospel, 9 for which I suffer hardship, even to the point of being chained like a criminal. But the word of God is not chained. 10 Therefore I endure everything for the sake of the elect, so that they may also obtain the salvation that is in Christ Jesus, with eternal glory. 11 The saying is sure:

If we have died with him, we will
    also live with him;
12 if we endure, we will also reign
    with him;
if we deny him, he will also
    deny us;
13 if we are faithless, he remains
    faithful—
for he cannot deny himself.

14 Remind them of this, and warn them before God*d* that they are to avoid wrangling over words, which does no good but only ruins those who are listening. 15 Do your best to present yourself to God as one approved by him, a worker who has no need to be ashamed, rightly explaining the word of truth. 16 Avoid profane chatter, for it will lead people into more and more impiety, 17 and their talk will spread like gangrene. Among them are Hymenaeus and Philetus, 18 who have swerved from the truth by claiming that the resurrection has already taken place. They are upsetting the faith of some. 19 But God's firm foundation stands, bearing this inscription: "The

Lord knows those who are his," and, "Let everyone who calls on the name of the Lord turn away from wickedness."

20 In a large house there are utensils not only of gold and silver but also of wood and clay, some for special use, some for ordinary. 21 All who cleanse themselves of the things I have mentioned*e* will become special utensils, dedicated and useful to the owner of the house, ready for every good work. 22 Shun youthful passions and pursue righteousness, faith, love, and peace, along with those who call on the Lord from a pure heart. 23 Have nothing to do with stupid and senseless controversies; you know that they breed quarrels. 24 And the Lord's servant*f* must not be quarrelsome but kindly to everyone, an apt teacher, patient, 25 correcting opponents with gentleness. God may perhaps grant that they will repent and come to know the truth, 26 and that they may escape from the snare of the devil, having been held captive by him to do his will.*g*

3 You must understand this, that in the last days distressing times will come. 2 For people will be lovers of themselves, lovers of money, boasters, arrogant, abusive, disobedient to their parents, ungrateful, unholy, 3 inhuman, implacable, slanderers, profligates, brutes, haters of good, 4 treacherous, reckless, swollen with conceit, lovers of pleasure rather than lovers of God, 5 holding to the outward form of godli-

d Other ancient authorities read *the Lord*
e Gk *of these things*   *f* Gk *slave*   *g* Or *by him, to do his* (that is, God's) *will*

---

farmer) having the same point: hold nothing back from commitment to your work. **9:** *The word ... is not chained*, Phil 1.12–14.
2.11: *The saying* is rhythmical in structure; it may be from an early Christian hymn. *We will also live with him*, Rom 6.5–11; Gal 6.14; Col 2.12. **12:** *If we deny him*, Mt 10.32–33; Mk 8.38; Lk 12.9. **13:** Rom 3.3–4.
2.14–4.5: The pastor and the flock. 2.15: The task of preaching demands work and discipline. **17:** *Hymenaeus*, 1 Tim 1.20. **19:** *Inscription*, involving quotations from Num 16.5

and Isa 26.13, one bearing on predestination, the other on free will.
2.20: *In a large house*, such as the church is (compare Rom 9.19–24). **25:** *Correcting with gentleness*, nothing is gained by becoming angry; patience and avoidance of controversy may lead some to repentance.
3.1: *The last days*, which have already begun (compare v. 5; Acts 2.16–17). *Distressing times*, Mt 24.4–5; Mk 13.22; 2 Thess 2.3–12. **2:** Rom 1.29–31. **5:** Having the outward *form*, but lacking the vital *power* (1 Cor 2.5; 4.19–

ness but denying its power. Avoid them! ⁶For among them are those who make their way into households and captivate silly women, overwhelmed by their sins and swayed by all kinds of desires, ⁷who are always being instructed and can never arrive at a knowledge of the truth. ⁸As Jannes and Jambres opposed Moses, so these people, of corrupt mind and counterfeit faith, also oppose the truth. ⁹But they will not make much progress, because, as in the case of those two men, *ʰ* their folly will become plain to everyone.

10 Now you have observed my teaching, my conduct, my aim in life, my faith, my patience, my love, my steadfastness, ¹¹my persecutions and suffering the things that happened to me in Antioch, Iconium, and Lystra. What persecutions I endured! Yet the Lord rescued me from all of them. ¹²Indeed, all who want to live a godly life in Christ Jesus will be persecuted. ¹³But wicked people and impostors will go from bad to worse, deceiving others and being deceived. ¹⁴But as for you, continue in what you have learned and firmly believed, knowing from whom you learned it, ¹⁵and how from childhood you have known the sacred writings that are able to instruct you for salvation through faith in Christ Jesus. ¹⁶All scripture is inspired by God and is *ⁱ* useful for teaching, for reproof, for correction, and for training in righteousness, ¹⁷so that everyone who belongs to God may be proficient, equipped for every good work.

4 In the presence of God and of Christ Jesus, who is to judge the living and the dead, and in view of his appearing and his kingdom, I solemnly urge you: ²proclaim the message; be persistent whether the time is favorable or unfavorable; convince, rebuke, and encourage, with the utmost patience in teaching. ³For the time is coming when people will not put up with sound doctrine, but having itching ears, they will accumulate for themselves teachers to suit their own desires, ⁴and will turn away from listening to the truth and wander away to myths. ⁵As for you, always be sober, endure suffering, do the work of an evangelist, carry out your ministry fully.

6 As for me, I am already being poured out as a libation, and the time of my departure has come. ⁷I have fought the good fight, I have finished the race, I have kept the faith. ⁸From now on there is reserved for me the crown of righteousness, which the Lord, the righteous judge, will give me on that day, and not only to me but also to all who have longed for his appearing.

9 Do your best to come to me soon, ¹⁰for Demas, in love with this present world, has deserted me and gone to Thessalonica; Crescens has gone to Galatia, *ʲ* Titus to Dalmatia. ¹¹Only Luke is with me. Get Mark and bring him with you, for he is useful in my ministry. ¹²I have sent Tychicus to Ephesus. ¹³When you come, bring the cloak that I left with Carpus at Troas, also the books, and above all the parchments. ¹⁴Alexander the coppersmith did me great harm; the Lord will pay him back for his deeds.

ʰ Gk lacks *two men inspired by God is also* authorities read *Gaul*
ⁱ Or *Every scripture*
ʲ Other ancient

20; 1 Thess 1.5; Titus 1.16). **6–7:** Love of novelty was a feature of life in ancient times also.

3.8: The names *Jannes* and *Jambres* are not given in Ex 7.11, but are supplied by Jewish tradition. **11:** Acts 13.14–52; 14.1–20; 16.1–5; 2 Cor 6.4–10. **15:** *The sacred writings,* the books of the Old Testament, which the church interpreted as pointing to Jesus Christ. **16:** *All scripture,* the Old Testament. *Inspired by God,* compare Gen 2.7 where God breathes life into the first human being. **17:** *Every good work,* including those mentioned in v. 16.

4.1: *His appearing,* Christ's second coming. *His kingdom,* the rule of Christ as king. **3:** 3.1. **4:** *Myths,* 1 Tim 1.4; 4.7. **5:** *Evangelist,* Acts 21.8; Eph 4.11.

4.6–22: **Concluding exhortations. 8:** Among Jews crowns or wreaths of leaves or of flowers were worn as symbols of joy and honor at feasts and weddings. The Greeks gave wreaths to the winners of athletic con-

15 You also must beware of him, for he strongly opposed our message.

16 At my first defense no one came to my support, but all deserted me. May it not be counted against them! 17 But the Lord stood by me and gave me strength, so that through me the message might be fully proclaimed and all the Gentiles might hear it. So I was rescued from the lion's mouth. 18 The Lord will rescue me from every evil attack and save me for his heavenly kingdom. To him be the glory forever and ever. Amen.

19 Greet Prisca and Aquila, and the household of Onesiphorus. 20 Erastus remained in Corinth; Trophimus I left ill in Miletus. 21 Do your best to come before winter. Eubulus sends greetings to you, as do Pudens and Linus and Claudia and all the brothers and sisters. *k*

22 The Lord be with your spirit. Grace be with you. *l*

k Gk *all the brothers*   *l* The Greek word for *you* here is plural. Other ancient authorities add *Amen*

---

tests (1 Cor 9.25). **14:** *Alexander,* Acts 19.33; 1 Tim 1.20.

4.17: *The lion's mouth,* a common Old Testament metaphor for a violent death (Ps 7.2; 17.12; 22.21). **19:** *Prisca and Aquila,* Acts 18.2, 18; Rom 16.3; 1 Cor 16.19. *Onesiphorus,* 1.16–17. **20:** *Trophimus,* Acts 20.4; 21.29. **21:** *Winter,* when navigation was suspended.

The Letter of Paul to
Titus

Although he is not mentioned in the Acts of the Apostles, Titus appears as an important companion of Paul in Galatians and in 2 Corinthians. In Gal 2.1–3 we learn that Titus accompanied Paul on the visit to Jerusalem when Paul's apostleship was recognized by the Jerusalem leaders (see also Acts 15.2, "some of the others" went with Paul and Barnabas). Titus was the focus of a vigorous debate about circumcision (since as a non-Jew he had not been circumcised), and Paul resisted all efforts made in demanding that he be circumcised.

After Paul had established the church in Corinth, and while he was absent, he twice sent Titus there in his effort to reestablish his leadership of that church. Titus's mission was important in effecting a reconciliation between Paul and the church in Corinth (2 Cor 7.6–7, 13–14; 8.6, 16–17; see the Introduction to 2 Corinthians). If Paul and Titus traveled to Crete as the letter to Titus presupposes, this may have been as part of the visit to Greece mentioned in Acts 20.3, or at a later time if Paul was released from the Roman imprisonment with which Acts 28 closes. The author of 2 Timothy mentions (4.10) that Titus had gone to Dalmatia, a region on the eastern side of the Adriatic Sea. In the present letter, Paul wrote to Titus giving directions for the administration of the new churches of Crete. If the author of Titus was writing in Paul's name, he drew upon the association of Paul and Titus to frame a series of directions for missionary supervision of new churches. (On date and authorship, see the Introduction to 1 Timothy).

The teaching of this letter is strongly parallel to that of 1 Timothy, and like that letter, it is principally made up of general teaching rather than being directed to an individual person. The Letter to Titus has three main topics, corresponding to its chapter divisions. The first (ch 1) sets forth what is required of church leaders ("elders" or "bishops") in the face of various false teachers and local shortcomings. As in both 1 and 2 Timothy, wrong teaching is seen to result in immoral behavior. The second division suggests the proper approach to different groups in the church: older men (2.1–2), older women (2.3–5), younger men (2.6–8), and slaves (2.9–10); the emphasis on the submissiveness of women and slaves is in accord with the patriarchal tradition of the time. This section concludes with a summary of what is expected of believers in view of God's grace (2.11–15). The third division (ch 3) unfolds this ethical program in terms that presuppose the continuing life of the church in the world and the responsibility of believers to be good citizens, admonishing the readers to avoid hatred and wrangling and to manifest the meekness, gentleness, obedience, and courtesy made possible by God's mercy in Christ.

1 Paul, a servant*a* of God and an apostle of Jesus Christ, for the sake of the faith of God's elect and the knowledge of the truth that is in accordance with godliness, 2 in the hope of eternal life that God, who never lies, promised before the ages began— 3 in due time he revealed his word through the proclamation with which I have been entrusted by the command of God our Savior,

4 To Titus, my loyal child in the faith we share:

Grace*b* and peace from God the Father and Christ Jesus our Savior.

5 I left you behind in Crete for this reason, so that you should put in order what remained to be done, and should appoint elders in every town, as I directed you: 6 someone who is blameless, married only once,*c* whose children are believers, not accused of debauchery and not rebellious. 7 For a bishop,*d* as God's steward, must be blameless; he must not be arrogant or quick-tempered or addicted to wine or violent or greedy for gain; 8 but he must be hospitable, a lover of goodness, prudent, upright, devout, and self-controlled. 9 He must have a firm grasp of the word that is trustworthy in accordance with the teaching, so that he may be able both to preach with sound doctrine and to refute those who contradict it.

10 There are also many rebellious people, idle talkers and deceivers, especially those of the circumcision; 11 they must be silenced, since they are upsetting whole families by teaching for sordid gain what it is not right to teach. 12 It was one of them, their very own prophet, who said,

"Cretans are always liars, vicious
brutes, lazy gluttons."

13 That testimony is true. For this reason rebuke them sharply, so that they may become sound in the faith, 14 not paying attention to Jewish myths or to commandments of those who reject the truth. 15 To the pure all things are pure, but to the corrupt and unbelieving nothing is pure. Their very minds and consciences are corrupted. 16 They profess to know God, but they deny him by their actions. They are detestable, disobedient, unfit for any good work.

2 But as for you, teach what is consistent with sound doctrine. 2 Tell the older men to be temperate, serious, prudent, and sound in faith, in love, and in endurance.

3 Likewise, tell the older women to be reverent in behavior, not to be slanderers or slaves to drink; they are to teach what is good, 4 so that they may encourage the young women to love their husbands, to love their children, 5 to be self-controlled, chaste, good managers of the household, kind, being submissive to their husbands, so that the word of God may not be discredited.

6 Likewise, urge the younger men to be self-controlled. 7 Show yourself in all

a Gk *slave*   *b* Other ancient authorities read
Grace, mercy,   *c* Gk *husband of one wife*
d Or *an overseer*

1.1–4: Salutation, see Rom 1.1–7 n. **1:** The church is *God's elect,* the chosen people. **2–3:** The good news of Christ was *promised before the ages began* (Lk 24.44; Acts 3.18; Rom 1.2) and revealed in *due time* (Mk 1.15; Gal 4.4); the stress is on God's initiative and sovereignty. *Who never lies,* contrast v. 12. **4:** *Grace and peace,* see 2 Thess 1.2 n.

1.5–16: Administration of the church in Crete. **5:** *Elder* and *bishop* (v. 7) are two terms for the same officer in the church. **7:** The *bishop* is pictured as *God's steward,* the servant who manages the affairs of a large household. **1.10:** *Those of the circumcision* maintained

that Christians are subject to the law of Moses and to Jewish traditions (Acts 15.1–29; see Introduction to Galatians). **12:** Quoted from the Cretan poet Epimenides (around 600 B.C.). In ancient parlance "to Cretanize" was to be a liar. **14:** *Jewish myths,* see 1 Tim 1.4 n.

2.1–3.11: The pastor and the flock. Advice is given on relationships to *older men* (v. 2), *older women* (vv. 3–5), *younger men* (vv. 6–8), *slaves* (vv. 9–10). **5:** *Submissive to their husbands,* see Eph 5.21 n. **7:** Titus himself is to set the example.

2.9: *Slaves,* 1 Cor 7.21–23; Eph 6.5–8; Col

respects a model of good works, and in your teaching show integrity, gravity, [8] and sound speech that cannot be censured; then any opponent will be put to shame, having nothing evil to say of us.

9 Tell slaves to be submissive to their masters and to give satisfaction in every respect; they are not to talk back, [10] not to pilfer, but to show complete and perfect fidelity, so that in everything they may be an ornament to the doctrine of God our Savior.

11 For the grace of God has appeared, bringing salvation to all,[e] [12] training us to renounce impiety and worldly passions, and in the present age to live lives that are self-controlled, upright, and godly, [13] while we wait for the blessed hope and the manifestation of the glory of our great God and Savior,[f] Jesus Christ. [14] He it is who gave himself for us that he might redeem us from all iniquity and purify for himself a people of his own who are zealous for good deeds.

15 Declare these things; exhort and reprove with all authority.[g] Let no one look down on you.

3 Remind them to be subject to rulers and authorities, to be obedient, to be ready for every good work, [2] to speak evil of no one, to avoid quarreling, to be gentle, and to show every courtesy to everyone. [3] For we ourselves were once foolish, disobedient, led astray, slaves to various passions and pleasures, passing our days in malice and envy, despicable, hating one another. [4] But when the goodness and loving kindness of God our Savior appeared, [5] he saved us, not because of any works of righteousness that we had done, but according to his mercy, through the water[h] of rebirth and renewal by the Holy Spirit. [6] This Spirit he poured out on us richly through Jesus Christ our Savior, [7] so that, having been justified by his grace, we might become heirs according to the hope of eternal life. [8] The saying is sure.

I desire that you insist on these things, so that those who have come to believe in God may be careful to devote themselves to good works; these things are excellent and profitable to everyone. [9] But avoid stupid controversies, genealogies, dissensions, and quarrels about the law, for they are unprofitable and worthless. [10] After a first and second admonition, have nothing more to do with anyone who causes divisions, [11] since you know that such a person is perverted and sinful, being self-condemned.

12 When I send Artemas to you, or Tychicus, do your best to come to me at Nicopolis, for I have decided to spend the winter there. [13] Make every effort to send Zenas the lawyer and Apollos on their way, and see that they lack nothing. [14] And let people learn to devote themselves to good works in order to meet urgent needs, so that they may not be unproductive.

15 All who are with me send greetings to you. Greet those who love us in the faith.

Grace be with all of you.[i]

e Or has appeared to all, bringing salvation
f Or of the great God and our Savior
g Gk commandment   *h Gk washing*   *i Other*
ancient authorities add *Amen*

3.22; 1 Tim 6.1–2; see Introduction to Philemon. **14:** *Who gave himself,* Mt 20.28; Mk 10.45; Gal 1.4. To *redeem . . . iniquity,* to ransom us from slavery to sin (Ps 130.8). *Purify . . . a people,* Ezek 37.23; 1 Pet 2.9.

3.1: *Be subject,* see 1 Tim 2.2 n. **5:** *The water of rebirth* combines two pictures descriptive of baptism: the washing away of sins (Acts 22.16; Eph 5.26), and rebirth, the beginning of a new life (Jn 3.5). *Renewal by the Holy Spirit,* the restoration of the relationship to God which was lost through sin (2 Cor 5.17). **9:** *Genealogies,* see 1 Tim 1.4 n. **10:** See 2 Jn 10 n.

3.12–15: Final instructions. 12: *Tychicus,* Acts 20.4; Eph 6.21; Col 4.7; 2 Tim 4.12. **13:** *Apollos,* Acts 18.24; 1 Cor 1.12; 16.12.

The Letter of Paul to
Philemon

The Letter to Philemon was to be delivered by Onesimus (v. 10), a slave who had emancipated himself and whom Paul was sending back to his master, as the law required. Although the letter has this specific purpose, it is not a private letter, for others are addressed as well as Philemon, including the whole church that meets in his house (v. 2). Paul sets the personal message and the question of how to treat the returning slave in the context of the life of the church, and thereby indirectly points to his own authority as apostle (v. 8) and as the one to whom Philemon owed his conversion to Christian faith (v. 19).

Why Onesimus had left is not stated. A slave who had deserted could be severely punished and even put to death. Evidently Onesimus had made his way to Paul, and had become a Christian while with him (v. 16). He had become very "useful" to Paul (v. 11), who suggests that he will now be the same to Philemon (in Greek Onesimus means "useful"); that is, Onesimus has changed and should return to a changed situation.

Paul does not address the general question of slavery as a social institution, nor does he discuss whether or not Onesimus should be set free. He expects Onesimus to be received back with forgiveness into the household of Philemon and at the same time to be received into the community of the church as a full and equal member (a "brother," v. 16). Elsewhere he speaks of slaves and free as one in Christ (Gal 3.28; Col 3.11).

The Letter to Philemon illustrates in a very concrete way how Paul applied his grasp of love or *agape* to the problem between Philemon and Onesimus. Paul's authority stands in the background, and he undertakes to be responsible for whatever Onesimus may owe (vv. 18–19). Yet Paul does not want to appeal to his own authority; he wants the decision to be Philemon's own choice. But Paul expects that the relation between master and slave will be set in the context of the church and transformed by the love that is active there.

Onesimus is identified elsewhere as coming from Colossae (Col 4.9), as is Archippus who is mentioned in the opening greeting of this letter (v. 2; Col 4.17). Hence Philemon and the church in his house were in Colossae. Since most of those who are greeted at the end of the letter are also mentioned in the close of Colossians, it is probable that the two letters were written at nearly the same time, if Paul was the author of Colossians, or that the author of Colossians had Philemon at hand, if Paul was not the author (see Introduction to Colossians).

Paul wrote from prison (vv. 1, 9), perhaps in Rome or Caesarea. But many scholars today think that Ephesus is the most likely to be the place where Paul wrote Philemon.

1 Paul, a prisoner of Christ Jesus, and Timothy our brother, *a*

To Philemon our dear friend and co-worker, [2]to Apphia our sister, *b* to Archippus our fellow soldier, and to the church in your house:

3 Grace to you and peace from God our Father and the Lord Jesus Christ.

4 When I remember you *c* in my prayers, I always thank my God [5]because I hear of your love for all the saints and your faith toward the Lord Jesus. [6]I pray that the sharing of your faith may become effective when you perceive all the good that we *d* may do for Christ. [7]I have indeed received much joy and encouragement from your love, because the hearts of the saints have been refreshed through you, my brother.

8 For this reason, though I am bold enough in Christ to command you to do your duty, [9]yet I would rather appeal to you on the basis of love—and I, Paul, do this as an old man, and now also as a prisoner of Christ Jesus. *e* [10]I am appealing to you for my child, Onesimus, whose father I have become during my imprisonment. [11]Formerly he was useless to you, but now he is indeed useful *f* both to you and to me. [12]I am sending him, that is, my own heart, back to you. [13]I wanted to keep him with me, so that he might be of service to me in your place during my imprisonment for the gospel; [14]but I preferred to do nothing without your consent, in order that your good

deed might be voluntary and not something forced. [15]Perhaps this is the reason he was separated from you for a while, so that you might have him back forever, [16]no longer as a slave but more than a slave, a beloved brother—especially to me but how much more to you, both in the flesh and in the Lord.

17 So if you consider me your partner, welcome him as you would welcome me. [18]If he has wronged you in any way, or owes you anything, charge that to my account. [19]I, Paul, am writing this with my own hand: I will repay it. I say nothing about your owing me even your own self. [20]Yes, brother, let me have this benefit from you in the Lord! Refresh my heart in Christ. [21]Confident of your obedience, I am writing to you, knowing that you will do even more than I say.

22 One thing more—prepare a guest room for me, for I am hoping through your prayers to be restored to you.

23 Epaphras, my fellow prisoner in Christ Jesus, sends greetings to you, *g* [24]and so do Mark, Aristarchus, Demas, and Luke, my fellow workers.

25 The grace of the Lord Jesus Christ be with your spirit. *h*

a Gk *the brother*   *b* Gk *the sister*   *c* From verse 4 through verse 21, *you* is singular   *d* Other ancient authorities read *you* (plural)   *e* Or *as an ambassador of Christ Jesus, and now also his prisoner*   *f* The name Onesimus means *useful* or (compare verse 20) *beneficial*   *g* Here *you* is singular   *h* Other ancient authorities add *Amen*

1–3: Salutation, see Rom 1.1–7 n. **1:** *Prisoner,* see Introduction. **2:** *Sister,* in the faith. *Archippus,* mentioned at the close of the Letter to Colossae (4.17). *The church in your house,* in the first century Christians met for worship in private homes (Rom 16.5; 1 Cor 16.19; Col 4.15). **3:** *Grace . . . peace,* see 2 Thess 1.2 n. **4–7: Thanksgiving,** see Rom 1.8–15 n. **7:** *Refreshed through* Philemon's charitable acts.

8–20: Paul's plea for Onesimus. 10: *My child* in the Christian faith (see Gal 4.19 n.). *Onesimus,* see note *f.* **14:** Although as an apostle Paul had the right to command (v. 8), he leaves the matter to Philemon's free choice.

15: *Was separated,* a tactful expression for "ran away." *Back forever,* suggesting God's providential oversight. **18:** Paul's promise to repay does not necessarily mean that Onesimus had taken anything from Philemon. **21:** *More than I say,* perhaps a gentle hint that Philemon should grant Onesimus his freedom, but more probably a courteous anticipation of Philemon's acceptance of Paul's letter.

21–25: Concluding hopes and greetings. 22: While Philemon's *guest,* Paul will be able to observe how Onesimus has been treated. **23:** *Epaphras,* Col 1.7; 4.12. **24:** All these individuals are referred to in Col 4.10, 14.

The Letter to the
Hebrews

This anonymous treatise combines a theological argument for the finality of salvation, achieved by Christ, with repeated exhortations, directed to those whose faith and practice are weakening, not to abandon that salvation (2.1; 10.32–36; 13.22). With a careful and closely-knit discussion, the unknown author moves with confidence step by step through an elaborate analysis. Three main points are emphasized; (a) The superiority of Jesus Christ to the prophets (1.1–3), to the angels (1.5–2.18), and to Moses himself (3.1–6); (b) The superiority of Christ's priesthood to the levitical priesthood (4.14–7.28); and (c) The superiority of Christ's sacrifice, offered in the heavenly sanctuary, to the many animal sacrifices offered on earth by the levitical priests (8.1–10.39).

Some scholars conclude that the author's argument for the superiority of Christ's sacrifice over those offered by the levitical priests indicates that the recipients were in danger of adopting or returning to Jewish ritual practices, and that therefore the letter must have been composed before the destruction of the Jerusalem temple in A.D. 70 had brought sacrificial ritual to an end. Although this conclusion cannot be definitively ruled out, it is not demanded, for both Jewish and Christian authors write about temple sacrifice after the cultus had ceased. In any case, Hebrews is not referring to the Jerusalem cultus, since the argument in the treatise is developed from references to the tent shrine of Exodus.

By mid-second century Alexandrian exegetes had placed Hebrews among the letters of Paul, though they recognized that it was so different in language and style from the Pauline correspondence that some special account of its authorship was required. Thus Clement thought that Luke had translated a Pauline letter written in Hebrew (though the presence of plays on words in Greek shows that it is not a translation), and Origen held that it was written by an unknown disciple of Paul's. In any case, 2.3–4 suggests that the author comes from a generation after that of the apostles.

The addressees had formerly been subject to persecution, imprisonment, and loss of property as a result of their faith. Having met all those trials with joy, they must continue in faith and good works until the coming of the Lord (10.32–39). This exhortation is followed by an encomium on faith that traces the faithful endurance of all the great heroes of the Hebrew Scriptures, who have their hopes fulfilled in the people redeemed by Christ's sacrifice (11.1–40). Jesus' own life is an example of faithful endurance that Christians should imitate (12.1–11).

The Letter to the Hebrews concludes with an exhortation to remember the exemplary lives of past leaders of the community (13.7) and to submit to its present leaders, who are responsible for the souls of those entrusted to them (13.17). An epistolary closing, modelled on the style of the Pauline letters (travel plans, benediction, appeal and greetings, farewell) ends the treatise (13.20–25). Although the author speaks of the work as a discourse or "word of exhortation" (5.11; 6.1; 13.22), this ending led to the common designation of Hebrews as a "letter."

1 Long ago God spoke to our ancestors in many and various ways by the prophets, 2but in these last days he has spoken to us by a Son,*a* whom he appointed heir of all things, through whom he also created the worlds. 3He is the reflection of God's glory and the exact imprint of God's very being, and he sustains*b* all things by his powerful word. When he had made purification for sins, he sat down at the right hand of the Majesty on high, 4having become as much superior to angels as the name he has inherited is more excellent than theirs.

5 For to which of the angels did God ever say,

"You are my Son;
today I have begotten you"?

Or again,

"I will be his Father,
and he will be my Son"?

6And again, when he brings the firstborn into the world, he says,

"Let all God's angels worship him."

7Of the angels he says,

"He makes his angels winds,
and his servants flames of fire."

8But of the Son he says,

"Your throne, O God,*c* is forever and ever,
and the righteous scepter is the scepter of your*d* kingdom.
9 You have loved righteousness and hated wickedness;
therefore God, your God, has anointed you

with the oil of gladness beyond your companions."

10And,

"In the beginning, Lord, you founded the earth,
and the heavens are the work of your hands;
11 they will perish, but you remain;
they will all wear out like clothing;
12 like a cloak you will roll them up,
and like clothing*e* they will be changed.
But you are the same,
and your years will never end."

13But to which of the angels has he ever said,

"Sit at my right hand
until I make your enemies a footstool for your feet"?

14Are not all angels*f* spirits in the divine service, sent to serve for the sake of those who are to inherit salvation?

2 Therefore we must pay greater attention to what we have heard, so that we do not drift away from it. 2For if the message declared through angels was valid, and every transgression or disobedience received a just penalty, 3how can we escape if we neglect so great a salvation? It was declared at first through the Lord, and it was attested to us by those who heard him, 4while God added his testimony by signs and wonders and

a Or *the Son*   *b* Or *bears along*   *c* Or *God is your throne*   *d* Other ancient authorities read *his*   *e* Other ancient authorities lack *like clothing*   *f* Gk *all of them*

1.1–4: The prologue, a stately declaration of the author's theme, namely the finality of salvation in Christ. Note the contrasts between *long ago* and *in these last days, to our ancestors* and *to us, by the prophets* and *by a Son;* yet it is the same God who spoke then and speaks now. God's *Son,* who is the exact counterpart of the Father (v. 3), not only had part in the creation (Jn 1.1–3), but continues to sustain *all things* (Col 1.15–17). Having accomplished his priestly work of *purification for sins* (v. 3), he was enthroned in divine splendor. Christ's sonship and priesthood are the basis of salvation.

1.5–2.18: The superiority of Christ to angels.
1.5–13: Seven quotations from the Old Testament. **5:** Ps 2.7 and 2 Sam 7.14. **6:** Deut 32.43 Septuagint, compare Ps 97.7. **7:** Ps 104.4. **8–9:** Ps 45.6–7. **10–12:** Ps 102.25–27. **13:** Ps 110.1. **14:** Unlike the divine Son, angels are *spirits,* who are *sent to serve.*
2.1–4: Warning against falling away. The superior status of the Son to angels authenticates his message and lays a greater obligation on those who *heard.* **2:** *The message declared through angels,* see Acts 7.53 n.; Gal 3.19 n.

various miracles, and by gifts of the Holy Spirit, distributed according to his will.

5 Now God[g] did not subject the coming world, about which we are speaking, to angels. [6]But someone has testified somewhere,

"What are human beings that you
are mindful of them,[h]
or mortals, that you care for
them?[i]

[7] You have made them for a little
while lower[j] than the
angels;
you have crowned them with
glory and honor,[k]

[8]   subjecting all things under
their feet."

Now in subjecting all things to them, God[g] left nothing outside their control. As it is, we do not yet see everything in subjection to them, [9]but we do see Jesus, who for a little while was made lower[l] than the angels, now crowned with glory and honor because of the suffering of death, so that by the grace of God[m] he might taste death for everyone.

10 It was fitting that God,[g] for whom and through whom all things exist, in bringing many children to glory, should make the pioneer of their salvation perfect through sufferings. [11]For the one who sanctifies and those who are sanctified all have one Father.[n] For this reason Jesus[g] is not ashamed to call them brothers and sisters,[o] [12]saying,

"I will proclaim your name to my
brothers and sisters,[o]
in the midst of the congregation
I will praise you."
[13]And again,
"I will put my trust in him."
And again,
"Here am I and the children
whom God has given me."

14 Since, therefore, the children share flesh and blood, he himself likewise shared the same things, so that through death he might destroy the one who has the power of death, that is, the devil, [15]and free those who all their lives were held in slavery by the fear of death. [16]For it is clear that he did not come to help angels, but the descendants of Abraham. [17]Therefore he had to become like his brothers and sisters[o] in every respect, so that he might be a merciful and faithful high priest in the service of God, to make a sacrifice of atonement for the sins of the people. [18]Because he himself was tested by what he suffered, he is able to help those who are being tested.

g Gk *he*   h Gk *What is man that you are mindful of him?*   i Gk *or the son of man that you care for him?* In the Hebrew of Psalm 8.4-6 both *man* and *son of man* refer to all humankind
j Or *them only a little lower*   k Other ancient authorities add *and set them over the works of your hands*   l Or *who was made a little lower*
m Other ancient authorities read *apart from God*
n Gk *are all of one*   o Gk *brothers*

2.5–18: The humiliation and exaltation of Jesus. Though Jesus, in his humiliation, was *for a little while . . . lower than the angels* (v. 9), his suffering was temporary and enabled him to *be a merciful and faithful high priest* (v. 17). **6–8:** Quoted from Ps 8.4–6; *for a little while* (v. 7) agrees with the Septuagint, not the Hebrew. **9:** Jesus' exaltation is the consequence of his humiliation (12.2; Phil 2.6–11). *Taste death,* a biblical expression for its bitterness (Isa 51.17; Jer 49.18; Mt 16.28). **2.10–18:** The high destiny of humankind described in Ps 8, is to be attained through the work of Christ. **10:** *It was fitting,* it was in accord with God's gracious plan of salvation. The expression, "to make perfect," is characteristic of this letter (5.9; 7.19, 28; 9.9; 10.1, 14; 11.40; 12.23) and means "to make complete, to bring to maturity." Christ as *pioneer* goes before and points out the way to his followers (12.2); the same Greek word is translated "Author" (Acts 3.15) and "Leader" (Acts 5.31). **11:** *The one who sanctifies,* Jesus Christ; *those who are sanctified,* see 10.10 n. *One Father,* God. **12:** Ps 22.22. **13:** Isa 8.17–18. **14–15:** The eternal Son became a human being in order to overcome the devil and to free humanity from death (Rom 6.23). **17:** As high priest the Son is both sympathetic (*merciful;* compare 5.2–3) and trustworthy (*faithful;* compare 3.2, 6), *to make a sacrifice of atonement* continually for *sins* that bring death and the fear of it to God's *people* (vv. 14–15). **18:** At Gethsemane, and elsewhere, Jesus was *tested* by death on the cross (5.7–8); therefore he can help those who *are being tested* by apostasy.

3 Therefore, brothers and sisters,[p] holy partners in a heavenly calling, consider that Jesus, the apostle and high priest of our confession, [2] was faithful to the one who appointed him, just as Moses also "was faithful in all[q] God's[r] house." [3] Yet Jesus[s] is worthy of more glory than Moses, just as the builder of a house has more honor than the house itself. [4] (For every house is built by someone, but the builder of all things is God.) [5] Now Moses was faithful in all God's[r] house as a servant, to testify to the things that would be spoken later. [6] Christ, however, was faithful over God's[r] house as a son, and we are his house if we hold firm[t] the confidence and the pride that belong to hope.

[7] Therefore, as the Holy Spirit says,
"Today, if you hear his voice,
[8] do not harden your hearts as in
    the rebellion,
  as on the day of testing in the
    wilderness,
[9] where your ancestors put me to
    the test,
  though they had seen my works
    [10] for forty years.
Therefore I was angry with that
    generation,
and I said, 'They always go astray
    in their hearts,
  and they have not known
    my ways.'
[11] As in my anger I swore,
  'They will not enter my rest.'"

[12] Take care, brothers and sisters,[p] that none of you may have an evil, unbelieving heart that turns away from the living God. [13] But exhort one another every day, as long as it is called "today," so that none of you may be hardened by the deceitfulness of sin. [14] For we have become partners of Christ, if only we hold our first confidence firm to the end. [15] As it is said,

"Today, if you hear his voice,
  do not harden your hearts as in
    the rebellion."

[16] Now who were they who heard and yet were rebellious? Was it not all those who left Egypt under the leadership of Moses? [17] But with whom was he angry forty years? Was it not those who sinned, whose bodies fell in the wilderness? [18] And to whom did he swear that they would not enter his rest, if not to those who were disobedient? [19] So we see that they were unable to enter because of unbelief.

4 Therefore, while the promise of entering his rest is still open, let us take care that none of you should seem to have failed to reach it. [2] For indeed the good news came to us just as to them; but the message they heard did not benefit them, because they were not united by

p Gk *brothers*   q Other ancient authorities lack *all*   r Gk *his*   s Gk *this one*   t Other ancient authorities add *to the end*

3.1–6: Christ is superior to Moses. 1: *Holy partners,* the readers are addressed in terms that remind them of their position and responsibility in sharing *in a heavenly calling,* as distinguished from the calling of Israel to an earthly Canaan. As *apostle, Jesus* is God's envoy to humankind; as *high priest,* he is their representative before God. **2:** The author first refers to what Jesus and Moses have in common, namely faithfulness to their respective tasks. **3–5:** As to position, Jesus (a son) is superior to Moses (a servant in God's house; Num 12.7). **6:** *We are his house,* the community of faithful believers.
3.7–4.13: Warning and exhortation, based on the fate of Israel in the wilderness (Ps 95.7–11).

3.8: *Rebellion* and *testing* are translations of the Hebrew names Meribah and Massah in Ps 95.8 (Ex 17.1–7; Num 20.1–13; Deut 33.8). **11:** *My rest,* the peaceful settlement of the promised land of Canaan (4.1–4). **12:** Jer 17.5–6. **13:** *The deceitfulness of sin,* disobedience (Num 14.1–4, 34, 41), leading to apostasy (v. 12). **16–18:** A series of questions (based on Num 14.1–35) suggests the lessons to be drawn from the failure of the wilderness generation.
4.1–13: The rest that God promised. 1: God's *promise,* which must be fulfilled, *remains* for faithful Christians to inherit (v. 6). **2:** *Those who listened,* Num 13.30–14.10. **3:** Ps 95.11. **4:** Gen 2.2. **5:** Ps 95.11. **6:** See v. 1 n. **7–8:** The possession of Canaan under *Joshua*

faith with those who listened." [3] For we who have believed enter that rest, just as God[v] has said,

"As in my anger I swore,
'They shall not enter my rest,'"

though his works were finished at the foundation of the world. [4] For in one place it speaks about the seventh day as follows, "And God rested on the seventh day from all his works." [5] And again in this place it says, "They shall not enter my rest." [6] Since therefore it remains open for some to enter it, and those who formerly received the good news failed to enter because of disobedience, [7] again he sets a certain day—"today"—saying through David much later, in the words already quoted,

"Today, if you hear his voice,
do not harden your hearts."

[8] For if Joshua had given them rest, God[v] would not speak later about another day. [9] So then, a sabbath rest still remains for the people of God; [10] for those who enter God's rest also cease from their labors as God did from his. [11] Let us therefore make every effort to enter that rest, so that no one may fall through such disobedience as theirs.

12 Indeed, the word of God is living and active, sharper than any two-edged sword, piercing until it divides soul from spirit, joints from marrow; it is able to judge the thoughts and intentions of the heart. [13] And before him no creature is hidden, but all are naked and laid bare to the eyes of the one to whom we must render an account.

14 Since, then, we have a great high priest who has passed through the heavens, Jesus, the Son of God, let us hold fast to our confession. [15] For we do not have a high priest who is unable to sympathize with our weaknesses, but we have one who in every respect has been tested[w] as we are, yet without sin. [16] Let us therefore approach the throne of grace with boldness, so that we may receive mercy and find grace to help in time of need.

5 Every high priest chosen from among mortals is put in charge of things pertaining to God on their behalf, to offer gifts and sacrifices for sins. [2] He is able to deal gently with the ignorant and wayward, since he himself is subject to weakness; [3] and because of this he must offer sacrifice for his own sins as well as for those of the people. [4] And one does not presume to take this honor, but takes it only when called by God, just as Aaron was.

5 So also Christ did not glorify himself in becoming a high priest, but was appointed by the one who said to him,

"You are my Son,
today I have begotten you";

[6] as he says also in another place,

"You are a priest forever,
according to the order of Melchizedek."

7 In the days of his flesh, Jesus[v] offered up prayers and supplications, with loud cries and tears, to the one who was

u Other ancient authorities read *it did not meet with faith in those who listened*   v Gk *he*
w Or *tempted*

---

was not the promised *rest,* otherwise *David* would not have spoken centuries later of a *rest* still remaining. **7:** Ps 95.7–8. **9–10:** *Sabbath rest,* points back to God's rest after the work of creation (Gen 2.2), and forward to rest as salvation in God's presence (11.16), completed at the end of time (10.25, 37–38; 12.26–29). **11–13:** An exhortation to diligence, reinforced by a reminder that *the word of God* can discern all secret *intentions of the heart.*

4.14–5.14: The theme of Jesus our high priest is resumed (compare 2.17–18); he has the two qualifications of a priest, divine appointment (5.4) and the ability to *sympathize with our weaknesses* (4.15).

4.15–16: Because Christ experienced real, human testing, he is able *to sympathize with our weaknesses.* **16:** At God's *throne of grace* humans *receive mercy* for past sins and *find grace* for present and future *need.*

5.1: *Gifts and sacrifices,* grain and animal sacrifices. **2:** The Old Testament provides no atoning sacrifice for deliberate and defiant sins (see Num 15.30 n.; Deut 17.12), but only for "unwitting" offenses committed by *the ignorant and wayward* (Lev ch 4). **4:** *Aaron,* Ex 28.1. **5:** Ps 2.7; Lk 3.22 note *z.* **6:** *The order of Melchizedek,* according to the rank which Melchizedek held (Ps 110.4; see 7.1–10 n.). **5.7–8:** Jesus' agonizing prayer in Geth-

able to save him from death, and he was heard because of his reverent submission. [8] Although he was a Son, he learned obedience through what he suffered; [9] and having been made perfect, he became the source of eternal salvation for all who obey him, [10] having been designated by God a high priest according to the order of Melchizedek.

[11] About this[x] we have much to say that is hard to explain, since you have become dull in understanding. [12] For though by this time you ought to be teachers, you need someone to teach you again the basic elements of the oracles of God. You need milk, not solid food; [13] for everyone who lives on milk, being still an infant, is unskilled in the word of righteousness. [14] But solid food is for the mature, for those whose faculties have been trained by practice to distinguish good from evil.

6 Therefore let us go on toward perfection,[y] leaving behind the basic teaching about Christ, and not laying again the foundation: repentance from dead works and faith toward God, [2] instruction about baptisms, laying on of hands, resurrection of the dead, and eternal judgment. [3] And we will do[z] this, if God permits. [4] For it is impossible to restore again to repentance those who have once been enlightened, and have tasted the heavenly gift, and have shared in the Holy Spirit, [5] and have tasted the goodness of the word of God and the powers of the age to come, [6] and then have fallen away, since on their own they are crucifying again the Son of God and are hold-

ing him up to contempt. [7] Ground that drinks up the rain falling on it repeatedly, and that produces a crop useful to those for whom it is cultivated, receives a blessing from God. [8] But if it produces thorns and thistles, it is worthless and on the verge of being cursed; its end is to be burned over.

[9] Even though we speak in this way, beloved, we are confident of better things in your case, things that belong to salvation. [10] For God is not unjust; he will not overlook your work and the love that you showed for his sake[a] in serving the saints, as you still do. [11] And we want each one of you to show the same diligence so as to realize the full assurance of hope to the very end, [12] so that you may not become sluggish, but imitators of those who through faith and patience inherit the promises.

[13] When God made a promise to Abraham, because he had no one greater by whom to swear, he swore by himself, [14] saying, "I will surely bless you and multiply you." [15] And thus Abraham,[b] having patiently endured, obtained the promise. [16] Human beings, of course, swear by someone greater than themselves, and an oath given as confirmation puts an end to all dispute. [17] In the same way, when God desired to show even more clearly to the heirs of the promise the unchangeable character of his purpose, he guaranteed it by an oath, [18] so that through two unchangeable things,

x Or *him*   y Or *toward maturity*   z Other ancient authorities read *let us do*   a Gk *for his name*   b Gk *he*

---

semane (Mk 14.32–42) *was heard* in the sense that *he learned obedience* by submitting to the divine will, which involved death. **9:** *Made perfect,* Jesus completed his divinely appointed discipline for priesthood (see 2.10 n.). *Eternal salvation,* not merely temporary deliverance, such as the levitical law provided (9.12). **10:** *Melchizedek,* see 7.1–10 n. **11–14:** The author deplores the spiritual immaturity of the readers. **12–14:** *Milk* stands for rudimentary teachings; *solid food,* more advanced teaching (1 Cor 3.1–2).

6.1–20: Exhortation and declaration of purpose. Despite misgivings as to the capac-

ity of the readers (5.11–14), and taking *basic* Christian *teaching* for granted (6.1–3), the author is resolved to proceed to more advanced subjects. Though it would be useless to address those who have apostatized from the faith (vv. 4–6), who are like *worthless ground* (vv. 7–8), this gloomy picture does not represent the condition of the readers (vv. 9–12), who are addressed for the first and only time as *beloved* (v. 9).

6.13–20: The certainty of God's promises. To encourage patient waiting (v. 12) the author mentions the example of Abraham (v. 15). **13–14:** Gen 22.16–17. **18:** *Two un-*

in which it is impossible that God would prove false, we who have taken refuge might be strongly encouraged to seize the hope set before us. 19 We have this hope, a sure and steadfast anchor of the soul, a hope that enters the inner shrine behind the curtain, 20 where Jesus, a forerunner on our behalf, has entered, having become a high priest forever according to the order of Melchizedek.

7 This "King Melchizedek of Salem, priest of the Most High God, met Abraham as he was returning from defeating the kings and blessed him"; 2 and to him Abraham apportioned "one-tenth of everything." His name, in the first place, means "king of righteousness"; next he is also king of Salem, that is, "king of peace." 3 Without father, without mother, without genealogy, having neither beginning of days nor end of life, but resembling the Son of God, he remains a priest forever.

4 See how great he is! Even*c* Abraham the patriarch gave him a tenth of the spoils. 5 And those descendants of Levi who receive the priestly office have a commandment in the law to collect tithes*d* from the people, that is, from their kindred,*e* though these also are descended from Abraham. 6 But this man, who does not belong to their ancestry, collected tithes*d* from Abraham and blessed him who had received the promises. 7 It is beyond dispute that the inferior is blessed by the superior. 8 In the one case, tithes are received by those who are mortal; in the other, by one of whom it is testified that he lives. 9 One might even

say that Levi himself, who receives tithes, paid tithes through Abraham, 10 for he was still in the loins of his ancestor when Melchizedek met him.

11 Now if perfection had been attainable through the levitical priesthood— for the people received the law under this priesthood—what further need would there have been to speak of another priest arising according to the order of Melchizedek, rather than one according to the order of Aaron? 12 For when there is a change in the priesthood, there is necessarily a change in the law as well. 13 Now the one of whom these things are spoken belonged to another tribe, from which no one has ever served at the altar. 14 For it is evident that our Lord was descended from Judah, and in connection with that tribe Moses said nothing about priests.

15 It is even more obvious when another priest arises, resembling Melchizedek, 16 one who has become a priest, not through a legal requirement concerning physical descent, but through the power of an indestructible life. 17 For it is attested of him,

"You are a priest forever,
   according to the order of
      Melchizedek."

18 There is, on the one hand, the abrogation of an earlier commandment because it was weak and ineffectual 19 (for the law made nothing perfect); there is, on the other hand, the introduction of a better hope, through which we approach God.

c Other ancient authorities lack *Even*   *d* Or *a tenth*   *e* Gk *brothers*

---

changeable things, God's promise and God's oath. **19**: *The inner shrine,* the Holy of Holies (9.3) of the tabernacle, *behind the curtain* or veil (Ex 26.31–35), restricted to the high priest alone (Lev 16.2). **20**: As *forerunner,* however, Jesus secures our unhindered access to God. *Melchizedek,* here the author picks up the argument of 5.10.

7.1–28: **The priesthood of Melchizedek and the levitical priesthood compared. 1–10**: From Gen 14.17–20 the author deduces that the mysterious priest-king, Melchizedek, was greater than either Abraham or his descendent, Levi. **3**: *Without father, without mother,* Melchizedek's ancestors, birth, and

death are not recorded in Scripture. **8**: *Those who are mortal,* the levitical priests. *One who lives,* Melchizedek, whose death is not recorded.

7.11–25: The levitical priesthood is inadequate, because it is provisional and temporary (vv. 11–14). On the other hand, a *priest . . . resembling Melchizedek* is eternal (Ps 110.4), and the office is neither inherited nor transmitted (vv. 15–19). (Among the Essene community at Qumran, Melchizedek was regarded as an angelic, heavenly figure and judge, who rescues the righteous.) Unlike the levitical priests, Jesus was appointed with a divine oath (Ps 110.4); furthermore, being immortal

20 This was confirmed with an oath; for others who became priests took their office without an oath, 21 but this one became a priest with an oath, because of the one who said to him,

"The Lord has sworn
   and will not change his mind,
'You are a priest forever' "—
22 accordingly Jesus has also become the guarantee of a better covenant.

23 Furthermore, the former priests were many in number, because they were prevented by death from continuing in office; 24 but he holds his priesthood permanently, because he continues forever. 25 Consequently he is able for all time to save^f those who approach God through him, since he always lives to make intercession for them.

26 For it was fitting that we should have such a high priest, holy, blameless, undefiled, separated from sinners, and exalted above the heavens. 27 Unlike the other^g high priests, he has no need to offer sacrifices day after day, first for his own sins, and then for those of the people; this he did once for all when he offered himself. 28 For the law appoints as high priests those who are subject to weakness, but the word of the oath, which came later than the law, appoints a Son who has been made perfect forever.

8 Now the main point in what we are saying is this: we have such a high priest, one who is seated at the right hand of the throne of the Majesty in the heavens, 2 a minister in the sanctuary and the true tent^h that the Lord, and not any mortal, has set up. 3 For every high priest is appointed to offer gifts and sacrifices; hence it is necessary for this priest also to have something to offer. 4 Now if he were on earth, he would not be a priest at all, since there are priests who offer gifts according to the law. 5 They offer worship in a sanctuary that is a sketch and shadow of the heavenly one; for Moses, when he was about to erect the tent,^h was warned, "See that you make everything according to the pattern that was shown you on the mountain." 6 But Jesusⁱ has now obtained a more excellent ministry, and to that degree he is the mediator of a better covenant, which has been enacted through better promises. 7 For if that first covenant had been faultless, there would have been no need to look for a second one.

8 God^j finds fault with them when he says:

"The days are surely coming, says
      the Lord,
   when I will establish a new
      covenant with the house
      of Israel
   and with the house of Judah;
9   not like the covenant that I made
         with their ancestors,
      on the day when I took them
         by the hand to lead them
         out of the land of Egypt;
   for they did not continue in
         my covenant,
   and so I had no concern for
         them, says the Lord.
10   This is the covenant that I will
         make with the house
         of Israel
      after those days, says the Lord:
   I will put my laws in their minds,
      and write them on their hearts,
   and I will be their God,
      and they shall be my people.
11   And they shall not teach
         one another

f Or *able to save completely*   g Gk lacks *other*
h Or *tabernacle*   i Gk *he*   j Gk *He*

Jesus *holds his priesthood permanently* (vv. 20–25).

7.26–28: Summary of the merits of our high priest, Jesus the Son of God.

8.1–13: The heavenly sanctuary and the new covenant. 1–6: The chief *point . . . is this, we have . . . a high priest* who is also the enthroned king, ministering on our behalf in the heavenly sanctuary, of which the earthly tabernacle erected by Moses was *a sketch* (v. 5; Ex 25.40).

8.7–12: The *new covenant,* of which Jeremiah prophesied (Jer 31.31–34), surpasses the *first covenant* in making available a new order

or say to each other, 'Know
the Lord,'
for they shall all know me,
from the least of them to
the greatest.
12 For I will be merciful toward their
iniquities,
and I will remember their sins
no more."
13 In speaking of "a new covenant," he
has made the first one obsolete. And
what is obsolete and growing old will
soon disappear.

9 Now even the first covenant had
regulations for worship and an
earthly sanctuary. 2 For a tent*k* was constructed, the first one, in which were the
lampstand, the table, and the bread of the
Presence;*l* this is called the Holy Place.
3 Behind the second curtain was a tent*k*
called the Holy of Holies. 4 In it stood the
golden altar of incense and the ark of the
covenant overlaid on all sides with gold,
in which there were a golden urn holding
the manna, and Aaron's rod that budded,
and the tablets of the covenant; 5 above it
were the cherubim of glory overshadowing the mercy seat. *m* Of these things we
cannot speak now in detail.

6 Such preparations having been
made, the priests go continually into the
first tent*k* to carry out their ritual duties;
7 but only the high priest goes into the
second, and he but once a year, and not
without taking the blood that he offers
for himself and for the sins committed
unintentionally by the people. 8 By this

the Holy Spirit indicates that the way
into the sanctuary has not yet been disclosed as long as the first tent*k* is still
standing. 9 This is a symbol*n* of the
present time, during which gifts and sacrifices are offered that cannot perfect the
conscience of the worshiper, 10 but deal
only with food and drink and various
baptisms, regulations for the body imposed until the time comes to set things
right.

11 But when Christ came as a high
priest of the good things that have
come,*o* then through the greater and
perfect*p* tent*k* (not made with hands,
that is, not of this creation), 12 he entered
once for all into the Holy Place, not with
the blood of goats and calves, but with
his own blood, thus obtaining eternal redemption. 13 For if the blood of goats and
bulls, with the sprinkling of the ashes of
a heifer, sanctifies those who have been
defiled so that their flesh is purified,
14 how much more will the blood of
Christ, who through the eternal Spirit*q*
offered himself without blemish to God,
purify our*r* conscience from dead works
to worship the living God!

15 For this reason he is the mediator
of a new covenant, so that those who are
called may receive the promised eternal

k Or *tabernacle*   *l* Gk *the presentation of the
loaves*   *m* Or *the place of atonement*
n Gk *parable*   *o* Other ancient authorities read
good things to come   *p* Gk *more perfect*   *q* Other
ancient authorities read *Holy Spirit*   *r* Other
ancient authorities read *your*

---

of perfection. **13**: Jeremiah shows that the old
covenant would not endure forever (Jer
31.31–32). **9.1–10**: **The ministry of the levitical
priests** (Ex 25.10–40). **2**: *The bread of the Presence,* Lev 24.5–9. **3**: Ex 26.31–33. **4**: *The gold-
en altar of incense* stood in the Holy Place (Ex
30.6). *Ark of the covenant,* the chest containing
the tablets of the Law (Ex 25.10–22). *Urn . . .
manna,* Ex 16.32–34. *Aaron's rod,* Num 17.1–
10. **7**: Lev ch 16. *Sins committed unintentionally
by the people,* see 5.2 n. **9–10**: *Perfect,* see
2.10 n. *The conscience,* levitical sacrifices cannot cleanse the inner guilt that results from
sin. **10**: *The time . . . to set things right,* the
period of the new covenant, inaugurated by
the death of Christ.

9.11–10.18: **Characteristics of the sacrifice of Christ.** The offering of the blood of
Christ is incomparably superior to sacrifices
of *the blood of goats and bulls* (9.13–14), because
(*a*) it is the life (Lev 17.11) of a person, the
Christ, not of irrational beasts (10.4); (*b*)
Christ was both priest and sacrificial victim
(9.12); (*c*) the offering was made *once for all*
(9.12, 25–26; 10.11–12); (*d*) it was made in the
heavenly *sanctuary* (9.24); (*e*) it was a voluntary act (10.9; compare Jn 10.17–18), accomplished through *the eternal Spirit* (9.14); and
therefore (*f*) the efficacy of his sacrifice remains *for all time* (10.14).

9.13: Lev 16.6, 16; Num 19.9, 17–18. **14**:
The eternal Spirit, probably Christ's spirit of
self-offering. **15**: Christ's death redeemed also

inheritance, because a death has occurred that redeems them from the transgressions under the first covenant. *s* 16 Where a will*s* is involved, the death of the one who made it must be established. 17 For a will*s* takes effect only at death, since it is not in force as long as the one who made it is alive. 18 Hence not even the first covenant was inaugurated without blood. 19 For when every commandment had been told to all the people by Moses in accordance with the law, he took the blood of calves and goats,*t* with water and scarlet wool and hyssop, and sprinkled both the scroll itself and all the people, 20 saying, "This is the blood of the covenant that God has ordained for you." 21 And in the same way he sprinkled with the blood both the tent*u* and all the vessels used in worship. 22 Indeed, under the law almost everything is purified with blood, and without the shedding of blood there is no forgiveness of sins.

23 Thus it was necessary for the sketches of the heavenly things to be purified with these rites, but the heavenly things themselves need better sacrifices than these. 24 For Christ did not enter a sanctuary made by human hands, a mere copy of the true one, but he entered into heaven itself, now to appear in the presence of God on our behalf. 25 Nor was it to offer himself again and again, as the high priest enters the Holy Place year after year with blood that is not his own; 26 for then he would have had to suffer again and again since the foundation of the world. But as it is, he has appeared once for all at the end of the age to remove sin by the sacrifice of himself. 27 And just as it is appointed for mortals to die once, and after that the judgment, 28 so Christ, having been offered once to bear the sins of many, will appear a sec-

ond time, not to deal with sin, but to save those who are eagerly waiting for him.

10 Since the law has only a shadow of the good things to come and not the true form of these realities, it*v* can never, by the same sacrifices that are continually offered year after year, make perfect those who approach. 2 Otherwise, would they not have ceased being offered, since the worshipers, cleansed once for all, would no longer have any consciousness of sin? 3 But in these sacrifices there is a reminder of sin year after year. 4 For it is impossible for the blood of bulls and goats to take away sins. 5 Consequently, when Christ*w* came into the world, he said,

"Sacrifices and offerings you have
        not desired,
    but a body you have prepared
        for me;
6 in burnt offerings and sin offerings
    you have taken no pleasure.
7 Then I said, 'See, God, I have
        come to do your will, O
        God'
    (in the scroll of the book*x* it is
        written of me)."

8 When he said above, "You have neither desired nor taken pleasure in sacrifices and offerings and burnt offerings and sin offerings" (these are offered according to the law), 9 then he added, "See, I have come to do your will." He abolishes the first in order to establish the second. 10 And it is by God's will*y* that we have been sanctified through the offering of the body of Jesus Christ once for all.

11 And every priest stands day after day at his service, offering again and

s The Greek word used here means both *covenant* and *will*   *t* Other ancient authorities lack *and goats*   *u* Or *tabernacle*   *v* Other ancient authorities read *they*   *w* Gk *he*   *x* Meaning of Gk uncertain   *y* Gk *by that will*

the Old Testament saints (11.39–40), and inaugurated the new covenant (Rom 3.24–25; 1 Cor 11.25).
9.15–17: See note *s*. **18–20**: Ex 24.6–8. **28**: *The sins of many*, Isa 53.12; Mk 10.45; Rom 5.19; Rev 7.9–10. *A second time*, Acts 1.10–11; Phil 3.20.
10.1: Insubstantial *shadow* is contrasted

with *the true form*. *Approach*, i.e. God (7.25). **2**: *Consciousness of sin*, see 9.9 n. **5–9**: Ps 40.6–8 Septuagint. **8**: 1 Sam 15.22; Ps 50.8–15; Isa 1.10–17; Jer 7.21–26; Hos 6.6. **10**: *Sanctified*, ceremonially cleansed and perfected through Christ's blood (10.29).
10.11–12: A priest stands, a king sits; Christ is both (1.3). **14**: *Sanctified*, see v. 10 n. **16–17**:

again the same sacrifices that can never take away sins. [12] But when Christ[z] had offered for all time a single sacrifice for sins, "he sat down at the right hand of God," [13] and since then has been waiting "until his enemies would be made a footstool for his feet." [14] For by a single offering he has perfected for all time those who are sanctified. [15] And the Holy Spirit also testifies to us, for after saying,

[16] "This is the covenant that I will
  make with them
    after those days, says the Lord:
  I will put my laws in their hearts,
    and I will write them on
      their minds,"

[17] he also adds,

  "I will remember[a] their sins and
    their lawless deeds no
      more."

[18] Where there is forgiveness of these, there is no longer any offering for sin.

19 Therefore, my friends,[b] since we have confidence to enter the sanctuary by the blood of Jesus, [20] by the new and living way that he opened for us through the curtain (that is, through his flesh), [21] and since we have a great priest over the house of God, [22] let us approach with a true heart in full assurance of faith, with our hearts sprinkled clean from an evil conscience and our bodies washed with pure water. [23] Let us hold fast to the confession of our hope without wavering, for he who has promised is faithful. [24] And let us consider how to provoke one another to love and good deeds, [25] not neglecting to meet together, as is the habit of some, but encouraging one another, and all the more as you see the Day approaching.

26 For if we willfully persist in sin after having received the knowledge of the truth, there no longer remains a sacrifice for sins, [27] but a fearful prospect of judgment, and a fury of fire that will consume the adversaries. [28] Anyone who has violated the law of Moses dies without mercy "on the testimony of two or three witnesses." [29] How much worse punishment do you think will be deserved by those who have spurned the Son of God, profaned the blood of the covenant by which they were sanctified, and outraged the Spirit of grace? [30] For we know the one who said, "Vengeance is mine, I will repay." And again, "The Lord will judge his people." [31] It is a fearful thing to fall into the hands of the living God.

32 But recall those earlier days when, after you had been enlightened, you endured a hard struggle with sufferings, [33] sometimes being publicly exposed to abuse and persecution, and sometimes being partners with those so treated. [34] For you had compassion for those who were in prison, and you cheerfully accepted the plundering of your possessions, knowing that you yourselves possessed something better and more lasting. [35] Do not, therefore, abandon that confidence of yours; it brings a great reward. [36] For you need endurance, so that when you have done the will of God, you may receive what was promised.

37 For yet "in a very little while,
    the one who is coming will
      come and will not delay;
38 but my righteous one will live
    by faith.
  My soul takes no pleasure in
    anyone who shrinks back."

[39] But we are not among those who shrink back and so are lost, but among those who have faith and so are saved.

11 Now faith is the assurance of things hoped for, the conviction

z Gk *this one*   a Gk *on their minds and I will
remember*   b Gk *Therefore, brothers*

The new covenant assures full and final remission of sins (Jer 31.33–34).
**10.19–39: Exhortations and warnings.
19–25:** Three privileges and duties of Christians: *let us approach* God in faith and worship (vv. 22 and 25); *let us hold fast* the public *confession of our hope* (v. 23); *let us consider* how we can help others in love (v. 24). **25:** *The Day* of Christ's second coming (see 9.28 n.).
10.26–31: The fate of the wilful sinner (see 5.2 n. and 6.4–6). **28:** Deut 17.2–6. **30:** Deut 32.35–36. **32–39:** Memory of past victories should inspire present *endurance*. **37:** Isa 26.20 Septuagint. **37–38:** Hab 2.3–4.

of things not seen. ²Indeed, by faith ᶜ our ancestors received approval. ³By faith we understand that the worlds were prepared by the word of God, so that what is seen was made from things that are not visible. ᵈ

4 By faith Abel offered to God a more acceptable ᵉ sacrifice than Cain's. Through this he received approval as righteous, God himself giving approval to his gifts; he died, but through his faith ᶠ he still speaks. ⁵By faith Enoch was taken so that he did not experience death; and "he was not found, because God had taken him." For it was attested before he was taken away that "he had pleased God." ⁶And without faith it is impossible to please God, for whoever would approach him must believe that he exists and that he rewards those who seek him. ⁷By faith Noah, warned by God about events as yet unseen, respected the warning and built an ark to save his household; by this he condemned the world and became an heir to the righteousness that is in accordance with faith.

8 By faith Abraham obeyed when he was called to set out for a place that he was to receive as an inheritance; and he set out, not knowing where he was going. ⁹By faith he stayed for a time in the land he had been promised, as in a foreign land, living in tents, as did Isaac and Jacob, who were heirs with him of the same promise. ¹⁰For he looked forward to the city that has foundations, whose architect and builder is God. ¹¹By faith he received power of procreation, even though he was too old—and Sarah herself was barren—because he considered him faithful who had promised. ᵍ ¹²Therefore from one person, and this one as good as dead, descendants were born, "as many as the stars of heaven and as the innumerable grains of sand by the seashore."

13 All of these died in faith without having received the promises, but from a distance they saw and greeted them. They confessed that they were strangers and foreigners on the earth, ¹⁴for people who speak in this way make it clear that they are seeking a homeland. ¹⁵If they had been thinking of the land that they had left behind, they would have had opportunity to return. ¹⁶But as it is, they desire a better country, that is, a heavenly one. Therefore God is not ashamed to be called their God; indeed, he has prepared a city for them.

17 By faith Abraham, when put to the test, offered up Isaac. He who had received the promises was ready to offer up his only son, ¹⁸of whom he had been told, "It is through Isaac that descendants shall be named for you." ¹⁹He considered the fact that God is able even to raise someone from the dead—and figuratively speaking, he did receive him back. ²⁰By faith Isaac invoked blessings for the future on Jacob and Esau. ²¹By faith Jacob, when dying, blessed each of the sons of Joseph, "bowing in worship over the top of his staff." ²²By faith Joseph, at the end of his life, made mention of the exodus of the Israelites and gave instructions about his burial. ʰ

23 By faith Moses was hidden by his parents for three months after his birth, because they saw that the child was beau-

---

c Gk by this   d Or *was not made out of visible things*   e Gk *greater*   f Gk *through it*
g Other ancient authorities read *By faith Sarah herself, though barren, received power to conceive, even when she was too old, because she considered him faithful who had promised.*   h Gk *his bones*

11.1–40: Roll call of heroes and heroines of faith, designed to reinforce the exhortation in 10.35–39. **1:** Instead of defining faith comprehensively, the author describes those aspects of it which bear upon the argument. **4:** Gen 4.3–10. **5:** Gen 5.21–24. **6:** A general axiom referring to the existence and the moral government of God. **7:** Gen 6.13–22.
11.8–9: Gen 12.1–8. **10:** *The city* is the heavenly Jerusalem (v. 16; Gal 4.26; Rev 21.2). *Foundations,* contrasted with tents (v. 9). **11:** Gen 17.19; 18.11–14; 21.2. **12:** Gen 15.5–6; 22.17; 32.12; Rom 4.19. **16:** *A city,* see v. 10 n. **17:** Gen 22.1–10. **18:** Gen 21.12. **19:** Abraham received Isaac *back* when he was told to offer a ram instead of his son (Gen 22.13). **20:** Gen 27.27–29, 39–40. **21:** Gen ch 48; 47.31 Septuagint. **22:** Gen 50.24–25; Ex 13.19.
11.23: Ex 2.2; 1.22. **24–25:** Ex 2.10–15. **26:**

tiful; and they were not afraid of the king's edict. *i* 24 By faith Moses, when he was grown up, refused to be called a son of Pharaoh's daughter, 25 choosing rather to share ill-treatment with the people of God than to enjoy the fleeting pleasures of sin. 26 He considered abuse suffered for the Christ *j* to be greater wealth than the treasures of Egypt, for he was looking ahead to the reward. 27 By faith he left Egypt, unafraid of the king's anger; for he persevered as though *k* he saw him who is invisible. 28 By faith he kept the Passover and the sprinkling of blood, so that the destroyer of the firstborn would not touch the firstborn of Israel. *l*

29 By faith the people passed through the Red Sea as if it were dry land, but when the Egyptians attempted to do so they were drowned. 30 By faith the walls of Jericho fell after they had been encircled for seven days. 31 By faith Rahab the prostitute did not perish with those who were disobedient, *m* because she had received the spies in peace.

32 And what more should I say? For time would fail me to tell of Gideon, Barak, Samson, Jephthah, of David and Samuel and the prophets— 33 who through faith conquered kingdoms, administered justice, obtained promises, shut the mouths of lions, 34 quenched raging fire, escaped the edge of the sword, won strength out of weakness, became mighty in war, put foreign armies to flight. 35 Women received their dead by resurrection. Others were tortured, refusing to accept release, in order to obtain a better resurrection. 36 Others

suffered mocking and flogging, and even chains and imprisonment. 37 They were stoned to death, they were sawn in two, *n* they were killed by the sword; they went about in skins of sheep and goats, destitute, persecuted, tormented— 38 of whom the world was not worthy. They wandered in deserts and mountains, and in caves and holes in the ground.

39 Yet all these, though they were commended for their faith, did not receive what was promised, 40 since God had provided something better so that they would not, apart from us, be made perfect.

12 Therefore, since we are surrounded by so great a cloud of witnesses, let us also lay aside every weight and the sin that clings so closely, *o* and let us run with perseverance the race that is set before us, 2 looking to Jesus the pioneer and perfecter of our faith, who for the sake of *p* the joy that was set before him endured the cross, disregarding its shame, and has taken his seat at the right hand of the throne of God.

3 Consider him who endured such hostility against himself from sinners, *q* so that you may not grow weary or lose heart. 4 In your struggle against sin you

---

i Other ancient authorities add *By faith Moses, when he was grown up, killed the Egyptian, because he observed the humiliation of his people* (Gk brothers)   *j* Or *the Messiah*   *k* Or *because*   *l* Gk *would not touch them*   *m* Or *unbelieving*   *n* Other ancient authorities add *they were tempted*   *o* Other ancient authorities read *sin that easily distracts*   *p* Or *who instead of*   *q* Other ancient authorities read *such hostility from sinners against themselves*

Abuse suffered for the Christ, 13.13 (see 1 Cor 10.3–4 n.). **27**: *Him who is invisible,* Jn 1.18; Col 1.15; 1 Tim 1.17; 6.16. **28**: *Sprinkling of blood,* on the lintels and doorposts. See Ex 12.21–30. **29**: Ex 14.21–31. **30**: Josh 6.12–21. **31**: Josh 2.1–21; 6.22–25.

11.32: Judg chs 6–8; 4–5; 13–16; 11–12; 1 Sam chs 16–30; 2 Sam chs 1–24; 1 Kings 1.1–2.11; 1 Sam 1–12; 15.1–16.13. **33**: *Lions,* Dan ch 6. **34**: *Fire,* Dan ch 3. **35**: 1 Kings 17.17–24; 2 Kings 4.25–37; 2 Macc 7.9, 14. **37**: 2 Chr 24.20–22; Jer 26.23; 2 Macc 5.27; 6.12–7.42. **38**: *Caves,* 2 Macc 6.11; 10.6. **40**: See 9.15 n.

12.1–29: Exhortations and warnings. 1–11: Suffering, its joy and discipline. **1**: *Witnesses,* those mentioned in ch 11. *Weight,* encumbrance. *Sin,* not a particular "besetting sin," but sin in general as a hindrance to running *the race* (1 Cor 9.24–27). **2**: *Pioneer and perfecter,* see 2.10 n. *Our faith,* literally "the faith." *For . . . the joy,* may also be translated "instead of the joy" (of unbroken life in heaven, Phil 2.6). **5–6**: Prov 3.11–12. **8**: *All . . . share,* namely, all children of God (v. 6). **9**: *Father of spirits,* God is the author of humanity's spiritual being (Num 16.22; Eccl 12.7; Zech 12.1).

have not yet resisted to the point of shedding your blood. 5 And you have forgotten the exhortation that addresses you as children—

"My child, do not regard lightly
        the discipline of the Lord,
    or lose heart when you are
        punished by him;
6 for the Lord disciplines those
        whom he loves,
    and chastises every child whom
        he accepts."

7 Endure trials for the sake of discipline. God is treating you as children; for what child is there whom a parent does not discipline? 8 If you do not have that discipline in which all children share, then you are illegitimate and not his children. 9 Moreover, we had human parents to discipline us, and we respected them. Should we not be even more willing to be subject to the Father of spirits and live? 10 For they disciplined us for a short time as seemed best to them, but he disciplines us for our good, in order that we may share his holiness. 11 Now, discipline always seems painful rather than pleasant at the time, but later it yields the peaceful fruit of righteousness to those who have been trained by it.

12 Therefore lift your drooping hands and strengthen your weak knees, 13 and make straight paths for your feet, so that what is lame may not be put out of joint, but rather be healed.

14 Pursue peace with everyone, and the holiness without which no one will see the Lord. 15 See to it that no one fails to obtain the grace of God; that no root of bitterness springs up and causes trouble, and through it many become defiled. 16 See to it that no one becomes like Esau, an immoral and godless person, who sold his birthright for a single meal. 17 You know that later, when he wanted to inherit the blessing, he was rejected, for he found no chance to repent,[r] even though he sought the blessing[s] with tears.

18 You have not come to something[t] that can be touched, a blazing fire, and darkness, and gloom, and a tempest, 19 and the sound of a trumpet, and a voice whose words made the hearers beg that not another word be spoken to them. 20 (For they could not endure the order that was given, "If even an animal touches the mountain, it shall be stoned to death." 21 Indeed, so terrifying was the sight that Moses said, "I tremble with fear.") 22 But you have come to Mount Zion and to the city of the living God, the heavenly Jerusalem, and to innumerable angels in festal gathering, 23 and to the assembly[u] of the firstborn who are enrolled in heaven, and to God the judge of all, and to the spirits of the righteous made perfect, 24 and to Jesus, the mediator of a new covenant, and to the sprinkled blood that speaks a better word than the blood of Abel.

25 See that you do not refuse the one who is speaking; for if they did not escape when they refused the one who warned them on earth, how much less will we escape if we reject the one who warns from heaven! 26 At that time his voice shook the earth; but now he has promised, "Yet once more I will shake not only the earth but also the heaven." 27 This phrase, "Yet once more," indicates the removal of what is shaken—

r Or no chance to change his father's mind   s Gk it
t Other ancient authorities read a mountain
u Or angels, and to the festal gathering 23 and assembly

12.12: Isa 35.3. **14–17**: The necessity of peace and purity. **15**: Deut 29.18 Septuagint. **16**: Gen 25.29–34. **17**: *Chance to repent*, or "chance to change his father's mind" (Gen 27.30–40).
12.18–29: The two covenants contrasted. **18–19**: *Something that can be touched*, Mount Sinai (Ex 19.12–22; 20.18–21; Deut 4.11–12; 5.22–27), **20**: Ex 19.12–13. **21**: Deut 9.19. **22**:
Mount Zion . . . the heavenly Jerusalem, contrast v. 18 and see 11.10 n. **24**: Abel's blood cried for vengeance (11.4; Gen 4.10); *the sprinkled blood* of Jesus, which established the new covenant (9.14; 13.20), *speaks . . . better* for forgiveness. **25**: *Who warns*, the Greek may also be translated "who is." **26**: Hag 2.6. **29**: Deut 4.24; 9.3; 2 Thess 1.7–8; compare Mt 3.12.

that is, created things—so that what cannot be shaken may remain. ²⁸Therefore, since we are receiving a kingdom that cannot be shaken, let us give thanks, by which we offer to God an acceptable worship with reverence and awe; ²⁹for indeed our God is a consuming fire.

13 Let mutual love continue. ²Do not neglect to show hospitality to strangers, for by doing that some have entertained angels without knowing it. ³Remember those who are in prison, as though you were in prison with them; those who are being tortured, as though you yourselves were being tortured.ᵛ ⁴Let marriage be held in honor by all, and let the marriage bed be kept undefiled; for God will judge fornicators and adulterers. ⁵Keep your lives free from the love of money, and be content with what you have; for he has said, "I will never leave you or forsake you." ⁶So we can say with confidence,

"The Lord is my helper;
    I will not be afraid.
What can anyone do to me?"

7 Remember your leaders, those who spoke the word of God to you; consider the outcome of their way of life, and imitate their faith. ⁸Jesus Christ is the same yesterday and today and forever. ⁹Do not be carried away by all kinds of strange teachings; for it is well for the heart to be strengthened by grace, not by regulations about food,ʷ which have not benefited those who observe them. ¹⁰We have an altar from which those who officiate in the tentˣ have no right to eat. ¹¹For the bodies of those animals whose blood is brought into the sanctuary by the high priest as a sacrifice for sin are burned outside the camp. ¹²Therefore Jesus also suffered outside the city gate in order to sanctify the people by his own blood. ¹³Let us then go to him outside the camp and bear the abuse he endured. ¹⁴For here we have no lasting city, but we are looking for the city that is to come. ¹⁵Through him, then, let us continually offer a sacrifice of praise to God, that is, the fruit of lips that confess his name. ¹⁶Do not neglect to do good and to share what you have, for such sacrifices are pleasing to God.

17 Obey your leaders and submit to them, for they are keeping watch over your souls and will give an account. Let them do this with joy and not with sighing—for that would be harmful to you.

18 Pray for us; we are sure that we have a clear conscience, desiring to act honorably in all things. ¹⁹I urge you all the more to do this, so that I may be restored to you very soon.

20 Now may the God of peace, who brought back from the dead our Lord Jesus, the great shepherd of the sheep, by the blood of the eternal covenant, ²¹make you complete in everything good so that you may do his will, working among usʸ that which is pleasing in his sight, through Jesus Christ, to whom be the glory forever and ever. Amen.

22 I appeal to you, brothers and sisters,ᶻ bear with my word of exhortation, for I have written to you briefly. ²³I want you to know that our brother Timothy has been set free; and if he comes in time, he will be with me when I see you. ²⁴Greet all your leaders and all the saints. Those from Italy send you greetings. ²⁵Grace be with all of you.ᵃ

---

v Gk *were in the body*   *w* Gk *not by foods*
x Or *tabernacle*   *y* Other ancient authorities
read *you*   *z* Gk *brothers*   *a* Other ancient
authorities add *Amen*

13.1–17: Concluding admonitions. 2: *Strangers,* Christian sisters and brothers from other places (see Rom 16.1–2 n.; 1 Pet 4.9 n.). *Angels,* Gen 18.1–8; 19.1–3. **3:** Mt 25.35–46. **5:** Josh 1.5. **6:** Ps 118.6. **9:** *Food,* 9.10. **11:** Lev 16.27. **12:** *City gate,* of Jerusalem. *Sanctify,* see 10.10 n. **14:** *The city that is to come,* i.e. the new Jerusalem; see 11.10 n.; 12.22 n. **15:** *A sacrifice of thanksgiving* (Lev 7.12; Ps 50.14, 23; Hos 14.2). **17:** *Leaders,* v. 7.

13.18–25: Personal messages (vv. 18–19), **benediction** (vv. 20–21), **and postscript** (vv. 22–25). **22:** *Briefly,* considering the extent of the subject.

The Letter of
James

Although it begins with an epistolatory greeting (1.1), James lacks other formal characteristics of a letter. The central section (2.1–3.12) contains three diatribe-style discussions (2.1–13, on partiality; 2.14–26, on faith and works; 3.1–12, on speech). This central section can also be seen as describing relationships within a community that is to embody the love-command. Smaller sections of wisdom sayings linked by catchword, and even single proverbs, surround the central section. James concludes with admonitions addressed to the community. The members are to await the coming of the Lord with patience (5.7–11), not to swear oaths (5.12; compare Mt 5.34–37), but to pray together (5.13). Elders of the community are to pray over the sick and to anoint them, for such prayer brings healing and forgiveness of sin (5.14–15). Sins within the community are to be dealt with by mutual confession, exhortation, and pastoral efforts to recall the sinner (5.16–19). Although Jesus is mentioned only twice (1.1; 2.1) and much of the exhortation can be paralleled in Jewish wisdom traditions, such directives in the conclusion of the letter suggest that James addresses a structured Christian community.

Like other wisdom materials, James speaks of tensions between rich and poor. The common social practice of favoring the wealthy and pushing aside the poor must be rejected (2.1–5). Christians have experienced some form of abuse by wealthy non-Christians (2.6–7). Wealthy merchants are warned against confidence that the future is assured (4.13–17). Those who gained their wealth by withholding just wages from workers will be condemned (5.1–6). Many of the sayings in James seem to echo sayings of Jesus in the synoptic tradition, though without directly invoking Jesus' authority. The mention, however, of "James" in the greeting (1.1) may place this teaching under the authority of James, the brother of the Lord (Gal 1.19; see Acts 12.17; 1 Cor 15.7; Gal 2.9).

The figure of James emerges in another early Christian context, which seems to be echoed in the letter, namely the dispute over "faith and works" (Jas 2.14–26; contrast Gal 2.1–14). James is not defending salvation through Christ against the Pauline understanding of "works" as the efforts of non-Jewish Christians to conform to Jewish rites of circumcision, kosher food laws, and religious feasts. Paul would certainly agree that Christian life should be expressed by deeds of charity (see Gal 5.6, "faith working through love"). Both Paul and James hold that love of one's neighbor fulfills the law (2.8; Gal 5.14). James appears to be opposed to a distorted and sloganizing Paulinism, which influenced Christians to neglect their obligations to aid their poverty-stricken and suffering brothers and sisters (2.15–17).

Although this elegantly composed treatise does not appear to have come from the James whom Paul met in Jerusalem, the author may have deliberately put this Christian wisdom-sermon under the patronage of that revered figure. Not only can James reformulate the wisdom teaching of Jesus, but James also has the authority to counter the false slogans about "faith and works" attributed to Paul.

1 James, a servant*a* of God and of the Lord Jesus Christ,

To the twelve tribes in the Dispersion: Greetings.

2 My brothers and sisters, *b* whenever you face trials of any kind, consider it nothing but joy, 3 because you know that the testing of your faith produces endurance; 4 and let endurance have its full effect, so that you may be mature and complete, lacking in nothing.

5 If any of you is lacking in wisdom, ask God, who gives to all generously and ungrudgingly, and it will be given you. 6 But ask in faith, never doubting, for the one who doubts is like a wave of the sea, driven and tossed by the wind; 7, 8 for the doubter, being double-minded and unstable in every way, must not expect to receive anything from the Lord.

9 Let the believer*c* who is lowly boast in being raised up, 10 and the rich in being brought low, because the rich will disappear like a flower in the field. 11 For the sun rises with its scorching heat and withers the field; its flower falls, and its beauty perishes. It is the same way with the rich; in the midst of a busy life, they will wither away.

12 Blessed is anyone who endures temptation. Such a one has stood the test and will receive the crown of life that the Lord*d* has promised to those who love him. 13 No one, when tempted, should say, "I am being tempted by God"; for God cannot be tempted by evil and he himself tempts no one. 14 But one is tempted by one's own desire, being lured and enticed by it; 15 then, when that desire has conceived, it gives birth to sin, and that sin, when it is fully grown, gives birth to death. 16 Do not be deceived, my beloved. *e*

17 Every generous act of giving, with every perfect gift, is from above, coming down from the Father of lights, with whom there is no variation or shadow due to change.*f* 18 In fulfillment of his own purpose he gave us birth by the word of truth, so that we would become a kind of first fruits of his creatures.

19 You must understand this, my beloved:*e* let everyone be quick to listen, slow to speak, slow to anger; 20 for your anger does not produce God's righteousness. 21 Therefore rid yourselves of all sordidness and rank growth of wickedness, and welcome with meekness the implanted word that has the power to save your souls.

22 But be doers of the word, and not merely hearers who deceive themselves. 23 For if any are hearers of the word and not doers, they are like those who look at themselves*g* in a mirror; 24 for they look at themselves and, on going away, immediately forget what they were like. 25 But those who look into the perfect law, the law of liberty, and persevere, being not hearers who forget but doers

a Gk *slave*   *b* Gk *brothers*   *c* Gk *brother*
d Gk *he*; other ancient authorities read *God*
e Gk *my beloved brothers*   *f* Other ancient authorities read *variation due to a shadow of turning*
g Gk *at the face of his birth*

1.1: Salutation. 1: *Dispersion* describes Jews scattered outside Palestine (see Jn 7.35). Its metaphorical use here testifies to the church's sense of being aliens in this world as well as the heir of Israel (Gal 4.21–31; Phil 3.20; 1 Pet 1.1).
1.2–18: The blessings of trials. Trials are a ground for rejoicing. **5:** *Wisdom,* see 3.13 n. **6:** *The sea,* Isa 57.20; Jer 49.23. **7–8:** *The doubter's* entire conduct reflects inconstancy of purpose (Mt 6.24). **10–11:** Isa 40.6–7.
1.12: *The crown of life,* see 2 Tim 4.8 n. **13:** Temptation is not from God, but from human passions. **17:** *Father of lights,* creator of the heavenly bodies (Gen 1.14–18; Ps 136.7).

18: The *first fruits* of harvest were frequently offered to God (Lev 23.10; Num 15.21; Deut 18.4).
1.19–27: True worship. 19–20: *Anger,* see Eph 4.26 n. **21:** *Rid yourselves,* strip off as dirty clothing. *The implanted word,* the gospel which has been received and is now growing. James uses a variety of expressions for the gospel: perfect law (v. 25), law of liberty (v. 25; 2.12), royal law (2.8). **22:** *Be doers . . . , not merely hearers,* Mt 7.24–27; Rom 2.13. **25:** *The perfect law,* the Jewish description of the Mosaic law, is here applied to the gospel, the "law" (1.27; 2.1–13) through which a believer obtains freedom. *Blessed,* Ps 1.1; Mt

who act—they will be blessed in their doing.

26 If any think they are religious, and do not bridle their tongues but deceive their hearts, their religion is worthless. 27Religion that is pure and undefiled before God, the Father, is this: to care for orphans and widows in their distress, and to keep oneself unstained by the world.

2 My brothers and sisters,*h* do you with your acts of favoritism really believe in our glorious Lord Jesus Christ?*i* 2For if a person with gold rings and in fine clothes comes into your assembly, and if a poor person in dirty clothes also comes in, 3and if you take notice of the one wearing the fine clothes and say, "Have a seat here, please," while to the one who is poor you say, "Stand there," or, "Sit at my feet,"*j* 4have you not made distinctions among yourselves, and become judges with evil thoughts? 5Listen, my beloved brothers and sisters.*k* Has not God chosen the poor in the world to be rich in faith and to be heirs of the kingdom that he has promised to those who love him? 6But you have dishonored the poor. Is it not the rich who oppress you? Is it not they who drag you into court? 7Is it not they who blaspheme the excellent name that was invoked over you?

8 You do well if you really fulfill the royal law according to the scripture, "You shall love your neighbor as yourself." 9But if you show partiality, you commit sin and are convicted by the law as transgressors. 10For whoever keeps the whole law but fails in one point has become accountable for all of it. 11For the one who said, "You shall not commit adultery," also said, "You shall not murder." Now if you do not commit adultery but if you murder, you have become a transgressor of the law. 12So speak and so act as those who are to be judged by the law of liberty. 13For judgment will be without mercy to anyone who has shown no mercy; mercy triumphs over judgment.

14 What good is it, my brothers and sisters,*k* if you say you have faith but do not have works? Can faith save you? 15If a brother or sister is naked and lacks daily food, 16and one of you says to them, "Go in peace; keep warm and eat your fill," and yet you do not supply their bodily needs, what is the good of that? 17So faith by itself, if it has no works, is dead.

18 But someone will say, "You have faith and I have works." Show me your faith apart from your works, and I by my works will show you my faith. 19You believe that God is one; you do well. Even the demons believe—and shudder. 20Do you want to be shown, you senseless person, that faith apart from works is barren? 21Was not our ancestor Abraham justified by works when he offered his son Isaac on the altar? 22You see that faith was active along with his works, and faith was brought to completion by the works. 23Thus the scripture was fulfilled that says, "Abraham believed God, and it was reckoned to him as righteousness," and he was called the friend of God. 24You see that a person is justified by works and not by faith alone. 25Like-

h Gk *My brothers*   *i* Or *hold the faith of our glorious Lord Jesus Christ without acts of favoritism*   *j* Gk *Sit under my footstool*   *k* Gk *brothers*

5.3–11. The word points to the happiness of the person who has God's favor. **26–27:** Religion consists of more than devotional exercises (Mt 25.35–36).
2.1–13: The respect due to the poor. 1: God shows no partiality; the disciple must not. **2:** *Assembly,* literally "synagogue." **5:** Lk 6.20; 1 Cor 1.26–29. **7:** *The excellent name* is that of Christ (see 2 Thess 1.12 n.).
2.8: Lev 19.18. **10:** *The law* forms a unity; therefore violating any one part breaks the entire law. **11:** Ex 20.13–14; Deut 5.17–18. **13:** *No mercy,* Mt 6.14–15; 18.21–35.
2.14–26: Faith and works. See Introduction. **14:** Faith without works is counterfeit; such faith cannot save (Mt 25.31–46; Gal 5.6). **19:** Demons know the truth about God (Deut 6.4), but are not saved. **21:** Gen 22.1–14. **23:** Gen 15.6 (compare Rom 4.3–25; Gal 3.6–14). *The friend of God,* 2 Chr 20.7; Isa 41.8. **25:**

wise, was not Rahab the prostitute also justified by works when she welcomed the messengers and sent them out by another road? 26 For just as the body without the spirit is dead, so faith without works is also dead.

3 Not many of you should become teachers, my brothers and sisters, *l* for you know that we who teach will be judged with greater strictness. 2 For all of us make many mistakes. Anyone who makes no mistakes in speaking is perfect, able to keep the whole body in check with a bridle. 3 If we put bits into the mouths of horses to make them obey us, we guide their whole bodies. 4 Or look at ships: though they are so large that it takes strong winds to drive them, yet they are guided by a very small rudder wherever the will of the pilot directs. 5 So also the tongue is a small member, yet it boasts of great exploits.

How great a forest is set ablaze by a small fire! 6 And the tongue is a fire. The tongue is placed among our members as a world of iniquity; it stains the whole body, sets on fire the cycle of nature, *m* and is itself set on fire by hell. *n* 7 For every species of beast and bird, of reptile and sea creature, can be tamed and has been tamed by the human species, 8 but no one can tame the tongue—a restless evil, full of deadly poison. 9 With it we bless the Lord and Father, and with it we curse those who are made in the likeness of God. 10 From the same mouth come blessing and cursing. My brothers and sisters, *o* this ought not to be so. 11 Does a spring pour forth from the same open-ing both fresh and brackish water? 12 Can a fig tree, my brothers and sisters, *p* yield olives, or a grapevine figs? No more can salt water yield fresh.

13 Who is wise and understanding among you? Show by your good life that your works are done with gentleness born of wisdom. 14 But if you have bitter envy and selfish ambition in your hearts, do not be boastful and false to the truth. 15 Such wisdom does not come down from above, but is earthly, unspiritual, devilish. 16 For where there is envy and selfish ambition, there will also be disorder and wickedness of every kind. 17 But the wisdom from above is first pure, then peaceable, gentle, willing to yield, full of mercy and good fruits, without a trace of partiality or hypocrisy. 18 And a harvest of righteousness is sown in peace for *q* those who make peace.

4 Those conflicts and disputes among you, where do they come from? Do they not come from your cravings that are at war within you? 2 You want something and do not have it; so you commit murder. And you covet *r* something and cannot obtain it; so you engage in disputes and conflicts. You do not have, because you do not ask. 3 You ask and do not receive, because you ask wrongly, in order to spend what you get on your pleasures. 4 Adulterers! Do you not know that friendship with the world is enmity with God? Therefore whoever wishes to be a friend of the world be-

l Gk *brothers*   *m* Or *wheel of birth*
n Gk *Gehenna*   *o* Gk *My brothers*   *p* Gk *my brothers*   *q* Or *by*   *r* Or *you murder and you covet*

Rahab, Josh 2.1–21. Abraham and Rahab represent two extremes: the friend of God and the prostitute; but both were justified by God. *Sent them out,* i.e. from Jericho. **26:** Where there are no works, faith is no more alive than a body without spirit.

3.1–18: True wisdom. Two besetting sins of the teacher are rebuked: intemperate speech (vv. 1–12) and arrogance (vv. 13–18). **5:** Sir 28.22. **12:** Mt 7.16.

3.13: True *wisdom* is not a human achievement, but is from God, and it reveals itself in a good life (v. 17). **14:** *Hearts,* see 1 Thess 3.13 n. **17:** Wisdom and godliness are intimately connected (Job 28.28; Sir 19.22–24). **18:** Isa 32.16–18; Mt 5.6, 9–10.

4.1–5.6: The contrast between godliness and worldliness. 4.1: *Cravings,* Rom 7.5–25; 1 Pet 2.11–12. **3:** Mt 7.7, 11. **4:** The prophets pictured the covenant as a marriage between God and Israel. Those who break covenant are *adulterers* (Isa 1.21; Jer 3.6–10; Hos 2.2). *Enmity with God,* Mt 6.24; Lk 16.13. **5:** For the quotation compare Ex 20.5; Deut

comes an enemy of God. ⁵Or do you suppose that it is for nothing that the scripture says, "God⁵ yearns jealously for the spirit that he has made to dwell in us"? ⁶But he gives all the more grace; therefore it says,

"God opposes the proud,
  but gives grace to the humble."
⁷Submit yourselves therefore to God. Resist the devil, and he will flee from you. ⁸Draw near to God, and he will draw near to you. Cleanse your hands, you sinners, and purify your hearts, you double-minded. ⁹Lament and mourn and weep. Let your laughter be turned into mourning and your joy into dejection. ¹⁰Humble yourselves before the Lord, and he will exalt you.

11 Do not speak evil against one another, brothers and sisters.¹ Whoever speaks evil against another or judges another, speaks evil against the law and judges the law; but if you judge the law, you are not a doer of the law but a judge. ¹²There is one lawgiver and judge who is able to save and to destroy. So who, then, are you to judge your neighbor?

13 Come now, you who say, "Today or tomorrow we will go to such and such a town and spend a year there, doing business and making money." ¹⁴Yet you do not even know what tomorrow will bring. What is your life? For you are a mist that appears for a little while and then vanishes. ¹⁵Instead you ought to say, "If the Lord wishes, we will live and do this or that." ¹⁶As it is, you boast in your arrogance; all such boasting is evil. ¹⁷Anyone, then, who knows the right thing to do and fails to do it, commits sin.

5 Come now, you rich people, weep and wail for the miseries that are coming to you. ²Your riches have rotted, and your clothes are moth-eaten. ³Your gold and silver have rusted, and their rust will be evidence against you, and it will eat your flesh like fire. You have laid up treasure" for the last days. ⁴Listen! The wages of the laborers who mowed your fields, which you kept back by fraud, cry out, and the cries of the harvesters have reached the ears of the Lord of hosts. ⁵You have lived on the earth in luxury and in pleasure; you have fattened your hearts in a day of slaughter. ⁶You have condemned and murdered the righteous one, who does not resist you.

7 Be patient, therefore, beloved,¹ until the coming of the Lord. The farmer waits for the precious crop from the earth, being patient with it until it receives the early and the late rains. ⁸You also must be patient. Strengthen your hearts, for the coming of the Lord is near.ᵛ ⁹Beloved,ʷ do not grumble against one another, so that you may not be judged. See, the Judge is standing at the doors! ¹⁰As an example of suffering and patience, beloved,¹ take the prophets who spoke in the name of the Lord. ¹¹Indeed we call blessed those who showed endurance. You have heard of the endurance of Job, and you have seen the purpose of the Lord, how the Lord is compassionate and merciful.

12 Above all, my beloved,¹ do not swear, either by heaven or by earth or by any other oath, but let your "Yes" be yes

s Gk He   t Gk brothers   u Or will eat your flesh, since you have stored up fire   v Or is at hand   w Gk Brothers

---

4.24; Zech 8.2. **6**: Prov 3.34 Septuagint; 1 Pet 5.5. **9**: Isa 32.11; Jer 4.13; Mic 2.4. **10**: Mt 23.12; 1 Pet 5.6. **12**: *One,* God.

4.13–17: The author rebukes the presumption of those who plan their future with complacency (Prov 27.1), forgetting that they live by the mercy of God. Human life is as evanescent as morning mist.

5.1–6: The rich are addressed as examples

of the folly of accumulating riches (Lk 12.16–20).

5.7–20: **Concluding encouragement. 7–9**: Three references to the coming of the Lord stand in contrast to the preceding passage; seeking riches is vain, but one who waits patiently for the Lord will be rewarded. **11**: Job 1.21–22; 2.10; Ps 111.4. **12**: Mt 5.34–37.

and your "No" be no, so that you may not fall under condemnation.

13 Are any among you suffering? They should pray. Are any cheerful? They should sing songs of praise. [14] Are any among you sick? They should call for the elders of the church and have them pray over them, anointing them with oil in the name of the Lord. [15] The prayer of faith will save the sick, and the Lord will raise them up; and anyone who has committed sins will be forgiven. [16] Therefore confess your sins to one another, and pray for one another, so that you may be healed. The prayer of the righteous is powerful and effective.

[17] Elijah was a human being like us, and he prayed fervently that it might not rain, and for three years and six months it did not rain on the earth. [18] Then he prayed again, and the heaven gave rain and the earth yielded its harvest.

19 My brothers and sisters,[x] if anyone among you wanders from the truth and is brought back by another, [20] you should know that whoever brings back a sinner from wandering will save the sinner's[y] soul from death and will cover a multitude of sins.

x Gk *My brothers*   y Gk *his*

5.13–15: This passage is cited by those churches that make a practice of anointing the sick (formerly called the sacrament of extreme unction by the Roman Catholic Church). **14:** *Oil* was a common medicinal remedy (Isa 1.6; Mk 6.13; see Lk 10.34 n.). Here it is invested with special significance through connection with the divine name (see 2 Thess 1.12 n.). **17:** 1 Kings 17.1; 18.1; Lk 4.25. **19:** God's *truth* is that which leads to righteousness. It is a closing indication of the practical character of the letter. **20:** *Cover . . . sins,* see 1 Pet 4.8 n.

The First Letter of
Peter

The First Letter of Peter is a pastoral exhortation from an elder in Rome (5.1; "Babylon" [5.13] was a common code-name for Rome in Jewish and Christian apocalyptic; see Rev 17.5–6). It was written to those having oversight of churches in Asia Minor (1.1; 5.2–3). The opening greeting claims Peter as author (1.1), while the epistolatory conclusion refers to Silvanus (5.12; associated with Paul's mission in 1 Thess 1.1). Some scholars hold that Peter had Silvanus dispatch this exhortation before his own martyrdom in Rome (ca. A.D. 64). Others find such an early date improbable.

The sociological setting of the communities assumes that their members are Gentiles, resident aliens and household slaves in rural Asia Minor. Outsiders recognize the Christian "name" as offensive (4.14, 16), and Christians must expect to endure persecution as well as milder forms of social ostracism and "name-calling" as a result of that identification. Christianity is not struggling to define itself in relationship to Jewish traditions. The letter assumes that the church is the true heir of the covenant (2.4–10). Christians must show by their lives that they reject the religious and moral ethos of a pagan culture to which they no longer belong (4.1–6).

The most vulnerable Christians, slaves of abusive masters and wives of non-believing husbands, are told to suffer abuse, to accept the authority of their husbands, to do what is right, not to be afraid (2.18–3.6). Such behavior may convert the persecutor. If it does not, it follows the pattern of Christ's own suffering and will also be rewarded. This characterization hardly fits the Petrine mission of Galatians and Acts, but is characteristic of Asia Minor at the end of the first century.

In order to encourage Christians who are suffering because their faith has called them to break with the social fabric of their community, the author reminds them that though they are "aliens" in this world, they belong to a new commonwealth, which is God's own household (1.2–5). Just as God chose Jesus to be the cornerstone of the new temple (2.6–7), so Christians are a "new race" (2.9), living stones of the spiritual house of God (2.5). God is not indifferent to their suffering. Jesus trusted God while being persecuted, bearing humanity's sins so that believers can now live in holiness (2.22–24). Therefore Christians are to continue their lives of mutual love and service, confident that God's judgment is at hand (4.7–11). They should not be surprised when they are tested by persecution (4.12). The devil tries to use suffering to destroy faith (4.8–9). By reminding the readers of Jesus' suffering and the suffering of countless fellow Christians throughout the world, the author assures them that they are not isolated in the struggle.

First Peter follows the typical pattern of a Pauline letter: greeting (1.1–2); thanksgiving (1.3–9); body of the letter, consisting of theological reflection on Christian identity (1.10–2.10), followed by exhortation (2.11–5.11); closing (5.12–14).

1 Peter, an apostle of Jesus Christ,
To the exiles of the Dispersion in Pontus, Galatia, Cappadocia, Asia, and Bithynia, ²who have been chosen and destined by God the Father and sanctified by the Spirit to be obedient to Jesus Christ and to be sprinkled with his blood:

May grace and peace be yours in abundance.

3 Blessed be the God and Father of our Lord Jesus Christ! By his great mercy he has given us a new birth into a living hope through the resurrection of Jesus Christ from the dead, ⁴and into an inheritance that is imperishable, undefiled, and unfading, kept in heaven for you, ⁵who are being protected by the power of God through faith for a salvation ready to be revealed in the last time. ⁶In this you rejoice, *ª* even if now for a little while you have had to suffer various trials, ⁷so that the genuineness of your faith—being more precious than gold that, though perishable, is tested by fire—may be found to result in praise and glory and honor when Jesus Christ is revealed. ⁸Although you have not seen*ᵇ* him, you love him; and even though you do not see him now, you believe in him and rejoice with an indescribable and glorious joy, ⁹for you are receiving the outcome of your faith, the salvation of your souls.

10 Concerning this salvation, the prophets who prophesied of the grace that was to be yours made careful search and inquiry, ¹¹inquiring about the person or time that the Spirit of Christ within them indicated when it testified in advance to the sufferings destined for Christ and the subsequent glory. ¹²It was revealed to them that they were serving not themselves but you, in regard to the things that have now been announced to you through those who brought you good news by the Holy Spirit sent from heaven—things into which angels long to look!

13 Therefore prepare your minds for action;*ᶜ* discipline yourselves; set all your hope on the grace that Jesus Christ will bring you when he is revealed. ¹⁴Like obedient children, do not be conformed to the desires that you formerly had in ignorance. ¹⁵Instead, as he who called you is holy, be holy yourselves in all your conduct; ¹⁶for it is written, "You shall be holy, for I am holy."

17 If you invoke as Father the one who judges all people impartially according to their deeds, live in reverent fear during the time of your exile. ¹⁸You know that you were ransomed from the futile ways inherited from your ancestors, not with perishable things like silver or gold, ¹⁹but with the precious blood of Christ, like that of a lamb without defect or blemish. ²⁰He was destined

a Or *Rejoice in this*   *b* Other ancient authorities read *known*   *c* Gk *gird up the loins of your mind*

1.1–2: Salutation, see Rom 1.1–7 n. **1:** *Exiles of the Dispersion,* see Jas 1.1 n. **2:** The transcendent origin of redemption, involving *God the Father,* the *Spirit,* and *Jesus Christ. Chosen and destined,* see Eph 1.4–10 n.; 1 Thess 1.4 n. *To be obedient to Jesus Christ* is the goal to which God's election and sanctification point (1.14, 22). *Sprinkled with his blood,* forgiveness is based on Christ's sacrificial death (see Mt 26.28 n.; Heb 9.11–10.18 n.). *Grace and peace,* see 2 Thess 1.2 n. **1.3–12: Rejoice in salvation.** *Living hope,* not forlorn or dead. *The resurrection of Jesus Christ* is the foundation of Christian hope, for in him God has shown what he intends to do for the faithful (1 Cor 15.20–28). **4:** *Inheritance* expresses the forward-looking character

of the Christian life; the greater part of the riches is yet to come. **5:** *Faith* puts one in God's keeping. *Salvation* is *ready to be revealed;* it is accomplished, but not fully manifest. **6:** God permits believers to suffer as proof of the quality of their faith (Ps 26.2; Jas 1.2–3). **10–12:** The prophets, and even angels, sought to understand what God was doing for the redemption of the faithful.

1.13–2.10: An appeal for holiness. **1.13:** *When he is revealed,* at the second coming. **14:** *Formerly had in ignorance,* Acts 17.30; Eph 4.17–18. **16:** Lev 11.44–45. **20:** Redemption through Christ was not a sudden whim of God, but was planned *before the foundation of the world* (Col 1.26).

before the foundation of the world, but was revealed at the end of the ages for your sake. 21 Through him you have come to trust in God, who raised him from the dead and gave him glory, so that your faith and hope are set on God.

22 Now that you have purified your souls by your obedience to the truth*d* so that you have genuine mutual love, love one another deeply*e* from the heart.*f* 23 You have been born anew, not of perishable but of imperishable seed, through the living and enduring word of God.*g* 24 For

"All flesh is like grass
    and all its glory like the flower
        of grass.
The grass withers,
    and the flower falls,
25 but the word of the Lord endures
        forever."

That word is the good news that was announced to you.

2 Rid yourselves, therefore, of all malice, and all guile, insincerity, envy, and all slander. 2 Like newborn infants, long for the pure, spiritual milk, so that by it you may grow into salvation— 3 if indeed you have tasted that the Lord is good.

4 Come to him, a living stone, though rejected by mortals yet chosen and precious in God's sight, and 5 like living stones, let yourselves be built*h* into a spiritual house, to be a holy priesthood, to offer spiritual sacrifices acceptable to God through Jesus Christ. 6 For it stands in scripture:

"See, I am laying in Zion a stone,
    a cornerstone chosen and
        precious;
and whoever believes in him*i* will
    not be put to shame."

7 To you then who believe, he is precious; but for those who do not believe,

"The stone that the builders
    rejected
has become the very head of
    the corner,"

8 and

"A stone that makes them
    stumble,
and a rock that makes them
    fall."

They stumble because they disobey the word, as they were destined to do.

9 But you are a chosen race, a royal priesthood, a holy nation, God's own people,*j* in order that you may proclaim the mighty acts of him who called you out of darkness into his marvelous light. 10 Once you were not a people,
    but now you are God's people;
once you had not received mercy,
    but now you have received
        mercy.

11 Beloved, I urge you as aliens and exiles to abstain from the desires of the flesh that wage war against the soul. 12 Conduct yourselves honorably among the Gentiles, so that, though they malign you as evildoers, they may see your hon-

d Other ancient authorities add *through the Spirit*
e Or *constantly*   *f* Other ancient authorities
read *a pure heart*   *g* Or *through the word of the
living and enduring God*   *h* Or *you yourselves are
being built*   *i* Or *it*   *j* Gk *a people for his
possession*

1.23–25: The Christian has been *born anew* by the creative *word of God*. Humans are frail, but the word that works in them *endures forever* (Isa 40.6–9).

2.2: *Pure, spiritual milk*, unadulterated nourishment supplied by the gospel (1 Thess 2.7–8). **3**: Ps 34.8. **4**: Ps 118.22; Isa 28.16; Mt 21.42. **5–8**: The images are mixed: a spiritual house being built of living stones, then a holy priesthood offering spiritual sacrifices. Christ is the rock of destiny: to those who believe, he is *chosen and precious*; to those who reject

him, he is *a rock that makes them fall.* **6**: Isa 28.16. **7**: Ps 118.22. **8**: Isa 8.14–15. **9–10**: As Gentiles they were *not a people* (Hos 2.23); now they are *God's own people*, chosen, holy, *a royal priesthood* (Ex 19.6). *Into his marvelous light*, in which God dwells (Isa 57.15; 1 Tim 6.16).

2.11–4.11: The obligation of Christians.
2.11: *That wage war*; Rom 7.23. **12**: *Malign you*, Christians were accused of committing immoralities during their secret meetings. *He comes to judge*, when God makes the innocence

orable deeds and glorify God when he comes to judge. *k*

13 For the Lord's sake accept the authority of every human institution, *l* whether of the emperor as supreme, 14 or of governors, as sent by him to punish those who do wrong and to praise those who do right. 15 For it is God's will that by doing right you should silence the ignorance of the foolish. 16 As servants *m* of God, live as free people, yet do not use your freedom as a pretext for evil. 17 Honor everyone. Love the family of believers. *n* Fear God. Honor the emperor.

18 Slaves, accept the authority of your masters with all deference, not only those who are kind and gentle but also those who are harsh. 19 For it is a credit to you if, being aware of God, you endure pain while suffering unjustly. 20 If you endure when you are beaten for doing wrong, what credit is that? But if you endure when you do right and suffer for it, you have God's approval. 21 For to this you have been called, because Christ also suffered for you, leaving you an example, so that you should follow in his steps.

22 "He committed no sin,
and no deceit was found in
his mouth."

23 When he was abused, he did not return abuse; when he suffered, he did not threaten; but he entrusted himself to the one who judges justly. 24 He himself bore our sins in his body on the cross, *o* so that, free from sins, we might live for righteousness; by his wounds *p* you have been healed. 25 For you were going astray like sheep, but now you have returned to the shepherd and guardian of your souls.

3 Wives, in the same way, accept the authority of your husbands, so that, even if some of them do not obey the word, they may be won over without a word by their wives' conduct, 2 when they see the purity and reverence of your lives. 3 Do not adorn yourselves outwardly by braiding your hair, and by wearing gold ornaments or fine clothing; 4 rather, let your adornment be the inner self with the lasting beauty of a gentle and quiet spirit, which is very precious in God's sight. 5 It was in this way long ago that the holy women who hoped in God used to adorn themselves by accepting the authority of their husbands. 6 Thus Sarah obeyed Abraham and called him lord. You have become her daughters as long as you do what is good and never let fears alarm you.

7 Husbands, in the same way, show consideration for your wives in your life together, paying honor to the woman as the weaker sex, *q* since they too are also heirs of the gracious gift of life—so that nothing may hinder your prayers.

8 Finally, all of you, have unity of spirit, sympathy, love for one another, a tender heart, and a humble mind. 9 Do not repay evil for evil or abuse for abuse; but, on the contrary, repay with a blessing. It is for this that you were called—that you might inherit a blessing. 10 For

"Those who desire life
and desire to see good days,
let them keep their tongues
from evil

k Gk *God on the day of visitation* *l* Or *every institution ordained for human beings* *m* Gk *slaves* *n* Gk *Love the brotherhood* *o* Or *carried up our sins in his body to the tree* *p* Gk *bruise* *q* Gk *vessel*

of the suffering faithful known. **13–17**: Respect for civil authority (Rom 13.1–7). **14**: *Governors,* of Roman provinces. **16**: Live as *servants of God,* only so can you *live as free people.* **17**: *Fear God. Honor the emperor,* Prov 24.21; Mt 22.21; Rom 13.7. **2.18–25**: Obedience to masters. The Christian slave must have respect even for harsh and unjust masters (Col 3.22–25; Eph 6.5–8). **21–25**: The example of Christ is set forth in language that echoes Isa 53.5–12. **25**:

Shepherd and guardian, synonymous terms, referring to Christ (5.4; Jn 10.11, 14; Heb 13.20).

3.1–7: Wives and husbands (Col 3.18). **1**: An unbelieving husband *may be won* (converted) by example (1 Cor 7.13–16) without argument. **4**: *Inner self,* Eph 3.16. **6**: Gen 18.12. **7**: 1 Thess 4.4.

3.8–12: Love for one another. **9**: Rom 12.17; 1 Thess 5.15. **10**: Ps 34.12–16. **13–17**: Patience in persecution (Mt 5.10–11). Be

and their lips from speaking
deceit;
11 let them turn away from evil
and do good;
let them seek peace and
pursue it.
12 For the eyes of the Lord are on
the righteous,
and his ears are open to
their prayer.
But the face of the Lord is against
those who do evil."

13 Now who will harm you if you are eager to do what is good? 14 But even if you do suffer for doing what is right, you are blessed. Do not fear what they fear,*r* and do not be intimidated, 15 but in your hearts sanctify Christ as Lord. Always be ready to make your defense to anyone who demands from you an accounting for the hope that is in you; 16 yet do it with gentleness and reverence.*s* Keep your conscience clear, so that, when you are maligned, those who abuse you for your good conduct in Christ may be put to shame. 17 For it is better to suffer for doing good, if suffering should be God's will, than to suffer for doing evil. 18 For Christ also suffered*t* for sins once for all, the righteous for the unrighteous, in order to bring you*u* to God. He was put to death in the flesh, but made alive in the spirit, 19 in which also he went and made a proclamation to the spirits in prison, 20 who in former times did not obey, when God waited patiently in the days of Noah, during the building of the ark, in which a few, that is, eight persons, were saved through water. 21 And baptism, which this prefigured, now

saves you—not as a removal of dirt from the body, but as an appeal to God for*v* a good conscience, through the resurrection of Jesus Christ, 22 who has gone into heaven and is at the right hand of God, with angels, authorities, and powers made subject to him.

4 Since therefore Christ suffered in the flesh,*w* arm yourselves also with the same intention (for whoever has suffered in the flesh has finished with sin), 2 so as to live for the rest of your earthly life*x* no longer by human desires but by the will of God. 3 You have already spent enough time in doing what the Gentiles like to do, living in licentiousness, passions, drunkenness, revels, carousing, and lawless idolatry. 4 They are surprised that you no longer join them in the same excesses of dissipation, and so they blaspheme.*y* 5 But they will have to give an accounting to him who stands ready to judge the living and the dead. 6 For this is the reason the gospel was proclaimed even to the dead, so that, though they had been judged in the flesh as everyone is judged, they might live in the spirit as God does.

7 The end of all things is near;*z* therefore be serious and discipline yourselves for the sake of your prayers. 8 Above all, maintain constant love for one another, for love covers a multitude of sins. 9 Be hospitable to one another without com-

r Gk *their fear*   *s* Or *respect*   *t* Other ancient authorities read *died*   *u* Other ancient authorities read *us*   *v* Or *a pledge to God from*   *w* Other ancient authorities add *for us*; others, *for you*   *x* Gk *rest of the time in the flesh*   *y* Or *they malign you*   *z* Or *is at hand*

---

ready to defend your hope as a Christian, but do so with gentleness and reverence (1 Cor 4.12–13).
3.18–22: The example of Christ (compare 2.21–25). **18:** *Put to death in the flesh* asserts that he really died. *Made alive in the spirit,* death did not hold him; he rose to new life (1 Cor 15.35–50). **19–20:** This difficult passage may mean that Christ announced his completed work in the realm of the dead to those who in Noah's day had been disobedient (compare 4.6). **21:** Mention of the flood (Gen chs 6–8) leads to a comparison with baptism: then wa-

ter destroyed; now in baptism it saves. **22:** *At the right hand of God,* the place of honor (Ps 110.1). *Authorities and powers,* angelic beings (see Col 1.16 n.).
4.1–6: Spiritual liberation through suffering. **1:** *Since therefore,* refers back to 3.18. **4:** *No longer join,* as you once did. **6:** *Proclaimed . . . to the dead,* see 3.19–20 n. **7–11:** Since the end is near, be sober stewards of God's diverse gifts (1 Cor 12.14–26). **8:** *Love covers . . . sins,* Prov 10.12; Lk 7.47; 1 Cor 13.7. **9:** *Be hospitable,* provide lodging for Christian travelers (see Heb 13.2 n.; 3 Jn 5–8 n.).

plaining. ¹⁰Like good stewards of the manifold grace of God, serve one another with whatever gift each of you has received. ¹¹Whoever speaks must do so as one speaking the very words of God; whoever serves must do so with the strength that God supplies, so that God may be glorified in all things through Jesus Christ. To him belong the glory and the power forever and ever. Amen.

12 Beloved, do not be surprised at the fiery ordeal that is taking place among you to test you, as though something strange were happening to you. ¹³But rejoice insofar as you are sharing Christ's sufferings, so that you may also be glad and shout for joy when his glory is revealed. ¹⁴If you are reviled for the name of Christ, you are blessed, because the spirit of glory,ᵃ which is the Spirit of God, is resting on you.ᵇ ¹⁵But let none of you suffer as a murderer, a thief, a criminal, or even as a mischief maker. ¹⁶Yet if any of you suffers as a Christian, do not consider it a disgrace, but glorify God because you bear this name. ¹⁷For the time has come for judgment to begin with the household of God; if it begins with us, what will be the end for those who do not obey the gospel of God? ¹⁸And

> "If it is hard for the righteous
>   to be saved,
> what will become of the
>   ungodly and the sinners?"

¹⁹Therefore, let those suffering in accordance with God's will entrust themselves to a faithful Creator, while continuing to do good.

5 Now as an elder myself and a witness of the sufferings of Christ, as well as one who shares in the glory to be revealed, I exhort the elders among you ²to tend the flock of God that is in your charge, exercising the oversight,ᶜ not under compulsion but willingly, as God would have you do itᵈ—not for sordid gain but eagerly. ³Do not lord it over those in your charge, but be examples to the flock. ⁴And when the chief shepherd appears, you will win the crown of glory that never fades away. ⁵In the same way, you who are younger must accept the authority of the elders.ᵉ And all of you must clothe yourselves with humility in your dealings with one another, for

> "God opposes the proud,
>   but gives grace to the humble."

6 Humble yourselves therefore under the mighty hand of God, so that he may exalt you in due time. ⁷Cast all your anxiety on him, because he cares for you. ⁸Discipline yourselves, keep alert.ᶠ Like a roaring lion your adversary the devil prowls around, looking for someone to devour. ⁹Resist him, steadfast in your faith, for you know that your brothers and sistersᵍ in all the world are undergoing the same kinds of suffering. ¹⁰And after you have suffered for a little while, the God of all grace, who has called you to his eternal glory in Christ, will himself restore, support, strengthen, and estab-

a Other ancient authorities add *and of power*
b Other ancient authorities add *On their part he is blasphemed, but on your part he is glorified*
c Other ancient authorities lack *exercising the oversight*   *d* Other ancient authorities lack *as God would have you do it*   *e* Or *of those who are older*   *f* Or *be vigilant*   *g* Gk *your brotherhood*

4.12–19: Recapitulation of previous exhortations. **12:** *Fiery ordeal,* persecutions (1.6–7). **13:** Rom 8.17; 2 Tim 2.12. **14:** 2.20; Isa 11.2. **16:** Phil 1.20. *Christian,* see Acts 11.26 n. **18:** Prov 11.31 Septuagint. **19:** 2.20.
5.1–14: Concluding exhortations and greetings. 1: *Elders,* church officials (Acts 14.23; 20.17–38; Jas 5.14). **3:** *Not . . . over,* Mk 10.42. **4:** *The chief shepherd,* Christ (see 2.25 n.). **5:** *Clothe yourselves,* humility is not natural to anyone; it must be put on like the

menial's apron. The quotation is from Prov 3.34. **7:** Ps 55.22. **8:** *Roaring,* the rage of hunger. **9:** *Resist,* the devil (compare Eph 6.11–18). **10:** God has *called;* he will also *strengthen.* **12:** *Silvanus,* probably identical with Silas (Acts 15.22, 40; 16.19; 17.4). **13:** *Babylon,* a cryptic designation of Rome (see Rev 17.1 n.). *My son,* in a spiritual sense (see Philem 10 n.). *Mark,* see Acts 12.12 n.; Col 4.10; Philem 24. **14:** *Kiss,* see Rom 16.16 n.

lish you. [11] To him be the power forever and ever. Amen.

12 Through Silvanus, whom I consider a faithful brother, I have written this short letter to encourage you and to testify that this is the true grace of God. Stand fast in it. [13] Your sister church[h] in Babylon, chosen together with you, sends you greetings; and so does my son Mark. [14] Greet one another with a kiss of love.

Peace to all of you who are in Christ. [i]

h Gk *She who is*   i Other ancient authorities add *Amen*

The Second Letter of
Peter

The Second Letter of Peter presents itself as the work of Simeon Peter (1.1; Simeon is Semitic orthography; see Acts 15.14 n.), whose death was predicted by Jesus (1.14). The author has witnessed the transfiguration, which is evidence that Jesus' promise to return in glory is no mere fable, as some allege (1.16). At the same time, identification with Peter, or even a disciple of Peter (like the author of 1 Peter), appears unlikely. Although the author claims to have written a previous letter (3.1), 2 Peter does not reflect the social setting of 1 Peter or its imagery for the new covenant community. Despite the author's claim to fellowship with Paul (3.15), the reference to misinterpretation of Paul's letters suggests that they have already been collected and are treated as "scriptures" in some churches (3.16)—clearly not the case during the lifetime of the apostles.

The author refers to "your apostles" as those who have transmitted the teaching of the prophets and the Lord, which the readers must take care to remember correctly (3.2). The reference to the transfiguration in 1.17–18 assumes an audience familiar with some form of the synoptic gospel tradition. In addition, 2 Peter has apparently incorporated sections of Jude (compare 2.1–8 with Jude 4–16), and uses the Stoic philosophical concept of the return of all things to fire at the end of the world (3.10). Thus, 2 Peter appears to have been composed late in the first century or early in the second century A.D. In this view, it is intended to meet the challenge of persons who are distorting the true apostolic tradition by denying that the world will be brought to an end by divine judgment (2.1–3; 3.3–4).

It should be borne in mind that in antiquity pseudonymous authorship was a widely accepted literary convention. Therefore the use of an apostle's name in reasserting his teaching was not regarded as dishonest but was merely a way of reminding the church of what it had received from God through that apostle.

From 1.12–14 it appears that the letter is cast as the "last testament" of the apostle. Like other testaments, the author seeks to preserve a legacy of correct teaching (1.15) against the turmoil created by false teachers in coming generations (3.3–4). These teachers are already active in the church. Since denunciations of the moral character of opponents were common in ancient polemic, the author's descriptions of the heretics' moral turpitude (2.2–14) should not be taken as evidence for their teaching. It does provide an opportunity to warn that divine judgment, which requires a high standard of ethical conduct (1.3–11; 2.11–21), will condemn such persons. The point of contention is clearly belief in the coming of judgment itself (3.10–12). The opposition may have argued that the failure of Christ to return (3.4), as well as the philosophical doctrine of the eternity of the world, disproves traditional belief in the parousia.

Second Peter embodies a sustained refutation of all such claims, drawing upon evidence of divine judgment in the Hebrew Scriptures, the testimony of the prophets, the transfiguration of Jesus witnessed by Peter himself, and even Paul's own writings. The author also advances logical arguments against their view; these arguments seek to show that the doctrine of the parousia is no myth or fable. Stoic cosmology taught that the universe would return to the fiery matter from which it emerged. As for the alleged delay of the parousia, the opponents fail to recognize that God's time is different from that of humans (3.8). Furthermore, God delays the judgment out of patience so that people may have opportunity to repent (3.9).

1 Simeon^a Peter, a servant^b and apostle of Jesus Christ,

To those who have received a faith as precious as ours through the righteousness of our God and Savior Jesus Christ:^c

2 May grace and peace be yours in abundance in the knowledge of God and of Jesus our Lord.

3 His divine power has given us everything needed for life and godliness, through the knowledge of him who called us by^d his own glory and goodness. 4 Thus he has given us, through these things, his precious and very great promises, so that through them you may escape from the corruption that is in the world because of lust, and may become participants of the divine nature. 5 For this very reason, you must make every effort to support your faith with goodness, and goodness with knowledge, 6 and knowledge with self-control, and self-control with endurance, and endurance with godliness, 7 and godliness with mutual^e affection, and mutual^e affection with love. 8 For if these things are yours and are increasing among you, they keep you from being ineffective and unfruitful in the knowledge of our Lord Jesus Christ. 9 For anyone who lacks these things is nearsighted and blind, and is forgetful of the cleansing of past sins. 10 Therefore, brothers and sisters,^f be all the more eager to confirm your call and election, for if you do this, you will never stumble. 11 For in this way, entry into the eternal kingdom of our Lord and Savior Jesus Christ will be richly provided for you.

12 Therefore I intend to keep on reminding you of these things, though you know them already and are established in the truth that has come to you. 13 I think it right, as long as I am in this body,^g to refresh your memory, 14 since I know that my death^h will come soon, as indeed our Lord Jesus Christ has made clear to me. 15 And I will make every effort so that after my departure you may be able at any time to recall these things.

16 For we did not follow cleverly devised myths when we made known to you the power and coming of our Lord Jesus Christ, but we had been eyewitnesses of his majesty. 17 For he received honor and glory from God the Father when that voice was conveyed to him by the Majestic Glory, saying, "This is my Son, my Beloved,ⁱ with whom I am well pleased." 18 We ourselves heard this voice come from heaven, while we were with him on the holy mountain.

19 So we have the prophetic message more fully confirmed. You will do well to be attentive to this as to a lamp shining in a dark place, until the day dawns and the morning star rises in your hearts. 20 First of all you must understand this, that no prophecy of scripture is a matter of one's own interpretation, 21 because no prophecy ever came by human will, but men and women moved by the Holy Spirit spoke from God.^j

a Other ancient authorities read *Simon*
b Gk *slave*   c Or *of our God and the Savior Jesus Christ*   d Other ancient authorities read *through*   e Gk *brotherly*   f Gk *brothers*   g Gk *tent*   h Gk *the putting off of my tent*
i Other ancient authorities read *my beloved Son*
j Other ancient authorities read *but moved by the Holy Spirit saints of God spoke*

1.1–2: Salutation, see Rom 1.1–7 n. **1:** *To those who have received,* i.e. Gentile converts who had not observed the Mosaic law. *God and Savior,* see Titus 2.13. **2:** *Knowledge* is one of the author's favorite words (1.3, 5, 6, 8; 2.20; 3.18). *Grace and peace,* see 2 Thess 1.2 n. **1.3–21: Exhortation to holiness.** Salvation is the result of God's goodness. One must hold firmly *his precious and very great promises,* and thus escape the corruption of the world. **5–7:** *Faith* should lead to Christian virtues (Rom 5.2–5, 1 Cor 13.13). **11:** *Kingdom,* Col 1.13.

1.16: Apostolic tradition is not a collection of *myths,* but is based upon the experience of *eyewitnesses.* **17–18:** The transfiguration confirms the tradition that Christ will return in glory (Mt 17.1–8; Mk 9.2–8; Lk 9.28–36). **19:** Prophecies also confirm the coming of the Lord. **20–21:** Since *prophecy* is inspired by the Spirit, its *interpretation* must be in accord with God's intention.

2 But false prophets also arose among the people, just as there will be false teachers among you, who will secretly bring in destructive opinions. They will even deny the Master who bought them—bringing swift destruction on themselves. 2 Even so, many will follow their licentious ways, and because of these teachers*k* the way of truth will be maligned. 3 And in their greed they will exploit you with deceptive words. Their condemnation, pronounced against them long ago, has not been idle, and their destruction is not asleep.

4 For if God did not spare the angels when they sinned, but cast them into hell*l* and committed them to chains*m* of deepest darkness to be kept until the judgment; 5 and if he did not spare the ancient world, even though he saved Noah, a herald of righteousness, with seven others, when he brought a flood on a world of the ungodly; 6 and if by turning the cities of Sodom and Gomorrah to ashes he condemned them to extinction*n* and made them an example of what is coming to the ungodly;*o* 7 and if he rescued Lot, a righteous man greatly distressed by the licentiousness of the lawless 8 (for that righteous man, living among them day after day, was tormented in his righteous soul by their lawless deeds that he saw and heard), 9 then the Lord knows how to rescue the godly from trial, and to keep the unrighteous under punishment until the day of judgment 10—especially those who indulge their flesh in depraved lust, and who despise authority.

Bold and willful, they are not afraid to slander the glorious ones,*p* 11 whereas angels, though greater in might and power, do not bring against them a slanderous judgment from the Lord.*q*

12 These people, however, are like irrational animals, mere creatures of instinct, born to be caught and killed. They slander what they do not understand, and when those creatures are destroyed,*r* they also will be destroyed, 13 suffering*s* the penalty for doing wrong. They count it a pleasure to revel in the daytime. They are blots and blemishes, reveling in their dissipation*t* while they feast with you. 14 They have eyes full of adultery, insatiable for sin. They entice unsteady souls. They have hearts trained in greed. Accursed children! 15 They have left the straight road and have gone astray, following the road of Balaam son of Bosor,*u* who loved the wages of doing wrong, 16 but was rebuked for his own transgression; a speechless donkey spoke with a human voice and restrained the prophet's madness.

17 These are waterless springs and mists driven by a storm; for them the deepest darkness has been reserved. 18 For they speak bombastic nonsense, and with licentious desires of the flesh they entice people who have just*v* escaped from those who live in error. 19 They promise them freedom, but they themselves are slaves of corruption; for people are slaves to whatever masters them. 20 For if, after they have escaped the defilements of the world through the

k Gk *because of them*   *l* Gk *Tartaros*
m Other ancient authorities read *pits*   *n* Other ancient authorities lack *to extinction*   *o* Other ancient authorities read *an example to those who were to be ungodly*   *p* Or *angels;* Gk *glories*
q Other ancient authorities read *before the Lord;* others lack the phrase   *r* Gk *in their destruction*
s Other ancient authorities read *receiving*
t Other ancient authorities read *love feasts*
u Other ancient authorities read *Beor*
v Other ancient authorities read *actually*

2.1–22: **Attack upon false teachers.** The language follows closely the Letter of Jude (vv. 4–16) in the denunciation of heretics. **1:** The future tense gives an atmosphere of prophecy; but in vv. 14–19 and 3.5 the present tense is used. False teachers are already present.
2.4–10a: God's past judgments indicate what is in store for these people. **5:** Gen 6.6–

8; 8.18. **6:** Gen 19.24. **7:** Gen 19.16, 29.
2.11–12: See Jude 9–10 n. **13–16:** The opponents are no better than pagans (see Rom 1.18–32; 1 Cor 6.9–11; Eph 4.17–19; Col 3.5–10; 1 Pet 4.3–4). **15:** Num 22.5, 7. **16:** Num 22.21, 23, 28, 30–31. *Slaves,* Jn 8.34; Rom 6.6, 16. **20:** *The last state,* Mt 12.45; Lk 11.26. **21:** *The holy commandment . . . passed on to them,* 1 Tim 6.14; Jude 3. **22:** Prov 26.11.

knowledge of our Lord and Savior Jesus Christ, they are again entangled in them and overpowered, the last state has become worse for them than the first. ²¹For it would have been better for them never to have known the way of righteousness than, after knowing it, to turn back from the holy commandment that was passed on to them. ²²It has happened to them according to the true proverb,

"The dog turns back to its
    own vomit,"

and,

"The sow is washed only to
    wallow in the mud."

3 This is now, beloved, the second letter I am writing to you; in them I am trying to arouse your sincere intention by reminding you ²that you should remember the words spoken in the past by the holy prophets, and the commandment of the Lord and Savior spoken through your apostles. ³First of all you must understand this, that in the last days scoffers will come, scoffing and indulging their own lusts ⁴and saying, "Where is the promise of his coming? For ever since our ancestors died, ʷ all things continue as they were from the beginning of creation!" ⁵They deliberately ignore this fact, that by the word of God heavens existed long ago and an earth was formed out of water and by means of water, ⁶through which the world of that time was deluged with water and perished.

⁷But by the same word the present heavens and earth have been reserved for fire, being kept until the day of judgment and destruction of the godless.

8 But do not ignore this one fact, beloved, that with the Lord one day is like a thousand years, and a thousand years are like one day. ⁹The Lord is not slow about his promise, as some think of slowness, but is patient with you, ˣ not wanting any to perish, but all to come to repentance. ¹⁰But the day of the Lord will come like a thief, and then the heavens will pass away with a loud noise, and the elements will be dissolved with fire, and the earth and everything that is done on it will be disclosed. ʸ

11 Since all these things are to be dissolved in this way, what sort of persons ought you to be in leading lives of holiness and godliness, ¹²waiting for and hasteningᶻ the coming of the day of God, because of which the heavens will be set ablaze and dissolved, and the elements will melt with fire? ¹³But, in accordance with his promise, we wait for new heavens and a new earth, where righteousness is at home.

14 Therefore, beloved, while you are waiting for these things, strive to be found by him at peace, without spot or

w Gk *our fathers fell asleep*   x Other ancient
authorities read *on your account*   y Other
ancient authorities read *will be burned up*
z Or *earnestly desiring*

3.1–18: The day of the Lord. Scoffers ridicule the hope of Christ's second coming. Delay is no proof that he will not come, for God does not measure time as mortals do. The delay shows God's patience and desire that sinners should repent. Because the day of the Lord will come as promised, Christians should await it in holiness and godliness. **3:** The coming of scoffers is a sign of *the last days* (see 2 Tim 3.1 n.). **4:** *Ever since our ancestors died* expresses the viewpoint of a later generation. The scoffers apparently argued that the universe is immutable, and therefore there would be no end. **5:** God's *word* is not just chatter; it is creative, and by it the world was brought into being (Gen 1.6–10). **6–7:** So far from being immutable the world has already been destroyed once, by the deluge (2.5; Gen 7.11), and, so the author affirms, it will be destroyed again, by fire. **3.8:** God's measure of time is cited from Ps 90.4. **10:** *Day of the Lord,* see Am 5.18–20 n.; Joel 2.28–32 n. *Like a thief,* Mt 24.43; Lk 12.39. **12:** In apocalyptic writings *fire* plays a prominent role in the destruction of the world. **13–14:** Freely quoted from Isa 65.17; 66.22. In Isaiah and Rev ch 21 the thought is not the destruction but the renewal and transformation of the universe. **14:** *By him,* i.e. Christ. **15–16:** The author knows several letters of Paul and equates them with *the other scriptures,* which confirm the tradition that Christians must lead blameless lives while awaiting Christ's coming.

blemish; 15 and regard the patience of our Lord as salvation. So also our beloved brother Paul wrote to you according to the wisdom given him, 16 speaking of this as he does in all his letters. There are some things in them hard to understand, which the ignorant and unstable twist to their own destruction, as they do the other scriptures. 17 You therefore, be-

loved, since you are forewarned, beware that you are not carried away with the error of the lawless and lose your own stability. 18 But grow in the grace and knowledge of our Lord and Savior Jesus Christ. To him be the glory both now and to the day of eternity. Amen. *a*

a Other ancient authorities lack *Amen*

The First Letter of
John

Although traditionally called a letter, 1 John is a treatise or sermon from an unknown teacher in the Johannine tradition to those in the community ("my little children," 2.1; "beloved," 2.7). Two letters from the same author indicate that the writer is an elder (2 Jn 1; 3 Jn 1) with authority in other Johannine communities. Travelling emissaries from the elder apparently maintained contact with a number of communities in the region (3 Jn 10). Consequently, the claim that 1 John represents the tradition that has been passed on by the official witnesses ("we" in 1.1–4) is made in the context of a network of Johannine churches.

By the end of the second century A.D., 1 John was commonly thought to have been written by the author of the Fourth Gospel, and 2 and 3 John by another member of the Johannine circle (see Eusebius, *Hist.* III. 39.4 for the distinction between John, the disciple, and John, the elder). The three letters, however, reflect a common community setting that is quite different from that which is implied by the gospel. In the gospel the Johannine community is a minority that had been excluded from the Jewish synagogue and still faces hostility from persons with ties to the synagogue (Jn 9.22; 12.42–43; 16.1–4a). The point of conflict lay in the Johannine claim that Jesus is the preexistent, divine Son of God, the only revelation of the Father (Jn 5.18; 8.23–30, 38, 52–59). First John is not concerned with the relationship of Christian faith to Jewish traditions but with the proper testimony about Jesus embodied in the Christian tradition itself (1.1–4). Its emphasis falls upon the physical reality of Jesus' coming in the flesh (1.1–3; 4.2). The author does not oppose Jewish claims to interpret the tradition of Moses, as the gospel does, but opposes the teaching of former members of the Johannine fellowship who have broken away (2.19; 4.1; 2 Jn 7).

Both the author and these secessionist teachers clearly appeal to a common Johannine tradition. The elder knows that people could easily be misled by the false teachers. The Holy Spirit, which members of the community have received, will confirm that the writer's teaching reflects the true gospel (2.20, 27; 4.13). The dissident teachers apparently emphasize the divine Jesus of the Gospel and deny the significance of Jesus' human reality and death on the cross as sacrifice for sin (1.7; 2.2; 3.16; 4.10; 5.6). The author challenges their claims to sinlessness by insisting that sinlessness is a gift that Christians receive from God through Christ (2.1–2). Persons who have violated the fundamental commandment of the Johannine tradition, the mutual love among Christians (Jn 13.35; 15.12), are not sinless (2.3–9; 3.22–23; 4.21; 5.2–3). Their apostasy and false teaching is a deadly sin (idolatry), which separates them from the communal prayer for forgiveness (5.15–17, 21) and makes them agents of the demonic attack on the faithful, which apocalyptic traditions had prophesied for the end-time (2.18, 22; 3.4–5; 4.1–5).

Claims to perfectionism, denial of the significance of Jesus' coming in the flesh, rejection of the saving power of Jesus' death, and schismatic preaching among established Christian communities are all features of second-century gnostic teaching. Since 1 John does not provide evidence of the peculiar teaching of the secessionist opponents, one cannot identify them with any known gnostic group. The conflict may have arisen over the true meaning of the Johannine Gospel prior to the emergence of well-defined gnostic groups. As part of the Christian canon, 1 John rejects any gnosticizing interpretation of the Fourth Gospel.

1 We declare to you what was from the beginning, what we have heard, what we have seen with our eyes, what we have looked at and touched with our hands, concerning the word of life— [2]this life was revealed, and we have seen it and testify to it, and declare to you the eternal life that was with the Father and was revealed to us— [3]we declare to you what we have seen and heard so that you also may have fellowship with us; and truly our fellowship is with the Father and with his Son Jesus Christ. [4]We are writing these things so that our[a] joy may be complete.

[5] This is the message we have heard from him and proclaim to you, that God is light and in him there is no darkness at all. [6]If we say that we have fellowship with him while we are walking in darkness, we lie and do not do what is true; [7]but if we walk in the light as he himself is in the light, we have fellowship with one another, and the blood of Jesus his Son cleanses us from all sin. [8]If we say that we have no sin, we deceive ourselves, and the truth is not in us. [9]If we confess our sins, he who is faithful and just will forgive us our sins and cleanse us from all unrighteousness. [10]If we say that we have not sinned, we make him a liar, and his word is not in us.

2 My little children, I am writing these things to you so that you may not sin. But if anyone does sin, we have an advocate with the Father, Jesus Christ the righteous; [2]and he is the atoning sacrifice for our sins, and not for ours only but also for the sins of the whole world.

[3] Now by this we may be sure that we know him, if we obey his commandments. [4]Whoever says, "I have come to know him," but does not obey his commandments, is a liar, and in such a person the truth does not exist; [5]but whoever obeys his word, truly in this person the love of God has reached perfection. By this we may be sure that we are in him: [6]whoever says, "I abide in him," ought to walk just as he walked.

[7] Beloved, I am writing you no new commandment, but an old commandment that you have had from the beginning; the old commandment is the word that you have heard. [8]Yet I am writing you a new commandment that is true in him and in you, because[b] the darkness is passing away and the true light is already shining. [9]Whoever says, "I am in the light," while hating a brother or sister,[c] is still in the darkness. [10]Whoever loves a brother or sister[d] lives in the light, and in such a person[e] there is no cause for stumbling. [11]But whoever hates another believer[f] is in the darkness, walks in the

---

a Other ancient authorities read *your*
b Or *that*   c Gk *hating a brother*   d Gk *loves a brother*   e Or *in it*   f Gk *hates a brother*

1.1–4: Introduction. 1–2: *From the beginning,* compare Gen 1.1; Jn 1.1. *We,* teachers in the Johannine community who preserve its traditions. *The word of life,* namely Christ, the word and source of life (Jn 1.14; 11.25; 14.6). *Heard, seen, looked at, touched* refer to the evidence of the senses, refuting the claim that Christ was not really human. **3–4:** The purpose of writing is the attainment of *fellowship* (oneness with Christ) and Christian joy (Jn 15.11; 17.13).

1.5–10: Right attitude toward sin. 5: *God is light,* absolute holiness without taint of evil. **6:** *Walking in darkness,* habitual and intentional evil conduct (Jn 3.19; 1 Jn 2.11). **8–9:** Denial of *sin* (guilt, sinful nature) is self-deception; confession brings forgiveness. **10:** Denial of sin contradicts God's declaration that no one is sinless. The secessionists are liars (2.4, 22; 4.20; Ps 14.1–2; 53.1–3).

2.1–6: Obedience. 1: *That you may not sin* is the ultimate goal of Christian living (Rom 6.11). *Advocate,* one who pleads the cause of another (Jn 14.16f.). **2:** *Atoning sacrifice,* 4.10. **3–5:** Obedience to God's *commandments* tests whether we know God, and measures the perfection (completeness) of our *love of God* (Jn 14.15, 21, 23; 15.10). **6:** Jesus is the pattern of obedience.

2.7–11: Love for one another. 8: The *commandment* to love, though old, is never obsolete or antiquated, but ever *new,* being the law of the new age and overcoming the *darkness* of evil (1.5; Jn 13.34; 15.12). **9–11:** Hatred of a *brother or sister* (fellow Christian) is incompatible with Christ's *light* (see Jn 8.12; 11.9–10; 12.35–36).

darkness, and does not know the way to go, because the darkness has brought on blindness. 12 I am writing to you, little children,
because your sins are forgiven on account of his name. 13 I am writing to you, fathers,
because you know him who is from the beginning.
I am writing to you, young people,
because you have conquered the evil one. 14 I write to you, children,
because you know the Father.
I write to you, fathers,
because you know him who is from the beginning.
I write to you, young people,
because you are strong
and the word of God abides in you,
and you have overcome the evil one.

15 Do not love the world or the things in the world. The love of the Father is not in those who love the world; 16for all that is in the world—the desire of the flesh, the desire of the eyes, the pride in riches—comes not from the Father but from the world. 17And the world and its desire are passing away, but those who do the will of God live forever.

18 Children, it is the last hour! As you have heard that antichrist is coming, so now many antichrists have come. From this we know that it is the last hour. 19They went out from us, but they did not belong to us; for if they had belonged to us, they would have remained with us. But by going out they made it plain that none of them belongs to us. 20But you have been anointed by the Holy One, and all of you have knowledge. *h* 21I write to you, not because you do not know the truth, but because you know it, and you know that no lie comes from the truth. 22Who is the liar but the one who denies that Jesus is the Christ?*i* This is the antichrist, the one who denies the Father and the Son. 23No one who denies the Son has the Father; everyone who confesses the Son has the Father also. 24Let what you heard from the beginning abide in you. If what you heard from the beginning abides in you, then you will abide in the Son and in the Father. 25And this is what he has promised us,*j* eternal life.

26 I write these things to you concerning those who would deceive you. 27As for you, the anointing that you received from him abides in you, and so you do not need anyone to teach you. But as his anointing teaches you about all things, and is true and is not a lie, and just as it has taught you, abide in him. *k*

28 And now, little children, abide in him, so that when he is revealed we may have confidence and not be put to shame before him at his coming.

29 If you know that he is righteous, you may be sure that everyone who does

g Or *the desire for it*   *h* Other ancient authorities read *you know all things*   *i* Or *the Messiah*   *j* Other ancient authorities read *you*   *k* Or *it*

2.12–14: **True relationship to God in Christ. 12**: *Little children,* the whole Christian group (2.1, 18, 28). **14**: *Fathers,* the aged. *Young people,* youth. The terms could refer to leaders and younger members of the community.
2.15–17: **True appraisal of the world. 15**: *The world,* all that is alienated from God (Jas 1.27). **16**: *Desire of the flesh,* unlawful physical gratification; *desire of the eyes,* sinful delights of mind or emotions; *pride in riches,* empty trust in possessions.

2.18–29: **Loyalty to the true faith. 18**: *Antichrist,* an expression coined by Johannine Christians meaning "opponent of Christ"; in this case, false teachers (4.3). **19**: They left the Christian fellowship because they were not really Christians. **20–21**: Christians have received the enlightening grace of the Holy Spirit, and therefore can discern *the truth* (v. 27; 1 Cor 2.15–16). **22–23**: The dissidents denied the human reality of Jesus (2 Jn 7). **27**: Jn 14.26. **28**: 4.17. **29**: 3.7–10; 4.7.

3 right has been born of him. ¹See what love the Father has given us, that we should be called children of God; and that is what we are. The reason the world does not know us is that it did not know him. ²Beloved, we are God's children now; what we will be has not yet been revealed. What we do know is this: when he[l] is revealed, we will be like him, for we will see him as he is. ³And all who have this hope in him purify themselves, just as he is pure.

4 Everyone who commits sin is guilty of lawlessness; sin is lawlessness. ⁵You know that he was revealed to take away sins, and in him there is no sin. ⁶No one who abides in him sins; no one who sins has either seen him or known him. ⁷Little children, let no one deceive you. Everyone who does what is right is righteous, just as he is righteous. ⁸Everyone who commits sin is a child of the devil; for the devil has been sinning from the beginning. The Son of God was revealed for this purpose, to destroy the works of the devil. ⁹Those who have been born of God do not sin, because God's seed abides in them;[m] they cannot sin, because they have been born of God. ¹⁰The children of God and the children of the devil are revealed in this way: all who do not do what is right are not from God, nor are those who do not love their brothers and sisters.[n]

11 For this is the message you have heard from the beginning, that we should love one another. ¹²We must not be like Cain who was from the evil one and murdered his brother. And why did he murder him? Because his own deeds were evil and his brother's righteous. ¹³Do not be astonished, brothers and sisters,[o] that the world hates you. ¹⁴We know that we have passed from death to life because we love one another. Whoever does not love abides in death. ¹⁵All who hate a brother or sister[n] are murderers, and you know that murderers do not have eternal life abiding in them. ¹⁶We know love by this, that he laid down his life for us—and we ought to lay down our lives for one another. ¹⁷How does God's love abide in anyone who has the world's goods and sees a brother or sister[p] in need and yet refuses help?

18 Little children, let us love, not in word or speech, but in truth and action. ¹⁹And by this we will know that we are from the truth and will reassure our hearts before him ²⁰whenever our hearts condemn us; for God is greater than our hearts, and he knows everything. ²¹Beloved, if our hearts do not condemn us, we have boldness before God; ²²and we receive from him whatever we ask, because we obey his commandments and do what pleases him.

23 And this is his commandment, that we should believe in the name of his Son Jesus Christ and love one another, just as he has commanded us. ²⁴All who obey his commandments abide in him,

l Or *it*   m Or *because the children of God abide in him*   n Gk *his brother*   o Gk *brothers*
p Gk *brother*

3.1–10: **Filial relation expressed in right conduct. 1–2**: God's *love* in making us *children* (Jn 1.12) progressively produces resemblance to God, here and hereafter. **3**: Hope of complete moral likeness to Christ motivates purity of life (Mt 5.8). **5**: Jn 1.29. **6**: *Sins,* habitually and constantly (3 Jn 11). **8**: Jn 8.44; Acts 13.10; Heb 2.14. **9**: 5.18.
3.11–18: **Love for one another. 11–12**: Abel's *righteous deeds* stirred Cain's hatred (Gen 4.8; Heb 11.4; Jude 11). **13**: *The world* likewise *hates* the church (Jn 15.18–19). **14–15**: Hatred of fellow Christians is equivalent to murder (compare Mt 5.21–22). **16–17**: Jesus, in contrast to taking life, *laid down his* life (Jn 15.13). He is our pattern. *To lay down our lives* may take the form of daily sacrifice for others *in need*. **18**: *Word,* intention without *action; speech,* hypocrisy, not *truth* (Jas 1.22).
3.19–24: **The Christian's assurance. 19**: *This,* the *love* described in v. 18. **20**: *God,* who *knows* everything, judges us by the abiding relation of love to others, rather than by our passing moods (Jn 21.17). **21–22**: *Boldness* in prayer results from obedience to God, and strengthens assurance (5.14). **23–24**: Belief *in the name of Jesus Christ,* Jesus bears God's name; belief makes people children of God (Jn 17.11, 12). *Love* (Jn 13.34; 15.17) is evidence of God's Spirit and presence (4.12–13).

and he abides in them. And by this we know that he abides in us, by the Spirit that he has given us.

4 Beloved, do not believe every spirit, but test the spirits to see whether they are from God; for many false prophets have gone out into the world. 2By this you know the Spirit of God: every spirit that confesses that Jesus Christ has come in the flesh is from God, 3and every spirit that does not confess Jesus*q* is not from God. And this is the spirit of the antichrist, of which you have heard that it is coming; and now it is already in the world. 4Little children, you are from God, and have conquered them; for the one who is in you is greater than the one who is in the world. 5They are from the world; therefore what they say is from the world, and the world listens to them. 6We are from God. Whoever knows God listens to us, and whoever is not from God does not listen to us. From this we know the spirit of truth and the spirit of error.

7 Beloved, let us love one another, because love is from God; everyone who loves is born of God and knows God. 8Whoever does not love does not know God, for God is love. 9God's love was revealed among us in this way: God sent his only Son into the world so that we might live through him. 10In this is love, not that we loved God but that he loved us and sent his Son to be the atoning sacrifice for our sins. 11Beloved, since God loved us so much, we also ought to love one another. 12No one has ever seen God; if we love one another, God lives in us, and his love is perfected in us.

13 By this we know that we abide in him and he in us, because he has given us of his Spirit. 14And we have seen and do testify that the Father has sent his Son as the Savior of the world. 15God abides in those who confess that Jesus is the Son of God, and they abide in God. 16So we have known and believe the love that God has for us.

God is love, and those who abide in love abide in God, and God abides in them. 17Love has been perfected among us in this: that we may have boldness on the day of judgment, because as he is, so are we in this world. 18There is no fear in love, but perfect love casts out fear; for fear has to do with punishment, and whoever fears has not reached perfection in love. 19We love*r* because he first loved us. 20Those who say, "I love God," and hate their brothers or sisters,*s* are liars; for those who do not love a brother or sister*t* whom they have seen, cannot love God whom they have not seen. 21The commandment we have from him is this: those who love God must love their brothers and sisters*s* also.

5 Everyone who believes that Jesus is the Christ*u* has been born of God, and everyone who loves the parent loves the child. 2By this we know that we love the children of God, when we love God and obey his commandments. 3For the love of God is this, that we obey his commandments. And his command-

q Other ancient authorities read *does away with Jesus* (Gk *dissolves Jesus*) r Other ancient authorities add *him*; others add *God* s Gk *brothers* t Gk *brother* u Or *the Messiah*

4.1–6: Discernment of truth and error. 1–2: Note the contrast between *spirits* (supernatural powers claimed by *false prophets*) and the Holy *Spirit* (Jn 16.13–15), which inspires the true confession about Christ (compare 1 Cor 12.1–3). **3:** 2 Jn 7. **4:** *The one who is in the world,* the devil (3.10; Jn 8.44; 12.31). **5:** *They,* the false prophets. **6:** *We,* the Christian teachers. *Whoever knows God* discriminates between *truth* and *error* (Jn 5.37–38; 8.47; 10.4–5).
4.7–21: The blessedness of love. 7–10: That *God is love* is seen in sending *his Son* as

the atoning sacrifice for our sins (2.2). **11–12:** To *love one another* is the only authentication that we know *God,* whom *no one has ever seen* (Jn 1.18). **13–18:** The Holy *Spirit* testifies that *Jesus,* God's *Son,* has revealed his *Father* as love. When his love is *perfected* (matured) in us, *fear* of *judgment* is allayed (2.28; 3.21). **19–21:** *Love* originates in *God.* Failure to *love* is visible evidence of a breach with the unseen God, and a violation of his *commandment.*
5.1–12: Victorious faith issuing in eternal life. **1:** Jn 8.42. **3:** *Love of God* involves obedience to his will. *Not burdensome,* Mt 11.30;

ments are not burdensome, [4]for whatever is born of God conquers the world. And this is the victory that conquers the world, our faith. [5]Who is it that conquers the world but the one who believes that Jesus is the Son of God?

[6] This is the one who came by water and blood, Jesus Christ, not with the water only but with the water and the blood. And the Spirit is the one that testifies, for the Spirit is the truth. [7]There are three that testify:[v] [8]the Spirit and the water and the blood, and these three agree. [9]If we receive human testimony, the testimony of God is greater; for this is the testimony of God that he has testified to his Son. [10]Those who believe in the Son of God have the testimony in their hearts. Those who do not believe in God[w] have made him a liar by not believing in the testimony that God has given concerning his Son. [11]And this is the testimony: God gave us eternal life, and this life is in his Son. [12]Whoever has the Son has life; whoever does not have the Son of God does not have life.

[13] I write these things to you who believe in the name of the Son of God, so that you may know that you have eternal life.

[14] And this is the boldness we have in him, that if we ask anything according to his will, he hears us. [15]And if we know that he hears us in whatever we ask, we know that we have obtained the requests made of him. [16]If you see your brother or sister[x] committing what is not a mortal sin, you will ask, and God[y] will give life to such a one—to those whose sin is not mortal. There is sin that is mortal; I do not say that you should pray about that. [17]All wrongdoing is sin, but there is sin that is not mortal.

[18] We know that those who are born of God do not sin, but the one who was born of God protects them, and the evil one does not touch them. [19]We know that we are God's children, and that the whole world lies under the power of the evil one. [20]And we know that the Son of God has come and has given us understanding so that we may know him who is true;[z] and we are in him who is true, in his Son Jesus Christ. He is the true God and eternal life.

[21] Little children, keep yourselves from idols.[a]

v A few other authorities read (with variations)
[7]There are three that testify in heaven, the Father, the Word, and the Holy Spirit, and these three are one. [8]And there are three that testify on earth:
w Other ancient authorities read in the Son
x Gk your brother   y Gk he   z Other ancient authorities read know the true God
a Other ancient authorities add Amen

Phil 4.13. **4**: Jn 16.33. *That conquers,* or "that has conquered." **6–8**: The Spirit's witness is to *the water* (Jesus' baptism) and to *the blood* (the cross); see Jn 19.34–35. **9**: Jn 5.32, 36; 8.18. **11–12**: *Eternal life* is *in* (living union with) Jesus, God's *Son,* and nowhere else (Jn 1.4; 3.36; 5.24–26; 6.57).

5.13–21: Conclusion. 13: Jn 17.3; 20.31. **14**: Jn 14.14–16; 15.16; 16.23–26. **15**: 3.22. **16**: *Sin that is mortal,* for unforgivable sin compare Mk 3.21; 8.38. Here the reference is probably to those who have left the Johannine community (1.2–3; 3.12–17); Christians are not to associate with such persons (2 Jn 10). **18–20**: *We know,* thrice repeated (compare v. 13). **18**: *Do not sin,* see 3.6 n. **19**: *World,* see 2.15 n. **20**: *The true God,* in contrast to the secessionists' view of God. **21**: *Idols,* any rival to God (1 Cor 10.14).

The Second Letter of
John

Language, literary style, and ideas, along with early church tradition, make it highly probable that this letter came from the same pen that produced 1 John. A copy of 1 John may have accompanied this letter, which instructs its addressees not to associate with the dissidents (vv. 9–11). The tone of the admonitions indicates that the author was well known to the readers and was one whose spiritual authority they acknowledged. Unlike 1 John, which appears to have been addressed to several churches, this letter was written to one specific church ("the elect lady," v. 1), probably one of the churches in Asia Minor.

The letter yields no definite evidence regarding date. The kinship of situation and ideas with those reflected in 1 John suggests a date near the end of the first century (see Introduction to 1 John).

The author alludes to the main teaching of 1 John (vv. 5–7), and adds a command not to show hospitality to false teachers. Anyone who does so is guilty of participation in their evil teaching (vv. 7–11).

1 The elder to the elect lady and her children, whom I love in the truth, and not only I but also all who know the truth, 2because of the truth that abides in us and will be with us forever:

3 Grace, mercy, and peace will be with us from God the Father and from*a* Jesus Christ, the Father's Son, in truth and love.

4 I was overjoyed to find some of your children walking in the truth, just as we have been commanded by the Father. 5But now, dear lady, I ask you, not as though I were writing you a new commandment, but one we have had from the beginning, let us love one another. 6And this is love, that we walk according to his commandments; this is the commandment just as you have heard it from the beginning—you must walk in it.

7 Many deceivers have gone out into the world, those who do not confess that Jesus Christ has come in the flesh; any such person is the deceiver and the antichrist! 8Be on your guard, so that you do not lose what we*b* have worked for, but may receive a full reward. 9Everyone who does not abide in the teaching of Christ, but goes beyond it, does not have God; whoever abides in the teaching has both the Father and the Son. 10Do not receive into the house or welcome anyone who comes to you and does not bring this teaching; 11for to welcome is to participate in the evil deeds of such a person.

12 Although I have much to write to you, I would rather not use paper and ink; instead I hope to come to you and talk with you face to face, so that our joy may be complete.

13 The children of your elect sister send you their greetings.*c*

a Other ancient authorities add *the Lord*
b Other ancient authorities read *you*   *c* Other ancient authorities add *Amen*

1: *Elect lady,* probably refers to a local church, the members of which are called *her children*. **2:** *The truth* about Jesus, in contrast to false teaching (v. 7). **4:** *Some,* not all; heresy has crept in. **5–6:** 1 Jn 2.7; 5.3; Jn 13.34. **7:** *Deceivers,* who taught that the Christ was not indissolubly united with the man Jesus (1 Jn 2.22), or that Jesus' body was not a real body of *flesh* and blood (1 Jn 4.2–3). **8:** *Full reward,* eternal life (Jn 20.31; 1 Jn 2.25; 5.11–12). **9:** *Goes beyond,* i.e. beyond the tradition about Jesus (1 Jn 1.1–4) and Jesus' own commandment of love (1 Jn 3.23; see 1 Jn 2.22–23 n.). **10–11:** Hospitality to false teachers incurs responsibility for their *evil deeds* (compare Titus 3.10; Rev 2.2). **13:** *Elect sister,* sister church (see v. 1 n.).

The Third Letter of
John

Of the three Johannine letters, this one alone is written to an individual. It reflects a period in the church's life when organization was loose and churches were bound together by letters from those in authority and by personal visits of their representatives and traveling missionaries. In the unknown church to which this letter was directed, a certain Diotrephes, who liked "to put himself first" (v. 9), had challenged the spiritual authority of "the elder," refusing to receive messengers from him and expelling "from the church" those who showed them hospitality (v. 10). The elder, therefore, writes to a respected and influential member of the church named Gaius (v. 1), who has cordially welcomed the messengers and provided for their further needs (vv. 5–6). He encourages Gaius to continue this practice, indicating that he will deal with Diotrephes on a subsequent visit (v. 10). Verse 11 echoes the exhortation of 1 Jn 2.6; 3.6, 10; 4.7. Hospitality is an application of the love-command (vv. 5–6). Verse 12 serves as a letter of recommendation for Demetrius, probably a missionary about to visit the area.

The authorship and date parallel those of 1 and 2 John (see Introductions to 1 and 2 John).

1 The elder to the beloved Gaius, whom I love in truth.

2 Beloved, I pray that all may go well with you and that you may be in good health, just as it is well with your soul. [3] I was overjoyed when some of the friends[a] arrived and testified to your faithfulness to the truth, namely how you walk in the truth. [4] I have no greater joy than this, to hear that my children are walking in the truth.

5 Beloved, you do faithfully whatever you do for the friends,[a] even though they are strangers to you; [6] they have testified to your love before the church. You will do well to send them on in a manner worthy of God; [7] for they began their journey for the sake of Christ,[b] accepting no support from non-believers.[c] [8] Therefore we ought to support such people, so that we may become co-workers with the truth.

9 I have written something to the church; but Diotrephes, who likes to put himself first, does not acknowledge our authority. [10] So if I come, I will call attention to what he is doing in spreading false charges against us. And not content with those charges, he refuses to welcome the friends,[a] and even prevents those who want to do so and expels them from the church.

11 Beloved, do not imitate what is evil but imitate what is good. Whoever does good is from God; whoever does evil has not seen God. [12] Everyone has testified favorably about Demetrius, and so has the truth itself. We also testify for him,[d] and you know that our testimony is true.

13 I have much to write to you, but I would rather not write with pen and ink; [14] instead I hope to see you soon, and we will talk together face to face.

15 Peace to you. The friends send you their greetings. Greet the friends there, each by name.

a Gk *brothers*   b Gk *for the sake of the name*
c Gk *the Gentiles*   d Gk lacks *for him*

1: *Gaius,* see Introduction. **3**: *Your faithfulness to the truth,* behavior contrasting with that of Diotrephes (vv. 9–10) in receiving emissaries from the elder. **5–8**: Commendation for, and encouragement to continue, rendering assistance to traveling missionaries, who were dependent on the church rather than on *non-believers.* Such support is true participation in spreading the gospel (contrast 2 Jn 10–11). **9–10**: *Written something,* probably a letter of apostolic counsel no longer extant. *Diotrephes,* see Introduction. Personal ambition rejects *authority. Friends,* emissaries of the elder (contrast 1 Jn 2.19). **11**: 1 Jn 3.6–10. **12**: *Demetrius,* perhaps the bearer of this letter. Such recommedations appear in the Pauline letters (Rom 16.1–2; 1 Cor 16.3; 2 Cor 3.1; Phil 2.25–30). **13**: 2 Jn 12. **15**: *Friends,* fellow-Christians.

The Letter of
Jude

Jude is a brief tract cast as a general letter that warns Christians against persons who cause divisions in the Christian community (vv. 4, 12, 19). The traditional identification of Jude with Judas, a brother of Jesus (Mt 13.55; Mk 6.3), is unlikely. The author is contending for the faith that had been "once for all entrusted to the saints" (v. 3). Readers are reminded that the apostles had predicted the coming of such false Christians in the last days (v. 18). Jude clearly originated in the post-apostolic age.

The readers should not be deceived by such false Christians. Their licentious way of life (vv. 4, 6–8, 12–13, 16) and divisiveness show that they are without the Holy Spirit and the love of God that leads to eternal life (vv. 19–21). Christians, however, should seek to rescue those who are weak, and even show mercy to persons whose evil lives they reject (vv. 22–23).

The central argument (vv. 5–16) includes examples of divine judgment that are drawn from both Hebrew Scriptures and Jewish apocalyptic. The latter provides accounts of the fate of the angels who fell from heaven and the prophecies of judgment associated with Enoch (vv. 6–15), as well as a tradition that the archangel Michael and the devil struggled over the body of Moses (v. 9). The author's exhortation is so general that the letter may not be directed against a particular group or crisis within the community. Perhaps the author composed this warning against the deceit of ungodly Christians in order to remind the readers that they must never relax their efforts to achieve holiness. The brief but trenchant letter concludes with a beautiful and moving doxology (vv. 24–25).

1 Jude,[a] a servant[b] of Jesus Christ and brother of James,

To those who are called, who are beloved[c] in[d] God the Father and kept safe for[d] Jesus Christ:

2 May mercy, peace, and love be yours in abundance.

3 Beloved, while eagerly preparing to write to you about the salvation we share, I find it necessary to write and appeal to you to contend for the faith that was once for all entrusted to the saints. [4]For certain intruders have stolen in among you, people who long ago were designated for this condemnation as ungodly, who pervert the grace of our God into licentiousness and deny our only Master and Lord, Jesus Christ.[e]

5 Now I desire to remind you, though you are fully informed, that the Lord, who once for all saved[f] a people out of the land of Egypt, afterward destroyed those who did not believe. [6]And the angels who did not keep their own position, but left their proper dwelling, he has kept in eternal chains in deepest darkness for the judgment of the great Day. [7]Likewise, Sodom and Gomorrah and the surrounding cities, which, in the same manner as they, indulged in sexual immorality and pursued unnatural lust,[g] serve as an example by undergoing a punishment of eternal fire.

8 Yet in the same way these dreamers also defile the flesh, reject authority, and slander the glorious ones.[h] [9]But when the archangel Michael contended with the devil and disputed about the body of Moses, he did not dare to bring a condemnation of slander[i] against him, but said, "The Lord rebuke you!" [10]But these people slander whatever they do not understand, and they are destroyed by those things that, like irrational animals, they know by instinct. [11]Woe to them! For they go the way of Cain, and abandon themselves to Balaam's error for the sake of gain, and perish in Korah's rebellion. [12]These are blemishes[j] on your love-feasts, while they feast with you without fear, feeding themselves.[k] They are waterless clouds carried along by the winds; autumn trees without fruit, twice dead, uprooted; [13]wild waves of the sea, casting up the foam of their own shame; wandering stars, for whom the deepest darkness has been reserved forever.

14 It was also about these that Enoch, in the seventh generation from Adam,

a Gk *Judas* b Gk *slave* c Other ancient authorities read *sanctified* d Or *by* e Or *the only Master and our Lord Jesus Christ* f Other ancient authorities read *though you were once for all fully informed, that Jesus* (or *Joshua*) *who saved* g Gk *went after other flesh* h Or *angels*; Gk *glories* i Or *condemnation for blasphemy* j Or *reefs* k Or *without fear. They are shepherds who care only for themselves*

1–2: Salutation, see 1 Pet 1.1–2; Rom 1.1–7 n. **1:** God has *called* (Rom 1.6–7; 8.28) the Christians, shown them his love, and keeps them for the coming of *Jesus Christ* (1 Thess 5.23). **3–4: Occasion of the letter.** The emergence of false teachers has stimulated the author to warn readers against those who pervert the gospel and deny Christ. **3:** *The faith,* a synonym for the gospel. *The saints,* see Rom 1.7 n. **5–16: False teachers. 5–7:** The Israelites in the wilderness (1 Cor 10.1–11; Heb 3.7–19), the rebellious angels (Gen 6.4; the legend of the fallen angels appears in 1 Enoch 6–8), and Sodom and Gomorrah (Gen ch 19) are illustrations of how God judges the disobedient. **8:** In spite of warnings, the heretics defy authority, revile angels (apparently those angels who are God's servants), and live licentiously. **9–10:** The author refers to a Jewish tradition according to which Michael, though provoked to anger by Satan (who had charged that Moses, being a murderer, was not worthy of burial), refrained from bringing *a condemnation of slander.* The heretics, however, slander spiritual truths they do not comprehend (1 Cor 2.7–16) and will perish because they follow animal passions. **11:** *The way of Cain,* Gen 4.9; Heb 11.4; 1 Jn 3.12. *Balaam,* Num chs 22–24; Rev 2.14. *Korah,* Num ch 16. **12:** Apparently the heretics are within the fellowship, even attending the love feasts (see 1 Cor 11.20–21 n., and 2 Pet 2.13 note *t*), and hence were the more dangerous. **14:** *See . . . ,* quoted from the book of Enoch, 1.9.

prophesied, saying, "See, the Lord is coming[1] with ten thousands of his holy ones, [15]to execute judgment on all, and to convict everyone of all the deeds of ungodliness that they have committed in such an ungodly way, and of all the harsh things that ungodly sinners have spoken against him." [16]These are grumblers and malcontents; they indulge their own lusts; they are bombastic in speech, flattering people to their own advantage.

[17] But you, beloved, must remember the predictions of the apostles of our Lord Jesus Christ; [18]for they said to you, "In the last time there will be scoffers, indulging their own ungodly lusts." [19]It is these worldly people, devoid of the Spirit, who are causing divisions. [20]But you, beloved, build yourselves up on your most holy faith; pray in the Holy Spirit; [21]keep yourselves in the love of God; look forward to the mercy of our Lord Jesus Christ that leads to[m] eternal life. [22]And have mercy on some who are wavering; [23]save others by snatching them out of the fire; and have mercy on still others with fear, hating even the tunic defiled by their bodies.[n]

[24] Now to him who is able to keep you from falling, and to make you stand without blemish in the presence of his glory with rejoicing, [25]to the only God our Savior, through Jesus Christ our Lord, be glory, majesty, power, and authority, before all time and now and forever. Amen.

l Gk *came*   *m* Gk *Christ to*   *n* Gk *by the flesh.* The Greek text of verses 22-23 is uncertain at several points

---

17–23: Exhortations. Recall the apostolic tradition, avoid the heretics, grow in faith, wait for the mercy of Jesus Christ, care for the erring (on excluding false Christians from fellowship, 1 Cor 5.9–11; 2 Thess 3.14–15; 2 Jn 9–11). **18:** *In the last time,* 1 Jn 2.18; 1 Pet 1.5. **19:** *Divisions* in the church (1 Cor 1.12–13). **20–21:** Note the Trinitarian reference. **24–25: Doxology.**

Apocalyptic Literature

Apocalyptic literature is a class of Jewish and Christian writings that first appeared about 250 B.C. and continued well into the opening centuries A.D. It frequently reflects a negative view of this world and expresses the hope for salvation in new creation or in another life. It served to comfort and encourage the faithful in difficult times.

Examples of apocalyptic literature are the following. In the Old Testament the second half of the book of Daniel (chs 7–12) has long been recognized as apocalyptic in genre. Other Old Testament writings also contain passages that resemble the eschatological symbolism of the apocalypses: e.g. Isaiah chs 24–27 ("the apocalypse of Isaiah"), a very late addition to the Isaianic collection; Ezekiel chs 38–39, depicting the final victory over eschatological enemies by God's people; Joel ch 2 and Zechariah chs 9–14. In the New Testament the leading example of apocalyptic genre is the book of Revelation; apocalyptic materials are present also in Mark ch 13 and its parallels in Matthew ch 24 and Luke ch 12. Among the books of the Apocrypha the work traditionally known as 2 Esdras contains as its central portion (chs 3–14) the "Ezra-Apocalypse." This section, which is believed to have been written in Hebrew by a Palestinian Jew at the close of the first century A.D., presents seven visions granted to Ezra in order to answer his questions about the disproportionate sufferings of Zion. Other examples of Jewish apocalyptic literature include 1 Enoch, 2 Enoch, Jubilees, and the Apocalypses of Abraham, of Baruch, and of Elijah. Among Christian apocryphal literature are the Apocalypses of Peter, of Paul, and the *Shepherd of Hermas.*

The word "apocalypse" is derived from the Greek word *apokalypsis,* meaning "disclosure," "unveiling," or "revelation." Apocalyptic revelations are of two kinds: (1) symbolic visions (e.g. Daniel 7–12; Revelation) and (2) otherworldly journeys (parts of 1 Enoch; 2 Enoch; 3 Baruch). In both kinds, there is a heavenly mediator (usually an angel, but sometimes Christ in the Christian apocalypses) who explains what the visionary sees. The otherworldly journeys have a stronger interest in cosmological matters than the visions do. All the apocalypses in the Bible and Apocrypha include symbolic visions. The symbolism is colorful. Gentile nations and institutions are represented as wild beasts, sometimes of composite make-up; for example, a leopard, with feet like a bear's, and a mouth like a lion's mouth (Rev 13.2). There is also sometimes an interest in numerology, whether in terms of cryptic reference to a person (as the number of the beast, 666, in Rev 13.18) or the duration of persecutions (a time, times, and half a time = three-and-a-half years) in Daniel (7.2–5) and Revelation (12.14).

The apocalyptic author often makes use of two literary devices that, though not confined to apocalyptic writings, are characteristic of many such writings. The first is that of secret books in which the seer is bidden to conceal these mysteries until the time of the End when they will be revealed to "the wise" as a sign that the End is soon at hand. In 2 Esdras, for example, Ezra is bidden to "keep the seventy [books] that were written last, in order to give them to the wise among your people. For in them is the spring of understanding, the fountain of wisdom, and the river of knowledge" (14.46f.; compare Dan 12.4, 9). It is noteworthy that the author of the New Testament Apocalypse is commanded not to seal up "the words of the prophecy of this book, for the time is near" (Rev 22.10). The second device is that of pseudonymity, whereby the author writes in the name of some honored person of antiquity, such as Enoch, Abraham, Moses, or Ezra. The intention is not to deceive but rather to strengthen the conviction that the apocalyptist is the transmitter of a long and authoritative tradition. The same device is followed in 2 Esdras and apocryphal Christian apocalypses, such as those of Peter and Paul, but not in the book of Revelation, where the writer uses his own name.

The content of the revelations invariably involves eschatology, that is, predictions about the final outcome of human affairs. To a degree, the eschatology resembles that of the prophets: apocalypses often predict the restoration of Israel as a messianic kingdom on earth. The eschatological kingdom plays an important part in the biblical apocalypses of Daniel and Revelation and in 2 Esdras. Apocalyptic eschatology is distinguished from prophecy, however, by the expectation of a judgment of the dead and of reward or punishment in a future life.

Apocalypticism is also characterized by dualism. There are two opposing personified forces in the universe, one good and the other evil. Yahweh is, of course, the supreme God. In Jewish apocalypticism the archangel Michael often functions as God's agent. Opposed to him is Belial, or Satan. The opposition of these supernatural figures is most clearly drawn in certain treatises among the Dead Sea Scrolls.

Dualism also modifies the eschatology of the apocalypses. There are two distinct and separate ages, the present age and the age to come. The second is not a natural development of the first, but is a new creation. At present, human history is under the domination of the power of evil; consequently it is evil and corrupt. God, however, has set a limit to the era of wickedness and will intervene at the appointed time to execute judgment.

Other features frequently found in apocalypses include the representation of present troubles as "birth-pangs" heralding the End (Mt 24.8; Mk 13.8). Sometimes this End is described in terms of political action and military struggle (the battle of "Armageddon"). At other times the conflict assumes cosmic proportions involving mysterious and awesome happenings on earth and in the heavens: earthquakes, famine, fearful celestial portents, and destruction by fire (Joel 2.30–32; Mt 24.29–31; Mk 13.24–26; Lk 21.25–27; Rev 6.12–17). In the final battle the powers of evil, together with the evil nations they represent, will be utterly destroyed (Rev 19.11–21). Then a new divine order will be established, when the End will be as the Beginning, and Paradise will be restored.

The revelance of apocalypticism has been variously estimated. On the one hand, to some it may seem to be hopelessly pessimistic. On the other hand, it has had a strong appeal as an uncomplicated explanation of the struggle between good and evil. In any case, one cannot deny that books such as Daniel and Revelation have sustained and strengthened believers, whether Jewish or Christian, amid experiences of trials and persecutions. The apocalyptists believed in God, and they, like the prophets, had faith in the divine control of history. That control, they were assured, was certain to be vindicated and displayed in the final climax of history, which was regarded solely and uniquely as the act of God.

The Revelation
to John

The book of Revelation, or Apocalypse, is a fitting close to the Scriptures of the Old and New Testaments, for its final chapters depict the consummation toward which the whole biblical message of redemption is focused. It may be described as an inspired picture-book that, by an accumulation of magnificent poetic imagery, makes a powerful appeal to the reader's imagination. Many of the details of its pictures are intended to contribute to the total impression, and are not to be isolated and interpreted with wooden literalism.

Through the centuries the Apocalypse has been the object of widely divergent systems of interpretation. It can be best understood when one takes into account the following considerations. (1) This book comprises the substance of real visions that repeat with kaleidoscopic variety certain great principles of God's just and merciful government of the whole creation. By centering attention on these principles, the church in all ages has been encouraged and sustained despite the fiercest antagonisms of both human and demonic foes. (2) The book is written in apocalyptic style, a recognized literary genre (see pp. 362–363 NT). It contains other elements as well, such as the seven letters in chs 2 and 3 and the several prophetic utterances scattered here and there throughout its pages, but its difficulty will be found to arise largely from our unfamiliarity with apocalyptic writings. (3) As an apocalypse, the message of the book is couched in symbolism, involving numbers, strange beasts, and other typical apocalyptic features. Throughout one must recognize that the author's descriptions are descriptions of the symbols, not of the reality conveyed by the symbol. (4) Although the key for understanding some of the symbols has been lost, in other cases a comparison with the prophetic symbolism of the Old Testament sheds light on the intended meaning. This is understandable in view of the author's frequent allusion to the Greek Septuagint translation of the Hebrew Scriptures; of the 404 verses in Revelation, some 275 include one or more allusions to passages in the Old Testament. (5) The structure of the book involves a series of parallel and yet ever-progressing sections; these bring before the reader, over and over again, but in climacteric form, the struggle of the church, and its victory over the world in the providence of God Almighty. There are probably seven of these sections, though only five are clearly marked. The plan of the whole is, then, something like the following: Prologue, 1.1–8; seven parallel sections divided at 3.22; 8.1; 11.19; 14.20; 16.21; and 19.21; Epilogue, 22.6–21.

Although parts of the book (e.g. ch 11) may have been reduced to writing before the fall of Jerusalem in A.D. 70, it is probable that the author, whose name is John (1.1, 4, 9; 22.8), put the book in its present form toward the close of the reign of the Emperor Domitian (A.D. 81–96). It was then that Domitian began to demand that his subjects address him as "Lord and God" and worship his image. For refusing to do so, many Christians were put to death (6.9; 13.15); others, like John (1.9), were exiled, and all were threatened. One reason for the author's couching his teaching in mysterious figures and extraordinary metaphors was to prevent the imperial police from recognizing that this book is a trumpet call to the persecuted, assuring them that, despite the worst that the Roman Empire could do, God reigns supreme, and Christ, who died and is alive forevermore (1.18), has the power to overcome all evil. And therefore John closes his book with the prayer, "Come, Lord Jesus!" (22.20).

1 The revelation of Jesus Christ, which God gave him to show his servants*a* what must soon take place; he made*b* it known by sending his angel to his servant*c* John, 2who testified to the word of God and to the testimony of Jesus Christ, even to all that he saw.

3 Blessed is the one who reads aloud the words of the prophecy, and blessed are those who hear and who keep what is written in it; for the time is near.

4 John to the seven churches that are in Asia:

Grace to you and peace from him who is and who was and who is to come, and from the seven spirits who are before his throne, 5and from Jesus Christ, the faithful witness, the firstborn of the dead, and the ruler of the kings of the earth.

To him who loves us and freed*d* us from our sins by his blood, 6and made*b* us to be a kingdom, priests serving*e* his God and Father, to him be glory and dominion forever and ever. Amen.

7 Look! He is coming with the
clouds;
every eye will see him,
even those who pierced him;
and on his account all the tribes
of the earth will wail.
So it is to be. Amen.

8 "I am the Alpha and the Omega," says the Lord God, who is and who was and who is to come, the Almighty.

9 I, John, your brother who share with you in Jesus the persecution and the kingdom and the patient endurance, was on the island called Patmos because of the word of God and the testimony of Jesus.*f* 10I was in the spirit*g* on the Lord's day, and I heard behind me a loud voice like a trumpet 11saying, "Write in a book what you see and send it to the seven churches, to Ephesus, to Smyrna, to Pergamum, to Thyatira, to Sardis, to Philadelphia, and to Laodicea."

12 Then I turned to see whose voice it was that spoke to me, and on turning I saw seven golden lampstands, 13and in the midst of the lampstands I saw one like the Son of Man, clothed with a long robe and with a golden sash across his chest. 14His head and his hair were white as white wool, white as snow; his eyes were like a flame of fire, 15his feet were like burnished bronze, refined as in a furnace, and his voice was like the sound of many waters. 16In his right hand he held seven stars, and from his mouth came a

a Gk *slaves*   *b* Gk *and he made*   *c* Gk *slave*
d Other ancient authorities read *washed*
e Gk *priests to*   *f* Or *testimony to Jesus*
g Or *in the Spirit*

1.1–3: The prologue. This revelation came from God through Jesus Christ and was communicated to John by an angel (referred to again in 22.16). **3:** *Blessed is . . .* , the first of seven beatitudes in Revelation (compare 14.13; 16.15; 19.9; 20.6; 22.7, 14) is pronounced upon the reader of this prophetic book in services of worship and upon the listening worshipers who heed its message. The solemn words, *the time is near*, provide a motive for obedience.

1.4–8: Introductory salutation to seven representative churches in the Roman province of Asia (in western Asia Minor). *Seven* suggests the idea of completeness and totality. **4:** *Grace . . . and peace*, see 2 Thess 1.2 n. The *seven spirits* are either angelic beings or, more likely, a symbolic reference to the manifold energies of the Spirit of God (Isa 11.2). **5:** *Faithful witness*, Jesus testifies to the truth (Jn 18.37). He *loves* continually; he *freed us*

once for all by his death as a sacrifice (Rom 6.10; Heb 7.27). **8:** *Alpha* and *Omega*, the first and last letters of the Greek alphabet (like our "A to Z"); hence, the beginning and end of all things (Isa 44.6).

1.9–20: Preparatory vision on *Patmos*, a rocky island, about ten by five miles, in the Aegean, where John had been exiled (see Introduction). **10:** *In the spirit*, in a state of prophetic illumination. *Lord's day*, Sunday.

1.12–16: In the midst of the churches (see v. 20) stands the exalted Christ, whose royalty, eternity, wisdom, and immutability are suggested by means of symbols; the effect is that of terrifying majesty (compare v. 17 with Isa 6.5). *Son of Man*, see Mk 2.10 n. *Sound of many waters*, Ezekiel also makes a similar comparison with the voice of the Almighty (Ezek 1.24; 43.2). *Two-edged sword*, the word of God (Heb 4.12). **18:** *Hades*, used here with its synonym *death*, is the abode of the dead;

sharp, two-edged sword, and his face was like the sun shining with full force.

17 When I saw him, I fell at his feet as though dead. But he placed his right hand on me, saying, "Do not be afraid; I am the first and the last, 18 and the living one. I was dead, and see, I am alive forever and ever; and I have the keys of Death and of Hades. 19 Now write what you have seen, what is, and what is to take place after this. 20 As for the mystery of the seven stars that you saw in my right hand, and the seven golden lampstands: the seven stars are the angels of the seven churches, and the seven lampstands are the seven churches.

2 "To the angel of the church in Ephesus write: These are the words of him who holds the seven stars in his right hand, who walks among the seven golden lampstands:

2 "I know your works, your toil and your patient endurance. I know that you cannot tolerate evildoers; you have tested those who claim to be apostles but are not, and have found them to be false. 3 I also know that you are enduring patiently and bearing up for the sake of my name, and that you have not grown weary. 4 But I have this against you, that you have abandoned the love you had at first. 5 Remember then from what you have fallen; repent, and do the works you did at first. If not, I will come to you and remove your lampstand from its place, unless you repent. 6 Yet this is to your credit: you hate the works of the Nicolaitans, which I also hate. 7 Let anyone who has an ear listen to what the Spirit is saying to the churches. To everyone who conquers, I will give permission to eat from the tree of life that is in the paradise of God.

8 "And to the angel of the church in Smyrna write: These are the words of the first and the last, who was dead and came to life:

9 "I know your affliction and your poverty, even though you are rich. I know the slander on the part of those who say that they are Jews and are not, but are a synagogue of Satan. 10 Do not fear what you are about to suffer. Beware, the devil is about to throw some of you into prison so that you may be tested, and for ten days you will have affliction. Be faithful until death, and I will give you the crown of life. 11 Let anyone who has an ear listen to what the Spirit is saying to the churches. Whoever conquers will not be harmed by the second death.

12 "And to the angel of the church in Pergamum write: These are the words of him who has the sharp two-edged sword:

13 "I know where you are living,

Christ has *the keys* to release those confined within its gates (Mt 16.18; Jn 5.25–29). **20:** *Mystery,* truth formerly hidden, but now to be revealed. Angels are assigned to *the seven churches,* as also to nations (Dan 10.13; 12.1) and individuals (Mt 18.10).

2.1–3.22: The letters to the seven churches, each containing an address, a descriptive phrase referring to the risen Lord, a commendation or condemnation of the church addressed, an admonition, and a concluding promise and exhortation to the faithful.

2.1–7: The first letter is appropriately sent to *Ephesus,* the most important city of proconsular Asia; Paul labored here in the 60s for two years and three months (Acts 19.8–10). **1:** Christ *walks among* his churches (1.20).

2: *Evildoers,* compare the warning to the Ephesian elders in Acts 20.29–30. **6:** To *hate* evil is the counterpart of the love of the good (Isa 61.8; Zech 8.17). The *Nicolaitans* (probably not connected with the Nicolaus of Acts 6.5) taught that Christians were free to eat food offered to idols and to practice immorality in the name of religion (v. 14). **7:** *What the Spirit is saying,* speaking through Christ to John. *Conquers,* a military term, suggesting continuous vigilance. *Tree of life,* Gen 2.9.

2.8–11: The second letter, sent to *Smyrna,* commends the church for its perseverance amid *affliction* and *poverty.* **8:** Isa 44.6. **9:** *Rich,* in spiritual things. **10:** *Ten days,* not a lengthy period (Dan 1.12). *Crown of life,* supreme blessedness. **11:** *The second death,* the final condemnation of sinners (20.14; Mt 10.28).

where Satan's throne is. Yet you are holding fast to my name, and you did not deny your faith in me[h] even in the days of Antipas my witness, my faithful one, who was killed among you, where Satan lives. 14 But I have a few things against you: you have some there who hold to the teaching of Balaam, who taught Balak to put a stumbling block before the people of Israel, so that they would eat food sacrificed to idols and practice fornication. 15 So you also have some who hold to the teaching of the Nicolaitans. 16 Repent then. If not, I will come to you soon and make war against them with the sword of my mouth. 17 Let anyone who has an ear listen to what the Spirit is saying to the churches. To everyone who conquers I will give some of the hidden manna, and I will give a white stone, and on the white stone is written a new name that no one knows except the one who receives it.

18 "And to the angel of the church in Thyatira write: These are the words of the Son of God, who has eyes like a flame of fire, and whose feet are like burnished bronze:

19 "I know your works—your love, faith, service, and patient endurance. I know that your last works are greater than the first. 20 But I have this against you: you tolerate that woman Jezebel, who calls herself a prophet and is teaching and beguiling my servants[i] to practice fornication and to eat food sacrificed to idols. 21 I gave her time to repent, but she refuses to repent of her fornication.

22 Beware, I am throwing her on a bed, and those who commit adultery with her I am throwing into great distress, unless they repent of her doings; 23 and I will strike her children dead. And all the churches will know that I am the one who searches minds and hearts, and I will give to each of you as your works deserve. 24 But to the rest of you in Thyatira, who do not hold this teaching, who have not learned what some call 'the deep things of Satan,' to you I say, I do not lay on you any other burden; 25 only hold fast to what you have until I come. 26 To everyone who conquers and continues to do my works to the end,

I will give authority over
    the nations;
27 to rule[j] them with an iron rod,
    as when clay pots are
        shattered—

28 even as I also received authority from my Father. To the one who conquers I will also give the morning star. 29 Let anyone who has an ear listen to what the Spirit is saying to the churches.

3 "And to the angel of the church in Sardis write: These are the words of him who has the seven spirits of God and the seven stars:

"I know your works; you have a name of being alive, but you are dead. 2 Wake up, and strengthen what remains and is on the point of death, for I have not found your works perfect in the sight of

h Or *deny my faith*   *i* Gk *slaves*   *j* Or *to shepherd*

2.12–17: The third letter is sent to *Pergamum*, a noted center of idolatrous worship. **12:** *Sharp two-edged sword*, 1.16. **14–15:** See v. 6; Num 25.1–2; 31.16. **17:** *Manna . . . name*, Ex 16.33–34; Ps 78.24; Isa 62.2. *White*, the color symbolizing victory and joy.

2.18–29: The fourth letter is sent to *Thyatira*, a commercial center renowned for its many trade guilds, which periodically sponsored idolatrous feasts. Lydia, one of Paul's converts, was a dealer in purple cloth from Thyatira (Acts 16.14). **18:** Dan 10.6. **20:** *Jezebel*, a false *prophet*, resembling Ahab's wicked and idolatrous queen (1 Kings 16.31; 2 Kings 9.22, 30). **22:** In the Old Testament idolatry is often called *adultery*, as an abandonment of God, the husband of his people (Deut 31.16; Judg 2.17; 1 Chr 5.25). **23:** *Her children*, those who follow her teachings. **24:** *Deep things of Satan*, a withering reference to heretical teachings (contrast 1 Cor 2.10). **26–27:** The conquerors will share in Christ's Messianic rule (Ps 2.8–9). **28:** *The morning star* is Christ himself (22.16). **29:** *Is saying*, in this revelation.

3.1–6: The fifth letter is sent to *Sardis*, notorious for its luxury and licentiousness. **1:** *Seven spirits*, see 1.4 n. *Seven stars*, 1.20. *You have a name*, its Christianity was only nomi-

my God. ³Remember then what you received and heard; obey it, and repent. If you do not wake up, I will come like a thief, and you will not know at what hour I will come to you. ⁴Yet you have still a few persons in Sardis who have not soiled their clothes; they will walk with me, dressed in white, for they are worthy. ⁵If you conquer, you will be clothed like them in white robes, and I will not blot your name out of the book of life; I will confess your name before my Father and before his angels. ⁶Let anyone who has an ear listen to what the Spirit is saying to the churches.

7 "And to the angel of the church in Philadelphia write:

These are the words of the holy
one, the true one,
who has the key of David,
who opens and no one will
shut,
who shuts and no one opens:

8 "I know your works. Look, I have set before you an open door, which no one is able to shut. I know that you have but little power, and yet you have kept my word and have not denied my name. ⁹I will make those of the synagogue of Satan who say that they are Jews and are not, but are lying—I will make them come and bow down before your feet, and they will learn that I have loved you. ¹⁰Because you have kept my word of patient endurance, I will keep you from the hour of trial that is coming on the whole world to test the inhabitants of the earth. ¹¹I am coming soon; hold fast to what you have, so that no one may seize your crown. ¹²If you conquer, I will make you a pillar in the temple of my God; you will never go out of it. I will write on you the name of my God, and the name of the city of my God, the new Jerusalem that comes down from my God out of heaven, and my own new name. ¹³Let anyone who has an ear listen to what the Spirit is saying to the churches.

14 "And to the angel of the church in Laodicea write: The words of the Amen, the faithful and true witness, the origin*ᵏ* of God's creation:

15 "I know your works; you are neither cold nor hot. I wish that you were either cold or hot. ¹⁶So, because you are lukewarm, and neither cold nor hot, I am about to spit you out of my mouth. ¹⁷For you say, 'I am rich, I have prospered, and I need nothing.' You do not realize that you are wretched, pitiable, poor, blind, and naked. ¹⁸Therefore I counsel you to buy from me gold refined by fire so that you may be rich; and white robes to clothe you and to keep the shame of your nakedness from being seen; and salve to anoint your eyes so that you may see. ¹⁹I reprove and discipline those whom I love. Be earnest, therefore, and repent. ²⁰Listen! I am standing at the door,

k Or *beginning*

nal. **3:** *Received,* the gospel. *Like a thief,* just when unexpected (16.15; Mt 24.42–44; 1 Thess 5.2). **4–5:** Those who have maintained spiritual purity will enjoy Christ's companionship here and will be acknowledged before God in heaven. **5:** *Book of life,* the register of God containing the names of the redeemed (13.8; 17.8; 20.12, 15; Ex 32.32; Ps 69.28; Dan 12.1; Mal 3.16; Lk 10.20). **3.7–13:** **The sixth letter** is sent to *Philadelphia,* a small town in Lydia. **7:** *Key of David,* a symbol of authority (Isa 22.22). **8:** *An open door,* of opportunity (1 Cor 16.9; 2 Cor 2.12). **9:** Isa 43.4; 60.14. **10:** The Philadelphian church, though feeble (v. 8), will be sustained during the coming persecution. **12:** *A pillar,* steadfast and permanent (Gal 2.9). *New Jerusalem,* 21.2.

3.14–22: **The seventh letter** is sent to *Laodicea,* a proud and wealthy city near Colossae (Col 4.13–16). **14:** *The Amen* is Jesus Christ (2 Cor 1.20). *The origin of God's creation,* Christ is the principle and source of all creation (Jn 1.3; Col 1.15, 18). **15–16:** Their lukewarm Christianity is nauseating. **17:** Complacent and self-satisfied, they are spiritually poverty-stricken (Hos 12.8). **18:** *Eyes,* Laodicea was noted for its manufacture of a medication for ophthalmic disorders. **19:** God's chastening has beneficent motives (Prov 3.12; Heb 12.5–11). **20:** An invitation to share the joys of the Messianic banquet in

knocking; if you hear my voice and open the door, I will come in to you and eat with you, and you with me. 21 To the one who conquers I will give a place with me on my throne, just as I myself conquered and sat down with my Father on his throne. 22 Let anyone who has an ear listen to what the Spirit is saying to the churches."

4 After this I looked, and there in heaven a door stood open! And the first voice, which I had heard speaking to me like a trumpet, said, "Come up here, and I will show you what must take place after this." 2 At once I was in the spirit, *l* and there in heaven stood a throne, with one seated on the throne! 3 And the one seated there looks like jasper and carnelian, and around the throne is a rainbow that looks like an emerald. 4 Around the throne are twenty-four thrones, and seated on the thrones are twenty-four elders, dressed in white robes, with golden crowns on their heads. 5 Coming from the throne are flashes of lightning, and rumblings and peals of thunder, and in front of the throne burn seven flaming torches, which are the seven spirits of God; 6 and in front of the throne there is something like a sea of glass, like crystal.

Around the throne, and on each side of the throne, are four living creatures, full of eyes in front and behind: 7 the first living creature like a lion, the second living creature like an ox, the third living creature with a face like a human face, and the fourth living creature like a fly-ing eagle. 8 And the four living creatures, each of them with six wings, are full of eyes all around and inside. Day and night without ceasing they sing,

"Holy, holy, holy,
the Lord God the Almighty,
who was and is and is to
come."

9 And whenever the living creatures give glory and honor and thanks to the one who is seated on the throne, who lives forever and ever, 10 the twenty-four elders fall before the one who is seated on the throne and worship the one who lives forever and ever; they cast their crowns before the throne, singing,

11 "You are worthy, our Lord
and God,
to receive glory and honor
and power,
for you created all things,
and by your will they existed
and were created."

5 Then I saw in the right hand of the one seated on the throne a scroll written on the inside and on the back, sealed *m* with seven seals; 2 and I saw a mighty angel proclaiming with a loud voice, "Who is worthy to open the scroll and break its seals?" 3 And no one in heaven or on earth or under the earth was able to open the scroll or to look into it. 4 And I began to weep bitterly because no one was found worthy to open the scroll or to look into it. 5 Then one of the elders

l Or *in the Spirit*   *m* Or *written on the inside, and sealed on the back*

---

the coming age (compare Mt 26.29). **21**: *A place with me*, a promise of reigning with Christ in glory (22.5; Lk 22.30).
4.1–5.14: **Vision of the glory of God and of the Lamb.**
4.1: *The first voice*, mentioned in 1.10. *Come up*, John was exalted in spirit, as was Ezekiel, to behold visions of God (Ezek 3.12; 8.3; 11.1). **2**: *A throne*, Ezek 1.26–28. **3**: The glory of the divine presence is described in terms of precious gems. **4**: *Twenty-four elders*, probably angelic beings of the heavenly court, symbolizing the twelve patriarchs of the Old Testament and the twelve apostles of the New Testament. **5**: *Flashes of lightning*, ex-pressive of the majesty of the Most High (Ex 19.16; Rev 11.19). *Seven spirits*, 1.4. **6**: *A sea of glass* suggests the distance between God and his creatures, even in heaven. *Four living creatures*, angelic beings representing humankind and all beasts (Ezek 1.5, 10). *Full of eyes*, symbolizing unceasing watchfulness. **8**: *Six wings . . . Holy, holy, holy*, Isa 6.2–3. **10**: *Cast their crowns*, acknowledging that all power comes from God. **11**: *They existed* in God's mind from all eternity.
5.1: *A scroll*, containing the fixed purposes of God for the future (Ezek 2.9–10). *Sealed*, therefore both unalterable and unknown to others. **3–5**: No created being is worthy to

said to me, "Do not weep. See, the Lion
of the tribe of Judah, the Root of David,
has conquered, so that he can open the
scroll and its seven seals."
6 Then I saw between the throne and
the four living creatures and among the
elders a Lamb standing as if it had been
slaughtered, having seven horns and sev-
en eyes, which are the seven spirits of
God sent out into all the earth. 7 He went
and took the scroll from the right hand
of the one who was seated on the throne.
8 When he had taken the scroll, the four
living creatures and the twenty-four elders fell before the Lamb, each holding a
harp and golden bowls full of incense,
which are the prayers of the saints.
9 They sing a new song:

"You are worthy to take the scroll
 and to open its seals,
for you were slaughtered and by
 your blood you ransomed
 for God
saints from[n] every tribe and
 language and people
 and nation;
10 you have made them to be a
 kingdom and priests
 serving[o] our God,
and they will reign on earth."

11 Then I looked, and I heard the
voice of many angels surrounding the
throne and the living creatures and the
elders; they numbered myriads of myriads and thousands of thousands, 12 singing with full voice,

"Worthy is the Lamb that was
 slaughtered
to receive power and wealth and
 wisdom and might
and honor and glory and
 blessing!"

13 Then I heard every creature in heaven
and on earth and under the earth and in
the sea, and all that is in them, singing,

"To the one seated on the throne
 and to the Lamb
be blessing and honor and glory
 and might
forever and ever!"

14 And the four living creatures said,
"Amen!" And the elders fell down and
worshiped.

6 Then I saw the Lamb open one of
the seven seals, and I heard one of
the four living creatures call out, as with
a voice of thunder, "Come!"[p] 2 I looked,
and there was a white horse! Its rider had
a bow; a crown was given to him, and he
came out conquering and to conquer.
3 When he opened the second seal, I
heard the second living creature call out,
"Come!"[p] 4 And out came[q] another
horse, bright red; its rider was permitted
to take peace from the earth, so that people would slaughter one another; and he
was given a great sword.
5 When he opened the third seal, I

n Gk *ransomed for God from* *o* Gk *priests to*
p Or *"Go!"* *q* Or *went*

---

carry out God's plan, only the Messianic king can do so; for his titles (*Lion . . . Root*), compare Gen 49.9–10; Isa 11.1, 10.

5.6–9: The *Lamb . . . slaughtered,* refers to Christ's sacrificial death, by which God's purposes contained in the scroll are accomplished (Isa 53.7; Jn 1.29, 36; 1 Pet 1.19). *Seven horns . . . seven eyes,* plenitude of power and insight. *Seven spirits,* 1.4. **8**: *The prayers of the saints* on earth are joined with the worship rendered to the *Lamb* by heavenly creatures. **9–10**: *A new song,* because Christ has inaugurated a new era (14.3). The Lamb is adored in terms similar to the adoration rendered to God (4.11). *A kingdom and priests,* the vocation promised to Israel (Ex 19.6; Isa 61.6) is extended to the church (1 Pet 2.9). **11–12**: The sevenfold praise of *myriads* in heaven honoring the sacrificial *Lamb.* **13**: Universal praise to the Creator and to the Redeemer as equal in majesty.

6.1–17: **The opening of the first six seals** of the scroll, and the pictorial enactment of what is written therein. **1–8**: *Come!* addressed successively to each of the four *riders* (Zech 6.1–3) who accomplish God's purposes. **2**: The *white horse* symbolizes a conquering power that none can resist (in 19.11–13 the reference is to Christ). **4**: The *red horse* symbolizes war and bloodshed. **5**: The *black horse* symbolizes famine, which follows upon war. **8**: The *pale green horse* symbolizes pestilence and death. *A fourth* part indicates wide but not total devastation. *Pestilence,* Ex 9.3;

heard the third living creature call out, "Come!"[r] I looked, and there was a black horse! Its rider held a pair of scales in his hand, 6 and I heard what seemed to be a voice in the midst of the four living creatures saying, "A quart of wheat for a day's pay,[s] and three quarts of barley for a day's pay,[s] but do not damage the olive oil and the wine!"

7 When he opened the fourth seal, I heard the voice of the fourth living creature call out, "Come!"[r] 8 I looked and there was a pale green horse! Its rider's name was Death, and Hades followed with him; they were given authority over a fourth of the earth, to kill with sword, famine, and pestilence, and by the wild animals of the earth.

9 When he opened the fifth seal, I saw under the altar the souls of those who had been slaughtered for the word of God and for the testimony they had given; 10 they cried out with a loud voice, "Sovereign Lord, holy and true, how long will it be before you judge and avenge our blood on the inhabitants of the earth?" 11 They were each given a white robe and told to rest a little longer, until the number would be complete both of their fellow servants[t] and of their brothers and sisters,[u] who were soon to be killed as they themselves had been killed.

12 When he opened the sixth seal, I looked, and there came a great earthquake; the sun became black as sackcloth, the full moon became like blood, 13 and the stars of the sky fell to the earth as the fig tree drops its winter fruit when shaken by a gale. 14 The sky vanished like a scroll rolling itself up, and every mountain and island was removed from its place. 15 Then the kings of the earth and the magnates and the generals and the rich and the powerful, and everyone, slave and free, hid in the caves and among the rocks of the mountains, 16 calling to the mountains and rocks, "Fall on us and hide us from the face of the one seated on the throne and from the wrath of the Lamb; 17 for the great day of their wrath has come, and who is able to stand?"

7 After this I saw four angels standing at the four corners of the earth, holding back the four winds of the earth so that no wind could blow on earth or sea or against any tree. 2 I saw another angel ascending from the rising of the sun, having the seal of the living God, and he called with a loud voice to the four angels who had been given power to damage earth and sea, 3 saying, "Do not damage the earth or the sea or the trees, until we have marked the servants[t] of our God with a seal on their foreheads."

4 And I heard the number of those who were sealed, one hundred forty-four thousand, sealed out of every tribe of the people of Israel:

5 From the tribe of Judah twelve
    thousand sealed,
from the tribe of Reuben twelve
    thousand,
from the tribe of Gad twelve thou-
    sand,

---

r Or "Go!"   s Gk *a denarius*   t Gk *slaves*
u Gk *brothers*

---

2 Sam 24.13; Ezek 5.12. *Wild animals,* Ezek 5.17; 29.5; 33.27.

6.9–17: The fifth and sixth seals describe the prayers of the martyrs in heaven (vv. 9–11) and their effects on earth (vv. 12–17). **9:** The souls of the martyrs are said to be *under the altar* because they had been martyred for the sake of Christ (Mt 24.9; Phil 2.17; 2 Tim 4.6). **10:** *How long?* is a cry for divine vindication. *Avenge our blood,* vengeance belongs to God (Rom 12.19). **11:** *White robe, see* 2.17 n. and 7.9 n. **12–14:** *The great earthquake* and cosmic catastrophes are not to be understood literally, but represent social upheavals and divine judgment on the Day of the Lord (Isa 34.4; Joel 2.30–31; Am 8.9). **15–17:** All classes of society seek to escape from God (Isa 2.10).

7.1–17: An interlude between the sixth and seventh seals: two visions (vv. 1–8 and vv. 9–17) which provide assurance that God's people are secure from the plagues and judgments. **3:** The seal (Ezek 9.4–6) marks those under God's protection; they are preserved through, not from, tribulations (1 Cor 10.13). **4:** The explicit number (144,000) symbolizes completeness; not one of the redeemed is missing.

6 from the tribe of Asher twelve
thousand,
from the tribe of Naphtali twelve
thousand,
from the tribe of Manasseh twelve
thousand,
7 from the tribe of Simeon twelve
thousand,
from the tribe of Levi twelve thousand,
from the tribe of Issachar twelve
thousand,
8 from the tribe of Zebulun twelve
thousand,
from the tribe of Joseph twelve
thousand,
from the tribe of Benjamin twelve
thousand sealed.

9 After this I looked, and there was a
great multitude that no one could count,
from every nation, from all tribes and
peoples and languages, standing before
the throne and before the Lamb, robed in
white, with palm branches in their
hands. 10 They cried out in a loud voice,
saying,

"Salvation belongs to our God
who is seated on the
throne, and to the Lamb!"

11 And all the angels stood around the
throne and around the elders and the four
living creatures, and they fell on their
faces before the throne and worshiped
God, 12 singing,

"Amen! Blessing and glory
and wisdom
and thanksgiving and honor
and power and might
be to our God forever and ever!
Amen."

13 Then one of the elders addressed
me, saying, "Who are these, robed in
white, and where have they come from?"

14 I said to him, "Sir, you are the one that
knows." Then he said to me, "These are
they who have come out of the great
ordeal; they have washed their robes and
made them white in the blood of the
Lamb.
15 For this reason they are before the
throne of God,
and worship him day and night
within his temple,
and the one who is seated on
the throne will shelter
them.
16 They will hunger no more, and
thirst no more;
the sun will not strike them,
nor any scorching heat;
17 for the Lamb at the center of the
throne will be their
shepherd,
and he will guide them to
springs of the water of life,
and God will wipe away every
tear from their eyes."

8 When the Lamb opened the seventh
seal, there was silence in heaven for
about half an hour. 2 And I saw the seven
angels who stand before God, and seven
trumpets were given to them.

3 Another angel with a golden censer
came and stood at the altar; he was given
a great quantity of incense to offer with
the prayers of all the saints on the golden
altar that is before the throne. 4 And the
smoke of the incense, with the prayers of
the saints, rose before God from the hand
of the angel. 5 Then the angel took the
censer and filled it with fire from the altar
and threw it on the earth; and there were
peals of thunder, rumblings, flashes of
lightning, and an earthquake.

6 Now the seven angels who had the

7.9: *A great multitude* of the redeemed, so
many they cannot be counted. *White* robes
and *palm branches* symbolize righteousness
and victory. **12**: A sevenfold ascription of
praise to God. **14**: *The blood of the Lamb*
cleanses from sin (Jn 1.29; 1 Jn 1.7). **15**: *Before
the throne of God*, in a favored position because
of their faithfulness. To *worship* God implies
activity in heaven. *Shelter them*, literally

"spread his tabernacle over them." **16**: Isa
49.10; Ps 121.6. **17**: A paradox, the *Lamb* is a
shepherd (Ps 23.1–2; Ezek 34.23–24; Jn 10.11).
Springs of . . . life, 21.6; 22.1, 17; Jn 4.10; 7.37.
Wipe away every tear, 21.4; Isa 25.8.

8.1–5: **The seventh seal** is opened. The
sight of the unsealed scroll leads to reverential
silence. **3**: *Incense . . . prayers,* Ps 141.2. **5**: Ezek
10.2.

seven trumpets made ready to blow them.

7 The first angel blew his trumpet, and there came hail and fire, mixed with blood, and they were hurled to the earth; and a third of the earth was burned up, and a third of the trees were burned up and all green grass was burned up.

8 The second angel blew his trumpet, and something like a great mountain, burning with fire, was thrown into the sea. 9 A third of the sea became blood, a third of the living creatures in the sea died, and a third of the ships were destroyed.

10 The third angel blew his trumpet, and a great star fell from heaven, blazing like a torch, and it fell on a third of the rivers and on the springs of water. 11 The name of the star is Wormwood. A third of the waters became wormwood, and many died from the water, because it was made bitter.

12 The fourth angel blew his trumpet, and a third of the sun was struck, and a third of the moon, and a third of the stars, so that a third of their light was darkened; a third of the day was kept from shining, and likewise the night.

13 Then I looked, and I heard an eagle crying with a loud voice as it flew in midheaven, "Woe, woe, woe to the inhabitants of the earth, at the blasts of the other trumpets that the three angels are about to blow!"

9 And the fifth angel blew his trumpet, and I saw a star that had fallen from heaven to earth, and he was given the key to the shaft of the bottomless pit;

2 he opened the shaft of the bottomless pit, and from the shaft rose smoke like the smoke of a great furnace, and the sun and the air were darkened with the smoke from the shaft. 3 Then from the smoke came locusts on the earth, and they were given authority like the authority of scorpions of the earth. 4 They were told not to damage the grass of the earth or any green growth or any tree, but only those people who do not have the seal of God on their foreheads. 5 They were allowed to torture them for five months, but not to kill them, and their torture was like the torture of a scorpion when it stings someone. 6 And in those days people will seek death but will not find it; they will long to die, but death will flee from them.

7 In appearance the locusts were like horses equipped for battle. On their heads were what looked like crowns of gold; their faces were like human faces, 8 their hair like women's hair, and their teeth like lions' teeth; 9 they had scales like iron breastplates, and the noise of their wings was like the noise of many chariots with horses rushing into battle. 10 They have tails like scorpions, with stingers, and in their tails is their power to harm people for five months. 11 They have as king over them the angel of the bottomless pit; his name in Hebrew is Abaddon, *v* and in Greek he is called Apollyon. *w*

12 The first woe has passed. There are still two woes to come.

v That is, *Destruction*   *w* That is, *Destroyer*

8.6–9.21: **The first six trumpets** introduce new convulsions of nature, in judgment upon the wicked (compare the plagues of Egypt, Ex 9.23, 25; 7.17–21; 10.21–23). These calamities are not consecutive nor are they to be taken literally.
8.11: *Wormwood,* a bitter drug symbolizing divine chastisements (Jer 9.15; 23.15). 13: *An eagle* announces impending judgment. *Woe . . . ,* thrice repeated because the last three plagues are especially grievous.
9.1–12: **The plague of demonic locusts,** which combine the terrors of evil spirits and of invading horsemen (probably Parthians). 1:

A star . . . fallen from heaven, one of the fallen angels, perhaps Satan himself (v. 11; Lk 10.18). 3: *Locusts,* Ex 10.12–15. 4: God's servants are not to be harmed (7.3). 5: *Five months,* the period from May to September, the life-cycle of a plague of locusts. 6: *Will seek death,* in utter despair because of the torments.
9.7–10: John apparently would have the reader understand the locusts to be of considerable size. 11: The name *Abaddon,* which means "Destruction," denotes the depths of Sheol (Job 26.6; see Prov 15.11 n.).

13 Then the sixth angel blew his trumpet, and I heard a voice from the four[x] horns of the golden altar before God, 14 saying to the sixth angel who had the trumpet, "Release the four angels who are bound at the great river Euphrates." 15 So the four angels were released, who had been held ready for the hour, the day, the month, and the year, to kill a third of humankind. 16 The number of the troops of cavalry was two hundred million; I heard their number. 17 And this was how I saw the horses in my vision: the riders wore breastplates the color of fire and of sapphire[y] and of sulfur; the heads of the horses were like lions' heads, and fire and smoke and sulfur came out of their mouths. 18 By these three plagues a third of humankind was killed, by the fire and smoke and sulfur coming out of their mouths. 19 For the power of the horses is in their mouths and in their tails; their tails are like serpents, having heads; and with them they inflict harm.

20 The rest of humankind, who were not killed by these plagues, did not repent of the works of their hands or give up worshiping demons and idols of gold and silver and bronze and stone and wood, which cannot see or hear or walk. 21 And they did not repent of their murders or their sorceries or their fornication or their thefts.

10 And I saw another mighty angel coming down from heaven, wrapped in a cloud, with a rainbow over his head; his face was like the sun, and his legs like pillars of fire. 2 He held a little scroll open in his hand. Setting his right foot on the sea and his left foot on the land, 3 he gave a great shout, like a lion roaring. And when he shouted, the seven thunders sounded. 4 And when the seven thunders had sounded, I was about to write, but I heard a voice from heaven saying, "Seal up what the seven thunders have said, and do not write it down." 5 Then the angel whom I saw standing on the sea and the land

raised his right hand to heaven
6   and swore by him who lives
        forever and ever,
who created heaven and what is in it, the earth and what is in it, and the sea and what is in it: "There will be no more delay, 7 but in the days when the seventh angel is to blow his trumpet, the mystery of God will be fulfilled, as he announced to his servants[z] the prophets."

8 Then the voice that I had heard from heaven spoke to me again, saying, "Go, take the scroll that is open in the hand of the angel who is standing on the sea and on the land." 9 So I went to the angel and told him to give me the little scroll; and he said to me, "Take it, and eat; it will be bitter to your stomach, but sweet as honey in your mouth." 10 So I took the little scroll from the hand of the angel and ate it; it was sweet as honey in my mouth, but when I had eaten it, my stomach was made bitter.

11 Then they said to me, "You must prophesy again about many peoples and nations and languages and kings."

11 Then I was given a measuring rod like a staff, and I was told, "Come and measure the temple of God and the altar and those who worship there, 2 but do not measure the court outside the temple; leave that out, for it is given over to the nations, and they will trample over the holy city for forty-two

x Other ancient authorities lack *four*
y Gk *hyacinth*   z Gk *slaves*

9.13–21: The plague of demonic cavalry. **14:** The *Euphrates* was at the eastern border of the Roman empire, where invasion by the Parthians was feared. **16–19:** As with the locusts, the details regarding the two hundred million cavalry are not to be understood literally. **20–21:** Compare the hardness of Pharaoh's heart despite the plagues (Ex 8.15, 19).

10.1–11.13: An interlude between the sixth and seventh trumpets; two visions (10.1–11 and 11.1–13) provide consolation and assurance for the seer and for his fellow-believers.

10.1–2: The smallness of the *little scroll* is emphasized in contrast to the *mighty angel* holding it. **6:** *No more delay* in the accomplishment of God's will. **7:** *The mystery of God,* Rom 16.25–26; Eph 1.9; 3.3–9; Col 1.26, 27. **8:** *The scroll,* which is not the sealed scroll of 5.1, is a special message from God to John. **9:**

months. ³And I will grant my two witnesses authority to prophesy for one thousand two hundred sixty days, wearing sackcloth."

4 These are the two olive trees and the two lampstands that stand before the Lord of the earth. ⁵And if anyone wants to harm them, fire pours from their mouth and consumes their foes; anyone who wants to harm them must be killed in this manner. ⁶They have authority to shut the sky, so that no rain may fall during the days of their prophesying, and they have authority over the waters to turn them into blood, and to strike the earth with every kind of plague, as often as they desire.

7 When they have finished their testimony, the beast that comes up from the bottomless pit will make war on them and conquer them and kill them, ⁸and their dead bodies will lie in the street of the great city that is prophetically *a* called Sodom and Egypt, where also their Lord was crucified. ⁹For three and a half days members of the peoples and tribes and languages and nations will gaze at their dead bodies and refuse to let them be placed in a tomb; ¹⁰and the inhabitants of the earth will gloat over them and celebrate and exchange presents, because these two prophets had been a torment to the inhabitants of the earth.

11 But after the three and a half days, the breath *b* of life from God entered them, and they stood on their feet, and those who saw them were terrified. ¹²Then they *c* heard a loud voice from heaven saying to them, "Come up here!" And they went up to heaven in a cloud while their enemies watched them. ¹³At that moment there was a great earthquake, and a tenth of the city fell; seven thousand people were killed in the earthquake, and the rest were terrified and gave glory to the God of heaven.

14 The second woe has passed. The third woe is coming very soon.

15 Then the seventh angel blew his trumpet, and there were loud voices in heaven, saying,

"The kingdom of the world has
    become the kingdom of
    our Lord
    and of his Messiah, *d*
    and he will reign forever and ever."

16 Then the twenty-four elders who sit on their thrones before God fell on their faces and worshiped God, ¹⁷singing,

"We give you thanks, Lord
    God Almighty,
    who are and who were,
    for you have taken your
    great power
    and begun to reign.
18 The nations raged,
    but your wrath has come,
    and the time for judging
    the dead,

a Or *allegorically*; Gk *spiritually*   *b* Or *the spirit*
c Other ancient authorities read *I*   *d* Gk *Christ*

Ezek 2.8; 3.1–3. **10**: *Sweet*, because it contains God's words; *bitter*, because it involves his terrible judgments. **11**: *You must*, in accord with the divine will, *prophesy again;* the second part of the book (chs 12–22) contains these prophecies.

11.1–13: The measuring of the temple and the two witnesses. 1: The sanctuary *and those who worship there* are measured with a view to their preservation (as in Zech 2.1–5; compare Ezek 40.3–42.20). **2**: *The court outside*, the court of the Gentiles. *The holy city*, Jerusalem (Mt 4.5; 27.53). *Forty-two months=* 1260 days=3½ years (Dan 7.25; 12.7). **3**: *My two witnesses*, unnamed but resembling Zerubbabel and Joshua (Zech 3.1–4.14) as well as Elijah (vv. 5–6; 2 Kings 1.10) and Moses (v. 6; Ex 7.17, 19). Clothed in *sackcloth*, a sign

that their prophecy was of repentance. **4**: *Two olive trees*, supplying the lamps in the temple (Zech 4.3–14). **6**: *Shut the sky*, cause a draught, as did Elijah (1 Kings 17.1). *Authority over the waters*, as did Moses and Aaron (Ex 7.17–21).

11.7: *The beast* (or "the monster") is the Antichrist (17.8; Dan 7.3, 7, 21). **8**: *The name Sodom* is applied to Jerusalem in Isa 1.10; Jer 23.14; Ezek 16.46–56. *Prophets had been a torment*, to those who have guilty consciences. **11–12**: The witnesses are resuscitated (compare Ezek 37.5, 10) and taken to heaven (2 Kings 2.11).

11.14–19: The seventh trumpet announces (v. 15) the consummation of God's kingdom (10.7). **17–18**: A song of triumph.

for rewarding your servants, [e]
   the prophets
and saints and all who fear
   your name,
both small and great,
and for destroying those who
   destroy the earth."

19 Then God's temple in heaven was opened, and the ark of his covenant was seen within his temple; and there were flashes of lightning, rumblings, peals of thunder, an earthquake, and heavy hail.

12 A great portent appeared in heaven: a woman clothed with the sun, with the moon under her feet, and on her head a crown of twelve stars. 2 She was pregnant and was crying out in birth pangs, in the agony of giving birth. 3 Then another portent appeared in heaven: a great red dragon, with seven heads and ten horns, and seven diadems on his heads. 4 His tail swept down a third of the stars of heaven and threw them to the earth. Then the dragon stood before the woman who was about to bear a child, so that he might devour her child as soon as it was born. 5 And she gave birth to a son, a male child, who is to rule [f] all the nations with a rod of iron. But her child was snatched away and taken to God and to his throne; 6 and the woman fled into the wilderness, where she has a place prepared by God, so that there she can be nourished for one thousand two hundred sixty days.

7 And war broke out in heaven; Michael and his angels fought against the dragon. The dragon and his angels fought back, 8 but they were defeated, and there was no longer any place for them in heaven. 9 The great dragon was thrown down, that ancient serpent, who is called the Devil and Satan, the deceiver of the whole world—he was thrown down to the earth, and his angels were thrown down with him.

10 Then I heard a loud voice in heaven, proclaiming,

"Now have come the salvation
   and the power
and the kingdom of our God
   and the authority of his
   Messiah, [g]
for the accuser of our comrades [h]
   has been thrown down,
who accuses them day and night
   before our God.
11 But they have conquered him by
   the blood of the Lamb
and by the word of their
   testimony,
for they did not cling to life even
   in the face of death.
12 Rejoice then, you heavens
   and those who dwell in them!
But woe to the earth and the sea,
   for the devil has come down
   to you
with great wrath,
   because he knows that his time
   is short!"

13 So when the dragon saw that he had been thrown down to the earth, he pursued [i] the woman who had given birth to the male child. 14 But the woman was given the two wings of the great eagle, so that she could fly from the serpent into the wilderness, to her place where she is nourished for a time, and times, and half a time. 15 Then from his mouth the serpent poured water like a river after the woman, to sweep her away with the flood. 16 But the earth came to the help of the woman; it opened

e Gk *slaves*   f Or *to shepherd*   g Gk *Christ*
h Gk *brothers*   i Or *persecuted*

12.1–17: The vision of the woman, the child, and the dragon portrays the conflict between Christ and Satan. **1:** The *woman* appears to be the heavenly representative of God's people, first as Israel (from whom Jesus the Messiah was born, v. 5), then as the Christian Church (which is persecuted by the dragon, v. 13). **3:** *Dragon,* identified in v.9 as *the Devil* or *Satan.* His enormous size and power are suggested in v. 4. **5:** Ps 2.9 identifies the *child* to be the Davidic Messiah. **6:** The church is sustained by God.

12.7: *Michael,* an archangel and the champion of Israel (Dan 10.13, 21; 12.1; Jude 9). *The dragon and his angels,* see Eph 6.12 n. **9:** Lk 10.18. **10:** *The accuser,* Job 1.9–11. **11:** *They have conquered,* through both divine and human effort. **12:** *Rejoice,* Ps 96.11; Isa 49.13. **16:**

its mouth and swallowed the river that the dragon had poured from his mouth. [17]Then the dragon was angry with the woman, and went off to make war on the rest of her children, those who keep the commandments of God and hold the testimony of Jesus.

18 Then the dragon[j] took his stand

13 on the sand of the seashore. [1]And I saw a beast rising out of the sea, having ten horns and seven heads; and on its horns were ten diadems, and on its heads were blasphemous names. [2]And the beast that I saw was like a leopard, its feet were like a bear's, and its mouth was like a lion's mouth. And the dragon gave it his power and his throne and great authority. [3]One of its heads seemed to have received a death-blow, but its mortal wound[k] had been healed. In amazement the whole earth followed the beast. [4]They worshiped the dragon, for he had given his authority to the beast, and they worshiped the beast, saying, "Who is like the beast, and who can fight against it?"

5 The beast was given a mouth uttering haughty and blasphemous words, and it was allowed to exercise authority for forty-two months. [6]It opened its mouth to utter blasphemies against God, blaspheming his name and his dwelling, that is, those who dwell in heaven. [7]Also it was allowed to make war on the saints and to conquer them.[l] It was given authority over every tribe and people and language and nation, [8]and all the inhabitants of the earth will worship it, everyone whose name has not been written from the foundation of the world in the book of life of the Lamb that was slaughtered.[m]

9 Let anyone who has an ear listen:
[10] If you are to be taken captive,
   into captivity you go;
   if you kill with the sword,
     with the sword you must be
       killed.
Here is a call for the endurance and faith of the saints.

11 Then I saw another beast that rose out of the earth; it had two horns like a lamb and it spoke like a dragon. [12]It exercises all the authority of the first beast on its behalf, and it makes the earth and its inhabitants worship the first beast, whose mortal wound[n] had been healed. [13]It performs great signs, even making fire come down from heaven to earth in the sight of all; [14]and by the signs that it is allowed to perform on behalf of the beast, it deceives the inhabitants of earth, telling them to make an image for the beast that had been wounded by the sword[o] and yet lived; [15]and it was allowed to give breath[p] to the image of the beast so that the image of the beast could even speak and cause those who would not worship the image of the beast to be killed. [16]Also it causes all, both small and great, both rich and poor, both free and slave, to be marked on the right hand or the forehead, [17]so that no one can buy or sell who does not have the mark, that is, the name of the beast or the number of its name. [18]This calls for wisdom: let anyone with understanding calculate the

j Gk *Then he*; other ancient authorities read *Then I stood*   k Gk *the plague of its death*   l Other ancient authorities lack this sentence
m Or *written in the book of life of the Lamb that was slaughtered from the foundation of the world*
n Gk *whose plague of its death*   o Or *that had received the plague of the sword*   p Or *spirit*

Aid comes from an unexpected quarter.

13.1–18: The two beasts. 1–10: The *beast* from *the sea* combines the powers of the four beasts of Dan ch 7 and represents the Roman empire, incited by the *dragon* (v. 2) to persecute *the saints* (v. 7). **5:** The sovereignty of God, even amid persecution, is implied by the use of passive verbs here and in vv. 7, 10, 14, 15 (compare 6.4; see 17.17 n.). *Forty-two months*, see 11.2 n. **8:** *Book of life,* see 3.5 n. **10:** Jer 15.2; Mt 26.52.

13.11–18. The *beast* from *the earth,* called the false prophet (19.20), enforces Emperor worship (v. 12), and produces *great signs* (v. 13) to deceive the people (v. 14; compare Deut 13.1–5; Mt 24.24; 2 Thess 2.9). **16–17:** *Marked* in imitation of the sealing of God's servants (7.2–4), resulting in economic boycott against the Christians. **18:** Since Hebrew and Greek letters have numerical equivalents, *the number of the beast* (666) is the sum of the separate letters of his name. Of countless ex-

number of the beast, for it is the number of a person. Its number is six hundred sixty-six. *q*

14 Then I looked, and there was the Lamb, standing on Mount Zion! And with him were one hundred forty-four thousand who had his name and his Father's name written on their foreheads. ²And I heard a voice from heaven like the sound of many waters and like the sound of loud thunder; the voice I heard was like the sound of harpists playing on their harps, ³and they sing a new song before the throne and before the four living creatures and before the elders. No one could learn that song except the one hundred forty-four thousand who have been redeemed from the earth. ⁴It is these who have not defiled themselves with women, for they are virgins; these follow the Lamb wherever he goes. They have been redeemed from humankind as first fruits for God and the Lamb, ⁵and in their mouth no lie was found; they are blameless.

6 Then I saw another angel flying in midheaven, with an eternal gospel to proclaim to those who live*ʳ* on the earth—to every nation and tribe and language and people. ⁷He said in a loud voice, "Fear God and give him glory, for the hour of his judgment has come; and worship him who made heaven and earth, the sea and the springs of water."

8 Then another angel, a second, followed, saying, "Fallen, fallen is Babylon the great! She has made all nations drink of the wine of the wrath of her fornication."

9 Then another angel, a third, followed them, crying with a loud voice, "Those who worship the beast and its image, and receive a mark on their foreheads or on their hands, ¹⁰they will also drink the wine of God's wrath, poured unmixed into the cup of his anger, and they will be tormented with fire and sulfur in the presence of the holy angels and in the presence of the Lamb. ¹¹And the smoke of their torment goes up forever and ever. There is no rest day or night for those who worship the beast and its image and for anyone who receives the mark of its name."

12 Here is a call for the endurance of the saints, those who keep the commandments of God and hold fast to the faith of*ˢ* Jesus.

13 And I heard a voice from heaven saying, "Write this: Blessed are the dead who from now on die in the Lord." "Yes," says the Spirit, "they will rest from their labors, for their deeds follow them."

14 Then I looked, and there was a white cloud, and seated on the cloud was one like the Son of Man, with a golden crown on his head, and a sharp sickle in his hand! ¹⁵Another angel came out of the temple, calling with a loud voice to the one who sat on the cloud, "Use your sickle and reap, for the hour to reap has

q Other ancient authorities read *six hundred sixteen   r* Gk *sit   s* Or *to their faith in*

---

planations, the most probable is *Neron Caesar* (in Hebrew letters), which, if spelled without the final *n,* also accounts for the variant reading, 616 (see note *q*).

14.1–20: An interlude of three visions intended to reassure the church amid trials and persecutions. **1–5:** The Lamb and those redeemed from the earth; as in 7.4, the one hundred forty-four thousand is a symbolic expression for the whole number of the faithful. **1:** *Mount Zion,* that is, the heavenly Zion (Heb 12.22). *Written on their foreheads,* 7.3; 22.4. **3:** *They,* the one hundred forty-four thousand. *New song,* compare 5.8–10. **4:** *They are virgins,* that is, chaste, in contrast to the devotees of the pagan cults (2.20–22).

First fruits, Ex 23.19; Jas 1.18. **6–13:** Three angels announce the coming judgment. **6:** God's *eternal* purpose is soon to be fulfilled (10.7). **7:** *Fear God,* 15.4; Jer 10.7; Acts 5.11. *The hour of his judgment,* Acts 10.42; 17.31; 24.25. **8:** An anticipation of ch 18. **10:** *Wine of God's wrath,* Jer 25.15–16; 51.7. *Unmixed,* not diluted. Wine was often mixed with water before being drunk. *Tormented with fire,* in Gehenna (see 19.20 n.; 20.10). **11:** *Smoke,* 18.9, 18–19.3.

14.13: *Blessed,* see 1.3 n. *Their deeds,* the fruits of Christian character (Gal 5.22–23). **14–20:** The final judgment of God (Joel 3.13), involving the ingathering of the saints by one like the Son of Man (vv. 14–16) and the assem-

and an artisan of any trade
    will be found in you no more;
and the sound of the millstone
    will be heard in you no more;
23 and the light of a lamp
    will shine in you no more;
and the voice of bridegroom
    and bride
    will be heard in you no more;
for your merchants were the
    magnates of the earth,
    and all nations were deceived by
    your sorcery.
24 And in you*f* was found the blood
    of prophets and of saints,
    and of all who have been
    slaughtered on earth."

19 After this I heard what seemed to be the loud voice of a great multitude in heaven, saying,
"Hallelujah!
Salvation and glory and power
    to our God,
2   for his judgments are true
    and just;
he has judged the great whore
    who corrupted the earth with
    her fornication,
and he has avenged on her the
    blood of his servants."*g*
3 Once more they said,
"Hallelujah!
The smoke goes up from her
    forever and ever."
4 And the twenty-four elders and the four living creatures fell down and worshiped God who is seated on the throne, saying,
"Amen. Hallelujah!"
5 And from the throne came a voice saying,
"Praise our God,
    all you his servants,*g*
and all who fear him,
    small and great."
6 Then I heard what seemed to be the

voice of a great multitude, like the sound of many waters and like the sound of mighty thunderpeals, crying out,
"Hallelujah!
For the Lord our God
    the Almighty reigns.
7 Let us rejoice and exult
    and give him the glory,
for the marriage of the Lamb
    has come,
and his bride has made
    herself ready;
8 to her it has been granted to
    be clothed
    with fine linen, bright and
    pure"—
for the fine linen is the righteous deeds of the saints.
9 And the angel said*h* to me, "Write this: Blessed are those who are invited to the marriage supper of the Lamb." And he said to me, "These are true words of God." 10 Then I fell down at his feet to worship him, but he said to me, "You must not do that! I am a fellow servant*i* with you and your comrades*j* who hold the testimony of Jesus.*k* Worship God! For the testimony of Jesus*k* is the spirit of prophecy."

11 Then I saw heaven opened, and there was a white horse! Its rider is called Faithful and True, and in righteousness he judges and makes war. 12 His eyes are like a flame of fire, and on his head are many diadems; and he has a name inscribed that no one knows but himself. 13 He is clothed in a robe dipped in*l* blood, and his name is called The Word of God. 14 And the armies of heaven, wearing fine linen, white and pure, were

f Gk *her*   g Gk *slaves*   h Gk *he said*
i Gk *slave*   j Gk *brothers*   k Or *to Jesus*
l Other ancient authorities read *sprinkled with*

19.1–10: Praises in heaven for the destruction of Rome (vv. 1–5) and for the marriage of the Lamb (vv. 6–9). **5:** Ps 115.13. **7:** *His bride,* the church (Eph 5.23–32). **8:** The church is holy as her members are holy (7.14; compare Mt 22.11). **9:** *Blessed,* see 1.3 n. **10:** An aside (compare 22.8–9).

19.11–21: The victory of Christ and his heavenly armies over the beast and his cohorts. **12:** *A name . . . that no one knows,* the greatness of Christ surpasses human knowledge (Mt 11.27). **13:** As revealer of God he is called *The Word of God* (Jn 1.1, 14). **14:** *Armies,* the angelic host (see Lk 2.13 note *j*). **15:** 1.16;

following him on white horses. [15]From his mouth comes a sharp sword with which to strike down the nations, and he will rule[m] them with a rod of iron; he will tread the wine press of the fury of the wrath of God the Almighty. [16]On his robe and on his thigh he has a name inscribed, "King of kings and Lord of lords."

17 Then I saw an angel standing in the sun, and with a loud voice he called to all the birds that fly in midheaven, "Come, gather for the great supper of God, [18]to eat the flesh of kings, the flesh of captains, the flesh of the mighty, the flesh of horses and their riders—flesh of all, both free and slave, both small and great." [19]Then I saw the beast and the kings of the earth with their armies gathered to make war against the rider on the horse and against his army. [20]And the beast was captured, and with it the false prophet who had performed in its presence the signs by which he deceived those who had received the mark of the beast and those who worshiped its image. These two were thrown alive into the lake of fire that burns with sulfur. [21]And the rest were killed by the sword of the rider on the horse, the sword that came from his mouth; and all the birds were gorged with their flesh.

20 Then I saw an angel coming down from heaven, holding in his hand the key to the bottomless pit and a great chain. [2]He seized the dragon, that ancient serpent, who is the Devil and Satan, and bound him for a thousand years, [3]and threw him into the pit, and locked and sealed it over him, so that he would deceive the nations no more, until the thousand years were ended. After that he must be let out for a little while.

4 Then I saw thrones, and those seated on them were given authority to judge. I also saw the souls of those who had been beheaded for their testimony to Jesus[n] and for the word of God. They had not worshiped the beast or its image and had not received its mark on their foreheads or their hands. They came to life and reigned with Christ a thousand years. [5](The rest of the dead did not come to life until the thousand years were ended.) This is the first resurrection. [6]Blessed and holy are those who share in the first resurrection. Over these the second death has no power, but they will be priests of God and of Christ, and they will reign with him a thousand years.

7 When the thousand years are ended, Satan will be released from his prison [8]and will come out to deceive the nations at the four corners of the earth, Gog and Magog, in order to gather them for battle; they are as numerous as the sands of the sea. [9]They marched up over the breadth of the earth and surrounded the camp of the saints and the beloved city.

m Or will shepherd   n Or for the testimony of Jesus

Ps 2.9. **16**: 17.14; Deut 10.17. **17–18**: *An angel summons the birds of prey* (Ezek 39.4, 17–20). **19–21**: The final battle between Christ and Antichrist (anticipated in 16.13–16). **20**: *The false prophet,* the second beast of 13.11–15. *The lake of fire* is Gehenna (see 14.10 n.; Mt 18.9 note *g*).

20.1–6: The binding of Satan and the reign of the martyrs. 1: *Chain,* 2 Pet 2.4; Jude 6. **2**: As other numerals in this book are to be understood symbolically and not literally, so this period of *a thousand years* represents the perfection and completion of the martyrs' reign with Christ, untroubled by Satan's wiles. The first limiting of the power of the evil one occurred during the ministry of the Seventy (Lk 10.18). One must beware of reading more into this passage than is warranted; e.g. nothing is said here about a reign on earth. **4**: *Thrones* of judgment (Dan 7.9, 22, 27; Mt 19.28; Lk 22.30). *Those . . . beheaded for their testimony,* martyrs (in 6.9–10 the souls of the martyrs, under the altar, cry for vengeance). *Its mark,* 13.16–17; 14.9. **6**: *Blessed,* see 1.3 n. *Priests,* 1.6; 5.10. *Second death,* see 2.11 n.

20.7–10: The loosing of Satan and the final conflict. *Gog and Magog,* Ezek chs 38–39. **9**: *The beloved city,* Jerusalem, symbol of the Church universal. *Fire,* 2 Kings 1.10–12. **10**: *Was thrown into the lake of fire,* the final overthrow of Satan.

And fire came down from heaven[o] and
consumed them. 10 And the devil who
had deceived them was thrown into the
lake of fire and sulfur, where the beast
and the false prophet were, and they will
be tormented day and night forever and
ever.

11 Then I saw a great white throne
and the one who sat on it; the earth and
the heaven fled from his presence, and no
place was found for them. 12 And I saw
the dead, great and small, standing before the throne, and books were opened.
Also another book was opened, the book
of life. And the dead were judged according to their works, as recorded in the
books. 13 And the sea gave up the dead
that were in it, Death and Hades gave up
the dead that were in them, and all were
judged according to what they had done.
14 Then Death and Hades were thrown
into the lake of fire. This is the second
death, the lake of fire; 15 and anyone
whose name was not found written in
the book of life was thrown into the lake
of fire.

21 Then I saw a new heaven and a
new earth; for the first heaven
and the first earth had passed away, and
the sea was no more. 2 And I saw the holy
city, the new Jerusalem, coming down
out of heaven from God, prepared as a
bride adorned for her husband. 3 And I
heard a loud voice from the throne saying,

"See, the home[p] of God is
among mortals.
He will dwell[p] with them as
their God;[q]
they will be his peoples,[r]
and God himself will be
with them;[s]
4 he will wipe every tear from
their eyes.
Death will be no more;
mourning and crying and pain
will be no more,
for the first things have
passed away."

5 And the one who was seated on the
throne said, "See, I am making all things
new." Also he said, "Write this, for these
words are trustworthy and true." 6 Then
he said to me, "It is done! I am the Alpha
and the Omega, the beginning and the
end. To the thirsty I will give water as a
gift from the spring of the water of life.
7 Those who conquer will inherit these
things, and I will be their God and they
will be my children. 8 But as for the cowardly, the faithless,[t] the polluted, the
murderers, the fornicators, the sorcerers,
the idolaters, and all liars, their place will
be in the lake that burns with fire and
sulfur, which is the second death."

9 Then one of the seven angels who
had the seven bowls full of the seven last
plagues came and said to me, "Come, I
will show you the bride, the wife of the

o Other ancient authorities read *from God, out of heaven,* or *out of heaven from God*
p Gk *tabernacle* *q* Other ancient authorities lack *as their God* *r* Other ancient authorities read *people* *s* Other ancient authorities add *and be their God* *t* Or *the unbelieving*

20.11–15: The final judgment. 12: *Books,* containing the record of what everyone has done (Dan 7.10). *Book of life,* see 3.5 n. *Judged according to their works,* Mt 16.27; Rom 2.6; 2 Cor 5.10. **13–14:** *Death and Hades* (the temporary abode of the dead) are personified. **14:** *The second death,* see 2.11 n.; Mt 25.41.

21.1–22.5: Vision of the new Jerusalem.

21.1–8: The renewal of creation. 1: The *new heaven and . . . new earth,* predicted by Isaiah (Isa 65.17; 66.22); all creation will be renewed, freed from imperfections, and transformed by the glory of God (Rom 8.19–21). *The sea,* a symbol of turbulence and unrest. **2:** *New Jerusalem,* the church (Gal 4.26). *Prepared as a bride,* suggesting all that is beautiful and lovely (19.7–9). **3:** Ezek 37.27. **4:** 7.16; Isa 25.8; 35.10. **5:** The speaker is the Lord God (1.8). **6:** *Alpha* and *Omega,* see 1.8 n. *Water of life,* Isa 55.1; Jn 4.13; 7.37. **7:** *Those who conquer,* compare the concluding words of each of the seven letters in chs 2 and 3. **8:** *The cowardly* and *the faithless* do not have faith enough to endure trials, and so fall away in time of persecution. *Sorcerers,* literally "poisoners," those dealing in philters and poisons (Acts 19.19). *The lake,* see 19.20 n. *The second death,* contrasted to the water of life in v. 6 (see 2.11 n.).

21.9–27: The measuring of the city. Since the city is symbolical, so are its measurements (all of which are multiples of 12).

Lamb." [10] And in the spirit[u] he carried me away to a great, high mountain and showed me the holy city Jerusalem coming down out of heaven from God. [11] It has the glory of God and a radiance like a very rare jewel, like jasper, clear as crystal. [12] It has a great, high wall with twelve gates, and at the gates twelve angels, and on the gates are inscribed the names of the twelve tribes of the Israelites; [13] on the east three gates, on the north three gates, on the south three gates, and on the west three gates. [14] And the wall of the city has twelve foundations, and on them are the twelve names of the twelve apostles of the Lamb.

15 The angel[v] who talked to me had a measuring rod of gold to measure the city and its gates and walls. [16] The city lies foursquare, its length the same as its width; and he measured the city with his rod, fifteen hundred miles;[w] its length and width and height are equal. [17] He also measured its wall, one hundred forty-four cubits[x] by human measurement, which the angel was using. [18] The wall is built of jasper, while the city is pure gold, clear as glass. [19] The foundations of the wall of the city are adorned with every jewel; the first was jasper, the second sapphire, the third agate, the fourth emerald, [20] the fifth onyx, the sixth carnelian, the seventh chrysolite, the eighth beryl, the ninth topaz, the tenth chrysoprase, the eleventh jacinth, the twelfth amethyst. [21] And the twelve gates are twelve pearls, each of the gates is a single pearl, and the street of the city is pure gold, transparent as glass.

22 I saw no temple in the city, for its temple is the Lord God the Almighty and the Lamb. [23] And the city has no need of sun or moon to shine on it, for the glory of God is its light, and its lamp is the Lamb. [24] The nations will walk by its light, and the kings of the earth will bring their glory into it. [25] Its gates will never be shut by day—and there will be no night there. [26] People will bring into it the glory and the honor of the nations. [27] But nothing unclean will enter it, nor anyone who practices abomination or falsehood, but only those who are written in the Lamb's book of life.

22 Then the angel[y] showed me the river of the water of life, bright as crystal, flowing from the throne of God and of the Lamb [2] through the middle of the street of the city. On either side of the river is the tree of life[z] with its twelve kinds of fruit, producing its fruit each month; and the leaves of the tree are for the healing of the nations. [3] Nothing accursed will be found there any more. But the throne of God and of the Lamb will be in it, and his servants[a] will worship him; [4] they will see his face, and his name will be on their foreheads. [5] And there will be no more night; they need no light of lamp or sun, for the Lord God will be

u Or *in the Spirit*   v Gk *He*   w Gk *twelve thousand stadia*   x That is, almost seventy-five yards   y Gk *he*   z Or *the Lamb*. [2] *In the middle of the street of the city, and on either side of the river, is the tree of life*   a Gk *slaves*

9: *The seven bowls,* see ch 16. 10: *In the spirit,* see 1.10 n. *High mountain,* Ezek 40.2. 11: *Jasper,* 4.3. 12: Ezek 48.30–34. 14: *Apostles,* Eph 2.20.
21.15–18: *The city* is represented as being a cube, symbol of perfection; its beauty and magnificence are suggested by the precious stones (Ex 28.17–21). 19: Isa 54.11–12. 21: *The street,* the paving of the streets.
21.22–23: *No temple . . . . no sun* are needed because the Presence and Glory of God pervades the entire community (Isa 24.23; 60.1, 19). *Its lamp is the Lamb,* Jn 8.12. 25: Gates are for protection from enemies; perpetually open gates symbolize perfect safety; see Isa 60.11. 27: *Book of life,* see 3.5 n.
22.1–5: **The river and the tree of life. 1:** *The river* of blessings from God (compare Gen 2.10; Ps 46.4; Ezek 47.1; Zech 14.8; Jn 4.10, 14). 2: *The tree,* used here generically of many trees *on either side of the river* (Ezek 47.12). 3: *The throne of God and of the Lamb* is one throne (Jn 10.30). 4: To *see his face* will be the crowning joy of heaven (Mt 5.8; 1 Jn 3.2; compare 1 Cor 13.12). *On their foreheads,* see 7.3 n.; contrast 13.16. 5: Those who *worship* God (v. 3) *will reign* with him in eternal triumph.

their light, and they will reign forever and ever.

6 And he said to me, "These words are trustworthy and true, for the Lord, the God of the spirits of the prophets, has sent his angel to show his servants*b* what must soon take place."

7 "See, I am coming soon! Blessed is the one who keeps the words of the prophecy of this book."

8 I, John, am the one who heard and saw these things. And when I heard and saw them, I fell down to worship at the feet of the angel who showed them to me; 9but he said to me, "You must not do that! I am a fellow servant*c* with you and your comrades*d* the prophets, and with those who keep the words of this book. Worship God!"

10 And he said to me, "Do not seal up the words of the prophecy of this book, for the time is near. 11Let the evildoer still do evil, and the filthy still be filthy, and the righteous still do right, and the holy still be holy."

12 "See, I am coming soon; my reward is with me, to repay according to everyone's work. 13I am the Alpha and the Omega, the first and the last, the beginning and the end."

14 Blessed are those who wash their robes,*e* so that they will have the right to the tree of life and may enter the city by the gates. 15Outside are the dogs and sorcerers and fornicators and murderers and idolaters, and everyone who loves and practices falsehood.

16 "It is I, Jesus, who sent my angel to you with this testimony for the churches. I am the root and the descendant of David, the bright morning star."

17 The Spirit and the bride say,
"Come."
And let everyone who hears say,
"Come."
And let everyone who is thirsty
come.
Let anyone who wishes take the
water of life as a gift.

18 I warn everyone who hears the words of the prophecy of this book: if anyone adds to them, God will add to that person the plagues described in this book; 19if anyone takes away from the words of the book of this prophecy, God will take away that person's share in the tree of life and in the holy city, which are described in this book.

20 The one who testifies to these things says, "Surely I am coming soon."

Amen. Come, Lord Jesus!

21 The grace of the Lord Jesus be with all the saints. Amen.*f*

b Gk *slaves*   *c* Gk *slave*   *d* Gk *brothers*
e Other ancient authorities read *do his commandments*   *f* Other ancient authorities lack *all*; others lack *the saints*; others lack *Amen*

22.6–21: Epilogue, consisting of warnings and exhortations. **6:** *He said*, an angel (v. 8), perhaps the one referred to in 1.1. *These words,* i.e. the contents of this book. *The God of the spirits of the prophets,* the one who inspired the spirits of the prophets. **7:** A parenthesis, reporting the words of Christ (compare 16.15). *Blessed,* see 1.3 n. **8–9:** 19.10. **10:** *Do not seal,* this book is intended for all to read. **11:** The end of the age is too near to allow time for change.
22.12: *My reward is with me,* Isa 40.10; Jer 17.10. **13:** Christ applies God's title to himself (see 1.8 n.). **14:** *Blessed,* see 1.3 n. *Wash,* 7.14. *The city,* heavenly Jerusalem. **15:** *Dogs,* impure, lascivious persons (Phil 3.2). *Sorcerers,* see 21.8 n. **16:** *Root . . . of David,* Isa 11.1, 10, Mt 1.1. *Morning star,* see 2.28 n. **17:** *The bride,* the church. *"Come,"* singular number in Greek, is addressed to Jesus, as v. 20. **18–19:** A solemn warning against perversion of the teachings of *this book* (compare Deut 4.2; 12.32). **20:** *The one who testifies,* Jesus Christ; see 1.2. **21:** A fitting conclusion to this book and to the Bible.

Modern Approaches to Biblical Study

Opening the Old Testament can be at first quite disconcerting to the average reader. If Moses wrote the Pentateuch (the Torah, see p. xxxv OT), he would appear to have done it in a very sporadic fashion, with laws and narrative put together in a puzzling mixture. And how was the book of Jeremiah assembled, with its personal sayings, the biography (Jer chs 37–45), and prose discourses? So many of the prophetical books have the character of pieces in an anthology, to which no indication of the date or life-setting have been provided. More basically, are we to think of the Old Testament as a *literary* work? In its final form it is literature, but how was it put together, especially since oral tradition (the prophets were *preachers*) must have played a role? These questions can also be addressed to the New Testament. How did the synoptic Gospels come to be written? What was the role of the early Christian community in the formation of the oral and written traditions about Jesus? It is problems such as these that gave rise to the scientific approach to the Bible in modern times—an approach that is manifested in the introductions and annotations of the present edition of the Bible.

I. LITERARY CRITICISM

During the nineteenth and twentieth centuries scholars attempted to answer these questions by the method of literary criticism. This approach aimed primarily at questions of authorship (did John write the Gospel attributed to him?), and the formation of the *written* documents (how were the J, E, P, and D documents (see pp. xxxv–xxxvi OT) put together to form the Pentateuch?). This method achieved many good results. For example, we can now better grasp the message of Isa 40–55 when we read these chapters in the light of the exilic period rather than as coming from the Isaiah of 700 B.C. It must be acknowledged, however, that new insights from archaeology and history have revealed definite limitations of this method. Among other things, it failed to take seriously the *oral* traditions that lay behind the final written form of a biblical book, and it did not adequately examine questions of the origin or life-setting of the individual units within a book. This task was taken over by the method called "form criticism."

II. FORM CRITICISM

The method of form criticism was first used by German scholars (Hermann Gunkel for the Old Testament; Martin Dibelius for the New Testament), who employed the term *Formegeschichte* (lit. "form history"). It was recognized that within the Bible certain definite literary genres or forms could be identified. Gunkel analyzed the Psalter into Hymns, Laments, Thanksgiving Songs, Liturgies, etc. The analysis rested upon a study of literary style, which was correlated with the life-setting (*Sitz im Leben*) of the individual Psalm. Often similar genres in the literature of Israel's neighbors, especially Egypt and Mesopotamia, served to suggest or confirm such analysis, which eventually was applied to all parts of the Bible.

The existence of various literary genres in the Bible had always been admitted, but they had not been studied in and for themselves or used as tools in the analysis of biblical literature. In a sense, they were taken for granted, just as we take for granted the several genres in our own literatures. Any newspaper presents its readers with many and various literary genres: advertisements, obituary notices, editorials, letters to the editor, human interest stories, the comics, etc. In all these cases fixed written forms can be recognized. Each has its own specific characteristics: certain expressions or formulae and perhaps even a given structure. Modern readers are oblivious to all these details. They are so attuned to the literary expressions of their own culture that they automatically "shift gears" as they move from one genre to another. But it is a different expe-

388

rience when they expose themselves to a literature as ancient and varied as the Bible.

The difference between the peoples of the ancient Semitic world, its way of thinking, and our own style, quickly dawns upon us. It would not do to exaggerate these differences, but the fact remains that the mode of expression of the people of biblical times is a challenge to our understanding. Thus, their way of writing history does not conform to ours, as we quickly learn from the story of David's rise in 1 Sam 16—2 Sam 5, or from the "throne succession" narrative in 2 Sam 9ff. Our task is to ascertain and respect the various literary types, whether historical, epic, saga, or legend, that occur in the Bible; otherwise we will fail to comprehend the literary riches preserved in the various parts of Scripture.

In contrast to an earlier generation of Bible readers, we are very much aware of the role played by oral tradition in the gradual formation of the Scriptures. Many of the biblical books have a long pre-history of oral transmission. The ancients retained in their memories far more than they wrote. Thus we read of Jeremiah being commissioned by the Lord to write "all the words that I have spoken to you . . . from the days of Josiah until today" (36.2). It would appear that it was only after a twenty-year period of preaching that Jeremiah resorted to writing. In his case, and in that of other prophets as well, it is easy to imagine that many of their listeners remembered certain sayings and even put them in writing. We know that Isaiah had a group of disciples (Isa 8.16). The wisdom-sayings were cultivated at the Jerusalem court (Prov 25.1), as was the case in Egypt. But many of these sayings doubtless were first coined among the common people. Similarly, the laws circulated by oral tradition before they became fixed in writing. They were proclaimed at the great feasts, such as that of the covenant renewal, and only later reduced to the codes we find in the Bible (Covenant [see Ex 20.22–23.33n.], Priestly, and Deuteronomic Codes). The stories of the Judges (Jephthah, Samson, Deborah, and others) circulated by word of mouth among the people concerned (in Transjordan and in Judah) before they were incorporated into a *book* such as Judges. The Gospels too have been analyzed in the light of the several units that went into their composition: the sayings of Jesus, the miracle stories, the Passion narratives, etc. Scholars came to recognize that some of these had their life-setting in the primitive Church. Such literary units reflect concerns that became particularly acute in the history of the community—such as the attitude to the syn-

agogue, the delay of the parousia (= second coming), etc.—and the gospel material was shaped to meet the needs of the growing Church. Thus, several parables reflect their new life-setting (see p. 400); they shift the emphasis which Jesus would have placed on the coming of the kingdom to the second coming of Christ (e.g., the parable of the ten bridesmaids, Mt 25.1–13).

The life-setting of any piece of literature is a particularly important element. To a given life-setting there corresponds a certain range of literary genres. Thus, Israelite liturgy is the general life-setting within which the Psalms were composed and utilized. It was in the temple and in association with various feasts that the Songs of Praise, Thanksgivings, Laments, etc., were used. And we are better able to understand the structures and motifs of a given genre if we can locate it within its proper life-setting. In everyday life we understand the life-setting implicit in a wedding invitation, in an obituary notice, etc. In a similar vein, we must be sensitive to the life-setting of a given prophetic oracle, or a given wisdom saying. It is also important to recognize that the life-setting changes as a biblical tradition is handed down within a community. This fact is especially important for understanding the Gospels. Thus, one may distinguish between the historical life-setting in Jesus' own experience, into which one of his sayings may fit (Mk 2.27, "the sabbath was made for humankind, and not humankind for the sabbath"), from the life-setting in the Christian community that applied one of the Lord's sayings to a particular situation in the life of the Church (Mt 28.19–20, "Go therefore and make disciples of all nations, baptizing them in the name of the Father and of the Son and of the Holy Spirit, and teaching them . . . ."). Furthermore, as will be indicated in the following section dealing with redaction criticism, there is a third life-setting—the place of, and emphasis given to, a particular section within a Gospel. Thus, one can see the different ways in which the tradition concerning John the Baptist is used in the Gospels of Matthew and Mark. Each evangelist has his own way of emphasizing the traditions about Jesus, and gives these a specific life-setting within his own Gospel.

III. REDACTION CRITICISM

The intensely analytical method of form criticism called forth another approach, synthetic in character. This is the method of "redaction criticism" (*Redaktionsgeschichte*, lit. "redaction history"), which studies the document in its final form from the theological point of view of the author who arranged, modified, and edited the traditional material that formed the raw material out of which the work was created. Such an editor may even have created new forms to express new meanings appropriate to new contexts. The method of redaction criticism is most clearly illustrated by the differences among the Synoptic Gospels, which show that Matthew's viewpoint goes beyond Mark's, and that Luke's is different from either—despite the similarities among them all that warrant for them the term "synoptic." Now there is another life-setting to take into consideration: the life-setting of the Gospel itself—how a given form is used in a Gospel.

A similar synthetic approach is applicable to the Old Testament. It is not enough to recognize a unit or given literary genre. One must also consider its function within the larger work of which it forms a part. Thus, the various episodes in the book of Judges have been brought together to contribute to a higher unity, the theme of which is clearly enunciated in ch 2 of that book: the cycle of apostasy, enslavement, repentance, and deliverance (Judg 2.16–20). As the author of this passage clearly implies, readers should understand from history that Israel was prosperous and free only when it was loyal to God. Indeed, the book of Judges itself is a part of a larger composition: Joshua, Judges, 1–2 Samuel, 1–2 Kings. This corpus has been given the name "Deuteronomistic History," because it interprets the history of Israel from Joshua to the Exile in the light of the teaching of the book of Deuteronomy (see "The Historical Books," pp. 267–269 OT). It is interesting, furthermore, to compare the treatment given to a particular unit by the Deuteronomistic historians with that accorded by the Chronicler in that historical narrative. In 2 Sam 24.1, the bare fact is recorded that "the anger of the Lord was kindled against Israel, and he incited David against them. . . ." And David is condemned for taking a census on this occasion. But in 1 Chr 21.1 we read, "Satan stood up against Israel, and incited David to count the people of Israel." The point of view of the redactor has shaped the text—hence the name, redaction criticism.

IV. TRANSMISSION
(OR TRADITION) HISTORY

Another term that is frequently used in this kind of study of the biblical text is "transmission history" (*Überlieferungsgeschichte;* also called "tradi-

tion history," *Traditionsgeschichte*). This means the recognition of the growth and gradual formation of the stories that have been handed down—from simple forms in the beginning, to large complexes that can be recognized in the books as we now have them. Thus, the manner in which the Yahwist writer in the J document has put together the material belonging to pre-history (Gen 2–11), and has also worked out the sequence of the patriarchal narratives (e.g., that of Abraham), can come alive by the study of the history of the transmission of a unit. And very often this kind of analysis lends itself to conclusions about the history of Israel and its beliefs, enabling us to date certain ideas and movements as very old or as relatively late.

V. OTHER WAYS OF
READING THE TEXT

The above methods of approaching the biblical text can be summarized under the term historical methodology, or historical criticism. The purpose is to span the years and get into the meaning of the text as it came to be written. But there are other ways of reading the text that have surfaced in the history of exegesis. The typological reading sought to find correspondences between persons (Adam and Christ; see Rom 5.14–21), or events (Exodus and return from Babylon; see Isa 43. 16–19). Allegory sought for hidden meanings behind the obvious sense of the text (Noah's ark and baptism; see 1 Pet 3.20–21). These time-honored methods still survive.

In recent times other ways of reading the text have been proposed and they deserve to be noticed here. For example, why should one not read the Pentateuch as a whole, and not merely as the product of four primary documents? What meaning does one derive from such an approach to the Torah? What was the "canonical shape" given to the book of Isaiah by joining together its three major parts (1–39; 40–55; 56–66)? What difference does it make that the original Luke/Acts (one work) was separated into two books in different places in the New Testament? This kind of "holistic" approach yields a different perspective for interpretation. In a similar way the addition of the New Testament to the Old creates a broader context of interpretation for the Christian; at the same time the Christian should be aware that the Old Testament can complete or correct a Christian understanding of the Bible (e.g., an exaggerated personal eschatology.)

Furthermore, modern literary critics have chided biblical scholars for composing commentaries that concentrate on the genesis of a biblical book. Why not approach the interpretation of such a work from the point of view of its final form, not its hypothetical pre-history? Repetitions can be deliberate and meaningful, and not just signs of separate documents (e.g., J or E in the Pentateuch) in a work. Modern tastes are beginning to make allowances for the different emphases in the styles of ancient writings.

The development of philosophical hermeneutics has also added further dimensions to biblical interpretation. No interpretation (of *any* text) takes place without presuppositions, of which we are often unaware. Historical methodology, too, has to make allowances for this. Contextual interpretation, i.e., interpretation from the situation in which either the reader or the writer begins, is inevitable. Hence many approaches have been recognized: liberationist, feminist, etc. Sophisticated methods of literary criticism have also been developed: structuralism, deconstruction, reader-response theory, etc. Perhaps the most practical rule for a Bible reader is to strive to become aware of his or her initial presuppositions in reading the Scriptures (and to recognize these also in the explanations offered by various interpreters). The long history of biblical interpretation teaches us valuable lessons about our own biases in reading a text.

VI. SUMMARY

Current biblical scholarship employs all the methods that we have briefly described. It is concerned to determine the given literary genre of a biblical book or most often a given section of it (form criticism). Further, it may be possible to discover how this genre was handed down (transmission history), and came to be incorporated by a collector or editor (redaction criticism) into a continuous narrative, and ultimately became part of a written document (literary criticism) The methods do not exclude one another; in fact, they are mutually complementary. But they are not merely methods; they are not intellectual exercises for an élite. Every perceptive reader of the Bible can profit from the several approaches. We shall try to illustrate the advantages in the case of the Decalogue.

The literary genre of the Decalogue can be described as apodictic law. As opposed to casuistic law (or "case law"), which stipulates a course of action in given circumstances, the apodictic law is a forceful command or prohibition. (Apodictic means "clear" or "incontrovertible" and

denotes foundational law from which there is generally no appeal.) There are several such laws in the Old Testament (for example, Ex 23.1–9; Lev 18.6–23). A motive clause may be appended to a law, as in the case of the sabbath law and honoring one's parents. Unfortunately, there is as yet no way of tracing the history of the development of the Decalogue, even though we can make many worthwhile observations about it. As is well known, there is more than one version of it: Ex 20.1–17, and Deut 5.6–22. A comparison of the two passages reveals something of the transmission history.

In general, the Decalogue in Deuteronomy is couched in language that is characteristic of this book: "as the Lord your God commanded you" (5.12–16; compare 4.5; 5.32); "that it may go well with you in the land that the Lord your God is giving you" (5.16, 29; 6.18). The familiar sabbath law is strengthened in Deuteronomy ("observe") as against Exodus ("remember"). A more striking difference is the motive clause: creation in six days and rest on the seventh (Ex 20.11), in contrast to the Exodus from Egypt (Deut 5.15). Originally, one may presume, there was no motive given; in the course of time, two different motives were offered.

The treatment of the second commandment is particularly interesting. The prohibition of a graven image (Ex 20.4; Deut 5.8) referred originally to an image of the Lord (see Ex 34.17). But both versions of the Decalogue, by adding the phrase "the likeness of anything . . . ," refer the prohibition to images of other gods—presumably at a time when *this* was the great temptation. The continuation, "you shall not bow down to them or serve them" refers now to the "other gods," and thus the second commandment stands in the shadow of the first; the prohibition of images (of the Lord) and of foreign gods is viewed as one and the same. The "false witness" in Ex 20.16 and Deut 5.20 calls for comment. In the Deuteronomy passage "false witness" is, more literally, "vain witness" (empty, and eventually, deceitful). One may reasonably surmise that the form in Exodus is the original. It is found frequently outside the Decalogue (for example, Ps 27.12; Prov 6.19 "lying witness"), whereas "vain witness" is found only in the Deuteronomic version of the Decalogue.

We may conclude, then, that the form of the Decalogue has had a complicated transmission history. What is suggested by the fact that it is found in two different books? The form in Exodus has the flavor of the editing given to the

Deuteronomistic history (Joshua to Kings): Egypt is "the house of slavery" (Ex 20.2); reverence for "the name of the LORD your God" (Ex 20.7). The import of such touches is to include the Decalogue within the sweep of Israel's salvation history as it is presented here. In Deuteronomy the role of the Decalogue is slightly different—it has become the sovereign will of God, mediated through Moses in a farewell speech. It is followed by the Deuteronomic code (Deut chs 12–26), which serves as a commentary on, or interpretation of, this Decalogue. But this life-setting of the Decalogue within the framework of the book of Deuteronomy is not necessarily the life-setting which the Decalogue had in the life of Israel. Here we are on somewhat uncertain ground. But it seems safe to say that *one* life-setting for the Decalogue has been the cult, perhaps at the covenant renewal ceremony, or the Feast of Tabernacles.

We have discussed the Decalogue merely as an example of how literary analysis of a biblical text can proceed. Basically the same approach can be used throughout the Old and New Testaments. And readers will find Bible reading more challenging, as well as more informative, if they study the text along these lines. They will understand how truly the Old Testament is a precipitate, as it were, from the life of Israel, the People of God. Similarly, the role of the Church in the formation of the New Testament, especially the Gospels, will stand out. We have not mentioned the Pauline Epistles, but readers will not fail to be impressed with the manner in which the "good news" is mediated by Paul, in contrast to the way in which it is presented in the four Gospels. In short, there is no book, or part of a book, in the Bible that does not have a literary form and a history behind its incorporation and transmission within the Scriptures. If we pay attention to this aspect of biblical literature, we shall be well on our way to understanding it.

Characteristics of Hebrew Poetry

I. POETRY IN THE BIBLE

Hebrew poetry has had a long and varied history, extending from the beginnings of Israel's life up to modern times. The earliest phases are represented by the poetical parts of the Old Testament (Hebrew Scriptures). These are, in relation to the total extent of the Old Testament, quite substantial: Psalms, Proverbs, Song of Solomon, Lamentations, almost the whole of Job, much of the prophetic literature, and many passages in the narrative literature (e.g., Gen ch 49; Ex 15.1–18; Deut chs 32 and 33; Judg 5.2–31). In all, this amounts to about one third of the Old Testament; but it is presumably only a fraction of the entire body of Hebrew poetry produced during the Old Testament period. The Hebrew poetic tradition is continued in the Apocrypha (Wisdom of Solomon, Sirach, and several passages in other books, e.g., Judith), and in parts of the Pseudepigrapha and of the literature of the Qumran sect (the Dead Sea Scrolls). In the New Testament much of the teaching of Jesus is expressed in poetic form (see "Literary Forms in the Gospels," §2); and there are passages of poetry in Lk chs 1–2 and in several other books, notably Revelation.

It is significant that the Hebrew poetic form is clear even when the poems have been transmitted in Greek. In translating from one language to another there is the double problem of reproducing both the spirit and the form of the original.

For technical reasons, and because of the dissimilarity between poetic structure in different languages, it may be difficult to reproduce the form satisfactorily. In order to preserve and communicate the spirit, it is often necessary in translation to adopt a different poetic form, or even to translate into rhythmical prose. The many attempts which have been made to render the *Iliad* and the *Odyssey* into English verse and prose show how formidable the undertaking can be. In some ways the translation of ancient Hebrew poetry presents a less intractable problem, because its formal characteristics are in the main simple and flexible. Increased knowledge of ancient Near Eastern literature, particularly the poetic texts from Ugarit in northern Syria, has enlarged our understanding of Hebrew poetry by providing parallels to its forms and techniques. Ps 29 is an excellent example of a Hebrew poem that reflects the structural characteristics and poetic techniques known to us from the Ugaritic texts. It has been suggested that it is an Israelite adaptation of a Canaanite hymn to Baal.

II. THE FORMS OF HEBREW POETRY

Although there are gaps in our knowledge of the forms of ancient Hebrew poetry, and important points are still open to dispute, its general structure is sufficiently clear. Three main formal characteristics have been attributed to it: §1. Parallelism; §2. Meter; §3. Strophic Arrangement.

1. Parallelism

This feature (sometimes referred to also as "thought rhyme") is a balance not only of form but also of the thought between successive members in a poem. The terminology applied to these members varies considerably, but in this article the poetic unit printed as a line in the New Revised Standard Version will be called a "stich" (Gr. *stichos,* "line"), and combinations of two and three of these units will be termed "distichs" and "tristichs" respectively. A distich is composed of two lines, and a tristich is composed of three. The following from Ps 19.1 is a distich, or couplet, in which such a balance of thought and form is evident:

The heavens are telling the glory of God;
and the firmament proclaims his handiwork.

As an example of a tristich we may take Ps 100.1:

Make a joyful noise to the LORD, all the earth.
Worship the LORD with gladness;
come into his presence with singing.

Parallelism is more than mere repetition of words or ideas in successive lines. The second line is a specification, often an intensification, of the first. It can be said to serve, sometimes with a third line, to complement the thought of the first line.

Parallelism appears in four main forms: (*a*) Synonymous, (*b*) Antithetic, (*c*) Formal, (*d*) Climactic; and it may be (*e*) Internal or External, and (*f*) Complete or Incomplete.

(*a*) In Synonymous Parallelism the same thought is expressed in successive stichs; for example,

The ox knows its owner,
and the donkey its master's crib . . . (Isa 1.3a)

and

But let justice roll down like waters,
and righteousness like an ever flowing stream.
(Am 5.24)

In these distichs, the second stich simply repeats the sense of the first in slightly different terms. Sometimes, however, the second stich gives more precise expression to the sense of the first, as in Ps 34.13:

Keep your tongue from evil,
and your lips from speaking deceit.

(*b*) In Antithetic Parallelism the thought expressed in the second stich is in contrast to that in the first; for example,

for the LORD watches over the way of
the righteous,
but the way of the wicked will perish.
(Ps 1.6)

and

A wise child makes a glad father,
but a foolish child is a mother's grief.
(Prov 10.1)

This type of parallelism is particularly characteristic of the Wisdom Literature. It also occurs frequently in the sayings of Jesus; for example,

A good tree cannot bear bad fruit,
nor can a bad tree bear good fruit. (Mt 7.18)

and

Those who find their life will lose it,
and those who lose their life for my sake
will find it. (Mt 10.39)

(*c*) Formal Parallelism, also called Synthetic Parallelism, contains neither repetition in different terms nor contrasted assertions. In it the thought

393

of the first stich is carried further and completed in the second; for example,

> The LORD looks down from heaven on humankind
> to see if there are any who are wise,
> who seek after God. (Ps 14.2)

Here there is a formal balance of clause with clause, but no such correspondence in content as we find in Synonymous and Antithetic Parallelism. The balance is emphasized by two factors: the clearly marked caesura or break between the stichs, and the correspondence in rhythm or meter (see below §2).

(*d*) In Climactic Parallelism the characteristics of Synonymous and Formal Parallelism are combined. The second stich echoes or repeats part of the first and also adds to it an element which carries forward or completes the sense; for example,

> Ascribe to the LORD, O heavenly beings,
> ascribe to the LORD glory and strength.
> (Ps 29.1)

and

> Pray to your Father who is in secret;
> and your Father who sees in secret will
> reward you. (Mt 6.6b)

(*e*) All the examples quoted above are instances of Internal Parallelism, in which the balance of form and thought is between the individual stichs within a distich or tristich. In External Parallelism there is balance not only *within* but *between* distichs. So in Isa 1.27–28:

> Zion shall be redeemed by justice,
> and those in her who repent, by
> righteousness.
> But rebels and sinners shall be destroyed
> together,
> and those who forsake the LORD shall
> be consumed.

There is here Internal Synonymous Parallelism *within* each of the two distichs and External Antithetic Parallelism *between* the distichs. Compare Isa 1.10:

> Hear the word of the LORD,
> you rulers of Sodom!
> Listen to the teaching of our God,
> you people of Gomorrah!

By contrast, here each distich has Internal Formal Parallelism, and there is External Parallelism between the distichs. External Parallelism gives scope

for still more elaborate patterns, as in Ps 30.8–10:

> To you, O LORD, I cried;
> and to the LORD I made supplication;
> "What profit is there in my death,
> if I go down to the Pit?
> Will the dust praise you?
> Will it tell of your faithfulness?
> Hear, O LORD, and be gracious to me!
> O LORD, be my helper!"

The first, third, and fourth distichs have Internal Synonymous Parallelism, but the second has Internal Formal Parallelism. There is External Synonymous Parallelism between the first and fourth distichs and between the second and third, giving the pattern a, b, b, a (chiasmus). The juxtaposition of the second and third distichs serves to emphasize the thought which they both express.

(*f*) Complete Parallelism occurs when each term in the first stich is matched by a corresponding term in the second, as in the following example from Psalm 146.2:

> I-will-praise the-LORD as-long-as-I-live;
> I-will-sing-praises to-my-God all-my-life-long.

(The hyphens link words that together represent a single term in the Hebrew.) In Incomplete Parallelism one or more terms in the first stich have no counterpart in the second, as in the following:

> Our-inheritance has-been-turned-over to-strangers,
> our-homes to-aliens. (Lam 5.2)

Here there is nothing in the second stich to correspond to "has-been-turned-over" in the first. Incomplete Parallelism may occur without compensation, as in the example just given, or with compensation, as in Lam 5.11:

> Women are-raped in-Zion,
> virgins in-the-towns-of Judah.

In this verse the second stich contains no counterpart to "are-raped" in the first, but the *two* terms "in-the-towns-of" and "Judah" in the second stich correspond to the *one* term "in-Zion" in the first.

These varied forms of parallelism provide almost innumerable possibilities of changing poetic patterns. One feature, however, remains constant. In a distich or tristich each stich is separated from what follows or precedes by an emphatic rhythmic break (a caesura); there may be a completion of the sense in what follows, but there is always a marked pause. The feature known as enjambement (running over without rhetorical pause from one line to another so that closely related words fall in different lines) is foreign to Hebrew poetry.

Even a superficial study of the poetic parts of the Bible makes it abundantly clear that parallelism is a formal characteristic of Hebrew poetry; and this is further borne out by the fact that the same feature also appears in other ancient Near Eastern literatures. There is, unfortunately, less clarity about the second formal characteristic, namely meter.

2. Meter

It is reasonable to suppose that parallelism is complemented by metrical or rhythmical patterns and that each stich is composed of a number of metrical units. But the evidence about the nature of the patterns of meter from ancient Near Eastern parallels and from writers such as Josephus, Origen, and Jerome is inadequate. Of the varied theories that have been advanced in modern times, the one that commands the widest agreement is that the meter is determined in terms of accented syllables and that the number of unaccented syllables is of no significance. This means that as a rule each word counts as one unit; but where a word has a secondary accent in addition to the main one, it counts as two units, and where a word loses its accent because of its close connection with the following word, it does not count as a unit. The metrical value of a stich is indicated by the number of stressed syllables that it contains. By far the most frequent metrical pattern is 3 + 3, that is, a distich with three stressed syllables in each stich. This is the characteristic meter of Hebrew proverbs; it abounds in the Wisdom Literature, but it is also very frequent in some parts of the prophetical books and in the Psalms. The following example is from Wisdom Literature:

A-mórtal, bórn of-wóman,
   féw-of dáys and-full-of-tróuble,
comes úp like-a-flówer, and-wíthers,
   flées like-a-shádow and does-not-
     lást. (Job 14.1–2)

The shorter 2+ 2 meter is well adapted to the expression of intense emotion and urgency; for example,

Céase to-do-évil,
   léarn to-do-góod;
séek jústice,
   réscue the-oppréssed,

defénd the-órphan,
   pléad for-the-widow. (Isa 1.16d–17)

The unequal metrical pattern 3 + 2 is known as the Qinah ("lament") meter, because it is the prevailing meter used in the book of Lamentations and occurs frequently in laments, though it is not confined to this type of composition. The unfulfilled expectation of a third beat in the second stich creates a peculiarly haunting effect. So in the two following passages:

Fállen, no-móre to-ríse,
   is-maíden Ísrael. (Am 5.2a)

Give-éar-to my-wórds, O-Lórd;
   give héed-to my-síghing.
Listen to-the-sóund-of my-crý,
   my-King and-my-Gód.
     (Ps 5.1–2b)

Other metrical patterns that occur less frequently are 4 + 4, 2+ 2 + 2, 3+ 3 + 3. The last of these is memorably illustrated in a familiar psalm:

Lift-up your-héads, O-gátes!
   and-be-lífted-up, O-áncient dóors!
   that-the-King-of glóry may-come-in.
Whó-is-the-King-of glóry?
   The-Lórd, stróng and-míghty,
   the-Lórd, míghty in-báttle. (Ps 24.7–8)

A Hebrew poem is not necessarily composed in a single meter throughout. Indeed, the variation of meter can be most effective. A good instance of this is Ps 46, in which 4 + 4 meter predominates (verses 1–2, 4–6, 8–10), with 3 + 3 distichs rounding off each section (verses 3, 7, 11), so that the metrical pattern suggests that the poem is constructed in three stanzas.

3. Strophic Arrangement

Ps 46 is by no means the only poem in the Old Testament in which a recurring refrain appears to indicate a grouping of distichs or tristichs into larger patterns resembling stanzas. For instance, in Pss 42 and 43 (which form one psalm), the following refrain occurs at 42.5, 11, and 43.5:

Why are you cast down, O my soul,
   and why are you disquieted within me?
Hope in God; for I shall again praise him,
   my help and my God.

Again in Isa 9.8–10.4 and 5.25–30, which are probably parts of a single poem, the following refrain appears to mark off recognizable stanzas:

For all this his anger has not turned away;
   his hand is stretched out still.
     (9.12b, 17c, 21b; 10.4h; 5.25c)

Ps 107.1–32 contains the double refrain,

Then they cried to the LORD in their trouble,
   and he delivered them from their distress; . . .
Let them thank the LORD for his steadfast love,
   for his wonderful works to humankind.

at verses 6, 8, 13, 15, 19, 21, 28, 31; but it does not mark off regular strophes.

Strophic arrangement is the least convincing of the formal characteristics attributed to Hebrew poetry. Some Hebrew poems do indeed fall into recognizable sections; but, with the exception of poems of a special type (see below, 3.§1), these sections are not strictly regular. Moreover, the term "strophe" is inappropriate, since it implies the recurrence of regular metrical patterns in a fixed number of poetic lines. All the indications are that such rigid structures were alien to ancient Hebrew poetry, which in meter and in the grouping of distichs and tristichs was as flexible as it manifestly was in the use of parallelism.

III. THE TECHNIQUES OF HEBREW POETRY

In addition to the structural forms of Hebrew poetry, certain technical devices were used to secure particular effects. These are: §1 Acrostic Patterns; §2 Alliteration; §3 Assonance; §4 Onomatopoeia; §5 Paranomasia.

1. Acrostic Patterns

The sole instances of deliberately contrived sequences of lines (see above, 2. §3) are acrostic poems. Acrostic patterns occur in varying degrees of complexity. In Ps 34.1–21 the successive verses begin with the letters of the Hebrew alphabet in the appropriate order (verse 22 is outside the pattern). Pss 9 and 10 form a single poem with an imperfectly preserved acrostic pattern in which pairs of verses begin with successive letters of the Hebrew alphabet. Varied acrostic patterns appear in Lam chs 1, 2, 3, and 4: in chs 1, 2, and 3 the distichs are grouped in threes (in ch 3 *all* the distichs in a given group begin with the same letter; in chs 1 and 2 only the first distich in each group) and in ch 4 the grouping is in pairs. The most elaborate acrostic poem in the Bible is Ps 119, in which there are twenty-two sections (corresponding to the successive letters of the alphabet) each consisting of eight distichs, and every distich in a given section begins with the same letter.

Various reasons have been suggested for the adoption of this somewhat exacting technical pattern: to serve as a mnemonic device; to exploit the supposed magical power of an alphabetic acrostic; or to express completeness of grief, aspiration, penitence, hope, or devotion. The last of these would be particularly appropriate to Ps 119 and the acrostics in the book of Lamentations.

2. Alliteration

Effect is sometimes gained by the juxtaposition of words or syllables that begin with the same consonant; for example, in Ps 122.6–7, where *sh* and *l* recur:

> *shaʿălu shĕlom Yĕrushalayim*
> *yishlayu ʾohăbhayikh*
> *yĕhi shalom bĕḥelekh*
> *shalwah bĕʾarmĕnothayikh.*

Similarly, in Ps 93.4 the occurrence of doubled consonants and particularly of *m,* brings out effectively the menacing reverberation of the flood waters:

> *miqqoloth mayim rabbim*
> *ʾaddir mimmishbĕre yam*
> *ʾaddir bammarom Yahweh.*

3. Assonance

There is no systematic use of rhyme in ancient Hebrew poetry; but the recurrence of the same vowel sound is often deliberately contrived. Where this occurs in suffixes such as *-enu* ("our," "us") and *-ehu* ("it"), something approximating to rhyme appears. Thus in Ps 90.17 we find

> *wihi noʿam ʾădhonay ʾĕlohenu ʿalenu*
> *umaʿăseh yadhenu konĕnah ʿalenu*
> *umaʿăseh yadhenu konĕnehu.*

In Isa 53.4–5 the echoing suffixes *-enu* ("our") and *-hu* ("him") produce a wailing effect and also, together with the emphatic repetition of the pronoun *huʾ* ("he") and the final *lanu* ("for us"), bring out the contrast between the Servant and those for whom he suffered.

4. Onomatopoeia

The use of words that sound like what they describe has already been illustrated from Ps 93.4 (see above, 3.§2). Similarly, in Isa 17. 12–13 a combination of doubled consonants and long vowels expresses the surge and thunder of the waves, to which the tumultuous onslaught of the nations is compared. Then, with a change of figure, there follows a description of the repulse and scattering of the nations like chaff or dust before a storm wind:

> *hōy hămōn ʿammīm rabbīm*
> *kahămōth yammīm yehĕmāyūn*

*ūsheʾōn lĕ ʾummīm
kishĕ ʾōn mayīm kabbīrīm yishshāʾun
lĕ ʾummīm kishĕ ʾōn mayīm rabbīm yishshāʾūn
wĕghāʿar bō wĕnās mimmerḥāq
wĕruddaph kemōṣ hārīm liphĕnē rūăḥ
ūkhĕghalgal liphnē sūphāh.*

Another outstanding example of onomatopoeia is the description in the Song of Deborah of the thudding hoofs of galloping horses:

*ʾaz ḥalĕmu ʾiqqĕbhe sus
middahăroth dahăroth ʾabbiraw* (Judg 5.22)

5. Paranomasia

A pointed play on words occurs more frequently in the Bible than is evident from an English rendering. In predicting national disaster Amos makes an effective play on the name "Gilgal" and the verb *galah* ("go into exile"):

ki haggilgal galoh yighleh.

Again, when Amos sees a basket of "summer fruit" (*quayiṣ*) it is a token that the 'end' (*qeṣ*) is coming on Israel (Am 8.12). Isaiah says that God looked for 'justice' (*mishpaṭ*) and 'righteousness' (*ṣĕdhaqah*) but found only 'bloodshed' (*mispaḥ*) and the 'cry' (*ṣĕʿaqah*) of the oppressed (Isa 5.7b). In Jer 31.16–19 the words "come back," "bring back," "turned away" all come from the same verb (*shubh*). It is probably impossible to reproduce in translation this sustained play on similar and contrasted meanings.

IV. CONCLUSION

The supreme quality of ancient Hebrew poetry is seen in the concentrated expression, within a limited compass, of experience, emotion, and aspiration. This is seen in ancient war songs and dirges, the love poems in the Song of Songs, many of the hymns, laments, and thanksgivings in the Psalter, and in the terse oracles of Amos and of Isaiah of Jerusalem. Apart from Job and (probably) Isa 40–55, there is no sustained poetic composition of any great length in the Old Testament, and nothing that is epic in character. Although the tradition of the Psalmists is continued in the later period, e.g., in the poems in Lk 1–2 and (with significant differences) the Qumran *Hymns of Thanksgiving,* it is on the whole the didactic and reflective notes that tend to predominate, as in Sirach and the Wisdom of Solomon. In the sayings of Jesus the form and spirit of classical prophecy and of Wisdom are fused in a new and creative manifestation of the Hebrew poetic genius.

Literary Forms in the Gospels

An analysis of the teaching of Jesus reported in the four Gospels reveals a variety of literary forms. Sometimes he conveyed his teaching by means of parables; at other times he used proverbs and plays on words (puns). Many passages in the Gospels are arranged in strophic, or poetic, form, and frequently one is struck by the vigorous, picturesque language by which the teaching is conveyed. Examples within each of these categories, considered in reverse order, include the following.

1. Picturesque Speech

Like other persons of the Near East, Jesus made use of striking contrasts and vivid metaphors. Using exaggerated and colorful expressions, he frequently drew attention to the ridiculous and the illogical behavior of the self-righteous. For example instead of saying in prosaic and commonplace terms that some people are inconsistent when judging others and themselves, Jesus put it thus:

Why do you see the speck in your neighbor's eye, but do not notice the log in your own eye? . . . You hypocrite, first take the log out of your own eye, and then you will see clearly to take the speck out of your neighbor's eye (Mt 7.3,5).

By taking into account the presence of picturesque expression in the Gospels the reader can sometimes avoid misinterpreting the meaning. For example, the hard saying preserved in the third Gospel, "Whoever comes to me and does not hate father and mother, wife and children, brothers and sisters, yes, and even life itself, cannot be my disciple" (Lk 14.26), must be understood in the light of the frequent use of overstatement as characteristic of the speech of Near Easterners. It is obvious that Jesus, so far from intending to increase the sum total of hatred in the world, states a principle in a startling, hyperbolic manner, and leaves it to his hearers to discover whatever qualifications are necessary in the light of his other pronouncements. The saying means that in order to be a follower of Jesus one must be prepared to choose between natural affection and loyalty to the Master. The same idea is expressed in Matthew's less rigorous version of Jesus' saying: "Whoever loves father or mother more than me is not worthy of me; and whoever loves son or daughter more than me is not worthy of me" (Mt 10.37).

One should, of course, be alert to the danger of diluting Jesus' teaching by finding overstatement in passages where it is not present. For example, Jesus' command to the rich man who inquired what he should do to inherit eternal life, "Sell all that you own and distribute the money to the poor, and you will have treasure in heaven; then come, follow me" (Lk 18.22), should not be discounted as exaggerated hyperbole, meaning merely, "Sell ten percent of what you own. . . ." The context makes it absolutely clear that the questioner as well as the disciples understood Jesus' words in their literal sense.

2. Poetic, Rhythmical Parallelism

Hebrew poetry, illustrated in the Old Testament Psalter, is characterized by parallelism of members. Sometimes the parallelism is synonymous and sometimes antithetic (see pp. 394–395).

In view of the frequency of Jesus' quotations from and allusions to the Psalms, it is not surprising that we find much of his teaching cast into the mold of Semitic poetry. Synonymous parallelism appears in the saying recorded in Lk 6.27–28:

> Love your enemies,
>> do good to those who hate you,
> bless those who curse you,
>> pray for those who abuse you.

Antithetic parallelism is illustrated by Mt 7.17–18:

> Every good tree bears good fruit,
>> but the bad tree bears bad fruit.
> A good tree cannot bear bad fruit,
>> nor can a bad tree bear good fruit.

Besides these two basic types of parallelism, several other kinds have been identified. What is called step parallelism, for example, occurs when the second line takes up a thought contained in the first line and, repeating it, makes it, as it were, a step toward the development of a further thought, which is the climax of the whole. An example of step parallelism is found in Lk 9.48 (the italics indicate the repeated member which serves as a step, and the vertical line stands before the climax):

> Whoever welcomes this child in my name
>> *welcomes me,*
> *and whoever welcomes me* |welcomes the one
>> who sent me.

For other passages that exhibit an elaborate rhythmical pattern, see Mt 6.19–21; 23.16–22; Mk 2.21–22; 9.43–48; Lk 11.31–32; 17.26–30.

3. Plays on Words

The Old Testament contains not a few instances of plays on words (for examples see p. 397 and the notes on Gen 11.9; Jer 1.11–12; Am 8.1–2). The text of the Gospels, which has been transmitted to us in Greek, contains more than one instance where the original Aramaic of Jesus' mother tongue probably involved a word-play. It is understandable that very few such puns in Aramaic could be reproduced in Greek. In one case, however, it happens that the Greek word *pneuma*, just as the Aramaic *rûhâ*, means both "wind" and "spirit." In Jn 3.8 Jesus is quoted as saying to Nicodemus, "The *pneuma* blows where it wills, and you hear the sound of it, but you do not know whence it comes or whither it goes; so it is with everyone who is born of the *pneuma*."

One of the most noteworthy of Jesus' sayings about the church involves a play on words. According to Mt 16.13ff. at Caesarea Philippi, in response to Jesus' question to his disciples who they thought he was, Simon Peter confessed, "You are the Messiah, the Son of the living God." After declaring that Peter had spoken this by divine revelation, Jesus retorted, "And I tell you, you are Peter [Greek *Petros*], and on this rock [Greek *petra*] I will build my church." In Jesus' mother tongue the play is even closer, for in Aramaic the word *kēphâ* serves as a proper name (Cephas) and also means "a rock, a stone." Jesus' statement therefore would have been, "And I tell you, you are *Kēphâ*, and on this *kēphâ* I will build my church" (there remains a difference in gender, for the common noun is feminine and the proper name is, of course, masculine; compare French *pierre* (f.), "a stone," and *Pierre* (m.), "Peter").

Another passage which probably involved a

pun is Mt 23.24, where the Greek text is unable to reproduce the jingle that is present in what is presumed to be the original Aramaic. In his condemnation of the inconsistency of certain scribes and Pharisees, Jesus reproached them for "straining out a gnat and swallowing a camel." Since in Aramaic the word for "gnat" or "louse" is *qalmâ* and the word for "camel" is *gamlâ*, the pun provides added piquancy to the picturesque speech used by Jesus: he is describing a punctilious Pharisee who, in view of Lev 11.41ff., which forbids the eating of what swarms or crawls on the earth, is careful to strain out a *qalmâ* that may have fallen into his food or wine, but is quite unconcerned over gulping down a whole *gamlâ!*

4. Proverbs

Every language has pithy sayings or maxims that express a truth crisply and forcefully. Because proverbs frequently express only one side of a truth, it happens that mutually contradictory proverbs may circulate, each of which is true when applied to the appropriate life-setting. The common saying, "Penny wise, pound foolish," correctly describes one who is scrupulous about small transactions, but is extravagant in great ones. On the other hand, the proverb, "Take care of the pennies, and the dollars will take care of themselves," is also true. More than once the Bible presents two proverbs that, though contradictory, are both true when applied to appropriate circumstances. In Prov 26.4 the writer cautions his reader, "Do not answer fools according to their folly, or you will be a fool yourself"; in the very next verse, however, he advises, "Answer fools according to their folly, or they will be wise in their own eyes." It is left to the reader to know when it is appropriate to heed one or the other of these two antithetical proverbs.

It is not surprising that Jesus sometimes cast his teaching in the form of proverbs. Since, however, these brief, salty sayings stress one side of a truth, they should not be exalted as maxims of inflexible conduct. On the contrary, one categorical statement must be interpreted in the light of another that may counsel the opposite of the first. For example, Jesus' command, "Do not judge, so that you may be not judged" (Mt 7.1), has sometimes been taken as a blanket prohibition against making judgments concerning right and wrong, good and evil. In the same context, however, the evangelist includes another of Jesus' pithy sayings, one which presupposes the necessity of forming judgments: "Do not give what is holy to dogs; and do not throw your pearls before

swine" (Mt 7.6). To obey this command against desecrating what is holy, one obviously must judge who is doggish and who is swinish. Spiritual prudence will know when it is appropriate to follow one precept and when it is appropriate to follow the other.

Similarly Jesus' proverb-like prohibiton, "Do not resist an evildoer" (Mt 5.39), is not to be taken to mean that his disciples are never to resist evil in any kind of way. In the light of Jesus' other teachings as well as his use of force to drive out the money-changers from the temple precincts (Mk 12.15), it is clear that the principle that he inculcates in this crisp maxim is non-retaliation for a malicious wrong inflicted by a personal enemy.

5. Parables

In all the teaching of Jesus there is no feature more striking than the parables. Although other religious teachers had made use of parabolic stories (see Judg 9.7–15; 2 Sam 12.1–6), in quantity and in excellence his parables are acknowledged to be outstanding. About sixty examples, from what was probably a larger number, have been preserved in the synoptic Gospels; these comprise more than one third of Jesus' recorded words. The fourth Gospel nowhere uses the word "parable," but it contains several parabolic sayings in the form of allegories (for example, Jn 10.1–18; 15.1–11).

The old definition of a parable as "an earthly story with a heavenly meaning" contains a certain amount of truth, but one must beware against seeking an elaborate allegorical meaning for every detail in a parable. That is, many details in Jesus' parables are present in order to make the story "live," and were not included primarily to instruct or edify the hearer. Defined more precisely, in Jesus' teaching a parable is a comparison drawn from nature or common experience in life and designed to illustrate some moral or religious truth, on the assumption that what is valid in one sphere is valid also in the other. The distinctions between parable and simile and metaphor are not easily defined. Often there is scarcely any difference, for all of them involve an aspect of comparison, but generally the metaphor and simile are short while the parable is more extended. "You are the salt of the earth" (Mt 5.13) is a metaphor; "Be wise as serpents" (Mt 10.16) is a simile; but "The kingdom of heaven is like yeast that a woman took and mixed in with three measures of flour until all of it was leavened" (Mt 13.33) is a parable.

The proper method of interpreting Jesus' parables is to make a thorough inquiry into the "life-setting" in his ministry when the parable was first uttered, and to seek out the chief point that, in that setting, it was intended to teach. In other words, To whom did Jesus speak the parable? and, Why did he speak it? Usually the details in a parable provide nothing more than the necessary background in order to make the story realistic, and are not to be assigned, point by point, special meanings in the manner of an allegory.

An analysis of Jesus' parables reveals that most of them are intended either *(a)* to portray a type of human character or disposition for warning or example, or *(b)* to reveal a principle of God's government of the world and humankind. In other words, Jesus' parables usually teach a certain kind of conduct that his hearers are to emulate or avoid (matters of ethics), or they disclose something of the character of God and his dealings with humankind (matters of theology). Examples of the former class of parables include The Two Builders (Mt 7.24–27), The Two Sons (Mt 21.28–32), The Pharisee and the Tax Collector (Lk 18.9–14), and The Good Samaritan (Lk 10.30–37); examples of the latter include the several parables concerning the Kingdom of Heaven (Mt ch 13; 20.1–15), The Seed Growing Secretly (Mk 4.26–29), The Great Supper (Lk 14.16–24), and The Lost Coin (Lk 15.8–10). Most parables of Jesus have two levels of meaning. One is the story itself, which usually reflects some aspect of daily life in the Near East. The other, deeper level of meaning (which may be paradoxical or surprising), is an open-ended invitation awaiting the hearer's response. In this respect the parable is not effective until the challenge inherent in the parable is freely accepted and acted upon.

Finally, it should be observed that when Christian teachers and evangelists retold Jesus' parables in the early church, they occasionally introduced small changes so as to apply the stories to new situations or to bring out the application more vividly. An example of the latter is the slight modification in the order of the wording in Matthew and Luke's retelling of the parable of The Wicked Tenants. According to Mk 12.8, when the owner of the vineyard sent his son to the tenants to get some of the fruit, they "killed him, and threw him out of the vineyard." Matthew and Luke, however, finding in the parable a parallel to what happened to Jesus when he was crucified *outside* the city walls, altered the sequence of the clauses so as to read, "they threw him out of the vineyard, and killed him" (Mt 21.39; Lk 20.15).

There was also a tendency to turn parables that Jesus addressed to the crowd, or to opponents, into parables for the disciples. For example, according to Luke (15.4–10) Jesus told the parable of the Lost Sheep as an answer to criticisms leveled against him by Pharisees and scribes (15.2). When Matthew recounts the same parable (18.12–14), however, it is no longer addressed to Jesus' opponents; it has now become part of Jesus' instruction to his disciples (18.1), that is, to the church, on the subject of how Christians are to relate to other Christians (see also 18.15–17).

In other cases the parables of Jesus were remembered long after the circumstances that gave rise to them had been forgotten. More than once, therefore, we find that the Evangelists, impressed by the sublimity of Jesus' teaching, recount his parables without mentioning the specific situation in which they were first narrated.

English Versions of the Bible

Prior to the sixteenth century, translations of the Bible into English were made from the Latin Vulgate instead of from the Hebrew or Greek, and were recorded only in manuscript copies.

To the Anglo-Saxon period belong the paraphrases of the Biblical narrative put into verse by Caedmon, herdsman for the abbey at Whitby (about A.D. 670); a version of the Psalms attributed to Aldhelm, bishop of Sherborne (640–709); a translation of the Gospel of John by the Venerable Bede (d. 735); portions of Exodus and the Acts of the Apostles, and some of the Psalms, by King Alfred (849–901); and a translation of the Heptateuch (Genesis through Judges) by Aelfric, abbot of Eynsham (955–1020). Four complete manuscripts and five fragmentary manuscripts remain of the Anglo-Saxon Gospels; these date from the eleventh to the thirteenth centuries. The famous Lindisfarne Gospels and the Rushworth Gospels are Latin manuscripts written toward the close of the seventh century, with interlinear Anglo-Saxon glosses (translations) inserted three centuries later. There were many Anglo-Saxon Psalters, with the Latin text and an interlinear Anglo-Saxon gloss.

To the Middle English period (1150–1500) belong the Ormulum, a metrical paraphrase of the Gospels, with interspersed moralizations; the Psalter of Richard Rolle (d. 1349), a prose version with commentary; and a prose version of the Psalms that has been attributed to William of Shoreham (about 1270–1350).

The first English versions of the entire Bible were the two associated with the work of John Wyclif, made by translation from the Latin Vulgate between 1380 and 1397. They were copied by hand, and there remain some one hundred and eighty manuscripts, mostly of the second version.

We do not know what part of the work upon the first version was done by Wyclif himself. But that is of no consequence; he inspired it all, including the making of the second version after his death in 1384. Both versions were made by scholars who were his immediate associates. Nicholas Hereford was largely responsible for the first version, which was completed before Wyclif's death. John Purvey, Wyclif's secretary, was responsible for the second version, which was completed by 1397.

The first version of the Wyclif Bible was a careful literal translation of the Latin Vulgate, with the English words following the order of the Latin words as closely as possible. Purvey gives a striking example of the mischief that may thus be wrought. In 1 Sam 2.10 the Latin has *Dominum formidabunt adversarii eius,* which the first version renders, "The Lord shulen drede the adversaries of hym." This follows the Latin word order, but ignores the Latin inflections which show that "adversaries" is the subject of the verb, that the verb for "shall dread" is plural, and that "the Lord" is its object. Purvey changes the translation to read, "Adversaries of the Lord shulen drede hym."

In the "General Prologue" to the second version Purvey states that it is best "to translate after the sentence and not only after the words, so that the sentence be as open, or opener, in English as Latin"—that is, so that the meaning be as clear, or clearer.

The sixteenth century brought the Bible in English to the common people as a printed book. Beginning with the first of Tyndale's translations in 1526, there appeared in rapid succession eight English versions of the Bible, culminating in the King James Version published in 1611.

The first English version of the Scriptures to be made by direct translation from the Hebrew and Greek was the work of William Tyndale. His New Testament, 1526, was followed by his translation of the Pentateuch in 1530 and of Jonah in 1531. In 1534 he issued a revision of his translation of the New Testament, and in 1535 *The New Testament yet once again corrected by William Tyndale.* These became the basis of all later revisions and the main source of the authorized versions of the New Testament in English.

Tyndale was bitterly opposed. He was accused of perverting the meaning of the Scriptures, and his New Testaments were ordered to be burned as "untrue translations," intended "for the advancement and setting forth of Luther's abominable heresies." He was finally betrayed into the hands of his enemies, and in October 1536 was executed and burned at the stake.

In 1535 appeared an English translation of the Bible by Miles Coverdale. This was the first complete Bible to be printed in English. It was not a direct translation from the original languages, but was based upon two Latin versions and upon the translations by Tyndale into English, and by Luther and Zwingli into German.

In 1537 a folio volume was published entitled *The Bible, which is all the Holy Scriptures, in which are contayned the Olde and Newe Testaments, truely and purely translated into Englysh, by Thomas Matthew.* "Thomas Matthew" was a pseudonym adopted by John Rogers, a friend of Tyndale, who took Tyndale's manuscript translations of the books of the Old Testament from Joshua to Second Chronicles, together with Tyndale's printed translations of the Pentateuch and the New Testament, and published them in this one volume, which he completed by adding Coverdale's version of the rest of the Old Testament and the Apocrypha.

In 1539 Richard Taverner, a layman and a lawyer, clerk of the signet to the king, published a revision of Matthew's Bible, one edition of which was issued in parts in order that people who could not afford to purchase the whole Bible might buy one or more parts. He was a good Greek scholar, and made some changes in the translation of the New Testament that have been kept in later versions.

Meanwhile at Paris, in early 1538, Miles Coverdale had begun a new revision of Matthew's Bible, for which he had been commissioned by Sir Thomas Cromwell, Secretary to King Henry VIII, and Vicar General. The Great Bible, as this was called, was published at London in April 1539. It was the first authorized English version, and a copy was ordered to be placed in every church. Until very recently, the Psalms in the *Book of Common Prayer* were from Coverdale's translation.

Under Queen Mary the printing of the English Bible ceased and its use in the churches was forbidden. Many English citizens sought refuge on the Continent, and a group of these at Geneva undertook the revision of the English Bible. The Geneva version appeared in 1560. It was set in Roman type instead of the old blackface, and it was a book easy to handle, instead of an unwieldy folio. It was the first English version to use numbered verses, each set off as a separate paragraph.

The Geneva Bible was never authorized, but it became at once the people's book, the household Bible of the English-speaking nations, and it held this place for three-quarters of a century. It was the Bible used by Shakespeare and John Bunyan, and it was the Bible of the Puritans who settled New England. Between 1560 and 1644 at least one hundred and forty editions of the Geneva Bible or New Testament were printed, and it lasted longer in competition with the King James Version than any other English version. This was the version of the Bible brought to America by the Pilgrims in 1620.

Queen Elizabeth renewed the injunction that a copy of "the whole Bible of the largest volume in English" be replaced in every church, and encouraged its reading. Because not enough copies of the Great Bible were available, Archbishop Parker proposed that the bishops themselves make a new revision of the English Bible. The resulting version, known as the Bishops' Bible, was published in 1568. It was authorized by Convocation, and its possession enjoined upon the churches. In 1572 another edition was published, with considerable revision of the New Testament. This revised edition, reprinted in 1602, became the basis of the revision under King James.

In 1582 an English translation of the New Testament was published at Rheims, made from the Latin by Roman Catholic scholars led by Gregory Martin, who had been trained at Oxford University. A similar translation was made of the Old Testament, but not published until 1609.

The distinctive characteristic of the Rhemish version of the New Testament is the closeness with which it adheres to the Latin.

On February 10, 1604, after a conference "for hearing and for the determining things pretended to be amiss in the church," King James I ordained: "That a translation be made of the whole Bible, as consonant as can be to the original Hebrew and Greek; and this to be set out and printed without any marginal notes, and only to be used in all churches of England in time of divine service." He appointed fifty-four men as translators, forty-eight of whom are named in the records that have come down to us. They worked in six companies, to each of which was assigned a section of the Bible. Two companies met at Oxford, two at Cambridge, and two at Westminster.

It was provided that each company would consider the work of each other company, and that differences would be resolved by correspondence if possible, and if not, be referred to the general meeting at the end. This was a meeting at London of a committee of six, made up of two representatives from the companies at each of the three centers, which devoted nine months to bringing together and finally editing the work. It was then seen through the press by Dr. Thomas Bilson, bishop of Winchester, and Dr. Myles Smith, of Oxford University; and Smith wrote an extended and informative preface, entitled "The Translators to the Reader." This version of the Bible, with a dedication to King James, was published in 1611.

It is a strange fact that no evidence has yet been found that the King James Version was ever authorized in the sense of being publicly sanctioned by Convocation or by Parliament. But it did not need that. Bishop Westcott, writing in the nineteenth century, said: "From the middle of the seventeenth century, the King's Bible has been the acknowledged Bible of the English-speaking nations throughout the world simply because it is the best. A revision which embodied the ripe fruits of nearly a century of labour, and appealed to the religious instinct of a great Christian people, gained by its own internal character a vital authority which could never have been secured by any edict of sovereign rulers."

An outstanding merit of the King James Version is the music of its cadences. The translators were experienced in the public reading of the Scriptures and in the conduct of public worship. Their choice of the final wording of a passage was often determined by a marvelously sure instinct for what would sound well when read aloud. Take as an example the successive translations of Prov 3.17, part of a discourse in praise of wisdom, where Coverdale, the Great Bible, and the Bishops' Bible agree in reading: "Her wayes are pleasant wayes and all her paths are peaceable."

The Geneva Bible has: "Her wayes are wayes of pleasure and all her paths prosperitie."

The King James Version gives to the verse a perfect melody: "Her wayes are wayes of pleasantnesse, and all her pathes are peace."

The English Bible owes more to William Tyndale than to any other man, not only because he was the first to translate the Bible from the original Hebrew and Greek, but because the basic structure of his translation has endured through all subsequent changes. It has been estimated that about sixty per cent of the text of the English Bible achieved its final literary form before the King James Version appeared, and that in the King James Version at least one-third of the New Testament is worded exactly as in Tyndale's New Testament, while the sentences of the remaining two-thirds follow the general pattern of the underlying structure of Tyndale's New Testament.

For two and a half centuries the King James Version maintained its place as the Authorized Version of the English-speaking peoples, without any serious consideration of its revision. But in the 1850's a movement toward revision began to gather strength, and in 1870 the Convocation of the Province of Canterbury appointed a Committee to undertake it. The Revised Version of the New Testament was published in 1881, of the Old Testament in 1885, and of the Apocrypha in 1895. The American Standard Version, a variant edition containing the renderings preferred by the American scholars who had cooperated in the work of revision, was published in 1901.

In 1928 the copyright of the American Standard Version was acquired by the International Council of Religious Education, and thus passed into the ownership of the churches of the United States and Canada that were associated in this Council through their boards of education and publication. The Council appointed a committee of scholars to have charge of the text of the American Standard Version and to undertake inquiry as to whether further revision was necessary. After more than two years of study and experimental work, this committee decided that there was need for a thorough revision of the version of 1901, which would stay as close to the Tyndale-King James tradition as it could in the light of present knowledge of the Hebrew and Greek texts and their meaning on the one hand, and present usage of English on the other.

In 1937 the revision was authorized by vote of the Council, which directed that the resulting version should "embody the best results of modern scholarship as to the meaning of the Scriptures, and express this meaning in English diction which is designed for use in public and private worship and preserves those qualities which have given to the King James Version a supreme place in English literature."

The Revised Standard Version of the New Testament was published in 1946, of the Old Testament in 1952, and of the Apocrypha in 1957. The publication of the Revised Standard Version of the Bible was authorized in 1951 by vote of the National Council of the Churches of Christ in the U.S.A.

The revised versions of 1881–1901 made permanent advances in their recognition of the principle that Hebrew poetry is to be translated as poetry, in characteristic parallelism, and in their giving due weight to the text of the ancient Greek manuscripts of the New Testament that had been discovered since 1611.

But the Revised New Testament of 1881, said Charles H. Spurgeon, the great London preacher, was "strong in Greek, weak in English." The revisers were literalists, especially in the New Testament. Their ideal of translation was a meticulous word-for-word reproduction of the Greek text in English words, using the same English word for a given Greek word whenever possible, leaving no Greek word without translation into a correspondent English word, following the order of the Greek words rather than the order natural to English, and attempting to translate the tenses and the definite article with a precision alien to English idioms. The result is that the Revised Version and the American Standard Version are distinctly "translation English."

The problem of the archaic language of the King James Version was ineffectively handled by the revisers, who changed some of the misleading words but actually increased the use of other archaic terms, such as aforetime, haply, holden, howbeit, would fain, must needs, peradventure. Where the King James Version reads, "this is the will of God in Christ Jesus concerning you" (1 Thess 5.18), the 1881–1901 revisers changed it to "this is the will of God in Christ Jesus to youward" (so also in Rom 8.18, Gal 5.10, Col 1.25). A full account of archaic words and obsolete meanings in the King James Version and the revised versions of 1881–1901 is given in *The Bible Word Book,* by Ronald Bridges and Luther A. Weigle.

It is one of the ironies of history that the King James Version remained unrevised for two hundred and sixty years, then was revised with the utmost care, but that almost immediately there began a period that was marked by discovery after discovery, widening our knowledge of the Bible text and its early history, and testing the results at which the scholars of 1881 had arrived by evidence with which they were totally unacquainted.

The Revised Standard Version took full account of the new knowledge of the history, geography, religions, and cultures of Bible lands, and of the rich new resources for understanding the vocabulary, grammar, and idioms of the Biblical and related languages. It also broke away from the literalism and mechanical exactitude of the revisions of 1881–1901, and returned to the basic structure and more natural cadence of the Tyndale-King James tradition.

Paraphrases and new translations, usually of the New Testament, have appeared from time to time, such as those by Henry Hammond, 1653; Richard Baxter, 1685; Daniel Mace, 1729; Edward Harwood, 1768; Andrews Norton, 1855; Leicester Sawyer, 1858. A group of scholars produced The Twentieth Century New Testament, 1898–1901; thereafter have come new translations by R.F. Weymouth, 1903; James Moffatt, 1913; Edgar J. Goodspeed, 1923; J.B. Phillips, 1947–1958 (2nd. ed. 1972); Charles Kingsley Williams, 1952; Hugh J. Schonfield, 1955 (2nd. ed. 1985); William F. Beck, 1963; William Barclay, 1968, 1969; Richmond Lattimore, 1982; and Heinz W. Cassirer, 1989.

Moffatt's translation of the Old Testament appeared in 1924, and a rendering of the Old Testament by J. M. Powis Smith and other scholars was published (1927) with the Goodspeed New Testament as *The Bible: An American Translation.* After Beck's death in 1966 several other Lutheran scholars finished his partially complete rendering of the Old Testament, and both parts of the Bible were issued in 1976. It is a dignified and contemporary rendering in mid-American English.

After the publication of the Revised Standard Version in 1952, the second half of the twentieth century saw the publication of many translations and revisions of the English Bible. In fact, between 1952 and 1990, when the New Revised Standard Version was published, no fewer than twenty-six different renderings of the complete English Bible were issued, with twenty-five additional translations and revisions of the New Testament. Among these the following Jewish, Protestant, and Catholic translations are of special note.

In 1955 the Jewish Publication Society, whose first translation of the Hebrew Scriptures had appeared in 1917, initiated a new translation of the Scriptures in English. The first part, published in 1962, was entitled *The Torah, The Five Books of Moses.* This was followed by *The Prophets (Nevi'im)* in 1978, and *The Writings (Kethuvim)* in 1982. These three volumes, with revisions, were brought together in 1985 under the title *Tanakh, A New Translation of the Holy Scriptures According to the Traditional Hebrew Text.* The rendering is supplied with useful footnotes of three kinds: textual, translational, and explanatory. The version is notable for its perceptive handling of the Hebrew vocabulary and syntax in contemporary English.

In the year that the RSV New Testament was published (1946), British Protestant Churches embarked on a totally new translation of the Scriptures, The New English Bible. Prior to its completion in 1970 a first edition of the New Testament appeared in 1961. At that time, C. H. Dodd, director of the enterprise and chairman of the New Testament translation panel, stated that this was "not another revision of an old version, but a genuinely new translation of the original, which should be frankly contemporary in vocabulary, idiom, style and rhythm—not to supersede the Authorized Version, but as a second version alongside it." The rendering is free and vigorous, tending at places to be periphrastic. Here and there the translators have rearranged the sequence of verses and sections of the text.

The first English Bible made from the original Hebrew and Greek languages that received Roman Catholic approval was The Revised Standard Version Catholic Edition, of which the New Testament was published in 1965 and the complete Bible in 1966. In it the Deuterocanonical Books, which comprise what others call the Apocrypha (but without 1 Esdras, 2 Esdras, and the Prayer of Manasseh), were placed among the Old Testament books in accord with Catholic usage. It was prepared by the Catholic Biblical Association of Great Britain with the consent of the Revised Standard Version Bible Committee and the Division of Christian Education of the National Council of the Churches of Christ in the U.S.A. There are no changes in the Old Testament text; the sixty-seven changes in the

New Testament, made for liturgical and theological reasons, are carefully noted in an appendix. A significant further advance toward a Common Bible was achieved with the publication of the RSV without any changes whatever in the Oxford Annotated Bible with the Apocrypha (1966), to which Richard Cardinal Cushing of Boston granted the imprimatur.

The first English translation of the Bible made from the original languages by Catholic scholars was The Jerusalem Bible, published in 1966. With close comparison with the Hebrew and Greek, it is based on the French translation made under the direction of l'École Biblique of Jerusalem by a committee of scholars headed by Père Roland de Vaux, O.P., and popularly known as La Bible de Jérusalem. The English translation was made by a British Committee headed by Alexander Jones, L.S.S., of Christ's College, Liverpool. The translation is well done and has been warmly received. At the same time it must be acknowledged that occasionally the French idiom intrudes. "Yahweh" is used instead of the surrogate, the LORD. The volume is provided with useful introductions and notes, and "thee," "thou," and "thine" disappear from the text.

The New American Bible, published September 30, 1970 (St. Jerome's Day in the ecclesiastical calendar), carries on its title page the statement that it is "translated from the original languages with critical use of all the ancient sources," that the translation is by members of the Catholic Biblical Association of America, and that it is sponsored by the Bishops' Committee of the Confraternity of Christian Doctrine. The version is the final outcome of a movement that began in response to the encyclical letter of Pope Pius XII, issued September 30, 1943, and urging translation of the Scriptures from the original languages. The work of translation was begun in 1944, and the first volume, Genesis to Ruth, was published on September 30, 1952. Later volumes appeared in due time, copyrighted by the Confraternity of Christian Doctrine. A few years before the end of the work, four Protestant scholars were added to the Committee. The completed work represents capable and dedicated scholarship and renders the Scriptures in modern American idiom.

Persons who have only a limited facility in the English language—such as immigrants, or children, or those with little education—find the traditional language of the King James Bible forbidding if not sometimes incomprehensible. It was chiefly for such readers that in 1966 the American Bible Society issued "Good News for Modern Man," a translation of the New Testament made by Robert G. Bratcher. Known as Today's English Version, the rendering uses simplified syntax and a limited vocabulary. The Old Testament (1976), prepared by a committee, was subsequently followed by a rendering of the Deuterocanonicals/Apocrypha (1979). The translators employ contemporary American English (although a British-Usage edition has also appeared) and adopt the translational practice of producing "dynamic equivalence," as opposed to "formal equivalence."

The popularity of the Good News Bible spurred the preparation of another modern-speech Bible. This is the New International Version, sponsored by the New York Bible Society (subsequently known as the New York International Bible Society). The New Testament was published in 1973 and the complete Bible in 1978. Less colloquial than Today's English Version, and more literal than the New English Bible, the version was prepared by twenty teams of translators in the USA, Canada, Australia, New Zealand, and Great Britain. It had the largest first printing ever for an English Bible.

Unlike the other versions mentioned here, the Living Bible (1971), prepared by Kenneth N. Taylor, is professedly a paraphrase of the Scriptures: "A restatement of the author's thought, using different words than he did." Its chatty style made it the best-selling book in 1972.

The latter part of the twentieth century saw the publication of a number of revisions of earlier translations. The New American Standard Version (1970), prepared by a group of anonymous scholars sponsored by the Lockman Foundation of California, is ostensibly the American Standard Version of 1901 purged of some of the archaisms taken over from the Revised Version of 1881–1884. A number of verses resting on doubtful manuscript authority have been reintroduced into the text from the margin. It remains a severely literalistic rendering.

The New King James Bible (1982) is the work of an international group of some 130 scholars, editors, and religious leaders who have sought to produce an English Bible that retains as much of the classic King James Version as possible, while at the same time bringing its English up-to-date. Changes have been introduced in word order, grammar, vocabulary, and spelling. It retains the practice of printing the numbered verses as separate units, rather than arranging the text in paragraphs. The chief deficiency of the New Testament results from the decision to follow the Greek text known as the Textus Receptus, despite the many scribal errors that this text embodies. A totally new feature, never used in the King James Bible or in English Bibles in general, is the capitalization of pronouns referring to the Deity. This introduces into English something that is foreign to the Hebrew and Greek originals.

The first stage of producing a revision of the New American Bible (mentioned above) was finished in 1986 when the second edition of the New Testament was issued. This is a totally new translation, done with Protestant cooperation. No longer is the First Epistle of John printed in strophic format similar to blank verse, and the reader will notice changes throughout made in the rendering, including an attempt to compensate in part for the masculine bias of the English language.

In 1973 a new edition of *La Bible de Jérusalem* was published, which incorporated the progress in scholarship over the two decades since the preparation of its first edition in 1955. The French revision was deemed important enough to warrant a completely new edition of the English-language Jerusalem Bible. This edition, entitled *The New Jerusalem Bible* (1985), was prepared under the supervision of Henry Wansbrough of Ampleforth Abbey, York. The translation of the biblical text was made directly from the Hebrew, Greek, or Aramaic, and paraphrase was avoided more rigorously than in the first edition. "Considerable efforts have been made, though not at all costs, to soften or avoid the inbuilt preferences of the English language . . . for the masculine [gender]." As in the earlier edition, extensive introductions and explanatory notes are provided throughout the volume.

In the early 1980s work began on a revision of the New English Bible, under the chairmanship of W. D. McHardy, and in 1989 the work of the revision panels was published under the title, *The Revised English Bible*. It is somewhat less idiosyncratic in the Old Testament than its predecessor. The revisers sought to insure that the style of English used is fluent and of appropriate dignity for liturgical use. In passages that address God the "you"-form replaces the traditional "thou"-form. Steps were taken in preferring more inclusive gender reference in place of male-oriented language "where that was possible without compro-

mising scholarly integrity of English style" (Preface).

The ecumenicity of our times is in various degrees represented in these newer translations, but perhaps most of all in the New Revised Standard Version. The committee that produced the Revised Standard Version is a continuing committee, holding meetings at regular intervals and having charge of the RSV text. In 1959 a few changes were authorized for subsequent printings of the Revised Standard Version, growing out of a study of suggestions sent to the committee by various readers. Meanwhile the committee became both international and ecumenical, with the appointment of Catholic and Protestant members from Great Britain, Canada, and the United States. Since 1946 an American Jewish scholar had been a member of the Old Testament section. More recently an Eastern Orthodox scholar was added to the committee.

In 1974 the Division of Christian Education of the National Council of Churches of Christ in the U.S.A. directed that the Standard Bible Committee undertake a revision of the RSV Bible, with the Apocrypha. Four mandates were given to the committee, namely, that necessary changes be made (1) in paragraph structure and punctuation; (2) in the elimination of archaisms while retaining the flavor of the Tyndale-King James Bible tradition; (3) in attaining a greater degree of accuracy, clarity, and euphony; and (4) in eliminating masculine-oriented language relating to people, so far as this could be done without distorting passages that reflect the historical situation of ancient patriarchal culture. Within the constraints set by the original text and the mandates given by the Division, the committee followed the maxim, "As literal as possible, as free as necessary." As a consequence, the New Revised Standard Version (1990) remains essentially a literal translation. Paraphrastic renderings were adopted only sparingly, and then chiefly to compensate for a deficiency in the English language— the lack of a common gender third person singular pronoun. The results may now be assessed in the present translation, which, unlike other English Bibles, contains all of the books that are regarded as authoritative by Protestant, Roman Catholic, and Eastern Orthodox churches.

Survey of the Geography, History, and Archaeology of the Bible Lands

I. GEOGRAPHY

1. Lands of the Bible

Physical geography is concerned with the land's feature, its climate, and ecology, whereas historical geography considers their influence on the people who inhabited the land. Physical and historical geography are complementary disciplines that furnish the background for biblical geography. The study of biblical geography gives rise to two immediate problems: the appropriate name of the biblical lands and their boundaries. Modern names and boundaries only add to the confusion; names change and boundaries fluctuate over time. Broadly speaking, modern Israel, Jordan, Lebanon, and Syria constitute the Lands of the Bible.

In addition to the Holy Land, two other names used to describe this geographical unit are Canaan and Palestine. Canaan, a term used somewhat loosely, is the common name in the Hebrew Bible for the Promised Land; it designates especially Phoenicia proper.

By an irony of history the name Palestine is derived from one of the Sea Peoples, the Philistines, who were Israel's archenemies. The Greek historian Herodotus first used the designation Palestine in the fifth century B.C., and the Romans later adopted it. Since Palestine is a broader term than Land of Israel (the modern Israeli designation), biblical scholars use Palestine as an inclusive geographical name, not as a modern political title. Palestine, consisting of Cisjordan on the west side of the Jordan River and Transjordan on the east, is a land bridge connecting Egypt to the south and Mesopotamia to the north. With the Amanus and Taurus mountains as its northern border, the desert to the south and east, and the Mediterranean Sea ("the Great Sea" in the Bible) as its western border, Palestine (Canaan) is a relatively small geographical unit, about 350 miles long and 60 miles wide. In terms of the traditional formula "from Dan to Beer-sheba," Palestine is only 150 miles long. (For these and other geographical details, consult the set of maps at the end of the volume.)

2. Geographical features of Palestine

Palestine may be divided into five main geographical regions, extending from north to south: the coastal plain, the central hills, the Rift Valley, the plateau of Jordan, and the deserts. These strips, varying in topography, have several subdivisions, which may be described as follows, though the Bible does not delineate boundaries clearly.

The coastal plain stretches about 125 miles from Lebanon to Gaza. The most important of the international highways, the Via Maris ("Way of the Sea"), runs the length of the Mediterranean coast. The Mount Carmel ridge, which rises above the harbor of modern Haifa, divides the coastal plain into a northern and southern region. Acco, about fourteen miles north of Haifa, was one of the important harbor cities in antiquity. The coastal plain widens at the plain of Sharon, lying between

Mount Carmel on the north and the Yarkon River on the south. Sharon is still forested and rich in agriculture. Dor and Joppa (modern Jaffa) are the chief harbors of Sharon.

The Philistines occupied the southern part of the fertile coastal plain, notable for barley and wheat. The city-states of Ashdod, Ashkelon, Gaza, Gath, and Ekron constitute the Philistine pentapolis. Ashdod is a half-mile inland from the Mediterranean coast. Ashkelon, the only Philistine city located on the seacoast, was the main Philistine harbor. Modern Gaza marks the site of the ancient city. Gath is unidentified, and Ekron is ten miles inland from Ashdod.

The Negeb (Hebrew for "dry"), a barren steppe, forms the southern border of Palestine. In modern terminology the Negeb extends to the Sinai peninsula. The Bible sometimes uses "Negeb" to refer to the south, designating the area around Beer-sheba (see Gen 20.1). In biblical times this dry, hot region had little economic significance. The Via Maris, connecting Egypt with Mesopotamia and Anatolia, passses through the Negeb. The two principal sites in the Negeb are Beer-sheba and Kadesh-barnea. Beer-sheba is the capital of the northern Negeb. Fifty miles south of Beer-sheba is Kadesh-barnea, where the Israelites encamped in the wilderness.

3. The Central Hills

The western highlands extend about 188 miles from northern Galilee to the Sinai. A great valley running from west to east divides Galilee into Upper and Lower Galilee. This region is rarely mentioned in the Old Testament; the New Testament relates that Jesus spent much of his early ministry in Lower Galilee. Both Jews and Christians relocated in Upper and Lower Galilee after A.D. 135 when this region flourished. The rolling country of Galilee, with the mountains of Upper Galilee rising more than 3,000 feet, was fertile and prosperous.

The central hills also include the plain of Megiddo, the valley of Jezreel, Mount Ephraim (at the center of the hill country), and the hill country of Judah. Between the hills of Galilee and Samaria is the plain of Jezreel (Esdraelon is its Greek name), which is also rich in agriculture. This important plain connects the Mediterranean coast with the Jordan Valley. Jezreel and Esdraelon are often identified; but, in fact, they are two distinct lowlands. The Jezreel Valley was the scene of several well-known battles in biblical history. The principal fortified cities on the west side of the Jezreel Valley are Jokneam, Megiddo, Taan-ach, and Ibleam. Megiddo is strategically located in the international route between Egypt and Mesopotamia.

Ephraim is a land of mountains and fertile valleys. Grapevines and olives are still cultivated on the slopes of the hills, while wheat and barley are planted in the valleys. The capital city of Samaria, situated 300 feet above the valley, is strategically located on the trade routes converging from all directions. Other cities in the Ephraim region are Shechem and Tirzah, both former capitals of the Northern Kingdom of Israel. Shechem controlled the pass between Mount Gerizim and Mount Ebal.

No sharp demarcation exists between the hill country of Ephraim and the hill country of Judah. The Southern Kingdom of Judah was more defensible than the Nothern Kingdom of Samaria because it was more isolated, but it lacks the physical attractiveness of the north. The hill country of Judah reaches altitudes of 3,345 feet. Jerusalem is located in the midst of the Judean hills, on the edge of the desert. Surrounded by the Kidron and Hinnom Valleys, Jerusalem stands 2,600 feet above sea level. Hebron and Bethlehem, well-known towns in biblical times, are situated in the hill country of Judah.

The Shephelah (Hebrew for "lowlands") separates the coastal plain from the central hills. Grains and vines flourish in the Shephelah. Situated between Philistia and the hill country of Judah, the Shephelah was protected by fortresses and served as the buffer between the Philistines and the Israelites. Among the fortress cities was Lachish, one of the most prominent cities in Judah, second only to Jerusalem.

4. The Rift Valley

This geological fault, called the Ghor in Arabic, is about ten miles wide; it is a unique feature of Palestine's physical geography. Between Mount Hermon in the north and Elath (Eilat) in the south this depression extends about 260 miles. The Jordan, the longest river in Palestine, follows the Rift Valley. If the Jordan River ran in a straight line between the Sea of Galilee and the Dead Sea, the distance covered would be 65 miles; in fact, the Jordan meanders more than twice that distance.

The Sea of Galilee (biblical Chinnereth), also known as the Lake of Tiberias, is a freshwater lake at the northern end of the Jordan Valley. Measuring thirteen miles long and eight miles wide, it is 700 feet below the level of the Mediterranean Sea. The Sea of Galilee is well known for plentiful fish and sudden storms. Though barely mentioned in the Old Testament, the Sea of Galilee was prominent in the early ministry of Jesus. The Gospels refer to Capernaum, Tiberias, and other towns in the vicinity of the Sea of Galilee.

The Jordan River flows into the Dead Sea, the surface of which, at 1300 feet below sea level, is the lowest spot of land on earth. Forty-eight miles long and ten miles wide, it lies between the hills of Moab on the east and the Judean hills on the west. There is no outlet from this body of water; instead, the inflow is balanced by evaporation. Since the mineral content of water is left behind when the water evaporates, the Dead Sea has a high mineral content. For this reason it is also known in the Bible as the Salt Sea. (Another name for it is Sea of the Plain.) Its high salt deposits and other chemicals in this lake preclude plant and animal life. Jericho, located seven miles northwest of the Dead Sea, is thought to be the oldest city of Palestine.

The term Arabah, connoting aridity, is sometimes used in the Bible with reference to the entire Rift Valley, extending from the Sea of Galilee to the Gulf of Aqabah at the Red Sea. Arabah may also be used, as in modern terminology, to designate simply the area from the Dead Sea to the Red Sea, a distance of about a hundred miles. The Arabah is noted for its copper mines.

5. The Plateau of Jordan

This zone is located east of the Jordan (Transjordan); it is the modern Hashemite Kingdom of Jordan. The King's Highway, running the length of Transjordan from Damascus to the Gulf of Aqabah, was a major international route in antiquity. Eastern tributaries emptying into the Jordan River or the Dead Sea have formed natural boundaries separating nations and peoples in the course of history. The principal rivers of Transjordan are the Yarmuk (not mentioned in the Bible), the Jabbok (modern Nahr es-Zerqa), the Arnon (Wadi el-Mojib), and the Brook Zered (Wadi el-Hesa).

Bashan is a fertile plateau northeast of the Jordan River. It is located north of Gilead and divided from it by the Yarmuk River. It is rich pasture land; wheat and barley are grown there; basalt is also found in Bashan. The hills of Gilead are located between Bashan to the north and Moab to the south. The Jabbok River divides Gilead into a northern and southern section. Gilead is a delightful region, abounding with forests, vines, olives, and grains.

Ammon is situated south of Gilead between the Jabbok and Arnon Rivers. The principal town of this region in Old Testament times was Rabbath-Ammon; in the third century B.C. it was renamed Philadelphia; today it is Amman, the capital of the modern kingdom of Jordan.

Moab is a small plateau located between Ammon and Edom. Moab's northern border is the Arnon, and the Wadi Zered is its southern boundary. Moab's economy depended on the wheat and barley crop, as well as on sheep and goats. Before entering the land of Canaan the Israelites camped in the plains of Moab. Among the principal towns in Moab are Medeba (modern Madeba) in the north, and Dibon (modern Dhiban) in the south.

Edom or Mount Seir (the Old Testament connects the two) is located south of the Dead Sea on both sides of the Wadi Arabah. The Brook Zered serves as the frontier between Moab and Edom. Edom extended to the Gulf of Aqabah. Important trade routes passed through Edom. This rugged, mountainous region (with some peaks over 5,000 feet) is notable for agriculture, commerce, and copper mines. Bozrah (modern Buseirah) was the principal city of northern Edom. After the Nabateans settled in Edom in the sixth century B.C., Petra, a city carved from sandstone, became the capital of the Nabatean kingdom. Ezion-geber and Elath were both situated in Edom, at the head of the Gulf of Aqabah, but their precise locations are disputed.

The deserts of Syria and Arabia constitute the final geographic zone of Palestine. Today most of this region lies in Saudi Arabia, as well as in eastern Jordan and Syria.

6. Climate and Rainfall

Palestine's location between the sea and the desert has a marked influence on the climate and rainfall of the country. There are great fluctuations in climate within small areas of Palestine. The climate is subtropical, with only two seasons: summer and winter. Between May and October the weather is warm, and there is almost no rainfall. Between October and April the weather is both cool and wet (see Deut 11.14). The early rains ("former rains"), coming in mid-October, soften the ground preparatory to plowing for the winter crop. The spring rains ("latter rains") occur usually in March and April when summer crops are planted. The amount of rainfall varies throughout the country; the annual rainfall at the Dead Sea is four inches; the annual rainfall in Jerusalem is about twenty-five inches. In addition to rainfall, dew is indispensable for watering the crops in the hot, dry months. Wadis (water beds that are usually dry except in the rainy season), springs, and cisterns also provide needed water.

The basis of the economy of Palestine, as in many neighboring countries, was agriculture. The Bible specifies the products of agriculture in Palestine: "a land of wheat and barley, of vines and fig trees and pomegranates, a land of olive trees and honey" (Deut 8.8).

7. Neighboring Peoples

Egypt and Mesopotamia were the two great empires in biblical times; they were also great rivals. Egypt (*misraim* in Hebrew), one of the great civilizations of antiquity, is located in the northeast corner of Africa, along the course of the Lower Nile River. Surrounded for the most part by desert, Egypt is bordered on the east by the Red Sea, and by Libya on the west. Lower Egypt comprises the Nile Delta, whereas Upper Egypt constitutes the remainder of the country south of Cairo. Egypt's route to Palestine passed through the Sinai wilderness, along the Mediterranean coast, and into the hill country. The peninsula of Sinai is triangular in shape, and lies between the Gulf of Suez and the Gulf of Aqabah. Mount Sinai (Horeb) is traditionally located at the southern end of the Sinai peninsula.

Egyptian involvement with Palestine began at some time in the Old Kingdom (approximately 2686–2181 B.C.); there is much evidence of connections in the Early Bronze Age (approximately 3300–2000 B.C.). Abundant biblical references document the interrelationship of Egypt and Palestine. This relationship was ambivalent, sometimes amicable, sometimes hostile. For the most part, Egypt's influence on Palestine was more indirect than direct.

Mesopotamia (the biblical name is Aram-naharayim, designating roughly "the land between the rivers"), constitutes the area of the Upper and Middle Euphrates and the Tigris Rivers. Ancient Mesopotamia was approximately coterminous with modern Iraq. The northern region of Mesopotamia was Assyria; the southern sector was Babylonia.

Assyria, situated in the Upper Mesopotamian plain (northern Iraq), was a mighty empire bent on territorial expansion. The beginnings of Assyria date from the second millennium B.C., but the empire enjoyed its greatest prominence in the Neo-Assyrian period (911–609 B.C.) when it controlled both provinces and vassal states, including, for a time, Egypt. Among the principal cities

of Assyria were Asshur, the first capital of the Assyrian empire, situated on the west bank of the Tigris; Calah (Nimrud), on the east bank of the Tigris; and Nineveh, the last capital of the Assyrian empire, located on the east bank of the Tigris, opposite modern Mosul in northern Iraq.

Babylonia in southern Iraq may designate both the region and its capital city. The Bible often refers to the region of southern Mesopotamia as Chaldea. The Neo-Babylonian empire (626–539 B.C.) is synonymous with Chaldea. The city of Babylon, which gave its name to the whole region, is located on the Euphrates River, about fifty miles south of modern Baghdad.

8. Phoenicia, Philistia, Aram, and Hittites

Occupying the land along the coast of the East Mediterranean, ancient Phoenicia was coextensive with modern Lebanon and the northern part of Palestine. The Greeks used the title "Phoenicia" to denote ancient Canaan. Among the Phoenician city-states were Tyre, Sidon, Arvad, and Byblos. For the most part, Phoenicia and Israel enjoyed a cordial and close relationship; the Phoenicians, for example, supplied both artisans and materials for the building of Solomon's temple. The economic base of Phoenicia was maritime trade.

Sometime after 1150 B.C., the Philistines settled on the southwest coast of Palestine, between Joppa and Gaza. They and the Israelites clashed constantly because both were bent on expansion of their territory.

Aram, a collection of city-states to the northeast of Israel, was also in constant conflict with its neighbors. Aram is usually equated with Aram-Damascus, the capital of modern Syria, located about sixty miles east of the East Mediterranean. Especially prominent from the tenth to the eighth century B.C., Damascus was an ancient and prosperous city, well located on the major trade routes.

The Hittites, an Indo-European people, established their kingdom during the second millennium B.C. in the central Anatolia plain. The Old Testament contains several references to the Hittites.

II. HISTORY

9. Nature of Biblical History

The Bible is not a textbook of ancient history. The biblical authors intended to tell the story of Israel, not to present a history of Israel in the modern sense. Writing from a theological perspective, these authors provided a religious interpretation of God's role in the events that shaped their destiny.

The historicity of the Pentateuch (see pp. xxxv–xxxvi OT) is a troublesome question. The literary tradition behind the patriarchal narratives and the other events related in the early books of the Bible is a complex subject. This is not to disparage the biblical accounts; it is rather to understand the nature of the biblical record. The Bible is not an eyewitness account of the early events of Israel's history. The sources for these events were compiled at a much later date than the events they describe. During generations of oral circulation the stories underwent considerable development. It is the function of scholars to reconstruct the history from available archaeological and literary evidence.

Old Testament history begins with Abraham, yet the period in which the patriarchs lived cannot be dated with certainty. Even the exodus from Egypt, the central event in Old Testament history, is surrounded with several difficulties of detail. The earliest reference to Israel appears on the Merneptah stele, dating from about 1207 B.C. This, then, marks the beginning of the actual history of Israel, so far as non-biblical sources are concerned.

10. The Patriarchal Period

A long period of oral transmission lies behind the episodes of the patriarchal period as related in Genesis. Conjecturally, this period may be dated to the first third of the second millennium (2000–1700 B.C.), inasmuch as the culture and customs reflected in the biblical story appear to match that epoch in ancient Near Eastern history.

Abraham, Isaac, Jacob, and Joseph stand at the center of the patriarchal narratives in Genesis. Despite the traditional term "patriarch," the spouses of the patriarchs also play a role in the Genesis narratives. The inclusive term "ancestors" then is more accurate than "patriarchs." The central theme of Genesis is God's promise to Abraham (Gen 12.3), meaning that God chose one people (Israel) through whom all others would be affected. Abraham and his wife Sarah migrated from Haran (in Turkey) to Canaan so as to fulfill the divine plan.

11. Exodus, Election, Covenant

These three events are the heart of the Old Testament; they are also the subject of the book of Exodus, which elaborates the meaning of these events in the life of Israel. Many details surrounding the exodus event are lacking, and consequently are disputed; for example the following: the name of the pharaoh of the exodus, the route of the exodus, the precise location of Sinai, and the origin of the personal name of Israel's God.

The Hebrews were slaves in Egypt, but they escaped under the leadership of Moses, who is the central figure in Israel's history from Exodus to Deuteronomy. Legendary elements surround the biblical account of Moses, but few would question his influence upon Israelite traditions.

12. The Israelite Settlement

According to the biblical account, the Israelites crossed into Canaan (Palestine) from Transjordan. The results of excavation and survey point to Iron Age I (1200–900 B.C.) as the period of the Israelite settlement in Canaan. The books of Joshua and Judges are the principal biblical sources for the settlement, although they furnish somewhat contradictory accounts. Describing the conquest very much as if it were a blitzkrieg, the book of Joshua gives the impression that the conquest was relatively rapid and definitive. On the contrary, other sources show that the Israelite settlement was a lengthy and complicated process. The book of Joshua, reflecting a variety of sources, was compiled and composed much later than the events described. With respect to at least three sites—Ai, Jericho, and Gibeon—archaeological evidence does not bear out the book of Joshua's account of a decisive assault. These cities appear to have been unoccupied at the time of the conquest (settlement). As Judges relates, the infiltration of Israelites into Canaan was gradual; it was brought to completion only with David and Solomon.

Israel's emergence in Canaan has been the object of intense archaeological investigation in recent times. Much more is known about settlement patterns in Canaan as a result of recent excavations and extensive site-surveys. Although consensus among scholars is still lacking, three models of settlement have been proposed: a decisive military assault (the traditional theory), a social movement from within (peasant revolt theory), and a process of peaceful infiltration. The most plausible explanation may be the last, as elaborated by the archaeologist I. Finkelstein on the basis of excavation and survey: namely, that Israel developed from pastoralist elements already residing in Canaan who settled in the hill country sometime around 1200 B.C.

The beginning of the Iron Age dates from 1200 B.C. when changes began to appear in architecture, weapons, and agriculture. This coincided with the period of the judges (Hebrew: *shophetim;* from about 1200 to 1050 B.C.). The judges were charismatic military leaders who delivered Israel from crises caused by the onslaughts of neighboring peoples, including the Philistines. Without a strong central government, the tribes were only loosely confederated. Such a situation led inevitably to anarchy, as the book of Judges attests: "In those days there was no king in Israel; all the people did what was right in their own eyes" (Judg 21.25).

The book of Judges, encompassing the period between Joshua and Samuel, is a valuable historical source for this barbarous period, even though Judges is a religious rather than an historical document. It is not easy to distinguish Canaanite and Israelite religious practices during this chaotic period when the neighboring peoples probably shared a common material culture.

13. The United Monarchy

With respect to the historicity of the Israelite monarchy from about 1020 B.C. onward, scholars are on firmer ground. Starting in this period, the biblical record and the archaeological evidence can be correlated with more assurance. Extrabiblical sources, including increased epigraphic remains, are available to clarify the biblical narrative.

Confronted with the anarchy of the period of judges, Israel had to take steps to establish a strong central rule if it was to survive the Philistine threat. The dominant figure at that time in Israel's history was Samuel. The two books of Samuel are a valuable historical resource, even though their narratives are more composite than continuous, and discrepancies appear in the accounts. The people (though not all) demanded that they be "like other nations" (1 Sam 8.20), that is, that they establish kingship, a new type of leadership. But kingship occasioned a serious theological crisis. Would it not compromise the position of the LORD, who was the sole king of Israel? The practical consideration of survival outweighed theology, and so the monarchy was established at a critical period in Israelite history. Kingship in Israel had to be different from that of Mesopotamia and Egypt, however, where the king was considered divine.

Saul became Israel's first king about 1020 B.C., and his major challenge was to contain the Philistines who were threatening Israel's existence. Little is known about the Philistines because documentary evidence is lacking; Philistine written texts are practically unknown to scholars. The

Philistines, one of the Sea Peoples, had migrated from the Aegean area and settled in southern Palestine after 1150 B.C., following their defeat at the hands of the Egyptians under Rameses III. Conflict between Israel and Philistia persisted until David's defeat of the Philistines, though it continued less intensely until the time of Hezekiah in the latter part of the eighth century B.C.

Saul's story dominates 1 Samuel (chs 13–31). More a warrior than a king, he had some military successes, but ultimately was unable to prevail over the Philistines, who drove him to suicide on Mount Gilboa. Saul was basically a good person, who was thrust into a role for which he was unprepared. Saul's life, especially his ultimate defeat by the Philistines, is full of pathos.

Saul and his successor David (1000–961 B.C.) are a study in contrasts. Saul was no match for David, with whom he had an ambivalent relationship. Saul, jealous of David, even tried to kill him. David's story is told principally in 2 Samuel, especially in the Succession Narrative (2 Sam chs 9–20; 1 Kings chs 1–2). This court history of David addresses the question of his successor to the throne. The disarming candor of the narrative, including the court intrigues, allows the reader to see the human side of David, who was one of Israel's greatest heroes.

As military leader and statesman, David united both Judah and Israel into one kingdom, and was victorious over the Philistines. He also expanded his empire to encompass territory from Eziongeber to Homs (in modern Syria), and from the Mediterranean to the Euphrates. For seven years David ruled over Judah from Hebron. After capturing Jerusalem from the Jebusites, David showed his diplomatic genius by establishing it as the capital of the kingdom. Jerusalem was ideally located for this purpose—on neutral territory and thereby acceptable to both the Northern Kingdom of Israel and the Southern Kingdom of Judah. From that time on, Jerusalem (specifically, the lower eastern ridge) was known as the City of David (Zion). By transferring the Ark of the Covenant to Jerusalem David made his city the religious center of the kingdom, in addition to its being the political capital. David ruled over the United Kingdom of Israel for thirty-three years.

David had planned to build the temple in Jerusalem, but was dissuaded by the prophet Nathan. David's son and successor, Solomon, built the temple and the other royal buildings just north of the original City of David. Nathan is best known for the dynastic oracle (2 Sam ch 7; see also Ps 89), describing the promissory covenant between the God of Israel and the Davidic dynasty.

The contents of 1 and 2 Kings, which are theological narratives, deal with Solomon and his achievements (1 Kings chs 1–11), the kingdoms of Israel and Judah (1 Kings ch 12 to 2 Kings ch 17), and the lone-surviving kingdom of Judah (2 Kings chs 18–25). The kings are judged on the basis of the Deuteronomic principle of purity of the Yahwist cult in Jerusalem. These narratives are ambivalent toward Solomon, a political rather than a military leader. He is praised for consolidating the kingdom by the establishment of administrative districts. Solomon also interjected himself onto the international scene by trade and commerce. He is also distinguished for his elaborate building projects, especially the splendid Jerusalem temple and adjacent palace complex. Solomon is also justly famous for the "Solomonic gates" built to fortify Hazor, Megiddo, and Gezer (1 Kings 9.15).

To finance his lavish building program, Solomon imposed heavy taxes and had recourse to forced labor, thereby alienating his subjects, especially those of the Northern Kingdom. At the same time, he fostered syncretism by introducing foreign religious practices. The United Kingdom of Israel was at the pinnacle of its wealth and power during Solomon's reign, but it was short-lived. The kingdom, united for seventy years, fell apart at Solomon's death in 922 B.C.

14. The Divided Kingdom

Egypt was in constant conflict with both the Northern and Southern Kingdoms; at the same time, internecine warfare persisted until the Northern Kingdom fell in 722/721 B.C. The Northern Kingdom (Israel) was larger and stronger than the Southern Kingdom (Judah), but the latter enjoyed more stability. Rehoboam (922–915 B.C.), Solomon's successor, could have deterred Israel from seceding, but his arrogance prevented him from conciliating the Northern Kingdom. Jeroboam I (922–901 B.C.), the first king of the North, set up royal sanctuaries at Dan (in the north) and Bethel (in the south) to discourage his subjects from worshipping at Jerusalem. He adorned these royal temples with golden bulls, probably to serve as pedestals upon which the invisible Yahweh was enthroned. Jeroboam's purpose probably was not idolatrous; nevertheless, he incurred the condemnation of the Jerusalem-based Deuteronomic historian (1 Kings ch 12). He established his first capital at Shechem, subsequently transferring it to Tirzah.

The next memorable ruler of the Northern Kingdom was Omri (876–869 B.C.), sometimes referred to as "David of the North"; he established a prestigious dynasty, and also expanded the kingdom. He set up his new capital at the strategic site of Samaria, which had been previously occupied by the Shomron family. Omri achieved stability for Israel through collaboration with Judah and a marriage alliance with Phoenicia. He arranged a political marriage between Jezebel, daughter of Ethbaal of Tyre, and his son Ahab (869–850 B.C.). The triple alliance of Israel, Judah, and Phoenicia was intended to hold Aram-Damascus in check. Israel's prosperity achieved under Omri continued through Ahab's reign.

Ahab was a great builder, as attested by the magnificent buildings at Samaria. Much of his reign was absorbed with successful wars against Aram-Damascus. Then in 853 B.C. Israel and Aram joined forces against Assyria at Qarqar on the Orontes. Ahab is reputed to have put two thousand chariots and ten thousand foot soldiers in battle against the armies of Shalmaneser III of Assyria. Ahab was killed subsequently in a campaign against the Arameans at Ramoth-gilead (in modern Jordan, though its location is disputed).

The biblical story focuses on the hostile encounter of Ahab and Jezebel with the prophet Elijah. Jezebel was as zealous for her Phoenician god Baal as Elijah ("Yahweh is my God") was for the God of Israel. But Elijah, an eerie, solitary figure, would tolerate no rival to Yahweh. By defeating the Baal prophets in a dramatic contest on Mount Carmel, Elijah proved that "the LORD indeed is God" (1 Kings 18.39).

During this period the Kingdom of Judah, overshadowed by the Northern Kingdom, was little more than a vassal state of Israel, partly due to its geographical isolation. King Jehoshaphat of Judah reigned twenty-five years (873–849 B.C.). Jehoram (Joram; 849–843 B.C.), his son and successor, maintained a close alliance with Israel during his reign. Ahaziah, the son of Jehoram and Athaliah, was king of Judah in 842 B.C. Athaliah (842–837 B.C.), the daughter of Ahab and Jezebel, usurped the throne and ruled in Judah upon the death of Ahaziah. She met her inglorious end by murder. Jehoash (Joash; 837–800 B.C.), the youngest son of Ahaziah, having escaped Athaliah's purge of the royal family, acceded to the Davidic throne.

15. Dynasty of Jehu (842–746 B.C.)

Jehu (842–815 B.C.), son of Jehoshaphat, established a dynasty that ruled Israel more than ninety years. Having usurped the throne, Jehu conducted a bloody crusade against the worship of Baal, resulting in the massacre of the whole house of Ahab, including the queen mother Jezebel. He put an end to the worship of the Tyrian Baal, destroying the Baal temple in Samaria. The prophet Elisha, among others, supported Jehu's rebellion. Jehu became a vassal to the Assyrian king Shalmaneser III (858–824 B.C.), paying tribute to win Assyrian favor against the Arameans, but he was not entirely successful. The Arameans continued to menace Israel for another half century. Jehu's submission to Shalmaneser III is memorialized on the famous Black Obelisk (discovered in Nimrud, and now in the British Museum). Jehoash (Joash; 802–786 B.C.) of Israel destroyed portions of the Jerusalem wall, and raided the palace treasures as well.

The greatest king of the Jehu dynasty was Jeroboam II (786–746 B.C.), who ruled when the Northern Kingdom was at the height of prosperity. Through territorial expansion Jeroboam II came close to restoring the boundaries of the Davidic empire. Many eighth-century Israelites acquired their wealth through exploitation and oppression of the poor. The prophets Amos and Hosea inveighed uncompromisingly against rampant social injustice, moral turpitude, and religious abuses. Amos aimed his attack against "the winter house as well as the summer house" and "the houses of ivory" (Am 3.15) of the affluent. Hosea indicted the Israelites for their practice of gross rites associated with fertility cults. The conduct of eighth-century Israel was a total contradiction of the covenant ideal.

In the same era Judah also was enjoying prosperity and stability under King Uzziah (Azariah; 783–742 B.C.), partly because Israel and Judah were at peace with each other. Internationally it was an ideal time for the Northern and Southern Kingdoms because Aram (Syria) and Assyria, absorbed in their own internal problems, were not serious threats. Uzziah fortified Jerusalem and strengthened the city walls. He also expanded the territory of Judah to the west, north, and east. His career was cut short when leprosy forced him to relinquish the throne to Jotham, who acted as regent. In 742 B.C., the year Uzziah died, Isaiah was called to be a prophet (Isa 6.1).

The international picture changed for the worse when Tiglath-pileser III (745–727 B.C.), also known as Pul (2 Kings 15.19), became king of Assyria. As founder of the Neo-Assyrian empire, this imperialist was devoted to the conquest of the west, including Syria, Israel, Philistia, Judah, and Transjordan. Forming a coalition against Assyrian advances, King Pekah of Israel and King Rezin of Damascus tried unsuccessfully to force Judah to join. Ahaz (735–715 B.C.), king of Judah at this time, ignored the advice of Isaiah to place absolute trust in Yahweh, and instead appealed to Tiglath-pileser III for help. He paid the price by becoming an Assyrian vassal. The upshot was that Assyria devastated Damascus in 733 B.C., made Galilee and Gilead Assyrian provinces, and reduced Israel to the status of vassal of Assyria.

The Assyrian siege of Samaria under Shalmaneser V began in 724 B.C., and the Northern Kingdom fell to Sargon II in 722/721 B.C. In accordance with Assyrian policy, inhabitants of Samaria were deported, and captives from elsewhere were relocated there. After the fall of Samaria a number of refugees from the Northern Kingdom migrated south and settled in Judah, including Jerusalem. The increase in population of Jerusalem accounts for the expansion of Jerusalem westward at that time. Samaria became an Assyrian province, known as Samarina.

16. The Age of Hezekiah (715–687 B.C.)

Hezekiah, son and successor of the weak Ahaz, was a vigorous leader. Like Ahaz, Hezekiah continued as a vassal of Assyria. Judah, for the most part, enjoyed great prosperity during the reign of Hezekiah. He may be best known for instituting cultic reforms: he abolished local shrines ("high places") and purified the Jerusalem temple, henceforth the exclusive center of worship. Isaiah, prominent during the reign of Hezekiah, exerted some influence on the king.

When Sargon II died in 705 B.C., Hezekiah revolted against Assyrian rule. In anticipation of an Assyrian military response, Hezekiah cut a 1,750-foot tunnel through solid rock to conduct the waters of the exposed Gihon spring to the pool of Siloam within the walls of Jerusalem. In 701 B.C. Sennacherib (704–681 B.C.), the new Assyrian king, invaded Philistia and Judah, suppressing the revolt of the west. He conquered forty-six cities surrounding Jerusalem, including Lachish. The wall reliefs in Sennacherib's palace at Nineveh (now in the British Museum) furnish an unusually vivid pictorial account of his siege and capture of Lachish. Jerusalem was spared when Hezekiah capitulated to Sennacherib by paying a heavy tribute. According to Sennacherib's description of the siege of Jerusalem, he shut up Hezekiah in Jerusalem "like a bird in a cage."

Manasseh (687–642 B.C.), Hezekiah's successor,

undid during his long reign the cultic reforms implemented by his father, and reintroduced idolatrous worship. Judah's vassaldom to Assyria continued through the reign of Manasseh. Assyria attained the zenith of its power during this era.

Like Hezekiah, Josiah (639–609 B.C.) inaugurated a sweeping religious reform, emphasizing the centralization of public worship in Jerusalem and the elimination of idolatory. This religious reform did not prove to be as effective as anticipated. Jeremiah, whose prophetic career extended roughly from 626 to 580 B.C., was disappointed in the reform, perhaps because it had more form than substance.

Josiah, sometimes compared with David, succeeded in expanding the territory of Judah, including, it would seem, much of the former Northern Kingdom of Israel. As the Assyrian empire was beginning to wane as the result of overextension, the Medes and the Babylonians emerged. In 612 B.C. Nineveh, the Assyrian capital, fell to the Babylonians (see Nahum). Josiah died in 609 B.C. at the age of thirty-nine, killed at the Megiddo pass (2 Chron 35.22–24), apparently attempting to prevent Egypt from bolstering the declining Assyrian empire in its struggle with Babylon.

During the last two decades of its history the Kingdom of Judah was caught in a power struggle between imperial Egypt and Babylonia, each striving to fill the power vacuum left by Assyria. The last three kings of Judah—Jehoiakim (609–598 B.C.), Jehoiachin (three months), and Zedekiah (597–587 B.C.)—were undistinguished. During Jehoiakim's reign in 605 B.C. the Babylonian king Nebuchadnezzar defeated the Egyptian pharaoh Neco in battle at Carchemish, placing Judah under the control of Babylonia. This was a turning point in Judah's history. Jehoiakim (see Jer 36) refused to listen to Jeremiah's prediction of the impending Babylonian destruction. In 597 B.C., Nebuchadnezzar besieged Jerusalem and deported Jehoiachin to Babylon, as well as thousands of leading citizens, among them Ezekiel, who became a prophet in Babylon. Nebuchadnezzar then appointed Zedekiah as a puppet in place of the exiled Jehoiachin. Jeremiah was a kind of adviser to Zedekiah, who sought his counsel but seldom followed it. Jeremiah, a realist, urged Zedekiah (Jer 37–38) not to rebel against Babylonia, but instead to capitulate. As a result, Jeremiah was accused of being a traitor. Zedekiah rebelled against the Babylonians, and in 587–586 B.C. Nebuchadnezzar attacked and destroyed Jerusalem and its temple, deporting many of its

inhabitants to Babylon. Zedekiah, the last king of Judah, was taken to Babylon, was imprisoned for treason, and died there. At the same time the fortified towns of Judah, including Lachish, were also destroyed.

17. The Babylonian Exile (586–539 B.C.)

After the captives of Judah were deported to Babylon, the Babylonians did not replace them with foreigners. Although the number of deportees is uncertain, the majority did not go into exile. In fact, some cultic practices continued to be observed at the site of the ruined Jerusalem temple (Jer 41.4–6). Meanwhile, the Babylonians and the Edomites (see Ps 137) sacked the land of Judah. Apparently, the Jews in exile prospered; among other considerations they were granted religious freedom.

Before the fall of Jerusalem, Jeremiah and Ezekiel uttered menacing oracles, to lead the people of Judah to repentance; after the fall, these prophets proclaimed oracles of hope: Jeremiah promised a new covenant (Jer 31.31–34), Ezekiel (37.40–48) a new beginning. Using the typology of the first exodus, the prophet Second Isaiah (Isa chs 40–55) also strove to restore the confidence of his people. Assuring them of imminent delivery from captivity, Second Isaiah declared that a new exodus would follow the demise of Babylon. Cyrus of Persia was God's "messiah" who would deliver the Jews (Isa 45.1–3).

In 539 B.C. at the battle of Opis on the Tigris, the Persians under Cyrus defeated the Babylonian army. Cyrus, a tolerant and enlightened leader, issued an edict liberating the Jews from captivity and permitting repatriation, though not all Jews were eager to return to Judah. The economic opportunities were greater in Babylon than in Jerusalem, which was in need of rebuilding. Cyrus also granted a certain amount of autonomy to Judah.

18. The Postexilic Period

The principal source of information for this period is in Ezra and Nehemiah. Cyrus appointed Sheshbazzar governor of the province of Judah, who led the first group of returning exiles back to Judah. Only in 520 B.C. did Zerubbabel, governor of the province of Judah, and Joshua, the high priest, join forces to undertake the rebuilding of the temple. With the prodding of the prophets Haggai and Zechariah, the temple was completed in 515 B.C. The Samaritans from the district of Samaria in the north offered assistance in rebuilding the temple, but Zerubbabel spurned them. Herein lies the root of the animosity, evident in the Gospels, between Jews and Samaritans. Later the Samaritans went on to build their own temple on Mount Gerizim, overlooking Shechem.

About 445 B.C. Nehemiah, who had been appointed governor of Jerusalem, came from Persia to rebuild the walls of Jerusalem. He met with stiff opposition from both outside and within Jerusalem. His opponents, opting for the status quo, may have felt threatened by a walled city. Nonetheless, Nehemiah completed the rebuilding of the walls in fifty-two days. He also repopulated Jerusalem. Instituting social and religious reforms, Nehemiah championed the cause of the poor, regulated tithing, enforced Sabbath observance, and forbade foreign marriages.

The priest Ezra came to Jerusalem from Babylon, perhaps after Nehemiah, as a religious leader in the company of a large number of exiles. He was skilled in the law of Moses. Some scholars describe his position as "secretary for Jewish affairs" (see Ezra 7.6–10). It was Ezra's function to enforce the observance of the Jewish law. Disturbed by the number of mixed marriages, Ezra fostered Jewish endogamy.

19. The Greek Period (333–63 B.C.)

In 334 B.C. Alexander the Great, king of Macedon, undertook the conquest of the Persian empire, including Anatolia, Phoenicia, Palestine, Egypt, and Mesopotamia. Persian rule came to an end in Judea when Alexander captured Jerusalem in 332 B.C.. The direct result of his victorious campaigns was hellenization, the imposition of Greek culture and language on the East. For the Jews, it meant the accommodation of Jewish faith to Greek practice. The Greek translation of the Old Testament (the Septuagint) and the composition of the New Testament in Greek are direct results of the Greek acculturation instituted by Alexander, and fostered by his successors.

After Alexander's death his vast empire was partitioned into four parts; two of these kingdoms competed for control of Palestine: Egypt, ruled by Ptolemy, and Asia (including Syria), ruled by Seleucus, both generals of Alexander. Ptolemy I captured Jerusalem in 320 B.C. With the Ptolemaic dynasty controlling Palestine for about a century, the Jewish community in Palestine underwent little change; it was a peaceful and prosperous period. The Ptolemies, and later the Seleucids, permitted the high priest to continue as both religious and civil head of Judea.

The Seleucids, also patrons of Hellenism, gained control of Palestine in 198 B.C. when Antiochus III (the Great; 223–187 B.C.) defeated the Ptolemies. Antiochus treated the Jews with consideration, providing financial assistance for the rebuilding of Jerusalem, and exempting them from taxes for a three-year period.

Whereas many Jews, especially the economically poor, opposed any form of hellenization, other Jews (Hellenistic Jews), especially those among the upper classes and the temple priesthood, were eager for assimilation. Embracing Greek culture willingly, those Jews became hellenizers. For example, the high priest Jason, who bought the office, transformed Jerusalem into a Hellenistic polis, a Greek city-state. Jason also built a gymnasium in Jerusalem under the patronage of Hermes and Hercules. The gymnasium was more than a coliseum; it became a social center in deliberate competition with the temple.

Antiochus IV (Epiphanes; 175–164 B.C.), who forbade Jewish religious practices, was the infamous Seleucid king against whom the Maccabees revolted. He erected in Jerusalem a citadel for his garrison; known as the Akra, it housed Syrian troops, and was at the same time the symbol of foreign rule. Attempting to destroy Judaism, Antiochus IV desecrated the temple by erecting an altar to Zeus. The Maccabean family (the Hasmoneans), including the priest Mattathias and his five sons, revolted in 167 B.C. against Antiochus IV when he mandated Jews to offer pagan sacrifice.

Judas Maccabeus, the third of the five sons, succeeded Mattathias; conducting guerrilla warfare against both opposing Jews and Syrian troops, he led his forces to victory, liberating Jerusalem (except for the Akra fortress) in 164 B.C. This resulted in the purification and rededication of the temple, memorialized in the festival of the Dedication (John 10.22; called Hanukkah today).

Upon the death of Judas in 160 B.C., Jonathan (160–143 B.C.), who usurped the high priesthood in 152 B.C., was chosen leader of the Jewish forces. Another brother, Simon (143–134 B.C.), succeeded Jonathan. He was recognized as high priest, commander, and ethnarch of the Jews. Simon was successful in capturing the Akra, and razing it. From the height of the Akra the Seleucids had been able to spy on the Jews during temple services. (For an account of the general history of the period see 1 and 2 Maccabees.)

20. The Hasmoneans

The Hasmonean dynasty traced its roots to the family of the priest Mattathias. The capital of the dynasty was Jerusalem, center of religious, political, and economic life. Excluding the Maccabees, the Hasmonean dynasty extended from 134 to 63 B.C., and embraced John Hyrcanus I (134–104 B.C.), Aristobulus I (104–103 B.C.), Alexander Jannaeus (103–76 B.C.), Salome Alexandra (76–69 B.C.), and Aristobulus II (69–63 B.C.). The Hasmoneans attained the summit of their power under John Hyrcanus I and his son, Alexander Jannaeus. (In editions of this book that include the Apocryphal/Deuterocanonical Books, see "Chronological Table of Rulers," pp. XIV-XV AP of Apocrypha.)

John Hyrcanus I, Simon's son, succeeded as high priest when his father was murdered; he was an exemplary ruler, often compared with David. John Hyrcanus I extended Jewish rule to Transjordan; he forcibly converted the Idumeans to Judaism, imposing Jewish ritual observances on them; he destroyed the Samaritan temple on Mount Gerizim. During this period, or shortly before, two rival parties or "sects"—the Pharisees and the Sadducees—emerged in the Jewish community. The Pharisees took issue with Hyrcanus' secular attitude; in turn, he supported the Sadducees.

The Pharisees (the name may mean "separated ones") were principally a lay group. Advocating a strict interpretation of Torah ("law" or, better, "teaching") the Pharisees accepted both the written Torah (in the Pentateuch) and the oral Torah (accumulated interpretations of the Torah). Included among their beliefs were the resurrection, the coming of a Messiah, and angels. They emphasized ritual purity, tithing, and Sabbath observance. The description of the Pharisees in the Gospels is somewhat tendentious, reflecting an early Christian polemic. Matthew ch 23, for example, is a litany of angry fulminations against (some of) the Pharisees.

The conservative Sadducees were opponents of the Pharisees. The name "Sadducees" may be derived from the priestly Zadokite family of David's time. Although it is difficult to get a coherent picture of the Sadducees, their membership appears to be priestly (not all priests were Sadducees) and aristocratic. The Sadducees supported the Hasmonean priest-kings; they were also influenced by Greek culture. Unlike the Pharisees, the Sadducees did not believe in the immortality of the soul, the general resurrection, or angels. The Sadducees disappeared after the fall of Jerusalem in A.D. 70.

Hyrcanus' son, Aristobulus, appropriated the title of king, and his successors followed suit.

The fact that he assumed a Greek name (*Aristobulus,* "excellent counselor") is an indication of Hellenistic influence.

During the long reign of Alexander Jannaeus, the Hasmoneans achieved considerable territorial expansion, approximating the boundaries of the ancient Davidic kingdom. A deep animosity developed between Alexander and the Pharisees, some of whom were put to death. Alexander's widow, Salome Alexandra, succeeded to the throne, and appointed her son, Hyrcanus II, as high priest. Unlike her husband, she favored the Pharisees, who were powerful in this period. At the death of Salome Alexandra, her two sons, Hyrcanus II and Aristobulus II, engaged in bitter rivalry. Aristobulus became king and high priest.

Ultimately, Pompey intervened with his Roman legions, capturing Jerusalem in 63 B.C.; he attacked the temple mount, entered the holy of holies, and destroyed the walls of the city. Aristobulus II was imprisoned in Rome, while Hyrcanus II ruled in Jerusalem as high priest, but not as king. Judea became part of the province of Syria. From 63 B.C. until the advent of Herod the Great in 37 B.C., the Hasmoneans ruled under the protection of Rome; in 37 B.C. the Herodians replaced the Hasmonean dynasty. It had been a period of great political intrigue.

21. The Herodian Period (63–4 B.C.)

Antipater II, the father of Herod the Great and Phasael, was a friend of Hyrcanus II. Because of their military support of Julius Caesar, both Hyrcanus II and Antipater were rewarded. Antipater was appointed procurator of Judea in 47 B.C., Herod was named governor of Galilee, and Phasael governor of Jerusalem. In 40 B.C. Herod was appointed king of the Jews by the Roman senate; in 37 B.C. he conquered the kingdom of Judea.

Herod the Great (37–4 B.C.) was king of the Jews for thirty-three years, ruling over Judea, Idumea, Perea, Galilee, and Jaffa. Though technically a Jew (as a native of Idumea), Herod was more Hellenist than Jewish; in fact, the Jews hated him. Herod was enamored of Greek culture. He was able to survive so long amidst intrigue and conspiracy because of his total loyalty to Rome, his ruthless repression of those who opposed him, and his unprincipled political maneuvering, not to mention his boundless ambition.

Herod's long reign marked an era of economic prosperity, when the Upper City of Jerusalem attained the zenith of its development; at the same time, the poor were living in the Lower City. As evidence of affluence, archaeologists have uncov ered the remains of luxurious houses, adorned with mosaic flooring, frescoes, and painted plaster. Herod may be remembered best for his grandiose building activities. The designation "Herodian" is synonymous with military architecture and massive masonry throughout Palestine. Herod's best-known building project was the Jerusalem temple, rebuilt to ingratiate himself with the Jews. Begun in 20 B.C. and dedicated ten years later, the temple was not actually completed until A.D. 63. To accommodate the newly enlarged temple, Herod doubled the size of the temple platform by filling in the surrounding valleys. Retaining walls, still standing, were built to support the temple esplanade. The most famous of these walls is called the *kotel* or "western wall," regarded as one of the holiest places in Judaism.

In the northwest corner of the Upper City of Jerusalem, near the present Jaffa Gate, Herod built a splendid, fortified palace, but nothing remains of its superstructure. At the north side of the palace stood three towers, named in memory of Phasael (Herod's brother), Hippicus (an unknown friend), and Miriamme (Herod's Hasmonean wife).

Herod also rebuilt the Baris (a Hasmonean fortress), situated at the northwest corner of the temple enclosure. He enlarged it with towers, naming it Antonia after his patron Mark Antony. This huge fortress served to protect the temple mount.

Herod also constructed fortress-palaces throughout the realm, notably the Alexandrium, north of Jericho; the Machaerus on the east shore of the Dead Sea, where, according to Josephus, Herod Antipas put John the Baptist to death; Masada on the western cliffs of the Dead Sea, first fortified by Alexander Jannaeus, where, according to Josephus, resisting Jewish Zealots committed suicide in A.D. 74 rather than be taken captive by the Romans; and the Herodion, southeast of Bethlehem, where Herod is supposedly buried, although no tomb has been unearthed. Herod magnificently rebuilt Caesarea, the seaport on the eastern coast of the Mediterranean, renaming it Caesarea Maritima in honor of Octavian (Caesar Augustus). Herod also constructed a splendid new city at Samaria, renaming it Sebaste (Greek-Latin "Augustus") in honor of Caesar Augustus.

22. The New Testament Period (4 B.C.–A.D. 30)

According to the New Testament, Jesus was born when Herod the Great was king and Caesar Au-

gustus was Roman emperor; the year was probably 6 B.C. At the death of Herod in 4 B.C., Augustus divided the kingdom among Herod's three sons. Archelaus (4 B.C.–A.D. 6) became ethnarch of Judea, Idumea, and Samaria. When he was removed for brutality, his territory became a Roman province ruled by a procurator. Herod Antipas (4 B.C.–A.D. 39), the most frequently mentioned Herod in the New Testament (and always designated as Herod), was tetrarch of Galilee and Perea. He established his capital at Tiberias on the western shore of the Sea of Galilee. John the Baptist reproached Herod Antipas for his marriage to Herodias, the mother of Salome, because Antipas was Herodias' half-uncle. At the instigation of the vindictive Herodias, Herod Antipas ordered the execution of John (Mk 6.17–29). He also had a part in the condemnation of Jesus.

Philip (4 B.C.–A.D. 34), was tetrarch of Batanea, Trachonitis, and Auranitis, place-names used in the Hellenistic period for the region of Bashan. Philip, a reputable ruler, rebuilt the old Greek city of Paneas (modern Baniyas) in the extreme north of Palestine, renaming it Caesarea Philippi, after the emperor and himself. This city is associated with Peter's confession (Mt 16.16). At Philip's death (A.D. 34) his region became part of the Roman province of Syria.

After the demise of Herod the Great, Jerusalem became a province of the Roman empire, ruled by Roman procurators residing in Caesarea. Pontius Pilate (A.D. 26–36) is the best-known of the Roman procurators of Judea. The preaching of John the Baptist and the ministry of Jesus took place during his governorship. Jewish sources underscore his cruel and insensitive nature. Lacking in respect for Jewish customs, he countered nonviolent protests with bloody responses. Despite his central role in the condemnation of Jesus, Pilate, according to the Gospels, thought Jesus innocent and tried to free him.

The four Gospels, the source for the life and especially the teachings of Jesus, are not intended as a biography in the modern sense (see pp. v–vi NT). On the basis of historical reconstruction, Jesus began his public ministry around A.D. 26, frequenting the Lower Galilee region (the territory of Herod Antipas). Three or four years later, Jesus moved south to Jerusalem to preach there. His public life is reckoned as extending two to three years. About A.D. 30 Roman soldiers put Jesus to death by crucifixion outside the walls of Jerusalem. According to the New Testament, Jesus rose from the dead and appeared to his followers.

23. The Post-Resurrection Period
(A.D. 30–135)

After Pentecost, the gospel was preached abroad, first to the Jews and then to the Gentiles; shortly thereafter, perhaps in A.D. 36, persecution followed. Paul the apostle was born a Hellenistic Jew at Tarsus of Cilicia (in Asia Minor) in the early first century A.D. After persecuting the nascent church, he converted to Christianity and made three famous missionary journeys, described in Acts. Thirteen New Testament letters are attributed to Paul.

Herod Agrippa I (A.D. 41–44), friend of Emperor Caligula and grandson of Herod the Great, was given the title of king and eventually ruled the former kingdom of Herod the Great. He persecuted the Christians and, according to Acts 12, executed James, the son of Zebedee, and also imprisoned Peter. Herod Agrippa's name is associated with the foundations of the "third wall," the northernmost wall of Jerusalem completed as late as the first Jewish revolt (A.D. 66–70). The three northern walls of Jerusalem are known from the writings of the Jewish historian Josephus. Because the archaeological evidence is ambiguous, the alignment of the "third wall" remains a problem.

Herod Agrippa II (called Agrippa in the New Testament), son of Herod Agrippa I, governed (from A.D. 53) the former tetrarchy of Philip, as well as parts of Galilee and Perea. He apparently had an incestuous relationship with his sister Bernice. Paul as a prisoner addressed Agrippa and Bernice at Caesarea (Acts 26.2–29). Agrippa observed Jewish laws and customs; at the beginning of the Jewish revolt in A.D. 66, however, he sided with Rome against the Jews.

24. The Jewish Revolts (A.D. 66–135)

The first revolt broke out in A.D. 66. At that time the Christian community abandoned Jerusalem and fled to Pella in Transjordan. The Jews had some success until the emperor Nero dispatched Vespasian as field commander. He moved against Jerusalem in A.D. 69; meanwhile having been acclaimed emperor, he returned to Rome before the actual siege of Jerusalem. His son Titus led the attack on Jerusalem from the north side, always the most vulnerable, and built a siege wall around the city. First, he razed the fortress Antonia, and then destroyed the temple by fire. In A.D. 70, after a siege lasting from April to September, Jerusalem was captured, and its walls destroyed.

The emperor Hadrian visited Jerusalem in A.D.

129–130; he rebuilt the city as a Roman military camp, renaming it Aelia Capitolina. He also erected a temple honoring Jupiter Capitolinus on the temple mount. The rebuilding and renaming of Jerusalem most likely sparked the second Jewish revolt. Simon ben Kosibah (Bar Cochba) led this revolt against Rome between A.D. 132 and 135. Sextus Julius Severus crushed the revolt only after the Jews had reoccupied Jerusalem. A decree of Hadrian forbade Jews henceforth to live in Jerusalem, or even to visit the city.

III. ARCHAEOLOGY

25. Archaeology and the Bible

Archaeology may be defined as the scientific study of the material remains of past human life and activities. The interrelationship of the Bible and archaeology is reflected in the title "biblical archaeology." It is the function of biblical archaeology to correlate archaeological evidence with the biblical record for the purpose of illuminating the biblical text. Although other archaeologists had preceded W. Flinders Petrie, he conducted the first systematic excavation in Palestine in 1890. Excavating at Tell el-Hesi on the northern edge of the Negeb, he laid the foundations for the scientific dating of sites in Palestine. Since that time, significant progress has been made in Near Eastern archaeology, and especially since 1960 with new developments in the method of excavation, as well as in the processing and interpreting of data.

For convenience, archaeologists refer to periods of time (called "ages") that are named from the material culture that was prevalent during them. This is determined by the artifacts that are found in excavations for the time periods. The periods discussed below are the following:

Early Bronze Age	3000–2100 B.C.
Middle Bronze Age	2100–1550 B.C.
Late Bronze Age	1550–1200 B.C.
Iron Age I	1200–900 B.C.
Iron Age II	900–600 B.C.
Iron Age III	600–300 B.C.

Iron Age III is also called the Persian period. After that era, the ages are named for the dominant political or cultural power. See the discussion below.

The archaeological periods in Palestine date as early as Paleolithic (Old Stone Age), but in biblical studies it is sufficient to begin with the Middle Bronze Age (2100–1550 B.C.). This was a period of power and prosperity in Palestine, characterized by imposing city walls, a simplified alphabet, and international trade. The biblical ancestors (patriarchs) are often associated with the Middle Bronze Age (or the Late Bronze Age), though the chronology of the Genesis narratives is not easy to determine. The Hyksos, the Asiatics who ruled in Egypt as the fifteenth and sixteenth dynasties (ca. 1667–1559 B.C.), belong to the Middle Bronze period. Excavation of Hyksos sites in Egypt, Mari documents from Syria, and Nuzi tablets from northwest Iraq are shedding light on the cultures surrounding the biblical peoples. Excavations at Ras-Shamra, the ancient city-state of Ugarit in northern Syria, have uncovered five major phases of occupation beginning in the seventh and sixth millennia. This site has illuminated the Early Bronze (3000–2100 B.C.) and Middle Bronze (2100–1550 B.C.) Ages, and especially the Late Bronze Age (below).

Hazor (Josh 11.10), located nine miles north of the Sea of Galilee, was the largest city in Palestine during the Middle Bronze period. At Dan on the northern border of Israel a huge mud-brick gateway, including a set of towers and an arch, has been unearthed intact. At Megiddo the earliest excavated city gate dates from the Middle Bronze Age. Taanach, Tirzah, and Shechem were important urban centers in this period. Gezer, with its massive fortifications and a "high place," reached its peak as a Canaanite city in Middle Bronze. In this period Lachish was a Hyksos settlement, protected by a glacis (a defensible slope) and a moat.

The Late Bronze Age (1550–1200 B.C.) was the era of Egyptian domination in Palestine. The archives at Amarna in Egypt, the capital of Pharaoh Amenhoptep IV (Akhenaton), have yielded correspondence from Canaanite kings to the Egyptian court describing conditions in Palestine. The Late Bronze period marked the golden age of Ugarit. The cuneiform (wedge-shaped writing) texts unearthed at Ugarit have been invaluable in clarifying the biblical text. The language of these cuneiform texts, called Ugaritic, bears a close resemblance to biblical Hebrew.

Despite scholarly disputes about details of the exodus, the event itself is usually dated in the Late Bronze period. Some date the Israelite settlement in Canaan to the Late Bronze Age, though that event too is fraught with historical problems. Excavations at Israelite villages of the Early Iron Age, including Ai and Radanna, are casting some light on the manner of Israelite settlement. At the farming villages in the central hill-country archaeologists have uncovered the remains of small, rectangular pillared houses, as well as agricultural terraces. From the material elements found at Iron Age I sites it is obvious that their occupants were not nomads recently arrived from the desert.

26. *Iron Age I (1200–900* B.C.)

The Iron Age is also known as the Israelite period. The introduction of iron metallurgy was one of the factors in the transition from the Bronze to the Iron Age. In the first half of the twelfth century B.C. the Philistines and other Sea Peoples settled in the East Mediterranean basin. The center of Philistine power was located along the southern coastal plain of Palestine (Ashdod, Ashkelon, Gaza, Gath, and Ekron), but not exclusively. It also extended north to Dor, and east to Beth-shan and Tell Deir Alla (Succoth?). Philistine pottery with its elaborate decoration testifies to the high level of culture achieved by the Philistines, despite the later pejorative connotation of their name. This pottery is painted in black and red on a white-slipped background.

Excavations at Ekron (Tel Miqne) and Ashkelon in the 1980s and 1990s add significantly to the history of the Philistines. Ekron, continuing as a Philistine city until the end of the seventh century B.C., is one of the largest Iron Age sites in modern Israel. In the seventh century B.C. Ekron was among the principal centers for production of olive oil in the ancient Near East. In 603 B.C. the Neo-Babylonians destroyed Ekron.

Ashkelon was one of the most important seaports in the East Mediterranean. Occupation at the site extended from about 2000 B.C. to A.D. 1500. The stele of Pharaoh Merneptah (ca. 1213–1204 B.C.) records the Egyptian conquest of Ashkelon. Indicative of the importance of this port city, Assyria, Egypt, and Neo-Babylonia successively conquered Ashkelon in the eighth and seventh centuries B.C.

Three superimposed Philistine temples, the earliest dating from the twelfth century B.C., have come to light at Tel Qasile, a prime Philistine site on the Yarkon river (within the modern city of Tel Aviv). Tel Batash, biblical Timnah, in the Sorek Valley, reveals continuous occupation from the Middle Bronze Age to the Persian period. This city played a prominent role in the Samson story. Philistine artifacts at Timnah indicate that the city prospered in Iron Age I.

Jerusalem may be the most excavated city in the world. Early excavations in Jerusalem took place from 1867, principally under British auspices. Between 1961 and 1967 Kenyon excavated stratigraphically in the City of David (Ophel hill). After 1967 Israeli archaeologists (principally, Mazar adjacent to the Temple Mount, Avigad in the Jewish Quarter, and Shiloh at the City of David) excavated Jerusalem intensively. The City of David, constituting only a small portion of modern Jerusalem, is located on the steep slope of the Old City's southeast spur. Kenyon discovered on the eastern ridge the Jebusite wall and adjacent tower (dating to 1800 B.C.) surrounding pre-Israelite Jerusalem. Continuing Kenyon's excavations, Shiloh uncovered twenty-five occupational strata, extending from the Chalcolithic period (4500–3200 B.C.) to medieval times. He also investigated Jerusalem's complex hydraulic system, including Hezekiah's 1,750-foot tunnel running under the city, the Siloam channel, and "Warren's shaft," a vertical shaft and connecting tunnels (named for Charles Warren, who discovered it in 1867). These three watercourses originated from the Gihon spring, Jerusalem's chief source of fresh water. Earlier archaeologists, dating "Warren's shaft" to the pre-Davidic Jebusite period, assumed that David's soldiers captured Jerusalem by entering the city through "Warren's shaft" (2 Sam 5). Shiloh, however, found no evidence for dating the shaft system before the Israelite occupation (in the tenth to ninth centuries B.C.).

Architectural remains from time of Solomon have also been uncovered in Palestine. Typical are the Solomonic casemate walls (a double line of fortification walls partitioned into compartments by cross walls), the impressive city gates with four entryways at Megiddo, Hazor, and Gezer (1 Kings 9.15), and the header-stretcher masonry (referring to the configuration of stones within a fortification wall). The tenth and ninth centuries B.C. were notable for hydraulic technology at several ancient cities; in addition to Jerusalem, archaeologists have uncovered water tunnels at Megiddo, Hazor, Gezer, and Gibeon. It was imperative to protect the water supply, especially during times of siege, by channeling it within the city walls.

Solomon undertook with Phoenician help the construction of the Jerusalem temple in the fourth year of his reign (ca. 960 B.C.), a project that took seven years. No remains of this temple have been unearthed; nevertheless a paper reconstruction is possible on the basis of the brief biblical description (1 Kings 6) and from comparison with extant temples in neighboring lands, especially the eighth-century B.C. temple at Tell Tainat in the Amuq valley (Syria). This temple and the Solomonic were long-room temples (the entrance on the short side, and the shrine at the opposite end),

with a tripartite architectural design consisting of the vestibule, the holy place (nave), and the holy of holies. The Jerusalem temple was relatively modest in size: about 105 feet long, 35 feet wide, and 52 feet high.

The temple and royal palace together constituted the administrative center of the Israelite kingdom. The temple, principally a religious center, also played a major role in the economic and political life of the kingdom.

The Gezer calendar, found at Tel Gezer in 1908, is the oldest inscription thus far found that relates to the Bible. This small limestone plaque, written in Hebrew in the late tenth century B.C., provides information about the agricultural cycle in Palestine.

27. *Iron Age II (900–600* B.C.*)*

This archaeological period is approximately coterminous with the divided kingdoms of Judah and Israel (922–586 B.C.). When the Northern Kingdom of Israel seceded under Jeroboam I, it had no permanent capital until Omri (876–869 B.C.) became king. He established the capital at Samaria (Shomron), and built an elaborate city virtually anew. This strategically located site, 300 feet above the surrounding area, has been excavated extensively. Samaria is a difficult site to interpret archaeologically because it has been destroyed and rebuilt several times, with the reuse of stone in the successive rebuilding of the city. After Omri and his son Ahab, the kings Jehu and Jeroboam II continued with the architectural development of the city.

Among the valuable artifacts from Samaria are ivory, pottery, and ostraca. More than five hundred fragments of ivory inlays, signs of the city's wealth, have been unearthed in the area of the royal quarter. Among ceramics of the ancient Near East, "Samaria ware" is highly prized; such vessels from Samaria are finely burnished with red slip (veneer). More than sixty ostraca (potsherds preserving writing in ink) were also uncovered. Dating from the reign of Jeroboam II in the eighth century B.C., these ostraca may have been tax receipts for the shipment of wine and oil, indicating Samaria's activity in commerce and trade. After the conquest of Samaria in 722/721 B.C., the Assyrians rebuilt the city as the headquarters of the whole administrative district.

The Siloam monumental inscription, the most important Hebrew inscription ever recovered, was found in 1880 carved on the east wall of Hezekiah's tunnel; it describes how two teams digging from opposite directions built the tunnel. Hezekiah, though unmentioned in the inscription, constructed this underground channel in Jerusalem in preparation for Sennacherib's siege in 701 B.C. (2 Chr 32.2–4, 30).

Another important discovery was a hoard in excess of two hundred ostraca at Arad; more than half are in Hebrew and dated to the monarchy; the remainder are in Aramaic from about 400 B.C.

At Ramat Rahel (Beth-haccherem?), situated south of Bethlehem, archaeologists excavated a royal citadel (ninth or eighth century B.C.) and a palace of one of the last kings of Judah, possibly Manasseh (687–642 B.C.). Khirbet Rabud, a large tell southwest of Hebron, is probably the site of the Levitical city of Debir, once thought to be identified with Tell Beit Mirsim. Debir was destroyed by Sennacherib, and rebuilt later. At Tel Goren in the En-gedi region, an oasis on the west shore of the Dead Sea, Israeli archaeologists unearthed an industrial installation dating to the reign of Josiah in the mid-seventh century B.C.

At the City of David (Jerusalem) archaeologists have found over fifty well-preserved clay bullae, used to seal papyri. These bullae, dating from the late seventh to the early sixth century B.C., may be from the royal archive of Jeremiah's time. Several of the names on the bullae are also found in Jer 36.

Beer-sheba, small in size, was the principal city of the northern Negeb. A strong fortification system combined with city-planning made Beersheba an impressive site. A street with buildings on each side encircled the city. An important discovery at Beer-sheba was a dismantled horned altar; some of its stones were found in a restored section of a storehouse wall, dating to the eighth century B.C. Three of the altar's horns were intact; the top of the fourth had been displaced. Archaeologists, rconstructing the altar, found that its original height was just over five feet. This altar may have been dismantled as a consequence of Hezekiah's reform, requiring worship to be centralized in Jerusalem.

Two Israelite forts located in the Negeb cast light on Iron Age II. The site of Ein el-Qudeirat, forty miles south of Beer-sheba, is considered to be Kadesh-barnea, the principal station during the Israelite sojourn in the wilderness. However, no Late Bronze or Early Iron Age remains that could relate to the exodus have been unearthed. The three successive fortresses found at the site date between the tenth and the sixth centuries B.C.

Kuntillet Ajrud, situated forty miles south of Ein el-Qudeirat, was a remote wayside shrine in northeastern Sinai. This southernmost outpost of the kingdom of Judah dates to about 800 B.C., when Amaziah was king of Judah and Joash king of Israel. More important than the buildings at the site are the inscriptions (dedications, prayers, requests, blessings) and accompanying drawings on the plaster walls and on two large storage jars. The interpretation of the inscriptions and the drawings is still uncertain.

Lachish in the foothills of the Hebron mountains was the second most important city of Judah. No site has illuminated more clearly Sennacherib's Assyrian assault on Judah in 701 B.C. Lachish, excavated in the 1930s and again in the 1970s and 1980s, is a textbook case in archaeology. The reliefs (now in the British Museum) adorning the walls of Sennacherib's palace at Nineveh, complemented by the Assyrian annals and the biblical record, depict in detail the siege of Lachish. In addition, they furnish military historians with valuable information about warfare in the Iron Age II period. A little more than a century after Sennacherib's siege of Lachish the Babylonian king Nebuchadnezzar destroyed the city as part of his conquest of Judah. The Lachish Letters, consisting of twenty-one ostraca, cast light on the city during the reign of King Zedekiah, shortly before it fell to the Babylonians.

Important Iron Age II sites are also found in the north of Israel. Dan, located at the foot of Mount Hermon, is the site of one of Jeroboam I's national sanctuaries in the Northern Kingdom. The excavators may have uncovered the sacred area or "high place" on the northwest side of the site. Jeroboam I may have built the square platform; the monumental steps were constructed during the reigns of Ahab and Jeroboam II.

Hazor, rebuilt during the Solomonic era, prospered during the reign of Jeroboam II. Israelite houses of this period are noteworthy: the excavators unearthed a four-room house with two stories, the second floor supported by pillars. Tiglath-pileser III destroyed Hazor in 732 B.C., as broken vessels in ash layers attest.

During Iron Age II the region east of the Jordan river (Transjordan) prospered. Extensive surveys of this region in the 1980s generated new data, adding considerably to the results of the earlier surveys of Glueck, who in the 1930s and the 1940s mapped, surveyed, and photographed Moab, Ammon, and Edom, lands practically unknown. The surveys and excavations in Transjordan have led to the discovery of new inscriptions that add considerably to the history of Transjordan.

With respect to the Ammonites, the following

written material has come to light: the Tell Siran bronze bottle inscription, the Amman citadel inscription, the Amman theatre inscription, and numerous seals. With regard to the Moabites, the Moabite Stone (Mesha stele), documenting the Moabite King Mesha's revolt against Israelite control in the ninth century B.C., was discovered in 1868 at Dibon (modern Dhiban), Mesha's capital.

A tell near the modern village of Hesban in Jordan was once thought to be the site of biblical Heshbon, capital of the Amorite king Sihon, formerly in the land of Moab. The lack of Bronze Age remains and the paucity of Early Iron Age artifacts, however, cast doubt on the identification of the site. Iron Age II is well represented here, including ostraca from the seventh and sixth centuries B.C.

Almost no Edomite writing has come to light, with the exception of some seventh-century B.C. jar stamps from Tell el-Kheleifeh and an Edomite ostracon with six lines of writing from Horvat Uza, a Judean fortress in the eastern Negeb. But the picture is changing with the discovery in the 1980s of an Edomite shrine, the first ever excavated. Surprisingly, the shrine, named Qitmit, is located in ancient Judah, and not in Edom. This cult center dates to about 586 B.C., when the Babylonians discovered Jerusalem. As this excavation continues into the 1990s, much more may be discovered.

28. Persian to Herodian Periods (539–4 B.C.)

Samaria was a provincial capital in the Persian era (sixth to fourth centuries B.C.), as it had been in the Assyrian period, but few Persian remains are in evidence at Samaria. During the Persian period the Upper City of Jerusalem was not resettled, the city being confined to the City of David and the temple mount. This restricted area was adequate for the small number of people living in Jerusalem at the time. New excavations underway in Palestine are more promising with respect to the Persian period.

Anafa in Upper Galilee was an affluent Hellenistic town from the mid-second century B.C. Molded glass vessels, fine red wares (terra sigillata), and elaborately decorated buildings attest to the wealth of this Greek settlement.

Concerning Jerusalem, Israeli excavators, concentrating on the southwest corner of the temple mount, as well as the area adjacent to the western and southern walls of the temple enclosure, have recovered remains dating from the Iron Age to the Ottoman era (A.D. 1516–1918), with most finds belonging to the Herodian period (37–4 B.C.). At the same time, excavations proceeding in the Jewish Quarter of the Old City unearthed remnants from the Hasmonean and Herodian periods. The artifacts reveal that during the Herodian period Jerusalem residents enjoyed considerable luxury. Beautiful stone objects (tables, bowls, cups, and purification jars) were recovered, as well as mosaics and frescoes.

29. New Testament Period

Archaeological evidence is far more extensive for the Old Testament than for the New. The New Testament encompasses a much shorter period; also, fewer New Testament excavations have been undertaken, though the situation is changing. The locations of many of the events in the life and ministry of Jesus are often uncertain; the sites currently associated with a number of them have been determined by pious tradition, not by scientific investigation.

The site of the famous Capernaum synagogue on the western shore of the Sea of Galilee is certain, but the date is in dispute. The Franciscan excavators favor the fourth or fifth century A.D. on the basis of coins found beneath the pavement, whereas the Israelis prefer a date in the late second or early third century A.D. The ruins of a more ancient synagogue, lying beneath the reconstructed limestone synagogue, may be the building where Jesus preached.

With respect to the trial and condemnation of Jesus, some scholars have conjectured that the Antonia fortress at the northwest corner of the temple enclosure was the site of the Roman praetorium where Pontius Pilate judged Jesus (Jn 18.28–19.16). It seems certain that the praetorium of Jesus' trial was located at Herod's palace, near the present Jaffa Gate in the northwest corner of the Upper City, which was the official residence of the Roman procurators (including Pontius Pilate) during their visits to Jerusalem. (A dedicatory inscription found in 1961 at Caesarea Maritima bears the name of Pontius Pilate.)

Some have conjectured that the broad stone pavement in the basement of the Sisters of Zion convent on the Via Dolorosa is the *lithostrothon* (stone pavement) mentioned in the trial of Jesus (Jn 19.13). Benoit demonstrated, however, that this stone pavement dates to the Roman emperor Hadrian (A.D. 117–138), and that the adjoining "Ecce Homo" arch (John 19.5) was the large central arch of Hadrian's east forum.

Concerning Josephus' three successive walls defending Jerusalem's northern side, the direction of the disputed second wall bears on the authenticity of the church of the Holy Sepulchre. The site of Jesus' burial had to have been located outside the city wall because burials were forbidden within the city. Recent excavations appear to confirm that at the time of Jesus' crucifixion the site of the Holy Sepulchre was located outside the second wall of Jerusalem.

Measures and Weights in the Bible

I. HEBREW MEASURES OF LENGTH

Hebrew	NRSV	Equivalence	U.S. Measures
'ammāh	cubit	2 spans	17.49 inches
zéreth	span	3 handbreadths	8.745 inches
ṭôphaḥ, ṭéphaḥ	handbreadth	4 fingers	2.915 inches
'eṣba'	finger		0.728 inch

The cubit described in Ezekiel (40.5; 43.13) is equal to seven (not six) handbreadths, namely 20.405 inches.

II. MEASURES OF LENGTH IN THE NEW TESTAMENT

Greek	NRSV	U.S. Measures
pēchus	cubit	about 1½ feet
orguia	fathom	about 72.44 inches
stadion	stadia, or the equivalent in miles	about 606 feet
milion	mile	about 4,854 feet

III. HEBREW MEASURES OF CAPACITY

DRY MEASURES

Hebrew	NRSV	Equivalence	U.S. Measures
ḥómer	homer	} 2 lethechs	6.524 bushels
kōr	measure, cor		
léthekh	lethech	5 ephahs	3.262 bushels
'êphāh	ephah, measure	3 seahs	20.878 quarts
se'āh	measure	3⅓ omers	6.959 quarts
'ómer	omer	1⅘ kabs	2.087 quarts
'issārôn	tenth part (of ephah)		
qabh	kab		1.159 quarts

LIQUID MEASURES

Hebrew	NRSV	Equivalence	U.S. Measures
kōr	measure, cor	10 baths	60.738 gallons
bath	bath	6 hins	6.073 gallons
hîn	hin	3 kabs	1.012 gallons
qabh	kab	4 logs	1.349 quarts
lōgh	log		0.674 pint

IV. MEASURES OF CAPACITY IN THE NEW TESTAMENT

Greek	NRSV	Equivalence	U.S. Measures
batos	measure	(Hebrew) bath	
koros	measure	(Hebrew) kōr	see Table III
saton	measure	(Hebrew) ṣe'āh	
metrētēs	measure		10.3 gallons
choinix	quart		0.98 dry quart
modios	bushel	(Latin) modius	7.68 dry quarts
xestēs	pot	(Latin) sextarius	0.96 dry pint, or
			1.12 fluid pints

V. HEBREW WEIGHTS

Hebrew	NRSV	Equivalence	U.S. Avoirdupois
kikkār	talent	60 minas	75.558 pounds
māneh	mina	50 shekels	20.148 ounces
shéqel	shekel	2 bekas	176.29 grains
pîm (or payim)	pim	⅔ shekel	117.52 grains
béqa'	beka, half a shekel	10 gerahs	88.14 grains
gērāh	gerah		8.81 grains

The practice of weighing unmarked ingots of metal used in commercial transactions prior to the invention of money explains that the names of the units of weight were used later as indications of value, and as names for monetary standards. There is, however, no direct relation between the shekel-weight and the weight of a shekel piece.

VI. WEIGHTS IN THE NEW TESTAMENT

Greek	NRSV	Equivalence	U.S. Avoirdupois
talenton	talent	(Hebrew) talent	see Table V
mna	pound	(Hebrew) mina	see Table V
litra	pound	(Latin) libra	0.719 pound

Index to the Annotations

The following index lists important names, institutions, and ideas that are mentioned in the annotations. In order to gain the fullest information, the Scriptural references as well as the annotation itself should be read, and all cross references should be consulted.

This index combines references from the Hebrew Scriptures, the New Testament, and the Apocryphal/Deuterocanonical Books. References to the Apocryphal/Deuterocanonical Books follow those to the other Scripture portions, in cases where a topic is treated in various places.

A separate Index to the Annotations of the Apocryphal/Deuterocanonical Books appears at the end of that section. In editions of this book that do not include the Apocryphal/Deuterocanonical Books, the reader should ignore references to them contained in this Index.

Page numbers are followed by letters indicating the section in which they are found, thus: OT = Old Testament, AP = Apocryphal/Deuterocanonical Books, NT = New Testament.

Aaron, Sir 44.19–47.11
Abaddon, Job 26.6; Rev 9.11
Abba, Rom 8.15
Abbot, George, p. viii AP
Abel, 4 Macc 18.11
Abiram, 4 Macc 2.17
Abraham, Sir 44.19–47.11; Song of Thr 12; 4 Macc 7.19; 13.12; 14.20
Abyss, the, Lk 8.31
Achior, Jdt 5.5–21
Acrostic, Pss 9–10; 25; 34; 37; 112; 119; 145; Intro to Lam, p. 1047 OT; Nah 1.2–11
Adam, creation of, 2 Esd 3.4–5
Adultery, Sir 23.16–27
Agriculture, Gen 3.17–19; 4.1–26; 9.20; 47.13–26
Ahikar, Tob 1.21; 11.18; 14.10
Alcimus, 1 Macc 7.5–25
Alexander Balas, 1 Macc 10.1–21
Alexander the Great, 1 Macc 1.1
Alexander Janneus, 1 Macc 14.41–43
Alexandria, 3 Macc 4.11
Alexandrian pseudepigrapha, p. xii AP
Alliances, foreign, 2 Kings 20.16–19; Ezek 23.5–10
Almsgiving, Tob 4.7–11; Sir 3.30–4.10
Almsgiving, right spirit in, Sir 18.15–18
Alpha and Omega, Rev 1.8
Altar, Ex 20.24–26; 27.1–8; Ezek 43.13–26
Altar, horns of the, 1 Kings 1.49–53

Amen-em-ope, Prov 22.17–24.22
Amorites, Gen 10.15–20; 1 Sam 7.13–14
Angel(s), Gen 16.7; 32.1–2; Judg 2.1; 2 Sam 24.16; Job 33.8–33; Isa 37.36; Dan 10.12–14; Acts 12.15; 1 Cor 11.10; Heb 1.14; Rev 1.20; 2 Esd 6.41; Tob 12.12–15; Pr Man 15b; 2 Macc 11.6
Anglo-Saxon poem of Judith, Jdt 10.21
Anoint, 1 Sam 10.1; Ps 2.2; see also Messiah
Anointed, Ps 151
Antichrist, 1 Jn 2.18; Rev 11.7; 19.19–21
Antioch, p. 341 AP
Antiochus III the Great, 1 Macc 1.10; 3 Macc 1.1–2
Antiochus Epiphanes, Dan 7.4–8; 8.9–14, 23–25; 11.21–45; Jn 10.22; 1 Macc 1.10; 3.30; 2 Macc 4.7; p. 285 AP; 4 Macc 1.11; 4.15–26; 17.24
Antiochus V Eupator, 1 Macc 3.33; 2 Macc 10.10–11
Antiochus VII Sidetes, 1 Macc 15.1–14
Anti-Semitism, Add Esth (Gk) 10.4–11.1
Apocalypse, Isaiah, Isa 24–27
Apocalypticism, Ezek 38.1–39.29; Intro to Dan, p. 1126 OT
Apocalyptic literature, p. v AP

Apocrypha, pp. iii AP
Apollos, Acts 18.24–28
Apollonius, governor of Samaria, 1 Macc 3.10
Apostasy, Num 25.1–18; Judg 2.6–3.6; 1 Kings 12.25–33; 14.1–16; 2 Kings 17.1–41; Jer 2.1–37; Ezek 20.1–44; 23.1–4; Hos 4.1–8.14; 1 Tim 4.1–16
Apostles, the twelve, Lk 6.12–16
Aramaic, 2 Kings 18.26; Esth 1.22; Isa 36.11; Jer 10.11; 44.1; Intro to Dan, p. 1126 OT; p. iv NT; Rom 8.15; pp. 411–415
Archaeology, pp. 419–424
Archelaus, Mt 2.22
Ark (of the Covenant), Ex 25.10–22; 33.8–9; Num 10.33–36; Deut 10.1–3,8; Josh 3.3; 1 Sam 3.3; 2 Sam 6.1–15; Ps 132; Jer 3.16–17; pp. 412–413
Arrogance, Sir 10.6–18
Arsinoë, 3 Macc 1.1
Art and the Apocrypha, p. x AP
Artaxerxes I, 1 Esd 2.16–30
Artaxerxes II, 2 Esd 1.3
Ascension of Christ, Jn 6.62–63; Acts 1.6–11
Asherah, 1 Kings 14.15; Jer 2.20–28; Ezek 6.1–14
Asmodeus, Tob 3.8
Assyrians, pp. 410, 414–415
Atargatis, 2 Macc 12.26
Atbash, Jer 25.26
Athens, Paul at, Acts 17.16–34

Alphonso October 19, 1994
Amy Jo August 15, 1981
Jennifer Denise December 26, 1982
Raquel Joan May 17, 1985
Thomas Micheal Jan 31, 1990
Megan Joy May 2, 1993

THE NEW OXFORD

BIBLE MAPS

Prepared by Oxford Cartographers
and based on the Oxford Bible Atlas.

MAP 1

The Land of Canaan
Abraham to Moses

GAD, etc. Tribes of Israel

EDOM, etc Kingdoms encountered by the Israelites in the 13th century, B.C.

• Cities mentioned in Numbers and Deuteronomy, but not in Genesis.

20 Miles

20 Kilometres

V · W · X · Y · Z

A R A M
(S Y R I A)

• Damascus

Mt. Hermon (Sirion, Senir)

Mt. Lebanon

• Sidon

• Ijon

Abel •
• Laish (Dan)
• Beth-anath?

MAACAH

• Kedesh

Tyre • Uzu
• Kanah
• Janoah

Achzib •

Acco •

N

A

GESHUR

• Aduru

R. Jordan
Sea of Chinnereth

ARGOB

• Karnaim
• Ashtaroth

B A S H A N

Golan •

Edrei

• Ramoth-gilead

H E S B O N

G R E A T   M A N A S S E H

• Ham

HAVVOTH-JAIR

• Yanoam

Arabah

Jordan

• Pehel (Pella)

Hazor •
• Merom

Chinnereth •
Madon •

Beth-yerah •
(Philoterio)
• Japhia

Anaharath?

Shunem •

• Beth-shean
Rehob •

• Tirzah
Mt. Ebal

Hannathon •
Achshaph ?
Shimron •

Jokneam •

Megiddo •

• Taanach
• Beth-haggan (En-gannim)
• Ibleam
• Dothan

Manasseh

country

Mt. Carmel

Aruna •
Migdal •
Gath of Sharon •
• Arubboth
• Yehem
• Socoh

Dor •

Sharon

of

T H E
G R E A T
S E A

(The Western Sea)

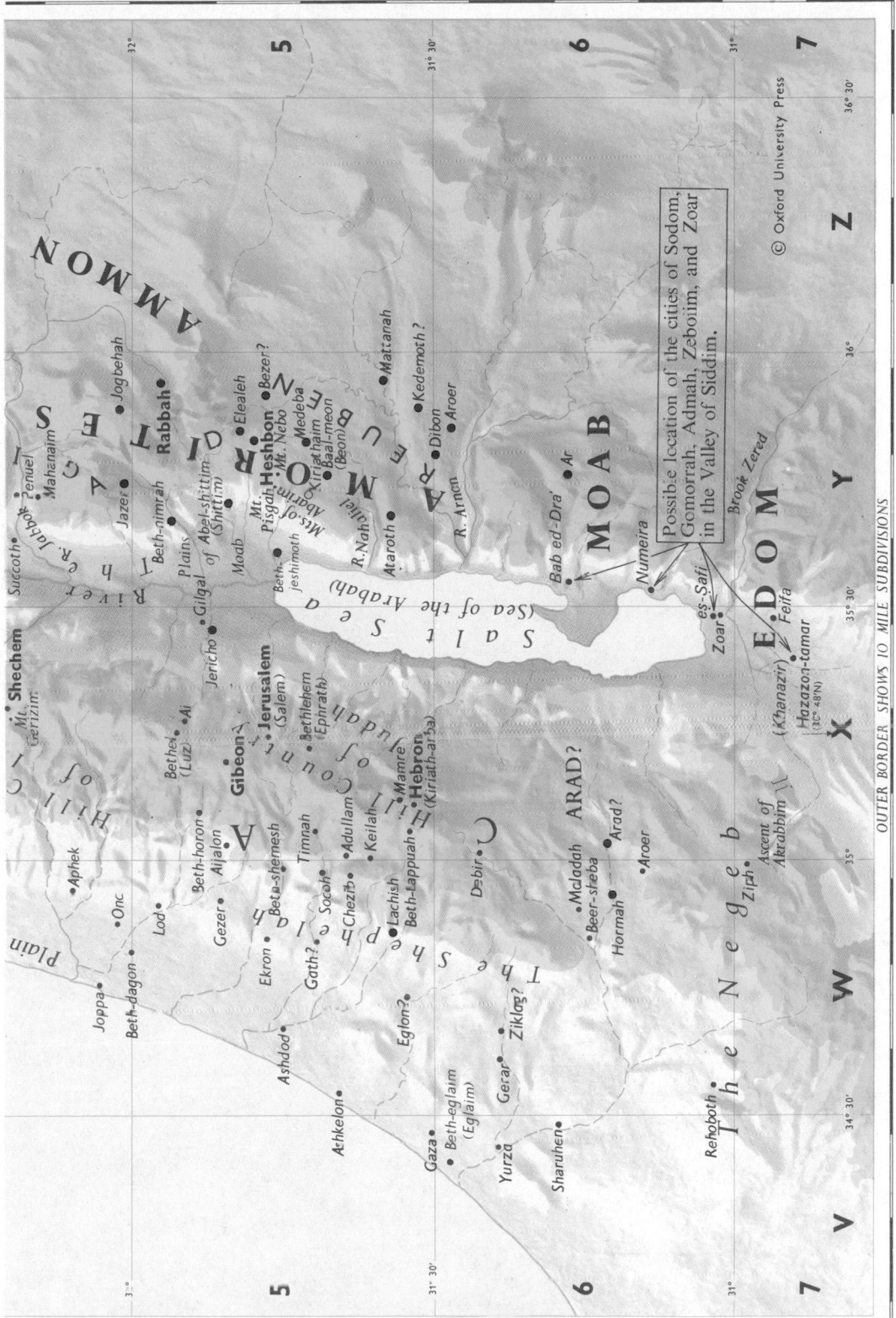

MAP 1

Possible location of the cities of Sodom, Gomorrah, Admah, Zeboiim, and Zoar in the Valley of Siddim.

© Oxford University Press

OUTER BORDER SHOWS 10 MILE SUBDIVISIONS

MAP 2

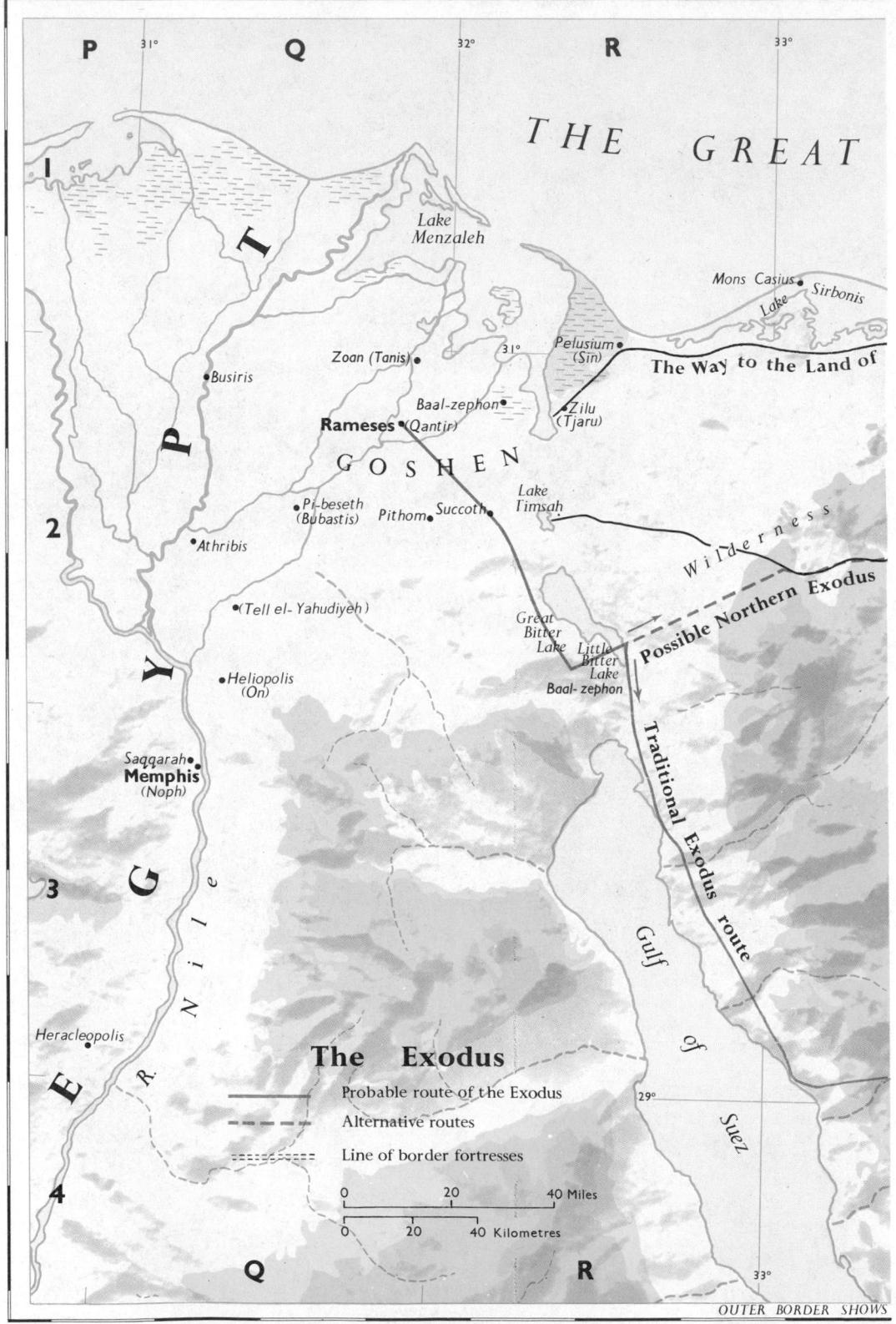

P   31°   Q   32°   R   33°

THE GREAT

1

Lake Menzaleh

Mons Casius • *Sirbonis*

Lake

• *Busiris*

Zoan (Tanis) •   31°   *Pelusium (Sin)* •

The Way to the Land of

Baal-zephon •   • *Zilu (Tjaru)*

Rameses • *(Qantir)*

G O S H E N

2

• *Pi-beseth (Bubastis)*   *Pithom* • *Succoth* •   *Lake Timsah*

Wilderness

• *Athribis*

Possible Northern Exodus

• *(Tell el-Yahudiyeh)*

Great Bitter Lake   *Little Bitter Lake*   *Baal-zephon*

• *Heliopolis (On)*

Traditional Exodus route

Saqqarah •

Memphis (Noph)

3

Gulf

R. Nile

of

Heracleopolis •

The Exodus

——————— Probable route of the Exodus

– – – – – – – Alternative routes

======== Line of border fortresses

Suez   29°

E   G   Y

0   20   40 Miles

0   20   40 Kilometres

4

Q   R   33°

OUTER BORDER SHOWS

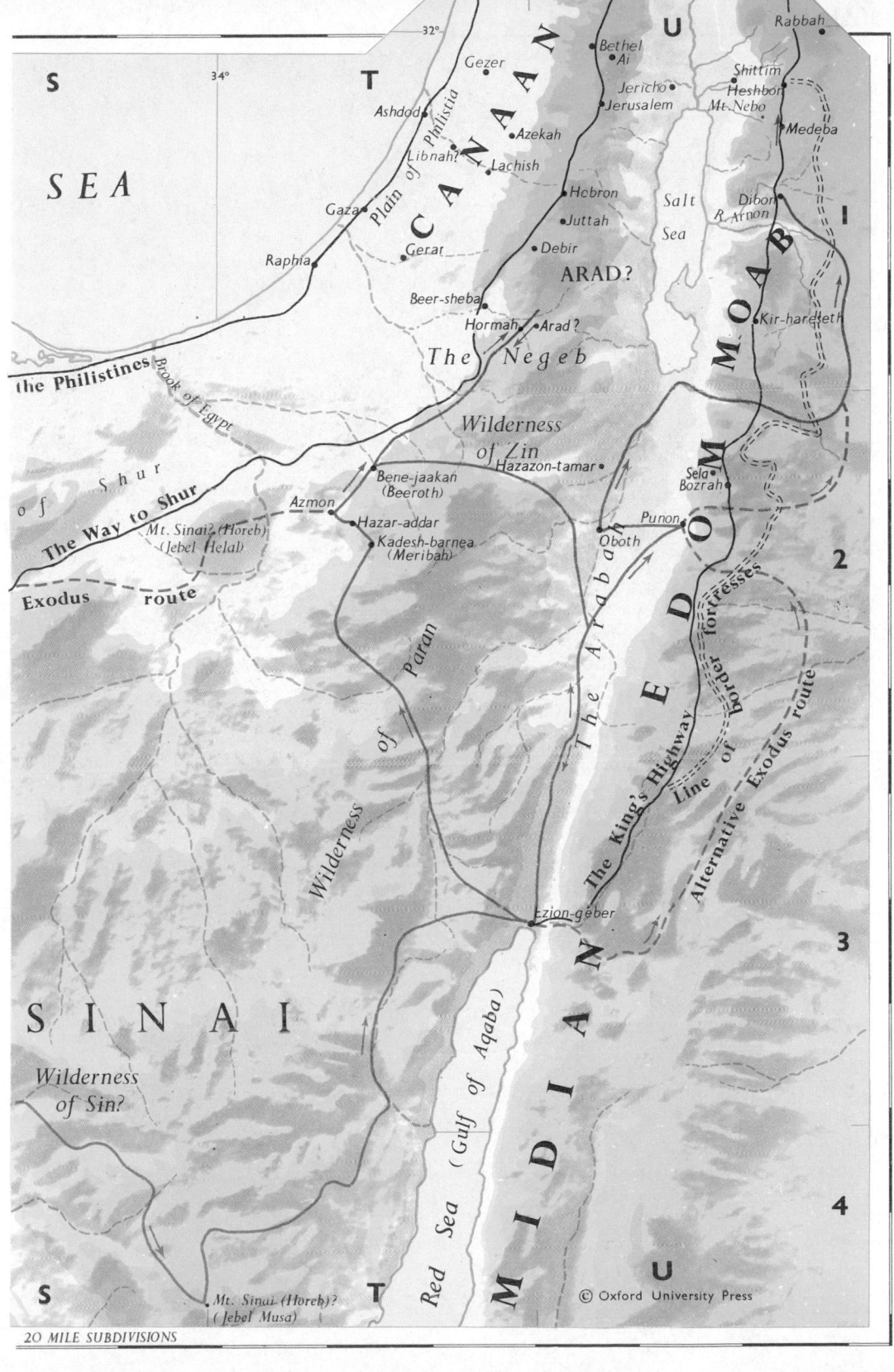

S

SEA

32°
34°

T

U

Rabbah

CANAAN

Bethel
Ai

Gezer

Shittim
Heshbon
Mt. Nebo

Ashdod

Jericho
Jerusalem

Azekah

Medeba

Libnah?
Lachish

Plain of Philistia

Gaza

Hebron

Salt
Sea

Dibon
R. Arnon

Juttah

Gerar

Debir

ARAD?

Raphia

Kir-hareteth

Beer-sheba

Hormah   Arad ?

MOAB

I

The   Negeb

the Philistines

Brook of Egypt

Wilderness

Sela
Bozrah

of Zin

Hazazon-tamar

of   Shur

Bene-jaakan
(Beeroth)

Punon

The Way to Shur

Azmon

Oboth

2

Exodus   route

Hazar-addar

EDOM

Mt. Sinai? (Horeb)
(Jebel Helal)

Kadesh-barnea
(Meribah)

Line of border fortresses

Alternative Exodus route

Wilderness

of

Paran

The Arabah

The King's Highway

Ezion-geber

3

SINAI

Wilderness
of Sin?

MIDIAN

Red Sea (Gulf of Aqaba)

4

S

T

U

Mt. Sinai (Horeb)?
(Jebel Musa)

© Oxford University Press

20 MILE SUBDIVISIONS

MAP 3

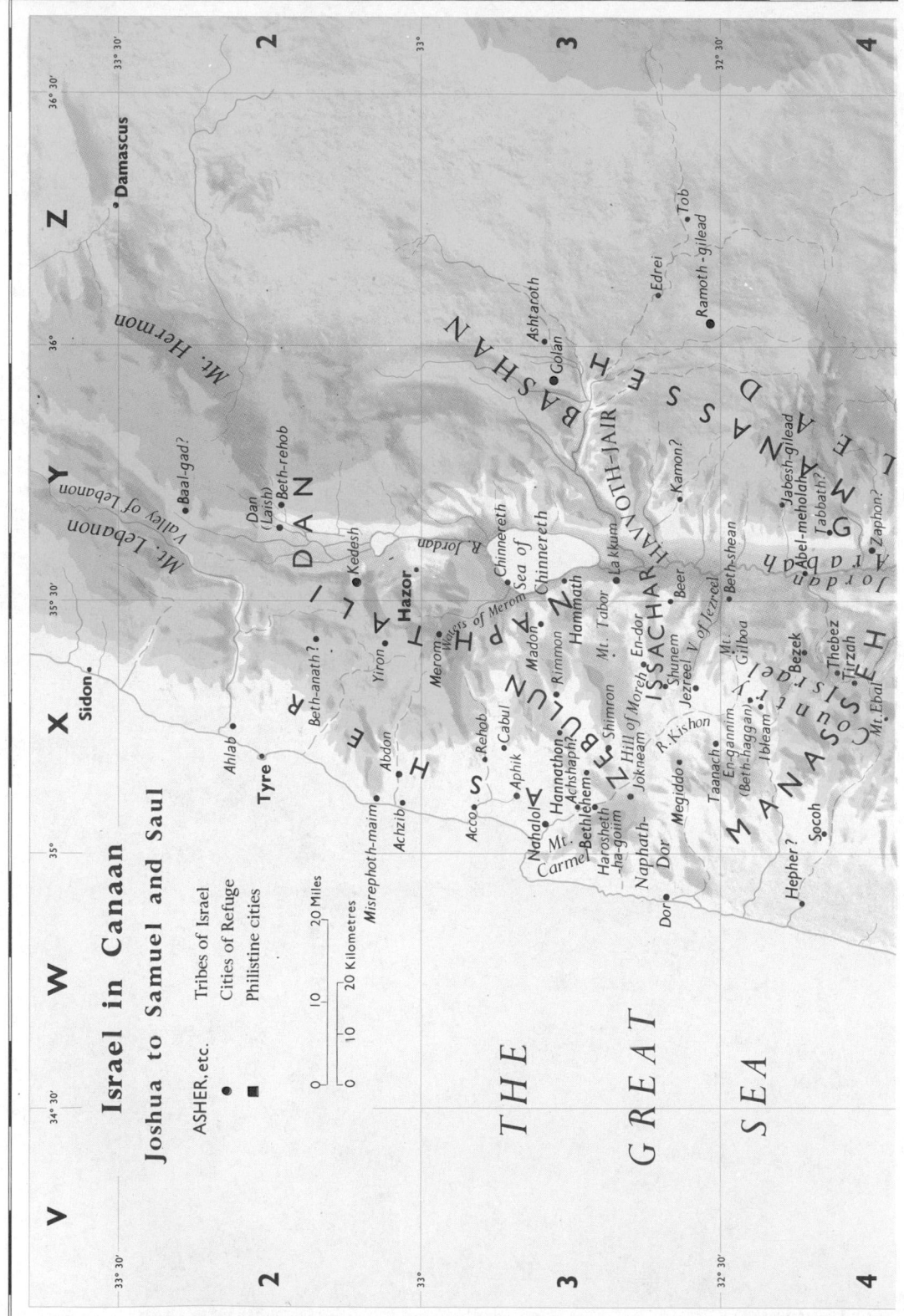

Israel in Canaan

Joshua to Samuel and Saul

ASHER, etc. Tribes of Israel
● Cities of Refuge
■ Philistine cities

20 Miles
20 Kilometres

Sidon

Damascus

Mt. Hermon

Valley of Lebanon

Mt. Lebanon

Baal-gad?

Beth-rehob

Dan (Laish)

Kedesh

Beth-anath?

Yiron

Hazor

Merom

Waters of Merom

Chinnereth

Sea of Chinnereth

R. Jordan

NAPHTALI

DAN

ASHER

Ahlab

Tyre

Abdon

Achzib

Misrephoth-maim

Acco

Rehob

Cabul

Aphik

Nahalol

Bethlehem

Mt. Carmel

Harosheth-ha-goim

Naphath-Dor

Dor

Megiddo

Taanach

Hannathon

Achshaph?

Shimron

Jokneam

ZEBULUN

Madon

Rimmon

Hammath

Mt. Tabor

En-dor

Hill of Moreh

Shunem

Beer

ISSACHAR

Jezreel

V. of Jezreel

Beth-shean

Mt. Gilboa

En-gannim (Beth-haggan)

Ibleam

Socoh

Hepher?

Bezek

Thebez

Tirzah

Mt. Ebal

MANASSEH

Hill Country of Israel

BASHAN

Ashtaroth

Golan

Edrei

Tob

Ramoth-gilead

HAVVOTH-JAIR

Kamon?

La kkum

Jabesh-gilead

Abel-meholah

Tabbath?

Zaphon?

MANASSEH

GILEAD

Arabah

Jordan

THE GREAT SEA

MAP 3

OUTER BORDER SHOWS 10 MILE SUBDIVISIONS

MAP 4

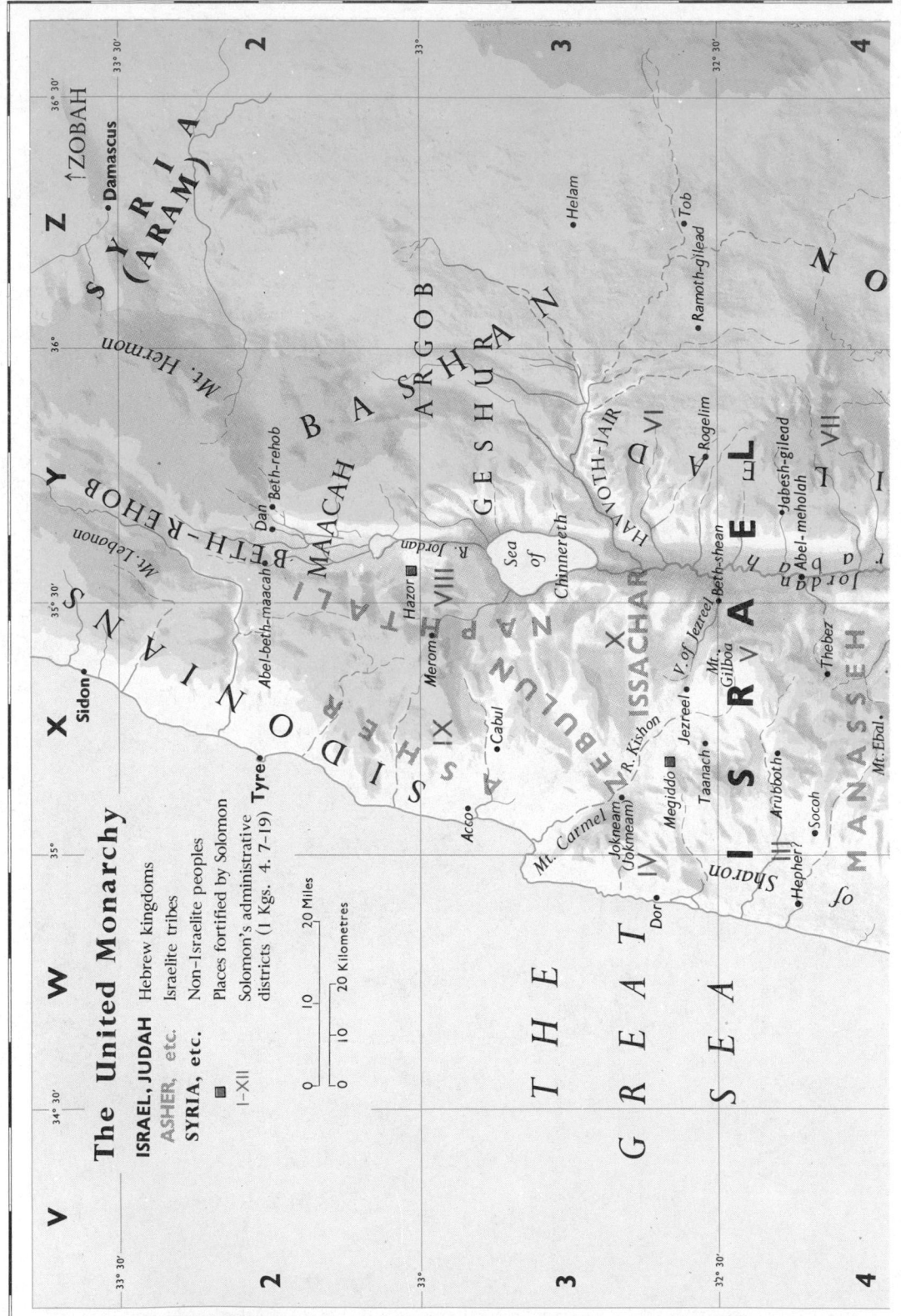

The United Monarchy

ISRAEL, JUDAH Hebrew kingdoms
ASHER, etc. Israelite tribes
SYRIA, etc. Non-Israelite peoples

■ Places fortified by Solomon

I–XII Solomon's administrative districts (1 Kgs. 4. 7–19)

0   10   20 Miles
0   10   20 Kilometres

THE   *GREAT*   *SEA*

Dor•

Mt. Carmel

Jokneam
(Jokmeam)•
R. Kishon

Acco•

•Cabul

Sharon

of MANASSEH

•Hepher?
•Socoh
Arubboth•

Megiddo ■
Taanach•

Jezreel•
Mt. Gilboa

V. of Jezreel

Beth-shean•

•Thebez

Mt. Ebal.

ISSACHAR

ZEBULUN

Merom•

Hazor ■
VIII

Sea of Chinnereth

R. Jordan

NAPHTALI

ASHER

SIDONIANS

Sidon•

Tyre•

Mt. Lebanon

Mt. Hermon

Dan•
•Beth-rehob

Abel-beth-maacah•

BETH-REHOB

MAACAH

BASHAN

ARGOB

GESHUR

HAVOTH-JAIR

VI

•Rogelim

•Ramoth-gilead

•Tob

•Helam

SYRIA
(ARAM)

↑ZOBAH
•Damascus

GILEAD

•Jabesh-gilead
•Abel-meholah

R. Jordan

•Jordan

VII

ISRAEL

III

IV

X

IX

XI

XII

MAP 4

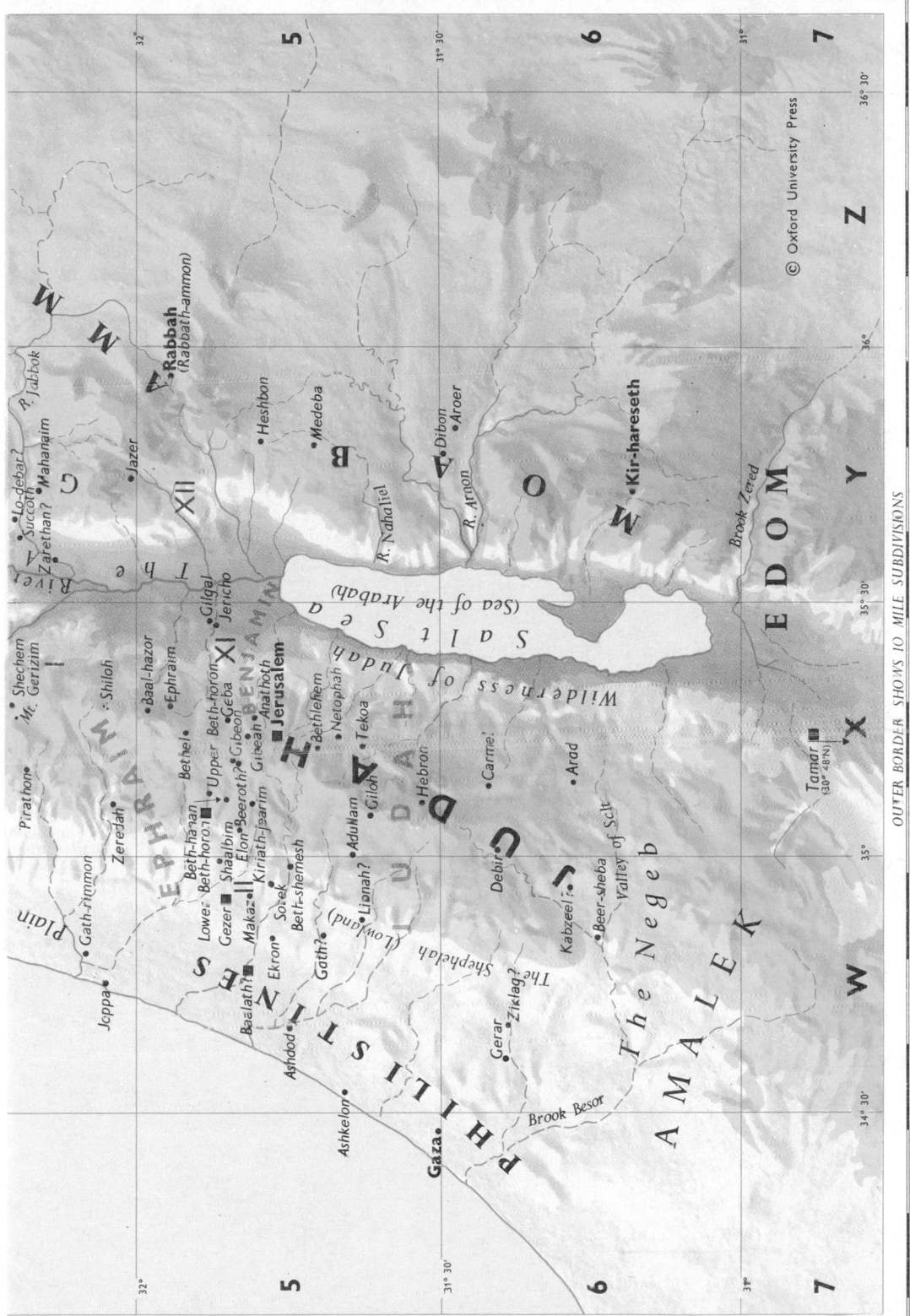

© Oxford University Press

OUTER BORDER SHOWS 10 MILE SUBDIVISIONS

Rabbah
(Rabbath-ammon)

Kir-hareseth

Heshbon
Medeba
Dibon
Aroer
R. Arnon
R. Nahaliel

Jazer

XII

The

Lo-debar?
Succoth
Zarethan?
Zerd
River

Mahanaim
R. Jabbok

AMMON

GAD

REUBEN

EDOM

Brook Zered

Salt Sea
(Sea of the Arabah)

Gilgal
Jericho

Wilderness of Judah

BENJAMIN

XI

Ephraim
Bethel
Baal-hazor
Shiloh
Shechem
Mt. Gerizim

EPHRAIM

Upper Beth-horon
Geba
Gibeah
Anathoth
Jerusalem
Bethlehem
Netophah
Tekoa

JUDAH

Gibeon
Beeroth?
Kiriath-jearim

Pirathon
Zeredah
Gath-rimmon
Joppa

Plain

Beth-haran
Lower Beth-horon
Gezer
Shaalbim
Makaz?
Elon
Ekron
Soreh
Beth-shemesh

PHILISTINES

Baalath?
Gath?
Ashdod

Ashkelon
Gaza

Libnah?
Adullam
Giloh
Hebron

Carmel?
Arad

Debir
Kabzeel?
Beer-sheba
Valley of Salt

The Shephelah (Lowland)

Gerar
Ziklag?
Brook Besor

The Negeb

AMALEK

Tamar
(30° 48′N)

EDOM

MAP 5

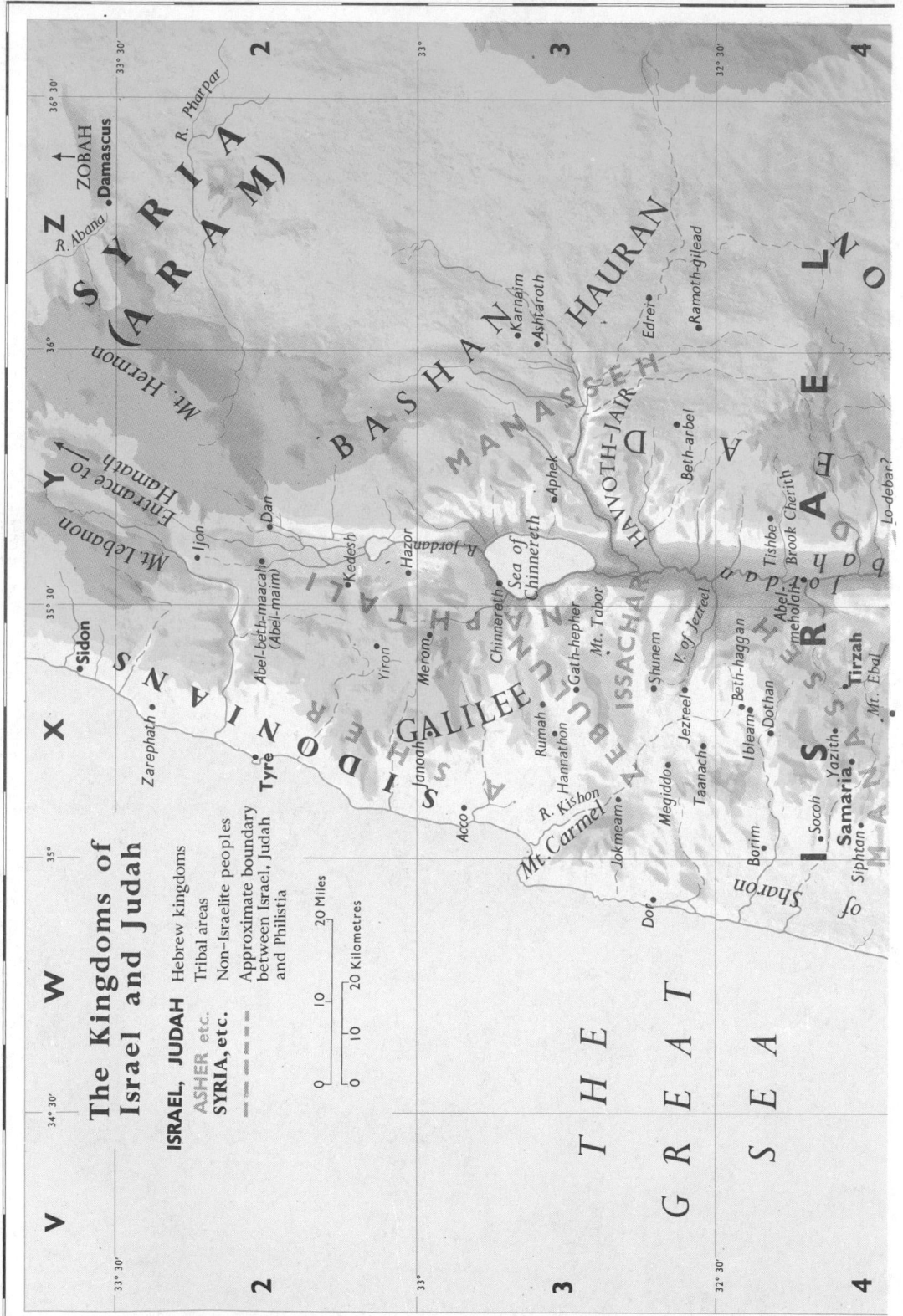

The Kingdoms of
Israel and Judah

ISRAEL, JUDAH Hebrew kingdoms
ASHER etc. Tribal areas
SYRIA, etc. Non-Israelite peoples

– – – – Approximate boundary
between Israel, Judah
and Philistia

20 Miles

20 Kilometres

SYRIA
(ARAM)
ZOBAH • Damascus
R. Abana
R. PharPar
Mt. Hermon
Entrance to Hamath
Mt. Lebanon

BASHAN
HAURAN
Karnaim
Ashtaroth
Edrei
Ramoth-gilead

MANASSEH
HAVVOTH-JAIR
Beth-ar'bel
Brook Cherith
Lo-debar?

Sidon
Zarephath
SIDONIANS
Tyre
Acco
Mt. Carmel
R. Kishon
Jokmeam
Megiddo
Taanach
Dor
Borim
of Sharon

Dan
Ijon
Kedesh
Hazor
Yiron
Abel-beth-maacah (Abel-maim)
Merom
NAPHTALI
Chinnereth
Sea of Chinnereth
Aphek
R. Jordan

GALILEE
Janoah
Rumah
Gath-hepher
Mt. Tabor
ZEBULUN
R. Hannathon
Shunem
V. of Jezreel
Jezreel
Ibleam
Dothan
ISSACHAR

Abel-meholah
GAD
Tishbe
Jordan

ISRAEL
MANASSEH
Socoh
Yazith
Samaria
Tirzah
Mt. Ebal
Siphtan

THE
GREAT
SEA

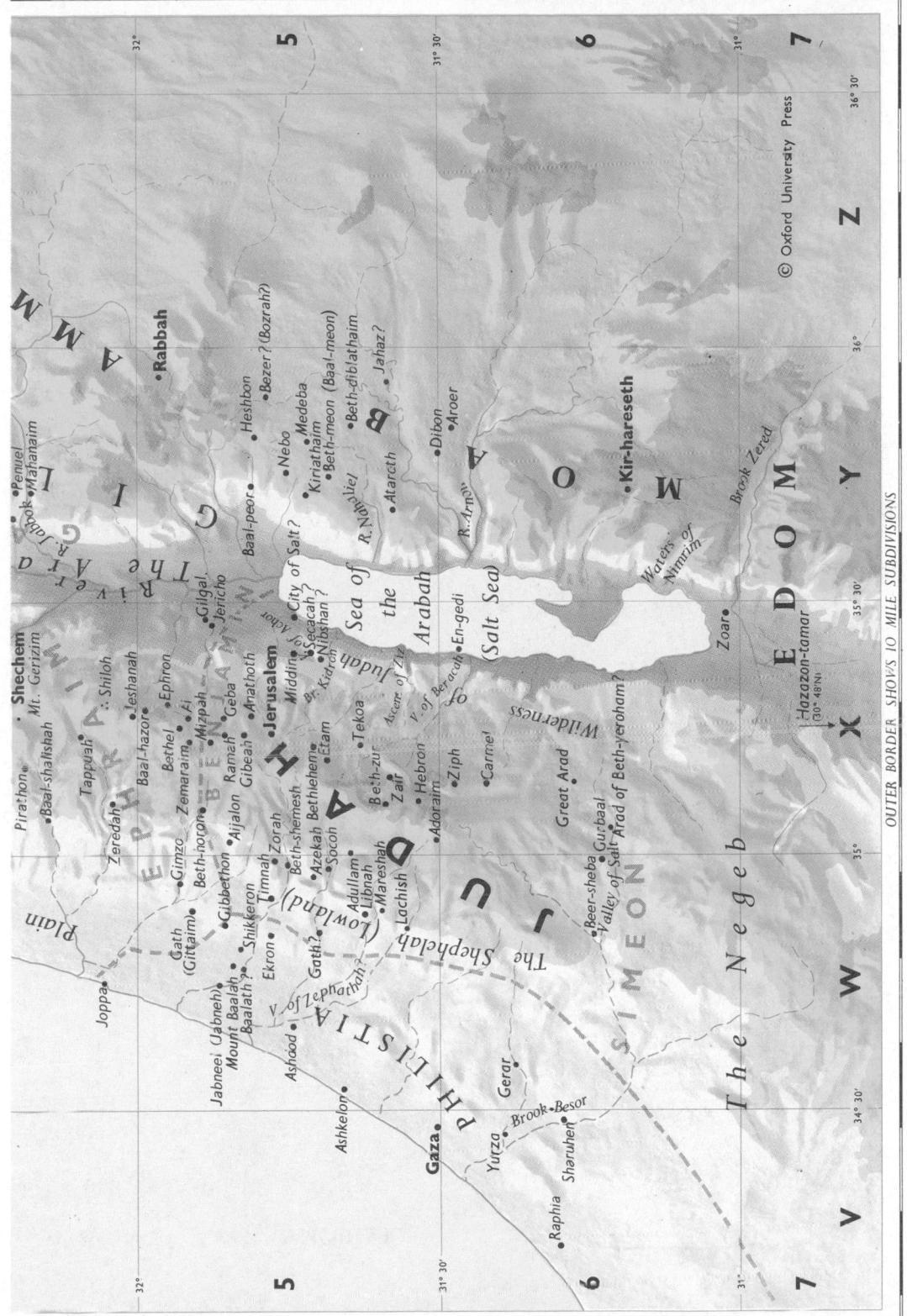

MAP 5

© Oxford University Press

OUTER BORDER SHOWS 10 MILE SUBDIVISIONS

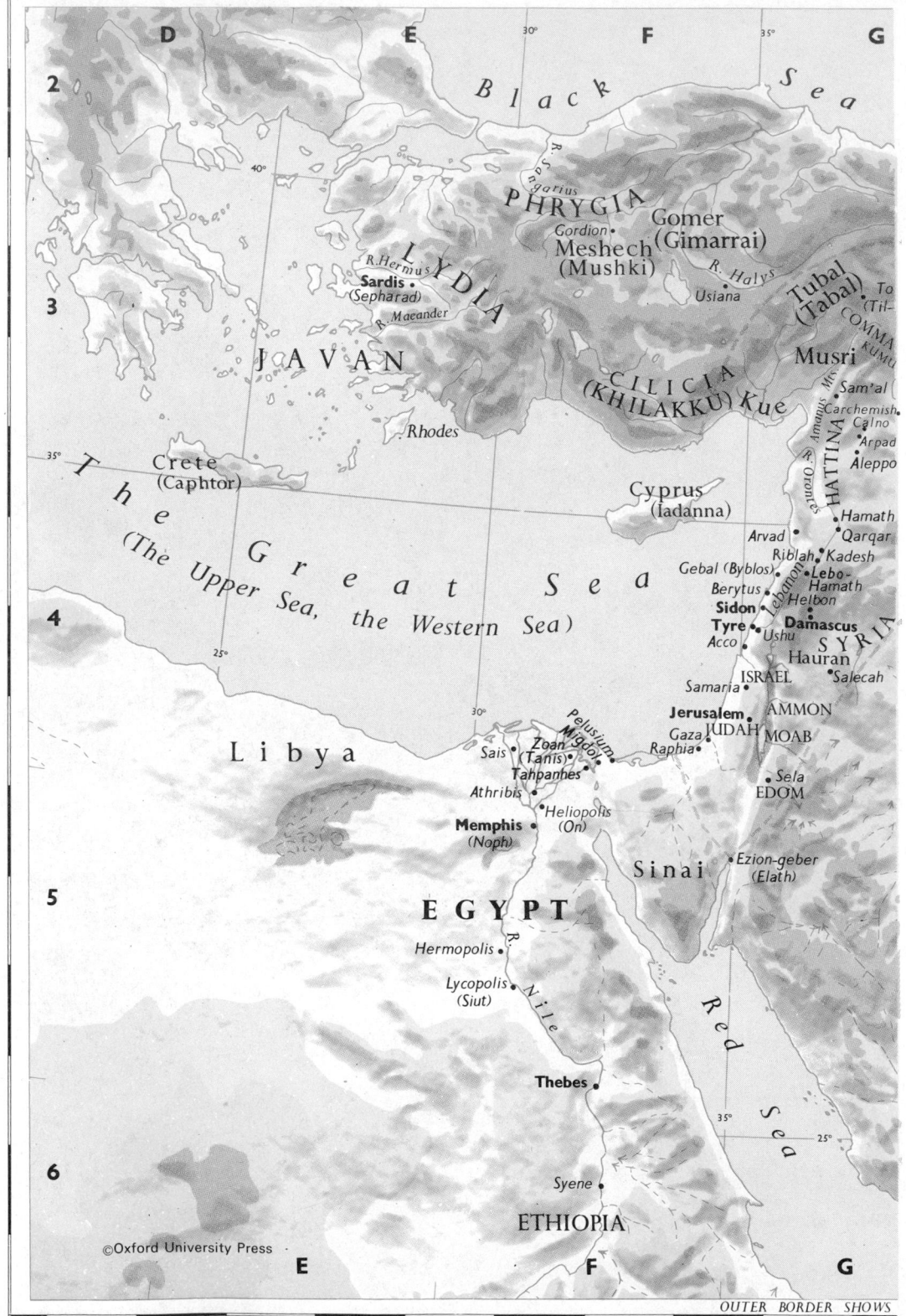

MAP 6

Black Sea

R. Sangarius

PHRYGIA

Gordion

Gomer
(Gimarrai)

Meshech
(Mushki)

R. Halys

Usiana

LYDIA

R. Hermus

Sardis
(Sepharad)

Tubal
(Tabal)

To (Til-

COMMA-
KUMU

R. Maeander

Musri

Sam'al

JAVAN

CILICIA
(KHILAKKU) Kue

Carchemish
Calno
Arpad
Aleppo

HATTINA

R. Orontes

Rhodes

Crete
(Caphtor)

Cyprus
(Iadanna)

Hamath
Qarqar

T h e
G r e a t
S e a

(The Upper Sea, the Western Sea)

Arvad

Riblah Kadesh

Gebal (Byblos)

Berytus

Lebo-
Hamath
Helbon

Sidon

Tyre

Damascus

Acco

Ushu

SYRIA

Hauran

Samaria

ISRAEL

Salecah

L i b y a

Sais

Pelusium
Migdol

Zoan
(Tanis)

Tahpanhes

Athribis

Jerusalem

Gaza

Raphia

AMMON

JUDAH MOAB

Sela

EDOM

Heliopolis
(On)

Memphis
(Noph)

S i n a i

Ezion-geber
(Elath)

E G Y P T

Hermopolis

R. Nile

Lycopolis
(Siut)

Red Sea

Thebes

Syene

ETHIOPIA

©Oxford University Press

MAP 6

H   **J**   **K**

Caspian Sea

ARARAT
(URARTU)

*garmah
garimmu)*
• *Milid
(Melitene)*
GENE
Khur
L. Van
Naïri
• *Turushpa
(Tuspar)*

L. Urmia
Minni
(Mannai)

A S S Y R I A

R. Balikh
• Haran   Gozan
Beth-eden
(Bit-adini)
• Tiphsah
• Rezeph

Dur-sharrukin
Nineveh
Calah
Asshur
Upper Zab
• *Arbela*
Lower Zab
• *Arrapkha*

MADAI
(MEDES)
• *Ecbatana*

• *Tadmor
(Tadmar)*

R. Habor
R. Euphrates
R. Tigris
R. Adhaim
R. Diyala

K e d a r
(Qidri)

Sippar •
Cuthah •
Babylon
Borsippa

*Pekod
(Puqudu)*

E L A M
• *Susa (Shushan)*

A
R
A
B
I
A

• *Nippur*

BABYLONIA
*Erech
(Uruk)* • *Larsa*
Ur •

The Lower (Eastern) Sea

• *Dumah*

• *Tema*

The Near East

in the time of the

Assyrian Empire

Approximate extent of Assyrian domination
in the latter part of the 8th. century.

(Later, under Esarhaddon (681–669), Assyria conquered Egypt.)

• *Dedan*

0   100   200 Miles

0   100   200 Kilometres

SHEBA
(SABA) OPHIR

H   **J**

© Oxford University Press

100 MILE SUBDIVISIONS

MAP 7

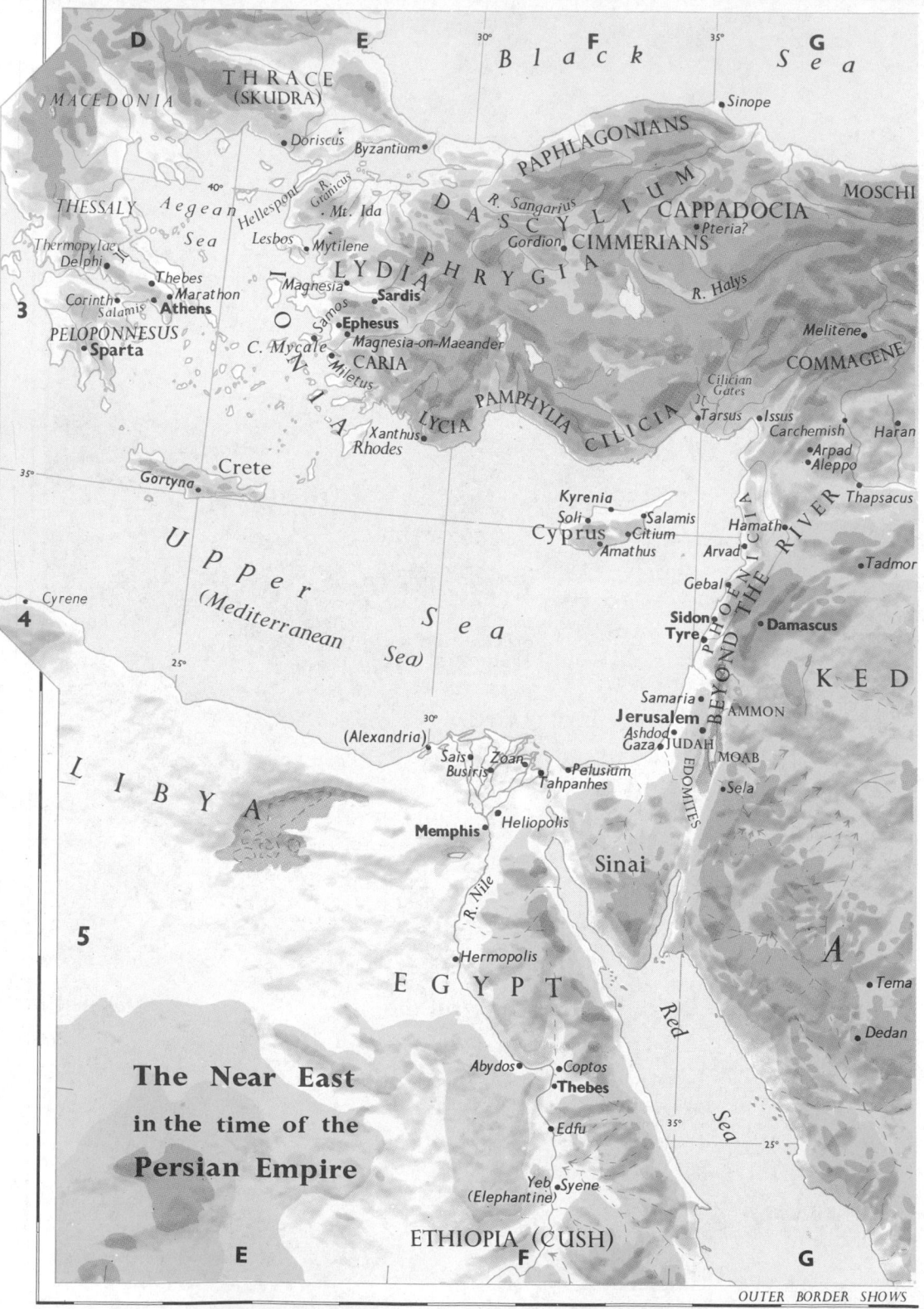

The Near East

in the time of the

Persian Empire

MAP 7

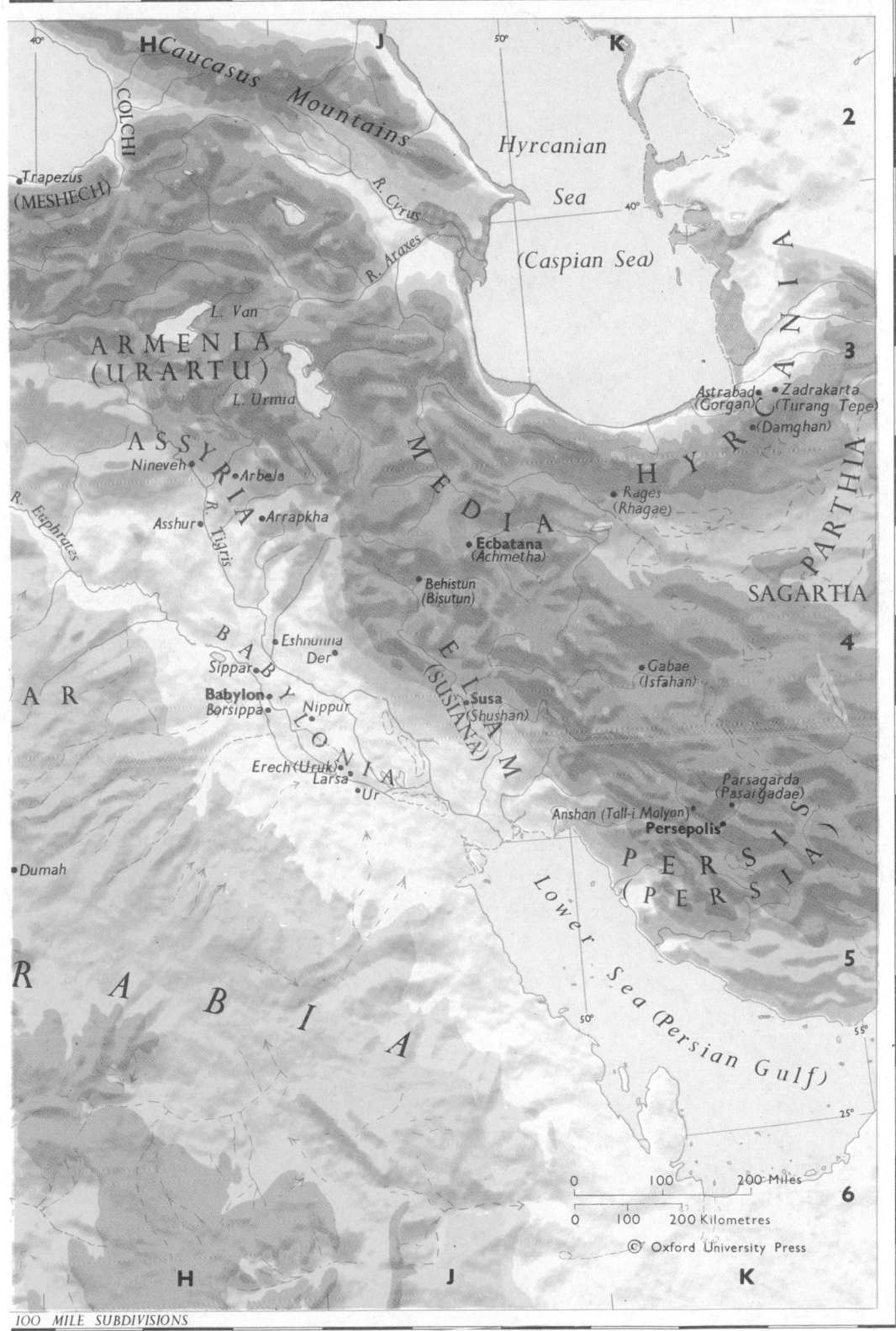

H   J   K

40°

Caucasus Mountains

COLCHI

Trapezus
(MESHECH)

R. Cyrus

R. Araxes

Hyrcanian

Sea

(Caspian Sea)

2

40°

3

L. Van

ARMENIA
(URARTU)

L. Urmia

HYRCANIA

Astrabad
(Gorgan)

Zadrakarta
(Turang Tepe)

(Damghan)

ASSYRIA

Nineveh

Arbela

Asshur

R. Tigris

Arrapkha

MEDIA

Rages
(Rhagae)

PARTHIA

SAGARTIA

Ecbatana
(Achmetha)

R. Euphrates

Behistun
(Bisutun)

BABYLONIA

Eshnunna
Der

Sippar

ELAM
(SUSIANA)

Susa
(Shushan)

Gabae
(Isfahan)

4

AR

Babylon
Borsippa

Nippur

Erech (Uruk)
Larsa

Ur

Parsagarda
(Pasargadae)

Anshan (Tall-i Malyan)
Persepolis

PERSIS
(PERSIA)

Dumah

ARABIA

Lower Sea (Persian Gulf)

50°

5

55°

25°

0   100   200 Miles

0   100   200 Kilometres

© Oxford University Press

6

H   J   K

100 MILE SUBDIVISIONS

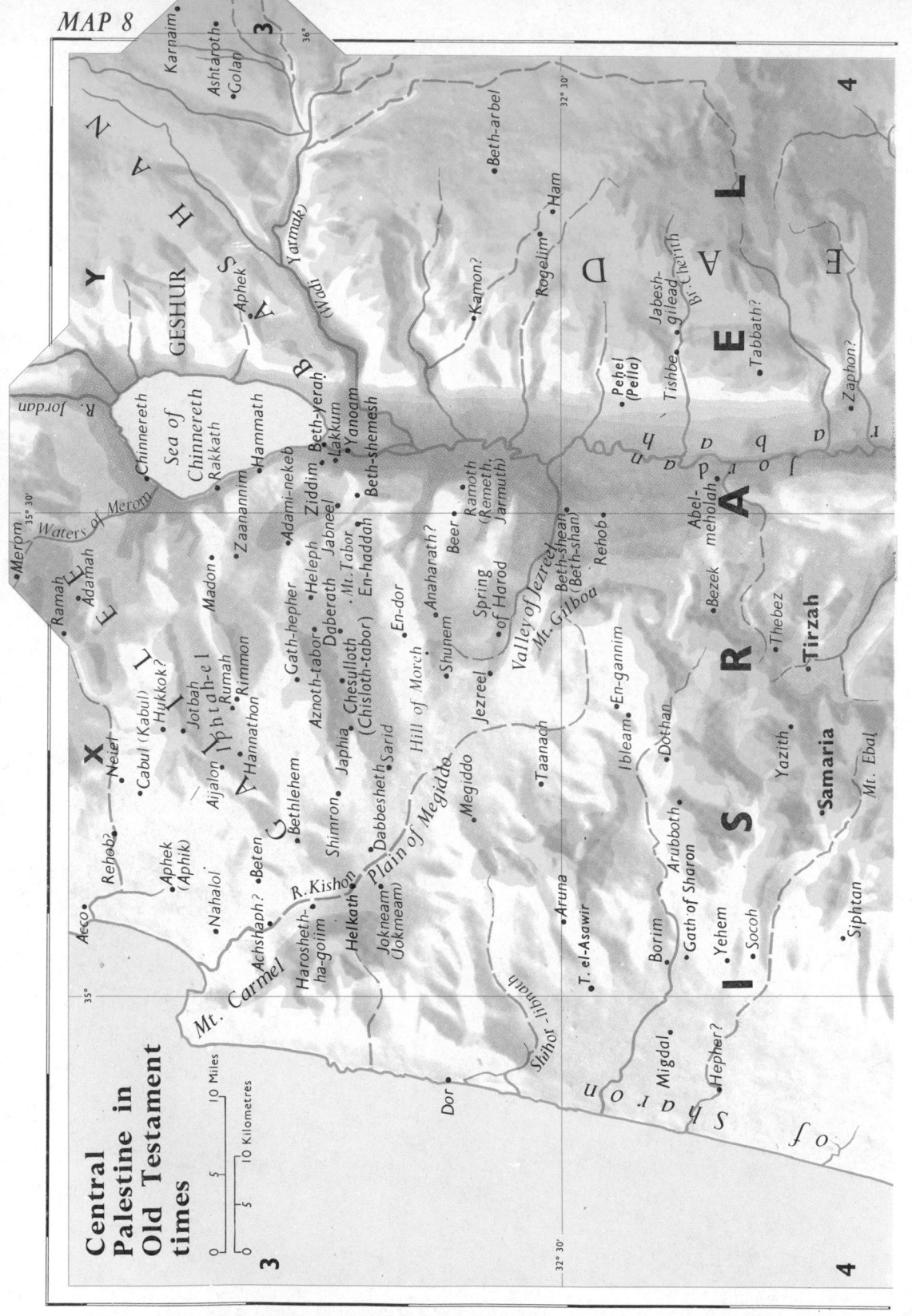

MAP 8

3   36°

4

32° 30'

Karnaim

Ashtaroth
Golan

Y N
A
H
S
GESHUR
A
S
B
E
Aphek
(Wadi) Yarmuk

Beth-arbel

Kamon?

Rogelim   Ham

Pehel
(Pella)   Tishbe
Jabesh-
gilead

D
L
A
E
G
Tabbath?
Zaphon?

R. Jordan

Merom

Sea of
Chinnereth
Rakkath   Chinnereth
Hammath

Waters of Merom
35° 30'

Merom
Ramah   Adamah

Beth-yerah
Beth-shemesh
Zaananim
Adami-nekeb   Ziddim
Lakkum
Yanoam   Hammath

E
Nahalol
Madon   Rumah
Rimmon

X
Neiel   Cabul (Kabul)
Hukkok?
Jotbah
Aijalon   Jiphtah-el
Hannathon

Y

Z
Rehob?
Aphek
(Aphik)   Beten
Bethlehem
Shimron
Japhia
Dabbesheth
Sarid

Gath-hepher
Aznoth-tabor   Heleph
Daberath
Chesulloth
(Chisloth-tabor)   Jabneel
Mt. Tabor
En-haddah

En-dor

Hill of Moreh
Shunem
Jezreel   Anaharath?
Beer
Spring
of Harod
Ramoth
(Remeth,
Jarmuth)

Valley of Jezreel   Beth-shean
(Beth-shan)
Rehob
Mt. Gilboa
Bezek

Abel-
meholah

Jordan

32° 30'

Acco

Achshaph?   R. Kishon
Harosheth-
ha-goiim
Helkath   Jokneam
(Jokmeam)   Plain of Megiddo
Megiddo
Taanach

Ibleam   Dothan
En-gannim

Arubboth
Gath of Sharon   Yazith
Samaria
Mt. Ebal

Tirzah
Thebez

I   S   R   A   E   L

Mt. Carmel

Shihor-libnath

Aruna
T. el-Asawir

Borim   Yehem
Socoh
Siphtan

Dor

Migdal   Hepher?

Sharon

of

35°

Central
Palestine in
Old Testament
times

10 Miles
5   10 Kilometres
0   5   0

3

4

MAP 8

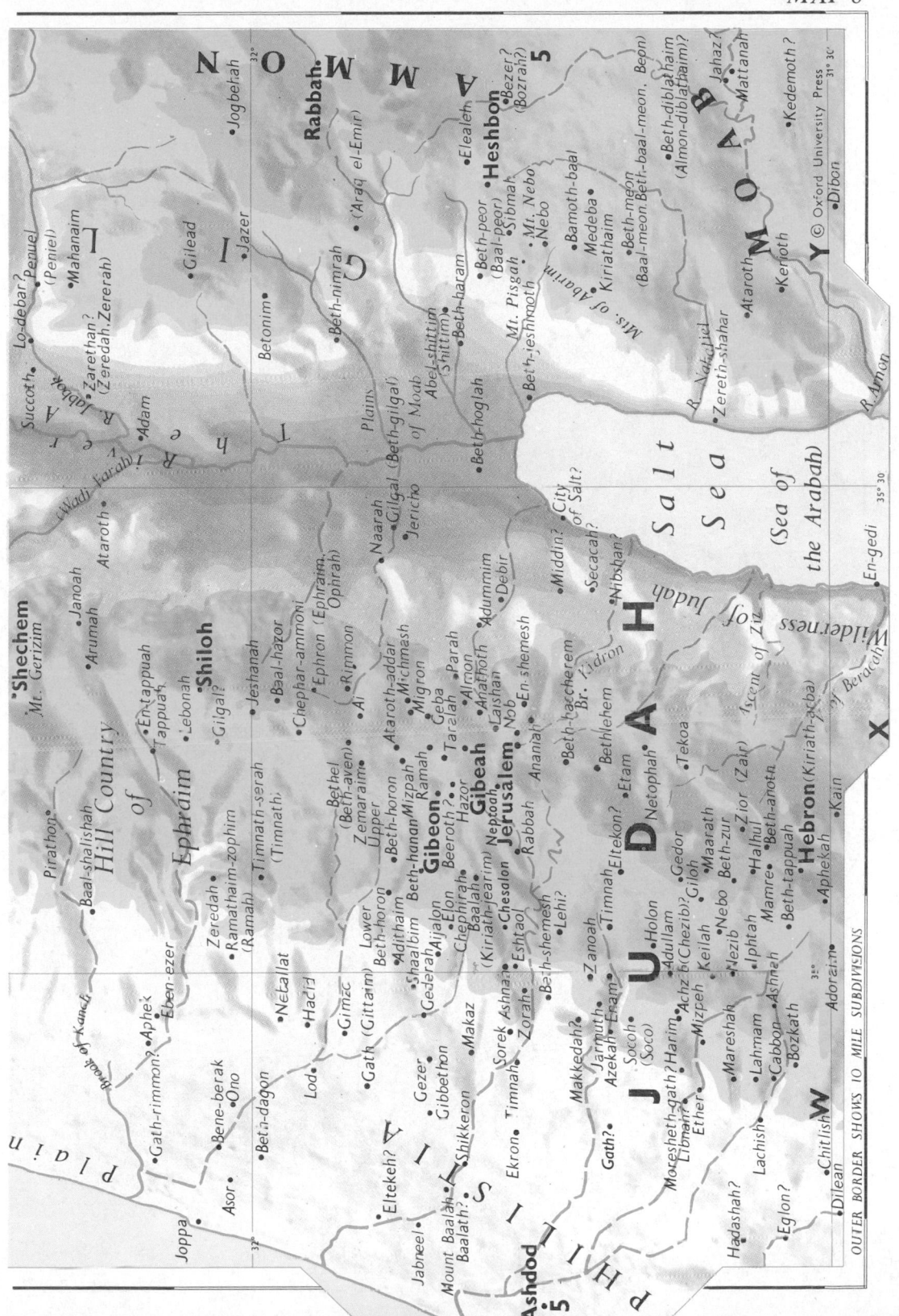

AMMON

Rabbah.

5

•Jogbehah

Mahanaim
(Peniel)
Lo-debar? •Penuel

Succoth•
Zarethan?
(Zeredah-Zererah)

•Adam

Gilead

•Jazer

(Araq el-Emir)

GILEAD

Beth-nimrah•

Betonim•

Abel-shittim
(Shittim)

Beth-haram•

Beth-peor•
(Baal-peor)

Mt. Pisgah.
Beth-jeshimoth•

Heshbon

Elealeh•

Sibmah•

•Baal-meon
Beth-meon.Beth-baal-meon.

Mt. Nebo
•Nebo

Bamoth-baal•
Medeba•

Kiriathaim•

Beth-diblathaim•
(Almon-diblathaim?)

Bezer?
(Bozrah?)

•Beon

Jahaz?
•Mattanah

•Kedemoth?

MOAB

Ataroth•

•Kerioth

•Dibon

The Plain of Moab

Gilgal (Beth-gilgal)

Beth-hoglah•

Mts. of Abarim

R. Nahaliel

Zereth-shahar•

R. Arnon

Salt Sea

(Sea of the Arabah)

City of Salt?

Middin?•

Secacah?•

Nibshan?•

Wilderness of Judah

Ziz

•En-gedi

Shechem
Mt. Gerizim

Janoah•

Arumah•

Hill Country of Ephraim

Shiloh

Lebonah•

Gilgal?•

Jeshanah•

Baal-hazor•

•Ephron (Ephraim, Ophrah)

Chephar-ammoni•

Naarah•

Gilgal•
Jericho•

En-tappuah
•Tappuah•

•Pirathon

Baal-shalishah•

Zeredah•

Ramathaim-zophim
(Ramah)

Timnath-serah
(Timnath)

Bethel
(Beth-aven)•

Zemaraim•
Upper Beth-horon•

Ai•

Rimmon•

Michmash•

Migron•

Geba•
Gibeah

Ataroth-addar•

Ramah•

Hazor•
Beeroth?•
Baalah (Kiriath-jearim)•

•Tar-elah

•Almon
•Anathoth
Nob•

•Parah

Adummim•

Debir•

Laishah•
En-shemesh•

JUDAH

Beth-haccherem•

Br. Kidron

Beth-anoth•

Beth-haccherem

Anatoth

Naaran
Mizpah•
Beth-hanan•

Chephirah•
Elon•
Ajjalon•
Shaalbim•
Gederah•

Makaz•

Lower Beth-horon•

Adithaim•

Beth-shemesh•

Chesalon (Kiriath-jearim)•

Eshtaol•
Zorah•

Lehi?•

Jerusalem

Rabbah•

Ananiah•

Bethlehem•

Netophah•

Etam•

Tekoa•

Zior (Zair)•

Halhul•

Mamre•
Beth-zur•

Hebron (Kiriath-arba)•

Ascent of

Valley of Berachah

Kain•

Aphekah•

Wadi Farah

Joppa

Asor•

Plain of Sharon

Brook of Kanah

Gath-rimmon?•

Bene-berak•
Ono•

Aphek•

Eben-ezer•

Beth-dagon•

Lod•

Neballat•

Hadid•

Gimzo•

Gezer•

Gath (Gittaim)•

Gibbethon•
Shikkeron•

Ekron•
Timnah•

Sorek•
Ashnah•

Eltekeh?•

Mount Baalah•
Baalath?•

Gath?•

Makkedah?•
Azekah•
Zanoah•

Jarmuth•
Eram•

Socoh (Soco)•

Zanoah•

Adullam•

Harim•
Achzib (Chezib)?•

Moresheth-gath?•
Libnah?•
Ether•

Keilah•

Gedor•
Nezib•
Iphtah•

Mareshah•

Lahmam•

Lachish•

Eglon?•

Dilean•

Chitlish•

Gath-zur•

Bozkath•

Cabbon•
Ashnah•

Beth-tappuah•

Maarath•

Holon•

Eltekon?•

Timnah•

Beth-shemesh•

Ashdod
5

PHILISTIA

Jabneel•

Hadashah?•

Gedor•
Nebo•

Adoraim•

Dumah•

PHILISTIA

MAP 9

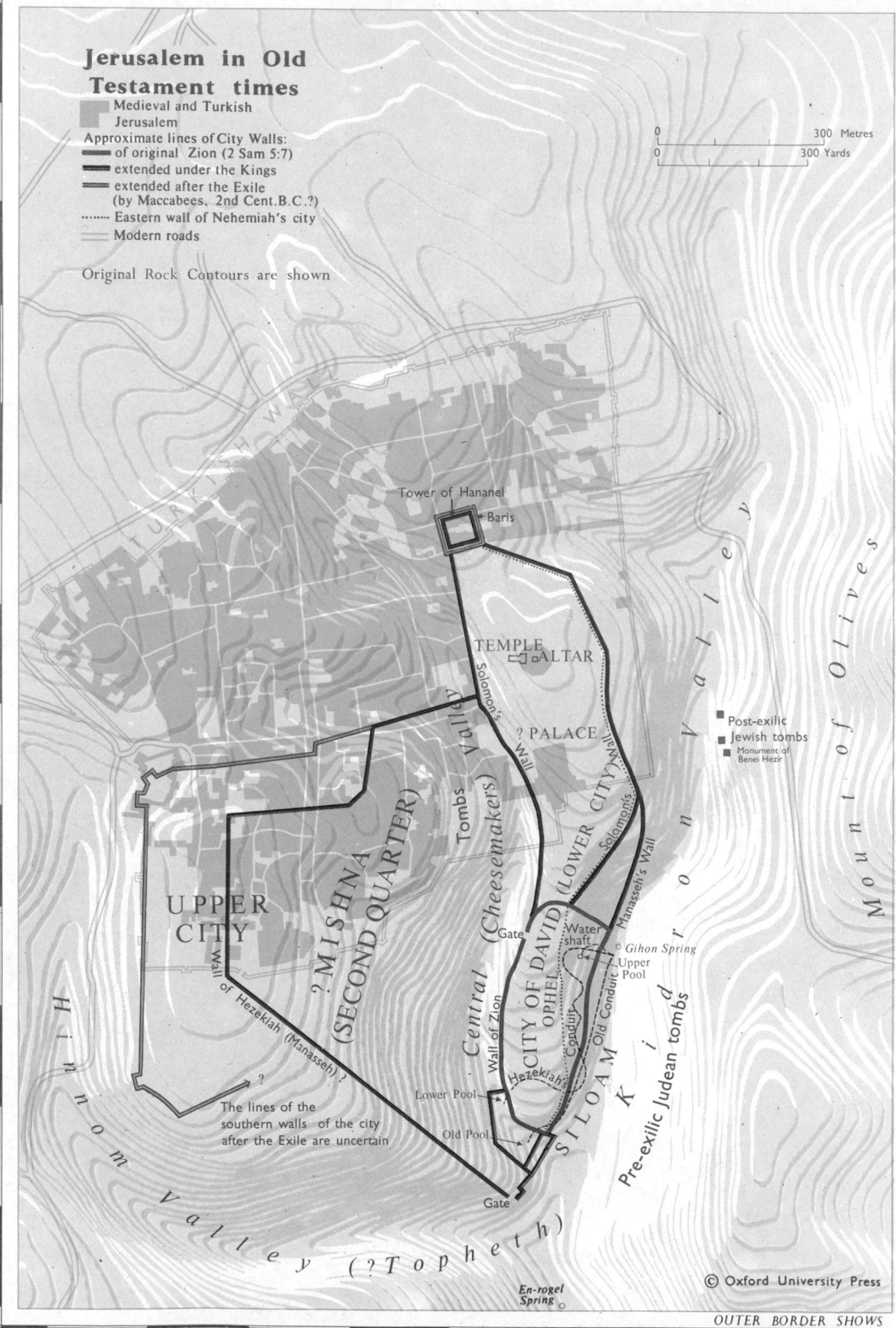

Jerusalem in Old Testament times

Medieval and Turkish Jerusalem

Approximate lines of City Walls:
- of original Zion (2 Sam 5:7)
- extended under the Kings
- extended after the Exile (by Maccabees, 2nd Cent.B.C.?)
- Eastern wall of Nehemiah's city
- Modern roads

Original Rock Contours are shown

0   300 Metres
0   300 Yards

Tower of Hananel
Baris

TEMPLE ☐ ALTAR

? PALACE

Post-exilic Jewish tombs
Monument of Benei Hezir

Solomon's Wall

Tombs

Solomon's Wall

(LOWER CITY)

Manasseh's Wall

Valley

Central (Cheesemakers) Valley

UPPER CITY

?MISHNA (SECOND QUARTER)?

Wall of Hezekiah (Manasseh)?

CITY OF DAVID
OPHEL

Gate

Water shaft

○ Gihon Spring
Upper Pool

Wall of Zion

Hezekiah's Conduit

Old Conduit

Mount of Olives

Kidron Valley

SILOAM

Pre-exilic Judean tombs

The lines of the southern walls of the city after the Exile are uncertain

Lower Pool

Old Pool

Gate

Hinnom valley (?Topheth)

En-rogel Spring

© Oxford University Press

MAP 9

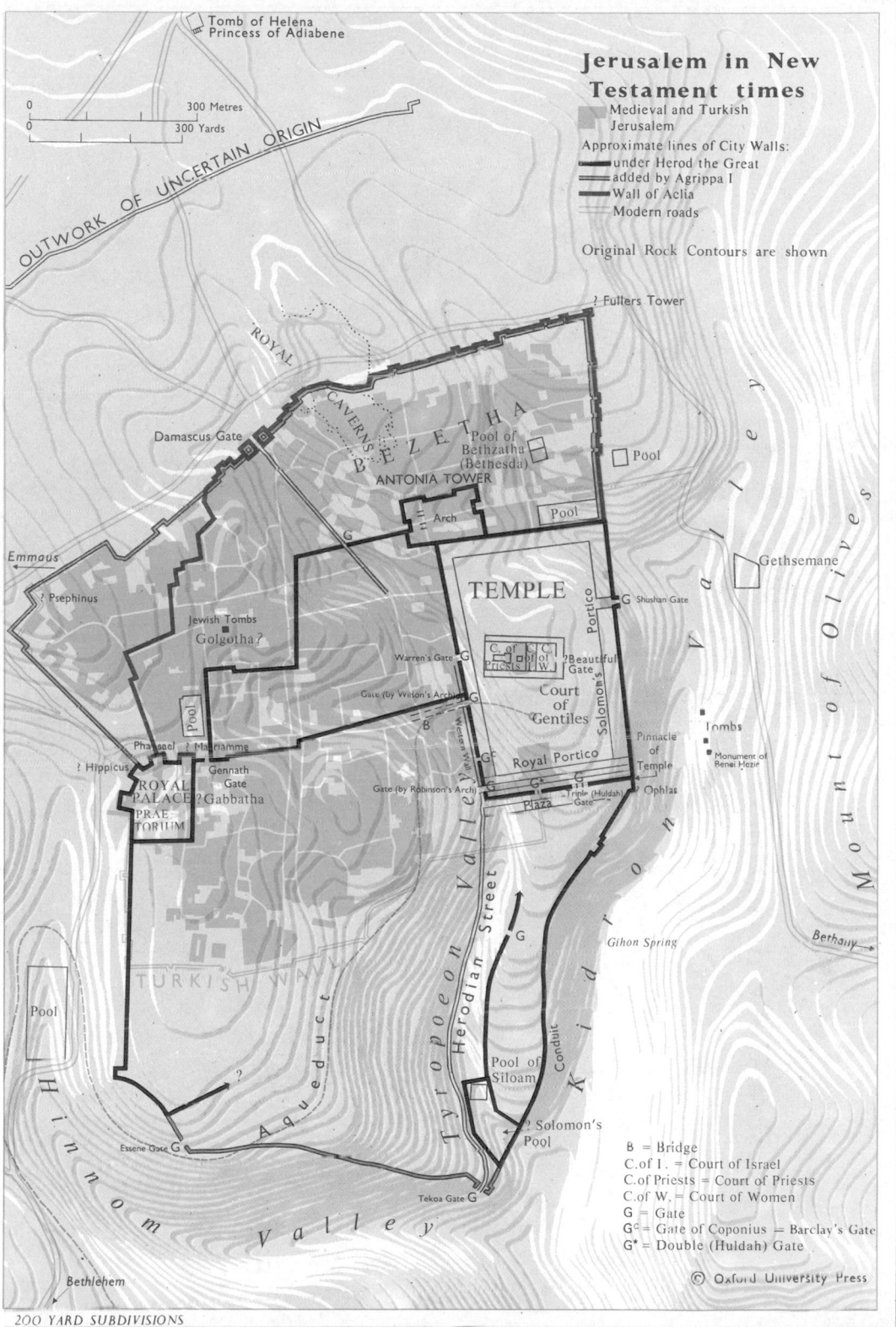

Tomb of Helena
Princess of Adiabene

OUTWORK OF UNCERTAIN ORIGIN

0   300 Metres
0   300 Yards

**Jerusalem in New
Testament times**

Medieval and Turkish
Jerusalem
Approximate lines of City Walls:
under Herod the Great
added by Agrippa I
Wall of Aelia
Modern roads

Original Rock Contours are shown

'ROYAL

CAVERNS

B E Z E T H A

? Fullers Tower

Damascus Gate

Pool of
Bethzatha
(Bethesda)

Pool

ANTONIA TOWER

Pool

Arch

Pool

Gethsemane

Emmaus

? Psephinus

Jewish Tombs
Golgotha ?

TEMPLE

Portico

Shushan Gate

C. of
Priests | C. | C.
I. | W.J
?Beautiful
Gate

Warren's Gate

Court
of
Gentiles

Solomon's
Portico

Gate (by Wilson's Arch)

Tombs

B

Monument of
Benei Hezir

Pinnacle
of
Temple

? Hippicus

Pool

Phasael ? Mariamme

Royal Portico

Gennath
Gate

ROYAL
PALACE ?Gabbatha
PRAE-
TORIUM

Gate (by Robinson's Arch)

Triple (Huldah)
Gate

Plaza

Ophlas

TURKISH WALL

Gihon Spring

Bethany

Pool

Aqueduct

Tyropoeon Valley

Herodian Street

Kidron Valley

Mount of Olives

Essene Gate G

Hinnom

Pool of
Siloam

Conduit

Valley

? Solomon's
Pool

Tekoa Gate G

B = Bridge
C.of I . = Court of Israel
C.of Priests = Court of Priests
C.of W. = Court of Women
G = Gate
G° = Gate of Coponius = Barclay's Gate
G* = Double (Huldah) Gate

Bethlehem

© Oxford University Press

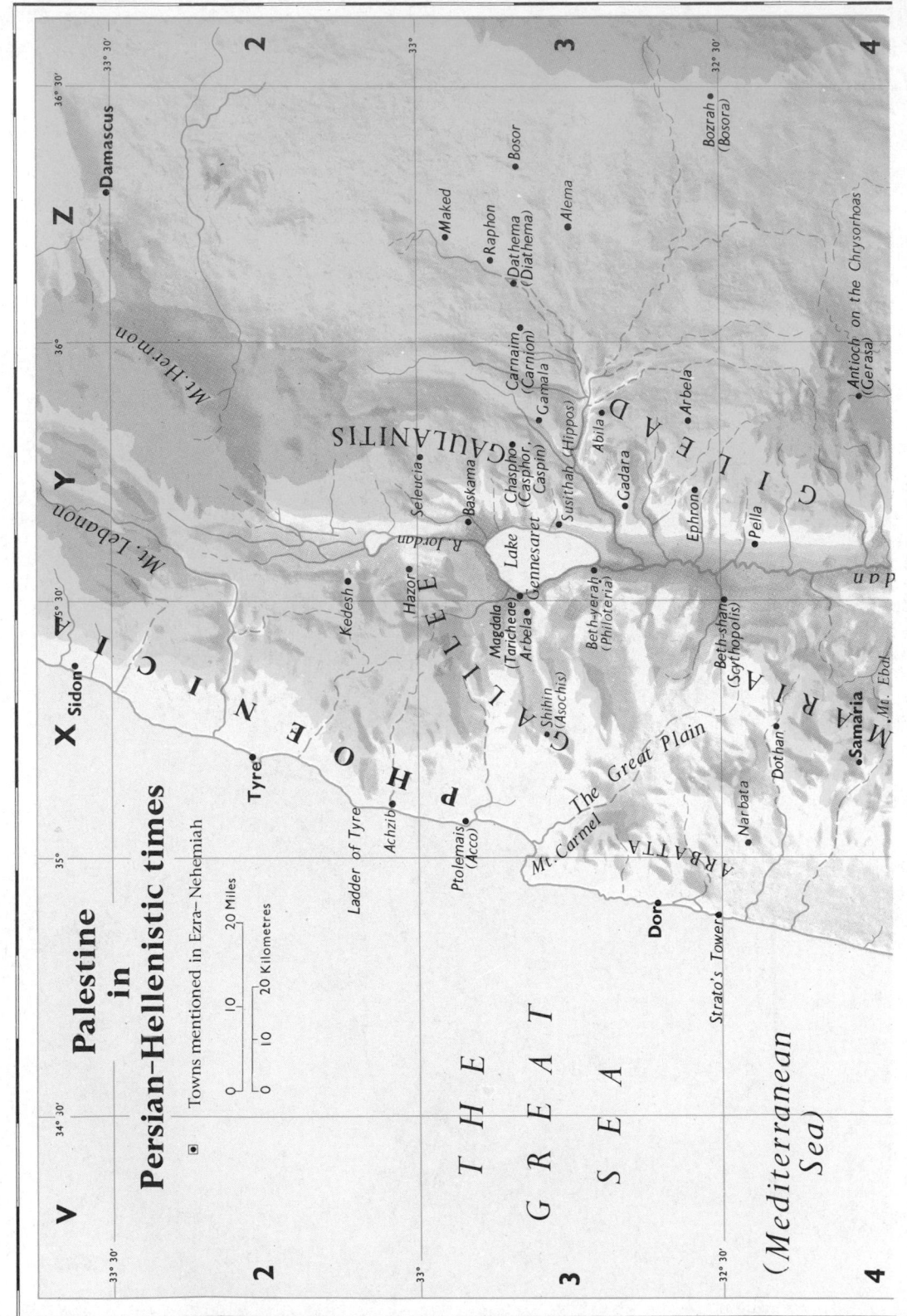

MAP 10

Palestine
in
Persian–Hellenistic times

▣ Towns mentioned in Ezra–Nehemiah

20 Miles
20 Kilometres

Damascus

Maked
Raphon
Dathema
(Diathema)
Bosor
Alema
Bozrah
(Bosora)

Antioch on the Chrysorhoas
(Gerasa)

Carnaim
(Carnion)
Gamala
Arbela
Abila
Susitah (Hippos)
Gadara
Ephron
Pella

GAULANITIS

Seleucia
Baskama
Chaspho,
Casphor,
Caspin)

R. Jordan

Lake
Gennesaret

Kedesh
Hazor
Magdala
(Tarichea)
Arbela

Beth-yerah
(Philoteria)

Beth-shan
(Scythopolis)

Shihin
(Asochis)

GALILEE

Mt. Hermon

Mt. Lebanon

Sidon

Tyre

Achzib

Ladder of Tyre

Ptolemais
(Acco)

Mt. Carmel

The Great Plain

Narbata

Dothan

Samaria

Mt. Ebal

Dor

Strato's Tower

ARBATTA

SAMARIA

GILEAD

PHOENICIA

THE
GREAT
SEA

(Mediterranean
Sea)

MAP 10

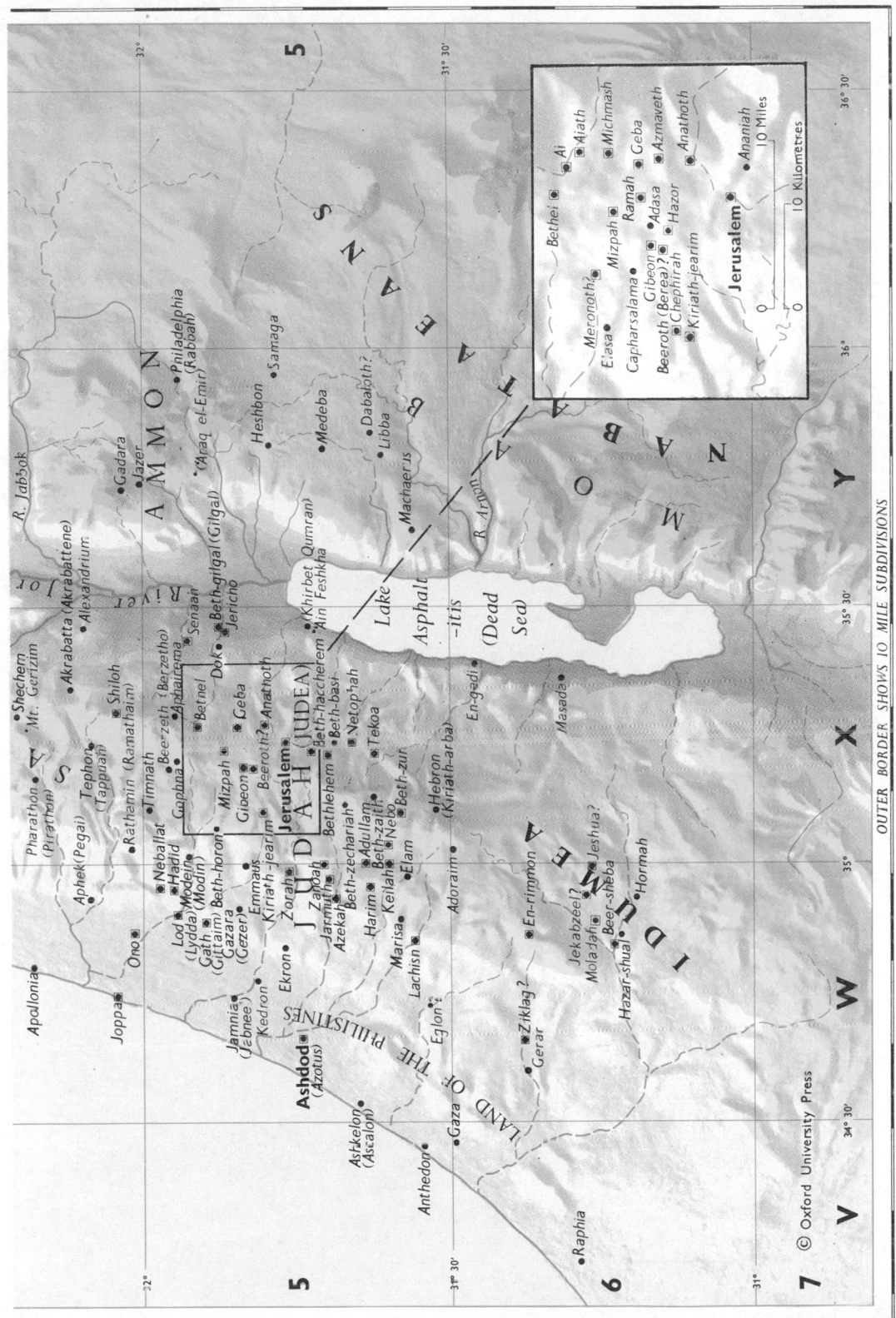

OUTER BORDER SHOWS 10 MILE SUBDIVISIONS

© Oxford University Press

MAP 11

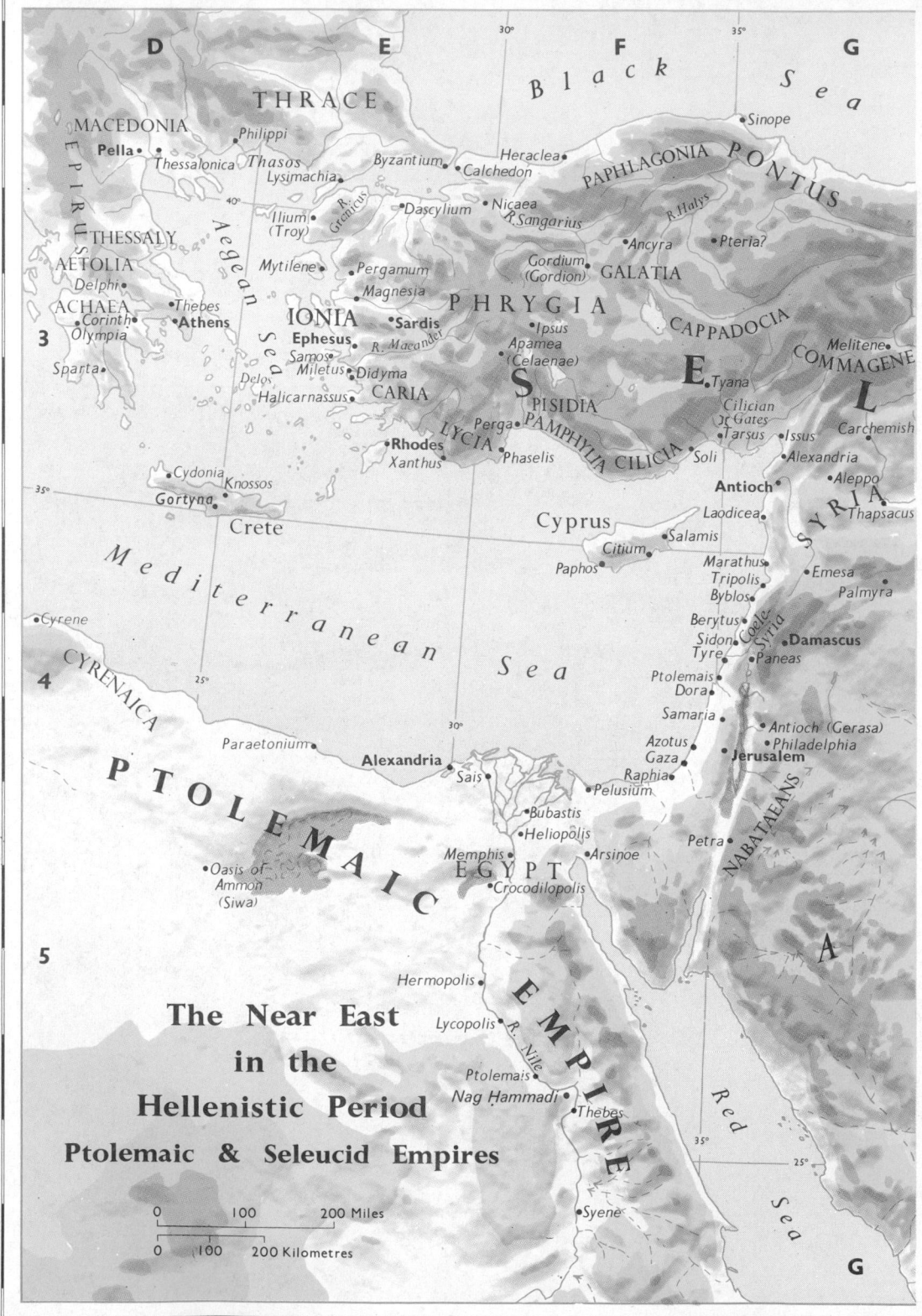

The Near East
in the
Hellenistic Period

Ptolemaic & Seleucid Empires

MAP 11

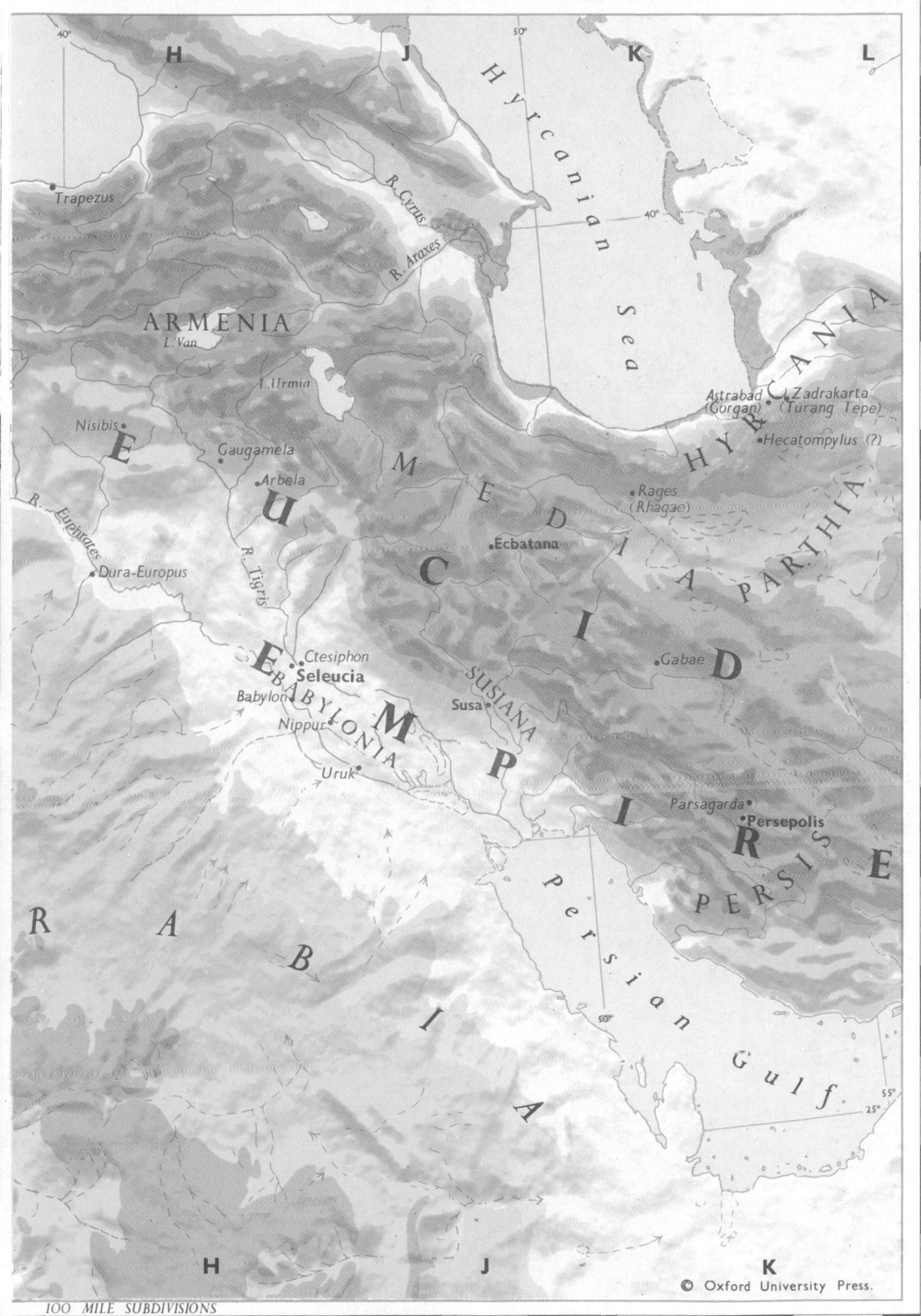

40°

H   J   50°   K   L

Hyrcanian Sea

•Trapezus

R. Cyrus

40°

R. Araxes

ARMENIA

L. Van

L. Urmia

HYRCANIA

Nisibis•

E

Gaugamela•

•Arbela

•Astrabad (Gorgan)   •Zadrakarta (Turang Tepe)

•Hecatompylus (?)

U

M

E

D

I

A

PARTHIA

R. Euphrates

•Rages (Rhagae)

•Ecbatana

C

R. Tigris

•Dura-Europus

•Gabae

D

•Ctesiphon

E

Seleucia

Babylon•

SUSIANA

M

•Susa

Nippur•

BABYLONIA

P

•Uruk

Parsagarda•

•Persepolis

R

I

PERSIS

E

R

A

B

Persian Gulf

I

A

50°

55°

25°

H   J   K

© Oxford University Press.

100 MILE SUBDIVISIONS

MAP 12

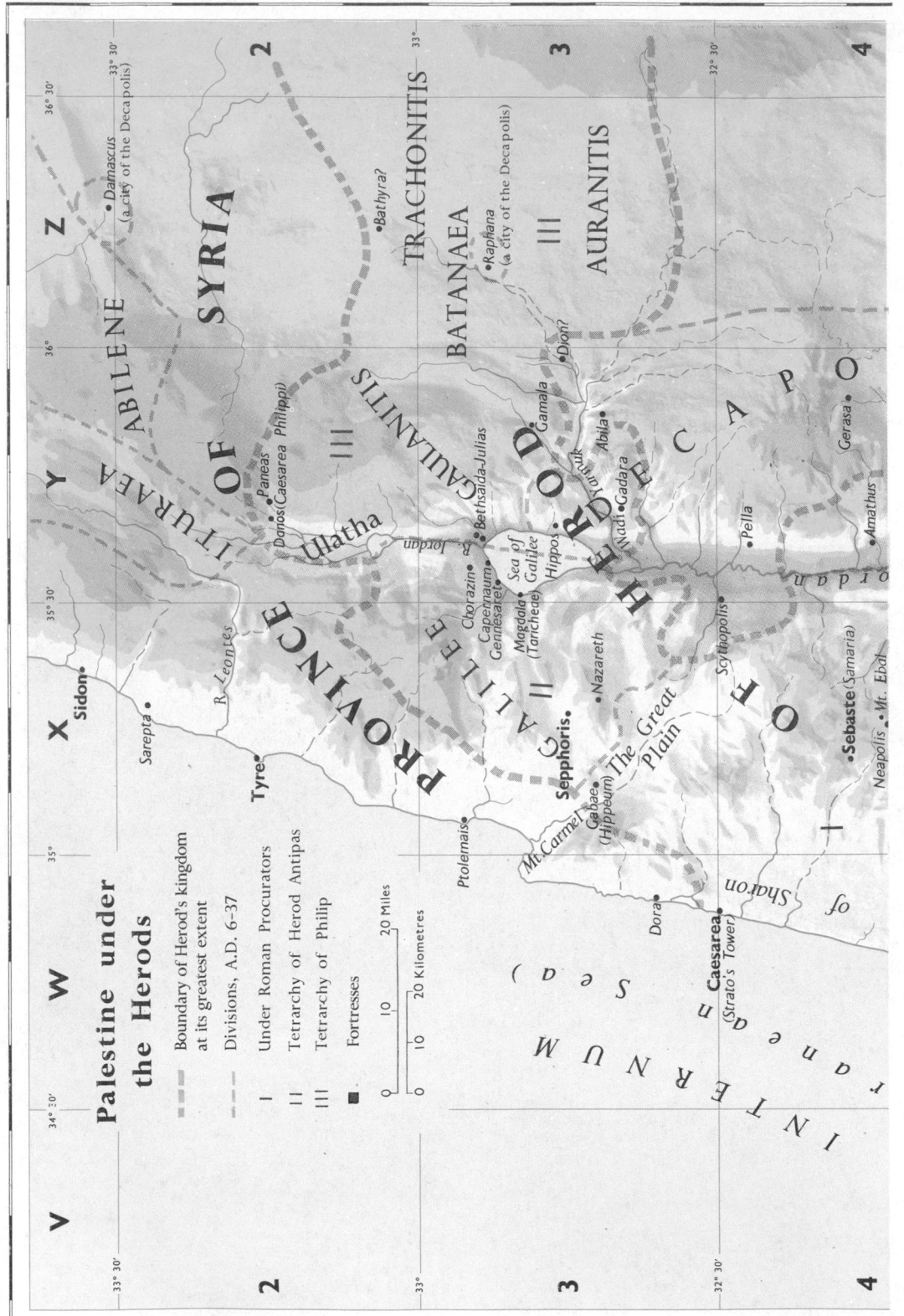

Palestine under the Herods

- — — Boundary of Herod's kingdom at its greatest extent
- — — Divisions, A.D. 6–37
- I   Under Roman Procurators
- II   Tetrarchy of Herod Antipas
- III   Tetrarchy of Philip
- ■   Fortresses

20 Miles
0   10   20

20 Kilometres
0   10   20

PROVINCE OF SYRIA

ITURAEA

ABILENE

•Damascus (a city of the Decapolis)

•Bathyra?

TRACHONITIS

BATANAEA

AURANITIS

•Raphana (a city of the Decapolis)

III

SILINANITIS

GAULANITIS

III

Paneas•
D丘ios (Caesarea Philippi)
Ulatha

R. Jordan

•Dion?

•Gamala

Bethsaida-Julias
Chorazin•
Capernaum•
Gennesaret•   Sea of Galilee
Magdala   Hippos•
(Tarichea)

Jamnia•
•Abila
•Gadara
Wadi

D E C A P O L I S

•Pella

Jordan

GALILEE II

•Nazareth

Sepphoris•

Gabae   The Great
(Hippeum)   Plain
Mt. Carmel•

Scythopolis•

•Gerasa

•Amathus

•Sebaste (Samaria)
Neapolis•   •Mt. Ebal

Ptolemais•

X   Sidon•

Sarepta•

R. Leontes

Tyre•

of Sharon

Dora•

Caesarea
(Strato's Tower)

M E D I T E R R A N E A N   S E A
(M A R E   I N T E R N U M)

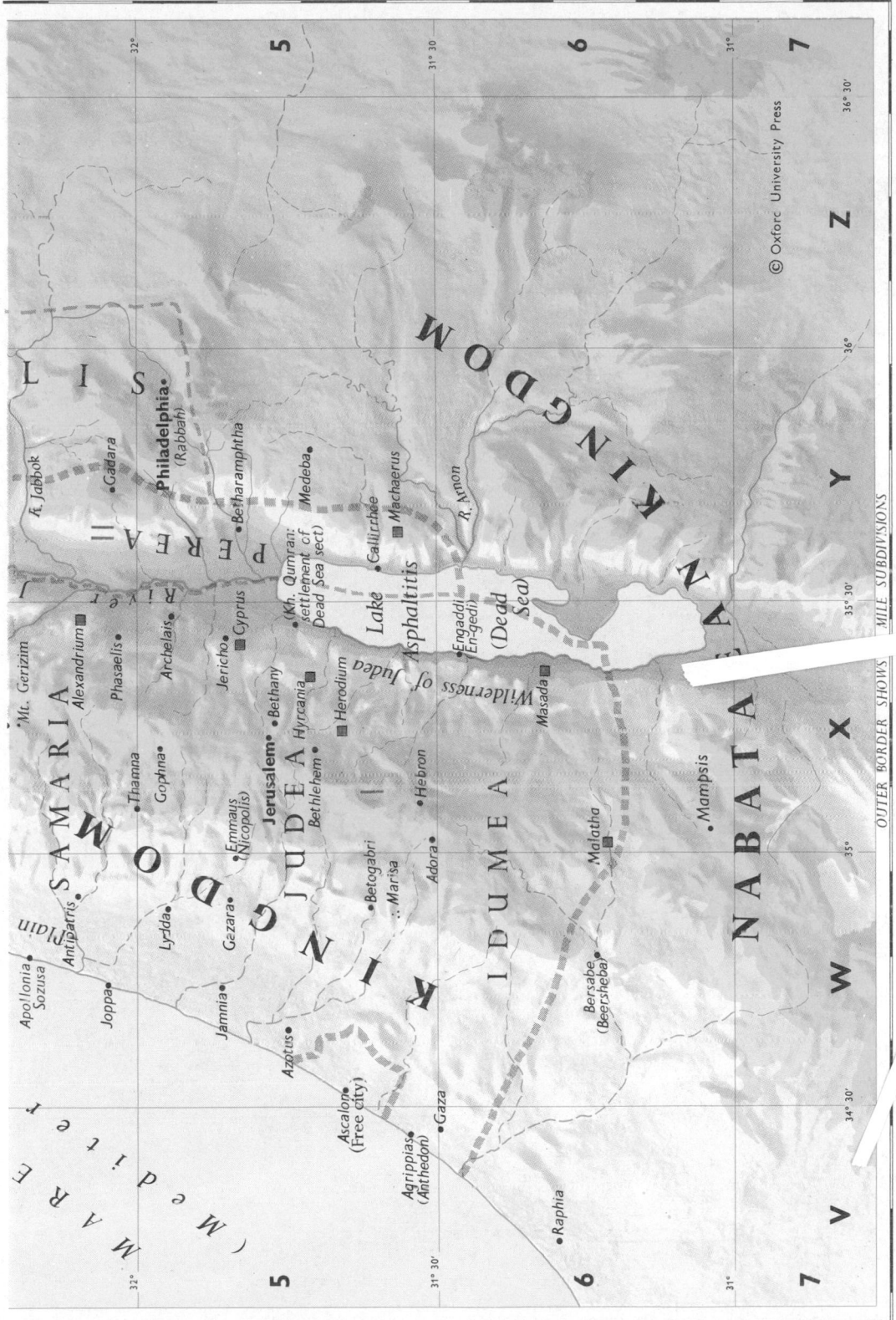

MAP 12

5

6

7

Z

© Oxford University Press

L I S

Philadelphia·
(Rabbah)

·Gadara

·Betharamphtha

Kh. Jabbok

·Medeba

P E R E A

Gallirrhoe

·Machaerus

R. Arnon

Kh. Qumran:
settlement of
(Dead Sea sect)

Cyprus·

Lake
Asphaltitis

·Jericho

·Engaddi
(En-gedi)

(Dead
Sea)

Mt. Gerizim·

Alexandrium■

S A M A R I A

·Phasaelis

·Archelais

Wilderness of Judea

Wilderness of Judea

Masada■

■Hyrcania

·Bethany

Herodium■

·Thamna

·Gophna

Jerusalem·

Emmaus·
(Nicopolis)

·Bethlehem

J U D E A

·Antipatris

·Lydda

·Gazara

·Hebron

·Betogabri

:Marisa

·Adora

I D U M E A

·Mampsis

N A B A T A

·Apollonia
Sozusa

Plain

·Joppa

K I N G D O M

·Jamnia

·Azotus

·Malatha

·Ascalon
(Free city)

·Gaza

·Bersabe
(Beersheba)

MARE

(Mediter

·Agrippias
(Anthedon)

·Raphia

W

X

Y

OUTER BORDER SHOWS

MILE SUBDIVISIONS

5

6

7

V

32°

31° 30'

31°

32°

31° 30'

31°

35°

34° 30'

35° 30'

36°

36° 30'

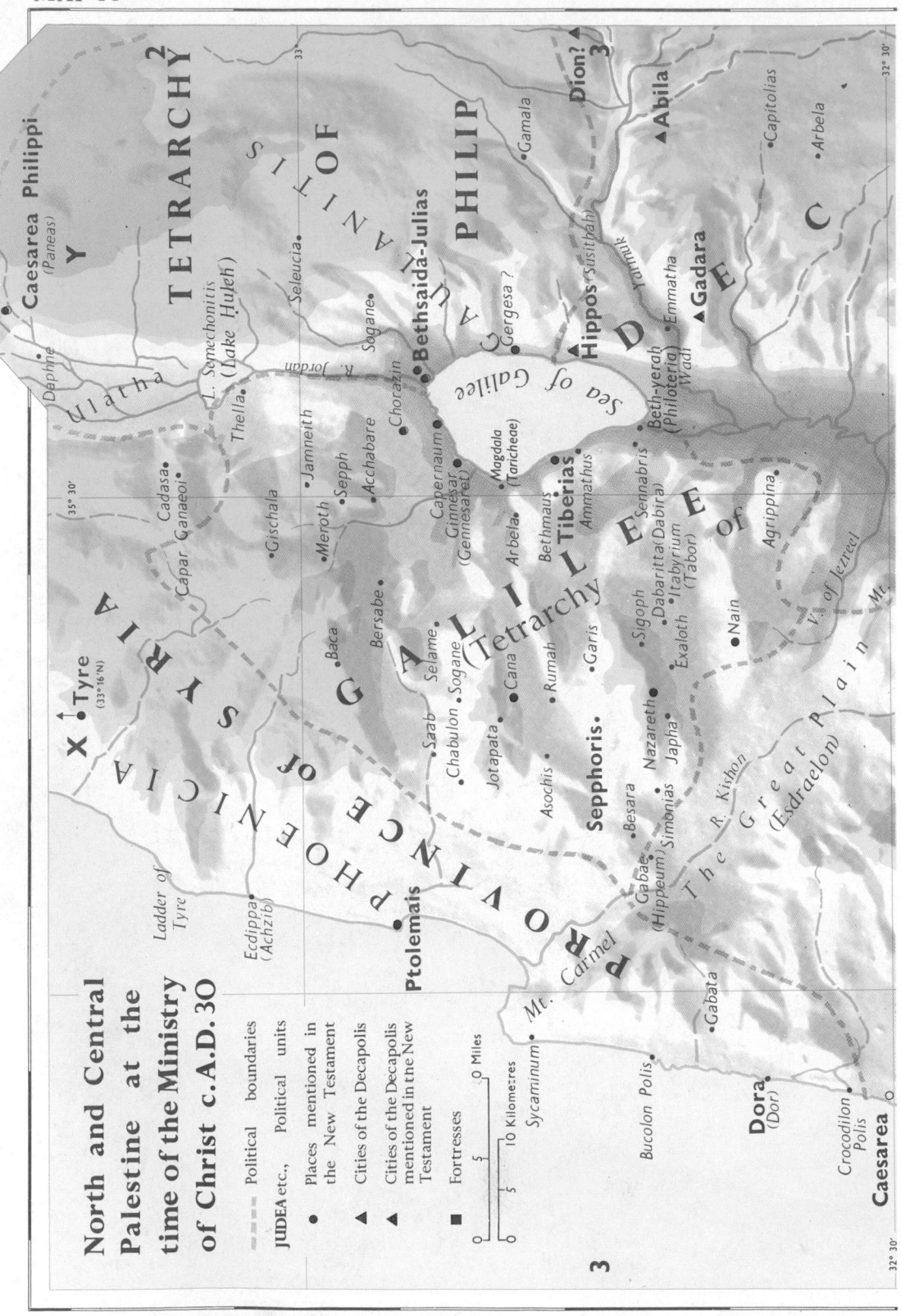

MAP 13

North and Central Palestine at the time of the Ministry of Christ c.A.D. 30

JUDEA etc., Political units

Political boundaries

• Places mentioned in the New Testament

▲ Cities of the Decapolis

▲ Cities of the Decapolis mentioned in the New Testament

■ Fortresses

Miles
Kilometres

PROVINCE OF SYRIA

PHOENICIA

PROVINCE of SYRIA

TETRARCHY OF PHILIP

Caesarea Philippi (Paneas)

Daphne

Ulatha

L. Semechonitis (Lake Huleh)

Thella

Seleucia

R. Jordan

GAULANITIS

Sogane

Bethsaida-Julias

Gamala

Dion?

Abila

Capitolias

Arbela

PHILIP

Chorazin

Gergesa?

Hippos (Susitha)

Yarmuk

Emmatha

Gadara

D E C

Capernaum

Magdala (Tarichede)

Beth-yerah (Philoteria) Wadi

Sea of Galilee

Ginnesar (Gennesaret)

Bethmaus

Sennabris

Tiberias

Ammathus

Tyre (33°16'N)

Cadasa

Capar Canaeoi

Gischala

Meroth

Sepph

Acchabare

Jamneith

Arbela

Rumah

Dabaritta (Dabira)

Itabyrium (Tabor)

Agrippina

V. of Jezreel

Mt.

Baca

Bersabe

Selame

Chabulon Sogane

Cana

Garis

Sigoph

Exaloth

Nain

Saab

Jotapata

G A L I L E E of

Tetrarchy

Asochis

Nazareth

Japha

G A L I L E E

Ladder of Tyre

Ecdippa (Achzib)

Sepphoris

Besara

Simonias

R. Kishon

The Great Plain (Esdraelon)

Ptolemais

Gabae (Hippeum)

Mt. Carmel

P R O

Gabata

Sycaminum

Bucolon Polis

Dora (Dor)

Crocodilon Polis

Caesarea

3

2

3

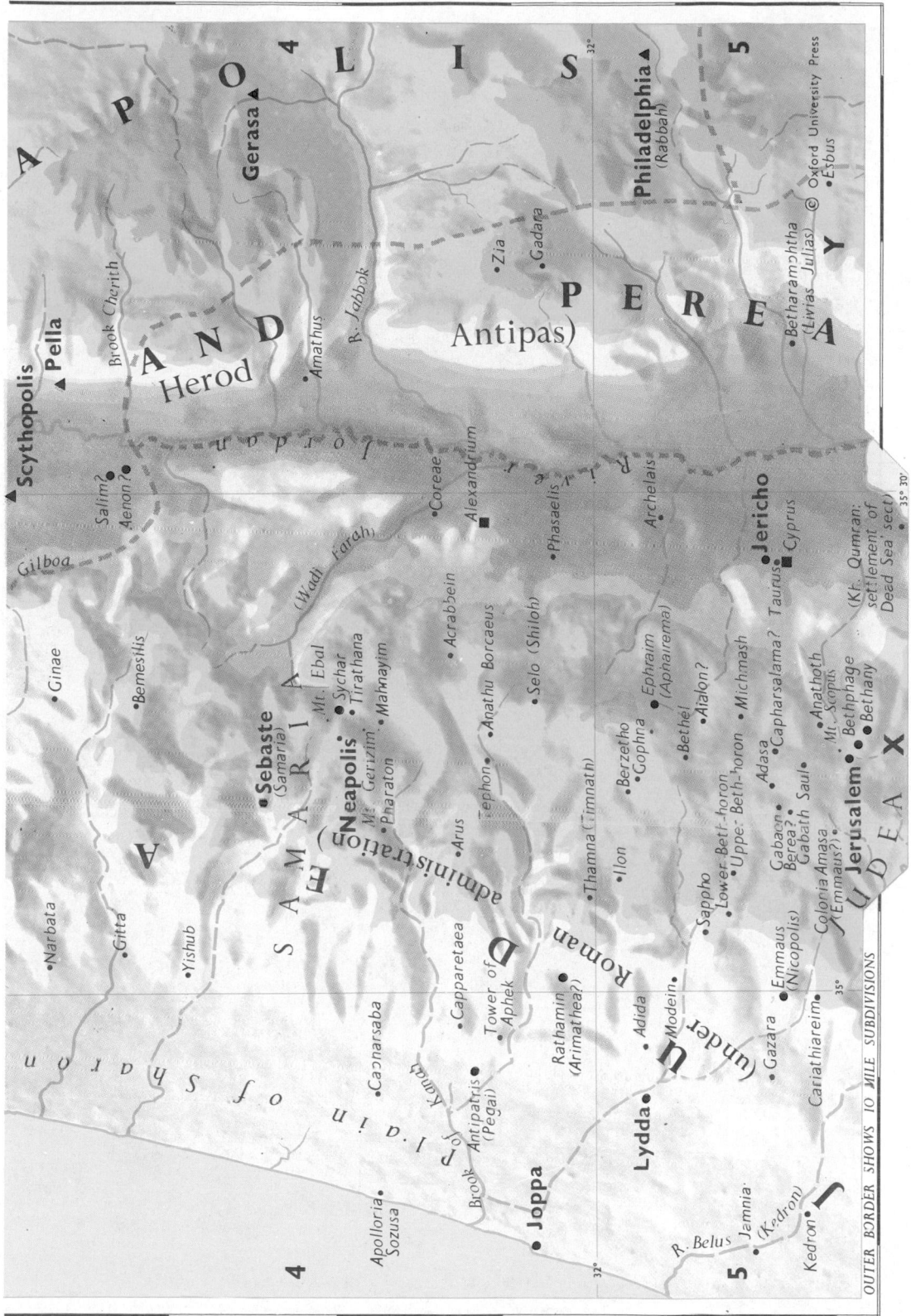

MAP 13

DECAPOLIS

Gerasa▲

5

Philadelphia▲
(Rabbah)

© Oxford University Press

•Esbus

AND

Pella▲

•Zia
•Gadara

P   E   R   E   A

Brook Cherith

R. Jabbok

•Amathus

R. Jordan

Herod

Antipas)

Betharamphtha
(Livias Julias)

35° 30'

Scythopolis▲

Salim?
Aenon?

•Coreae

•Alexandrium

•Phasaelis

Archelais

Jericho

Cyprus

(Kh. Qumran:
settlement of
Dead Sea sect)

Gilboa

(Wadi Farah)

•Acrabbein

•Selo (Shiloh)

•Ginae

•Bemeshis

Mt. Ebal

Sychar

Tirathana

•Anathu Borcaeus

Ephram
(Aphairema)

•Michmash

Taurus•

•Anathoth

Mt. Scopus

Sebaste•
(Samaria)

Neapolis

Mt. Gerizim

Malwayim

Pharaton

Berzetho

Gophna•

Bethel•

Ajalon?

Adasa•

Capharsalama?

Bethphage•

•Bethany

Jerusalem•

J U D E A

S   A   M   A   R   I   A

Arus

Tephon•

•Thamna (Timnath)

•Ilon

Sappho•

Lower-Beth-horon

Uppe-Beth-horon•

Gabath Saul•

Gabaon•
Berea?

Colonia Amasa•

(Roman administration)

•Narbata

•Gitta

•Yishub

Capparetaea•

Tower of
Aphek

Rathamin
(Arimathea?)

Modein•

(under Roman administration)

Emmaus•
(Nicopolis)

Colonia Amasa•
(Emmaus?)

35°

Plain of Sharon

Apolloria•
Sozusa

Caparsaba•

Kana

Brook

Antipatris•
(Pegai)

•Adida

Lydda•

Gazara•

Cariathiareim•

Joppa

Jamnia•

R. Belus

Kedron•

(Kedron)

Kedron•

4

5

MAP 14

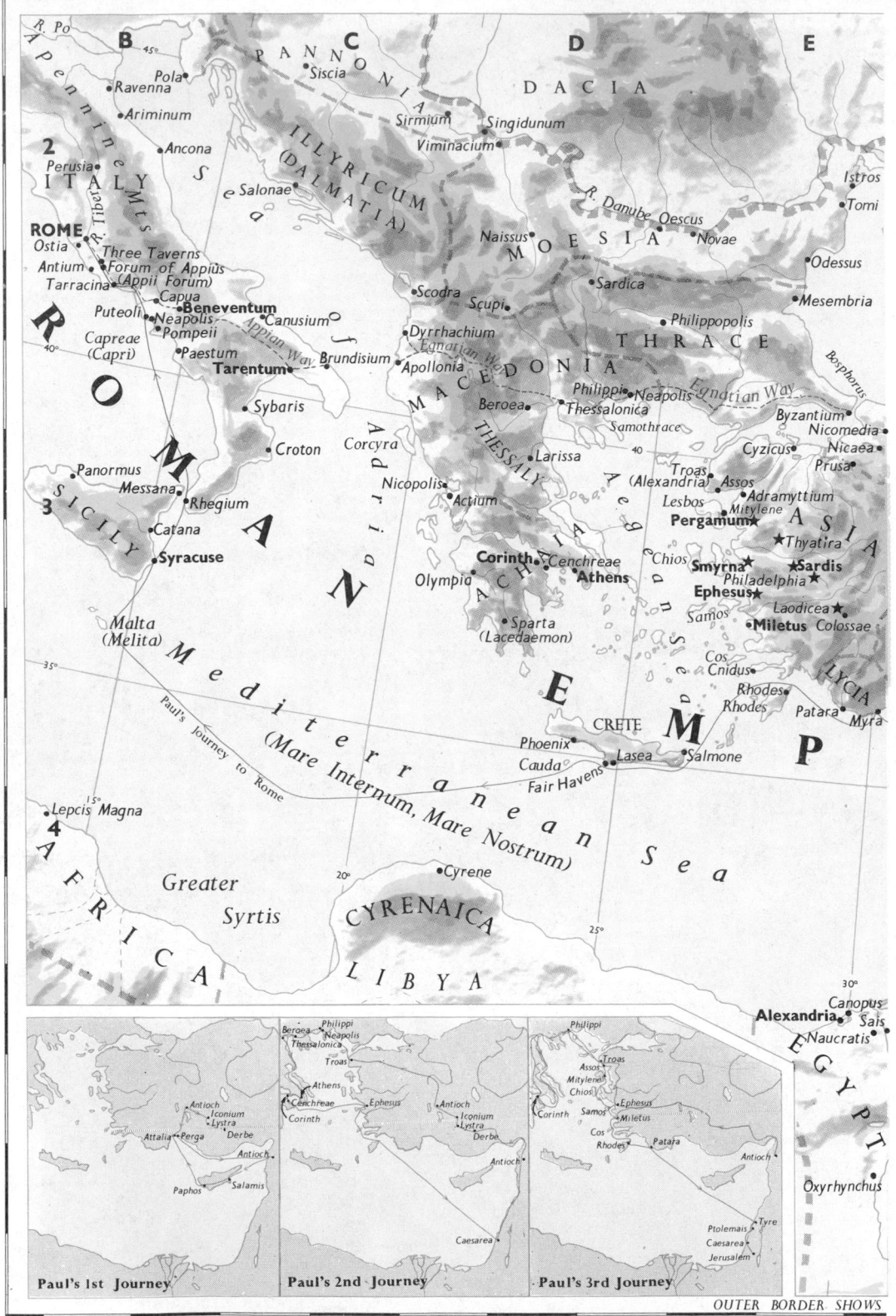

R. Po

B 45° C D E

Pola
Ravenna
PANNONIA
Siscia
DACIA

Ariminum
ILLYRICUM
(DALMATIA)
Sirmium
Singidunum

2 Ancona
Viminacium

Perusia
ITALY
R. Danube
Oescus
Istros

Salonae
Naissus
Novae
Tomi

ROME
MOESIA

Ostia
Three Taverns
Scodra
Sardica
Odessus

Antium
Forum of Appius
(Appii Forum)
Scupi
Philippopolis
Mesembria

Tarracina
Capua
THRACE
Bosphorus

Puteoli
Beneventum
Canusium
Dyrrhachium
Egnatian Way
Byzantium

Capreae
(Capri)
Neapolis
Pompeii
Brundisium
Apollonia
Philippi
Neapolis
Egnatian Way
Nicomedia

40°
Paestum
Tarentum
MACEDONIA
Beroea
Thessalonica
Samothrace
Cyzicus
Prusa
Nicaea

Sybaris
THESSALY
Larissa
40
Troas
(Alexandria)
Assos
Adramyttium

Croton
Corcyra
Nicopolis
Actium
Lesbos
Mitylene
Pergamum
ASIA

Panormus
Messana
ACHAIA
Corinth
Cenchreae
Athens
Chios
Smyrna
Thyatira
Sardis

Rhegium
Olympia
Samos
Ephesus
Philadelphia

3 SICILY
Catana
Sparta
(Lacedaemon)
Miletus
Laodicea
Colossae

Syracuse
Cos
Cnidus
LYCIA

Malta
(Melita)
Rhodes
Rhodes
Patara
Myra

35°
Mediterranean
CRETE
EM
P

Paul's Journey to Rome
Phoenix
Cauda
Lasea
Salmone

(Mare Internum, Mare Nostrum)
Fair Havens
Sea

15°
Lepcis Magna

4 AFRICA
Greater
Syrtis
Cyrene
20°
CYRENAICA
25°

LIBYA
30°
Alexandria
Canopus
Sais

EGYPT
Naucratis

Oxyrhynchus

Paul's 1st Journey | Paul's 2nd Journey | Paul's 3rd Journey

Antioch
Iconium
Lystra
Derbe
Attalia
Perga
Antioch
Paphos
Salamis

Beroea
Philippi
Neapolis
Thessalonica
Troas
Athens
Cenchreae
Corinth
Ephesus
Antioch
Iconium
Lystra
Derbe
Antioch
Caesarea

Philippi
Assos
Mitylene
Chios
Samos
Ephesus
Miletus
Cos
Rhodes
Patara
Corinth
Antioch
Ptolemais
Tyre
Caesarea
Jerusalem

OUTER BORDER SHOWS

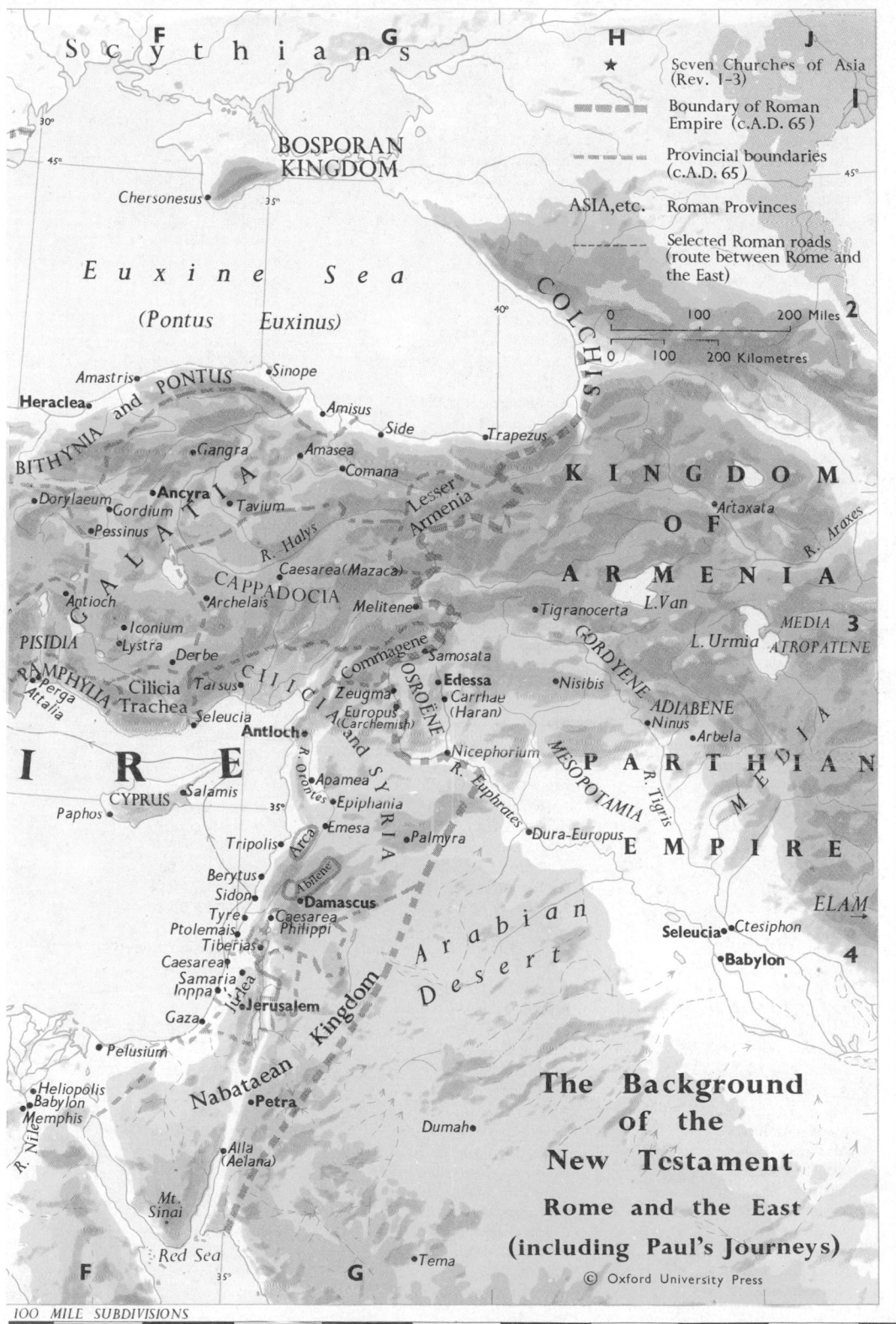

MAP14

S c y t h i a n s

F · G · H · J · I

★ Seven Churches of Asia (Rev. 1–3)

▬ ▬ ▬ Boundary of Roman Empire (c.A.D. 65)

– · – · – Provincial boundaries (c.A.D. 65)

ASIA, etc. Roman Provinces

– – – – – Selected Roman roads (route between Rome and the East)

BOSPORAN KINGDOM

Chersonesus 35°

E u x i n e S e a

(Pontus Euxinus)

COLCHIS

30° 45° 45°

40°

0 — 100 — 200 Miles **2**

0 — 100 — 200 Kilometres

Amastris · PONTUS · Sinope

Heraclea · BITHYNIA and PONTUS

Amisus · Side · Trapezus

Giangra · Amasea · Comana

Dorylaeum · Ancyra · Tavium

Gordium · GALATIA · Lesser Armenia

Pessinus · R. Halys

Antioch · Archelais · CAPPADOCIA · Caesarea (Mazaca)

PISIDIA · Iconium · Lystra · Melitene

PAMPHYLIA · Perga · Attalia · Derbe · CILICIA · Commagene · Samosata

Cilicia Trachea · Tarsus · Zeugma · OSROENE · Edessa

Seleucia · Europus (Carchemish) · Carrhae (Haran)

KINGDOM OF ARMENIA

Artaxata · R. Araxes

Tigranocerta · L. Van

L. Urmia · MEDIA ATROPATENE **3**

GORDYENE · Nisibis · ADIABENE · Ninus · Arbela

I R E · CYPRUS · Salamis · Antioch · R. Orontes · Apamea · Nicephorium

Paphos · Epiphania · Emesa · R. Euphrates · Dura-Europus

MESOPOTAMIA · R. Tigris · P A R T H I A N · M E D I A · E M P I R E

Tripolis · Arca · Palmyra

Berytus · Abilene · Damascus

Sidon · Caesarea Philippi

Tyre · Ptolemais · Tiberias

Caesarea · Samaria · Judea · Joppa · Jerusalem

Gaza

ELAM

Seleucia · Ctesiphon · Babylon **4**

A r a b i a n D e s e r t

Nabataean Kingdom

Pelusium

Heliopolis · Babylon · Memphis

Petra · Dumah

R. Nile

Alla (Aelana)

Mt. Sinai

Red Sea 35°

Tema

The Background of the New Testament

Rome and the East

(including Paul's Journeys)

© Oxford University Press

F · G

100 MILE SUBDIVISIONS

INDEX TO MAPS

Babylonia 6,7,11, **J4**
Baca 13, **X3**
Balikh, R. 6, **G3**
Bamoth-baal 8, **Y5**
Baris 9
Bashan 1,3,4,5,8, **Y3**
Baskama 10, **Y3**
Batanaca 12, **Z3**
Bathyra 12, **Z2**
Beautiful Gate 9
Beer 3,8, **X3**
Beeroth: Benjamin 3,8, **X5**
Beeroth (Bene-jaakan) 2, **T2**
Beeroth (Berea) 10, **X5** & inset
Beer-sheba 1,3,4,5,10,12, **W6**
   also 2, **T1**
Beerzeth (Berzetho) 10, **X5**
Behistun 7, **J4**
Belus, R. 13, **W5**
Bemesilis 13, **X4**
Bene-berak 8, **W4**
Bene-jaakan 2, **T2**
Beneventum 14, **B2**
Benjamin 3,4,5, **X5**
Beon 1,8, **Y5**
Beracah, V. of 5,8, **X6**
Berea 10, **X5** & inset; 13, **X5**
Beroea 14, **D2**
Bersabe (Beersheba) 12, **W6**
Bersabe: Galilee 13, **X3**
Berytus 6,11,14, **G4**
Berzetho 10, **X4**; 13, **X5**
Besara 13, **X3**
Besor, Brook 3,4,5, **V6**
Beten 8, **X3**
Beth-anath 1,3, **X2**
Beth-anoth 8, **X5**
Bethany 12,13, **X5**
Betharamphtha 12,13, **Y5**
Beth-arbel 5,8, **Y3**
Beth-aven 8, **X5**
Beth-baal-meon 3,8, **Y5**
Beth-basi 10, **X5**
Beth-dagon 1,3,8, **W5**
Beth-diblathaim 5,8, **Y5**
Beth-eden 6, **G3**
Beth-eglaim 1, **V6**
Bethel 1,3,4,5,8,13 **X5**; 10, **X5** &
   inset
Bethesda, Pool of 9
Beth-gilgal 8, **X5**
Beth-gilgal (Gilgal) 10, **X5**
Beth-haccherem 8,10, **X5**
Beth-haggan 1,3,5, **X4**
Beth-hanan 4, **X5**
Beth-haram 8, **Y5**
Beth-hoglah 8, **Y5**
Beth-horon 10, **X5**
Beth-horon, Upper & Lower
   1,5, **X5**
Beth-jeshimoth 1,3,8, **Y5**
Bethlehem: Galilee 3,8, **X3**
Bethlehem: Judah
   1,3,4,5,8,10,12, **X5**

Bethmaus 13, **X3**
Beth-meon 5,8, **Y5**
Beth-nimrah 1,3,8, **Y5**
Beth-peor 3,8, **Y5**
Bethphage 13, **X5**
Beth-rehob: region 4, **Y2**
Beth-rehob: town 3,4, **Y2**
Bethsaida-Julias 12,13, **Y3**
Beth-shan (Beth-shean) 8, **X4**
Beth-shan (Scythopolis) 10, **Y4**
Beth-shean 1,3,4,8, **Y4**
Beth-shemesh: Issachar 8, **Y3**
Beth-shemesh: Judah
   1,3,4,5,8, **W5**
Beth-tappuah 8, **X5**
Bethul (Bethuel) 3, **W6**
Beth yerah 1,8,11, **Y3**
Beth-yerah (Philoteria) 10, **Y3**
Beth-zachariah 10, **X5**
Beth-zaith 10, **X5**
Bethzatha, Pool of 9
Beth-zur 3,5,8,10, **X5**
Betogabri 12, **W5**
Betonim 3,8, **Y5**
Beyond the River 7, **G4**
Bezek 3,8, **X4**
Bezer 1,3,5,8, **Y5**
Bezetha 9
Bisitun 7, **J4**
Bit-adini 6, **G3**
Bithynia & Pontus 14, **F2**
Black Sea 6,7,11, **F2**
Borim 5,8, **X4**
Borsippa 6,7, **H4**
Bosor 10, **Z3**
Bosora 10, **Z3**
Bosphorus 14, **E2**
Bosporan Kingdom 14, **G1**
Bozkath 8, **W5**
Bozrah (Bosora) 10, **Z3**
Bozrah: Edom 2, **U2**
Bozrah: Moab 5,8, **Y5**
Brundisium 14, **C2**
Bubastis 2, **Q2**; 11, **F4**
Bucolon Polis 13, **W3**
Busiris 2,7, **Q2**
Byblos 6,11, **G4**
Byzantium 7,11,14, **E2**

Cabbon 8, **W5**
Cabul 3,4,8, **X3**
Cadasa 13, **Y2**
Caesarea: Palestine 12,13, **W3**
   also 14, **F4**
Caesarea (Mazaca) 14, **G3**
Caesarea Philippi 12, **Y2**
   also 14, **G4**
Calah 6, **H3**
Calchedon 11, **E2**
Callirrhoe 12, **Y5**
Calno 6, **G3**
Cana 13, **X3**
Canaan 1 –; 2, **T1**
Canopus 14, **F4**

Canusium 14, **C2**
Capar Ganaeoi 13, **Y2**
Capernaum 12,13, **Y3**
Capharsaba 13, **W4**
Capharsalama 10, inset; 13, **X5**
Caphtor (Crete) 6, **D3**
Capitolias 13, **Y3**
Cappadocia 7,14, **G3**; 11, **F3**
Capparetaea 13, **W4**
Capreae (Capri) 14, **B2**
Capua 14, **B2**
Carchemish 6,11,14, **G3**
Caria 7,11, **E3**
Cariathiareim 13, **W5**
Carmel: town 4,5, **X6**
Carmel, Mt. 1,3,4,5,8,10,12,13, **X3**
Carnaim (Carnion) 10, **Z3**
Carrhae (Haran) 14, **G3**
Casphor 10, **Y3**
Caspian Sea 6,7, **K3**
Caspin 10, **Y3**,
Caucasus Mts. 7, **J2**
Cauda 14, **D4**
Celaenae 11, **F3**
Cenchreae 14, **D3**
Chabulon 13, **X3**
Chaspo (Casphor, Caspin) 10, **Y3**
Chephar-ammoni 8, **X5**
Chephirah 3,8, **X5**; 10, inset
Cherith, Brook 5,8,13, **Y4**
Chersonesus 14, **F2**
Chesulloth 8, **X3**
Chesalon 8, **X5**
Chezib 1,8, **W5**
Chinnereth 1,3,5,8, **Y3**
Chinnereth, Sea of
   1,3,4,5,8, **Y3**
Chios 14, **E3**
Chisloth-tabor 8, **X3**
Chitlish 8, **W5**
Chorazin 12,13, **Y3**
Cilicia 6,7,11, **F3**
Cilicia & Syria 14, **G3**
Cilicia Trachea 14, **F3**
Cilician Gates 11, **G3**, 7, **F3**
Cimmerians 7, **F3**
Citium 7,11, **F4**
City of David 9
City of Salt 5,8, **X5**
Cnidus 14, **E3**
Coele Syria 11, **G4**
Colchi 7, **H2**
Colchis 14, **H2**
Colonia Amasa 13, **X5**
Colossae 14, **E3**
Comana 14, **G2**
Commagene 6,7,11,14, **G3**
Coponius, Gate of 9
Coptos 7, **F5**
Corcyra 14, **C3**
Coreae 13, **X4**
Corinth 7,11,14, **D3**
Cos 14, **E3**
Court of Gentiles 9